Finite Element
Bibliography

Finite Element Bibliography

Compiled by

Douglas Norrie and Gerard de Vries
University of Calgary, Calgary, Alberta, Canada

IFI/PLENUM · NEW YORK-WASHINGTON-LONDON

Library of Congress Cataloging in Publication Data

Norrie, D   H
   Finite element bibliography.

   Includes indexes.
   1. Finite element method—Bibliography. I. De Vries, Gerard, joint
author. II. Title.
Z5853.F55N67   [TA347.F5] 016.62'0004'2   76-45615
ISBN-13: 978-1-4684-1386-1   e-ISBN-13: 978-1-4684-1384-7
DOI: 10.1007/978-1-4684-1384-7

© 1976 IFI/Plenum Data Company

Softcover reprint of the hardcover 1st edition 1976

A Division of Plenum Publishing Corporation
227 West 17th Street, New York, N.Y. 10011

Preface

This bibliography had its inception in 1967, when the compilers first attempted a comprehensive coverage of the finite element literature using both manual and computer-based retrieval. Initially, the data base was stored on a card index, but this was subsequently transferred to punched cards and magnetic tape. Computer processing was adopted at an early stage to derive the three index formats from the data base.

Over the subsequent years, several versions of the bibliography were produced, with two of these being made available in a report form to other researchers. From the widespread interest in these documents, it became evident that there was a need for a commercially-available *comprehensive bibliography* in this area. A major effort was undertaken to revise, update, and extend the data base, resulting in this present volume.

The bibliography covers the period 1956—1975 primarily, although some earlier publications of historical interest are included. The citations are not restricted to the English language and documents are listed in many languages and from diverse places of origin. All publication formats were accepted, so that references will be found to books, monographs, journal papers and articles, theses, dissertations, reports, surveys, and the like.

Although it is known that this bibliography is not complete, it does include 7115 citations and is believed to reference the larger portion of the literature published in this area. The compilers apologize to the authors of publications which are not included and would appreciate omissions (or errors) being brought to their attention so that they may be incorporated (or corrected) in future editions. Some references are included which are not specific to the finite element method but are closely related, such as matrix structural analysis, computational procedure, and automatic mesh generation.

Many, but not all, of the citations have been checked from the original publications. Those from secondary sources can only be as accurate as the listing or cross-reference from which they were taken. In spite of care in standardizing, merging, and processing, no doubt errors have crept in which are to be regretted.

The following, without whose efforts this bibliography would not have been completed, deserve special appreciation: John Roberts (development of the bibliographic computer programs); Roger Ingles (modification of the programs and computer processing); Carol Weaver, Evelyn Arason, and Beatrice Norrie (verification and standardization of citations, and conversion to machine-readable form). The assistance of Betty Ann Maylor and Susan Milner is also gratefully acknowledged. Finally, our debt to Oldrich Standera, Head of Information Systems, University of Calgary Library, for support and counsel must similarly be recorded.

This project was pursued in association with research on the finite element method supported by the National Research Council of Canada (Grants A4192 and A7432), and the assistance received from this source is gratefully acknowledged.

Douglas Norrie
Gerard de Vries

July, 1976

List of Contents and Explanatory Notes

AUTHOR LISTING (page 1)

Publications are listed under the name of the author alphabetically. Where there is more than one author, the document is listed separately under the name of each author. Cross-reference to the Citation Listing is through the citation number.

KEYWORD LISTING (page 119)

References are indexed under each significant word in the title. However, because of their large number, listings under the keywords FINITE and ELEMENT, have been deleted. The citation number allows cross-reference to the Citation Listing, where a full bibliographic description will be found.

In the alphabetization, hyphened prefixes are treated as words, so that there are separate entries for, e.g., NON-LINEAR (page 300) and NONLINEAR (page 301). Also, there are separate entries for the singular and plural forms of nouns, often with many other keywords intervening, e.g., MODE (page 291) and MODES (page 294). The user is advised to explore all possible variants.

CITATION LISTING (page 451)

Author(s), title, source, and publication date are given for each document, indexed under the citation number. The first two digits of this number identify the year of publication.

Note: Where a document may otherwise be difficult to obtain, the NTIS number has been added, if known, to the end of the citation. Microfiche or photocopies of these documents can be purchased from the National Technical Information Service, U.S. Department of Commerce, 5285 Port Royal Rd., Springfield, Virginia 22151, U.S.A.

Author Listing

AALAMI, B.	LARGE DEFLECTION OF ELASTIC PLATES UNDER PATCH LOADING	72-0120
AALAMI, B.	ROTATIONAL STIFFNESS OF CONCRETE SLABS	74-0660
AAMODT, B.	APPLICATION OF THE FINITE ELEMENT METHOD TO PROBLEMS IN LINEAR FRACTURE MECHANICS	74-0903
AAMODT, B.	CALCULATION OF STRESS INTENSITY FACTORS AND FATIGUE CRACK PROPAGATION OF SEMI-ELLIPTICAL, PART-THROU	73-0360
AAMODT, B.	EFFICIENT FORMULATIONS OF THE FINITE ELEMENT METHOD IN LINEAR AND NONLINEAR FRACTURE MECHANICS	75-0403
AAMODT, B.	ELASTO-PLASTIC ANALYSIS USING AN EFFICIENT FORMULATION OF THE FINITE ELEMENT METHOD	75-0012
AAMODT, B.	FINITE ELEMENT ANALYSIS OF CRACK PROPAGATION IN THREE-DIMENSIONAL SOLIDS UNDER CYCLING LOADING	73-0529
AAMODT, B.	FINITE ELEMENT ANALYSIS OF CRACK PROPAGATION IN 3-DIMENSIONAL SOLIDS UNDER CYCLIC LOADING	74-0479
AAMODT, B.	NUMERICAL TECHNIQUES IN LINEAR AND NONLINEAR FRACTURE MECHANICS	75-0260
AAMODT, B.	PROPAGATION OF ELLIPTICAL SURFACE CRACKS AND NONLINEAR FRACTURE MECHANICS BY THE FINITE ELEMENT METH	74-0906
AAMODT, B.	SHELL ANALYSIS BY SESAM-69	70-0249
AASEN, E.	APPLICATION OF THE FINITE ELEMENT METHOD TO MACHINERY	74-0198
AASOKA, K.	SIMULATION ON THE TENSILE BEHAVIOUR OF TWO-PHASE ALLOY CONTAINING A FIBER UNDER TENSILE LOAD	73-0542
ABBAS, B. A. H.	DYNAMIC STABILITY OF TIMOSHENKO BEAMS BY FINITE ELEMENT METHOD	75-0194
ABBAS, B. A. H.	DYNAMIC STABILITY OF TIMOSHENKO BEAMS BY FINITE ELEMENT METHOD	75-0422
ABBAS, B. A. H.	FINITE ELEMENT MODEL FOR DYNAMIC ANALYSIS OF TIMOSHENKO BEAM	75-0275
ABBASI, J.	PLANE STRESS REINFORCED CONCRETE FINITE ELEMENTS	74-0108
ABBASI, J. A.	REINFORCED CONCRETE - A FINITE-ELEMENT FORMULATION	74-0929
ABDEL-HAMIDABDE	A FINITE ELEMENT METHOD FOR RANDOM FOLDED PLATE STRUCTURES (EINE FINITE E.EMENT METHODE ZUR BERECHNU	73-0787
ABDELRAOUF, M.	FINITE ELEMENT ANALYSIS OF SHELL-TYPE STRUCTURES	71-0789
ABDELRAOUG, M.	FINITE-ELEMENT ANALYSIS OF BRIDGE DECKS	74-0672
ABE, T.	ELASTIC DEFORMATION OF POLYCRYSTALLINE METAL - 1. INFLUENCE OF GRAIN SHAPE ON DISTRIBUTIONS OF STRES	72-0463
ABE, T.	ELASTIC DEFORMATION OF POLYCRYSTALLINE METAL - 2 EFFECT OF SLENDERNESS OF GRAINS ON DEFORMATION MODE	73-0275
ABE, T.	ELASTIC DEFORMATION OF POLYCRYSTALLINE METALS - 3	74-0361
ABE, T.	STRESS DISTRIBUTION UNDER UNIAXIAL ELASTIC PLANE STRAIN COMPRESSION WITH FRICTION	74-0526
ABEL, J.	COMPARISON OF FINITE ELEMENTS FOR PLATE BENDING	72-0625
ABEL, J. F.	BUCKLING OF COOLING-TOWER SHELLS: BIFURCATION RESULTS	75-0384
ABEL, J. F.	BUCKLING OF COOLING-TOWER SHELLS: STATE-OF-THE-ART	75-0385
ABEL, J. F.	FINITE ELEMENT GALERKIN METHOD FOR ANALYSIS OF FLOW IN FRACTURED POROUS MEDIA	74-0066
ABEL, J. F.	INTRODUCTION TO THE FINITE ELEMENT METHOD	72-0362
ABEL, J. F.	PHOTOELASTIC AND FINITE-ELEMENT ANALYSIS OF A QUADRIPARTITE VAULT	73-0119
ABEL, J. F.	STATIC AND DYNAMIC FINITE ELEMENT ANALYSIS OF SANDWICH STRUCTURES	68-0003
ABEL, J. F.	STRESSES AROUND FLEXIBLE ELLIPTIC PIPES	73-0215
ABIRU, H.	DYNAMIC ANALYSIS OF THE GROUND USING THE THREE DIMENSIONAL FINITE ELEMENT	73-0293
ABOU-AYYASH, A.	A SUMMARY OF DISCRETE-ELEMENT METHODS OF ANALYSIS FOR PAVEMENT SLABS	72-0674
ABSHER, R. G.	ANALYSIS OF FEEDBACK CONTROL STABILIZED VIBRATION BY THE FINITE ELEMENT METHOD	74-1188
ABSI, E.	COMPARISON OF EQUIVALENCE AND FINITE ELEMENT METHODS	75-0483
ABSI, E.	FINITE ELEMENT METHOD	69-0036
ABSI, E.	THEORY OF EQUIVALENCE AND ITS APPLICATION TO VARIOUS PROBLEMS IN ELASTICITY AND DESIGN OF SKEW BRIDG	72-0470
ABU-SHUMAYS, I.	ADJOINING APPROPRIATE SINGULAR ELEMENTS TO TRANSPORT THEORY COMPUTATIONS	74-0946
ABU-SHUMAYS, I.	FINITE ELEMENTS IN NEUTRON TRANSPORT THEORY	73-0961
ABU-SHUMAYS, I.	SINGULAR ELEMENTS IN VARIATIONAL AND FINITE ELEMENT TRANSPORT COMPUTATIONS	73-0902
ACHUTARAMAYYA,	FRACTURE SURFACE ENERGIES BY FINITE ELEMENT STRESS ANALYSIS	73-0915
ADACHI, J.	ON THE DYNAMIC BUCKLING OF SHELLS OF REVOLUTION	73-1102
ADALI, S.	FINITE ELEMENT SOLUTION OF A NON-LINEAR DIFFERENTIAL EQUATION USING KACHANOV'S AND SUCCESSIVE APPROX	75-0108
ADAMEK, J. R.	AN AUTOMATIC MESH GENERATOR USING TWO AND THREE DIMENSIONAL ISOPARAMETRIC ELEMENTS	73-0066
ADAMS, D. F.	ELASTOPLASTIC CRACK PROPAGATION IN A TRANSVERSELY LOADED UNIDIRECTIONAL COMPOSITE	74-0217
ADAMS, D. F.	HIGH-PERFORMANCE COMPOSITE MATERIALS FOR VEHICLE CONSTRUCTION: A FINITE ELEMENT COMPUTER PROGRAM FOR	73-0767
ADAMS, D. F.	INELASTIC ANALYSIS OF A UNIDIRECTIONAL COMPOSITE SUBJECTED TO TRANSVERSE NORMAL LOADING	70-0015
ADAMS, D. F.	MICROMECHANICAL ANALYSIS OF CRACK PROPAGATION IN AN ELASTOPLASTIC COMPOSITE MATERIAL	74-0565
ADAMS, D. F.	PRACTICAL PROBLEMS ASSOCIATED WITH THE APPLICATION OF THE FINITE ELEMENT METHOD TO COMPOSITE MATERIA	74-0483
ADAMS, D. G.	AUTOMOBILE PANEL SWEEP STIFFNESS ANALYSIS	74-0175
ADAMS, D. G.	HIGH STRENGTH MATERIALS AND VEHICLE WEIGHT REDUCTION ANALYSIS	75-0157
ADAMS, F. D.	INTERFERENCE-FIT FASTENER DISPLACEMENT MEASUREMENT BY SPECKLE PHOTOGRAPHY	74-1178
ADAMS, R. D.	STRESS ANALYSIS OF ADHESIVE-BONDED LAP JOINTS	74-0548
ADAMS, R. R.	FINITE ELEMENT STRESS ANALYSIS OF PROSTHETIC TOOTH IMPLANTS	75-0073
ADDICOTT, G. W.	INTERPRETATION OF FINITE ELEMENT STRESSES ACCORDING TO ASME SECTION 111	75-0655
ADELMAN, H. M.	A FINITE ELEMENT FOR THERMAL STRESS ANALYSIS OF SHELLS OF REVOLUTION	74-0798
ADELMAN, H. M.	A METHOD FOR COMPUTATION OF VIBRATION MODES AND FREQUENCIES OF ORTHOTROPIC THIN SHELLS OF REVOLUTION	69-0037
ADELMAN, H. M.	ACCURACY OF MODAL STRESS CALCULATIONS BY FINITE ELEMENT METHOD	70-0223
ADELMAN, H. M.	CALCULATION OF TEMPERATURE DISTRIBUTIONS IN THIN SHELLS OF REVOLUTION BY THE FINITE ELEMENT METHOD	71-0018
ADELMAN, H. M.	INCLUSION OF TRANSVERSE SHEAR DEFORMATION IN FINITE ELEMENT DISPLACEMENT FORMULATIONS	74-0662
ADEY, R. A.	EFFICIENT METHOD FOR SOLUTION OF VISCOELASTIC PROBLEMS	73-0232
ADEY, R. A.	FINITE ELEMENT SOLUTION FOR EFFLUENT DISPERSION	74-1371
ADHAM, S.	REINFORCED CONCRETE ON CONSTITUTIVE RELATIONS	75-0753
ADINI, A.	ANALYSIS OF PLATE BENDING BY THE FINITE ELEMENT METHOD	60-0003
ADINI, A.	ANALYSIS OF SHELL STRUCTURES BY THE FINITE ELEMENT METHOD	61-0008
ADLER, H.	THE USE OF FINITE ELEMENT METHODS IN HEAT FLOW ANALYSIS	68-0000
ADLER, D.	NUMERICAL CALCULATION OF MERIDIONAL FLOW FIELD IN TURBOMACHINES USING FINITE ELEMENTS METHOD	74-0481
ADLER, W. F.	STRESS ANALYSIS OF COLDWORKED FASTENER HOLES	74-0783
AGARWAL, A. C.	DIFFERENTIAL EXPANSION IN ELASTIC LAMINATES	73-0190
AGARWAL, B. D.	ELASTIC-PLASTIC FINITE ELEMENT ANALYSIS OF SHORT FIBRE COMPOSITES	74-0138
AGARWAL, B. D.	MICROMECHANICS ANALYSIS OF COMPOSITE MATERIALS USING FINITE ELEMENT METHODS	72-0770
AGARWAL, B. D.	THEORETICAL STUDY OF THE EFFECT OF THE INTERFACE ON COMPOSITE TOUGHNESS	73-0386
AGARWAL, B. D.	THREE-DIMENSIONAL FINITE ELEMENT ANALYSIS OF SPHERICAL PARTICLE COMPOSITES	74-0137
AGARWAL, R. K.	STRESSES AND DISPLACEMENTS AROUND A CIRCULAR TUNNEL IN A THREE LAYER MEDIUM	69-0043
AGARWAL, S. L.	EXPERIMENTAL VERIFICATION OF DISCRETE-ELEMENT SOLUTIONS FOR PLATES AND PAVEMENT SLABS	70-0411
AGRAWAL, G. L.	ELAD PROGRAM MANUAL, A COMPUTER PROGRAM FOR THE ELASTIC ANALYSIS OF DOME STRUCTURES AND AXISYMMETRIC	69-0064
AGRAWAL, G. L.	EPAD, A COMPUTER PROGRAM FOR THE ELASTIC-PLASTIC ANALYSIS OF DOME STRUCTURES AND ARBITRARY SOLIDS OF	70-0036
AGRAWAL, K. M.	ANALYSIS OF ELASTIC SHELLS OF REVOLUTION WITH MEMBRANE AND FLEXURE STRESSES UNDER ARBITRARY LOADING	68-0289
AGRAWAL, K. M.	FINITE ELEMENT ANALYSIS OF HYPERBOLIC PARABOLOID SHELLS	71-0245
AGRAWAL, K. M.	FINITE ELEMENT ANALYSIS OF HYPERBOLIC PARABOLOID SHELLS	72-0887
AGRAWAL, P. K.	FINITE ELEMENT ANALYSIS FOR THE SEISMIC STABILITY OF EARTH STRUCTURES	74-1106
AGRAWAL, P. K.	FINITE ELEMENT TREATMENT OF SOIL-STRUCTURE INTERACTION PROBLEM FOR NUCLEAR POWER PLANT UNDER SEISMIC	73-0834
AGUIRRE-RAMIREZ	A NUMERICAL SOLUTION OF THE BOLTZMANN EQUATION BY THE FINITE ELEMENT TECHNIQUE	70-0095
AGUIRRE-RAMIREZ	FINITE ELEMENT TECHNIQUE APPLIED TO HEAT CONDUCTION IN SOLIDS WITH TEMPERATURE DEPENDENT THERMAL CON	69-0004
AHMAD, R.	ON THE STRESS COMPUTATION IN FINITE ELEMENT MODELS BASED UPON DISPLACEMENT APPROXIMATIONS	74-0598
AHMAD, S.	A SIMPLE MATRIX-VECTOR HANDLING SCHEME FOR THREE-DIMENSIONAL AND SHELL ANALYSIS	70-0317
AHMAD, S.	AN ASSUMED STRESS APPROACH TO REFINED ISOPARAMETRIC ELEMENTS IN THREE DIMENSIONS	74-0421
AHMAD, S.	ANALYSIS OF THICK AND THIN SHELL STRUCTURE BY CURVED FINITE ELEMENTS	70-0016
AHMAD, S.	CURVED FINITE ELEMENTS IN THE ANALYSIS OF SOLID SHELL AND PLATE STRUCTURES	69-0357
AHMAD, S.	CURVED THICK SHELL AND MEMBRANE ELEMENTS WITH PARTICULAR REFERENCE TO AXISYMMETRIC PROBLEMS	68-0005
AHMAD, S.	FINITE ELEMENT METHOD IN STRESS ANALYSIS	69-0413
AHMAD, S.	FINITE-ELEMENT METHODS FOR INHOMOGENEOUS WAVEGUIDES	69-0039
AHMAD, S.	ISO-PARAMETRIC AND ASSOCIATED ELEMENT FAMILIES FOR TWO AND THREE DIMENSIONAL ANALYSIS	69-0285
AHMAD, S.	PSEUDO ISOPARAMETRIC FINITE ELEMENTS FOR SHELL AND PLATE ANALYSIS	69-0040
AHMAD, S.	VIBRATION OF THICK, CURVED, SHELLS WITH PARTICULAR REFERENCE TO TURBINE BLADES	70-0017
AHMED, K. M.	APPLICATIONS OF CURVED FINITE ELEMENT TO SANDWICH BEAMS USING SHEAR STRAIN FORMULATION	74-0523
AHMED, K. M.	APPLICATIONS OF CURVED FINITE ELEMENTS TO SANDWICH SHELLS USING SHEAR STRAIN FORMULATION	75-0433
AHMED, K. M.	DYNAMIC ANALYSIS OF SANDWICH BEAMS	72-0586
AHMED, K. M.	FREE VIBRATION OF CURVED SANDWICH BEAMS BY METHOD OF FINITE ELEMENTS	71-0351

3

AHMED, K. M.	FREE VIBRATIONS OF CURVED SANDWICH BEAMS	71-0729
AHMED, K. M.	ON THE ENERGY FUNCTIONS FOR CURVED SANDWICH PLATES	71-0587
AHMED, K. M.	STATIC ANALYSIS OF SANDWICH STRUCTURES BY THE FINITE ELEMENT METHOD	71-0730
AHMED, K. M.	STATIC AND DYNAMIC ANALYSIS OF SANDWICH STRUCTURES BY METHOD OF FINITE ELEMENTS	71-0352
AHMED, K. M.	THE EFFECT OF IN PLANE STATIC STRESSES ON THE VIBRATION CHARACTERISTICS OF CURVED SANDWICH PLATES	71-0588
AHMED, S.	FINITE ELEMENT METHODS FOR INHOMOGENEOUS WAVEGUIDES	69-0498
AHMED, S.	FINITE-ELEMENT METHOD FOR WAVEGUIDE PROBLEMS	68-0006
AHMED, S.	WAVEGUIDE SOLUTIONS BY THE FINITE ELEMENT METHOD	69-0011
AHMED, S. R.	BERECHNUNG DES REIBUNGSLOSEN STROEMUNGSFELDES VON DREIDIMENSIONALEN AUFTRIEBSBEHAFTETENTRAGFLUEGELN	73-0533
AHN, C. S.	MATHEMATICAL MODEL OF HEATED EFFLUENTS IN COASTAL WATER	73-0921
AICHER, W.	OSCILLATION STABILITY OF INTERNALLY AND EXTERNALLY NOTCHED PLANE BARS OF EQUAL FORM FACTOR FROM 3943	73-0745
AKAGI, T.	FINITE ELEMENT ANALYSIS OF CONSOLIDATION OF UNSATURATED SOIL	74-0082
AKAGI, T.	VISCOELASTIC ANALYSIS OF P C CONTAINMENT USING FINITE ELEMENT METHOD	73-0034
AKAGI, T.	VISCOELASTIC ANALYSIS OF PRESTRESSED CONCRETE SHELL OF REVOLUTION	73-0464
AKAY, H. U.	EFFECT OF HIGHER ORDER TERMS IN CERTAIN NONLINEAR FINITE ELEMENT MODELS	73-0125
AKAY, H. U.	LINEAR BUCKLING AND GEOMETRICALLY NONLINEAR ANALYSIS OF PLANAR PLATE-STIFFENER TYPE STRUCTURES BY TH	74-1097
AKAY, H. U.	ON THE EFFECT OF APPROXIMATING HIGHER ORDER TERMS IN CERTAIN FINITE ELEMENT MODELS OF NONLINEAR STRU	74-0603
AKBARZADEH, A.	STRESS ANALYSIS OF COMPOSITE MATERIALS BY THE HYBRID FINITE ELEMENT METHOD	75-0799
AKESSON, B. A.	FINITE ELEMENT METHOD	68-0205
AKHRAS, G.	AN AUTOMATIC RELABELING ALGORITHM FOR BANDWIDTH MINIMIZATION	75-0763
AKIMOTO, M.	ON THE STATES OF STRESS AND DEFORMATION AROUND THE BOTTOM OF A CIRCULAR VERTICAL SHAFT	69-0323
AKIN, J. E.	A FINITE ELEMENT COLLOCATION SOLUTION OF DIFFERENTIAL EQUATIONS	72-0050
AKIN, J. E.	A LEAST SQUARES FINITE ELEMENT SOLUTION OF NONLINEAR OPERATORS	72-0060
AKIN, J. E.	ELEMENT RESEQUENCING FOR FRONTAL SOLUTIONS	75-0132
AKIN, J. E.	FINITE ELEMENT APPLICATIONS OF THE LEAST SQUARES METHOD	72-0052
AKIN, J. E.	THE FINITE ELEMENT METHOD - A BIBLIOGRAPHY OF ITS THEORY AND APPLICATIONS	71-0510
AKTAN, A. E.	RC COLUMN EARTHQUAKE RESPONSE IN TWO DIMENSIONS	74-0651
AKYUZ, A.	NATURAL COORDINATE SYSTEMS, AN AUTOMATIC INPUT DATA GENERATION SCHEME FOR A FINITE ELEMENT METHOD	70-0018
AKYUZ, F. A.	AN AUTOMATIC NODE RELABELLING SCHEME FOR BANDWIDTH MINIMIZATION OF STIFFNESS MATRICES	68-0256
AKYUZ, F. A.	ELAS - A GENERAL PURPOSE COMPUTER PROGRAM FOR THE EQUILIBRIUM PROBLEMS OF LINEAR STRUCTURES	68-0158
AKYUZ, F. A.	FEDGE - A GENERAL PURPOSE COMPUTER PROGRAM FOR FINITE ELEMENT DATA GENERATION. VOLUME 2 - PROGRAM MA	69-0492
AKYUZ, F. A.	FEDGE - A GENERAL-PURPOSE COMPUTER PROGRAM FOR FINITE ELEMENT DATA GENERATION. VOLUME 1 - USER'S MAN	69-0491
AKYUZ, F. A.	ON THE SOLUTION OF TWO-DIMENSIONAL PROBLEMS OF ELASTOPLASTICITY BY THE FINITE ELEMENT AND DIRECT STI	67-0012
AKYUZ, F. A.	SOLUTION OF NONLINEAR PROBLEMS OF ELASTOPLASTICITY BY THE FINITE ELEMENT METHOD	68-0007
AKYUZ, F. A.	VISCEL A GENERAL PURPOSE COMPUTER PROGRAM FOR ANALYSIS OF LINEAR VISCOELASTIC STRUCTURES, VOL. 2	72-0789
AL-ABDULLA, J.	SOME PLANE QUADRILATERAL HYBRID FINITE ELEMENTS	69-0082
AL-ADEEB, A. M.	A CURVED CIRCULAR CYLINDRICAL SHELL FINITE ELEMENT FOR HIGHER ORDER SHELL THEORY	69-0528
AL-DABBAGH, A.	FINITE ELEMENT METHOD FOR WOOD MECHANICS	71-0294
AL-DABBAGH, A.	FINITE ELEMENT STRESS ANALYSIS FOR ANISOTROPIC SOLIDS	70-0478
AL-HASHIMI, K.	FRACTURE IN SOIL CEMENT USING THE FINITE ELEMENT METHOD	74-0434
AL-HUSSAINI, M.	ANALYSIS OF PLANE STRAIN TEST USING THE FINITE ELEMENT METHOD	72-1002
AL-MAHAIDI, R.	COUPLED SHEAR WALL ANALYSIS BY LAGRANGE MULTIPLIERS	75-0202
AL-MAHAIDI, R.	COUPLED SHEAR WALLS. IMPROVED METHODS FOR ELASTIC ANALYSIS	74-0818
AL-MASHIDANI, G	FINITE ELEMENT APPROACH TO WATERSHED RUNOFF	74-0234
AL-MASHIDANI, G	AN ANALYSIS OF TWO-DIMENSIONAL SURFACE RUN-OFF BY FINITE ELEMENTS	74-1336
AL-MASHIDANI, G	DISPERSION IN STEADY UNSTEADY SEEPAGE FLOWFIELD	74-0072
AL-MASHIDANI, G	FINITE ELEMENT SOLUTIONS OF THE SHALLOW WATER EQUATIONS SURFACE RUN OFF	74-0046
AL-MASHIDANI, G	FINITE ELEMENT SOLUTION OF BOUSSINESQ'S EQUATION FOR UNSTEADY GROUNDWATER FLOW	74-0459
AL-NAJAFI, A. M	FREE VIBRATION OF THIN CYLINDRICAL SHELLS WITH A DISCONTINUITY IN THE THICKNESS	69-0439
AL-NAJAFI, A. M	FREE VIBRATION OF RING STIFFENED CYLINDRICAL SHELLS	70-0172
AL-RIFAIE, W. N	EXPERIMENTAL AND THEORETICAL INVESTIGATION OF THE BEHAVIOR OF BOX GIRDERS CURVED IN PLAN	75-0659
ALANI, K. G.	LEAST SQUARES FINITE ELEMENT ANALYSIS OF TWO-DIMENSIONAL LAMINAR BOUNDARY-LAYER EQUATION WITH SEPARA	75-0796
ALAYLIOGLU, H.	DEVELOPMENT AND APPLICATION OF HYBRID STRESS FINITE ELEMENTS TO AN AUTOMOTIVE STRUCTURE	74-1221
ALBRECHT, B.	WAVE PROPAGATION IN A THIN HOLLOW CONE BY A FINITE ELEMENT METHOD	72-0045
ALBRECHT, H.	FESTIGKEITSUNTERSUCHUNGEN UND-BERECHNUNGEN AN PUMPENSPIRALGEHAEUSEN. (INVESTIGATIONS AND CALCULATION	75-0238
ALBUQUERQUE, F.	A BEAM ELEMENT FOR USE WITH THE SEMILOOF SHELL ELEMENT	73-0620
ALDABBAGH, A.	FINITE ELEMENT METHOD FOR WOOD MECHANICS	71-0613
ALDSTEDT, E.	SESAM A PROGRAMMING SYSTEM FOR FINITE ELEMENT PROBLEMS	68-0276
ALDSTEDT, E.	SHELL ANALYSIS USING PLANAR TRIANGULAR ELEMENTS	69-0041
ALEWINE, R. W.	STUDY OF 1964 ALASKAN EARTHQUAKE USING FINITE ELEMENT AND GENERALIZED INVERSION TECHNIQUES	72-0003
ALEXANDER, J. H	3-D SEISMIC CODE FOR ILLIAC-1V	73-0868
ALEXANDER, J. M	PRELIMINARY INVESTIGATION OF THE DIE-LESS DRAWING OF TITANIUM AND SOME STEELS	75-0770
ALEXANDER, R. M	DYNAMIC STRAINS IN A FOUR BAR MECHANISM	73-0677
ALEXANDER, S. S	DYNAMIC FINITE ELEMENT METHOD CALCULATION OF ELASTIC WAVE PROPAGATION IN TWO DIMENSIONAL STRUCTURES	74-0166
ALEXANDER, S. S	EFFECTS OF SOURCE STRUCTURE AND SITE RESPONSE ON GROUND MOTION IN THE NEAR FIELD OF EARTHQUAKES	73-0732
ALEXANDER, S. S	INVERSION TECHNIQUE FOR STATIC FINITE ELEMENT METHOD	74-0170
ALI, R.	APPLICATION OF FINITE ELEMENT TECHNIQUES TO THE ANALYSIS OF AN AUTOMOBILE STRUCTURE	70-0136
ALI, R.	COMPUTER AIDED DESIGN APPLIED TO A MODEL OF A CHASSIS TYPE STRUCTURE USING FINITE ELEMENT TECHNIQUES	69-0421
ALI, R.	DYNAMIC ANALYSIS OF AN AUTOMOBILE CHASSIS FRAME	70-0135
ALI, R.	THE DEVELOPMENT OF A SERIES OF HYBRID-STRESS FINITE ELEMENTS	75-0335
ALLAN, T.	APPLICATION OF FINITE ELEMENT ANALYSIS TO HYDRODYNAMIC AND EXTERNALLY PRESSURIZED POCKET BEARINGS	72-0295
ALLAN, T.	THE EFFECT OF DEFORMATION ON THE BEHAVIOUR OF HYDRODYNAMIC JOURNAL BEARINGS	71-0009
ALLEN, F. C.	DEVELOPMENT OF A GRAPHITE HORIZONTAL STABILIZER	73-0931
ALLEN, J. J.	SIGNIFICANCE OF VARIABLY CONFINED TRIAXIAL TESTING ROADBUILDING MATERIALS	74-0554
ALLEN, J. J.	THE EFFECT OF STRESS HISTORY ON THE RESILIENT RESPONSE OF SOILS	73-0914
ALLEN, J. M.	NASTRAN ANALYSIS OF A TURBINE BLADE AND COMPARISON WITH TEST AND FIELD DATA	75-0367
ALLEN, M.	NASTRAN EXPERIENCES OF FORT WORTH OPERATION, CONVAIR AEROSPACE DIVISION OF GENERAL DYNAMICS	71-0019
ALLEN, M.	STRESS AND STRAIN DISTRIBUTION IN THE VICINITY OF INTERFERENCE FIT FASTENERS	73-0363
ALLEN, V. D.	DIRECT AND INVERSE HEAT CONDUCTION IN ROCK MATERIALS USING THE FINITE ELEMENT METHOD	74-1098
ALLGOOD, J. R.	BALANCED DESIGN AND FINITE-ELEMENT ANALYSIS OF CULVERTS.	72-0462
ALLGOOD, J. R.	RESPONSE OF BURIED CAPSULES IN THE HIGH-OVERPRESSURE REGION	70-0093
ALLGOOD, J. R.	SUMMARY OF SOIL-STRUCTURE INTERACTION	72-0817
ALLIK, H.	FINITE ELEMENT FORMULATION FOR PIEZOELECTRIC CONTINUA BY A VARIATIONAL THEOREM	68-0230
ALLIK, H.	FINITE ELEMENT METHOD FOR PIEZOELECTRIC VIBRATION	70-0012
ALLIK, H.	FINITE ELEMENTS FOR COMPRESSIBLE AND INCOMPRESSIBLE CONTINUA	69-0133
ALLIK, H.	VIBRATIONAL RESPONSE OF SONAR TRANSDUCERS USING PIEZO-ELECTRIC FINITE ELEMENTS	74-0522
ALLMAN, D. J.	CALCULATION OF THE ELASTIC BUCKLING LOADS OF THIN FLAT REINFORCED PLATES USING TRIANGULAR FINITE ELE	73-1024
ALLMAN, D. J.	CALCULATION OF THE ELASTIC BUCKLING LOADS OF THIN FLAT REINFORCED PLATES USING TRIANGULAR FINITE ELE	75-0177
ALLMAN, D. J.	FINITE ELEMENT ANALYSIS OF PLATE BUCKLING USING A MIXED VARIATIONAL PRINCIPLE	71-0020
ALLMAN, D. J.	TRIANGULAR FINITE ELEMENTS FOR PLATE BENDING WITH CONSTANT AND LINEARLY VARYING BENDING MOMENTS	70-0019
ALLMENDINGER, E	THE EVOLUTION OF SHELL STRUCTURES IN OCEAN TECHNOLOGY	69-0042
ALLOUARD, Y.	NUMERICAL STUDY OF TRANSITORY OR NONLINEAR WAVES	74-0031
ALLWOOD, R. J.	A POLYGONAL FINITE ELEMENT FOR PLATE BENDING PROBLEMS USING THE ASSUMED STRESS APPROACH	69-0043
ALLWOOD, R. J.	MATRIX METHODS OF STRUCTURAL ANALYSIS	71-0620
ALMOND, J.	FORMAT, A GENERAL PURPOSE MATRIX ANALYSIS SYSTEM IN FORTRAN	68-0266
ALMROTH, B. O.	EXTENSIONS TO THE STAGS COMPUTER CODE	73-0364
ALMROTH, B. O.	EXTENSIONS TO THE STAGS COMPUTER CODE	73-0917
ALMROTH, B. O.	FINITE-DIFFERENCE ENERGY METHOD FOR NONLINEAR SHELL ANALYSIS	71-0399
ALPAY, I. B.	ON RESPONSE OF INITIALLY STRESSED STRUCTURES TO RANDOM EXCITATIONS	73-0278
ALSPAUGH, W.	OPTIMUM CONFIGURATIONAL AND DIMENSIONAL DESIGN OF TRUSS STRUCTURES	74-0588
ALTMAN, W.	STABILITY OF PLATES USING A MIXED FINITE ELEMENT FORMULATION	74-0218
ALUJEVIC, A.	APPLICATION OF THREE-DIMENSIONAL FINITE ELEMENTS FOR ANALYSIS OF IRRADIATION INDUCED STRESSES IN NUC	72-0078

4

ALUJEVIC, A.	FINITE ELEMENT MATRIX DISPLACEMENT METHOD FOR TIME DEPENDENT STRESS ANALYSIS OF REACTOR MATERIALS IN	71-0674
ALUJEVIC, A.	FUEL ELEMENT STRESSES AT DISCONTINUITIES AND INTERACTIONS BY FINITE ELEMENTS IN TWO DIMENSIONS	73-0083
ALUJEVIC, A.	SOME MODELS FOR THE ANALYSIS OF STRESSES IN A TUBULAR FUEL PIN FOR HIGH TEMPERATURE REACTORS BY FINI	73-0816
ALWAY, G. G.	AN ALGORITHM FOR REDUCING THE BANDWIDTH OF A MATRIX OF SYMMETRICAL CONFIGURATION	65-0067
AMBA-RAO, C. L.	VIBRATION OF BEAMS WITH OVERHANGS	73-0488
AMEND, J. H. 11	A FINITE ELEMENT ANALYSIS OF DISSOLVED OXYGEN DRAWDOWN AND SULFATE PRODUCTION IN STRIP MINE SPOIL DA	75-0795
AMENIYA, E.	FLAT FINITE ELEMENT FOR THIN SHELL ANALYSIS DERIVED BY ASSUMED STRESS APPROACH	73-0147
AMMAR, A. R.	FINITE ELEMENT ANALYSIS OF METAL DECK SHEAR DIAPHRAGMS	74-0128
ANAND, S. C.	A FINITE ELEMENT MODEL FOR PLASTICITY PROBLEMS	74-1225
ANAND, S. C.	ELASTIC-PLASTIC ANALYSIS OF A WHEEL ROLLING ON A RIGID TRACK	74-1040
ANAND, S. C.	FINITE ELEMENT ANALYSIS OF ELASTIC-PLASTIC PLANE STRESS PROBLEMS BASED UPON TRESCA YIELD CRITERION	70-0179
ANAND, S. C.	FINITE ELEMENT STRESS ANALYSIS OF PROSTHETIC TOOTH IMPLANTS	75-0073
ANAND, S. C.	NONLINEAR ANALYSIS OF A ROLLING CONTACT PROBLEM BY FINITE ELEMENTS	74-1216
ANAND, S. C.	ON TRESCA YIELD CRITERION IN FINITE ELEMENT ANALYSIS OF PLANE STRESS PROBLEMS	73-0314
ANAND, S. C.	SHAKEDOWN OF ROLLING WHEEL UNDER HUB LOADING	74-0663
ANAND, S. C.	STRESS DISTRIBUTIONS AROUND SHALLOW BURIED RIGID PIPES	74-0142
ANAND, S. C.	STRUCTURAL INELASTICITY. V11. ELASTIC-PLASTIC ANALYSIS OF A WHEEL ROLLING ON A RIGID TRACK	73-0886
ANAND, S. C.	STRUCTURAL INELASTICITY. V1. A FINITE ELEMENT METHOD FOR PLASTICITY PROBLEMS	73-0887
ANANEV, I. N.	SOLUTION OF THE PROBLEM OF PROCESSING METALS BY PRESSURE USING THE METHOD OF FINITE ELEMENTS	74-1363
ANDERHEGGEN, E.	A CONFORMING TRIANGULAR FINITE ELEMENT PLATE BENDING SOLUTION	70-0020
ANDERHEGGEN, E.	FINITE ELEMENT LIMIT ANALYSIS USING LINEAR PROGRAMMING.	72-0461
ANDERHEGGEN, E.	FINITE ELEMENT PLATE BENDING EQUILIBRIUM ANALYSIS	69-0045
ANDERHEGGEN, E.	PROGRAMME ZUR METHODE DER FINITEN ELEMETE	69-0416
ANDERHEGGEN, E.	STARR-PLASTISCHE TRAGLASTBERECHNUNGEN MITTELS DER METHODE DER FINITEN ELEMENTE	71-0635
ANDERSEN, C. M.	USE OF COMPUTERIZED SYMBOLIC INTEGRATION IN FINITE ELEMENT DEVELOPMENT	74-0520
ANDERSEN, C. M.	USE OF GROUP-THEORETIC METHODS IN THE DEVELOPMENT OF NONLINEAR SHELL FINITE ELEMENTS	71-0704
ANDERSEN, D. W.	LAPLACIAN ELECTROSTATIC FIELD CALCULATIONS BY FINITE ELEMENTS WITH AUTOMATIC GRID GENERATION	73-0084
ANDERSEN, K. H.	ANALYSIS OF GEOTECHNICAL PROBLEMS BY MEANS OF THE FINITE ELEMENTS METHOD	72-0376
ANDERSEN, O. W.	TRANSFORMER LEAKAGE FLUX PROGRAM BASED ON FINITE ELEMENT METHOD	73-0000
ANDERSEN, S. I.	THE SCANDINAVIAN PORV MODEL PROJECT: STRESS CALCULATION AND EXPERIMENTAL VERIFICATION	71-0021
ANDERSON, C. A.	CAVITY STABILITY: FINITE ELEMENT ANALYSIS FOR STEADY AND TRANSIENT CREEP	74-1082
ANDERSON, C. A.	SPONTANEOUS IGNITION - FINITE ELEMENT SOLUTIONS FOR STEADY AND TRANSIENT CONDITIONS	74-0353
ANDERSON, C. A.	TSAAS: FINITE ELEMENT THERMAL AND STRESS ANALYSIS OF AXISYMMETRIC SOLIDS WITH ORTHOTROPIC TEMPERATUR	74-0886
ANDERSON, C. M.	MIXED ISOPARAMETRIC LAMINATED COMPOSITE SHELL ELEMENTS	75-0450
ANDERSON, C. M.	USE OF SYMBOLIC MANIPULATION IN DEVELOPMENT OF TWO-DIMENSIONAL FINITE ELEMENTS	74-0103
ANDERSON, D. A.	MODELING OF GAS TURBINE ENGINE COMPRESSOR BLADES FOR VIBRATION ANALYSIS	75-0508
ANDERSON, D. L.	FINITE ELEMENT METHOD: A GALERKIN APPROACH	71-0451
ANDERSON, D. M.	FRACTURE TOUGHNESS PARAMETERS AND ELASTIC-PLASTIC ANALYSIS OF NON-MODERATE FRACTURE CONDITIONS USING	73-0258
ANDERSON, D. T.	DYNAMIC ANALYSIS OF A CAR CHASSIS FRAME USING FINITE ELEMENT METHOD	72-0265
ANDERSON, F. A.	FOUNDATION TESTING FOR AUBURN DAM	70-0119
ANDERSON, G. P.	FINITE ELEMENT IN ADHESION ANALYSES	73-0355
ANDERSON, G. P.	FINITE ELEMENT IN NUMERICAL ANALYSIS OF ADHESIVE FRACTURE	73-0085
ANDERSON, G. P.	USE OF FINITE ELEMENT COMPUTER PROGRAMS IN FRACTURE MECHANICS	71-0022
ANDERSON, H.	A GENERAL METHOD OF ANALYSIS FOR TWO-DIMENSIONAL PILE GROUPS	71-0298
ANDERSON, J. C.	FINITE ELEMENT APPROACH TO HYDROTHERMAL ANALYSIS OF SMALL LAKES	72-0363
ANDERSON, J. C.	HYDROTHERMAL ANALYSIS BY FINITE ELEMENT METHOD	72-0023
ANDERSON, J. C.	INELASTIC RESPONSE OF NUCLEAR PIPING SUBJECTED TO RUPTURE FORCES	75-0515
ANDERSON, J. C.	SEISMIC BEHAVIOUR OF FRAMED TUBES	75-0306
ANDERSON, J. C.	SEISMIC RESPONSE EFFECTS ON EMBEDDED STRUCTURES	72-0902
ANDERSON, J. C.	TRANSIENT HYDROTHERMAL ANALYSIS OF SMALL LAKES	73-0445
ANDERSON, J. M.	CASE BOND STRESS CALCULATIONS FOR FLAPPED CYLINDRICAL ANALOGS OF SOLID PROPELLANT ROCKET MOTORS. BON	72-0842
ANDERSON, J. M.	FINITE ELEMENT ANALYSIS OF CRACKED STRUCTURES SUBJECTED TO SHOCK LOADS	75-0258
ANDERSON, R. G.	A FINITE ELEMENT EIGENVALUE SYSTEM	68-0192
ANDERSON, R. G.	THE BEHAVIOUR OF SQUARE SHEAR WEBS HAVING A CIRCULAR HOLE	67-0108
ANDERSON, R. G.	VIBRATION AND STABILITY OF PLATES USING FINITE ELEMENTS	68-0244
ANDERSON, R. G.	VIBRATION OF THICK, CURVED, SHELLS WITH PARTICULAR REFERENCE TO TURBINE BLADES	70-0017
ANDERSON, R. V.	FUEL ROD FINITE ELEMENT STRESS ANALYSIS: SEGMENTED FLIGHT REACTOR FUEL VERSUS S8DR FUEL	71-0798
ANDERSON, S. I.	ULTIMATE LOAD BEHAVIOUR OF PCRV TOP CLOSURES. THEORETICAL AND EXPERIMENTAL INVESTIGATIONS	73-0835
ANDERSON, T. L.	THREE-DIMENSIONAL STRESS ANALYSES OF MITERED ELBOWS	75-0650
ANDERSON, W. E.	FRACTURE MECHANICS OF TINY CRACKS NEAR FASTENERS	74-0735
ANDERSON, W. J.	WHIRLING CABLE SUBJECTED TO VISCOUS DRAG	74-0457
ANDERSSON, H.	FINITE ELEMENT REPRESENTATION OF STABLE CRACK-GROWTH	73-0086
ANDERSSON, H.	FINITE ELEMENT TREATMENT OF A UNIFORMLY MOVING ELASTIC-PLASTIC CRACK TIP	74-0470
ANDERSSON, H.	FINITE-ELEMENT ANALYSIS OF A FRACTURE TOUGHNESS TEST SPECIMEN IN NON-LINEAR RANGE	72-0093
ANDERSSON, H.	STEADILY GROWING ELASTIC-PLASTIC CRACK TIP IN A FINITE ELEMENT TREATMENT	73-0087
ANDERTON, G. L.	ANALYSIS OF THE 747 AIRCRAFT WING BODY INTERSECTION	68-0275
ANDO, R.	SIMPLIFIED HYBRID DISPLACEMENT METHOD FOR LINEAR FINITE ELEMENT ANALYSIS OF GENERAL SHELLS	74-0614
ANDO, Y.	A FINITE ELEMENT METHOD FOR INITIAL VALUE PROBLEMS	71-0145
ANDO, Y.	A FINITE ELEMENT METHOD FOR FRIEDRICH'S SYMMETRIC POSITIVE SYSTEMS	71-0146
ANDO, Y.	ANALYSIS FOR CRACKS IN SHIP STRUCTURE USING THE FINITE ELEMENT METHOD WITH THE CONCEPT OF CRACK SUB-	74-0248
ANDO, Y.	ANALYSIS OF ELASTIC-PLASTIC CREEP BUCKLING OF AXISYMMETRIC SHELLS BY FINITE-ELEMENT METHOD	75-0355
ANDO, Y.	ANALYSIS OF LARGE ELASTIC-PLASTIC CREEP DEFORMATIONS OF AXISYMMETRIC SHELLS BY THE FINITE ELEMENT ME	74-0563
ANDO, Y.	APPLICATION OF FINITE ELEMENT METHOD TO AXISYMMETRIC BUCKLING OF SHALLOW SPHERICAL SHELLS UNDER EXTE	73-0111
ANDO, Y.	APPLICATION OF SIMPLIFIED HYBRID DISPLACEMENT METHOD TO PLATE AND SHELL PROBLEMS	73-1037
ANDO, Y.	APPLICATION OF THE FINITE ELEMENT METHOD TO THE ANALYSIS OF CYLINDRICAL SHELLS	73-0828
ANDO, Y.	CONVERGENCE OF A MIXED FINITE ELEMENT SCHEME FOR PLATE BENDING	73-0023
ANDO, Y.	CONVERGENCE OF LUMPED FINITE ELEMENT SCHEMES FOR SELECTED INITIAL VALUE PROBLEMS	73-1026
ANDO, Y.	CONVERGENCE OF SIMPLIFIED HYBRID DISPLACEMENT METHOD FOR PLATE BENDING	72-0412
ANDO, Y.	FINITE ELEMENT CREEP BUCKLING ANALYSIS OF VISCO-ELASTIC SHALLOW SPHERICAL SHELL	75-0159
ANDO, Y.	FINITE ELEMENT LARGE DEFLECTION ANALYSIS OF ELASTIC-PLASTIC SHELLS OF REVOLUTION SUBJECTED TO AXISYM	72-0137
ANDO, Y.	FINITE ELEMENT SOLUTIONS FOR PLATE BENDING PROBLEMS BY SIMPLIFIED HYBRID DISPLACEMENT METHOD	72-0780
ANDO, Y.	LOW CYCLE FATIGUE STRENGTH OF PIPING COMPONENTS	73-0263
ANDO, Y.	LUMPED FINITE ELEMENT BASES FOR BEAMS AND PLATES	72-0263
ANDO, Y.	NEW VARIATIONAL FUNCTIONAL FOR THE FINITE-ELEMENT METHOD AND ITS APPLICATION TO PLATE SHELL PROBLEMS	72-0108
ANDO, Y.	ON THE CONVERGENCE OF FINITE ELEMENT SCHEMES BASED ON NON-CONFORMING AND SIMPLIFIED HYBRID DISPLACEM	72-0841
ANDO, Y.	SIMPLIFIED HYBRID DISPLACEMENT METHOD APPLIED TO PLATE BUCKLING PROBLEMS	72-0781
ANDO, Y.	SNAP-THROUGH BUCKLING ANALYSIS OF SHELLS OF REVOLUTION USING INCREMENTAL STIFFNESS MATRICES	74-1349
ANDO, Y.	SOME FINITE ELEMENT SOLUTIONS FOR PLATE BENDING PROBLEMS BY SIMPLIFIED HYBRID DISPLACEMENT METHOD	72-0268
ANDO, Y.	STRESS DISTRIBUTIONS IN THIN-WALLED INTERSECTING CYLINDRICAL SHELLS SUBJECTED TO INTERNAL PRESSURE A	71-0023
ANDO, Y.	THE APPLICATION OF THE FINITE ELEMENT METHOD TO THE ANALYSIS OF FRACTURE OF CYLINDRICAL SHELLS	73-0728
ANDO, Y.	THE FINITE ELEMENT CALCULATION OF STRESS INTENSITY FACTORS USING SUPERPOSITION	75-0886
ANDO, Y.	THREE-DIMENSIONAL FINITE ELEMENT METHOD OF THERMOELASTOPLASTICITY WITH CREEP EFFECT	73-0837
ANDRESEN, K.	BERECHNUNG DES KEGELDRUCKVERSUCHES MIT EINER FINITEN-ELEMENT-METHODE. (CALCULATION OF THE CONE INDENT	75-0187
ANDRESEN, K.	CALCULATION OF CONE INDENTATION TEST USING A FINITE-ELEMENT METHOD	75-0438
ANDREWS, W. R.	ON THE USE OF FINITE ELEMENTS AND COMPLIANCE CALCULATIONS FOR DETERMINING CRACK TIP STRESS INTENSITY	69-0448
ANEJA, I. K.	EXPERIMENTAL AND ANALYTICAL INVESTIGATION OF A HORIZONTALLY CURVED BOX-BEAM HIGHWAY BRIDGE MODEL	71-0024
ANEJA, I. K.	MODEL ANALYSIS OF CURVED BOX-BEAM HIGHWAY BRIDGE	71-0282
ANG, A. H.	DISCRETE MODEL ANALYSIS OF ELASTIC-PLASTIC PLATES	68-0009
ANGELOPOULOS, T	DIE LOSUNG NICHT-LINEARER PROBLEME NACH DER METHODE DER FINITEN ELEMENTE	74-0673
ANGELOPOULOS, T	A GENERAL METHOD FOR THE SHAPE FINDING OF LIGHTWEIGHT TENSION STRUCTURES	74-1217
ANGELOPOULOS, T	DYNAMIC RESPONSE BY LARGE STEP INTEGRATION	73-0567
ANGELOPOULOS, T	EIN VERFAHREN DUR DIE FORMFINDUNG VEON BELIEBIGEN VORGESPANNTEN NETZWERKKONSTRUKTIONEN	72-0641

ANGELOPOULOS, T	FINITE ELEMENT DIFFICULTIES IN NONLINEAR AND INCOMPRESSIBLE MATERIAL	74-1242
ANGELOPOULOS, T	LINEAR AND NONLINEAR OSCILLATIONS IN MECHANICS	74-0677
ANGELOPOULOS, T	NONLINEAR OSCILLATIONS USING THE FINITE ELEMENT TECHNIQUE	73-0568
ANGELOPOULOS, T	THEORIE, PROGRAMMENTWICKLUNG UND ERFAHRUNG ANVORGESPANNTEN NETZWERKKONSTRUKTIONEN	72-0640
ANON	A FINITE ELEMENT METHOD FOR THE PLASTIC BUCKLING ANALYSIS OF PLATES	69-0390
ANON	ANALYSIS FOR CRACKS IN SHIP STRUCTURES USING THE FINITE-ELEMENT METHOD WITH THE CONCEPT OF CRACK SUB	74-1367
ANON	ASKA - AUTOMATIC SYSTEM FOR KINEMATIC ANALYSIS	65-0059
ANON	AXISYMMETRIC RING AND PLATE STRESS ANALYSIS PROGRAM - USERS GUIDE	70-0333
ANON	BOX CAR BODY BOLSTER PROBED IN STRESS STUDY	74-0654
ANON	BULLETIN DE L ASSOCIATION TECHNIQUE MARITIME ET AERONAUTIQUE (BULLETIN OF THE TECHNICAL MARITIME AND	74-1108
ANON	CALCULATION OF ELASTIC-PLASTIC DEFORMATION BY THE FINITE ELEMENT METHOD OF NUMERICAL ANALYSIS	71-0793
ANON	CALCULATION OF STRESS INTENSITY FACTORS USING THE FINITE ELEMENT METHOD WITH CRACKED ELEMENTS	70-0475
ANON	CATCHING WAVES WITH COMPUTERS	74-0280
ANON	COMPUTATIONAL METHODS FOR INVISCID AND VISCOUS TWO-AND-THREE DIMENSIONAL FLOW FIELDS	75-0716
ANON	CONF. ON MATRIX METHODS IN STRUCTURAL MECHANICS	68-0284
ANON	CONFERENCE ON APPLICATION OF FINITE ELEMENT METHODS TO STRESS ANALYSIS PROBLEMS IN NUCLEAR ENGINEERI	71-0774
ANON	CONFERENCE ON MATRIX METHODS IN STRUCTURAL MECHANICS	65-0065
ANON	CONFERENCE ON MATRIX METHODS IN STRUCTURAL MECHANICS	71-0773
ANON	DESIGN OPTIMIZATION IN UNDERGROUND COAL SYSTEMS	74-1095
ANON	DEVELOPMENTS IN DISCRETE ELEMENT FINITE DEFLECTION STRUCTURAL ANALYSIS BY FUNCTION MINIMIZATION	68-0243
ANON	DEVELOPMENTS IN STRUCTURAL ANALYSIS AND SYNTHESIS BY FUNCTION MINIMIZATION	71-0384
ANON	EAC/EASE USER INFORMATION MANUAL	70-0320
ANON	FINITE ELEMENT AND CUMULATIVE DAMAGE ANALYSIS OF A KEYHOLE TEST SPECIMEN	75-0774
ANON	FINITE ELEMENT COMPUTER ANALYSIS OF KOCHMUS HULL 520.	70-0293
ANON	FINITE ELEMENT COMPUTER ANALYSIS OF MITSUBISHI HULL 1670	70-0294
ANON	FINITE ELEMENT METHOD FOR LINEAR FRACTURE MECHANICS	74-1359
ANON	FINITE ELEMENT METHODS	75-0876
ANON	FINITE ELEMENT SOLUTION OF STEADY STATE POTENTIAL FLOW PROBLEMS	70-0480
ANON	FINITE ELEMENTS FOR DETERMINATION OF CRACK-TIP ELASTIC STRESS INTENSITY FACTORS	71-0785
ANON	FINITE-ELEMENT METHOD FOR CALCULATING THE CONF-INDENTATION TEST	75-0813
ANON	FRACTURE TOUGHNESS PARAMETERS AND ELASTIC-PLASTIC ANALYSIS OF NON-MODERATE FRACTURE CONDITIONS USING	73-1126
ANON	GENERAL PURPOSE PROGRAM FOR SHELL STRUCTURES: ISTRAN-S	75-0511
ANON	LARGE STRUCTURES FOR MANNED SPACECRAFT: MATHEMATICAL ANALYSIS, DESIGN, CONSTRUCTION AND TESTS	74-1327
ANON	MACHINERY SPACE DEFLECTION AND VIBRATION ANALYSES USING FINITE ELEMENT METHODS	72-0474
ANON	MRI/STARDYNE USER INFORMATION MANUAL	73-0541
ANON	NASTRAN: USER'S EXPERIENCES	75-0675
ANON	NAVY-NASTRAN COLLOQUIM AT WASHINGTON, D. C.	72-0669
ANON	NOTE ON THE FINITE ELEMENT METHOD IN LINEAR FRACTURE MECHANICS	70-0474
ANON	PROGRAM TO REDUCE THE SIZE OF STRUCTURAL MATRICES	72-0912
ANON	REPORT OF INTERAGENCY CONFERENCE ON COORDINATION OF RESEARCH ACTIVITIES	73-1100
ANON	REPORT ON POST CONF. MEETING ON COMPUTATIONAL ASPECTS OF ASPECTS OF FINITE ELEMENT METHOD ISD STUTTG	72-0475
ANON	RUBBER IN ENGINEERING	73-0367
ANON	SHIP STRUCTURAL ANALYSIS IN LLOYD'S REGISTER OF SHIPPING	75-0285
ANON	SNAP DYNAMICS	72-0911
ANON	SOLVING THE PROBLEM OF PROCESSING METALS BY PRESSURE THROUGH THE METHOD OF FINITE ELEMENTS	74-1354
ANON	SOME ASPECTS OF NON-LINEAR FINITE ELEMENT COMPUTER PROGRAMS	75-0778
ANON	SP-222 NASTRAN THEORETICAL MANUAL	70-0387
ANON	SP-222 NASTRAN USERS MANUAL	70-0386
ANON	STIFFNESS DERIVATIVE FINITE ELEMENT TECHNIQUE FOR DETERMINATION OF CRACK TIP STRESS INTENSITY FACTOR	74-1358
ANON	STRESS ANALYSIS FOR CLEAN LEAN DESIGN	71-0386
ANON	STRUCTURAL VIBRATION - A MATRIX APPROACH	71-0385
ANON	THREE-DIMENSIONAL FINITE ELEMENT ANALYSIS FOR FRACTURE MECHANICS	72-0921
ANON	TRANSVERSE STRENGTH AT FORE AND AFT SECTION OF TANKER CARGO HOLD	74-0618
ANON	USER INFORMATION MANUAL	73-0543
ANON	USING FINITE ELEMENTS IN AUTOMOTIVE DESIGN	74-0104
ANON.	THERMOMECHANICAL ANALYSIS OF THE WELDING PROCESS USING THE FINITE ELEMENT METHOD	75-0776
ANQUEZ, L.	VIBRATIONS OF TANKS PARTIALLY FILLED WITH LIQUIDS	74-0089
ANSOURIAN, P.	APPLICATION OF METHOD OF FINITE ELEMENTS TO ANALYSIS OF COMPOSITE FLOOR SYSTEMS	75-0418
ANSOURIAN, P.	FACTOR OF SAFETY APPROACH FOR EVALUATING SEISMIC STABILITY OF SLOPES	75-0645
ANTES, H.	UEBER PAUSCHALE FEHLERSCHRANKEN FUER APPROXIMATIONEN VON SCHALENVERFORMUNGEN. (GLOBAL BOUNDARY CONDI	75-0292
AOKI, I.	APPLICATION OF FINITE-ELEMENT METHOD AND MOIRE TECHNIQUE TO SHEET-METAL FORMING PROBLEMS	71-0784
AOYAMA, S.	FINITE ELEMENT ANALYSIS OF SOME MECHANICAL PROBLEMS OF HOLLOW GRAVITY DAM	70-0470
AOYAMA, T.	DYNAMIC BEHAVIOR OF HYDROSTATIC THRUST BEARING	75-0189
APERGHIS, G. G.	SLAB BRIDGE DESIGN AND DRAWING. AN AUTOMATED PROCESS	70-0347
APOSTAL, M. C.	DEVELOPMENT OF AN ANISOTROPIC SINGULARITY FINITE ELEMENT UTILIZING THE HYBRID DISPLACEMENT METHOD	74-1099
APOSTOL, R. T.	SOLUTION OF A TWO DIMENSIONAL DIFFUSION CONVECTION EQUATION DESCRIBING THE MASS TRANSFER OF A NEUTRA	74-0037
APPA RAO, T. V.	AN ASSUMED STRESS HYBRID FINITE ELEMENT MODEL FOR THE ANALYSIS OF AN AXISYMMETRIC THICK WALLED PRESS	71-0025
APPA, K.	FINITE ELEMENT APPROACH TO THE INTEGRATED POTENTIAL FORMULATION OF GENERAL UNSTEADY SUPERSONIC AERODO	73-0910
APPA, K.	KINEMATICALLY CONSISTENT UNSTEADY AERODYNAMIC COEFFICIENTS IN SUPERSONIC FLOW	73-0316
APPERT, K.	A USEFUL FORM OF THE VARIATIONAL PRINCIPLE OF IDEAL MHD FOR ONE-DIMENSIONAL NUMERICAL STABILITY CALC	73-0746
APPERT, K.	APPLICATION OF METHOD OF FINITE ELEMENTS TO ANALYSIS OF COMPOSITE FLOOR SYSTEMS	75-0418
APPERT, K.	DETERMINATION OF EIGENMODES OF HOMOGENEOUS PLASMA WITH ELLIPTICAL SECTION USING BI-DIMENSIONAL FINIT	74-0553
APPERT, K.	EIGENMODES OF CYLINDRICAL PLASMA USING FINITE ELEMENT METHOD	74-0174
APPERT, K.	NEW FINITE ELEMENT APPROACH TO THE NORMAL MODE ANALYSIS IN MAGNETOHYDRODYNAMICS	74-0876
APPERT, K.	NUMERICAL STUDY OF EIGEN OSCILLATIONS OF A SCREW PINCH WITH FINITE ELEMENTS	74-0159
APPERT, K.	NUMERICAL STUDY OF PROPER MODES OF A HOMOGENEOUS CYLINDRICAL PLASMA BY METHOD OF FINITE ELEMENTS	74-0160
APPERT, K.	THALIA - ONE-DIMENSIONAL MAGNETOHYDRODYNAMIC STABILITY PROGRAM USING METHOD OF FINITE-ELEMENTS	75-0458
APPERT, K.	THALIA: A ONE-DIMENSIONAL MAGNETOHYDRODYNAMIC STABILITY PROGRAM USING THE METHOD OF FINITE ELEMENTS	74-1085
APPERT, K.	USEFUL FORM OF THE VARIATIONAL PRINCIPLE OF IDEAL MHD FOR ONE-DIMENSIONAL NUMERICAL STABILITY CALCUL	73-0946
APRAHAMIAN, R.	HOLOGRAPHIC INSTRUMENTATION OF TURBINE BLADES	73-0466
ARAI, H.	ANALYSIS ON THE LARGE DEFORMATION OF PLATE STRUCTURES	74-0435
ARAI, H.	APPLICATION OF FINITE ELEMENT METHOD TO THE STUDY OF STRENGTH OF CAR BODY STRUCTURE AND WHEEL	74-0281
ARAI, H.	FINITE ELEMENT SIMULATION OF ENERGY-TRAPPED ELECTROMECHANICAL RESONATORS	75-0142
ARAI, K.	ANALYSIS OF OUT-OF-PLANE STRESS BY THE SIX NODES, FOURTH ORDER ACCURACY TRIANGULAR ELEMENTS	69-0375
ARAI, K.	FINITE ELEMENT METHOD WITH CONSIDERATION TO THE CONTINUITY OF NORMAL SLOPE	72-0752
ARAI, K.	SHELL ANALYSIS BY THE FINITE ELEMENT METHOD	69-0376
ARAL, K.	DATA GENERATION FOR A 3-D FINITE ELEMENT SYSTEM	74-0361
ARAL, K.	DATA GENERATION FOR A 3-D FINITE ELEMENT SYSTEM	74-0475
ARAL, M. M.	APPLICATION OF FINITE ELEMENT ANALYSIS IN FLUID MECHANICS	71-0026
ARAL, M. M.	FINITE ELEMENT GALERKIN METHOD SOLUTIONS TO SELECTED ELLIPTIC AND PARABOLIC DIFFERENTIAL EQUATIONS	71-0027
ARAL, M. M.	FINITE ELEMENT SOLUTION OF PARTIAL DIFFERENTIAL EQUATIONS FEMAC COMPUTER PROGRAM	73-0648
ARAL, M. M.	FINITE ELEMENT SOLUTIONS OF SELECTED PARTIAL DIFFERENTIAL EQUATIONS "FEMAC COMPUTER PROGRAM"	74-0327
ARAL, M. M.	FINITE ELEMENT SOLUTION OF PROBLEMS WITH UNKNOWN AND MOVING BOUNDARIES	74-0460
ARAL, M. M.	FINITE ELEMENT SOLUTION OF A NON-LINEAR DIFFERENTIAL EQUATION USING KACHANOV'S AND SUCCESSIVE APPROX	75-0108
ARALDSEN, P. O.	EXAMPLE OF LARGE-SCALE STRUCTURAL ANALYSIS. COMPARISON BETWEEN FINITE ELEMENT CALCULATION AND FULL S	74-0177
ARALDSEN, P. O.	GENERAL DESCRIPTION OF SESAM-69, SUPER ELEMENT STRUCTURAL ANALYSIS (PROGRAM) MODULES	71-0612
ARALDSEN, P. O.	ORSAM - A PROGRAMMING SYSTEM FOR THE FINITE ELEMENT METHOD	73-0528
ARALDSEN, P. O.	SESAM 69 A GENERAL PURPOSE FINITE ELEMENT METHOD PROGRAM CDC/NASTRAN APPLICATIONS MANUAL	72-0552
ARALDSEN, P. O.	SESAM-69 - A GENERAL PURPOSE FINITE ELEMENT METHOD PROGRAM	74-0183
ARALDSEN, P. O.	THE APPLICATION OF THE SUPER ELEMENT METHOD IN ANALYSIS AND DESIGN OF SHIP STRUCTURES AND MACHINERY	72-0657
ARANY, I.	AN IMPROVED METHOD FOR REDUCING THE BANDWIDTH OF SPARSE SYMMETRIC MATRICES	72-0990
ARAV, F.	THEORETICAL AND EXPERIMENTAL INVESTIGATION ON A STEAM GENERATOR TUBE PLATE	69-0424

ARBIC, B. J.   NUMERICAL SOLUTION OF SOME NONSTATIONARY PARTIAL DIFFERENTIAL EQUATIONS: THE FINITE ELEMENT METHOD A   72-0930
ARBON, P. F.   FINITE ELEMENT STUDY OF COMPOSITE ACTION IN WALLS   69-0291
ARCHAMBE, C. B.   DISPLACEMENT AND STRAIN FIELDS FROM UNDERGROUND EXPLOSIONS USING ANALYTICAL AND FINITE ELEMENT METHO   72-0587
ARCHER, J. S.   A STIFFNESS MATRIX METHOD OF NATURAL MODE ANALYSIS   58-0002
ARCHER, J. S.   A TWO-DIMENSIONAL AXISYMMETRIC SHELL FLUID MODEL FOR LAUNCH VEHICLE LONGITUDINAL DYNAMIC ANALYSIS   69-0046
ARCHER, J. S.   CONSISTENT MASS MATRIX FOR DISTRIBUTED SYSTEMS   63-0001
ARCHER, J. S.   CONSISTENT MATRIX FORMULATIONS FOR STRUCTURAL ANALYSIS USING INFLUENCE COEFFICIENT TECHNIQUES   64-0017
ARCHER, J. S.   DYNAMIC ANALYSIS   75-0171
ARCHER, J. S.   IMPROVED ANALYTIC LONGITUDINAL RESPONSE ANALYSIS FOR AXISYMMETRIC LAUNCH VEHICLES, COMPUTER PROGRAM   65-0002
ARCHER, J. S.   IMPROVED LINEAR AXISYMMETRIC SHELL FLUID MODEL FOR LAUNCH VEHICLE LONGITUDINAL RESPONSE ANALYSIS   65-0044
ARCHER, J. S.   WAVEGUIDE SOLUTIONS BY THE FINITE ELEMENT METHOD   69-0011
ARGYRIS, J. H.   A FULLY COMPATIBLE PLATE BENDING ELEMENT   67-0142
ARGYRIS, J. H.   A GENERAL METHOD FOR THE SHAPE FINDING OF LIGHTWEIGHT TENSION STRUCTURES   74-1217
ARGYRIS, J. H.   A LOOK INTO THE FUTURE: HOW COMPUTERS WILL INFLUENCE ENGINEERING   67-0139
ARGYRIS, J. H.   A SEQUEL TO TECHNICAL NOTE 14 ON THE TUBA FAMILY OF PLATE ELEMENTS   68-0015
ARGYRIS, J. H.   A SEQUEL TO TECHNICAL NOTE 13 - THE CURVED TETRAHEDRONAL AND TRIANGULAR ELEMENTS TEC AND TRIC FOR TH   69-0048
ARGYRIS, J. H.   A SEQUEL TO TECHNICAL NOTE 15: THE SHEBA FAMILY OF SHELL ELEMENTS FOR THE MATRIX DISPLACEMENT METHOD   69-0444
ARGYRIS, J. H.   A SURVEY OF THE APPLICATION OF FINITE ELEMENT METHODS TO STRESS ANALYSIS PROBLEMS WITH PARTICULAR EM   71-0632
ARGYRIS, J. H.   A TAPERED THICKNESS TRIM 6 ELEMENT FOR THE MATRIX DISPLACEMENT METHOD   66-0007
ARGYRIS, J. H.   ANWENDUNG DER METHODEN DER FINITEN ELEMENTE MIT DEM PROGRAMMSYSTEM ASKA UNTER BESONDERER BERUCKSICHT   71-0624
ARGYRIS, J. H.   APPLICATION OF NATURAL KINEMATIC MODES TO MATRIX ANALYSIS: SMALL AND LARGE DISPLACEMENTS AND STRAINS   67-0138
ARGYRIS, J. H.   APPLICATION OF THE MATRIX DISPLACEMENT METHOD TO THE ANALYSIS OF PRESSURE VESSELS   69-0051
ARGYRIS, J. H.   APPLICATIONS OF FINITE ELEMENTS IN SPACE AND TIME   72-0147
ARGYRIS, J. H.   ARBITRARY QUADRILATERAL SPAR WEBS FOR THE MATRIX DISPLACEMENT METHOD   66-0003
ARGYRIS, J. H.   ASKA AUTOMATIC SYSTEM FOR KINEMATIC ANALYSIS   69-0307
ARGYRIS, J. H.   ASKA, SOFTWARE ENGINEERING FOR THE SEVENTIES IN STRUCTURAL ANALYSIS   73-0704
ARGYRIS, J. H.   ASPECTS OF THE FINITE ELEMENT METHOD AS APPLIED TO AERO-SPACE STRUCTURES   72-0649
ARGYRIS, J. H.   AUTOMATIC SYSTEM FOR KINEMATIC ANALYSIS, NASKA   65-0054
ARGYRIS, J. H.   BERECHNUNG RAUMLICHER SPANNUNGSVERTEILUNGEN IN STAUDAMMEN NACH DER MATRIZENVERSCHIEBUNGSMETHODE   67-0003
ARGYRIS, J. H.   BERECHNUNG VON SPANNUNGSKONZENTRATIONEN AN TRAGWERKEN UNTER THERMO-MECHANISCHER BELASTUNG   73-0686
ARGYRIS, J. H.   BERECHNUNG VORGESPANNTER NETZWERKE   70-0399
ARGYRIS, J. H.   BOOLEAN MATRICES AS A STANDARD FACILITY IN A MATRIX CODE   65-0057
ARGYRIS, J. H.   CALCULATION OF STRESS CONCENTRATIONS ON SUPPORT STRUCTURES UNDER THERMOMECHANICAL LOADING (BERECHNUN   73-0796
ARGYRIS, J. H.   CALCULATIONS FOR PRESTRESSED CONCRETE PRESSURE VESSELS   71-0643
ARGYRIS, J. H.   COMPUTER AIDED STRUCTURAL ANALYSIS: THE MACHINE INDEPENDENT SYSTEM ASKA   70-0397
ARGYRIS, J. H.   CONTINUA AND DISCONTINUA   66-0008
ARGYRIS, J. H.   DESIGN AND CALCULATION WITH FINITE ELEMENTS   74-0161
ARGYRIS, J. H.   DIE ANWENDUNG DER MATRIZENVERSCHIEBUNGSMETHODE ZUM BERECHNEN VON SPANNUNGSKONZENTRATIONEN   68-0208
ARGYRIS, J. H.   DIE ANWENDUNG VON INTERAKTIVER COMPUTER-GRAPHIK BEI DER BERECHNUNG VON TRAGWERKEN. (APPLICATION OF I   72-0102
ARGYRIS, J. H.   DIE BERECHNUNG ACHSENSY METRISCHER TRAGWERKE UNTER ACHSENSYMETRISCHER UND NIGHT-SYMMETRISHCER BELAST   69-0452
ARGYRIS, J. H.   DIE ELASTO-PLASTISCHE BERECHNUNG VON ALLGEMEINEN TRAGWERKEN UND KONTINUA   69-0443
ARGYRIS, J. H.   DIE LOSUNG NICHT-LINEARER PROBLEME NACH DER METHODE DER FINITEN ELEMENTE   74-0673
ARGYRIS, J. H.   DIE MATRIZENTHEORIE DER STATIK   57-0003
ARGYRIS, J. H.   DIE METHODE DER ENDLICHEN ELEMENTE UND IHRE ANWENDUNG IM MASCHINENBAU   71-0628
ARGYRIS, J. H.   DIE VERWENDUNG DER MATRIZENVERSCHIEBUNGSMETHODE ZUR BERECHNUNG VON SPANNUNGSKOMPONENTEN   67-0140
ARGYRIS, J. H.   DIE VERWENDUNG DER MATRIZENVERSCHIEBUNGSMETHODE ZUR BERECHNUNG VON SPANNUNGSKONZENTRATIONEN   70-0400
ARGYRIS, J. H.   DISCUSSION ON FINITE ELEMENT ANALYSIS OF PRESTRESSED CONCRETE REACTOR VESSELS   72-0648
ARGYRIS, J. H.   DYNAMIC RESPONSE BY LARGE STEP INTEGRATION   73-0567
ARGYRIS, J. H.   EIN VERFAHREN DUR DIE FORMFINDUNG VEON BELIEBIGEN VORGESPANNTEN NETZWERKKONSTRUKTIONEN   72-0641
ARGYRIS, J. H.   EINIGE ENTWICKLUNGEN IN DER MATRIZEN-VERSCHIEBUNGSMETHODE   69-0451
ARGYRIS, J. H.   ELASTIC-PLASTIC ANALYSIS OF THREE-DIMENSIONAL MEDIA   66-0002
ARGYRIS, J. H.   ELASTO-PLASTIC MATRIX DISPLACEMENT ANALYSIS OF THREE-DIMENSIONAL CONTINUA   66-0004
ARGYRIS, J. H.   ENERGY THEOREMS AND STRUCTURAL ANALYSIS   54-0001
ARGYRIS, J. H.   ENERGY THEOREMS AND STRUCTURAL ANALYSIS   67-0136
ARGYRIS, J. H.   FINITE ELEMENT ANALYSIS OF THERMOMECHANICAL PROBLEMS   71-0030
ARGYRIS, J. H.   FINITE ELEMENT DIFFICULTIES IN NONLINEAR AND INCOMPRESSIBLE MATERIAL   74-1242
ARGYRIS, J. H.   FINITE ELEMENT FORMULATION OF THE INCOMPRESSIBLE LUBRICATION PROBLEM   70-0022
ARGYRIS, J. H.   FINITE ELEMENT THEORY OF PLATES AND SHELLS INCLUDING TRANSVERSE SHEAR STRAIN EFFECTS   70-0395
ARGYRIS, J. H.   FINITE ELEMENT ULTIMATE LOAD ANALYSIS OF THREE DIMENSIONAL CONCRETE STRUCTURES   74-0672
ARGYRIS, J. H.   FINITE ELEMENTE ZUR BERECHNUNG VON SPANNBETONREAKTORDRUCKBEHALTERN   73-0702
ARGYRIS, J. H.   FINITE ELEMENTS FOR CALCULATION OF PRESTRESSED CONCRETE REACTOR PRESSURE VESSELS (FINITE ELEMENTE ZU   73-0797
ARGYRIS, J. H.   FINITE ELEMENTS IN TIME AND SPACE   69-0442
ARGYRIS, J. H.   FINITE ELEMENTS IN TIME AND SPACE   69-0450
ARGYRIS, J. H.   FINITE ELEMENTS METHOD IN DESIGN OF PRESTRESSED REACTOR CONCRETE CONSTRUCTION   73-0550
ARGYRIS, J. H.   FINITE TIME-SPACE ELEMENTS APPLIED TO NONLINEAR PROBLEMS   70-0398
ARGYRIS, J. H.   FINITE-ELEMENT ANALYSIS OF SLOW INCOMPRESSIBLE VISCOUS FLUID MOTION   74-0154
ARGYRIS, J. H.   GENERAL TREATMENT OF STRUCTURAL MODIFICATIONS   72-0166
ARGYRIS, J. H.   GEOMETRIC NONLINEARITY AND THE FINITE ELEMENT DISPLACEMENT METHOD   74-0679
ARGYRIS, J. H.   GROPRECHENANLAGEN IN DER FORSCHUNG   58-0264
ARGYRIS, J. H.   INITIAL STRAINS IN THE MATRIX FORCE METHOD OF STRUCTURAL ANALYSIS   60-0008
ARGYRIS, J. H.   INTERACTIVE COMPUTER GRAPHICS IN STRUCTURAL ANALYSIS: ONLINE 72   72-0639
ARGYRIS, J. H.   INTRODUCTION TO FINITE ELEMENT ANALYSIS   68-0218
ARGYRIS, J. H.   KONSTRUKTION UND BERECHNUNG MIT DER METHODE FINITE ELEMENTE   73-0707
ARGYRIS, J. H.   KONSTRUKTION UND BERECHNUNG MIT FINITEN ELEMENTEN   74-0678
ARGYRIS, J. H.   LARGE DEFLECTION ANALYSIS OF PRESTRESSED NETWORKS   72-0169
ARGYRIS, J. H.   LARGE NATURAL STRAINS AND SOME SPECIAL DIFFICULTIES DUE TO NONLINEARITY AND INCOMPRESSIBILITY IN FIN   74-0528
ARGYRIS, J. H.   LARGE NATURAL STRAINS AND SOME SPECIAL DIFFICULTIES DUE TO NONLINEARITY AND INCOMPRESSIBILITY IN FIN   74-1027
ARGYRIS, J. H.   LINEAR AND NONLINEAR OSCILLATIONS IN MECHANICS   74-0677
ARGYRIS, J. H.   LINEAR METHODS OF STRUCTURAL ANALYSIS   71-0028
ARGYRIS, J. H.   MATERIAL NONLINEARITIES IN THE FINITE ELEMENT ANALYSIS   74-0647
ARGYRIS, J. H.   MATRIX ANALYSIS OF SHELLS - SMALL AND LARGE DISPLACEMENTS   68-0016
ARGYRIS, J. H.   MATRIX ANALYSIS OF THREE-DIMENSIONAL ELASTIC MEDIA: SMALL AND LARGE DISPLACEMENTS   65-0003
ARGYRIS, J. H.   MATRIX DISPLACEMENT ANALYSIS OF ANISOTROPIC SHELLS BY TRIANGULAR ELEMENTS   65-0005
ARGYRIS, J. H.   MATRIX DISPLACEMENT ANALYSIS OF PLATES AND SHELLS   66-0010
ARGYRIS, J. H.   MATRIX DISPLACEMENT ANALYSIS OF PLATES AND SHELLS, PART 11, GENERAL FORMULATION OF THE LINEAR THEORY   71-0621
ARGYRIS, J. H.   MATRIX DISPLACEMENT ANALYSIS OF SHELLS AND PLATES INCLUDING TRANSVERSE SHEAR STRAIN EFFECTS   72-0643
ARGYRIS, J. H.   MATRIX METHODS OF STRUCTURAL ANALYSIS   63-0014
ARGYRIS, J. H.   MATRIX METHODS OF STRUCTURAL ANALYSIS. A PRECIS OF RECENT DEVELOPMENTS   63-0019
ARGYRIS, J. H.   MATRIZENTHEORIE DER STATIK   60-0009
ARGYRIS, J. H.   MEMBRANE PARALLELOGRAM ELEMENT WITH LINEARLY VARYING EDGE STRAIN FOR THE MATRIX DISPLACEMENT METHOD   66-0006
ARGYRIS, J. H.   METHODS OF ELASTO-PLASTIC ANALYSIS   71-0617
ARGYRIS, J. H.   METHODS OF ELASTOPLASTIC ANALYSIS   69-0343
ARGYRIS, J. H.   MODERN DEVELOPMENTS IN THE STRESS ANALYSIS OF PRESSURE VESSELS   69-0201
ARGYRIS, J. H.   MODERN FUSELAGE ANALYSIS AND THE ELASTIC AIRCRAFT   63-0023
ARGYRIS, J. H.   NATURAL FACTOR FORMULATION OF THE STIFFNESS FOR THE MATRIX DISPLACEMENT METHOD   75-0057
ARGYRIS, J. H.   NATURAL GEOMETRY OF SURFACE WITH SPECIFIC REFERENCE TO THE MATRIX DISPLACEMENT ANALYSIS OF SHELLS, 1   73-0709
ARGYRIS, J. H.   NONLINEAR METHODS OF STRUCTURAL ANALYSIS   71-0029
ARGYRIS, J. H.   NONLINEAR OSCILLATIONS USING THE FINITE ELEMENT TECHNIQUE   73-0568
ARGYRIS, J. H.   NOTE ON THE THEORY OF AIRCRAFT STRUCTURAL ANALYSIS   59-0005
ARGYRIS, J. H.   ON THE APPLICATION OF THE SHEBA SHELL ELEMENT   72-0651
ARGYRIS, J. H.   ON THE MATRIX THEORY OF STRUCTURES   60-0007
ARGYRIS, J. H.   ON THE REDUCTION OF NUMERICAL ERROR IN THE MATRIX DISPLACEMENT METHOD   74-0417

7

ARGYRIS, J. H.	PARAMETRISCHE KERBSPANNUNGSUNTERSUCHUNG AM ELASTISCHEN KERN.(PARAMETRICAL INVESTIGATION OF NOTCH STR	72-0208
ARGYRIS, J. H.	PARAMETRISCHE UNTERSUCHUNGEN VON KERBEN MIT DER MATRIZENVERSCHIEBUNGSMETHODE UND BILDSCHIRMUNTERSTUT	72-0638
ARGYRIS, J. H.	PART 11: LUMINA AND HEXE ELEMENTS	68-0261
ARGYRIS, J. H.	POTENTIAL FLOW ANALYSIS BY FINITE ELEMENTS	72-0004
ARGYRIS, J. H.	REANALYSIS FOR LIMITED STRUCTURAL DESIGN MODIFICATIONS	72-0650
ARGYRIS, J. H.	RECENT ADVANCES IN MATRIX METHODS OF STRUCTURAL ANALYSIS	64-0016
ARGYRIS, J. H.	RECENT DEVELOPMENTS OF MATRIX THEORY OF STRUCTURES	59-0006
ARGYRIS, J. H.	RECENT DEVELOPMENTS IN FINITE ELEMENT ANALYSIS OF PRESTRESSED CONCRETE REACTOR VESSELS	74-0469
ARGYRIS, J. H.	RECENT DEVELOPMENTS IN THE FINITE ELEMENT ANALYSIS OF PRE-STRESSED CONCRETE REACTOR VESSELS	74-0582
ARGYRIS, J. H.	REINFORCED FIELDS OF TRIANGULAR ELEMENTS WITH LINEARLY VARYING STRAIN: EFFECT OF INITIAL STRAINS	65-0004
ARGYRIS, J. H.	SHAPE DETERMINATION OF PREFABRICATED PRESTRESSED NETWORKS	73-0889
ARGYRIS, J. H.	SOME ASPECTS OF FINITE ELEMENT TECHNIQUES	69-0340
ARGYRIS, J. H.	SOME CONSIDERATION FOR THE EVALUATION OF THE FINITE ELEMENT MODELS	73-0710
ARGYRIS, J. H.	SOME CONSIDERATIONS FOR THE EVALUATION OF FINITE ELEMENT MODELS	73-0706
ARGYRIS, J. H.	SOME CONSIDERATIONS FOR EVALUATION OF FINITE ELEMENT MODELS	74-0471
ARGYRIS, J. H.	SOME CONTRIBUTIONS TO NONLINEAR SOLID MECHANICS	73-0705
ARGYRIS, J. H.	SOME GENERAL CONSIDERATIONS ON THE NATURAL MODE TECHNIQUE	69-0049
ARGYRIS, J. H.	SOME GENERAL CONSIDERATIONS ON THE NATURAL MODE TECHNIQUE, PART II: LARGE DISPLACEMENTS	69-0445
ARGYRIS, J. H.	SOME NEW ELEMENTS FOR THE MATRIX DISPLACEMENT METHOD	68-0017
ARGYRIS, J. H.	SOME RESULTS ON THE FREE-FREE OSCILLATIONS OF AIRCRAFT TYPE STRUCTURES	65-0007
ARGYRIS, J. H.	SPEKTRALANALYSE VON ROTORBLATTERN NACH DER MATRIZENVERSCHIEBUNGSMETHODE UND MATRIZENKRAFTEMETHODE	66-0123
ARGYRIS, J. H.	STATIC AND DYNAMIC ELASTO-PLASTIC ANALYSIS BY THE METHOD OF FINITE ELEMENTS IN SPACE AND TIME	72-0642
ARGYRIS, J. H.	STRING EIN PROGRAMMSYSTEM FUR DIE MASCHINENUNABHANGIGE BEARBEITUNG VON ZEICHENKETTEN IN FORTRAN	68-0265
ARGYRIS, J. H.	STRUCTURAL ANALYSIS BY THE MATRIX FORCE METHOD WITH APPLICATION TO AIRCRAFT WINGS	56-0004
ARGYRIS, J. H.	STRUCTURAL ANALYSIS BY PROBLEM ORIENTED LANGUAGES	71-0623
ARGYRIS, J. H.	TETRAHEDRON ELEMENTS WITH LINEARLY VARYING STRAIN FOR THE MATRIX DISPLACEMENT METHOD	65-0006
ARGYRIS, J. H.	THE CHANGES IN THE (TWO-DIMENSIONAL) STRESS ROUNDED CORNERS, CAUSED BY A VARYING INTERNAL PRESSURE A	67-0141
ARGYRIS, J. H.	THE CURVED TETRAHEDRONAL AND TRIANGULAR ELEMENTS TEC AND TRIC FOR THE MATRIX DISPLACEMENT METHOD	69-0047
ARGYRIS, J. H.	THE ELASTO-PLASTIC CALCULATION OF GENERAL STRUCTURES AND CONTINUA	68-0207
ARGYRIS, J. H.	THE FINITE ELEMENT METHOD APPLIED TO FLUID MECHANICS	74-0676
ARGYRIS, J. H.	THE FUGA FAMILY FOR FOLDED PLATE STRUCTURES	69-0333
ARGYRIS, J. H.	THE FUGA FAMILY OF ELEMENTS FOR FOLDED PLATE STRUCTURES A FURTHER SEQUEL TO TECHNICAL NOTE 14	69-0446
ARGYRIS, J. H.	THE HERMES 8 ELEMENT FOR THE MATRIX DISPLACEMENT METHOD	68-0011
ARGYRIS, J. H.	THE IMPACT OF THE DIGITAL COMPUTER ON STRUCTURAL ANALYSIS	69-0447
ARGYRIS, J. H.	THE IMPACT OF THE DIGITAL COMPUTER ON THE ENGINEERING SCIENCES - PART I	70-0003
ARGYRIS, J. H.	THE INCOMPRESSIBLE LUBRICATION PROBLEM	69-0022
ARGYRIS, J. H.	THE LUMINA ELEMENT FOR THE MATRIX DISPLACEMENT METHOD	68-0010
ARGYRIS, J. H.	THE MATRIX ANALYSIS OF STRUCTURES WITH CUT-OUTS AND MODIFICATIONS	56-0003
ARGYRIS, J. H.	THE MATRIX FORCE METHOD OF STRUCTURAL ANALYSIS AND SOME NEW APPLICATIONS	56-0005
ARGYRIS, J. H.	THE MATRIX THEORY OF STATICS	57-0002
ARGYRIS, J. H.	THE NATURAL FACTOR APPROACH IN THE DISPLACEMENT METHOD - FOUNDATION AND FURTHER DEVELOPMENTS	75-0111
ARGYRIS, J. H.	THE SHEBA FAMILY OF SHELL ELEMENTS FOR THE MATRIX DISPLACEMENT METHOD	68-0014
ARGYRIS, J. H.	THE SHEBA FAMILY OF SHELL ELEMENTS FOR THE MATRIX DISPLACEMENT METHOD. 111. LARGE DISPLACEMENTS	69-0499
ARGYRIS, J. H.	THE TET 20 AND TEA 8 ELEMENTS FOR THE MATRIX DISPLACEMENT METHOD	68-0012
ARGYRIS, J. H.	THE THERMO-ELASTO-PLASTIC ANALYSIS OF A RE-ENTRY BODY TYPE COMPOSITE SOLID STRUCTURE USING A NEW FIN	66-0093
ARGYRIS, J. H.	THE TRIAX 6 ELEMENT FOR AXISYMMETRIC ANALYSIS BY THE MATRIX DISPLACEMENT METHOD: PART 1, FOUNDATIONS	66-0124
ARGYRIS, J. H.	THE TUBA FAMILY OF PLATE ELEMENTS FOR THE MATRIX DISPLACEMENT METHOD	68-0013
ARGYRIS, J. H.	THE VALIDITY OF THE INITIAL STRAIN CONCEPT	61-0010
ARGYRIS, J. H.	THEORIE, PROGRAMMENTWICKLUNG UND ERFAHRUNG ANVORGESPANNTEN NETZWERKKONSTRUKTIONEN	72-0640
ARGYRIS, J. H.	THERMOMECHANICAL ANALYSIS OF STRUCTURES	71-0618
ARGYRIS, J. H.	THERMOMECHANICAL ANALYSIS OF STRUCTURES	73-0368
ARGYRIS, J. H.	THERMOMECHANICAL CREEP OF AGEING CONCRETE - A UNIFIED APPROACH	74-0607
ARGYRIS, J. H.	THREE DIMENSIONAL ANISOTROPIC AND INHOMOGENEOUS ELASTIC MEDIA MATRIX ANALYSIS FOR SMALL AND LARGE DI	65-0008
ARGYRIS, J. H.	THREE DIMENSIONAL ANALYSIS OF TWO ARCH DAMS BY THE FINITE ELEMENT METHOD	68-0018
ARGYRIS, J. H.	THREE DIMENSIONAL ELASTIC AND ELASTO-PLASTIC ANALYSIS OF REACTOR PRESSURE VESSELS	69-0453
ARGYRIS, J. H.	TRIANGULAR ELEMENTS WITH LINEARLY VARYING STRAIN FOR THE MATRIX DISPLACEMENT METHOD	65-0053
ARGYRIS, J. H.	TRIAX 6 ELEMENT FOR AXISYMMETRIC ANALYSIS BY THE MATRIX DISPLACEMENT METHOD	66-0009
ARGYRIS, J. H.	TWO AND THREE DIMENSIONAL POTENTIAL FLOW BY THE METHOD OF SINGULARITIES	69-0449
ARGYRIS, J. H.	TWO AND THREE DIMENSIONAL POTENTIAL FLOW ANALYSIS BY THE METHOD OF SINGULARITIES	70-0357
ARGYRIS, J. H.	TWO AND THREE-DIMENSIONAL FLOW USING FINITE ELEMENTS	69-0008
ARGYRIS, J. H.	TWO-AND THREE-DIMENSIONAL FLOW ANALYSIS USING FINITE ELEMENTS	70-0021
ARIATHURAI, C.	A FINITE ELEMENT MODEL FOR SEDIMENT TRANSPORT IN ESTUARIES	74-1102
ARKALEV, B. A.	SOLVING A NONLINEAR PROBLEM OF NONSTEADY THERMAL CONDUCTIVITY IN SOLID BODIES OF ARBITRARY FORM	68-0019
ARLETT, P. L.	APPLICATION OF FINITE ELEMENTS TO THE SOLUTION OF HELMHOLT'S EQUATION	68-0020
ARLETT, P. L.	SOLUTION OF THREE-DIMENSIONAL FIELD PROBLEMS BY THE FINITE ELEMENT METHOD	67-0091
ARMAN, A.	EFFECT OF DENSIFICATION ON THE ENGINEERING CHARACTERISTICS OF ORGANIC SOILS - VOL. 1 AND 2.	73-0805
ARMAN, A.	NUMERICAL ANALYSIS OF EMBANKMENTS OVER SOFT SOILS	75-0719
ARMAN, A.	PHOTOELASTIC AND FINITE ELEMENT ANALYSIS OF EMBANKMENTS CONSTRUCTED OVER SOFT SOILS	70-0151
ARMEN, H.	A FINITE ELEMENT METHOD FOR THE PLASTIC BENDING ANALYSIS OF STRUCTURES	68-0021
ARMEN, H.	COMPUTER PROGRAMS FOR THE PLASTIC ANALYSIS OF STRUCTURES USING DISCRETE ELEMENT METHODS	67-0004
ARMEN, H.	DISCRETE ELEMENT METHODS FOR THE PLASTIC ANALYSIS OF STRUCTURES SUBJECTED TO CYCLIC LOADING	70-0023
ARMEN, H.	ELASTIC AND PLASTIC INTERLAMINAR SHEAR DEFORMATION IN LAMINATED COMPOSITES UNDER GENERALIZED PLANE	71-0158
ARMEN, H.	FINITE ELEMENT ANALYSIS OF STRUCTURES IN THE PLASTIC RANGE	71-0031
ARMEN, H.	FINITE ELEMENT METHOD FOR THE PLASTIC BENDING ANALYSIS OF STRUCTURES	68-0206
ARMEN, H.	NONLINEAR ANALYSIS OF STRUCTURES	74-0363
ARMEN, H.	NONLINEAR BEHAVIOR OF SHELLS OF REVOLUTION UNDER CYCLIC LOADING	73-0242
ARMEN, H.	NONLINEAR CRACK ANALYSIS WITH FINITE ELEMENTS	73-0369
ARMEN, H.	PLANS - A FINITE-ELEMENT PROGRAM FOR NONLINEAR ANALYSIS OF STRUCTURES	75-0011
ARMEN, H.	PLASTICITY	70-0509
ARMEN, H.	PLASTICITY THEORY AND FINITE ELEMENT APPLICATIONS	72-0318
ARMEN, H.	VARIABLE TIME STEP METHOD FOR DETERMINING PLASTIC STRESS REFLECTIONS FROM BOUNDARIES	74-0641
ARMEN, H. A.	EVALUATION OF NUMERICAL TIME INTEGRATION METHODS AS APPLIED TO ELASTIC-PLASTIC DYNAMIC PROBLEMS INVO	74-0858
ARMEN, H. A. JR	PLASTIC ANALYSIS	74-0859
ARMEN, H. JR.	APPLICATION OF A SUBSTRUCTURING TECHNIQUE TO THE PROBLEM OF CRACK EXTENSION AND CLOSURE	74-0772
ARMEN, H. JR.	ELASTIC-PLASTIC ANALYSIS OF A PROPAGATING CRACK UNDER CYCLIC LOADING	75-0857
ARMEN, H. JR.	FEASIBILITY OF USING FINITE ELEMENTS IN ANALYSIS OF SECOND BREAKDOWN IN SEMICONDUCTOR DEVICES	73-0874
ARMENAKAS, A. E	LINEAR AND NONLINEAR ANALYSIS OF SHELLS	72-0832
ARMIT, A. P.	ICON: THE INTERACTIVE CREATION OF NASTRAN DATA. A SYSTEM DESCRIPTION	75-0274
ARMSTRONG, W. H	ANALYSIS OF NONLINEAR DYNAMIC, COUPLED THERMOVISCOELASTICITY PROBLEMS BY THE FINITE ELEMENT METHOD	71-0527
ARMSTRONG, W. H	ANALYSIS OF NONLINEAR THERMOVISCOELASTICITY PROBLEMS BY THE FINITE ELEMENT METHOD	71-0528
ARNAOUTI, C.	YIELD CRITERION FOR REINFORCED CONCRETE SLABS	73-0985
AROCKIASAMY, M.	LIMIT ANALYSIS OF REINFORCED CONCRETE SLABS BY FINITE ELEMENT METHOD	71-0802
ARORA, J. S.	ON IMPROVING EFFICIENCY OF AN ALGORITHM FOR STRUCTURAL OPTIMIZATION AND A USERS MANUAL FOR PROGRAM T	74-0778
ARRIYAVAT, P.	ANALYSIS OF NONLINEAR REINFORCED CONCRETE FRAME BY FINITE ELEMENT METHOD	74-1104
ARTHUR, I. P.	PRESTRESSED CONCRETE PRESSURE VESSELS	69-0052
ARYA, B.	APPLICATION OF THE FINITE ELEMENT METHOD TO THE NONLINEAR ANALYSIS OF NUCLEAR REACTOR FUEL BEHAVIOR	74-1246
ARYA, S. K.	FINITE ELEMENTS FORMULATED BY THE WEIGHTED DISCRETE LEAST SQUARES METHOD	74-0200
ARYA, S. K.	USE OF THE LEAST SQUARES CRITERION IN THE FINITE ELEMENT FORMULATION	73-0270
ARYGYRIS, J. H.	ASPECTS OF FINITE ELEMENT METHOD APPLIED TO AEROSPACE STRUCTURES	73-0703
ARYGYRIS, J. H.	OSCILLATION STABILITY OF INTERNALLY AND EXTERNALLY NOTCHED PLANE BARS OF EQUAL FORM FACTOR FROM 3943	73-0745
ASADA, Y.	APPROACH TO INVESTIGATE NOTCH EFFECT ON LOW-CYCLE FATIGUE WITH FINITE ELEMENT	73-0509
ASAOKA, K.	SIMULATION ON THE INITIATION AND GROWTH OF DUCTILE FRACTURE VOIDS	75-0836

8

ASHBY, M. F.	NON-UNIFORM FLOW OF POLYCRYSTALS BY GRAIN-BOUNDARY SLIDING ACCOMMODATED BY POWER-LAW CREEP	75-0562
ASHER, G. W.	CALCULATION OF AXISYMMETRIC LONGITUDINAL MODES FOR FLUID-ELASTIC TANK-ULLAGE GAS SYSTEMS AND COMPARI	64-0019
ASHIE, I.	COMPUTER BASED ANALYSES OF THE RESPONSE OF BOX TYPE STRUCTURES TO RANDOM PRESSURES	73-0227
ASHWELL, D. G.	A COMPARISON OF CURVED BEAM FINITE ELEMENTS WHEN USED IN VIBRATION PROBLEMS	71-0227
ASHWELL, D. G.	A STIFFNESS MATRIX FOR SHALLOW SHELL FINITE ELEMENTS	69-0229
ASHWELL, D. G.	APPLICATION OF CURVED FINITE ELEMENTS TO CIRCULAR ARCHES	71-0360
ASHWELL, D. G.	BEHAVIOR WITH DIMINISHING CURVATURE OF STRAIN-BASED ARCH FINITE ELEMENTS	73-0088
ASHWELL, D. G.	CORRECTED ASSESSMENT OF CYLINDRICAL SHELL FINITE ELEMENT OF BOGNER, FOX AND SCHMIT WHEN APPLIED TO A	73-0024
ASHWELL, D. G.	FINITE ELEMENT ANALYSIS OF A BROAD ARCH BRIDGE WITH A SLAB DECK	71-0724
ASHWELL, D. G.	FINITE ELEMENT CALCULATION OF STRESS DISTRIBUTIONS IN ARCHES	74-0094
ASHWELL, D. G.	FURTHER STUDIES IN THE APPLICATION OF CURVED FINITE ELEMENTS TO CIRCULAR ARCHES	71-0644
ASHWELL, D. G.	LIMITATIONS OF CERTAIN CURVED FINITE ELEMENTS WHEN APPLIED TO ARCHES	71-0279
ASHWELL, D. G.	NEW CYLINDRICAL SHELL FINITE ELEMENT BASED ON SIMPLE INDEPENDENT STRAIN FUNCTIONS	72-0094
ASHWELL, D. G.	POST-BUCKLING ANALYSIS OF SLIGHTLY CURVED PLATES BY THE FINITE ELEMENT METHOD	69-0365
ASHWELL, D. G.	THE USE OF FINITE ELEMENT MID-INCREMENT STIFFNESS MATRICES IN THE POST BUCKLING ANALYSIS OF IMPERFEC	71-0383
ASKAR, A.	FINITE ELEMENT METHOD FOR BOUND STATE CALCULATIONS IN QUANTUM MECHANICS	75-0005
ATCHISON, D. L.	FINITE ELEMENT SOLUTION OF THE INTERACTION OF A PLANE ACOUSTIC BLAST WAVE AND A CYLINDRIC STRUCTURE	74-0790
ATCHLEY, C. E.	USE OF FINITE ELEMENT ANALYSIS IN FRACTURE MECHANICS	74-0837
ATHANASIU, C.	IN-SITU HORIZONTAL STRESSES IN LATERITE BY FINITE ELEMENT ANALYSIS	73-1048
ATKINS, J. T.	CONTROLLING OPEN-PIT SLOPE FAILURES AT SHIRLEY BASIN	73-0149
ATKINSON, B.	APPLICATION OF THE FINITE ELEMENT METHOD TO CREEPING FLOW PROBLEMS	70-0024
ATKINSON, B.	LOW REYNOLDS NUMBER DEVELOPING FLOW	69-0020
ATKINSON, P.	PREDICTION OF MARINE PROPELLER DISTORTION AND STRESSES USING A SUPERPARAMETRIC THICK SHELL FINITE EL	73-0370
ATLURI, S.	A NEW ASSUMED STRESS HYBRID FINITE ELEMENT MODEL FOR SOLID CONTINUA	71-0371
ATLURI, S.	ASSUMED STRESS HYBRID FINITE ELEMENT MODEL FOR LINEAR ELASTODYNAMIC ANALYSIS	73-0089
ATLURI, S.	MECHANICS OF BRAIN TISSUE FRAGILITY	73-0214
ATLURI, S.	NONLINEAR HYBRID STRESS FINITE ELEMENT ANALYSIS OF LAMINATED SHELLS	74-1350
ATLURI, S.	NONLINEAR HYBRID STRESS FINITE ELEMENT ANALYSIS OF PNEUMATIC TIRES	75-0029
ATLURI, S.	ON THE HYBRID STRESS FINITE ELEMENT MODEL FOR INCREMENTAL ANALYSIS OF LARGE DEFLECTION PROBLEMS	75-0371
ATLURI, S.	STATIC ANALYSIS OF SHELLS OF REVOLUTION USING DOUBLY-CURVED QUADRILATERAL ELEMENTS DERIVED FROM ALTE	69-0368
ATLURI, S.	THEORETICAL FORMULATION OF FINITE ELEMENT METHODS IN LINEAR ELASTIC ANALYSIS OF GENERAL SHELLS	72-0496
ATLURI, S. N.	APPLICATION OF AN ASSUMED DISPLACEMENT HYBRID FINITE ELEMENT PROCEDURE TO TWO-DIMENSIONAL PROBLEMS I	74-0729
ATLURI, S. N.	ASSUMED DISPLACEMENT HYBRID FINITE ELEMENT MODEL FOR LINEAR FRACTURE MECHANICS	75-0146
ATLURI, S. N.	BRAIN TISSUE FRAGILITY - A FINITE STRAIN ANALYSIS BY A HYBRID FINITE-ELEMENT METHOD	75-0162
ATLURI, S. N.	BRAIN TISSUE FRAGILITY - A FINITE STRAIN ANALYSIS BY A HYBRID FINITE-ELEMENT METHOD	75-0243
ATLURI, S. N.	BRAIN-TISSUE FRAGILITY - FINITE STRAIN ANALYSIS BY A HYBRID FINITE-ELEMENT METHOD	75-0631
ATLURI, S. N.	FINITE ELEMENT PERTURBATION ANALYSIS OF NONLINEAR DYNAMIC RESPONSE OF ELASTIC CONTINUA	74-0452
ATLURI, S. N.	FRACTURE MECHANICS APPLICATION OF AN ASSUMED DISPLACEMENT HYBRID FINITE ELEMENT PROCEDURE	75-0212
ATLURI, S. N.	THREE-DIMENSIONAL LINEAR FRACTURE MECHANICS ANALYSIS BY A DISPLACEMENT-HYBRID FINITE-ELEMENT MODEL	75-0884
ATTAR-HASSAN, G	A REFINED FINITE ELEMENT ANALYSIS OF FOLDED PLATE STRUCTURES	71-0671
ATTEIA, M.	FONCTIONS SPLINES ET METHODE D ELEMENTS FINIS (SPLINE FUNCTIONS AND FINITE ELEMENT METHOD)	75-0846
ATTEIA, M.	SPLINE FUNCTIONS AND FINITE-ELEMENT METHOD	75-0430
ATTIA, A. V.	FINITE METHOD AS AN ASPECT OF THE PRINCIPLE OF MAXIMUM UNIFORMITY: NEW HYDRODYNAMICAL RAMIFA	75-0009
AUCONI, F.	MODAL RESPONSE ANALYSIS OF CONSTRAINED FINITE-ELEMENTS ELASTIC CONTINUUM	75-0767
AUGUST, A.	THE BEHAVIOUR OF ADVANCED FILAMENTARY COMPOSITE PLATES WITH CUTOUTS	70-0025
AUGUSTUS, K. J.	A FINITE ELEMENT SOLUTION OF THE TRANSIENT MAGNETOHYDRODYNAMIC COUETTE FLOW	74-1103
AUSTIN, B. A.	APPLICATION OF FRACTURE MECHANICS TO FAILURE OF THICK WALLED CYLINDERS	71-0511
AUSTIN, F.	VIBRATION AND STRESS ANALYSIS OF SOFT-BONDED SHUTTLE INSULATION TILES. MODAL ANALYSIS WITH COMPACT W	74-0708
AUTHOR	ANALYSIS OF LAYERED SYSTEMS SUBJECTED TO WHEEL LOADS	72-0868
AUTHOR	CYCLIC-STRESS ANALYSIS OF NOTCHES FOR SUPERSONIC TRANSPORT CONDITIONS	74-0834
AUTHOR	DYNAMICS OF SUBMARINE STRUCTURES. VOLUME 1	68-0251
AUTHOR	FINITE ELEMENT ANALYSIS OF MINE STRUCTURES	72-0687
AUTHOR	M.I.T. TEST SECTION INSTRUMENTATION, MASSACHUSETTS BAY TRANSPORTATION AUTHORITY, HAYMARKET-NORTH EXT	72-0673
AUTHOR	NASTRAN USERS EXPERIENCES	72-0753
AUTHOR	SINGLE-TRACK SUBWAY ENVIRONMENTAL SIMULATION MODEL. PHASE 1	71-0693
AUTHOR	THE ROLE OF MECHANICS IN DESIGN - STRUCTURAL JOINTS	74-0770
AUTHOR	THREE DIMENSIONAL SUPERTANKER COMPUTER PROGRAM	71-0661
AUTHOR	USE OF SPARSITY IN THE SOLUTION OF FINITE ELEMENT SYSTEMS OF EQUATIONS BY GAUSSIAN ELIMINATION-TYPE	74-0868
AUTHORS	ARMY NUMERICAL ANALYSIS CONFERENCE	71-0707
AUTHORS	COMPUTATIONAL TECHNIQUES AS AN AID IN PHYSICAL METALLURGY	71-0554
AUTHORS	CONFERENCE ON MATRIX METHODS IN STRUCTURAL MECHANICS	69-0470
AUTHORS	JOURNAL OF RESEARCH OF THE NATIONAL BUREAU OF STANDARDS: C. ENGINEERING AND INSTRUMENTATION	71-0664
AUTHORS	NASTRAN USERS EXPERIENCES	71-0713
AUTHORS	NAVY-NASTRAN COLLOQUIUM, WASHINGTON, D.C.	70-0403
AUTHORS	NSRDC: NASTRAN COLLOQUIUM, WASHINGTON, D.C.	70-0404
AUTHORS	PROCEEDINGS OF THE NAVY-NASTRAN COLLOQUIUM (4TH) HELD AT NAVY SHIP RESEARCH AND DEVELOPMENT CENTER,	73-0888
AUTHORS	PROCEEDINGS OF THE 5TH NAVY-NASTRAN COLLOQUIUM, HELD AT NAVAL SHIP RESEARCH AND DEVELOPMENT CENTER,	74-0709
AUTHORS	SYMPOSIUM ON ACOUSTIC FATIGUE	73-0898
AUTHORS	THE ROLE OF MECHANICS IN DESIGN-STRUCTURAL JOINTS	74-0767
AUTHORS	TRANSACTIONS OF ARMY MATHEMATICIANS RESEARCH CONFERENCE	73-0925
AUYER, S. E.	FINITE-ELEMENT MODELING OF PIEZOELECTRIC CERAMIC HYDROPHONES	75-0595
AVENT, R. R.	FIELD ANALYSIS FOR PREFABRICATED RIBBED CYLINDRICAL SHELLS	73-0899
AWBI, H. B.	A FINITE ELEMENT SOLUTION OF THE CONFINED FLOW OVER A CIRCULAR CYLINDER	74-0023
AXELSON, K.	ANISOTROPIC WORK HARDENING IN METALS. FINITE ELEMENT APPLICATION TO PLANE STRESS AND PLANE STRAIN	73-0950
AYALA-MILIAN, G	SOLUTION OF WAVE PROPAGATION PROBLEMS IN SATURATED MEDIUM	72-0784
AYATOLLAHI, M.	FLOW IN ROCKS WITH DEFORMABLE FRACTURES	74-0065
AYORINDE, O. A.	FINITE ELEMENT LARGE DEFORMATION ANALYSIS OF A TAPERED AORTA	75-0856
AYRES, D. J.	COMPANY TRAINING OF ENGINEERING IN STRUCTURAL ANALYSIS COMPUTER PROGRAM USAGE	72-0460
AYRES, D. J.	CREEP DATA ACQUISITION AND APPLICATION TO REACTOR COMPONENT DESIGN	72-0113
AYRES, D. J.	ELASTIC-PLASTIC AND CREEP ANALYSIS VIA THE MARC 2 FINITE ELEMENT PROGRAM	71-0032
AYRES, D. J.	FINITE ELEMENT ANALYSIS OF STRUCTURAL INTEGRITY OF A REACTOR PRESSURE VESSEL DURING EMERGENCY CORE C	70-0026
AYRES, D. J.	THREE-DIMENSIONAL ELASTIC ANALYSIS OF SEMI-ELLIPTICAL SURFACE CRACKS SUBJECT TO THERMAL SHOCK	75-0256
AZAR, J. J.	PIPELINE STRESS ANALYSIS	75-0078
AZEVEDO, M. C.	COMPARISON OF ANALYTICAL AND EXPERIMENTAL RESULTS IN THE DESIGN OF A HOLLOW GRAVITY DAM	67-0109
AZIZ, A.	LECTURES ON MATHEMATICAL FOUNDATIONS OF THE FINITE ELEMENT METHOD	72-0777
AZIZ, A. K.	ANGLE CONDITION IN THE FINITE ELEMENT METHOD	74-1083
AZIZ, A. K.	FINITE ELEMENT METHOD FOR NON SMOOTH DOMAINS AND COEFFICIENTS	72-0504
AZIZ, A. K.	ON NUMERICAL SOLUTIONS OF EQUATIONS OF HYPERBOLIC-ELLIPTIC TYPE	74-0741
AZIZ, A. K.	ON THE ANGLE CONDITION IN THE FINITE ELEMENT METHOD	74-0907
AZIZ, A. K.	ONE PARAMETER FAMILIES OF VARIATIONAL PRINCIPLES	72-0503
AZIZ, A. K.	PRELIMINARY REMARKS	72-0497
AZIZ, A. K.	PROPERTIES OF SOLUTIONS OF ELLIPTIC BOUNDARY VALUE PROBLEMS	72-0499
AZIZ, A. K.	RATE OF CONVERGENCE OF THE FINITE ELEMENT METHOD	72-0502
AZIZ, A. K.	SURVEY LECTURES ON THE MATHEMATICAL FOUNDATIONS OF THE FINITE ELEMENT METHOD	72-0670
AZIZ, A. K.	THE EIGENVALUE PROBLEM	72-0506
AZIZ, A. K.	THE FINITE ELEMENT METHOD FOR TIME DEPENDENT PROBLEMS	72-0507
AZIZ, A. K.	THE FUNDAMENTAL NOTIONS	72-0498
AZIZ, A. K.	THE MATHEMATICAL FOUNDATIONS OF THE FINITE ELEMENT METHOD - WITH APPLICATIONS TO PARTIAL DIFFERENTIA	73-0286
AZIZ, A. K.	THE PROBLEMS OF PERTURBATIONS IN THE FINITE ELEMENT METHOD	72-0505
AZIZ, A. K.	THEORY OF APPROXIMATION	72-0500
AZIZ, A. K.	VARIATIONAL PRINCIPLES	72-0501
BABA, K.	STRESS ANALYSIS OF FRAME STRUCTURE CONNECTED WITH PLATES	74-0265

9

BABCOCK, C. O.	A METHOD FOR THE PREDICTION OF STRESSES IN AN ISOTROPIC INCLUSION OR OREBODY OF IRREGULAR SHAPE	72-0827
BABCOCK, C. O.	A NEW METHOD OF ANALYSIS TO OBTAIN EXACT SOLUTIONS FOR STRESSES AND STRAINS IN CIRCULAR INCLUSIONS	74-1063
BABCOCK, C. O.	GEOMETRIC METHOD FOR THE PREDICTION OF STRESSES IN INCLUSIONS, OREBODIES, AND MINING SYSTEMS	74-0219
BABCOCK, C. O.	STRESS DISTRIBUTION AND ROCK FRACTURE ZONES IN THE ROOF OF LONGWALL FACE IN COAL MINE	73-0247
BABU, P. V.	FREQUENCY ANALYSIS OF ORTHOTROPIC CIRCULAR CYLINDRICAL PANELS BY THE FINITE STRIP METHOD	73-0274
BABUSKA, I.	A REMARK ON THE FINITE ELEMENT METHOD	70-0300
BABUSKA, I.	ANGLE CONDITION IN THE FINITE ELEMENT METHOD	74-1083
BABUSKA, I.	COMPUTATION OF DERIVATIVES IN THE FINITE ELEMENT METHOD	70-0304
BABUSKA, I.	CONNECTION BETWEEN THE FINITE DIFFERENCE LIKE METHODS AND THE METHODS BASED ON INITIAL VALUE PROBLEM	74-0877
BABUSKA, I.	ERROR BOUNDS FOR FINITE ELEMENT METHOD	71-0033
BABUSKA, I.	FINITE ELEMENT METHOD FOR ELLIPTIC EQUATIONS WITH DISCONTINUOUS COEFFICIENTS	70-0193
BABUSKA, I.	FINITE ELEMENT METHOD FOR DOMAINS WITH CORNERS	70-0244
BABUSKA, I.	FINITE ELEMENT METHOD WITH PENALTY	71-0512
BABUSKA, I.	FINITE ELEMENT METHOD FOR INFINITE DOMAINS	72-0288
BABUSKA, I.	FINITE ELEMENT METHOD FOR NON SMOOTH DOMAINS AND COEFFICIENTS	72-0504
BABUSKA, I.	FINITE ELEMENT METHOD WITH LAGRANGIAN MULTIPLIERS	73-0046
BABUSKA, I.	FINITE ELEMENT SCHEME FOR DOMAINS WITH CORNERS	72-0261
BABUSKA, I.	LECTURES ON MATHEMATICAL FOUNDATIONS OF THE FINITE ELEMENT METHOD	72-0777
BABUSKA, I.	MATHEMATICAL AND COMPUTATIONAL PROBLEMS IN REACTOR CALCULATIONS	73-0806
BABUSKA, I.	MATHEMATICAL AND COMPUTATIONAL PROBLEMS IN REACTOR CALCULATIONS	73-0973
BABUSKA, I.	METHOD OF WEAK ELEMENTS	74-1084
BABUSKA, I.	NONCONFORMING ELEMENTS IN THE FINITE ELEMENT METHOD	72-0341
BABUSKA, I.	NONCONFORMING ELEMENTS IN FINITE ELEMENT METHOD WITH PENALTY	73-0056
BABUSKA, I.	NONUNIFORM ERROR ESTIMATES FOR THE FINITE ELEMENT METHOD	74-0890
BABUSKA, I.	NUMERICAL EXPERIMENT WITH FINITE ELEMENT I	70-0303
BABUSKA, I.	NUMERICAL SOLUTION OF PARTIAL DIFFERENTIAL EQUATIONS	73-0968
BABUSKA, I.	ON THE ANGLE CONDITION IN THE FINITE ELEMENT METHOD	74-0907
BABUSKA, I.	ONE PARAMETER FAMILIES OF VARIATIONAL PRINCIPLES	72-0503
BABUSKA, I.	PRELIMINARY REMARKS	72-0497
BABUSKA, I.	PROBLEMS OF SELF ADAPTIVENESS IN THE FINITE ELEMENT METHOD	75-0097
BABUSKA, I.	PROPERTIES OF SOLUTIONS OF ELLIPTIC BOUNDARY VALUE PROBLEMS	72-0499
BABUSKA, I.	RATE OF CONVERGENCE OF THE FINITE ELEMENT METHOD	72-0502
BABUSKA, I.	SELF-ADAPTIVE REFINEMENTS IN THE FINITE ELEMENT METHOD	75-0712
BABUSKA, I.	SOLUTION OF PROBLEMS WITH INTERFACES AND SINGULARITIES	74-0822
BABUSKA, I.	SOLUTION OF PROBLEMS WITH INTERFACES AND SINGULARITIES	74-0891
BABUSKA, I.	SURVEY LECTURES ON THE MATHEMATICAL FOUNDATIONS OF THE FINITE ELEMENT METHOD	72-0670
BABUSKA, I.	THE EIGENVALUE PROBLEM	72-0506
BABUSKA, I.	THE FINITE ELEMENT METHOD FOR ELLIPTIC DIFFERENTIAL EQUATIONS	70-0027
BABUSKA, I.	THE FINITE ELEMENT METHOD FOR TIME DEPENDENT PROBLEMS	72-0507
BABUSKA, I.	THE FINITE ELEMENT METHOD WITH LAGRANGIAN MULTIPLIERS	72-0901
BABUSKA, I.	THE FINITE ELEMENT METHOD WITH PENALTY	73-0967
BABUSKA, I.	THE FUNDAMENTAL NOTIONS	72-0498
BABUSKA, I.	THE MATHEMATICAL FOUNDATIONS OF THE FINITE ELEMENT METHOD - WITH APPLICATIONS TO PARTIAL DIFFERENTIA	73-0286
BABUSKA, I.	THE PROBLEMS OF PERTURBATIONS IN THE FINITE ELEMENT METHOD	72-0505
BABUSKA, I.	THE RATE OF CONVERGENCE FOR THE FINITE ELEMENT METHOD	71-0034
BABUSKA, I.	THEORY OF APPROXIMATION	72-0500
BABUSKA, I.	VARIATIONAL PRINCIPLES	72-0501
BACHER, A. E. S	STRUCTURAL BEHAVIOR OF A FLEXIBLE METAL CULVERT UNDER A DEEP EARTH EMBANKMENT USING METHOD A BACKFIL	73-0739
BACK, N.	PRESSURE DISTRIBUTION AND DEFORMATIONS OF MACHINED COMPONENTS IN CONTACT.	73-0372
BACK, P. A. A.	THE SEISMIC DESIGN STUDY OF A DOUBLE CURVATURE ARCH DAM	69-0053
BACKLUND, J.	FINITE ELEMENT ANALYSIS OF NONLINEAR STRUCTURES	73-0975
BACKLUND, J.	FINITE ELEMENT ANALYSIS OF ELASTO-PLASTIC SHELLS	74-0249
BACKLUND, J.	LIMIT ANALYSIS OF REINFORCED CONCRETE SLABS BY A FINITE ELEMENT METHOD	72-0985
BACKLUND, J.	MIXED FINITE ELEMENT ANALYSIS OF PLATES IN BENDING. SMALL DEFLECTION THEORY OF ELASTIC AND ELASTO-PL	71-0783
BACKLUND, J.	MIXED FINITE ELEMENT ANALYSIS OF ELASTO-PLASTIC PLATES IN BENDING	72-0135
BACKUS, W. E.	SOME APPLICATIONS OF FINITE ELEMENT ANALYSIS TO SHELL BUCKLING PREDICTION	68-0023
BADER, R. M.	MATRIX METHODS IN STRUCTURAL MECHANICS	66-0065
BAGCHI, D. K.	BUCKLING OF PLATE GIRDER WEBS UNDER PARTIAL EDGE LOADINGS	70-0283
BAGCI, C.	COMPUTER METHOD FOR COMPUTING CRITICAL SPEEDS OF NON-UNIFORM SHAFTS GEARED SYSTEMS AND CURVED ASSEMB	73-0675
BAGCI, C.	FORCE AND TORQUE ANALYSIS OF CONSTRAINED SPACE MECHANISMS AND PLANE MECHANISMS WITH OFFSET LINKS BY	74-0569
BAGCI, C.	MATRIX-DISPLACEMENT DIRECT-ELEMENT METHOD FOR FORCE AND TORQUE ANALYSIS OF DETERMINATE AND INDETERMI	74-0531
BAGUELIN, F.	ETUDE DE LA CAPACITE PORTANTE DES PIEUX PAR LA METHODE DES ELEMENTS FINIS. INFLUENCE DE LA DILATANCE	75-0624
BAGUELIN, F.	LA CAPACITE PORTANTE DES PIEUX. (BEARING CAPACITY OF PILES)	75-0463
BAHGAT, B. M.	GENERAL FINITE ELEMENT VIBRATIONAL ANALYSIS OF PLANAR MECHANISMS	74-1107
BAHRANI, A. K.	APPLICATION OF FINITE ELEMENTS TO THE SOLUTION OF HELMHOLT'S EQUATION	68-0020
BAHRANI, A. K.	COMMENT ON "SOLUTION OF LAPLACE'S EQUATION IN A REGION WITH DIELECTRIC INTERFACES OF ARBITRARY SHAPE	70-0238
BAHRANI, A. K.	SOLUTION OF THREE-DIMENSIONAL FIELD PROBLEMS BY THE FINITE ELEMENT METHOD	67-0091
BAI, K. J.	A VARIATIONAL METHOD IN POTENTIAL FLOWS WITH A FREE SURFACE	72-0723
BAI, K. J.	DIFFRACTION OF OBLIQUE WAVES BY AN INFINITE CYLINDER	75-0605
BAINBRIDGE, C.	DESIGN APPRAISAL OF OFFSHORE PLATFORMS OF WELDED TUBULAR CONSTRUCTION	74-1165
BAKER, A. J.	A FINITE ELEMENT COMPUTATIONAL THEORY FOR THE MECHANICS AND THERMODYNAMICS OF A VISCOUS, COMPRESSIBL	71-0787
BAKER, A. J.	A FINITE ELEMENT SOLUTION ALGORITHM FOR THE NAVIER-STOKES EQUATIONS	74-0784
BAKER, A. J.	A HIGHLY STABLE EXPLICIT INTEGRATION TECHNIQUE FOR COMPUTATIONAL CONTINUUM MECHANICS	74-1277
BAKER, A. J.	A NUMERICAL SOLUTION TECHNIQUE FOR A CLASS OF TWO-DIMENSIONAL PROBLEMS IN FLUID DYNAMICS FORMULATED	69-0019
BAKER, A. J.	FINITE ELEMENT COMPUTATIONAL THEORY FOR THREE-DIMENSIONAL BOUNDARY LAYER FLOW	72-0632
BAKER, A. J.	FINITE ELEMENT SOLUTION FOR ENERGY CONSERVATION USING A HIGHLY STABLE EXPLICIT INTEGRATION ALGORITHM	72-0633
BAKER, A. J.	FINITE ELEMENT SOLUTION ALGORITHM FOR VISCOUS INCOMPRESSIBLE FLUID DYNAMICS	73-0063
BAKER, A. J.	FINITE ELEMENT SOLUTION ALGORITHM FOR INCOMPRESSIBLE FLUID DYNAMICS	74-0006
BAKER, A. J.	FINITE ELEMENT SOLUTION THEORY FOR THREE-DIMENSIONAL BOUNDARY FLOWS	74-0579
BAKER, A. J.	FINITE ELEMENT THEORY FOR THE MECHANICS AND THERMODYNAMICS OF A VISCOUS, COMPRESSIBLE MULTI-SPECIE F	71-0005
BAKER, A. J.	NONLINEAR INITIAL-BOUNDARY VALUE SOLUTIONS BY THE FINITE ELEMENT METHOD	74-0812
BAKER, A. J.	NUMERICAL SOLUTION TO THE DYNAMICS OF VISCOUS FLUID FLOW BY A FINITE ELEMENT ALGORITHM: A FIRST STEP	71-0004
BAKER, A. J.	NUMERICAL SOLUTION TO THE DYNAMICS OF VISCOUS FLUID FLOW BY A FINITE ELEMENT ALGORITHM: A FIRST STEP	71-0791
BAKER, A. J.	PREDICTIONS IN ENVIRONMENTAL HYDRODYNAMICS USING FINITE ELEMENT METHOD 1, THEORETICAL DEVELOPMENT	75-0003
BAKER, A. J.	PREDICTIONS IN ENVIRONMENTAL HYDRODYNAMICS USING THE FINITE ELEMENT METHOD 2, APPLICATIONS	75-0064
BAKER, A. J.	THE FINITE ELEMENT METHOD IN LOW SPEED AERODYNAMICS	75-0702
BAKER, G. A.	FINITE ELEMENT - GALERKIN METHODS FOR HYPERBOLIC EQUATIONS WITH DISCONTINUOUS COEFFICIENTS	73-0726
BAKER, G. A.	FINITE-ELEMENT METHOD FOR 1ST ORDER HYPERBOLIC EQUATIONS	75-0590
BAKER, L. E.	APPLICATION OF ELASTO-PLASTIC ANALYSIS IN ROCK MECHANICS BY FINITE ELEMENT METHOD	71-0349
BAKER, O. C.	A COMPARISON OF FINITE ELEMENT AND FOURIER SERIES SOLUTIONS AS APPLIED TO RADIALLY LOADED CIRCULAR R	68-0024
BAKHREBAH, S. A	FINITE ELEMENT ANALYSIS OF INTERSECTING CYLINDERS	73-0866
BAKHREBAH, S. A	FINITE ELEMENT SOLUTION OF NORMALLY INTERSECTING CYLINDERS	73-0723
BAKHSHANDEHPOUR	FINITE ELEMENT SOLUTION FOR AXISYMMETRIC TRANSIENT THERMAL STRESSES	72-0823
BALAAM, N. P.	FINITE ELEMENT ANALYSIS OF THE EFFECTS OF INSTALLATION ON PILE LOAD-SETTLEMENT BEHAVIOUR	74-0763
BALAKRISHNA RAO	EVALUATION OF RIGID PAVEMENTS BY NONDESTRUCTIVE TESTS.	72-0459
BALAS, J.	VELKE PRIEHYBY DOSIEK METODOU KONECNYCH PRVKOV (LARGE DEFLECTIONS OF PLATES ANALYZED BY FINITE ELEME	73-0373
BALASUBRAMANIAN	AN ITERATIVE FINITE ELEMENT METHOD OF VISCOUS FLOW	74-1353
BALASUBRAMANIAN	APPLICATION OF THE PSEUDO-FUNCTIONAL FINITE ELEMENT METHOD TO VISCOUS FLOW PROBLEMS	74-1183
BALASUBRAMANIAN	FINITE ELEMENT STRESS ANALYSIS OF TURBINE BLADE HOOKS	70-0479
BALDUR, R.	AXISYMMETRIC FINITE ELEMENT IN NATURAL COORDINATES	75-0826
BALDWIN, J.	SHAPE FUNCTION SUBROUTING FOR AN ISOPARAMETRIC THIN PLATE ELEMENT	73-0618
BALDWIN, J. J.	ANALYSIS OF SPACE CLOSING SPRINGS IN ORTHODONTICS	74-0236

10

BALIGH, M. M.	NUMERICAL STUDY OF UNIAXIAL AND TRIAXIAL ROCK COMPRESSION TESTS	73-0896
BALL, R. E.	A COMPARISON OF SEVERAL COMPUTER SOLUTIONS TO THREE STRUCTURAL SHELL ANALYSIS PROBLEMS	73-0884
BALL, R. E.	A COMPARISON OF SEVERAL COMPUTER SOLUTIONS TO THREE STRUCTURAL SHELL ANALYSIS PROBLEMS	73-0907
BALL, R. E.	AIRCRAFT FUEL TANK VULNERABILITY TO HYDRAULIC RAM: MODIFICATTION OF THE NORTHROP FINITE ELEMENT COMP	74-0724
BALMER, H.	ASKA 105 - A TECHNICAL MANUAL FOR THE DIRECT SOLUTION AND INVERSION PACKAGE	68-0255
BALMER, H.	ASKA 105 A TECHNICAL MANUAL AND PROGRAMMING THEORY FOR THE ELEMENTS, FLA, TRIM, SPARTA AND TET	68-0254
BALMER, H.	ELASTOPLASTIC AND CREEP ANALYSIS WITH THE ASKA PROGRAM SYSTEM	74-0863
BALMER, H.	FINITE ELEMENT ANALYSIS OF THERMOMECHANICAL PROBLEMS	71-0030
BALMER, H.	GENERAL BEAM ELEMENT FOR ASKA	69-0458
BALMER, H.	MATERIAL NONLINEARITIES	72-0647
BALMER, J. H.	MATERIAL NONLINEARITIES IN THE FINITE ELEMENT ANALYSIS	74-0647
BALSARA, J. P.	VIBRATION CHARACTERISTICS OF THE NORTH FORK DAM MODEL	74-0845
BALTRUKONIS, J.	DYNAMICS OF SUBMARINE STRUCTURES, PART 2	68-0025
BALTRUKONIS, J.	DYNAMICS OF SUBMARINE STRUCTURES, PART 1.	69-0054
BALUCH, M. H.	FINITE ELEMENT ANALYSIS OF A THICK WALLED MICROPOLAR CYLINDER LOADED AXISYMMETRICALLY	73-0847
BALUCH, M. H.	FINITE ELEMENT APPROACH TO PLANE MICROELASTICITY	72-0205
BALUCH, M. H.	FINITE ELEMENT APPROACH TO BENDING OF MICROPOLAR PLATES	74-0255
BAMFORD, R. M.	EFFICIENT SOLUTION OF LOAD-DEFLECTING EQUATIONS	69-0396
BAMFORD, W. H.	REACTOR COOLANT PUMP FLYWHEELS OVERSPEED EVALUATION	74-0506
BANAVALKAR, P.	ANALYSIS OF THIN-SHELL HYPERBOLIC PARABOLOID SHELLS	72-0121
BANDLER, J. W.	IEEE-G-MTT INTERNATIONAL MICROWAVE SYMPOSIUM, 1973: APPLICATIONS IN THE 70'S	73-0374
BANERJEE, S.	APPLICATION OF THE FINITE ELEMENT METHOD TO THE NONLINEAR ANALYSIS OF NUCLEAR REACTOR FUEL BEHAVIOR	74-1246
BANGASH, Y.	CIRCUMFERENTIAL PRESTRESSING LOAD ANALYSIS DUE TO WIRE/STRAND WINDING SYSTEMS ON PRESTRESSED CONCRET	74-1202
BANGASH, Y.	RIDDLE OF BONDED AND UNBONDED TENDONS IN PRESTRESSED CONCRETE REACTOR VESSELS	74-0519
BANGASH, Y.	STRUCTURAL DESIGN OF CIRCULAR DIVERSION, POWER AND PRESSURE TUNNELS	71-0315
BANKS, D. C.	APPLICATION OF FINITE ELEMENT METHOD IN DETERMINING STABILITY OF CRATER SLOPES, PRELIMINARY REPORT	68-0026
BANKS, D. C.	APPLICATION OF FINITE ELEMENT METHOD IN DETERMINING STABILITY OF CRATER SLOPES	72-0888
BARACH, D.	A FINITE-ELEMENT APPROACH TO SCATTERING FROM ELASTIC SPHERICAL SHELLS	71-0702
BARACH, D.	A FINITE-ELEMENT APPROACH TO THE ANALYSIS OF TANGENTIALLY POLARIZED PIEZOELECTRIC-CERAMIC FREE-FLOOD	74-0802
BARACH, D.	APPLICATIONS OF THE FINITE-ELEMENT METHOD AND COMPUTER GRAPHICS. (A VIBRATIONAL ANALYSIS OF A SONAR	72-0729
BARACH, D.	FINITE ELEMENT ANALYSIS OF ACOUSTICALLY RADIATING STRUCTURES WITH APPLICATIONS TO SONAR TRANSDUCERS	74-0524
BARACH, D.	FINITE-ELEMENT APPROACH TO ACOUSTIC SCATTERING FROM ELASTIC STRUCTURES	75-0069
BARACH, D.	VIBRATIONAL ANALYSIS OF THE FLEXING HEAD OF THE BQS-6 TRANSDUCER	71-0708
BARACH, D. R.	FINITE ELEMENT ANALYSIS OF ACOUSTICALLY RADIATING STRUCTURES WITH APPLICATIONS TO SONAR TRANSDUCERS	73-0075
BARACH, D. R.	FINITE ELEMENT APPROACH TO ACOUSTIC RADIATION FROM ELASTIC STRUCTURES	74-0121
BARADAT, Y.	UTILISATION DE MODELES MATHEMATIQUES MULTIDIMENSIONNELS POUR L'ETUDE DES ECOULEMENTS DANS LES DIGUES	73-0381
BARBERA, F. J.	MATHEMATICAL MODELING OF SPINNING ELASTIC BODIES FOR MODAL ANALYSIS	73-0441
BARCLAY, R. G.	NONLINEAR CONSIDERATION OF GRAVITY IN A STIFFNESS TEST OF A WEAK STRUCTURE AT SMALL STRAINS	71-0035
BAREISS, E. H.	ADJOINING APPROPRIATE SINGULAR ELEMENTS TO TRANSPORT THEORY COMPUTATIONS	74-0946
BAREISS, E. H.	FINITE ELEMENTS IN NEUTRON TRANSPORT THEORY	73-0961
BARFIELD, W. D.	NUMERICAL METHOD FOR GENERATING ORTHOGONAL CURVILINEAR MESHES	74-0443
BARFIS, E. H.	SINGULAR ELEMENTS IN VARIATIONAL AND FINITE ELEMENT TRANSPORT COMPUTATIONS	73-0902
BARKER, R. M.	A FINITE ELEMENT ANALYSIS INCLUDING TRANSVERSE SHEAR EFFECTS FOR APPLICATIONS TO LAMINATED PLATES	71-0216
BARKER, R. M.	A THEOREM FOR OPTIMUM IDEALIZATIONS IN FINITE-ELEMENT ANALYSIS	72-0791
BARKER, R. M.	ANALYSIS OF BONDED JOINTS IN VEHICULAR STRUCTURES	73-0375
BARKER, R. M.	FINITE ELEMENT ANALYSIS OF BENDING-EXTENSIONAL COUPLING IN LAMINATED COMPOSITES	70-0292
BARKER, R. M.	FINITE ELEMENT BENDING ANALYSIS OF REISSNER PLATES	70-0214
BARKER, R. M.	SINGER: A COMPUTER CODE FOR GENERAL ANALYSIS OF TWO-DIMENSIONAL CONCRETE STRUCTURES. VOL. 11. PROGRA	75-0723
BARKER, R. M.	SINGER: A COMPUTER CODE FOR GENERAL ANALYSIS OF TWO-DIMENSIONAL REINFORCED CONCRETE STRUCTURES. VOLU	75-0724
BARKER, R. M.	SINGER: A COMPUTER CODE FOR GENERAL ANALYSIS TO TWO-DIMENSIONAL CONCRETE STRUCTURES. VOLUME 4. DEMON	75-0730
BARKER, R. M.	SINGER: A COMPUTER CODE FOR GENERAL ANALYSIS OF TWO-DIMENSIONAL REINFORCED CONCRETE STRUCTURES. VOLU	75-0731
BARKER, R. M.	STRESS CONCENTRATIONS NEAR A DISCONTINUITY IN FIBROUS COMPOSITES	71-0316
BARKER, R. M.	THEOREM FOR OPTIMUM FINITE-ELEMENT IDEALIZATIONS	73-0255
BARKER, R. M.	THREE DIMENSIONAL ANALYSIS OF STRESS CONCENTRATIONS NEAR HOLES IN LAMINATED COMPOSITES	72-0713
BARKER, R. M.	THREE-DIMENSIONAL ANALYSIS FOR THE STRESS DISTRIBUTION NEAR CIRCULAR HOLES IN LAMINATED COMPOSITES	74-0704
BARKER, R. M.	THREE-DIMENSIONAL FINITE ELEMENT ANALYSIS OF LAMINATED COMPOSITES	72-0345
BARKER, R. M.	THREE-DIMENSIONAL FINITE-ELEMENT COMPUTER PROGRAM - USERS GUIDE	74-0755
BARKSDALE, R. D	NONLINEAR THEORY FOR PREDICTING THE PERFORMANCE OF FLEXIBLE HIGHWAY PAVEMENTS	70-0143
BARLA, G.	DIRECT TENSILE TESTING OF ANISOTROPIC ROCKS	74-0989
BARLA, G.	INDIRECT TENSILE TESTING OF ANISOTROPIC ROCKS	73-0376
BARLA, G.	STRESSES AND DISPLACEMENTS AROUND TWO ADJACENT CIRCULAR OPENINGS NEAR TO THE GROUND SURFACE	74-0993
BARLOW, J.	A STIFFNESS MATRIX FOR A CURVED MEMBRANE SHELL	69-0055
BARLOW, J.	COMMENT ON MATRICES FOR THE DIRECT STIFFNESS METHOD BY R. J. MELOSH	64-0022
BARLOW, J.	FINITE ELEMENTS FOR IMPROVED SKIN STRINGER STRUCTURAL ANALYSIS	70-0297
BARLOW, J.	INERTIA LOADING IN FINITE-ELEMENT ANALYSIS OF STRUCTURES SUBJECT TO COMPOUND MOTION	70-0351
BARLOW, J.	ON THE AUTOMATIC SELECTION OF INDEPENDENT DISPLACEMENT VARIABLES IN GENERAL PURPOSE FINITE ELEMENT C	73-0676
BARNARD, A. J.	A SANDWICH PLATE FINITE ELEMENT	72-0077
BARNARD, A. J.	NONLINEAR RESPONSE OF PLATES TO IMPULSIVE LOADING USING HYBRID FINITE ELEMENTS	74-1018
BARNARD, A. J.	THE FINITE ELEMENT METHOD APPLIED TO THE ANALYSIS OF SANDWICH PLATE AND SHELL STRUCTURES	72-0824
BARNARD, AL. J.	THE ELASTO-PLASTIC AND LARGE-DISPLACEMENT RESPONSE OF PLATES TO BLAST LOADING	73-0731
BARNES, J. J.	2-DIMENSIONAL SIMULATION OF SEMICONDUCTOR-DEVICES USING FINITE ELEMENT METHODS	75-0548
BARNES, J. J.	TWO-DIMENSIONAL FINITE ELEMENT SIMULATION OF SEMICONDUCTOR DEVICES	74-0260
BARNHART, D. F.	STRESS ANALYSIS OF INDUSTRIAL COMPONENTS WITH PLASTIC AND FINITE ELEMENT MODELS	74-0512
BARNHILL, R. E.	BLENDING FUNCTION FINITE ELEMENTS FOR CURVED BOUNDARIES	75-0093
BARNHILL, R. E.	COMPUTABLE ERROR BOUNDS FOR THE FINITE ELEMENT METHOD FOR ELLIPTIC BOUNDARY VALUE PROBLEMS	74-1219
BARNHILL, R. E.	ERROR ANALYSIS OF FINITE ELEMENT METHODS WITH TRIANGLES FOR ELLIPTIC BOUNDARY VALUE PROBLEMS	73-0072
BARNHILL, R. E.	FINITE ELEMENT METHODS FOR ELLIPTIC BOUNDARY VALUE PROBLEMS CONTAINING SINGULARITIES	72-0360
BARNHILL, R. E.	FUNCTIONAL ANALYSIS AND THE FINITE ELEMENT METHOD APPLIED TO ELLIPTIC BOUNDARY VALUE PROBLEMS	72-0056
BARNHILL, R. E.	SARD KERNEL THEOREMS ON TRIANGULAR AND RECTANGULAR DOMAINS WITH EXTENSIONS AND APPLICATIONS TO FINIT	72-0361
BARNHILL, R. E.	THE EXTENSION AND APPLICATION OF SARD KERNEL THEOREMS TO COMPUTE FINITE ELEMENT ERROR BOUNDS	72-0519
BARNWELL, U. L.	FINITE ELEMENT SOLUTION FOR WAVE PROPAGATION IN LAYERED MEDIA CAUSED BY A NUCLEAR DETONATION	67-0016
BARON, M. L.	MATHEMATICAL MODELS FOR GEOLOGICAL MATERIALS FOR WAVE PROPAGATION STUDIES	71-0740
BARONET, C. N.	ELASTOPLASTIC INDENTATION OF HALF-SPACE BY AN INFINITELY LONG RIGID CIRCULAR CYLINDER	71-0395
BARONET, C. N.	THE ELASTO-PLASTIC INDENTATION OF A HALF-SPACE BY A RIGID SPHERE	71-0113
BARRACO, A.	APPLICATION DE LA METHODE DES ELEMENTS FINIS AU CALCUL DES PLAQUES FLECHIES (APPLICATION OF THE FINI	73-0019
BARRON, G. E.	FINITE ELEMENT AND CUMULATIVE DAMAGE ANALYSIS OF A KEYHOLE TEST SPECIMEN	75-0163
BARRON, G. E.	THE USE OF ELASTIC-PLASTIC FINITE ELEMENT ANALYSIS IN THE CALCULATION OF CUMULATIVE FATIGUE DAMAGE	74-0314
BARROW, H.	THE FINITE ELEMENT SOLUTION OF FLOW AND HEAT TRANSFER IN ELLIPTICAL DUCTS	74-0039
BARROW, H.	THE PREDICTION OF FULLY DEVELOPED TURBULENT FLOW IN DUCTS BY THE FINITE ELEMENT METHOD	75-0117
BARRY, J. M.	APPLICATION OF THE FINITE ELEMENT METHOD TO ELLIPTIC PROBLEMS	74-0570
BARRY, J. M.	THREE DIMENSIONAL STRESS ANALYSIS OF PENETRATIONS IN THICK WALLED PRESSURE VESSELS	74-0422
BARRY, W. R.	FINITE ELEMENT ANALYSIS OF CASING THREADS	75-0330
BARRY, W. R.	FINITE-ELEMENT ANALYSIS OF CASING THREADS	75-0417
BARRY, W. R.	STRUCTURAL ANALYSIS OF PETROCHEMICAL EQUIPMENT BY THE FINITE ELEMENT METHOD	73-0377
BARSOUM, R. S.	A FINITE ELEMENT FORMULATION FOR THE GENERAL STABILITY ANALYSIS OF THIN WALLED MEMBERS	70-0337
BARSOUM, R. S.	APPLICATION OF QUADRATIC ISOPARAMETRIC FINITE ELEMENTS IN LINEAR FRACTURE MECHANICS	74-0521
BARSOUM, R. S.	DISCUSSION OF PROCEEDINGS, PAPER 6612, ELASTIC BUCKLING OF TWO-COMPONENT COLUMNS	70-0321
BARSOUM, R. S.	FINITE ELEMENT ANALYSIS OF TORSIONAL AND TORSIONAL-FLEXURAL STABILITY PROBLEMS	70-0028
BARSOUM, R. S.	FINITE ELEMENT METHOD APPLIED TO THE PROBLEM OF STABILITY OF A NON-CONSERVATIVE SYSTEM	71-0291
BARSOUM, R. S.	FURTHER APPLICATION OF QUADRATIC ISOPARAMETRIC FINITE-ELEMENTS TO LINEAR FRACTURE MECHANICS OF PLATE	75-0639
BARSOUM, R. S.	USE OF ISOPARAMETRIC FINITE-ELEMENTS IN LINEAR FRACTURE MECHANICS	75-0352
BART, J.	KOMBINATION EINES INTEGRALGLEICHUNGS VERFAHRENS MIT METHODE DER FINITEN ELEMENTEN ZUR BERECHNUNG EBE	74-0856

11

BARTEL, D. L.	FINITE ELEMENT ANALYSIS OF A BONE-PLATE-SCREW SYSTEM	74-0920
BARTEL, L. C.	COMPUTER SIMULATION OF AN IN SITU SHALE-OIL RETORT	74-1059
BARTELDS, B.	FINITE ELEMENT ANALYSIS OF SANDWICH PANELS	70-0449
BARTELDS, G.	APPLICATION OF THE FINITE ELEMENT DISPLACEMENT METHOD TO PROBLEMS OF ELASTOPLASTIC DEFORMATION	72-0199
BARTELDS, G.	FINITE ELEMENT ANALYSIS OF SANDWICH PANELS	71-0662
BARTELDS, G.	SANDWICH AND COMPOSITE ELEMENTS	75-0136
BARTEN, H. J.	STRESS AND VIBRATION ANALYSIS OF TURBOPUMP INDUCER BLADES BY FINITE ELEMENTS	71-0363
BARTEN, H. J.	STRESSES AND VIBRATION ANALYSIS OF INDUCER BLADES USING FINITE ELEMENT TECHNIQUE	70-0305
BARTHOLOME, G.	FRACTURE AND SAFETY ANALYSIS OF NUCLEAR PRESSURE VESSELS	73-0257
BARUCH, M.	INCLUSION OF TRANSVERSE-SHEAR DEFORMATION IN FINITE-ELEMENT DISPLACEMENT FORMULATIONS - COMMENT	75-0406
BARWELL, F. T.	LUBRICATION PROBLEMS - THE SELECTION OF MATHEMATICAL MODELS	75-0881
BARWELL, F. T.	THE ROLE OF NUMERICAL METHODS IN THE STUDY OF LUBRICATION	74-0042
BASA, J.	ASPECTS OF MECHANICAL BEHAVIOR OF ROCK UNDER STATIC AND CYCLIC LOADING. PART A: MECHANICAL BEHAVIOR	72-0762
BASA, J. E.	STATIC ANALYSIS OF STIFFENED SHELLS BY THE FINITE ELEMENT METHOD	71-0803
BASAS, J.	ASPECTS OF MECHANICAL BEHAVIOR OF ROCK UNDER STATIC AND CYCLIC LOADING. PART A. MECHANICAL BEHAVIOR	72-0761
BASSENGE, CH.	NUMERISCHE BERECHNUNG INSTATIONAERER TEMPERATURFELDER UND THERMISCH BEDINGTER VERFORMUNGEN VON WERKZ	75-0847
BASTING, W. J.	PREDICTION OF TEMPERATURE DISTRIBUTIONS IN CARGOES WITH THE AID OF A COMPUTER PROGRAM USING THE METH	74-1215
BASU, A. K.	NEW LIGHT ON THE NAYAK ALPHA TECHNIQUE	73-0614
BATDORF, N.	THE ROLE OF COMPUTER GRAPHICS IN THE STRUCTURAL DESIGN PROCESS	68-0215
BATDORF, S. B.	A STATISTICAL THEORY FOR THE FRACTURE OF BRITTLE STRUCTURES SUBJECTED TO NONUNIFORM POLYAXIAL STRESS	73-0928
BATES, C. J.	A COMPUTATIONAL TECHNIQUE FOR THE EFFICIENT HANDLING OF LARGE MATRICES	73-1083
BATHE, K. J.	ANALYSIS OF PLATES IN BENDING USING LARGE FINITE ELEMENTS	70-0185
BATHE, K. J.	ANALYSIS OF PLATES SUBJECTED TO IN-PLANE FORCES USING LARGE FINITE ELEMENTS	70-0271
BATHE, K. J.	COMPUTER PROGRAM FOR STATIC AND DYNAMIC ANALYSIS OF LINEAR STRUCTURAL SYSTEMS	72-0683
BATHE, K. J.	FINITE ELEMENT ANALYSIS OF DYNAMIC RESPONSE	72-0681
BATHE, K. J.	FINITE ELEMENT ANALYSIS OF LINEAR AND NONLINEAR HEAT TRANSFER	74-0626
BATHE, K. J.	FINITE ELEMENT FORMULATIONS FOR LARGE DISPLACEMENT AND LARGE STRAIN ANALYSIS	73-0457
BATHE, K. J.	NONSAP - A GENERAL FINITE ELEMENT PROGRAM FOR NONLINEAR DYNAMIC ANALYSIS OF COMPLEX STRUCTURES	73-0740
BATHE, K. J.	NONSAP - A NONLINEAR STRUCTURAL ANALYSIS PROGRAM	74-1006
BATHE, K. J.	NONSAP: A STRUCTURAL ANALYSIS PROGRAM FOR STATIC AND DYNAMIC RESPONSE OF NONLINEAR SYSTEMS	74-0848
BATHE, K. J.	SAP IV. A STRUCTURAL ANALYSIS PROGRAM FOR STATIC AND DYNAMIC RESPONSE OF LINEAR SYSTEMS	73-0650
BATHE, K. J.	SOLUTION METHODS FOR EIGENVALUE PROBLEMS IN STRUCTURAL MECHANICS	73-1095
BATHE, K. J.	STATIC AND DYNAMIC GEOMETRIC AND MATERIAL NONLINEAR ANALYSIS	74-1302
BATHE, K. J.	THEORETICAL BASIS FOR CBL/NONSAP, A NONLINEAR STRUCTURAL ANALYSIS PROGRAM	74-0740
BATHE, K. J.	USERS MANUAL FOR CEL/NONSAP, A NONLINEAR STRUCTURAL ANALYSIS PROGRAM	74-0706
BATHE, K. M.	FINITE ELEMENT FORMULATIONS FOR LARGE DEFORMATION DYNAMIC ANALYSIS	75-0564
BATOZ, J. L.	BUCKLING OF DEEP SHELLS	73-0849
BATOZ, J. L.	DEVELOPMENT OF TWO SIMPLE SHELL ELEMENTS	72-0676
BATOZ, J. L.	FINITE ELEMENT LARGE DEFLECTION ANALYSIS OF SHALLOW SHELLS	75-0356
BATOZ, J. L.	LARGE DEFLECTION ANALYSIS OF ARBITRARY SHALLOW SHELLS BY THE FINITE ELEMENT METHOD	73-0937
BATSON, G. B.	INFLATION FORMING OF STEEL FIBER-REINFORCED CONCRETE DOMES	74-0705
BATT, J. R.	MAGIC 111: AN AUTOMATED GENERAL PURPOSE SYSTEM FOR STRUCTURAL ANALYSIS VOLUME 1. ENGINEER'S MANUAL	72-0727
BAUDENDISTEL, M	EFFECT OF DISCONTINUITIES ON STRESSES AND DEFORMATIONS OF ROCK AROUND A TUNNEL. (EINFLUSS VON DISKON	70-0170
BAUDENDISTEL, M	INTERACTION OF ROCK AND TUNNEL LINING	73-1057
BAUDOUIN, M.	LARGE DISPLACEMENT FORMULATION FOR ELASTIC BODIES	70-0429
BAUER, A. M.	COMOC THERMAL ANALYSIS VARIANT, USER'S MANUAL	72-0701
BAUSINGER, R.	ELASTOSTATIK-ELEMENTMETHODE (ESEM)	69-0415
BAYER, D. M.	A FINITE ELEMENT TECHNIQUE FOR THE IDEALIZATION OF CONTINUOUS SHELL STRUCTURES	68-0290
BAYLOR, J. L.	TRANAL: A 3-D FINITE ELEMENT CODE FOR TRANSIENT NONLINEAR ANALYSIS	74-0707
BAZANT, E.	ANWENDUNG DER FINITE-ELEMENT-METHODE BEI DER BERECHNUNG VON REAKTORDRUCKBEHAELTERN	71-0036
BAZANT, E.	APPLICATION OF THE FINITE ELEMENT METHOD TO THE CALCULATION OF REACTOR PRESSURE VESSELS	71-0645
BAZANT, Z. P.	CREEP AND SHRINKAGE LAW FOR CONCRETE AT VARIABLE HUMIDITY	74-0668
BAZANT, Z. P.	FINITE-ELEMENT FOR BUCKLING OF CURVED BEAMS AND SHELLS WITH SHEAR	75-0607
BAZANT, Z. P.	INCREMENTAL FINITE ELEMENT MATRICES	74-0354
BAZANT, Z. P.	LARGE DEFLECTION SPATIAL BUCKLING OF THIN WALLED BEAMS AND FRAMES	73-1085
BAZANT, Z. P.	MATRIX DIFFERENTIAL EQUATION AND HIGHER-ORDER NUMERICAL METHODS FOR PROBLEMS OF NON-LINEAR CREEP, VI	72-0198
BAZANT, Z. P.	THREE-DIMENSIONAL HARMONIC FUNCTIONS NEAR TERMINATION OR INTERSECTION OF GRADIENT SINGULARITY LINES:	72-0178
BAZELEY, G. P.	TRIANGULAR ELEMENTS IN PLATE BENDING CONFORMING AND NON-CONFORMING SOLUTIONS	66-0011
BEADLE, C. W.	RANDOM VIBRATION RESPONSE OF CANTILEVER PLATES USING THE FINITE ELEMENT METHOD	70-0181
BEADLE, C. W.	STRENGTH DISTRIBUTIONS OF SINGLE FILAMENTS	75-0305
BEARD, R. M.	STRUCTURAL RESPONSE OF UNSTIFFENED TOROIDAL SHELLS	69-0198
BEAUFAIT, F. W.	HIGH-ORDER FINITE ELEMENT TECHNIQUE FOR PRISMATIC AND NONPRISMATIC FOLDED PLATE STRUCTURES	74-1130
BEAUMONT, C.	CRUSTAL STRUCTURE FROM SURFACE LOAD TILTS USING A FINITE ELEMENT MODEL	72-0005
BEAVERS, J. E.	A HIGH-ORDER FINITE ELEMENT FOR PRISMATIC AND NONPRISMATIC FOLDED PLATE STRUCTURES	75-0889
BEAVERS, J. E.	HIGH-ORDER FINITE ELEMENT TECHNIQUE FOR PRISMATIC AND NONPRISMATIC FOLDED PLATE STRUCTURES	74-1130
BECK, H.	ZUSAMMENWIRKEN VON EINZELNEN FERTIGTEILEN ALS GROSSFLAECHIGE SCHEIBE (INTERACTION OF INDIVIDUAL PREF	73-0552
BECKER, E. B.	A NUMERICAL SOLUTION OF A CLASS OF PROBLEMS OF FINITE ELASTIC DEFORMATION	66-0097
BECKER, E. B.	APPLICATION OF THE FINITE ELEMENT METHOD TO STRESS ANALYSIS OF SOLID PROPELLANT ROCKET GRAINS	65-0009
BECKER, E. B.	APPLICATION OF THE FINITE ELEMENT METHOD TO HEAT CONDUCTION IN SOLIDS	67-0005
BECKER, E. B.	COMPUTATIONALLY EFFICIENT FINITE ELEMENT ANALYSIS OF VISCOUS FLOW PROBLEMS	74-1050
BECKER, E. B.	COMPUTATIONALLY EFFICIENT FINITE ELEMENT ANALYSIS OF VISCOUS FLOW PROBLEMS	74-1258
BECKER, E. B.	FINITE ELEMENT ANALYSIS OF VISCOUS, INCOMPRESSIBLE FLUID FLOW	75-0760
BECKER, E. B.	FINITE ELEMENT STRESS FORMULATION FOR WAVE PROPAGATION	73-1043
BECKER, E. B.	FINITE ELEMENT STRESS FORMULATION FOR DYNAMIC ELASTIC-PLASTIC ANALYSIS	75-0484
BECKER, E. B.	FINITE-ELEMENT ANALYSIS OF STRUCTURAL PROBLEMS	74-0606
BECKER, E. B.	INVESTIGATION OF TECHNIQUES OF THREE-DIMENSIONAL FINITE ELEMENT STRESS ANALYSIS	70-0029
BECKER, E. B.	LARGE DISPLACEMENT ANALYSIS OF THIN PLATES	73-0261
BECKER, E. B.	ON APPLICATION OF THE FINITE ELEMENT METHOD TO LIMIT ANALYSIS	74-1237
BECKER, E. B.	ORGANIZATION AND FUNCTIONAL PURPOSE OF A FINITE ELEMENT COMPUTER PROGRAM	74-0362
BECKER, E. B.	SHOCK TUNNEL TESTS OF PRELOADED AND ARCHED WALL PANELS FINITE ELEMENT STRESS FORMULATION FOR WAVE PR	74-0892
BECKER, E. B.	SOME STRESS INTENSITY CALCULATIONS USING FINITE ELEMENTS	74-0423
BECKER, E. B.	THE APPLICATION OF FINITE ELEMENT METHODS TO THE SOLUTION OF STRESS AND DIFFUSION PROBLEMS IN CONTIN	68-0034
BECKER, J. M.	FIRES-RC A COMPUTER PROGRAM FOR THE FIRE RESPONSE OF STRUCTURES-REINFORCED CONCRETE FRAMES	74-0747
BECKER, J. M.	FIRES-T. A COMPUTER PROGRAM FOR THE FIRE RESPONSE OF STRUCTURES - THERMAL	74-0850
BECKERS, P.	COMPUTER INTEGRATION OF CONGRUENTLY TRANSFORMED MATRICES	71-0580
BECKERS, P.	DELINQUENT FINITE ELEMENTS FOR SHELL IDEALIZATION	75-0478
BECKERS, P.	DIGITAL COMPUTATION OF STRESSES AND DEFLECTIONS IN A BOX BEAM	67-0073
BECKERS, P.	DIGITAL COMPUTATION OF STRESSES AND DEFLECTIONS IN A BOX BEAM - A PERFORMANCE COMPARISON BETWEEN FIN	69-0355
BECKERS, P.	IMPROVEMENTS OF FINITE ELEMENT SOLUTIONS FOR STRUCTURAL AND NON-STRUCTURAL APPLICATIONS	71-0229
BECKERS, P.	IMPROVEMENTS OF FINITE ELEMENT SOLUTIONS FOR STRUCTURAL AND NONSTRUCTURAL APPLICATIONS	72-0671
BECKERS, P.	LES FONCTIONS DE TENSION DANS LA METHODE DES ELEMENTS FINIS	73-1113
BECKETT, R. E.	FINITE-ELEMENT METHOD APPLIED TO HEAT CONDUCTION IN SOLIDS WITH NONLINEAR BOUNDARY CONDITIONS	73-0005
BECKETT, V.	ANALYSIS OF LOADING AND THERMAL EFFECTS ON FUELLED GRAPHITE BRICKS FOR A HIGH-TEMPERATURE GAS COOLED	71-0085
BEDFORD, K. W.	A NUMERICAL INVESTIGATION OF STABLY STRATIFIED, WIND DRIVEN CAVITY FLOW BY THE FINITE ELEMENT METHOD	74-1372
BEDFORD, K. W.	CONVECTIVE TRANSPORT FINITE ELEMENT ANALOG	75-0289
BEECKMAN, R.	A FINITE ELEMENT ANALYSIS OF SHELLS OF REVOLUTION	73-0759
BEECKMAN, R.	A FINITE ELEMENT ANALYSIS OF BODIES OF REVOLUTION. A COMPARISON BETWEEN CURRENTLY USED TYPES OF ELEM	73-1319
BEER, G.	TRANSIENT HEAT FLOW IN SOLIDS	74-0461
BEEUWKES, R.	A NUMERICAL AND EXPERIMENTAL INVESTIGATION OF THE USE OF J-INTEGRALS	72-0858
BEGIS, D.	APPLICATION DE LA METHODE DES ELEMENTS FINIS A LA RESOLUTION D'UN PROBLEME DE DOMAINE OPTIMAL	74-0412
BEGIS, D.	APPLICATION OF FINITE-ELEMENT METHOD TO APPROXIMATION OF OPTIMAL DOMAIN PROBLEM - SOLUTION METHODS F	75-0349
BEGLEY, J. A.	HEAVY SECTION STEEL TECHNOLOGY PROGRAM TECHNICAL OR PROGRAMMATIC MANUSCRIPT NO. 25. VARIABLE THICKNE	73-0948

BEGLINGER, V.	EFFECTS OF SHEAR DEFORMATION, ROTARY INERTIA, AND ELASTICITY OF THE SUPPORT ON THE RESONANCE FREQUEN	75-0369
BEISINGER, Z. E	CHILES: A FINITE ELEMENT COMPUTER PROGRAM THAT CALCULATES THE INTENSITIES OF LINEAR ELASTIC SINGULAR	73-0959
BEISINGER, Z. E	SLADE D: A COMPUTER PROGRAM FOR THE DYNAMIC ANALYSIS OF THIN SHELLS	73-0964
BEISINGER, Z. E	SLADE, A COMPUTER PROGRAM FOR THE STATIC ANALYSIS OF THIN SHELLS	70-0079
BEISINGER, Z. E	THE ANALYSIS OF THIN SHELLS WITH TRANSVERSE SHEAR STRAINS BY THE FINITE ELEMENT METHOD	68-0098
BEISINGER, Z. E	THE ANALYSIS OF THIN SHELLS BY THE FINITE ELEMENT METHOD	70-0078
BEISINGER, Z. E	THE TRANSIENT DYNAMIC ANALYSIS OF THIN SHELLS BY THE FINITE ELEMENT METHODS	71-0142
BEISSNER, R. E.	DEFORMATION MECHANICS OF DEEP SURFACE FLAW CRACKS	72-0785
BELKUNE, R. M.	FREE VIBRATIONS OF PLATE-BEAM SYSTEMS	73-0322
BELL, C. G.	EARTHQUAKE-RESISTANT DESIGN OF ENGINEERING STRUCTURES	73-0284
BELL, J. C.	NON-LINEAR ANALYSIS OF REINFORCED CONCRETE SLABS	72-0191
BELL, K.	A REFINED TRIANGULAR PLATE BENDING FINITE ELEMENT	69-0056
BELL, K.	ANALYSIS OF THIN PLATES IN BENDING USING TRIANGULAR FINITE ELEMENTS	68-0213
BELL, K.	FINITE ELEMENT METHODS IN STRESS ANALYSIS	69-0126
BELL, K.	ON THE QUINTIC TRIANGULAR PLATE BENDING ELEMENT	72-0677
BELL, K.	ORSAM - A PROGRAMMING SYSTEM FOR THE FINITE ELEMENT METHOD	73-0528
BELL, K.	TRIANGULAR PLATE BENDING ELEMENTS	69-0057
BELL, R.	AN ENGINEERING APPROACH TO FINITE ELEMENT ANALYSIS OF NUCLEAR COMPONENTS	73-1054
BELL, R.	ENGINEERING APPROACH TO FINITE ELEMENT ANALYSIS OF NUCLEAR COMPONENTS	73-0839
BELLEVAUX, C.	APPLICATIONS OF FINITE ELEMENT METHODS IN FLUID DYNAMICS	75-0184
BELYTSCHKO, T.	CONVECTED COORDINATES FOR TRANSIENT NONLINEAR FINITE-ELEMENT ANALYSIS	73-0850
BELYTSCHKO, T.	DYNAMIC RESPONSE OF FAST-REACTOR CORE SUBASSEMBLIES	74-0540
BELYTSCHKO, T.	EXPLICIT INTEGRATION - STABILITY, SOLUTION PROPERTIES, COST	75-0298
BELYTSCHKO, T.	FINITE ELEMENT ANALYSIS, AN AUTOMOBILE ENGINEER'S TOOL	74-0238
BELYTSCHKO, T.	FINITE ELEMENT ANALYSIS OF TRANSIENT NONLINEAR STRUCTURAL BEHAVIOR	75-0297
BELYTSCHKO, T.	FINITE ELEMENT APPROACH TO HYDROTHERMAL ANALYSIS OF SMALL LAKES	72-0363
BELYTSCHKO, T.	FINITE ELEMENT APPROACH TO HYDRODYNAMICS AND MESH STABILIZATION	74-1298
BELYTSCHKO, T.	FINITE ELEMENT APPROACH TO PRESSURE WAVE ATTENUATION BY REACTOR FUEL SUBASSEMBLIES	75-0522
BELYTSCHKO, T.	FINITE ELEMENT METHOD FOR ELASTIC PLASTIC PLATES	72-0167
BELYTSCHKO, T.	FINITE ELEMENT METHOD FOR A SOLID ENCLOSING AN INVISCID INCOMPRESSIBLE FLUID	73-0090
BELYTSCHKO, T.	FINITE ELEMENT STRESS ANALYSIS OF AN INTERVERTEBRAL DISC	74-0244
BELYTSCHKO, T.	FINITE ELEMENT STUDY OF PRESSURE WAVE ATTENUATION BY REACTOR FUEL SUBASSEMBLIES	75-0215
BELYTSCHKO, T.	FINITE ELEMENTS FOR AXISYMMETRIC SOLIDS UNDER ARBITRARY LOADINGS WITH NODES ON ORIGIN,	72-0251
BELYTSCHKO, T.	FINITE-ELEMENT ANALYSIS OF STRUCTURAL RESPONSE IN HCDA	75-0407
BELYTSCHKO, T.	LARGE DISPLACEMENT, NONLINEAR TRANSIENT ANALYSIS BY FINITE ELEMENTS	74-0324
BELYTSCHKO, T.	LOWER BOUND LIMIT ANALYSIS OF PLANE STRESS PROBLEMS BY A FINITE ELEMENT METHOD	70-0428
BELYTSCHKO, T.	NON-LINEAR TRANSIENT FINITE ELEMENT ANALYSIS WITH CONVECTED CO-ORDINATES	73-0378
BELYTSCHKO, T.	NONLINEAR FINITE-ELEMENT METHOD FOR PLATES AND ITS APPLICATION TO DYNAMIC RESPONSE OF REACTOR FUEL S	74-0670
BELYTSCHKO, T.	NONLINEAR TRANSIENT ANALYSIS OF SHELLS AND SOLIDS OF REVOLUTION BY CONVECTED ELEMENTS	73-0519
BELYTSCHKO, T.	NONLINEAR TRANSIENT ANALYSIS OF SHELLS AND SOLIDS OF REVOLUTION BY CONVECTED ELEMENTS	74-0476
BELYTSCHKO, T.	NUMERICAL METHODS FOR THE LIMIT ANALYSIS OF PLATES	68-0181
BELYTSCHKO, T.	NUMERICAL STRESS ANALYSIS OF INTERVERTEBRAL DISK	72-0125
BELYTSCHKO, T.	ON THE UNCONDITIONAL STABILITY OF AN IMPLICIT ALGORITHM FOR NONLINEAR STRUCTURAL DYNAMICS	75-0825
BELYTSCHKO, T.	PLANE STRESS LIMIT ANALYSIS BY FINITE ELEMENTS	70-0255
BELYTSCHKO, T.	PLANE STRESS SHAKEDOWN ANALYSIS BY FINITE ELEMENTS	72-0266
BELYTSCHKO, T.	RESPONSE OF REACTOR CORE SUBASSEMBLIES TO IMPULSIVE LOADINGS	73-0076
BELYTSCHKO, T.	NONLINEAR BEHAVIOR OF THE HUMAN INTERVERTEBRAL DISC UNDER AXIAL LOADING	74-0919
BELYTSCHKO, T.	NONLINEAR FINITE-ELEMENT FORMULATION FOR TRANSIENT ANALYSIS OF THREE DIMENSIONAL THIN STRUCTURES	74-0874
BELYTSCHKO, T.	STATIC THREE DIMENSIONAL ANALYSIS OF ELEVATED STEEL TRANSPORTATION STRUCTURES	74-1187
BENFIELD, W. A.	VIBRATION ANALYSIS OF STRUCTURES BY COMPONENT MODE SUBSTITUTION	71-0303
BENJAMIN, C.	DETERMINATION DU CHAMP DES CONTRAINTES DANS LE CAISSON DU REACTEUR DE BUGEY L'	67-0121
BENKATARAMANA,	STABILITY OF MODERATELY THICK RECTANGULAR PLATES USING A HIGH PRECISION TRIANGULAR FINITE ELEMENT	75-0334
BENNETT, J. G.	A STUDY OF A DYNAMICAL FINITE ELEMENT ANALYSIS FOR APPLICATION TO AXIAL WAVE PROPAGATION PROBLEMS IN	71-0589
BENNETT, P.	NATURAL FREQUENCIES AND MODE SHAPES OF TRANSFORMER CORES	65-0048
BENNETT, R. D.	NASTRAN DIFFERENTIAL STIFFNESS ANALYSIS OF AN AIRCRAFT CANOPY	71-0037
BENTHIEN, G. W.	ON THE SEMI-DISCRETE GALERKIN METHOD FOR HYPERBOLIC PROBLEMS AND ITS APPLICATION TO PROBLEMS IN ELAS	72-0457
BENZLEY, S. E.	CHILES: A FINITE ELEMENT COMPUTER PROGRAM THAT CALCULATES THE INTENSITIES OF LINEAR ELASTIC SINGULAR	73-0959
BENZLEY, S. E.	DYNAMIC SHELL THEORY COUPLING THICKNESS STRESS WAVE EFFECTS WITH GROSS STRUCTURAL RESPONSE	73-0379
BENZLEY, S. E.	REPRESENTATION OF SINGULARITIES WITH ISOPARAMETRIC FINITE ELEMENTS	74-1075
BENZLEY, S. E.	SHELL DEFORMATION UNDER TOOL LOADS REPRESENTATION OF SINGULARITIES WITH ISOPARAMETRIC FINITE ELEMENT	74-0365
BENZLEY, S. E.	TACOS: A FINITE ELEMENT COMPUTER PROGRAM FOR THE TRANSIENT ANALYSIS OF CYLINDRICAL OBESE SHELLS	72-0588
BENZLEY, S. E.	TRANSIENT ANALYSIS OF THICK CYLINDRICAL SHELLS	71-0590
BERARD, G. P.	APPLICATION OF THE FINITE ELEMENT TECHNIQUE TO COMPRESSIBLE FLOW PROBLEMS	70-0311
BERARD, G. P.	APPLICATION OF THE FINITE ELEMENT TECHNIQUE TO COMPRESSIBLE FLOW PROBLEMS	73-0003
BERARD, G. P.	APPLICATION OF THE FINITE ELEMENT TECHNIQUE TO COMPRESSIBLE FLOW PROBLEMS	74-0090
BEREDUGO, Y. O.	VERTICAL VIBRATION OF EMBEDDED FOOTINGS	72-0231
BERG, B. A.	STACUSS 1: A DISCRETE-ELEMENT PROGRAM FOR THE STATIC ANALYSIS OF SINGLE-LAYER CURVED STIFFENED SHELL	69-0160
BERG, B. A.	THE TRANSIENT LINEAR ELASTIC RESPONSE ANALYSIS COMPLEX THIN SHELLS OF REVOLUTION SUBJECTED TO ARBITR	70-0080
BERG, C. A.	FATIGUE OF GRAPHITE FIBRE-REINFORCED EPOXY IN COMPRESSION.	72-0383
BERG, C. A.	FATIGUE OF PRENOTCHED GRAPHITE FIBER COMPOSITES IN COMPRESSION	72-0215
BERG, S.	NONLINEAR FINITE ELEMENT ANALYSIS OF REINFORCED CONCRETE PLATES	73-0530
BERG, S.	NONLINEAR FINITE ELEMENT ANALYSIS OF REINFORCED CONCRETE PLATES	73-0845
BERGAN, P. G.	A COMPARATIVE STUDY OF DIFFERENT NUMERICAL SOLUTION TECHNIQUES AS APPLIED TO A NONLINEAR STRUCTURAL	73-1117
BERGAN, P. G.	A NEW APPROACH FOR DERIVING GOOD ELEMENT STIFFNESS MATRICES	75-0128
BERGAN, P. G.	ANALYSIS OF STIFFENED PLATES USING THE FINITE ELEMENT METHOD	70-0030
BERGAN, P. G.	CALCULATION OF STRESS INTENSITY FACTORS AND FATIGUE CRACK PROPAGATION OF SEMI-ELLIPTICAL, PART-THROU	73-0360
BERGAN, P. G.	CONVERGENCE CRITERIA FOR ITERATIVE PROCESSES	72-0664
BERGAN, P. G.	ELASTO-PLASTIC ANALYSIS OF PLATES USING THE FINITE ELEMENT METHOD	71-0038
BERGAN, P. G.	FINITE ELEMENT ANALYSIS OF CRACK PROPAGATION IN THREE-DIMENSIONAL SOLIDS UNDER CYCLING LOADING	73-0529
BERGAN, P. G.	FINITE ELEMENT ANALYSIS OF CRACK PROPAGATION IN 3-DIMENSIONAL SOLIDS UNDER CYCLIC LOADING	74-0479
BERGAN, P. G.	HIGHER-ORDER FINITE ELEMENT FOR PLANE STRESS	68-0089
BERGAN, P. G.	LARGE DEFLECTION ANALYSIS OF PLATES AND SHALLOW SHELLS USING THE FINITE ELEMENT METHOD	73-0989
BERGAN, P. G.	NON-LINEAR ANALYSIS OF PLATES CONSIDERING GEOMETRIC AND MATERIAL EFFECTS	71-0702
BERGAN, P. G.	NONLINEAR FINITE ELEMENT ANALYSIS OF REINFORCED CONCRETE PLATES	73-0845
BERGAN, P. G.	NUMERICAL TECHNIQUES IN LINEAR AND NONLINEAR FRACTURE MECHANICS	75-0260
BERGAN, P. G.	PLANE STRESS ANALYSIS USING THE FINITE ELEMENT METHOD. TRIANGULAR FINITE ELEMENT WITH SIX PARAMETERS	67-0106
BERGAN, P. G.	PROPAGATION OF ELLIPTICAL SURFACE CRACKS AND NONLINEAR FRACTURE MECHANICS BY THE FINITE ELEMENT METH	74-0906
BERGAN, P. G.	SESAM A PROGRAMMING SYSTEM FOR FINITE ELEMENT PROBLEMS	68-0276
BERGER, A.	APPROXIMATE BOUNDARY CONDITIONS IN THE FINITE ELEMENT METHOD	72-0995
BERGER, A. E.	(L SUP 2) ERROR ESTIMATES FOR FINITE ELEMENTS WITH INTERPOLATED BOUNDARY CONDITIONS	73-0885
BERGER, A. E.	ERROR ESTIMATES FOR THE FINITE ELEMENT METHOD	72-0686
BERGER, A. E.	L2 ERROR ESTIMATES FOR FINITE ELEMENTS WITH INTERPOLATED BOUNDARY CONDITIONS	73-0194
BERGER, A. E.	NUMERICAL SOLUTION OF A DIFFUSION CONSUMPTION PROBLEM WITH A FREE BOUNDARY	74-0835
BERGER, A. E.	USE OF FINITE ELEMENT METHOD IN A MARKER AND CELL PROGRAM FOR SOLVING TIME-DEPENDENT FLUID FLOW PROB	75-0043
BERGER, A. E..	TWO TYPES OF PIECEWISE QUADRATIC SPACES AND THEIR ORDER OF ACCURACY OF POISSON'S EQUATION	72-0520
BERGER, D.	A USEFUL FORM OF THE VARIATIONAL PRINCIPLE OF IDEAL MHD FOR ONE-DIMENSIONAL NUMERICAL STABILITY CALC	73-0746
BERGER, D.	BIDIMENSIONAL COMPUTER-PROGRAM FOR CALCULATION OF GROWTH-RATES OF INCLUDED AXIALLY SYMMETRICAL PLASM	75-0420
BERGER, D.	DETERMINATION OF EIGENMODES OF HOMOGENEOUS PLASMA WITH ELLIPTICAL SECTION USING BI-DIMENSIONAL FINIT	74-0553
BERGER, D.	EIGENMODES OF CYLINDRICAL PLASMA USING FINITE ELEMENT METHOD	74-0174
BERGER, D.	NEW FINITE ELEMENT APPROACH TO THE NORMAL MODE ANALYSIS IN MAGNETOHYDRODYNAMICS	74-0876
BERGER, D.	NUMERICAL STUDY OF EIGEN OSCILLATIONS OF A SCREW PINCH WITH FINITE ELEMENTS	74-0159
BERGER, D.	NUMERICAL STUDY OF PROPER MODES OF A HOMOGENEOUS CYLINDRICAL PLASMA BY METHOD OF FINITE ELEMENTS	74-0160

BERGER, D.	THALIA - ONE-DIMENSIONAL MAGNETOHYDRODYNAMIC STABILITY PROGRAM USING METHOD OF FINITE-ELEMENTS	75-0458
BERGER, D.	THALIA: A ONE-DIMENSIONAL MAGNETOHYDRODYNAMIC STABILITY PROGRAM USING THE METHOD OF FINITE ELEMENTS	74-1085
BERGER, D.	USEFUL FORM OF THE VARIATIONAL PRINCIPLE OF IDEAL MHD FOR ONE-DIMENSIONAL NUMERICAL STABILITY CALCU.	73-0946
BERGER, H.	ON A SPECTRAL PROBLEM IN VIBRATION MECHANICS: COMPUTATION OF ELASTIC TANKS PARTIALLY FILLED WITH LIQ	75-0267
BERGER, H.	UNE METHODE DE CALCUL PAR ELEMENTS FINIS DES MOUVEMENTS DE LIQUIDES DANS DES RESERVOIRS RIGIDES OU D	74-1171
BERGER, H.	VIBRATIONS OF TANKS PARTIALLY FILLED WITH LIQUIDS	74-0089
BERGERON, W. J.	FINITE ELEMENT ANALYSIS OF SALT PILLAR MODELS	68-0291
BERGMAN, H. W.	ASKA, A LARGE SCALE SOFTWARE SYSTEM FOR FINITE ELEMENT ANALYSIS	69-0331
BERKE, L.	A REVIEW OF SOME ASPECTS OF STRUCTURAL OPTIMIZATION	75-0467
BERKE, L.	AUTOMATED METHOD DEFLECTION AND STABILITY ANALYSIS OF THREE DIMENSIONAL BAR STRUCTURES	69-0393
BERKE, L.	AUTOMATED METHOD FOR THE LARGE DEFLECTION AND INSTABILITY ANALYSIS OF THREE-DIMENSIONAL TRUSS AND FR	66-0105
BERKE, L.	OPTIM 11: A MAGIC COMPATIBLE LARGE-SCALE AUTOMATED MINIMUM WEIGHT DESIGN PROGRAM. VOLUME 1. ENGINEER	74-1096
BERKE, L.	REVIEW OF OPTIMALITY CRITERIA APPROACHES TO STRUCTURAL OPTIMIZATION	74-0652
BERKE, L.	THE FINITE ELEMENT ANALYSIS OF ELASTIC-PLASTIC THICK-WALLED TUBES	74-0768
BERKHOFF, J. C.	LINEAR WAVE PROPAGATION PROBLEMS AND THE FINITE ELEMENT METHOD	75-0395
BERKHOFF, J. C.	COMPUTATION OF WAVE PROPAGATION BY MEANS OF THE FINITE ELEMENT METHOD	74-0032
BERKHOFF, J. C.	LINEAR WAVE PROPAGATION PROBLEMS AND THE FINITE ELEMENT METHOD	74-1340
BERKOVIC, M.	FREE-FREE STRUCTURES IN FINITE ELEMENT ANALYSIS	70-0442
BERKOVIC, M.	GENERAL MEMBRANE ISOPARAMETRIC ELEMENTS	75-0442
BERKOVIC, M.	HYBRID FINITE ELEMENTS IN THE PLANE STRESS PROBLEM	74-1071
BERKOVIC, M.	THE CALCULATION OF STATIC AND DYNAMIC CHARACTERISTICS OF FLIGHT VEHICLE STRUCTURES BY THE FINITE ELE	73-101R
BERKOWITZ, L.	SIMPLIFIED MINIMUM INTERNAL ENERGY OF DEFORMATION POSTULATE AND ITS APPLICATION TO METALFORMING PROC	74-1155
BERLAMONT, J.	THE PROGRAM " ASSEN " FOR STEADY FREE SURFACE FLOWS	74-0075
BERMAN, I.	CREEP BENDING TESTS OF LARGE WELD OVERLAY AND BASE METAL SPECIMENS AT 1000 DEGREE F. (537.7 DEG. C)	75-0521
BERMAN, I.	EVALUATION OF THE COMPUTER PROGRAM CREEP-BLAST (CDC VERSION)	74-0684
BERNDT, P.	IMPORTANCE OF VARIATIONAL CALCULUS IN FLUID MECHANICS AND ITS APPLICATION TO COMPRESSIBLE POTENTIAL	73-1103
BERNHARDT, K.	DISCUSSION OF A PAPER BY F. W. WILLIAMS	73-1078
BERNSTEIN, M.	ANALYTICAL AND EXPERIMENTAL INVESTIGATION OF A 1/8 SCALE DYNAMIC MODEL OF THE SHUTTLE ORBITER. VOLUM	74-1186
BERNSTEIN, M.	DEVELOPMENT OF TECHNOLOGY FOR MODELING OF A 1/8-SCALE DYNAMIC MODEL OF THE SHUTTLE SOLID ROCKET BOOS	74-0703
BERRY, G. F.	TRANSIENT RESPONSE OF NON-UNIFORM, NON-HOMOGENEOUS BEAMS	73-0162
BERRY, G. F.	TRANSIENT RESPONSE OF ELASTIC, VISCO-PLASTIC BEAMS	73-0380
BERRY, J. T.	APPLICATION OF THE DOUBLE LIGAMENT TENSILE TEST IN DEFORMATION PROCESSING	75-0827
BERRY, V. L.	A NASTRAN VIBRATION MODEL OF THE AH-1G HELICOPTER AIRFRAME. VOLUME 1	74-1338
BERRY, V. L.	A NASTRAN VIBRATION MODEL OF THE AH-1G HELICOPTER AIRFRAME. VOLUME 11	74-1339
BERT, C. W.	FINITE-ELEMENT ANALYSIS FOR FREE VIBRATION OF GENERAL ANISOTROPIC LAMINATED THIN SHELLS	72-0618
BERTELS, A. W.	APPLICATION OF THE FINITE ELEMENT METHOD TO THE NONLINEAR ANALYSIS OF NUCLEAR REACTOR FUEL BEHAVIOR	74-1246
BERTELS, A. W.	IMPROVED APPROXIMATION OF CONSTITUTIVE ELASTO-PLASTIC STRESS STRAIN RELATIONSHIP FOR FINITE ELEMENT	74-0566
BERTELS, A. W.	PROPAGATION AND OPENING OF A THROUGH CRACK IN PIPE SUBJECT TO COMBINED CYCLIC THERMOMECHANICAL LOADI	75-0516
BERTSCH, O. L.	THREE-DIMENSIONAL FINITE ELEMENT ANALYSIS OF A STEAM GENERATOR CHANNEL-HEAD-COMPLEX - COMPARISON OF	75-0653
BERWANGER, C.	THERMAL STRESSES IN STEEL-CONCRETE COMPOSITE BRIDGES	75-0154
BESANT, C. B.	INEXPENSIVE, INTERACTIVE COMPUTER-AIDED DESIGN APPLIED TO FINITE ELEMENT STRESS ANALYSIS	74-0592
BESSELING, J. F	MATRIX ANALYSIS OF CREEP AND PLASTICITY PROBLEMS	65-0001
BESSELING, J. F	NONLINEAR ANALYSIS	75-0627
BESSELING, J. F	POST-BUCKLING AND NON-LINEAR ANALYSIS BY FINITE ELEMENT METHOD AS A SUPPLEMENT TO A LINEAR ANALYSIS	75-0505
BEST, B.	AN INVESTIGATION INTO THE USE OF FINITE ELEMENT METHODS FOR ANALYSING STRESS DISTRIBUTIONS IN BLOCK	71-0530
BEST, B.	ANALYSIS OF NONLINEAR PROBLEMS IN ROCK MECHANICS WITH PARTICULAR REFERENCE TO JOINTED ROCK SYSTEMS	70-0282
BEST, B.	SOME NON LINEAR PROBLEMS IN SOIL AND ROCK MECHANICS - FINITE ELEMENT SOLUTION	69-0363
BEST, G. C.	A FORMULA FOR CERTAIN TYPES OF STIFFNESS MATRICES OF STRUCTURAL ELEMENTS	63-0003
BEST, G. C.	A GENERAL FORMULA FOR STIFFNESS MATRICES OF STRUCTURAL ELEMENTS	63-0002
BEST, G. C.	HELPFUL FORMULAS FOR INTEGRATING POLYNOMIALS IN THREE DIMENSIONS	64-0031
BEST, G. C.	VIBRATION ANALYSIS OF A CANTILEVERED SQUARE PLATE BY THE STIFFNESS MATRIX METHOD	66-0012
BESUNER, P. M.	APPLICATION OF THE TWO-DIMENSIONAL INTEGRAL EQUATION METHOD TO ENGINEERING PROBLEMS	75-0377
BETTES, R. S.	ELASTIC-PLASTIC BEHAVIOR OF A LONGITUDINAL SEMI-ELLIPTIC CRACK IN A THICK PRESSURE VESSEL	73-0952
BETTESS, P.	A NOTE ON SOME VARIATIONAL STATEMENTS FOR THE SLOW FLOW OF A NAVIER-POISSON FLUID	74-1054
BEUSSEM, W. R.	HIGH STRENGTH CERAMICS	72-0878
BEVILACQUA, L.	VARIATIONAL PRINCIPLE FOR THE LAPLACE'S OPERATOR WITH APPLICATION IN THE TORSION OF COMPOSITE RODS	74-1044
BHANDARI, D. R.	A LARGE DEFORMATION ANALYSIS OF CRYSTALLINE ELASTIC-VISCOPLASTIC MATERIALS	73-0602
BHANDARI, D. R.	A NEW APPROACH TO THE FINITE-ELEMENT FORMULATION AND SOLUTION OF A CLASS OF PROBLEMS IN COUPLED THER	72-0732
BHANDARI, D. R.	A NEW APPROACH TO THE FINITE-ELEMENT FORMULATION AND SOLUTION OF A CLASS OF PROBLEMS IN COUPLED THER	73-0600
BHANDARI, D. R.	GENERAL MIXED FINITE ELEMENT METHODS OF NONLINEAR CONTINUA	72-0831
BHANDARI, D. R.	GENERAL MIXED FINITE ELEMENT METHODS OF NONLINEAR CONTINUA	73-0091
BHANDARI, D. R.	NEW APPROACH TO FINITE ELEMENT FORMULATION AND SOLUTION OF A CLASS OF PROBLEMS IN COUPLED THERMOELAS	73-0013
BHATT, P.	EFFECT OF BEAM-SHEARWALL JUNCTION DEFORMATIONS ON THE FLEXIBILITY OF THE CONNECTING BEAMS	73-0213
BHAUMIK, A.	REINFORCED CONCRETE ON CONSTITUTIVE RELATIONS	75-0753
BHAUNIK, A. K.	SPHERICAL WAVES IN INELASTIC MATERIAL	70-0075
BHUTWALA, J. N.	STEADY STATE AND TRANSIENT HEAT CONDUCTION AND STRESS ANALYSES OF TWO AND THREE DIMENSIONAL STRUCTUR	71-0675
BIAREZ, J.	APPLICATION DES CALCULUS DE FONDATIONS A L'EVALUATION DU COMPORTEMENT DES CHAUSSEES EM DASH 1, 2. (A	71-0429
BICHAT, B.	A GENERAL METHOD FOR THE SHAPE FINDING OF LIGHTWEIGHT TENSION STRUCTURES	74-1217
BICHAT, B.	FINITE ELEMENT DIFFICULTIES IN NONLINEAR AND INCOMPRESSIBLE MATERIAL	74-1242
BICHAT, B.	LINEAR AND NONLINEAR OSCILLATIONS IN MECHANICS	74-0677
BICKOVSKI, V.	STATIC AND DYNAMIC ANALYSIS OF AN ARCH DAM	74-1101
BIELAK, C.	DYNAMICS OF BUILDING - SOIL INTERACTION	72-0856
BIELAK, J.	A SIMPLIFIED VERSION OF GURTIN'S VARIATIONAL PRINCIPLE	74-1244
BIENIEK, M. P.	TRANAL: A 3-D FINITE ELEMENT CODE FOR TRANSIENT NONLINEAR ANALYSIS	74-0707
BIFFLE, J. H.	FINITE ELEMENT ANALYSIS FOR WAVE PROPAGATION IN ELASTIC-PLASTIC SOLIDS	73-0990
BIFFLE, J. H.	FINITE ELEMENT STRESS FORMULATION FOR DYNAMIC ELASTIC-PLASTIC ANALYSIS	74-0484
BIFFLE, J. H.	INTERACTIVE GRAPHICS AIDED STRUCTURAL ANALYSIS	74-0867
BIGGS, J. M.	IMPACT OF COMPUTERS ON THE PRACTICE OF STRUCTURAL ENGINEERING IN CONCRETE	72-0415
BILEK, Z.	APLIKACE MKP NA RESENI PROBLEMU DIFUZE V KOVECH. (APPLICATION OF THE FINITE ELEMENT METHOD TO THE SO	74-0609
BILEK, Z.	APPLICATION OF THE FINITE ELEMENT METHOD IN METAL DIFFUSION	74-1366
BILGEN, E.	ON THE FINITE ELEMENT FORMULATION OF NAVIER-STOKES EQUATIONS	74-0720
BILLERBECK, W.	ADVANCED INTERCONNECT SYSTEMS FOR LIGHTWEIGHT SOLAR ARRAYS	73-0696
BILLERBECK, W.	DEVELOPMENT OF ADVANCED INTERCONNECTORS FOR SOLAR CELLS	74-0371
BILLINGTON, A.	ON THE FICTION OF LEADING EDGE SUCTION	71-0712
BILLINGTON, D.	BUCKLING OF CYLINDRICAL SHELLS BY WIND PRESSURE	74-0642
BILLINGTON, D.	THERMAL LOADING OF THIN-SHELL CONCRETE COOLING TOWERS	74-0667
BINDSEIL, H.	UEBER DAS VERHALTEN VON SPANNEBETONDRUCKBEHAELTRERN IM GRENZTRAGFAEHIGKEITSZUSTAND UND DESSEN EINFLU	73-0855
BINGHAM, C. M.	UNUSUAL FOUNDATIONS OVER THE MERSEY TUNNEL	70-0197
BIOT, M. A.	MECHANICS AND THERMOMECHANICS	73-0929
BIOT, M. A.	NONLINEAR THERMOELASTICITY, IRREVERSIBLE THERMODYNAMICS AND ELASTIC INSTABILITY	73-0762
BIRCHFIELD, E.	EVALUATION OF COMPOSITE SHEAR PANELS WITH CUTOUTS	72-0615
BIRCHLER, W. D.	FINITE ELEMENT MODELS FOR SHELLS OF REVOLUTION	68-0027
BIRCHLER, W. D.	USING INTERACTIVE GRAPHICS FOR THE PREPARATION AND MANAGEMENT OF FINITE ELEMENT DATA	74-0176
BIRD, T. S.	EVALUATION OF ATTENUATION FROM LOSSLESS TRIANGULAR-FINITE-ELEMENT SOLUTIONS FOR INHOMOGENEOUSLY FILL	73-0059
BIRKEMOE, P. C.	HIGH STRENGTH BOLTS SUBJECT TO TENSION AND PRYING	74-0201
BIRKHOFF, G.	COMPATIBLE TRIANGULAR FINITE ELEMENTS	74-0330
BIRKHOFF, G.	FINITE ELEMENT METHODS IN REACTOR DESIGN CALCULATIONS	71-0571
BIRKHOFF, G.	HIGHER-ORDER LINEAR FINITE ELEMENT METHODS	74-0826
BIRKHOFF, G.	PIECEWISE ANALYTIC INTERPOLATION AND APPROXIMATION IN TRIANGULATED POLYGONS	72-0508
BIRKHOFF, G.	PIECEWISE HERMITE INTERPOLATION IN ONE AND TWO VARIABLES WITH APPLICATIONS TO PARTIAL DIFFERENTIAL E	68-0201
BIRKHOFF, G.	PIECEWISE POLYNOMIAL INTERPOLATION AND APPROXIMATION	65-0058
BIRKHOFF, G.	RAYLEIGH-RITZ APPROXIMATION OF PIECEWISE CUBIC POLYNOMIALS	66-0087

BIRKHOFF, G.	TRICUBIC POLYNOMIAL INTERPOLATION	71-0377
BISHIO, P. T.	PREDICTION OF STATIC AND FATIGUE DAMAGE AND CRACK PROPAGATION IN COMPOSITE MATERIALS	75-0643
BISWAS, J. K.	THREE-DIMENSIONAL ANALYSIS OF SHEAR WALL BUILDINGS TO LATERAL LOAD	74-0220
BIWAS, A. K.	PROCEEDINGS OF INTERNATIONAL ASSOCIATION FOR HYDRAULIC RESEARCH AND PROCEEDINGS OF INTERNATIONAL SYM	73-0534
BIZRI, H.	FIRES-T. A COMPUTER PROGRAM FOR THE FIRE RESPONSE OF STRUCTURES - THERMAL	74-0850
BIZRI, H.	STRUCTURAL CAPACITY OF REINFORCED CONCRETE COLUMNS SUBJECTED TO FIRE INDUCED THERMAL GRADIENTS	73-0913
BJERTNAES, L. T	MEASUREMENT OF THE WATER DIFFUSIVITY OF POROUS MATERIALS	74-1138
BLAAUWENDRAAD,	FINITE ELEMENT METHOD. BASIC PRINCIPLES AND FORMULATION	69-0328
BLAAUWENDRAAD,	METHODS AND POSSIBILITIES FOR ELECTRONIC COMPUTATION OF PLATES IN FLEXURE	69-0329
BLACK, D. C.	CERTAIN FINITE ELEMENTS FOR PLATE BENDING PROBLEMS	74-0395
BLACK, D. C.	DJAL ANALYSIS OF COLUMN SUPPORTED SLABS	74-0399
BLACK, D. C.	ON THE BENDING STIFFNESS OF SLABS IN CROSS-WALLED STRUCTURES	74-0398
BLACK, D. C.	SUBSTRUCTURE ANALYSIS OF MULTISTOREY FLAT SLAB BUILDINGS	74-0394
BLACK, D. E.	DYNAMIC STABILITY OF CYLINDRICAL SHELLS BY FINITE ELEMENTS	68-0292
BLACKBURN, C. L	PROCEDURE FOR EFFICIENT GENERATING, CHECKING, AND DISPLAYING NASTRAN INPUT AND OUTPUT DATA FOR ANALY	71-0098
BLACKBURN, W. S	AN APPLICATION OF LEAST SQUARES FINITE ELEMENT METHODS TO TWO DIMENSIONAL STEADY INVISCID COMPRESSIB	75-0114
BLACKBURN, W. S	CALCULATION OF STRESS INTENSITY FACTORS AT CRACK TIPS USING SPECIAL FINITE ELEMENTS	72-0071
BLACKBURN, W. S	CALCULATION OF STRESS INTENSITY FACTORS FOR COMBINED TENSILE AND SHEAR LOADING	75-0205
BLACKBURN, W. S	CALCULATION OF STRESS INTENSITY FACTORS IN TWO AND THREE DIMENSIONS USING FINITE ELEMENTS	75-0254
BLACKKETTER, D.	DYNAMIC RESPONSE OF THE HUMAN ARM AS A COMPOSITE STRUCTURE	72-0611
BLACKLOCK, J. R	FINITE ELEMENT ANALYSIS OF INELASTIC STRUCTURES	69-0224
BLACKLOCK, J. R	PLANE STRESS FINITE ELEMENTS FOR NONLINEAR ANALYSIS	69-0529
BLACKMAN, K.	ELEMENT STIFFNESS MATRICES FOR CONSTANT AND LINEAR STRESS - TRIANGULAR, CONSTANT AND LINEAR STRESS T	67-0006
BLACKSTONE, W.	DISCONTINUITY STRESS DECAY IN OFFSHORE PLATFORM JOINTS	74-0491
BLAIR, P.	FINITE ELEMENT SOLUTION OF THE FLOW AROUND A TWO-DIMENSIONAL DESCOID	69-0298
BLAKE, W.	A METHOD FOR THE PREDICTION OF STRESSES IN AN ISOTROPIC INCLUSION OR OREBODY OF IRREGULAR SHAPE	72-0827
BLAKE, W.	APPLICATION OF THE FINITE ELEMENT METHOD OF ANALYSIS IN SOLVING BOUNDARY VALUE PROBLEMS IN ROCK MECH	66-0013
BLAKE, W.	DESTRESSING TEST AT THE GALENA MINE, WALLACE, IDAHO	72-0126
BLAKER, B.	VIBRATION OF SUBMERGED STRUCTURES AS COMPUTED BY THE FINITE ELEMENT METHOD	75-0227
BLAKEY, L. H.	DYNAMIC RESPONSE OF REINFORCED OPENINGS IN ROCK	71-0311
BLAKEY, L. H.	FINITE ELEMENT TECHNIQUES IN ROCK MECHANICS	68-0028
BLANC, G.	UTILISATION DE MODELES MATHEMATIQUES MULTIDIMENSIONNELS POUR L'ETUDE DES ECOULEMENTS DANS LES DIGUES	73-0381
BLAND, D. R.	SOLUTIONS OF LAPLACE'S EQUATIONS	61-0002
BLATZ, P.	APPLICATION OF FINITE ELASTICITY THEORY TO THE DEFORMATION OF RUBBERS	62-0005
BLEICH, H. H.	COMPARISON OF THE RESULTS OF THE DYNAMIC RESPONSE OF CYLINDRICAL SHELLS BY CHARACTERISTIC AND BY FIN	74-0781
BLEW, J. M.	DYNAMIC TESTING AND COMPUTER ANALYSIS OF AUTOMOTIVE FRAMES	72-0089
BLOKHOV, V. G.	ISSLEDOVANIE PROCHNOSTI I ZHESTKOSTI PLASTIN, NAGRUZHENNYKH V SVOEI PLOSKOSTI, METHODOM KONECHNYKH E	73-0382
BLOKHOV, V. V.	INVESTIGATION OF FLEXURAL STIFFNESS OF RECTANGULAR PLATES BY THE FINITE ELEMENT METHOD	73-0991
BLOMFIELD, J. A	STRESS ANALYSIS OF CURVED PIPES WITH END RESTRAINTS	75-0667
BLUM, R. E.	CALCULATING STRESSES IN A PRESSURE VESSEL WITH FILL TUBE - COMPARISON OF THE FINITE ELEMENT AND PHOT	75-0782
BOCK, H.	COMPUTER SIMULATION OF SECOND ORDER FAULTS	71-0010
BODIG, J.	A RATIONAL ANALYSIS AND DESIGN PROCEDURE FOR WOOD JOIST FLOOR SYSTEMS	74-0730
BODIG, J.	FINITE ELEMENT METHOD FOR WOOD MECHANICS	71-0613
BODNER, S. R.	FINITE ELEMENT ANALYSIS FOR THE TIME-DEPENDENT INELASTIC MATERIAL BEHAVIOR	75-0749
BOECKER, H.	DIGITAL CALCULATION OF ELECTRIC FIELDS IN METAL CLAD INSTALLATIONS	73-0212
BOEHMER, J. W.	CONSOLIDATION OF A LAYER UNDER A STRIP LOAD	72-0222
BOEHMER, J. W.	PLANE STRAIN CONSOLIDATION BY FINITE ELEMENTS	69-0471
BOEHMER, J. W.	PLANE STRAIN CONSOLIDATION BY FINITE ELEMENTS	70-0043
BOEHMER, J. W.	THE EFFECTS OF SOIL PARAMETERS AND BOUNDARY CONDITIONS ON THE CONSOLIDATION OF AN ELASTIC LAYER	70-0042
BOEKENBRINK, D.	INFLUENCE OF COOLING, FREE-STREAM TURBULENCE AND SURFACE ROUGHNESS ON THE AERODYNAMIC BEHAVIOR OF CA	74-0970
BOGDEN, F. J.	VERIFICATION AND QUALIFICATION OF A 3-D FINITE ELEMENT COMPUTER ANALYSIS	71-0487
BOGNER, F. K.	A CYLINDRICAL SHELL DISCRETE ELEMENT	67-0007
BOGNER, F. K.	DISCUSSION OF A SHORT COMMUNICATION BY S. GOPALACHARYULU	74-0404
BOGNER, F. K.	FINITE DEFLECTION STRUCTURAL ANALYSIS USING PLATE AND CYLINDRICAL SHELL DISCRETE ELEMENTS	68-0193
BOGNER, F. K.	FINITE DEFLECTION, DISCRETE ELEMENT ANALYSIS OF SHELLS	68-0029
BOGNER, F. K.	THE GENERATION OF INTERELEMENT, COMPATIBLE STIFFNESS AND MASS MATRICES BY THE USE OF INTERPOLATION F	66-0014
BOHM, G. J.	ANALYTICAL PROBLEMS ASSOCIATED WITH CORE SUPPORT STRUCTURES OF PWR	71-0039
BOHM, G. J.	STRUCTURALLY COMPATIBLE FLUID FINITE-ELEMENT FOR SOLID FLUID INTERACTION STUDIES	75-0344
BOISSERIE, J. M	APPLICATION DE LA METHODE DES ELEMENTS FINIS - UN PROCEDE DE SOUS-ASSEMBLAGE	74-0405
BOISSERIE, J. M	DETERMINATION DU CHAMP DES CONTRAINTES DANS LE CAISSON DU REACTEUR DE BUGEY L'	67-0121
BOISSERIE, J. M	GENERATION OF TWO- AND THREE-DIMENSIONAL FINITE ELEMENTS	71-0040
BOLEY, B. A.	STRESSES AND DEFORMATIONS IN MELTING PLATES	69-0476
BOLLAND, G. B.	FINITE ELEMENT FOR TORSION OF THIN WALLED OPEN TUBES	71-0308
BOLLAND, G. B.	FLEXIBILITY AND STIFFNESS MATRICES FOR AN OPEN-TUBE WARPING CONSTRAINT FINITE ELEMENT	72-0076
BOLLAND, G. B.	FLEXIBILITY MATRIX FOR AN EXACT TORSION-BENDING FINITE ELEMENT	72-0214
BOLLETER, U.	EFFECTS OF SHEAR DEFORMATION, ROTARY INERTIA, AND ELASTICITY OF THE SUPPORT ON THE RESONANCE FREQUEN	75-0369
BOLSTAD, J. H.	EMPIRICAL INVESTIGATION OF REORDERING AND DATA MANAGEMENT FOR FINITE ELEMENT SYSTEMS OF EQUATIONS	73-0944
BOLT, B. A.	LOVE WAVE DISPERSION THROUGH IRREGULAR STRUCTURES USING FINITE-ELEMENT ALGORITHMS	75-0431
BOLT, S. E.	EXPERIMENTAL AND FINITE ELEMENT STRESS ANALYSIS OF A THIN SHELLED CYLINDER TO CYLINDER MODEL	71-0066
BOLT, S. E.	THEORETICAL AND EXPERIMENTAL STRESS ANALYSES OF ORNL THIN-SHELL CYLINDER-TO-CYLINDER MODEL 3	75-0690
BOLT, S. E.	THEORETICAL AND EXPERIMENTAL STRESS ANALYSES OF ORNL THIN-SHELL CYLINDER-TO-CYLINDER MODEL 4	75-0695
BOLTSINIS, J. S	MATERIAL NONLINEARITIES IN THE FINITE ELEMENT ANALYSIS	74-0647
BOMBA, J. G.	PIPELINE STRESS ANALYSIS	75-0078
BOMBICH, A. A.	ULTIMATE STRAIN CAPACITY AND TEMPERATURE RISE STUDIES, TRUMBULL POND DAM	72-0705
BON, C.	FINITE ELEMENT ANALYSIS OF SUPERSONIC PANEL FLUTTER	73-0481
BON, C.	ON THE NUMERICAL SOLUTION OF LARGE EIGENVALUE PROBLEMS ARISING IN PANEL FLUTTER ANALYSIS BY THE FINI	74-0942
BOND, D.	OPTIMUM DESIGN OF CONCRETE STRUCTURES	74-0979
BOND, T. J.	COMPARISON OF SOME CURVED TWO DIMENSIONAL FINITE ELEMENTS	73-0383
BOND, T. J.	RING FINITE ELEMENTS FOR AXISYMMETRIC AND NON-AXISYMMETRIC THIN SHELL ANALYSIS	72-0922
BOND, T. J.	SOME CONSIDERATIONS OF THE FINITE ELEMENT METHOD IN STRESS ANALYSIS	72-0997
BONNEROT, R.	A SECOND ORDER FINITE ELEMENT METHOD FOR THE ONE-DIMENSIONAL STEFAN PROBLEM	74-1222
BONNEROT, R.	NUMERICAL-SOLUTION OF EULERIAN EQUATIONS OF COMPRESSIBLE FLOW BY A FINITE-ELEMENT METHOD WHICH FOLLO	75-0535
BONNES, G.	ANALYSE DES BOILES MINCES PAR ELEMENTS FINIS COURBES	69-0524
BONNES, G.	CURVED TRIANGULAR ELEMENTS FOR THE ANALYSIS OF SHELLS	68-0030
BONNES, G.	FINITE ELEMENT ANALYSIS OF SHELLS UNDER NONUNIFORM SNOW LOADS	69-0113
BOOKER, J. F.	APPLICATION OF FINITE ELEMENT METHODS TO LUBRICATION - ENGINEERING APPROACH	72-0006
BOOKER, J. F.	APPLICATION OF FINITE ELEMENT METHODS TO LUBRICATION - SQUEEZE FILMS BETWEEN POROUS SURFACES	75-0764
BOOKER, J. F.	FINITE ELEMENT MODEL FOR DISTRIBUTED PARAMETER TURBOMOTOR SYSTEMS	71-0332
BOOKER, J. R.	A NUMERICAL METHOD FOR THE SOLUTION OF BIOT'S CONSOLIDATION THEORY	72-0688
BOOKER, J. R.	AN INVESTIGATION OF THE STABILITY OF NUMERICAL SOLUTIONS OF BIOT'S EQUATIONS OF CONSOLIDATION	74-0746
BOOKER, J. R.	EFFECT OF SEEPAGE ON EMBANKMENT DEFORMATIONS DUE TO WATER LOADING	74-1329
BOOKER, J. R.	EMBANKMENT DEFORMATIONS DUE TO WATER LOADS	73-0782
BOOKER, J. R.	FINITE ELEMENT ANALYSIS OF THE EFFECTS OF INSTALLATION ON PILE LOAD-SETTLEMENT BEHAVIOUR	74-0763
BOOKER, J. R.	THE ECONOMICAL SOLUTION OF ELASTIC PROBLEMS FOR A RANGE OF POISSONS RATIO	74-0786
BOOKER, J. R.	THE NUMERICAL SOLUTION OF VISCO-ELASTIC PROBLEMS USING LAPLACE TRANSFORMS	74-0436
BOOKER, J. R.	THE SIGNIFICANCE OF THE RATE OF PLASTIC WORK IN ELASTO-PLASTIC ANALYSIS	74-0437
BOOKER, J. R.	THE SIGNIFICANCE OF THE RATE OF PLASTIC WORK IN ELASTO-PLASTIC ANALYSIS	74-0764
BOON, J. D. 111	INVESTIGATION OF THE WATER TABLE IN A TIDAL BEACH	71-0648
BOONLUALOHR, P.	ELASTIC PLASTIC ANALYSIS OF SHALLOW FOUNDATIONS	74-0446
BOONLUALOHR, P.	FINITE ELEMENT ANALYSIS OF CONSOLIDATION PROBLEM	74-0081
BOORMAN, R.	THE ANALYSIS OF FLEXIBLE BULKHEADS IN SANDS	74-0369
BORDAS, J.	MATRIX ANALYSIS OF MULTILAYERED AND SANDWICH SHELLS BY FINITE ELEMENT METHOD	73-0101

15

BORESI, A. P.	CREEP OF A UNIAXIAL METAL MATRIX COMPOSITE SUBJECTED TO AXIAL AND NORMAL LATERAL LOADS	71-0041
BORESI, A. P.	KINDS OF CONVERGENCE AND IMPROVED CONVERGENCE OF CONFORMING FINITE ELEMENT SOLUTIONS IN PLATE BENDIN	70-0323
BORING, L. M.	MATRIX DISPLACEMENT ANALYSIS OF SHELLS WITH AN AXISYMMETRIC MIDSURFACE BUT HAVING CIRCUMFERENTIALLY	68-0031
BORONKAY, T. G.	ANALYSIS AND DESIGN OF MULTIPLE INPUT FLEXIBLE LINK MECHANISMS	70-0354
BOROWIEC, Z.	STATICAL CALCULATIONS OF PLANE AND SPACE BAR STRUCTURES AS WELL AS STUMPY FRAMEWORKS USING A MODIFIE	73-0747
BOROWSKI, V. J.	FINITE ELEMENT DYNAMIC ANALYSIS OF AN AUTOMOTIVE FRAME	73-0256
BOS, H. D.	DYNAMIC MODELING OF HALF-PLANE BY FINITE ELEMENTS	74-0953
BOSHKOV, S. H.	STRESSES AND DISPLACEMENTS AROUND A CIRCULAR TUNNEL IN A THREE LAYER MEDIUM	69-0038
BOSSHARD, W.	A FULLY COMPATIBLE PLATE BENDING ELEMENT	67-0142
BOSSHAPD, W.	A NEW, FULLY COMPATIBLE FINITE ELEMENT FOR PLATE BENDING EIN NEUES VOLLVERTRAGLICHES ENDLICHES ELEME	68-0032
BOSSONEY, C.	FINITE ELEMENT METHOD APPLIED TO CALCULATIONS OF NON-COPLANAR PLATE SYSTEMS	74-0241
BOTKIN, M. E.	WELDED JOINT STIFFNESS OBTAINED USING SOLID FINITE ELEMENTS	75-0190
BOTMAN, M.	SUBSTRUCTURES ANALYSIS OF IMPELLER VIBRATION MODES	75-0283
BOTTERUD, O. H.	FORMAT, A GENERAL PURPOSE MATRIX ANALYSIS SYSTEM IN FORTRAN	68-0266
BOUGHTON, N. O.	ELASTIC ANALYSIS FOR BEHAVIOR OF ROCKFILL	70-0032
BOUJOT, J.	ON A SPECTRAL PROBLEM IN VIBRATION MECHANICS: COMPUTATION OF ELASTIC TANKS PARTIALLY FILLED WITH LIQ	75-0267
BOULIN, M.	APPLICATION DES CALCULUS DE FONDATIONS A L'EVALUATION DU COMPORTEMENT DES CHAUSSEES EM DASH 1, 2, (A	71-0429
BOUSQUET, R. D.	APPLICATION OF INTERACTIVE GRAPHICS TO THE NUMERICAL METHODS USED IN STRUCTURAL ANALYSIS	73-0384
BOUSQUET, R. D.	THE DEVELOPMENT OF COMPUTER GRAPHICS FOR LARGE SCALE FINITE ELEMENT CODES	71-0042
BOUWKAMP, J. G.	STRUCTURAL BEHAVIOR OF A TWO SPAN REINFORCED CONCRETE BOX GIRDER BRIDGE MODEL. VOLUME 11. REDUCTION,	71-0659
BOVE, B.	A FINITE ELEMENT PROCEDURE FOR NUMERICAL OPTIMIZATION IN MECHANICAL DESIGN	74-1218
BOWIE, O. L.	STRESS ANALYSIS OF CONFIGURATIONS INVOLVING SMALL FILLETS	75-0361
BOWLEY, W. W.	FINITE ELEMENT ANALYSIS OF GENERAL FLUID FLOW PROBLEMS	71-0536
BOYD, T. J.	NUMERICAL STUDY OF HYDROGMAGNETIC STABILITY USING FINITE ELEMENT METHOD	73-0092
BOYD, T. J.	STUDIES OF HYDROMAGNETIC STABILITY	74-0035
BOYD, T. J. M.	HYDROMAGNETIC STABILITY STUDIES USING THE FINITE ELEMENT METHOD	75-0878
BOYD, W. W.	APPLICATION OF FINITE ELEMENT METHOD TO MASS TRANSPORT LIMITED EPITAXIAL GROWTH PROCESSES	74-0488
BOYD, W. W.	FINITE ELEMENT METHOD APPLIED TO IDEAL GASEOUS MIXTURES.	72-0393
BOZICH, W. F.	AN EVALUATION OF CURRENT NASTRAN DISCRETE ELEMENT MODELS FOR MONOCOQUE AND SEMI-MONOCOQUE STRUCTURES	71-0236
BOZICH, W. F.	MATRIX METHODS IN STRUCTURAL MECHANICS	66-0065
BRADFORD, E.	FINITE-ELEMENT CALCULATION OF PRESSURE HISTORY IN A GAS FIELD	75-0546
BRADFORD, L. G.	ELASTODYNAMIC BEHAVIOR OF LAMINATED ORTHOTROPIC PLATES UNDER INITIAL STRESS	75-0249
BRADLEY, P. G.	APPLICATION OF FINITE-ELEMENT THEORY TO AIRPLANE CONFIGURATIONS	71-0458
BRADLEY, W. A.	A COMPARISON OF FIRST AND SECOND ORDER AXIALLY SYMMETRIC FINITE ELEMENTS	73-0997
BRADLEY, W. B.	CRACK BRANCHING IN HOMALITE-100 SHEETS	72-0843
BRADLEY, W. B.	FRACTURE DYNAMICS OF HOMALITE-100 SHEETS	72-0809
BRADSHAW, J. C.	SINGER: A COMPUTER CODE FOR GENERAL ANALYSIS TO TWO-DIMENSIONAL CONCRETE STRUCTURES. VOLUME 4. DEMON	75-0730
BRADY, B. T.	EFFECTS OF INSERTS ON THE ELASTIC BEHAVIOUR OF CYLINDRICAL MATERIALS LOADED BETWEEN ROUGH END-PLATES	71-0318
BRADY, B. T.	EXACT SOLUTION TO THE RADIALLY END CONSTRAINED CIRCULAR CYLINDER UNDER TRIAXIAL LOADING	71-0433
BRADY, B. T.	THE USE OF ROCK BOLTS OR WIRE ROPE TO INCREASE THE STRENGTH OF FRACTURED MODEL PILLARS	71-0722
BRAMBLE, J. H.	APPROXIMATION OF STEKLOV EIGENVALUES OF NON-SELFADJOINT SECOND ORDER ELLIPTIC OPERATORS	72-0509
BRAMBLE, J. H.	BOUNDS FOR A CLASS OF LINEAR FUNCTIONALS WITH APPLICATIONS TO HERMITE INTERPOLATION	71-0582
BRAMBLE, J. H.	INTERIOR MAXIMUM NORM ESTIMATES FOR SOME SIMPLE FINITE ELEMENT METHODS	74-1023
BRAMBLE, J. H.	SURVEY OF SOME FINITE ELEMENT METHOD PROPOSED FOR TREATING DIRICHLET PROBLEM	75-0612
BRAMBLE, J. H.	TRIANGULAR ELEMENTS IN THE FINITE ELEMENT METHOD	70-0194
BRAMLETTE, T. T	A FINITE ELEMENT SOLUTION TECHNIQUE FOR THE BOLTZMANN EQUATION	69-0026
BRAMLETTE, T. T	FINITE ELEMENT ANALYSIS OF THE UNSTEADY, LINEARIZED BOLTZMANN EQUATION	70-0033
BRAMLETTE, T. T	FINITE ELEMENT SOLUTIONS TO DIFFERENTIAL EQUATIONS	70-0084
BRAMLETTE, T. T	PLANE POISEUILLE FLOW OF A RAREFIED GAS BASED ON FINITE ELEMENT METHOD	71-0575
BRAMLETTE, T. T	FINITE ELEMENT ANALYSIS OF THE LINEARIZED BGK BOLTZMANN EQUATION.	72-0408
BRANCA, T. R.	CREEP OF A UNIAXIAL METAL MATRIX COMPOSITE SUBJECTED TO AXIAL AND NORMAL LATERAL LOADS	71-0041
BRANCA, T. R.	CREEP OF A UNIAXIAL METAL MATRIX COMPOSITE SUBJECTED TO AXIAL AND NORMAL LATERAL LOADS	71-0727
BRANCA, T. R.	ON THE DETERMINATION OF EFFECTIVE ELASTIC-PLASTIC PROPER TIES FOR THE EQUIVALENT SOLID PLATE ANALYSI	74-0494
BRAND, L.	INTRODUCTION TO FINITE ELEMENT METHODS OF STRUCTURAL ANALYSIS	68-0231
BRANDES, H.	SPANNUNGEN IN MEMBRANWAENDEN DURCH WAERMEBEANSPRUCHUNG UND MECHANISCHE BELASTUNG. (STRESSES IN MEMBR	75-0066
BRANDES, K.	A DISCRETE LEAST SQUARES AND FINITE ELEMENT METHOD	69-0322
BRANDES, K.	ZUR SYSTEMATIK NUMERISCHER NAEHERUNGSVERFAHREN IN DER STATIK UND DYNAMIK DER KONSTRUKTIONEN (METHODS	72-0557
BRANTLY, E.	ANALYSIS OF FRAMES WITH SHEAR WALLS WITH FINITE ELEMENTS	69-0500
BRASFIELD, R.L.	MATRIX-DISPLACEMENT DIRECT-ELEMENT METHOD FOR FORCE AND TORQUE ANALYSIS OF DETERMINATE AND INDETERMI	74-0531
BRASHERS, M. R.	FINITE ELEMENT ANALYSIS OF TRANSONIC FLOW	74-1223
BRATANOW, T.	ANALYSIS OF MOVING BODY PROBLEMS IN AERODYNAMICS	74-0024
BRATANOW, T.	ANALYSIS OF THREE-DIMENSIONAL UNSTEADY VISCOUS FLOW AROUND OSCILLATING WINGS	74-0982
BRATANOW, T.	APPLICATIONS OF FINITE ELEMENT METHOD IN UNSTEADY AERODYNAMICS	74-0105
BRATANOW, T.	COMPUTATIONAL CONSIDERATIONS IN APPLICATION OF THE FINITE ELEMENT METHOD FOR ANALYSIS OF UNSTEADY FL	73-0068
BRATANOW, T.	FINITE ELEMENT ANALYSIS OF UNSTEADY INCOMPRESSIBLE FLOW AROUND AN OSCILLATING OBSTACLE OF ARBITRARY	73-0067
BRATANOW, T.	FINITE ELEMENT ANALYSIS AND COMPUTER GRAPHICS VISUALIZATION OF FLOW AROUND PITCHING AND PLUNGING AIR	73-0385
BRATANOW, T.	NUMERICAL CALCULATIONS OF VELOCITY AND PRESSURE DISTRIBUTION AROUND OSCILLATING AIRFOILS	74-0847
BRATANOW, T.	SENSITIVITY ANALYSIS OF TORSIONAL VIBRATION CHARACTERISTICS OF HELICOPTER ROTOR BLADES. PART 1: STRU	74-0846
BRATANOW, T.	SUITABILITY OF THE FINITE ELEMENT METHOD FOR ANALYSIS OF UNSTEADY FLOW AROUND OSCILLATING AIRFOILS	73-0069
BRATONOW, T.	SENSITIVITY OF ROTOR BLADE VIBRATION CHARACTERISTICS TO TORSIONAL OSCILLATIONS	74-0611
BRAUCHLI, H. J.	A NOTE ON ACCURACY AND CONVERGENCE OF FINITE ELEMENT APPROXIMATIONS	71-0565
BRAUCHLI, H. J.	CONJUGATE APPROXIMATION IN FINITE ELEMENT ANALYSIS	71-0043
BRAUCHLI, H. J.	ON THE CALCULATION OF CONSISTENT STRESS DISTRIBUTIONS IN FINITE ELEMENT APPROXIMATIONS	71-0191
BRAUCHLI, J. H.	THEMIS FINAL REPORT. VOL. 1. BASIC RESEARCH IN METHODS OF APPROXIMATION IN THE NONLINEAR MECHANICS O	74-0723
BRAUER, J. R.	SATURATED MAGNETIC ENERGY FUNCTIONAL FOR FINITE-ELEMENT ANALYSIS OF ELECTRIC MACHINES	75-0636
BRAUER, J. R.	SIMPLE EQUATIONS FOR THE MAGNETIZATION AND RELUCTIVITY CURVES OF STEEL	75-0156
BRAUN, K. A.	DYNAN LECTURE NOTES WITH COMPUTATIONAL EXAMPLES	71-0581
BREBBIA, C.	A COMPARISON OF RECENT SHALLOW SHELL FINITE-ELEMENT ANALYSES	70-0232
BREBBIA, C.	EFFICIENT METHOD FOR SOLUTION OF VISCOELASTIC PROBLEMS	73-0232
BREBBIA, C.	FINITE ELEMENT TECHNIQUES IN STRUCTURAL MECHANICS	70-0116
BREBBIA, C.	GEOMETRICALLY NONLINEAR FINITE-ELEMENT ANALYSIS	69-0332
BREBBIA, C.	INTEGRATION OF AREA AND VOLUME COORDINATES IN FINITE-ELEMENT METHOD	69-0370
BREBBIA, C.	ON THE PARABOLIC VELAROIDAL SHELL	69-0314
BREBBIA, C.	STATICS AND DYNAMICS OF HYDROSTATICALLY LOADED SHELLS BY FINITE ELEMENT METHOD	71-0044
BREBBIA, C.	STIFFNESS MATRIX FOR SHALLOW RECTANGULAR SHELL ELEMENT	67-0014
BREBBIA, C.	THE FINITE ELEMENT METHOD IN SHELL ANALYSIS	69-0058
BREBBIA, C. A.	DYNAMIC BEHAVIOR OF STEEL FOUNDATIONS FOR TURBO-ALTERNATORS	71-0837
BREBBIA, C. A.	FINITE ELEMENT MODELLING TECHNIQUES FOR THE ANALYSIS OF BUILDING VIBRATIONS	72-0839
BREBBIA, C. A.	FINITE ELEMENT SOLUTION FOR EFFLUENT DISPERSION	74-1371
BREBBIA, C. A.	FINITE-ELEMENT SOLUTION OF NAVIER-STOKES EQUATIONS FOR TRANSIENT 2-DIMENSIONAL INCOMPRESSIBLE FLOW	75-0134
BREBBIA, C. A.	FUNDAMENTALS OF FINITE ELEMENT TECHNIQUES FOR STRUCTURAL ENGINEERS	73-1050
BREBBIA, C. A.	NUMERICAL METHODS IN FLUID DYNAMICS	73-0689
BREBBIA, C. A.	RATE OF CHANGE OF DYNAMIC RESPONSE	72-1004
BREBBIA, C. A.	SOLUTION OF WAVE PROPAGATION PROBLEMS IN SATURATED MEDIUM	72-0784
BREBBIA, C. A.	SOME APPLICATIONS OF FINITE ELEMENTS FOR FLOW PROBLEMS	73-0698
BREIG, W. F.	V-BELT LIFE PREDICTION AND POWER RATING	75-0838
BREKKE, T.	A MODEL FOR THE MECHANICS OF JOINTED ROCK	68-0079
BREKKE, T. L.	ANALYSIS AND MODEL STUDIES OF UNDERGROUND OPENINGS IN JOINTED ROCK. RESEARCH ON ROCK BOLT REINFORCEM	71-0750
BRESLER, B.	FIRES-RC. A COMPUTER PROGRAM FOR THE FIRE RESPONSE OF STRUCTURES-REINFORCED CONCRETE FRAMES	74-0747
BRESLER, B.	FIRES-T. A COMPUTER PROGRAM FOR THE FIRE RESPONSE OF STRUCTURES - THERMAL	74-0850
BRESLER, E.	FINITE ELEMENT ANALYSIS OF 2-DIMENSIONAL FLOW IN SOILS CONSIDERING WATER UPTAKE BY ROOTS .1. THEORY	75-0137
BRESLER, E.	FINITE ELEMENT ANALYSIS OF 2-DIMENSIONAL FLOW IN SOILS CONSIDERING WATER UPTAKE BY ROOTS .2. FIELD A	75-0147

16

BRIDWELL, R. J.	NONLINEAR IN-PLANE BENDING OF A PLANE STRESS-STRAIN FINITE ELEMENT MODEL		74-0211
BRIDWELL, R. J.	STABILITY ANALYSIS OF AN EXPERIMENTALLY DEFORMED SINGLE LAYER OF INDIANA LIMESTONE USING FINITE ELEM		74-0212
BRIGGES, D. A.	MATHEMATICAL MODELS OF THE MASSACHUSETTS BAY. PART 1. FINITE ELEMENT MODELING OF TWO-DIMENSIONAL HYD		73-0800
BRIGGS, L. L.	RAY-EFFECT MITIGATION IN DISCRETE ORDINATE-LIKE ANGULAR FINITE-ELEMENT APPROXIMATIONS IN NEUTRON-TRA		75-0496
BRILLA, J.	A GENERALIZATION OF FINITE ELEMENT METHOD TO DYNAMIC VISCOELASTIC ANALYSIS		75-0106
BRILLA, J.	FINITE ELEMENT METHOD IN LINEAR VISCOELASTICITY		74-0213
BRILLA, J.	RIESENIE VAZKOPRUZNYCH ANIZOTROPICKYCH DOASK METODOU KONECNYCH PRVKOV (SOLUTION OF VISCOELASTIC ANIS		74-0477
BRILLANT, J.	SPECIAL STUDY ON THE STABILITY OF A DAM SPILLWAY		72-0469
BRINK, F. I. A.	STRESS ANALYSIS OF A TUBULAR CROSS-JOINT WITHOUT ANY INTERNAL STIFFENING FOR OFFSHORE STRUCTURES		74-0549
BRINKLEY, F. W.	TRIPLET: A TWO-DIMENSIONAL, MULTIGROUP, TRIANGULAR MESH, PLANAR GEOMETRY, EXPLICIT TRANSPORT CODE		73-0955
BRINSON, H. F.	APPLICATIONS OF TRILINEAR COORDINATES TO SOME PROBLEMS IN PLANE ELASTICITY		72-0728
BRISBANE, J. J.	APPLICATION OF THE FINITE ELEMENT METHOD TO STRESS ANALYSIS OF SOLID PROPELLANT ROCKET GRAINS		65-0009
BRISBANE, J. J.	FINITE ELEMENT STRESS ANALYSIS OF ANISOTROPIC BODIES		68-0033
BRISBANE, J. J.	HEAT CONDUCTION AND STRESS ANALYSIS OF SOLID PROPELLANT ROCKET MOTOR NOZZLES		69-0059
BRISBANE, J. J.	INVESTIGATION OF TECHNIQUES OF THREE-DIMENSIONAL FINITE ELEMENT STRESS ANALYSIS		70-0029
BRISBANE, J. J.	ROHM AND HAAS THERMOSTRUCTURAL ANALYSIS SYSTEM, THERMAL STRUCTURAL ANALYSIS PROGRAMS - A SURVEY AND		72-0007
BRISBANE, J. J.	THE APPLICATION OF FINITE ELEMENT METHODS TO THE SOLUTION OF STRESS AND DIFFUSION PROBLEMS IN CONTIN		68-0034
BRISTEAU, M.	FINITE ELEMENT ANALYSIS OF THE UNSTEADY FLOW OF A VISCOUS PLASTIC FLUID IN A CYLINDRICAL PIPE		74-0055
BRITTEN, S. S.	EIGENVALUE BOUNDS FOR FINITE ELEMENT MODELS		72-0570
BRITTEN, S. S.	EIGENVALUE BOUNDS FOR FINITE ELEMENT MODELS		72-0932
BROCCI, R. A.	ANALYSIS OF AXISYMMETRIC LINEAR HEAT CONDUCTION PROBLEMS BY FINITE ELEMENT METHOD		69-0015
BROCKLEBANK, M.	LOW REYNOLDS NUMBER DEVELOPING FLOW		69-0020
BROELMANN, J.	TRAGLAST VON T-TRAEGERN. EXPERIMENT UND F.E. BERECHNUNGEN (ULTIMATE STRENGTH OF T-BEAMS. EXPERIMENTS		74-0935
BROERE, I.	USE OF THE FINITE ELEMENT METHOD IN EXAMINING ELASTIC BUCKLING OF THIN PLATES OF ANY SHAPE		75-0153
BROGAN, F. A.	EXTENSIONS TO THE STAGS COMPUTER CODE		73-0364
BROGAN, F. A.	EXTENSIONS TO THE STAGS COMPUTER CODE		73-0917
BROGAN, F. A.	PROBLEM CONTROLLED GRID GENERATION FOR THE NUMERICAL SOLUTION OF PARTIAL DIFFERENTIAL EQUATIONS		75-0725
BROLLIAR, R. H.	NASTRAN BUCKLING ANALYSIS OF A LARGE STIFFENED CYLINDRICAL SHELL WITH A CUTOUT		71-0045
BROMBOLICH, L.	HIGH-PRECISION CURVED SHELL FINITE ELEMENT		71-0422
BROMBOLICH, L.	FINITE ELEMENT ANALYSIS OF SHELLS OF REVOLUTION BY MINIMIZATION OF THE POTENTIAL ENERGY FUNCTIONAL		69-0354
BROMBOLICH, L.	FINITE ELEMENT ANALYSIS OF SHELLS OF REVOLUTION BY MINIMIZATION OF THE POTENTIAL ENERGY FUNCTIONAL		70-0481
BROMBOLICH, L.	HIGH PRECISION FINITE ELEMENT ANALYSIS OF SHELLS OF REVOLUTION		71-0524
BRON, J.	MIXED QUADRILATERAL ELEMENTS FOR BENDING		72-0623
BRONLUND, O.	THE NATURAL FACTOR APPROACH IN THE DISPLACEMENT METHOD - FOUNDATION AND FURTHER DEVELOPMENTS		75-0111
BRONLUND, O. E.	A SURVEY OF THE APPLICATION OF FINITE ELEMENT METHODS TO STRESS ANALYSIS PROBLEMS WITH PARTICULAR EM		71-0632
BRONLUND, O. E.	ANWENDUNG DER METHODEN DER FINITEN ELEMENTE MIT DEM PROGRAMMSYSTEM ASKA UNTER BESONDERER BERUCKSICHT		71-0624
BRONLUND, O. E.	ASKA, SOFTWARE ENGINEERING FOR THE SEVENTIES IN STRUCTURAL ANALYSIS		73-0704
BRONLUND, O. E.	COMPUTER AIDED STRUCTURAL ANALYSIS: THE MACHINE INDEPENDENT SYSTEM ASKA		70-0397
BRONLUND, O. E.	DYNAN LECTURE NOTES WITH COMPUTATIONAL EXAMPLES		71-0581
BRONLUND, O. E.	EIGENVALUES OF LARGE MATRICES		69-0341
BRONLUND, O. E.	EINIGE VERFAHREN ZUR BERECHNUNG VON EIGENWERTEN UND EIGENVEKTOREN VON NICHT HERMITESCHEN MATRIZEN U		72-0646
BRONLUND, O. E.	NATURAL FACTOR FORMULATION OF THE STIFFNESS FOR THE MATRIX DISPLACEMENT METHOD		75-0057
BRONLUND, O. E.	THE CHANGES IN THE (TWO-DIMENSIONAL) STRESS ROUNDED CORNERS, CAUSED BY A VARYING INTERNAL PRESSURE A		67-0141
BROOKS, D.F.	COMPUTER SYSTEM FOR ANALYSIS OF LARGE FRAMEWORKS		67-0122
BROTTON, D.M.	COMPUTER SYSTEM FOR ANALYSIS OF LARGE FRAMEWORKS		67-0122
BROUGHTON, P.	OFFSHORE GRAVITY BASED OIL PRODUCTION PLATFORM INTERACTION WITH THE SEA BED		75-0221
BROUTMAN, L. J.	THEORETICAL STUDY OF THE EFFECT OF THE INTERFACE ON COMPOSITE TOUGHNESS		73-0386
BROUTMAN, L. J.	THREE-DIMENSIONAL FINITE ELEMENT ANALYSIS OF SPHERICAL PARTICLE COMPOSITES		74-0137
BROWN, A. J.	FUNDAMENTAL STUDY OF UNDERWATER WELDING		72-0382
BROWN, C. A.	VISCOELASTIC PRESPONSE OF THE CORNEO-SCLERAL SHELL UNDER TONOMETER LOADING		75-0855
BROWN, C. B.	AUTOMATIC EMBANKMENT ANALYSIS. EQUILIBRIUM AND INSTABILITY CONDITIONS		66-0015
BROWN, C. B.	DARCY FLOW SOLUTIONS WITH A FREE SURFACE		67-0081
BROWN, C. B.	FORCES ON RIGID CULVERTS UNDER HIGH FILLS		67-0000
BROWN, E. T.	CONTROLLED FAILURE OF HOLLOW ROCK CYLINDERS IN UNIAXIAL COMPRESSION		72-0196
BROWN, J. E.	FINITE ELEMENT SOLUTION TO DYNAMIC STABILITY OF BARS		68-0035
BROWN, P. T.	FINITE ELEMENT ANALYSIS OF PILE-SUPPORTED PAVEMENT STRUCTURES		75-0145
BROWN, R. C. JR	COMPUTER ANALYSIS OF SEGMENTALLY ERECTED BRIDGES		75-0172
BROWN, R. E.	A MULTI-LAYERED FINITE ELEMENT MODEL FOR PREDICTING MINE SUBSIDENCE		68-0293
BROWN, S. F.	DYNAMIC RESPONSE OF MODEL PAVEMENT STRUCTURE		72-0226
BROWN, S. JR.	COMPARISON OF THE NUMERICAL INTEGRATION TECHNIQUE AND THE FINITE ELEMENT METHOD IN THE ANALYSIS OF T		73-0848
BROWN, T. G.	FINITE STRIP ANALYSIS OF SKEW SLABS		72-0369
BROWN, T. G.	SEMI-ANALYTIC SOLUTION OF SKEW BOX GIRDER BRIDGES		75-0578
BROWNE, B. T.	EFFICIENT TECHNIQUES FOR FINITE ELEMENT ANALYSIS OF ELECTRIC MACHINES		73-0134
BROWNING, R. V.	TSAAS: FINITE ELEMENT THERMAL AND STRESS ANALYSIS OF AXISYMMETRIC SOLIDS WITH ORTHOTROPIC TEMPERATUR		74-0886
BROZ, P.	STABILITA STEN V PLASTICKEM OBORU (STABILITY IN WEBPLATES IN PLASTIC RANGE)		73-0387
BRUCE, R. W.	LARGE DISPLACEMENT, NONLINEAR TRANSIENT ANALYSIS BY FINITE ELEMENTS		74-0324
BRUCE, R. W.	NONLINEAR TRANSIENT ANALYSIS OF AUTOMOTIVE STRUCTURES BY FINITE ELEMENTS		75-0301
BRUCH, J. C.	AN APPLICATION OF LEAST SQUARES TO ONE-DIMENSIONAL TRANSIENT PROBLEMS		74-0347
BRUCH, J. C.	ELECTRO-OSMOSIS AND THE CONTAMINATION OF UNDERGROUND FLUIDS		73-0609
BRUCH, J. C.	FINITE ELEMENT SOLUTION OF UNSTEADY AND UNSATURATED FLOW IN POROUS MEDIA		72-0064
BRUCH, J. C.	FINITE ELEMENT SOLUTION OF TWO DIMENSIONAL UNSTEADY AND UNSATURATED FLOW IN POROUS MEDIA		75-0113
BRUCH, J. C.	FINITE ELEMENT WEIGHTED RESIDUAL SOLUTION TO ONE DIMENSIONAL FIELD PROBLEMS		73-0388
BRUCH, J. C.	NONLINEAR EQUATION OF UNSTEADY GROUND-WATER FLOW		73-0262
BRUCH, J. C.	TRANSIENT TWO DIMENSIONAL HEAT CONDUCTION PROBLEMS SOLVED BY THE FINITE ELEMENT METHOD		74-0364
BRUCH, J. C. J	TRANSIENT TWO-DIMENSIONAL PROBLEMS UTILIZING THE LEAST SQUARES ALGORITHM		75-0219
BRUCH, J. C. JR	DISPERSION IN STEADY UNSTEADY SEEPAGE FLOWFIELD		74-0072
BRUCH, J. C. JR	SOLUTION OF A FREE-SURFACE BOUNDARY VALUE PROBLEM USING AN INVERSE FORMULATION AND THE FINITE ELEMEN		75-0466
BRUMAN, Z. I.	METHODS OF FINITE ELEMENTS IN ANALYSES OF CONTINUOUS AND COMBINED CONSTRUCTIONS		74-0789
BRUNKEN, J. E.	A NASTRAN VIBRATION MODEL OF THE AH-IG HELICOPTER AIRFRAME. VOLUME 1		74-1338
BRUNKEN, J. E.	A NASTRAN VIBRATION MODEL OF THE AH-IG HELICOPTER AIRFRAME. VOLUME 11		74-1339
BRUSA, L.	COMPUTATIONAL METHOD FOR DYNAMIC ANALYSIS OF STRUCTURES WITH LARGE SIZE STIFFNESS AND MASS MATRICES.		73-0830
BRUSA, L.	TWO-DIMENSIONAL FINITE ELEMENT COMPUTER CODES FOR STRESS ANALYSIS AND FIELD PROBLEMS		73-0055
BRUSA, L.	USE OF PAS-1 FINITE ELEMENT COMPUTER SYSTEM FOR STATIC AND DYNAMIC ANALYSIS OF 2-DIMENSIONAL AND 3-D		75-0070
BRYAN, J. J.	ANALYSIS OF THERMAL STRESSES AND METAL MOVEMENT DURING WELDING 1. ANALYTICAL STUDY		74-1026
BRYON, C. W.	FINITE ELEMENT ANALYSIS OF BENDING-EXTENSIONAL COUPLING IN LAMINATED COMPOSITES		70-0292
BRYSON, J. W.	THEORETICAL AND EXPERIMENTAL STRESS ANALYSES OF ORNL THIN-SHELL CYLINDER-TO-CYLINDER MODEL 3		75-0690
BRYSON, J. W.	THEORETICAL AND EXPERIMENTAL STRESS ANALYSES OF ORNL THIN-SHELL CYLINDER-TO-CYLINDER MODEL 4		75-0695
BUBECK, R. B.	TRANSIENT THERMAL STRESS ANALYSIS OF COMPOSITE STRUCTURES INCLUDING CONTINUOUSLY VARYING PROPERTIES		75-0681
BUCCIARELLI, L.	MODE SHAPES AND FREQUENCIES BY FINITE ELEMENT METHOD USING CONSISTENT AND LUMPED MASSES		71-0610
BUCHAKJIAN, L.	FINITE-ELEMENT MODELING OF DIRECTIONAL SOLIDIFICATION OF SUPER ALLOY EUTECTICS		75-0350
BUCHALET, C.	FINITE-ELEMENT ELASTIC-PLASTIC ANALYSIS OF RESIDUAL STRESSES DUE TO CLAD WELDING IN REACTOR VESSELS		75-0779
BUCHANAN, G. R.	STRESS-HYBRID FINITE-ELEMENT		75-0619
BUCHTA, R.	KOLBENKUEHLUNG (PISTON COOLING)		75-0266
BUCK, K. E.	DIE BERECHNUNG ACHSENSY METRISCHER TRAGWERKE UNTER ACHSENSYMETRISCHER UND NIGHT-SYMMETRISHCER BELAST		69-0452
BUCK, K. E.	DISCUSSION ON FINITE ELEMENT ANALYSIS OF PRESTRESSED CONCRETE REACTOR VESSELS		72-0648
BUCK, K. E.	EINIGE ALLGEMEINE PROGRAMMSYSTEME FUR FINITE ELEMENTE		73-0581
BUCK, K. E.	EINIGE ENTWICKLUNGEN IN DER MATRIZEN-VERSCHIEBUNGSMETHODE		69-0451
BUCK, K. E.	FINITE ELEMENTS FOR AXISYMMETRIC SOLIDS UNDER ARBITRARY LOADINGS WITH NODES ON ORIGIN		73-0094
BUCK, K. E.	LINEAR METHODS OF STRUCTURAL ANALYSIS		71-0028
BUCK, K. E.	MATRIX DISPLACEMENT ANALYSIS OF PLATES AND SHELLS. PART 11. GENERAL FORMULATION OF THE LINEAR THEORY		71-0621
BUCK, K. E.	MODERN DEVELOPMENTS IN THE STRESS ANALYSIS OF PRESSURE VESSELS		69-0201
BUCK, K. E.	NONLINEAR METHODS OF STRUCTURAL ANALYSIS		71-0029

BUCK, K. E.	PRACTICAL APPLICATION OF METHOD OF FINITE ELEMENTS	74-0162
BUCK, K. E.	SOME ASPECTS OF FINITE ELEMENT TECHNIQUES	69-0340
BUCK, K. E.	SOME NEW ELEMENTS FOR THE MATRIX DISPLACEMENT METHOD	68-0017
BUDD, W. I. H.	REDUCTION GEAR DAMAGES RELATED TO EXTERNAL INFLUENCES	75-0307
BUDIANSKY, B.	SLOSHING OF LIQUID IN CIRCULAR CHANNELS AND SPHERICAL TANKS	60-0005
BUELL, W. R.	APPLICATION OF THE FINITE ELEMENT METHOD TO PREDICT STATIC AND DYNAMIC RESPONSE OF AN UNSHROUDED CEN	71-0574
BUELL, W. R.	MATHEMATICAL PROGRAMMING	72-0100
BUELL, W. R.	MESH GENERATION - SURVEY	73-0048
BUERHOP, H.	ZUR BERECHNUNG DER BIEGESTEIFIGKEIT ABGESETZTER STAEBE UND WELLEN UNTER ANWENDUNG VON FINITEN ELEMEN	75-0481
BUESSEM, W. R.	TWO-DIMENSIONAL FINITE ELEMENT MODEL OF A CERAMIC BODY	73-0010
BUFLER, H.	CALCULATION OF PLATES BY MEANS OF METHOD OF FINITE ELEMENTS	70-0345
BUFLER, H.	PLATE MEASUREMENTS USING FINITE ELEMENTS	70-0229
BUG, G.	FINITE ELEMENT SOLUTION OF THE FLOW AROUND A TWO-DIMENSIONAL DESCOID	69-0298
BUGROV, A. K.	METOD KONECHNYKH ELEMENTOV V RASCHETAKH KONSOLIDATSII VODONASYSHCHENNYKH GRUNTOV. (FINITE ELEMENT ME	75-0469
BUGROV, A. K.	SOLUTION OF THE MIXED PROBLEM OF THE THEORY OF ELASTICITY AND THE THEORY OF PLASTICITY OF SOILS	74-1136
BUHLER, J.	EXPERIMENTAL AND THEORETICAL STUDIES OF HULL-DECKHOUSE INTERACTION - 2-LEVEL DECKHOUSE	71-0672
BUHLMEIER, J.	DYNAN LECTURE NOTES WITH COMPUTATIONAL EXAMPLES	71-0581
BUHLMEIER, J.	EINIGE VERFAHREN ZUR BERECHNUNG VON EIGENWERTEN UND EIGENVEKTOREN VON NICHT HERMITESCHEN MATRIZEN J	72-0646
BUHLMEIER, J.	F 104 G FINAL MOTIVE: GROUND VIBRATION CALCULATIONS FOR THE STARFIGHTER F 104 G	69-0456
BUHLMEIER, J.	MALEN DIMENSIONIERUNG VON TRAGWERKEN	74-0683
BULLIS, S.	ON THE CONVERGENCE OF THE FINITE ELEMENT SOLUTION IN A NONLINEAR HEAT CONDUCTION PROBLEM	74-1272
BURDEKIN, M.	PRESSURE DISTRIBUTION AND DEFORMATIONS OF MACHINED COMPONENTS IN CONTACT.	73-0372
BURLAND, J. B.	OBSERVED AND PREDICTED DEFORMATIONS IN A LARGE EMBANKMENT DAM DURING CONSTRUCTION	71-0301
BURMAN, B. C.	DEVELOPMENT OF A NUMERICAL MODEL FOR DISCONTINUA	74-0961
BURMAN, Z. I.	ON THE THEORY OF CALCULATION OF THE OVERALL STRENGTH OF A FUSELAGE BY THE METHOD OF FINITE ELEMENTS	72-0692
BURMAN, Z. I.	PROBLEM OF USING THE METHOD OF FINITE ELEMENTS TO CONSTRUCT AN ALGORITHM FOR CALCULATING A THIN-WALL	72-0851
BURNETT, E. F.	INFLUENCE OF JOINTS IN PANELIZED STRUCTURAL SYSTEMS	72-0204
BURNHAM, M. W.	SHELL DEFORMATION UNDER TOOL LOADS REPRESENTATION OF SINGULARITIES WITH ISOPARAMETRIC FINITE ELEMENT	74-0365
BURNS, B. P.	STRESS ANALYSIS OF 175 MM PROJECTILE, HE, M437	71-0736
BURNS, N. H.	COMPUTER ANALYSIS OF SEGMENTALLY ERECTED BRIDGES	75-0172
BURNS, T. A.	AEROELASTIC LOADS PREDICTIONS USING FINITE-ELEMENT AERODYNAMICS	75-0815
BURRIDGE, H.	THEORETICAL COMPUTATIONS ON RIDGE ACOUSTIC SURFACE WAVES USING THE FINITE-ELEMENT METHOD	71-0011
BUPSTEIN, S. Z.	ADVANCES IN NUMERICAL FLUID DYNAMICS	73-0813
BUSBY, H. R.	DYNAMIC STABILITY OF A NONLINEAR BEAM SUBJECTED TO BOTH LONGITUDINAL AND TRANSVERSE EXCITATION	73-0553
BUSBY, H. R., JR	NON-LINEAR RESPONSE OF A BEAM TO PERIODIC LOADING	72-0407
BUSBY, H. R., JR	RESPONSE OF NONLINEAR BEAM TO RANDOM EXCITATION	73-0285
BUSCH, R. A.	DETERMINING SEEPAGE CHARACTERISTICS OF MILL-TAILING DAMS BY THE FINITE ELEMENT METHOD	71-0477
BUSCH, R.A.	SEEPAGE-ENVIRONMENTAL ANALYSIS OF THE SLIME ZONE OF A TAILINGS POND	74-0513
BUSH, B. A.	MATHEMATICAL PROGRAMMING	72-0100
BUSH, B. A.	MESH GENERATION - SURVEY	73-0048
BUSH, D. I.	DYNAMIC RESPONSE OF MODEL PAVEMENT STRUCTURE	72-0226
BUSHNELL, D.	ANALYSIS OF BUCKLING AND VIBRATION OF RING-STIFFENED, SEGMENTED SHELLS OF REVOLUTION	69-0496
BUSHNELL, D.	ANALYSIS OF RING-STIFFENED SHELLS OF REVOLUTION UNDER COMBINED THERMAL AND MECHANICAL LOADING	71-0342
BUSHNELL, D.	COMPUTER ANALYSIS OF SHELL STRUCTURES	69-0060
BUSHNELL, D.	ENERGY APPROACHES TO FINITE-DIFFERENCE AND FINITE ELEMENT METHODS	71-0046
BUSHNELL, D.	FINITE-DIFFERENCE ENERGY METHOD FOR NONLINEAR SHELL ANALYSIS	71-0399
BUSHNELL, D.	STRESS, STABILITY AND VIBRATION OF COMPLEX SHELLS OF REVOLUTION: ANALYSIS AND USERS MANUAL FOR BOSOR	69-0441
BUSSE, L.	ACCURATE CALCULATION OF STRESSES IN BEAMS, DISKS AND PLATES BY MEANS OF FINITE ELEMENT METHODS AND I	70-0133
BUSSE, L.	SCHWINGUNGEN ZYLINDRISCHER SCHRAUBENFEDERN (VIBRATIONS OF CYLINDRICAL COIL SPRINGS)	74-0331
BUSTAMANTE, M.	LA CAPACITE PORTANTE DES PIEUX. (BEARING CAPACITY OF PILES)	75-0463
BUTLIN, G.	A COMPATIBLE TRIANGULAR PLATE BENDING FINITE ELEMENT	70-0385
BUTLIN, G. A.	A COMPATIBLE TRIANGULAR PLATE BENDING FINITE ELEMENT	68-0209
BUTLIN, G. A.	A STUDY OF FINITE ELEMENTS APPLIED TO PLATE FLEXURE	66-0016
BUTLIN, G. A.	MAN MACHINE INTERACTIVE STRUCTURAL ANALYSIS AS A PRELIMINARY DESIGN AID	69-0293
BUTLIN, G. A.	THE FINITE ELEMENT METHOD APPLIED TO PLATE FLEXURE	66-0095
BUTMAN, Z. I.	SOME RESULTS OF FUSELAGE ANALYSIS BY THE FINITE ELEMENT METHOD USING A COMPUTER	73-0535
BUTTER, U.	FINITE ELEMENT METHOD FOR STRESS ANALYSIS OF BODIES OF REVOLUTION SUBJECTED TO AXISYMMETRIC LOADING	71-0409
BUTTERFIELD, R.	DYNAMIC ANALYSIS OF SOIL - STRUCTURE SYSTEM USING FINITE ELEMENT METHOD	67-0001
BUTTSTAEDT, K.	DER EINSATZ DER FINIT-ELEMENT-METHODE IN PRESSENBAU (APPLICATION OF THE FINITE ELEMENT METHOD FOR TH	74-0615
BUTURLA, E. M.	ERROR BOUNDS IN AN OPTIMIZATION PROBLEM USING FINITE-ELEMENT METHOD	73-0049
BUTURLA, E. M.	EVALUATION OF THE ERROR BOUNDS IN AN OPTIMIZATION PROBLEM USING THE FINITE-ELEMENT METHOD	73-0211
BUTURLA, E. M.	ON SPARSE MATRICES IN FINITE ELEMENT WAVE PROPAGATION	73-0290
BUTURLA, E. M.	SPARSE MATRICES IN FINITE ELEMENT PROGRAM DEVELOPMENT	73-0337
BUTURLA, E. M.	TWO-DIMENSIONAL FINITE ELEMENT ANALYSIS OF SEMI-CONDUCTOR STEADY STATE TRANSPORT EQUATIONS	74-1224
BUYUKOZTURK, O.	DEFORMATION AND FRACTURE OF PARTICULATE COMPOSITE	72-0221
BUYUKOZTURK, O.	STRENGTH OF REINFORCED CONCRETE CHAMBERS UNDER EXTERNAL PRESSURE	75-0293
BUYUKOZTURK, O.	STRENGTH OF REINFORCED CONCRETE CHAMBERS UNDER EXTERNAL PRESSURE	75-0512
BUYUKOZTURK, O.	STRESS STRAIN RESPONSE AND FRACTURE OF A CONCRETE MODEL IN BIAXIAL LOADING	71-0295
BUYUKOZTURK, O.	WATER IMPACT ANALYSIS OF SPACE SHUTTLE SOLID ROCKET BY THE FINITE ELEMENT METHOD	74-0760
BYERS, N. R.	ANALYSIS OF A STUB END BY FINITE ELEMENT METHOD	72-0210
BYKAT, A.	AUTOMATIC TRIANGULATION OF TWO DIMENSIONAL REGIONS	72-0697
BYRAM, K. V.	MATHEMATICAL MODEL OF THE COLUMBIA RIVER FROM THE PACIFIC OCEAN TO BONNEVILLE DAM. PART II: INPUT-OU	70-0417
BYRAN, K. V.	MATHEMATICAL MODEL OF THE COLUMBIA RIVER FROM THE PACIFIC OCEAN TO BONNEVILLE DAM. PART 1. THEORY, P	69-0464
BYSKOV, E.	CALCULATION OF STRESS INTENSITY FACTORS USING FINITE ELEMENT METHOD WITH CRACKED ELEMENTS	70-0224
BYSKOV, E.	ELEMENTMETODEN OG DEFORMATIONSMETODEN (FINITE ELEMENT AND DEFORMATION METHODS IN STRESS ANALYSIS)	74-0580
BYSKOV, E.	FINITE ELEMENT METHOD AND DENDIXEN-OSTENFELD SLOPE DEFLECTION METHOD	74-0557
BYSKOV, E.	TWO NEARLY POLYGONAL HOLES, MATHEMATICAL CRACK PROBLEMS	69-0061
CABAYAN, H. S.	EFFICIENT TECHNIQUES FOR FINITE ELEMENT ANALYSIS OF ELECTRIC MACHINES	73-0134
CALDER, C. A.	APPLICATIONS OF PULSED LASER INDUCED STRESS WAVES	73-0956
CALLABRESI, M.	GNATS: A FINITE ELEMENT COMPUTER PROGRAM FOR THE GENERAL NONLINEAR ANALYSIS OF TWO-DIMENSIONAL STRUC	74-0875
CALLABRESI, M.	ORGANIZATION OF GNATS: A GENERAL NONLINEAR ANALYSIS COMPUTER PROGRAM FOR TWO-DIMENSIONAL STRUCTURES	74-0871
CALLABRESI, M.	SASL - A FINITE ELEMENT CODE FOR THE STATIC ANALYSIS OF AXISYMMETRIC AND PLANE SOLIDS SUBJECTED TO A	72-0779
CALLADINE, C. R	NEW FINITE ELEMENT METHOD FOR ANALYZING SYMMETRICALLY LOADED THIN SHELLS OF REVOLUTION	73-0389
CALLAWAY, R. J.	MATHEMATICAL MODEL OF THE COLUMBIA RIVER FROM THE PACIFIC OCEAN TO BONNEVILLE DAM. PART 1. THEORY, P	69-0464
CALLAWAY, R. J.	MATHEMATICAL MODEL OF THE COLUMBIA RIVER FROM THE PACIFIC OCEAN TO BONNEVILLE DAM. PART II: INPUT-OU	70-0417
CALZOLARI, P. U	FINITE DIFFERENCE AND FINITE ELEMENT METHODS FOR THE SEMICONDUCTOR FLOW EQUATIONS	74-0028
CALZONA, R.	ANALISI NUMERICA DEL COMPORTAMENTO ELASTICO DI UNA STRUTTURA BIDIMENSIONALE PIANA. (NUMERICAL ANALYS	74-1029
CAMBEFORT, H.	SPECIAL STUDY ON THE STABILITY OF A DAM SPILLWAY	72-0469
CAMBURN, G. L.	THE EFFECT OF GRAVITY UPON THE MELT-THROUGH TIME OF A SOLID SUBJECTED TO A HIGH INTENSITY LASER	72-0693
CAMERON, I. G.	A DYNAMIC ELASTOPLASTIC ANALYSIS OF THIN SHELLS OF REVOLUTION UNDER ASYMMETRIC MECHANICAL AND THERMA	72-0068
CAMPBELL, D. M.	A FINITE-ELEMENT ANALYSIS FOR THIN SHELLS	68-0087
CAMPBELL, F. S.	DEVELOPMENTS IN DISCRETE ELEMENT FINITE DEFLECTION STRUCTURAL ANALYSIS BY FUNCTION MINIMIZATION	68-0144
CAMPBELL, J. S.	A PENALTY FUNCTION APPROACH TO THE MINIMIZATION OF QUADRATIC FUNCTIONALS IN FINITE ELEMENT ANALYSIS	74-0418
CAMPBELL, J. S.	FINESSE - A FINITE ELEMENT SYSTEM, PART V FINITE ELEMENT DESIGN - DERIVATIVES FOR STRUCTURAL SHAPE D	72-0963
CAMPBELL, J. S.	FINITE ELEMENT STRESS ANALYSIS OF PLANE AND AXISYMMETRIC STRUCTURES	69-0062
CAMPBELL, J. S.	FINITE ELEMENTS FOR THE REPRESENTATION OF PLANE AND AXISYMMETRIC STRUCTURAL REINFORCEMENT	68-0037
CAMPBELL, J. S.	LOCAL AND GLOBAL SMOOTHING OF DISCONTINUOUS FINITE ELEMENT FUNCTIONS USING A LEAST SQUARES METHOD	74-0378
CAMPBELL, J. S.	STRESS ANALYSIS BY THE FINITE ELEMENT DISPLACEMENT METHOD	68-0036
CAMPBELL, J. S.	THREE DIMENSIONAL STRESS ANALYSIS	70-0086
CAMPBELL, R. D.	INELASTIC ANALYSIS OF A BRANCH SHELL JUNCTION	74-0682
CANDOGAN, A.	CYCLIC LOADING AND CRACK PROPAGATION - AN ELASTOPLASTIC FINITE ELEMENT STUDY	74-1214
CANDOLFO, G.	ON THE SOLUTION OF PLANE STRESS PROBLEMS BY FINITE ELEMENTS COMPUTER PROGRAMS	73-0978

CANNON, M.	ADAPTABILITY OF COMPUTER GRAPHICS TO STUDIES OF ATHEROSCLEROSIS PATHOGENESIS	72-0206
CANTATORE, W. P	NUMERICAL SOLUTION OF THE STEADY INTERFACE IN A CONFINED COASTAL AQUIFER	74-1007
CANTATORE, W. P	THE NUMERICAL SOLUTION OF THE STEADY INTERFACE IN A COASTAL CONFINED AQUIFER	73-0711
CANTIN, G.	A CURVED CYLINDRICAL SHELL DISCRETE ELEMENT	67-0009
CANTIN, G.	A CURVED FINITE ELEMENT FOR CYLINDRICAL SHELLS	68-0294
CANTIN, G.	A CURVED, CYLINDRICAL SHELL, FINITE ELEMENT	68-0039
CANTIN, G.	A REFINED CURVED CYLINDRICAL SHELL FINITE ELEMENT	68-0038
CANTIN, G.	AN EQUATION SOLVER OF VERY LARGE CAPACITY	71-0677
CANTIN, G.	RIGID BODY MOTIONS AND EQUILIBRIUM IN FINITE ELEMENTS	74-0102
CANTIN, G.	RIGID BODY MOTIONS IN CURVED FINITE ELEMENTS	70-0035
CANTIN, G.	THREE DIMENSIONAL FINITE ELEMENT STUDIES. PART ONE: SERVICE ROUTINES	72-0721
CAPEY, E. C.	A RELAXATION PROCEDURE FOR COMPUTING STRESSES IN BUILT-UP STRUCTURES	69-0063
CAPPELLI, A. P.	ELAD PROGRAM MANUAL, A COMPUTER PROGRAM FOR THE ELASTIC ANALYSIS OF DOME STRUCTURES AND AXISYMMETRIC	69-0064
CAPPELLI, A. P.	EPAD, A COMPUTER PROGRAM FOR THE ELASTIC-PLASTIC ANALYSIS OF DOME STRUCTURES AND ARBITRARY SOLIDS OF	70-0036
CAPPELLI, A. P.	THEORETICAL ANALYSIS OF THICK DOME STRUCTURES AND ARBITRARY SOLIDS OF REVOLUTION SUBJECTED TO UNSYMM	70-0037
CAPRILI, M.	SPLINE INTERPOLATION TECHNIQUES FOR VARIATIONAL METHODS	73-1079
CAPURSO, M.	ON THE EXTREMAL PROPERTIES OF THE SOLUTION IN DYNAMICS OF RIGID VISCOPLASTIC BODIES ALLOWING FOR LAR	73-1080
CARCY, G. F.	PERTURBATION VARIATIONAL SOLUTION OF COMPRESSIBLE FLOWS	74-0025
CARD, C. C. H.	FINITE ELEMENT APPLICATIONS TO NEWTONIAN FLUIDS	68-0197
CARD, C. C. H.	LOW REYNOLDS NUMBER DEVELOPING FLOW	69-0020
CARD, C. C. M.	APPLICATION OF THE FINITE ELEMENT METHOD TO CREEPING FLOW PROBLEMS	70-0024
CAREY, G. F.	A COMPUTER PROGRAMME FOR ELASTIC PLASTIC CALCULATIONS WITH DUAL PERTURBATION EXPANSION AND VARIATION	75-0882
CAREY, G. F.	A VARIATIONAL FORMULATION OF THE COMPRESSIBLE THROUGHFLOW PROBLEM	74-1343
CAREY, G. F.	A VARIATIONAL FORMULATION OF THE COMPRESSIBLE THROUGHFLOW PROBLEM	75-0021
CAREY, G. F.	AUTOMATED MESH REFINEMENT STRATEGIES FOR FINITE ELEMENT COMPUTATIONS	75-0877
CAREY, G. F.	AXISYMMETRIC SOLUTION OF ELASTO PLASTIC PROBLEMS BY FINITE ELEMENT METHODS	69-0361
CAREY, G. F.	BASIS FUNCTIONS IN FINITE ELEMENT THEORY AND APPLICATION	74-0419
CAREY, G. F.	COMPUTER MODELLING OF PLUG SEEDLING SURVIVAL	75-0176
CAREY, G. F.	ELASTO PLASTIC ANALYSIS OF PLANE STRUCTURES BY THE FINITE ELEMENT DISPLACEMENT METHOD	70-0299
CAREY, G. F.	EXPERIENCE WITH FINITE ELEMENT COMPUTATIONS FOR NONLINEAR STRUCTURAL PROBLEMS	74-1239
CAREY, G. F.	INTRODUCTION TO FINITE ELEMENT ANALYSIS	72-0358
CAREY, G. F.	ORTHOGONAL COLLOCATION ON FINITE ELEMENTS	75-0364
CAREY, G. F.	PERTURBATION-VARIATIONAL SOLUTION OF COMPRESSIBLE FLOWS	74-1088
CAREY, G. F.	STUDIES IN FINITE-ELEMENT ANALYSIS	74-1193
CAREY, G. F.	UNIFIED APPROACH TO THREE FINITE ELEMENT THEORIES FOR GEOMETRIC NONLINEARITY	74-0599
CAREY, J. J.	EVALUATION OF A CLASS OF METHODS FOR BOUNDING STEADY-STATE CREEP DEFORMATION	74-0892
CARL, E. J.	HOW FINITE ELEMENT METHODS ARE INTRODUCED IN LARGE AND SMALL ORGANIZATIONS	74-0366
CARLSEN, C. A.	INDETS - INTEGRATED DESIGN OF TANKER STRUCTURES	75-0524
CARLSON, M.	A FINITE ELEMENT APPROACH TO SCATTERING FROM ELASTIC SPHERICAL SHELLS	74-0702
CARLSSON, A. J.	INFLUENCE OF NON-SINGULAR STRESS TERMS AND SPECIMEN GEOMETRY ON SMALL SCALE YIELDING AT CRACK TIPS I	73-0203
CARLTON, C. H.	CASE LINER BOND ANALYSIS	74-0817
CAPMEAN, I. C.	VISCOPLASTICITY, PLASTICITY AND CREEP, A UNIFIED NUMERICAL SOLUTION APPROACH	74-1205
CARMICHAEL, G.	ADVANCES IN THE ANALYSIS OF PRESTRESSED CONCRETE PRESSURE VESSELS	71-0162
CARMICHAEL, G.	APPLICATION OF FRACTURE MECHANICS TO PRESTRESSED CONCRETE PRESSURE VESSELS	73-0228
CARMICHAEL, G.	THE APPLICATION OF THREE-DIMENSIONAL FINITE ELEMENTS TO THE ANALYSIS OF PODDED BOILER TYPE PRESTRESS	71-0047
CARMICHAEL, R.	RECENT EXPERIENCE IN USING FINITE ELEMENT METHODS FOR THE SOLUTION OF PROBLEMS IN AERODYNAMIC INTERF	70-0339
CARMIGNANI, C.	CHARACTERISTICS OF A HYBRID CALCULATION SYSTEM BASED ON THE FINITE ELEMENT METHOD. (CARATTERISTICHE	72-0715
CARMIGNANI, C.	COMPUTER SYSTEM FOR THE ANALYSIS OF LARGE SCALE PROBLEMS IN SOLID, FLUID AND THERMODYNAMICS	73-0554
CARMIGNANI, C.	ELASTIC ANALYSIS OF CRACKED THIN SHELLS BY THE FINITE ELEMENT METHOD	71-0048
CARMIGNANI, C.	FOURIER ANALYSIS OF TRANSIENT PROBLEMS BY THE FINITE ELEMENT METHOD	74-1331
CARMIGNANI, C.	FUNCTIONAL MINIMIZATION IN NONLINEAR SOLID MECHANICS	74-1251
CARMIGNANI, C.	ON THE RESOLUTION OF NONLINEAR ELASTIC AND ELASTOPLASTIC PROBLEMS BY THE METHOD OF FINITE ELEMENTS.	74-1326
CARMIGNANI, F.	STRUCTURAL DYNAMICS BY FINITE ELEMENTS: MODAL AND FOURIER ANALYSIS	73-0836
CARNAHAN, N. F.	ANCHOR-PILE DESIGN FOR OCEAN-FLOOR ENVIRONMENTS USING FINITE-ELEMENT ANALYSIS	75-0459
CARNEGIE, W.	FINITE ELEMENT ANALYSIS OF COUPLED VIBRATION OF TAPERED TWISTED BLADES	72-0766
CARPENTER, C. J	FINITE-ELEMENT NETWORK MODELS AND THEIR APPLICATION TO EDDY-CURRENT PROBLEMS	75-0143
CARPENTER, L. R	POLYMER-IMPREGNATED CONCRETE TUNNEL SUPPORT AND LINING	73-0736
CARPENTER, S. H	PAVEMENT CRACKING IN WEST TEXAS DUE TO FREEZE-THAW CYCLING	75-0276
CARPENTER, W. C	ANALYSIS OF PLATES ON ELASTIC FOUNDATIONS	73-0684
CARPENTER, W. C	AUTOMATED SOLUTION OF TIME DEPENDENT PROBLEMS	72-0084
CARPENTER, W. C	FINITE ELEMENT ANALYSIS OF BONDED CONNECTIONS	73-1096
CARR, A. J.	A REFINED FINITE ELEMENT ANALYSIS OF THIN SHELL STRUCTURES	67-0101
CARR, A. J.	A REFINED FINITE ELEMENT ANALYSIS OF THIN SHELL STRUCTURES INCLUDING DYNAMIC LOADINGS	67-0157
CARR, A. J.	DYNAMIC BEHAVIOUR OF SHELL ROOFS	69-0228
CARR, A. J.	FINITE ELEMENT ANALYSIS OF DYNAMIC SHELL BEHAVIOUR	67-0013
CARRARA, S.	MATRIX AND INTERFACE STRESSES IN A DISCONTINUOUS FIBER COMPOSITE MODEL	68-0040
CARRETT, R. J.	DEVELOPMENT OF A DESIGN FOR A PRECAST CONCRETE BRIDGE BEAM OF J-SECTION	73-0392
CARROLL, W. E.	A THEOREM FOR OPTIMUM IDEALIZATIONS IN FINITE-ELEMENT ANALYSIS	72-0791
CARROLL, W. E.	RAMIFICATIONS OF OPTIMUM IDEALIZATION GEOMETRY IN DISCRETE ELEMENT ANALYSIS	75-0182
CARROLL, W. E.	THEOREM FOR OPTIMUM FINITE-ELEMENT IDEALIZATIONS	73-0255
CARSLON, A. D.	FINITE ELEMENT STRESS ANALYSIS OF A MULTICONDUCTOR ARMOR CABLE	73-0095
CARSON, J.	AN ANALYSIS OF FINITE ELEMENT CODE TORT 11	75-0752
CARSON, W. G.	PLATE BUCKLING ANALYSIS USING A FULLY COMPATIBLE FINITE ELEMENT	69-0065
CARSON, W. W.	EVALUATION OF THE USE OF FINITE ELEMENT METHOD IN COMPUTING TEMPERATURE	69-0017
CARSON, W. W.	EVALUATION OF USE OF THE FINITE-ELEMENT METHOD IN COMPUTATION OF TEMPERATURE	71-0467
CARTER, C.	DIAGONAL TENSILE STRENGTH OF BRICKWORK	70-0158
CARTER, J. P.	EFFECT OF SEEPAGE ON EMBANKMENT DEFORMATIONS DUE TO WATER LOADING	74-1329
CARTWRIGHT, K.	ANALYSIS OF LIQUID-WASTE INJECTION WELLS IN ILLINOIS BY MATHEMATICAL MODELS	74-0518
CARUSO, S.	INTERNAL SUPPORTING STRUCTURES OF PRESSURE TUBES TYPE REACTOR CORE, ANALYTICAL MODEL FOR STATIC ANAL	71-0049
CARUSO, S.	INTERNAL SUPPORTING STRUCTURES OF THE (CIRENE) REACTOR CORE - RESULTS OF STATIC STRESS ANALYSIS FOR	73-0818
CARVER, D. R.	STRESS CONCENTRATION FACTORS FOR BONDED LAP JOINTS	71-0842
CASARELLA, M. J	A SURVEY OF ANALYTICAL METHODS FOR DYNAMIC SIMULATION OF CAB	73-0936
CASCIARO, R.	DISCUSSION ON "PLANE STRESS LIMIT ANALYSIS BY FINITE ELEMENTS"	71-0792
CASCIARO, R.	MIXED FINITE ELEMENT MODELS IN LIMIT ANALYSIS	74-1314
CASE, W. R. JR.	FINITE ELEMENT MODELLING OF THREE-DIMENSIONAL STRUCTURES	68-0267
CASSELL, A. C.	THE SEISMIC DESIGN STUDY OF A DOUBLE CURVATURE ARCH DAM	69-0053
CASTLETON, G. A	FIELD TESTS OF A STEEL-COMPOSITE BOX-GIRDER BRIDGE	75-0633
CASTLETON, G. A	LOAD DISTRIBUTION IN A COMPOSITE STEEL BOX GIRDER BRIDGE	73-0765
CASWELL, B.	SOLUTION OF VISCOUS INCOMPRESSIBLE JET AND FREE-SURFACE FLOWS USING FINITE ELEMENT METHODS	74-0357
CATALOGLU, A.	STRESS ANALYSIS OF THE HUMAN AORTIC VALVE	73-0166
CATALOGLU, A.	THE FINITE ELEMENT STRESS ANALYSIS OF THE HUMAN AORTIC VALVE TOWARD THE DEVELOPMENT OF A PROSTHETIC	74-1201
CATHERINES, D.	ACCURACY OF MODAL STRESS CALCULATIONS BY FINITE ELEMENT METHOD	70-0223
CATHERINES, D.	CALCULATION OF TEMPERATURE DISTRIBUTIONS IN THIN SHELLS OF REVOLUTION BY THE FINITE ELEMENT METHOD	71-0018
CAVENDISH, J. C	AUTOMATIC TRIANGULATION OF ARBITRARY PLANAR DOMAINS FOR THE FINITE ELEMENT METHOD	74-0480
CAVENDISH, J. C	LOCAL MESH REFINEMENT USING RECTANGULAR BLENDED FINITE-ELEMENTS	75-0452
CAYWOOD, W. C.	GENERAL PURPOSE COMPUTER PROGRAM FOR THE DYNAMIC SIMULATION OF VEHICLE GUIDEWAY INTERACTIONS	73-0260
CAZACU, M. D.	THEORETICAL AND EXPERIMENTAL CONSIDERATIONS ON FLOW OF VISCOUS LIQUIDS PAST PLATE AT SMALL AND AVERA	70-0144
CECCHI, M. M.	ERROR ESTIMATES FOR FINITE ELEMENT METHOD FOR HEAT TRANSFER PROBLEMS IN THE SPACE-TIME DOMAIN	75-0104
CECCI, M. M.	A FINITE ELEMENT SOLUTION OF THE STABILITY OF SUPERPOSED FLUIDS	74-1226
CELESTINI, A. C	APPLICATION OF FINITE ELEMENT ANALYSIS TO COPLANAR WAVEGUIDE	73-0924
CELLA, A.	A FINITE ELEMENT SOLUTION OF THE STABILITY OF SUPERPOSED FLUIDS	74-1226
CELLA, A.	AN EXTENDED THEORY FOR THE FINITE ELEMENT METHOD	72-0350

CELLA, A.	APPROXIMATION TECHNIQUES IN THE FINITE ELEMENT METHOD	72-0700
CELLA, A.	COMPUTER SYSTEM FOR THE ANALYSIS OF LARGE SCALE PROBLEMS IN SOLID, FLUID AND THERMODYNAMICS	73-0554
CELLA, A.	ELASTIC ANALYSIS OF CRACKED THIN SHELLS BY THE FINITE ELEMENT METHOD	71-0048
CELLA, A.	FUNCTIONAL MINIMIZATION IN NONLINEAR SOLID MECHANICS	74-1251
CELLA, A.	ON THE ACCURACY AND NUMERICAL STABILITY OF THE FINITE ELEMENT APPROXIMATION FOR PARABOLIC AND HYPERB	75-0102
CELLA, A.	ON THE RESOLUTION OF NONLINEAR ELASTIC AND ELASTOPLASTIC PROBLEMS BY THE METHOD OF FINITE ELEMENTS.	74-1326
CELLA, A.	SPLINE INTERPOLATION TECHNIQUES FOR VARIATIONAL METHODS	73-1079
CELLA, A.	STRUCTURAL DYNAMICS BY FINITE ELEMENTS: MODAL AND FOURIER ANALYSIS	73-0836
CERVENKA, V.	INELASTIC FINITE ELEMENT ANALYSIS OF REINFORCED CONCRETE PANELS UNDER IN-PLANE LOADS	70-0482
CESARI, E.	VERIFICATION OF THE PFC CONTAINMENT VESSEL FOR SOME OPERATING CONDITIONS BY USE OF THE FINITE ELEMEN	74-0883
CESARI, F.	DESCRIPTION AND INSTRUCTION FOR USE OF THE SAFE-SHELL. B FINITE ELEMENT PROGRAM IN THE IBM/360 VERSI	73-0945
CETINER, A.	ELEMENT STIFFNESS MATRIX GENERATOR	71-0106
CHABERT, G.	EVALUATION OF STRESSES AND DEFLECTION OF SPUR GEAR TEETH UNDER STRAIN	72-0115
CHACOUR, S.	A HIGH PRECISION AXISYMMETRIC TRIANGULAR ELEMENT USED IN THE ANALYSIS OF HYDRAULIC TURBINE COMPONENT	70-0038
CHACOUR, S.	DANUTA THREE DIMENSIONAL FINITE ELEMENT PROGRAM USED IN ANALYSIS OF TURBOMACHINERY	72-0296
CHAINOV, N. D.	METODY RASCHETNOGO OPREDELENIYA TEMPERATURNYKH NAPRYAZHENII V KRYSHKAKH TSILINDROV DVIGATELEI VNUTRE	74-0179
CHAKRABARTI, G.	A DISCRETE LEAST-SQUARES FINITE ELEMENT METHOD FOR TRANSIENT HEAT CONDUCTION PROBLEMS	74-1191
CHAKRABARTI, P.	A COMPUTER PROGRAM FOR EARTHQUAKE ANALYSIS OF DAMS	70-0469
CHAKRABARTI, P.	EARTHQUAKE ANALYSIS OF GRAVITY DAMS INCLUDING HYDRODYNAMIC INTERACTION	73-0350
CHAKRABARTI, P.	EARTHQUAKE RESPONSE OF GRAVITY DAMS INCLUDING RESERVOIR INTERACTION EFFECTS	72-0675
CHAKRABARTI, P.	THE EARTHQUAKE EXPERIENCE AT KOYNA DAM AND STRESSES IN CONCRETE GRAVITY DAMS	72-0589
CHAKRABARTI, S.	ANALYSIS OF SIMPLY SUPPORTED THICK, LAYERED PLATES	71-0428
CHAKRABARTI, S.	APPROXIMATIONS IN FINITE-ELEMENT HEAT CONDUCTION ANALYSIS	74-1167
CHAKRABARTI, S.	FREE VIBRATION OF THICK, LAYERED RECTANGULAR PLATES BY FINITE LAYER METHOD	72-0440
CHAKRABARTI, S.	TRIGONOMETRIC FUNCTION REPRESENTATIONS FOR RECTANGULAR PLATE BENDING ELEMENTS	71-0765
CHAKRABAS, S.	THREE DIMENSIONAL FINITE ELEMENT METHOD FOR PLATE BENDING	74-0243
CHALOUPKA, A. B	A COMPUTER PROGRAM FOR THE ANALYSIS OF TWO-DIMENSIONAL HEAT CONDUCTION USING THE FINITE ELEMENT TECH	69-0066
CHAMIS, C. C.	ANALYTICAL DISPLACEMENTS AND VIBRATIONS OF CANTILEVERED UNSYMMETRIC FIBER COMPOSITE LAMINATES	75-0736
CHAMIS, C. C.	MECHANICS OF LOAD TRANSFER AT THE FIBER/MATRIX INTERFACE	72-0879
CHAN, A. S. L.	ANALYSIS OF COOLING TOWERS BY THE MATRIX FINITE ELEMENT METHOD- 2	70-0148
CHAN, A. S. L.	ANALYSIS OF RADIAL FLOW IMPELLERS BY THE MATRIX FINITE ELEMENT METHOD	71-0412
CHAN, A. S. L.	APPLICATIONS OF FINITE ELEMENTS IN SPACE AND TIME	72-0147
CHAN, A. S. L.	STATIC AND DYNAMIC ELASTO-PLASTIC ANALYSIS BY THE METHOD OF FINITE ELEMENTS IN SPACE AND TIME	72-0642
CHAN, C. F.	A DYNAMIC PHOTOELASTIC ANALYSIS OF DYNAMIC TEAR TEST SPECIMEN	74-0736
CHAN, H. C.	STATIC AND DYNAMIC ANALYSIS OF MULTILAYERED SANDWICH PLATES	72-0493
CHAN, H. S.	MATHEMATICAL MODEL FOR CLOSED HEAD IMPACT	74-0934
CHAN, H. S. Y.	COLLAPSE LOAD OF REINFORCED CONCRETE PLATES.	72-0406
CHAN, H. T.	TEST OF VALIDITY OF DARCY'S LAW FOR A CLAY SOIL	74-0070
CHAN, S. C.	NONLINEAR ANALYSIS OF ELASTIC FRAMED STRUCTURES	68-0222
CHAN, S. K.	ANALYSIS OF STRESSES IN PRESSURIZED WELDED PIPE IN THE CREEP RANGE	71-0404
CHAN, S. K.	FINITE ELEMENT METHOD IN LINEAR FRACTURE MECHANICS	70-0476
CHAN, S. K.	FINITE ELEMENT METHOD FOR CONTACT PROBLEMS OF SOLID BODIES - 2.	71-0435
CHAN, S. K.	ON THE FINITE ELEMENT METHOD IN LINEAR FRACTURE MECHANICS	69-0032
CHAN, S. S. M.	A CASE STUDY OF IN-SITU ROCK DEFORMATION BEHAVIOR FOR THE DESIGN OF GROUND SUPPORT SYSTEM	72-0668
CHAN, S. T. K.	FINITE ELEMENT ANALYSIS OF IRROTATIONAL FLOWS OF AN IDEAL FLUID	71-0806
CHAN, S. T. K.	FINITE ELEMENT ANALYSIS OF TRANSONIC FLOW	74-1223
CHAN, S. T. K.	FINITE ELEMENT CIRCULATION ANALYSIS OF VARIABLE-DEPTH SHALLOW LAKES	73-0727
CHAN, S. T. K.	FREE-SURFACE IDEAL FLUID FLOWS BY FINITE ELEMENTS	73-0254
CHAN, S. T. K.	HIGHER-ORDER FINITE ELEMENT ANALYSIS OF LAKE CIRCULATION	73-1002
CHAN, S. T. K.	VARIATIONAL APPROACH TO THE LIFTING SURFACE PROBLEM	73-0779
CHAN, Y. K.	A COMPATIBLE ELEMENT FOR BUCKLING ANALYSIS OF SKEW PLATES WITH PARALLEL EDGES	73-0328
CHAN, Y. K.	A CONFORMING QUADRILATERAL ELEMENT FOR BUCKLING ANALYSIS OF STIFFENED PLATES	73-0613
CHANDRASEKARAN,	INFLUENCE OF RESERVOIR WATER ON EARTHQUAKE RESPONSE OF EARTH DAMS	70-0375
CHANDRASEKARAN,	STATIC AND DYNAMIC ANALYSIS OF AN ARCH DAM	74-1101
CHANDRASEKARAN,	INTERACTIVE ANALYSIS OF A RAFTED MULTISTOREY SPACE FRAME RESSTING ON AN INHOMOGENEOUS CLAY STRATUM	74-0447
CHANE, H. L.	SABOR CS-2 USER'S MANUAL	70-0408
CHANG, B.	STRESS CONCENTRATIONS IN HULL SHAPES FROM SURFACE DISCONTINUITIES	69-0130
CHANG, C.	ANALYSIS OF SOIL MOVEMENT AROUND A DEEP EXCAVATION	70-0039
CHANG, C.	NONLINEAR ANALYSIS OF STRESS AND STRAIN IN SOILS	70-0050
CHANG, C. H.	PART 1 - TASK 3.2.9. BENDING AND BUCKLING OF A ONE-DIMENSIONAL MODEL SIMULATING A CONICAL SHELL. PAR	71-0745
CHANG, C. Y.	ANALYTICAL METHODS FOR PREDICTING SUBSIDENCE ABOVE SOLUTION-MINED CAVITIES	74-1145
CHANG, C. Y.	DEVELOPMENT AND APPLICATIONS OF THEORETICAL METHODS FOR EVALUATING STABILITY OF OPENINGS IN ROCK	73-0791
CHANG, C. Y.	TIME-DEPENDENT ANALYSIS TO PREDICT CLOSURE IN SALT CAVITIES	74-1149
CHANG, C-Y.	A THEORETICAL METHOD FOR EVALUATING STABILITY OF OPENINGS IN ROCK	72-0877
CHANG, C-Y.	DEVELOPMENT AND APPLICATIONS OF A GENERAL COMPUTER PROGRAM FOR EVALUATING STABILITY OF OPENINGS IN R	74-0994
CHANG, C-Y.	FINITE ELEMENT ANALYSES OF SOIL MOVEMENTS CAUSED BY DEEP EXCAVATION AND DEWATERING	69-0530
CHANG, D. C.	EFFECTS OF FLEXIBLE CONNECTIONS ON BODY STRUCTURAL RESPONSE	74-0140
CHANG, D. C.	PLANE STRESS AND PLANE STRAIN ANALYSIS BY FINITE ELEMENT METHOD	70-0388
CHANG, G. C.	FINITE ELEMENT APPROACH TO WAVE MOTIONS IN ELASTIC-PLASTIC CONTINUA	68-0041
CHANG, G. W.	VERIFICATION OF FRACTURE MECHANICS CONCEPTS TO PREDICT CRACKING OF FLEXIBLE PAVEMENTS	73-0772
CHANG, H. K.	A MODEL STUDY OF GAS DIFFUSION IN ALVEOLAR SACS	73-0029
CHANG, J. C.	STRESSES AND DEFORMATIONS IN JAIL GULCH EMBANKMENT	73-0391
CHANG, P. Y.	SIMPLE APPROACH TO THE STRENGTH ANALYSIS OF TANKERS	71-0319
CHANG, P. Y.	STRUCTURAL ANALYSIS OF LONGITUDINALLY FRAMED SHIPS	72-0436
CHANG, P. Y.	TANKER TRANSVERSE STRENGTH ANALYSIS: USER'S MANUAL	72-0750
CHANG, S. K.	SCATTERING BY BURIED OBSTACLES	74-0739
CHANG, S. K.	SCATTERING BY DIELECTRIC CYLINDERS	74-1025
CHANG, T. Y.	AN APPROXIMATE METHOD FOR THERMO-VISCOELASTIC STRESS ANALYSIS	66-0100
CHANG, T. Y.	APPROXIMATE SOLUTION IN LINEAR VISCOELASTICITY	66-0101
CHANG, T. Y.	ELASTIC PLASTIC DEFORMATION OF CYLINDRICAL PRESSURE VESSELS UNDER CYCLIC LOADING	74-0604
CHANG, T. Y.	NONLINEAR CREEP ANALYSIS AT ELEVATED TEMPERATURE	71-0050
CHANG, T. Y.	STRESS ANALYSIS OF TWO-DIMENSIONAL PROBLEMS UNDER SIMULTANEOUS CREEP AND PLASTICITY	71-0220
CHANG, T. Y.	VISCOELASTIC RESPONSE OF GRAPHITE MATERIALS IN IRRADIATION ENVIRONMENTS	70-0314
CHANGLIANG, V.	EARTHQUAKE FINITE-ELEMENT ANALYSIS OF STRUCTURE-FOUNDATION SYSTEMS	75-0410
CHAO, H.	STRESS-WAVE PROPAGATION AND DYNAMIC EFFECTS IN SHELLS (SPADES) MICROMECHANICAL STRESS-WAVE RESPONSE	72-0851
CHAO, R.	LONG TERM CREEP CLOSURE OF SOLUTION CAVITY SYSTEM	74-1150
CHAPEAU, C.	IMMEDIATE SETTLEMENTS OF THREE TEST EMBANKMENTS ON CHAMPLAIN CLAY	74-0275
CHAPLIN, E. C.	DEVELOPMENT OF A DESIGN FOR A PRECAST CONCRETE BRIDGE BEAM OF U-SECTION	73-0392
CHAR, C. V.	FINITE ELEMENT ANALYSIS OF CREEP IN ROCK-SALT PILLAR MODELS	72-0935
CHARAP, S. H.	NUMERICAL METHOD FOR INVESTIGATING THE PROPAGATION OF SURFACE WAVES ON DISSIPATIVE GUIDES	74-0613
CHARGIN, A. K.	MAGNET FIELD AND FORCE BY FINITE ELEMENT TECHNIQUES	69-0035
CHARI, M. V.	ANALYSIS OF TURBOALTERNATOR MAGNETIC FIELDS BY FINITE ELEMENTS	71-0460
CHARI, M. V.	FINITE ELEMENT ANALYSIS OF MAGNETICALLY SATURATED D-C MACHINES	71-0566
CHARI, M. V.	FINITE-ELEMENT ANALYSIS OF NONLINEAR MAGNETIC FIELDS IN ELECTRIC MACHINES	70-0484
CHARI, M. V.	FINITE-ELEMENT SOLUTION OF EDDY-CURRENT PROBLEM IN MAGNETIC STRUCTURES	74-0106
CHARI, M. V.	NONLINEAR FINITE ELEMENT SOLUTION OF ELECTRICAL MACHINES UNDER NO-LOAD AND FULL-LOAD CONDITIONS	74-0472
CHARLES, A.	CONSTRUCTIONAL DEFORMATIONS IN ROCKFILL DAM	73-0281
CHARLES, J. A.	PERFORMANCE OF CULVERT UNDER WINSCAR DAM	75-0638
CHARLTON, J. A.	SOLUTION OF A TWO DIMENSIONAL DIFFUSION CONVECTION EQUATION DESCRIBING THE MASS TRANSFER OF A NEUTRA	74-0037
CHARLWOOD, R. G	THERMO-PLASTIC CONSTITUTIVE RELATIONS AND A VARIATIONAL SOLUTION TECHNIQUE USING FINITE ELEMENTS	70-0486
CHARNLEY, C. J.	INTERNATIONAL MACHINE TOOL DESIGN AND RESEARCH CONFERENCE, 12TH PROCEEDINGS, 1971.	72-0224
CHARPENET, G.	THE TRIDIMENSIONAL THERMOELASTIC COMPUTER CODE TITUS	71-0156

CHATANI, A.	SOME CONSIDERATIONS ON THE DYNAMIC STRESS ANALYSIS OF A STRIP BY FINITE ELEMENT METHOD	71-0591
CHATTERJEE, A.	A MIXED FINITE ELEMENT FORMULATION FOR PLATE PROBLEMS	72-0556
CHATTOPADHYAY,	APPLICATION OF FINITE ELEMENT METHOD TO CONTINUUM MECHANICS PROBLEMS	71-0051
CHATTOPADHYAY,	FINITE ELEMENT LARGE DEFLECTION ANALYSIS OF SHALLOW SHELLS	75-0356
CHATTOT, J. J.	COUPLAGE ENTRE L ECOULEMENT AUTOUR D UN ARRIERE-CORPS ET LE JET PROPULSIF EN THEORIE DE FLUIDE PARFA	74-1137
CHAUDHARY, V. P	FINITE ELEMENT ANALYSIS OF AXISYMMETRICAL SHELLS OF COMPOSITE MATERIAL	70-0483
CHAUMP, L. E.	MODAL SENSITIVITY STUDY OF A CONICAL RE-ENTRY VEHICLE SUBJECTED TO AERODYNAMICALLY INDUCED ACOUSTIC	69-0431
CHAUVIN, G. A.	POST-TENSIONED CONCRETE NUCLEAR STRUCTURES	70-0040
CHEBAN, W. G.	EXCITATION OF PLANE THERMOELASTIC WAVES IN A LAYER WITH A FINITE RATE OF HEAT PROPAGATION	74-0803
CHELAPATI, C. V	PROBABILISTIC ASSESSMENT OF AIRCRAFT HAZARD FOR NUCLEAR POWER PLANTS	72-0235
CHEN, A. C. T.	CONSTITUTIVE EQUATIONS AND PUNCH-INDENTATION OF CONCRETE	75-0286
CHEN, A. C. T.	CONSTITUTIVE RELATIONS OF CONCRETE AND PUNCH-INDENTATION PROBLEMS	73-0877
CHEN, A. C. T.	CONSTITUTIVE RELATIONS FOR CONCRETE	75-0541
CHEN, A. T. F.	MESH GENERATING COMPUTER PROGRAM FOR FINITE ELEMENT ANALYSES OF SIMPLE SLOPES IN HORIZONTALLY LAYERE	71-0748
CHEN, C. H. S.	APPLICATION OF FINITE ELEMENT METHOD TO PNEUMATIC TYRE	75-0119
CHEN, C. J.	ERROR ANALYSIS OF A REMOTE TRANSIENT HEAT FLUX SENSOR	74-0753
CHEN, H. S.	HYBRID-ELEMENT METHOD FOR WATER WAVES	75-0808
CHEN, H. S.	OSCILLATIONS AND WAVE FORCES IN A MAN-MADE HARBOUR IN THE OPEN SEA	74-0151
CHEN, H. S.	OSCILLATIONS AND WAVE FORCES IN AN OFFSHORE HARBOR - APPLICATIONS OF HYBRID FINITE ELEMENT METHOD TO	74-1024
CHEN, H. Y.	A COMPUTER PROGRAM FOR THE STATIC AND DYNAMIC FINITE ELEMENT ANALYSIS OF AXISYMMETRIC THIN SHELLS	66-0017
CHEN, J. C.	FINITE ELEMENT FORMULATION FOR LINEAR THERMOVISCO ELASTIC MATERIALS	69-0351
CHEN, J. C. K	FINITE ELEMENT LARGE DEFORMATION ANALYSIS OF A TAPERED AORTA	75-0856
CHEN, J. T.	APPLICATION OF THE FINITE ELEMENT METHOD TO STRESS DISTRIBUTION IN SOIL MEDIA	67-0075
CHEN, K.	STIFFNESS ANALYSIS OF SHEET METAL SHELLS UNDER CONCENTRATED LOADS	74-0325
CHEN, K. C.	COMPUTER IMPLEMENTATION OF THE CONSTRAINT METHOD	75-0001
CHEN, K. C.	CONFORMING FINITE ELEMENTS BASED ON COMPLETE POLYNOMIALS	74-0302
CHEN, K. C.	HIGH PRECISION FINITE ELEMENTS FOR PLANE ELASTIC PROBLEMS	72-0862
CHEN, L. K-C.	A FINITE ELEMENT FORMULATION FOR LARGE DEFLECTION ELASTIC STRUCTURAL ANALYSIS APPLIED TO PLATE BENDI	70-0485
CHEN, L. T.	A FINITE ELEMENT ANALYSIS FOR STEADY AND OSCILLATORY SUBSONIC FLOW AROUND COMPLEX CONFIGURATIONS	74-1317
CHEN, L. T.	A FINITE-ELEMENT ANALYSIS FOR STEADY-AND OSCILLATORY SUPERSONIC FLOWS AROUND COMPLEX CONFIGURATIONS	74-1316
CHEN, M-C.	SEISMIC RESPONSE OF HIGHWAY BRIDGES	75-0879
CHEN, P. C. T.	CONSTITUTIVE MATRIX AND FINITE ELEMENT FORMULATION FOR A DEFORMATION THEORY OF PLASTICITY	71-0700
CHEN, P. C. T.	ELASTIC-PLASTIC SOLUTION OF A TWO-DIMENSIONAL TUBE PROBLEM BY THE FINITE ELEMENT METHOD	74-0769
CHEN, P. C. T.	THE FINITE ELEMENT ANALYSIS OF ELASTIC-PLASTIC THICK-WALLED TUBES	74-0768
CHEN, P. E.	MECHANICAL PROPERTIES OF TAPE COMPOSITES	69-0317
CHEN, P. E.	STRESS ANALYSIS OF RIBBON REINFORCED COMPOSITES	70-0221
CHEN, P. E.	STRESS ANALYSIS OF TAPE REINFORCED COMPOSITES	69-0068
CHEN, P. E.	STRESS FIELDS AROUND MULTIPLE INCLUSIONS	69-0067
CHEN, P. E.	TRANSVERSE PROPERTIES OF FIBROUS COMPOSITES	69-0069
CHEN, P. E.	TRANSVERSE PROPERTIES OF UNIDIRECTIONAL ALUMINUM MATRIX FIBROUS COMPOSITES	69-0167
CHEN, P. Y. P.	DISCONTINUOUS FINITE ELEMENTS IN THERMAL ANALYSIS	70-0256
CHEN, P.C.	FINITE ELEMENT ANALYSIS OF ELASTIC PLASTIC THICK-WALLED TUBES	74-0368
CHEN, R.M.M.	SOLVING A CLASS OF LARGE SPARSE LINEAR SYSTEMS OF EQUATIONS BY PARTITIONING	73-0580
CHEN, S. F.	THE AIRFOIL PROBLEM VIA THE FINITE ELEMENT METHOD	73-1167
CHEN, W. F.	CONSTITUTIVE EQUATIONS AND PUNCH-INDENTATION OF CONCRETE	75-0286
CHEN, W. F.	CONSTITUTIVE RELATIONS OF CONCRETE AND PUNCH-INDENTATION PROBLEMS	73-0877
CHEN, W. F.	ELASTIC-PLASTIC LARGE DEFORMATION RESPONSE OF CLAY TO FOOTING LOADS	74-0699
CHEN, W-F.	CONSTITUTIVE RELATIONS FOR CONCRETE	75-0541
CHEN, Y.	FINITE ELEMENT ANALYSIS OF AN AXISYMMETRICALLY LOADED ORTHOTROPIC SHELL OF REVOLUTION.	72-0438
CHEN, Y.	SOME RESULTS IN THE FINITE ELEMENT MODELING OF THIN PLATES	66-0018
CHEN, Y-N.	PREDICTION OF LONG TERM DEFORMATION OF A COMPACTED COHESIVE SOIL EMBANKMENT OVER A SOFT FOUNDATION	74-0727
CHENAULT, D. W.	FINITE ELEMENT SOLUTION FOR ADDED MASS AND DAMPING	74-0018
CHENG, F. Y.	DYNAMIC RESPONSE OF NONLINEAR SPACE FRAMES BY FINITE ELEMENT METHODS	71-0584
CHENG, F. Y.	FINITE ELEMENT ANALYSIS OF STRUCTURAL INSTABILITY BY ASSOCIATION OF PULSATING EXCITATIONS	74-0453
CHENG, J. S.	BRAIN-TISSUE FRAGILITY - FINITE STRAIN ANALYSIS BY A HYBRID FINITE-ELEMENT METHOD	75-0631
CHENG, K. Y.	NOTE ON MINIMIZING THE BANDWIDTH OF SPARSE, SYMMETRIC MATRICES	73-1107
CHENG, P.	FINITE ELEMENT ANALYSIS OF LAMINAR CONVECTIVE HEAT TRANSFER IN VERTICAL DUCTS WITH ARBITRARY CROSS-S	75-0555
CHENG, P. Y.	DYNAMICS OF PRISMATIC AND OPEN SECTION MEMBER GRIDS	69-0070
CHENG, R. C.	TRANSIENT LAMINAR FLOW IN DUCTS OF ARBITRARY CROSS-SECTION BY FINITE ELEMENT METHODS	72-0013
CHENG, R. C-N.	FINITE-ELEMENT ANALYSIS OF FLOW IN BLOOD VESSEL WITH ARBITRARY CROSS-SECTION	72-0936
CHENG, R. T.	A MODEL STUDY OF GAS DIFFUSION IN ALVEOLAR SACS	73-0029
CHENG, R. T.	FINITE ELEMENT SOLUTION OF VISCOUS FLOW IN AN ARBITRARY INTERNAL PASSAGE	71-0461
CHENG, R. T.	NUMERICAL INVESTIGATIONS OF LAKE CIRCULATION AROUND ISLANDS BY THE FINITE ELEMENT METHOD	72-0233
CHENG, R. T.	NUMERICAL SOLUTION OF NAVIER-STOKES EQUATIONS BY FINITE ELEMENT METHOD	72-0008
CHENG, R. T.	NUMERICAL SOLUTION OF THE VISCOUS FLOW IN THE ENTRANCE REGION OF PARALLEL PLATES	73-0022
CHENG, R. T.	ON SEAWATER ENCROACHMENT IN COASTAL AQUIFERS	74-0577
CHENG, R. T.	ON THE ACCURACY OF CERTAIN C DEGREE CONTINUOUS FINITE ELEMENT REPRESENTATIONS	74-0367
CHENG, R. T.	ON THE SOLUTION OF TRANSIENT FREE-SURFACE FLOW PROBLEMS IN POROUS MEDIA BY THE FINITE ELEMENT METHOD	73-1019
CHENG, R. T.	ON THE STUDY OF CONVECTIVE DISPERSION EQUATION	74-0005
CHENG, R. T.	TRANSIENT FREE-SURFACE FLOW IN POROUS MEDIA BY FINITE ELEMENT METHOD	72-0297
CHENG, R. T.	WIND DRIVEN LAKE CIRCULATION BY THE FINITE ELEMENT METHOD	70-0041
CHENG, R. T. S.	A STUDY OF CONVECTIVE-DISPERSION EQUATION BY ISOPARAMETRIC FINITE ELEMENTS. TRANSPORT OF POLLUTANTS	75-0773
CHENG, R. T-S.	FINITE ELEMENT MODELING OF FLOW THROUGH POROUS MEDIA	75-0706
CHENG, R. T-S.	STUDY OF CONVECTIVE-DISPERSION EQUATION BY ISOPARAMETRIC FINITE ELEMENTS	75-0074
CHENG, R. T-S.	STUDY OF FLUID MOVEMENTS THROUGH CAUSEWAY	75-0083
CHENG, W. C.	CREEP DEFORMATION OF A MULTI-STOREY STRUCTURE IN A FIRE	74-1197
CHENG, W. K.	ANALYSIS OF CASTELLATED BEAMS BY THE FINITE ELEMENT METHOD	72-0620
CHENG, W-C.	COMPUTER ANALYSIS OF STEEL FRAME IN FIRE	75-0173
CHENG, Y. F.	FACTORS OF STRESS CONCENTRATION AT A HOLE IN FINITE STRIPS UNDER BENDING	72-0744
CHENG, Y. F.	ON MAXIMUM FILLET STRESSES IN BREECH RING	72-0736
CHENG, Y. K.	DEFLECTION ANALYSIS OF EXPANDED OPEN-WEB STEEL BEAMS	74-0224
CHERN, J. M.	INELASTIC ANALYSIS OF A STRAIGHT TUBE UNDER COMBINED BENDING, PRESSURE AND THERMAL LOADS	75-0218
CHERN, J. M.	INELASTIC ANALYSIS OF A STRAIGHT TUBE UNDER COMBINED BENDING, PRESSURE AND THERMAL LOADS	75-0513
CHERNUKA, M. W.	APPLICATION OF THE HIGH PRECISION TRIANGULAR PLATE BENDING ELEMENT TO PROBLEMS WITH CURVED BOUNDARIE	69-0071
CHERNUKA, M. W.	ESTIMATES OF THE CREEP RUPTURE LIFETIME OF STRUCTURES USING THE FINITE ELEMENT METHOD	75-0833
CHERNUKA, M. W.	FINITE ELEMENT ANALYSIS OF PLATES WITH CURVED EDGES.	72-0173
CHESNOKOV, S. S	A MODIFIED FINITE ELEMENT METHOD FOR CALCULATION OF VIBRATION OF THIN PLATES	72-0598
CHESNOKOV, S. S	APPLICATION OF FINITE ELEMENT METHODS FOR CALCULATION OF TRANSVERSE OSCILLATIONS IN ONE DIMENSIONAL	71-0599
CHESNOKOV, S. S	COMPARISON OF CALCULATION BY THE FINITE ELEMENT METHOD WITH EXPERIMENT AND AN EXAMPLE OF A DYNAMIC S	72-0597
CHESNOKOV, S. S	MODEL OF ELASTIC PLATE FROM FINITE ELEMENTS IN A SUPERSONIC FLOW	75-0038
CHESNOKOV, S. S	MODIFICIROVANNYJ KONECHNYJ ELEMENT DIJA RASCHIOTA KOLEBANIJ TONKIKH PLASTIN	72-0891
CHESNOKOV, S. S	REDUCTION OF THE DEGREES OF FREEDOM IN SOLVING DYNAMIC PROBLEMS BY THE FINITE ELEMENT METHOD	73-0647
CHESNOKOV, S. S	STUDY ON THE VIBRATION OF THIN PLATES IN A GAS STREAM BY THE FINITE ELEMENT METHOD	74-0454
CHEUNG, J. S. T	COMPUTATION OF POST-YIELD BEHAVIOUR IN NOTCH-BEND AND TENSION TESTPIECES.	72-0244
CHEUNG, M. S.	A THEORETICAL AND EXPERIMENTAL STUDY ON THE FREE VIBRATION OF CURVED SLAB BRIDGES	71-0496
CHEUNG, M. S.	ANALYSIS OF CURVED BOX GIRDER BRIDGES BY FINITE STRIP METHOD	71-0485
CHEUNG, M. S.	ANALYSIS OF SLAB AND GIRDER BRIDGES BY THE FINITE STRIP METHOD	70-0210
CHEUNG, M. S.	ANALYSIS OF SLAB BEAM BRIDGES	69-0077
CHEUNG, M. S.	FLEXURAL VIBRATIONS OF RECTANGULAR AND OTHER POLYGONAL PLATES	71-0417
CHEUNG, M. S.	FREE VIBRATION OF CURVED AND STRAIGHT BEAM SLAB OR BOX GIRDER BRIDGES	72-0487
CHEUNG, M. S.	FREE VIBRATION OF ORTHOTROPIC VARIABLE THICKNESS, CONTINUOUS RECTANGULAR PLATES	70-0270
CHEUNG, M. S.	NATURAL VIBRATIONS OF THIN, FLAT WALLED STRUCTURES WITH DIFFERENT BOUNDARY CONDITIONS	71-0509

CHEUNG, M. S.	STATIC AND DYNAMIC BEHAVIOUR OF RECTANGULAR PLATES USING HIGHER ORDER FINITE STRIPS	72-0458
CHEUNG, M. S.	VIBRATION ANALYSIS OF CYLINDRICAL PANELS	72-0485
CHEUNG, R. T.	NUMERICAL INVESTIGATION OF LAKE CIRCULATION AROUND ISLANDS BY THE FINITE ELEMENT METHOD	72-0337
CHEUNG, Y. K.	A HIGHER ORDER IN-PLANE PARALLELOGRAM ELEMENT AND ITS APPLICATION TO SKEWED CURVED BOX GIRDER BRIDGE	71-0002
CHEUNG, Y. K.	A HIGHER ORDER TRIANGULAR FINITE ELEMENT FOR THE SOLUTION OF FIELD PROBLEMS IN ORTHOTROPIC MEDIA	72-0370
CHEUNG, Y. K.	A THEORETICAL AND EXPERIMENTAL STUDY ON THE FREE VIBRATION OF CURVED SLAB BRIDGES	71-0496
CHEUNG, Y. K.	A TRIANGULAR ELEMENT FOR BENDING AND VIBRATION OF MULTILAYER SANDWICH PLATES	72-0489
CHEUNG, Y. K.	AN ADVANCED ANALYSIS OF RIGID PAVEMENTS	72-0417
CHEUNG, Y. K.	ANALYSIS OF ARCH DAMS USING A SPACE-FRAME MODEL	73-0176
CHEUNG, Y. K.	ANALYSIS OF BOX GIRDER BRIDGES BY FINITE STRIP METHOD	69-0050
CHEUNG, Y. K.	ANALYSIS OF CIRCULAR AND ANGULAR LAMINATED THICK PLATES	74-0426
CHEUNG, Y. K.	ANALYSIS OF CORNER SUPPORTED SKEW SLABS	68-0203
CHEUNG, Y. K.	ANALYSIS OF CURVED BOX GIRDER BRIDGES BY FINITE STRIP METHOD	71-0485
CHEUNG, Y. K.	ANALYSIS OF LONG RECTANGULAR TANKS	70-0276
CHEUNG, Y. K.	ANALYSIS OF ORTHOTROPIC FOLDED PLATES	70-0272
CHEUNG, Y. K.	ANALYSIS OF RECTANGULAR TANKS BY FINITE ELEMENT TECHNIQUES	67-0044
CHEUNG, Y. K.	ANALYSIS OF SIMPLY SUPPORTED THICK, LAYERED PLATES	71-0428
CHEUNG, Y. K.	ANALYSIS OF SLAB AND GIRDER BRIDGES BY THE FINITE STRIP METHOD	70-0210
CHEUNG, Y. K.	ANALYSIS OF SLAB BEAM BRIDGES	69-0077
CHEUNG, Y. K.	APPLICATION OF THE FINITE ELEMENT METHOD TO PROBLEMS OF ROCK MECHANICS	66-0088
CHEUNG, Y. K.	ARCH DAMS ANALYSED BY A LINEAR FINITE ELEMENT SHELL THE FINITE STRIP METHOD IN THE ANALYSIS OF ELAST	68-0189
CHEUNG, Y. K.	AUTOMATIC PREPARATION OF DATA CARDS FOR THE FINITE ELEMENT METHOD	66-0086
CHEUNG, Y. K.	BENDING AND VIBRATION OF MULTILAYER SANDWICH BEAMS AND PLATES	73-0272
CHEUNG, Y. K.	BENDING MOMENTS IN LONG WALL TANKS	67-0110
CHEUNG, Y. K.	BUTTRESS DAMS ON COMPLEX ROCK FOUNDATIONS	64-0013
CHEUNG, Y. K.	COMPARISON OF ANALYTICAL AND EXPERIMENTAL RESULTS IN THE DESIGN OF A HOLLOW GRAVITY DAM	67-0109
CHEUNG, Y. K.	DYNAMIC ANALYSIS OF CIRCULAR AND SECTOR THICK, LAYERED PLATES	75-0198
CHEUNG, Y. K.	EFFECT OF OPENINGS ON STRESSES IN RIGID PAVEMENTS	73-0344
CHEUNG, Y. K.	FINITE ELEMENT ANALYSIS OF SKEW, CURVED BOX-GIRDER BRIDGE	70-0261
CHEUNG, Y. K.	FINITE ELEMENT ANALYSIS OF SKEW BOX GIRDER BRIDGES	72-0209
CHEUNG, Y. K.	FINITE ELEMENT ANALYSIS OF CABLE STAYED BRIDGES	72-0494
CHEUNG, Y. K.	FINITE ELEMENT ANALYSIS OF AXISYMMETRIC MULTILAYER SANDWICH PLATES AND SHELLS	74-0101
CHEUNG, Y. K.	FINITE ELEMENT ANALYSIS OF MULTILAYER SANDWICH PLATES AND SHELLS	74-0425
CHEUNG, Y. K.	FINITE ELEMENT AND FINITE STRIP ANALYSES OF BRIDGES	72-0490
CHEUNG, Y. K.	FINITE ELEMENT METHOD OF ANALYSIS FOR ARCH DAMS SHELLS AND COMPARISON WITH FINITE DIFFERENCE PROCEDU	65-0036
CHEUNG, Y. K.	FINITE ELEMENT METHOD APPLIED TO THE SOLUTION OF HEAT CONDUCTION PROBLEMS	69-0009
CHEUNG, Y. K.	FINITE ELEMENT METHOD AND FINITE STRIP METHOD	73-0347
CHEUNG, Y. K.	FINITE ELEMENT OF SKEW VAULT BRIDGES	72-0434
CHEUNG, Y. K.	FINITE ELEMENTS IN THE SOLUTION OF FIELD PROBLEMS	65-0034
CHEUNG, Y. K.	FINITE STRIP METHOD OF ANALYSIS OF ELASTIC SLABS	68-0190
CHEUNG, Y. K.	FLEXURAL VIBRATIONS OF RECTANGULAR AND OTHER POLYGONAL PLATES	71-0417
CHEUNG, Y. K.	FOLDED PLATE STRUCTURES BY THE FINITE STRIP METHOD	69-0359
CHEUNG, Y. K.	FREE VIBRATION OF CURVED AND STRAIGHT BEAM SLAB OR BOX GIRDER BRIDGES	72-0487
CHEUNG, Y. K.	FREE VIBRATION OF ORTHOTROPIC VARIABLE THICKNESS, CONTINUOUS RECTANGULAR PLATES	70-0270
CHEUNG, Y. K.	FREE VIBRATION OF THICK, LAYERED RECTANGULAR PLATES BY FINITE LAYER METHOD	72-0440
CHEUNG, Y. K.	FREE VIBRATION OF THICK, LAYERED CYLINDERS HAVING FINITE LENGTH WITH VARIOUS BOUNDARY CONDITIONS	72-0491
CHEUNG, Y. K.	FREQUENCY ANALYSIS OF CERTAIN SINGLE AND CONTINUOUS SPAN BRIDGES	71-0337
CHEUNG, Y. K.	GENERAL METHOD OF ANALYSIS BASED ON FINITE ELEMENTS	68-0277
CHEUNG, Y. K.	GROUNDWATER PREDICTIONS FOR A HIGHWAY CUT USING FINITE ELEMENTS	73-0346
CHEUNG, Y. K.	NATURAL VIBRATIONS OF THIN, FLAT WALLED STRUCTURES WITH DIFFERENT BOUNDARY CONDITIONS	71-0509
CHEUNG, Y. K.	NEW FINITE ELEMENTS WITH APPLICATION TO BOX GIRDER BRIDGES.	72-0465
CHEUNG, Y. K.	ORTHOTROPIC FLAT PLATES SUPPORTED ON COLUMNS ELONGATED IN PLANE	72-0492
CHEUNG, Y. K.	ORTHOTROPIC RIGHT BRIDGES BY THE FINITE STRIP METHOD	69-0028
CHEUNG, Y. K.	PLATE AND SHELL PROBLEMS, FINITE ELEMENT DISPLACEMENT APPROACH	66-0005
CHEUNG, Y. K.	PLATES AND BEAMS ON ELASTIC FOUNDATIONS: LINEAR AND NONLINEAR BEHAVIOUR	68-0042
CHEUNG, Y. K.	PLATES AND TANKS ON ELASTIC FOUNDATIONS, AN APPLICATION OF FINITE ELEMENT METHOD	65-0010
CHEUNG, Y. K.	SLAB BRIDGES WITH ARBITRARY SHAPE AND SUPPORT CONDITIONS - A GENERAL METHOD OF ANALYSIS BASED ON FIN	68-0233
CHEUNG, Y. K.	SOLUTION OF ANISOTROPIC SEEPAGE PROBLEMS BY FINITE ELEMENTS	66-0080
CHEUNG, Y. K.	STABILITY ANALYSIS OF MULTILAYER SANDWICH STRUCTRES	73-0345
CHEUNG, Y. K.	STATIC AND DYNAMIC ANALYSIS OF MULTILAYERED SANDWICH PLATES	72-0493
CHEUNG, Y. K.	STATIC AND DYNAMIC BEHAVIOUR OF RECTANGULAR PLATES USING HIGHER ORDER FINITE STRIPS	72-0458
CHEUNG, Y. K.	STEADY STATE AND TRANSIENT GROUNDWATER FLOW IN SLOPES	70-0273
CHEUNG, Y. K.	STRESS ANALYSIS BY THE FINITE ELEMENT METHOD - THERMAL EFFECTS	67-0022
CHEUNG, Y. K.	STRESSES DUE TO CONCENTRATED LOADS IN RESTRAINED BENDING TESTS	67-0037
CHEUNG, Y. K.	STRESSES IN ANISOTROPIC MEDIA WITH PARTICULAR REFERENCE TO PROBLEMS OF ROCK MECHANICS	66-0079
CHEUNG, Y. K.	STRESSES IN ANISOTROPIC MEDIA OF ROCK MECHANICS	66-0127
CHEUNG, Y. K.	STRESSES IN BUTTRESS DAMS	65-0035
CHEUNG, R. K.	STRESSES IN SHAFTS	67-0104
CHEUNG, Y. K.	THE ANALYSIS OF COOLING TOWER RING BEAMS	68-0063
CHEUNG, Y. K.	THE ANALYSIS OF CURVILINEAR ORTHOTROPIC CURVED BRIDGE DECKS	69-0162
CHEUNG, Y. K.	THE APPLICATION OF HIGHER ORDER FINITE ELEMENTS TO THREE DIMENSIONAL HEAT CONDUCTION PROBLEMS	70-0266
CHEUNG, Y. K.	THE BEHAVIOUR OF SQUARE SHEAR WEBS HAVING A CIRCULAR HOLE	67-0108
CHEUNG, Y. K.	THE FINITE ELEMENT METHOD FOR ANALYSIS OF ELASTIC ISOTROPIC AND ORTHOTROPIC SLABS	64-0014
CHEUNG, Y. K.	THE FINITE ELEMENT METHOD IN STRUCTURAL AND CONTINUUM MECHANICS	67-0093
CHEUNG, Y. K.	THE FINITE STRIP METHOD IN THE ANALYSIS OF ELASTIC PLATES WITH TWO SIMPLY SUPPORTED ENDS	68-0068
CHEUNG, Y. K.	THE FINITE STRIP METHOD AND ITS APPLICATIONS	73-0642
CHEUNG, Y. K.	THREE DIMENSIONAL FINITE ELEMENT METHOD FOR PLATE BENDING	74-0243
CHEUNG, Y. K.	THREE-DIMENSIONAL FIELD PROBLEMS BY HIGHER ORDER FINITE ELEMENTS	70-0257
CHEUNG, Y. K.	TRIANGULAR ELEMENT FOR MULTILAYER SANDWICH PLATES	72-0581
CHEUNG, Y. K.	TRIANGULAR ELEMENTS IN PLATE BENDING CONFORMING AND NON-CONFORMING SOLUTIONS	66-0011
CHEUNG, Y. K.	TWO AND THREE DIMENSIONAL GROUNDWATER SEEPAGE BY FINITE ELEMENTS	74-0463
CHEUNG, Y. K.	UNSTEADY RADIAL FLOW IN GAS RESERVOIRS BY FINITE ELEMENTS	74-0462
CHEUNG, Y. K.	VIBRATION ANALYSIS OF CYLINDRICAL PANELS	72-0485
CHEW, B.	DIFFUSION OF HYDROGEN IN FILLET WELDS	75-0373
CHI CHEONG POON	RESERVOIR SIMULATION BY FINITE ELEMENT METHOD	73-0062
CHI, H. M.	COMPUTATIONAL PROCEDURES FOR INELASTIC STRESS ANALYSIS	73-0859
CHI, H-M.	COMPUTATIONAL PROCEDURES FOR INELASTIC FINITE-ELEMENT ANALYSIS	72-0933
CHIAPETTA, R. L	DYNAMIC RESPONSE OF SOIL/CONCRETE INTERFACES AT HIGH PRESSURE	74-0836
CHICUREL, R.	SPACE TIME FINITE ELEMENTS IN DYNAMIC PROBLEMS	72-0402
CHIU, S. T.	A NUMERICAL AND EXPERIMENTAL INVESTIGATION OF THE USE OF J-INTEGRALS	72-0858
CHIU, S. T.	CRACK BRANCHING IN HOMALITE-100 SHEETS	72-0843
CHIU, S. T.	NUMERICAL AND EXPERIMENTAL INVESTIGATION ON THE USE OF J-INTEGRAL	73-0245
CHOI, C-K.	NONCONFORMING FINITE ELEMENT ANALYSIS OF SHELLS	75-0539
CHOI, C-K.	USE OF NONCONFORMING MODES IN FINITE ELEMENT ANALYSIS OF PLATES AND SHELLS	73-0865
CHOO, Y-I.	A SURVEY OF ANALYTICAL METHODS FOR DYNAMIC SIMULATION OF CAB	73-0936
CHOPPA, A. K.	EARTHQUAKE ANALYSIS OF EARTH DAMS	69-0072
CHOPPA, A. K.	EARTHQUAKE BEHAVIOUR OF RESERVOIR-DAM SYSTEMS	68-0043
CHOPPA, A. K.	EARTHQUAKE RESPONSE OF EARTH DAMS	67-0010
CHOPPA, A. K.	EARTHQUAKE RESPONSE OF CONCRETE GRAVITY DAMS	70-0042
CHOPPA, A. K.	THE IMPORTANCE OF THE VERTICAL COMPONENT OF EARTHQUAKE MOTIONS	66-0019
CHOPRA, A. K.	A COMPUTER PROGRAM FOR EARTHQUAKE ANALYSIS OF DAMS	70-0469
CHOPRA, A. K.	DYNAMICS OF EARTH DAMS WITH FOUNDATION INTERACTION	71-0343

22

CHOPRA, A. K.	EARTHQUAKE ANALYSIS OF GRAVITY DAMS INCLUDING HYDRODYNAMIC INTERACTION	73-0350
CHOPRA, A. K.	EARTHQUAKE ANALYSIS OF STRUCTURE-FOUNDATION SYSTEMS	73-0876
CHOPRA, A. K.	EARTHQUAKE ANALYSIS OF AXISYMMETRIC TOWERS PARTIALLY SUBMERGED IN WATER	75-0076
CHOPRA, A. K.	EARTHQUAKE FINITE ELEMENT ANALYSIS OF STRUCTURE-FOUNDATION SYSTEMS	74-0558
CHOPRA, A. K.	EARTHQUAKE RESPONSE OF GRAVITY DAMS INCLUDING RESERVOIR INTERACTION EFFECTS	72-0675
CHOPRA, A. K.	EARTHQUAKE STRESS ANALYSIS IN EARTH DAMS	65-0043
CHOPRA, A. K.	THE EARTHQUAKE EXPERIENCE AT KOYNA DAM AND STRESSES IN CONCRETE GRAVITY DAMS	72-0589
CHOPRA, P. S.	A FINITE ELEMENT CONTRIBUTION TO FRACTURE MECHANICS	72-0320
CHOPRA, P. S.	FINITE ELEMENT FRACTURE MECHANICS ANALYSIS OF CREEP RUPTURE OF FUEL ELEMENT CLADDING	74-0929
CHOPRA, P. S.	FINITE ELEMENT INVESTIGATION OF SECONDARY CRACKING DURING BRITTLE FRACTURE	71-0804
CHOPRA, P. S.	INFLUENCE OF DEFECTS ON FUEL ELEMENT CLADDING STRESSES	73-0863
CHOPRA, P. S.	SIGNIFICANCE OF SYMMETRY IN MULTIPLE FRACTURE ANALYSIS	74-0869
CHOU, C-K.	IMPROVEMENT IN PLATE MODELING AND PLATE DESIGN BY THE FINITE ELEMENT METHOD	75-0791
CHOU, S. I.	MEAN-SQUARE ERRORS OF FINITE ELEMENT APPROXIMATIONS ON LINEAR ELASTOSTATICS, ELASTODYNAMICS, AND THE	74-0107
CHOU, Y. T.	DESIGN METHOD FOR FLEXIBLE AIRFIELD PAVEMENTS	74-1174
CHOUDHURY, J. R	ELASTIC ANALYSIS OF SPATIAL SYSTEMS OF INTERCONNECTED SHEAR WALLS AND FRAMES	74-1195
CHOW, H. Y.	FINITE ELEMENT SOLUTION OF AXISYMMETRICAL DYNAMIC PROBLEMS OF SHELLS OF REVOLUTION	66-0020
CHOW, T. S.	COMPUTING WITH SPARSE MATRICES	73-0926
CHOW, T. Y.	VERIFICATION OF A HYBRID GENERAL PURPOSE SHELL AND BODY OF REVOLUTION COMPUTER PROGRAM	71-0333
CHOWDHURY, P. C	FLUID FINITE ELEMENTS FOR ADDED-MASS CALCULATIONS	72-0009
CHOWDHURY, P. C	FREE VIBRATIONS OF FLUID-BORNE STRUCTURES: INVESTIGATIONS ON A SIMPLE MODEL	74-0976
CHOWDHURY, R. N	FINITE ELEMENT SOLUTION FOR QUANTITY OF STEADY SEEPAGE	71-0824
CHOWDHURY, R. N	GENERALIZED STEADY STATE FIELD PROBLEM AND ITS SOLUTION	74-0059
CHOWDHURY, R. N	THE APPLICATION OF THE FINITE ELEMENT METHOD TO SEEPAGE AND STRESS-DEFORMATION PROBLEMS IN SOIL MECH	70-0267
CHRIST, G.	NEUESTE ERFAHRUNGEN BEI DER RECHNERISCHEN UND KONSTRUKTIVEN AUSLEGUNG KOMPLIZIERTER SCHALENSYSTEME	71-0052
CHRISTENSEN, H.	EXPERIENCES FROM STRESS ANALYSIS OF AXI-SYMMETRIC PROBLEMS IN MACHINE DESIGN	69-0247
CHRISTENSEN, H.	FINITE-ELEMENT ANALYSIS OF AXISYMMETRIC ROTORS	69-0244
CHRISTIAN, J. T	A FINITE ELEMENT FORMULATION FOR MATERIALS WITH SPECIFIED CHANGE OF VOLUME	68-0044
CHRISTIAN, J. T	CONSOLIDATION OF A LAYER UNDER A STRIP LOAD	72-0222
CHRISTIAN, J. T	FINITE ELEMENT PROGRAM FEECON FOR UNDRAINED DEFORMATION ANALYSES OF GRANULAR EMBANKMENTS ON SOFT CLA	72-0875
CHRISTIAN, J. T	INITIATION OF FAILURE IN SLOPES IN OVERCONSOLIDATED CLAYS AND CLAY SHALES.	70-0186
CHRISTIAN, J. T	PLANE STRAIN CONSOLIDATION BY FINITE ELEMENTS	69-0471
CHRISTIAN, J. T	PLANE STRAIN CONSOLIDATION BY FINITE ELEMENTS	70-0043
CHRISTIAN, J. T	THE EFFECTS OF SOIL PARAMETERS AND BOUNDARY CONDITIONS ON THE CONSOLIDATION OF AN ELASTIC LAYER	70-0422
CHRISTIAN, J. T	UNDRAINED STRESS DISTRIBUTION BY NUMERICAL METHODS	68-0045
CHRISTIAN, J. T	UNDRAINED VISCO-ELASTIC ANALYSIS OF SOIL DEFORMATION	74-0424
CHRISTIANO, P.	FINITE DEFORMATION OF BIAXIALLY LOADED COLUMNS	72-0021
CHRISTIANO, P.	STIFFNESS COEFFICIENTS FOR EMBEDDED FOOTINGS	75-0540
CHRISTIANO, P.	COMPLIANCES OF LAYERED ELASTIC SYSTEMS	74-1204
CHRISTIANO, P.	COMPUTER SIMULATION OF LAMINATED ROOF REINFORCED WITH GROUTED BOLTS	73-1045
CHRISTIANO, P.	OPTIMAL DESIGN FOR PRESCRIBED BUCKLING LOADS	74-0568
CHRISTIANO, P.	SNAP-THROUGH BUCKLING OF RETICULATED SHELLS	75-0640
CHRISTIANSEN, H	APPLICATIONS OF CONTINUOUS TONE COMPUTER-GENERATED IMAGES IN STRUCTURAL MECHANICS	74-1035
CHRISTIANSEN, H	COMPUTER GENERATED DISPLAYS OF STRUCTURES IN VIBRATION	74-1305
CHRISTIANSEN, H	DISPLAYS OF KINEMATIC AND ELASTIC SYSTEMS	71-0053
CHRISTIANSEN, H	STRUCTURAL ANALYSIS AND MATRIX INTERPRETIVE SYSTEM (SAMIS)	66-0098
CHU, S. C.	ELASTIC PLASTIC DEFORMATION OF CYLINDRICAL PRESSURE VESSELS UNDER CYCLIC LOADING	72-0604
CHU, S. C.	FINITE ELEMENT METHOD APPLIED TO TRANSIENT TWO-DIMENSIONAL HEAT TRANSFER WITH CONVECTION AND RADIATI	70-0127
CHU, S. C.	FINITE-ELEMENT METHOD APPLIED TO HEAT CONDUCTION IN SOLIDS WITH NONLINEAR BOUNDARY CONDITIONS	73-0005
CHU, S. C.	STABILITY AND OSCILLATION CHARACTERISTICS OF FINITE ELEMENT, FINITE DIFFERENCE AND WEIGHTED RESIDUAL	73-0139
CHU, S. L.	ANALYSIS AND DESIGN CAPABILITY OF THE STRUDL PROGRAM	71-0054
CHU, S. L.	FINITE ELEMENT TREATMENT OF SOIL-STRUCTURE INTERACTION PROBLEM FOR NUCLEAR POWER PLANT UNDER SEISMIC	73-0834
CHU, S-C.	APPLICATION OF THE FINITE-ELEMENT METHOD TO HEAT-TRANSFER PROBLEMS. PART 11. TRANSIENT TWO-DIMENSION	71-0759
CHU, S-C.	APPLICATION OF THE FINITE-ELEMENT METHOD TO HEAT-TRANSFER PROBLEMS. PART 1. FINITE-ELEMENT METHOD AP	71-0760
CHU, S-C.	DISCRETE VARIABLE METHOD APPLIED TO TRANSIENT HEAT CONDUCTION PROBLEMS	74-1163
CHU, T. C.	FINITE ELEMENT ANALYSIS OF TRANSLATIONAL SHELLS	72-0348
CHU, T. Y.	FINITE ELEMENT SOLUTION OF THE STEADY STATE COMPRESSIBLE LUBRICATION PROBLEM	70-0007
CHU, W. H.	A COMPARISON OF SOME FINITE ELEMENT AND FINITE-DIFFERENCE METHODS FOR A SIMPLE SLOSHING PROBLEM	71-0055
CHUGH, Y. P.	APPLICATION OF FINITE ELEMENT ANALYSIS TO UNDERGROUND STORAGE OF NATURAL GAS	70-0198
CHUGH, Y. P.	FINITE ELEMENT STUDY OF PRESSURIZED CAVITIES IN GEOLOGIC MEDIA	71-0805
CHULSOO, Y. J.	AUTOMATIC GENERATION OF FINITE ELEMENT MESH WITHIN A BANDWIDTH AND AN EFFICIENT SOLUTION OF THE BAND	73-0338
CHUM K. H.	BUCKLING OF OPEN CYLINDRICAL SHELLS	67-0115
CHUNG, T. J.	A FINITE ELEMENT ANALYSIS OF TRANSIENT RAREFIED GAS FLOW	73-0604
CHUNG, T. J.	A NEW APPROACH TO THE FINITE-ELEMENT FORMULATION AND SOLUTION OF A CLASS OF PROBLEMS IN COUPLED THER	72-0732
CHUNG, T. J.	A NEW APPROACH TO THE FINITE-ELEMENT FORMULATION AND SOLUTION OF A CLASS OF PROBLEMS IN COUPLED THER	73-0600
CHUNG, T. J.	A SURVEY OF ANALYTICAL METHODS FOR DYNAMIC SIMULATION OF CAB	73-0936
CHUNG, T. J.	ANALYSIS OF NONLINEAR THERMOELASTIC AND THERMOPLASTIC BEHAVIOUR OF SOLIDS OF REVOLUTION BY THE FINIT	71-0192
CHUNG, T. J.	ANALYSIS OF RAREFIED GAS FLOW THROUGH AN ARBITRARY CROSS SECTION BY THE FINITE ELEMENT METHOD	73-0365
CHUNG, T. J.	ANALYSIS OF VISCOELASTOPLASTIC STRUCTURAL BEHAVIOUR OF ANISOTROPIC SHELLS BY THE FINITE ELEMENT METH	71-0056
CHUNG, T. J.	CONVERGENCE AND STABILITY OF NONLINEAR FINITE-ELEMENT EQUATIONS	75-0441
CHUNG, T. J.	DYNAMIC ANALYSIS OF VISCOELASTOPLASTIC ANISOTROPIC SHELLS	73-0253
CHUNG, T. J.	DYNAMIC ELASTOPLASTIC RESPONSE OF GEOMETRICALLY NONLINEAR ARBITRARY SHELLS OF REVOLUTION UNDER IMPUL	70-0098
CHUNG, T. J.	DYNAMIC RESPONSE OF ARBITRARY SHELLS OF REVOLUTION UNDER COMBINED IMPULSIVE NONISOTHERMAL AND IRREGU	70-0365
CHUNG, T. J.	INCREMENTAL THEORY OF THREE DIMENSIONAL TRANSIENT THERMOELASTOPLASTICITY - FORMULATION AND SOLUTION	70-0555
CHUNG, T. J.	INCREMENTAL THERMOMECHANICAL THEORY OF VISCOELASTOPLASTIC SOLIDS AND SOLUTION BY FINITE ELEMENTS	73-0861
CHUNG, T. J.	NEW APPROACH TO FINITE ELEMENT FORMULATION AND SOLUTION OF A CLASS OF PROBLEMS IN COUPLED THERMOELAS	73-0013
CHUNG, T. J.	STATIC AND DYNAMIC ANALYSIS OF VISCOELASTOPLASTIC FIBER-REINFORCED COMPOSITE SHELLS IN MISSILE STRUC	73-0651
CHUNG, T. J.	STEADY STATE AND TRANSIENT HEAT CONDUCTION AND STRESS ANALYSES OF TWO AND THREE DIMENSIONAL STRUCTUR	71-0675
CHUNG, T. J.	THE FINITE ELEMENT ANALYSIS OF TRANSIENT RAREFIED GAS FLOW	74-0027
CHUNG, T. J.	THE FINITE ELEMENT DYNAMIC STABILITY ANALYSIS OF THIN SHELLS SUBMERGED IN FLUIDS	71-0057
CHUNG, T. J.	THEMIS FINAL REPORT, VOL. II. NONLINEAR MECHANICS IN SOLIDS AND STRUCTURES	74-0722
CHUNG, T. J.	THEMIS FINAL REPORT, VOL. 1. BASIC RESEARCH IN METHODS OF APPROXIMATION IN THE NONLINEAR MECHANICS O	74-0723
CHUNG, T. J.	THERMOMECHANICAL RESPONSE OF INELASTIC FIBRE COMPOSITES	75-0235
CHURCH, K.	A GENERAL QUADRILATERAL PLATE ELEMENT	73-0807
CHURCH, K.	GENERAL QUADRILATERAL PLATE ELEMENT	75-0339
CHURCH, K.	IMPROVED GENERAL QUADRILATERAL PLATE ELEMENTS	75-0036
CHWIRUT, D. J.	ANALYSIS OF COMPOSITE REINFORCED CUTOUTS AND CRACKS	74-0824
CHWIRUT, D. J.	COMPOSITE-OVERLAY REINFORCEMENT OF CUTOUTS AND CRACKS IN METAL SHEET	73-0878
CIACCI, R.	COMPUTATIONAL METHOD FOR DYNAMIC ANALYSIS OF STRUCTURES WITH LARGE SIZE STIFFNESS AND MASS MATRICES.	73-0830
CIACCI, R.	TWO-DIMENSIONAL FINITE ELEMENT COMPUTER CODES FOR STRESS ANALYSIS AND FIELD PROBLEMS	73-0055
CIACCI, R.	USE OF PAS-1 FINITE ELEMENT COMPUTER SYSTEM FOR STATIC AND DYNAMIC ANALYSIS OF 2-DIMENSIONAL AND 3-D	75-0070
CIARLET, P. G.	MULTIPOINT TAYLOR FORMULAS AND APPLICATIONS TO THE FINITE ELEMENT METHOD	74-0446
CIARLET, P. G.	A MIXED FINITE ELEMENT METHOD FOR THE BIHARMONIC EQUATION	74-1227
CIARLET, P. G.	CONFORMING AND NONCONFORMING FINITE ELEMENT METHODS FOR SOLVING THE PLATE PROBLEM	73-0621
CIARLET, P. G.	CONFORMING FINITE ELEMENT METHOD FOR THE SHELL PROBLEM	75-0096
CIARLET, P. G.	CONVERGENCE OF CONFORMING FINITE-ELEMENT METHODS FOR SHELL PROBLEMS	75-0494
CIARLET, P. G.	DUAL ITERATIVE TECHNIQUES FOR SOLVING A FINITE ELEMENT APPROXIMATION OF THE BIHARMONIC EQUATION	75-0480
CIARLET, P. G.	GENERAL LAGRANGE AND HERMITE INTERPOLATION IN R**N WITH APPLICATIONS TO FINITE ELEMENT METHODS	72-0110
CIARLET, P. G.	INTERPOLATION THEORY OVER CURVED ELEMENTS WITH APPLICATION TO THE FINITE ELEMENT METHOD	72-0356
CIARLET, P. G.	LA METHODE DES ELEMENTS FINIS POUR LES PROBLEMES AUX LIMITES ELLIPTIQUES	74-1228
CIARLET, P. G.	LAGRANGE INTERPOLATION ON CURVED FINITE ELEMENTS	72-0163
CIARLET, P. G.	MAXIMUM PRINCIPLE AND UNIFORM CONVERGENCE FOR THE FINITE ELEMENT METHOD	73-0852

CIARLET, P. G.	NON-CONFORMING FINITE ELEMENT METHODS AND THE PATCH TEST	73-0640
CIARLET, P. G.	NUMERICAL METHODS OF HIGH-ORDER ACCURACY FOR NONLINEAR BOUNDARY VALUE PROBLEMS	68-0194
CIARLET, P. G.	ORDERS OF CONVERGENCE IN FINITE ELEMENT METHODS	72-0057
CIARLET, P. G.	QUELQUES METHODES D'ELEMENTS FINIS POUR DE PROBLEME D'UNE PLAQUE ENCASTREE	74-0406
CIARLET, P. G.	SUR L ELEMENT DE CLOUGH ET TOCHER (CLOUGH-TOCHER ELEMENT)	74-1022
CIARLET, P. G.	THE COMBINED EFFECT TO CURVED BOUNDARIES AND NUMERICAL INTEGRATION IN ISOPARAMETRIC FINITE ELEMENT M	72-0357
CIARLET, P. W.	USE OF FINITE ELEMENT METHOD FOR SHELL PROBLEMS	75-0598
CIAVALDINI, J.	ANALYSE NUMERIQUE D'UN PROBLEME DE STEFAN A DEUX PHASES PAR UNE METHODE D'ELEMENTS FINIS. (NUMERICAL	75-0387
CIAVALDINI, J.	FINITE-ELEMENT METHOD TO COMPUTE STATIONARY SUBCRITICAL FLOWS	75-0439
CIAVALDINI, J.	NUMERICAL ANALYSIS OF 2-PHASE STEFAN PROBLEM BY METHOD OF FINITE ELEMENTS	75-0616
CIAVALDINI, J.	SUR L ELEMENT DE FRAEIJS DE VEUBEKE ET SANDER (FRAEIJS DE VEUBEKE-SANDER ELEMENT)	74-0924
CICALA, P.	LINEAR SHELL THEORY FOR RAPID VARIATION SOLUTIONS	71-0406
CIMENT, M.	NUMERICAL SOLUTION OF A DIFFUSION CONSUMPTION PROBLEM WITH A FREE BOUNDARY	74-0835
CLARK, G. B.	THERMAL-MECHANICAL FRAGMENTATION OF HARD ROCK - FIELD TESTS	74-0998
CLARK, J. L.	GROUNDWATER PREDICTIONS FOR A HIGHWAY CUT USING FINITE ELEMENTS	73-0346
CLARK, J. N.	STRESS ANALYSIS OF INDUSTRIAL COMPONENTS WITH PLASTIC AND FINITE ELEMENT MODELS	74-0512
CLARKE, P. W.	CONIDA: A FINITE ELEMENT PROGRAM FOR THE STRESS ANALYSIS OF AXISYMMETRIC THIN SHELLS	74-1087
CLARKSON, B. L.	APPLICATION OF FINITE ELEMENT METHOD TO THE DYNAMIC ANALYSIS OF TALL STRUCTURES	71-0111
CLARKSON, B. L.	COMPUTER BASED ANALYSES OF THE RESPONSE OF BOX TYPE STRUCTURES TO RANDOM PRESSURES	73-0227
CLARY, R. R.	MEASURED AND PREDICTED LONGITUDINAL VIBRATIONS OF A LIQUID PROPELLANT TWO-STAGE LAUNCH VEHICLE	70-0366
CLARY, R. R.	PRACTICAL ANALYSIS OF PLATE VIBRATIONS USING NASTRAN	71-0058
CLEMENT, P.	APPROXIMATION BY FINITE-ELEMENT FUNCTIONS USING LOCAL REGULARIZATION	75-0436
CLEMENT, P.	ON THE RIGID DISPLACEMENT CONDITION.	72-0405
CLEMENT, PH.	APPROXIMATION BY FINITE ELEMENT FUNCTIONS USING REGULARIZATION	74-1229
CLEMENT, S.	INFLUENCE OF JOINTS IN PANELIZED STRUCTURAL SYSTEMS	72-0204
CLINARD, J. A.	COMPARISON OF TYPICAL INELASTIC ANALYSIS PREDICTIONS WITH BENCHMARK PROBLEM EXPERIMENTAL RESULTS	75-0696
CLOUGH, G. W.	ANALYTICAL PROBLEMS IN MODELING SLURRY WALL CONSTRUCTION	73-0764
CLOUGH, G. W.	FINITE ELEMENT ANALYSES OF PORT ALLEN AND OLD RIVER LOCKS	69-0203
CLOUGH, G. W.	FINITE ELEMENT ANALYSES OF SOIL-STRUCTURE INTERACTION IN U-FRAME LOCKS	69-0531
CLOUGH, G. W.	FINITE ELEMENT ANALYSIS OF PORT ALLEN LOCK	71-0293
CLOUGH, G. W.	FINITE ELEMENT ANALYSES OF RETAINING WALL BEHAVIOUR	71-0320
CLOUGH, G. W.	PERFORMANCE OF TIED-BACK WALLS IN CLAY	74-0665
CLOUGH, G. W.	SUBTERRANEAN GUIDEWAY RESPONSES TO LEVITATION VEHICLE LOADING	74-0927
CLOUGH, G. W.	TEMPORARY EXCAVATION IN VARVED CLAY	75-0049
CLOUGH, R. W.	A COMPUTER PROGRAM FOR ANALYSIS OF STIFFENED PLATES UNDER COMBINED INPLANE AND LATERAL LOADS	71-0703
CLOUGH, R. W.	A CURVED CYLINDRICAL SHELL DISCRETE ELEMENT	67-0009
CLOUGH, R. W.	A CURVED, CYLINDRICAL SHELL, FINITE ELEMENT	68-0039
CLOUGH, R. W.	A FINITE ELEMENT APPROXIMATION FOR THE ANALYSIS OF THIN SHELLS	68-0046
CLOUGH, R. W.	A REFINED CURVED CYLINDRICAL SHELL FINITE ELEMENT	68-0038
CLOUGH, R. W.	A REFINED QUADRILATERAL ELEMENT FOR ANALYSIS OF PLATE BENDING	68-0047
CLOUGH, R. W.	ANALYSIS OF EMBANKMENT STRESSES AND DEFORMATIONS	67-0012
CLOUGH, R. W.	ANALYSIS OF PLATE BENDING BY THE FINITE ELEMENT METHOD	60-0003
CLOUGH, R. W.	ANALYSIS OF STIFFENED PLATES USING THE FINITE ELEMENT METHOD	70-0030
CLOUGH, R. W.	ANALYSIS OF STRUCTURAL VIBRATIONS AND DYNAMIC RESPONSE	71-0059
CLOUGH, R. W.	ANALYSIS OF THIN ARCH DAMS BY THE FINITE ELEMENT METHOD	64-0001
CLOUGH, R. W.	APPLICATION OF THE FINITE ELEMENT METHOD	75-0035
CLOUGH, R. W.	AREAS OF APPLICATION OF THE FINITE ELEMENT. METHOD	74-0180
CLOUGH, R. W.	BASIC PRINCIPLES OF STRUCTURAL DYNAMICS	71-0505
CLOUGH, R. W.	COMPARISON OF THREE DIMENSIONAL FINITE ELEMENTS	69-0073
CLOUGH, R. W.	CONSTRUCTION STRESSES IN DWORSHAK DAM	65-0029
CLOUGH, R. W.	CONVERGENCE CRITERIA FOR ITERATIVE PROCESSES	72-0664
CLOUGH, R. W.	DYNAMIC BEHAVIOUR OF SHELL ROOFS	69-0228
CLOUGH, R. W.	DYNAMIC FINITE ELEMENT ANALYSIS OF ARBITRARY THIN SHELLS	71-0061
CLOUGH, R. W.	DYNAMIC MATRIX ANALYSIS OF FRAMED STRUCTURES	62-0012
CLOUGH, J. W.	DYNAMIC RESPONSE BY A STEP BY STEP ANALYSIS	62-0014
CLOUGH, R. W.	EARTHQUAKE ANALYSIS OF EARTH DAMS	69-0072
CLOUGH, R. W.	EARTHQUAKE STRESS ANALYSIS IN EARTH DAMS	65-0043
CLOUGH, R. W.	ELASTO-PLASTIC ANALYSIS OF PLATES USING THE FINITE ELEMENT METHOD	71-0038
CLOUGH, R. W.	EXPLICIT ADDITION OF RIGID-BODY MOTIONS IN CURVED FINITE ELEMENTS	73-0037
CLOUGH, R. W.	FINITE ELEMENT ANALYSIS OF AXISYMMETRIC SOLIDS	65-0011
CLOUGH, R. W.	FINITE ELEMENT ANALYSIS OF DYNAMIC SHELL BEHAVIOUR	67-0013
CLOUGH, R. W.	FINITE ELEMENT ANALYSIS OF TUBULAR JOINTS: A REPORT ON A FEASIBILITY STUDY	67-0147
CLOUGH, R. W.	FINITE ELEMENT ANALYSIS OF ARBITRARY THIN SHELLS	70-0329
CLOUGH, R. W.	FINITE ELEMENT ANALYSIS OF TUBULAR K JOINTS	70-0427
CLOUGH, R. W.	FINITE ELEMENT ANALYSIS OF ARCH DAMS AND FOUNDATIONS	71-0668
CLOUGH, R. W.	FINITE ELEMENT ANALYSIS OF DYNAMIC RESPONSE	72-0681
CLOUGH, R. W.	FINITE ELEMENT APPLICATIONS IN THE CHARACTERIZATION OF ELASTIC SOLIDS	71-0138
CLOUGH, R. W.	FINITE ELEMENT STIFFNESS MATRICES FOR ANALYSIS OF PLATE BENDING	66-0021
CLOUGH, R. W.	IMPROVED NUMERICAL INTEGRATION OF THICK SHELL FINITE ELEMENTS	71-0396
CLOUGH, R. W.	LARGE DEFLECTION ANALYSIS OF PLATES AND SHALLOW SHELLS USING THE FINITE ELEMENT METHOD	73-0989
CLOUGH, R. W.	NUMERICAL INTEGRATION OF THE EQUATION OF MOTION	71-0507
CLOUGH, R. W.	RECENT ADVANCES IN COMPUTATIONAL METHODS IN STRUCTURAL MECHANICS AND DESIGN.	72-0575
CLOUGH, R. W.	SOME ASPECTS OF RECENT CONTRIBUTIONS TO THE MATHEMATICAL THEORY OF FINITE ELEMENTS	73-0601
CLOUGH, R. W.	STIFFNESS AND DEFLECTION ANALYSIS OF COMPLEX STRUCTURES	56-0001
CLOUGH, R. W.	STRESS ANALYSIS OF A GRAVITY DAM BY THE FINITE ELEMENT METHOD	63-0004
CLOUGH, R. W.	STRESS ANALYSIS OF B16.9 TEES BY THE FINITE ELEMENT METHOD	71-0060
CLOUGH, R. W.	STRESS ANALYSIS OF B16.9 TEES BY THE FINITE ELEMENT METHOD, A PROGRESS REPORT	71-0215
CLOUGH, R. W.	THE FINITE ELEMENT IN PLANE STRESS ANALYSIS	60-0001
CLOUGH, R. W.	THE FINITE ELEMENT METHOD IN STRUCTURAL MECHANICS	65-0012
CLOUGH, R. W.	THE FINITE ELEMENT METHOD IN SOLID MECHANICS	70-0295
CLOUGH, R. W.	THE STRESS DISTRIBUTION OF NORFOLK DAM	62-0001
CLOUGH, R. W.	VIBRATION ANALYSIS OF FINITE ELEMENT SYSTEMS	71-0506
COALE, C. W.	INFLUENCE OF ASCENT HEATING ON THE SEPARATION DYNAMICS OF A SPACECRAFT FAIRING	72-0906
COATES, D. F.	ANALYSIS OF ROCK SLOPES, USING THE FINITE ELEMENT METHOD	70-0184
COATES, D. F.	DEVELOPMENT AND USE OF COMPUTER PROGRAMS FOR FINITE ELEMENT ANALYSIS. (LA MISE AU POINT ET L'UTILIZA	69-0289
COATES, D. F.	NOTE ON THE STRESS CONCENTRATIONS AT THE END OF A CYLINDRICAL HOLE	70-0139
COATES, D. F.	UNDERGROUND MEASUREMENTS IN A STEEPLY DIPPING OREBODY	68-0085
COCO, R. H.	STIFFNESS MATRIX FOR CURVED TAPERED AND STRAIGHT TAPERED SHELL STIFFENERS	69-0074
COFFIN, G. K.	FINITE ELEMENT ANALYSIS OF OPEN SHELLS OF REVOLUTION UNDER NONSYMMETRICAL LOADING	68-0295
COFFIN, G. K.	STRESSES AND DEFLECTIONS IN COOLING TOWER SHELLS DUE TO WIND LOADING	68-0088
COGGON, J. H.	ELECTROMAGNETIC AND ELECTRICAL MODELING BY THE FINITE ELEMENT METHOD	71-0462
COHEN, E.	IMPROVED DEFORMATION FUNCTIONS FOR THE FINITE ELEMENT ANALYSIS OF BEAM SYSTEMS	69-0075
COHRON, G. T.	CONCEPT FEASIBILITY COMBAT TRACKED VEHICLE SIGNATURE DUPLICATOR	64-0028
COLE, J. A.	IDENTIFICATION OF AQUIFER CONSTANTS: TESTS OF TWO INVERSE METHODS	74-0051
COLE, P. P.	BUCKLING OF COOLING-TOWER SHELLS: BIFURCATION RESULTS	75-0384
COLE, P. P.	BUCKLING OF COOLING-TOWER SHELLS: STATE-OF-THE-ART	75-0385
COLE, R. T.	EVALUATION OF COMPOSITE SHEAR PANELS WITH CUTOUTS	72-0615
COLEMAN, G. E.	FINITE ELEMENT ANALYSIS OF SOIL COMPACTION	74-0931
COLEMAN, W.	APPLICATION OF FINITE ELEMENTS TO THE ANALYSIS OF GEAR TOOTH STRESSES	72-0116
COLLATZ, L.	METHODS FOR SOLUTION OF PARTIAL DIFFERENTIAL EQUATIONS	73-0521
COLLEN, M. J. P	A SIMPLE FINITE ELEMENT METHOD FOR METEOROLOGICAL PROBLEMS	73-1121
COLLIER, W. D.	HEATRAN, A FINITE ELEMENT CODE FOR HEAT TRANSFER PROBLEMS	69-0076

24

COLLIER, W. D.	TRESS - A FINITE ELEMENT CODE FOR STRESS ANALYSIS PROBLEMS	71-0715
COLLINS, A. G.	FINITE-ELEMENT METHOD FOR WATER-DISTRIBUTION NETWORKS	75-0269
COLLINS, J. D.	STATISTICAL IDENTIFICATION OF STRUCTURES	74-0141
COLLINS, R. J.	BANDWIDTH REDUCTION BY AUTOMATIC RENUMBERING	73-0993
COLLINS, T. P.	APPLICATION OF THE COLLISION-IMPARTED VELOCITY METHOD FOR ANALYZING THE RESPONSES OF CONTAINMENT AND	73-0770
COLVILLE, J.	A GENERAL SOLUTION OF THE VON KARMAN PLATE EQUATIONS BY THE FINITE ELEMENT METHOD	70-0487
COLVILLE, J.	PLANE STRESS REINFORCED CONCRETE FINITE ELEMENTS	74-0108
COMARTIN, C. D.	ANALYSIS AND DESIGN OF SKEW BOX GIRDER BRIDGES	72-0659
COMINI, C.	FINITE ELEMENT SOLUTION OF NON-LINEAR HEAT CONDUCTION PROBLEMS WITH SPECIAL REFERENCE TO PHASE CHANG	74-0348
COMSA, R.	ASUPRA ANALIZEI PRIN METODA ELEMENTULUI FINIT A STARILOR DE EFORTURI SI DEFORMATII DIN MASIVELE DE P	70-0471
COMSTOCK, C.	SOLVING THE LINEAR BALANCE EQUATION GLOBALLY	74-0810
CONLEY, W. H. J	CONSISTENT STRESSES FOR THE FINITE ELEMENT STIFFNESS METHOD	72-0698
CONNACHER, N.	SAMECS STRUCTURAL ANALYSIS SYSTEM - USER'S DOCUMENT	69-0495
CONNACHER, N. E	ANALYSIS OF THE 747 AIRCRAFT WING BODY INTERSECTION	68-0275
CONNELL, D. H.	PERFORMANCE OF AN EMBANKMENT CONSTRUCTED ON VARVED CLAY	73-0743
CONNOLLY, P. L.	APPLICATION OF SAMIS TO THE ANALYSIS OF A FRONT SUSPENSION UPPER ARM	70-0260
CONNOR, J.	A SURVEY OF FINITE-ELEMENT METHODS IN CONTINUUM MECHANICS	72-0751
CONNOR, J.	A TRIANGULAR FLAT PLATE BENDING ELEMENT	68-0252
CONNOR, J.	COMPUTER AIDED TEACHING OF THE FINITE ELEMENT DISPLACEMENT METHOD	69-0430
CONNOR, J.	MIXED MODELS FOR PLATES	70-0384
CONNOR, J. J.	A MIXED FINITE ELEMENT SHELL FORMULATION	71-0063
CONNOR, J. J.	A MIXED FINITE-ELEMENT SHALLOW SHELL FORMULATION	69-0078
CONNOR, J. J.	ANALYSIS OF GEOMETRICALLY NONLINEAR PLATES AND SHELLS BY THE FINITE ELEMENT DISPLACEMENT METHOD	68-0050
CONNOR, J. J.	APPLICATION TO PLANE STRESS AND PLANE STRAIN--LINEAR ELASTIC MATERIAL	68-0048
CONNOR, J. J.	AUTOMATIC GENERATION OF FINITE ELEMENT MATRICES	71-0165
CONNOR, J. J.	BENDING OF THIN PLATES - DISPLACEMENT MODELS	68-0049
CONNOR, J. J.	FINITE ELEMENT MODEL OF TWO LAYER COASTAL CIRCULATION	74-1192
CONNOR, J. J.	FINITE ELEMENT MODELLING OF HYDRODYNAMIC CIRCULATION	74-1370
CONNOR, J. J.	FINITE ELEMENTS AND DYNAMIC VISCOELASTICITY	71-0479
CONNOR, J. J.	FUNDAMENTALS OF FINITE ELEMENT TECHNIQUES FOR STRUCTURAL ENGINEERS	73-1050
CONNOR, J. J.	GEOMETRICALLY NONLINEAR FINITE-ELEMENT ANALYSIS	69-0332
CONNOR, J. J.	MATHEMATICAL MODELS OF THE MASSACHUSETTS BAY. PART 1. FINITE ELEMENT MODELING OF TWO-DIMENSIONAL HYD	73-0800
CONNOR, J. J.	NONLINEAR ANALYSIS OF ELASTIC FRAMED STRUCTURES	68-0222
CONNOR, J. J.	NUMERICAL METHODS IN FLUID DYNAMICS	73-0689
CONNOR, J. J.	STIFFNESS MATRIX FOR SHALLOW RECTANGULAR SHELL ELEMENT	67-0014
CONSTANTINO, C.	ANALYSIS OF SOIL-STRUCTURE INTERACTION EFFECTS UNDER SEISMIC EXCITATION	71-0064
CONSTANTINO, C.	DYNAMIC RESPONSE ON NONLINEAR MEDIA AT LARGE STRAINS	72-0237
CONSTANTINO, C.	FINITE ELEMENT APPROACH TO STRESS WAVE PROBLEMS	67-0015
CONSTANTINO, C.	FINITE ELEMENT SOLUTION FOR WAVE PROPAGATION IN LAYERED MEDIA CAUSED BY A NUCLEAR DETONATION	67-0016
CONSTANTOPOULOS	INITIATION OF FAILURE IN SLOPES IN OVERCONSOLIDATED CLAYS AND CLAY SHALES.	70-0186
CONTRACTOR, D.	FINITE-ELEMENT SIMULATION OF FLOOD HYDROGRAPHS	75-0632
CONWAY, J. C.	FINITE ELEMENT TECHNIQUES APPLIED TO CRACKS INTERACTING WITH SELECTED SINGULARITIES	75-0196
COOK D. J.	TENSILE CREEP ANALYSIS OF CONCRETE STRUCTURES	74-0439
COOK, E. L.	FINITE ELEMENT STRUCTURAL ANALYSIS USING COMPUTERS	70-0216
COOK, R. D.	A NOTE ON CERTAIN INCOMPATIBLE ELEMENTS	73-0994
COOK, R. D.	AVOIDANCE OF PARASITIC SHEAR IN PLANE ELEMENT	75-0383
COOK, R. D.	COMMENT ON DISCRETE ELEMENT IDEALIZATION OF AN INCOMPRESSIBLE LIQUID FOR VIBRATION ANALYSIS	73-0694
COOK, R. D.	CONCEPTS AND APPLICATIONS OF FINITE ELEMENT ANALYSIS	74-1373
COOK, R. D.	EFFECT OF HIGHER ORDER TERMS IN CERTAIN NONLINEAR FINITE ELEMENT MODELS - COMMENT	74-0473
COOK, R. D.	EIGENVALUE PROBLEMS WITH A ' MIXED ' PLATE ELEMENT	69-0081
COOK, R. D.	FINITE ELEMENT BUCKLING ANALYSIS OF HOMOGENEOUS AND SANDWICH PLATES	75-0613
COOK, R. D.	IMPROVED TWO-DIMENSIONAL FINITE ELEMENT	74-0355
COOK, R. D.	MORE ON REDUCED INTEGRATION AND ISOPARAMETRIC ELEMENTS	72-0548
COOK, R. D.	OBSERVATIONS REGARDING ASSUMED STRESS HYBRID PLATE ELEMENTS	74-0370
COOK, R. D.	SOME ELEMENTS FOR ANALYSIS OF PLATE BENDING	72-0626
COOK, R. D.	SOME MODIFICATIONS OF AN ISOPARAMETRIC SHELL ELEMENT	73-1010
COOK, R. D.	SOME PLANE QUADRILATERAL HYBRID FINITE ELEMENTS	69-0082
COOK, R. D.	STRAIN RESULTANTS IN CERTAIN FINITE ELEMENTS	69-0079
COOK, R. D.	TWO HYBRID ELEMENTS FOR ANALYSIS OF THICK, THIN AND SANDWICH PLATES	72-0544
COOK, W. A.	BODY ORIENTED (NATURAL) CO-ORDINATES FOR GENERATING THREE-DIMENSIONAL MESHES	74-0908
COOK, W. A.	THREE-DIMENSIONAL GRAIN STRESS ANALYSIS USING THE FINITE ELEMENT METHOD	71-0732
COOK, W. A.	THREE-DIMENSIONAL MESH GENERATOR FOR FINITE ELEMENT COMPUTER CODES	74-0864
COOK, W. A.	THREE-DIMENSIONAL PHOTOELASTIC AND FINITE-ELEMENT ANALYSIS OF A PROPELLANT GRAIN	72-0031
COOK, W. L.	AUTOMATED INPUT DATA PREPARATION FOR NASTRAN	69-0088
COOK, W. L.	FINITE ELEMENT MODELING OF A CIRCULAR RING USING HALF AND QUARTER SYMMETRY	66-0119
COOK, W. L.	NASTRAN-GAP, AN INTEGRATED APPROACH TO THE ANALYSIS OF RADIATION PATTERNS OF DISTORTED REFLECTORS	71-0062
COOLEY, R. L.	A FINITE DIFFERENCE METHOD FOR UNSTEADY FLOW IN VARIABLY SATURATED POROUS MEDIA	71-0586
COOLEY, R. L.	FINITE ELEMENT SOLUTIONS FOR THE EQUATIONS OF GROUND-WATER FLOW	74-1341
COOLEY, R. L.	HYDROLOGIC ENGINEERING METHODS FOR WATER RESOURCES DEVELOPMENT. VOLUME 10. PRINCIPLES OF GROUND-WATE	72-0696
COOMER, J. T.	FINITE-ELEMENT ANALYSIS OF PERIODIC-STRUCTURES	73-1129
COPE, R. J.	ANALYSIS OF MULTI-CELL BRIDGES WITHOUT TRANSVERSE DIAPHRAGMS, A FINITE ELEMENT APPROACH	69-0235
COPE, R. J.	THE ANALYSIS OF SKEW BRIDGE DECKS, A NEW FINITE ELEMENT APPROACH	69-0234
CORDS, H.	THERMAL AND STRESS ANALYSIS ON PRISMATIC NUCLEAR FUEL ELEMENTS	75-0120
CORMEAU, I. C.	VISCO-PLASTICITY SOLUTION BY FINITE ELEMENT PROCESS	72-0048
CORMEAU, I. C.	VISCO-PLASTICITY, PLASTICITY AND CREEP IN ELASTIC SOLIDS, A UNIFIED NUMERICAL SOLUTION APPROACH	74-1279
CORNELL, D. C.	CAPE-2D, A COMPUTER PROGRAM FOR THE STRESS ANALYSIS OF PLANE AND AXISYMMETRIC COMPOSITE STRUCTURES.	69-0083
CORNELL, D. C.	ITERATIVE TECHNIQUE FOR THE STRESS ANALYSIS OF PCR STRUCTURES USING THE FINITE ELEMENT METHOD	64-0026
CORNELL, D. C.	SAFE PCRS, A COMPUTER PROGRAM FOR THE STRESS ANALYSIS OF COMPOSITES BODIES OF REVOLUTION. INPUT INST	65-0013
CORNELL, D. C.	SAFE SHELL, A COMPUTER PROGRAM FOR THE STRESS ANALYSIS OF THIN SHELL BODIES OF REVOLUTION. A USER'S	67-0018
CORNELL, D. C.	SAFE-AXISYM, A COMPUTER PROGRAM FOR THE DESIGN OF COMPOSITE BODIES OF REVOLUTION. INPUT INSTRUCTIONS	65-0014
CORNELL, D. C.	SAFE-PLANE, A COMPUTER PROGRAM FOR THE STRESS ANALYSIS AND DESIGN OF TWO-DIMENSIONAL COMPOSITE BODIE	67-0017
CORNES, G. M. M	A POLYGONAL FINITE ELEMENT FOR PLATE BENDING PROBLEMS USING THE ASSUMED STRESS APPROACH	69-0043
CORNETT, D.	ON THE NUMERICAL SOLUTION OF A CLASS OF PROBLEMS IN A LINEAR FIRST STRAIN-GRADIENT THEORY OF ELASTIC	70-0318
COROTIS, R. B.	ANALYTICAL AND EXPERIMENTAL EVALUATION OF MODULUS OF SOIL REACTION	74-1168
CORP, E. L.	A FINITE ELEMENT MODEL FOR STABILITY ANALYSIS OF MINE WASTE EMBANKMENTS SUBJECTED TO SEEPAGE	74-1166
CORP, E. L.	ELASTIC-PLASTIC STABILITY ANALYSIS OF MINE-WASTE EMBANKMENTS	75-0835
CORRADI, L.	INELASTIC ANALYSIS OF REINFORCED CONCRETE FRAMES	74-0587
CORRADI, L.	LINEAR PROGRAMMING APPROACH TO SHAKEDOWN ANALYSIS OF STRUCTURES	74-0181
CORUM, J. M.	A THREE-DIMENSIONAL FINITE ELEMENT ANALYSIS OF A PRESTRESSED CONCRETE REACTOR VESSEL MODEL	69-0084
CORUM, J. M.	ANALYTIC INVESTIGATION OF COMPACT REINFORCEMENT FOR RADIAL NOZZLES IN SPHERICAL SHELLS	71-0107
CORUM, J. M.	COMPARISON OF TYPICAL INELASTIC ANALYSIS PREDICTIONS WITH BENCHMARK PROBLEM EXPERIMENTAL RESULTS	75-0696
CORUM, J. M.	EFFECT OF FILLETS ON STRESS CONCENTRATIONS IN CYLINDRICAL SHELLS WITH STEP CHANGES IN OUTSIDE DIAMET	71-0108
CORUM, J. M.	EXPERIMENTAL AND FINITE ELEMENT STRESS ANALYSIS OF A THIN SHELLED CYLINDER TO CYLINDER MODEL	71-0066
CORUM, J. M.	EXPERIMENTAL ELASTIC STRESS ANALYSES OF CYLINDER-TO-CYLINDER SHELL MODELS AND COMPARISONS WITH THEOR	71-0065
CORUM, J. M.	THEORETICAL AND EXPERIMENTAL STRESS ANALYSES OF ORNL THIN-SHELL CYLINDER-TO-CYLINDER MODEL 3	75-0690
CORUM, J. M.	USE OF SMALL MODELS IN DESIGN AND ANALYSIS OF PRESTRESSED-CONCRETE REACTOR VESSELS	70-0433
CORUM, J. M.	USE OF SMALL MODELS IN DESIGN AND ANALYSIS OF PRESTRESSED-CONCRETE REACTOR VESSELS	71-0426
CORUM, J. M.	ELASTIC-PLASTIC-CREEP ANALYSIS OF THERMAL RATCHETTING IN STRAIGHT PIPE AND COMPARISONS WITH TEST RES	73-0393
COST, T. L.	ANALYSIS OF THE BIAXIAL STRIP AND SHEAR LAP TESTS FOR SOLID PROPELLANT CHARACTERIZATION	67-0019
COST, T. L.	BASIC RESEARCH IN METHODS OF APPROXIMATION IN THE NONLINEAR MECHANICS OF SOLIDS AND STRUCTURES, VOLU	74-1091
COST, T. L.	THEMIS FINAL REPORT, VOL. 1, BASIC RESEARCH IN METHODS OF APPROXIMATION IN THE NONLINEAR MECHANICS O	74-0723
COST, T. L.	THERMOMECHANICAL DISSIPATION ANALYSIS OF THERMOVISCOELASTIC SOLIDS BY FINITE ELEMENTS	72-0403

COST, TH. L.	AUTOMATED EVALUATION OF INTERCEPTOR ROCKET MOTOR DESIGNS UNDER COMBINED OPERATIONAL AND NUCLEAR EFFE	74-0698
COSTANTINO, C.	INFLUENCE OF SOIL/STRUCTURE INTERACTION PARAMETERS ON FLOOR RESPONSE SPECTRA	73-0833
COSTELLO, M. G.	SOLUTION PROCEDURE FOR NONLINEAR STRUCTURAL PROBLEMS	73-0425
COTTRELL, P. E.	TWO-DIMENSIONAL FINITE ELEMENT ANALYSIS OF SEMI-CONDUCTOR STEADY STATE TRANSPORT EQUATIONS	74-1224
COUDERT, J. F.	NUMERICAL STUDY OF TRANSITORY OR NONLINEAR WAVES	74-0031
COURANT, R.	VARIATIONAL METHODS FOR THE SOLUTION OF PROBLEMS OF EQUILIBRIUM AND VIBRATION	43-0001
COURBON, J.	CALCULATION OF THE BRIDGE DECKS BY USING NUMERICAL METHODS, (LE CALCUL DES TABLIERS DE PONTS PAR DES	71-0411
COURTADE, R. M.	AUTOMATIC CALCULATION OF THIN STRUCTURES	73-0201
COURTADE, R. M.	CHARACTERISATION ELASTIQUE D'UN STRATIFIE RESINE-VERRE PAR LA METHODE DES DEPLACEMENTS (DETERMINATIO	75-0272
COURTNEY, R. L.	FINITE ELEMENT MODELING STUDIES IN THE NORMAL-MODE METHOD AND NORMAL-MODE SYNTHESIS	71-0067
COURTNEY, R. L.	NASTRAN MODELING STUDIES IN THE NORMAL-MODE AND NORMAL-MODE SYNTHESIS	71-0069
COURTNEY, R. L.	STRUCTURAL ANALYSIS AND DESIGN OPTIMIZATION OF SPACECRAFT USING NASTRAN	71-0068
COVARRUBIAS, S.	CRACKING OF EARTH AND ROCKFILL DAMS, A THEORETICAL INVESTIGATION BY MEANS OF THE FINITE ELEMENT METH	69-0085
COVARRUBIAS, S.	CRACKING OF EARTH AND ROCKFILL DAMS	69-0294
COVARRUBIAS, S.	CRACKING OF EARTH AND ROCKFILL DAMS: COMPARISON OF OBSERVED AND THEORETICAL TENSILE STRAINS IN THE C	69-0699
COWAN, W. C.	POLYMER-IMPREGNATED CONCRETE TUNNEL SUPPORT AND LINING	73-0736
COWDREY, D. R.	APPLICATION OF FINITE ELEMENT METHOD TO VIBRATIONS OF QUARTZ PLATES	74-0345
COWDREY, D. R.	FINITE ELEMENT CALCULATIONS RELEVANT TO AS-CUT QUARTZ RESONATORS	73-0672
COWDREY, D. R.	VARIATION WITH TEMPERATURE OF RESONANT FREQUENCIES OF ANISOTROPIC PLATES	74-1046
COWLEY, A.	FINITE ELEMENT METHOD FOR MACHINE TOOL STRUCTURAL ANALYSIS	71-0136
COWLEY, A.	FINITE ELEMENT METHOD FOR THE ANALYSIS OF MACHINE TOOL STRUCTURES.	72-0448
COWPER, G. R.	A HIGH PRECISION TRIANGULAR PLATE BENDING ELEMENT	68-0052
COWPER, G. R.	A SHALLOW SHELL FINITE ELEMENT OF TRIANGULAR SHAPE	70-0044
COWPER, G. R.	APPLICATION OF THE HIGH PRECISION TRIANGULAR PLATE BENDING ELEMENT TO PROBLEMS WITH CURVED BOUNDARIE	69-0071
COWPER, G. R.	COMPARISON OF TWO HIGH-PRECISION TRIANGULAR FINITE ELEMENTS FOR ARBITRARY DEEP SHELLS	71-0071
COWPER, G. R.	CURSHL: A HIGH-PRECISION FINITE ELEMENT FOR SHELLS OF ARBITRARY SHAPE	71-0430
COWPER, G. R.	FINITE ELEMENT ANALYSIS OF PLATES WITH CURVED EDGES.	72-0173
COWPER, G. R.	FORMULATION OF A NEW TRIANGULAR PLATE BENDING ELEMENT	68-0051
COWPER, G. R.	GAUSSIAN QUADRATURE FORMULAS FOR TRIANGLES	73-0906
COWPER, G. R.	NEW DEVELOPMENTS IN THE FINITE ELEMENT ANALYSIS OF SHELLS	69-0334
COWPER, G. R.	REFINED FINITE ELEMENTS FOR FOLDED PLATE STRUCTURES	72-0574
COWPER, G. R.	STATIC AND DYNAMIC APPLICATIONS OF A HIGH-PRECISION TRIANGULAR PLATE BENDING ELEMENT	69-0086
COWPER, G. R.	STRESS ANALYSIS OF MULTI-CELLULAR CAISSONS	74-0338
COWPER, G. R.	STRUCTURAL DESIGN AND ANALYSIS OF THE SPACE SHUTTLE	70-0336
COWPER, G. R.	VARIATIONAL PROCEDURES AND CONVERGENCE OF FINITE ELEMENT METHODS	71-0070
COX, H. L.	VIBRATION OF MISSILES	61-0007
CRAGGS, A.	ACOUSTIC FINITE ELEMENT APPROACH FOR STUDYING BOUNDARY FLEXIBILITY AND SOUND TRANSMISSION BETWEEN IR	73-0096
CRAGGS, A.	COMPUTATION OF THE RESPONSE OF COUPLED PLATE ACOUSTIC SYSTEMS USING PLATE FINITE ELEMENTS AND ACOUST	71-0072
CRAGGS, A.	SOME PRACTICAL APPLICATIONS OF ACOUSTIC FINITE ELEMENTS	75-0414
CRAGGS, A.	TRANSIENT RESPONSE OF A COUPLED PLATE-ACOUSTIC SYSTEM USING PLATE AND ACOUSTIC FINITE ELEMENTS	71-0463
CRAGGS, A.	TRANSIENT VIBRATION ANALYSIS OF LINEAR SYSTEMS USING TRANSITION MATRICES	68-0053
CRAGGS, A.	USE OF SIMPLE THREE-DIMENSIONAL ACOUSTIC FINITE ELEMENTS FOR DETERMINING NATURAL MODES AND FREQUENCI	72-0010
CRAIG, R. G.	FINITE ELEMENT STRESS ANALYSIS OF A RESTORED AXISYMMETRIC FIRST MOLAR	73-0053
CRAIG, R. G.	PHOTOELASTIC AND FINITE ELEMENT STRESS ANALYSIS OF A RESTORED AXISYMMETRIC FIRST MOLAR	73-0142
CRAIG, R. G.	OPTIMIZATION OF A SUPERSONIC PANEL SUBJECT TO A FLUTTER CONSTRAINT, A FINITE ELEMENT SOLUTION	71-0073
CRASTAN, V.	APPLICATION OF AN ' ELEMENT-METHOD ' PROGRAM TO SOLVING TEMPERATURE DISTRIBUTION PROBLEMS (ANWENDUNG	71-0464
CRASTAN, V.	FINITE ELEMENT SOLUTION OF THE THREE DIMENSIONAL FLOW PROBLEMS AND OF REYNOLD'S EQUATION FOR INCOMPR	72-0011
CRASTAN, V.	GENERALIZATION OF THE ELEMENT METHOD	71-0646
CRAWFORD, J.	A STRUCTURAL EVALUATION OF RAPID METHODS OF BACKFILLING FOR BOMB DAMAGE REPAIR - PHASE 11	75-0682
CRAWFORD, J.	FINITE-ELEMENT ANALYSIS OF PAVEMENT STRUCTURES USING AFPAV CODE (NONLINEAR ELASTIC ANALYSIS)	75-0748
CRAWFORD, J. E.	A TECHNIQUE FOR CONNECTING BEAM ELEMENTS TO A PLATE MODEL OF A COMPLICATED BOX SECTION	74-0271
CRAWFORD, J. E.	FINITE ELEMENT ANALYSIS OF PILE-SUPPORTED PAVEMENT STRUCTURES	74-0157
CRAWFORD, J. E.	FINITE ELEMENT ANALYSES OF QUICK CAMP MODULE	75-0017
CRAWFORD, J. E.	STRESS ANALYSIS OF A CONICAL ACRYLIC VIEWPORT	70-0106
CRAWFORD, J. E.	STRUCTURAL RESPONSE OF UNSTIFFENED TOROIDAL SHELLS	69-0198
CRAWFORD, P. B.	FINITE ELEMENT METHOD IN PETROLEUM RESERVOIR ANALYSIS	72-0049
CRAWFORD, P. B.	RELIABILITY STUDY OF LINEAR VARIATION TRIANGULAR FINITE ELEMENT	72-0281
CRAWLEY, J. T.	NORMAL MODE ANALYSIS OF A FUZE SUPPORT STRUCTURE USING NASTRAN, PART 1	72-0853
CREAGER, M.	DEVELOPMENT OF A CRACKED FINITE ELEMENT	70-0279
CREGO, H. L.	FUNDAMENTAL VIBRATIONAL FREQUENCY BY A 3D - FEM PROCEDURE	75-0710
CREWS, J. H.	A METHOD FOR DETERMINING LOCAL ELASTOPLASTIC STRESS AND STRAIN IN METALLURGICALLY BONDED NOTCHED LAM	74-0686
CREWS, J. H. JR	AN ELASTOPLASTIC ANALYSIS OF A UNIAXIALLY LOADED SHEET WITH AN INTERFERENCE - FIT BOLT	74-0687
CREWS, J. H. JR	ANALYTICAL AND EXPERIMENTAL INVESTIGATION OF FATIGUE IN A SHEET SPECIMEN WITH AN INTERFERENCE-FIT BO	75-0701
CRISFIELD, M. A	A NOTE ON THE ELASTIC ANALYSIS OF BOX-GIRDER DIAPHRAGMS USING FINITE ELEMENTS	74-1230
CRISFIELD, M. A	AN ITERATIVE IMPROVEMENT FOR NON-CONFORMING BENDING ELEMENTS	75-0844
CRISFIELD, M. A	COLLAPSE ANALYSIS OF BOX-GIRDER COMPONENTS USING FINITE ELEMENTS	74-0951
CRISFIELD, M. A	COMBINED MATERIAL AND GEOMETRIC NONLINEARITY FOR THIN STEEL PLATES	75-0397
CRISFIELD, M. A	ELASTIC-PLASTIC FINITE ELEMENT ANALYSIS OF PLATES - DISCUSSION	75-0139
CRISFIELD, M. A	FINITE ELEMENT METHODS FOR ANALYSIS OF MULTICELLULAR STRUCTURES	72-0561
CRISFIELD, M. A	FULL-RANGE ANALYSIS OF STEEL PLATES AND STIFFENED PLATING UNDER UNIAXIAL COMPRESSION	75-0803
CRISFIELD, M. A	LARGE-DEFLECTION ELASTO-PLASTIC BUCKLING ANALYSIS OF PLATES USING FINITE ELEMENTS	73-0394
CRISFIELD, M. A	MULTI-CELL BRIDGE DECKS - MODEL TESTS AND THEORETICAL ANALYSIS	72-0381
CRISFIELD, M. A	SOME APPROXIMATIONS IN THE NONLINEAR ANALYSIS OF RECTANGULAR PLATES USING FINITE ELEMENTS	74-1231
CRISWELL, M. E.	A RATIONAL ANALYSIS AND DESIGN PROCEDURE FOR WOOD JOIST FLOOR SYSTEMS	74-0730
CROCKER, M. J.	PREDICTION OF TRANSMISSION LOSS IN MUFFLERS BY THE FINITE ELEMENT METHOD	75-0160
CROCKER, T. J.	TECHNIQUE OF MEASURING INITIAL DEFORMATION AROUND AN OPENING	71-0331
CROLL, J. C. A.	THE TREATMENT OF NATURAL BOUNDARY CONDITIONS IN THE FINITE ELEMENT AND FINITE DIFFERENCE METHODS	73-1098
CROLL, J. G. A.	THE FINITE DIFFERENCE AND LOCALIZED RITZ METHODS	71-0447
CRONK, M.	AN INTERACTIVE COMPUTER GRAPHICS PROGRAM FOR NASTRAN	71-0074
CRONKHITE, J. D	A NASTRAN VIBRATION MODEL OF THE AH-IG HELICOPTER AIRFRAME. VOLUME 1	74-1338
CRONKHITE, J. D	A NASTRAN VIBRATION MODEL OF THE AH-IG HELICOPTER AIRFRAME. VOLUME 11	74-1339
CRONSHAGEN, E.	COMPUTER PROGRAMS FOR SIMULATING PHYSICAL PHENOMENA IN THREE SPACE DIMENSIONS AND TIME	74-0866
CROSE, J. G.	A STATISTICAL THEORY FOR THE FRACTURE OF BRITTLE STRUCTURES SUBJECTED TO NONUNIFORM POLYAXIAL STRESS	73-0928
CROSE, J. G.	ASAAS: ASYMMETRIC STRESS ANALYSIS OF AXISYMMETRIC SOLIDS WITH ORTHOTROPIC, TEMPERATURE-DEPENDENT MAT	74-0696
CROSE, J. G.	BANDWIDTH MINIMIZATION OF STIFFNESS MATRICES	70-0045
CROSE, J. G.	FINITE ELEMENT STRESS ANALYSIS OF POROUS MEDIA	69-0440
CROSE, J. G.	SAAS 111: FINITE ELEMENT STRESS ANALYSIS OF AXISYMMETRIC AND PLANE SOLIDS WITH DIFFERENT ORTHOTROPIC	71-0741
CROSE, J. G.	STRESS ANALYSIS OF AXISYMMETRIC SOLIDS WITH ASYMMETRIC PROPERTIES	72-0140
CROSE, J. G.	THERMAL STRESS ANALYSIS OF REENTRY VEHICLE NOSETIPS AT ANGLE OF ATTACK	72-0456
CROSE, J. G.	THERMAL STRESS ANALYSIS OF RE-ENTRY VEHICLE NOSETIPS AT ANGLE OF ATTACK	72-0852
CROSS, J. G.	SAAS II: FINITE ELEMENT STRESS ANALYSIS OF AXISYMMETRIC SOLIDS WITH ORTHOTROPIC, TEMPERATURE DEPENDE	68-0096
CROSSMAN, F. W.	CREEP OF B/AL COMPOSITES AS INFLUENCED BY RESIDUAL STRESSES, BOND STRENGTH, AND FIBER PACKING GEOMET	74-1208
CROSSMAN, F. W.	FAILURE MECHANISMS IN COMPOSITE SYSTEMS	75-0688
CROSSMAN, F. W.	INTERFACE FAILURES IN COMPOSITES	74-1209
CROSSMAN, F. W.	NON-UNIFORM FLOW OF POLYCRYSTALS BY GRAIN-BOUNDARY SLIDING ACCOMMODATED BY POWER-LAW CREEP	75-0562
CROSSMAN, F. W.	COMPUTER SIMULATION OF THE DEFORMATION OF COMPOSITE AND POLYCRYSTALLINE MATERIALS BY FINITE ELEMENT	72-0937
CROUCH, S. L.	ANALYSIS OF ROCK MASS DEFORMATIONS DUE TO EXCAVATIONS	73-0395
CROUCH, T. G.	GRIP: AN INFORMATION SYSTEM FOR DESIGN AUTOMATION	73-0210
CROUZEIX, M.	ELEMENTS FINIS ET PROBLEMES ELLIPTIQUES DEGENERES (FINITE ELEMENTS AND DEGENERATE ELLIPTIC PROBLEMS)	73-0582
CROUZEIX, M.	FINITE ELEMENTS AND DEGENERATED ELLIPTIC PROBLEMS	73-0357
CROUZEIX, M.	NUMERICAL SOLUTION BY METHODS OF FINITE ELEMENTS OF DEGENERATED ELLIPTICAL PROBLEMS	72-0284
CROUZEIX, M.	RESOLUTION NUMERIQUE PAR DES METHODES D ELEMENTS FINIS DE PROBLEMES ELLIPTIQUES DEGENERES	72-0702

26

CROUZET-PASCAL,	DOUBLY CURVED TRIANGULAR FINITE ELEMENTS FOR SHELLS OF ARBITRARY SHAPE	73-0881
CROUZIEX, M.	CONFORMING AND NONCONFORMING FINITE ELEMENT METHODS FOR SOLVING THE STATIONARY STOKES EQUATIONS - 1	73-0390
CRUSE, T. A.	APPLICATION OF THE BOUNDARY-INTEGRAL EQUATION METHOD TO THREE DIMENSIONAL STRESS ANALYSIS	73-0252
CRUZ., L. V.	A FINITE ELEMENT APPROACH TO PLATE BENDING PROBLEMS	68-0296
CSAGOLY, P.	FINITE ELEMENT ANALYSIS OF PRESTRESSED CONCRETE VOIDED BRIDGE DECKS	73-0106
CSENDES, Z.	FINITE ELEMENT SOLUTION OF DIELECTRIC LOADED WAVEGUIDES	70-0206
CSENDES, Z. J.	FINITE ELEMENT METHOD FOR THE GENERAL SOLUTION OF ORDINARY DIFFERENTIAL EQUATIONS	75-0473
CUBAUD, J. C.	APPLICATIONS DE LA METHODE DE GAUSS-SEIDEL A LA RESOLUTION DES PROBLEMS D'ELASTICITE. (APPLICATION D	72-0145
CUBAUD, J. C.	PHOTOELASTIC DETERMINATION OF THE STRESSES IN THE FOUNDATION OF THE NUCLEAR-POWER STATION BUGEY 2	75-0834
CUBITT, N. J.	APPLICATION OF THE FINITE ELEMENT METHOD TO CASES REQUIRING THE COMBINATION OF ELEMENTS POSSESSING D	72-0580
CULHAM, W. E.	NUMERICAL METHODS FOR TIME DEPENDENT NONLINEAR BOUNDARY VALUE PROBLEMS	70-0296
CULLEN, M. J.	INTEGRATIONS OF PRIMITIVE EQUATIONS ON A SPHERE USING FINITE ELEMENT METHOD	74-0649
CULLEN, M. J. P	INTEGRATIONS OF THE PRIMITIVE EQUATIONS ON A SPHERE USING THE FINITE ELEMENT METHOD	74-1347
CULLEN, M. J. P	SOLUTION OF WATER PREDICTION PROBLEMS BY FINITE ELEMENT METHODS	74-1345
CULLEN, T. M.	CREEP DATA ACQUISITION AND APPLICATION TO REACTOR COMPONENT DESIGN	72-0113
CULLIMORE, M. S	DISTRIBUTION OF THE CLAMPING PRESSURE IN FRICTION-GRIP BOLTED JOINTS	74-0250
CULVER, C.	INSTABILITY OF HORIZONTALLY CURVED MEMBERS. FLANGE BUCKLING STUDIES	69-0480
CULVILLE, J.	LARGE DISPLACEMENT ANALYSIS OF THIN PLATES	73-0261
CUNNINGHAM, F.	VIBRATION CHARACTERISTICS OF FREE THIN CYLINDRICAL SHELLS	73-0396
CUNNINGHAM, A.	UNSTEADY SUBSONIC COLLOCATION METHOD FOR WINGS WITH AND WITHOUT CONTROL SURFACES	74-0012
CUPPS, F. J.	USE OF SINGULARITY PROGRAMMING IN FINITE ELEMENT CALCULATIONS OF ELASTIC STRESS INTENSITY FACTORS. P	75-0252
CURTIN, D. J.	DEVELOPMENT OF ADVANCED INTERCONNECTORS FOR SOLAR CELLS	74-0371
CURTIN, J.	ADVANCED INTERCONNECT SYSTEMS FOR LIGHTWEIGHT SOLAR ARRAYS	73-0696
CURTIS, A. J.	APPLICATION OF RESPONSE LIMITING TO FINITE-ELEMENT STRUCTURAL-ANALYSIS PROGRAMS	75-0424
CURTIS, A. J.	ON THE APPLICATION OF RESPONSE LIMITING TO FINITE ELEMENT STRUCTURAL ANALYSIS PROGRAMS	75-0192
CURTIS, A. J.	ON THE APPLICATION OF RESPONSE LIMITING TO FINITE ELEMENT STRUCTURAL ANALYSIS PROGRAMS	75-0822
CURTIS, R.	STARDYNE STRUCTURAL ANALYSIS SYSTEM, ANALYTICAL CAPABILITY MANUAL	68-0235
CUTHILL, E.	REDUCING THE BANDWIDTH OF SPARSE SYMMETRIC MATRICES	69-0469
CUTHILL, E. H.	REDUCING THE BANDWIDTH OF SPARSE SYMMETRIC MATRICES	69-0502
CUTHILL, E. H.	SEVERAL STRATEGIES FOR REDUCING THE BANDWIDTH OF MATRICES	72-0996
CYR, N. A.	FINITE ELEMENT ELASTIC CREEP ANALYSIS OF TWO DIMENSIONAL CONTINUUM WITH TEMPERATURE DEPENDEN	73-0226
CYR, N. A.	FINITE ELEMENT THERMOELASTOPLASTIC ANALYSIS	72-0139
CYR, N. A.	TRIANGULAR FINITE ELEMENT FOR CYLINDRICAL SHELLS	70-0488
CYRUS, N. J.	ACCURACY AND CONVERGENCE OF FINITE ELEMENT APPROXIMATION	68-0228
CYRUS, N. J.	ACCURACY OF FINITE ELEMENT APPROXIMATIONS TO STRUCTURAL PROBLEMS	70-0438
CZENDES, Z. J.	FINPLT; A FINITE-ELEMENT FIELD-PLOTTING PROGRAM	72-0299
CZENDES, Z. J.	NUMERICAL SOLUTION OF DIELECTRIC LOADED WAVEGUIDES: I. FINITE-ELEMENT ANALYSIS	70-0199
CZENDES, Z. J.	ON LOCAL APPROXIMATIONS OF OPERATOR EQUATIONS	72-0063
D'APPOLONIA, D.	INITIAL SETTLEMENT OF STRUCTURES ON CLAY	71-0304
DACKO, M.	OBLICZANIE KONSTRUKCJI TOROIDALNEGO ZBIORNIKA NA WODE. (STATICAL ANALYSIS OF A TOROIDAL WATER RESERV	75-0211
DAEMEN, J. J. K	RATIONAL DESIGN OF TUNNEL SUPPORTS: TUNNEL SUPPORT LOADING CAUSED BY ROCK FAILURE	75-0687
DAHL, H. D.	A FINITE ELEMENT MODEL FOR ANISOTROPIC YIELDING IN GRAVITY LOADED ROCK	69-0012
DAI, P. K.	EFFECTS OF RISE TIME AND DAMPING ON FINITE ELEMENT ANALYSIS OF RESPONSE OF STRUCTURES	71-0087
DAILEY, G.	COMPUTER PROGRAMS FOR PLOTTING FINITE ELEMENT PATTERNS OF TWO-AND THREE-DIMENSIONAL STRUCTURES	70-0416
DAILEY, G.	GENERAL PURPOSE COMPUTER PROGRAM FOR THE DYNAMIC SIMULATION OF VEHICLE GUIDEWAY INTERACTIONS	73-0260
DAILEY, J. E.	A NUMERICAL MODEL OF TRANSIENT WATER QUALITY IN A ONE-DIMENSIONAL ESTUARY BASED ON THE FINITE ELEMEN	74-1369
DAILEY, J. E.	NJMERICAL MODEL FOR THE PREDICTION OF TRANSIENT WATER QUALITY IN ESTUARY NETWORKS	72-0635
DALEY, G. C.	OPTIMIZATION OF TENSION LEVEL AND STINGER LENGTH FOR OFFSHORE PIPELINE INSTALLATION	74-0504
DALY, P.	DIRECT METHOD OF OBTAINING CAPACITANCE FROM FINITE-ELEMENT MATRICES	72-0289
DALY, P.	EXACT FINITE ELEMENT SOLUTIONS TO THE HELMHOLTZ EQUATIONS	73-0652
DALY, P.	FINITE ELEMENT COUPLING MATRICES	69-0300
DALY, P.	FINITE ELEMENT METHODS FOR INHOMOGENEOUS WAVEGUIDES	69-0498
DALY, P.	FINITE ELEMENT SOLUTIONS FOR AN EQUILATERAL TRIANGLE	74-1281
DALY, P.	FINITE ELEMENTS FOR FIELD PROBLEMS IN CYLINDRICAL CO-ORDINATES	73-0171
DALY, P.	FINITE-ELEMENT METHODS FOR INHOMOGENEOUS WAVEGUIDES	69-0039
DALY, P.	HYBRID MODE ANALYSIS OF MICROSTRIP BY FINITE ELEMENT METHODS	71-0465
DALY, P.	POLAR GEOMETRY WAVEGUIDES BY FINITE-ELEMENT METHOD	74-0109
DALY, P.	PROPAGATION IN ELLIPTICAL AND PARABOLIC WAVEGUIDES	74-1173
DALY, P.	SINGULARITIES IN ELECTROMAGNETIC TRANSMISSION LINES	72-0072
DALY, P.	WAVEGUIDE SOLUTIONS BY THE FINITE ELEMENT METHOD	69-0011
DAME, R. E.	APPROXIMATION AND BOUNDS FOR EIGENVALUES OF VARIOUS DOMAINS USING THE INTERMEDIATE PROBLEM TECHNIQUE	72-0711
DAMLE, P. S.	ON THE NUMERICAL SOLUTION OF FREE BOUNDARY PROBLEMS	75-0577
DAMLE, S. K.	A FINITE ELEMENT ANALYSIS FOR DYNAMIC BEHAVIOR OF STIFFENED PLATES	69-0392
DAMLE, S. K.	RESPONSE OF RECTANGULAR PLATES TO MOVING LOADS BY A FINITE ELEMENT PROCEDURE	69-0515
DAMLE, S. K.	VIBRATION OF FOUR POINT-SUPPORTED PLATES BY A FINITE ELEMENT METHOD	71-0592
DANA, J. R.	STRESS CONCENTRATIONS NEAR HOLES IN LAMINATES	74-0329
DANA, J. R.	THREE DIMENSIONAL ANALYSIS OF STRESS CONCENTRATIONS NEAR HOLES IN LAMINATED COMPOSITES	72-0713
DANA, J. R.	THREE DIMENSIONAL FINITE ELEMENT ANALYSIS OF THICK LAMINATED COMPOSITES - INCLUDING INTERLAMINAR AND	73-0882
DANA, J. R.	THREE-DIMENSIONAL ANALYSIS FOR THE STRESS DISTRIBUTION NEAR CIRCULAR HOLES IN LAMINATED COMPOSITES	74-0793
DANA, J. R.	THREE-DIMENSIONAL FINITE ELEMENT ANALYSIS OF LAMINATED COMPOSITES	72-0345
DANA, J. R.	THREE-DIMENSIONAL FINITE-ELEMENT COMPUTER PROGRAM - USERS GUIDE	72-0755
DANA, J. R.	THREE-DIMENSIONAL FINITE-ELEMENT ANALYSIS OF THICK LAMINATED COMPOSITES - INCLUDING INTERLAMINAR AND	74-1151
DANCE, S. H.	RESPONSE OF NUCLEAR STRUCTURAL SYSTEMS TO TRANSIENT AND RANDOM EXCITATIONS. USING BOTH DETERMINISTIC	74-1047
DANHOF, R. H.	SECTOR FINITE ELEMENTS IN THE THEORY OF PLANE ELASTICITY	71-0766
DANIEL, D. E.	STRESS-STRAIN PROPERTIES OF COMPACTED CLAYS	74-0650
DANIEL, I. M.	DEFORMATION AND FAILURE OF BORON-EPOXY PLATE WITH CIRCULAR HOLE	72-0536
DANIEL, I. M.	MECHANICAL BEHAVIOR OF A GRAPHITE-EPOXY LAMINATE CONTAINING A HOLE	72-0614
DANIEL, I. M.	STRESS AND FAILURE ANALYSIS OF A GLASS-EPOXY COMPOSITE PLATE WITH A CIRCULAR HOLE	73-0276
DANIEU, D. J.	MINUTEMAN STV STRESS ANALYSIS AND TESTING	74-0809
DANTS, J.	FINITE ELEMENT NUMERICAL MODELLING OF FLOW AND DISPERSION IN ESTUARIES	73-0070
DAOUD, N. A-H.	FORCE AND MOMENTS ON ASYMMETRIC AND YAWED BODIES IN A FREE SURFACE	73-0908
DARBOIS, F.	USE OF FINITE ELEMENTS IN THE ANALYSIS OF A HYDROTHERMAL DOUBLET	74-0049
DARBY, J. B. JR	THREE-DIMENSIONAL MECHANICAL STRESSES IN TOROIDAL MAGNETS FOR CONTROLLED THERMONUCLEAR REACTORS	73-0669
DARIO, N. P.	A COMPARISON OF FINITE ELEMENT AND FINITE DIFFERENCE METHODS IN ELASTOSTATIC PROBLEMS	69-0273
DARIO, N. P.	A COMPARISON OF FIRST AND SECOND ORDER AXIALLY SYMMETRIC FINITE ELEMENTS	73-0997
DARWIN, D.	INELASTIC MODEL FOR CYCLIC BIAXIAL LOADING OF REINFORCED CONCRETE	74-0503
DAS, P. C.	MULTI-CELL BRIDGE DECKS - MODEL TESTS AND THEORETICAL ANALYSIS	72-0381
DASGUPTA, G.	WAVE PROPAGATION IN FLUIDS	74-0034
DASGUPTA, S.	BENDING PROBLEM OF AXIALLY CONSTRAINED BEAMS	73-0397
DASGUPTA, S. P.	FINITE ELEMENT SOLUTION TO SOME PLANE PROBLEMS IN SOIL DYNAMICS	74-0448
DATTAGUPU, B.	DETERMINATION OF STRESSES DUE TO DISCONTINUITIES IN FINITE PLATES OF ISOTROPIC AND ORTHOTROPIC MATER	73-1052
DAUGHTRIDGE, A.	GENERAL ANALYSIS OF FOLDED PLATES BY THE FINITE ELEMENT METHOD	69-0200
DAVIDS, N.	DOUBLE STRESS-WAVE DISCONTINUITIES IN FINITE, SHEAR-CORRECTED BEAMS AND PLATES	68-0210
DAVIDS, N.	DYNAMICAL FINITE ELEMENT ANALYSIS FOR ELASTIC WAVES IN BEAMS AND PLATES	68-0211
DAVIDS, N.	EFFECTS OF TURBULENCE ON FLOOD FLOW EXPLORED BY FINITE ELEMENT ANALYSIS.	72-0410
DAVIDS, N.	RADIAL INERTIA EFFECTS ON AN IDEALLY PLASTIC CIRCULAR PLATE UNDER IMPULSIVE AXIAL COMPRESSION	69-0324
DAVIDS, N.	THE DAMPED TRANSIENT BEHAVIOR OF FINITE BEAMS AND PLATES	69-0520
DAVIDS, N.	TRANSIENT LAMINAR FLOW IN DUCTS OF ARBITRARY CROSS-SECTION BY FINITE ELEMENT METHODS	72-0013
DAVIDSON, B. J.	A FINITE ELEMENT APPROACH TO STABILITY ANALYSIS IN FRAMES	74-0455
DAVIDSON, H. L.	ELASTIC-PLASTIC LARGE DEFORMATION RESPONSE OF CLAY TO FOOTING LOADS	74-0699
DAVIES, G.	NEW DIFFERENCE BASED FINITE-ELEMENT METHOD	75-0444
DAVIES, J. D.	ANALYSIS OF CORNER SUPPORTED SKEW SLABS	68-0203

DAVIES, J. D.	ANALYSIS OF CORNER SUPPORTED RECTANGULAR SLABS	70-0046
DAVIES, J. D.	ANALYSIS OF LONG RECTANGULAR TANKS	70-0276
DAVIES, J. D.	ANALYSIS OF RECTANGULAR TANKS BY FINITE ELEMENT TECHNIQUES	67-0044
DAVIES, J. D.	ANALYSIS OF SLABS WITH EDGE BEAMS	71-0075
DAVIES, J. D.	BENDING MOMENTS IN LONG WALL TANKS	67-0110
DAVIES, J. D.	EXPLICIT TRIANGULAR BENDING ELEMENT MATRIX	74-0288
DAVIES, J. D.	FINITE ELEMENT BENDING ANALYSIS OF MULTILAYER PLATES	74-0266
DAVIES, J. D.	INFLUENCE CHARACTERISTICS FOR SLAB BRIDGES	71-0188
DAVIES, J. D.	LIMITATIONS OF THE FINITE ELEMENT METHOD IN STRUCTURAL ANALYSIS	73-0420
DAVIES, J. D.	SIMPLE FINITE ELEMENT SOLUTION FOR PLATES OF HOMOGENEOUS, SANDWICH AND CELLULAR CONSTRUCTION	75-0140
DAVIES, J. D.	STRESSES DUE TO CONCENTRATED LOADS IN RESTRAINED BENDING TESTS	67-0097
DAVIES, J. D.	THE ANALYSIS OF COOLING TOWER RING BEAMS	68-0063
DAVIES, L. G.	LEAKAGE THROUGH CRACKS IN LNG TANKAGE	74-0749
DAVIS, C. G.	EFFECT OF INITIAL STATE FORCES ON THE HYDROELASTIC VIBRATION OF TUBES	73-0589
DAVIS, C. S.	ILLUSTRATIONS OF AUTOMOTIVE FINITE ELEMENT MODELS - STATICS	74-0373
DAVIS, D. B.	STRESSES IN CONTINUOUS CONCRETE BOX GIRDER BRIDGES	71-0233
DAVIS, E. H.	THE SIGNIFICANCE OF THE RATE OF PLASTIC WORK IN ELASTO-PLASTIC ANALYSIS	74-0437
DAVIS, E. H.	THE SIGNIFICANCE OF THE RATE OF PLASTIC WORK IN ELASTO-PLASTIC ANALYSIS	74-0764
DAVIS, J.	A FINITE ELEMENT MODEL OF TIDES IN ESTUARIES	74-0044
DAVIS, J.	A NUMERICAL MODEL OF DISPERSION IN ESTUARIES	74-0045
DAVIS, J. M.	IDENTIFICATION OF AQUIFER CONSTANTS: TESTS OF TWO INVERSE METHODS	74-0051
DAVIS, J. M.	NUMERICAL MODEL OF DISPERSION IN ESTUARIES	74-0905
DAVIS, J. M.	TIDAL AND LONG WAVE PROPAGATION - A FINITE ELEMENT APPROACH	73-1059
DAVIS, J. M.	TIDAL PROPAGATION AND DISPERSION IN ESTUARIES	75-0390
DAVIS, M. E.	FINITE BOUNDARY CORRECTIONS TO THE COPLANAR WAVEGUIDE ANALYSIS	73-0398
DAVIS, R.	CONSTANT CURVATURE BEAM FINITE ELEMENTS FOR IN-PLANE VIBRATION	72-0254
DAVIS, R.	CURVED BEAM FINITE ELEMENTS FOR COUPLED BENDING AND TORSIONAL VIBRATION	72-0720
DAVIS, R. E.	FIELD TESTS OF A STEEL-COMPOSITE BOX-GIRDER BRIDGE	75-0633
DAVIS, R. E.	LOAD DISTRIBUTION IN A COMPOSITE STEEL BOX GIRDER BRIDGE	73-0765
DAVIS, R. E. SR	STRUCTURAL BEHAVIOR OF A FLEXIBLE METAL CULVERT UNDER A DEEP EARTH EMBANKMENT USING METHOD A BACKFIL	73-0739
DAVIS, R. L.	A TIMOSHENKO BEAM ELEMENT	72-0590
DAVIS, R. L.	FINITE ELEMENT ANALYSIS OF PRESSURE VESSELS	71-0348
DAVIS, R. L.	FORMING OF 7075-T6 ALUMINUM IN HIGH PRESSURE ENVIRONMENTS	71-0322
DAVIS, R. L.	HOW FINITE ELEMENT METHODS IMPROVE THE DESIGN CYCLE	74-0372
DAVIS, R. L.	USER'S MANUAL FOR UMR FINITE ELEMENT COMPUTER PROGRAM - STRATA	69-0429
DAWE, D. J.	A FINITE ELEMENT APPROACH TO PLATE VIBRATION PROBLEMS	65-0015
DAWE, D. J.	A FINITE-DEFLECTION ANALYSIS OF SHALLOW ARCHES BY THE DISCRETE ELEMENT METHOD	71-0077
DAWE, D. J.	APPLICATION OF THE DISCRETE ELEMENT METHOD TO THE BUCKLING ANALYSIS OF RECTANGULAR PLATES UNDER ARBI	68-0054
DAWE, D. J.	CURVED FINITE ELEMENTS IN THE ANALYSIS OF SHELL STRUCTURES	71-0076
DAWE, D. J.	CURVED FINITE ELEMENTS FOR THE ANALYSIS OF SHALLOW AND DEEP ARCHES	74-0332
DAWE, D. J.	FREE VIBRATIONS OF FINITE ELEMENT PLATES SUBJECTED TO COMPLEX MIDDLE-PLANE FORCE SYSTEMS	73-0097
DAWE, D. J.	NUMERICAL STUDIES USING CIRCULAR ARCH FINITE ELEMENTS	74-0600
DAWE, D. J.	ON ASSUMED DISPLACEMENTS FOR THE RECTANGULAR PLATE BENDING ELEMENT	67-0020
DAWE, D. J.	PARALLELOGRAM ELEMENT IN THE SOLUTION OF RHOMBIC CANTILEVER PLATE BENDING	66-0022
DAWE, D. J.	RIGID-BODY MOTIONS AND STRAIN-DISPLACEMENT EQUATIONS OF CURVED SHELL FINITE ELEMENTS	72-0267
DAWE, D. J.	SHELL ANALYSIS USING A SIMPLE FACET ELEMENT	72-0986
DAWKINS, W. P.	A DISCRETE-ELEMENT METHOD OF ANALYSIS FOR COMBINED BENDING AND SHEAR DEFORMATION OF A BEAM	69-0494
DAWKINS, W. P.	FINITE ELEMENT FOR REINFORCED CONCRETE FRAME STUDY	74-0110
DAY, C. P.	TRANSIENT ELASTOHYDRODYNAMIC LUBRICATION BY FINITE ELEMENT METHODS	74-0087
DAY, S. M.	FINITE-ELEMENT TREATMENT OF QUADRUPOLE GROUND MOTIONS IN AXISYMMETRIC EARTH STRUCTURES	75-0434
DAYARATNAM, P.	INTERACTION OF FRAME AND SHEAR WALL WITH OPENINGS	74-1190
DE AMORIM, A. M	FREE-SURFACE IDEAL FLUID FLOWS BY FINITE ELEMENTS	74-0125
DE ANDRADE, J.	LINEAR AND NON LINEAR ANALYSIS OF SHELLS OF REVOLUTION WITH ASYMMETRICAL STIFFNESS PROPERTIES	68-0151
DE ARANTES E OL	A GENERAL THEORY OF VARIATIONAL METHODS WITH APPLICATION TO FINITE ELEMENTS	72-0897
DE ARANTES E OL	A THEORY OF VARIATIONAL METHODS AND ITS APPLICATION TO FINITE ELEMENTS	73-0639
DE ARANTES E OL	COMPLETENESS AND CONVERGENCE IN THE FINITE ELEMENT METHOD	68-0279
DE ARANTES E OL	CONVERGENCE AND ACCURACY IN THE FINITE ELEMENT METHOD	75-0630
DE ARANTES E OL	CONVERGENCE OF FINITE ELEMENT SOLUTIONS IN VISCOPLASTICITY	74-0628
DE ARANTES E OL	CONVERGENCE OF FINITE ELEMENT SOLUTIONS IN VISCOUS FLOW PROBLEMS	75-0337
DE ARANTES E OL	MATHEMATICAL FOUNDATION OF THE FINITE ELEMENT METHOD	67-0126
DE ARANTES E OL	MATHEMATICAL THEORY OF LINEAR STRUCTURES	71-0498
DE ARANTES E OL	RESULTS ON THE CONVERGENCE OF THE FINITE ELEMENT METHOD IN STRUCTURAL AND NONSTRUCTURAL CASES	74-0416
DE ARANTES E OL	THEORETICAL FOUNDATIONS ON THE FINITE ELEMENT METHOD	68-0055
DE ARANTES E OL	THEORETICAL FOUNDATIONS OF THE FINITE ELEMENT METHOD (DISCUSSION)	70-0074
DE ARANTES E OL	THEORETICAL FOUNDATIONS OF THE FINITE ELEMENT METHOD	70-0453
DE ARANTES E OL	COMPLETENESS AND CONVERGENCE IN THE FINITE ELEMENT METHOD	68-0056
DE BOOR, C.	PIECEWISE POLYNOMIAL INTERPOLATION AND APPROXIMATION	65-0058
DE COEN, J.	INFLUENCE OF THE UNPRECISION OF EXECUTION ON THE BEHAVIOUR OF A SPHERICAL SHELL WITH CONCENTRATED LO	73-0802
DE DONATO, O.	A MODIFIED GRADIENT METHOD FOR FINITE ELEMENT ELASTOPLASTIC ANALYSIS BY QUADRATIC PROGRAMMING	73-0998
DE DONATO, O.	FINITE ELEMENT ELASTOPLASTIC ANALYSIS OF UNDERGROUND OPENINGS BY QUADRATIC PROGRAMMING	74-1288
DE DONATO, O.	INELASTIC ANALYSIS OF REINFORCED CONCRETE FRAMES	74-0587
DE FRAINE, J.	PRAKTISCHE RESULTATEN. DOEL EN NUT VAN HET ONTWERPEN VAN MECHANISCHE ELEMENTEN MET BEHULP VAN COMPUT	73-0400
DE KONING, A. U	A COMPUTER PROGRAMME FOR ELASTIC-PLASTIC CALCULATIONS WITH USE OF TRIM-6 FINITE ELEMENTS	72-0993
DE KONING, A. U	A MATHEMATICAL THEORY OF PLASTICITY FOR STRAIN HARDENING MATERIALS EXHIBITING INITIAL ANISOTROPY	72-0660
DE KONING, A. U	RESULT OF CALCULATIONS WITH TRIM 6 AND TRIAX 6 ELASTIC-PLASTIC ELEMENTS	73-1122
DE KONING, A. U	THEORETICAL AND EXPERIMENTAL ANALYSIS OF ELASTOPLASTIC DEFORMATION IN A UNIAXIALLY LOADED STRIP OF 2	73-0720
DE LA CRUZ, R.	THEORETICAL BASIS OF THE BOREHOLE DEEPENING METHOD OF ABSOLUTE STRESS MEASUREMENT	70-0047
DE ROUVRAY, A.	ANALYSIS AND MODEL STUDIES OF UNDERGROUND OPENINGS IN JOINTED ROCK. RESEARCH ON ROCK BOLT REINFORCEM	71-0438
DE ROUVRAY, A.	ANALYSIS AND MODEL STUDIES OF UNDERGROUND OPENINGS IN JOINTED ROCK. RESEARCH ON ROCK BOLT REINFORCEM	71-0750
DE ROUVRAY, A.	FINITE ELEMENT ANALYSIS OF CRACK INITIATION IN A BLOCK MODEL EXPERIMENT	72-0455
DE VAHL DAVIS,	APPLICATION OF THE FINITE ELEMENT METHOD TO CONVECTION HEAT TRANSFER BETWEEN PARALLEL PLANES	71-0244
DE VRIES, G.	A FINITE ELEMENT BIBLIOGRAPHY - PART I: AUTHOR LISTING	74-0284
DE VRIES, G.	A FINITE ELEMENT BIBLIOGRAPHY - PART II: KEYWORD LISTING	74-0285
DE VRIES, G.	A FINITE ELEMENT BIBLIOGRAPHY - PART III: CITATION LISTING	74-0286
DE VRIES, G.	A LAGRANGIAN FINITE ELEMENT SOLUTION TO UNSTEADY FLOW	74-1264
DE VRIES, G.	APPLICATION OF FINITE ELEMENT METHODS IN FLUID DYNAMICS	71-0369
DE VRIES, G.	APPLICATION OF THE FINITE ELEMENT METHOD TO UNSTEADY FLOW PROBLEMS	70-0310
DE VRIES, G.	APPLICATION OF THE FINITE ELEMENT TECHNIQUE TO COMPRESSIBLE FLOW PROBLEMS	70-0311
DE VRIES, G.	APPLICATION OF THE FINITE ELEMENT TECHNIQUE TO COMPRESSIBLE FLOW PROBLEMS	73-0003
DE VRIES, G.	APPLICATION OF THE FINITE ELEMENT TECHNIQUE TO COMPRESSIBLE FLOW PROBLEMS	74-0090
DE VRIES, G.	APPLICATION OF THE PSEUDO-FUNCTIONAL FINITE ELEMENT METHOD TO VISCO-PLASTIC TORSION	73-0002
DE VRIES, G.	APPLICATION OF THE PSEUDO-FUNCTIONAL FINITE ELEMENT METHOD TO NON-LINEAR PROBLEMS	74-0282
DE VRIES, G.	APPLICATION OF THE PSEUDO-FUNCTIONAL FINITE ELEMENT METHOD TO VISCOUS FLOW PROBLEMS	74-1183
DE VRIES, G.	APPLICATION OF THE PSEUDO-FUNCTIONAL FINITE ELEMENT METHOD TO COMPRESSIBLE FLOW PROBLEMS	74-1360
DE VRIES, G.	NON-STRUCTURAL APPLICATIONS OF THE FINITE ELEMENT METHOD	73-0330
DE VRIES, G.	THE APPLICATION OF THE FINITE ELEMENT TECHNIQUE TO POTENTIAL FLOW PROBLEMS: PART I	69-0087
DE VRIES, G.	THE APPLICATION OF THE FINITE ELEMENT TECHNIQUE TO POTENTIAL FLOW PROBLEMS: PART II	69-0395
DE VRIES, G.	THE APPLICATION OF THE FINITE ELEMENT TECHNIQUE TO POTENTIAL FLOW PROBLEMS	71-0370
DE VRIES, G.	THE APPLICATION OF THE PSEUDO-FUNCTIONAL FINITE ELEMENT METHOD TO VISCOUS FLOW PROBLEMS	74-0287
DE VRIES, G.	THE FINITE ELEMENT METHOD - FUNDAMENTALS AND APPLICATIONS	73-0001
DE VRIES, G.	THE PSEUDO FUNCTIONAL FINITE ELEMENT METHOD	74-0004
DE VRIES, K. L.	FINITE ELEMENT IN NUMERICAL ANALYSIS OF ADHESIVE FRACTURE	73-0085

DE WINDT, P.	EURCYL: A COMPUTER PROGRAM TO GENERATE FINITE ELEMENT MESHES FOR CYLINDER-CYLINDER INTERSECTIONS	73-0962
DE WINDT, P.	EURCYL: A PROGRAM TO GENERATE FINITE ELEMENT MESHES FOR PRESSURE VESSEL NOZZLES	74-1086
DE, A. K.	ANALYSIS OF CLAMPED CIRCULAR PLATES OF VARIABLE THICKNESS SUBJECTED TO UNIFORMLY DISTRIBUTED LOAD BY	73-0583
DE, A. K.	INTRODUCTION TO FINITE ELEMENT METHOD AND ITS APPLICATION TO MACHINE TOOL STRUCTURES	72-0112
DE, A. K.	SOME DESIGN ASPECTS OF IC ENGINE PISTON HEADS	73-0578
DEAK, A.	APPLICATION OF THE SMOOTH-SURFACE INTERPOLATION TO THE FINITE ELEMENT ANALYSIS	69-0088
DEAK, A. L.	NONLINEAR HYBRID STRESS FINITE ELEMENT ANALYSIS OF LAMINATED SHELLS	74-1350
DEAK, A. L.	NONLINEAR HYBRID STRESS FINITE ELEMENT ANALYSIS OF PNEUMATIC TIRES	75-0029
DEAK, A. L.	THE APPLICATION OF THE SMOOTH-SURFACE INTERPOLATION TECHNIQUE TO THE FINITE-ELEMENT ANALYSIS OF THIN	66-0120
DEAN, D. L.	FIELD ANALYSIS FOR PREFABRICATED RIBBED CYLINDRICAL SHELLS	73-0899
DEAN, W.	CORRELATION OF FINITE ELEMENT STRUCTURAL DYNAMIC ANALYSIS WITH MEASURED FREE VIBRATION CHARACTERISTI	74-0486
DEARIEN, J. A.	STRAP - A COMPUTER CODE FOR THE STATIC AND DYNAMIC ANALYSIS OF REACTOR STRUCTURAL SYSTEMS	71-0078
DEARIEN, J. A.	STRAP: COMPUTER CODE FOR STATIC AND DYNAMIC STRUCTURAL ANALYSIS AND STUDIES MADE USING THE CODE	70-0409
DEARRIAGA, F. J	SAFE-CREEP: A COMPUTER PROGRAM FOR THE VISCOELASTIC ANALYSIS OF AXISYMMETRIC AND PLANE CONCRETE STRU	67-0148
DEASON, D. H.	A FINITE ELEMENT TECHNIQUE FOR ANALYZING FOLDED PLATE TYPE STRUCTURES WITH SKEWED SUPPORTS	68-0525
DEB NATH, J. M.	USE OF HIGHER ORDER DISPLACEMENT FUNCTIONS IN THE FREE VIBRATION ANALYSIS OF SHELLS OF REVOLUTION HA	74-1055
DEBNATH, J. M.	A COMPARISON OF RECENT SHALLOW SHELL FINITE-ELEMENT ANALYSES	70-0232
DEBOOR, C.	RAYLEIGH-RITZ APPROXIMATION OF PIECEWISE CUBIC POLYNOMIALS	66-0087
DECKER, R. W.	FINITE-ELEMENT MODELING OF SURFACE DEFORMATION ASSOCIATED WITH VOLCANISM	75-0596
DECRETON, M. C.	ANALYSIS OF OPEN STRUCTURES BY FINITE ELEMENT RESOLUTION OF EQUIVALENT CLOSED-BOUNDARY PROBLEM	74-0111
DECRETON, M. C.	CALCUL DES CHAMPS ELECTROMAGNETIQUES PAR LA METHODE DES ELEMENTS FINIS (COMPUTATION OF ELECTROMAGNET	73-0399
DECRETON, M. C.	INTEGRAL FINITE-ELEMENT APPROACH FOR SOLVING LAPLACE EQUATION	73-0121
DECRETON, M. C.	NUMERICAL ANALYSIS OF THE LINE CAPACITANCE AND CROSSTALK FACTOR FOR INSULATED WIRE PAIRS	74-0685
DEDONATO, O.	FINITE ELEMENT ELASTOPLASTIC ANALYSIS BY QUADRATIC PROGRAMMING: THE MULTISTAGE METHOD	73-0842
DEERE, D. U.	ELASTO-PLASTIC ANALYSIS OF UNDERGROUND OPENINGS BY THE FINITE ELEMENT METHOD	66-0067
DEHART, R. C.	DISCONTINUITY STRESS DECAY IN OFFSHORE PLATFORM JOINTS	74-0491
DEHART, R. C.	PENETRATION IN SHELLS UNDER EXTERNAL PRESSURE	70-0237
DEHART, R. C.	USE OF FRACTURE MECHANICS AND FINITE ELEMENT TECHNIQUES FOR DETERMINING THE SAFETY OF OFFSHORE STRUC	74-1211
DEHLEN, G. L.	EFFECT OF NONLINEAR MATERIAL RESPONSE ON THE BEHAVIOUR OF PAVEMENTS UNDER TRAFFIC	70-0171
DEIBEL, W. T.	EFFECTS OF SHOE FORCE GEOMETRY ON HEAVY DUTY INTERNAL SHOE BRAKE PERFORMANCE	68-0219
DEICKE, K.	KOLBENKUEHLUNG (PISTON COOLING)	75-0266
DEIST, F. H.	COMPUTER APPLICATIONS IN ROCK MECHANICS	72-0190
DEIST, F. H.	FINITE ELEMENT METHOD	69-0386
DEL GUIDICE, C.	FINITE ELEMENT SOLUTION OF NON-LINEAR HEAT CONDUCTION PROBLEMS WITH SPECIAL REFERENCE TO PHASE CHANG	74-0348
DELPAK, R.	AXISYMMETRIC VIBRATION OF SHELLS OF REVOLUTION BY THE FINITE ELEMENT METHOD	75-0767
DELPUGLIA, A.	CALCULATION METHOD FOR THE STUDY OF CRACKED RIGID STRUCTURES (METODO DI CALCOLO PER LO STUDIO DI STR	71-0637
DELVES, L. M.	VARIATIONAL SOLUTION OF TWO DIMENSIONAL FLUID FLOW PROBLEMS	73-1014
DEMEY, G.	FINITE-ELEMENT METHOD FOR POTENTIAL CALCULATIONS IN A HALL PLATE	75-0454
DEMICHELE, D. W	THREE-DIMENSIONAL MECHANICAL STRESSES IN TOROIDAL MAGNETS FOR CONTROLLED THERMONUCLEAR REACTORS	73-0669
DEMIRCHYAN, K.	RASHET TREKHMERNYKH MAGNITNYKH POLEI METODOM KONECHNYKH ELEMENTOV. (CALCULATION OF THREE-DIMENSIONAL	75-0766
DEMIRCHYAN, K.	REALIZATSIYA METODA KONECHNYKH ELEMENTOV NA EVM DLYA RASCHETA DVUMERNYKH ELEKTRICHESKIKH I MAGNITNYK	74-0251
DEMITRACK, G.	FINITE ELEMENT STRESS ANALYSIS OF ARTILLERY PROJECTILES	73-0432
DEMS, K.	APPLICATION OF FINITE ELEMENTS FOR SOLVING THE TWO-DIMENSIONAL ELASTICITY PROBLEM BY MEANS OF THE	75-0482
DEMS, K.	APPLICATION OF HERMITE POLYNOMIALS TO THE DETERMINATION OF THE STIFFNESS MATRIX OF PLATE BY FINITE E	71-0365
DEMS, K.	WIELOSTOPNIOWA SYNTEZA MACIERZY SZTYWNOSCI (MULTI-STAGE SYNTHESIS OF THE STIFFNESS MATRICES)	73-0668
DENDY, J. E. JR	A METHOD TO GALERKIN TYPE ACHIEVING OPTIMUM L(2) ACCURACY FOR FIRST ORDER HYPERBOLICS AND EQUATIONS	72-0521
DENKE, P. H.	A GENERAL DIGITAL COMPUTER ANALYSIS OF STATICALLY INDETERMINATE STRUCTURES	59-0007
DENKE, P. H.	DIGITAL ANALYSIS OF NON-LINEAR STRUCTURES BY THE FORCE METHOD	64-0023
DENKE, P. H.	MATRIX DIFFERENCE EQUATION ANALYSIS OF VIBRATING PERIODIC STRUCTURES	75-0151
DENKE, P. H.	MATRIX METHODS OF AEROSPACE STRUCTURAL ANALYSIS	68-0198
DENKE, P. H.	THE MATRIC SOLUTION OF CERTAIN NONLINEAR PROBLEMS IN STRUCTURAL ANALYSIS	56-0002
DENKE, P. H.	THE MATRIX DEFORMATION METHOD OF STRUCTURAL ANALYSIS	71-0079
DENKMANN, W. J.	ANALYSIS OF STRUCTURAL-ACOUSTIC INTERACTIONS IN METAL-CERAMIC TRANSDUCERS	72-0850
DENKMANN, W. J.	ANALYSIS OF STRUCTURAL ACOUSTIC INTERACTIONS IN METAL CERAMIC TRANSDUCERS	73-0401
DENKMANN, W. J.	FINITE-ELEMENT ANALYSIS OF SANDWICH SHELLS	73-1130
DENNIS, R. W.	THEORETICAL PREDICTION OF DISC BRAKE TEMPERATURES AND A COMPARISON WITH EXPERIMENTAL DATA	72-0306
DEPAULIS, A.	FUNCTIONAL MINIMIZATION IN NONLINEAR SOLID MECHANICS	74-1251
DEPAULIS, A.	ON THE RESOLUTION OF NONLINEAR ELASTIC AND ELASTOPLASTIC PROBLEMS BY THE METHOD OF FINITE ELEMENTS.	74-1326
DEPPE, L.	FINITE ELEMENT METHOD APPLIED TO NEUTRON DIFFUSION PROBLEMS	73-0977
DEPPE, L. O.	APPLICATION OF FINITE ELEMENT METHOD TO 2-DIMENSIONAL DIFFUSION PROBLEMS	73-0098
DEPRES, J.	THIN SHELL THEORY IN CARTESIAN COORDINATES - METHODS FOR FINITE ELEMENTS	70-0359
DERBIDGE, T. C.	ANALYSIS OF THERMAL STRESSES IN CORED CERAMIC BRICKS USED IN HYPERSONIC WIND TUNNEL HEATERS	72-0655
DERNER, W. J.	HOLLOW-ENDED ROLLER. A SOLUTION FOR IMPROVING FATIGUE LIFE IN ASYMMETRICALLY LOADED CYLINDRICAL ROLL	71-0393
DESAI, C.	COMPARISON OF FINITE ELEMENTS FOR PLATE BENDING	72-0625
DESAI, C. S.	A CONSISTENT FINITE ELEMENT TECHNIQUE FOR WORKSOFTENING BEHAVIOUR	74-1233
DESAI, C. S.	ANALYSIS OF CIRCULAR FOOTINGS ON LAYERED SOILS	70-0281
DESAI, C. S.	ANALYSIS OF PILE-SUPPORTED GRAVITY LOCK	74-0585
DESAI, C. S.	APPROXIMATE SOLUTION FOR UNCONFINED SEEPAGE	73-0348
DESAI, C. S.	EVALUATION AND ANALYSES OF SOME FINITE ELEMENT AND FINITE DIFFERENCE PROCEDURES FOR TIME-DEPENDENT P	75-0745
DESAI, C. S.	EVALUATION OF SOME NUMERICAL SCHEMES FOR CONSOLIDATION	73-0349
DESAI, C. S.	EVALUATION OF TWO FINITE ELEMENT FORMULATIONS FOR ONE-DIMENSIONAL CONSOLIDATION	72-0453
DESAI, C. S.	FINITE ELEMENT ANALYSIS OF UNCONFINED FLOW WITH A VARIABLE TIME-STEP PROCEDURE	74-0776
DESAI, C. S.	FINITE ELEMENT METHODS FOR FLOW IN POROUS MEDIA	74-0057
DESAI, C. S.	FINITE ELEMENT PROCEDURES FOR SEEPAGE ANALYSIS USING AN ISOPARAMETRIC ELEMENT	72-0366
DESAI, C. S.	INTRODUCTION TO THE FINITE ELEMENT METHOD	72-0362
DESAI, C. S.	LOAD DEFORMATION ANALYSIS OF DEEP PILE FOUNDATION	72-0931
DESAI, C. S.	NONLINEAR ANALYSES USING SPLINE FUNCTIONS	71-0466
DESAI, C. S.	NUMERICAL DESIGN-ANALYSIS FOR PILES IN SANDS	74-0333
DESAI, C. S.	ON APPLICATION OF THE FINITE ELEMENT METHOD TO LIMIT ANALYSIS CONSISTENT FINITE ELEMENT TECHNIQUE FO	74-1234
DESAI, C. S.	OVERVIEW, TRENDS AND PROJECTIONS: THEORY AND APPLICATIONS OF THE FINITE ELEMENT METHOD IN GEOTECHNIC	72-0988
DESAI, C. S.	SEEPAGE ANALYSIS OF EARTH BANKS UNDER DRAWDOWN	72-0014
DESAI, C. S.	SEEPAGE IN MISSISSIPPI RIVER BANKS. REPORT 1. ANALYSIS OF TRANSIENT SEEPAGE USING VISCOUS FLOW MODEL	70-0405
DESAI, C. S.	SEEPAGE IN MISSISSIPPI RIVER BANKS. REPORT 1. ANALYSIS OF TRANSIENT SEEPAGE USING A VISCOUS-FLOW MOD	73-0909
DESAI, C. S.	SOLUTION OF STRESS-DEFORMATION PROBLEMS IN SOIL AND ROCK MECHANICS USING FINITE ELEMENT METHODS	69-0272
DESAI, C. S.	SOME NUMERICAL PROCEDURES FOR ANALYSIS OF ONE DIMENSIONAL CONSOLIDATION	72-0928
DESAI, C. S.	STABILITY CRITERIA FOR TWO FINITE ELEMENT SCHEMES FOR PARABOLIC EQUATION	75-0568
DESAI, C. S.	THEORY AND APPLICATION OF THE FINITE ELEMENT METHOD IN GEOTECHNICAL ENGINEERING	72-0704
DESAI, C. S.	ULTIMATE CAPACITY OF CIRCULAR FOOTINGS ON LAYERED SOILS	70-0280
DESAI, C. S.	UNCONFINED TRANSIENT SEEPAGE IN SLOPING BANKS	71-0378
DESAI, S.	USE OF AUXILIARY FUNCTIONS IN THE FINITE ELEMENT METHOD	72-0938
DESALVO, G. J.	APPLICATION OF FRACTURE MECHANICS TO THERMALLY STRESSED CYLINDERS USING A FINITE ELEMENT METHOD	69-0089
DESALVO, G. J.	STRESS INTENSITY FACTORS FOR HEXAGONAL FRACTURE TOUGHNESS SPECIMENS	71-0799
DESANTIS, D.	MAGIC, AN AUTOMATED GENERAL PURPOSE SYSTEM FOR STRUCTURAL ANALYSIS. VOLUME III. PROGRAMMER'S MANUAL	69-0090
DESCHAMPS, L. C	TANKER TRANSVERSE STRENGTH ANALYSIS: USER'S MANUAL	72-0750
DESCLOUX, J.	ON FINITE ELEMENT MATRICES	72-0103
DESCLOUX, J.	ON THE RIGID DISPLACEMENT CONDITION	72-0405
DESCOEUDRES, F.	ANALYSE TRIDIMENSIONELLE DE LA STABILITE D'UN TUNNEL AU VOISINAGE DU FRONT DE TAILLE DANS UNE ROCHE	74-0995
DESILVA, B. M.	HYBRID FINITE ELEMENT OPTIMIZATION PROGRAMS FOR DISCS USING A DYNAMICS DESIGN TECHNOLOGY	74-0691
DESROCHERS, L.	THERMAL SLEEVE DESIGN FOR PRESSURE VESSELS	74-1129
DEVERALL, L. I.	CASE LINER BOND ANALYSIS	74-0817
DEVERALL, L. I.	COMPARISON OF NUMERICAL METHODS FOR DETERMINING STRESS INTENSITY FACTORS	72-0176
DEVERALL, L. I.	FRACTURE OF ROCKET GRAIN MODELS SUBJECTED TO COOLDOWN: A STUDY OF TEMPERATURE-INITIATED FRACTURES OF	74-0536

29

DEVOS, J. E.	FINITE ELEMENT SOLUTION OF THE THREE DIMENSIONAL FLOW PROBLEMS AND OF REYNOLD'S EQUATION FOR INCOMPR	72-0011
DEVRIES, K. L.	FINITE ELEMENT IN ADHESION ANALYSES	73-0355
DEVRIES, K. L.	MECHANICS OF FRACTURE IN ADHESIVE JOINTS	72-0691
DEWAAL, J. F.	STRESS CONCENTRATION FACTORS FOR A FATIGUE TEST SPECIMEN WITH A U-TYPE NOTCH IN ONE EDGE BY FINITE E	74-1344
DEZFULIAN, H.	SEISMIC RESPONSE OF SOIL DEPOSITS UNDERLAIN BY SLOPING ROCK BOUNDARIES	69-0484
DHALLA, A. K.	DIRECT FLEXIBILITY FINITE ELEMENT ELASTOPLASTIC ANALYSIS	71-0095
DHALLA, A. K.	INELASTIC ANALYSIS AND SATISFACTION OF DESIGN CRITERIA OF A HIGH TEMPERATURE COMPONENT	75-0265
DHATT, G.	AN AUTOMATIC RELABELING ALGORITHM FOR BANDWIDTH MINIMIZATION	75-0763
DHATT, G.	AN EFFICIENT TRIANGULAR SHELL ELEMENT	70-0382
DHATT, G.	BUCKLING OF DEEP SHELLS	73-0849
DHATT, G.	DEVELOPMENT OF TWO SIMPLE SHELL ELEMENTS	72-0676
DHATT, G.	INSTABILITY OF THIN SHELLS BY THE FINITE ELEMENT METHOD	70-0187
DHATT, G.	MIXED QUADRILATERAL ELEMENTS FOR BENDING	72-0623
DHATT, G. S.	AN EFFICIENT TRIANGULAR SHELL ELEMENT	70-0441
DHATT, G. S.	CURVED TRIANGULAR ELEMENTS FOR THE ANALYSIS OF SHELLS	68-0030
DHATT, G. S.	FINITE ELEMENT ANALYSIS OF CONTAINMENT VESSELS	71-0080
DHATT, G. S.	INSTABILITY OF THIN SHELLS BY THE FINITE ELEMENT METHOD	70-0477
DHATT, G. S.	LARGE DEFLECTION ANALYSIS OF ARBITRARY SHALLOW SHELLS BY THE FINITE ELEMENT METHOD	73-0937
DHATT, G. S.	NUMERICAL ANALYSIS OF THIN SHELLS BY CURVED TRIANGULAR ELEMENTS BASED ON DISCRETE - KIRCHHOFF HYPOTH	69-0091
DHILLON, B. S.	CONVERGENCE OF EIGENVALUE SOLUTIONS IN CONFORMING PLATE BENDING FINITE ELEMENTS	72-0195
DHILLON, B. S.	TRIANGULAR FINITE ELEMENTS FOR THE BENDING ANALYSIS OF THICK ELASTIC PLATES	70-0466
DHILLON, B. S.	TRIANGULAR THICK PLATE BENDING ELEMENTS	71-0167
DHIR, S. K.	IMPROVED METHOD FOR OBTAINING THE GENERAL-DISPLACEMENT FIELD FROM A HOLOGRAPHIC INTERFEROGRAM	72-0015
DHOOPAR, B. L.	MEMBRANE ANALOGY FOR ANISOTROPIC CABLE NETWORKS	74-0233
DI CARLO, A.	DISCUSSION ON "PLANE STRESS LIMIT ANALYSIS BY FINITE ELEMENTS"	71-0792
DI CARLO, A.	FINITE ELEMENT SIMULATION OF THERMALLY INDUCED FLOW FIELDS	74-1235
DI CARLO, A.	MIXED FINITE ELEMENT MODELS IN LIMIT ANALYSIS	74-1314
DI CARLO, A.	NUMERICAL TECHNIQUES FOR CONVECTION/DIFFUSION PROBLEMS	75-0123
DI MONACO, A.	PROCESS FOR THE AUTOMATIC GENERATION OF FLAT TRIANGULAR LATTICES FOR APPLICATION TO THE FINITE ELEME	71-0366
DI MONACO, A.	STUDIO DI CAMPI ELETTRICI E MAGNETICI STANZIONARI CON IL METODO DEGLI ELEMENTI FINITI. APPLICAZIONE	75-0207
DI NAPOLI, A.	EDDY CURRENTS SCREENING PROBLEMS SOLVED BY MEANS OF FINITE ELEMENT DIFFERENCE AND FINITE ELEMENT MET	75-0552
DI PASQUALE, S.	ENERGY FORMS IN THE FINITE ELEMENT TECHNIQUES	72-0680
DIAB, B.	CONTRIBUTION A L'ETUDE DU COMPLEXE BARRAGE-FONDATION ET EXPLICATION DES SEISMES DUS AU REMPLISSAGE D	72-0554
DIANA, G.	SULLA DETERMINAZIONE DELLE FREQUENZE PROPRIE DI ALBERI SU PIU SUPPORTI ELASTICI. (DETERMINATION OF F	74-0916
DIBAJ, M.	EARTHQUAKE ANALYSIS OF EARTH DAMS	69-0072
DIBAJ, M.	RESPONSE OF EARTH DAMS TO TRAVELLING SEISMIC WAVES	69-0092
DIBAJ, M.	THE EFFECTS OF THIN CORES AND NONHOMOGENEITY ON DYNAMIC RESPONSE OF EARTH DAMS	67-0021
DIBENEDETTO, A.	TRANSVERSE PROPERTIES OF UNIDIRECTIONAL ALUMINUM MATRIX FIBROUS COMPOSITES	69-0167
DICELLO, J. A.	HIGH STRENGTH MATERIALS AND VEHICLE WEIGHT REDUCTION ANALYSIS	75-0157
DICKINSON, S. M	CLOUGH-TOCHER TRIANGULAR PLATE BENDING ELEMENT IN VIBRATION	69-0093
DICKINSON, S. M	FREE VIBRATION OF MODEL OF CAR BODY	71-0312
DICKSON, J. N.	DEVELOPMENT OF AN UNDERSTANDING OF THE FATIGUE PHENOMENA OF BONDED AND BOLTED JOINTS IN ADVANCED FIL	72-0798
DIEMONT, W.	THERMAL AND STRESS ANALYSIS ON PRISMATIC NUCLEAR FUEL ELEMENTS	75-0120
DIETERICH, D. A	FINITE ELEMENT COMPUTER PROGRAM FOR PREDICTING THE NONLINEAR STATIC AND DYNAMIC BEHAVIOR OF VISCOELA	75-0191
DIETERICH, J. H	EARTHQUAKE TRIGGERING BY FLUID INJECTION AT RANGELY, COLORADO	72-0634
DIETERICH, J. H	FINITE-ELEMENT MODELING OF SURFACE DEFORMATION ASSOCIATED WITH VOLCANISM	75-0596
DIETERICH, J. H	SLOW FINITE DEFORMATIONS OF VISCOUS SOLIDS	69-0423
DIETRICH, D. E.	NONLINEAR ANALYSIS OF ARBITRARY, HYPERELASTIC MEMBRANE SHELLS BY THE FINITE-ELEMENT METHOD	73-1132
DIETRICH, D. E.	THREE DIMENSIONAL INELASTIC ANALYSIS OF A SHELL APPURTENANCE PER SECTION 111 - APPENDIX F.	74-0716
DIETRICH, G.	DYNAN LECTURE NOTES WITH COMPUTATIONAL EXAMPLES	71-0581
DIETRICH, G.	SIMULTANEOUS EIGENVECTOR ITERATION APPLIED TO LARGE SYSTEMS	74-0680
DIETRICH, R.	BERECHNUNGSMETHODEN ZUR ERMITTLUNG VON BELASTUNGSBEDINGUNGEN UND GRENZTRAGFAEHIGKEITSVERHALTEN VON K	73-0858
DIETZ, C. G.	DEVELOPMENT OF A GRAPHITE HORIZONTAL STABILIZER	73-0991
DILL, E. H.	LARGE DEFLECTIONS OF STRUCTURES SUBJECT TO HEATING AND EXTERNAL LOADS	60-0002
DILLON, E. C.	APPLICATION OF THE FAST FOURIER TRANSFORM TO LINEAR SYSTEMS IN CIVIL ENGINEERING	73-1119
DILLON, E. C.	SOLUTION BOUNDS IN SOME STRESS PROBLEMS BY THE HYPERCIRCLE AND FINITE ELEMENT METHODS	72-0706
DILWORTH, T. E.	HINGE	71-0634
DIMAROGONAS, A.	GENERAL METHOD FOR STABILITY ANALYSIS OF ROTATING SHAFTS	75-0528
DIMITRIOU, C.	FINITE ELEMENT METHOD	69-0386
DIMITRIOU, C.	STRESSES IN SHORT THICK COMPOUND CYLINDERS	69-0385
DIMMER, R.	ESTIMATES OF THE CREEP RUPTURE LIFETIME OF STRUCTURES USING THE FINITE ELEMENT METHOD	75-0833
DIMOCK, R. R.	CORRELATION BETWEEN MODEL STUDY AND FINITE ELEMENT ANALYSIS IN AN INVESTIGATION OF SLOPE STABILITY	72-0227
DIMONACO, A.	ANALYSIS OF STEADY ELECTRICAL AND MAGNETIC-FIELDS BY FINITE-ELEMENT METHOD - APPLICATION TO TRANSFOR	75-0550
DINIS DA GAMA,	APPLICATION OF THE FINITE ELEMENT METHOD TO THE ANALYSIS OF ROCK SAMPLES UNDER UNIAXIAL COMPRESSION	70-0182
DINYOVSZKY, P.	FINITE-ELEMENT TECHNIQUES FOR THE DYNAMIC ANALYSIS OF UNSYMMETRICALLY LAMINATED PLATES	73-1131
DISNEY, R. K.	CURRENT FINITE ELEMENT ANALYSIS PRACTICES	73-0118
DISTEFANO, N.	DYNAMIC PROGRAMMING APPROACH TO THE FORMULATION AND SOLUTION OF FINITE ELEMENT EQUATIONS	75-0059
DISTEFANO, N.	INVARIANT IMBEDDING AND SOLUTION OF FINITE ELEMENT EQUATIONS	74-0158
DITSWORTH, G. R	MATHEMATICAL MODEL OF THE COLUMBIA RIVER FROM THE PACIFIC OCEAN TO BONNEVILLE DAM. PART 1. THEORY, P	69-0464
DITTMAR, S.	STRESS DETERMINATION IN 2 CONDENSED HEAT EXCHANGER BAFFLE CAPS BY METHOD OF FINITE ELEMENTS AND A ST	74-0661
DIXIT, V. D.	ANALYSIS OF A SKEW SLAB-BEAM SYSTEM BY FINITE ELEMENT METHOD	73-0233
DIXON, J. D.	STRUCTURAL DESIGN DATA FOR CONCRETE DRIFT LININGS IN BLOCK CAVING	73-0773
DIXON, J. R.	ANALYSIS OF SHIP STRUCTURES BY THE FINITE ELEMENT METHOD	69-0336
DIXON, J. R.	DETERMINATION OF ENERGY RELEASE RATES AND STRESS - INTENSITY FACTORS BY THE FINITE-ELEMENT METHOD	72-0183
DIXON, J. R.	STRESS INTENSITY FACTORS CALCULATED GENERALLY BY FINITE ELEMENT TECHNIQUE	69-0369
DJUBEK, J.	BOX SHAPED GIRDERS. (NOSNIKY UZAVRETEHO PRIEREZU)	70-0169
DJUBEK, J.	GEOMETRICKY NELINEARNE ULOHY PLOCHYCH SKRUPIN RIESENE PRIRASTKOVOU METODOU KONECNYCH PRVKOV. (GEOMET	74-1032
DJURIC, M. P.	NEW METHOD OF BULKHEAD ANALYSIS IN AIRCRAFT STRUCTURES	74-0334
DLUGACH, M. I.	AN APPLICATION OF THE FINITE ELEMENT METHOD TO THE CALCULATION OF CYLINDRICAL SHELLS WITH RECTANGULA	74-1236
DLUGACH, M. I.	ISSLEDOVANIE NAPRYAZHENNOGO SOSTOYANIYA REBRISTYKH ISILINDRICHESKIKH OBOLOCHEK S PRYAMOUGOL'NYMI OTV	74-0944
DLUGACH, M. I.	METOD KONECHNYKH ELEMENTOV V PRIMENENII K RASCHETU TSILINDRICHESKIKH OBOLOCHEK S PRYAMOUGOL'NYMI OTV	73-0536
DOBOVISEK, B.	DESIGN OF PRISMATIC SHELLS	72-0404
DOBYNS, A. L.	A TETRA-CORE STRESS ANALYSIS MODEL	72-0845
DOCTORS, L. J.	AN APPLICATION OF THE FINITE ELEMENT TECHNIQUE TO BOUNDARY VALUE PROBLEMS OF POTENTIAL FLOW	70-0002
DODDS, B. J.	COMPARISON OF SOME PLATE BENDING FINITE ELEMENTS	72-0413
DOERR, K.	BERECHNUNG VON STAHLBETONSCHEIBEN IM ZUSTAND II BEI ANNAHME EINES WIRKLICHKEITSNAHEN WERKSTOFFVERHAL	74-0978
DOGGERT, R. V.	A DESIGN STUDY FOR THE INCORPORATION OF AEROELASTIC CAPABILITY INTO NASTRAN	71-0112
DOHERTY, W. P.	STRESS ANALYSIS OF AXISYMMETRIC SOLIDS UTILIZING HIGHER ORDER QUADRILATERAL FINITE ELEMENTS	69-0356
DOKAINISH, M. A	DEFLECTION AND VIBRATION ANALYSIS OF A FLEXIBLE FAN BLADE	73-0556
DOKAINISH, M. A	MODEL TEST AND ANALYSIS OF OBLIQUE FOLDED PLATE STRUCTURE	73-0230
DOKAINISH, M. A	NEW APPROACH FOR PLATE VIBRATIONS. COMBINATION OF TRANSFER MATRIX AND FINITE ELEMENT TECHNIQUE	71-0434
DOKAINISH, M. A	PSEUDO STATIC DEFORMATION AND FREQUENCIES OF ROTATING TURBOMACHINERY BLADES	72-0239
DOKAINISH, M. A	THEORETICAL AND EXPERIMENTAL VIBRATION ANALYSIS OF AN OBLIQUE SPACE FRAME	70-0460
DOKUMACI, E.	FINITE ELEMENT ANALYSIS OF COUPLED VIBRATION OF TAPERED TWISTED BLADES	72-0766
DOKUMACI, E.	IMPROVED FINITE ELEMENTS FOR VIBRATION ANALYSIS OF TAPERED BEAMS	73-0033
DOKUMACI, E.	SIMPLE FINITE ELEMENTS FOR PRE-TWISTED BLADING VIBRATION	74-0239
DOLTSINIS J.	ELASTOPLASTIC AND CREEP ANALYSIS WITH THE ASKA PROGRAM SYSTEM	74-0863
DOLTSINIS, J.	ASPECTS OF FINITE ELEMENT METHOD APPLIED TO AEROSPACE STRUCTURES	73-0703
DOLTSINIS, J.	FINITE ELEMENT ANALYSIS OF THERMOMECHANICAL PROBLEMS	71-0030
DOLTSINIS, J.	MATERIAL NONLINEARITIES	72-0647
DOLTSINIS, J.	RECENT DEVELOPMENTS IN THE FINITE ELEMENT ANALYSIS OF PRE-STRESSED CONCRETE REACTOR VESSELS	74-0582
DOLTSINIS, J. S	ASPECTS OF THE FINITE ELEMENT METHOD AS APPLIED TO AERO-SPACE STRUCTURES	72-0649

DOMASCHUK, L.	NONLINEAR SETTLEMENT ANALYSIS BY FINITE ELEMENT	75-0486
DOMINGUEZ, R. F	NUMERICAL SOLUTIONS FOR DETERMINING WAVE-INDUCED PRESSURE DISTRIBUTIONS AROUND BURIED PIPELINES	74-1334
DOMINGUEZ, R. F	MECHANICS OF CABLE MOORING SYSTEMS. VOLUME IV. A COMPUTER PROGRAM FOR ANALYZING THE STEADY STATE RES	72-0653
DOMINGUEZ, R. F	MECHANICS OF CABLE MOORING SYSTEMS. VOLUME 1. THREE DIMENSIONAL RESPONSE OF DEEP WATER MOORING LINES	72-0654
DOMINGUEZ, R. F	MECHANICS OF CABLE MOORING SYSTEMS. VOLUME V11. THE STEADY-STATE BEHAVIOR OF A PYRAMID ARRAY SYSTEM	73-0724
DOMINGUEZ, R. F	MECHANICS OF CABLE MOORING SYSTEMS. VOLUME V1. A COMPUTER PROGRAM FOR ANALYZING THE STEADY STATE CON	73-0725
DONALDSON, I. G	GALERKIN-FINITE ELEMENT ANALYSIS OF HYDROTHERMAL SYSTEM AT WAIRAKEI, NEW-ZEALAND	75-0601
DONEA, J.	ANALYSE DES CONTRAINTES DAN DES STRUCTURES COMPLEXES EN PRESENCE DE FLUAGE, DE VARIATIONS DE DIMENSI	71-0081
DONEA, J.	CODE TAFEST: NUMERICAL SOLUTION TO TRANSIENT HEAT-CONDUCTION PROBLEMS USING FINITE ELEMENTS IN SPACE	74-0901
DONEA, J.	CREEP ANALYSIS OF TRANSVERSELY ISOTROPIC BODIES SUBJECTED TO TIME DEPENDENT LOADING	73-0170
DONEA, J.	FINITE ELEMENT ANALYSIS OF STEADY-STATE NONLINEAR HEAT TRANSFER PROBLEMS	74-0572
DONEA, J.	FINITE ELEMENTS IN THE SOLUTION OF ELECTROMAGNETIC INDUCTION PROBLEMS	74-0252
DONEA, J.	FINITE-ELEMENT ANALYSIS OF RADIATION-DAMAGE STRESSES IN THE GRAPHITE OF MATRIX FUEL ELEMENTS	70-0048
DONEA, J.	ON THE ACCURACY OF FINITE ELEMENT SOLUTIONS TO THE TRANSIENT HEAT CONDUCTION EQUATION	74-0182
DONG, S. B.	ANALYSIS OF LAMINATED SHELL FOR REVOLUTION	66-0023
DONG, S. B.	ARBITRARILY SHAPED INHOMOGENEOUS OPTICAL FIBER OR INTEGRATED OPTICAL WAVEGUIDES	75-0372
DONG, S. B.	EFFECT OF TRANSVERSE SHEAR DEFORMATION IN VIBRATIONS OF PLANAR STRUCTURES COMPOSED OF BEAM TYPE ELEM	73-0402
DONG, S. B.	ELASTODYNAMIC BEHAVIOR OF LAMINATED ORTHOTROPIC PLATES UNDER INITIAL STRESS	75-0249
DONG, S. B.	EMBANKMENT ANALYSIS AND FIELD CORRELATION	68-0057
DONG, S. B.	FINITE ELEMENT ANALYSIS FOR SOLIDS OF REVOLUTION	68-0153
DONG, S. B.	GLOBAL-LOCAL FINITE ELEMENT STRUCTURAL ANALYSIS, ANALYTICAL CONSIDERATIONS/ELEMENT FORMULATION	74-1090
DONG, S. B.	NATURAL VIBRATIONS OF LAMINATED ORTHOTROPIC SHELLS OF REVOLUTION	7C-0367
DONG, S. B.	STABILITY ANALYSIS OF STRUCTURES BY A REDUCED SYSTEM OF GENERALIZED COORDINATES	70-0168
DORN, L.	EXPERIMENTS IN FORMING AND JOINTING OF TITANIUM AND TITANIUM ALLOYS FOR AIRCRAFT AND SPACE VEHICLES	70-0134
DORNQUAST, P.	PROBLEMS OF LONGITUDINAL STRENGTH IN THE CONSTRUCTION OF VERY LARGE TANK TRUCKS	73-0251
DOROSZ, S.	FUNKCJA KSZTALTU NIESPELNIAPACA WARUNKOW CIAGLOSCI W ANALIZIE PLYT METODA ELEMENTOW SKONCZONYCH. (SH	75-0263
DOUGHERTY, C. S	ANALYSIS OF THE 747 AIRCRAFT WING BODY INTERSECTION	68-0275
DOUGLAS, A.	BOUNDARIES AND FRACTURES IN FINITE ELEMENT MODELS OF GEOLOGICAL STRUCTURES	73-0352
DOUGLAS, A.	FINITE ELEMENTS FOR GEOLOGICAL MODELLING	73-0049
DOUGLAS, F.	THE NASTRAN PROGRAMMER'S MANUAL	72-0480
DOUGLAS, J.	A FINITE ELEMENT COLLOCATION METHOD FOR THE HEAT EQUATION	72-0488
DOUGLAS, J.	A FINITE ELEMENT COLLOCATION METHOD FOR QUASILINEAR PARABOLIC EQUATIONS	73-1000
DOUGLAS, J.	A SUPERCONVERGENCE RESULT FOR THE APPROXIMATE SOLUTION OF THE HEAT EQUATION BY A COLLOCATION METHOD	72-0510
DOUGLAS, J.	FINITE ELEMENT COLLOCATION METHOD FOR QUASILINEAR PARABOLIC EQUATIONS	73-0047
DOUGLAS, J.	GALERKIN METHODS FOR PARABOLIC EQUATIONS	70-0448
DOUGLAS, J.	GALERKIN METHODS FOR PARABOLIC EQUATIONS WITH NONLINEAR BOUNDARY CONDITIONS	73-1023
DOUGLAS, J.	LOCATION OF POINTS AT WHICH FINITE ELEMENT SOLUTIONS ARE VERY ACCURATE	74-0112
DOUGLAS, J.	SOME SUPERCONVERGENCE RESULTS FOR AN H-GALERKIN PROCEDURE FOR THE HEAT EQUATION	74-0411
DOUGLAS, J.	STABILITY IN LQ OF L2-PROJECTION INTO FINITE ELEMENT FUNCTION SPACES	75-0071
DOUGLAS, M. R.	AN APPLICATION OF FINITE ELEMENT TECHNIQUE IN BRIDGE DESIGN	70-0225
DOUGLAS, M. R.	FINITE ELEMENT PROGRAMS FOR SLAB BRIDGE DESIGN	71-0082
DOVE, R. C.	WAVE PROPAGATION IN A THIN HOLLOW CONE BY A FINITE ELEMENT METHOD	72-0045
DOVEY, H. H.	COMPUTER PROGRAM FOR STATIC AND DYNAMIC ANALYSIS OF LINEAR STRUCTURAL SYSTEMS	72-0683
DOVEY, H. H.	EXTENSION OF THREE DIMENSIONAL ANALYSIS TO SHELL STRUCTURES USING THE FINITE ELEMENT IDEALIZATION	74-0849
DOWLING, A. R.	THREE-DIMENSIONAL CRACK ANALYSIS APPLIED TO AN LWR NOZZLE-CYLINDER INTERSECTION	75-0165
DOWLING, P. J.	SHEAR LAG IN STEEL BOX GIRDER BRIDGES	75-0310
DOYLE, D. M.	STIFFNESS OF STEPPED BARS	74-0711
DOYLE, W. S.	EXPLICIT TRIANGULAR BENDING ELEMENT MATRIX	74-0288
DRAKE, L. A.	RAYLEIGH WAVES AT A CONTINENTAL BOUNDARY BY FINITE ELEMENT METHOD	72-0016
DRAKE, L. A.	THE PROPAGATION OF LOVE WAVES ACROSS NONHORIZONTALLY LAYERED STRUCTURES	71-0600
DRAPER, J. K.	INADEQUACY OF NODAL CONNECTIONS IN A STIFFNESS SOLUTION FOR PLATE BENDING	65-0021
DRAPER, K.	LAGRANGE MULTIPLIER TECHNIQUES IN STRUCTURAL ANALYSIS	65-0045
DRAPER, K.	STRUCTURAL EIGENVALUE PROBLEMS: ELIMINATION OF UNWANTED VARIABLES	65-0050
DRNEVICH, V. P.	CONSTRAINED AND SHEAR MODULI FOR FINITE ELEMENTS	74-0774
DRNEVICH, V. P.	CONSTRAINED AND SHEAR MODULI FOR FINITE ELEMENTS	75-0228
DROZD, K.	ANALYSIS AND MODEL STUDIES OF UNDERGROUND OPENINGS IN JOINTED ROCK. RESEARCH ON ROCK BOLT REINFORCEM	71-0750
DRUCKER, D. C.	EFFECTS OF GEOMETRY CHANGE ON ESSENTIAL FEATURES OF INELASTIC BEHAVIOUR	73-0448
DRYSDALE, W. H.	AN ANALYSIS OF FINITE ELEMENT CODE TORT 11	75-0752
DSTAF'EV, V. A.	DETERMINATION OF THE STRESS-STRAINED STATE WHEN CUTTING METAL WITH A WEDGE	73-0237
DU PREEZ, R. J.	ANALYSIS OF BOX GIRDERS BY FOLDED PLATE AND STRIP METHODS	73-0982
DU PREEZ, R. J.	ANALYSIS OF SLAB BRIDGES BY FINITE STRIP METHOD	73-0687
DU PREEZ, R. J.	REVIEW OF BASIC THEORY AND METHODS OF ANALYSIS OF CONTINUOUS STRUCTURES	73-0697
DUBOIS, J.	DUPLICATION OF DILATANT BEHAVIOR IN THE ANALYSIS OF JOINTED ROCKS	72-0757
DUCKER, W. L.	ANALYSIS FOR THE DESIGN OF BEAM TESTING APPARATUS	75-0527
DUDDECK, H.	DER GERISSENE ZUSTAND ZWEISEITIG GELAGERTER PLATTEN UNTER EINZELLASTEN - NICHTLINEARE BERECHNUNG MIT	75-0858
DUDGEON, C. R.	OPTIMAL DESIGN OF WATER WELLS IN UNCONSOLIDATED SEDIMENTS	74-1182
DUDGEON, C. R.	TECHNIQUES FOR HANDLING INPUT DATA FOR FINITE ELEMENT ANALYSIS OF REGIONAL GROUNDWATER FLOW	75-0806
DUGUID, J. O.	FINITE ELEMENT GALERKIN METHOD FOR ANALYSIS OF FLOW IN FRACTURED POROUS MEDIA	74-0066
DUGUID, J. O.	WATER FLOW THROUGH SATURATED-UNSATURATED POROUS MEDIA USING FINITE-ELEMENT-GALERKIN METHODS	74-0879
DUGUID, J. O.	WATER MOVEMENT THROUGH SATURATED-UNSATURATED POROUS MEDIA: A FINITE-ELEMENT GALERKIN MODEL	75-0761
DUHAU, R.	MATRICE DE RIGIDITE D'UN ELEMENT DE POUTRE SOLLICITE A LA TORSION-FLEXION (STIFFNESS MATRIX OF A BEA	73-0442
DUKES, T. P.	ANALYSIS OF SHIP STRUCTURES BY THE FINITE ELEMENT METHOD	69-0336
DUKES, W. H.	BRITTLE MATERIALS: A DESIGN CHALLENGE	75-0769
DULLAGE, C.	ANALYSIS OF NONLINEAR PROBLEMS IN ROCK MECHANICS WITH PARTICULAR REFERENCE TO JOINTED ROCK SYSTEMS	70-0282
DUMAS, G.	ELASTOPLASTIC INDENTATION OF HALF-SPACE BY AN INFINITELY LONG RIGID CIRCULAR CYLINDER	71-0395
DUMITRESCU, D.	THEORETICAL AND EXEPIMENTAL CONSIDERATIONS ON FLOW OF VISCOUS LIQUIDS PAST PLATE AT SMALL AND AVERA	70-0144
DUNCAN, J. L.	THREE-DIMENSIONAL FINITE ELEMENT ANALYSES OF DAMS	73-0202
DUNCAN, J. M.	ANALYSIS OF SOIL MOVEMENT AROUND A DEEP EXCAVATION	70-0039
DUNCAN, J. M.	APPLICATIONS OF THE FINITE ELEMENT METHOD IN GEOTECHNICAL ENGINEERING. PROCEEDINGS OF THE SYMPOSIUM	73-0273
DUNCAN, J. M.	DEVELOPMENT OF FAILURE AROUND EXCAVATED SLOPES	70-0051
DUNCAN, J. M.	EARTH PRESSURES ON STRUCTURES DUE TO FAULT MOVEMENT	73-0403
DUNCAN, J. M.	EFFECT OF RESERVOIR FILLING ON STRESSES AND MOVEMENT IN EARTH AND ROCKFILL DAMS. A REPORT OF AN INVE	72-0836
DUNCAN, J. M.	FINITE ELEMENT ANALYSIS OF PAVEMENTS	68-0058
DUNCAN, J. M.	FINITE ELEMENT ANALYSES OF SLOPES IN SOIL	68-0061
DUNCAN, J. M.	FINITE ELEMENT ANALYSES OF SLOPES IN JOINTED ROCK	68-0269
DUNCAN, J. M.	FINITE ELEMENT ANALYSES OF PORT ALLEN AND OLD RIVER LOCKS	69-0203
DUNCAN, J. M.	FINITE ELEMENT ANALYSIS OF PORT ALLEN LOCK	71-0293
DUNCAN, J. M.	FINITE ELEMENT ANALYSIS OF RETAINING WALL BEHAVIOUR	71-0320
DUNCAN, J. M.	FINITE ELEMENT ANALYSIS OF TRANSVERSE CRACKING IN LOW-EMBANKMENT DAMS	74-0696
DUNCAN, J. M.	HYDRAULIC FRACTURING IN ZONED EARTH AND ROCKFILL DAMS: A REPORT OF AN INVESTIGATION	73-0873
DUNCAN, J. M.	NONLINEAR ANALYSIS OF STRESS AND STRAIN IN SOILS	70-0050
DUNCAN, J. M.	SEEPAGE AND GROUNDWATER EFFECTS ASSOCIATED WITH EXPLOSIVE CRATERING	72-0656
DUNCAN, J. M.	SLOPES IN STIFF FISSURED CLAYS AND SHALES	69-0094
DUNCAN, J. M.	STRESSES AND MOVEMENTS IN OROVILLE DAM	72-0200
DUNCAN, J. M.	THREE-DIMENSIONAL FINITE ELEMENT ANALYSIS OF DAMS	71-0733
DUNCAN, W.	EFFECTIVENESS OF EXCESSIVE NODAL CONTINUITIES IN THE FINITE ELEMENT ANALYSIS OF THIN RECTANGULAR AND	70-0346
DUNCAN, W.	INCREMENTAL NUMERICAL SOLUTION OF A SIMPLE DEFORMATION PROBLEM IN SOIL MECHANICS	70-0380
DUNCAN, W.	THE EFFECTIVENESS OF EXCESSIVE NODAL CONTINUITIES IN THE FINITE ELEMENT ANALYSIS OF THIN RECTANGULAR	70-0105
DUNG, L.	POWER SERIES EXPANSION OF THE GENERAL STIFFNESS MATRIX FOR BEAM ELEMENTS	75-0543
DUNGAR, R.	THE ANALYSIS OF ARCH DAMS UNDER STATIC LOADING BY THE FINITE ELEMENT METHOD	68-0059
DUNGAR, R.	THE SEISMIC DESIGN STUDY OF A DOUBLE CURVATURE ARCH DAM	69-0053
DUNGAR, R.	TRIANGULAR FINITE-ELEMENTS OF VARIABLE THICKNESS AND THEIR APPLICATION TO PLATE AND SHELL PROBLEMS	69-0095

DUNGAR, R.	VIBRATIONS OF PLATE AND SHELL STRUCTURES USING TRIANGULAR FINITE ELEMENTS	67-0023
DUNGLAS, J.	NOUVELLE CONCEPTION DES DRAINS DANS LES BARRAGES EN TERRE HOMOGENE DE PETITE ET MOYENNE DIMENSIONS (	73-0404
DUNHAM, R. S.	A FINITE ELEMENT APPLICATION OF THE HELLINGER-REISSNER VARIATIONAL THEOREM	68-0060
DUNHAM, R. S.	BOTTOM STABILITY UNDER WATER INDUCED LOADING	72-0373
DUNHAM, R. S.	FINITE ELEMENT ANALYSIS OF AXISYMMETRIC SOLIDS WITH ARBITRARY LOADINGS	67-0145
DUNHAM, R. S.	ON APPLICATION OF THE FINITE ELEMENT METHOD TO LIMIT ANALYSIS	74-1237
DUNHAM, R. S.	ORGANIZATION AND FUNCTIONAL PURPOSE OF A FINITE ELEMENT COMPUTER PROGRAM	74-0362
DUNHAM, R. S.	SOME STRESS INTENSITY CALCULATIONS USING FINITE ELEMENTS	74-0423
DUNLAP, W. A.	NUMERICAL SOLUTIONS FOR DETERMINING WAVE-INDUCED PRESSURE DISTRIBUTIONS AROUND BURIED PIPELINES	74-1334
DUNLOP, P.	ANALYSES OF SLOPES BY THE FINITE ELEMENT METHOD	68-0297
DUNLOP, P.	DEVELOPMENT OF FAILURE AROUND EXCAVATED SLOPES	70-0051
DUNLOP, P.	FINITE ELEMENT ANALYSIS OF SLOPES IN SOIL	68-0061
DUNLOP, P.	SLOPES IN STIFF FISSURED CLAYS AND SHALES	69-0094
DUNN, J. C.	RE-ENTRY THERMAL ANALYSIS OF VARIABLE THICKNESS SPHERICAL VEHICLES.	72-0468
DUNN, R. J. JR.	AN ERROR EXPANSION FOR A FINITE-ELEMENT SOLUTION OF A TWO-POINT BOUNDARY-VALUE PROBLEM	75-0801
DUNN, S. F. JR.	A STUDY OF INTERNAL STRESSES IN STATICALLY DEFORMED PNEUMATIC TIRES. A MATHEMATICAL MODEL FOR THE PN	70-0421
DUNNE, P. C.	COMPLETE POLYNOMIAL DISPLACEMENT FIELDS FOR FINITE ELEMENT METHODS	68-0062
DUNNE, P. C.	DIE LOSUNG NICHT-LINEARER PROBLEME NACH DER METHODE DER FINITEN ELEMENTE	74-0673
DUNNE, P. C.	DYNAMIC RESPONSE BY LARGE STEP INTEGRATION	73-0567
DUNNE, P. C.	FINITE ELEMENT DIFFICULTIES IN NONLINEAR AND INCOMPRESSIBLE MATERIAL	74-1242
DUNNE, P. C.	LARGE NATURAL STRAINS AND SOME SPECIAL DIFFICULTIES DUE TO NONLINEARITY AND INCOMPRESSIBILITY IN FIN	74-0528
DUNNE, P. C.	LINEAR AND NONLINEAR OSCILLATIONS IN MECHANICS	74-0677
DUNNE, P. C.	NONLINEAR OSCILLATIONS USING THE FINITE ELEMENT TECHNIQUE	73-0568
DUNNE, P. C.	SHAPE DETERMINATION OF PREFABRICATED PRESTRESSED NETWORKS	73-0889
DUNNE, P. C.	SOME CONTRIBUTIONS TO NONLINEAR SOLID MECHANICS	73-0705
DUNNE, P. C.	SOME CONTRIBUTIONS TO NONLINEAR SOLID MECHANICS	74-0674
DUNNE, P. C.	THE FINITE ELEMENT METHOD APPLIED TO FLUID MECHANICS	74-0676
DUNNE, P.C.	LARGE NATURAL STRAINS AND SOME SPECIAL DIFFICULTIES DUE TO NONLINEARITY AND INCOMPRESSIBILITY IN FIN	74-1027
DUNPHEY, F. P.	PARAMETRIC STUDIES OF LAYERED PAVEMENT SYSTEMS USING AFPAV CODE WITH FAST EQUATION SOLVER	75-0747
DUNS, C. S.	DYNAMIC ANALYSIS OF SOIL - STRUCTURE SYSTEM USING FINITE ELEMENT METHOD	67-0001
DUNS, C. S.	VERTICAL RESPONSE OF A RIGID BASE TO DYNAMIC LOADING	69-0316
DUPONT, T.	A FINITE ELEMENT COLLOCATION METHOD FOR THE HEAT EQUATION	72-0488
DUPONT, T.	A FINITE ELEMENT COLLOCATION METHOD FOR QUASILINEAR PARABOLIC EQUATIONS	73-1000
DUPONT, T.	ADI METHODS FOR FINITE ELEMENT SYSTEMS	71-0652
DUPONT, T.	FINITE ELEMENT COLLOCATION METHOD FOR QUASILINEAR PARABOLIC EQUATIONS	73-0047
DUPONT, T.	GALERKIN METHODS FOR PARABOLIC EQUATIONS	70-0448
DUPONT, T.	GALERKIN METHODS FOR PARABOLIC EQUATIONS WITH NONLINEAR BOUNDARY CONDITIONS	73-1023
DUPONT, T.	MODELING WAVE PROPAGATION WITH FINITE ELEMENT METHODS	75-0015
DUPONT, T.	SOME L(2) ERROR ESTIMATES FOR PARABOLIC GALERKIN METHODS	72-0511
DUPONT, T.	SOME SUPERCONVERGENCE RESULTS FOR AN H-GALERKIN PROCEDURE FOR THE HEAT EQUATION	74-0411
DUPONT, T.	STABILITY IN LQ OF L2-PROJECTION INTO FINITE ELEMENT FUNCTION SPACES	75-0016
DUPREE, D. M.	OPTIM 11: A (MAGIC) COMPATIBLE LARGE SCALE AUTOMATED MINIMUM WEIGHT DESIGN PROGRAM - 1. ENGINEERS AN	74-1048
DUPREE, D. M.	OPTIM 11: A MAGIC COMPATIBLE LARGE-SCALE AUTOMATED MINIMUM WEIGHT DESIGN PROGRAM. VOLUME 1. ENGINEER	74-1096
DUPREE, D. M.	STRESS ANALYSIS OF COLDWORKED FASTENER HOLES	74-0783
DUPUIS, G.	A CURVED FINITE ELEMENT FOR THIN ELASTIC SHELLS	69-0486
DUPUIS, G.	APPLICATION OF RITZ'S METHOD TO THIN ELASTIC SHELL ANALYSIS	71-0083
DUPUIS, G.	CURVED FINITE ELEMENT FOR THIN ELASTIC SHELLS	70-0173
DUPUIS, G.	ELEMENTS FINIS RAFFINES EN ELASTICITE BIDIMENSIONNELLE	69-0411
DUPUIS, G.	FINITE ELEMENT WITH A HIGH DEGREE OF REGULARITY	70-0052
DUPUIS, G.	REFINED FINITE ELEMENTS IN 2-DIMENSIONAL ELASTICITY	69-0320
DUPUIS, G. A.	INCREMENTAL FINITE ELEMENT ANALYSIS OF LARGE ELASTIC DEFORMATION PROBLEMS	71-0753
DUPUIS, G. A.	NONLINEAR MATERIAL AND GEOMETRIC BEHAVIOUR OF SHELL STRUCTURES	71-0084
DURBIN, C. F.	APPLICATION, TESTING, AND EVALUATION OF DOUBLY CURVED SHELL ELEMENTS FOR DYNAMIC ANALYSIS. PART III.	69-0256
DURFEE, R. L.	LEAKAGE THROUGH CRACKS IN LNG TANKAGE	74-0749
DUTTON, R. D.	APPLICATION OF NASTRAN TO THE ANALYSIS OF A SPACE SHUTTLE ORBITER STRUCTURE	71-0099
DUVALL, F. W.	FINITE ELEMENT STRUCTURAL ANALYSIS AS APPLIED TO AN AUTOMOTIVE DOOR STRUCTURE	74-0310
DUVALL, W. I.	THE USE OF ROCK BOLTS OR WIRE ROPE TO INCREASE THE STRENGTH OF FRACTURED MODEL PILLARS	71-0722
DVORAK, G. J.	PLASTICITY THEORY OF FIBROUS COMPOSITES - PART 11	73-0750
DVORAK, G. J.	PLASTICITY THEORY OF FIBROUS COMPOSITES - PART 1	73-0751
DVORAK, G. J.	YIELDING IN UNIDIRECTIONAL COMPOSITES UNDER EXTERNAL LOADS AND TEMPERATURE CHANGES.	73-0169
DWYER, M. G.	FINITE ELEMENT ANALYSES OF SALT DOMES WITH STORED HOT WASTES	74-1148
DWYER, W. J.	AN IMPROVED AUTOMATED STRUCTURAL OPTIMIZATION PROGRAM	74-0732
DWYER, W. J.	FINITE ELEMENT MODELING AND OPTIMIZATION OF AEROSPACE STRUCTURES	72-0416
DWYER, W. J.	FINITE ELEMENT MODELING AND OPTIMIZATION OF AEROSPACE STRUCTURES	72-0734
DYER, D. F.	FLUID MECHANICS AND FINITE ELEMENTS	72-0707
EAGLE, J.	ANALYSIS OF LOADING AND THERMAL EFFECTS ON FUELLED GRAPHITE BRICKS FOR A HIGH-TEMPERATURE GAS COOLED	71-0085
EBEL, H.	DIE BEULLASTEN DIAGONAL VERSTEIFTER RECHTECKPLATTEN (BUCKLING LOADS IN DIAGONALLY STIFFENED RECTANGU	73-0405
EBERHARDT, A. C	A FINITE ELEMENT APPROACH TO THE DYNAMIC ANALYSIS OF CONTINUOUS HIGHWAY BRIDGES	72-0591
EBERHARDT, A. C	AIRCRAFT-PAVEMENT INTERACTION STUDIES, PHASE 1: A FINITE ELEMENT MODEL OF A JOINTED CONCRETE PAVEMEN	73-0894
EBERT, L. J.	AN ANALYTIC METHOD FOR PREDICTING THE FLOW AND FRACTURE IN NOTCHED FIBER COMPOSITE MATERIALS	75-0689
EBERT, L. J.	FINITE ELEMENT ANALYSIS SYSTEM FOR THE MECHANICAL BEHAVIOR OF ORIENTED FIBER COMPOSITE MATERIALS UND	73-0869
EBNER, A. M.	A THEORETICAL AND NUMERICAL COMPARISON OF ELASTIC NONLINEAR FINITE ELEMENT METHODS	72-0709
EBNER, A. M.	THEORETICAL AND NUMERICAL COMPARISON OF ELASTIC NONLINEAR FINITE ELEMENT METHODS.	72-0452
ECER, A.	ANALYSIS OF THREE-DIMENSIONAL UNSTEADY VISCOUS FLOW AROUND OSCILLATING WINGS	74-0982
ECER, A.	APPLICATIONS OF FINITE ELEMENT METHOD IN UNSTEADY AERODYNAMICS	74-0105
ECER, A.	COMPUTATIONAL CONSIDERATIONS IN APPLICATION OF THE FINITE ELEMENT METHOD FOR ANALYSIS OF UNSTEADY FL	73-0068
ECER, A.	FINITE ELEMENT ANALYSIS OF POST-BUCKLING BEHAVIOR	71-0808
ECER, A.	FINITE ELEMENT ANALYSIS OF UNSTEADY INCOMPRESSIBLE FLOW AROUND AN OSCILLATING OBSTACLE OF ARBITRARY	73-0067
ECER, A.	FINITE ELEMENT ANALYSIS OF POSTBUCKLING BEHAVIOR OF STRUCTURES	73-0099
ECER, A.	FINITE ELEMENT ANALYSIS AND COMPUTER GRAPHICS VISUALIZATION OF FLOW AROUND PITCHING AND PLUNGING AIR	73-0385
ECER, A.	NUMERICAL CALCULATIONS OF VELOCITY AND PRESSURE DISTRIBUTION AROUND OSCILLATING AIRFOILS	74-0847
ECER, A.	SENSITIVITY ANALYSIS OF TORSIONAL VIBRATION CHARACTERISTICS OF HELICOPTER ROTOR BLADES. PART 1: STRU	74-0846
ECER, A.	SENSITIVITY OF ROTOR BLADE VIBRATION CHARACTERISTICS TO TORSIONAL OSCILLATIONS	74-0611
ECER, A.	SUITABILITY OF THE FINITE ELEMENT METHOD FOR ANALYSIS OF UNSTEADY FLOW AROUND OSCILLATING AIRFOILS	73-0069
ECK, D. L.	ECONOMICAL DESIGN ANALYSIS OF PISTONS FOR INTERNAL COMBUSTION ENGINES	73-0188
ECKHART, J. B.	DISTRIBUTION OF THE CLAMPING PRESSURE IN FRICTION-GRIP BOLTED JOINTS	74-0250
EDGE, F. W.	TWO-DIMENSIONAL AUTOMATIC MESH GENERATION FOR STRUCTURAL ANALYSIS	70-0058
EDWARDS, A. D.	ANALYSIS OF CONCRETE BOX GIRDERS BEFORE AND AFTER CRACKING	72-0247
EDWARDS, B. A.	CHARACTERIZATION OF ELASTIC SOLIDS USING FINITE ELEMENT METHODS	72-0712
EDWARDS, G.	HYBRID PLATE BENDING ELEMENTS	72-0792
EFIMOV, Y. N.	REALIZATSIYA METODA KONECHNYKH ELEMENTOV NA EVM DLYA RASCHETA DVUMERNYKH ELEKTRICHESKIKH I MAGNITNYK	74-0251
EGEBERG, J. L.	MESHGEN, A COMPUTER CODE FOR AUTOMATIC FINITE ELEMENT MESH GENERATION	69-0096
EGELAND, O.	APPLICATION OF FINITE ELEMENT TECHNIQUES TO PLASTICITY PROBLEMS	69-0097
EGELAND, O.	GENERAL DESCRIPTION OF SESAM-69, SUPER ELEMENT STRUCTURAL ANALYSIS (PROGRAM) MODULES	71-0612
EGELAND, O.	PROGRAM SPECIFICATION NV 300. GENERAL FINITE ELEMENT PROGRAM FOR 2- AND 3-DIMENSIONAL MEMBRANE STRUC	68-0184
EGELAND, O.	SESAM-69 - A GENERAL PURPOSE FINITE ELEMENT METHOD PROGRAM	74-0183
EGGERS, H.	CONTINUUM CALCULATION FROM ELASTIC PLASTIC REINFORCED MATERIAL USING A FINITE ELEMENT METHOD (DIE BE	72-0658
EGGERS, H.	VARIATIONAL PRINCIPLES FOR ELASTOPLASTIC CONTINUA	75-0837
EHLE, P. E.	DEVELOPMENT OF A FINITE ELEMENT APPROACH FOR APPROXIMATE ANALYSIS OF UNSTEADY COMPRESSIBLE FLUID FLO	72-0828
EHLERS, K. D.	BERECHNUNG INSTATIONARER GRUND - UND SICKERWASSERSTROMUNGEN MIT FREIER OBERFLACHE NACH DER METHODE F	71-0541
EHRLICH, L. W.	SOME DIFFERENCE SCHEMES FOR THE BIHARMONIC EQUATION	75-0204
EIBL, J.	FINITE ELEMENTS IN CURVILINEAR COORDINATES - SECOND ORDER SURFACES	73-0753

EIDE, G. R.	MATRIX DIFFERENCE EQUATION ANALYSIS OF VIBRATING PERIODIC STRUCTURES	75-0151
EIDE, G. R.	THE MATRIX DEFORMATION METHOD OF STRUCTURAL ANALYSIS	71-0079
EIDELBERG, B. E	APPLICATION OF FINITE ELEMENT METHODS TO LUBRICATION: SQUEEZE FILMS BETWEEN POROUS SURFACES	75-0764
EIDELBERG, B. E	FINITE ELEMENT ANALYSIS OF LUBRICATION IN NATURAL JOINTS	74-0086
EIDSON, R. L.	ANALYSIS OF VISCOELASTOPLASTIC STRUCTURAL BEHAVIOUR OF ANISOTROPIC SHELLS BY THE FINITE ELEMENT METH	75-0056
EIDSON, R. L.	DYNAMIC ANALYSIS OF VISCOELASTOPLASTIC ANISOTROPIC SHELLS	73-0253
EIDSON, R. L.	DYNAMIC ELASTOPLASTIC RESPONSE OF GEOMETRICALLY NONLINEAR ARBITRARY SHELLS OF REVOLUTION UNDER IMPUL	70-0098
EIDSON, R. L.	STATIC AND DYNAMIC ANALYSIS OF VISCOELASTOPLASTIC FIBER-REINFORCED COMPOSITE SHELLS IN MISSILE STRUC	73-0651
EIGENHEER, LUIZ	EMBANKMENT DAM FOR THE MARIMBONDO HYDRO PROJECT	75-0270
EISENBERG, M. A	ON FINITE ELEMENT INTEGRATION IN NATURAL COORDINATES	73-0683
EISENSTAT, S. C	COMPUTATIONAL ASPECTS OF THE FINITE ELEMENT METHOD	72-0512
EISENSTEIN, Z.	FINITE ELEMENTS METHOD IN SOIL MECHANICS (METODA KONECNYCH PRVKU V MECHANICE SEMIN).	71-0468
EL NIMEIRI, M.	LARGE DEFLECTION SPATIAL BUCKLING OF THIN WALLED BEAMS AND FRAMES	73-1085
EL-ADAWY NASSEF	MODIFIED APPROACH FOR ESTIMATING THE CRACKING MOMENT OF REINFORCED CONCRETE BEAMS	75-0485
EL-GAALY, M. A.	FAILURE OF THIN-WALLED MEMBERS UNDER PATCH LOADINGS	72-0230
ELBATOUTI, A.	FINITE ELEMENT ANALYSIS OF CONTAINER SHIPS	72-0372
ELBATOUTI, A. M	STRUCTURAL ANALYSIS OF SL-7 CONTAINERSHIP UNDER COMBINED LOADING OF VERTICAL, LATERAL AND TORSIONAL	74-0734
ELDER, A. S.	STRESS ANALYSIS OF 175 MM PROJECTILE, HE, M437	71-0736
ELDRIDGE, C. M.	ANALYSIS OF SHELLS OF REVOLUTION BY A CURVED FINITE ELEMENT	75-0680
ELIAS, Z. M.	DUALITY IN FINITE ELEMENT METHODS	67-0143
ELIAS, Z. M.	DUALITY IN FINITE ELEMENTS METHODS	68-0065
ELIAS, Z. M.	DYNAMIC ANALYSIS OF FRAME STRUCTURES BY THE FORCE METHOD	72-0313
ELIAS, Z. M.	MIXED FINITE ELEMENT METHOD FOR AXISYMMETRIC SHELLS	72-0149
ELIAS, Z. M.	VARIATIONAL FORMULATIONS AND FINITE ELEMENT METHOD FOR THIN ELASTIC SHELLS AND PLATES	68-0064
ELLINGTON, J. P	SIMPLE TRIANGULAR MEMBRANE FINITE ELEMENT	74-0881
ELLIOTT, G. L.	CALCULATION OF NATURAL FREQUENCIES OF HEAT EXCHANGER TUBES	73-0665
ELLIOTT, G. L.	PARAMETRIC ANALYSIS OF CONTACT CHATTER IN RELAYS	74-0487
ELLIS, J. A.	STRESS AND STRAIN DISTRIBUTION IN THE VICINITY OF INTERFERENCE FIT FASTENERS	73-0363
ELMEGED, A. B.	MACRO-APPROACH FOR RIBBED AND GRID PLATE SYSTEMS	75-0079
ELMS, D. G.	NON-LINEAR ANALYSIS OF REINFORCED CONCRETE SLABS	72-0191
ELNIMEIRI, M.	FINITE-ELEMENT FOR BUCKLING OF CURVED BEAMS AND SHELLS WITH SHEAR	75-0607
ELSAIE, A. M.	VIBRATION OF SQUARE PLATES HAVING VARIOUS EDGE CONDITIONS AND LOADINGS USING A REFINED FINITE ELEMEN	75-0024
ELSTE, P.	SCHWINGUNGSMAESSIGE AUSLEGUNG DER BESCHAUFELUNG AXIALER TURBOMASCHINEN (DESIGN OF DISPLACEMENT-FREE	74-0253
ELTAWILA, F. M.	AN OPTIMUM IDEALIZATION IN FINITE ELEMENT METHOD FOR A TWO-GROUP NEUTRON DIFFUSION PROBLEM	74-1143
ELTER, C.	STRESS ANALYSIS OF BWR PRESSURE VESSEL WITH BERSAFE AND FLHE	73-0406
EMERY, A.	CALCULATION OF HARMONIC FIELDS IN WHICH SINGULARITIES ARE PRESENT USING FINITE ELEMENT AND FINITE DI	73-0979
EMERY, A. F.	EVALUATION OF THE USE OF FINITE ELEMENT METHOD IN COMPUTING TEMPERATURE	69-0017
EMERY, A. F.	EVALUATION OF USE OF THE FINITE-ELEMENT METHOD IN COMPUTATION OF TEMPERATURE	71-0467
EMERY, A. F.	THERMAL-STRESS CONCENTRATION CAUSED BY STRUCTURAL DISCONTINUITIES	69-0308
EMERY, A. F.	USE OF SINGULARITY PROGRAMMING IN FINITE-DIFFERENCE AND FINITE ELEMENT COMPUTATIONS OF TEMPERATURE	73-0100
EMERY, A. F.	USE OF SINGULARITY PROGRAMMING IN FINITE ELEMENT CALCULATIONS OF ELASTIC STRESS INTENSITY FACTORS. P	75-0252
EMERY, A. F.	VISCOELASTIC PRESPONSE OF THE CORNEO-SCLERAL SHELL UNDER TONOMETER LOADING	75-0855
EMERY, J. J.	FINITE ELEMENT ANALYSIS OF CREEP PROBLEMS IN SOIL MECHANICS	71-0809
EMORI, R. I.	STUDY ON AUTOMOBILE CRASH WORTHINESS	70-0217
ENDA, Y.	ANALYSIS OF THIN-WALLED CURVED BEAMS BY THE TRANSFER MATRIX METHOD	72-0335
ENDERBY, L. R.	FINITE-ELEMENT STRESS ANALYSIS OF A MULTIPIECE PISTON	69-0274
ENDERBY, L. R.	FINITE-ELEMENT STRESS ANALYSIS OF A NON-LINEAR PROBLEM. A CONNECTING-ROD EYE LOADED BY MEANS OF A PI	70-0239
ENGLAND, G. L.	THE CALCULATION OF TIME-DEPENDENT STRESSES IN PCPVS BY AN APPROXIMATE METHOD	71-0086
ENGLAND, G. L.	TWO-DIMENSIONAL TIME-DEPENDENT ANALYSIS OF PERFORATED END CAPS FOR NUCLEAR REACTOR PRESSURE VESSELS	71-0239
ENGLAND, R.	FINITE-ELEMENT CODE TO SIMULATE ANISOTROPIC-PLASMA DIFFUSION IN LINEAR MULTIPOLES	75-0400
ENGLAND, R.	SEMIDISCRETE GALERKIN TECHNIQUES WITH TIME INTERPOLATION AND SPLITTING UP FOR PLASMA SIMULATION	75-0121
ENGLAND, W. G.	CALCULATION OF TEMPERATURE DISTRIBUTION WITHIN A MELTING-FREEZING MATERIAL USING FINITE ELEMENT TECH	72-0043
ENGRAND, D.	MATRIX ANALYSIS OF MULTILAYERED AND SANDWICH SHELLS BY FINITE ELEMENT METHOD	73-0101
EPPINK, R. T.	ACCURACY OF FINITE ELEMENT APPROXIMATIONS TO STRUCTURAL PROBLEMS	70-0438
EPSTEIN, H. I.	NONLINEAR ANALYSIS OF RETICULATED SPACE TRUSSES	75-0287
EPSTEIN, H. I.	SNAP-THROUGH BUCKLING OF RETICULATED SHELLS	75-0640
ERGATOUDIS, I.	CURVED ISOPARAMETRIC QUADRILATERAL ELEMENTS FOR FINITE ANALYSIS	68-0008
ERGATOUDIS, J.	FINITE ELEMENT METHOD IN STRESS ANALYSIS	69-0413
ERGATOUDIS, J.	ISO-PARAMETRIC AND ASSOCIATED ELEMENT FAMILIES FOR TWO AND THREE DIMENSIONAL ANALYSIS	69-0285
ERGATOUDIS, J.	ISOPARAMETRIC FINITE ELEMENTS IN TWO AND THREE-DIMENSIONAL ANALYSIS	68-0237
ERGATOUDIS, J.	QUADRILATERAL ELEMENTS IN PLANE ANALYSIS	66-0085
ERGATOUDIS, J.	THREE-DIMENSIONAL STRESS ANALYSIS OF ARCH DAMS BY THE FINITE ELEMENT METHOD: PART I, ARCH DAM NO. 1.	66-0099
ERGATOUDIS, J.	COMPLETE POLYNOMIAL DISPLACEMENT FIELDS FOR FINITE ELEMENT METHODS	68-0092
ERGATOUDIS, J.	CURVED ISOPARAMETRIC QUADRILATERAL ELEMENTS FOR FINITE ELEMENT ANALYSIS	68-0067
ERGATOUDIS, J.	THREE DIMENSIONAL ANALYSIS OF ARCH DAMS AND THEIR FOUNDATIONS	68-0066
ERICKSON, L. B.	NASTRAN ANALYSIS OF A TURBINE BLADE AND COMPARISON WITH TEST AND FIELD DATA	75-0367
ERICSON, W. A.	AN INVESTIGATION OF INITIAL YIELD SURFACES FOR UNIDIRECTIONAL REINFORCED COMPOSITES	72-0837
ESCAPID, V.	INFLUENCE OF CONSTRUCTION PROCEDURES ON GROUND PRESSURES AROUND A TUNNEL	74-0997
ESSELINK, L.	USING INTERACTIVE GRAPHICS FOR THE PREPARATION AND MANAGEMENT OF FINITE ELEMENT DATA	74-0176
ETTLES, C.	APPLICATION OF A FINITE ELEMENT METHOD TO EVALUATION OF OIL WHIRL CHARACTERISTICS	74-0590
ETTLES, C.	FINITE ELEMENT METHOD FOR THE CALCULATION OF LOCUS PATHS IN DYNAMICALLY LOADED BEARINGS	73-0487
ETTLES, C. M. M	MECHANISM OF PARALLEL SURFACE THRUST BEARING	75-0322
EVANS, D. J.	THE ANALYSIS AND APPLICATION OF SPARSE MATRIX ALGORITHMS IN THE FINITE ELEMENT METHOD	72-0079
EVANS, H. R.	A FINITE ELEMENT SOLUTION FOR FOLDED PLATE STRUCTURES	67-0072
EVANS, H. R.	EXPERIMENTAL AND THEORETICAL INVESTIGATION OF THE BEHAVIOR OF BOX GIRDERS CURVED IN PLAN	75-0659
EVANS, J. D.	INVESTIGATION OF THE FINITE ELEMENT METHOD FOR DETERMINING GREEN'S FUNCTIONS	75-0751
EVANS, W. T.	PREDICTION OF SLOW CRACK GROWTH BY A FINITE-ELEMENT METHOD	75-0416
EVENSEN, D. A.	LARGE STRAIN ELASTO-PLASTIC FINITE ELEMENT ANALYSIS	71-0121
EVENSEN, D. A.	PURE MOMENT LOADING OF AXISYMMETRIC FINITE ELEMENT MODELS	73-1007
EVENSEN, D. A.	VIBRATION PROBLEMS USING ISOPARAMETRIC SHELL ELEMENTS	72-0549
EVENSEN, H.	CALCULATION AND DESIGN OF JOINTS MADE FROM COMPOSITE MATERIALS	70-0156
EVER, A.	ANALYSIS OF MOVING BODY PROBLEMS IN AERODYNAMICS	74-0024
EVERETT, J.	BENTONITE PILES IN DURBAN	75-0797
EVERLING, G.	FORECASTING THE ROCK PRESSURE AND ITS EFFECT ON THE LONGWALL FACES AND ROADWAYS IN THE BITUMINOUS CO	74-0374
EVERLING, G.	VORAUSSAGE DES GEBIRGSDRUCKES UND SEINER AUSWIRKUNGEN AUF STREBEN UND STRECKEN IM STEINKOHLENBERGBAU	73-0679
EVERSTINE, G. C	A SURVEY OF PRE AND POSTPROCESSORS FOR NASTRAN	74-0714
EVERSTINE, G. C	THE BANDIT COMPUTER PROGRAM FOR THE REDUCTION OF MATRIX BANDWIDTH FOR NASTRAN	72-0481
EWANG, R.	SOIL-STRUCTURE INTERACTION ANALYSES FOR EVALUATING SEISMIC RESPONSE	74-0745
EWING, D. J. F.	RULES GOVERNING THE NUMBER OF NODES AND ELEMENTS IN A FINITE ELEMENT MESH	70-0053
EZZUDDIN, Z. Y.	NUMERICAL SOLUTIONS OF NONLINEAR PLASMA EQUATIONS BY THE FINITE-ELEMENT METHOD	75-0671
FABIAN, P. J.	DESIGNING BY FINITE ELEMENTS CATCHING ON IN AUTOMOTIVE ENGINEERING	74-0113
FACCIOLI, E.	A FINITE ELEMENT LINEAR PROGRAMMING METHOD FOR THE LIMIT ANALYSIS OF THIN PLATES	73-1064
FAHERTY, K. F.	AN ANALYSIS OF A REINFORCED AND A PRESTRESSED CONCRETE BEAM BY FINITE-ELEMENT METHOD	72-0940
FAIR, G. S.	STRUCTURE-MEDIUM INTERACTION AND DESIGN PROCEDURES STUDY. VOLUME IV. FEAT CODE USER'S GUIDE	69-0265
FAIRBAIRN, I. S	GRIP: AN INFORMATION SYSTEM FOR DESIGN AUTOMATION	73-0210
FAIRCLOTH, J. M	FINITE-ELEMENT ANALYSIS OF AN ELECTRODE-NIPPLE JOINT	75-0843
FAIRHURST, C.	COMPUTER SIMULATION OF LAMINATED ROOF REINFORCED WITH GROUTED BOLTS	73-1045
FAIRHURST, C.	ESTIMATION OF THE MECHANICAL PROPERTIES OF ROCK MASSES	73-0935
FAIRHURST, C. F	ANALYSIS OF ROCK MASS DEFORMATIONS DUE TO EXCAVATIONS	73-0395
FAIRWEATHER, G.	RICHARDSON EXTRAPOLATION FOR PARABOLIC GALERKIN METHODS	72-0522
FAIZ, A.	PAVEMENT DESIGN AND CONTINUOUSLY REINFORCED CONCRETE PAVEMENT PERFORMANCE	73-0749
FAKHROD, E. A.	FINITE ELEMENT SOLUTION OF HEAT TRANSMISSION IN STEEL INGOTS	72-0311

33

FALBY, W. E.	MATRIX ANALYSIS METHODS FOR INELASTIC STRUCTURES	65-0051
FALBY, W. E.	MATRIX ANALYSIS METHODS FOR ANISOTROPIC INELASTIC STRUCTURES	66-0104
FALCONER, R.	STARDYNE STRUCTURAL ANALYSIS SYSTEM, ANALYTICAL CAPABILITY MANUAL	68-0235
FALK, S.	DAS VERFAHREN VON RAYLEIGH-RITZ MIT HERMITESCHEN INTERPOLATIONSPOLYNOMEN	63-0018
FAM, A.	FINITE ELEMENT SCHEME FOR BOX BRIDGE ANALYSIS	75-0666
FAM, A. R. M.	ANALYSIS OF CURVED BOX GIRDER BRIDGES	73-0294
FANELLI, M.	FINITE ELEMENT ANALYSIS OF PRESTRESSED CONCRETE PRESSURE VESSELS	74-1289
FANELLI, M.	FINITE ELEMENT STUDY OF THE TRIAXIAL STRESS STATE AROUND AN INSPECTION TUNNEL IN AN ARCH DAM	74-0831
FANELLI, M.	IL METODO DEGLI ELEMENTI FINITI: POSSIBILITA DI APPLICAZIONE A PROBLEMI DI INTERESSE DEGLI ELETTROTE	75-0200
FANG, C. S.	GROUNDWATER FLOW IN A SANDY TIDAL BEACH: II. 2-DIMENSIONAL FINITE ELEMENT ANALYSIS	72-0300
FANG, C. S.	GROUNDWATER FLOWN IN A SANDY TIDAL BEACH .1. ONE-DIMENSIONAL FINITE ELEMENT ANALYSIS	71-0567
FANG, C. S.	INVESTIGATION OF THE WATER TABLE IN A TIDAL BEACH	71-0648
FARAH, J. W.	FINITE ELEMENT STRESS ANALYSIS OF A RESTORED AXISYMMETRIC FIRST MOLAR	73-0053
FARAH, J. W.	PHOTOELASTIC AND FINITE ELEMENT STRESS ANALYSIS OF A RESTORED AXISYMMETRIC FIRST MOLAR	73-0142
FARAH, J. W.	STRESS ANALYSIS OF FIRST MOLARS WITH FULL CROWN PREPARATIONS BY THREE-DIMENSIONAL PHOTOELASTICITY AND	72-0939
FARGETTE, F.	PRACTICAL METHOD FOR COMPUTER CALCULATION OF STRUCTURES	69-0098
FARHI, L. E.	A MODEL STUDY OF GAS DIFFUSION IN ALVEOLAR SACS	73-0029
FAPHOOMAND, I.	A NONLINEAR FINITE ELEMENT CODE FOR ANALYZING THE BLAST RESPONSE OF UNDERGROUND STRUCTURES	70-0054
FARHOOMAND, I.	NONLINEAR DYNAMIC ANALYSIS OF COMPLEX STRUCTURES	73-0193
FARNUM, P.	COMPUTER MODELLING OF PLUG SEEDLING SURVIVAL	75-0176
FARRADAY, R. V.	GALERKIN FINITE ELEMENT SOLUTIONS FOR POLLUTION PROBLEMS IN PARTIALLY MIXED ESTUARIES	74-0047
FARRADAY, R. V.	NUMERICAL ANALYSIS OF PLASTICITY IN SOILS	73-0611
FARRADAY, R. V.	RAYLEIGH-RITZ AND GALERKIN FINITE ELEMENTS FOR DIFFUSION-CONVECTION PROBLEMS	73-0017
FARRELL, J. J.	EFFECTS OF RISE TIME AND DAMPING ON FINITE ELEMENT ANALYSIS OF RESPONSE OF STRUCTURES	71-0087
FARRIS, R. J.	DEVELOPMENT OF A SOLID ROCKET PROPELLANT NONLINEAR CONSTITUTTIVE THEORY	75-0722
FARVOLDEN, R. N	SENSITIVITY ANALYSIS OF INPUT PARAMETERS IN NUMERICAL MODELING OF STEADY STATE REGIONAL GROUNDWATER	74-0926
FARZIN, M. H.	EVALUATION OF STRESS CELL PERFORMANCE	74-0664
FAULKES, K. A.	A FINITE ELEMENT PLATE BENDING PROGRAMME	72-0579
FAUSAK, L. E.	INVESTIGATION OF THE WATER TABLE IN A TIDAL BEACH	71-0648
FAUST, G.	FINITE ELEMENT ULTIMATE LOAD ANALYSIS OF THREE DIMENSIONAL CONCRETE STRUCTURES	74-0672
FAUST, G.	FINITE ELEMENTE ZUR BERECHNUNG VON SPANNBETONREAKTORDRUCKBEHALTERN	73-0702
FAUST, G.	FINITE ELEMENTS FOR CALCULATION OF PRESTRESSED CONCRETE REACTOR PRESSURE VESSELS (FINITE ELEMENTE ZU	73-0797
FAUST, G.	FINITE ELEMENTS METHOD IN DESIGN OF PRESTRESSED REACTOR CONCRETE CONSTRUCTION	73-0550
FAUST, G.	RECENT DEVELOPMENTS IN FINITE ELEMENT ANALYSIS OF PRESTRESSED CONCRETE REACTOR VESSELS	74-0469
FAUST, G.	RECENT DEVELOPMENTS IN THE FINITE ELEMENT ANALYSIS OF PRE-STRESSED CONCRETE REACTOR VESSELS	74-0582
FAWCETT, D. J.	GENERATION OF FINITE ELEMENT MODELS VIA COMPUTER GRAPHICS	74-0194
FAWKES, A. J.	RULES GOVERNING THE NUMBER OF NODES AND ELEMENTS IN A FINITE ELEMENT MESH	70-0053
FAWKES, A. J.	SOME NONCONFORMING VARIANTS ON THE TRILINEAR ISOPARAMETRIC BRICK FOR LINEAR ELASTIC PROBLEMS	72-0531
FAZIO, P. P.	SANDWICH PLATE STRUCTURE ANALYSIS BY FINITE ELEMENT	74-0163
FEDDERSEN, C. E	EXPERIMENTAL AND THEORETICAL INVESTIGATION OF PLANE STRESS FRACTURE OF 2024 – T351 ALUMINUM ALLOY	70-0147
FEDDES, R. A.	FINITE ELEMENT ANALYSIS OF 2-DIMENSIONAL FLOW IN SOILS CONSIDERING WATER UPTAKE BY ROOTS .1. THEORY	75-0137
FEDDES, R. A.	FINITE ELEMENT ANALYSIS OF 2-DIMENSIONAL FLOW IN SOILS CONSIDERING WATER UPTAKE BY ROOTS .2. FIELD A	75-0147
FEDOROVSKUU, V.	RIGID PLATE ON NONLINEARLY DEFORMABLE COHESIVE BASE (PLANE PROBLEM)	75-0264
FEESER, L. J.	RESPONSE OF RECTANGULAR PLATES TO MOVING LOADS BY A FINITE ELEMENT PROCEDURE	69-0515
FEESER, L. J.	VIBRATION OF FOUR POINT-SUPPORTED PLATES BY A FINITE ELEMENT METHOD	71-0592
FEIJOO, R.	VARIATIONAL PRINCIPLE FOR THE LAPLACE'S OPERATOR WITH APPLICATION IN THE TORSION OF COMPOSITE RODS	74-1044
FEIKEN, R. W.	FLEXURE OF A BEAM SUPPORTED ON AN ELASTIC FOUNDATION	72-0871
FELIPPA, C. A.	A REFINED QUADRILATERAL ELEMENT FOR ANALYSIS OF PLATE BENDING	68-0047
FELIPPA, C. A.	AN ALPHANUMERIC FINITE ELEMENT MESH PLOTTER	72-0543
FELIPPA, C. A.	ANALYSIS OF ELEMENT STIFFNESS	67-0117
FELIPPA, C. A.	BASIS FOR FORMULATION OF FINITE ELEMENT MODELS	72-0273
FELIPPA, C. A.	COMPUTATIONAL TECHNIQUES IN FINITE ELEMENT ANALYSIS OF STRUCTURAL PROBLEMS	71-0556
FELIPPA, C. A.	COMPUTER IMPLEMENTATION OF NONLINEAR FINITE ELEMENT ANALYSIS	73-0407
FELIPPA, C. A.	DATA-STRUCTURES IN FINITE ELEMENT ANALYSIS	74-0114
FELIPPA, C. A.	FINITE ELEMENT ANALYSIS OF THREE-DIMENSIONAL CABLE STRUCTURES	74-1275
FELIPPA, C. A.	FINITE ELEMENT AND FINITE DIFFERENCE ENERGY TECHNIQUES FOR THE NUMERICAL SOLUTION OF PARTIAL DIFFERE	73-0570
FELIPPA, C. A.	INCREMENTAL FINITE ELEMENT MATRICES	74-0556
FELIPPA, C. A.	PROBLEM CONTROLLED GRID GENERATION FOR THE NUMERICAL SOLUTION OF PARTIAL DIFFERENTIAL EQUATIONS	75-0725
FELIPPA, C. A.	REFINED FINITE ELEMENT ANALYSIS OF LINEAR AND NONLINEAR TWO-DIMENSIONAL STRUCTURES	66-0024
FELIPPA, C. A.	SOLUTION OF LINEAR EQUATIONS WITH SKYLINE-STORED SYMMETRIC MATRIX	75-0662
FELIPPA, C. A.	THE FINITE ELEMENT METHOD IN SOLID MECHANICS	70-0295
FELIPPA, C. A.	VARIABLE GRID FINITE DIFFERENCE-ELEMENT SOLUTION OF ELLIPTIC PARTIAL DIFFERENTIAL EQUATIONS	74-0862
FENG, C. C.	APPLICATION OF FLOWGRAPHS TO FREE VIBRATION OF STRUCTURES	68-0196
FENG, G. C.	DYNAMICS OF A FLEXIBLE BULKHEAD AND CONTAINED FLUID	73-0082
FENG, G. C.	STUDY OF PROPELLANT DYNAMICS IN A SHUTTLE TYPE LAUNCH VEHICLE	72-0699
FENG, M.	TRANSVERSE LOADING OF UNIDIRECTIONAL FIBER COMPOSITES	69-0119
FENNER, D.N.	DUGDALE MODEL SOLUTIONS FOR A SINGLE EDGE CRACKED PLATE	74-0221
FENNER, R. T.	FINITE ELEMENT ANALYSIS OF TWO-DIMENSIONAL SLOW NON-NEWTONIAN FLOWS	72-0035
FENNER, R. T.	FINITE ELEMENT ANALYSIS OF SLOW NON-NEWTONIAN CHANNEL FLOW	72-0308
FENSEL, P. A.	AN AXISYMMETRIC FINITE ELEMENT ANALYSIS OF THE MECHANICAL AND THERMAL STRESSES IN BRAKE DRUMS	74-0311
FENTON, D. L.	ECONOMICAL SOLUTION TECHNIQUES FOR LOAD DEFLECTION EQUATIONS	74-0414
FENTON, D. L.	THE FINITE ELEMENT METHOD - A BIBLIOGRAPHY OF ITS THEORY AND APPLICATIONS	71-0510
FENTON, P. H.	STRUCTURAL ANALYSIS AND DESIGN OF A CATAMARAN CROSS-STRUCTURE BY FINITE ELEMENT METHOD	73-0052
FENTON, R. G.	APPLICATION OF A REFINED PLATE BENDING ELEMENT TO BUCKLING PROBLEMS	74-0937
FENTON, R. G.	FINITE ELEMENT SOLUTION OF EXTRUSION OF ELASTO-PLASTIC WORK HARDENING MATERIALS	73-1127
FENTON, R. G.	VIBRATION OF SQUARE PLATES HAVING VARIOUS EDGE CONDITIONS AND LOADINGS USING A REFINED FINITE ELEMEN	75-0024
FERRANTE, A. J.	ICES-STRUDL-11	68-0257
FERRARA, S.	SOME REMARKS ON INFINITE VENEZIANO REPRESENTATIONS	69-0472
FERREIRA, M. J.	COMPARISON OF ANALYTICAL AND EXPERIMENTAL RESULTS IN THE DESIGN OF A HOLLOW GRAVITY DAM	67-0109
FERREIRA, S.	TRANSIENT NATURAL CONVECTION COOLING OF A VERTICAL CIRCULAR CYLINDER	74-1169
FERRELL, C. S.	DYNAMIC BEHAVIOR OF ECCENTRICALLY STIFFENED PLATES	71-0796
FERRING, M.	THE FINITE ELEMENT METHOD APPLIED TO COMPRESSIBLE CASCADE FLOW	75-0179
FERRISS, D. H.	SOLUTION OF MIXED BOUNDARY VALUE PROBLEMS IN FINITE DOMAINS BY FOURIERS METHOD	72-0790
FERRITTO, J. M.	DYNAMIC RESPONSE OF A CYLINDER BURIED IN AN EARTH BEAM - RESULTS OF A FINITE ELEMENT ANALYSIS	72-0592
FESSLER, H.	PREDICTION OF THE CREEP BEHAVIOUR OF A FLANGED JOINT	74-0757
FETTE, H.	CURVED FINITE ELEMENTS IN SHELL CALCULATIONS	70-0227
FETTE, H.	FINITE ELEMENTE FUR GEKRUEMMTE FLAECHENTRAGWERKE. (FINITE ELEMENTS SOLUTIONS FOR CURVED SHELL STRUCT	71-0391
FILCEK, H.	DER SPANNUNGS UND BEANSPRUCHUNGSZUSTAND DES BODENMATERIALS IN DER NACHBARSCHAFT EINER BODENBOESCHUNG	75-0371
FILSTRUP, A. W.	FINITE ELEMENT ANALYSIS OF A GAS TURBINE BLADE	74-0971
FINE, D. S.	STIFFNESS ANALYSIS OF SHEET METAL SHELLS UNDER CONCENTRATED LOADS	74-0325
FINE, J.	EFFECT OF SHAPE AND HETEROGENEITY OF ROCK SAMPLES ON THEIR COMPRESSIVE STRENGTH	69-0315
FINLAYSON, B. A	ORTHOGONAL COLLOCATION ON FINITE ELEMENTS	75-0364
FINLAYSON, B. A	WEIGHTED RESIDUAL METHODS AND THEIR RELATION TO FINITE ELEMENT METHODS IN FLOW PROBLEMS	74-0003
FINN, W. D.	DYNAMIC RESPONSE OF EARTH DAMS	67-0024
FINN, W. D.	DYNAMICS OF GRAVITY DAM-RESERVOIR SYSTEMS	73-0225
FINN, W. D.	FINITE-ELEMENT ANALYSIS OF SEEPAGE THROUGH DAMS	67-0025
FINN, W. D.	STATIC AND DYNAMIC STRESSES IN SLOPES	66-0025
FINN, W. D.	STATIC AND SEISMIC BEHAVIOUR OF AN EARTH DAM	67-0026
FINN, W. D. L.	FINITE ELEMENT ANALYSIS OF SEEPAGE	67-0130
FIRKINS, N. L.	STEEL PLATE ANALYSIS BY FINITE ELEMENTS	69-0100
FIRMIN, A.	ANALYSIS OF COOLING TOWERS BY THE MATRIX FINITE ELEMENT METHOD- 2	70-0148
FISCHER, D.	NEUESTE ERFAHRUNGEN BEI DER RECHNERISCHEN UND KONSTRUKTIVEN AUSLEGUNG KOMPLIZIERTER SCHALENSYSTEME	71-0052

FISCHER, F. D.	ZUR LOESUNG DES KONTAKTPROBLEMS ELASTISCHER KOERPER MIT AUSGEDEHNTER KONTAKTFLAECHE DURCH QUADRATISC	74-1135
FISCHL, C. F.	FLOW BETWEEN LINES OF A GIVEN PRESSURE	74-0020
FISHER, C. R.	FORMULATION AND EXPERIMENTAL VERIFICATION OF AN AXISYMMETRIC FINITE-ELEMENT STRUCTURAL ANALYSIS	71-0354
FISHER, I.	THE MULTIDIMENSIONAL STEFAN PROBLEM: A FINITE ELEMENT APPROACH	74-0464
FISHER, S. M.	HARBOUR OSCILLATION - A NUMERICAL TREATMENT FOR UNDAMPED NATURAL MODES	70-0208
FISHKAA, G.	COMPUTER CALCULATIONS OF STRESSES IN AXISYMMETRIC THERMALLY LOADED COMPONENTS	68-0069
FISHMAN, H. M.	A THREE-DIMENSIONAL FINITE ELEMENT COMPUTER CODE FOR THE ANALYSIS OF COMPLEX STRUCTURES	71-0275
FITZGERALD, J.	HANDBOOK FOR THE ENGINEERING STRUCTURAL ANALYSIS OF SOLID PROPELLANTS	71-0642
FITZGERALD, J.	APPLICATION OF THE FAST FOURIER TRANSFORM TO LINEAR SYSTEMS IN CIVIL ENGINEERING	73-1119
FIX, G.	HYBRID FINITE-ELEMENT METHODS	75-0445
FIX, G. F.	AN ANALYSIS OF THE FINITE ELEMENT METHOD	73-0071
FIX, G. F.	FINITE ELEMENT APPROXIMATIONS TO TIME-DEPENDENT PROBLEMS	72-0269
FIX, G. J.	EFFECTS OF QUADRATURE ERRORS IN FINITE ELEMENT APPROXIMATION OF STEADY STATE, EIGENVALUE, AND PARABO	72-0343
FIX, G. J.	EIGENVALUE APPROXIMATION BY FINITE ELEMENT METHOD	73-0039
FIX, G. J.	FINITE ELEMENT APPROXIMATIONS TO PARABOLIC PROBLEMS	71-0448
FIX, G. J.	FINITE ELEMENT METHOD FOR NONLINEAR HYPERBOLIC EQUATIONS	74-1238
FIX, G. J.	FINITE ELEMENT MODELS FOR OCEAN CIRCULATION PROBLEMS	75-0206
FIX, G. J.	FINITE ELEMENTS AND SCIENTIFIC COMPUTING	75-0042
FIX, G. J.	FOURIER ANALYSIS OF THE FINITE ELEMENT METHOD IN RITZ-GALERKIN THEORY	69-0297
FIX, G. J.	HIGHER-ORDER LINEAR FINITE ELEMENT METHODS	74-0826
FIX, G. J.	ITERATIVE METHODS FOR FINITE ELEMENT SYSTEMS	70-0196
FIX, G. J.	ON FINITE ELEMENT APPROXIMATIONS ON TIME DEPENDENT PROBLEMS	72-0528
FIX, G. J.	ON THE CONVERGENCE OF SOR ITERATIONS FOR FINITE ELEMENT APPROXIMATIONS TO ELLIPTIC BOUNDARY VALUE PR	71-0449
FIX, G. J.	ON THE EFFECTS OF QUADRATURE ERRORS IN THE FINITE ELEMENT METHOD	72-0258
FIX, G. J.	USE OF SINGULAR FUNCTIONS WITH FINITE ELEMENT APPROXIMATIONS	73-0102
FJELD, S. A.	THREE DIMENSIONAL THEORY OF ELASTICITY	69-0099
FLACHSBART, B.	ICES-STRUDL-11	68-0257
FLAIR, G. S.	FINITE ELEMENT APPROACH TO WAVE MOTIONS IN ELASTIC-PLASTIC CONTINUA	68-0041
FLANAGAN, R. F.	ANALYSIS OF BEHAVIOR OF EDWARD HYATT POWER PLANT	75-0055
FLATABO, N.	TRANSIENT HEAT CONDUCTION PROBLEMS IN POWER CABLES SOLVED BY FINITE ELEMENT METHOD	73-0007
FLEISCHER, C. C	FREE VIBRATION OF A CURVED BEAM	71-0606
FLEISCHER, C. C	VIBRATION OF CURVED STRUCTURES USING QUADRILATERAL FINITE ELEMENTS	72-0074
FLEISCHER, C. C	VIBRATION OF MULTISUPPORTED CURVED BEAMS	74-1297
FLEMING, J. M.	DIESEL ENGINE COMPONENT DESIGN USING THE FINITE ELEMENT METHOD AND INTERACTIVE GRAPHICS	74-0237
FLEMMING, M.	PRACTICAL APPLICATION OF FINITE ELEMENT METHODS IN THE DESIGN OF STRUCTURES	69-0420
FLINN, E. A.	STUDY OF NUMERICAL METHODS IN SEISMIC PROBLEMS	72-0738
FLYNN, P. L.	AN ANALYTIC METHOD FOR PREDICTING THE FLOW AND FRACTURE IN NOTCHED FIBER COMPOSITE MATERIALS	75-0689
FOGGIA, A.	FINITE ELEMENT SOLUTION OF SATURATED TRAVELLING MAGNETIC FIELD PROBLEMS	75-0600
FONDER, G.	STUDIES IN DOUBLY CURVED SHELL ELEMENTS	74-0133
FONDER, G. A.	EXPLICIT ADDITION OF RIGID-BODY MOTIONS IN CURVED FINITE ELEMENTS	73-0037
FOOS, R. C.	MODELING AN INPUT-OUTPUT GEOKINETIC SYSTEM UTILIZING A FINITE ELEMENT APPROACH	75-0746
FORD, J. L.	A METHOD FOR MATERIAL CHARACTERIZATION EMPLOYING THE FINITE-ELEMENT TECHNIQUE WITH APPLICATION TO HO	72-0941
FORD, R.	A COMPATIBLE TRIANGULAR PLATE BENDING FINITE ELEMENT	68-0209
FORD, R.	A COMPATIBLE TRIANGULAR PLATE BENDING FINITE ELEMENT	70-0385
FORREST, J. B.	A COMPUTER MODEL FOR PREDICTING THE LOAD-DEFLECTION RESPONSE OF EXPEDIENT SOIL SURFACING	73-0880
FORREST, J. B.	A STRUCTURAL EVALUATION OF RAPID METHODS OF BACKFILLING FOR BOMB DAMAGE REPAIR	74-0821
FORREST, J. B.	A STRUCTURAL EVALUATION OF RAPID METHODS OF BACKFILLING FOR BOMB DAMAGE REPAIR - PHASE 11	75-0682
FORREST, J. B.	LAYERED PAVEMENT SYSTEMS. PART 1. LAYERED SYSTEM DESIGN. PART 11. FATIGUE OF PLAIN CONCRETE	72-0867
FORSBERG, K.	AN EVALUATION OF FINITE DIFFERENCE AND FINITE ELEMENT TECHNIQUES FOR ANALYSIS OF GENERAL SHELLS	70-0254
FORSYTH, R. A.	STRESSES AND DEFORMATIONS IN JAIL GULCH EMBANKMENT	73-0391
FORTIN	CALCUL NUMERIQUE DES ECOULEMENTS DES FLUIDES DE BINGHAM ET DES FLUIDES NEWTONIENS INCOMPRESSIBLES PA	72-0349
FOST, R. B.	A FINITE ELEMENT ANALYSIS OF SHOCK AND FINITE-AMPLITUDE WAVES IN ONE-DIMENSIONAL HYPERELASTIC BODIES	72-0731
FOST, R. B.	A NOTE ON THE ANALYSIS OF NONLINEAR DYNAMICS OF ELASTIC MEMBRANES BY THE FINITE ELEMENT METHOD	74-0341
FOST, R. B.	CONVERGENCE, ACCURACY AND STABILITY OF FINITE ELEMENT APPROXIMATIONS OF A CLASS OF NON-LINEAR HYPERB	73-0159
FOST, R. B.	THEMIS FINAL REPORT, VOL. IV. RECENT DEVELOPMENTS IN THE NUMERICAL SOLUTION OF CERTAIN PROBLEMS IN N	74-0721
FOSTER, E. P.	EXPERIMENTAL AND FINITE ELEMENT ANALYSIS OF CABLE ROOF STRUCTURES INCLUDING PRECAST PANELS	74-1139
FOSTER, W. A.	A METHOD FOR THE TRANSIENT NONLINEAR COUPLED THERMOELASTIC ANALYSIS OF AXISYMMETRIC BODIES USING FIN	74-1140
FOTHERGILL, J.	PREDICTION OF LONG-TERM STRESS RANGES: USER'S MANUAL - BRIDGE LOAD GENERATOR	73-0737
FOTHERGILL, J.	PREDICTION OF LONG-TERM STRESS RANGES: USER'S MANUAL - BRIDGE DYNAMIC STRESS ANALYSIS	74-0807
FOTHERGILL, P.	PREDICTION OF LONG-TERM STRESS RANGES: USER'S MANUAL - BRIDGE LOAD GENERATOR	73-0737
FOTHERGILL, P.	PREDICTION OF LONG-TERM STRESS RANGES: USER'S MANUAL - BRIDGE DYNAMIC STRESS ANALYSIS	74-0807
FOURNEY, W. G.	LEAKAGE THROUGH CRACKS IN LNG TANKAGE	74-0749
FOURNEY, W. L.	NORMAL MODE ANALYSIS OF A FUZE SUPPORT STRUCTURE USING NASTRAN. PART 1	72-0853
FOWLER, J.	VIBRATION CHARACTERISTICS OF THE NORTH FORK DAM MODEL	74-0845
FOWLER, J. H.	FINITE-ELEMENT ANALYSIS OF HIGH-PRESSURE, LARGE VALVE GATES	75-0329
FOWLER, J. H.	FINITE-ELEMENT ANALYSIS OF HIGH-PRESSURE LARGE VALVE GATES	75-0413
FOWLER, J. N.	ELASTIC-PLASTIC DISCRETE ELEMENT ANALYSIS OF AXISYMMETRICALLY LOADED SHELLS OF REVOLUTION	67-0099
FOWLKES, C. W.	FINITE ELEMENT ANALYSIS OF CRACKS IN REGION OF A BI-MATERIAL INTERFACE	74-0164
FOX, R. L.	A CYLINDRICAL SHELL DISCRETE ELEMENT	67-0007
FOX, R. L.	AUTOMATED DESIGN OPTIMIZATION OF SUPERSONIC AIRPLANE WING STRUCTURES UNDER DYNAMIC CONSTRAINTS	72-0178
FOX, R. L.	DEVELOPMENTS IN DISCRETE ELEMENT FINITE DEFLECTION STRUCTURAL ANALYSIS BY FUNCTION MINIMIZATION	68-0144
FOX, R. L.	DEVELOPMENTS IN STRUCTURAL ANALYSIS BY DIRECT ENERGY MINIMIZATION	68-0070
FOX, R. L.	FINITE DEFLECTION STRUCTURAL ANALYSIS USING PLATE AND CYLINDRICAL SHELL DISCRETE ELEMENTS	68-0193
FOX, R. L.	THE GENERATION OF INTERELEMENT, COMPATIBLE STIFFNESS AND MASS MATRICES BY THE USE OF INTERPOLATION F	66-0014
FOYE, R. L.	INELASTIC MICROMECHANICS OF CURING STRESSES IN COMPOSITES	75-0284
FOYE, R. L.	THEORETICAL POST YIELDING BEHAVIOUR OF COMPOSITE LAMINATES - 1. INELASTIC MICROMECHANICS	73-0168
FOYE, R. L.	THEORETICAL POST YIELDING BEHAVIOUR OF COMPOSITE LAMINATES -2.	73-0408
FRAEIJS DE VEUB	A CONFORMING FINITE ELEMENT FOR PLATE BENDING	68-0182
FRAEIJS DE VEUB	BASIS OF A WELL CONDITIONED FORCE PROGRAM FOR EQUILIBRIUM MODELS VIA THE SOUTHWELL SLAB ANALOGIES	67-0111
FRAEIJS DE VEUB	BENDING AND STRETCHING OF PLATES, SPECIAL MODELS FOR UPPER AND LOWER BOUNDS	66-0026
FRAEIJS DE VEUB	DIFFUSIVE EQUILIBRIUM MODELS	73-1112
FRAEIJS DE VEUB	DISPLACEMENT AND EQUILIBRIUM MODELS IN THE FINITE ELEMENT METHOD	65-0016
FRAEIJS DE VEUB	DUAL ANALYSIS BY FINITE ELEMENTS: LINEAR AND NONLINEAR APPLICATIONS	72-0476
FRAEIJS DE VEUB	DUAL ANALYSIS FOR HEAT CONDUCTION PROBLEMS BY FINITE ELEMENTS	72-0401
FRAEIJS DE VEUB	DUALITY BETWEEN DISPLACEMENT AND EQUILIBRIUM METHOD WITH A VIEW TO OBTAINING UPPER AND LOWER BOUNDS	62-0013
FRAEIJS DE VEUB	DUALITY IN STRUCTURAL ANALYSIS BY FINITE ELEMENTS, STATIC GEOMETRIC ANALOGIES, THE DUAL PRINCIPLES O	71-0583
FRAEIJS DE VEUB	ELASTIC EQUILIBRIUM MODELS, STRESS FUNCTIONS, AND DUALITY NON LINEARITIES AND STABILITY. FIVE LECTUR	73-0631
FRAEIJS DE VEUB	FINITE ELEMENTS METHOD IN AEROSPACE ENGINEERING PROBLEMS	74-0409
FRAEIJS DE VEUB	IMPROVEMENTS OF FINITE ELEMENT SOLUTIONS FOR STRUCTURAL AND NONSTRUCTURAL APPLICATIONS	72-0671
FRAEIJS DE VEUB	LARGE DISPLACEMENT FORMULATION FOR ELASTIC BODIES	70-0429
FRAEIJS DE VEUB	NATRUAL STRAINS AND STRESSES FOR TRAPEZOIDAL STRUCTURES ANALYSIS	69-0101
FRAEIJS DE VEUB	NEW VARIATIONAL PRINCIPLE FOR FINITE ELASTIC DISPLACEMENTS	72-0454
FRAEIJS DE VEUB	QUALITY IN STRUCTURAL ANALYSIS BY FINITE ELEMENTS	71-0500
FRAEIJS DE VEUB	STABILITY ANALYSIS BY FINITE ELEMENTS	70-0056
FRAEIJS DE VEUB	STATIC-GEOMETRIC ANALOGIES AND FINITE ELEMENT MODELS	71-0499
FRAEIJS DE VEUB	STRAIN-ENERGY BOUNDS IN FINITE-ELEMENT ANALYSIS BY SLAB ANALOGY	67-0027
FRAEIJS DE VEUB	STRESS FUNCTION APPROACH	75-0016
FRAEIJS DE VEUB	THE DUAL PRINCIPLE OF ELASTODYNAMICS FINITE ELEMENT APPLICATIONS	71-0501
FRAEIJS DE VEUB	THE SECOND VARIAION TEST WITH ALGEBRAIC AND DIFFERENTIAL CONTRASTS	69-0342
FRAEIJS DE VEUB	UPPER AND LOWER BOUNDS IN MATRIX STRUCTURAL ANALYSIS	64-0018
FRAEIJS DE VEUB	UPPER AND LOWER BOUNDS TO STRUCTURAL DEFORMATIONS BY DUAL ANALYSIS IN FINITE ELEMENTS	67-0074
FRAEIJS DE VEUB	VARIATIONAL PRINCIPLES AND THE PATCH TEST	74-1282

FRAEIJS DE VEUB	AN EQUILIBRIUM MODEL FOR PLATE BENDING	68-0072
FRAEIJS DE VEUB	BIFURCATION OF SPACE FRAMES	70-0055
FRANCAVILLA, A.	A NOTE ON NUMERICAL COMPUTATION OF ELASTIC CONTACT PROBLEMS	75-0621
FRANCE, P. N.	ON SOME FREE SURFACE ON TRANSIENT FLOW PROBLEMS OF SEEPAGE AND IRROTATIONAL FLOW	72-0070
FRANCE, P. W.	FINITE ELEMENT ANALYSIS OF UNCONFINED SEEPAGE PROBLEMS	71-0767
FRANCE, P. W.	FINITE ELEMENT ANALYSIS OF THREE-DIMENSIONAL GROUNDWATER FLOW PROBLEMS	74-0254
FRANCE, P. W.	NUMERICAL ANALYSIS OF FREE SURFACE SEEPAGE PROBLEMS	71-0469
FRANCE, P. W.	THE DETERMINATION OF A WATER/GAS INTERFACE USING FINITE ELEMENTS	74-0346
FRANCHI, A.	A MODIFIED GRADIENT METHOD FOR FINITE ELEMENT ELASTOPLASTIC ANALYSIS BY QUADRATIC PROGRAMMING	73-0998
FRANCHI, A.	FINITE ELEMENT ELASTOPLASTIC ANALYSIS OF UNDERGROUND OPENINGS BY QUADRATIC PROGRAMMING	74-1288
FRANCIS, E. C.	CASE LINER BOND ANALYSIS	74-0817
FRANCIS, E. G.	FRACTURE OF ROCKET GRAIN MODELS SUBJECTED TO COOLDOWN: A STUDY OF TEMPERATURE-INITIATED FRACTURES OF	74-0536
FRANCIS, P. H.	DEFORMATION MECHANICS OF DEEP SURFACE FLAW CRACKS	72-0785
FRANCKEN, A. J.	THEORETICAL AND EXPERIMENTAL INVESTIGATION ON A STEAM GENERATOR TUBE PLATE	69-0424
FRANECK, H.	BERECHNUNG VON LAUFRAEDERN MIT DER METHODE DER FINITEN ELEMENTE (DESIGN OF ROTATING WHEELS BY THE FI	74-0547
FRANECK, H.	COMPUTATION OF IMPELLERS BY METHOD OF FINITE ELEMENTS	74-0646
FRANECK, H.	PROBLEMS OF CROSS-LINKAGE IN USING METHOD OF FINITE ELEMENTS	74-0165
FRANK, K. H.	THE FATIGUE STRENGTH OF FILLET WELDED CONNECTIONS	71-0684
FRANK, R.	ETUDE DE LA CAPACITE PORTANTE DES PIEUX PAR LA METHODE DES ELEMENTS FINIS. INFLUENCE DE LA DILATANCE	75-0624
FRANK, R.	ETUDE THEORIQUE DU COMPORTEMENT DES PIEUX SOUS CHARGE VERTICALE. INTRODUCTION DE LA DILATANCE. (THEO	75-0309
FRANKE, H. P.	FEM2D: PROGRAM FOR SOLVING THE TWO-DIMENSIONAL DIFFUSION EQUATION BY THE METHOD OF FINITE ELEMENTS	74-1089
FRANKE, H. P.	FINITE ELEMENTS VERSUS FINITE DIFFERENCES, A COMPARISON OF THE TWO METHODS FOR THE SOLUTION OF THE D	73-1055
FRANKE, H. P.	ON THE APPLICATION OF THE FINITE ELEMENT METHOD IN REACTOR PHYSICS	75-0122
FRANKE, H. P.	2-DIMENSIONAL AND 3-DIMENSIONAL REACTOR-PHYSICS CALCULATIONS WITH FINITE-ELEMENT METHOD	75-0423
FRANKLIN, H. A.	FINITE ELEMENT STUDY OF REINFORCED CONCRETE BEAMS WITH DIAGONAL TENSION CRACKS	74-1308
FRANKLIN, H. A.	NONLINEAR ANALYSIS OF REINFORCED CONCRETE FRAMES AND PANELS	70-0436
FRANKUS, A.	LOCAL COMPLIANCE PROBLEM OF LUNG ELASTICITY	75-0660
FRANKUS, A.	THEORY FOR DISTORTION STUDIES OF LUNG PARENCHYMA BASED ON ALVEOLAR MEMBRANE PROPERTIES	74-0222
FRANQUES, J. T.	TWO-DIMENSIONAL ANALYSIS OF BACKWATER AT BRIDGES	74-0184
FRANQUES, J. T.	A FINITE ELEMENT MODEL FOR TWO-DIMENSIONAL STEADY FLOW THROUGH CONTRACTIONS IN NATURAL CHANNELS	71-0810
FRASER, R. A.	FINITE ELEMENT ANALYSIS OF A PLATE ON A LAYERED CROSS-ANISOTROPIC FOUNDATION	74-0451
FRAZIER, G.	FINITE-ELEMENT TREATMENT OF QUADRUPOLE GROUND MOTIONS IN AXISYMMETRIC EARTH STRUCTURES	75-0434
FRAZIER, G. A.	VIBRATIONAL CHARACTERISTICS OF THREE-DIMENSIONAL SOLIDS, WITH APPLICATIONS TO EARTH DAMS	69-0432
FRAZIER, G. A.	3-D SEISMIC CODE FOR ILLIAC-1V	73-0868
FRAZIER, G. A.	3-D STRESS WAVE CODE FOR THE ILLTAC 1V	74-0780
FREDERCIK, D. L	RAPID CASE DESIGN USING FINITE ELEMENT STRESS ANALYSIS AND NASTRAN	73-0730
FREDERICK, C. O	TWO-DIMENSIONAL AUTOMATIC MESH GENERATION FOR STRUCTURAL ANALYSIS	70-0058
FREESE, C. E.	COLLOCATION AND FINITE ELEMENTS - A COMBINED METHOD	73-0734
FREESE, C. E.	STRESS ANALYSIS OF CONFIGURATIONS INVOLVING SMALL FILLETS	75-0361
FREMOND, M.	DUAL FORMULATION FOR POTENTIAL ENERGY AND COMPLEMENTARY ENERGY APPLICATION TO THE FINITE ELEMENT MET	72-0062
FREMOND, M.	FORMULATIONS DUALES DES ENERGIES POTENTIELLES ET COMPLEMENTAIRES. APPLICATION A LA METHODE DES ELEME	71-0752
FREMOND, M.	LA METHODE DES ELEMENTS FINIS	72-0724
FREMOND, M.	LA METHODE FRONTALE POUR LA RESOLUTION DES SYSTEMES LINEARESS. (FRONTAL METHOD FOR SOLVING LINEAR SY	73-1097
FRENCH, S.	PLANE STRESS ELEMENT OF VARYING THICKNESS APPLIED TO END BLOCK ANALYSIS	74-0396
FRENCH, S.	SINGLE ELEMENT PANEL FOR FLAT PLATE STRUCTURES	75-0279
FRENCH, S. B.	A COMPATIBLE MEMBRANE QUADRILATERAL ELEMENT	74-0483
FRENCH, S. B.	DISCRETE-ELEMENT MODELLING AND SENSITIVITY STUDIES OF STATICALLY LOADED SINGLE-LAYER AND BOUNDED DOU	70-0129
FRENCH, S. E.	FEABL (FINITE ELEMENT ANALYSIS BASIC LIBRARY) USER'S GUIDE	72-0748
FRIDLEY, R. B.	ANALYSIS OF VISCOELASTIC CONTACT STRESSES IN AGRICULTURAL PRODUCTS USING A FINITE ELEMENT METHOD	74-1199
FRIDLEY, R. B.	COMPUTER ANALYSIS OF FRUIT DETACHMENT DURING TREE SHAKING	74-1162
FRIDLEY, R. B.	COMPUTER SIMULATION OF VIBRATION OF WHOLE TREE SYSTEMS USING FINITE ELEMENT METHODS	74-1161
FRIDLEY, R. B.	SIMULATION OF VIBRATION OF WHOLE TREE SYSTEMS USING FINITE ELEMENTS	75-0765
FRIED, I.	A COMPUTATIONAL PROCEDURE FOR GRADIENT ITERATIVE TECHNIQUES IN THE FINITE ELEMENT METHOD	68-0229
FRIED, I.	A FULLY COMPATIBLE PLATE BENDING ELEMENT	67-0142
FRIED, I.	A GRADIENT COMPUTATIONAL PROCEDURE FOR THE SOLUTION OF LARGE PROBLEMS ARISING FROM THE FINITE ELEMEN	70-0059
FRIED, I.	ACCURACY AND CONDITION OF CURVED (ISOPARAMETRIC) FINITE ELEMENTS	73-0057
FRIED, I.	ACCUPACY OF COMPLEX FINITE ELEMENTS	72-0291
FRIED, I.	ACCURACY OF FINITE ELEMENT EIGENPROBLEMS	71-0088
FRIED, I.	BASIC COMPUTATIONAL PROBLEMS IN THE FINITE ELEMENT ANALYSIS OF SHELLS	71-0678
FRIED, I.	BEST FINITE ELEMENTS DISTRIBUTION AROUND A SINGULARITY	72-0252
FRIED, I.	BOUNDARY AND INTERIOR APPROXIMATION ERRORS IN THE FINITE ELEMENT METHOD	73-0209
FRIED, I.	BOUNDS ON EXTERNAL EIGENVALUES OF FINITE ELEMENT STIFFNESS AND MASS MATRICES AND THEIR SPECTRAL COND	72-0256
FRIED, I.	BOUNDS ON THE SPECTRAL AND MAXIMUM NORMS OF THE FINITE ELEMENT STIFFNESS, FLEXIBILITY AND MASS MATRI	73-0409
FRIED, I.	COMMENT ON THE PAPER "POSSIBLE LOSS OF ACCURACY IN CURVED (ISOPARAMETRIC) FINITE ELEMENTS" BY HENSHE	72-0279
FRIED, I.	CONDENSATION OF FINITE ELEMENT EIGENPROBLEMS	72-0168
FRIED, I.	CONDITION OF FINITE ELEMENT MATRICES GENERATED FROM NONUNIFORM MESHES	72-0104
FRIED, I.	DISCRETIZATION AND COMPUTATIONAL ERRORS IN HIGHER-ORDER FINITE ELEMENTS	71-0089
FRIED, I.	DISCRETIZATION AND ROUND-OFF ERROR IN THE FINITE ELEMENT ANALYSIS OF ELLIPTIC BOUNDARY VALUE PROBLEM	71-0514
FRIED, I.	FINITE ELEMENT ANALYSIS OF PROBLEMS FORMULATED BY AN INTEGRAL EQUATION: APPLICATION TO POTENTIAL FLO	68-0191
FRIED, I.	FINITE ELEMENT ANALYSIS OF INCOMPRESSIBLE MATERIAL BY RESIDUAL ENERGY BALANCING	74-1041
FRIED, I.	FINITE ELEMENT ANALYSIS OF THIN ELASTIC SHELLS WITH RESIDUAL ENERGY BALANCING AND THE ROLE OF THE RI	75-0034
FRIED, I.	FINITE ELEMENT METHOD IN FLUID DYNAMICS AND HEAT TRANSFER	67-0151
FRIED, I.	FINITE ELEMENT METHOD - ACCURACY AT A POINT	74-0240
FRIED, I.	FINITE-ELEMENT ANALYSIS OF TIME-DEPENDENT PHENOMENA	69-0104
FRIED, I.	FINITE-ELEMENT ANALYSIS OF THIN ELASTIC SHELLS WITH RESIDUAL ENERGY BALANCING AND ROLE OF RIGID BODY	75-0048
FRIED, I.	GRADIENT METHODS FOR FINITE ELEMENT EIGENPROBLEMS	69-0103
FRIED, I.	INFLUENCE OF POISSON'S RATIO ON THE CONDITION OF THE FINITE ELEMENT STIFFNESS MATRIX	71-0519
FRIED, I.	MONOTON FINITE-ELEMENT MATRICES AND THEIR COMPUTED CONDITION NUMBERS	75-0820
FRIED, I.	MORE ON GRADIENT ITERATIVE METHODS IN FINITE-ELEMENT ANALYSIS	69-0102
FRIED, I.	NOTES ON THE FINITE ELEMENT ANALYSIS OF THE AXISYMMETRIC ELASTIC SOLID	74-0185
FRIED, I.	NUMERICAL INTEGRATION IN THE FINITE ELEMENT METHOD	74-0576
FRIED, I.	NUMERICAL RESULTS FROM THE APPLICATION OF GRADIENT ITERATIVE TECHNIQUES TO THE FINITE ELEMENT VIBRAT	72-0138
FRIED, I.	OPTIMAL GRADIENT MINIMIZATION SCHEME FOR FINITE ELEMENT EIGENPROBLEMS	72-0292
FRIED, I.	PERTURBATION ERRORS IN FINITE ELEMENT METHOD	72-0277
FRIED, I.	RESIDUAL ENERGY BALANCING TECHNIQUE IN THE GENERATION OF PLATE BENDING FINITE ELEMENTS	74-0594
FRIED, I.	SHAPE FUNCTIONS AND ACCURACY OF ARCH FINITE ELEMENTS	73-0038
FRIED, I.	SHEAR IN C AND C PLATE BENDING ELEMENTS	73-0692
FRIED, I.	SOME ASPECTS OF THE NATURAL COORDINATE SYSTEM IN THE FINITE-ELEMENT METHOD	69-0105
FRIED, I.	SOME NEW ELEMENTS FOR THE MATRIX DISPLACEMENT METHOD	68-0017
FRIED, I.	THE CHANGES IN THE (TWO-DIMENSIONAL) STRESS ROUNDED CORNERS, CAUSED BY A VARYING INTERNAL PRESSURE A	67-0141
FRIED, I.	THE LUMINA ELEMENT FOR THE MATRIX DISPLACEMENT METHOD	68-0010
FRIED, I.	THE L2 AND L∞ CONDITION NUMBERS OF THE FINITE ELEMENT MATRICES AND THE POINTWISE CONVERGENCE OF THE	72-0061
FRIEDMAN, E.	DIRECT ITERATION METHOD FOR THE INCORPORATION OF PHASE CHANGE IN FINITE ELEMENT HEAT CONDUCTION PROG	74-0894
FRIEDMAN, E.	STRAIN CONCENTRATION ANALYSIS USING THE FINITE ELEMENT METHOD	71-0090
FRIEDMAN, E.	STRESSES AND DEFORMATIONS IN MELTING PLATES	69-0476
FRIEDMAN, E.	THERMOMECHANICAL ANALYSIS OF THE WELDING PROCESS USING THE FINITE ELEMENT METHOD	75-0213
FRIEDMAN, E.	THERMOMECHANICAL ANALYSIS OF THE WELDING PROCESS USING THE FINITE ELEMENT METHOD	75-0518
FRIEDMAN, M.	INTEGRAL FINITE-ELEMENT APPROACH FOR SOLVING LAPLACE EQUATION	73-0121
FRIEDRICH, C. M	BESTRAN: A TECHNIQUE FOR PERFORMING STRUCTURAL ANALYSES	75-0759
FRIEDRICH, C. M	SEAL-SHELL-2, A COMPUTER PROGRAM FOR THE STRESS ANALYSIS OF A THICK SHELL OF REVOLUTION WITH AXISYMM	69-0107
FRIK, G.	DYNAN LECTURE NOTES WITH COMPUTATIONAL EXAMPLES	71-0581
FRILEY, J. R.	FINITE ELEMENT APPLICATIONS OF A PRECISION TRIANGULAR PLATE BENDING ELEMENT TO LINEAR BOUNDARY VALUE	75-0788

36

FRIMANN CLAUSEN	ANALYSIS OF GEOTECHNICAL PROBLEMS BY MEANS OF THE FINITE ELEMENTS METHOD	72-0376
FRIND, E. O.	APPLICATION OF GALERKIN'S PROCEDURE TO AQUIFER ANALYSIS	72-0636
FRIND, E. O.	FINITE ELEMENTS IN THE SOLUTION OF THE INVERSE PROBLEM IN GROUNDWATER FLOW	74-0063
FRIND, E. O.	FUNCTIONAL COEFFICIENTS IN THE ANALYSIS OF GROUNDWATER FLOW USING FINITE ELEMENTS	73-0077
FRISCH-FAY, R.	LINEARIZED LARGE DEFLECTION PROBLEM	73-0685
FRISCH-FAY, R.	STABILITY OF MASONRY PIERS	75-0248
FRITZ, K.	NONSTEADY PYROMETRIC CALCULATIONS AS THE BASIS FOR STRESS ANALYSIS OF STEAM GENERATORS AND HEAT EXCH	72-0451
FROEHLICH, R.	REVIEW OF CURRENT PROBLEMS FOR MULTIDIMENSIONAL REACTOR STATICS CALCULATIONS	73-0960
FROIDEVAUX, H.	A FINITE ELEMENT METHOD FOR THE RESOLUTION OF NON-LINEAR BOUNDARY-VALUE PROBLEMS	75-0107
FROIER, M.	THE RECTANGULAR PLANE STRESS ELEMENT BY TURNER, PIAN AND WILSON	74-1265
FROST, G. R.	MODIFICATIONS TO ARL COMPUTER PROGRAMS USED FOR DESIGN OF AXIAL COMPRESSOR AIRFOILS	74-0508
FRUEHAUF, H.	ENTWURF UND SPANNUNGSBERECHNUNG EINES SPANNBETON DRUCKBEHAELTERS AUS FERTIGTEILEN. (DESIGN AND STRES	73-0856
FRYE, J. W.	AN ANALYSIS OF SOME FACTORS THAT AFFECT HYDROPHONE SENSITIVITY	72-0739
FU, C. C.	A METHOD FOR THE NUMERICAL INTEGRATION OF THE EQUATIONS OF MOTION ARISING FROM A FINITE ELEMENT ANAL	70-0061
FU, C. C.	ON THE STABILITY OF EXPLICIT METHODS FOR THE NUMERICAL INTEGRATION OF THE EQUATIONS OF MOTION IN FIN	72-0172
FUCHS, H. O.	SELF-STRESS CONCENTRATIONS	71-0317
FUEHRING, H.	PARAMETRISCHE KERBSPANNUNGSUNTERSUCHUNGEN AN DER LOCHSCHEIBE MIT DER METHODE DER FINITEN ELEMENTE. (	75-0603
FUERSTE, W.	BERECHNUNGSMETHODEN ZUR ERMITTLUNG VON BELASTUNGSBEDINGUNGEN UND GRENZTRAGFAEHIGKEITSVERHALTEN VON K	73-0858
FUHRING, H.	A DISCUSSION - ON THE DENSITY OF FINITE ELEMENT MATRICES	73-1066
FUHRING, H.	THE APPLICATION OF NODE-ELEMENT RULES FOR FORECASTING PROBLEMS IN THE GENERATION OF FINITE ELEMENT M	75-0614
FUJII, H.	FINITE ELEMENT GALERKIN METHOD FOR MIXED INITIAL-BOUNDARY VALUE PROBLEMS IN ELASTICITY THEORY	71-0520
FUJII, H.	FINITE ELEMENT SCHEMES: STABILITY AND CONVERGENCE	72-0302
FUJII, H.	SOME REMARKS ON FINITE ELEMENT ANALYSIS OF TIME DEPENDENT FIELD PROBLEMS	73-0307
FUJII, H.	STABILITY OF FINITE ELEMENT SCHEMES FOR VIBRATION PROBLEMS OF THE EQUATION OF ELASTICITY THEORY	71-0521
FUJII, T.	AUTOMATED DATA GENERATION AND LARGE MATRIX OPERATION FOR 3 DIMENSIONAL PLATE STRUCTURE ANALYSIS	73-0339
FUJIMURA, T.	FINITE ELEMENT METHOD FOR SOLVING NEUTRON TRANSPORT PROBLEMS IN TWO-DIMENSIONAL CYLINDRICAL GEOMETRY	74-1081
FUJINO, S.	ANALYSIS OF HYDROSTATIC EXTRUSION BY THE FINITE ELEMENT METHOD	72-0020
FUJINO, T.	ANALYSIS OF COMPRESSIBLE HYDRODYNAMICS BY THE FINITE ELEMENT METHOD	71-0196
FUJINO, T.	ANALYSIS OF CONTINUUM BY THE FINITE ELEMENT METHOD	69-0352
FUJINO, T.	ANALYSIS OF HYDRODYNAMIC PROBLEMS BY THE FINITE ELEMENT METHOD	69-0027
FUJINO, T.	ANALYSIS OF HYDRODYNAMIC AND PLATE STRUCTURES PROBLEMS BY THE FINITE ELEMENT METHOD	71-0091
FUJINO, T.	ANALYSIS OF OUT-OF-PLANE STRESS BY THE SIX NODES, FOURTH ORDER ACCURACY TRIANGULAR ELEMENTS	69-0375
FUJINO, T.	APPLICATION OF FINITE ELEMENT METHOD TO THE PROBLEM OF HEAT CONDUCTION AND FLUID MECHANICS	72-0486
FUJINO, T.	SELF ADJOINT DIFFERENTIAL EQUATIONS AND VARIATIONAL PRINCIPLE - 1.	71-0388
FUJINO, T.	SELF ADJOINT DIFFERENTIAL EQUATIONS AND VARIATIONAL PRINCIPLE - 2.	72-0106
FUJINO, T.	SHELL ANALYSIS BY THE FINITE ELEMENT METHOD	69-0376
FUJINO, T.	STATIC STRUCTURAL ANALYSIS OF SUSPENSION BRIDGES	66-0110
FUJINO, T.	THE HEAT CONDUCTION AND THERMAL STRESS ANALYSIS BY THE FINITE ELEMENT METHOD	68-0073
FUJINO, T.	VARIATIONAL PRINCIPLE OF LINEAR DIFFERENTIAL EQUATIONS	72-0107
FUJISAWA, N.	ANALYSIS OF MAGNETIC FIELDS BY FINITE ELEMENT METHOD	72-0126
FUJITA, Y.	A METHOD OF FINITE DEFLECTION ELASTIC PLASTIC ANALYSIS OF FRAMED STRUCTURES SUBJECTED TO LARGE AXIAL	69-0239
FUJITANI, Y.	A DIFFUSION ANALYSIS OF ION IN ELECTROLYSIS BY FINITE ELEMENT METHOD	73-0296
FUJITANI, Y.	SHEAR DEFORMATION ANALYSIS OF BEAMS AND PLATES BY FINITE ELEMENT METHOD	73-0873
FUKII, T.	STUDY ON FINITE ELEMENT METHOD FOR STRUCTURAL ANALYSIS - ISTRAN/PL (INH STRUCTURE ANALYSIS PLATE STR	73-0167
FUKUDA, J.	AUTOMATIC MESH GENERATION FOR FINITE ELEMENT ANALYSIS	72-0328
FUKUDA, J.	NEW APPROACHES TO THE FINITE ELEMENT ANALYSIS	71-0241
FUKUDA, J.	ON THE APPLICATIONS OF COLLOCATION OF ANALYTIC SOLUTION TO FINITE ELEMENT METHOD	69-0384
FUKUDA, S.	COMPUTER PREDICTION OF FATIGUE CRACK PROPAGATION UNDER RANDOM LOADING	73-0851
FULLARD, K.	CALCULATION OF THERMAL STRESSES IN A CYLINDER-CYLINDER INTERSECTION BY MEANS OF FINITE ELEMENTS	73-0571
FULLARD, K.	EXPERIENCE IN THE USE OF THE CONSTANT STRESS TRIANGLE FOR FINITE ELEMENT STRESS ANALYSIS	69-0407
FULLARD, K.	FREQUENCY TUNING OF COMPRESSOR ROTOR BLADES	66-0112
FULLARD, K.	THE COMPUTATION OF TEMPERATURE DISTRIBUTIONS AND THERMAL STRESSES USING FINITE ELEMENT TECHNIQUES	71-0092
FULLARD, K.	THERMAL STRESSES IN A NOZZLE-DRUM INTERSECTION	73-0824
FULLER, D. D.	STABILITY OF TILTING-PAD JOURNAL BEARINGS	71-0687
FULTON, R. E.	ACCURACY AND CONVERGENC OF FINITE ELEMENT APPROXIMATION	68-0228
FULTON, R. E.	ACCURACY OF FINITE ELEMENT APPROXIMATIONS TO STRUCTURAL PROBLEMS	70-0438
FULTON, R. E.	APPLICATION OF COMPUTER-AIDED AIRCRAFT DESIGN IN A MULTIDISCIPLINARY ENVIRONMENT	73-0527
FULTON, R. E.	CONVERGENCE OF THE NASTRAN PLATE ELEMENTS FOR SHELL STRESS ANALYSIS	71-0261
FULTON, R. E.	IMPACT OF CDC STAR-100 COMPUTER ON FINITE ELEMENT SYSTEMS	75-0039
FULTON, R. E.	THE ACCURACY OF THE FINITE ELEMENT METHODS IN CONTINUUM PROBLEMS	66-0027
FUNAOKA, K.	ON STRENGTH OF TRANSVERSE WEBS AND THEIR JOINTS OF LARGE TANKER	69-0327
FUNG, Y. C.	SLOW PARTICULATE VISCOUS FLOW IN CHANNELS AND TUBES, APPLICATION TO BIOMECHANICS	71-0248
FUNG, Y. C.	SLOW VISCOUS FLOW AND ITS APPLICATION TO BIOMECHANICS	71-0619
FUNG, Y. C.	THIN-SHELL STRUCTURES: THEORY, EXPERIMENT, AND DESIGN	74-0828
FUNNELL, W. R.	SIMULATING BEHAVIOR OF EARDRUM BY FINITE-ELEMENT METHOD	74-0511
FUNNELL, W. R.	MODELING EARDRUM AS A DOUBLY CURVED SHELL USING FINITE-ELEMENT METHOD	75-0047
FUREY, M.	STRUCTURAL EVALUATION OF CANDIDATE DESIGNS FOR THE LARGE SPACE TELESCOPE PRIMARY MIRROR	75-0708
FURRY, R. B.	LINEAR AND NONLINEAR FINITE ELEMENT ANALYSIS OF WOOD STRUCTURAL MEMBERS	73-0117
FURUIKE, T.	THE EFFECT OF AN INITIAL AXISYMMETRIC IMPERFECTION ON THE NATURAL FREQUENCIES OF CONICAL SHELLS	70-0368
FURUIKE, T.	VEHICLE STRUCTURAL ANALYSIS, PRE AND POST PROCESSOR DATA HANDLING	71-0093
GABRIELSE, S. E	FUEL ROD DEFORMATIONS - A TIME DEPENDENT FINITE ELEMENT ANALYSIS ON AXISYMMETRIC FUEL RODS, INCLUDIN	70-0414
GABRIELSE, S. E	HEAVY SECTION STEEL TECHNOLOGY PROGRAM. TECHNICAL REPORT NO. 4, OCTOBER 1969. A TWO-DIMENSIONAL ELAS	69-0474
GABRIELSEN, B.	STRUCTURAL RESPONSE AND LOADING OF WALL PANELS	71-0735
GABRIELSEN, B.	SHOCK TUNNEL TESTS OF PRELOADED AND ARCHED WALL PANELS FINITE ELEMENT STRESS FORMULATION FOR WAVE PR	73-0892
GABRIELSON, V.	FEMESH: A FINITE ELEMENT CODE PREPROCESSOR	73-0949
GABRIELSON, V.	GRAPHICS APPLICATIONS FOR FINITE ELEMENT CODE PROCESSING	74-0893
GAGGERO, A.	ON THE SOLUTION OF PLANE STRESS PROBLEMS BY FINITE ELEMENTS COMPUTER PROGRAMS	73-0978
GAINS, J. H.	TRANSVERSE VIBRATIONS OF CANTILEVER BARS OF VARIABLE CROSS SECTIONS	66-0001
GAJOWNICZEK, S.	OBLICZANIE KONSTRUKCJI TOROIDALNEGO ZBIORNIKA NA WODE. (STATICAL ANALYSIS OF A TOROIDAL WATER RESERV	75-0211
GAL-OR, B.	GENERAL VARIATIONAL ANALYSIS OF HYDRODYNAMIC, THERMAL AND DIFFUSIONAL BOUNDARY LAYERS	70-0191
GALANTE, J. O.	FINITE ELEMENT STRESS ANALYSIS OF AN INTERVERTEBRAL DISC	74-0244
GALE, J. E.	FLOW IN ROCKS WITH DEFORMABLE FRACTURES	74-0065
GALLAGHER, J. P	SPECTRUM TRUNCATION AND DAMAGE TOLERANCE STUDY ASSOCIATED WITH THE C-5A OUTBOARD PYLON AFT TRUSS LUG	74-0535
GALLAGHER, R. H	A CORRELATION STUDY OF METHODS OF MATRIX STRUCTURAL ANALYSIS	69-0110
GALLAGHER, P. H	A DISCRETE ELEMENT PROCEDURE FOR THIN-SHELL INSTABILITY ANALYSIS	67-0029
GALLAGHER, R. H	A METHOD OF LIMIT POINT CALCULATION IN FINITE ELEMENT STRUCTURAL ANALYSIS	72-0786
GALLAGHER, R. H	A PROCEDURE FOR FINITE ELEMENT PLATE AND SHELL PRE AND POST BUCKLING ANALYSIS	71-0096
GALLAGHER, R. H	A SURVEY OF FINITE ELEMENT FRAMEWORK STABILITY ANALYSIS	69-0418
GALLAGHER, R. H	A TRIANGULAR THIN SHELL FINITE ELEMENT: NONLINEAR ANALYSIS	75-0697
GALLAGHER, R. H	A TRIANGULAR THIN SHELL FINITE ELEMENT: LINEAR ANALYSIS	75-0698
GALLAGHER, R. H	ANALYSIS OF PLATE AND SHELL STRUCTURES	69-0109
GALLAGHER, R. H	APPLICATIONS OF FINITE ELEMENT ANALYSIS	72-0330
GALLAGHER, R. H	COMMENTS ON DERIVATION OF ELEMENT STIFFNESS MATRICES BY T. H. H. PIAN	65-0060
GALLAGHER, R. H	COMPUTATIONAL METHODS IN NUCLEAR REACTOR STRUCTURAL DESIGN FOR HIGH TEMPERATURE APPLICATIONS: AN INT	73-1001
GALLAGHER, R. H	DEVELOPMENT OF ADVANCED STRUCTURAL OPTIMIZATION PROGRAMS AND THEIR APPLICATION TO LARGE ORDER STRAIN	66-0028
GALLAGHER, R. H	DIRECT FLEXIBILITY FINITE ELEMENT ELASTOPLASTIC ANALYSIS	71-0095
GALLAGHER, R. H	DISCRETE ELEMENT APPROACH TO STRUCTURAL INSTABILITY ANALYSIS	63-0005
GALLAGHER, R. H	DISCRETE ELEMENT PROCEDURE FOR THIN SHELL INSTABILITY ANALYSIS	67-0028
GALLAGHER, R. H	EFFICIENT SOLUTION PROCESSES FOR FINITE ELEMENT ANALYSIS OF TRANSIENT HEAT CONDUCTION	69-0014
GALLAGHER, R. H	ELASTIC DEFORMATION OF LIGHTWEIGHT MIRRORS METHOD OF LIMIT POINT CALCULATION IN FINITE ELEMENT STRUC	72-0232
GALLAGHER, R. H	ELASTIC INSTABILITY PREDICTIONS FOR DOUBLY-CURVED SHELLS	68-0075
GALLAGHER, R. H	ENHANCEMENTS OF THE FINITE ELEMENT METHOD THROUGH MULTIDISCIPLINARY APPLICATIONS	75-0842

GALLAGHER, R. H	FINITE ELEMENT ANALYSIS OF PLATE AND SHELL STRUCTURES	69-0360
GALLAGHER, R. H	FINITE ELEMENT ANALYSIS OF TORSIONAL AND TORSIONAL-FLEXURAL STABILITY PROBLEMS	70-0028
GALLAGHER, R. H	FINITE ELEMENT ANALYSIS IN BRITTLE MATERIAL DESIGN	70-0360
GALLAGHER, R. H	FINITE ELEMENT ANALYSIS OF GEOMETRICALLY NONLINEAR PROBLEMS	73-0308
GALLAGHER, R. H	FINITE ELEMENT ANALYSIS: FUNDAMENTALS	75-0772
GALLAGHER, R. H	FINITE ELEMENT CIRCULATION ANALYSIS OF VARIABLE-DEPTH SHALLOW LAKES	73-0727
GALLAGHER, R. H	FINITE ELEMENT LAKE CIRCULATION AND THERMAL ANALYSIS	74-0043
GALLAGHER, R. H	FINITE ELEMENT METHOD IN ELASTIC INSTABILITY ANALYSIS	69-0108
GALLAGHER, R. H	FINITE ELEMENT METHOD IN SHELL STABILITY ANALYSIS	73-0250
GALLAGHER, R. H	FINITE ELEMENT METHOD IN PLATE AND SHELL INSTABILITY ANALYSIS	73-0983
GALLAGHER, R. H	FINITE ELEMENT METHODS IN FLOW PROBLEMS	74-0752
GALLAGHER, R. H	FINITE ELEMENT PROCEDURE FOR NONLINEAR PREBUCKLING AND INITIAL POSTBUCKLING ANALYSIS	72-0170
GALLAGHER, R. H	GENERAL POTENTIAL ENERGY AND COMPLEMENTARY ENERGY MODELS BASED ON STRESS PARAMETERS	74-1240
GALLAGHER, R. H	GEOMETRICALLY NONLINEAR FINITE ELEMENT ANALYSIS	72-0347
GALLAGHER, R. H	HIGHER-ORDER FINITE ELEMENT ANALYSIS OF LAKE CIRCULATION	73-1002
GALLAGHER, R. H	LARGE SCALE COMPUTER PROGRAMS FOR STRUCTURAL ANALYSIS	70-0211
GALLAGHER, R. H	MATRIX DYNAMIC AND INSTABILITY ANALYSIS WITH NON-UNIFORM ELEMENTS	70-0062
GALLAGHER, R. H	PERTURBATION PROCEDURES IN NONLINEAR FINITE ELEMENT ANALYSIS	74-1241
GALLAGHER, R. H	PLATES AND SHELLS, INELASTIC ANALYSIS, TIME DEPENDENT INELASTICITY, FIVE LECTURES	73-0633
GALLAGHER, R. H	RECENT ADVANCES IN MATRIX METHODS IN STRUCTURAL ANALYSIS AND DESIGN	70-0290
GALLAGHER, R. H	SHEAR BUCKLING OF SQUARE PERFORATED PLATES	74-0713
GALLAGHER, R. H	SHELL ELEMENTS	75-0545
GALLAGHER, R. H	STABILITY OF PLATES USING THE FINITE ELEMENT METHOD	67-0030
GALLAGHER, R. H	STIFFNESS MATRIX FOR SHALLOW RECTANGULAR SHELL ELEMENT	68-0074
GALLAGHER, R. H.	STRESS ANALYSIS IN HEATED, COMPLEX SHAPES	62-0002
GALLAGHER, R. H	SURVEY AND EVALUATION OF THE FINITE ELEMENT METHOD IN LINEAR FRACTURE MECHANICS ANALYSIS	71-0094
GALLAGHER, R. H	THE FINITE ELEMENT METHOD IN ELASTIC INSTABILITY ANALYSIS	69-0339
GALLAGHER, R. H	THE FINITE ELEMENT METHOD OF THIN-SHELL STABILITY ANALYSIS	72-0725
GALLAGHER, R. H	THE FINITE ELEMENT METHOD IN PLATE AND SHELL STABILITY ANALYSIS	73-0644
GALLAGHER, R. H	THE USE OF FINITE ELEMENT METHODS IN HEAT FLOW ANALYSIS	68-0004
GALLAGHER, R. H	THEORY AND PRACTICE IN FINITE ELEMENT STRUCTURAL ANALYSIS	73-0700
GALLAGHER, R. H	TRENDS AND DIRECTIONS IN THE APPLICATIONS OF NUMERICAL ANALYSIS.	71-0667
GALLAGHER, R. H	TRIANGULAR THIN SHELL FINITE ELEMENT: LINEAR ANALYSIS	75-0602
GALLAGHER, R. H	TRIANGULAR THIN SHELL FINITE ELEMENT: NONLINEAR ANALYSIS	75-0668
GALLETLY, G. D.	FREE VIBRATIONS OF CYLINDRICAL SHELLS WITH VARIOUS END CLOSURES	74-0571
GALLETLY, R. D.	THE COMPARISON OF THE RESPONSE OF A HIGHWAY BRIDGE TO UNIFORM GROUND SHOCK AND MOVING GROUND EXCITAT	72-0737
GALLO, A. M.	MAGIC II. AN AUTOMATED GENERAL PURPOSE SYSTEM FOR STRUCTURAL ANALYSIS	71-0495
GALLO, A. M.	MAGIC 11: AN AUTOMATED GENERAL PURPOSE SYSTEM FOR STRUCTURAL ANALYSIS. VOLUME 111. PROGRAMMER'S MANU	71-0756
GALLO, A. M.	MAGIC 111: AN AUTOMATED GENERAL PURPOSE SYSTEM FOR STRUCTURAL ANALYSIS. VOLUME 111. PROGRAMMER'S MAN	72-0726
GAMBLE, W. L.	THICK WALLED MULTIPLE OPENING REINFORCED CONCRETE CONDUITS	72-0666
GAMBOLATI, G.	DIAGONALLY DOMINANT MATRICES FOR THE FINITE ELEMENT METHOD IN HYDROLOGY	73-0410
GAMBOLATI, G.	MATHEMATICAL SIMULATION OF THE SUBSIDENCE OF VENICE 1 THEORY	73-0992
GAMBOLATI, G.	OPTIMIZZAZIONE DI RETICOLI TRIANGOLARI PER IL METODO DEGLI ELEMENTI FINITI IN IDROLOGIA	72-0987
GAMBOLATI, G.	USE OF THE OVER-RELAXATION TECHNIQUE IN THE SIMULATION OF LARGE GROUNDWATER BASINS BY THE FINITE ELE	75-0606
GANGADHARAN, A.	DESIGN AND ANALYSIS EXPERIENCE WITH A LIQUID METAL HEAT EXCHANGER FOR PFTF SERVICE	72-0772
GANGAL, M. B.	FINITE ELEMENT METHOD IN STRESS ANALYSIS	71-0576
GANGAL, M. B.	STRESS ANALYSIS OF PRESSURIZED CYLINDRICAL BORE IN RECTANGULAR BLOCK	69-0312
GANGAL, M. D.	DIRECT FINITE ELEMENT ANALYSIS OF ELASTIC CONTACT PROBLEMS	72-0733
GANGAL, M. D.	FINITE ELEMENT ANALYSIS OF ELASTIC CONTACT PROBLEM	72-0550
GANGARAO, H. V.	MACRO-APPROACH FOR RIBBED AND GRID PLATE SYSTEMS	75-0079
GANTAYAT, A. N.	STRESS ANALYSIS OF B16.9 TEES BY THE FINITE ELEMENT METHOD	71-0060
GANTAYAT, A. N.	STRESS ANALYSIS OF B16.9 TEES BY THE FINITE ELEMENT METHOD, A PROGRESS REPORT	71-0215
GANTAYAT, A. N.	STRESS ANALYSIS OF TWO JOINTS BY THE FINITE-ELEMENT METHOD	72-0942
GANTAYAT, A. N.	THREE-DIMENSIONAL BAR CELL FOR ELASTIC STRESS ANALYSIS	70-0219
GAONKAR, G. H.	AVERAGE ELASTIC PROPERTIES AND STRESS FIELDS FOR COMPOSITE CONTINUA	69-0106
GARDIOL, F. E.	NUMERICAL ANALYSIS OF THE LINE CAPACITANCE AND CROSSTALK FACTOR FOR INSULATED WIRE PAIRS	74-0685
GARDNER, G. A.	NUMERICAL STUDY OF HYDROGMAGNETIC STABILITY USING FINITE ELEMENT METHOD	73-0092
GARDNER, G. A.	STUDIES OF HYDROMAGNETIC STABILITY	74-0035
GARDNER, G. A.	STUDY OF THE MHD INSTABILITIES OF CYLINDRICAL BELT PINCH USING THE FINITE ELEMENT METHOD	74-0700
GARDNER, L. R.	NUMERICAL STUDY OF HYDROGMAGNETIC STABILITY USING FINITE ELEMENT METHOD	73-0092
GARDNER, L. R.	HYDROMAGNETIC STABILITY STUDIES USING THE FINITE ELEMENT METHOD	75-0878
GARDNER, L. R.	STUDIES OF HYDROMAGNETIC STABILITY	74-0035
GARDNER, L. R.	STUDY OF THE MHD INSTABILITIES OF CYLINDRICAL BELT PINCH USING THE FINITE ELEMENT METHOD	74-0700
GARDNER, N. A.	COMPARISON OF FINITE-ELEMENT AND EXPERIMENTAL STUDIES ON DEFORMATIONS OF ZIRCONIUM NOTCHED BEND SPEC	71-0359
GARDNER, W. S.	ANALYSIS OF LOAD-BEARING FILLS OVER SOFT SUBSOILS	71-0330
GARG, R. P.	EFFECT OF OPENINGS ON THE LATERAL STIFFNESS OF INFILLED FRAMES	71-0300
GARG, V. K.	A FINITE ELEMENT MODEL FOR PLASTICITY PROBLEMS	74-1225
GARG, V. K.	ELASTIC-PLASTIC ANALYSIS OF A WHEEL ROLLING ON A RIGID TRACK	74-1040
GARG, V. K.	SHAKEDOWN OF ROLLING WHEEL UNDER HUB LOADING	74-0663
GARG, V. K.	STRUCTURAL INELASTICITY. V11. ELASTIC-PLASTIC ANALYSIS OF A WHEEL ROLLING ON A RIGID TRACK	73-0886
GARG, V. K.	STRUCTURAL INELASTICITY. V1. A FINITE ELEMENT METHOD FOR PLASTICITY PROBLEMS	73-0887
GARGESA, G.	NEW DESIGN FACTOR FOR A SHRINK FITTED ASSEMBLY	75-0167
GARLANGER, J. E	PERFORMANCE OF AN EMBANKMENT CONSTRUCTED ON VARVED CLAY	73-0743
GARNER, E. R.	DYNAMIC RESPONSE OF THE HUMAN ARM AS A COMPOSITE STRUCTURE	72-0611
GARNER, R. W.	FINITE ELEMENT SOLUTION OF COUPLED ELECTROKINETIC AND HYDRODYNAMIC FLOW IN POROUS MEDIA	72-0398
GARNET, H.	DOUBLY CURVED TRIANGULAR FINITE ELEMENTS FOR SHELLS OF ARBITRARY SHAPE	73-0881
GARNET, H.	EVALUATION OF NUMERICAL TIME INTEGRATION METHODS AS APPLIED TO ELASTIC-PLASTIC DYNAMIC PROBLEMS INVO	74-0858
GARNET, H.	VARIABLE TIME STEP METHOD FOR DETERMINING PLASTIC STRESS REFLECTIONS FROM BOUNDARIES	75-0641
GARNIER, H.	DEFORMABILITY OF THE HULL STEEL-WORK AND DEFORMATIONS OF THE ENGINE-ROOM OF LARGE TANKERS	74-0532
GARTLING, D. K.	COMPUTATIONALLY EFFICIENT FINITE ELEMENT ANALYSIS OF VISCOUS FLOW PROBLEMS	74-1050
GARTLING, D. K.	COMPUTATIONALLY EFFICIENT FINITE ELEMENT ANALYSIS OF VISCOUS FLOW PROBLEMS	74-1258
GARTLING, D. K.	FINITE ELEMENT ANALYSIS OF VISCOUS, INCOMPRESSIBLE FLUID FLOW	75-0674
GARTLING, D. K.	FINITE ELEMENT ANALYSIS OF VISCOUS, INCOMPRESSIBLE FLUID FLOW	75-0760
GARVEY, S. J.	THE QUADRILATERAL SHEAR PANEL	51-0002
GASCH, R.	UNBALANCE GENERATED VIBRATIONS AND STABILITY OF TURBINE ROTORS	73-0224
GASCHEN, J. P.	THE CALCULATION OF THREE-DIMENSIONAL TEMPERATURE DISTRIBUTIONS AND THERMAL STRESSES USING FINITE ELE	71-0187
GASS, N.	EIN NEUES VARIATIONSPRINZIP MIT ANWENDUNG AUF SCHWINGUNGEN FLACHER SCHALEN NACH DER METHODE DER FINI	73-0563
GASS, N.	NEW HYBRID CYLINDRICAL SHELL FINITE ELEMENT	72-0092
GASS, N.	NEW VARIATIONAL PRINCIPLE WITH APPLICATION TO VIBRATIONS OF SHALLOW SHELLS USING FINITE ELEMENTS MET	73-0354
GASS, N.	SOME TWO-FIELD VARIATIONAL PRINCIPLES FOR NONLINEAR DEFORMATION ANALYSIS OF SHELLS	75-0294
GASS, N.	VIBRATION OF CYLINDRICAL SHELLS BY HYBRID FINITE ELEMENT METHOD	72-0253
GATES, R. H.	APPLICATIONS OF THE FINITE ELEMENT METHOD IN GEOTECHNICAL ENGINEERING, PROCEEDINGS OF THE SYMPOSIUM	73-0273
GATES, R. H.	INELASTIC ANALYSIS OF SLOPES BY THE FINITE ELEMENT METHOD	68-0298
GATES, R. M.	EFFECT OF DAMPING ON EXCITABILITY OF HIGH-ORDER NORMAL MODES	75-0709
GAUDU, R.	USE OF FINITE ELEMENTS IN THE ANALYSIS OF A HYDROTHERMAL DOUBLET	74-0049
GAUKROGER, D. R	THE SEISMIC DESIGN STUDY OF A DOUBLE CURVATURE ARCH DAM	69-0053
GAWRONSKI, W.	ANALIZA DRGAN WYMUSZONYCH ZLOZONYCH UKLADOW LINIOWYCH METODA SZTYWNYCH ELEMENTOW SKONCZONYCH (ANALYS	72-0246
GAWRONSKI, W.	EFFECT OF SYSTEM PARAMETER VARIATIONS ON NATURAL FREQUENCIES	75-0661
GAWRONSKI, W.	METHOD SZTYWNYCH ELEMENTOW SKONCZONYCH W OBLICZENIACH KONSTRUCJI OKRETOWYCH. (STIFF FINITE ELEMENT M	74-0923
GAYLORD, C. E.	RESEARCH TO IMPROVE TUNNEL SUPPORT SYSTEMS	74-0777
GAZDA, I. W.	FINITE-ELEMENT ANALYSIS OF AN ELECTRODE-NIPPLE JOINT	75-0843
GEBHARDT, L.	APPLICATION OF THE FINITE ELEMENT STIFFNESS METHOD TO NONCONSERVATIVE STABILITY PROBLEMS	67-0158

GEDGE, M. R.	RESPONSE OF STRUCTURE TO THE PSEUDO-SOUND FIELD OF A JET (USING A COMBINED CONTINUUM AND FINITE ELEM	67-0146
GEERTSMA, J.	FINITE-ELEMENT ANALYSIS OF SHALLOW TEMPERATURE ANOMALIES	71-0471
GEFFEN, N.	FINITE ELEMENTS FOR FLUID DYNAMICS. VARIATIONAL FORMULATIONS FOR NONLINEAR, INITIAL-VALUE PROBLEMS	74-0710
GEISS, R. H.	MECHANICS OF DEFORMABLE SOLIDS - EURASIAN COMMUNIST COUNTRIES	73-0794
GEISSLER, H.	FINITE ELEMENT ANALYSIS, AN AUTOMOBILE ENGINEER'S TOOL	74-0238
GELDER, D.	SOLUTION OF THE COMPRESSIBLE FLOW EQUATIONS	71-0003
GELLATLY, R. A.	A REVIEW OF SOME ASPECTS OF STRUCTURAL OPTIMIZATION	75-0467
GELLATLY, R. A.	DEVELOPMENT OF ADVANCED STRUCTURAL OPTIMIZATION PROGRAMS AND THEIR APPLICATION TO LARGE ORDER STRAIN	66-0028
GELLATLY, R. A.	DISCRETE ELEMENT PROCEDURE FOR THIN SHELL INSTABILITY ANALYSIS	67-0028
GELLATLY, R. A.	OPTIM 11: A (MAGIC) COMPATIBLE LARGE SCALE AUTOMATED MINIMUM WEIGHT DESIGN PROGRAM - 1. ENGINEERS AN	74-1048
GELLATLY, R. A.	OPTIM 11: A MAGIC COMPATIBLE LARGE-SCALE AUTOMATED MINIMUM WEIGHT DESIGN PROGRAM. VOLUME 1. ENGINEER	74-1096
GELLERT, M.	FINITE ELEMENT EQUILIBRIUM MODEL FOR GENERAL BODY FORCE DISTRIBUTION	75-0268
GELLERT, M.	GENERATION OF FINITE ELEMENTS BY EULER'S VARIATIONAL METHOD	72-0596
GENALIS, P.	ELASTIC STRENGTH OF PROPELLERS: AN ANALYSIS BY MATRIX METHODS	70-0468
GENALIS, P.	THE APPLICATION OF THE FINITE ELEMENT METHOD TO MULTIHULLED SHIP STRUCTURAL ANALYSIS	72-0868
GENCO, M.	INTERNAL SUPPORTING STRUCTURES OF THE (CIRENE) REACTOR CORE - RESULTS OF STATIC STRESS ANALYSIS FOR	73-0818
GEORGE, A.	NESTED DISSECTION OF A REGULAR FINITE ELEMENT MESH	73-0042
GEORGE, A.	SPARSE MATRIX TECHNIQUES IN DIRECT SOLUTION OF FINITE ELEMENT EQUATIONS	74-0115
GEORGE, A.	SURVEY OF SPARSE MATRIX METHODS IN THE DIRECT SOLUTION OF FINITE ELEMENT EQUATIONS	73-0572
GEORGE, J.	ON THE NUMERICAL SOLUTION OF FREE BOUNDARY PROBLEMS	75-0577
GEORGE, J. A.	BLOCK ELIMINATION ON FINITE ELEMENT SYSTEMS OF EQUATIONS	72-0530
GEORGE, J. A.	COMPUTER IMPLEMENTATION OF THE FINITE ELEMENT METHOD	71-0376
GEORGE, J. A.	ON THE DENSITY OF FINITE ELEMENT MATRICES	72-0545
GEORGE, W.	CALCULATION OF THERMAL STRESSES IN TUBULAR CERAMIC COMPONENTS	73-0670
GEORGESON, D. L	ENVIRONMENTAL CONSIDERATIONS OF DAM CONSTRUCTION AND OPERATION IN SEISMICALLY ACTIVE URBAN AREAS	73-0470
GEORGIADIS, E.	COMPUTER APPLICATIONS IN ROCK MECHANICS	72-0190
GEORGOPOULOS, G	A FORTRAN PROGRAM FOR CALCULATING THE FREQUENCIES AND MODE SHAPES OF AN ARBITRARY SHAPE CYLINDRICAL	72-0594
GEORGOPOULOS, G	A FORTRAN PROGRAM FOR CALCULATING THE FREQUENCIES AND MODE SHAPES OF A GUIDED CLAMPED, GUIDED PINNED	72-0890
GEORGOPOULOS, G	DEVELOPMENT OF A FINITE ELEMENT MODEL FOR THE CLASS V FLEXTENSIONAL UNDERWATER TRANSDUCER SHELL	71-0754
GEORGOPOULOS, G	DEVELOPMENT OF A FINITE ELEMENT MODEL FOR THE CLASS 5 FLEXTENSIONAL UNDERWATER TRANSDUCER SHELL	71-0840
GEORGOPOULOS, G	DEVELOPMENT OF A FINITE ELEMENT MODEL FOR THE CLASS IV FLEXTENSIONAL UNDERWATER TRANSDUCER SHELL	72-0593
GEORGOPOULOS, G	FREQUENCY COEFFICIENTS OF A SPHERICAL SHELL FOR VARYING THICKNESS AND OPENING ANGLE WITH GUIDED PINN	72-0595
GEORGOPOULOS, G	VIBRATIONAL ANALYSIS OF GENERAL SHELLS UTILIZING FINITE-ELEMENT TECHNIQUES	73-1133
GERADIN, M.	ANALYSE DYNAMIQUE DUALE DES STRUCTURES PAR LA METHODE DES ELEMENTS FINIS	73-0623
GERADIN, M.	ERROR BOUNDS FOR EIGENVALUE ANALYSIS BY ELIMINATION OF VARIABLES	71-0594
GERADIN, M.	MODAL ANALYSIS OF A SATELLITE PRIMARY STRUCTURE USING A FINITE ELEMENT PROCEDURE	73-0624
GERADIN, M.	ON THE NUMERICAL SOLUTION OF LARGE EIGENVALUE PROBLEMS ARISING IN PANEL FLUTTER ANALYSIS BY THE FINI	74-0942
GERADIN, M.	THE COMPUTATIONAL EFFICIENCY OF A NEW MINIMIZATION ALGORITHM FOR EIGENVALUE ANALYSIS	71-0595
GERARD, R.	FINITE ELEMENT SOLUTION FOR FLOW IN NONCIRCULAR CONDUITS	74-0116
GERBERT, B. G.	PRESSURE DISTRIBUTION AND BELT DEFORMATION IN V-BELT DRIVES	74-1028
GERDEEN, J. C.	LARGE DEFLECTION ANALYSIS OF ELASTIC-PLASTIC SHELLS OF REVOLUTION	70-0507
GERGELY, P.	ANALYSIS OF THIN-SHELL HYPERBOLIC PARABOLOID SHELLS	72-0121
GERIJ, J.	FINITE-ELEMENT METHOD IN CYCLIC THERMO-ELASTOPLASTICITY	73-1111
GERIJ, J.	THERMO-ELASTIC-PLASTIC CYCLIC ANALYSIS BY FINITE ELEMENT METHOD	73-1003
GERINROZE, J.	ISOPARAMETRIC FINITE-ELEMENT METHODS FOR 2-DIMENSIONAL TRANSPORT CALCULATIONS	75-0354
GERLACH, H. D.	FINITE ELEMENT METHOD IN ROLLING BEARING ENGINEERING	73-0054
GERRISH, R. W.	PERIODIC HEAT CONDUCTION IN A TWO PHASE, TWO DIMENSIONAL SOLID DOMAIN	74-1164
GERSTENKORN, G.	APPLICATION OF THE DIRECT STIFFNESS METHOD TO PLANE PROBLEMS INVOLVING LARGE, TIME-DEPENDENT DEFORMA	66-0108
GERSTENKORN, G.	STRUCTURAL ANALYSIS OF AN ARTERIOLE BY THE DIRECT STIFFNESS METHOD	66-0109
GERSTNER, R. W.	STRESS ANALYSIS AND SPECIAL PROBLEMS OF PRESTRESSED DAMS	61-0006
GERSTNER, R. W.	THE METHOD OF INTERFACE STRESS ADJUSTMENT AND ITS USES IN SOME PLANE ELASTICITY PROBLEMS	61-0005
GHABOUSSI, J.	DYNAMIC STRESS ANALYSIS OF POROUS ELASTIC SOLIDS SATURATED WITH COMPRESSIBLE FLUIDS	71-0651
GHABOUSSI, J.	FINITE ELEMENT FOR ROCK JOINTS AND INTERFACES	73-0411
GHABOUSSI, J.	FLOW OF COMPRESSIBLE FLUID IN POROUS ELASTIC MEDIA	73-1004
GHABOUSSI, J.	SEISMIC ANALYSIS OF EARTH DAM RESERVOIR SYSTEMS	73-0412
GHABOUSSI, J.	TUNNEL DESIGN CONSIDERATIONS: ANALYSIS OF MEDIUM-SUPPORT INTERACTION	74-1062
GHABOUSSI, J.	VARIATIONAL FORMULATION OF DYNAMICS OF FLUID-SATURATED POROUS ELASTIC SOLIDS	72-0017
GHADIALI, N. D.	ELASTIC-PLASTIC ANALYSIS OF INTERFERENCE FIT FASTENERS	74-1177
GHALI, A.	ANALYSIS OF PLATES IN BENDING USING LARGE FINITE ELEMENTS	70-0185
GHALI, A.	ANALYSIS OF PLATES SUBJECTED TO IN-PLANE FORCES USING LARGE FINITE ELEMENTS	70-0271
GHALI, A.	ANALYSIS OF SLAB AND GIRDER BRIDGES BY THE FINITE STRIP METHOD	70-0210
GHALI, A.	CONVERGENCE OF SEMIANALYTICAL SOLUTION OF PLATES	73-0497
GHALI, A.	DIAPHRAGMS IN SINGLE- AND DOUBLE-CELL BOX GIRDER BRIDGES WITH VARYING ANGLE OF SKEW	72-0130
GHALI, A.	FINITE ELEMENT ANALYSIS OF SKEW, CURVED BOX-GIRDER BRIDGE	70-0241
GHALI, A.	FINITE ELEMENT ANALYSIS OF SKEW BOX GIRDER BRIDGES	72-0209
GHALI, A.	FINITE ELEMENT ANALYSIS OF PERFORATED SHELLS	72-0228
GHALI, A.	FINITE ELEMENT OF SKEW VAULT BRIDGES	72-0434
GHALI, A.	FINITE STRIP ANALYSIS OF SKEW SLABS	72-0369
GHALI, A.	LE CALCUL DES TABLIERS DE PONTS PAR DES METHODES NUMERIQUES	72-0197
GHALI, A.	SEMI-ANALYTIC SOLUTION OF SKEW BOX GIRDER BRIDGES	75-0578
GHALI, A.	STRUCTURAL ANALYSIS: A UNIFIED CLASSICAL AND MATRIX APPROACH	72-0055
GHAZVINIAN, B.	NONCONSERVATIVE STABILITY OF PLATES BY FINITE ELEMENT METHOD	75-0789
GHERI, G.	SPLINE INTERPOLATION TECHNIQUES FOR VARIATIONAL METHODS	73-1079
GHISTA, D. N.	FINITE ELEMENT ANALYSES IN BIOMECHANICS	72-0741
GHISTA, D. N.	FINITE ELEMENT ANALYSIS OF HUMAN CARDIAC STRUCTURES	74-0438
GHOSH, K.	ANCHOR ZONE STRESS ANALYSIS BY FINITE ELEMENT METHOD	68-0147
GHOSH, S.	DYNAMIC STRESS ANALYSIS OF AXISYMMETRIC STRUCTURES UNDER ARBITRARY LOADING	69-0433
GIANNINI, M.	A CURVED ELEMENT APPROXIMATION IN THE ANALYSIS OF AXISYMMETRIC THIN SHELLS	70-0063
GIBBS, N. E.	AN ALGORITHM FOR REDUCING THE BANDWIDTH AND PROFILE OF A SPARSE MATRIX	74-0792
GIBSON, W.	DEVELOPMENTS IN DISCRETE ELEMENT FINITE DEFLECTION STRUCTURAL ANALYSIS BY FUNCTION MINIMIZATION	68-0144
GIBSON, W. C.	FESTRAN: A FINITE ELEMENT COMPUTER PROGRAM	71-0811
GIDDA, J.	FINITE ELEMENT ELASTOPLASTIC ANALYSIS OF UNDERGROUND OPENINGS BY QUADRATIC PROGRAMMING	74-1288
GIENCKE, E.	A SIMPLE MIXED METHOD FOR PLATE AND SHELL PROBLEMS	73-1005
GIENCKE, E.	SIMPLE "MIXED" METHOD FOR PLATE AND SHELL PROBLEMS	74-0952
GIENCKE, E.	SIMPLE FINITE ELEMENT METHOD FOR STRESS CONCENTRATION PROBLEMS IN PRESSURE VESSELS	71-0669
GIESELBERG, K.	ELASTIC DEFORMATION OF SQUARE EXTRUSION DIES DUE TO LOAD AND TEMPERATURE	75-0581
GIESKE, R. K.	NODAL ANALYSIS OF THE MATED SPACE SHUTTLE CONFIGURATION	71-0097
GIFFORD, L. N.	FINITE ELEMENT ANALYSIS FOR ARBITRARY AXISYMMETRIC STRUCTURES ORTHOTROPIC, TEMPERATURE-DEPENDENT MAT	68-0076
GIFFORD, L. N.	FINITE ELEMENT FRACTURE MECHANICS ANALYSIS OF TWO-DIMENSIONAL AND AXISYMMETRIC ELASTIC AND ELASTIC-P	74-1092
GIGLIOTTI, M. F	FINITE-ELEMENT MODELING OF DIRECTIONAL SOLIDIFICATION OF SUPER ALLOY EUTECTICS	75-0350
GIGNAC, D. A.	A COMPARATIVE STUDY OF SEVERAL CORE STORAGE SCHEMES FOR LARGE SPARSE POSITIVE DEFINITE MATRICES WITH	72-0578
GILBERT, D. E.	THE FINITE ELEMENT SOLUTION OF FLOW AND HEAT TRANSFER IN ELLIPTICAL DUCTS	74-0039
GILES, G. L.	APPLICATION OF NASTRAN TO THE ANALYSIS OF A SPACE SHUTTLE ORBITER STRUCTURE	71-0099
GILES, G. L.	PROCEDURE FOR EFFICIENT GENERATING, CHECKING, AND DISPLAYING NASTRAN INPUT AND OUTPUT DATA FOR ANALY	71-0098
GILL, J. A.	COMPUTER-AIDED DESIGN OF SHELL STRUCTURES USING THE FINITE ELEMENT METHOD	72-0998
GILL, J. I.	SHELLS, PLATES AND MEMBRANES - A COMPUTER-AIDED DESIGN SYSTEM	74-1159
GILL, P. A. T.	AUTOMATED SOLUTION OF TIME DEPENDENT PROBLEMS	72-0084
GILL, S. S.	LOAD TRANSFER MECHANISM FOR PIERS SOCKETED IN HARD SOILS OR ROCK	74-0969
GILLESPIE, D. C	AN INVESTIGATION OF THE EFFECTS OF SQUARE CUTOUTS ON THE NATURAL FREQUENCIES AND MODE SHAPES OF RECT	71-0638
GILLHAM, R. W.	SENSITIVITY ANALYSIS OF INPUT PARAMETERS IN NUMERICAL MODELING OF STEADY STATE REGIONAL GROUNDWATER	74-0926
GILMAN, J. D.	THREE-DIMENSIONAL ANALYSIS OF REACTOR PRESSURE VESSEL NOZZLES	71-0221
GIMM, W.	BERECHNUNG UND BEURTEILUNG DER STANDSICHERHEIT VON ABBAUHOHLRAEUMEN IN STEILEN GANGERZLAGERSTAETTEN	74-0597

GIMM, W.	NEUARTIGE ANWENDUNG GEOMECHANISCHER UNTERSUCHUNGSMETHODEN ZUR BEURTEILUNG DER ABBAUWIRKUNGEN IM GANG	75-0530
GINESU, F.	ELASTO-PLASTIC ANALYSIS OF A ROTATING MODEL CONICAL TURBINE DISC	75-0281
GINTER, L. E.	STATISTICAL STATE OF STRESS STUDIED BY GRID ANALYSIS: "NUMERICAL METHODS OF ANALYSIS"	49-0001
GIORDANA, F.	CALCULATION OF THE NATURAL FREQUENCIES AND THE PRINCIPAL MODES OF HELICOPTER BLADES	72-0907
GIRARD, A.	MODAL ANALYSIS OF A SATELLITE PRIMARY STRUCTURE USING A FINITE ELEMENT PROCEDURE	73-0624
GIRAULT, V.	THEORY OF A FINITE DIFFERENCE METHOD ON IRREGULAR NETWORKS	74-0375
GIRIJAVALLABHAN	ANALYSIS OF SHEAR WALLS WITH OPENINGS	69-0111
GIRIJAVALLABHAN	ANALYSIS OF SHEAR WALLS BY FINITE ELEMENT METHOD	69-0112
GIRIJAVALLABHAN	APPLICATION OF THE FINITE ELEMENT METHOD TO PROBLEMS IN SOIL AND ROCK MECHANICS	67-0118
GIRIJAVALLABHAN	FINITE ELEMENT METHOD FOR PROBLEMS IN SOIL MECHANICS	68-0078
GIRIJAVALLABHAN	FINITE-ELEMENT METHOD APPLIED TO SOME PROBLEMS IN SOIL MECHANICS	68-0077
GIROUX, Y. M.	CURVED TRIANGULAR ELEMENTS FOR THE ANALYSIS OF SHELLS	68-0030
GIROUX, Y. M.	FINITE ELEMENT ANALYSIS OF SHELLS UNDER NONUNIFORM SNOW LOADS	69-0113
GIULIANI, S.	ANALYSE DES CONTRAINTES DAN DES STRUCTURES COMPLEXES EN PRESENCE DE FLUAGE, DE VARIATIONS DE DIMENSI	71-0081
GIULIANI, S.	CODE TAFEST: NUMERICAL SOLUTION TO TRANSIENT HEAT-CONDUCTION PROBLEMS USING FINITE ELEMENTS IN SPACE	74-0901
GIULIANI, S.	CREEP ANALYSIS OF TRANSVERSELY ISOTROPIC BODIES SUBJECTED TO TIME DEPENDENT LOADING	73-0170
GIULIANI, S.	FINITE ELEMENT ANALYSIS OF STEADY-STATE NONLINEAR HEAT TRANSFER PROBLEMS	74-0572
GIULIANI, S.	FINITE ELEMENTS IN THE SOLUTION OF ELECTROMAGNETIC INDUCTION PROBLEMS	74-0252
GIUSEPPETTI, G.	ANALYSIS OF STEADY ELECTRICAL AND MAGNETIC-FIELDS BY FINITE-ELEMENT METHOD - APPLICATION TO TRANSFOR	75-0550
GIUSEPPETTI, G.	FINITE ELEMENT STUDY OF THE TRIAXIAL STRESS STATE AROUND AN INSPECTION TUNNEL IN AN ARCH DAM	74-0831
GIUSEPPETTI, G.	STUDIO DI CAMPI ELETTRICI E MAGNETICI STANZIONARI CON IL METODO DEGLI ELEMENTI FINITI. APPLICAZIONE	75-0207
GJELLEC, P.	CALCUL DE MASSIFS EN ELASTOPLASTICITE APPLICATIONS A LA MECANIQUE DES ROCHES ET DES SOLS. (APPLICATI	74-0223
GLADWELL, G.	A TRIANGULAR PLATE BENDING ELEMENT FOR CONTACT PROBLEMS	73-0693
GLADWELL, G. M.	FINITE ELEMENT ANALYSIS OF AXISYMMETRIC VIBRATIONS OF CYLINDERS	72-0255
GLADWELL, G. M.	A FINITE ELEMENT METHOD FOR ACOUSTICS	65-0017
GLADWELL, G. M.	A VARIATIONAL FORMULATION OF DAMPED ACOUSTO-STRUCTURAL PROBLEMS	66-0133
GLADWELL, G. M.	ERRORS IN SHELL FINITE-ELEMENT MODELS FOR VIBRATION OF CIRCULAR-CYLINDERS	75-0203
GLADWELL, G. M.	FINITE ELEMENT ANALYSIS OF FLEXURE - TYPE VIBRATORS WITH ELECTROSTRICTIVE TRANSDUCERS	70-0338
GLADWELL, G. M.	VARIATIONAL FINITE ELEMENT CALCULATION OF THE ACOUSTIC RESPONSE OF A RECTANGULAR PANEL	71-0100
GLADWELL, G. M.	VIBRATION ANALYSIS OF AXISYMMETRIC RESONATORS	75-0199
GLANCY, J. J.	DETERMINATION OF THE LOWEST CRITICAL BUCKLING STRESS OF VARIOUS SHAPED PLATES BY FINITE-ELEMENTS	73-1134
GLANVILLE, J. I	PREFABRICATED PRETENSIONED LIGHTWEIGHT CONCRETE FOLDED PLATE	73-0808
GLASER, F.	A FINITE ELEMENT PROGRAM FOR DESIGN AND ANALYSIS OF THREE-DIMENSIONAL FLAT SPRINGS	73-0738
GLOSS, R. J.	TRANSPIRATION AND FILM COOLING BOUNDARY LAYER COMPUTER PROGRAM. VOLUME 2, COMPUTER PROGRAM AND USER'	71-0655
GLOUDEMAN, J. F	ASPECTS OF FINITE ELEMENT METHOD APPLIED TO AEROSPACE STRUCTURES	73-0703
GLOUDEMAN, J. F	ASPECTS OF THE FINITE ELEMENT METHOD AS APPLIED TO AERO-SPACE STRUCTURES	72-0649
GLOUDEMAN, J. F	FINITE TIME-SPACE ELEMENTS APPLIED TO NONLINEAR PROBLEMS	70-0398
GLOUDEMAN, J. F	NUMERICAL CONDITIONING OF STIFFNESS MATRIX FORMULATIONS FOR FRAME STRUCTURES	68-0022
GLOUDEMAN, J. F	SOME ASPECTS OF FINITE ELEMENT TECHNIQUES	69-0340
GLOUDEMAN, J. F	THE GROWTH OF INTERACTIVE GRAPHICS AS AN ENGINEERING TOOL.	69-0347
GLOVER, P.	DYNAMIC FINITE ELEMENT METHOD CALCULATION OF ELASTIC WAVE PROPAGATION IN TWO DIMENSIONAL STRUCTURES	74-0166
GLOVER, P.	EFFECTS OF SOURCE STRUCTURE AND SITE RESPONSE ON GROUND MOTION IN THE NEAR FIELD OF EARTHQUAKES	73-0732
GLOVER, P.	INVERSION TECHNIQUE FOR STATIC FINITE ELEMENT METHOD	74-0170
GLOWINSKI, R.	ANALYSE NUMERIQUE DU CHAMP MAGNETIQUE D'UN ALTERNATEUR PAR ELEMENTS FINIS ET SURRELAXATION PONCTUELL	74-0625
GLOWINSKI, R.	APPLICATION DE LAT METHODE DES ELEMENTS FINIS A LA RESOLUTION D'UN PROBLEME DE DOMAINE OPTIMAL	74-0412
GLOWINSKI, R.	APPLICATION OF FINITE-ELEMENT METHOD TO APPROXIMATION OF OPTIMAL DOMAIN PROBLEM - SOLUTION METHODS =	75-0349
GLOWINSKI, R.	APPROXIMATION BY FIRST ORDER FINITE ELEMENTS AND SOLUTION BY PENALTY AND DUALITY OF NONLINEAR DIRICH	74-0245
GLOWINSKI, R.	APPROXIMATIONS EXTERNES, PAR ELEMENTS FINIS DE LAGRANGE D ORDRE, UN ET DEUX, DU PROBLEME DE DIRICHL=	72-0495
GLOWINSKI, R.	DUAL ITERATIVE TECHNIQUES FOR SOLVING A FINITE ELEMENT APPROXIMATION OF THE BIHARMONIC EQUATION	75-0480
GLOWINSKI, R.	ETUDE NUMERIQUE DU CHAMP MAGNETIQUE DANS UN ALTERNATEUR TETRAPOLAIRE PAR LA METHODE DES ELEMENTS FIN	74-0413
GLOWINSKI, R.	FINITE ELEMENT ANALYSIS OF THE UNSTEADY FLOW OF A VISCOUS PLASTIC FLUID IN A CYLINDRICAL PIPE	74-0055
GLOWINSKI, R.	METHOD FOR EXTERNAL APPROXIMATION USING FINITE ELEMENTS FOR DELTA - 2 AND ITERATIVE METHOD FOR SOLVI	72-0282
GLOWINSKI, R.	METHOD FOR EXTERNAL APPROXIMATION USING FINITE ELEMENTS FOR DELTA -2 AND ITERATIVE METHOD FOR SOLVIN	72-0283
GLOWINSKI, R.	SUR L APPROXIMATION, PAR ELEMENTS FINIS D ORDRE UN, ET LA RESOLUTION, PAR PENALISATION-DUALITE, D UN	75-0845
GLOWINSKI, R.	SURE UNE METHODE D'APPROXIMATION EXTERNE PAR ELEMENTS FINIS D ORDRE DEUX	75-0742
GLYNN, T. E.	EXPERIMENTAL AND THEORETICAL INVESTIGATION OF PAVEMENT DEFLECTIONS	69-0154
GOANKAR, G. H.	AVERAGE ELASTIC PROPERTIES AND STRESS FIELDS FOR COMPOSITE MATERIALS	69-0435
GOBLE, G. G.	DEVELOPMENTS IN DISCRETE ELEMENT FINITE DEFLECTION STRUCTURAL ANALYSIS BY FUNCTION MINIMIZATION	68-0144
GODBOLD, C. W.	CONVERGENCE CHARACTERISTICS OF CYLINDRICAL SHELL FINITE ELEMENTS	72-0943
GODBOLE, A. N.	VISCOUS, INCOMPRESSIBLE FLOW WITH SPECIAL REFFRENCE TO NON-NEWTONIAN (PLASTIC) FLUIDS	74-0053
GODBOLE, P. N.	APPLICATION OF FINITE ELEMENT METHOD TO PLANE STRAIN EXTRUSION PROCESSES	74-1243
GODBOLE, P. N.	CREEPING FLOW IN RECTANGULAR DUCTS BY FINITE ELEMENT METHOD	75-0569
GODBOLE, P. N.	ELASTOPLASTIC ANALYSIS OF A CANTILEVER BEAM	73-0984
GODBOLE, P. N.	FINITE ELEMENT ANALYSIS OF STEADY FLOW OF NON-NEWTONIAN FLUIDS	74-0465
GODBOLE, P. N.	FLOW OF PLASTIC AND VISCO-PLASTIC SOLIDS WITH SPECIAL REFERENCE TO EXTRUSION AND FORMING PROCESSES	74-0210
GODBOLE, P. N.	PENALTY FUNCTION APPROACH TO PROBLEMS OF PLASTIC FLOW OF METALS WITH LARGE SURFACE DEFORMATIONS	75-0282
GODDEN, W. G.	STRUCTURAL MODEL STUDIES OF LARGE PIPE BIFURCATION	73-0283
GODFREY, D. A.	DYNAMIC ANALYSIS OF AXISYMMETRIC SHELLS UNDER ARBITRARY TRANSIENT PRESSURES.	72-0467
GODFREY, D. A.	STRUCTURAL ANALYSIS OF AIRCRAFT IMPACT ON A NUCLEAR CONTAINMENT VESSEL AND ASSOCIATED STRUCTURES	69-0279
GOEL, J. J.	A CURVED FINITE ELEMENT FOR THIN ELASTIC SHELLS	69-0486
GOEL, J. J.	CURVED FINITE ELEMENT FOR THIN ELASTIC SHELLS	70-0173
GOEL, J. J.	ELEMENTS FINIS RAFFINES EN ELASTICITE BIDIMENSIONNELLE	69-0411
GOEL, J. J.	FINITE ELEMENT WITH A HIGH DEGREE OF REGULARITY	70-0052
GOEL, J. J.	REFINED FINITE ELEMENTS IN 2-DIMENSIONAL ELASTICITY	69-0320
GOFFI, L.	DIRECT TENSILE TESTING OF ANISOTROPIC ROCKS	74-0989
GOLAS, J.	NUMERYCZNE ROZWIAZANIE PROBLEMU ORGAN WLASNYCH NA PRZYKLADZIE PLYT. (NUMERICAL SOLUTION OF THE PROBL	75-0296
GOLAS, J.	ZASTOSOWANIE ITERACJI SEIDLA W METODZIE ELEMENTOW SKONCZONYCCH NA PRZYKLADZIE OBLICZEN STATYCZNYCH P	74-1031
GOLD, L. M.	ANALYTICAL PERFORMANCE OF STEEL, ALUMINUM AND PLASTIC FOR A NEW DESIGN HIGH PRESSURE THIN-WALLED CAR	74-0756
GOLD, L. M.	FINITE ELEMENT ANALYSIS OF A MULTI-COMPONENT KINETIC ENERGY PROJECTILE	74-1180
GOLD, L. M.	FINITE ELEMENT ANALYSIS OF A MULTI-COMPONENT KINETIC ENERGY PROJECTILE	74-0754
GOLD, L. M.	FINITE ELEMENT DESIGN STUDY FOR 30-MM CARTRIDGE CASE	74-0376
GOLDBERG, D. R.	FINITE ELEMENT STRUCTURAL ANALYSIS USING COMPUTERS	70-0216
GOLDBERG, J. E.	ANALYSIS OF MODERATELY THICK FLANGED REACTOR PRESSURE VESSELS	73-0821
GOLDBERG, J. E.	ANALYSIS OF MODERATELY THICK FLANGED REACTOR VESSELS	75-0320
GOLDBERG, J. E.	BOUSSINESQ PROBLEM OF PLANE MICROPOLAR ELASTICITY	74-0197
GOLDBERG, J. E.	FINITE ELEMENT APPROACH TO PLANE MICROELASTICITY	72-0205
GOLDBERG, J. E.	FINITE ELEMENT APPROACH TO BENDING OF MICROPOLAR PLATES	74-0255
GOLDEN, M. E.	DEFINIT - A NEW ELEMENT DEFINITION CAPABILITY FOR NASTRAN: USER'S MANUAL	73-0738
GOLDEN, M. E.	NEW ELEMENT DEFINITION CAPABILITY FOR NASTRAN	71-0125
GOLDHAMMER, M.	METHODS FOR PREDICTING THE AERODYNAMIC AND STABILITY AND CONTROL CHARACTERISTICS OF STOL AIRCRAFT. V	73-0718
GOLDHAMMER, M.	METHODS FOR PREDICTING THE AERODYNAMIC AND STABILITY AND CONTROL CHARACTERISTICS OF STOL AIRCRAFT. V	73-0719
GOLDHAMMER, M.	METHODS OF PREDICTING THE AERODYNAMIC AND STABILITY AND CONTROL CHARACTERISTICS OF STOL AIRCRAFT. VO	73-0717
GOLDIN, A. L.	BEHAVIOUR OF THE ROCK MASS AT THE SITE OF A ROCKFILL DAM	74-0992
GOLDSBERRY, F.	RELIABILITY STUDY OF LINEAR VARIATION TRIANGULAR FINITE ELEMENT	72-0281
GOLDSMITH, W.	IMPACT ON A TRANSVERSELY ANISOTROPIC HALF-SPACE	74-0950
GOLDSMITH, W.	WAVE PROPAGATION IN ANISOTROPIC ROCKS	72-0801
GOLDSMITH, W.	WAVE PROPAGATION IN ANISOTROPIC ROCKS	72-0873
GOLDSMITH, W.	WAVE PROPAGATION IN ANISOTROPIC ROCKS	73-0922
GOLDSTEIN, M.	TENUE DES STRUCTURES DE GRAPHITE VIS-A-VIS DES SOLLICITATIONS RENCONTREES DANS UN COEUR DE REACTEUR	73-0817
GOLEMBESKI, T.	STRUCTURAL ANALYSIS OF A THREE-DIMENSIONAL PERFORATED PLATE BY THE FINITE ELEMENT METHOD WITH TEST C	74-0945
GOLIA, C.	LA TECNICA DEGLI ELEMENTI FINITI PER PROBLEMI DI LUBRIFICAZIONE INCOMPRESSIBILE. (FINITE ELEMENT MET	75-0471
GONCALVES MORGA	SOBRE TORRES ESPIADAS - METODO DE CALCULO E UM EXEMPLO DE APLICACAO. (METHOD AND AN EXAMPLE OF CALCU	74-0915

40

GONG, C.	PSEUDO 3-DIMENSIONAL FINITE-ELEMENT FORMULATION FOR ELASTOSTATIC PROBLEMS AND ITS GEOPHYSICAL APPLIC	75-0597
GONG, C.	PSEUDO-THREE-DIMENSIONAL FINITE ELEMENT FORMULATION FOR ELASTOSTATIC PROBLEMS	74-0167
GONSHOR, H.	THE RING OF FINITE ELEMENTS IN A NON-STANDARD MODEL OF THE REALS	71-0450
GONTAROVSKII, V	APPLICATION OF THE METHOD OF FINITE ELEMENTS TO CALCULATIONS OF THE STRESSED AND STRAINED STATE OF I	75-0775
GONTAROVSKII, V	METHOD OF FINITE ELEMENTS FOR CALCULATION OF HETEROGENEOUS BODIES OF REVOLUTION IN ELASTO-PLASTIC ME	75-0144
GOODELL, R. A.	ANALYTICAL AND EXPERIMENTAL INVESTIGATION OF A NOZZLE TO CYLINDRICAL SHELL JUNCTION	74-0638
GOODHEIM, H.	FINITE ELEMENT ANALYSIS OF SEVERAL SWAGE MANDREL DESIGNS	74-0794
GOODMAN, J. R.	A RATIONAL ANALYSIS AND DESIGN PROCEDURE FOR WOOD JOIST FLOOR SYSTEMS	74-0730
GOODMAN, J. R.	FINITE ELEMENT ANALYSIS OF LAYERED WOOD SYSTEMS	75-0288
GOODMAN, J. R.	FINITE ELEMENT METHOD FOR WOOD MECHANICS	71-0294
GOODMAN, J. R.	FINITE ELEMENT METHOD FOR WOOD MECHANICS	71-0613
GOODMAN, R. E.	A DESIGN PROCEDURE FOR HIGH CUTS IN JOINTED HARD ROCK. THREE DIMENSIONAL SOLUTIONS. VOLUME 1	71-0656
GOODMAN, R. E.	A MODEL FOR THE MECHANICS OF JOINTED ROCK	68-0079
GOODMAN, R. E.	ANALYSIS AND MODEL STUDIES OF UNDERGROUND OPENINGS IN JOINTED ROCK. RESEARCH ON ROCK BOLT REINFORCEM	71-0438
GOODMAN, R. E.	ANALYSIS AND MODEL STUDIES OF UNDERGROUND OPENINGS IN JOINTED ROCK. RESEARCH ON ROCK BOLT REINFORCEM	71-0750
GOODMAN, R. E.	DUPLICATION OF DILATANT BEHAVIOR IN THE ANALYSIS OF JOINTED ROCKS	72-0757
GOODMAN, R. E.	FINITE ELEMENT ANALYSES OF SLOPES IN JOINTED ROCK	68-0269
GOODMAN, R. E.	FINITE ELEMENT ANALYSIS OF CRACK INITIATION IN A BLOCK MODEL EXPERIMENT	72-0455
GOODMAN, R. E.	FINITE ELEMENT STUDIES OF PILE DRIVER TUNNELS INCLUDING CONSIDERATIONS OF SUPPORT REQUIREMENTS	72-0679
GOODMAN, R. E.	NUMERICAL ANALYSES OF DEFORMABILITY TESTS IN JOINTED ROCK. ' JOINT PERTURBATION ' AND ' NO TENSION '	71-0439
GOODMAN, R. E.	NUMERICAL AND PHYSICAL MODELLING OF REINFORCEMENT SYSTEMS FOR TUNNELS IN JOINTED ROCK	73-0872
GOODMAN, R. E.	ON THE DISTRIBUTION OF STRESSES AROUND CIRCULAR TUNNELS IN NON-HOMOGENEOUS ROCKS	66-0029
GOODMAN, R. E.	THEORETICAL BASIS OF THE BOREHOLE DEEPENING METHOD OF ABSOLUTE STRESS MEASUREMENT	70-0047
GOOS, G.	FORMULATION AND APPLICATION OF CERTAIN PRIMAL AND MIXED FINITE ELEMENT MODELS OF FINITE DEFORMATIONS	74-0344
GOPALACHARYULU,	A HIGHER ORDER CONFORMING RECTANGULAR ELEMENT	73-0690
GOPALACHARYULU,	THE DEVELOPMENT OF A SERIES OF HYBRID-STRESS FINITE ELEMENTS	75-0335
GOPINATH, A.	ON LOCAL APPROXIMATIONS OF OPERATOR EQUATIONS	72-0063
GORDON, J. L.	A FINITE ELEMENT ANALYSIS OF THE PLANE STRAIN SHEET DRAWING PROBLEM	71-0812
GORDON, J. L.	FINITE ELEMENT ANALYSIS OF THE PLANE STRAIN DRAWING PROBLEM	74-1154
GORDON, W. J.	CONSTRUCTION OF CURVILINEAR COORDINATE SYSTEMS AND APPLICATIONS TO MESH GENERATION	74-1295
GORDON, W. J.	DISCRETIZATION ERROR BOUNDS FOR TRANSFINITE ELEMENTS	72-0533
GORDON, W. J.	GEOMETRIC ASPECTS OF THE FINITE ELEMENT METHOD	72-0523
GORDON, W. J.	GEOMETRIC ASPECTS OF THE FINITE ELEMENT METHOD: CONSTRUCTION OF CURVILINEAR COORDINATE SYSTEMS AND T	72-0534
GORDON, W. J.	TRANSFINITE ELEMENT METHODS: BLENDING FUNCTION INTERPOLATION OVER ARBITRARY CURVED ELEMENT DOMAINS	73-1006
GORECKI, A.	ANALYSIS OF LONG RECTANGULAR TANKS	70-0276
GORHOLT, W.	DESIGN OF A COMPOSITE STEEL AND CONCRETE CLOSURE FOR A PRESTRESSED CONCRETE REACTOR VESSEL.	72-0187
GOSSAIN, D. M.	DEFLECTION AND VIBRATION ANALYSIS OF A FLEXIBLE FAN BLADE	73-0556
GOTOH, M.	FINITE ELEMENT ANALYSIS OF GENERAL DEFORMATION OF SHEET METALS	74-0501
GOTOH, M.	PLASTICITY OF POLYCRYSTALS AND GENERAL DISCUSSIONS ON PLASTICITY THEORY WITH IT	75-0312
GOUDREAU, G. L.	EVALUATION OF DISCRETE METHODS FOR THE LINEAR DYNAMIC RESPONSE OF ELASTIC AND VISCOELASTIC SOLIDS	70-0439
GOULD, H. H.	AREAS OF CONTACT AND PRESSURE DISTRIBUTION IN BOLTED JOINTS	71-0329
GOULD, J. P.	MOVEMENTS OF ARTICULATED CONDUITS UNDER EARTH DAMS ON COMPRESSIBLE FOUNDATIONS	73-0477
GOULD, L.	FINITE ELEMENT ANALYSIS OF SHELLS OF REVOLUTION BY MINIMIZATION OF THE POTENTIAL ENERGY FUNCTIONAL	69-0114
GOULD, P. L.	CRITERIA FOR FINITE ELEMENT DISCRETIZATION OF SHELLS OF REVOLUTION	73-0152
GOULD, P. L.	CURVED ROTATIONAL SHELL ELEMENTS BY THE CONSTRAINT METHOD	72-0924
GOULD, P. L.	DYNAMIC ANALYSIS OF COLUMN-SUPPORTED HYPERBOLOIDAL SHELLS	74-0188
GOULD, P. L.	FINITE ELEMENT ANALYSIS OF SHELLS OF REVOLUTION BY MINIMIZATION OF THE POTENTIAL ENERGY FUNCTIONAL	69-0354
GOULD, P. L.	FREE VIBRATION OF SHELLS OF REVOLUTION USING FINITE ELEMENT METHOD	74-0232
GOULD, P. L.	HIGH PRECISION FINITE ELEMENT ANALYSIS OF SHELLS OF REVOLUTION	71-0524
GOULD, P. L.	HIGH-PRECISION CURVED SHELL FINITE ELEMENT	71-0422
GOULD, P. L.	HIGH-PRECISION PLATE AND SHELL FINITE ELEMENTS	71-0437
GOULD, P. L.	REFINED MIXED METHOD FINITE ELEMENTS FOR SHELLS OF REVOLUTION	71-0101
GOULD, P. L.	STRESS ANALYSIS OF THE HUMAN AORTIC VALVE	73-0166
GOULD, P. L.	STRESS RESULTANTS IN SHELLS BY AN AUGMENTED FINITE ELEMENT METHOD	73-0585
GOULD, PHILLIP	APPROXIMATE ANALYSIS OF PREFABRICATED TOWER STRUCTURES USING THIN SHELL FINITE ELEMENTS	73-0893
GOVIL, A. K.	SHEAR LAG ANALYSIS OF RECTANGULAR FULL-WIDTH TUBE JUNCTIONS	72-0087
GOWAN, M. J.	EFFECT OF THE VARIATION OF CONSISTENCY IN RESIDUAL SOILS ON LOAD TRANSFER BY BORED PILES	75-0812
GRABER, P.-W.	BEITRAG ZUR SIMULATION NICHTSTATIONARER HORIZONTAL-EBENER GRUNDWASSERSTROMUNGSFELDER AUF DER GRUNDLA	74-0088
GRADOWCZYK, M.	VISCOELASTIC STRESS ANALYSIS	68-0080
GRAFF, W. J.	ANALYSIS OF TUBULAR T-CONNECTIONS BY THE FINITE ELEMENT METHOD: COMPARISON WITH EXPERIMENTS	72-0380
GRAFFI, S.	FINITE DIFFERENCE AND FINITE ELEMENT METHODS FOR THE SEMICONDUCTOR FLOW EQUATIONS	74-0028
GRAFTON, P. E.	ANALYSIS OF AXISYMMETRIC SHELLS BY THE DIRECT STIFFNESS METHOD	63-0006
GRAGGS, A.	AN ACOUSTIC FINITE ELEMENT APPROACH FOR STUDYING BOUNDARY FLEXIBILITY AND SOUND TRANSMISSION BETWEEN	73-1087
GRANDLE, R. E.	NODAL ANALYSIS OF A NINE BAY SKIN-STRINGER PANEL	71-0102
GRANEY, L.	RESPONSE OF UNCONFINED AQUIFERS TO PUMPING	74-0073
GRANT, G. N. C.	HYBRID FINITE ELEMENT OPTIMIZATION PROGRAMS FOR DISCS USING A DYNAMICS DESIGN TECHNOLOGY	74-0691
GRANTHAM, P.	COMPUTER PROGRAMS FOR SIMULATING PHYSICAL PHENOMENA IN THREE SPACE DIMENSIONS AND TIME	74-0866
GRAVAS, J.	TOUR MAINE-MONTPARNASSE QUELQUES PROBLEMES DE STRUCTURE (SPECIFIC PROBLEMS ENCOUNTERED IN THE CONSTR	74-0539
GRAVES SMITH, T	A VARIATIONAL METHOD FOR LARGE DEFLECTION ELASTO-PLASTIC THEORY	69-0115
GRAY, D.	A THREE-DIMENSIONAL FINITE ELEMENT COMPUTER CODE FOR THE ANALYSIS OF COMPLEX STRUCTURES	71-0275
GRAY, I.	THE APPLICATION OF FINITE ELEMENT ANALYSIS TO THE STUDY OF TWO-DIMENSIONAL CRACKED BODIES	72-0695
GRAY, W. G.	GALERKIN APPROXIMATION OF THE TIME DERIVATIVE IN THE FINITE ELEMENT ANALYSIS	73-0573
GRAY, W. G.	GALERKIN APPROXIMATION OF THE TIME DERIVATIVE IN THE FINITE ELEMENT ANALYSIS OF GROUNDWATER FLOW	74-0658
GRAY, W. G.	GALERKIN-FINITE ELEMENT TECHNIQUE FOR CALCULATING TRANSIENT POSITION OF SALTWATER FRONT	75-0135
GRAY, W. H.	COMPARISON OF THE FINITE ELEMENT AND FINITE DIFFERENCE METHODS FOR THE ANALYSIS OF STEADY TWO DIMENS	75-0223
GRAY, W. H.	THIN WINDOWS FOR GASEOUS AND LIQUID TARGETS: AN OPTIMIZATION PROCEDURE	75-0741
GRECO, A.	USE OF PAS-1 FINITE ELEMENT COMPUTER SYSTEM FOR STATIC AND DYNAMIC ANALYSIS OF 2-DIMENSIONAL AND 3-D	75-0070
GREEN, D. R.	INTERACTION OF SOLID SHEAR WALLS AND THEIR SUPPORTING STRUCTURES	72-0450
GREEN, P.	POST-TENSIONED SLAB BRIDGES ON ISOLATED SUPPORTS	75-0084
GREENBAUM, G. A	CREEP ANALYSIS OF AXISYMMETRIC BODIES	68-0081
GREENBAUM, G. A	LARGE STRAIN ELASTO-PLASTIC FINITE ELEMENT ANALYSIS	71-0121
GREENBAUM, G. A	PURE MOMENT LOADING OF AXISYMMETRIC FINITE ELEMENT MODELS	73-1007
GREENE, B. E.	ANALYSIS OF AXISYMMETRIC SOLIDS BY THE DIRECT STIFFNESS METHOD	66-0049
GREENE, B. E.	APPLICATION OF GENERALIZED CONSTRAINTS IN THE STIFFNESS METHOD OF STRUCTURAL ANALYSIS	66-0030
GREENE, B. E.	APPLICATION OF THE STIFFNESS METHOD TO THE ANALYSIS OF SHELL STRUCTURES	61-0004
GREENE, B. E.	APPLICATION, TESTING AND EVALUATION OF DOUBLY CURVED SHELL ELEMENTS FOR DYNAMIC ANALYSIS. PART II. F	69-0117
GREENE, B. E.	APPLICATION, TESTING, AND EVALUATION OF DOUBLY CURVED SHELL ELEMENTS FOR DYNAMIC ANALYSIS. PART 1. T	69-0246
GREENE, B. E.	ASTRA-BOEING'S ADVANCED STRUCTURAL ANALYZER	69-0118
GREENE, B. E.	DYNAMIC ANALYSIS OF SHELLS USING DOUBLY-CURVED FINITE ELEMENTS	68-0082
GREENE, B. E.	GENERALIZED VARIATIONAL PRINCIPLES IN THE FINITE ELEMENT METHOD	69-0116
GREENE, B. E.	ON THE APPLICATION OF GENERALIZED VARIATIONAL PRINCIPLES IN THE FINITE ELEMENT METHOD	68-0083
GREENE, B. E.	STIFFNESS MATRIX FOR BENDING OF A RECTANGULAR PLATE ELEMENT WITH INITIAL MEMBRANE STRESSES	62-0006
GREENFIELD, H.	ADAPTABILITY OF COMPUTER GRAPHICS TO STUDIES OF ATHEROSCLEROSIS PATHOGENESIS	72-0206
GREENING, T. A.	ADVANCED WIRE-WOUND TUNGSTEN NOZZLES	66-0031
GREENSTREET, W.	EFFECT OF FILLETS ON STRESS CONCENTRATIONS IN CYLINDRICAL SHELLS WITH STEP CHANGES IN OUTSIDE DIAMET	71-0108
GREENSTREET, W.	EXPERIMENTAL AND FINITE ELEMENT STRESS ANALYSIS OF A THIN SHELLED CYLINDER TO CYLINDER MODEL	71-0066
GREENSTREET, W.	EXPERIMENTAL ELASTIC STRESS ANALYSES OF CYLINDER-TO-CYLINDER SHELL MODELS AND COMPARISONS WITH THEOR	71-0065
GREER, G. G.	MECHANICS OF CABLE MOORING SYSTEMS. VOLUME VII. THE STEADY-STATE BEHAVIOR OF A PYRAMID ARRAY SYSTEM	73-0724
GREER, G. G.	MECHANICS OF CABLE MOORING SYSTEMS. VOLUME VI. A COMPUTER PROGRAM FOR ANALYZING THE STEADY STATE CON	73-0725
GREGORY, J. A.	FINITE ELEMENT ERROR BOUNDS	75-0100
GREGORY, J. A.	LOCAL MESH REFINEMENT WITH FINITE ELEMENTS FOR ELLIPTIC PROBLEMS	73-0073
GREGORY, J. A.	SARD KERNEL THEOREMS ON TRIANGULAR AND RECTANGULAR DOMAINS WITH EXTENSIONS AND APPLICATIONS TO FINIT	72-0361

41

GREGORY, J. A.	THE EXTENSION AND APPLICATION OF SARD KERNEL THEOREMS TO COMPUTE FINITE ELEMENT ERROR BOUNDS	72-0519
GREGORY, M. S.	MEASUREMENT OF SLOPE CHANGE ON CYLINDRICAL SHELL SURFACES	73-0439
GREIGER, I.	INGA - INTERACTIVE GRAPHIC ANALYSIS	74-0681
GREIGER, I.	KONSTRUKTION UND BERECHNUNG MIT DER METHODE FINITE ELEMENTE	73-0707
GREIMANN, L. F.	FINITE ELEMENT ANALYSIS OF PLATE BENDING WITH TRANSVERSE SHEAR DEFORMATION	70-0109
GREIMANN, L. F.	PENETRATION IN SHELLS UNDER EXTERNAL PRESSURE	70-0237
GRESTE, O.	A COMPUTER PROGRAM FOR THE ANALYSIS OF TUBULAR K JOINTS	69-0488
GRESTE, O.	FINITE ELEMENT ANALYSIS OF TUBULAR JOINTS: A REPORT ON A FEASIBILITY STUDY	67-0147
GRESTE, O.	FINITE ELEMENT ANALYSIS OF TUBULAR K JOINTS	70-0427
GRIEGER, I.	A SURVEY OF THE APPLICATION OF FINITE ELEMENT METHODS TO STRESS ANALYSIS PROBLEMS WITH PARTICULAR EM	71-0632
GRIEGER, I.	ASKA - EINE PROGRAMMIERSPRACHE ZUR STATISCHEN UND DYNAMISCHEN ANALYSE	69-0457
GRIEGER, I.	BERECHNUNG VON SPANNUNGSKONZENTRATIONEN AN TRAGWERKEN UNTER THERMO-MECHANISCHER BELASTUNG	73-0686
GRIEGER, I.	CALCULATION OF STRESS CONCENTRATIONS ON SUPPORT STRUCTURES UNDER THERMOMECHANICAL LOADING (BERECHNUN	73-0796
GRIFGER, I.	COMPUTER AIDED DESIGN AND CONSTRUCTION BY MEANS OF CURVED SURFACE AND SOLID ELEMENTS	73-0165
GRIEGER, I.	DESIGN AND CALCULATION WITH FINITE ELEMENTS	74-0161
GRIEGER, I.	DIE ANWENDUNG VON INTERAKTIVER COMPUTER-GRAPHIK BEI DER BERECHNUNG VON TRAGWERKEN. (APPLICATION OF I	72-0102
GRIEGER, I.	DIE METHODE DER ENDLICHEN ELEMENTE UND IHRE ANWENDUNG IM MASCHINENBAU	71-0628
GRIEGER, I.	DISCUSSION ON FINITE ELEMENT ANALYSIS OF PRESTRESSED CONCRETE REACTOR VESSELS	72-0648
GRIEGER, I.	INGA - INTERAKTIVE GRAPHISCHE ANALYSE BENUTZERHAND	73-0701
GRIEGER, I.	INTERACTIVE COMPUTER GRAPHICS IN STRUCTURAL ANALYSIS: ONLINE 72	72-0639
GRIEGER, I.	KONSTRUKTION UND BERECHNUNG MIT FINITEN ELEMENTEN	74-0678
GRIEGER, I.	PARAMETRISCHE UNTERSUCHUNGEN VON KERBEN MIT DER MATRIZENVERSCHIEBUNGSMETHODE UND BILDSCHIRMUNTERSTUT	72-0638
GRIEGER, I.	STRUCTURAL ANALYSIS BY PROBLEM ORIENTED LANGUAGES	71-0623
GRIEGER, I.	THREE DIMENSIONAL ELASTIC AND ELASTO-PLASTIC ANALYSIS OF REACTOR PRESSURE VESSELS	69-0453
GRIESBACH, T. J	FINITE ELEMENT ANALYSIS SYSTEM FOR THE MECHANICAL BEHAVIOR OF ORIENTED FIBER COMPOSITE MATERIALS UND	73-0869
GRIFFIN, D. F.	LAYERED PAVEMENT SYSTEMS. PART 1. LAYERED SYSTEM DESIGN. PART 11. FATIQUE OF PLAIN CONCRETE	72-0867
GRIFFIN, D. S.	PLASTIC COLLAPSE LOADS FOR PIPE ELBOWS USING INELASTIC ANALYSIS	73-0941
GRIFFIN, D. S.	PLASTIC COLLAPSE LOADS FOR PIPE ELBOWS USING INELASTIC ANALYSIS	74-0507
GRIFFIN, D. S.	THE QUALIFICATION OF SOLUTION METHODS	71-0549
GRIFFIN, D. S.	THE VERIFICATION AND ACCEPTANCE OF COMPUTER PROGRAMS FOR DESIGN ANALYSIS	70-0334
GRIFFIS, C. A.	FINITE ELEMENT ANALYSIS OF NOTCHED TENSILE SPECIMENS IN PLANE STRESS	71-0557
GRIFFITH, W. I.	INTERFERENCE-FIT FASTENER DISPLACEMENT MEASUREMENT BY SPECKLE PHOTOGRAPHY	74-1178
GRIFFITHS, J. R	ELASTIC-PLASTIC STRESS ANALYSIS FOR A NOTCHED BAR IN PLANE STRAIN BENDING	71-0410
GRIFFITHS, J. R	RULES GOVERNING THE NUMBER OF NODES AND ELEMENTS IN A FINITE ELEMENT MESH	70-0053
GRIFFITHS, J. R	STRESS INTENSITY FACTORS FOR CRACKS IN A PLATE CONTAINING A HOLE AND IN A SPINNING DISC	73-0467
GRIFFITHS, J. R	STRESSES IN A PARTLY YIELDED NOTCHED BAR - AN ASSESSMENT OF THREE ALTERNATIVE PROGRAMS	73-0267
GRILLO, A. F.	SOME REMARKS ON INFINITE VENEZIANO REPRESENTATIONS	69-0472
GRIMM, A. F.	DEFORMATION OF THE END DIASTOLIC LEFT VENTRICLE, COMPARISON OF THEORY WITH EXPERIMENT	71-0131
GRIMM, A. F.	FINITE-ELEMENT MODEL FOR THE MECHANICAL BEHAVIOUR OF THE LEFT VENTRICLE, PREDICTION OF DEFORMATION I	72-0304
GRIN, R.	STRUCTURAL EVALUATION OF CANDIDATE DESIGNS FOR THE LARGE SPACE TELESCOPE PRIMARY MIRROR	75-0708
GRINDLEY, R. E.	INEXPENSIVE, INTERACTIVE COMPUTER-AIDED DESIGN APPLIED TO FINITE ELEMENT STRESS ANALYSIS	74-0592
GRINSHPAN, Z.	FINITE-ELEMENT MODEL FOR POSSIBLE ISOSTATIC REBOUND IN GRAND CANYON	75-0037
GRISHAM, A. F.	THE BOEING SST PROTYPE INTERNAL LOADS ANALYSIS SYSTEM AND PROCEDURES	71-0103
GROOMS, D. W.	FINITE ELEMENTS IN STRUCTURAL ANALYSIS	75-0739
GROOMS, D. W.	NUMERICAL METHODS IN FLUID FLOW PROBLEMS. VOL. 2.	75-0714
GROOMS, D. W.	NUMERICAL METHODS IN FLUID FLOW PROBLEMS. VOL. 1	75-0715
GROOMS, D. W.	STRUCTURAL MECHANICS SOFTWARE. A BIBLIOGRAPHY WITH ABSTRACTS	74-0842
GROOMS, H. R.	SPACE SHUTTLE, THE NEED FOR SUBSTRUCTURING	71-0104
GROOMS, H. R.	ALGORITHM FOR MATRIX BANDWIDTH REDUCTION	72-0895
GROS, G.	TENUE DES STRUCTURES DE GRAPHITE VIS-A-VIS DES SOLLICITATIONS RENCONTREES DANS UN COEUR DE REACTEUR	73-0817
GROSS, B.	EVALUATION OF THE USE OF A SINGULARITY ELEMENT IN FINITE ELEMENT ANALYSIS OF CENTER-CRACKED PLATES	72-0870
GROSS, H.	CLASSICAL AND NATURAL CONDUCTION AND DEFORMATION MODES CONCEPTS FOR AXISYMMETRIC VARIATIONAL METHODS	72-1003
GROSS, H.	THERMOVISCOELASTIC AXISYMMETRIC STRESS ANALYSIS BY FINITE ELEMENT AND FINITE DIFFERENCE COMPUTER TEC	71-0105
GROSSKURTH, J.	SHEAR BUCKLING OF SQUARE PERFORATED PLATES	74-0713
GROTH, P.	METHODE DER ELASTOSTATIK PROGRAMMIERUNG UND ANWENDUNG	70-0381
GROTKOP, G.	DIE BERECHNUNG VON FLACHWASSERWELLEN NACH DER METHODE DER FINITEN ELEMENTE	72-0359
GROTKOP, G.	FINITE ELEMENT ANALYSIS OF LONG-PERIOD WATER WAVES	73-1008
GRUBBS, B. R.	A FINITE-ELEMENT METHOD OF SOLUTION FOR STRUCTURAL FRAMES	67-0133
GRUBER, R.	A USEFUL FORM OF THE VARIATIONAL PRINCIPLE OF IDEAL MHD FOR ONE-DIMENSIONAL NUMERICAL STABILITY CALC	73-0746
GRUBER, R.	BIDIMENSIONAL COMPUTER-PROGRAM FOR CALCULATION OF GROWTH-RATES OF INCLUDED AXIALLY SYMMETRICAL PLASM	75-0420
GRUBER, R.	DETERMINATION OF EIGENMODES OF HOMOGENEOUS PLASMA WITH ELLIPTICAL SECTION USING BI-DIMENSIONAL FINIT	74-0553
GRUBER, R.	EIGENMODES OF CYLINDRICAL PLASMA USING FINITE ELEMENT METHOD	74-0174
GRUBER, R.	NEW FINITE ELEMENT APPROACH TO THE NORMAL MODE ANALYSIS IN MAGNETOHYDRODYNAMICS	74-0876
GRUBER, R.	NJMERICAL STUDY OF EIGEN OSCILLATIONS OF A SCREW PINCH WITH FINITE ELEMENTS	74-0159
GRUBER, R.	NUMERICAL STUDY OF PROPER MODES OF A HOMOGENEOUS CYLINDRICAL PLASMA BY METHOD OF FINITE ELEMENTS	74-0160
GRUBER, R.	THALIA - ONE-DIMENSIONAL MAGNETOHYDRODYNAMIC STABILITY PROGRAM USING METHOD OF FINITE-ELEMENTS	75-0458
GRUBER, R.	THALIA: A ONE-DIMENSIONAL MAGNETOHYDRODYNAMIC STABILITY PROGRAM USING THE METHOD OF FINITE ELEMENTS	74-1085
GRUBER, R.	USEFUL FORM OF THE VARIATIONAL PRINCIPLE OF IDEAL MHD FOR ONE-DIMENSIONAL NUMERICAL STABILITY CALCUL	73-0946
GRUBITZSCH, W.	APPROXIMATION METHOD FOR COMPUTATION OF NATURAL FREQUENCIES OF SPIRAL SECTIONS	71-0775
GRUNDY, P.	STATISTICAL ASPECTS OF THE STABILITY OF COMPOUND SYSTEMS WITH BILINEAR ELEMENTS	73-0986
GRYCZMANSKI, M.	GENERATION OF SERENDIPEAN SHAPE FUNCTIONS IN FINITE ELEMENT METHOD	74-1012
GRYCZMANSKI, M.	SCALAR ALGORITHM OF DETERMINATION OF STIFFNESS MATRIX AND OF ANOTHER FINITE ELEMENT CHARACTERISTIC	74-1013
GUBBELS, M. H.	ANALYSIS OF AN IN-PILE REACTOR TUBE	75-0193
GUELLEC, P.	CALCULATION OF FLOWS IN POROUS MEDIA BY THE FINITE ELEMENT METHOD	70-0473
GUELLEC, P.	LES PLAQUES ET LES COQUES MINCES OU EPAISSES (THIN AND THICK PLATES AND SHELLS)	71-0432
GUERRA, F. M.	ORGANIZATION AND FUNCTIONAL PURPOSE OF A FINITE ELEMENT COMPUTER PROGRAM	74-0362
GUERRINI, M.	A FINITE ELEMENT PROCEDURE FOR NUMERICAL OPTIMIZATION IN MECHANICAL DESIGN	74-1218
GUGLIELMO, F. D	METHODE DES ELEMENTS FINIS; UNE FAMILE D'APPROXIMATION DES ESPACES DE SOBOLEV PAR LES TRANSLATES DE	70-0301
GUILINGE, W. H.	FINITE ELEMENT METHOD	71-0570
GULATI, S.	USE OF SINGULAR FUNCTIONS WITH FINITE ELEMENT APPROXIMATIONS	73-0102
GUNDERSON, R. H	ELEMENT STIFFNESS MATRIX GENERATOR	71-0106
GUPALP, Y. U. P	DIFFUSION ON SOLID SPHERICAL PARTICLE IN VISCOUS FLUID FLOW AT FINITE VALUES OF RE NUMBERS	69-0419
GUPTA, A. K.	A METHOD OF COMPUTING NUMERICALLY INTEGRATED STIFFNESS MATRICES	72-0546
GUPTA, A. K.	ELASTO-PLASTIC ANALYSIS OF THREE-DIMENSIONAL STRUCTURES USING THE ISOPARAMETRIC ELEMENT	72-0248
GUPTA, A. K.	3-DIMENSIONAL ISOPARAMETRIC ELEMENT IN CURVILINEAR COORDINATES	72-0098
GUPTA, B. P.	SOME RECENT FINITE ELEMENT APPROACHES IN ANALYSIS OF PLASTIC IMPACT	69-0299
GUPTA, K. K.	DYNAMIC RESPONSE ANALYSIS OF GEOMETRICALLY NON-LINEAR STRUCTURES SUBJECTED TO HIGH IMPACT	72-0555
GUPTA, K. K.	FREE VIBRATION ANALYSIS OF SPINNING STRUCTURAL SYSTEMS	73-1077
GUPTA, K. K.	MATHEMATICAL MODELLING AND STUCTURAL ANALYSIS OF THE MANDIBLE	72-0124
GUPTA, K. K.	ON A COMBINED STURM SEQUENCE AND INVERSE ITERATION TECHNIQUE FOR EIGENPROBLEM SOLUTION OF SPINNING S	73-0413
GUPTA, K. K.	RECENT ADVANCES IN NUMERICAL ANALYSIS OF STRUCTURAL EIGENVALUE PROBLEMS	73-0316
GUPTA, K. K.	SOLUTION OF EIGENVALUE PROBLEMS BY STURM SEQUENCE METHOD.	72-0400
GUPTA, K. K.	VIBRATION OF FRAMES AND OTHER STRUCTURES WITH BANDED STIFFNESS MATRIX	70-0322
GUPTA, K. K.	VISCEL A GENERAL PURPOSE COMPUTER PROGRAM FOR ANALYSIS OF LINEAR VISCOELASTIC STRUCTURES. VOL. 2	72-0789
GUPTA, M. M.	SOME DIFFERENCE SCHEMES FOR THE BIHARMONIC EQUATION	75-0204
GUPTA, S. K. S.	A FINITE SOLUTION OF PUCHER'S EQUATION	75-0782
GUPTAK, D. C.	FINITE ELEMENT ANALYSIS FOR THE SEISMIC STABILITY OF EARTH STRUCTURES	74-1106
GUREGHIAN, A. B	CALCULATION OF STEADY-STATE WATER-TABLE HEIGHTS IN DRAINED SOILS BY MEANS OF FINITE-ELEMENT METHOD	75-0457
GUREGHIAN, A. B	STUDY BY THE FINITE-ELEMENT METHOD OF THE INFLUENCE OF FRACTURES IN CONFINED AQUIFERS.	75-0604
GUREGHIAN, A. B	THE USE OF THE FINITE ELEMENT METHOD IN OBTAINING SOIL-WATER PRESSURE AND SOIL-WATER CONTENT DISTRIB	75-0112
GURFINKEL, G.	SEISMIC BEHAVIOUR OF FRAMED TUBES	75-0306
GURLAND, J.	COMBINED MACROSCOPIC AND MICROSCOPIC APPROACH TO THE FRACTURE OF METALS	74-0897

GURTIN, M. E.	GALERKIN METHOD AS APPLIED TO PROBLEMS IN VISCOELASTICITY	74-1039
GURTIN, M. E.	ON THE SEMI-DISCRETE GALERKIN METHOD FOR HYPERBOLIC PROBLEMS AND ITS APPLICATION TO PROBLEMS IN ELAS	72-0457
GURTIN, M. E.	VARIATIONAL PRINCIPLES FOR LINEAR INITIAL VALUE PROBLEMS	64-0029
GUTHER, H.	SEEPAGE INTO AN EXCAVATION IN A MEDIUM POSSESSING STRESS-DEPENDENT PERMEABILITY	72-0629
GUTZWILLER, M.	FINITE ELEMENT FOR REINFORCED CONCRETE FRAME STUDY	73-0204
GUTZWILLER, M.	FINITE ELEMENT FOR REINFORCED CONCRETE FRAME STUDY - CLOSURE	75-0087
GUY, A. W.	EFFECT OF 2450-MHZ RADIATION ON THE RABBIT EYE	75-0365
GUYAN, R. J.	DISTRIBUTED MASS MATRIX FOR PLATE ELEMENTS IN BENDING	65-0018
GUYAN, R. J.	HYDROELASTIC ANALYSIS OF AXISYMMETRIC SYSTEMS BY A FINITE ELEMENT METHOD	68-0084
GUYAN, R. J.	REDUCTION OF STIFFNESS AND MASS MATRICES	65-0037
GUYMON, G. L.	A FINITE ELEMENT SOLUTION OF THE ONE DIMENSIONAL DIFFUSION CONVECTION EQUATION	70-0258
GUYMON, G. L.	A GENERAL NUMERICAL SOLUTION OF THE TWO-DIMENSIONAL DIFFUSION-CONVECTION EQUATION BY THE FINITE ELEM	70-0004
GUYMON, G. L.	APPLICATION OF FINITE ELEMENT METHOD TO GENERAL NUMERICAL SOLUTION OF 2-DIMENSIONAL DIFFUSION-CONVEC	70-0269
GUYMON, G. L.	APPLICATION OF THE FINITE-ELEMENT METHOD FOR SIMULATION OF SURFACE WATER TRANSPORT PROBLEMS	72-0826
GUYMON, G. L.	COUPLED HEAT AND MOISTURE TRANSPORT MODEL FOR ARCTIC SOILS	74-0578
GUYMON, G. L.	FINITE ELEMENT SOLUTION OF DIFFUSION-CONVECTION EQUATION	72-0018
GUYMON, G. L.	FINITE ELEMENT SOLUTION FOR GENERAL FLUID MOTION	73-0249
GUYMON, G. L.	MATHEMATICAL MODELING OF MOVEMENT OF DISSOLVED CONSTITUENTS IN GROUND WATER AQUIFERS BY THE FINITE E	70-0490
GWALTNEY, R. C.	ANALYTIC INVESTIGATION OF COMPACT REINFORCEMENT FOR RADIAL NOZZLES IN SPHERICAL SHELLS	71-0107
GWALTNEY, R. C.	EFFECT OF FILLETS ON STRESS CONCENTRATIONS IN CYLINDRICAL SHELLS WITH STEP CHANGES IN OUTSIDE DIAMET	71-0108
GWALTNEY, R. C.	THEORETICAL AND EXPERIMENTAL STRESS ANALYSES OF ORNL THIN-SHELL CYLINDER-TO-CYLINDER MODEL 3	75-0690
GWALTNEY, R. C.	THEORETICAL AND EXPERIMENTAL STRESS ANALYSES OF ORNL THIN-SHELL CYLINDER-TO-CYLINDER MODEL 4	75-0695
GWALTNEY, W. L.	EXPERIMENTAL AND FINITE ELEMENT STRESS ANALYSIS OF A THIN SHELLED CYLINDER TO CYLINDER MODEL	71-0066
HA, H. K.	ANALYSIS OF THREE-DIMENSIONAL ORTHOTROPIC SANDWICH PLATE STRUCTURES BY FINITE-ELEMENT METHOD	72-0945
HA, K. H.	FINITE ELEMENT ANALYSIS OF SPATIAL SANDWICH PLATE STRUCTURES	74-1132
HA, K. H.	SANDWICH PLATE STRUCTURE ANALYSIS BY FINITE ELEMENT	74-0163
HAALAND, E.	TESTRAN PROGRAMS, A TOOL FOR TEMPERATURE AND STRESS ANALYSIS	73-0414
HAASE, M.	GEOMETRIC NONLINEARITY AND THE FINITE ELEMENT DISPLACEMENT METHOD	74-0679
HAASE, M.	NATURAL GEOMETRY OF SURFACE WITH SPECIFIC REFERENCE TO THE MATRIX DISPLACEMENT ANALYSIS OF SHELLS, 1	74-0709
HABASHI, W. G.	A STUDY OF THE FINITE ELEMENT METHOD FOR AERODYNAMIC APPLICATIONS	75-0793
HABERCOM, G. E.	MATHEMATICAL ANALYSIS OF STRESS CRACKS. VOLUME 2.	75-0742
HABERCOM, G. E.	MATHEMATICAL ANALYSIS OF STRESS CRACKS. VOLUME 1	75-0743
HABERCOM, G. E.	STRESS CRACK PHENOMENA	75-0744
HABERCOM, G. E.	TUNNEL CONSTRUCTION - A BIBLIOGRAPHY WITH ABSTRACTS	74-0748
HABERL, G.	FINITE ELEMENTS SOLUTION FOR TORSION RIGIDITY AND SHEAR CENTER OF ARBITRARY CROSS-SECTIONS	74-0168
HABIP, L. M.	STRUCTURAL EFFECTS OF CONFINED DYNAMIC LOADS	72-0151
HACKETT, R. M.	APPLICATION OF FINITE ELEMENT METHODS IN CIVIL ENGINEERING	69-0031
HACKNEY, J. R.	PROBLEMS IN BUILDING AND MANIPULATING LARGE STIFFNESS MATRICES	71-0596
HADCOCK, R.	THE BEHAVIOUR OF ADVANCED FILAMENTARY COMPOSITE PLATES WITH CUTOUTS	70-0025
HADDADIN, M. J.	MATS AND COMBINED FOOTINGS. ANALYSIS BY THE FINITE ELEMENT METHOD	71-0289
HADDOW, J. B.	A NUMERICAL METHOD FOR THE ANALYSIS OF LONGITUDINAL ELASTIC PLASTIC STRESS WAVE PROPAGATION	73-0526
HADEDANK, G.	FREQUENCY-DEPENDENT SHAPE FUNCTIONS FOR FREE AND FORCED STRUCTURAL VIBRATION CALCULATION	73-1091
HADJIAN, A. H.	DYNAMIC MODELING OF HALF-PLANE BY FINITE ELEMENTS	74-0953
HADJIAN, A. H.	SIMPLIFIED SOIL-STRUCTURE INTERACTION ANALYSIS WITH STRAIN DEPENDENT SOIL PROPERTIES	74-1052
HADJIAN, A. H.	SOIL-STRUCTURE INTERACTION - CONTINUUM OR FINITE ELEMENT	74-0954
HAENER, J.	TRANSVERSE LOADING OF UNIDIRECTIONAL FIBER COMPOSITES	69-0119
HAFTKA, R. T.	A KOITER TYPE METHOD FOR FINITE ELEMENT ANALYSIS OF NONLINEAR STRUCTURAL BEHAVIOUR	70-0064
HAFTKA, R. T.	A KOITER-TYPE METHOD FOR FINITE ELEMENT ANALYSIS OF NONLINEAR STRUCTURAL BEHAVIOR	70-0493
HAFTKA, R. T.	ADAPTION OF KOITER'S METHOD TO FINITE ELEMENT ANALYSIS OF SNAP-THROUGH BUCKLING BEHAVIOUR	71-0109
HAFTKA, R. T.	AUTOMATED PROCEDURE FOR DESIGN OF WING STRUCTURES TO SATISFY STRENGTH AND FLUTTER REQUIREMENTS	73-0867
HAFTKA, R. T.	EFFECT OF OUT-OF-PLANENESS OF MEMBRANE QUADRILATERAL FINITE ELEMENTS	73-0031
HAFTKA, R. T.	PROGRESS IN NONLINEAR FINITE ELEMENT ANALYSIS USING ASYMPTOTIC SOLUTION TECHNIQUES	72-0316
HAGEDOORN, A. H	AN ELASTIC FINITE-ELEMENT ANALYSIS OF THE BUCKLING OF ECCENTRICALLY STIFFENED PLATES AND SHELLS	73-1136
HAGEN, R. L.	SYMMETRICAL BENDING OF CIRCULAR PLATES USING FINITE ELEMENTS	72-0428
HAGER, W. W.	FREE BOUNDARIES AND FINITE-ELEMENTS IN ONE DIMENSION	75-0447
HAGER, W. W.	FREE BOUNDARIES AND FINITE-ELEMENTS IN ONE DIMENSION	75-0587
HAGGENMACHER, G	A CURVED BEAM ELEMENT	68-0202
HAGGENMACHER, G	BASIS FOR ELEMENT INTERCHANGEABILITY IN FINITE ELEMENT PROGRAMS	71-0224
HAGGENMACHER, G	ELEMENTS FOR FINITE ELEMENT STRUCTURAL ANALYSIS PROGRAMS	71-0335
HAGGENMACHER, G	EVALUATION OF CALAC SEMI-MONOCOQUE MEMBRANE ELEMENT	70-0278
HAGGENMACHER, G	OPTIMIZATION OF REDUNDANCY SELECTION IN FINITE-ELEMENT FORCE METHOD	70-0228
HAGGENMACHER, G	SOME NEW DEVELOPMENTS IN MATRIX FORCE ANALYSIS	71-0225
HAGGENMACHER, G	THE USE OF NASTRAN IN THE ANALYSIS OF LARGE COMPLEX AIRFRAME STRUCTURES, MODELING TECHNIQUES AND ORG	71-0110
HAGGENMACHER, G	WARPED SEMI-MONOCOQUE QUADRILATERAL MEMBRANE ELEMENT FOR THE FINITE ELEMENT METHODS	70-0195
HAGUE, M. I.	CONVERGENCE OF THE FINITE-ELEMENT METHOD UNDER THE CONSTRAINT OF MATERIAL INCOMPRESSIBILITY	73-1135
HAHN, G. T.	FAST FRACTURE RESISTANCE AND CRACK ARREST IN STRUCTURAL STEELS	73-0766
HAHN, O.	UNTERSUCHUNGEN ZUM FESTIGKEITSVERHALTEN VON METALLKLEBUNGEN IM HINBLICK AUF DIE ERSTELLUNG EINES PRA	74-0289
HAISLER, W. E.	A COMPARISON OF CURRENT WORK-HARDENING MODELS USED IN THE ANALYSIS OF PLASTIC DEFORMATIONS	73-0769
HAISLER, W. E.	A FINITE ELEMENT PROGRAM FOR THE DYNAMIC LARGE DEFLECTION, ELASTIC-PLASTIC ANALYSIS OF STIFFENED SHE	72-0908
HAISLER, W. E.	COMPUTATION AND SOLUTION PROCEDURE FOR NONLINEAR ANALYSIS BY COMBINED FINITE ELEMENT-FINITE DIFFEREN	72-0571
HAISLER, W. E.	DYNAPLAS: A FINITE ELEMENT PROGRAM FOR THE DYNAMIC, LARGE DEFLECTION, ELASTIC-PLASTIC ANALYSIS OF ST	72-0773
HAISLER, W. E.	EVALUATION OF SOLUTION PROCEDURES FOR MATERIAL AND/OR GEOMETRICALLY NONLINEAR STRUCTURAL ANALYSIS	73-0362
HAISLER, W. E.	FAMSOR, A FINITE ELEMENT PROGRAM FOR THE FREQUENCIES AND MODE SHAPES OF SHELLS OF REVOLUTION	70-0090
HAISLER, W. E.	FORMULATION, COMPUTATION, AND SOLUTION PROCEDURES FOR MATERIAL AND/OR GEOMETRIC NONLINEAR STRUCTURAL	70-0346
HAISLER, W. E.	GEOMETRICALLY NONLINEAR ANALYSIS BY THE DIRECT STIFFNESS METHOD	70-0331
HAISLER, W. E.	GEOMETRICALLY NONLINEAR STRUCTURAL ANALYSIS BY THE DIRECT STIFFNESS METHOD	71-0560
HAISLER, W. E.	LARGE DEFLECTION ELASTIC-PLASTIC DYNAMIC RESPONSE OF STIFFENED SHELLS OF REVOLUTION	72-0608
HAISLER, W. E.	LARGE DEFLECTION ELASTIC-PLASTIC DYNAMIC RESPONSE OF STIFFENED SHELLS OF REVOLUTION	74-0496
HAISLER, W. E.	NONLINEAR ANALYSIS OF SHELLS OF REVOLUTION BY THE MATRIX DISPLACEMENT METHOD	67-0079
HAISLER, W. E.	NONLINEAR FINITE ELEMENT ANALYSIS INCLUDING HIGHER ORDER STRAIN ENERGY TERMS	70-0065
HAISLER, W. E.	RIGID BODY DISPLACEMENTS OF CURVED ELEMENTS IN THE ANALYSIS OF SHELLS BY THE MATRIX-DISPLACEMENT MET	67-0031
HAISLER, W. E.	SAMMSOR III: A FINITE ELEMENT PROGRAM TO DETERMINE STIFFNESS AND MASS MATRICES OF RING-STIFFENED SHE	72-0774
HAISLER, W. E.	STATIC GEOMETRIC AND MATERIAL NONLINEAR ANALYSIS	72-0314
HAISLER, W. E.	SURVEY OF SOLUTION PROCEDURES FOR NONLINEAR STATIC AND DYNAMIC ANALYSES	74-0307
HAISLER, W. E.	SURVEY OF STATIC GEOMETRIC AND MATERIAL NONLINEAR ANALYSIS BY THE FINITE ELEMENT METHOD	72-0783
HAJIDJAFARI, S.	NUMERICAL PREDICTION OF TRACER CONCENTRATION IN AQUIFERS WITH TRANSIENT GROUNDWATER FLOW BY FINITE-E	75-0610
HALAT, W.	DER SPANNUNGS UND BEANSPRUCHUNGSZUSTAND DES BODENMATERIALS IN DER NACHBARSCHAFT EINER BODENBOESCHUNG	75-0371
HALIBURTON, T.	A FINITE ELEMENT METHOD OF SOLUTION FOR LINEARLY ELASTIC BEAM-COLUMNS	66-0117
HALIBURTON, T.	FINITE ELEMENT ANALYSIS OF STRUCTURAL FRAMES	67-0032
HALINA, H.	NUMERICAL DETERMINATION OF STRESSES AND DEFORMATIONS IN ROCK TAKING INTO ACCOUNT DISCONTINUITIES	70-0066
HALL, A. S.	FINITE ELEMENT TECHNIQUES - A SHORT COURSE OF FUNDAMENTALS AND APPLICATIONS	73-0612
HALL, C. A.	BICUBIC INTERPOLATION OVER TRIANGLES	69-0367
HALL, C. A.	CONSTRUCTION OF CURVILINEAR COORDINATE SYSTEMS AND APPLICATIONS TO MESH GENERATION	74-1295
HALL, C. A.	DISCRETIZATION ERROR BOUNDS FOR TRANSFINITE ELEMENTS	72-0533
HALL, C. A.	GEOMETRIC ASPECTS OF THE FINITE ELEMENT METHOD	72-0523
HALL, C. A.	GEOMETRIC ASPECTS OF THE FINITE ELEMENT METHOD: CONSTRUCTION OF CURVILINEAR COORDINATE SYSTEMS AND T	72-0534
HALL, C. A.	TRANSFINITE ELEMENT METHODS: BLENDING FUNCTION INTERPOLATION OVER ARBITRARY CURVED ELEMENT DOMAINS	73-1006
HALL, D. B.	BENDING ELEMENTS FOR PLATE AND SHELL NETWORKS	67-0043
HALL, D. G.	AN ANALYSIS OF THE LARGE DEFLECTIONS OF BEAMS USING THE RAYLEIGH-RITZ FINITE ELEMENT METHOD	68-0162
HALL, E. J.	ICES-STRUDL-II	68-0257
HALL, U.	THE FINITE ELEMENT METHOD APPLIED TO COMPRESSIBLE CASCADE FLOW	75-0179
HALLWACHS, D. A	AN INVESTIGATION INTO THE METHODS OF DETECTING DEWETTED ZONES IN SOLID ROCKET PROPELLANTS USING HOLO	73-0890

43

HALSEY, N.	NON-METALLIC ANTENNA-SUPPORT MATERIALS	73-0716
HALTEMAN, E. K.	PERIODIC HEAT CONDUCTION IN A TWO PHASE, TWO DIMENSIONAL SOLID DOMAIN	74-1164
HALTINER, G. J.	RECENT ADVANCES IN NUMERICAL WEATHER PREDICTION	74-0787
HAMANN, W. C.	FINITE ELEMENT METHODS IN PRODUCT DESIGN	74-0117
HAMANN, W. C.	HOW FINITE ELEMENT METHODS ARE INTRODUCED IN LARGE AND SMALL ORGANIZATIONS	74-0366
HAMANN, W. C.	INTERFACING FINITE ELEMENT METHODS WITH PRODUCT DESIGN ENGINEERING	73-0189
HAMELIN, P.	CHARACTERISATION ELASTIQUE D'UN STRATIFIE RESINE-VERRE PAR LA METHODE DES DEPLACEMENTS (DETERMINATIO	75-0272
HAMID, M. S.	FINITE ELEMENT ANALYSIS OF HUMAN CARDIAC STRUCTURES	74-0438
HAMID, M. S.	FINITE ELEMENT METHOD FOR PREDICTION OF CRACK BEHAVIOUR	70-0342
HAMILTON, J.	STRUCTURAL EVALUATION OF CANDIDATE DESIGNS FOR THE LARGE SPACE TELESCOPE PRIMARY MIRROR	75-0708
HAMMEL, D. J.	AN APPLICATION OF THE FINITE ELEMENT METHOD FOR ROCK SLOPE STABILITY ANALYSIS	71-0813
HAMPHREYS, J. D	PERFORMANCE OF CULVERT UNDER WINSCAR DAM	75-0638
HANAI, S.	APPLICATION OF FINITE-ELEMENT METHOD AND MOIRE TECHNIQUE TO SHEET-METAL FORMING PROBLEMS	71-0784
HAND, F. R.	NONLINEAR LAYERED ANALYSIS OF RC PLATES AND SHELLS	73-0208
HAND, F. R. JR.	NONLINEAR LAYERED FINITE-ELEMENT ANALYSIS OF REINFORCED-CONCRETE PLATES AND SHELLS	72-0946
HANDA, K. N.	ANALYSIS OF IN-PLANE VIBRATION OF SHEAR WALLS BY A FINITE ELEMENT METHOD	72-0090
HANDA, K. N.	ANALYSIS OF INPLANE VIBRATION OF BOX-TYPE STRUCTURES BY A FINITE ELEMENT METHOD	72-0091
HANDA, K. N.	APPLICATION OF FINITE ELEMENT METHOD TO THE DYNAMIC ANALYSIS OF TALL STRUCTURES	71-0111
HANEKE, R.	TEMPERATURVERLAUF IM QUERSCHNITT GESCHWEISSTER ROHRWAENDE BEI JEBERWEIGEND KONVEKTIVEM WAERMEUEBERGA	75-0628
HANGAI, Y.	PERTURBATION METHOD IN THE ANALYSIS OF GEOMETRICALLY NON LINEAR AND STABILITY PROBLEMS	72-0321
HANLEIN, S. L.	NONLINEAR CONSIDERATION OF GRAVITY IN A STIFFNESS TEST OF A WEAK STRUCTURE AT SMALL STRAINS	71-0035
HANRATTY, T. J.	NUMERICAL SOLUTION FOR THE FLOW AROUND A CYLINDER AT REYNOLDS NUMBERS OF 40, 200 AND 500	69-0378
HANS, D.	STRESS ANALYSIS OF TUBULAR JOINTS FOR OFFSHORE STRUCTURES	73-0207
HANSEN, H. R.	HULL AND SUPERSTRUCTURE VIBRATIONS DESIGN CALCULATIONS BY FINITE ELEMENTS	75-0556
HANSEN, H. R.	SOME EXAMPLES OF THE APPLICATION OF THE FINITE ELEMENT METHOD SHIP STRUCTURES	74-0189
HANSEN, K.	FINITE ELEMENT METHOD APPLIED TO NEUTRON DIFFUSION PROBLEMS	73-0977
HANSEN, K. F.	ANNUAL PROGRESS REPORT, FY1973	73-0969
HANSEN, K. F.	APPLICATION OF FINITE ELEMENT METHOD TO 2-DIMENSIONAL DIFFUSION PROBLEMS	73-0098
HANSEN, K. F.	FINITE ELEMENT METHODS FOR THE NEUTRON DIFFUSION EQUATIONS	71-0372
HANSEN, K. F.	FINITE ELEMENT METHODS FOR SPACE-TIME REACTOR ANALYSIS	71-0673
HANSEN, K. F.	FINITE ELEMENT METHODS FOR REACTOR ANALYSIS	73-0108
HANSEN, K. F.	FINITE ELEMENT METHODS IN REACTOR PHYSICS ANALYSIS	75-0020
HANSEN, K. F.	FINITE ELEMENT SOLUTIONS FOR MULTIREGION PROBLEMS	73-1021
HANSEN, K. F.	FINITE-ELEMENT METHODS IN REACTOR-PHYSICS ANALYSIS	75-0502
HANSEN, S. D.	ANALYSIS OF THE 747 AIRCRAFT WING BODY INTERSECTION	68-0275
HANSEN, S. D.	LARGE SCALE ANALYSIS OF CURRENT AIRCRAFT	70-0383
HANSEN, S. D.	SAMECS STRUCTURAL ANALYSIS SYSTEM - USER'S DOCUMENT	69-0495
HANSLIAN, J.	SOLUTION OF STRESSES AND TRANSFORMATIONS OF EARTH DAMS BY THE METHOD OF FINITE ELEMENTS	69-0426
HANSON, D. E.	THREE-DIMENSIONAL ELASTO-STATIC PROBLEMS USING ISOPARAMETRIC FINITE ELEMENTS	71-0751
HANSSEN, L.	A NEW APPROACH FOR DERIVING GOOD ELEMENT STIFFNESS MATRICES	75-0128
HANSTEEN, H.	FINITE ELEMENT DISPLACEMENT ANALYSIS OF PLATE BENDING BASED ON RECTANGULAR ELEMENTS	66-0032
HANSTEEN, O. E.	A CONICAL ELEMENT FOR DISPLACEMENT ANALYSIS OF AXI-SYMMETRIC SHELLS	69-0120
HANSTEEN, O. E.	ANALYSIS OF STRESS DISTRIBUTION IN SHEAR WALLS BY THE FINITE ELEMENT DISPLACEMENT METHOD	67-0033
HANSTEEN, O. E.	FINITE ELEMENT METHODS AS APPLICATIONS OF VARIATIONAL PRINCIPLES	69-0121
HANSTEEN, O. E.	ORSAM - A PROGRAMMING SYSTEM FOR THE FINITE ELEMENT METHOD	73-0528
HAQUE, M. I.	HIGH ORDER FINITE ELEMENT FOR COMPLETELY INCOMPRESSIBLE CREEPING FLOW	73-0144
HAQUE, M. N.	TENSILE CREEP ANALYSIS OF CONCRETE STRUCTURES	74-0439
HARBERL, G.	EINE FINITE-ELEMENTE-LOESUNG FUR DIE TORSIONSSTEIFIGKEIT UND DEN SCHUBMITTELPUNKT BELIEBIGER QUERSCH	74-0256
HARBORD, R.	COMPUTATION OF SHELLS WITH FINITE DISPLACEMENTS: MIXED FINITE ELEMENTS (BERECHNUNG VON SCHALEN MIT E	73-0788
HARDCASTLE, J.	SEEPAGE AND GROUNDWATER EFFECTS ASSOCIATED WITH EXPLOSIVE CRATERING	72-0656
HARDER, R. L.	A DESIGN STUDY FOR THE INCORPORATION OF AEROELASTIC CAPABILITY INTO NASTRAN	71-0112
HARDER, R. L.	KERNEL FUNCTION FOR NONPLANAR OSCILLATING SURFACES IN SUPERSONIC FLOW	74-0472
HARDY, C.	THE ELASTO-PLASTIC INDENTATION OF A HALF-SPACE BY A RIGID SPHERE	71-0113
HARDY, C. H.	ELASTIC ANALYSIS OF A SKULL	73-0415
HARDY, H. R.	APPLICATION OF FINITE ELEMENT ANALYSIS TO UNDERGROUND STORAGE OF NATURAL GAS	70-0198
HARDY, R. H.	A HIGH-ORDER FINITE ELEMENT FOR TWO-DIMENSIONAL CRACK PROBLEMS	74-1128
HAREN, R.	FINITE ELEMENT ANALYSIS OF ELECTROMAGNETIC FIELD PROBLEMS	74-0080
HAREN, R. J.	THREE DIMENSIONAL FINITE ELEMENT APPROXIMATION IN ELECTROMAGNETIC EXPLORATION	74-0358
HARKEGARD, G.	APPLICATION OF FINITE ELEMENT METHOD TO CYCLIC LOADING OF ELASTIC-PLASTIC STRUCTURES CONTAINING EFFE	73-0103
HARKEGARD, G.	ON THE FINITE ELEMENT ANALYSIS OF CRACK AND INCLUSION PROBLEMS IN ELASTIC-PLASTIC MATERIALS	74-0227
HARKEGARD, G.	THEORETICAL STUDY OF THE INFLUENCE OF INCLUSIONS UPON THE INITIATION AND GROWTH OF FATIGUE CRACKS IN	71-0394
HARKRIDER, D. G	SEISMOLOGY AND ACOUSTIC-GRAVITY WAVES	73-0744
HARLEMAN, D. R.	A NUMERICAL MODEL OF TRANSIENT WATER QUALITY IN A ONE-DIMENSIONAL ESTUARY BASED ON THE FINITE ELEMEN	74-1369
HARLEMAN, D. R.	FINITE ELEMENT MODEL FOR TRANSIENT TWO-LAYER COOLING POND BEHAVIOR	75-0465
HARLEMAN, D. R.	NUMERICAL MODEL FOR THE PREDICTION OF TRANSIENT WATER QUALITY IN ESTUARY NETWORKS	72-0635
HARMINAS, J.	FORMULATION AND APPLICATION OF CERTAIN PRIMAL AND MIXED FINITE ELEMENT MODELS OF FINITE DEFORMATIONS	74-0344
HARMON, M. B.	CONICAL SEGMENT METHOD FOR ANALYZING OPEN CROWN SHELLS OF REVOLUTION FOR EDGE LOADING	63-0007
HARMON, T. G.	SOME STRUCTURAL PROBLEMS - STANDARD OIL OF INDIANA BUILDING	73-0180
HARNAGE, D.	EVALUATION OF RIGID PAVEMENTS BY NONDESTRUCTIVE TESTS.	72-0459
HARRENSTINE, H.	THE EVOLUTION OF SHELL STRUCTURES IN OCEAN TECHNOLOGY	69-0042
HARRIS, H. G.	ANALYTICAL AND EXPERIMENTAL INVESTIGATION OF A 1/8 SCALE DYNAMIC MODEL OF THE SHUTTLE ORBITER. VOLUM	74-1186
HARRIS, H. G.	ELASTIC PLASTIC BUCKLING OF STIFFENED RECTANGULAR PLATES	69-0122
HARRIS, J. G.	STRESS AND DEFLECTION ANALYSIS OF MECHANICALLY FASTENED JOINTS	74-0067
HARRIS, P. J.	FURTHER DEVELOPMENTS IN THE USE OF LINEAR DISPLACEMENT FUNCTIONS IN THE FINITE ELEMENT ANALYSIS OF O	71-0344
HARRISON, D. G.	A HIGH ORDER TRIANGULAR FIELD ELEMENTS	71-0535
HARRISON, D. G.	A HIGHER ORDER TRIANGULAR FINITE ELEMENT FOR THE SOLUTION OF FIELD PROBLEMS IN ORTHOTROPIC MEDIA	72-0370
HARRISON, H. B.	NON-LINEAR ELASTIC ANALYSIS OF UNGUYED TOWERS AND STACKS	74-1157
HARRISON, H. T.	A VARIATIONAL METHOD FOR FREE BOUNDARY PROBLEMS	75-0105
HARRISON, N. L.	STRAIN ENERGY RELEASE RATES FOR TURNING CRACKS	72-0146
HARRISON, N. L.	THE STRESSES IN AN ADHESIVE LAYER	72-0910
HARRISON, R. L.	EXPLICIT TRIANGULAR BENDING ELEMENT MATRIX	74-0288
HARRISON, W.	GROUNDWATER FLOWN IN A SANDY TIDAL BEACH .1. ONE-DIMENSIONAL FINITE ELEMENT ANALYSIS	71-0567
HARRISON, W.	INVESTIGATION OF THE WATER TABLE IN A TIDAL BEACH	71-0648
HARRISON, W. J.	THE STRESSES IN AN ADHESIVE LAYER	72-0910
HARROLD, A. J.	REMOVAL OF TRUNCATION ERROR IN FINITE ELEMENT ANALYSIS	75-0131
HARRYSSON, C.	EXPERIMENTELL UNDERSOKNING AV ELEMENTFOGARS HALLFASTHETSEGENSKAPER (EXPERIMENTAL INVESTIGATION OF ST	74-0192
HARSH, J. F.	HYDROLOGIC ENGINEERING METHODS FOR WATER RESOURCES DEVELOPMENT. VOLUME 10. PRINCIPLES OF GROUND-WATE	72-0696
HART, G. C.	STATISTICAL IDENTIFICATION OF STRUCTURES	74-0141
HART, G. C.	TREATMENT OF RANDOMNESS IN FINITE ELEMENT MODELING	70-0353
HARTIG, D.	HERMITEAN RECTANGLE FINITE ELEMENT FAMILY	73-1009
HARTIG, D.	USE OF RECTANGULAR FINITE ELEMENTS	73-0104
HARTLEY, G. A.	ARCH DAM CONSTRUCTION STRESSES DUE TO HYDROSTATIC AND GRAVITY LOADS	75-0658
HARTLEY, G. A.	VOGT BOUNDARY FOR FINITE ELEMENT ARCH DAM ANALYSIS	74-0156
HARTMAN, J. P.	FINITE ELEMENT PARAMETRIC STUDY OF VERTICAL STRAIN INFLUENCE FACTORS AND THE PRESSUREMETER TEST TO E	74-1127
HARTMANN, B.	MESY - EIN PROGRAMMSYSTEM ZUR UNTERSUCHUNG VON TRAGWERKEN. (STRUCTURAL ANALYSIS BY MESY PROGRAMMING	75-0360
HARTRANFT, R. J	THREE-DIMENSIONAL STRESS ANALYSIS OF A FINITE SLAB CONTAINING A TRANSVERSE CENTRAL CRACK	75-0732
HARTUNG, R. F.	A COMPARISON OF SEVERAL COMPUTER SOLUTIONS TO THREE STRUCTURAL SHELL ANALYSIS PROBLEMS	73-0907
HARTUNG, R. F.	AN ASSESSMENT OF CURRENT CAPABILITY FOR COMPUTER ANALYSIS OF SHELL STRUCTURES	71-0706
HARTUNG, R. F.	AXISYMMETRIC VIBRATION OF CONICAL SHELLS	70-0350
HARTUNG, R. F.	COMPUTER ORIENTED ANALYSIS OF SHELL STRUCTURES	71-0559
HARTUNG, R. F.	COMPUTER ORIENTED ANALYSIS OF SHELL STRUCTURES	71-0762
HARTZ, B. J.	A FINITE ELEMENT CONTRIBUTION TO FRACTURE MECHANICS	72-0320

44

HARTZ, B. J.	AN EQUILIBRIUM STRESS FIELD MODEL FOR FINITE ELEMENT SOLUTIONS OF TWO-DIMENSIONAL ELASTIC PROBLEMS	68-0163
HARTZ, B. J.	EQUILIBRIUM STRESS FIELD MODELS IN FINITE ELEMENTS	67-0011
HARTZ, B. J.	FINITE ELEMENT ANALYSIS OF PLYWOOD PLATES	67-0052
HARTZ, B. J.	FINITE ELEMENT FORMULATION OF GEOMETRICALLY NONLINEAR PROBLEMS OF ELASTICITY	71-0114
HARTZ, B. J.	FINITE ELEMENT MATRIX FORMULATION OF POST-BUCKLING STABILITY AND IMPERFECTION SENSITIVITY	70-0246
HARTZ, B. J.	HIGHER-ORDER FINITE ELEMENT FOR PLANE STRESS	67-0082
HARTZ, B. J.	MATRIX FORMULATION OF STRUCTURAL STABILITY PROBLEMS	65-0039
HARTZ, B. J.	STABILITY OF PLATES USING THE FINITE ELEMENT METHOD	66-0081
HARTZMAN, M.	APPROXIMATE METHOD FOR CALCULATING THE RESPONSE OF EMPLACEMENT STRUCTURES SUBJECTED TO GROUND SHOCK	73-0654
HARTZMAN, M.	ELASTIC-PLASTIC STRESS-STRAIN RELATIONS FOR MATERIALS WITH DIFFERENT YIELD STRENGTHS IN TENSION AND	72-0764
HARTZMAN, M.	ELASTIC-PLASTIC STRESS-STRAIN RELATIONS USED IN CODE EPSOLA	73-0784
HARTZMAN, M.	NONLINEAR DYNAMIC ANALYSIS OF AXISYMMETRIC SOLIDS BY THE FINITE ELEMENT METHOD	73-0942
HARTZMAN, M.	NONLINEAR DYNAMIC ANALYSIS OF AXISYMMETRIC SOLIDS BY THE FINITE ELEMENT METHOD	74-0510
HARTZMAN, M.	NONLINEAR DYNAMICS OF SOLIDS BY FINITE ELEMENT METHODS	69-0490
HARTZMAN, M.	NONLINEAR DYNAMICS OF SOLIDS BY THE FINITE ELEMENT METHOD	71-0116
HARTZMAN, M.	STATIC STRESS ANALYSIS OF AXISYMMETRIC SOLIDS WITH MATERIAL AND GEOMETRIC NONLINEARITIES BY THE FINI	73-0958
HARVEY, J.	TRIANGULAR PLATE BENDING ELEMENT WITH ENFORCED COMPATIBILITY	71-0593
HARVEY, J. W.	CONSTRAINT TECHNIQUE IN FINITE ELEMENT ANALYSIS	70-0491
HASEGAWA, T.	FINITE ELEMENT ANALYSIS OF SOME MECHANICAL PROBLEMS OF HOLLOW GRAVITY DAM	70-0470
HASHIMOTO, K.	ANALYSIS OF LONGITUDINAL WAVE IN A BAR EXHIBITING VISCO-ELASTIC BEHAVIOR OR WITH MATERIAL DAMPING	71-0780
HASHIMOTO, R.	THE APPLICATION OF FINITE ELEMENT METHOD TO THE LIFTING SURFACE PROBLEM	75-0868
HASHIMOTO, R.	STRESSES IN A ROTATING DISK WITH CENTRAL AND NONCENTRAL HOLES (STRESS ANALYSIS BY THE FINITE ELEMENT	74-1337
HASKELL, W. E.	GEOMETRIC NONLINEAR ANALYSIS OF THIN PLATES BY FINITE ELEMENTS	70-0492
HASLUM, K.	FINITE ELEMENTS IN TRANSVERSE FRAME ANALYSIS	72-0182
HASSELIN, G.	USE OF THE ' TITUS ' FINITE ELEMENT CODE FOR THE NUCLEAR STRUCTURES AND INTERNAL STRESS-ANALYSIS	73-0840
HASSLER, C. R.	MEASUREMENTS OF HEALING AT AN OSTEOTOMY IN A RABBIT CALVARIUM: THE INFLUENCE OF APPLIED COMPRESSIVE	74-0975
HASSLER, G. R.	FINITE ELEMENT MODEL SYSTEM OF A RABBIT CALVARIUM: AN INVESTIGATION OF OPTIMAL COMPRESSIVE LOADING	73-0206
HASSON, C.	EMBANKMENT ANALYSIS AND FIELD CORRELATION	68-0057
HASTING, R. F.	A COMPARISON OF SEVERAL COMPUTER SOLUTIONS TO THREE STRUCTURAL SHELL ANALYSIS PROBLEMS	73-0884
HASUIKE, A.	STUDY ON FOIL BEARINGS - AN APPLICATION TO TAPE MEMORY DEVICES	75-0311
HATLESTAD, B.	ORSAM - A PROGRAMMING SYSTEM FOR THE FINITE ELEMENT METHOD	73-0528
HATT, F.	ANALYSIS OF BONDED JOINTS IN VEHICULAR STRUCTURES	73-0375
HATT, F.	THREE-DIMENSIONAL FINITE ELEMENT ANALYSIS INCLUDING ELASTIC-PLASTIC MATERIAL RESPONSE	74-1126
HATTORI, T.	STRESS AND VIBRATION ANALYSIS ON THE SYSTEM OF FRAME AND FOUNDATION	69-0218
HAUG, E.	FINITE ELEMENT ANALYSIS OF NONLINEAR MEMBRANE STRUCTURES	72-0909
HAUG, E.	FINITE ELEMENT ANALYSIS OF NONLINEAR MEMBRANE STRUCTURES	72-0944
HAUG, E.	FINITE ELEMENT ANALYSIS OF INFLATABLE SHELLS	74-0559
HAUPT, R. S.	PERFORMANCE OF EARTH AND EARTH-SUPPORTED STRUCTURES	72-0449
HAUSER, L. E.	CALCULATION OF TEMPERATURE DISTRIBUTION WITHIN A MELTING-FREEZING MATERIAL USING FINITE ELEMENT TECH	72-0043
HAVER, S.	STOCHASTIC DYNAMIC RESPONSE ANALYSIS OF OFFSHORE PLATFORMS, WITH PARTICULAR REFERENCE TO GRAVITY-TYP	75-0220
HAVILAND, J. K.	A METHOD OF ANALYSIS OF LINE STRUCTURES BY TRANSFER MATRICES DERIVED FROM FINITE ELEMENTS	74-0808
HAWARD, R. N.	YIELDING OF A TWO DIMENSIONAL VOID ASSEMBLY IN AN ORGANIC GLASS	73-0416
HAWS, E. T.	FOUNDATION INSTRUMENTATION FOR THE NATIONAL WESTMINISTER BANK TOWER	74-1147
HAY, J. K.	AVOIDING DANGEROUS AND COSTLY FAN FAILURES	75-0061
HAY, J. K.	DYNAMIC TESTING AND COMPUTER ANALYSIS OF AUTOMOTIVE FRAMES	72-0089
HAYAMIZU, Y.	A SEISMIC DESIGN OF NUCLEAR REACTOR BUILDING. STRESS ANALYSIS AND STIFFNESS EVALUATION OF THE ENTIRE	71-0251
HAYASHI, H.	APPLICATION OF FINITE ELEMENT METHOD TO HYDRODYNAMIC LUBRICATION PROBLEMS (PART 1, "INFINITE-WIDTH B	71-0537
HAYASHI, H.	APPLICATION OF FINITE ELEMENT METHOD TO HYDRODYNAMIC LUBRICATION PROBLEMS (PART 2, " FINITE WIDTH BE	71-0538
HAYASHI, H.	FINITE ELEMENT METHOD IN HYDRODYNAMIC LUBRICATION	72-0301
HAYASHI, I.	STRESS DISTRIBUTION NEAR A BROKEN POINT OF FILAMENT IN UNIDIRECTIONALLY FIBER REINFORCED COMPOSITES	73-0811
HAYASHI, N.	ON THE EXPLICIT REPRESENTATION OF TRANSFER-MATRICES OF CURVED BEAMS AND THEIR APPLICATIONS	75-0565
HAYATDAVOUDI, A	ASPECTS OF MECHANICAL BEHAVIOR OF ROCK UNDER STATIC AND CYCLIC LOADING. PART A. MECHANICAL BEHAVIOR	72-0761
HAYCOCKS, C.	CENTRIFUGAL MODEL STUDY AND FINITE ELEMENT ANALYSIS OF STRESS NEAR A FAULT	67-0159
HAYES, D. J.	APPLICATION OF FINITE ELEMENT TECHNIQUES TO POST-YIELD ANALYSIS OF PROPOSED STANDARD 3 POINT BEND FR	74-0118
HAYES, D. J.	DETERMINATION OF UPPER BOUNDS FOR PROBLEMS IN PLANE STRESS USING FINITE ELEMENT TECHNIQUES	67-0034
HAYHURST, D. R.	ESTIMATES OF THE CREEP RUPTURE LIFETIME OF STRUCTURES USING THE FINITE ELEMENT METHOD	75-0833
HAZZENZAHL, W.	THIN WINDOWS FOR GASEOUS AND LIQUID TARGETS: AN OPTIMIZATION PROCEDURE	75-0741
HEAD, J. L.	APPLICATION OF THREE-DIMENSIONAL FINITE ELEMENTS FOR ANALYSIS OF IRRADIATION INDUCED STRESSES IN NUC	72-0078
HEAD, J. L.	FINITE ELEMENT ANALYSIS OF STRESSES IN FUEL ELEMENTS OF A HIGH-TEMPERATURE GAS-COOLED REACTOR	71-0115
HEAD, J. L.	FINITE ELEMENT MATRIX DISPLACEMENT METHOD FOR TIME DEPENDENT STRESS ANALYSIS OF REACTOR MATERIALS IN	71-0674
HEAD, J. L.	FINITE ELEMENT STRESS ANALYSIS OF INTERACTING FUEL PELLET AND CANNING UNDER FAST REACTOR CONDITION	73-0814
HEAD, J. L.	FUEL ELEMENT STRESSES AT DISCONTINUITIES AND INTERACTIONS BY FINITE ELEMENTS IN TWO DIMENSIONS	73-0083
HEAD, J. L.	SOME MODELS FOR THE ANALYSIS OF STRESSES IN A TUBULAR FUEL PIN FOR HIGH TEMPERATURE REACTORS BY FINI	73-0816
HEAP, B. R.	ALGORITHMS FOR THE PRODUCTION OF CONTOUR MAPS OVER AN IRREGULAR TRIANGULAR MESH	72-0825
HEATON, H. S.	ACCURACY, STABILITY AND OSCILLATION CHARACTERISTICS OF FINITE ELEMENT METHOD FOR SOLVING HEAT CONDUC	69-0007
HECHMER, J. L.	THERMAL MECHANICAL ANALYSIS PROCEDURE FOR A SEMICIRCULAR PLATE WITH CLAMPED STRAIGHT EDGE	75-0651
HECHT, M.	STUDY OF THE FLEXURAL BEHAVIOUR OF CASTINGS	74-0377
HEDGES, J. L.	APPLICATION OF FINITE ELEMENT TECHNIQUES TO THE ANALYSIS OF AN AUTOMOBILE STRUCTURE	70-0136
HEDGES, J. L.	COMPUTER AIDED DESIGN APPLIED TO A MODEL OF A CHASSIS TYPE STRUCTURE USING FINITE ELEMENT TECHNIQUES	69-0421
HEDGES, J. L.	DYNAMIC ANALYSIS OF AN AUTOMOBILE CHASSIS FRAME	70-0135
HEDGES, J. L.	STRESS ANALYSIS OF AN AUTOMOBILE CHASSIS FRAME	70-0137
HEDLEY, D. G. F	UNDERGROUND MEASUREMENTS IN A STEEPLY DIPPING OREBODY	68-0085
HEER, E.	FINITE ELEMENT FORMULATION FOR LINEAR THERMOVISCO ELASTIC MATERIALS	69-0351
HEIDARI, M.	ANALYSIS OF LIQUID-WASTE INJECTION WELLS IN ILLINOIS BY MATHEMATICAL MODELS	74-0518
HEIDELBERG, S.	SASL - A FINITE ELEMENT CODE FOR THE STATIC ANALYSIS OF AXISYMMETRIC AND PLANE SOLIDS SUBJECTED TO A	72-0779
HEIFETZ, J. H.	FINITE DEFORMATIONS IN ELASTIC-PLASTIC MEDIA SUBJECTED TO IMPULSIVE LOADINGS BY THE FINITE ELEMENT M	72-0947
HEIFITZ, J. H.	DYNAMIC RESPONSE ON NONLINEAR MEDIA AT LARGE STRAINS	72-0237
HEINRICH, F.	BERECHNUNG UND BEURTEILUNG DER STANDSICHERHEIT VON ABBAUHOHLRAEUMEN IN STEILEN GANGERZLAGERSTAETTEN	74-0597
HEINRICH, F.	NEUARTIGE ANWENDUNG GEOMECHANISCHER UNTERSUCHUNGSMETHODEN ZUR BEURTEILUNG DER ABBAUWIRKUNGEN IM GANG	75-0530
HEINS, R. W.	ASPECTS OF MECHANICAL BEHAVIOR OF ROCK UNDER STATIC AND CYCLIC LOADING. PART A. MECHANICAL BEHAVIOR	72-0761
HEINS, R. W.	ASPECTS OF MECHANICAL BEHAVIOR OF ROCK UNDER STATIC AND CYCLIC LOADING. PART A: MECHANICAL BEHAVIOR	72-0762
HEINS, R. W.	NEW HORIZONS IN ROCK MECHANICS	73-0446
HEISE, U.	COMBINED APPLICATION OF FINITE ELEMENT METHODS AND RICHARDSON EXTRAPOLATION TO THE TORSION PROBLEM	73-0066
HEISE, U.	COMPILED APPLICATION OF FINITE ELEMENT METHODS AND RICHARDSON EXTRAPOLATION TO THE TORSION PROBLEM	73-1012
HEISE, U.	EINE FINITE-ELEMENT-METHODE MIT KREISRINGSEKTORFORMIGEN ELEMENTEN	73-1115
HEISE, U.	FINITE ELEMENT ANALYSIS IN POLAR CO-ORDINATES OF THE SAINT VENANT TORSION PROBLEM	74-0500
HEISE, U.	FINITE-ELEMENT METHOD WITH ELEMENTS WITH CIRCULAR RING SECTOR FORM	73-0105
HEITNER, K. L.	NUMERICAL MODEL FOR TSUNAMI RUN-UP	70-0200
HELBING, T. C.	A STUDY OF THE PLANE STRESS OR STRAIN FINITE ELEMENT ANALYSIS FOR SOLUTION OF STRESS DISTRIBUTION IN	69-0487
HELIOT, J.	USE OF THE ' TITUS ' FINITE ELEMENT CODE FOR THE NUCLEAR STRUCTURES AND INTERNAL STRESS-ANALYSIS	73-0840
HELLAN, K.	ANALYSIS OF ELASTIC PLATES BY A SIMPLIFIED FINITE ELEMENT METHOD	67-0156
HELLAN, K.	ANALYSIS OF RECTANGULAR PLATES IN TRANSIENT CREEP BENDING	69-0425
HELLAN, K.	ON THE UNITY OF THE CONSTANT STRAIN-CONSTANT MOMENT FINITE ELEMENT METHODS	73-0164
HELLE, E.	FORMULATION AND EVALUATION OF A TOROIDAL RING DISCRETE ELEMENT	66-0082
HELLE, E.	FORMULATION AND EVALUATION OF A TRIANGULAR CROSS-SECTION RING DISCRETE ELEMENT	66-0083
HELLEN, T. K.	A FRONT SOLUTION FOR FINITE ELEMENT TECHNIQUES	69-0412
HELLEN, T. K.	ANALYSIS OF ELASTIC PLATES IN FLEXURE BY A SIMPLIFIED FINITE ELEMENT METHOD	67-0035
HELLEN, T. K.	BERSAFE - A COMPUTER SYSTEM FOR STRESS ANALYSIS BY FINITE ELEMENTS	70-0212
HELLEN, T. K.	BERSAFE FINITE ELEMENT SYSTEM	74-0190
HELLEN, T. K.	CALCULATION OF STRESS INTENSITY FACTORS FOR COMBINED TENSILE AND SHEAR LOADING	75-0205
HELLEN, T. K.	CALCULATION OF STRESS INTENSITY FACTORS IN TWO AND THREE DIMENSIONS USING FINITE ELEMENTS	75-0254
HELLEN, T. K.	ECONOMICAL COMPUTER TECHNIQUES FOR SOLID FINITE ELEMENTS	68-0086

45

HELLEN, T. K.	FINITE ELEMENT CALCULATION OF STRESS INTENSITY FACTORS USING ENERGY TECHNIQUES	73-0825
HELLEN, T. K.	FINITE ELEMENT TECHNIQUES IN FRACTURE MECHANICS	72-0746
HELLEN, T. K.	NUMERICAL INTEGRATION CONSIDERATIONS IN TWO AND THREE DIMENSIONAL ISOPARAMETRIC FINITE ELEMENTS	75-0130
HELLEN, T. K.	ON THE METHOD OF VIRTUAL CRACK EXTENSIONS	75-0237
HELLEN, T. K.	SOLID WEDGE FINITE ELEMENT FOR STRESS ANALYSIS	69-0124
HELLEN, T. K.	THE APPLICATION OF THREE-DIMENSIONAL FINITE ELEMENTS TO A CYLINDER-CYLINDER INTERSECTION	70-0068
HELLEN, T. K.	THE APPLICATION OF THE BERSAFE FINITE ELEMENT SYSTEM TO NUCLEAR DESIGN PROBLEMS	71-0117
HELLEN, T. K.	THE USE OF FINITE ELEMENT TECHNIQUES IN FRACTURE MECHANICS	72-0926
HELLEN, T. K.	THREE-DIMENSIONAL CRACK ANALYSIS APPLIED TO AN LWR NOZZLE-CYLINDER INTERSECTION	75-0165
HELLINGS, J. E.	COMPARISON OF PRACTICAL METHODS OF ANALYSING ELASTIC LATERALLY LOADED PLATES	71-0425
HELPS, J. D.	DIRECT METHOD OF OBTAINING CAPACITANCE FROM FINITE-ELEMENT MATRICES	72-0289
HELPS, J. D.	EXACT FINITE ELEMENT SOLUTIONS TO THE HELMHOLTZ EQUATIONS	73-0652
HENDERSON, H. L	PRESSURE VESSEL CERTIFICATION BASED ON FRACTURE MECHANICS TECHNOLOGY	74-0948
HENDERSON, N. C	ANALYSIS OF STRENGTH AND RESIDUAL STRESSES IN FILAMENT REINFORCED ALUMINUM CYLINDERS	75-0517
HENDRIE, J. M.	FINITE ELEMENT APPROACH TO POLYMER CONCRETE BRIDGE DECK DESIGN ANALYSIS	72-0918
HENDRON, A. J.	RESEARCH TO IMPROVE TUNNEL SUPPORT SYSTEMS	74-0777
HENDRON, A. J.	LINER-MEDIUM INTERACTION IN TUNNELS	75-0052
HENDRY, J. A.	VARIATIONAL SOLUTION OF TWO DIMENSIONAL FLUID FLOW PROBLEMS	73-1014
HENNART, J. P.	COMPARISON OF EXTRAPOLATION TECHNIQUES WITH HIGH ORDER FINITE ELEMENT METHODS FOR DIFFUSION EQUATION	73-0586
HENNART, J. P.	EXTRAPOLATED FINITE ELEMENT PROCEDURES OVER CARTESIAN PRODUCT DOMAINS	75-0044
HENNART, J. P.	FINITE-ELEMENT CODE TO SIMULATE ANISOTROPIC-PLASMA DIFFUSION IN LINEAR MULTIPOLES	75-0409
HENNART, J. P.	FINITE-ELEMENT METHODS FOR REACTOR ANALYSIS - COMMENT	75-0009
HENNART, J. P.	FINITE-ELEMENT PLASMA-DIFFUSION CODE FOR ARBITRARY TOROIDAL GEOMETRY	75-0500
HENNART, J. P.	SEMIDISCRETE GALERKIN TECHNIQUES WITH TIME INTERPOLATION AND SPLITTING UP FOR PLASMA SIMULATION	75-0121
HENNART, J. P.	SINGULARITIES IN THE FINITE ELEMENT APPROXIMATION OF TWO DIMENSIONAL DIFFUSION PROBLEMS	73-1015
HENNING, G.	ZUR GENAUEN BERECHNUNG KONSTRUKTIV ORTHOTROPER PLATTEN (EXACT CALCULATION OF STRUCTURALLY ORTHOTROPI	72-0188
HENNRICH, C. H.	SUBSTRUCTURE ANALYSIS TECHNIQUES AND AUTOMATION	73-0597
HENRICHSEN, L.	VISCOELASTIC FINITE ELEMENTS	74-0440
HENRY, A. F.	ANNUAL PROGRESS REPORT, FY1973	73-0969
HENRY, M. F.	FINITE-ELEMENT MODELING OF DIRECTIONAL SOLIDIFICATION OF SUPER ALLOY EUTECTICS	75-0350
HENRY, R.	VIBRATION ANALYSIS OF ROTATING COMPRESSOR BLADES	73-0417
HENRY, T. P.	2-DIMENSIONAL FINITE-ELEMENT METHOD FOR INTEGRAL NEUTRON-TRANSPORT CALCULATIONS	75-0443
HENRYWOOD, R. K	ANALYSIS OF RADIAL FLOW IMPELLERS BY THE MATRIX FINITE ELEMENT METHOD	71-0412
HENRYWOOD, R. K	DESIGN, DEVELOPMENT, DOCUMENTATION AND SUPPORT OF A MAJOR FINITE ELEMENT SYSTEM	73-0205
HENSCHEL, R.	CREEP BENDING TESTS OF LARGE WELD OVERLAY AND BASE METAL SPECIMENS AT 1000 DEGREE F. (537.7 DEG. C)	75-0521
HENSHAW, D. H.	TWO-DIMENSIONAL AIRFOIL ANALYSIS USING A REFINED FINITE ELEMENT TECHNIQUE	73-0587
HENSHELL, R. D.	A NEW FAMILY OF CURVILINEAR PLATE BENDING ELEMENTS FOR VIBRATION AND STABILITY	72-0645
HENSHELL, R. D.	A NEW FAMILY OF CURVILINEAR PLATE BENDING ELEMENTS FOR VIBRATION AND STABILITY	72-0999
HENSHELL, R. D.	A NEW HYBRID CYLINDRICAL SHELL FINITE ELEMENT	71-0118
HENSHELL, R. D.	A TIMOSHENKO BEAM ELEMENT	72-0590
HENSHELL, R. D.	AUTOMATIC MASTERS FOR EIGENVALUE ECONOMIZATION	75-0557
HENSHELL, R. D.	CLOUGH-TOCHER TRIANGULAR PLATE BENDING ELEMENT IN VIBRATION	69-0093
HENSHELL, R. D.	COMPARISON OF RECENT SHALLOW SHELL FINITE ELEMENT ANALYSES	71-0362
HENSHELL, R. D.	CONSTANT CURVATURE BEAM FINITE ELEMENTS FOR IN-PLANE VIBRATION	72-0254
HENSHELL, R. D.	CRACK TIP FINITE ELEMENTS ARE UNNECESSARY	73-1016
HENSHELL, R. D.	CRACK TIP FINITE ELEMENTS ARE UNNECESSARY	75-0472
HENSHELL, R. D.	CURVED BEAM FINITE ELEMENTS FOR COUPLED BENDING AND TORSIONAL VIBRATION	72-0720
HENSHELL, R. D.	FINITE ELEMENT TECHNIQUES APPLIED TO THE VIBRATIONS OF PLATE SYSTEMS	66-0033
HENSHELL, R. D.	HYBRID PLATE BENDING ELEMENTS	72-0792
HENSHELL, R. D.	NATURAL FREQUENCIES AND MODE SHAPES OF TRANSFORMER CORES	65-0048
HENSHELL, R. D.	ON HYBRID FINITE ELEMENTS	72-0069
HENSHELL, R. D.	POSSIBLE LOSS OF ACCURACY IN CURVED FINITE ELEMENTS	72-0278
HENSHELL, R. D.	RING FINITE ELEMENTS FOR AXISYMMETRIC AND NON-AXISYMMETRIC THIN SHELL ANALYSIS	72-0922
HENSHELL, R. D.	THEORETICAL BASIS FOR HYBRID FINITE ELEMENTS IN DYNAMIC PROBLEMS	75-0489
HENSHELL, R. D.	TRANSMISSION OF VIBRATION IN BEAM SYSTEMS	69-0405
HENSLEY, E. K.	EVALUATION OF DAMAGE TOLERANCE IN AIRCRAFT STRUCTURES	75-0559
HENZE, E.	PRACTICAL FINITE ELEMENT METHOD OF FAILURE PREDICTION FOR COMPOSITE MATERIAL STRUCTURES	75-0644
HENZI, A. N.	WEAPON-MOUNT INTERFACE STUDY - TRIPOD MOUNT M3	72-0846
HERAKOVICH, C.	PLASTIC STRAINS ASSOCIATED WITH TORSION OF NONCIRCULAR BARS	72-0177
HERAKOVICH, C.	STRAIN HARDENING TORSION OF SOLID BARS AS A MINIMIZATION PROBLEM	70-0174
HERMANN, L. R.	MIXED FORMULATIONS FOR FINITE ELEMENT SHELL ANALYSIS	71-0222
HERNES, M. W.	ASTRA-BOEING'S ADVANCED STRUCTURAL ANALYZER	69-0118
HERNESS, E. D.	A CRITICAL REVIEW OF NASTRAN	71-0246
HERRERA, I.	A SIMPLIFIED VERSION OF GURTIN'S VARIATIONAL PRINCIPLE	74-1244
HERRMAN, L. R.	A GENERAL NUMERICAL SOLUTION OF THE TWO-DIMENSIONAL DIFFUSION-CONVECTION EQUATION BY THE FINITE ELEM	70-0004
HERRMANN, G.	PERIODIC SHOCK EXCITATION OF ELASTIC STRUCTURES	69-0163
HERRMANN, L. R.	A BENDING ANALYSIS FOR PLATES	66-0035
HERRMANN, L. R.	A FINITE ELEMENT ANALYSIS FOR CYLINDRICAL SHELLS AND PLATES	66-0034
HERRMANN, L. R.	A FINITE ELEMENT COMPUTER PROGRAM FOR THE PREDICTION OF THE BEHAVIOR OF REINFORCED AND PRESTRESSED C	72-0799
HERRMANN, L. R.	A FINITE-ELEMENT ANALYSIS FOR THIN SHELLS	68-0087
HERRMANN, L. R.	DEVELOPMENT OF A SOLID ROCKET PROPELLANT NONLINEAR CONSTITUTTIVE THEORY	75-0722
HERRMANN, L. R.	EFFICIENCY EVALUATION OF A TWO-DIMENSIONAL INCOMPATIBLE FINITE ELEMENT	73-0988
HERRMANN, L. R.	ELASTIC TORSIONAL ANALYSIS OF IRREGULAR SHAPES	65-0020
HERRMANN, L. R.	ELASTICITY EQUATIONS FOR INCOMPRESSIBLE AND NEARLY INCOMPRESSIBLE MATERIALS BY A VARIATIONAL THEOREM	65-0019
HERRMANN, L. R.	FINITE-ELEMENT BENDING ANALYSIS FOR PLATES	67-0036
HERRMANN, L. R.	INTERPRETATION OF FINITE ELEMENT PROCEDURE AS STRESS ERROR MINIMIZATION PROCEDURE	72-0984
HERRMANN, L. R.	ON A VARIATIONAL THEOREM FOR INCOMPRESSIBLE AND NEARLY INCOMPRESSIBLE ORTHOTROPIC ELASTICITY	68-0249
HERSCOVITCH, P.	FINITE ELEMENT CALCULATION OF LEFT VENTRICULAR WALL STRESS IN NORMAL AND DISEASE STATES	73-0588
HERTING, D. N.	ACOUSTIC ANALYSIS OF SOLID ROCKET MOTOR CAVITIES BY A FINITE ELEMENT METHOD	71-0119
HERTLING, J.	NUMERICAL TREATMENT OF ALGEBRAIC INTEGRAL EQUATIONS BY VARIATIONAL METHODS	75-0374
HESS, J. L.	CALCULATION OF POTENTIAL FLOW ABOUT ARBITRARY THREE-DIMENSIONAL LIFTING BODIES	72-0722
HESSEL, J. J.	SOLUTION OF AUTOMOTIVE STRUCTURAL PROBLEMS USING THE FINITE ELEMENT METHOD AND COMPUTER GRAPHICS	71-0341
HEUBNER, K. H.	FINITE ELEMENT ANALYSIS OF FLUID FILM LUBRICATION - A SURVEY	74-0041
HEUZE, F.	ANALYSIS AND MODEL STUDIES OF UNDERGROUND OPENINGS IN JOINTED ROCK. RESEARCH ON ROCK BOLT REINFORCEM	71-0750
HEUZE, F. E.	A DESIGN PROCEDURE FOR HIGH CUTS IN JOINTED HARD ROCK. THREE DIMENSIONAL SOLUTIONS. VOLUME 1	71-0656
HEUZE, F. E.	FINITE ELEMENT STUDIES OF PILE DRIVER TUNNELS INCLUDING CONSIDERATIONS OF SUPPORT REQUIREMENTS	72-0679
HEUZE, F. E.	NUMERICAL ANALYSES OF DEFORMABILITY TESTS IN JOINTED ROCK. ' JOINT PERTURBATION ' AND ' NO TENSION '	71-0439
HEUZE, F. E.	NUMERICAL AND PHYSICAL MODELLING OF REINFORCEMENT SYSTEMS FOR TUNNELS IN JOINTED ROCK	73-0872
HEWITT, D. R.	PLANETARY EXPLORER STRUCTURAL CONCEPT EVALUATION MODEL USING NASTRAN	71-0742
HIBBERT, P. D.	BOUNDING PLANE STRESS SOLUTIONS BY FINITE ELEMENTS	66-0055
HIBBERT, P. D.	IMPACT OF COMPUTERS ON THE PRACTICE OF STRUCTURAL ENGINEERING IN CONCRETE	72-0414
HIBBERT, P. D.	STRESS CONCENTRATIONS IN TIMBER BEAMS	67-0038
HIBBITT, H. D.	A FINITE ELEMENT FORMULATION FOR PROBLEMS OF LARGE STRAIN AND LARGE DISPLACEMENT	70-0069
HIBBITT, H. D.	A NUMERICAL THERMO-MECHANICAL MODEL FOR THE WELDING AND SUBSEQUENT LOADING OF A FABRICATED STRUCTURE	72-0847
HIBBITT, H. D.	ELASTIC PLASTIC AND CREEP ANALYSIS OF PIPELINES BY FINITE ELEMENTS	73-0419
HIBBITT, H. D.	GENERAL PURPOSE PROGRAMS FOR NONLINEAR FINITE ELEMENT ANALYSIS	70-0247
HIBBITT, H. D.	HYBRID FINITE ELEMENT ANALYSIS WITH PARTICULAR REFERENCE TO AXISYMMETRIC STRUCTURES	70-0248
HIBBITT, H. D.	MARC SERIES OF PROGRAMS	72-0154
HIBBITT, H. D.	NONLINEAR MATERIAL AND GEOMETRIC BEHAVIOUR OF SHELL STRUCTURES	71-0084
HIBBITT, H. D.	NUMERICAL, THERMO-MECHANICAL MODEL FOR THE WELDING AND SUBSEQUENT LOADING OF A FABRICATED STRUCTURE	73-0418
HIBBITT, H. D.	SPECIAL STRUCTURAL ELEMENTS FOR PIPING ANALYSIS	74-0502
HIBBITT, H. D.	THERMAL AND MECHANICAL ANALYSIS OF WELDED STRUCTURES	73-0819

HIBBITT, H. D.	WATER IMPACT ANALYSIS OF SPACE SHUTTLE SOLID ROCKET BY THE FINITE ELEMENT METHOD	74-0760
HIBINO, S.	STABILITY OF FINITE ELEMENT SCHEMES FOR VIBRATION PROBLEMS OF THE EQUATION OF ELASTICITY THEORY	71-0521
HICKS, G. W.	A FINITE ELEMENT ANALYSIS OF THE ELASTIC BEHAVIOR OF BEAM COLUMNS	66-0122
HICKS, G. W. JR	A FINITE ELEMENT ANALYSIS FOR THE STRENGTH OF BEAM COLUMNS	66-0037
HICKS, G. W. JR	A GENERAL TECHNIQUE FOR BUCKLING ANALYSIS WITH A FINITE ELEMENT SOLUTION APPLICATION	66-0036
HICKS, G. W. JR	FINITE ELEMENT ELASTIC BUCKLING ANALYSIS	67-0107
HIDAKA, M.	ON THE DEVELOPMENT OF THE "PASSAGE" PROGRAM	73-0342
HIGGINS, J. E.	RAPID CASE DESIGN USING FINITE ELEMENT STRESS ANALYSIS AND NASTRAN	73-0730
HIGHT, T. K.	DETAILED STRUCTURAL ANALYSIS OF THE HUMAN TIBIA	75-0379
HIGUCHI, M.	NUMERICAL ANALYSIS OF SHIP STRUCTURAL ELEMENTS BY MEANS OF FINITE ELEMENT METHOD AND PHOTO-ELASTIC E	69-0125
HIGUCHI, M.	STRESS ANALYSIS IN HULL STRUCTURAL MEMBERS BY PHOTOELASTIC EXPERIMENTS AND FINITE ELEMENT METHODS -	70-0341
HIJINK, J. A. W	ON THE DESIGN OF DIE-SETS	75-0582
HILBER, H.	A FULLY COMPATIBLE PLATE BENDING ELEMENT	67-0142
HILBER, H. M.	SOME NEW ELEMENTS FOR THE MATRIX DISPLACEMENT METHOD	68-0017
HILBERT, S. H.	BOUNDS FOR A CLASS OF LINEAR FUNCTIONALS WITH APPLICATIONS TO HERMITE INTERPOLATION	71-0582
HILDEBRAND, R.	FINAL REPORT ON NASTRAN/GRAPHICS, 1971 FORD DOOR PROJECT	73-0595
HILL, D. W.	STRESSES AND DEFLECTIONS IN COOLING TOWER SHELLS DUE TO WIND LOADING	68-0088
HILL, J. H.	FEASIBILITY OF USING FINITE ELEMENTS IN ANALYSIS OF SECOND BREAKDOWN IN SEMICONDUCTOR DEVICES	73-0874
HILL, J. L.	EFFECT OF INITIAL STATE FORCES ON THE HYDROELASTIC VIBRATION OF TUBES	73-0589
HILL, J. L.	THERMOMECHANICAL DISSIPATION ANALYSIS OF THERMOVISCOELASTIC SOLIDS BY FINITE ELEMENTS	72-0403
HILL, J. R. M.	SUPPORT PERFORMANCE OF HYDRAULIC BACKFILL	74-0257
HILL, T. R.	LINEAR DISCONTINUOUS FINITE-ELEMENT SCHEME IN TIME-DEPENDENT TRANSPORT CALCULATIONS	75-0498
HILL, T. R.	TRIANGULAR MESH METHODS FOR THE NEUTRON TRANSPORT EQUATION	73-1062
HILL, T. R.	TRIPLET: A TWO-DIMENSIONAL, MULTIGROUP, TRIANGULAR MESH, PLANAR GEOMETRY, EXPLICIT TRANSPORT CODE	73-0955
HILTON, P. D.	THREE-DIMENSIONAL STRESS ANALYSIS OF A FINITE SLAB CONTAINING A TRANSVERSE CENTRAL CRACK	75-0732
HILTON, P. D.	APPLICATIONS OF THE FINITE ELEMENT METHOD TO THE CALCULATIONS OF STRESS INTENSITY FACTORS	72-0900
HILTON, P. D.	ELASTIC-PLASTIC ANALYSIS FOR CRACKED MEMBERS	75-0519
HILTON, P. D.	FINITE ELEMENT FRACTURE MECHANICS ANALYSIS OF TWO-DIMENSIONAL AND AXISYMMETRIC ELASTIC AND ELASTIC-P	74-1092
HILTON, P. D.	FINITE ELEMENT METHOD FOR CALCULATING STRESS INTENSITY FACTORS AND ITS APPLICATION TO COMPOSITES	74-1000
HILTON, P. D.	PLASTIC INTENSITY FACTORS FOR CRACKED PLATES	71-0164
HINATA, M.	NUMERICAL SOLUTION OF PLATEAUS PROBLEM BY A FINITE ELEMENT METHOD	74-0119
HINDUJA, S.	FINITE ELEMENT METHOD FOR MACHINE TOOL STRUCTURAL ANALYSIS	71-0136
HINDUJA, S.	FINITE ELEMENT METHOD FOR THE ANALYSIS OF MACHINE TOOL STRUCTURES.	72-0448
HINTON, E.	DYNAMIC TRANSIENT NON-LINEAR BEHAVIOUR OF THICK AND THIN PLATES	75-0125
HINTON, E.	FINITE ELEMENT ANALYSIS OF PLATES ALLOWING FOR TRANSVERSE SHEAR DEFORMATION EFFECTS	73-0996
HINTON, E.	FREE VIBRATION AND TRANSIENT RESPONSE OF THICK AND THIN PLATES USING THE FINITE ELEMENT METHOD	74-0517
HINTON, E.	LEAST SQUARES ANALYSIS USING FINITE ELEMENTS	68-0287
HINTON, E.	LEAST SQUARES SMOOTHING OF EXPERIMENTAL DATA USING FINITE ELEMENTS	68-0224
HINTON, E.	LIMITATIONS OF THE FINITE ELEMENT METHOD IN STRUCTURAL ANALYSIS	73-0420
HINTON, E.	LOCAL AND GLOBAL SMOOTHING OF DISCONTINUOUS FINITE ELEMENT FUNCTIONS USING A LEAST SQUARES METHOD	74-0378
HINTON, E.	LOCAL LEAST SQUARES STRESS SMOOTHING FOR PARABOLIC ISOPARAMETRIC ELEMENTS	75-0593
HINTON, E.	SIMPLE FINITE ELEMENT SOLUTION FOR PLATES OF HOMOGENEOUS, SANDWICH AND CELLULAR CONSTRUCTION	75-0140
HINTZ, R. M.	ANALYTICAL METHODS IN COMPONENT MODAL SYNTHESIS	75-0824
HIRAI, T.	HYBRID EFFECT ON BENDING BEHAVIORS OF CFRP-AL LAMINATE	74-1213
HIRATA, T.	COMPUTING SYSTEM FOR STRUCTURAL ANALYSIS OF CAR BODIES	72-0324
HIRIART, G.	ANALYSIS OF CURVED TARGET-TYPE THRUST REVERSERS	75-0152
HIRIART, G.	FINITE ELEMENT ANALYSIS OF JET IMPINGEMENT ON AXISYMMETRIC CURVED DEFLECTORS	75-0396
HIRIART, G.	JET IMPINGEMENT ON AXISYMMETRIC CURVED DEFLECTORS	74-0017
HIRSCH, C.	A FINITE ELEMENT METHOD FOR FLOW CALCULATIONS IN TURBOMACHINES	74-0726
HIRSCH, C.	FINITE-ELEMENT METHOD FOR AXISYMMETRIC FLOW COMPUTATION IN A TURBOMACHINE	75-0359
HIRSCH, CH.	THE MERIDIONAL THROUGH-FLOW CALCULATION IN AN AXIAL FLOW MACHINE BY THE FINITE ELEMENT METHOD	75-0115
HIRST, M. J. S.	THE SOLUTION OF STRUCTURAL EQUILIBRIUM EQUATIONS BY THE CONJUGATE GRADIENT METHOD WITH PARTICULAR RE	71-0268
HIRT, M.	MEASUREMENT OF ACTUAL DEDENDUM STRAINS, AND THE INFLUENCE OF DEDENDUM RADIUS ON STRESSES AND DEDENDU	72-0118
HISCHBERG, M. H	A GENERAL APPROACH TO THE PRACTICAL SOLUTION OF CREEP PROBLEMS	59-0003
HITCHINGS, D.	RESPONSE OF NUCLEAR STRUCTURAL SYSTEMS TO TRANSIENT AND RANDOM EXCITATIONS, USING BOTH DETERMINISTIC	74-1047
HO, B. P. C.	APPLYING FINITE ELEMENT ANALYSIS	70-0150
HO, B. P. C.	SEISMIC ANALYSIS OF A THREE-PHASE BREAKER INSTALLATION	74-0622
HO, M. K.	FINITE ELEMENT ANALYSIS OF SOIL DEFORMATIONS	69-0134
HO, P. K.	ANALYSIS OF PLATE STRUCTURES BY A DUAL FINITE ELEMENT METHOD	69-0482
HOA, V. S.	THERMAL STRESS ANALYSIS OF PLATES AND SHALLOW SHELLS BY HYBRID FINITE-ELEMENT METHOD	74-0485
HOAGLAND, R. G.	FAST FRACTURE RESISTANCE AND CRACK ARREST IN STRUCTURAL STEELS	73-0766
HOBBS, G. K.	METHODS OF MODELING AND ANALYZING VISCOELASTICALLY DAMPED STRUCTURES	71-0473
HOBBS, G. K.	TRANSIENT DYNAMIC RESPONSE OF VISCOELASTIC STRUCTURES	72-0234
HOBBS, R.	FINITE ELEMENT ANALYSIS OF CENTRIFUGED AND BUILT-UP SLOPES	74-0415
HOBBS, R.	FINITE ELEMENT ANALYSIS OF CENTRIFUGED AND BUILT UP SLOPES	74-1376
HOBBS, R.	NUMERICAL AND PHYSICAL MODELLING	74-0393
HODGE, P. G.	A NEW MODEL FOR ELASTIC-PLASTIC TRUSSES	74-0805
HODGE, P. G.	A SIMPLE FINITE ELEMENT MODEL FOR ELASTIC-PLASTIC PLATE BENDING	72-0889
HODGE, P. G.	CONSISTENT FINITE ELEMENT MODEL FOR TWO DIMENSIONAL CONTINUUM	70-0361
HODGE, P. G.	FINITE ELEMENT METHODS IN PLASTICITY	74-0801
HODGE, P. G.	NUMERICAL METHODS FOR THE LIMIT ANALYSIS OF PLATES	68-0181
HODGE, P. G.	PLANE STRESS LIMIT ANALYSIS BY FINITE ELEMENTS	70-0255
HODGE, P. G.	SIMPLE FINITE ELEMENT MODEL FOR ELASTIC-PLASTIC PLATE BENDING.	72-0447
HODGE, P. G. JR	A FINITE ELEMENT MODEL FOR PLASTICITY PROBLEMS	74-1225
HODGE, P. G. JR	COMPUTER SOLUTIONS OF PLASTICITY PROBLEMS	72-0829
HODGE, P. G. JR	LOWER BOUND LIMIT ANALYSIS OF PLANE STRESS PROBLEMS BY A FINITE ELEMENT METHOD	70-0428
HODGE, P. G. JR	SOME APPLICATIONS OF THE PRINCIPLE OF VIRTUAL WORK	73-0930
HODGE, P. G. JR	STRUCTURAL INELASTICITY. V11. ELASTIC-PLASTIC ANALYSIS OF A WHEEL ROLLING ON A RIGID TRACK	73-0886
HODGE, P. G. JR	STRUCTURAL INELASTICITY. V1. A FINITE ELEMENT METHOD FOR PLASTICITY PROBLEMS	73-0887
HOEG, K.	EFFECTIVE STRESS-STRAIN-STRENGTH MODEL FOR SOILS	75-0056
HOEG, K.	FINITE ELEMENT ANALYSIS OF STRAIN-SOFTENING CLAY	72-0028
HOFFMAN, A. J.	COMPLEXITY BOUNDS FOR REGULAR FINITE DIFFERENCE AND FINITE ELEMENT GRIDS	73-0043
HOFFMANN, A.	ANALYSIS OF SHELLS OF ANY FORM IN THE PLASTIC REGION BY THE FINITE ELEMENT METHOD	73-0940
HOFFMANN, A.	PLASTIC ANALYSIS OF SHELLS BY FINITE ELEMENTS	73-1017
HOFMANN, H. H.	EINFACHES NAEHERUNGSVERFAHREN FUR DIE GRENZTRAGFAEHIGKEITSANLYSE VON SPANNBETON-REAKTORDRUCKBEHAELTE	73-0854
HOFMANN, H. H.	LOAD LIMIT ANALYSIS, COMBINATION OF STATISTICAL METHODS (MONTE CARLO METHOD) AND FINITE ELEMENT METH	71-0768
HOFMANN, H. H.	PRESTRESSED CONCRETE REACTOR PRESSURE VESSEL FAILURE ANALYSIS, THE INCORPORATION OF STATISTICAL METH	71-0120
HOFMEISTER, L.	LARGE STRAIN ELASTO-PLASTIC FINITE ELEMENT ANALYSIS	71-0121
HOFMEISTER, L.	PURE MOMENT LOADING OF AXISYMMETRIC FINITE ELEMENT MODELS	73-1007
HOFMEISTER, L.	VIBRATION PROBLEMS USING ISOPARAMETRIC SHELL ELEMENTS	72-0549
HOGAN, J. C.	DEEP SEA ENGINEERING	71-0739
HOGGE, M.	FAMILLE D ELEMENTS FINIS DE COQUE PLAN CINEMATIQUEMENT ADMISSIBLES POUR L'ANALYSE DES STRUCTURES	74-0488
HOGGE, M. A.	COMPARISON OF SOME FORMULATIONS IN THERMAL ANALYSIS OF STRUCTURES	75-0888
HOGGE, M. A.	DUAL ANALYSIS FOR HEAT CONDUCTION PROBLEMS BY FINITE ELEMENTS	72-0401
HOLAND, I.	APPLICATION OF FINITE ELEMENT METHODS TO STRESS ANALYSIS	73-0524
HOLAND, I.	FINITE ELEMENT METHODS IN STRESS ANALYSIS	69-0126
HOLAND, I.	FINITE ELEMENTS FOR THE COMPUTATIONS OF HYDRODYNAMIC MASS	69-0346
HOLAND, I.	FUNDAMENTALS OF THE FINITE ELEMENT METHOD	74-0191
HOLAND, I.	HIGHER-ORDER FINITE ELEMENT FOR PLANE STRESS	68-0089
HOLAND, I.	INCORPORATION OF FOUNDATION DEFORMATION IN A FINITE ELEMENT SHELL ANALYSIS	68-0227
HOLAND, I.	MEMBRANE AND PLATE BENDING ELEMENTS	75-0405
HOLAND, I.	NUMERICAL ACCURACY AND ROUNDING-OFF ERRORS	69-0348

47

HOLAND, I.	STIFFNESS MATRICES FOR PLATE BENDING ELEMENTS	69-0129
HOLAND, I.	THE FINITE ELEMENT METHOD IN PLANE STRESS ANALYSIS	69-0127
HOLAND, I.	THE FINITE ELEMENT METHOD IN PLATE BUCKLING	69-0128
HOLEHOUSE, I.	SONIC FATIGUE OF DIFFUSION BOUNDED TITANIUM SANDWICH STRUCTURES	72-0290
HOLISTER, G. S.	STRESS ANALYSIS	65-0055
HOLLINGSHEAD, G	PREDICTION OF UNDRAINED MOVEMENTS CAUSED BY EMBANKMENTS ON MUSKEG	70-0070
HOLLOWAY, G. F.	THE INELASTIC BEHAVIOUR OF STRUCTURES SUBJECTED TO CYCLIC, THERMAL AND MECHANICAL STRESSING CONDITIO	60-0004
HOLLOWAY, M.	LOAD DEFORMATION ANALYSIS OF DEEP PILE FOUNDATION	72-0931
HOLMES, N.	EXPLICIT INTEGRATION - STABILITY, SOLUTION PROPERTIES, COST	75-0298
HOLMES, P.	YIELD CRITERION FOR REINFORCED CONCRETE SLABS	73-0985
HOLUSA, L.	CALCULATION OF PLATE OF CONSTANT THICKNESS BY FINITE ELEMENT METHOD	69-0227
HOLUSA, L.	THE FINITE ELEMENT METHOD	70-0327
HOLZER, S. M.	SINGER: A COMPUTER CODE FOR GENERAL ANALYSIS OF TWO-DIMENSIONAL CONCRETE STRUCTURES. VOL. 11. PROGRA	75-0723
HOLZER, S. M.	SINGER: A COMPUTER CODE FOR GENERAL ANALYSIS OF TWO-DIMENSIONAL REINFORCED CONCRETE STRUCTURES. VOLU	75-0724
HOLZER, S. M.	SINGER: A COMPUTER CODE FOR GENERAL ANALYSIS TO TWO-DIMENSIONAL CONCRETE STRUCTURES. VOLUME 4. DEMON	75-0730
HOLZER, S. M.	SINGER: A COMPUTER CODE FOR GENERAL ANALYSIS OF TWO-DIMENSIONAL REINFORCED CONCRETE STRUCTURES. VOLU	75-0731
HOLZLOEHNER, U.	DIE BERECHNUNG VON KONSOLIDATIONSVORGAENGEN MIT DER FINITE-ELEMENT-METHODE, (CALCULATION OF CONSOLID	71-0012
HOLZLOEHNER, U.	FINITE ELEMENT ANALYSIS FOR TIME-DEPENDENT PROBLEMS	74-0192
HOMMEL, G.	COMMENT ON LIMITATIONS OF CERTAIN FINITE ELEMENTS WHEN APPLIED TO ARCHES - FURTHER STUDIES IN APPLIC	72-0097
HONG, J. H.	NONLINEAR DYNAMIC ANALYSIS OF SHELLS OF REVOLUTION BY MATRIX DISPLACEMENT METHOD	70-0110
HOOD, P.	A NUMERICAL SOLUTION OF THE NAVIER-STOKES EQUATIONS USING THE FINITE ELEMENT TECHNIQUE	73-0061
HOOD, P.	FINITE ELEMENT FORMULATION WITH REFERENCE TO FLUID DYNAMICS	74-1245
HOOD, P.	NAVIER STOKES EQUATIONS USING MIXED INTERPOLATION	74-0014
HOOD, P.	THE SOLUTION OF THE NAVIER STOKES EQUATIONS	74-1220
HOOKER, R. J.	NATURAL FREQUENCIES OF BOX-TYPE STRUCTURES BY A FINITE-ELEMENT METHOD	74-0193
HOOLEY, R. F.	COMPARISON AND EVALUATION OF STIFFNESS MATRICES	66-0045
HOOPER, J. A.	ANALYSIS OF A CIRCULAR RAFT IN ADHESIVE CONTACT WITH A THICK ELASTIC LAYER	74-1152
HOOPER, J. A.	ELASTIC SETTLEMENT OF A CIRCULAR RAFT IN ADHESIVE CONTACT WITH A TRANSVERSELY ISOTROPIC MEDIUM	75-0828
HOOPER, J. A.	FIELD INSTRUMENTATION FOR LONG TERM MEASUREMENT OF PILE LOAD AND RAFT CONTACT PRESSURE	73-0150
HOOPER, J. A.	OBSERVATIONS ON THE BEHAVIOUR OF A PILED-RAFT FOUNDATION ON LONDON CLAY.	73-0421
HOOSON, R. E.	STRESS AND DEFLECTION ANALYSIS OF MECHANICALLY FASTENED JOINTS	70-0067
HOPKINS, A. S.	THE MOTION OF INTERCONNECTED FLEXIBLE BODIES	75-0727
HOPPE, V.	FINITE ELEMENTS WITH HARMONIC INTERPOLATION FUNCTIONS	72-0058
HOPPE, V.	FINITE ELEMENTS WITH HARMONIC INTERPOLATION FUNCTIONS	73-1065
HOPPE, V.	SPECIFICATION OF GEOMETRICAL PARAMETERS FOR ELEMENTS WITH CUBIC MAPPING FUNCTIONS	73-1082
HOPPER, A. T.	ELASTIC-PLASTIC ANALYSIS OF INTERFERENCE FIT FASTENERS	74-1177
HOPPER, A. T.	HIGH-ORDER FINITE ELEMENT ANALYSIS FOR TRANSIENT TEMPERATURE ANALYSIS OF INHOMOGENEOUS MATERIALS	69-0301
HOPPER, A. T.	TRANSFORMATIONS FOR ELLIPTICAL HOLES APPLIED TO FINITE ELEMENT ANALYSIS	74-0132
HORAK, M.	CALCULATION OF THE CRITICAL LOAD OF ELASTIC THIN WALLED BARS BY THE METHOD OF FINITE ELEMENTS	72-0446
HORI, Y.	STUDY ON FOIL BEARINGS - AN APPLICATION TO TAPE MEMORY DEVICES	75-0311
HORII, K.	LARGE STRAIN, ELASTO-PLASTIC NUMERICAL ANALYSIS BY MEANS OF FINITE ELEMENT METHOD	71-0790
HORII, K.	NUMERICAL ANALYSIS OF VISCO-ELASTIC STRUCTURES BY THE FINITE ELEMENT METHOD	70-0177
HORIKAMI, K.	FINITE ELEMENT METHOD FOR SOLVING NEUTRON TRANSPORT PROBLEMS IN TWO-DIMENSIONAL CYLINDRICAL GEOMETRY	74-1081
HORIKAWA, J.	ELASTIC AND PLASTIC STRESS DISTRIBUTION ON BASE PLATES IN THE NEIGHBOURHOOD OF FILLET WELDS IN LAP J	70-0175
HORINO, F. G.	THE USE OF ROCK BOLTS OR WIRE ROPE TO INCREASE THE STRENGTH OF FRACTURED MODEL PILLARS	71-0722
HORIOKA, M.	STRESS ANALYSIS IN HULL STRUCTURAL MEMBERS BY PHOTOELASTIC EXPERIMENTS AND FINITE ELEMENT METHODS -	70-0341
HORITA, T.	APPLICATION OF FINITE-ELEMENT METHOD AND MOIRE TECHNIQUE TO SHEET-METAL FORMING PROBLEMS	71-0784
HORLOCK, J. H.	FINITE ELEMENT METHODS IN FLOW PROBLEMS	74-0643
HORLOCK, J. H.	COMPUTATION OF FLOWS IN TURBOMACHINES	74-0076
HORN, H. R. JR.	CONICAL FINITE ELEMENT STRESS ANALYSIS OF AN AXISYMMETRIC TANK ON AN ELASTIC FOUNDATION	70-0494
HORNBERGER, G.	NUMERICAL METHODS IN SUBSURFACE HYDROLOGY: WITH AN INTRODUCTION TO THE FINITE ELEMENT METHOD	71-0800
HORNBUCKLE, J.	STRUCTURAL OPTIMIZATION USING THE FINITE ELEMENT METHOD APPLIED TO A BEAM	75-0233
HORWATH, J. K.	STRUCTURAL AND SYSTEM MODELS	75-0669
HOSAIN, M. U.	ANALYSIS OF CASTELLATED BEAMS BY THE FINITE ELEMENT METHOD	72-0620
HOSAIN, M. U.	DEFLECTION ANALYSIS OF EXPANDED OPEN-WEB STEEL BEAMS	74-0224
HOSHIYA, S.	FLEXUAL RIGIDITY OF A THIN WALLED BUILD-UP ROTOR FOR JET ENGINE. MEASUREMENT BY STATIC LOAD TEST AND	73-0756
HOSKINS, E. R.	A DEEP HOLE STRESS MEASUREMENT DEVICE	74-0804
HOSOMUPA, T.	LARGE DEFORMATION AND CRITICAL LOADS ANALYSIS OF FRAMED STRUCTURES	74-1293
HOSSAIN, Q. A.	NONLINEAR FINITE ELEMENT ANALYSIS OF REINFORCED CONCRETE FRAMES	74-1124
HOTOS, G. K.	METHODS OF MODELING AND ANALYZING VISCOELASTICALLY DAMPED STRUCTURES	71-0836
HOTTA, Y.	ELASTO-PLASTIC ANALYSIS OF AUTOMOBILE BODY STRUCTURE BY THE FINITE ELEMENT METHOD	74-0196
HOUDE, J.	A STUDY OF BEHAVIOUR OF REINFORCED CONCRETE ELEMENTS USING FINITE ELEMENTS	70-0335
HOUDE, J.	FINITE ELEMENT ANALYSIS OF SHEAR STRENGTH OF REINFORCED CONCRETE BEAMS	74-1306
HOUDE, J.	STUDY OF FORCE DISPLACEMENT RELATIONSHIPS FOR THE FINITE ELEMENT ANALYSIS OF REINFORCED CONCRETE	73-0645
HOUSNER, G. W.	NUMERICAL MODEL FOR TSUNAMI RUN-UP	70-0200
HOUSTIS, E. N.	FINITE ELEMENT METHODS FOR SOLVING INITIAL BOUNDARY VALUE PROBLEMS	74-1125
HOWELL, L. J.	POWER SPECTRAL DENSITY ANALYSIS OF VEHICLE VIBRATION USING THE NASTRAN COMPUTER PROGRAM	74-0318
HOWLETT, J. T.	APPLICATIONS OF NASTRAN TO COUPLED STRUCTURAL AND HYDRODYNAMIC RESPONSES IN AIRCRAFT HYDRAULIC SYSTE	71-0744
HOYAUX, B.	ROCK FAILURE AROUND A CIRCULAR OPENING IN A GRAVITY FIELD WITH TECTONIC FORCES	70-0071
HOYAUX, B.	STRESS DISTRIBUTION DUE TO GRAVITY IN A VERTICAL ROCK BANK	68-0090
HOYT, P.	STRESS CONCENTRATIONS IN HULL SHAPES FROM SURFACE DISCONTINUITIES	69-0130
HRDINA, J.	THREE DIMENSIONAL STRESS ANALYSIS OF PENETRATIONS IN THICK WALLED PRESSURE VESSELS	74-0422
HRENNIKOFF, A.	PRECISION OF FINITE ELEMENT METHOD IN PLANE STRESS	69-0295
HRENNIKOFF, A.	SOLUTION OF PROBLEMS IN ELASTICITY BY THE FRAMEWORK METHOD	41-0001
HRENNIKOFF, A.	THE FINITE ELEMENT METHOD IN APPLICATION TO PLANE STRESS	68-0117
HRENNIKOFF, A.	THREE-DIMENSIONAL BAR CELL FOR ELASTIC STRESS ANALYSIS	70-0219
HRIBAR, J. A.	ELASTIC ANALYSIS OF A REFRACTORY SPRUNG ARCH	74-0386
HRIBAR, J. A.	STEADY STATE IN SOIL SUBJECTED TO SEVERAL COUPLED FORCES: A FINITE ELEMENT FORMULATION	70-0207
HRUDA, R. F.	VIBRATION ANALYSIS OF STRUCTURES BY COMPONENT MODE SUBSTITUTION	71-0307
HRUDEY, T. M.	A FINITE ELEMENT PROCEDURE FOR PLATES WITH CURVED BOUNDARIES	75-0030
HRUDEY, T. M.	A FINITE ELEMENT PROCEDURE FOR PLATES WITH CURVED BOUNDARIES	75-0750
HRUDEY, T. M.	APPLICATION OF A REFINED SHALLOW SHELL FINITE ELEMENT TO GEOMETRICALLY NONLINEAR STRUCTURES	73-0358
HRUDEY, T. M.	CREEP BENDING ANALYSIS OF PLATES BY THE FINITE ELEMENT METHOD	71-0415
HRUDEY, T. M.	FINITE ELEMENT VIBRATION ANALYSIS OF MULTICONDUCTOR ELECTRICAL TRANSMISSION LINES	73-0590
HRUDEY, T. M.	REFINED FINITE ELEMENTS FOR FOLDED PLATE STRUCTURES	72-0574
HSICH, M. S.	FINITE ELEMENT SOLUTION OF TWO DIMENSIONAL EXTERIOR FIELD PROBLEMS	71-0491
HSIEH, B. J.	CONVECTED COORDINATES FOR TRANSIENT NONLINEAR FINITE-ELEMENT ANALYSIS	73-0850
HSIEH, B. J.	NON-LINEAR TRANSIENT FINITE ELEMENT ANALYSIS WITH CONVECTED CO-ORDINATES	73-0378
HSIEH, B. J.	NONLINEAR TRANSIENT ANALYSIS OF SHELLS AND SOLIDS OF REVOLUTION BY CONVECTED ELEMENTS	73-0519
HSIEH, B. J.	NONLINEAR TRANSIENT ANALYSIS OF SHELLS AND SOLIDS OF REVOLUTION BY CONVECTED ELEMENTS	74-0476
HSIUNG, H. C. H	DYNAMIC ANALYSIS OF HYDROELASTIC SYSTEMS USING THE FINITE-ELEMENT METHOD	73-1137
HSU, C. C.	GALERKIN METHOD FOR A CLASS OF STEADY, 2-DIMENSIONAL, INCOMPRESSIBLE, LAMINAR BOUNDARY-LAYER FLOWS	75-0437
HSU, M. B.	BLOWDOWN EFFECTS ON A FUEL ROD	75-0575
HSU, M. B.	COUPLED CONVECTIVE AND CONDUCTIVE HEAT TRANSFER BY FINITE ELEMENT METHODS	74-0050
HSU, M. B.	EVALUATION OF THE COMPUTER PROGRAM CREEP-BLAST (CDC VERSION)	74-0684
HSU, M. B.	FETE PROGRAM PACKAGE	72-0161
HSU, M. B.	INELASTIC ANALYSIS OF TUBESHEETS BY THE FINITE ELEMENT METHOD	75-0652
HSU, T. R.	APPLICATION OF THE FINITE ELEMENT METHOD TO THE NONLINEAR ANALYSIS OF NUCLEAR REACTOR FUEL BEHAVIOR	74-1246
HSU, T. R.	APPLICATION OF THE SINGLE-EDGE NOTCHED BEAM TO FRACTURE TOUGHNESS TESTING OF CERAMICS	74-0623
HSU, T. R.	ESTABLISHMENT OF A VALID FRACTURE TOUGHNESS SPECIMEN GEOMETRY USING A FINITE ELEMENT ANALYSIS	74-0135
HSU, T. R.	FULMOD - AN INELASTIC ANALYSIS PROGRAM TO PREDICT THE OPERATING BEHAVIOUR OF CANDU FUEL ELEMENTS	75-0574
HSU, T. R.	IMPROVED APPROXIMATION OF CONSTITUTIVE ELASTO-PLASTIC STRESS STRAIN RELATIONSHIP FOR FINITE ELEMENT	74-0566

HSU, T. R.	PROPAGATION AND OPENING OF A THROUGH CRACK IN PIPE SUBJECT TO COMBINED CYCLIC THERMOMECHANICAL LOADI	75-0516
HSU, T. R.	TRANSIENT THERMAL STRESS TECHNIQUES IN NUCLEAR REACTORS. APPLIED MECHANICS ASPECTS OF NUCLEAR EFFECT	71-0474
HSU, T-M.	DEVELOPMENT OF AN UNDERSTANDING OF THE FATIQUE PHENOMENA OF BONDED AND BOLTED JOINTS IN ADVANCED FIL	72-0798
HSUEH, T.	LARGE DISPLACEMENT ANALYSIS OF ASIXYMMETRIC SHELLS	69-0276
HSUEH, T.	STABILITY ANALYSIS OF AXISYMMETRIC SHELLS	69-0131
HU, M-H.	STUDY OF FLUID MOVEMENTS THROUGH CAUSEWAY	75-0083
HU, W-L.	CRACK TIP STRAIN - A COMPARISON OF FEM CALCULATIONS AND MEASUREMENTS	75-0721
HUA, H. M.	A FINITE ELEMENT METHOD FOR CALCULATING AERODYNAMIC COEFFICIENTS OF A SUBSONIC AIRPLANE	73-0655
HUANG, A. B.	A NON VARIATIONAL FINITE ELEMENT ANALYSIS FOR THE NAVIER STOKES EQUATIONS	74-0010
HUANG, C. T.	ON SOLUTION OF PLANE STRAIN CONSOLIDATION PROBLEMS BY FINITE ELEMENT METHOD	71-0532
HUANG, J-C.	ANALYSIS OF SHELLS OF REVOLUTION BY A CURVED FINITE ELEMENT	71-0680
HUANG, P. C.	MISSILE BODY INPUT GENERATOR (BING), A NASTRAN PRE-PROCESSOR: THEORETICAL DEVELOPMENT, USER'S MANUAL	75-0717
HUANG, P. C.	PLANFORM INPUT GENERATOR (PING), A NASTRAN PRE-PROCESSOR FOR LIFTING SURFACES-THEORETICAL DEVELOPMEN	73-0776
HUANG, P. C.	PLANFORM INPUT GENERATOR (PING), A NASTRAN PROCESSOR FOR LIFTING SURFACES-THEORETICAL DEVELOPMENT, U	73-0780
HUANG, S. L.	STATIC AND DYNAMIC ANALYSIS OF F-14A BORON HORIZONTAL STABILIZER	71-0123
HUANG, S. N.	TRANSVERSE IMPACT PROBLEMS BY HIGHER-ORDER BEAM FINITE-ELEMENTS	75-0819
HUANG, S. W.	APPLICATION OF GRIFFITH'S THEORY TO ANALYSIS OF PROGRESSIVE FRACTURE	75-0150
HUANG, S-W.	FINITE-ELEMENT ANALYSIS OF FRACTURE PROPAGATION IN TWO-DIMENSIONAL ELASTIC BRITTLE SOLIDS	72-0948
HUANG, T. C.	COUPLED RESPONSE OF SPATIAL VIBRATORY STRUCTURES MOUNTED TO ISOTROPIC PLATE ELEMENTS	71-0228
HUANG, T. C.	ELASTODYNAMICS OF COMPLEX STRUCTURAL SYSTEMS	71-0563
HUANG, T. S.	CURVED TRIANGULAR FINITE ELEMENTS FOR THE ANALYSIS OF THICK ELASTIC PLATES AND SHELLS	71-0814
HUANG, Y. H.	ANALYSIS OF UNSTEADY FLOW TOWARD AN ARTESIAN WELL BY 3-DIMENSIONAL FINITE ELEMENTS	74-0351
HUANG, Y. H.	ANALYSIS OF UNSTEADY FLOW TOWARD ARTESIAN WELLS BY THREE-DIMENSIONAL FINITE ELEMENTS	74-0742
HUANG, Y. H.	ANALYSIS OF UNSTEADY FLOW TOWARD AN ARTESIAN WELL BY THREE-DIMENSIONAL FINITE ELEMENTS	74-1351
HUANG, Y. H.	FINITE ELEMENT ANALYSIS OF NONLINEAR SOIL MEDIA	69-0132
HUANG, Y. H.	FINITE ELEMENT ANALYSIS OF SLABS ON ELASTIC SOLIDS	74-0120
HUANG, Y. H.	FINITE-ELEMENT ANALYSIS OF CONCRETE SLABS AND ITS IMPLICATIONS FOR RIGID PAVEMENT DESIGN	73-0538
HUANYEN, L.	FINITE ELEMENT ANALYSIS OF TIME-DEPENDENT PUMPING TEST DATA IN AN UNCONFINED AQUIFER	75-0497
HUBBARD, B.	NUMERICAL SOLUTION OF PARTIAL DIFFERENTIAL EQUATIONS, 11	70-0072
HUBBARD, P. P.	DYNAMIC RESPONSE OF ELASTIC HEAD MODELS	75-0378
HUBER, A. E.	EINFACHES NAEHERUNGSVERFAHREN FUR DIE GRENZTRAGFAEHIGKEITSANLYSE VON SPANNBETON-REAKTORDRUCKBEHAELTE	73-0854
HUCK, P. J.	DYNAMIC RESPONSE OF SOIL/CONCRETE INTERFACES AT HIGH PRESSURE	74-0836
HUDSON, J. A.	CONTROLLED FAILURE OF HOLLOW ROCK CYLINDERS IN UNIAXIAL COMPRESSION	72-0196
HUDSON, W. R.	A SUMMARY OF DISCRETE-ELEMENT METHODS OF ANALYSIS FOR PAVEMENT SLABS	72-0674
HUDSON, W. R.	EXPERIMENTAL VERIFICATION OF DISCRETE-ELEMENT SOLUTIONS FOR PLATES AND PAVEMENT SLABS	70-0411
HUDSPITH, I. B.	THE ASAS ANALYSIS OF THE LYNX	74-0932
HUEBNER, K. H.	FINITE ELEMENT ANALYSIS OF FLUID FILM LUBRICATION SURVEY	75-0880
HUEBNER, K. H.	APPLICATION OF FINITE ELEMENT METHODS TO LUBRICATION - ENGINEERING APPPROACH	72-0006
HUEBNER, K. H.	APPLICATION OF FINITE ELEMENT METHODS TO THERMOHYDRODYNAMIC LUBRICATION	73-0074
HUEBNER, K. H.	FINITE ELEMENT ANALYSIS OF CONTINUUM PROBLEMS AND ITS APPLICATION TO THE INCOMPRESSIBLE LUBRICATION	71-0539
HUEBNER, K. H.	FINITE ELEMENT ANALYSIS OF FLUID FILM LUBRICATION - A SURVEY	74-1100
HUEBNER, K. H.	FINITE ELEMENT METHOD - STRESS ANALYSIS AND MUCH MORE	74-0095
HUEBNER, K. H.	SOLUTION OF THE ELASTOHYDRODYNAMIC FINITE JOURNAL BEARING PROBLEM	72-0033
HUEBNER, K. H.	THE FINITE ELEMENT METHOD FOR ENGINEERS	75-0887
HUFF, R. D.	THE INELASTIC BEHAVIOUR OF STRUCTURES SUBJECTED TO CYCLIC, THERMAL AND MECHANICAL STRESSING CONDITIO	60-0004
HUFFERD, W. L.	HANDBOOK FOR THE ENGINEERING STRUCTURAL ANALYSIS OF SOLID PROPELLANTS	71-0642
HUGGINS, M. W.	DIFFERENTIAL EXPANSION IN ELASTIC LAMINATES	73-0190
HUGHES, O. F.	SOME CONSIDERATIONS REGARDING STRUCTURAL OPTIMIZATION AND FINITE ELEMENT ANALYSIS	74-0424
HUGHES, T. J.	FINITE ELEMENT FORMULATION AND SOLUTION OF CONTACT-IMPACT PROBLEMS IN CONTINUUM MECHANICS	74-0806
HUGHES, T. J.	FINITE ELEMENT FORMULATION AND SOLUTION OF CONTACT-IMPACT PROBLEMS IN CONTINUUM MECHANICS-II	75-0740
HUGHES, T. J. R	FINITE ELEMENT FORMULATION FOR PIEZOELECTRIC CONTINUA BY A VARIATIONAL THEOREM	68-0230
HUGHES, T. J. R	FINITE ELEMENT METHOD FOR PIEZOELECTRIC VIBRATION	70-0012
HUGHES, T. J. R	FINITE ELEMENTS FOR COMPRESSIBLE AND INCOMPRESSIBLE CONTINUA	69-0133
HUIZER, A.	A THEORETICAL AND EXPERIMENTAL STUDY ON THE FREE VIBRATION OF CURVED SLAB BRIDGES	71-0496
HUMPHERSON, J.	FINITE ELEMENT ANALYSIS OF STRESSES IN SHAFTS DUE TO INTERFERENCE-FIT HUBS	69-0353
HUMPHESON, C.	APPLICATIONS OF ELECTRO-OSMOSIS TO GROUND-WATER FLOW PROBLEMS	75-0336
HUMPHESON, C.	FINITE ELEMENT ANALYSIS OF ELECTRO-OSMOTIC DEWATERING	73-0592
HUMPHESON, C.	NUMERICAL ANALYSIS OF ELECTRO-OSMOTIC FLOW IN SOILS	73-0444
HUMPHREY, A. T.	APPLICATION OF FINITE ELEMENT TECHNIQUES TO STRESS AND STRUCTURAL ANALYSIS	70-0362
HUMPHREY, A. T.	FINITE ELEMENT ANALYSIS OF AN EXPANDED I - SECTION BEAM AND AN AXISYMMETRIC FLANGED CYLINDER	68-0002
HUMPHRIES, R. W	APPROACH TO ANALYSIS OF ARCH DAMS IN WIDE VALLEYS	73-0153
HUNG, N. D.	DUALITY IN THE ANALYSIS OF SHELLS BY THE FINITE ELEMENT METHOD	71-0124
HUNG, N. D.	NATRUAL STRAINS AND STRESSES FOR TRAPEZOIDAL STRUCTURES ANALYSIS	69-0101
HUNNISETT, S. F	MIXED-DISPLACEMENT FINITE-ELEMENT ANALYSIS WITH PARTICULAR APPLICATION USING PLANE-STRESS TRIANGLES	72-0245
HUNSAKER, B.	COMPARISON OF THE CAPABILITY OF OUR HARDENING RULES TO PREDICT A MATERIAL'S PLASTIC BEHAVIOR	75-0523
HUNSAKER, B. JR	A COMPARISON OF CURRENT WORK-HARDENING MODELS USED IN THE ANALYSIS OF PLASTIC DEFORMATIONS	73-0769
HUNT, C.	ON A VARIATIONAL INEQUALITY AND ITS APPROXIMATION, IN THE THEORY OF SEMICONDUCTORS	75-0829
HUNT, D. A.	DISCRETE ELEMENT IDEALIZATION OF AN INCOMPRESSIBLE LIQUID FOR VIBRATION ANALYSIS	70-0391
HUNT, D. A.	DISCRETE ELEMENT STRUCTURAL THEORY OF FLUIDS	70-0009
HUNT, D. A.	DYNAMIC ANALYSIS OF A FLEXIBLE VEHICLE MOVING ALONG A FLEXIBLE SUPPORT	71-0597
HUNT, G. W.	COMPARATIVE PERTURBATION STUDIES OF THE ELASTICA	69-0255
HUNT, J. T.	APPLICATIONS OF THE FINITE-ELEMENT METHOD AND COMPUTER GRAPHICS. (A VIBRATIONAL ANALYSIS OF A SONAR	72-0729
HUNT, J. T.	FINITE ELEMENT ANALYSIS OF ACOUSTICALLY RADIATING STRUCTURES WITH APPLICATIONS TO SONAR TRANSDUCERS	73-0075
HUNT, J. T.	FINITE ELEMENT APPROACH TO ACOUSTIC RADIATION FROM ELASTIC STRUCTURES	74-0121
HUNT, J. T.	FINITE-ELEMENT APPROACH TO ACOUSTIC SCATTERING FROM ELASTIC STRUCTURES	75-0069
HUNT, J. T.	VIBRATIONAL ANALYSIS OF THE FLEXING HEAD OF THE BOS-6 TRANSDUCER	71-0708
HUNT, J. T.	VIBRATIONAL RESPONSE OF SONAR TRANSDUCERS USING PIEZO-ELECTRIC FINITE ELEMENTS	74-0269
HUNT, M. O.	PREDICTION OF ELASTIC CONSTANTS OF PARTICLEBOARD	74-0259
HUNTER, A. R.	PHOTOELASTIC AND FINITE ELEMENT INVESTIGATION OF A NONSYMMETRICAL PLUG HATCH CONFIGURATION	73-0025
HUNTER, A. R.	STRESS DISTRIBUTIONS AROUND PLUG MATCHES WITH ANCHOR BLOCKS	69-0135
HUNTER, D. T.	LARGE DEFLECTION ANALYSIS OF ELASTIC-PLASTIC SHELLS OF REVOLUTION	70-0507
HURBAN, J. M.	ANALYSIS OF BAND PRESSURE IN A GUN TUBE USING FINITE ELEMENTS AND SHELL PUSHING DATA	72-0793
HURBAN, J. M.	STRESS ANALYSIS OF 175 MM PROJECTILE, HE, M437	71-0736
HURR, R. T.	MODELLING GROUNDWATER FLOW BY THE FINITE ELEMENT METHOD	72-0637
HURST, C. J.	CONDUCTION THROUGH DROPLETS DURING DROPWISE CONDENSATION	72-0609
HURST, H. T.	RELATIONSHIP BETWEEN FINITE ELEMENT AND MEASURED JOINT-FLOOR CONNECTION RIGIDITIES	75-0738
HURTADO, J.	RHEOLOGICAL THEORY OF STRENGTH OF MATERIALS	69-0313
HURTY, C. H.	DYNAMIC ANALYSIS OF STRUCTURAL SYSTEMS USING COMPONENT MODES	64-0024
HURWITZ, M. M.	DEFINIT - A NEW ELEMENT DEFINITION CAPABILITY FOR NASTRAN: USER'S MANUAL	73-0738
HURWITZ, M. M.	NEW ELEMENT DEFINITION CAPABILITY FOR NASTRAN	71-0125
HURWITZ, M. M.	PROGRAMMER'S MANUAL ADDITIONS AND DEMONSTRATION PROBLEMS FOR A THERMOSTRUCTURAL CAPABILITY FOR NASTR	75-0734
HURWITZ, M. M.	THERMAL AND STRESS ANALYSIS BY NASTRAN	72-0153
HUS, M.	THERMAL AND STRESS ANALYSIS OF A FUEL ROD	75-0713
HUSAIN, H. M.	GENERALISED MATRIX FORCE AND DISPLACEMENT METHODS FOR LINEAR STRUCTURAL ANALYSIS	65-0064
HUSAIN, H. M.	PLANE FRAMEWORK METHODS FOR PLATES IN EXTENSION	66-0126
HUSSEY, M. J. L	ON THE CONSTRUCTION AND USE OF FINITE ELEMENTS	70-0265
HUSTON, R. L.	AN ANALYSIS OF GENERAL CHAIN SYSTEMS	72-0787
HUSTON, R. L.	ON CONSTRAINT EQUATIONS - A NEW APPROACH	74-1017
HUTCHINSON, J.	DEVELOPMENT OF A SOLID ROCKET PROPELLANT NONLINEAR CONSTITUTTIVE THEORY	75-0722
HUTCHINSON, J.	DYNAMIC SHELL THEORY COUPLING THICKNESS STRESS WAVE EFFECTS WITH GROSS STRUCTURAL RESPONSE	73-0379
HUTCHINSON, J.	NONLINEAR DYNAMICS OF SOLIDS BY THE FINITE ELEMENT METHOD	71-0116
HUTCHINSON, J.	USE OF TRIGONOMETRIC TERMS IN FINITE ELEMENT METHOD WITH APPLICATION TO VIBRATING MEMBRANES	74-0127

HUTCHINSON, J.	PLASTIC INTENSITY FACTORS FOR CRACKED PLATES	71-0164
HUTCHINSON, P.	CALCULATION OF AREAL RAINFALL USING FINITE ELEMENT TECHNIQUES WITH ALTITUDINAL CORRECTIONS	72-0344
HUTCHINSON, R.	DESIGN METHOD FOR FLEXIBLE AIRFIELD PAVEMENTS	74-1174
HUTHER, M.	DYNAMIC AND STOCHASTIC ASPECTS IN THE STRUCTURAL ANALYSIS OF VLCCS AND ULCCS	74-0298
HUTT, J. M.	DYNAMIC STABILITY OF PLATES BY FINITE ELEMENTS	71-0573
HUTTON, A. G.	ON FLOW NEAR SINGULAR POINTS OF A WALL BOUNDARY	74-0008
HUTTON, M.	SIMPLIFICATION OF HIGH ORDER MECHANICAL SYSTEMS USING THE ROUTH APPROXIMATION	75-0853
HUTTON, S. G.	FINITE ELEMENT METHOD: A GALERKIN APPROACH	71-0451
HUTULA D. N.	MESH 3D: A THREE-DIMENSIONAL FINITE ELEMENT MESH GENERATOR PROGRAM FOR EIGHT-NODE ISOPARAMETRIC ELEM	73-0972
HUTULA, D. N.	APACHE: A THREE-DIMENSIONAL FINITE ELEMENT PROGRAM FOR STEADY-STATE OR TRANSIENT HEAT CONDUCTION ANA	73-0971
HUTULA, D. N.	MATUS: A THREE-DIMENSIONAL FINITE ELEMENT PROGRAM FOR SMALL-STRAIN ELASTIC ANALYSIS	73-0970
HUYAKORN, P.	FINITE ELEMENT SOLUTION OF TWO-REGIME FLOW TOWARDS WELLS	74-1352
HUYAKORN, P. S.	FINITE ELEMENT SOLUTIONS OF TRANSIENT, TWO-REGIME FLOW TOWARDS WELLS	74-0466
HUYAKORN, P. S.	TECHNIQUES FOR HANDLING INPUT DATA FOR FINITE ELEMENT ANALYSIS OF REGIONAL GROUNDWATER FLOW	75-0806
HWANG, C. T.	APPLICATION OF THE FINITE ELEMENT METHOD TO CONSOLIDATION PROBLEMS	72-0421
HWANG, C. T.	FINITE ELEMENT ANALYSIS OF SOIL DEFORMATIONS	69-0134
HWANG, C. T.	FINITE ELEMENT METHOD FOR SOIL DEFORMATIONS	70-0495
HWANG, C. T.	NONLINEAR ACOUSTIC RESPONSE ANALYSIS OF PLATES USING THE FINITE ELEMENT METHOD	72-0223
HWANG, J. H.	FINITE ELEMENT ANALYSIS OF THE MAGNETIC FIELD DISTRIBUTION INSIDE A ROTATING FERROMAGNETIC BAR	74-1037
HWANG, J. H.	FINITE ELEMENT MODELING OF MAGNETIC FIELD DEFECT INTERACTIONS	75-0068
HWANG, J. H.	PREDICTING MAGNETIC LEAKAGE FIELDS BY FINITE ELEMENT TECHNIQUES	75-0046
HWANG, R.	LUSH: A COMPUTER PROGRAM FOR COMPLEX RESPONSE ANALYSIS OF SOIL-STRUCTURE SYSTEMS	74-0744
HWANG, R.	QUAD-4: A COMPUTER PROGRAM FOR EVALUATING THE SEISMIC RESPONSE OF SOIL STRUCTURES BY VARIABLE DAMPIN	73-0771
HYLARIDES, S.	FINITE ELEMENT TECHNIQUE IN SHIP VIBRATION ANALYSIS	68-0180
HYLARIDES, S.	RECENT DEVELOPMENTS IN HULL AND SHAFT VIBRATION ANALYSIS	70-0176
HYLARIDES, S.	TRANSVERSE VIBRATIONS OF SHIP'S PROPULSION SYSTEMS. PART 1: THEORETICAL ANALYSIS	74-1323
HYLARIDES, S.	TRANSVERSE VIBRATIONS OF SHIP'S PROPULSION SYSTEMS - 1 THEORETICAL ANALYSIS	75-0291
HYMAN, B. I.	EFFECT OF FIBRE DISTRIBUTION ON STRESS AND STRAIN CONCENTRATION AT HOLES IN COMPOSITE PLATES	72-0616
ICE, M. W.	NASTRAN USER INTERFACES, AUTOMATED INPUT INNOVATIONS	71-0126
ICEMAN, K. R.	A FINITE ELEMENT MODEL FOR TWO DIMENSIONAL FLOW	74-0015
ICEMAN, K. R.	A FINITE ELEMENT SOLUTION FOR TWO-DIMENSIONAL STRATIFIED FLOW PROBLEMS	75-0391
ICHIKAWA, T.	SOME CONSIDERATIONS ON METHODS FOR FLUTTER ANALYSIS	69-0136
IDELSOHN, S.	ANALYSE STATIQUE ET DYNAMIQUE DES COQUES PAR LA MÉTHODE DES ELEMENTS FINIS	74-0509
IDING, R.	APPLICATION OF EXTENDED VARIATIONAL PRINCIPLES TO FINITE ELEMENT ANALYSIS	72-0529
IDING, R. H.	IDENTIFICATION OF NONLINEAR MATERIALS BY FINITE ELEMENT METHODS	73-0927
IDING, R. H.	IDENTIFICATION OF NONLINEAR ELASTIC SOLIDS BY A FINITE ELEMENT METHOD	74-0530
IDING, R. H.	NONSAP: A STRUCTURAL ANALYSIS PROGRAM FOR STATIC AND DYNAMIC RESPONSE OF NONLINEAR SYSTEMS	74-0848
IDRISS, I. H.	STUDIES OF SEISMIC RESPONSE OF CLAY BANKS	71-0833
IDRISS, I. M.	DYNAMIC ANALYSIS OF THE SLIDE IN THE LOWER SAN FERNANDO DAM DURING THE EARTHQUAKE OF FEBRUARY 9, 197	75-0277
IDRISS, I. M.	FINITE ELEMENT ANALYSIS FOR SEISMIC RESPONSE OF EARTH DAMS	68-0091
IDRISS, I. M.	QUAD-4: A COMPUTER PROGRAM FOR EVALUATING THE SEISMIC RESPONSE OF SOIL STRUCTURES BY VARIABLE DAMPIN	73-0771
IDRISS, I. M.	RESPONSE OF EARTH BANKS DURING EARTHQUAKES	67-0039
IDRISS, I. M.	SEISMIC RESPONSE BY VARIABLE DAMPING FINITE ELEMENTS	74-0143
IDRISS, I. M.	STATIC STRESSES BY LINEAR AND NONLINEAR METHODS	75-0278
IGHARO, P.	ANALYSIS OF PLANE STRAIN CONSOLIDATION IN SATURATED CLAYS BY FINITE ELEMENT METHOD	71-0815
IGUTI, F.	VIBRATIONS OF GRIDS BY THE FINITE ELEMENT METHOD	73-1089
IIDA, K.	FINITE ELEMENT LARGE DEFLECTION ANALYSIS OF ELASTIC-PLASTIC SHELLS OF REVOLUTION SUBJECTED TO AXISYM	72-0137
IIDA, S.	APPLICATION OF THE METHOD OF FINITE ELEMENT ANALYSIS TO TWO-DIMENSIONAL PROBLEMS IN FRACTURE MECHANI	68-0100
IIDA, Y.	APPLICATION OF FINITE ELEMENT METHOD TO SOIL-FOUNDATION INTERACTION ANALYSES	74-1080
IJAM, A.	COUPLED CONVECTIVE/CONDUCTIVE HEAT TRANSFER INCLUDING VELOCITY FIELD EVALUATION	75-0116
IKEGAWA, M.	FINITE ELEMENT ANALYSIS OF FLUID MOTION IN A CONTAINER	74-0079
IKEGAWA, M.	FINITE ELEMENT ANALYSIS OF POTENTIAL FLOW PROBLEMS	75-0869
IKEGAWA, M.	FINITE ELEMENT METHOD APPLIED TO ANALYSIS OF FLOW OVER A SPILLWAY CREST	73-0300
IKEGAWA, M.	FINITE ELEMENT TECHNIQUE IN LIFTING SURFACE PROBLEMS	74-0022
IKEGAWA, M.	LIFTING SURFACE PROBLEMS ANALYSIS	73-0333
IKEGAWA, M.	SOME APPLICATIONS OF THE FINITE ELEMENT METHOD TO FLUID MECHANICS	74-1274
IKENOUCHI, M.	AN APPROXIMATE NUMERICAL SOLUTION OF THE NAVIER STOKES EQUATIONS BY GALERKIN METHOD	74-0011
IMAMASA, J.	EXPERIMENTAL STUDY OF FLEXIBILITY FACTORS AND STRESSES OF WELDING ELBOWS WITH END EFFECTS	73-0422
IMAMASA, J.	STRESS ANALYSIS PROGRAM BY FINITE ELEMENT METHOD FOR RADIAL FLOW IMPELLERS	74-1016
IMBERT, J. F.	MODAL ANALYSIS OF A SATELLITE PRIMARY STRUCTURE USING A FINITE ELEMENT PROCEDURE	73-0624
INASAKI, I.	DYNAMIC BEHAVIOR OF HYDROSTATIC THRUST BEARING	75-0189
INGRAM, W. B.	A FINITE ELEMENT METHOD FOR BENDING ANALYSIS OF LAYERED STRUCTURAL SYSTEMS	67-0134
INNAURATO, N.	INDIRECT TENSILE TESTING OF ANISOTROPIC ROCKS	73-0376
INOUE, T.	ANALYSIS OF ELASTO-PLASTIC THERMAL STRESSES BY THE FINITE ELEMENT METHOD	70-0142
INOUE, T.	THERMAL ELASTIC PLASTIC STRESS ANALYSIS BY THE FINITE ELEMENT METHOD	69-0391
IQBAL, M. A.	INTERACTION STRESSES IN COMPOSITE SYSTEMS	70-0215
IQBAL, S.	FINITE ELEMENT ANALYSIS OF ANISOTROPIC PLATES USING 0-19 ELEMENT: PART 11. INSTRUCTIONS FOR USERS AN	74-0731
IREMONGER, M. J	PLASTIC FLOW AND FAILURE OF DISCONTINUOUS FIBRE COMPOSITE MATERIALS	70-0157
IROBE, M.	FINITE ELEMENT ANALYSIS OF CONSOLIDATION OF UNSATURATED SOIL	74-0082
IRONS, B.	THE EVOLUTION OF THE ISOPARAMETRIC ELEMENTS	75-0425
IRONS, B. M.	A BOUND THEOREM IN EIGENVALUES AND ITS PRACTICAL APPLICATIONS	71-0128
IRONS, B. M.	A CONFORMING QUARTIC TRIANGULAR ELEMENT FOR PLATE BENDING	69-0137
IRONS, B. M.	A FRONTAL SOLUTION PROGRAM FOR FINITE ELEMENT ANALYSIS	70-0073
IRONS, B. M.	A FURTHER MODIFICATION TO AHMAD'S SHELL ELEMENT	73-0619
IRONS, B. M.	A SIMPLE MATRIX-VECTOR HANDLING SCHEME FOR THREE-DIMENSIONAL AND SHELL ANALYSIS	70-0317
IRONS, B. M.	A TECHNIQUE FOR DEGENERATING BRICK-TYPE ISOPARAMETRIC ELEMENTS USING HIERARCHICAL MIDSIDE NODES	74-0403
IRONS, B. M.	AN ASSUMED STRESS APPROACH TO REFINED ISOPARAMETRIC ELEMENTS IN THREE DIMENSIONS	74-0421
IRONS, B. M.	AN ASSUMED STRESS VERSION OF THE WILSON 8-NODE ISOPARAMETRIC BRICK	72-0532
IRONS, B. M.	ANALYSIS OF THICK AND THIN SHELL STRUCTURE BY CURVED FINITE ELEMENTS	70-0016
IRONS, B. M.	APPLICATION OF THE FINITE ELEMENT METHOD TO CREEPING FLOW PROBLEMS	70-0024
IRONS, B. M.	COMMENT ON A HIGHER ORDER CONFORMING RECTANGULAR PLATE ELEMENT	73-0999
IRONS, B. M.	COMMENT ON MATRICES FOR THE DIRECT STIFFNESS METHOD BY R. J. MELOSH	64-0022
IRONS, B. M.	COMMENT ON STIFFNESS MATRICES FOR SECTOR ELEMENTS	70-0233
IRONS, B. M.	COMPLETE POLYNOMIAL DISPLACEMENT FIELDS FOR FINITE ELEMENT METHODS	68-0092
IRONS, B. M.	COMPUTATIONAL TECHNIQUES, NONLINEAR EQUATIONS, ISOPARAMETRIC ELEMENTS, FIVE LECTURES	73-0634
IRONS, B. M.	COMPUTER TECHNIQUE: THE FRONT SOLUTION	72-0631
IRONS, B. M.	CURVED ISOPARAMETRIC QUADRILATERAL ELEMENTS FOR FINITE ELEMENT ANALYSIS	68-0008
IRONS, B. M.	CURVED THICK SHELL AND MEMBRANE ELEMENTS WITH PARTICULAR REFERENCE TO AXISYMMETRIC PROBLEMS	68-0005
IRONS, B. M.	DISCUSSION OF FINITE ELEMENT TECHNIQUES IN STRUCTURAL MECHANICS	70-0263
IRONS, B. M.	DISTRIBUTED MASS MATRIX FOR PLATE ELEMENT BENDING (DISCUSSION)	66-0038
IRONS, B. M.	ECONOMICAL COMPUTER TECHNIQUES FOR NUMERICALLY INTEGRATED FINITE ELEMENTS	69-0138
IRONS, B. M.	EIGENVALUE ECONOMISERS IN VIBRATION PROBLEMS	63-0013
IRONS, B. M.	ENGINEERING APPLICATIONS OF NUMERICAL INTEGRATION IN STIFFNESS METHOD	66-0040
IRONS, B. M.	EQUATION SOLVING ALGORITHMS FOR THE FINITE ELEMENT METHOD	71-0127
IRONS, B. M.	EXPERIENCE WITH THE PATCH TEST FOR CONVERGENCE OF FINITE ELEMENTS	72-0365
IRONS, B. M.	FINITE ELEMENT METHOD IN STRESS ANALYSIS	69-0413
IRONS, B. M.	INADEQUACY OF NODAL CONNECTIONS IN A STIFFNESS SOLUTION FOR PLATE BENDING	65-0021
IRONS, B. M.	INTRODUCTION TO SHEARS DEFORMATION INTO A THIN PLATE DISPLACEMENTS FORMULATIONS	73-0617
IRONS, B. M.	ISO-PARAMETRIC AND ASSOCIATED ELEMENT FAMILIES FOR TWO AND THREE DIMENSIONAL ANALYSIS	69-0285
IRONS, B. M.	ISOPARAMETRIC AND OTHER NUMERICALLY INTEGRATED ELEMENTS	71-0272
IRONS, B. M.	ISOPARAMETRIC FINITE ELEMENT SYSTEM: A NEW CONCEPT IN FINITE ELEMENT ANALYSIS	68-0223
IRONS, B. M.	LAGRANGE MULTIPLIER TECHNIQUES IN STRUCTURAL ANALYSIS	65-0045

IRONS, B. M.	LEAST SQUARES SMOOTHING OF EXPERIMENTAL DATA USING FINITE ELEMENTS	68-0224
IRONS, B. M.	LEAST SQUARES SURFACE FITTING BY FINITE ELEMENTS AND AN APPLICATION TO STRESS SMOOTHING	67-0154
IRONS, B. M.	MATRIX ITERATION AND ACCELERATION PROCESSES IN FINITE ELEMENT PROBLEMS OF STRUCTURAL MECHANICS	70-0132
IRONS, B. M.	NATURAL FREQUENCIES OF COMPLEX FREE OR SUBMERGED STRUCTURES BY THE FINITE ELEMENT METHOD	65-0026
IRONS, B. M.	NUMERICAL INTEGRATION APPLIED TO FINITE ELEMENT METHODS	66-0039
IRONS, B. M.	NUMERICAL INTEGRATION APPLIED TO FINITE ELEMENT METHODS	70-0390
IRONS, B. M.	PLATE BENDING WITH NONCONFORMING ELEMENT: FINE-MESH CONVERGENCE TEST	65-0049
IRONS, B. M.	QUADRATURE RULES FOR BRICK BASED FINITE ELEMENTS	71-0531
IRONS, B. M.	ROLE OF PART-INVERSION IN FLUID-STRUCTURE PROBLEMS WITH MIXED VARIABLES	70-0209
IRONS, B. M.	ROUNDOFF CRITERIA IN DIRECT STIFFNESS SOLUTIONS	68-0204
IRONS, B. M.	SHAPE FUNCTION FORMULATIONS FOR ELEMENTS OTHER THAN DISPLACEMENT MODELS	72-0472
IRONS, B. M.	SHAPE FUNCTION SUBROUTINE FOR AN ISOPARAMETRIC THIN PLATE ELEMENT	73-0618
IRONS, B. M.	STRUCTURAL EIGENVALUE PROBLEMS: ELIMINATION OF UNWANTED VARIABLES	65-0050
IRONS, B. M.	TESTING AND ASSESSING FINITE ELEMENTS BY AN EIGENVALUE TECHNIQUE	68-0188
IRONS, B. M.	THE CONJUGATE NEWTON METHOD	73-0560
IRONS, B. M.	THE ISOPARAMETRIC FINITE ELEMENT SYSTEM - A NEW CONCEPT IN FINITE ELEMENT ANALYSIS	69-0139
IRONS, B. M.	THE PATCH TEST	74-0339
IRONS, B. M.	THE SEMI-LOOF SHELL ELEMENT	74-0629
IRONS, B. M.	THE SEMILOOF SHELL ELEMENT	75-0536
IRONS, B. M.	THE SUPERPATCH THEOREM AND OTHER PROPOSITIONS RELATING TO THE PATCH TEST	75-0026
IRONS, B. M.	THEORETICAL FOUNDATIONS ON THE FINITE ELEMENT METHOD	68-0055
IRONS, B. M.	THEORETICAL FOUNDATIONS OF THE FINITE ELEMENT METHOD (DISCUSSION)	70-0074
IRONS, B. M.	THREE DIMENSIONAL STRESS ANALYSIS	70-0086
IRONS, B. M.	THREE-DIMENSIONAL STRESS ANALYSIS OF ARCH DAMS BY THE FINITE ELEMENT METHOD: PART I, ARCH DAM NO. 1,	66-0099
IRONS, B. M.	TRIANGULAR ELEMENTS IN PLATE BENDING CONFORMING AND NON-CONFORMING SOLUTIONS	66-0011
IRONS, B. M.	UN NOVEL ELEMENT DE COQUES GENERALES	74-0407
IRONS, B. M.	VIBRATION AND STABILITY OF PLATES USING FINITE ELEMENTS	68-0244
IRVING, J.	ADVANCES IN THE ANALYSIS OF PRESTRESSED CONCRETE PRESSURE VESSELS	71-0162
ISAACS, L. T.	CURVED CUBIC TRIANGULAR FINITE ELEMENT FOR POTENTIAL FLOW PROBLEMS	73-0423
ISAKSON, G.	A FINITE-ELEMENT METHOD FOR THE PLASTIC BUCKLING ANALYSIS OF PLATES	69-0214
ISAKSON, G.	DISCRETE ELEMENT METHODS FOR THE PLASTIC ANALYSIS OF STRUCTURES SUBJECTED TO CYCLIC LOADING	70-0023
ISAKSON, G.	DISCRETE-ELEMENT METHODS FOR THE PLASTIC ANALYSIS OF STRUCTURES	67-0040
ISAKSON, G.	DISCRETE-ELEMENT PLASTIC ANALYSIS OF STRUCTURES IN A STATE OF MODIFIED PLANE STRAIN	69-0140
ISAKSON, G.	FINITE ELEMENT ANALYSIS OF INTERLAMINAR SHEAR FIBROUS COMPOSITES	71-0129
ISAKSON, G.	PLASTICITY	70-0509
ISEKI, H.	FINITE ELEMENT METHOD OF ANALYSIS OF THE HYDROSTATIC BULGING OF A SHEET METAL - 1.	74-0958
ISENBERG, J.	ANALYTIC MODELING OF ROCK-STRUCTURE INTERACTION	72-0808
ISENBERG, J.	ANALYTIC MODELING OF ROCK-STRUCTURE INTERACTION. VOLUME 3. COMPUTER PROGRAM	73-0918
ISENBERG, J.	ANALYTIC MODELING OF ROCK-STRUCTURE INTERACTION. VOLUME 2. USERS GUIDE FOR A COMPUTER PROGRAM	73-0919
ISENBERG, J.	ANALYTIC MODELING OF ROCK-STRUCTURE INTERACTION. VOLUME 1	73-0920
ISENBERG, J.	EFFECTS OF FOUNDATION ROTATION ON SEISMIC INERTIA FORCES OF NUCLEAR POWER PLANT STRUCTURES	73-0177
ISENBERG, J.	RESPONSE OF STRUCTURES TO COMBINED BLAST EFFECTS	73-0424
ISENBERG, J.	SPHERICAL WAVES IN INELASTIC MATERIAL	70-0075
ISHIHARA, K.	ANALYSIS OF ACOUSTIC FIELD IN IRREGULARLY SHAPED ROOMS BY FINITE ELEMENT METHOD	73-0133
ISHIHARA, K.	FINITE ELEMENT ANALYSIS OF IRREGULARLY SHAPED ACOUSTIC FIELD	71-0711
ISHIJIMA, Y.	SIMULATION OF PROGRESSIVE FAILURE IN ROCK AROUND MINING EXCAVATIONS	70-0146
ISHIJIMA, Y.	THE APPLICATION OF THE FINITE ELEMENT METHOD TO FRACTURE MECHANICS	72-0331
ISHINO, B.	VIBRATION ANALYSIS OF AN INTERNAL REFERENCE UNIT	69-0176
ISIDA, M.	ARBITRARY SYMMETRIC LOADING PROBLEMS OF CENTRALLY CRACKED RECTANGULAR PLATES	75-0318
ISIDA, M.	SOME EXAMINATIONS ON AVAILABILITY OF FINITE ELEMENT METHODS IN LINEAR FRACTURE MECHANICS	73-0112
ISSACS, L. T.	FINITE ELEMENT METHODS: TWO-DIMENSIONAL SEEPAGE WITH A FREE SURFACE	71-0475
ISSHIKI, M.	VARIATIONAL PRINCIPLES AND DUALISTIC SCHEME FOR INTERSECTION PROBLEMS IN ELASTICITY	69-0305
ITO, H.	ACTUAL STRESS OF HIGH HEAD PUMP-TURBINE RUNNER	75-0567
ITO, K.	FINITE ELEMENT ANALYSIS OF STEADY FLUID AND METAL FLOW	74-0054
ITO, K.	FINITE ELEMENT ANALYSIS OF STEADY FLUID AND METAL FLOW	75-0178
ITO, K.	FINITE ELEMENT FORMULATION FOR SOLUTION OF STEADY FLOW OF INCOMPRESSIBLE VISCOUS FLUID	73-0299
ITOH, S.	ANALYSIS OF MAGNETIC FIELDS BY FINITE ELEMENT METHOD	73-0126
ITOH, Y.	FINITE ELEMENT ANALYSIS OF CONSOLIDATION OF UNSATURATED SOIL	74-0082
IUZZOLINI, H. J	PARAMETRIC STUDIES OF LAYERED PAVEMENT SYSTEMS USING AFPAV CODE WITH FAST EQUATION SOLVER	75-0747
IVANOV, Y. I.	CALCULATION OF THE REINFORCED THIN-WALLED STRUCTURES BY THE METHOD OF FINITE ELEMENTS (RASCHET PODKR	75-0685
IVANOV, Y. I.	CALCULATION OF THE REINFORCED THIN-WALLED CONSTRUCTION BY THE FINITE ELEMENT	75-0756
IVASHCHENKO, N.	METODY RASCHETNOGO OPREDELENIYA TEMPERATURNYKH NAPRYAZHENII V KRYSHKAKH TSILINDROV DVIGATELEI VNUTRE	74-0179
IVASHCHENKO, N.	OPREDELENIE STATSIONARNYKH TEMPERATURNYKH POLEI V DETALYAKH DVIGATELEI VNUTRENNEGO SGORANIYA METODOM	73-0454
IVASHCHENKO, N.	RASCHET POLEI DEFORMATSII I NAPRYAZHENII V DETALYAKH DVIGATELEI VNUTRENNEGO SGORANIYA (CALCULATION O	73-0465
IVERSEN, P. A.	COMPUTER CALCULATIONS OF STRESSES IN AXISYMMETRIC THERMALLY LOADED COMPONENTS	68-0069
IVERSEN, P. A.	SOME ASPECTS OF THE FINITE ELEMENT METHOD IN TWO-DIMENSIONAL PROBLEMS	69-0141
IWAKI, T.	FINITE ELEMENT ANALYSIS OF THIN-WALLED STRUCTURES BASED ON THE MODERN ENGINEERING THEORY OF BEAMS	71-0140
IWAKI, T.	ON THE PHYSICAL MEANINGS OF THE MATRICES DERIVED FROM FINITE ELEMENT METHOD OF HEAT CONDUCTION PROBL	73-0297
IWAMOTO, J.	STRESS DISTRIBUTION NEAR A BROKEN POINT OF FILAMENT IN UNIDIRECTIONALLY FIBER REINFORCED COMPOSITES	73-0811
IWASAKI, H.	CALCULATION OF RESIDUAL STRESSES IN PLASTIC DEFORMATION PROCESSES	72-0241
IWASAKI, T.	EIGENVALUE ANALYSIS BY FINITE ELEMENT METHODS	69-0377
IWATA, K.	ANALYSIS OF HYDROSTATIC EXTRUSION BY THE FINITE ELEMENT METHOD	72-0020
IWATA, K.	FINITE ELEMENT ANALYSIS OF THERMO-VISCOELASTIC PROBLEMS	71-0526
IWATA, K.	FINITE ELEMENT ANALYSIS OF THERMO-VISCOELASTIC PROBLEMS	71-0781
IWATA, K.	INSTABILITY ANALYSIS BY THE FINITE ELEMENT METHOD	75-0180
IWATA, K.	LARGE DEFORMATION AND CRITICAL LOADS ANALYSIS OF FRAMED STRUCTURES	74-1293
IWATA, K.	NONLINEAR ANALYSIS BY THE FINITE ELEMENT METHOD AND SOME EXPOSITORY EXAMPLES	73-0309
IWATA, K.	NONLINEAR ANALYSIS OF BEAMS AND SHELLS	73-1074
IYENGAR, N. G.	FREE VIBRATION OF A SIMPLY SUPPORTED BEAM WITH NONLINEAR MATERIAL PROPERTIES	71-0689
IYENGAR, N. G.	EFFECTS OF MATERIAL NONLINEARITIES ON VIBRATION	73-0323
IYENGAR, N. G.	NON-LINEAR EFFECTS OF HIGH TEMPERATURE ON THE VIBRATION OF A SIMPLY SUPPORTED BEAM WITH A CENTRAL MA	73-0838
IYER, M. S.	ANALYSIS OF A PRESSURE-VESSEL JUNCTION BY THE FINITE-ELEMENT METHOD	72-0949
IZUMI, H.	ON ANALYSIS OF LARGE DEFORMATION PROBLEMS OF BEAM	71-0660
JABBOUR, K. N.	NORMAL MODE ANALYSIS OF THE RADIO ASTRONOMY EXLPORER BOOMS AND SPACECRAFT	71-0130
JABLON, C.	COMPARAISON DES DIVERSES MODELISATIONS NUMERIQUES D'UN SYSTEME MAGNETIQUE SATURABLE. (COMPARISON OF	75-0295
JACHENS, R. C.	PSEUDO 3-DIMENSIONAL FINITE-ELEMENT FORMULATION FOR ELASTOSTATIC PROBLEMS AND ITS GEOPHYSICAL APPLIC	75-0597
JACHENS, R. C.	PSEUDO-THREE-DIMENSIONAL FINITE ELEMENT FORMULATION FOR ELASTOSTATIC PROBLEMS	74-0167
JACOBS, H. R.	INFLUENCE OF INTERLAMINATE FLAWS ON TRANSIENT TEMPERATURE DISTRIBUTIONS	73-0223
JACOBS, L. D.	FINITE-ELEMENT ANALYSIS OF COMPLEX PANEL RESPONSE TO RANDOM LOADS	68-0093
JACOBS, L. D.	RANDOM-VIBRATION ANALYSIS SYSTEM FOR COMPLEX STRUCTURES. PART 1, ENGINEERING USER'S GUIDE	68-0106
JACQUIN, J. C.	FINITE ELEMENT SOLUTIONS FOR MULTIREGION PROBLEMS	73-1021
JAEGER, L. G.	NEAR-CRACK DEFORMATION FIELDS IN STRAIN-HARDENING MATERIAL	74-0938
JAEGER, L. G.	STRUCTURAL ANALYSIS OF TALL BUILDINGS HAVING IRREGULARLY POSITIONED SHEAR WALLS	73-0282
JAEGER, L. G.	THE ROLE OF FINITE DEFORMATION ANALYSIS IN PLANE STRESS AND STRAIN FRACTURES	74-0322
JAENSSON, B. O.	DETERMINATION OF YOUNG'S MODULUS AND POISSON'S RATIO FOR WC-CO ALLOYS BY THE FINITE ELEMENT METHOD	72-0142
JAGANNATHAN, D.	NONLINEAR ANALYSIS OF RETICULATED SPACE TRUSSES	75-0287
JAGANNATHAN, D.	SNAP-THROUGH BUCKLING OF RETICULATED SHELLS	75-0640
JAHN, F.	NONSTEADY PYROMETRIC CALCULATIONS AS THE BASIS FOR STRESS ANALYSIS OF STEAM GENERATORS AND HEAT EXCH	72-0451
JAIN, R. K.	NAVIGATION LOCK STRUCTURE ANALYSIS BY FINITE ELEMENT METHOD	71-0816
JAMES, J. H.	ACOUSITC FINITE ELEMENT ANALYSIS OF AXISYMMETRIC FLUID REGIONS	73-0763
JAMET, P.	A SECOND ORDER FINITE ELEMENT METHOD FOR THE ONE-DIMENSIONAL STEFAN PROBLEM	74-1222
JAMET, P.	CONVERGENCE AND ERROR ESTIMATES FOR FINITE ELEMENT APPROXIMATIONS OF STATIONARY NAVIER-STOKES EQUATI	74-0122

JAMET, P.	ERROR ESTIMATE FOR QUADRILATERAL ISOPARAMETRIC FINITE-ELEMENTS OF TYPE 01 WHICH CAN DEGENERATE INTO	75-0401
JAMET, P.	ESTIMATIONS D'ERREUR POUR DES ELEMENTS FINIS DROITS PRESQUE DEGENERES	74-1247
JAMET, P.	NUMERICAL SOLUTION OF THE STATIONARY NAVIER-STOKES EQUATIONS BY FINITE ELEMENT METHODS	74-0408
JAMET, P.	NUMERICAL-SOLUTION OF EULERIAN EQUATIONS OF COMPRESSIBLE FLOW BY A FINITE-ELEMENT METHOD WHICH FOLLO	75-0535
JAMOVSKY, V.	ELLIPTIC BOUNDARY VALUE PROBLEMS WITH NONVARIATIONAL PERTURBATION AND THE FINITE ELEMENT METHOD	73-1022
JAN, H. Y.	STRUCTURAL ANALYSIS OF SL-7 CONTAINERSHIP UNDER COMBINED LOADING OF VERTICAL, LATERAL AND TORSIONAL	74-0734
JAN, S. F.	IMPACT ON COMPLEX MECHANICAL STRUCTURES	71-0598
JANSEN, R.	FINITE-ELEMENT CURRENT-DENSITY REPRESENTATION IN NUMERICAL-SOLUTION OF MICROSTRIP PROBLEMS	75-0400
JANSSEN, J. D.	DETERMINATION OF THE APPLICABILITY OF SOME THEORIES ABOUT STRENGTH AND STIFFNESS OF THIN-WALLED BEAM	69-0318
JANSSEN, J. D.	POSSIBILITIES OF THE METHOD OF FINITE ELEMENTS IN THE CALCULATION OF THE STRENGTH AND STIFFNESS OF T	70-0344
JANSSEN, T. L.	SHELL INSTABILITY ANALYSIS APPLIED TO A RADOME	74-0261
JANZ, R. F.	DEFORMATION OF THE DIASTOLIC LEFT VENTRICLE - 2. NONLINEAR GEOMETRIC EFFECTS	74-0974
JANZ, R. F.	DEFORMATION OF THE END DIASTOLIC LEFT VENTRICLE, COMPARISON OF THEORY WITH EXPERIMENT	71-0131
JANZ, R. F.	FINITE-ELEMENT MODEL FOR THE MECHANICAL BEHAVIOUR OF THE LEFT VENTRICLE, PREDICTION OF DEFORMATION I	72-0304
JARAUSCH, R.	FESTIGKEITSUNTERSUCHUNGEN UND-BERECHNUNGEN AN PUMPENSPIRALGEHAEUSEN. (INVESTIGATIONS AND CALCULATION	75-0238
JATEGAONKAR, R.	SHEAR STRESSES IN BEAMS WITH DISCONTINUITY IN DEPTH	75-0487
JAVANDEL, I.	A METHOD OF ANALYZING TRANSIENT FLUID FLOW IN MULTILAYERED AQUIFERS	69-0142
JAVANDEL, I.	APPLICATION OF THE FINITE ELEMENT METHOD TO TRANSIENT FLOW IN POROUS MEDIA	68-0094
JAVANDEL, I.	USE OF THE FINITE ELEMENT METHOD IN SOLVING TRANSIENT FLOW PROBLEMS IN AQUIFER SYSTEMS	68-0167
JAYAWARDENA, A.	FINITE-ELEMENT APPROACH TO WATERSHED RUNOFF- DISCUSSION	75-0402
JEAN, C.	USE OF HYBRID COMPUTATION FOR SOLVING TRANSIENT FLOW PROBLEM IN POROUS MEDIA BY FINITE ELEMENT METHO	74-0068
JENKINS, J. F.	THE FINITE ELEMENT DYNAMIC STABILITY ANALYSIS OF THIN SHELLS SUBMERGED IN FLUIDS	71-0057
JENKINS, T. B.	EFFECT OF ELASTIC MODULAR RATIO AND ANISOTROPY ON THE STRESS FIELD IN A TWO-DIMENSIONAL MATRIX INTO	71-0647
JENKINS, W. M.	MATRIX AND DIGITAL COMPUTER METHODS IN STRUCTURAL ANALYSIS	69-0143
JENNEQUIN, G.	IS THE COMPUTATION OF NOISE LEVEL INSIDE A CAR FEASIBLE	71-0631
JENNING, A.	NATURAL VIBRATION OF A FREE SURFACE	62-0008
JENNINGS, A.	A COMPACT STORAGE SCHEME FOR THE SOLUTION OF SYMMETRICAL LINEAR SIMULTANEOUS EQUATIONS	66-0116
JENNINGS, A.	AN ITERATIVE METHOD FOR LARGE SYSTEMS OF LINEAR STRUCTURAL EQUATIONS	73-1020
JENNINGS, P. C.	DYNAMICS OF BUILDING - SOIL INTERACTION	72-0856
JENSEN, F. R.	THREE-DIMENSIONAL FINITE-ELEMENT ELASTIC ANALYSIS OF PRISMATIC SOLIDS	73-1138
JENSEN, P. S.	PROBLEM CONTROLLED GRID GENERATION FOR THE NUMERICAL SOLUTION OF PARTIAL DIFFERENTIAL EQUATIONS	75-0725
JENSEN, P. S.	VARIABLE GRID FINITE DIFFERENCE-ELEMENT SOLUTION OF ELLIPTIC PARTIAL DIFFERENTIAL EQUATIONS	74-0862
JENSEN, W. R.	MATRIX ANALYSIS METHODS FOR INELASTIC STRUCTURES	65-0051
JENSEN, W. R.	MATRIX ANALYSIS METHODS FOR ANISOTROPIC INELASTIC STRUCTURES	66-0104
JERGER, E. W.	DEEP SEA ENGINEERING	71-0739
JERRAM, K.	A FRACTURE MECHANICS APPROACH TO THE PROPAGATION OF FATIGUE CRACKS IN REACTOR COMPONENTS BY ACOUSTIC	71-0132
JERRAM, K.	APPLICATION OF FRACTURE MECHANICS TO PRESTRESSED CONCRETE PRESSURE VESSELS	73-0228
JERRAM, K.	EXPERIMENTAL AND THEORETICAL FATIGUE-CRACK PROPAGATION LIVES OF VARIOUSLY NOTCHED PLATES	74-0602
JERRAM, K.	FINITE ELEMENT TECHNIQUES IN FRACTURE MECHANICS	72-0746
JERRAM, K.	FINITE-ELEMENT TECHNIQUES IN FRACTURE MECHANICS	75-0780
JERRAM, K.	THE USE OF FINITE ELEMENT TECHNIQUES IN FRACTURE MECHANICS	72-0926
JESPERSE, D. C.	ARAKAWAS METHOD IS A FINITE-ELEMENT METHOD	74-0633
JEZERNIK, A.	FINITE ELEMENT ANALYSIS OF STRESSES IN FUEL ELEMENTS OF A HIGH-TEMPERATURE GAS-COOLED REACTOR	71-0115
JEZERNIK, A.	FINITE ELEMENT MATRIX DISPLACEMENT METHOD FOR TIME DEPENDENT STRESS ANALYSIS OF REACTOR MATERIALS IN	71-0674
JEZERNIK, A.	LARGE USER-ORIENTATED SYSTEMS OF PROGRAMS FOR STRUCTURAL ANALYSIS AND DESIGN	74-0482
JEZERNIK, A.	THE COMPARISON OF ITERATIVE AND DIRECT SOLUTION TECHNIQUES IN THE ANALYSIS OF TIME-DEPENDENT STRESS	72-0080
JIMMA, T.	FINITE ELEMENT METHOD OF ANALYSIS OF THE HYDROSTATIC BULGING OF A SHEET METAL - 1.	74-0958
JIMMA, T.	NON-UNIFORM DEFORMATION OF MATERIAL IN AXIALLY SYMMETRIC COMPRESSION CAUSED BY FRICTION. PART 3, THE	72-0923
JINOCH, J.	APLIKACE MKP NA RESENI PROBLEMU DIFUZE V KOVECH. (APPLICATION OF THE FINITE ELEMENT METHOD TO THE SO	74-0609
JINOCH, J.	APPLICATION OF THE FINITE ELEMENT METHOD IN METAL DIFFUSION	74-1366
JIROUSEK, J.	BENDING OF CYLINDRICAL SHELLS - SOLUTION USING CYLINDRICAL STRIP ELEMENTS - CONTRIBUTION TO FINITE	70-0363
JIROUSEK, J.	CALCUL DES PONTS BIAIS A POUTRES MULTIPLES SANS ENTRETOISES PAR LA METHODE DES ELEMENTS (CALCULATION	72-0184
JIROUSEK, J.	FINITE ELEMENT ANALYSIS OF CYLINDRICAL SHELLS. SOLUTION USING RECTANGULAR CIRCULAR CYLINDRICAL ELEME	70-0178
JOBSON, D. A.	DETERMINATION OF STRESSES AT NODAL POINTS FROM FINITE ELEMENT SOLUTIONS TO ELASTIC PROBLEMS	71-0307
JOBSON, D. A.	PIECE-WISE CONTINUOUS FINITE ELEMENT FOR THE STRESS ANALYSIS OF PLATES AND SHELLS	75-0758
JOFRIET, J. C.	FINITE ELEMENT ANALYSIS OF PRESTRESSED CONCRETE VOIDED BRIDGE DECKS	73-0106
JOFRIET, J. C.	PATTERN LOADING ON REINFORCED CONCRETE FLAT PLATES	71-0290
JOFRIET, J. C.	SHORT TERM DEFLECTIONS OF CONCRETE FLAT PLATES	73-0277
JOHANSSON, P.	LOADS AND RESPONSE OF LARGE DUCTED PROPELLER SYSTEMS	73-0810
JOHNS, J. T.	ECONOMICAL DESIGN ANALYSIS OF PISTONS FOR INTERNAL COMBUSTION ENGINES	73-0188
JOHNSEN, T. L.	DYNAN LECTURE NOTES WITH COMPUTATIONAL EXAMPLES	71-0581
JOHNSEN, T. L.	SIMULTANEOUS EIGENVECTOR ITERATION APPLIED TO LARGE SYSTEMS	74-0680
JOHNSEN, TH.	DYNAN USERS MANUAL	71-0625
JOHNSON, C.	APPLICATIONS OF THE FINITE-ELEMENT METHOD AND COMPUTER GRAPHICS. (A VIBRATIONAL ANALYSIS OF A SONAR	72-0729
JOHNSON, C.	CONVERGENCE OF A MIXED FINITE-ELEMENT METHOD FOR PLATE BENDING PROBLEMS	73-0107
JOHNSON, C.	CONVERGENCE OF ANOTHER MIXED FINITE ELEMENT METHOD FOR PLATE BENDING PROBLEMS	72-0367
JOHNSON, C.	ERROR ESTIMATES FOR A FINITE-ELEMENT APPROXIMATION OF A MINIMAL SURFACE	75-0491
JOHNSON, C.	VIBRATIONAL ANALYSIS OF THE FLEXING HEAD OF THE BQS-6 TRANSDUCER	71-0708
JOHNSON, C. O.	V-BELT LIFE PREDICTION AND POWER RATING	75-0838
JOHNSON, C. P.	A COMPUTER PROGRAM FOR THE ANALYSIS OF THIN SHELLS	67-0041
JOHNSON, C. P.	A FINITE ELEMENT APPROXIMATION FOR THE ANALYSIS OF THIN SHELLS	68-0046
JOHNSON, C. P.	BEAM BUCKLING BY FINITE ELEMENT PROCEDURE	74-0123
JOHNSON, C. P.	EFFECT OF HIGHER ORDER TERMS IN CERTAIN NONLINEAR FINITE ELEMENT MODELS	73-0125
JOHNSON, C. P.	FINITE ELEMENT ANALYSIS OF ARBITRARY THIN SHELLS	70-0329
JOHNSON, C. P.	ON THE EFFECT OF APPROXIMATING HIGHER ORDER TERMS IN CERTAIN FINITE ELEMENT MODELS OF NONLINEAR STRU	73-0603
JOHNSON, C. P.	THE ANALYSIS OF THIN SHELLS BY A FINITE ELEMENT PROCEDURE	67-0123
JOHNSON, E. H.	OPTIMIZATION OF STRUCTURES UNDERGOING HARMONIC OR STOCHASTIC EXCITATION	75-0711
JOHNSON, F. R.	FINITE ELEMENT ANALYSES OF QUICK CAMP MODULE	75-0017
JOHNSON, G. R.	STIFFNESS COEFFICIENTS FOR EMBEDDED FOOTINGS	75-0540
JOHNSON, J. A.	A TECHNIQUE FOR THE SETTLEMENT AND STRESS ANALYSIS OF MAT FOUNDATIONS USING THE FINITE ELEMENT METHO	72-0950
JOHNSON, J. P.	MATRIX METHODS IN STRUCTURAL MECHANICS	66-0065
JOHNSON, J. P.	RICHARDSON EXTRAPOLATION FOR PARABOLIC GALERKIN METHODS	72-0522
JOHNSON, K.	STRESS ANALYSIS OF LAMINATED CFRP PLATES CONTAINING A HOLE	74-1175
JOHNSON, L. D.	ANALYSIS OF PILE-SUPPORTED GRAVITY LOCK	74-0585
JOHNSON, L. D.	EVALUATION OF SOME NUMERICAL SCHEMES FOR CONSOLIDATION	73-0349
JOHNSON, L. D.	EVALUATION OF TWO FINITE ELEMENT FORMULATIONS FOR ONE-DIMENSIONAL CONSOLIDATION	72-0453
JOHNSON, L. D.	SOME NUMERICAL PROCEDURES FOR ANALYSIS OF ONE DIMENSIONAL CONSOLIDATION	72-0928
JOHNSON, M. W.	CONVERGENCE OF THE FINITE ELEMENT METHOD IN THE THEORY OF ELASTICITY	68-0095
JOHNSON, N. E.	THE COMPARISON OF THE RESPONSE OF A HIGHWAY BRIDGE TO UNIFORM GROUND SHOCK AND MOVING GROUND EXCITAT	72-0737
JOHNSON, P. W.	MECHANICS OF DEFORMABLE SOLIDS - EURASIAN COMMUNIST COUNTRIES	73-0794
JOHNSON, R. L.	FINITE-ELEMENT METHOD FOR WATER-DISTRIBUTION NETWORKS	75-0269
JOHNSON, R. L.	BEHAVIOR OF A BI-MATERIAL CYLINDER INCLUDING THE EFFECTS OF FRACTURE	69-0466
JONES, A. T.	FRACTURE TOUGHNESS TESTING WITH SECTIONS OF CYLINDERS	74-0999
JONES, A. T.	LOAD POINT COMPLIANCE OF THE CHARPY IMPACT SPECIMEN	73-0775
JONES, A. T.	RANDOM VIBRATION RESPONSE OF CANTILEVER PLATES USING THE FINITE ELEMENT METHOD	70-0181
JONES, B. D.	CONCEPT FEASIBILITY COMBAT TRACKED VEHICLE SIGNATURE DUPLICATOR	64-0028
JONES, C. E.	ADVANCED SUBSTRUCTURING TECHNIQUES	71-0838
JONES, C. E.	DYNAMICS OF A FLEXIBLE BULKHEAD AND CONTAINED FLUID	73-0082
JONES, C. E.	STUDY OF PROPELLANT DYNAMICS IN A SHUTTLE TYPE LAUNCH VEHICLE	72-0699
JONES, C. E.	SUBSTRUCTURING TECHNIQUES	71-0692
JONES, C. E.	VIBRATIONAL CHARACTERISTICS OF SPACE FRAMES	69-0434
JONES, C. J. F.	COMPARISON BETWEEN THEORETICAL AND MEASURED EARTH PRESSURES ACTING ON A LARGE MOTORWAY RETAINING WAL	74-0917

JONES, C. J. F.	EARTH PRESSURES AGAINST THE ABUTMENTS AND WING WALLS OF STANDARD MOTORWAY BRIDGES	75-0629
JONES, D. P.	FINITE ELEMENT ANALYSIS OF PERFORATED PLATES CONTAINING TRIANGULAR PENETRATION PATTERNS OF 5 AND 10	75-0216
JONES, D. P.	FINITE ELEMENT ANALYSIS OF PERFORATED PLATES CONTAINING TRIANGULAR PENETRATION PATTERNS OF 5 AND 10	75-0520
JONES, D. P.	INFLUENCE OF CRACK CLOSURE AND ELASTO-PLASTIC FLOW ON THE BENDING OF A CRACKED PLATE	75-0735
JONES, G. K.	THE RESPONSE OF SHELLS TO DISTRIBUTED RANDOM LOADS USING NASTRAN	71-0134
JONES, L.	LARGE DISPLACEMENT ANALYSIS OF ASIXYMMETRIC SHELLS	69-0276
JONES, L.	UNIFICATION OF THE FINITE ELEMENT AND CLASSICAL RITZ METHODS	73-1139
JONES, P. J.	COMPARISON OF STRUCTURAL LOADS; STATIC VERSUS DYNAMIC	70-0464
JONES, R. E.	A GENERALIZATION OF THE DIRECT-STIFFNESS METHOD OF STRUCTURAL ANALYSIS	64-0002
JONES, R. E.	A SELF-ORGANIZING MESH GENERATION PROGRAM	74-1051
JONES, R. E.	A SELF-ORGANIZING MESH GENERATION PROGRAM	74-1249
JONES, R. E.	A SURVEY OF THE ANALYSIS OF SHELLS BY THE DISPLACEMENT METHOD	66-0041
JONES, R. E.	ANALYSIS OF AXISYMMETRIC SOLIDS BY THE DIRECT STIFFNESS METHOD	66-0049
JONES, R. E.	APPLICATION, TESTING AND EVALUATION OF DOUBLY CURVED SHELL ELEMENTS FOR DYNAMIC ANALYSIS. PART II. F	69-0117
JONES, R. E.	APPLICATION, TESTING, AND EVALUATION OF DOUBLY CURVED SHELL ELEMENTS FOR DYNAMIC ANALYSIS. PART 1, T	69-0246
JONES, R. E.	DIRECT STIFFNESS METHOD OF ANALYSIS OF SHELLS OF REVOLUTION UTILIZING CURVED ELEMENTS	66-0042
JONES, R. E.	DYNAMIC ANALYSIS OF SHELLS USING DOUBLY-CURVED FINITE ELEMENTS	68-0082
JONES, R. E.	ELASTIC WAVE PROPAGATION, A COMPARISON BETWEEN FINITE ELEMENT PREDICTIONS AND EXACT SOLUTION	67-0076
JONES, R. E.	GENERALIZED VARIATIONAL PRINCIPLES IN THE FINITE ELEMENT METHOD	69-0116
JONES, R. E.	ON THE APPLICATION OF GENERALIZED VARIATIONAL PRINCIPLES IN THE FINITE ELEMENT METHOD	68-0083
JONES, R. E.	QMESH: A SELF-ORGANIZING MESH GENERATION PROGRAM	74-0872
JONES, R. E.	QMESH: A SELF-ORGANIZING MESH GENERATION PROGRAM	74-0899
JONES, R. E.	SHELL AND PLATE ANALYSIS BY FINITE ELEMENTS	73-0163
JONES, R. E.	STRUCTURE-MEDIUM INTERACTION AND DESIGN PROCEDURES STUDY. VOLUME I. ANALYSIS METHOD, THEORY, VERIFIC	69-0166
JONES, R. F.	USERS MANUAL FOR QMESH, A SELF-ORGANIZING MESH GENERATION PROGRAM	74-0880
JONES, R. F.	INCREMENTAL SOLUTION PROCEDURE FOR NONLINEAR STRUCTURAL DYNAMICS PROBLEMS	75-0656
JONES, R. F.	SOLUTION PROCEDURE FOR NONLINEAR STRUCTURAL PROBLEMS	73-0425
JONES, R. M.	FINITE ELEMENT STRESS ANALYSIS OF AXISYMMETRIC SOLIDS WITH ORTHOTROPIC TEMPERATURE DEPENDENT MATERIA	67-0089
JONES, R. M.	NEW MATERIAL MODEL FOR THE NONLINEAR BIAXIAL BEHAVIOR OF ATJ-S GRAPHITE	75-0063
JONES, R. M.	PROCEEDINGS OF THE NOSETIP STRESS ANALYSIS TECHNICAL INTERCHANGE MEETING	67-0042
JONES, R. M.	SAAS II: FINITE ELEMENT STRESS ANALYSIS OF AXISYMMETRIC SOLIDS WITH ORTHOTROPIC, TEMPERATURE DEPENDE	68-0096
JONES, T. C.	ADAPTATION OF NASTRAN TO THE ANALYSIS OF THE VIKING SPACE VEHICLE	71-0135
JORDAN, S.	FORMULATION AND EVALUATION OF A TRAPEZOIDAL CROSS-SECTION RING DISCRETE ELEMENT	66-0084
JORDAN, S.	MAGIC II: AN AUTOMATED GENERAL PURPOSE SYSTEM FOR STRUCTURAL ANALYSIS	71-0469
JORDAN, S.	MAGIC 11: AN AUTOMATED GENERAL PURPOSE SYSTEM FOR STRUCTURAL ANALYSIS. VOLUME 1. ENGINEER'S MANUAL (	71-0757
JORDAN, S.	MAGIC 111: AN AUTOMATED GENERAL PURPOSE SYSTEM FOR STRUCTURAL ANALYSIS VOLUME 1. ENGINEER'S MANUAL	72-0727
JORDAN, S.	MAGIC, AN AUTOMATED GENERAL PURPOSE SYSTEM FOR STRUCTURAL ANALYSIS, VOLUME II. USER'S MANUAL	69-0144
JORDAN, S.	MAGIC, AN AUTOMATED GENERAL PURPOSE SYSTEM FOR STRUCTURAL ANALYSIS	69-0173
JORDAN, S.	STRESSES ANALYSIS USING FINITE ELEMENT METHODS	70-0349
JORDAN, W. B.	IMPROVED CALCULATION OF ELASTO-PLASTIC SYSTEMS BY THE FINITE ELEMENT METHOD	70-0076
JORDAN, W. B.	THE PLANE ISOPARAMETRIC STRUCTURAL ELEMENT	70-0262
JORRIET, J. C.	FINITE ELEMENT ANALYSIS OF REINFORCED CONCRETE SLABS	71-0133
JOSEPH, J. A.	ACOUSTIC ANALYSIS OF SOLID ROCKET MOTOR CAVITIES BY A FINITE ELEMENT METHOD	71-0119
JOSEPH, J. A.	MSC/NASTRAN APPLICATION MANUAL	72-0567
JOUNG, K. S.	APPLYING FINITE ELEMENT ANALYSIS	70-0150
JOUNG, K. S.	SEISMIC ANALYSIS OF A THREE-PHASE BREAKER INSTALLATION	74-0622
JOUNG, K. S.	STRUCTURAL ANALYSIS OF MAST BY FINITE ELEMENT TECHNIQUE	74-0385
JUDAH, O. M.	FINITE-ELEMENT SIMULATION OF FLOOD HYDROGRAPHS	75-0632
JUDAH, O. M.	SIMULATION OF RUNOFF HYDROGRAPHS FROM NATURAL WATERSHEDS BY FINITE-ELEMENT METHOD	73-1140
JUDD, W. R.	STRAIN DISTRIBUTION AROUND UNDERGROUND OPENINGS. STATISTICAL RELATIONSHIPS FOR CERTAIN ROCK PROPERTI	71-0710
JUIGNET, N.	CALCULATION BY FINITE ELEMENT TECHNIQUE OF STEADY-STATE OR TRANSIENT HEAT-FLOW CONDUCTION	74-0637
JULLIEN, J. F.	PHOTOELASTIC DETERMINATION OF THE STRESSES IN THE FOUNDATION OF THE NUCLEAR-POWER STATION BUGEY 2	75-0834
JUNGELS, P.	DISPLACEMENT AND STRAIN FIELDS FROM UNDERGROUND EXPLOSIONS USING ANALYTICAL AND FINITE ELEMENT METHO	72-0587
JUNGELS, P.	STUDY OF 1964 ALASKAN EARTHQUAKE USING FINITE ELEMENT, AND GENERALIZED INVERSION TECHNIQUES	72-0003
JUNKINS, J. L.	FINITE ELEMENT MODEL OF EARTHS GRAVITY FIELD	74-0516
JUNKINS, J. L.	INVESTIGATION OF FINITE ELEMENT REPRESENTATIONS OF THE GEOPOTENTIAL	75-0290
KABAILA, A. P.	A COMPATIBLE ELEMENT FOR BUCKLING ANALYSIS OF SKEW PLATES WITH PARALLEL EDGES	73-0328
KABAILA, A. P.	A COMPATIBLE MEMBRANE QUADRILATERAL ELEMENT	72-0483
KABAILA, A. P.	A COMPATIBLE QUADRILATERAL PLATE BENDING ELEMENT	72-0482
KABAILA, A. P.	A CONFORMING QUADRILATERAL ELEMENT FOR BUCKLING ANALYSIS OF STIFFENED PLATES	73-0613
KABAILA, A. P.	A FINITE ELEMENT PLATE BENDING PROGRAMME	72-0579
KABAILA, A. P.	AN EQUILIBRIUM MEMBRANE QUADRILATERAL ELEMENT	72-0484
KABAILA, A. P.	BIFURCATION OF ECCENTRICALLY LOADED MEMBERS AND OF STIFFENED PLATES	70-0057
KABAILA, A. P.	BIFURCATION OF SPACE FRAMES	70-0055
KABAILA, A. P.	CERTAIN FINITE ELEMENTS FOR PLATE BENDING PROBLEMS	74-0395
KABAILA, A. P.	COMPUTER INTEGRATION OF CONGRUENTLY TRANSFORMED MATRICES	71-0580
KABAILA, A. P.	COMPUTER INTEGRATION FOR FINITE ELEMENTS	73-0329
KABAILA, A. P.	DUAL ANALYSIS OF COLUMN SUPPORTED SLABS	74-0399
KABAILA, A. P.	FINITE ELEMENT METHODS IN ENGINEERING	74-0400
KABAILA, A. P.	FINITE ELEMENT TECHNIQUES - A SHORT COURSE OF FUNDAMENTALS AND APPLICATIONS	73-0612
KABAILA, A. P.	INTEGRATION FOR THE FINITE ELEMENTS - AN ALGORITHM IN ORTHOGONAL AXES	73-0295
KABAILA, A. P.	MESH GRADING TECHNIQUES FOR COMPATIBLE AND EQUILIBRIUM ELEMENTS	74-0397
KABAILA, A. P.	ON THE BENDING STIFFNESS OF SLABS IN CROSS-WALLED STRUCTURES	74-0398
KABAILA, A. P.	PLANE STRESS ELEMENT OF VARYING THICKNESS APPLIED TO END BLOCK ANALYSIS	74-0396
KABAILA, A. P.	SINGLE ELEMENT PANEL FOR FLAT PLATE STRUCTURES	75-0279
KABAILA, A. P.	STABILITY ANALYSIS BY FINITE ELEMENTS	70-0056
KABAILA, A. P.	SUBSTRUCTURE ANALYSIS OF MULTISTOREY FLAT SLAB BUILDINGS	74-0394
KACENA, W. J.	COMPARISON OF STRUCTURAL LOADS; STATIC VERSUS DYNAMIC	70-0464
KACHROD, P. N.	MODIFICATION OF LAYERED THEORY FOR PAVEMENT DESIGN ANALYSIS	73-0574
KACZKOWSKI, Z.	METHOD OF FINITE SPACE-TIME ELEMENTS IN DYNAMICS OF STRUCTURES	75-0823
KADEN, S. T.	PRAKTISCHE BEISPIELE DER DIGITALEN BERECHNUNG VON GRUNDWASSERSTROMUNGS PROBLEMEN NACH DER METHODE EN	73-0081
KADEN, ST.	APPLICATION OF FINITE ELEMENT METHODS IN GEOHYDRAULICS	74-0078
KADEN, ST.	AUFBEREITUNG MATHEMATISCHER MODELLE FUR DIE DIGITALE BERECHNUNG VON GRUNDWASSERSTROMUNGSPROBLEMEN NA	71-0543
KADEN, ST.	BEITRAG ZUR SIMULATION NICHTSTATIONAERER HORIZONTAL-EBENER GRUNDWASSERSTROMUNGSFELDER AUF DER GRUNDLA	74-0088
KADEN, ST.	GRUNDLAGEN DER DIGITALEN LOSUNG VON GRUNDWASSERSTROMUNGSPROBLEMEN NACH DER METHODE ENDLICHER ELEMENT	71-0542
KADHAMOH, S. K.	REINVESTIGATION OF BUCKLING OF SHELLS OF REVOLUTION BY A REFINED FINITE ELEMENT	74-0097
KADO, F.	ON THE APPLICATION OF THE FINITE ELEMENT METHOD TO THE VIBRATION ANALYSIS OF SPACE PLATE STRUCTURE	72-0630
KAEMPF, H.	INFLUENCE OF DEFECTS ON FUEL ELEMENT CLADDING STRESSES	73-0863
KAERCHER, H. J.	FINITE ELEMENTS ON THE BASIS OF CONTINUUM MECHANICS	75-0234
KAGANOVSKAYA, S	RIGID PLATE ON NONLINEARLY DEFORMABLE COHESIVE BASE (PLANE PROBLEM)	75-0264
KAGAWA, Y.	FINITE ELEMENT ANALYSIS OF FLEXURE - TYPE VIBRATORS WITH ELECTROSTRICTIVE TRANSDUCERS	70-0338
KAGAWA, Y.	FINITE ELEMENT SIMULATION OF TWO-DIMENSIONAL ELECTROMECHANICAL RESONATORS	74-1014
KAGAWA, Y.	FINITE ELEMENT SIMULATION OF ENERGY-TRAPPED ELECTROMECHANICAL RESONATORS	75-0142
KAGAWA, Y.	FINITE ELEMENT SIMULATION OF ELECTROMECHANICAL RESONATORS WITH ARBITRARY ELECTRODES	75-0492
KAGAWA, Y.	FINITE ELEMENT SIMULATION OF ACOUSTIC FILTERS WITH VARIABLE CIRCULAR CROSS SECTION	75-0871
KAGAWA, Y.	NEW APPROACH TO ANALYSIS AND DESIGN OF ELECTROMECHANICAL FILTERS BY FINITE-ELEMENT TECHNIQUE	71-0476
KAHL, M. R.	EFFECTS OF LOCAL STRENGTH VARIATIONS ON PLASTIC ENCLAVE DEVELOPMENT AT CRACK TIPS	72-0822
KAHLERT, F.	NUMERISCHE BERECHNUNG INSTATIONAERER TEMPERATURFELDER UND THERMISCH BEDINGTER VERFORMUNGEN VON WERKZ	75-0847
KAHN, L. F.	INFLUENCE OF END-CLOSURE STIFFNESS ON BEHAVIOR OF CONCRETE CYLINDRICAL HULLS SUBJECTED TO HYDROSTATI	71-0725
KAITZ, M. J.	A COMPUTER PROGRAM TO DISPLAY THE DYNAMIC STRESS ANALYSIS OF STRUCTURES	70-0077
KAJITA, T.	FINITE ELEMENT ANALYSIS ON UNIAXIAL COMPRESSIVE STRENGTH OF CYLINDRICAL BRITTLE SPECIMEN	70-0155
KAJITA, T.	FINITE ELEMENT ANALYSIS OF CABLE STAYED BRIDGES	72-0494

KAJITA, T.	STUDY ON STATE OF STRESS AND DEFORMATION OF CYLINDRICAL SPECIMENS OF BRITTLE MATERIAL UNDER UNIAXIAL	69-0152
KAKIMI, T.	LARGE DEFORMATION AND CRITICAL LOADS ANALYSIS OF FRAMED STRUCTURES	74-1293
KAKINO, Y.	RESIDUAL STRESS PRODUCED BY METAL CUTTING	71-0414
KALAJDZIC, M.	DOSTIGNUCA I TENDENCIJE U ISPITIVANJIMA ALATNIH MASINA (ACHIEVEMENTS AND TENDENCIES IN MACHINE TOOLS	73-0426
KALAJDZIC, M.	FINITE ELEMENT METHOD FOR DETERMINATION OF MACHINE TOOL STRUCTURES	72-0095
KALKANI, E. C.	MESH GENERATION PROGRAM FOR HIGHWAY EXCAVATION CUTS	74-1250
KAMAL, S. A.	ANALYSIS OF MULTI-AXIAL ANISOTROPIC CREEP, ROOF RESIN BOLTS, AND BED SEPARATION IN ROCK STRUCTURES U	74-1122
KAMAT, M. P.	EFFECT OF SHEAR DEFORMATIONS AND ROTARY INERTIA ON OPTIMUM BEAM FREQUENCIES	75-0231
KAMAT, M. P.	OPTIMAL BEAM FREQUENCIES BY THE FINITE ELEMENT DISPLACEMENT METHOD	73-0187
KAMAT, M. P.	STRONGEST COLUMN BY FINITE ELEMENT DISPLACEMENT METHOD	73-0186
KAMEI, A.	SIZE OF PLASTIC ZONE AT TIP OF A CRACK IN PLANE STRAIN SLATE BY FINITE ELEMENT METHOD	73-0006
KAMEL, H.	AUTOMATIC SYSTEM FOR KINEMATIC ANALYSIS, NASKA	65-0054
KAMEL, H. A.	AN INTERACTIVE GRAPHICS SHIP DESIGN ORIENTED PROGRAM PACKAGE FOR TIME SHARING SYSTEMS	73-0341
KAMEL, H. A.	ANALYSIS OF LARGE CLOSED SHIP STRUCTURES: APPLICATIONS TO TANKER STRUCTURAL DESIGN	70-0289
KAMEL, H. A.	APPLICATION OF THE FINITE ELEMENT METHOD TO SHIP STRUCTURES	71-0403
KAMEL, H. A.	AUTOMATIC ANALYSIS OF FUSELAGES AND PROBLEMS OF CONDITIONING	64-0027
KAMEL, H. A.	AUTOMATIC MESH GENERATION IN TWO AND THREE DIMENSIONAL INTERCONNECTED DOMAINS	70-0251
KAMEL, H. A.	BOOLEAN MATRICES AS A STANDARD FACILITY IN A MATRIX CODE	65-0057
KAMEL, H. A.	COMPUTATIONAL STABILITY IN NONLINEAR DISCRETE STRUCTURAL SYSTEMS	69-0145
KAMEL, H. A.	GIFTS SYSTEM	74-1060
KAMEL, H. A.	GIFTS SYSTEM FOR THE PDP-15 AND OTHER CONSIDERATIONS IN DEVELOPING GRAPHICS AND CAPABILITIES FOR FIN	73-0671
KAMEL, H. A.	MATRIX METHODS OF STRUCTURAL ANALYSIS	63-0014
KAMEL, H. A.	SOME DEVELOPMENTS IN THE ANALYSIS OF COMPLEX SHIP STRUCTURES	72-0332
KAMEL, W. H.	MATRIX METHODS OF STRUCTURAL ANALYSIS, A PRECIS OF RECENT DEVELOPMENTS	63-0019
KAMEMURA, K.	ELASTO-VISCOPLASTIC FINITE ELEMENT ANALYSIS BY PERTURBATION METHOD	74-1271
KAMESWARA RAO,	FINITE ELEMENT SOLUTION TO SOME PLANE PROBLEMS IN SOIL DYNAMICS	74-0448
KAMMINGA, W.	FINITE-ELEMENT SOLUTIONS FOR DEVICES WITH PERMANENT MAGNETS	75-0464
KAN, D.	MESH AND CONTOUR PLOT FOR TRIANGLE AND ISOPARAMETRIC ELEMENTS	70-0445
KAN, D. K-Y.	EQUATION SOLVING ALGORITHMS FOR THE FINITE ELEMENT METHOD	71-0127
KANAKARA, K.	GALERKIN FINITE ELEMENT ANALYSIS OF A UNIFORM BEAM CARRYING A CONCENTRATED MASS AND ROTARY INERTIA W	74-0552
KANDA, T.	STRESS ANALYSIS OF SELF-SEALING TYPE PISTON HEAD BY FINITE-ELEMENT METHOD	72-0259
KANDA, T.	STRESS ANALYSIS OF THE "SELF-SEALING" TYPE PISTON HEAD BY THE FINITE ELEMENT METHOD	72-0617
KANDAOUROFF, W.	TOUR MAINE-MONTPARNASSE QUELQUES PROBLEMES DE STRUCTURE (SPECIFIC PROBLEMS ENCOUNTERED IN THE CONSTR	74-0539
KANDEL, W.	NOTCH STRESSES AT THE LOAD CARRYING ELASTIC CORE	73-0266
KANDIDOV, V. P.	A MODIFIED FINITE ELEMENT METHOD FOR CALCULATION OF VIBRATION OF THIN PLATES	72-0598
KANDIDOV, V. P.	APPLICATION OF FINITE ELEMENT METHODS FOR CALUCLATION OF TRANSVERSE OSCILLATIONS IN ONE DIMENSIONAL	71-0599
KANDIDOV, V. P.	COMPARISON OF CALCULATION BY THE FINITE ELEMENT METHOD WITH EXPERIMENT AND AN EXAMPLE OF A DYNAMIC S	72-0597
KANDIDOV, V. P.	MODEL OF ELASTIC PLATE FROM FINITE ELEMENTS IN A SUPERSONIC FLOW	75-0038
KANDIDOV, V. P.	MODIFICIROVANNYJ KONECHNYJ ELEMENT DIJA RASCHIOTA KOLEBANIJ TONKIKH PLASTIN	72-0891
KANDIDOV, V. P.	ON THE CONVERGENCE OF THE METHOD OF FINITE ELEMENTS IN THE ANALYSIS OF MEMBRANE DYNAMICS	72-0537
KANDIDOV, V. P.	RASCHIOT USTOICHIVOSTI IZGIBNO-KRUTILNYKH KOLEBANIJ KRULA V DOZVUKOVOM POTOKE METODOM KONECHNYKH ELE	72-0903
KANDIDOV, V. P.	REDUCTION OF THE DEGREES OF FREEDOM IN SOLVING DYNAMIC PROBLEMS BY THE FINITE ELEMENT METHOD	73-0647
KANDIDOV, V. P.	STABILITY ANALYSIS OF THE TORSIONAL BENDING VIBRATIONS OF WING IN A SUBSONIC FLOW BY METHOD OF THE F	75-0755
KANDIDOV, V. P.	STUDY ON THE VIBRATION OF THIN PLATES IN A GAS STREAM BY THE FINITE ELEMENT METHOD	74-0454
KANG, C. M.	FINITE ELEMENT METHODS FOR THE NEUTRON DIFFUSION EQUATIONS	71-0372
KANG, C. M.	FINITE ELEMENT METHODS FOR SPACE-TIME REACTOR ANALYSIS	71-0673
KANG, C. M.	FINITE ELEMENT METHODS FOR REACTOR ANALYSIS	73-0108
KANG, C. M.	FINITE ELEMENT METHODS IN REACTOR PHYSICS ANALYSIS	75-0020
KANG, C. M.	FINITE-ELEMENT METHODS IN REACTOR-PHYSICS ANALYSIS	75-0502
KANG, S. I.	ON THE DETERMINATION OF EFFECTIVE MODULI OF COMPOSITE MATERIALS BY A THREE-DIMENSIONAL FINITE ELEMEN	74-1123
KANNINEN, M. F.	FAST FRACTURE RESISTANCE AND CRACK ARREST IN STRUCTURAL STEELS	73-0766
KAO, D. W.	FINITE ELEMENT ANALYSIS OF BUCKLING AND POST-BUCKLING BEHAVIORS OF ARCHES WITH GEOMETRIC IMPERFECTIO	73-0186
KAO, D-W.	INCREMENTAL VARIATIONAL METHOD FOR THE LARGE DISPLACEMENT ANALYSIS OF SHELLS WITH GEOMETRIC IMPERFEC	72-0894
KAO, R.	A GENERAL FINITE DIFFERENCE METHOD FOR ARBITRARY MESHES	72-0883
KAO, R.	APPLICATION OF HILL FUNCTIONS TO CIRCULAR PLATE PROBLEMS	75-0572
KAO, R.	APPLICATIONS OF HILL FUNCTIONS (APPLIED MATHEMATIC'S FINITE ELEMENT) TO APPLIED MECHANICS PROBLEMS	72-0884
KAPER, H. G.	APPLICATION OF FINITE ELEMENT TECHNIQUES FOR THE NUMERICAL SOLUTION OF THE NEUTRON TRANSPORT AND DIF	71-0013
KAPER, H. G.	APPLICATIONS OF FINITE ELEMENT METHODS IN REACTOR MATHEMATICS - NUMERICAL SOLUTION OF THE NEUTRON DI	72-0535
KAPER, H. G.	APPLICATIONS OF FINITE ELEMENT METHODS IN REACTOR MATHEMATICS. NUMERICAL SOLUTION OF THE NEUTRON TRA	74-0853
KAPER, H. G.	CONSTRUCTION OF A FINITE ELEMENT APPROXIMATION WHICH CROSSES MATERIAL INTERFACES	74-0896
KAPER, H. G.	EMPIRICAL INVESTIGATION OF REORDERING AND DATA MANAGEMENT FOR FINITE ELEMENT SYSTEMS OF EQUATIONS	73-0944
KAPER, H. G.	THE USE OF INTERPOLATORY POLYNOMIALS FOR A FINITE ELEMENT SOLUTION OF THE MULTIGROUP DIFFUSION EQUAT	72-0524
KAPER, H. G.	TIMING COMPARISON STUDY FOR SOME HIGH ORDER FINITE ELEMENT PROCEDURES AND A LOW ORDER FINITE DIFFERE	72-0767
KAPKOWSKI, J.	A FINITE ELEMENT STUDY OF ELASTIC-PLASTIC STRESS DISTRIBUTIONS IN NOTCHED SPECIMENS UNDER TENSION	68-0177
KAPOOR, M. P.	INTEGRATED SEQUENTIAL SOLVER FOR LARGE MATRIX EQUATIONS	74-0276
KAPPEL, V. V.	VIBRATION OF STIFFENED PLATES USING FINITE-ELEMENTS	72-0951
KAPUR, K. K.	METHOD FOR ESTIMATING RESPONSE OF PAYLOAD SECONDARY STRUCTURES TO RANDOM EXCITATION	68-0232
KAPUR, K. K.	PREDICTION OF PLATE VIBRATIONS USING A CONSISTENT MASS MATRIX	66-0043
KAPUR, K. K.	STABILITY OF PLATES USING THE FINITE ELEMENT METHOD	66-0081
KAPUR, K. K.	VIBRATIONS OF A TIMOSHENKO BEAM, USING FINITE-ELEMENT APPROACH	66-0044
KAPUR, S.	THE ROLE OF COMPUTER GRAPHICS IN THE STRUCTURAL DESIGN PROCESS	68-0215
KARCHER, H. J.	APPLICATION OF A CONJUGATED-GRADIENT-PROCESS IN FINITE-ELEMENT-CALCULATION OF NON-LINEAR STRUCTURES	74-0214
KARCHER, H. J.	TWO DUAL FINITE ELEMENT METHODS DERIVED FROM EXTENDED VARIATIONAL PRINCIPLE	73-0109
KARIAPPA, S.	THE MATRIX EIGENVALUE PROBLEM FOR DAMPED STRUCTURAL VIBRATION	66-0121
KARIAPPA, V.	APPLICATION OF MATRIX DISPLACEMENT METHODS IN THE STUDY OF PANEL FLUTTER	69-0406
KARIAPPA, V.	FLUTTER OF SKEW PANELS BY THE MATRIX DISPLACEMENT APPROACH	70-0201
KARIAPPA, V.	FURTHER DEVELOPMENTS IN CONSISTENT UNSTEADY SUPERSONIC AERODYNAMIC COEFFICIENTS	72-0559
KARIAPPA, V.	KINEMATICALLY CONSISTENT UNSTEADY AERODYNAMIC COEFFICIENTS IN SUPERSONIC FLOW	70-0014
KARLAK, R. F.	CREEP OF B/AL COMPOSITES AS INFLUENCED BY RESIDUAL STRESSES, BOND STRENGTH, AND FIBER PACKING GEOMET	74-1208
KARLAK, R. F.	FAILURE MECHANISMS IN COMPOSITE SYSTEMS	75-0688
KARLAK, R. F.	INTERFACE FAILURES IN COMPOSITES	74-1209
KAPLSSON, B. I.	PRESTRESSED CONCRETE DEEP SLABS WITH OPENINGS	73-0427
KARNES, R. N.	A USER-ORIENTED PROGRAM FOR CRASH DYNAMICS	74-0321
KARWOSKI, W.	NEW HORIZONS IN ROCK MECHANICS	73-0428
KARWOSKI, W. J.	THEORETICAL INVESTIGATION OF ROCK AND SUPPORT INTERACTION - DEVELOP MORE RATIONAL DESIGN METHODS AND	74-0782
KASHEF, A. I.	MANAGEMENT OF RETARDATION OF SALT WATER INTRUSION IN COASTAL AQUIFERS	75-0692
KASHEF, A-A. I.	COMPARATIVE STUDY OF FRESH-SALT WATER INTERFACES USING FINITE ELEMENT AND SIMPLE APPROACHES	75-0468
KASHIWAGI, T.	A RADIATIVE IGNITION MODEL OF A SOLID FUEL	74-1312
KASPER, R. G.	FINITE ELEMENT STRESS ANALYSIS OF A MULTICONDUCTOR ARMOR CABLE	73-0095
KASPERSKI, Z.	NUMERYCZNE ROZWIAZANIE PROBLEMU DRGAN WLASNYCH NA PRZYKLADZIE PLYT. (NUMERICAL SOLUTION OF THE PROBL	75-0296
KASPERSKI, Z.	ZASTOSOWANIE ITERACJI SEIDLA W METODZIE ELEMENTOW SKONCZONYCH NA PRZYKLADZIE OBLICZEN STATYCZNYCH P	74-1031
KASSOS, T.	LINEAR EQUALITY CONSTRAINTS IN FINITE ELEMENT APPROXIMATION	75-0475
KASTIL, H.	HYBRIDE FINITE VERFAHREN FUER LINEARE UND NICHTLINEARE INSTATIONAERE WAERMELEITPROBLEME (HYBRID FINI	73-0484
KATAYAMA, D.	FINITE ELEMENT APPLICATION IN LAKE CIRCULATION	75-0493
KATHIRESAN, K.	THREE-DIMENSIONAL LINEAR FRACTURE MECHANICS ANALYSIS BY A DISPLACEMENT-HYBRID FINITE-ELEMENT MODEL	75-0884
KATO, B.	STRENGTH OF TRANSVERSE FILLET WELDED JOINTS	74-0144
KATO, S.	NONLINEAR DYNAMIC ANALYSIS OF TORSIONAL SHELLS BY COMBINED USE OF FINITE ELEMENT AND MODE SUPERPOSIT	75-0859
KATONA, M. G.	DEVELOPMENT OF FINITE ELEMENT HEAD INJURY MODEL	75-0506
KATONA, M. G.	ICE ENGINEERING: VISCOELASTIC FINITE ELEMENT FORMULATION	74-0861
KATONA, M. G.	LAYERED PAVEMENT SYSTEMS. PART 1. LAYERED SYSTEM DESIGN. PART 11. FATIGUE OF PLAIN CONCRETE	72-0867
KATOW, M. S.	STATIC ANALYSIS OF THE 64-M 210-FT ANTENNA REFLECTION STRUCTURE	71-0137
KATZ, I. M.	NODAL VARIABLES FOR ARBITRARY ORDER FINITE ELEMENTS	75-0002

KATZ, I. N.	THE CONSTRAINT METHOD FOR FINITE ELEMENT STRESS ANALYSIS	75-0432
KAUFFMANN, E. M	ANALYSIS OF FATIGUE OF PAVING MIXTURES FROM THE FRACTURE MECHANICS VIEWPOINT	72-0540
KAUFFMANN, E. M	VERIFICATION OF FRACTURE MECHANICS CONCEPTS TO PREDICT CRACKING OF FLEXIBLE PAVEMENTS	73-0772
KAUFMAN, A.	ANALYSIS OF THE EFFECTS ON LIFE OF LEADING-EDGE HOLES IN AN AIRFOIL SUBJECTED TO ARBITRARY SPANWISE	75-0700
KAUFMAN, S.	AXISYMMETRIC SHELLS	69-0481
KAUFMAN, S.	BENDING ELEMENTS FOR PLATE AND SHELL NETWORKS	67-0043
KAUFMAN, S.	FINITE ELEMENT METHOD FOR DETERMINATION OF THERMAL STRESSES INTERACTIVE GRAPHICS FOR FINITE ELEMENT	74-0155
KAUFMAN, S.	FINITE-ELEMENT METHOD FOR THE DETERMINATION OF THERMAL STRESSES IN ANISOTROPIC SOLIDS OF REVOLUTION	74-0262
KAUSEL, E.	DYNAMIC ANALYSIS OF FOOTINGS ON LAYERED MEDIA	75-0586
KAUSEL, E.	EARTHQUAKE FINITE-ELEMENT ANALYSIS OF STRUCTURE-FOUNDATION SYSTEMS	75-0410
KAUSEL, E.	FORCED VIBRATIONS OF CIRCULAR FOUNDATIONS ON LAYERED MEDIA	74-0762
KAVANAGH, K. T.	APPROXIMATE ALGORITHM FOR THE REANALYSIS OF STRUCTURES BY THE FINITE ELEMENT METHOD.	72-0445
KAVANAGH, K. T.	EXTENSION OF CLASSICAL EXPERIMENTAL TECHNIQUES FOR CHARACTERIZING COMPOSITE - MATERIAL BEHAVIOUR	72-0088
KAVANAGH, K. T.	FINITE ELEMENT APPLICATIONS IN THE CHARACTERIZATION OF ELASTIC SOLIDS	71-0138
KAVANAGH, K. T.	THE FINITE ELEMENT ANALYSIS OF PHYSICALLY AND KINEMATICALLY NONLINEAR ELASTIC SOLIDS	69-0146
KAVARI, K.	ON THE DIMENSIONING OF UNDERGROUND STRUCTURES	69-0147
KAVLIE, D.	A COMPUTER PROGRAM FOR ANALYSIS OF STIFFENED PLATES UNDER COMBINED INPLANE AND LATERAL LOADS	71-0703
KAVLIE, D.	INDETS - INTEGRATED DESIGN OF TANKER STRUCTURES	75-0524
KAWAHAPA, M.	ANALYSIS OF STEADY INCOMPRESSIBLE VISCOUS FLOW	74-0013
KAWAHARA, M.	ANALYSIS OF VISCOELASTIC FLUID BY FINITE ELEMENT METHOD	75-0865
KAWAHARA, M.	ELASTO-VISCOPLASTIC FINITE ELEMENT ANALYSIS BY PERTURBATION METHOD	74-1271
KAWAHARA, M.	FINITE ELEMENT ANALYSIS OF VISCOUS FLOW BY STREAM FUNCTION	75-0526
KAWAHARA, M.	FINITE ELEMENT ANALYSIS OF INCOMPRESSIBLE MICROPOLAR VISCOUS FLUID FLOW	75-0866
KAWAHARA, M.	LARGE STRAIN, ELASTO-PLASTIC NUMERICAL ANALYSIS BY MEANS OF FINITE ELEMENT METHOD	71-0790
KAWAHARA, M.	LARGE STRAIN, VISCOELASTIC NUMERICAL ANALYSIS BY MEANS OF FINITE ELEMENT METHOD	72-0768
KAWAHARA, M.	LARGE STRAIN, VISCOELASTIC AND ELASTO-VISCOPLASTIC NUMERICAL-ANALYSIS BY MEANS OF FINITE-ELEMENT MET	75-0435
KAWAHARA, M.	LARGE STRAIN, VISCOELASTIC AND ELASTO VISCOPLASTIC NUMERICAL ANALYSIS BY MEANS OF THE FINITE ELEMENT	75-0563
KAWAHARA, M.	NUMERICAL ANALYSIS OF VISCO-ELASTIC STRUCTURES BY THE FINITE ELEMENT METHOD	70-0177
KAWAHARA, M.	SPARSE MATRICES IN FINITE ELEMENT PROGRAM DEVELOPMENT	73-0337
KAWAHARA, M.	STEADY FLOW ANALYSIS OF INCOMPRESSIBLE VISCOUS FLOW BY THE FINITE ELEMENT METHOD	73-0332
KAWAI, S.	APPLICATION OF THE FINITE ELEMENT METHOD TO SURFACE WAVE ANALYSIS	74-0913
KAWAI, S.	FINITE ELEMENT ANALYSIS OF WAVE PROPAGATION IN OPEN CHANNELS	75-0570
KAWAI, T.	A DIFFUSION ANALYSIS OF ION IN ELECTROLYSIS BY FINITE ELEMENT METHOD	73-0296
KAWAI, T.	A METHOD OF FINITE DEFLECTION ELASTIC PLASTIC ANALYSIS OF FRAMED STRUCTURES SUBJECTED TO LARGE AXIAL	69-0239
KAWAI, T.	A METHOD OF SOLUTION FOR THE COMPLICATED BUCKLING PROBLEMS OF ELASTIC PLATES WITH COMBINED USE OF RA	68-0097
KAWAI, T.	A TORSIONAL STRENGTH ANALYSIS ON THE CONTAINER SHIP BY MEANS OF THE FINITE ELEMENT PROCEDURE AND FUL	71-0243
KAWAI, T.	ANALYSIS OF LARGE DEFLECTION OF PLATES BY THE FINITE ELEMENT METHOD	69-0148
KAWAI, T.	ANALYSIS OF THE ELASTIC PLASTIC PROBLEMS BY THE MATRIX DISPLACEMENT METHOD	68-0173
KAWAI, T.	APPLICATION OF FINITE ELEMENT METHODS TO SHIP STRUCTURES	73-0429
KAWAI, T.	APPLICATION OF THE FINITE ELEMENT METHOD TO PRESSURE VESSEL DESIGN (JAPANESE)	70-0446
KAWAI, T.	FINITE ELEMENT ANALYSIS ON THE TORSION OF A BAR WITH UNIFORM CROSS SECTION	68-0186
KAWAI, T.	FINITE ELEMENT ANALYSIS OF THE GEOMETRICALLY NONLINEAR PROBLEMS	69-0149
KAWAI, T.	FINITE ELEMENT ANALYSIS OF SHELL STRUCTURES IN JAPAN	69-0151
KAWAI, T.	FINITE ELEMENT ANALYSIS OF THIN-WALLED STRUCTURES BASED ON THE MODERN ENGINEERING THEORY OF BEAMS	71-0140
KAWAI, T.	FINITE ELEMENT ANALYSIS OF A WING STRUCTURE	72-0333
KAWAI, T.	MATRIX METHOD OF ANALYSIS OF SHIP STRUCTURES 3RD REPORT	68-0225
KAWAI, T.	ON THE DEVELOPMENT OF THE "PASSAGE" PROGRAM	73-0342
KAWAI, T.	ON THE FINITE ELEMENT ANALYSIS OF DIFFUSION PROBLEMS	73-0331
KAWAI, T.	ON THE FINITE ELEMENT ANALYSIS OF THIN-WALLED STRUCTURES WITH APPLICATION TO STRUCTURAL PROBLEMS IN	73-0832
KAWAI, T.	ON THE STATES OF STRESS AND DEFORMATION OF CYLINDRICAL SPECIMENS OF BRITTLE MATERIAL UNDER UNIAXIAL	69-0150
KAWAI, T.	RECENT DEVELOPMENTS OF FINITE ELEMENT METHODS IN JAPAN	75-0226
KAWAI, T.	SHEAR DEFORMATION ANALYSIS OF BEAMS AND PLATES BY FINITE ELEMENT METHOD	75-0873
KAWAI, T.	SOME PROGRAM DEVELOPMENTS OF MATRIX STRUCTURAL ANALYSIS IN JAPAN	72-0322
KAWAMATA, S.	ANALYSIS OF CABLE NETS IN MIXED FORMULATION	73-0311
KAWAMATA, S.	MATRIX ANALYSIS OF A COMPOSITE STRUCTURE OF PLATE AND LINE ELEMENTS	71-0250
KAWAMATA, S.	PERTURBATION METHOD IN THE ANALYSIS OF GEOMETRICALLY NON LINEAR AND STABILITY PROBLEMS	72-0321
KAWAMO, K.	A TEST ON CONVERGENCE OF EIGENSOLUTIONS BY SIMULTANEOUS ITERATION METHOD	73-0317
KAWAMOTO, T.	FINITE ELEMENT ANALYSIS ON UNIAXIAL COMPRESSIVE STRENGTH OF CYLINDRICAL BRITTLE SPECIMEN	70-0155
KAWAMOTO, T.	ON THE STATES OF STRESS AND DEFORMATION AROUND THE BOTTOM OF A CIRCULAR VERTICAL SHAFT	69-0323
KAWAMOTO, T.	STUDY ON STATE OF STRESS AND DEFORMATION OF CYLINDRICAL SPECIMENS OF BRITTLE MATERIAL UNDER UNIAXIAL	69-0152
KAWASAKI, H.	CALCULATION OF THE VIRTUAL MASS, MOVEMENT AND INERTIA BY THE FINITE ELEMENT METHOD	68-0221
KAWASHIMA, K.	ANALYSIS OF STRESSES IN SKEWED PLATES BY THE METHOD OF FINITE ELEMENTS	73-0248
KAWASHIMA, M.	A STUDY ON THE STRESS ANALYSIS OF BEAM PLATE STRUCTURE	69-0404
KAWASHIMA, M.	PRACTICAL ANALYSIS OF STEEL STRUCTURES USING HIGHER-ORDER ELEMENTS	72-0326
KAWAZURA, A.	BANDED MATRIX TECHNIQUES AND SOLUTIONS OF LINEAR SIMULTANEOUS EQUATIONS OF FINITE ELEMENT ANALYSES	70-0302
KAY, S.	AN EXPERIMENTAL AND THEORETICAL STUDY OF HOLLOW CYLINDER TESTS ON SAND	71-0008
KAY, S.	STRESS ANALYSIS OF CONTRACTIVE OR DILATIVE SOIL	71-0579
KAYE, G.	A DATA-GENERATION PROCEDURE FOR THE FINITE ELEMENT ANALYSIS OF PLANE FRAMES AND GRILLAGES	68-0179
KEALY, C. D.	DETERMINING SEEPAGE CHARACTERISTICS OF MILL-TAILING DAMS BY THE FINITE ELEMENT METHOD	71-0477
KEALY, C.D.	SEEPAGE-ENVIRONMENTAL ANALYSIS OF THE SLIME ZONE OF A TAILINGS POND	74-0513
KEAYS, R. H.	ANALYSIS OF AN ORTHOTROPIC PLATE WITH TWO OPPOSITE EDGES SIMPLY SUPPORTED AND THE OTHER TWO FREE	73-0741
KEEGSTRA, P. N.	FINITE ELEMENT STRESS ANALYSIS OF INTERACTING FUEL PELLET AND CANNING UNDER FAST REACTOR CONDITION	73-0814
KEITH, H. D.	FINITE ELEMENT ANALYSIS OF PRESSURE VESSELS	71-0348
KEITH, H. D.	USER'S MANUAL FOR UMP FINITE ELEMENT COMPUTER PROGRAM - STRATA	69-0429
KELLEY, B. E.	FINITE ELEMENT FORMULATION OF GENERAL THERMO-ELASTICITY PROBLEMS	71-0001
KELLOGG, R. B.	HIGHER ORDER SINGULARITIES FOR INTERFACE PROBLEMS	72-0513
KELLOGG, R. B.	MATHEMATICAL AND COMPUTATIONAL PROBLEMS IN REACTOR CALCULATIONS	73-0806
KELLOGG, R. B.	MATHEMATICAL AND COMPUTATIONAL PROBLEMS IN REACTOR CALCULATIONS	73-0973
KELLOGG, R. B.	NONUNIFORM ERROR ESTIMATES FOR THE FINITE ELEMENT METHOD	74-0890
KELLOGG, R. B.	SOME MATHEMATICAL ASPECTS OF FINITE ELEMENT CALCULATIONS	73-0110
KELLY, B. M.	THE NASTRAN CONTOUR PLOTTER	72-0703
KELSEY, S.	ENERGY THEOREMS AND STRUCTURAL ANALYSIS	54-0001
KELSEY, S.	FINITE ELEMENT METHODS IN CIVIL ENGINEERING	72-0336
KELSEY, S.	MATRIX METHODS OF STRUCTURAL ANALYSIS	63-0014
KELSEY, S.	MATRIX METHODS OF STRUCTURAL ANALYSIS. A PRECIS OF RECENT DEVELOPMENTS	63-0019
KELSEY, S.	MODERN FUSELAGE ANALYSIS AND THE ELASTIC AIRCRAFT	63-0023
KELSEY, S.	ON THE MATRIX THEORY OF STRUCTURES	60-0007
KELSEY, S.	PROBABILISTIC BUCKLING BEHAVIOR OF STRUCTURES WITH RANDOM LACK-OF-FIT AND GEOMETRIC IMPERFECTIONS	73-0862
KELSEY, S.	STRUCTURAL ANALYSIS BY THE MATRIX FORCE METHOD WITH APPLICATION TO AIRCRAFT WINGS	56-0004
KELSEY, S.	THE APPLICATION OF DYNAMIC RELAXATION TO THE FINITE ELEMENT METHOD OF STRUCTURAL ANALYSIS	68-0108
KELSEY, S.	THE CONDITION OF SOME FINITE ELEMENT COEFFICIENT MATRICES	70-0393
KELSEY, S.	THE MATRIX FORCE METHOD OF STRUCTURAL ANALYSIS AND SOME NEW APPLICATIONS	56-0005
KELSEY, S.	THE VALIDITY OF THE INITIAL STRAIN CONCEPT	61-0010
KELSEY, S.	TRIANGULAR PLATE BENDING ELEMENT WITH ENFORCED COMPATIBILITY	71-0593
KEMAIRE, M.	AUTOMATIC CALCULATION OF THIN STRUCTURES	73-0201
KEMP, K. O.	COMPARISON OF FINITE ELEMENT AND UNIQUE LIMIT ANALYSIS SOLUTIONS FOR CERTAIN REINFORCED CONCRETE SLA	69-0184
KENDALL, D. P.	EXPERIMENTAL AND ANALYTICAL STRAINS IN A EDGE-CRACKED SHEET	69-0483
KENDALL, J. M.	BLOWDOWN EFFECTS ON A FUEL ROD	75-0575
KENDRICK, R. F.	PREDICTING THE BLOCK CAVING BEHAVIOUR OF OREBODIES	69-0044
KENDRICK, S.	STRUCTURAL DESIGN OF SUPERTANKERS	70-0152
KENINGSBERG, I.	CH-53A FLEXIBLE FRAME VIBRATION ANALYSIS/TEST CORRELATION	73-1093
KENINGSBERG, I.	CORRELATION OF FINITE ELEMENT STRUCTURAL DYNAMIC ANALYSIS WITH MEASURED FREE VIBRATION CHARACTERISTI	74-0486

KENNEDY, F. E.	A THERMAL, THERMOELASTIC, AND WEAR ANALYSIS OF HIGH-ENERGY DISK BRAKES	74-0839
KENNEDY, F. E.	THERMAL, THERMOELASTIC, AND WEAR SIMULATION OF A HIGH-ENERGY SLIDING CONTACT PROBLEM	73-0430
KENNEDY, F. E.	ANALYSIS OF NONLINEAR CONTACT PROBLEMS BY THE FINITE-ELEMENT METHOD	72-0952
KENNEDY, F. F.	ELASTO-PLASTIC INDENTATION OF A LAYERED MEDIUM	74-0225
KENNEDY, J. M.	DYNAMIC RESPONSE OF FAST-REACTOR CORE SUBASSEMBLIES	74-0540
KENNEDY, J. M.	FINITE ELEMENT APPROACH TO PRESSURE WAVE ATTENUATION BY REACTOR FUEL SUBASSEMBLIES	75-0522
KENNEDY, J. M.	FINITE ELEMENT STUDY OF PRESSURE WAVE ATTENUATION BY REACTOR FUEL SUBASSEMBLIES	75-0215
KENNEDY, J. M.	FINITE-ELEMENT ANALYSIS OF STRUCTURAL RESPONSE IN HCDA	75-0407
KENNEDY, J. M.	RESPONSE OF REACTOR CORE SUBASSEMBLIES TO IMPULSIVE LOADINGS	73-0076
KENNEDY, R. P.	PROBABILISTIC ASSESSMENT OF AIRCRAFT HAZARD FOR NUCLEAR POWER PLANTS	72-0235
KENNER, P. M.	STRUCTURAL ANALYSIS OF A PARAWING DURING DEPLOYMENT	72-0560
KENNER, V. H.	DYNAMIC LOADING OF A FLUID-FILLED SPHERICAL SHELL	72-0444
KENNEY, T. C.	TEST OF VALIDITY OF DARCY'S LAW FOR A CLAY SOIL	74-0070
KERFOOT, R. P.	WOOD SHEAR PANELS BONDED WITH FLEXIBLE ADHESIVES	75-0086
KERR, R. I.	SOME PROBLEMS IN THE DISCRETE ELEMENT REPRESENTATION OF AIRCRAFT STRUCTURES	64-0021
KERTESZ, T. J.	INFLUENCE OF ASCENT HEATING ON THE SEPARATION DYNAMICS OF A SPACECRAFT FAIRING	72-0906
KESHAVARZ, M. H	OPTIMAL DESIGN OF WATER WELLS IN UNCONSOLIDATED SEDIMENTS	74-1182
KESLER, C. E.	DETERMINATION OF VOLUME CHANGE STRESSES IN PLAIN AND REINFORCED CONCRETE USING FINITE ELEMENT ANALYS	69-0477
KESLER, C. E.	RESEARCH TO IMPROVE TUNNEL SUPPORT SYSTEMS	74-0777
KESLER, C. E.	STUDIES TO IMPROVE TUNNEL SUPPORT SYSTEMS	74-0957
KETCHMAN, J.	APPLICATION OF THE FINITE-ELEMENT METHOD TO TOWED CABLE DYNAMICS	73-1141
KETCHMAN, J.	APPLICATION OF THE FINITE ELEMENT METHOD TO TOWED CABLE DYNAMICS	75-0809
KEUNING, D. H.	APPLICATION OF FINITE ELEMENT METHOD WITH SECTIONAL LINEARIZATION TO FLOW PROBLEMS	75-0611
KEVICZKY, L.	NONLINEAR STRUCTURES FOR SYSTEM IDENTIFICATION	74-1030
KEY, J. E.	A NOTE ON THE ANALYSIS OF NONLINEAR DYNAMICS OF ELASTIC MEMBRANES BY THE FINITE ELEMENT METHOD	74-0341
KEY, J. E.	A NOTE ON THE FINITE DEFORMATIONS OF A THICK ELASTIC SHELL OF REVOLUTION	71-0193
KEY, J. E.	ANALYSIS OF FINITE DEFORMATIONS OF ELASTIC SOLIDS BY THE FINITE ELEMENT METHOD	70-0096
KEY, J. E.	ANALYSIS OF NONLINEAR THERMOELASTIC AND THERMOPLASTIC BEHAVIOUR OF SOLIDS OF REVOLUTION BY THE FINIT	71-0192
KEY, J. E.	COMPUTER PROGRAM FOR SOLUTION OF LARGE, SPARSE, UNSYMMETRIC SYSTEMS OF LINEAR EQUATIONS	72-0719
KEY, J. E.	NOTE ON THE ANALYSIS OF NONLINEAR DYNAMICS OF ELASTIC MEMBRANES BY THE FINITE ELEMENT METHOD	74-0230
KEY, J. E.	NUMERICAL ANALYSIS FOR FINITE AXISYMMETRIC DEFORMATIONS OF INCOMPRESSIBLE ELASTIC SOLIDS OF REVOLUTI	70-0094
KEY, J. E.	ON THE EFFECT OF THE FORM OF THE STRAIN ENERGY FUNCTION ON THE SOLUTION OF A BOUNDARY-VALUE PROBLEM	72-0429
KEY, J. E.	ON THE NUMERICAL SOLUTION OF CERTAIN PROBLEMS IN FINITE ELASTICITY BY THE FINITE ELEMENT METHOD	75-0542
KEY, J. E.	SOLUTION OF NONLINEAR EQUATIONS BY EXPLICIT TIME INTEGRATION - A SHORT COMMUNICATION	73-0606
KEY, J. E.	STATIC, STABILITY, AND DYNAMIC ANALYSIS OF SHELLS OF REVOLUTION BY NUMERICAL INTEGRATION - A COMPARI	74-0389
KEY, J. E.	THEMIS FINAL REPORT, VOL. IV. RECENT DEVELOPMENTS IN THE NUMERICAL SOLUTION OF CERTAIN PROBLEMS IN N	74-0721
KEY, J. E.	USERS MANUAL FOR DIGITAL COMPUTER PROGRAM DK5 NUMERICAL ANALYSIS OF FINITE STRAINS AND DISPLACEMENTS	72-0718
KEY, J. E.	USERS' MANUAL FOR DIGITAL COMPUTER PROGRAM DK1. NUMERICAL ANALYSIS OF FINITE AXISYMMETRIC DEFORMATIO	72-0861
KEY, J. N.	ON SOME GENERALIZATIONS OF THE INCREMENTAL STIFFNESS RELATIONS FOR FINITE DEFORMATIONS OF COMPRESSIB	71-0189
KEY, S. W.	A CONVERGENCE INVESTIGATION OF THE DIRECT STIFFNESS METHOD	66-0102
KEY, S. W.	ANALYSIS OF THIN SHELLS WITH A DOUBLY CURVED ARBITRARY QUADRILATERAL FINITE ELEMENT.	72-0443
KEY, S. W.	COMPARISON OF FINITE ELEMENT AND FINITE DIFFERENCE METHODS	71-0143
KEY, S. W.	FINITE ELEMENT PROCEDURE FOR THE LARGE DEFORMATION DYNAMIC RESPONSE OF AXISYMMETRIC SOLIDS	74-0527
KEY, S. W.	FINITE ELEMENT PROCEDURE FOR THE LARGE DEFORMATION DYNAMIC RESPONSE OF AXISYMMETRIC SOLIDS	74-0967
KEY, S. W.	FINITE ELEMENT PROCEDURE FOR LARGE DEFORMATION DYNAMIC RESPONSE OF AXISYMMETRIC SOLIDS	75-0013
KEY, S. W.	SLADE D: A COMPUTER PROGRAM FOR THE DYNAMIC ANALYSIS OF THIN SHELLS	73-0964
KEY, S. W.	SLADE, A COMPUTER PROGRAM FOR THE STATIC ANALYSIS OF THIN SHELLS	70-0079
KEY, S. W.	THE ANALYSIS OF THIN SHELLS WITH TRANSVERSE SHEAR STRAINS BY THE FINITE ELEMENT METHOD	68-0098
KEY, S. W.	THE ANALYSIS OF THIN SHELLS BY THE FINITE ELEMENT METHOD	70-0078
KEY, S. W.	THE TRANSIENT DYNAMIC ANALYSIS OF THIN SHELLS BY THE FINITE ELEMENT METHODS	71-0142
KEY, S. W.	TRANSIENT SHELL RESPONSE BY NUMERICAL TIME INTEGRATION	72-0305
KEY, S. W.	VARIATIONAL PRINCIPLE FOR INCOMPRESSIBLE AND NEARLY-INCOMPRESSIBLE ANISOTROPIC ELASTICITY	69-0319
KFOURI, A. P.	ELASTIC-PLASTIC FINITE ELEMENT ANALYSIS OF CRACK TIP FIELDS UNDER BIAXIAL LOADING CONDITIONS	74-0296
KFOURI, A. P.	STRESS, DISPLACEMENT, LINE INTEGRAL AND CLOSURE ENERGY DETERMINATIONS OF CRACK TIP STRESS INTENSITY	74-0621
KFOURI, A. P.	STRESSES IN A PARTLY YIELDED NOTCHED BAR - AN ASSESSMENT OF THREE ALTERNATIVE PROGRAMS	73-0267
KHABBAZ, G. R.	DYNAMIC BEHAVIOR OF LIQUIDS IN ELASTIC TANKS	71-0141
KHAL, R.	COMPUTER STRESS ANALYSIS	68-0099
KHALIL, T. B.	DYNAMIC RESPONSE OF ELASTIC HEAD MODELS	75-0378
KHALIL, T. B.	WAVE PROPAGATION IN FLUIDS	74-0034
KHAN, A. Q.	GEOMETRICALLY NONLINEAR BEHAVIOUR OF THIN-WALLED MEMBERS USING FINITE-ELEMENTS	73-1142
KHAN, M. H.	ANALYSE TRIDIMENSIONALE DU COMPORTEMENT NON-LINEAIRE D'UN CAISSON DE REACTEUR NUCLEAIRE EN BETON PRE	71-0144
KHAN, M. H.	FINITE ELEMENT METHOD IN NONLINEAR DOMAIN, (METHODE DES ELEMENTS FINIS DANS LE DOMAINE NON LINEAIRE)	71-0288
KHAN, M. H.	NONLINEAR ANALYSIS OF PRESTRESSED CONCRETE FAST REACTOR PRESSURE VESSEL, (ANALYSE NON LINEAIRE DES C	71-0145
KHANNA, J.	COMPARISON AND EVALUATION OF STIFFNESS MATRICES	66-0045
KHANNA, J.	CRITERION FOR SELECTING STIFFNESS MATRICES	65-0061
KHANNA, J.	DYNAMIC RESPONSE OF EARTH DAMS	67-0024
KHANNA, J.	SOME INVESTIGATIONS INTO THE FINITE ELEMENT METHOD WITH SPECIAL REFERENCE TO PLANE STRESS	66-0128
KHANNA, S. K.	MODIFICATION OF LAYERED THEORY FOR PAVEMENT DESIGN ANALYSIS	73-0574
KHARKHURIM, I.	APPLICATION OF THE METHOD OF FINITE ELEMENTS TO THE CALCULATION OF SHIP STRUCTURES (ISPOLZOVANIE MET	73-0933
KHATUA, T. P.	A TRIANGULAR ELEMENT FOR BENDING AND VIBRATION OF MULTILAYER SANDWICH PLATES	72-0489
KHATUA, T. P.	BENDING AND VIBRATION OF MULTILAYER SANDWICH BEAMS AND PLATES	73-0272
KHATUA, T. P.	FINITE ELEMENT ANALYSIS OF AXISYMMETRIC MULTILAYER SANDWICH PLATES AND SHELLS	74-0101
KHATUA, T. P.	FINITE ELEMENT ANALYSIS OF MULTILAYER SANDWICH PLATES AND SHELLS	74-0425
KHATUA, T. P.	FINITE-ELEMENT ANALYSIS OF MULTILAYER SANDWICH STRUCTURES	72-0953
KHATUA, T. P.	STABILITY ANALYSIS OF MULTILAYER SANDWICH STRUCTRES	73-0345
KHATUA, T. P.	TRIANGULAR ELEMENT FOR MULTILAYER SANDWICH PLATES	72-0581
KHEGAZI, S. K.	APPLICATION OF THE METHOD OF FINITE ELEMENTS TO THE CALCULATION OF SHIP STRUCTURES (ISPOLZOVANIE MET	73-0933
KHLYBOV, E. P.	ON THE CONVERGENCE OF THE METHOD OF FINITE ELEMENTS IN THE ANALYSIS OF MEMBRANE DYNAMICS	72-0537
KHOJASTEH-BAKHT	ANALYSIS OF ELASTIC-PLASTIC SHELLS OF REVOLUTION UNDER AXI-SYMMETRIC LOADING BY THE FINITE ELEMENT M	67-0102
KHOJASTEH-BAKHT	ANALYSIS OF ELASTIC-PLASTIC SHELLS OF REVOLUTION	70-0220
KHOJASTEH-BAKHT	BENDING OF CIRCULAR PLATES OF HARDENING MATERIAL	67-0112
KHOJASTEH-BAKHT	COMPUTER PROGRAM FOR ELASTIC-PLASTIC ANALYSIS OF AXISYMMETRICALLY LOADED SHELLS OF REVOLUTION	68-0239
KHOJASTEH-BAKHT	ELASTIC-PLASTIC ANALYSIS OF SOME PRESSURE VESSEL HEADS	69-0379
KHOT, N. S.	APPLICATION OF OPTIMALLY CRITERION TO FIBER-REINFORCED COMPOSITES	73-0431
KHOT, N. S.	DESIGN OF OPTIMUM STRUCTURES FOR DYNAMIC LOADS	71-0255
KHOT, N. S.	OPTIMIZATION OF STRUCTURES FOR STRENGTH AND STABILITY REQUIREMENTS	73-0575
KHOT, N. S.	OPTIMUM DESIGN OF ADVANCED COMPOSITE STRUCTURES FOR STATIC LOADS	72-0613
KHOZEIMEH, K.	COMPUTER ANALYSIS OF RIGID FRAMES	71-0401
KHOZEIMEH, K.	INELASTIC RESPONCE OF FRAMES TO DYNAMIC LOADS	71-0611
KI, M. K.	MESH GENERATING COMPUTER PROGRAM FOR FINITE ELEMENT ANALYSES OF SIMPLE SLOPES IN HORIZONTALLY LAYERE	71-0748
KIDBIRIGE, S. L	A FINITE ELEMENT SOLUTION OF TIME DEPENDENT FIELD PROBLEMS USING A REDUCED VARIATIONAL FORMULATION	69-0504
KIDDER, R. L.	QUADRILATERAL FINITE ELEMENT FOR PLANE STRESS AND PLANE STRAIN	70-0496
KIDYBINSKI, A.	STRESS DISTRIBUTION AND ROCK FRACTURE ZONES IN THE ROOF OF LONGWALL FACE IN COAL MINE	73-0247
KIEFER, B. V.	THREE-DIMENSIONAL STRESS ANALYSIS OF A FINITE SLAB CONTAINING A TRANSVERSE CENTRAL CRACK	75-0732
KIERNER, G.	OSCILLATION STABILITY OF INTERNALLY AND EXTERNALLY NOTCHED PLANE BARS OF EQUAL FORM FACTOR FROM 3943	73-0745
KIERSKY, L. B.	APPLICATION, TESTING, AND EVALUATION OF DOUBLY CURVED SHELL ELEMENTS FOR DYNAMIC ANALYSIS, PART III.	69-0256
KIESBAUER, H. T	DYNAN LECTURE NOTES WITH COMPUTATIONAL EXAMPLES	71-0581
KIESS, D. W.	A CURVILINEAR SHELL FINITE ELEMENT	69-0153
KIHARA, T.	A DIFFUSION ANALYSIS OF ION IN ELECTROLYSIS BY FINITE ELEMENT METHOD	73-0296
KIKUCHI, F.	A FINITE ELEMENT METHOD FOR INITIAL VALUE PROBLEMS	71-0145
KIKUCHI, F.	A FINITE ELEMENT METHOD FOR FRIEDRICH'S SYMMETRIC POSITIVE SYSTEMS	71-0146
KIKUCHI, F.	APPLICATION OF FINITE ELEMENT METHOD TO AXISYMMETRIC BUCKLING OF SHALLOW SPHERICAL SHELLS UNDER EXTE	73-0111

56

KIKUCHI, F.	APPLICATION OF SIMPLIFIED HYBRID DISPLACEMENT METHOD TO PLATE AND SHELL PROBLEMS	73-1037
KIKUCHI, F.	CONVERGENCE OF A MIXED FINITE ELEMENT SCHEME FOR PLATE BENDING	73-0023
KIKUCHI, F.	CONVERGENCE OF LUMPED FINITE ELEMENT SCHEMES FOR SELECTED INITIAL VALUE PROBLEMS	73-1026
KIKUCHI, F.	CONVERGENCE OF SIMPLIFIED HYBRID DISPLACEMENT METHOD FOR PLATE BENDING	72-0412
KIKUCHI, F.	FINITE ELEMENT ANALYSIS OF VIBRATION OF THIN ELASTIC PLATES BY SIMPLIFIED HYBRID DISPLACEMENT METHOD	73-0565
KIKUCHI, F.	FINITE ELEMENT METHOD FOR PLATE BENDING ANALYSIS BY DECOMPOSITION OF DIFFERENTIAL OPERATOR	71-0558
KIKUCHI, F.	FINITE ELEMENT METHOD FOR NON-SELF-ADJOINT PROBLEMS	73-0271
KIKUCHI, F.	FINITE ELEMENT SOLUTIONS FOR PLATE BENDING PROBLEMS BY SIMPLIFIED HYBRID DISPLACEMENT METHOD	72-0780
KIKUCHI, F.	FINITE-ELEMENT SCHEME BASED ON DISCRETE KIRCHHOFF ASSUMPTION	75-0544
KIKUCHI, F.	LUMPED FINITE ELEMENT BASES FOR BEAMS AND PLATES	72-0263
KIKUCHI, F.	NEW VARIATIONAL FUNCTIONAL FOR THE FINITE-ELEMENT METHOD AND ITS APPLICATION TO PLATE SHELL PROBLEMS	72-0108
KIKUCHI, F.	ON THE CONVERGENCE OF FINITE ELEMENT SCHEMES BASED ON NON-CONFORMING AND SIMPLIFIED HYBRID DISPLACEM	73-0841
KIKUCHI, F.	ON THE VALIDITY OF AN APPROXIMATION AVAILABLE IN THE FINITE ELEMENT SHELL ANALYSIS	75-0663
KIKUCHI, F.	ON THE VALIDITY OF THE FINITE ELEMENT ANALYSIS OF CIRCULAR ARCHES REPRESENTED BY AN ASSEMBLAGE OF BE	75-0479
KIKUCHI, F.	RECTANGULAR FINITE ELEMENT FOR PLATE BENDING ANALYSIS BASED ON HELLINGER-REISSNERS VARIATIONAL PRINC	71-0545
KIKUCHI, F.	SHORT TERM DEFLECTIONS OF CONCRETE FLAT PLATES	73-0277
KIKUCHI, F.	SIMPLIFIED HYBRID DISPLACEMENT METHOD APPLIED TO PLATE BUCKLING PROBLEMS	72-0781
KIKUSHITA, F.	SIMPLIFIED HYBRID DISPLACEMENT METHOD FOR LINEAR FINITE ELEMENT ANALYSIS OF GENERAL SHELLS	74-0614
KIKUCHI, F.	SOME CONSIDERATIONS ON THE CONVERGENCE OF HYBRID STRESS METHOD	73-0303
KIKUCHI, F.	SOME FINITE ELEMENT SOLUTIONS FOR PLATE BENDING PROBLEMS BY SIMPLIFIED HYBRID DISPLACEMENT METHOD	72-0268
KIKUCHI, F.	STRESS DISTRIBUTIONS IN THIN-WALLED INTERSECTING CYLINDRICAL SHELLS SUBJECTED TO INTERNAL PRESSURE A	71-0023
KIKUCHI, F.	THE HYBRID DISPLACEMENT METHOD APPLIED TO PLATE AND SHELL PROBLEMS	75-0089
KIKUCHI, F.	THEORY AND EXAMPLES OF PARTIAL APPROXIMATION IN FINITE-ELEMENT METHOD	75-0353
KIKUCHI, F.	TRIANGULAR FINITE ELEMENT FOR PLATE BENDING ANALYSIS WITH TRANSVERSE SHEAR DEFORMATION	71-0555
KIM, H. W.	A STUDY OF FREE-VIBRATIONS OF THIN ELASTIC SHELLS SUBJECTED TO INITIAL LOADS USING FINITE-ELEMENTS	73-1143
KIM, H. W.	ASYMMETRICAL BENDING AND VIBRATION OF A CONICAL SHELL FINITE ELEMENT	74-0085
KIM, L. P.	RASCHIOT USTOICHIVOSTI IZGIBNO-KRUTILNYKH KOLEBANIJ KPULA V DOZVUKOVOM POTOKE METODOM KONECHNYKH ELE	72-0903
KIM, L. P.	STABILITY ANALYSIS OF THE TORSIONAL BENDING VIBRATIONS OF WING IN A SUBSONIC FLOW BY METHOD OF THE F	75-0755
KIM, Y. J.	AN EFFICIENT ITERATIVE PROCEDURE FOR USE WITH THE FINITE-ELEMENT METHOD	73-1145
KIM, Y. S.	CORRELATION BETWEEN MODEL STUDY AND FINITE ELEMENT ANALYSIS IN AN INVESTIGATION OF SLOPE STABILITY	72-0227
KIMSCH, M.	FRACTURE AND SAFETY ANALYSIS OF NUCLEAR PRESSURE VESSELS	73-0257
KIMURA, L. A.	DIGITAL COMPUTER PROGRAMS FOR STEADY-STATE OR TRANSIENT TEMPERATURE ANALYSIS OF PLANE AND AXISYMMETR	72-0915
KIMURA, N.	AN APPROXIMATE NUMERICAL SOLUTION OF THE NAVIER STOKES EQUATIONS BY GALERKIN METHOD	74-0011
KING, G. J. W.	FINITE ELEMENT SOLUTION FOR QUANTITY OF STEADY SEEPAGE	71-0824
KING, G. J. W.	INTERACTIVE ANALYSIS OF A RAFTED MULTISTOREY SPACE FRAME RESTING ON AN INHOMOGENEOUS CLAY STRATUM	74-0447
KING, I. P.	A FINITE ELEMENT MODEL FOR TWO DIMENSIONAL FLOW	74-0015
KING, I. P.	A FINITE ELEMENT SOLUTION FOR TWO-DIMENSIONAL DENSITY STRATIFIED FLOW	73-0911
KING, I. P.	A FINITE ELEMENT SOLUTION FOR TWO-DIMENSIONAL STRATIFIED FLOW PROBLEMS	75-0391
KING, I. P.	A NUMERICAL METHOD OF VISCO-ELASTIC STRESS ANALYSIS	68-0175
KING, I. P.	AN AUTOMATIC RECORDING SCHEME FOR SIMULTANEOUS EQUATIONS DERIVED FROM NETWORK SYSTEMS	70-0285
KING, I. P.	ARCH DAMS ANALYSED BY A LINEAR FINITE ELEMENT SHELL THE FINITE STRIP METHOD IN THE ANALYSIS OF ELAST	68-0189
KING, I. P.	AUTOMATIC EMBANKMENT ANALYSIS. EQUILIBRIUM AND INSTABILITY CONDITIONS	66-0015
KING, I. P.	ELASTIC-PLASTIC ANALYSIS OF TWO-DIMENSIONAL STRESS SYSTEMS BY THE FINITE ELEMENT METHOD	67-0049
KING, I. P.	ELASTO-PLASTIC SOLUTIONS OF ENGINEERING PROBLEMS ' INITIAL STRESS ' FINITE ELEMENT APPROACH	69-0282
KING, I. P.	FINITE ELEMENT ANALYSIS OF TWO-DIMENSIONAL TIME-DEPENDENT STRESS PROBLEMS	65-0022
KING, I. P.	FINITE ELEMENT SOLUTION SYSTEMS (FESS).	67-0098
KING, I. P.	GENERAL METHOD OF ANALYSIS BASED ON FINITE ELEMENTS	68-0277
KING, I. P.	ON THE FINITE ELEMENT ANALYSIS OF TWO-DIMENSIONAL STRESS PROBLEMS WITH TIME DEPENDENT PROPERTIES	65-0052
KING, I. P.	SLAB BRIDGES WITH ARBITRARY SHAPE AND SUPPORT CONDITIONS - A GENERAL METHOD OF ANALYSIS BASED ON FIN	68-0233
KING, I. P.	STRESS ANALYSIS OF ROCK AS A NO-TENSION MATERIAL	68-0174
KING, I. P.	TRIANGULAR MESH GENERATION PROGRAM	67-0127
KING, I. P., ET	FINITE ELEMENT MODEL FOR TWO-DIMENSIONAL DENSITY STRATIFIED FLOW	73-0064
KING, J. T.	QUASIOPTIMAL FINITE-ELEMENT METHOD FOR ELLIPTIC INTERFACE PROBLEMS	75-0427
KING, P. H.	NONLINEAR CONSIDERATION OF GRAVITY IN A STIFFNESS TEST OF A WEAK STRUCTURE AT SMALL STRAINS	71-0035
KING, R. G.	A FINITE ELEMENT ANALYSIS OF WAVE PROPAGATION IN HUMAN SPINE	73-0792
KING, W. W.	ELASTODYNAMICS OF CRACKED STRUCTURES USING FINITE ELEMENTS	74-0625
KINITAKE, M.	ANALYSIS OF SLOPE STABILITY BY FINITE ELEMENT METHOD AND ITS APPLICATION PROGRAM. SOIL MECHANICS	72-0919
KINOSHITA, H.	ANALYSIS OF DISLOCATION STRESS FIELD BY FINITE ELEMENT METHOD	75-0862
KINSNER, W.	ITERATIVE APPROACH TO FINITE-ELEMENT METHOD IN FIELD PROBLEMS	74-0124
KIRCHHOFF, L. W	COMPUTER GRAPHICS FOR 3-D FINITE ELEMENT MODELS	74-0290
KIRICHEVSKII, V	INVESTIGATION OF THE STRESS-STRAINED STATE OF SPHERICAL SHELLS WITH AN ECCENTRIC CUTOUT ON THE BASIS	73-0222
KIRIOKA, K.	ANALYSIS OF BODY STRUCTURES - 2.	71-0339
KIRIOKA, K.	COMPUTING SYSTEM FOR STRUCTURAL ANALYSIS OF CAR BODIES	72-0324
KIRIOKA, K.	ELASTO-PLASTIC ANALYSIS OF AUTOMOBILE BODY STRUCTURE BY THE FINITE ELEMENT METHOD	74-0196
KIRKHOPE, J.	DISCUSSION OF "FINITE ELEMENT BENDING ANALYSIS OF REISSNER PLATES"	71-0832
KIRKHOPE, J.	VIBRATION ANALYSIS OF AXIAL FLOW TURBINE DISKS USING FINITE ELEMENTS	75-0195
KIRKHOPE, J.	VIBRATION ANALYSIS OF AXIAL-FLOW TURBINE DISKS USING FINITE-ELEMENTS	75-0415
KIRKHOPE, J.	VIBRATION OF CIRCULAR AND ANNULAR PLATES USING FINITE ELEMENTS	72-0194
KIRKLAND, J. L.	DYNAMIC FINITE ELEMENT ANALYSIS OF AXI-SYMMETRIC STRESS WAVE PROPAGATION IN SOIL-FILLED BIN	72-0622
KIRKLAND, J. L.	FINITE ELEMENT ANALYSIS OF BURIED CYLINDERS	71-0817
KIRKLAND, J. L.	FUNDAMENTAL STUDIES OF MEDIUM-STRUCTURE INTERACTION. REPORT 1. FINITE ELEMENT ANALYSIS OF BURIED CYL	72-0854
KIRKWOOD, W. F.	CALCULATING STRESSES IN A PRESSURE VESSEL WITH FILL TUBE - COMPARISON OF THE FINITE ELEMENT AND PHOT	72-0782
KIRWAN, R. W.	EXPERIMENTAL AND THEORETICAL INVESTIGATION OF PAVEMENT DEFLECTIONS	69-0154
KISHORE, B. R.	CHARACTERISTICS OF FINITE-ELEMENTS FOR ANALYSIS OF BEAMS SUBJECTED TO IMPACT LOAD	73-1144
KITAGAWA, H.	AN INCREMENTAL FINITE ELEMENT ANALYSIS OF TWO DIMENSIONAL LARGE STRAIN AND LARGE DISPLACEMENT PROBLE	72-0338
KITAGAWA, H.	INCREMENTAL THEORY OF LARGE STRAIN AND LARGE DISPLACEMENT PROBLEMS AND ITS FINITE ELEMENT FORMULATIO	72-0274
KITAGAWA, H.	SOME EXAMINATIONS ON AVAILABILITY OF FINITE ELEMENT METHODS IN LINEAR FRACTURE MECHANICS	73-0112
KITIPORNCHAI, S	BUCKLING OF INELASTIC I-BEAMS UNDER MOMENT GRADIENT	74-0765
KIUSALAAS, J.	OPTIMAL DESIGN OF STRUCTURES WITH BUCKLING CONSTRAINTS	73-0246
KIYONO, T.	NUMERICAL SOLUTION OF PLATEAUS PROBLEM BY A FINITE ELEMENT METHOD	74-0119
KLAMECKI, B. E.	INCIPIENT CHIP FORMATION IN METAL CUTTING - A THREE- DIMENSION FINITE-ELEMENT ANALYSIS	73-1146
KLAWITTER, J. J	FINITE ELEMENT STRESS ANALYSIS OF PROSTHETIC TOOTH IMPLANTS	75-0073
KLEEMANN, P. H.	THE GENERALIZED INVERSE - SOME PROPERTIES. A COMPUTATIONAL ALGORITHM AND A STRUCTURAL APPLICATION	68-0248
KLEIBER, M.	FUNKCJA KSZTALTU NIESPELNIAPACA WARUNKOW CIAGLOSCI W ANALIZIE PLYT METODA ELEMENTOW SKONCZONYCH. (SH	75-0263
KLEIBER, M.	LAGRANGIAN AND EULERIAN FINITE ELEMENT FORMULATION FOR LARGE STRAIN ELASTO-PLASTICITY	75-0209
KLEIN, L.	TRANSVERSE VIBRATIONS OF NON-UNIFORM BEAMS	74-1058
KLEIN, S.	A STATIC AND DYNAMIC FINITE ELEMENT SHELL ANALYSIS WITH EXPERIMENTAL VERIFICATION	69-0156
KLEIN, S.	A STUDY OF THE MATRIX DISPLACEMENT METHOD AS APPLIED TO SHELLS OF REVOLUTION	66-0111
KLEIN, S.	APPLICATION OF MATRIX DISPLACEMENT METHOD TO LINEAR ELASTIC ANALYSIS OF SHELLS OF REVOLUTION	65-0025
KLEIN, S.	DISCUSSION ON FINITE ELEMENT SOLUTION FOR AXISYMMETRICAL SHELLS	65-0062
KLEIN, S.	DYNAMIC ANALYSIS OF A COMPLEX SHELL STRUCTURE UNDER IMPULSIVE ASYMMETRIC LOADING	68-0226
KLEIN, S.	MATRIX ANALYSIS OF SHELL STRUCTURES	64-0003
KLEIN, S.	NONLINEAR DYNAMIC ANALYSIS OF SHELLS OF REVOLUTION WITH ASYMMETRIC PROPERTIES BY THE FINITE ELEMENT	75-0217
KLEIN, S.	NONLINEAR DYNAMIC ANALYSIS OF SHELLS OF REVOLUTION WITH ASYMMETRIC PROPERTIES BY THE FINITE ELEMENT	75-0657
KLEIN, S.	SABOR 1, A FORTRAN PROGRAM FOR THE LINEAR ELASTIC AXISYMMETRIC LOADING BY USING THE MATRIX DISPLACEM	65-0027
KLEIN, S.	STATIC ANALYSIS OF A COMPLEX SHELL STRUCTURE UNDER ASYMMETRIC LOADING	67-0120
KLEIN, S.	STRUCTURAL ANALYSIS AND EXPERIMENTAL CORRELATION OF A COMPLEX SHELL UNDER ASYMMETRIC LOADS	69-0155
KLEIN, S.	THE LINEAR ELASTIC DYNAMIC ANALYSIS OF SHELLS OF REVOLUTION BY THE MATRIX DISPLACEMENT METHOD	66-0046
KLEIN, S.	THE NONLINEAR DYNAMIC ANALYSIS OF SHELLS OF REVOLUTION WITH ASYMMETRIC PROPERTIES BY THE FINITE ELEM	71-0818
KLEIN, S. A.	A STATIC AND DYNAMIC FINITE ELEMENT SHELL ANALYSIS WITH EXPERIMENTAL VERIFICATION	71-0769
KLINEBERG, J. M	THE NUMERICAL CALCULATION OF LAMINAR BOUNDARY-LAYER SEPARATION	74-0785
KLINGNER, R.	ELASTIC LATERAL BUCKLING OF STEEL BEAMS	70-0364

KLINK, W. H.	FINITE AND DISCONNECTED SUBGROUPS OF SU(3) AND THEIR APPLICATION TO THE ELEMENTARY PARTICLE SPECTRUM	66-0129
KLOSTERMAN, A.	COMBINING EXPERIMENTAL AND ANALYTICAL TECHNIQUES FOR DYNAMIC SYSTEM ANALYSIS	73-0321
KLOSTERMAN, A.	DYNAMIC SIMULATION OF COMPLEX SYSTEMS UTILIZING EXPERIMENTAL AND ANALYTICAL TECHNIQUES	75-0864
KLOSTERMAN, A.	USING NASTRAN FOR DYNAMIC ANALYSIS OF VEHICLE SYSTEMS	74-0316
KLOTZ, J. A.	EFFECT OF PERFORATION DAMAGE ON WELL PRODUCTIVITY	74-0936
KNAPP, R.	NUMERICAL ANALYSIS OF COOLING TOWER PLUMES AND FOG INCIDENCE	74-0040
KNAPP, R. H.	FINITE-ELEMENT NONLINEAR BUCKLING ANALYSIS OF A PSEUDO-CYLINDRICAL CONCAVE POLYHEDRAL SHELL UNDER EX	73-1147
KNAPP, R. H.	NUMERICAL AND EXPERIMENTAL ANALYSIS OF A PSEUDOCYLINDRICAL SHELL	74-1133
KNAUSS, W. G.	CASE LINER BOND ANALYSIS	74-0817
KNIGHT, C. J.	A VARIATIONAL FORMULATION OF THE COMPRESSIBLE THROUGHFLOW PROBLEM	75-0021
KNIGHT, W. A.	INTERNATIONAL MACHINE TOOL DESIGN AND RESEARCH CONFERENCE, 13TH PROCEEDINGS, 1972	73-0504
KNITTEL, M. R.	A FINITE ELEMENT APPROACH TO SCATTERING FROM ELASTIC SPHERICAL SHELLS	74-0702
KNITTEL, M. R.	A FINITE-ELEMENT APPROACH TO THE ANALYSIS OF TANGENTIALLY POLARIZED PIEZOELECTRIC-CERAMIC FREE-FLOOD	74-0802
KNITTEL, M. R.	FINITE ELEMENT ANALYSIS OF ACOUSTICALLY RADIATING STRUCTURES WITH APPLICATIONS TO SONAR TRANSDUCERS	74-0524
KNITTEL, M. R.	FINITE ELEMENT APPROACH TO ACOUSTIC RADIATION FROM ELASTIC STRUCTURES	74-0121
KNITTEL, M. R.	FINITE-ELEMENT APPROACH TO ACOUSTIC SCATTERING FROM ELASTIC STRUCTURES	75-0069
KNOELL, A. C.	MATHEMATICAL MODELLING AND STUCTURAL ANALYSIS OF THE MANDIBLE	72-0124
KNOEPFEL, H.	FINITE ELEMENT LIMIT ANALYSIS USING LINEAR PROGRAMMING	72-0461
KNOTHE, K.	VARIATIONAL PRINCIPLES OF ELASTOSTATICS WITH RELAXED CONTINUITY REQUIREMENTS AND THEIR APPLICABILITY	74-0758
KNOWLES, N. C.	THE FINITE ELEMENT METHOD APPLIED TO THE STATIC AND DYNAMIC BUCKLING OF CIRCULAR ARCHES	70-0497
KNUDSON, R. M.	PERFORMANCE OF STRUCTURAL WOOD MEMBERS EXPOSED TO FIRE	75-0060
KNUTELSKY, B. G	FINITE ELEMENT STRESS ANALYSIS OF ARTILLERY PROJECTILES	73-0432
KO, W. L.	APPLICATION OF FINITE ELASTICITY THEORY TO THE DEFORMATION OF RUBBERS	62-0005
KOBAYASHI, A. S	A DYNAMIC PHOTOELASTIC ANALYSIS OF DYNAMIC TEAR TEST SPECIMEN	74-0736
KOBAYASHI, A. S	A MATHEMATICAL MODEL OF THE CORNEO-SCLERAL SHELL	71-0150
KOBAYASHI, A. S	A NUMERICAL AND EXPERIMENTAL INVESTIGATION OF THE USE OF J-INTEGRALS	72-0858
KOBAYASHI, A. S	AN INVESTIGATION ON THE PLASTICITY MAGNIFICATION FACTOR IN DEEP SURFACE FLOWS	74-0766
KOBAYASHI, A. S	ANALYSES OF AXISYMMETRIC UPSETTING AND PLANE STRAIN SIDE PRESSING OF SOLID CYLINDERS BY THE FINITE E	70-0083
KOBAYASHI, A. S	ANALYSIS OF BIOLOGICAL STRUCTURES	71-0148
KOBAYASHI, A. S	ANALYSIS OF THE CORNEO-SCLERAL SHELL BY THE METHOD OF DIRECT STIFFNESS	71-0151
KOBAYASHI, A. S	ANALYTICAL PREDICTION OF DEFECTS OCCURRENCE IN SIMPLE AND COMPLEX	69-0157
KOBAYASHI, A. S	APPLICATION OF AN ASSUMED DISPLACEMENT HYBRID FINITE ELEMENT PROCEDURE TO TWO-DIMENSIONAL PROBLEMS I	74-0729
KOBAYASHI, A. S	APPLICATION OF FINITE ELEMENT ANALYSIS METHOD TO TWO-DIMENSIONAL PROBLEMS IN FRACTURE MECHANICS	69-0033
KOBAYASHI, A. S	APPLICATION OF THE DIRECT STIFFNESS METHOD TO PROBLEMS INVOLVING LARGE, TIME-DEPENDENT DEFORMA	66-0108
KOBAYASHI, A. S	APPLICATION OF THE METHOD OF FINITE ELEMENT ANALYSIS TO TWO-DIMENSIONAL PROBLEMS IN FRACTURE MECHANI	68-0100
KOBAYASHI, A. S	ASSUMED DISPLACEMENT HYBRID FINITE ELEMENT MODEL FOR LINEAR FRACTURE MECHANICS	75-0146
KOBAYASHI, A. S	BRAIN TISSUE FRAGILITY - A FINITE STRAIN ANALYSIS BY A HYBRID FINITE-ELEMENT METHOD	75-0162
KOBAYASHI, A. S	BRAIN TISSUE FRAGILITY - A FINITE STRAIN ANALYSIS BY A HYBRID FINITE-ELEMENT METHOD	75-0243
KOBAYASHI, A. S	BRAIN-TISSUE FRAGILITY - FINITE STRAIN ANALYSIS BY A HYBRID FINITE-ELEMENT METHOD	75-0631
KOBAYASHI, A. S	CRACK BRANCHING IN HOMALITE-100 SHEETS	72-0843
KOBAYASHI, A. S	CRACK BRANCHING IN HOMALITE-100 SHEETS	74-0291
KOBAYASHI, A. S	CRACK PROPAGATION AND ARREST IN IMPACTED PLATES	72-0599
KOBAYASHI, A. S	DEFORMATION OF THE ARTERIAL VASA VASORUM AT NORMAL AND HYPERTENSIVE ARTERIAL PRESSURE	71-0149
KOBAYASHI, A. S	ELASTOPLASTIC ANALYSIS OF PLANE STRAIN AND AXISYMMETRIC FLAT PUNCH INDENTATION BY THE FINITE ELEMENT	70-0343
KOBAYASHI, A. S	FINITE ELEMENT ANALYSIS AND EXPERIMENTAL INVESTIGATION OF SHEET METAL STRETCHING	72-0229
KOBAYASHI, A. S	FINITE ELEMENT LARGE DEFORMATION ANALYSIS OF A TAPERED AORTA	75-0856
KOBAYASHI, A. S	FRACTURE DYNAMICS OF HOMALITE-100 SHEETS	72-0809
KOBAYASHI, A. S	FRACTURE MECHANICS APPLICATION OF AN ASSUMED DISPLACEMENT HYBRID FINITE ELEMENT PROCEDURE	75-0212
KOBAYASHI, A. S	MATHEMATICAL MODEL OF THE CORNEO-SCLERAL SHELL AS APPLIED TO INTRAOCULAR PRESSURE VOLUME RELATIONS A	72-0471
KOBAYASHI, A. S	MECHANICS OF BRAIN TISSUE FRAGILITY	73-0214
KOBAYASHI, A. S	MECHANICS OF MACKAY-MARG TONOMETER	73-0433
KOBAYASHI, A. S	NUMERICAL AND EXPERIMENTAL INVESTIGATION ON THE USE OF J-INTEGRAL	73-0245
KOBAYASHI, A. S	PHOTOELASTIC INVESTIGATION ON THE CRACK-ARREST CAPABILITY OF A HOLE	72-0086
KOBAYASHI, A. S	PHOTOELASTIC INVESTIGATION ON THE CRACK-ARREST CAPABILITY OF A PRETENSIONED STIFFENED PLATE	75-0088
KOBAYASHI, A. S	STRESS INTENSITY FACTOR FOR A STRAIGHT CRACK APPROACHING A HOLE	69-0158
KOBAYASHI, A. S	STRESS INTENSITY FACTOR OF A SURFACE CRACK IN A PRESSURIZED CYLINDER	75-0255
KOBAYASHI, A. S	STRUCTURAL ANALYSIS OF AN ARTERIOLE BY THE DIRECT STIFFNESS METHOD	66-0109
KOBAYASHI, A. S	STRUCTURAL RESPONSE OF RELAXED AND CONSTRICTED ARTERIOLES	69-0200
KOBAYASHI, A. S	THE EFFECT OF END RESTRAINT CONDITION ON THE FRACTURE MECHANISM AND STRENGTH OF ROCK-LIKE MATERIALS	72-0134
KOBAYASHI, A. S	THREE-DIMENSIONAL LINEAR FRACTURE MECHANICS ANALYSIS BY A DISPLACEMENT-HYBRID FINITE-ELEMENT MODEL	75-0884
KOBAYASHI, A. S	VISCOELASTIC PRESPONSE OF THE CORNEO-SCLERAL SHELL UNDER TONOMETER LOADING	75-0855
KOBAYASHI, S.	DEFORMATION MODE OF VOID-GROWTH AND COALESCENCE IN THE PROCESS OF DUCTILE FRACTURE	75-0683
KOBISKE, M.	FINITE ELEMENT ANALYSIS OF UNSTEADY INCOMPRESSIBLE FLOW AROUND AN OSCILLATING OBSTACLE OF ARBITRARY	73-0067
KOBISKE, M.	NUMERICAL CALCULATIONS OF VELOCITY AND PRESSURE DISTRIBUTION AROUND OSCILLATING AIRFOILS	74-0847
KOBORI, K.	DYNAMIC RESPONSE ANALYSIS OF THE IRREGULAR GROUND BY MEANS OF THE FINITE ELEMENT METHOD	75-0860
KOBZA, W.	METODA ELEMENTOW SKONSZONYCH ROZWIAZWANIA OSIOWO-SYMETRYCZNEGO PRZEWODNICTWA CIEPLA (FINITE-ELEMENT	73-0434
KOBZA, W.	MULTIANGULAR ELEMENTS IN PLATE BENDING	73-0435
KOEHLER, W.	BERECHNUNG INSTATIONAERER TEMPERATUREFELDER MIT FINITEN ELEMENTEN IN RAUM-UND ZEITDIMENSION. (CALCUL	74-0987
KOEHLER, W.	CALCULATION OF TRANSIENT TEMPERATURE FIELDS WITH FINITE ELEMENTS IN SPACE AND TIME DIMENSIONS	74-0379
KOENEMAN, J. B.	BENDING PROBLEM OF AXIALLY CONSTRAINED BEAMS	73-0397
KOENIG, H. A.	DOUBLE STRESS-WAVE DISCONTINUITIES IN FINITE, SHEAR-CORRECTED BEAMS AND PLATES	68-0210
KOENIG, H. A.	DYNAMICAL FINITE ELEMENT ANALYSIS FOR ELASTIC WAVES IN BEAMS AND PLATES	68-0211
KOENIG, H. A.	ELASTO-PLASTIC CYCLIC ANALYSIS OF STRUCTURAL MEMBERS	72-0202
KOENIG, H. A.	MULTIPLE STRESS WAVE DISCONTINUITIES IN FINITE BEAMS AND PLATES	67-0114
KOENIG, H. A.	THE DAMPED TRANSIENT BEHAVIOR OF FINITE BEAMS AND PLATES	69-0520
KOENIG, H. A.	TRANSIENT RESPONSE OF NON-UNIFORM, NON-HOMOGENEOUS BEAMS	73-0162
KOENIG, H. A.	TRANSIENT RESPONSE OF ELASTIC, VISCO-PLASTIC BEAMS	73-0380
KOHNKE, P. C.	A FINITE ELEMENT ANALYSIS OF ECCENTRICALLY STIFFENED-CIRCULAR CYLINDRICAL SHELLS	69-0159
KOHNKE, P. C.	ANALYSIS OF ECCENTRICALLY STIFFENED CYLINDRICAL SHELLS	72-0217
KOICHI, A.	FINITE ELEMENT METHOD WITH CONSIDERATION TO THE CONTINUITY OF NORMAL SLOPE	72-0813
KOKUSHO, S.	TWO DIMENSIONAL NON LINEAR ANALYSIS OF REINFORCED CONCRETE MEMBERS BY THE FINITE ELEMENT METHOD, PAR	71-0210
KOLAR, V.	DESIGN OF TWO AND THREE DIMENSIONAL STRUCTURES BY THE FINITE ELEMENT METHOD (IN GERMAN)	74-1364
KOLAR, V.	EFFICIENT FINITE ELEMENT ANALYSIS OF RECTANGULAR AND SKEW LAMINATED PLATES	73-0436
KOLAR, V.	INFLUENCE FUNCTIONS IN FINITE ELEMENT METHOD	70-0226
KOLAR, V.	STRESS ANALYSIS OF TWO AND THREE DIMENSIONAL STRUCTURES BY THE FINITE ELEMENT METHOD	72-0794
KOLAR, V.	TECHNICAL, PHYSICAL AND MATHEMATICAL PRINCIPLES OF THE FINITE ELEMENT METHOD	71-0676
KOLAR, V.	THE INFLUENCE OF DIVISION ON THE RESULTS IN THE FINITE ELEMENT METHOD	70-0447
KOLATA, G. B.	FINITE ELEMENT METHOD - MATHEMATICAL REVIVAL	74-0169
KOMORIDA, H.	APPLICATION OF FINITE-ELEMENT METHOD AND MOIRE TECHNIQUE TO SHEET-METAL FORMING PROBLEMS	71-0784
KOMORIYA, H.	ENERGY-DEPENDENT FINITE-ELEMENT METHOD FOR FEW-GROUP DIFFUSION EQUATIONS	75-0499
KONDERLA, P.	SOLUTION OF A SINGLE HYPERBOLOID SHELL BY FINITE ELEMENT METHOD	73-0151
KONG, F. K.	MODIFIED FINITE ELEMENT METHOD APPLIED TO R C DEEP BEAMS	73-0476
KONIECZNY, S.	MULTI-ELEMENT SPATIAL STRUCTURES	73-0768
KONIG, M.	ELASTOPLASTIC AND CREEP ANALYSIS WITH THE ASKA PROGRAM SYSTEM	74-0863
KONIG, M.	GEOMETRIC NONLINEARITY AND THE FINITE ELEMENT DISPLACEMENT METHOD	74-0679
KONO, I.	ANALYSIS OF INTERFACE PROBLEM IN GROUNDWATER FLOW BY FINITE ELEMENT METHOD	74-0467
KONO, I.	ANALYSIS OF THE SEEPAGE THROUGH AN EARTH DAM BY FINITE ELEMENT METHOD	72-0585
KONO, I.	FINITE ELEMENT ANALYSIS OF INTERFACE PROBLEM IN GROUND-WATER FLOW	74-0497
KONO, I.	FINITE ELEMENT ANALYSIS OF A NONSTEADY GROUNDWATER SEEPAGE PROBLEM DAM, DRAINAGE	74-1375
KONRAD, A.	ANALYSIS OF TRANSFORMER LEAKAGE PHENOMENA BY HIGH-ORDER FINITE ELEMENTS	73-0135
KONRAD, A.	AXISYMMETRIC TRIANGULAR ELEMENT FOR THE SCALAR HELMHOLTZ EQUATION	73-0688
KONRAD, A.	FINITE ELEMENT PROGRAM PACKAGE FOR AXISYMMETRIC SCALAR FIELD PROBLEMS	73-0113

KONRAD, A.	FINITE ELEMENT PROGRAM PACKAGE FOR AXISYMMETRIC VECTOR FIELD PROBLEMS	75-0138
KONRAD, A.	LINEAR ACCELERATOR CAVITY FIELD CALCULATION BY FINITE ELEMENT	73-0008
KONRAD, A.	SCALAR FINITE-ELEMENT PROGRAM PACKAGE FOR TWO-DIMENSIONAL FIELD PROBLEMS	71-0478
KONRAD, A.	TRIANGULAR FINITE ELEMENTS FOR THE GENERALIZED BESSEL EQUATION OF ORDER M	73-1094
KONRATH, E. J.	SUBSTRUCTURE ANALYSIS TECHNIQUES AND AUTOMATION	73-0597
KOOPMANN, G. H.	BUILDING VIBRATIONS AND ACOUSTICS	73-0656
KOPLICK, B.	FINITE ELEMENT APPROACH TO POLYMER CONCRETE BRIDGE DECK DESIGN ANALYSIS	72-0918
KOPLIK, B.	ANALYSIS OF BRIDGE DECKS USING POLYMER IMPREGNATED CONCRETE	73-0673
KORMAN, T.	BENDING OF MICROPOLAR PLATES	74-0263
KORMAN, T.	BOUSSINESQ PROBLEM OF PLANE MICROPOLAR ELASTICITY	74-0197
KORMAN, T.	FINITE ELEMENT ANALYSIS OF A THICK WALLED MICROPOLAR CYLINDER LOADED AXISYMMETRICALLY	73-0847
KORMAN, T.	SHELL STRUCTURES AND CLIMATIC INFLUENCES	72-0399
KORNEEV, V. G.	CONCERNING THE FINITE ELEMENT METHOD FOR THE SOLUTION OF PROBLEMS OF ELASTIC EQUILIBRIUM	71-0516
KORNEEV, V. G.	THE COMPARISON OF THE FINITE ELEMENT METHOD WITH THE VARIATIONAL-DIFFERENCE METHOD IN THE THEORY OF	67-0096
KOSKO, E.	A HIGH PRECISION TRIANGULAR PLATE BENDING ELEMENT	68-0052
KOSKO, E.	FORMULATION OF A NEW TRIANGULAR PLATE BENDING ELEMENT	68-0051
KOSKO, E.	STATIC AND DYNAMIC APPLICATIONS OF A HIGH-PRECISION TRIANGULAR PLATE BENDING ELEMENT	69-0086
KOSKO, E.	STRUCTURAL DESIGN AND ANALYSIS OF THE SPACE SHUTTLE	70-0336
KOSSOWSKI, R.	APPLICATION OF THE FINITE ELEMENT METHOD FOR SOLVING ORDINARY LINEAR DIFFERENTIAL EQUATION WITH BOUN	74-0084
KOST, G.	NONLINEAR DYNAMIC ANALYSIS OF FRAMES WITH FILLER PANELS	74-0226
KOSTEM, C.	FINITE ELEMENT ANALYSIS OF PLATES AND ECCENTRICALLY STIFFENED PLATES	73-0691
KOSTEM, C. N.	ANALYTICAL MODELING OF BEAM-SLAB BRIDGES	74-1131
KOSTEM, C. N.	DYNAMIC ANALYSIS OF BEAM-SLAB HIGHWAY BRIDGES	73-0199
KOSTEM, C. N.	THE INELASTIC ANALYSIS OF REINFORCED AND PRESTRESSED CONCRETE BEAMS	72-0667
KOSTEM, C. N.	THE INELASTIC ANALYSIS OF REINFORCED CONCRETE SLABS	74-1333
KOSTEM, C. N.	USER'S MANUAL FOR PROGRAM BEAM	73-1105
KOSTEM, C. N.	USERS MANUAL FOR PROGRAM BEAM	73-0714
KOSTRO, C.	FINITE ELEMENT ANALYSIS OF HYPERBOLIC PARABOLOID SHELLS	71-0245
KOSTRO, G.	FINITE ELEMENT ANALYSIS OF HYPERBOLIC PARABOLOID SHELLS	72-0887
KOTANCHIK, J. J	DISCRETE-ELEMENT STATIC ANALYSIS OF BONDED, DOUBLE-LAYER, BRANCHED, THIN SHELLS OF REVOLUTION, PART	68-0102
KOTANCHIK, J. J	DISCRETE-ELEMENT STATIC ANALYSIS OF BONDED, DOUBLE-LAYER, BRANCHED, THIN SHELLS OF REVOLUTION, PART	70-0081
KOTANCHIK, J. J	PROGRESS REPORT ON DISCRETE-ELEMENT ELASTIC AND ELASTIC-PLASTIC ANALYSIS OF SHELLS OF REVOLUTION SUB	68-0170
KOTANCHIK, J. J	STACUSS 1: A DISCRETE-ELEMENT PROGRAM FOR THE STATIC ANALYSIS OF SINGLE-LAYER CURVED STIFFENED SHELL	69-0160
KOTANCHIK, J. J	THE TRANSIENT LINEAR ELASTIC RESPONSE ANALYSIS COMPLEX THIN SHELLS OF REVOLUTION SUBJECTED TO ARBITR	70-0080
KOTCHERGENKO, I	FREE-SURFACE IDEAL FLUID FLOWS BY FINITE ELEMENTS	74-0125
KOTERAS, J. R.	A HYBRID FINITE ELEMENT FOR DETERMINATION OF STRESS INTENSITY FACTORS AT A CRACK TIP	73-0643
KOTHAWALA, K. S	FINAL REPORT ON NASTRAN/GRAPHICS, 1971 FORD DOOR PROJECT	73-0595
KOUKAL, S.	PIECEWISE POLYNOMIAL INTERPOLATIONS AND THEIR APPLICATIONS TO PARTIAL DIFFERENTIAL EQUATIONS	70-0458
KOUKAL, S.	PIECEWISE POLYNOMIAL INTERPOLATIONS IN THE FINITE ELEMENT METHOD	73-1028
KOUNO, T.	A CONSIDERATION ON THE FINITE ELEMENT METHOD FOR CALCULATING STRESS INTENSITY FACTOR KI USING THE J	75-0863
KOVACS, W. D.	STUDIES OF SEISMIC RESPONSE OF CLAY BANKS	74-0833
KOVAL'CHUK, N.	ISSLEDOVANIE NAPRYAZHENNOGO SOSTOYANIYA REBRISTYKH ISILINDRICHESKIKH OBOLOCHEK S PRYAMOUGOL'NYMI OTV	74-0944
KOVAL'CHUK, N.	METOD KONECHNYKH ELEMENTOV V PRIMENENII K RASCHETU TSILINDRICHESKIKH OBOLOCHEK S PRYAMOUGOL'NYMI OTV	73-0536
KOVALCHUK, N. V	AN APPLICATION OF THE FINITE ELEMENT METHOD TO THE CALCULATION OF CYLINDRICAL SHELLS WITH RECTANGULA	74-1236
KOWALIK, J. S.	COMPUTING WITH SPARSE MATRICES	73-0926
KOZAKOFF, D. J.	THREE DIMENSIONAL NONLINEAR MAGNETIC FIELD BOUNDARY VALUE PROBLEM AND ITS NUMERICAL SOLUTION	70-0060
KOZLOV, I. A.	APPLICATION OF THE METHOD OF FINITE ELEMENTS TO CALCULATIONS OF THE STRESSED AND STRAINED STATE OF I	75-0775
KOZLOV, I. A.	METHOD OF FINITE ELEMENTS FOR CALCULATION OF HETEROGENEOUS BODIES OF REVOLUTION IN ELASTO-PLASTIC ME	75-0144
KRAFT, L. M.	ANALYTICAL EVALUATION OF STRESS-STRAIN TEST DATA	72-0442
KRAHULA, J. L.	A FINITE ELEMENT SOLUTION FOR SAINT-VENANT TORSION	69-0161
KRAHULA, J. L.	ANALYSIS OF BENT AND TWISTED BARS USING THE FINITE ELEMENT METHOD	67-0045
KRAHULA, J. L.	USE OF FOURIER SERIES IN THE FINITE ELEMENT METHOD	68-0103
KRAJCINOVIC, D.	A CONSISTENT DISCRETE ELEMENT TECHNIQUE FOR THIN-WALLED ASSEMBLAGES	68-0104
KRAJCINOVIC, D.	COUPLED VIBRATIONS OF THIN-WALLED BEAMS OF OPEN SECTION USING FINITE ELEMENT METHOD	71-0568
KRAJCINOVIC, D.	PERIODIC SHOCK EXCITATION OF ELASTIC STRUCTURES	69-0163
KRAJCINOVIC, D.	PRISMATIC SHELLS WITH HEXAGONAL CROSS-SECTION	74-1043
KRASNIKOV, N. D	RASCHET METODOM KONECHNYKH ELEMENTOV DINAMICHESKIKH KHARAKTEERISTIK ZEMLYANOI PLOTINY SOVMESTNO S OS	73-0437
KRATOCHVIL, J.	A SIMPLE ALGORITHM FOR THE STIFFNESS MATRIX OF TRIANGULAR PLATE BENDING ELEMENTS	71-0152
KRATOCHVIL, J.	CALCULATION OF PLATE OF CONSTANT THICKNESS BY FINITE ELEMENT METHOD	69-0227
KRATOCHVIL, J.	CALCULATION OF RECTANGULAR PLATES BY THE FINITE ELEMENT METHOD	69-0506
KRATOCHVIL, J.	DESIGN OF TWO AND THREE DIMENSIONAL STRUCTURES BY THE FINITE ELEMENT METHOD (IN GERMAN)	74-1364
KRATOCHVIL, J.	SOLUTION OF STRESSES AND TRANSFORMATIONS OF EARTH DAMS BY THE METHOD OF FINITE ELEMENTS	69-0426
KRATOCHVIL, J.	STRESS ANALYSIS OF TWO AND THREE DIMENSIONAL STRUCTURES BY THE FINITE ELEMENT METHOD	72-0794
KRATOCHVIL, J.	TECHNICAL, PHYSICAL AND MATHEMATICAL PRINCIPLES OF THE FINITE ELEMENT METHOD	71-0676
KRATOCHVIL, J.	THE FINITE ELEMENT METHOD	70-0327
KRATOCHVIL, J.	THREE-DIMENSIONAL ISOPARAMETRIC ELEMENTS AND THEIR USE FOR STATICAL ANALYSIS OF MASSIVE WATER-POWER	71-0364
KRAUS, H.	ANALYSIS OF BOTTOM ENTRY NOZZLES FOR ATMOSPHERIC STORAGE TANKS	75-0331
KRAUS, H.	SOFTWARE USER: EDUCATION AND QUALIFICATION	72-0441
KRAUSE, W.	THE MBB-UFE SUBSONIC PANEL METHOD. PART 2 - THE LIFT-CONNECTED DISPLACEMENT PROBLEM IN COMPRESSIBLE	72-0865
KREIG, R. D.	TRANSIENT SHELL RESPONSE BY NUMERICAL TIME INTEGRATION	72-0305
KREISS, H. O.	ADVANCES IN NUMERICAL FLUID DYNAMICS	73-0813
KRIEG, R. D.	COMPARISON OF FINITE ELEMENT AND FINITE DIFFERENCE METHODS	71-0143
KRILE, T. F.	FINITE ELEMENT ANALYSIS OF ELASTIC WAVE SCATTERING FROM DISCONTINUITIES	73-0547
KRILE, T. F.	2-DIMENSIONAL FINITE-ELEMENT ANALYSIS OF ELECTROELASTIC NORMAL MODES OF FERROELECTRIC DISKS AND SHEL	75-0594
KRILE, T. F.	3-DIMENSIONAL FINITE-ELEMENT ANALYSIS OF ELECTROELASTIC NORMAL MODES OF A TANGENTIALLY POLARIZED, SE	75-0591
KRIMERMA, Y.	NUMERICAL CALCULATION OF MERIDIONAL FLOW FIELD IN TURBOMACHINES USING FINITE ELEMENTS METHOD	74-0481
KRISHNA MURTY,	A COMPARATIVE STUDY OF THE CONSISTENT AND SIMPLIFIED FINITE ELEMENTS ANALYSIS OF EIGENVALUE PROBLEMS	70-0034
KRISHNA MURTY,	A FINITE ELEMENT ANALOGUE OF THE MODIFIED RAYLEIGH-RITZ METHOD FOR VIBRATION PROBLEMS	72-0478
KRISHNA MURTY,	A POWERFUL HYBRID METHOD IN FINITE ELEMENT ANALYSIS	71-0218
KRISHNA MURTY,	ASSESSMENT OF ACCURACIES OF FINITE ELEMENT EIGENVALUES	72-0075
KRISHNA MURTY,	ASSESSMENT OF ACCURACIES OF FINITE ELEMENT EIGENVALUES	72-0747
KRISHNA MURTY,	BOUNDS FOR EIGENVALUES IN SOME VIBRATION AND STABILITY PROBLEMS.	72-0390
KRISHNA MURTY,	FINITE ELEMENT ANALOGUE OF THE MODIFIED RAYLEIGH RITZ METHOD FOR VIBRATION PROBLEMS	73-0181
KRISHNA MURTY,	GALERKIN FINITE ELEMENT METHOD FOR VIBRATION PROBLEMS	73-0659
KRISHNA MURTY,	SECTOR ELEMENTS FOR MATRIX DISPLACEMENT ANALYSIS	73-1049
KRISHNA MURTY,	SPECIAL FINITE ELEMENTS FOR THE ANALYSIS OF STRESS CONCENTRATIONS AND SINGULARITIES	71-0219
KRISHNA, JAI	A SEISMIC STRENGTH OF KOLKEWADI DAM	73-0197
KRISHNAMURTHY,	BUCKLING OF OPEN CYLINDRICAL SHELLS	67-0115
KRISHNAMURTHY,	A THREE-DIMENSIONAL FINITE ELEMENT ANALYSIS OF A PRESTRESSED CONCRETE REACTOR VESSEL MODEL	69-0084
KRISHNAMURTHY,	ANALYTICAL EVALUATION OF STRESS-STRAIN TEST DATA.	72-0442
KRISHNAMURTHY,	ERROR ESTIMATES FOR CONFORMING FINITE ELEMENT SOLUTIONS	74-0940
KRISHNAMURTHY,	THEORETICAL AND NUMERICAL CONVERGENCE RATES FOR SOME CONFORMING FINITE ELEMENTS	74-0941
KRISHNAMURTHY,	THREE-DIMENSIONAL FINITE ELEMENT ANALYSIS OF THICK-WALLED PIPE-NOZZLE JUNCTIONS WITH CURVED TRANSITI	71-0153
KRIZEK, B. J.	MIMIC SOURCE APPROACH TO VISCOELASTIC ANALYSIS	69-0164
KRIZEK, R. J.	EVALUATION OF STRESS CELL PERFORMANCE	74-0664
KRIZEK, R. J.	GRAVITY FLOW TO EXCAVATIONS AND DRAINAGE TRENCHES IN LAYERED AQUIFERS	71-0014
KROENKE, W. C.	ANALYSIS OF NONLINEAR CONCRETE FRAMES USING SMALL GYPSUM MODELS AND THE FINITE ELEMENT METHOD	70-0498
KROENKE, W. C.	CLASSIFICATION OF FINITE ELEMENT STRESSES ACCORDING TO ASME SECTION 111 STRESS CATEGORIES	74-0610
KROENKE, W. C.	FINITE ELEMENT FOR REINFORCED CONCRETE FRAME STUDY	73-0204
KROENKE, W. C.	FINITE ELEMENT FOR REINFORCED CONCRETE FRAME STUDY - CLOSURE	75-0087
KROENKE, W. C.	INTERPRETATION OF FINITE ELEMENT STRESSES ACCORDING TO ASME SECTION 111	75-0655
KROKOSKY, E. M.	FINITE ELEMENT PREDICTION OF FIREPROOFING EFFECTIVENESS	72-0047

KROKOSKY, E. M.	INTERACTION STRESSES IN COMPOSITE SYSTEMS	70-0215
KROSS, D.	FINITE ELEMENT ANALYSIS OF THIN SHELLS	70-0457
KROSS, D. A.	A STUDY OF STIFFNESS MATRICES FOR THE ANALYSIS OF FLAT PLATES	68-0105
KROSS, D. A.	ANALYSIS OF GENERAL COUPLED THERMOELASTICITY PROBLEMS BY THE FINITE ELEMENT METHOD	68-0129
KROSS, D. A.	FINITE-ELEMENT ANALYSIS OF THIN SHELLS	68-0164
KROUSKOP, T. A.	A FINITE ELEMENT MODEL FOR PREDICTING LONG BONE FRACTURE FIXATION	71-0819
KRUEGER, R. F.	EFFECT OF PERFORATION DAMAGE ON WELL PRODUCTIVITY	74-0936
KRUSZEWSKI, J.	ALGORYTM OBLICZEN DRGAN WYMUSZONYCH METODA SZTYWNYCH ELEMENTOW SKONCZONYCH (ALGORITHM FOR CALCULATIN	74-0581
KRUSZEWSKI, J.	METHOD SZTYWNYCH ELEMENTOW SKONCZONYCH W OBLICZENIACH KONSTRUCJI OKRETOWYCH. (STIFF FINITE ELEMENT M	74-0923
KRUSZEWSKI, J.	OBLICZANIE DRGAN METODA SZTYWNYCH ELEMENTOW SKONCZCNYCH. (CALCULATING VIBRATIONS BY THE RIGID-FINITE	75-0649
KRUSZWESKI, J.	ANALIZA DRGAN WYMUSZONYCH ZLOZONYCH UKLADOW LINIOWYCH METODA SZTYWNYCH ELEMENTOW SKONCZONYCH (ANALYS	72-0246
KUBERT, B. R.	DEFORMATION OF THE DIASTOLIC LEFT VENTRICLE - 2. NONLINEAR GEOMETRIC EFFECTS	74-0974
KUBITZA, W. K.	A LARGE DEFORMATION ANALYSIS OF CRYSTALLINE ELASTIC-VISCOPLASTIC MATERIALS	73-0602
KUBITZA, W. K.	NUMERICAL ANALYSIS OF NONLINEAR PNEUMATIC STRUCTURES	67-0060
KUDO, H.	STRESS ANALYSIS IN SOME COLD FORGING TOOL COMPONENTS AND THEIR DESIGN GUIDES	74-0589
KUDO, K.	LOCAL POTENTIAL APPROACH TO FINITE ELEMENT METHOD IN UNSTEADY VISCOUS INCOMPRESSIBLE FLUID FLOW	73-0298
KUESEL, T. R.	SETTLEMENTS AND STRENGTHENING OF SOFT CLAY ACCELERATED BY SAND DRAINS	73-0438
KUESSNER, H. G.	A COMPARISON OF METHODS USED IN FLUTTER RESEARCH	72-0600
KUFEL, W.	OCENA DOKLADNOSCI ROZWIAZAN PROBLEMOW BRZEGOWYCH SPREZYSTYCH CIAL DYSKERTYZOWANYCH. (ASSESSING THE A	74-0921
KUHLEMEYER, R.	FINITE DYNAMIC MODEL FOR INFINITE MEDIA	69-0171
KUHLEMEYER, R.	VERTICAL VIBRATIONS OF FOOTINGS EMBEDDED IN LAYERED MEDIA	69-0436
KUHN, H. A.	SIMPLIFIED MINIMUM INTERNAL ENERGY OF DEFORMATION POSTULATE AND ITS APPLICATION TO METALFORMING PROC	74-1155
KULAK, R. F.	A STUDY OF INTERVERTEBRAL DISC MECHANICS BY THE FINITE ELEMENT METHOD	74-1121
KULAK, R. F.	FINITE ELEMENT METHOD FOR A SOLID ENCLOSING AN INVISCID INCOMPRESSIBLE FLUID	73-0090
KULAK, R. F.	FINITE ELEMENT STRESS ANALYSIS OF AN INTERVERTEBRAL DISC	74-0244
KULAK, R. F.	NONLINEAR BEHAVIOR OF THE HUMAN INTERVERTEBRAL DISC UNDER AXIAL LOADING	74-0919
KULAK, R. F.	NUMERICAL STRESS ANALYSIS OF INTERVERTEBRAL DISK	72-0125
KULHAWY, F. H.	ACCURACY OF EQUILIBRIUM SLOPE STABILITY ANALYSIS	73-0517
KULHAWY, F. H.	ANALYSIS OF BEHAVIOR OF EDWARD HYATT POWER PLANT	75-0055
KULHAWY, F. H.	ANALYSIS OF UNDERGROUND OPENINGS IN ROCK BY FINITE ELEMENT METHODS	73-0895
KULHAWY, F. H.	FINITE ELEMENT ANALYSIS OF THE BEHAVIOR OF EMBANKMENTS	70-0499
KULHAWY, F. H.	FINITE ELEMENT MODELING CRITERIA FOR UNDERGROUND OPENINGS IN ROCK	74-0630
KULHAWY, F. H.	STRESS AND DISPLACEMENTS AROUND OPENINGS IN ROCK CONTAINING AN INELASTIC DISCONTINUITY	75-0170
KULHAWY, F. H.	STRESSES AND DISPLACEMENTS AROUND OPENINGS IN HOMOGENEOUS ROCK	75-0168
KULHAWY, F. H.	STRESSES AND DISPLACEMENTS AROUND OPENINGS IN ROCK CONTAINING AN ELASTIC DISCONTINUITY	75-0169
KULHAWY, F. H.	STRESSES AND MOVEMENTS IN OROVILLE DAM	72-0200
KULICKI, J. M.	THE INELASTIC ANALYSIS OF REINFORCED AND PRESTRESSED CONCRETE BEAMS	72-0667
KULICKI, J. M.	THE INELASTIC ANALYSIS OF REINFORCED CONCRETE SLABS	74-1333
KULICKI, J. M.	USER'S MANUAL FOR PROGRAM BEAM	73-1105
KULIKOV, Y. A.	TREUGOL'NY I ELEMENT DLYA KRUGOVOI TSILINDRICHESKOI OBOLOCHKI (TRIANGULAR FINITE ELEMENT FOR CIRCULA	73-0493
KUMAR, V.	FINITE ELEMENT ANALYSIS FOR THE SEISMIC STABILITY OF EARTH STRUCTURES	74-1106
KUMBLE, R. G.	APPLICATION OF THE DOUBLE LIGAMENT TENSILE TEST IN DEFORMATION PROCESSING	75-0827
KUNITA, J.	ANALYSIS OF CABLE NETS IN MIXED FORMULATION	73-0311
KUNOO, K.	OPTIMUM CONFIGURATIONAL AND DIMENSIONAL DESIGN OF TRUSS STRUCTURES	74-0588
KUNOO, K.	OPTIMUM CONFIGURATION OF TRUSS STRUCTURES	74-1315
KUO, C. T.	UNSYMMETRIC BUCKLING OF SHELLS OF REVOLUTION BY FINITE ELEMENT METHOD	70-0500
KUO, J. T.	PSEUDO 3-DIMENSIONAL FINITE-ELEMENT FORMULATION FOR ELASTOSTATIC PROBLEMS AND ITS GEOPHYSICAL APPLIC	75-0597
KUO, J. T.	PSEUDO-THREE-DIMENSIONAL FINITE ELEMENT FORMULATION FOR ELASTOSTATIC PROBLEMS	74-0167
KUO, P. S.	FINITE ELEMENT APPROACH TO THE VIBRATION ANALYSIS OF ELASTIC DISKS ON A FLEXIBLE SHAFT	75-0368
KUO, W. C.	SCATTERING BY BURIED OBSTACLES	74-0739
KUPTA, K. K.	DYNAMIC RESPONSE ANALYSIS OF GEOMETRICALLY NON-LINEAR STRUCTURES SUBJECTED TO HIGH IMPACT	72-0193
KURAKATA, Y.	FINITE DISPLACEMENT BEAM THEORY	75-0529
KURATA, T.	ANALYSIS OF VEHICLE BODY STRUCTURE BY THE FINITE ELEMENT METHOD	71-0324
KURIBAYA, E.	APPLICATION OF FINITE ELEMENT METHOD TO SOIL-FOUNDATION INTERACTION ANALYSES	74-1080
KURIBAYASHI, E.	EIGENVALUE ANALYSIS BY FINITE ELEMENT METHODS	69-0377
KURODA, M.	SOME EXAMINATIONS ON AVAILABILITY OF FINITE ELEMENT METHODS IN LINEAR FRACTURE MECHANICS	73-0112
KURODA, T.	EXPERIMENTAL AND THEORETICAL STUDY ON A 1/10TH SCALE MODEL PCPV	73-0829
KURODA, T.	STRESS ANALYSIS OF A PRESTRESSED CONCRETE NUCLEAR PRESSURE VESSEL BY THE FINITE ELEMENT METHOD USING	71-0154
KUROKAWA, Y.	STUDIES ON THE DETERMINATION OF SEEPAGE-LINE OF FILL DAM BY THE METHOD OF FINITE ELEMENT. WATER FLOW	74-1348
KUSAKA, K.	LARGE DEFLECTION OF CANTILEVER BEAMS	71-0649
KUSUMOTO, S.	METHOD OF EVALUATION OF THREE-DIMENSIONAL STRESS INTENSITY FACTOR USING THE FINITE ELEMENT METHOD	75-0316
KUSUMOTO, S.	NOTCH EFFECT IN LOW CYCLE FATIGUE - REPORT 1: LOW-CYCLE FATIGUE STRENGTH OF 0.48 PERCENT CARBON STEE	73-1090
KUSUMOTO, S.	THE SIMPLE METHOD OF EVALUATION OF STRESS INTENSITY FACTOR USING THE FINITE ELEMENT METHOD	74-0829
KUUSINEN, L. R.	ACOUSTIC ANALYSIS OF SOLID ROCKET MOTOR CAVITIES BY A FINITE-ELEMENT METHOD	71-0119
KUZANEK, J. F.	FINITE ELEMENT MODELING OF A CANTILVER BEAM	67-0132
KUZIN, V. I.	FINITE-DIFFERENCE AND FINITE-ELEMENT METHODS IN GLOBAL WIND-DRIVEN OCEAN CIRCULATION	75-0345
KVITKA, A. L.	CHISLENNOE RESHENIE PLOSKIKH ZADACH TEPLOPROVODNOSTI DLYA OBLASTEI SLOZHNOI FORMY (NUMERICAL SOLUTIO	74-0659
KVITKA, A. L.	ISSLEDOVANIE NAPRYAZHENNO-DEFORMIROVANNOGO SOSTOYANIYA SIMMETRICHNYKH DISKOV SO STUPITSEI. (INVESTIG	74-0965
KVITKA, A. L.	K VOPROSU OB AVTOMATIZATSII ZADANIYA INFORMATSII V METODE KONECHNYKH ELEMENTOV. (PROBLEM OF AUTOMATI	75-0382
KWAN, C.	ANALYSIS OF TUBULAR T-CONNECTIONS BY THE FINITE ELEMENT METHOD: COMPARISON WITH EXPERIMENTS	72-0380
KWAN, G.	LIMITATIONS ON THE APPLICATION OF EFFECTIVE BREADTH CONCEPT IN BOX GIRDER DESIGN	74-0493
KWOK, W. L.	ANALYSIS OF CIRCULAR AND ANGULAR LAMINATED THICK PLATES	74-0426
KWOK, W. L.	DYNAMIC ANALYSIS OF CIRCULAR AND SECTOR THICK, LAYERED PLATES	75-0198
KWOK, W. L.	FINITE-ELEMENT ANALYSIS OF SKEW SANDWICH PLATES	75-0185
KYTE, W. S.	DIFFUSION OF HYDROGEN IN FILLET WELDS	75-0373
LA-ROCHELLE, P.	STRESSES AND DEFORMATIONS OF VERTICAL SLOPES IN ELASTO-PLASTIC ROCKS	70-0114
LABANTI, L.	DESCRIPTION AND INSTRUCTION FOR USE OF THE SAFE-SHELL. B FINITE ELEMENT PROGRAM IN THE IBM/360 VERSI	73-0945
LABBENS, R. C.	EFFICIENT FINITE-ELEMENT METHODS FOR STRESS INTENSITY FACTORS USING WEIGHT FUNCTIONS	74-0411
LABRUJERE, TH.	FINITE ELEMENT CALCULATION OF AXISYMMETRIC SUBCRITICAL COMPRESSIBLE FLOW	74-1064
LABRUJERE, TH.	PROPOSALS FOR THE APPLICATION OF THE FINITE ELEMENT TECHNIQUE TO SOME AIRCRAFT AERODYNAMICAL PROBLEM	72-0628
LABRUJERE, TH.E	FINITE ELEMENT SOLUTION OF THE INCOMPRESSIBLE FLOW OVER AN AIRFOIL IN A NON UNIFORM STREAM	74-0432
LACASSE, S. M.	BEHAVIOR OF EMBANKMENTS ON NEW LISKHEARD VARVED CLAY	73-0721
LACHANCE, P.	LINEAR AND NON-LINEAR ANALYSIS OF THIN SHALLOW SHELLS BY MIXED FINITE ELEMENTS	75-0665
LACHAT, J. C.	ANALYSIS OF CONTINUA BY THE FINITE ELEMENT AND INTEGRAL EQUATION METHOD	75-0370
LACHAT, J. E.	SECOND GENERATION BOUNDARY INTEGRAL EQUATION PROGRAM FOR THREE-DIMENSIONAL ELASTIC ANALYSIS	75-0376
LADANYI, B.	ANALYSE DE LA STABILITY DES TALUS ROCHEUX PAR UNE METHODE GENERALISEE DE L EQUILIBRE LIMITE (ROCK SL	74-1019
LADANYI, B.	ROCK FAILURE AROUND A CIRCULAR OPENING IN A GRAVITY FIELD WITH TECTONIC FORCES	70-0071
LADANYI, B.	STRESS DISTRIBUTION DUE TO GRAVITY IN A VERTICAL ROCK BANK	68-0090
LADD, C. C.	BEHAVIOR OF EMBANKMENTS ON NEW LISKHEARD VARVED CLAY	73-0721
LADD, C. C.	FINITE ELEMENT PROGRAM FEECON FOR UNDRAINED DEFORMATION ANALYSES OF GRANULAR EMBANKMENTS ON SOFT CLA	72-0875
LADD, C. C.	PERFOPMANCE OF AN EMBANKMENT CONSTRUCTED ON VARVED CLAY	73-0743
LADKANY, S. G.	OBSERVATIONS REGARDING ASSUMED STRESS HYBRID PLATE ELEMENTS	74-0370
LAEVASTU, T.	HYDRODYNAMICAL-NUMERICAL MODELS FOR COASTAL WATERS AND OPEN OCEAN AREAS	75-0571
LAGASSE, P. E.	FINITE ELEMENT ANALYSIS OF DISPERSION-FREE GUIDES FOR ELASTIC SURFACE WAVES	73-0009
LAGASSE, P. E.	FINITE ELEMENT ANALYSIS OF PIEZOELECTRIC ELASTIC WAVEGUIDES	73-0115
LAGASSE, P. E.	HIGHER-ORDER FINITE-ELEMENT ANALYSIS OF TOPOGRAPHIC GUIDES SUPPORTING ELASTIC SURFACE WAVES	73-0114
LAGER, J. R.	DESIGN OF LOW-THERMAL-DISTORTION LST METERING STRUCTURE	75-0840
LAGERQUIST, D.	FINITE-ELEMENT ANALYSIS OF COMPLEX PANEL RESPONSE TO RANDOM LOADS	68-0093
LAGERQUIST, D.	RANDOM-VIBRATION ANALYSIS SYSTEM FOR COMPLEX STRUCTURES. PART 1, ENGINEERING USER'S GUIDE	68-0106
LAGNADO, I.	ANALYSIS AND REDESIGN OF SONAR PIEZOELECTRIC RESONATORS USING THE FINITE ELEMENT METHOD.	72-0374
LAGUE, G.	AXISYMMETRIC FINITE ELEMENT IN NATURAL COORDINATES	75-0826
LAHELD, P.	DESIGN OF NEW SHIP TYPES BY MEANS OF FINITE ELEMENT METHOD	72-0562

LAI, N. W.	NUMERICAL SOLUTIONS FOR DETERMINING WAVE-INDUCED PRESSURE DISTRIBUTIONS AROUND BURIED PIPELINES	74-1334
LAING, I. E.	TENSION ZONES IN DECKED ROCKFILL DAMS	73-0540
LAING, V. C.	DYNAMIC RESPONSE OF STRUCTURES IN LAYERED SOILS	74-0761
LAKIS, A. A.	DYNAMICAL ANALYSIS OF AXIALLY NON-UNIFORM, THIN CYLINDRICAL SHELLS. PART 1 - MATRIX FORMULATION	70-0413
LAKIS, A. A.	DYNAMICAL ANALYSIS OF AXIALLY NONUNIFORM, THIN CYLINDRICAL SHELLS. PART 2 FREE VIBRATION	71-0716
LAKIS, A. A.	FREE VIBRATION AND RESPONSE TO RANDOM PRESSURE FIELD OF ANISOTROPIC THIN CYLINDRICAL SHELLS	74-0830
LAKIS, A. A.	FREE VIBRATION OF CYLINDRICAL SHELLS PARTIALLY FILLED WITH LIQUID	71-0839
LAKSHMIKANTHAM,	STRESSES AROUND HOLES IN STIFFENED COMPOSITE PANELS USING LAURENT-SERIES AND FINITE ELEMENT METHODS	74-1252
LAKSHMIKANTHAM,	STRESSES AROUND HOLES IN STIFFENED COMPOSITE PANELS USING LAURENT-SERIES AND FINITE ELEMENT METHODS	75-0733
LAKSHMINARAYANA	DISCRETIZATION-DISPERSION IN FINITE ELEMENT MODELLING OF WAVE PROPAGATION IN SOLIDS	74-1253
LAKSHMINARAYANA	ON THE DEVELOPMENT OF A REFINED FINITE-ELEMENT FREE FROM DISCRETIZATION - DISPERSION FOR ELASTIC WAV	73-1148
LALANNE, M.	VIBRATION ANALYSIS OF ROTATING COMPRESSOR BLADES	73-0417
LALANNE, M.	VIBRATION ANALYSIS OF ROTATING TURBINE BLADES	74-0959
LAM, P.	MEASUREMENT OF SLOPE CHANGE ON CYLINDRICAL SHELL SURFACES	73-0439
LAM, T. L.	DESIGN AND ANALYSIS OF A CERAMIC STATOR VANE	75-0366
LAMA, R. D.	APPLICATION OF NEW AUSTRIAN TUNNELLING METHOD IN DIFFICULT BUILTOVER AREAS IN FRANKFURT/MAIN METRO	73-1025
LAMBERT, A.	CRUSTAL STRUCTURE FROM SURFACE LOAD TILTS USING A FINITE ELEMENT MODEL	72-0005
LAMMERS, S. J.	SOLUTION OF AUTOMOTIVE STRUCTURAL PROBLEMS USING THE FINITE ELEMENT METHOD AND COMPUTER GRAPHICS	71-0341
LANG, T. E.	DISCRETIZATION-DISPERSION IN FINITE ELEMENT MODELLING OF WAVE PROPAGATION IN SOLIDS	74-1253
LANG, T. E.	FINITE ELEMENT MATRIX FORMULATION OF POST-BUCKLING STABILITY AND IMPERFECTION SENSITIVITY	70-0246
LANG, T. E.	MODIFIED POTENTIAL ENERGY MASS REPRESENTATION FOR FREQUENCY PREDICTION	66-0053
LANG, T. E.	POST-BUCKLING RESPONSE OF STRUCTURES USING THE FINITE ELEMENT METHOD	69-0414
LANG, T. E.	STRUCTURAL ANALYSIS AND MATRIX INTERPRETIVE SYSTEM (SAMIS) USER REPORT	67-0125
LANGBALLE, M.	APPLICATION OF THE FINITE ELEMENT METHOD TO MACHINERY	74-0198
LANGE, K.	ELASTIC DEFORMATION OF SQUARE EXTRUSION DIES DUE TO LOAD AND TEMPERATURE	75-0581
LANGEFORS, B.	STRUCTURAL ANALYSIS OF SWEPT-BACK WINGS BY MATRIX-TRANSFORMATION	51-0001
LANGLAND, R. T.	NONDETERMINISTIC PROBLEMS IN STRUCTURAL MECHANICS	71-0746
LANGLAND, R. T.	PROBABILISTIC THEORY OF STRUCTURES USING FINITE ELEMENTS	71-0820
LANNOY, F. G.	DISPLACEMENT METHOD IN STRUCTURAL ANALYSIS AS APPLIED TO RECTANGULAR SHELL BEAMS	72-0411
LANSBERRY, C. R	A FULLY COMPATIBLE ANNULAR SEGMENT FINITE ELEMENT	72-0582
LANSBERRY, C. R	A FULLY COMPATIBLE ANNULAR SEGMENT FINITE ELEMENT	72-0663
LANSBERRY, C. R	HORIZONTALLY CURVED HIGHWAY BRIDGES. A FULLY COMPATIBLE ANNULAR SEGMENT FINITE ELEMENT	72-0834
LANSING, W.	A METHOD FOR REDUCING THE ANALYSIS OF COMPLEX REDUNDANT STRUCTURES TO A ROUTINE PROCEDURE	52-0002
LANSING, W.	MATRIX ANALYSIS METHODS FOR INELASTIC STRUCTURES	65-0051
LAPSHIN, IU. G.	APPLICATION OF THE FINITE ELEMENT METHOD TO CALCULATE WOOD TENSION	75-0768
LARDER, R. A.	APPLICATION OF THREE-DIMENSIONAL FINITE ELEMENT ANALYSES TO THE MICROMECHANICS OF FIBROUS COMPOSITE	74-1079
LARDER, R. A.	STOCHASTIC FINITE ELEMENT SIMULATION OF THE NONLINEAR STRUCTURAL RESPONSE OF FIBROUS COMPOSITE MATER	75-0762
LARDER, R. A.	STRENGTH DISTRIBUTIONS OF SINGLE FILAMENTS	75-0305
LARDER, R. A.	THE STOCHASTIC FINITE ELEMENT SIMULATION OF THE NONLINEAR STRUCTURAL RESPONSE OF FIBROUS COMPOSITE M	75-0798
LARKIN, L. A.	ELASTIC-PLASTIC PLATE BENDING WITH CONSTANT CURVATURE ELEMENTS	74-0308
LAROCK, B. E.	FREE-SURFACE IDEAL FLUID FLOWS BY FINITE ELEMENTS	73-0254
LARRABEE, R. D.	THERMAL LOADING OF THIN-SHELL CONCRETE COOLING TOWERS	74-0667
LARSEN, K.	ON THE CONVERGENCE OF SOR ITERATIONS FOR FINITE ELEMENT APPROXIMATIONS TO ELLIPTIC BOUNDARY VALUE PR	71-0449
LARSEN, P. K.	ELASTIC-PLASTIC ANALYSIS OF THICK WALLED PRESSURE VESSELS WITH SHARP DISCONTINUITIES	71-0155
LARSEN, P. K.	NOTE ON INCREMENTAL EQUILIBRIUM EQUATIONS AND APPROXIMATE CONSTITUTIVE RELATIONS IN LARGE INELASTIC	74-0984
LARSSON, S. G.	INFLUENCE OF NON-SINGULAR STRESS TERMS AND SPECIMEN GEOMETRY ON SMALL SCALE YIELDING AT CRACK TIPS I	73-0203
LARSSON, S. G.	ON THE FINITE ELEMENT ANALYSIS OF CRACK AND INCLUSION PROBLEMS IN ELASTIC-PLASTIC MATERIALS	74-0227
LASCAUX, P.	APPLICATION OF THE FINITE ELEMENTS METHOD IN TWO-DIMENSIONAL HYDRODYNAMICS USING THE LAGRANGE VAPIAB	73-0976
LASCAUX, P.	APPLICATION OF THE FINITE ELEMENT METHOD TO TWO DIMENSIONAL LAGRANGIAN HYDRODYNAMICS	74-0016
LASCAUX, P.	SOME NONCONFORMING FINITE ELEMENTS FOR THE PLATE BENDING PROBLEM	75-0324
LASHKARI, M.	EFFECT OF LATERAL PRESSURE ON NATURAL FREQUENCIES OF HYPERBOLOIDAL SHELLS OF REVOLUTION	71-0278
LASHKARI, M.	VIBRATIONS OF PRESSURE LOADED HYPERBOLOIDAL SHELLS	72-0874
LASHKARI, M.	VIBRATIONS OF SEGMENTED SHELLS	73-0161
LASKARIS, T. E.	FINITE ELEMENT ANALYSIS OF SEVERAL COMPRESSIBLE AND INCOMPRESSIBLE VISCOUS FLOW PROBLEMS	74-1119
LASKARIS, T. E.	FINITE-ELEMENT ANALYSIS OF COMPRESSIBLE AND INCOMPRESSIBLE VISCOUS-FLOW AND HEAT-TRANSFER PROBLEMS	75-0347
LASKER, G.	STRUCTURAL DYNAMIC PROPERTIES OF TACTICAL MISSILE JOINTS - PHASE 3	74-0811
LASRY, S. J.	A HYBRID-ELEMENT APPROACH TO CRACK PROBLEMS IN PLANE ELASTICITY	73-0569
LASRY, S. J.	A SUPER-ELEMENT FOR CRACK ANALYSIS	73-0754
LASRY, S. J.	DERIVATION OF CRACK ELEMENT-STIFFNESS MATRIX BY THE COMPLEX VARIABLE APPROACH	73-0870
LASSKER, A. J.	NONLINEAR BEHAVIOR OF REINFORCED CONCRETE BEAMS BY THE FINITE-ELEMENT METHOD	72-0954
LASZLO, C. A.	MODELING EARDRUM AS A DOUBLY CURVED SHELL USING FINITE-ELEMENT METHOD	75-0047
LASZLO, C. A.	SIMULATING BEHAVIOR OF EARDRUM BY FINITE-ELEMENT METHOD	74-0511
LATHROP, K. D.	NEW TRANSPORT METHODS AND CODES	74-0878
LATHROP, K. D.	TRANSPORT THEORY NUMERICAL METHODS	72-0769
LATHROP, K. D.	TRIPLET: A TWO-DIMENSIONAL, MULTIGROUP, TRIANGULAR MESH, PLANAR GEOMETRY, EXPLICIT TRANSPORT CODE	73-0955
LATZKO, D. G. H	DESIGN ASPECTS OF HIGH DUTY PRESSURE COMPONENTS	74-0199
LAU, P. C. M.	INTERPRETATION OF COMPLEX STRESS ANALYSIS BY MULTIPLE REGRESSION	74-1203
LAUDIERO, F.	FINITE ELEMENT INCREMENTAL ANALYSIS OF ELASTOPLASTIC PLATE BENDING	73-0557
LAUNAY, P.	THE TRIDIMENSIONAL THERMOELASTIC COMPUTER CODE TITUS	71-0156
LAURA, P. A.	NUMERICAL EXPERIMENTS ON THE DETERMINATION OF STRESS CONCENTRATION FACTORS	74-0292
LAURA, P. A.	USE OF TRIGONOMETRIC TERMS IN FINITE ELEMENT METHOD WITH APPLICATION TO VIBRATING MEMBRANES	74-0555
LAURA, P. A. A.	COMPARISON OF ANALYTICAL AND NUMERICAL SOLUTIONS IN HEAT CONDUCTION PROBLEMS	74-1170
LAURSEN, J. I.	DYNAMIC MATRIX ANALYSIS OF FRAMED STRUCTURES	62-0012
LAURSEN, M. F.	FINITE ELEMENT EQUILIBRIUM MODEL FOR GENERAL BODY FORCE DISTRIBUTION	75-0268
LAUTERBACH, G.	FINITE ELEMENT SOLUTION FOR SAINT-VENANT BENDING	71-0346
LAVENGOOD, R. E	STRESS FIELDS AROUND MULTIPLE INCLUSIONS	69-0067
LAWRENCE, C.	A MATHEMATICAL MODEL OF THE CORNEO-SCLERAL SHELL	71-0150
LAWRENCE, C.	ANALYSIS OF THE CORNEO-SCLERAL SHELL BY THE METHOD OF DIRECT STIFFNESS	71-0151
LAWRENCE, D. J.	FINITE ELEMENT ANALYSIS. MODELLING AND COMPUTER GRAPHICS	74-1292
LAWRENCE, H. H.	EFFECTS OF SHOE FORCE GEOMETRY ON HEAVY DUTY INTERNAL SHOE BRAKE PERFORMANCE	68-0219
LAWRENCE, K. L.	DYNAMIC STRAINS IN A FOUR BAR MECHANISM	73-0677
LAWRENCE, K. L.	TWISTED BEAM ELEMENT MATRICES FOR BENDING	70-0082
LAWRENCE, S. J.	FINITE ELEMENT ANALYSIS OF CLAD STRUCTURES	73-0353
LAWTON, C. W.	STRAIN CONCENTRATION BEHAVIOUR IN A NOTCHED ROUND BAR SUBJECTED TO CREEP	73-0440
LAWTON, R. G.	AYER HEAT CONDUCTION COMPUTER PROGRAM	74-0885
LAWTON, R. G.	PLACID: A GENERAL FINITE ELEMENT COMPUTER PROGRAM FOR STRESS ANALYSIS OF PLANE AND AXISYMMETRIC SOLI	74-0884
LAZEAR, D. J.	SOME APPLICATIONS OF FINITE ELEMENTS TO MODEL GEOLOGICAL STRUCTURES	73-0143
LEAF, G. K.	APPLICATION OF FINITE ELEMENT TECHNIQUES FOR THE NUMERICAL SOLUTION OF THE NEUTRON TRANSPORT AND DIF	71-0013
LEAF, G. K.	APPLICATIONS OF FINITE ELEMENT METHODS IN REACTOR MATHEMATICS - NUMERICAL SOLUTION OF THE NEUTRON DI	72-0535
LEAF, G. K.	APPLICATIONS OF FINITE ELEMENT METHODS IN REACTOR MATHEMATICS. NUMERICAL SOLUTION OF THE NEUTRON TRA	74-0853
LEAF, G. K.	COMPUTATIONAL ANALYSIS AND EVALUATION OF THE FINITE ELEMENT METHOD FOR A CLASS OF NUCLEAR REACTOR CO	74-0595
LEAF, G. K.	CONSTRUCTION OF A FINITE ELEMENT APPROXIMATION WHICH CROSSES MATERIAL INTERFACES	74-0896
LEAF, G. K.	EMPIRICAL INVESTIGATION OF REORDERING AND DATA MANAGEMENT FOR FINITE ELEMENT SYSTEMS OF EQUATIONS	73-0944
LEAF, G. K.	THE USE OF INTERPOLATORY POLYNOMIALS FOR A FINITE ELEMENT SOLUTION OF THE MULTIGROUP DIFFUSION EQUAT	72-0524
LEAF, G. K.	TIMING COMPARISON STUDY FOR SOME HIGH ORDER FINITE ELEMENT PROCEDURES AND A LOW ORDER FINITE DIFFERE	73-0396
LEANHARDT, D. E	VIBRATION CHARACTERISTICS OF FREE THIN CYLINDRICAL SHELLS	73-0396
LECKIE, F. A.	A STUDY OF FINITE ELEMENTS APPLIED TO PLATE FLEXURE	66-0016
LECKIE, F. A.	THE EFFECT OF LUMPED PARAMETERS ON BEAM FREQUENCIES	63-0012
LEDERER, E.	FORMAT. A GENERAL PURPOSE MATRIX ANALYSIS SYSTEM IN FORTRAN	68-0266
LEE, C. H.	ANALYSES OF AXISYMMETRIC UPSETTING AND PLANE STRAIN SIDE PRESSING OF SOLID CYLINDERS BY THE FINITE E	70-0083
LEE, C. H.	ANALYSES OF AXYSYMMETRIC UPSETTING AND PLANE-STRAIN SIDE-PRESSING OF SOLID CYLINDERS BY THE FINITE E	71-0794
LEE, C. H.	ANALYTICAL PREDICTION OF DEFECTS OCCURRENCE IN SIMPLE AND COMPLEX	69-0157

LEE, C. H.	CALCULATION OF RESIDUAL STRESSES IN PLASTIC DEFORMATION PROCESSES	72-0241
LEE, C. H.	ELASTOPLASTIC ANALYSIS OF PLANE STRAIN AND AXISYMMETRIC FLAT PUNCH INDENTATION BY THE FINITE ELEMENT	70-0343
LEE, C. H.	FINITE ELEMENT METHOD FOR TRANSIENT LINEAR VISCOUS FLOW PROBLEMS	74-0474
LEE, C. H.	MATRIX DYNAMIC AND INSTABILITY ANALYSIS WITH NON-UNIFORM ELEMENTS	70-0062
LEE, C. H.	ON SEAWATER ENCROACHMENT IN COASTAL AQUIFERS	74-0577
LEE, E. H.	DYNAMIC ANALYSIS OF THE DEFORMATION OF STRUCTURES AND OF METAL FORMING	74-0823
LEE, E. H.	DYNAMIC ANALYSIS OF STRUCTURAL DEFORMATION AND METAL FORMING	75-0058
LEE, G. C.	DERIVATION OF STIFFNESS MATRICES FOR PROBLEMS IN PLANE ELASTICITY BY GALERKIN'S METHOD	69-0250
LEE, G. C.	DERIVATION OF THE STIFFNESS MATRIX FOR PLATES BY GALERKIN METHOD	69-0402
LEE, G. C.	FINITE ELEMENT SOLUTION TO AN ELASTICA PROBLEM OF BEAMS	70-0112
LEE, G. C.	LOCAL COMPLIANCE PROBLEM OF LUNG ELASTICITY	75-0660
LEE, G. C.	STIFFNESS METHODS FOR PLATES BY GALERKIN'S METHOD	69-0249
LEE, G. C.	STRAIN ENERGY FUNCTIONS AND MATHEMATICAL MODELS OF LUNG ELASTICITY	73-0591
LEE, G. C.	THEORY FOR DISTORTION STUDIES OF LUNG PARENCHYMA BASED ON ALVEOLAR MEMBRANE PROPERTIES	74-0222
LEE, H.	FINITE BEAM ELEMENTS FOR DYNAMIC ANALYSIS	70-0372
LEE, H. Y.	PREDICTION OF LONG-TERM STRESS RANGES: USER'S MANUAL - BRIDGE LOAD GENERATOR	73-0737
LEE, H. Y.	PREDICTION OF LONG-TERM STRESS RANGES: USER'S MANUAL - BRIDGE DYNAMIC STRESS ANALYSIS	74-0807
LEE, H-P.	USING THE NASTRAN THERMAL ANALYZER TO SIMULATE A FLIGHT SCIENTIFIC INSTRUMENT PACKAGE	74-1011
LEE, I. K.	ELASTIC PLASTIC ANALYSIS OF SHALLOW FOUNDATIONS	74-0446
LEE, I. K.	FINITE ELEMENT ANALYSIS OF CONSOLIDATION PROBLEM	74-0081
LEE, J. C.	COMPUTER IMPLEMENTATION OF THE CONSTRAINT METHOD	75-0001
LEE, K. K.	HYDRAULIC FRACTURING IN ZONED EARTH AND ROCKFILL DAMS: A REPORT OF AN INVESTIGATION	73-0873
LEE, K. L.	FACTOR OF SAFETY APPROACH FOR EVALUATING SEISMIC STABILITY OF SLOPES	75-0645
LEE, K. L.	STATIC STRESSES BY LINEAR AND NONLINEAR METHODS	75-0278
LEE, K. N.	FINITE ELEMENT FORMULATIONS FOR ELASTIC PLATES BY GENERAL VARIATIONAL STATEMENTS WITH DISCONTINUOUS	73-0559
LEE, K. N.	FINITE ELEMENT FORMULATIONS FOR ELASTIC PLATES BY GENERAL VARIATIONAL STATEMENTS WITH DISCONTINUOUS	73-1124
LEE, K. N.	FINITE ELEMENT METHODS IN ELASTICITY	74-1118
LEE, K. N.	MIXED VARIATIONAL PRINCIPLES, FINITE ELEMENTS AND FINITE ELASTICITY	74-1254
LEE, K. N.	THE CONDITION OF SOME FINITE ELEMENT COEFFICIENT MATRICES	70-0393
LEE, K. P.	GENERALIZED STIFFNESS MATRIX OF A CURVED-BEAM ELEMENT	69-0165
LEE, L.	RESPONSE OF STRUCTURES TO COMBINED BLAST EFFECTS	73-0424
LEE, L. H. N.	INCREMENTAL VARIATIONAL METHOD FOR THE LARGE DISPLACEMENT ANALYSIS OF SHELLS WITH GEOMETRIC IMPERFEC	72-0894
LEE, R. H.	FINITE ELEMENT FOR REINFORCED CONCRETE FRAME STUDY - CLOSURE	75-0087
LEE, S. L.	FINITE ELEMENT ANALYSIS OF ELASTIC-PLASTIC PLANE STRESS PROBLEMS BASED UPON TRESCA YIELD CRITERION	70-0179
LEE, S. L.	ON TRESCA YIELD CRITERION IN FINITE ELEMENT ANALYSIS OF PLANE STRESS PROBLEMS	73-0314
LEE, S. W.	AN ASSUMED STRESS HYBRID FINITE ELEMENT FOR THREE-DIMENSIONAL PLASTIC STRUCTURAL ANALYSIS	74-0717
LEE, S. Y.	MASS LOADING EFFECTS ON VIBRATION OF RING AND SHELL STRUCTURES	68-0245
LEE, S. Y.	RADIAL AND AXIAL TRANSMISSIBILITY CHARACTERISTICS OF TYPICAL ROCKET VEHICLE STRUCTURE TECHNICAL SUMM	69-0475
LEE, W.	THERMAL AND STRESS ANALYSIS OF A FUEL ROD	75-0713
LEECH, A.	THE COMPARISON OF ITERATIVE AND DIRECT SOLUTION TECHNIQUES IN THE ANALYSIS OF TIME-DEPENDENT STRESS	72-0080
LEECH, W. J.	DISCRETE VARIABLE METHOD APPLIED TO TRANSIENT HEAT CONDUCTION PROBLEMS	74-1163
LEFEBVRE DU PRE	FRONT DISPLACEMENT MODEL IN A FRACTURED RESERVOIR	74-0071
LEFEBVRE, G.	EARTH PRESSURES ON STRUCTURES DUE TO FAULT MOVEMENT	73-0403
LEFEBVRE, G.	THREE-DIMENSIONAL BEHAVIOR OF A CENTRAL CORE DAM	72-0714
LEFEBVRE, G.	THREE-DIMENSIONAL FINITE ELEMENT ANALYSIS OF DAMS	71-0733
LEFEBVRE, G.	THREE-DIMENSIONAL FINITE ELEMENT ANALYSES OF DAMS	73-0202
LEFEBVRE, G. D.	FINITE ELEMENT ANALYSIS OF TRANSVERSE CRACKING IN LOW-EMBANKMENT DAMS	74-0696
LEGGERI, B.	ENERGY FORMS IN THE FINITE ELEMENT TECHNIQUES	72-0680
LEHMAN, G. M.	DEVELOPMENT OF A GRAPHITE HORIZONTAL STABILIZER	73-0931
LEHNHOFF, T. F.	INFLUENCE OF TEMPERATURE DEPENDENT PROPERTIES ON THERMAL ROCK FRAGMENTATION	75-0222
LEHNHOFF, T. F.	THERMAL-MECHANICAL FRAGMENTATION OF HARD ROCK - FIELD TESTS	74-0998
LEIBER, P.	FINITE ELEMENT METHOD AS AN ASPECT OF THE PRINCIPLE OF MAXIMUM UNIFORMITY: NEW HYDRODYNAMICAL RAMIFA	74-0009
LEIBOWITZ, R. C	VIBROACOUSTIC RESPONSE OF TURBULENCE EXCITED THIN RECTANGULAR FINITE PLATES IN HEAVY AND FLUID MEDIA	74-1309
LEIMBACH, K. R.	STUDY OF HYPERVELOCITY METEOROID IMPACT ON ORBITAL SPACE STATIONS	73-0905
LEISTNER, H. G.	ELASTIC WAVE PROPAGATION, A COMPARISON BETWEEN FINITE ELEMENT PREDICTIONS AND EXACT SOLUTION	67-0015
LEISTNER, H. G.	STRUCTURE-MEDIUM INTERACTION AND DESIGN PROCEDURES STUDY. VOLUME I. ANALYSIS METHOD, THEORY, VERIFIC	69-0166
LEITERER, R.	NUMERICAL INTEGRATION METHODS IN FINITE ELEMENTS	74-0841
LEITNER, F.	DESIGN OF TWO AND THREE DIMENSIONAL STRUCTURES BY THE FINITE ELEMENT METHOD (IN GERMAN)	74-1364
LEITNER, F.	STRESS ANALYSIS OF TWO AND THREE DIMENSIONAL STRUCTURES BY THE FINITE ELEMENT METHOD	72-0794
LEITNER, F.	THREE-DIMENSIONAL ISOPARAMETRIC ELEMENTS AND THEIR USE FOR STATICAL ANALYSIS OF MASSIVE WATER-POWER	71-0364
LEMAIRE, M.	APPLICATIONS DE LA METHODE DE GAUSS-SEIDEL A LA RESOLUTION DES PROBLEMS D'ELASTICITE. (APPLICATION O	72-0145
LEMAIRE, M.	MATRICE DE RIGIDITE D'UN ELEMENT DE POUTRE SOLLICITE A LA TORSION-FLEXION (STIFFNESS MATRIX OF A BEA	73-0442
LEMAIRE, M. H.	DEPLACEMENTS DANS LES STRUCTURES MINCES A FAIBLE RIGIDITE GEOMETRIQUE-DETERMINATION NUMERIQUE (NUMER	72-0213
LEMKE, H. U.	ICON: THE INTERACTIVE CREATION OF NASTRAN DATA, A SYSTEM DESCRIPTION	75-0274
LEMMON, E. C.	ACCURACY, STABILITY AND OSCILLATION CHARACTERISTICS OF FINITE ELEMENT METHOD FOR SOLVING HEAT CONDUC	69-0064
LEMMON, E. C.	COUPLE - A PSEUDO FINITE ELEMENT CONDUCTION CODE FOR THE TRANSIENT THERMAL RESPONE OF AXISYMMETRIC O	72-0778
LEMMON, E. C.	FINITE CONDUCTANCE ELEMENT METHOD OF CONDUCTION HEAT TRANSFER	73-0963
LEMPP, W.	BERECHNUNG VON SPANNUNGEN UND TEMPERATURFELDERN IN FLOSSENWAENDEN UND AUFGESCHWEISSIEN BAUTFILEN. (C	74-0544
LENCOVA, B.	FINITE ELEMENT METHOD IN ELECTRON OPTICS	75-0081
LENGYEL, P.	TEMPERATURE RISE IN THE HIGH SPEED COMPRESSION OF RIGHT CYLINDRICAL BILLETS	74-1153
LENNOX, W. C.	FINITE ELEMENT PERMAFROST THAW SETTLEMENT MODEL	74-0543
LENNOX, W. C.	TWO-DIMENSIONAL HEATED PIPELINE IN PERMAFROST PIPELINES	74-0541
LEONARD, J. W.	COMPRESSIBLE FLOW BY THE FINITE ELEMENT METHOD	69-0417
LEONARD, J. W.	CURVED FINITE ELEMENT APPROXIMATION TO NONLINEAR CABLES	72-0379
LEONARD, J. W.	FINITE ELEMENT ANALYSIS OF THE UNSTEADY, LINEARIZED BOLTZMANN EQUATION	70-0033
LEONARD, J. W.	FINITE ELEMENT ANALYSIS OF PERTURBED COMPRESSIBLE FLOW	72-0001
LEONARD, J. W.	FINITE ELEMENT ANALYSIS OF THE LINEARIZED BGK BOLTZMANN EQUATION.	72-0408
LEONARD, J. W.	FINITE ELEMENT ANALYSIS OF INFLATABLE SHELLS	73-0241
LEONARD, J. W.	FINITE ELEMENT SOLUTIONS TO DIFFERENTIAL EQUATIONS	70-0084
LEONARD, J. W.	GALERKIN FINITE ELEMENT FORMULATION FOR INCOMPRESSIBLE FLOW	70-0006
LEONARD, J. W.	LINEARIZED COMPRESSIBLE FLOW BY THE FINITE ELEMENT METHOD	69-0025
LEONARD, J. W.	NONLINEAR DYNAMICS OF CABLES WITH LOW INITIAL TENSION	72-0740
LEONARD, J. W.	STRONGLY CURVED FINITE ELEMENT FOR SHELL ANALYSIS	73-0244
LEONARD, J. W.	3-D FINITE ELEMENT MODEL FOR LAKE CIRCULATION	71-0157
LEONARD, R.	COMPUTER AIDED DESIGN OF THERMALLY LOADED AXISYMMETRIC DIESEL ENGINE COMPONENTS	74-0596
LEONE, S. G.	NUMERICAL STUDY OF DAMPING IN VISCOELASTIC SANDWICH BEAMS	73-0443
LEONIDAS, E.	A GENERAL PURPOSE THREE DIMENSIONAL STRESS ANALYSIS PROGRAM	71-0686
LESAINT, P.	CONTINUOUS AND DISCONTINUOUS FINITE ELEMENT METHODS FOR SOLVING THE TRANSPORT EQUATION	75-0099
LESAINT, P.	FINITE ELEMENT METHODS FOR SYMMETRIC HYPERBOLIC EQUATIONS	73-0116
LESAINT, P.	FINITE ELEMENT METHODS FOR THE TRANSPORT EQUATION	74-0925
LESAINT, P.	ISOPARAMETRIC FINITE-ELEMENT METHODS FOR 2-DIMENSIONAL TRANSPORT CALCULATIONS	75-0354
LESAINT, P.	ON A FINITE ELEMENT METHOD FOR SOLVING THE NEUTRON TRANSPORT EQUATION	74-1255
LESAINT, P.	ON A FINITE ELEMENT METHOD FOR SOLVING THE NEUTRON TRANSPORT EQUATION	74-1355
LESAINT, P.	RAYLEIGH RITZ GALERKIN METHODS FOR SYMMETRIC POSITIVE DIFFERENTIAL EQUATIONS APPLICATION TO FINITE E	73-1030
LESAINT, P.	SOME NONCONFORMING FINITE ELEMENTS FOR THE PLATE BENDING PROBLEM	75-0324
LESTER, H. C.	A FINITE ELEMENT FOR THERMAL STRESS ANALYSIS OF SHELLS OF REVOLUTION	73-0798
LESTINGI, J.	BLOCKING TECHNIQUE FOR LARGE SCALE STRUCTURAL ANALYSIS	73-0243
LESTINGI, J.	COMPARISON OF THE NUMERICAL INTEGRATION TECHNIQUE AND THE FINITE ELEMENT METHOD IN THE ANALYSIS OF T	73-0848
LEVENTHAL, S. H	ON NUMERICAL SOLUTIONS OF EQUATIONS OF HYPERBOLIC-ELLIPTIC TYPE	74-0741
LEVERENZ, R. K.	FINITE ELEMENT STRESS ANALYSIS OF A CRACK IN A BI-MATERIAL PLATE	72-0165
LEVERENZ, R. K.	INTERACTIVE GRAPHICS - AN AID FOR DEVELOPING FINITE ELEMENT MODELS	74-0139
LEVERENZ, R. K.	USING INTERACTIVE GRAPHICS FOR THE PREPARATION AND MANAGEMENT OF FINITE ELEMENT DATA	74-0176

LEVINE, H.	NONLINEAR CRACK ANALYSIS WITH FINITE ELEMENTS	73-0875
LEVINE, H. S.	A FINITE ELEMENT METHOD FOR THE PLASTIC BENDING ANALYSIS OF STRUCTURES	68-0021
LEVINE, H. S.	FINITE ELEMENT ANALYSIS OF STRUCTURES IN THE PLASTIC RANGE	71-0031
LEVINE, H. S.	FINITE ELEMENT METHOD FOR THE PLASTIC BENDING ANALYSIS OF STRUCTURES	68-0206
LEVINE, H. S.	NONLINEAR ANALYSIS OF STRUCTURES	74-0363
LEVINE, H. S.	PLANS - A FINITE ELEMENT PROGRAM FOR NONLINEAR ANALYSIS OF STRUCTURES	74-0966
LEVINE, H. S.	PLANS - A FINITE-ELEMENT PROGRAM FOR NONLINEAR ANALYSIS OF STRUCTURES	75-0011
LEVINE, H. S.	PLASTICITY	70-0509
LEVINE, H. S.	PLASTICITY THEORY AND FINITE ELEMENT APPLICATIONS	72-0318
LEVINE, H. W.	NONLINEAR BEHAVIOR OF SHELLS OF REVOLUTION UNDER CYCLIC LOADING	73-0242
LEVY, A.	A THREE DIMENSIONAL "VARIABLE NODE" ISOPARAMETRIC SOLID ELEMENT	74-0779
LEVY, A.	DEVELOPMENT OF TECHNOLOGY FOR MODELING OF A 1/8-SCALE DYNAMIC MODEL OF THE SHUTTLE SOLID ROCKET BOOS	74-0703
LEVY, A.	ELASTIC AND PLASTIC INTERLAMINAR SHEAR DEFORMATION IN LAMINATED COMPOSITES UNDER GENERALIZED PLANE	71-0158
LEVY, A.	FINITE ELEMENT ANALYSIS OF INTERLAMINAR SHEAR FIBROUS COMPOSITES	71-0129
LEVY, A.	PLANS - A FINITE-ELEMENT PROGRAM FOR NONLINEAR ANALYSIS OF STRUCTURES	75-0011
LEVY, A.	VIBRATION AND STRESS ANALYSIS OF SOFT-BONDED SHUTTLE INSULATION TILES, MODAL ANALYSIS WITH COMPACT W	74-0708
LEVY, L. J.	EXPERIMENTAL AND THEORETICAL STUDIES OF HULL-DECKHOUSE INTERACTION - 2-LEVEL DECKHOUSE	71-0672
LEVY, M. J.	DEVELOPMENT OF FINITE ELEMENT COMPUTATIONAL METHODS FOR THREE DIMENSIONAL ELASTIC PLASTIC CRACK ANAL	70-0356
LEVY, N.	APPLICATION OF FINITE ELEMENT METHODS TO LARGE SCALE ELASTIC PLASTIC PROBLEMS OF FRACTURE MECHANICS	69-0372
LEVY, N.	APPLICATION OF THE FINITE ELEMENT METHOD TO 2 AND 3 DIMENSIONAL ELASTIC-PLASTIC PROBLEMS OF FRACTURE	70-0501
LEVY, N.	FINITE ELEMENT FORMULATION OF STRESS STRAIN MATRIX FOR AN ELASTICALLY PERFECTLY PLASTIC TRESCA MATER	70-0275
LEVY, N.	GENERAL PURPOSE PROGRAMS FOR NONLINEAR FINITE ELEMENT ANALYSIS	70-0247
LEVY, N.	PROGRESS IN THREE-DIMENSIONAL ELASTIC-PLASTIC STRESS ANALYSIS FOR FRACTURE MECHANICS	71-0314
LEVY, N.	SMALL SCALE YIELDING NEAR A CRACK IN PLANE STRAIN-FINITE ELEMENT ANALYSIS	71-0355
LEVY, N.	THREE DIMENSIONAL ELASTIC PLASTIC ANALYSIS FOR FRACTURE MECHANICS - 2.	71-0358
LEVY, N.	THREE-DIMENSIONAL ELASTIC-PLASTIC STRESS AND STRAIN ANALYSIS FOR FRACTURE MECHANICS. PHASE 1. SIMPLE	70-0198
LEVY, R.	A NASTRAN POSTPROCESSOR FOR STRUCTURAL MODIFICATION REANALYSIS	71-0160
LEVY, R.	GUYAN REDUCTION SOLUTIONS RECYCLED FOR IMPROVED ACCURACY	71-0159
LEVY, R.	NONPROPORTIONAL LOADING LIMIT FOR STRUCTURES	74-0228
LEVY, R.	SAVINGS IN NASTRAN DECOMPOSITION TIME BY SEQUENCING TO REDUCE ACTIVE COLUMNS	71-0161
LEVY, S.	COMPUTATION OF INFLUENCE COEFFICIENTS FOR AIRCRAFT STRUCTURES WITH DISCONTINUITIES AND SWEEPBACK	47-0002
LEVY, S.	NUMERICAL CONDITIONING OF STIFFNESS MATRIX FORMULATIONS FOR FRAME STRUCTURES	68-0022
LEVY, S.	THREE-DIMENSIONAL STUDY OF NUCLEAR FUEL ROD BEHAVIOUR DURING STARTUP	73-0815
LEVY, S.	THREE-DIMENSIONAL STUDY OF NUCLEAR FUEL ROD BEHAVIOR DURING STARTUP	74-1003
LEW, G. T.	A THREE DIMENSIONAL SOLUTION OF THE TRANSIENT FIELD PROBLEM USING ISOPARAMETRIC FINITE ELEMENTS	72-0601
LEW, T. K.	A COMPUTER MODEL FOR PREDICTING THE LOAD-DEFLECTION RESPONSE OF EXPEDIENT SOIL SURFACING	73-0880
LEW, T. K.	EVALUATION OF A THREE-DIMENSIONAL STRESS CELL FOR GRANULAR SOILS	72-0882
LEWIS, D. C.	HYDROLOGIC ENGINEERING METHODS FOR WATER RESOURCES DEVELOPMENT. VOLUME 10. PRINCIPLES OF GROUND-WATE	72-0696
LEWIS, D. J.	ADVANCES IN THE ANALYSIS OF PRESTRESSED CONCRETE PRESSURE VESSELS	71-0162
LEWIS, E. E.	A FINITE ELEMENT TREATMENT OF NEUTRON DIFFUSION	71-0373
LEWIS, E. E.	APPLICATION OF FINITE ELEMENT METHOD TO MULTIGROUP NEUTRON DIFFUSION EQUATION	72-0310
LEWIS, E. E.	APPLICATION OF PHASE-SPACE FINITE ELEMENTS TO ONE DIMENSIONAL NEUTRON TRANSPORT EQUATION	73-0123
LEWIS, E. E.	APPLICATION OF PHASE SPACE FINITE ELEMENTS TO THE TWO DIMENSIONAL NEUTRON TRANSPORT EQUATION IN X-Y	73-0453
LEWIS, E. E.	DUAL FINITE ELEMENT METHODS FOR NEUTRON DIFFUSION	71-0489
LEWIS, E. E.	FINITE ELEMENT INTEGRAL NEUTRON TRANSPORT	73-1032
LEWIS, E. E.	FINITE ELEMENT NEUTRON DIFFUSION IN CURVILINEAR COORDINATES	72-0039
LEWIS, E. E.	QUADRATIC FINITE ELEMENT IN NEUTRON TRANSPORT	73-1039
LEWIS, E. E.	RAY-EFFECT MITIGATION IN DISCRETE ORDINATE-LIKE ANGULAR FINITE-ELEMENT APPROXIMATIONS IN NEUTRON-TRA	75-0496
LEWIS, E. E.	SPACE FINITE-ELEMENTS 2-DIMENSIONAL MULTIGROUP TRANSPORT CALCULATIONS USING PHASE-SPACE FINITE-ELEME	75-0504
LEWIS, E. E.	TRANSPORT SOLUTIONS USING FINITE ELEMENTS IN SPACE-ANGLE PHASE SPACE	71-0480
LEWIS, E. E.	TWO-DIMENSIONAL TRANSPORT CALCULATIONS USING PHASE-SPACE FINITE ELEMENTS	72-0027
LEWIS, E. E.	2-DIMENSIONAL FINITE-ELEMENT METHOD FOR INTEGRAL NEUTRON-TRANSPORT CALCULATIONS	75-0443
LEWIS, G. M.	CONTINUUM AND FINITE ELEMENT BRANCHING STUDIES OF THE CIRCULAR PLATE.	72-0427
LEWIS, R. W.	A FINITE ELEMENT APPROACH TO TWO-PHASE FLOW IN POROUS MEDIA	75-0392
LEWIS, R. W.	A FINITE ELEMENT FORMULATION OF HEAT CONDUCTION AND HEAT AND MASS TRANSFER PROBLEMS	74-0350
LEWIS, R. W.	A TWO-DIMENSIONAL FINITE ELEMENT - LEAST SQUARES SOLUTION	75-0025
LEWIS, R. W.	AN ANALYSIS OF VARIOUS TIME STEPPING SCHEMES FOR INITIAL VALUE PROBLEMS	73-0607
LEWIS, R. W.	AN APPLICATION OF LEAST SQUARES TO ONE-DIMENSIONAL TRANSIENT PROBLEMS	74-0347
LEWIS, R. W.	AN APPLICATION OF THE FINITE ELEMENT METHOD TO TRANSIENT INTERFACE PROBLEMS IN POROUS MEDIA	73-0608
LEWIS, R. W.	APPLICATIONS OF ELECTRO-OSMOSIS TO GROUND-WATER FLOW PROBLEMS	75-0336
LEWIS, R. W.	COMPARISON OF TIME MARCHING SCHEMES FOR THE TRANSIENT HEAT CONDUCTION EQUATION	75-0477
LEWIS, R. W.	ELECTRO-OSMOSIS AND THE CONTAMINATION OF UNDERGROUND FLUIDS	73-0609
LEWIS, R. W.	FINITE ELEMENT ANALYSIS OF ELECTRO-OSMOTIC DEWATERING	73-0592
LEWIS, R. W.	FINITE ELEMENT ANALYSIS OF THE MOTION OF A GAS-LIQUID INTERFACE IN A POROUS MEDIUM	75-0672
LEWIS, R. W.	FINITE ELEMENT SOLUTION OF COUPLED ELECTROKINETIC AND HYDRODYNAMIC FLOW IN POROUS MEDIA	72-0398
LEWIS, R. W.	FINITE ELEMENT SOLUTION OF NON-LINEAR HEAT CONDUCTION PROBLEMS WITH SPECIAL REFERENCE TO PHASE CHANG	74-0348
LEWIS, R. W.	FINITE ELEMENTS FOR TWO PHASE FLOW IN POROUS MEDIA	74-0060
LEWIS, R. W.	NUMERICAL ANALYSIS OF ELECTRO-OSMOTIC FLOW IN SOILS	73-0444
LEWIS, R. W.	POROUS MEDIA - SOME PROBLEMS IN CIVIL AND PETROLEUM ENGINEERING	74-0349
LEWIS, R. W.	THE DETERMINATION OF A WATER/GAS INTERFACE USING FINITE ELEMENTS	74-0346
LEWIS, R. W.	TRANSIENT TWO-DIMENSIONAL PROBLEMS UTILIZING THE LEAST SQUARES ALGORITHM	75-0219
LEWIS, T. B.	STRESS ANALYSIS OF RIBBON REINFORCED COMPOSITES	70-0221
LEWIS, T. B.	STRESS ANALYSIS OF TAPE REINFORCED COMPOSITES	69-0068
LI, C. Y.	ON THE SOLUTION OF TRANSIENT FREE-SURFACE FLOW PROBLEMS IN POROUS MEDIA BY THE FINITE ELEMENT METHOD	73-1019
LI, C.	FINITE ELEMENT ANALYSIS OF INFLATABLE SHELL	73-0241
LI, C.	STRONGLY CURVED FINITE ELEMENT FOR SHELL ANALYSIS	73-0244
LI, C. Y.	FINITE ELEMENT ANALYSIS OF SLUICE GATE FLOW	71-0540
LI, C-T.	ANALYSIS OF INFLATABLE SHELLS BY THE FINITE ELEMENT METHOD	71-0822
LI, P.	ERROR ANALYSIS OF A REMOTE TRANSIENT HEAT FLUX SENSOR	74-0753
LIAM FINN, W. D	FINITE ELEMENT ANALYSIS OF SEEPAGE THROUGH DAMS	67-0152
LIANIS, G.	LARGE DISPLACEMENT ANALYSIS OF VISCOELASTIC BEAMS AND FRAMES BY FINITE-ELEMENT METHOD	74-0644
LIAU, A. C.	MECHANICS OF CABLE MOORING SYSTEMS. VOLUME V11. THE STEADY-STATE BEHAVIOR OF A PYRAMID ARRAY SYSTEM	73-0724
LIAU, A. C.	MECHANICS OF CABLE MOORING SYSTEMS. VOL V1. A COMPUTER PROGRAM FOR ANALYZING THE STEADY STATE CON	73-0725
LIAW, C. Y.	EARTHQUAKE ANALYSIS OF AXISYMMETRIC TOWERS PARTIALLY SUBMERGED IN WATER	75-0076
LIBER, T.	DYNAMIC RESPONSE OF SOIL/CONCRETE INTERFACES AT HIGH PRESSURE	74-0836
LICHARDUS, S.	ORTHOTROPIC FLAT PLATES SUPPORTED ON COLUMNS ELONGATED IN PLANE	72-0492
LIDA, M.	DETERMINATION DU CHAMP DES CONTRAINTES DANS LE CAISSON DU REACTEUR DE BUGEY L'	67-0121
LIDA, S.	APPLICATION OF FINITE ELEMENT ANALYSIS METHOD TO TWO-DIMENSIONAL PROBLEMS IN FRACTURE MECHANICS	69-0033
LIEN, S.	APPLICATION OF FINITE ELEMENT ANALYSIS TO CERAMIC COMPONENTS	75-0040
LIEN, S.	A PROCEDURE FOR FINITE ELEMENT PLATE AND SHELL PRE AND POST BUCKLING ANALYSIS	71-0096
LIEN, S-Y.	FINITE ELEMENT ELASTIC THIN SHELL PRE-BUCKLING AND POST-BUCKLING ANALYSIS	71-0821
LIFSHITZ, J. M.	ELASTIC-PLASTIC FINITE ELEMENT ANALYSIS OF SHORT FIBRE COMPOSITES	74-0178
LIGGETT, J. A.	CONVECTIVE TRANSPORT FINITE ELEMENT ANALOG	75-0289
LIGGETT, J. A.	FINITE ELEMENT CIRCULATION ANALYSIS OF VARIABLE-DEPTH SHALLOW LAKES	73-0727
LIGHT, M. F.	PREDICTION OF SLOW CRACK GROWTH BY A FINITE-ELEMENT METHOD	75-0416
LIKINS, P. W.	ANALYTICAL DYNAMICS AND NONRIGID SPACECRAFT SIMULATION	74-0771
LIKINS, P. W.	DYNAMIC ANALYSIS OF A SYSTEM OF HINGE-CONNECTED RIGID BODIES WITH NONRIGID APPENDAGES	74-0852
LIKINS, P. W.	FINITE ELEMENT APPENDAGE EQUATIONS FOR HYBRID COORDINATE DYNAMIC ANALYSIS	72-0022
LIKINS, P. W.	GEOMETRIC STIFFNESS CHARACTERISTICS OF A ROTATING ELASTIC APPENDAGE	74-1296
LIKINS, P. W.	MATHEMATICAL MODELING OF SPINNING ELASTIC BODIES FOR MODAL ANALYSIS	73-0441
LILJEDAHL, J. B	FINITE ELEMENT METHOD OF PREDICTING STRESS DISTRIBUTIONS AND SOIL DEFORMATION UNDER A TRACTIVE DEVIC	69-0389
LILJEDAHL, J. B	THE FINITE ELEMENT METHOD FOR PREDICTING STRESS DISTRIBUTION AND SOIL DEFORMATION UNDER A TRACTIVE D	71-0801

LIN, C. W.	SHELL STRUCTURES AND CLIMATIC INFLUENCES.	72-0389
LIN, C-S.	FINITE ELEMENT STUDY OF A REINFORCED CONCRETE CYLINDRICAL SHELL THROUGH ELASTIC, CRACKING, AND ULTIM	75-0317
LIN, C-S.	NONLINEAR ANALYSIS OF RC SHELLS OF GENERAL FORM	75-0053
LIN, E. S.	FINITE ELEMENT METHOD FOR INCOMPRESSIBLE SLOW VISCOUS FLOW WITH A FREE SURFACE	69-0303
LIN, F. T.	THE FINITE ELEMENT ANALYSIS OF LAMINATED COMPOSITES	71-0731
LIN, F. T.	THE FINITE ELEMENT ANALYSIS OF LAMINATED COMPOSITES	72-0955
LIN, F. T.	THREE-DIMENSIONAL FINITE ELEMENT ANALYSIS OF LAMINATED COMPOSITES	72-0345
LIN, F-T.	THE ELASTIC STRESS ANALYSIS OF A BI-MATERIAL PLATE WITH A CRACK NORMAL TO THE INTERFACES	72-0819
LIN, J. C.	EFFECT OF 2450-MHZ RADIATION ON THE RABBIT EYE	75-0365
LIN, J. M.	TRANSVERSE PROPERTIES OF FIBROUS COMPOSITES	69-0069
LIN, J. M.	TRANSVERSE PROPERTIES OF UNIDIRECTIONAL ALUMINUM MATRIX FIBROUS COMPOSITES	69-0167
LIN, T. H.	INITIAL YIELD SURFACE OF A UNIDIRECTIONALLY REINFORCED COMPOSITE	71-0280
LIN, Y. K.	APPLICATIONS OF THE ASSUMED-STRESS HYBRID METHOD TO FINITE ELEMENT ANALYSIS IN FRACTURE MECHANICS	75-0495
LINDBERG, G. M.	COMPARISON OF TWO HIGH-PRECISION TRIANGULAR FINITE ELEMENTS FOR ARBITRARY DEEP SHELLS	71-0071
LINDBERG, G. M.	FINITE ELEMENT DYNAMIC ANALYSIS OF SHALLOW SHELL STRUCTURES	70-0164
LINDBERG, G. M.	A FINITE CYLINDRICAL SHELL ELEMENT AND THE VIBRATIONS OF A CURVED FAN BLADE	68-0131
LINDBERG, G. M.	A HIGH PRECISION TRIANGULAR PLATE BENDING ELEMENT	68-0052
LINDBERG, G. M.	A HIGH-PRECISION TRIANGULAR CYLINDRICAL SHELL FINITE ELEMENT	71-0163
LINDBERG, G. M.	A SHALLOW SHELL FINITE ELEMENT OF TRIANGULAR SHAPE	70-0044
LINDBERG, G. M.	ACCURATE FINITE ELEMENT MODELLING OF FLAT AND CURVED STIFFENED PANELS	72-0538
LINDBERG, G. M.	ANNULAR AND CIRCULAR SECTOR FINITE ELEMENTS FOR PLATE BENDING	70-0100
LINDBERG, G. M.	APPLICATION OF THE HIGH PRECISION TRIANGULAR PLATE BENDING ELEMENT TO PROBLEMS WITH CURVED BOUNDARIE	69-0071
LINDBERG, G. M.	CLOSED FORM, FINITE ELEMENT SOLUTIONS FOR PLATE VIBRATIONS	69-0168
LINDBERG, G. M.	CONVERGENCE STUDIES OF EIGENVALUE SOLUTIONS USING TWO FINITE PLATE ELEMENTS	70-0085
LINDBERG, G. M.	DYNAMIC ANALYSIS OF SHALLOW SHELLS WITH A DOUBLY-CURVED TRIANGULAR FINITE ELEMENT	71-0353
LINDBERG, G. M.	FINITE ELEMENT VIBRATION ANALYSIS OF MULTICONDUCTOR ELECTRICAL TRANSMISSION LINES	73-0590
LINDBERG, G. M.	FINITE ELEMENTS FOR PLATE BENDING	70-0325
LINDBERG, G. M.	FINITE PLATE-BENDING ELEMENTS IN POLAR COORDINATES	68-0234
LINDBERG, G. M.	FREE VIBRATIONS AND RANDOM RESPONSE OF AN INTEGRALLY-STIFFENED PANEL	70-0138
LINDBERG, G. M.	JET NOISE EXCITATION OF AN INTEGRALLY STIFFENED PANEL	71-0546
LINDBERG, G. M.	LUMPED PARAMETER METHODS APPLIED TO ELASTIC VIBRATIONS	63-0016
LINDBERG, G. M.	NEW DEVELOPMENTS IN THE FINITE ELEMENT ANALYSIS OF SHELLS	69-0334
LINDBERG, G. M.	REFINED FINITE ELEMENTS FOR FOLDED PLATE STRUCTURES	72-0574
LINDBERG, G. M.	STATIC AND DYNAMIC APPLICATIONS OF A HIGH-PRECISION TRIANGULAR PLATE BENDING ELEMENT	69-0086
LINDBERG, G. M.	STRESS ANALYSIS OF MULTI-CELLULAR CAISSONS	74-0338
LINDBERG, G. M.	STRUCTURAL DESIGN AND ANALYSIS OF THE SPACE SHUTTLE	70-0336
LINDBERG, G. M.	THE EFFECT OF LUMPED PARAMETERS ON BEAM FREQUENCIES	63-0012
LINDBERG, G. M.	THE VIBRATION OF STEPPED CANTILEVERED PLATES	67-0129
LINDBERG, G. M.	VIBRATION ANALYSIS OF CANTILEVERED CURVED PLATES USING A NEW CYLINDRICAL SHELL FINITE ELEMENT	68-0130
LINDBERG, G. M.	VIBRATION MODES AND RANDOM RESPONSE OF A MULTI-BAY PANEL SYSTEM USING FINITE ELEMENTS	67-0128
LINDBERG, G. M.	VIBRATION OF NON-UNIFORM BEAMS	63-0021
LINDEMAN, A. J.	APPLICATION OF FINITE ELEMENT TECHNIQUES FOR THE NUMERICAL SOLUTION OF THE NEUTRON TRANSPORT AND DIF	71-0013
LINDEMAN, A. J.	APPLICATIONS OF FINITE ELEMENT METHODS IN REACTOR MATHEMATICS - NUMERICAL SOLUTION OF THE NEUTRON DI	72-0535
LINDEMAN, A. J.	APPLICATIONS OF FINITE ELEMENT METHODS IN REACTOR MATHEMATICS. NUMERICAL SOLUTION OF THE NEUTRON TRA	74-0853
LINDEMAN, A. J.	CONSTRUCTION OF A FINITE ELEMENT APPROXIMATION WHICH CROSSES MATERIAL INTERFACES	74-0896
LINDEMAN, A. J.	EMPIRICAL INVESTIGATION OF REORDERING AND DATA MANAGEMENT FOR FINITE ELEMENT SYSTEMS OF EQUATIONS	73-0944
LINDEMAN, A. J.	THE USE OF INTERPOLATORY POLYNOMIALS FOR A FINITE ELEMENT SOLUTION OF THE MULTIGROUP DIFFUSION EQUAT	74-0524
LINDEMAN, A. J.	TIMING COMPARISON STUDY FOR SOME HIGH ORDER FINITE ELEMENT PROCEDURES AND A LOW ORDER FINITE DIFFERE	72-0767
LINDEMAN, J.	COMPUTATIONAL ANALYSIS AND EVALUATION OF THE FINITE ELEMENT METHOD FOR A CLASS OF NUCLEAR REACTOR CO	74-0595
LINDEMAN, R. A.	FINITE ELEMENT COMPUTER PROGRAM FOR THE SOLUTION OF NONLINEAR AXISYMMETRIC CONTACT PROBLEMS WITH INT	74-1094
LINDLEY, P. B.	ENERGY FOR CRACK GROWTH IN MODEL RUBBER COMPONENTS	72-0568
LINDLEY, P. B.	FINITE-ELEMENT PROGRAMME FOR THE PLANE-STRAIN ANALYSIS OF RUBBER	75-0362
LINDLEY, P. B.	PLANE STRESS ANALYSIS OF RUBBER AT HIGH STRAINS USING FINITE ELEMENTS	71-0680
LINDLEY, P. B.	SOME APPLICATIONS OF FINITE ELEMENT ANALYSIS IN RUBBER SPRING DESIGN	72-0539
LINDSEY, G. H.	COMPARISON OF NUMERICAL METHODS FOR DETERMINING STRESS INTENSITY FACTORS	72-0176
LINDSEY, G. H.	PLANE STRESS CRACK TRAJECTORIES IN LOW MODULUS MATERIALS	75-0646
LING, F. E.	THERMAL, THERMOELASTIC, AND WEAR SIMULATION OF A HIGH-ENERGY SLIDING CONTACT PROBLEM	73-0430
LING, F. F.	A THERMAL, THERMOELASTIC, AND WEAR ANALYSIS OF HIGH-ENERGY DISK BRAKES	74-0879
LING, F. F.	ELASTO-PLASTIC INDENTATION OF A LAYERED MEDIUM	74-0225
LINK, J.	ZUR WIRKLICHKEITSNAHEN BERECHNUNG VON STAHLBETONPLATTEN MIT DER FINITE-ELEMENT-METHODE. (APPROXIMATE	75-0832
LINK, M.	STABILITY INVESTIGATIONS OF PLATES USING FINITE ELEMENT METHOD BASED ON AN EXPANDED VARIATION PRINCI	71-0663
LINK, M.	ZUR BERECHNUNG EINFACH-SYMMETRISCHER I-TRAEGER NACH DER THEORIE - 2. ORDNUNG UNTER BERUECKSICHTIGUNG	75-0512
LINK, M.	ZUR BERECHNUNG VON PLATTEN NACH DER THEORIE ZWEITER. ORDNUNG MIT HILFE EINES HYBRIDEN DEFORMATIONS M	73-0558
LINO, N.	STUDY ON FRACTURE TOUGHNESS EVALUATION BY ELASTO-PLASTIC ANALYSIS AROUND NOTCHES	74-0619
LINZER, V.	BERECHNUNG VON SPANNUNGEN UND TEMPERATURFELDERN IN FLOSSENWAENDEN UND AUFGESCHWEISSTEN BAUTEILEN. (C	74-0544
LIPINSKI, J.	APPLICATION OF FINITE DIFFERENCES FOR SOLVING THE TWO-DIMENSIONAL ELASTICITY PROBLEM BY MEANS OF THE	75-0482
LIPINSKI, J.	METODA ELEMENTOW SKONSZONYCH ROZWIAZYWANIA OSIOWO-SYMETRYCZNEGO PRZEWODNICTWA CIEPLA (FINITE-ELEMENT	75-0434
LIPP, R.	FINITE ELEMENT SOLUTION FOR AXISYMMETRIC SHELLS	65-0023
LIPPARD, D. C.	FOUNDATION INSTRUMENTATION FOR THE NATIONAL WESTMINISTER BANK TOWER	74-1147
LIPPMANN, M. J.	TWO DIMENSIONAL STOCHASTIC MODEL OF A HETEROGENEOUS GEOLOGIC SYSTEM	73-0715
LIS, S. J.	WEAPON-MOUNT INTERFACE STUDY - TRIPOD MOUNT M3	72-0846
LIU, D.	APPLICATION OF THE FINITE ELEMENT METHOD TO SHIP STRUCTURES	71-0403
LIU, D.	SOME DEVELOPMENTS IN THE ANALYSIS OF COMPLEX SHIP STRUCTURES	72-0332
LIU, D.	STRUCTURAL ANALYSIS OF SL-7 CONTAINERSHIP UNDER COMBINED LOADING OF VERTICAL, LATERAL AND TORSIONAL	74-0734
LIU, H. W.	CRACK TIP STRAIN - A COMPARISON OF FEM CALCULATIONS AND MEASUREMENTS	75-0721
LIU, S. C.	APPLICATION OF RESPONSE-BOUND METHOD IN SHOCK AND VIBRATION ANALYSIS OF TELEPHONE STRUCTURES	74-0655
LIU, S-N.	AN ANALYSIS OF THE PHYSICAL BEHAVIOR OF NONWOVEN FIBROUS MATERIAL BY THE FINITE-ELEMENT METHOD	73-1149
LIU, T. C. Y.	BIAXIAL STRESS-STRAIN RELATIONS FOR CONCRETE	72-0203
LIU, W-H.	COMPARATIVE ANALYSIS OF THE CUTHILL-MCKEE AND THE REVERSE CUTHILL-MCKEE ORDERING ALGORITHMS FOR SPAR	73-1099
LIU, Y.	A FINITE ELEMENT ANALYSIS OF WAVE PROPAGATION IN HUMAN SPINE	73-0792
LIVOLANT, M.	ANALYSIS OF SHELLS OF ANY FORM IN THE PLASTIC REGION BY THE FINITE ELEMENT METHOD	73-0940
LIVOLANT, M.	PLASTIC ANALYSIS OF SHELLS BY FINITE ELEMENT METHODS	73-1017
LIYEOS, J. G.	MASS LOADING EFFECTS ON VIBRATION OF RING AND SHELL STRUCTURES	68-0245
LLOYD, J. R.	ANALYSIS MODELS FOR PLATE ELEMENTS IN THE MEMBRANE STATE OF STRESS	67-0046
LLOYD, T.	EFFECT OF THE VARIATION OF CONSISTENCY IN RESIDUAL SOILS ON LOAD TRANSFER BY BORED PILES	75-0812
LO, K. S.	FINITE SEGMENT ANALYSIS OF FOLDED PLATES	69-0169
LO, K. V.	STRESSES AND DEFORMATIONS OF VERTICAL SLOPES IN ELASTO-PLASTIC ROCKS	70-0114
LO, K. Y.	EVALUATION OF THE STABILITY OF NATURAL SLOPES IN PLASTIC CHAMPLAIN CLAYS	74-0264
LOBITZ, D. W.	MESH GENERATION CODE AND A CONTOUR PLOTTING ROUTINE	72-0802
LOBITZ, D. W.	NUMERICAL SUFFICIENCY TEST FOR MONOTONIC CONVERGENCE OF FINITE ELEMENT MODELS	75-0620
LOBITZ, D. W.	ON A NUMERICAL SUFFICIENCY TEST FOR MONOTONIC CONVERGENCE OF FINITE ELEMENT MODELS	75-0533
LOCHNER, N.	MATRIX DISPLACEMENT ANALYSIS OF PLATES AND SHELLS. PART 11. GENERAL FORMULATION OF THE LINEAR THEORY	71-0621
LOCHNER, N.	ON THE APPLICATION OF THE SHEBA SHELL ELEMENT	72-0651
LOCHNER, N. M.	SPEKTRALANALYSE VON ROTORBLAETTERN NACH DERMATRIZENVERS CHIEBUNGSMETHODE UND MATRIZENKRAFTEMETHODE	66-0123
LOCK, A. C.	ALGORITHM FOR THE LARGE DEFLECTION GEOMETRICALLY NON-LINEAR BEHAVIOUR OF PLANE AND CURVED STRUCTURES	72-0083
LOCK, A. C.	CURVED. CYLINDRICAL SHELL. FINITE ELEMENT	72-0096
LOCK, A. C.	LARGE AMPLITUDE. FREE VIBRATION GEOMETRICALLY NONLINEAR CIRCULAR ARCHES	72-0606
LOCK, A. C.	LARGE DEFLECTION GEOMETRICALLY NON-LINEAR FINITE ELEMENT ANALYSIS OF CIRCULAR ARCHES	73-0021
LOCK, A. C.	THE APPLICATION OF FINITE ELEMENTS TO THE LARGE DEFLECTION GEOMETRICALLY NONLINEAR BEHAVIOR OF CYLIN	72-0815
LOCKE, B.	AN ANALYSIS OF AN INTERSECTION PROBLEM USING THE SEMILOOP SHELL ELEMENT	74-0401
LODEN, W. A.	A DOUBLY CURVED TRIANGULAR SHELL ELEMENT	68-0149
LODEN, W. A.	A STUDY OF MATRIX ANALYSIS METHODS FOR INELASTIC STRUCTURES	63-0022

LODEN, W. A.	AXISYMMETRIC VIBRATION OF CONICAL SHELLS	70-0350
LODEN, W. A.	ELEMENT STIFFNESS MATRICES OF THICK-WALLED ORTHOTROPIC SHELLS WITH APPLICATIONS	69-0261
LOENDORF, D. D.	NASTRAN AS ANALYSIS TOOL IN A STRUCTURAL DESIGN OPTIMIZATION PROCESS	71-0235
LOF, C. J.	COMPLIANCE OF A TAPERED DCB SPECIMEN - CONFIGURATIONS BY A FINITE ELEMENT METHOD	74-0545
LOF, C. J.	FINITE ELEMENT CALCULATIONS OF THE COMPLIANCE OF A TAPERED DOUBLE CANTILEVER BEAM SPECIMEN FOR DIFFE	72-0665
LOGCHER J.	NONLINEAR ANALYSIS OF ELASTIC FRAMED STRUCTURES	68-0222
LOGCHER, R. D.	A GENERAL SYSTEM FOR STRUCTURAL FINITE ELEMENT ANALYSIS	68-0107
LOGCHER, R. D.	COMPUTER - AIDED DESIGN - STANDARD OIL OF INDIANA BUILDING	73-0182
LOGCHER, R. D.	ICES STRUDL-II ENGINEERING USERS MANUAL. VOLS. 1, 11, 111	69-0398
LOGCHER, R. D.	ICES-STRUDL-11	68-0257
LOGCHER, R. D.	IMPACT OF COMPUTERS ON THE PRACTICE OF STRUCTURAL ENGINEERING IN CONCRETE	72-0415
LOGCHER, R. D.	STRUDL-A COMPUTER SYSTEM FOR STRUCTURAL DESIGN	66-0118
LOGCHER, R. D.	THE DEVELOPMENT OF ICES STRUDL	72-0327
LOMACKY, O.	EVALUATION OF FINITE ELEMENT METHODS FOR THE COMPUTATION OF ELASTIC STRESS INTENSITY FACTORS	72-0249
LOMACKY, O.	FINITE ELEMENT FRACTURE MECHANICS ANALYSIS OF TWO-DIMENSIONAL AND AXISYMMETRIC ELASTIC AND ELASTIC-P	74-1092
LOMAX, R.	2-DIMENSIONAL SIMULATION OF SEMICONDUCTOR-DEVICES USING FINITE ELEMENT METHODS	75-0548
LOMAX, R. J.	TWO-DIMENSIONAL FINITE ELEMENT SIMULATION OF SEMICONDUCTOR DEVICES	74-0260
LONG, A. E.	PUNCHING STRENGTH OF SLABS, A FLEXURAL APPROACH USING FINITE ELEMENTS	74-1307
LONGENBAUGH, R.	FINITE ELEMENT METHOD FOR HYDRODYNAMIC DISPERISON EQUATION WITH MIXED PARTIAL DERIVATIVES	74-0030
LONGREE, W. D.	DESIGN, ANALYSIS AND CONSTRUCTION OF THE RESEARCH PLATFORM (NORDSEE)	75-0261
LONKAR, S.	BENDING AND TORSION OF THIN-WALLED STRAIGHT AND CURVED BEAMS WITH VARIABLE OPEN CROSS-SECTION	69-0493
LOOF, H. W.	THE ECONOMICAL COMPUTATION OF STIFFNESS OF LARGE STRUCTURAL ELEMENTS	66-0090
LOOV, R. E.	THE DETERMINATION OF STRESSES AND DEFORMATIONS OF REINFORCED CONCRETE AFTER CRACKING	64-0004
LOPEZ, L. A.	DISCRETE MODEL ANALYSIS OF ELASTIC-PLASTIC PLATES	68-0009
LOPEZ, M. L.	JET-WING LIFTING - SURFACE THEORY USING ELEMENTARY VORTEX DISTRIBUTIONS	75-0245
LOPEZ, M. L.	METHODS FOR PREDICTING THE AERODYNAMIC AND STABILITY AND CONTROL CHARACTERISTICS OF STOL AIRCRAFT. V	73-0719
LOPEZ, M. L.	METHODS OF PREDICTING THE AERODYNAMIC AND STABILITY AND CONTROL CHARACTERISTICS OF STOL AIRCRAFT. VO	73-0717
LORD, H. W.	ELASTIC BENDING OF CIRCULAR PLATES OF VARIABLE THICKNESS. AN ANALYTICAL AND EXPERIMENTAL STUDY	70-0163
LORD, W.	FINITE ELEMENT ANALYSIS OF THE MAGNETIC FIELD DISTRIBUTION INSIDE A ROTATING FERROMAGNETIC BAR	74-1037
LORD, W.	FINITE ELEMENT MODELING OF MAGNETIC FIELD DEFECT INTERACTIONS	75-0068
LORD, W.	PREDICTING MAGNETIC LEAKAGE FIELDS BY FINITE ELEMENT TECHNIQUES	75-0046
LORIOT DE ROUVR	NONLINEAR FINITE ELEMENT ANALYSIS OF MULTI JOINTED ROCK MASSES	72-0956
LOTT, L. J.	DETERMINATION OF VOLUME CHANGE STRESSES IN PLAIN AND REINFORCED CONCRETE USING FINITE ELEMENT ANALYS	69-0477
LOU, Y. K.	APPLICATION OF THE FINITE ELEMENT METHOD TO TOWED CABLE DYNAMICS	75-0809
LOUDIERE, D.	NOUVELLE CONCEPTION DES DRAINS DANS LES BARRAGES EN TERRE HOMOGENE DE PETITE ET MOYENNE DIMENSIONS (	73-0404
LOVATT, J. D.	AN ENGINEERING APPROACH TO FINITE ELEMENT ANALYSIS OF NUCLEAR COMPONENTS	73-1054
LOVERHER, A.	FINITE ELEMENT METHOD FOR DETERMINATION OF THERMAL STRESSES INTERACTIVE GRAPHICS FOR FINITE ELEMENT	74-0155
LOVERHER, A.	FINITE ELEMENT MODEL DATA CHECKOUT WITH INTERACTIVE GRAPHICS	74-0195
LOZIUK, L. A.	FINITE ELEMENT APPROACH TO HYDROTHERMAL ANALYSIS OF SMALL LAKES	72-0363
LOZIUK, L. A.	HYDROTHERMAL ANALYSIS BY FINITE ELEMENT METHOD	72-0023
LOZIUK, L. A.	TRANSIENT HYDROTHERMAL ANALYSIS OF SMALL LAKES	73-0445
LU, Z. A.	FINITE ELEMENT SOLUTION FOR THIN SHELLS OF REVOLUTION	63-0024
LU, Z. A.	FINITE ELEMENT SOLUTION FOR THIN SHELLS OF REVOLUTION	64-0005
LU, Z. A.	FINITE ELEMENT SOLUTION FOR AXISYMMETRIC SHELLS	64-0010
LUBKIN, J. L.	STRUCTURAL MODELING OF A WELDED JOINT TYPICAL FOR VEHICLE FRAMES	73-0539
LUBKIN, J. L.	THE FLEXIBILITY OF A TUBULAR WELDED JOINT IN A VEHICLE FRAME	74-0209
LUBKIN, J. L.	WELDED JOINT STIFFNESS OBTAINED USING SOLID FINITE ELEMENTS	75-0190
LUCKNER, L.	APPLICATION OF FINITE ELEMENT METHODS IN GEOHYDRAULICS	74-0078
LUCKNER, L.	AUFBEREITUNG MATHEMATISCHER MODELLE FUR DIE DIGITALE BERECHNUNG VON GRUNDWASSERSTROMUNGSPROBLEMEN NA	71-0543
LUCKNER, L.	BEITRAG ZUR SIMULATION NICHTSTATIONARER HORIZONTAL-EVENER GRUNDWASSERSTROMUNGSFELDER AUF DER GRUNDLA	74-0088
LUCKNER, L.	GRUNDLAGEN DER DIGITALEN LOSUNG VON GRUNDWASSERSTROMUNGSPROBLEMEN NACH DER METHODE ENDLICHER ELEMENT	71-0542
LUCKNER, L.	PRAKTISCHE BEISPIELE DER DIGITALEN BERECHNUNG VON GRUNDWASSERSTROMUNGS PROBLEMEN NACH DER METHODE EN	73-0081
LUCO, J. E.	DYNAMIC MODELING OF HALF-PLANE BY FINITE ELEMENTS	74-0953
LUCO, J. E.	SOIL-STRUCTURE INTERACTION - CONTINUUM OR FINITE ELEMENT	74-0954
LUEHR, H-P.	ORGANIZATION TECHNIQUES OF THE FINITE ELEMENT METHOD FOR REDUCING CORE STORAGE REQUIREMENTS	74-1120
LUFT, R. W.	AUTOMATIC GENERATION OF FINITE ELEMENT MATRICES	71-0165
LUIR, H. P.	ORGANIZATION TECHNIQUES OF THE FINITE ELEMENT METHOD FOR REDUCING CORE STORAGE REQUIREMENTS	74-0074
LUK, C. H.	ASSUMED STRESS HYBRID FINITE ELEMENT METHOD FOR FRACTURE MECHANICS AND ELASTIC-PLASTIC ANALYSIS	72-0584
LUK, C. H.	ELASTIC CRACK ANALYSIS BY A FINITE ELEMENT HYBRID METHOD	71-0212
LUK, C. H.	ELASTIC PLASTIC ANALYSIS BY ASSUMED STRESS HYBRID MODEL	74-0443
LUK, C. H.	FINITE ELEMENT ANALYSIS FOR LIQUID SLOSHING PROBLEMS	69-0029
LUKAS, I. L.	CURVED BOUNDARY ELEMENTS. GENERAL FORMS OF POLYNOMIAL MAPPINGS	74-1256
LUKAS, L.	CONTRIBUTION TO SOLUTION OF BODIES WITH CURVED BOUNDARY BY FINITE ELEMENT METHOD	72-1001
LUKAS, L.	CURVED BOUNDARY ELEMENTS FOR FINITE ELEMENT METHOD	72-0859
LUKAS, L.	STRESS SEPARATION IN PHOTOELASTICIMETRIC MEASUREMENT BY THE FINITE ELEMENT METHOD	71-0807
LUKASHENKO, V.	SOME RESULTS OF FUSELAGE ANALYSIS BY THE FINITE ELEMENT METHOD USING A COMPUTER	73-0535
LUMING, H.	LUMPED PARAMETER APPROACH TO AXISYMMETRIC DYNAMIC SOIL DEFORMATIONS	69-0170
LUNDER, C. A.	DERIVATION OF A STIFFNESS MATRIX FOR A RIGHT TRIANGULAR PLATE IN BENDING AND SUBJECTED TO INITIAL ST	62-0007
LUNDGREN, H. R.	BUCKLING OF MULTILAYER PLATES BY FINITE ELEMENTS	67-0103
LUNDQUIST, R. G	ASPECTS OF MECHANICAL BEHAVIOR OF ROCK UNDER STATIC AND CYCLIC LOADING. PART A: MECHANICAL BEHAVIOR	72-0762
LUNDQUIST, R. G	NEW HORIZONS IN ROCK MECHANICS	73-0446
LUNG, M.	FINITE ELEMENT PROCEDURE FOR ANALYSIS OF METAL FORMING PROCESSES	74-0380
LUNG, M.	USE OF METHOD OF FINITE ELEMENTS FOR STUDY OF STATIONARY METAL FORMING TECHNIQUES	72-0024
LUNNISS, R. C.	AN APPLICATION OF FINITE ELEMENT TECHNIQUE IN BRIDGE DESIGN	70-0225
LUTHIN, J. N.	COUPLED HEAT AND MOISTURE TRANSPORT MODEL FOR ARCTIC SOILS	74-0578
LUTTRELL, A. L.	PRESSURE VESSEL CERTIFICATION BASED ON FRACTURE MECHANICS TECHNOLOGY	74-0948
LUTZ, L. A.	ANALYSIS OF STRESSES IN CONCRETE NEAR A REINFORCING BAR DUE TO BOND AND TRANSVERSE CRACKING	70-0351
LUXMOORE, A.	PREDICTION OF SLOW CRACK GROWTH BY A FINITE-ELEMENT METHOD	75-0416
LUXMOORE, A. R.	COMPARISON OF FINITE-ELEMENT AND EXPERIMENTAL STUDIES ON DEFORMATIONS OF ZIRCONIUM NOTCHED BEND SPEC	71-0359
LYALL, J. S.	MODELLING HEAT FLOW FROM UNDERGROUND POWER CABLES	74-1158
LYNCH, F.	A FINITE ELEMENT METHOD OF VISCOELASTIC STRESS ANALYSES WITH APPLICATION TO ROLLING CONTACT PROBLEMS	69-0330
LYNCH, F. D.	DYNAMIC RESPONSE OF A CONSTRAINED FIBROUS SYSTEM SUBJECTED TO TRANSVERSE IMPACT. II. A MECHANICAL MO	74-0369
LYNCH, R. D.	THE APPLICATION OF DYNAMIC RELAXATION TO THE FINITE ELEMENT METHOD OF STRUCTURAL ANALYSIS	68-0108
LYNESS, J. F.	FINITE ELEMENT ANALYSIS OF THE STEADY FLOW OF NON NEWTONIAN FLUIDS THROUGH PARALLEL SIDED CONDUITS	74-0056
LYNESS, J. F.	FINITE ELEMENT ANALYSIS OF ENGINEERING SYSTEMS GOVERNED BY A NON-LINEAR QUASI-HARMONIC EQUATION	75-0664
LYNESS, J. F.	INVESTIGATION OF BOND FAILURE IN FIBRE-REINFORCED MATERIALS BY THE FINITE ELEMENT METHOD	72-0034
LYNN, P. O.	KINDS OF CONVERGENCE AND IMPROVED CONVERGENCE OF CONFORMING FINITE ELEMENT SOLUTIONS IN PLATE BENDIN	70-0323
LYNN, P. P.	A LEAST SQUARES FINITE ELEMENT ANALYSIS OF NONLINEAR BIOPHYSICAL DIFFUSION-KINETICS EQUATIONS	74-1257
LYNN, P. P.	A METHOD TO GENERATE BOTH UPPER AND LOWER BOUNDS TO PLATE EIGENVALUES BY CONFORMING DISPLACEMENT FIN	71-0166
LYNN, P. P.	CONVERGENCE OF EIGENVALUE SOLUTIONS IN CONFORMING PLATE BENDING FINITE ELEMENTS	72-0195
LYNN, P. P.	FINITE ELEMENT ANALYSIS OF PLATE BENDING WITH TRANSVERSE SHEAR DEFORMATION	70-0109
LYNN, P. P.	FINITE ELEMENTS FORMULATED BY THE WEIGHTED DISCRETE LEAST SQUARES METHOD	74-0200
LYNN, P. P.	LEAST SQUARES FINITE ELEMENT ANALYSIS OF LAMINAR BOUNDARY LAYER FLOWS	74-0616
LYNN, P. P.	TRIANGULAR THICK PLATE BENDING ELEMENTS	71-0167
LYNN, P. P.	USE OF THE LEAST SQUARES CRITERION IN THE FINITE ELEMENT FORMULATION	73-0270
LYSMER, J.	FINITE DYNAMIC MODEL FOR INFINITE MEDIA	69-0171
LYSMER, J.	LUSH: A COMPUTER PROGRAM FOR COMPLEX RESPONSE ANALYSIS OF SOIL-STRUCTURE SYSTEMS	74-0744
LYSMER, J.	QUAD-4: A COMPUTER PROGRAM FOR EVALUATING THE SEISMIC RESPONSE OF SOIL STRUCTURES BY VARIABLE DAMPIN	73-0771
LYSMER, J.	SOIL-STRUCTURE INTERACTION ANALYSES FOR EVALUATING SEISMIC RESPONSE	74-0745
LYSMER, J.	THE FINITE ELEMENT METHOD AND ITS APPLICATION TO SOIL DYNAMIC PROBLEMS	68-0288
LYSMER, J.	THE PROPAGATION OF LOVE WAVES ACROSS NONHORIZONTALLY LAYERED STRUCTURES	71-0600
LYTTON, R. L.	PAVEMENT CRACKING IN WEST TEXAS DUE TO FREEZE-THAW CYCLING	75-0276

65

LYTTON, R. L.	STABILITY CRITERIA FOR TWO FINITE ELEMENT SCHEMES FOR PARABOLIC EQUATION	75-0568
MA, J. H.	STRESS ANALYSIS OF COMPLEX SHIP COMPONENTS BY A NUMERICAL PROCEDURE USING CURVED FINITE ELEMENTS	73-0879
MA, P. S.	APPLICATION OF THE FINITE ELEMENT METHOD TO PROBLEMS OF TEMPERATURE DISTRIBUTION ANALYSIS FOR BODIES	74-1117
MA, S. P. K.	ANALYSIS OF CORNER SUPPORTED SKEW SLABS	68-0203
MAARSINGH, R. A	PROPOSALS FOR THE APPLICATION OF THE FINITE ELEMENT TECHNIQUE TO SOME AIRCRAFT AERODYNAMICAL PROBLEM	72-0628
MACBAIN, J. C.	VIBRATORY BEHAVIOR OF TWISTED CANTILEVERED PLATES	75-0507
MACCHI, A.	TEMPERATURES AND STRESSES IN A BOILER MEMBRANE WALL TUBE	74-1053
MACIOTTA, R.	SIGNIFICANT APPLICATIONS OF FINITE ELEMENTS AND FINITE DIFFERENCES METHODS TO SOLVE PROBLEMS OF THER	73-0864
MACK, L. R.	FINITE ELEMENT METHOD FOR INCOMPRESSIBLE SLOW VISCOUS FLOW WITH A FREE SURFACE	69-0303
MACLAUGHLIN, T.	STRESS CONCENTRATIONS NEAR A DISCONTINUITY IN FIBROUS COMPOSITES	71-0316
MACLEAD, I. A.	NEW RECTANGULAR FINITE ELEMENT FOR SHEAR WALL ANALYSIS	69-0362
MACNEAL, R. H.	A HYBRID METHOD OF COMPONENT MODE SYNTHESIS	71-0681
MACNEAL, R. H.	COMPUTERIZED SUBSTRUCTURE ANALYSIS	75-0419
MACNEAL, R. H.	DYNAMIC STRUCTURAL ANALYSIS WITH THE NASTRAN COMPUTER PROGRAM	70-0213
MACNEAL, R. H.	SOME ORGANIZATIONAL ASPECTS OF NASTRAN	73-1033
MACNEAL, R. H.	SOME ORGANIZATIONAL ASPECTS OF NASTRAN	74-1005
MACNEAL, R. H.	THE NASTRAN COMPUTER PROGRAM FOR STRUCTURAL ANALYSIS	71-0682
MACNEAL, R. H.	THE NASTRAN THEORETICAL MANUAL	70-0298
MADABUSHI, V. J	FINITE ELEMENT SOLUTION OF SATURABLE MAGNETIC FIELD PROBLEMS	70-0205
MADDEN, R.	ANALYSIS OF STRESSES AND DEFLECTIONS IN FRAME SUPPORTED TENTS	74-0689
MADDUX, G. E.	MAGIC, AN AUTOMATED GENERAL PURPOSE SYSTEM FOR STRUCTURAL ANALYSIS, VOLUME II, USER'S MANUAL	69-0144
MADSEN, O. S.	MATHEMATICAL MODELS OF THE MASSACHUSETTS BAY. PART 1. FINITE ELEMENT MODELING OF TWO-DIMENSIONAL HYD	73-0800
MAEDA, Y.	RECENT TRENDS IN STUDIES AT INTERNATIONAL CONFERENCE ON STRUCTURAL ENGINEERING	71-0347
MAESTRELLO, K.	RESPONSE OF STRUCTURE TO THE PSEUDO-SOUND FIELD OF A JET (USING A COMBINED CONTINUUM AND FINITE ELEM	67-0146
MAESTRELLO, L.	TEST RESULTS FROM THE BOUNDARY LAYER FACILITY - RESPONSE OF STRUCTURE TO THE PSEUDO-SOUND FIELD OF A	68-0262
MAGARA, E.	ANALYSIS OF CABLE NETS IN MIXED FORMULATION	73-0311
MAGHSOOD, J.	FINITE ELEMENT ANALYSIS OF WOOD STRUCTURAL MEMBERS	70-0180
MAGHSOOD, J.	FINITE ELEMENT ANALYSIS OF WOOD BEAMS	70-0503
MAGHSOOD, J.	LINEAR AND NONLINEAR FINITE ELEMENT ANALYSIS OF WOOD STRUCTURAL MEMBERS	73-0117
MAHAPATRA, B. C	TANGENT STIFFNESS MATRIX FOR SPACE FRAME MEMBERS	69-0397
MAHATA, P. C.	NEW TRIANGULAR ELEMENT FOR FINITE DIFFERENCE SOLUTION OF AXISYMMETRIC CONDUCTION PROBLEMS IN CYLINDR	74-0381
MAHMOUD, M. H.	SILAGE - SILO INTERACTION USING MATERIAL CHARACTERIZATION AND FINITE ELEMENT ANALYSIS	75-0792
MAHOMED, M.	APPLICATION OF THE FINITE ELEMENT TECHNIQUE TO THE STUDY OF AREAS OF HIGH STRESS GRADIENTS WITHIN A	73-0259
MAHRENHOLTZ, O.	FINITE ELEMENT PROCEDURE FOR ANALYSIS OF METAL FORMING PROCESSES	74-0380
MAHTAB, M. A.	THREE DIMENSIONAL FINITE ELEMENT ANALYSIS OF JOINTED ROCK SLOPES	70-0502
MAI, M.	F 104 G FINAL MOTIVE: GROUND VIBRATION CALCULATIONS FOR THE STARFIGHTER F 104 G	69-0456
MAI, M.	MALEN DIMENSIONIERUNG VON TRAGWERKEN	74-0683
MAI, M.	TURBAN VERSION 1.0. EIN SPEZIELLES PROGRAMM ZUR LINEAREN UND NICHTLINEAREN STATISCHEN UND DYNAMISCHE	71-0622
MAIDEN, D. E.	APPLICATION OF FINITE ELEMENT ANALYSIS METHOD TO TWO-DIMENSIONAL PROBLEMS IN FRACTURE MECHANICS	69-0033
MAIDEN, D. E.	APPLICATION OF THE METHOD OF FINITE ELEMENT ANALYSIS TO TWO-DIMENSIONAL PROBLEMS IN FRACTURE MECHANI	68-0100
MAIDEN, D. E.	STRESS INTENSITY FACTOR FOR A STRAIGHT CRACK APPROACHING A HOLE	69-0158
MAIER, G.	EFFECTS OF GEOMETRY CHANGE ON ESSENTIAL FEATURES OF INELASTIC BEHAVIOUR	73-0448
MAIER, G.	FINITE ELEMENT APPROACH TO OPTIMAL DESIGN OF PLASTIC STRUCTURES IN PLANE STRESS.	72-0397
MAIER, G.	FINITE ELEMENT ELASTOPLASTIC ANALYSIS BY QUADRATIC PROGRAMMING: THE MULTISTAGE METHOD	73-0842
MAIER, G.	SHAKEDOWN THEORY IN PERFECT ELASTOPLASTICITY WITH ASSOCIATED AND NON ASSOCIATED FLOW LAWS - A FINITE	69-0172
MAILLE, M.	APPLICATIONS OF FINITE ELEMENT METHODS IN FLUID DYNAMICS	75-0184
MAINS, R. M.	COMPARISONS OF CONSISTENT MASS MATRIX SCHEMES	69-0437
MAJID, K. I.	DESIGN OF INCLINED TIED ARCH RAILWAY BRIDGES OVER THE M56	71-0297
MAJID, K. I.	THE EFFECT OF COMPOSITE ACTION ON THE ELASTOPLASTIC ANALYSIS OF COMPLETE BUILDING STRUCTURES	74-0441
MAJIDZADEH, K.	ANALYSIS OF FATIGUE OF PAVING MIXTURES FROM THE FRACTURE MECHANICS VIEWPOINT	72-0540
MAJIDZADEH, K.	VERIFICATION OF FRACTURE MECHANICS CONCEPTS TO PREDICT CRACKING OF FLEXIBLE PAVEMENTS	73-0772
MAK, C. K.	CREEP DEFORMATION OF A MULTI-STOREY STRUCTURE IN A FIRE	74-1197
MAK, C. K.	FINITE ELEMENT ANALYSIS OF BUCKLING AND POST-BUCKLING BEHAVIORS OF ARCHES WITH GEOMETRIC IMPERFECTIO	73-0186
MAK, C. K.	INCREMENTAL VARIATIONAL METHOD FOR THE LARGE DISPLACEMENT ANALYSIS OF SHELLS WITH GEOMETRIC IMPERFEC	72-0894
MAK, C. K.	PROBABILISTIC BUCKLING BEHAVIOR OF STRUCTURES WITH RANDOM LACK-OF-FIT AND GEOMETRIC IMPERFECTIONS	73-0862
MAK, C. K. K.	THE CONDITION OF SOME FINITE ELEMENT COEFFICIENT MATRICES	70-0393
MAK, S. W.	SUBSTRUCTURES ANALYSIS OF IMPELLER VIBRATION MODES	75-0283
MAKI, A. C.	FINITE ELEMENT TECHNIQUES FOR ORTHOTROPIC PLANE STRESS AND ORTHOTROPIC PLATE ANALYSIS	68-0274
MAKOJU, J. O.	RING FINITE ELEMENTS FOR AXISYMMETRIC AND NON-AXISYMMETRIC THIN SHELL ANALYSIS	72-0922
MAKSIMOVIC, M.	OPTIMUM POSITION OF THE CENTRAL CLAY CORE OF A ROCKFILL DAM IN RESPECT TO ARCHING AND HYDRAULIC FRAC	73-0447
MALANG, S.	HETRAP: A HEAT TRANSFER ANALYSIS PROGRAM	74-0870
MALCOM, D. J.	SHEAR LAG IN STIFFENED BOX GIRDERS	70-0243
MALE, D. J.	FINITE ELEMENT STUDY OF COMPOSITE ACTION IN WALLS	69-0291
MALEJANNAKIS, G	GEOMETRIC NONLINEARITY AND THE FINITE ELEMENT DISPLACEMENT METHOD	74-0679
MALEJANNAKIS, G	ANWENDUNG DER MATRIZENVERSCHIEBUNGSMETHODE AUF ERZWUNGENE SCHWINGUNGEN (SCHWINGUNGSPROBLEME) (ARBEIT	71-0627
MALEJANNAKIS, G	DYNAN LECTURE NOTES WITH COMPUTATIONAL EXAMPLES	71-0581
MALEJANNAKIS, G	NATURAL GEOMETRY OF SURFACE WITH SPECIFIC REFERENCE TO THE MATRIX DISPLACEMENT ANALYSIS OF SHELLS, 1	73-0709
MALHOTRA, V. M.	IMPACT OF COMPUTERS ON THE PRACTICE OF STRUCTURAL ENGINEERING IN CONCRETE	72-0414
MALINA, H.	BERECHNUNG VON SPANNUNGSUMLAGERUNGEN IN FELS UND BODEN MIT HILFE DER ELEMENTENMETHODE	69-0364
MALINA, H.	EFFECT OF DISCONTINUITIES ON STRESSES AND DEFORMATIONS OF ROCK AROUND A TUNNEL. (EINFLUSS VON DISKON	70-0170
MALINA, H.	NUMERICAL DETERMINATION OF STRESSES AND DEFORMATIONS IN ROCK TAKING INTO ACCOUNT DISCONTINUITIES	70-0167
MALLETT, R. H.	A FINITE ELEMENT SOLUTION TECHNIQUE FOR THE BOLTZMANN EQUATION	69-0026
MALLETT, R. H.	A KOITER TYPE METHOD FOR FINITE ELEMENT ANALYSIS OF NONLINEAR STRUCTURAL BEHAVIOUR	70-0064
MALLETT, R. H.	ADAPTION OF KOITER'S METHOD TO FINITE ELEMENT ANALYSIS OF SNAP-THROUGH BUCKLING BEHAVIOUR	71-0109
MALLETT, R. H.	AUTOMATED LARGE DEFLECTION AND STABILITY ANALYSIS OF THREE DIMENSIONAL BAR STRUCTURES	69-0393
MALLETT, R. H.	AUTOMATED METHOD FOR THE LARGE DEFLECTION AND INSTABILITY ANALYSIS OF THREE-DIMENSIONAL TRUSS AND FR	66-0105
MALLETT, R. H.	CURRENT FINITE ELEMENT ANALYSIS PRACTICES	73-0118
MALLETT, R. H.	DISCRETE ELEMENT PROCEDURE FOR THIN SHELL INSTABILITY ANALYSIS	67-0028
MALLETT, R. H.	EFFICIENT SOLUTION PROCESSES FOR FINITE ELEMENT ANALYSIS OF TRANSIENT HEAT CONDUCTION	69-0014
MALLETT, R. H.	ELASTIC-PLASTIC ANALYSIS OF FLAT PLATES BY THE FINITE ELEMENT METHOD	68-0111
MALLETT, R. H.	FINITE ELEMENT ANALYSIS OF NONLINEAR STRUCTURES	68-0109
MALLETT, R. H.	FORMULATION AND EVALUATION OF A TOROIDAL RING DISCRETE ELEMENT	66-0082
MALLETT, R. H.	FORMULATION AND EVALUATION OF A TRAPEZOIDAL CROSS-SECTION RING DISCRETE ELEMENT	66-0084
MALLETT, R. H.	MAGIC, AN AUTOMATED GENERAL PURPOSE SYSTEM FOR STRUCTURAL ANALYSIS, VOLUME II, USER'S MANUAL	69-0144
MALLETT, R. H.	MAGIC, AN AUTOMATED GENERAL PURPOSE SYSTEM FOR STRUCTURAL ANALYSIS	69-0173
MALLETT, R. H.	PROGRESS IN NONLINEAR FINITE ELEMENT ANALYSIS USING ASYMPTOTIC SOLUTION TECHNIQUES	72-0316
MALLETT, R. H.	STRESSES ANALYSIS USING FINITE ELEMENT METHODS	70-0349
MALLETT, R. L.	DYNAMIC ANALYSIS OF THE DEFORMATION OF STRUCTURES AND OF METAL FORMING	74-0823
MALLETT, R. L.	DYNAMIC ANALYSIS OF STRUCTURAL DEFORMATION AND METAL FORMING	75-0058
MALLICK, D. V.	EFFECT OF OPENINGS ON THE LATERAL STIFFNESS OF INFILLED FRAMES	71-0300
MALNIG, H. W.	FINITE ELEMENTE ZUR BERECHNUNG DER WAERMEUEBERTRAGUNG FUER EBENE UND AXISYMMETRISCHE PROBLEME (FINIT	73-0449
MALONE, D. W.	FINITE ELEMENTS AND DYNAMIC VISCOELASTICITY	71-0479
MALONE, J. P.	A COMPUTER PROGRAM FOR THE ANALYSIS OF LINEARLY ELASTIC PLANE-STRESS, PLANE-STRAIN PROBLEMS	68-0110
MALONEY, J. G.	STRUCTURAL DYNAMIC PROPERTIES OF TACTICAL MISSILE JOINTS - PHASE 1	70-0423
MALONEY, J. G.	STRUCTURAL DYNAMIC PROPERTIES OF TACTICAL MISSILE JOINTS - PHASE 3	74-0811
MALTBAEK, J. C.	VIBRATIONS OF A NON-UNIFORM BEAM	68-0246
MALVERN, L. E.	ON FINITE ELEMENT INTEGRATION IN NATURAL COORDINATES	73-0683
MALVICK, A. J.	ELASTIC DEFORMATION OF LIGHTWEIGHT MIRRORS METHOD OF LIMIT POINT CALCULATION IN FINITE ELEMENT STRUC	72-0232
MAMET, J. C.	ANALYSIS OF SHEAR WALLS BY FINITE ELEMENTS	71-0569
MAMET, J. C.	HYPAR DISPLACEMENT FIELDS FOR RECTANGULAR FINITE ELEMENTS	73-0240
MAMET, J-C.	FINITE-ELEMENT ANALYSIS OF TALL BUILDINGS	72-0959
MAMORU, H.	APPLICATION OF PROGRAM FATIQUE TEST TO MEMBER JOINTS OF HULLS	72-0814

MANCUSO, P.	LIMIT ANALYSIS OF FRAME SYSTEMS STIFFENED BY PANELS	
MANDELL, J. F.	FRACTURE OF GRAPHITE FIBER REINFORCED COMPOSITES	74-0584
MANDELL, J. F.	STRESS INTENSITY FACTORS FOR ANISOTROPIC FRACTURE TEST SPECIMENS OF SEVERAL GEOMETRIES	74-0695
MANG, H. A.	ANALYSIS OF DOUBLE CORRUGATED SHELL STRUCTURES BY THE FINITE ELEMENT METHOD	74-0293
MANGARELLA, P.	NUMERICAL ANALYSIS OF COOLING TOWER PLUMES AND FOG INCIDENCE	74-1116
MANHARDT, P. D.	FINITE ELEMENT SOLUTION FOR ENERGY CONSERVATION USING A HIGHLY STABLE EXPLICIT INTEGRATION ALGORITHM	74-0040
MANHARDT, P. D.	THE FINITE ELEMENT METHOD IN LOW SPEED AERODYNAMICS	72-0633
MANI, M. K.	EFFECTS OF TURBULENCE ON FLOOD FLOW EXPLORED BY FINITE ELEMENT ANALYSIS.	75-0702
MANJOINE, M. J.	ANALYSIS OF STRESSES IN PRESSURIZED WELDED PIPE IN THE CREEP RANGE	72-0410
MANOR, H.	FLEXURE ANALYSIS OF AXIALLY-LOADED BEAM BY FINITE-ELEMENT METHOD WITH INTEGRAL PARAMETERS	71-0404
MANOR, H.	USE OF INTEGRAL PARAMETERS IN ROD PROBLEMS	75-0462
MANSFIELD, L.	A " LOCAL " BASIS OF GENERALIZED SPLINES OVER RIGHT TRIANGLES DETERMINED FROM A NONUNIFORM PARTITION	74-1194
MANSFIELD, L.	HIGHER ORDER COMPATIBLE TRIANGULAR FINITE ELEMENTS	72-0525
MANSON, S. S.	A GENERAL APPROACH TO THE PRACTICAL SOLUTION OF CREEP PROBLEMS	74-0152
MANSON, S. S.	PRACTICAL SOLUTION OF PLASTIC DEFORMATION PROBLEMS IN THE ELASTIC PLASTIC RANGE	59-0003
MANSOUR, A.	STRUCTURAL ANALYSIS AND DESIGN OF A CATAMARAN CROSS-STRUCTURE BY FINITE ELEMENT METHOD	59-0004
MANUEL, R. F.	FINITE ELEMENT ANALYSIS OF A DIAGONAL CRACKING BEHAVIOUR IN REINFORCED CONCRETE	73-0052
MAPLE, J. A.	A STUDY OF FINITE ELEMENTS FOR THE ANALYSIS OF SANDWICH PLATES	73-0462
MAR, G. W.	APPLICATIONS OF THE ASSUMED-STRESS HYBRID METHOD TO FINITE ELEMENT ANALYSIS IN FRACTURE MECHANICS	69-0503
MARCAL, P. V.	A COMPARATIVE STUDY OF NUMERICAL METHODS OF ELASTIC-PLASTIC ANALYSIS	75-0495
MARCAL, P. V.	A FINITE ELEMENT FORMULATION FOR PROBLEMS OF LARGE STRAIN AND LARGE DISPLACEMENT	67-0047
MARCAL, P. V.	A NUMERICAL THERMO-MECHANICAL MODEL FOR THE WELDING AND SUBSEQUENT LOADING OF A FABRICATED STRUCTURE	70-0069
MARCAL, P. V.	COMPARATIVE STUDY OF NUMERICAL METHODS OF ELASTIC-PLASTIC ANALYSIS	72-0847
MARCAL, P. V.	DETERMINATION OF UPPER BOUNDS FOR PROBLEMS IN PLANE STRESS USING FINITE ELEMENT TECHNIQUES	67-0153
MARCAL, P. V.	DEVELOPMENT OF FINITE ELEMENT COMPUTATIONAL METHODS FOR THREE DIMENSIONAL ELASTIC PLASTIC CRACK ANAL	67-0034
MARCAL, P. V.	EFFECT OF INITIAL DISPLACEMENT ON PROBLEM OF LARGE DEFLECTION AND STABILITY	70-0356
MARCAL, P. V.	ELASTIC ANALYSIS OF A SKULL	67-0117
MARCAL, P. V.	ELASTIC PLASTIC ANALYSIS OF PRESSURE VESSEL COMPONENTS	73-0415
MARCAL, P. V.	ELASTIC-PLASTIC ANALYSIS OF TWO-DIMENSIONAL STRESS SYSTEMS BY THE FINITE ELEMENT METHOD	69-0380
MARCAL, P. V.	ELASTIC-PLASTIC ANALYSIS OF FLAT PLATES BY THE FINITE ELEMENT METHOD	67-0049
MARCAL, P. V.	ELASTIC-PLASTIC BEHAVIOR OF A LONGITUDINAL SEMI-ELLIPTIC CRACK IN A THICK PRESSURE VESSEL	68-0111
MARCAL, P. V.	FINITE ELEMENT ALGORITHM FOR THE DETERMINATION OF DYNAMIC BUCKLING	73-0952
MARCAL, P. V.	FINITE ELEMENT ANALYSIS OF NONLINEAR STRUCTURES	75-0302
MARCAL, P. V.	FINITE ELEMENT ANALYSIS OF COMBINED PROBLEMS ON NONLINEAR MATERIAL AND GEOMETRIC BEHAVIOUR	68-0109
MARCAL, P. V.	FINITE ELEMENT ANALYSIS WITH MATERIAL NONLINEARITIES: THEORY AND PRACTICE	69-0175
MARCAL, P. V.	FINITE ELEMENT ANALYSIS OF COMBINED PROBLEMS OF MATERIAL AND GEOMETRIC BEHAVIOUR	69-0459
MARCAL, P. V.	FINITE ELEMENT ANALYSIS WITH MATERIAL NONLINEARITIES, THEORY AND PRACTICE	69-0508
MARCAL, P. V.	FINITE-ELEMENT ANALYSIS WITH MATERIAL NONLINEARITIES	71-0168
MARCAL, P. V.	GENERAL PURPOSE PROGRAMS FOR NONLINEAR FINITE ELEMENT ANALYSIS	69-0177
MARCAL, P. V.	GENERAL PURPOSE PROGRAM FOR FINITE ELEMENT ANALYSIS WITH REFERENCE TO INTERACTION BETWEEN SOFTWARE A	70-0247
MARCAL, P. V.	HYBRID FINITE ELEMENT ANALYSIS WITH PARTICULAR REFERENCE TO AXISYMMETRIC STRUCTURES	73-0334
MARCAL, P. V.	IN-VACUO MODAL DYNAMIC RESPONSE OF THE HUMAN SKULL	70-0248
MARCAL, P. V.	INCREMENTAL STIFFNESS METHOD FOR FINITE ELEMENT ANALYSIS OF THE NONLINEAR DYNAMIC PROBLEM	73-0460
MARCAL, P. V.	LARGE DEFLECTION ANALYSIS OF ELASTIC-PLASTIC SHELLS OF REVOLUTION	71-0525
MARCAL, P. V.	LARGE DEFLECTION ANALYSIS OF ELASTIC-PLASTIC PLATES AND SHELLS	69-0174
MARCAL, P. V.	LARGE STRAIN, LARGE DISPLACEMENT ANALYSIS	69-0178
MARCAL, P. V.	MARC-CDC USER INFORMATION MANUALS	71-0508
MARCAL, P. V.	NONLINEAR ANALYSIS	71-0562
MARCAL, P. V.	NONLINEAR MATERIAL AND GEOMETRIC BEHAVIOUR OF SHELL STRUCTURES	71-0169
MARCAL, P. V.	NUMERICAL THERMO-MECHANICAL MODEL FOR THE WELDING AND SUBSEQUENT LOADING OF A FABRICATED STRUCTURE	71-0084
MARCAL, P. V.	ON GENERAL PURPOSE PROGRAMS FOR FINTIE ELEMENT ANALYSIS WITH SPECIAL REFERENCE TO GEOMETRIC AND MATE	73-0418
MARCAL, P. V.	OPTIMALITY CRITERIA IN STRUCTURAL DESIGN	70-0087
MARCAL, P. V.	OPTIMIZATION OF FINITE ELEMENT GRIDS BASED ON MINIMUM POTENTIAL ENERGY	71-0758
MARCAL, P. V.	OPTIMIZATION OF FINITE ELEMENT GRIDS BASED ON MINIMUM POTENTIAL ENERGY	71-0747
MARCAL, P. V.	PROGRESS IN THREE-DIMENSIONAL ELASTIC-PLASTIC STRESS ANALYSIS FOR FRACTURE MECHANICS	72-0109
MARCAL, P. V.	SOME CURRENT FINITE-ELEMENT MODELS FOR MATERIAL BEHAVIOR	71-0314
MARCAL, P. V.	STRENGTH OF REINFORCED CONCRETE CHAMBERS UNDER EXTERNAL PRESSURE	75-0549
MARCAL, P. V.	STRENGTH OF REINFORCED CONCRETE CHAMBERS UNDER EXTERNAL PRESSURF	75-0293
MARCAL, P. V.	SURVEY OF GENERAL PURPOSE PROGRAMS FOR FINITE ELEMENT ANALYSIS	75-0512
MARCAL, P. V.	THE EFFECT OF INITIAL DISPLACEMENTS OF PROBLEMS OF LARGE DEFLECTION AND STABILITY	72-0323
MARCAL, P. V.	THREE DIMENSIONAL ELASTIC PLASTIC ANALYSIS FOR FRACTURE MECHANICS - 2.	67-0048
MARCAL, P. V.	THREE-DIMENSIONAL ELASTIC-PLASTIC STRESS AND STRAIN ANALYSIS FOR FRACTURE MECHANICS. PHASE 1. SIMPLE	71-0358
MARCAL, P. V.	THREE-DIMENSIONAL FINITE ELEMENT ANALYSIS FOR FRACTURE MECHANICS	70-0188
MARCHERTAS, A.	NONLINEAR FINITE-ELEMENT METHOD FOR PLATES AND ITS APPLICATION TO DYNAMIC RESPONSE OF REACTOR FUEL S	72-0162
MARCHERTAS, A.	NONLINEAR FINITE-ELEMENT FORMULATION FOR TRANSIENT ANALYSIS OF THREE DIMENSIONAL THIN STRUCTURES	74-0670
MARCHUK, G. I.	FINITE-DIFFERENCE AND FINITE-ELEMENT METHODS IN GLOBAL WIND-DRIVEN OCEAN CIRCULATION	74-0874
MARCUS, L. A.	FATIGUE CRACK INITIATION IN A BORON EPOXY PLATE WITH A CIRCULAR HOLE	75-0345
MARECZEK, G.	FINITE-ELEMENT ANALYSIS OF SLOW INCOMPRESSIBLE VISCOUS FLUID MOTION	74-0820
MARECZEK, G.	POTENTIAL FLOW ANALYSIS BY FINITE ELEMENTS	74-0154
MARECZEK, G.	SOME NEW ELEMENTS FOR THE MATRIX DISPLACEMENT METHOD	72-0004
MARECZEK, G.	THERMOMECHANICAL ANALYSIS OF STRUCTURES	68-0017
MARECZEK, G.	THERMOMECHANICAL ANALYSIS OF STRUCTURES	71-0618
MARECZEK, G.	TWO AND THREE-DIMENSIONAL FLOW USING FINITE ELEMENTS	73-0368
MARECZEK, G.	TWO-AND THREE-DIMENSIONAL FLOW ANALYSIS USING FINITE ELEMENTS	69-0008
MAREK, I.	FINITE ELEMENT IN THE THEORY OF SH-WAVE PROPAGATION	70-0021
MARGASS, B.	LOADS AND RESPONSE OF LARGE DUCTED PROPELLER SYSTEMS	73-1034
MARGERIE, J. C.	STUDY OF THE FLEXURAL BEHAVIOUR OF CASTINGS	73-0810
MARGOLIAS, D. S	FREE VIBRATIONS OF PRESSURE LOADED PARABOLOIDAL SHELLS	74-0377
MARGOLIAS, D. S	INEXTENSIONAL VIBRATIONS OF PARABOLOIDAL SHELLS OF REVOLUTION. DOMES AND SHELLS	71-0601
MARGOLIAS, D. S	VIBRATIONS OF PRESSURE LOADED HYPERBOLOIDAL SHELLS	71-0427
MARGOLIS, S. G.	FINITE ELEMENT METHODS FOR UNSTEADY COMPRESSIBLE GAS DYNAMICS	72-0874
MARGUERRE, K.	BERECHNUNG DER ERDEBENSCHWINGUNGEN VON STRUKTUREN MIT DER FINITE ELEMENT METHODE. MECHANISCHE MODELL	75-0045
MARGUERRE, K.	CALCULATION OF VIBRATIONS OF STRUCTURES CAUSED BY EARTHQUAKES USING THE FINITE ELEMENT METHOD	71-0170
MARGUERRE, K.	UBER DIE ANWENDUNG DER ENERGETISCHEN METHODE AUF STABILITATSPROBLEME	71-0690
MARIASUBE, R.	FINITE ELEMENT METHOD FOR AXISYMMETRIC JETS OF FLUID WITH GRAVITY FORCE	38-0001
MARINSHAW, S. A	PLANE STRESS CRACK TRAJECTORIES IN LOW MODULUS MATERIALS	75-0428
MARION, R.	ISOFINEL: ISOPARAMETRIC FINITE ELEMENT CODE FOR ELASTIC ANALYSIS OF TWO-DIMENSIONAL BODIES	75-0646
MARK, R.	PHOTOELASTIC AND FINITE-ELEMENT ANALYSIS OF A QUADRIPARTITE VAULT	75-0676
MARK, R.	PHOTOELASTIC INVESTIGATION OF STRESS CONCENTRATIONS IN SPHERE-CYLINDER TRANSITION REGIONS: INCLUDING	73-0119
MARK, R.	STRESSES AROUND FLEXIBLE ELLIPTIC PIPES	68-0258
MARLOWE, D. E.	NON-METALLIC ANTENNA-SUPPORT MATERIALS	73-0215
MARR, W. A. JR.	FINITE ELEMENT ANALYSIS OF ELASTO-PLASTIC SOILS. REPORT NO. 4. FINITE ELEMENT ANALYSIS OF ELASTO-PLA	73-0716
MARROCCO, A.	ANALYSE NUMERIQUE DU CHAMP MAGNETIQUE D'UN ALTERNATEUR PAR ELEMENTS FINIS ET SURRELAXATION PONCTUELL	72-0788
MARROCCO, A.	APPROXIMATION BY FIRST ORDER FINITE ELEMENTS AND SOLUTION BY PENALTY AND DUALITY OF NONLINEAR DIRICH	73-0625
MARROCCO, A.	ETUDE NUMERIQUE DU CHAMP MAGNETIQUE DANS UN ALTERNATEUR TETRAPOLAIRE PAR LA METHODE DES ELEMENTS FIN	74-0245
MARROCCO, A.	SUR L APPROXIMATION, PAR ELEMENTS FINIS D ORDRE UN, ET LA RESOLUTION, PAR PENALISATION-DUALITE, D UN	74-0413
MARSHALL, J.	ON THE NUMERICAL TREATMENT OF THE NAVIER-STOKES EQUATIONS FOR AN INCOMPRESSIBLE FLUID	75-0845
MARSHALL, J.	AN INVESTIGATION INTO THE ECONOMICS OF USING COMPLEX ELEMENTS IN PLANE STRESS ANALYSIS	73-0809
MARSHALL, J.	SHEAR STRESSES IN BEAMS WITH DISCONTINUITY IN DEPTH	74-0427
MARSHALL, J. A.	AN EXACT BOUNDARY TECHNIQUE FOR IMPROVED ACCURACY IN THE FINITE ELEMENT METHOD	75-0487
MARSHALL, J. A.	MATCHING OF ESSENTIAL BOUNDARY CONDITIONS IN THE FINITE ELEMENT METHOD	73-1035
MARTIN, B. W.	CONTRIBUTION TO NUMERICAL-SOLUTION OF DEVELOPING LAMINAR-FLOW IN ENTRANCE REGION OF CONCENTRIC ANNUL	74-1260
		75-0448

MARTIN, C. W.	AN ITERATIVE IMPROVEMENT FOR FINITE ELEMENT ANALYSIS	74-0428
MARTIN, C. W.	COMPUTER WORD LENGTH AND SOLUTION ACCURACY. EXPERIENCE WITH ELAS AND THE IBM 360	73-0335
MARTIN, G. P.	AN INVESTIGATION OF THE DYNAMIC RESPONSE CHARACTERISTICS OF THE BON TEMPE DAM	66-0047
MARTIN, G. R.	DYNAMIC RESPONSE OF COHESIVE EARTH DAMS TO EARTHQUAKES	67-0131
MARTIN, H.	SPANNUNGEN IN MEMBRANWAENDEN DURCH WAERMEBEANSPRUCHUNG UND MECHANISCHE BELASTUNG. (STRESSES IN MEMBR	75-0066
MARTIN, H.	ZUSAMMENHANG ZWISCHEN OBERFLAECHENBESCHAFFENHEIT, VERBUND UND SPRENGWIRKUNG VON BEWEHRUNGSSTAEHLEN U	75-0450
MARTIN, H. C.	A SURVEY OF FINITE ELEMENT FORMULATIONS OF GEOMETRICALLY NONLINEAR PROBLEMS	71-0171
MARTIN, H. C.	DERIVATION OF STIFFNESS MATRICES FOR THE ANALYSIS OF LARGE DEFLECTION AND STABILITY PROBLEMS	65-0066
MARTIN, H. C.	FINITE ELEMENT ANALYSIS OF FLUID FLOWS	68-0112
MARTIN, H. C.	FINITE ELEMENT FORMULATIONS OF GEOMETRICALLY NONLINEAR PROBLEMS	69-0180
MARTIN, H. C.	FINITE ELEMENTS AND THE ANALYSIS OF GEOMETRICALLY NONLINEAR PROBLEMS	69-0179
MARTIN, H. C.	FURTHER DEVELOPMENT AND APPLICATIONS OF THE STIFFNESS METHOD	64-0020
MARTIN, H. C.	INTRODUCTION TO FINITE ELEMENT ANALYSIS	72-0358
MARTIN, H. C.	LARGE DEFLECTIONS OF STRUCTURES SUBJECT TO HEATING AND EXTERNAL LOADS	60-0002
MARTIN, H. C.	ON THE DERIVATION OF STIFFNESS MATRICES FOR THE ANALYSIS OF LARGE DEFLECTION AND STABILITY PROBLEMS	66-0050
MARTIN, H. C.	PLANE ELASTICITY PROBLEMS AND THE DIRECT STIFFNESS METHOD	61-0009
MARTIN, H. C.	STIFFNESS AND DEFLECTION ANALYSIS OF COMPLEX STRUCTURES	56-0001
MARTIN, H. C.	STIFFNESS MATRIX FOR A TRIANGULAR SANDWICH ELEMENT IN BENDING	67-0050
MARTIN, H. C.	TRUSS ANALYSIS BY STIFFNESS CONSIDERATIONS	58-0001
MARTIN, J. F.	CYCLIC MECHANICAL TESTS AND AN APPROPRIATE ANALYTICAL STRESS-STRAIN MODEL FOR A36 STEEL	74-0815
MARTIN, J. G.	FINITE-ELEMENT CODE TO SIMULATE ANISOTROPIC-PLASMA DIFFUSION IN LINEAR MULTIPOLES	75-0409
MARTIN, J. G.	FINITE-ELEMENT PLASMA-DIFFUSION CODE FOR ARBITRARY TOROIDAL GEOMETRY	75-0500
MARTIN, J. G.	SEMIDISCRETE GALERKIN TECHNIQUES WITH TIME INTERPOLATION AND SPLITTING UP FOR PLASMA SIMULATION	75-0121
MARTIN, M. S.	COMPLEXITY BOUNDS FOR REGULAR FINITE DIFFERENCE AND FINITE ELEMENT GRIDS	72-0872
MARTIN, M. S.	COMPLEXITY BOUNDS FOR REGULAR FINITE DIFFERENCE AND FINITE ELEMENT GRIDS	73-0043
MARTIN, S. W.	AN ALGORITHM FOR REDUCING THE BANDWIDTH OF A MATRIX OF SYMMETRICAL CONFIGURATION	65-0067
MARTIN, W. C.	REMOVAL OF TRUNCATION ERROR IN FINITE ELEMENT ANALYSIS	75-0131
MARTIN, W. R.	CONVERGENCE OF FINITE-ELEMENT METHOD IN NEUTRON-TRANSPORT	75-0408
MARTINEZ, J. E.	EFFICIENT COMPUTER ANALYSIS OF LARGE THREE-DIMENSIONAL PIPING SYSTEMS BY THE FINITE ELEMENT METHOD	72-0378
MARTINEZ, J. E.	NONLINEAR DYNAMIC ANALYSIS OF SHELLS OF REVOLUTION BY MATRIX DISPLACEMENT METHOD	70-0110
MARTZ, J. W.	AVOIDING DANGEROUS AND COSTLY FAN FAILURES	75-0061
MARTZ, J. W.	IMPROVED TECHNIQUES FOR DYNAMIC ANALYSIS OF EARTHMOVING EQUIPMENT	74-0294
MARUYAMA, K.	STRESS ANALYSIS OF A BOLT-NUT JOINT BY THE FINITE ELEMENT METHOD AND THE COPPER-ELECTROPLATING METHO	73-0220
MARUYAMA, K.	STRESS ANALYSIS OF A BOLT-NUT JOINT BY FINITE ELEMENT METHOD AND COPPER-ELECTROPLATING METHOD - 2. S	74-0153
MASAAKI, M.	APPLICATION OF PROGRAM FATIQUE TEST TO MEMBER JOINTS OF HULLS	72-0814
MASAHIRO, M.	APPLICATION OF PROGRAM FATIQUE TEST TO MEMBER JOINTS OF HULLS	72-0814
MASER, K. R.	AN ANALYSIS OF THE SMALL-SCALE STRENGTH TESTING OF ICE	72-0981
MASER, K. R.	ELASTIC ANALYSIS OF LARGE SPACEBOUND MIRRORS	70-0519
MASLENNIKOV, A.	VYUZITI METODY KONECNYCH PRVKU PRO PROJEKTOVANI KOMBINOVANYCH KONSTRUCKCI (FINITE ELEMENT METHOD IDE	73-0667
MASLIYAH, J. H.	FULLY DEVELOPED VISCOUS FLOW IN INTERNALLY FINNED TUBES	75-0648
MASON, J. B.	ANALYSIS OF THERMALLY INDUCED STRUCTURAL VIBRATIONS OF FINITE ELEMENT TECHNIQUES	68-0272
MASON, J. B.	FINITE ELEMENT ANALYSIS OF COUPLED IRREVERSIBLE VECTOR PROCESSES	71-0701
MASON, J. B.	NASTRAN STRUCTURAL/THERMAL ANALYZER	72-0152
MASON, J. B.	THE NASTRAN HYDROELASTIC ANALYZER	71-0738
MASON, J. B.	THERMAL DEFORMATION PREDICTION IN ORBITING OBSERVATORY STRUCTURES	69-0292
MASON, P. W.	ANALYTICAL AND EXPERIMENTAL INVESTIGATION OF A 1/8 SCALE DYNAMIC MODEL OF THE SHUTTLE ORBITER. VOLUM	74-1186
MASON, P. W.	DEVELOPMENT OF TECHNOLOGY FOR MODELING OF A 1/8-SCALE DYNAMIC MODEL OF THE SHUTTLE SOLID ROCKET BOOS	74-0703
MASON, V.	ON THE USE OF RECTANGULAR FINITE ELEMENTS	67-0162
MASON, V.	RECTANGULAR FINITE ELEMENTS FOR ANALYSIS OF PLATE VIBRATIONS	68-0113
MASON, V.	THE FINITE ELEMENT TECHNIQUE APPLIED TO AN ACOUSTIC DAMPING PROBLEM	66-0048
MASON, V.	VARIATIONAL FINITE ELEMENT CALCULATION OF THE ACOUSTIC RESPONSE OF A RECTANGULAR PANEL	71-0100
MASON, W. E.	MIXED FORMULATIONS FOR FINITE ELEMENT SHELL ANALYSIS	71-0222
MASON, W. E. JR	FINITE ELEMENT ANALYSIS OF CYLINDRICAL SHELLS UTILIZING CURVED ELEMENTS CONTAINING RIGID BODY DEFORM	71-0823
MASON, WM. E. J	ELASTIC SHEAR ANALYSIS OF GENERAL PRISMATIC BEAMS	68-0114
MASSARD, J. M.	ELEMENT STIFFNESS MATRICES OF THICK-WALLED ORTHOTROPIC SHELLS WITH APPLICATIONS	69-0261
MASSINI, G.	DESCRIPTION OF THE IBM-360 VERSION OF THE FINITE ELEMENTS COMPUTER PROGRAM SAFE-3D	72-0775
MASSINI, G.	REBO: COMPUTER CODE FOR THE GENERATION OF A THREE-DIMENSIONAL FINITE-ELEMENT GRID OF THE NOZZLE-CYLI	74-1310
MASSINI, G.	VERIFICATION OF THE FINITE ELEMENT COMPUTER CODE SAFE-2D	74-0873
MASSMANN, P.	STRUCTURAL RESPONSE TO IMPACT DAMAGE	75-0325
MASSON, J. C.	INVESTIGATION ON FREE AND FORCED VIBRATIONS OF A LONG TANKER WITH OVERLAPPING PROPELLER ARRANGEMENT	73-0981
MASSONNET, C. E	NUMERICAL USE OF INTEGRAL PROCEDURES	65-0041
MASTERS, A. E.	DYNAMIC RESPONSE OF ARBITRARY SHELLS OF REVOLUTION UNDER COMBINED IMPULSIVE NONISOTHERMAL AND IRREGU	70-0365
MASTERSON, D. M	PUNCHING STRENGTH OF SLABS, A FLEXURAL APPROACH USING FINITE ELEMENTS	74-1307
MASTROGEORGOPOU	FINITE-ELEMENT ANALYSIS OF ORTHOTROPIC PLATES WITH ECCENTRIC STIFFENERS	72-0958
MASUBUCHI, K.	THERMAL ANALYSIS OF M552 EXPERIMENT FOR MATERIALS PROCESSING IN SPACE	74-0704
MASUBUCHI, K.	THERMAL ANALYSIS OF M551 EXPERIMENT FOR MATERIALS PROCESSING IN SPACE	74-0725
MATE, D. P.	COMPUTER ANALYSIS OF A RAILROAD FREIGHT CAR BOLSTER UTILIZING THE FINITE ELEMENT METHOD	71-0283
MATERNA, A.	RESENI MAZNIHO STAVU ZELEZOBETONOVYCH DESEK (PLASTIC ANALYSIS OF REINFORCED CONCRETE PLATES)	74-0495
MATERNA, A.	STATICKE RESENI FYZIKALNE NELINEARNIHO CHOVANI ZELEZOBETO- NOVYCH DESEK (ANALYSIS OF PHYSICALLY NONL	74-0514
MATHERS, M. D.	NONLINEAR FINITE ELEMENT ANALYSIS OF LAMINATED COMPOSITE SHELLS	74-0751
MATHES, H. B.	METHODS FOR DETERMINING CHARACTERISTICS OF ACOUSTIC WAVES IN ROCKET MOTORS	75-0075
MATHEWS, D.	FEASIBILITY OF USING FINITE ELEMENTS IN ANALYSIS OF SECOND BREAKDOWN IN SEMICONDUCTOR DEVICES	73-0874
MATLOCK, H.	A COMPUTER PROGRAM TO ANALYZE BEAM-COLUMN UNDER MOVEABLE LOADS	68-0268
MATLOCK, H.	A DISCRETE-ELEMENT METHOD OF MULTIPLE-LOADING ANALYSIS FOR TWO-WAY BRIDGE FLOOR SLABS	70-0435
MATLOCK, H.	A FINITE ELEMENT METHOD OF SOLUTION FOR LINEARLY ELASTIC BEAM-COLUMNS	66-0117
MATLOCK, H.	A FINITE ELEMENT METHOD FOR BENDING ANALYSIS OF LAYERED STRUCTURAL SYSTEMS	67-0134
MATLOCK, H.	A FINITE-ELEMENT METHOD OF SOLUTION FOR STRUCTURAL FRAMES	67-0133
MATLOCK, H.	A FINITE-ELEMENT METHOD OF ANALYSIS FOR COMPOSITE BEAMS	68-0263
MATLOCK, H.	FINITE ELEMENT ANALYSIS OF STRUCTURAL FRAMES	67-0032
MATLOCK, H.	FINITE-ELEMENT ANALYSIS OF BRIDGE DECKS	72-0672
MATRA, J. P. JR	MISSILE BODY INPUT GENERATOR (BING), A NASTRAN PRE-PROCESSOR: THEORETICAL DEVELOPMENT, USER'S MANUAL	75-0717
MATRA, J. P. JR	PLANFORM INPUT GENERATOR (PING), A NASTRAN PRE-PROCESSOR FOR LIFTING SURFACES-THEORETICAL DEVELOPMEN	73-0776
MATRA, J. P. JR	PLANFORM INPUT GENERATOR (PING), A NASTRAN PROCESSOR FOR LIFTING SURFACES-THEORETICAL DEVELOPMENT, U	73-0780
MATSUBARA, S.	STRESS ANALYSIS IN SOME COLD FORGING TOOL COMPONENTS AND THEIR DESIGN GUIDES	74-0589
MATSUI, T.	NEW FINITE-ELEMENT SCHEME FOR INSTABILITY ANALYSIS OF THIN SHELLS	75-0357
MATSUISHI, M.	ELASTIC-PLASTIC BUCKLING OF PLATES USING THE FINITE ELEMENT METHOD	67-0094
MATSUISHI, M.	ELASTIC-PLASTIC BUCKLING OF PLATES BY FINITE ELEMENT METHOD	69-0253
MATSUISKI, M.	ANALYSIS OF ELASTIC-PLASTIC BUCKLING OF PLATES BY THE FINITE ELEMENT METHOD	69-0253
MATSUMOTO, G. Y	COUPLED, NONCONSERVATIVE STABILITY-FINITE ELEMENT	72-0220
MATSUMOTO, G. Y	FINITE ELEMENT ANALYSIS OF PROBLEMS IN NONCONSERVATIVE ELASTIC STABILITY	72-0957
MATSUMOTO, K.	APPLICATION OF FINITE ELEMENT METHOD TO ADDED VIRTUAL MASS OF SHIP HULL VIBRATION	70-0463
MATSUMOTO, K.	VIBRATION OF AN ELASTIC BODY IMMERSED IN A FLUID	72-0312
MATSUOKA, O.	NEW FINITE-ELEMENT SCHEME FOR INSTABILITY ANALYSIS OF THIN SHELLS	75-0357
MATSUOKA, T.	STRESS ANALYSIS OF FRAME STRUCTURE CONNECTED WITH PLATES	74-0265
MATSUSHIMA, Y.	HYSTERETIC DAMPING SYSTEMS APPLIED TO SOIL STRUCTURE INTERACTION BY FINITE ELEMENT METHOD	73-0663
MATSUSHITA, T.	ANALYTICAL STUDY OF THE PROBLEM OF EVAPORATION AND CONDENSATION USING THE FINITE ELEMENT METHOD	74-0960
MATSUSUE, K.	STRESSES IN A ROTATING DISK WITH CENTRAL AND NONCENTRAL HOLES (STRESS ANALYSIS BY THE FINITE ELEMENT	74-1337
MATSUURA, Y.	CALCULATION OF THE VIRTUAL MASS, MOVEMENT AND INERTIA BY THE FINITE ELEMENT METHOD	68-0221
MATTHEWS, A. T.	COMPARISON OF THE RESULTS OF THE DYNAMIC RESPONSE OF CYLINDRICAL SHELLS BY CHARACTERISTIC AND BY FIN	74-0781
MATTHEWS, F. L.	FINITE ELEMENT DISPLACEMENT ANALYSIS OF A LUNG	72-0026
MATTHEWS, J. B.	HYDRODYNAMICAL-NUMERICAL MODELS FOR COASTAL WATERS AND OPEN OCEAN AREAS	75-0571
MAU, S. T.	A METHOD OF LIMIT POINT CALCULATION IN FINITE ELEMENT STRUCTURAL ANALYSIS	72-0786

MAU, S. T.	A PROCEDURE FOR FINITE ELEMENT PLATE AND SHELL PRE AND POST BUCKLING ANALYSIS	71-0096
MAU, S. T.	CALCULATION OF MECHANICAL IMPEDANCE BY FINITE ELEMENT HYBRID MODEL	74-0126
MAU, S. T.	DERIVATION OF GEOMETRIC STIFFNESS AND MASS MATRICES FOR FINITE ELEMENT HYBRID MODELS	74-0844
MAU, S. T.	ELASTIC DEFORMATION OF LIGHTWEIGHT MIRRORS METHOD OF LIMIT POINT CALCULATION IN FINITE ELEMENT STRUC	72-0232
MAU, S. T.	FINITE ELEMENT PROCEDURE FOR NONLINEAR PREBUCKLING AND INITIAL POSTBUCKLING ANALYSIS	72-0170
MAU, S. T.	FINITE ELEMENT SOLUTIONS FOR LAMINATED THICK PLATES	72-0143
MAU, S. T.	SOME RECENT STUDIES IN ASSUMED STRESS HYBRID MODELS	72-0262
MAU, S. T.	STATIC, VIBRATION, AND THERMAL STRESS ANALYSES OF LAMINATED PLATE AND SHELLS BY THE HYBRID-STRESS FI	72-0602
MAU, S. T.	VIBRATION ANALYSIS OF LAMINATED PLATES AND SHELLS BY A HYBRID STRESS ELEMENT	73-0120
MAURIZI, M. J.	USE OF TRIGONOMETRIC TERMS IN FINITE ELEMENT METHOD WITH APPLICATION TO VIBRATING MEMBRANES	74-0555
MAURY, J. F.	PRACTICAL METHOD FOR COMPUTER CALCULATION OF STRUCTURES	69-0098
MAWENYA, A. S.	EXPLICIT TRIANGULAR BENDING ELEMENT MATRIX	74-0288
MAWENYA, A. S.	FINITE ELEMENT BENDING ANALYSIS OF MULTILAYER PLATES	74-0266
MAY, B.	FINITE ELEMENTE FUR GEKRUEMMTE FLAECHENTRAGWERKE. (FINITE ELEMENTS SOLUTIONS FOR CURVED SHELL STRUCT	71-0391
MAY, G. W.	CRACKING ANALYSIS OF REINFORCED CONCRETE PLATES	75-0085
MAYER, P.	SOLUTION OF ANISOTROPIC SEEPAGE PROBLEMS BY FINITE ELEMENTS	66-0080
MAYER, P. G.	FINITE ELEMENT GALERKIN METHOD SOLUTIONS TO SELECTED ELLIPTIC AND PARABOLIC DIFFERENTIAL EQUATIONS	71-0027
MAYER, P. G.	THE FATE OF RADIONUCLIDES IN A GROUNDWATER ENVIRONMENT	74-0061
MAYERJAK, R. J.	FATIQUE STRENGTH OF LUGS CONTAINING LINERS. VOLUME II. COMPUTER PROGRAM USED FOR ANALYSIS	70-0418
MCALLISTER, G.	VARIATIONAL FORMULATION FOR A CLASS OF FREE BOUNDARY PROBLEMS ARISING IN HYDRODYNAMIC LUBRICATION	75-0531
MCAULAY, A. D.	NUMERICAL METHOD FOR INVESTIGATING THE PROPAGATION OF SURFACE WAVES ON DISSIPATIVE GUIDES	74-0613
MCBEAN, R. P.	ANALYSIS OF STIFFENED PLATES BY THE FINITE ELEMENT METHOD	68-0299
MCBEAN, R. P.	FINITE ELEMENT ANALYSIS OF STIFFENED PLATES	68-0240
MCCABE, J. T.	STABILITY OF TILTING-PAD JOURNAL BEARINGS	71-0687
MCCABE, M. W.	AN INTERACTIVE GRAPHICS SHIP DESIGN ORIENTED PROGRAM PACKAGE FOR TIME SHARING SYSTEMS	73-0341
MCCABE, M. W.	GIFTS SYSTEM	74-1060
MCCABE, M. W.	GIFTS SYSTEM FOR THE PDP-15 AND OTHER CONSIDERATIONS IN DEVELOPING GRAPHICS AND CAPABILITIES FOR FIN	73-0671
MCCABE, M. W.	SOME DEVELOPMENTS IN THE ANALYSIS OF COMPLEX SHIP STRUCTURES	72-0332
MCCALLEY, R. B.	ROTARY INERTIA CORRECTION FOR MASS MATRICES	63-0011
MCCALLEY, R. B.	MASS MATRIX FOR A PRISMATIC BEAM SEGMENT	68-0116
MCCALLEY, R. B.	SHOCK ANALYSIS BY MATRIX METHODS	67-0051
MCCALLION, H.	IMPROVED DEFORMATION FUNCTIONS FOR THE FINITE ELEMENT ANALYSIS OF BEAM SYSTEMS	69-0075
MCCALLION, H.	NATURAL FREQUENCIES AND MODE SHAPES OF TRANSFORMER CORES	65-0048
MCCALLION, H.	THE NATURAL FREQUENCIES OF PORTAL FRAMES II	65-0047
MCCALLUM, R. B.	A FEASIBILITY STUDY OF THE ANALYSIS OF ASYMMETRIC SHELL BY THE DISCRETE-ELEMENT METHOD	68-0115
MCCANN, M.	ANALYSIS OF THE DYNAMICS OF THIN PRIMARY MIRRORS FOR LARGE ASTRONOMICAL TELESCOPES	73-1084
MCCARTHY, J. B.	INELASTIC ANALYSIS OF A BRANCH SHELL JUNCTION	74-0682
MCCHESNEY, R. S	IMPROVED FINITE-ELEMENT ANALYSIS OF NUCLEAR-CONTAINMENT SHELL STRUCTURES	73-0239
MCCLEARY, L.	APPLICATIONS OF THE FINITE-ELEMENT METHOD AND COMPUTER GRAPHICS. (A VIBRATIONAL ANALYSIS OF A SONAR	72-0729
MCCLELLAND, W.	COMBINING EXPERIMENTAL AND ANALYTICAL TECHNIQUES FOR DYNAMIC SYSTEM ANALYSIS	73-0321
MCCLELLAND, W.	DYNAMIC SIMULATION OF COMPLEX SYSTEMS UTILIZING EXPERIMENTAL AND ANALYTICAL TECHNIQUES	75-0864
MCCLELLAND, W.	IMPROVED TECHNIQUES FOR DYNAMIC ANALYSIS OF EARTHMOVING EQUIPMENT	74-0294
MCCLELLAND, W.	USING NASTRAN FOR DYNAMIC ANALYSIS OF VEHICLE SYSTEMS	74-0316
MCCOMAS, R. L.	NASTRAN, NASA'S GENERAL PURPOSE STRUCTURAL ANALYSIS PROGRAM	70-0370
MCCONNELEE, J.	APPLICATIONS OF FINITE ELEMENT STRESS ANALYSIS AND STRESS-STRAIN PROPERTIES IN DETERMINING NOTCH FAT	73-1114
MCCONNELLEE, J.	APPLICATION OF FINITE ELEMENT ELASTIC-PLASTIC STRESS ANALYSIS TO NOTCHED FATIQUE SPECIMEN BEHAVIOUR	71-0185
MCCORMICK, C. W	HARDWARE AND SOFTWARE CONSIDERATIONS IN GENERAL PURPOSE PROGRAMS FOR FINITE ELEMENT ANALYSIS	70-0251
MCCORMICK, C. W	REVIEW OF NASTRAN DEVELOPMENT RELATIVE TO EFFICIENCY OF EXECUTION	73-0938
MCCORMICK, C. W	SHELL ANALYSIS WITH LARGE GENERAL PURPOSE PROGRAMS	71-0402
MCCORMICK, C. W	SPARSE MATRIX OPERATIONS IN NASTRAN	73-0336
MCCORMICK, C. W	THE NASTRAN COMPUTER PROGRAM FOR STRUCTURAL ANALYSIS	69-0183
MCCORMICK, C. W	THE NASTRAN COMPUTER PROGRAM FOR STRUCTURAL ANALYSIS	71-0682
MCCORMICK, C. W	THE NASTRAN PROGRAM FOR STRUCTURAL ANALYSIS	72-0325
MCCORMICK, C. W	THE NASTRAN USER'S MANUAL	72-0479
MCCORQUODALE, J	FINITE ELEMENT ANALYSIS OF NON-DARCY FLOW	70-0268
MCCORQUODALE, J	FINITE ELEMENT ANALYSIS OF SLUICE GATE FLOW	71-0540
MCCORQUODALE, J	NON-DARCY FLOW SOLVED BY FINITE ELEMENT ANALYSIS	69-0366
MCCORQUODALE, J	NUMERICAL METHODS FOR UNSTEADY NON-DARCY FLOW	74-0062
MCCORQUODALE, J	FINITE ELEMENT METHOD APPLIED TO UNSTEADY NON-DARCY FLOW	70-0088
MCCORQUODALE, J	VARIATIONAL APPROACH TO NON-DARCY FLOW	70-0089
MCCOWAN, D. W.	DYNAMIC FINITE ELEMENT CALCULATION OF ELASTIC WAVE PROPAGATION IN TWO DIMENSIONAL STRUCTURES	74-0166
MCCOWAN, D. W.	DYNAMIC FINITE ELEMENT ANALYSIS WITH APPLICATIONS TO SEISMOLOGICAL PROBLEMS	75-0785
MCCOWAN, D. W.	EFFECTS OF SOURCE STRUCTURE AND SITE RESPONSE ON GROUND MOTION IN THE NEAR FIELD OF EARTHQUAKES	73-0732
MCCOWAN, D. W.	INVERSION TECHNIQUE FOR STATIC FINITE ELEMENT METHOD	74-0170
MCCOWAN, D. W.	SLAMCODE: FINITE ELEMENT STRESS WAVE ANALYSIS	72-0603
MCCOWAN, D. W.	STUDY OF NUMERICAL METHODS IN SEISMIC PROBLEMS	72-0738
MCCULLERS, L. A	AUTOMATED STRUCTURAL DESIGN AND ANALYSIS OF ADVANCED COMPOSITE WING MODELS	73-0221
MCCULLERS, L. A	COMPOSITE WING FOR TRANSONIC IMPROVEMENT. VOLUME 11. ADVANCED ANALYSIS EVALUATION	71-0657
MCCUTCHEON, J.	A STUDY OF BEHAVIOUR OF REINFORCED CONCRETE BEAMS USING FINITE ELEMENTS	70-0335
MCDONALD, B. H.	FINITE-ELEMENT SOLUTION OF UNBOUNDED FIELD PROBLEMS	72-0025
MCDONALD, B. H.	INTEGRAL FINITE-ELEMENT APPROACH FOR SOLVING LAPLACE EQUATION	73-0121
MCDONALD, D.	CONFORMING RECTANGULAR AND TRIANGULAR PLATE BENDING ELEMENTS	71-0720
MCDONALD, M. M.	ELASTIC-PLASTIC STABILITY ANALYSIS OF MINE-WASTE EMBANKMENTS	75-0835
MCDONALD, M. M.	SUPPORT PERFORMANCE OF HYDRAULIC BACKFILL	74-0257
MCDONOUGH, J. R	NASTRAN PLOTTING CAPABILITIES	71-0172
MCELROY, M.	SAMECS STRUCTURAL ANALYSIS SYSTEM - USER'S DOCUMENT	69-0495
MCEVILY, A. J.	ON THE ROLE OF DEFECTS IN CRACK INITIATION IN WELDED STRUCTURES	74-1142
MCGARRY, F.	MATRIX AND INTERFACE STRESSES IN A DISCONTINUOUS FIBER COMPOSITE MODEL	68-0040
MCGARRY, F.	STRESS INTENSITY FACTORS FOR ANISOTROPIC FRACTURE TEST SPECIMENS OF SEVERAL GEOMETRIES	74-0293
MCGARRY, F. J.	FRACTURE OF GRAPHITE FIBER REINFORCED COMPOSITES	74-0695
MCGEE, T. M.	AN INVESTIGATION OF MICRO-BUCKLING USING PHOTOELASTIC AND FINITE-ELEMENT METHODS	72-0960
MCGEORGE, P.	REFINED CRACKED CONCRETE ANALYSIS OF CONCRETE CONTAINMENT STRUCTURES SUBJECT TO OPERATIONAL AND ENVI	74-0943
MCGLEEN, J.	FINITE ELEMENT ANALYSIS OF PLYWOOD PLATES	67-0052
MCGOVERN, D. J.	THE APPLICATION OF GRADIENT MINIMIZATION METHODS AND HIGHER ORDER DISCRETE ELEMENTS TO SHELL BUCKLIN	71-0237
MCHENRY, D.	A LATTICE ANALOGY FOR THE SOLUTION OF PLANE STRESS	43-0002
MCHENRY, H. I.	EVALUATION OF DAMAGE TOLERANCE IN AIRCRAFT STRUCTURES	75-0559
MCKEE, J.	REDUCING THE BANDWIDTH OF SPARSE SYMMETRIC MATRICES	69-0469
MCKEE, J. M.	A SURVEY OF PRE AND POSTPROCESSORS FOR NASTRAN	74-0714
MCKEE, J. M.	REDUCING THE BANDWIDTH OF SPARSE SYMMETRIC MATRICES	69-0502
MCKEOWN, J. J.	QUASI-LINEAR PROGRAMMING ALGORITHM FOR OPTIMIZING FIBRE-REINFORCED STRUCTURES OF FIXED STIFFNESS	75-0225
MCKINLEY, T. K.	THERMAL STRESS ANALYSIS OF REENTRY VEHICLE NOSETIPS AT ANGLE OF ATTACK	72-0456
MCKINLEY, T. K.	THERMAL STRESS ANALYSIS OF RE-ENTRY VEHICLE NOSETIPS AT ANGLE OF ATTACK	72-0852
MCKINNEY, J.	DEVELOPMENT OF AN UNDERSTANDING OF THE FATIGUE PHENOMENA OF BONDED AND BOLTED JOINTS IN ADVANCED FIL	73-0798
MCKINNIS, J. A.	MEASUREMENT OF THERMAL EXPANSION IN SOLID PROPELLANTS	73-0916
MCKINSTRY, H. A	HIGH STRENGTH CERAMICS	72-0878
MCKINSTRY, H. A	TWO-DIMENSIONAL FINITE ELEMENT MODEL OF A CERAMIC BODY	73-0010
MCKNIGHT, R. L.	FINITE ELEMENT CYCLIC THERMOPLASTICITY ANALYSIS BY THE METHOD OF SUBVOLUMES	75-0787
MCLARNAN, C. W.	OPTIMAL SYNTHESIS OF FLEXIBLE LINK MECHANISMS WITH LARGE STATIC DEFLECTIONS	74-0525
MCLAY, R. W.	A SPECIAL VARIATIONAL PRINCIPLE FOR THE FINITE-ELEMENT METHOD	69-0181
MCLAY, R. W.	ANALYSIS OF AXISYMMETRIC SOLIDS BY THE DIRECT STIFFNESS METHOD	66-0049
MCLAY, R. W.	ANALYSIS OF FEEDBACK CONTROL STABILIZED VIBRATION BY THE FINITE ELEMENT METHOD	74-1188
MCLAY, R. W.	APPLICATION, TESTING AND EVALUATION OF DOUBLY CURVED SHELL ELEMENTS FOR DYNAMIC ANALYSIS, PART II. F	69-0117

MCLAY, R. W.	CERTAIN APPROXIMATIONS IN FINITE ELEMENT METHOD	71-0577
MCLAY, R. W.	COMPLETENESS AND CONVERGENCE PROPERTIES OF FINITE ELEMENT DISPLACEMENT FUNCTIONS., A GENERAL TREATME	67-0053
MCLAY, R. W.	CONVERGENCE OF THE FINITE ELEMENT METHOD IN THE THEORY OF ELASTICITY	68-0095
MCLAY, R. W.	DYNAMIC ANALYSIS OF SHELLS USING DOUBLY-CURVED FINITE ELEMENTS	68-0082
MCLAY, R. W.	ERROR BOUNDS IN AN OPTIMIZATION PROBLEM USING FINITE-ELEMENT METHOD	73-0049
MCLAY, R. W.	EVALUATION OF THE ERROR BOUNDS IN AN OPTIMIZATION PROBLEM USING THE FINITE-ELEMENT METHOD	73-0211
MCLAY, R. W.	GENERALIZED VARIATIONAL PRINCIPLES IN THE FINITE ELEMENT METHOD	69-0116
MCLAY, R. W.	NATURAL BOUNDARY CONDITIONS IN FINITE-ELEMENT METHOD	72-0276
MCLAY, R. W.	ON CERTAIN APPROXIMATIONS IN THE FINITE ELEMENT METHOD	71-0515
MCLAY, R. W.	ON SPARSE MATRICES IN FINITE ELEMENT WAVE PROPAGATION	73-0290
MCLAY, R. W.	ON THE APPLICATION OF GENERALIZED VARIATIONAL PRINCIPLES IN THE FINITE ELEMENT METHOD	68-0083
MCLAY, R. W.	ON THE CONVERGENCE OF THE FINITE ELEMENT SOLUTION IN A NONLINEAR HEAT CONDUCTION PROBLEM	74-1272
MCLAY, R. W.	SPARSE MATRICES IN FINITE ELEMENT PROGRAM DEVELOPMENT	73-0337
MCLEAN, F. G.	GRAVITY FLOW TO EXCAVATIONS AND DRAINAGE TRENCHES IN LAYERED AQUIFERS	71-0014
MCLEAN, F. G.	THE KANSAS TEST TRACK. NON-CONVENTIONAL TRACK STRUCTURES. DESIGN REPORT	72-0797
MCLEOD, R.	CURVED ELEMENTS IN THE FINITE ELEMENT METHOD	74-1261
MCLEOD, R.	OVERCOMING LOSS OF ACCURACY WHEN USING CURVED FINITE ELEMENTS	75-0092
MCLEOD, R. J. Y	BASIS FUNCTIONS FOR CURVED ELEMENTS IN THE FINITE ELEMENT METHOD	72-0807
MCLEOD, R. J. Y	USE OF PARABOLIC ARCS IN MATCHING CURVED BOUNDARIES IN FINITE-ELEMENT METHOD	75-0589
MCLEOD, R. Y. J	THE CONSTRUCTION OF BASIS FUNCTIONS FOR CURVED ELEMENTS IN THE FINITE ELEMENT METHOD	72-0558
MCLEON, I. A.	NEW RECTANGULAR FINITE ELEMENT FOR SHEAR WALL ANALYSIS	69-0182
MCMAHAN, L. L.	DEVELOPMENT AND APPLICATION OF THE DIRECT STIFFNESS METHOD FOR OUT-OF-PLANE BENDING USING A TRIANGUL	62-0009
MCMAHON, A.	A SIMPLE FINITE ELEMENT MODEL FOR ELASTIC-PLASTIC PLATE BENDING	72-0889
MCMAHON, A. A.	SIMPLE FINITE ELEMENT MODEL FOR ELASTIC-PLASTIC PLATE BENDING.	72-0447
MCMAHON, B. K.	PREDICTING THE BLOCK CAVING BEHAVIOUR OF OREBODIES	69-0044
MCMEEKING, P. M	EFFICIENT FINITE ELEMENT METHODS FOR STRESS INTENSITY FACTOR	75-0148
MCMEEKING, R. M	EFFICIENT FINITE-ELEMENT METHODS FOR STRESS INTENSITY FACTORS USING WEIGHT FUNCTIONS - REPLY	75-0421
MCMEEKING, R. N	FINITE ELEMENT FORMULATIONS FOR PROBLEMS OF LARGE ELASTIC-PLASTIC DEFORMATION	74-0796
MCMICHAEL, C. L	RESERVOIR SIMULATION BY GALERKIN'S METHOD	73-0662
MCMILLAN, C. M.	ANALYSIS OF SLAB BRIDGES BY FINITE ELEMENTS	73-0695
MCMILLAN, C. M.	BENTONITE PILES IN DURBAN	72-0797
MCMUNN, J. C.	FINITE ELEMENT ANALYSIS OF STATIC AND DYNAMIC STRESSES AROUND SUDDENLY PUNCHED HOLES IN PLATES AND S	73-1088
MCNAMARA, J. F.	ANALYTICAL MODEL FOR DETERMINING ENERGY DISSIPATION IN DYNAMICALLY LOADED STRUCTURES	75-0328
MCNAMARA, J. F.	INCREMENTAL STIFFNESS METHOD FOR FINITE ELEMENT ANALYSIS OF THE NONLINEAR DYNAMIC PROBLEM	71-0525
MCNAMARA, J. F.	ISOTROPIC-KINEMATIC HARDENING MODEL FOR ELASTIC-PLASTIC CYCLIC STRUCTURAL ANALYSIS	75-0210
MCNAMARA, J. F.	NUMERICAL SOLUTION SCHEMES FOR HIGHLY NONLINEAR STATIC STRUCTURAL BEHAVIOR	75-0679
MCNAMARA, J. F.	SOLUTION SCHEMES FOR PROBLEMS OF NONLINEAR STRUCTURAL DYNAMICS	74-0505
MCNAMARA, S. F.	NONLINEAR MATERIAL AND GEOMETRIC BEHAVIOUR OF SHELL STRUCTURES	71-0084
MCNEAL, R. H.	A DESIGN STUDY FOR THE INCORPORATION OF AEROELASTIC CAPABILITY INTO NASTRAN	71-0112
MCNEAL, R. H.	ACOUSTIC ANALYSIS OF SOLID ROCKET MOTOR CAVITIES BY A FINITE ELEMENT METHOD	71-0119
MCNEAL, R. H.	THE NASTRAN COMPUTER PROGRAM FOR STRUCTURAL ANALYSIS	69-0183
MCNEAL, R. H.	THE NASTRAN COMPUTER PROGRAM FOR STRUCTURAL ANALYSIS	71-0173
MCNEICE, G. M.	A STRAIN ENERGY BASIS FOR STUDIES OF ELEMENT STIFFNESS MATRICES	72-0621
MCNEICE, G. M.	A VARIATIONAL APPROACH TO GRID OPTIMIZATION IN THE FINITE ELEMENT METHOD	72-0572
MCNEICE, G. M.	AN ELASTIC-PLASTIC FINITE ELEMENT ANALYSIS FOR PLATES WITH EDGE BEAMS	69-0185
MCNEICE, G. M.	APPLICATION OF GRID SELECTION PROCEDURES FOR IMPROVED FINITE ELEMENT STRESS ANALYSIS	74-0326
MCNEICE, G. M.	ARCH DAM CONSTRUCTION STRESSES DUE TO HYDROSTATIC AND GRAVITY LOADS	75-0658
MCNEICE, G. M.	COMPARISON OF FINITE ELEMENT AND UNIQUE LIMIT ANALYSIS SOLUTIONS FOR CERTAIN REINFORCED CONCRETE SLA	69-0184
MCNEICE, G. M.	FINITE ELEMENT ANALYSIS OF REINFORCED CONCRETE SLABS	71-0133
MCNEICE, G. M.	FINITE ELEMENT ANALYSIS OF FINITE SIZED PLATES BOUNDED TO AN ELASTIC HALF SPACE	72-0880
MCNEICE, G. M.	FINITE ELEMENT ANALYSIS OF PRESTRESSED CONCRETE VOIDED BRIDGE DECKS	73-0106
MCNEICE, G. M.	FINITE ELEMENT STUDIES OF FEMORAL ENDOPROSTHESES FOR HIP RECONSTRUCTION	75-0381
MCNEICE, G. M.	FURTHER LIMITATIONS ON GENERAL HEXAHEDRON FINITE ELEMENTS WITH DERIVATIVE DEGREES OF FREEDOM.	72-0395
MCNEICE, G. M.	GUIDELINES FOR SELECTING FINITE ELEMENT GRIDS BASED ON AN OPTIMIZATION STUDY	74-0303
MCNEICE, G. M.	MIXED-DISPLACEMENT FINITE-ELEMENT ANALYSIS WITH PARTICULAR APPLICATION USING PLANE-STRESS TRIANGLES	72-0245
MCNEICE, G. M.	OPTIMIZATION OF FINITE ELEMENT GRIDS BASED ON MINIMUM POTENTIAL ENERGY	71-0747
MCNEICE, G. M.	OPTIMIZATION OF FINITE ELEMENT GRIDS BASED ON MINIMUM POTENTIAL ENERGY	72-0109
MCNEICE, G. M.	PATTERN LOADING ON REINFORCED CONCRETE FLAT PLATES	71-0290
MCNEICE, G. M.	PROCEDURES FOR SELECTING NEAR OPTIMUM FINITE ELEMENT GRIDS	73-0537
MCNEICE, G. M.	VOGT BOUNDARY FOR FINITE ELEMENT ARCH DAM ANALYSIS	74-0156
MCRAE, G. J.	FINITE ELEMENT SOLUTIONS OF THE WAVE EQUATION	74-0216
MCVERRY, G. H.	FINITE-ELEMENT CALCULATION OF PRESSURE HISTORY IN A GAS FIELD	75-0546
MCVEY, J. R.	NX BOREHOLE JACK MODULUS DETERMINATIONS IN HOMOGENEOUS, ISOTROPIC, ELASTIC MATERIALS	74-0295
MCWHORTER, L. B	FAMSOR, A FINITE ELEMENT PROGRAM FOR THE FREQUENCIES AND MODE SHAPES OF SHELLS OF REVOLUTION	70-0090
MEACHAM, H. W.	ASSESSMENT OF DESIGN TOOLS AND CRITERIA FOR URBAN RAIL TRACK STRUCTURES. VOLUME 11. AT-GRADE SLAB TR	74-0813
MEAK, J. L.	REDUNDANT STRUCTURE CONCEPTS IN FINITE ELEMENT ANALYSES	75-0626
MEBANE, P. M.	AN IMPROVED SHELL OF REVOLUTION ELEMENT UTILISING CUBIC DISPLACEMENT FUNCTIONS	70-0451
MEBANE, P. M.	IMPLICIT RIGID BODY MOTION IN CURVED FINITE ELEMENTS	71-0174
MEDAGLIA, J.	ELASTIC PLATE SPALLATION	72-0682
MEDAGLIA, J. M	INVESTIGATION OF ELASTIC PLATE SPALLATION BY FINITE ELEMENTS	74-0382
MEDLAND, I. C.	A FINITE ELEMENT APPROACH TO STABILITY ANALYSIS IN FRAMES	74-0455
MEDLAND, I. C.	THE MULTIDIMENSIONAL STEFAN PROBLEM: A FINITE ELEMENT APPROACH	74-0464
MEDWELL, O.	FINITE ELEMENT METHOD APPLIED TO THE SOLUTION OF HEAT CONDUCTION PROBLEMS	69-0009
MEEK, J. L.	APPLICATION OF FINITE ELEMENT METHOD TO PLANE STRAIN EXTRUSION PROCESSES	74-1243
MEEK, J. L.	AXISYMMETRIC SOLUTION OF ELASTO PLASTIC PROBLEMS BY FINITE ELEMENT METHODS	69-0361
MEEK, J. L.	COMPARISON OF SOME PLATE BENDING FINITE ELEMENTS	72-0413
MEEK, J. L.	ELASTOPLASTIC ANALYSIS OF A CANTILEVER BEAM	73-0984
MEEK, J. L.	EXCAVATION IN ROCK, AN APPRECIATION OF THE FINITE ELEMENT METHOD OF ANALYSIS	73-0313
MEEK, J. L.	FIELD PROBLEMS SOLUTIONS BY FINITE METHODS	68-0118
MEEK, J. L.	MATRIX STRUCTURAL ANALYSIS	71-0614
MEEK, J. L.	MEMBRANE ANALYSIS OF UNFRAMED HIGH RISE BUILDINGS	73-0451
MEEK, J. L.	REDUNDANT STRUCTURE CONCEPTS IN FINITE ELEMENT ANALYSES	75-0338
MEEK, J. L.	THE FINITE ELEMENT METHOD IN STRUCTURAL MECHANICS	66-0051
MEEK, J. L.	TRANSIENT HEAT FLOW IN SOLIDS	74-0461
MEERWALD, K.	ENTWURF UND SPANNUNGSBERECHNUNG EINES SPANNBETON DRUCKBEHAELTERS AUS FERTIGTEILEN. (DESIGN AND STRES	73-0856
MEGAHED, M. M.	AXISYMMETRIC ELEMENT ANALYSIS USING ANALYTICAL COMPUTING	75-0333
MEGARD, G.	ANALYSIS OF THIN SHELLS USING CURVED FINITE ELEMENTS	69-0523
MEGARD, G.	PLANAR AND CURVED SHELL ELEMENTS	69-0186
MEHLHORN, G.	BERECHNUNG VON STAHLBETONSCHEIBEN IM ZUSTAND II BEI ANNAHME EINES WIRKLICHKEITSNAHEN WERKSTOFFVERHAL	74-0978
MEHLHORN, G.	ZUM TRAGVERHALTEN VON WAENDEN AUS FERTIGTEILTAFELN (BEHAVIOR OF LARGE PANEL SHEAR WALLS)	74-0299
MEHLHORN, G.	ZUSAMMENWIRKEN VON EINZELNEN FERTIGTEILEN ALS GROSSFLAECHIGE SCHEIBE (INTERACTION OF INDIVIDUAL PREF	73-0552
MEHLHORN, G.	ZUR WIRKLICHKEITSNAHEN BERECHNUNG VON STAHLBETONPLATTEN MIT DER FINITE-ELEMENT-METHODE. (APPROXIMATE	75-0832
MEHRAIN, M.	FINITE ELEMENT ANALYSIS OF SKEW COMPOSITE GIRDER BRIDGES	67-0150
MEHRAIN, M.	FINITE ELEMENT ANALYSIS OF SKEW COMPOSITE PLATES	67-0160
MEHRAIN, M.	ROTATIONAL STIFFNESS OF CONCRETE SLABS	74-0660
MEHROTRA, B. L.	ANALYSIS OF THREE-DIMENSIONAL THIN WALLED STRUCTURES	69-0188
MEHROTRA, B. L.	FINITE ELEMENT ANALYSIS OF THIN SHELLS (DISCUSSION)	69-0187
MEHROTRA, B. L.	SHEAR LAG ANALYSIS OF RECTANGULAR FULL-WIDTH TUBE JUNCTIONS	72-0087
MEHTA, H. S.	FINITE ELEMENT ANALYSIS AND EXPERIMENTAL INVESTIGATION OF SHEET METAL STRETCHING	72-0229
MEHTA, K. C.	STRESS STRAIN RELATIONSHIP FROM COMPRESSION TESTS ON NONLINEAR MATERIALS	69-0189
MEHTA, Y. B.	FORCE AND TORQUE ANALYSIS OF CONSTRAINED SPACE MECHANISMS AND PLANE MECHANISMS WITH OFFSET LINKS BY	74-0569
MEI, C.	ANALYSIS AND DESIGN OF MULTIPLE INPUT FLEXIBLE LINK MECHANISMS	70-0354

MEI, C.	COUPLED VIBRATIONS OF THIN-WALLED BEAMS OF OPEN SECTION USING FINITE ELEMENT METHOD	70-0230
MEI, C.	FINITE ELEMENT ANALYSIS OF FREE VIBRATION AND STABILITY OF STRUCTURES SUBJECTED TO INITIAL STRESS	68-0300
MEI, C.	FINITE ELEMENT ANALYSIS OF NONLINEAR VIBRATION OF BEAM COLUMNS	73-0028
MEI, C.	FINITE ELEMENT DISPLACEMENT METHOD FOR LARGE AMPLITUDE FREE FLEXURAL VIBRATIONS OF BEAMS AND PLATES	73-0185
MEI, C.	FREE VIBRATIONS OF CIRCULAR MEMBRANES UNDER ARBITRARY TENSION BY THE FINITE ELEMENT METHOD	69-0190
MEI, C.	FREE VIBRATIONS OF FINITE ELEMENT PLATES SUBJECTED TO COMPLEX MIDDLE-PLANE FORCE SYSTEMS	72-0280
MEI, C.	FREE VIBRATIONS OF FINITE ELEMENT PLATES SUBJECTED TO COMPLEX MIDDLE-PLANE FORCE SYSTEMS - REPLY	73-0122
MEI, C.	OSCILLATIONS AND WAVE FORCES IN A MAN-MADE HARBOUR IN THE OPEN SEA	74-0151
MEI, C. C.	HYBRID-ELEMENT METHOD FOR WATER WAVES	75-0808
MEI, C. C.	OSCILLATIONS AND WAVE FORCES IN AN OFFSHORE HARBOR - APPLICATIONS OF HYBRID FINITE ELEMENT METHOD TO	74-1024
MEI, K. K.	APPLICATION OF A UNIMOMENT TECHNIQUE TO A BICONICAL ANTENNA WITH INHOMOGENEOUS DIELECTRIC LOADING	75-0375
MEI, K. K.	SCATTERING BY BURIED OBSTACLES	74-0739
MEI, K. K.	SCATTERING BY DIELECTRIC CYLINDERS	74-1025
MEI, K. K.	UNIMOMENT METHOD OF SOLVING ANTENNA AND SCATTERING PROBLEMS	74-1020
MEIJERS, P.	ELASTIC PLASTIC DEFORMATION OF THICK WALLED CYLINDERS	69-0427
MEIJERS, P.	NUMERICAL HULL VIBRATION ANALYSIS OF A FAR EAST CONTAINER SHIP	74-1325
MEIJERS, P.	OPTIMIZATION OF STRUCTURAL DESIGN 1	73-0904
MEIJERS, P.	REVIEW OF THE ASKA PROGRAM	71-0175
MEISSNER, C. J.	DIRECT BEAM STIFFNESS MATRIX CALCULATIONS INCLUDING SHEAR EFFECTS	62-0003
MEISSNER, H.	NUMERICAL COMPUTATION OF BOUNDARY VALUE PROBLEMS IN SOIL MECHANICS	73-0269
MEISSNER, H.	PARAMETER EINES ELASTO-PLASTISCHEN STOFFANSATZES FUER KOERNIGE ERDSTOFFE (DETERMINATION OF PARAMETER	74-0657
MEISSNER, U.	A MIXED FINITE ELEMENT MODEL FOR USE IN POTENTIAL FLOW PROBLEMS	73-0078
MEISSNER, U.	FEHLERBETRACHTUNGEN AUFGRUND DER METHODE DER GEWICHTETEN RESIDUEN FUR DAS VERFAHREN DER FINITEN ELEM	72-0368
MEISSNER, U.	GENERALIZED VARIATIONAL PRINCIPLES FOR USE IN FLOW PROBLEMS THROUGH POROUS MEDIA	74-0067
MELENDEZ, L.	FINITE-ELEMENT CODE TO SIMULATE ANISOTROPIC-PLASMA DIFFUSION IN LINEAR MULTIPOLES	75-0409
MELENDEZ, L.	SEMIDISCRETE GALERKIN TECHNIQUES WITH TIME INTERPOLATION AND SPLITTING UP FOR PLASMA SIMULATION	75-0121
MELFI, D.	3-D FINITE ELEMENT MODEL FOR LAKE CIRCULATION	71-0157
MELKES, F.	THE FINITE ELEMENT METHOD FOR NONLINEAR PROBLEMS	70-0245
MELLER, E.	EXTENSIONS TO THE STAGS COMPUTER CODE	73-0917
MELLIERE, R. A.	A FINITE ELEMENT METHOD FOR GEOMETRICALLY NONLINEAR LARGE DISPLACEMENT PROBLEMS IN THIN, ELASTIC PLA	69-0005
MELLO, R. M.	PLASTIC COLLAPSE LOADS FOR PIPE ELBOWS USING INELASTIC ANALYSIS	73-0941
MELLO, R. M.	PLASTIC COLLAPSE LOADS FOR PIPE ELBOWS USING INELASTIC ANALYSIS	74-0507
MELLO, R. M.	SIMPLIFIED INELASTIC (PLASTIC AND CREEP) ANALYSIS OF PIPE ELBOWS SUBJECTED TO INPLANE AND OUT-OF-PLA	75-0654
MELLO, R. M.	SOME APPLICATIONS OF FINITE ELEMENT ANALYSIS TO SHELL BUCKLING PREDICTION	68-0023
MELOSH, R. J.	A FLAT TRIANGULAR SHELL ELEMENT STIFFNESS MATRIX	66-0052
MELOSH, R. J.	A STIFFNESS MATRIX FOR THE ANALYSIS OF THIN PLATES IN BENDING	61-0001
MELOSH, R. J.	BASIS OF DERIVATION OF MATRICES FOR THE DIRECT STIFFNESS METHOD	63-0008
MELOSH, R. J.	BEHAVIOR OF TRIANGULAR SHELL ELEMENT STIFFNESS MATRICES ASSOCIATED WITH POLYHEDRAL DEFLECTION DISTRI	68-0159
MELOSH, R. J.	CHARACTERISTICS OF MANIPULATION ERRORS IN SOLVING LOAD DEFLECTION EQUATIONS	70-0355
MELOSH, R. J.	COMPUTATIONAL TECHNIQUES FOR FINITE ELEMENT ANALYSIS	73-1036
MELOSH, R. J.	DEVELOPMENT OF THE STIFFNESS METHOD TO DEFINE BOUNDS ON ELASTIC BEHAVIOUR OF STRUCTURES	62-0011
MELOSH, R. J.	EFFICIENT SOLUTION OF LOAD-DEFLECTING EQUATIONS	69-0396
MELOSH, R. J.	FINITE ELEMENT ANALYSIS OF AUTOMOBILE STRUCTURES	74-0309
MELOSH, R. J.	INHERITED ERROR IN FINITE ELEMENT ANALYSES OF STRUCTURES	73-0452
MELOSH, R. J.	LARGE DEFLECTIONS OF STRUCTURES SUBJECT TO HEATING AND EXTERNAL LOADS	60-0002
MELOSH, R. J.	MANIPULATION ERRORS IN FINITE ELEMENT ANALYSIS	68-0253
MELOSH, R. J.	MANIPULATION ERRORS IN FINITE ELEMENT ANALYSIS OF STRUCTURES	69-0349
MELOSH, R. J.	MANIPULATION ERRORS IN THE FINITE ELEMENT ANALYSES	71-0176
MELOSH, R. J.	MANIPULATION ERRORS IN COMPUTER SOLUTION OF CRITICAL SIZE STRUCTURAL EQUATIONS	71-0367
MELOSH, R. J.	MATRIX METHODS OF STRUCTURAL ANALYSIS	62-0004
MELOSH, R. J.	MODIFIED POTENTIAL ENERGY MASS REPRESENTATION FOR FREQUENCY PREDICTION	66-0053
MELOSH, R. J.	NONPROPORTIONAL LOADING LIMIT FOR STRUCTURES	74-0228
MELOSH, R. J.	NUMERICAL ANALYSIS OF AUTOMOBILE STRUCTURES	72-0334
MELOSH, R. J.	NUMERICAL SUFFICIENCY TEST FOR MONOTONIC CONVERGENCE OF FINITE ELEMENT MODELS	75-0620
MELOSH, R. J.	ON A NUMERICAL SUFFICIENCY TEST FOR MONOTONIC CONVERGENCE OF FINITE ELEMENT MODELS	75-0533
MELOSH, R. J.	SINGER: A COMPUTER CODE FOR GENERAL ANALYSIS OF TWO-DIMENSIONAL CONCRETE STRUCTURES. VOL. 11. PROGRA	75-0723
MELOSH, R. J.	SINGER: A COMPUTER CODE FOR GENERAL ANALYSIS OF TWO-DIMENSIONAL REINFORCED CONCRETE STRUCTURES. VOLU	75-0724
MELOSH, R. J.	SINGER: A COMPUTER CODE FOR GENERAL ANALYSIS TO TWO-DIMENSIONAL CONCRETE STRUCTURES. VOLUME 4. DEMON	75-0730
MELOSH, R. J.	SINGER: A COMPUTER CODE FOR GENERAL ANALYSIS OF TWO-DIMENSIONAL REINFORCED CONCRETE STRUCTURES. VOLU	75-0731
MELOSH, R. J.	STATUS REPORT ON COMPUTATIONAL TECHNIQUES FOR FINITE ELEMENT ANALYSES	74-0242
MELOSH, R. J.	STRUCTURAL ANALYSIS OF SOLIDS	63-0009
MELOSH, R. J.	STRUCTURAL ANALYSIS AND MATRIX INTERPRETIVE SYSTEM (SAMIS)	66-0098
MELOSH, R. J.	THE OPTIMUM APPROACH TO ANALYSIS OF ELASTIC CONTINUA	71-0177
MELOSH, R. J.	UNBOUNDED ERRORS IN NUMERICAL ANALYSIS OF STRUCTURES	74-1210
MELVILLE, J. G.	FINITE-ELEMENT ANALYSIS OF TAPERED VISCOUS FLOW	72-0961
MENDELSON, A.	A GENERAL APPROACH TO THE PRACTICAL SOLUTION OF CREEP PROBLEMS	59-0003
MENDELSON, A.	EVALUATION OF THE USE OF A SINGULARITY ELEMENT IN FINITE ELEMENT ANALYSIS OF CENTER-CRACKED PLATES	72-0870
MENTEL, T. J.	STUDY AND DEVELOPMENT OF SIMPLE MATRIX METHODS OF INELASTIC STRUCTURES	66-0115
MENZEL, R.	PROCEDURE FOR SOLVING LINEAR EQUATION SYSTEMS WITH SPARSELY OCCUPIED COEFFICIENT MATRICES ASSOCIATED	74-0352
MERA, A.	STATIC AND HEAT TRANSFER ANALYSIS OF A SPHERE-CONE INTERSECTION IN A NUCLEAR CONTAINMENT VESSEL	75-0166
MERATI, J. K.	FINITE ELEMENT LARGE DEFORMATION ANALYSIS OF A TAPERED AORTA	75-0856
MERCER, J. W.	FINITE ELEMENT ANALYSIS OF HYDROTHERMAL SYSTEMS	74-0046
MERCER, J. W.	FINITE ELEMENT APPROACH TO THE MODELING OF HYDROTHERMAL SYSTEMS	73-0011
MERCER, J. W.	GALERKIN-FINITE ELEMENT ANALYSIS OF HYDROTHERMAL SYSTEM AT WAIRAKEI, NEW-ZEALAND	75-0601
MERCER, J. W. J	GALERKIN FINITE-ELEMENT SIMULATION OF A GEOTHERMAL RESERVOIR	73-0900
MERCHANT, D. H.	EFFECT OF DAMPING ON EXCITABILITY OF HIGH-ORDER NORMAL MODES	75-0709
MERCHANT, J. K.	MODAL SENSITIVITY STUDY OF A CONICAL RE-ENTRY VEHICLE SUBJECTED TO AERODYNAMICALLY INDUCED ACOUSTIC	69-0431
MERCIER, B.	FINITE ELEMENT APPROXIMATION, AND SOLUTION BY A PENALTY-DUALITY ALGORITHM OF AN ELASTO-PLASTIC PROBL	75-0041
MERCIER, B.	NUMERICAL SOLUTION OF THE BIHARMONIC PROBLEM BY MIXED FINITE OF CLASS C-0	73-0803
MERCKX, K. R.	CALCULATIONAL PROCEDURE FOR DETERMINING CREEP COLLAPSE OF LWR FUEL RODS	74-1001
MERIMAN, P. A.	AN ANNULAR SEGMENT FINITE ELEMENT FOR PLATE BENDING	71-0285
MERRETT, G.	ESTABLISHMENT OF A VALID FRACTURE TOUGHNESS SPECIMEN GEOMETRY USING A FINITE ELEMENT ANALYSIS	74-0135
MERRIFIELD, B.	A COMPUTER PROGRAM FOR THE FINITE ELEMENT ANALYSIS OF A CLASS OF PLATE BENDING PROBLEMS	69-0191
MERRIFIELD, B.	FORTRAN SUBROUTINES FOR FINITE ELEMENT ANALYSIS	72-0694
MERRIFIELD, B.	ON THE CONFORMING CUBIC TRIANGULAR ELEMENT FOR PLATE BENDING	72-0439
MERRIMAN, P. A.	FINITE ELEMENT ANALYSIS OF BRIDGES CURVED IN PLAN	71-0547
MERWIN, J. E.	SOLUTION OF NONLINEAR PROBLEMS OF ELASTOPLASTICITY BY THE FINITE ELEMENT METHOD	68-0007
MERZ, K. L.	NONLINEAR SEISMIC ANALYSIS OF CURRENT DIVIDER SUPPORT SYSTEM FOR PACIFIC DC INTERTIE	73-0200
MESSIER, R. H.	FINITE ELEMENT ALGORITHM FOR THE DETERMINATION OF DYNAMIC BUCKLING	75-0302
MESZTENYI, C.	SELF-ADAPTIVE REFINEMENTS IN THE FINITE ELEMENT METHOD	75-0712
MEYER, C.	ANALYSIS OF CURVED FOLDED PLATE STRUCTURES	70-0431
MEYER, C.	ANALYSIS OF CURVED FOLDED PLATE STRUCTURES	71-0139
MEYER, C.	COMPUTER PROGRAM FOR PRISMATIC FOLDED PLATES BY PLATE AND BEAM ELEMENTS	70-0371
MEYER, G. H.	FINITE ELEMENT SOLUTION OF DEGENERATE INTERFACE PROBLEMS	75-0101
MEYER, P. M.	DATA GENERATION FOR FINITE-ELEMENT STRUCTURAL ANALYSIS OF THREE-DIMENSIONAL NAVAL SHIP STRUCTURES	75-0707
MEYER, P. M.	DATA GENERATOR FOR THE IDEALIZATION FOR FINITE ELEMENT STRUCTURAL ANALYSIS OF NAVAL SHIP FLAT PLATED	72-0860
MEYER, R. R.	CONICAL SEGMENT METHOD FOR ANALYZING OPEN CROWN SHELLS OF REVOLUTION FOR EDGE LOADING	63-0007
MEYER, T. O.	NX BOREHOLE JACK MODULUS DETERMINATIONS IN HOMOGENEOUS, ISOTROPIC, ELASTIC MATERIALS	74-0295
MEYERS, J. F.	BUILDING THREE NEW MILLION-HORSEPOWER HYDRAULIC TURBINES FOR GRAND COULEE	74-0981
MEYERS, V. J.	PLASTIC SANDWICH BUCKLE-SHELL STRUCTURES	72-0119
MICHAIL, M. G.	FINITE ELEMENT ANALYSIS OF SKEW SANDWICH PLATES	72-0810
MIESSEN, W.	ANWENDUNG DER METHODE FINITER ELEMENTE BEI DER ANALYSE DES DYNAMISCHEN VERHALTENS GEDAEMPFTER WERKZE	75-0580

71

MIGITA, M.	APPLICATION OF FINITE ELEMENT METHOD TO HYDRODYNAMIC LUBRICATION PROBLEMS (PART 1, "INFINITE-WIDTH B	71-0537
MIHALCEA, A.	COMPORTAREA ELASTICA A DIGURILOR DE PAMINT SOLICITE SEISMIC (ELASTIC BEHAVIOR OF THE EARTH DAMS SEI	74-1185
MIKI, T.	STRESS ANALYSIS PROGRAM BY FINITE ELEMENT METHOD FOR RADIAL FLOW IMPELLERS	74-1016
MIKIC, B. B.	AREAS OF CONTACT AND PRESSURE DISTRIBUTION IN BOLTED JOINTS	71-0329
MIKKELSON, P. T	COMPUTERIZED ANALYSIS OF NON-ISOTROPIC STRUCTURES	69-0192
MIKKOLA, M.	MATERIAL BEHAVIOUR CHARACTERISTICS FOR REINFORCED CONCRETE SHELLS STRESSED BEYOND THE ELASTIC RANGE	70-0166
MIKKOLA, M.	ON THE CONVERGENCE OF THE FINITE ELEMENT METHOD	71-0513
MIKSCH, M.	APPLICATION OF THE FINITE ELEMENT METHOD FOR THE SAFETY EVALUATION OF REACTOR COMPONENTS	73-0853
MIKSCH, M.	STATIC AND HEAT TRANSFER ANALYSIS OF A SPHERE-CONE INTERSECTION IN A NUCLEAR CONTAINMENT VESSEL	70-0166
MILEIKOVSKII, I	THE DEVELOPMENT OF APPLIED METHODS IN PROBLEMS OF STATIC CALCULATION OF THIN-WALLED THREE-DIMENSIONA	73-0934
MILES, G. A.	A CURVED ELEMENT APPROXIMATION IN THE ANALYSIS OF AXISYMMETRIC THIN SHELLS	70-0063
MILES, G. A.	AN ECONOMICAL METHOD FOR DETERMINING THE SMALLEST EIGENVALUES OF LARGE LINEAR SYSTEMS	71-0551
MILES, G. A.	FINITE-ELEMENT METHOD AND ITS APPLICATIONS IN AN ENGINEERING LABORATORY	70-0202
MILLER, C. J.	INFLUENCE OF SOIL/STRUCTURE INTERACTION PARAMETERS ON FLOOR RESPONSE SPECTRA	73-0833
MILLER, C. J.	LIGHT GAGE STEEL INFILL PANELS IN MULTISTORY STEEL FRAMES	74-0229
MILLER, D. G.	TSAAS: FINITE ELEMENT THERMAL AND STRESS ANALYSIS OF AXISYMMETRIC SOLIDS WITH ORTHOTROPIC TEMPERATUR	74-0886
MILLER, D. W.	TRANSPORT SOLUTIONS USING FINITE ELEMENTS IN SPACE-ANGLE PHASE SPACE	71-0480
MILLER, K. J.	AN ELASTIC-PLASTIC FINITE ELEMENT ANALYSIS OF CRACK TIP FIELDS UNDER BIAXIAL LOADING CONDITIONS	73-0760
MILLER, K. J.	ELASTIC-PLASTIC FINITE ELEMENT ANALYSIS OF CRACK TIP FIELDS UNDER BIAXIAL LOADING CONDITIONS	74-0296
MILLER, K. J.	STRESS, DISPLACEMENT, LINE INTEGRAL AND CLOSURE ENERGY DETERMINATIONS OF CRACK TIP STRESS INTENSITY	74-0621
MILLER, M. C.	LARGE USER-ORIENTATED SYSTEMS OF PROGRAMS FOR STRUCTURAL ANALYSIS AND DESIGN	74-0482
MILLER, M. F.	DESIGN, MANUFACTURE, DEVELOPMENT, TEST, AND EVALUATION OF BORON/ALUMINUM STRUCTURAL COMPONENTS FOR S	74-0798
MILLER, R. D.	APPLICATION OF FINITE-ELEMENT THEORY TO AIRPLANE CONFIGURATIONS	71-0458
MILLER, R. E.	LARGE SCALE ANALYSIS OF CURRENT AIRCRAFT	70-0383
MILLER, R. W.	CYCLIC FATIGUE ANALYSIS OF ROCKET THRUST CHAMBERS. VOLUME 2: ATTITUDE CONTROL THRUSTER HIGH CYCLE FA	74-0693
MILLER, R. W.	CYCLIC FATIGUE ANALYSIS OF ROCKET THRUST CHAMBERS. VOLUME 1: OFHC COPPER CHAMBER LOW CYCLE FATIGUE	74-0694
MILLER, R. W.	RETSCP: A COMPUTER PROGRAM FOR ANALYSIS OF ROCKET ENGINE THERMAL STRAINS WITH CYCLIC PLASTICITY	74-0692
MILLER, W. E.	FINITE ELEMENT INTEGRAL NEUTRON TRANSPORT	73-1032
MILLER, W. F.	APPLICATION OF PHASE-SPACE FINITE ELEMENTS TO ONE DIMENSIONAL NEUTRON TRANSPORT EQUATION	73-0123
MILLER, W. F.	APPLICATION OF PHASE SPACE FINITE ELEMENTS TO THE TWO DIMENSIONAL NEUTRON TRANSPORT EQUATION IN X-Y	73-1039
MILLER, W. F.	QUADRATIC FINITE ELEMENT IN NEUTRON TRANSPORT	73-1039
MILLER, W. F.	RAY-EFFECT MITIGATION IN DISCRETE ORDINATE-LIKE ANGULAR FINITE-ELEMENT APPROXIMATIONS IN NEUTRON-TRA	75-0496
MILLER, W. F.	TWO-DIMENSIONAL TRANSPORT CALCULATIONS USING PHASE-SPACE FINITE ELEMENTS	72-0027
MILLER, W. F.	2-DIMENSIONAL FINITE-ELEMENT METHOD FOR INTEGRAL NEUTRON-TRANSPORT CALCULATIONS	75-0443
MILLER, W. F. J	APPLICATION OF FINITE-ELEMENTS TO ONE-DIMENSIONAL AND TWO-DIMENSIONAL NEUTRON-TRANSPORT	73-1150
MILLOY, J. A.	A VARIATIONAL FORMULATION FOR DISCONTINUOUS FIELDS IN LINEARLY ELASTIC FINITE ELEMENT APPLICATIONS	72-0962
MILLS, B.	APPLIED FINITE ELEMENT DISPLACEMENT ANALYSIS OF A SIMPLE INTEGRAL CONSTRUCTION VEHICLE BODY SHELL	72-0264
MILLS, B.	DYNAMIC ANALYSIS OF A CAR CHASSIS FRAME USING FINITE ELEMENT METHOD	72-0265
MILNE, R. D.	APPLICATION OF INTEGRAL EQUATIONS TO FLUID FLOWS IN UNBOUNDED REGIONS	74-0029
MILNER, M.	NATURAL FREQUENCIES AND MODE SHAPES OF TRANSFORMER CORES	65-0048
MILSTED, M. G.	USE OF TRIGONOMETRIC TERMS IN FINITE ELEMENT METHOD WITH APPLICATION TO VIBRATING MEMBRANES	74-0127
MINAMI, H. M.	PRESSURE VESSEL WALL THICKNESS LIMITS BASED ON THERMAL STRESS RACHETTING AND CREEP FATIGUE INTERACTI	74-0968
MINAMI, T.	ELASTIC-PLASTIC EARTHQUAKE RESPONSE OF SOIL-BUILDING SYSTEMS	72-0717
MINEAR, J. W.	FINITE ELEMENT MODELS OF FRACTURE	73-0012
MINICH, M. D.	ANALYTICAL DISPLACEMENTS AND VIBRATIONS OF CANTILEVERED UNSYMMETRIC FIBER COMPOSITE LAMINATES	75-0736
MIRSKY, I.	IN VIVO STRESSES IN THE HUMAN LEFT VENTRICULAR WALL: ANALYSIS ACCOUNTING FOR THE IRREGULAR THREE DIM	72-0136
MIRZA, F. A.	A HYBRID FINITE ELEMENT METHOD FOR TWO-DIMENSIONAL NAVIER STOKES EQUATIONS	75-0027
MIRZA, M. S.	A STUDY OF BEHAVIOUR OF REINFORCED CONCRETE ELEMENTS USING FINITE ELEMENTS	70-0335
MIRZA, M. S.	FINITE ELEMENT ANALYSIS OF SHEAR STRENGTH OF REINFORCED CONCRETE BEAMS	74-1306
MIRZA, W. H.	DYNAMIC BEHAVIOR OF IN-LINE SHEAR WALLS CONNECTED BY FLOOR SLABS	72-0689
MIRZA, W. H.	FINITE ELEMENT MODELLING TECHNIQUES FOR THE ANALYSIS OF BUILDING VIBRATIONS	72-0839
MIRZA, W. H.	ON THE VIBRATION OF POINT-SUPPORTED PLATES	71-0602
MIRZAD, S. S.	CONTRIBUTION A L'ANALYSE DES VIBRATIONS LIBRES D'UNE STRUCTURE. APPLICATION AUX AUBAGES DE TURBOMACH	75-0804
MISEL, J.	STARDYNE STRUCTURAL ANALYSIS SYSTEM, ANALYTICAL CAPABILITY MANUAL	68-0235
MISHU, F.	AN INVESTIGATION INTO THE ECONOMICS OF USING COMPLEX ELEMENTS IN PLANE STRESS ANALYSIS	74-0427
MISRA, N.	BEHAVIOR OF ISOLATED SHEAR WALLS SUBJECTED TO LATERAL LOADS	75-0704
MISTREE, F.	SOME CONSIDERATIONS REGARDING STRUCTURAL OPTIMIZATION AND FINITE ELEMENT ANALYSIS	74-0424
MISTRY, J.	FREE VIBRATIONS OF CYLINDRICAL SHELLS WITH VARIOUS END CLOSURES	74-0571
MISTRY, J.	THE FINITE ELEMENT SOLUTION OF FLOW AND HEAT TRANSFER IN ELLIPTICAL DUCTS	74-0039
MISTRY, J.	THE PREDICTION OF FULLY DEVELOPED TURBULENT FLOW IN DUCTS BY THE FINITE ELEMENT METHOD	75-0117
MITCHELL, A. R.	AN EXACT BOUNDARY TECHNIQUE FOR IMPROVED ACCURACY IN THE FINITE ELEMENT METHOD	73-1035
MITCHELL, A. R.	BASIC FUNCTIONS FOR CURVED ELEMENTS IN THE MATHEMATICAL THEORY OF FINITE ELEMENTS	75-0091
MITCHELL, A. R.	CORNER SINGULARITIES IN ELLIPTIC PROBLEMS BY FINITE ELEMENT METHODS	71-0456
MITCHELL, A. R.	CURVED BOUNDARIES IN THE FINITE ELEMENT METHOD	73-0638
MITCHELL, A. R.	CURVED ELEMENTS IN THE FINITE ELEMENT METHOD	74-1261
MITCHELL, A. R.	ELEMENT TYPES AND BASE FUNCTIONS	73-0522
MITCHELL, A. R.	FORBIDDEN SHAPES IN THE FINITE ELEMENT METHOD	74-0705
MITCHELL, A. R.	INTRODUCTION TO THE MATHEMATICS OF THE FINITE ELEMENT METHOD	72-0054
MITCHELL, A. R.	MATCHING OF ESSENTIAL BOUNDARY CONDITIONS IN THE FINITE ELEMENT METHOD	74-1260
MITCHELL, A. R.	THE CONSTRUCTION OF BASIC FUNCTIONS FOR CURVED ELEMENTS IN THE FINITE ELEMENT METHOD	72-0558
MITCHELL, A. R.	THE FINITE ELEMENT METHOD	74-1259
MITCHELL, A. R.	USE OF PARABOLIC ARCS IN MATCHING CURVED BOUNDARIES IN FINITE-ELEMENT METHOD	75-0589
MITCHELL, A. R.	VARIATIONAL PRINCIPLES AND FINITE ELEMENT METHOD IN PARTIAL DIFFERENTIAL EQUATIONS	71-0452
MITCHELL, G. C.	ACCURACY AND OPTIMIZATION IN THE ANALYSIS OF SHIP STRUCTURES, USING FINITE ELEMENT METHODS	71-0418
MITCHELL, G. C.	PRACTICAL CONSIDERATIONS IN THE APPLICATION OF FINITE ELEMENT TECHNIQUES TO SHIP STRUCTURES	69-0338
MITCHELL, J. K.	ANALYSIS OF LOAD-BEARING FILLS OVER SOFT SUBSOILS	71-0330
MITCHELL, J. K.	SEEPAGE AND GROUNDWATER EFFECTS ASSOCIATED WITH EXPLOSIVE CRATERING	72-0656
MITCHELL, J. K.	STRESS - DEFORMATION PREDICTION IN CEMENT - TREATED SOIL PAVEMENTS	71-0286
MITCHELL, R. A.	ANALYSIS OF COMPOSITE REINFORCED CUTOUTS AND CRACKS	74-0824
MITCHELL, R. A.	COMPOSITE-OVERLAY REINFORCEMENT OF CUTOUTS AND CRACKS IN METAL SHEET	73-0878
MITCHELL, R. A.	FORMULATION AND EXPERIMENTAL VERIFICATION OF AN AXISYMMETRIC FINITE-ELEMENT STRUCTURAL ANALYSIS	71-0354
MITCHELL, R. A.	HIGH STRENGTH END FITTINGS FOR FRP ROD AND ROPE	74-0605
MITCHELL, R. A.	NON-METALLIC ANTENNA-SUPPORT MATERIALS	73-0716
MITKEVICH, V. M.	CONSTRUCTION OF NORMAL DISPLACEMENT FUNCTIONS OF TRIANGULAR FINITE ELEMENT OF PLATES AND SHELLS	72-0563
MITSUAKI, N.	APPLICATION OF PROGRAM FATIQUE TEST TO MEMBER JOINTS OF HULLS	72-0814
MITTELMANN, H.	FINITE ELEMENT VERFAHREN BEI QUASILINEAREN ELLIPTISCHEN RANDWERTPROBLEM	74-1262
MITTELMANN, H.	NICHTLINEARE DIRICHLET PROBLEME UND EINFACHE FINITE ELEMENT VERFAHREN	74-1263
MITTELMANN, H.	STABILITAT BEI DER METHODE DER FINITEN ELEMENTE FUR QUASILINEARE ELLIPTISCHE RANDWERTPROBLEME	74-1283
MIURA, H.	AUTOMATED DESIGN OPTIMIZATION OF SUPERSONIC AIRPLANE WING STRUCTURES UNDER DYNAMIC CONSTRAINTS	72-0178
MIURA, I.	SIMULATION ON THE INITIATION AND GROWTH OF DUCTILE FRACTURE VOIDS	75-0836
MIURA, I.	SIMULATION ON THE TENSILE BEHAVIOUR OF TWO-PHASE ALLOY CONTAINING A FIBER UNDER TENSILE LOAD	73-0542
MIYACHI, T.	FLEXURAL RIGIDITY OF A THIN WALLED BUILD-UP ROTOR FOR JET ENGINE. MEASUREMENT BY STATIC LOAD TEST AND	73-0756
MIYAMOTO, H.	ANALYSIS OF CRACK PROPAGATION IN WELDED STRUCTURES	74-1144
MIYAMOTO, H.	ANALYSIS OF STRESS AND STRAIN DISTRIBUTIONS AT CRACK TIP BY FINITE ELEMENT METHOD	69-0193
MIYAMOTO, H.	APPLICATION OF FINITE ELEMENT METHOD TO FRACTURE MECHANICS	71-0178
MIYAMOTO, H.	COMPUTER PREDICTION OF FATIGUE CRACK PROPAGATION UNDER RANDOM LOADING	73-0851
MIYAMOTO, H.	EFFECT OF GRAIN BOUNDARY FOR THE ELASTIC BEHAVIOR OF AL AND CU CRYSTALS (ANALYSIS BY THE FINITE ELEM	74-0947
MIYAMOTO, H.	ELASTIC AND PLASTIC ANALYSIS BY FINITE ELEMENT METHOD	68-0119
MIYAMOTO, H.	ELASTIC PLASTIC ANALYSIS OF THREE DIMENSIONAL PROBLEMS BY THE FINITE ELEMENT METHOD	69-0387
MIYAMOTO, H.	SIMULATION OF ELASTIC MODULUS AND POISSON'S RATIO OF SPHEROIDAL GRAPHITE CAST IRON	74-0955
MIYAMOTO, H.	STRESS AND STRAIN DISTRIBUTIONS AT THE ROOT OF CRACKS	69-0290
MIYAMOTO, H.	THE APPLICATION OF THE FINITE ELEMENT METHOD TO FRACTURE MECHANICS	72-0331

MIYASHITA, T.	NONLINEAR ANALYSIS OF REINFORCED CONCRETE BUILDINGS	73-0324
MIYATA, H.	METHOD OF EVALUATION OF THREE-DIMENSIONAL STRESS INTENSITY FACTOR USING THE FINITE ELEMENT METHOD	75-0316
MIYATA, H.	THE SIMPLE METHOD OF EVALUATION OF STRESS INTENSITY FACTOR USING THE FINITE ELEMENT METHOD	74-0829
MIYAZAKI, N.	ANALYSIS OF ELASTIC-PLASTIC CREEP BUCKLING OF AXISYMMETRIC SHELLS BY FINITE-ELEMENT METHOD	75-0355
MIYAZAKI, N.	ANALYSIS OF LARGE ELASTIC-PLASTIC CREEP DEFORMATIONS OF AXISYMMETRIC SHELLS BY THE FINITE ELEMENT ME	74-0563
MIYAZAKI, N.	FINITE ELEMENT CREEP BUCKLING ANALYSIS OF VISCO-ELASTIC SHALLOW SPHERICAL SHELL	75-0159
MIYOSHI, T.	ANALYSIS OF CRACK PROPAGATION IN WELDED STRUCTURES	74-1144
MIYOSHI, T.	CONVERGENCE OF FINITE ELEMENT SOLUTIONS REPRESENTED BY A NON-CONFORMING BASIS	72-0812
MIYOSHI, T.	ELASTIC AND PLASTIC ANALYSIS BY FINITE ELEMENT METHOD	68-0119
MIYOSHI, T.	ELASTIC PLASTIC ANALYSIS OF THREE DIMENSIONAL PROBLEMS BY THE FINITE ELEMENT METHOD	69-0387
MIYOSHI, T.	FINITE ELEMENT METHOD OF MIXED TYPE AND ITS CONVERGENCE IN LINEAR SHELL PROBLEMS	73-1041
MIYOSHI, T.	NONLINEAR ANALYSIS OF CRACKED BODY BY FINITE ELEMENT METHOD	75-0315
MIYOSHI, T.	STRESS AND STRAIN DISTRIBUTIONS AT THE ROOT OF CRACKS	69-0290
MIYOSHI, T.	STUDY ON J//I//C FRACTURE CRITERION	74-1034
MIYOSHI, T.	THE APPLICATION OF THE FINITE ELEMENT METHOD TO FRACTURE MECHANICS	72-0331
MIZERNYUK, G. N	OPREDELENIE STATSIONARNYKH TEMPERATURNYKH POLEI V DETALYAKH DVIGATELEI VNUTRENNEGO SGORANIYA METODOM	73-0454
MIZUMACHI, W.	ANALYSIS FOR ATOMIC POWER PLANT BY FINITE ELEMENT METHOD	71-0198
MIZUMACHI, W.	APPLICATION OF FINITE ELEMENT METHOD TO THE THREE DIMENSIONAL STRESS AND VIBRATION ANALYSIS OF BWR P	71-0180
MIZUMACHI, W.	APPLICATION OF FINITE ELEMENT METHOD TO BOILING WATER REACTOR POWER PLANT (JAPANESE)	72-0816
MIZUMACHI, W.	FINITE ELEMENT ANALYSIS IN NUCLEAR POWER PLANTS (JAPANESE)	72-0685
MIZUMACHI, W.	NON-LINEAR THERMAL STRESS ANALYSIS FOR NUCLEAR POWER PLANT BY FINITE ELEMENT METHOD	71-0181
MLEJNEK, H. P.	BERECHNUNG EINES EINACHSIGEN STAHLDRAHT-KUNSTSTOFFVERBUNDES UNTER EINSATZ DER FINIT-ELEMENT METHODE	72-0644
MLEJNEK, H. P.	F 104 G FINAL MOTIVE: GROUND VIBRATION CALCULATIONS FOR THE STARFIGHTER F 104 G	69-0456
MLEJNEK, H. P.	MALEN DIMENSIONIERUNG VON TRAGWERKEN	74-0683
MLEJNEK, H. P.	TURBAN VERSION 1.0. EIN SPEZIELLES PROGRAMM ZUR LINEAREN UND NICHTLINEAREN STATISCHEN UND DYNAMISCHE	71-0622
MO, O.	ELASTO-PLASTIC ANALYSIS USING AN EFFICIENT FORMULATION OF THE FINITE ELEMENT METHOD	75-0012
MOAN, T.	ANALYSIS OF STIFFENED PLATES CONSIDERING NONLINEAR MATERIAL AND GEOMETRIC BEHAVIOUR	74-0327
MOAN, T.	EXPERIENCES WITH ORTHOGONAL POLYNOMIALS AND "BEST" NUMERICAL INTEGRATION FORMULAS ON A TRIANGLE; WIT	74-0977
MOAN, T.	FINITE ELEMENT STRESS FIELD SOLUTION OF THE PROBLEM OF SAINT VENANT TORSION	73-1042
MOAN, T.	NOTE ON THE CONVERGENCE OF FINITE ELEMENT APPROXIMATIONS FOR PROBLEMS FORMULATED IN CURVILINEAR COOR	74-0267
MOAN, T.	ON SHELL EFFECTS ON FERRO-CEMENT VESSELS	73-0455
MOAN, T.	ON THE LOCAL DISTRIBUTION OF ERRORS BY FINITE ELEMENT APPROXIMATIONS	73-0304
MOAN, T.	STOCHASTIC DYNAMIC RESPONSE ANALYSIS OF OFFSHORE PLATFORMS, WITH PARTICULAR REFERENCE TO GRAVITY-TYP	75-0220
MOAN, T.	THE FINITE ELEMENT METHOD IN PLATE BUCKLING	69-0128
MOCK, B.	LUMPING OF FINITE ELEMENT MATRICES	74-0645
MODENA, C.	DISTRIBUZIONI OTTIMALI DI ARMATURE DI PRECOMPRESSIONE IN TRAVI-PARETE DI CALCESTRUZZO. (OPTIMAL DIST	75-0560
MOE, J.	FINITE ELEMENT TECHNIQUES IN SHIP STRUCTURE DESIGN IN FINITE ELEMENT TECHNIQUES	69-0335
MOE, J.	MODEL EXPERIMENTS AND FINITE ELEMENT ANALYSIS OF STRESS IN AN OPEN SHIP	66-0054
MOE, J.	THE FINITE ELEMENT TECHNIQUE - A NEW TOOL IN STRUCTURAL ANALYSIS	69-0195
MOEHLING, C.	COMPUTER ANALYSIS OF A RAILROAD FREIGHT CAR BOLSTER UTILIZING THE FINITE ELEMENT METHOD	71-0283
MOFFATT, K. R.	SHEAR LAG IN STEEL BOX GIRDER BRIDGES	75-0310
MOHAMMED, M. A.	VIBRATION ANALYSIS OF PLATES, SHELLS AND BLADED RIMMED DISCS USING THE TRANSFER-MATRIX-FINITE-ELEMEN	73-1151
MOHAN, A.	FINITE-ELEMENT ANALYSIS OF HEAT FLOW AROUND BURIED PIPES	73-1153
MOHITPOUR, M.	TEMPERATURE RISE IN THE HIGH SPEED COMPRESSION OF RIGHT CYLINDRICAL BILLETS	74-1153
MOHR, C. L.	NONLINEAR STRESS ANALYSIS OF INDETERMINATE STRUCTURAL ELEMENTS SUBJECTED TO BIAXIAL STRESS STATES US	73-1152
MOHRA, B.	RESEARCH TO IMPROVE TUNNEL SUPPORT SYSTEMS	74-0777
MOHRAZ, B.	A METHOD OF COMPUTING NUMERICALLY INTEGRATED STIFFNESS MATRICES	72-0546
MOHRAZ, B.	ANALYSIS OF SHALLOW SHELL STRUCTURES BY A DISCRETE ELEMENT SYSTEM	66-0113
MOHRAZ, B.	ELASTO-PLASTIC ANALYSIS OF THREE-DIMENSIONAL STRUCTURES USING THE ISOPARAMETRIC ELEMENT	72-0248
MOHRAZ, B.	FINITE ELEMENT METHOD FOR THE ANALYSIS OF NOZZLE OPENINGS IN SHELLS OF REVOLUTION	69-0254
MOHRAZ, B.	LINER-MEDIUM INTERACTION IN TUNNELS	75-0052
MOHRAZ, B.	LUMPED PARAMETER ELEMENT FOR THE ANALYSIS OF HYPERBOLIC PARABOLOID SHELLS	72-0148
MOHRAZ, B.	NONLINEAR ANALYSIS OF PLANAR REINFORCED CONCRETE STRUCTURES	74-0533
MOHRAZ, B.	THICK WALLED MULTIPLE OPENING REINFORCED CONCRETE CONDUITS	72-0666
MOHRAZ, B.	3-DIMENSIONAL ISOPARAMETRIC ELEMENT IN CURVILINEAR COORDINATES	72-0098
MOJTAHEDI, S.	FINITE ELEMENT ANALYSIS OF ARCH DAMS AND FOUNDATIONS	71-0668
MOKRY, M.	INTEGRAL EQUATION METHOD FOR SUBSONIC FLOW PAST AIRFOILS IN VENTILATED WIND TUNNELS	75-0065
MOLHOEK, F. A.	DETERMINATION OF THE INITIAL BUCKLING STRESS OF FLAT PLATES USING THE DISPLACEMENT METHOD	69-0465
MOLLENDORF, J.	APPLICABILITY OF APPROXIMATE AND EXACT TRANSIENT HEAT TRANSFER ANALYSES TO HEATING PROCESSES USED TO	75-0244
MOLZ, F. J.	NUMERICAL METHODS IN SUBSURFACE HYDROLOGY: WITH AN INTRODUCTION TO THE FINITE ELEMENT METHOD	71-0800
MONAHAN, J.	NATURAL FREQUENCIES AND MODE SHAPES OF PLATES WITH INTERIOR CUT-OUTS	70-0402
MONDKAR, D. P.	LARGE CAPACITY EQUATION SOLVER FOR STRUCTURAL ANALYSIS	74-0983
MONDKAR, D. P.	STATIC AND DYNAMIC ANALYSIS OF NONLINEAR STRUCTURES	75-0718
MONEY, H. A.	THE APPLICATION OF THREE-DIMENSIONAL FINITE ELEMENTS TO A CYLINDER-CYLINDER INTERSECTION	70-0068
MONFORTON, G. R	DISCRETE ELEMENT FINITE DISPLACEMENT ANALYSIS OF ANISOTROPIC SANDWICH SHELLS	70-0504
MONFORTON, G. R	FINITE DEFLECTION DISCRETE ELEMENT ANALYSIS OF SANDWICH PLATES AND CYLINDRICAL SHELLS WITH LAMINATED	69-0236
MONFORTON, G. R	FINITE ELEMENT ANALYSIS OF SKEW PLATES IN BENDING	68-0120
MONFORTON, G. R	FINITE ELEMENT ANALYSIS OF SANDWICH PLATES AND CYLINDRICAL SHELLS WITH LAMINATED FACES	68-0121
MONFORTON, G. R	FINITE ELEMENT ANALYSIS OF SKEW SANDWICH PLATES	72-0810
MONISMITH, C. L	EFFECT OF NONLINEAR MATERIAL RESPONSE ON THE BEHAVIOUR OF PAVEMENTS UNDER TRAFFIC	70-0171
MONISMITH, C. L	FATIGUE CRACK FORMATION AND PROPAGATION IN PAVEMENTS CONTAINING SOIL-CEMENT BASES.	72-0435
MONISMITH, C. L	FINITE ELEMENT ANALYSIS OF PAVEMENTS	68-0058
MONISMITH, C. L	FRACTURE (ULTIMATE STRENGTH) ANALYSES OF ASPHALT PAVEMENT LAYERS RESULTING FROM TRAFFIC LOADING	73-0548
MONTI, R.	PREPROCESSORS FOR GENERAL PURPOSE FINITE ELEMENT PROGRAMS	74-1033
MOOK, D. T.	NUMERICAL-PERTURBATION METHOD FOR THE NONLINEAR ANALYSIS OF STRUCTURAL VIBRATIONS	74-0499
MOOLEY, R. F.	BOUNDING PLANE STRESS SOLUTIONS BY FINITE ELEMENTS	66-0055
MOOLEY, R. F.	STRESS CONCENTRATIONS IN TIMBER BEAMS	67-0038
MORAND, H. J. P	INTERNAL PRESSURE EFFECTS ON THE VIBRATION OF PARTIALLY FILLED ELASTIC TANKS	75-0241
MORANDI CECCHI,	A NUMERICAL STUDY OF NONLINEAR INSTABILITY BY MEANS OF AN EXTENDED FINITE ELEMENT METHOD	74-0632
MORANDI CECCHI,	AN EXTENDED THEORY FOR THE FINITE ELEMENT METHOD	72-0350
MORDFIN, L.	NON-METALLIC ANTENNA-SUPPORT MATERIALS	73-0716
MOREL, A.	PHOTOELASTIC DETERMINATION OF THE STRESSES IN THE FOUNDATION OF THE NUCLEAR-POWER STATION BUGEY 2	75-0834
MORGAN, B. J.	THE APPLICATION OF THE FINITE ELEMENT METHOD TO A PROBLEM IN STRESS WAVE PROPAGATION	69-0505
MORGAN, J. D.	DYNAMICS OF CRACKED STRUCTURES USING FINITE ELEMENTS	74-1115
MORGAN, J. D.	ELASTODYNAMICS OF CRACKED STRUCTURES USING FINITE ELEMENTS	74-0625
MORGAN, L. A.	DIFFERENCE AND ELEMENT METHODS FOR POTENTIAL FLOW	74-0019
MORGAN, S.	THEORETICAL PREDICTION OF DISC BRAKE TEMPERATURES AND A COMPARISON WITH EXPERIMENTAL DATA	72-0306
MORGENSTERN, N.	ON SOLUTION OF PLANE STRAIN CONSOLIDATION PROBLEMS BY FINITE ELEMENT METHOD	71-0532
MORGENSTERN, N.	SEEPAGE INTO AN EXCAVATION IN A MEDIUM POSSESSING STRESS-DEPENDENT PERMEABILITY	72-0629
MORGHEM, F.T.	BENDING OF MICROPOLAR PLATES	74-0263
MORI, K.	CODE FOR STRESS ANALYSIS BY THE THREE DIMENSIONAL FINITE ELEMENT METHOD (JAPANESE)	69-0509
MORI, K.	EXAMPLES OF STRESS ANALYSIS OF STRUCTURES BY THE FINITE ELEMENT METHOD. (JAPANESE)	70-0450
MORIGUCHI, K.	ACTUAL STRESS OF HIGH HEAD PUMP-TURBINE RUNNER	75-0567
MORIHARA, H.	NUMERICAL SOLUTION OF THE VISCOUS FLOW IN THE ENTRANCE REGION OF PARALLEL PLATES	73-0022
MORINO, L.	A FINITE ELEMENT ANALYSIS FOR STEADY AND OSCILLATORY SUBSONIC FLOW AROUND COMPLEX CONFIGURATIONS	74-1317
MORINO, L.	A FINITE-ELEMENT ANALYSIS FOR STEADY-AND OSCILLATORY SUPERSONIC FLOWS AROUND COMPLEX CONFIGURATIONS	74-1316
MORINO, L.	A FINITE-ELEMENT FORMULATION FOR SUBSONIC FLOWS AROUND COMPLEX CONFIGURATIONS	73-0742
MORINO, L.	A FINITE-ELEMENT METHOD FOR LIFTING SURFACES IN STEADY INCOMPRESSIBLE SUBSONIC FLOW	74-1321
MORINO, L.	SUBSONIC AND SUPERSONIC INDICIAL AERODYNAMICS AND AERODYNAMIC TRANSFER FUNCTION FOR COMPLEX CONFIGUR	74-1320
MORITA, J.	STRESS ANALYSIS OF THIN PLATE STRUCTURE BY THE FINITE ELEMENT METHOD	69-0403
MORITA, K.	STRENGTH OF TRANSVERSE FILLET WELDED JOINTS	74-0144
MORITA, Y.	STRENGTH OF RAILROAD WHEELS - 3	71-0306

MORLEY, L. S. D	A FINITE ELEMENT APPLICATION OF THE MODIFIED RAYLEIGH-RITZ METHOD	70-0091
MORLEY, L. S. D	A TRIANGULAR EQUILIBRIUM ELEMENT WITH LINEARLY VARYING BENDING MOMENTS FOR PLATE BENDING PROBLEMS	67-0054
MORLEY, L. S. D	EXTENDED INTERPOLATION IN FINITE ELEMENT ANALYSIS	71-0183
MORLEY, L. S. D	FINITE ELEMENT SOLUTION OF BOUNDARY-VALUE PROBLEMS WITH NON-REMOVABLE SINGULARITIES	73-0124
MORLEY, L. S. D	ON THE CONFORMING CUBIC TRIANGULAR ELEMENT FOR PLATE BENDING	72-0439
MORLEY, L. S. D	THE CONSTANT BENDING MOMENT PLATE BENDING ELEMENT	71-0182
MORLEY, L. S. D	THE TRIANGULAR EQUILIBRIUM ELEMENT IN THE SOLUTION OF PLATE BENDING PROBLEMS	68-0122
MOROIANU, A.	COMPORTAREA ELASTICA A DIGURILOR DE PAMINT SOLICIATE SEISMIC (ELASTIC BEHAVIOR OF THE EARTH DAMS SEI	74-1185
MORRIS, A. J.	DEFICIENCY IN CURRENT FINITE ELEMENTS FOR THIN SHELL APPLICATIONS	73-0184
MORRIS, P.	STRUCTURAL RESPONSE AND LOADING OF WALL PANELS	71-0735
MORRIS, W. F.	A FINITE ELEMENT APPROACH TO THE DETERMINATION OF THE DYNAMIC RESPONSE OF BEAMS SUPPORTING MOVING MA	74-1114
MORTENSTERN, N.	APPLICATION OF THE FINITE ELEMENT METHOD TO CONSOLIDATION PROBLEMS	72-0421
MORTON, K. W.	STABILITY AND ACCURACY OF NUMERICAL APPROXIMATIONS TO TIME DEPENDENT FLOWS	75-0183
MOSCO, U.	ONE-SIDED APPROXIMATION AND VARIATIONAL INEQUALITIES	74-0335
MOTE, C. D.	COUPLED, NONCONSERVATIVE STABILITY-FINITE ELEMENT	72-0220
MOTE, C. D.	FORMULATION OF DISCRETE ELEMENT MODELS FOR STRESS AND VIBRATION ANALYSIS OF PLATES	70-0328
MOTE, C. D.	GLOBAL-LOCAL FINITE ELEMENTS	71-0184
MOTE, C. D.	NONCONSERVATIVE STABILITY BY FINITE ELEMENT	71-0572
MOTE, C. D.	SOLUTION OF MIXED BOUNDARY VALUE PROBLEMS WITH LOCAL ERROR BOUND BY THE FINITE ELEMENT METHOD	73-0795
MOTE, C. D.	STABILITY CONTROL ANALYSIS OR ROTATING PLATES BY FINITE ELEMENT. EMPHASIS ON SLOTS AND HOLES	71-0284
MOTE, C. D.	UNSYMMETRICAL TRANSIENT HEAT CONDUCTION: ROTATING DISK APPLICATIONS	70-0005
MOUSSA, M. M.	DYNAMIC ANALYSIS OF AXISYMMETRIC SHELLS UNDER ARBITRARY TRANSIENT PRESSURES.	72-0467
MOWATT, G. A.	STRENGTH OF SHIP STRUCTURAL ELEMENTS	75-0051
MOWBRAY, D. F.	A NOTE ON THE FINITE ELEMENT METHOD IN LINEAR FRACTURE MECHANICS	70-0489
MOWBRAY, D. F.	APPLICATION OF FINITE ELEMENT ELASTIC-PLASTIC STRESS ANALYSIS TO NOTCHED FATIGUE SPECIMEN BEHAVIOUR	71-0185
MOWBRAY, D. F.	APPLICATIONS OF FINITE ELEMENT STRESS ANALYSIS AND STRESS-STRAIN PROPERTIES IN DETERMINING NOTCH FAT	73-1114
MOWBRAY, D. F.	NOTE ON STRESS AND STRAIN REDISTRIBUTION IN A NOTCHED PLATE SPECIMEN DURING CYCLIC LOADING	69-0194
MOWBRAY, D. F.	ON THE USE OF FINITE ELEMENTS AND COMPLIANCE CALCULATIONS FOR DETERMINING CRACK TIP STRESS INTENSITY	69-0448
MUELLER, W.	DIE BERECHNUNG DER BEWEGUNGEN UND SPANNUNGEN DES GEBIRGES VOM ABBAU BIS ZUR TAGESOBERFLAECHE NACH DE	70-0456
MUFTI, A. A.	A STUDY OF BEHAVIOUR OF REINFORCED CONCRETE ELEMENTS USING FINITE ELEMENTS	70-0335
MUFTI, A. A.	ANALYSIS OF THREE-DIMENSIONAL THIN WALLED STRUCTURES	69-0188
MUFTI, A. A.	FINITE ELEMENT ANALYSIS OF THIN SHELLS (DISCUSSION)	69-0187
MUFTI, A. A.	FINITE ELEMENT CALCULATION OF LEFT VENTRICULAR WALL STRESS IN NORMAL AND DISEASE STATES	73-0588
MUFTI, A. A.	FURTHER DEVELOPMENTS IN THE USE OF LINEAR DISPLACEMENT FUNCTIONS IN THE FINITE ELEMENT ANALYSIS OF O	71-0344
MUFTI, A. A.	HYPAR DISPLACEMENT FIELDS FOR RECTANGULAR FINITE ELEMENTS	73-0240
MUFTI, A. A.	MATRIX ANALYSIS OF THIN SHELLS USING FINITE ELEMENTS	69-0527
MUFTI, A. A.	STRUCTURAL ANALYSIS OF TALL BUILDINGS HAVING IRREGULARLY POSITIONED SHEAR WALLS	73-0282
MUHA, T. J. JR.	COMPARISON OF THEORETICAL AND EXPERIMENTAL SHEAR STRESS IN THE ADHESIVE LAYER OF A LAP JOINT MODEL	74-1176
MUIR, W. E.	SELECTION OF LOCATIONS FOR LONG TERM STORAGE OF WHEAT	75-0323
MUKHERJEE, A. K	ANCHOR ZONE STRESS ANALYSIS BY FINITE ELEMENT METHOD	68-0147
MUKHERJI, B.	CRACK PROPAGATION IN SEA ICE - FINITE ELEMENT APPROACH	72-0029
MUKHERJI, B.	NOTE ON RECTANGULAR FINITE ELEMENTS WITH IN-PLANE FORCES	75-0080
MUKHERJI, B.	STRESS INTENSITY FACTORS FOR A NOTCHED BEAM ON AN ELASTIC FOUNDATION	75-0566
MULINAZZI, T. E	GEOMETRIC HIGHWAY AND CULVERT DESIGN	74-1342
MULLINS, M.	PROBABILISTIC EVALUATION OF THE RELATIVE BEHAVIOUR OF TUNGSTEN CARBIDE COMPONENTS IN STRUCTURAL APPL	75-0579
MULLINS, W. M.	STRESS ANALYSIS OF PARACHUTES USING FINITE ELEMENTS	71-0481
MUNCH, K. V.	A COLLECTION OF PROBLEMS SOLVED WITH ASKA	70-0401
MUNFAKH, G. A.	EFFECT OF DENSIFICATION ON THE ENGINEERING CHARACTERISTICS OF ORGANIC SOILS - VOL. 1 AND 2.	73-0805
MUNROE, J. A.	FINITE ELEMENT ANALYSIS OF AN AXISYMMETRICALLY LOADED ORTHOTROPIC SHELL OF REVOLUTION.	72-0438
MUPAKI, T.	FINITE ELEMENT ANALYSIS OF THIN-WALLED STRUCTURES BASED ON THE MODERN ENGINEERING THEORY OF BEAMS	71-0140
MURAKAMI, H.	STRESS ANALYSIS OF A PRESTRESSED CONCRETE NUCLEAR PRESSURE VESSEL BY THE FINITE ELEMENT METHOD USING	71-0154
MURAKAMI, Y.	NOTCH EFFECT IN LOW CYCLE FATIGUE - REPORT 1: LOW-CYCLE FATIGUE STRENGTH OF 0.48 PERCENT CARBON STEE	73-1090
MURAKAMI, Y.	NOTCH EFFECT IN LOW-CYCLE FATIGUE - 2. EFFECT OF STRESS CHANGE AND MEAN STRESS	75-0537
MURAKI, T.	ANALYSIS OF THERMAL STRESSES AND METAL MOVEMENT DURING WELDING 1. ANALYTICAL STUDY	74-1026
MURAKI, T.	ON THE FINITE ELEMENT ANALYSIS OF THIN-WALLED STRUCTURES WITH APPLICATION TO STRUCTURAL PROBLEMS IN	74-0832
MURAKI, T.	THERMAL ANALYSIS OF M552 EXPERIMENT FOR MATERIALS PROCESSING IN SPACE	74-0704
MURAKI, T.	THERMAL ANALYSIS OF M551 EXPERIMENT FOR MATERIALS PROCESSING IN SPACE	74-0725
MURFIN, W. B.	EFFECT OF GEOLOGIC IRREGULARITIES OF SEISMIC RESPONSE	70-0898
MURNEN, G. J.	ANALYSIS AND DESIGN OF AXISYMMETRIC REINFORCEMENT ARCUND A CIRCULAR HOLE IN A THIN FLAT PLATE BY THE	74-1113
MUROTA, T.	ELASTIC, PLASTIC BEHAVIOUR OF ROTO - 1	73-0458
MUROTA, T.	NON-UNIFORM DEFORMATION OF A BLOCK IN PLANE-STRAIN COMPRESSION CAUSED BY FRICTION. PART 4, ANALYSIS	71-0786
MUROTA, T.	NON-UNIFORM DEFORMATION OF MATERIAL IN AXIALLY SYMMETRIC COMPRESSION CAUSED BY FRICTION. PART 3, THE	72-0923
MUROTA, T.	ON THE NON-UNIFORM DEFORMATION OF A BLOCK IN PLANE-STRAIN COMPRESSION CAUSED BY FRICTION -3	71-0328
MUROTA, T.	ON THE NONUNIFORM DEFORMATION OF MATERIAL IN AXIALLY SYMMETRIC COMPRESSION CAUSED BY FRICTION - 2	71-0321
MURPHY, D. J.	TEMPORARY EXCAVATION IN VARVED CLAY	75-0049
MURPHY, W. D.	CUBIC SPLINE GALERKIN APPROXIMATIONS TO PARABOLIC SYSTEMS WITH COUPLED NON-LINEAR BOUNDARY CONDITION	75-0232
MURRAY, D. W.	AN APPROXIMATE NON-LINEAR ANALYSIS OF THIN PLATES	68-0286
MURRAY, D. W.	APPLICATION OF THE FINITE ELEMENT METHOD TO CONSOLIDATION PROBLEMS	72-0421
MURRAY, D. W.	COUPLED LOCAL BUCKLING IN WIDE-FLANGE BEAM-COLUMNS	73-0235
MURRAY, D. W.	FINITE ELEMENT LARGE DEFLECTION ANALYSIS OF THIN-WALLED BEAMS	74-0444
MURRAY, D. W.	FINITE ELEMENT POST BUCKLING ANALYSIS OF THIN ELASTIC PLATES	68-0123
MURRAY, D. W.	FINITE ELEMENT SOLUTION OF INELASTIC BEAM EQUATIONS	73-0234
MURRAY, D. W.	FINITE-ELEMENT LARGE DEFLECTION ANALYSIS OF PLATES	69-0196
MURRAY, D. W.	INCREMENTAL FINITE ELEMENT MATRICES	73-0475
MURRAY, D. W.	INCREMENTAL FINITE-ELEMENT MATRICES	75-0635
MURRAY, D. W.	TECHNIQUE FOR FORMULATING BEAM EQUATIONS	75-0584
MURRAY, D. W.	TIME DEPENDENT REINFORCED CONCRETE SLAB DEFLECTIONS	74-0586
MURRAY, K. H.	COMMENTS ON CONVERGENCE OF FINITE ELEMENT SOLUTIONS	70-0234
MURRAY, K. H.	FINITE ELEMENT ANALYSIS OF DIAMETRAL TEST OF POLYMER MOLDINGS	73-0129
MURRAY, K. H.	FINITE ELEMENT ANALYSIS OF A COMPOSITE MATERIAL INTERFACE	73-0777
MURRAY, M. T.	STRUCTURAL DYNAMICS PROGRAMS	70-0412
MURRAY, R. C.	SEISMIC EFFECTS ON A PROPOSED UNDERGROUND REACTOR FACILITY	73-0953
MURTHA, R. N.	REDUCTION OF STRESSES IN BURIED STRUCTURES	71-0698
MURTHY, A. V. K	NONLINEAR VIBRATIONS OF NONUNIFORM BEAMS WITH CONCENTRATED MASSES	74-1300
MURTHY, K. N.	OPTIMAL DESIGN FOR PRESCRIBED BUCKLING LOADS	74-0568
MURTHY, P. N.	EFFECTS OF MATERIAL NONLINEARITIES ON VIBRATION	73-0323
MURTHY, P. N.	FREE VIBRATION OF A SIMPLY SUPPORTED BEAM WITH NONLINEAR MATERIAL PROPERTIES	71-0689
MURTHY, P. N.	NON-LINEAR EFFECTS OF HIGH TEMPERATURE ON THE VIBRATION OF A SIMPLY SUPPORTED BEAM WITH A CENTRAL MA	73-0838
MURTHY, P. N.	STUDIES ON TWO PHASE COMPOSITE MATERIALS THROUGH COMPUTER SIMULATION	73-0315
MURTHY, T. V. G	VIBRATION OF RECTANGULAR PLATES WITH MIXED BOUNDARY CONDITIONS	73-1086
MURTY, A. V.	AN INVERSE FORM OF RAMBERG - OSGOOD FORMULA	69-0479
MURTY, A. V.	STABILITY ANALYSIS OF COLUMNS USING FINITE ELEMENTS	69-0478
MURTY, A. V. K.	A COMPARATIVE STUDY OF THE CONSISTENT AND SIMPLIFIED FINITE ELEMENT ANALYSES OF EIGENVALUE PROBLEMS	70-0373
MURTY, A. V. K.	GALERKIN FINITE ELEMENT METHOD FOR VIBRATION PROBLEMS	73-0026
MURTY, A.V.K.	FINITE ELEMENT MODELLING OF NATURAL VIBRATION PROBLEMS	73-0320
MURTY, V. V. N.	A FINITE ELEMENT MODEL FOR MISCIBLE DISPLACEMENT IN GROUNDWATER AQUIFERS	75-0783
MUSKA, N. M.	MECHANICS OF CABLE MOORING SYSTEMS. VOLUME IV. A COMPUTER PROGRAM FOR ANALYZING THE STEADY STATE RES	72-0653
MUTO, K.	ELASTIC-PLASTIC ANALYSIS OF REINFORCED CONCRETE MEMBERS BY FEM	73-0292
MUTO, K.	NON LINEAR ANALYSIS OF REINFORCED CONCRETE MEMBERS BY FINITE ELEMENT METHOD, PART 1: ANALYSIS, PART	71-0313
MUTO, K.	NONLINEAR ANALYSIS OF REINFORCED CONCRETE BUILDINGS	73-0324
MUYAMOTO, H.	EFFECTS OF A PEAK OVERLOAD ON FATIGUE CRACK PROPAGATION RATE-PLANE STRESS ANALYSIS OF CRACKED PLATE	74-1365
MYERS, G. E.	ANALYTIC METHODS IN CONDUCTION HEAT TRANSFER	71-0186

MYERS, G. E.	STABILITY AND OSCILLATION CHARACTERISTICS OF FINITE ELEMENT, FINITE-DIFFERENCE, AND WEIGHTED-RESIDUA	75-0615
MYKTOW, W. J.	MATRIX METHODS IN STRUCTURAL MECHANICS	66-0065
NABERHAUS, J. D	AUTOMATED STRUCTURAL DESIGN AND ANALYSIS OF ADVANCED COMPOSITE WING MODELS	73-0221
NABERHAUS, J. D	COMPOSITE WING FOR TRANSONIC IMPROVEMENT. VOLUME 11. ADVANCED ANALYSIS EVALUATION	71-0657
NACCI, V. A.	GROUNDWATER FLOW IN PARTIALLY SATURATED SOILS	72-0745
NACHBAR, W.	A KOITER TYPE METHOD FOR FINITE ELEMENT ANALYSIS OF NONLINEAR STRUCTURAL BEHAVIOUR	70-0064
NACHBAR, W.	ADAPTION OF KOITER'S METHOD TO FINITE ELEMENT ANALYSIS OF SNAP-THROUGH BUCKLING BEHAVIOUR	71-0109
NAEHRIG, TH. K.	THE CALCULATION OF THREE-DIMENSIONAL TEMPERATURE DISTRIBUTIONS AND THERMAL STRESSES USING FINITE ELE	71-0187
NAG, D. K.	PLATES AND BEAMS ON ELASTIC FOUNDATIONS: LINEAR AND NONLINEAR BEHAVIOUR	68-0042
NAGAI, K.	GENERAL PURPOSE PROGRAM OF PLANE STRESS ANALYSIS BY FINITE ELEMENT METHOD, AND ITS APPLICATIONS	72-0186
NAGAI, K.	3-DIMENSIONAL ANALYSIS OF HULL STRUCTURE BY FINITE ELEMENT METHOD	73-0160
NAGAI, Y.	ANALYSIS OF LONGITUDINAL WAVE IN A BAR EXHIBITING VISCO-ELASTIC BEHAVIOR OR WITH MATERIAL DAMPING	71-0780
NAGAI, Y.	ANALYSIS OF ONE-DIMENSIONAL STRESS WAVE BY THE FINITE ELEMENT METHOD	71-0523
NAGAI, Y.	FINITE ELEMENT ANALYSIS OF LOAD CELL RESPONSE IN HIGH SPEED TENSILE TESTING	73-1071
NAGAMATO, R.	ON STRENGTH OF TRANSVERSE WEBS AND THEIR JOINTS OF LARGE TANKER	69-0327
NAGAMATSU, A.	ELASTIC, PLASTIC BEHAVIOUR OF ROTO - 1	73-0458
NAGAMATSU, A.	NON-UNIFORM DEFORMATION OF A BLOCK IN PLANE-STRAIN COMPRESSION CAUSED BY FRICTION. PART 4, ANALYSIS	71-0786
NAGAMATSU, A.	NON-UNIFORM DEFORMATION OF MATERIAL IN AXIALLY SYMMETRIC COMPRESSION CAUSED BY FRICTION. PART 3, THE	72-0923
NAGAMATSU, A.	ON THE NON-UNIFORM DEFORMATION OF A BLOCK IN PLANE-STRAIN COMPRESSION CAUSED BY FRICTION -3	71-0328
NAGAMATSU, A.	ON THE NONUNIFORM DEFORMATION OF MATERIAL IN AXIALLY SYMMETRIC COMPRESSION CAUSED BY FRICTION - 2	71-0321
NAGAMATSU, A.	STRESSES IN A ROTATING DISK WITH CENTRAL AND NONCENTRAL HOLES (STRESS ANALYSIS BY THE FINITE ELEMENT	74-1337
NAGAMOTO, R.	ON THE STRESS ANALYSIS AT JOINTS OF DEEP BEAMS	67-0055
NAGANO, H.	SOME TECHNIQUES FOR FINITE ELEMENT ANALYSIS	71-0195
NAGARAJ, V. T.	ROTOR BLADE VIBRATIONS BY GALERKIN FINITE-ELEMENT METHOD	75-0389
NAGARAJAN, S.	ELASTIC PLASTIC DYNAMIC ANALYSIS OF AXISYMMETRIC SOLIDS	73-0657
NAGARAJAN, S.	NON-LINEAR DYNAMIC ANALYSIS OF AXISYMMETRIC SHELLS	75-0474
NAGARAJAN, S.	NONLINEAR ELASTIC-VISCOPLASTIC ANALYSIS USING THE FINITE ELEMENT METHOD	74-0750
NAGARAJAN, S.	NONLINEAR FINITE ELEMENT DYNAMIC ANALYSIS OF AXISYMMETRIC SOLIDS	74-0775
NAGARAJAN, S.	PLASTIC AND VISCOPLASTIC ANALYSIS OF AXISYMMETRIC SHELLS	75-0246
NAGATO, K.	DYNAMIC ELASTIC-PLASTIC ANALYSIS OF AXISYMMETRIC STRUCTURES BY FINITE ELEMENT METHOD COMPUTER CODE	73-0844
NAGATO, K.	EXAMPLES OF STRESS ANALYSIS OF STRUCTURES BY THE FINITE ELEMENT METHOD, (JAPANESE)	70-0450
NAGEL, D. A.	DETAILED STRUCTURAL ANALYSIS OF THE HUMAN TIBIA	75-0379
NAGTEGAAL, J. C	ON NUMERICALLY ACCURATE FINITE ELEMENT SOLUTIONS IN THE FULLY PLASTIC RANGE	74-0529
NAGTEGAAL, J. C	ON NUMERICALLY ACCURATE FINITE ELEMENTS	74-0797
NAGY, A.	DEFORMATION MECHANICS OF DEEP SURFACE FLAW CRACKS	72-0785
NAGY, D. A.	GEOMETRIC NONLINEARITY AND THE FINITE ELEMENT DISPLACEMENT METHOD	74-0679
NAGY, L. I.	EVALUATION OF NASTRAN FROM THE THE USER'S POINT OF VIEW	73-0596
NAGY, L. I.	FINITE ELEMENT METHOD IN AUTOMOBILE DESIGN	73-0326
NAGY, L. I.	STATIC ANALYSIS VIA SUBSTRUCTURING OF AN EXPERIMENTAL VEHICLE FRONT-END BODY STRUCTURE	74-0313
NAHAVANDI, A. N	STRUCTURALLY COMPATIBLE FLUID FINITE-ELEMENT FOR SOLID FLUID INTERACTION STUDIES	75-0344
NAIK, T.	ASPECTS OF MECHANICAL BEHAVIOR OF ROCK UNDER STATIC AND CYCLIC LOADING. PART A: MECHANICAL BEHAVIOR	72-0762
NAIK, T. B.	ANALYSIS OF FOUNDATIONS ON A SEMI-INFINITE ELASTIC MASS UNDER DYNAMIC LOADS	72-0763
NAIR, K.	A THEORETICAL METHOD FOR EVALUATING STABILITY OF OPENINGS IN ROCK	72-0877
NAIR, K.	ANALYTICAL METHODS FOR PREDICTING SUBSIDENCE ABOVE SOLUTION-MINED CAVITIES	74-1145
NAIR, K.	DEVELOPMENT AND APPLICATIONS OF THEORETICAL METHODS FOR EVALUATING STABILITY OF OPENINGS IN ROCK	73-0791
NAIR, K.	DEVELOPMENT AND APPLICATIONS OF A GENERAL COMPUTER PROGRAM FOR EVALUATING STABILITY OF OPENINGS IN R	74-0994
NAIR, K.	TIME-DEPENDENT ANALYSIS TO PREDICT CLOSURE IN SALT CAVITIES	74-1149
NAIR, P.	UNIMOD - APPLICATIONS ORIENTED FINITE ELEMENT SCHEME FOR THE ANALYSIS OF FRACTURE MECHANICS PROBLEMS	74-1356
NAIR, R. S.	HIGH STRENGTH BOLTS SUBJECT TO TENSION AND PRYING	74-0201
NAKAGAKI, M.	APPLICATION OF AN ASSUMED DISPLACEMENT HYBRID FINITE ELEMENT PROCEDURE TO TWO-DIMENSIONAL PROBLEMS I	74-0729
NAKAGAKI, N.	ASSUMED DISPLACEMENT HYBRID FINITE ELEMENT MODEL FOR LINEAR FRACTURE MECHANICS	75-0146
NAKAGAWA, K.	ANALYSIS OF STEADY INCOMPRESSIBLE VISCOUS FLOW	74-0013
NAKAGAWA, K.	STEADY FLOW ANALYSIS OF INCOMPRESSIBLE VISCOUS FLOW BY THE FINITE ELEMENT METHOD	73-0332
NAKAGAWA, K.	THE EFFECT OF END RESTRAINT CONDITION ON THE FRACTURE MECHANISM AND STRENGTH OF ROCK-LIKE MATERIALS	72-0134
NAKAGIRI, S.	ANALYSIS OF SAINT-VENANT TORSION PROBLEM BY A HYBRID STRESS MODEL	69-0277
NAKAGIRI, S.	ELASTIC PLASTIC ANALYSIS OF SAINT-VENANT TORSION PROBLEM BY A HYBRID STRESS MODEL	72-0385
NAKAGIRI, S.	PLASTIC STRESS STRAIN MATRIX OF ANISOTROPIC MATERIALS AND ITS APPLICATION	69-0383
NAKAHARA, Y.	FINITE ELEMENT METHOD FOR SOLVING NEUTRON TRANSPORT PROBLEMS IN TWO-DIMENSIONAL CYLINDRICAL GEOMETRY	74-1081
NAKAI, Y.	EXAMPLES OF STRESS ANALYSIS OF STRUCTURES BY THE FINITE ELEMENT METHOD, (JAPANESE)	70-0450
NAKAMURA, S.	ITERATIVE SOLUTIONS FOR THE FINITE ELEMENT METHOD	72-0820
NAKAMURA, S.	STUDY OF ITERATIVE SOLUTIONS FOR FINITE ELEMENT METHOD	71-0453
NAKAMURA, T.	ELASTIC-PLASTIC ANALYSIS OF FRAMED STRUCTURES	69-0197
NAKAO, K.	A STUDY ON THE STRESS ANALYSIS OF BEAM PLATE STRUCTURE	69-0404
NAKAO, Y.	CONSIDERATION ON THE SEISMIC RESPONSE ANALYSIS OF UNDERGROUND STRUCTURES	75-0470
NAKAO, Y.	DYNAMIC ANALYSIS OF THE GROUND USING THE THREE DIMENSIONAL FINITE ELEMENT	73-0293
NAKAO, Y.	ELASTO-PLASTIC SEISMIC RESPONSE ANALYSIS OF FRAMED STRUCTURES BY MODE SUPERPOSITION	73-0325
NAKAO, Y.	ON THE APPLICATION OF THE FINITE ELEMENT METHOD TO THE VIBRATION ANALYSIS OF SPACE PLATE STRUCTURE	72-0630
NAKAO, Y.	PRACTICAL ANALYSIS OF STEEL STRUCTURES USING HIGHER-ORDER ELEMENTS	72-0326
NAKAO, Y.	THREE DIMENSIONAL SEISMIC ANALYSIS OF STRUCTURE-GROUND SYSTEMS BY THE FINITE ELEMENT METHOD	74-0268
NALLUSWAMI, M.	FINITE ELEMENT METHOD FOR HYDRODYNAMIC DISPERSION EQUATION WITH MIXED PARTIAL DERIVATIVES	72-0030
NAM, C. H.	A FINITE-ELEMENT STUDY ON THE NONLINEAR BEHAVIOR OF REINFORCED CONCRETE FLEXURAL MEMBERS DUE TO SHOR	73-1154
NAM, C. H.	FINITE ELEMENT ANALYSIS OF CONCRETE BEAMS	74-0560
NAMAROS, T. C.	LOAD POINT COMPLIANCE OF THE CHARPY IMPACT SPECIMEN	73-0775
NANDAKUMAR, K.	FULLY DEVELOPED VISCOUS FLOW IN INTERNALLY FINNED TUBES	75-0648
NAPOLITANO, L.	FINITE ELEMENT METHODS IN FLUID-DYNAMICS	73-0060
NAPOLITANO, L.	FUNCTIONAL ANALYSIS DERIVATION AND GENERALIZATION OF HYBRID VARIATIONAL METHODS	73-1104
NAPOLITANO, L.	PREPROCESSORS FOR GENERAL PURPOSE FINITE ELEMENT PROGRAMS	74-1033
NARASIMHA, R. R	GALERKIN FINITE ELEMENT SOLUTION FOR THE STABILITY OF CANTILEVER COLUMNS SUBJECTED TO TANGENTIAL LOA	75-0534
NARAYANA, R.	DEPENDENCE OF PLATE-BENDING FINITE ELEMENT DEFLECTIONS AND EIGENVALUES ON POISSONS RATIO	74-0583
NARAYANA, R.	INCLUSION OF TRANSVERSE SHEAR DEFORMATION IN FINITE ELEMENT DISPLACEMENT FORMULATIONS	74-0662
NARAYANA, R.	NEW TRIANGULAR PLATE-BENDING FINITE ELEMENT WITH TRANSVERSE SHEAR FLEXIBILITY	74-0627
NARAYANAN, J.	A NUMERICAL MODEL FOR THE SIMULATION OF TWO DIMENSIONAL CONVECTIVE DISPERSION IN SHALLOW ESTUARIES	74-0052
NARAYANASWAMI,	DEPENDENCE OF PLATE-BENDING FINITE ELEMENT DEFLECTIONS AND EIGENVALUES ON POISSONS RATIO	74-0567
NARAYANASWAMI,	HIGHER ORDER FINITE ELEMENT METHOD FOR STATIC AND DYNAMIC ANALYSIS OF SHEAR WALLS	71-0368
NARAYANASWAMI,	NEW TRIANGULAR AND QUADRILATERAL PLATE-BENDING FINITE ELEMENTS	74-0833
NARAYANASWAMI,	NEW TRIANGULAR PLATE-BENDING FINITE ELEMENT WITH TRANSVERSE SHEAR FLEXIBILITY	74-0964
NARUOKA, M.	ANALYSIS OF STRESSES IN SKEWED PLATES BY THE METHOD OF FINITE ELEMENTS	73-0248
NASCH, L.	VELKE PRIEHYBY DOSIEK METODOU KONECNYCH PRVKOV (LARGE DEFLECTIONS OF PLATES ANALYZED BY FINITE ELEME	73-0373
NASH, J. K. T.	PERFORMANCE OF CULVERT UNDER WINSCAR DAM	75-0638
NASIR, G.	INSTABILITY OF HORIZONTALLY CURVED MEMBERS. FLANGE BUCKLING STUDIES	69-0480
NASRELDIN, H. A	INTERACTIVE FINITE ELEMENT DATA GENERATION FOR THREE-DIMENSIONAL PLATE STRUCTURES	74-0909
NASRELDIN, H. A	INTERACTIVE FINITE ELEMENT DATA GENERATION USING THE SUB-STRUCTURING TECHNIQUE	74-1291
NASSER, M. S.	NUMERICAL METHODS FOR UNSTEADY NON-DARCY FLOW	74-0062
NASSIF, N.	FINITE ELEMENT APPROXIMATIONS TO TIME-DEPENDENT PROBLEMS	72-0269
NASSIF, N.	FINITE ELEMENT METHOD FOR TIME DEPENDENT PROBLEMS	72-0821
NASSIF, N.	ON FINITE ELEMENT APPROXIMATIONS ON TIME DEPENDENT PROBLEMS	72-0528
NASSIF, N. R.	ON A VARIATIONAL INEQUALITY AND ITS APPROXIMATION, IN THE THEORY OF SEMICONDUCTORS	75-0289
NASSIF, N. R.	ON THE DISCRETIZATION OF THE TIME VARIABLE FOR PARABOLIC PARTIAL DIFFERENTIAL EQUATIONS	75-0103
NATARAJA, M. S.	FINITE-ELEMENT SOLUTION OF STRESSES AND DISPLACEMENTS IN A SOIL-CULVERT SYSTEM	73-1156
NATARAJAN, R.	STRESS ANALYSIS OF CURVED PIPES WITH END RESTRAINTS	75-0667
NATARIUS, Y. I.	INVESTIGATION OF THE EFFECT OF FOUNDATION YIELD ON THE FREQUENCY AND SHAPE OF INHERENT VIBRATIONS OF	72-0419
NATH, B.	FINITE ELEMENT - ANALOGUE METHOD FOR DETERMINING THE DYNAMIC CHARACTERISTICS OF AN ARCH DAM-RESERVOI	73-0678

NATH, B.	FUNDAMENTALS OF FINITE ELEMENT METHODS FOR ENGINEERS	74-0719
NATH, B.	HYDRODYNAMIC PRESSURES ON HIGH DAMS DUE TO VERTICAL EARTHQUAKE MOTIONS	69-0438
NATH, J. M.	FREE VIBRATION OF DOUBLY CURVED RECTANGULAR PLATES INCLUDING THE PRESENCE OF MEMBRANE STRESSES WITH	69-0468
NATH, P.	NATURAL FREQUENCIES OF COMPLEX FREE OR SUBMERGED STRUCTURES BY THE FINITE ELEMENT METHOD	65-0026
NATH, P.	STRESSES DUE TO CONCENTRATED LOADS IN RESTRAINED BENDING TESTS	67-0097
NATH, P.	TENSILE CRACK PROPAGATION IN REINFORCED CONCRETE BEAMS BY THE FINITE ELEMENT TECHNIQUES	69-0264
NATHAN, N. D.	FINITE ELEMENT FORMULATION OF GEOMETRICALLY NONLINEAR PROBLEMS OF ELASTICITY	71-0114
NATTERER, F.	POINTWISE CONVERGENCE OF FINITE-ELEMENTS	75-0821
NAVARATNA, D. R	A STUDY OF MATRIX ANALYSIS METHODS FOR INELASTIC STRUCTURES	63-0022
NAVARATNA, D. R	ANALYSIS OF ELASTIC STABILITY OF SHELLS OF REVOLUTION BY THE FINITE ELEMENT METHOD	67-0056
NAVARATNA, D. R	APPLICATION OF MATRIX DISPLACEMENT METHOD TO LINEAR ELASTIC ANALYSIS OF SHELLS OF REVOLUTION	65-0025
NAVARATNA, D. R	COMPUTATION OF STRESS RESULTANTS IN FINITE ELEMENT ANALYSIS	66-0056
NAVARATNA, D. R	ELASTIC STABILITY OF SHELLS OF REVOLUTION BY THE VARIATIONAL APPROACH USING DISCRETE ELEMENTS	66-0106
NAVARATNA, D. R	IMPROVEMENTS ON THE ANALYSIS OF SHELLS OF REVOLUTION BY THE MATRIX DISPLACEMENT METHOD	66-0070
NAVARATNA, D. R	SABOR 1, A FORTRAN PROGRAM FOR THE LINEAR ELASTIC AXISYMMETRIC LOADING BY USING THE MATRIX DISPLACEM	65-0027
NAVARATNA, D. R	STABILITY ANALYSIS OF SHELLS OF REVOLUTION BY THE FINITE ELEMENT METHOD	68-0124
NAVEH, A.	GENERATION OF FINITE ELEMENTS BY EULER'S VARIATIONAL METHOD	72-0596
NAY, R. A.	AN ALTERNATIVE FOR THE FINITE-ELEMENT METHOD	73-1155
NAYAK, A. L.	FINITE ELEMENT ANALYSIS OF LAMINAR CONVECTIVE HEAT TRANSFER IN VERTICAL DUCTS WITH ARBITRARY CROSS-S	75-0555
NAYAK, G. C.	A GENERAL APPROACH TO PROBLEMS OF LARGE DEFORMATION, AND PLASTICITY USING ISOPARAMETRIC ELEMENT	71-0273
NAYAK, G. C.	ELASTO-PLASTIC STRESS ANALYSIS. A GENERALIZATION FOR VARIOUS CONSTITUTIVE RELATIONS INCLUDING STRAIN	72-0547
NAYAK, G. C.	FINITE ELEMENT METHODS IN ANALYSIS OF REACTOR VESSELS	72-0271
NAYAK, G. C.	INFLUENCE CHARACTERISTICS FOR SLAB BRIDGES	71-0188
NAYAK, G. C.	PLASTICITY AND LARGE DEFORMATION PROBLEMS BY FINITE ELEMENT METHOD	71-0529
NAYAK, G. C.	STRESSES IN A PARTLY YIELDED NOTCHED BAR - AN ASSESSMENT OF THREE ALTERNATIVE PROGRAMS	73-0267
NAYAK, G. C.	STRESSES IN A PARTLY YIELDED NOTCHED BAR EM DASH AN ASSESSMENT OF THREE ALTERNATIVE PROGRAMS	73-0268
NAYAK, G. C.	THE ALPHA CONSTANT-STIFFNESS METHOD FOR THE ANALYSIS OF NONLINEAR PROBLEMS	72-0566
NAYFEH, A. H.	NUMERICAL-PERTURBATION METHOD FOR THE NONLINEAR ANALYSIS OF STRUCTURAL VIBRATIONS	74-0499
NAYLOR, D. J.	ADAPTATION OF CRITICAL STATE SOIL MECHANICS FOR USE IN FINITE ELEMENTS	71-0534
NAYLOR, D. J.	DISCUSSION ON THE ADAPTION OF CRITICAL STATE SOLID MECHANICS THEORY FOR USE IN FINITE ELEMENTS	71-0007
NAYLOR, D. J.	FINITE ELEMENT STUDIES OF SOILS AND POROUS MEDIA	71-0504
NAYLOR, D. J.	SETTLEMENT ANALYSIS OF A STRIP FOOTING USING A CRITICAL STATE SOIL MODEL IN CONJUNCTION WITH FINITE	71-0533
NAYLOR, D. J.	STRESSES IN NEARLY INCOMPRESSIBLE MATERIALS BY FINITE ELEMENTS WITH APPLICATION TO THE CALCULATION O	74-0383
NCCORMICK, C. W	COMPUTERIZED SUBSTRUCTURE ANALYSIS	75-0419
NEALE, B. K.	A NEW HYBRID CYLINDRICAL SHELL FINITE ELEMENT	71-0118
NEALE, B. K.	BERGEN, AN AUTOMATIC FINITE ELEMENT MESH GENERATION PROGRAM FOR ARBITRARY STRUCTURES	72-0706
NEALE, B. K.	COMPARISON OF RECENT SHALLOW SHELL FINITE ELEMENT ANALYSES	71-0362
NEALE, B. K.	FINITE ELEMENT CRACK ANALYSIS USING THE J-INTEGRAL METHOD	73-1106
NEALE, B. K.	FINITE ELEMENT CRACK TIP MODELING WHEN EVALUATING J-INTEGRAL	75-0637
NEALE, B. K.	FINITE ELEMENT MESH GENERATION PROGRAM FOR ARBITRARY TWO-AND THREE-DIMENSIONAL STRUCTURES	73-0966
NEALE, B. K.	HYBRID PLATE BENDING ELEMENTS	72-0792
NEDELEC, J. C.	SUR L ELEMENT DE FRAEIJS DE VEUBEKE ET SANDER (FRAEIJS DE VEUBEKE-SANDER ELEMENT)	74-0924
NEDELEC, J. C.	UNE METHODE VARIATIONNELLE ELEMENTS FINIS POUR LA RESULUTION NUMERIQUE UN PROBLEME EXTERIEUR DANS R*	73-0576
NEDELEC, J. C.	VARIATION METHOD ON FINITE ELEMENTS FOR NUMERICAL RESOLUTION TO PROBLEMS WITH EXTERNAL R3	73-0359
NEDOMA, J.	FINITE ELEMENT IN THE THEORY OF SH-WAVE PROPAGATION	73-1034
NEEDLEMAN, A.	NUMERICAL STUDY OF UNIAXIAL COMPRESSION IN CIRCULAR ELASTIC-PLASTIC COLUMNS	73-0238
NEGRIN, M. P.	A FINITE ELEMENT APPROACH TO PHYSIOLOGICAL TRANSPORT PHENOMENA BY CILIARY AND PERISTALTIC PUMPING	72-0353
NEILLE, D. S.	COMPARISON OF PRACTICAL METHODS OF ANALYSING ELASTIC LATERALLY LOADED PLATES	71-0425
NEIS, V. V.	ANALYSIS OF CASTELLATED BEAMS BY THE FINITE ELEMENT METHOD	72-0620
NEKI, I.	GENERAL PURPOSE PROGRAM OF PLANE STRESS ANALYSIS BY FINITE ELEMENT METHOD, AND ITS APPLICATIONS	72-0186
NEKI, I.	3-DIMENSIONAL ANALYSIS OF HULL STRUCTURE BY FINITE ELEMENT METHOD	73-0160
NELSON, D. A. R	NEW MATERIAL MODEL FOR THE NONLINEAR BIAXIAL BEHAVIOR OF ATJ-S GRAPHITE	75-0063
NELSON, I.	MATHEMATICAL MODELS FOR GEOLOGICAL MATERIALS FOR WAVE PROPAGATION STUDIES	71-0740
NELSON, J. M.	THREE-DIMENSIONAL PHOTOELASTIC AND FINITE-ELEMENT ANALYSIS OF A PROPELLANT GRAIN	72-0031
NELSON, M. E.	EIGENVALUE BOUNDS FOR FINITE ELEMENT MODELS	72-0570
NELSON, M. E.	THE USE OF CONDENSATION TECHNIQUES FOR SOLVING DYNAMICS PROBLEMS	74-0320
NELSON, N. W.	ANALYSIS OF BOTTOM ENTRY NOZZLES FOR ATMOSPHERIC STORAGE TANKS	75-0331
NEMAT-NASSER, S	CONTINUUM BASES FOR CONSISTENT NUMERICAL FORMULATIONS OF FINITE STRAINS IN ELASTIC AND INELASTIC STR	75-0299
NEMAT-NASSER, S	FINITE ELEMENT FORMULATIONS FOR ELASTIC PLATES BY GENERAL VARIATIONAL STATEMENTS WITH DISCONTINUOUS	73-0559
NEMAT-NASSER, S	FINITE ELEMENT FORMULATIONS FOR ELASTIC PLATES BY GENERAL VARIATIONAL STATEMENTS WITH DISCONTINUOUS	73-1124
NEMAT-NASSER, S	GENERAL VARIATIONAL PRINCIPLES IN NONLINEAR AND LINEAR ELASTICITY WITH APPLICATIONS	72-0914
NEMAT-NASSER, S	MIXED VARIATIONAL PRINCIPLES, FINITE STRAINS AND FINITE ELASTICITY	74-1254
NEMEC, I.	EFFICIENT FINITE ELEMENT ANALYSIS OF RECTANGULAR AND SKEW LAMINATED PLATES	73-0436
NEMETHY, A.	RIESENIE VAZKOPRUZNYCH ANIZOTROPICKYCH DOASK METODOU KONECNYCH PRVKOV (SOLUTION OF VISCOELASTIC ANIS	74-0477
NENCIONI, S.	ENERGY FORMS IN THE FINITE ELEMENT TECHNIQUES	72-0680
NERLI, G.	CALCULATION OF AXISYMMETRICAL OSCILLATION OF FLEXIBLE TURBINE DISCS. (CALCOLO DELLE OSCILLAZIONI ASS	74-1328
NERLI, G.	STUDY OF UMBRELLA OSCILLATIONS OF TURBINE DISKS OF ARBITRARY PROFILE BY THE FINITE ELEMENT METHOD	75-0332
NESSIM, A.	A COMPARISON BETWEEN TWO-HIGH-PRECISION FINITE ELEMENTS	75-0033
NETHERCOT, D. A	FINITE ELEMENT SOLUTIONS FOR BUCKLING OF COLUMNS AND BEAMS	71-0361
NETHERCOT, D. A	INELASTIC BUCKLING OF STEEL BEAMS UNDER NON UNIFORM MOMENT	75-0149
NEU, T. F.	FINITE-ELEMENT ANALYSIS OF EDGE EFFECTS IN ANGLE-PLY COMPOSITE LAMINATES	74-0843
NEUBERT, V. H.	FINITE BEAM ELEMENTS FOR DYNAMIC ANALYSIS	70-0372
NEUBERT, V. H.	FINITE ELEMENT FOR SHOCK ANALYSIS	71-0603
NEUBERT, V. H.	FINITE ELEMENT FOR TIMOSHENKO BEAM BASED ON MECHANICAL IMPEDANCE	71-0607
NEUBERT, V. H.	RELATIONSHIP OF SUBSTRUCTURING TO IMPROVEMENT OF DYNAMIC MODELING	71-0323
NEUBERT, V. H.	SHOCK ANALYSIS OF STRUCTURAL SYSTEMS	70-0426
NEUMAN, S. P.	ANALYSIS OF NONSTEADY FLOW WITH A FREE SURFACE USING FINITE ELEMENT METHOD	71-0482
NEUMAN, S. P.	FINITE ELEMENT ANALYSIS OF 2-DIMENSIONAL FLOW IN SOILS CONSIDERING WATER UPTAKE BY ROOTS .1. THEORY	75-0137
NEUMAN, S. P.	FINITE ELEMENT ANALYSIS OF 2-DIMENSIONAL FLOW IN SOILS CONSIDERING WATER UPTAKE BY ROOTS .2. FIELD A	75-0147
NEUMAN, S. P.	FINITE ELEMENT COMPUTER PROGRAMS FOR FLOW IN SATURATED UNSATURATED POROUS MEDIA	72-0364
NEUMAN, S. P.	FINITE ELEMENT METHOD OF ANALYZING STEADY SEEPAGE WITH A FREE SURFACE	70-0092
NEUMAN, S. P.	GALERKIN APPROACH TO UNSATURATED FLOW IN SOILS	74-0058
NEUMAN, S. P.	GALERKIN APPROACH TO SATURATED-UNSATURATED FLOW IN POROUS MEDIA	75-0393
NEUMAN, S. P.	SATURATED-UNSATURATED SEEPAGE BY FINITE ELEMENTS	73-0459
NEUMAN, S. P.	THEORY OF FLOW IN A CONFINED TWO AQUIFER SYSTEM	69-0382
NEUMAN, S. P.	TRANSIENT FLOW OF GROUNDWATER TO WELLS IN MULTIPLE-AQUIFER SYSTEMS	69-0501
NEUMAN, S. P.	USE OF THE FINITE ELEMENT METHOD IN SOLVING TRANSIENT FLOW PROBLEMS IN AQUIFER SYSTEMS	68-0167
NEUMAN, S. P.	VARIATIONAL PRINCIPLES FOR FLUID FLOW IN POROUS MEDIA	71-0844
NEVILL, G. E.	STRUCTURAL OPTIMIZATION USING THE FINITE ELEMENT METHOD APPLIED TO A BEAM	75-0233
NEVILLE, A. M.	STRUCTURAL ANALYSIS: A UNIFIED CLASSICAL AND MATRIX APPROACH	72-0055
NEWMAN, J. C. J	A FINITE ELEMENT ANALYSIS OF FATIGUE CRACK CLOSURE	74-0759
NEWMAN, J. C. J	ELASTIC-PLASTIC ANALYSIS OF A PROPAGATING CRACK UNDER CYCLIC LOADING	75-0857
NEWMAN, J. C. R	FINITE ELEMENT ANALYSIS OF FATIGUE CRACK PROPAGATION - INCLUDING THE EFFECTS OF CRACK CLOSURE	74-1112
NEWMAN, M.	FAST MODAL EXTRACTION IN NASTRAN VIA THE FEER COMPUTER PROGRAM	73-0356
NEWMAN, M.	FINITE ELEMENT ANALYSIS FOR THE TIME-DEPENDENT INELASTIC MATERIAL BEHAVIOR	75-0749
NEWMAN, M.	HABEAS- A STRUCTURAL DYNAMICS ANALYSIS SYSTEM	70-0121
NEWMARK, N. M.	A METHOD OF COMPUTATION FOR STRUCTURAL DYNAMICS	59-0002
NEWSOM, C. D.	A FINITE ELEMENT APPROACH FOR THE ANALYSIS OF RANDOMLY EXCITED COMPLEX ELASTIC STRUCTURES	67-0057
NEWSON, C. D.	FINITE-ELEMENT FORMULATION OF MIXED AND STRESS VARIATIONAL PRINCIPLES FOR STRUCTURAL DYNAMIC ANALYSI	72-0964
NEWTON, C. A.	STUDY OF NUMERICAL METHODS IN SEISMIC PROBLEMS	72-0738
NEWTON, R. E.	COUPLED VIBRATIONS OF A STRUCTURE SUBMERGED IN A COMPRESSIBLE FLUID	69-0001
NEWTON, R. E.	COUPLED VIBRATIONS OF A STRUCTURE SUBMERGED IN A COMPRESSIBLE FLUID	69-0507

NEWTON, R. E.	DEGENERATION OF BRICK-TYPE ISOPARAMETRIC ELEMENTS	73-1031
NEWTON, R. E.	FINITE ELEMENT ANALYSIS OF TWO-DIMENSIONAL ADDED MASS AND DAMPING	75-0394
NEWTON, R. E.	FINITE ELEMENT SOLUTION FOR ADDED MASS AND DAMPING	74-0018
NEWTON, R. F.	PLATE BUCKLING ANALYSIS USING A FULLY COMPATIBLE FINITE ELEMENT	69-0065
NEY, R. A.	AN ALTERNATIVE FOR THE FINITE ELEMENT METHOD	72-0833
NEYLAN, A. J.	DESIGN OF A COMPOSITE STEEL AND CONCRETE CLOSURE FOR A PRESTRESSED CONCRETE REACTOR VESSEL.	72-0187
NG, B. L.	USING INTERACTIVE GRAPHICS FOR THE PREPARATION AND MANAGEMENT OF FINITE ELEMENT DATA	74-0176
NG, S. S. F.	FINITE-ELEMENT ANALYSIS OF SKEW SANDWICH PLATES	75-0185
NGO, D.	FINITE ELEMENT ANALYSIS OF REINFORCED CONCRETE BEAMS	67-0058
NGUYEN, D. H.	FINITE ELEMENT SOLUTIONS OF A NONLINEAR REACTOR DYNAMICS PROBLEM	74-1268
NGUYEN, D. H.	SPACE-TIME SOLUTIONS OF NONLINEAR REACTOR DYNAMICS BY FINITE-ELEMENT METHOD	75-0503
NGUYEN, H. D.	DIGITAL COMPUTATION OF STRESSES AND DEFLECTIONS IN A BOX BEAM	67-0073
NGUYEN, H. D.	DIGITAL COMPUTATION OF STRESSES AND DEFLECTIONS IN A BOX BEAM - A PERFORMANCE COMPARISON BETWEEN FIN	69-0355
NICHOLS, C. S.	A FINITE ELEMENT APPROACH TO SCATTERING FROM ELASTIC SPHERICAL SHELLS	74-0702
NICHOLS, C. S.	FINITE ELEMENT ANALYSIS OF ACOUSTICALLY RADIATING STRUCTURES WITH APPLICATIONS TO SONAR TRANSDUCERS	74-0524
NICHOLS, C. S.	FINITE-ELEMENT APPROACH TO ACOUSTIC SCATTERING FROM ELASTIC STRUCTURES	75-0069
NICHOLSON, D. E	GRAVITY FLOW BIN DESIGN FOR LUNAR SOIL	72-0240
NICKEL, R. E.	CONVERGENCE OF CONSISTENTLY DERIVED TIMOSHENKO BEAM FINITE ELEMENTS	72-0396
NICKELL, R. E.	ANALYSIS OF STRUCTURAL-ACOUSTIC INTERACTIONS IN METAL-CERAMIC TRANSDUCERS	72-0850
NICKELL, R. E.	ANALYSIS OF STRUCTURAL ACOUSTIC INTERACTIONS IN METAL CERAMIC TRANSDUCERS	73-0401
NICKELL, R. E.	APPLICATION OF THE FINITE ELEMENT METHOD TO HEAT CONDUCTION ANALYSIS	66-0076
NICKELL, R. E.	APPROXIMATE SOLUTIONS IN LINEAR COUPLED THERMOELASTICITY	68-0220
NICKELL, R. E.	COMPUTER PROGRAM CONSTRUCTION AND MAINTENANCE - FUTURE OF CENTRALIZED FINITE ELEMENT ACTIVITY	75-0014
NICKELL, R. E.	COUPLED CONVECTIVE AND CONDUCTIVE HEAT TRANSFER BY FINITE ELEMENT METHODS	74-0050
NICKELL, R. E.	FINITE ELEMENT ANALYSIS OF AXISYMMETRIC SOLIDS WITH ARBITRARY LOADINGS	67-0145
NICKELL, R. E.	FINITE ELEMENT METHODS FOR THE SOLUTION OF SOME INCOMPRESSIBLE NON-NEWTONIAN FLUID MECHANICS PROBLEM	75-0224
NICKELL, R. E.	FINITE-ELEMENT METHODS FOR COUPLED FIELD-STRESS PROBLEMS	74-0635
NICKELL, R. E.	IN-VACUO MODAL DYNAMIC RESPONSE OF THE HUMAN SKULL	73-0460
NICKELL, R. E.	RE-ENTRY THERMAL ANALYSIS OF VARIABLE THICKNESS SPHERICAL VEHICLES.	72-0468
NICKELL, R. E.	SOLUTION OF VISCOUS INCOMPRESSIBLE JET AND FREE-SURFACE FLOWS USING FINITE ELEMENT METHODS	74-0357
NICKELL, R. E.	STRESS WAVE ANALYSIS IN LAYERED THERMOVISCOELASTIC MATERIALS BY THE EXTENDED RITZ METHOD, VOLUME II	68-0125
NICKELL, R. E.	THERMAL AND MECHANICAL ANALYSIS OF WELDED STRUCTURES	73-0819
NICKELL, R. E.	USER EDUCATION: THE FINITE ELEMENT SHORT COURSE	72-0437
NICKERSON, E.	A FINITE ELEMENT PROGRAM FOR DESIGN AND ANALYSIS OF THREE-DIMENSIONAL FLAT SPRINGS	74-0738
NICOLAIDES, R.	CLASS OF FINITE ELEMENTS GENERATED BY LAGRANGE INTERPOLATION - 2.	73-0044
NICOLAIDES, R.	ON A CLASS OF FINITE ELEMENTS GENERATED BY LAGRANGE INTERPOLATION.	72-0101
NICOLAIDES, R.	ON LAGRANGIAN INTERPOLATION IN N VARIABLES	70-0452
NIEHOFF, D.	SIMPLIFIED SOIL-STRUCTURE INTERACTION ANALYSIS WITH STRAIN DEPENDENT SOIL PROPERTIES	74-1052
NIFLSEN, J. P.	AFPAV COMPUTER CODE FOR STRUCTURAL ANALYSIS OF AIRFIELD PAVEMENTS	75-0684
NIELSEN, L. E.	MECHANICAL PROPERTIES OF TAPE COMPOSITES	69-0317
NIELSEN, L. O.	ELEMENTMETODEN OG DEFORMATIONSMETODEN (FINITE ELEMENT AND DEFORMATION METHODS IN STRESS ANALYSIS)	74-0580
NIELSEN, L. O.	FINITE ELEMENT METHOD AND DENDIXEN-OSTENFELD SLOPE DEFLECTION METHOD	74-0557
NIELSEN, L. O.	SPAENDINGSHYBRIDE FINITE ELEMENTER TIL SVINGNINGSPROBLEMER. (STRESS HYBRID FINITE ELEMENT METHOD FOR	75-0208
NIELSEN, R.	SIMPLE APPROACH TO THE STRENGTH ANALYSIS OF TANKERS	71-0319
NIELSEN, R.	TANKER TRANSVERSE STRENGTH ANALYSIS: USER'S MANUAL	72-0750
NIELSON, H. B.	A FINITE ELEMENT METHOD FOR CALCULATING OPEN CHANNEL FLOW	69-0510
NIELSON, R.	STRUCTURAL ANALYSIS OF LONGITUDINALLY FRAMED SHIPS	72-0436
NIISEKI, S.	SOME APPLICATIONS OF TOPOLOGICAL CONSIDERATION AND WEIGHT MATRIX METHOD TO FINITE ELEMENT ANALYSIS	73-0305
NIISEKI, S.	STUDY ON WEIGHT MATRIX IN STRUCTURAL ANALYSIS	75-0062
NIKITENKO, V. I	CHISLENNYI METOD RASCHETA SOBSTVENNYKH I VYNUZHDENNYKH KOLEBANII SOSTAVNYKH OBOLOCHECHNYKH KONSTRUKT	73-0461
NILFOROUSH, J.	EFFECT OF FIBRE DISTRIBUTION ON STRESS AND STRAIN CONCENTRATION AT HOLES IN COMPOSITE PLATES	72-0616
NILSON, A. H.	ANALYSIS OF LIGHT GATE STEEL SHEAR DIAPHRAGMS	73-0733
NILSON, A. H.	BIAXIAL STRESS-STRAIN RELATIONS FOR CONCRETE	72-0203
NILSON, A. H.	COUPLED SHEAR WALL ANALYSIS BY LAGRANGE MULTIPLIERS	75-0202
NILSON, A. H.	COUPLED SHEAR WALLS, IMPROVED METHODS FOR ELASTIC ANALYSIS	74-0818
NILSON, A. H.	DEFORMATION AND FRACTURE OF PARTICULATE COMPOSITE	72-0221
NILSON, A. H.	FINITE ELEMENT ANALYSIS OF REINFORCED CONCRETE	67-0161
NILSON, A. H.	FINITE ELEMENT ANALYSIS OF METAL DECK SHEAR DIAPHRAGMS	74-0128
NILSON, A. H.	NON-LINEAR FINITE ELEMENT ANALYSIS OF REINFORCED CONCRETE BY THE FINITE ELEMENT METHOD	68-0241
NILSON, A. H.	NONLINEAR ANALYSIS OF REINFORCED CONCRETE BY THE FINITE ELEMENT METHOD	68-0126
NILSON, A. H.	STRESS STRAIN RESPONSE AND FRACTURE OF A CONCRETE MODEL IN BIAXIAL LOADING	71-0295
NILSSON, L.	THE RECTANGULAR PLANE STRESS ELEMENT BY TURNER, PIAN AND WILSON	74-1265
NISHIMJRA, M.	TRANSIENT TEMPERATURE RESPONSE OF COMPOSITE SLABS	74-1038
NISHIMURA, T.	FINITE ELEMENT ANALYSIS OF COMPOSITE MATERIAL BEHAVIOURS FROM MECHANICAL PROPERTIES OF CONSTITUENT M	75-0181
NISHINO, F.	FINITE DISPLACEMENT BEAM THEORY	75-0529
NISHIOKA, K.	STRENGTH OF RAILROAD WHEELS - 3	71-0306
NISHIOKA, T.	ELASTIC-PLASTIC FINITE ELEMENT ANALYSIS USING SUPERPOSITION	75-0314
NISHIOKA, T.	ELASTIC-PLASTIC FINITE-ELEMENT ANALYSIS USING SUPERPOSITION	75-0429
NISHIOKA, T.	FINITE ELEMENT CALCULATION OF STRESS INTENSITY FACTORS USING SUPERPOSITION	75-0251
NISHIOKA, T.	THE FINITE ELEMENT CALCULATION OF STRESS INTENSITY FACTORS USING SUPERPOSITION	75-0886
NISHIYAMA, K.	A SEISMIC DESIGN OF NUCLEAR REACTOR BUILDING. STRESS ANALYSIS AND STIFFNESS EVALUATION OF THE ENTIRE	71-0254
NISHIYAMA, T.	STUDIES ON THE DETERMINATION OF SEEPAGE-LINE OF FILL DAM BY THE METHOD OF FINITE ELEMENT. WATER FLOW	74-1348
NISITANI, H.	NOTCH EFFECT IN LOW-CYCLE FATIGUE - 2. EFFECT OF STRESS CHANGE AND MEAN STRESS	75-0537
NITSCHE, J. A.	NON PROJECTIONAL METHODS	73-0632
NITSCHE, J. A.	ON DIRICHLET PROBLEMS USING SUBSPACES WITH NEARLY ZERO BOUNDARY CONDITIONS	72-0514
NIXON, C. D.	FINITE ELEMENT ANALYSIS OF A DIAGONAL CRACKING BEHAVIOUR IN REINFORCED CONCRETE	73-0462
NOBARI, E. S.	EFFECT OF RESERVOIR FILLING ON STRESSES AND MOVEMENT IN EARTH AND ROCKFILL DAMS. A REPORT OF AN INVE	72-0836
NOBARI, E. S.	HYDRAULIC FRACTURING IN ZONED EARTH AND ROCKFILL DAMS: A REPORT OF AN INVESTIGATION	73-0873
NOBLE, B.	FINITE ELEMENT METHODS FOR INITIAL VALUE PROBLEMS	72-0059
NOBLE, C.	STRAP: COMPUTER CODE FOR STATIC AND DYNAMIC STRUCTURAL ANALYSIS AND STUDIES MADE USING THE CODE	70-0409
NOOR, A. K.	HYPERMATRIX SCHEME FOR FINITE-ELEMENT SYSTEMS ON CDC STAR-100 COMPUTER	75-0818
NOOR, A. K.	IMPACT OF CDC STAR-100 COMPUTER ON FINITE ELEMENT SYSTEMS	75-0039
NOOR, A. K.	MIXED ISOPARAMETRIC LAMINATED COMPOSITE SHELL ELEMENTS	75-0450
NOOR, A. K.	NONLINEAR FINITE ELEMENT ANALYSIS OF LAMINATED COMPOSITE SHELLS	74-0751
NOOR, A. K.	USE OF GROUP-THEORETIC METHODS IN THE DEVELOPMENT OF NONLINEAR SHELL FINITE ELEMENTS	71-0704
NOOR, A. K.	USE OF SYMBOLIC MANIPULATION IN DEVELOPMENT OF TWO-DIMENSIONAL FINITE ELEMENTS	74-0103
NOORISHAD, J.	INFLUENCE OF FLUID INJECTION ON THE STATE OF STRESS IN THE EARTH'S CRUST	72-0624
NOORISHAD, J.	SEEPAGE IN FRACTURED ROCK: FINITE ELEMENT ANALYSIS OF ROCK MASS BEHAVIOR UNDER COUPLED ACTION OF BOD	71-0825
NOPPEN, R.	BERECHNUNG DER ELASTIZITATSEIGENSCHAFTEN VON MASCHINENBAUTEILEN NACH DER METHODE FINITER ELEMENTE	73-1046
NOPPEN, R.	EVALUATION OF COMPUTER AIDED METHODS FOR THE DESIGN OF MACHINE TOOL STRUCTURES	73-0577
NOPPEN, R.	FINEL, UNIVERSELLES PROGRAMMSYSTEM ZUR BERECHNUNG DER ELASTISCHEN EIGENSCHAFTEN VON MASCHINENBAUTEIL	72-0132
NORDELL, W. J.	STRUCTURAL RESPONSE OF UNSTIFFENED TOROIDAL SHELLS	69-0198
NORRIE, D. H.	A FINITE ELEMENT BIBLIOGRAPHY - PART I: AUTHOR LISTING	74-0284
NORRIE, D. H.	A FINITE ELEMENT BIBLIOGRAPHY - PART II: KEYWORD LISTING	74-0285
NORRIE, D. H.	A FINITE ELEMENT BIBLIOGRAPHY - PART III: CITATION LISTING	74-0286
NORRIE, D. H.	A LAGRANGIAN FINITE ELEMENT SOLUTION TO UNSTEADY FLOW	74-1264
NORRIE, D. H.	APPLICATION OF FINITE ELEMENT METHODS IN FLUID DYNAMICS	71-0369
NORRIE, D. H.	APPLICATION OF THE FINITE ELEMENT METHOD TO UNSTEADY FLOW PROBLEMS	70-0310
NORRIE, D. H.	APPLICATION OF THE FINITE ELEMENT TECHNIQUE TO COMPRESSIBLE FLOW PROBLEMS	70-0311
NORRIE, D. H.	APPLICATION OF THE FINITE ELEMENT TECHNIQUE TO COMPRESSIBLE FLOW PROBLEMS	73-0003
NORRIE, D. H.	APPLICATION OF THE FINITE ELEMENT TECHNIQUE TO COMPRESSIBLE FLOW PROBLEMS	74-0090
NORRIE, D. H.	APPLICATION OF THE PSEUDO-FUNCTIONAL FINITE ELEMENT METHOD TO VISCO-PLASTIC TORSION	73-0002

NORRIE, D. H.	APPLICATION OF THE PSEUDO-FUNCTIONAL FINITE ELEMENT METHOD TO NON-LINEAR PROBLEMS	74-0282
NORRIE, D. H.	APPLICATION OF THE PSEUDO-FUNCTIONAL FINITE ELEMENT METHOD TO VISCOUS FLOW PROBLEMS	74-1183
NORRIE, D. H.	APPLICATION OF THE PSEUDO-FUNCTIONAL FINITE ELEMENT METHOD TO COMPRESSIBLE FLOW PROBLEMS	74-1360
NORRIE, D. H.	NON-STRUCTURAL APPLICATIONS OF THE FINITE ELEMENT METHOD	73-0330
NORRIE, D. H.	THE APPLICATION OF THE FINITE ELEMENT TECHNIQUE TO POTENTIAL FLOW PROBLEMS: PART I	69-0087
NORRIE, D. H.	THE APPLICATION OF THE FINITE ELEMENT TECHNIQUE TO POTENTIAL FLOW PROBLEMS: PART II	69-0395
NORRIE, D. H.	THE APPLICATION OF THE FINITE ELEMENT TECHNIQUE TO POTENTIAL FLOW PROBLEMS	71-0370
NORRIE, D. H.	THE APPLICATION OF THE PSEUDO-FUNCTIONAL FINITE ELEMENT METHOD TO VISCOUS FLOW PROBLEMS	74-0287
NORRIE, D. H.	THE FINITE ELEMENT METHOD - FUNDAMENTALS AND APPLICATIONS	73-0001
NORRIE, D. H.	THE PSEUDO FUNCTIONAL FINITE ELEMENT METHOD	74-0004
NORRIS, D. M. J	TENSILE FIELD CALCULATION ABOUT A PRESSURIZED CAVITY: MONERO	72-0684
NORRIS, D. M. J	DYNAMIC LOADS ON GROUTED EMPLACEMENT SYSTEMS	70-0406
NORRIS, V.	THE DETERMINATION OF A WATER/GAS INTERFACE USING FINITE ELEMENTS	74-0346
NORRIS, V. A.	AN APPLICATION OF THE FINITE ELEMENT METHOD TO TRANSIENT INTERFACE PROBLEMS IN POROUS MEDIA	73-0608
NORRIS, V. A.	AN INTRODUCTION TO THE FINITE ELEMENT ANALYSIS OF THE MOTION OF A GAS-LIQUID INTERFACE IN A POROUS M	72-0904
NORRIS, V. A.	FINITE ELEMENT ANALYSIS OF THE MOTION OF A GAS-LIQUID INTERFACE IN A POROUS MEDIUM	75-0672
NORTH, E. L.	PAPA, STRUCTURAL ANALYSIS OF PLATES AND SHELLS USING TRAPEZOIDAL AND TRIANGULAR PLATE ELEMENTS	67-0059
NORTON, W. R.	A FINITE ELEMENT MODEL FOR TWO DIMENSIONAL FLOW	74-0015
NORTON, W. R.	A FINITE ELEMENT SOLUTION FOR TWO-DIMENSIONAL DENSITY STRATIFIED FLOW	73-0911
NORTON, W. R.	A FINITE ELEMENT SOLUTION FOR TWO-DIMENSIONAL STRATIFIED FLOW PROBLEMS	75-0391
NORTON, W. R.	FINITE ELEMENT MODEL FOR LOWER GRANITE RESERVOIR	73-0065
NORVILLE, C. C.	APPLIED FINITE ELEMENT DISPLACEMENT ANALYSIS OF A SIMPLE INTEGRAL CONSTRUCTION VEHICLE BODY SHELL	72-0264
NORVILLE, C. C.	STRESS ANALYSIS OF AN AUTOMOBILE CHASSIS FRAME	70-0137
NOSSEIR, S.	PROBABILISTIC MODEL FOR MATERIAL STRENGTH VARIATION AND SIZE EFFECT	72-0876
NOSSEIR, T. A.	FREE VIBRATION OF MODEL OF CAR BODY	71-0312
NOVAK, M.	VERTICAL VIBRATION OF EMBEDDED FOOTINGS	72-0231
NOZUE, Y.	NOTCH EFFECT IN LOW-CYCLE FATIGUE. CONSIDERATIONS ON NOTCH EFFECT IN LOW-CYCLE FATIGUE WITH FINITE-E	75-0343
NYASHIN, YU. I.	SOLUTION OF THE PROBLEM OF PROCESSING METALS BY PRESSURE USING THE METHOD OF FINITE ELEMENTS	74-1363
O'BRIEN, D. J.	NATURAL FREQUENCIES OF BOX-TYPE STRUCTURES BY A FINITE-ELEMENT METHOD	74-0193
O'BRIEN, M.	SOLUTION BOUNDS IN SOME STRESS PROBLEMS BY THE HYPERCIRCLE AND FINITE ELEMENT METHODS	72-0706
O'BRIEN, M. A.	APPLICATION OF THE FAST FOURIER TRANSFORM TO LINEAR SYSTEMS IN CIVIL ENGINEERING	73-1119
O'CALLAGHAN, J.	ANALYSIS OF STRESSES AND DEFLECTIONS IN FRAME SUPPORTED TENTS	74-0689
O'CALLAGHAN, J.	NUMERICAL SOLUTION OF THE ELASTOHYDRODYNAMIC LUBRICATION PROBLEM USING FINITE ELEMENTS	72-0042
O'CARROLL, M. J	A VARIATIONAL METHOD FOR FREE BOUNDARY PROBLEMS	75-0105
O'CARROLL, M. J	DIFFERENCE AND ELEMENT METHODS FOR POTENTIAL FLOW	74-0019
O'CARROLL, M. J	UNIFORM CONVERGENCE OF VARIATIONAL FINITE ELEMENT AND RITZ METHODS	74-0420
O'CONNELL, J. M	STRESS CONCENTRATION FACTORS AROUND A PENETRATION IN A STRIP OR SLAB OF FINITE DIMENSIONS	69-0409
O'CONNOR, B. A.	GALERKIN FINITE ELEMENT SOLUTIONS FOR POLLUTION PROBLEMS IN PARTIALLY MIXED ESTUARIES	74-0047
O'CONNOR, B. A.	NUMERICAL ANALYSIS OF PLASTICITY IN SOILS	73-0611
O'CONNOR, B. A.	RAYLEIGH-RITZ AND GALERKIN FINITE ELEMENTS FOR DIFFUSION-CONVECTION PROBLEMS	73-0017
O'CONNOR, J. J.	FINITE ELEMENT ANALYSIS OF ALTERNATIVE AXIAL LOADING OF AN ELASTIC PLATE PRESSED BETWEEN TWO ELASTIC	71-0357
O'DONNELL, W. J	COMPARISON OF STRUCTURAL TEST RESULTS WITH PREDICTIONS OF FINITE ELEMENT ANALYSIS	73-0158
O'HARA, G. P.	FINITE ELEMENT ANALYSIS OF SEVERAL SWAGE MANDREL DESIGNS	74-0794
O'HARA, P.	FINITE ELEMENT ANALYSIS OF THREADED CONNECTIONS	74-1179
OAKBERG, R. G.	ANALYSIS OF FRAMES WITH SHEAR WALLS BY FINITE ELEMENTS	69-0199
OATES, G. C.	A VARIATIONAL FORMULATION OF THE COMPRESSIBLE THROUGHFLOW PROBLEM	74-1343
OATES, G. C.	A VARIATIONAL FORMULATION OF THE COMPRESSIBLE THROUGHFLOW PROBLEM	75-0021
OBAID, P.	THREE DIMENSIONAL FINITE ELEMENT ANALYSIS	72-0965
OCH, F.	EINE FINITE-ELEMENTE-LOESUNG FUR DIE TORSIONSSTEIFIGKEIT UND DEN SCHUBMITTELPUNKT BELIEBIGER QUERSCH	74-0256
OCH, F.	FINITE ELEMENTS SOLUTION FOR TORSION RIGIDITY AND SHEAR CENTER OF ARBITRARY CROSS-SECTIONS	74-0168
OCHIAI, S-I.	FORMULATION OF STRESS STRAIN INDUCED DIFFUSION OF HYDROGEN AND ITS SOLUTION BY COMPUTER-AIDED FINITE	75-0326
ODA, E.	METHOD OF ANALYZING NON-DARCY FLOW PROBLEMS BY FINITE ELEMENTS	75-0158
ODA, J.	EFFECT OF GRAIN BOUNDARY FOR THE ELASTIC BEHAVIOR OF AL AND CU CRYSTALS (ANALYSIS BY THE FINITE ELEM	74-0947
ODA, J.	SIMULATION OF ELASTIC MODULUS AND POISSON'S RATIO OF SPHEROIDAL GRAPHITE CAST IRON	74-0955
ODELLO, R. J.	RESPONSE OF BURIED CAPSULES IN THE HIGH-OVERPRESSURE REGION	70-0093
ODEN, J. T.	A FINITE ELEMENT ANALOGUE OF THE NAVIER-STOKES EQUATION	70-0099
ODEN, J. T.	A FINITE ELEMENT ANALYSIS OF SHOCK AND FINITE-AMPLITUDE WAVES IN ONE-DIMENSIONAL HYPERELASTIC BODIES	72-0731
ODEN, J. T.	A FINITE ELEMENT ANALYSIS OF TRANSIENT RAREFIED GAS FLOW	73-0604
ODEN, J. T.	A GENERAL THEORY OF FINITE ELEMENTS	69-0034
ODEN, J. T.	A GENERALIZATION OF THE FINITE ELEMENT CONCEPT AND ITS APPLICATION TO A CLASS OF PROBLEMS IN NONLINE	68-0212
ODEN, J. T.	A LARGE DEFORMATION ANALYSIS OF CRYSTALLINE ELASTIC-VISCOPLASTIC MATERIALS	73-0602
ODEN, J. T.	A NEW APPROACH TO THE FINITE-ELEMENT FORMULATION AND SOLUTION OF A CLASS OF PROBLEMS IN COUPLED THER	72-0732
ODEN, J. T.	A NEW APPROACH TO THE FINITE-ELEMENT FORMULATION AND SOLUTION OF A CLASS OF PROBLEMS IN COUPLED THER	73-0600
ODEN, J. T.	A NOTE ON ACCURACY AND CONVERGENCE OF FINITE ELEMENT APPROXIMATIONS	71-0565
ODEN, J. T.	A NOTE ON THE ANALYSIS OF NONLINEAR DYNAMICS OF ELASTIC MEMBRANES BY THE FINITE ELEMENT METHOD	74-0341
ODEN, J. T.	A NOTE ON THE FINITE DEFORMATIONS OF A THICK ELASTIC SHELL OF REVOLUTION	71-0193
ODEN, J. T.	A NUMERICAL SOLUTION OF THE BOLTZMANN EQUATION BY THE FINITE ELEMENT TECHNIQUE	70-0095
ODEN, J. T.	A THEORY OF MIXED FINITE ELEMENT APPROXIMATIONS OF NON-SELF-ADJOINT BOUNDARY-VALUE PROBLEMS	74-0304
ODEN, J. T.	A THEORY OF MIXED FINITE ELEMENT APPROXIMATIONS OF NON-SELF-ADJOINT BOUNDARY VALUE PROBLEMS	74-0715
ODEN, J. T.	ACCURACY AND CONVERGENCE OF FINITE ELEMENT GALERKIN APPROXIMATIONS OF TIME DEPENDENT PROBLEMS WITH E	73-0987
ODEN, J. T.	ACCURACY AND CONVERGENCE OF FINITE ELEMENT/GALERKIN APPROXIMATIONS OF TIME DEPENDENT PROBLEMS WITH E	75-0006
ODEN, J. T.	AN IMBEDDING TECHNIQUE FOR THE GENERATION OF WEAK-WEAK FINITE ELEMENT APPROXIMATIONS OF LINEAR AND N	72-0577
ODEN, J. T.	ANALYSIS OF FINITE DEFORMATIONS OF ELASTIC SOLIDS BY THE FINITE ELEMENT METHOD	70-0096
ODEN, J. T.	ANALYSIS OF FLOW OF VISCOUS FLUIDS BY FINITE-ELEMENT METHOD	72-0032
ODEN, J. T.	ANALYSIS OF GENERAL COUPLED THERMOELASTICITY PROBLEMS BY THE FINITE ELEMENT METHOD	68-0129
ODEN, J. T.	ANALYSIS OF LARGE DEFORMATIONS OF ELASTIC MEMBRANES BY THE FINITE ELEMENT METHOD	66-0057
ODEN, J. T.	ANALYSIS OF NONLINEAR THERMOELASTIC AND THERMOPLASTIC BEHAVIOUR OF SOLIDS OF REVOLUTION BY THE FINIT	71-0192
ODEN, J. T.	ANALYSIS OF NONLINEAR DYNAMIC, COUPLED THERMOVISCOELASTICITY PROBLEMS BY THE FINITE ELEMENT METHOD	71-0527
ODEN, J. T.	ANALYSIS OF RAREFIED GAS FLOW THROUGH AN ARBITRARY CROSS SECTION BY THE FINITE ELEMENT METHOD	73-0365
ODEN, J. T.	APPROXIMATIONS AND NUMERICAL ANALYSIS OF FINITE DEFORMATIONS OF ELASTIC SOLIDS	74-0343
ODEN, J. T.	BASIC RESEARCH IN METHODS OF APPROXIMATION IN THE NONLINEAR MECHANICS OF SOLIDS AND STRUCTURES	72-0857
ODEN, J. T.	CALCULATION OF GEOMETRIC STIFFNESS MATRICES FOR COMPLEX STRUCTURES	66-0058
ODEN, J. T.	CALCULATION OF STIFFNESS MATRICES FOR FINITE ELEMENTS OF THIN SHELLS OF ARBITRARY SHAPE	68-0127
ODEN, J. T.	COMPUTATIONAL METHODS IN NONLINEAR MECHANICS	74-1232
ODEN, J. T.	CONJUGATE APPROXIMATION FUNCTIONS IN FINITE ELEMENT ANALYSIS	71-0043
ODEN, J. T.	CONVERGENCE OF MIXED FINITE ELEMENT APPROXIMATIONS OF A CLASS OF LINEAR-BOUNDARY VALUE PROBLEMS	73-0599
ODEN, J. T.	CONVERGENCE, ACCURACY AND STABILITY OF FINITE ELEMENT APPROXIMATIONS OF A CLASS OF NON-LINEAR HYPERB	73-0159
ODEN, J. T.	DISCONTINUOUS FINITE-ELEMENT APPROXIMATIONS FOR ANALYSIS OF SHOCK-WAVES IN NONLINEARLY ELASTIC-MATER	75-0451
ODEN, J. T.	DYNAMIC ELASTOPLASTIC RESPONSE OF GEOMETRICALLY NONLINEAR ARBITRARY SHELLS OF REVOLUTION UNDER IMPUL	70-0098
ODEN, J. T.	EFFECT OF HIGHER ORDER TERMS IN CERTAIN NONLINEAR FINITE ELEMENT MODELS	73-0125
ODEN, J. T.	EVALUATION AND ANALYSES OF SOME FINITE ELEMENT AND FINITE DIFFERENCE PROCEDURES FOR TIME-DEPENDENT P	75-0745
ODEN, J. T.	FINITE DEFORMATION OF ELASTIC MEMBRANES BY THE FINITE ELEMENT METHOD	67-0105
ODEN, J. T.	FINITE DEFORMATION OF ELASTIC PLATES, SHELLS, AND MEMBRANES BY THE FINITE ELEMENT METHOD	72-0805
ODEN, J. T.	FINITE ELEMENT ANALOGUE OF NAVIER-STOKES EQUATIONS	69-0002
ODEN, J. T.	FINITE ELEMENT ANALYSIS OF NONLINEAR PROBLEMS IN THE DYNAMICAL THEORY OF COUPLED THERMOELASTICITY	69-0202
ODEN, J. T.	FINITE ELEMENT ANALYSIS OF NONLINEAR STRUCTURES	69-0205
ODEN, J. T.	FINITE ELEMENT ANALYSIS OF THIN SHELLS	70-0457
ODEN, J. T.	FINITE ELEMENT ANALYSIS OF FINITE DEFORMATIONS OF PLATES, SHELLS, AND MEMBRANES	72-0578
ODEN, J. T.	FINITE ELEMENT ANALYSIS OF WAVES IN NONLINEAR MATERIALS	75-0110
ODEN, J. T.	FINITE ELEMENT APPLICATIONS IN LINEAR AND NONLINEAR THERMOVISCOELASTICITY	67-0065
ODEN, J. T.	FINITE ELEMENT APPLICATIONS IN FLUID DYNAMICS	69-0003
ODEN, J. T.	FINITE ELEMENT APPLICATIONS IN NONLINEAR STRUCTURAL ANALYSIS	69-0204

ODEN, J. T.	FINITE ELEMENT APPLICATIONS IN MATHEMATICAL PHYSICS	
ODEN, J. T.	FINITE ELEMENT APPROXIMATION OF A CLASS OF NONLINEAR OPERATORS	72-0067
ODEN, J. T.	FINITE ELEMENT APPROXIMATIONS IN NONLINEAR THERMOVISCOELASTICITY	71-0454
ODEN, J. T.	FINITE ELEMENT APPROXIMATIONS IN NONLINEAR ELASTICITY	71-0459
ODEN, J. T.	FINITE ELEMENT FORMULATION OF FINITE DEFORMATIONS AND IRREVERSIBLE THERMODYNAMICS OF NONLINEAR CONTI	71-0470
ODEN, J. T.	FINITE ELEMENT FORMULATION OF GENERAL THERMO-ELASTICITY PROBLEMS	69-0207
ODEN, J. T.	FINITE ELEMENT FORMULATIONS OF PROBLEMS OF FINITE DEFORMATION AND IRREVERSIBLE THERMODYNAMICS OF NONL	71-0001
ODEN, J. T.	FINITE ELEMENT LARGE DEFLECTION ANALYSIS OF PLATES	71-0190
ODEN, J. T.	FINITE ELEMENT METHODS IN FLOW PROBLEMS	71-0497
ODEN, J. T.	FINITE ELEMENT MODELS OF NONLINEAR OPERATOR EQUATIONS	69-0208
ODEN, J. T.	FINITE ELEMENT TECHNIQUE APPLIED TO HEAT CONDUCTION IN SOLIDS WITH TEMPERATURE DEPENDENT THERMAL CON	74-0752
ODEN, J. T.	FINITE ELEMENTS IN NONLINEAR CONTINUA	71-0199
ODEN, J. T.	FINITE PLANE STRAIN OF INCOMPRESSIBLE ELASTIC SOLIDS BY THE FINITE ELEMENT METHOD	69-0004
ODEN, J. T.	FINITE STRAINS AND DISPLACEMENTS OF ELASTIC MEMBRANES BY THE FINITE ELEMENT METHOD	71-0200
ODEN, J. T.	FINITE-ELEMENT ANALYSIS OF THIN SHELLS	68-0128
ODEN, J. T.		67-0062
ODEN, J. T.	FORMULATION AND APPLICATION OF CERTAIN PRIMAL AND MIXED FINITE ELEMENT MODELS OF FINITE DEFORMATIONS	68-0164
ODEN, J. T.	FORMULATION OF GENERAL DISCRETE MODELS OF THERMOMECHANICAL BEHAVIOR OF MATERIALS WITH MEMORY	74-0344
ODEN, J. T.	GENERAL MIXED FINITE ELEMENT METHODS OF NONLINEAR CONTINUA	69-0381
ODEN, J. T.	GENERAL MIXED FINITE ELEMENT METHODS OF NONLINEAR CONTINUA	72-0831
ODEN, J. T.	GENERALIZED CONJUGATE FUNCTIONS FOR MIXED FINITE ELEMENT APPROXIMATIONS OF BOUNDARY VALUE PROBLEMS	73-0091
ODEN, J. T.	LECTURES ON FINITE ELEMENT METHODS IN CONTINUUM MECHANICS	72-0515
ODEN, J. T.	MATHEMATICAL ASPECTS OF FINITE ELEMENT APPROXIMATIONS IN THEORETICAL MECHANICS	71-0714
ODEN, J. T.	MATHEMATICAL FEATURES OF FINITE ELEMENT APPROXIMATIONS OF CERTAIN FLOW PROBLEMS WITH EMPHASIS ON CON	75-0852
ODEN, J. T.	MIXED CONJUGATE FINITE ELEMENT APPROXIMATIONS OF LINEAR OPERATORS	74-0002
ODEN, J. T.	MIXED FINITE-ELEMENT APPROXIMATIONS OF LINEAR BOUNDARY-VALUE PROBLEMS	72-0085
ODEN, J. T.	NEW APPROACH TO FINITE ELEMENT FORMULATION AND SOLUTION OF A CLASS OF PROBLEMS IN COUPLED THERMOELAS	75-0453
ODEN, J. T.	NOTE ON AN APPROXIMATE METHOD FOR COMPUTING NONCONSERVATIVE GENERALIZED FORCES ON FINITELY DEFORMED	73-0013
ODEN, J. T.	NOTE ON AN APPROXIMATE METHOD FOR COMPUTING CONSISTENT CONJUGATE STRESSES IN ELASTIC FINITE ELEMENTS	70-0097
ODEN, J. T.	NOTE ON THE ACCURACY AND CONVERGENCE OF FINITE ELEMENT APPROXIMATIONS OF THE CONVECTION EQUATION	73-0050
ODEN, J. T.	NOTE ON THE ANALYSIS OF NONLINEAR DYNAMICS OF ELASTIC MEMBRANES BY THE FINITE ELEMENT METHOD	75-0490
ODEN, J. T.	NUMERICAL ANALYSIS FOR FINITE AXISYMMETRIC DEFORMATIONS OF INCOMPRESSIBLE ELASTIC SOLIDS OF REVOLUTI	74-0230
ODEN, J. T.	NUMERICAL ANALYSIS OF NONLINEAR PNEUMATIC STRUCTURES	70-0094
ODEN, J. T.	NUMERICAL FORMULATION OF A CLASS OF PROBLEMS IN NONLINEAR VISCOELASTICITY	67-0060
ODEN, J. T.	NUMERICAL FORMULATION OF NONLINEAR ELASTICITY PROBLEMS	67-0061
ODEN, J. T.	ON A GENERALIZATION OF THE FINITE ELEMENT CONCEPT AND ITS APPLICATION TO A CLASS OF PROBLEMS IN NONL	67-0063
ODEN, J. T.	ON COMPLEMENTARY-DUAL VARIATIONAL PRINCIPLES IN MATHEMATICAL PHYSICS	69-0206
ODEN, J. T.	ON MIXED FINITE ELEMENT APPROXIMATIONS	74-0342
ODEN, J. T.	ON SOME GENERALIZATIONS OF THE INCREMENTAL STIFFNESS RELATIONS FOR FINITE DEFORMATIONS OF COMPRESSIB	74-0701
ODEN, J. T.	ON THE APPLICATION OF THE FINITE ELEMENT METHOD TO A CLASS OF PROBLEMS IN FLUID DYNAMICS	71-0189
ODEN, J. T.	ON THE CALCULATION OF CONSISTENT STRESS DISTRIBUTIONS IN FINITE ELEMENT APPROXIMATIONS	69-0400
ODEN, J. T.	ON THE EFFECT OF APPROXIMATING HIGHER ORDER TERMS IN CERTAIN FINITE ELEMENT MODELS OF NONLINEAR STRU	71-0191
ODEN, J. T.	ON THE EFFECT OF THE FORM OF THE STRAIN ENERGY FUNCTION ON THE SOLUTION OF A BOUNDARY-VALUE PROBLEM	73-0603
ODEN, J. T.	ON THE NUMERICAL SOLUTION OF A CLASS OF PROBLEMS IN A LINEAR FIRST STRAIN-GRADIENT THEORY OF ELASTIC	72-0429
ODEN, J. T.	ON THE NUMERICAL SOLUTION OF A CLASS OF NONLINEAR PROBLEMS IN DYNAMIC COUPLED THERMOELASTICITY	70-0318
ODEN, J. T.	RECENT ADVANCES IN COMPUTATIONAL METHODS IN STRUCTURAL MECHANICS AND DESIGN.	71-0441
ODEN, J. T.	RECENT ADVANCES IN MATRIX METHODS IN STRUCTURAL ANALYSIS AND DESIGN	72-0575
ODEN, J. T.	SOLUTION OF NONLINEAR EQUATIONS BY EXPLICIT TIME INTEGRATION — A SHORT COMMUNICATION	70-0290
ODEN, J. T.	SOME ASPECTS OF RECENT CONTRIBUTIONS TO THE MATHEMATICAL THEORY OF FINITE ELEMENTS	73-0606
ODEN, J. T.	SOME ASPECTS OF RECENT CONTRIBUTIONS TO THE MATHEMATICAL THEORY OF FINITE ELEMENTS	72-0811
ODEN, J. T.	SOME ASPECTS OF THE MATHEMATICAL THEORY OF FINITE ELEMENTS	73-0601
ODEN, J. T.	SOME CONTRIBUTIONS TO THE MATHEMATICAL THEORY OF MIXED FINITE ELEMENT APPROXIMATIONS	72-0243
ODEN, J. T.	SOME DISTRIBUTIONAL AND CONVERGENCE PROPERTIES OF THE FINITE ELEMENT METHOD, WITH APPLICATIONS IN NO	73-0302
ODEN, J. T.	SOME OBSERVATIONS ON PROPERTIES OF CERTAIN MIXED FINITE ELEMENT APPROXIMATIONS	73-0598
ODEN, J. T.	SOME RECENT DEVELOPMENTS IN THE THEORY AND APPLICATION OF FINITE ELEMENTS TO MIXED PROBLEMS AND SHOC	75-0617
ODEN, J. T.	SOME RESULTS OF FINITE ELEMENT APPLICATIONS IN FINITE ELASTICITY	73-0637
ODEN, J. T.	SOME RESULTS OF FINITE ELEMENT APPLICATIONS ON NONLINEAR ELASTICITY	73-0183
ODEN, J. T.	STABILITY AND POSTBUCKLING BEHAVIOUR OF HYPERELASTIC BODIES AT FINITE STRAIN BY THE FINITE ELEMENT M	73-0605
ODEN, J. T.	STRUCTURAL ANALYSIS OF AERODYNAMIC DECELERATION SYSTEMS BY THE FINITE ELEMENT METHOD	73-0562
ODEN, J. T.	THE FINITE ELEMENT ANALYSIS OF TRANSIENT RAREFIED GAS FLOW	67-0064
ODEN, J. T.	THE FINITE ELEMENT METHOD IN FLUID MECHANICS	74-0027
ODEN, J. T.	THE FINITE ELEMENT METHOD IN MATHEMATICAL PHYSICS	71-0197
ODEN, J. T.	THEMIS FINAL REPORT, VOL. IV, RECENT DEVELOPMENTS IN THE NUMERICAL SOLUTION OF CERTAIN PROBLEMS IN N	73-0289
ODEN, J. T.	THEMIS FINAL REPORT, VOL. 1, BASIC RESEARCH IN METHODS OF APPROXIMATION IN THE NONLINEAR MECHANICS O	74-0721
ODEN, J. T.	THEORY OF CONJUGATE PROJECTIONS IN FINITE ELEMENT ANALYSIS	74-0723
ODEN, J. T.	VARIATIONAL PRINCIPLES, NONLINEAR OPERATORS, AND FINITE ELEMENT APPLICATIONS	71-0194
ODLAND, J. R.	SIMPLIFIED CALCULATION OF THE INTERACTION FORCES BETWEEN HULL AND TANK SYSTEM FOR A MOSS ROSENBERG T	72-0576
OGAWA, A.	A STRUCTURAL ANALYSIS OF CYLINDER-CONE-CYLINDER SHELLS BY THE FINITE ELEMENT METHOD	75-0561
OGAWA, A.	FLEXURAL RIGIDITY OF A THIN WALLED BUILD-UP ROTOR FOR JET ENGINE. MEASUREMENT BY STATIC LOAD TEST AND	72-0796
OGLESBY, J. J.	EVALUATION OF FINITE ELEMENT METHODS FOR THE COMPUTATION OF ELASTIC STRESS INTENSITY FACTORS	73-0756
OGURA, K.	ANALYTICAL APPROACH TO THE NON-PROPAGATING CRACK PROBLEM USING THE FINITE ELEMENT METHOD	72-0249
OGURA, K.	CYCLIC ANALYSIS OF A PROPAGATING CRACK AND ITS CORRELATION WITH FATIGUE CRACK GROWTH	71-0389
OGURA, K.	ELASTIC PLASTIC ANALYSIS OF NOTCHED SPECIMENS USING THE FINITE ELEMENT METHOD	75-0188
OGURA, K.	FINITE ELEMENT APPROACH TO THE CRITICAL CYCLIC STRESS REQUIRED TO PROPAGATE A CRACK	70-0140
OGURA, N.	THE FINITE ELEMENT CALCULATION OF STRESS INTENSITY FACTORS USING SUPERPOSITION	73-0463
OH, H. L.	THE COMPUTATION OF TEARING ENERGY OF NICKED RUBBER STRIPS IN EXTENSION	75-0886
OH, K. P.	A UNIFIED TREATMENT OF THICK AND THIN FILM ELASTOHYDRODYNAMIC PROBLEMS USING HIGHER ORDER ELEMENT PR	74-0315
OH, K. P.	APPLICATION OF A GRADIENT METHOD IN FINITE ELEMENT SOLUTION OF ELASTOHYDRODYNAMIC PROBLEMS	75-0890
OH, K. P.	SOLUTION OF THE ELASTOHYDRODYNAMIC FINITE JOURNAL BEARING PROBLEM	74-0129
OH, S. I.	DEFORMATION MODE OF VOID-GROWTH AND COALESCENCE IN THE PROCESS OF DUCTILE FRACTURE	72-0033
OHAYON, R.	INTERNAL PRESSURE EFFECTS ON THE VIBRATION OF PARTIALLY FILLED ELASTIC TANKS	75-0683
OHAYON, R.	UNE METHODE DE CALCUL PAR ELEMENTS FINIS DES MOUVEMENTS DE LIQUIDES DANS DES RESERVOIRS RIGIDES OU D	75-0241
OHAYON, R.	VIBRATIONS OF TANKS PARTIALLY FILLED WITH LIQUIDS	74-1171
OHCHI, Y.	STRESS ANALYSIS FOR STRUCTURES UNDER CONSTRUCTION	74-0089
OHISUBO, T.	FINITE ELEMENT ANALYSIS OF STEADY FLUID AND METAL FLOW	69-0209
OHJI, K.	ANALYTICAL APPROACH TO THE NON-PROPAGATING CRACK PROBLEM USING THE FINITE ELEMENT METHOD	74-0054
OHJI, K.	CYCLIC ANALYSIS OF A PROPAGATING CRACK AND ITS CORRELATION WITH FATIGUE CRACK GROWTH	71-0389
OHJI, K.	ELASTIC PLASTIC ANALYSIS OF NOTCHED SPECIMENS USING THE FINITE ELEMENT METHOD	75-0188
OHJI, K.	FINITE ELEMENT APPROACH TO THE CRITICAL CYCLIC STRESS REQUIRED TO PROPAGATE A CRACK	70-0140
OHKUBO, Y.	ANALYSIS OF BODY STRUCTURES - 2.	73-0463
OHMORI, N.	NONLINEAR ANALYSIS OF REINFORCED CONCRETE BUILDINGS	71-0339
OHNAKA, I.	CALCULATION OF SOLIDIFICATION OF MOLTEN METAL BY FINITE-ELEMENT METHOD	73-0324
OHNISHI, S.	EXPERIMENTAL AND THEORETICAL STUDY ON A 1/10TH SCALE MODEL PCPV	74-1357
OHNISHI, T.	APPLICATION OF FINITE ELEMENT SOLUTION TECHNIQUE TO NEUTRON DIFFUSION AND TRANSPORT EQUATIONS	73-0829
OHNISHI, T.	FINITE ELEMENT METHOD APPLIED TO REACTOR PHYSICS PROBLEMS	71-0015
OHNISHI, T.	FINITE ELEMENT METHOD FOR SOLVING NEUTRON TRANSPORT PROBLEMS IN TWO-DIMENSIONAL CYLINDRICAL GEOMETRY	71-0483
OHNISHI, T.	FINITE ELEMENT SOLUTION TECHNIQUE FOR NEUTRON TRANSPORT EQUATION	74-1081
OHNISHI, T.	STUDY OF ITERATIVE SOLUTIONS FOR FINITE ELEMENT METHOD	72-0730
OHNO, T.	VISCOELASTIC ANALYSIS OF P C CONTAINMENT USING FINITE ELEMENT METHOD	71-0453
OHNO, T.	VISCOELASTIC ANALYSIS OF PRESTRESSED CONCRETE SHELL OF REVOLUTION	73-0034
OHSAKA, H.	STEADY FLOW ANALYSIS OF INCOMPRESSIBLE VISCOUS FLOW BY THE FINITE ELEMENT METHOD	73-0464
OHSAKA, K.	STATIC STRUCTURAL ANALYSIS OF SUSPENSION BRIDGES	73-0332
		66-0110

OHSAKA, K.	THE HEAT CONDUCTION AND THERMAL STRESS ANALYSIS BY THE FINITE ELEMENT METHOD	68-0073
OHTA, M.	ANALYSIS OF HYDROMAGNETIC PLASMA STABILITY BY FINITE-ELEMENT METHOD	72-0307
OHTA, M.	ANALYSIS OF MAGNETIC FIELDS BY FINITE ELEMENT METHOD	73-0126
OHTA, M.	NUMERICAL ANALYSIS OF MAGNETOHYDRODYNAMIC INSTABILITIES BY FINITE ELEMENT METHOD	72-0041
OHTANI, R.	CREEP STRESS AND STRAIN ANALYSIS OF NOTCHED PLATES AND BARS BY MEANS OF THE FINITE ELEMENT METHOD	70-0141
OHTE, S.	FINITE ELEMENT ANALYSIS OF ELASTIC CONTACT PROBLEMS	73-0219
OHTMER, O.	AUTOMATIC GENERATION AND CONDENSATION OF LARGE MASS AND STIFFNESS MATRICES	72-0466
OHTMER, O.	CALCULATION OF TWO AND THREE DIMENSIONAL FLOW PROBLEMS ACCORDING TO SUB-STRUCTURE THEORY WITH ISOPAR	74-0215
OHTSUBO, H.	FINITE ELEMENT FORMULATION FOR SOLUTION OF STEADY FLOW OF INCOMPRESSIBLE VISCOUS FLUID	73-0299
OHTSUBO, H.	A METHOD OF ELASTIC-PLASTIC ANALYSIS OF LARGELY DEFORMED PLATE PROBLEMS	72-0319
OHTSUBO, H.	A METHOD OF SOLUTION FOR THE COMPLICATED BUCKLING PROBLEMS OF ELASTIC PLATES WITH COMBINED USE OF RA	68-0097
OHTSUBO, H.	ON THE STATES OF STRESS AND DEFORMATION OF CYLINDRICAL SPECIMENS OF BRITTLE MATERIAL UNDER UNIAXIAL	69-0150
OHTSUBO, T.	FINITE ELEMENT ANALYSIS OF STEADY FLUID AND METAL FLOW	75-0178
OHYA, H.	APPLICATION OF FINITE ELEMENT METHOD TO AXISYMMETRIC BUCKLING OF SHALLOW SPHERICAL SHELLS UNDER EXTE	72-0111
OHYA, H. G.	SNAP-THROUGH BUCKLING ANALYSIS OF SHELLS OF REVOLUTION USING INCREMENTAL STIFFNESS MATRICES	74-1349
OHYAMA, H.	ANALYSIS OF TENSION STRUCTURES BY MATRIX METHODS	67-0119
OJALVO, I. U.	IMPROVED THERMAL STRESS DETERMINATION BY FINITE ELEMENT METHODS	74-0247
OJALVO, I. U.	STRESS AND DEFLECTION ANALYSIS OF MECHANICALLY FASTENED JOINTS	70-0067
OJALVO, I. U.	VIBRATION AND STRESS ANALYSIS OF SOFT-BONDED SHUTTLE INSULATION TILES. MODAL ANALYSIS WITH COMPACT W	74-0708
OKABAYASHI, K.	THE APPLICATION OF THE FINITE ELEMENT METHOD TO THE ANALYSIS OF FRACTURE OF CYLINDRICAL SHELLS	73-0728
OKABE, T.	ON THE DEVELOPMENT OF THE "PASSAGE" PROGRAM	73-0342
OKAMOTO, T.	FINITE ELEMENT ANALYSIS OF VISCOUS FLOW BY STREAM FUNCTION	75-0526
OKAMURA, H.	ERROR ANALYSIS OF VARIOUS FINITE ELEMENT PATTERNS	73-0510
OKUBO, N.	COMPUTER AIDED DESIGN OF MACHINE TOOL STRUCTURE	74-1189
OKUBO, N.	COMPUTERUNTERSTUETZTES SYSTEM FUER DIE CONSTRUKTION DES WERKZEUGMASCHINENAUFBAUS (COMPUTER-AIDED SYS	73-0482
OKUBO, Y.	ANALYSIS OF VEHICLE BODY STRUCTURE BY THE FINITE ELEMENT METHOD	71-0324
OKUDA, S.	FINITE ELEMENT SIMULATION OF ENERGY-TRAPPED ELECTROMECHANICAL RESONATORS	75-0142
OKUMURA, T.	ELASTIC AND PLASTIC STRESS DISTRIBUTION ON BASE PLATES IN THE NEIGHBOURHOOD OF FILLET WELDS IN LAP J	70-0175
OKUSHIMA, K.	RESIDUAL STRESS PRODUCED BY METAL CUTTING	71-0414
OLINE, L.	ELASTIC PLATE SPALLATION	72-0682
OLINE, L. W.	INVESTIGATION OF ELASTIC PLATE SPALLATION BY FINITE ELEMENTS	74-0382
OLIVER, L. R.	V-BELT LIFE PREDICTION AND POWER RATING	75-0838
OLIVER, W. A.	FINITE ELEMENT ANALYSIS OF THE PIEZOELECTRIC TOPOGRAPHIC WAVEGUIDE	74-1111
OLSEN, M. D.	FINITE ELEMENTS FOR PLATE BENDING	70-0325
OLSON, D. R.	CONDUCTION THROUGH DROPLETS DURING DROPWISE CONDENSATION	72-0019
OLSON, J. P.	PERFORMANCE OF EARTH AND EARTH-SUPPORTED STRUCTURES	72-0449
OLSON, M.	ANALYSIS OF ARBITRARY SHELLS USING SHALLOW SHELL FINITE ELEMENTS	71-0717
OLSON, M. D.	A CONSISTENT FINITE ELEMENT METHOD FOR RANDOM RESPONSE PROBLEMS	71-0202
OLSON, M. D.	A CONSISTENT FINITE ELEMENT METHOD FOR RANDOM RESPONSE PROBLEMS	72-0896
OLSON, M. D.	A FINITE CYLINDRICAL SHELL ELEMENT AND THE VIBRATIONS OF A CURVED FAN BLADE	68-0131
OLSON, M. D.	A HIGH PRECISION TRIANGULAR PLATE BENDING ELEMENT	68-0052
OLSON, M. D.	A HIGH-PRECISION TRIANGULAR CYLINDRICAL SHELL FINITE ELEMENT	71-0163
OLSON, M. D.	A HYBRID FINITE ELEMENT METHOD FOR TWO-DIMENSIONAL NAVIER STOKES EQUATIONS	75-0027
OLSON, M. D.	A SHALLOW SHELL FINITE ELEMENT OF TRIANGULAR SHAPE	70-0044
OLSON, M. D.	A VARIATIONAL FINITE ELEMENT METHOD FOR TWO DIMENSIONAL STEADY VISCOUS FLOWS	72-0355
OLSON, M. D.	ANALYSIS OF ARBITRARY SHELLS USING SHALLOW SHELL FINITE ELEMENTS	74-1374
OLSON, M. D.	ANNULAR AND CIRCULAR SECTOR FINITE ELEMENTS FOR PLATE BENDING	70-0100
OLSON, M. D.	APPLICATION OF THE HIGH PRECISION TRIANGULAR PLATE BENDING ELEMENT TO PROBLEMS WITH CURVED BOUNDARIE	69-0071
OLSON, M. D.	CLOSED FORM, FINITE ELEMENT SOLUTIONS FOR PLATE VIBRATIONS	69-0168
OLSON, M. D.	COMPARISON OF TWO HIGH-PRECISION TRIANGULAR FINITE ELEMENTS FOR ARBITRARY DEEP SHELLS	71-0071
OLSON, M. D.	COMPATIBILITY	75-0781
OLSON, M. D.	CONVERGENCE STUDIES OF EIGENVALUE SOLUTIONS USING TWO FINITE PLATE ELEMENTS	70-0085
OLSON, M. D.	DYNAMIC ANALYSIS OF SHALLOW SHELLS WITH A DOUBLY-CURVED TRIANGULAR FINITE ELEMENT	71-0353
OLSON, M. D.	FINITE ELEMENT DYNAMIC ANALYSIS OF SHALLOW SHELL STRUCTURES	70-0164
OLSON, M. D.	FINITE ELEMENTS APPLIED TO PANEL FLUTTER	67-0066
OLSON, M. D.	FINITE PLATE-BENDING ELEMENTS IN POLAR COORDINATES	68-0234
OLSON, M. D.	FORMULATION OF A VARIATIONAL PRINCIPLE - FINITE ELEMENT METHOD FOR VISCOUS FLOW	72-0354
OLSON, M. D.	FREE VIBRATIONS AND RANDOM RESPONSE OF AN INTEGRALLY-STIFFENED PANEL	70-0138
OLSON, M. D.	JET NOISE EXCITATION OF AN INTEGRALLY STIFFENED PANEL	71-0546
OLSON, M. D.	NEW DEVELOPMENTS IN THE FINITE ELEMENT ANALYSIS OF SHELLS	69-0334
OLSON, M. D.	ON APPLYING FINITE ELEMENTS TO PANEL FLUTTER	67-0135
OLSON, M. D.	SOME FLUTTER SOLUTIONS USING FINITE ELEMENTS	70-0011
OLSON, M. D.	STATIC AND DYNAMIC APPLICATIONS OF A HIGH-PRECISION TRIANGULAR PLATE BENDING ELEMENT	69-0086
OLSON, M. D.	STRUCTURAL DESIGN AND ANALYSIS OF THE SPACE SHUTTLE	70-0336
OLSON, M. D.	THE VIBRATION OF STEPPED CANTILEVERED PLATES	67-0129
OLSON, M. D.	VARIATIONAL FINITE ELEMENT METHODS FOR TWO DIMENSIONAL AND AXISYMMETRIC NAVIER STOKES EQUATIONS	74-0012
OLSON, M. D.	VIBRATION ANALYSIS OF CANTILEVERED CURVED PLATES USING A NEW CYLINDRICAL SHELL FINITE ELEMENT	68-0130
OLSON, M. D.	VIBRATION MODES AND RANDOM RESPONSE OF A MULTI-BAY PANEL SYSTEM USING FINITE ELEMENTS	67-0128
OLSON, R. E.	STRESS-STRAIN PROPERTIES OF COMPACTED CLAYS	74-0650
OMOTE, T.	FINITE ELEMENT SIMULATION OF ACOUSTIC FILTERS WITH VARIABLE CIRCULAR CROSS SECTION	75-0871
ON, F. J.	MATHEMATICAL MODELING VIA DIRECT USE OF VIBRATION DATA	69-0310
ONAT, E. T.	SLOW FINITE DEFORMATIONS OF VISCOUS SOLIDS	69-0423
ONDRACEK, E.	A NUMERICAL METHOD FOR THE ANALYSIS OF LONGITUDINAL ELASTIC PLASTIC STRESS WAVE PROPAGATION	73-0526
ONEILL, K.	PHOTOELASTIC AND FINITE-ELEMENT ANALYSIS OF A QUADRIPARTITE VAULT	73-0119
ONG, J.	AUTOMATIC MASTERS FOR EIGENVALUE ECONOMIZATION	75-0557
OPSTEEGH, J. D.	THE PREDICTION OF REGULAR WAVE FORCES ON FLOATING BODIES WITH THE FINITE ELEMENT METHOD (IN DUTCH)	71-0743
OPTIZ, H.	EVALUATION OF COMPUTER AIDED METHODS FOR THE DESIGN OF MACHINE TOOL STRUCTURES	73-0577
ORLIN, A. S.	RASCHET POLEI DEFORMATSII I NAPRYAZHENII V DETALYAKH DVIGATELEI VNUTRENNEGO SGORANIYA (CALCULATION O	74-0465
ORLOB, G. T.	A FINITE ELEMENT SOLUTION FOR TWO-DIMENSIONAL DENSITY STRATIFIED FLOW	73-0911
ORME, D. H.	CONSTRUCTION SELF-WEIGHT STRESSES OF PRECAST SHEAR WALLS	71-0203
ORR, F. M.	MENISCI IN ARRAYS OF CYLINDERS - NUMERICAL-SIMULATION BY FINITE-ELEMENTS	75-0449
ORRINGER, O.	APPLICATIONS OF THE ASSUMED-STRESS HYBRID METHOD TO FINITE ELEMENT ANALYSIS IN FRACTURE MECHANICS	75-0495
ORRINGER, O.	FEABL (FINITE ELEMENT ANALYSIS BASIC LIBRARY) USER'S GUIDE	72-0748
ORRIS, R. M.	A FINITE ELEMENT STUDY OF THE VIBRATION OF SKIN-RIB STRUCTURES	74-0537
ORRIS, R. M.	FINITE ELEMENT STUDY OF VIBRATION OF TRAPEZOIDAL PLATES	73-0035
ORRIS, R. M.	FINITE ELEMENT STUDY OF HARMONIC WAVE PROPAGATION IN PERIODIC STRUCTURES	74-0130
ORRIS, R. M.	RANDOM RESPONSE OF PERIODIC STRUCTURES BY A FINITE-ELEMENT TECHNIQUE	75-0426
ORTEGA, M. A.	APPLICATION OF THE STIFFNESS METHOD TO A SNAP-THROUGH PROBLEM	60-0006
ORTLEPP, W. D.	FAILURE OF THE CONCRETE LINING IN AN INCLINED SHAFT AT GREAT DEPTH.	74-0996
OSAKADA, K.	ANALYSIS OF HYDROSTATIC EXTRUSION BY THE FINITE ELEMENT METHOD	72-0020
OSBORN, J. E.	APPROXIMATION OF STEKLOV EIGENVALUES OF NON-SELFADJOINT SECOND ORDER ELLIPTIC OPERATORS	72-0509
OSHIER, E. H.	A DEEP HOLE STRESS MEASUREMENT DEVICE	74-0804
OSIAS, J. R.	FINITE ELASTO-PLASTIC DEFORMATION - 1. THEORY AND NUMERICAL EXAMPLES	74-0202
OSIAS, J. R.	FINITE ELEMENT ANALYSIS OF TRANSIENT NONLINEAR STRUCTURAL BEHAVIOR	75-0297
OSIAS, J. R.	USER'S GUIDE FOR ANALYSIS OF FINITE ELASTOPLASTIC DEFORMATION: THE FIPDEF AND FIPAX PROGRAMS FOR THE	74-0799
OSTERBERG, J. O	LOAD TRANSFER MECHANISM FOR PIERS SOCKETTED IN HARD SOILS OR ROCK	74-0969
OSTERGREN, W. J	SMALL SCALE YIELDING NEAR A CRACK IN PLANE STRAIN-FINITE ELEMENT ANALYSIS	71-0355
OSTERMANN, H.	FORM FACTORS OF FLAT BARS FOR FATIGUE STRENGTH TESTS. (FORMZAHLEN VON FLACHSTAEBEN FUER SCHWINGFESTI	71-0641
OSTROFF, A. J.	ANALYSIS OF THE DYNAMICS OF THIN PRIMARY MIRRORS FOR LARGE ASTRONOMICAL TELESCOPES	73-1084
OTSUBO, H.	A METHOD OF FINITE DEFLECTION ELASTIC PLASTIC ANALYSIS OF FRAMED STRUCTURES SUBJECTED TO LARGE AXIAL	69-0239
OTTAVIANI, M.	THREE-DIMENSIONAL FINITE ELEMENT ANALYSIS OF VERTICALLY LOADED PILE GROUPS	75-0239
OTTENS, H. H.	COMPLIANCE OF A TAPERED DCB SPECIMEN - CONFIGURATIONS BY A FINITE ELEMENT METHOD	74-0545

OTTENS, H. H.	FINITE ELEMENT ANALYSIS OF SANDWICH PANELS	70-0449
OTTENS, H. H.	FINITE ELEMENT ANALYSIS OF SANDWICH PANELS	71-0662
OTTOSEN, N. S.	ULTIMATE LOAD BEHAVIOUR OF PCRV TOP CLOSURES. THEORETICAL AND EXPERIMENTAL INVESTIGATIONS	73-0835
OTTOVIANI, M.	STRESSES AND DISPLACEMENTS AROUND TWO ADJACENT CIRCULAR OPENINGS NEAR TO THE GROUND SURFACE	74-0993
OUDENHOVEN, M.	A METHOD FOR THE PREDICTION OF STRESSES IN AN ISOTROPIC INCLUSION OR OREBODY OF IRREGULAR SHAPE	72-0827
OVEROYE, K. R.	HOLOGRAPHIC INSTRUMENTATION OF TURBINE BLADES	73-0466
OVUNC, B.	AN ITERATION METHOD FOR THE NONLINEAR BUCKLING OF FRAME STRUCTURES	66-0107
OVUNC, B.	NONLINEAR ANALYSIS OF ELASTIC SHELLS OF REVOLUTION UNDER AXISYMMETRIC LOADING	70-0161
OVUNC, B. A.	ANALYSIS OF BUILDINGS ON ELASTIC MEDIUM UNDER DYNAMIC OR SEISMIC LOADS	74-0456
OWEN, D. R. J.	ANALYSIS OF VISCOPLASTIC EFFECTS IN PRESSURE VESSELS BY THE FINITE ELEMENT METHOD	74-0573
OWEN, D. R. J.	DYNAMIC TRANSIENT NON-LINEAR BEHAVIOUR OF THICK AND THIN PLATES	75-0125
OWEN, D. R. J.	ELASTIC-PLASTIC STRESS ANALYSIS FOR A NOTCHED BAR IN PLANE STRAIN BENDING	71-0410
OWEN, D. R. J.	FESS AND FINESSE PROGRAMS	72-0158
OWEN, D. R. J.	FINITE ELEMENT ANALYSIS OF THE STEADY FLOW OF NON NEWTONIAN FLUIDS THROUGH PARALLEL SIDED CONDUITS	74-0056
OWEN, D. R. J.	FINITE ELEMENT ANALYSIS OF ELASTO-PLASTIC MATERIALS BY USE OF DISLOCATION DIPOLE SYSTEMS	74-0269
OWEN, D. R. J.	FINITE ELEMENT ANALYSIS OF NON-LINEAR COMPOSITE MATERIALS BY USE OF OVERLAY SYSTEMS	74-0928
OWEN, D. R. J.	FINITE ELEMENT ANALYSIS OF ENGINEERING SYSTEMS GOVERNED BY A NON-LINEAR QUASI-HARMONIC EQUATION	75-0664
OWEN, D. R. J.	FINITE ELEMENT METHOD IN ANALYSIS OF REINFORCED AND PRESTRESSED CONCRETE STRUCTURES	71-0270
OWEN, D. R. J.	FINITE ELEMENT METHODS IN ANALYSIS OF REACTOR VESSELS	72-0271
OWEN, D. R. J.	INVESTIGATION OF BOND FAILURE IN FIBRE-REINFORCED MATERIALS BY THE FINITE ELEMENT METHOD	72-0034
OWEN, D. R. J.	LEAST SQUARE-FINITE ELEMENT FOR ELASTO-STATIC PROBLEMS	74-0279
OWEN, D. R. J.	SHORT COMMUNICATIONS: TORSION OF AXI-SYMMETRIC SOLIDS OF VARIABLE DIAMETER - INCLUDING ACCELERATION	74-0203
OWEN, D. R. J.	STRESS INTENSITY FACTORS FOR CRACKS IN A PLATE CONTAINING A HOLE AND IN A SPINNING DISC	73-0467
OWEN, D. R. J.	STRESSES IN A PARTLY YIELDED NOTCHED BAR - AN ASSESSMENT OF THREE ALTERNATIVE PROGRAMS	73-0267
OWEN, D. R. J.	STRESSES IN A PARTLY YIELDED NOTCHED BAR EM DASH AN ASSESSMENT OF THREE ALTERNATIVE PROGRAMS	73-0268
OWEN, D. R. J.	THREE-DIMENSIONAL ELASTO-PLASTIC FINITE ELEMENT ANALYSIS	75-0236
OWEN, D. R. J.	YIELDING OF A TWO DIMENSIONAL VOID ASSEMBLY IN AN ORGANIC GLASS	73-0416
OWEN, M. J.	PREDICTION OF STATIC AND FATIGUE DAMAGE AND CRACK PROPAGATION IN COMPOSITE MATERIALS	75-0643
OWEN, M. J.	STRESS ANALYSIS OF LAMINATED CFRP PLATES CONTAINING A HOLE	74-1175
OWENS, D. R. J.	FINITE ELEMENT ANALYSIS OF SOME CONCRETE NONLINEARITIES. THEORY AND EXAMPLES	74-1287
OWENS, G. E.	MECHANICS OF CABLE MOORING SYSTEMS. VOLUME 1. THREE DIMENSIONAL RESPONSE OF DEEP WATER MOORING LINES	72-0654
OWENS, R. H.	APPLICATIONS OF NUMERICAL ANALYSIS	71-0755
OZAWA, Y.	ELASTO-PLASTIC FINITE ELEMENT ANALYSIS OF SOIL DEFORMATION	73-1157
OZDEMIR, H.	STATIC AND DYNAMIC GEOMETRIC AND MATERIAL NONLINEAR ANALYSIS	74-1302
PADLOG, J.	DISCRETE ELEMENT APPROACH TO STRUCTURAL INSTABILITY ANALYSIS	63-0005
PADLOG, J.	DISCRETE ELEMENT PROCEDURE FOR THIN SHELL INSTABILITY ANALYSIS	67-0028
PADLOG, J.	THE INELASTIC BEHAVIOUR OF STRUCTURES SUBJECTED TO CYCLIC, THERMAL AND MECHANICAL STRESSING CONDITIO	60-0004
PADOVAN, J.	QUASI-ANALYTICAL FINITE ELEMENT PROCEDURES FOR AXISYMMETRIC ANISOTROPIC SHELLS AND SOLIDS	74-0297
PADOVAN, J.	SEMI-ANALYTICAL FINITE ELEMENT PROCEDURE FOR CONDUCTION IN ANISOTROPIC AXISYMMETRIC SOLIDS	74-0270
PADOVAN, J.	STEADY CONDUCTION OF HEAT IN LINEAR AND NONLINEAR FULLY ANISOTROPIC MEDIA BY FINITE ELEMENTS	74-0359
PADUART, A.	CONTINUOUS CONCEPT OF A SPACE GRID	73-0755
PAGAY, S. N.	FINITE ELEMENT FOR REINFORCED CONCRETE FRAME STUDY - SPACE DISCUSSION	74-0246
PAGE, E. A.	STUDY OF NUMERICAL METHODS IN SEISMIC PROBLEMS	72-0738
PAGENKOPF, J. R	EVALUATION OF TECHNIQUES FOR NUMERICAL CALCULATION OF STORM SURGES	75-0554
PAGENKOPF, J. R	NUMERICAL CALCULATION OF STORM SURGES: AN EVALUATION OF TECHNIQUES	75-0461
PAI, D. H.	INELASTIC ANALYSIS OF A STRAIGHT TUBE UNDER COMBINED BENDING, PRESSURE AND THERMAL LOADS	75-0218
PAI, D. H.	INELASTIC ANALYSIS OF A STRAIGHT TUBE UNDER COMBINED BENDING, PRESSURE AND THERMAL LOADS	75-0513
PAI, D. H.	INELASTIC ANALYSIS OF TUBESHEETS BY THE FINITE ELEMENT METHOD	75-0652
PAIDOUSSIS, M.	DYNAMICAL ANALYSIS OF AXIALLY NON-UNIFORM, THIN CYLINDRICAL SHELLS. PART 1 - MATRIX FORMULATION	70-0413
PAIDOUSSIS, M.	FREE VIBRATION OF CYLINDRICAL SHELLS PARTIALLY FILLED WITH LIQUID	71-0839
PAINE, A. A.	DEVELOPMENT AND APPLICATIONS OF SUPERSONIC UNSTEADY CONSISTENT AERODYNAMICS FOR INTERFERING PARALLEL	72-0716
PAKSTYS, M. JP.	DYNAMIC SUBSTRUCTURES METHOD FOR SHOCK ANALYSIS	68-0285
PALACOL, E. L.	ANISOTROPIC PARAMETRIC PLATE DISCRETE ELEMENTS	73-1047
PALACOL, E. L.	MANIPULATION ERRORS IN FINITE ELEMENT ANALYSIS OF STRUCTURES	69-0349
PALACOL, E. L.	STRESS-WAVE PROPAGATION AND DYNAMIC EFFECTS IN SHELLS (SPADES) MICROMECHANICAL STRESS-WAVE RESPONSE	72-0851
PALEKAR, M. G.	NUMERICAL SOLUTION OF 2-POINT BOUNDARY VALUE PROBLEMS	74-1141
PALIT, K.	FINITE ELEMENT ANALYSIS OF TWO-DIMENSIONAL SLOW NON-NEWTONIAN FLOWS	72-0035
PALIT, K.	FINITE ELEMENT ANALYSIS OF SLOW NON-NEWTONIAN CHANNEL FLOW	72-0308
PALKA, L. C.	INVESTIGATION OF VENEER CUTTING BY ELASTIC FINITE ELEMENT MODELS	74-1110
PALMBERG, J. O.	ON THERMO-ELASTO-HYDRODYNAMIC FLUID FILM BEARINGS	75-0532
PALMER, J. H.	CALCULATION OF AXISYMMETRIC LONGITUDINAL MODES FOR FLUID-ELASTIC TANK-ULLAGE GAS SYSTEMS AND COMPARI	64-0019
PALMERTON, J. B	APPLICATION OF FINITE ELEMENT METHOD IN DETERMINING STABILITY OF CRATER SLOPES, PRELIMINARY REPORT	68-0026
PALMERTON, J. B	APPLICATION OF FINITE ELEMENT METHOD IN DETERMINING STABILITY OF CRATER SLOPES	72-0888
PALMERTON, J. B	PRELIMINARY FINITE ELEMENT ANALYSIS, ATCHAFALAYA BASIN PROTECTION LEVEES, TEST SECTION 3	69-0463
PALMERTON, J. B	THREE-DIMENSIONAL BEHAVIOR OF A CENTRAL CORE DAM	72-0714
PALMIERI, I.	DESCRIPTION AND INSTRUCTION FOR USE OF THE SAFE-SHELL. B FINITE ELEMENT PROGRAM IN THE IBM/360 VERSI	73-0945
PALMISANO, P. R	NORMAL MODE ANALYSIS OF A FUZE SUPPORT STRUCTURE USING NASTRAN. PART 1	72-0853
PAN, Y. S.	THREE-DIMENSIONAL STRESS ANALYSES OF MITERED ELBOWS	75-0650
PANAK, J. J.	A DISCRETE-ELEMENT METHOD OF MULTIPLE-LOADING ANALYSIS FOR TWO-WAY BRIDGE FLOOR SLABS	70-0435
PANAK, J. J.	A DISCRETE-ELEMENT METHOD OF ANALYSIS FOR ORTHOGONAL SLAB AND GRID BRIDGE FLOOR SYSTEMS	72-0690
PANAK, J. J.	DEVELOPMENT OF METHODS FOR COMPUTER SIMULATION OF BEAM-COLUMNS AND GRID-BEAM AND SLAB SYSTEMS	73-1101
PANDARINATHAN,	COMPARISON OF SOME FINITE-ELEMENT MODELS FOR CIRCULAR CYLINDRICAL SHELLS	73-1158
PAO, C.	MISSILE BODY INPUT GENERATOR (BING), A NASTRAN PRE-PROCESSOR; THEORETICAL DEVELOPMENT, USER'S MANUAL	75-0717
PAO, Y. C.	FINITE ELEMENT ANALYSIS OF LEFT VENTRICULAR MYOCARDIAL STRESSES	74-0669
PAO, Y. C.	ON COMPUTATIONS INVOLVING STIFFNESS MATRICES STORED IN RECTANGULAR FORM	75-0576
PAPADOPULOUS, S	FUNCTIONAL COEFFICIENTS IN THE ANALYSIS OF GROUNDWATER FLOW USING FINITE ELEMENTS	73-0077
PAPAIOANNOU, S.	FINITE ELEMENT METHOD FOR CALCULATING STRESS INTENSITY FACTORS AND ITS APPLICATION TO COMPOSITES	74-1000
PAPANTONOPOULOS	ANALYSE DE LA STABILITY DES TALUS ROCHEUX PAR UNE METHODE GENERALISEE DE L EQUILIBRE LIMITE (ROCK SL	74-1019
PARAMASIVAN, P.	DISCRETE METHODS IN THE ANALYSIS OF PLATE SYSTEMS	71-0835
PARDOEN, G. C.	ASYMMETRIC BENDING OF CIRCULAR PLATES USING THE FINITE ELEMENT METHOD	75-0175
PARDOEN, G. C.	DEFLECTION FUNCTION FOR THE ASYMMETRICAL BENDING OF CIRCULAR PLATES	73-0468
PARDOEN, G. C.	STATIC AND DYNAMIC BUCKLING OF THIN-WALLED COLUMNS USING FINITE ELEMENTS	70-0505
PARDOEN, G. C.	STATIC, VIBRATION AND BUCKLING ANALYSIS OF AXISYMMETRIC CIRCULAR PLATES USING FINITE ELEMENTS	73-0157
PARDOEN, G. C.	SYMMETRICAL BENDING OF CIRCULAR PLATES USING FINITE ELEMENTS	72-0423
PARDOEN, G. C.	VIBRATION AND BUCKLING ANALYSIS OF AXISYMMETRIC POLAR ORTHOTROPIC CIRCULAR PLATES	74-0574
PARDUE, R. M.	ELEMENT RESEQUENCING FOR FRONTAL SOLUTIONS	75-0132
PAREKH, C.	THERMAL STRESS AND TEMPERATURE FIELDS - A FINITE-ELEMENT APPROACH	71-0782
PAREKH, C. J.	A FINITE ELEMENT SOLUTION OF TIME DEPENDENT FIELD PROBLEMS	67-0155
PAREKH, C. J.	ANALYSIS OF SLABS WITH EDGE BEAMS	71-0075
PAREKH, C. J.	APPLICATION OF FINITE ELEMENTS TO HEAT CONDUCTION PROBLEMS INVOLVING LATENT HEAT	73-0520
PAREKH, C. J.	ARCH DAMS ANALYSED BY A LINEAR FINITE ELEMENT SHELL THE FINITE STRIP METHOD IN THE ANALYSIS OF ELAST	68-0189
PAREKH, C. J.	FINITE ELEMENT PROGRAMS FOR SLAB BRIDGE DESIGN	71-0082
PAREKH, C. J.	FINITE ELEMENT SOLUTION SYSTEM	69-0358
PAREKH, C. J.	NUMERICAL ANALYSIS OF FREE SURFACE SEEPAGE PROBLEMS	71-0469
PAREKH, C. J.	THREE DIMENSIONAL ANALYSIS OF BUILDINGS COMPOSED OF FLOOR AND WALL PANELS	71-0387
PAREKH, C. J.	TRANSIENT FIELD PROBLEMS: TWO-DIMENSIONAL AND THREE-DIMENSIONAL ANALYSIS BY ISO-PARAMETRIC FINITE EL	70-0001
PAREKH, R. J.	STARS - A COMPUTER CODE FOR STRUCTURAL DYNAMIC ANALYSIS OF REACTOR SYSTEM	71-0204
PARFITT, V. R.	FINITE ELEMENT STRESS ANALYSIS OF THIN SHELLS	65-0024
PARIKH, S. K.	ANALYSIS OF EARTHEN DAMS BY THE METHOD OF FINITE ELEMENTS	70-0454
PARIS, P. C.	EFFICIENT FINITE ELEMENT METHODS FOR STRESS INTENSITY FACTOR	75-0148
PARIS, P. C.	EFFICIENT FINITE-ELEMENT METHODS FOR STRESS INTENSITY FACTORS USING WEIGHT FUNCTIONS - REPLY	75-0421
PARISEAU, W. G.	FINITE ELEMENT ANALYSES OF ELASTIC-PLASTIC PROBLEMS IN THE MECHANICS OF GEOLOGIC MEDIA: AN OVERVIEW	70-0203

PARISEAU, W. G.	GRAVITY FLOW BIN DESIGN FOR LUNAR SOIL	72-0240
PARISEAU, W. G.	GRAVITY FLOW OF POWDER IN A LUNAR ENVIRONMENT. PT. 2-ANALYSIS OF FLOW INITIATION	71-0484
PARISEAU, W. G.	INFLUENCE OF ROCK PROPERTIES VARIABILITY ON MINE OPENING STABILITY ANALYSIS	74-1021
PARISEAU, W. G.	ROCK MECHANICS AND RISK IN OPEN PIT MINING	73-0544
PARK, A. C.	FINITE ELEMENT SOLUTION OF LOW BOND NUMBER SLOSHING	75-0728
PARK, K. C.	AN INTERACTIVE HYBRID TECHNIQUE FOR CRASHWORTHY DESIGN OF COMPLEX VEHICULAR STRUCTURAL SYSTEMS	74-0317
PARK, K. C.	TRANSIENT RESPONSE OF INELASTICALLY CONSTRAINED RIGID BODY SYSTEMS	73-0291
PARK, S. W.	FINITE ELEMENT STRUCTURAL ANALYSIS AS APPLIED TO AN AUTOMOTIVE DOOR STRUCTURE	74-0310
PARKER, J. V.	PRACTICAL APPLICATIONS OF FINITE ELEMENT METHODS IN THE STRESS ANALYSIS OF THIN SHELL PRESSURE VESSE	71-0205
PARKING, A. K.	FIELD SOLUTIONS FOR TURBULENT SEEPAGE FLOW	71-0206
PARKS, D. M.	ON NUMERICALLY ACCURATE FINITE ELEMENT SOLUTIONS IN THE FULLY PLASTIC RANGE	74-0529
PARKS, D. M.	ON NUMERICALLY ACCURATE FINITE ELEMENTS	74-0797
PARKS, D. M.	STIFFNESS DERIVATIVE FINITE ELEMENT TECHNIQUE FOR DETERMINATION OF CRACK TIP STRESS INTENSITY FACTOR	74-0956
PARMELEE, R. A.	ANALYTICAL AND EXPERIMENTAL EVALUATION OF MODULUS OF SOIL REACTION	74-1168
PARMERTER, R. R	STRESS INTENSITY FACTORS FOR A NOTCHED BEAM ON AN ELASTIC FOUNDATION	75-0566
PARR, C. H.	ANALYSIS OF THE BIAXIAL STRIP AND SHEAR LAP TESTS FOR SOLID PROPELLANT CHARACTERIZATION	67-0019
PARR, C. H.	APPLICATION OF THE FINITE ELEMENT METHOD TO HEAT CONDUCTION IN SOLIDS	67-0005
PARR, C. H.	THE APPLICATION OF FINITE ELEMENT METHODS TO THE SOLUTION OF STRESS AND DIFFUSION PROBLEMS IN CONTIN	68-0034
PARRISH, D.	NONLINEAR FINITE ELEMENT FOLD MODEL	73-0014
PARSONS, B.	A METHOD FOR DETERMINING THE SURFACE CONTACT STRESSES RESULTING FROM INTERFERENCE FITS	70-0101
PARSONS, B.	ANALYSIS OF AXISYMMETRIC COMPOSITE BODIES BY THE FINITE ELEMENT METHOD	68-0242
PARSONS, B.	FINITE ELEMENT ANALYSIS OF ELASTIC CONTACT PROBLEMS USING DIFFERENTIAL DISPLACEMENTS	70-0126
PARSONS, B.	RADIAL DISPLACEMENTS OF A HOLLOW SHAFT SUBJECTED TO A UNIFORM BAND OF EXTERNAL PRESSURE	72-0242
PARSONS, B.	TRAPEZOIDAL FINITE ELEMENTS: THEIR DERIVATION AND USE FOR AXISYMMETRIC ROTATING BODIES	73-0192
PASHA, M. A.	CONTROLLING OPEN-PIT SLOPE FAILURES AT SHIRLEY BASIN	73-0149
PASKALOV, T.	STATIC AND DYNAMIC ANALYSIS OF AN ARCH DAM	74-1101
PASSERELLO, C.	AN ANALYSIS OF GENERAL CHAIN SYSTEMS	72-0787
PASSERELLO, C.	ON CONSTRAINT EQUATIONS - A NEW APPROACH	74-1017
PASTERNACK, S.	ANALYTICAL FINITE ELEMENT SIMULATION MODEL FOR STRUCTURAL CRASHWORTHINESS PREDICTION	73-0778
PATEL, J. S.	COMPLEX EIGENVALUE SOLUTION TO A SPINNING SKYLAB PROBLEM	71-0207
PATEL, K. S.	COMPUTER - AIDED DESIGN - STANDARD OIL OF INDIANA BUILDING	73-0182
PATEL, M. R.	ROCK FRAGMENTATION BY SUBSURFACE THERMAL INCLUSIONS - A FINITE ELEMENT STUDY	73-1159
PATENT, P. D.	LEAST SQUARE POLYNOMIAL SPLINE APPROXIMATION	72-0526
PATHAK, D. V.	ANALYSIS OF MODERATELY THICK FLANGED REACTOR VESSELS	75-0320
PATIL, B. S.	HARBOUR OSCILLATION - A NUMERICAL TREATMENT FOR UNDAMPED NATURAL MODES	70-0208
PATIL, S. S.	SPECIAL FINITE ELEMENTS FOR FLAW AND DISCONTINUITY STRESSES IN COMPOSITES	73-0846
PATRICK, G. E.	FINITE DEFLECTIONS, BUCKLING AND POSTBUCKLING OF AN ARCH	69-0123
PATTERSON, C.	A CLASS OF FUNCTIONALS GIVING IMPROVED CONVERGENCE WITH FINITE ELEMENTS	72-0553
PATTERSON, C.	AN EVALUATION OF FINITE ELEMENTS WITH IMPROVED ASSURED CONVERGENCE FOR PLANE STRESS AND PLANE STRAIN	74-0429
PATTERSON, C.	IMPROVED FINITE ELEMENT METHOD FOR STRESS CONCENTRATIONS	73-0127
PATTERSON, C.	SUFFICIENT CONDITIONS FOR CONVERGENCE IN THE FINITE ELEMENT METHOD FOR ANY SOLUTION OF FINITE ENERGY	72-0065
PATTERSON, G. J	FINITE ELEMENT ANALYSIS OF HYPERBOLOIDAL PRESTRESSED CONCRETE CONTAINMENT VESSELS	69-0526
PATTERSON, J. F	APPLICATION OF FINITE ELEMENT METHODS FOR THE ANALYSIS OF THERMAL CREEP, IRRADIATION INDUCED CREEP,	71-0242
PATTERSON, J. F	DESIGN ANALYSIS FOR LMFBR CORE RESTRAINT	71-0208
PATTON, P. C.	A LOOK INTO THE FUTURE: HOW COMPUTERS WILL INFLUENCE ENGINEERING	67-0139
PATTON, P. C.	THE MATRIX EIGENVALUE PROBLEM FOR DAMPED STRUCTURAL VIBRATION	66-0121
PAUL, B.	FINITE ELEMENT METHOD IN STRESS ANALYSIS	71-0576
PAUL, B.	SHIFTING ROLES OF COMPUTERS, EXPERIMENTS AND ANALYSIS IN APPLIED MECHANICS	71-0604
PAUL, B.	STRESS ANALYSIS OF PRESSURIZED CYLINDRICAL BORE IN RECTANGULAR BLOCK	69-0312
PAUL, S. L.	FINITE ELEMENT SOLUTION OF NORMALLY INTERSECTING CYLINDERS	73-0723
PAUL, S. L.	RESEARCH TO IMPROVE TUNNEL SUPPORT SYSTEMS	74-0777
PAUL, S. L.	STUDIES TO IMPROVE TUNNEL SUPPORT SYSTEMS	74-0957
PAULEY, K. E.	ELAD PROGRAM MANUAL, A COMPUTER PROGRAM FOR THE ELASTIC ANALYSIS OF DOME STRUCTURES AND AXISYMMETRIC	69-0064
PAULLING, J. R.	APPLICATION OF FINITE ELEMENT COMPUTATIONS TO SOME SHIP STRUCTURAL PROBLEMS	69-0337
PAULLING, J. R.	APPLICATIONS OF FINITE ELEMENT METHOD TO SHIP STRUCTURAL ANALYSIS	69-0210
PAULLING, J. R.	HULL-DECKHOUSE INTERACTION BY FINITE ELEMENT CALCULATIONS	68-0132
PAULLING, J. R.	THE ANALYSIS OF COMPLEX SHIP STRUCTURES BY THE FINITE ELEMENT TECHNIQUE	64-0006
PAULSEN, W. C.	FINITE ELEMENT STRESS ANALYSIS, PART 2	71-0209
PAULSEN, W. C.	FINITE ELEMENT STRESS ANALYSIS PART 1	71-0440
PAWSEY, S. F.	IMPROVED NUMERICAL INTEGRATION OF THICK SHELL FINITE ELEMENTS	71-0396
PAWSEY, S. F.	THE ANALYSIS OF MODERATELY THICK TO THIN SHELLS BY THE FINITE ELEMENT METHOD	70-0315
PAYER, H. G.	DESIGN, ANALYSIS AND CONSTRUCTION OF THE RESEARCH PLATFORM (NORDSEE)	75-0261
PAYER, H. G.	HULL-DECKHOUSE INTERACTION BY FINITE ELEMENT CALCULATIONS	68-0132
PAYER, H. G.	TRAGLAST VON T-TRAEGERN. EXPERIMENT UND F.E. BERECHNUNGEN (ULTIMATE STRENGTH OF T-BEAMS. EXPERIMENTS	74-0935
PAZ, M.	MATHEMATICAL OBSERVATIONS IN STRUCTURAL DYNAMICS	73-0156
PAZ, M.	POWER SERIES EXPANSION OF THE GENERAL STIFFNESS MATRIX FOR BEAM ELEMENTS	75-0543
PEAKE, D. J.	TECHNICAL EVALUATION REPORT ON AGARD SPECIALISTS	71-0737
PEANO, A. G.	NODAL VARIABLES FOR ARBITRARY ORDER FINITE ELEMENTS	75-0002
PEANO, A. G.	THE CONSTRAINT METHOD FOR FINITE ELEMENT STRESS ANALYSIS	75-0432
PEARCE, B. R.	EVALUATION OF TECHNIQUES FOR NUMERICAL CALCULATION OF STORM SURGES	75-0554
PEARCE, B. R.	NUMERICAL CALCULATION OF STORM SURGES: AN EVALUATION OF TECHNIQUES	75-0461
PECK, J. C.	STRESS-WAVE PROPAGATION AND DYNAMIC EFFECTS IN SHELLS (SPADES) MICROMECHANICAL STRESS-WAVE RESPONSE	72-0851
PECKNOLD, D. A.	FINITE ELEMENT ANALYSIS OF SKEWED SHALLOW SHELLS	69-0211
PECKNOLD, D. A.	NONLINEAR LAYERED ANALYSIS OF RC PLATES AND SHELLS	73-0208
PECKNOLD, D. A.	RC COLUMN EARTHQUAKE RESPONSE IN TWO DIMENSIONS	74-0651
PECKNOLD, D. A.	FINITE ELEMENT ANALYSIS OF SKEWED SHALLOW SHELLS	68-0250
PECKNOLD, D. A.	INELASTIC MODEL FOR CYCLIC BIAXIAL LOADING OF REINFORCED CONCRETE	74-0503
PECQUET, R. A.	A FINITE-ELEMENT PROGRAM FOR THE ANALYSIS OF EMBANKMENTS OVER SOFT, SATURATED SOILS INCLUDING CONSOL	73-1160
PECQUET, R. A.	NUMERICAL ANALYSIS OF EMBANKMENTS OVER SOFT SOILS	75-0719
PEDERSEN, B.	APPLICATION OF THE FINITE ELEMENT METHOD TO OFF-SHORE STRUCTURES	74-0206
PEDERSEN, P.	AXISYMMETRIC ELEMENT ANALYSIS USING ANALYTICAL COMPUTING	75-0333
PEDERSEN, P.	SOME PROPERTIES OF LINEAR STRAIN TRIANGLES AND OPTIMAL FINITE ELEMENT MODELS	73-1044
PEDRIDO, R. R.	STRUCTURALLY COMPATIBLE FLUID FINITE-ELEMENT FOR SOLID FLUID INTERACTION STUDIES	75-0344
PEDRO, J. O.	AUTOMATIC PREPARATION OF DATA CARDS FOR THE FINITE ELEMENT METHOD	66-0086
PEDRO, J. O.	COMPARISON OF ANALYTICAL AND EXPERIMENTAL RESULTS IN THE DESIGN OF A HOLLOW GRAVITY DAM	67-0109
PEEL, M.	A FINITE ELEMENT PROGRAM FOR DESIGN AND ANALYSIS OF THREE-DIMENSIONAL FLAT SPRINGS	74-0738
PELLEGRIND, A.	SURFACE FOOTINGS ON SOFT ROCKS	74-0990
PENDLETON, R. L	FORMING OF 7075-T6 ALUMINUM IN HIGH PRESSURE ENVIRONMENTS	71-0322
PENMAN, A.	CONSTRUCTIONAL DEFORMATIONS IN ROCKFILL DAM	73-0281
PENMAN, A. D. M	OBSERVED AND PREDICTED DEFORMATIONS IN A LARGE EMBANKMENT DAM DURING CONSTRUCTION	71-0301
PENMAN, A. D. M	PERFORMANCE OF CULVERT UNDER WINSCAR DAM	75-0638
PENZIEN, J.	EARTHQUAKE ANALYSIS OF EARTH DAMS	69-0072
PENZIEN, J.	FINITE ELEMENT SOLUTION FOR THIN SHELLS OF REVOLUTION	63-0024
PENZIEN, J.	FINITE ELEMENT SOLUTION FOR THIN SHELLS OF REVOLUTION	64-0005
PENZIEN, J.	FINITE ELEMENT SOLUTION FOR AXISYMMETRIC SHELLS	64-0010
PENZIEN, J.	RESPONSE OF EARTH DAMS TO TRAVELLING SEISMIC WAVES	69-0092
PENZIEN, J.	SEISMIC RESPONSE OF HIGHWAY BRIDGES	75-0879
PENZIEN, J.	THE EFFECTS OF THIN CORES AND NONHOMOGENEITY ON DYNAMIC RESPONSE OF EARTH DAMS	67-0021
PEPINO, A.	DESCRIPTION OF THE IBM-360 VERSION OF THE FINITE ELEMENTS COMPUTER PROGRAM SAFE-3D	72-0775
PEPPIATT, N. A.	STRESS ANALYSIS OF ADHESIVE-BONDED LAP JOINTS	74-0548
PEPPIN, R. J.	FINITE ELEMENT SOLUTION FOR SAINT-VENANT BENDING	71-0346
PERCY, J. H.	A STUDY OF MATRIX ANALYSIS METHODS FOR INELASTIC STRUCTURES	63-0022

PERCY, J. H.	APPLICATION OF MATRIX DISPLACEMENT METHOD TO LINEAR ELASTIC ANALYSIS OF SHELLS OF REVOLUTION	65-0025
PERCY, J. H.	QUADRILATERAL FINITE ELEMENT IN ELASTIC-PLASTIC PLANE STRESS ANALYSIS	67-0067
PERCY, J. H.	SABOR 1. A FORTRAN PROGRAM FOR THE LINEAR ELASTIC AXISYMMETRIC LOADING BY USING THE MATRIX DISPLACEM	65-0027
PERCY, M. J.	DIESEL ENGINE COMPONENT DESIGN USING THE FINITE ELEMENT METHOD AND INTERACTIVE GRAPHICS	74-0237
PERIARD, A. R.	USING INTERACTIVE GRAPHICS FOR THE PREPARATION AND MANAGEMENT OF FINITE ELEMENT DATA	74-0176
PERIAUX, J.	THREE DIMENSIONAL ANALYSIS OF COMPRESSIBLE POTENTIAL FLOWS	75-0625
PERKINS, H. J.	COMPUTATION OF FLOWS IN TURBOMACHINES	74-0076
PERLMAN, A. B.	NUMERICAL STUDY OF DAMPING IN VISCOELASTIC SANDWICH BEAMS	73-0443
PERLOFF, W. H.	FINITE ELEMENT METHOD OF PREDICTING STRESS DISTRIBUTIONS AND SOIL DEFORMATION UNDER A TRACTIVE DEVIC	69-0389
PERLOFF, W. H.	STRAIN DISTRIBUTION AROUND UNDERGROUND OPENINGS	69-0212
PERLOFF, W. H.	STRAIN DISTRIBUTION AROUND UNDERGROUND OPENINGS. STATISTICAL RELATIONSHIPS FOR CERTAIN ROCK PROPERTI	71-0710
PERLOFF, W. H.	STRAIN DISTRIBUTION AROUND UNDERGROUND OPENINGS. FINITE ELEMENT ANALYSIS OF JOINTED SYSTEMS	71-0838
PERLOFF, W. H.	THE FINITE ELEMENT METHOD FOR PREDICTING STRESS DISTRIBUTION AND SOIL DEFORMATION UNDER A TRACTIVE D	71-0801
PERNG, D. Y. P.	ON THE ASYMPTOTICALLY SPHERICAL DEFORMATIONS OF ARBITRARY MEMBRANES OF REVOLUTION FIXED ALONG AN EDG	71-0640
PERRELL, C. S.	DYNAMIC BEHAVIOUR OF ECCENTRICALLY STIFFENED PLATES	71-0605
PERRIAUX, J.	THREE DIMENSIONAL ANALYSIS OF COMPRESSIBLE POTENTIAL FLOWS WITH THE FINITE ELEMENT METHOD	73-0058
PERRIAUX, J.	TWO AND THREE DIMENSIONAL ANALYSIS OF SUBSONIC COMPRESSIBLE POTENTIAL FLOWS IN DUCTS AND AROUND LIFT	74-0026
PERRONE, N.	A GENERAL FINITE DIFFERENCE METHOD FOR ARBITRARY MESHES	72-0883
PERRY, P. M.	ROCK MECHANICS STUDIES FOR A HIGH CUT AT A NUCLEAR POWER STATION IN PENNSYLVANIA	70-0122
PERSON, P.	CORRELATION BETWEEN PRESSURE AND THERMAL TRANSIENT STRESS LEVELS DERIVED BY THEORY AND EXPERIMENT ON	73-0469
PERSSON, B.	THE FINITE ELEMENT METHOD APPLIED TO COMPRESSIBLE CASCADE FLOW	75-0179
PERUMALSWAMI, P	DYNAMICS OF EARTH DAMS WITH FOUNDATION INTERACTION	71-0343
PERUMPRAL, J. V	FINITE ELEMENT ANALYSIS OF SOIL COMPACTION	74-0931
PERUMPRAL, J. V	FINITE ELEMENT METHOD OF PREDICTING STRESS DISTRIBUTIONS AND SOIL DEFORMATION UNDER A TRACTIVE DEVIC	69-0389
PERUMPRAL, J. V	THE FINITE ELEMENT METHOD FOR PREDICTING STRESS DISTRIBUTION AND SOIL DEFORMATION UNDER A TRACTIVE D	69-0522
PERUMPRAL, J. V	THE FINITE ELEMENT METHOD FOR PREDICTING STRESS DISTRIBUTION AND SOIL DEFORMATION UNDER A TRACTIVE D	71-0801
PESTEL, E.	DYNAMIC STIFFNESS MATRIX FORMULATION BY MEANS OF HERMITIAN POLYNOMIALS	66-0059
PETERS, D. H.	SHOCK LOADS ON PIPING SYSTEMS	72-0760
PETERS, J. C.	NUMERICAL ANALYSIS OF FREE SURFACE SEEPAGE PROBLEMS	71-0469
PETERSEN, C. M.	3-D SEISMIC CODE FOR ILLIAC-1V	73-0868
PETERSEN, C. M.	3-D STRESS WAVE CODE FOR THE ILLTAC 1V	74-0780
PETERSEN, H. T.	EXTENSIONS TO THE STAGS COMPUTER CODE	73-0917
PETERSEN, W.	APPLICATION OF FINITE ELEMENT METHOD TO PREDICT STATIC RESPONSE OF AUTOMOTIVE BODY STRUCTURES	71-0340
PETERSON, A. F.	STUDY OF NUMERICAL METHODS IN SEISMIC PROBLEMS	72-0738
PETERSON, F. E.	COMPUTER PROGRAM FOR STATIC AND DYNAMIC ANALYSIS OF LINEAR STRUCTURAL SYSTEMS	72-0683
PETERSON, F. E.	FINITE ELEMENT ANALYSIS OF LINEAR AND NONLINEAR HEAT TRANSFER	74-0626
PETERSON, F. E.	SAP IV. A STRUCTURAL ANALYSIS PROGRAM FOR STATIC AND DYNAMIC RESPONSE OF LINEAR SYSTEMS	73-0650
PETERSON, F. E.	THEORETICAL BASIS FOR CBL/NONSAP, A NONLINEAR STRUCTURAL ANALYSIS PROGRAM	74-0740
PETERSON, F. E.	USERS MANUAL FOR CEL/NONSAP, A NONLINEAR STRUCTURAL ANALYSIS PROGRAM	74-0706
PETERSON, W. S.	DYNAMIC ANALYSIS OF BEAM-SLAB HIGHWAY BRIDGES	73-0199
PETERSON, W. S.	THE INELASTIC ANALYSIS OF REINFORCED CONCRETE SLABS	74-1333
PETINER, A.	EFFICIENT COMPUTER ANALYSIS OF LARGE THREE-DIMENSIONAL PIPING SYSTEMS BY THE FINITE ELEMENT METHOD	72-0378
PETRASOVITS, G.	SETTLEMENT ANALYSIS OF DRIVEN PILES	75-0319
PETROF, R. C.	ELASTO-PLASTIC ANALYSIS OF STRESS IN A GAS-TURBINE WHEEL	74-0323
PETROF, R. C.	METHOD FOR THE COMPUTATION OF CYCLIC STRESSES IN A GAS-TURBINE WHEEL	74-0072
PETROVSKI, J.	STATIC AND DYNAMIC ANALYSIS OF AN ARCH DAM	74-1101
PETTERSEN, E.	FINITE ELEMENTS IN TRANSVERSE FRAME ANALYSIS	72-0182
PETYT, M.	DYNAMIC ANALYSIS OF STRUCTURES USING THE RANK FORCE METHOD	71-0553
PETYT, M.	DYNAMIC BEHAVIOR OF IN-LINE SHEAR WALLS CONNECTED BY FLOOR SLABS	72-0689
PETYT, M.	FINITE ELEMENT METHODS FOR THE RESPONSE OF STRUCTURES TO RANDOM EXCITATIONS	75-0230
PETYT, M.	FINITE ELEMENT MODELLING TECHNIQUES FOR THE ANALYSIS OF BUILDING VIBRATIONS	72-0839
PETYT, M.	FINITE ELEMENT STUDY OF VIBRATION OF TRAPEZOIDAL PLATES	73-0035
PETYT, M.	FINITE ELEMENT STUDY OF HARMONIC WAVE PROPAGATION IN PERIODIC STRUCTURES	74-0130
PETYT, M.	FINITE ELEMENT VIBRATION ANALYSIS OF CRACKED PLATES IN TENSION	68-0133
PETYT, M.	FREE VIBRATION OF A CURVED BEAM	71-0606
PETYT, M.	FREE VIBRATION OF DOUBLY CURVED RECTANGULAR PLATES INCLUDING THE PRESENCE OF MEMBRANE STRESSES WITH	69-0468
PETYT, M.	ON THE VIBRATION OF POINT-SUPPORTED PLATES	71-0602
PETYT, M.	RANDOM RESPONSE OF PERIODIC STRUCTURES BY A FINITE-ELEMENT TECHNIQUE	75-0426
PETYT, M.	STRUCTURAL VIBRATION ANALYSIS USING TRIANGULAR FINITE ELEMENTS	66-0060
PETYT, M.	THE APPLICATION OF FINITE ELEMENT TECHNIQUES TO PLATE AND SHELL PROBLEMS	65-0028
PETYT, M.	THE VIBRATION CHARACTERISTICS OF A TENSIONED PLATE CONTAINING A FATIGUE CRACK	68-0134
PETYT, M.	VIBRATION OF CURVED STRUCTURES USING QUADRILATERAL FINITE ELEMENTS	72-0074
PETYT, M.	VIBRATION OF MULTISUPPORTED CURVED BEAMS	74-1297
PEYRONNE, C.	ETUDE THEORIQUE DU COMPORTEMENT D'UNE DALLE DE BETON. (THEORETICAL STUDY OF THE BEHAVIOR OF A CONCRE	75-0273
PEYROT, A. H.	MULTIDEGREE DYNAMIC ANALYSIS OF TALL BUILDINGS SUBJECTED TO WIND AS A STOCHASTIC PROCESS	74-1196
PFEIFER, C. G.	EVALUATION OF A THREE-DIMENSIONAL STRESS ANALYSIS PROGRAM	72-0743
PHILIPPOPOULOS,	SOME DEVELOPMENTS IN THE ANALYSIS OF COMPLEX SHIP STRUCTURES	72-0332
PHILLIPS, D. V.	AN AUTOMATIC MESH GENERATION SCHEME FOR PLANE AND CURVED SURFACES BY ISOPARAMETRIC COORDINATES	71-0271
PHILLIPS, D. V.	FINITE ELEMENT ANALYSIS OF SOME CONCRETE NONLINEARITIES. THEORY AND EXAMPLES	74-1287
PHILLIPS, D. V.	FINITE ELEMENT METHOD IN ANALYSIS OF REINFORCED AND PRESTRESSED CONCRETE STRUCTURES	71-0270
PHILLIPS, D. V.	FINITE ELEMENT METHODS IN ANALYSIS OF REACTOR VESSELS	72-0271
PHILLIPS, D. V.	TECHNIQUES IN TRIANGULAR MESH GENERATION FLAT AND CURVED SURFACES	69-0394
PHILLIPS, G.	FORBIDDEN SHAPES IN THE FINITE ELEMENT METHOD	71-0705
PHILLIPS, G.	NUMERICAL ANALYSIS	73-0628
PHILLIPS, R. V.	ENVIRONMENTAL CONSIDERATIONS OF DAM CONSTRUCTION AND OPERATION IN SEISMICALLY ACTIVE URBAN AREAS	73-0470
PHU, N. D.	A FINITE ELEMENT METHOD FOR THREE DIMENSIONAL ANALYSIS OF METAL FORMING	74-1109
PI, W. S.	NONLINEAR ACOUSTIC RESPONSE ANALYSIS OF PLATES USING THE FINITE ELEMENT METHOD	72-0223
PIAN, T. H.	HYBRID MODELS IN NUMERICAL AND COMPUTER METHODS	73-0682
PIAN, T. H.	NONLINEAR CREEP ANALYSIS BY ASSUMED STRESS FINITE ELEMENT METHODS	74-0631
PIAN, T. H. H.	A HYBRID-ELEMENT APPROACH TO CRACK PROBLEMS IN PLANE ELASTICITY	73-0569
PIAN, T. H. H.	A VARIATIONAL PRINCIPLE AND THE CONVERGENCE OF A FINITE-ELEMENT METHOD BASED ON ASSUMED STRESS DISTR	69-0260
PIAN, T. H. H.	APPLICATION OF FINITE ELEMENT METHOD TO MIXED-MODE FRACTURE	73-1123
PIAN, T. H. H.	APPLICATION OF MATRIX DISPLACEMENT METHOD TO LINEAR ELASTIC ANALYSIS OF SHELLS OF REVOLUTION	65-0025
PIAN, T. H. H.	APPLICATION OF THE SMOOTH-SURFACE INTERPOLATION TO THE FINITE ELEMENT ANALYSIS	69-0088
PIAN, T. H. H.	BASIS OF FINITE ELEMENT METHODS FOR SOLID CONTINUA	69-0213
PIAN, T. H. H.	BOUNDS TO THE INFLUENCE COEFFICIENTS BY THE ASSUMED STRESS METHOD	70-0165
PIAN, T. H. H.	CRACK ELEMENTS	75-0558
PIAN, T. H. H.	DERIVATION OF ELEMENT STIFFNESS MATRICES BY ASSUMED STRESS DISTRIBUTIONS	64-0007
PIAN, T. H. H.	DERIVATION OF ELEMENT STIFFNESS MATRICES	64-0008
PIAN, T. H. H.	DERIVATION OF GEOMETRIC STIFFNESS AND MASS MATRICES FOR FINITE ELEMENT HYBRID MODELS	74-0844
PIAN, T. H. H.	ELASTIC CRACK ANALYSIS BY A FINITE ELEMENT HYBRID METHOD	71-0212
PIAN, T. H. H.	ELASTIC PLASTIC ANALYSIS BY ASSUMED STRESS HYBRID MODEL	74-0443
PIAN, T. H. H.	ELEMENT STIFFNESS MATRICES FOR BOUNDARY COMPATIBILITY AND FOR PRESCRIBED BOUNDARY STRESSES	66-0061
PIAN, T. H. H.	FINITE ELEMENT FORMULATION BY VARIATIONAL PRINCIPLES WITH RELAXED CONTINUITY REQUIREMENTS	72-0516
PIAN, T. H. H.	FINITE ELEMENT METHODS BY DIFFERENT VARIATIONAL PRINCIPALS IN ELASTICITY	70-0407
PIAN, T. H. H.	FINITE ELEMENT METHODS BY VARIATIONAL PRINCIPLES WITH RELAXED CONTINUITY REQUIREMENTS	72-0342
PIAN, T. H. H.	FINITE ELEMENT METHODS IN CONTINUUM MECHANICS.	72-0377
PIAN, T. H. H.	FINITE ELEMENT STIFFNESS METHODS BY DIFFERENT VARIATIONAL PRINCIPLES IN ELASTICITY	70-0455
PIAN, T. H. H.	FORMULATIONS OF FINITE ELEMENT METHODS FOR SOLID CONTINUA	71-0679
PIAN, T. H. H.	FOUNDATIONS OF FINITE ELEMENT METHODS FOR SOLID CONTINUA	71-0213
PIAN, T. H. H.	HYBRID MODELS	71-0211
PIAN, T. H. H.	HYBRID-ELEMENT APPROACH TO CRACK PROBLEMS IN PLANE ELASTICITY	73-0508

PIAN, T. H. H.	IMPROVEMENTS ON THE ANALYSIS OF SHELLS OF REVOLUTION BY THE MATRIX DISPLACEMENT METHOD	66-0070
PIAN, T. H. H.	INTERACTIVE PROGRAM IN DESIGN AND ANALYSIS OF COMPOSITE MATERIALS	74-0854
PIAN, T. H. H.	LINEAR DYNAMIC ANALYSES OF LAMINATED PLATES AND SHELLS BY THE HYBRID-STRESS FINITE-ELEMENT METHOD	73-0786
PIAN, T. H. H.	MODE SHAPES AND FREQUENCIES BY FINITE ELEMENT METHOD USING CONSISTENT AND LUMPED MASSES	71-0610
PIAN, T. H. H.	NONLINEAR ANALYSIS BY ASSUMED STRESS HYBRID MODELS	74-1004
PIAN, T. H. H.	NONLINEAR CREEP ANALYSIS BY ASSUMED STRESS FINITE ELEMENT METHODS	74-0922
PIAN, T. H. H.	ON THE CONVERGENCE OF THE FINITE ELEMENT METHOD FOR PROBLEMS WITH SINGULARITY	73-0173
PIAN, T. H. H.	PLATE BENDING ANALYSIS USING ASSUMED STRESS METHODS	68-0135
PIAN, T. H. H.	RATIONALIZATION IN DERIVING ELEMENT STIFFNESS MATRIX BY ASSUMED STRESS APPROACH	68-0138
PIAN, T. H. H.	SOME RECENT STUDIES IN ASSUMED STRESS HYBRID MODELS	72-0262
PIAN, T. H. H.	STABILITY ANALYSIS OF SHELLS OF REVOLUTION	68-0136
PIAN, T. H. H.	STATIC AND DYNAMIC NONLINEAR STRUCTURAL PROBLEMS	72-0848
PIAN, T. H. H.	THE APPLICATION OF THE SMOOTH-SURFACE INTERPOLATION TECHNIQUE TO THE FINITE-ELEMENT ANALYSIS OF THIN	66-0120
PIAN, T. H. H.	THE CONVERGENCE OF FINITE ELEMENT METHOD IN SOLVING LINEAR ELASTIC PROBLEMS	67-0083
PIAN, T. H. H.	THEORETICAL FORMULATION OF FINITE ELEMENT METHODS IN LINEAR ELASTIC ANALYSIS OF GENERAL SHELLS	72-0496
PIAN, T. H. H.	VARIATIONAL FORMULATION OF FINITE DISPLACEMENT ANALYSIS	70-0284
PIAN, T. H. H.	VARIATIONAL FORMULATIONS OF NUMERICAL METHODS	70-0392
PIAN, T. H. H.	VARIATIONAL FORMULATIONS OF NUMERICAL METHODS IN SOLID CONTINUA	71-0670
PIAN, T. H. H.	VARIATIONAL PRINCIPLES AND THEIR APPLICATION TO FINITE ELEMENT METHODS	68-0137
PIAN, T. H. H.	VIBRATION ANALYSIS OF LAMINATED PLATES AND SHELLS BY A HYBRID STRESS ELEMENT	73-0120
PICASSO, B.	ELASTO-PLASTIC ANALYSIS OF A ROTATING MODEL CONICAL TURBINE DISC	75-0281
PICHUMANI, R.	FINITE ELEMENT ANALYSIS OF PAVEMENT STRUCTURES USING AFPAV CODE LINEAR ELASTIC ANALYSIS	73-0658
PICHUMANI, R.	FINITE-ELEMENT ANALYSIS OF PILE-SUPPORTED PAVEMENT STRUCTURES	74-0157
PICHUMANI, R.	FINITE-ELEMENT ANALYSIS OF PAVEMENT STRUCTURES USING AFPAV CODE (NONLINEAR ELASTIC ANALYSIS)	75-0748
PICHUMANI, R.	PARAMETRIC STUDIES OF LAYERED PAVEMENT SYSTEMS USING AFPAV CODE WITH FAST EQUATION SOLVER	75-0747
PICK, R. J.	CALCULATION OF NATURAL FREQUENCIES OF HEAT EXCHANGER TUBES	73-0665
PICKARD, J.	FORMAT-FORTRAN MATRIX ABSTRACTION TECHNIQUE	68-0214
PICKETT, R. M.	AUTOMATED STRUCTURAL SYNTHESIS USING A RE NUMBER OF DESIGN COORDINATES	73-0584
PIENING, M.	DESH, EIN EINFACHES FLACHES SCHALENELEMENT UEBER DREIECKIGEM GRUNDRISS. (DESH, A SIMPLE TRIANGULAR S	75-0304
PIERINI, G.	FINITE DIFFERENCE AND FINITE ELEMENT METHODS FOR THE SEMICONDUCTOR FLOW EQUATIONS	74-0028
PIETRAS, F.	OCENA DOKLADNOSCI ROZWIAZAN PROBLEMOW BRZEGOWYCH SPREZYSTYCH CIAL DYSKERTYZOWANYCH. (ASSESSING THE A	74-0921
PIFKO, A. B.	ELASTIC PLASTIC BUCKLING OF STIFFENED RECTANGULAR PLATES	69-0122
PIFKO, A.	NONLINEAR CRACK ANALYSIS WITH FINITE ELEMENTS	73-0875
PIFKO, A.	PLANS - A FINITE ELEMENT PROGRAM FOR NONLINEAR ANALYSIS OF STRUCTURES	74-0966
PIFKO, A.	PLANS - A FINITE-ELEMENT PROGRAM FOR NONLINEAR ANALYSIS OF STRUCTURES	75-0011
PIFKO, A. B.	A FINITE ELEMENT METHOD FOR THE PLASTIC BENDING ANALYSIS OF STRUCTURES	68-0021
PIFKO, A. B.	A FINITE-ELEMENT METHOD FOR THE PLASTIC BUCKLING ANALYSIS OF PLATES	69-0214
PIFKO, A. B.	COMPUTER PROGRAMS FOR THE PLASTIC ANALYSIS OF STRUCTURES USING DISCRETE ELEMENT METHODS	67-0004
PIFKO, A. B.	DISCRETE ELEMENT METHODS FOR THE PLASTIC ANALYSIS OF STRUCTURES SUBJECTED TO CYCLIC LOADING	70-0023
PIFKO, A. B.	FINITE ELEMENT ANALYSIS OF STRUCTURES IN THE PLASTIC RANGE	71-0031
PIFKO, A. B.	FINITE ELEMENT METHOD FOR THE PLASTIC BENDING ANALYSIS OF STRUCTURES	68-0206
PIFKO, A. B.	PLASTICITY	70-0509
PIFKO, A. B.	PLASTICITY THEORY AND FINITE ELEMENT APPLICATIONS	72-0318
PIH, H.	THREE DIMENSIONAL SCATTERED-LIGHT STRESS ANALYSIS OF DISCONTINUOUS FIBER-REINFORCED COMPOSITES	73-0218
PILKEY, W.	STRUCTURAL MECHANICS COMPUTER PROGRAMS	74-0902
PILKEY, W. D.	A METHOD OF ANALYSIS OF LINE STRUCTURES BY TRANSFER MATRICES DERIVED FROM FINITE ELEMENTS	74-0808
PINCEMIN, M.	INTEGRATED PROGRAM FOR THE DYNAMIC STRUCTURAL CALCULATION OF MOBIL OFFSHORE UNITS	74-0490
PINDER, G. F.	APPLICATION OF GALERKIN'S PROCEDURE TO AQUIFER ANALYSIS	72-0636
PINDER, G. F.	FINITE ELEMENT ANALYSIS OF HYDROTHERMAL SYSTEMS	74-0048
PINDER, G. F.	FINITE ELEMENT APPROACH TO THE MODELING OF HYDROTHERMAL SYSTEMS	73-0011
PINDER, G. F.	FINITE ELEMENTS IN THE SOLUTION OF THE INVERSE PROBLEM IN GROUNDWATER FLOW	74-0063
PINDER, G. F.	FUNCTIONAL COEFFICIENTS IN THE ANALYSIS OF GROUNDWATER FLOW USING FINITE ELEMENTS	73-0070
PINDER, G. F.	GALERKIN APPROXIMATION OF THE TIME DERIVATIVE IN THE FINITE ELEMENT ANALYSIS	73-0573
PINDER, G. F.	GALERKIN APPROXIMATION OF THE TIME DERIVATIVE IN THE FINITE ELEMENT ANALYSIS OF GROUNDWATER FLOW	74-0658
PINDER, G. F.	GALERKIN FINITE-ELEMENT SIMULATION OF A GEOTHERMAL RESERVOIR	73-0900
PINDER, G. F.	GALERKIN-FINITE ELEMENT SIMULATION OF GROUNDWATER CONTAMINATION ON LONG-ISLAND, NEW-YORK	73-0093
PINDER, G. F.	GALERKIN-FINITE ELEMENT TECHNIQUE FOR CALCULATING TRANSIENT POSITION OF SALTWATER FRONT	75-0135
PINDER, G. F.	GALERKIN-FINITE ELEMENT ANALYSIS OF HYDROTHERMAL SYSTEM AT WAIRAKEI, NEW-ZEALAND	75-0601
PINDER, G. F.	SIMULATION OF GROUND WATER CONTAMINATION USING A GALERKIN FINITE ELEMENT TECHNIQUE	74-1362
PINI, F.	APPPOXIMATION BY FINITE ELEMENT FUNCTIONS USING REGULARIZATION	74-1229
PINKNEY, R. B.	CYCLIC PLASTIC ANALYSIS OF STRUCTURAL STEEL JOINTS	73-0799
PINSON, L. D.	A TWO-DIMENSIONAL AXISYMMETRIC SHELL FLUID MODEL FOR LAUNCH VEHICLE LONGITUDINAL DYNAMIC ANALYSIS	69-0046
PINSON, L. D.	ADAPTATION OF NASTRAN TO THE ANALYSIS OF THE VIKING SPACE VEHICLE	71-0135
PINSON, L. D.	EVALUATION OF A FINITE ELEMENT ANALYSIS FOR THE LONGITUDINAL VIBRATIONS OF LIQUID-PROPELLANT LAUNCH	70-0204
PINSON, L. D.	VIBRATION CHARACTERISTICS OF THIN, PRESSURIZED SHELLS OF REVOLUTION PARTIALLY FILLED WITH LIQUID	72-0604
PINTUS, L. V.	APPLICATION OF THE FINITE ELEMENT METHOD TO CALCULATE WOOD TENSION	75-0768
PIPANO, A.	FAST MODAL EXTRACTION IN NASTRAN VIA THE FEER COMPUTER PROGRAM	73-0634
PISKUN, V. V.	SOLUTION OF A THREE DIMENSIONAL AXISYMMETRICAL PROBLEM OF THERMOPLASTICITY AS APPLIED TO THICK TURBI	74-0384
PISTER, K.	ON A VARIATIONAL THEOREM FOR INCOMPRESSIBLE AND NEARLY INCOMPRESSIBLE ORTHOTROPIC ELASTICITY	68-0249
PISTER, K. S.	A FINITE ELEMENT APPLICATION OF THE HELLINGER-REISSNER VARIATIONAL THEOREM	68-0060
PISTER, K. S.	CONTINUUM MECHANICS - A BRIEF REVIEW	71-0407
PISTER, K. S.	IDENTIFICATION OF NONLINEAR ELASTIC SOLIDS BY A FINITE ELEMENT METHOD	74-0530
PISTER, K. S.	THERMOMECHANICAL ANALYSIS OF VISCOELASTIC SOLIDS	70-0240
PISTER, K. S.	THERMOMECHANICAL CREEP OF AGEING CONCRETE - A UNIFIED APPROACH	74-0607
PITKARAN, J.	FINITE ELEMENT SOLUTION OF MULTIGROUP TRANSPORT EQUATION IN 2 DIMENSIONAL GEOMETRIES	74-0648
PITKARANTA, J.	COMPUTATIONAL EXPERIMENTATION ON FINITE ELEMENT METHOD IN BARE SLAB CRITICALITY CALCULATIONS	73-0015
PITKARANTA, J.	FINITE ELEMENT ANALYSIS OF SOME CRITICAL FAST ASSEMBLIES	73-0128
PITKARANTA, J.	NON-SELF-ADJOINT VARIATIONAL PROCEDURE FOR FINITE-ELEMENT APPROXIMATION OF TRANSPORT EQUATION	75-0072
PITTR, J.	BERECHNUNG INSTATIONAERER TEMPERATUREFELDER MIT FINITEN ELEMENTEN IN RAUM-UND ZEITDIMENSION. (CALCUL	74-0987
PITTR, J.	CALCULATION OF TRANSIENT TEMPERATURE FIELDS WITH FINITE ELEMENTS IN SPACE AND TIME DIMENSIONS	74-0379
PIVA, R.	FINITE ELEMENT SIMULATION OF THERMALLY INDUCED FLOW FIELDS	74-1235
PIVA, R.	NUMERICAL TECHNIQUES FOR CONVECTION/DIFFUSION PROBLEMS	75-0123
PIZIALI, R. L.	DYNAMIC TORSIONAL RESPONSE OF THE HUMAN LEG RELATIVE TO SKIING INJURIES	74-0471
PLANCHARD, J.	UNE METHODE VARIATIONNELLE ELEMENTS FINIS POUR LA RESULUTION NUMERIQUE UN PROBLEME EXTERIEUR DANS R*	73-0576
PLANCHARD, J.	VARIATION METHOD ON FINITE ELEMENTS FOR NUMERICAL RESOLUTION TO PROBLEMS WITH EXTERNAL R3	73-0359
PLANEIX, J. M.	APPLICATION OF THE FINITE ELEMENT METHOD TO THE CALCULATION OF SHIP HULLS	71-0431
PLANEIX, J. M.	DYNAMIC AND STOCHASTIC ASPECTS IN THE STRUCTURAL ANALYSIS OF VLCCS AND ULCCS	74-0298
PLANFIX, J. M.	INTEGRATED PROGRAM FOR THE DYNAMIC STRUCTURAL CALCULATION OF MOBIL OFFSHORE UNITS	74-0490
PLANK, R. J.	BUCKLING UNDER COMBINED LOADING OF THIN FLAT-WALLED STRUCTURES BY A COMPLEX FINITE STRIP METHOD	74-0788
PLANT, R. E.	FINITE ELEMENT ANALYSIS OF A BONE-PLATE-SCREW SYSTEM	74-0920
PLECK, M. H.	GRAPHICAL DISPLAY OF COMPUTER SIMULATED UNBALANCED ROTOR RESPONSE	71-0486
PLETTA, D. H.	RELATIONSHIP BETWEEN FINITE ELEMENT AND MEASURED JOINT-FLOOR CONNECTION RIGIDITIES	75-0738
PLUMMER, F. B.	A NEW LOOK AT STRUCTURAL ENERGY DISSIPATION	74-0816
POBEDRIA, B. E.	SOME METHODS OF SOLVING PROBLEMS OF NON-LINEAR THERMOVISCOELASTICITY	74-1207
POCESKI, A.	MIXED FINITE ELEMENT METHOD FOR BENDING OF PLATES	75-0229
POCHA, J. J.	STUDY OF THE ACOUSTIC EXCITATION OF STRUCTURES	74-1332
POE, J.	ON THE NUMERICAL SOLUTION OF A CLASS OF NONLINEAR PROBLEMS IN DYNAMIC COUPLED THERMOELASTICITY	71-0441
POLENSEK, A.	STATIC AND DYNAMIC ANALYSIS OF WOOD-JOIST FLOORS BY THE FINITE-ELEMENT METHOD	73-1161
POLHEMUS, J. F.	USE OF FOURIER SERIES IN THE FINITE ELEMENT METHOD	68-0103
POLIZZOTTO, C.	OPTIMUM PLASTIC DESIGN FOR MULTIPLE SETS OF LOADS	74-0980
POLVANICH, N.	STRESS INTENSITY FACTOR OF A SURFACE CRACK IN A PRESSURIZED CYLINDER	75-0255
POMERANTZ, M. A	HORIZONS IN THE STRUCTURAL USE OF ADVANCED MATERIALS - THRAUSTICS	70-0415

84

PONTER, A. R. S	APPLICATION OF DUAL MINIMUM THEOREMS TO THE FINITE ELEMENT SOLUTION OF POTENTIAL PROBLEMS WITH SPECI	72-0036
POOK, L. P.	STRESS INTENSITY FACTORS CALCULATED GENERALLY BY FINITE ELEMENT TECHNIQUE	69-0369
POOLE, W. G. JR	AN ALGORITHM FOR REDUCING THE BANDWITH AND PROFILE OF A SPARSE MATRIX	74-0792
POPE, G.	A DISCRETE ELEMENT METHOD FOR ANALYSIS OF PLANE ELASTIC-PLASTIC STRESS PROBLEMS	65-0063
POPE, G. G.	A DISCRETE ELEMENT METHOD FOR THE ANALYSIS OF PLANE ELASTO-PLASTIC STRESS PROBLEMS	66-0062
POPE, G. G.	OPTIMUM DESIGN OF STRESSED SKIN STRUCTURES	73-0472
POPE, G. G.	THE APPLICATION OF THE MATRIX DISPLACEMENT METHOD IN PLANE ELASTO-PLASTIC STRESS PROBLEMS	64-0009
POPOV, E. P.	A REFINED CURVED ELEMENT FOR THIN SHELLS OF REVOLUTION	71-0214
POPOV, E. P.	ANALYSES OF SHELLS OF REVOLUTION BY FINITE ELEMENTS	67-0149
POPOV, E. P.	ANALYSIS OF ELASTIC-PLASTIC SHELLS OF REVOLUTION	70-0220
POPOV, E. P.	BENDING OF CIRCULAR PLATES OF HARDENING MATERIAL	67-0112
POPOV, E. P.	ELASTIC PLASTIC DYNAMIC ANALYSIS OF AXISYMMETRIC SOLIDS	73-0657
POPOV, E. P.	ELASTIC-PLASTIC ANALYSIS OF SOME PRESSURE VESSEL HEADS	69-0379
POPOV, E. P.	ELASTIC-PLASTIC ANALYSIS OF THICK WALLED PRESSURE VESSELS WITH SHARP DISCONTINUITIES	71-0155
POPOV, E. P.	FINITE ELEMENT SOLUTION FOR THIN SHELLS OF REVOLUTION	63-0024
POPOV, E. P.	FINITE ELEMENT SOLUTION FOR THIN SHELLS OF REVOLUTION	64-0005
POPOV, E. P.	FINITE ELEMENT SOLUTION FOR AXISYMMETRIC SHELLS	64-0010
POPOV, E. P.	FINITE ELEMENT SOLUTION OF AXISYMMETRICAL DYNAMIC PROBLEMS OF SHELLS OF REVOLUTION	66-0020
POPOV, E. P.	INCREMENTAL ANALYSIS OF LARGE DEFLECTIONS OF SHELLS OF REVOLUTION	71-0296
POPOV, E. P.	INELASTIC BUCKLING ANALYSIS OF PIPES SUBJECTED TO INTERNAL PRESSURE, FLEXURE AND AXIAL LOADING	74-0515
POPOV, E. P.	LARGE DEFLECTION ELASTIC-PLASTIC ANALYSIS OF SHELLS OF REVOLUTION	71-0263
POPOV, E. P.	LINEAR AND NONLINEAR STATIC ANALYSIS OF AXISYMMETRICALLY LOADED THIN SHELLS OF REVOLUTION	69-0215
POPOV, E. P.	NON-LINEAR DYNAMIC ANALYSIS OF AXISYMMETRIC SHELLS	75-0474
POPOV, E. P.	NONLINEAR BUCKLING ANALYSIS OF SANDWICH ARCHES	71-0292
POPOV, E. P.	NONLINEAR ELASTIC-VISCOPLASTIC ANALYSIS USING THE FINITE ELEMENT METHOD	74-0750
POPOV, E. P.	NONLINEAR FINITE ELEMENT ANALYSIS OF SANDWICH SHELLS OF REVOLUTION	72-0180
POPOV, E. P.	NONLINEAR FINITE ELEMENT DYNAMIC ANALYSIS OF AXISYMMETRIC SOLIDS	74-0775
POPOV, E. P.	NOTE ON INCREMENTAL EQUILIBRIUM EQUATIONS AND APPROXIMATE CONSTITUTIVE RELATIONS IN LARGE INELASTIC	74-0984
POPOV, E. P.	PLASTIC AND VISCOPLASTIC ANALYSIS OF AXISYMMETRIC SHELLS	75-0246
POPOV, E. P.	STATIC AND DYNAMIC FINITE ELEMENT ANALYSIS OF SANDWICH STRUCTURES	68-0003
POPOVIC, J. R.	IEEE-G-MTT INTERNATIONAL MICROWAVE SYMPOSIUM, 1973: APPLICATIONS IN THE 70'S	73-0374
POPOVICI, L.	ASJPRA ANALIZEI PRIN METODA ELEMENTULUI FINIT A STARILOR DE EFORTURI SI DEFORMATII DIN MASIVELE DE P	74-0471
POPPLEWELL, N.	CONFORMING RECTANGULAR AND TRIANGULAR PLATE BENDING ELEMENTS	71-0720
POPPLEWELL, N.	RESPONSE OF BOX-LIKE STRUCTURES TO WEAK EXPLOSIONS	75-0197
POPPLEWELL, N.	VIBRATION OF A BOX-TYPE STRUCTURE - 1	71-0405
POPPLEWELL, N.	VIBRATION OF A BOX-TYPE STRUCTURE - 2.	71-0420
POSTNOV, V. A.	APPLICATION OF THE METHOD OF FINITE ELEMENTS TO THE CALCULATION OF SHIP STRUCTURES (ISPOLZOVANIE MET	73-0933
POTASH, R. L.	SECOND ORDER THEORY OF OSCILLATING CYLINDERS	70-0430
POTTS, J. S.	HABEAS- A STRUCTURAL DYNAMICS ANALYSIS SYSTEM	70-0121
POULOS, H. G.	EFFECT OF SEEPAGE ON EMBANKMENT DEFORMATIONS DUE TO WATER LOADING	74-1329
POULOS, H. G.	FINITE ELEMENT ANALYSIS OF THE EFFECTS OF INSTALLATION ON PILE LOAD-SETTLEMENT BEHAVIOUR	74-0763
POULOS, H. G.	FINITE ELEMENT ANALYSIS OF PILE-SUPPORTED PAVEMENT STRUCTURES	75-0145
POULOS, H. G.	INITIAL SETTLEMENT OF STRUCTURES ON CLAY	71-0304
POWARS, C. A.	BASELINE SOLUTIONS FOR THE SMOOTH WALL THERMOCHEMICAL ABLATION RESPONSE OF GRAPHITE AND CARBON PHENO	71-0728
POWELL, G.	ELASTIC LATERAL BUCKLING OF STEEL BEAMS	70-0364
POWELL, G. H.	COMPUTATIONAL PROCEDURES FOR INELASTIC STRESS ANALYSIS	73-0859
POWELL, G. H.	FINITE ELEMENT ANALYSIS OF NONLINEAR MEMBRANE STRUCTURES	72-0909
POWELL, G. H.	FINITE ELEMENT ANALYSIS OF ELASTO-PLASTIC TEE JOINTS	74-1313
POWELL, G. H.	LARGE CAPACITY EQUATION SOLVER FOR STRUCTURAL ANALYSIS	74-0983
POWELL, G. H.	STATIC AND DYNAMIC ANALYSIS OF NONLINEAR STRUCTURES	75-0718
POWELL, G. H.	STRESS ANALYSIS OF B16.9 TEES BY THE FINITE ELEMENT METHOD	71-0060
POWELL, G. H.	STRESS ANALYSIS OF B16.9 TEES BY THE FINITE ELEMENT METHOD, A PROGRESS REPORT	71-0215
POWER, C. M.	ICES-STRUDL-11	68-0257
POWERS, C. A.	ANALYSIS OF THERMAL STRESSES IN CORED CERAMIC BRICKS USED IN HYPERSONIC WIND TUNNEL HEATERS	72-0655
PRABHU, K. S.	AN IMPROVED MATRIX DISPLACEMENT METHOD FOR VIBRATION OF NONUNIFORM BEAMS	71-0841
PRACHUKTAM, S.	BLOCKING TECHNIQUE FOR LARGE SCALE STRUCTURAL ANALYSIS	73-0243
PRAGER, W.	APPROXIMATION IN ELASTICITY BASED ON THE CONCEPT OF FUNCTION SPACE	47-0001
PRAGER, W.	COMPARISON OF EQUIVALENCE AND FINITE ELEMENT METHODS	75-0483
PRAGER, W.	OPTIMALITY CRITERIA IN STRUCTURAL DESIGN	71-0758
PRAKASH, A.	FINITE ELEMENT ANALYSIS OF ELASTO-PLASTIC MATERIALS BY USE OF DISLOCATION DIPOLE SYSTEMS	74-0269
PRAKASH, A.	FINITE ELEMENT ANALYSIS OF NON-LINEAR COMPOSITE MATERIALS BY USE OF OVERLAY SYSTEMS	74-0928
PRAKASO, B.	AXISYMMETRIC VIBRATIONS OF LINEARLY TAPERED ANNULAR PLATES	74-1266
PRASAD, K. S. R	GALERKIN FINITE ELEMENT METHOD FOR VIBRATION PROBLEMS	73-0659
PRASAD, K.S.R.K	A FINITE ELEMENT ANALOGUE OF THE MODIFIED RAYLEIGH-RITZ METHOD FOR VIBRATION PROBLEMS	72-0478
PRASAD, K.S.R.K	FINITE ELEMENT ANALOGUE OF THE MODIFIED RAYLEIGH RITZ METHOD FOR VIBRATION PROBLEMS	73-0181
PRASAD, K.S.R.K	FINITE ELEMENT MODELLING OF NATURAL VIBRATION PROBLEMS	73-0320
PRATO, C. A.	A MIXED FINITE ELEMENT METHOD FOR THIN SHELLS ANALYSIS	68-0182
PRATO, C. A.	COMMENT ON EXACT SOLUTION OF CERTAIN PROBLEMS BY FINITE ELEMENT METHOD	69-0371
PRATO, C. A.	SHELL FINITE ELEMENT METHOD VIA REISSNER'S PRINCIPLE	69-0216
PRAUSE, R. H.	ASSESSMENT OF DESIGN TOOLS AND CRITERIA FOR URBAN RAIL TRACK STRUCTURES, VOLUME 11. AT-GRADE SLAB TR	74-0813
PRESCOTT, P.	INTERPRETATION OF COMPLEX STRESS ANALYSIS BY MULTIPLE REGRESSION	74-1203
PRETORIUS, P. C	FATIGUE CRACK FORMATION AND PROPAGATION IN PAVEMENTS CONTAINING SOIL-CEMENT BASES.	72-0435
PREVOST, J. H.	EFFECTIVE STRESS-STRAIN-STRENGTH MODEL FOR SOILS	75-0056
PRICE, E. W.	METHODS FOR DETERMINING CHARACTERISTICS OF ACOUSTIC WAVES IN ROCKET MOTORS	75-0075
PRICE, H. L.	FINITE ELEMENT ANALYSIS OF DIAMETRAL TEST OF POLYMER MOLDINGS	73-0129
PRICE, H. S.	ERROR BOUNDS FOR SEMIDISCRETE GALERKIN APPROXIMATIONS OF PARABOLIC PROBLEMS WITH APPLICATIONS TO PET	70-0291
PRICE, J.	ELASTIC DEFLECTION IN THIN WALLED CONES ROTATING ABOUT THEIR AXES OF SYMMETRY	73-0545
PRIETO, L.	AN INCORE FINITE ELEMENT PROGRAM FOR RESEARCH AND TEACHING PURPOSE	74-0666
PRINCE, J. F.	FINITE ELEMENT ANALYSIS OF GENERAL FLUID FLOW PROBLEMS	71-0536
PRINCE, N.	MATRIX ANALYSIS METHODS FOR ANISOTROPIC INELASTIC STRUCTURES	66-0104
PRINCE, N.	ON FURTHER APPLICATION OF THE FINITE ELEMENT METHOD TO THREE-DIMENSIONAL ELASTIC ANALYSIS	70-0252
PRINCE, N.	STRUCTURAL ANALYSIS OF SHELL INTERSECTIONS	69-0217
PRIVITZER, E.	SOME FACTORS AFFECTING DENTAL IMPLANT DESIGN	74-1146
PROCHAZKA, P.	STABILITA STEN V PLASTICKEM OBORU (STABILITY IN WEBPLATES IN PLASTIC RANGE)	73-0387
PROMSIT, S.	DYNAMIC STABILITY OF CONICAL SHELLS BY FINITE ELEMENTS	70-0467
PROTHEROE, S. J	BERSAFE FINITE ELEMENT SYSTEM	74-0190
PROZAN, R. J.	STUDY OF HYPERVELOCITY METEOROID IMPACT ON ORBITAL SPACE STATIONS	73-0905
PRYOR, C. W.	A FINITE ELEMENT ANALYSIS INCLUDING TRANSVERSE SHEAR EFFECTS FOR APPLICATIONS TO LAMINATED PLATES	71-0216
PRYOR, C. W. JR	FINITE ELEMENT ANALYSIS OF LAMINATED ANISOTROPIC PLATES INCLUDING TRANSVERSE SHEAR DEFORMATIONS	70-0506
PRYOR, C. W. JR	FINITE ELEMENT BENDING ANALYSIS OF REISSNER PLATES	70-0214
PRYOR, C. W. JR	THREE DIMENSIONAL ANALYSIS OF STRESS CONCENTRATIONS NEAR HOLES IN LAMINATED COMPOSITES	72-0713
PRZEMIENIECKI,	DISCRETE-ELEMENT METHODS FOR STABILITY ANALYSIS OF COMPLEX STRUCTURES	68-0140
PRZEMIENIECKI,	EQUIVALENT MASS MATRICES FOR RECTANGULAR PLATES IN BENDING	66-0063
PRZEMIENIECKI,	FINITE ELEMENT ANALYSIS OF FREQUENCY SPECTRA FOR ELASTIC WAVEGUIDES	75-0247
PRZEMIENIECKI,	FINITE ELEMENT STRUCTURAL ANALYSIS OF LOCAL INSTABILITY	72-0179
PRZEMIENIECKI,	INFLUENCE OF HIGHER ORDER TERMS IN THE LARGE DEFLECTION ANALYSIS OF FRAMEWORKS	68-0259
PRZEMIENIECKI,	MATRIX ANALYSIS OF LOCAL INSTABILITY IN PLATES, STIFFENED PANELS AND COLUMNS	72-0542
PRZEMIENIECKI,	MATRIX METHODS IN STRUCTURAL MECHANICS	66-0065
PRZEMIENIECKI,	QUADRATIC MATRIX EQUATIONS FOR DETERMINING VIBRATION MODES AND FREQUENCIES OF CONTINUOUS ELASTIC SYS	66-0064
PRZEMIENIECKI,	STABILITY ANALYSIS OF COMPLEX STRUCTURES USING DISCRETE ELEMENT TECHNIQUE	67-0068
PRZEMIENIECKI,	TETRAHEDRON ELEMENTS IN THE MATRIX FORCE METHOD OF STRUCTURAL ANALYSIS	64-0011
PRZEMIENIECKI,	THEORY OF MATRIX STRUCTURAL ANALYSIS	68-0139

PRZEMIENIECKI,	TRIANGULAR PLATE ELEMENTS IN THE MATRIX FORCE METHOD OF STRUCTURAL ANALYSIS	63-0010
PRZEMIENIECKI,	FINITE ELEMENT STRUCTURAL ANALYSIS OF LOCAL INSTABILITY - REPLY	74-0099
PUCCINELLI, E.	VIEW FACTOR COMPUTER PROGRAM (PROGRAM VIEW) USER'S MANUAL	73-0789
PUGH, J. W.	STRUCTURAL MODEL FOR THE MECHANICAL BEHAVIOUR OF TRABECULAR BONE	73-0473
PULMANO, V. A.	A COMPATIBLE MEMBRANE QUADRILATERAL ELEMENT	72-0483
PULMANO, V. A.	A COMPATIBLE PLATE BENDING ANNULAR SECTOR FINITE ELEMENT OF NONUNIFORM THICKNESS	74-0430
PULMANO, V. A.	A COMPATIBLE QUADRILATERAL PLATE BENDING ELEMENT	72-0482
PULMANO, V. A.	CERTAIN FINITE ELEMENTS FOR PLATE BENDING PROBLEMS	74-0395
PULMANO, V. A.	DUAL ANALYSIS OF COLUMN SUPPORTED SLABS	74-0399
PULMANO, V. A.	FINITE ELEMENT METHODS IN ENGINEERING	74-0400
PULMANO, V. A.	ON THE BENDING STIFFNESS OF SLABS IN CROSS-WALLED STRUCTURES	74-0398
PULMANO, V. A.	PLANE STRESS ELEMENT OF VARYING THICKNESS APPLIED TO END BLOCK ANALYSIS	74-0396
PULMANO, V. A.	SUBSTRUCTURE ANALYSIS OF MULTISTOREY FLAT SLAB BUILDINGS	74-0394
PULMANO, V. A.	TAPERED PLANE STRESS ANNULAR SECTOR FINITE ELEMENT	75-0538
PUPPO, A.	CALCULATION AND DESIGN OF JOINTS MADE FROM COMPOSITE MATERIALS	70-0156
PUPPO, A.	TRANSVERSE LOADING OF UNIDIRECTIONAL FIBER COMPOSITES	69-0119
PURDY, D. M.	DEVELOPMENT OF A GRAPHITE HORIZONTAL STABILIZER	73-0931
PURDY, D. M.	DISCRETE ELEMENT ANALYSIS OF COMPOSITE STRUCTURES	75-0201
PURDY, M. D.	INFLUENCE OF HIGHER ORDER TERMS IN THE LARGE DEFLECTION ANALYSIS OF FRAMEWORKS	68-0259
PURUSHOTHAMAN,	STUDIES ON TWO PHASE COMPOSITE MATERIALS THROUGH COMPUTER SIMULATION	73-0315
PUTTAIAH, G.	RADIAL INERTIA EFFECTS ON AN IDEALLY PLASTIC CIRCULAR PLATE UNDER IMPULSIVE AXIAL COMPRESSION	69-0324
PUTTAIAH, G.	SOME RECENT FINITE ELEMENT APPROACHES IN ANALYSIS OF PLASTIC IMPACT	69-0299
PUTTER, S.	BEAM BENDING FINITE ELEMENT CONTAINING INTEGRATION PARAMETERS	71-0179
PUTTER, S.	FLEXURE ANALYSIS OF AXIALLY-LOADED BEAM BY FINITE-ELEMENT METHOD WITH INTEGRAL PARAMETERS	75-0462
PUTTER, S.	USE OF INTEGRAL PARAMETERS IN ROD PROBLEMS	74-1194
PUTZEYS, J.	STRESS ANALYSIS OF A GRAPHITE FUEL ELEMENT INCLUDING RADIATION CREEP AND DIMENSIONAL CHANGES	71-0217
PUTZEYS, J.	STRESS ANALYSIS OF PRESSURE VESSEL NOZZLES	73-0823
QUAASS, S. T.	THREE DIMENSIONAL STRESS ANALYSIS OF PENETRATIONS IN THICK WALLED PRESSURE VESSELS	74-0422
QUAST, J.	APPLICATION OF FINITE ELEMENT METHODS IN GEOHYDRAULICS	74-0078
QUAST, J.	PRAKTISCHE BEISPIELE DER DIGITALEN BERECHNUNG VON GRUNDWASSERSTROMUNGS PROBLEMEN NACH DER METHODE EN	73-0081
RABINS, M. J.	SIMPLIFICATION OF HIGH ORDER MECHANICAL SYSTEMS USING THE ROUTH APPROXIMATION	75-0853
RABIZADEH, R. O	DYNAMIC ANALYSIS OF CURVED BOX-GIRDER BRIDGES	75-0280
RADAJ, D.	COMPARATIVE CONSIDERATION OF PARAMETRIC STRESS CONCENTRATION INVESTIGATIONS WITH FINITE ELEMENTS AND	72-0423
RADAJ, D.	FINITE-ELEMENT METHOD FOR DETERMINING STRENGTH OF WELDED STRUCTURES	72-0272
RADAJ, D.	ACCURACY OF THE FINITE ELEMENT ANALYSIS FOR THE ELASTIC PLATE WITH A CIRCULAR HOLE	73-0155
RADAJ, D.	ANWENDUNG DES FINITE ELEMENT VERFAHRENS SUR BESTIMMUNG DER FESTIGKERIT VON SCHWEISSKONSTRUKTIONEN. (	72-0185
RADAJ, D.	BERECHNUNG DER SCHWEISSEIGENSPANNUNGEN UND SCHWEISSFORMAENDERUNGEN MIT ELASTISCH-PLASTISCHEN FINITEN	75-0271
RADAJ, D.	COMPARATIVE OBSERVATION ON PARAMETRIC STRESS STUDIES WITH FINITE ELEMENTS WITH COMPLEX STRESS FUNCTI	72-0250
RADAJ, D.	COMPUTER-UNTERSTUETZTES KONSTRUIEREN, PRAKTISCHE BEDEUTUNG DER FINIT-ELEMENT-VERFAHREN	71-0552
RADAJ, D.	EINSATZ DES COMPUTERS BEI DER ENTWICKLUNG EINER PANZERWANNE (THE USE OF A COMPUTER FOR DEVELOPING A	72-0211
RADAJ, D.	FESTIGKEITSNACHWEISE	74-0258
RADAJ, D.	FINITE ELEMENT ANALYSIS, AN AUTOMOBILE ENGINEER'S TOOL	74-0238
RADAJ, D.	FORCES AND DEFORMATIONS AT THE LATERAL STIFFENER OF THIN WALLED U-BARS	73-0236
RADAJ, D.	KERBSPANNUNGSANALYSE NACH DER FINITE-ELEMENT-METHODE AN KEHLNAEHTEN. (NOTCH EFFECT ANALYSIS USING TH	75-0673
RADAJ, D.	NOTCH STRESSES AT THE LOAD CARRYING ELASTIC CORE	73-0266
RADAJ, D.	PARAMETRISCHE KERBSPANNUNGSUNTERSUCHUNG AM ELASTISCHEN KERN.(PARAMETRICAL INVESTIGATION OF NOTCH STR	72-0208
RADAJ, D.	VOLLSTAENDIGE SPANNUNGS-DEHNUNGS-TEMPERATURAENDERUNGSBEZIEHUNG FUER DIE SCHWEISSEIGENSPANNUNGSBERECH	75-0807
RADHAKRISHNAN,	ANALYSIS OF PLANE STRAIN TEST USING THE FINITE ELEMENT METHOD	72-1002
RADHAKRISHNAN,	ANALYSIS OF STRESS AND STRAIN DISTRIBUTIONS IN TRIAXIAL TESTS USING THE METHOD OF FINITE ELEMENTS	73-0903
RADHAKRISHNAN,	SOIL RESISTANCE MOVEMENT RELATIONSHIPS FOR LATERALLY LOADED PILES IN CLAYS AT GREAT DEPTHS	70-0162
RADHAKRISHNAN,	SOLUTION OF SOME PLANE STRAIN PROBLEMS IN SOIL MECHANICS USING THE METHOD OF FINITE ELEMENTS	69-0521
RADHAMOH, S. K.	BUCKLING OF SHELLS BY FINITE ELEMENT METHOD	74-0091
RAESTAD, A. E.	MARINE PROPELLER BLADES STRESSES AND DEFORMATIONS WITH GIVEN, FROZEN, HYDRODYNAMIC LOADING	73-0474
RAFINEJA, P.	CURVILINEAR FINITE ELEMENTS FOR 2-DIMENSIONAL SATURABLE MAGNETIC FIELDS	74-0562
RAGHAVA, R. S.	THEORETICAL AND EXPERIMENTAL VIBRATION ANALYSIS OF AN OBLIQUE SPACE FRAME	70-0460
RAGLE, R.	STARDYNE STRUCTURAL ANALYSIS SYSTEM, ANALYTICAL CAPABILITY MANUAL	68-0235
RAHE, A. E.	DEVELOPMENT OF A FINITE ELEMENT APPROACH FOR APPROXIMATE ANALYSIS OF UNSTEADY COMPRESSIBLE FLUID FLO	72-0828
RAHMAN, K. M. K	THEORETICAL STUDY OF THE SEQUENCE OF FAILURE IN PRECAST PANEL SHEAR WALLS	73-0490
RAI,	FINITE ELEMENT ANALYSIS OF ANISOTROPIC PLATES USING Q-19 ELEMENT: PART 11. INSTRUCTIONS FOR USERS AN	74-0731
RAI, I. S.	FINITE ELEMENT ANALYSIS OF ANISOTROPIC PLATES USING Q-19 ELEMENT. PART 1. THEORETICAL DEVELOPMENT	74-0728
RAI, I. S.	FINITE ELEMENT ANALYSIS OF ANISOTROPIC PLATES USING Q-19 ELEMENT - 1, 2	74-0985
RAIZER, V. D.	THE DEVELOPMENT OF APPLIED METHODS IN PROBLEMS OF STATIC CALCULATION OF THIN-WALLED THREE-DIMENSIONA	73-0934
RAJAIAH, K.	DETERMINATION OF STRESSES DUE TO DISCONTINUITIES IN FINITE PLATES OF ISOTROPIC AND ORTHOTROPIC MATER	73-1052
RAJASEKARAN, S.	COUPLED LOCAL BUCKLING IN WIDE-FLANGE BEAM-COLUMNS	72-0235
RAJASEKARAN, S.	FINITE ELEMENT ANALYSIS OF THIN-WALLED MEMBERS OF OPEN SECTION	72-0966
RAJASEKARAN, S.	FINITE ELEMENT ANALYSIS OF REINFORCED CONCRETE	74-0356
RAJASEKARAN, S.	FINITE ELEMENT LARGE DEFLECTION ANALYSIS OF THIN-WALLED BEAMS	74-0444
RAJASEKARAN, S.	FINITE ELEMENT SOLUTION OF INELASTIC BEAM EQUATIONS	73-0234
RAJASEKARAN, S.	INCREMENTAL FINITE ELEMENT MATRICES	73-0475
RAJASEKARAN, S.	INCREMENTAL FINITE-ELEMENT MATRICES	75-0635
RAJASEKARAN, S.	TECHNIQUE FOR FORMULATING BEAM EQUATIONS	75-0584
RAJENDRA, R.	INFLUENCE OF JOINTS IN PANELIZED STRUCTURAL SYSTEMS	72-0204
RAJU, I. S.	A COMPARATIVE STUDY OF VARIABLE AND CONSTANT THICKNESS HIGH PRECISION TRIANGULAR PLATE BENDING ELEME	74-0857
RAJU, I. S.	A POWERFUL HYBRID METHOD IN FINITE ELEMENT ANALYSIS	71-0218
RAJU, I. S.	AN INHOMOGENEOUS, TAPERED FINITE ELEMENT	73-1038
RAJU, I. S.	BUCKLING OF SHELLS BY FINITE ELEMENT METHOD	74-0091
RAJU, I. S.	DISCONTINUITY STRESS ANALYSIS OF PRESSURE-VESSELS USING FINITE-ELEMENT METHOD	75-0346
RAJU, I. S.	FINITE ELEMENT ANALYSES IN BIOMECHANICS	72-0741
RAJU, I. S.	FINITE ELEMENT ANALYSIS OF STRESS CONCENTRATIONS AND SINGULARITIES	72-0841
RAJU, I. S.	REINVESTIGATION OF BUCKLING OF SHELLS OF REVOLUTION BY A REFINED FINITE ELEMENT	74-0097
RAJU, I. S.	SECTOR ELEMENTS FOR MATRIX DISPLACEMENT ANALYSIS	73-1049
RAJU, I. S.	SPECIAL FINITE ELEMENTS FOR THE ANALYSIS OF STRESS CONCENTRATIONS AND SINGULARITIES	71-0219
RAJU, I. S.	STIFFNESS MATRICES FOR SECTOR ELEMENTS	69-0219
RAJU, I. S.	VIBRATION OF RECTANGULAR PLATES WITH MIXED BOUNDARY CONDITIONS	73-1086
RAJU, K. K.	CALCULATION OF NON-LINEAR AXISYMMETRIC VIBRATIONS OF THIN SHELLS OF REVOLUTION BY A FINITE ELEMENT M	75-0067
RAJU, S. P. G.	A MATRIX METHOD FOR VIBRATION AND STABILITY PROBLEMS	66-0091
RAJU, S. P. G.	SOME APPLICATIONS OF MATRIX FORCE METHOD	66-0092
RALEIGH, C. B.	EARTHQUAKE TRIGGERING BY FLUID INJECTION AT RANGELY, COLORADO	72-0634
RALSTON, T. D.	GALERKIN METHOD AS APPLIED TO PROBLEMS IN VISCOELASTICITY	74-1039
RAMAKRISHNAN, V	BEHAVIOR OF ISOLATED SHEAR WALLS SUBJECTED TO LATERAL LOADS	75-0704
RAMAKRISHNAN, V	COMPARISON OF THE THREE-DIMENSIONAL ANALYSIS OF A SHEAR WALL BUILDING BY TWO DIFFERENT METHODS	75-0705
RAMAMURTHY, T.	BOUNDS AND ERROR CONTROL FOR EIGENVALUES	73-0319
RAMAMURTI, V.	APPLICATION OF SIMULTANEOUS ITERATION METHOD TO TORSIONAL VIBRATION PROBLEMS	73-0660
RAMANATHA, I.	IN-SITU HORIZONTAL STRESSES IN LATERITE BY FINITE ELEMENT ANALYSIS	73-1048
RAMANI, D. T.	PLASTIC SANDWICH BUCKLE-SHELL STRUCTURES	72-0119
RAMANI, D. T.	THE FINITE ELEMENT ANALYSIS OF SANDWICH BUCKLE-SHELL STRUCTURES	72-0968
RAMANI, D. T.	TRIANGULAR THICK SHELL FINITE ELEMENT FOR THE THREE-DIMENSIONAL STRESS ANALYSIS OF PRESSURE VESSELS	73-0860
RAMASWAMY, G. S	SECTOR ELEMENT FOR THIN PLATE FLEXURE	70-0174
RAMESH, C. K.	ANALYSIS OF A SKEW SLAB-BEAM SYSTEM BY FINITE ELEMENT METHOD	73-0233
RAMESH, C. K.	FREE VIBRATIONS OF PLATE-BEAM SYSTEMS	73-0322
RAMEY, G. E.	A METHOD TO GENERATE BOTH UPPER AND LOWER BOUNDS TO PLATE EIGENVALUES BY CONFORMING DISPLACEMENT FIN	71-0166
RAMEY, G. E.	A STUDY OF THE CONVERGENCE CHARACTERISTICS AND ACCURACY OF CONFORMING FINITE ELEMENT SOLUTIONS	72-0967

86

RAMEY, G. E.	ERROR ESTIMATES FOR CONFORMING FINITE ELEMENT SOLUTIONS	74-0940
RAMEY, G. E.	SOME EFFECTS OF SYSTEM IDEALIZATIONS, SINGULARITIES AND MESH PATTERNS ON FINITE ELEMENT SOLUTIONS	74-0939
RAMEY, G. E.	THE BEHAVIOR OF STATISTICALLY HETEROGENEOUS EXCAVATED EARTH SLOPES	74-1330
RAMEY, G. E.	THEORETICAL AND NUMERICAL CONVERGENCE RATES FOR SOME CONFORMING FINITE ELEMENTS	74-0941
RAMEY, M. R.	A FINITE ELEMENT COMPUTER PROGRAM FOR THE PREDICTION OF THE BEHAVIOR OF REINFORCED AND PRESTRESSED C	72-0799
RAMEY, M. R.	REINFORCING REQUIREMENTS FOR CONCRETE BEAMS WITH LARGE WEB OPENINGS	73-0198
RAMEY, M. R.	VIERENDEEL BENT CAPS	72-0758
RAMM, E.	FINITE ELEMENT FORMULATIONS FOR LARGE DISPLACEMENT AND LARGE STRAIN ANALYSIS	73-0457
RAMM, E.	FINITE ELEMENT FORMULATIONS FOR LARGE DEFORMATION DYNAMIC ANALYSIS	75-0564
RAMSDEN, J. N.	MASS CONDENSATION, A SEMI-AUTOMATIC METHOD FOR REDUCING THE SIZE OF VIBRATION PROBLEMS	69-0220
RAMSDEN, V. S.	APPLICATION OF THE FINITE ELEMENT METHOD TO EDDY CURRENT PROBLEMS	74-1156
RAMSEY, J. W. J	FINITE ELEMENT ANALYSIS FOR BUCKLING OF SEMI-RIGID SPACE FRAMES	68-0270
RAMSEY, J. W. J	STRESS CONCENTRATION FACTORS FOR CIRCULAR, REINFORCED PENETRATIONS IN PRESSURIZED CYLINDRICAL SHELLS	75-0726
RAMSTAD, H.	CONVERGENCE AND NUMERICAL ACCURACY WITH SPECIAL REFERENCE TO PLATE BENDING	69-0221
RAMSTAD, H.	FINITE ELEMENT METHODS IN STRESS ANALYSIS	69-0410
RANBIR, S.	FINITE ELEMENT ANALYSIS OF ANISOTROPIC PLATES USING 0-19 ELEMENT: PART 11. INSTRUCTIONS FOR USERS AN	74-0731
RAND, R. A.	OPTIMUM DESIGN OF COMPOSITE SHELLS SUBJECT TO NATURAL FREQUENCY CONSTRAINTS	73-0154
RANGAIAH, V. P.	FINITE ELEMENT FOR TIMOSHENKO BEAM BASED ON MECHANICAL IMPEDANCE	71-0607
RANGAIAH, V. P.	SHOCK ANALYSIS OF STRUCTURAL SYSTEMS	70-0426
RANGASWAMY, T.	FINITE ELEMENT ANALYSIS OF ORTHOGONAL FRAME-PLATE INTERACTION	70-0508
RANKEN, R. E.	TUNNEL DESIGN CONSIDERATIONS: ANALYSIS OF MEDIUM-SUPPORT INTERACTION	74-1062
RANKIN, D.	MAGNETOTELLURIC RESPONSE OF A 2-DIMENSIONAL SLOPING CONTACT BY FINITE ELEMENT METHOD	73-0130
RANKIN, D.	MAGNETOTELLURIC RESPONSE OF LATERALLY INHOMOGENEOUS AND ANISOTROPIC MEDIA	75-0850
RAO, J. K. S.	STUDIES ON TWO PHASE COMPOSITE MATERIALS THROUGH COMPUTER SIMULATION	73-0315
RAO, A. K.	A FINITE-ELEMENT ANALOGUE OF THE MODIFIED RAYLEIGH-RITZ METHOD FOR VIBRATION PROBLEMS	72-0478
RAO, A. K.	A MATRIX METHOD FOR VIBRATION AND STABILITY PROBLEMS	66-0091
RAO, A. K.	A POWERFUL HYBRID METHOD IN FINITE ELEMENT ANALYSIS	71-0218
RAO, A. K.	BOUNDS AND ERROR CONTROL FOR EIGENVALUES	73-0319
RAO, A. K.	DETERMINATION OF STRESSES DUE TO DISCONTINUITIES IN FINITE PLATES OF ISOTROPIC AND ORTHOTROPIC MATER	73-1052
RAO, A. K.	FINITE ELEMENT MODELLING OF NATURAL VIBRATION PROBLEMS	73-0320
RAO, A. K.	REVIEW OF CONTINUUM, FINITE ELEMENT AND HYBRID TECHNIQUES IN THE ANALYSIS OF STRESS CONCENTRATIONS I	73-1051
RAO, A. K.	REVIEW OF CONTINUUM, FINITE ELEMENT AND HYBRID TECHNIQUES IN ANALYSIS OF STRESS CONCENTRATIONS IN ST	74-0914
RAO, A. K.	SECTOR ELEMENTS FOR MATRIX DISPLACEMENT ANALYSIS	73-1049
RAO, A. K.	SPECIAL FINITE ELEMENTS FOR THE ANALYSIS OF STRESS CONCENTRATIONS AND SINGULARITIES	71-0219
RAO, A. K.	SPECIAL FINITE ELEMENTS FOR FLAW AND DISCONTINUITY STRESSES IN COMPOSITES	73-0846
RAO, A. K.	STIFFNESS MATRICES FOR SECTOR ELEMENTS	69-0219
RAO, A. K.	STRESS CONCENTRATIONS AND SINGULARITIES AT INTERFACE CORNERS	71-0419
RAO, G. V.	A COMPARATIVE STUDY OF THE CONSISTENT AND SIMPLIFIED FINITE ELEMENT ANALYSES OF EIGENVALUE PROBLEMS	70-0373
RAO, G. V.	AN INVERSE FORM OF RAMBERG - OSGOOD FORMULA	69-0479
RAO, G. V.	BOUNDS AND ERROR CONTROL FOR EIGENVALUES	73-0319
RAO, G. V.	CALCULATION OF NON-LINEAR AXISYMMETRIC VIBRATIONS OF THIN SHELLS OF REVOLUTION BY A FINITE ELEMENT M	75-0067
RAO, G. V.	DISCONTINUITY STRESS ANALYSIS OF PRESSURE-VESSELS USING FINITE-ELEMENT METHOD	75-0346
RAO, G. V.	FINITE ELEMENT MODELLING OF NATURAL VIBRATION PROBLEMS	73-0320
RAO, G. V.	FINITE-ELEMENT ANALYSIS OF MODERATELY THICK SHELLS	75-0588
RAO, G. V.	GALERKIN FINITE-ELEMENT SOLUTION FOR STABILITY OF CANTILIVER COLUMNS SUBJECTED TO TANGENTIAL LOADS	75-0618
RAO, G. V.	REINVESTIGATION OF BUCKLING OF SHELLS OF REVOLUTION BY A REFINED FINITE ELEMENT	74-0097
RAO, G. V.	STABILITY ANALYSIS OF COLUMNS USING FINITE ELEMENTS	69-0478
RAO, G. V.	VIBRATION OF RECTANGULAR PLATES WITH MIXED BOUNDARY CONDITIONS	73-1086
RAO, J. K. S.	DISCRETE METHODS IN THE ANALYSIS OF PLATE SYSTEMS	71-0835
RAO, M. S.	PLASTICITY THEORY OF FIBROUS COMPOSITES - PART 11	73-0750
RAO, M. S. M.	PLASTICITY THEORY OF FIBROUS COMPOSITES - PART 1	73-0751
RAO, M. S. M.	YIELDING IN UNIDIRECTIONAL COMPOSITES UNDER EXTERNAL LOADS AND TEMPERATURE CHANGES.	73-0169
RAO, R. V. N.	GALERKIN FINITE-ELEMENT SOLUTION FOR STABILITY OF CANTILIVER COLUMNS SUBJECTED TO TANGENTIAL LOADS	75-0618
RAO, S. S.	AUTOMATED OPTIMUM DESIGN OF AIRCRAFT WINGS TO SATISFY STRENGTH, STABILITY, FREQUENCY AND FLUTTER REQ	72-0605
RAO, S. S.	FINITE ELEMENT FLUTTER ANALYSIS OF MULTI-WEB WING STRUCTURES	75-0004
RAO, S. S.	OPTIMIZATION OF COMPLEX STRUCTURES TO SATISFY STATIC, DYNAMIC AND AEROELASTIC REQUIREMENTS	74-0272
RAPHAEL, J. M.	CONSTRUCTION STRESSES IN DWORSHAK DAM	65-0029
RAPHAEL, J. M.	FINITE ELEMENT ANALYSIS OF ARCH DAMS AND FOUNDATIONS	71-0668
RAPPAZ, J.	NUMERICAL STUDY OF EIGEN OSCILLATIONS OF A SCREW PINCH WITH FINITE ELEMENTS	74-0159
RAPPAZ, J.	NUMERICAL STUDY OF PROPER MODES OF A HOMOGENEOUS CYLINDRICAL PLASMA BY METHOD OF FINITE ELEMENTS	74-0160
RAQUET, E.	EINSATZ MODERNER RECHENVERFAHREN ZUR ERMITTLUNG DER SPANNUNGEN UND VERFORMUNGEN BEI RADSAETZEN. (USE	75-0186
RAQUET, E.	UNTERSUCHUNGEN ZUR SPURMASSTABILITAET BEI RADSAETZEN MIT KLOTZGEBREMSTEN VOLLRAEDERN (INVESTIGATIONS	75-0810
RAS, G. V.	BUCKLING OF SHELLS BY FINITE ELEMENT METHOD	74-0091
RASHID, Y.	SOLUTION OF ELASTO-STATIC BOUNDARY VALUE PROBLEMS BY THE FINITE ELEMENT METHOD	64-0025
RASHID, Y. R.	ANALYSIS OF AXISYMMETRIC COMPOSITE STRUCTURES BY THE FINITE ELEMENT METHOD	66-0066
RASHID, Y. R.	ANALYSIS OF NUCLEAR FUEL ROD LOCALIZED STRESSES	73-0708
RASHID, Y. R.	COMPUTATIONAL METHODS IN SOLID MECHANICS AND STRESS ANALYSIS	69-0006
RASHID, Y. R.	FINITE ELEMENT ANALYSIS OF AXISYMMETRIC SOLIDS	65-0011
RASHID, Y. R.	FINITE ELEMENT ANALYSIS OF AXISYMMETRIC COMPOSITE STRUCTURES	65-0030
RASHID, Y. R.	MATHEMATICAL MODELING AND ANALYSIS OF FUEL RODS	74-0930
RASHID, Y. R.	NONLINEAR ANALYSIS OF TWO-DIMENSIONAL PROBLEMS IN CONCRETE CREEP	71-0299
RASHID, Y. R.	NONLINEAR CREEP ANALYSIS AT ELEVATED TEMPERATURE	71-0050
RASHID, Y. R.	ON FURTHER APPLICATION OF THE FINITE ELEMENT METHOD TO THREE-DIMENSIONAL ELASTIC ANALYSIS	70-0252
RASHID, Y. R.	PART 1. THEORY REPORT FOR CREEP-PLAST COMPUTER PROGRAM: ANALYSIS OF TWO-DIMENSIONAL PROBLEMS UNDER S	72-0765
RASHID, Y. R.	PRESSURE VESSELS ANALYSIS BY FINITE ELEMENT TECHNIQUES	67-0069
RASHID, Y. R.	PRESTRESSED CONCRETE REACTOR VESSEL ANALYTICAL METHODS DEVELOPMENTS	68-0141
RASHID, Y. R.	SAFE-COLAPS: A FINITE ELEMENT PROGRAM FOR THE CREEP COLLAPSE ANALYSIS OF NUCLEAR FUEL RODS	75-0573
RASHID, Y. R.	SAFE-CREEP: A COMPUTER PROGRAM FOR THE VISCOELASTIC ANALYSIS OF AXISYMMETRIC AND PLANE CONCRETE STRU	67-0148
RASHID, Y. R.	SOLUTION OF ELASTO-STATIC BOUNDARY VALUE PROBLEMS BY THE FINITE ELEMENT METHOD	64-0030
RASHID, Y. R.	STRESS ANALYSIS OF TWO-DIMENSIONAL PROBLEMS UNDER SIMULTANEOUS CREEP AND PLASTICITY	71-0220
RASHID, Y. R.	STRUCTURAL ANALYSIS OF SHELL INTERSECTIONS	69-0217
RASHID, Y. R.	THREE-DIMENSIONAL ANALYSIS OF ELASTIC SOLIDS I. ANALYSIS PROCEDURE	69-0222
RASHID, Y. R.	THREE-DIMENSIONAL ANALYSIS OF REACTOR PRESSURE VESSEL NOZZLES	71-0221
RASHID, Y. R.	THREE-DIMENSIONAL APPLICATION OF THE FINITE ELEMENT METHOD	70-0250
RASHID, Y. R.	VISCOELASTIC RESPONSE OF GRAPHITE MATERIALS IN IRRADIATION ENVIRONMENTS	70-0314
RASMUSSEN, H.	UNIFORM CONVERGENCE OF VARIATIONAL FINITE ELEMENT AND RITZ METHODS	74-0420
RATTI, U.	EDDY CURRENTS SCREENING PROBLEMS SOLVED BY MEANS OF FINITE ELEMENT DIFFERENCE AND FINITE ELEMENT MET	75-0552
RATTI, U.	SU LA DETERMINAZIONE DEI CAMPI MAGNETICI LINEARI E NON LINEARI MEDIANTE L APPLICAZIONE DEI METODI DE	74-1061
RAUHUT, J. B.	A FINITE-ELEMENT METHOD FOR ANALYSIS OF ANCHORED BULKHEADS AND ANCHOR WALLS	66-0130
RAUSCHENPLAT, H	CLOSURE ANALYSIS USING SELF ENERGIZING, DOUBLE CONICAL GASKETS	74-0675
RAVENSCRAFT, R.	PARAMETRIC STUDY OF A BEAM WITH A COMPOUND SIDE-BRANCH RESONATOR AS A DEVICE TO EVALUATE PRELIMINARY	70-0465
RAVERA, R. J.	RESPONSE OF SIMPLY SUPPORTED TIMOSHENKO BEAMS	69-0489
RAVIART, P. A.	A MIXED FINITE ELEMENT METHOD FOR THE BIHARMONIC EQUATION	74-1227
RAVIART, P. A.	CONFORMING AND NONCONFORMING FINITE ELEMENT METHODS FOR SOLVING THE STATIONARY STOKES EQUATIONS - 1	74-0390
RAVIART, P. A.	CONVERGENCE AND ERROR ESTIMATES FOR FINITE ELEMENT APPROXIMATIONS OF STATIONARY NAVIER-STOKES EQUATI	74-0122
RAVIART, P. A.	GENERAL LAGRANGE AND HERMITE INTERPOLATION IN R**N WITH APPLICATIONS TO FINITE ELEMENT METHODS	72-0110
RAVIART, P. A.	HYBRID FINITE ELEMENT METHODS FOR SOLVING 2ND ORDER ELLIPTIC EQUATIONS	74-1267
RAVIART, P. A.	INTERPOLATION THEORY OVER CURVED ELEMENTS WITH APPLICATION TO THE FINITE ELEMENT METHOD	72-0356
RAVIART, P. A.	LA METHODE DES ELEMENTS FINIS POUR LES PROBLEMES AUX LIMITES ELLIPTIQUES	74-1228
RAVIART, P. A.	LAGRANGE INTERPOLATION ON CURVED FINITE ELEMENTS	72-0163
RAVIART, P. A.	MAXIMUM PRINCIPLE AND UNIFORM CONVERGENCE FOR THE FINITE ELEMENT METHOD	73-0852
RAVIART, P. A.	METHODE DES ELEMENTS FINIS	73-1053

87

RAVIART, P. A.	NUMERICAL SOLUTION OF THE STATIONARY NAVIER-STOKES EQUATIONS BY FINITE ELEMENT METHODS	74-0408
RAVIART, P. A.	ON A FINITE ELEMENT METHOD FOR SOLVING THE NEUTRON TRANSPORT EQUATION	74-1255
RAVIART, P. A.	ON A FINITE ELEMENT METHOD FOR SOLVING THE NEUTRON TRANSPORT EQUATION	74-1355
RAVIART, P. A.	THE COMBINED EFFECT TO CURVED BOUNDARIES AND NUMERICAL INTEGRATION IN ISOPARAMETRIC FINITE ELEMENT M	72-0357
RAVIART, P. A.	THE USE OF NUMERICAL INTEGRATION IN FINITE ELEMENT METHODS FOR SOLVING PARABOLIC EQUATIONS	72-0351
RAVINGER, J.	GEOMETRICKY NELINEARNE ULOHY PLOCHYCH SKRUPIN RIESENE PRIRASTKOVOU METODOU KONECNYCH PRVKOV. (GEOMET	74-1032
RAWTANI, S.	NATURAL FREQUENCIES OF WIDE CHORD COMPRESSOR BLADES	74-0231
RAWTANI, S.	PSEUDO STATIC DEFORMATION AND FREQUENCIES OF ROTATING TURBOMACHINERY BLADES	72-0239
RAWTANI, S.	THE EFFECT OF CAMBER ON THE NATURAL FREQUENCIES OF LOW ASPECT RATIO TURBOMACHINERY BLADES	73-1092
RAY, G.	FINITE ELEMENT ANALYSIS METHODS IN DYNAMICS OF PULSATILE FLOW	69-0519
RAYMOND, G.	PREDICTION OF UNDRAINED MOVEMENTS CAUSED BY EMBANKMENTS ON MUSKEG	70-0070
RAYMOND, G. P.	FOUNDATION FAILURE OF NEW LISKEARD EMBANKMENT	73-0549
RAYMOND, L.	USE OF FINITE ELEMENT ANALYSIS IN FRACTURE MECHANICS	74-0837
RAZZAQUE, A.	A CONFORMING TRIANGULAR ELEMENT IN THE ANALYSIS OF THIN AND SEMI-THICK PLATES IN BENDING	70-0548
RAZZAQUE, A.	A FURTHER MODIFICATION TO AHMAD'S SHELL ELEMENT	73-0619
RAZZAQUE, A.	EXPERIENCE WITH THE PATCH TEST FOR CONVERGENCE OF FINITE ELEMENTS	72-0365
RAZZAQUE, A.	FINITE ELEMENT ANALYSIS OF PLATES AND SHELLS	72-0409
RAZZAQUE, A.	FINITE ELEMENT ANALYSIS OF PLATES ALLOWING FOR TRANSVERSE SHEAR DEFORMATION EFFECTS	73-0996
RAZZAQUE, A.	INTRODUCTION TO SHEARS DEFORMATION INTO A THIN PLATE DISPLACEMENTS FORMULATIONS	73-0617
RAZZAQUE, A.	PROGRAM FOR TRIANGULAR BENDING ELEMENTS WITH DERIVATIVES SMOOTHING	72-0473
RAZZAQUE, A.	SHAPE FUNCTION FORMULATIONS FOR ELEMENTS OTHER THAN DISPLACEMENT MODELS	72-0472
RAZZAQUE, A.	SHAPE FUNCTION SUBROUTINE FOR AN ISOPARAMETRIC THIN PLATE ELEMENT	73-0618
RAZZAQUE, A.	SIMPLE FINITE ELEMENT SOLUTION FOR PLATES OF HOMOGENEOUS, SANDWICH AND CELLULAR CONSTRUCTION	75-0140
RAZZAQUE, A.	THE EVOLUTION OF THE ISOPARAMETRIC ELEMENTS	75-0425
RECKE, H. G.	BERECHNUNG VON LAUFRAEDERN MIT DER METHODE DER FINITEN ELEMENTE (DESIGN OF ROTATING WHEELS BY THE FI	74-0547
RECKE, H. G.	COMPUTATION OF IMPELLERS BY METHOD OF FINITE ELEMENTS	74-0646
RECKE, H. G.	PROBLEMS OF CROSS-LINKAGE IN USING METHOD OF FINITE ELEMENTS	74-0165
RECKER, W. W.	NONLINEAR DYNAMICS OF CABLES WITH LOW INITIAL TENSION	72-0740
RED, W. E.	FINITE ELEMENT ANALYSIS OF STATIC AND DYNAMIC STRESSES AROUND SUDDENLY PUNCHED HOLES IN PLATES AND S	73-1088
REDDAWAY, A. R.	RESPONSE OF STRUCTURE TO THE PSEUDO-SOUND FIELD OF A JET (USING A COMBINED CONTINUUM AND FINITE ELEM	67-0044
REDDI M. M.	DYPLAS, A FINITE ELEMENT DYNAMIC ELASTIC-PLASTIC LARGE DEFORMATION ANALYSIS PROGRAM	74-0306
REDDI, M. M.	A THREE-DIMENSIONAL FINITE ELEMENT COMPUTER CODE FOR THE ANALYSIS OF COMPLEX STRUCTURES	71-0275
REDDI, M. M.	ANALYSIS OF ASYMMETRIC STIFFENED SHELL TYPE STRUCTURES BY FINITE ELEMENT METHOD II. SIX DEGREE-OF-FR	69-0373
REDDI, M. M.	FINITE ELEMENT SOLUTION OF THE INCOMPRESSIBLE LUBRICATION PROBLEM	69-0021
REDDI, M. M.	FINITE ELEMENT SOLUTION OF THE STEADY STATE COMPRESSIBLE LUBRICATION PROBLEM	70-0007
REDDI, M. M.	FINITE ELEMENT SOLUTION OF LUBRICATION PROBLEMS	71-0683
REDDY, D. P.	MAGAZINE HEADWALL RESPONSE TO EXPLOSIVE BLAST	74-0754
REDDY, D. V.	FREQUENCY ANALYSIS OF CERTAIN SINGLE AND CONTINUOUS SPAN BRIDGES	71-0337
REDDY, D. V.	FREQUENCY ANALYSIS OF SKEW ORTHOTROPIC PLATES BY THE FINITE STRIP METHOD	71-0413
REDDY, D. V.	FREQUENCY ANALYSIS OF ORTHOTROPIC CIRCULAR CYLINDRICAL PANELS BY THE FINITE STRIP METHOD	73-0274
REDDY, D. V.	USE OF FINITE ELEMENT METHOD IN THE DYNAMICS OF PLATES AND SHELLS	70-0313
REDDY, I. K.	MAGNETOTELLURIC RESPONSE OF A 2-DIMENSIONAL SLOPING CONTACT BY FINITE ELEMENT METHOD	73-0130
REDDY, I. K.	MAGNETOTELLURIC RESPONSE OF LATERALLY INHOMOGENEOUS AND ANISOTROPIC MEDIA	75-0850
REDDY, J. N.	A MATHEMATICAL THEORY OF COMPLEMENTARY-DUAL VARIATIONAL PRINCIPLES AND MIXED FINITE ELEMENT APPROXIM	73-0804
REDDY, J. N.	ACCURACY AND CONVERGENCE OF MIXED FINITE-ELEMENT APPROXIMATIONS OF THIN BARS, MEMBRANES, AND PLATES	73-1120
REDDY, J. N.	CONVERGENCE OF MIXED FINITE ELEMENT APPROXIMATIONS OF A CLASS OF LINEAR-BOUNDARY VALUE PROBLEMS	73-0599
REDDY, J. N.	MIXED CONJUGATE FINITE ELEMENT APPROXIMATIONS OF LINEAR OPERATORS	72-0085
REDDY, J. N.	MIXED FINITE-ELEMENT APPROXIMATIONS OF LINEAR BOUNDARY-VALUE PROBLEMS	75-0453
REDDY, J. N.	NOTE ON AN APPROXIMATE METHOD FOR COMPUTING CONSISTENT CONJUGATE STRESSES IN ELASTIC FINITE ELEMENTS	73-0050
REDDY, J. N.	ON COMPLEMENTARY-DUAL VARIATIONAL PRINCIPLES IN MATHEMATICAL PHYSICS	74-0342
REDDY, J. N.	ON MIXED FINITE ELEMENT APPROXIMATIONS	74-0701
REDDY, J. N.	SOME MATHEMATICAL PROPERTIES OF CERTAIN MIXED GALERKIN APPROXIMATIONS IN NONLINEAR ELASTICITY	74-1248
REDDY, J. N.	SOME OBSERVATIONS ON PROPERTIES OF CERTAIN MIXED FINITE ELEMENT APPROXIMATIONS	75-0617
REDDY, J. N.	THEMIS FINAL REPORT, VOL. IV. RECENT DEVELOPMENTS IN THE NUMERICAL SOLUTION OF CERTAIN PROBLEMS IN N	74-0721
REDSHAW, S. C.	THREE DIMENSIONAL ANALYSIS OF TWO ARCH DAMS BY THE FINITE ELEMENT METHOD	68-0018
REDWOOD, R. G.	ANALYSIS OF THREE-DIMENSIONAL THIN WALLED STRUCTURES	69-0188
REDWOOD, R. G.	SHEAR LAG IN STIFFENED BOX GIRDERS	70-0243
REED, W. H.	TRIANGULAR MESH METHODS FOR THE NEUTRON TRANSPORT EQUATION	73-1062
REED, W. H.	TRIPLET: A TWO-DIMENSIONAL, MULTIGROUP, TRIANGULAR MESH, PLANAR GEOMETRY, EXPLICIT TRANSPORT CODE	73-0955
REEFMAN, R. J.	PRINCIPLES OF THE FINITE ELEMENT METHOD	75-0303
REEFMAN, R. J.	TEMPERATURE CALCULATIONS USING THE NASTRAN FINITE ELEMENT PROGRAMME	75-0642
REESE, L. C.	ANALYSIS OF CIRCULAR FOOTINGS ON LAYERED SOILS	70-0281
REESE, L. C.	BEHAVIOUR OF STRIP FOOTINGS ON LAYERED COHESIVE SOILS	69-0223
REESE, L. C.	FINITE ELEMENT METHOD APPLIED TO PROBLEMS IN STRESSES AND DEFORMATION OF SOIL	67-0085
REESE, L. C.	FINITE ELEMENT METHOD FOR PROBLEMS IN SOIL MECHANICS	68-0078
REESE, L. C.	SOIL RESISTANCE MOVEMENT RELATIONSHIPS FOR LATERALLY LOADED PILES IN CLAYS AT GREAT DEPTHS	70-0162
REESE, L. C.	USE OF THE FINITE ELEMENT IN FOUNDATION ENGINEERING	72-0422
REESE, L. C.	FINITE-ELEMENT METHOD APPLIED TO SOME PROBLEMS IN SOIL MECHANICS	68-0077
REEVES, G. N.	A FINITE ELEMENT STUDY OF THE PENETRATION OF A SOIL HALF-SPACE BY A RIGID PROJECTILE	69-0518
REEVES, M.	WATER FLOW THROUGH SATURATED-UNSATURATED POROUS MEDIA USING FINITE-ELEMENT-GALERKIN METHODS	74-0879
REEVES, M.	WATER MOVEMENT THROUGH SATURATED-UNSATURATED POROUS MEDIA: A FINITE-ELEMENT GALERKIN MODEL	75-0761
REICH, M.	ANALYSIS OF BRIDGE DECKS USING POLYMER IMPREGNATED CONCRETE	73-0673
REICH, M.	FINITE ELEMENT APPROACH TO POLYMER CONCRETE BRIDGE DECK DESIGN ANALYSIS	72-0918
REICHERT, K.	DIGITAL CALCULATION OF ELECTRIC FIELDS IN METAL CLAD INSTALLATIONS	73-0212
REID, J. K.	A DISCUSSION ON A MODIFIED CONJUGATE GRADIENT METHOD BY W.L. WOOD	74-0402
REID, J. K.	FORTRAN SUBROUTINES FOR THE SOLUTION OF LAPLACE'S EQUATION OVER A GENERAL REGION IN TWO DIMENSIONS	70-0425
REID, J. K.	ON THE CONSTRUCTION AND CONVERGENCE OF A FINITE ELEMENT SOLUTION OF LAPLACE'S EQUATION	72-0340
REIFSNIDER, K.	EFFECTS OF LOCAL STRENGTH VARIATIONS ON PLASTIC ENCLAVE DEVELOPMENT AT CRACK TIPS	72-0822
REIGSTAD, G. H.	FINITE ELEMENT ANALYSIS OF DRIVEN PILES	74-1105
REIMER, R. B.	DECONVOLUTION OF SEISMIC RESPONSE FOR LINEAR SYSTEMS	73-0774
REINHOUDT, J. P	ON THE STABILITY OF ROTOR-AND-BEARING SYSTEMS AND ON THE CALCULATION OF SLIDING BEARINGS	72-0037
REJALI, H. M.	DETERMINATION OF VOLUME CHANGE STRESSES IN PLAIN AND REINFORCED CONCRETE USING FINITE ELEMENT ANALYS	69-0477
REJALI, H. M.	USER'S GUIDE FOR THE COMPUTER PROGRAM DESCRIBED IN TAM REPORT NO. 680 "DETERMINATION OF VOLUME CHANG	69-0473
REMEDIOS, E. E.	ENGINEERING APPROACH TO FINITE ELEMENT ANALYSIS OF NUCLEAR COMPONENTS	73-0839
REMEDIOS, F. E.	AN ENGINEERING APPROACH TO FINITE ELEMENT ANALYSIS OF NUCLEAR COMPONENTS	73-1054
REMINGTON, P. J	ANALYSIS OF STRESSES AND DEFLECTIONS IN FRAME SUPPORTED TENTS	74-0689
REMSON, I.	NUMERICAL METHODS IN SUBSURFACE HYDROLOGY: WITH AN INTRODUCTION TO THE FINITE ELEMENT METHOD	71-0800
RENICK, J. D.	ANALYSIS OF THE MUTUAL INDUCTANCE PARTICLE VELOCIMETER	74-0712
REPNAU, T.	HIGH-PERFORMANCE COMPOSITE MATERIALS FOR VEHICLE CONSTRUCTION: A FINITE ELEMENT COMPUTER PROGRAM FOR	73-0767
RESTAD, K.	INVESTIGATION ON FREE AND FORCED VIBRATIONS OF A LONG TANKER WITH OVERLAPPING PROPELLER ARRANGEMENT	73-0981
REYER, E.	UEBER DIE PRAKTISCHE ANWENDUNG VON FINITEN ELEMENTEN IN DER PLATTENSTATIK (APPLICATION OF THE FINITE	75-0805
REYES, F.	FINITE ELEMENT FOR THE INTERACTION OF FRAMES AND SHEAR WALLS	73-0681
REYES, J. A.	COMPARISON OF ANALYTICAL AND NUMERICAL SOLUTIONS IN HEAT CONDUCTION PROBLEMS	74-1170
REYES, J. A.	NUMERICAL EXPERIMENTS ON THE DETERMINATION OF STRESS CONCENTRATION FACTORS	74-0292
REYES, S. F.	ELASTO-PLASTIC ANALYSIS OF UNDERGROUND OPENINGS BY THE FINITE ELEMENT METHOD	66-0067
REYNEN, J.	EURCYL: A COMPUTER PROGRAM TO GENERATE FINITE ELEMENT MESHES FOR CYLINDER-CYLINDER INTERSECTIONS	73-0962
REYNEN, J.	EURCYL: A PROGRAM TO GENERATE FINITE ELEMENT MESHES FOR PRESSURE VESSEL NOZZLES	74-1086
REYNEN, J.	ON THE USE OF FINITE ELEMENTS IN FRACTURE ANALYSIS OF PRESSURE VESSEL COMPONENTS	75-0514
REYNEN, J.	STRESS ANALYSIS OF A GRAPHITE FUEL ELEMENT INCLUDING RADIATION CREEP AND DIMENSIONAL CHANGES	71-0217
REYNEN, J.	STRESS ANALYSIS OF PRESSURE VESSEL NOZZLES	73-0823
REYNOLDS, D. T.	STRESS ANALYSIS OF PARACHUTES USING FINITE ELEMENTS	71-0481
REYNOLDS, P. J.	PROCEEDINGS OF INTERNATIONAL ASSOCIATION FOR HYDRAULIC RESEARCH AND PROCEEDINGS OF INTERNATIONAL SYM	73-0534

REYNOLDS, R. J.	PRODUCTION OF MOTION PICTURES OF A CONTINUOUSLY DEFORMING STRUCTURE USING NASTRAN	71-0223
RHEINBOLDT, W.	SELF-ADAPTIVE REFINEMENTS IN THE FINITE ELEMENT METHOD	75-0712
RHEINBOLDT, W.	ON THE SOLUTION OF SETS OF NONLINEAR EQUATIONS ARISING IN THE APPLICATION OF FINITE ELEMENT METHODS	75-0127
RIBEIRO, A. A.	COMPARISON OF ANALYTICAL AND EXPERIMENTAL RESULTS IN THE DESIGN OF A HOLLOW GRAVITY DAM	67-0109
RICARD, A.	CALCUL DE MASSIFS EN ELASTOPLASTICITE APPLICATIONS A LA MECANIQUE DES ROCHES ET DES SOLS. (APPLICATI	74-0223
RICCARDELLA, P.	COMBINED ANALYTICAL - EXPERIMENTAL FRACTURE STUDY OF THE TWO LEADING THEORIES OF ELASTIC-PLASTIC FRA	73-0951
RICCARDELLA, P.	REACTOR COOLANT PUMP FLYWHEELS OVERSPEED EVALUATION	74-0506
RICCIONI, R.	FINITE ELEMENT ANALYSIS OF PRESTRESSED CONCRETE PRESSURE VESSELS	74-1289
RICE, J. R.	A FINITE ELEMENT FORMULATION FOR PROBLEMS OF LARGE STRAIN AND LARGE DISPLACEMENT	70-0069
RICE, J. R.	COMBINED MACROSCOPIC AND MICROSCOPIC APPROACH TO THE FRACTURE OF METALS	74-0897
RICE, J. R.	COMPUTATIONAL FRACTURE MECHANICS	71-0564
RICE, J. R.	DEVELOPMENT OF FINITE ELEMENT COMPUTATIONAL METHODS FOR THREE DIMENSIONAL ELASTIC PLASTIC CRACK ANAL	70-0356
RICE, J. R.	FINITE ELEMENT FORMULATIONS FOR PROBLEMS OF LARGE ELASTIC-PLASTIC DEFORMATION	74-0796
RICE, J. R.	LIMITATIONS TO THE SMALL SCALE YIELDING APPROXIMATION FOR CRACK TIP PLASTICITY	74-0145
RICE, J. R.	NUMERICAL ANALYSIS	70-0432
RICE, J. R.	ON NUMERICALLY ACCURATE FINITE ELEMENTS	74-0797
RICE, J. R.	SMALL SCALE YIELDING NEAR A CRACK IN PLANE STRAIN-FINITE ELEMENT ANALYSIS	71-0355
RICHARD, R. M.	ELASTIC DEFORMATION OF LIGHTWEIGHT MIRRORS METHOD OF LIMIT POINT CALCULATION IN FINITE ELEMENT STRUC	72-0232
RICHARD, R. M.	FINITE ELEMENT ANALYSIS OF INELASTIC STRUCTURES	69-0224
RICHARDS, A. M.	FINITE ELEMENT ANALYSIS OF DISCONTINUOUS SHELLS	68-0301
RICHARDS, B. G.	THE USE OF THE FINITE ELEMENT METHOD IN THE SOLUTION OF THE FLOW EQUATIONS IN SOILS	74-0449
RICHARDS, B. G.	THEORETICAL TRANSIENT BEHAVIOUR OF SATURATED AND UNSATURATED SOILS UNDER LOAD AND CHANGING MOISTURE	73-0901
RICHARDS, D. J.	FINITE ELEMENT TECHNIQUES FOR SOLUTION OF POISSON'S EQUATION	71-0494
RICHARDS, D. J.	FINITE-ELEMENT SOLUTIONS WITHIN CURVED BOUNDARIES	72-0038
RICHARDS, J. A.	WOOD SHEAR PANELS BONDED WITH FLEXIBLE ADHESIVES	75-0086
RICHARDSON, J.	A DERIVATION OF PERTURBED EQUATIONS OF MOTION AND HEAT CONDUCTION OF A THERMOELASTIC SOLID	72-0735
RICHARDSON, J.	MODAL RESPONSE OF FREE ROCKETS UNDER THRUST LOADING	74-0855
RICHARDSON, P.	USE OF FINITE ELEMENT METHODS IN SOLUTION OF TRANSIENT HEAT CONDUCTION PROBLEMS	69-0016
RICKETTS, R. E.	LOCAL LEAST SQUARES STRESS SMOOTHING FOR PARABOLIC ISOPARAMETRIC ELEMENTS	75-0593
RICKS, R. G.	USE OF THE FINITE ELEMENT DAMPED FORCED RESPONSE STRAIN ENERGY DISTRIBUTION FOR VIBRATION REDUCTION	72-0759
RIDGWAY, S.	APPROXIMATE BOUNDARY CONDITIONS IN THE FINITE ELEMENT METHOD	72-0995
RIEGER, N.F.	THE NATURAL FREQUENCIES OF PORTAL FRAMES II	65-0047
RIGBY, G. L.	A STRAIN ENERGY BASIS FOR STUDIES OF ELEMENT STIFFNESS MATRICES	72-0621
RIGBY, G. L.	FURTHER LIMITATIONS ON GENERAL HEXAHEDRON FINITE ELEMENTS WITH DERIVATIVE DEGREES OF FREEDOM.	72-0395
RIGBY, G. L.	MIXED DISPLACEMENT 3D ISOPARAMETRIC FINITE ELEMENTS AND A STRAIN ENERGY BASIS FOR STUDIES OF ELEMENT	70-0286
RIGSBY, D. M.	ON THE NUMERICAL SOLUTION OF A CLASS OF PROBLEMS IN A LINEAR FIRST STRAIN-GRADIENT THEORY OF ELASTIC	70-0318
RIIKONEN, I.	ANALYSIS OF THICK-WALLED PRESSURE VESSELS USING THE CONSTANT STRAIN AXISYMMETRIC FINITE ELEMENT	71-0653
RINDAL, R. A.	BASELINE SOLUTIONS FOR THE SMOOTH WALL THERMOCHEMICAL ABLATION RESPONSE OF GRAPHITE AND CARBON PHENO	71-0728
RING, G. J.	THE SIGNIFICANCE OF THE RATE OF PLASTIC WORK IN ELASTO-PLASTIC ANALYSIS	74-0437
RISH, R. F.	ANALYSIS OF CYLINDRICAL SHELL ROOFS WITH POST TENSIONED EDGEBEAMS	74-1042
RITMAN, E. L.	FINITE ELEMENT ANALYSIS OF LEFT VENTRICULAR MYOCARDIAL STRESSES	74-0669
RIVAS, A. P.	MENISCI IN ARRAYS OF CYLINDERS - NUMERICAL-SIMULATION BY FINITE-ELEMENTS	75-0449
RIVELLO, R. M.	NASTRAN: USER EXPERIENCE WITH FOUR EXAMPLE PROBLEMS	72-0804
RIVIER, M. L.	FINITE-ELEMENT SOLUTIONS OF PLASMA EQUILIBRIUM	75-0455
RIZZO, P. C.	COMPLIANCES OF LAYERED ELASTIC SYSTEMS	74-1204
RIZZO, R. R.	A FINITE ELEMENT ANALYSIS OF LAMINATED ANISOTROPIC TUBES	70-0319
ROBBINS, K.	ANCHORAGE ZONE STRESSES IN AXIALLY POST-TENSIONED I-SECTION MEMBERS WITH END BLOCKS	71-0305
ROBERTS, E. JR.	RELABELING OF FINITE ELEMENT MESHES USING A RANDOM PROCESS	72-0754
ROBERTS, K. V.	THALIA - ONE-DIMENSIONAL MAGNETOHYDRODYNAMIC STABILITY PROGRAM USING METHOD OF FINITE-ELEMENTS	75-0458
ROBERTS, K. V.	THALIA: A ONE-DIMENSIONAL MAGNETOHYDRODYNAMIC STABILITY PROGRAM USING THE METHOD OF FINITE ELEMENTS	74-1085
ROBERTS, T. A.	DYNAMIC STABILITY ANALYSIS BY THE CONJUGATE GRADIENT METHOD	72-0749
ROBERTS, T. M.	APPLICATION OF CURVED FINITE ELEMENTS TO CIRCULAR ARCHES	71-0360
ROBERTS, T. M.	FURTHER STUDIES IN THE APPLICATION OF CURVED FINITE ELEMENTS TO CIRCULAR ARCHES	71-0644
ROBERTS, T. M.	POST-BUCKLING ANALYSIS OF SLIGHTLY CURVED PLATES BY THE FINITE ELEMENT METHOD	69-0365
ROBERTS, T. M.	THE USE OF FINITE ELEMENT MID-INCREMENT STIFFNESS MATRICES IN THE POST BUCKLING ANALYSIS OF IMPERFEC	71-0383
ROBICHAUD, L. P	CURVED TRIANGULAR ELEMENTS FOR THE ANALYSIS OF SHELLS	68-0030
ROBINS, P. J.	MODIFIED FINITE ELEMENT METHOD APPLIED TO R C DEEP BEAMS	73-0476
ROBINSON, J.	A CURVED BEAM ELEMENT	68-0202
ROBINSON, J.	A DERIVATION PROCEDURE FOR THE DYNAMIC FLEXIBILITY MATRIX OF A TRIANGULAR BENDING ELEMENT	70-0103
ROBINSON, J.	A GENERAL EQUILIBRIUM BEAM ELEMENT WITH OFFSETS	73-0351
ROBINSON, J.	A GENERAL EQUILIBRIUM BENDING ELEMENT	73-0699
ROBINSON, J.	A GENERAL EQUILIBRIUM AXIAL ELEMENT	73-0729
ROBINSON, J.	A GENERAL EQUILIBRIUM TORQUE ELEMENT	74-0187
ROBINSON, J.	A MODE AMPLITUDE TECHNIQUE FOR GENERAL FINITE ELEMENT EQUILIBRIUM	74-0283
ROBINSON, J.	A SIX NODE TRIANGULAR EQUILIBRIUM (CONSTANT STRESS) MEMBRANE ELEMENT	73-0783
ROBINSON, J.	A SIX NODE TRIANGULAR SHEAR PANEL FOR EQUILIBRIUM MODELS	73-1011
ROBINSON, J.	A SYNTHESIZED WARPED QUADRILATERAL ELEMENT GIVING SIMPLIFIED INPUT FOR EQUILIBRIUM MODELS	73-1013
ROBINSON, J.	A TRIANGULAR EQUILIBRIUM (LINEAR STRESS) MEMBRANE ELEMENT	75-0883
ROBINSON, J.	A WARPED QUADRILATERAL STRAIN MEMBRANE ELEMENTS	75-0141
ROBINSON, J.	A WARPED QUADRILATERAL EQUILIBRIUM MEMBRANE ELEMENT	75-0885
ROBINSON, J.	ASAS - A COMMERCIAL FINITE ELEMENT SYSTEM WITH DISPLACEMENT AND FORCE CONTINUITY OPTIONS	73-0947
ROBINSON, J.	AUTOMATIC EXTRACTION OF RIGID BODY MODES FROM STRESS AND STRAIN ELEMENTS	72-0913
ROBINSON, J.	BASIS FOR ELEMENT INTERCHANGEABILITY IN FINITE ELEMENT PROGRAMS	71-0224
ROBINSON, J.	BASIS FOR ISOPARAMETRIC STRESS ELEMENTS	73-0532
ROBINSON, J.	DYNAMIC ANALYSIS OF STRUCTURES USING THE RANK FORCE METHOD	71-0553
ROBINSON, J.	EFFECT OF OUT-OF-PLANENESS OF MEMBRANE QUADRILATERAL FINITE ELEMENTS	73-0031
ROBINSON, J.	EIGENVALUES OF COLLINEAR BEAM STRUCTURES USING FINITE ELEMENT TECHNIQUES AND VARIOUS DYNAMIC REPRESE	67-0124
ROBINSON, J.	ELASTIC AND EQUILIBRIUM MATRICES OF GENERAL SEMI-MONOCOQUE MEMBRANE ELEMENTS	70-0159
ROBINSON, J.	ELEMENT STATIC FLEXIBILITY AND STIFFNESS MATRICES FOR THE FINITE ELEMENT ANALYSIS OF BEAM, PLATE, AN	70-0070
ROBINSON, J.	ELEMENTS FOR FINITE ELEMENT STRUCTURAL ANALYSIS PROGRAMS	71-0335
ROBINSON, J.	EQUILIBRIUM MODELS	75-0008
ROBINSON, J.	EVALUATION OF CALAC SEMI-MONOCOQUE MEMBRANE ELEMENT	70-0278
ROBINSON, J.	FLAT QUADRILATERAL EQUILIBRIUM MEMBRANE ELEMENTS	73-1116
ROBINSON, J.	INTEGRATED THEORY OF FINITE ELEMENT METHODS	73-0531
ROBINSON, J.	OPTIMIZATION OF REDUNDANCY SELECTION IN FINITE-ELEMENT FORCE METHOD	70-0228
ROBINSON, J.	PRACTICAL APPLICATION OF A COMPUTERIZED STRUCTURAL ANALYSIS SYSTEM WHICH ADOPTS A FINITE ELEMENT TEC	69-0225
ROBINSON, J.	SOME NEW DEVELOPMENTS IN MATRIX FORCE ANALYSIS	71-0225
ROBINSON, J.	STATIC FLEXIBILITY MATRICES FOR A RECTANGULAR PLATE ELEMENT IN BENDING, TWISTING, AND SHEAR. UNIFORM	68-0217
ROBINSON, J.	THREE NODE ISOPARAMETRIC STRESS AXIAL ELEMENT	72-0992
ROBINSON, J.	WARPED SEMI-MONOCOQUE QUADRILATERAL MEMBRANE ELEMENT FOR THE FINITE ELEMENT METHODS	70-0195
ROBINSON, J. C.	SOLUTION OF DIFFUSION AND P1 FINITE-ELEMENT EQUATIONS BY ITERATION	75-0404
ROBUTTI, G.	FINITE ELEMENT ANALYSIS OF PRESTRESSED CONCRETE PRESSURE VESSELS	74-1289
ROCHE, R.	ANALYSIS OF SHELLS OF ANY FORM IN THE PLASTIC REGION BY THE FINITE ELEMENT METHOD	73-0940
ROCHE, R.	PLASTIC ANALYSIS OF SHELLS BY FINITE ELEMENT METHODS	73-1017
ROCHE, R. V.	INELASTIC ANALYSIS AND SATISFACTION OF DESIGN CRITERIA OF A HIGH TEMPERATURE COMPONENT	75-0265
ROCK, A. O.	ENGINEERING OPERATIONS REPORT: NERVA THIN SHELL FINITE ELEMENT STRESS ANALYSIS METHODOLOGY COMPUTER	72-0916
ROCK, A. O.	ENGINEERING OPERATIONS REPORT: NERVA AXISYMMETRIC AND PLANE STRUCTURES FINITE ELEMENT STRESS ANALYSI	72-0917
ROCK, T.	FREE VIBRATION AND TRANSIENT RESPONSE OF THICK AND THIN PLATES USING THE FINITE ELEMENT METHOD	74-0517
ROCKENHAUSER, W	PRESSURE VESSELS ANALYSIS BY FINITE ELEMENT TECHNIQUES	67-0069
ROCKEY, K. C.	A FINITE ELEMENT SOLUTION FOR FOLDED PLATE STRUCTURES	67-0072
ROCKEY, K. C.	BUCKLING OF PLATE GIRDER WEBS UNDER PARTIAL EDGE LOADINGS	70-0283
ROCKEY, K. C.	FAILURE OF THIN-WALLED MEMBERS UNDER PATCH LOADINGS	72-0230

ROCKEY, K. C.	FINITE ELEMENT SOLUTIONS FOR BUCKLING OF COLUMNS AND BEAMS	71-0361
ROCKEY, K. C.	THE BEHAVIOUR OF SQUARE SHEAR WEBS HAVING A CIRCULAR HOLE	67-0108
ROCKEY, K. C. E	THE FINITE ELEMENT METHOD: A BASIC INTRODUCTION	75-0019
ROCKWELL, R. D.	COMPUTER-AIDED INPUT/OUTPUT FOR USE WITH FINITE ELEMENT METHOD OF STRUCTURAL ANALYSIS	70-0330
ROCKWELL, R. D.	COMPUTER-AIDED INPUT/OUTPUT FOR THE USE WITH FINITE ELEMENT STRUCTURAL ANALYSES	73-0923
RODABAUGH, E. C	INFLUENCE OF END-EFFECTS ON STRESSES AND FLEXIBILITY OF A PIPING ELBOW WITH IN-PLANE MOMENT	74-0534
RODABAUGH, E. C	SIMPLIFIED INELASTIC HIGH TEMPERATURE STRUCTURAL ANALYSIS OF MODERATELY COMPLEX SPATIALLY THREE DIME	74-0445
RODABAUGH, E. C	SPECIAL PURPOSE COMPUTER PROGRAMS	70-0154
RODATZ, W.	WECHSELWIRKUNG ZWISCHEN DEFORMATION UND DURCHSTROIMUNG IM KLUEFTIGEN, ANISOTROPEN GEBIRGE. (INTERACT	72-0612
RODDEN, W. P.	KERNEL FUNCTION FOR NONPLANAR OSCILLATING SURFACES IN SUPERSONIC FLOW	71-0472
RODDIS, R.	THE APPLICATION OF FINITE ELEMENT ANALYSIS TO THE STUDY OF TWO-DIMENSIONAL CRACKED BODIES	72-0695
RODERICK, J. E.	INCREMENTAL SOLUTION PROCEDURE FOR NONLINEAR STRUCTURAL DYNAMICS PROBLEMS	75-0656
RODRIGUES, J. C. W	FINITE ELEMENT ANALYSIS OF THE SURFACE DEFORMATION DUE TO A UNIFORM LOADING ON A LAYER OF GIBSON SOI	75-0240
RODRIGUES, J. S	NODE NUMBERING OPTIMIZATION IN STRUCTURAL ANALYSIS	75-0077
RODRIGUEZ, L.	FINITE ELEMENT NONLINEAR ANALYSIS FOR PLATES AND SHALLOW SHELLS	68-0216
ROEHRLE, M. D.	DETERMINATION OF STRESSES AND DEFORMATIONS ON PISTONS BY MEANS OF COMPUTER PROGRAMS AND PHOTOELASTIC	75-0553
ROESSET, J. M.	AUTOMATIC GENERATION OF FINITE ELEMENT MATRICES	71-0165
ROESSET, J. M.	DYNAMIC ANALYSIS OF FOOTINGS ON LAYERED MEDIA	75-0586
ROESSET, J. M.	SOME STRUCTURAL PROBLEMS - STANDARD OIL OF INDIANA BUILDING	73-0180
ROGERS, C. R.	FINITE ELEMENT APPLICATION TO STRAIN GAGE DATA REDUCTION	73-0561
ROGERS, J. C. W	NUMERICAL SOLUTION OF A DIFFUSION CONSUMPTION PROBLEM WITH A FREE BOUNDARY	74-0835
ROGERS, J. L. J	A FINITE ELEMENT FOR THERMAL STRESS ANALYSIS OF SHELLS OF REVOLUTION	73-0798
ROGERS, P. H.	SHIP (SIMPLIFIED-HELMHOLTZ-INTEGRAL PROGRAM): A FAST COMPUTER PROGRAM FOR CALCULATING THE ACOUSTIC R	72-0835
ROHDE, S. M.	A UNIFIED TREATMENT OF THICK AND THIN FILM ELASTOHYDRODYNAMIC PROBLEMS USING HIGHER ORDER ELEMENT PR	75-0890
ROHDE, S. M.	FINITE ELEMENT OPTIMIZATION OF FINITE STEPPED SLIDER BEARING PROFILES	74-0131
ROHDE, S. M.	HIGHER ORDER FINITE ELEMENT METHODS FOR THE SOLUTION OF COMPRESSIBLE POROUS BEARING PROBLEMS	75-0341
ROHDE, S. M.	VARIATIONAL FORMULATION FOR A CLASS OF FREE BOUNDARY PROBLEMS ARISING IN HYDRODYNAMIC LUBRICATION	75-0531
ROHWER, K.	DESH, EIN EINFACHES FLACHES SCHALENELEMENT UEBER DREIECKIGEM GRUNDRISS. (DESH, A SIMPLE TRIANGULAR S	75-0304
ROLL, F.	EXPERIMENTAL AND ANALYTICAL INVESTIGATION OF A HORIZONTALLY CURVED BOX-BEAM HIGHWAY BRIDGE MODEL	71-0024
ROLL, F.	MODEL ANALYSIS OF CURVED BOX-BEAM HIGHWAY BRIDGE	71-0282
ROMERO, V. E.	EPAD, A COMPUTER PROGRAM FOR THE ELASTIC-PLASTIC ANALYSIS OF DOME STRUCTURES AND ARBITRARY SOLIDS OF	70-0036
ROMSTAD, K. M.	A FINITE ELEMENT COMPUTER PROGRAM FOR THE PREDICTION OF THE BEHAVIOR OF REINFORCED AND PRESTRESSED C	72-0799
ROMSTAD, K. M.	NUMERICAL BIAXIAL CHARACTERIZATION FOR CONCRETE	74-0641
RONEY, B. D.	APPLICATION OF COMPUTER GRAPHICS TO FINITE ELEMENT MESH GENERATION	73-0546
ROPCHAN, D. M.	STRUCTURAL ANALYSIS OF A COAL MINE OPENING IN ELASTIC, MULTILAYERED MATERIAL	74-0235
ROPCHAN, D. M.	STRUCTURAL ANALYSIS OF A COAL MINE OPENING IN ELASTIC, MULTILAYERED MATERIAL	74-0840
ROREN, E. M. Q.	FINITE ELEMENT ANALYSIS OF SHIP STRUCTURES	69-0226
ROREN, E. M. Q.	IMPACT OF FINITE ELEMENT TECHNIQUES ON PRACTICAL DESIGN OF SHIP STRUCTURES	69-0344
ROREN, E. M. Q.	TRANSVERSE STRENGTH OF TANKERS, FINITE ELEMENT APPLICATIONS	68-0142
ROSANOFF, R.	NUMERICAL CONDITIONING OF STIFFNESS MATRIX FORMULATIONS FOR FRAME STRUCTURES	68-0022
ROSE, D. J.	AUTOMATIC NESTED DISSECTION	74-0860
ROSE, D. J.	COMPLEXITY BOUNDS FOR REGULAR FINITE DIFFERENCE AND FINITE ELEMENT GRIDS	72-0872
ROSE, D. J.	COMPLEXITY BOUNDS FOR REGULAR FINITE DIFFERENCE AND FINITE ELEMENT GRIDS	73-0043
ROSE, R. M.	STRUCTURAL MODEL FOR THE MECHANICAL BEHAVIOUR OF TRABECULAR BONE	73-0473
ROSEN, R.	FINITE ELEMENT ANALYSIS - INTRODUCTION	74-0100
ROSEN, R.	INTRODUCTION TO FINITE ELEMENT ANALYSIS	73-0179
ROSEN, R.	STARDYNE STRUCTURAL ANALYSIS SYSTEM, ANALYTICAL CAPABILITY MANUAL	68-0235
ROSENFIELD, A.	FAST FRACTURE RESISTANCE AND CRACK ARREST IN STRUCTURAL STEELS	73-0766
ROSENZWEIG, M.	FINITE ELEMENT SCHEME FOR DOMAINS WITH CORNERS	72-0261
ROSKAM, J.	STEADY STATE EQUATIONS OF MOTION, EQUILIBRIUM SHAPE AND STABILITY DERIVATIVES OF ELASTIC AIRPLANES E	74-0541
ROSS, A. L.	DESIGNING WITH THREE DIRECTIONAL COMPOSITES	74-0204
ROSS, C. T.	ANALYSIS OF STRUCTURES USING FINITE ELEMENT TECHNIQUES	71-0356
ROSS, C. T. F.	FINITE ELEMENTS FOR THE VIBRATION OF CONES AND CYLINDERS	75-0340
ROSS, G. A.	RIGID PAVEMENT STRESSES UNDER AIRCRAFT LOADING	71-0423
ROSS, W.	MAGAZINE HEADWALL RESPONSE TO EXPLOSIVE BLAST	74-0754
ROSSETTOS, J. N	FINITE ELEMENT ANALYSIS OF VIBRATION AND FLUTTER OF CANTILEVER ANISOTROPIC PLATES	74-0484
ROSSETTOS, J. N	FINITE-ELEMENT ANALYSIS OF VIBRATION AND FLUTTER OF CANTILEVER ANISOTROPIC PLATES	74-0933
ROSSETTOS, J. N	MODULAR APPROACH TO STRUCTURAL SIMULATION FOR VEHICLE CRASHWORTHINESS PREDICTION	75-0737
ROSSI, F. A.	CALCULATION METHOD FOR THE STUDY OF CRACKED RIGID STRUCTURES (METODO DI CALCOLO PER LO STUDIO DI STR	71-0637
ROSSOW, E. C.	A FINITE ELEMENT TREATMENT OF NEUTRON DIFFUSION	71-0373
ROSSOW, E. C.	APPLICATION OF FINITE ELEMENT METHOD TO MULTIGROUP NEUTRON DIFFUSION EQUATION	72-0310
ROSSOW, E. C.	APPLICATION OF PHASE-SPACE FINITE ELEMENTS TO ONE DIMENSIONAL NEUTRON TRANSPORT EQUATION	73-0123
ROSSOW, E. C.	DUAL FINITE ELEMENT METHODS FOR NEUTRON DIFFUSION	71-0489
ROSSOW, E. C.	FINITE ELEMENT NEUTRON DIFFUSION IN CURVILINEAR COORDINATES	72-0039
ROSSOW, E. C.	TRANSPORT SOLUTIONS USING FINITE ELEMENTS IN SPACE-ANGLE PHASE SPACE	71-0480
ROSSOW, E. C.	TWO-DIMENSIONAL TRANSPORT CALCULATIONS USING PHASE-SPACE FINITE ELEMENTS	72-0027
ROSSOW, M. P.	A FINITE-ELEMENT APPROACH TO OPTIMAL STRUCTURAL DESIGN	73-1162
ROSSOW, M. P.	COMPUTER IMPLEMENTATION OF THE CONSTRAINT METHOD	75-0001
ROSSOW, M. P.	FINITE ELEMENT METHOD FOR OPTIMAL DESIGN OF VARIABLE THICKNESS SHEETS	73-0131
ROSSOW, M. P.	LEAST-SQUARES VARIATIONAL PRINCIPLE FOR FINITE-ELEMENT APPLICATIONS	75-0412
ROSSOW, M. P.	THE CONSTRAINT METHOD FOR FINITE ELEMENT STRESS ANALYSIS	75-0432
ROTH, S.	PRACTICAL FINITE ELEMENT METHOD OF FAILURE PREDICTION FOR COMPOSITE MATERIAL STRUCTURES	75-0644
ROTH, W.	FACTOR OF SAFETY APPROACH FOR EVALUATING SEISMIC STABILITY OF SLOPES	75-0645
ROUCH, K. E.	STRUCTURAL ANALYSIS OF MAST BY FINITE ELEMENT TECHNIQUE	74-0385
ROUSSOPO, A. A.	APPLICATION OF FINITE ELEMENT METHOD TO A MICROZONATION STUDY OF SALONICA REGION	73-0610
ROWAN, J. C.	AEROELASTIC LOADS PREDICTIONS USING FINITE-ELEMENT AERODYNAMICS	75-0815
ROWAN, W. H.	APPLICATION OF FINITE ELEMENT METHODS IN CIVIL ENGINEERING	69-0031
ROWE, G. H.	MATRIX DISPLACEMENT METHODS IN FRACTURE MECHANICS ANALYSIS OF REACTOR VESSELS	71-0226
ROWE, W. M.	STRESS ANALYSIS AND DESIGN OF SILICON SOLAR CELL ARRAYS AND RELATED MATERIAL PROPERTIES	72-0840
ROWLANDS, R. E.	DEFORMATION AND FAILURE OF BORON-EPOXY PLATE WITH CIRCULAR HOLE	72-0536
ROWLANDS, R. E.	MECHANICAL BEHAVIOR OF A GRAPHITE-EPOXY LAMINATE CONTAINING A HOLE	72-0614
ROWLANDS, R. E.	STRESS AND FAILURE ANALYSIS OF A GLASS-EPOXY COMPOSITE PLATE WITH A CIRCULAR HOLE	73-0276
ROY, J.	A STUDY OF NUMERICAL ERROR IN STRUCTURAL SYSTEMS	70-0394
ROY, J.	AN AUTOMATIC SYSTEM FOR KINEMATIC ANALYSIS ASKA PART 1	71-0626
ROY, J. R.	AN AUTOMATIC SYSTEM FOR KINEMATIC ANALYSIS, ASKA PART 1	70-0396
ROY, J. R.	DISCUSSION OF A PAPER BY F. W. WILLIAMS	73-1078
ROY, J. R.	FINITE ELEMENTE ZUR BERECHNUNG VON SPANNBETONREAKTORDRUCKBEHAELTERN	73-0702
ROY, J. R.	FINITE ELEMENTS FOR CALCULATION OF PRESTRESSED CONCRETE REACTOR PRESSURE VESSELS (FINITE ELEMENTE ZU	73-0797
ROY, J. R.	GENERAL TREATMENT OF STRUCTURAL MODIFICATIONS	72-0166
ROY, J. R.	ON THE REDUCTION OF NUMERICAL ERROR IN THE MATRIX DISPLACEMENT METHOD	74-0417
ROY, J. R.	REANALYSIS FOR LIMITED STRUCTURAL DESIGN MODIFICATIONS	72-0650
ROY, S. K.	APPLICATION OF THE FINITE ELEMENT METHOD TO CASES REQUIRING THE COMBINATION OF ELEMENTS POSSESSING D	72-0580
ROYLANCE, D.	BALLISTIC IMPACT OF TEXTILE STRUCTURES.	72-0433
ROYLANCE, D. K.	WAVE PROPAGATION IN A VISCOELASTIC FIBER SUBJECTED TO TRANSVERSE IMPACT	71-0639
ROYSTER, L. H.	A FORTRAN PROGRAM FOR CALCULATING THE FREQUENCIES AND MODE SHAPES OF AN ARBITRARY SHAPE CYLINDRICAL	72-0594
ROYSTER, L. H.	A FORTRAN PROGRAM FOR CALCULATING THE FREQUENCIES AND MODE SHAPES OF A GUIDED CLAMPED, GUIDED PINNED	72-0890
ROYSTER, L. H.	DEVELOPMENT OF A FINITE ELEMENT MODEL FOR THE CLASS V FLEXTENSIONAL UNDERWATER TRANSDUCER SHELL	71-0754
ROYSTER, L. H.	DEVELOPMENT OF A FINITE ELEMENT MODEL FOR THE CLASS 5 FLEXTENSIONAL UNDERWATER TRANSDUCER SHELL	71-0840
ROYSTER, L. H.	DEVELOPMENT OF A FINITE ELEMENT MODEL FOR THE CLASS IV FLEXTENSIONAL UNDERWATER TRANSDUCER SHELL	72-0593
ROZENDAL, D. B.	FINITE ELEMENT ANALYSIS OF PRESTRESSED FOLDED PLATE STRUCTURES	74-1066
RUBIN, C. P.	A TWO-DIMENSIONAL AXISYMMETRIC SHELL FLUID MODEL FOR LAUNCH VEHICLE LONGITUDINAL DYNAMIC ANALYSIS	69-0046
RUBIN, C. P.	IMPROVED ANALYTIC LONGITUDINAL RESPONSE ANALYSIS FOR AXISYMMETRIC LAUNCH VEHICLES, COMPUTER PROGRAM	65-0002

RUBIN, C. P.	IMPROVED LINEAR AXISYMMETRIC SHELL FLUID MODEL FOR LAUNCH VEHICLE LONGITUDINAL RESPONSE ANALYSIS	65-0044
RUBIN, C. P.	MINIMUM WEIGHT DESIGN OF COMPLEX STRUCTURES SUBJECT TO A FREQUENCY CONSTRAINT	70-0374
RUBIN, H.	STATIC AND DYNAMIC ANALYSIS OF F-14A BORON HORIZONTAL STABILIZER	71-0123
RUBINSTEIN, M.	AUTOMATED STRUCTURAL SYNTHESIS USING A RE NUMBER OF DESIGN COORDINATES	73-0584
RUBINSTEIN, M.	CREEP ANALYSIS OF AXISYMMETRIC BODIES	68-0081
RUBINSTEIN, M.	DYNAMIC STABILITY ANALYSIS BY THE CONJUGATE GRADIENT METHOD	72-0749
RUBINSTEIN, M.	ELASTIC-PLASTIC ANALYSIS BY QUADRATIC PROGRAMMING	72-0238
RUCKER, C. E.	NODAL ANALYSIS OF A NINE BAY SKIN-STRINGER PANEL	71-0102
RUDDER, F. F. J	SKIN-STRINGER PANEL NORMAL MODE RESPONSE EXPERIMENTAL DATA	71-0691
RUDDER, F. F. J	STUDY OF EFFECTS OF DESIGN DETAILS ON STRUCTURAL RESPONSE TO ACOUSTIC EXCITATION	72-0864
RUDY DU PREEZ,	BERECHNUNG PRISMATISCHER FLAECHENTRAGWERKE MIT HILFE VON FINITEN STREIFENELEMENTEN. (CALCULATION OF	72-0133
RUGGIERI, G.	SULLA DETERMINAZIONE DELLE FREQUENZE PROPRIE DI ALBERI SU PIU SUPPORTI ELASTICI. (DETERMINATION OF F	74-0916
RUGGLES, V. L.	MINUTEMAN STV STRESS ANALYSIS AND TESTING	74-0809
RUGGLES, V. L.	USE OF FINITE ELEMENT COMPUTER PROGRAMS IN FRACTURE MECHANICS	71-0022
RUHL, R. L.	DYNAMICS OF DISTRIBUTED PARAMETER ROTOR SYSTEMS: TRANSFER MATRIX AND FINITE ELEMENT TECHNIQUES	70-0461
RUHL, R. L.	FINITE ELEMENT MODEL FOR DISTRIBUTED PARAMETER TURBOMOTOR SYSTEMS	71-0332
RUHL, R. L.	GRAPHICAL DISPLAY OF COMPUTER SIMULATED UNBALANCED ROTOR RESPONSE	71-0486
RUIZ, P.	FINITE ELEMENT FOR THE INTERACTION OF FRAMES AND SHEAR WALLS	73-0681
RUMSEY, T. R.	ANALYSIS OF VISCOELASTIC CONTACT STRESSES IN AGRICULTURAL PRODUCTS USING A FINITE ELEMENT METHOD	74-0634
RUMSEY, T. R.	ANALYSIS OF VISCOELASTIC CONTACT STRESSES IN AGRICULTURAL PRODUCTS USING A FINITE ELEMENT METHOD	74-1199
RUNNER, T. A.	AN INVESTIGATION OF THE BENDING OF SYMMETRIC PLATES UNDER SYMMETRIC TRANSVERSE LOADING	72-0662
RUSER, J. R.	APPLICATION OF THE THREE-DIMENSIONAL FINITE ELEMENT METHOD TO SOME PROBLEMS IN SOIL MECHANICS	70-0510
RUSHNER, R. F.	STRUCTURAL ANALYSIS OF AN ARTERIOLE BY THE DIRECT STIFFNESS METHOD	66-0109
RUSHTON, K. R.	SOLUTION OF ANISOTROPIC SEEPAGE BY FINITE ELEMENTS	67-0071
RUSSELL, J. J.	WHIRLING CABLE SUBJECTED TO VISCOUS DRAG	74-0457
RUTLEDGE, J. C.	TIED-BACK RETAINING WALLS	75-0308
RUTLEDGE, P. C.	MOVEMENTS OF ARTICULATED CONDUITS UNDER EARTH DAMS ON COMPRESSIBLE FOUNDATIONS	73-0477
RYAN, M. O.	THICK WALLED MULTIPLE OPENING REINFORCED CONCRETE CONDUITS	72-0666
RYAZANTSEV, Y.	DIFFUSION ON SOLID SPHERICAL PARTICLE IN VISCOUS FLUID FLOW AT FINITE VALUES OF RE NUMBERS	69-0419
RYBICKI, E. E.	APPROXIMATE THREE-DIMENSIONAL SOLUTIONS FOR SYMMETRIC LAMINATES UNDER INPLANE LOADING	71-0392
RYBICKI, E. F.	AN INCREMENTAL COMPLEMENTARY ENERGY METHOD OF NONLINEAR STRESS ANALYSIS	69-0281
RYBICKI, E. F.	APPROXIMATE THREE-DIMENSIONAL STRESS SOLUTIONS FOR SOME LAMINATED PLATE PROBLEMS USING A HIGHER ORDE	70-0419
RYBICKI, E. F.	FINITE ELEMENT MODEL SYSTEM OF A RABBIT CALVARIUM: AN INVESTIGATION OF OPTIMAL COMPRESSIVE LOADING	73-0206
RYBICKI, E. F.	HIGH-ORDER FINITE ELEMENT ANALYSIS FOR TRANSIENT TEMPERATURE ANALYSIS OF INHOMOGENEOUS MATERIALS	69-0301
RYBICKI, E. F.	MATHEMATICAL AND EXPERIMENTAL STUDIES ON THE MECHANICS OF PLATED TRANSVERSE FRACTURES	74-0656
RYBICKI, E. F.	MEASUREMENTS OF HEALING AT AN OSTEOTOMY IN A RABBIT CALVARIUM: THE INFLUENCE OF APPLIED COMPRESSIVE	74-0975
RYBICKI, E. F.	ON THE MATHEMATICAL ANALYSIS OF STRESS IN THE HUMAN FEMUR	72-0207
RYBICKI, E. F.	TRANSFORMATIONS FOR ELLIPTICAL HOLES APPLIED TO FINITE ELEMENT ANALYSIS	74-0132
RYDZEWSKI, J. R	APPROACH TO ANALYSIS OF ARCH DAMS IN WIDE VALLEYS	73-0153
RYNN, P. G.	PROGRESS IN COMPUTERIZED STRUCTURAL ANALYSIS OF LARGE VESSELS	74-0949
RYU, J.	FINITE ELEMENT METHOD TO ELECTROMAGNETIC INDUCTION PROBLEMS IN GEOPHYSICS	72-0309
SABANATHAN, S.	STATICS AND DYNAMICS OF HYDROSTATICALLY LOADED SHELLS BY FINITE ELEMENT METHOD	71-0044
SABBAGH, H. A.	FINITE ELEMENT ANALYSIS OF ELASTIC WAVE SCATTERING FROM DISCONTINUITIES	73-0547
SABBAGH, H. A.	2-DIMENSIONAL FINITE-ELEMENT ANALYSIS OF FERROELECTRIC NORMAL MODES OF FERROELECTRIC DISKS AND SHEL	75-0594
SABBAGH, H. A.	3-DIMENSIONAL FINITE-ELEMENT ANALYSIS OF ELECTROELASTIC NORMAL MODES OF A TANGENTIALLY POLARIZED, SF	75-0591
SABINA, F. J.	THEORETICAL COMPUTATIONS ON RIDGE ACOUSTIC SURFACE WAVES USING THE FINITE-ELEMENT METHOD	71-0011
SABIR, A. B.	A COMPARISON OF CURVED BEAM FINITE ELEMENTS WHEN USED IN VIBRATION PROBLEMS	71-0227
SABIR, A. B.	A STIFFNESS MATRIX FOR SHALLOW SHELL FINITE ELEMENTS	69-0229
SABIR, A. B.	ALGORITHM FOR THE LARGE DEFLECTION GEOMETRICALLY NON-LINEAR BEHAVIOUR OF PLANE AND CURVED STRUCTURES	72-0083
SABIR, A. B.	AN EXTENSION OF THE SHALLOW TO THE NON-SHALLOW STIFFNESS MATRIX FOR A CYLINDRICAL SHELL FINITE ELEME	00-0104
SABIR, A. B.	APPLICATION OF CURVED FINITE ELEMENTS TO CIRCULAR ARCHES	71-0360
SABIR, A. B.	APPLICATION OF THE FINITE ELEMENT METHOD TO THE BUCKLING OF RECTANGULAR PLATES AND PLATES ON ELASTIC	73-0478
SABIR, A. B.	APPLICATION OF THE FINITE ELEMENT METHOD OF ANALYSIS TO HYPERBOLIC PARABOLOIDS WITH EDGE BEAMS	75-0250
SABIR, A. B.	CORRECTED ASSESSMENT OF CYLINDRICAL SHELL FINITE ELEMENT OF BOGNER, FOX AND SCHMIT WHEN APPLIED TO A	73-0024
SABIR, A. B.	CURVED, CYLINDRICAL SHELL, FINITE ELEMENT	72-0096
SABIR, A. B.	FINITE ELEMENT ANALYSIS OF A BROAD ARCH BRIDGE WITH A SLAB DECK	71-0724
SABIR, A. B.	FINITE ELEMENT CALCULATION OF STRESS DISTRIBUTIONS IN ARCHES	74-0094
SABIR, A. B.	FURTHER STUDIES IN THE APPLICATION OF CURVED FINITE ELEMENTS TO CIRCULAR ARCHES	71-0644
SABIR, A. B.	LARGE AMPLITUDE, FREE VIBRATION GEOMETRICALLY NONLINEAR CIRCULAR ARCHES	72-0606
SABIR, A. B.	LARGE DEFLECTION GEOMETRICALLY NON-LINEAR FINITE ELEMENT ANALYSIS OF CIRCULAR ARCHES	73-0021
SABIR, A. B.	LIMITATIONS OF CERTAIN CURVED FINITE ELEMENTS WHEN APPLIED TO ARCHES	71-0279
SABIR, A. B.	NEW CYLINDRICAL SHELL FINITE ELEMENT BASED ON SIMPLE INDEPENDENT STRAIN FUNCTIONS	72-0094
SABIR, A. B.	STABILITY INVESTIGATION OF A CYLINDRICAL SHELL WITH FINITE ELEMENTS	72-0127
SABIR, R. B.	STIFFNESS MATRICES FOR THE GENERAL DEFORMATION (OUT-OF-PLANE AND INPLANE) OF CURVED BEAM MEMBERS BAS	75-0124
SABIR, R. A.	THE APPLICATION OF FINITE ELEMENTS TO THE LARGE DEFLECTION GEOMETRICALLY NONLINEAR BEHAVIOR OF CYLIN	72-0815
SABLE, W. W.	THE DEVELOPMENT OF COMPUTER GRAPHICS FOR LARGE SCALE FINITE ELEMENT CODES	71-0042
SABLE, W. W.	THE DEVELOPMENT OF LARGE SCALE DIGITAL COMPUTER CODES FOR PRODUCTION STRUCTURAL ANALYSIS	71-0721
SABODASH, P. F.	EXCITATION OF PLANE THERMOELASTIC WAVES IN A LAYER WITH A FINITE RATE OF HEAT PROPAGATION	74-0803
SABONNADIERE, J	FINITE ELEMENT SOLUTION OF SATURATED TRAVELLING MAGNETIC FIELD PROBLEMS	75-0600
SABOTA, J.	VYPOCET KRUHOVYCH LOMENICOVYCH KONSTRUKCII METODCU KONECNYCH PASOV. (CALCULATION OF CIRCULAR FOLDED-	74-0612
SACHS, R. L.	A FINITE ELEMENT ANALYSIS OF A DOUBLY CURVED GOTHIC VAULT	71-0827
SACK, R. L.	COMPUTATIONAL STABILITY IN NONLINEAR DISCRETE STRUCTURAL SYSTEMS	69-0145
SACKMAN, J. J.	APPROXIMATE SOLUTIONS IN LINEAR COUPLED THERMOELASTICITY	68-0220
SACKMAN, J. L.	FINITE ELEMENT FORMULATION AND SOLUTION OF CONTACT-IMPACT PROBLEMS IN CONTINUUM MECHANICS	74-0806
SACKMAN, J. L.	FINITE ELEMENT FORMULATION AND SOLUTION OF CONTACT-IMPACT PROBLEMS IN CONTINUUM MECHANICS-II	75-0740
SACKMAN, J. L.	WAVE PROPAGATION IN ANISOTROPIC ROCKS	72-0873
SACKMAN, J. L.	WAVE PROPAGATION IN ANISOTROPIC ROCKS	73-0922
SACZALSKI, K.	STRUCTURAL MECHANICS COMPUTER PROGRAMS	74-0902
SACZALSKI, K. J	AN INTERACTIVE HYBRID TECHNIQUE FOR CRASHWORTHY DESIGN OF COMPLEX VEHICULAR STRUCTURAL SYSTEMS	74-0317
SACZALSKI, K. J	COUPLED RESPONSE OF SPATIAL VIBRATORY STRUCTURES MOUNTED TO ISOTROPIC PLATE ELEMENTS	71-0228
SACZALSKI, K. J	ELASTODYNAMICS OF COMPLEX STRUCTURAL SYSTEMS	71-0563
SACZALSKI, K. J	MODELING AND ANALYSIS TECHNIQUES FOR PREDICTION OF STRUCTURAL AND BIODYNAMIC CRASH-IMPACT RESPONSE	75-0300
SACZALSKI, K. J	TRANSIENT RESPONSE OF INELASTICALLY CONSTRAINED RIGID BODY SYSTEMS	73-0291
SADAT, M. M.	A QUADRILATERAL PLANE STRESS FINITE ELEMENT WITH QUADRATIC PROPERTIES	72-0969
SAEED MIRZA, M.	NONLINEAR FINITE ELEMENT ANALYSIS OF REINFORCED CONCRETE STRUCTURES	74-0442
SAFAR, M. M.	COMPARATIVE STUDY OF FRESH-SALT WATER INTERFACES USING FINITE ELEMENT AND SIMPLE APPROACHES	75-0468
SAGASETA, G.	INFLUENCE OF CONSTRUCTION PROCEDURES ON GROUND PRESSURES AROUND A TUNNEL	74-0997
SAGAWA, K.	AUTOMATIC MESH GENERATION FOR THREE DIMENSIONAL STRUCTURES BASED ON THEIR THREE VIEWS	73-0340
SAGHERA, S. S.	EFFECTS OF SHEAR WALLS ON DYNAMIC RESPONSE OF FRAMES	69-0230
SAHA, S. K.	SOME DESIGN ASPECTS OF IC ENGINE PISTON HEADS	73-0578
SAINI, S. S.	A SEISMIC STRENGTH OF KOLKEWADI DAM	73-0197
SAINI, S. S.	INFLUENCE OF RESERVOIR WATER ON EARTHQUAKE RESPONSE OF EARTH DAMS	70-0375
SAITO, G. E.	STRESS ANALYSIS OF HUMAN TEETH	73-0505
SAITO, T.	DESIGN OF CONCRETE TRACK SLAB ON ELASTIC FOUNDATION USING THE FINITE ELEMENT METHOD	73-0479
SAITO, T.	STRESS ANALYSIS OF CONCRETE TRACK SLABS ON AN ELASTIC FOUNDATION BY THE FINITE ELEMENT METHOD	74-1049
SAKAI, F.	APPLICATION OF THE FINITE ELEMENT METHOD TO SURFACE WAVE ANALYSIS	74-0913
SAKAI, F.	EQUIVALENCY BETWEEN FINITE ELEMENT METHOD AND FINITE DIFFERENCE METHOD AND A NEW DISCRETIZATION APPR	73-0480
SAKAI, F.	FINITE ELEMENT ANALYSIS OF SURFACE WAVE PROBLEMS USING INTEGRAL EQUATIONS	74-0624
SAKAI, F.	FINITE ELEMENT ANALYSIS OF WAVE PROPAGATION IN OPEN CHANNELS	75-0570
SAKAI, F.	FINITE ELEMENT METHOD FOR WAVE GUIDE PROBLEMS	75-0872
SAKAI, K.	STUDY ON FRACTURE TOUGHNESS EVALUATION BY ELASTO-PLASTIC ANALYSIS AROUND NOTCHES	74-0619
SAKAI, T.	A UNIFIED THEORY OF VARIATIONAL PRINCIPLES IN MECHANICS	75-0155

91

SAKENEM E.	NONLINEAR CRACK ANALYSIS WITH FINITE ELEMENTS	73-0875
SAKHAROV, A. S.	INVESTIGATION OF THE STRESS-STRAINED STATE OF SPHERICAL SHELLS WITH AN ECCENTRIC CUTOUT ON THE BASIS	73-0222
SAKURAI, T.	A FINITE ELEMENT SIMULATION OF MECHANICAL PROPERTIES OF COMPOSITE MATERIALS	74-1294
SAKURAI, T.	A STUDY ON COMPOSITE STRUCTURES EMPLOYING DIFFERENT TYPES OF FINITE ELEMENTS	69-0231
SAKURAI, T.	ANALYSIS OF THE ELASTIC PLASTIC PROBLEMS BY THE MATRIX DISPLACEMENT METHOD	68-0173
SAKURAI, T.	STRESS STRAIN MATRIX AND ITS APPLICATIONS FOR THE SOLUTION OF ELASTIC-PLASTIC PROBLEMS BY THE FINITE	68-0282
SALAAM, U.	FINITE ELEMENTS WITH RELAXED CONTINUITY - AN INVESTIGATION OF THE VARIATIONAL BASIS AND NUMERICAL PF	74-1065
SALAAM, U.	NUMERICAL PERFORMANCE OF A FINITE ELEMENT GALERKIN PROCEDURE	75-0022
SALAH EL-DIN, A	MODIFIED APPROACH FOR ESTIMATING THE CRACKING MOMENT OF REINFORCED CONCRETE BEAMS	75-0485
SALAM, A. E.	DYNAMIC STABILITY OF PLATES BY FINITE ELEMENTS	71-0573
SALAM, Y. M.	FRACTURE (ULTIMATE STRENGTH) ANALYSES OF ASPHALT PAVEMENT LAYERS RESULTING FROM TRAFFIC LOADING	73-0548
SALAMA, A. M.	STRESS ANALYSIS AND DESIGN OF SILICON SOLAR CELL ARRAYS AND RELATED MATERIAL PROPERTIES	72-0840
SALAMA, M.	FATIGUE OF GRAPHITE FIBRE-REINFORCED EPOXY IN COMPRESSION.	72-0383
SALAMA, M.	FATIGUE OF PRENOTCHED GRAPHITE FIBER COMPOSITES IN COMPRESSION	72-0215
SALAMON, P. F.	LEAST SQUARE OPTIMAL APPROXIMATION SIMULATION TECHNIQUES	73-0871
SALANI, H.	A FINITE-ELEMENT METHOD FOR TRANSVERSE VIBRATIONS OF BEAMS AND PLATES	67-0144
SALEM, M. H.	DATA GENERATION FOR A 3-D FINITE ELEMENT SYSTEM	74-0475
SALEM, M. H.	NONLINEAR ANALYSIS OF PLANAR REINFORCED CONCRETE STRUCTURES	74-0533
SALEM, M. H.	THICK WALLED MULTIPLE OPENING REINFORCED CONCRETE CONDUITS	72-0666
SALEME, E.	NONLINEAR CRACK ANALYSIS WITH FINITE ELEMENTS	73-0369
SALINAS, D.	INITIAL YIELD SURFACE OF A UNIDIRECTIONALLY REINFORCED COMPOSITE	71-0280
SALINAS, D.	FINITE ELEMENT SOLUTIONS OF A NONLINEAR REACTOR DYNAMICS PROBLEM	74-1268
SALMON, C. G.	FINITE ELEMENT ANALYSIS OF CONCRETE BEAMS	74-0560
SALMON, M.	AN APPLICATION OF THE FINITE ELEMENT METHOD TO ELASTIC-PLASTIC PROBLEMS OF PLANE STRESS	70-0031
SALOMON, D.	USE OF FINITE ELEMENTS IN THE ANALYSIS OF A HYDROTHERMAL DOUBLET	74-0049
SALONEN, E. M.	A FRAMEWORK METHOD FOR STRETCHED PLATES	71-0697
SALONEN, E. M.	A RECTANGULAR PLATE BENDING ELEMENTS THE USE OF WHICH IS EQUIVALENT TO THE USE OF THE FINITE DIFFERE	71-0304
SALONEN, E. M.	THREE-DIMENSIONAL ELASTO-PLASTIC FINITE ELEMENT ANALYSIS	75-0236
SALUS, W. L.	ASTRA-BOEING'S ADVANCED STRUCTURAL ANALYZER	69-0118
SAMARTIN, A.	DYNAMIC PROGRAMMING APPROACH TO THE FORMULATION AND SOLUTION OF FINITE ELEMENT EQUATIONS	75-0059
SAMLER, R. G.	VIBRATIONS OF TAPERED AND CURVED MEMBERS BY FINITE ELEMENT METHODS	69-0517
SAMUELSON, A.	MIXED FINITE ELEMENT METHODS IN THEORY AND APPLICATIONS	73-1108
SAMUELSON, A. C	FINITE-ELEMENT ANALYSES OF INITIAL ELASTIC DEFORMATION OF VERTICALLY DEFORMED CRYSTAL BLOCKS	72-0970
SAMUELSON, A. C	ON THE APPLICATION OF FINITE-ELEMENT TECHNIQUES TO PROBLEMS CONCERNING POTENTIAL DISTRIBUTION AND ST	69-0268
SAMUELSSON, A.	THE RECTANGULAR PLANE STRESS ELEMENT BY TURNER, PIAN AND WILSON	74-1265
SANBONGI, S.	A TORSIONAL STRENGTH ANALYSIS ON THE CONTAINER SHIP BY MEANS OF THE FINITE ELEMENT PROCEDURE AND FUL	71-0243
SANBONGI, S.	APPLICATION OF FINITE ELEMENT METHOD TO ANALYSIS OF AIRCRAFT WING STRUCTURES	69-0232
SANBONGI, S.	APPLICATIONS OF THE FINITE ELEMENT METHOD TO BOX BEAMS	68-0273
SANBONGI, S.	FINITE DIFFERENCE SOLUTIONS FOR LARGE DEFORMATIONS OF CYLINDRICAL SHELLS - A COMPARISON WITH FINITE	72-0270
SANCHEZGUTIERRE	FINITE-ELEMENT PLASMA-DIFFUSION CODE FOR ARBITRARY TOROIDAL GEOMETRY	75-0500
SANDAY, S. C.	DESIGN AND ANALYSIS OF A CERAMIC STATOR VANE	75-0366
SANDER, C.	BORNES SUPERIEURES ET INFERIEURES DANS L'ANALYSE MATRICIELLE DES PLAQUES EN FLEXION TORSION	64-0015
SANDER, G.	APPLICATION OF THE DUAL ANALYSIS PRINCIPLE	70-0306
SANDER, G.	APPLICATIONS DE LA METHODE DES ELEMENTS FINIS A LA FLEXION DES PLAQUES	69-0511
SANDER, G.	DELINQUENT FINITE ELEMENTS FOR SHELL IDEALIZATION	75-0478
SANDER, G.	DIGITAL COMPUTATION OF STRESSES AND DEFLECTIONS IN A BOX BEAM	67-0073
SANDER, G.	DIGITAL COMPUTATION OF STRESSES AND DEFLECTIONS IN A BOX BEAM - A PERFORMANCE COMPARISON BETWEEN FIN	69-0355
SANDER, G.	DUAL ANALYSIS BY FINITE ELEMENTS: LINEAR AND NONLINEAR APPLICATIONS	72-0476
SANDER, G.	FINITE ELEMENT ANALYSIS OF SUPERSONIC PANEL FLUTTER	73-0481
SANDER, G.	IMPROVEMENTS OF FINITE ELEMENT SOLUTIONS FOR STRUCTURAL AND NON-STRUCTURAL APPLICATIONS	71-0229
SANDER, G.	IMPROVEMENTS OF FINITE ELEMENT SOLUTIONS FOR STRUCTURAL AND NONSTRUCTURAL APPLICATIONS	72-0671
SANDER, G.	UPPER AND LOWER BOUNDS TO STRUCTURAL DEFORMATIONS BY DUAL ANALYSIS IN FINITE ELEMENTS	67-0074
SANDERSE, A. A.	A COMPUTER PROGRAMME FOR ELASTIC-PLASTIC CALCULATIONS WITH USE OF TRIM-6 FINITE ELEMENTS	72-0993
SANDHU, R. S.	AN APPLICATION OF THE FINITE ELEMENT METHOD TO ELASTIC-PLASTIC PROBLEMS OF PLANE STRESS	70-0031
SANDHU, R. S.	APPLICATION OF ELASTO-PLASTIC ANALYSIS IN ROCK MECHANICS BY FINITE ELEMENT METHOD	71-0349
SANDHU, R. S.	APPLICATION OF GRIFFITH'S THEORY TO ANALYSIS OF PROGRESSIVE FRACTURE	75-0150
SANDHU, R. S.	FINITE ELEMENT ANALYSIS OF SEEPAGE IN ELASTIC MEDIA	69-0024
SANDHU, R. S.	FINITE ELEMENT ANALYSIS OF LAND SUBSIDENCE	69-0233
SANDHU, R. S.	FINITE ELEMENT ANALYSIS OF CONSOLIDATION IN CREEP	72-0420
SANDHU, R. S.	FINITE ELEMENT ANALYSIS OF ANISOTROPIC PLATES USING Q-19 ELEMENT. PART 1. THEORETICAL DEVELOPMENT	74-0728
SANDHU, R. S.	FINITE ELEMENT ANALYSIS OF UNCONFINED FLOW WITH A VARIABLE TIME-STEP PROCEDURE	74-0776
SANDHU, R. S.	FINITE ELEMENT ANALYSIS OF ANISOTROPIC PLATES USING Q-19 ELEMENT - 1, 2	74-0985
SANDHU, R. S.	NEW HORIZONS IN ROCK MECHANICS	73-0428
SANDHU, R. S.	NUMERICAL PERFORMANCE OF A FINITE ELEMENT GALERKIN PROCEDURE	75-0022
SANDHU, R. S.	STRIP FOOTING ANALYSIS	72-0755
SANDHU, R. S.	VARIABLE TIME STEP ANALYSIS OF UNCONFINED SEEPAGE	74-0064
SANDIDGE, D. W.	STABILITY AND POSTBUCKLING BEHAVIOUR OF HYPERELASTIC BODIES AT FINITE STRAIN BY THE FINITE ELEMENT M	73-0562
SANDIDGE, D. W.	STABILITY AND POSTBUCKLING BEHAVIOR OF HYPERELASTIC BODIES AT FINITE STRAIN BY THE FINITE ELEMENT ME	74-0838
SANDLER, I.	MATHEMATICAL MODELS FOR GEOLOGICAL MATERIALS FOR WAVE PROPAGATION STUDIES	71-0740
SANKAR, A.	CURRENT COMPUTATION ON COMPLEX STRUCTURES BY FINITE-ELEMENT METHOD	75-0456
SANTI, R. L.	A FINITE ELEMENT METHOD FOR OPTIMUM BEAM DESIGN	71-0685
SAPPER, E.	ON THE APPLICATION OF THE FINITE ELEMENT METHOD IN REACTOR PHYSICS	75-0122
SAPPER, E.	2-DIMENSIONAL AND 3-DIMENSIONAL REACTOR-PHYSICS CALCULATIONS WITH FINITE-ELEMENT METHOD	75-0423
SARAZIN, A. C.	FINITE ELEMENT DYNAMIC ANALYSIS OF SHALLOW SHELL STRUCTURES	70-0164
SARGIOUS, M.	AN ADVANCED ANALYSIS OF RIGID PAVEMENTS	72-0417
SARGIOUS, M.	PRINCIPAL STRESSES AT THE INTERMEDIATE SUPPORT OF PRESTRESSED CONCRETE CONTINUOUS BEAMS	70-0352
SARGIOUS, M.	RIGID PAVEMENT STRESSES UNDER AIRCRAFT LOADING	71-0423
SARGIOUS, M.	STEP AND LOADS EFFECT ON STRESSES IN PRESTRESSED CONCRETE SHORT BRACKETS	71-0334
SARGIOUS, M. A.	EFFECT OF OPENINGS ON STRESSES IN RIGID PAVEMENTS	73-0344
SARPKAYA, T.	ANALYSIS OF CURVED TARGET-TYPE THRUST REVERSERS	75-0151
SARPKAYA, T.	FINITE ELEMENT ANALYSIS OF JET IMPINGEMENT ON AXISYMMETRIC CURVED DEFLECTORS	75-0390
SARPKAYA, T.	JET IMPINGEMENT ON AXISYMMETRIC CURVED DEFLECTORS	74-0017
SARSTEN, A.	COMPUTER CALCULATIONS OF STRESSES IN AXISYMMETRIC THERMALLY LOADED COMPONENTS	68-0069
SARTORY, W. K.	COMPARISON OF TYPICAL INELASTIC ANALYSIS PREDICTIONS WITH BENCHMARK PROBLEM EXPERIMENTAL RESULTS	75-0690
SARTORY, W. K.	ELASTIC-PLASTIC-CREEP ANALYSIS OF THERMAL RATCHETTING IN STRAIGHT PIPE AND COMPARISONS WITH TEST RES	73-0391
SASAKI, N.	CONSIDERATION ON THE SEISMIC RESPONSE ANALYSIS OF UNDERGROUND STRUCTURES	75-0471
SASAKI, N.	DYNAMIC ANALYSIS OF THE GROUND USING THE THREE DIMENSIONAL FINITE ELEMENT	73-0291
SASAKI, N.	ON THE APPLICATION OF THE FINITE ELEMENT METHOD TO THE VIBRATION ANALYSIS OF SPACE PLATE STRUCTURE	72-0631
SASAKI, N.	THREE DIMENSIONAL SEISMIC ANALYSIS OF STRUCTURE-GROUND SYSTEMS BY THE FINITE ELEMENT METHOD	74-0262
SASAKI, T.	EXAMPLES OF STRESS ANALYSIS OF STRUCTURES BY THE FINITE ELEMENT METHOD. (JAPANESE)	70-0453
SASAKI, T.	STUDIES ON THE DETERMINATION OF SEEPAGE-LINE OF FILL DAM BY THE METHOD OF FINITE ELEMENT. WATER FLOW	74-1341
SASAOKA, G.	ANALYSIS OF CONTACT PROBLEMS BY THE MATRIX METHOD	69-0381
SATA, T.	ANALYSIS OF THERMAL DEFORMATION OF MACHINE TOOL BY THE FINITE ELEMENT METHOD.	72-0431
SATA, T.	COMPUTERUNTERSTUETZTES SYSTEM FUER DIE CONSTRUCTION DES WERKZEUGMASCHINENAUFBAUS (COMPUTER-AIDED SYS	73-0481
SATA, T.	DYNAMIC ANALYSIS OF MACHINE TOOL STRUCTURES BY THE FINITE ELEMENT METHOD	71-0417
SATAKE, M.	SOME APPLICATIONS OF TOPOLOGICAL CONSIDERATION AND WEIGHT MATRIX METHOD TO FINITE ELEMENT ANALYSIS	73-0302
SATAKE, M.	STUDY ON WEIGHT MATRIX IN STRUCTURAL ANALYSIS	75-0061
SATO, H.	ON VIBRATION ANALYSIS OF RECTANGULAR CANTILEVER PLATES WITH A STEPPED CROSS-SECTION BY THE FINITE EL	71-0601
SATO, N.	ANALYSIS OF THERMAL DEFORMATION OF MACHINE TOOL BY THE FINITE ELEMENT METHOD.	72-0431
SATO, T.	EFFECT OF TIME-DEPENDENT MATERIAL PROPERTIES ON DYNAMIC RESPONSE	74-0905
SATO, T.	FINITE DEFORMATION OF ELASTIC MEMBRANES BY THE FINITE ELEMENT METHOD	67-0101
SATO, T.	FINITE STRAINS AND DISPLACEMENTS OF ELASTIC MEMBRANES BY THE FINITE ELEMENT METHOD	67-0061

SATO, T.	STRUCTURAL ANALYSIS OF AERODYNAMIC DECELERATION SYSTEMS BY THE FINITE ELEMENT METHOD	67-0064
SATOH, K.	RECENT TRENDS OF RESEARCH INTO RESTRAINT STRESSES AND STRAINS IN RELATION TO WELD CRACKING	73-0483
SATOH, K.	TENDANCES ACTUELLES DES RECHERCHES SUR LES CONTRAINTES ET DEFORMATIONS DUES AU BRIDAGE EN RELATION A	74-0273
SATTAR, S. A.	GAS TURBINE ENGINE DISK CYCLIC LIFE PREDICTION	75-0509
SATTERLY, G. T.	GEOMETRIC HIGHWAY AND CULVERT DESIGN	74-1342
SAUER, G.	APPLICATION OF NEW AUSTRIAN TUNNELLING METHOD IN DIFFICULT BUILTOVER AREAS IN FRANKFURT/MAIN METRO	73-1025
SAUGY, B.	ANALYSE TRIDIMENSIONALE DU COMPORTEMENT NON-LINEAIRE D'UN CAISSON DE REACTEUR NUCLEAIRE EN BETON PRE	71-0144
SAUGY, B.	APPLICATION OF NONLINEAR ANALYSIS TO SOLID STRUCTURES	72-0212
SAUGY, B.	NONLINEAR ANALYSIS OF PRESTRESSED CONCRETE FAST REACTOR PRESSURE VESSEL, (ANALYSE NON LINEAIRE DES C	70-0145
SAUGY, B.	THREE DIMENSIONAL RUPTURE ANALYSIS OF A PRESTRESSED CONCRETE PRESSURE VESSEL INCLUDING CREEP EFFECTS	73-0826
SAUGY, B.	THREE DIMENSIONAL RUPTURE ANALYSIS OF A PRESTRESSED CONCRETE PRESSURE VESSEL INCLUDING CREEP EFFECTS	74-0538
SAUL, W. E.	MULTIDEGREE DYNAMIC ANALYSIS OF TALL BUILDINGS SUBJECTED TO WIND AS A STOCHASTIC PROCESS	74-1196
SAVCHENKO, V. G	SOLUTION OF A THREE DIMENSIONAL AXISYMMETRICAL PROBLEM OF THERMOPLASTICITY AS APPLIED TO THICK TURBI	74-0384
SAW, C. B.	AUTOMATIC NODAL TRIANGULATION FOR FINITE ELEMENTS	73-0178
SAW, C. B.	LINEAR ELASTIC FINITE ELEMENT ANALYSIS OF MASONRY WALLS ON BUILDINGS	74-1015
SAWADA, T.	FINITE ELEMENT ANALYSIS OF SOME MECHANICAL PROBLEMS OF HOLLOW GRAVITY DAM	70-0470
SAWADA, Y.	STRESS ANALYSIS OF SELF-SEALING TYPE PISTON HEAD BY FINITE-ELEMENT METHOD	72-0259
SAWADA, Y.	STRESS ANALYSIS OF THE "SELF-SEALING" TYPE PISTON HEAD BY THE FINITE ELEMENT METHOD	72-0617
SAWKO, F.	AN ANNULAR SEGMENT FINITE ELEMENT FOR PLATE BENDING	71-0285
SAWKO, F.	ANALYSIS OF MULTI-CELL BRIDGES WITHOUT TRANSVERSE DIAPHRAGMS, A FINITE ELEMENT APPROACH	69-0235
SAWKO, F.	ANALYSIS OF SPINE BEAM BRIDGES USING FINITE ELEMENTS	70-0236
SAWKO, F.	FINITE ELEMENT ANALYSIS OF BRIDGES CURVED IN PLAN	71-0547
SAWKO, F.	THE ANALYSIS OF SKEW BRIDGE DECKS, A NEW FINITE ELEMENT APPROACH	69-0234
SAWYER, S. G.	ANALYSIS OF BAND PRESSURE IN A GUN TUBE USING FINITE ELEMENTS AND SHELL PUSHING DATA	72-0793
SAWYER, S. G.	BRLESC FINITE ELEMENT PROGRAM FOR AXISYMMETRIC, PLANE STRAIN, AND PLANE STRESS, ORTHOTROPIC SOLIDS W	71-0749
SAXE, H. C.	THE APPLICATION OF DYNAMIC RELAXATION TO THE FINITE ELEMENT METHOD OF STRUCTURAL ANALYSIS	68-0108
SAXTON, H. J.	LOAD POINT COMPLIANCE OF THE CHARPY IMPACT SPECIMEN	73-0775
SAYEGH, A. F.	ELASTIC-PLASTIC ANALYSIS BY QUADRATIC PROGRAMMING	72-0238
SAYER, R.	THE ROLE OF COMPUTER GRAPHICS IN THE STRUCTURAL DESIGN PROCESS	68-0215
SAYLES, F. H.	IN SITU CREEP ANALYSIS OF ROOM IN FROZEN SOIL	72-0128
SCAIFFE, B. K.	STUDIES IN NUMERICAL ANALYSIS	74-1269
SCANLON, A.	TIME DEPENDENT REINFORCED CONCRETE SLAB DEFLECTIONS	74-0586
SCAVUZZO, R. J.	EFFECTS OF FOUNDATION ROTATION ON SEISMIC INERTIA FORCES OF NUCLEAR POWER PLANT STRUCTURES	73-0177
SCHACHT, C. A.	ELASTIC ANALYSIS OF A REFRACTORY SPRUNG ARCH	74-0386
SCHACHT, C. A.	THERMO-STRUCTURAL CAPABILITY OF ANSYS	72-0155
SCHACHT, C. A.	VERIFICATION AND QUALIFICATION OF A 3-D FINITE ELEMENT COMPUTER ANALYSIS	71-0487
SCHAEFER, H.	CONTRIBUTION TO THE SOLUTION OF CONTACT PROBLEMS WITH THE AID OF BOND ELEMENTS	75-0321
SCHAEFER, H.	EINE EINFACHE KONSTRUKTION VON KOORDINATENFUNKTIONEN FUR DIE NUMERISCHE LOSUNG ZWEIDIMENSIONALER RAN	66-0134
SCHAEFER, H.	STRUCTURAL MECHANICS COMPUTER PROGRAMS	74-0902
SCHAEFER, H.	ZUR WIRKLICHKEITSNAHEN BERECHNUNG VON STAHLBETONPLATTEN MIT DER FINITE-ELEMENT-METHODE. (APPROXIMATE	75-0832
SCHALK, M.	BERECHNUNG DER ERDEBENSCHWINGUNGEN VON STRUKTUREN MIT DER FINITE ELEMENT METHODE, MECHANISCHE MODELL	71-0170
SCHALK, M.	CALCULATION OF VIBRATIONS OF STRUCTURES CAUSED BY EARTHQUAKES USING THE FINITE ELEMENT METHOD	71-0690
SCHALLER, R. J.	APPLICATION OF FINITE ELEMENT ANALYSIS TO CERAMIC COMPONENTS	75-0040
SCHALLER, R. J.	CERAMIC ROTATING BLADES: SOME CRITICAL DESIGN PARAMETERS FOR GAS TURBINE APPLICATIONS	74-0146
SCHAPERY, R. A.	DEVELOPMENT OF A SOLID ROCKET PROPELLANT NONLINEAR CONSTITUTTIVE THEORY	75-0722
SCHARPF, D. W.	SPEKTRALANALYSE VON ROTORBLATTERN NACH DERMATRIZENVERS CHIEBUNGSMETHODE UND MATRIZENKRAFTEMFHODE	66-0123
SCHARPF, D. W.	A COMPUTATIONAL PROCEDURE FOR GRADIENT ITERATIVE TECHNIQUES IN THE FINITE ELEMENT METHOD	68-0229
SCHARPF, D. W.	A SEQUEL TO TECHNICAL NOTE 13 - THE CURVED TETRAHEDRONAL AND TRIANGULAR ELEMENTS TEC AND TRIC FOR TH	69-0048
SCHARPF, D. W.	A SEQUEL TO TECHNICAL NOTE 15: THE SHEBA FAMILY OF SHELL ELEMENTS FOR THE MATRIX DISPLACEMENT METHOD	69-0444
SCHARPF, D. W.	BERECHNUNG VORGESPANNTER NETZWERKE	70-0399
SCHARPF, D. W.	DIE ELASTO-PLASTISCHE BERECHNUNG VON ALLGEMEINEN TRAGWERKEN UND KONTINUA	69-0443
SCHARPF, D. W.	EINIGE ALLGEMEINE PROGRAMMSYSTEME FUR FINITE ELEMENTE	73-0581
SCHARPF, D. W.	ELASTO-PLASTIC ANALYSIS OF METAL SHEETS WITH A CENTRAL CRACK	67-0137
SCHARPF, D. W.	FINITE ELEMENT FORMULATION OF THE INCOMPRESSIBLE LUBRICATION PROBLEM	70-0022
SCHARPF, D. W.	FINITE ELEMENT THEORY OF PLATES AND SHELLS INCLUDING TRANSVERSE SHEAR STRAIN EFFECTS	70-0395
SCHARPF, D. W.	FINITE ELEMENTS IN TIME AND SPACE	69-0442
SCHARPF, D. W.	FINITE ELEMENTS IN TIME AND SPACE	69-0450
SCHARPF, D. W.	LARGE DEFLECTION ANALYSIS OF PRESTRESSED NETWORKS	72-0169
SCHARPF, D. W.	LINEAR METHODS OF STRUCTURAL ANALYSIS	71-0028
SCHARPF, D. W.	MATRIX DISPLACEMENT ANALYSIS OF PLATES AND SHELLS, PART 11. GENERAL FORMULATION OF THE LINEAR THEORY	71-0621
SCHARPF, D. W.	MATRIX DISPLACEMENT ANALYSIS OF SHELLS AND PLATES INCLUDING TRANSVERSE SHEAR STRAIN EFFECTS	72-0643
SCHARPF, D. W.	METHODS OF ELASTO-PLASTIC ANALYSIS	71-0617
SCHARPF, D. W.	METHODS OF ELASTOPLASTIC ANALYSIS	69-0343
SCHARPF, D. W.	NONLINEAR METHODS OF STRUCTURAL ANALYSIS	71-0029
SCHARPF, D. W.	SOME ASPECTS OF FINITE ELEMENT TECHNIQUES	69-0340
SCHARPF, D. W.	SOME GENERAL CONSIDERATIONS ON THE NATURAL MODE TECHNIQUE	69-0049
SCHARPF, D. W.	SOME GENERAL CONSIDERATIONS ON THE NATURAL MODE TECHNIQUE, PART II: LARGE DISPLACEMENTS	69-0445
SCHARPF, D. W.	SOME NEW ELEMENTS FOR THE MATRIX DISPLACEMENT METHOD	68-0017
SCHARPF, D. W.	THE CURVED TETRAHEDRONAL AND TRIANGULAR ELEMENTS TEC AND TRIC FOR THE MATRIX DISPLACEMENT METHOD	69-0047
SCHARPF, D. W.	THE ELASTO-PLASTIC CALCULATION OF GENERAL STRUCTURES AND CONTINUA	68-0207
SCHARPF, D. W.	THE FUGA FAMILY FOR FOLDED PLATE STRUCTURES	69-0333
SCHARPF, D. W.	THE FUGA FAMILY OF ELEMENTS FOR FOLDED PLATE STRUCTURES A FURTHER SEQUEL TO TECHNICAL NOTE 14	69-0446
SCHARPF, D. W.	THE INCOMPRESSIBLE LUBRICATION PROBLEM	69-0022
SCHARPF, D. W.	THE SHEBA FAMILY OF SHELL ELEMENTS FOR THE MATRIX DISPLACEMENT METHOD	68-0014
SCHARPF, D. W.	THE SHEBA FAMILY OF SHELL ELEMENTS FOR THE MATRIX DISPLACEMENT METHOD, 111, LARGE DISPLACEMENTS	69-0499
SCHARPF, D. W.	TWO AND THREE DIMENSIONAL POTENTIAL FLOW BY THE METHOD OF SINGULARITIES	69-0449
SCHARPF, D. W.	TWO AND THREE DIMENSIONAL POTENTIAL FLOW ANALYSIS BY THE METHOD OF SINGULARITIES	70-0357
SCHARPF, D. W.	TWO AND THREE-DIMENSIONAL FLOW USING FINITE ELEMENTS	69-0008
SCHARPF, D. W.	TWO-AND THREE-DIMENSIONAL FLOW ANALYSIS USING FINITE ELEMENTS	70-0021
SCHATZ, A. H.	SOME INTERIOR SUPER CONVERGENCE ESTIMATES FOR SPLINE APPROXIMATION	73-0641
SCHAUER, D. A.	COMPUTER CODE TO GENERATE FINITE ELEMENT MESH FOR PROBLEMS OF COMPLEX AXISYMMETRIC GEOMETRY	74-0360
SCHAUER, D. A.	ELFED A COMPUTER CODE TO GENERATE A FINITE ELEMENT MESH FOR PROBLEMS OF COMPLEX AXISYMMETRIC GEOMETR	74-1270
SCHAUER, D. A.	ELFED: A COMPUTER CODE TO GENERATE FINITE ELEMENT MESH FOR PROBLEMS OF COMPLEX AXISYMMETRIC GEOMETRY	73-0943
SCHEER, D. D.	STRESS AND VIBRATION ANALYSIS OF TURBOPUMP INDUCER BLADES BY FINITE ELEMENTS	71-0363
SCHEER, D. D.	STRESSES AND VIBRATION ANALYSIS OF INDUCER BLADES USING FINITE ELEMENT TECHNIQUE	70-0305
SCHELLER, J. D.	INFLUENCE OF TEMPERATURE DEPENDENT PROPERTIES ON THERMAL ROCK FRAGMENTATION	75-0222
SCHELLER, J. D.	SIMPLIFIED INELASTIC (PLASTIC AND CREEP) ANALYSIS OF PIPE ELBOWS SUBJECTED TO INPLANE AND OUT-OF-PLA	75-0654
SCHEURENBRAND,	STRESS AND VIBRATION ANALYSIS OF TURBOPUMP INDUCER BLADES BY FINITE ELEMENTS	71-0363
SCHEURENBRAND,	STRESSES AND VIBRATION ANALYSIS OF INDUCER BLADES USING FINITE ELEMENT TECHNIQUE	70-0305
SCHIFF, M.	ANALYSIS AND REDESIGN OF SONAR PIEZOELECTRIC RESONATORS USING THE FINITE ELEMENT METHOD.	72-0374
SCHIFFMAN, R. L	PORTABILITY AND ADAPTABILITY OF CIVIL ENGINEERING SOFTWARE	73-0343
SCHIFFMAN, R. L	THE EFFICIENT USE OF COMPUTER RESOURCES	72-0099
SCHIFFNER, K.	COMPARIOSN OF RESULTS OBTAINED FOR THE PROBLEM OF SYMMETRIC FLUID OSCILLATIONS IN TANKS BY THE FINIT	71-0761
SCHIFFNER, K.	HYBRIDE FINITE VERFAHREN FUER LINEARE UND NICHTLINEARE INSTATIONAERE WAERMELEITPROBLEME (HYBRID FINI	73-0484
SCHIFFNER, K.	VERGLEICH NUMERISCHER LOESUNGEN NACH DEM DIFFERENZENVERFAHREN UND NACH DER FINITE-ELEMENT-METHODE AM	69-0016
SCHILLING, C. H	APPLICATION OF MATRIX METHODS IN STRUCTURAL THEORY	69-0454
SCHILLING, U.	CALCULATION OF THE FREE OSCILLATIONS OF A LIQUID IN AXISYMMETRIC MOTIONLESS CONTAINERS OF ARBITRARY	74-0387
SCHKADE, A. F.	A REFINED AXISYMMETRIC FINITE ELEMENT FOR THE ANALYSIS OF NEARLY INCOMPRESSIBLE SOLIDS	70-0309
SCHKADE, A. F.	INVESTIGATION OF TECHNIQUES OF THREE-DIMENSIONAL FINITE ELEMENT STRESS ANALYSIS	70-0029
SCHLEGEL, W. A.	A MATHEMATICAL MODEL OF THE CORNEO-SCLERAL SHELL	71-0150
SCHLEGEL, W. A.	ANALYSIS OF THE CORNEO-SCLERAL SHELL BY THE METHOD OF DIRECT STIFFNESS	71-0151
SCHMID, G.	INCOMPRESSIBLE FLOW IN MULTIPLY CONNECTED REGIONS	74-0904

SCHMID, G.	THE HARMONIC FINITE ELEMENT MODEL FOR USE IN FIELD PROBLEMS	70-0389
SCHMIDT, B.	SETTLEMENTS AND STRENGTHENING OF SOFT CLAY ACCELERATED BY SAND DRAINS	73-0438
SCHMIDT, E.	SPLINES AND FINITE ELEMENTS	73-0629
SCHMIDT, F. A.	FEM2D: PROGRAM FOR SOLVING THE TWO-DIMENSIONAL DIFFUSION EQUATION BY THE METHOD OF FINITE ELEMENTS	74-1089
SCHMIDT, F. A.	FINITE ELEMENTS VERSUS FINITE DIFFERENCES, A COMPARISON OF THE TWO METHODS FOR THE SOLUTION OF THE D	73-1055
SCHMIDT, F. A.	ON THE APPLICATION OF THE FINITE ELEMENT METHOD IN REACTOR PHYSICS	75-0122
SCHMIDT, F. A.	2-DIMENSIONAL AND 3-DIMENSIONAL REACTOR-PHYSICS CALCULATIONS WITH FINITE-ELEMENT METHOD	75-0423
SCHMIDT, G.	METHOD FOR FINITE ELEMENTS AS SPECIAL CASE OF METHOD OF WEIGHTED RESIDUALS	72-0141
SCHMIDT, G. H.	APPLICATION OF THE FINITE ELEMENT METHOD TO THE EXTENSIONAL VIBRATIONS OF PIEZOELECTRIC PLATES	72-0073
SCHMIDT, J. H.	DYNAMIC ANALYSIS OF STIFFENED PANEL STRUCTURES	71-0522
SCHMIDT, J. H.	THEORETICAL AND EXPERIMENTAL ANALYSIS OF STIFFENED PANEL UNDER DYNAMIC CONDITIONS	70-0130
SCHMIDT, W. F.	A COMPARISON AND EVALUATION OF INCREMENTAL FINITE ELEMENT FORMULATIONS	75-0023
SCHMIDT, W. F.	PROJECTIVE METHOD APPLIED TO THREE-DIMENSIONAL ELASTICITY EQUATIONS	74-0617
SCHMIT, L. A.	A CYLINDRICAL SHELL DISCRETE ELEMENT	67-0007
SCHMIT, L. A.	A DISCRETE ELEMENT STRESS AND DISPLACEMENT ANALYSIS OF ELASTOPLASTIC PLATES	70-0108
SCHMIT, L. A.	AN INCREMENTAL COMPLEMENTARY ENERGY METHOD OF NONLINEAR STRESS ANALYSIS	69-0281
SCHMIT, L. A.	DEVELOPMENTS IN DISCRETE ELEMENT FINITE DEFLECTION STRUCTURAL ANALYSIS BY FUNCTION MINIMIZATION	68-0144
SCHMIT, L. A.	FINITE DEFLECTION DISCRETE ELEMENT ANALYSIS OF SANDWICH PLATES AND CYLINDRICAL SHELLS WITH LAMINATED	69-0236
SCHMIT, L. A.	FINITE DEFLECTION STRUCTURAL ANALYSIS USING PLATE AND SHELL DISCRETE ELEMENTS	68-0143
SCHMIT, L. A.	FINITE DEFLECTION STRUCTURAL ANALYSIS USING PLATE AND CYLINDRICAL SHELL DISCRETE ELEMENTS	68-0193
SCHMIT, L. A.	FINITE ELEMENT ANALYSIS OF SKEW PLATES IN BENDING	68-0120
SCHMIT, L. A.	FINITE ELEMENT ANALYSIS OF SANDWICH PLATES AND CYLINDRICAL SHELLS WITH LAMINATED FACES	68-0121
SCHMIT, L. A.	STRUCTURAL SYNTHESIS, 1959-1969, A DECADE OF PROGRESS	69-0237
SCHMIT, L. A.	THE GENERATION OF INTERELEMENT, COMPATIBLE STIFFNESS AND MASS MATRICES BY THE USE OF INTERPOLATION F	66-0014
SCHMIT, L. C. J	FRAME OPTIMIZATION INCLUDING FREQUENCY CONSTRAINTS	75-0082
SCHMITT, K.	APPLICATIONS OF VARIATIONAL EQUATIONS TO ORDINARY AND PARTIAL DIFFERENTIAL EQUATIONS - MULTIPLE SOLU	75-0010
SCHMITT, K. H.	NUMERICAL RESULTS FROM THE APPLICATION OF GRADIENT ITERATIVE TECHNIQUES TO THE FINITE ELEMENT VIBRAT	72-0138
SCHMITT, W.	APPLICATION OF THE FINITE ELEMENT METHOD FOR THE SAFETY EVALUATION OF REACTOR COMPONENTS	73-0853
SCHMITZ, R. P.	APPLICATION OF SMALL AND LARGE-DISPLACEMENT THEORY TO A LARGE FLEXIBLE SOLAR ARRAY AND COMPARISON WI	71-0231
SCHMITZ, R. P.	STRUCTURAL DYNAMIC ANALYSIS OF ELECTRONIC ASSEMBLIES USING NASTRAN RESTART/FORMAT CHANGE CAPABILITY	71-0230
SCHMOHL, H. P.	MESSVERFAHREN RECHENVERFAHREN ZUR ERMITTLUNG DREIACHSIGER EIGENSPANNUNGSZUSTAENDE	73-0507
SCHMUGAR, K. L.	VIBRATORY WEAR OF FUEL RODS	75-0771
SCHNEIDER, G. E	FINITE ELEMENT FORMULATION OF THE HEAT CONDUCTION EQUATION IN GENERAL ORTHOGONAL CURVILINEAR COORDIN	75-0848
SCHNELL, W.	MODERN TECHNIQUES FOR CALCULATING STRESSES IN TANKS, AND THEIR LIMITATIONS, (MODERNE VERFAHREN ZUR S	70-0153
SCHNIEWIND, A.	PERFORMANCE OF STRUCTURAL WOOD MEMBERS EXPOSED TO FIRE	75-0060
SCHNOBRICH, W.	FINITE ELEMENT ANALYSIS OF TRANSLATIONAL SHELLS	72-0348
SCHNOBRICH, W.	A FINITE ELEMENT ANALYSIS OF ECCENTRICALLY STIFFENED-CIRCULAR CYLINDRICAL SHELLS	69-0159
SCHNOBRICH, W.	ANALYSIS OF ECCENTRICALLY STIFFENED CYLINDRICAL SHELLS	72-0217
SCHNOBRICH, W.	ANALYSIS OF HIPPED ROOF HYPERBOLIC PARABOLOID STRUCTURES	72-0216
SCHNOBRICH, W.	ANALYSIS OF HYPERBOLIC PARABOLOID SHELLS	70-0324
SCHNOBRICH, W.	ANALYSIS OF SHALLOW SHELL STRUCTURES BY A DISCRETE ELEMENT SYSTEM	66-0113
SCHNOBRICH, W.	FINITE ELEMENT ANALYSIS OF SKEWED SHALLOW SHELLS	68-0250
SCHNOBRICH, W.	FINITE ELEMENT ANALYSIS OF SKEWED SHALLOW SHELLS	69-0211
SCHNOBRICH, W.	FINITE ELEMENT ANALYSIS OF REINFORCED CONCRETE	73-0494
SCHNOBRICH, W.	FINITE ELEMENT ANALYSIS OF INTERSECTING CYLINDERS	73-0866
SCHNOBRICH, W.	FINITE ELEMENT METHOD FOR THE ANALYSIS OF NOZZLE OPENINGS IN SHELLS OF REVOLUTION	69-0254
SCHNOBRICH, W.	FINITE ELEMENT SOLUTION OF NORMALLY INTERSECTING CYLINDERS	73-0723
SCHNOBRICH, W.	MATERIAL BEHAVIOUR CHARACTERISTICS FOR REINFORCED CONCRETE SHELLS STRESSED BEYOND THE ELASTIC RANGE	70-0166
SCHNOBRICH, W.	NONCONFORMING FINITE ELEMENT ANALYSIS OF SHELLS	75-0539
SCHNOBRICH, W.	NUMERICAL PROCEDURES FOR THE DETERMINATION OF THE BEHAVIOUR OF A SHEAR WALL FRAME SYSTEM	73-0216
SCHNOBRICH, W.	STRESSES IN GABLE ROOF H. P. SHELLS.	72-0394
SCHNOBRICH, W.	USE OF NONCONFORMING MODES IN FINITE ELEMENT ANALYSIS OF PLATES AND SHELLS	73-0865
SCHNURR, N. M.	COMPARISON OF THE FINITE ELEMENT AND FINITE DIFFERENCE METHODS FOR THE ANALYSIS OF STEADY TWO DIMENS	75-0223
SCHOEBERLE, D.	ON THE UNCONDITIONAL STABILITY OF AN IMPLICIT ALGORITHM FOR NONLINEAR STRUCTURAL DYNAMICS	75-0825
SCHOENSTER, J.	MEASURED AND CALCULATED VIBRATION PROPERTIES OF RING-STIFFENED HONEY-COMB CYLINDERS	71-0834
SCHOLES, A.	THE PIECEWISE LINEAR ANALYSIS OF TWO CONNECTED STRUCTURES INCLUDING THE EFFECT OF CLEARANCE AT THE C	71-0379
SCHOLES, A.	THE STRESS ANALYSIS OF AXISYMMETRIC SHELLS USING A PLANE FRAME ANALYSIS PROGRAM	69-0238
SCHOMBURG, U.	THE FINITE ELEMENT METHOD AND LOCAL BOUNDS FOR BOUNDARY VALUE PROBLEMS OF ELASTIC STRUCTURES	73-1056
SCHRADER, K. H.	MESY - EIN PROGRAMMSYSTEM ZUR UNTERSUCHUNG VON TRAGWERKEN. (STRUCTURAL ANALYSIS BY MESY PROGRAMMING	75-0360
SCHREM, E.	AN AUTOMATIC SYSTEM FOR KINEMATIC ANALYSIS, ASKA PART 1	70-0396
SCHREM, E.	AN AUTOMATIC SYSTEM FOR KINEMATIC ANALYSIS ASKA PART 1	71-0626
SCHREM, E.	ANWENDUNG DER METHODEN DER FINITEN ELEMENTE MIT DEM PROGRAMMSYSTEM ASKA UNTER BESONDERER BERUCKSICHT	71-0624
SCHREM, E.	ASKA A COMPUTER SYSTEM FOR STRUCTURAL ENGINEERS	69-0345
SCHREM, E.	ASKA, SOFTWARE ENGINEERING FOR THE SEVENTIES IN STRUCTURAL ANALYSIS	73-0704
SCHREM, E.	COMPUTER IMPLEMENTATION OF THE FINITE ELEMENT PROCEDURE	71-0633
SCHREM, E.	COMPUTER IMPLEMENTATION OF THE FINITE ELEMENT PROCEDURE	71-0650
SCHREM, E.	DEVELOPMENT AND MAINTENANCE OF LARGE FINITE ELEMENT SOFTWARE SYSTEMS	74-0865
SCHREM, E.	EINIGE ALLGEMEINE PROGRAMMSYSTEME FUR FINITE ELEMENTE	73-0581
SCHREM, E.	FINITE ELEMENT SOFTWARE IN THE NEXT DECADE	75-0259
SCHREM, E.	IMPLEMENTATION OF THE FINITE ELEMENT PROCEDURE	71-0201
SCHREM, E.	STRING EIN PROGRAMMSYSTEM FUR DIE MASCHINENUNABHANGIGE BEARBEITUNG VON ZEICHENKETTEN IN FORTRAN	68-0265
SCHREM, E.	STRUCTURAL ANALYSIS BY PROBLEM ORIENTED LANGUAGES	71-0623
SCHREM, E.	AN AUTOMATIC SYSTEM FOR KINEMATIC ANALYSIS ASKA HIGH SPEED COMPUTING OF ELASTIC STRUCTURES	70-0358
SCHREYER, H. L.	LOWER BOUNDS TO COLUMN BUCKLING LOADS	73-0485
SCHRODER, J. J.	COMPARISON OF FINITE-DIFFERENCE AND FINITE-ELEMENT SOLUTION TECHNIQUE IN TRANSIENT HEAT-CONDUCTION	75-0348
SCHROEDER, E. A	THERMAL AND STRESS ANALYSIS BY NASTRAN	72-0153
SCHUBERT, L.	ZUR BERECHNUNG, BEMESSUNG UND BEWEHRUNG RECHTECKIGER PLATTEN KONSTNATER DICKE MIT OEFFNUNGEN ODER AU	73-0664
SCHULTCHEN, E.	INTERACTIVE SOLUTION OF LINEAR EQUATIONS IN FINITE ELEMENT ANALYSIS	73-1163
SCHULTE, D.	APPLICATION OF FINITE ELEMENT METHOD - STRESS CALCULATIONS ON 90 DEGREES NOZZLES OF THIN-WALLED CYLI	71-0253
SCHULTZ, A. B.	FINITE ELEMENT STRESS ANALYSIS OF AN INTERVERTEBRAL DISC	74-0244
SCHULTZ, C. C.	COMBINED ELASTIC-PLASTIC CREEP ANALYSIS OF TWO-DIMENSIONAL BODIES	71-0232
SCHULTZ, M. H.	COMPARISON OF FINITE ELEMENT METHODS FOR A MODEL PROBLEM	74-0134
SCHULTZ, M. H.	COMPUTATIONAL ASPECTS OF THE FINITE ELEMENT METHOD	72-0512
SCHULTZ, M. H.	NUMERICAL METHODS OF HIGH-ORDER ACCURACY FOR NONLINEAR BOUNDARY VALUE PROBLEMS	68-0194
SCHULTZ, M. H.	PIECEWISE HERMITE INTERPOLATION IN ONE AND TWO VARIABLES WITH APPLICATIONS TO PARTIAL DIFFERENTIAL E	68-0201
SCHULTZ, R. F.	ANALYSIS OF A STUB END BY FINITE ELEMENT METHOD	72-0210
SCHULZE, K. W.	FINITE ELEMENT ANALYSIS OF LONG WAVES IN OPEN CHANNEL SYSTEMS	74-0033
SCHUMANN, W.	ON SHELLS OF CONSTANT STRENGTH	72-0431
SCHUMANN, W.	SOME REMARKS CONCERNING HETEROGENEOUS ANISOTROPIC PLATES.	72-0430
SCHUSTER, R. L.	ELASTIC-PLASTIC STABILITY ANALYSIS OF MINE-WASTE EMBANKMENTS	75-0835
SCHWAEGLER, R.	THE ANALYSIS OF CLASSICAL PLATE BENDING PROBLEMS USING AN EQUILIBRIUM FINITE ELEMENT MODEL	68-0302
SCHWARZ, H. R.	EIGENVALUE PROBLEMS (A - LAMBDA B) X EQUALS O FOR SYMMETRIC MATRICES OF HIGH ORDER	74-0205
SCHWERZLER, D.	A TECHNIQUE FOR CONNECTING BEAM ELEMENTS TO A PLATE MODEL OF A COMPLICATED BOX SECTION	74-0271
SCHWING, H.	ZUM TRAGVERHALTEN VON WAENDEN AUS FERTIGTEILTAFELN (BEHAVIOR OF LARGE PANEL SHEAR WALLS)	74-0299
SCIARRA, J. J.	COUPLED ROTOR-AIRFRAME VIBRATION PREDICTION METHODS	74-0498
SCIARRA, J. J.	USE OF THE FINITE ELEMENT DAMPED FORCED RESPONSE STRAIN ENERGY DISTRIBUTION FOR VIBRATION REDUCTION	72-0759
SCIARRA, J. J.	VIBRATION REDUCTION BY USING BOTH THE FINITE ELEMENT STRAIN ENERGY DISTRIBUTION AND MOBILITY TECHNIQ	74-1303
SCORDELIS, A. C	ANALYSIS AND DESIGN OF SKEW BOX GIRDER BRIDGES	72-0659
SCORDELIS, A. C	ANALYSIS OF CURVED FOLDED PLATE STRUCTURES	70-0431
SCORDELIS, A. C	ANALYSIS OF CURVED FOLDED PLATE STRUCTURES	71-0139
SCORDELIS, A. C	ANALYSIS OF ORTHOTROPIC FOLDED PLATES WITH ECCENTRIC STIFFENERS	70-0437
SCORDELIS, A. C	ANALYTICAL AND EXPERIMENTAL STUDIES OF MULTI-CELL CONCRETE BOX GIRDER BRIDGES	74-1134

SCORDELIS, A. C	CELLULAR STRUCTURES OF ARBITRARY PLAN GEOMETRY	72-0201
SCORDELIS, A. C	COMPUTER PROGRAM FOR PRISMATIC FOLDED PLATES BY PLATE AND BEAM ELEMENTS	70-0371
SCORDELIS, A. C	FINITE ELEMENT ANALYSIS OF REINFORCED CONCRETE BEAMS	67-0058
SCORDELIS, A. C	FINITE ELEMENT ANALYSIS OF REINFORCED CONCRETE STRUCTURES	72-0051
SCORDELIS, A. C	FINITE ELEMENT STUDY OF A REINFORCED CONCRETE CYLINDRICAL SHELL THROUGH ELASTIC, CRACKING, AND ULTIM	75-0317
SCORDELIS, A. C	FINITE SEGMENT ANALYSIS OF FOLDED PLATES	69-0169
SCORDELIS, A. C	NONLINEAR ANALYSIS OF RC SHELLS OF GENERAL FORM	75-0053
SCORDELIS, A. C	STRESSES IN CONTINUOUS CONCRETE BOX GIRDER BRIDGES	71-0233
SCORDELIS, A. C	STRUCTURAL BEHAVIOR OF A TWO SPAN REINFORCED CONCRETE BOX GIRDER BRIDGE MODEL. VOLUME 11. REDUCTION,	71-0659
SCOTT, F. C.	A QUARTIC, TWO DIMENSIONAL ISOPARAMETRIC ELEMENT	68-0187
SCOTT, F. C.	FINITE ELEMENT METHOD IN STRESS ANALYSIS	69-0413
SCOTT, F. C.	ISO-PARAMETRIC AND ASSOCIATED ELEMENT FAMILIES FOR TWO AND THREE DIMENSIONAL ANALYSIS	69-0285
SCOTT, F. C.	LOCAL LEAST SQUARES STRESS SMOOTHING FOR PARABOLIC ISOPARAMETRIC ELEMENTS	75-0593
SCOTT, F. C.	THREE DIMENSIONAL STRESS ANALYSIS	70-0086
SCOTT, N. R.	DIRECT MEASUREMENT OF BOND SLIP IN REINFORCED CONCRETE	69-0428
SCOTT, N. R.	FINITE ELEMENT ANALYSIS OF WOOD STRUCTURAL MEMBERS	70-0180
SCOTT, N. R.	LINEAR AND NONLINEAR FINITE ELEMENT ANALYSIS OF WOOD STRUCTURAL MEMBERS	73-0117
SCOTT, R.	ACCURATE FINITE ELEMENT APPROXIMATION FOR PROBLEMS WITH SINGULAR DATA	75-0098
SCOTT, R.	FINITE ELEMENT CONVERGENCE FOR SINGULAR DATA	73-0079
SCOTT, R.	FINITE ELEMENT TECHNIQUES FOR CURVED BOUNDARIES	73-0622
SCOTT, R.	INTERPOLATED BOUNDARY CONDITIONS IN THE FINITE ELEMENT METHOD	75-0388
SCOTT, R.	SUBSPACES WITH ACCURATELY INTERPOLATED BOUNDARY CONDITIONS	72-0527
SCOTT, V. H.	A GENERAL NUMERICAL SOLUTION OF THE TWO-DIMENSIONAL DIFFUSION-CONVECTION EQUATION BY THE FINITE ELEM	70-0004
SCOTT, V. H.	APPLICATION OF FINITE ELEMENT METHOD TO GENERAL NUMERICAL SOLUTION OF 2-DIMENSIONAL DIFFUSION-CONVEC	70-0269
SCOTT, W. D.	FRACTURE SURFACE ENERGIES BY FINITE ELEMENT STRESS ANALYSIS	73-0915
SCOTTO, F. L.	THIN-WALLED PRESTRESSED CONCRETE PRESSURE VESSEL FOR HIGH TEMPERATURE REACTORS: EXPERIMENTAL INVESTI	73-0857
SCRIVEN, L. E.	MENISCI IN ARRAYS OF CYLINDERS - NUMERICAL-SIMULATION BY FINITE-ELEMENTS	75-0449
SEBASTIAN, J. D	A USER-ORIENTED PROGRAM FOR CRASH DYNAMICS	74-0321
SECHLER, E. E.	THIN-SHELL STRUCTURES: THEORY, EXPERIMENT, AND DESIGN	74-0828
SECOR, G. A.	CONVERGENCE OF CONSISTENTLY DERIVED TIMOSHENKO BEAM FINITE ELEMENTS	72-0396
SECREST, B. G.	APPLICATION OF FINITE ELEMENT METHOD TO MASS TRANSPORT LIMITED EPITAXIAL GROWTH PROCESSES	71-0488
SECREST, B. G.	FINITE ELEMENT METHOD APPLIED TO IDEAL GASEOUS MIXTURES.	72-0393
SEED, B. H.	SEISMIC RESPONSE OF SOIL DEPOSITS UNDERLAIN BY SLOPING ROCK BOUNDARIES	69-0484
SEED, H. B.	AN INVESTIGATION OF THE DYNAMIC RESPONSE CHARACTERISTICS OF THE BON TEMPE DAM	66-0047
SEED, H. B.	DYNAMIC ANALYSIS OF THE SLIDE IN THE LOWER SAN FERNANDO DAM DURING THE EARTHQUAKE OF FEBRUARY 9, 197	75-0277
SEED, H. B.	EARTHQUAKE ANALYSIS OF EARTH DAMS	69-0072
SEED, H. B.	FINITE ELEMENT ANALYSIS OF SLOPES IN SOIL	68-0061
SEED, H. B.	LUSH: A COMPUTER PROGRAM FOR COMPLEX RESPONSE ANALYSIS OF SOIL-STRUCTURE SYSTEMS	74-0744
SEED, H. B.	QUAD-4: A COMPUTER PROGRAM FOR EVALUATING THE SEISMIC RESPONSE OF SOIL STRUCTURES BY VARIABLE DAMPIN	73-0771
SEED, H. B.	SEISMIC RESPONSE BY VARIABLE DAMPING FINITE ELEMENTS	74-0143
SEED, H. B.	SOIL-STRUCTURE INTERACTION ANALYSES FOR EVALUATING SEISMIC RESPONSE	74-0745
SEED, H. B.	STUDIES OF SEISMIC RESPONSE OF CLAY BANKS	71-0833
SEED, M. B.	RESPONSE OF EARTH BANKS DURING EARTHQUAKES	67-0039
SEGAL, V. M.	INVESTIGATION OF THE STATIONARY FLOW OF A RIGID PLASTIC MATERIAL BY THE NUMERICAL FINITE ELEMENT MET	73-0231
SEGAL, V. M.	NUMERICAL INVESTIGATION OF AXI-SYMMETRIC VISCOPLASTIC FLOWS BY FINITE-ELEMENT PROCEDURE	73-0132
SEGERLIND, L. J	FINITE ELEMENT METHOD IN FOOD ENGINEERING	74-1200
SEGERLIND, L. J	FINITE ELEMENTS AND THE AG ENGINEER	75-0647
SEGETH, K.	NUMERICAL EXPERIMENT WITH FINITE ELEMENT I	70-0303
SEGETHOVA, J.	NUMERICAL EXPERIMENT WITH FINITE ELEMENT I	70-0303
SEGOL, G.	GALERKIN-FINITE ELEMENT TECHNIQUE FOR CALCULATING TRANSIENT POSITION OF SALTWATER FRONT	75-0135
SEGUCHI, Y.	ELASTIC-PLASTIC PROBLEMS OF COMPRESSIBLE STRAIGHT BAR UNDER COMBINED LOADING CONDITION	75-0313
SEGUCHI, Y.	INCREMENTAL THEORY OF LARGE STRAIN AND LARGE DISPLACEMENT PROBLEMS AND ITS FINITE ELEMENT FORMULATIO	72-0274
SEGUI, W. T.	COMPUTER PROGRAMS FOR THE SOLUTION OF SYSTEMS OF LINEAR ALGEBRAIC EQUATIONS	73-1040
SEGUI, W. T.	FINITE ELEMENT ANALYSIS OF STRUCTURES WITH SURFACES OF ARBITRARY CURVATURE	70-0102
SEIDE, P.	ACCURACY OF SOME NUMERICAL METHODS FOR COLUMN BUCKLING	75-0583
SEIREG, A.	COMPUTER SIMULATION OF DYNAMIC STRESS, DEFORMATION, AND FRACTURE OF GEAR TEETH	72-0114
SEIREG, A.	FINITE ELEMENT BASED PROCEDURE FOR SIMULATING THE TRANSIENT RESPONSE AND FAILURE OF A TWO-DIMENSIONA	72-0123
SEITELMAN, L. H	SOME PRACTICAL SOLUTION TECHNIQUES FOR FINITE ELEMENT ANALYSIS	72-0082
SEJNOHA, R.	FAST FRACTURE RESISTANCE AND CRACK ARREST IN STRUCTURAL STEELS	73-0766
SELBY, R. A.	THREE-DIMENSIONAL THERMAL AND STRESS ANALYSIS OF A PIPING TEE	75-0703
SELNA, L.	FEATHER: FINITE ELEMENT ANALYSIS FOR THREE-DIMENSIONAL ELASTIC RESPONSE	72-0607
SELNA, L.	DYNAMIC ANALYSIS OF AUTOMOTIVE STRUCTURAL SYSTEMS	70-0340
SELNA, L. G.	FINITE ELEMENT ANALYSIS OF DENTAL STRUCTURES - AXISYMMETRIC AND PLANE STRESS IDEALIZATIONS	75-0525
SELNA, L. G.	NATURAL VIBRATIONS OF LAMINATED ORTHOTROPIC SHELLS OF REVOLUTION	70-0367
SELTZER, S. M.	COMPLEX EIGENVALUE SOLUTION TO A SPINNING SKYLAB PROBLEM	71-0207
SEMENOV, V. V.	CALCULATION OF DISPLACEMENTS AND STRESSES IN ANISOTROPIC ROCKS BY USING METHOD OF FINITE ELEMENTS	73-0265
SEMENZA, L. A.	A FINITE ELEMENT TREATMENT OF NEUTRON DIFFUSION	71-0373
SEMENZA, L. A.	APPLICATION OF FINITE ELEMENT METHOD TO MULTIGROUP NEUTRON DIFFUSION EQUATION	72-0310
SEMENZA, L. A.	APPLICATION OF THE FINITE ELEMENT METHOD TO TWO DIMENSIONAL MULTIGROUP NEUTRON DIFFUSION	72-0971
SEMENZA, L. A.	DUAL FINITE ELEMENT METHODS FOR NEUTRON DIFFUSION	71-0489
SEMENZA, L. A.	FINITE ELEMENT NEUTRON DIFFUSION IN CURVILINEAR COORDINATES	72-0039
SEMSARZADEH, G.	AN INTERACTIVE COMPUTER-AIDED MODELING SYSTEM FOR THE DISCRETIZATION OF BRIDGE STRUCTURES FOR FINITE	73-1165
SEN GUPTA, S. R	FINITE ELEMENT APPLICATIONS OF THE LEAST SQUARES METHOD	72-0052
SEN, S. K.	CRITERIA FOR FINITE ELEMENT DISCRETIZATION OF SHELLS OF REVOLUTION	73-0152
SEN, S. K.	DYNAMIC ANALYSIS OF COLUMN-SUPPORTED HYPERBOLOIDAL SHELLS	74-0188
SEN, S. K.	FREE VIBRATION OF SHELLS OF REVOLUTION USING FINITE ELEMENT METHOD	74-0232
SEN, S. K.	HIGH-PRECISION FINITE-ELEMENT ANALYSIS OF SHELLS OF REVOLUTION	73-1164
SEN, S. K.	REFINED MIXED METHOD FINITE ELEMENTS FOR SHELLS OF REVOLUTION	71-0101
SEQUIT, W. T.	COMPUTER PROGRAMS FOR THE SOLUTION OF SYSTEMS OF LINEAR ALGEBRAIC EQUATIONS	73-0932
SERCHIA, L.	DESCRIPTION OF THE IBM-360 VERSION OF THE FINITE ELEMENTS COMPUTER PROGRAM SAFE-3D	72-0775
SERPANOS, J. E.	FEATHER: FINITE ELEMENT ANALYSIS FOR THREE-DIMENSIONAL ELASTIC RESPONSE	72-0607
SERVICE, K.G.	BOUNDARIES AND FRACTURES IN FINITE ELEMENT MODELS OF GEOLOGICAL STRUCTURES	73-0352
SETHURATHNAM, A	INTERACTION OF FRAME AND SHEAR WALL WITH OPENINGS	74-1190
SETLUR, A. V.	A MIXED FINITE ELEMENT FORMULATION FOR PLATE PROBLEMS	72-0556
SETLUR, A. V.	APPLICATION OF FINITE ELEMENT METHOD TO CONTINUUM MECHANICS PROBLEMS	71-0051
SETO, H.	ANALYSIS OF WATER WAVE PROBLEMS BY THE METHOD OF SUPERPOSITION OF ANALYTICAL AND FINITE ELEMENT SOLU	75-0870
SETO, H.	FUNDAMENTAL STUDIES ON STEADY SHIP WAVE PROBLEMS BY THE FINITE ELEMENT METHOD	74-0564
SETOGUCHI, K.	STRUCTURAL ANALYSIS OF WELDED WALL FOR BOILER FURNACE	74-0653
SEVAK, N. M.	OPTIMAL SYNTHESIS OF FLEXIBLE LINK MECHANISMS WITH LARGE STATIC DEFLECTIONS	74-0525
SEVERN, P. T.	THE SEISMIC DESIGN STUDY OF A DOUBLE CURVATURE ARCH DAM	69-0053
SEVERN, R. T.	DOUBLE CURVATURE SHELL FINITE-ELEMENT AND ITS USE IN THE DYNAMIC ANALYSIS OF COOLING TOWERS AND OTHE	70-0312
SEVERN, R. T.	INCLUSION OF SHEAR DEFLECTION IN THE STIFFNESS MATRIX OF A BEAM ELEMENT	70-0307
SEVERN, R. T.	PLATE VIBRATION PROBLEMS USING THE FINITE ELEMENT METHOD WITH ASSUMED STRESS DISTRIBUTION AND TRIANG	66-0068
SEVERN, R. T.	THE ANALYSIS OF ARCH DAMS UNDER STATIC LOADING BY THE FINITE ELEMENT METHOD	68-0059
SEVERN, R. T.	THE FINITE ELEMENT METHOD FOR FLEXURE OF SLABS WHEN STRESS DISTRIBUTIONS ARE ASSUMED	66-0069
SEVERN, R. T.	TRIANGULAR FINITE-ELEMENTS OF VARIABLE THICKNESS AND THEIR APPLICATION TO PLATE AND SHELL PROBLEMS	69-0095
SEWELL, E. G.	AUTOMATIC GENERATION OF TRIANGULATIONS FOR PIECEWISE POLYNOMIAL APPROXIMATION	73-1118
SEWELL, E. G.	FINITE ELEMENT SOLUTION OF DEGENERATE INTERFACE PROBLEMS	75-0101
SEWELL, G.	AN ADAPTIVE COMPUTER PROGRAM FOR DIV. (P(X,Y) GRAD U) EQUALS F(X,Y,U) IN A POLYGONAL REGION	75-0133
SEWELL, M. J.	GOVERNING EQUATIONS AND EXTREMUM PRINCIPLES OF ELASTICITY AND PLASTICITY GENERATED FROM A SINGLE FUN	73-0666
SEXSMITH, R.	LINEAR AND NONLINEAR FINITE ELEMENT ANALYSIS OF WOOD STRUCTURAL MEMBERS	73-0117
SHABAIK, A. H.	ANALYTICAL PREDICTION OF DEFECTS OCCURRENCE IN SIMPLE AND COMPLEX	69-0157

SHAH, A. H.	PREFABRICATED PRETENSIONED LIGHTWEIGHT CONCRETE FOLDED PLATE	73-0808
SHAH, R. C.	EXPLORATORY INVESTIGATION OF RAPID CRACK PROPAGATION AND CRACK ARREST	73-0757
SHALLON, S. M.	A COMPUTER PROGRAM TO DISPLAY THE DYNAMIC STRESS ANALYSIS OF STRUCTURES	70-0077
SHANHOLTZ, V. O	FINITE-ELEMENT SIMULATION OF FLOOD HYDROGRAPHS	75-0632
SHANKAR, N. J.	A NUMERICAL MODEL FOR THE SIMULATION OF TWO DIMENSIONAL CONVECTIVE DISPERSION IN SHALLOW ESTUARIES	74-0052
SHANNON, R. W.	APPLICATION OF FRACTURE MECHANICS TO FAILURE OF THICK WALLED CYLINDERS	71-0511
SHANNON, R. W.	CRACK GROWTH MONITORING BY STRAIN SENSING	73-0486
SHANNON, R. W.	STRESS INTENSITY FACTORS FOR THICK-WALLED CYLINDERS	74-0300
SHANTARAM, D.	DYNAMIC TRANSIENT NON-LINEAR BEHAVIOUR OF THICK AND THIN PLATES	75-0125
SHANTHAKUMAR, P	ROTOR BLADE VIBRATIONS BY GALERKIN FINITE-ELEMENT METHOD	75-0389
SHARIFI, P.	A REFINED CURVED ELEMENT FOR THIN SHELLS OF REVOLUTION	71-0214
SHARIFI, P.	COMPUTER IMPLEMENTATION OF NONLINEAR FINITE ELEMENT ANALYSIS	73-0407
SHARIFI, P.	INELASTIC BUCKLING ANALYSIS OF PIPES SUBJECTED TO INTERNAL PRESSURE, FLEXURE AND AXIAL LOADING	74-0515
SHARIFI, P.	NONLINEAR BUCKLING ANALYSIS OF SANDWICH ARCHES	71-0292
SHARIFI, P.	NONLINEAR BUCKLING ANALYSIS OF COMPOSITE SHELLS	74-0973
SHARIFI, P.	NONLINEAR BUCKLING ANALYSIS OF COMPOSITE SHELLS	75-0214
SHARIFI, P.	NONLINEAR FINITE ELEMENT ANALYSIS OF SANDWICH SHELLS OF REVOLUTION	72-0180
SHARIFI, P.	NONLINEAR FINITE ELEMENT AND ITS APPLICATION TO LARGE SCALE STRUCTURAL PROBLEMS	75-0386
SHARIFI, P.	NONLINEAR THERMO ELASTIC PLASTIC AND CREEP ANALYSIS BY FINITE ELEMENT METHOD	74-0092
SHARIFI, P.	REFINED FINITE ELEMENT ANALYSIS OF ELASTIC-PLASTIC THIN SHELLS OF REVOLUTION	69-0497
SHARMA, S. K.	ANALYTICAL MODEL FOR DETERMINING ENERGY DISSIPATION IN DYNAMICALLY LOADED STRUCTURES	75-0328
SHARMA, S. K.	ISOTROPIC-KINEMATIC HARDENING MODEL FOR ELASTIC-PLASTIC CYCLIC STRUCTURAL ANALYSIS	75-0210
SHARMAN, P. W.	NONLINEAR RESPONSE OF PLATES TO IMPULSIVE LOADING USING HYBRID FINITE ELEMENTS	74-1018
SHARMAN, P. W.	THE DEVELOPMENT OF A SERIES OF HYBRID-STRESS FINITE ELEMENTS	75-0335
SHARMAN, P. W.	THE ELASTO-PLASTIC AND LARGE-DISPLACEMENT RESPONSE OF PLATES TO BLAST LOADING	73-0731
SHARPE, W. N. J	COMPARISON OF THEORETICAL AND EXPERIMENTAL SHEAR STRESS IN THE ADHESIVE LAYER OF A LAP JOINT MODEL	74-1176
SHATOFF, H. D.	THREE-DIMENSIONAL ELASTIC FINITE ELEMENT ANALYSIS USING GRADIENT DEGREES OF FREEDOM	73-0843
SHATROVSKAYA, G	SOLVING A NONLINEAR PROBLEM OF NONSTEADY THERMAL CONDUCTIVITY IN SOLID BODIES OF ARBITRARY FORM	68-0019
SHAUG, J. C.	SOLUTION OF A FREE-SURFACE BOUNDARY VALUE PROBLEM USING AN INVERSE FORMULATION AND THE FINITE ELEMEN	75-0466
SHAW, D. W.	APPLICATION OF FINITE ELEMENT METHOD TO MASS TRANSPORT LIMITED EPITAXIAL GROWTH PROCESSES	71-0488
SHAW, H.	NONLINEAR ANALYSIS OF A ROLLING CONTACT PROBLEM BY FINITE ELEMENTS	74-1216
SHAW, K. G.	CRACK TIP FINITE ELEMENTS ARE UNNECESSARY	73-1016
SHAW, K. G.	CRACK TIP FINITE ELEMENTS ARE UNNECESSARY	75-0472
SHEARS, M.	STATIC AND DYNAMIC BEHAVIOUR OF GUYED MASTS	68-0271
SHEFFLER, A. W.	A COMPACT FINITE-ELEMENT METHOD FOR THE ANALYSIS OF ANISOTROPIC LAMINATED PLATES	73-1166
SHELLY, P.	APPLICATION OF A FINITE ELEMENT METHOD TO EVALUATION OF OIL WHIRL CHARACTERISTICS	74-0590
SHELLY, P. D.	FINITE ELEMENT METHOD FOR THE CALCULATION OF LOCUS PATHS IN DYNAMICALLY LOADED BEARINGS	73-0487
SHELTON, M. T.	STRUCTURAL DYNAMIC PROPERTIES OF TACTICAL MISSILE JOINTS - PHASE 1	70-0423
SHELTON, M. T.	STRUCTURAL DYNAMIC PROPERTIES OF TACTICAL MISSILE JOINTS - PHASE 3	74-0811
SHEN, C. C.	JET-WING LIFTING - SURFACE THEORY USING ELEMENTARY VORTEX DISTRIBUTIONS	75-0245
SHEN, C. C.	METHODS FOR PREDICTING THE AERODYNAMIC AND STABILITY AND CONTROL CHARACTERISTICS OF STOL AIRCRAFT. V	73-0719
SHEN, C. N.	OPTIMUM DESIGN OF COMPOSITE SHELLS SUBJECT TO NATURAL FREQUENCY CONSTRAINTS	73-0154
SHEN, S. F.	AN AERODYNAMICIST LOOKS AT THE FINITE ELEMENT METHOD	74-0021
SHEN, S. F.	VARIATIONAL APPROACH TO THE LIFTING SURFACE PROBLEM	73-0779
SHERIF, M. A.	APPLICATION OF THE FINITE ELEMENT METHOD TO STRESS DISTRIBUTION IN SOIL MEDIA	67-0075
SHERLOCK, J. E.	DYNAMIC SIMULATION OF COMPLEX SYSTEMS UTILIZING EXPERIMENTAL AND ANALYTICAL TECHNIQUES	75-0864
SHERMAN, A. H.	COMPARATIVE ANALYSIS OF THE CUTHILL-MCKEE AND THE REVERSE CUTHILL-MCKEE ORDERING ALGORITHMS FOR SPAR	73-1099
SHERMAN, C. H.	IMPEDANCE OF INTERSTITIAL GAPS IN TRANSDUCER ARRAYS	70-0410
SHERMAN, W. C.	UNCONFINED TRANSIENT SEEPAGE IN SLOPING BANKS	71-0378
SHIBATA, H.	DYNAMIC ANALYSIS OF MECHANICAL STRUCTURES	71-0381
SHIDA, S.	THE SIMPLE METHOD OF EVALUATION OF STRESS INTENSITY FACTOR USING THE FINITE ELEMENT METHOD	74-0829
SHIEH, W. Y. J.	ANALYSIS OF PLATE BENDING BY TRIANGULAR ELEMENTS	68-0145
SHIEH, W. Y-J.	ANALYSIS OF PLATE AND SHELL STRUCTURES BY TRIANGULAR FINITE ELEMENTS	68-0303
SHIH, C. F.	J INTEGRAL ESTIMATES FOR STRAIN HARDENING MATERIALS USING FULLY PLASTIC SOLUTIONS	74-0690
SHIH, C. F.	SMALL SCALE YIELDING ANALYSIS OF MIXED MODE PLANE STRAIN CRACK PROBLEMS	73-0801
SHIH, P.	LOWER BOUNDS TO COLUMN BUCKLING LOADS	73-0485
SHILLINGBURG, H	FINITE ELEMENT ANALYSIS OF DENTAL STRUCTURES - AXISYMMETRIC AND PLANE STRESS IDEALIZATIONS	75-0525
SHIMASAKI, M.	NUMERICAL SOLUTION OF PLATEAUS PROBLEM BY A FINITE ELEMENT METHOD	74-0119
SHIMIZU, H.	NONLINEAR ANALYSIS OF REINFORCED CONCRETE BUILDINGS	73-0324
SHIMIZU, S.	A CONSIDERATION ON THE FINITE ELEMENT METHOD FOR CALCULATING STRESS INTENSITY FACTOR KI USING THE J	75-0863
SHIMIZU, S.	FREE VIBRATION ANALYSIS OF STIFFENED PLATES	72-0303
SHIMOMURA, Y.	ANALYSIS OF HYDROMAGNETIC PLASMA STABILITY BY FINITE-ELEMENT METHOD	72-0307
SHIMOMURA, Y.	NUMERICAL ANALYSIS OF MAGNETOHYDRODYNAMIC INSTABILITIES BY FINITE ELEMENT METHOD	72-0041
SHIN, Y. K.	A COMPUTER METHOD FOR THE SECOND ORDER ELASTIC ANALYSIS OF PLANE FRAMED STRUCTURES	73-0310
SHINDO, A.	ELASTIC-PLASTIC PROBLEMS OF COMPRESSIBLE STRAIGHT BAR UNDER COMBINED LOADING CONDITION	75-0313
SHINOZUKA, M.	PROBABILISTIC FORMULATION FOR ANALYTICAL MODELING OC CONCRETE STRUCTURES	71-0709
SHINOZUKA, M.	PROBABILISTIC MODELING OF CONCRETE STRUCTURES	72-0236
SHINOZUKA, M.	PROBABILISTIC MODEL FOR MATERIAL STRENGTH VARIATION AND SIZE EFFECT	72-0876
SHIOYA, S.	MATRIX ANALYSIS OF A COMPOSITE STRUCTURE OF PLATE AND LINE ELEMENTS	71-0250
SHIPLEY, S. A.	ELASTIC WAVE PROPAGATION, A COMPARISON BETWEEN FINITE ELEMENT PREDICTIONS AND EXACT SOLUTION	67-0076
SHIRAI, K.	FINITE ELEMENT SIMULATION OF ENERGY-TRAPPED ELECTROMECHANICAL RESONATORS	75-0142
SHIRATORI, M.	ANALYSIS OF DISLOCATION STRESS FIELD BY FINITE ELEMENT METHOD	75-0862
SHIRATORI, M.	NONLINEAR ANALYSIS OF CRACKED BODY BY FINITE ELEMENT METHOD	75-0315
SHIRATORI, M.	STUDY ON J//I//C FRACTURE CRITERION	74-1034
SHIRATORI, M.	THE APPLICATION OF THE FINITE ELEMENT METHOD TO FRACTURE MECHANICS	72-0331
SHORE, S.	A FULLY COMPATIBLE ANNULAR SEGMENT FINITE ELEMENT	72-0582
SHORE, S.	A FULLY COMPATIBLE ANNULAR SEGMENT FINITE ELEMENT	72-0663
SHORE, S.	DYNAMIC ANALYSIS OF CURVED BOX-GIRDER BRIDGES	75-0280
SHORE, S.	HORIZONTALLY CURVED HIGHWAY BRIDGES. A FULLY COMPATIBLE ANNULAR SEGMENT FINITE ELEMENT	72-0834
SHORE, S.	INTERACTIVE TECHNIQUES WITH GRAPHICAL OUTPUT FOR BRIDGE ANALYSES	75-0622
SHOTTER, B. A.	NEW APPROACH TO GEAR TOOTH ROOT STRESSES	72-0117
SHUBINSKI, R. P	DYNAMIC MATRIX ANALYSIS OF FRAMED STRUCTURES	62-0012
SHUGAR, T. A.	A STRUCTURAL EVALUATION OF RAPID METHODS OF BACKFILLING FOR BOMB DAMAGE REPAIR	74-0821
SHUGAR, T. A.	DEVELOPMENT OF FINITE ELEMENT HEAD INJURY MODEL	75-0506
SHUGAR, T. A.	STRUCTURAL ANALYSIS OF SYMMETRICAL PENSTOCK BIFURCATIONS BY THE FINITE ELEMENT METHOD	71-0310
SHUKER, W. A.	A SURVEY OF HEAT CONDUCTION COMPUTER PROGRAMS	71-0017
SHUKU, T.	ANALYSIS OF ACOUSTIC FIELD IN IRREGULARLY SHAPED ROOMS BY FINITE ELEMENT METHOD	73-0133
SHUKU, T.	FINITE ELEMENT ANALYSIS OF IRREGULARLY SHAPED ACOUSTIC FIELD	71-0711
SHULL, H. E.	HIGH STRENGTH CERAMICS	72-0878
SHULL, H. E.	TWO-DIMENSIONAL FINITE ELEMENT MODEL OF A CERAMIC BODY	73-0010
SHUM, Y. M.	USE OF FINITE ELEMENT METHODS IN SOLUTION OF TRANSIENT HEAT CONDUCTION PROBLEMS	69-0016
SHUM, Y. M.	USE OF FINITE-ELEMENT METHOD IN THE SOLUTION OF DIFFUSION-CONVECTION EQUATIONS	71-0490
SIDDALL, J. N.	MODEL TEST AND ANALYSIS OF OBLIQUE FOLDED PLATE STRUCTURE	73-0230
SIDDALL, J. N.	THEORETICAL AND EXPERIMENTAL VIBRATION ANALYSIS OF AN OBLIQUE SPACE FRAME	70-0460
SIDDALL, W. F.	FINITE ELEMENT ANALYSIS OF STRUCTURAL INTEGRITY OF A REACTOR PRESSURE VESSEL DURING EMERGENCY CORE C	70-0026
SIDNAY, S. E.	THE APPLICATION OF FINITE ELEMENT ANALYSIS TO THE STUDY OF TWO-DIMENSIONAL CRACKED BODIES	72-0695
SIEKMANN, J.	CALCULATION OF THE FREE OSCILLATIONS OF A LIQUID IN AXISYMMETRIC MOTIONLESS CONTAINERS OF ARBITRARY	74-0387
SIGVALDASON, O.	FINITE-ELEMENT ANALYSIS OF AN ELECTRODE-NIPPLE JOINT	75-0843
SIH, G. C.	APPLICATIONS OF THE FINITE ELEMENT METHOD TO THE CALCULATIONS OF STRESS INTENSITY FACTORS	72-0900
SIH, G. C.	THREE-DIMENSIONAL STRESS ANALYSIS OF A FINITE SLAB CONTAINING A TRANSVERSE CENTRAL CRACK	75-0732
SIKARSKIE, D. L	PHOTOELASTIC AND FINITE ELEMENT STRESS ANALYSIS OF A RESTORED AXISYMMETRIC FIRST MOLAR	73-0142
SIKORA, J. P.	IMPROVED METHOD FOR OBTAINING THE GENERAL-DISPLACEMENT FIELD FROM A HOLOGRAPHIC INTERFEROGRAM	72-0015

SILVENNOINEN, P	COMPUTATIONAL EXPERIMENTATION ON FINITE ELEMENT METHOD IN BARE SLAB CRITICALITY CALCULATIONS	73-0015
SILVENNOINEN, P	CURVED LOW ORDER ANGULAR ELEMENTS IN TWO DIMENSIONAL NEUTRON TRANSPORT PROBLEMS	74-1322
SILVENNOINEN, P	FINITE ELEMENT ANALYSIS OF SOME CRITICAL FAST ASSEMBLIES	73-0128
SILVER, M. L.	STATIC THREE DIMENSIONAL ANALYSIS OF ELEVATED STEEL TRANSPORTATION STRUCTURES	74-1187
SILVER, P.	ON LOCAL APPROXIMATIONS OF OPERATOR EQUATIONS	72-0063
SILVESTER, P.	A GENERAL HIGH-ORDER FINITE-ELEMENT WAVEGUIDE ANALYSIS PROGRAM	69-0512
SILVESTER, P.	ANALYSIS OF TRANSFORMER LEAKAGE PHENOMENA BY HIGH-ORDER FINITE ELEMENTS	73-0135
SILVESTER, P.	ANALYSIS OF TURBOALTERNATOR MAGNETIC FIELDS BY FINITE ELEMENTS	71-0460
SILVESTER, P.	AXISYMMETRIC TRIANGULAR ELEMENT FOR THE SCALAR HELMHOLTZ EQUATION	73-0688
SILVESTER, P.	CURVILINEAR FINITE ELEMENTS FOR 2-DIMENSIONAL SATURABLE MAGNETIC FIELDS	74-0562
SILVESTER, P.	EFFICIENT TECHNIQUES FOR FINITE ELEMENT ANALYSIS OF ELECTRIC MACHINES	73-0134
SILVESTER, P.	FINITE ELEMENT ANALYSIS OF MAGNETICALLY SATURATED D-C MACHINES	71-0566
SILVESTER, P.	FINITE ELEMENT ANALYSIS OF PLANAR MICROWAVE NETWORKS	73-0016
SILVESTER, P.	FINITE ELEMENT PROGRAM PACKAGE FOR AXISYMMETRIC SCALAR FIELD PROBLEMS	73-0113
SILVESTER, P.	FINITE ELEMENT PROGRAM PACKAGE FOR AXISYMMETRIC VECTOR FIELD PROBLEMS	75-0138
SILVESTER, P.	FINITE ELEMENT SOLUTION OF THE HOMOGENEOUS WAVEGUIDE PROBLEMS	69-0010
SILVESTER, P.	FINITE ELEMENT SOLUTION OF SATURABLE MAGNETIC FIELD PROBLEMS	70-0205
SILVESTER, P.	FINITE ELEMENT SOLUTION OF DIELECTRIC LOADED WAVEGUIDES	70-0206
SILVESTER, P.	FINITE ELEMENT SOLUTION OF TWO DIMENSIONAL EXTERIOR FIELD PROBLEMS	71-0491
SILVESTER, P.	FINITE ELEMENT SOLUTION OF SATURATED TRAVELLING MAGNETIC FIELD PROBLEMS	75-0600
SILVESTER, P.	FINPLT; A FINITE-ELEMENT FIELD-PLOTTING PROGRAM	72-0299
SILVESTER, P.	HIGH ORDER POLYNOMIAL TRIANGULAR FINITE ELEMENTS FOR POTENTIAL PROBLEMS	69-0240
SILVESTER, P.	NUMERICAL SOLUTION OF DIELECTRIC LOADED WAVEGUIDES: I. FINITE-ELEMENT ANALYSIS	70-0199
SILVESTER, P.	SCALAR FINITE-ELEMENT PROGRAM PACKAGE FOR TWO-DIMENSIONAL FIELD PROBLEMS	71-0478
SILVESTER, P.	TETRAHEDRAL POLYNOMIAL FINITE ELEMENTS FOR THE HELMHOLTZ EQUATION.	72-0392
SILVESTER, P.	TRIANGULAR FINITE ELEMENTS FOR THE GENERALIZED BESSEL EQUATION OF ORDER M	73-1094
SIMITSES, G. J.	OPTIMAL BEAM FREQUENCIES BY THE FINITE ELEMENT DISPLACEMENT METHOD	73-0187
SIMITSES, G. J.	STRONGEST COLUMN BY FINITE ELEMENT DISPLACEMENT METHOD	73-0136
SIMKINS, T. E.	STRUCTURAL RESPONSE TO MOVING PROJECTILE MASS BY THE FINITE ELEMENT METHOD	75-0693
SIMON, B. J.	APPLICATION OF FINITE ELEMENT ANALYSIS METHOD TO TWO-DIMENSIONAL PROBLEMS IN FRACTURE MECHANICS	69-0033
SIMON, B. J.	APPLICATION OF THE METHOD OF FINITE ELEMENT ANALYSIS TO TWO-DIMENSIONAL PROBLEMS IN FRACTURE MECHANI	68-0100
SIMON, B. R.	DEFORMATION OF THE ARTERIAL VASA VASORUM AT NORMAL AND HYPERTENSIVE ARTERIAL PRESSURE	71-0149
SIMON, B. R.	FINITE ELEMENT AND EXPERIMENTAL ANALYSIS OF AN INTERNAL FIXATION PLATE MODEL	75-0380
SIMON, R. M.	ANALYSIS OF EMBANKMENT CONSTRUCTION BY FINITE-ELEMENTS	72-0972
SIMON, R. M.	FINITE ELEMENT PROGRAM FEECON FOR UNDRAINED DEFORMATION ANALYSES OF GRANULAR EMBANKMENTS ON SOFT CLA	72-0875
SIMONEN, F. A.	ANALYSIS OF STRENGTH AND RESIDUAL STRESSES IN FILAMENT REINFORCED ALUMINUM CYLINDERS	75-0517
SIMONEN, F. A.	EXPERIMENTAL AND THEORETICAL INVESTIGATION OF PLANE STRESS FRACTURE OF 2024 - T351 ALUMINUM ALLOY	70-0147
SIMONEN, F. A.	LARGE DEFLECTION ANALYSIS OF ELASTIC-PLASTIC SHELLS OF REVOLUTION	70-0507
SIMONEN, F. A.	MATHEMATICAL AND EXPERIMENTAL STUDIES ON THE MECHANICS OF PLATED TRANSVERSE FRACTURES	74-0656
SIMONEN, F. A.	ON THE MATHEMATICAL ANALYSIS OF STRESS IN THE HUMAN FEMUR	72-0207
SIMONS, JR. F.	THREE DIMENSIONAL NONLINEAR MAGNETIC FIELD BOUNDARY VALUE PROBLEM AND ITS NUMERICAL SOLUTION	70-0060
SIMONS, N. E.	FINITE ELEMENT ANALYSIS OF THE SURFACE DEFORMATION DUE TO A UNIFORM LOADING ON A LAYER OF GIBSON SOI	75-0240
SIMPSON, B.	KRON'S METHOD: AN ALGORITHM FOR THE EIGENVALUE ANALYSIS OF LARGE-SCALE STRUCTURAL SYSTEMS	73-0758
SIMPSON, B.	THE USES OF FINITE ELEMENT TECHNIQUES IN SOIL MECHANICS WITH PARTICULAR REFERENCE TO DEFORMATION PLA	72-0002
SIMPSON, F. M.	APPLICATION OF THE FINITE ELEMENT METHOD TO PREDICT STATIC AND DYNAMIC RESPONSE OF AN UNSHROUDED CEN	71-0574
SIMPSON, L. A.	APPLICATION OF THE SINGLE-EDGE NOTCHED BEAM TO FRACTURE TOUGHNESS TESTING OF CERAMICS	74-0623
SIMPSON, L. A.	ESTABLISHMENT OF A VALID FRACTURE TOUGHNESS SPECIMEN GEOMETRY USING A FINITE ELEMENT ANALYSIS	74-0135
SIMS, F. A.	COMPARISON BETWEEN THEORETICAL AND MEASURED EARTH PRESSURES ACTING ON A LARGE MOTORWAY RETAINING WAL	74-0917
SIMS, F. A.	EARTH PRESSURES AGAINST THE ABUTMENTS AND WING WALLS OF STANDARD MOTORWAY BRIDGES	75-0629
SINGA-RAO, K.	VIBRATION OF BEAMS WITH OVERHANGS	73-0488
SINGH, A. K.	INELASTIC RESPONSE OF NUCLEAR PIPING SUBJECTED TO RUPTURE FORCES	75-0515
SINGH, B.	COMPUTER SIMULATION OF LAMINATED ROOF REINFORCED WITH GROUTED BOLTS	73-1045
SINGH, B.	CONTINUUM CHARACTERIZATION OF JOINTED ROCK MASSES - 1, 2	73-0196
SINGH, B. P.	MEMBRANE ANALOGY FOR ANISOTROPIC CABLE NETWORKS	74-0233
SINGH, C.	ON CRACKS IN WELDS AND WELDED STRUCTURES	71-0688
SINGH, J. N.	FINITE ELEMENT ANALYSIS OF MERCURY SLOSH IN THE SOLAR ELECTRIC PROPULSION STAGE	75-0757
SINGH, M. M.	DYNAMIC RESPONSE OF SOIL/CONCRETE INTERFACES AT HIGH PRESSURE	74-0836
SINGH, R. D.	DEVELOPMENT AND APPLICATIONS OF THEORETICAL METHODS FOR EVALUATING STABILITY OF OPENINGS IN ROCK	73-0791
SINGH, R. D.	MECHANICAL CHARACTERIZATION AND FINITE-ELEMENT ANALYSIS OF ELASTIC-PLASTIC, WORK-HARDENING SOILS	72-0974
SINGH, R. P.	FINITE ELEMENT METHOD IN FOOD ENGINEERING	74-1200
SINGH, S.	SECTOR ELEMENT FOR THIN PLATE FLEXURE	72-0174
SINGH, S.	STRUCTURAL DESIGN OF NUCLEAR PLANT FACILITIES	73-0812
SINGHAL, A. C.	THE USE OF FINITE ELEMENT METHODS FOR STATIC AND DYNAMIC ANALYSES OF COMPLEX SHELLS	68-0247
SINGHAL, A. C.	775 SELECTED REFERENCES ON THE FINITE ELEMENT METHOD AND MATRIX METHODS OF STRUCTURAL ANALYSIS	69-0241
SINGLA, O. P.	ANALOG SUBROUTINES FOR THE FINITE ELEMENT METHOD	72-0652
SISODIYA, R. G.	A HIGHER ORDER IN-PLANE PARALLELOGRAM ELEMENT AND ITS APPLICATION TO SKEWED CURVED BOX GIRDER BRIDGE	71-0002
SISODIYA, R. G.	DIAPHRAGMS IN SINGLE- AND DOUBLE-CELL BOX GIRDER BRIDGES WITH VARYING ANGLE OF SKEW	72-0130
SISODIYA, R. G.	FINITE ELEMENT ANALYSIS OF SKEW, CURVED BOX-GIRDER BRIDGE	70-0241
SISODIYA, R. G.	FINITE ELEMENT ANALYSIS OF SKEW BOX GIRDER BRIDGES	72-0209
SISODIYA, R. G.	FINITE ELEMENT OF SKEW VAULT BRIDGES	72-0434
SISODIYA, R. G.	FINITE-ELEMENT ANALYSIS OF BRIDGES	72-0973
SISODIYA, R. G.	NEW FINITE ELEMENTS WITH APPLICATION TO BOX GIRDER BRIDGES.	72-0465
SKAAR, K. T.	HULL AND SUPERSTRUCTURE VIBRATIONS DESIGN CALCULATIONS BY FINITE ELEMENTS	75-0556
SKAAR, K. T.	ON THE FINITE ELEMENT STRESS ANALYSIS OF OIL TANKER STRUCTURES	74-0606
SKALSKI, S. C.	OPTIM 11: (MAGIC) COMPATIBLE LARGE SCALE AUTOMATED MINIMUM WEIGHT DESIGN PROGRAM - V 2, PROGRAMMER'S	74-1036
SKALSKI, S. C.	OPTIM 11: MAGIC COMPATIBLE LARGE SCALE AUTOMATED MINIMUM WEIGHT DESIGN PROGRAM. VOLUME 11. PROGRAMME	74-1093
SKATTUM, K.S.	MODELING TECHNIQUES OF THIN-WALLED BEAMS WITH OPEN CROSS SECTIONS	72-0573
SKERMER, N. A.	MICA DAM EMBANKMENT STRESS ANALYSIS	75-0054
SKIBA, E.	A FINITE ELEMENT SOLUTION OF GENERAL FLUID DYNAMICS PROBLEMS - NATURAL CONVECTION IN RECTANGULAR CAV	70-0008
SKJOLINGSTAD, L	ANALYSIS OF ARCH DAMS USING A SPACE-FRAME MODEL	73-0176
SKJOLINGSTAD, L	GROUNDWATER PREDICTIONS FOR A HIGHWAY CUT USING FINITE ELEMENTS	73-0346
SKJOLINGSTAD, L	STEADY STATE AND TRANSIENT GROUNDWATER FLOW IN SLOPES	70-0273
SKJOLINGSTAD, L	THE APPLICATION OF HIGHER ORDER FINITE ELEMENTS TO THREE DIMENSIONAL HEAT CONDUCTION PROBLEMS	70-0266
SKJOLINGSTAD, L	THREE-DIMENSIONAL FIELD PROBLEMS BY HIGHER ORDER FINITE ELEMENTS	70-0257
SKJOLINGSTAD, L	THREE-DIMENSIONAL STRESS ANALYSIS AND FIELD PROBLEMS BY THE FINITE ELEMENT METHOD	70-0183
SKJOLINGSTAD, L	TRANSIENT AND NONLINEAR-ANALYSIS OF FIELD PROBLEMS BY FINITE ELEMENTS	72-0975
SKJOLINGSTAD, L	TWO AND THREE DIMENSIONAL GROUNDWATER SEEPAGE BY FINITE ELEMENTS	74-0463
SKJOLINGSTAD, L	UNSTEADY RADIAL FLOW IN GAS RESERVOIRS BY FINITE ELEMENTS	74-0462
SKOGH, J.	PHOTOELASTIC AND FINITE ELEMENT INVESTIGATION OF A NONSYMMETRICAL PLUG HATCH CONFIGURATION	73-0025
SKOGH, J.	POSTBUCKLING BEHAVIOUR OF A SECTION REPRESENTATIVE OF THE B-1 AFT INTERMEDIATE FUSELAGE	73-0489
SKOGH, J.	STRESS DISTRIBUTIONS AROUND PLUG MATCHES WITH ANCHOR BLOCKS	69-0135
SLAGTER, W.	FAST REACTOR PROGRAMME. FINITE ELEMENT TEMPERATURE ANALYSIS IN FUEL PINS	73-0939
SLEBIR, E. J.	FOUNDATION TESTING FOR AUBURN DAM	70-0119
SLETTEN, R.	APPLICATION OF THE FINITE ELEMENT METHOD TO OFF-SHORE STRUCTURES	74-0206
SLOT, T.	NOTE ON STRESS AND STRAIN REDISTRIBUTION IN A NOTCHED PLATE SPECIMEN DURING CYCLIC LOADING	69-0194
SLOT, T.	ON THE DETERMINATION OF EFFECTIVE ELASTIC-PLASTIC PROPERTIES FOR THE EQUIVALENT SOLID PLATE ANALYSI	74-0494
SLUPEK, B. A.	METHOD FOR ESTIMATING RESPONSE OF PAYLOAD SECONDARY STRUCTURES TO RANDOM EXCITATION	68-0232
SLYPER, H. A.	DEVELOPMENT OF EXPLICIT STIFFNESS AND MASS MATRICES FOR A TRIANGULAR PLATE ELEMENT	69-0242
SMALL, J. C.	AN INVESTIGATION OF THE STABILITY OF NUMERICAL SOLUTIONS OF BIOT'S EQUATIONS OF CONSOLIDATION	74-0746
SMALL, J. C.	THE ECONOMICAL SOLUTION OF ELASTIC PROBLEMS FOR A RANGE OF POISSONS RATIO	74-0786
SMALL, J. C.	THE NUMERICAL SOLUTION OF VISCO-ELASTIC PROBLEMS USING LAPLACE TRANSFORMS	74-0436
SMALL, R. J.	POST-TENSIONED CONCRETE NUCLEAR STRUCTURES	70-0040

SMEDLEY, G. P.	DESIGN APPRAISAL OF OFFSHORE PLATFORMS OF WELDED TUBULAR CONSTRUCTION	74-1165
SMIRAK, S.	RESENI MAZNIHO STAVU ZELEZOBETONOVYCH DESEK (PLASTIC ANALYSIS OF REINFORCED CONCRETE PLATES)	74-0495
SMIRAK, S.	STATICKE RESENI FYZIKALNE NELINEARNIHO CHOVANI ZELEZOBETO- NOVYCH DESEK (ANALYSIS OF PHYSICALLY NONL	74-0514
SMIRNOFF, T. P.	STEADY STATE IN SOIL SUBJECTED TO SEVERAL COUPLED FORCES: A FINITE ELEMENT FORMULATION	70-0207
SMITH, B. S.	DIAGONAL TENSILE STRENGTH OF BRICKWORK	70-0158
SMITH, B. S.	ELASTIC ANALYSIS OF SPATIAL SYSTEMS OF INTERCONNECTED SHEAR WALLS AND FRAMES	74-1195
SMITH, B. S.	THEORETICAL STUDY OF THE SEQUENCE OF FAILURE IN PRECAST PANEL SHEAR WALLS	73-0490
SMITH, C. S.	BUCKLING PROBLEMS IN THE DESIGN OF FIBREGLASS-REINFORCED PLASTIC SHIPS	72-0164
SMITH, C. S.	PRACTICAL CONSIDERATIONS IN THE APPLICATION OF FINITE ELEMENT TECHNIQUES TO SHIP STRUCTURES	69-0338
SMITH, C. V.	FINITE ELEMENT MODEL WITH APPLICATIONS TO BUILDINGS K-33 AND K-31	70-0420
SMITH, C. V.	STRONGEST COLUMN BY FINITE ELEMENT DISPLACEMENT METHOD	73-0136
SMITH, C. V.	THE FATE OF RADIONUCLIDES IN A GROUNDWATER ENVIRONMENT	74-0061
SMITH, C. V. (J	FINITE ELEMENT GALERKIN METHOD SOLUTIONS TO SELECTED ELLIPTIC AND PARABOLIC DIFFERENTIAL EQUATIONS	71-0027
SMITH, D. A. JR	FINITE ELEMENT ANALYSIS OF THE FORCED OSCILLATION OF SHIP HULL FORMS	74-0814
SMITH, D. A. JR	FINITE ELEMENT SOLUTION FOR ADDED MASS AND DAMPING	74-0018
SMITH, E. T.	A FINITE ELEMENT PLATE BENDING PROGRAMME	72-0579
SMITH, F. W.	ELASTIC STRESSES IN LAYERED SNOW PACKS	72-0925
SMITH, G. C.	FINITE ELEMENT APPROACH TO THE INTEGRATED POTENTIAL FORMULATION OF GENERAL UNSTEADY SUPERSONIC AEROD	73-0910
SMITH, G. C. C.	FURTHER DEVELOPMENTS IN CONSISTENT UNSTEADY SUPERSONIC AERODYNAMIC COEFFICIENTS	72-0559
SMITH, G. L.	AUTOMOTIVE USE OF FINITE ELEMENT METHODS - INTRODUCTION AND OVERVIEW	74-0388
SMITH, G. L.	USING FINITE ELEMENTS IN AUTOMOTIVE DESIGN	74-1299
SMITH, H. S.	SHELL STRUCTURES AND CLIMATIC INFLUENCES	72-0399
SMITH, H. W.	FINITE ELEMENT AIRPLANE COST ANALYSIS	74-0301
SMITH, I. M.	A FINITE ELEMENT ANALYSIS FOR ' MODERATELY THICK ' RECTANGULAR PLATES IN BENDING	68-0146
SMITH, I. M.	A FINITE ELEMENT APPROACH TO ELASTIC SOIL-STRUCTURE INTERACTION	70-0013
SMITH, I. M.	EFFECTIVENESS OF EXCESSIVE NODAL CONTINUITIES IN THE FINITE ELEMENT ANALYSIS OF THIN RECTANGULAR AND	70-0346
SMITH, I. M.	FINITE ELEMENT ANALYSIS OF CENTRIFUGED AND BUILT-UP SLOPES	70-0415
SMITH, I. M.	FINITE ELEMENT ANALYSIS OF CENTRIFUGED AND BUILT UP SLOPES	74-1376
SMITH, I. M.	GALERKIN FINITE ELEMENT SOLUTIONS FOR POLLUTION PROBLEMS IN PARTIALLY MIXED ESTUARIES	74-0047
SMITH, I. M.	INCREMENTAL NUMERICAL SOLUTION OF A SIMPLE DEFORMATION PROBLEM IN SOIL MECHANICS	70-0380
SMITH, I. M.	NUMERICAL ANALYSIS OF PLASTICITY IN SOILS	73-0611
SMITH, I. M.	NUMERICAL AND PHYSICAL MODELLING	74-0393
SMITH, I. M.	PLANE PLASTIC DEFORMATION OF SOIL,	71-0006
SMITH, I. M.	RAYLEIGH-RITZ AND GALERKIN FINITE ELEMENTS FOR DIFFUSION-CONVECTION PROBLEMS	73-0017
SMITH, I. M.	RESPONSE OF UNCONFINED AQUIFERS TO PUMPING	74-0073
SMITH, I. M.	STRESS ANALYSIS OF CONTRACTIVE OR DILATIVE SOIL	71-0579
SMITH, I. M.	THE ANALYSIS OF FLEXIBLE BULKHEADS IN SANDS	71-0579
SMITH, I. M.	THE EFFECTIVENESS OF EXCESSIVE NODAL CONTINUITIES IN THE FINITE ELEMENT ANALYSIS OF THIN RECTANGULAR	70-0105
SMITH, J. E.	USE OF SMALL MODELS IN DESIGN AND ANALYSIS OF PRESTRESSED-CONCRETE REACTOR VESSELS	70-0433
SMITH, J. H.	NONLINEAR BEAM AND PLATE ELEMENTS	72-0866
SMITH, J. M.	LOW REYNOLDS NUMBER DEVELOPING FLOW	69-0020
SMITH, L. J.	STRESS CONCENTRATION FACTORS FOR CIRCULAR NOTCHES	74-0791
SMITH, P. D.	FINITE ELEMENT INCREMENTAL ELASTIC-PLASTIC ANALYSIS OF PRESSURE VESSELS	70-0253
SMITH, P. G.	AUTOMATIC DESIGN OF SHELL STRUCTURES	71-0336
SMITH, Q.	A COMPUTER PROGRAM FOR THE ANALYSIS OF THIN SHELLS	67-0041
SMITH, R. A.	EXPERIMENTAL AND THEORETICAL FATIGUE-CRACK PROPAGATION LIVES OF VARIOUSLY NOTCHED PLATES	74-0602
SMITH, R. G.	AUTOMATIC NODAL TRIANGULATION FOR FINITE ELEMENTS	73-0178
SMITH, R. G.	CONSTRUCTION SELF-WEIGHT STRESSES OF PRECAST SHEAR WALLS	71-0203
SMITH, R. G.	MATRIX ANALYSIS OF BEAM COLUMNS BY THE METHOD OF FINITE ELEMENTS	64-0012
SMITH, R. R.	APPLICATIONS OF THE FINITE-ELEMENT METHOD AND COMPUTER GRAPHICS. (A VIBRATIONAL ANALYSIS OF A SONAR	72-0729
SMITH, R. R.	FINITE ELEMENT ANALYSIS OF ACOUSTICALLY RADIATING STRUCTURES WITH APPLICATIONS TO SONAR TRANSDUCERS	73-0075
SMITH, R. R.	FINITE ELEMENT ANALYSIS OF ACOUSTICALLY RADIATING STRUCTURES WITH APPLICATIONS TO SONAR TRANSDUCERS	74-0524
SMITH, S. L.	FINITE-ELEMENT SOLUTION OF NAVIER-STOKES EQUATIONS FOR TRANSIENT 2-DIMENSIONAL INCOMPRESSIBLE FLOW	75-0134
SMITH, W. D.	APPLICATION OF FINITE-ELEMENT ANALYSIS TO BODY WAVE-PROPAGATION PROBLEMS	75-0446
SMITH, W. W.	A SPECIAL NASTRAN PROGRAM FOR INPUT CHECKING AND UNDEFORMED STRUCTURE PLOTTING	71-0234
SMOLDEREN, J. J	NUMERICAL METHODS IN FLUID DYNAMICS	72-0844
SMYTH, W. F.	AN IMPROVED METHOD FOR REDUCING THE BANDWIDTH OF SPARSE SYMMETRIC MATRICES	72-0990
SNELL, C.	NEW DIFFERENCE BASED FINITE-ELEMENT METHOD	75-0444
SNOEY, M. R.	STRESS ANALYSIS OF A CONICAL ACRYLIC VIEWPORT	70-0106
SNOW, D. W.	APPLICATION OF THE TWO-DIMENSIONAL INTEGRAL EQUATION METHOD TO ENGINEERING PROBLEMS	75-0377
SNOWDON, J. C.	TRANSVERSE VIBRATION OF BEAMS WITH INTERNAL DAMPING, ROTATORY INERTIA AND SHEAR	63-0020
SNYDER, R. E.	A STABILITY ANALYSIS OF CYLINDRICAL PANELS USING A FINITE ELEMENT FORMULATION	71-0636
SNYDER, R. E.	VIKING DYNAMICS - AN OVERVIEW	74-1304
SOANE, A. J. M.	UNUSUAL FOUNDATIONS OVER THE MERSEY TUNNEL	70-0197
SOARES, C. A.	FINITE ELEMENT ANALYSIS OF ROTATING SHELLS	74-0096
SOBIESZCZANSKI,	APPLICATION OF COMPUTER-AIDED AIRCRAFT DESIGN IN A MULTIDISCIPLINARY ENVIRONMENT	73-0527
SOBIESZCZANSKI,	EVALUATION OF ALGORITHMS FOR STRUCTURAL MODIFICATION	69-0243
SOBIESZCZANSKI,	ON THE ACCURACY OF THE TAYLOR APPROXIMATION FOR STRUCTURAL RESIZING	74-0147
SOBIESZCZANSKI,	NASTRAN AS ANALYSIS TOOL IN A STRUCTURAL DESIGN OPTIMIZATION PROCESS	71-0235
SOBOLESKI, J. J	NASTRAN PLOTTING AT A REMOTE TERMINAL	74-0186
SODERLAND, H. W	UNDERGROUND MEASUREMENTS IN A STEEPLY DIPPING OREBODY	68-0085
SODHI, D. S.	MIXED FINITE ELEMENT METHOD FOR VIBRATION ANALYSIS OF SHALLOW SHELLS	73-0593
SODHI, D. S.	ON THE GENERALIZATION OF STRESS FUNCTION PROCEDURE FOR DYNAMIC ANALYSIS OF PLATES	73-0661
SOEDEL, W.	CONTACT OF AN INFLATED TOROIDAL MEMBRANE WITH A FLAT SURFACE AS AN APPROACH TO THE TIRE DEFLECTION P	75-0164
SOFUE, Y.	FLEXURAL RIGIDITY OF A THIN WALLED BUILD-UP ROTOR FOR JET ENGINE. MEASUREMENT BY STATIC LOAD TEST AND	73-0756
SOGGE, R. L.	FINITE ELEMENT ANALYSIS OF ANCHORED BULKHEAD BEHAVIOR	74-1068
SOKAL, Y. J.	NATURAL PERIODS OF INFILLED FRAMES	72-0122
SOKOLOV, V. F.	CHISLENNYI METOD RASCHETA SOBSTVENNYKH I VYNUZHDENNYKH KOLEBANII SOSTAVNYKH OBOLOCHECHNYKH KONSTRUKT	73-0461
SOLIMAN, J. I.	FINITE ELEMENT SOLUTION OF HEAT TRANSMISSION IN STEEL INGOTS	72-0311
SOLNYSHKIN, N.	RASHET TREKHMERNYKH MAGNITNYKE POLEI METODOM KONECHNYKH ELEMENTOV. (CALCULATION OF THREE-DIMENSIONAL	75-0766
SOLTIS, L. A.	FINITE DEFORMATION OF BIAXIALLY LOADED COLUMNS	72-0021
SOM, P. K.	ANCHOR ZONE STRESS ANALYSIS BY FINITE ELEMENT METHOD	68-0147
SOMASHEKAR, B.	APPLICATION OF MATRIX DISPLACEMENT METHODS IN THE STUDY OF PANEL FLUTTER	69-0406
SOMASHEKAR, B.	FLUTTER OF SKEW PANELS BY THE MATRIX DISPLACEMENT APPROACH	70-0201
SOMERS, A. E.	SINGER: A COMPUTER CODE FOR GENERAL ANALYSIS OF TWO-DIMENSIONAL REINFORCED CONCRETE STRUCTURES. VOLU	75-0724
SOMERTON, W. H.	ROCK MECHANICS - THEORY AND PRACTICE	70-0107
SOMERVAILLE, I.	A COMPATIBLE QUADRILATERAL PLATE BENDING ELEMENT	72-0482
SOMERVAILLE, I.	A FAMILY OF EQUILIBRIUM PLATE BENDING ELEMENTS	74-0431
SOMERVAILLE, I.	A TECHNIQUE FOR MESH GRADING APPLIED TO CONFORMING PLATE BENDING FINITE ELEMENTS	73-1058
SOMERVAILLE, I.	AN EQUILIBRIUM MEMBRANE QUADRILATERAL ELEMENT	72-0484
SOMERVAILLE, I.	MESH GRADING TECHNIQUES FOR COMPATIBLE AND EQUILIBRIUM ELEMENTS	74-0397
SOMOGYI, D.	ON THE APPLICATION OF THE FINITE ELEMENT METHOD TO A CLASS OF PROBLEMS IN FLUID DYNAMICS	69-0400
SON, J. S.	NUMERICAL SOLUTION FOR THE FLOW AROUND A CYLINDER AT REYNOLDS NUMBERS OF 40, 200 AND 500	69-0378
SONNENFE, S. L.	ANALYSIS OF UNSTEADY FLOW TOWARD AN ARTESIAN WELL BY 3-DIMENSIONAL FINITE ELEMENTS	74-0351
SONNENFELD, S.	ANALYSIS OF UNSTEADY FLOW TOWARD AN ARTESIAN WELL BY THREE-DIMENSIONAL FINITE ELEMENTS	74-1351
SONNOGYI, D.	FINITE ELEMENT APPLICATIONS IN FLUID DYNAMICS	69-0003
SONTVEDT, T.	LOADS AND RESPONSE OF LARGE DUCTED PROPELLER SYSTEMS	73-0810
SONTVEDT, T.	MARINE PROPELLER BLADES STRESSES AND DEFORMATIONS WITH GIVEN, FROZEN, HYDRODYNAMIC LOADING	73-0474
SONTVEDT, T.	PROPELLER BLADE STRESSES, APPLICATION OF FINITE ELEMENT METHODS	74-0207
SOOSAAR, K.	APPLICATION OF FINITE ELEMENT TECHNIQUE TO DESIGN AND MANUFACTURE OF ASTRONOMICAL MIRRORS	69-0302
SOOSAAR, K.	DESIGN OF OPTICAL MIRROR STRUCTURES	71-0658
SOOSAAR, K.	ELASTIC ANALYSIS OF LARGE SPACEBOUND MIRRORS	72-0619

SOOSAAR, K.	STRUCTURAL EVALUATION OF CANDIDATE DESIGNS FOR THE LARGE SPACE TELESCOPE PRIMARY MIRROR	75-0708
SOREIDE, T.	ANALYSIS OF STIFFENED PLATES CONSIDERING NONLINEAR MATERIAL AND GEOMETRIC BEHAVIOUR	75-0327
SOREIDE, T. H.	A COMPARATIVE STUDY OF DIFFERENT NUMERICAL SOLUTION TECHNIQUES AS APPLIED TO A NONLINEAR STRUCTURAL	73-1117
SORENSEN, E. P.	ELASTIC PLASTIC AND CREEP ANALYSIS OF PIPELINES BY FINITE ELEMENTS	73-0419
SORENSEN, E. P.	WATER IMPACT ANALYSIS OF SPACE SHUTTLE SOLID ROCKET BY THE FINITE ELEMENT METHOD	74-0760
SORENSEN, J. P.	SOLUTION OF TRANSPORT PROBLEMS BY USE OF TRIAL FUNCTION TECHNIQUES	72-0991
SORENSEN, M.	A SURVEY OF THE APPLICATION OF FINITE ELEMENT METHODS TO STRESS ANALYSIS PROBLEMS WITH PARTICULAR EM	71-0632
SORENSEN, M.	A TECHNICAL MANUAL FOR THE BEAM ELEMENTS	69-0455
SORENSEN, M.	APPLICATION OF MATRIX METHODS IN STRUCTURAL THEORY	69-0454
SORENSEN, M.	ASKA 105 - A PROBLEM ORIENTED LANGUAGE FOR STRUCTURAL ENGINEERS, VOLUME 11, APPLICATION	69-0461
SORENSEN, M.	ASKA 105, A PROBLEM ORIENTED LANGUAGE FOR STRUCTURAL ENGINEERS, VOLUME 1, DESIGN	69-0460
SORENSEN, M.	COMPUTER AIDED STRUCTURAL ANALYSIS: THE MACHINE INDEPENDENT SYSTEM ASKA	70-0397
SORENSEN, M.	DIE METHODE DER ENDLICHEN ELEMENTE UND IHRE ANWENDUNG IM MASCHINENBAU	71-0628
SORENSON, M.	ASKA 105 - A TECHNICAL MANUAL FOR THE DIRECT SOLUTION AND INVERSION PACKAGE	68-0255
SORENSON, M.	ASKA 105 A TECHNICAL MANUAL AND PROGRAMMING THEORY FOR THE ELEMENTS, FLA, TRIM, SPARTA AND TET	68-0254
SOUTO SILVEIRA,	EMBANKMENT DAM FOR THE MARIMBONDO HYDRO PROJECT	75-0270
SOVA, J. A.	A METHOD FOR DETERMINING LOCAL ELASTOPLASTIC STRESS AND STRAIN IN METALLURGICALLY BONDED NOTCHED LAM	74-0686
SOZEN, M. A.	PRESTRESSED CONCRETE DEEP SLABS WITH OPENINGS	73-0427
SPAAS, H. A. C.	DETERMINATION OF STRESS CONCENTRATIONS AT NOZZLE TO CYLINDER INTERSECTIONS AND COMPARISON WITH EXPER	73-0491
SPAAS, H.A.C.M.	ON THE STRESS DISTRIBUTION IN NOZZLE TO CYLINDER INTERSECTIONS FOR SMALL DIAMETER RATIOS	72-0426
SPANNAGEL, D. W	STRUCTURAL BEHAVIOR OF A FLEXIBLE METAL CULVERT UNDER A DEEP EARTH EMBANKMENT USING METHOD A BACKFIL	73-0739
SPANNAGEL, D. W	TRANSVERSE REINFORCEMENT OF CONTINUOUS, TROUGH-SHAPED CONCRETE RAILROAD STRUCTURES. AN IMPLEMENTATIO	71-0719
SPEARE, J. C.	DISCRETE-ELEMENT STATIC ANALYSIS OF CORE-STIFFENED SHELLS OF REVOLUTION FOR ASYMMETRIC MECHANICAL LO	68-0148
SPEARS, R. K.	FINITE ELEMENT VISCOELASTIC STRESS ANALYSIS STUDY USING THE GENERALIZED KELVIN SOLID MODEL	73-0957
SPEELPENNING, B	THE ELEMENT-CHAIN CONCEPT IN FINITE ELEMENT BASED PROGRAMMING	72-0934
SPENCER, A. M.	ELASTIC DEFLECTION IN THIN WALLED CONES ROTATING ABOUT THEIR AXES OF SYMMETRY	73-0545
SPENCER, R. W.	POLYMER-IMPREGNATED CONCRETE TUNNEL SUPPORT AND LINING	73-0736
SPIER, E. E.	DESIGN, MANUFACTURE, DEVELOPMENT, TEST, AND EVALUATION OF BORON/ALUMINUM STRUCTURAL COMPONENTS FOR S	74-0798
SPIER, E. E.	ON BUCKLING OF UNIDIRECTIONAL BORON-ALUMINUM STIFFENERS - A CAUTION TO DESIGNERS	75-0634
SPILKER, R. L.	A FINITE ELEMENT MODEL FOR LAMINATED PLATES INCLUDING TRANSVERSE SHEAR DEFORMATION	72-0818
SPILKER, R. L.	A STUDY OF ELASTIC-PLASTIC ANALYSIS BY THE ASSUMED-STRESS HYBRID FINITE-ELEMENT MODEL, WITH APPLICAT	74-0718
SPILKER, R. L.	ELASTIC PLASTIC ANALYSIS BY ASSUMED STRESS HYBRID MODEL	74-0443
SPINDEL, J. E.	DESIGN OF INCLINED TIED ARCH RAILWAY BRIDGES OVER THE M56	71-0297
SPOKOWSKI, R. W	FINITE ELEMENT ANALYSIS OF REINFORCED CONCRETE MEMBERS	72-0583
SPOONER, J. B.	A HISTORY OF THE FINITE ELEMENT METHOD	75-0398
SPOONER, J. B.	ASAS - A COMMERCIAL FINITE ELEMENT SYSTEM WITH DISPLACEMENT AND FORCE CONTINUITY OPTIONS	73-0947
SPOONER, J. B.	DIE ELASTO-PLASTISCHE BERECHNUNG VON ALLGEMEINEN TRAGWERKEN UND KONTINUA	69-0443
SPOONER, J. B.	DIE VERWENDUNG DER MATRIZENVERSCHIEBUNGSMETHODE ZUR BERECHNUNG VON SPANNUNGSKOMPONENTEN	67-0140
SPOONER, J. B.	DIE VERWENDUNG DER MATRIZENVERSCHIEBUNGSMETHODE ZUR BERECHNUNG VON SPANNUNGSKONZENTRATIONEN	70-0400
SPOONER, J. B.	FINITE ELEMENT ANALYSIS, MODELLING AND COMPUTER GRAPHICS	74-1292
SPOONER, J. B.	THE CHANGES IN THE (TWO-DIMENSIONAL) STRESS ROUNDED CORNERS, CAUSED BY A VARYING INTERNAL PRESSURE A	67-0141
SPOONER, J. H.	DIE ANWENDUNG DER MATRIZENVERSCHIEBUNGSMETHODE ZUM BERECHNEN VON SPANNUNGSKONZENTRATIONEN	68-0208
SPOONER, J. H.	THE ELASTO-PLASTIC CALCULATION OF GENERAL STRUCTURES AND CONTINUA	68-0207
SPOONER, J. H.	THE THERMO-ELASTO-PLASTIC ANALYSIS OF A RE-ENTRY BODY TYPE COMPOSITE SOLID STRUCTURE USING A NEW FIN	66-0093
SPREEUW, E.	FIESTA - FINITE ELEMENT STRESS AND TEMPERATURE ANALYSIS. A DESCRIPTION OF THE THEORY USED FOR THE CO	71-0718
SPREEUW, E.	FINITE ELEMENT COMPUTER PROGRAMS FOR HEAT CONDUCTION PROBLEMS	73-0525
SPREEUW, E.	PRINCIPLES OF THE FINITE ELEMENT METHOD	75-0303
SRAWLEY, J. E.	EVALUATION OF THE USE OF A SINGULARITY ELEMENT IN FINITE ELEMENT ANALYSIS OF CENTER-CRACKED PLATES	72-0870
SRINATHA, H. R.	DISCONTINUITY STRESS ANALYSIS OF PRESSURE-VESSELS USING FINITE-ELEMENT METHOD	75-0346
SRINIVASAN, R.	STRUCTURAL DESIGN OF NUCLEAR PLANT FACILITIES	73-0812
SRINIVASAN, S.	INTERNATIONAL MACHINE TOOL DESIGN AND RESEARCH CONFERENCE, 12TH PROCEEDINGS, 1971.	72-0224
SRIVASTAVA, N.	ANALYSIS OF PNEUMATIC SHELLS WITH OR WITHOUT CABLE FORMULATION	74-0591
STAATS, J. R.	DYNAMIC LOADS ON GROUTED EMPLACEMENT SYSTEMS	70-0406
STABERG, L. G.	MECHANICS OF MACKAY-MARG TONOMETER	73-0433
STACEY, T. R.	THREE DIMENSIONAL FINITE ELEMENT STRESS ANALYSIS APPLIED TO TWO PROBLEMS IN ROCK MECHANICS	72-0189
STACEY, T. R.	THREE-DIMENSIONAL CONSIDERATION OF THE STRESSES SURROUNDING OPEN-PIT MINE SLOPES	73-0492
STADTER, J. T.	CONPLOT: A GENERAL PURPOSE CONTOUR PLOTTING PROGRAM	74-0887
STAGG, K. G.	ANALYSIS OF NONLINEAR PROBLEMS IN ROCK MECHANICS WITH PARTICULAR REFERENCE TO JOINTED ROCK SYSTEMS	70-0282
STAGG, K. G.	STRESSES IN ANISOTROPIC MEDIA WITH PARTICULAR REFERENCE TO PROBLEMS OF ROCK MECHANICS	66-0079
STAGG, K. G.	STRESSES IN ANISOTROPIC MEDIA OF ROCK MECHANICS	66-0127
STAGG, K. G.	THERMAL STRESS AND TEMPERATURE FIELDS - A FINITE-ELEMENT APPROACH	71-0782
STAHL, P. A.	EFFECTS OF SPECIMEN RADIUS ON THE STRESS STATE NEAR A ROOF BOLT ANCHOR: A FINITE ELEMENT DETERMINATI	73-0229
STALEY, J. R.	COUPLED ROTOR-AIRFRAME VIBRATION PREDICTION METHODS	74-0498
STALK, G.	APPLICATIONS OF THE ASSUMED-STRESS HYBRID METHOD TO FINITE ELEMENT ANALYSIS IN FRACTURE MECHANICS	75-0495
STALNAKER, H. D	SPECTRUM TRUNCATION AND DAMAGE TOLERANCE STUDY ASSOCIATED WITH THE C-5A OUTBOARD PYLON AFT TRUSS LUG	74-0535
STANTON, E. L.	A DISCRETE ELEMENT STRESS AND DISPLACEMENT ANALYSIS OF ELASTOPLASTIC PLATES	70-0108
STANTON, E. L.	AN EVALUATION OF CURRENT NASTRAN DISCRETE ELEMENT MODELS FOR MONOCOQUE AND SEMI-MONOCOQUE STRUCTURES	71-0236
STANTON, E. L.	ANISOTROPIC PARAMETRIC PLATE DISCRETE ELEMENTS	73-1047
STANTON, E. L.	DEVELOPMENTS IN DISCRETE ELEMENT FINITE DEFLECTION STRUCTURAL ANALYSIS BY FUNCTION MINIMIZATION	58-0144
STANTON, E. L.	DEVELOPMENTS IN STRUCTURAL ANALYSIS BY DIRECT ENERGY MINIMIZATION	68-0070
STANTON, E. L.	THE APPLICATION OF GRADIENT MINIMIZATION METHODS AND HIGHER ORDER DISCRETE ELEMENTS TO SHELL BUCKLIN	71-0237
STARNES, J. H.	VIBRATION STUDIES OF A FLAT PLATE AND BUILT-UP WING	71-0238
STASENKO, I. V.	TREUGOL'NY I ELEMENT DLYA KRUGOVOI TSILINDRICHESKOI OBOLOCHKI (TRIANGULAR FINITE ELEMENT FOR CIRCULA	73-0493
STAUB, J. A.	FUNDAMENTAL STUDY OF UNDERWATER WELDING	72-0382
STEARNS, B. K.	SPARSE MATRICES IN FINITE ELEMENT PROGRAM DEVELOPMENT	73-0337
STECCO, S.	A FINITE ELEMENT PROCEDURE FOR NUMERICAL OPTIMIZATION IN MECHANICAL DESIGN	74-1218
STEFANKO, R.	FINITE ELEMENT OF UNDERGROUND STRESSES UTILIZING STOCHASTICALLY SIMULATED MATERIAL PROPERTIES	70-0111
STEFANOU, G. D.	STRESS DISTRIBUTION IN THE LIGAMENT ZONE OF PERFORATED CIRCULAR DISCS WITH SPECIAL REFERENCE TO THE	73-0827
STEFANOU, G. D.	TWO-DIMENSIONAL TIME-DEPENDENT ANALYSIS OF PERFORATED END CAPS FOR NUCLEAR REACTOR PRESSURE VESSELS	71-0239
STEGER, J. L.	THE NUMERICAL CALCULATION OF LAMINAR BOUNDARY-LAYER SEPARATION	74-0785
STEIN, E.	CALCULATION OF PLATES BY MEANS OF METHOD OF FINITE ELEMENTS	70-0345
STEIN, E.	EINIGE ALLGEMEINE PROGRAMMSYSTEME FUR FINITE ELEMENTE	73-0581
STEIN, E.	ON THE STRESS COMPUTATION IN FINITE ELEMENT MODELS BASED UPON DISPLACEMENT APPROXIMATIONS	74-0598
STEIN, E.	PLATE MEASUREMENTS USING FINITE ELEMENTS	70-0229
STEINCHE, W.	PROCEDURE FOR SOLVING LINEAR EQUATION SYSTEMS WITH SPARSELY OCCUPIED COEFFICIENT MATRICES ASSOCIATED	74-0352
STEINMUELLER, G	RESTRICTIONS IN THE APPLICATION OF AUTOMATIC GENERATION SCHEMES BY ISOPARAMETRIC CO-ORDINATES	74-1377
STELZER, F.	PRACTICAL HANDLING OF FINITE-ELEMENT METHOD 2	75-0547
STELZER, F.	PRACTICAL HANDLING OF FINITE-ELEMENT METHOD 1	75-0551
STENSCH, W.	VOGT BOUNDARY FOR FINITE ELEMENT ARCH DAM ANALYSIS	74-0156
STEPHENSON, J.	FINITE-ELEMENT PROCEDURES FOR PROBLEMS WITH CURVED BOUNDARIES	72-0977
STERN, M.	CALCULATION OF STRESS INTENSITY FACTORS IN ANISOTROPIC MATERIALS BY A CONTOUR INTEGRAL METHOD	75-0257
STERN, M.	SOME STRESS INTENSITY CALCULATIONS USING FINITE ELEMENTS	74-0423
STERN, M.	STRONGLY PLANAR AC-LATTICES IN WHICH IDEAL OF FINITE-ELEMENTS IS STANDARD	75-0816
STERN, P.	NUMERICAL ANALYSIS OF SHELL STRUCTURES	67-0077
STERN, P.	POSTBUCKLING BEHAVIOUR OF A SECTION REPRESENTATIVE OF THE B-1 AFT INTERMEDIATE FUSELAGE	73-0489
STETSON, M.	FINITE ELEMENT STUDY OF THE ELASTIC BEHAVIOR OF PLANE FRAMES WITH FILLER WALLS	71-0828
STEURY, R. L.	FINITE ELEMENT DYNAMIC ANALYSIS OF AN AUTOMOTIVE FRAME	73-0256
STEVEN, G. P.	AUTOMATIC GENERATION OF FINITE ELEMENTS	74-1335
STEVEN, G. P.	FINITE ELEMENT ANALYSIS OF A HYDRODYNAMIC BEARING	74-0773
STEVEN, G. P.	SOLUTION OF FLUID DYNAMIC PROBLEMS BY FINTE ELEMENTS	74-1184
STEVEN, G. P.	STRESS DIFFUSION IN THICK STIFFENED PANELS	74-0620
STEVENSON, M. G	USING THE FINITE ELEMENT METHOD TO DETERMINE TEMPERATURE DISTRIBUTIONS IN ORTHOGONAL MACHINING	74-1212

STEWARD, D. V.	ON AN APPROACH TO TECHNIQUES FOR THE ANALYSIS OF THE STRUCTURES OF LARGE SYSTEMS OF EQUATIONS	62-0015
STIANSEN, S. G.	FINITE ELEMENT ANALYSIS OF CONTAINER SHIPS	72-0372
STIBOR, G. S.	THREE-DIMENSIONAL PHOTOELASTIC AND FINITE-ELEMENT ANALYSIS OF A PROPELLANT GRAIN	72-0031
STIBOR, G. S.	USE OF FINITE ELEMENT COMPUTER PROGRAMS IN FRACTURE MECHANICS	71-0022
STICKLER, D. C.	ANALYSIS OF STRUCTURAL-ACOUSTIC INTERACTIONS IN METAL-CERAMIC TRANSDUCERS	72-0850
STIEDA, C. K. A	A COMPUTER PROGRAM FOR THE CALCULATION OF TORSION FACTORS AND SHEAR STRESS FACTORS FOR BOX SECTIONS	69-0485
STIFLER, J. F.	LOVE WAVE DISPERSION THROUGH IRREGULAR STRUCTURES USING FINITE-ELEMENT ALGORITHMS	75-0431
STINCHCOMB, W.	FATIGUE CRACK INITIATION IN A BORON EPOXY PLATE WITH A CIRCULAR HOLE	74-0820
STOCKMEYER, P.	AN ALGORITHM FOR REDUCING THE BANDWIDTH AND PROFILE OF A SPARSE MATRIX	74-0792
STODDART, W. C.	THE FINITE ELEMENT METHOD - A BIBLIOGRAPHY OF ITS THEORY AND APPLICATIONS	71-0510
STODDART, W. C.	TRANSIENT RESPONSE OF LINEAR ELASTIC STRUCTURES DETERMINED BY THE MATRIX EXPONENTAL METHOD	71-0240
STOKER, J. R.	FINITE ELEMENTS	74-0274
STOKER, J. R.	FINITE ELEMENTS	74-0478
STOKER, J. R.	MASS CONDENSATION, A SEMI-AUTOMATIC METHOD FOR REDUCING THE SIZE OF VIBRATION PROBLEMS	69-0220
STONE, B. J.	APPLICATION OF COMPUTER GRAPHICS TO FINITE ELEMENT MESH GENERATION	73-0546
STONE, G. O.	COUPLING MATRICES FOR HIGH ORDER FINITE ELEMENT ANALYSIS OF ACOUSTIC - WAVE PROPAGATION	72-0040
STONE, G. O.	HIGH-ORDER FINITE ELEMENTS FOR INHOMOGENEOUS ACOUSTIC GUIDING STRUCTURES	73-0137
STORAASLI, O. O	ON THE ACCURACY OF THE TAYLOR APPROXIMATION FOR STRUCTURE RESIZING	74-0147
STORDAHL, H.	EXPERIENCES FROM STRESS ANALYSIS OF AXI-SYMMETRIC PROBLEMS IN MACHINE DESIGN	69-0247
STORDAHL, H.	FINITE-ELEMENT ANALYSIS OF AXISYMMETRIC ROTORS	69-0244
STORM, J. H.	A FINITE-ELEMENT MODEL TO SIMULATE THE NONLINEAR RESPONSE OF REINFORCED CONCRETE FRAMES WITH MASONRY	73-1168
STOVALL, R. E.	APPLICATION OF A UNIMOMENT TECHNIQUE TO A BICONICAL ANTENNA WITH INHOMOGENEOUS DIELECTRIC LOADING	75-0375
STOWE, R. L.	PROPAGATION OF FAILURE IN A CIRCULAR CYLINDER OF ROCK SUBJECTED TO A COMPRESSIVE FORCE	73-0793
STOWELL, N. A.	FINITE ELEMENT ANALYSIS OF A MULTI-COMPONENT KINETIC ENERGY PROJECTILE	74-1180
STOWELL, N. A.	FINITE ELEMENT ANALYSIS OF A MULTI-COMPONENT KINETIC ENERGY PROJECTILE	75-0754
STOYKOVICH, M.	EVALUATION OF TORSIONAL EFFECTS IN THE SEISMIC ANALYSIS OF SYMMETRIC AND UNSYMMETRIC NUCLEAR POWER P	73-0175
STRAAYER, J. W.	BOPACE 3-D (THE BOEING PLASTIC ANALYSIS CAPABILITY FOR 3-DIMENSIONAL SOLIDS USING ISOPARAMETRIC FINI	75-0699
STRAAYER, J. W.	EFFECT OF DAMPING ON EXCITABILITY OF HIGH-ORDER NORMAL MODES	75-0709
STRANDNESS, D.	DEFORMATION OF THE ARTERIAL VASA VASORUM AT NORMAL AND HYPERTENSIVE ARTERIAL PRESSURE	71-0149
STRANG, G.	AN ANALYSIS OF THE FINITE ELEMENT METHOD	73-0071
STRANG, G.	APPROXIMATE BOUNDARY CONDITIONS IN THE FINITE ELEMENT METHOD	72-0995
STRANG, G.	APPROXIMATION IN FINITE ELEMENT METHOD	72-0293
STRANG, G.	FINITE ELEMENT METHOD	71-0455
STRANG, G.	FOURIER ANALYSIS OF THE FINITE ELEMENT METHOD IN RITZ-GALERKIN THEORY	69-0297
STRANG, G.	FREE BOUNDARIES AND FINITE-ELEMENTS IN ONE DIMENSION	75-0447
STRANG, G.	FREE BOUNDARIES AND FINITE-ELEMENTS IN ONE DIMENSION	75-0587
STRANG, G.	ONE-SIDED APPROXIMATION AND VARIATIONAL INEQUALITIES	74-0335
STRANG, G.	ONE-SIDED APPROXIMATION AND PLATE BENDING	74-0336
STRANG, G.	PIECEWISE POLYNOMIALS AND FINITE ELEMENT METHOD	73-0041
STRANG, G.	THE FINITE ELEMENT METHOD AND APPROXIMATION THEORY	71-0380
STRANG, G.	THE FINITE ELEMENT METHOD-LINEAR AND NONLINEAR APPLICATIONS	74-0337
STRANG, G.	VARIATIONAL CRIMES IN THE FINITE ELEMENT METHOD	72-0517
STRANG, W. G.	SURVEY OF UNDERLYING THEORY, APPROXIMATION BY PIECEWISE POLYNOMIALS, ILLEGAL ELEMENTS AND THE PATCH	73-0635
STRANNIGAN, J.	DETERMINATION OF ENERGY RELEASE RATES AND STRESS - INTENSITY FACTORS BY THE FINITE-ELEMENT METHOD	72-0183
STRATONOVA, M.	RASCHET KILEBANII PLASTIN SLOZHNOI FORMY S NEODNORODNYMI GRANICHNYMI USLOVIYAMI. (CALCULATION OF VIB	74-0988
STRAUB, K.	ASPECTS OF FINITE ELEMENT METHOD APPLIED TO AEROSPACE STRUCTURES	73-0703
STRAUB, K.	ASPECTS OF THE FINITE ELEMENT METHOD AS APPLIED TO AERO-SPACE STRUCTURES	72-0649
STRAUB, K.	DYNAN LECTURE NOTES WITH COMPUTATIONAL EXAMPLES	71-0581
STRAUB, K.	DYNAN USERS MANUEL	71-0625
STRAUSS, A. M.	HEAT TRANSFER WITH NONLINEAR BOUNDARY CONDITIONS VIA A VARIATIONAL PRINCIPLE	71-0443
STRICKLAND, G.	A DOUBLY CURVED TRIANGULAR SHELL ELEMENT	68-0149
STRICKLIN, J. A	A COMPARISON OF CURRENT WORK-HARDENING MODELS USED IN THE ANALYSIS OF PLASTIC DEFORMATIONS	73-0769
STRICKLIN, J. A	A FINITE ELEMENT PROGRAM FOR THE DYNAMIC LARGE DEFLECTION, ELASTIC-PLASTIC ANALYSIS OF STIFFENED SHE	72-0908
STRICKLIN, J. A	A RAPIDLY CONVERGING TRIANGULAR PLATE ELEMENT	69-0245
STRICKLIN, J. A	COMPUTATION AND SOLUTION PROCEDURE FOR NONLINEAR ANALYSIS BY COMBINED FINITE ELEMENT-FINITE DIFFEREN	72-0571
STRICKLIN, J. A	COMPUTATION OF STRESS RESULTANTS FROM ELEMENT STIFFNESS MATRICES	66-0125
STRICKLIN, J. A	CONSISTENT STIFFNESS MATRICES IN THE ANALYSIS OF SHELLS	67-0078
STRICKLIN, J. A	CURVED FINITE ELEMENTS BY METHODS OF INITIAL STRAINS	73-0030
STRICKLIN, J. A	DYNAPLAS: A FINITE ELEMENT PROGRAM FOR THE DYNAMIC, LARGE DEFLECTION, ELASTIC-PLASTIC ANALYSIS OF ST	72-0773
STRICKLIN, J. A	EVALUATION OF SOLUTION PROCEDURES FOR MATERIAL AND/OR GEOMETRICALLY NONLINEAR STRUCTURAL ANALYSIS	73-0362
STRICKLIN, J. A	FORMULATION, COMPUTATION, AND SOLUTION PROCEDURES FOR MATERIAL AND/OR GEOMETRIC NONLINEAR STRUCTURAL	72-0346
STRICKLIN, J. A	GEOMETRICALLY NONLINEAR ANALYSIS BY THE DIRECT STIFFNESS METHOD	70-0331
STRICKLIN, J. A	GEOMETRICALLY NONLINEAR STRUCTURAL ANALYSIS BY THE DIRECT STIFFNESS METHOD	71-0560
STRICKLIN, J. A	IMPLICIT RIGID BODY MOTION IN CURVED FINITE ELEMENTS	71-0174
STRICKLIN, N. A	IMPROVEMENTS ON THE ANALYSIS OF SHELLS OF REVOLUTION BY THE MATRIX DISPLACEMENT METHOD	66-0070
STRICKLIN, J. A	INTERACTION OF AREA COORDINATES IN MATRIX STRUCTURAL ANALYSIS	68-0150
STRICKLIN, J. A	LARGE DEFLECTION ELASTIC-PLASTIC DYNAMIC RESPONSE OF STIFFENED SHELLS OF REVOLUTION	72-0608
STRICKLIN, J. A	LARGE DEFLECTION ELASTIC-PLASTIC DYNAMIC RESPONSE OF STIFFENED SHELLS OF REVOLUTION	74-0496
STRICKLIN, J. A	LINEAR AND NON LINEAR ANALYSIS OF SHELLS OF REVOLUTION WITH ASYMMETRICAL STIFFNESS PROPERTIES	68-0151
STRICKLIN, J. A	NONLINEAR ANALYSIS OF SHELLS OF REVOLUTION BY THE MATRIX DISPLACEMENT METHOD	67-0079
STRICKLIN, J. A	NONLINEAR DYNAMIC ANALYSIS OF SHELLS OF REVOLUTION BY MATRIX DISPLACEMENT METHOD	70-0110
STRICKLIN, J. A	NONLINEAR FINITE ELEMENT ANALYSIS INCLUDING HIGHER ORDER STRAIN ENERGY TERMS	70-0065
STRICKLIN, J. A	NUMERICAL METHODS FOR THE SOLUTION OF NONLINEAR PROBLEMS IN STRUCTURAL ANALYSIS	73-0506
STRICKLIN, J. A	RIGID BODY DISPLACEMENTS OF CURVED ELEMENTS IN THE ANALYSIS OF SHELLS BY THE MATRIX-DISPLACEMENT MET	67-0031
STRICKLIN, J. A	SAMMSOR 111: A FINITE ELEMENT PROGRAM TO DETERMINE STIFFNESS AND MASS MATRICES OF RING-STIFFENED SHE	72-0774
STRICKLIN, J. A	STATIC GEOMETRIC AND MATERIAL NONLINEAR ANALYSIS	72-0314
STRICKLIN, J. A	SURVEY OF SOLUTION PROCEDURES FOR NONLINEAR STATIC AND DYNAMIC ANALYSES	74-0307
STRICKLIN, J. A	SURVEY OF STATIC GEOMETRIC AND MATERIAL NONLINEAR ANALYSIS BY THE FINITE ELEMENT METHOD	72-0783
STROEBEL, G. J.	DYNAMIC COUPLING IN AN UNSYMMETRICALLY TAPERED BEAM	74-1301
STROEVE, A.	ACOUSTIC RESPONSE OF A SPACECRAFT SOLAR PANEL USING NORMAL MODE METHOD	69-0422
STROH, G. B.	EFFECTS OF SHOE FORCE GEOMETRY ON HEAVY DUTY INTERNAL SHOE BRAKE PERFORMANCE	68-0219
STROMBERG, D. D	STRUCTURAL RESPONSE OF RELAXED AND CONSTRICTED ARTERIOLES	68-0200
STROME, D. R.	A SURVEY OF THE ANALYSIS OF SHELLS BY THE DISPLACEMENT METHOD	66-0041
STROME, D. R.	ANALYSIS OF AXISYMMETRIC SHELLS BY THE DIRECT STIFFNESS METHOD	63-0006
STROME, D. R.	APPLICATION OF THE STIFFNESS METHOD TO THE ANALYSIS OF SHELL STRUCTURES	61-0004
STROME, D. R.	APPLICATION, TESTING, AND EVALUATION OF DOUBLY CURVED SHELL ELEMENTS FOR DYNAMIC ANALYSIS. PART 1, T	69-0246
STROME, D. R.	ASTRA-BOEING'S ADVANCED STRUCTURAL ANALYZER	69-0118
STROME, D. R.	DIRECT STIFFNESS METHOD OF ANALYSIS OF SHELLS OF REVOLUTION UTILIZING CURVED ELEMENTS	66-0042
STROME, D. R.	DYNAMIC ANALYSIS OF SHELLS USING DOUBLY-CURVED FINITE ELEMENTS	68-0082
STROME, D. R.	GENERALIZED VARIATIONAL PRINCIPLES IN THE FINITE ELEMENT METHOD	69-0116
STROME, D. R.	ON THE APPLICATION OF GENERALIZED VARIATIONAL PRINCIPLES IN THE FINITE ELEMENT METHOD	68-0083
STROVER, E. M.	THE PIECEWISE LINEAR ANALYSIS OF TWO CONNECTED STRUCTURES INCLUDING THE EFFECT OF CLEARANCE AT THE C	71-0379
STROVER, E. M.	THE STRESS ANALYSIS OF AXISYMMETRIC SHELLS USING A PLANE FRAME ANALYSIS PROGRAM	69-0238
STRUIK, J. H. A	APPLICATIONS OF FINITE-ELEMENT ANALYSIS TO NONLINEAR PLANE STRESS PROBLEMS	72-0976
STUART, P. M.	ELASTIC-PLASTIC BEHAVIOR OF A LONGITUDINAL SEMI-ELLIPTIC CRACK IN A THICK PRESSURE VESSEL	73-0952
STUEMKE, W.	CONTRIBUTIONS TO FLUID MECHANICS AND FLIGHT MECHANICS	74-1311
STUMPP, W.	BERECHNUNGSBEISPIELE UND ERFAHRUNGEN MIT DER METHODE DER FINITEN ELEMENTE IN MASCHINENBAU. (EXAMPLES	71-0390
STURGIS, J. D.	A FINITE ELEMENT ANALYSIS OF AN AIRCRAFT LANDING MAT ON AN ELASTIC FOUNDATION	68-0280
STURGUL, J. R.	FINITE-ELEMENT MODEL FOR POSSIBLE ISOSTATIC REBOUND IN GRAND CANYON	75-0037
STURGUL, J. R.	SOME APPLICATIONS OF FINITE ELEMENTS TO MODEL GEOLOGICAL STRUCTURES	73-0143
STURMAN, G. M.	STRUDL-A COMPUTER SYSTEM FOR STRUCTURAL DESIGN	66-0118
SU, C. L.	VARIATIONAL PRINCIPLES FOR A RAREFIED GAS FLOW PAST AND ARBITRARY THREE-DIMENSIONAL BODY	70-0189

SU, Y. L.	FINITE ELEMENT OF UNDERGROUND STRESSES UTILIZING STOCHASTICALLY SIMULATED MATERIAL PROPERTIES	70-0111
SUBE, R. M.	COUPLAGE ENTRE L ECOULEMENT AUTOUR D UN ARRIERE-CORPS ET LE JET PROPULSIF EN THEORIE DE FLUIDE PARFA	74-1137
SUCIU, E.	A FINITE-ELEMENT METHOD FOR LIFTING SURFACES IN STEADY INCOMPRESSIBLE SUBSONIC FLOW	74-1321
SUCIU, E. O.	A FINITE ELEMENT ANALYSIS FOR STEADY AND OSCILLATORY SUBSONIC FLOW AROUND COMPLEX CONFIGURATIONS	74-1317
SUDDARTH, S. K.	PREDICTION OF ELASTIC CONSTANTS OF PARTICLEBOARD	74-0259
SUGANO, T.	NONLINEAR ANALYSIS OF REINFORCED CONCRETE BUILDINGS	73-0324
SUGIYAMA, S.	NUMERICAL ANALYSIS ON STRESS DISTRIBUTION IN PARTIAL COMPRESSED SPECIMEN (HARDBOARD) BY FINITE ELEME	72-0920
SUGIYAMA, S.	TRANSIENT TEMPERATURE RESPONSE OF COMPOSITE SLABS	74-1038
SUHARA, J.	AUTOMATIC MESH GENERATION FOR FINITE ELEMENT ANALYSIS	72-0328
SUHARA, J.	NEW APPROACHES TO THE FINITE ELEMENT ANALYSIS	71-0241
SUHARA, J.	ON THE APPLICATIONS OF COLLOCATION OF ANALYTIC SOLUTION TO FINITE ELEMENT METHOD	69-0384
SUHARA, T.	STUDY OF MOTION AND STRENGTH OF FLOATING MARINE STRUCTURES IN WAVES	74-0489
SUIDAN, M.	FINITE ELEMENT ANALYSIS OF REINFORCED CONCRETE	73-0494
SUITS, K. W.	NONLINEAR ANALYSES OF THE STRESS FIELD NEAR A CRACK OR HOLE IN A PLATE	75-0694
SULLIVAN, B. R.	ULTIMATE STRAIN CAPACITY AND TEMPERATURE RISE STUDIES, TRUMBULL POND DAM	72-0705
SUMI, Y.	FINITE ELEMENT TREATMENT OF SINGULARITIES OF BOUNDARY VALUE PROBLEMS AND ITS APPLICATION TO ANALYSIS	73-0306
SUMPTER, J. D.	FRACTURE ANALYSIS IN AREAS OF HIGH NOMINAL STRAIN	73-0495
SUN, C. L.	THREE-DIMENSIONAL FINITE ELEMENT ANALYSIS OF A STEAM GENERATOR CHANNEL-HEAD-COMPLEX - COMPARISON OF	75-0653
SUN, C. T.	AXIAL-FLEXURAL VIBRATION OF FRAMEWORKS USING FINITE-ELEMENT APPROACH	73-0051
SUN, C. T.	CONTINUUM APPROACH TOWARD DYNAMICS OF GRIDWORKS	73-0174
SUN, C. T.	FINITE ELEMENTS FOR VIBRATION OF FRAMED SHEAR WALLS	73-0036
SUN, C. T.	TRANSVERSE IMPACT PROBLEMS BY HIGHER-ORDER BEAM FINITE-ELEMENTS	75-0819
SUN, M. C.	COMPUTER PROGRAM FOR GENERATING FINITE ELEMENT MODELS OF MINE STRUCTURES	71-0493
SUN, M. C.	COMPUTER PROGRAM FOR PIT SLOPE STABILITY ANALYSIS BY THE FINITE ELEMENT STRESS ANALYSIS AND LIMITING	72-0225
SUN, M. C.	SLOPE STABILITY ANALYSIS BY THE FINITE ELEMENT STRESS ANALYSIS AND LIMITING EQUILIBRIUM METHOD	70-0222
SUN, M. C.	STRUCTURAL ANALYSIS OF A COAL MINE OPENING IN ELASTIC, MULTILAYERED MATERIAL	74-0840
SUNADA, D. K.	FINITE ELEMENT METHOD FOR HYDRODYNAMIC DISPERISON EQUATION WITH MIXED PARTIAL DERIVATIVES	72-0030
SUNDARARAJAN, C	RESPONSE OF RECTANGULAR ORTHOTROPIC PLATES TO RANDOM EXCITATIONS	73-0496
SUNDARARAMAIAH,	FINITE ELEMENT ANALYSIS OF VIBRATIONS OF INITIALLY STRESSED THIN SHELLS OF REVOLUTION	74-1056
SUNDSTROM, B. O	DETERMINATION OF YOUNG'S MODULUS AND POISSON'S RATIO FOR WC-CO ALLOYS BY THE FINITE ELEMENT METHOD	72-0142
SUNDSTROM, B.	ENERGY CONDITION FOR INITIATION OF INTERFACIAL MICROCRACKS AT INCLUSIONS	74-0546
SUNDT, C. V.	GAS TURBINE ENGINE DISK CYCLIC LIFE PREDICTION	75-0509
SUNLEY, V. K.	COMPUTER GRAPHICS FOR STRUCTURAL ANALYSIS	74-0832
SUNLEY, V. K.	FINITE ELEMENT ANALYSIS OF AN EXPANDED I - SECTION BEAM AND AN AXISYMMETRIC FLANGED CYLINDER	68-0002
SURYOUTOMO, H.	CURVED ROTATIONAL SHELL ELEMENTS BY THE CONSTRAINT METHOD	72-0924
SUTHERLAND, W.	A FINITE ELEMENT COMPUTER CODE, AXICAP FOR CREEP ANALYSIS	69-0248
SUTHERLAND, W.	ANALYSIS OF AXISYMMETRIC SOLIDS BY THE DIRECT STIFFNESS METHOD	66-0049
SUTLIFF, D. R.	THREE DIMENSIONAL SCATTERED-LIGHT STRESS ANALYSIS OF DISCONTINUOUS FIBER-REINFORCED COMPOSITES	73-0218
SUTPHIN, D. L.	MARK5C/DAISY SERIES OF PROGRAMS	72-0160
SUZUKI, K.	SIMULATION OF PROGRESSIVE FAILURE IN ROCK AROUND MINING EXCAVATIONS	70-0146
SVALBONAS, V.	LINEAR AND NONLINEAR ANALYSIS OF SHELLS	72-0832
SVALBONAS, V.	STATIC, STABILITY, AND DYNAMIC ANALYSIS OF SHELLS OF REVOLUTION BY NUMERICAL INTEGRATION - A COMPARI	74-0389
SVEC, O. J.	A TRIANGULAR PLATE BENDING ELEMENT FOR CONTACT PROBLEMS	73-0693
SVEC, O. J.	FINITE ELEMENT ANALYSIS OF FINITE SIZED PLATES BOUNDED TO AN ELASTIC HALF SPACE	72-0880
SVEC, O. J.	FINITE ELEMENT APPROACH TO THE CONTACT PROBLEM OF A PLATE ON THE ELASTIC HALF SPACE	72-0830
SVEC, O. J.	UNBONDED CONTACT PROBLEM OF A PLATE ON THE ELASTIC HALF SPACE	74-0208
SVEC, O. T.	REFINED HEAT CONDUCTION FINITE ELEMENT FOR PERMAFROST ANALYSIS	73-0594
SVED, G.	FINITE ELEMENT ANALYSIS OF CLAD STRUCTURES	73-0353
SVENSSON, N. L.	STRESS ANALYSIS OF HUMAN FEMUR	73-0327
SVIRID, G. P.	INVESTIGATION OF THE STATIONARY FLOW OF A RIGID PLASTIC MATERIAL BY THE NUMERICAL FINITE ELEMENT MET	73-0231
SVIRID, G. P.	NUMERICAL INVESTIGATION OF AXI-SYMMETRIC VISCOPLASTIC FLOWS BY FINITE-ELEMENT PROCEDURE	73-0132
SWADA, T.	ANALYSIS OF SEEPAGE THROUGH DAMS AND IN FOUNDATION BY FINITE ELEMENT METHOD	73-1110
SWANEY, T. G.	SHELL INSTABILITY ANALYSIS APPLIED TO A RADOME	74-0261
SWANNELL, J. H.	A FINITE ELEMENT SOLUTION OF THE CONFINED FLOW OVER A CIRCULAR CYLINDER	74-0023
SWANNELL, J. H.	COMPARISON OF SOME CURVED TWO DIMENSIONAL FINITE ELEMENTS	73-0383
SWANNELL, J. H.	PREDICTION OF THE CREEP BEHAVIOUR OF A FLANGED JOINT	74-0757
SWANNELL, P.	THE AUTOMATIC COMPUTATION OF THE NATURAL FREQUENCIES OF STRUCTURAL FRAMES USING AN EXACT MATRIX TECH	73-0318
SWANSON, J. A.	ANSYS (ENGINEERING ANALYSIS SYSTEM) PROGRAM	72-0156
SWANSON, J. A.	ANSYS USER'S MANUAL	71-0561
SWANSON, J. A.	APPLICATION OF FINITE ELEMENT METHODS FOR THE ANALYSIS OF THERMAL CREEP, IRRADIATION INDUCED CREEP,	71-0242
SWANSON, J. A.	FEATS, A COMPUTER PROGRAM FOR THE FINITE ELEMENT THERMAL STRESS ANALYSIS OF PLANE OR AXISYMMETRIC SO	68-0152
SWANSON, S. R.	FINITE-ELEMENT SOLUTION FOR A CRACKED TWO-LAYERED ELASTIC CYLINDER	71-0397
SWARTS, J. D.	DEVELOPMENTS IN DISCRETE ELEMENT FINITE DEFLECTION STRUCTURAL ANALYSIS BY FUNCTION MINIMIZATION	68-0144
SWARTZ, B K.	CREATION AND COMPARISON OF FINITE DIFFERENCE ANALOGS OF SOME FINITE ELEMENT SCHEMES	74-0888
SWARTZ, B.	COMPARATIVE EFFICIENCY OF CERTAIN FINITE ELEMENT AND FINITE DIFFERENCE METHODS FOR A HYPERBOLIC PROB	73-0965
SWARTZ, B.	RAYLEIGH-RITZ APPROXIMATION OF PIECEWISE CUBIC POLYNOMIALS	66-0087
SWARTZ, B.	RELATIVE EFFICIENCY OF FINITE DIFFERENCE AND FINITE ELEMENT METHODS - 1. HYPERBOLIC PROBLEMS AND SPL	74-0986
SWARTZ, B.	THE COMPARATIVE EFFICIENCY OF CERTAIN FINITE ELEMENT AND FINITE DIFFERENCE SCHEMES FOR A HYPERBOLIC	74-0910
SWARTZ, B.	THE RELATION BETWEEN THE GALERKIN AND COLLOCATION METHODS USING SMOOTH SPLINES	74-0912
SWARTZ, B.	THE RELATIVE EFFICIENCY OF FINITE DIFFERENCE AND FINITE ELEMENT METHODS. 1. HYPERBOLIC PROBLEMS AND	74-0911
SWEC, L. F. JR.	REFINED CRACKED CONCRETE ANALYSIS OF CONCRETE CONTAINMENT STRUCTURES SUBJECT TO OPERATIONAL AND ENVI	74-0943
SWEDLOW, J. L.	COMBINED ANALYTICAL - EXPERIMENTAL FRACTURE STUDY OF THE TWO LEADING THEORIES OF ELASTIC-PLASTIC FRA	73-0951
SWEDLOW, J. L.	ELASTIC PLASTIC CRACKED PLATES IN PLANE STRAIN	69-0321
SWEDLOW, J. L.	ELASTO PLASTIC STRESS IN CRACKED PLATES	65-0042
SWEDLOW, J. L.	EXPERIMENTAL AND ANALYTICAL STRAINS IN A EDGE-CRACKED SHEET	69-0483
SWEDLOW, J. L.	FINITE ELASTO-PLASTIC DEFORMATION - 1. THEORY AND NUMERICAL EXAMPLES	74-0202
SWEDLOW, J. L.	INFLUENCE OF CRACK CLOSURE AND ELASTO-PLASTIC FLOW ON THE BENDING OF A CRACKED PLATE	75-0735
SWEDLOW, J. L.	PROCEDURE FOR SOLVING PROBLEMS OF ELASTO-PLASTIC FLOW	73-0217
SWEDLOW, J. L.	STIFFNESS ANALYSIS OF ELASTIC PLASTIC PLATES	65-0040
SWEEN, J. (JR.)	NONLINEAR CONSIDERATION OF GRAVITY IN A STIFFNESS TEST OF A WEAK STRUCTURE AT SMALL STRAINS	71-0035
SWEETSUR, G.	LAMELLAR TEARING IN WELDED GIRDERS	70-0218
SWIFT, G. W.	RELATIONSHIP BETWEEN FINITE ELEMENT AND MEASURED JOINT-FLOOR CONNECTION RIGIDITIES	75-0738
SWITZER, C. Z.	FINITE-ELEMENT ANALYSIS OF AN ELECTRODE-NIPPLE JOINT	75-0843
SWITZKY, H.	MINIMUM WEIGHT DESIGN OF FINITE ELEMENT STRUCTURES	74-0148
SWOFFORD, D. P.	A METHOD FOR THE THERMAL ANALYSIS OF SPACECRAFT INCLUDING ALL MULTIPLE REFLECTIONS AND SHADING AMONG	70-0424
SWOLFS, H. S.	STABILITY ANALYSIS OF AN EXPERIMENTALLY DEFORMED SINGLE LAYER OF INDIANA LIMESTONE USING FINITE ELEM	74-0212
SYKES, J. F.	FINITE ELEMENT PERMAFROST THAW SETTLEMENT MODEL	74-0543
SYKES, J. F.	TWO-DIMENSIONAL HEATED PIPELINE IN PERMAFROST PIPELINES	74-0541
SYLVESTER, R. J	THE LINEAR ELASTIC DYNAMIC ANALYSIS OF SHELLS OF REVOLUTION BY THE MATRIX DISPLACEMENT METHOD	66-0046
SYMKO, Y.	THERMAL STRESSES IN STEEL-CONCRETE COMPOSITE BRIDGES	75-0154
SYNGE, J. L.	APPROXIMATION IN ELASTICITY BASED ON THE CONCEPT OF FUNCTION SPACE	47-0001
SYNGE, J. L.	THE HYPERCIRCLE IN MATHEMATICAL PHYSICS	57-0001
SYNGE, J. L.	THE HYPERCIRCLE METHOD	74-1284
SYNGE, J. L.	TRIANGULATION IN THE HYPERCIRCLE METHOD FOR PLANE PROBLEMS	52-0001
SYRMAKEZ, C. A.	APPLICATION OF FINITE ELEMENT METHOD TO A MICROZONATION STUDY OF SALONICA REGION	73-0610
SZABO, B. A.	CONFORMING FINITE ELEMENTS BASED ON COMPLETE POLYNOMIALS	74-0302
SZABO, B. A.	CURVED ROTATIONAL SHELL ELEMENTS BY THE CONSTRAINT METHOD	72-0924
SZABO, B. A.	DERIVATION OF STIFFNESS MATRICES FOR PROBLEMS IN PLANE ELASTICITY BY GALERKIN'S METHOD	69-0250
SZABO, B. A.	DERIVATION OF THE STIFFNESS MATRIX FOR PLATES BY GALERKIN METHOD	69-0402
SZABO, B. A.	HIGH-PRECISION PLATE AND SHELL FINITE ELEMENTS	71-0437
SZABO, B. A.	LINEAR EQUALITY CONSTRAINTS IN FINITE ELEMENT APPROXIMATION	75-0475
SZABO, B. A.	NODAL VARIABLES FOR ARBITRARY ORDER FINITE ELEMENTS	75-0002

SZABO, B. A.	STIFFNESS METHODS FOR PLATES BY GALERKIN'S METHOD	69-0249
SZABO, B. A.	THE CONSTRAINT METHOD - A NEW FINITE ELEMENT TECHNIQUE	73-1027
SZABO, B. A.	THE CONSTRAINT METHOD FOR FINITE ELEMENT STRESS ANALYSIS	75-0432
SZABO, B. A.	THE QUADRATIC PROGRAMMING APPROACH TO THE FINITE ELEMENT METHOD	73-0831
SZILARD, R.	DERIVATION OF COMPATIBLE STIFFNESS MATRICES FOR LARGE DISCRETE ELEMENTS	67-0080
SZILARD, R.	ESTIMATING MATRIX DISPLACEMENT SOLUTIONS OF TWO-DIMENSIONAL PROBLEMS BY LARGE ELEMENT TECHNIQUE	70-0160
SZILARD, R.	THE EVOLUTION OF SHELL STRUCTURES IN OCEAN TECHNOLOGY	69-0042
SZIMMAT, J.	FINITE ELEMENT ULTIMATE LOAD ANALYSIS OF THREE DIMENSIONAL CONCRETE STRUCTURES	74-0672
SZIMMAT, J.	FINITE ELEMENTE ZUR BERECHNUNG VON SPANNBETONREAKTORDRUCKBEHALTERN	73-0702
SZIMMAT, J.	FINITE ELEMENTS FOR CALCULATION OF PRESTRESSED CONCRETE REACTOR PRESSURE VESSELS (FINITE ELEMENTE ZU	73-0797
SZIMMAT, J.	RECENT DEVELOPMENTS IN THE FINITE ELEMENT ANALYSIS OF PRE-STRESSED CONCRETE REACTOR VESSELS	74-0582
SZMELTER, J.	THE ENERGY METHOD OF NETWORKS OF ARBITRARY SHAPE IN PROBLEMS OF THE THEORY OF ELASTICITY	59-0001
SZODA, L.	AN IMPROVED METHOD FOR REDUCING THE BANDWIDTH OF SPARSE SYMMETRIC MATRICES	72-0990
TABANDEH, N.	CONVERGENCE OF THE FINITE ELEMENT METHOD	70-0377
TABARROK, B.	VIBRATION OF CYLINDRICAL SHELLS BY A HYBRID FINITE ELEMENT METHOD	72-0795
TABARROK, B.	APPLICATION OF A REFINED PLATE BENDING ELEMENT TO BUCKLING PROBLEMS	74-0937
TABARROK, B.	EIN NEUES VARIATIONSPRINZIP MIT ANWENDUNG AUF SCHWINGUNGEN FLACHER SCHALEN NACH DER METHODE DER FINI	73-0563
TABARROK, B.	NEW HYBRID CYLINDRICAL SHELL FINITE ELEMENT	72-0092
TABARROK, B.	NEW VARIATIONAL PRINCIPLE WITH APPLICATION TO VIBRATIONS OF SHALLOW SHELLS USING FINITE ELEMENTS MET	73-0354
TABARROK, B.	ON SOME ASPECTS OF CONTINUUM MECHANICS	73-0626
TABARROK, B.	ON THE GENERALIZATION OF STRESS FUNCTION PROCEDURE FOR DYNAMIC ANALYSIS OF PLATES	73-0661
TABARROK, B.	THERMAL STRESS ANALYSIS OF PLATES AND SHALLOW SHELLS BY HYBRID FINITE-ELEMENT METHOD	74-0485
TABARROK, B.	VARIATIONAL PRINCIPLE FOR THE DYNAMIC ANALYSIS OF CONTINUA BY HYBRID FINITE ELEMENT METHOD	71-0492
TABARROK, B.	VIBRATION OF CYLINDRICAL SHELLS BY HYBRID FINITE ELEMENT METHOD	72-0253
TABARROK, B.	VIBRATION OF SQUARE PLATES HAVING VARIOUS EDGE CONDITIONS AND LOADINGS USING A REFINED FINITE ELEMEN	75-0024
TACKE, G.	UNTERSUCHUNGEN ZUR SPURMASSTABILITAET BEI RADSAETZEN MIT KLOTZGEBREMSTEN VOLLRAEDERN (INVESTIGATIONS	75-0810
TADA, Y.	FINITE ELEMENT ANALYSIS OF A WING STRUCTURE	72-0333
TADA, Y.	FINITE ELEMENT SOLUTION TO AN ELASTICA PROBLEM OF BEAMS	70-0112
TADA, Y.	LARGE DEFLECTION OF CANTILEVER BEAMS	71-0649
TADROS, G.	STEP AND LOADS EFFECT ON STRESSES IN PRESTRESSED CONCRETE SHORT BRACKETS	71-0334
TADROS, G. S.	CONVERGENCE OF SEMIANALYTICAL SOLUTION OF PLATES	73-0497
TAGNFORS, H.	GENERAL SOLUTION ROUTINES FOR SYMMETRIC EQUATION SYSTEMS	75-0129
TAHBILDAR, A.	THE FINITE ELEMENT METHOD IN SHELL ANALYSIS	69-0058
TAHBILDAR, U. C	DYNAMIC ANALYSIS OF SHELLS USING CURVED FINITE ELEMENTS	69-0296
TAHBILDAR, U. C	EARTHQUAKE RESPONSE OF ARCH DAMS	70-0113
TAHBILDAR, U. C	FINITE ELEMENT ANALYSES OF SHELL RESPONSE FOR ARBITRARY EXCITATIONS	70-0348
TAHBILDAR, U. C	FINITE ELEMENT ANALYSIS OF AXISYMMETRIC VIBRATIONS OF CYLINDERS	72-0255
TAHBILDAR, U. C	STATICS AND DYNAMICS OF HYDROSTATICALLY LOADED SHELLS BY FINITE ELEMENT METHOD	71-0044
TAHIANI, C.	ANALYSE DES VOILES MINCES DANS LES DOMAINES LINEAIRES ET GEOMETRIQUEMENT NONLINEAIRES PAR LA METHODE	71-0122
TAHIANI, C.	LINEAR AND NON-LINEAR ANALYSIS OF THIN SHALLOW SHELLS BY MIXED FINITE ELEMENTS	75-0665
TAIG, I. C.	AUTOMATED STRESS ANALYSIS USING SUBSTRUCTURES	66-0071
TAIG, I. C.	MODELLING AND INTERPRETATION OF RESULTS	75-0399
TAIG, I. C.	SOME PROBLEMS IN THE DISCRETE ELEMENT REPRESENTATION OF AIRCRAFT STRUCTURES	64-0021
TAIG, I. C.	STRUCTURAL ANALYSIS BY THE MATRIX DISPLACEMENT METHOD	61-0003
TAIRA, S.	ANALYSIS OF ELASTO-PLASTIC THERMAL STRESSES BY THE FINITE ELEMENT METHOD	70-0142
TAIRA, S.	CREEP STRESS AND STRAIN ANALYSIS OF NOTCHED PLATES AND BARS BY MEANS OF THE FINITE ELEMENT METHOD	70-0141
TAIRA, S.	THERMAL ELASTIC PLASTIC STRESS ANALYSIS BY THE FINITE ELEMENT METHOD	69-0391
TAKABATAKE, H.	DYNAMIC RESPONSE OF VISCOELASTIC MATERIALS TO SINUSOIDALLY VARYING LOADINGS	73-1072
TAKABATAKE, H.	DYNAMIC RESPONSE OF VISCOELASTIC MATERIALS	73-1075
TAKABATAKE, H.	EFFECT OF TIME-DEPENDENT MATERIAL PROPERTIES ON DYNAMIC RESPONSE	74-0900
TAKACS, R. R.	A FINITE ELEMENT PROGRAM FOR FREE FORM SHELLS	75-0786
TAKAHASHI, S. K	BALANCED DESIGN AND FINITE-ELEMENT ANALYSIS OF CULVERTS.	72-0452
TAKAHASHI, S. K	EFFECT OF BACKPACKING AND INTERNAL PRESSURIZATION ON STRESSES TRANSMITTED TO BURIED CYLINDERS	73-0897
TAKAHASHI, S. K	FINITE ELEMENT ANALYSIS FOR SOLIDS OF REVOLUTION	68-0153
TAKAHASHI, S. K	PHOTOELASTIC INVESTIGATION OF STRESS CONCENTRATIONS IN SPHERE-CYLINDER TRANSITION REGIONS: INCLUDING	68-0258
TAKAHASHI, S. K	REDUCTION OF STRESSES IN BURIED STRUCTURES	71-0698
TAKANO, S.	ELASTO-PLASTIC SEISMIC RESPONSE ANALYSIS OF FRAMED STRUCTURES BY MODE SUPERPOSITION	73-0325
TAKARZ, F. J.	USER'S MANUAL FOR NTEDYNS - A NONLINEAR FINITE ELEMENT CODE FOR ANALYZING THE BLAST RESPONSE OF UNDE	71-0695
TAKASHIMA, N.	DYNAMIC ANALYSIS OF MACHINE TOOL STRUCTURES BY THE FINITE ELEMENT METHOD	71-0416
TAKATSUKA, K.	ANALYSIS OF ELASTIC-PLASTIC BEHAVIOR OF BEAMS AND COLUMNS BY PLANE FINITE ELEMENT MODEL	73-1073
TAKATSUKA, K.	ANALYSIS OF SAINT-VENANT TORSION PROBLEM BY A HYBRID STRESS MODEL	69-0277
TAKATSUKA, K.	ELASTIC PLASTIC ANALYSIS OF SAINT-VENANT TORSION PROBLEM BY A HYBRID STRESS MODEL	72-0385
TAKATSUKA, K.	ELASTIC-PLASTIC ANALYSIS OF BEAMS WITH UNIFORM CROSS-SECTION UNDER COMBINED LOADINGS	71-0777
TAKATSUKA, K.	ELASTIC-PLASTIC ANALYSIS OF A BAR WITH UNIFORM CROSS-SECTION UNDER COMBINED LOADINGS 1 - COMBINATION	71-0778
TAKATSUKA, K.	ELASTIC-PLASTIC ANALYSIS OF BEAMS WITH UNIFORM CROSS-SECTION UNDER COMBINED LOADINGS (2)	73-1070
TAKATSUKA, K.	NONLINEAR ANALYSIS BY THE FINITE ELEMENT METHOD AND SOME EXPOSITORY EXAMPLES	73-0309
TAKE, H.	A FINITE ELEMENT SIMULATION OF MECHANICAL PROPERTIES OF COMPOSITE MATERIALS	74-1294
TAKEDA, T.	ANALYSIS OF HYDROMAGNETIC PLASMA STABILITY BY FINITE-ELEMENT METHOD	72-0307
TAKEDA, T.	NUMERICAL ANALYSIS OF MAGNETOHYDRODYNAMIC INSTABILITIES BY FINITE ELEMENT METHOD	72-0041
TAKEMOTO, H.	SOME MODIFICATIONS OF AN ISOPARAMETRIC SHELL ELEMENT	73-1010
TAKENAKA, Y.	EXAMPLES OF STRESS ANALYSIS OF STRUCTURES BY THE FINITE ELEMENT METHOD. (JAPANESE)	70-0450
TAKETO, K.	A TEST ON CONVERGENCE OF EIGENVALUES BY SIMULTANEOUS ITERATION METHOD	73-0317
TAKEUCHI, N.	ANALYSIS OF VISCOELASTIC FLUID BY FINITE ELEMENT METHOD	75-0865
TAKEUCHI, Y.	COMPUTER AIDED DESIGN OF MACHINE TOOL STRUCTURE	74-1189
TALBOT, R. J.	ANALYSIS OF THE DYNAMICS OF ELASTIC STRUCTURES BY THE FINITE ELEMENT METHOD	72-0893
TALBOT, R. J.	FINITE ELEMENT ANALYSIS OF FREQUENCY SPECTRA FOR ELASTIC WAVEGUIDES	75-0247
TALL, L.	LINEAR STABILITY ANALYSIS OF BEAM-COLUMNS	73-0499
TALL, L.	ON CRACKS IN WELDS AND WELDED STRUCTURES	71-0688
TAMANO, T.	FINITE ELEMENT ANALYSIS OF STEADY FLUID AND METAL FLOW	74-0054
TAMANO, T.	FINITE ELEMENT ANALYSIS OF STEADY FLUID AND METAL FLOW	75-0178
TAMULY-PHUKAN,	STRESSES AND DEFORMATIONS OF VERTICAL SLOPES IN ELASTO-PLASTIC ROCKS	70-0114
TANAKA, H.	OPTIMUM DESIGN OF FRAMED STRUCTURES	69-0251
TANAKA, K.	ANALYSIS OF SLOPE STABILITY BY FINITE ELEMENT METHOD AND ITS APPLICATION PROGRAM. SOIL MECHANICS	72-0919
TANAKA, M.	DEVELOPMENT AND EVALUATION OF A TRIANGULAR THIN SHELL ELEMENT BASED UPON THE HYBRID ASSUMED STRESS M	69-0252
TANAKA, N.	A TORSIONAL STRENGTH ANALYSIS ON THE CONTAINER SHIP BY MEANS OF THE FINITE ELEMENT PROCEDURE AND FUL	71-0243
TANAKA, N.	FINITE ELEMENT ANALYSIS OF THIN-WALLED STRUCTURES BASED ON THE MODERN ENGINEERING THEORY OF BEAMS	71-0140
TANAKA, T.	FINITE ELEMENT APPLICATION IN LAKE CIRCULATION	75-0493
TANAKA, T.	HYBRID EFFECT ON BENDING BEHAVIORS OF CFRP-AL LAMINATE	74-1213
TANG, H. T.	SAFE-COLAPS: A FINITE ELEMENT PROGRAM FOR THE CREEP COLLAPSE ANALYSIS OF NUCLEAR FUEL RODS	75-0573
TANG, S. C.	ELASTO-PLASTIC ANALYSIS OF STRESS IN A GAS-TURBINE WHEEL	74-0323
TANG, S. C.	METHOD FOR THE COMPUTATION OF CYCLIC STRESSES IN A GAS-TURBINE WHEEL	74-0972
TANG, S. S.	MASS LOADING EFFECTS ON VIBRATION OF RING AND SHELL STRUCTURES	68-0245
TANI, M.	STUDY ON AUTOMOBILE CRASH WORTHINESS	70-0217
TANIGUCHI, S.	A THERMO-ELASTIC ANALYSIS OF THE PARALLEL SURFACE THRUST BEARING	74-1069
TANIGUCHI, S.	MECHANISM OF PARALLEL SURFACE THRUST BEARING	75-0322
TANIGUCHI, S.	THERMO-ELASTIC ANALYSIS OF THE PARALLEL SURFACE THRUST BEARING	74-1172
TANKERSLEY, D.	A DISCRETE-ELEMENT METHOD OF ANALYSIS FOR COMBINED BENDING AND SHEAR DEFORMATION OF A BEAM	69-0494
TANNER, R. I.	DIE-SWELL RECONSIDERED: SOME NUMERICAL SOLUTIONS USING A FINITE ELEMENT PROGRAM	73-0498
TANNER, R. I.	FINITE ELEMENT METHODS FOR THE SOLUTION OF SOME INCOMPRESSIBLE NON-NEWTONIAN FLUID MECHANICS PROBLEM	75-0224
TANNER, R. I.	SOLUTION OF VISCOUS INCOMPRESSIBLE JET AND FREE-SURFACE FLOWS USING FINITE ELEMENT METHODS	74-0357
TARNOWSKI, J.	ZASTOSOWANIE METODY SZTYWNYCH ELEMENTOW SKONCZONYCH DO OBLICZEN ORGAN WALOW OKRETOYCH Z UWZGLEDNIENI	74-0962
TARPY, T. S. JR	A TECHNIQUE FOR THE FINITE ELEMENT ANALYSIS OF LARGE COMPLEX STRUCTURAL SYSTEMS	69-0514

TASAI, F.	STUDY OF MOTION AND STRENGTH OF FLOATING MARINE STRUCTURES IN WAVES	74-0489
TATTERSHALL, D.	REINFORCING REQUIREMENTS FOR CONCRETE BEAMS WITH LARGE WEB OPENINGS	73-0198
TAVENAS, F. A.	IMMEDIATE SETTLEMENTS OF THREE TEST EMBANKMENTS ON CHAMPLAIN CLAY	74-0275
TAY, A. O.	APPLICATION OF THE FINITE ELEMENT METHOD TO CONVECTION HEAT TRANSFER BETWEEN PARALLEL PLANES	71-0244
TAY, A. O.	USING THE FINITE ELEMENT METHOD TO DETERMINE TEMPERATURE DISTRIBUTIONS IN ORTHOGONAL MACHINING	74-1212
TAYLOR, C.	A FINITE ELEMENT MODEL OF TIDES IN ESTUARIES	74-0044
TAYLOR, C.	A NUMERICAL MODEL OF DISPERSION IN ESTUARIES	74-0045
TAYLOR, C.	A NUMERICAL SOLUTION OF THE NAVIER-STOKES EQUATIONS USING THE FINITE ELEMENT TECHNIQUE	73-0061
TAYLOR, C.	AN ANALYSIS OF TWO-DIMENSIONAL SURFACE RUN-OFF BY FINITE ELEMENTS	74-1336
TAYLOR, C.	COUPLED CONVECTIVE/CONDUCTIVE HEAT TRANSFER INCLUDING VELOCITY FIELD EVALUATION	75-0116
TAYLOR, C.	DISPERSION IN STEADY UNSTEADY SEEPAGE FLOWFIELD	74-0072
TAYLOR, C.	FINITE ELEMENT APPROACH TO WATERSHED RUNOFF	74-0234
TAYLOR, C.	FINITE ELEMENT METHODS IN FLOW PROBLEMS	74-0752
TAYLOR, C.	FINITE ELEMENT NUMERICAL MODELLING OF FLOW AND DISPERSION IN ESTUARIES	73-0070
TAYLOR, C.	FINITE ELEMENT SOLUTIONS OF THE SHALLOW WATER EQUATIONS SURFACE RUN OFF	74-0046
TAYLOR, C.	FINITE ELEMENT SOLUTION OF BOUSSINESQ'S EQUATION FOR UNSTEADY GROUNDWATER FLOW	74-0459
TAYLOR, C.	HARBOUR OSCILLATION - A NUMERICAL TREATMENT FOR UNDAMPED NATURAL MODES	70-0208
TAYLOR, C.	NAVIER STOKES EQUATIONS USING MIXED INTERPOLATION	74-0014
TAYLOR, C.	NUMERICAL ANALYSIS OF FREE SURFACE SEEPAGE PROBLEMS	71-0469
TAYLOR, C.	NUMERICAL MODEL OF DISPERSION IN ESTUARIES	74-0905
TAYLOR, C.	NUMERICAL SOLUTION OF THE ELASTOHYDRODYNAMIC LUBRICATION PROBLEM USING FINITE ELEMENTS	72-0042
TAYLOR, C.	ON SOME FREE SURFACE ON TRANSIENT FLOW PROBLEMS OF SEEPAGE AND IRROTATIONAL FLOW	72-0070
TAYLOR, C.	TIDAL AND LONG WAVE PROPAGATION - A FINITE ELEMENT APPROACH	73-1059
TAYLOR, C.	TIDAL PROPAGATION AND DISPERSION IN ESTUARIES	75-0390
TAYLOR, C.	WEIGHTED RESIDUAL PROCESSES IN FINITE ELEMENT WITH PARTICULAR REFERENCE TO SOME TRANSIENT AND COUPLE	71-0503
TAYLOR, E. G.	ESTIMATIONS OF THE EFFECT OF DESIGN CHANGES ON THE MODES AND FREQUENCIES OF VIBRATING STRUCTURAL ELE	74-0819
TAYLOR, J. E.	FINITE ELEMENT METHOD FOR OPTIMAL DESIGN OF VARIABLE THICKNESS SHEETS	73-0131
TAYLOR, M. A.	A FINITE ELEMENT COMPUTER PROGRAM FOR THE PREDICTION OF THE BEHAVIOR OF REINFORCED AND PRESTRESSED C	72-0799
TAYLOR, M. A.	NUMERICAL BIAXIAL CHARACTERIZATION FOR CONCRETE	74-0641
TAYLOR, P.	PLATE VIBRATION PROBLEMS USING THE FINITE ELEMENT METHOD WITH ASSUMED STRESS DISTRIBUTION AND TRIANG	66-0068
TAYLOR, P.	THE FINITE ELEMENT METHOD FOR FLEXURE OF SLABS WHEN STRESS DISTRIBUTIONS ARE ASSUMED	66-0069
TAYLOR, R. F.	DESIGN OF OPTIMUM STRUCTURES FOR DYNAMIC LOADS	71-0255
TAYLOR, R. L.	A MODEL FOR THE MECHANICS OF JOINTED ROCK	68-0079
TAYLOR, R. L.	AN APPROXIMATE METHOD FOR THERMO-VISCOELASTIC STRESS ANALYSIS	66-0100
TAYLOR, R. L.	APPLICATION OF EXTENDED VARIATIONAL PRINCIPLES TO FINITE ELEMENT ANALYSIS	72-0529
TAYLOR, R. L.	DARCY FLOW SOLUTIONS WITH A FREE SURFACE	67-0081
TAYLOR, R. L.	DISCUSSION OF "FINITE ELEMENT BENDING ANALYSIS OF REISSNER PLATES"	71-0826
TAYLOR, R. L.	FINITE ELEMENT FORMULATION AND SOLUTION OF CONTACT-IMPACT PROBLEMS IN CONTINUUM MECHANICS	74-0806
TAYLOR, R. L.	FINITE ELEMENT FORMULATION AND SOLUTION OF CONTACT-IMPACT PROBLEMS IN CONTINUUM MECHANICS-II	75-0740
TAYLOR, R. L.	FLOW IN ROCKS WITH DEFORMABLE FRACTURES	74-0065
TAYLOR, R. L.	IMPACT OF FINITE ELEMENT METHOD ON STRUCTURAL ENGINEERING	75-0007
TAYLOR, R. L.	INCOMPATIBLE DISPLACEMENT MODELS	71-0262
TAYLOR, R. L.	ON A VARIATIONAL THEOREM FOR INCOMPRESSIBLE AND NEARLY INCOMPRESSIBLE ORTHOTROPIC ELASTICITY	68-0249
TAYLOR, R. L.	ON COMPLETENESS OF SHAPE FUNCTIONS FOR FINITE ELEMENT ANALYSIS	72-0105
TAYLOR, R. L.	ON THE COMPLETENESS OF SHAPE FUNCTIONS FOR FINITE ELEMENT ANALYSIS	72-0885
TAYLOR, R. L.	REDUCED INTEGRATION TECHNIQUE IN GENERAL ANALYSIS OF PLATES AND SHELLS	71-0374
TAYLOR, R. L.	STRESS ANALYSIS OF AXISYMMETRIC SOLIDS UTILIZING HIGHER ORDER QUADRILATERAL FINITE ELEMENTS	69-0356
TAYLOR, R. L.	THERMOMECHANICAL ANALYSIS OF VISCOELASTIC SOLIDS	70-0240
TAYLOR, R. L.	WAVE PROPAGATION IN FLUIDS	74-0034
TAYLOR, S.	COMPUTER METHODS FOR THE STRUCTURAL ANALYSIS OF MACHINE TOOLS	69-0325
TAYLOR, S.	COMPUTERS AND MECHANICAL DESIGN	73-0195
TAYLOR, T. P.	A COMPUTER PROGRAM TO ANALYZE BEAM-COLUMN UNDER MOVEABLE LOADS	68-0268
TAYLOR, T. P.	A FINITE-ELEMENT METHOD OF ANALYSIS FOR COMPOSITE BEAMS	68-0263
TEBEDGE, N.	APPLICATIONS OF THE FINITE-ELEMENT METHOD TO BEAM-COLUMN PROBLEMS	72-0978
TEBEDGE, N.	ETUDES DES POUTRES ET POTEAUX PAR LA METHODE DES ELEMENTS FINIS. (STABILITY ANALYSIS OF BEAMS AND C	74-1008
TEBEDGE, N.	LINEAR STABILITY ANALYSIS OF BEAM-COLUMNS	73-0499
TELEGA, J. J.	METODA ELEMENTOW SKONSZONYCH W MECHANICE GRUNTOW I MECHANICE GOROTWORU (FINITE ELEMENT METHOD IN SOI	73-0500
TEMAM, R.	FINITE ELEMENT METHODS IN FLUID FLOW	75-0729
TEMAM, R.	NUMERICAL ANALYSIS	73-1060
TEMPLIER, A.	EFFECT OF SHAPE AND HETEROGENEITY OF ROCK SAMPLES ON THEIR COMPRESSIVE STRENGTH	69-0315
TENERELLI, D. J	THERMOSTRUCTURAL DESIGN CONSIDERATIONS TO ACHIEVE THE LARGE SPACE TELESCOPE LINE-OF-SIGHT REQUIREMEN	75-0839
TERAMAE, T.	APPLICATION OF FINITE ELEMENT METHOD TO HYDRO-DYNAMICS	73-0301
TERAZAWA, K.	ELASTIC-PLASTIC BUCKLING OF PLATES USING THE FINITE ELEMENT METHOD	67-0094
TERAZAWA, K.	ELASTIC-PLASTIC BUCKLING OF PLATES BY FINITE ELEMENT METHOD	69-0253
TESK, J. A.	STRESS DISTRIBUTION IN BONE ARISING FROM LOADING ON ENDOSTEAL DENTAL IMPLANTS	73-0501
TETER, R. D.	FINITE ELEMENT ELASTIC PLASTIC CREEP ANALYSIS OF TWO DIMENSIONAL CONTINUUM WITH TEMPERATURE DEPENDEN	73-0226
TETER, R. D.	FINITE ELEMENT THERMOELASTOPLASTIC ANALYSIS	72-0139
TETSURO, K.	APPLICATION OF PROGRAM FATIQUE TEST TO MEMBER JOINTS OF HULLS	72-0814
TEUSCHER, L. H.	CALCULATION OF TEMPERATURE DISTRIBUTION WITHIN A MELTING-FREEZING MATERIAL USING FINITE ELEMENT TECH	72-0043
TEZCAN, S. S.	AN ITERATION METHOD FOR THE NONLINEAR BUCKLING OF FRAME STRUCTURES	66-0107
TEZCAN, S. S.	FINITE ELEMENT ANALYSIS OF HYPERBOLIC PARABOLOID SHELLS	71-0245
TEZCAN, S. S.	FINITE ELEMENT ANALYSIS OF HYPERBOLIC PARABOLOID SHELLS	72-0887
TEZCAN, S. S.	TANGENT STIFFNESS MATRIX FOR SPACE FRAME MEMBERS	69-0397
THAMM, B. R.	ANWENDUNG DER FINITE-ELEMENT-METHODE ZUR BERECHNUNG VON SPANNUNGEN IN WASSERGESAETTIGTEN BOEDEN (FIN	73-0502
THANGAM BABU, P	FREQUENCY ANALYSIS OF SKEW ORTHOTROPIC PLATES BY THE FINITE STRIP METHOD	71-0413
THATCHER, R. W.	THE THEORY AND APPLICATION OF THE FINITE ELEMENT METHOD	71-0770
THEIMERT, P. H.	DER EINSATZ DER FINIT-ELEMENT-METHODE IN PRESSENBAU (APPLICATION OF THE FINITE ELEMENT METHOD FOR TH	74-0615
THELEN, J. F.	COMPUTERIZED STRUCTURAL ANALYSIS OF THE WORLD'S LARGEST LIGHT-GAGE STEEL PRIMARY STRUCTURAL SYSTEM	73-0503
THIERAUF, G.	ELASTIC-PLASTIC DEFORMATIONS OF FLEXURALLY STIFF FRAMEWORKS FROM SECOND ORDER STRESS THEORY	73-1061
THIERAUF, G.	STRUCTURAL OPTIMIZATION USING THE FORCE METHOD	75-0440
THIRRIOT, C.	USE OF FINITE ELEMENTS IN THE ANALYSIS OF A HYDROTHERMAL DOUBLET	75-0441
THIRRIOT, C.	USE OF HYBRID COMPUTATION FOR SOLVING TRANSIENT FLOW PROBLEM IN POROUS MEDIA BY FINITE ELEMENT METHO	74-0068
THOMANN, G. E.	AEROELASTIC PROBLEMS OF LOW ASPECT RATIO WINGS (PART 1)	56-0006
THOMAS, D. H.	INTERACTIVE PROCEDURE FOR CURVILINEAR COORDINATION OF A PLANAR REGION WITH APPLICATIONS TO FINITE-EL	73-0040
THOMAS, D. L.	TIMOSHENKO BEAM FINITE ELEMENTS	73-0032
THOMAS, D. L.	USE OF STRAIGHT BEAM FINITE ELEMENTS FOR ANALYSIS OF VIBRATIONS OF CURVED BEAMS	73-0616
THOMAS, D. T.	FUNCTIONAL APPROXIMATIONS FOR SOLVING BOUNDARY VALUE PROBLEMS BY COMPUTER	69-0013
THOMAS, E. A.	DIGITAL COMPUTER PROGRAMS FOR STEADY-STATE OR TRANSIENT TEMPERATURE ANALYSIS OF PLANE AND AXISYMMETR	72-0915
THOMAS, G. R.	A TRIANGULAR THIN SHELL FINITE ELEMENT: NONLINEAR ANALYSIS	75-0697
THOMAS, G. R.	A TRIANGULAR THIN SHELL FINITE ELEMENT: LINEAR ANALYSIS	75-0698
THOMAS, G. R.	COMPUTATIONAL EFFICIENCY ASPECTS OF THE FINITE ELEMENT NONLINEAR ANALYSIS OF SHELLS	75-0031
THOMAS, G. R.	NONLINEAR FINITE-ELEMENT ANALYSIS OF THIN SHELLS	73-1169
THOMAS, G. R.	SHEAR BUCKLING OF SQUARE PERFORATED PLATES	74-0713
THOMAS, G. R.	TRIANGULAR THIN SHELL FINITE ELEMENT: LINEAR ANALYSIS	75-0602
THOMAS, G. R.	TRIANGULAR THIN SHELL FINITE ELEMENT: NONLINEAR ANALYSIS	75-0668
THOMAS, G. W.	RESERVOIR SIMULATION BY GALERKIN'S METHOD	73-0662
THOMAS, J.	DYNAMIC STABILITY OF TIMOSHENKO BEAMS BY FINITE ELEMENT METHOD	75-0194
THOMAS, J.	DYNAMIC STABILITY OF TIMOSHENKO BEAMS BY FINITE ELEMENT METHOD	75-0422
THOMAS, J.	FINITE ELEMENT ANALYSIS OF COUPLED VIBRATION OF TAPERED TWISTED BLADES	72-0766
THOMAS, J.	FINITE ELEMENT ANALYSIS OF ROTATING SHELLS	74-0096
THOMAS, J.	FINITE ELEMENT MODEL FOR DYNAMIC ANALYSIS OF TIMOSHENKO BEAM	75-0275
THOMAS, J.	IMPROVED FINITE ELEMENTS FOR VIBRATION ANALYSIS OF TAPERED BEAMS	73-0033

THOMAS, J.	SIMPLE FINITE ELEMENTS FOR PRE-TWISTED BLADING VIBRATION	74-0239
THOMAS, J. M.	A FINITE ELEMENT APPROACH TO THE STRUCTURAL INSTABILITY OF BEAM COLUMNS, FRAMES, AND ARCHES	70-0434
THOMAS, J. M.	ELEMENTS FINIS ET PROBLEMES ELLIPTIQUES DEGENERES (FINITE ELEMENTS AND DEGENERATE ELLIPTIC PROBLEMS)	73-0582
THOMAS, J. M.	FINITE ELEMENTS AND DEGENERATED ELLIPTIC PROBLEMS	73-0357
THOMAS, J. M.	RESOLUTION NUMERIQUE PAR DES METHODES D ELEMENTS FINIS DE PROBLEMES ELLIPTIQUES DEGENERES	72-0702
THOMAS, M. D.	INTERNATIONAL MACHINE TOOL DESIGN AND RESEARCH CONFERENCE, 13TH PROCEEDINGS, 1972	73-0504
THOMEE, V.	ERROR ESTIMATES FOR A FINITE-ELEMENT APPROXIMATION OF A MINIMAL SURFACE	75-0491
THOMEE, V.	INTERIOR MAXIMUM NORM ESTIMATES FOR SOME SIMPLE FINITE ELEMENT METHODS	74-1023
THOMEE, V.	SPLINE APPROXIMATION AND DIFFERENCE SCHEMES FOR THE HEAT EQUATION	72-0518
THOMOPOULOS, N.	DYNAMIC RESPONSE OF SOIL/CONCRETE INTERFACES AT HIGH PRESSURE	74-0836
THOMPSON, D. G.	FINITE ELEMENT METHOD FOR CALCULATING STRESS INTENSITY FACTORS FOR CRACKED PLATES IN BENDING	71-0424
THOMPSON, D. S.	FINITE ELEMENT ANALYSIS OF THE FLOW THROUGH A CASCADE OF AEROFOILS	73-0761
THOMPSON, D. S.	FLOW THROUGH A CASCADE OF AEROFOILS	74-0077
THOMPSON, E. G.	AVERAGE AND COMPLETE INCOMPRESSIBILITY IN THE FINITE ELEMENT METHOD	75-0342
THOMPSON, E. G.	FINITE ELEMENT ANALYSIS OF LAYERED WOOD SYSTEMS	75-0288
THOMPSON, E. G.	FINITE ELEMENT METHOD FOR INCOMPRESSIBLE SLOW VISCOUS FLOW WITH A FREE SURFACE	69-0303
THOMPSON, E. G.	HIGH ORDER FINITE ELEMENT FOR COMPLETELY INCOMPRESSIBLE CREEPING FLOW	73-0144
THOMPSON, E. G.	IN SITU CREEP ANALYSIS OF ROOM IN FROZEN SOIL	72-0128
THOMPSON, G. L.	COMPUTATION OF VIBRATION MODAL SHAPES AND NATURAL FREQUENCIES OF A GENERAL ORTHOTROPIC LAMINATED, TH	71-0309
THOMPSON, G. L.	FINITE-ELEMENT ANALYSIS FOR FREE VIBRATION OF GENERAL ANISOTROPIC LAMINATED THIN SHELLS	72-0618
THOMPSON, G. V.	CLOSURE ANALYSIS USING SELF ENERGIZING, DOUBLE CONICAL GASKETS	74-0675
THOMPSON, J. E.	FUTURE DEVELOPMENTS IN STRUCTURAL ANALYSIS	74-0390
THOMPSON, J. E.	VEHICLE CRUSH PREDICTION USING FINITE-ELEMENT TECHNIQUES	73-0280
THOMPSON, J. J.	DISCONTINUOUS FINITE ELEMENTS IN THERMAL ANALYSIS	70-0256
THOMPSON, J. M.	COMPARATIVE PERTURBATION STUDIES OF THE ELASTICA	69-0255
THOMPSON, J. M.	CONTINUUM AND FINITE ELEMENT BRANCHING STUDIES OF THE CIRCULAR PLATE.	72-0427
THOMPSON, M. R.	SIGNIFICANCE OF VARIABLY CONFINED TRIAXIAL TESTING ROADBUILDING MATERIALS	74-0554
THOMS, R. L.	FINITE ELEMENT ANALYSES OF SALT DOMES WITH STORED HOT WASTES	74-1148
THOMS, R. L.	NUMERICAL ANALYSIS OF EMBANKMENTS OVER SOFT SOILS	75-0719
THOMS, R. L.	PHOTOELASTIC AND FINITE ELEMENT ANALYSIS OF EMBANKMENTS CONSTRUCTED OVER SOFT SOILS	75-0151
THORNE, R. G.	APPLICATION, TESTING, AND EVALUATION OF DOUBLY CURVED SHELL ELEMENTS FOR DYNAMIC ANALYSIS, PART III.	69-0256
THORNTON, C. H.	COMPUTERIZED STRUCTURAL ANALYSIS OF THE WORLD'S LARGEST LIGHT-GAGE STEEL PRIMARY STRUCTURAL SYSTEM	73-0503
THORNTON, E. A.	A NASTRAN CORRELATION STUDY FOR VIBRATIONS OF A CROSS-STIFFENED SHIP'S DECK	72-0756
THORNTON, E. A.	APPLICATION OF NASTRAN TO A SPACE SHUTTLE DYNAMICS MODEL	71-0247
THORNTON, E. A.	VIBRATION ANALYSIS OF A 1/15-SCALE DYNAMIC MODAL OF A SPACE SHUTTLE CONFIGURATION	71-0609
THRESHER, R. W.	STRESS ANALYSIS OF HUMAN TEETH	73-0505
THROSBY, P. W.	A FINITE ELEMENT APPROACH TO SURFACE DEFINITION	69-0030
THUBOI, Y.	ANALYSIS OF TENSION STRUCTURES BY MATRIX METHODS	67-0119
THURSTON, G. A.	ON THE INTERRELATION BETWEEN STABILITY AND COMPUTATIONS	73-0674
TICOZZI, C.	TWO-DIMENSIONAL FINITE ELEMENT COMPUTER CODES FOR STRESS ANALYSIS AND FIELD PROBLEMS	73-0055
TICOZZI, C.	USE OF PAS-1 FINITE ELEMENT COMPUTER SYSTEM FOR STATIC AND DYNAMIC ANALYSIS OF 2-DIMENSIONAL AND 3-D	75-0070
TIEU, A. K.	NUMERICAL SIMULATION OF FINITE-WIDTH THRUST BEARINGS, TAKING INTO ACCOUNT VISCOSITY VARIATION WITH T	75-0623
TIEU, A. K.	OIL-FILM TEMPERATURE DISTRIBUTION IN AN INFINITELY WIDE SLIDER BEARING: AN APPLICATION OF THE FINITE	73-1128
TIEU, A. K.	RESEARCH NOTE: A THREE-DIMENSIONAL OIL FILM TEMPERATURE DISTRIBUTION IN TILTING THRUST BEARINGS	74-0608
TILLERSON, J. R	A TREATISE OF NONLINEAR FINITE ELEMENT ANALYSIS	73-1063
TILLERSON, J. R	CURVED FINITE ELEMENTS BY METHODS OF INITIAL STRAINS	73-0248
TILLERSON, J. R	DYNASOR 11 – A FINITE ELEMENT PROGRAM FOR THE DYNAMIC NONLINEAR ANALYSIS OF SHELLS OF REVOLUTION (US	71-0666
TILLERSON, J. R	NONLINEAR DYNAMIC ANALYSIS OF SHELLS OF REVOLUTION BY MATRIX DISPLACEMENT METHOD	70-0110
TILLERSON, J. R	NUMERICAL METHODS FOR THE SOLUTION OF NONLINEAR PROBLEMS IN STRUCTURAL ANALYSIS	73-0506
TILLERSON, J. R	STATIC GEOMETRIC AND MATERIAL NONLINEAR ANALYSIS	72-0314
TILLERSON, J. R	SURVEY OF STATIC GEOMETRIC AND MATERIAL NONLINEAR ANALYSIS BY THE FINITE ELEMENT METHOD	72-0783
TINAWI, R.	A COMPARISON BETWEEN TWO-HIGH-PRECISION FINITE ELEMENTS	75-0033
TINAWI, R. A.	A STUDY OF VARIOUS IDEALIZATIONS FOR WING STRUCTURES AND NUMERICAL PROCEDURES INVOLVED USING MATRIX	69-0257
TINAWI, R. A.	ANISOTROPIC TAPERED ELEMENTS USING DISPLACEMENT MODELS	72-0391
TING, E. C.	NUMERICAL PROCEDURE FOR THE ANALYSIS OF STRUCTURES SUPPORTED BY VISCOELASTIC FOUNDATIONS	73-0564
TING, T. C. T.	DYNAMIC ANALYSIS OF THE DEFORMATION OF STRUCTURES AND OF METAL FORMING	74-0823
TING, V. C.	INFLUENCE OF INTERLAMINAR FLAWS ON TRANSIENT TEMPERATURE DISTRIBUTIONS	73-0223
TINGLEFF, O.	A METHOD FOR SIMULATING CONCENTRATED FORCES AND LOCAL REINFORCEMENTS IN STRESS COMPUTATIONS	72-0081
TIROSH, J.	EFFECT OF PLASTICITY AND CRACK BLUNTING ON THE STRESS DISTRIBUTION IN ORTHOTROPIC COMPOSITE MATERIAL	73-1123
TISCHLEP, V. A.	DESIGN OF OPTIMUM STRUCTURES FOR DYNAMIC LOADS	71-0255
TIWARI, S. K.	THEORETICAL AND EXPERIMENTAL VIBRATION ANALYSIS OF AN OBLIQUE SPACE FRAME	70-0460
TOBIAS, S. A.	COMPUTER METHODS FOR THE STRUCTURAL ANALYSIS OF MACHINE TOOLS	69-0325
TOCHER, J. L.	A CRITICAL REVIEW OF NASTRAN	71-0246
TOCHER, J. L.	A TECHNICAL EVALUATION OF THE NASTRAN COMPUTER PROGRAM	71-0550
TOCHER, J. L.	A USER-ORIENTED PROGRAM FOR CRASH DYNAMICS	74-0321
TOCHER, J. L.	ANALYSIS OF PLATE BENDING USING TRIANGULAR ELEMENTS	62-0010
TOCHER, J. L.	ANALYSIS OF THIN ARCH DAMS BY THE FINITE ELEMENT METHOD	64-0001
TOCHER, J. L.	FINITE ELEMENT ANALYSIS OF PLASTICITY PROBLEMS	68-0154
TOCHER, J. L.	FINITE ELEMENT STIFFNESS MATRICES FOR ANALYSIS OF PLATE BENDING	66-0021
TOCHER, J. L.	HIGHER-ORDER FINITE ELEMENT FOR PLANE STRESS	67-0082
TOCHER, J. L.	NON-LINEAR MATERIAL ANALYSIS WITH FINITE ELEMENTS AND INCREMENTAL METHODS	68-0199
TODD, M.	THE FINITE ELEMENT METHOD APPLIED TO THIN SHELLS AND BOX STRUCTURES	70-0444
TODD, P. H.	SELF-STRESS CONCENTRATIONS	71-0317
TOENSHOFF, H. K	MESSVERFAHREN RECHENVERFAHREN ZUR ERMITTLUNG DREIACHSIGER EIGENSPANNUNGSZUSTAENDE	73-0507
TOEWS, N. A.	COMPUTER PROGRAM FOR AUTOMATIC MESH GENERATION FOR THE 2-D FINITE ELEMENT PROGRAM WILAX	73-0565
TOKARZ, F. J.	INPUT SOIL PROPERTIES FOR THE FINITE ELEMENT CODE NTEDYNS	71-0694
TOKARZ, F. J.	SEISMIC EFFECTS ON A PROPOSED UNDERGROUND REACTOR FACILITY	73-0953
TOKUDA, N.	A NOTE ON CONVERGENCE OF FINITE ELEMENT SOLUTIONS	71-0252
TOKUDA, N.	DETERMINATION OF STRESS INTENSITY FACTORS IN CRACKED PLATES BY THE FINITE ELEMENT METHOD	73-0148
TOKUDA, N.	FINITE ELEMENT TREATMENT OF SINGULARITIES OF BOUNDARY VALUE PROBLEMS AND ITS APPLICATION TO ANALYSIS	73-0306
TOKUDA, N.	STRESS INTENSITY FACTORS IN PLATE STRUCTURES CALCULATED BY THE FINITE ELEMENT METHOD (IN JAPANESE)	71-0788
TOKUDA, N.	THE ANALYSIS OF STRESS DISTRIBUTION AROUND THE TIP OF A CRACK BY FINITE ELEMENT METHOD WITH THE AID	71-0517
TOLEFSON, D. C.	INTRODUCTION TO FINITE ELEMENT METHODS OF STRUCTURAL ANALYSIS	68-0231
TOLKACHNIK, S.	INVESTIGATION OF THE EFFECT OF FOUNDATION YIELD ON THE FREQUENCY AND SHAPE OF INHERENT VIBRATIONS OF	72-0419
TOLMACH, G. U.	RASCHET KILEBANII PLASTIN SLOZHNOI FORMY S NEODNORODNYMI GRANICHNYMI USLOVIYAMI. (CALCULATION OF VIB	74-0988
TOLMAN, F.	THE ELEMENT-CHAIN CONCEPT IN FINITE ELEMENT BASED PROGRAMMING	72-0934
TOMINO, H.	STRESS AND VIBRATION ANALYSIS ON THE SYSTEM OF FRAME AND FOUNDATION	69-0218
TOMITA, M.	AN INCREMENTAL FINITE ELEMENT ANALYSIS OF TWO DIMENSIONAL LARGE STRAIN AND LARGE DISPLACEMENT PROBLE	72-0338
TOMITA, Y.	INCREMENTAL THEORY OF LARGE STRAIN AND LARGE DISPLACEMENT PROBLEMS AND ITS FINITE ELEMENT FORMULATIO	72-0274
TOMLINSON, E. T	SOLUTION OF DIFFUSION AND P1 FINITE-ELEMENT EQUATIONS BY ITERATION	75-0404
TOMLINSON, G. R	COMPUTER AIDED DESIGN OF THERMALLY LOADED AXISYMMETRIC DIESEL ENGINE COMPONENTS	74-0596
TONG, P.	A FINITE ELEMENT METHOD FOR LOW REYNOLDS NUMBER FLOW AND ITS APPLICATION TO BIOMECHANICS	69-0304
TONG, P.	A HYBRID-ELEMENT APPROACH TO CRACK PROBLEMS IN PLANE ELASTICITY	73-0569
TONG, P.	A NEW FINITE ELEMENT MODEL FOR SOLID CONTINUA	68-0156
TONG, P.	A SUPER-ELEMENT FOR CRACK ANALYSIS	73-0754
TONG, P.	A VARIATIONAL PRINCIPLE AND THE CONVERGENCE OF A FINITE-ELEMENT METHOD BASED ON ASSUMED STRESS DISTR	69-0260
TONG, P.	AN ASSUMED STRESS HYBRID FINITE ELEMENT METHOD FOR AN INCOMPRESSIBLE AND NEAR INCOMPRESSIBLE MATERIA	69-0259
TONG, P.	ANALYSIS OF PERISTALTIC PUMPING	72-0044
TONG, P.	APPLICATION OF FINITE ELEMENT METHOD TO MIXED-MODE FRACTURE	73-1123
TONG, P.	APPLICATIONS OF THE ASSUMED-STRESS HYBRID METHOD TO FINITE ELEMENT ANALYSIS IN FRACTURE MECHANICS	75-0495
TONG, P.	BASIS OF FINITE ELEMENT METHODS FOR SOLID CONTINUA	69-0213
TONG, P.	BOUNDS TO THE INFLUENCE COEFFICIENTS BY THE ASSUMED STRESS METHOD	70-0165

TONG, P.	CALCULATION OF MECHANICAL IMPEDANCE BY FINITE ELEMENT HYBRID MODEL	74-0126
TONG, P.	DERIVATION OF GEOMETRIC STIFFNESS AND MASS MATRICES FOR FINITE ELEMENT HYBRID MODELS	74-0844
TONG, P.	ELASTIC CRACK ANALYSIS BY A FINITE ELEMENT HYBRID METHOD	71-0212
TONG, P.	ELASTIC PLASTIC ANALYSIS BY ASSUMED STRESS HYBRID MODEL	74-0443
TONG, P.	EXACT SOLUTION OF CERTAIN PROBLEMS BY THE FINITE-ELEMENT METHOD	69-0258
TONG, P.	FINITE ELEMENT ANALYSIS OF VIBRATION AND FLUTTER OF CANTILEVER ANISOTROPIC PLATES	74-0484
TONG, P.	FINITE ELEMENT METHODS FOR FLUID PROBLEMS	72-0352
TONG, P.	FINITE ELEMENT METHODS IN CONTINUUM MECHANICS.	72-0377
TONG, P.	FINITE ELEMENT SOLUTIONS FOR LAMINATED THICK PLATES	72-0143
TONG, P.	FINITE ELEMENT SOLUTION OF THE WIND DRIVEN CURRENTS AND ITS MASS TRANSPORT IN LAKES	74-1368
TONG, P.	FINITE-ELEMENT ANALYSIS OF VIBRATION AND FLUTTER OF CANTILEVER ANISOTROPIC PLATES	74-0933
TONG, P.	HYBRID-ELEMENT APPROACH TO CRACK PROBLEMS IN PLANE ELASTICITY	73-0508
TONG, P.	LIQUID SLOSHING IN AN ELASTIC CONTAINER	66-0096
TONG, P.	MODE SHAPES AND FREQUENCIES BY FINITE ELEMENT METHOD USING CONSISTENT AND LUMPED MASSES	71-0610
TONG, P.	MODULAR APPROACH TO STRUCTURAL SIMULATION FOR VEHICLE CRASHWORTHINESS PREDICTION	75-0737
TONG, P.	NEW DISPLACEMENT HYBRID FINITE ELEMENT MODELS FOR SOLID CONTINUA	70-0115
TONG, P.	ON THE CONVERGENCE OF FINITE ELEMENT METHODS	68-0155
TONG, P.	ON THE CONVERGENCE OF THE FINITE ELEMENT METHOD FOR PROBLEMS WITH SINGULARITY	73-0173
TONG, P.	ON THE DYNAMIC BUCKLING OF SHELLS OF REVOLUTION	73-1102
TONG, P.	ON THE NUMERICAL PROBLEMS BY THE FINITE ELEMENT METHODS	70-0264
TONG, P.	ON THE SOLUTION OF THE NAVIER-STOKES EQUATION IN TWO DIMENSIONAL AND AXIAL SYMMETRIC PROBLEMS	74-0007
TONG, P.	RATIONALIZATION IN DERIVING ELEMENT STIFFNESS MATRIX BY ASSUMED STRESS APPROACH	68-0138
TONG, P.	SLOW PARTICULATE VISCOUS FLOW IN CHANNELS AND TUBES. APPLICATION TO BIOMECHANICS	71-0248
TONG, P.	SLOW VISCOUS FLOW AND ITS APPLICATION TO BIOMECHANICS	71-0619
TONG, P.	STATIC AND DYNAMIC NONLINEAR STRUCTURAL PROBLEMS	72-0848
TONG, P.	STRESSES AROUND HOLES IN STIFFENED COMPOSITE PANELS USING LAURENT-SERIES AND FINITE ELEMENT METHODS	74-1252
TONG, P.	STRESSES AROUND HOLES IN STIFFENED COMPOSITE PANELS USING LAURENT-SERIES AND FINITE ELEMENT METHODS	75-0733
TONG, P.	THE CONVERGENCE OF FINITE ELEMENT METHOD IN SOLVING LINEAR ELASTIC PROBLEMS	67-0083
TONG, P.	THE FINITE ELEMENT METHOD FOR FLUID FLOW	69-0018
TONG, P.	THE FINITE ELEMENT METHOD IN FLUID FLOW ANALYSIS	71-0249
TONG, P.	VARIATIONAL FORMULATION OF FINITE DISPLACEMENT ANALYSIS	70-0284
TONG, P.	VIBRATION ANALYSIS OF LAMINATED PLATES AND SHELLS BY A HYBRID STRESS ELEMENT	73-0120
TONG, T. C.	CURRENT COMPUTATION ON COMPLEX STRUCTURES BY FINITE-ELEMENT METHOD	75-0456
TONNESSEN, A.	MODEL EXPERIMENTS AND FINITE ELEMENT ANALYSIS OF STRESS IN AN OPEN SHIP	66-0054
TONTINI, G.	ANALYSIS OF STEADY ELECTRICAL AND MAGNETIC-FIELDS BY FINITE-ELEMENT METHOD - APPLICATION TO TRANSFOR	75-0550
TOO, J.	REDUCED INTEGRATION TECHNIQUE IN GENERAL ANALYSIS OF PLATES AND SHELLS	71-0374
TOO, J. J. M.	FULMOD - AN INELASTIC ANALYSIS PROGRAM TO PREDICT THE OPERATING BEHAVIOUR OF CANDU FUEL ELEMENTS	75-0472
TOO, J. J. M.	ON THE FINITE ELEMENT FORMULATION OF NAVIER-STOKES EQUATIONS	74-0720
TOO, J. M.	TWO DIMENSION, PLATE, SHELL AND FINITE PRISM ISOPARAMETRIC ELEMENTS AND THEIR APPLICATIONS	71-0772
TOPCU, A.	STRUCTURAL OPTIMIZATION USING THE FORCE METHOD	75-0440
TOPP, L. J.	STIFFNESS AND DEFLECTION ANALYSIS OF COMPLEX STRUCTURES	56-0001
TORBE, I.	A CRUCIFORM ELEMENT FOR SURFACES OF INFLATABLE STRUCTURES	75-0118
TORBE, I.	A GENERAL QUADRILATERAL PLATE ELEMENT	73-0807
TORBE, I.	GENERAL QUADRILATERAL PLATE ELEMENT	75-0339
TORBE, I.	IMPROVED GENERAL QUADRILATERAL PLATE ELEMENTS	75-0036
TORDION, G. V.	THE ELASTO-PLASTIC INDENTATION OF A HALF-SPACE BY A RIGID SPHERE	71-0113
TORIDIS, T. C.	COMPUTER ANALYSIS OF RIGID FRAMES	71-0401
TORIDIS, T. G.	DYNAMIC ANALYSIS OF FRAME AND PLATE STRUCTURES WITH GEOMETRIC AND MATERIAL NONLINEARITIES	73-0649
TORIDIS, T. G.	INELASTIC RESPONCE OF FRAMES TO DYNAMIC LOADS	71-0611
TORIDIS, T. G.	SOME RESULTS IN THE FINITE ELEMENT MODELING OF THIN PLATES	66-0018
TORRE, E. D.	ITERATIVE APPROACH TO FINITE-ELEMENT METHOD IN FIELD PROBLEMS	74-0124
TORVIK, P. J.	A NUMERICAL PROCEDURE FOR TWO DIMENSIONAL HEATING AND MELTING CALCULATIONS, WITH APPLICATIONS TO LAS	72-0708
TOTTENHAM, H.	DYNAMIC ANALYSIS OF SHELLS USING CURVED FINITE ELEMENTS	69-0296
TOTTENHAM, H.	EARTHQUAKE RESPONSE OF ARCH DAMS	70-0113
TOTTENHAM, H.	FINITE ELEMENT ANALYSES OF SHELL RESPONSE FOR ARBITRARY EXCITATIONS	70-0348
TOTTENHAM, H.	FINITE ELEMENT MODELLING TECHNIQUES FOR THE ANALYSIS OF BUILDING VIBRATIONS	72-0839
TOTTENHAM, H.	FINITE ELEMENT TECHNIQUES IN STRUCTURAL MECHANICS	70-0116
TOTTENHAM, H.	STATICS AND DYNAMICS OF HYDROSTATICALLY LOADED SHELLS BY FINITE ELEMENT METHOD	71-0044
TOTTENHAM, H.	THE FINITE ELEMENT METHOD IN SHELL ANALYSIS	69-0058
TOURNAIRE, M.	COMPARAISON DES DIVERSES MODELISATIONS NUMERIQUES D'UN SYSTEME MAGNETIQUE SATURABLE. (COMPARISON OF	75-0295
TOURNEMINE, G.	FINITE-ELEMENT METHOD TO COMPUTE STATIONARY SUBCRITICAL FLOWS	75-0439
TOVAR DE LEMOS,	CONVERGENCE OF FINITE ELEMENT SOLUTIONS IN VISCOELASTICITY	74-0628
TOWNLEY, C. H.	APPLICATION OF ELASTIC PLASTIC ANALYSIS TO THE DESIGN OF PRESSURE VESSELS	68-0157
TRACEY, D. M.	COMPUTATIONAL FRACTURE MECHANICS	71-0564
TRACEY, D. M.	FINITE ELEMENTS FOR DETERMINATION OF CRACK TIP ELASTIC STRESS INTENSITY FACTORS	71-0436
TRACEY, D. M.	FINITE ELEMENTS FOR THREE-DIMENSIONAL ELASTIC CRACK ANALYSIS	74-0136
TRACEY, D. M.	ON THE FRACTURE MECHANICS ANALYSIS OF ELASTIC-PLASTIC MATERIALS USING THE FINITE ELEMENT METHOD	73-1067
TRAHAIR, N. S.	BUCKLING OF INELASTIC I-BEAMS UNDER MOMENT GRADIENT	74-0765
TRAHAIR, N. S.	WARPING AND DISTORTION AT I-SECTION JOINTS	73-0722
TRALLI, A.	FINITE ELEMENT INCREMENTAL ANALYSIS OF ELASTOPLASTIC PLATE BENDING	73-0557
TRANTINA, G. G.	COMBINED MODE CRACK EXTENSION IN ADHESIVE JOINTS	72-0111
TRANTINA, G. G.	FRACTURE MECHANICS APPROACH TO ADHESIVE JOINTS	72-0181
TRATINA, G. G.	FRACTURE MECHANICS APPROACH TO ADHESIVE JOINTS	71-0726
TRAUBENIK, M.	STATIC THREE DIMENSIONAL ANALYSIS OF ELEVATED STEEL TRANSFORMATION STRUCTURES	74-1187
TREHARNE, G.	A BOUND THEOREM IN EIGENVALUES AND ITS PRACTICAL APPLICATIONS	71-0128
TREHARNE, G.	APPLICATIONS OF THE FINITE ELEMENT METHOD TO THE STRESS ANALYSIS OF MATERIALS SUBJECT TO CREEP	71-0548
TREYBIG, H. J.	A SUMMARY OF DISCRETE-ELEMENT METHODS OF ANALYSIS FOR PAVEMENT SLABS	72-0674
TRIANDAF, G. E.	FINITE ELEMENT ANALYSIS OF PILE-SUPPORTED PAVEMENT STRUCTURES	74-0157
TRIKHA, D. N.	ANALYSIS OF CONCRETE BOX GIRDERS BEFORE AND AFTER CRACKING	72-0247
TRIPP, L.	APPLICATION, TESTING, AND EVALUATION OF DOUBLY CURVED SHELL ELEMENTS FOR DYNAMIC ANALYSIS, PART III.	69-0256
TROITSKII, A. P	RASCHET METODOM KONECHNYKH ELEMENTOV DINAMICHESKIKH KHARAKTERISTIK ZEMLYANOI PLOTINY SOVMESTNO S OS	73-0437
TROITSKY, A. P.	BEHAVIOUR OF THE ROCK MASS AT THE SITE OF A ROCKFILL DAM	74-0992
TROMPETTTE, P.	VIBRATION ANALYSIS OF ROTATING TURBINE BLADES	74-0959
TROSCH, J.	TRANSIENT TWO DIMENSIONAL UNCONFINED AQUIFER PROBLEMS	74-0069
TROYON, F.	EIGENMODES OF CYLINDRICAL PLASMA USING FINITE ELEMENT METHOD	74-0174
TROYON, F.	NUMERICAL STUDY OF EIGEN OSCILLATIONS OF A SCREW PINCH WITH FINITE ELEMENTS	74-0159
TROYON, F.	NUMERICAL STUDY OF PROPER MODES OF A HOMOGENEOUS CYLINDRICAL PLASMA BY METHOD OF FINITE ELEMENTS	74-0160
TROYON, F.	THALIA - ONE-DIMENSIONAL MAGNETOHYDRODYNAMIC STABILITY PROGRAM USING METHOD OF FINITE-ELEMENTS	75-0458
TROYON, F.	THALIA: A ONE-DIMENSIONAL MAGNETOHYDRODYNAMIC STABILITY PROGRAM USING THE METHOD OF FINITE ELEMENTS	74-1085
TS'AO, H. S.	MAGAZINE HEADWALL RESPONSE TO EXPLOSIVE BLAST	74-0754
TSAI, C. T.	THE CONSTRAINT METHOD - A NEW FINITE ELEMENT TECHNIQUE	73-1027
TSAI, C-T.	THE QUADRATIC PROGRAMMING APPROACH TO THE FINITE ELEMENT METHOD	73-0831
TSAI, H. C.	DYPLAS, A FINITE ELEMENT DYNAMIC ELASTIC-PLASTIC LARGE DEFORMATION ANALYSIS PROGRAM	74-0306
TSAI, N. C.	SOIL-STRUCTURE INTERACTION - CONTINUUM OR FINITE ELEMENT	74-0954
TSENBERG, J.	REINFORCED CONCRETE ON CONSTITUTIVE RELATIONS	75-0753
TSENG, M. T.	FINITE ELEMENT MODEL FOR BACKWATER COMPUTATION	75-0851
TSENG, W-S.	SEISMIC RESPONSE OF HIGHWAY BRIDGES	75-0879
TSIPOURAS, P.	RECTANGULAR AND TRIANGULAR FINITE ELEMENTS FOR POISSON'S EQUATION	75-0802
TSOU, H. M.	STRAIN DISTRIBUTION IN AN ALUMINA BICRYSTAL STUDIES BY THE FINITE ELEMENT METHOD AND X-RAY DIFFRACTI	74-1070
TSUBOI, Y.	MATRIX ANALYSIS OF A COMPOSITE STRUCTURE OF PLATE AND LINE ELEMENTS	71-0250
TSUI, E. Y. W.	ELEMENT STIFFNESS MATRICES OF THICK-WALLED ORTHOTROPIC SHELLS WITH APPLICATIONS	69-0261
TSUI, Y.	PERFORMANCE OF TIED-BACK WALLS IN CLAY	74-0665

TSUKIOKA, Y.	FINITE ELEMENT ANALYSIS OF SURFACE WAVE PROBLEMS USING INTEGRAL EQUATIONS	74-0624
TSURUSAKI, K.	RANDOM VIBRATION ANALYSIS SYSTEM FOR COMPLEX STRUCTURES, PART 2, COMPUTER PROGRAM DESCRIPTION	69-0262
TSUSHIMA, Y.	A SEISMIC DESIGN OF NUCLEAR REACTOR BUILDING, STRESS ANALYSIS AND STIFFNESS EVALUATION OF THE ENTIRE	71-0251
TSUTA, T.	FINITE ELEMENT ANALYSIS OF CONTACT PROBLEMS	73-0312
TU, R. K-C.	FREE BOUNDARY POTENTIAL FLOW USING FINITE ELEMENTS	71-0829
TUBA, I. S.	A METHOD OF ELASTIC-PLASTIC PLANE STRESS AND PLANE STRAIN ANALYSIS	66-0114
TUBA, I. S.	FINITE ELEMENT METHOD IN LINEAR FRACTURE MECHANICS	70-0476
TUBA, I. S.	FINITE ELEMENT METHOD FOR CONTACT PROBLEMS OF SOLID BODIES - 2.	71-0435
TUBA, I. S.	ON THE FINITE ELEMENT METHOD IN LINEAR FRACTURE MECHANICS	69-0032
TUCCHIO, M. A.	AN ANALYSIS OF SOME FACTORS THAT AFFECT HYDROPHONE SENSITIVITY	72-0739
TUCCHIO, M. A.	FINITE ELEMENT STRESS ANALYSIS OF A MULTICONDUCTOR ARMOR CABLE	73-0095
TUFF, A. D.	AN ITERATIVE METHOD FOR LARGE SYSTEMS OF LINEAR STRUCTURAL EQUATIONS	73-1020
TUKIOKA, Y.	FINITE ELEMENT METHOD FOR WAVE GUIDE PROBLEMS	75-0872
TULLOCH, H. A.	CLOSED FORM, FINITE ELEMENT SOLUTIONS FOR PLATE VIBRATIONS	69-0168
TULLOCH, H. A.	FINITE PLATE-BENDING ELEMENTS IN POLAR COORDINATES	68-0234
TUNG, C.	WIND DRIVEN LAKE CIRCULATION BY THE FINITE ELEMENT METHOD	70-0041
TUOMINEN, J.	CURVED LOW ORDER ANGULAR ELEMENTS IN TWO DIMENSIONAL NEUTRON TRANSPORT PROBLEMS	74-1322
TURCKE, D. J.	A VARIATIONAL APPROACH TO GRID OPTIMIZATION IN THE FINITE ELEMENT METHOD	72-0572
TURCKE, D. J.	APPLICATION OF GRID SELECTION PROCEDURES FOR IMPROVED FINITE ELEMENT STRESS ANALYSIS	74-0326
TURCKE, D. J.	FURTHER DEVELOPMENTS IN GRID SELECTION PROCEDURES IN THE FINITE ELEMENT METHOD	75-0028
TURCKE, D. J.	GUIDELINES FOR SELECTING FINITE ELEMENT GRIDS BASED ON AN OPTIMIZATION STUDY	74-0303
TURCKE, D. J.	OPTIMUM MESH CONFIGURATIONS IN THE FINITE ELEMENT METHOD	74-1346
TURCKE, D. J.	PROCEDURES FOR SELECTING NEAR OPTIMUM FINITE ELEMENT GRIDS	73-0537
TURGAY, H. M.	FATIGUE CRACK INITIATION IN A BORON EPOXY PLATE WITH A CIRCULAR HOLE	74-0820
TURKSTRA, C.	FINITE ELEMENT SCHEME FOR BOX BRIDGE ANALYSIS	75-0666
TURKSTRA, C. J.	ANALYSIS OF CURVED BOX GIRDER BRIDGES	73-0294
TURNBELL, R. C.	THE KANSAS TEST TRACK. NON-CONVENTIONAL TRACK STRUCTURES. DESIGN REPORT	72-0797
TURNER, A. B.	FORTRAN SUBROUTINES FOR THE SOLUTION OF LAPLACE'S EQUATION OVER A GENERAL REGION IN TWO DIMENSIONS	70-0425
TURNER, C. E.	APPLICATION OF FINITE ELEMENT TECHNIQUES TO POST-YIELD ANALYSIS OF PROPOSED STANDARD 3 POINT BEND FR	74-0118
TURNER, C. E.	COMPUTATION OF POST-YIELD BEHAVIOUR IN NOTCH-BEND AND TENSION TESTPIECES.	72-0244
TURNER, C. E.	FRACTURE ANALYSIS IN AREAS OF HIGH NOMINAL STRAIN	73-0495
TURNER, L. J.	MEASURED AND PREDICTED LONGITUDINAL VIBRATIONS OF A LIQUID PROPELLANT TWO-STAGE LAUNCH VEHICLE	70-0366
TURNER, M. J.	FURTHER DEVELOPMENT AND APPLICATIONS OF THE STIFFNESS METHOD	64-0020
TURNER, M. J.	LARGE DEFLECTIONS OF STRUCTURES SUBJECT TO HEATING AND EXTERNAL LOADS	60-0002
TURNER, M. J.	OPTIMIZATION OF STRUCTURES TO SATISFY FLUTTER REQUIREMENTS	69-0311
TURNER, M. J.	STIFFNESS AND DEFLECTION ANALYSIS OF COMPLEX STRUCTURES	56-0001
TURNER, T. W.	PRELIMINARY INVESTIGATION OF THE DIE-LESS DRAWING OF TITANIUM AND SOME STEELS	75-0770
TURNEY, M.	COMPUTER GRAPHICS FOR STRUCTURAL ANALYSIS	74-0832
TVEIT, A.	MEASUREMENT OF THE WATER DIFFUSIVITY OF POROUS MATERIALS	74-1138
TWIGG, D. W.	A USER-ORIENTED PROGRAM FOR CRASH DYNAMICS	74-0321
UCCIFERRO, J. J	A THEORETICAL AND NUMERICAL COMPARISON OF ELASTIC NONLINEAR FINITE ELEMENT METHODS	72-0709
UCCIFERRO, J. J	ELASTIC NONLINEAR FINITE ELEMENT ANALYSIS OF PLANAR STRUCTURES	71-0830
UCCIFERRO, J. J	THEORETICAL AND NUMERICAL COMPARISON OF ELASTIC NONLINEAR FINITE ELEMENT METHODS.	72-0452
UDAKA, T.	LUSH: A COMPUTER PROGRAM FOR COMPLEX RESPONSE ANALYSIS OF SOIL-STRUCTURE SYSTEMS	74-0744
UDOGUCHI, T.	APPROACH TO INVESTIGATE NOTCH EFFECT ON LOW-CYCLE FATIGUE WITH FINITE ELEMENT	73-0509
UDOGUCHI, T.	ERROR ANALYSIS OF VARIOUS FINITE ELEMENT PATTERNS	73-0510
UDOGUCHI, T.	NOTCH EFFECT IN LOW-CYCLE FATIGUE. CONSIDERATIONS ON NOTCH EFFECT IN LOW-CYCLE FATIGUE WITH FINITE-E	75-0343
UEDA, Y.	THERMAL ELASTIC PLASTIC STRESS ANALYSIS BY THE FINITE ELEMENT METHOD	69-0391
UEDA, Y.	A NEW MEASURING METHOD OF RESIDUAL STRESSES WITH THE AID OF FINITE ELEMENT METHOD AND RELIABILITY OF	75-0875
UEDA, Y.	ANALYSIS OF ELASTIC-PLASTIC BUCKLING OF PLATES BY THE FINITE ELEMENT METHOD	69-0263
UEDA, Y.	ANALYSIS OF THERMAL ELASTIC PLASTIC STRESS AND STRAIN DURING WELDING BY FINITE ELEMENT METHOD	71-0382
UEDA, Y.	ANALYSIS OF WELDING STRESS RELIEVING BY ANNEALING BASED ON FINITE ELEMENT METHOD	75-0777
UEDA, Y.	APPLICATION OF FINITE ELEMENT METHOD TO PROBLEMS OF NON-LINEAR BEHAVIOR WITH CREEP	75-0811
UEDA, Y.	ELASTIC-PLASTIC BUCKLING OF PLATES USING THE FINITE ELEMENT METHOD	67-0094
UEDA, Y.	ELASTIC-PLASTIC BUCKLING OF PLATES BY FINITE ELEMENT METHOD	69-0253
UEDA, Y.	RECENT TRENDS OF RESEARCH INTO RESTRAINT STRESSES AND STRAINS IN RELATION TO WELD CRACKING	73-0483
UEDA, Y.	TENDANCES ACTUELLES DES RECHERCHES SUR LES CONTRAINTES ET DEFORMATIONS DUES AU BRIDAGE EN RELATION A	74-0273
UEDA, Y.	THERMAL NONLINEAR BEHAVIOUR OF STRUCTURES	72-0317
UEHARA, S.	COMPUTATIONAL ASPECTS OF FINITE ELEMENT METHOD FOR A FRAME STRUCTURE	72-0175
UJIHARA, B. H.	HYDROELASTIC ANALYSIS OF AXISYMMETRIC SYSTEMS BY A FINITE ELEMENT METHOD	68-0084
UKAI, S.	SOLUTION OF MULTI-DIMENSIONAL NEUTRON TRANSPORT EQUATION BY FINITE ELEMENT METHOD	72-0564
UKHOV, S. B.	CALCULATION OF DISPLACEMENTS AND STRESSES IN ANISOTROPIC ROCKS BY USING METHOD OF FINITE ELEMENTS	73-0265
UKHOV, S. B.	POSSIBILITY OF USING FINITE ELEMENT METHOD IN THE ANALYSIS OF JOINT PERFORMANCE OF THE HYDRAULIC STR	72-0418
ULDRICH, E. D.	STRAP: COMPUTER CODE FOR STATIC AND DYNAMIC STRUCTURAL ANALYSIS AND STUDIES MADE USING THE CODE	70-0409
UNDERHILL, D. A	STRUCTURAL DYNAMIC PROPERTIES OF TACTICAL MISSILE JOINTS - PHASE 1	70-0423
UNDERHILL, D. A	STRUCTURAL DYNAMIC PROPERTIES OF TACTICAL MISSILE JOINTS - PHASE 3	74-0811
UNDERWOOD, J. H	EXPERIMENTAL AND ANALYTICAL STRAINS IN A EDGE-CRACKED SHEET	69-0483
UNNY, T. E.	A FINITE ELEMENT SOLUTION FOR A CLASS OF TWO DIMENSIONAL VISCOUS FLUID FLOW DYNAMICS PROBLEMS	71-0544
URA, T.	SIMULATION OF FINITE ELEMENT METHOD FOR LOADING PROCESS OF GRAIN CARGO AND INTERACTION WITH SHIP'S S	75-0161
URAGAMI, K.	EXPERIMENTAL STUDY OF FLEXIBILITY FACTORS AND STRESSES OF WELDING ELBOWS WITH END EFFECTS	73-0422
URAL, O.	FINITE ELEMENT METHOD - A VERSATILE TOOL FOR CIVIL ENGINEERING	72-0627
URAL, O.	FINITE ELEMENT METHOD AND ROCK MECHANICS	73-0551
URAL, O.	FINITE ELEMENT METHOD: BASIC CONCEPTS AND APPLICATIONS	73-0646
URAL, O.	FINITE ELEMENT METHOD AND COMPUTER PROGRAMS	74-0450
URAL, O.	MATRIX METHODS AND USE OF COMPUTERS IN STRUCTURAL ENGINEERING	71-0615
URAL, OKTAY	ANALYSIS OF TOTALLY SUBMERGED THIN SHEELS BY FINITE ELEMENT TECHNIQUE	71-0616
USUKI, S.	APPLICATIONS OF FINITE ELEMENT METHODS TO UNSTEADY VISCOUS FLOW AROUND A BOX-GIRDER BRIDGE OSCILLATI	74-0827
USUKI, S.	FINITE ELEMENT METHOD VIA LOCAL POTENTIAL	74-0038
USUKI, S.	LOCAL POTENTIAL APPROACH TO FINITE ELEMENT METHOD IN UNSTEADY VISCOUS INCOMPRESSIBLE FLUID FLOW	73-0298
UTKU, S.	AN ALTERNATIVE FOR THE FINITE ELEMENT METHOD	72-0833
UTKU, S.	AN AUTOMATIC NODE RELABELLING SCHEME FOR BANDWIDTH MINIMIZATION OF STIFFNESS MATRICES	68-0256
UTKU, S.	BEHAVIOR OF TRIANGULAR SHELL ELEMENT STIFFNESS MATRICES ASSOCIATED WITH POLYHEDRAL DEFLECTION DISTRI	68-0159
UTKU, S.	BOUNDING SOLUTIONS OF PLATE BENDING PROBLEMS	67-0084
UTKU, S.	COMPUTATION OF STRESSES IN TRIANGULAR FINITE ELEMENTS	66-0072
UTKU, S.	ELAS - A GENERAL PURPOSE COMPUTER PROGRAM FOR THE EQUILIBRIUM PROBLEMS OF LINEAR STRUCTURES	68-0158
UTKU, S.	ELAS: A GENERAL-PURPOSE COMPUTER PROGRAM FOR THE EQUILIBRIUM PROBLEMS OF LINEAR STRUCTURES. VOLUME 2	69-0462
UTKU, S.	EXPLICIT EXPRESSIONS FOR TRIANGULAR TOWA ELEMENT STIFFNESS MATRIX	68-0160
UTKU, S.	FINITE ELEMENT METHOD FOR BLOCK ADJUSTMENT PROBLEMS OF PHOTOGRAMMETRY	73-0264
UTKU, S.	ON DERIVATION OF STIFFNESS MATRICES WITH C0 ROTATION FIELDS FOR PLATES AND SHELLS	71-0254
UTKU, S.	ON RESPONSE OF INITIALLY STRESSED STRUCTURES TO RANDOM EXCITATIONS	73-0278
UTKU, S.	ON SMALL VIBRATIONS AND PERTURBATIONS OF FLEXIBLE BODIES UNDERGOING ARBITRARY NOMINAL MOTION	70-0376
UTKU, S.	SOLUTION OF LARGE SYSTEMS OF EQUATIONS	75-0608
UTKU, S.	STIFFNESS MATRICES FOR THIN TRIANGULAR ELEMENTS OF NONZERO GAUSSIAN CURVATURE	66-0073
UTKU, S.	SYSTEMATIC SUBSTRUCTURING	74-0895
VAAGE, B.	LOADS AND RESPONSE OF LARGE DUCTED PROPELLER SYSTEMS	73-0810
VACHARAJITTIPHA	WARPING AND DISTORTION AT I-SECTION JOINTS	73-0722
VAIL, C. F.	ILLUSTRATIONS OF AUTOMOTIVE FINITE ELEMENT MODELS - DYNAMICS	74-0391
VAIL, C.F.	A MODAL SYNTHESIS TECHNIQUE FOR DETERMINING DYNAMIC PROPERTIES FOR A STRUCTURE FOR MASS AND STIFFNES	74-0319
VAISH, A. K.	EARTHQUAKE ANALYSIS OF STRUCTURE-FOUNDATION SYSTEMS	73-0876
VAISH, A. K.	EARTHQUAKE FINITE ELEMENT ANALYSIS OF STRUCTURE-FOUNDATION SYSTEMS	74-0558
VAITYS, R. P.	WEAPON-MOUNT INTERFACE STUDY - TRIPOD MOUNT M3	72-0846
VALACHOVIC, R.	APPLICATION OF FRACTURE MECHANICS TO GRAPHITE UNDER COMPLEX STRESS CONDITIONS	72-0771

106

VALATHUR, M.	WAVE PROPAGATION IN A THIN HOLLOW CONE BY A FINITE ELEMENT METHOD	72-0045
VALENTE, G.	DISCUSSION ON "PLANE STRESS LIMIT ANALYSIS BY FINITE ELEMENTS"	71-0792
VALENTIN, R. A.	EVALUATION OF A CLASS OF METHODS FOR BOUNDING STEADY-STATE CREEP DEFORMATION	74-0892
VALID, R.	VIBRATIONS OF TANKS PARTIALLY FILLED WITH LIQUIDS	74-0089
VALLABHAN, C. V	ANALYSIS FOR THE DESIGN OF BEAM TESTING APPARATUS	75-0527
VALLABHAN, C. V	FINITE ELEMENT METHOD APPLIED TO PROBLEMS IN STRESSES AND DEFORMATION OF SOIL	67-0085
VALLIANOS, G.	DYNAN LECTURE NOTES WITH COMPUTATIONAL EXAMPLES	71-0581
VALLIAPPAN, P.	NONLINEAR SETTLEMENT ANALYSIS BY FINITE ELEMENT	75-0486
VALLIAPPAN, P.	NONLINEAR STRESS DEFORMATION ANALYSES OF LAKE AGASSIZ CLAYS USING FINITE ELEMENT METHOD	74-1072
VALLIAPPAN, S.	ANALYSIS OF REAL STRUCTURES FOR CREEP, PLASTICITY AND OTHER COMPLEX CONSTITUTIVE LAWS.	69-0287
VALLIAPPAN, S.	ELASTIC PLASTIC ANALYSIS OF SHALLOW FOUNDATIONS	74-0446
VALLIAPPAN, S.	ELASTO-PLASTIC SOLUTIONS OF ENGINEERING PROBLEMS ' INITIAL STRESS ' FINITE ELEMENT APPROACH	69-0282
VALLIAPPAN, S.	FINITE ELEMENT ANALYSIS OF ELECTROMAGNETIC FIELD PROBLEMS	74-0080
VALLIAPPAN, S.	FINITE ELEMENT ANALYSIS OF CONSOLIDATION PROBLEM	74-0081
VALLIAPPAN, S.	NON-LINEAR STRESS ANALYSIS OF TWO DIMENSIONAL PROBLEMS WITH SPECIAL REFERENCE TO ROCK AND SOIL MECHA	68-0183
VALLIAPPAN, S.	STRESS ANALYSIS OF HUMAN FEMUR	73-0327
VALLIAPPAN, S.	STRESS ANALYSIS OF ROCK AS A NO-TENSION MATERIAL	68-0174
VALLIAPPAN, S.	TENSILE CRACK PROPAGATION IN REINFORCED CONCRETE BEAMS BY THE FINITE ELEMENT TECHNIQUES	69-0254
VALLIAPPAN, S.	TENSILE CREEP ANALYSIS OF CONCRETE STRUCTURES	74-0439
VALLIAPPAN, S.	USE OF COMPUTERS IN WELDING TECHNOLOGY	74-1160
VAN BATEN, T. J	DETERMINATION OF THE INITIAL BUCKLING STRESS OF FLAT PLATES USING THE DISPLACEMENT METHOD	69-0465
VAN BOMMEL, P.	PRACTICAL APPLICATION OF THE FINITE ELEMENT METHOD AND SOME NUMERICAL ASPECTS	69-0309
VAN CAMPEN, D.	ON THE STRESS DISTRIBUTION IN NOZZLE TO CYLINDER CONNECTIONS FOR SMALL DIAMETER RATIOS	72-0426
VAN CAMPEN, D.	SOME APPLICATIONS OF ENGINEERING MECHANICS IN PRESSURE VESSEL ANALYSIS	72-0892
VAN COUVERING,	STARDYNE STRUCTURAL ANALYSIS SYSTEM, ANALYTICAL CAPABILITY MANUAL	68-0235
VAN DER KROGT,	STRESS ANALYSIS OF A TUBULAR CROSS-JOINT WITHOUT ANY INTERNAL STIFFENING FOR OFFSHORE STRUCTURES	74-0549
VAN DER REE, H.	PREDICTION OF TEMPERATURE DISTRIBUTIONS IN CARGOES WITH THE AID OF A COMPUTER PROGRAM USING THE METH	74-1215
VAN DER VOOREN,	FINITE ELEMENT CALCULATION OF AXISYMMETRIC SUBCRITICAL COMPRESSIBLE FLOW	74-1064
VAN DER VOOREN,	FINITE ELEMENT SOLUTION OF THE INCOMPRESSIBLE FLOW OVER AN AIRFOIL IN A NON UNIFORM STREAM	74-0432
VAN DER WILT, M	A NUMERICAL APPROACH TO THE STUDY OF IRREGULAR SHIP MOTIONS	74-0030
VAN DER WOLF, A	ON THE DESIGN OF DIE-SETS	75-0582
VAN LAETHEM, M.	INFLUENCE OF THE UNPRECISION OF EXECUTION ON THE BEHAVIOUR OF A SPHERICAL SHELL WITH CONCENTRATED LO	73-0802
VAN LEEUWEN, D.	A FINITE ELEMENT ANALYSIS OF SHELLS OF REVOLUTION	73-0759
VAN POSSEN, D.	COMBINED ELASTIC-PLASTIC CREEP ANALYSIS OF TWO-DIMENSIONAL BODIES	71-0232
VAN SPIEGEL, E.	ON THE NUMERICAL TREATMENT OF THE NAVIER-STOKES EQUATIONS FOR AN INCOMPRESSIBLE FLUID	73-0809
VAN WINSEN, F.H	MODERNE VERFAHREN ZUR UNTERSUCHUNG DER STEIFIGKEITS-UND FESTIGKEITSVERHALTNISSE VON RAHMENKORPERN. I	69-0401
VANBUREN, W.	FUEL ROD DEFORMATIONS - A TIME DEPENDENT FINITE ELEMENT ANALYSIS ON AXISYMMETRIC FUEL RODS, INCLUDIN	70-0414
VANBUREN, W.	HEAVY SECTION STEEL TECHNOLOGY PROGRAM. TECHNICAL REPORT NO. 4, OCTOBER 1969. A TWO-DIMENSIONAL ELAS	69-0474
VANBUREN, W.	STRESS ANALYSIS OF A CERAMIC BLADE DESIGN SUBJECTED TO CENTRIFUGAL AND THERMAL LOADS	73-0790
VANDERBILT, M.	A RATIONAL ANALYSIS AND DESIGN PROCEDURE FOR WOOD JOIST FLOOR SYSTEMS	74-0730
VANDERLINDEN, J	STRUCTURE-MEDIUM INTERACTION AND DESIGN PROCEDURES STUDY. VOLUME IV. FEAT CODE USER'S GUIDE	69-0265
VANDERLINDEN, J	STRUCTURE- MEDIUM INTERACTION PHENOMENOLOGY	69-0863
VANDERVORST, M.	USE OF FINITE ELEMENT METHOD IN A MARKER AND CELL PROGRAM FOR SOLVING TIME-DEPENDENT FLUID FLOW PROB	75-0043
VANDHAN, C. P.	INTEGRATED SEQUENTIAL SOLVER FOR LARGE MATRIX EQUATIONS	74-0276
VANLEEUWEN, D.	A FINITE ELEMENT ANALYSIS OF BODIES OF REVOLUTION. A COMPARISON BETWEEN CURRENTLY USED TYPES OF ELEM	74-1319
VANN, W. P.	A FINITE ELEMENT TENSOR APPROACH TO PLATE BUCKLING AND POST BUCKLING	73-0883
VARADARAJAN, A.	STRESS-PATH DEPENDENT DEFORMATION MODULI OF CLAY	75-0050
VARANASI, S. R.	ELASTO PLASTIC ANALYSIS OF PLANE STRUCTURES BY THE FINITE ELEMENT DISPLACEMENT METHOD	70-0299
VARANASI, S. R.	EXPERIENCE WITH FINITE ELEMENT COMPUTATIONS FOR NONLINEAR STRUCTURAL PROBLEMS	74-1239
VARANASI, S. R.	FINITE ELEMENT ANALYSIS OF PLASTICITY PROBLEMS	68-0154
VARANASI, S. R.	NON-LINEAR MATERIAL ANALYSIS WITH FINITE ELEMENTS AND INCREMENTAL METHODS	68-0199
VARGA, R. S.	ERROR BOUNDS FOR SEMIDISCRETE GALERKIN APPROXIMATIONS OF PARABOLIC PROBLEMS WITH APPLICATIONS TO PET	70-0291
VARGA, R. S.	EXTENSIONS OF THE SUCCESSIVE OVER-RELAXATION THEORY WITH APPLICATIONS TO FINITE ELEMENT APPROXIMATIO	73-0980
VARGA, R. S.	HERMITE INTERPOLATION TYPE RITZ METHODS FOR TWO POINT BOUNDARY VALUE PROBLEMS	66-0089
VARGA, R. S.	NUMERICAL METHODS FOR TIME DEPENDENT NONLINEAR BOUNDARY VALUE PROBLEMS	70-0296
VARGA, R. S.	NUMERICAL METHODS OF HIGH-ORDER ACCURACY FOR NONLINEAR BOUNDARY VALUE PROBLEMS	68-0194
VARGA, R. S.	PIECEWISE HERMITE INTERPOLATION IN ONE AND TWO VARIABLES WITH APPLICATIONS TO PARTIAL DIFFERENTIAL E	68-0201
VAROGLU, E.	DYNAMICS OF GRAVITY DAM-RESERVOIR SYSTEMS	73-0225
VARSHNEY, R. S.	DAMS ON ROCKS OF VARYING ELASTICITY	74-0671
VARVASINSKAYA,	SOLVING NONLINEAR PROBLEM OF NONSTEADY THERMAL CONDUCTIVITY IN SOLID BODIES OF ARBITRARY FORM	68-0019
VASARHELYI, D.	TESTS OF GUSSET PLATE MODELS	71-0338
VAUGHAN, D. K.	A COMPARISON OF CURRENT WORK-HARDENING MODELS USED IN THE ANALYSIS OF PLASTIC DEFORMATIONS	73-0769
VAUGHAN, D. K.	COMPARISON OF THE CAPABILITY OF OUR HARDENING RULES TO PREDICT A MATERIAL'S PLASTIC BEHAVIOR	75-0523
VAUGHAN, G. N.	CALCULATION OF THERMAL STRESSES IN TUBULAR CERAMIC COMPONENTS	73-0670
VAWTER, D.	ANALYSIS OF PERISTALTIC PUMPING	72-0044
VEIDINGE, L.	ORDER OF CONVERGENCE OF A FINITE ELEMENT SCHEME	74-0550
VEIDINGER, L.	ORDER OF CONVERGENCE OF A FINITE-ELEMENT METHOD IN REGIONS WITH CURVED BOUNDARIES	75-0817
VELEBIT, M.	FINITE ELEMENT METHOD FOR ELASTIC PLASTIC PLATES	72-0167
VELJOVICH, W.	PRESSURE VESSEL WALL THICKNESS LIMITS BASED ON THERMAL STRESS RACHETTING AND CREEP FATIGUE INTERACTI	74-0968
VENANCIO-FILHO,	STABILITY OF PLATES USING A MIXED FINITE ELEMENT FORMULATION	74-0218
VENANCIO-FILHO,	VIBRATIONS OF GRIDS BY THE FINITE ELEMENT METHOD	73-1089
VENART, J. E. S	THE APPLICATION OF HIGHER ORDER FINITE ELEMENTS TO THREE DIMENSIONAL HEAT CONDUCTION PROBLEMS	70-0266
VENKATARAMAN, N	DETERMINATION OF STRESSES DUE TO DISCONTINUITIES IN FINITE PLATES OF ISOTROPIC AND ORTHOTROPIC MATER	73-1052
VENKATARAMANA,	FINITE-ELEMENT ANALYSIS OF MODERATELY THICK SHELLS	75-0588
VENKATARAMANA,	VIBRATIONS OF THICK PLATES USING A HIGH PRECISION TRIANGULAR ELEMENT	74-1002
VENKATASUBBU, S	FINITE ELEMENT ANALYSIS OF CONTAINMENT VESSELS	71-0080
VENKATASUBBU, S	THERMAL SLEEVE DESIGN FOR PRESSURE VESSELS	74-1129
VENKATES, G.	GALERKIN FINITE ELEMENT ANALYSIS OF A UNIFORM BEAM CARRYING A CONCENTRATED MASS AND ROTARY INERTIA W	74-0552
VENKATESWARA RA	A COMPARATIVE STUDY OF THE CONSISTENT AND SIMPLIFIED FINITE ELEMENTS ANALYSIS OF EIGENVALUE PROBLEMS	70-0034
VENKATESWARA RA	AN INHOMOGENEOUS, TAPERED FINITE ELEMENT	73-1038
VENKATESWARA RA	ASSESSMENT OF ACCURACIES OF FINITE ELEMENT EIGENVALUES	72-0075
VENKATESWARA RA	ASSESSMENT OF ACCURACIES OF FINITE ELEMENT EIGENVALUES	72-0747
VENKATESWARA RA	BOUNDS FOR EIGENVALUES IN SOME VIBRATION AND STABILITY PROBLEMS.	72-0390
VENKATESWARA RA	STABILITY OF MODERATELY THICK RECTANGULAR PLATES USING A HIGH PRECISION TRIANGULAR FINITE ELEMENT	75-0334
VENKATESWARA, G	AXISYMMETRIC VIBRATIONS OF LINEARLY TAPERED ANNULAR PLATES	74-1266
VENKATESWARA, G	CONICAL SHELL FINITE ELEMENT	74-0593
VENKATESWARA, R	A COMPARATIVE STUDY OF VARIABLE AND CONSTANT THICKNESS HIGH PRECISION TRIANGULAR PLATE BENDING ELEME	74-0857
VENKATESWARA, R	FINITE ELEMENT ANALYSIS OF VIBRATIONS OF INITIALLY STRESSED THIN SHELLS OF REVOLUTION	74-1056
VENKATESWARA, R	GALERKIN FINITE ELEMENT SOLUTION FOR THE STABILITY OF CANTILEVER COLUMNS SUBJECTED TO TANGENTIAL LOA	75-0534
VENKATESWARA, R	VIBRATIONS OF THICK PLATES USING A HIGH PRECISION TRIANGULAR ELEMENT	74-1002
VENKATESWARA, R	FINITE ELEMENT ANALYSIS OF VIBRATION AND STABILITY PROBLEMS	72-0477
VENKAYYA, V. B.	APPLICATION OF OPTIMALLY CRITERION TO FIBER-REINFORCED COMPOSITES	73-0431
VENKAYYA, V. B.	DESIGN OF OPTIMUM STRUCTURES FOR DYNAMIC LOADS	71-0255
VENKAYYA, V. B.	ITERATIVE METHOD FOR THE ANALYSIS OF LARGE STRUCTURAL SYSTEMS	68-0236
VENKAYYA, V. B.	OPTIMIZATION OF STRUCTURES FOR STRENGTH AND STABILITY REQUIREMENTS	73-0575
VENKAYYA, V. B.	OPTIMUM DESIGN OF ADVANCED COMPOSITE STRUCTURES FOR STATIC LOADS	72-0613
VENKAYYA, V. B.	REVIEW OF OPTIMALITY CRITERIA APPROACHES TO STRUCTURAL OPTIMIZATION	74-0652
VENTER, P.	PROBABILISTIC EVALUATION OF THE RELATIVE BEHAVIOUR OF TUNGSTEN CARBIDE COMPONENTS IN STRUCTURAL APPL	75-0579
VENTER, R. D.	APPLICATION OF THE FINITE ELEMENT TECHNIQUE TO THE STUDY OF AREAS OF HIGH STRESS GRADIENTS WITHIN A	73-0259
VERMA, M. K.	NONLINEAR VIBRATIONS OF NONUNIFORM BEAMS WITH CONCENTRATED MASSES	74-1300
VERMA, V. K.	FINITE ELEMENT ANALYSIS OF NONLINEAR CABLE REINFORCED MEMBRANES	74-1073
VERNER, E.	A FINITE ELEMENT APPROACH TO TWO-PHASE FLOW IN POROUS MEDIA	75-0392

VERNER, E. A.	FINITE ELEMENT STRESS FORMULATION FOR WAVE PROPAGATION	71-0831
VERNER, E. A.	FINITE ELEMENT STRESS FORMULATION FOR WAVE PROPAGATION	73-1043
VERNER, E. A.	FINITE ELEMENTS FOR TWO PHASE FLOW IN POROUS MEDIA	74-0060
VERNER, E. A.	SHOCK TUNNEL TESTS OF PRELOADED AND ARCHED WALL PANELS FINITE ELEMENT STRESS FORMULATION FOR WAVE PR	73-0892
VERONDA, D. R.	BUCKLING OF SEGMENTS OF TOROIDAL SHELLS	73-0515
VERONDA, D. R.	STABILITY OF HYPERBOLOIDAL SHELLS	75-0488
VERONDA, D. R.	STABILITY OF PRESSURIZED HYPERBOLOIDAL SHELLS	75-0585
VERRUIJT, A.	GROUNDWATER FLOW - THE FINITE ELEMENT METHOD	68-0260
VERRUIJT, A.	SOLUTION OF TRANSIENT GROUNDWATER FLOW PROBLEMS BY FINITE ELEMENT METHOD	72-0046
VERRUIJT, A.	THEORY OF GROUNDWATER FLOW	70-0117
VERYARD, D. A.	PROBLEMS ASSOCIATED WITH THE CONVERGENCE OF ISOPARAMETRIC AND MIXOPARAMETRIC FINITE ELEMENTS	71-0375
VESTRONI, F.	ANALISI NUMERICA DEL COMPORTAMENTO ELASTICO DI UNA STRUTTURA BIDIMENSIONALE PIANA. (NUMERICAL ANALYS	74-1029
VICK, J. L.	THE USE OF EIGEN-DERIVATIVES TO PREDICT THE EFFECT OF DESIGN CHANGES UPON THE DYNAMIC CHARACTERISTIC	74-0737
VIDOUSE, F.	DETERMINATION OF THE TECHNICAL CONSTANTS OF LAMINATES IN OBLIQUE DIRECTIONS. THEORY AND EXPERIMENTAL	73-0748
VIGNESWAPAN, K.	THE FINITE-ELEMENT METHOD AS APPLIED TO SHALLOW FOUNDATIONS	74-1074
VIGNON, B.	USE OF VARIATIONAL PRINCIPLE IN STUDY ON RELATIVISTIC BOLTZMANN EQUATION	71-0442
VIJAY, D. K.	ERRORS IN SHELL FINITE-ELEMENT MODELS FOR VIBRATION OF CIRCULAR-CYLINDERS	75-0203
VIJAY, D. K.	VIBRATION ANALYSIS OF AXISYMMETRIC RESONATORS	75-0199
VILLAGGIO, P.	PROPERTIES OF STABILITY AND MONOTONY WITH THE FINITE ELEMENT METHOD	69-0266
VINJE, T.	CALCULATION OF THE CHARACTERISTIC FREQUENCIES OF FLUID-FILLED SPHERICAL SHELLS	73-0511
VINSON, T. J.	APPLICATION OF THE FINITE ELEMENT METHOD FOR ANALYSIS AND OPTIMIZATION OF SUBMARINE STRUCTURES	71-0350
VINSON, T. J.	MARK5C/DAISY SERIES OF PROGRAMS	72-0160
VINSON, T. J.	THE DEVELOPMENT OF COMPUTER GRAPHICS FOR LARGE SCALE FINITE ELEMENT CODES	71-0042
VISON, T. J.	THE DEVELOPMENT OF LARGE SCALE DIGITAL COMPUTER CODES FOR PRODUCTION STRUCTURAL ANALYSIS	71-0721
VISSER, C.	APPLICATION OF FINITE ELEMENT ANALYSIS TO CERAMIC COMPONENTS	75-0040
VISSER, C.	FINITE ELEMENT ANALYSIS OF HEAT CONDUCTION IN ANISOTROPIC SOLIDS	68-0161
VISSER, C.	FUEL ROD DEFORMATIONS - A TIME DEPENDENT FINITE ELEMENT ANALYSIS ON AXISYMMETRIC FUEL RODS, INCLUDIN	70-0414
VISSER, C.	HEAVY SECTION STEEL TECHNOLOGY PROGRAM. TECHNICAL REPORT NO. 4, OCTOBER 1969. A TWO-DIMENSIONAL ELAS	69-0474
VISSER, C.	STRESS ANALYSIS OF A CERAMIC BLADE DESIGN SUBJECTED TO CENTRIFUGAL AND THERMAL LOADS	73-0790
VISSER, C.	THE APPROXIMATE ANALYSIS OF THIN SHELLS BY THE FINITE ELEMENT METHOD	66-0131
VISSER, K.	THE APPLICATION OF A CURVED MIXED-TYPE SHELL ELEMENT	70-0440
VISSER, M.	THE FINITE ELEMENT METHOD IN DEFORMATION AND HEAT CONDUCTION PROBLEMS	68-0238
VISSER, R. J.	CERAMIC ROTATING BLADES: SOME CRITICAL DESIGN PARAMETERS FOR GAS TURBINE APPLICATIONS	74-0146
VISSER, W.	A FINITE ELEMENT METHOD FOR THE DETERMINATION OF NON-STATIONARY TEMPERATURE DISTRIBUTIONS AND THERMA	66-0074
VISSER, W.	A NUMERICAL APPROACH TO THE STUDY OF IRREGULAR SHIP MOTIONS	74-0030
VISSER, W.	A REFINED MIXED-TYPE PLATE BENDING ELEMENT	69-0267
VISSER, W.	HEAT TRANSFER PROBLEMS AND TEMPERATURE STRESSES	69-0326
VISSER, W.	ON THE STRUCTURAL DESIGN OF TUBULAR JOINTS	74-0492
VISSER, W.	THE APPLICATION OF A CURVED MIXED TYPE SHELL ELEMENT	70-0118
VITIELLO, E.	A FINITE ELEMENT LINEAR PROGRAMMING METHOD FOR THE LIMIT ANALYSIS OF THIN PLATES	73-1064
VOGEL, W.	ELASTO-PLASTIC CYCLIC ANALYSIS OF STRUCTURAL MEMBERS	72-0202
VOGEL, W. H.	SHOCK ANALYSIS OF STRUCTURAL SYSTEMS	70-0426
VOIGHT, B	FINITE ELEMENT ANALYSES OF ELASTIC-PLASTIC PROBLEMS IN THE MECHANICS OF GEOLOGIC MEDIA: AN OVERVIEW	70-0203
VOIGHT, B.	ON THE APPLICATION OF FINITE-ELEMENT TECHNIQUES TO PROBLEMS CONCERNING POTENTIAL DISTRIBUTION AND ST	69-0268
VOIGT, S.	PROGRAM DESIGN BY A MULTIDISCIPLINARY TEAM	75-0854
VOIGT, S. J.	HYPERMATRIX SCHEME FOR FINITE-ELEMENT SYSTEMS ON CDC STAR-100 COMPUTER	75-0818
VOLCY, G.	DEFORMABILITY OF THE HULL STEEL-WORK AND DEFORMATIONS OF THE ENGINE-ROOM OF LARGE TANKERS	74-0532
VOLCY, G. C.	INVESTIGATION ON FREE AND FORCED VIBRATIONS OF A LONG TANKER WITH OVERLAPPING PROPELLER ARRANGEMENT	73-0981
VOLCY, G. C.	REDUCTION GEAR DAMAGES RELATED TO EXTERNAL INFLUENCES	75-0307
VOLKER, R. E.	NON-LINEAR FLOW IN POROUS MEDIA BY FINITE ELEMENTS	69-0023
VOLKER, R. E.	NUMERICAL DETERMINATION OF SEA WATER INTRUSION IN AQUIFERS	74-1181
VOLKER, R. E.	NUMERICAL SOLUTION OF THE STEADY INTERFACE IN A CONFINED COASTAL AQUIFER	74-1007
VOLTERRA, E.	TRANSVERSE VIBRATIONS OF CANTILEVER BARS OF VARIABLE CROSS SECTIONS	66-0001
VON FUCHS, G.	ASKA A COMPUTER SYSTEM FOR STRUCTURAL ENGINEERS	69-0345
VON FUCHS, G.	ASKA SERIES OF PROGRAMS	72-0157
VON FUCHS, G.	STRING EIN PROGRAMMSYSTEM FUR DIE MASCHINENUNABHANGIGE BEARBEITUNG VON ZEICHENKETTEN IN FORTRAN	68-0265
VON RIESEMANN,	A FINITE ELEMENT PROGRAM FOR THE DYNAMIC LARGE DEFLECTION, ELASTIC-PLASTIC ANALYSIS OF STIFFENED SHE	72-0908
VON RIESEMANN,	ANALYSIS OF AN IN-PILE REACTOR TUBE	75-0193
VON RIESEMANN,	COMPUTATION AND SOLUTION PROCEDURE FOR NONLINEAR ANALYSIS BY COMBINED FINITE ELEMENT-FINITE DIFFEREN	72-0571
VON RIESEMANN,	DYNAPLAS: A FINITE ELEMENT PROGRAM FOR THE DYNAMIC, LARGE DEFLECTION, ELASTIC-PLASTIC ANALYSIS OF ST	72-0773
VON RIESEMANN,	GEOMETRICALLY NONLINEAR ANALYSIS BY THE DIRECT STIFFNESS METHOD	70-0331
VON RIESEMANN,	GEOMETRICALLY NONLINEAR STRUCTURAL ANALYSIS BY THE DIRECT STIFFNESS METHOD	71-0560
VON RIESEMANN,	LARGE DEFLECTION ELASTIC-PLASTIC DYNAMIC RESPONSE OF STIFFENED SHELLS OF REVOLUTION	72-0608
VON RIESEMANN,	SAMMSOR 111: A FINITE ELEMENT PROGRAM TO DETERMINE STIFFNESS AND MASS MATRICES OF RING-STIFFENED SHE	72-0774
VON RIESEMANN,	STATIC GEOMETRIC AND MATERIAL NONLINEAR ANALYSIS	72-0314
VON RIESEMANN,	SURVEY OF STATIC GEOMETRIC AND MATERIAL NONLINEAR ANALYSIS BY THE FINITE ELEMENT METHOD	72-0783
VON RIESEMANN,	LARGE DEFLECTIONS OF ELASTIC BEAMS AND PLATES USING THE FINITE ELEMENT METHOD	68-0101
VON RUSMAN, W.	FORMULATION, COMPUTATION, AND SOLUTION PROCEDURES FOR MATERIAL AND/OR GEOMETRIC NONLINEAR STRUCTURAL	72-0346
VONDY, D. R.	SOLUTION OF DIFFUSION AND P1 FINITE-ELEMENT EQUATIONS BY ITERATION	75-0404
VOOREN, J. V. D	PROPOSALS FOR THE APPLICATION OF THE FINITE ELEMENT TECHNIQUE TO SOME AIRCRAFT AERODYNAMICAL PROBLEM	72-0628
VOROSHKO, P. P.	CHISLENNOE RESHENIE PLOSKIKH ZADACH TEPLOPROVODNOSTI DLYA OBLASTEI SLOZHNOI FORMY (NUMERICAL SOLUTIO	74-0659
VOROSHKO, P. P.	ISSLEDOVANIE NAPRYAZHENNO-DEFORMIROVANNOGO SOSTOYANIYA SIMMETRICHNYKH DISKOV SO STUPITSEI. (INVESTIG	74-0965
VOROSHKO, P. P.	K VOPROSU OB AVTOMATIZATSII ZADANIYA INFORMATSII V METODE KONECHNYKH ELEMENTOV. (PROBLEM OF AUTOMATI	74-0382
VOS, R. F.	FINITE ELEMENT ANALYSIS OF PLATE BUCKLING AND POST-BUCKLING	70-0332
VOS, R. G.	A FINITE ELEMENT TENSOR APPROACH TO PLATE BUCKLING AND POST BUCKLING	73-0883
VOS, R. G.	BOPACE 3-D (THE BOEING PLASTIC ANALYSIS CAPABILITY FOR 3-DIMENSIONAL SOLIDS USING ISOPARAMETRIC FINI	75-0699
VOS, R. G.	FINITE ELEMENT SOLUTION OF NONLINEAR STRUCTURES BY PERTURBATION TECHNIQUE	74-1273
VOS, R. G.	GENERALIZATION OF PLATE FINITE ELEMENTS TO SHELLS	72-0171
VOUILLON, C. C.	THE TRIDIMENSIONAL THERMOELASTIC COMPUTER CODE TITUS	71-0156
VUCKY, K. E.	A SEQUEL TO TECHNICAL NOTE 14 ON THE TUBA FAMILY OF PLATE ELEMENTS	68-0015
VUJANOVIC, B.	HEAT TRANSFER WITH NONLINEAR BOUNDARY CONDITIONS VIA A VARIATIONAL PRINCIPLE	71-0443
VYSLOUKH, V. A.	MODEL OF ELASTIC PLATE FROM FINITE ELEMENTS IN A SUPERSONIC FLOW	75-0038
VYSLOUKH, V. A.	REDUCTION OF THE DEGREES OF FREEDOM IN SOLVING DYNAMIC PROBLEMS BY THE FINITE ELEMENT METHOD	73-0647
VYSLOUKH, V. A.	STUDY ON THE VIBRATION OF THIN PLATES IN A GAS STREAM BY THE FINITE ELEMENT METHOD	74-0454
WAAS, G.	EARTH VIBRATION EFFECTS AND ABATEMENT FOR MILITARY FACILITIES. REPORT 3. ANALYSIS METHOD FOR FOOTING	72-0806
WAAS, G.	FINITE ELEMENT MODELING	73-0018
WACHOWSKI, A.	FINITE ELEMENT SOLUTION FOR WAVE PROPAGATION IN LAYERED MEDIA CAUSED BY A NUCLEAR DETONATION	67-0016
WACHSPRESS, E.	FORBIDDEN SHAPES IN THE FINITE ELEMENT METHOD	71-0705
WACHSPRESS, E.	ALGEBRAIC-GEOMETRY FOUNDATIONS FOR FINITE-ELEMENT COMPUTATION	73-0974
WADA, H.	STRUCTURAL ANALYSIS OF WELDED WALL FOR BOILER FURNACE	74-0653
WADA, S.	APPLICATION OF FINITE ELEMENT METHOD TO HYDRODYNAMIC LUBRICATION PROBLEMS (PART 1, "INFINITE-WIDTH B	71-0537
WADA, S.	APPLICATION OF FINITE ELEMENT METHOD TO HYDRODYNAMIC LUBRICATION PROBLEMS (PART 2, " FINITE WIDTH BE	71-0538
WADA, S.	FINITE ELEMENT METHOD IN HYDRODYNAMIC LUBRICATION	72-0301
WADDELL, G. G.	TECHNIQUE OF MEASURING INITIAL DEFORMATION AROUND AN OPENING	74-0331
WADDELL, J.	ASSESSMENT OF DESIGN TOOLS AND CRITERIA FOR URBAN RAIL TRACK STRUCTURES. VOLUME 11. AT-GRADE SLAB TR	74-0813
WADDOUPS, M. E.	COMPOSITE WING FOR TRANSONIC IMPROVEMENT. VOLUME 11. ADVANCED ANALYSIS EVALUATION	71-0657
WADE, B. G.	CRACK BRANCHING IN HOMALITE-100 SHEETS	72-0843
WADE, B. G.	CRACK BRANCHING IN HOMALITE-100 SHEETS	74-0291
WADE, B. G.	CRACK PROPAGATION AND ARREST IN IMPACTED PLATES	72-0599
WADE, B. G.	FRACTURE DYNAMICS OF HOMALITE-100 SHEETS	72-0809
WADE, B. G.	PHOTOELASTIC INVESTIGATION ON THE CRACK-ARREST CAPABILITY OF A HOLE	72-0086
WADE, B. G.	PHOTOELASTIC INVESTIGATION ON THE CRACK-ARREST CAPABILITY OF A PRETENSIONED STIFFENED PLATE	75-0088

WADE, L. V.   EFFECTS OF SPECIMEN RADIUS ON THE STRESS STATE NEAR A ROOF BOLT ANCHOR: A FINITE ELEMENT DETERMINATI   73-0229
WADE, L. V.   FINITE ELEMENT PREDICTION OF FIREPROOFING EFFECTIVENESS   72-0047
WADE, L. V.   USE OF THE FINITE ELEMENT TECHNIQUE TO DETERMINE THE FIRE RESISTANCE CAPABILITIES OF INSULATED STRUC   71-0147
WADLEIGH, K. H.   APPLICATION OF FINITE ELEMENT METHODS TO COMPLETE AUTOMOBILE STRUCTURAL DESIGN EVALUATION   74-0312
WAGEMANN, G.   BERECHNUNG VON SPANNUNGSKONZENTRATIONEN AN TRAGWERKEN UNTER THERMO-MECHANISCHER BELASTUNG   73-0686
WAGEMANNM G.   CALCULATION OF STRESS CONCENTRATIONS ON SUPPORT STRUCTURES UNDER THERMOMECHANICAL LOADING (BERECHNUN   73-0796
WAGNER, R.   THERMAL STRESSES IN OPEN CYLINDRICAL SHELLS OF ARBITRARY BOUNDARY CONDITIONS BY A FINITE STRIP METHO   72-0371
WAGNER, R. J.   A FINITE ELEMENT DISPLACEMENT APPROACH FOR THE ELASTO-PLASTIC ANALYSIS OF THIN CYLINDRICAL SHELLS   74-1077
WAGNER, R. J.   ELASTIC-PLASTIC FORMULATION FOR A CYLINDRICAL-SHELL FINITE-ELEMENT   75-0841
WAGSCHAL, C.   MULTIPOINT TAYLOR FORMULAS AND APPLICATIONS TO THE FINITE ELEMENT METHOD   71-0446
WAHLA, M. I.   DIRECT MEASUREMENT OF BOND SLIP IN REINFORCED CONCRETE   69-0428
WAHLBIN, L.   STABILITY IN LQ OF L2-PROJECTION INTO FINITE ELEMENT FUNCTION SPACES   75-0071
WAHLSTROM, S.   SOLID ELEMENT REQUIREMENT FOR NASTRAN   71-0256
WAIT, R.   A FINITE ELEMENT METHOD FOR THREE DIMENSIONAL FUNCTION APPROXIMATION   71-0654
WAIT, R.   CORNER SINGULARITIES IN ELLIPTIC PROBLEMS BY FINITE ELEMENT METHODS   71-0456
WAIT, R.   FINITE-ELEMENT-TYPE SOLUTION OF INTEGRAL EQUATIONS   73-0995
WAIT, R.   THE FINITE ELEMENT METHOD IN PARTIAL DIFFERENTIAL EQUATIONS   70-0456
WAKESUGI, T.   BANDED MATRIX TECHNIQUES AND SOLUTION OF LINEAR SIMULTANEOUS EQUATIONS OF FINITE ELEMENT ANALYSES   70-0302
WAKOFF, G. I.   USE OF SINGULAR FUNCTIONS WITH FINITE ELEMENT APPROXIMATIONS   73-0102
WALDNER, K.   STRESS ANALYSIS OF BWR PRESSURE VESSEL WITH BERSAFE AND FLHE   73-0406
WALKER, A. C.   A NONLINEAR FINITE ELEMENT ANALYSIS OF SHALLOW CIRCULAR ARCHES   69-0269
WALKER, A. C.   AN ANALYSIS OF THE LARGE DEFLECTIONS OF BEAMS USING THE RAYLEIGH-RITZ FINITE ELEMENT METHOD   68-0162
WALKER, A. C.   THE FINITE DIFFERENCE AND LOCALIZED RITZ METHODS   71-0447
WALKER, R. E.   DYNAMIC FINITE ELEMENT ANALYSIS OF AXI-SYMMETRIC STRESS WAVE PROPAGATION IN SOIL-FILLED BIN   72-0622
WALKER, R. E.   FUNDAMENTAL STUDIES OF MEDIUM-STRUCTURE INTERACTION. REPORT 1. FINITE ELEMENT ANALYSIS OF BURIED CYL   72-0854
WALKER, R. E.   VIBRATION CHARACTERISTICS OF THE NORTH FORK DAM MODEL   74-0845
WALKER, T. J.   QUANTITATIVE STRAIN-AND-STRESS STATE CRITERION FOR FAILURE IN THE VICINITY OF SHARP CRACKS   74-0639
WALKER, W. J.   STRUCTURE- MEDIUM INTERACTION PHENOMENOLOGY   72-0963
WALKER, W. J.   STRUCTURE-MEDIUM INTERACTION AND DESIGN PROCEDURES STUDY. VOLUME I. ANALYSIS METHOD, THEORY, VERIFIC   69-0166
WALL, S.   SAVINGS IN NASTRAN DECOMPOSITION TIME BY SEQUENCING TO REDUCE ACTIVE COLUMNS   71-0161
WALLACE, C. D.   MATRIX ANALYSIS OF AXISYMMETRIC SHELLS UNDER GENERAL LOADING   66-0075
WALLACE, D. B.   A COMPUTER BASED PROCEDURE FOR PREDICTING THE TRANSIENT RESPONSE AND FAILURE OF A TWO-DIMENSIONAL CO   71-0630
WALLACE, D. B.   COMPUTER SIMULATION OF DYNAMIC STRESS, DEFORMATION, AND FRACTURE OF GEAR TEETH   72-0114
WALLACE, D. B.   FINITE ELEMENT BASED PROCEDURE FOR SIMULATING THE TRANSIENT RESPONSE AND FAILURE OF A TWO-DIMENSIONA   72-0123
WALLACE, F. S.   RANDOM VIBRATION ANALYSIS SYSTEM FOR COMPLEX STRUCTURES, PART 2, COMPUTER PROGRAM DESCRIPTION   69-0262
WALLACE, J. A.   FOUNDATION TESTING FOR AUBURN DAM   70-0119
WALLACE, J. A.   STRUCTURAL FINITE ELEMENT ANALYSIS AIDED BY COMPUTER GRAPHICS   74-0304
WALLACE, P. W.   A FINITE-ELEMENT, PLANAR-FLOW MODEL OF CAMAS PRAIRIE, IDAHO   72-0980
WALLERSTEIN, D.   GENERAL LINEAR GEOMETRIC MATRIX FOR A FULLY COMPATIBLE FINITE ELEMENT   72-0294
WALLERSTEIN, D.   THERMAL DEFORMATION VECTOR FOR A BILINEAR TEMPERATURE DISTRIBUTION IN AN ANISOTROPIC QUADRILATERAL M   75-0670
WALLEY, W. J.   CALCULATION OF AREAL RAINFALL USING FINITE ELEMENT TECHNIQUES WITH ALTITUDINAL CORRECTIONS   72-0344
WALLNER, M.   THEORIE UND EXPERIMENT ZU EINEM NEUEN INJEKTIONSVERFAHREN. (THEORY AND EXPERIMENT OF A NEW GROUTING   74-0991
WALSER, A.   POST-TENSIONED CONCRETE NUCLEAR STRUCTURES   70-0040
WALSH, J. E.   FINITE DIFFERENCE AND FINITE ELEMENT METHODS OF APPROXIMATION   71-0457
WALSH, J. E.   RESPONSE OF UNCONFINED AQUIFERS TO PUMPING   74-0073
WALSH, P. F.   COMPUTATION OF STRESS INTENSITY FACTORS BY A SPECIAL FINITE ELEMENT TECHNIQUE   71-0287
WALSH, P. F.   CRACKS, NOTCHES AND FINITE ELEMENTS   75-0253
WALSH, P. F.   NUMERICAL ANALYSIS IN ORTHOTROPIC LINEAR FRACTURE MECHANICS   73-0512
WALTERS, D.   A NEW FAMILY OF CURVILINEAR PLATE BENDING ELEMENTS FOR VIBRATION AND STABILITY   72-0645
WALTERS, D.   A NEW FAMILY OF CURVILINEAR PLATE BENDING ELEMENTS FOR VIBRATION AND STABILITY   72-0999
WALTERS, D.   POSSIBLE LOSS OF ACCURACY IN CURVED FINITE ELEMENTS   72-0278
WALTERS, W. F.   ENERGY-DEPENDENT FINITE-ELEMENT METHOD FOR FEW-GROUP DIFFUSION EQUATIONS   75-0499
WALTON, D.   FINITE-ELEMENT METHOD APPLIED TO PREDICTING FATIGUE-CRACK GROWTH   73-0513
WALTON, W. C.   ACCURACY OF MODAL STRESS CALCULATIONS BY FINITE ELEMENT METHOD   70-0223
WALZ, J. E.   ACCURACY AND CONVERGENCE OF FINITE ELEMENT APPROXIMATION   68-0228
WALZ, J. E.   ACCURACY OF FINITE ELEMENT APPROXIMATIONS TO STRUCTURAL PROBLEMS   70-0120
WALZ, J. E.   ACCURACY OF FINITE ELEMENT APPROXIMATIONS TO STRUCTURAL PROBLEMS   70-0438
WANCHOO, M. K.   CRACKING ANALYSIS OF REINFORCED CONCRETE PLATES   75-0085
WANCHOO, M. K.   POST-ELASTIC FINITE-ELEMENT ANALYSIS OF REINFORCED PLATES   72-0979
WANG, S. N.   GROUNDWATER FLOWN IN A SANDY TIDAL BEACH .1. ONE-DIMENSIONAL FINITE ELEMENT ANALYSIS   71-0567
WANG, C. F.   VARIATIONAL PRINCIPLE OF STEADY-STATE TRANSPORT PROCESSES   70-0190
WANG, C. Y-J.   FRACTURE MECHANICS FOR AN INTERFACIAL CRACK BETWEEN ADHESIVELY BONDED DISSIMILAR MATERIALS   72-0869
WANG, E. D.   COMPUTER PROGRAM FOR GENERATING FINITE ELEMENT MODELS OF MINE STRUCTURES   71-0493
WANG, F. D.   COMPUTER PROGRAM FOR PIT SLOPE STABILITY ANALYSIS BY THE FINITE ELEMENT STRESS ANALYSIS AND LIMITING   72-0225
WANG, F. D.   SLOPE STABILITY ANALYSIS BY THE FINITE ELEMENT STRESS ANALYSIS AND LIMITING EQUILIBRIUM METHOD   70-0222
WANG, F. D.   STRUCTURAL ANALYSIS OF A COAL MINE OPENING IN ELASTIC, MULTILAYERED MATERIAL   74-0235
WANG, F. D.   STRUCTURAL ANALYSIS OF A COAL MINE OPENING IN ELASTIC, MULTILAYERED MATERIAL   74-0840
WANG, G.   ON BUCKLING OF UNIDIRECTIONAL BORON-ALUMINUM STIFFENERS - A CAUTION TO DESIGNERS   75-0634
WANG, H-P   MODELING AN OCEAN POND. A TWO-DIMENSIONAL, FINITE ELEMENT HYDRODYNAMIC MODEL OF NINIGRET POND, CHARL   75-0686
WANG, J.   FINITE ELEMENT MODELLING OF HYDRODYNAMIC CIRCULATION   74-1370
WANG, J. D.   FINITE ELEMENT MODEL OF TWO LAYER COASTAL CIRCULATION   74-1192
WANG, J. D.   MATHEMATICAL MODELING OF NEAR COASTAL CIRCULATION   75-0720
WANG, J. D.   MATHEMATICAL MODELS OF THE MASSACHUSETTS BAY. PART 1. FINITE ELEMENT MODELING OF TWO-DIMENSIONAL HYD   73-0800
WANG, J. T. S.   SHELL STRUCTURES AND CLIMATIC INFLUENCES   72-0389
WANG, J-K.   BIT PENETRATION INTO ROCK - A FINITE ELEMENT STUDY   75-0800
WANG, M. C.   GROUNDWATER FLOW IN PARTIALLY SATURATED SOILS   72-0745
WANG, M. C.   STRESS - DEFORMATION PREDICTION IN CEMENT - TREATED SOIL PAVEMENTS   71-0286
WANG, M. S.   A STUDY OF CONVECTIVE-DISPERSION EQUATION BY ISOPARAMETRIC FINITE ELEMENTS. TRANSPORT OF POLLUTANTS   75-0773
WANG, M-H.   STUDY OF CONVECTIVE-DISPERSION EQUATION BY ISOPARAMETRIC FINITE ELEMENTS   75-0074
WANG, N.   FINITE ELEMENT ANALYSIS OF CUT-GROWTH IN SHEETS OF HIGHLY ELASTIC MATERIALS   73-0514
WANG, N.   THE COMPUTATION OF TEARING ENERGY OF NICKED RUBBER STRIPS IN EXTENSION   74-0315
WANG, P.   MINIMUM WEIGHT DESIGN OF FINITE ELEMENT STRUCTURES   74-0148
WANG, R. C-J.   REFINED FINITE ELEMENT STABILITY ANALYSIS OF THIN-WALLED MEMBERS   75-0784
WANG, S. K.   AN ADVANCED ANALYSIS OF RIGID PAVEMENTS   72-0417
WANG, S. K.   EFFECT OF OPENINGS ON STRESSES IN RIGID PAVEMENTS   73-0344
WANG, S. N.   GROUNDWATER FLOW IN A SANDY TIDAL BEACH: II. 2-DIMENSIONAL FINITE ELEMENT ANALYSIS   72-0300
WANG, S. N.   HABEAS- A STRUCTURAL DYNAMICS ANALYSIS SYSTEM   70-0121
WANG, S. N.   INVESTIGATION OF THE WATER TABLE IN A TIDAL BEACH   71-0648
WANG, S. S.   FRACTURE OF GRAPHITE FIBER REINFORCED COMPOSITES   74-0695
WANG, S. T.   FINITE-ELEMENT ANALYSIS OF CONCRETE SLABS AND ITS IMPLICATIONS FOR RIGID PAVEMENT DESIGN   73-0538
WANG, Y. J.   FINITE ELEMENT OF UNDERGROUND STRESSES UTILIZING STOCHASTICALLY SIMULATED MATERIAL PROPERTIES   70-0111
WANG, Y. S.   BUCKLING OF CYLINDRICAL SHELLS BY WIND PRESSURE   74-0642
WARBURTON, G. B   A NEW FAMILY OF CURVILINEAR PLATE BENDING ELEMENTS FOR VIBRATION AND STABILITY   72-0645
WARBURTON, G. B   A NEW FAMILY OF CURVILINEAR PLATE BENDING ELEMENTS FOR VIBRATION AND STABILITY   72-0999
WARBURTON, G. B   A NEW HYBRID CYLINDRICAL SHELL FINITE ELEMENT   71-0118
WARBURTON, G. B   A TIMOSHENKO BEAM ELEMENT   72-0590
WARBURTON, G. B   COMPARISON OF RECENT SHALLOW SHELL FINITE ELEMENT ANALYSES   71-0362
WARBURTON, G. B   CONSTANT CURVATURE BEAM FINITE ELEMENTS FOR IN-PLANE VIBRATION   72-0254
WARBURTON, G. B   CURVED BEAM FINITE ELEMENTS FOR COUPLED BENDING AND TORSIONAL VIBRATION   72-0720
WARBURTON, G. B   FINITE ELEMENT TECHNIQUES APPLIED TO THE VIBRATIONS OF PLATE SYSTEMS   66-0033
WARBURTON, G. B   FREE VIBRATION OF RING STIFFENED CYCLINDRICAL SHELLS   70-0172
WARBURTON, G. B   FREE VIBRATION OF THIN CYLINDRICAL SHELLS WITH A DISCONTINUITY IN THE THICKNESS   69-0439

WARBURTON, G. B	POSSIBLE LOSS OF ACCURACY IN CURVED FINITE ELEMENTS	72-0278
WARBURTON, G. B	TRANSMISSION OF VIBRATION IN BEAM SYSTEMS	69-0405
WARD, C.	MODES OF MOTION OF A FINITE-ELEMENT BRAIN MODEL	75-0592
WARD, C. C.	A DYNAMIC FINITE ELEMENT MODEL OF THE HUMAN BRAIN	74-1076
WARD, G.	PROPELLER-EXCITED VIBRATION WITH PARTICULAR REFERENCE TO FULL-SCALE MEASUREMENTS	75-0018
WARDLE, L. J.	FINITE ELEMENT ANALYSIS OF A PLATE ON A LAYERED CROSS-ANISOTROPIC FOUNDATION	74-0451
WARNER, D. M.	FINITE ELEMENT SOLUTION OF LOW BOND NUMBER SLOSHING	74-0728
WARNKE, E.	FINITE ELEMENTS FOR CALCULATION OF PRESTRESSED CONCRETE REACTOR PRESSURE VESSELS (FINITE ELEMENTE ZU	73-0797
WARNKE, E. P.	FINITE ELEMENT ULTIMATE LOAD ANALYSIS OF THREE DIMENSIONAL CONCRETE STRUCTURES	74-0672
WARNKE, E. P.	FINITE ELEMENTE ZUR BERECHNUNG VON SPANNBETONREAKTORDRUCKBEHALTERN	73-0702
WARNKE, P.	ROTATIONSSYMMETRISCHE BERECHNUNG DES 1:5 THTR SPANNBETON-BEHAL-BEHAL TERMODELLS	71-0629
WARREN, G. E.	PHOTOELASTIC TESTS ON COMPLEX INTERSECTIONS OF REINFORCED THIN-SHELL CYLINDERS	75-0363
WARREN, J. W. L	CREEP OF FABRICATED STRUCTURES	68-0001
WARZEE, G.	A FINITE ELEMENT METHOD FOR FLOW CALCULATIONS IN TURBOMACHINES	74-0726
WARZEE, G.	APPLICATION OF LAPLACE TRANSFORM AND FINITE ELEMENT METHOD TO SOLUTION OF NONSTATIONARY THERMAL COND	74-0171
WARZEE, G.	FINITE ELEMENT ANALYSIS OF TRANSIENT HEAT CONDUCTION. APPLICATION OF THE WEIGHTED RESIDUAL PROCESS	74-0277
WARZEE, G.	FINITE ELEMENT METHOD AND LAPLACE TRANSFORM COMPARATIVE SOLUTIONS OF TRANSIENT HEAT CONDUCTION PROBL	74-0575
WARZEE, G.	FINITE-ELEMENT METHOD FOR AXISYMMETRIC FLOW COMPUTATION IN A TURBOMACHINE	75-0359
WARZEE, G.	THE MERIDIONAL THROUGH-FLOW CALCULATION IN AN AXIAL FLOW MACHINE BY THE FINITE ELEMENT METHOD	75-0115
WASHIZU, K.	FINITE ELEMENT METHOD APPLIED TO ANALYSIS OF FLOW OVER A SPILLWAY CREST	73-0300
WASHIZU, K.	FINITE ELEMENT TECHNIQUE IN LIFTING SURFACE PROBLEMS	74-0022
WASHIZU, K.	LIFTING SURFACE PROBLEMS ANALYSIS	73-0333
WASHIZU, K.	ON THE ROLE OF VARIATIONAL PRINCIPLES AND METHOD OF WEIGHTED RESIDUALS IN THE FINITE ELEMENT METHODS	74-0551
WASHIZU, K.	SOME APPLICATIONS OF THE FINITE ELEMENT METHOD TO FLUID MECHANICS	74-1274
WASHIZU, K.	SOME CONSIDERATIONS ON BASIC THEORY FOR THE FINITE ELEMENT METHOD	72-0257
WASHIZU, K.	SOME REMARKS ON BASIC THEORY FOR FINITE ELEMENT METHOD	71-0257
WASSON, N. F.	METHODS FOR PREDICTING THE AERODYNAMIC AND STABILITY AND CONTROL CHARACTERISTICS OF STOL AIRCRAFT. V	73-0718
WASTI, S. T.	STRUCTURAL BEHAVIOR OF A TWO SPAN REINFORCED CONCRETE BOX GIRDER BRIDGE MODEL. VOLUME 11. REDUCTION,	71-0659
WATANABE, H.	ANALYSIS OF NONELASTIC AND NONLINEAR VIBRATION OF ROCK DAMS USING FINITE ELEMENT METHOD	73-0680
WATANABE, M.	FINITE ELEMENT MODEL FOR TRANSIENT TWO-LAYER COOLING POND BEHAVIOR	75-0465
WATKINS, D.	ANALYSIS	73-0627
WATKINS, D. J.	SEEPAGE AND GROUNDWATER EFFECTS ASSOCIATED WITH EXPLOSIVE CRATERING	72-0656
WATKINS, D. S.	BLENDING FUNCTIONS AND FINITE ELEMENTS	74-0889
WATKINS, D. S.	SOME CONFORMING, RECTANGULAR, PLATE ELEMENTS	75-0094
WATKINSON, K. W	THE MIDCOHV WEIGHT AND BALANCE COMPUTER PROGRAM (WTBAL)	74-0697
WATSON, J. O.	ANALYSIS OF CONTINUA BY THE FINITE ELEMENT AND INTEGRAL EQUATION METHOD	75-0370
WATSON, J. O.	SECOND GENERATION BOUNDARY INTEGRAL EQUATION PROGRAM FOR THREE-DIMENSIONAL ELASTIC ANALYSIS	75-0376
WATSON, M.	A NUMERICAL METHOD OF VISCO-ELASTIC STRESS ANALYSIS	68-0175
WATSON, M.	ORTHOTROPIC FLAT PLATES SUPPORTED ON COLUMNS ELONGATED IN PLANE	72-0492
WATSON, M.	SOME CREEP EFFECTS IN STRESS ANALYSIS WITH PARTICULAR REFERENCE TO CONCRETE PRESSURE VESSELS	69-0284
WATSON, M.	STEADY STATE AND TRANSIENT GROUNDWATER FLOW IN SLOPES	70-0273
WATSON, M.	STRESS ANALYSIS BY THE FINITE ELEMENT METHOD - THERMAL EFFECTS	67-0022
WATWOOD, V. B.	AN EQUILIBRIUM STRESS FIELD MODEL FOR FINITE ELEMENT SOLUTIONS OF TWO-DIMENSIONAL ELASTIC PROBLEMS	68-0163
WATWOOD, V. B.	EQUILIBRIUM STRESS FIELD MODELS IN FINITE ELEMENTS	67-0011
WATWOOD, V. B.	FINITE ELEMENT METHOD FOR PREDICTION OF CRACK BEHAVIOUR	70-0342
WATWOOD, V. B.	RELATIONSHIPS BETWEEN MATRIX METHODS IN FINITE ELEMENTS	67-0086
WATWOOD, V. B.	THERMAL AND STRESS ANALYSIS PROGRAM - HYBOS	72-0509
WATWOOD, V. B.	A STUDY OF THE EQUILIBRIUM MODEL FOR USE IN FINITE ELEMENT ANALYSIS	66-0132
WEAVER, D. S.	A FINITE ELEMENT SOLUTION FOR A CLASS OF TWO DIMENSIONAL VISCOUS FLUID FLOW DYNAMICS PROBLEMS	71-0544
WEAVER, J. R. W	RESPONSE OF BEAMS AND PLATES TO MOVING LOADS	71-0277
WEAVER, W.	NONLINEAR DYNAMIC ANALYSIS OF FRAMES WITH FILLER PANELS	74-0226
WEBBER, J. P. H	EXPERIMENTAL AND THEORETICAL BENDING DEFLECTIONS OF LAMINATED PLATES	72-0144
WEBBER, J. P. H	STRESS ANALYSIS IN VISCOELASTIC BODIES USING FINITE ELEMENTS AND A CORRESPONDENCE RULE WITH ELASTICI	69-0270
WEBBER, J. P. H	THERMO-ELASTIC ANALYSIS OF RECTANGULAR PLATES IN PLANE STRESS BY THE FINITE-ELEMENT DISPLACEMENT MET	67-0087
WEBER, J.	ELASTO-PLASTIC ANALYSIS OF METAL SHEETS WITH A CENTRAL CRACK	67-0137
WEBMAN, K. M.	VIBRATIONAL RESPONSE OF SONAR TRANSDUCERS USING PIEZO-ELECTRIC FINITE ELEMENTS	74-0522
WEBSTER, J. A.	MATRIX ANALYSIS OF BEAM COLUMNS BY THE METHOD OF FINITE ELEMENTS	64-0012
WEBSTER, J. J.	ACCURACY OF FINITE-ELEMENT SOLUTIONS FOR MODAL CHARACTERISTICS OF SHELLS OF REVOLUTION	70-0231
WEBSTER, J. J.	FREE VIBRATION OF SHELLS OF REVOLUTION USING RING ELEMENTS	67-0088
WEBSTER, R. L.	NONLINEAR STATIC AND DYNAMIC RESPONSE OF UNDERWATER CABLE STRUCTURES USING THE FINITE ELEMENT METHOD	75-0460
WECK, M.	ANWENDUNG DER METHODE FINITER ELEMENTE BEI DER ANALYSE DES DYNAMISCHEN VERHALTENS GEDAEMPFTER WERKZE	75-0580
WEEKS, G. B.	PART 1 - TASK 3.2.9. BENDING AND BUCKLING OF A ONE-DIMENSIONAL MODEL SIMULATING A CONICAL SHELL. PAR	71-0745
WEEKS, G. E.	AUTOMATED EVALUATION OF INTERCEPTOR ROCKET MOTOR DESIGNS UNDER COMBINED OPERATIONAL AND NUCLEAR EFFE	74-0698
WEEKS, G. E.	BASIC RESEARCH IN METHODS OF APPROXIMATION IN THE NONLINEAR MECHANICS OF SOLIDS AND STRUCTURES. VOLU	74-1091
WEEKS, G. E.	FINITE ELEMENT MODEL FOR SHELLS BASED ON THE DISCRETE KIRCHHOFF HYPOTHESIS.	72-0388
WEEKS, G. R.	THEMIS FINAL REPORT. VOL. 1. BASIC RESEARCH IN METHODS OF APPROXIMATION IN THE NONLINEAR MECHANICS O	74-0723
WEGMULLER, A.	FINITE ELEMENT ANALYSIS OF PLATES AND ECCENTRICALLY STIFFENED PLATES	73-0691
WEGMULLER, A. W	ELASTIC PLASTIC FINITE ELEMENT ANALYSIS OF PLATES	74-0093
WEGMULLER, A. W	ELASTIC-PLASTIC FINITE ELEMENT ANALYSIS OF PLATES - DISCUSSION	75-0139
WEGMULLER, A. W	ELASTIC-PLASTIC FINITE ELEMENT ANALYSIS OF PLATES	75-0599
WEGMULLER, A. W	FINITE ELEMENT ANALYSIS OF ELASTIC-PLASTIC PLATES AND ECCENTRICALLY STIFFENED PLATES	71-0585
WEGMULLER, A. W	FULL RANGE ANALYSIS OF ECCENTRICALLY STIFFENED PLATES	74-0149
WEGMULLER, A. W	INELASTIC RESPONSE OF PLATES AND ECCENTRICALLY STIFFENED PLATES	74-0458
WEGMULLER, A. W	REFINED PLATE BENDING FINITE ELEMENT	74-1045
WEGNER, R.	DER GERISSENE ZUSTAND ZWEISEITIG GELAGERTER PLATTEN UNTER EINZELLASTEN - NICHTLINEARE BERECHNUNG MIT	75-0858
WEHLE, L. B.	A METHOD FOR REDUCING THE ANALYSIS OF COMPLEX REDUNDANT STRUCTURES TO A ROUTINE PROCEDURE	52-0002
WEI, B. C. F.	BALANCED STRESSES IN POST-YIELDED MULTI-MATERIAL STRUCTURAL JOINTS	69-0271
WEIHS, D.	GENERAL VARIATIONAL ANALYSIS OF HYDRODYNAMIC, THERMAL AND DIFFUSIONAL BOUNDARY LAYERS	70-0191
WEIKEL, R. C.	APPLICATION OF THE STIFFNESS METHOD TO THE ANALYSIS OF SHELL STRUCTURES	61-0004
WEIKEL, R. C.	FURTHER DEVELOPMENT AND APPLICATIONS OF THE STIFFNESS METHOD	64-0020
WEILER, F. C.	ANALYSIS OF THERMAL STRESSES IN CORED CERAMIC BRICKS USED IN HYPERSONIC WIND TUNNEL HEATERS	72-0655
WEILL, L. M.	FRONT DISPLACEMENT MODEL IN A FRACTURED RESERVOIR	74-0071
WEINBERG, H. F.	VARIATIONAL PROPERTIES OF STEADY FALL IN STOKES FLOW	72-0285
WEINGARTEN, V.	BUCKLING OF SEGMENTS OF TOROIDAL SHELLS	73-0515
WEINGARTEN, V.	DYNAMIC STABILITY OF A NONLINEAR BEAM SUBJECTED TO BOTH LONGITUDINAL AND TRANSVERSE EXCITATION	73-0553
WEINGARTEN, V.	EFFECT OF LATERAL PRESSURE ON NATURAL FREQUENCIES OF HYPERBOLOIDAL SHELLS OF REVOLUTION	71-0278
WEINGARTEN, V.	FREE VIBRATIONS OF PRESSURE LOADED PARABOLOIDAL SHELLS	71-0601
WEINGARTEN, V.	INEXTENSIONAL VIBRATIONS OF PARABOLOIDAL SHELLS OF REVOLUTION. DOMES AND SHELLS	71-0427
WEINGARTEN, V.	NON-LINEAR RESPONSE OF A BEAM TO PERIODIC LOADING.	72-0407
WEINGARTEN, V.	RESPONSE OF NONLINEAR BEAM TO RANDOM EXCITATION	73-0285
WEINGARTEN, V.	STABILITY OF HYPERBOLOIDAL SHELLS	75-0488
WEINGARTEN, V.	STABILITY OF PRESSURIZED HYPERBOLOIDAL SHELLS	75-0585
WEINGARTEN, V.	VIBRATIONS OF PRESSURE LOADED HYPERBOLOIDAL SHELLS	72-0874
WEINGARTEN, V.	VIBRATIONS OF SEGMENTED SHELLS	73-0161
WEINSTEIN, A. M	FINITE ELEMENT STRESS ANALYSIS OF PROSTHETIC TOOTH IMPLANTS	75-0073
WEINSTEIN, A. S	FINITE ELEMENT ANALYSIS OF THE PLANE STRAIN DRAWING PROBLEM	74-1154
WEISS, R. O.	CONPLOT: A GENERAL PURPOSE CONTOUR PLOTTING PROGRAM	74-0887
WEISSGERBER, V.	HYBRIDE SCHALENELEMENTE MIT EINER ANWENDUNG AUF WENDELSCHALEN. (HYBRID SHELL ELEMENTS IN APPLICATION	75-0609
WEISSHAAR, T. A	AN APPLICATION OF FINITE ELEMENT METHODS TO PANEL FLUTTER OPTIMIZATION	71-0258
WEISSHAAR, T. A	PANEL FLUTTER OPTIMIZATION - REFINED FINITE-ELEMENT APPROACH	75-0358
WELCH, P. W.	HYDROELASTIC ANALYSIS OF AXISYMMETRIC SYSTEMS BY A FINITE ELEMENT METHOD	68-0084
WELCH, R. E.	LARGE DISPLACEMENT, NONLINEAR TRANSIENT ANALYSIS BY FINITE ELEMENTS	74-0324

110

WELCH, R. E.	NONLINEAR TRANSIENT ANALYSIS OF AUTOMOTIVE STRUCTURES BY FINITE ELEMENTS	75-0301
WELCH, R. E.	THREE DIMENSIONAL STRUCTURES/MEDIA INTERACTION, DYNAX DEVELOPMENT AND APPLICATIONS.	70-0378
WELLFORD, L. C.	ACCURACY AND CONVERGENCE OF FINITE ELEMENT GALERKIN APPROXIMATIONS OF TIME DEPENDENT PROBLEMS WITH E	73-0987
WELLFORD, L. C.	ACCURACY AND CONVERGENCE OF FINITE ELEMENT/GALERKIN APPROXIMATIONS OF TIME DEPENDENT PROBLEMS WITH E	75-0006
WELLFORD, L. C.	ANALYSIS OF FLOW OF VISCOUS FLUIDS BY FINITE-ELEMENT METHOD	72-0032
WELLFORD, L. C.	DISCONTINUOUS FINITE-ELEMENT APPROXIMATIONS FOR ANALYSIS OF SHOCK-WAVES IN NONLINEARLY ELASTIC-MATER	75-0451
WELLFORD, L. C.	MATHEMATICAL FEATURES OF FINITE ELEMENT APPROXIMATIONS OF CERTAIN FLOW PROBLEMS WITH EMPHASIS ON CON	74-0002
WELLFORD, L. C.	THEMIS FINAL REPORT, VOL. IV. RECENT DEVELOPMENTS IN THE NUMERICAL SOLUTION OF CERTAIN PROBLEMS IN N	74-0721
WELLFORD, L. C.	NOTE ON THE ACCURACY AND CONVERGENCE OF FINITE ELEMENT APPROXIMATIONS OF THE CONVECTION EQUATION	75-0490
WELLS, A.	ICES-STRUDL-11	68-0257
WELLS, A. A.	FRACTURE CONTROL: PAST, PRESENT AND FUTURE	73-0516
WEMPER, G.	DISCRETE APPROXIMATIONS OF ELASTIC-PLASTIC BODIES BY VARIATIONAL METHODS	72-0710
WEMPNER, G.	FINITE ELEMENT ANALYSIS OF THIN SHELLS	70-0457
WEMPNER, G. A.	DISCRETE APPROXIMATIONS RELATED TO NONLINEAR THEORIES OF SOLIDS	71-0259
WEMPNER, G. A.	FINITE DEFLECTIONS, BUCKLING AND POSTBUCKLING OF AN ARCH	69-0123
WEMPNER, G. A.	FINITE ELEMENTS, FINITE ROTATIONS AND SMALL STRAINS OF FLEXIBLE SHELLS	69-0516
WEMPNER, G. A.	FINITE-DIFFERENCES VIA FINITE-ELEMENTS	71-0665
WEMPNER, G. A.	FINITE-ELEMENT ANALYSIS OF THIN SHELLS	68-0164
WEMPNER, G. A.	NEW CONCEPTS FOR FINITE ELEMENTS OF SHELLS	68-0195
WEN, K. S.	FINITE ELEMENT METHOD AS AN ASPECT OF THE PRINCIPLE OF MAXIMUM UNIFORMITY: NEW HYDRODYNAMICAL RAMIFA	74-0009
WENDROFF, B.	COMPARATIVE EFFICIENCY OF CERTAIN FINITE ELEMENT AND FINITE DIFFERENCE METHODS FOR A HYPERBOLIC PROB	73-0965
WENDROFF, B.	RAYLEIGH-RITZ APPROXIMATION OF PIECEWISE CUBIC POLYNOMIALS	66-0087
WENDROFF, B.	RELATIVE EFFICIENCY OF FINITE DIFFERENCE AND FINITE ELEMENT METHODS - 1. HYPERBOLIC PROBLEMS AND SPL	74-0986
WENDROFF, B.	THE COMPARATIVE EFFICIENCY OF CERTAIN FINITE ELEMENT AND FINITE DIFFERENCE SCHEMES FOR A HYPERBOLIC	74-0910
WENDROFF, B.	THE RELATION BETWEEN THE GALERKIN AND COLLOCATION METHODS USING SMOOTH SPLINES	74-0912
WENDROFF, B.	THE RELATIVE EFFICIENCY OF FINITE DIFFERENCE AND FINITE ELEMENT METHODS. 1. HYPERBOLIC PROBLEMS AND	74-0911
WENGER, N. C.	VARIATIONAL PRINCIPLE FOR MAGNETOHYDRODYNAMIC CHANNEL FLOW	70-0192
WENNERSTROM, H.	FINITE ELEMENT ANALYSIS OF ELASTO-PLASTIC SHELLS	74-0249
WERNER, S. D.	ANALYSIS OF DYNAMIC RESPONSE OF SAFEGUARD STRUCTURES	70-0462
WEST, A. J.	LOAD POINT COMPLIANCE OF THE CHARPY IMPACT SPECIMEN	73-0775
WEST, A. S.	DERIVATION OF COMPATIBLE STIFFNESS MATRICES FOR LARGE DISCRETE ELEMENTS	67-0080
WEST, J. B.	FINITE ELEMENT DISPLACEMENT ANALYSIS OF A LUNG	72-0026
WEST, L. J.	ROCK MECHANICS STUDIES FOR A HIGH CUT AT A NUCLEAR POWER STATION IN PENNSYLVANIA	70-0122
WESTBROOK, D. R	A VARIATIONAL PRINCIPLE WITH APPLICATIONS IN FINITE ELEMENTS	74-1285
WESTBROOK, D. R	THREE DIMENSIONAL FINITE ELEMENT METHOD FOR PLATE BENDING	74-0243
WESTBROOK, D. R	VARIATIONAL AND WEIGHTED RESIDUAL METHODS	73-0630
WESTBY, O.	APPLICATION OF THE FINITE ELEMENT METHOD ON RESIDUAL STRESSES AND DEFORMATIONS IN WELDS	68-0178
WESTMANN, B. A.	STRESS ANALYSIS BY FINITE ELEMENTS	68-0165
WESTMANN, R. A.	EMBANKMENT ANALYSIS AND FIELD CORRELATION	68-0057
WESTPHAL, J. L.	ADDITION OF AN ARBITRARY BODY ANALYSIS CAPABILITY TO THE BOEING TEA 236 FINITE ELEMENT COMPUTER PROG	73-0785
WETENKAMP, H. R	THERMAL STRESS DEVELOPED IN S PLATE, STRAIGHT PLATE, AND DEEP DISH WHEELS	73-0172
WEVERS, L. J.	TRANSVERSE VIBRATIONS OF SHIP'S PROPULSION SYSTEMS. PART 2: EXPERIMENTAL ANALYSIS	74-1324
WEXLER, A.	FINITE ELEMENT TECHNIQUES FOR SOLUTION OF POISSON'S EQUATION	71-0494
WEXLER, A.	FINITE-ELEMENT FIELD ANALYSIS OF AN INHOMOGENEOUS ANISOTROPIC RELUCTANCE MACHINE ROTOR	73-0020
WEXLER, A.	FINITE-ELEMENT SOLUTION OF UNBOUNDED FIELD PROBLEMS	72-0025
WEXLER, A.	FINITE-ELEMENT SOLUTIONS WITHIN CURVED BOUNDARIES	72-0038
WEXLER, A.	INTEGRAL FINITE-ELEMENT APPROACH FOR SOLVING LAPLACE EQUATION	73-0121
WHANG, B.	A FINITE ELEMENT ANALYSIS OF LAMINATED ORTHOTROPIC PLATES AND SHALLOW SHELLS	70-0123
WHANG, B.	ELASTO-PLASTIC ANALYSIS OF ORTHOTROPIC PLATES AND SHELLS	68-0166
WHANG, B.	ELASTO-PLASTIC ORTHOTROPIC PLATES AND SHELLS	69-0467
WHANG, B.	LAMINATED ORTHOTROPIC PLATES AND SHALLOW SHELLS	71-0400
WHANG, B.	THERMAL STRESS ANALYSIS BY THE FINITE ELEMENT METHOD	70-0124
WHEELER, M. F.	SOME SUPERCONVERGENCE RESULTS FOR AN H-GALERKIN PROCEDURE FOR THE HEAT EQUATION	74-0411
WHETSONE, W. D.	SUBSTRUCTURING TECHNIQUES	71-0692
WHETSTONE, W. D	COMPUTER ANALYSIS OF LARGE LINEAR FRAMES	69-0408
WHETSTONE, W. D	VIBRATIONAL CHARACTERISTICS OF SPACE FRAMES	69-0434
WHITE, D. J.	FINITE ELEMENT ANALYSIS OF STRESSES IN SHAFTS DUE TO INTERFERENCE-FIT HUBS	69-0353
WHITE, D. J.	FINITE-ELEMENT METHOD AND ITS APPLICATIONS IN AN ENGINEERING LABORATORY	70-0202
WHITE, D. J.	FINITE-ELEMENT STRESS ANALYSIS OF A MULTIPIECE PISTON	69-0274
WHITE, D. J.	FINITE-ELEMENT STRESS ANALYSIS OF A NON-LINEAR PROBLEM. A CONNECTING-ROD EYE LOADED BY MEANS OF A PI	70-0239
WHITE, J. A.	OH - 58A PROPULSION SYSTEM VIBRATION INVESTIGATION	74-0733
WHITE, J. K.	A FINITE ELEMENT DETERMINISTIC CATCHMENT MODEL	74-1361
WHITE, J. K.	FINITE-ELEMENT APPROACH TO WATERSHED RUNOFF- DISCUSSION	75-0402
WHITE, J. L.	FINITE ELEMENTS IN LINEAR VISCOELASTICITY	68-0281
WHITE, R. N.	SHEAR BUCKLING OF SQUARE PERFORATED PLATES	74-0713
WHITEMAN, J.	THE MATHEMATICS OF FINITE ELEMENTS AND APPLICATIONS	73-0288
WHITEMAN, J. R.	A BIBLIOGRAPHY FOR FINITE ELEMENT METHODS	72-0339
WHITEMAN, J. R.	AN INTRODUCTION TO VARIATIONAL AND FINITE ELEMENT METHODS	72-0994
WHITEMAN, J. R.	COMPUTABLE ERROR BOUNDS FOR THE FINITE ELEMENT METHOD FOR ELLIPTIC BOUNDARY VALUE PROBLEMS	74-1219
WHITEMAN, J. R.	CONFORMING FINITE ELEMENT METHODS FOR THE CLAMPED PLATE PROBLEM	75-0174
WHITEMAN, J. R.	ERROR ANALYSIS OF FINITE ELEMENT METHODS WITH TRIANGLES FOR ELLIPTIC BOUNDARY VALUE PROBLEMS	73-0072
WHITEMAN, J. R.	FINITE ELEMENT METHODS FOR ELLIPTIC BOUNDARY VALUE PROBLEMS CONTAINING SINGULARITIES	72-0360
WHITEMAN, J. R.	FUNCTIONAL ANALYSIS AND THE FINITE ELEMENT METHOD APPLIED TO ELLIPTIC BOUNDARY VALUE PROBLEMS	72-0056
WHITEMAN, J. R.	LAGRANGIAN FINITE ELEMENT AND FINITE DIFFERENCE METHODS FOR POISSON PROBLEMS	74-1276
WHITEMAN, J. R.	LOCAL MESH REFINEMENT WITH FINITE ELEMENTS FOR ELLIPTIC PROBLEMS	73-0073
WHITEMAN, J. R.	NUMERICAL SOLUTION OF STEADY STATE DIFFUSION PROBLEMS CONTAINING SINGULARITIES	74-0306
WHITEMAN, J. R.	SOME ASPECTS OF THE MATHEMATICS OF FINITE ELEMENTS	75-0090
WHITEMAN, J. R.	THE EXTENSION AND APPLICATION OF SARD KERNEL THEOREMS TO COMPUTE FINITE ELEMENT ERROR BOUNDS	72-0519
WHITESIDE, J.	ELASTIC AND PLASTIC INTERLAMINAR SHEAR DEFORMATION IN LAMINATED COMPOSITES UNDER GENERALIZED PLANE	71-0158
WHITESIDE, J. B	THE BEHAVIOR OF ADVANCED FILAMENTARY COMPOSITE PLATES WITH CUTOUTS	73-0891
WHITESTONE, W.	COMPARISON OF MEMBRANE FINITE ELEMENT FORMULATIONS	70-0287
WHITTEN, G. F.	AUTOMATIC NESTED DISSECTION	74-0860
WIANCKO, B. E.	APACHE: A THREE-DIMENSIONAL FINITE ELEMENT PROGRAM FOR STEADY-STATE OR TRANSIENT HEAT CONDUCTION ANA	73-0971
WIANCKO, B. E.	MATUS: A THREE-DIMENSIONAL FINITE ELEMENT PROGRAM FOR SMALL-STRAIN ELASTIC ANALYSIS	73-0970
WIBEL, A.	PARAMETER EINES ELASTO-PLASTISCHEN STOFFANSATZES FUER KOERNIGE ERDSTOFFE (DETERMINATION OF PARAMETER	74-0657
WIBEL, A. R.	NUMERICAL COMPUTATION OF BOUNDARY VALUE PROBLEMS IN SOIL MECHANICS	73-0269
WIBERG, N.	GENERAL SOLUTION ROUTINES FOR SYMMETRIC EQUATION SYSTEMS	75-0129
WICKS, T. M.	ON APPLICATION OF THE FINITE ELEMENT METHOD TO LIMIT ANALYSIS	74-1237
WIDERA, O.	SOME FACTORS AFFECTING DENTAL IMPLANT DESIGN	74-1146
WIDERA, O.	STRESS DISTRIBUTION IN BONE ARISING FROM LOADING ON ENDOSTEAL DENTAL IMPLANTS	73-0501
WIEDERHIELM, C.	DEFORMATION OF THE ARTERIAL VASA VASORUM AT NORMAL AND HYPERTENSIVE ARTERIAL PRESSURE	71-0149
WIEDERHIELM, C.	STRUCTURAL ANALYSIS OF AN ARTERIOLE BY THE DIRECT STIFFNESS METHOD	66-0109
WIEDERHIELM, C.	STRUCTURAL RESPONSE OF RELAXED AND CONSTRICTED ARTERIOLES	68-0200
WIESNER, T. J.	FINITE ELEMENT ANALYSIS OF PILE-SUPPORTED PAVEMENT STRUCTURES	75-0145
WIGGERT, D. C.	NUMERICAL PREDICTION OF TRACER CONCENTRATION IN AQUIFERS WITH TRANSIENT GROUNDWATER FLOW BY FINITE-E	75-0610
WIGGERTS, D. C.	TWO-DIMENSIONAL FINITE ELEMENT MODELING OF TRANSIENT FLOW IN REGIONAL AQUIFER SYSTEMS. ANALYSIS OF G	74-0743
WIGHT, R. B.	THREE-DIMENSIONAL THERMAL AND STRESS ANALYSIS OF A PIPING TEE	75-0703
WILCOX, L.	APPLICATION OF FINITE ELEMENTS TO THE ANALYSIS OF GEAR TOOTH STRESSES	72-0116
WILCOX, P. R.	CALCULATION OF STRESS CONCENTRATION DUE TO A HYPERBOLOID CAVITY IN A THIN PLATE USING NASTRAN	71-0260
WILCOX, W. W.	APPLICATIONS OF PULSED LASER INDUCED STRESS WAVES	73-0956
WILDE, A.	BALLISTIC IMPACT OF TEXTILE STRUCTURES	72-0433
WILKINSON, J. P	THREE-DIMENSIONAL STUDY OF NUCLEAR FUEL ROD BEHAVIOUR DURING STARTUP	73-0815

WILKINSON, J. P	THREE-DIMENSIONAL STUDY OF NUCLEAR FUEL ROD BEHAVIOR DURING STARTUP	74-1003
WILKINSON, M. T	CONVERGENCE OF THE NASTRAN PLATE ELEMENTS FOR SHELL STRESS ANALYSIS	71-0261
WILL, G.	A TRIANGULAR FLAT PLATE BENDING ELEMENT	68-0252
WILL, G.	COMPUTER AIDED TEACHING OF THE FINITE ELEMENT DISPLACEMENT METHOD	69-0430
WILL, G. T.	A MIXED FINITE ELEMENT SHELL FORMULATION	71-0063
WILL, G. T.	A MIXED FINITE-ELEMENT SHALLOW SHELL FORMULATION	69-0078
WILL, K. M.	BEAM BUCKLING BY FINITE ELEMENT PROCEDURE	74-0123
WILLAM, K. J.	RECENT DEVELOPMENTS IN THE FINITE ELEMENT ANALYSIS OF PRE-STRESSED CONCRETE REACTOR VESSELS	74-0582
WILLAM, K.	ASPECTS OF FINITE ELEMENT METHOD APPLIED TO AEROSPACE STRUCTURES	73-0703
WILLAM, K.	DISCUSSION ON FINITE ELEMENT ANALYSIS OF PRESTRESSED CONCRETE REACTOR VESSELS	72-0648
WILLAM, K.	FINITE ELEMENT ANALYSIS OF THERMOMECHANICAL PROBLEMS	71-0030
WILLAM, K.	FINITE ELEMENTE ZUR BERECHNUNG VON SPANNBETONREAKTORDRUCKBEHALTERN	73-0702
WILLAM, K.	LINEAR METHODS OF STRUCTURAL ANALYSIS	71-0028
WILLAM, K.	NONLINEAR METHODS OF STRUCTURAL ANALYSIS	71-0029
WILLAM, K.	RECENT DEVELOPMENTS IN FINITE ELEMENT ANALYSIS OF PRESTRESSED CONCRETE REACTOR VESSELS	74-0469
WILLAM, K.	ROTATIONSSYMMETRISCHE BERECHNUNG DES 1:5 THTR SPANNBETON-BEHAL-BEHAL TERMODELLS	71-0629
WILLAM, K.	SOME CONSIDERATIONS FOR THE EVALUATION OF FINITE ELEMENT MODELS	73-0706
WILLAM, K.	SOME CONSIDERATIONS FOR EVALUATION OF FINITE ELEMENT MODELS	74-0471
WILLAM, K. J.	ANALYSIS OF ORTHOTROPIC FOLDED PLATES WITH ECCENTRIC STIFFENERS	70-0437
WILLAM, K. J.	ASPECTS OF THE FINITE ELEMENT METHOD AS APPLIED TO AERO-SPACE STRUCTURES	72-0649
WILLAM, K. J.	FINITE ELEMENT ULTIMATE LOAD ANALYSIS OF THREE DIMENSIONAL CONCRETE STRUCTURES	74-0672
WILLAM, K. J.	SOME CONSIDERATION FOR THE EVALUATION OF THE FINITE ELEMENT MODELS	73-0710
WILLIAM, K. P.	CALCULATIONS FOR PRESTRESSED CONCRETE PRESSURE VESSELS	71-0643
WILLIAM, K.	FINITE ELEMENT ANALYSIS OF CELLULAR STRUCTURES	69-0350
WILLIAM, K. J.	CELLULAR STRUCTURES OF ARBITRARY PLAN GEOMETRY	72-0201
WILLIAM, K. J.	THERMOMECHANICAL CREEP OF AGEING CONCRETE - A UNIFIED APPROACH	74-0607
WILLIAMS, E. W.	FINITE BOUNDARY CORRECTIONS TO THE COPLANAR WAVEGUIDE ANALYSIS	73-0398
WILLIAMS, F. W.	COMPARISON BETWEEN SPARSE STIFFNESS MATRIX AND SUB-STRUCTURE METHODS	73-1068
WILLIAMS, J. A.	THERMAL-STRESS CONCENTRATION CAUSED BY STRUCTURAL DISCONTINUITIES	69-0308
WILLIAMS, M. L.	ELASTO PLASTIC STRESS IN CRACKED PLATES	65-0042
WILLIAMS, M. L.	FINITE ELEMENT IN ADHESION ANALYSES	73-0355
WILLIAMS, M. L.	FINITE ELEMENT IN NUMERICAL ANALYSIS OF ADHESIVE FRACTURE	73-0085
WILLIAMS, M. L.	MECHANICS OF FRACTURE IN ADHESIVE JOINTS	72-0691
WILLIAMS, R.	APPLICATIONS OF TRILINEAR COORDINATES TO SOME PROBLEMS IN PLANE ELASTICITY	72-0728
WILLIAMS, R.	ELASTIC AND EQUILIBRIUM MATRICES OF GENERAL SEMI-MONOCOGUE MEMBRANE ELEMENTS	70-0159
WILLIAMS, R. D.	THE KANSAS TEST TRACK. NON-CONVENTIONAL TRACK STRUCTURES. DESIGN REPORT	72-0797
WILLIAMS, R. T.	RECENT ADVANCES IN NUMERICAL WEATHER PREDICTION	74-0787
WILLIANS, G. M.	SURVEY OF STRUCTURAL ANALYSIS SYSTEMS	74-1290
WILLIS, J. P.	APPLICATION OF FINITE ELEMENT METHOD TO VIBRATIONS OF QUARTZ PLATES	74-0345
WILLIS, J. R.	FINITE ELEMENT CALCULATIONS RELEVANT TO AS-CUT QUARTZ RESONATORS	73-0672
WILLIS, J. R.	VARIATION WITH TEMPERATURE OF RESONANT FREQUENCIES OF ANISOTROPIC PLATES	74-1046
WILLOUGHBY, R.	A SURVEY OF SPARSE MATRIX TECHNOLOGY	72-0989
WILLS, C. M. R.	NASTRAN - A FINITE ELEMENT PROGRAM FOR STRUCTURAL ANALYSIS	72-0131
WILLSHARE, G. T	PROPELLER-EXCITED VIBRATION WITH PARTICULAR REFERENCE TO FULL-SCALE MEASUREMENTS	75-0018
WILSON, B. W.	HARBOUR OSCILLATION - A NUMERICAL TREATMENT FOR UNDAMPED NATURAL MODES	70-0208
WILSON, C. R.	STEADY STATE FLOW IN RIGID NETWORKS OF FRACTURES	74-0278
WILSON, E. A.	A DESIGN ORIENTED APPROACH TO CREEP AND PLASTICITY IN FINITE ELEMENT PROGRAMS	70-0125
WILSON, E. A.	A METHOD FOR DETERMINING THE SURFACE CONTACT STRESSES RESULTING FROM INTERFERENCE FITS	70-0101
WILSON, E. A.	A NONLINEAR FINITE ELEMENT CODE FOR ANALYZING THE ELAST RESPONSE OF UNDERGROUND STRUCTURES	70-0054
WILSON, E. A.	DESIGN ORIENTED APPROACH TO CREEP AND PLASTICITY IN FINITE ELEMENT PROGRAMS	71-0281
WILSON, E. A.	END DISPLACEMENTS OF SEMI-INFINITE CYLINDERS DUE TO ANNULAR LOADINGS	69-0399
WILSON, E. A.	FINITE ELEMENT ANALYSIS OF ELASTIC CONTACT PROBLEMS USING DIFFERENTIAL DISPLACEMENTS	70-0124
WILSON, E. A.	LARGE DISPLACEMENT ANALYSIS OF ASIXYMMETRIC SHELLS	69-0276
WILSON, E. A.	RADIAL DISPLACEMENTS OF A HOLLOW SHAFT SUBJECTED TO A UNIFORM BAND OF EXTERNAL PRESSURE	72-0242
WILSON, E. A.	TRAPEZOIDAL FINITE ELEMENTS: THEIR DERIVATION AND USE FOR AXISYMMETRIC ROTATING BODIES	73-0192
WILSON, E. L.	A DIGITAL COMPUTER PROGRAM FOR THE FINITE ELEMENT ANALYSIS OF SOLIDS WITH NONLINEAR MATERIAL PROPERT	65-0032
WILSON, E. L.	AN APPROXIMATE NON-LINEAR ANALYSIS OF THIN PLATES	68-0286
WILSON, E. L.	APPLICATION OF THE FINITE ELEMENT METHOD TO HEAT CONDUCTION ANALYSIS	66-0076
WILSON, E. L.	AUTOMATIC DESIGN OF SHELL STRUCTURES	71-0336
WILSON, E. L.	COMPUTER PROGRAM FOR STATIC AND DYNAMIC ANALYSIS OF LINEAR STRUCTURAL SYSTEMS	72-0683
WILSON, E. L.	DYNAMIC FINITE ELEMENT ANALYSIS OF ARBITRARY THIN SHELLS	71-0061
WILSON, E. L.	DYNAMIC RESPONSE BY A STEP BY STEP ANALYSIS	62-0014
WILSON, E. L.	ELASTIC DYNAMIC RESPONSE OF AXISYMMETRIC STRUCTURES	69-0275
WILSON, E. L.	FINITE ELEMENT ANALYSIS OF TWO DIMENSIONAL STRUCTURES	63-0015
WILSON, E. L.	FINITE ELEMENT ANALYSIS OF PAVEMENTS	68-0058
WILSON, E. L.	FINITE ELEMENT ANALYSIS OF SEEPAGE IN ELASTIC MEDIA	69-0024
WILSON, E. L.	FINITE ELEMENT ANALYSIS OF LAND SUBSIDENCE	69-0233
WILSON, E. L.	FINITE ELEMENT ANALYSIS OF LINEAR AND NONLINEAR HEAT TRANSFER	74-0626
WILSON, E. L.	FINITE ELEMENT FOR ROCK JOINTS AND INTERFACES	73-0411
WILSON, E. L.	FINITE ELEMENT FORMULATIONS FOR LARGE DISPLACEMENT AND LARGE STRAIN ANALYSIS	73-0457
WILSON, E. L.	FINITE ELEMENT FORMULATIONS FOR LARGE DEFORMATION DYNAMIC ANALYSIS	75-0564
WILSON, E. L.	FINITE ELEMENT POST BUCKLING ANALYSIS OF THIN ELASTIC PLATES	68-0123
WILSON, E. L.	FINITE ELEMENT STRESS ANALYSIS OF AXISYMMETRIC SOLIDS WITH ORTHOTPOPIC TEMPERATURE DEPENDENT MATERIA	67-0089
WILSON, E. L.	FINITE-ELEMENT LARGE DEFLECTION ANALYSIS OF PLATES	69-0196
WILSON, E. L.	FLOW OF COMPRESSIBLE FLUID IN POROUS ELASTIC MEDIA	73-1004
WILSON, E. L.	INCOMPATIBLE DISPLACEMENT MODELS	71-0262
WILSON, E. L.	NONLINEAR DYNAMIC ANALYSIS OF COMPLEX STRUCTURES	73-0193
WILSON, E. L.	NONSAP - A GENERAL FINITE ELEMENT PROGRAM FOR NONLINEAR DYNAMIC ANALYSIS OF COMPLEX STRUCTURES	73-0740
WILSON, E. L.	NONSAP - A NONLINEAR STRUCTURAL ANALYSIS PROGRAM	74-1006
WILSON, E. L.	NONSAP: A STRUCTURAL ANALYSIS PROGRAM FOR STATIC AND DYNAMIC RESPONSE OF NONLINEAR SYSTEMS	74-0848
WILSON, E. L.	SAP IV. A STRUCTURAL ANALYSIS PROGRAM FOR STATIC AND DYNAMIC RESPONSE OF LINEAR SYSTEMS	73-0650
WILSON, E. L.	SAP-A GENERAL STRUCTURAL ANALYSIS PROGRAM FOR LINEAR SYSTEMS	72-0329
WILSON, E. L.	SEISMIC ANALYSIS OF EARTH DAM RESERVOIR SYSTEMS	73-0412
WILSON, E. L.	SOLID SAP. A STATIC ANALYSIS PROGRAM FOR THREE DIMENSIONAL SOLID STRUCTURES	72-0855
WILSON, E. L.	SOLUTION METHODS FOR EIGENVALUE PROBLEMS IN STRUCTURAL MECHANICS	73-1095
WILSON, E. L.	STABILITY ANALYSIS OF AXISYMMETRIC SHELLS	69-0131
WILSON, E. L.	STATIC AND DYNAMIC GEOMETRIC AND MATERIAL NONLINEAR ANALYSIS	74-1302
WILSON, E. L.	STRESS ANALYSIS OF A GRAVITY DAM BY THE FINITE ELEMENT METHOD	63-0004
WILSON, E. L.	STRESS ANALYSIS OF AXISYMMETRIC SOLIDS UTILIZING HIGHER ORDER QUADRILATERAL FINITE ELEMENTS	69-0356
WILSON, E. L.	STRUCTURAL ANALYSIS OF AXISYMMETRIC SOLIDS	65-0031
WILSON, E. L.	STRUCTURAL ANALYSIS OF AXISYMMETRIC SOLIDS	65-0056
WILSON, E. L.	THE STATIC CONDENSATION ALGORITHM	74-0328
WILSON, E. L.	VARIATIONAL FORMULATION OF DYNAMICS OF FLUID-SATURATED POROUS ELASTIC SOLIDS	72-0017
WILSON, G. J.	APPLICATION OF THE FINITE ELEMENT METHOD TO THE VIBRATION ANALYSIS OF AXIAL FLOW TURBINES	73-1125
WILSON, G. J.	DISCUSSION OF "FINITE ELEMENT BENDING ANALYSIS OF REISSNER PLATES"	71-0832
WILSON, G. J.	VIBRATION ANALYSIS OF AXIAL FLOW TURBINE DISKS USING FINITE ELEMENTS	75-0195
WILSON, G. J.	VIBRATION ANALYSIS OF AXIAL-FLOW TURBINE DISKS USING FINITE-ELEMENTS	75-0415
WILSON, G. J.	VIBRATION OF CIRCULAR AND ANNULAR PLATES USING FINITE ELEMENTS	72-0194
WILSON, H. B.	FEASIBILITY OF USING FINITE ELEMENTS IN ANALYSIS OF SECOND BREAKDOWN IN SEMICONDUCTOR DEVICES	73-0874
WILSON, H. B.	MODAL RESPONSE OF FREE ROCKETS UNDER THRUST LOADING	74-0855
WILSON, I. H.	CRUCIFORM SPECIMENS FOR BIAXIAL FATIGUE TESTS - AN INVESTIGATION USING FINITE-ELEMENT ANALYSIS AND P	71-0795
WILSON, I. H.	TWO-DIMENSIONAL PHOTOELASTIC STUDY OF STEAM-TURBINE CASING FLANGES WITH INTERFACE RELIEF	74-0603

WILSON, J. D.	THERMAL STRESSES IN THIN CYLINDRICAL SHELLS	73-0822
WILSON, J. F.	SUBTERRANEAN GUIDEWAY RESPONSES TO LEVITATION VEHICLE LOADING	74-0927
WILSON, J. F.	THE FINITE ELEMENT ANALYSIS OF THIN ELASTIC SHELLS	73-1076
WILSON, J. L.	INTERACTIVE TECHNIQUES WITH GRAPHICAL OUTPUT FOR BRIDGE ANALYSES	75-0622
WILSON, J. M.	TIMOSHENKO BEAM FINITE ELEMENTS	73-0032
WILSON, M.	USE OF COMPUTERS IN WELDING TECHNOLOGY	74-1160
WILSON, N. E.	FINITE ELEMENT ANALYSIS OF SOIL DEFORMATIONS	69-0134
WILSON, R. R.	DYNAMIC BEHAVIOR OF STEEL FOUNDATIONS FOR TURBO-ALTERNATORS	71-0837
WILSON, R. R.	RATE OF CHANGE OF DYNAMIC RESPONSE	72-1004
WILSON, R. R.	TIMOSHENKO BEAM FINITE ELEMENTS	73-0032
WILSON, R. R.	USE OF STRAIGHT BEAM FINITE ELEMENTS FOR ANALYSIS OF VIBRATIONS OF CURVED BEAMS	73-0616
WILSON, W. K.	CRACK TIP FINITE ELEMENTS FOR PLANE ELASTICITY	71-0764
WILSON, W. K.	FINITE ELEMENT METHOD FOR CALCULATING STRESS INTENSITY FACTORS FOR CRACKED PLATES IN BENDING	71-0424
WILSON, W. K.	FINITE ELEMENT METHODS FOR ELASTIC BODIES CONTAINING CRACKS	73-1029
WILSON, W. K.	FUEL ROD DEFORMATIONS - A TIME DEPENDENT FINITE ELEMENT ANALYSIS ON AXISYMMETRIC FUEL RODS, INCLUDIN	70-0414
WILSON, W. K.	ON THE FINITE ELEMENT METHOD IN LINEAR FRACTURE MECHANICS	69-0032
WILSON, W.K.	HEAVY SECTION STEEL TECHNOLOGY PROGRAM TECHNICAL OR PROGRAMMATIC MANUSCRIPT NO. 25. VARIABLE THICKNE	73-0948
WILTON, C.	SHOCK TUNNEL TESTS OF PRELOADED AND ARCHED WALL PANELS FINITE ELEMENT STRESS FORMULATION FOR WAVE PR	73-0892
WINFREY, R. C.	MULTIDEGREE-OF-FREEDOM ELASTIC SYSTEMS HAVING MULTIPLE CLEARANCES	73-0653
WINK, J. H.	APPLICATION OF FINITE ELEMENT METHOD FOR THE DETAILED ANALYSIS OF HATCH CORNER STRESSES	73-0735
WINKEL, B. V.	STRESS AND DEFORMATION ANALYSIS OF IRRADIATION INDUCED SWELLING	74-0305
WINKEL, B. V.	STRESS AND DEFORMATION ANALYSIS OF IRRADIATION INDUCED SWELLING	74-0882
WINNICKI, R. T.	FINITE-ELEMENT MODELING OF PIEZOELECTRIC CERAMIC HYDROPHONES	75-0595
WINSLOW, A. M.	AN IRREGULAR TRIANGLE MESH GENERATOR	64-0032
WINSLOW, A. M.	EQUIPOTENTIAL ZONING OF TWO-DIMENSIONAL MESHES	63-0025
WINSLOW, A. M.	EQUIPOTENTIAL ZONING OF TWO-DIMENSIONAL MESHES	64-0033
WINSLOW, A. M.	NUMERICAL SOLUTION OF THE QUASI LINEAR POISSON EQUATION IN A NON-UNIFORM TRIANGLE MESH	57-0090
WINTER, H.	MEASUREMENT OF ACTUAL DEDENDUM STRAINS, AND THE INFLUENCE OF DEDENDUM RADIUS ON STRESSES AND DEDENDU	72-0118
WISSMANN, J. W.	NONLINEAR STRUCTURAL ANALYSIS: TENSOR FORMULATION	65-0046
WISSMANN, J. W.	NUMERISCHE BERECHNUNG NICHTLINEARER ELASTISCHER KOERPER	63-0017
WITHERSPOON, P.	A METHOD OF ANALYZING TRANSIENT FLUID FLOW IN MULTILAYERED AQUIFERS	69-0142
WITHERSPOON, P.	ANALYSIS OF NONSTEADY FLOW WITH A FREE SURFACE USING FINITE ELEMENT METHOD	71-0482
WITHERSPOON, P.	APPLICATION OF THE FINITE ELEMENT METHOD TO TRANSIENT FLOW IN POROUS MEDIA	68-0094
WITHERSPOON, P.	FINITE ELEMENT METHOD OF ANALYZING STEADY SEEPAGE WITH A FREE SURFACE	70-0092
WITHERSPOON, P.	FLOW IN ROCKS WITH DEFORMABLE FRACTURES	74-0065
WITHERSPOON, P.	SEEPAGE AND GROUNDWATER EFFECTS ASSOCIATED WITH EXPLOSIVE CRATERING	72-0656
WITHERSPOON, P.	STEADY STATE FLOW IN RIGID NETWORKS OF FRACTURES	74-0278
WITHERSPOON, P.	TRANSIENT FLOW OF GROUNDWATER TO WELLS IN MULTIPLE-AQUIFER SYSTEMS	69-0501
WITHERSPOON, P.	USE OF THE FINITE ELEMENT METHOD IN SOLVING TRANSIENT FLOW PROBLEMS IN AQUIFER SYSTEMS	68-0167
WITHERSPOON, P.	VARIATIONAL PRINCIPLES FOR FLUID FLOW IN POROUS MEDIA	71-0844
WITHERSPOON, P.	INFLUENCE OF FLUID INJECTION ON THE STATE OF STRESS IN THE EARTH'S CRUST	72-0624
WITHUM, D.	BERECHNUNG VON PLATTEN NACH DEM RITZ' SCHEN VERFAHREN MIT HILFE DREIECKFOERMIGER MASCHENNETZE	66-0094
WITMER, E. A.	A FEASIBILITY STUDY OF THE ANALYSIS OF ASYMMETRIC SHELL BY THE DISCRETE-ELEMENT METHOD	68-0115
WITMER, E. A.	APPLICATION OF THE COLLISION-IMPARTED VELOCITY METHOD FOR ANALYZING THE RESPONSES OF CONTAINMENT AND	73-0770
WITMER, E. A.	COMPUTER PROGRAM JET 3 TO CALCULATE THE LARGE ELASTIC PLASTIC DYNAMICALLY INDUCED DEFORMATIONS OF FR	72-0609
WITMER, E. A.	DISCRETE-ELEMENT ANALYSIS OF SHELLS OF REVOLUTION	68-0168
WITMER, E. A.	DISCRETE-ELEMENT MODELLING AND SENSITIVITY STUDIES OF STATICALLY LOADED SINGLE-LAYER AND BONDED DOU	70-0129
WITMER, E. A.	DYNAMIC RESPONSES OF CYLINDRICAL SHELLS INCLUDING GEOMETRIC AND MATERIAL NONLINEARITIES	74-0150
WITMER, E. A.	ELASTIC-PLASTIC AND THERMAL STRESS ANALYSIS	68-0169
WITMER, E. A.	FINITE ELEMENT ANALYSIS OF LARGE TRANSIENT ELASTIC-PLASTIC DEFORMATIONS OF SIMPLE STRUCTURES, WITH A	72-0849
WITMER, E. A.	FINITE-ELEMENT ANALYSIS OF LARGE ELASTIC-PLASTIC TRANSIENT DEFORMATIONS OF SIMPLE STRUCTURES	71-0326
WITMER, E. A.	FINITE-ELEMENT PREDICTIONS OF TRANSIENT ELASTIC-PLASTIC LARGE DEFLECTIONS OF STIFFENED AND/OR UNSTIF	74-0825
WITMER, E. A.	NONLINEAR TRANSIENT RESPONSES OF STRUCTURES BY SPATIAL FINITE-ELEMENT METHOD	74-0138
WITMER, E. A.	PROGRESS REPORT ON DISCRETE-ELEMENT ELASTIC AND ELASTIC-PLASTIC ANALYSIS OF SHELLS OF REVOLUTION SUB	68-0170
WITMER, E. A.	STATIC, VIBRATION, AND THERMAL STRESS ANALYSES OF LAMINATED PLATE AND SHELLS BY THE HYBRID-STRESS FI	72-0602
WITMER, E. A.	THE TRANSIENT LINEAR ELASTIC RESPONSE ANALYSIS COMPLEX THIN SHELLS OF REVOLUTION SUBJECTED TO ARBITR	70-0080
WITMER, E. A.	THEORETICAL AND EXPERIMENTAL STUDIES OF TRANSIENT ELASTIC-PLASTIC LARGE DEFLECTIONS OF GEOMETRICALLY	75-0830
WITMER, E. A.	THEORETICAL AND EXPERIMENTAL STUDIES OF TRANSIENT ELASTIC-PLASTIC LARGE DEFLECTIONS OF GEOMETRICALLY	75-0831
WITTBRODT, E.	ALGORYTM OBLICZEN ORGAN WYMUSZONYCH METODA SZTYWNYCH ELEMENTOW SKONCZONYCH (ALGORITHM FOR CALCULATIN	74-0581
WITTBRODT, E.	HYBRYDOWA METODA ELEMENTOW SKONCZONYCH W ZASTOSOWANIU DO OBLICZEN ORGAN URZADZEN OKRETOWYCH. (HYBRID	74-0963
WITTBRODT, E.	OBLICZANIE ORGAN METODA SZTYWNYCH ELEMENTOW SKONCZONYCH. (CALCULATING VIBRATIONS BY THE RIGID-FINITE	75-0649
WITTKE, W.	APPLICATION OF THE FINITE ELEMENT METHOD TO THE DESIGN OF UNDERGROUND BULKHEADS	72-0387
WITTKE, W.	METHODS OF CALCULATION OF THREE-DIMENSIONAL PROBLEMS OF PERCOLATION OF FISSURED ROCK BY FINITE ELEME	70-0472
WITTKE, W.	THEORIE UND EXPERIMENT ZU EINEM NEUEN INJEKTIONSVERFAHREN. (THEORY AND EXPERIMENT OF A NEW GROUTING	74-0991
WITTKE, W.	WECHSELWIRKUNG ZWISCHEN DEFORMATION UND DURCHSTROEMUNG IM KLUEFTIGEN, ANISOTROPEN GEBIRGE. (INTERACT	72-0612
WITTRICK, W. H.	BUCKLING UNDER COMBINED LOADING OF THIN FLAT-WALLED STRUCTURES BY A COMPLEX FINITE STRIP METHOD	74-0788
WITTRICK, W. H.	GENERAL SINUSOIDAL STIFFNESS MATRICES FOR BUCKLING AND VIBRATION ANALYSES OF THIN FLAT-WALLED STRUCT	68-0278
WITTRICK, W. H.	STRESS SINGULARITIES AT DIAPHRAGM-WEB-FLANGE JUNCTIONS IN BOX GIRDERS	75-0242
WITTRICK, W.H.	FINITE ELEMENT STRUCTURAL ANALYSIS OF LOCAL INSTABILITY	74-0098
WIXOM, J. A.	INTERACTIVE PROCEDURE FOR CURVILINEAR COORDINATION OF A PLANAR REGION WITH APPLICATIONS TO FINITE-EL	73-0040
WOELFEL, H.	CALCULATION OF VIBRATIONS OF STRUCTURES CAUSED BY EARTHQUAKES USING THE FINITE ELEMENT METHOD	71-0690
WOHLEN, R. L.	FINITE ELEMENT SOLUTION OF LOW BOND NUMBER SLOSHING	75-0728
WOLBERG, J. R.	A FINITE ELEMENT PROGRAM FOR DESIGN AND ANALYSIS OF THREE-DIMENSIONAL FLAT SPRINGS	74-0738
WOLDEMARIAM, Y.	DISTRIBUTION OF STRESS AROUND A TUNNEL CAVITY, USING THE FINITE ELEMENT METHOD	71-0843
WOLF, J. A.	EFFECT OF TRANSVERSE SHEAR DEFORMATION IN VIBRATIONS OF PLANAR STRUCTURES COMPOSED OF BEAM TYPE ELEM	73-0402
WOLF, J. A.	GENERALIZED HYBRID STRESS FINITE-ELEMENT MODELS	73-0027
WOLF, J. A.	NATURAL FREQUENCIES OF CIRCULAR ARCHES	71-0302
WOLF, J. A.	STABILITY ANALYSIS OF STRUCTURES BY A REDUCED SYSTEM OF GENERALIZED COORDINATES	70-0168
WOLF, J. P.	ALTERNATE HYBRID STRESS FINITE ELEMENT MODELS	75-0476
WOLF, J. P.	COMPUTER-DEVELOPED STATICAL METHODS	69-0374
WOLF, J. P.	DAS FLAECHENTRAGWERKSPROGRAM VON STRIP. (STRIP PROGRAM FOR PLATE STRUCTURES)	72-0150
WOLF, J. P.	GENERALIZED STRESS MODELS FOR FINITE ELEMENT ANALYSIS	74-1067
WOLF, J. P.	STRIP PROGRAM FOR DESIGNING LOAD BEARING SURFACE STRUCTURES (PROGRAMME STRIP POUR LE CALCUL DES STRU	71-0408
WOLF, J. P.	SYSTEMATIC ENFORCEMENT OF STRESS BOUNDARY CONDITIONS IN THE ASSUMED STRESS HYBRID MODEL BASED ON THE	71-0771
WOLFEL, H.	BERECHNUNG VON ERDBEBENSCHWINGUNGEN VON STRUKTUREN MIT DER FINITE ELEMENT METHODE. MECHANISCHE MODELL	71-0170
WONG, P. S.	SPHERICAL WAVES IN INELASTIC MATERIAL	70-0075
WONG, Y. C.	TWO-DIMENSIONAL AUTOMATIC MESH GENERATION FOR STRUCTURAL ANALYSIS	70-0058
WOO, S. L. Y.	A MATHEMATICAL MODEL OF THE CORNEO-SCLERAL SHELL	71-0150
WOO, S. L. Y.	ANALYSIS OF BIOLOGICAL STRUCTURES	71-0148
WOO, S. L. Y.	ANALYSIS OF THE CORNEO-SCLERAL SHELL BY THE METHOD OF DIRECT STIFFNESS	71-0151
WOO, S. L. Y.	FINITE ELEMENT AND EXPERIMENTAL ANALYSIS OF AN INTERNAL FIXATION PLATE MODEL	75-0780
WOO, S. L. Y.	MATHEMATICAL MODEL OF THE CORNEO-SCLERAL SHELL AS APPLIED TO INTRAOCULAR PRESSURE VOLUME RELATIONS A	72-0471
WOO, S. L. Y.	STRUCTURAL RESPONSE OF RELAXED AND CONSTRICTED ARTERIOLES	68-0200
WOOD, J. E.	ANALYSIS OF NUCLEAR FUEL ROD LOCALIZED STRESSES	73-0708
WOOD, R.	STRESS ANALYSIS OF HUMAN FEMUR	73-0327
WOOD, W. G.	PLASTIC FLOW AND FAILURE OF DISCONTINUOUS FIBRE COMPOSITE MATERIALS	70-0157
WOOD, W. L.	COMPARISON OF TIME MARCHING SCHEMES FOR THE TRANSIENT HEAT CONDUCTION EQUATION	75-0477
WOOD, W. L.	NOTE ON A MODIFIED CONJUGATE GRADIENT METHOD	73-0615
WOOD, W. L.	SIMPLE TEACHING EXAMPLES FOR FINITE ELEMENT STRESS ANALYSIS	72-0898
WOODING, R. A.	FINITE-ELEMENT CALCULATION OF PRESSURE HISTORY IN A GAS FIELD	75-0546
WOODMAN, N. J.	DOUBLE CURVATURE SHELL FINITE-ELEMENT AND ITS USE IN THE DYNAMIC ANALYSIS OF COOLING TOWERS AND OTHE	70-0312
WOODMAN, N. J.	FINITE-ELEMENT METHOD APPLIED TO PREDICTING FATIGUE-CRACK GROWTH	73-0513

WOODS, E. H.	FINITE ELEMENT ANALYSIS OF LEFT VENTRICULAR MYOCARDIAL STRESSES	74-0669
WOODWARD, R. J.	ANALYSIS OF EMBANKMENT STRESSES AND DEFORMATIONS	67-0012
WOOL, M. R.	BASELINE SOLUTIONS FOR THE SMOOTH WALL THERMOCHEMICAL ABLATION RESPONSE OF GRAPHITE AND CARBON PHENO	71-0728
WOOLEY, G. R.	A FINITE-ELEMENT ANALYSIS OF STEADY, TWO-DIMENSIONAL, INCOMPRESSIBLE, LAMINAR FLOW	72-0981
WOOLEY, G. R.	STRESS CONCENTRATION FACTORS FOR BONDED LAP JOINTS	71-0842
WOOLEY, R. M.	COMPOSITE-OVERLAY REINFORCEMENT OF CUTOUTS AND CRACKS IN METAL SHEET	73-0878
WOOLLEY, R. M.	ANALYSIS OF COMPOSITE REINFORCED CUTOUTS AND CRACKS	74-0824
WOOLLEY, R. M.	FORMULATION AND EXPERIMENTAL VERIFICATION OF AN AXISYMMETRIC FINITE-ELEMENT STRUCTURAL ANALYSIS	71-0354
WOOLLEY, R. M.	HIGH STRENGTH END FITTINGS FOR FRP ROD AND ROPE	74-0605
WORKMAN, G. H.	DYNAMIC RESPONSE OF STRUCTURES SUBJECTED TO TIME-DEPENDENT BOUNDARY CONDITIONS USING THE FINITE ELEM	72-0129
WORKMAN, G. H.	SIMPLIFIED INELASTIC HIGH TEMPERATURE STRUCTURAL ANALYSIS OF MODERATELY COMPLEX SPATIALLY THREE DIME	74-0445
WOZNIAK, C. Z.	NONLINEAR MECHANICS OF CONSTRAINED MATERIAL CONTINUA: FOUNDATIONS OF THE THEORY	74-0640
WOZNIAK, CZ.	BASIC CONCEPTS OF THE MECHANICS OF DISCRETIZED BODIES WITH AN INTRODUCTION TO DISCRETE ELEMENT CALCU	73-0146
WOZNIAK, CZ.	EXACT THEORY OF SPATIAL STRUCTURES	73-0781
WOZNIAK, CZ.	MULTI-ELEMENT SPATIAL STRUCTURES	73-0768
WRIGHT, F. D.	DESIGN OF ROOF BOLT PATTERNS FOR JOINTED ROCK	74-1318
WRIGHT, G. C.	AN ECONOMICAL METHOD FOR DETERMINING THE SMALLEST EIGENVALUES OF LARGE LINEAR SYSTEMS	71-0551
WRIGHT, J. P.	FINITE ELEMENT ANALYSIS OF ALTERNATIVE AXIAL LOADING OF AN ELASTIC PLATE PRESSED BETWEEN TWO ELASTIC	71-0357
WRIGHT, J. P.	TRANAL: A 3-D FINITE ELEMENT CODE FOR TRANSIENT NONLINEAR ANALYSIS	74-0707
WRIGHT, M. B.	THERMAL STRESSES IN FACT REACTOR BOILER TUBES AT THE SODIUM/GAS INTERFACE	73-0820
WRIGHT, P. K.	FINITE ELEMENT ANALYSIS SYSTEM FOR THE MECHANICAL BEHAVIOR OF ORIENTED FIBER COMPOSITE MATERIALS UND	73-0869
WRIGHT, P. M.	APPLICATION OF FLOWGRAPHS TO FREE VIBRATION OF STRUCTURES	68-0196
WRIGHT, S. G.	ACCURACY OF EQUILIBRIUM SLOPE STABILITY ANALYSIS	73-0517
WRIGHT, S. G.	BOTTOM STABILITY UNDER WATER INDUCED LOADING	72-0373
WRIGHT, W. B.	INFLUENCE OF END-EFFECTS ON STRESSES AND FLEXIBILITY OF A PIPING ELBOW WITH IN-PLANE MOMENT	74-0534
WU, C. I.	FREE VIBRATION OF THICK, LAYERED CYLINDERS HAVING FINITE LENGTH WITH VARIOUS BOUNDARY CONDITIONS	72-0491
WU, C-H.	ON THE ASYMPTOTICALLY SPHERICAL DEFORMATIONS OF ARBITRARY MEMBRANES OF REVOLUTION FIXED ALONG AN EDG	71-0640
WU, J. C.	DIVISION OF COMPUTATION FIELD FOR THE FINITE ELEMENT METHOD	74-0083
WU, J. C.	INTEGRAL REPRESENTATION OF FIELD VARIABLES FOR THE FINITE ELEMENT SOLUTION OF VISCOUS FLOW PROBLEMS	74-0468
WU, J. J.	A THERMAL, THERMOELASTIC, AND WEAR ANALYSIS OF HIGH-ENERGY DISK BRAKES	74-0839
WU, J. J.	A UNIFIED FINITE ELEMENT APPROACH TO COLUMNS SUBJECTED TO NONCONSERVATIVE LOADS, WITH DAMPING EFFECT	74-0795
WU, J. J.	APPLICATION OF THE FINITE ELEMENT METHOD TO NONCONSERVATIVE STABILITY PROBLEMS WITH DAMPING	72-0610
WU, J. J.	ON THE NUMERICAL CONVERGENCE OF MATRIX EIGENVALUE PROBLEMS DUE TO CONSTRAINT CONDITIONS	74-1057
WU, J. J.	SOLUMN INSTABILITY UNDER NONCONSERVATIVE FORCES, WITH INTERNAL AND EXTERNAL DAMPING FINITE ELEMENT U	73-0566
WU, J. N. C.	THERMAL MECHANICAL ANALYSIS PROCEDURE FOR A SEMICIRCULAR PLATE WITH CLAMPED STRAIGHT EDGE	75-0651
WU, R. W.	COMPUTER PROGRAM JET 3 TO CALCULATE THE LARGE ELASTIC PLASTIC DYNAMICALLY INDUCED DEFORMATIONS OF FR	72-0609
WU, R. W.	DYNAMIC PLASTIC LARGE DEFLECTIONS OF CLAMPED RECTANGULAR PLATE	74-0392
WU, R. W.	FINITE ELEMENT ANALYSIS OF LARGE TRANSIENT ELASTIC-PLASTIC DEFORMATIONS OF SIMPLE STRUCTURES, WITH A	72-0849
WU, R. W.	NONLINEAR TRANSIENT RESPONSES OF STRUCTURES BY SPATIAL FINITE-ELEMENT METHOD	73-0138
WU, R. W. H.	DYNAMIC RESPONSES OF CYLINDRICAL SHELLS INCLUDING GEOMETRIC AND MATERIAL NONLINEARITIES	74-0150
WU, R. W. H.	FINITE-ELEMENT ANALYSIS OF LARGE ELASTIC-PLASTIC TRANSIENT DEFORMATIONS OF SIMPLE STRUCTURES	71-0326
WU, R. W. H.	FINITE-ELEMENT PREDICTIONS OF TRANSIENT ELASTIC-PLASTIC LARGE DEFLECTIONS OF STIFFENED AND/OR UNSTIF	74-0825
WU, R. W. H.	THEORETICAL AND EXPERIMENTAL STUDIES OF TRANSIENT ELASTIC-PLASTIC LARGE DEFLECTIONS OF GEOMETRICALLY	75-0830
WU, R. W. H.	THEORETICAL AND EXPERIMENTAL STUDIES OF TRANSIENT ELASTIC-PLASTIC LARGE DEFLECTIONS OF GEOMETRICALLY	75-0831
WU, S. J.	ANALYSIS OF UNSTEADY FLOW TOWARD ARTESIAN WELLS BY THREE-DIMENSIONAL FINITE ELEMENTS	74-0742
WU, S. T.	A FINITE ELEMENT ANALYSIS OF TRANSIENT RAREFIED GAS FLOW	73-0604
WU, S. T.	ANALYSIS OF RAREFIED GAS FLOW THROUGH AN ARBITRARY CROSS SECTION BY THE FINITE ELEMENT METHOD	73-0365
WU, S. T.	CREEP AND SHRINKAGE LAW FOR CONCRETE AT VARIABLE HUMIDITY	74-0668
WU, S. T.	THE FINITE ELEMENT ANALYSIS OF TRANSIENT RAREFIED GAS FLOW	74-0027
WU, S. T.	UNSTEADY MHD DUCT FLOW BY THE FINITE ELEMENT METHOD	73-0263
WU, T. H.	INVESTIGATION OF SOIL-STRUCTURE INTERACTION: FLEXIBLE CULVERT PIPES	72-0661
WU, T. H.	LONG-TERM DEFORMATION IN CLAY	73-0713
WU, T. S.	A NUMERICAL SOLUTION OF THE BOLTZMANN EQUATION BY THE FINITE ELEMENT TECHNIQUE	70-0095
WUETHRICH, W.	ON SHELLS OF CONSTANT STRENGTH	72-0431
WUETHRICH, W.	SOME REMARKS CONCERNING HETEROGENEOUS ANISOTROPIC PLATES.	72-0430
WURMNEST, W.	ANTRAS COMPUTER PROGRAM FOR LINEAR STRESS ANALYSIS BY FINITE ELEMENT METHOD	74-0601
WURMNEST, W.	DAS PROGRAMMSYSTEM ANTRAS EDV-SYSTEM ZUR BERECHNUNG BELIEBIGER STATISCHER SYSTEME NACH DER METHODE D	74-0918
WURMNEST, W.	FESTIGKEITSUNTERSUCHUNG MIT DER METHODE DER FINITEN ELEMENTE. (STRENGTH OF THE MATERIALS INVESTIGATE	74-1009
WYATT, M. J.	NEW DIFFERENCE BASED FINITE-ELEMENT METHOD	75-0444
WYATT, P. J.	COMPARISON OF FINITE-ELEMENT AND EXPERIMENTAL STUDIES ON DEFORMATIONS OF ZIRCONIUM NOTCHED BEND SPEC	71-0359
WYNN, J. C.	SOME APPLICATIONS OF FINITE ELEMENTS TO MODEL GEOLOGICAL STRUCTURES	73-0143
YACIUK, G.	SELECTION OF LOCATIONS FOR LONG TERM STORAGE OF WHEAT	75-0323
YAGAWA, G	SNAP-THROUGH BUCKLING ANALYSIS OF SHELLS OF REVOLUTION USING INCREMENTAL STIFFNESS MATRICES	74-1349
YAGAWA, G.	A NEW APPROACH TO THE FINITE-ELEMENT FORMULATION AND SOLUTION OF A CLASS OF PROBLEMS IN COUPLED THER	72-0732
YAGAWA, G.	A NEW APPROACH TO THE FINITE-ELEMENT FORMULATION AND SOLUTION OF A CLASS OF PROBLEMS IN COUPLED THER	73-0600
YAGAWA, G.	ANALYSIS OF ELASTIC-PLASTIC CREEP BUCKLING OF AXISYMMETRIC SHELLS BY FINITE-ELEMENT METHOD	73-0355
YAGAWA, G.	ANALYSIS OF LARGE ELASTIC-PLASTIC CREEP DEFORMATIONS OF AXISYMMETRIC SHELLS BY THE FINITE ELEMENT ME	74-0563
YAGAWA, G.	APPLICATION OF FINITE ELEMENT DYNAMIC ANALYSIS TO FRACTURE MECHANICS	75-0861
YAGAWA, G.	APPLICATION OF THE FINITE ELEMENT METHOD TO THE ANALYSIS OF CYLINDRICAL SHELLS	73-0828
YAGAWA, G.	ELASTIC-PLASTIC FINITE ELEMENT ANALYSIS USING SUPERPOSITION	75-0314
YAGAWA, G.	ELASTIC-PLASTIC FINITE ELEMENT ANALYSIS USING SUPERPOSITION	75-0429
YAGAWA, G.	FINITE ELEMENT CALCULATION OF STRESS INTENSITY FACTORS USING SUPERPOSITION	75-0251
YAGAWA, G.	FINITE ELEMENT CREEP BUCKLING ANALYSIS OF VISCO-ELASTIC SHALLOW SPHERICAL SHELL	75-0159
YAGAWA, G.	INCREMENTAL THEORY OF THREE DIMENSIONAL TRANSIENT THERMOELASTOPLASTICITY - FORMULATION AND SOLUTION	73-0555
YAGAWA, G.	INCREMENTAL THERMOMECHANICAL THEORY OF VISCOELASTOPLASTIC SOLIDS AND SOLUTION BY FINITE ELEMENTS	73-0861
YAGAWA, G.	LOW CYCLE FATIGUE STRENGTH OF PIPING COMPONENTS	73-0366
YAGAWA, G.	NEW APPROACH TO FINITE ELEMENT FORMULATION AND SOLUTION OF A CLASS OF PROBLEMS IN COUPLED THERMOELAS	73-0013
YAGAWA, G.	STRESS DISTRIBUTIONS IN THIN-WALLED INTERSECTING CYLINDRICAL SHELLS SUBJECTED TO INTERNAL PRESSURE A	71-0023
YAGAWA, G.	THE APPLICATION OF THE FINITE ELEMENT METHOD TO THE ANALYSIS OF FRACTURE OF CYLINDRICAL SHELLS	73-0728
YAGAWA, G.	THE FINITE ELEMENT CALCULATION OF STRESS INTENSITY FACTORS USING SUPERPOSITION	75-0886
YAGAWA, G.	THREE-DIMENSIONAL FINITE ELEMENT METHOD OF THERMOELASTOPLASTICITY WITH CREEP EFFECT	73-0837
YAGHMAI, S.	INCREMENTAL ANALYSIS OF LARGE DEFLECTIONS OF SHELLS OF REVOLUTION	71-0296
YAGHMAI, S.	LARGE DEFLECTION ELASTIC-PLASTIC ANALYSIS OF SHELLS OF REVOLUTION	71-0263
YAGHMAI, S.	LINEAR AND NONLINEAR STATIC ANALYSIS OF AXISYMMETRICALLY LOADED THIN SHELLS OF REVOLUTION	69-0215
YAGHMAI, S. K.	BENDING OF CIRCULAR PLATES OF HARDENING MATERIAL	67-0112
YAGISHITA, S.	TIDAL SIMULATION BY TIME SPACE ELEMENT	75-0867
YAGUCHI, M.	STRESS ANALYSIS BY SUPERPOSITION OF THE ANALYTICAL SOLUTION AND FINITE ELEMENT SOLUTION	72-1000
YAHATA, S.	SPACE SHUTTLE. THE NEED FOR SUBSTRUCTURING	71-0104
YAHATA, S.	VEHICLE STRUCTURAL ANALYSIS, PRE AND POST PROCESSOR DATA HANDLING	71-0093
YAHR, G. T.	APPLICATION OF FRACTURE MECHANICS TO GRAPHITE UNDER COMPLEX STRESS CONDITIONS	72-0771
YALAMANCHILI, R	ACCURACY, STABILITY, AND OSCILLATION CHARACTERISTICS OF TRANSIENT TWO-DIMENSIONAL HEAT CONDUCTION	75-0849
YALAMANCHILI, R	AN EXTENSION OF FINITE-ELEMENT METHOD TO INCLUDE NONLINEAR BOUNDARY CONDITIONS	75-0032
YALAMANCHILI, R	STABILITY AND OSCILLATION CHARACTERISTICS OF FINITE ELEMENT, FINITE DIFFERENCE AND WEIGHTED RESIDUAL	73-0139
YALAMANCHILI, R	FINITE ELEMENT METHOD APPLIED TO TRANSIENT TWO-DIMENSIONAL HEAT TRANSFER WITH CONVECTION AND RADIATI	70-0127
YALAMANCHILI, R	APPLICATION OF THE FINITE-ELEMENT METHOD TO HEAT-TRANSFER PROBLEMS. PART 11. TRANSIENT TWO-DIMENSION	71-0759
YALAMANCHILI, R	APPLICATION OF THE FINITE-ELEMENT METHOD TO HEAT-TRANSFER PROBLEMS. PART 1. FINITE-ELEMENT AP	71-0760
YAM, L. C. P.	CONFORMING FINITE ELEMENT WITH INTERNAL EQUILIBRIUM FOR TWO DIMENSIONAL PROBLEMS	72-0386
YAMABUCHI, T.	FINITE ELEMENT SIMULATION OF TWO-DIMENSIONAL ELECTROMECHANICAL RESONATORS	74-1014
YAMABUCHI, T.	FINITE ELEMENT SIMULATION OF ELECTROMECHANICAL RESONATORS WITH ARBITRARY ELECTRODES	75-0492
YAMADA, Y.	ANALYSIS OF CONTACT PROBLEMS BY THE MATRIX METHOD	69-0388
YAMADA, Y.	ANALYSIS OF ONE-DIMENSIONAL STRESS WAVE BY THE FINITE ELEMENT METHOD	71-0523
YAMADA, Y.	ANALYSIS OF SAINT-VENANT TORSION PROBLEM BY A HYBRID STRESS MODEL	69-0277

114

YAMADA, Y.	ANALYSIS OF THE ELASTIC PLASTIC PROBLEMS BY THE MATRIX DISPLACEMENT METHOD	68-0173
YAMADA, Y.	ELASTIC PLASTIC ANALYSIS OF SAINT-VENANT TORSION PROBLEM BY A HYBRID STRESS MODEL	72-0385
YAMADA, Y.	FINITE DIFFERENCE SOLUTIONS FOR LARGE DEFORMATIONS OF CYLINDRICAL SHELLS - A COMPARISON WITH FINITE	72-0270
YAMADA, Y.	FINITE ELEMENT ANALYSIS OF THERMO-VISCOELASTIC PROBLEMS	71-0526
YAMADA, Y.	FINITE ELEMENT ANALYSIS OF STEADY FLUID AND METAL FLOW	74-0054
YAMADA, Y.	FINITE ELEMENT ANALYSIS OF COMPOSITE MATERIAL BEHAVIORS FROM MECHANICAL PROPERTIES OF CONSTITUENT M	75-0874
YAMADA, Y.	FINITE ELEMENT FORMULATION FOR SOLUTION OF STEADY FLOW OF INCOMPRESSIBLE VISCOUS FLUID	73-0299
YAMADA, Y.	INCREMENTAL FORMULATIONS FOR PROBLEMS WITH GEOMETRIC AND MATERIAL NONLINEARITIES	72-0315
YAMADA, Y.	NONLINEAR ANALYSIS BY THE FINITE ELEMENT METHOD AND SOME EXPOSITORY EXAMPLES	73-0309
YAMADA, Y.	PLASTIC STRESS STRAIN MATRIX OF ANISOTROPIC MATERIALS AND ITS APPLICATION	69-0383
YAMADA, Y.	PLASTIC STRESS-STRAIN MATRIX AND ITS APPLICATION FOR THE SOLUTION OF ELASTIC-PLASTIC PROBLEMS BY THE	68-0171
YAMADA, Y.	RECENT ADVANCES IN MATRIX METHODS IN STRUCTURAL ANALYSIS AND DESIGN	70-0290
YAMADA, Y.	RECENT DEVELOPMENTS IN MATRIX DISPLACEMENT METHOD FOR ELASTIC-PLASTIC PROBLEMS IN JAPAN	71-0265
YAMADA, Y.	STIFFNESS MATRIX IN PLASTIC-ELASTIC PROBLEMS OF CONTINUA	67-0116
YAMADA, Y.	STRESS STRAIN MATRIX AND ITS APPLICATIONS FOR THE SOLUTION OF ELASTIC-PLASTIC PROBLEMS BY THE FINITE	68-0282
YAMADA, Y.	THE STATUS AND OUTLOOK FOR THE SOLUTION OF NONLINEAR PROBLEMS BY THE FINITE ELEMENT METHOD	70-0259
YAMADA, YOSHIAK	A FINITE ELEMENT SIMULATION OF MECHANICAL PROPERTIES OF COMPOSITE MATERIALS	74-1294
YAMADA, YOSHIAK	ANALYSIS OF COMPRESSIBLE HYDRODYNAMICS BY THE FINITE ELEMENT METHOD	71-0196
YAMADA, YOSHIAK	ANALYSIS OF ELASTIC-PLASTIC BEHAVIOR OF BEAMS AND COLUMNS BY PLANE FINITE ELEMENT MODEL	73-1073
YAMADA, YOSHIAK	ANALYSIS OF LONGITUDINAL WAVE IN A BAR EXHIBITING VISCO-ELASTIC BEHAVIOR OR WITH MATERIAL DAMPING	71-0780
YAMADA, YOSHIAK	DYNAMIC RESPONSE OF VISCOELASTIC MATERIALS TO SINUSOIDALLY VARYING LOADINGS	73-1072
YAMADA, YOSHIAK	DYNAMIC RESPONSE OF VISCOELASTIC MATERIALS	73-1075
YAMADA, YOSHIAK	EFFECT OF TIME-DEPENDENT MATERIAL PROPERTIES ON DYNAMIC RESPONSE	74-0900
YAMADA, YOSHIAK	ELASTIC-PLASTIC ANALYSIS OF BEAMS WITH UNIFORM CROSS-SECTION UNDER COMBINED LOADINGS	71-0777
YAMADA, YOSHIAK	ELASTIC-PLASTIC ANALYSIS OF A BAR WITH UNIFORM CROSS-SECTION UNDER COMBINED LOADINGS 1 - COMBINATION	71-0778
YAMADA, YOSHIAK	ELASTIC-PLASTIC ANALYSIS OF AXISYMMETRIC PLATES AND SHELLS	71-0779
YAMADA, YOSHIAK	ELASTIC-PLASTIC ANALYSIS OF BEAMS WITH UNIFORM CROSS-SECTION UNDER COMBINED LOADINGS (2)	73-1070
YAMADA, YOSHIAK	FINITE ELEMENT ANALYSIS OF THERMO-VISCOELASTIC PROBLEMS	71-0526
YAMADA, YOSHIAK	FINITE ELEMENT ANALYSIS OF LOAD CELL RESPONSE IN HIGH SPEED TENSILE TESTING	73-1071
YAMADA, YOSHIAK	FINITE ELEMENT ANALYSIS OF STEADY FLUID AND METAL FLOW	75-0178
YAMADA, YOSHIAK	FINITE ELEMENT ANALYSIS OF COMPOSITE MATERIAL BEHAVIOURS FROM MECHANICAL PROPERTIES OF CONSTITUENT M	75-0181
YAMADA, YOSHIAK	INCREMENTAL SOLUTION OF AXISYMMETRIC PLATE AND SHELL FINITE DEFORMATION	71-0776
YAMADA, YOSHIAK	INSTABILITY ANALYSIS BY THE FINITE ELEMENT METHOD	75-0180
YAMADA, YOSHIAK	LARGE DEFORMATION AND CRITICAL LOADS ANALYSIS OF FRAMED STRUCTURES	74-1293
YAMADA, YOSHIAK	MATRIX METHODS IN MECHANICS OF MATERIALS	70-0459
YAMADA, YOSHIAK	NONLINEAR ANALYSIS OF BEAMS AND SHELLS	73-1074
YAMADA, YOSHIAK	PLASTICITY AND VISCOELASTICITY	72-0905
YAMADA, YOSHIAK	THEORY AND PRACTICE IN FINITE ELEMENT STRUCTURAL ANALYSIS	73-0700
YAMADA, YOSHIKA	DYNAMIC ANALYSIS OF CIVIL ENGINEERING STRUCTURES	71-0264
YAMADA, YOSHIKA	DYNAMIC ANALYSIS OF STRUCTURES	72-0298
YAMAGAMI, T.	ANALYSIS OF QUICK SAND PHENOMENON BY FINITE ELEMENT METHOD	73-0191
YAMAGAMI, T.	METHOD OF ANALYZING NON-DARCY FLOW PROBLEMS BY FINITE ELEMENTS	75-0158
YAMAGUCHI, H.	STUDY OF ELASTIC-PLASTIC MATRIX ANALYSIS FOR FOUNDATION AND REINFORCED CONCRETE IN JAPAN	69-0278
YAMAJI, S.	FINITE ELEMENT ANALYSIS OF CONTACT PROBLEMS	73-0312
YAMAKAWA, T.	ANALYSIS OF THERMAL ELASTIC PLASTIC STRESS AND STRAIN DURING WELDING BY FINITE ELEMENT METHOD	71-0382
YAMAKAWA, T.	THERMAL NONLINEAR BEHAVIOUR OF STRUCTURES	72-0317
YAMAMOTO, S.	STRESS ANALYSIS OF A PRESTRESSED CONCRETE NUCLEAR PRESSURE VESSEL BY THE FINITE ELEMENT METHOD USING	71-0154
YAMAMOTO, Y.	A FORMULATION OF MATRIX DISPLACEMENT METHOD	66-0103
YAMAMOTO, Y.	A NOTE ON CONVERGENCE OF FINITE ELEMENT SOLUTIONS	71-0252
YAMAMOTO, Y.	ANALYSIS OF WATER WAVE PROBLEMS BY THE METHOD OF SUPERPOSITION OF ANALYTICAL AND FINITE ELEMENT SOLU	75-0870
YAMAMOTO, Y.	DETERMINATION OF STRESS INTENSITY FACTORS IN CRACKED PLATES BY THE FINITE ELEMENT METHOD	73-0148
YAMAMOTO, Y.	FINITE ELEMENT APPROACHES WITH THE AID OF ANALYTIC SOLUTIONS	71-0266
YAMAMOTO, Y.	FINITE ELEMENT TREATMENT OF SINGULARITIES OF BOUNDARY VALUE PROBLEMS AND ITS APPLICATION TO ANALYSIS	73-0306
YAMAMOTO, Y.	FUNDAMENTAL STUDIES ON STEADY SHIP WAVE PROBLEMS BY THE FINITE ELEMENT METHOD	74-0564
YAMAMOTO, Y.	RATE OF CONVERGENCE FOR THE ITERATIVE APPROACH IN ELASTIC-PLASTIC ANALYSIS OF CONTINUA	73-0518
YAMAMOTO, Y.	RECENT ADVANCES IN COMPUTATIONAL METHODS IN STRUCTURAL MECHANICS AND DESIGN.	72-0575
YAMAMOTO, Y.	SIMULATION OF FINITE ELEMENT METHOD FOR LOADING PROCESS OF GRAIN CARGO AND INTERACTION WITH SHIP'S S	75-0161
YAMAMOTO, Y.	SOME ASPECTS OF RECENT CONTRIBUTIONS TO THE MATHEMATICAL THEORY OF FINITE ELEMENTS	73-0601
YAMAMOTO, Y.	SOME CONSIDERATIONS ON ROUND-OFF ERRORS OF THE FINITE ELEMENT METHOD	72-0260
YAMAMOTO, Y.	SOME PROBLEMS ON THE BASIC THEORY OF THE FINITE ELEMENT METHODS	74-0561
YAMAMOTO, Y.	STRESS INTENSITY FACTORS IN PLATE STRUCTURES CALCULATED BY THE FINITE ELEMENT METHOD (IN JAPANESE)	71-0788
YAMAMOTO, Y.	THE ANALYSIS OF STRESS DISTRIBUTION AROUND THE TIP OF A CRACK BY FINITE ELEMENT METHOD WITH THE AID	71-0517
YAMAMOTO, Y.	VARIATIONAL PRINCIPLES AND DUALISTIC SCHEME FOR INTERSECTION PROBLEMS IN ELASTICITY	69-0305
YAMASHITA, M.	A STUDY ON THE STRESS ANALYSIS OF BEAM PLATE STRUCTURE	69-0404
YAMOMOTO, Y.	STRESS ANALYSIS BY SUPERPOSITION OF THE ANALYTICAL SOLUTION AND FINITE ELEMENT SOLUTION	72-1000
YANG, H. T. Y.	A FINITE ELEMENT STRESS ANALYSIS OF THE VERTICAL BUTTRESSES OF A NUCLEAR CONTAINMENT VESSEL	69-0280
YANG, H. T. Y.	ANALYSIS OF MODERATELY THICK FLANGED REACTOR PRESSURE VESSELS	73-0821
YANG, H. T. Y.	ANALYSIS OF SPACE CLOSING SPRINGS IN ORTHODONTICS	74-0236
YANG, H. T. Y.	FLEXIBLE PLATE FINITE ELEMENT ON ELASTIC FOUNDATION	70-0128
YANG, H. T. Y.	NON-LINEAR ANALYSIS OF THE ROTATIONAL STIFFNESS OF A NUCLEAR STEAM GENERATOR FOUNDATION	69-0286
YANG, H. T. Y.	NUMERICAL PROCEDURE FOR THE ANALYSIS OF STRUCTURES SUPPORTED BY VISCOELASTIC FOUNDATIONS	73-0564
YANG, H. T. Y.	STRUCTURAL ANALYSIS OF AIRCRAFT IMPACT ON A NUCLEAR CONTAINMENT VESSEL AND ASSOCIATED STRUCTURES	69-0279
YANG, H. T-Y.	A FINITE ELEMENT FORMULATION FOR STABILITY ANALYSIS OF DOUBLY CURVED THIN SHELL STRUCTURES	69-0513
YANG, R. Y.	ELASTIC SNAP-TRHOUGH ANALYSIS OF CURVED PLATES USING DISCRETE ELEMENTS	72-0927
YANG, S. K.	CONDITION OF FINITE ELEMENT MATRICES	74-1078
YANG, T. Y.	A MATRIX DISPLACEMENT METHOD ON PRE AND POST BUCKLING ANALYSIS OF LINERS FOR REACTOR VESSELS	71-0267
YANG, T. Y.	ASYMMETRICAL BENDING AND VIBRATION OF A CONICAL SHELL FINITE ELEMENT	74-0085
YANG, T. Y.	AXIAL-FLEXURAL VIBRATION OF FRAMEWORKS USING FINITE-ELEMENT APPROACH	73-0051
YANG, T. Y.	CONTINUUM APPROACH TOWARD DYNAMICS OF GRIDWORKS	73-0174
YANG, T. Y.	ELASTIC SNAP THROUGH ANALYSIS OF CURVED PLATES USING DISCRETE ELEMENTS	72-0899
YANG, T. Y.	ELASTIC-PLASTIC FORMULATION FOR A CYLINDRICAL-SHELL FINITE-ELEMENT	75-0841
YANG, T. Y.	FINITE DISPLACEMENT PLATE FLEXURE BY THE USE OF MATRIX INCREMENTAL APPROACH	72-0384
YANG, T. Y.	FINITE ELEMENT ANALYSIS OF PLATES ON A TWO PARAMETER FOUNDATION MODEL.	72-0425
YANG, T. Y.	FINITE ELEMENT PROCEDURE FOR LARGE DEFLECTION ANALYSIS OF PLATES WITH INITIAL DEFLECTIONS	71-0327
YANG, T. Y.	FINITE ELEMENT PROCEDURE FOR POSTBUCKLING ANALYSIS OF INITIALLY CURVED PLATES	71-0421
YANG, T. Y.	FINITE ELEMENTS FOR VIBRATION OF FRAMED SHEAR WALLS	73-0036
YANG, T. Y.	FLUTTER OF FLAT FINITE-ELEMENT PANELS IN A SUPERSONIC POTENTIAL FLOW	75-0814
YANG, T. Y.	FREE VIBRATIONS OF FINITE ELEMENT PLATES SUBJECTED TO COMPLEX MIDDLE-PLANE FORCE SYSTEMS - REPLY	73-0122
YANG, T. Y.	HIGH ORDER RECTANGULAR SHALLOW SHELL FINITE ELEMENT	73-0279
YANG, T. Y.	LARGE DISPLACEMENT ANALYSIS OF VISCOELASTIC BEAMS AND FRAMES BY FINITE-ELEMENT METHOD	74-0644
YANG, W. H.	DYNAMIC ANALYSIS OF THE DEFORMATION OF STRUCTURES AND OF METAL FORMING	74-0823
YANG, W. H.	STIFFNESS ANALYSIS OF ELASTIC PLASTIC PLATES	65-0040
YANG, W. M.	ELASTO PLASTIC STRESS IN CRACKED PLATES	65-0042
YANG, Y. C.	RECORD CONCRETE SHELL ROOF FEATURED IN COLLISEUM	72-0218
YANIV, S.	FINITE ELEMENTS FOR FLUID DYNAMICS. VARIATIONAL FORMULATIONS FOR NONLINEAR, INITIAL-VALUE PROBLEMS	74-0710
YANNITELL, D. M	TWO-DIMENSIONAL ANALYSIS OF BACKWATER AT BRIDGES	74-0184
YANT, H. W.	DESIGN ANALYSIS FOR LMFBR CORE RESTRAINT	71-0208
YAD, F. C-C.	ANALYSIS OF SHELLS BY THE FINITE ELEMENT DISPLACEMENT METHOD	68-0185
YASAKA, T.	METHOD OF MINIMUM WEIGHT DESIGN WITH REQUIREMENTS IMPOSED ON STRESSES AND NATURAL FREQUENCIES	70-0149
YASHIMURA, N.	STRESS STRAIN MATRIX AND ITS APPLICATIONS FOR THE SOLUTION OF ELASTIC-PLASTIC PROBLEMS BY THE FINITE	68-0282
YASUI, R. K.	STRESS ANALYSIS AND DESIGN OF SILICON SOLAR CELL ARRAYS AND RELATED MATERIAL PROPERTIES	72-0840
YASUNAMI, K.	STRESS ANALYSIS OF SELF-SEALING TYPE PISTON HEAD BY FINITE-ELEMENT METHOD	72-0259

YASUNAMI, K.	STRESS ANALYSIS OF THE "SELF-SEALING" TYPE PISTON HEAD BY THE FINITE ELEMENT METHOD	72-0617
YATES, D. N.	APPLICATION OF THE FINITE ELEMENT METHOD FOR ANALYSIS AND OPTIMIZATION OF SUBMARINE STRUCTURES	71-0350
YATES, D. N.	NONLINEAR FINITE ELEMENT AND ITS APPLICATION TO LARGE SCALE STRUCTURAL PROBLEMS	75-0386
YATES, D. N.	NONLINEAR THERMO ELASTIC PLASTIC AND CREEP ANALYSIS BY FINITE ELEMENT METHOD	74-0092
YATES, D. N.	THE DEVELOPMENT OF COMPUTER GRAPHICS FOR LARGE SCALE FINITE ELEMENT CODES	71-0042
YATES, D. N.	THE DEVELOPMENT OF LARGE SCALE DIGITAL COMPUTER CODES FOR PRODUCTION STRUCTURAL ANALYSIS	71-0721
YATT, B. J.	UNDRAINED VISCO-ELASTIC ANALYSIS OF SOIL DEFORMATION	72-0424
YAZICI, A.	FINITE ELEMENT SOLUTION OF PROBLEMS WITH UNKNOWN AND MOVING BOUNDARIES	74-0460
YEE, F.	DIGITAL COMPUTER PROGRAMS FOR STEADY-STATE OR TRANSIENT TEMPERATURE ANALYSIS OF PLANE AND AXISYMMETR	72-0915
YEE, W. S.	LATERAL RESISTANCE AND DEFLECTION OF VERTICAL PILES	73-0912
YEGHIAYAN, R. P	DISCRETE-ELEMENT MODELLING AND SENSITIVITY STUDIES OF STATICALLY LOADED SINGLE-LAYER AND BOUNDED DOU	70-0129
YEGHIAYAN, R. P	THE TRANSIENT LINEAR ELASTIC RESPONSE ANALYSIS COMPLEX THIN SHELLS OF REVOLUTION SUBJECTED TO ARBITR	70-0080
YEH, C.	ARBITRARILY SHAPED INHOMOGENEOUS OPTICAL FIBER OR INTEGRATED OPTICAL WAVEGUIDES	75-0372
YEH, C.	NONLINEAR DYNAMIC ANALYSIS OF COOLING TOWER.	72-0219
YEH, C. H.	LARGE DEFLECTION DYNAMIC ANALYSIS OF THIN SHELLS USING FINITE ELEMENT METHOD	70-0379
YEN, C. L.	COMPARISON OF MEMBRANE FINITE ELEMENT FORMULATIONS	70-0287
YEN, C. L.	COMPARISON OF PLATE BENDING FINITE ELEMENT FORMULATIONS	70-0288
YEO, M. F.	A MORE EFFICIENT FRONT SOLUTION: ALLOCATING ASSEMBLY LOCATIONS BY LONGEVITY CONSIDERATIONS	73-1069
YETTRAM, A. L.	ANCHORAGE ZONE STRESSES IN AXIALLY POST-TENSIONED I-SECTION MEMBERS WITH END BLOCKS	71-0305
YETTRAM, A. L.	GENERALISED MATRIX FORCE AND DISPLACEMENT METHODS FOR LINEAR STRUCTURAL ANALYSIS	65-0064
YETTRAM, A. L.	PLANE FRAMEWORK METHODS FOR PLATES IN EXTENSION	66-0126
YETTRAM, A. L.	THE SOLUTION OF STRUCTURAL EQUILIBRIUM EQUATIONS BY THE CONJUGATE GRADIENT METHOD WITH PARTICULAR RE	71-0268
YEW, C. H.	ON APPLICATION OF THE FINITE ELEMENT METHOD TO LIMIT ANALYSIS	74-1237
YIH, C. S.	ADVANCES IN APPLIED MECHANICS, VOLUME 12.	72-0375
YILMAZ, C.	ULTIMATE STRENGTH OF BOX GIRDERS BY FINITE ELEMENT METHOD	75-0794
YIN, F. C. P.	INTERFACE CORE DISCRETE-ELEMENT FOR STIFFENED SHELLS OF REVOLUTION	67-0092
YOKOBORI, T.	SIZE OF PLASTIC ZONE AT TIP OF A CRACK IN PLANE STRAIN SLATE BY FINITE ELEMENT METHOD	73-0006
YOKOO, Y. ET AL	FINITE ELEMENT METHOD APPLIED TO BIOT'S CONSOLIDATION THEORY	71-0518
YOKOUCHI, Y.	ANALYSIS OF CONTACT PROBLEMS BY THE MATRIX METHOD	69-0388
YOKOUCHI, Y.	ELASTIC-PLASTIC ANALYSIS OF AXISYMMETRIC PLATES AND SHELLS	71-0779
YOKOUCHI, Y.	FINITE DIFFERENCE SOLUTIONS FOR LARGE DEFORMATIONS OF CYLINDRICAL SHELLS - A COMPARISON WITH FINITE	72-0270
YOKOUCHI, Y.	FINITE ELEMENT ANALYSIS OF STEADY FLUID AND METAL FLOW	74-0054
YOKOUCHI, Y.	FINITE ELEMENT ANALYSIS OF STEADY FLUID AND METAL FLOW	75-0178
YOKOUCHI, Y.	FINITE ELEMENT ANALYSIS OF COMPOSITE MATERIAL BEHAVIOURS FROM MECHANICAL PROPERTIES OF CONSTITUENT M	75-0181
YOKOUCHI, Y.	INCREMENTAL SOLUTION OF AXISYMMETRIC PLATE AND SHELL FINITE DEFORMATION	71-0776
YOKOUCHI, Y.	PLASTIC STRESS STRAIN MATRIX OF ANISOTROPIC MATERIALS AND ITS APPLICATION	69-0383
YONAIYAMA, M.	STRESSES IN A ROTATING DISK WITH CENTRAL AND NONCENTRAL HOLES (STRESS ANALYSIS BY THE FINITE ELEMENT	74-1337
YOON, K. E.	STRUCTURAL ANALYSIS OF A THREE-DIMENSIONAL PERFORATED PLATE BY THE FINITE ELEMENT METHOD WITH TEST C	74-0945
YOSHIDA, D. M.	RESPONSE OF BEAMS AND PLATES TO MOVING LOADS	71-0277
YOSHIDA, I.	EVALUATION OF STRESSES IN SOIL UNDER RIGID WHEELS BY FINITE ELEMENT METHOD	73-1109
YOSHIDA, Y.	A HYBRID STRESS ELEMENT FOR THIN SHELL ANALYSIS	74-0433
YOSHIDA, Y.	DISCRETE TRIANGULAR APPROXIMATION OF MOMENT AND DISPLACEMENT SURFACES FOR PLATE BENDING	71-0269
YOSHIDA, Y.	EQUIVALENT FINITE ELEMENTS ON DIFFERENT BASES	72-0287
YOSHIDA, Y.	FLAT FINITE ELEMENT FOR THIN SHELL ANALYSIS DERIVED BY ASSUMED STRESS APPROACH	73-0147
YOSHIDA, Y.	ON THE VARIATIONAL PRINCIPLES IN THE FINITE ELEMENT METHOD	74-1010
YOSHIKAWA, K. K	VARIATIONAL SOLUTIONS BY USE OF STEPWISE CONSTANT FUNCTIONS: ONE LINEAR CASE	72-0286
YOSHIKAWA, M.	NUMERICAL ANALYSIS OF MAGNETOHYDRODYNAMIC INSTABILITIES BY FINITE ELEMENT METHOD	72-0041
YOSHIKI, M.	ON THE DEVELOPMENT OF THE "PASSAGE" PROGRAM	73-0342
YOSHIKII, M.	MATRIX METHOD OF ANALYSIS OF SHIP STRUCTURES 3RD REPORT	68-0225
YOSHIMURA, N.	ANALYSIS OF LARGE DEFLECTION OF PLATES BY THE FINITE ELEMENT METHOD	69-0148
YOSHIMURA, N.	ANALYSIS OF STEADY INCOMPRESSIBLE VISCOUS FLOW	74-0013
YOSHIMURA, N.	ANALYSIS OF THE ELASTIC PLASTIC PROBLEMS BY THE MATRIX DISPLACEMENT METHOD	68-0173
YOSHIMURA, N.	FINITE ELEMENT ANALYSIS ON THE TORSION OF A BAR WITH UNIFORM CROSS SECTION	68-0186
YOSHIMURA, N.	MATRIX METHOD OF ANALYSIS OF SHIP STRUCTURES 3RD REPORT	68-0225
YOSHIMURA, N.	STEADY FLOW ANALYSIS OF INCOMPRESSIBLE VISCOUS FLOW BY THE FINITE ELEMENT METHOD	73-0332
YOSHIMURA, N.	STRESS ANALYSIS OF THIN PLATE STRUCTURE BY THE FINITE ELEMENT METHOD	69-0403
YOSHIMURA, N.	TIDAL SIMULATION BY TIME SPACE ELEMENT	75-0867
YOSHINAGA, S.	FORMULATION OF STRESS STRAIN INDUCED DIFFUSION OF HYDROGEN AND ITS SOLUTION BY COMPUTER-AIDED FINITE	75-0326
YOSHITAKE, Y.	ANALYSIS OF SEEPAGE THROUGH DAMS AND IN FOUNDATION BY FINITE ELEMENT METHOD	73-1110
YOUNG, C-I. J.	PREDICTION OF TRANSMISSION LOSS IN MUFFLERS BY THE FINITE ELEMENT METHOD	75-0160
YOUNG, D. M.	ITERATIVE SOLUTION OF LINEAR SYSTEMS ARISING FROM FINITE ELEMENT TECHNIQUES	75-0126
YOUNG, J. P.	MATHEMATICAL MODELING VIA DIRECT USE OF VIBRATION DATA	69-0310
YOUNG, J. W.	CRASH: A COMPUTER SIMULATION OF NONLINEAR TRANSIENT RESPONSE OF STRUCTURES	72-0569
YOUNG, R. C.	GNATS: A FINITE ELEMENT COMPUTER PROGRAM FOR THE GENERAL NONLINEAR ANALYSIS OF TWO-DIMENSIONAL STRUC	74-0875
YOUNG, R. C.	LOCAL ERROR BOUNDS IN BOUNDARY VALUE PROBLEMS USING FINITE ELEMENT DISCRETIZATION	72-0982
YOUNG, R. C.	ORGANIZATION OF GNATS: A GENERAL NONLINEAR ANALYSIS COMPUTER PROGRAM FOR TWO-DIMENSIONAL STRUCTURES	74-0871
YOUNG, R. C.	SOLUTION OF MIXED BOUNDARY VALUE PROBLEMS WITH LOCAL ERROR BOUND BY THE FINITE ELEMENT METHOD	73-0795
YOUNG, V. Y. C.	FINITE ELEMENT ANALYSIS OF TRANSONIC FLOW	74-1223
YOUNG, V.Y.C.	A NON VARIATIONAL FINITE ELEMENT ANALYSIS FOR THE NAVIER STOKES EQUATIONS	74-0010
YOUNGS, E.	THE USE OF THE FINITE ELEMENT METHOD IN OBTAINING SOIL-WATER PRESSURE AND SOIL-WATER CONTENT DISTRIB	75-0112
YOUNGS, E. G.	CALCULATION OF STEADY-STATE WATER-TABLE HEIGHTS IN DRAINED SOILS BY MEANS OF FINITE-ELEMENT METHOD	75-0457
YOUSEF, S. S.	ELASTIC BENDING OF CIRCULAR PLATES OF VARIABLE THICKNESS. AN ANALYTICAL AND EXPERIMENTAL STUDY	70-0163
YOUSSEF, A.	THE ROLE OF FINITE DEFORMATION ANALYSIS IN PLANE STRESS AND STRAIN FRACTURES	74-0322
YPEREZ, L. P.	A GENERAL CALCULATION OF THE STRUCTURAL LOADS ON A SHIP IN A SEAWAY	71-0345
YU, C. W.	TWO-DIMENSIONAL TIME-DEPENDENT ANALYSIS OF PERFORATED END CAPS FOR NUCLEAR REACTOR PRESSURE VESSELS	71-0239
YU, J. C. M.	APPLICATION OF CONFORMAL MAPPING AND VARIATIONAL METHOD TO THE STUDY OF HEAT CONDUCTION IN POLYGONAL	71-0444
YU, Y. S.	ANALYSIS OF ROCK SLOPES, USING THE FINITE ELEMENT METHOD	70-0184
YU, Y. S.	COMPUTER PROGRAM FOR AUTOMATIC MESH GENERATION FOR THE 2-D FINITE ELEMENT PROGRAM WILAX	73-0565
YU, Y. S.	DEVELOPMENT AND USE OF COMPUTER PROGRAMS FOR FINITE ELEMENT ANALYSIS, (LA MISE AU POINT ET L'UTILIZA	69-0289
YU, Y. S.	NOTE ON THE STRESS CONCENTRATIONS AT THE END OF A CYLINDRICAL HOLE	70-0139
YU, Y. Y.	DYNAMIC INSTABILITY OF DUCTS CONVEYING FLUID	70-0691
YUAN, Y. C.	SPACE FINITE-ELEMENTS 2-DIMENSIONAL MULTIGROUP TRANSPORT CALCULATIONS USING PHASE-SPACE FINITE-ELEME	75-0504
YUDHBIR,	STRESS-PATH DEPENDENT DEFORMATION MODULI OF CLAY	75-0050
YUFIN, S. A.	RASCHET NAPRYAZHENIL L PEREMESHCHENII V SVODE PODZEMNOGO MASHINNOGO ZALA GES I V OKRUZHAYUSHCHEM SKA	74-0542
YUHARA, T.	A METHOD OF FINITE DEFLECTION ELASTIC PLASTIC ANALYSIS OF FRAMED STRUCTURES SUBJECTED TO LARGE AXIAL	69-0239
YUKI, T.	AUTOMATED DATA GENERATION AND LARGE MATRIX OPERATION FOR 3 DIMENSIONAL PLATE STRUCTURE ANALYSIS	73-0339
YUKI, T.	STUDY ON FINITE ELEMENT METHOD FOR STRUCTURAL ANALYSIS - ISTRAN/PL (INH STRUCTURE ANALYSIS PLATE STR	73-0167
YUNG, C.	COMPUTER ANALYSIS OF FRUIT DETACHMENT DURING TREE SHAKING	74-1162
YUNG, C.	COMPUTER SIMULATION OF VIBRATION OF WHOLE TREE SYSTEMS USING FINITE ELEMENT METHODS	74-1161
YUNG, C.	SIMULATION OF VIBRATION OF WHOLE TREE SYSTEMS USING FINITE ELEMENTS	75-0765
YUNG, CH.	VIBRATION STUDY OF COMPLETE TREE SYSTEMS USING FINITE ELEMENT METHOD	75-0790
YURKOVICH, R. N	DYNAMIC ANALYSIS OF STIFFENED PANEL STRUCTURES	71-0522
YURKOVICH, R. N	THEORETICAL AND EXPERIMENTAL ANALYSIS OF STIFFENED PANEL UNDER DYNAMIC CONDITIONS	70-0130
YUUKI, R.	SOME EXAMINATIONS ON AVAILABILITY OF FINITE ELEMENT METHODS IN LINEAR FRACTURE MECHANICS	73-0112
YUZUGULLU, O.	FINITE-ELEMENT APPROACH FOR THE PREDICTION OF INELASTIC BEHAVIOR OF SHEAR WALL-FRAME SYSTEMS	72-0983
YUZUGULLU, O.	NUMERICAL PROCEDURES FOR THE DETERMINATION OF THE BEHAVIOUR OF A SHEAR WALL FRAME SYSTEM	73-0216
ZACCARELLI, P.	DESCRIPTION OF THE IBM-360 VERSION OF THE FINITE ELEMENTS COMPUTER PROGRAM SAFE-3D	72-0775
ZACCARELLI, P.	VERIFICATION OF THE FINITE ELEMENT COMPUTER CODE SAFE-2D	74-0873
ZAFIRAKOPOULOS,	FORCES AND DEFORMATIONS AT THE LATERAL STIFFENER OF THIN WALLED U-BARS	73-0236
ZAHARY, G.	UNDERGROUND MEASUREMENTS IN A STEEPLY DIPPING OREBODY	68-0085
ZAK, A. R.	DEVELOPMENT OF A GUN-TURRET MODEL AND OF THE EQUATIONS FOR COUPLING IT WITH A HELICOPTER FINITE-ELEM	73-0752
ZAK, A. R.	ELASTIC ANALYSIS OF CYLINDRICAL CONFIGURATIONS WITH STRESS SINGULARITIES	71-0325

ZAK, A. R.	FINITE ELEMENT COMPUTER PROGRAM FOR THE ANALYSIS OF LAYERED ORTHOTROPIC STRUCTURES	74-0800
ZAK, A. R.	THEORETICAL AND EXPERIMENTAL ANALYSIS OF STIFFENED PANEL UNDER DYNAMIC CONDITIONS	70-0130
ZALESAK, J.	ANALYTICAL AND EXPERIMENTAL INVESTIGATION OF A 1/8 SCALE DYNAMIC MODEL OF THE SHUTTLE ORBITER. VOLUM	74-1186
ZALESAK, J.	DEVELOPMENT OF TECHNOLOGY FOR MODELING OF A 1/8-SCALE DYNAMIC MODEL OF THE SHUTTLE SOLID ROCKET BOOS	74-0703
ZALESAK, J.	MODAL COUPLING PROCEDURES ADAPTED TO NASTRAN ANALYSIS OF THE 1/8 - SCALE SHUTTLE STRUCTURAL DYNAMICS	75-0677
ZALESAK, J.	MODAL COUPLING PROCEDURES ADAPTED TO NASTRAN ANALYSIS OF THE 1/8-SCALE SHUTTLE STRUCTURAL DYNAMICS M	75-0678
ZALESNYI, V. B.	FINITE-DIFFERENCE AND FINITE-ELEMENT METHODS IN GLOBAL WIND-DRIVEN OCEAN CIRCULATION	75-0345
ZANGHIRELLA, E.	PWR DEVELOPMENT FOR NUCLEAR PROPULSION: VIBRATION ANALYSIS OF NUCLEAR POWER PLANT	71-0797
ZANGS, L.	CALCULATION OF THERMALLY INDUCED STRAINS IN MACHINE TOOL ELEMENTS	72-0464
ZAPHIR, Z.	FINITE ELEMENT ANALYSIS FOR THE TIME-DEPENDENT INELASTIC MATERIAL BEHAVIOR	75-0749
ZAPIR, Z.	STRUCTURAL ANALYSIS OF ARAVA FUSELAGE USING FINITE ELEMENT METHOD	71-0578
ZAR, J. L.	THE USE OF A LASER FOR ARPA MILITARY GEOPHYSICS PROGRAM (ROCK MECHANICS AND RAPID EXCAVATION)	72-0800
ZARGHAMEE, M. S	MINIMUM WEIGHT DESIGN WITH STABILITY CONSTRAINT	70-0274
ZARGHAMER, M. S	DIFFERENTIAL OF VARIATIONAL FORMS	71-0445
ZAVELANI-ROSSI,	FINITE ELEMENT TECHNIQUES IN PLANE LIMIT PROBLEMS	74-1206
ZAVELANI, A.	LINEAR PROGRAMMING APPROACH TO SHAKEDOWN ANALYSIS OF STRUCTURES	74-0181
ZAVENLANI-ROSSI	FINITE ELEMENT APPROACH TO OPTIMAL DESIGN OF PLASTIC STRUCTURES IN PLANE STRESS.	72-0397
ZEHLEIN, H.	ASSESSMENT OF COMMERCIALLY AVAILABLE FINITE ELEMENT CODES FOR THE DYNAMIC ANALYSIS OF STRUCTURES	73-0954
ZEHLEIN, H.	EINSATZ DES COMPUTERS BEI DER ENTWICKLUNG EINER PANZERWANNE (THE USE OF A COMPUTER FOR DEVELOPING A	73-0211
ZEILER, S. M.	APACHE: A THREE-DIMENSIONAL FINITE ELEMENT PROGRAM FOR STEADY-STATE OR TRANSIENT HEAT CONDUCTION ANA	73-0971
ZEILER, S. M.	MESH 3D: A THREE-DIMENSIONAL FINITE ELEMENT MESH GENERATOR PROGRAM FOR EIGHT-NODE ISOPARAMETRIC ELEM	73-0972
ZELAZNY, S. W.	PREDICTIONS IN ENVIRONMENTAL HYDRODYNAMICS USING THE FINITE ELEMENT METHOD 2. APPLICATIONS	75-0064
ZENISEK, A.	A SIMPLE ALGORITHM FOR THE STIFFNESS MATRIX OF TRIANGULAR PLATE BENDING ELEMENTS	71-0152
ZENISEK, A.	CALCULATION OF RECTANGULAR PLATES BY THE FINITE ELEMENT METHOD	69-0506
ZENISEK, A.	CONVERGENCE OF A FINITE ELEMENT PROCEDURE FOR SOLVING BOUNDARY VALUE PROBLEMS OF THE FOURTH ORDER	70-0131
ZENISEK, A.	DESIGN OF TWO AND THREE DIMENSIONAL STRUCTURES BY THE FINITE ELEMENT METHOD (IN GERMAN)	74-1364
ZENISEK, A.	INTERPOLATION POLYNOMIALS ON THE TRIANGLE	70-0235
ZENISEK, A.	INTERPOLATION POLYNOMIALS ON THE TRIANGLE ELEMENT	70-0326
ZENISEK, A.	MATHEMATICAL ASPECTS OF THE FINITE ELEMENT METHOD	71-0723
ZENISEK, A.	SOME ELEMENT TYPES AND TRIAL FUNCTIONS IN THE FINITE ELEMENT METHOD	70-0242
ZENISEK, A.	STRESS ANALYSIS OF TWO AND THREE DIMENSIONAL STRUCTURES BY THE FINITE ELEMENT METHOD	72-0794
ZENISEK, A.	TECHNICAL, PHYSICAL AND MATHEMATICAL PRINCIPLES OF THE FINITE ELEMENT METHOD	71-0676
ZENISEK, A.	THE FINITE ELEMENT METHOD	70-0327
ZIEGLER, F.	PLANAR FINITE ELEMENT AT RANDOM TEMPERATURE	73-0140
ZIENKIEWICZ, O.	A SIMPLE MATRIX-VECTOR HANDLING SCHEME FOR THREE-DIMENSIONAL AND SHELL ANALYSIS	70-0317
ZIENKIEWICZ, O.	A FINITE ELEMENT APPROACH TO TWO-PHASE FLOW IN POROUS MEDIA	75-0392
ZIENKIEWICZ, O.	A GENERAL APPROACH TO PROBLEMS OF LARGE DEFORMATION, AND PLASTICITY USING ISOPARAMETRIC ELEMENT	71-0273
ZIENKIEWICZ, O.	A NOTE ON NUMERICAL COMPUTATION OF ELASTIC CONTACT PROBLEMS	75-0621
ZIENKIEWICZ, O.	A NUMERICAL METHOD OF VISCO-ELASTIC STRESS ANALYSIS	68-0175
ZIENKIEWICZ, O.	ADAPTATION OF CRITICAL STATE SOIL MECHANICS FOR USE IN FINITE ELEMENTS	71-0534
ZIENKIEWICZ, O.	AN ANALYSIS OF VARIOUS TIME STEPPING SCHEMES FOR INITIAL VALUE PROBLEMS	73-0607
ZIENKIEWICZ, O.	AN AUTOMATIC MESH GENERATION SCHEME FOR PLANE AND CURVED SURFACES BY ISOPARAMETRIC COORDINATES	71-0271
ZIENKIEWICZ, O.	ANALYSIS OF NONLINEAR PROBLEMS IN ROCK MECHANICS WITH PARTICULAR REFERENCE TO JOINTED ROCK SYSTEMS	70-0282
ZIENKIEWICZ, O.	ANALYSIS OF REAL STRUCTURES FOR CREEP, PLASTICITY AND OTHER COMPLEX CONSTITUTIVE LAWS.	69-0287
ZIENKIEWICZ, O.	ANALYSIS OF SLABS WITH EDGE BEAMS	71-0075
ZIENKIEWICZ, O.	ANALYSIS OF THICK AND THIN SHELL STRUCTURE BY CURVED FINITE ELEMENTS	70-0016
ZIENKIEWICZ, O.	ANALYSIS OF VISCOPLASTIC EFFECTS IN PRESSURE VESSELS BY THE FINITE ELEMENT METHOD	74-0573
ZIENKIEWICZ, O.	APPLICATION OF FINITE ELEMENTS TO THE SOLUTION OF HELMHOLT'S EQUATION	68-0020
ZIENKIEWICZ, O.	APPLICATION OF FINITE ELEMENTS TO HEAT CONDUCTION PROBLEMS INVOLVING LATENT HEAT	73-0520
ZIENKIEWICZ, O.	ARCH DAMS ANALYSED BY A LINEAR FINITE ELEMENT SHELL THE FINITE STRIP METHOD IN THE ANALYSIS OF ELAST	68-0189
ZIENKIEWICZ, O.	BUTTRESS DAMS ON COMPLEX ROCK FOUNDATIONS	64-0013
ZIENKIEWICZ, O.	COMPARISON OF ANALYTICAL AND EXPERIMENTAL RESULTS IN THE DESIGN OF A HOLLOW GRAVITY DAM	57-0109
ZIENKIEWICZ, O.	COMPLETE POLYNOMIAL DISPLACEMENT FIELDS FOR FINITE ELEMENT METHODS	68-0092
ZIENKIEWICZ, O.	CONSTRAINED VARIATIONAL PRINCIPLES AND PENALTY FUNCTION METHODS IN FINITE ELEMENT ANALYSIS	72-0551
ZIENKIEWICZ, O.	CONTINUUM MECHANICS AS AN APPROACH TO ROCK MASS PROBLEMS	68-0172
ZIENKIEWICZ, O.	COUPLED VIBRATIONS OF A STRUCTURE SUBMERGED IN A COMPRESSIBLE FLUID	69-0507
ZIENKIEWICZ, O.	CURVED ISOPARAMETRIC QUADRILATERAL ELEMENTS FOR FINITE ANALYSIS	68-0008
ZIENKIEWICZ, O.	CURVED THICK SHELL AND MEMBRANE ELEMENTS WITH PARTICULAR REFERENCE TO AXISYMMETRIC PROBLEMS	68-0005
ZIENKIEWICZ, O.	DISCUSSION ON THE ADAPTION OF CRITICAL STATE SOLID MECHANICS THEORY FOR USE IN FINITE ELEMENTS	71-0007
ZIENKIEWICZ, O.	EARTHQUAKE BEHAVIOUR OF RESERVOIR-DAM SYSTEMS (DISCUSSION)	69-0283
ZIENKIEWICZ, O.	ELASTIC TORSIONAL ANALYSIS OR IRREGULAR SHAPES	66-0078
ZIENKIEWICZ, O.	ELASTO-PLASTIC SOLUTIONS OF ENGINEERING PROBLEMS * INITIAL STRESS * FINITE ELEMENT APPROACH	69-0282
ZIENKIEWICZ, O.	ELASTO-PLASTIC STRESS ANALYSIS. A GENERALIZATION FOR VARIOUS CONSTITUTIVE RELATIONS INCLUDING STRAIN	72-0547
ZIENKIEWICZ, O.	ELASTOPLASTIC ANALYSIS OF A CANTILEVER BEAM	73-0984
ZIENKIEWICZ, O.	FINITE ELEMENT ANALYSIS OF THE STEADY FLOW OF NON NEWTONIAN FLUIDS THROUGH PARALLEL SIDED CONDUITS	74-0056
ZIENKIEWICZ, O.	FINITE ELEMENT ANALYSIS OF STEADY FLOW OF NON-NEWTONIAN FLUIDS	74-0465
ZIENKIEWICZ, O.	FINITE ELEMENT ANALYSIS OF SOME CONCRETE NONLINEARITIES. THEORY AND EXAMPLES	74-1287
ZIENKIEWICZ, O.	FINITE ELEMENT METHOD OF ANALYSIS FOR ARCH DAMS SHELLS AND COMPARISON WITH FINITE DIFFERENCE PROCEDU	65-0036
ZIENKIEWICZ, O.	FINITE ELEMENT METHOD IN STRESS ANALYSIS	69-0413
ZIENKIEWICZ, O.	FINITE ELEMENT METHOD IN ANALYSIS OF REINFORCED AND PRESTRESSED CONCRETE STRUCTURES	71-0270
ZIENKIEWICZ, O.	FINITE ELEMENT METHODS IN ANALYSIS OF REACTOR VESSELS	72-0271
ZIENKIEWICZ, O.	FINITE ELEMENT METHODS IN FLOW PROBLEMS	74-0001
ZIENKIEWICZ, O.	FINITE ELEMENT METHODS IN FLOW PROBLEMS	74-0752
ZIENKIEWICZ, O.	FINITE ELEMENT PROCEDURES IN THE SOLUTION OF PLATE AND SHELL PROBLEMS	65-0038
ZIENKIEWICZ, O.	FINITE ELEMENT PROGRAMS FOR SLAB BRIDGE DESIGN	71-0082
ZIENKIEWICZ, O.	FINITE ELEMENT SOLUTION OF NON-LINEAR HEAT CONDUCTION PROBLEMS WITH SPECIAL REFERENCE TO PHASE CHANG	74-0348
ZIENKIEWICZ, O.	FINITE ELEMENT STUDIES OF SOILS AND POROUS MEDIA	71-0504
ZIENKIEWICZ, O.	FINITE ELEMENTS - THE BACKGROUND STORY	72-0053
ZIENKIEWICZ, O.	FINITE ELEMENTS FOR TWO PHASE FLOW IN POROUS MEDIA	74-0060
ZIENKIEWICZ, O.	FINITE ELEMENTS IN THE SOLUTION OF FIELD PROBLEMS	65-0034
ZIENKIEWICZ, O.	FLOW OF PLASTIC AND VISCO-PLASTIC SOLIDS WITH SPECIAL REFERENCE TO EXTRUSION AND FORMING PROCESSES	74-0210
ZIENKIEWICZ, O.	GENERAL METHOD OF ANALYSIS BASED ON FINITE ELEMENTS	68-0277
ZIENKIEWICZ, O.	HARBOUR OSCILLATION - A NUMERICAL TREATMENT FOR UNDAMPED NATURAL MODES	70-0208
ZIENKIEWICZ, O.	ISO-PARAMETRIC AND ASSOCIATED ELEMENT FAMILIES FOR TWO AND THREE DIMENSIONAL ANALYSIS	69-0285
ZIENKIEWICZ, O.	ISOPARAMETRIC AND ALLIED NUMERICALLY INTEGRATED ELEMENTS - A REVIEW	71-0763
ZIENKIEWICZ, O.	ISOPARAMETRIC AND OTHER NUMERICALLY INTEGRATED ELEMENTS	71-0272
ZIENKIEWICZ, O.	ISOPARAMETRIC ELEMENT FORMS IN FINITE ELEMENT ANALYSIS	71-0502
ZIENKIEWICZ, O.	ISOPARAMETRIC FINITE ELEMENT SYSTEM: A NEW CONCEPT IN FINITE ELEMENT ANALYSIS	68-0223
ZIENKIEWICZ, O.	LEAST SQUARE-FINITE ELEMENT FOR ELASTO-STATIC PROBLEMS	74-0279
ZIENKIEWICZ, O.	MATRIX ITERATION AND ACCELERATION PROCESSES IN FINITE ELEMENT PROBLEMS OF STRUCTURAL MECHANICS	70-0132
ZIENKIEWICZ, O.	NATURAL FREQUENCIES OF COMPLEX FREE OR SUBMERGED STRUCTURES BY THE FINITE ELEMENT METHOD	65-0026
ZIENKIEWICZ, O.	NEWTONIAN AND NON-NEWTONIAN VISCOUS FLOW	75-0109
ZIENKIEWICZ, O.	ON SOME FREE SURFACE ON TRANSIENT FLOW PROBLEMS OF SEEPAGE AND IRROTATIONAL FLOW	72-0070
ZIENKIEWICZ, O.	PENALTY FUNCTION APPROACH TO PROBLEMS OF PLASTIC FLOW OF METALS WITH LARGE SURFACE DEFORMATIONS	75-0282
ZIENKIEWICZ, O.	PLATE AND SHELL PROBLEMS. FINITE ELEMENT DISPLACEMENT APPROACH	66-0005
ZIENKIEWICZ, O.	PLATES AND TANKS ON ELASTIC FOUNDATIONS, AN APPLICATION OF FINITE ELEMENT METHOD	65-0010
ZIENKIEWICZ, O.	POROUS MEDIA - SOME PROBLEMS IN CIVIL AND PETROLEUM ENGINEERING	74-0349
ZIENKIEWICZ, O.	PRACTICAL HINTS ON HOW TO UNDERSTAND, MODEL, USE AND INTERPRET FINITE ELEMENTS	73-0141
ZIENKIEWICZ, O.	PROBLEMS IN ROCK MECHANICS	65-0033
ZIENKIEWICZ, O.	REDUCED INTEGRATION TECHNIQUE IN GENERAL ANALYSIS OF PLATES AND SHELLS	71-0374
ZIENKIEWICZ, O.	SETTLEMENT ANALYSIS OF A STRIP FOOTING USING A CRITICAL STATE SOIL MODEL IN CONJUNCTION WITH FINITE	71-0533
ZIENKIEWICZ, O.	SHORT COMMUNICATIONS: TORSION OF AXI-SYMMETRIC SOLIDS OF VARIABLE DIAMETER - INCLUDING ACCELERATION	74-0203

117

ZIENKIEWICZ, O.	SIMPLE FINITE ELEMENT SOLUTION FOR PLATES OF HOMOGENEOUS, SANDWICH AND CELLULAR CONSTRUCTION	75-0140
ZIENKIEWICZ, O.	SLAB BRIDGES WITH ARBITRARY SHAPE AND SUPPORT CONDITIONS - A GENERAL METHOD OF ANALYSIS BASED ON FIN	68-0233
ZIENKIEWICZ, O.	SOLUTION OF ANISOTROPIC SEEPAGE PROBLEMS BY FINITE ELEMENTS	66-0080
ZIENKIEWICZ, O.	SOLUTION OF THREE-DIMENSIONAL FIELD PROBLEMS BY THE FINITE ELEMENT METHOD	67-0091
ZIENKIEWICZ, O.	SOME CREEP EFFECTS IN STRESS ANALYSIS WITH PARTICULAR REFERENCE TO CONCRETE PRESSURE VESSELS	69-0284
ZIENKIEWICZ, O.	SOME LINEAR AND NON-LINEAR PROBLEMS IN FLUID MECHANICS. FINITE ELEMENT METHOD FORMULATION	73-0523
ZIENKIEWICZ, O.	SOME NON LINEAR PROBLEMS IN SOIL AND ROCK MECHANICS - FINITE ELEMENT SOLUTION	69-0363
ZIENKIEWICZ, O.	SPONTANEOUS IGNITION - FINITE ELEMENT SOLUTIONS FOR STEADY AND TRANSIENT CONDITIONS	74-0353
ZIENKIEWICZ, O.	STRESS ANALYSIS	65-0055
ZIENKIEWICZ, O.	STRESS ANALYSIS AND SPECIAL PROBLEMS OF PRESTRESSED DAMS	61-0006
ZIENKIEWICZ, O.	STRESS ANALYSIS BY THE FINITE ELEMENT METHOD - THERMAL EFFECTS	67-0022
ZIENKIEWICZ, O.	STRESS ANALYSIS OF ROCK AS A NO-TENSION MATERIAL	68-0174
ZIENKIEWICZ, O.	STRESSES IN ANISOTROPIC MEDIA WITH PARTICULAR REFERENCE TO PROBLEMS OF ROCK MECHANICS	66-0079
ZIENKIEWICZ, O.	STRESSES IN ANISOTROPIC MEDIA OF ROCK MECHANICS	66-0127
ZIENKIEWICZ, O.	STRESSES IN BUTTRESS DAMS	65-0035
ZIENKIEWICZ, O.	STRESSES IN SHAFTS	67-0104
ZIENKIEWICZ, O.	THE ALPHA CONSTANT-STIFFNESS METHOD FOR THE ANALYSIS OF NONLINEAR PROBLEMS	72-0566
ZIENKIEWICZ, O.	THE FINITE ELEMENT METHOD FOR ANALYSIS OF ELASTIC ISOTROPIC AND ORTHOTROPIC SLABS	64-0014
ZIENKIEWICZ, O.	THE FINITE ELEMENT METHOD AND ITS APPLICATION IN VIBRATION ANALYSIS	66-0077
ZIENKIEWICZ, O.	THE FINITE ELEMENT METHOD IN STRUCTURAL AND CONTINUUM MECHANICS	67-0093
ZIENKIEWICZ, O.	THE FINITE ELEMENT METHOD IN VIBRATION ANALYSIS	67-0095
ZIENKIEWICZ, O.	THE FINITE ELEMENT METHOD: FROM INTUITION TO GENERALITY	70-0010
ZIENKIEWICZ, O.	THE FINITE ELEMENT METHOD IN ENGINEERING SCIENCE	71-0274
ZIENKIEWICZ, O.	THE FINITE ELEMENT METHOD AND BOUNDARY SOLUTION PROCEDURES AS GENERAL APPROXIMATION METHODS FOR FIEL	75-0262
ZIENKIEWICZ, O.	THE ISOPARAMETRIC FINITE ELEMENT SYSTEM - A NEW CONCEPT IN FINITE ELEMENT ANALYSIS	69-0139
ZIENKIEWICZ, O.	THE METHOD OF INTERFACE STRESS ADJUSTMENT AND ITS USES IN SOME PLANE ELASTICITY PROBLEMS	61-0005
ZIENKIEWICZ, O.	THEORETICAL FOUNDATIONS ON THE FINITE ELEMENT METHOD	68-0055
ZIENKIEWICZ, O.	THEORETICAL FOUNDATIONS OF THE FINITE ELEMENT METHOD (DISCUSSION)	70-0074
ZIENKIEWICZ, O.	THERMAL STRESS AND TEMPERATURE FIELDS - A FINITE-ELEMENT APPROACH	71-0782
ZIENKIEWICZ, O.	THREE DIMENSIONAL ANALYSIS OF BUILDINGS COMPOSED OF FLOOR AND WALL PANELS	71-0387
ZIENKIEWICZ, O.	THREE DIMENSIONAL STRESS ANALYSIS	70-0086
ZIENKIEWICZ, O.	THREE-DIMENSIONAL STRESS ANALYSIS OF ARCH DAMS BY THE FINITE ELEMENT METHOD: PART I, ARCH DAM NO. 1,	66-0099
ZIENKIEWICZ, O.	TRANSIENT FIELD PROBLEMS: TWO-DIMENSIONAL AND THREE-DIMENSIONAL ANALYSIS BY ISO-PARAMETRIC FINITE EL	70-0001
ZIENKIEWICZ, O.	TRANSIENT RESPONSE OF A COUPLED PLATE-ACOUSTIC SYSTEM USING PLATE AND ACOUSTIC FINITE ELEMENTS	72-0275
ZIENKIEWICZ, O.	TRIANGULAR ELEMENTS IN PLATE BENDING CONFORMING AND NON-CONFORMING SOLUTIONS	66-0011
ZIENKIEWICZ, O.	TRIANGULAR MESH GENERATION PROGRAM	67-0037
ZIENKIEWICZ, O.	VIBRATION AND STABILITY OF PLATES USING FINITE ELEMENTS	68-0244
ZIENKIEWICZ, O.	VIBRATION OF THICK, CURVED, SHELLS WITH PARTICULAR REFERENCE TO TURBINE BLADES	70-0017
ZIENKIEWICZ, O.	VISCO-PLASTICITY AND PLASTICITY-AN ALTERNATIVE FOR FINITE ELEMENT SOLUTION OF MATERIAL NONLINEARITIE	74-0410
ZIENKIEWICZ, O.	VISCO-PLASTICITY SOLUTION BY FINITE ELEMENT PROCESS	72-0048
ZIENKIEWICZ, O.	VISCO-PLASTICITY, PLASTICITY AND CREEP IN ELASTIC SOLIDS. A UNIFIED NUMERICAL SOLUTION APPROACH	74-1279
ZIENKIEWICZ, O.	VISCOPLASTICITY, PLASTICITY AND CREEP. A UNIFIED NUMERICAL SOLUTION APPROACH	74-1205
ZIENKIEWICZ, O.	VISCOPLASTICITY, PLASTICITY AND PLASTIC FLOW	74-1278
ZIENKIEWICZ, O.	VISCOUS, INCOMPRESSIBLE FLOW WITH SPECIAL REFERENCE TO NON-NEWTONIAN (PLASTIC) FLUIDS	74-0053
ZIENKIEWICZ, O.	WEIGHTED RESIDUAL PROCESSES IN FINITE ELEMENT WITH PARTICULAR REFERENCE TO SOME TRANSIENT AND COUPLE	71-0507
ZIENKIEWICZ, O.	WHY FINITE ELEMENTS	74-1198
ZIKE, J. H.	DESIGN OF PROPELLANT FINS SUBJECTED TO AXIAL ACCELERATION	71-0734
ZIMMER, A	ELASTOSTATIK-ELEMENTMETHODE (ESEM)	69-0415
ZIMMER, A.	FINITE ELEMENT ANALYSIS, AN AUTOMOBILE ENGINEER'S TOOL	74-0238
ZIMMER, A.	MODERNE VERFAHREN ZUR UNTERSUCHUNG DER STEIFIGKEITS-UND FESTIGKEITSVERHALTNISSE VON RAHMENKORPERN, I	69-0401
ZIMMER, R. A.	ANCHOR-PILE DESIGN FOR OCEAN-FLOOR ENVIRONMENTS USING FINITE-ELEMENT ANALYSIS	75-0459
ZIMMERMAN, K. L	PLOTS - THE PLOTTING SUBROUTINE INCORPORATED IN THE BALLISTIC RESEARCH LABORATORIES ELECTRONIC SCIEN	72-0803
ZIMMERMANN, T.	ANALYSE TRIDIMENSIONALE DU COMPORTEMENT NON-LINEAIRE D'UN CAISSON DE REACTEUR NUCLEAIRE EN BETON PRE	71-0144
ZIMMERMANN, T.	THREE DIMENSIONAL RUPTURE ANALYSIS OF A PRESTRESSED CONCRETE PRESSURE VESSEL INCLUDING CREEP EFFECTS	73-0826
ZIMMERMANN, T.	THREE DIMENSIONAL RUPTURE ANALYSIS OF A PRESTRESSED CONCRETE PRESSURE VESSEL INCLUDING CREEP EFFECTS	74-0538
ZIMMERMANN, T.	APPLICATION OF NONLINEAR ANALYSIS TO SOLID STRUCTURES	72-0212
ZINGONE, G.	LIMIT ANALYSIS OF FRAME SYSTEMS STIFFENED BY PANELS	74-0584
ZITZER, H. H.	CASE LINER BOND ANALYSIS	74-0817
ZLAMAL, M.	A FINITE ELEMENT PROCEDURE OF THE SECOND ORDER ACCURACY	70-0277
ZLAMAL, M.	A REMARK ON THE SERENDIPITY FAMILY	73-1081
ZLAMAL, M.	A SIMPLE ALGORITHM FOR THE STIFFNESS MATRIX OF TRIANGULAR PLATE BENDING ELEMENTS	71-0152
ZLAMAL, M.	CONVERGENCE OF A FINITE ELEMENT PROCEDURE FOR SOLVING BOUNDARY VALUE PROBLEMS OF THE FOURTH ORDER	70-0131
ZLAMAL, M.	CURVED ELEMENTS IN FINITE ELEMENT METHOD. 1.	73-0045
ZLAMAL, M.	CURVED ELEMENTS IN FINITE ELEMENT METHOD - 2	74-0172
ZLAMAL, M.	CURVED ELEMENTS, ERROR BOUNDS, TIME DEPENDENT PROBLEMS FIVE LECTURES	73-0636
ZLAMAL, M.	FINITE ELEMENT METHODS FOR PARABOLIC EQUATIONS	74-0173
ZLAMAL, M.	FINITE ELEMENT METHODS FOR PARABOLIC EQUATIONS	74-1280
ZLAMAL, M.	FINITE ELEMENT METHODS IN HEAT CONDUCTION PROBLEMS	75-0095
ZLAMAL, M.	FINITE-ELEMENT MULTISTEP DISCRETIZATIONS OF PARABOLIC BOUNDARY-VALUE PROBLEMS	75-0501
ZLAMAL, M.	MATHEMATICAL ASPECTS OF THE FINITE ELEMENT METHOD	71-0723
ZLAMAL, M.	NONCONFORMING ELEMENTS IN THE FINITE ELEMENT METHOD	72-0341
ZLAMAL, M.	NONCONFORMING ELEMENTS IN FINITE ELEMENT METHOD WITH PENALTY	73-0056
ZLAMAL, M.	ON SOME FINITE ELEMENT PROCEDURES FOR SOLVING SECOND ORDER BOUNDARY VALUE PROBLEMS	69-0288
ZLAMAL, M.	ON THE FINITE ELEMENT METHOD	68-0176
ZLAMAL, M.	SOME RECENT ADVANCES IN THE MATHEMATICS OF FINITE ELEMENTS	73-0287
ZLAMAL, M.	TECHNICAL, PHYSICAL AND MATHEMATICAL PRINCIPLES OF THE FINITE ELEMENT METHOD	71-0676
ZLAMAL, M.	THE FINITE ELEMENT METHOD	70-0327
ZLAMAL, M.	THE FINITE ELEMENT METHOD IN DOMAINS WITH CURVED BOUNDARIES	73-0080
ZLAMAL, M.	TRIANGULAR ELEMENTS IN THE FINITE ELEMENT METHOD	70-0194
ZLAMAL, M.	UNCONDITIONALLY STABLE FINITE ELEMENT SCHEMES FOR PARABOLIC EQUATIONS	74-1286
ZOLTI, E.	REBO: COMPUTER CODE FOR THE GENERATION OF A THREE-DIMENSIONAL FINITE-ELEMENT GRID OF THE NOZZLE-CYLI	74-1310
ZOROWSKI, C. F.	A STUDY OF INTERNAL STRESSES IN STATICALLY DEFORMED PNEUMATIC TIRES. A MATHEMATICAL MODEL FOR THE PN	70-0421
ZOROWSKI, C.F.	MATHEMATICAL PREDICTION OF DYNAMIC TIRE BEHAVIOR	73-0579
ZUDANS, Z.	A THREE-DIMENSIONAL FINITE ELEMENT COMPUTER CODE FOR THE ANALYSIS OF COMPLEX STRUCTURES	71-0275
ZUDANS, Z.	ANALYSIS OF ASYMMETRIC STIFFENED SHELL TYPE STRUCTURES BY THE FINITE ELEMENT METHOD: I FLAT RECTANGU	68-0283
ZUDANS, Z.	ANALYSIS OF ASYMMETRIC STIFFENED SHELL TYPE STRUCTURES BY FINITE ELEMENT METHOD II. SIX DEGREE-OF-FR	69-0373
ZUDANS, Z.	ANALYSIS OF ELASTICALLY COUPLED SHELLS ON ELASTIC FOUNDATION BY HYBRID METHOD	71-0276
ZUDANS, Z.	DYPLAS, A FINITE ELEMENT DYNAMIC ELASTIC-PLASTIC LARGE DEFORMATION ANALYSIS PROGRAM	74-0306
ZUDANS, Z.	FINITE ELEMENT INCREMENTAL ELASTIC-PLASTIC ANALYSIS OF PRESSURE VESSELS	70-0253
ZUDANS, Z.	SURVEY OF ADVANCED STRUCTURAL DESIGN ANALYSIS TECHNIQUES	69-0306
ZUDANS, Z.	VERIFICATION OF A HYBRID GENERAL PURPOSE SHELL AND BODY OF REVOLUTION COMPUTER PROGRAM	71-0333
ZUK, J.	COMPRESSIBLE SEAL FLOW ANALYSIS USING THE FINITE ELEMENT METHOD WITH GALERKIN SOLUTION TECHNIQUE	74-0688
ZULL, L.	FINITE ELEMENT METHOD IN PETROLEUM RESERVOIR ANALYSIS	72-0049
ZUNDERDORP, H.	STRESS ANALYSIS OF TUBULAR JOINTS FOR OFFSHORE STRUCTURES	73-0207
ZUNGUO, H.	FINITE ELEMENT ANALYSIS OF TIME-DEPENDENT PUMPING TEST DATA IN AN UNCONFINED AQUIFER	75-0497
ZWINGENBERGER,	NEUESTE ERFAHRUNGEN BEI DER RECHNERISCHEN UND KONSTRUKTIVEN AUSLEGUNG KOMPLIZIERTER SCHALENSYSTEME	71-0052
ZYVOLOSKI, G.	FINITE ELEMENT SOLUTION OF UNSTEADY AND UNSATURATED FLOW IN POROUS MEDIA	72-0064
ZYVOLOSKI, G.	FINITE ELEMENT SOLUTION OF TWO DIMENSIONAL UNSTEADY AND UNSATURATED FLOW IN POROUS MEDIA	75-0113
ZYVOLOSKI, G.	FINITE ELEMENT WEIGHTED RESIDUAL SOLUTION TO ONE DIMENSIONAL FIELD PROBLEMS	73-0388
ZYVOLOSKI, G.	FINITE ELEMENT WEIGHTED RESIDUAL SOLUTION OF ONE-DIMENSIONAL, UNSTEADY AND UNSATURATED FLOWS IN PORO	73-0712
ZYVOLOSKI, G.	TRANSIENT TWO DIMENSIONAL HEAT CONDUCTION PROBLEMS SOLVED BY THE FINITE ELEMENT METHOD	74-0364

Keyword Listing

ABATEMENT	EARTH VIBRATION EFFECTS AND ABATEMENT FOR MILITARY FACILITIES. REPORT 3. ANALYSIS METHOD FOR FOOTING	72-0806
ABBAU	DIE BERECHNUNG DER BEWEGUNGEN UND SPANNUNGEN DES GEBIRGES VOM ABBAU BIS ZUR TAGESOBERFLAECHE NACH DE	73-0456
ABBAUHOHLRAEUME	BERECHNUNG UND BEURTEILUNG DER STANDSICHERHEIT VON ABBAUHOHLRAEUMEN IN STEILEN GANGERZLAGERSTAETTEN	74-0597
ABBAUWIRKUNGEN	NEUARTIGE ANWENDUNG GEOMECHANISCHER UNTERSUCHUNGSMETHODEN ZUR BEURTEILUNG DER ABBAUWIRKUNGEN IM GANG	75-0530
ABGESETZTER	ZUR BERECHNUNG DER BIEGESTEIFIGKEIT ABGESETZTER STAEBE UND WELLEN UNTER ANWENDUNG VON FINITEN ELEMEN	75-0481
ABLATION	BASELINE SOLUTIONS FOR THE SMOOTH WALL THERMOCHEMICAL ABLATION RESPONSE OF GRAPHITE AND CARBON PHENO	71-0728
ABSOLUTE	THEORETICAL BASIS OF THE BOREHOLE DEEPENING METHOD OF ABSOLUTE STRESS MEASUREMENT	70-0047
ABSTRACTION	FORMAT-FORTRAN MATRIX ABSTRACTION TECHNIQUE	68-0214
ABSTRACTS	STRUCTURAL MECHANICS SOFTWARE. A BIBLIOGRAPHY WITH ABSTRACTS	74-0842
ABSTRACTS	TUNNEL CONSTRUCTION - A BIBLIOGRAPHY WITH ABSTRACTS	74-0748
ABUTMENTS	EARTH PRESSURES AGAINST THE ABUTMENTS AND WING WALLS OF STANDARD MOTORWAY BRIDGES	75-0629
AC-LATTICES	STRONGLY PLANAR AC-LATTICES IN WHICH IDEAL OF FINITE-ELEMENTS IS STANDARD	75-0816
ACCELERATION	DESIGN OF PROPELLANT FINS SUBJECTED TO AXIAL ACCELERATION	71-0734
ACCELERATION	MATRIX ITERATION AND ACCELERATION PROCESSES IN FINITE ELEMENT PROBLEMS OF STRUCTURAL MECHANICS	70-0132
ACCELERATION	SHORT COMMUNICATIONS: TORSION OF AXI-SYMMETRIC SOLIDS OF VARIABLE DIAMETER - INCLUDING ACCELERATION	74-0203
ACCELERATOR	LINEAR ACCELERATOR CAVITY FIELD CALCULATION BY FINITE ELEMENT	73-0008
ACCEPTANCE	THE VERIFICATION AND ACCEPTANCE OF COMPUTER PROGRAMS FOR DESIGN ANALYSIS	70-0334
ACCURACIES	ASSESSMENT OF ACCURACIES OF FINITE ELEMENT EIGENVALUES	72-0075
ACCURACIES	ASSESSMENT OF ACCURACIES OF FINITE ELEMENT EIGENVALUES	72-0747
ACCURACY	A FINITE ELEMENT PROCEDURE OF THE SECOND ORDER ACCURACY	70-0277
ACCURACY	A METHOD TO GALERKIN TYPE ACHIEVING OPTIMUM L(2) ACCURACY FOR FIRST ORDER HYPERBOLICS AND EQUATIONS	72-0521
ACCURACY	A NOTE ON ACCURACY AND CONVERGENCE OF FINITE ELEMENT APPROXIMATIONS	71-0565
ACCURACY	A STUDY OF THE CONVERGENCE CHARACTERISTICS AND ACCURACY OF CONFORMING FINITE ELEMENT SOLUTIONS	72-0967
ACCURACY	ACCURACY AND CONDITION OF CURVED (ISOPARAMETRIC) FINITE ELEMENTS	73-0057
ACCURACY	ACCURACY AND CONVERGENCE OF FINITE ELEMENT APPROXIMATION	68-0228
ACCURACY	ACCURACY AND CONVERGENCE OF FINITE ELEMENT GALERKIN APPROXIMATIONS OF TIME DEPENDENT PROBLEMS WITH E	73-0987
ACCURACY	ACCURACY AND CONVERGENCE OF MIXED FINITE-ELEMENT APPROXIMATIONS OF THIN BARS, MEMBRANES, AND PLATES	73-1120
ACCURACY	ACCURACY AND CONVERGENCE OF FINITE ELEMENT/GALERKIN APPROXIMATIONS OF TIME DEPENDENT PROBLEMS WITH E	75-0006
ACCURACY	ACCURACY AND OPTIMIZATION IN THE ANALYSIS OF SHIP STRUCTURES. USING FINITE ELEMENT METHODS	71-0418
ACCURACY	ACCURACY OF COMPLEX FINITE ELEMENTS	72-0291
ACCURACY	ACCURACY OF EQUILIBRIUM SLOPE STABILITY ANALYSIS	73-0517
ACCURACY	ACCURACY OF FINITE ELEMENT APPROXIMATIONS TO STRUCTURAL PROBLEMS	70-0120
ACCURACY	ACCURACY OF FINITE ELEMENT APPROXIMATIONS TO STRUCTURAL PROBLEMS	70-0438
ACCURACY	ACCURACY OF FINITE ELEMENT EIGENPROBLEMS	71-0088
ACCURACY	ACCURACY OF FINITE-ELEMENT SOLUTIONS FOR MODAL CHARACTERISTICS OF SHELLS OF REVOLUTION	70-0231
ACCURACY	ACCURACY OF MODAL STRESS CALCULATIONS BY FINITE ELEMENT METHOD	70-0223
ACCURACY	ACCURACY OF SOME NUMERICAL METHODS FOR COLUMN BUCKLING	75-0583
ACCURACY	ACCURACY OF THE FINITE ELEMENT ANALYSIS FOR THE ELASTIC PLATE WITH A CIRCULAR HOLE	73-0155
ACCURACY	ACCURACY, STABILITY AND OSCILLATION CHARACTERISTICS OF FINITE ELEMENT METHOD FOR SOLVING HEAT CONDUC	69-0007
ACCURACY	ACCURACY, STABILITY, AND OSCILLATION CHARACTERISTICS OF TRANSIENT TWO-DIMENSIONAL HEAT CONDUCTION	75-0849
ACCURACY	AN EXACT BOUNDARY TECHNIQUE FOR IMPROVED ACCURACY IN THE FINITE ELEMENT METHOD	73-1035
ACCURACY	ANALYSIS OF OUT-OF-PLANE STRESS BY THE SIX NODES, FOURTH ORDER ACCURACY TRIANGULAR ELEMENTS	69-0375
ACCURACY	COMMENT ON THE PAPER "POSSIBLE LOSS OF ACCURACY IN CURVED (ISOPARAMETRIC) FINITE ELEMENTS" BY HENSHE	72-0279
ACCURACY	COMPUTER WORD LENGTH AND SOLUTION ACCURACY. EXPERIENCE WITH ELAS AND THE IBM 360	73-0335
ACCURACY	CONVERGENCE AND ACCURACY IN THE FINITE ELEMENT METHOD	75-0630
ACCURACY	CONVERGENCE AND NUMERICAL ACCURACY WITH SPECIAL REFERENCE TO PLATE BENDING	69-0221
ACCURACY	CONVERGENCE, ACCURACY AND STABILITY OF FINITE ELEMENT APPROXIMATIONS OF A CLASS OF NON-LINEAR HYPERB	73-0159
ACCURACY	FINITE ELEMENT METHOD - ACCURACY AT A POINT	74-0240
ACCURACY	GUYAN REDUCTION SOLUTIONS RECYCLED FOR IMPROVED ACCURACY	71-0159
ACCURACY	NOTE ON THE ACCURACY AND CONVERGENCE OF FINITE ELEMENT APPROXIMATIONS OF THE CONVECTION EQUATION	75-0490
ACCURACY	NUMERICAL ACCURACY AND ROUNDING-OFF ERRORS	69-0348
ACCURACY	NUMERICAL METHODS OF HIGH-ORDER ACCURACY FOR NONLINEAR BOUNDARY VALUE PROBLEMS	68-0194
ACCURACY	OCENA DOKLADNOSCI ROZWIAZAN PROBLEMOW BRZEGOWYCH SPREZYSTYCH CIAL DYSKERTYZOWANYCH. (ASSESSING THE A	70-0921
ACCURACY	ON THE ACCURACY AND NUMERICAL STABILITY OF THE FINITE ELEMENT APPROXIMATION FOR PARABOLIC AND HYPERB	75-0102
ACCURACY	ON THE ACCURACY OF CERTAIN C DEGREE CONTINUOUS FINITE ELEMENT REPRESENTATIONS	74-0367
ACCURACY	ON THE ACCURACY OF FINITE ELEMENT SOLUTIONS TO THE TRANSIENT HEAT CONDUCTION EQUATION	74-0182
ACCURACY	ON THE ACCURACY OF THE TAYLOR APPROXIMATION FOR STRUCTURE RESIZING	74-0147
ACCURACY	OVERCOMING LOSS OF ACCURACY WHEN USING CURVED FINITE ELEMENTS	75-0092
ACCURACY	POSSIBLE LOSS OF ACCURACY IN CURVED FINITE ELEMENTS	72-0278
ACCURACY	SHAPE FUNCTIONS AND ACCURACY OF ARCH FINITE ELEMENTS	73-0038
ACCURACY	STABILITY AND ACCURACY OF NUMERICAL APPROXIMATIONS TO TIME DEPENDENT FLOWS	75-0183
ACCURACY	THE ACCURACY OF THE FINITE ELEMENT METHODS IN CONTINUUM PROBLEMS	66-0027
ACCURACY	TWO TYPES OF PIECEWISE QUADRATIC SPACES AND THEIR ORDER OF ACCURACY FOR POISSON'S EQUATION	72-0520
ACCURATE	ACCURATE CALCULATION OF STRESSES IN BEAMS, DISKS AND PLATES BY MEANS OF FINITE ELEMENT METHODS AND !	70-0133
ACCURATE	ACCURATE FINITE ELEMENT MODELLING OF FLAT AND CURVED STIFFENED PANELS	72-0538
ACCURATE	ACCURATE FINITE ELEMENT APPROXIMATION FOR PROBLEMS WITH SINGULAR DATA	75-0098
ACCURATE	LOCATION OF POINTS AT WHICH FINITE ELEMENT SOLUTIONS ARE VERY ACCURATE	74-0112
ACCURATE	ON NUMERICALLY ACCURATE FINITE ELEMENT SOLUTIONS IN THE FULLY PLASTIC RANGE	74-0529
ACCURATE	ON NUMERICALLY ACCURATE FINITE ELEMENTS	74-0797
ACCURATELY	SUBSPACES WITH ACCURATELY INTERPOLATED BOUNDARY CONDITIONS	72-0527
ACHSENSY	DIE BERECHNUNG ACHSENSY METRISCHER TRAGWERKE UNTER ACHSENSYMETRISCHER UND NIGHT-SYMMETRISHCER BELAST	69-0452
ACHSENSYMETRISC	DIE BERECHNUNG ACHSENSY METRISCHER TRAGWERKE UNTER ACHSENSYMETRISCHER UND NIGHT-SYMMETRISHCER BELAST	69-0452
ACOUSITC	ACOUSITC FINITE ELEMENT ANALYSIS OF AXISYMMETRIC FLUID REGIONS	73-0763
ACOUSTIC	ACOUSTIC ANALYSIS OF SOLID ROCKET MOTOR CAVITIES BY A FINITE ELEMENT METHOD	73-0119
ACOUSTIC	ACOUSTIC FINITE ELEMENT APPROACH FOR STUDYING BOUNDARY FLEXIBILITY AND SOUND TRANSMISSION BETWEEN IR	73-0096
ACOUSTIC	ACOUSTIC RESPONSE OF A SPACECRAFT SOLAR PANEL USING NORMAL MODE METHOD	69-0422
ACOUSTIC	AN ACOUSTIC FINITE ELEMENT APPROACH FOR STUDYING BOUNDARY FLEXIBILITY AND SOUND TRANSMISSION BETWEEN	73-1087
ACOUSTIC	ANALYSIS OF ACOUSTIC FIELD IN IRREGULARLY SHAPED ROOMS BY FINITE ELEMENT METHOD	73-0133
ACOUSTIC	ANALYSIS OF STRUCTURAL ACOUSTIC INTERACTIONS IN METAL CERAMIC TRANSDUCERS	73-0401
ACOUSTIC	COMPUTATION OF THE RESPONSE OF COUPLED PLATE ACOUSTIC SYSTEMS USING PLATE FINITE ELEMENTS AND ACOUST	71-0072
ACOUSTIC	COMPUTATION OF THE RESPONSE OF COUPLED PLATE ACOUSTIC SYSTEMS USING PLATE FINITE ELEMENTS AND ACOUST	71-0072
ACOUSTIC	COUPLING MATRICES FOR HIGH ORDER FINITE ELEMENT ANALYSIS OF ACOUSTIC - WAVE PROPAGATION	72-0040
ACOUSTIC	FINITE ELEMENT ANALYSIS OF IRREGULARLY SHAPED ACOUSTIC FIELD	71-0711
ACOUSTIC	FINITE ELEMENT APPROACH TO ACOUSTIC RADIATION FROM ELASTIC STRUCTURES	74-0121
ACOUSTIC	FINITE ELEMENT SIMULATION OF ACOUSTIC FILTERS WITH VARIABLE CIRCULAR CROSS SECTION	75-0871
ACOUSTIC	FINITE ELEMENT SOLUTION OF THE INTERACTION OF A PLANE ACOUSTIC BLAST WAVE AND A CYLINDRIC STRUCTURE	74-0790
ACOUSTIC	FINITE-ELEMENT APPROACH TO ACOUSTIC SCATTERING FROM ELASTIC STRUCTURES	75-0069
ACOUSTIC	HIGH-ORDER FINITE ELEMENTS FOR INHOMOGENEOUS ACOUSTIC GUIDING STRUCTURES	73-0137
ACOUSTIC	METHODS FOR DETERMINING CHARACTERISTICS OF ACOUSTIC WAVES IN ROCKET MOTORS	75-0075
ACOUSTIC	MODAL SENSITIVITY STUDY OF A CONICAL RE-ENTRY VEHICLE SUBJECTED TO AERODYNAMICALLY INDUCED ACOUSTIC	69-0431
ACOUSTIC	NONLINEAR ACOUSTIC RESPONSE ANALYSIS OF PLATES USING THE FINITE ELEMENT METHOD	72-0223
ACOUSTIC	SHIP (SIMPLIFIED-HELMHOLTZ-INTEGRAL PROGRAM): A FAST COMPUTER PROGRAM FOR CALCULATING THE ACOUSTIC R	72-0835
ACOUSTIC	SOME PRACTICAL APPLICATIONS OF ACOUSTIC FINITE ELEMENTS	75-0414
ACOUSTIC	STUDY OF EFFECTS OF DESIGN DETAILS ON STRUCTURAL RESPONSE TO ACOUSTIC EXCITATION	72-0864
ACOUSTIC	STUDY OF THE ACOUSTIC EXCITATION OF STRUCTURES	74-1332
ACOUSTIC	SYMPOSIUM ON ACOUSTIC FATIGUE	73-0898
ACOUSTIC	THE FINITE ELEMENT TECHNIQUE APPLIED TO AN ACOUSTIC DAMPING PROBLEM	66-0048
ACOUSTIC	THEORETICAL COMPUTATIONS ON RIDGE ACOUSTIC SURFACE WAVES USING THE FINITE-ELEMENT METHOD	71-0011
ACOUSTIC	TRANSIENT RESPONSE OF A COUPLED PLATE-ACOUSTIC SYSTEM USING PLATE AND ACOUSTIC FINITE ELEMENTS	71-0463
ACOUSTIC	TRANSIENT RESPONSE OF A COUPLED PLATE-ACOUSTIC SYSTEM USING PLATE AND ACOUSTIC FINITE ELEMENTS	72-0275
ACOUSTIC	USE OF SIMPLE THREE-DIMENSIONAL ACOUSTIC FINITE ELEMENTS FOR DETERMINING NATURAL MODES AND FREQUENCI	72-0010

ACOUSTIC	VARIATIONAL FINITE ELEMENT CALCULATION OF THE ACOUSTIC RESPONSE OF A RECTANGULAR PANEL	71-0100
ACOUSTIC-GRAVIT.	SEISMOLOGY AND ACOUSTIC-GRAVITY WAVES	73-0744
ACOUSTICALLY	A FRACTURE MECHANICS APPROACH TO THE PROPAGATION OF FATIGUE CRACKS IN REACTOR COMPONENTS BY ACOUSTIC	71-0132
ACOUSTICALLY	FINITE ELEMENT ANALYSIS OF ACOUSTICALLY RADIATING STRUCTURES WITH APPLICATIONS TO SONAR TRANSDUCERS	73-0075
ACOUSTICALLY	FINITE ELEMENT ANALYSIS OF ACOUSTICALLY RADIATING STRUCTURES WITH APPLICATIONS TO SONAR TRANSDUCERS	74-0524
ACOUSTICS	A FINITE ELEMENT METHOD FOR ACOUSTICS	65-0017
ACOUSTICS	BUILDING VIBRATIONS AND ACOUSTICS	73-0656
ACOUSTO-STRUCTU	A VARIATIONAL FORMULATION OF DAMPED ACOUSTO-STRUCTURAL PROBLEMS	66-0133
ACRYLIC	STRESS ANALYSIS OF A CONICAL ACRYLIC VIEWPORT	70-0106
ACTIVITIES	REPORT OF INTERAGENCY CONFERENCE ON COORDINATION OF RESEARCH ACTIVITIES	73-1100
ACTIVITY	COMPUTER PROGRAM CONSTRUCTION AND MAINTENANCE - FUTURE OF CENTRALIZED FINITE ELEMENT ACTIVITY	75-0014
ADAPTED	MODAL COUPLING PROCEDURES ADAPTED TO NASTRAN ANALYSIS OF THE 1/8 - SCALE SHUTTLE STRUCTURAL DYNAMICS	75-0677
ADAPTED	MODAL COUPLING PROCEDURES ADAPTED TO NASTRAN ANALYSIS OF THE 1/8-SCALE SHUTTLE STRUCTURAL DYNAMICS M	75-0678
ADAPTING	PROBLEM OF USING THE METHOD OF FINITE ELEMENTS TO CONSTRUCT AN ALGORITHM FOR CALCULATING A THIN-WALL	74-0851
ADAPTIVE	AN ADAPTIVE COMPUTER PROGRAM FOR DIV. (P(X,Y) GRAD U) EQUALS F(X,Y,U) IN A POLYGONAL REGION	75-0133
ADAPTIVENESS	PROBLEMS OF SELF ADAPTIVENESS IN THE FINITE ELEMENT METHOD	75-0097
ADDED	APPLICATION OF FINITE ELEMENT METHOD TO ADDED VIRTUAL MASS OF SHIP HULL VIBRATION	70-0463
ADDED	FINITE ELEMENT ANALYSIS OF TWO-DIMENSIONAL ADDED MASS AND DAMPING	75-0394
ADDED	FINITE ELEMENT SOLUTION FOR ADDED MASS AND DAMPING	74-0018
ADDED-MASS	FLUID FINITE ELEMENTS FOR ADDED-MASS CALCULATIONS	72-0009
ADDENDUM	MAGIC 11: AN AUTOMATED GENERAL PURPOSE SYSTEM FOR STRUCTURAL ANALYSIS. VOLUME 1. ENGINEER'S MANUAL (	71-0757
ADDITIONS	PROGRAMMER'S MANUAL ADDITIONS AND DEMONSTRATION PROBLEMS FOR A THERMOSTRUCTURAL CAPABILITY FOR NASTR	75-0734
ADHESION	FINITE ELEMENT IN ADHESION ANALYSES	73-0355
ADHESIVE	ANALYSIS OF A CIRCULAR RAFT IN ADHESIVE CONTACT WITH A THICK ELASTIC LAYER	74-1152
ADHESIVE	COMBINED MODE CRACK EXTENSION IN ADHESIVE JOINTS	72-0111
ADHESIVE	COMPARISON OF THEORETICAL AND EXPERIMENTAL SHEAR STRESS IN THE ADHESIVE LAYER OF A LAP JOINT MODEL	74-1176
ADHESIVE	ELASTIC SETTLEMENT OF A CIRCULAR RAFT IN ADHESIVE CONTACT WITH A TRANSVERSELY ISOTROPIC MEDIUM	75-0828
ADHESIVE	FINITE ELEMENT IN NUMERICAL ANALYSIS OF ADHESIVE FRACTURE	73-0085
ADHESIVE	FRACTURE MECHANICS APPROACH TO ADHESIVE JOINTS	71-0726
ADHESIVE	FRACTURE MECHANICS APPROACH TO ADHESIVE JOINTS	72-0181
ADHESIVE	MECHANICS OF FRACTURE IN ADHESIVE JOINTS	72-0691
ADHESIVE	THE STRESSES IN AN ADHESIVE LAYER	72-0910
ADHESIVE-BONDED	STRESS ANALYSIS OF ADHESIVE-BONDED LAP JOINTS	74-0548
ADHESIVELY	FRACTURE MECHANICS FOR AN INTERFACIAL CRACK BETWEEN ADHESIVELY BONDED DISSIMILAR MATERIALS	72-0869
ADHESIVES	WOOD SHEAR PANELS BONDED WITH FLEXIBLE ADHESIVES	75-0086
ADI	ADI METHODS FOR FINITE ELEMENT SYSTEMS	71-0652
ADJACENT	STRESSES AND DISPLACEMENTS AROUND TWO ADJACENT CIRCULAR OPENINGS NEAR TO THE GROUND SURFACE	74-0993
ADJOINING	ADJOINING APPROPRIATE SINGULAR ELEMENTS TO TRANSPORT THEORY COMPUTATIONS	74-0946
ADJOINT	SELF ADJOINT DIFFERENTIAL EQUATIONS AND VARIATIONAL PRINCIPLE - 1.	71-0388
ADJOINT	SELF ADJOINT DIFFERENTIAL EQUATIONS AND VARIATIONAL PRINCIPLE - 2.	72-0106
ADJOINT	COLUMN INSTABILITY UNDER NONCONSERVATIVE FORCES, WITH INTERNAL AND EXTERNAL DAMPING FINITE ELEMENT U	73-0566
ADJUSTMENT	FINITE ELEMENT METHOD FOR BLOCK ADJUSTMENT PROBLEMS OF PHOTOGRAMMETRY	73-0264
ADJUSTMENT	THE METHOD OF INTERFACE STRESS ADJUSTMENT AND ITS USES IN SOME PLANE ELASTICITY PROBLEMS	61-0005
ADMISSIBLES	FAMILLE D ELEMENTS FINIS DE COQUE PLAN CINEMATIQUEMENT ADMISSIBLES POUR L'ANALYSE DES STRUCTURES	74-0488
AERO-SPACE	ASPECTS OF THE FINITE ELEMENT METHOD AS APPLIED TO AERO-SPACE STRUCTURES	72-0694
AERODYNAMIC	A FINITE ELEMENT METHOD FOR CALCULATING AERODYNAMIC COEFFICIENTS OF A SUBSONIC AIRPLANE	73-0655
AERODYNAMIC	A STUDY OF THE FINITE ELEMENT METHOD FOR AERODYNAMIC APPLICATIONS	75-0793
AERODYNAMIC	FURTHER DEVELOPMENTS IN CONSISTENT UNSTEADY SUPERSONIC AERODYNAMIC COEFFICIENTS	72-0559
AERODYNAMIC	INFLUENCE OF COOLING, FREE-STREAM TURBULENCE AND SURFACE ROUGHNESS ON THE AERODYNAMIC BEHAVIOR OF CA	74-0970
AERODYNAMIC	KINEMATICALLY CONSISTENT UNSTEADY AERODYNAMIC COEFFICIENTS IN SUPERSONIC FLOW	70-0014
AERODYNAMIC	KINEMATICALLY CONSISTENT UNSTEADY AERODYNAMIC COEFFICIENTS IN SUPERSONIC FLOW	70-0316
AERODYNAMIC	METHODS FOR PREDICTING THE AERODYNAMIC AND STABILITY AND CONTROL CHARACTERISTICS OF STOL AIRCRAFT. V	73-0718
AERODYNAMIC	METHODS FOR PREDICTING THE AERODYNAMIC AND STABILITY AND CONTROL CHARACTERISTICS OF STOL AIRCRAFT. V	73-0718
AERODYNAMIC	METHODS FOR PREDICTING THE AERODYNAMIC AND STABILITY AND CONTROL CHARACTERISTICS OF STOL AIRCRAFT. V	73-0719
AERODYNAMIC	METHODS OF PREDICTING THE AERODYNAMIC AND STABILITY AND CONTROL CHARACTERISTICS OF STOL AIRCRAFT. VO	73-0717
AERODYNAMIC	RECENT EXPERIENCE IN USING FINITE ELEMENT METHODS FOR THE SOLUTION OF PROBLEMS IN AERODYNAMIC INTERF	70-0339
AERODYNAMIC	STRUCTURAL ANALYSIS OF AERODYNAMIC DECELERATION SYSTEMS BY THE FINITE ELEMENT METHOD	67-0064
AERODYNAMIC	SUBSONIC AND SUPERSONIC INDICIAL AERODYNAMICS AND AERODYNAMIC TRANSFER FUNCTION FOR COMPLEX CONFIGUR	74-1320
AERODYNAMICAL	PROPOSALS FOR THE APPLICATION OF THE FINITE ELEMENT TECHNIQUE TO SOME AIRCRAFT AERODYNAMICAL PROBLEM	72-0628
AERODYNAMICALLY	MODAL SENSITIVITY STUDY OF A CONICAL RE-ENTRY VEHICLE SUBJECTED TO AERODYNAMICALLY INDUCED ACOUSTIC	69-0431
AERODYNAMICIST	AN AERODYNAMICIST LOOKS AT THE FINITE ELEMENT METHOD	74-0021
AERODYNAMICS	AEROELASTIC LOADS PREDICTIONS USING FINITE-ELEMENT AERODYNAMICS	75-0815
AERODYNAMICS	ANALYSIS OF MOVING BODY PROBLEMS IN AERODYNAMICS	74-0024
AERODYNAMICS	APPLICATIONS OF FINITE ELEMENT METHOD IN UNSTEADY AERODYNAMICS	74-0105
AERODYNAMICS	DEVELOPMENT AND APPLICATIONS OF SUPERSONIC UNSTEADY CONSISTENT AERODYNAMICS FOR INTERFERING PARALLEL	72-0716
AERODYNAMICS	FINITE ELEMENT APPROACH TO THE INTEGRATED POTENTIAL FORMULATION OF GENERAL UNSTEADY SUPERSONIC AEROD	73-0910
AERODYNAMICS	SUBSONIC AND SUPERSONIC INDICIAL AERODYNAMICS AND AERODYNAMIC TRANSFER FUNCTION FOR COMPLEX CONFIGUR	74-1320
AERODYNAMICS	THE FINITE ELEMENT METHOD IN LOW SPEED AERODYNAMICS	75-0702
AEROELASTIC	A DESIGN STUDY FOR THE INCORPORATION OF AEROELASTIC CAPABILITY INTO NASTRAN	71-0112
AEROELASTIC	AEROELASTIC LOADS PREDICTIONS USING FINITE-ELEMENT AERODYNAMICS	75-0815
AEROELASTIC	AEROELASTIC PROBLEMS OF LOW ASPECT RATIO WINGS (PART 1)	56-0006
AEROELASTIC	OPTIMIZATION OF COMPLEX STRUCTURES TO SATISFY STATIC, DYNAMIC AND AEROELASTIC REQUIREMENTS	74-0272
AEROFOILS	FINITE ELEMENT ANALYSIS OF THE FLOW THROUGH A CASCADE OF AEROFOILS	73-0761
AEROFOILS	FLOW THROUGH A CASCADE OF AEROFOILS	74-0077
AERONAUTICS	BULLETIN DE L ASSOCIATION TECHNIQUE MARITIME ET AERONAUTIQUE (BULLETIN OF THE TECHNICAL MARITIME AND	74-1108
AERONAUTIQUE	BULLETIN DE L ASSOCIATION TECHNIQUE MARITIME ET AERONAUTIQUE (BULLETIN OF THE TECHNICAL MARITIME AND	74-1108
AEROSPACE	ASPECTS OF FINITE ELEMENT METHOD APPLIED TO AEROSPACE STRUCTURES	73-0703
AEROSPACE	FINITE ELEMENT MODELING AND OPTIMIZATION OF AEROSPACE STRUCTURES	72-0416
AEROSPACE	FINITE ELEMENT MODELING AND OPTIMIZATION OF AEROSPACE STRUCTURES	72-0734
AEROSPACE	FINITE ELEMENTS METHOD IN AEROSPACE ENGINEERING PROBLEMS	74-0409
AEROSPACE	MATRIX METHODS OF AEROSPACE STRUCTURAL ANALYSIS	68-0198
AEROSPACE	NASTRAN EXPERIENCES OF FORT WORTH OPERATION, CONVAIR AEROSPACE DIVISION OF GENERAL DYNAMICS	71-0019
AEROSPACE	PROCEDURE FOR EFFICIENT GENERATING, CHECKING, AND DISPLAYING NASTRAN INPUT AND OUTPUT DATA FOR ANALY	71-0098
AFPAV	AFPAV COMPUTER CODE FOR STRUCTURAL ANALYSIS OF AIRFIELD PAVEMENTS	75-0684
AFPAV	FINITE ELEMENT ANALYSIS OF PAVEMENT STRUCTURES USING AFPAV CODE LINEAR ELASTIC ANALYSIS	73-0658
AFPAV	FINITE-ELEMENT ANALYSIS OF PAVEMENT STRUCTURES USING AFPAV CODE (NONLINEAR ELASTIC ANALYSIS)	75-0748
AFPAV	PARAMETRIC STUDIES OF LAYERED PAVEMENT SYSTEMS USING AFPAV CODE WITH FAST EQUATION SOLVER	75-0747
AFTERBODIES	COUPLAGE ENTRE L ECOULEMENT AUTOUR D UN ARRIERE-CORPS ET LE JET PROPULSIF EN THEORIE DE FLUIDE PARFA	74-1137
AGARD	TECHNICAL EVALUATION REPORT ON AGARD SPECIALISTS	71-0737
AGASSIZ	NONLINEAR STRESS DEFORMATION ANALYSES OF LAKE AGASSIZ CLAYS USING FINITE ELEMENT METHOD	74-1072
AGEING	THERMOMECHANICAL CREEP OF AGEING CONCRETE - A UNIFIED APPROACH	74-0607
AGRICULTURAL	ANALYSIS OF VISCOELASTIC CONTACT STRESSES IN AGRICULTURAL PRODUCTS USING A FINITE ELEMENT METHOD	74-0634
AGRICULTURAL	ANALYSIS OF VISCOELASTIC CONTACT STRESSES IN AGRICULTURAL PRODUCTS USING A FINITE ELEMENT METHOD	74-1199
AH-IG	A NASTRAN VIBRATION MODEL OF THE AH-IG HELICOPTER AIRFRAME. VOLUME 1	74-1338
AH-IG	A NASTRAN VIBRATION MODEL OF THE AH-IG HELICOPTER AIRFRAME. VOLUME 11	74-1339
AHMAD'S	A FURTHER MODIFICATION TO AHMAD'S SHELL ELEMENT	73-0619
AIRCRAFT	A FINITE ELEMENT ANALYSIS OF AN AIRCRAFT LANDING MAT ON AN ELASTIC FOUNDATION	68-0280
AIRCRAFT	AIRCRAFT FUEL TANK VULNERABILITY TO HYDRAULIC RAM: MODIFICATION OF THE NORTHROP FINITE ELEMENT COMP	74-0724
AIRCRAFT	ANALYSIS OF THE 747 AIRCRAFT WING BODY INTERSECTION	68-0275
AIRCRAFT	APPLICATION OF COMPUTER-AIDED AIRCRAFT DESIGN IN A MULTIDISCIPLINARY ENVIRONMENT	73-0527
AIRCRAFT	APPLICATION OF FINITE ELEMENT METHOD TO ANALYSIS OF AIRCRAFT WING STRUCTURES	69-0232
AIRCRAFT	APPLICATIONS OF NASTRAN TO COUPLED STRUCTURAL AND HYDRODYNAMIC RESPONSES IN AIRCRAFT HYDRAULIC SYSTE	71-0744

AIRCRAFT	AUTOMATED OPTIMUM DESIGN OF AIRCRAFT WINGS TO SATISFY STRENGTH, STABILITY, FREQUENCY AND FLUTTER REQ	72-0605
AIRCRAFT	BERECHNUNGSMETHODEN ZUR ERMITTLUNG VON BELASTUNGSBEDINGUNGEN UND GRENZTRAGFAEHIGKEITSVERHALTEN VON K	73-0858
AIRCRAFT	COMPUTATION OF INFLUENCE COEFFICIENTS FOR AIRCRAFT STRUCTURES WITH DISCONTINUITIES AND SWEEPBACK	47-0002
AIRCRAFT	EVALUATION OF DAMAGE TOLERANCE IN AIRCRAFT STRUCTURES	75-0559
AIRCRAFT	EXPERIMENTS IN FORMING AND JOINTING OF TITANIUM AND TITANIUM ALLOYS FOR AIRCRAFT AND SPACE VEHICLES	70-0134
AIRCRAFT	LARGE SCALE ANALYSIS OF CURRENT AIRCRAFT	70-0383
AIRCRAFT	METHODS FOR PREDICTING THE AERODYNAMIC AND STABILITY AND CONTROL CHARACTERISTICS OF STOL AIRCRAFT. V	73-0718
AIRCRAFT	METHODS FOR PREDICTING THE AERODYNAMIC AND STABILITY AND CONTROL CHARACTERISTICS OF STOL AIRCRAFT. V	73-0719
AIRCRAFT	METHODS OF PREDICTING THE AERODYNAMIC AND STABILITY AND CONTROL CHARACTERISTICS OF STOL AIRCRAFT. VO	73-0717
AIRCRAFT	MODERN FUSELAGE ANALYSIS AND THE ELASTIC AIRCRAFT	63-0023
AIRCRAFT	NASTRAN DIFFERENTIAL STIFFNESS ANALYSIS OF AN AIRCRAFT CANOPY	71-0037
AIRCRAFT	NEW METHOD OF BULKHEAD ANALYSIS IN AIRCRAFT STRUCTURES	74-0334
AIRCRAFT	NOTE ON THE THEORY OF AIRCRAFT STRUCTURAL ANALYSIS	59-0005
AIRCRAFT	PROBABILISTIC ASSESSMENT OF AIRCRAFT HAZARD FOR NUCLEAR POWER PLANTS	72-0235
AIRCRAFT	PROPOSALS FOR THE APPLICATION OF THE FINITE ELEMENT TECHNIQUE TO SOME AIRCRAFT AERODYNAMICAL PROBLEM	72-0628
AIRCRAFT	RIGID PAVEMENT STRESSES UNDER AIRCRAFT LOADING	71-0423
AIRCRAFT	SOME PROBLEMS IN THE DISCRETE ELEMENT REPRESENTATION OF AIRCRAFT STRUCTURES	64-0021
AIRCRAFT	SOME RESULTS ON THE FREE-FREE OSCILLATIONS OF AIRCRAFT TYPE STRUCTURES	65-0007
AIRCRAFT	STRUCTURAL ANALYSIS BY THE MATRIX FORCE METHOD WITH APPLICATION TO AIRCRAFT WINGS	56-0004
AIRCRAFT	STRUCTURAL ANALYSIS OF AIRCRAFT IMPACT ON A NUCLEAR CONTAINMENT VESSEL AND ASSOCIATED STRUCTURES	69-0279
AIRCRAFT-PAVEME	AIRCRAFT-PAVEMENT INTERACTION STUDIES, PHASE 1: A FINITE ELEMENT MODEL OF A JOINTED CONCRETE PAVEMEN	73-0894
AIRCRAFT-PAVEME	AIRCRAFT-PAVEMENT INTERACTION STUDIES, PHASE 1: A FINITE ELEMENT MODEL OF A JOINTED CONCRETE PAVEMEN	73-0894
AIRFIELD	AFPAV COMPUTER CODE FOR STRUCTURAL ANALYSIS OF AIRFIELD PAVEMENTS	75-0684
AIRFIELD	DESIGN METHOD FOR FLEXIBLE AIRFIELD PAVEMENTS	74-1174
AIRFOIL	ANALYSIS OF THE EFFECTS ON LIFE OF LEADING-EDGE HOLES IN AN AIRFOIL SUBJECTED TO ARBITRARY SPANWISE	75-0700
AIRFOIL	FINITE ELEMENT SOLUTION OF THE INCOMPRESSIBLE FLOW OVER AN AIRFOIL IN A NON UNIFORM STREAM	74-0432
AIRFOIL	THE AIRFOIL PROBLEM VIA THE FINITE ELEMENT METHOD	73-1167
AIRFOIL	TWO-DIMENSIONAL AIRFOIL ANALYSIS USING A REFINED FINITE ELEMENT TECHNIQUE	74-0587
AIRFOILS	COMPUTATIONAL CONSIDERATIONS IN APPLICATION OF THE FINITE ELEMENT METHOD FOR ANALYSIS OF UNSTEADY FL	73-0068
AIRFOILS	FINITE ELEMENT ANALYSIS AND COMPUTER GRAPHICS VISUALIZATION OF FLOW AROUND PITCHING AND PLUNGING AIR	73-0385
AIRFOILS	INTEGRAL EQUATION METHOD FOR SUBSONIC FLOW PAST AIRFOILS IN VENTILATED WIND TUNNELS	75-0065
AIRFOILS	MODIFICATIONS TO ARL COMPUTER PROGRAMS USED FOR DESIGN OF AXIAL COMPRESSOR AIRFOILS	74-0508
AIRFOILS	NUMERICAL CALCULATIONS OF VELOCITY AND PRESSURE DISTRIBUTION AROUND OSCILLATING AIRFOILS	74-0847
AIRFOILS	SUITABILITY OF THE FINITE ELEMENT METHOD FOR ANALYSIS OF UNSTEADY FLOW AROUND OSCILLATING AIRFOILS	73-0069
AIRFRAME	A NASTRAN VIBRATION MODEL OF THE AH-IG HELICOPTER AIRFRAME. VOLUME 1	74-1338
AIRFRAME	A NASTRAN VIBRATION MODEL OF THE AH-IG HELICOPTER AIRFRAME. VOLUME 11	74-1339
AIRFRAME	THE USE OF NASTRAN IN THE ANALYSIS OF LARGE COMPLEX AIRFRAME STRUCTURES, MODELING TECHNIQUES AND ORG	71-0110
AIRPLANE	A FINITE ELEMENT METHOD FOR CALCULATING AERODYNAMIC COEFFICIENTS OF A SUBSONIC AIRPLANE	73-0655
AIRPLANE	APPLICATION OF FINITE-ELEMENT THEORY TO AIRPLANE CONFIGURATIONS	71-0458
AIRPLANE	AUTOMATED DESIGN OPTIMIZATION OF SUPERSONIC AIRPLANE WING STRUCTURES UNDER DYNAMIC CONSTRAINTS	72-0178
AIRPLANE	FINITE ELEMENT AIRPLANE COST ANALYSIS	74-0301
AIRPLANES	STEADY STATE EQUATIONS OF MOTION, EQUILIBRIUM SHAPE AND STABILITY DERIVATIVES OF ELASTIC AIRPLANES E	72-0541
ALASKAN	STUDY OF 1964 ALASKAN EARTHQUAKE USING FINITE ELEMENT AND GENERALIZED INVERSION TECHNIQUES	72-0003
ALATNIH	DOSTIGNUCA I TENDENCIJE U ISPITIVANJIMA ALATNIH MASINA (ACHIEVEMENTS AND TENDENCIES IN MACHINE TOOLS	74-0426
ALBERI	SULLA DETERMINAZIONE DELLE FREQUENZE PROPRIE DI ALBERI SU PIU SUPPORTI ELASTICI. (DETERMINATION OF F	74-0916
ALGEBRAIC	COMPUTER PROGRAMS FOR THE SOLUTION OF SYSTEMS OF LINEAR ALGEBRAIC EQUATIONS	73-0932
ALGEBRAIC	COMPUTER PROGRAMS FOR THE SOLUTION OF SYSTEMS OF LINEAR ALGEBRAIC EQUATIONS	73-1040
ALGEBRAIC	NUMERICAL TREATMENT OF ALGEBRAIC INTEGRAL EQUATIONS BY VARIATIONAL METHODS	75-0374
ALGEBRAIC	THE SECOND VARIATON TEST WITH ALGEBRAIC AND DIFFERENTIAL CONTRASTS	69-0342
ALGEBRAIC-GEOME	ALGEBRAIC-GEOMETRY FOUNDATIONS FOR FINITE-ELEMENT COMPUTATION	73-0974
ALGORITHM	A COMPARATIVE STUDY OF SEVERAL CORE STORAGE SCHEMES FOR LARGE SPARSE POSITIVE DEFINITE MATRICES WITH	72-0678
ALGORITHM	A FINITE ELEMENT SOLUTION ALGORITHM FOR THE NAVIER-STOKES EQUATIONS	74-0784
ALGORITHM	A SIMPLE ALGORITHM FOR THE STIFFNESS MATRIX OF TRIANGULAR PLATE BENDING ELEMENTS	71-0152
ALGORITHM	ALGORITHM FOR MATRIX BANDWIDTH REDUCTION	72-0895
ALGORITHM	ALGORITHM FOR THE LARGE DEFLECTION GEOMETRICALLY NON-LINEAR BEHAVIOUR OF PLANE AND CURVED STRUCTURES	72-0083
ALGORITHM	ALGORYTM OBLICZEN ORGAN WYMUSZONYCH METODA SZTYWNYCH ELEMENTOW SKONCZONYCH (ALGORITHM FOR CALCULATIN	74-0581
ALGORITHM	AN ALGORITHM FOR REDUCING THE BANDWIDTH OF A MATRIX OF SYMMETRICAL CONFIGURATION	65-0067
ALGORITHM	AN ALGORITHM FOR REDUCING THE BANDWITH AND PROFILE OF A SPARSE MATRIX	74-0792
ALGORITHM	AN AUTOMATIC RELABELING ALGORITHM FOR BANDWIDTH MINIMIZATION	75-0763
ALGORITHM	APPROXIMATE ALGORITHM FOR THE REANALYSIS OF STRUCTURES BY THE FINITE ELEMENT METHOD.	72-0445
ALGORITHM	FINITE ELEMENT ALGORITHM FOR THE DETERMINATION OF DYNAMIC BUCKLING	75-0302
ALGORITHM	FINITE ELEMENT APPROXIMATION, AND SOLUTION BY A PENALTY-DUALITY ALGORITHM OF AN ELASTO-PLASTIC PROBL	75-0041
ALGORITHM	FINITE ELEMENT SOLUTION FOR ENERGY CONSERVATION USING A HIGHLY STABLE EXPLICIT INTEGRATION ALGORITHM	72-0633
ALGORITHM	FINITE ELEMENT SOLUTION ALGORITHM FOR VISCOUS INCOMPRESSIBLE FLUID DYNAMICS	73-0063
ALGORITHM	FINITE ELEMENT SOLUTION ALGORITHM FOR INCOMPRESSIBLE FLUID DYNAMICS	74-0006
ALGORITHM	INTEGRATION FOR THE FINITE ELEMENTS - AN ALGORITHM IN ORTHOGONAL AXES	73-0295
ALGORITHM	KRON'S METHOD: AN ALGORITHM FOR THE EIGENVALUE ANALYSIS OF LARGE-SCALE STRUCTURAL SYSTEMS	73-0758
ALGORITHM	NUMERICAL SOLUTION TO THE DYNAMICS OF VISCOUS FLUID FLOW BY A FINITE ELEMENT ALGORITHM: A FIRST STEP	71-0004
ALGORITHM	NUMERICAL SOLUTION TO THE DYNAMICS OF VISCOUS FLUID FLOW BY A FINITE ELEMENT ALGORITHM: A FIRST STEP	71-0791
ALGORITHM	ON IMPROVING EFFICIENCY OF AN ALGORITHM FOR STRUCTURAL OPTIMIZATION AND A USERS MANUAL FOR PROGRAM T	74-0778
ALGORITHM	ON THE UNCONDITIONAL STABILITY OF AN IMPLICIT ALGORITHM FOR NONLINEAR STRUCTURAL DYNAMICS	75-0825
ALGORITHM	PROBLEM OF USING THE METHOD OF FINITE ELEMENTS TO CONSTRUCT AN ALGORITHM FOR CALCULATING A THIN-WALL	74-0851
ALGORITHM	QUASI-LINEAR PROGRAMMING ALGORITHM FOR OPTIMIZING FIBRE-REINFORCED STRUCTURES OF FIXED STIFFNESS	75-0225
ALGORITHM	SCALAR ALGORITHM OF DETERMINATION OF STIFFNESS MATRIX AND OF ANOTHER FINITE ELEMENT CHARACTERISTIC	74-1013
ALGORITHM	THE COMPUTATIONAL EFFICIENCY OF A NEW MINIMIZATION ALGORITHM FOR EIGENVALUE ANALYSIS	71-0595
ALGORITHM	THE GENERALIZED INVERSE - SOME PROPERTIES, A COMPUTATIONAL ALGORITHM AND A STRUCTURAL APPLICATION	68-0248
ALGORITHM	THE STATIC CONDENSATION ALGORITHM	74-0328
ALGORITHM	TRANSIENT TWO-DIMENSIONAL PROBLEMS UTILIZING THE LEAST SQUARES ALGORITHM	75-0219
ALGORITHMS	ALGORITHMS FOR THE PRODUCTION OF CONTOUR MAPS OVER AN IRREGULAR TRIANGULAR MESH	72-0825
ALGORITHMS	COMPARATIVE ANALYSIS OF THE CUTHILL-MCKEE AND THE REVERSE CUTHILL-MCKEE ORDERING ALGORITHMS FOR SPAR	73-1099
ALGORITHMS	EQUATION SOLVING ALGORITHMS FOR THE FINITE ELEMENT METHOD	71-0127
ALGORITHMS	EVALUATION OF ALGORITHMS FOR STRUCTURAL MODIFICATION	69-0243
ALGORITHMS	LOVE WAVE DISPERSION THROUGH IRREGULAR STRUCTURES USING FINITE-ELEMENT ALGORITHMS	75-0431
ALGORITHMS	THE ANALYSIS AND APPLICATION OF SPARSE MATRIX ALGORITHMS IN THE FINITE ELEMENT METHOD	72-0079
ALGORYTM	ALGORYTM OBLICZEN ORGAN WYMUSZONYCH METODA SZTYWNYCH ELEMENTOW SKONCZONYCH (ALGORITHM FOR CALCULATIN	74-0581
ALLEN	FINITE ELEMENT ANALYSES OF PORT ALLEN AND OLD RIVER LOCKS	69-0203
ALLEN	FINITE ELEMENT ANALYSIS OF PORT ALLEN LOCK	71-0293
ALLGEMEINE	EINIGE ALLGEMEINE PROGRAMMSYSTEME FUR FINITE ELEMENTE	73-0581
ALLGEMEINEN	DIE ELASTO-PLASTISCHE BERECHNUNG VON ALLGEMEINEN TRAGWERKEN UND KONTINUA	69-0443
ALLOCATING	A MORE EFFICIENT FRONT SOLUTION: ALLOCATING ASSEMBLY LOCATIONS BY LONGEVITY CONSIDERATIONS	73-1069
ALLOY	EXPERIMENTAL AND THEORETICAL INVESTIGATION OF PLANE STRESS FRACTURE OF 2024 - T351 ALUMINUM ALLOY	70-0147
ALLOY	FINITE-ELEMENT MODELING OF DIRECTIONAL SOLIDIFICATION OF SUPER ALLOY EUTECTICS	75-0350
ALLOY	SIMULATION ON THE TENSILE BEHAVIOUR OF TWO-PHASE ALLOY CONTAINING A FIBER UNDER TENSILE LOAD	73-0542
ALLOYS	DETERMINATION OF YOUNG'S MODULUS AND POISSON'S RATIO FOR WC-CO ALLOYS BY THE FINITE ELEMENT METHOD	72-0142
ALLOYS	EXPERIMENTS IN FORMING AND JOINTING OF TITANIUM AND TITANIUM ALLOYS FOR AIRCRAFT AND SPACE VEHICLES	70-0134
ALPHA	NEW LIGHT ON THE NAYAK ALPHA TECHNIQUE	73-0614
ALPHA	THE ALPHA CONSTANT-STIFFNESS METHOD FOR THE ANALYSIS OF NONLINEAR PROBLEMS	72-0566
ALPHANUMERIC	AN ALPHANUMERIC FINITE ELEMENT MESH PLOTTER	72-0543
ALTERNATEUR	ANALYSE NUMERIQUE DU CHAMP MAGNETIQUE D'UN ALTERNATEUR PAR ELEMENTS FINIS ET SURRELAXATION PONCTUELL	73-0625
ALTERNATEUR	ETUDE NUMERIQUE DU CHAMP MAGNETIQUE DANS UN ALTERNATEUR TETRAPOLAIRE PAR LA METHODE DES ELEMENTS FIN	74-0413
ALTITUDINAL	CALCULATION OF AREAL RAINFALL USING FINITE ELEMENT TECHNIQUES WITH ALTITUDINAL CORRECTIONS	72-0344
ALUMINA	STRAIN DISTRIBUTION IN AN ALUMINA BICRYSTAL STUDIES BY THE FINITE ELEMENT METHOD AND X-RAY DIFFRACTI	74-1070

123

ALUMINUM	ANALYSIS OF STRENGTH AND RESIDUAL STRESSES IN FILAMENT REINFORCED ALUMINUM CYLINDERS	75-0517
ALUMINUM	ANALYTICAL PERFORMANCE OF STEEL, ALUMINUM AND PLASTIC FOR A NEW DESIGN HIGH PRESSURE THIN-WALLED CAR	74-0756
ALUMINUM	DESIGN, MANUFACTURE, DEVELOPMENT, TEST, AND EVALUATION OF BORON/ALUMINUM STRUCTURAL COMPONENTS FOR S	74-0798
ALUMINUM	EXPERIMENTAL AND THEORETICAL INVESTIGATION OF PLANE STRESS FRACTURE OF 2024 - T351 ALUMINUM ALLOY	70-0147
ALUMINUM	FORMING OF 7075-T6 ALUMINUM IN HIGH PRESSURE ENVIRONMENTS	71-0322
ALUMINUM	TRANSVERSE PROPERTIES OF UNIDIRECTIONAL ALUMINUM MATRIX FIBROUS COMPOSITES	69-0167
ALVEOLAR	A MODEL STUDY OF GAS DIFFUSION IN ALVEOLAR SACS	73-0029
ALVEOLAR	THEORY FOR DISTORTION STUDIES OF LUNG PARENCHYMA BASED ON ALVEOLAR MEMBRANE PROPERTIES	74-0222
AMPLITUDE	A MODE AMPLITUDE TECHNIQUE FOR GENERAL FINITE ELEMENT EQUILIBRIUM	74-0283
AMPLITUDE	FINITE ELEMENT DISPLACEMENT METHOD FOR LARGE AMPLITUDE FREE FLEXURAL VIBRATIONS OF BEAMS AND PLATES	73-0185
AMPLITUDE	LARGE AMPLITUDE, FREE VIBRATION GEOMETRICALLY NONLINEAR CIRCULAR ARCHES	72-0606
ANALOG	ANALOG SUBROUTINES FOR THE FINITE ELEMENT METHOD	72-0652
ANALOG	CONVECTIVE TRANSPORT FINITE ELEMENT ANALOG	75-0289
ANALOGIES	BASIS OF A WELL CONDITIONED FORCE PROGRAM FOR EQUILIBRIUM MODELS VIA THE SOUTHWELL SLAB ANALOGIES	67-0111
ANALOGIES	DUALITY IN STRUCTURAL ANALYSIS BY FINITE ELEMENTS, STATIC GEOMETRIC ANALOGIES, THE DUAL PRINCIPLES O	71-0583
ANALOGIES	STATIC-GEOMETRIC ANALOGIES AND FINITE ELEMENT MODELS	71-0499
ANALOGS	CASE BOND STRESS CALCULATIONS FOR FLAPPED CYLINDRICAL ANALOGS OF SOLID PROPELLANT ROCKET MOTORS. BON	72-0842
ANALOGS	CREATION AND COMPARISON OF FINITE DIFFERENCE ANALOGS OF SOME FINITE ELEMENT SCHEMES	74-0888
ANALOGUE	A FINITE ELEMENT ANALOGUE OF THE NAVIER-STOKES EQUATION	70-0099
ANALOGUE	A FINITE ELEMENT ANALOGUE OF THE MODIFIED RAYLEIGH-RITZ METHOD FOR VIBRATION PROBLEMS	72-0478
ANALOGUE	FINITE ELEMENT - ANALOGUE METHOD FOR DETERMINING THE DYNAMIC CHARACTERISTICS OF AN ARCH DAM-RESERVOI	73-0678
ANALOGUE	FINITE ELEMENT ANALOGUE OF NAVIER-STOKES EQUATIONS	69-0002
ANALOGUE	FINITE ELEMENT ANALOGUE OF THE MODIFIED RAYLEIGH RITZ METHOD FOR VIBRATION PROBLEMS	73-0181
ANALOGY	A LATTICE ANALOGY FOR THE SOLUTION OF PLANE STRESS	43-0002
ANALOGY	MEMBRANE ANALOGY FOR ANISOTROPIC CABLE NETWORKS	74-0233
ANALOGY	STRAIN-ENERGY BOUNDS IN FINITE-ELEMENT ANALYSIS BY SLAB ANALOGY	67-0027
ANALYZER	ASTRA-BOEING'S ADVANCED STRUCTURAL ANALYZER	69-0118
ANALYZER	NASTRAN STRUCTURAL/THERMAL ANALYZER	72-0152
ANALYZER	THE NASTRAN HYDROELASTIC ANALYZER	71-0738
ANALYZER	USING THE NASTRAN THERMAL ANALYZER TO SIMULATE A FLIGHT SCIENTIFIC INSTRUMENT PACKAGE	74-1011
ANCHOR	A FINITE-ELEMENT METHOD FOR ANALYSIS OF ANCHORED BULKHEADS AND ANCHOR WALLS	66-0130
ANCHOR	ANCHOR ZONE STRESS ANALYSIS BY FINITE ELEMENT METHOD	68-0147
ANCHOR	EFFECTS OF SPECIMEN RADIUS ON THE STRESS STATE NEAR A ROOF BOLT ANCHOR: A FINITE ELEMENT DETERMINATI	73-0229
ANCHOR	STRESS DISTRIBUTIONS AROUND PLUG MATCHES WITH ANCHOR BLOCKS	69-0135
ANCHOR-PILE	ANCHOR-PILE DESIGN FOR OCEAN-FLOOR ENVIRONMENTS USING FINITE-ELEMENT ANALYSIS	75-0459
ANCHORAGE	ANCHORAGE ZONE STRESSES IN AXIALLY POST-TENSIONED I-SECTION MEMBERS WITH END BLOCKS	71-0305
ANCHORED	A FINITE-ELEMENT METHOD FOR ANALYSIS OF ANCHORED BULKHEADS AND ANCHOR WALLS	66-0130
ANCHORED	FINITE ELEMENT ANALYSIS OF ANCHORED BULKHEAD BEHAVIOR	74-1068
ANGLE-PLY	FINITE-ELEMENT ANALYSIS OF EDGE EFFECTS IN ANGLE-PLY COMPOSITE LAMINATES	74-0843
ANGLES	MATRIX-DISPLACEMENT DIRECT-ELEMENT METHOD FOR FORCE AND TORQUE ANALYSIS OF DETERMINATE AND INDETERMI	74-0531
ANGULAR	ANALYSIS OF CIRCULAR AND ANGULAR LAMINATED THICK PLATES	74-0426
ANGULAR	CURVED LOW ORDER ANGULAR ELEMENTS IN TWO DIMENSIONAL NEUTRON TRANSPORT PROBLEMS	74-1322
ANGULAR	RAY-EFFECT MITIGATION IN DISCRETE ORDINATE-LIKE ANGULAR FINITE-ELEMENT APPROXIMATIONS IN NEUTRON-TRA	75-0496
ANISOTROPEN	WECHSELWIRKUNG ZWISCHEN DEFORMATION UND DURCHSTROEMUNG IM KLUEFTIGEN, ANISOTROPEN GEBIRGE. (INTERACT	72-0612
ANISOTROPIC	A COMPACT FINITE-ELEMENT METHOD FOR THE ANALYSIS OF ANISOTROPIC LAMINATED PLATES	73-1166
ANISOTROPIC	A FINITE ELEMENT ANALYSIS OF LAMINATED ANISOTROPIC TUBES	70-0319
ANISOTROPIC	A FINITE ELEMENT MODEL FOR ANISOTROPIC YIELDING IN GRAVITY LOADED ROCK	69-0012
ANISOTROPIC	ANALYSIS OF MULTI-AXIAL ANISOTROPIC CREEP, ROOF RESIN BOLTS, AND BED SEPARATION IN ROCK STRUCTURES U	74-1122
ANISOTROPIC	ANALYSIS OF VISCOELASTOPLASTIC STRUCTURAL BEHAVIOUR OF ANISOTROPIC SHELLS BY THE FINITE ELEMENT METH	71-0056
ANISOTROPIC	ANISOTROPIC PARAMETRIC PLATE DISCRETE ELEMENTS	73-1047
ANISOTROPIC	ANISOTROPIC TAPERED ELEMENTS USING DISPLACEMENT MODELS	72-0391
ANISOTROPIC	ANISOTROPIC WORK HARDENING IN METALS. FINITE ELEMENT APPLICATION TO PLANE STRESS AND PLANE STRAIN	73-0950
ANISOTROPIC	CALCULATION OF DISPLACEMENTS AND STRESSES IN ANISOTROPIC ROCKS BY USING METHOD OF FINITE ELEMENTS	73-0265
ANISOTROPIC	CALCULATION OF STRESS INTENSITY FACTORS IN ANISOTROPIC MATERIALS BY A CONTOUR INTEGRAL METHOD	75-0257
ANISOTROPIC	COUPLE - A PSEUDO FINITE ELEMENT CONDUCTION CODE FOR THE TRANSIENT THERMAL RESPONSE OF AXISYMMETRIC O	72-0778
ANISOTROPIC	DEVELOPMENT OF AN ANISOTROPIC SINGULARITY FINITE ELEMENT UTILIZING THE HYBRID DISPLACEMENT METHOD	74-1099
ANISOTROPIC	DIRECT TENSILE TESTING OF ANISOTROPIC ROCKS	74-0989
ANISOTROPIC	DISCRETE ELEMENT FINITE DISPLACEMENT ANALYSIS OF ANISOTROPIC SANDWICH SHELLS	70-0504
ANISOTROPIC	DYNAMIC ANALYSIS OF VISCOELASTOPLASTIC ANISOTROPIC SHELLS	73-0253
ANISOTROPIC	FINITE ELEMENT ANALYSIS OF HEAT CONDUCTION IN ANISOTROPIC SOLIDS	68-0161
ANISOTROPIC	FINITE ELEMENT ANALYSIS OF LAMINATED ANISOTROPIC PLATES INCLUDING TRANSVERSE SHEAR DEFORMATIONS	70-0506
ANISOTROPIC	FINITE ELEMENT ANALYSIS OF VIBRATION AND FLUTTER OF CANTILEVER ANISOTROPIC PLATES	74-0484
ANISOTROPIC	FINITE ELEMENT ANALYSIS OF ANISOTROPIC PLATES USING Q-19 ELEMENT. PART 1. THEORETICAL DEVELOPMENT	74-0728
ANISOTROPIC	FINITE ELEMENT ANALYSIS OF ANISOTROPIC PLATES USING Q-19 ELEMENT: PART 11. INSTRUCTIONS FOR USERS AN	74-0731
ANISOTROPIC	FINITE ELEMENT ANALYSIS OF ANISOTROPIC PLATES USING Q-19 ELEMENT - 1, 2	74-0985
ANISOTROPIC	FINITE ELEMENT METHOD FOR DETERMINATION OF THERMAL STRESSES INTERACTIVE GRAPHICS FOR FINITE ELEMENT	74-0155
ANISOTROPIC	FINITE ELEMENT STRESS ANALYSIS OF ANISOTROPIC BODIES	68-0033
ANISOTROPIC	FINITE ELEMENT STRESS ANALYSIS FOR ANISOTROPIC SOLIDS	70-0478
ANISOTROPIC	FINITE-ELEMENT ANALYSIS FOR FREE VIBRATION OF GENERAL ANISOTROPIC LAMINATED THIN SHELLS	72-0618
ANISOTROPIC	FINITE-ELEMENT ANALYSIS OF VIBRATION AND FLUTTER OF CANTILEVER ANISOTROPIC PLATES	74-0933
ANISOTROPIC	FINITE-ELEMENT FIELD ANALYSIS OF AN INHOMOGENEOUS ANISOTROPIC RELUCTANCE MACHINE ROTOR	73-0020
ANISOTROPIC	FINITE-ELEMENT METHOD FOR THE DETERMINATION OF THERMAL STRESSES IN ANISOTROPIC SOLIDS OF REVOLUTION	74-0262
ANISOTROPIC	FREE VIBRATION AND RESPONSE TO RANDOM PRESSURE FIELD OF ANISOTROPIC THIN CYLINDRICAL SHELLS	74-0830
ANISOTROPIC	IMPACT ON A TRANSVERSELY ANISOTROPIC HALF-SPACE	74-0950
ANISOTROPIC	INDIRECT TENSILE TESTING OF ANISOTROPIC ROCKS	73-0376
ANISOTROPIC	MAGNETOTELLURIC RESPONSE OF LATERALLY INHOMOGENEOUS AND ANISOTROPIC MEDIA	75-0850
ANISOTROPIC	MATRIX ANALYSIS METHODS FOR ANISOTROPIC INELASTIC STRUCTURES	66-0104
ANISOTROPIC	MATRIX DISPLACEMENT ANALYSIS OF ANISOTROPIC SHELLS BY TRIANGULAR ELEMENTS	65-0005
ANISOTROPIC	MEMBRANE ANALOGY FOR ANISOTROPIC CABLE NETWORKS	74-0233
ANISOTROPIC	PLASTIC STRESS STRAIN MATRIX OF ANISOTROPIC MATERIALS AND ITS APPLICATION	69-0383
ANISOTROPIC	QUASI-ANALYTICAL FINITE ELEMENT PROCEDURES FOR AXISYMMETRIC ANISOTROPIC SHELLS AND SOLIDS	74-0297
ANISOTROPIC	RIESENIE VAZKOPRUZNYCH ANIZOTROPICKYCH DOASK METODOU KONECNYCH PRVKOV (SOLUTION OF VISCOELASTIC ANIS	74-0477
ANISOTROPIC	SEMI-ANALYTICAL FINITE ELEMENT PROCEDURE FOR CONDUCTION IN ANISOTROPIC AXISYMMETRIC SOLIDS	74-0270
ANISOTROPIC	SOLUTION OF ANISOTROPIC SEEPAGE PROBLEMS BY FINITE ELEMENTS	66-0080
ANISOTROPIC	SOLUTION OF ANISOTROPIC SEEPAGE BY FINITE ELEMENTS	67-0071
ANISOTROPIC	SOME REMARKS CONCERNING HETEROGENEOUS ANISOTROPIC PLATES.	72-0430
ANISOTROPIC	STEADY CONDUCTION OF HEAT IN LINEAR AND NONLINEAR FULLY ANISOTROPIC MEDIA BY FINITE ELEMENTS	74-0359
ANISOTROPIC	STRESS INTENSITY FACTORS FOR ANISOTROPIC FRACTURE TEST SPECIMENS OF SEVERAL GEOMETRIES	74-0293
ANISOTROPIC	STRESSES IN ANISOTROPIC MEDIA WITH PARTICULAR REFERENCE TO PROBLEMS OF ROCK MECHANICS	66-0079
ANISOTROPIC	STRESSES IN ANISOTROPIC MEDIA OF ROCK MECHANICS	66-0127
ANISOTROPIC	THERMAL DEFORMATION VECTOR FOR A BILINEAR TEMPERATURE DISTRIBUTION IN AN ANISOTROPIC QUADRILATERAL M	75-0670
ANISOTROPIC	THREE DIMENSIONAL ANISOTROPIC AND INHOMOGENEOUS ELASTIC MEDIA MATRIX ANALYSIS FOR SMALL AND LARGE DI	65-0008
ANISOTROPIC	VARIATION WITH TEMPERATURE OF RESONANT FREQUENCIES OF ANISOTROPIC PLATES	74-1046
ANISOTROPIC	VARIATIONAL PRINCIPLE FOR INCOMPRESSIBLE AND NEARLY-INCOMPRESSIBLE ANISOTROPIC ELASTICITY	69-0319
ANISOTROPIC	WAVE PROPAGATION IN ANISOTROPIC ROCKS	72-0801
ANISOTROPIC	WAVE PROPAGATION IN ANISOTROPIC ROCKS	72-0873
ANISOTROPIC	WAVE PROPAGATION IN ANISOTROPIC ROCKS	73-0922
ANISOTROPIC	WECHSELWIRKUNG ZWISCHEN DEFORMATION UND DURCHSTROIMUNG IM KLUEFTIGEN, ANISOTROPEN GEBIRGE. (INTERACT	72-0612
ANISOTROPIC-PLA	FINITE-ELEMENT CODE TO SIMULATE ANISOTROPIC-PLASMA DIFFUSION IN LINEAR MULTIPOLES	75-0409
ANISOTROPY	A MATHEMATICAL THEORY OF PLASTICITY FOR STRAIN HARDENING MATERIALS EXHIBITING INITIAL ANISOTROPY	72-0660
ANISOTROPY	EFFECT OF ELASTIC MODULAR RATIO AND ANISOTROPY ON THE STRESS FIELD IN A TWO-DIMENSIONAL MATRIX INTO	71-0647
ANIZOTROPICKYCH	RIESENIE VAZKOPRUZNYCH ANIZOTROPICKYCH DOASK METODOU KONECNYCH PRVKOV (SOLUTION OF VISCOELASTIC ANIS	74-0477

124

ANLEITUNG	TURBAN VERSION 1.0. EIN SPEZIELLES PROGRAMM ZUR LINEAREN UND NICHTLINEAREN STATISCHEN UND DYNAMISCHE	71-0622
ANNAHME	BERECHNUNG VON STAHLBETONSCHEIBEN IM ZUSTAND II BEI ANNAHME EINES WIRKLICHKEITSNAHEN WERKSTOFFVERHAL	74-0978
ANNEALING	ANALYSIS OF WELDING STRESS RELIEVING BY ANNEALING BASED ON FINITE ELEMENT METHOD	75-0777
ANNULAR	A COMPATIBLE PLATE BENDING ANNULAR SECTOR FINITE ELEMENT OF NONUNIFORM THICKNESS	74-0430
ANNULAR	A FULLY COMPATIBLE ANNULAR SEGMENT FINITE ELFMENT	72-0582
ANNULAR	A FULLY COMPATIBLE ANNULAR SEGMENT FINITE ELEMENT	72-0663
ANNULAR	AN ANNULAR SEGMENT FINITE ELEMENT FOR PLATE BENDING	71-0285
ANNULAR	ANNULAR AND CIRCULAR SECTOR FINITE ELEMENTS FOR PLATE BENDING	70-0100
ANNULAR	AXISYMMETRIC VIBRATIONS OF LINEARLY TAPERED ANNULAR PLATES	74-1266
ANNULAR	END DISPLACEMENTS OF SEMI-INFINITE CYLINDERS DUE TO ANNULAR LOADINGS	69-0399
ANNULAR	HORIZONTALLY CURVED HIGHWAY BRIDGES. A FULLY COMPATIBLE ANNULAR SEGMENT FINITE ELEMENT	72-0834
ANNULAR	TAPERED PLANE STRESS ANNULAR SECTOR FINITE ELEMENT	75-0538
ANNULAR	VIBRATION OF CIRCULAR AND ANNULAR PLATES USING FINITE ELEMENTS	72-0194
ANNULI	CONTRIBUTION TO NUMERICAL-SOLUTION OF DEVELOPING LAMINAR-FLOW IN ENTRANCE REGION OF CONCENTRIC ANNUL	75-0448
ANOMALIES	FINITE-ELEMENT ANALYSIS OF SHALLOW TEMPERATURE ANOMALIES	71-0471
ANSYS	ANSYS (ENGINEERING ANALYSIS SYSTEM) PROGRAM	72-0156
ANSYS	ANSYS USER'S MANUAL	71-0561
ANSYS	THERMO-STRUCTURAL CAPABILITY OF ANSYS	72-0155
ANTENNA	APPLICATION OF A UNIMOMENT TECHNIQUE TO A BICONICAL ANTENNA WITH INHOMOGENEOUS DIELECTRIC LOADING	75-0375
ANTENNA	STATIC ANALYSIS OF THE 64-M 210-FT ANTENNA REFLECTION STRUCTURE	71-0137
ANTENNA	UNIMOMENT METHOD OF SOLVING ANTENNA AND SCATTERING PROBLEMS	74-1020
ANTENNA-SUPPORT	NON-METALLIC ANTENNA-SUPPORT MATERIALS	73-0716
ANTRAS	ANTRAS COMPUTER PROGRAM FOR LINEAR STRESS ANALYSIS BY FINITE ELEMENT METHOD	74-0601
ANTRAS	DAS PROGRAMMSYSTEM ANTRAS EDV-SYSTEM ZUR BERECHNUNG BELIEBIGER STATISCHER SYSTEME NACH DER METHODE D	74-0918
ANTRAS	DAS PROGRAMMSYSTEM ANTRAS EDV-SYSTEM ZUR BERECHNUNG BELIEBIGER STATISCHER SYSTEME NACH DER METHODE D	74-0918
ANVORGESPANNTEN	THEORIE, PROGRAMMENTWICKLUNG UND ERFAHRUNG ANVORGESPANNTEN NETZWERKKONSTRUKTIONEN	72-0640
AORTA	FINITE ELEMENT LARGE DEFORMATION ANALYSIS OF A TAPERED AORTA	75-0856
AORTIC	STRESS ANALYSIS OF THE HUMAN AORTIC VALVE	73-0166
AORTIC	THE FINITE ELEMENT STRESS ANALYSIS OF THE HUMAN AORTIC VALVE TOWARD THE DEVELOPMENT OF A PROSTHETIC	74-1201
AORTIC	THE FINITE ELEMENT STRESS ANALYSIS OF THE HUMAN AORTIC VALVE TOWARD THE DEVELOPMENT OF A PROSTHETIC	74-1201
APACHE	APACHE: A THREE-DIMENSIONAL FINITE ELEMENT PROGRAM FOR STEADY-STATE OR TRANSIENT HEAT CONDUCTION ANA	73-0971
APPENDAGE	FINITE ELEMENT APPENDAGE EQUATIONS FOR HYBRID COORDINATE DYNAMIC ANALYSIS	72-0022
APPENDAGE	GEOMETRIC STIFFNESS CHARACTERISTICS OF A ROTATING ELASTIC APPENDAGE	74-1296
APPENDAGES	DYNAMIC ANALYSIS OF A SYSTEM OF HINGE-CONNECTED RIGID BODIES WITH NONRIGID APPENDAGES	74-0852
APPENDIX	THREE DIMENSIONAL INELASTIC ANALYSIS OF A SHELL APPURTENANCE PER SECTION 111 - APPENDIX F.	74-0716
APPLANATION	MATHEMATICAL MODEL OF THE CORNEO-SCLERAL SHELL AS APPLIED TO INTRAOCULAR PRESSURE VOLUME RELATIONS A	72-0471
APPROXIMATE	A SUPERCONVERGENCE RESULT FOR THE APPROXIMATE SOLUTION OF THE HEAT EQUATION BY A COLLOCATION METHOD	72-0510
APPROXIMATE	AN APPROXIMATE METHOD FOR THERMO-VISCOELASTIC STRESS ANALYSIS	66-0100
APPROXIMATE	AN APPROXIMATE NON-LINEAR ANALYSIS OF THIN PLATES	68-0286
APPROXIMATE	AN APPROXIMATE NUMERICAL SOLUTION OF THE NAVIER STOKES EQUATIONS BY GALERKIN METHOD	74-0011
APPROXIMATE	APPLICABILITY OF APPROXIMATE AND EXACT TRANSIENT HEAT TRANSFER ANALYSES TO HEATING PROCESSES USED TO	75-0244
APPROXIMATE	APPROXIMATE ALGORITHM FOR THE REANALYSIS OF STRUCTURES BY THE FINITE ELEMENT METHOD.	72-0445
APPROXIMATE	APPROXIMATE ANALYSIS OF PREFABRICATED TOWER STRUCTURES USING THIN SHELL FINITE ELEMENTS	73-0893
APPROXIMATE	APPROXIMATE BOUNDARY CONDITIONS IN THE FINITE ELEMENT METHOD	72-0995
APPROXIMATE	APPROXIMATE METHOD FOR CALCULATING THE RESPONSE OF EMPLACEMENT STRUCTURES SUBJECTED TO GROUND SHOCK	73-0654
APPROXIMATE	APPROXIMATE SOLUTION IN LINEAR VISCOELASTICITY	66-0101
APPROXIMATE	APPROXIMATE SOLUTIONS IN LINEAR COUPLED THERMOELASTICITY	68-0220
APPROXIMATE	APPROXIMATE SOLUTION FOR UNCONFINED SEEPAGE	73-0348
APPROXIMATE	APPROXIMATE THREE-DIMENSIONAL STRESS SOLUTIONS FOR SOME LAMINATED PLATE PROBLEMS USING A HIGHER ORDE	70-0419
APPROXIMATE	APPROXIMATE THREE-DIMENSIONAL SOLUTIONS FOR SYMMETRIC LAMINATES UNDER INPLANE LOADING	71-0392
APPROXIMATE	DEVELOPMENT OF A FINITE ELEMENT APPROACH FOR APPROXIMATE ANALYSIS OF UNSTEADY COMPRESSIBLE FLUID FLO	72-0828
APPROXIMATE	EINFACHES NAEHERUNGSVERFAHREN FUR DIE GRENZTRAGFAEHIGKEITSANLYSE VON SPANNBETON-PEAKTORDRUCKBEHAELTE	73-0854
APPROXIMATE	NOTE ON AN APPROXIMATE METHOD FOR COMPUTING NONCONSERVATIVE GENERALIZED FORCES ON FINITELY DEFORMED	70-0097
APPROXIMATE	NOTE ON AN APPROXIMATE METHOD FOR COMPUTING CONSISTENT CONJUGATE STRESSES IN ELASTIC FINITE ELEMENTS	73-0050
APPROXIMATE	NOTE ON INCREMENTAL EQUILIBRIUM EQUATIONS AND APPROXIMATE CONSTITUTIVE RELATIONS IN LARGE INELASTIC	74-0984
APPROXIMATE	THE APPROXIMATE ANALYSIS OF THIN SHELLS BY THE FINITE ELEMENT METHOD	66-0131
APPROXIMATE	THE CALCULATION OF TIME-DEPENDENT STRESSES IN PCPVS BY AN APPROXIMATE METHOD	71-0086
APPROXIMATE	ZJR WIRKLICHKEITSNAHEN BERECHNUNG VON STAHLBETONPLATTEN MIT DER FINITE-ELEMENT-METHODE. (APPROXIMATE	75-0832
APPROXIMATED	APPLICATION OF FINITE-ELEMENT METHOD TO APPROXIMATION OF OPTIMAL DOMAIN PROBLEM - SOLUTION METHODS F	75-0349
APPROXIMATING	ON THE EFFECT OF APPROXIMATING HIGHER ORDER TERMS IN CERTAIN FINITE ELEMENT MODELS OF NONLINEAR STRU	73-0603
APPROXIMATION	A CURVED ELEMENT APPROXIMATION IN THE ANALYSIS OF AXISYMMETRIC THIN SHELLS	70-0063
APPROXIMATION	A FINITE ELEMENT APPROXIMATION FOR THE ANALYSIS OF THIN SHELLS	68-0046
APPROXIMATION	A FINITE ELEMENT METHOD FOR THREE DIMENSIONAL FUNCTION APPROXIMATION	71-0654
APPROXIMATION	ACCURACY AND CONVERGENCE OF FINITE ELEMENT APPROXIMATION	68-0228
APPROXIMATION	ACCURATE FINITE ELEMENT APPROXIMATION FOR PROBLEMS WITH SINGULAR DATA	75-0098
APPROXIMATION	APPLICATION OF FINITE-ELEMENT METHOD TO APPROXIMATION OF OPTIMAL DOMAIN PROBLEM - SOLUTION METHODS F	75-0349
APPROXIMATION	APPROXIMATION AND BOUNDS FOR EIGENVALUES OF VARIOUS DOMAINS USING THE INTERMEDIATE PROBLEM TECHNIQUE	72-0711
APPROXIMATION	APPROXIMATION BY FINITE ELEMENT FUNCTIONS USING REGULARIZATION	74-1229
APPROXIMATION	APPROXIMATION BY FINITE-ELEMENT FUNCTIONS USING LOCAL REGULARIZATION	75-0436
APPROXIMATION	APPROXIMATION BY FIRST ORDER FINITE ELEMENTS AND SOLUTION BY PENALTY AND DUALITY OF NONLINEAR DIRICH	74-0245
APPROXIMATION	APPROXIMATION IN ELASTICITY BASED ON THE CONCEPT OF FUNCTION SPACE	47-0001
APPROXIMATION	APPROXIMATION IN FINITE ELEMENT METHOD	72-0293
APPROXIMATION	APPROXIMATION METHOD FOR COMPUTATION OF NATURAL FREQUENCIES OF SPIRAL SECTIONS	71-0775
APPROXIMATION	APPROXIMATION OF STEKLOV EIGENVALUES OF NON-SELFADJOINT SECOND ORDER ELLIPTIC OPERATORS	72-0509
APPROXIMATION	APPROXIMATION TECHNIQUES IN THE FINITE ELEMENT METHOD	72-0700
APPROXIMATION	AUTOMATIC GENERATION OF TRIANGULATIONS FOR PIECEWISE POLYNOMIAL APPROXIMATION	73-1118
APPROXIMATION	BASIC RESEARCH IN METHODS OF APPROXIMATION IN THE NONLINEAR MECHANICS OF SOLIDS AND STRUCTURES	72-0857
APPROXIMATION	BASIC RESEARCH IN METHODS OF APPROXIMATION IN THE NONLINEAR MECHANICS OF SOLIDS AND STRUCTURES, VOLU	74-1091
APPROXIMATION	BOUNDARY AND INTERIOR APPROXIMATION ERRORS IN THE FINITE ELEMENT METHOD	73-0209
APPROXIMATION	CONJUGATE APPROXIMATION FUNCTIONS IN FINITE ELEMENT ANALYSIS	71-0043
APPROXIMATION	CONSTRUCTION OF A FINITE ELEMENT APPROXIMATION WHICH CROSSES MATERIAL INTERFACES	74-0896
APPROXIMATION	CURVED FINITE ELEMENT APPROXIMATION TO NONLINEAR CABLES	72-0379
APPROXIMATION	DISCRETE TRIANGULAR APPROXIMATION OF MOMENT AND DISPLACEMENT SURFACES FOR PLATE BENDING	71-0269
APPROXIMATION	DUAL ITERATIVE TECHNIQUES FOR SOLVING A FINITE ELEMENT APPROXIMATION OF THE BIHARMONIC EQUATION	75-0480
APPROXIMATION	EFFECTS OF QUADRATURE ERRORS IN FINITE ELEMENT APPROXIMATION OF STEADY STATE, EIGENVALUE, AND PARABO	72-0343
APPROXIMATION	EIGENVALUE APPROXIMATION BY FINITE ELEMENT METHOD	73-0039
APPROXIMATION	ERROR ESTIMATES FOR A FINITE-ELEMENT APPROXIMATION OF A MINIMAL SURFACE	75-0491
APPROXIMATION	FINITE DIFFERENCE AND FINITE ELEMENT METHODS OF APPROXIMATION	71-0457
APPROXIMATION	FINITE ELEMENT APPROXIMATION OF A CLASS OF NONLINEAR OPERATORS	71-0454
APPROXIMATION	FINITE ELEMENT APPROXIMATION, AND SOLUTION BY A PENALTY-DUALITY ALGORITHM OF AN ELASTO-PLASTIC PROBL	75-0041
APPROXIMATION	FINITE ELEMENT SOLUTION OF A NON-LINEAR DIFFERENTIAL EQUATION USING KACHANOV'S AND SUCCESSIVE APPROX	75-0108
APPROXIMATION	GALERKIN APPROXIMATION OF THE TIME DERIVATIVE IN THE FINITE ELEMENT ANALYSIS	73-0573
APPROXIMATION	GALERKIN APPROXIMATION OF THE TIME DERIVATIVE IN THE FINITE ELEMENT ANALYSIS OF GROUNDWATER FLOW	74-0658
APPROXIMATION	IMPROVED APPROXIMATION OF CONSTITUTIVE ELASTO-PLASTIC STRESS STRAIN RELATIONSHIP FOR FINITE ELEMENT	74-0566
APPROXIMATION	LEAST SQUARE OPTIMAL APPROXIMATION SIMULATION TECHNIQUES	73-0871
APPROXIMATION	LEAST SQUARE POLYNOMIAL SPLINE APPROXIMATION	72-0526
APPROXIMATION	LIMITATIONS TO THE SMALL SCALE YIELDING APPROXIMATION FOR CRACK TIP PLASTICITY	74-0145
APPROXIMATION	LINEAR EQUALITY CONSTRAINTS IN FINITE ELEMENT APPROXIMATION	75-0475
APPROXIMATION	METHOD FOR EXTERNAL APPROXIMATION USING FINITE ELEMENTS FOR DELTA - 2 AND ITERATIVE METHOD FOR SOLVI	72-0282
APPROXIMATION	METHOD FOR EXTERNAL APPROXIMATION USING FINITE ELEMENTS FOR DELTA - 2 AND ITERATIVE METHOD FOR SOLVI	72-0282
APPROXIMATION	METHOD FOR EXTERNAL APPROXIMATION USING FINITE ELEMENTS FOR DELTA -2 AND ITERATIVE METHOD FOR SOLVIN	72-0283
APPROXIMATION	METHOD FOR EXTERNAL APPROXIMATION USING FINITE ELEMENTS FOR DELTA -2 AND ITERATIVE METHOD FOR SOLVIN	72-0283

APPROXIMATION	NON-SELF-ADJOINT VARIATIONAL PROCEDURE FOR FINITE-ELEMENT APPROXIMATION OF TRANSPORT EQUATION	75-0072
APPROXIMATION	ON A VARIATIONAL INEQUALITY AND ITS APPROXIMATION, IN THE THEORY OF SEMICONDUCTORS	75-0829
APPROXIMATION	ON THE ACCURACY AND NUMERICAL STABILITY OF THE FINITE ELEMENT APPROXIMATION FOR PARABOLIC AND HYPERB	75-0102
APPROXIMATION	ON THE ACCURACY OF THE TAYLOR APPROXIMATION FOR STRUCTURE RESIZING	74-0147
APPROXIMATION	ON THE VALIDITY OF AN APPROXIMATION AVAILABLE IN THE FINITE ELEMENT SHELL ANALYSIS	75-0663
APPROXIMATION	ONE-SIDED APPPROXIMATION AND VARIATIONAL INEQUALITIES	74-0335
APPROXIMATION	ONE-SIDED APPROXIMATION AND PLATE BENDING	74-0336
APPROXIMATION	PIECEWISE ANALYTIC INTERPOLATION AND APPROXIMATION IN TRIANGULATED POLYGONS	72-0508
APPROXIMATION	PIECEWISE POLYNOMIAL INTERPOLATION AND APPROXIMATION	65-0058
APPROXIMATION	RAYLEIGH-RITZ APPROXIMATION OF PIECEWISE CUBIC POLYNOMIALS	66-0087
APPROXIMATION	SIMPLIFICATION OF HIGH ORDER MECHANICAL SYSTEMS USING THE ROUTH APPROXIMATION	75-0853
APPROXIMATION	SINGULARITIES IN THE FINITE ELEMENT APPROXIMATION OF TWO DIMENSIONAL DIFFUSION PROBLEMS	73-1015
APPROXIMATION	SOME INTERIOR SUPER CONVERGENCE ESTIMATES FOR SPLINE APPROXIMATION	73-0641
APPROXIMATION	SPLINE APPROXIMATION AND DIFFERENCE SCHEMES FOR THE HEAT EQUATION	72-0518
APPROXIMATION	SUR L APPROXIMATION, PAR ELEMENTS FINIS D ORDRE UN, ET LA RESOLUTION, PAR PENALISATION-DUALITE, D UN	75-0845
APPROXIMATION	SUR L APPROXIMATION, PAR ELEMENTS FINIS D ORDRE UN, ET LA RESOLUTION, PAR PENALISATION-DUALITE, D UN	75-0845
APPROXIMATION	SURVEY OF UNDERLYING THEORY, APPROXIMATION BY PIECEWISE POLYNOMIALS, ILLEGAL ELEMENTS AND THE PATCH	73-0635
APPROXIMATION	THE FINITE ELEMENT METHOD AND APPROXIMATION THEORY	71-0380
APPROXIMATION	THE FINITE ELEMENT METHOD AND BOUNDARY SOLUTION PROCEDURES AS GENERAL APPROXIMATION METHODS FOR FIEL	75-0262
APPROXIMATION	THEMIS FINAL REPORT, VOL. 1, BASIC RESEARCH IN METHODS OF APPROXIMATION IN THE NONLINEAR MECHANICS O	74-0723
APPROXIMATION	THEORY AND EXAMPLES OF PARTIAL APPROXIMATION IN FINITE-ELEMENT METHOD	75-0353
APPROXIMATION	THEORY OF APPROXIMATION	72-0500
APPROXIMATION	THREE DIMENSIONAL FINITE ELEMENT APPROXIMATION IN ELECTROMAGNETIC EXPLORATION	74-0358
APPROXIMATION	TIMING COMPARISON STUDY FOR SOME HIGH ORDER FINITE ELEMENT PROCEDURES AND A LOW ORDER FINITE DIFFERE	72-0767
APPROXIMATION	UEBER PAUSCHALE FEHLERSCHRANKEN FUER APPROXIMATIONEN VON SCHALENVERFORMUNGEN. (GLOBAL BOUNDARY CONDI	75-0292
APPROXIMATIONEN	UEBER PAUSCHALE FEHLERSCHRANKEN FUER APPROXIMATIONEN VON SCHALENVERFORMUNGEN. (GLOBAL BOUNDARY CONDI	75-0292
APPROXIMATIONS	A MATHEMATICAL THEORY OF COMPLEMENTARY-DUAL VARIATIONAL PRINCIPLES AND MIXED FINITE ELEMENT APPROXIM	73-0804
APPROXIMATIONS	A NOTE ON ACCURACY AND CONVERGENCE OF FINITE ELEMENT APPROXIMATIONS	71-0565
APPROXIMATIONS	A THEORY OF MIXED FINITE ELEMENT APPROXIMATIONS OF NON-SELF-ADJOINT BOUNDARY-VALUE PROBLEMS	74-0340
APPROXIMATIONS	A THEORY OF MIXED FINITE ELEMENT APPROXIMATIONS OF NON-SELF-ADJOINT BOUNDARY VALUE PROBLEMS	74-0715
APPROXIMATIONS	ACCURACY AND CONVERGENCE OF FINITE ELEMENT GALERKIN APPROXIMATIONS OF TIME DEPENDENT PROBLEMS WITH E	73-0987
APPROXIMATIONS	ACCURACY AND CONVERGENCE OF MIXED FINITE-ELEMENT APPROXIMATIONS OF THIN BARS, MEMBRANES, AND PLATES	73-1120
APPROXIMATIONS	ACCURACY AND CONVERGENCE OF FINITE ELEMENT/GALERKIN APPROXIMATIONS OF TIME DEPENDENT PROBLEMS WITH E	75-0006
APPROXIMATIONS	ACCURACY OF FINITE ELEMENT APPROXIMATIONS TO STRUCTURAL PROBLEMS	70-0120
APPROXIMATIONS	ACCURACY OF FINITE ELEMENT APPPROXIMATIONS TO STRUCTURAL PROBLEMS	70-0438
APPROXIMATIONS	AN IMBEDDING TECHNIQUE FOR THE GENERATION OF WEAK-WEAK FINITE ELEMENT APPROXIMATIONS OF LINEAR AND N	72-0577
APPROXIMATIONS	APPROXIMATIONS AND NUMERICAL ANALYSIS OF FINITE DEFORMATIONS OF ELASTIC SOLIDS	74-0343
APPROXIMATIONS	APPROXIMATIONS EXTERNES, PAR ELEMENTS FINIS DE LAGRANGE D ORDRE, UN ET DEUX, DU PROBLEME DE DIRICHLE	72-0495
APPROXIMATIONS	APPROXIMATIONS IN FINITE-ELEMENT HEAT CONDUCTION ANALYSIS	74-1167
APPROXIMATIONS	CERTAIN APPROXIMATIONS IN FINITE ELEMENT METHOD	71-0577
APPROXIMATIONS	CONVERGENCE AND ERROR ESTIMATES FOR FINITE ELEMENT APPROXIMATIONS OF STATIONARY NAVIER-STOKES EQUATI	74-0122
APPROXIMATIONS	CONVERGENCE OF MIXED FINITE ELEMENT APPROXIMATIONS OF A CLASS OF LINEAR-BOUNDARY VALUE PROBLEMS	73-0599
APPROXIMATIONS	CONVERGENCE, ACCURACY AND STABILITY OF FINITE ELEMENT APPROXIMATIONS OF A CLASS OF NON-LINEAR HYPERB	73-0159
APPROXIMATIONS	CUBIC SPLINE GALERKIN APPROXIMATIONS TO PARABOLIC SYSTEMS WITH COUPLED NON-LINEAR BOUNDARY CONDITION	75-0232
APPROXIMATIONS	DISCONTINUOUS FINITE-ELEMENT APPROXIMATIONS FOR ANALYSIS OF SHOCK-WAVES IN NONLINEARLY ELASTIC-MATER	75-0451
APPROXIMATIONS	DISCRETE APPROXIMATIONS RELATED TO NONLINEAR THEORIES OF SOLIDS	71-0259
APPROXIMATIONS	DISCRETE APPROXIMATIONS OF ELASTIC-PLASTIC BODIES BY VARIATIONAL METHODS	72-0710
APPROXIMATIONS	ERROR BOUNDS FOR SEMIDISCRETE GALERKIN APPROXIMATIONS OF PARABOLIC PROBLEMS WITH APPLICATIONS TO PET	70-0291
APPROXIMATIONS	EXPERIENCES WITH ORTHOGONAL POLYNOMIALS AND "BEST" NUMERICAL INTEGRATION FORMULAS ON A TRIANGLE; WIT	74-0977
APPROXIMATIONS	EXTENSIONS OF THE SUCCESSIVE OVER-RELAXATION THEORY WITH APPLICATIONS TO FINITE ELEMENT APPROXIMATIO	73-0980
APPROXIMATIONS	FINITE ELEMENT APPROXIMATIONS TO PARABOLIC PROBLEMS	71-0448
APPROXIMATIONS	FINITE ELEMENT APPROXIMATIONS IN NONLINEAR THERMOVISCOELASTICITY	71-0459
APPROXIMATIONS	FINITE ELEMENT APPROXIMATIONS IN NONLINEAR ELASTICITY	71-0470
APPROXIMATIONS	FINITE ELEMENT APPROXIMATIONS TO TIME-DEPENDENT PROBLEMS	72-0269
APPROXIMATIONS	FUNCTIONAL APPROXIMATIONS FOR SOLVING BOUNDARY VALUE PROBLEMS BY COMPUTER	69-0013
APPROXIMATIONS	GENERALIZED CONJUGATE FUNCTIONS FOR MIXED FINITE ELEMENT APPROXIMATIONS OF BOUNDARY VALUE PROBLEMS	72-0515
APPROXIMATIONS	MATHEMATICAL ASPECTS OF FINITE ELEMENT APPROXIMATIONS IN THEORETICAL MECHANICS	75-0852
APPROXIMATIONS	MATHEMATICAL FEATURES OF FINITE ELEMENT APPROXIMATIONS OF CERTAIN FLOW PROBLEMS WITH EMPHASIS ON CON	74-0002
APPROXIMATIONS	MEAN-SQUARE ERRORS OF FINITE ELEMENT APPROXIMATIONS ON LINEAR ELASTOSTATICS, ELASTODYNAMICS, AND THE	74-0107
APPROXIMATIONS	MIXED CONJUGATE FINITE ELEMENT APPROXIMATIONS OF LINEAR OPERATORS	72-0085
APPROXIMATIONS	MIXED FINITE-ELEMENT APPROXIMATIONS OF LINEAR BOUNDARY-VALUE PROBLEMS	75-0453
APPROXIMATIONS	NOTE ON THE ACCURACY AND CONVERGENCE OF FINITE ELEMENT APPROXIMATIONS OF THE CONVECTION EQUATION	75-0490
APPROXIMATIONS	NOTE ON THE CONVERGENCE OF FINITE ELEMENT APPROXIMATIONS FOR PROBLEMS FORMULATED IN CURVILINEAR COOR	74-0267
APPROXIMATIONS	ON CERTAIN APPROXIMATIONS IN THE FINITE ELEMENT METHOD	71-0515
APPROXIMATIONS	ON FINITE ELEMENT APPROXIMATIONS ON TIME DEPENDENT PROBLEMS	72-0528
APPROXIMATIONS	ON LOCAL APPROXIMATIONS OF OPERATOR EQUATIONS	72-0063
APPROXIMATIONS	ON MIXED FINITE ELEMENT APPROXIMATIONS	74-0701
APPROXIMATIONS	ON THE CALCULATION OF CONSISTENT STRESS DISTRIBUTIONS IN FINITE ELEMENT APPROXIMATIONS	71-0191
APPROXIMATIONS	ON THE CONVERGENCE OF SOR ITERATIONS FOR FINITE ELEMENT APPROXIMATIONS TO ELLIPTIC BOUNDARY VALUE PR	71-0449
APPROXIMATIONS	ON THE LOCAL DISTRIBUTION OF ERRORS BY FINITE ELEMENT APPROXIMATIONS	73-0304
APPROXIMATIONS	ON THE STRESS COMPUTATION IN FINITE ELEMENT MODELS BASED UPON DISPLACEMENT APPROXIMATIONS	74-0598
APPROXIMATIONS	RAY-EFFECT MITIGATION IN DISCRETE ORDINATE-LIKE ANGULAR FINITE-ELEMENT APPROXIMATIONS IN NEUTRON-TRA	75-0496
APPROXIMATIONS	SOME APPROXIMATIONS IN THE NONLINEAR ANALYSIS OF RECTANGULAR PLATES USING FINITE ELEMENTS	74-1231
APPROXIMATIONS	SOME CONTRIBUTIONS TO THE MATHEMATICAL THEORY OF MIXED FINITE ELEMENT APPROXIMATIONS	73-0302
APPROXIMATIONS	SOME MATHEMATICAL PROPERTIES OF CERTAIN MIXED GALERKIN APPROXIMATIONS IN NONLINEAR ELASTICITY	74-1248
APPROXIMATIONS	SOME OBSERVATIONS ON PROPERTIES OF CERTAIN MIXED FINITE ELEMENT APPROXIMATIONS	75-0617
APPROXIMATIONS	STABILITY AND ACCURACY OF NUMERICAL APPROXIMATIONS TO TIME DEPENDENT FLOWS	75-0183
APPROXIMATIONS	USE OF SINGULAR FUNCTIONS WITH FINITE ELEMENT APPROXIMATIONS	73-0102
APPROXIMATIONS	ZUR SYSTEMATIK NUMERISCHER NAEHERUNGSVERFAHREN IN DER STATIK UND DYNAMIK DER KONSTRUKTIONEN (METHODS	75-0557
APPURTENANCE	THREE DIMENSIONAL INELASTIC ANALYSIS OF A SHELL APPURTENANCE PER SECTION 111 - APPENDIX F.	74-0716
AQUIFER	APPLICATION OF GALERKIN'S PROCEDURE TO AQUIFER ANALYSIS	72-0636
AQUIFER	FINITE ELEMENT ANALYSIS OF TIME-DEPENDENT PUMPING TEST DATA IN AN UNCONFINED AQUIFER	75-0497
AQUIFER	IDENTIFICATION OF AQUIFER CONSTANTS: TESTS OF TWO INVERSE METHODS	74-0051
AQUIFER	NUMERICAL SOLUTION OF THE STEADY INTERFACE IN A CONFINED COASTAL AQUIFER	74-1007
AQUIFER	THE NUMERICAL SOLUTION OF THE STEADY INTERFACE IN A COASTAL CONFINED AQUIFER	73-0711
AQUIFER	THEORY OF FLOW IN A CONFINED TWO AQUIFER SYSTEM	69-0382
AQUIFER	TRANSIENT TWO DIMENSIONAL UNCONFINED AQUIFER PROBLEMS	74-0069
AQUIFER	TWO-DIMENSIONAL FINITE ELEMENT MODELING OF TRANSIENT FLOW IN REGIONAL AQUIFER SYSTEMS. ANALYSIS OF G	74-0743
AQUIFER	USE OF THE FINITE ELEMENT METHOD IN SOLVING TRANSIENT FLOW PROBLEMS IN AQUIFER SYSTEMS	68-0167
AQUIFERS	A FINITE ELEMENT MODEL FOR MISCIBLE DISPLACEMENT IN GROUNDWATER AQUIFERS	75-0783
AQUIFERS	A METHOD OF ANALYZING TRANSIENT FLUID FLOW IN MULTILAYERED AQUIFERS	69-0142
AQUIFERS	GRAVITY FLOW TO EXCAVATIONS AND DRAINAGE TRENCHES IN LAYERED AQUIFERS	71-0014
AQUIFERS	MANAGEMENT OF RETARDATION OF SALT WATER INTRUSION IN COASTAL AQUIFERS	75-0692
AQUIFERS	MATHEMATICAL MODELING OF MOVEMENT OF DISSOLVED CONSTITUENTS IN GROUND WATER AQUIFERS BY THE FINITE E	70-0490
AQUIFERS	NUMERICAL DETERMINATION OF SEA WATER INTRUSION IN AQUIFERS	74-1181
AQUIFERS	NUMERICAL PREDICTION OF TRACER CONCENTRATION IN AQUIFERS WITH TRANSIENT GROUNDWATER FLOW BY FINITE-E	75-0610
AQUIFERS	ON SEAWATER ENCROACHMENT IN COASTAL AQUIFERS	74-0577
AQUIFERS	RESPONSE OF UNCONFINED AQUIFERS TO PUMPING	74-0073
AQUIFERS	STUDY BY THE FINITE-ELEMENT METHOD OF THE INFLUENCE OF FRACTURES IN CONFINED AQUIFERS.	75-0604
ARAKAWAS	ARAKAWAS METHOD IS A FINITE-ELEMENT METHOD	74-0633
ARAVA	STRUCTURAL ANALYSIS OF ARAVA FUSELAGE USING FINITE ELEMENT METHOD	71-0578
ARBEITSTITEL	ANWENDUNG DER MATRIZENVERSCHIEBUNGSMETHODE AUF ERZWUNGENE SCHWINGUNGEN (SCHWINGUNGSPROBLEME) (ARBEIT	71-0627

ARBITRARILY	ARBITRARILY SHAPED INHOMOGENEOUS OPTICAL FIBER OR INTEGRATED OPTICAL WAVEGUIDES	75-0372
ARBITRARY	A FORTRAN PROGRAM FOR CALCULATING THE FREQUENCIES AND MODE SHAPES OF AN ARBITRARY SHAPE CYLINDRICAL	72-0594
ARBITRARY	A GENERAL FINITE DIFFERENCE METHOD FOR ARBITRARY MESHES	72-0883
ARBITRARY	ADDITION OF AN ARBITRARY BODY ANALYSIS CAPABILITY TO THE BOEING TEA 236 FINITE ELEMENT COMPUTER PROG	73-0785
ARBITRARY	ANALYSIS OF ARBITRARY SHELLS USING SHALLOW SHELL FINITE ELEMENTS	71-0717
ARBITRARY	ANALYSIS OF ARBITRARY SHELLS USING SHALLOW SHELL FINITE ELEMENTS	74-1374
ARBITRARY	ANALYSIS OF ELASTIC SHELLS OF REVOLUTION WITH MEMBRANE AND FLEXURE STRESSES UNDER ARBITRARY LOADING	69-0289
ARBITRARY	ANALYSIS OF RAREFIED GAS FLOW THROUGH AN ARBITRARY CROSS SECTION BY THE FINITE ELEMENT METHOD	73-0365
ARBITRARY	ANALYSIS OF THE EFFECTS ON LIFE OF LEADING-EDGE HOLES IN AN AIRFOIL SUBJECTED TO ARBITRARY SPANWISE	75-0700
ARBITRARY	ANALYSIS OF THIN SHELLS WITH A DOUBLY CURVED ARBITRARY QUADRILATERAL FINITE ELEMENT.	72-0443
ARBITRARY	APPLICATION OF THE DISCRETE ELEMENT METHOD TO THE BUCKLING ANALYSIS OF RECTANGULAR PLATES UNDER ARBI	68-0054
ARBITRARY	ARBITRARY QUADRILATERAL SPAR WEBS FOR THE MATRIX DISPLACEMENT METHOD	66-0003
ARBITRARY	ARBITRARY SYMMETRIC LOADING PROBLEMS OF CENTRALLY CRACKED RECTANGULAR PLATES	75-0318
ARBITRARY	AUTOMATIC TRIANGULATION OF ARBITRARY PLANAR DOMAINS FOR THE FINITE ELEMENT METHOD	74-0480
ARBITRARY	BERGEN, AN AUTOMATIC FINITE ELEMENT MESH GENERATION PROGRAM FOR ARBITRARY STRUCTURES	72-0776
ARBITRARY	CALCULATION OF POTENTIAL FLOW ABOUT ARBITRARY THREE-DIMENSIONAL LIFTING BODIES	72-0722
ARBITRARY	CALCULATION OF STIFFNESS MATRICES FOR FINITE ELEMENTS OF THIN SHELLS OF ARBITRARY SHAPE	68-0127
ARBITRARY	CALCULATION OF THE FREE OSCILLATIONS OF A LIQUID IN AXISYMMETRIC MOTIONLESS CONTAINERS OF ARBITRARY	74-0387
ARBITRARY	CELLULAR STRUCTURES OF ARBITRARY PLAN GEOMETRY	72-0201
ARBITRARY	COMMENT ON "SOLUTION OF LAPLACE'S EQUATION IN A REGION WITH DIELECTRIC INTERFACES OF ARBITRARY SHAPE	70-0238
ARBITRARY	COMPARISON OF TWO HIGH-PRECISION TRIANGULAR FINITE ELEMENTS FOR ARBITRARY DEEP SHELLS	71-0071
ARBITRARY	CURSHL: A HIGH-PRECISION FINITE ELEMENT FOR SHELLS OF ARBITRARY SHAPE	71-0430
ARBITRARY	DOUBLY CURVED TRIANGULAR FINITE ELEMENTS FOR SHELLS OF ARBITRARY SHAPE	73-0881
ARBITRARY	DYNAMIC ANALYSIS OF AXISYMMETRIC SHELLS UNDER ARBITRARY TRANSIENT PRESSURES.	72-0467
ARBITRARY	DYNAMIC ELASTOPLASTIC RESPONSE OF GEOMETRICALLY NONLINEAR ARBITRARY SHELLS OF REVOLUTION UNDER IMPUL	70-0098
ARBITRARY	DYNAMIC FINITE ELEMENT ANALYSIS OF ARBITRARY THIN SHELLS	71-0061
ARBITRARY	DYNAMIC RESPONSE OF ARBITRARY SHELLS OF REVOLUTION UNDER COMBINED IMPULSIVE NONISOTHERMAL AND IRREGU	70-0365
ARBITRARY	DYNAMIC STRESS ANALYSIS OF AXISYMMETRIC STRUCTURES UNDER ARBITRARY LOADING	69-0433
ARBITRARY	EPAD, A COMPUTER PROGRAM FOR THE ELASTIC-PLASTIC ANALYSIS OF DOME STRUCTURES AND ARBITRARY SOLIDS OF	70-0036
ARBITRARY	FINITE ELEMENT ANALYSIS OF AXISYMMETRIC SOLIDS WITH ARBITRARY LOADINGS	67-0145
ARBITRARY	FINITE ELEMENT ANALYSIS FOR ARBITRARY AXISYMMETRIC STRUCTURES ORTHOTROPIC, TEMPERATURE-DEPENDENT MAT	68-0076
ARBITRARY	FINITE ELEMENT ANALYSIS OF STRUCTURES WITH SURFACES OF ARBITRARY CURVATURE	70-0102
ARBITRARY	FINITE ELEMENT ANALYSIS OF ARBITRARY THIN SHELLS	70-0329
ARBITRARY	FINITE ELEMENT ANALYSES OF SHELL RESPONSE FOR ARBITRARY EXCITATIONS	70-0348
ARBITRARY	FINITE ELEMENT ANALYSIS OF UNSTEADY INCOMPRESSIBLE FLOW AROUND AN OSCILLATING OBSTACLE OF ARBITRARY	73-0067
ARBITRARY	FINITE ELEMENT ANALYSIS OF LAMINAR CONVECTIVE HEAT TRANSFER IN VERTICAL DUCTS WITH ARBITRARY CROSS-S	75-0555
ARBITRARY	FINITE ELEMENT MESH GENERATION PROGRAM FOR ARBITRARY TWO-AND THREE-DIMENSIONAL STRUCTURES	73-0966
ARBITRARY	FINITE ELEMENT SIMULATION OF ELECTROMECHANICAL RESONATORS WITH ARBITRARY ELECTRODES	75-0492
ARBITRARY	FINITE ELEMENT SOLUTION OF VISCOUS FLOW IN AN ARBITRARY INTERNAL PASSAGE	71-0461
ARBITRARY	FINITE ELEMENTS FOR AXISYMMETRIC SOLIDS UNDER ARBITRARY LOADINGS WITH NODES ON ORIGIN,	72-0251
ARBITRARY	FINITE ELEMENTS FOR AXISYMMETRIC SOLIDS UNDER ARBITRARY LOADINGS WITH NODES ON ORIGIN	73-0094
ARBITRARY	FINITE ELEMENTS SOLUTION FOR TORSION RIGIDITY AND SHEAR CENTER OF ARBITRARY CROSS-SECTIONS	74-0168
ARBITRARY	FINITE-ELEMENT ANALYSIS OF FLOW IN BLOOD VESSEL WITH ARBITRARY CROSS-SECTION	72-0936
ARBITRARY	FINITE-ELEMENT PLASMA-DIFFUSION CODE FOR ARBITRARY TOROIDAL GEOMETRY	75-0500
ARBITRARY	FREE VIBRATIONS OF CIRCULAR MEMBRANES UNDER ARBITRARY TENSION BY THE FINITE ELEMENT METHOD	69-0190
ARBITRARY	LARGE DEFLECTION ANALYSIS OF ARBITRARY SHALLOW SHELLS BY THE FINITE ELEMENT METHOD	73-0937
ARBITRARY	MATRIX-DISPLACEMENT DIRECT-ELEMENT METHOD FOR FORCE AND TORQUE ANALYSIS OF DETERMINATE AND INDETERMI	74-0531
ARBITRARY	NODAL VARIABLES FOR ARBITRARY ORDER FINITE ELEMENTS	75-0002
ARBITRARY	NONLINEAR ANALYSIS OF ARBITRARY, HYPERELASTIC MEMBRANE SHELLS BY THE FINITE-ELEMENT METHOD	73-1132
ARBITRARY	ON SMALL VIBRATIONS AND PERTURBATIONS OF FLEXIBLE BODIES UNDERGOING ARBITRARY NOMINAL MOTION	70-0376
ARBITRARY	ON THE ASYMPTOTICALLY SPHERICAL DEFORMATIONS OF ARBITRARY MEMBRANES OF REVOLUTION FIXED ALONG AN EDG	71-0640
ARBITRARY	SLAB BRIDGES WITH ARBITRARY SHAPE AND SUPPORT CONDITIONS - A GENERAL METHOD OF ANALYSIS BASED ON FIN	68-0233
ARBITRARY	SOLVING A NONLINEAR PROBLEM OF NONSTEADY THERMAL CONDUCTIVITY IN SOLID BODIES OF ARBITRARY FORM	68-0019
ARBITRARY	STUDY OF UMBRELLA OSCILLATIONS OF TURBINE DISKS OF ARBITRARY PROFILE BY THE FINITE ELEMENT METHOD	75-0332
ARBITRARY	THE ENERGY METHOD OF NETWORKS OF ARBITRARY SHAPE IN PROBLEMS OF THE THEORY OF ELASTICITY	59-0001
ARBITRARY	THE TRANSIENT LINEAR ELASTIC RESPONSE ANALYSIS COMPLEX THIN SHELLS OF REVOLUTION SUBJECTED TO ARBITR	70-0080
ARBITRARY	THEORETICAL ANALYSIS OF THICK DOME STRUCTURES AND ARBITRARY SOLIDS OF REVOLUTION SUBJECTED TO UNSYMM	70-0037
ARBITRARY	THERMAL STRESSES IN OPEN CYLINDRICAL SHELLS OF ARBITRARY BOUNDARY CONDITIONS BY A FINITE STRIP METHO	72-0371
ARBITRARY	TRANSFINITE ELEMENT METHODS: BLENDING FUNCTION INTERPOLATION OVER ARBITRARY CURVED ELEMENT DOMAINS	73-1006
ARBITRARY	TRANSIENT LAMINAR FLOW IN DUCTS OF ARBITRARY CROSS-SECTION BY FINITE ELEMENT METHODS	72-0013
ARBITRARY	VARIATIONAL PRINCIPLES FOR A RAREFIED GAS FLOW PAST AND ARBITRARY THREE-DIMENSIONAL BODY	70-0189
ARCH	ANALYSIS OF ARCH DAMS USING A SPACE-FRAME MODEL	73-0176
ARCH	ANALYSIS OF THIN ARCH DAMS BY THE FINITE ELEMENT METHOD	64-0001
ARCH	APPROACH TO ANALYSIS OF ARCH DAMS IN WIDE VALLEYS	73-0153
ARCH	ARCH DAM CONSTRUCTION STRESSES DUE TO HYDROSTATIC AND GRAVITY LOADS	75-0658
ARCH	ARCH DAMS ANALYSED BY A LINEAR FINITE ELEMENT SHELL THE FINITE STRIP METHOD IN THE ANALYSIS OF ELAST	68-0189
ARCH	BEHAVIOR WITH DIMINISHING CURVATURE OF STRAIN-BASED ARCH FINITE ELEMENTS	73-0088
ARCH	DESIGN OF INCLINED TIED ARCH RAILWAY BRIDGES OVER THE M56	71-0297
ARCH	EARTHQUAKE RESPONSE OF ARCH DAMS	70-0113
ARCH	ELASTIC ANALYSIS OF A REFRACTORY SPRUNG ARCH	74-0386
ARCH	FINITE DEFLECTIONS, BUCKLING AND POSTBUCKLING OF AN ARCH	69-0123
ARCH	FINITE ELEMENT - ANALOGUE METHOD FOR DETERMINING THE DYNAMIC CHARACTERISTICS OF AN ARCH DAM-RESERVOI	73-0678
ARCH	FINITE ELEMENT ANALYSIS OF ARCH DAMS AND FOUNDATIONS	71-0668
ARCH	FINITE ELEMENT ANALYSIS OF A BROAD ARCH BRIDGE WITH A SLAB DECK	71-0724
ARCH	FINITE ELEMENT METHOD OF ANALYSIS FOR ARCH DAMS SHELLS AND COMPARISON WITH FINITE DIFFERENCE PROCEDU	65-0036
ARCH	FINITE ELEMENT STUDY OF THE TRIAXIAL STRESS STATE AROUND AN INSPECTION TUNNEL IN AN ARCH DAM	74-0831
ARCH	NUMERICAL STUDIES USING CIRCULAR ARCH FINITE ELEMENTS	74-0600
ARCH	SHAPE FUNCTIONS AND ACCURACY OF ARCH FINITE ELEMENTS	73-0038
ARCH	STATIC AND DYNAMIC ANALYSIS OF AN ARCH DAM	74-1101
ARCH	THE ANALYSIS OF ARCH DAMS UNDER STATIC LOADING BY THE FINITE ELEMENT METHOD	68-0059
ARCH	THE SEISMIC DESIGN STUDY OF A DOUBLE CURVATURE ARCH DAM	69-0053
ARCH	THREE DIMENSIONAL ANALYSIS OF TWO ARCH DAMS BY THE FINITE ELEMENT METHOD	68-0018
ARCH	THREE DIMENSIONAL ANALYSIS OF ARCH DAMS AND THEIR FOUNDATIONS	68-0066
ARCH	THREE-DIMENSIONAL STRESS ANALYSIS OF ARCH DAMS BY THE FINITE ELEMENT METHOD: PART I, ARCH DAM NO. 1,	66-0099
ARCH	THREE-DIMENSIONAL STRESS ANALYSIS OF ARCH DAMS BY THE FINITE ELEMENT METHOD: PART I, ARCH DAM NO. 1,	66-0099
ARCH	THREE-DIMENSIONAL STRESS ANALYSIS OF ARCH DAMS BY THE FINITE ELEMENT METHOD: PART I, ARCH DAM NO. 1,	66-0099
ARCH	VOGT BOUNDARY FOR FINITE ELEMENT ARCH DAM ANALYSIS	74-0156
ARCHED	SHOCK TUNNEL TESTS OF PRELOADED AND ARCHED WALL PANELS FINITE ELEMENT STRESS FORMULATION FOR WAVE PR	73-0892
ARCHES	A FINITE ELEMENT APPROACH TO THE STRUCTURAL INSTABILITY OF BEAM COLUMNS, FRAMES, AND ARCHES	70-0434
ARCHES	A FINITE-DEFLECTION ANALYSIS OF SHALLOW ARCHES BY THE DISCRETE ELEMENT METHOD	71-0077
ARCHES	A NONLINEAR FINITE ELEMENT ANALYSIS OF SHALLOW CIRCULAR ARCHES	69-0269
ARCHES	APPLICATION OF CURVED FINITE ELEMENTS TO CIRCULAR ARCHES	71-0360
ARCHES	COMMENT ON LIMITATIONS OF CERTAIN FINITE ELEMENTS WHEN APPLIED TO ARCHES - FURTHER STUDIES IN APPLIC	72-0097
ARCHES	COMMENT ON LIMITATIONS OF CERTAIN FINITE ELEMENTS WHEN APPLIED TO ARCHES - FURTHER STUDIES IN APPLIC	72-0097
ARCHES	CORRECTED ASSESSMENT OF CYLINDRICAL SHELL FINITE ELEMENT OF BOGNER, FOX AND SCHMIT WHEN APPLIED TO A	73-0024
ARCHES	CURVED FINITE ELEMENTS FOR THE ANALYSIS OF SHALLOW AND DEEP ARCHES	74-0332
ARCHES	FINITE ELEMENT ANALYSIS OF BUCKLING AND POST-BUCKLING BEHAVIORS OF ARCHES WITH GEOMETRIC IMPERFECTIO	73-0186
ARCHES	FINITE ELEMENT CALCULATION OF STRESS DISTRIBUTIONS IN ARCHES	74-0094
ARCHES	FURTHER STUDIES IN THE APPLICATION OF CURVED FINITE ELEMENTS TO CIRCULAR ARCHES	71-0644
ARCHES	LARGE AMPLITUDE, FREE VIBRATION GEOMETRICALLY NONLINEAR CIRCULAR ARCHES	72-0606
ARCHES	LARGE DEFLECTION GEOMETRICALLY NON-LINEAR FINITE ELEMENT ANALYSIS OF CIRCULAR ARCHES	73-0021
ARCHES	LIMITATIONS OF CERTAIN CURVED FINITE ELEMENTS WHEN APPLIED TO ARCHES	71-0279

127

ARCHES	NATURAL FREQUENCIES OF CIRCULAR ARCHES	71-0302
ARCHES	NONLINEAR BUCKLING ANALYSIS OF SANDWICH ARCHES	71-0292
ARCHES	ON THE VALIDITY OF THE FINITE ELEMENT ANALYSIS OF CIRCULAR ARCHES REPRESENTED BY AN ASSEMBLAGE OF BE	75-0479
ARCHES	THE FINITE ELEMENT METHOD APPLIED TO THE STATIC AND DYNAMIC BUCKLING OF CIRCULAR ARCHES	70-0497
ARCHING	OPTIMUM POSITION OF THE CENTRAL CLAY CORE OF A ROCKFILL DAM IN RESPECT TO ARCHING AND HYDRAULIC FRAC	73-0447
ARCS	USE OF PARABOLIC ARCS IN MATCHING CURVED BOUNDARIES IN FINITE-ELEMENT METHOD	75-0589
ARCTIC	COUPLED HEAT AND MOISTURE TRANSPORT MODEL FOR ARCTIC SOILS	74-0578
AREA	INTEGRATION OF AREA AND VOLUME COORDINATES IN FINITE-ELEMENT METHOD	69-0370
AREA	INTERACTION OF AREA COORDINATES IN MATRIX STRUCTURAL ANALYSIS	68-0150
AREAL	CALCULATION OF AREAL RAINFALL USING FINITE ELEMENT TECHNIQUES WITH ALTITUDINAL CORRECTIONS	72-0344
ARL	MODIFICATIONS TO ARL COMPUTER PROGRAMS USED FOR DESIGN OF AXIAL COMPRESSOR AIRFOILS	74-0508
ARM	APPLICATION OF SAMIS TO THE ANALYSIS OF A FRONT SUSPENSION UPPER ARM	70-0260
ARM	DYNAMIC RESPONSE OF THE HUMAN ARM AS A COMPOSITE STRUCTURE	72-0611
ARMATURE	DISTRIBUZIONI OTTIMALI DI ARMATURE DI PRECOMPRESSIONE IN TRAVI-PARETE DI CALCESTRUZZO. (OPTIMAL DIST	75-0560
ARMOR	FINITE ELEMENT STRESS ANALYSIS OF A MULTICONDUCTOR ARMOR CABLE	73-0095
ARMY	ARMY NUMERICAL ANALYSIS CONFERENCE	71-0707
ARMY	TRANSACTIONS OF ARMY MATHEMATICIANS RESEARCH CONFERENCE	73-0925
ARPA	THE USE OF A LASER FOR ARPA MILITARY GEOPHYSICS PROGRAM (ROCK MECHANICS AND RAPID EXCAVATION)	72-0800
ARRANGEMENT	INVESTIGATION ON FREE AND FORCED VIBRATIONS OF A LONG TANKER WITH OVERLAPPING PROPELLER ARRANGEMENT	73-0691
ARRAY	APPLICATION OF SMALL AND LARGE-DISPLACEMENT THEORY TO A LARGE FLEXIBLE SOLAR ARRAY AND COMPARISON WI	71-0231
ARRAY	MECHANICS OF CABLE MOORING SYSTEMS. VOLUME V11. THE STEADY-STATE BEHAVIOR OF A PYRAMID ARRAY SYSTEM	73-0724
ARRAY	MECHANICS OF CABLE MOORING SYSTEMS. VOLUME V1. A COMPUTER PROGRAM FOR ANALYZING THE STEADY STATE CON	73-0725
ARRAYS	ADVANCED INTERCONNECT SYSTEMS FOR LIGHTWEIGHT SOLAR ARRAYS	73-0696
ARRAYS	IMPEDANCE OF INTERSTITIAL GAPS IN TRANSDUCER ARRAYS	70-0410
ARRAYS	MECHANICS OF CABLE MOORING SYSTEMS. VOLUME IV. A COMPUTER PROGRAM FOR ANALYZING THE STEADY STATE RES	72-0653
ARRAYS	MENISCI IN ARRAYS OF CYLINDERS - NUMERICAL-SIMULATION BY FINITE-ELEMENTS	75-0449
ARRAYS	STRESS ANALYSIS AND DESIGN OF SILICON SOLAR CELL ARRAYS AND RELATED MATERIAL PROPERTIES	72-0840
ARREST	CRACK PROPAGATION AND ARREST IN IMPACTED PLATES	72-0599
ARREST	EXPLORATORY INVESTIGATION OF RAPID CRACK PROPAGATION AND CRACK ARREST	73-0757
ARREST	FAST FRACTURE RESISTANCE AND CRACK ARREST IN STRUCTURAL STEELS	73-0766
ARRIERE-CORPS	COUPLAGE ENTRE L ECOULEMENT AUTOUR D UN ARRIERE-CORPS ET LE JET PROPULSIF EN THEORIE DE FLUIDE PARFA	74-1137
ARTERIAL	DEFORMATION OF THE ARTERIAL VASA VASORUM AT NORMAL AND HYPERTENSIVE ARTERIAL PRESSURE	71-0149
ARTERIAL	DEFORMATION OF THE ARTERIAL VASA VASORUM AT NORMAL AND HYPERTENSIVE ARTERIAL PRESSURE	71-0149
ARTERIOLE	STRUCTURAL ANALYSIS OF AN ARTERIOLE BY THE DIRECT STIFFNESS METHOD	66-0109
ARTERIOLES	STRUCTURAL RESPONSE OF RELAXED AND CONSTRICTED ARTERIOLES	68-0200
ARTESIAN	ANALYSIS OF UNSTEADY FLOW TOWARD AN ARTESIAN WELL BY 3-DIMENSIONAL FINITE ELEMENTS	74-0351
ARTESIAN	ANALYSIS OF UNSTEADY FLOW TOWARD ARTESIAN WELLS BY THREE-DIMENSIONAL FINITE ELEMENTS	74-0742
ARTESIAN	ANALYSIS OF UNSTEADY FLOW TOWARD AN ARTESIAN WELL BY THREE-DIMENSIONAL FINITE ELEMENTS	74-1351
ARTICULATED	MOVEMENTS OF ARTICULATED CONDUITS UNDER EARTH DAMS ON COMPRESSIBLE FOUNDATIONS	73-0477
ARTILLERY	FINITE ELEMENT STRESS ANALYSIS OF ARTILLERY PROJECTILES	73-0432
AS-CUT	FINITE ELEMENT CALCULATIONS RELEVANT TO AS-CUT QUARTZ RESONATORS	73-0672
ASAAS	ASAAS: ASYMMETRIC STRESS ANALYSIS OF AXISYMMETRIC SOLIDS WITH ORTHOTROPIC, TEMPERATURE-DEPENDENT MAT	71-0696
ASAS	ASAS - A COMMERCIAL FINITE ELEMENT SYSTEM WITH DISPLACEMENT AND FORCE CONTINUITY OPTIONS	73-0947
ASAS	THE ASAS ANALYSIS OF THE LYNX	74-0932
ASCENT	INFLUENCE OF ASCENT HEATING ON THE SEPARATION DYNAMICS OF A SPACECRAFT FAIRING	72-0906
ASIXYMMETRIC	LARGE DISPLACEMENT ANALYSIS OF ASIXYMMETRIC SHELLS	69-0276
ASKA	A COLLECTION OF PROBLEMS SOLVED WITH ASKA	70-0401
ASKA	AN AUTOMATIC SYSTEM FOR KINEMATIC ANALYSIS ASKA HIGH SPEED COMPUTING OF ELASTIC STRUCTURES	70-0358
ASKA	AN AUTOMATIC SYSTEM FOR KINEMATIC ANALYSIS, ASKA PART 1	70-0396
ASKA	AN AUTOMATIC SYSTEM FOR KINEMATIC ANALYSIS ASKA PART 1	71-0626
ASKA	ANWENDUNG DER METHODEN DER FINITEN ELEMENTE MIT DEM PROGRAMMSYSTEM ASKA UNTER BESONDERER BERUCKSICHT	71-0624
ASKA	ASKA - AUTOMATIC SYSTEM FOR KINEMATIC ANALYSIS	65-0059
ASKA	ASKA - EINE PROGRAMMIERSPRACHE ZUR STATISCHEN UND DYNAMISCHEN ANALYSE	69-0457
ASKA	ASKA A COMPUTER SYSTEM FOR STRUCTURAL ENGINEERS	69-0345
ASKA	ASKA AUTOMATIC SYSTEM FOR KINEMATIC ANALYSIS	69-0307
ASKA	ASKA SERIES OF PROGRAMS	72-0157
ASKA	ASKA 105 - A PROBLEM ORIENTED LANGUAGE FOR STRUCTURAL ENGINEERS, VOLUME 11. APPLICATION	69-0461
ASKA	ASKA 105 - A TECHNICAL MANUAL FOR THE DIRECT SOLUTION AND INVERSION PACKAGE	68-0255
ASKA	ASKA 105 A TECHNICAL MANUAL AND PROGRAMMING THEORY FOR THE ELEMENTS, FLA, TRIM, SPARTA AND TET	68-0254
ASKA	ASKA 105, A PROBLEM ORIENTED LANGUAGE FOR STRUCTURAL ENGINEERS, VOLUME 1. DESIGN	69-0460
ASKA	ASKA, A LARGE SCALE SOFTWARE SYSTEM FOR FINITE ELEMENT ANALYSIS	69-0331
ASKA	ASKA, SOFTWARE ENGINEERING FOR THE SEVENTIES IN STRUCTURAL ANALYSIS	73-0704
ASKA	COMPUTER AIDED STRUCTURAL ANALYSIS: THE MACHINE INDEPENDENT SYSTEM ASKA	70-0397
ASKA	ELASTOPLASTIC AND CREEP ANALYSIS WITH THE ASKA PROGRAM SYSTEM	74-0863
ASKA	GENERAL BEAM ELEMENT FOR ASKA	69-0458
ASKA	REVIEW OF THE ASKA PROGRAM	71-0175
ASME	CLASSIFICATION OF FINITE ELEMENT STRESSES ACCORDING TO ASME SECTION 111 STRESS CATEGORIES	74-0610
ASME	INTERPRETATION OF FINITE ELEMENT STRESSES ACCORDING TO ASME SECTION 111	75-0655
ASPECT	AEROELASTIC PROBLEMS OF LOW ASPECT RATIO WINGS (PART 1)	56-0006
ASPECT	FINITE ELEMENT METHOD AS AN ASPECT OF THE PRINCIPLE OF MAXIMUM UNIFORMITY: NEW HYDRODYNAMICAL RAMIFA	74-0009
ASPECT	THE EFFECT OF CAMBER ON THE NATURAL FREQUENCIES OF LOW ASPECT RATIO TURBOMACHINERY BLADES	73-1092
ASPHALT	FRACTURE (ULTIMATE STRENGTH) ANALYSES OF ASPHALT PAVEMENT LAYERS RESULTING FROM TRAFFIC LOADING	73-0548
ASSEMBLAGE	ON THE VALIDITY OF THE FINITE ELEMENT ANALYSIS OF CIRCULAR ARCHES REPRESENTED BY AN ASSEMBLAGE OF BE	75-0479
ASSEMBLAGES	A CONSISTENT DISCRETE ELEMENT TECHNIQUE FOR THIN-WALLED ASSEMBLAGES	68-0104
ASSEMBLIES	AUTOMATED METHOD FOR THE LARGE DEFLECTION AND INSTABILITY ANALYSIS OF THREE-DIMENSIONAL TRUSS AND FR	66-0105
ASSEMBLIES	COMPUTER METHOD FOR COMPUTING CRITICAL SPEEDS OF NON-UNIFORM SHAFTS GEARED SYSTEMS AND CURVED ASSEMB	73-0675
ASSEMBLIES	FINITE ELEMENT ANALYSIS OF SOME CRITICAL FAST ASSEMBLIES	73-0128
ASSEMBLIES	STRUCTURAL DYNAMIC ANALYSIS OF ELECTRONIC ASSEMBLIES USING NASTRAN RESTART/FORMAT CHANGE CAPABILITY	71-0230
ASSEMBLY	A MORE EFFICIENT FRONT SOLUTION: ALLOCATING ASSEMBLY LOCATIONS BY LONGEVITY CONSIDERATIONS	73-1069
ASSEMBLY	NEW DESIGN FACTOR FOR A SHRINK FITTED ASSEMBLY	75-0167
ASSEMBLY	YIELDING OF A TWO DIMENSIONAL VOID ASSEMBLY IN AN ORGANIC GLASS	73-0416
ASSEN	THE PROGRAM " ASSEN " FOR STEADY FREE SURFACE FLOWS	74-0075
ASSIALSIMMETRIC	CALCULATION OF AXISYMMETRICAL OSCILLATION OF FLEXIBLE TURBINE DISCS. (CALCOLO DELLE OSCILLAZIONI ASS	74-1328
ASSOCIATION	BULLETIN DE L ASSOCIATION TECHNIQUE MARITIME ET AERONAUTIQUE (BULLETIN OF THE TECHNICAL MARITIME AND	74-1108
ASSOCIATION	FINITE ELEMENT ANALYSIS OF STRUCTURAL INSTABILITY BY ASSOCIATION OF PULSATING EXCITATIONS	74-0453
ASSOCIATION	PROCEEDINGS OF INTERNATIONAL ASSOCIATION FOR HYDRAULIC RESEARCH AND PROCEEDINGS OF INTERNATIONAL SYM	73-0534
ASSUMED-STRESS	A STUDY OF ELASTIC-PLASTIC ANALYSIS BY THE ASSUMED-STRESS HYBRID FINITE-ELEMENT MODEL, WITH APPLICAT	74-0718
ASSUMED-STRESS	APPLICATIONS OF THE ASSUMED-STRESS HYBRID METHOD TO FINITE ELEMENT ANALYSIS IN FRACTURE MECHANICS	75-0495
ASSYMMETRIC	PROGRESS REPORT ON DISCRETE-ELEMENT ELASTIC AND ELASTIC-PLASTIC ANALYSIS OF SHELLS OF REVOLUTION SUB	68-0170
ASTRA-BOEING'S	ASTRA-BOEING'S ADVANCED STRUCTURAL ANALYZER	69-0118
ASTRONOMICAL	ANALYSIS OF THE DYNAMICS OF THIN PRIMARY MIRRORS FOR LARGE ASTRONOMICAL TELESCOPES	73-1084
ASTRONOMICAL	APPLICATION OF FINITE ELEMENT TECHNIQUE TO DESIGN AND MANUFACTURE OF ASTRONOMICAL MIRRORS	69-0302
ASTRONOMY	NORMAL MODE ANALYSIS OF THE RADIO ASTRONOMY EXLPORER BOOMS AND SPACECRAFT	71-0130
ASYMMETRIC	A DYNAMIC ELASTOPLASTIC ANALYSIS OF THIN SHELLS OF REVOLUTION UNDER ASYMMETRIC MECHANICAL AND THERMA	72-0068
ASYMMETRIC	A FEASIBILITY STUDY OF THE ANALYSIS OF ASYMMETRIC SHELL BY THE DISCRETE-ELEMENT METHOD	68-0115
ASYMMETRIC	ANALYSIS OF ASYMMETRIC STIFFENED SHELL TYPE STRUCTURES BY THE FINITE ELEMENT METHOD: I FLAT RECTANGU	68-0283
ASYMMETRIC	ANALYSIS OF ASYMMETRIC STIFFENED SHELL TYPE STRUCTURES BY FINITE ELEMENT METHOD II. SIX DEGREE-OF-FR	69-0373
ASYMMETRIC	ASAAS: ASYMMETRIC STRESS ANALYSIS OF AXISYMMETRIC SOLIDS WITH ORTHOTROPIC, TEMPERATURE-DEPENDENT MAT	71-0696
ASYMMETRIC	ASYMMETRIC BENDING OF CIRCULAR PLATES USING THE FINITE ELEMENT METHOD	75-0725
ASYMMETRIC	DISCRETE-ELEMENT STATIC ANALYSIS OF CORE-STIFFENED SHELLS OF REVOLUTION FOR ASYMMETRIC MECHANICAL LO	68-0148
ASYMMETRIC	DYNAMIC ANALYSIS OF A COMPLEX SHELL STRUCTURE UNDER IMPULSIVE ASYMMETRIC LOADING	68-0226
ASYMMETRIC	FORCE AND MOMENTS ON ASYMMETRIC AND YAWED BODIES IN A FREE SURFACE	73-0908

ASYMMETRIC	NONLINEAR DYNAMIC ANALYSIS OF SHELLS OF REVOLUTION WITH ASYMMETRIC PROPERTIES BY THE FINITE ELEMENT	75-0217
ASYMMETRIC	NONLINEAR DYNAMIC ANALYSIS OF SHELLS OF REVOLUTION WITH ASYMMETRIC PROPERTIES BY THE FINITE ELEMENT	75-0657
ASYMMETRIC	STATIC ANALYSIS OF A COMPLEX SHELL STRUCTURE UNDER ASYMMETRIC LOADING	67-0120
ASYMMETRIC	STRESS ANALYSIS OF AXISYMMETRIC SOLIDS WITH ASYMMETRIC PROPERTIES	72-0140
ASYMMETRIC	STRUCTURAL ANALYSIS AND EXPERIMENTAL CORRELATION OF A COMPLEX SHELL UNDER ASYMMETRIC LOADS	69-0155
ASYMMETRIC	THE NONLINEAR DYNAMIC ANALYSIS OF SHELLS OF REVOLUTION WITH ASYMMETRIC PROPERTIES BY THE FINITE ELEM	71-0818
ASYMMETRICAL	ASYMMETRICAL BENDING AND VIBRATION OF A CONICAL SHELL FINITE ELEMENT	74-0085
ASYMMETRICAL	CORRELATION BETWEEN PRESSURE AND THERMAL TRANSIENT STRESS LEVELS DERIVED BY THEORY AND EXPERIMENT ON	73-0469
ASYMMETRICAL	DEFLECTION FUNCTION FOR THE ASYMMETRICAL BENDING OF CIRCULAR PLATES	73-0468
ASYMMETRICAL	LINEAR AND NON LINEAR ANALYSIS OF SHELLS OF REVOLUTION WITH ASYMMETRICAL STIFFNESS PROPERTIES	68-0151
ASYMMETRICALLY	HOLLOW-ENDED ROLLER, A SOLUTION FOR IMPROVING FATIGUE LIFE IN ASYMMETRICALLY LOADED CYLINDRICAL ROLL	71-0393
ASYMPTOTIC	PROGRESS IN NONLINEAR FINITE ELEMENT ANALYSIS USING ASYMPTOTIC SOLUTION TECHNIQUES	72-0316
ASYMPTOTICALLY	ON THE ASYMPTOTICALLY SPHERICAL DEFORMATIONS OF ARBITRARY MEMBRANES OF REVOLUTION FIXED ALONG AN EDG	71-0640
AT-GRADE	ASSESSMENT OF DESIGN TOOLS AND CRITERIA FOR URBAN RAIL TRACK STRUCTURES, VOLUME 11. AT-GRADE SLAB TR	74-0813
ATCHAFALAYA	PRELIMINARY FINITE ELEMENT ANALYSIS, ATCHAFALAYA BASIN PROTECTION LEVEES, TEST SECTION 3	69-0463
ATHEROSCLEROSIS	ADAPTABILITY OF COMPUTER GRAPHICS TO STUDIES OF ATHEROSCLEROSIS PATHOGENESIS	72-0206
ATJ-S	NEW MATERIAL MODEL FOR THE NONLINEAR BIAXIAL BEHAVIOR OF ATJ-S GRAPHITE	75-0063
ATMOSPHERIC	ANALYSIS OF BOTTOM ENTRY NOZZLES FOR ATMOSPHERIC STORAGE TANKS	75-0331
ATOMIC	ANALYSIS FOR ATOMIC POWER PLANT BY FINITE ELEMENT METHOD	71-0198
ATTACK	THERMAL STRESS ANALYSIS OF REENTRY VEHICLE NOSETIPS AT ANGLE OF ATTACK	72-0456
ATTACK	THERMAL STRESS ANALYSIS OF RE-ENTRY VEHICLE NOSETIPS AT ANGLE OF ATTACK	72-0852
ATTENUATION	EVALUATION OF ATTENUATION FROM LOSSLESS TRIANGULAR-FINITE-ELEMENT SOLUTIONS FOR INHOMOGENEOUSLY FILL	73-0059
ATTENUATION	FINITE ELEMENT APPROACH TO PRESSURE WAVE ATTENUATION BY REACTOR FUEL SUBASSEMBLIES	75-0522
ATTENUATION	FINITE ELEMENT STUDY OF PRESSURE WAVE ATTENUATION BY REACTOR FUEL SUBASSEMBLIES	75-0215
ATTITUDE	CYCLIC FATIGUE ANALYSIS OF ROCKET THRUST CHAMBERS, VOLUME 2: ATTITUDE CONTROL THRUSTER HIGH CYCLE FA	74-0693
ATTRAVERSO	ON THE RESOLUTION OF NONLINEAR ELASTIC AND ELASTOPLASTIC PROBLEMS BY THE METHOD OF FINITE ELEMENTS.	74-1326
AUBAGES	CONTRIBUTION A L'ANALYSE DES VIBRATIONS LIBRES D'UNE STRUCTURE. APPLICATION AUX AUBAGES DE TURBOMACH	75-0804
AUBURN	FOUNDATION TESTING FOR AUBURN DAM	70-0119
AUFGESCHWEISSIE	BERECHNUNG VON SPANNUNGEN UND TEMPERATURFELDERN IN FLOSSENWAENDEN UND AUFGESCHWEISSIEN BAUTEILEN. (C	70-0544
AUFTRIEBSBEHAFT	BERECHNUNG DES REIBUNGSLOSEN STROEMUNGSFELDES VON DREIDIMENSIONALEN AUFTRIEBSBEHAFTETENTRAGFLUEGELN	73-0533
AUGMENTED	STRESS RESULTANTS IN SHELLS BY AN AUGMENTED FINITE ELEMENT METHOD	73-0585
AUSGEDEHNTER	ZUR LOESUNG DES KONTAKTPROBLEMS ELASTISCHER KOERPER MIT AUSGEDEHNTER KONTAKTFLAECHE DURCH QUADRATISC	74-1135
AUSLEGUNG	NEUESTE ERFAHRUNGEN BEI DER RECHNERISCHEN UND KONSTRUKTIVEN AUSLEGUNG KOMPLIZIERTER SCHALENSYSTEME	71-0052
AUSLEGUNG	SCHWINGUNGSMAESSIGE AUSLEGUNG DER BESCHAUFELUNG AXIALER TURBOMASCHINEN (DESIGN OF DISPLACEMENT-FREE	74-0253
AUSQIRKUNGEN	VORAUSSAGE DES GEBIRGSDRUCKES UND SEINER AUSQIRKUNGEN AUF STREBEN UND STRECKEN IM STEINKOHLENBERGBAU	73-0679
AUSSENGEKERBTER	OSCILLATION STABILITY OF INTERNALLY AND EXTERNALLY NOTCHED PLANE BARS OF EQUAL FORM FACTOR FROM 3943	73-0745
AUSSEREN	BERECHNUNGSMETHODEN ZUR ERMITTLUNG VON BELASTUNGSBEDINGUNGEN UND GRENZTRAGFAEHIGKEITSVERHALTEN VON K	73-0858
AUSSERGEWEOHNLI	BERECHNUNGSMETHODEN ZUR ERMITTLUNG VON BELASTUNGSBEDINGUNGEN UND GRENZTRAGFAEHIGKEITSVERHALTEN VON K	73-0858
AUSSPARUNGEN	ZUR BEPECHNUNG, BEMESSUNG UND BEWEHRUNG RECHTECKIGER PLATTEN KONSTNATER DICKE MIT OEFFNUNGEN ODER AU	73-0664
AUSTRIAN	APPLICATION OF NEW AUSTRIAN TUNNELLING METHOD IN DIFFICULT BUILTOVER AREAS IN FRANKFURT/MAIN METRO	73-1025
AUTHORITY	M.I.T. TEST SECTION INSTRUMENTATION, MASSACHUSETTS BAY TRANSPORTATION AUTHORITY, HAYMARKET-NORTH EXT	72-0673
AUTOMATED	AN IMPROVED AUTOMATED STRUCTURAL OPTIMIZATION PROGRAM	74-0732
AUTOMATED	AUTOMATED DATA GENERATION AND LARGE MATRIX OPERATION FOR 3 DIMENSIONAL PLATE STRUCTURE ANALYSIS	73-0339
AUTOMATED	AUTOMATED DESIGN OPTIMIZATION OF SUPERSONIC AIRPLANE WING STRUCTURES UNDER DYNAMIC CONSTRAINTS	72-0178
AUTOMATED	AUTOMATED EVALUATION OF INTERCEPTOR ROCKET MOTOR DESIGNS UNDER COMBINED OPERATIONAL AND NUCLEAR EFFE	74-0698
AUTOMATED	AUTOMATED INPUT DATA PREPARATION FOR NASTRAN	69-0080
AUTOMATED	AUTOMATED LARGE DEFLECTION AND STABILITY ANALYSIS OF THREE DIMENSIONAL BAR STRUCTURES	69-0393
AUTOMATED	AUTOMATED MESH REFINEMENT STRATEGIES FOR FINITE ELEMENT COMPUTATIONS	75-0877
AUTOMATED	AUTOMATED METHOD FOR THE LARGE DEFLECTION AND INSTABILITY ANALYSIS OF THREE-DIMENSIONAL TRUSS AND FR	66-0105
AUTOMATED	AUTOMATED OPTIMUM DESIGN OF AIRCRAFT WINGS TO SATISFY STRENGTH, STABILITY, FREQUENCY AND FLUTTER REQ	72-0605
AUTOMATED	AUTOMATED PROCEDURE FOR DESIGN OF WING STRUCTURES TO SATISFY STRENGTH AND FLUTTER REQUIREMENTS	73-0867
AUTOMATED	AUTOMATED SOLUTION OF TIME DEPENDENT PROBLEMS	72-0084
AUTOMATED	AUTOMATED STRESS ANALYSIS USING SUBSTRUCTURES	66-0071
AUTOMATED	AUTOMATED STRUCTURAL DESIGN AND ANALYSIS OF ADVANCED COMPOSITE WING MODELS	73-0221
AUTOMATED	AUTOMATED STRUCTURAL SYNTHESIS USING A RE NUMBER OF DESIGN COORDINATES	73-0584
AUTOMATED	MAGIC II. AN AUTOMATED GENERAL PURPOSE SYSTEM FOR STRUCTURAL ANALYSIS	71-0495
AUTOMATED	MAGIC 11: AN AUTOMATED GENERAL PURPOSE SYSTEM FOR STRUCTURAL ANALYSIS. VOLUME 111. PROGRAMMER'S MANU	71-0756
AUTOMATED	MAGIC 11: AN AUTOMATED GENERAL PURPOSE SYSTEM FOR STRUCTURAL ANALYSIS. VOLUME 1. ENGINEER'S MANUAL (	71-0757
AUTOMATED	MAGIC 111: AN AUTOMATED GENERAL PURPOSE SYSTEM FOR STRUCTURAL ANALYSIS. VOLUME 111. PROGRAMMER'S MAN	72-0726
AUTOMATED	MAGIC 111: AN AUTOMATED GENERAL PURPOSE SYSTEM FOR STRUCTURAL ANALYSIS VOLUME 1. ENGINEER'S MANUAL	72-0727
AUTOMATED	MAGIC, AN AUTOMATED GENERAL PURPOSE SYSTEM FOR STRUCTURAL ANALYSIS, VOLUME III, PROGRAMMER'S MANUAL	69-0090
AUTOMATED	MAGIC, AN AUTOMATED GENERAL PURPOSE SYSTEM FOR STRUCTURAL ANALYSIS, VOLUME II, USER'S MANUAL	69-0144
AUTOMATED	MAGIC, AN AUTOMATED GENERAL PURPOSE SYSTEM FOR STRUCTURAL ANALYSIS	69-0173
AUTOMATED	NASTRAN USER INTERFACES, AUTOMATED INPUT INNOVATIONS	71-0126
AUTOMATED	OPTIM 11: (MAGIC) COMPATIBLE LARGE SCALE AUTOMATED MINIMUM WEIGHT DESIGN PROGRAM - V 2. PROGRAMMER'S	74-1036
AUTOMATED	OPTIM 11: A (MAGIC) COMPATIBLE LARGE SCALE AUTOMATED MINIMUM WEIGHT DESIGN PROGRAM - 1. ENGINEERS AN	74-1048
AUTOMATED	OPTIM 11: A MAGIC COMPATIBLE LARGE-SCALE AUTOMATED MINIMUM WEIGHT DESIGN PROGRAM. VOLUME 1. ENGINEER	74-1096
AUTOMATED	OPTIM 11: MAGIC COMPATIBLE LARGE SCALE AUTOMATED MINIMUM WEIGHT DESIGN PROGRAM. VOLUME 11. PROGRAMME	74-1093
AUTOMATED	SLAB BRIDGE DESIGN AND DRAWING. AN AUTOMATED PROCESS	70-0347
AUTOMATIC	AN AUTOMATIC MESH GENERATION SCHEME FOR PLANE AND CURVED SURFACES BY ISOPARAMETRIC COORDINATES	71-0271
AUTOMATIC	AN AUTOMATIC MESH GENERATOR USING TWO AND THREE DIMENSIONAL ISOPARAMETRIC ELEMENTS	73-0066
AUTOMATIC	AN AUTOMATIC NODE RELABELLING SCHEME FOR BANDWIDTH MINIMIZATION OF STIFFNESS MATRICES	68-0256
AUTOMATIC	AN AUTOMATIC RECORDING SCHEME FOR SIMULTANEOUS EQUATIONS DERIVED FROM NETWORK SYSTEMS	70-0285
AUTOMATIC	AN AUTOMATIC RELABELING ALGORITHM FOR BANDWIDTH MINIMIZATION	75-0763
AUTOMATIC	AN AUTOMATIC SYSTEM FOR KINEMATIC ANALYSIS ASKA HIGH SPEED COMPUTING OF ELASTIC STRUCTURES	70-0358
AUTOMATIC	AN AUTOMATIC SYSTEM FOR KINEMATIC ANALYSIS. ASKA PART 1	70-0396
AUTOMATIC	AN AUTOMATIC SYSTEM FOR KINEMATIC ANALYSIS ASKA PART 1	71-0626
AUTOMATIC	ASKA - AUTOMATIC SYSTEM FOR KINEMATIC ANALYSIS	65-0059
AUTOMATIC	ASKA AUTOMATIC SYSTEM FOR KINEMATIC ANALYSIS	69-0307
AUTOMATIC	AUTOMATIC ANALYSIS OF FUSELAGES AND PROBLEMS OF CONDITIONING	64-0027
AUTOMATIC	AUTOMATIC CALCULATION OF THIN STRUCTURES	73-0201
AUTOMATIC	AUTOMATIC DESIGN OF SHELL STRUCTURES	71-0336
AUTOMATIC	AUTOMATIC EMBANKMENT ANALYSIS. EQUILIBRIUM AND INSTABILITY CONDITIONS	66-0015
AUTOMATIC	AUTOMATIC EXTRACTION OF RIGID BODY MODES FROM STRESS AND STRAIN ELEMENTS	72-0913
AUTOMATIC	AUTOMATIC GENERATION OF FINITE ELEMENT MATRICES	71-0165
AUTOMATIC	AUTOMATIC GENERATION AND CONDENSATION OF LARGE MASS AND STIFFNESS MATRICES	72-0466
AUTOMATIC	AUTOMATIC GENERATION OF FINITE ELEMENT MESH WITHIN A BANDWIDTH AND AN EFFICIENT SOLUTION OF THE BAND	73-0338
AUTOMATIC	AUTOMATIC GENERATION OF TRIANGULATIONS FOR PIECEWISE POLYNOMIAL APPROXIMATION	73-1118
AUTOMATIC	AUTOMATIC GENERATION OF FINITE ELEMENTS	74-1335
AUTOMATIC	AUTOMATIC MASTERS FOR EIGENVALUE ECONOMIZATION	75-0557
AUTOMATIC	AUTOMATIC MESH GENERATION IN TWO AND THREE DIMENSIONAL INTERCONNECTED DOMAINS	70-0251
AUTOMATIC	AUTOMATIC MESH GENERATION FOR FINITE ELEMENT ANALYSIS	72-0328
AUTOMATIC	AUTOMATIC MESH GENERATION FOR THREE DIMENSIONAL STRUCTURES BASED ON THEIR THREE VIEWS	73-0340
AUTOMATIC	AUTOMATIC NESTED DISSECTION	74-0860
AUTOMATIC	AUTOMATIC NODAL TRIANGULATION FOR FINITE ELEMENTS	73-0178
AUTOMATIC	AUTOMATIC PREPARATION OF DATA CARDS FOR THE FINITE ELEMENT METHOD	66-0086
AUTOMATIC	AUTOMATIC SYSTEM FOR KINEMATIC ANALYSIS, NASKA	65-0054
AUTOMATIC	AUTOMATIC TRIANGULATION OF TWO DIMENSIONAL REGIONS	72-0697
AUTOMATIC	AUTOMATIC TRIANGULATION OF ARBITRARY PLANAR DOMAINS FOR THE FINITE ELEMENT METHOD	74-0480
AUTOMATIC	BANDWIDTH REDUCTION BY AUTOMATIC RENUMBERING	73-0993
AUTOMATIC	BERGEN, AN AUTOMATIC FINITE ELEMENT MESH GENERATION PROGRAM FOR ARBITRARY STRUCTURES	72-0776
AUTOMATIC	COMPUTER PROGRAM FOR AUTOMATIC MESH GENERATION FOR THE 2-D FINITE ELEMENT PROGRAM WILAX	73-0565

129

AUTOMATIC	LAPLACIAN ELECTROSTATIC FIELD CALCULATIONS BY FINITE ELEMENTS WITH AUTOMATIC GRID GENERATION	73-0084
AUTOMATIC	MESHGEN, A COMPUTER CODE FOR AUTOMATIC FINITE ELEMENT MESH GENERATION	69-0096
AUTOMATIC	NATURAL COORDINATE SYSTEMS, AN AUTOMATIC INPUT DATA GENERATION SCHEME FOR A FINITE ELEMENT METHOD	70-0018
AUTOMATIC	ON THE AUTOMATIC SELECTION OF INDEPENDENT DISPLACEMENT VARIABLES IN GENERAL PURPOSE FINITE ELEMENT C	73-0676
AUTOMATIC	PROCESS FOR THE AUTOMATIC GENERATION OF FLAT TRIANGULAR LATTICES FOR APPLICATION TO THE FINITE ELEME	71-0366
AUTOMATIC	RESTRICTIONS IN THE APPLICATION OF AUTOMATIC GENERATION SCHEMES BY ISOPARAMETRIC CO-ORDINATES	74-1377
AUTOMATIC	THE AUTOMATIC COMPUTATION OF THE NATURAL FREQUENCIES OF STRUCTURAL FRAMES USING AN EXACT MATRIX TECH	73-0318
AUTOMATIC	TWO-DIMENSIONAL AUTOMATIC MESH GENERATION FOR STRUCTURAL ANALYSIS	70-0058
AUTOMATICA	PROCESS FOR THE AUTOMATIC GENERATION OF FLAT TRIANGULAR LATTICES FOR APPLICATION TO THE FINITE ELEME	71-0366
AUTOMATION	GRIP: AN INFORMATION SYSTEM FOR DESIGN AUTOMATION	73-0210
AUTOMATION	K VOPROSU OB AVTOMATIZATSII ZADANIYA INFORMATSII V METODE KONECHNYKH ELEMENTOV. (PROBLEM OF AUTOMATI	75-0382
AUTOMATION	SUBSTRUCTURE ANALYSIS TECHNIQUES AND AUTOMATION	73-0597
AUTOMOBILE	APPLICATION OF FINITE ELEMENT TECHNIQUES TO THE ANALYSIS OF AN AUTOMOBILE STRUCTURE	70-0136
AUTOMOBILE	APPLICATION OF FINITE ELEMENT METHODS TO COMPLETE AUTOMOBILE STRUCTURAL DESIGN EVALUATION	74-0312
AUTOMOBILE	AUTOMOBILE PANEL SWEEP STIFFNESS ANALYSIS	74-0175
AUTOMOBILE	DYNAMIC ANALYSIS OF AN AUTOMOBILE CHASSIS FRAME	70-0135
AUTOMOBILE	ELASTO-PLASTIC ANALYSIS OF AUTOMOBILE BODY STRUCTURE BY THE FINITE ELEMENT METHOD	74-0196
AUTOMOBILE	FINITE ELEMENT ANALYSIS, AN AUTOMOBILE ENGINEER'S TOOL	74-0238
AUTOMOBILE	FINITE ELEMENT ANALYSIS OF AUTOMOBILE STRUCTURES	74-0309
AUTOMOBILE	FINITE ELEMENT METHOD IN AUTOMOBILE DESIGN	73-0326
AUTOMOBILE	NUMERICAL ANALYSIS OF AUTOMOBILE STRUCTURES	72-0334
AUTOMOBILE	STRESS ANALYSIS OF AN AUTOMOBILE CHASSIS FRAME	70-0137
AUTOMOBILE	STUDY ON AUTOMOBILE CRASH WORTHINESS	70-0217
AUTOMOTIVE	APPLICATION OF FINITE ELEMENT METHOD TO PREDICT STATIC RESPONSE OF AUTOMOTIVE BODY STRUCTURES	71-0340
AUTOMOTIVE	AUTOMOTIVE USE OF FINITE ELEMENT METHODS - INTRODUCTION AND OVERVIEW	74-0388
AUTOMOTIVE	DESIGNING BY FINITE ELEMENTS CATCHING ON IN AUTOMOTIVE ENGINEERING	74-0113
AUTOMOTIVE	DEVELOPMENT AND APPLICATION OF HYBRID STRESS FINITE ELEMENTS TO AN AUTOMOTIVE STRUCTURE	74-1221
AUTOMOTIVE	DYNAMIC ANALYSIS OF AUTOMOTIVE STRUCTURAL SYSTEMS	70-0340
AUTOMOTIVE	DYNAMIC TESTING AND COMPUTER ANALYSIS OF AUTOMOTIVE FRAMES	72-0089
AUTOMOTIVE	FINITE ELEMENT DYNAMIC ANALYSIS OF AN AUTOMOTIVE FRAME	73-0256
AUTOMOTIVE	FINITE ELEMENT STRUCTURAL ANALYSIS AS APPLIED TO AN AUTOMOTIVE DOOR STRUCTURE	74-0310
AUTOMOTIVE	ILLUSTRATIONS OF AUTOMOTIVE FINITE ELEMENT MODELS - STATICS	74-0373
AUTOMOTIVE	ILLUSTRATIONS OF AUTOMOTIVE FINITE ELEMENT MODELS - DYNAMICS	74-0391
AUTOMOTIVE	NONLINEAR TRANSIENT ANALYSIS OF AUTOMOTIVE STRUCTURES BY FINITE ELEMENTS	75-0301
AUTOMOTIVE	SOLUTION OF AUTOMOTIVE STRUCTURAL PROBLEMS USING THE FINITE ELEMENT METHOD AND COMPUTER GRAPHICS	71-0341
AUTOMOTIVE	USING FINITE ELEMENTS IN AUTOMOTIVE DESIGN	74-0104
AUTOMOTIVE	USING FINITE ELEMENTS IN AUTOMOTIVE DESIGN	74-1299
AUXILIARY	USE OF AUXILIARY FUNCTIONS IN THE FINITE ELEMENT METHOD	72-0938
AVTOMATIZATSII	K VOPROSU OB AVTOMATIZATSII ZADANIYA INFORMATSII V METODE KONECHNYKH ELEMENTOV. (PROBLEM OF AUTOMATI	75-0382
AXES	ELASTIC DEFLECTION IN THIN WALLED CONES ROTATING ABOUT THEIR AXES OF SYMMETRY	73-0545
AXES	INTEGRATION FOR THE FINITE ELEMENTS - AN ALGORITHM IN ORTHOGONAL AXES	73-0295
AXI-SYMMETRIC	A CONICAL ELEMENT FOR DISPLACEMENT ANALYSIS OF AXI-SYMMETRIC SHELLS	69-0120
AXI-SYMMETRIC	ANALYSIS OF ELASTIC-PLASTIC SHELLS OF REVOLUTION UNDER AXI-SYMMETRIC LOADING BY THE FINITE ELEMENT M	67-0102
AXI-SYMMETRIC	DYNAMIC FINITE ELEMENT ANALYSIS OF AXI-SYMMETRIC STRESS WAVE PROPAGATION IN SOIL-FILLED BIN	72-0622
AXI-SYMMETRIC	EXPERIENCES FROM STRESS ANALYSIS OF AXI-SYMMETRIC PROBLEMS IN MACHINE DESIGN	69-0247
AXI-SYMMETRIC	METODA ELEMENTOW SKONSZONYCH ROZWIAZYWANIA OSIOWO-SYMETRYCZNEGO PRZEWODNICTWA CIEPLA (FINITE-ELEMENT	73-0434
AXI-SYMMETRIC	NUMERICAL INVESTIGATION OF AXI-SYMMETRIC VISCOPLASTIC FLOWS BY FINITE-ELEMENT PROCEDURE	73-0132
AXI-SYMMETRIC	SHORT COMMUNICATIONS: TORSION OF AXI-SYMMETRIC SOLIDS OF VARIABLE DIAMETER - INCLUDING ACCELERATION	74-0203
AXIAL	A GENERAL EQUILIBRIUM AXIAL ELEMENT	73-0729
AXIAL	A METHOD OF FINITE DEFLECTION ELASTIC PLASTIC ANALYSIS OF FRAMED STRUCTURES SUBJECTED TO LARGE AXIAL	69-0239
AXIAL	A STUDY OF A DYNAMICAL FINITE ELEMENT ANALYSIS FOR APPLICATION TO AXIAL WAVE PROPAGATION PROBLEMS IN	71-0589
AXIAL	APPLICATION OF THE FINITE ELEMENT METHOD TO THE VIBRATION ANALYSIS OF AXIAL FLOW TURBINES	73-1125
AXIAL	CREEP OF A UNIAXIAL METAL MATRIX COMPOSITE SUBJECTED TO AXIAL AND NORMAL LATERAL LOADS	71-0041
AXIAL	CREEP OF A UNIAXIAL METAL MATRIX COMPOSITE SUBJECTED TO AXIAL AND NORMAL LATERAL LOADS	71-0727
AXIAL	DESIGN OF PROPELLANT FINS SUBJECTED TO AXIAL ACCELERATION	71-0734
AXIAL	ELASTIC-PLASTIC ANALYSIS OF A BAR WITH UNIFORM CROSS-SECTION UNDER COMBINED LOADINGS 1 - COMBINATION	71-0778
AXIAL	FINITE ELEMENT ANALYSIS OF ALTERNATIVE AXIAL LOADING OF AN ELASTIC PLATE PRESSED BETWEEN TWO ELASTIC	71-0357
AXIAL	INELASTIC BUCKLING ANALYSIS OF PIPES SUBJECTED TO INTERNAL PRESSURE, FLEXURE AND AXIAL LOADING	74-0515
AXIAL	MODIFICATIONS TO ARL COMPUTER PROGRAMS USED FOR DESIGN OF AXIAL COMPRESSOR AIRFOILS	74-0508
AXIAL	NONLINEAR BEHAVIOR OF THE HUMAN INTERVERTEBRAL DISC UNDER AXIAL LOADING	74-0919
AXIAL	ON THE SOLUTION OF THE NAVIER-STOKES EQUATION IN TWO DIMENSIONAL AND AXIAL SYMMETRIC PROBLEMS	74-0007
AXIAL	RADIAL AND AXIAL TRANSMISSIBILITY CHARACTERISTICS OF TYPICAL ROCKET VEHICLE STRUCTURE TECHNICAL SUMM	69-0475
AXIAL	RADIAL INERTIA EFFECTS ON AN IDEALLY PLASTIC CIRCULAR PLATE UNDER IMPULSIVE AXIAL COMPRESSION	69-0324
AXIAL	SCHWINGUNGSMAESSIGE AUSLEGUNG DER BESCHAUFELUNG AXIALER TURBOMASCHINEN (DESIGN OF DISPLACEMENT-FREE	74-0253
AXIAL	THE MERIDIONAL THROUGH-FLOW CALCULATION IN AN AXIAL FLOW MACHINE BY THE FINITE ELEMENT METHOD	75-0115
AXIAL	THREE NODE ISOPARAMETRIC STRESS AXIAL ELEMENT	72-0992
AXIAL	VIBRATION ANALYSIS OF AXIAL FLOW TURBINE DISKS USING FINITE ELEMENTS	75-0195
AXIAL-FLEXURAL	AXIAL-FLEXURAL VIBRATION OF FRAMEWORKS USING FINITE-ELEMENT APPROACH	73-0051
AXIAL-FLOW	VIBRATION ANALYSIS OF AXIAL-FLOW TURBINE DISKS USING FINITE-ELEMENTS	75-0415
AXIALER	SCHWINGUNGSMAESSIGE AUSLEGUNG DER BESCHAUFELUNG AXIALER TURBOMASCHINEN (DESIGN OF DISPLACEMENT-FREE	74-0253
AXIALLY	A COMPARISON OF FIRST AND SECOND ORDER AXIALLY SYMMETRIC FINITE ELEMENTS	73-0997
AXIALLY	ANCHORAGE ZONE STRESSES IN AXIALLY POST-TENSIONED I-SECTION MEMBERS WITH END BLOCKS	71-0305
AXIALLY	BENDING PROBLEM OF AXIALLY CONSTRAINED BEAMS	73-0397
AXIALLY	BIDIMENSIONAL COMPUTER-PROGRAM FOR CALCULATION OF GROWTH-RATES OF INCLUDED AXIALLY SYMMETRICAL PLASM	74-0420
AXIALLY	DYNAMICAL ANALYSIS OF AXIALLY NON-UNIFORM, THIN CYLINDRICAL SHELLS. PART 1 - MATRIX FORMULATION	70-0413
AXIALLY	DYNAMICAL ANALYSIS OF AXIALLY NONUNIFORM, THIN CYLINDRICAL SHELLS. PART 2 FREE VIBRATION	71-0716
AXIALLY	NON-UNIFORM DEFORMATION OF MATERIAL IN AXIALLY SYMMETRIC COMPRESSION CAUSED BY FRICTION. PART 3, THE	72-0923
AXIALLY	ON THE NONUNIFORM DEFORMATION OF MATERIAL IN AXIALLY SYMMETRIC COMPRESSION CAUSED BY FRICTION - 2	71-0321
AXIALLY	THE THERMO-ELASTO-PLASTIC ANALYSIS OF A RE-ENTRY BODY TYPE COMPOSITE SOLID STRUCTURE USING A NEW FIN	66-0093
AXIALLY-LOADED	FLEXURE ANALYSIS OF AXIALLY-LOADED BEAM BY FINITE-ELEMENT METHOD WITH INTEGRAL PARAMETERS	75-0462
AXICAP	A FINITE ELEMENT COMPUTER CODE, AXICAP FOR CREEP ANALYSIS	69-0248
AXISYMETRIC	AXISYMETRIC TRIANGULAR ELEMENT FOR THE SCALAR HELMHOLTZ EQUATION	73-0688
AXISYMMETRIC	A COMPUTER PROGRAM FOR THE STATIC AND DYNAMIC FINITE ELEMENT ANALYSIS OF AXISYMMETRIC THIN SHELLS	66-0017
AXISYMMETRIC	A CURVED ELEMENT APPPOXIMATION IN THE ANALYSIS OF AXISYMMETRIC THIN SHELLS	70-0063
AXISYMMETRIC	A HIGH PRECISION AXISYMMETRIC TRIANGULAR ELEMENT USED IN THE ANALYSIS OF HYDRAULIC TURBINE COMPONENT	70-0038
AXISYMMETRIC	A METHOD FOR THE TRANSIENT NONLINEAR COUPLED THERMOELASTIC ANALYSIS OF AXISYMMETRIC BODIES USING FIN	74-1140
AXISYMMETRIC	A REFINED AXISYMMETRIC FINITE ELEMENT FOR THE ANALYSIS OF NEARLY INCOMPRESSIBLE SOLIDS	70-0309
AXISYMMETRIC	A TWO-DIMENSIONAL AXISYMMETRIC SHELL FLUID MODEL FOR LAUNCH VEHICLE LONGITUDINAL DYNAMIC ANALYSIS	69-0046
AXISYMMETRIC	ACOUSITC FINITE ELEMENT ANALYSIS OF AXISYMMETRIC FLUID REGIONS	73-0763
AXISYMMETRIC	AN ASSUMED STRESS HYBRID FINITE ELEMENT MODEL FOR THE ANALYSIS OF AN AXISYMMETRIC THICK WALLED PRESS	71-0025
AXISYMMETRIC	AN AXISYMMETRIC FINITE ELEMENT ANALYSIS OF THE MECHANICAL AND THERMAL STRESSES IN BRAKE DRUMS	74-0311
AXISYMMETRIC	ANALYSES OF AXISYMMETRIC UPSETTING AND PLANE STRAIN SIDE PRESSING OF SOLID CYLINDERS BY THE FINITE E	70-0083
AXISYMMETRIC	ANALYSIS AND DESIGN OF AXISYMMETRIC REINFORCEMENT AROUND A CIRCULAR HOLE IN A THIN FLAT PLATE BY THE	74-1113
AXISYMMETRIC	ANALYSIS OF AXISYMMETRIC SHELLS BY THE DIRECT STIFFNESS METHOD	63-0006
AXISYMMETRIC	ANALYSIS OF AXISYMMETRIC SOLIDS BY THE DIRECT STIFFNESS METHOD	66-0049
AXISYMMETRIC	ANALYSIS OF AXISYMMETRIC COMPOSITE STRUCTURES BY THE FINITE ELEMENT METHOD	66-0066
AXISYMMETRIC	ANALYSIS OF AXISYMMETRIC COMPOSITE BODIES BY THE FINITE ELEMENT METHOD	68-0242
AXISYMMETRIC	ANALYSIS OF AXISYMMETRIC LINEAR HEAT CONDUCTION PROBLEMS BY FINITE ELEMENT METHOD	69-0015
AXISYMMETRIC	ANALYSIS OF ELASTIC-PLASTIC CREEP BUCKLING OF AXISYMMETRIC SHELLS BY FINITE-ELEMENT METHOD	75-0355
AXISYMMETRIC	ANALYSIS OF LARGE ELASTIC-PLASTIC CREEP DEFORMATIONS OF AXISYMMETRIC SHELLS BY THE FINITE ELEMENT ME	74-0563
AXISYMMETRIC	ANALYSIS OF THICK-WALLED PRESSURE VESSELS USING THE CONSTANT STRAIN AXISYMMETRIC FINITE ELEMENT	71-0653
AXISYMMETRIC	APPLICATION OF FINITE ELEMENT METHOD TO AXISYMMETRIC BUCKLING OF SHALLOW SPHERICAL SHELLS UNDER EXTE	73-0111

AXISYMMETRIC	ASAAS: ASYMMETRIC STRESS ANALYSIS OF AXISYMMETRIC SOLIDS WITH ORTHOTROPIC, TEMPERATURE-DEPENDENT MAT	71-0696
AXISYMMETRIC	AXISYMMETRIC ELEMENT ANALYSIS USING ANALYTICAL COMPUTING	75-0333
AXISYMMETRIC	AXISYMMETRIC FINITE ELEMENT IN NATURAL COORDINATES	75-0826
AXISYMMETRIC	AXISYMMETRIC RING AND PLATE STRESS ANALYSIS PROGRAM - USERS GUIDE	70-0333
AXISYMMETRIC	AXISYMMETRIC SHELLS	69-0481
AXISYMMETRIC	AXISYMMETRIC SOLUTION OF ELASTO PLASTIC PROBLEMS BY FINITE ELEMENT METHODS	69-0361
AXISYMMETRIC	AXISYMMETRIC VIBRATION OF SHELLS OF REVOLUTION BY THE FINITE ELEMENT METHOD	67-0100
AXISYMMETRIC	AXISYMMETRIC VIBRATION OF CONICAL SHELLS	70-0350
AXISYMMETRIC	AXISYMMETRIC VIBRATIONS OF LINEARLY TAPERED ANNULAR PLATES	74-1266
AXISYMMETRIC	BRLESC FINITE ELEMENT PROGRAM FOR AXISYMMETRIC, PLANE STRAIN, AND PLANE STRESS, ORTHOTROPIC SOLIDS W	71-0749
AXISYMMETRIC	CALCULATION OF AXISYMMETRIC LONGITUDINAL MODES FOR FLUID-ELASTIC TANK-ULLAGE GAS SYSTEMS AND COMPARI	64-0019
AXISYMMETRIC	CALCULATION OF NON-LINEAR AXISYMMETRIC VIBRATIONS OF THIN SHELLS OF REVOLUTION BY A FINITE ELEMENT M	75-0067
AXISYMMETRIC	CALCULATION OF THE FREE OSCILLATIONS OF A LIQUID IN AXISYMMETRIC MOTIONLESS CONTAINERS OF ARBITRARY	74-0387
AXISYMMETRIC	CAPE-2D, A COMPUTER PROGRAM FOR THE STRESS ANALYSIS OF PLANE AND AXISYMMETRIC COMPOSITE STRUCTURES.	69-0083
AXISYMMETRIC	CLASSICAL AND NATURAL CONDUCTION AND DEFORMATION MODES CONCEPTS FOR AXISYMMETRIC VARIATIONAL METHODS	72-1003
AXISYMMETRIC	COMPARISON OF THE RESULTS OF THE DYNAMIC RESPONSE OF CYLINDRICAL SHELLS BY CHARACTERISTIC AND BY FIN	74-0781
AXISYMMETRIC	COMPUTER AIDED DESIGN OF THERMALLY LOADED AXISYMMETRIC DIESEL ENGINE COMPONENTS	74-0596
AXISYMMETRIC	COMPUTER CALCULATIONS OF STRESSES IN AXISYMMETRIC THERMALLY LOADED COMPONENTS	68-0069
AXISYMMETRIC	COMPUTER CODE TO GENERATE FINITE ELEMENT MESH FOR PROBLEMS OF COMPLEX AXISYMMETRIC GEOMETRY	74-0360
AXISYMMETRIC	CONICAL FINITE ELEMENT STRESS ANALYSIS OF AN AXISYMMETRIC TANK ON AN ELASTIC FOUNDATION	70-0494
AXISYMMETRIC	CONIDA: A FINITE ELEMENT PROGRAM FOR THE STRESS ANALYSIS OF AXISYMMETRIC THIN SHELLS	74-1087
AXISYMMETRIC	COUPLE - A PSEUDO FINITE ELEMENT CONDUCTION CODE FOR THE TRANSIENT THERMAL RESPONE OF AXISYMMETRIC O	72-0778
AXISYMMETRIC	CREEP ANALYSIS OF AXISYMMETRIC BODIES	68-0081
AXISYMMETRIC	CURVED THICK SHELL AND MEMBRANE ELEMENTS WITH PARTICULAR REFERENCE TO AXISYMMETRIC PROBLEMS	68-0005
AXISYMMETRIC	DIGITAL COMPUTER PROGRAMS FOR STEADY-STATE OR TRANSIENT TEMPERATURE ANALYSIS OF PLANE AND AXISYMMETR	72-0915
AXISYMMETRIC	DYNAMIC ANALYSIS OF AXISYMMETRIC SHELLS UNDER ARBITRARY TRANSIENT PRESSURES.	72-0467
AXISYMMETRIC	DYNAMIC ELASTIC-PLASTIC ANALYSIS OF AXISYMMETRIC STRUCTURES BY FINITE ELEMENT METHOD COMPUTER CODE	73-0844
AXISYMMETRIC	DYNAMIC STRESS ANALYSIS OF AXISYMMETRIC STRUCTURES UNDER ARBITRARY LOADING	69-0433
AXISYMMETRIC	EARTHQUAKE ANALYSIS OF AXISYMMETRIC TOWERS PARTIALLY SUBMERGED IN WATER	75-0076
AXISYMMETRIC	ELAD PROGRAM MANUAL, A COMPUTER PROGRAM FOR THE ELASTIC ANALYSIS OF DOME STRUCTURES AND AXISYMMETRIC	69-0064
AXISYMMETRIC	ELASTIC DYNAMIC RESPONSE OF AXISYMMETRIC STRUCTURES	69-0275
AXISYMMETRIC	ELASTIC PLASTIC DYNAMIC ANALYSIS OF AXISYMMETRIC SOLIDS	73-0657
AXISYMMETRIC	ELASTIC-PLASTIC ANALYSIS OF AXISYMMETRIC PLATES AND SHELLS	71-0779
AXISYMMETRIC	ELASTOPLASTIC ANALYSIS OF PLANE STRAIN AND AXISYMMETRIC FLAT PUNCH INDENTATION BY THE FINITE ELEMENT	70-0343
AXISYMMETRIC	ELFED A COMPUTER CODE TO GENERATE A FINITE ELEMENT MESH FOR PROBLEMS OF COMPLEX AXISYMMETRIC GEOMETR	74-1270
AXISYMMETRIC	ELFED: A COMPUTER CODE TO GENERATE FINITE ELEMENT MESH FOR PROBLEMS OF COMPLEX AXISYMMETRIC GEOMETRY	73-0943
AXISYMMETRIC	ENGINEERING OPERATIONS REPORT: NERVA AXISYMMETRIC AND PLANE STRUCTURES FINITE ELEMENT STRESS ANALYSI	72-0917
AXISYMMETRIC	FEATS, A COMPUTER PROGRAM FOR THE FINITE ELEMENT THERMAL STRESS ANALYSIS OF PLANE OR AXISYMMETRIC SO	68-0152
AXISYMMETRIC	FINITE ELEMENT ANALYSIS OF AXISYMMETRIC SOLIDS	65-0011
AXISYMMETRIC	FINITE ELEMENT ANALYSIS OF AXISYMMETRIC COMPOSITE STRUCTURES	65-0030
AXISYMMETRIC	FINITE ELEMENT ANALYSIS OF AXISYMMETRIC SOLIDS WITH ARBITRARY LOADINGS	67-0145
AXISYMMETRIC	FINITE ELEMENT ANALYSIS OF AN EXPANDED I - SECTION BEAM AND AN AXISYMMETRIC FLANGED CYLINDER	68-0002
AXISYMMETRIC	FINITE ELEMENT ANALYSIS FOR ARBITRARY AXISYMMETRIC STRUCTURES ORTHOTROPIC, TEMPERATURE-DEPENDENT MAT	68-0076
AXISYMMETRIC	FINITE ELEMENT ANALYSIS OF AXISYMMETRIC VIBRATIONS OF CYLINDERS	72-0255
AXISYMMETRIC	FINITE ELEMENT ANALYSIS OF AXISYMMETRIC MULTILAYER SANDWICH PLATES AND SHELLS	74-0101
AXISYMMETRIC	FINITE ELEMENT ANALYSIS OF JET IMPINGEMENT ON AXISYMMETRIC CURVED DEFLECTORS	75-0396
AXISYMMETRIC	FINITE ELEMENT ANALYSIS OF DENTAL STRUCTURES - AXISYMMETRIC AND PLANE STRESS IDEALIZATIONS	75-0525
AXISYMMETRIC	FINITE ELEMENT CALCULATION OF AXISYMMETRIC SUBCRITICAL COMPRESSIBLE FLOW	74-1064
AXISYMMETRIC	FINITE ELEMENT COMPUTER PROGRAM FOR THE SOLUTION OF NONLINEAR AXISYMMETRIC CONTACT PROBLEMS WITH INT	74-1094
AXISYMMETRIC	FINITE ELEMENT FRACTURE MECHANICS ANALYSIS OF TWO-DIMENSIONAL AND AXISYMMETRIC ELASTIC AND ELASTIC-P	74-1092
AXISYMMETRIC	FINITE ELEMENT LARGE DEFLECTION ANALYSIS OF ELASTIC-PLASTIC SHELLS OF REVOLUTION SUBJECTED TO AXISYM	72-0137
AXISYMMETRIC	FINITE ELEMENT METHOD FOR STRESS ANALYSIS OF BODIES OF REVOLUTION SUBJECTED TO AXISYMMETRIC LOADING	71-0409
AXISYMMETRIC	FINITE ELEMENT METHOD FOR AXISYMMETRIC JETS OF FLUID WITH GRAVITY FORCE	75-0428
AXISYMMETRIC	FINITE ELEMENT PROCEDURE FOR THE LARGE DEFORMATION DYNAMIC RESPONSE OF AXISYMMETRIC SOLIDS	74-0527
AXISYMMETRIC	FINITE ELEMENT PROCEDURE FOR THE LARGE DEFORMATION DYNAMIC RESPONSE OF AXISYMMETRIC SOLIDS	74-0967
AXISYMMETRIC	FINITE ELEMENT PROCEDURE FOR LARGE DEFORMATION DYNAMIC RESPONSE OF AXISYMMETRIC SOLIDS	75-0013
AXISYMMETRIC	FINITE ELEMENT PROGRAM PACKAGE FOR AXISYMMETRIC SCALAR FIELD PROBLEMS	73-0113
AXISYMMETRIC	FINITE ELEMENT PROGRAM PACKAGE FOR AXISYMMETRIC VECTOR FIELD PROBLEMS	75-0138
AXISYMMETRIC	FINITE ELEMENT SOLUTION FOR AXISYMMETRIC SHELLS	64-0010
AXISYMMETRIC	FINITE ELEMENT SOLUTION FOR AXISYMMETRIC SHELLS	65-0023
AXISYMMETRIC	FINITE ELEMENT SOLUTION FOR AXISYMMETRIC TRANSIENT THERMAL STRESSES	72-0823
AXISYMMETRIC	FINITE ELEMENT STRESS ANALYSIS OF AXISYMMETRIC SOLIDS WITH ORTHOTROPIC TEMPERATURE DEPENDENT MATERIA	67-0089
AXISYMMETRIC	FINITE ELEMENT STRESS ANALYSIS OF PLANE AND AXISYMMETRIC STRUCTURES	69-0062
AXISYMMETRIC	FINITE ELEMENT STRESS ANALYSIS OF A RESTORED AXISYMMETRIC FIRST MOLAR	73-0053
AXISYMMETRIC	FINITE ELEMENTE ZUR BERECHNUNG DER WAERMEUEBERTRAGUNG FUER EBENE UND AXISYMMETRISCHE PROBLEME (FINIT	73-0449
AXISYMMETRIC	FINITE ELEMENTS FOR THE REPRESENTATION OF PLANE AND AXISYMMETRIC STRUCTURAL REINFORCEMENT	68-0037
AXISYMMETRIC	FINITE ELEMENTS FOR AXISYMMETRIC SOLIDS UNDER ARBITRARY LOADINGS WITH NODES ON ORIGIN.	72-0251
AXISYMMETRIC	FINITE ELEMENTS FOR AXISYMMETRIC SOLIDS UNDER ARBITRARY LOADINGS WITH NODES ON ORIGIN	73-0094
AXISYMMETRIC	FINITE-ELEMENT ANALYSIS OF AXISYMMETRIC ROTORS	69-0244
AXISYMMETRIC	FINITE-ELEMENT METHOD FOR AXISYMMETRIC FLOW COMPUTATION IN A TURBOMACHINE	75-0359
AXISYMMETRIC	FINITE-ELEMENT TREATMENT OF QUADRUPOLE GROUND MOTIONS IN AXISYMMETRIC EARTH STRUCTURES	75-0434
AXISYMMETRIC	FORMULATION AND EXPERIMENTAL VERIFICATION OF AN AXISYMMETRIC FINITE-ELEMENT STRUCTURAL ANALYSIS	71-0354
AXISYMMETRIC	FUEL ROD DEFORMATIONS - A TIME DEPENDENT FINITE ELEMENT ANALYSIS ON AXISYMMETRIC FUEL RODS, INCLUDIN	70-0414
AXISYMMETRIC	HYBRID FINITE ELEMENT ANALYSIS WITH PARTICULAR REFERENCE TO AXISYMMETRIC STRUCTURES	70-0248
AXISYMMETRIC	HYDROELASTIC ANALYSIS OF AXISYMMETRIC SYSTEMS BY A FINITE ELEMENT METHOD	68-0084
AXISYMMETRIC	IMPROVED ANALYTIC LONGITUDINAL RESPONSE ANALYSIS FOR AXISYMMETRIC LAUNCH VEHICLES, COMPUTER PROGRAM	65-0002
AXISYMMETRIC	IMPROVED LINEAR AXISYMMETRIC SHELL FLUID MODEL FOR LAUNCH VEHICLE LONGITUDINAL RESPONSE ANALYSIS	65-0044
AXISYMMETRIC	INCREMENTAL SOLUTION OF AXISYMMETRIC PLATE AND SHELL FINITE DEFORMATION	71-0776
AXISYMMETRIC	JET IMPINGEMENT ON AXISYMMETRIC CURVED DEFLECTORS	74-0017
AXISYMMETRIC	LUMPED PARAMETER APPROACH TO AXISYMMETRIC DYNAMIC SOIL DEFORMATIONS	69-0170
AXISYMMETRIC	MATRIX ANALYSIS OF AXISYMMETRIC SHELLS UNDER GENERAL LOADING	66-0075
AXISYMMETRIC	MATRIX DISPLACEMENT ANALYSIS OF SHELLS WITH AN AXISYMMETRIC MIDSURFACE BUT HAVING CIRCUMFERENTIALLY	68-0031
AXISYMMETRIC	MIXED FINITE ELEMENT METHOD FOR AXISYMMETRIC SHELLS	72-0149
AXISYMMETRIC	NEW TRIANGULAR ELEMENT FOR FINITE DIFFERENCE SOLUTION OF AXISYMMETRIC CONDUCTION PROBLEMS IN CYLINDR	74-0381
AXISYMMETRIC	NON-LINEAR DYNAMIC ANALYSIS OF AXISYMMETRIC SHELLS	75-0474
AXISYMMETRIC	NONLINEAR ANALYSIS OF ELASTIC SHELLS OF REVOLUTION UNDER AXISYMMETRIC LOADING	70-0161
AXISYMMETRIC	NONLINEAR DYNAMIC ANALYSIS OF AXISYMMETRIC SOLIDS BY THE FINITE ELEMENT METHOD	73-0942
AXISYMMETRIC	NONLINEAR DYNAMIC ANALYSIS OF AXISYMMETRIC SOLIDS BY THE FINITE ELEMENT METHOD	74-0510
AXISYMMETRIC	NONLINEAR FINITE ELEMENT DYNAMIC ANALYSIS OF AXISYMMETRIC SOLIDS	74-0775
AXISYMMETRIC	NOTES ON THE FINITE ELEMENT ANALYSIS OF THE AXISYMMETRIC ELASTIC SOLID	74-0185
AXISYMMETRIC	NUMERICAL ANALYSIS FOR FINITE AXISYMMETRIC DEFORMATIONS OF INCOMPRESSIBLE ELASTIC SOLIDS OF REVOLUTI	70-0094
AXISYMMETRIC	PART 1 - TASK 3.2.9. BENDING AND BUCKLING OF A ONE-DIMENSIONAL MODEL SIMULATING A CONICAL SHELL. PAR	71-0745
AXISYMMETRIC	PHOTOELASTIC AND FINITE ELEMENT STRESS ANALYSIS OF A RESTORED AXISYMMETRIC FIRST MOLAR	73-0142
AXISYMMETRIC	PLACID: A GENERAL FINITE ELEMENT COMPUTER PROGRAM FOR STRESS ANALYSIS OF PLANE AND AXISYMMETRIC SOLI	74-0884
AXISYMMETRIC	PLASTIC AND VISCOPLASTIC ANALYSIS OF AXISYMMETRIC SHELLS	75-0246
AXISYMMETRIC	PROGRESS REPORT ON DISCRETE-ELEMENT ELASTIC AND ELASTIC-PLASTIC ANALYSIS OF SHELLS OF REVOLUTION SUB	68-0170
AXISYMMETRIC	PURE MOMENT LOADING OF AXISYMMETRIC FINITE ELEMENT MODELS	73-1007
AXISYMMETRIC	QUASI-ANALYTICAL FINITE ELEMENT PROCEDURES FOR AXISYMMETRIC ANISOTROPIC SHELLS AND SOLIDS	74-0297
AXISYMMETRIC	RING FINITE ELEMENTS FOR AXISYMMETRIC AND NON-AXISYMMETRIC THIN SHELL ANALYSIS	72-0922
AXISYMMETRIC	SAAS II: FINITE ELEMENT STRESS ANALYSIS OF AXISYMMETRIC SOLIDS WITH ORTHOTROPIC, TEMPERATURE DEPENDE	68-0096
AXISYMMETRIC	SAAS 111: FINITE ELEMENT STRESS ANALYSIS OF AXISYMMETRIC AND PLANE SOLIDS WITH DIFFERENT ORTHOTROPIC	71-0741
AXISYMMETRIC	SABOR 1, A FORTRAN PROGRAM FOR THE LINEAR ELASTIC AXISYMMETRIC LOADING BY USING THE MATRIX DISPLACEM	65-0027

131

AXISYMMETRIC	SAFE-CREEP: A COMPUTER PROGRAM FOR THE VISCOELASTIC ANALYSIS OF AXISYMMETRIC AND PLANE CONCRETE STRU	67-0148
AXISYMMETRIC	SASL - A FINITE ELEMENT CODE FOR THE STATIC ANALYSIS OF AXISYMMETRIC AND PLANE SOLIDS SUBJECTED TO A	72-0779
AXISYMMETRIC	SASL - A FINITE ELEMENT CODE FOR THE STATIC ANALYSIS OF AXISYMMETRIC AND PLANE SOLIDS SUBJECTED TO A	72-0779
AXISYMMETRIC	SEAL-SHELL-2, A COMPUTER PROGRAM FOR THE STRESS ANALYSIS OF A THICK SHELL OF REVOLUTION WITH AXISYMM	69-0107
AXISYMMETRIC	SEMI-ANALYTICAL FINITE ELEMENT PROCEDURE FOR CONDUCTION IN ANISOTROPIC AXISYMMETRIC SOLIDS	74-0270
AXISYMMETRIC	STABILITY ANALYSIS OF AXISYMMETRIC SHELLS	69-0131
AXISYMMETRIC	STATIC STRESS ANALYSIS OF AXISYMMETRIC SOLIDS WITH MATERIAL AND GEOMETRIC NONLINEARITIES BY THE FINI	73-0958
AXISYMMETRIC	STATIC, VIBRATION AND BUCKLING ANALYSIS OF AXISYMMETRIC CIRCULAR PLATES USING FINITE ELEMENTS	73-0157
AXISYMMETRIC	STRESS ANALYSIS OF AXISYMMETRIC SOLIDS UTILIZING HIGHER ORDER QUADRILATERAL FINITE ELEMENTS	69-0356
AXISYMMETRIC	STRESS ANALYSIS OF AXISYMMETRIC SOLIDS WITH ASYMMETRIC PROPERTIES	72-0140
AXISYMMETRIC	STRUCTURAL ANALYSIS OF AXISYMMETRIC SOLIDS	65-0031
AXISYMMETRIC	STRUCTURAL ANALYSIS OF AXISYMMETRIC SOLIDS	65-0056
AXISYMMETRIC	THE EFFECT OF AN INITIAL AXISYMMETRIC IMPERFECTION ON THE NATURAL FREQUENCIES OF CONICAL SHELLS	70-0368
AXISYMMETRIC	THE STRESS ANALYSIS OF AXISYMMETRIC SHELLS USING A PLANE FRAME ANALYSIS PROGRAM	69-0238
AXISYMMETRIC	THE TRIAX 6 ELEMENT FOR AXISYMMETRIC ANALYSIS BY THE MATRIX DISPLACEMENT METHOD: PART 1, FOUNDATIONS	66-0124
AXISYMMETRIC	THERMOVISCOELASTIC AXISYMMETRIC STRESS ANALYSIS BY FINITE ELEMENT AND FINITE DIFFERENCE COMPUTER TEC	71-0105
AXISYMMETRIC	TRAPEZOIDAL FINITE ELEMENTS: THEIR DERIVATION AND USE FOR AXISYMMETRIC ROTATING BODIES	73-0192
AXISYMMETRIC	TRIAX 6 ELEMENT FOR AXISYMMETRIC ANALYSIS BY THE MATRIX DISPLACEMENT METHOD	66-0009
AXISYMMETRIC	TSAAS: FINITE ELEMENT THERMAL AND STRESS ANALYSIS OF AXISYMMETRIC SOLIDS WITH ORTHOTROPIC TEMPERATUR	74-0886
AXISYMMETRIC	USE OF SINGULARITY PROGRAMMING IN FINITE ELEMENT CALCULATIONS OF ELASTIC STRESS INTENSITY FACTORS. P	75-0252
AXISYMMETRIC	USERS' MANUAL FOR DIGITAL COMPUTER PROGRAM OK1. NUMERICAL ANALYSIS OF FINITE AXISYMMETRIC DEFORMATIO	72-0861
AXISYMMETRIC	VARIATIONAL FINITE ELEMENT METHODS FOR TWO DIMENSIONAL AND AXISYMMETRIC NAVIER STOKES EQUATIONS	74-0012
AXISYMMETRIC	VIBRATION ANALYSIS OF AXISYMMETRIC RESONATORS	75-0199
AXISYMMETRIC	VIBRATION AND BUCKLING ANALYSIS OF AXISYMMETRIC POLAR ORTHOTROPIC CIRCULAR PLATES	74-0574
AXISYMMETRICAL	CALCULATION OF AXISYMMETRICAL OSCILLATION OF FLEXIBLE TURBINE DISCS. (CALCOLO DELLE OSCILLAZIONI ASS	74-1328
AXISYMMETRICAL	DISCUSSION ON FINITE ELEMENT SOLUTION FOR AXISYMMETRICAL SHELLS	65-0062
AXISYMMETRICAL	FINITE ELEMENT ANALYSIS OF AXISYMMETRICAL SHELLS OF COMPOSITE MATERIAL	70-0483
AXISYMMETRICAL	FINITE ELEMENT SOLUTION OF AXISYMMETRICAL DYNAMIC PROBLEMS OF SHELLS OF REVOLUTION	66-0020
AXISYMMETRICAL	SOLUTION OF A THREE DIMENSIONAL AXISYMMETRICAL PROBLEM OF THERMOPLASTICITY AS APPLIED TO THICK TURBI	74-0384
AXISYMMETRICALL	COMPUTER PROGRAM FOR ELASTIC-PLASTIC ANALYSIS OF AXISYMMETRICALLY LOADED SHELLS OF REVOLUTION	68-0239
AXISYMMETRICALL	ELASTIC-PLASTIC DISCRETE ELEMENT ANALYSIS OF AXISYMMETRICALLY LOADED SHELLS OF REVOLUTION	67-0099
AXISYMMETRICALL	FINITE ELEMENT ANALYSIS OF AN AXISYMMETRICALLY LOADED ORTHOTROPIC SHELL OF REVOLUTION.	72-0438
AXISYMMETRICALL	FINITE ELEMENT ANALYSIS OF A THICK WALLED MICROPOLAR CYLINDER LOADED AXISYMMETRICALLY	73-0847
AXISYMMETRICALL	LINEAR AND NONLINEAR STATIC ANALYSIS OF AXISYMMETRICALLY LOADED THIN SHELLS OF REVOLUTION	69-0215
AXISYMMETRISCHE	FINITE ELEMENTE ZUR BERECHNUNG DER WAERMEUEBERTRAGUNG FUER EBENE UND AXISYMMETRISCHE PROBLEME (FINIT	73-0449
AXYSYMMETRIC	ANALYSES OF AXYSYMMETRIC UPSETTING AND PLANE-STRAIN SIDE-PRESSING OF SOLID CYLINDERS BY THE FINITE E	71-0794
AYER	AYER HEAT CONDUCTION COMPUTER PROGRAM	74-0885
A36	CYCLIC MECHANICAL TESTS AND AN APPROPRIATE ANALYTICAL STRESS-STRAIN MODEL FOR A36 STEEL	74-0815
BACKFILL	STRUCTURAL BEHAVIOR OF A FLEXIBLE METAL CULVERT UNDER A DEEP EARTH EMBANKMENT USING METHOD A BACKFIL	73-0739
BACKFILL	SUPPORT PERFORMANCE OF HYDRAULIC BACKFILL	74-0257
BACKFILLING	A STRUCTURAL EVALUATION OF RAPID METHODS OF BACKFILLING FOR BOMB DAMAGE REPAIR	74-0821
BACKFILLING	A STRUCTURAL EVALUATION OF RAPID METHODS OF BACKFILLING FOR BOMB DAMAGE REPAIR - PHASE 11	75-0682
BACKPACKING	EFFECT OF BACKPACKING AND INTERNAL PRESSURIZATION ON STRESSES TRANSMITTED TO BURIED CYLINDERS	73-0897
BACKWATER	FINITE ELEMENT MODEL FOR BACKWATER COMPUTATION	75-0851
BACKWATER	TWO-DIMENSIONAL ANALYSIS OF BACKWATER AT BRIDGES	74-0184
BAFFLE	STRESS DETERMINATION IN 2 CONDENSER HEAT EXCHANGER BAFFLE CAPS BY METHOD OF FINITE ELEMENTS AND A ST	74-0661
BALANCE	SOLVING THE LINEAR BALANCE EQUATION GLOBALLY	74-0810
BALANCE	THE MIDCOHV WEIGHT AND BALANCE COMPUTER PROGRAM (WTBAL)	74-0697
BALANCED	BALANCED DESIGN AND FINITE-ELEMENT ANALYSIS OF CULVERTS.	72-0462
BALANCED	BALANCED STRESSES IN POST-YIELDED MULTI-MATERIAL STRUCTURAL JOINTS	69-0271
BALANCING	FINITE ELEMENT ANALYSIS OF INCOMPRESSIBLE MATERIAL BY RESIDUAL ENERGY BALANCING	74-1041
BALANCING	FINITE ELEMENT ANALYSIS OF THIN ELASTIC SHELLS WITH RESIDUAL ENERGY BALANCING AND THE ROLE OF THE RI	75-0034
BALANCING	FINITE-ELEMENT ANALYSIS OF THIN ELASTIC SHELLS WITH RESIDUAL ENERGY BALANCING AND ROLE OF RIGID BODY	75-0048
BALANCING	RESIDUAL ENERGY BALANCING TECHNIQUE IN THE GENERATION OF PLATE BENDING FINITE ELEMENTS	74-0594
BALKEN	ACCURATE CALCULATION OF STRESSES IN BEAMS, DISKS AND PLATES BY MEANS OF FINITE ELEMENT METHODS AND I	70-0133
BALLISTIC	BALLISTIC IMPACT OF TEXTILE STRUCTURES.	72-0433
BALLISTIC	PLOTS - THE PLOTTING SUBROUTINE INCORPORATED IN THE BALLISTIC RESEARCH LABORATORIES ELECTRONIC SCIEN	72-0803
BAND	ANALYSIS OF BAND PRESSURE IN A GUN TUBE USING FINITE ELEMENTS AND SHELL PUSHING DATA	72-0793
BAND	RADIAL DISPLACEMENTS OF A HOLLOW SHAFT SUBJECTED TO A UNIFORM BAND OF EXTERNAL PRESSURE	72-0242
BANDED	BANDED MATRIX TECHNIQUES AND SOLUTIONS OF LINEAR SIMULTANEOUS EQUATIONS OF FINITE ELEMENT ANALYSES	70-0302
BANDED	VIBRATION OF FRAMES AND OTHER STRUCTURES WITH BANDED STIFFNESS MATRIX	70-0322
BANDIT	THE BANDIT COMPUTER PROGRAM FOR THE REDUCTION OF MATRIX BANDWIDTH FOR NASTRAN	72-0481
BANDMATRIX	AUTOMATIC GENERATION OF FINITE ELEMENT MESH WITHIN A BANDWIDTH AND AN EFFICIENT SOLUTION OF THE BAND	73-0338
BANDWIDTH	ALGORITHM FOR MATRIX BANDWIDTH REDUCTION	72-0895
BANDWIDTH	AN ALGORITHM FOR REDUCING THE BANDWIDTH OF A MATRIX OF SYMMETRICAL CONFIGURATION	65-0067
BANDWIDTH	AN AUTOMATIC NODE RELABELLING SCHEME FOR BANDWIDTH MINIMIZATION OF STIFFNESS MATRICES	68-0256
BANDWIDTH	AN AUTOMATIC RELABELING ALGORITHM FOR BANDWIDTH MINIMIZATION	75-0763
BANDWIDTH	AN IMPROVED METHOD FOR REDUCING THE BANDWIDTH OF SPARSE SYMMETRIC MATRICES	72-0990
BANDWIDTH	AUTOMATIC GENERATION OF FINITE ELEMENT MESH WITHIN A BANDWIDTH AND AN EFFICIENT SOLUTION OF THE BAND	73-0338
BANDWIDTH	BANDWIDTH MINIMIZATION OF STIFFNESS MATRICES	70-0549
BANDWIDTH	BANDWIDTH REDUCTION BY AUTOMATIC RENUMBERING	73-0993
BANDWIDTH	NOTE ON MINIMIZING THE BANDWIDTH OF SPARSE, SYMMETRIC MATRICES	73-1107
BANDWIDTH	REDUCING THE BANDWIDTH OF SPARSE SYMMETRIC MATRICES	69-0469
BANDWIDTH	REDUCING THE BANDWIDTH OF SPARSE SYMMETRIC MATRICES	69-0502
BANDWIDTH	SEVERAL STRATEGIES FOR REDUCING THE BANDWIDTH OF MATRICES	72-0996
BANDWIDTH	THE BANDIT COMPUTER PROGRAM FOR THE REDUCTION OF MATRIX BANDWIDTH FOR NASTRAN	72-0481
BANDWITH	AN ALGORITHM FOR REDUCING THE BANDWITH AND PROFILE OF A SPARSE MATRIX	74-0792
BANK	FOUNDATION INSTRUMENTATION FOR THE NATIONAL WESTMINISTER BANK TOWER	74-1147
BANK	STRESS DISTRIBUTION DUE TO GRAVITY IN A VERTICAL ROCK BANK	68-0090
BANKS	RESPONSE OF EARTH BANKS DURING EARTHQUAKES	67-0039
BANKS	SEEPAGE ANALYSIS OF EARTH BANKS UNDER DRAWDOWN	72-0014
BANKS	SEEPAGE IN MISSISSIPPI RIVER BANKS. REPORT 1, ANALYSIS OF TRANSIENT SEEPAGE USING VISCOUS FLOW MODEL	70-0405
BANKS	SEEPAGE IN MISSISSIPPI RIVER BANKS. REPORT 1. ANALYSIS OF TRANSIENT SEEPAGE USING A VISCOUS-FLOW MOD	73-0909
BANKS	STUDIES OF SEISMIC RESPONSE OF CLAY BANKS	71-0833
BANKS	UNCONFINED TRANSIENT SEEPAGE IN SLOPING BANKS	71-0378
BAR	ANALYSIS OF LONGITUDINAL WAVE IN A BAR EXHIBITING VISCO-ELASTIC BEHAVIOR OR WITH MATERIAL DAMPING	71-0780
BAR	ANALYSIS OF STRESSES IN CONCRETE NEAR A REINFORCING BAR DUE TO BOND AND TRANSVERSE CRACKING	70-0351
BAR	AUTOMATED LARGE DEFLECTION AND STABILITY ANALYSIS OF THREE DIMENSIONAL BAR STRUCTURES	69-0393
BAR	DYNAMIC STRAINS IN A FOUR BAR MECHANISM	73-0677
BAR	ELASTIC-PLASTIC ANALYSIS OF A BAR WITH UNIFORM CROSS-SECTION UNDER COMBINED LOADINGS 1 - COMBINATION	71-0778
BAR	ELASTIC-PLASTIC PROBLEMS OF COMPRESSIBLE STRAIGHT BAR UNDER COMBINED LOADING CONDITION	75-0313
BAR	ELASTIC-PLASTIC STRESS ANALYSIS FOR A NOTCHED BAR IN PLANE STRAIN BENDING	71-0410
BAR	FINITE ELEMENT ANALYSIS ON THE TORSION OF A BAR WITH UNIFORM CROSS SECTION	68-0186
BAR	FINITE ELEMENT ANALYSIS OF THE MAGNETIC FIELD DISTRIBUTION INSIDE A ROTATING FERROMAGNETIC BAR	74-1037
BAR	STATICAL CALCULATIONS OF PLANE AND SPACE BAR STRUCTURES AS WELL AS STUMPY FRAMEWORKS USING A MODIFIE	73-0747
BAR	STATICAL CALCULATIONS OF PLANE AND SPACE BAR STRUCTURES AS WELL AS STUMPY FRAMEWORKS USING A MODIFIE	73-0747
BAR	STRAIN CONCENTRATION BEHAVIOUR IN A NOTCHED ROUND BAR SUBJECTED TO CREEP	73-0440
BAR	STRESSES IN A PARTLY YIELDED NOTCHED BAR - AN ASSESSMENT OF THREE ALTERNATIVE PROGRAMS	73-0267
BAR	STRESSES IN A PARTLY YIELDED NOTCHED BAR EM DASH AN ASSESSMENT OF THREE ALTERNATIVE PROGRAMS	73-0268
BAR	THREE-DIMENSIONAL BAR CELL FOR ELASTIC STRESS ANALYSIS	70-0219
BARE	COMPUTATIONAL EXPERIMENTATION ON FINITE ELEMENT METHOD IN BARE SLAB CRITICALITY CALCULATIONS	73-0015
BARRAGE-FONDATI	CONTRIBUTION A L'ETUDE DU COMPLEXE BARRAGE-FONDATION ET EXPLICATION DES SEISMES DUS AU REMPLISSAGE D	72-0554

132

BARRAGES	NOUVELLE CONCEPTION DES DRAINS DANS LES BARRAGES EN TERRE HOMOGENE DE PETITE ET MOYENNE DIMENSIONS (	73-0404
BARRAGES	UTILISATION DE MODELES MATHEMATIQUES MULTIDIMENSIONNELS POUR L'ETUDE DES ECOULEMENTS DANS LES DIGUES	73-0381
BARS	ACCURACY AND CONVERGENCE OF MIXED FINITE-ELEMENT APPROXIMATIONS OF THIN BARS, MEMBRANES, AND PLATES	73-1120
BARS	ANALYSIS OF BENT AND TWISTED BARS USING THE FINITE ELEMENT METHOD	67-0045
BARS	CALCULATION OF THE CRITICAL LOAD OF ELASTIC THIN WALLED BARS BY THE METHOD OF FINITE ELEMENTS	72-0446
BARS	CREEP STRESS AND STRAIN ANALYSIS OF NOTCHED PLATES AND BARS BY MEANS OF THE FINITE ELEMENT METHOD	70-0141
BARS	FINITE ELEMENT SOLUTION TO DYNAMIC STABILITY OF BARS	68-0035
BARS	FORM FACTORS OF FLAT BARS FOR FATIGUE STRENGTH TESTS. (FORMZAHLEN VON FLACHSTAEBEN FUER SCHWINGFESTI	71-0641
BARS	OSCILLATION STABILITY OF INTERNALLY AND EXTERNALLY NOTCHED PLANE BARS OF EQUAL FORM FACTOR FROM 3943	73-0745
BARS	PLASTIC STRAINS ASSOCIATED WITH TORSION OF NONCIRCULAR BARS	72-0177
BARS	STIFFNESS OF STEPPED BARS	74-0711
BARS	STRAIN HARDENING TORSION OF SOLID BARS AS A MINIMIZATION PROBLEM	70-0174
BARS	TRANSVERSE VIBRATIONS OF CANTILEVER BARS OF VARIABLE CROSS SECTIONS	66-0001
BASATO	CHARACTERISTICS OF A HYBRID CALCULATION SYSTEM BASED ON THE FINITE ELEMENT METHOD. (CARATTERISTICHE	72-0715
BASE	CREEP BENDING TESTS OF LARGE WELD OVERLAY AND BASE METAL SPECIMENS AT 1000 DEGREE F. (537.7 DEG. C)	75-0521
BASE	ELASTIC AND PLASTIC STRESS DISTRIBUTION ON BASE PLATES IN THE NEIGHBOURHOOD OF FILLET WELDS IN LAP J	70-0175
BASE	ELEMENT TYPES AND BASE FUNCTIONS	73-0522
BASE	FINITE ELEMENT ANALYSIS OF THE SURFACE DEFORMATION DUE TO A UNIFORM LOADING ON A LAYER OF GIBSON SOI	75-0240
BASE	RIGID PLATE ON NONLINEARLY DEFORMABLE COHESIVE BASE (PLANE PROBLEM)	75-0264
BASE	VERTICAL RESPONSE OF A RIGID BASE TO DYNAMIC LOADING	69-0316
BASELINE	BASELINE SOLUTIONS FOR THE SMOOTH WALL THERMOCHEMICAL ABLATION RESPONSE OF GRAPHITE AND CARBON PHENO	71-0728
BASES	CONTINUUM BASES FOR CONSISTENT NUMERICAL FORMULATIONS OF FINITE STRAINS IN ELASTIC AND INELASTIC STR	75-0299
BASES	EQUIVALENT FINITE ELEMENTS ON DIFFERENT BASES	72-0287
BASES	FATIGUE CRACK FORMATION AND PROPAGATION IN PAVEMENTS CONTAINING SOIL-CEMENT BASES.	72-0435
BASES	LUMPED FINITE ELEMENT BASES FOR BEAMS AND PLATES	72-0263
BASIN	CONTROLLING OPEN-PIT SLOPE FAILURES AT SHIRLEY BASIN	73-0149
BASIN	PRELIMINARY FINITE ELEMENT ANALYSIS, ATCHAFALAYA BASIN PROTECTION LEVEES, TEST SECTION 3	69-0463
BASINS	MATHEMATICAL MODELS OF THE MASSACHUSETTS BAY. PART 1. FINITE ELEMENT MODELING OF TWO-DIMENSIONAL HYD	73-0800
BASINS	USE OF THE OVER-RELAXATION TECHNIQUE IN THE SIMULATION OF LARGE GROUNDWATER BASINS BY THE FINITE ELE	75-0606
BAUTEILEN	BERECHNUNG VON SPANNUNGEN UND TEMPERATURFELDERN IN FLOSSENWAENDEN UND AUFGESCHWEISSTEN BAUTEILEN. (C	74-0544
BAUWERKSSTRUKTJ	BERECHNUNGSMETHODEN ZUR ERMITTLUNG VON BELASTUNGSBEDINGUNGEN UND GRENZTRAGFAEHIGKEITSVERHALTEN VON K	73-0858
BAY	M.I.T. TEST SECTION INSTRUMENTATION, MASSACHUSETTS BAY TRANSPORTATION AUTHORITY, HAYMARKET-NORTH EXT	72-0673
BAY	MATHEMATICAL MODELS OF THE MASSACHUSETTS BAY. PART 1. FINITE ELEMENT MODELING OF TWO-DIMENSIONAL HYD	73-0800
BAY	NODAL ANALYSIS OF A NINE BAY SKIN-STRINGER PANEL	71-0102
BEACH	GROUNDWATER FLOW IN A SANDY TIDAL BEACH: II. 2-DIMENSIONAL FINITE ELEMENT ANALYSIS	72-0300
BEACH	GROUNDWATER FLOW IN A SANDY TIDAL BEACH .1. ONE-DIMENSIONAL FINITE ELEMENT ANALYSIS	71-0567
BEACH	INVESTIGATION OF THE WATER TABLE IN A TIDAL BEACH	71-0648
BEACTOP	ANALYSIS OF LOADING AND THERMAL EFFECTS ON FUELLED GRAPHITE BRICKS FOR A HIGH-TEMPERATURE GAS COOLED	71-0085
BEAM	A BEAM ELEMENT FOR USE WITH THE SEMILOOF SHELL ELEMENT	73-0620
BEAM	A COMPARISON OF CURVED BEAM FINITE ELEMENTS WHEN USED IN VIBRATION PROBLEMS	71-0227
BEAM	A CURVED BEAM ELEMENT	68-0202
BEAM	A DISCRETE-ELEMENT METHOD OF ANALYSIS FOR COMBINED BENDING AND SHEAR DEFORMATION OF A BEAM	69-0494
BEAM	A FINITE ELEMENT ANALYSIS FOR THE STRENGTH OF BEAM COLUMNS	66-0037
BEAM	A FINITE ELEMENT ANALYSIS OF THE ELASTIC BEHAVIOR OF BEAM COLUMNS	66-0122
BEAM	A FINITE ELEMENT APPROACH TO THE STRUCTURAL INSTABILITY OF BEAM COLUMNS, FRAMES, AND ARCHES	70-0434
BEAM	A FINITE ELEMENT METHOD FOR OPTIMUM BEAM DESIGN	71-0685
BEAM	A GENERAL EQUILIBRIUM BEAM ELEMENT WITH OFFSETS	73-0351
BEAM	A STUDY ON THE STRESS ANALYSIS OF BEAM PLATE STRUCTURE	69-0404
BEAM	A TECHNICAL MANUAL FOR THE BEAM ELEMENTS	69-0455
BEAM	A TECHNIQUE FOR CONNECTING BEAM ELEMENTS TO A PLATE MODEL OF A COMPLICATED BOX SECTION	74-0271
BEAM	A TIMOSHENKO BEAM ELEMENT	72-0590
BEAM	AN ANALYSIS OF A REINFORCED AND A PRESTRESSED CONCRETE BEAM BY FINITE-ELEMENT METHOD	72-0940
BEAM	ANALYSIS FOR THE DESIGN OF BEAM TESTING APPARATUS	75-0527
BEAM	ANALYSIS OF SLAB BEAM BRIDGES	69-0077
BEAM	ANALYSIS OF SPINE BEAM BRIDGES USING FINITE ELEMENTS	70-0236
BEAM	APPLICATION OF THE SINGLE-EDGE NOTCHED BEAM TO FRACTURE TOUGHNESS TESTING OF CERAMICS	74-0623
BEAM	BEAM BENDING FINITE ELEMENT CONTAINING INTEGRATION PARAMETERS	71-0179
BEAM	BEAM BUCKLING BY FINITE ELEMENT PROCEDURE	74-0123
BEAM	COMPUTER PROGRAM FOR PRISMATIC FOLDED PLATES BY PLATE AND BEAM ELEMENTS	70-0371
BEAM	CONSTANT CURVATURE BEAM FINITE ELEMENTS FOR IN-PLANE VIBRATION	72-0254
BEAM	CONVERGENCE OF CONSISTENTLY DERIVED TIMOSHENKO BEAM FINITE ELEMENTS	72-0396
BEAM	CURVED BEAM FINITE ELEMENTS FOR COUPLED BENDING AND TORSIONAL VIBRATION	72-0720
BEAM	DESIGN, MANUFACTURE, DEVELOPMENT, TEST, AND EVALUATION OF BORON/ALUMINUM STRUCTURAL COMPONENTS FOR S	74-0798
BEAM	DEVELOPMENT OF A DESIGN FOR A PRECAST CONCRETE BRIDGE BEAM OF U-SECTION	73-0392
BEAM	DIGITAL COMPUTATION OF STRESSES AND DEFLECTIONS IN A BOX BEAM	67-0073
BEAM	DIGITAL COMPUTATION OF STRESSES AND DEFLECTIONS IN A BOX BEAM - A PERFORMANCE COMPARISON BETWEEN FIN	69-0355
BEAM	DIRECT BEAM STIFFNESS MATRIX CALCULATIONS INCLUDING SHEAR EFFECTS	62-0003
BEAM	DYNAMIC COUPLING IN AN UNSYMMETRICALLY TAPERED BEAM	74-1301
BEAM	DYNAMIC RESPONSE OF A CYLINDER BURIED IN AN EARTH BEAM - RESULTS OF A FINITE ELEMENT ANALYSIS	72-0592
BEAM	DYNAMIC STABILITY OF A NONLINEAR BEAM SUBJECTED TO BOTH LONGITUDINAL AND TRANSVERSE EXCITATION	73-0553
BEAM	EFFECT OF SHEAR DEFORMATIONS AND ROTARY INERTIA ON OPTIMUM BEAM FREQUENCIES	75-0231
BEAM	EFFECT OF TRANSVERSE SHEAR DEFORMATION IN VIBRATIONS OF PLANAR STRUCTURES COMPOSED OF BEAM TYPE ELEM	73-0402
BEAM	EIGENVALUES OF COLLINEAR BEAM STRUCTURES USING FINITE ELEMENT TECHNIQUES AND VARIOUS DYNAMIC REPRESE	67-0124
BEAM	ELASTOPLASTIC ANALYSIS OF A CANTILEVER BEAM	73-0984
BEAM	ELEMENT STATIC FLEXIBILITY AND STIFFNESS MATRICES FOR THE FINITE ELEMENT ANALYSIS OF BEAM, PLATE, AN	67-0070
BEAM	FINITE BEAM ELEMENTS FOR DYNAMIC ANALYSIS	70-0372
BEAM	FINITE DISPLACEMENT BEAM THEORY	75-0529
BEAM	FINITE ELEMENT ANALYSIS OF AN EXPANDED I - SECTION BEAM AND AN AXISYMMETRIC FLANGED CYLINDER	68-0002
BEAM	FINITE ELEMENT ANALYSIS OF NONLINEAR VIBRATION OF BEAM COLUMNS	73-0028
BEAM	FINITE ELEMENT CALCULATIONS OF THE COMPLIANCE OF A TAPERED DOUBLE CANTILEVER BEAM SPECIMEN FOR DIFFE	72-0665
BEAM	FINITE ELEMENT FOR TIMOSHENKO BEAM BASED ON MECHANICAL IMPEDANCE	71-0607
BEAM	FINITE ELEMENT MODELING OF A CANTILVER BEAM	67-0132
BEAM	FINITE ELEMENT MODEL FOR DYNAMIC ANALYSIS OF TIMOSHENKO BEAM	75-0275
BEAM	FINITE ELEMENT SOLUTION OF INELASTIC BEAM EQUATIONS	73-0234
BEAM	FLEXURE ANALYSIS OF AXIALLY-LOADED BEAM BY FINITE-ELEMENT METHOD WITH INTEGRAL PARAMETERS	75-0462
BEAM	FLEXURE OF A BEAM SUPPORTED ON AN ELASTIC FOUNDATION	72-0871
BEAM	FREE VIBRATION OF A CURVED BEAM	71-0606
BEAM	FREE VIBRATION OF A SIMPLY SUPPORTED BEAM WITH NONLINEAR MATERIAL PROPERTIES	71-0689
BEAM	FREE VIBRATION OF CURVED AND STRAIGHT BEAM SLAB OR BOX GIRDER BRIDGES	72-0487
BEAM	GALERKIN FINITE ELEMENT ANALYSIS OF A UNIFORM BEAM CARRYING A CONCENTRATED MASS AND ROTARY INERTIA W	74-0552
BEAM	GENERAL BEAM ELEMENT FOR ASKA	69-0458
BEAM	IMPROVED DEFORMATION FUNCTIONS FOR THE FINITE ELEMENT ANALYSIS OF BEAM SYSTEMS	69-0075
BEAM	INCLUSION OF SHEAR DEFLECTION IN THE STIFFNESS MATRIX OF A BEAM ELEMENT	70-0307
BEAM	MASS MATRIX FOR A PRISMATIC BEAM SEGMENT	68-0116
BEAM	MATRICE DE RIGIDITE D'UN ELEMENT DE POUTRE SOLLICITE A LA TORSION-FLEXION (STIFFNESS MATRIX OF A BEA	73-0442
BEAM	MATRIX ANALYSIS OF BEAM COLUMNS BY THE METHOD OF FINITE ELEMENTS	64-0012
BEAM	NON-LINEAR EFFECTS OF HIGH TEMPERATURE ON THE VIBRATION OF A SIMPLY SUPPORTED BEAM WITH A CENTRAL MA	73-0838
BEAM	NON-LINEAR RESPONSE OF A BEAM TO PERIODIC LOADING.	72-0407
BEAM	NONLINEAR BEAM AND PLATE ELEMENTS	72-0866
BEAM	ON ANALYSIS OF LARGE DEFORMATION PROBLEMS OF BEAM	71-0660
BEAM	ON THE VALIDITY OF THE FINITE ELEMENT ANALYSIS OF CIRCULAR ARCHES REPRESENTED BY AN ASSEMBLAGE OF BE	75-0479
BEAM	OPTIMAL BEAM FREQUENCIES BY THE FINITE ELEMENT DISPLACEMENT METHOD	73-0187

BEAM	PARAMETRIC STUDY OF A BEAM WITH A COMPOUND SIDE-BRANCH RESONATOR AS A DEVICE TO EVALUATE PRELIMINARY	70-0465
BEAM	POWER SERIES EXPANSION OF THE GENERAL STIFFNESS MATRIX FOR BEAM ELEMENTS	75-0543
BEAM	RESPONSE OF NONLINEAR BEAM TO RANDOM EXCITATION	73-0285
BEAM	STIFFNESS MATRICES FOR THE GENERAL DEFORMATION (OUT-OF-PLANE AND INPLANE) OF CURVED BEAM MEMBERS BAS	75-0124
BEAM	STRESS INTENSITY FACTORS FOR A NOTCHED BEAM ON AN ELASTIC FOUNDATION	75-0566
BEAM	STRUCTURAL OPTIMIZATION USING THE FINITE ELEMENT METHOD APPLIED TO A BEAM	75-0233
BEAM	TECHNIQUE FOR FORMULATING BEAM EQUATIONS	75-0584
BEAM	THE EFFECT OF LUMPED PARAMETERS ON BEAM FREQUENCIES	63-0012
BEAM	TIMOSHENKO BEAM FINITE ELEMENTS	73-0032
BEAM	TRANSMISSION OF VIBRATION IN BEAM SYSTEMS	69-0405
BEAM	TRANSVERSE IMPACT PROBLEMS BY HIGHER-ORDER BEAM FINITE-ELEMENTS	75-0819
BEAM	TWISTED BEAM ELEMENT MATRICES FOR BENDING	70-0082
BEAM	USE OF STRAIGHT BEAM FINITE ELEMENTS FOR ANALYSIS OF VIBRATIONS OF CURVED BEAMS	73-0616
BEAM	USER'S MANUAL FOR PROGRAM BEAM	73-1105
BEAM	USERS MANUAL FOR PROGRAM BEAM	73-0714
BEAM	VIBRATIONS OF A NON-UNIFORM BEAM	68-0246
BEAM	VIBRATIONS OF A TIMOSHENKO BEAM, USING FINITE-ELEMENT APPROACH	66-0044
BEAM	ZUR BERECHNUNG EINFACH-SYMMETRISCHER I-TRAEGER NACH DER THEORIE - 2. ORDNUNG UNTER BERUECKSICHTIGUNG	75-0510
BEAM-COLUMN	A COMPUTER PROGRAM TO ANALYZE BEAM-COLUMN UNDER MOVEABLE LOADS	68-0268
BEAM-COLUMN	APPLICATIONS OF THE FINITE-ELEMENT METHOD TO BEAM-COLUMN PROBLEMS	72-0978
BEAM-COLUMN	A FINITE ELEMENT METHOD OF SOLUTION FOR LINEARLY ELASTIC BEAM-COLUMNS	66-0117
BEAM-COLUMNS	COUPLED LOCAL BUCKLING IN WIDE-FLANGE BEAM-COLUMNS	73-0235
BEAM-COLUMNS	DEVELOPMENT OF METHODS FOR COMPUTER SIMULATION OF BEAM-COLUMNS AND GRID-BEAM AND SLAB SYSTEMS	73-1101
BEAM-COLUMNS	LINEAR STABILITY ANALYSIS OF BEAM-COLUMNS	73-0499
BEAM-SHEARWALL	EFFECT OF BEAM-SHEARWALL JUNCTION DEFORMATIONS ON THE FLEXIBILITY OF THE CONNECTING BEAMS	73-0213
BEAM-SLAB	ANALYTICAL MODELING OF BEAM-SLAB BRIDGES	74-1131
BEAM-SLAB	DYNAMIC ANALYSIS OF BEAM-SLAB HIGHWAY BRIDGES	73-0199
BEAMS	A FINITE ELEMENT APPROACH TO THE DETERMINATION OF THE DYNAMIC RESPONSE OF BEAMS SUPPORTING MOVING MA	74-1114
BEAMS	A FINITE-ELEMENT METHOD FOR TRANSVERSE VIBRATIONS OF BEAMS AND PLATES	67-0144
BEAMS	A FINITE-ELEMENT METHOD OF ANALYSIS FOR COMPOSITE BEAMS	68-0263
BEAMS	ACCURATE CALCULATION OF STRESSES IN BEAMS, DISKS AND PLATES BY MEANS OF FINITE ELEMENT METHODS AND I	70-0133
BEAMS	AN ANALYSIS OF THE LARGE DEFLECTIONS OF BEAMS USING THE RAYLEIGH-RITZ FINITE ELEMENT METHOD	68-0162
BEAMS	AN ELASTIC-PLASTIC FINITE ELEMENT ANALYSIS FOR PLATES WITH EDGE BEAMS	69-0185
BEAMS	AN IMPROVED MATRIX DISPLACEMENT METHOD FOR VIBRATION OF NONUNIFORM BEAMS	71-0841
BEAMS	ANALYSIS OF CASTELLATED BEAMS BY THE FINITE ELEMENT METHOD	72-0620
BEAMS	ANALYSIS OF ELASTIC-PLASTIC BEHAVIOR OF BEAMS AND COLUMNS BY PLANE FINITE ELEMENT MODEL	73-1073
BEAMS	ANALYSIS OF SLABS WITH EDGE BEAMS	71-0075
BEAMS	ANALYSIS OF THIN-WALLED CURVED BEAMS BY THE TRANSFER MATRIX METHOD	72-0335
BEAMS	APPLICATION OF THE FINITE ELEMENT METHOD OF ANALYSIS TO HYPERBOLIC PARABOLOIDS WITH EDGE BEAMS	75-0250
BEAMS	APPLICATIONS OF CURVED FINITE ELEMENT TO SANDWICH BEAMS USING SHEAR STRAIN FORMULATION	74-0523
BEAMS	APPLICATIONS OF THE FINITE ELEMENT METHOD TO BOX BEAMS	68-0273
BEAMS	BENDING AND TORSION OF THIN-WALLED STRAIGHT AND CURVED BEAMS WITH VARIABLE OPEN CROSS-SECTION	69-0493
BEAMS	BENDING AND VIBRATION OF MULTILAYER SANDWICH BEAMS AND PLATES	73-0272
BEAMS	BENDING PROBLEM OF AXIALLY CONSTRAINED BEAMS	73-0397
BEAMS	CHARACTERISTICS OF FINITE-ELEMENTS FOR ANALYSIS OF BEAMS SUBJECTED TO IMPACT LOAD	73-1144
BEAMS	COUPLED VIBRATIONS OF THIN-WALLED BEAMS OF OPEN SECTION USING FINITE ELEMENT METHOD	70-0230
BEAMS	COUPLED VIBRATIONS OF THIN-WALLED BEAMS OF OPEN SECTION USING FINITE ELEMENT METHOD	71-0568
BEAMS	DEFLECTION ANALYSIS OF EXPANDED OPEN-WEB STEEL BEAMS	74-0224
BEAMS	DETERMINATION OF THE APPLICABILITY OF SOME THEORIES ABOUT STRENGTH AND STIFFNESS OF THIN-WALLED BEAM	69-0318
BEAMS	DISPLACEMENT METHOD IN STRUCTURAL ANALYSIS AS APPLIED TO RECTANGULAR SHELL BEAMS	72-0411
BEAMS	DOUBLE STRESS-WAVE DISCONTINUITIES IN FINITE, SHEAR-CORRECTED BEAMS AND PLATES	68-0210
BEAMS	DYNAMIC ANALYSIS OF SANDWICH BEAMS	72-0586
BEAMS	DYNAMIC STABILITY OF TIMOSHENKO BEAMS BY FINITE ELEMENT METHOD	75-0194
BEAMS	DYNAMIC STABILITY OF TIMOSHENKO BEAMS BY FINITE ELEMENT METHOD	75-0422
BEAMS	DYNAMICAL FINITE ELEMENT ANALYSIS FOR ELASTIC WAVES IN BEAMS AND PLATES	68-0211
BEAMS	EFFECT OF BEAM-SHEARWALL JUNCTION DEFORMATIONS ON THE FLEXIBILITY OF THE CONNECTING BEAMS	73-0213
BEAMS	EFFECTS OF SHEAR DEFORMATION, ROTARY INERTIA, AND ELASTICITY OF THE SUPPORT ON THE RESONANCE FREQUEN	75-0369
BEAMS	ELASTIC LATERAL BUCKLING OF STEEL BEAMS	70-0364
BEAMS	ELASTIC SHEAR ANALYSIS OF GENERAL PRISMATIC BEAMS	68-0114
BEAMS	ELASTIC-PLASTIC ANALYSIS OF BEAMS WITH UNIFORM CROSS-SECTION UNDER COMBINED LOADINGS	71-0777
BEAMS	ELASTIC-PLASTIC ANALYSIS OF BEAMS WITH UNIFORM CROSS-SECTION UNDER COMBINED LOADINGS (2)	73-1070
BEAMS	ETUDES DES POUTRES ET POTEAUX PAR LA METHODE DES ELEMENTS FINIS. (STABILITY ANALYSIS OF BEAMS AND C	74-1008
BEAMS	FINITE ELEMENT ANALYSIS OF REINFORCED CONCRETE BEAMS	67-0058
BEAMS	FINITE ELEMENT ANALYSIS OF WOOD BEAMS	70-0503
BEAMS	FINITE ELEMENT ANALYSIS OF THIN-WALLED STRUCTURES BASED ON THE MODERN ENGINEERING THEORY OF BEAMS	71-0140
BEAMS	FINITE ELEMENT ANALYSIS OF CONCRETE BEAMS	74-0560
BEAMS	FINITE ELEMENT ANALYSIS OF SHEAR STRENGTH OF REINFORCED CONCRETE BEAMS	74-1306
BEAMS	FINITE ELEMENT DISPLACEMENT METHOD FOR LARGE AMPLITUDE FREE FLEXURAL VIBRATIONS OF BEAMS AND PLATES	73-0185
BEAMS	FINITE ELEMENT LARGE DEFLECTION ANALYSIS OF THIN-WALLED BEAMS	74-0444
BEAMS	FINITE ELEMENT SOLUTION TO AN ELASTICA PROBLEM OF BEAMS	70-0112
BEAMS	FINITE ELEMENT SOLUTIONS FOR BUCKLING OF COLUMNS AND BEAMS	71-0361
BEAMS	FINITE ELEMENT STUDY OF REINFORCED CONCRETE BEAMS WITH DIAGONAL TENSION CRACKS	74-1308
BEAMS	FINITE-ELEMENT FOR BUCKLING OF CURVED BEAMS AND SHELLS WITH SHEAR	75-0607
BEAMS	FREE VIBRATION OF CURVED SANDWICH BEAMS BY METHOD OF FINITE ELEMENTS	71-0351
BEAMS	FREE VIBRATIONS OF CURVED SANDWICH BEAMS	71-0729
BEAMS	IMPROVED FINITE ELEMENTS FOR VIBRATION ANALYSIS OF TAPERED BEAMS	73-0033
BEAMS	INELASTIC BUCKLING OF STEEL BEAMS UNDER NON UNIFORM MOMENT	75-0149
BEAMS	LARGE DEFLECTION OF CANTILEVER BEAMS	71-0649
BEAMS	LARGE DEFLECTION SPATIAL BUCKLING OF THIN WALLED BEAMS AND FRAMES	73-1085
BEAMS	LARGE DEFLECTIONS OF ELASTIC BEAMS AND PLATES USING THE FINITE ELEMENT METHOD	68-0101
BEAMS	LARGE DISPLACEMENT ANALYSIS OF VISCOELASTIC BEAMS AND FRAMES BY FINITE-ELEMENT METHOD	74-0644
BEAMS	LUMPED FINITE ELEMENT BASES FOR BEAMS AND PLATES	72-0263
BEAMS	MODELING TECHNIQUES OF THIN-WALLED BEAMS WITH OPEN CROSS SECTIONS	72-0573
BEAMS	MODIFIED APPROACH FOR ESTIMATING THE CRACKING MOMENT OF REINFORCED CONCRETE BEAMS	75-0485
BEAMS	MODIFIED FINITE ELEMENT METHOD APPLIED TO R C DEEP BEAMS	73-0476
BEAMS	MULTIPLE STRESS WAVE DISCONTINUITIES IN FINITE BEAMS AND PLATES	67-0114
BEAMS	NON LINEAR ANALYSIS OF REINFORCED CONCRETE MEMBERS BY FINITE ELEMENT METHOD. PART 1: ANALYSIS, PART	71-0313
BEAMS	NONLINEAR ANALYSIS OF BEAMS AND SHELLS	73-1074
BEAMS	NONLINEAR BEHAVIOR OF REINFORCED CONCRETE BEAMS BY THE FINITE-ELEMENT METHOD	72-0954
BEAMS	NONLINEAR VIBRATIONS OF NONUNIFORM BEAMS WITH CONCENTRATED MASSES	74-1300
BEAMS	NUMERICAL STUDY OF DAMPING IN VISCOELASTIC SANDWICH BEAMS	73-0443
BEAMS	ON THE EXPLICIT REPRESENTATION OF TRANSFER-MATRICES OF CURVED BEAMS AND THEIR APPLICATIONS	75-0565
BEAMS	ON THE STRESS ANALYSIS AT JOINTS OF DEEP BEAMS	67-0055
BEAMS	PLATES AND BEAMS ON ELASTIC FOUNDATIONS: LINEAR AND NONLINEAR BEHAVIOUR	68-0042
BEAMS	POSSIBILITIES OF THE METHOD OF FINITE ELEMENTS IN THE CALCULATION OF THE STRENGTH AND STIFFNESS OF T	70-0344
BEAMS	PRINCIPAL STRESSES AT THE INTERMEDIATE SUPPORT OF PRESTRESSED CONCRETE CONTINUOUS BEAMS	70-0352
BEAMS	REINFORCING REQUIREMENTS FOR CONCRETE BEAMS WITH LARGE WEB OPENINGS	73-0198
BEAMS	RESPONSE OF BEAMS AND PLATES TO MOVING LOADS	71-0277
BEAMS	RESPONSE OF SIMPLY SUPPORTED TIMOSHENKO BEAMS	69-0489
BEAMS	SHEAR DEFORMATION ANALYSIS OF BEAMS AND PLATES BY FINITE ELEMENT METHOD	75-0873
BEAMS	SHEAR STRESSES IN BEAMS WITH DISCONTINUITY IN DEPTH	75-0487
BEAMS	STRESS CONCENTRATIONS IN TIMBER BEAMS	67-0038

BEAMS	TENSILE CRACK PROPAGATION IN REINFORCED CONCRETE BEAMS BY THE FINITE ELEMENT TECHNIQUES	69-0264
BEAMS	THE ANALYSIS OF COOLING TOWER RING BEAMS	68-0063
BEAMS	THE DAMPED TRANSIENT BEHAVIOR OF FINITE BEAMS AND PLATES	69-0520
BEAMS	THE INELASTIC ANALYSIS OF REINFORCED AND PRESTRESSED CONCRETE BEAMS	72-0667
BEAMS	TRANSIENT RESPONSE OF NON-UNIFORM, NON-HOMOGENEOUS BEAMS	73-0162
BEAMS	TRANSIENT RESPONSE OF ELASTIC, VISCO-PLASTIC BEAMS	73-0380
BEAMS	TRANSVERSE VIBRATION OF BEAMS WITH INTERNAL DAMPING, ROTATORY INERTIA AND SHEAR	63-0020
BEAMS	TRANSVERSE VIBRATIONS OF NON-UNIFORM BEAMS	74-1058
BEAMS	USE OF STRAIGHT BEAM FINITE ELEMENTS FOR ANALYSIS OF VIBRATIONS OF CURVED BEAMS	73-0616
BEAMS	VIBRATION OF BEAMS WITH OVERHANGS	73-0488
BEAMS	VIBRATION OF MULTISUPPORTED CURVED BEAMS	74-1297
BEAMS	VIBRATION OF NON-UNIFORM BEAMS	63-0021
BEANSPRUCHUNGSZ	DER SPANNUNGS UND BEANSPRUCHUNGSZUSTAND DES BODENMATERIALS IN DER NACHBARSCHAFT EINER BODENBOESCHUNG	75-0371
BEARBEITUNG	STRING EIN PROGRAMMSYSTEM FUR DIE MASCHINENUNABHANGIGE BEARBEITUNG VON ZEICHENKETTEN IN FORTRAN	68-0265
BEARING	A THERMO-ELASTIC ANALYSIS OF THE PARALLEL SURFACE THRUST BEARING	74-1069
BEARING	DYNAMIC BEHAVIOR OF HYDROSTATIC THRUST BEARING	75-0189
BEARING	ETUDE DE LA CAPACITE PORTANTE DES PIEUX PAR LA METHODE DES ELEMENTS FINIS. INFLUENCE DE LA DILATANCE	75-0624
BEARING	FINITE ELEMENT ANALYSIS OF A HYDRODYNAMIC BEARING	74-0773
BEARING	FINITE ELEMENT METHOD IN ROLLING BEARING ENGINEERING	73-0054
BEARING	FINITE ELEMENT OPTIMIZATION OF FINITE STEPPED SLIDER BEARING PROFILES	74-0131
BEARING	HIGHER ORDER FINITE ELEMENT METHODS FOR THE SOLUTION OF COMPRESSIBLE POROUS BEARING PROBLEMS	75-0341
BEARING	LA CAPACITE PORTANTE DES PIEUX. (BEARING CAPACITY OF PILES)	75-0463
BEARING	MECHANISM OF PARALLEL SURFACE THRUST BEARING	75-0322
BEARING	OIL-FILM TEMPERATURE DISTRIBUTION IN AN INFINITELY WIDE SLIDER BEARING: AN APPLICATION OF THE FINITE	73-1128
BEARING	SOLUTION OF THE ELASTOHYDRODYNAMIC FINITE JOURNAL BEARING PROBLEM	72-0033
BEARING	STRIP PROGRAM FOR DESIGNING LOAD BEARING SURFACE STRUCTURES (PROGRAMME STRIP POUR LE CALCUL DES STRU	71-0408
BEARING	THERMO-ELASTIC ANALYSIS OF THE PARALLEL SURFACE THRUST BEARING	74-1172
BEARINGS	APPLICATION OF FINITE ELEMENT METHOD TO HYDRODYNAMIC LUBRICATION PROBLEMS (PART 1, "INFINITE-WIDTH B	71-0537
BEARINGS	APPLICATION OF FINITE ELEMENT METHOD TO HYDRODYNAMIC LUBRICATION PROBLEMS (PART 2, " FINITE WIDTH BE	71-0538
BEARINGS	APPLICATION OF FINITE ELEMENT ANALYSIS TO HYDRODYNAMIC AND EXTERNALLY PRESSURIZED POCKET BEARINGS	72-0295
BEARINGS	FINITE ELEMENT METHOD FOR THE CALCULATION OF LOCUS PATHS IN DYNAMICALLY LOADED BEARINGS	73-0487
BEARINGS	HOLLOW-ENDED ROLLER. A SOLUTION FOR IMPROVING FATIGUE LIFE IN ASYMMETRICALLY LOADED CYLINDRICAL ROLL	71-0393
BEARINGS	NUMERICAL SIMULATION OF FINITE-WIDTH THRUST BEARINGS, TAKING INTO ACCOUNT VISCOSITY VARIATION WITH T	75-0623
BEARINGS	ON THE STABILITY OF ROTOR-AND-BEARING SYSTEMS AND ON THE CALCULATION OF SLIDING BEARINGS	72-0037
BEARINGS	ON THERMO-ELASTO-HYDRODYNAMIC FLUID FILM BEARINGS	75-0532
BEARINGS	RESEARCH NOTE: A THREE-DIMENSIONAL OIL FILM TEMPERATURE DISTRIBUTION IN TILTING THRUST BEARINGS	74-0608
BEARINGS	STABILITY OF TILTING-PAD JOURNAL BEARINGS	71-0687
BEARINGS	STUDY ON FOIL BEARINGS - AN APPLICATION TO TAPE MEMORY DEVICES	75-0311
BEARINGS	SULLA DETERMINAZIONE DELLE FREQUENZE PROPRIE DI ALBERI SU PIU SUPPORTI ELASTICI. (DETERMINATION OF F	74-0916
BEARINGS	THE EFFECT OF DEFORMATION ON THE BEHAVIOUR OF HYDRODYNAMIC JOURNAL BEARINGS	71-0009
BED	ANALYSIS OF MULTI-AXIAL ANISOTROPIC CREEP, ROOF RESIN BOLTS, AND BED SEPARATION IN ROCK STRUCTURES U	74-1122
BED	OFFSHORE GRAVITY BASED OIL PRODUCTION PLATFORM INTERACTION WITH THE SEA BED	75-0221
BEDEUTUNG	COMPUTER-UNTERSTUTZTES KONSTRUIEREN, PRAKTISCHE BEDEUTUNG DER FINIT-ELEMENT-VERFAHREN	71-0552
BEDEUTUNG	IMPORTANCE OF VARIATIONAL CALCULUS IN FLUID MECHANICS AND ITS APPLICATION TO COMPRESSIBLE POTENTIAL	71-1103
BEDINGTER	NUMERISCHE BERECHNUNG INSTATIONAERER TEMPERATURFELDER UND THERMISCH BEDINGTER VERFORMUNGEN VON WERKZ	75-0847
BEHAELTER	UEBER DAS VERHALTEN VON SPANNBETONDRUCKBEHAELTERRN IM GRENZTRAGFAEHIGKEITSZUSTAND UND DESSEN EINFLU	73-0855
BEHULP	PRAKTISCHE RESULTATEN, DOEL EN NUT VAN HET ONTWERPEN VAN MECHANISCHE ELEMENTEN MET BEHULP VAN COMPUT	73-0400
BELASTUNGSBEDIN	BERECHNUNGSMETHODEN ZUR ERMITTLUNG VON BELASTUNGSBEDINGUNGEN UND GRENZTRAGFAEHIGKEITSVERHALTEN VON K	73-0858
BELIEBIGEN	EIN VERFAHREN DUR DIE FORMFINDUNG VEON BELIEBIGEN VORGESPANNTEN NETZWERKKONSTRUKTIONEN	72-0641
BELIEBIGER	A FINITE ELEMENT METHOD FOR RANDOM FOLDED PLATE STRUCTURES (EINE FINITE ELEMENT METHODE ZUR BERECHNU	73-0787
BELIEBIGER	DAS PROGRAMMSYSTEM ANTRAS EDV-SYSTEM ZUR BERECHNUNG BELIEBIGER STATISCHER SYSTEME NACH DER METHODE D	74-0918
BELIEBIGER	EINE FINITE-ELEMENTE-LOESUNG FUR DIE TORSIONSSTEIFIGKEIT UND DEN SCHUBMITTELPUNKT BELIEBIGER QUERSCH	74-0256
BELT	PRESSURE DISTRIBUTION AND BELT DEFORMATION IN V-BELT DRIVES	74-1028
BELT	STUDY OF THE MHD INSTABILITIES OF CYLINDRICAL BELT PINCH USING THE FINITE ELEMENT METHOD	74-0700
BEMESSUNG	ZUR BERECHNUNG, BEMESSUNG UND BEWEHRUNG RECHTECKIGER PLATTEN KONSTNATER DICKE MIT OEFFNUNGEN ODER AU	73-0664
BENCHMARK	COMPARISON OF TYPICAL INELASTIC ANALYSIS PREDICTIONS WITH BENCHMARK PROBLEM EXPERIMENTAL RESULTS	75-0696
BEND	APPLICATION OF FINITE ELEMENT TECHNIQUES TO POST-YIELD ANALYSIS OF PROPOSED STANDARD 3 POINT BEND FR	74-0118
BEND	COMPARISON OF FINITE-ELEMENT AND EXPERIMENTAL STUDIES ON DEFORMATIONS OF ZIRCONIUM NOTCHED BEND SPEC	71-0359
BEND	HEAVY SECTION STEEL TECHNOLOGY PROGRAM TECHNICAL OR PROGRAMMATIC MANUSCRIPT NO. 25. VARIABLE THICKNE	73-0948
BENDING	A BENDING ANALYSIS FOR PLATES	66-0035
BENDING	A COMPARATIVE STUDY OF VARIABLE AND CONSTANT THICKNESS HIGH PRECISION TRIANGULAR PLATE BENDING ELEME	74-0857
BENDING	A COMPATIBLE PLATE BENDING ANNULAR SECTOR FINITE ELEMENT OF NONUNIFORM THICKNESS	74-0430
BENDING	A COMPATIBLE QUADRILATERAL PLATE BENDING ELEMENT	72-0482
BENDING	A COMPATIBLE TRIANGULAR PLATE BENDING FINITE ELEMENT	68-0209
BENDING	A COMPATIBLE TRIANGULAR PLATE BENDING FINITE ELEMENT	70-0385
BENDING	A COMPUTER PROGRAM FOR THE FINITE ELEMENT ANALYSIS OF A CLASS OF PLATE BENDING PROBLEMS	69-0191
BENDING	A CONFORMING FINITE ELEMENT FOR PLATE BENDING	68-0071
BENDING	A CONFORMING QUARTIC TRIANGULAR ELEMENT FOR PLATE BENDING	69-0137
BENDING	A CONFORMING TRIANGULAR FINITE ELEMENT PLATE BENDING SOLUTION	70-0020
BENDING	A CONFORMING TRIANGULAR ELEMENT IN THE ANALYSIS OF THIN AND SEMI-THICK PLATES IN BENDING	70-0308
BENDING	A DERIVATION PROCEDURE FOR THE DYNAMIC FLEXIBILITY MATRIX OF A TRIANGULAR BENDING ELEMENT	70-0103
BENDING	A DISCRETE-ELEMENT METHOD OF ANALYSIS FOR COMBINED BENDING AND SHEAR DEFORMATION OF A BEAM	69-0494
BENDING	A FAMILY OF EQUILIBRIUM PLATE BENDING ELEMENTS	74-0431
BENDING	A FINITE ELEMENT ANALYSIS FOR ' MODERATELY THICK ' RECTANGULAR PLATES IN BENDING	68-0146
BENDING	A FINITE ELEMENT APPROACH TO PLATE BENDING PROBLEMS	68-0296
BENDING	A FINITE ELEMENT FORMULATION FOR LARGE DEFLECTION ELASTIC STRUCTURAL ANALYSIS APPLIED TO PLATE BENDI	70-0485
BENDING	A FINITE ELEMENT METHOD FOR BENDING ANALYSIS OF LAYERED STRUCTURAL SYSTEMS	67-0134
BENDING	A FINITE ELEMENT METHOD FOR THE PLASTIC BENDING ANALYSIS OF STRUCTURES	68-0021
BENDING	A FINITE ELEMENT PLATE BENDING PROGRAMME	72-0579
BENDING	A FULLY COMPATIBLE PLATE BENDING ELEMENT	67-0142
BENDING	A GENERAL EQUILIBRIUM BENDING ELEMENT	73-0699
BENDING	A HIGH PRECISION TRIANGULAR PLATE BENDING ELEMENT	68-0052
BENDING	A NEW FAMILY OF CURVILINEAR PLATE BENDING ELEMENTS FOR VIBRATION AND STABILITY	72-0645
BENDING	A NEW FAMILY OF CURVILINEAR PLATE BENDING ELEMENTS FOR VIBRATION AND STABILITY	72-0999
BENDING	A NEW, FULLY COMPATIBLE FINITE ELEMENT FOR PLATE BENDING EIN NEUES VOLLVERTRAGLICHES ENDLICHES ELEME	68-0032
BENDING	A POLYGONAL FINITE ELEMENT FOR PLATE BENDING PROBLEMS USING THE ASSUMED STRESS APPROACH	69-0043
BENDING	A RECTANGULAR PLATE BENDING ELEMENTS THE USE OF WHICH IS EQUIVALENT TO THE USE OF THE FINITE DIFFERE	71-0398
BENDING	A REFINED MIXED-TYPE PLATE BENDING ELEMENT	69-0267
BENDING	A REFINED QUADRILATERAL ELEMENT FOR ANALYSIS OF PLATE BENDING	68-0047
BENDING	A REFINED TRIANGULAR PLATE BENDING FINITE ELEMENT	69-0056
BENDING	A SIMPLE ALGORITHM FOR THE STIFFNESS MATRIX OF TRIANGULAR PLATE BENDING ELEMENTS	71-0152
BENDING	A SIMPLE FINITE ELEMENT MODEL FOR ELASTIC-PLASTIC PLATE BENDING	72-0889
BENDING	A STIFFNESS MATRIX FOR THE ANALYSIS OF THIN PLATES IN BENDING	61-0001
BENDING	A TECHNIQUE FOR MESH GRADING APPLIED TO CONFORMING PLATE BENDING FINITE ELEMENTS	73-1058
BENDING	A TRIANGULAR ELEMENT FOR BENDING AND VIBRATION OF MULTILAYER SANDWICH PLATES	72-0489
BENDING	A TRIANGULAR EQUILIBRIUM ELEMENT WITH LINEARLY VARYING BENDING MOMENTS FOR PLATE BENDING PROBLEMS	67-0054
BENDING	A TRIANGULAR EQUILIBRIUM ELEMENT WITH LINEARLY VARYING BENDING MOMENTS FOR PLATE BENDING PROBLEMS	67-0054
BENDING	A TRIANGULAR FLAT PLATE BENDING ELEMENT	68-0252
BENDING	A TRIANGULAR PLATE BENDING ELEMENT FOR CONTACT PROBLEMS	73-0693
BENDING	AN ANNULAR SEGMENT FINITE ELEMENT FOR PLATE BENDING	71-0285
BENDING	AN EQUILIBRIUM MODEL FOR PLATE BENDING	68-0072
BENDING	AN INVESTIGATION OF THE BENDING OF SYMMETRIC PLATES UNDER SYMMETRIC TRANSVERSE LOADING	72-0662

BENDING	AN ITERATIVE IMPROVEMENT FOR NON-CONFORMING BENDING ELEMENTS	75-0844
BENDING	ANALYSIS OF PLATE BENDING BY THE FINITE ELEMENT METHOD	60-0003
BENDING	ANALYSIS OF PLATE BENDING USING TRIANGULAR ELEMENTS	62-0010
BENDING	ANALYSIS OF PLATE BENDING BY TRIANGULAR ELEMENTS	68-0145
BENDING	ANALYSIS OF PLATES IN BENDING USING LARGE FINITE ELEMENTS	70-0185
BENDING	ANALYSIS OF RECTANGULAR PLATES IN TRANSIENT CREEP BENDING	69-0425
BENDING	ANALYSIS OF THIN PLATES IN BENDING USING TRIANGULAR FINITE ELEMENTS	68-0213
BENDING	ANNULAR AND CIRCULAR SECTOR FINITE ELEMENTS FOR PLATE BENDING	70-0100
BENDING	APPLICATION DE LA METHODE DES ELEMENTS FINIS AU CALCUL DES PLAQUES FLECHIES (APPLICATION OF THE FINI	73-0019
BENDING	APPLICATION OF A REFINED PLATE BENDING ELEMENT TO BUCKLING PROBLEMS	74-0937
BENDING	APPLICATION OF THE HIGH PRECISION TRIANGULAR PLATE BENDING ELEMENT TO PROBLEMS WITH CURVED BOUNDARIE	69-0071
BENDING	ASYMMETRIC BENDING OF CIRCULAR PLATES USING THE FINITE ELEMENT METHOD	75-0175
BENDING	ASYMMETRICAL BENDING AND VIBRATION OF A CONICAL SHELL FINITE ELEMENT	74-0085
BENDING	BEAM BENDING FINITE ELEMENT CONTAINING INTEGRATION PARAMETERS	71-0179
BENDING	BENDING AND STRETCHING OF PLATES, SPECIAL MODELS FOR UPPER AND LOWER BOUNDS	66-0026
BENDING	BENDING AND TORSION OF THIN-WALLED STRAIGHT AND CURVED BEAMS WITH VARIABLE OPEN CROSS-SECTION	69-0493
BENDING	BENDING AND VIBRATION OF MULTILAYER SANDWICH BEAMS AND PLATES	73-0272
BENDING	BENDING ELEMENTS FOR PLATE AND SHELL NETWORKS	67-0043
BENDING	BENDING MOMENTS IN LONG WALL TANKS	67-0110
BENDING	BENDING OF CIRCULAR PLATES OF HARDENING MATERIAL	67-0112
BENDING	BENDING OF CYLINDRICAL SHELLS - SOLUTION USING CYCLINDRICAL STRIP ELEMENTS - CONTRIBUTION TO FINITE	70-0363
BENDING	BENDING OF MICROPOLAR PLATES	74-0263
BENDING	BENDING OF THIN PLATES - DISPLACEMENT MODELS	68-0049
BENDING	BENDING PROBLEM OF AXIALLY CONSTRAINED BEAMS	73-0397
BENDING	BOUNDING SOLUTIONS OF PLATE BENDING PROBLEMS	67-0084
BENDING	CERTAIN FINITE ELEMENTS FOR PLATE BENDING PROBLEMS	74-0395
BENDING	CLOUGH-TOCHER TRIANGULAR PLATE BENDING ELEMENT IN VIBRATION	69-0093
BENDING	COMPARISON OF FINITE ELEMENTS FOR PLATE BENDING	72-0625
BENDING	COMPARISON OF PLATE BENDING FINITE ELEMENT FORMULATIONS	70-0288
BENDING	COMPARISON OF SOME PLATE BENDING FINITE ELEMENTS	72-0413
BENDING	CONFORMING RECTANGULAR AND TRIANGULAR PLATE BENDING ELEMENTS	71-0720
BENDING	CONVERGENCE AND NUMERICAL ACCURACY WITH SPECIAL REFERENCE TO PLATE BENDING	69-0221
BENDING	CONVERGENCE OF A MIXED FINITE ELEMENT SCHEME FOR PLATE BENDING	73-0023
BENDING	CONVERGENCE OF A MIXED FINITE-ELEMENT METHOD FOR PLATE BENDING PROBLEMS	73-0107
BENDING	CONVERGENCE OF ANOTHER MIXED FINITE ELEMENT METHOD FOR PLATE BENDING PROBLEMS	72-0367
BENDING	CONVERGENCE OF EIGENVALUE SOLUTIONS IN CONFORMING PLATE BENDING FINITE ELEMENTS	72-0195
BENDING	CONVERGENCE OF SIMPLIFIED HYBRID DISPLACEMENT METHOD FOR PLATE BENDING	72-0412
BENDING	CREEP BENDING ANALYSIS OF PLATES BY THE FINITE ELEMENT METHOD	71-0415
BENDING	CREEP BENDING TESTS OF LARGE WELD OVERLAY AND BASE METAL SPECIMENS AT 1000 DEGREE F. (537.7 DEG. C)	75-0521
BENDING	CURVED BEAM FINITE ELEMENTS FOR COUPLED BENDING AND TORSIONAL VIBRATION	72-0720
BENDING	DEFLECTION FUNCTION FOR THE ASYMMETRICAL BENDING OF CIRCULAR PLATES	73-0468
BENDING	DERIVATION OF A STIFFNESS MATRIX FOR A RIGHT TRIANGULAR PLATE IN BENDING AND SUBJECTED TO INITIAL ST	62-0007
BENDING	DEVELOPMENT AND APPLICATION OF THE DIRECT STIFFNESS METHOD FOR OUT-OF-PLANE BENDING USING A TRIANGUL	62-0009
BENDING	DISCRETE TRIANGULAR APPROXIMATION OF MOMENT AND DISPLACEMENT SURFACES FOR PLATE BENDING	71-0269
BENDING	DISCUSSION OF "FINITE ELEMENT BENDING ANALYSIS OF REISSNER PLATES"	71-0826
BENDING	DISCUSSION OF "FINITE ELEMENT BENDING ANALYSIS OF REISSNER PLATES"	71-0832
BENDING	DISTRIBUTED MASS MATRIX FOR PLATE ELEMENTS IN BENDING	65-0018
BENDING	DISTRIBUTED MASS MATRIX FOR PLATE ELEMENT BENDING (DISCUSSION)	66-0038
BENDING	EFFECTIVENESS OF EXCESSIVE NODAL CONTINUITIES IN THE FINITE ELEMENT ANALYSIS OF THIN RECTANGULAR AND	70-0346
BENDING	ELASTIC BENDING OF CIRCULAR PLATES OF VARIABLE THICKNESS. AN ANALYTICAL AND EXPERIMENTAL STUDY	70-0163
BENDING	ELASTIC-PLASTIC PLATE BENDING WITH CONSTANT CURVATURE ELEMENTS	74-0308
BENDING	ELASTIC-PLASTIC STRESS ANALYSIS FOR A NOTCHED BAR IN PLANE STRAIN BENDING	71-0410
BENDING	EQUIVALENT MASS MATRICES FOR RECTANGULAR PLATES IN BENDING	66-0063
BENDING	EXPERIMENTAL AND THEORETICAL BENDING DEFLECTIONS OF LAMINATED PLATES	72-0144
BENDING	EXPLICIT TRIANGULAR BENDING ELEMENT MATRIX	74-0288
BENDING	FACTORS OF STRESS CONCENTRATION AT A HOLE IN FINITE STRIPS UNDER BENDING	72-0744
BENDING	FINITE ELEMENT ANALYSIS OF SKEW PLATES IN BENDING	68-0120
BENDING	FINITE ELEMENT ANALYSIS OF PLATE BENDING WITH TRANSVERSE SHEAR DEFORMATION	70-0109
BENDING	FINITE ELEMENT APPLICATIONS OF A PRECISION TRIANGULAR PLATE BENDING ELEMENT TO LINEAR BOUNDARY VALUE	75-0788
BENDING	FINITE ELEMENT APPROACH TO BENDING OF MICROPOLAR PLATES	74-0255
BENDING	FINITE ELEMENT BENDING ANALYSIS OF REISSNER PLATES	70-0214
BENDING	FINITE ELEMENT BENDING ANALYSIS OF MULTILAYER PLATES	74-0266
BENDING	FINITE ELEMENT DISPLACEMENT ANALYSIS OF PLATE BENDING BASED ON RECTANGULAR ELEMENTS	66-0032
BENDING	FINITE ELEMENT INCREMENTAL ANALYSIS OF ELASTOPLASTIC PLATE BENDING	73-0557
BENDING	FINITE ELEMENT METHOD FOR THE PLASTIC BENDING ANALYSIS OF STRUCTURES	68-0206
BENDING	FINITE ELEMENT METHOD FOR CALCULATING STRESS INTENSITY FACTORS FOR CRACKED PLATES IN BENDING	71-0424
BENDING	FINITE ELEMENT METHOD FOR PLATE BENDING ANALYSIS BY DECOMPOSITION OF DIFFERENTIAL OPERATOR	71-0558
BENDING	FINITE ELEMENT PLATE BENDING EQUILIBRIUM ANALYSIS	69-0045
BENDING	FINITE ELEMENT SOLUTION FOR SAINT-VENANT BENDING	71-0346
BENDING	FINITE ELEMENT SOLUTIONS FOR PLATE BENDING PROBLEMS BY SIMPLIFIED HYBRID DISPLACEMENT METHOD	72-0780
BENDING	FINITE ELEMENT STIFFNESS MATRICES FOR ANALYSIS OF PLATE BENDING	66-0021
BENDING	FINITE ELEMENTS FOR PLATE BENDING	70-0325
BENDING	FINITE-ELEMENT BENDING ANALYSIS FOR PLATES	67-0036
BENDING	FORMULATION OF A NEW TRIANGULAR PLATE BENDING ELEMENT	68-0051
BENDING	FURTHER APPLICATION OF QUADRATIC ISOPARAMETRIC FINITE-ELEMENTS TO LINEAR FRACTURE MECHANICS OF PLATE	75-0639
BENDING	HYBRID EFFECT ON BENDING BEHAVIORS OF CFRP-AL LAMINATE	74-1213
BENDING	HYBRID PLATE BENDING ELEMENTS	72-0792
BENDING	INADEQUACY OF NODAL CONNECTIONS IN A STIFFNESS SOLUTION FOR PLATE BENDING	65-0021
BENDING	INELASTIC ANALYSIS OF A STRAIGHT TUBE UNDER COMBINED BENDING, PRESSURE AND THERMAL LOADS	75-0218
BENDING	INELASTIC ANALYSIS OF A STRAIGHT TUBE UNDER COMBINED BENDING, PRESSURE AND THERMAL LOADS	75-0513
BENDING	INFLUENCE OF CRACK CLOSURE AND ELASTO-PLASTIC FLOW ON THE BENDING OF A CRACKED PLATE	75-0735
BENDING	KINDS OF CONVERGENCE AND IMPROVED CONVERGENCE OF CONFORMING FINITE ELEMENT SOLUTIONS IN PLATE BENDIN	70-0323
BENDING	MATRICE DE RIGIDITE D'UN ELEMENT DE POUTRE SOLLICITE A LA TORSION-FLEXION (STIFFNESS MATRIX OF A BEA	73-0442
BENDING	MEMBRANE AND PLATE BENDING ELEMENTS	75-0405
BENDING	MIXED FINITE ELEMENT ANALYSIS OF PLATES IN BENDING, SMALL DEFLECTION THEORY OF ELASTIC AND ELASTO-PL	71-0783
BENDING	MIXED FINITE ELEMENT ANALYSIS OF ELASTO-PLASTIC PLATES IN BENDING	72-0135
BENDING	MIXED FINITE ELEMENT METHOD FOR BENDING OF PLATES	75-0229
BENDING	MIXED QUADRILATERAL ELEMENTS FOR BENDING	72-0623
BENDING	MULTIANGULAR ELEMENTS IN PLATE BENDING	73-0435
BENDING	NONLINEAR IN-PLANE BENDING OF A PLANE STRESS-STRAIN FINITE ELEMENT MODEL	74-0211
BENDING	ON ASSUMED DISPLACEMENTS FOR THE RECTANGULAR PLATE BENDING ELEMENT	67-0020
BENDING	ON THE BENDING STIFFNESS OF SLABS IN CROSS-WALLED STRUCTURES	74-0398
BENDING	ON THE CONFORMING CUBIC TRIANGULAR ELEMENT FOR PLATE BENDING	72-0439
BENDING	ON THE QUINTIC TRIANGULAR PLATE BENDING ELEMENT	72-0677
BENDING	ONE-SIDED APPROXIMATION AND PLATE BENDING	74-0336
BENDING	PARALLELOGRAM ELEMENT IN THE SOLUTION OF RHOMBIC CANTILEVER PLATE BENDING	66-0022
BENDING	PART 1 - TASK 3.2.9. BENDING AND BUCKLING OF A ONE-DIMENSIONAL MODEL SIMULATING A CONICAL SHELL. PAR	71-0745
BENDING	PLATE BENDING ANALYSIS USING ASSUMED STRESS METHODS	68-0135
BENDING	PLATE BENDING WITH NONCONFORMING ELEMENT: FINE-MESH CONVERGENCE TEST	65-0049
BENDING	PROGRAM FOR TRIANGULAR BENDING ELEMENTS WITH DERIVATIVES SMOOTHING	72-0473
BENDING	RECTANGULAR FINITE ELEMENT FOR PLATE BENDING ANALYSIS BASED ON HELLINGER-REISSNERS VARIATIONAL PRINC	71-0545
BENDING	REFINED PLATE BENDING FINITE ELEMENT	74-1045

BENDING	RESIDUAL ENERGY BALANCING TECHNIQUE IN THE GENERATION OF PLATE BENDING FINITE ELEMENTS	74-0594
BENDING	SHEAR IN C AND C PLATE BENDING ELEMENTS	73-0692
BENDING	SIMPLE FINITE ELEMENT MODEL FOR ELASTIC-PLASTIC PLATE BENDING.	72-0447
BENDING	SIMPLIFIED INELASTIC (PLASTIC AND CREEP) ANALYSIS OF PIPE ELBOWS SUBJECTED TO INPLANE AND OUT-OF-PLA	75-0654
BENDING	SOME ELEMENTS FOR ANALYSIS OF PLATE BENDING	72-0626
BENDING	SOME FINITE ELEMENT SOLUTIONS FOR PLATE BENDING PROBLEMS BY SIMPLIFIED HYBRID DISPLACEMENT METHOD	72-0268
BENDING	SOME NONCONFORMING FINITE ELEMENTS FOR THE PLATE BENDING PROBLEM	75-0324
BENDING	STABILITY ANALYSIS OF THE TORSIONAL BENDING VIBRATIONS OF WING IN A SUBSONIC FLOW BY METHOD OF THE F	75-0755
BENDING	STATIC AND DYNAMIC APPLICATIONS OF A HIGH-PRECISION TRIANGULAR PLATE BENDING ELEMENT	69-0086
BENDING	STATIC FLEXIBILITY MATRICES FOR A RECTANGULAR PLATE ELEMENT IN BENDING, TWISTING, AND SHEAR. UNIFORM	68-0217
BENDING	STIFFNESS MATRICES FOR PLATE BENDING ELEMENTS	69-0129
BENDING	STIFFNESS MATRIX FOR BENDING OF A RECTANGULAR PLATE ELEMENT WITH INITIAL MEMBRANE STRESSES	62-0006
BENDING	STIFFNESS MATRIX FOR A TRIANGULAR SANDWICH ELEMENT IN BENDING	67-0050
BENDING	STRESSES DUE TO CONCENTRATED LOADS IN RESTRAINED BENDING TESTS	67-0097
BENDING	SYMMETRICAL BENDING OF CIRCULAR PLATES USING FINITE ELEMENTS	72-0428
BENDING	THE ANALYSIS OF CLASSICAL PLATE BENDING PROBLEMS USING AN EQUILIBRIUM FINITE ELEMENT MODEL	68-0302
BENDING	THE CONSTANT BENDING MOMENT PLATE BENDING ELEMENT	71-0182
BENDING	THE CONSTANT BENDING MOMENT PLATE BENDING ELEMENT	71-0182
BENDING	THE EFFECTIVENESS OF EXCESSIVE NODAL CONTINUITIES IN THE FINITE ELEMENT ANALYSIS OF THIN RECTANGULAR	70-0105
BENDING	THE TRIANGULAR EQUILIBRIUM ELEMENT IN THE SOLUTION OF PLATE BENDING PROBLEMS	68-0122
BENDING	THREE DIMENSIONAL FINITE ELEMENT METHOD FOR PLATE BENDING	74-0243
BENDING	TRIANGULAR ELEMENTS IN PLATE BENDING CONFORMING AND NON-CONFORMING SOLUTIONS	66-0011
BENDING	TRIANGULAR FINITE ELEMENTS FOR PLATE BENDING WITH CONSTANT AND LINEARLY VARYING BENDING MOMENTS	70-0019
BENDING	TRIANGULAR FINITE ELEMENTS FOR PLATE BENDING WITH CONSTANT AND LINEARLY VARYING BENDING MOMENTS	70-0019
BENDING	TRIANGULAR FINITE ELEMENTS FOR THE BENDING ANALYSIS OF THICK ELASTIC PLATES	70-0466
BENDING	TRIANGULAR FINITE ELEMENT FOR PLATE BENDING ANALYSIS WITH TRANSVERSE SHEAR DEFORMATION	71-0555
BENDING	TRIANGULAR PLATE BENDING ELEMENTS	69-0057
BENDING	TRIANGULAR PLATE BENDING ELEMENT WITH ENFORCED COMPATIBILITY	71-0593
BENDING	TRIANGULAR THICK PLATE BENDING ELEMENTS	71-0167
BENDING	TRIGONOMETRIC FUNCTION REPRESENTATIONS FOR RECTANGULAR PLATE BENDING ELEMENTS	71-0765
BENDING	TWISTED BEAM ELEMENT MATRICES FOR BENDING	70-0082
BENDING	ZUR BERECHNUNG DER BIEGESTEIFIGKEIT ABGESETZTER STAEBE UND WELLEN UNTER ANWENDUNG VON FINITEN ELEMEN	75-0481
BENDING-EXTENSI	FINITE ELEMENT ANALYSIS OF BENDING-EXTENSIONAL COUPLING IN LAMINATED COMPOSITES	70-0292
BENT	ANALYSIS OF BENT AND TWISTED BARS USING THE FINITE ELEMENT METHOD	67-0045
BENT	VIERENDEEL BENT CAPS	72-0758
BENTONITE	BENTONITE PILES IN DURBAN	75-0797
BENUTZER	TURBAN VERSION 1.0. EIN SPEZIELLES PROGRAMM ZUR LINEAREN UND NICHTLINEAREN STATISCHEN UND DYNAMISCHE	71-0622
BENUTZERHAND	INGA - INTERAKTIVE GRAPHISCHE ANALYSE BENUTZERHAND	73-0701
BERECHNUNGEN	TRAGLAST VON T-TRAEGERN. EXPERIMENT UND F.E. BERECHNUNGEN (ULTIMATE STRENGTH OF T-BEAMS. EXPERIMENTS	74-0935
BERECHNUNGSBEIS	BERECHNUNGSBEISPIELE UND ERFAHRUNGEN MIT DER METHODE DER FINITEN ELEMENTE IN MASCHINENBAU. (EXAMPLES	71-0390
BERECHNUNGSMETH	BERECHNUNGSMETHODEN ZUR ERMITTLUNG VON BELASTUNGSBEDINGUNGEN UND GRENZTRAGFAEHIGKEITSVERHALTEN VON K	73-0858
BERECHUNG	UEBER DAS VERHALTEN VON SPANNEBETONDRUCKBEHAELTRERN IM GRENZTRAGFAEHIGKEITSZUSTAND UND DESSEN EINFUL	73-0855
BERGEN	BERGEN, AN AUTOMATIC FINITE ELEMENT MESH GENERATION PROGRAM FOR ARBITRARY STRUCTURES	72-0776
BERSAFE	BERSAFE - A COMPUTER SYSTEM FOR STRESS ANALYSIS BY FINITE ELEMENTS	70-0212
BERSAFE	BERSAFE FINITE ELEMENT SYSTEM	74-0190
BERSAFE	STRESS ANALYSIS OF BWR PRESSURE VESSEL WITH BERSAFE AND FLHE	73-0406
BERSAFE	THE APPLICATION OF THE BERSAFE FINITE ELEMENT SYSTEM TO NUCLEAR DESIGN PROBLEMS	71-0117
BERUCKSICHTIGUN	ANWENDUNG DER METHODEN DER FINITEN ELEMENTE MIT DEM PROGRAMMSYSTEM ASKA UNTER BESONDERER BERUCKSICHT	71-0624
BERUCKSICHTIGUN	EINIGE VERFAHREN ZUR BERECHNUNG VON EIGENWERTEN UND EIGENVEKTOREN VON NICHT HERMITESCHEN MATRIZEN U	72-0646
BERUECKSICHTIGU	ZUR BERECHNUNG EINFACH-SYMMETRISCHER I-TRAEGER NACH DER THEORIE - 2. ORDNUNG UNTER BERUECKSICHTIGUNG	75-0510
BESCHAUFELUNG	SCHWINGUNGSMAESSIGE AUSLEGUNG DER BESCHAUFELUNG AXIALER TURBOMASCHINEN (DESIGN OF DISPLACEMENT-FREE	74-0253
BESONDERER	ANWENDUNG DER METHODEN DER FINITEN ELEMENTE MIT DEM PROGRAMMSYSTEM ASKA UNTER BESONDERER BERUCKSICHT	71-0624
BESONDERER	EINIGE VERFAHREN ZUR BERECHNUNG VON EIGENWERTEN UND EIGENVEKTOREN VON NICHT HERMITESCHEN MATRIZEN U	72-0646
BESSEL	TRIANGULAR FINITE ELEMENTS FOR THE GENERALIZED BESSEL EQUATION OF ORDER M	73-1094
BESTRAN	BESTRAN: A TECHNIQUE FOR PERFORMING STRUCTURAL ANALYSES	75-0759
BETHESDA	PROCEEDINGS OF THE 5TH NAVY-NASTRAN COLLOQUIUM, HELD AT NAVAL SHIP RESEARCH AND DEVELOPMENT CENTER,	74-0709
BETON	ANALYSE TRIDIMENSIONALE DU COMPORTEMENT NON-LINEAIRE D'UN CAISSON DE REACTEUR NUCLEAIRE EN BETON PRE	71-0144
BETON	ETUDE THEORIQUE DU COMPORTEMENT D'UNE DALLE DE BETON. (THEORETICAL STUDY OF THE BEHAVIOR OF A CONCRE	75-0273
BETON	NONLINEAR ANALYSIS OF PRESTRESSED CONCRETE FAST REACTOR PRESSURE VESSEL. (ANALYSE NON LINEAIRE DES C	70-0145
BEULLASTEN	DIE BEULLASTEN DIAGONAL VERSTEIFTER RECHTECKPLATTEN (BUCKLING LOADS IN DIAGONALLY STIFFENED RECTANGU	73-0405
BEURTEILUNG	BERECHNUNG UND BEURTEILUNG VON DER STANDSICHERHEIT VON ABBAUHOHLRAEUMEN IN STEILEN GANGERZLAGERSTAETTEN	74-0597
BEURTEILUNG	NEUARTIGE ANWENDUNG GEOMECHANISCHER UNTERSUCHUNGSMETHODEN ZUR BEURTEILUNG DER ABBAUWIRKUNGEN IM GANG	75-0530
BEWEGUNGEN	DIE BERECHNUNG DER BEWEGUNGEN UND SPANNUNGEN DES GEBIRGES VOM ABBAU BIS ZUR TAGESOBERFLAECHE NACH DE	73-0456
BEWEHRUNG	ZUR BERECHNUNG, BEMESSUNG UND BEWEHRUNG RECHTECKIGER PLATTEN KONSTANTER DICKE MIT OEFFNUNGEN ODER AU	73-0664
BEWEHRUNGSSTAEH	ZUSAMMENHANG ZWISCHEN OBERFLAECHENBESCHAFFENHEIT, VERBUND UND SPRENGWIRKUNG VON BEWEHRUNGSSTAEHLEN U	73-0450
BEZUGNAHME	UEBER DAS VERHALTEN VON SPANNEBETONDRUCKBEHAELTRERN IM GRENZTRAGFAEHIGKEITSZUSTAND UND DESSEN EINFUL	73-0855
BGK	FINITE ELEMENT ANALYSIS OF THE LINEARIZED BGK BOLTZMANN EQUATION.	72-0408
BI-DIMENSIONAL	DETERMINATION OF EIGENMODES OF HOMOGENEOUS PLASMA WITH ELLIPTICAL SECTION USING BI-DIMENSIONAL FINIT	74-0553
BI-MATERIAL	BEHAVIOR OF A BI-MATERIAL CYLINDER INCLUDING THE EFFECTS OF FRACTURE	69-0466
BI-MATERIAL	FINITE ELEMENT ANALYSIS OF CRACKS IN REGION OF A BI-MATERIAL INTERFACE	74-0164
BI-MATERIAL	FINITE ELEMENT STRESS ANALYSIS OF A CRACK IN A BI-MATERIAL PLATE	72-0165
BI-MATERIAL	THE ELASTIC STRESS ANALYSIS OF A BI-MATERIAL PLATE WITH A CRACK NORMAL TO THE INTERFACES	72-0819
BIAXIAL	AN ELASTIC-PLASTIC FINITE ELEMENT ANALYSIS OF CRACK TIP FIELDS UNDER BIAXIAL LOADING CONDITIONS	73-0760
BIAXIAL	ANALYSIS OF THE BIAXIAL STRIP AND SHEAR LAP TESTS FOR SOLID PROPELLANT CHARACTERIZATION	67-0019
BIAXIAL	BIAXIAL STRESS-STRAIN RELATIONS FOR CONCRETE	72-0203
BIAXIAL	CRUCIFORM SPECIMENS FOR BIAXIAL FATIGUE TESTS - AN INVESTIGATION USING FINITE-ELEMENT ANALYSIS AND P	71-0795
BIAXIAL	ELASTIC-PLASTIC FINITE ELEMENT ANALYSIS OF CRACK TIP FIELDS UNDER BIAXIAL LOADING CONDITIONS	74-0296
BIAXIAL	INELASTIC MODEL FOR CYCLIC BIAXIAL LOADING OF REINFORCED CONCRETE	74-0503
BIAXIAL	NEW MATERIAL MODEL FOR THE NONLINEAR BIAXIAL BEHAVIOR OF ATJ-S GRAPHITE	75-0063
BIAXIAL	NONLINEAR STRESS ANALYSIS OF INDETERMINATE STRUCTURAL ELEMENTS SUBJECTED TO BIAXIAL STRESS STATES US	73-1152
BIAXIAL	NUMERICAL BIAXIAL CHARACTERIZATION FOR CONCRETE	74-0641
BIAXIAL	STRESS STRAIN RESPONSE AND FRACTURE OF A CONCRETE MODEL IN BIAXIAL LOADING	71-0295
BIAXIALLY	FINITE DEFORMATION OF BIAXIALLY LOADED COLUMNS	72-0021
BIBLIOGRAPHY	A BIBLIOGRAPHY FOR FINITE ELEMENT METHODS	72-0339
BIBLIOGRAPHY	A FINITE ELEMENT BIBLIOGRAPHY - PART I: AUTHOR LISTING	74-0284
BIBLIOGRAPHY	A FINITE ELEMENT BIBLIOGRAPHY - PART II: KEYWORD LISTING	74-0285
BIBLIOGRAPHY	A FINITE ELEMENT BIBLIOGRAPHY - PART III: CITATION LISTING	74-0286
BIBLIOGRAPHY	STRUCTURAL MECHANICS SOFTWARE. A BIBLIOGRAPHY WITH ABSTRACTS	74-0842
BIBLIOGRAPHY	THE FINITE ELEMENT METHOD - A BIBLIOGRAPHY OF ITS THEORY AND APPLICATIONS	71-0510
BIBLIOGRAPHY	TUNNEL CONSTRUCTION - A BIBLIOGRAPHY WITH ABSTRACTS	74-0748
BICONICAL	APPLICATION OF A UNIMOMENT TECHNIQUE TO A BICONICAL ANTENNA WITH INHOMOGENEOUS DIELECTRIC LOADING	75-0375
BICRYSTAL	STRAIN DISTRIBUTION IN AN ALUMINA BICRYSTAL STUDIES BY THE FINITE ELEMENT METHOD AND X-RAY DIFFRACTI	74-1070
BICUBIC	BICUBIC INTERPOLATION OVER TRIANGLES	69-0367
BIDIMENSIONAL	BIDIMENSIONAL COMPUTER-PROGRAM FOR CALCULATION OF GROWTH-RATES OF INCLUDED AXIALLY SYMMETRICAL PLASM	75-0420
BIDIMENSIONALE	ANALISI NUMERICA DEL COMPORTAMENTO ELASTICO DI UNA STRUTTURA BIDIMENSIONALE PIANA. (NUMERICAL ANALYS	74-1029
BIDIMENSIONNELL	ELEMENTS FINIS RAFFINES EN ELASTICITE BIDIMENSIONNELLE	69-0411
BIEGESTEIFIGKEI	ZUR BERECHNUNG DER BIEGESTEIFIGKEIT ABGESETZTER STAEBE UND WELLEN UNTER ANWENDUNG VON FINITEN ELEMEN	75-0481
BIFURCATION	BIFURCATION OF ECCENTRICALLY LOADED MEMBERS AND OF STIFFENED PLATES	70-0057
BIFURCATION	BIFURCATION OF SPACE FRAMES	70-0055
BIFURCATION	BUCKLING OF COOLING-TOWER SHELLS: BIFURCATION RESULTS	75-0384
BIFURCATION	STRUCTURAL MODEL STUDIES OF LARGE PIPE BIFURCATION	73-0283
BIFURCATIONS	STRUCTURAL ANALYSIS OF SYMMETRICAL PENSTOCK BIFURCATIONS BY THE FINITE ELEMENT METHOD	71-0310

137

BIHARMONIC	A MIXED FINITE ELEMENT METHOD FOR THE BIHARMONIC EQUATION	74-1227
BIHARMONIC	DUAL ITERATIVE TECHNIQUES FOR SOLVING A FINITE ELEMENT APPROXIMATION OF THE BIHARMONIC EQUATION	75-0480
BIHARMONIC	NUMERICAL SOLUTION OF THE BIHARMONIC PROBLEM BY MIXED FINITE OF CLASS C-0	73-0803
BIHARMONIC	SOME DIFFERENCE SCHEMES FOR THE BIHARMONIC EQUATION	75-0204
BIHARMONIQUE	APPROXIMATIONS EXTERNES, PAR ELEMENTS FINIS DE LAGRANGE D ORDRE, UN ET DEUX, DU PROBLEME DE DIRICHLE	72-0495
BILATERALLY	DER GEPISSENE ZUSTAND ZWEISEITIG GELAGERTER PLATTEN UNTER EINZELLASTEN - NICHTLINEARE BERECHNUNG MIT	75-0858
BILDSCHIRMUNTER	PARAMETRISCHE UNTERSUCHUNGEN VON KERBEN MIT DER MATRIZENVERSCHIEBUNGSMETHODE UND BILDSCHIRMUNTERSTUT	72-0638
BILINEAR	STATISTICAL ASPECTS OF THE STABILITY OF COMPOUND SYSTEMS WITH BILINEAR ELEMENTS	73-0986
BILINEAR	THERMAL DEFORMATION VECTOR FOR A BILINEAR TEMPERATURE DISTRIBUTION IN AN ANISOTROPIC QUADRILATERAL M	75-0670
BILLETS	TEMPERATURE RISE IN THE HIGH SPEED COMPRESSION OF RIGHT CYLINDRICAL BILLETS	74-1153
BIN	DYNAMIC FINITE ELEMENT ANALYSIS OF AXI-SYMMETRIC STRESS WAVE PROPAGATION IN SOIL-FILLED BIN	72-0622
BIN	GRAVITY FLOW BIN DESIGN FOR LUNAR SOIL	72-0240
BING	MISSILE BODY INPUT GENERATOR (BING), A NASTRAN PRE-PROCESSOR: THEORETICAL DEVELOPMENT, USER'S MANUAL	75-0717
BINGHAM	CALCUL NUMERIQUE DES ECOULEMENTS DES FLUIDES DE BINGHAM ET DES FLUIDES NEWTONIENS INCOMPRESSIBLES PA	72-0349
BIODYNAMIC	MODELING AND ANALYSIS TECHNIQUES FOR PREDICTION OF STRUCTURAL AND BIODYNAMIC CRASH-IMPACT RESPONSE	75-0020
BIOLOGICAL	ANALYSIS OF BIOLOGICAL STRUCTURES	71-0148
BIOMECHANICS	A FINITE ELEMENT METHOD FOR LOW REYNOLDS NUMBER FLOW AND ITS APPLICATION TO BIOMECHANICS	69-0304
BIOMECHANICS	FINITE ELEMENT ANALYSES IN BIOMECHANICS	72-0741
BIOMECHANICS	SLOW PARTICULATE VISCOUS FLOW IN CHANNELS AND TUBES, APPLICATION TO BIOMECHANICS	71-0248
BIOMECHANICS	SLOW VISCOUS FLOW AND ITS APPLICATION TO BIOMECHANICS	71-0619
BIOPHYSICAL	A LEAST SQUARES FINITE ELEMENT ANALYSIS OF NONLINEAR BIOPHYSICAL DIFFUSION-KINETICS EQUATIONS	74-1257
BIOT'S	A NUMERICAL METHOD FOR THE SOLUTION OF BIOT'S CONSOLIDATION THEORY	72-0688
BIOT'S	AN INVESTIGATION OF THE STABILITY OF NUMERICAL SOLUTIONS OF BIOT'S EQUATIONS OF CONSOLIDATION	74-0746
BIOT'S	FINITE ELEMENT METHOD APPLIED TO BIOT'S CONSOLIDATION THEORY	71-0518
BIT	BIT PENETRATION INTO ROCK - A FINITE ELEMENT STUDY	75-0800
BITUMINOUS	FORECASTING THE ROCK PRESSURE AND ITS EFFECT ON THE LONGWALL FACES AND ROADWAYS IN THE BITUMINOUS CO	74-0374
BLADE	A FINITE CYLINDRICAL SHELL ELEMENT AND THE VIBRATIONS OF A CURVED FAN BLADE	68-0131
BLADE	APPLICATION OF THE FINITE ELEMENT METHOD TO PREDICT STATIC AND DYNAMIC RESPONSE OF AN UNSHROUDED CEN	71-0574
BLADE	DEFLECTION AND VIBRATION ANALYSIS OF A FLEXIBLE FAN BLADE	73-0556
BLADE	FINITE ELEMENT ANALYSIS OF A GAS TURBINE BLADE	74-0971
BLADE	FINITE ELEMENT STRESS ANALYSIS OF TURBINE BLADE HOOKS	70-0479
BLADE	NASTRAN ANALYSIS OF A TURBINE BLADE AND COMPARISON WITH TEST AND FIELD DATA	75-0367
BLADE	PROPELLER BLADE STRESSES. APPLICATION OF FINITE ELEMENT METHODS	74-0207
BLADE	ROTOR BLADE VIBRATIONS BY GALERKIN FINITE-ELEMENT METHOD	75-0389
BLADE	SENSITIVITY OF ROTOR BLADE VIBRATION CHARACTERISTICS TO TORSIONAL OSCILLATIONS	74-0611
BLADE	STRESS ANALYSIS OF A CERAMIC BLADE DESIGN SUBJECTED TO CENTRIFUGAL AND THERMAL LOADS	73-0790
BLADED	VIBRATION ANALYSIS OF PLATES, SHELLS AND BLADED RIMMED DISCS USING THE TRANSFER-MATRIX-FINITE-ELEMEN	73-1151
BLADES	CALCULATION OF THE NATURAL FREQUENCIES AND THE PRINCIPAL MODES OF HELICOPTER BLADES	72-0907
BLADES	CERAMIC ROTATING BLADES: SOME CRITICAL DESIGN PARAMETERS FOR GAS TURBINE APPLICATIONS	74-0146
BLADES	FINITE ELEMENT ANALYSIS OF COUPLED VIBRATION OF TAPERED TWISTED BLADES	72-0766
BLADES	FREQUENCY TUNING OF COMPRESSOR ROTOR BLADES	66-0112
BLADES	HOLOGRAPHIC INSTRUMENTATION OF TURBINE BLADES	73-0466
BLADES	MARINE PROPELLER BLADES STRESSES AND DEFORMATIONS WITH GIVEN, FROZEN, HYDRODYNAMIC LOADING	73-0474
BLADES	MODELING OF GAS TURBINE ENGINE COMPRESSOR BLADES FOR VIBRATION ANALYSIS	75-0508
BLADES	NATURAL FREQUENCIES OF WIDE CHORD COMPRESSOR BLADES	74-0231
BLADES	PSEUDO STATIC DEFORMATION AND FREQUENCIES OF ROTATING TURBOMACHINERY BLADES	72-0239
BLADES	SCHWINGUNGSMAESSIGE AUSLEGUNG DER BESCHAUFELUNG AXIALER TURBOMASCHINEN (DESIGN OF DISPLACEMENT-FREE	74-0253
BLADES	SENSITIVITY ANALYSIS OF TORSIONAL VIBRATION CHARACTERISTICS OF HELICOPTER ROTOR BLADES. PART 1: STRU	74-0846
BLADES	STRESS AND VIBRATION ANALYSIS OF TURBOPUMP INDUCER BLADES BY FINITE ELEMENTS	71-0363
BLADES	STRESSES AND VIBRATION ANALYSIS OF INDUCER BLADES USING FINITE ELEMENT TECHNIQUE	70-0305
BLADES	THE EFFECT OF CAMBER ON THE NATURAL FREQUENCIES OF LOW ASPECT RATIO TURBOMACHINERY BLADES	73-1092
BLADES	VIBRATION ANALYSIS OF ROTATING COMPRESSOR BLADES	73-0417
BLADES	VIBRATION ANALYSIS OF ROTATING TURBINE BLADES	74-0959
BLADES	VIBRATION OF THICK, CURVED, SHELLS WITH PARTICULAR REFERENCE TO TURBINE BLADES	70-0017
BLADING	SIMPLE FINITE ELEMENTS FOR PRE-TWISTED BLADING VIBRATION	74-0239
BLADINGS	CONTRIBUTION A L'ANALYSE DES VIBRATIONS LIBRES D'UNE STRUCTURE. APPLICATION AUX AUBAGES DE TURBOMACH	75-0804
BLAST	A NONLINEAR FINITE ELEMENT CODE FOR ANALYZING THE BLAST RESPONSE OF UNDERGROUND STRUCTURES	70-0054
BLAST	FINITE ELEMENT SOLUTION OF THE INTERACTION OF A PLANE ACOUSTIC BLAST WAVE AND A CYLINDRIC STRUCTURE	74-0790
BLAST	MAGAZINE HEADWALL RESPONSE TO EXPLOSIVE BLAST	74-0754
BLAST	RESPONSE OF STRUCTURES TO COMBINED BLAST EFFECTS	73-0424
BLAST	THE ELASTO-PLASTIC AND LARGE-DISPLACEMENT RESPONSE OF PLATES TO BLAST LOADING	73-0731
BLAST	USER'S MANUAL FOR NTEDYNS - A NONLINEAR FINITE ELEMENT CODE FOR ANALYZING THE BLAST RESPONSE OF UNDE	71-0695
BLENDED	LOCAL MESH REFINEMENT USING RECTANGULAR BLENDED FINITE-ELEMENTS	75-0452
BLENDING	BLENDING FUNCTION FINITE ELEMENTS FOR CURVED BOUNDARIES	75-0093
BLENDING	BLENDING FUNCTIONS AND FINITE ELEMENTS	74-0889
BLENDING	TRANSFINITE ELEMENT METHODS: BLENDING FUNCTION INTERPOLATION OVER ARBITRARY CURVED ELEMENT DOMAINS	73-1006
BLOCH-FOURIER	STRESS-WAVE PROPAGATION AND DYNAMIC EFFECTS IN SHELLS (SPADES) MICROMECHANICAL STRESS-WAVE RESPONSE	72-0851
BLOCK	AN INVESTIGATION INTO THE USE OF FINITE ELEMENT METHODS FOR ANALYSING STRESS DISTRIBUTIONS IN BLOCK	71-0530
BLOCK	BLOCK ELIMINATION ON FINITE ELEMENT SYSTEMS OF EQUATIONS	72-0530
BLOCK	FINITE ELEMENT ANALYSIS OF CRACK INITIATION IN A BLOCK MODEL EXPERIMENT	72-0455
BLOCK	FINITE ELEMENT METHOD FOR BLOCK ADJUSTMENT PROBLEMS OF PHOTOGRAMMETRY	73-0264
BLOCK	NON-UNIFORM DEFORMATION OF A BLOCK IN PLANE-STRAIN COMPRESSION CAUSED BY FRICTION. PART 4, ANALYSIS	71-0786
BLOCK	ON THE NON-UNIFORM DEFORMATION OF A BLOCK IN PLANE-STRAIN COMPRESSION CAUSED BY FRICTION -3	71-0328
BLOCK	PLANE STRESS ELEMENT OF VARYING THICKNESS APPLIED TO END BLOCK ANALYSIS	74-0396
BLOCK	PREDICTING THE BLOCK CAVING BEHAVIOUR OF OREBODIES	69-0044
BLOCK	STRESS ANALYSIS OF PRESSURIZED CYLINDRICAL BORE IN RECTANGULAR BLOCK	69-0312
BLOCK	STRUCTURAL DESIGN DATA FOR CONCRETE DRIFT LININGS IN BLOCK CAVING	73-0773
BLOCK-BRAKED	UNTERSUCHUNGEN ZUR SPURMASSTABILITAET BEI RADSAETZEN MIT KLOTZGEBREMSTEN VOLLRAEDERN (INVESTIGATIONS	75-0810
BLOCKING	BLOCKING TECHNIQUE FOR LARGE SCALE STRUCTURAL ANALYSIS	73-0243
BLOCKS	ANCHORAGE ZONE STRESSES IN AXIALLY POST-TENSIONED I-SECTION MEMBERS WITH END BLOCKS	71-0305
BLOCKS	FINITE ELEMENT ANALYSIS OF ALTERNATIVE AXIAL LOADING OF AN ELASTIC PLATE PRESSED BETWEEN TWO ELASTIC	71-0357
BLOCKS	FINITE-ELEMENT ANALYSES OF INITIAL ELASTIC DEFORMATION OF VERTICALLY DEFORMED CRYSTAL BLOCKS	72-0970
BLOCKS	STRESS DISTRIBUTIONS AROUND PLUG MATCHES WITH ANCHOR BLOCKS	69-0135
BLOOD	FINITE-ELEMENT ANALYSIS OF FLOW IN BLOOD VESSEL WITH ARBITRARY CROSS-SECTION	72-0936
BLOWDOWN	BLOWDOWN EFFECTS ON A FUEL ROD	75-0575
BLUNTING	EFFECT OF PLASTICITY AND CRACK BLUNTING ON THE STRESS DISTRIBUTION IN ORTHOTROPIC COMPOSITE MATERIAL	73-0145
BOARDS	APPLICABILITY OF APPROXIMATE AND EXACT TRANSIENT HEAT TRANSFER ANALYSES TO HEATING PROCESSES USED TO	75-0244
BODENBOESCHUNG	DER SPANNUNGS UND BEANSPRUCHUNGSZUSTAND DES BODENMATERIALS IN DER NACHBARSCHAFT EINER BODENBOESCHUNG	75-0371
BODENMATERIALS	DER SPANNUNGS UND BEANSPRUCHUNGSZUSTAND DES BODENMATERIALS IN DER NACHBARSCHAFT EINER BODENBOESCHUNG	75-0371
BODIES	A FINITE ELEMENT ANALYSIS OF SHOCK AND FINITE-AMPLITUDE WAVES IN ONE-DIMENSIONAL HYPERELASTIC BODIES	72-0731
BODIES	A FINITE ELEMENT ANALYSIS OF BODIES OF REVOLUTION. A COMPARISON BETWEEN CURRENTLY USED TYPES OF ELEM	74-1319
BODIES	A METHOD FOR THE TRANSIENT NONLINEAR COUPLED THERMOELASTIC ANALYSIS OF AXISYMMETRIC BODIES USING FIN	74-1140
BODIES	ACCURATE CALCULATION OF STRESSES IN BEAMS, DISKS AND PLATES BY MEANS OF FINITE ELEMENT METHODS AND I	70-0133
BODIES	ANALYSIS OF AXISYMMETRIC COMPOSITE BODIES BY THE FINITE ELEMENT METHOD	68-0242
BODIES	APPLICATION OF THE FINITE ELEMENT METHOD TO PROBLEMS OF TEMPERATURE DISTRIBUTION ANALYSIS FOR BODIES	74-1117
BODIES	BASIC CONCEPTS OF THE MECHANICS OF DISCRETIZED BODIES WITH AN INTRODUCTION TO DISCRETE ELEMENT CALCU	74-0146
BODIES	CALCULATION OF POTENTIAL FLOW ABOUT ARBITRARY THREE-DIMENSIONAL LIFTING BODIES	72-0722
BODIES	COMBINED ELASTIC-PLASTIC CREEP ANALYSIS OF TWO-DIMENSIONAL BODIES	71-0232
BODIES	COMPUTING SYSTEM FOR STRUCTURAL ANALYSIS OF CAR BODIES	72-0324
BODIES	CONTRIBUTION TO SOLUTION OF BODIES WITH CURVED BOUNDARY BY FINITE ELEMENT METHOD	72-1001
BODIES	CREEP ANALYSIS OF AXISYMMETRIC BODIES	68-0081
BODIES	CREEP ANALYSIS OF TRANSVERSELY ISOTROPIC BODIES SUBJECTED TO TIME DEPENDENT LOADING	73-0170

BODIES	DIGITAL COMPUTER PROGRAMS FOR STEADY-STATE OR TRANSIENT TEMPERATURE ANALYSIS OF PLANE AND AXISYMMETR	72-0915
BODIES	DISCRETE APPROXIMATIONS OF ELASTIC-PLASTIC BODIES BY VARIATIONAL METHODS	72-0710
BODIES	DYNAMIC ANALYSIS OF A SYSTEM OF HINGE-CONNECTED RIGID BODIES WITH NONRIGID APPENDAGES	74-0852
BODIES	FINITE ELEMENT METHOD FOR STRESS ANALYSIS OF BODIES OF REVOLUTION SUBJECTED TO AXISYMMETRIC LOADING	71-0409
BODIES	FINITE ELEMENT METHOD FOR CONTACT PROBLEMS OF SOLID BODIES - 2.	71-0435
BODIES	FINITE ELEMENT METHODS FOR ELASTIC BODIES CONTAINING CRACKS	73-1029
BODIES	FINITE ELEMENT STRESS ANALYSIS OF ANISOTROPIC BODIES	68-0033
BODIES	FORCE AND MOMENTS ON ASYMMETRIC AND YAWED BODIES IN A FREE SURFACE	73-0908
BODIES	FORMULATION AND APPLICATION OF CERTAIN PRIMAL AND MIXED FINITE ELEMENT MODELS OF FINITE DEFORMATIONS	74-0344
BODIES	ISOFINEL: ISOPARAMETRIC FINITE ELEMENT CODE FOR ELASTIC ANALYSIS OF TWO-DIMENSIONAL BODIES	75-0676
BODIES	LARGE DISPLACEMENT FORMULATION FOR ELASTIC BODIES	70-0429
BODIES	MATHEMATICAL MODELING OF SPINNING ELASTIC BODIES FOR MODAL ANALYSIS	73-0441
BODIES	METHOD OF FINITE ELEMENTS FOR CALCULATION OF HETEROGENEOUS BODIES OF REVOLUTION IN ELASTO-PLASTIC ME	75-0144
BODIES	ON SMALL VIBRATIONS AND PERTURBATIONS OF FLEXIBLE BODIES UNDERGOING ARBITRARY NOMINAL MOTION	70-0376
BODIES	ON THE EXTREMAL PROPERTIES OF THE SOLUTION IN DYNAMICS OF RIGID VISCOPLASTIC BODIES ALLOWING FOR LAR	73-1080
BODIES	SAFE PCPS, A COMPUTER PROGRAM FOR THE STRESS ANALYSIS OF COMPOSITES BODIES OF REVOLUTION. INPUT INST	65-0013
BODIES	SAFE SHELL, A COMPUTER PROGRAM FOR THE STRESS ANALYSIS OF THIN SHELL BODIES OF REVOLUTION. A USER'S	67-0018
BODIES	SAFE-AXISYM, A COMPUTER PROGRAM FOR THE DESIGN OF COMPOSITE BODIES OF REVOLUTION. INPUT INSTRUCTIONS	65-0014
BODIES	SAFE-PLANE, A COMPUTER PROGRAM FOR THE STRESS ANALYSIS AND DESIGN OF TWO-DIMENSIONAL COMPOSITE BODIE	67-0017
BODIES	SOLVING A NONLINEAR PROBLEM OF NONSTEADY THERMAL CONDUCTIVITY IN SOLID BODIES OF ARBITRARY FORM	68-0019
BODIES	STABILITY AND POSTBUCKLING BEHAVIOUR OF HYPERELASTIC BODIES AT FINITE STRAIN BY THE FINITE ELEMENT M	73-0562
BODIES	STABILITY AND POSTBUCKLING BEHAVIOR OF HYPERELASTIC BODIES AT FINITE STRAIN BY THE FINITE ELEMENT ME	74-0838
BODIES	STRESS ANALYSIS IN VISCOELASTIC BODIES USING FINITE ELEMENTS AND A CORRESPONDENCE RULE WITH ELASTICI	69-0270
BODIES	THE APPLICATION OF FINITE ELEMENT ANALYSIS TO THE STUDY OF TWO-DIMENSIONAL CRACKED BODIES	72-0695
BODIES	THE MOTION OF INTERCONNECTED FLEXIBLE BODIES	75-0727
BODIES	THE PREDICTION OF REGULAR WAVE FORCES ON FLOATING BODIES WITH THE FINITE ELEMENT METHOD (IN DUTCH)	71-0743
BODIES	TRAPEZOIDAL FINITE ELEMENTS: THEIR DERIVATION AND USE FOR AXISYMMETRIC ROTATING BODIES	73-0192
BODIES	TWO AND THREE DIMENSIONAL ANALYSIS OF SUBSONIC COMPRESSIBLE POTENTIAL FLOWS IN DUCTS AND AROUND LIFT	74-0024
BODIES	ZUR LOESUNG DES KONTAKTPROBLEMS ELASTISCHER KOERPER MIT AUSGEDEHNTER KONTAKTFLAECHE DURCH QUADRATISC	74-1135
BODY	ADDITION OF AN ARBITRARY BODY ANALYSIS CAPABILITY TO THE BOEING TEA 236 FINITE ELEMENT COMPUTER PROG	73-0785
BODY	ANALYSIS OF BODY STRUCTURES - 2.	71-0339
BODY	ANALYSIS OF MOVING BODY PROBLEMS IN AERODYNAMICS	74-0024
BODY	ANALYSIS OF THE 747 AIRCRAFT WING BODY INTERSECTION	68-0275
BODY	ANALYSIS OF VEHICLE BODY STRUCTURE BY THE FINITE ELEMENT METHOD	71-0324
BODY	APPLICATION OF FINITE ELEMENT METHOD TO PREDICT STATIC RESPONSE OF AUTOMOTIVE BODY STRUCTURES	71-0340
BODY	APPLICATION OF FINITE ELEMENT METHOD TO THE STUDY OF STRENGTH OF CAR BODY STRUCTURE AND WHEEL	74-0281
BODY	APPLICATION OF FINITE-ELEMENT ANALYSIS TO BODY WAVE-PROPAGATION PROBLEMS	75-0446
BODY	APPLICATION OF THE FINITE ELEMENT TECHNIQUE TO THE STUDY OF AREAS OF HIGH STRESS GRADIENTS WITHIN A	73-0259
BODY	APPLIED FINITE ELEMENT DISPLACEMENT ANALYSIS OF A SINGLE INTEGRAL CONSTRUCTION VEHICLE BODY SHELL	72-0264
BODY	AUTOMATIC EXTRACTION OF RIGID BODY MODES FROM STRESS AND STRAIN ELEMENTS	72-0913
BODY	BODY ORIENTED (NATURAL) CO-ORDINATES FOR GENERATING THREE-DIMENSIONAL MESHES	74-0908
BODY	BOX CAR BODY BOLSTER PROBED IN STRESS STUDY	74-0654
BODY	EFFECTS OF FLEXIBLE CONNECTIONS ON BODY STRUCTURAL RESPONSE	74-0140
BODY	ELASTO-PLASTIC ANALYSIS OF AUTOMOBILE BODY STRUCTURE BY THE FINITE ELEMENT METHOD	74-0196
BODY	FINITE ELEMENT ANALYSIS OF CYLINDRICAL SHELLS UTILIZING CURVED ELEMENTS CONTAINING RIGID BODY DEFORM	71-0823
BODY	FINITE ELEMENT ANALYSIS OF THIN ELASTIC SHELLS WITH RESIDUAL ENERGY BALANCING AND THE ROLE OF THE RI	75-0034
BODY	FINITE ELEMENT EQUILIBRIUM MODEL FOR GENERAL BODY FORCE DISTRIBUTION	75-0268
BODY	FINITE-ELEMENT ANALYSIS OF THIN ELASTIC SHELLS WITH RESIDUAL ENERGY BALANCING AND ROLE OF RIGID BODY	75-0048
BODY	FREE VIBRATION OF MODEL OF CAR BODY	71-0312
BODY	IMPLICIT RIGID BODY MOTION IN CURVED FINITE ELEMENTS	71-0174
BODY	MISSILE BODY INPUT GENERATOR (BING), A NASTRAN PRE-PROCESSOR; THEORETICAL DEVELOPMENT, USER'S MANUAL	75-0717
BODY	NONLINEAR ANALYSIS OF CRACKED BODY BY FINITE ELEMENT METHOD	75-0315
BODY	RIGID BODY DISPLACEMENTS OF CURVED ELEMENTS IN THE ANALYSIS OF SHELLS BY THE MATRIX-DISPLACEMENT MET	67-0031
BODY	RIGID BODY MOTIONS AND EQUILIBRIUM IN FINITE ELEMENTS	74-0102
BODY	RIGID BODY MOTIONS IN CURVED FINITE ELEMENTS	70-0035
BODY	SEEPAGE IN FRACTURED ROCK: FINITE ELEMENT ANALYSIS OF ROCK MASS BEHAVIOR UNDER COUPLED ACTION OF BOD	71-0825
BODY	STATIC ANALYSIS VIA SUBSTRUCTURING OF AN EXPERIMENTAL VEHICLE FRONT-END BODY STRUCTURE	74-0313
BODY	THE THERMO-ELASTO-PLASTIC ANALYSIS OF A RE-ENTRY BODY TYPE COMPOSITE SOLID STRUCTURE USING A NEW FIN	66-0093
BODY	TRANSIENT RESPONSE OF INELASTICALLY CONSTRAINED RIGID BODY SYSTEMS	73-0291
BODY	TWO-DIMENSIONAL FINITE ELEMENT MODEL OF A CERAMIC BODY	73-0010
BODY	VARIATIONAL PRINCIPLES FOR A RAREFIED GAS FLOW PAST AND ARBITRARY THREE-DIMENSIONAL BODY	70-0189
BODY	VERIFICATION OF A HYBRID GENERAL PURPOSE SHELL AND BODY OF REVOLUTION COMPUTER PROGRAM	71-0333
BODY	VIBRATION OF AN ELASTIC BODY IMMERSED IN A FLUID	72-0312
BOEDEN	ANWENDUNG DER FINITE-ELEMENT-METHODE ZUR BERECHNUNG VON SPANNUNGEN IN WASSERGESAETTIGTEN BOEDEN (FIN	73-0502
BOEING	ADDITION OF AN ARBITRARY BODY ANALYSIS CAPABILITY TO THE BOEING TEA 236 FINITE ELEMENT COMPUTER PROG	73-0785
BOEING	BOPACE 3-D (THE BOEING PLASTIC ANALYSIS CAPABILITY FOR 3-DIMENSIONAL SOLIDS USING ISOPARAMETRIC FINI	75-0699
BOEING	THE BOEING SST PROTYPE INTERNAL LOADS ANALYSIS SYSTEM AND PROCEDURES	71-0103
BOGNER	CORRECTED ASSESSMENT OF CYLINDRICAL SHELL FINITE ELEMENT OF BOGNER, FOX AND SCHMIT WHEN APPLIED TO A	73-0024
BOILER	STRUCTURAL ANALYSIS OF WELDED WALL FOR BOILER FURNACE	74-0653
BOILER	TEMPERATURES AND STRESSES IN A BOILER MEMBRANE WALL TUBE	74-1053
BOILER	THE APPLICATION OF THREE-DIMENSIONAL FINITE ELEMENTS TO THE ANALYSIS OF PODDED BOILER TYPE PRESTRESS	71-0047
BOILER	THERMAL STRESSES IN FACT REACTOR BOILER TUBES AT THE SODIUM/GAS INTERFACE	73-0820
BOILES	ANALYSE DES BOILES MINCES PAR ELEMENTS FINIS COURBES	69-0524
BOILING	APPLICATION OF FINITE ELEMENT METHOD TO BOILING WATER REACTOR POWER PLANT (JAPANESE)	72-0816
BOLSTER	BOX CAR BODY BOLSTER PROBED IN STRESS STUDY	74-0654
BOLSTER	COMPUTER ANALYSIS OF A RAILROAD FREIGHT CAR BOLSTER UTILIZING THE FINITE ELEMENT METHOD	71-0283
BOLT	AN ELASTOPLASTIC ANALYSIS OF A UNIAXIALLY LOADED SHEET WITH AN INTERFERENCE - FIT BOLT	74-0687
BOLT	ANALYSIS AND MODEL STUDIES OF UNDERGROUND OPENINGS IN JOINTED ROCK. RESEARCH ON ROCK BOLT REINFORCEM	71-0438
BOLT	ANALYSIS AND MODEL STUDIES OF UNDERGROUND OPENINGS IN JOINTED ROCK. RESEARCH ON ROCK BOLT REINFORCEM	71-0750
BOLT	ANALYTICAL AND EXPERIMENTAL INVESTIGATION OF FATIGUE IN A SHEET SPECIMEN WITH AN INTERFERENCE-FIT BO	75-0701
BOLT	DESIGN OF ROOF BOLT PATTERNS FOR JOINTED ROCK	74-1318
BOLT	EFFECTS OF SPECIMEN RADIUS ON THE STRESS STATE NEAR A ROOF BOLT ANCHOR: A FINITE ELEMENT DETERMINATI	73-0229
BOLT	STRESS ANALYSIS OF A BOLT-NUT JOINT BY FINITE ELEMENT METHOD AND COPPER-ELECTROPLATING METHOD - 2. S	74-0153
BOLT	STRESS ANALYSIS OF A BOLT-NUT JOINT BY THE FINITE ELEMENT METHOD AND THE COPPER-ELECTROPLATING METHO	73-0220
BOLT-NUT	STRESS ANALYSIS OF A BOLT-NUT JOINT BY FINITE ELEMENT METHOD AND COPPER-ELECTROPLATING METHOD - 2. S	74-0153
BOLTED	AREAS OF CONTACT AND PRESSURE DISTRIBUTION IN BOLTED JOINTS	71-0329
BOLTED	DEVELOPMENT OF AN UNDERSTANDING OF THE FATIGUE PHENOMENA OF BONDED AND BOLTED JOINTS IN ADVANCED FIL	72-0798
BOLTED	DISTRIBUTION OF THE CLAMPING PRESSURE IN FRICTION-GRIP BOLTED JOINTS	74-0250
BOLTS	ANALYSIS OF MULTI-AXIAL ANISOTROPIC CREEP, ROOF RESIN BOLTS, AND BED SEPARATION IN ROCK STRUCTURES U	74-1122
BOLTS	COMPUTER SIMULATION OF LAMINATED ROOF REINFORCED WITH GROUTED BOLTS	73-1045
BOLTS	HIGH STRENGTH BOLTS SUBJECT TO TENSION AND PRYING	74-0201
BOLTS	THE USE OF ROCK BOLTS OR WIRE ROPE TO INCREASE THE STRENGTH OF FRACTURED MODEL PILLARS	71-0722
BOLTZMANN	A FINITE ELEMENT SOLUTION TECHNIQUE FOR THE BOLTZMANN EQUATION	69-0026
BOLTZMANN	A NUMERICAL SOLUTION OF THE BOLTZMANN EQUATION BY THE FINITE ELEMENT TECHNIQUE	70-0095
BOLTZMANN	FINITE ELEMENT ANALYSIS OF THE UNSTEADY, LINEARIZED BOLTZMANN EQUATION	70-0033
BOLTZMANN	FINITE ELEMENT ANALYSIS OF THE LINEARIZED BGK BOLTZMANN EQUATION.	72-0408
BOLTZMANN	USE OF VARIATIONAL PRINCIPLE IN STUDY ON RELATIVISTIC BOLTZMANN EQUATION	71-0442
BOMB	A FINITE ELEMENT EVALUATION OF RAPID METHODS OF BACKFILLING FOR BOMB DAMAGE REPAIR	74-0821
BOMB	A STRUCTURAL EVALUATION OF RAPID METHODS OF BACKFILLING FOR BOMB DAMAGE REPAIR - PHASE 11	75-0682
BOND	ANALYSIS OF STRESSES IN CONCRETE NEAR A REINFORCING BAR DUE TO BOND AND TRANSVERSE CRACKING	70-0351
BOND	CASE BOND STRESS CALCULATIONS FOR FLAPPED CYLINDRICAL ANALOGS OF SOLID PROPELLANT ROCKET MOTORS. BON	72-0842
BOND	CASE LINER BOND ANALYSIS	74-0817

BOND	CONTRIBUTION TO THE SOLUTION OF CONTACT PROBLEMS WITH THE AID OF BOND ELEMENTS	75-0321
BOND	CREEP OF B/AL COMPOSITES AS INFLUENCED BY RESIDUAL STRESSES, BOND STRENGTH, AND FIBER PACKING GEOMET	74-1208
BOND	DIRECT MEASUREMENT OF BOND SLIP IN REINFORCED CONCRETE	69-0428
BOND	FINITE ELEMENT SOLUTION OF LOW BOND NUMBER SLOSHING	75-0728
BOND	INVESTIGATION OF BOND FAILURE IN FIBRE-REINFORCED MATERIALS BY THE FINITE ELEMENT METHOD	72-0034
BOND	ZUSAMMENHANG ZWISCHEN OBERFLAECHENBESCHAFFENHEIT, VERBUND UND SPRENGWIRKUNG VON BEWEHRUNGSSTAEHLEN U	73-0450
BONDED	A METHOD FOR DETERMINING LOCAL ELASTOPLASTIC STRESS AND STRAIN IN METALLURGICALLY BONDED NOTCHED LAM	74-0686
BONDED	ANALYSIS OF BONDED JOINTS IN VEHICULAR STRUCTURES	73-0375
BONDED	DEVELOPMENT OF AN UNDERSTANDING OF THE FATIGUE PHENOMENA OF BONDED AND BOLTED JOINTS IN ADVANCED FIL	72-0798
BONDED	DISCRETE-ELEMENT STATIC ANALYSIS OF BONDED, DOUBLE-LAYER, BRANCHED, THIN SHELLS OF REVOLUTION, PART	68-0102
BONDED	DISCRETE-ELEMENT STATIC ANALYSIS OF BONDED, DOUBLE-LAYER, BRANCHED, THIN SHELLS OF REVOLUTION, PART	70-0081
BONDED	FINITE ELEMENT ANALYSIS OF BONDED CONNECTIONS	73-1096
BONDED	FRACTURE MECHANICS FOR AN INTERFACIAL CRACK BETWEEN ADHESIVELY BONDED DISSIMILAR MATERIALS	72-0869
BONDED	RIDDLE OF BONDED AND UNBONDED TENDONS IN PRESTRESSED CONCRETE REACTOR VESSELS	74-0519
BONDED	STRESS CONCENTRATION FACTORS FOR BONDED LAP JOINTS	71-0842
BONDED	UNTERSUCHUNGEN ZUM FESTIGKEITSVERHALTEN VON METALLKLEBUNGEN IM HINBLICK AUF DIE ERSTELLUNG EINES PRA	74-0289
BONDED	WOOD SHEAR PANELS BONDED WITH FLEXIBLE ADHESIVES	75-0086
BONDLINE	CASE BOND STRESS CALCULATIONS FOR FLAPPED CYLINDRICAL ANALOGS OF SOLID PROPELLANT ROCKET MOTORS. BON	72-0842
BONDS	OCENA DOKLADNOSCI ROZWIAZAN PROBLEMOW BRZEGOWYCH SPREZYSTYCH CIAL DYSKERTYZOWANYCH. (ASSESSING THE A	74-0921
BONE	A FINITE ELEMENT MODEL FOR PREDICTING LONG BONE FRACTURE FIXATION	71-0819
BONE	STRESS DISTRIBUTION IN BONE ARISING FROM LOADING ON ENDOSTEAL DENTAL IMPLANTS	73-0501
BONE	STRUCTURAL MODEL FOR THE MECHANICAL BEHAVIOUR OF TRABECULAR BONE	73-0473
BONE-PLATE-SCRE	FINITE ELEMENT ANALYSIS OF A BONE-PLATE-SCREW SYSTEM	74-0920
BONNEVILLE	MATHEMATICAL MODEL OF THE COLUMBIA RIVER FROM THE PACIFIC OCEAN TO BONNEVILLE DAM. PART 1. THEORY, P	69-0464
BONNEVILLE	MATHEMATICAL MODEL OF THE COLUMBIA RIVER FROM THE PACIFIC OCEAN TO BONNEVILLE DAM. PART II: INPUT-OU	70-0417
BOOLEAN	BOOLEAN MATRICES AS A STANDARD FACILITY IN A MATRIX CODE	65-0057
BOOMS	NORMAL MODE ANALYSIS OF THE RADIO ASTRONOMY EXPLORER BOOMS AND SPACECRAFT	71-0130
BOOSTER	DEVELOPMENT OF TECHNOLOGY FOR MODELING OF A 1/8-SCALE DYNAMIC MODEL OF THE SHUTTLE SOLID ROCKET BOOS	74-0703
BOPACE	BOPACE 3-D (THE BOEING PLASTIC ANALYSIS CAPABILITY FOR 3-DIMENSIONAL SOLIDS USING ISOPARAMETRIC FINI	75-0699
BORE	SIGNIFICANT APPLICATIONS OF FINITE ELEMENTS AND FINITE DIFFERENCES METHODS TO SOLVE PROBLEMS OF THER	73-0864
BORE	STRESS ANALYSIS OF PRESSURIZED CYLINDRICAL BORE IN RECTANGULAR BLOCK	69-0312
BORED	EFFECT OF THE VARIATION OF CONSISTENCY IN RESIDUAL SOILS ON LOAD TRANSFER BY BORED PILES	75-0812
BOREHOLE	NX BOREHOLE JACK MODULUS DETERMINATIONS IN HOMOGENEOUS, ISOTROPIC, ELASTIC MATERIALS	74-0259
BOREHOLE	THEORETICAL BASIS OF THE BOREHOLE DEEPENING METHOD OF ABSOLUTE STRESS MEASUREMENT	70-0047
BORON	DESIGN, MANUFACTURE, DEVELOPMENT, TEST, AND EVALUATION OF BORON/ALUMINUM STRUCTURAL COMPONENTS FOR S	74-0798
BORON	FATIGUE CRACK INITIATION IN A BORON EPOXY PLATE WITH A CIRCULAR HOLE	74-0820
BORON	STATIC AND DYNAMIC ANALYSIS OF F-14A BORON HORIZONTAL STABILIZER	71-0123
BORON-ALUMINUM	ON BUCKLING OF UNIDIRECTIONAL BORON-ALUMINUM STIFFENERS - A CAUTION TO DESIGNERS	75-0634
BORON-EPOXY	DEFORMATION AND FAILURE OF BORON-EPOXY PLATE WITH CIRCULAR HOLE	72-0536
BOS-6	VIBRATIONAL ANALYSIS OF THE FLEXING HEAD OF THE BOS-6 TRANSDUCER	71-0708
BOSOR	STRESS, STABILITY AND VIBRATION OF COMPLEX SHELLS OF REVOLUTION: ANALYSIS AND USERS MANUAL FOR BOSOR	69-0441
BOTTOM	ANALYSIS OF BOTTOM ENTRY NOZZLES FOR ATMOSPHERIC STORAGE TANKS	75-0331
BOTTOM	BOTTOM STABILITY UNDER WATER INDUCED LOADING	72-0373
BOTTOM	ON THE STATES OF STRESS AND DEFORMATION AROUND THE BOTTOM OF A CIRCULAR VERTICAL SHAFT	69-0323
BOUND	A BOUND THEOREM IN EIGENVALUES AND ITS PRACTICAL APPLICATIONS	71-0128
BOUND	FINITE ELEMENT METHOD FOR BOUND STATE CALCULATIONS IN QUANTUM MECHANICS	75-0005
BOUND	LOWER BOUND LIMIT ANALYSIS OF PLANE STRESS PROBLEMS BY A FINITE ELEMENT METHOD	70-0428
BOUND	SOLUTION OF MIXED BOUNDARY VALUE PROBLEMS WITH LOCAL ERROR BOUND BY THE FINITE ELEMENT METHOD	73-0795
BOUNDARIES	A FINITE ELEMENT PROCEDURE FOR PLATES WITH CURVED BOUNDARIES	75-0030
BOUNDARIES	A FINITE ELEMENT PROCEDURE FOR PLATES WITH CURVED BOUNDARIES	75-0750
BOUNDARIES	APPLICATION OF THE HIGH PRECISION TRIANGULAR PLATE BENDING ELEMENT TO PROBLEMS WITH CURVED BOUNDARIE	69-0071
BOUNDARIES	BLENDING FUNCTION FINITE ELEMENTS FOR CURVED BOUNDARIES	75-0093
BOUNDARIES	BOUNDARIES AND FRACTURES IN FINITE ELEMENT MODELS OF GEOLOGICAL STRUCTURES	73-0352
BOUNDARIES	CURVED BOUNDARIES IN THE FINITE ELEMENT METHOD	73-0638
BOUNDARIES	FINITE ELEMENT SOLUTION OF PROBLEMS WITH UNKNOWN AND MOVING BOUNDARIES	74-0460
BOUNDARIES	FINITE ELEMENT TECHNIQUES FOR CURVED BOUNDARIES	73-0622
BOUNDARIES	FINITE-ELEMENT PROCEDURES FOR PROBLEMS WITH CURVED BOUNDARIES	72-0977
BOUNDARIES	FINITE-ELEMENT SOLUTIONS WITHIN CURVED BOUNDARIES	72-0038
BOUNDARIES	FREE BOUNDARIES AND FINITE-ELEMENTS IN ONE DIMENSION	75-0447
BOUNDARIES	FREE BOUNDARIES AND FINITE-ELEMENTS IN ONE DIMENSION	75-0587
BOUNDARIES	ORDER OF CONVERGENCE OF A FINITE-ELEMENT METHOD IN REGIONS WITH CURVED BOUNDARIES	75-0817
BOUNDARIES	SEISMIC RESPONSE OF SOIL DEPOSITS UNDERLAIN BY SLOPING ROCK BOUNDARIES	69-0484
BOUNDARIES	THE COMBINED EFFECT TO CURVED BOUNDARIES AND NUMERICAL INTEGRATION IN ISOPARAMETRIC FINITE ELEMENT M	72-0357
BOUNDARIES	THE FINITE ELEMENT METHOD IN DOMAINS WITH CURVED BOUNDARIES	73-0080
BOUNDARIES	USE OF PARABOLIC ARCS IN MATCHING CURVED BOUNDARIES IN FINITE-ELEMENT METHOD	75-0589
BOUNDARIES	VARIABLE TIME STEP METHOD FOR DETERMINING PLASTIC STRESS REFLECTIONS FROM BOUNDARIES	75-0641
BOUNDARY	(L SUP 2) ERROR ESTIMATES FOR FINITE ELEMENTS WITH INTERPOLATED BOUNDARY CONDITIONS	73-0885
BOUNDARY	A MATHEMATICAL THEORY OF COMPLEMENTARY-DUAL VARIATIONAL PRINCIPLES AND MIXED FINITE ELEMENT APPROXIM	74-0804
BOUNDARY	A THEORY OF MIXED FINITE ELEMENT APPROXIMATIONS OF NON-SELF-ADJOINT BOUNDARY VALUE PROBLEMS	74-0715
BOUNDARY	A VARIATIONAL METHOD FOR FREE BOUNDARY PROBLEMS	75-0105
BOUNDARY	ACOUSTIC FINITE ELEMENT APPROACH FOR STUDYING BOUNDARY FLEXIBILITY AND SOUND TRANSMISSION BETWEEN IR	73-0096
BOUNDARY	AN ACOUSTIC FINITE ELEMENT APPROACH FOR STUDYING BOUNDARY FLEXIBILITY AND SOUND TRANSMISSION BETWEEN	73-1087
BOUNDARY	AN APPLICATION OF THE FINITE ELEMENT TECHNIQUE TO VISCOUS FLOW PROBLEMS OF POTENTIAL FLOW	70-0002
BOUNDARY	AN EXACT BOUNDARY TECHNIQUE FOR IMPROVED ACCURACY IN THE FINITE ELEMENT METHOD	73-1035
BOUNDARY	AN EXTENSION OF FINITE-ELEMENT METHOD TO INCLUDE NONLINEAR BOUNDARY CONDITIONS	75-0032
BOUNDARY	APPLICATION OF THE FINITE ELEMENT METHOD OF ANALYSIS IN SOLVING BOUNDARY VALUE PROBLEMS IN ROCK MECH	66-0013
BOUNDARY	APPLICATION OF THE FINITE-ELEMENT METHOD TO HEAT-TRANSFER PROBLEMS. PART 11. TRANSIENT TWO-DIMENSION	71-0759
BOUNDARY	APPLICATION OF THE FINITE-ELEMENT METHOD TO HEAT-TRANSFER PROBLEMS. PART 1. FINITE-ELEMENT METHOD AP	71-0760
BOUNDARY	APPLICATION OF THE FINITE-ELEMENT METHOD FOR SOLVING ORDINARY LINEAR DIFFERENTIAL EQUATION WITH BOUN	74-0084
BOUNDARY	APPLICATIONS OF VARIATIONAL EQUATIONS TO ORDINARY AND PARTIAL DIFFERENTIAL EQUATIONS - MULTIPLE SOLU	75-0010
BOUNDARY	APPROXIMATE BOUNDARY CONDITIONS IN THE FINITE ELEMENT METHOD	72-0995
BOUNDARY	BOUNDARY AND INTERIOR APPROXIMATION ERRORS IN THE FINITE ELEMENT METHOD	73-0209
BOUNDARY	COMPUTABLE ERROR BOUNDS FOR THE FINITE ELEMENT METHOD FOR ELLIPTIC BOUNDARY VALUE PROBLEMS	74-1219
BOUNDARY	CONTRIBUTION TO SOLUTION OF BODIES WITH CURVED BOUNDARY BY FINITE ELEMENT METHOD	72-1001
BOUNDARY	CONVERGENCE OF A FINITE ELEMENT PROCEDURE FOR SOLVING BOUNDARY VALUE PROBLEMS OF THE FOURTH ORDER	70-0131
BOUNDARY	CUBIC SPLINE GALERKIN APPROXIMATIONS TO PARABOLIC SYSTEMS WITH COUPLED NON-LINEAR BOUNDARY CONDITION	75-0232
BOUNDARY	CURVED BOUNDARY ELEMENTS FOR FINITE ELEMENT METHOD	72-0859
BOUNDARY	CURVED BOUNDARY ELEMENTS, GENERAL FORMS OF POLYNOMIAL MAPPINGS	74-1256
BOUNDARY	DISCRETIZATION AND ROUND-OFF ERROR IN THE FINITE ELEMENT ANALYSIS OF ELLIPTIC BOUNDARY VALUE PROBLEM	71-0514
BOUNDARY	DYNAMIC RESPONSE OF STRUCTURES SUBJECTED TO TIME-DEPENDENT BOUNDARY CONDITIONS USING THE FINITE ELEM	72-0129
BOUNDARY	EFFECT OF GRAIN BOUNDARY FOR THE ELASTIC BEHAVIOR OF AL AND CU CRYSTALS (ANALYSIS BY THE FINITE ELEM	74-0947
BOUNDARY	ELEMENT STIFFNESS MATRICES FOR BOUNDARY COMPATIBILITY AND FOR PRESCRIBED BOUNDARY STRESSES	66-0061
BOUNDARY	ELEMENT STIFFNESS MATRICES FOR BOUNDARY COMPATIBILITY AND FOR PRESCRIBED BOUNDARY STRESSES	66-0061
BOUNDARY	ELLIPTIC BOUNDARY VALUE PROBLEMS WITH NONVARIATIONAL PERTURBATION AND THE FINITE ELEMENT METHOD	73-1022
BOUNDARY	ERROR ANALYSIS OF FINITE ELEMENT METHODS WITH TRIANGLES FOR ELLIPTIC BOUNDARY VALUE PROBLEMS	73-0072
BOUNDARY	FINITE BOUNDARY CORRECTIONS TO THE COPLANAR WAVEGUIDE ANALYSIS	73-0398
BOUNDARY	FINITE ELEMENT APPLICATIONS OF A PRECISION TRIANGULAR PLATE BENDING ELEMENT TO LINEAR BOUNDARY VALUE	75-0788
BOUNDARY	FINITE ELEMENT COMPUTATIONAL THEORY FOR THREE-DIMENSIONAL BOUNDARY LAYER FLOW	72-0632
BOUNDARY	FINITE ELEMENT FORMULATIONS OF NONLINEAR BOUNDARY VALUE PROBLEMS	71-0497
BOUNDARY	FINITE ELEMENT METHOD APPLIED TO TRANSIENT TWO-DIMENSIONAL HEAT TRANSFER WITH CONVECTION AND RADIATI	70-0127
BOUNDARY	FINITE ELEMENT METHODS FOR ELLIPTIC BOUNDARY VALUE PROBLEMS CONTAINING SINGULARITIES	72-0360
BOUNDARY	FINITE ELEMENT METHODS FOR SOLVING INITIAL BOUNDARY VALUE PROBLEMS	74-1125

BOUNDARY	FINITE ELEMENT SOLUTION THEORY FOR THREE-DIMENSIONAL BOUNDARY FLOWS	74-0579
BOUNDARY	FINITE ELEMENT TREATMENT OF SINGULARITIES OF BOUNDARY VALUE PROBLEMS AND ITS APPLICATION TO ANALYSIS	73-0306
BOUNDARY	FINITE-ELEMENT METHOD APPLIED TO HEAT CONDUCTION IN SOLIDS WITH NONLINEAR BOUNDARY CONDITIONS	73-0005
BOUNDARY	FREE BOUNDARY POTENTIAL FLOW USING FINITE ELEMENTS	71-0829
BOUNDARY	FREE VIBRATION OF THICK, LAYERED CYLINDERS HAVING FINITE LENGTH WITH VARIOUS BOUNDARY CONDITIONS	72-0491
BOUNDARY	FREQUENCY COEFFICIENTS OF A SPHERICAL SHELL FOR VARYING THICKNESS AND OPENING ANGLE WITH GUIDED PINN	72-0595
BOUNDARY	FUNCTIONAL ANALYSIS AND THE FINITE ELEMENT METHOD APPLIED TO ELLIPTIC BOUNDARY VALUE PROBLEMS	72-0056
BOUNDARY	FUNCTIONAL APPROXIMATIONS FOR SOLVING BOUNDARY VALUE PROBLEMS BY COMPUTER	69-0013
BOUNDARY	GALERKIN METHODS FOR PARABOLIC EQUATIONS WITH NONLINEAR BOUNDARY CONDITIONS	73-1023
BOUNDARY	GENERAL VARIATIONAL ANALYSIS OF HYDRODYNAMIC, THERMAL AND DIFFUSIONAL BOUNDARY LAYERS	70-0191
BOUNDARY	GENERALIZED CONJUGATE FUNCTIONS FOR MIXED FINITE ELEMENT APPROXIMATIONS OF BOUNDARY VALUE PROBLEMS	72-0515
BOUNDARY	HEAT TRANSFER WITH NONLINEAR BOUNDARY CONDITIONS VIA A VARIATIONAL PRINCIPLE	71-0443
BOUNDARY	HERMITE INTERPOLATION TYPE RITZ METHODS FOR TWO POINT BOUNDARY VALUE PROBLEMS	66-0089
BOUNDARY	INTERPOLATED BOUNDARY CONDITIONS IN THE FINITE ELEMENT METHOD	75-0388
BOUNDARY	LEAST SQUARES FINITE ELEMENT ANALYSIS OF LAMINAR BOUNDARY LAYER FLOWS	74-0616
BOUNDARY	LOCAL ERROR BOUNDS IN BOUNDARY VALUE PROBLEMS USING FINITE ELEMENT DISCRETIZATION	72-0982
BOUNDARY	L2 ERROR ESTIMATES FOR FINITE ELEMENTS WITH INTERPOLATED BOUNDARY CONDITIONS	73-0194
BOUNDARY	MATCHING OF ESSENTIAL BOUNDARY CONDITIONS IN THE FINITE ELEMENT METHOD	74-1260
BOUNDARY	NATURAL BOUNDARY CONDITIONS IN FINITE-ELEMENT METHOD	72-0276
BOUNDARY	NATURAL VIBRATIONS OF THIN, FLAT WALLED STRUCTURES WITH DIFFERENT BOUNDARY CONDITIONS	71-0509
BOUNDARY	NUMERICAL COMPUTATION OF BOUNDARY VALUE PROBLEMS IN SOIL MECHANICS	73-0269
BOUNDARY	NUMERICAL METHODS FOR TIME DEPENDENT NONLINEAR BOUNDARY VALUE PROBLEMS	70-0296
BOUNDARY	NUMERICAL METHODS OF HIGH-ORDER ACCURACY FOR NONLINEAR BOUNDARY VALUE PROBLEMS	68-0194
BOUNDARY	NUMERICAL SOLUTION OF A DIFFUSION CONSUMPTION PROBLEM WITH A FREE BOUNDARY	74-0835
BOUNDARY.	NUMERICAL SOLUTION OF 2-POINT BOUNDARY VALUE PROBLEMS	74-1141
BOUNDARY	NUMERICAL-SOLUTION OF EULERIAN EQUATIONS OF COMPRESSIBLE FLOW BY A FINITE-ELEMENT METHOD WHICH FOLLO	75-0535
BOUNDARY	OCENA DOKLADNOSCI ROZWIAZAN PROBLEMOW BRZEGOWYCH SPREZYSTYCH CIAL DYSKERTYZOWANYCH. (ASSESSING THE A	74-0921
BOUNDARY	ON DIRICHLET PROBLEMS USING SUBSPACES WITH NEARLY ZERO BOUNDARY CONDITIONS	72-0514
BOUNDARY	ON FLOW NEAR SINGULAR POINTS OF A WALL BOUNDARY	74-0008
BOUNDARY	ON SOME FINITE ELEMENT PROCEDURES FOR SOLVING SECOND ORDER BOUNDARY VALUE PROBLEMS	69-0288
BOUNDARY	ON THE ASYMPTOTICALLY SPHERICAL DEFORMATIONS OF ARBITRARY MEMBRANES OF REVOLUTION FIXED ALONG AN EDG	71-0640
BOUNDARY	ON THE CONVERGENCE OF SOR ITERATIONS FOR FINITE ELEMENT APPROXIMATIONS TO ELLIPTIC BOUNDARY VALUE PR	71-0449
BOUNDARY	ON THE NUMERICAL SOLUTION OF FREE BOUNDARY PROBLEMS	75-0577
BOUNDARY	PROPERTIES OF SOLUTIONS OF ELLIPTIC BOUNDARY VALUE PROBLEMS	72-0499
BOUNDARY	RASCHET KILEBANII PLASTIN SLOZHNDI FORMY S NEODNORODNYMI GRANICHNYMI USLOVIYAMI. (CALCULATION OF VIB	74-0988
BOUNDARY	RAYLEIGH WAVES AT A CONTINENTAL BOUNDARY BY FINITE ELEMENT METHOD	72-0016
BOUNDARY	SECOND GENERATION BOUNDARY INTEGRAL EQUATION PROGRAM FOR THREE-DIMENSIONAL ELASTIC ANALYSIS	75-0376
BOUNDARY	SOLUTION OF A FREE-SURFACE BOUNDARY VALUE PROBLEM USING AN INVERSE FORMULATION AND THE FINITE ELEMEN	75-0466
BOUNDARY	SOLUTION OF ELASTO-STATIC BOUNDARY VALUE PROBLEMS BY THE FINITE ELEMENT METHOD	64-0025
BOUNDARY	SOLUTION OF ELASTO-STATIC BOUNDARY VALUE PROBLEMS BY THE FINITE ELEMENT METHOD	64-0030
BOUNDARY	SOLUTION OF MIXED BOUNDARY VALUE PROBLEMS IN FINITE DOMAINS BY FOURIERS METHOD	72-0790
BOUNDARY	SOLUTION OF MIXED BOUNDARY VALUE PROBLEMS WITH LOCAL ERROR BOUND BY THE FINITE ELEMENT METHOD	73-0795
BOUNDARY	SUBSPACES WITH ACCURATELY INTERPOLATED BOUNDARY CONDITIONS	72-0527
BOUNDARY	SYSTEMATIC ENFORCEMENT OF STRESS BOUNDARY CONDITIONS IN THE ASSUMED STRESS HYBRID MODEL BASED ON THE	71-0771
BOUNDARY	TEST RESULTS FROM THE BOUNDARY LAYER FACILITY - RESPONSE OF STRUCTURE TO THE PSEUDO-SOUND FIELD OF A	68-0262
BOUNDARY	THE EFFECTS OF SOIL PARAMETERS AND BOUNDARY CONDITIONS ON THE CONSOLIDATION OF AN ELASTIC LAYER	70-0422
BOUNDARY	THE FINITE ELEMENT METHOD AND LOCAL BOUNDS FOR BOUNDARY VALUE PROBLEMS OF ELASTIC STRUCTURES	73-1056
BOUNDARY	THE FINITE ELEMENT METHOD AND BOUNDARY SOLUTION PROCEDURES AS GENERAL APPROXIMATION METHODS FOR FIEL	75-0262
BOUNDARY	THE TREATMENT OF NATURAL BOUNDARY CONDITIONS IN THE FINITE ELEMENT AND FINITE DIFFERENCE METHODS	73-1098
BOUNDARY	THERMAL STRESSES IN OPEN CYLINDRICAL SHELLS OF ARBITRARY BOUNDARY CONDITIONS BY A FINITE STRIP METHO	72-0371
BOUNDARY	THREE DIMENSIONAL FINITE ELEMENT ANALYSIS OF THICK LAMINATED COMPOSITES - INCLUDING INTERLAMINAR AND	73-0882
BOUNDARY	THREE DIMENSIONAL NONLINEAR MAGNETIC FIELD BOUNDARY VALUE PROBLEM AND ITS NUMERICAL SOLUTION	70-0060
BOUNDARY	THREE-DIMENSIONAL FINITE ELEMENT ANALYSIS OF THICK LAMINATED COMPOSITES - INCLUDING INTERLAMINAR AND	74-1151
BOUNDARY	TRANSPIRATION AND FILM COOLING BOUNDARY LAYER COMPUTER PROGRAM. VOLUME 2, COMPUTER PROGRAM AND USER!	71-0655
BOUNDARY	UEBER PAUSCHALE FEHLERSCHRANKEN FUER APPROXIMATIONEN VON SCHALENVERFORMUNGEN. (GLOBAL BOUNDARY CONDI	75-0292
BOUNDARY	VARIATIONAL FORMULATION FOR A CLASS OF FREE BOUNDARY PROBLEMS ARISING IN HYDRODYNAMIC LUBRICATION	75-0531
BOUNDARY	VIBRATION OF RECTANGULAR PLATES WITH MIXED BOUNDARY CONDITIONS	73-1086
BOUNDARY	VOGT BOUNDARY FOR FINITE ELEMENT ARCH DAM ANALYSIS	74-0156
BOUNDARY-INTEGR	APPLICATION OF THE BOUNDARY-INTEGRAL EQUATION METHOD TO THREE DIMENSIONAL STRESS ANALYSIS	73-0252
BOUNDARY-LAYER	GALERKIN METHOD FOR A CLASS OF STEADY, 2-DIMENSIONAL, INCOMPRESSIBLE, LAMINAR BOUNDARY-LAYER FLOWS	75-0437
BOUNDARY-LAYER	LEAST SQUARES FINITE ELEMENT ANALYSIS OF TWO-DIMENSIONAL LAMINAR BOUNDARY-LAYER EQUATION WITH SEPARA	75-0796
BOUNDARY-LAYER	THE NUMERICAL CALCULATION OF LAMINAR BOUNDARY-LAYER SEPARATION	74-0785
BOUNDARY-VALUE	A FINITE ELEMENT METHOD FOR THE RESOLUTION OF NON-LINEAR BOUNDARY-VALUE PROBLEMS	75-0107
BOUNDARY-VALUE	A THEORY OF MIXED FINITE ELEMENT APPROXIMATIONS OF NON-SELF-ADJOINT BOUNDARY-VALUE PROBLEMS	74-0340
BOUNDARY-VALUE	AN ERROR EXPANSION FOR A FINITE-ELEMENT SOLUTION OF A TWO-POINT BOUNDARY-VALUE PROBLEM	75-0801
BOUNDARY-VALUE	FINITE ELEMENT SOLUTION OF BOUNDARY-VALUE PROBLEMS WITH NON-REMOVABLE SINGULARITIES	73-0124
BOUNDARY-VALUE	FINITE-ELEMENT MULTISTEP DISCRETIZATIONS OF PARABOLIC BOUNDARY-VALUE PROBLEMS	75-0501
BOUNDARY-VALUE	MIXED FINITE-ELEMENT APPROXIMATIONS OF LINEAR BOUNDARY-VALUE PROBLEMS	75-0453
BOUNDARY-VALUE	ON THE EFFECT OF THE FORM OF THE STRAIN ENERGY FUNCTION ON THE SOLUTION OF A BOUNDARY-VALUE PROBLEM	74-0429
BOUNDED	DISCRETE-ELEMENT MODELLING AND SENSITIVITY STUDIES OF STATICALLY LOADED SINGLE-LAYER AND BOUNDED DOU	70-0129
BOUNDED	FINITE ELEMENT ANALYSIS OF FINITE SIZED PLATES BOUNDED TO AN ELASTIC HALF SPACE	72-0880
BOUNDED	SONIC FATIGUE OF DIFFUSION BONDED TITANIUM SANDWICH STRUCTURES	72-0290
BOUNDING	BOUNDING PLANE STRESS SOLUTIONS BY FINITE ELEMENTS	66-0055
BOUNDING	BOUNDING SOLUTIONS OF PLATE BENDING PROBLEMS	67-0084
BOUNDING	EVALUATION OF A CLASS OF METHODS FOR BOUNDING STEADY-STATE CREEP DEFORMATION	74-0892
BOUNDS	A METHOD TO GENERATE BOTH UPPER AND LOWER BOUNDS TO PLATE EIGENVALUES BY CONFORMING DISPLACEMENT FIN	71-0166
BOUNDS	APPROXIMATION AND BOUNDS FOR EIGENVALUES OF VARIOUS DOMAINS USING THE INTERMEDIATE PROBLEM TECHNIQUE	72-0711
BOUNDS	BENDING AND STRETCHING OF PLATES, SPECIAL MODELS FOR UPPER AND LOWER BOUNDS	66-0026
BOUNDS	BOUNDS AND ERROR CONTROL FOR EIGENVALUES	73-0319
BOUNDS	BOUNDS FOR A CLASS OF LINEAR FUNCTIONALS WITH APPLICATIONS TO HERMITE INTERPOLATION	71-0582
BOUNDS	BOUNDS FOR EIGENVALUES IN SOME VIBRATION AND STABILITY PROBLEMS.	72-0390
BOUNDS	BOUNDS ON EXTERNAL EIGENVALUES OF FINITE ELEMENT STIFFNESS AND MASS MATRICES AND THEIR SPECTRAL COND	72-0256
BOUNDS	BOUNDS ON THE SPECTRAL AND MAXIMUM NORMS OF THE FINITE ELEMENT STIFFNESS, FLEXIBILITY AND MASS MATRI	73-0409
BOUNDS	BOUNDS TO THE INFLUENCE COEFFICIENTS BY THE ASSUMED STRESS METHOD	70-0165
BOUNDS	COMPLEXITY BOUNDS FOR REGULAR FINITE DIFFERENCE AND FINITE ELEMENT GRIDS	72-0872
BOUNDS	COMPLEXITY BOUNDS FOR REGULAR FINITE DIFFERENCE AND FINITE ELEMENT GRIDS	73-0043
BOUNDS	COMPUTABLE ERROR BOUNDS FOR THE FINITE ELEMENT METHOD FOR ELLIPTIC BOUNDARY VALUE PROBLEMS	74-1219
BOUNDS	CURVED ELEMENTS, ERROR BOUNDS, TIME DEPENDENT PROBLEMS FIVE LECTURES	73-0636
BOUNDS	DETERMINATION OF UPPER BOUNDS FOR PROBLEMS IN PLANE STRESS USING FINITE ELEMENT TECHNIQUES	67-0034
BOUNDS	DEVELOPMENT OF THE STIFFNESS METHOD TO DEFINE BOUNDS ON ELASTIC BEHAVIOUR OF STRUCTURES	62-0011
BOUNDS	DISCRETIZATION ERROR BOUNDS FOR TRANSFINITE ELEMENTS	72-0533
BOUNDS	DUALITY BETWEEN DISPLACEMENT AND EQUILIBRIUM METHOD WITH A VIEW TO OBTAINING UPPER AND LOWER BOUNDS	62-0013
BOUNDS	EIGENVALUE BOUNDS FOR FINITE ELEMENT MODELS	72-0570
BOUNDS	EIGENVALUE BOUNDS FOR FINITE ELEMENT MODELS	72-0932
BOUNDS	ERROR BOUNDS FOR EIGENVALUE ANALYSIS BY ELIMINATION OF VARIABLES	71-0594
BOUNDS	ERROR BOUNDS FOR FINITE ELEMENT METHOD	71-0033
BOUNDS	ERROR BOUNDS FOR SEMIDISCRETE GALERKIN APPROXIMATIONS OF PARABOLIC PROBLEMS WITH APPLICATIONS TO PET	70-0291
BOUNDS	ERROR BOUNDS IN AN OPTIMIZATION PROBLEM USING FINITE-ELEMENT METHOD	73-0049
BOUNDS	EVALUATION OF THE ERROR BOUNDS IN AN OPTIMIZATION PROBLEM USING THE FINITE-ELEMENT METHOD	73-0211
BOUNDS	FINITE ELEMENT ERROR BOUNDS	75-0100
BOUNDS	LOCAL ERROR BOUNDS IN BOUNDARY VALUE PROBLEMS USING FINITE ELEMENT DISCRETIZATION	72-0982
BOUNDS	LOWER BOUNDS TO COLUMN BUCKLING LOADS	73-0485

BOUNDS	SARD KERNEL THEOREMS ON TRIANGULAR AND RECTANGULAR DOMAINS WITH EXTENSIONS AND APPLICATIONS TO FINIT	72-0361
BOUNDS	SOLUTION BOUNDS IN SOME STRESS PROBLEMS BY THE HYPERCIRCLE AND FINITE ELEMENT METHODS	72-0706
BOUNDS	STRAIN-ENERGY BOUNDS IN FINITE-ELEMENT ANALYSIS BY SLAB ANALOGY	67-0027
BOUNDS	THE EXTENSION AND APPLICATION OF SARD KERNEL THEOREMS TO COMPUTE FINITE ELEMENT ERROR BOUNDS	72-0519
BOUNDS	THE FINITE ELEMENT METHOD AND LOCAL BOUNDS FOR BOUNDARY VALUE PROBLEMS OF ELASTIC STRUCTURES	73-1056
BOUNDS	UPPER AND LOWER BOUNDS IN MATRIX STRUCTURAL ANALYSIS	64-0018
BOUNDS	UPPER AND LOWER BOUNDS TO STRUCTURAL DEFORMATIONS BY DUAL ANALYSIS IN FINITE ELEMENTS	67-0074
BOUSSINESQ	BOUSSINESQ PROBLEM OF PLANE MICROPOLAR ELASTICITY	74-0197
BOUSSINESQ'S	FINITE ELEMENT SOLUTION OF BOUSSINESQ'S EQUATION FOR UNSTEADY GROUNDWATER FLOW	74-0459
BOX	A COMPUTER PROGRAM FOR THE CALCULATION OF TORSION FACTORS AND SHEAR STRESS FACTORS FOR BOX SECTIONS	69-0485
BOX	A HIGHER ORDER IN-PLANE PARALLELOGRAM ELEMENT AND ITS APPLICATION TO SKEWED CURVED BOX GIRDER BRIDGE	71-0002
BOX	A TECHNIQUE FOR CONNECTING BEAM ELEMENTS TO A PLATE MODEL OF A COMPLICATED BOX SECTION	74-0271
BOX	ANALYSIS AND DESIGN OF SKEW BOX GIRDER BRIDGES	72-0659
BOX	ANALYSIS OF BOX GIRDER BRIDGES BY FINITE STRIP METHOD	69-0050
BOX	ANALYSIS OF BOX GIRDERS BY FOLDED PLATE AND STRIP METHODS	73-0982
BOX	ANALYSIS OF CONCRETE BOX GIRDERS BEFORE AND AFTER CRACKING	72-0247
BOX	ANALYSIS OF CURVED BOX GIRDER BRIDGES BY FINITE STRIP METHOD	71-0485
BOX	ANALYSIS OF CURVED BOX GIRDER BRIDGES	73-0294
BOX	ANALYTICAL AND EXPERIMENTAL STUDIES OF MULTI-CELL CONCRETE BOX GIRDER BRIDGES	74-1134
BOX	APPLICATIONS OF THE FINITE ELEMENT METHOD TO BOX BEAMS	68-0273
BOX	BOX CAR BODY BOLSTER PROBED IN STRESS STUDY	74-0654
BOX	BOX SHAPED GIRDERS. (NOSNIKY UZAVRETEHO PRIEREZU)	70-0169
BOX	COMPUTER BASED ANALYSES OF THE RESPONSE OF BOX TYPE STRUCTURES TO RANDOM PRESSURES	73-0227
BOX	DIAPHRAGMS IN SINGLE- AND DOUBLE-CELL BOX GIRDER BRIDGES WITH VARYING ANGLE OF SKEW	72-0130
BOX	DIGITAL COMPUTATION OF STRESSES AND DEFLECTIONS IN A BOX BEAM	67-0073
BOX	DIGITAL COMPUTATION OF STRESSES AND DEFLECTIONS IN A BOX BEAM - A PERFORMANCE COMPARISON BETWEEN FIN	69-0355
BOX	EXPERIMENTAL AND THEORETICAL INVESTIGATION OF THE BEHAVIOR OF BOX GIRDERS CURVED IN PLAN	75-0659
BOX	FINITE ELEMENT ANALYSIS OF SKEW BOX GIRDER BRIDGES	72-0209
BOX	FINITE ELEMENT SCHEME FOR BOX BRIDGE ANALYSIS	75-0666
BOX	FREE VIBRATION OF CURVED AND STRAIGHT BEAM SLAB OR BOX GIRDER BRIDGES	72-0487
BOX	LIMITATIONS ON THE APPLICATION OF EFFECTIVE BREADTH CONCEPT IN BOX GIRDER DESIGN	74-0493
BOX	LOAD DISTRIBUTION IN A COMPOSITE STEEL BOX GIRDER BRIDGE	73-0765
BOX	NEW FINITE ELEMENTS WITH APPLICATION TO BOX GIRDER BRIDGES.	72-0465
BOX	SEMI-ANALYTIC SOLUTION OF SKEW BOX GIRDER BRIDGES	75-0578
BOX	SHEAR LAG IN STEEL BOX GIRDER BRIDGES	75-0310
BOX	SHEAR LAG IN STIFFENED BOX GIRDERS	70-0243
BOX	STRESS SINGULARITIES AT DIAPHRAGM-WEB-FLANGE JUNCTIONS IN BOX GIRDERS	75-0242
BOX	STRESSES IN CONTINUOUS CONCRETE BOX GIRDER BRIDGES	71-0233
BOX	STRUCTURAL BEHAVIOR OF A TWO SPAN REINFORCED CONCRETE BOX GIRDER BRIDGE MODEL. VOLUME 11. REDUCTION,	71-0659
BOX	THE FINITE ELEMENT METHOD APPLIED TO THIN SHELLS AND BOX STRUCTURES	70-0444
BOX	ULTIMATE STRENGTH OF BOX GIRDERS BY FINITE ELEMENT METHOD	75-0794
BOX-BEAM	EXPERIMENTAL AND ANALYTICAL INVESTIGATION OF A HORIZONTALLY CURVED BOX-BEAM HIGHWAY BRIDGE MODEL	71-0024
BOX-BEAM	MODEL ANALYSIS OF CURVED BOX-BEAM HIGHWAY BRIDGE	71-0282
BOX-GIRDER	A NOTE ON THE ELASTIC ANALYSIS OF BOX-GIRDER DIAPHRAGMS USING FINITE ELEMENTS	74-1230
BOX-GIRDER	APPLICATIONS OF FINITE ELEMENT METHODS TO UNSTEADY VISCOUS FLOW AROUND A BOX-GIRDER BRIDGE OSCILLATI	74-0827
BOX-GIRDER	COLLAPSE ANALYSIS OF BOX-GIRDER COMPONENTS USING FINITE ELEMENTS	74-0951
BOX-GIRDER	DYNAMIC ANALYSIS OF CURVED BOX-GIRDER BRIDGES	75-0280
BOX-GIRDER	FIELD TESTS OF A STEEL-COMPOSITE BOX-GIRDER BRIDGE	75-0633
BOX-GIRDER	FINITE ELEMENT ANALYSIS OF SKEW, CURVED BOX-GIRDER BRIDGE	70-0241
BOX-LIKE	RESPONSE OF BOX-LIKE STRUCTURES TO WEAK EXPLOSIONS	75-0197
BOX-TYPE	ANALYSIS OF INPLANE VIBRATION OF BOX-TYPE STRUCTURES BY A FINITE ELEMENT METHOD	72-0091
BOX-TYPE	NATURAL FREQUENCIES OF BOX-TYPE STRUCTURES BY A FINITE-ELEMENT METHOD	74-0193
BOX-TYPE	VIBRATION OF A BOX-TYPE STRUCTURE - 1	71-0405
BOX-TYPE	VIBRATION OF A BOX-TYPE STRUCTURE - 2.	71-0420
BR-1	AIRCRAFT FUEL TANK VULNERABILITY TO HYDRAULIC RAM: MODIFICATTION OF THE NORTHROP FINITE ELEMENT COMP	74-0724
BR-1HR	AIRCRAFT FUEL TANK VULNERABILITY TO HYDRAULIC RAM: MODIFICATTION OF THE NORTHROP FINITE ELEMENT COMP	74-0724
BRACINGS	CALCUL DES PONTS BIAIS A POUTRES MULTIPLES SANS ENTRETOISES PAR LA METHODE DES ELEMENTS (CALCULATION	72-0184
BRACKETS	STEP AND LOADS EFFECT ON STRESSES IN PRESTRESSED CONCRETE SHORT BRACKETS	71-0334
BRAIN	A DYNAMIC FINITE ELEMENT MODEL OF THE HUMAN BRAIN	74-1076
BRAIN	BRAIN TISSUE FRAGILITY - A FINITE STRAIN ANALYSIS BY A HYBRID FINITE-ELEMENT METHOD	75-0162
BRAIN	BRAIN TISSUE FRAGILITY - A FINITE STRAIN ANALYSIS BY A HYBRID FINITE-ELEMENT METHOD	75-0243
BRAIN	MECHANICS OF BRAIN TISSUE FRAGILITY	73-0214
BRAIN	MODES OF MOTION OF A FINITE-ELEMENT BRAIN MODEL	75-0592
BRAIN-TISSUE	BRAIN-TISSUE FRAGILITY - FINITE STRAIN ANALYSIS BY A HYBRID FINITE-ELEMENT METHOD	75-0631
BRAKE	AN AXISYMMETRIC FINITE ELEMENT ANALYSIS OF THE MECHANICAL AND THERMAL STRESSES IN BRAKE DRUMS	74-0311
BRAKE	EFFECTS OF SHOE FORCE GEOMETRY ON HEAVY DUTY INTERNAL SHOE BRAKE PERFORMANCE	68-0219
BRAKE	THEORETICAL PREDICTION OF DISC BRAKE TEMPERATURES AND A COMPARISON WITH EXPERIMENTAL DATA	72-0306
BRAKES	A THERMAL, THERMOELASTIC, AND WEAR ANALYSIS OF HIGH-ENERGY DISK BRAKES	74-0839
BRANCH	INELASTIC ANALYSIS OF A BRANCH SHELL JUNCTION	74-0682
BRANCHED	DISCRETE-ELEMENT STATIC ANALYSIS OF BONDED, DOUBLE-LAYER, BRANCHED, THIN SHELLS OF REVOLUTION, PART	68-0102
BRANCHED	DISCRETE-ELEMENT STATIC ANALYSIS OF BONDED, DOUBLE-LAYER, BRANCHED, THIN SHELLS OF REVOLUTION, PART	70-0081
BRANCHING	CONTINUUM AND FINITE ELEMENT BRANCHING STUDIES OF THE CIRCULAR PLATE.	72-0427
BRANCHING	CRACK BRANCHING IN HOMALITE-100 SHEETS	72-0843
BRANCHING	CRACK BRANCHING IN HOMALITE-100 SHEETS	74-0291
BREADTH	LIMITATIONS ON THE APPLICATION OF EFFECTIVE BREADTH CONCEPT IN BOX GIRDER DESIGN	74-0493
BREAKDOWN	FEASIBILITY OF USING FINITE ELEMENTS IN ANALYSIS OF SECOND BREAKDOWN IN SEMICONDUCTOR DEVICES	73-0874
BREAKER	SEISMIC ANALYSIS OF A THREE-PHASE BREAKER INSTALLATION	74-0622
BREECH	ON MAXIMUM FILLET STRESSES IN BREECH RING	72-0736
BRICK	AN ASSUMED STRESS VERSION OF THE WILSON 8-NODE ISOPARAMETRIC BRICK	72-0532
BRICK	QUADRATURE RULES FOR BRICK BASED FINITE ELEMENTS	71-0531
BRICK	SOME NONCONFORMING VARIANTS ON THE TRILINEAR ISOPARAMETRIC BRICK FOR LINEAR ELASTIC PROBLEMS	72-0531
BRICK-TYPE	A TECHNIQUE FOR DEGENERATING BRICK-TYPE ISOPARAMETRIC ELEMENTS USING HIERARCHICAL MIDSIDE NODES	74-0403
BRICK-TYPE	DEGENERATION OF BRICK-TYPE ISOPARAMETRIC ELEMENTS	73-1031
BRICKS	ANALYSIS OF LOADING AND THERMAL EFFECTS ON FUELLED GRAPHITE BRICKS FOR A HIGH-TEMPERATURE GAS COOLED	71-0085
BRICKS	ANALYSIS OF THERMAL STRESSES IN CORED CERAMIC BRICKS USED IN HYPERSONIC WIND TUNNEL HEATERS	72-0655
BRICKWORK	DIAGONAL TENSILE STRENGTH OF BRICKWORK	70-0158
BRIDAGE	TENDANCES ACTUELLES DES RECHERCHES SUR LES CONTRAINTES ET DEFORMATIONS DUES AU BRIDAGE EN RELATION A	74-0273
BRIDGE	A DISCRETE-ELEMENT METHOD OF MULTIPLE-LOADING ANALYSIS FOR TWO-WAY BRIDGE FLOOR SLABS	70-0435
BRIDGE	A DISCRETE-ELEMENT METHOD OF ANALYSIS FOR ORTHOGONAL SLAB AND GRID BRIDGE FLOOR SYSTEMS	72-0690
BRIDGE	AN APPLICATION OF FINITE ELEMENT TECHNIQUE IN BRIDGE DESIGN	70-0225
BRIDGE	AN INTERACTIVE COMPUTER-AIDED MODELING SYSTEM FOR THE DISCRETIZATION OF BRIDGE STRUCTURES FOR FINITE	73-1165
BRIDGE	ANALYSIS OF BRIDGE DECKS USING POLYMER IMPREGNATED CONCRETE	73-0673
BRIDGE	APPLICATIONS OF FINITE ELEMENT METHODS TO UNSTEADY VISCOUS FLOW AROUND A BOX-GIRDER BRIDGE OSCILLATI	74-0827
BRIDGE	CALCULATION OF THE BRIDGE DECKS BY USING NUMERICAL METHODS. (LE CALCUL DES TABLIERS DE PONTS PAR DES	71-0411
BRIDGE	DEVELOPMENT OF A DESIGN FOR A PRECAST CONCRETE BRIDGE BEAM OF U-SECTION	73-0392
BRIDGE	EXPERIMENTAL AND ANALYTICAL INVESTIGATION OF A HORIZONTALLY CURVED BOX-BEAM HIGHWAY BRIDGE MODEL	71-0024
BRIDGE	FIELD TESTS OF A STEEL-COMPOSITE BOX-GIRDER BRIDGE	75-0633
BRIDGE	FINITE ELEMENT ANALYSIS OF SKEW, CURVED BOX-GIRDER BRIDGE	70-0241
BRIDGE	FINITE ELEMENT ANALYSIS OF A BROAD ARCH BRIDGE WITH A SLAB DECK	71-0724
BRIDGE	FINITE ELEMENT ANALYSIS OF PRESTRESSED CONCRETE VOIDED BRIDGE DECKS	73-0106
BRIDGE	FINITE ELEMENT APPROACH TO POLYMER CONCRETE BRIDGE DECK DESIGN ANALYSIS	72-0918
BRIDGE	FINITE ELEMENT PROGRAMS FOR SLAB BRIDGE DESIGN	71-0082

BRIDGE	FINITE ELEMENT SCHEME FOR BOX BRIDGE ANALYSIS	75-0666
BRIDGE	FINITE-ELEMENT ANALYSIS OF BRIDGE DECKS	72-0672
BRIDGE	INTERACTIVE TECHNIQUES WITH GRAPHICAL OUTPUT FOR BRIDGE ANALYSES	75-0622
BRIDGE	LOAD DISTRIBUTION IN A COMPOSITE STEEL BOX GIRDER BRIDGE	73-0765
BRIDGE	MODEL ANALYSIS OF CURVED BOX-BEAM HIGHWAY BRIDGE	71-0282
BRIDGE	MULTI-CELL BRIDGE DECKS - MODEL TESTS AND THEORETICAL ANALYSIS	72-0381
BRIDGE	PREDICTION OF LONG-TERM STRESS RANGES: USER'S MANUAL - BRIDGE LOAD GENERATOR	73-0737
BRIDGE	PREDICTION OF LONG-TERM STRESS RANGES: USER'S MANUAL - BRIDGE DYNAMIC STRESS ANALYSIS	74-0807
BRIDGE	SLAB BRIDGE DESIGN AND DRAWING. AN AUTOMATED PROCESS	70-0347
BRIDGE	STRUCTURAL BEHAVIOR OF A TWO SPAN REINFORCED CONCRETE BOX GIRDER BRIDGE MODEL. VOLUME 11. REDUCTION,	71-0659
BRIDGE	THE ANALYSIS OF CURVILINEAR ORTHOTROPIC CURVED BRIDGE DECKS	69-0162
BRIDGE	THE ANALYSIS OF SKEW BRIDGE DECKS, A NEW FINITE ELEMENT APPROACH	69-0234
BRIDGE	THE COMPARISON OF THE RESPONSE OF A HIGHWAY BRIDGE TO UNIFORM GROUND SHOCK AND MOVING GROUND EXCITAT	72-0737
BRIDGES	A FINITE ELEMENT APPROACH TO THE DYNAMIC ANALYSIS OF CONTINUOUS HIGHWAY BRIDGES	72-0591
BRIDGES	A HIGHER ORDER IN-PLANE PARALLELOGRAM ELEMENT AND ITS APPLICATION TO SKEWED CURVED BOX GIRDER BRIDGE	71-0002
BRIDGES	A THEORETICAL AND EXPERIMENTAL STUDY ON THE FREE VIBRATION OF CURVED SLAB BRIDGES	71-0496
BRIDGES	ANALYSIS AND DESIGN OF SKEW BOX GIRDER BRIDGES	72-0659
BRIDGES	ANALYSIS OF BOX GIRDER BRIDGES BY FINITE STRIP METHOD	69-0050
BRIDGES	ANALYSIS OF CURVED BOX GIRDER BRIDGES BY FINITE STRIP METHOD	71-0485
BRIDGES	ANALYSIS OF CURVED BOX GIRDER BRIDGES	73-0294
BRIDGES	ANALYSIS OF MULTI-CELL BRIDGES WITHOUT TRANSVERSE DIAPHRAGMS, A FINITE ELEMENT APPROACH	69-0235
BRIDGES	ANALYSIS OF SLAB AND GIRDER BRIDGES BY THE FINITE STRIP METHOD	70-0210
BRIDGES	ANALYSIS OF SLAB BEAM BRIDGES	69-0077
BRIDGES	ANALYSIS OF SLAB BRIDGES BY FINITE STRIP METHOD	73-0687
BRIDGES	ANALYSIS OF SLAB BRIDGES BY FINITE ELEMENTS	73-0695
BRIDGES	ANALYSIS OF SPINE BEAM BRIDGES USING FINITE ELEMENTS	70-0236
BRIDGES	ANALYTICAL AND EXPERIMENTAL STUDIES OF MULTI-CELL CONCRETE BOX GIRDER BRIDGES	74-1134
BRIDGES	ANALYTICAL MODELING OF BEAM-SLAB BRIDGES	74-1131
BRIDGES	CALCUL DES PONTS BIAIS A POUTRES MULTIPLES SANS ENTRETOISES PAR LA METHODE DES ELEMENTS (CALCULATION	72-0184
BRIDGES	COMPUTER ANALYSIS OF SEGMENTALLY ERECTED BRIDGES	75-0172
BRIDGES	DESIGN OF INCLINED TIED ARCH RAILWAY BRIDGES OVER THE M56	71-0297
BRIDGES	DIAPHRAGMS IN SINGLE- AND DOUBLE-CELL BOX GIRDER BRIDGES WITH VARYING ANGLE OF SKEW	72-0130
BRIDGES	DYNAMIC ANALYSIS OF BEAM-SLAB HIGHWAY BRIDGES	73-0199
BRIDGES	DYNAMIC ANALYSIS OF CURVED BOX-GIRDER BRIDGES	75-0280
BRIDGES	EARTH PRESSURES AGAINST THE ABUTMENTS AND WING WALLS OF STANDARD MOTORWAY BRIDGES	75-0629
BRIDGES	FINITE ELEMENT ANALYSIS OF SKEW COMPOSITE GIRDER BRIDGES	67-0150
BRIDGES	FINITE ELEMENT ANALYSIS OF BRIDGES CURVED IN PLAN	71-0547
BRIDGES	FINITE ELEMENT ANALYSIS OF SKEW BOX GIRDER BRIDGES	72-0209
BRIDGES	FINITE ELEMENT ANALYSIS OF CABLE STAYED BRIDGES	72-0494
BRIDGES	FINITE ELEMENT AND FINITE STRIP ANALYSES OF BRIDGES	72-0490
BRIDGES	FINITE ELEMENT OF SKEW VAULT BRIDGES	72-0434
BRIDGES	FINITE-ELEMENT ANALYSIS OF BRIDGES	72-0973
BRIDGES	FREE VIBRATION OF CURVED AND STRAIGHT BEAM SLAB OR BOX GIRDER BRIDGES	72-0487
BRIDGES	FREQUENCY ANALYSIS OF CERTAIN SINGLE AND CONTINUOUS SPAN BRIDGES	71-0337
BRIDGES	HORIZONTALLY CURVED HIGHWAY BRIDGES. A FULLY COMPATIBLE ANNULAR SEGMENT FINITE ELEMENT	72-0834
BRIDGES	INFLUENCE CHARACTERISTICS FOR SLAB BRIDGES	71-0188
BRIDGES	NEW FINITE ELEMENTS WITH APPLICATION TO BOX GIRDER BRIDGES.	72-0465
BRIDGES	ORTHOTROPIC RIGHT BRIDGES BY THE FINITE STRIP METHOD	69-0028
BRIDGES	POST-TENSIONED SLAB BRIDGES ON ISOLATED SUPPORTS	75-0084
BRIDGES	SEISMIC RESPONSE OF HIGHWAY BRIDGES	75-0879
BRIDGES	SEMI-ANALYTIC SOLUTION OF SKEW BOX GIRDER BRIDGES	75-0578
BRIDGES	SHEAR LAG IN STEEL BOX GIRDER BRIDGES	75-0310
BRIDGES	SLAB BRIDGES WITH ARBITRARY SHAPE AND SUPPORT CONDITIONS - A GENERAL METHOD OF ANALYSIS BASED ON FIN	68-0233
BRIDGES	STATIC STRUCTURAL ANALYSIS OF SUSPENSION BRIDGES	66-0110
BRIDGES	STRESSES IN CONTINUOUS CONCRETE BOX GIRDER BRIDGES	71-0233
BRIDGES	THEORY OF EQUIVALENCE AND ITS APPLICATION TO VARIOUS PROBLEMS IN ELASTICITY AND DESIGN OF SKEW BRIDG	72-0470
BRIDGES	THERMAL STRESSES IN STEEL-CONCRETE COMPOSITE BRIDGES	75-0154
BRIDGES	TWO-DIMENSIONAL ANALYSIS OF BACKWATER AT BRIDGES	74-0184
BRITTLE	A STATISTICAL THEORY FOR THE FRACTURE OF BRITTLE STRUCTURES SUBJECTED TO NONUNIFORM POLYAXIAL STRESS	73-0928
BRITTLE	BRITTLE MATERIALS: A DESIGN CHALLENGE	75-0769
BRITTLE	FINITE ELEMENT ANALYSIS ON UNIAXIAL COMPRESSIVE STRENGTH OF CYLINDRICAL BRITTLE SPECIMEN	70-0155
BRITTLE	FINITE ELEMENT ANALYSIS IN BRITTLE MATERIAL DESIGN	70-0360
BRITTLE	FINITE ELEMENT INVESTIGATION OF SECONDARY CRACKING DURING BRITTLE FRACTURE	71-0804
BRITTLE	FINITE-ELEMENT ANALYSIS OF FRACTURE PROPAGATION IN TWO-DIMENSIONAL ELASTIC BRITTLE SOLIDS	72-0948
BRITTLE	ON THE STATES OF STRESS AND DEFORMATION OF CYLINDRICAL SPECIMENS OF BRITTLE MATERIAL UNDER UNIAXIAL	69-0150
BRITTLE	STUDY ON STATE OF STRESS AND DEFORMATION OF CYLINDRICAL SPECIMENS OF BRITTLE MATERIAL UNDER UNIAXIAL	69-0152
BRLESC	BRLESC FINITE ELEMENT PROGRAM FOR AXISYMMETRIC, PLANE STRAIN, AND PLANE STRESS, ORTHOTROPIC SOLIDS W	71-0749
BRLESC	PLOTS - THE PLOTTING SUBROUTINE INCORPORATED IN THE BALLISTIC RESEARCH LABORATORIES ELECTRONIC SCIEN	72-0803
BROAD	FINITE ELEMENT ANALYSIS OF A BROAD ARCH BRIDGE WITH A SLAB DECK	71-0724
BROKEN	STRESS DISTRIBUTION NEAR A BROKEN POINT OF FILAMENT IN UNIDIRECTIONALLY FIBER REINFORCED COMPOSITES	73-0811
BRZEGOWYCH	OCENA DOKLADNOSCI ROZWIAZAN PROBLEMOW BRZEGOWYCH SPREZYSTYCH CIAL DYSKERTYZOWANYCH. (ASSESSING THE A	74-0921
BUCKLE-SHELL	PLASTIC SANDWICH BUCKLE-SHELL STRUCTURES	72-0119
BUCKLE-SHELL	THE FINITE ELEMENT ANALYSIS OF SANDWICH BUCKLE-SHELL STRUCTURES	72-0968
BUCKLING	A COMPATIBLE ELEMENT FOR BUCKLING ANALYSIS OF SKEW PLATES WITH PARALLEL EDGES	73-0328
BUCKLING	A CONFORMING QUADRILATERAL ELEMENT FOR BUCKLING ANALYSIS OF STIFFENED PLATES	73-0613
BUCKLING	A FINITE ELEMENT METHOD FOR THE PLASTIC BUCKLING ANALYSIS OF PLATES	69-0390
BUCKLING	A FINITE ELEMENT TENSOR APPROACH TO PLATE BUCKLING AND POST BUCKLING	73-0883
BUCKLING	A FINITE ELEMENT TENSOR APPROACH TO PLATE BUCKLING AND POST BUCKLING	73-0883
BUCKLING	A FINITE-ELEMENT METHOD FOR THE PLASTIC BUCKLING ANALYSIS OF PLATES	69-0214
BUCKLING	A GENERAL TECHNIQUE FOR BUCKLING ANALYSIS WITH A FINITE ELEMENT SOLUTION APPLICATION	66-0036
BUCKLING	A MATRIX DISPLACEMENT METHOD ON PRE AND POST BUCKLING ANALYSIS OF LINERS FOR REACTOR VESSELS	71-0267
BUCKLING	A METHOD OF SOLUTION FOR THE COMPLICATED BUCKLING PROBLEMS OF ELASTIC PLATES WITH COMBINED USE OF RA	68-0097
BUCKLING	A PROCEDURE FOR FINITE ELEMENT PLATE AND SHELL PRE AND POST BUCKLING ANALYSIS	71-0096
BUCKLING	ACCURACY OF SOME NUMERICAL METHODS FOR COLUMN BUCKLING	75-0583
BUCKLING	ADAPTION OF KOITER'S METHOD TO FINITE ELEMENT ANALYSIS OF SNAP-THROUGH BUCKLING BEHAVIOUR	71-0109
BUCKLING	AN ELASTIC FINITE-ELEMENT ANALYSIS OF THE BUCKLING OF ECCENTRICALLY STIFFENED PLATES AND SHELLS	73-1136
BUCKLING	AN ITERATION METHOD FOR THE NONLINEAR BUCKLING OF FRAME STRUCTURES	66-0107
BUCKLING	ANALYSIS OF BUCKLING AND VIBRATION OF RING-STIFFENED, SEGMENTED SHELLS OF REVOLUTION	69-0496
BUCKLING	ANALYSIS OF ELASTIC-PLASTIC BUCKLING OF PLATES BY THE FINITE ELEMENT METHOD	69-0263
BUCKLING	ANALYSIS OF ELASTIC-PLASTIC CREEP BUCKLING OF AXISYMMETRIC SHELLS BY FINITE-ELEMENT METHOD	75-0355
BUCKLING	APPLICATION OF A REFINED PLATE BENDING ELEMENT TO BUCKLING PROBLEMS	74-0937
BUCKLING	APPLICATION OF FINITE ELEMENT METHOD TO AXISYMMETRIC BUCKLING OF SHALLOW SPHERICAL SHELLS UNDER EXTE	73-0111
BUCKLING	APPLICATION OF THE DISCRETE ELEMENT METHOD TO THE BUCKLING ANALYSIS OF RECTANGULAR PLATES UNDER ARBI	68-0054
BUCKLING	APPLICATION OF THE FINITE ELEMENT METHOD TO THE BUCKLING OF RECTANGULAR PLATES AND PLATES ON ELASTIC	73-0478
BUCKLING	BEAM BUCKLING BY FINITE ELEMENT PROCEDURE	74-0123
BUCKLING	BUCKLING OF COOLING-TOWER SHELLS: BIFURCATION RESULTS	75-0384
BUCKLING	BUCKLING OF COOLING-TOWER SHELLS: STATE-OF-THE-ART	75-0385
BUCKLING	BUCKLING OF CYLINDRICAL SHELLS BY WIND PRESSURE	74-0642
BUCKLING	BUCKLING OF DEEP SHELLS	73-0849
BUCKLING	BUCKLING OF INELASTIC I-BEAMS UNDER MOMENT GRADIENT	74-0765
BUCKLING	BUCKLING OF MULTILAYER PLATES BY FINITE ELEMENTS	67-0103
BUCKLING	BUCKLING OF OPEN CYLINDRICAL SHELLS	67-0115

BUCKLING	BUCKLING OF PLATE GIRDER WEBS UNDER PARTIAL EDGE LOADINGS	70-0283
BUCKLING	BUCKLING OF SEGMENTS OF TOROIDAL SHELLS	73-0515
BUCKLING	BUCKLING OF SHELLS BY FINITE ELEMENT METHOD	74-0091
BUCKLING	BUCKLING PROBLEMS IN THE DESIGN OF FIBREGLASS-REINFORCED PLASTIC SHIPS	72-0164
BUCKLING	BUCKLING UNDER COMBINED LOADING OF THIN FLAT-WALLED STRUCTURES BY A COMPLEX FINITE STRIP METHOD	74-0788
BUCKLING	CALCULATION OF THE ELASTIC BUCKLING LOADS OF THIN FLAT REINFORCED PLATES USING TRIANGULAR FINITE ELE	73-1024
BUCKLING	CALCULATION OF THE ELASTIC BUCKLING LOADS OF THIN FLAT REINFORCED PLATES USING TRIANGULAR FINITE ELE	75-0177
BUCKLING	COUPLED LOCAL BUCKLING IN WIDE-FLANGE BEAM-COLUMNS	73-0235
BUCKLING	DETERMINATION OF THE INITIAL BUCKLING STRESS OF FLAT PLATES USING THE DISPLACEMENT METHOD	69-0465
BUCKLING	DETERMINATION OF THE LOWEST CRITICAL BUCKLING STRESS OF VARIOUS SHAPED PLATES BY FINITE-ELEMENTS	73-1134
BUCKLING	DIE BEULLASTEN DIAGONAL VERSTEIFTER RECHTECKPLATTEN (BUCKLING LOADS IN DIAGONALLY STIFFENED RECTANGU	73-0405
BUCKLING	DISCUSSION OF PROCEEDINGS, PAPER 6612, ELASTIC BUCKLING OF TWO-COMPONENT COLUMNS	70-0321
BUCKLING	ELASTIC LATERAL BUCKLING OF STEEL BEAMS	70-0364
BUCKLING	ELASTIC PLASTIC BUCKLING OF STIFFENED RECTANGULAR PLATES	69-0122
BUCKLING	ELASTIC-PLASTIC BUCKLING OF PLATES USING THE FINITE ELEMENT METHOD	67-0094
BUCKLING	ELASTIC-PLASTIC BUCKLING OF PLATES BY FINITE ELEMENT METHOD	69-0253
BUCKLING	FINITE DEFLECTIONS, BUCKLING AND POSTBUCKLING OF AN ARCH	69-0123
BUCKLING	FINITE ELEMENT ALGORITHM FOR THE DETERMINATION OF DYNAMIC BUCKLING	75-0302
BUCKLING	FINITE ELEMENT ANALYSIS FOR BUCKLING OF SEMI-RIGID SPACE FRAMES	68-0270
BUCKLING	FINITE ELEMENT ANALYSIS OF PLATE BUCKLING AND POST-BUCKLING	70-0332
BUCKLING	FINITE ELEMENT ANALYSIS OF PLATE BUCKLING USING A MIXED VARIATIONAL PRINCIPLE	71-0020
BUCKLING	FINITE ELEMENT ANALYSIS OF BUCKLING AND POST-BUCKLING BEHAVIORS OF ARCHES WITH GEOMETRIC IMPERFECTIO	73-0186
BUCKLING	FINITE ELEMENT BUCKLING ANALYSIS OF HOMOGENEOUS AND SANDWICH PLATES	75-0613
BUCKLING	FINITE ELEMENT CREEP BUCKLING ANALYSIS OF VISCO-ELASTIC SHALLOW SPHERICAL SHELL	75-0159
BUCKLING	FINITE ELEMENT ELASTIC BUCKLING ANALYSIS	67-0107
BUCKLING	FINITE ELEMENT POST BUCKLING ANALYSIS OF THIN ELASTIC PLATES	68-0127
BUCKLING	FINITE ELEMENT SOLUTIONS FOR BUCKLING OF COLUMNS AND BEAMS	71-0361
BUCKLING	FINITE-ELEMENT FOR BUCKLING OF CURVED BEAMS AND SHELLS WITH SHEAR	75-0607
BUCKLING	FINITE-ELEMENT NONLINEAR BUCKLING ANALYSIS OF A PSEUDO-CYLINDRICAL CONCAVE POLYHEDRAL SHELL UNDER EX	73-1147
BUCKLING	GENERAL SINUSOIDAL STIFFNESS MATRICES FOR BUCKLING AND VIBRATION ANALYSES OF THIN FLAT-WALLED STRUCT	68-0278
BUCKLING	INELASTIC BUCKLING ANALYSIS OF PIPES SUBJECTED TO INTERNAL PRESSURE, FLEXURE AND AXIAL LOADING	74-0515
BUCKLING	INELASTIC BUCKLING OF STEEL BEAMS UNDER NON UNIFORM MOMENT	75-0149
BUCKLING	INSTABILITY OF HORIZONTALLY CURVED MEMBERS. FLANGE BUCKLING STUDIES	69-0480
BUCKLING	LARGE DEFLECTION SPATIAL BUCKLING OF THIN WALLED BEAMS AND FRAMES	73-1085
BUCKLING	LARGE-DEFLECTION ELASTO-PLASTIC BUCKLING ANALYSIS OF PLATES USING FINITE ELEMENTS	73-0394
BUCKLING	LINEAR BUCKLING AND GEOMETRICALLY NONLINEAR ANALYSIS OF PLANAR PLATE-STIFFENER TYPE STRUCTURES BY TH	74-1097
BUCKLING	LOWER BOUNDS TO COLUMN BUCKLING LOADS	73-0485
BUCKLING	NASTRAN BUCKLING ANALYSIS OF A LARGE STIFFENED CYLINDRICAL SHELL WITH A CUTOUT	71-0045
BUCKLING	NONLINEAR BUCKLING ANALYSIS OF SANDWICH ARCHES	71-0292
BUCKLING	NONLINEAR BUCKLING ANALYSIS OF COMPOSITE SHELLS	74-0973
BUCKLING	NONLINEAR BUCKLING ANALYSIS OF COMPOSITE SHELLS	75-0214
BUCKLING	ON BUCKLING OF UNIDIRECTIONAL BORON-ALUMINUM STIFFENERS - A CAUTION TO DESIGNERS	75-0634
BUCKLING	ON THE DYNAMIC BUCKLING OF SHELLS OF REVOLUTION	73-1102
BUCKLING	OPTIMAL DESIGN FOR PRESCRIBED BUCKLING LOADS	74-0568
BUCKLING	OPTIMAL DESIGN OF STRUCTURES WITH BUCKLING CONSTRAINTS	73-0246
BUCKLING	PART 1 - TASK 3.2.9. BENDING AND BUCKLING OF A ONE-DIMENSIONAL MODEL SIMULATING A CONICAL SHELL. PAR	71-0745
BUCKLING	PLATE BUCKLING ANALYSIS USING A FULLY COMPATIBLE FINITE ELEMENT	69-0065
BUCKLING	PROBABILISTIC BUCKLING BEHAVIOR OF STRUCTURES WITH RANDOM LACK-OF-FIT AND GEOMETRIC IMPERFECTIONS	73-0862
BUCKLING	REINVESTIGATION OF BUCKLING OF SHELLS OF REVOLUTION BY A REFINED FINITE ELEMENT	74-0097
BUCKLING	SHEAR BUCKLING OF SQUARE PERFORATED PLATES	74-0713
BUCKLING	SIMPLIFIED HYBRID DISPLACEMENT METHOD APPLIED TO PLATE BUCKLING PROBLEMS	72-0781
BUCKLING	SNAP-THROUGH BUCKLING ANALYSIS OF SHELLS OF REVOLUTION USING INCREMENTAL STIFFNESS MATRICES	74-1349
BUCKLING	SNAP-THROUGH BUCKLING OF RETICULATED SHELLS	75-0640
BUCKLING	SOME APPLICATIONS OF FINITE ELEMENT ANALYSIS TO SHELL BUCKLING PREDICTION	68-0023
BUCKLING	STATIC AND DYNAMIC BUCKLING OF THIN-WALLED COLUMNS USING FINITE ELEMENTS	70-0505
BUCKLING	STATIC, VIBRATION AND BUCKLING ANALYSIS OF AXISYMMETRIC CIRCULAR PLATES USING FINITE ELEMENTS	73-0157
BUCKLING	THE APPLICATION OF GRADIENT MINIMIZATION METHODS AND HIGHER ORDER DISCRETE ELEMENTS TO SHELL BUCKLIN	71-0237
BUCKLING	THE FINITE ELEMENT METHOD IN PLATE BUCKLING	69-0128
BUCKLING	THE FINITE ELEMENT METHOD APPLIED TO THE STATIC AND DYNAMIC BUCKLING OF CIRCULAR ARCHES	70-0497
BUCKLING	THE USE OF FINITE ELEMENT MID-INCREMENT STIFFNESS MATRICES IN THE POST BUCKLING ANALYSIS OF IMPERFEC	71-0383
BUCKLING	UNSYMMETRIC BUCKLING OF SHELLS OF REVOLUTION BY FINITE ELEMENT METHOD	70-0500
BUCKLING	USE OF THE FINITE ELEMENT METHOD IN EXAMINING ELASTIC BUCKLING OF THIN PLATES OF ANY SHAPE	75-0153
BUCKLING	VIBRATION AND BUCKLING ANALYSIS OF AXISYMMETRIC POLAR ORTHOTROPIC CIRCULAR PLATES	74-0574
BUGEY	DETERMINATION DU CHAMP DES CONTRAINTES DANS LE CAISSON DU REACTEUR DE BUGEY L'	67-0121
BUGEY	PHOTOELASTIC DETERMINATION OF THE STRESSES IN THE FOUNDATION OF THE NUCLEAR-POWER STATION BUGEY 2	70-0834
BUILD-UP	FLEXURAL RIGIDITY OF A THIN WALLED BUILD-UP ROTOR FOR JET ENGINE. MEASUREMENT BY STATIC LOAD TEST AND	73-0756
BUILDING	A SEISMIC DESIGN OF NUCLEAR REACTOR BUILDING, STRESS ANALYSIS AND STIFFNESS EVALUATION OF THE ENTIRE	71-0251
BUILDING	A SEISMIC DESIGN OF NUCLEAR REACTOR BUILDING, STRESS ANALYSIS AND STIFFNESS EVALUATION OF THE ENTIRE	71-0251
BUILDING	BERECHNUNGSMETHODEN ZUR ERMITTLUNG VON BELASTUNGSBEDINGUNGEN UND GRENZTRAGFAEHIGKEITSVERHALTEN VON K	73-0858
BUILDING	BUILDING THREE NEW MILLION-HORSEPOWER HYDRAULIC TURBINES FOR GRAND COULEE	74-0981
BUILDING	BUILDING VIBRATIONS AND ACOUSTICS	73-0656
BUILDING	COMPARISON OF THE THREE-DIMENSIONAL ANALYSIS OF A SHEAR WALL BUILDING BY TWO DIFFERENT METHODS	75-0705
BUILDING	COMPUTER - AIDED DESIGN - STANDARD OIL OF INDIANA BUILDING	73-0182
BUILDING	DYNAMICS OF BUILDING - SOIL INTERACTION	72-0856
BUILDING	FINITE ELEMENT MODELLING TECHNIQUES FOR THE ANALYSIS OF BUILDING VIBRATIONS	72-0839
BUILDING	PROBLEMS IN BUILDING AND MANIPULATING LARGE STIFFNESS MATRICES	71-0596
BUILDING	SOME STRUCTURAL PROBLEMS - STANDARD OIL OF INDIANA BUILDING	73-0180
BUILDING	THE EFFECT OF COMPOSITE ACTION ON THE ELASTOPLASTIC ANALYSIS OF COMPLETE BUILDING STRUCTURES	74-0441
BUILDINGS	ANALYSIS OF BUILDINGS ON ELASTIC MEDIUM UNDER DYNAMIC OR SEISMIC LOADS	74-0456
BUILDINGS	FINITE ELEMENT MODEL WITH APPLICATIONS TO BUILDINGS K-33 AND K-31	70-0420
BUILDINGS	FINITE-ELEMENT ANALYSIS OF TALL BUILDINGS	72-0959
BUILDINGS	LINEAR ELASTIC FINITE ELEMENT ANALYSIS OF MASONRY WALLS ON BUILDINGS	74-1015
BUILDINGS	MEMBRANE ANALYSIS OF UNFRAMED HIGH RISE BUILDINGS	73-0451
BUILDINGS	MULTIDEGREE DYNAMIC ANALYSIS OF TALL BUILDINGS SUBJECTED TO WIND AS A STOCHASTIC PROCESS	74-1196
BUILDINGS	NONLINEAR ANALYSIS OF REINFORCED CONCRETE BUILDINGS	73-0324
BUILDINGS	STRUCTURAL ANALYSIS OF TALL BUILDINGS HAVING IRREGULARLY POSITIONED SHEAR WALLS	73-0282
BUILDINGS	SUBSTRUCTURE ANALYSIS OF MULTISTOREY FLAT SLAB BUILDINGS	74-0394
BUILDINGS	THREE DIMENSIONAL ANALYSIS OF BUILDINGS COMPOSED OF FLOOR AND WALL PANELS	71-0387
BUILDINGS	THREE-DIMENSIONAL ANALYSIS OF SHEAR WALL BUILDINGS TO LATERAL LOAD	74-0220
BUILT	FINITE ELEMENT ANALYSIS OF CENTRIFUGED AND BUILT UP SLOPES	74-1376
BUILT-UP	A RELAXATION PROCEDURE FOR COMPUTING STRESSES IN BUILT-UP STRUCTURES	69-0063
BUILT-UP	FINITE ELEMENT ANALYSIS OF CENTRIFUGED AND BUILT-UP SLOPES	74-0415
BUILT-UP	VIBRATION STUDIES OF A FLAT PLATE AND BUILT-UP WING	71-0238
BUILTOVER	APPLICATION OF NEW AUSTRIAN TUNNELLING METHOD IN DIFFICULT BUILTOVER AREAS IN FRANKFURT/MAIN METRO	73-1025
BULGING	FINITE ELEMENT METHOD OF ANALYSIS OF THE HYDROSTATIC BULGING OF A SHEET METAL - 1.	74-0958
BULKHEAD	DYNAMICS OF A FLEXIBLE BULKHEAD AND CONTAINED FLUID	73-0082
BULKHEAD	FINITE ELEMENT ANALYSIS OF ANCHORED BULKHEAD BEHAVIOR	74-1068
BULKHEAD	NEW METHOD OF BULKHEAD ANALYSIS IN AIRCRAFT STRUCTURES	74-0334
BULKHEADS	A FINITE-ELEMENT METHOD FOR ANALYSIS OF ANCHORED BULKHEADS AND ANCHOR WALLS	66-0130
BULKHEADS	APPLICATION OF THE FINITE ELEMENT METHOD TO THE DESIGN OF UNDERGROUND BULKHEADS	72-0387
BULKHEADS	THE ANALYSIS OF FLEXIBLE BULKHEADS IN SANDS	74-0369
BULLETIN	BULLETIN DE L ASSOCIATION TECHNIQUE MARITIME ET AERONAUTIQUE (BULLETIN OF THE TECHNICAL MARITIME AND	74-1108

BULLETIN	BULLETIN DE L ASSOCIATION TECHNIQUE MARITIME ET AERONAUTIQUE (BULLETIN OF THE TECHNICAL MARITIME AND	74-1108
BUOYANT	SOLUTION OF A TWO DIMENSIONAL DIFFUSION CONVECTION EQUATION DESCRIBING THE MASS TRANSFER OF A NEUTRA	74-0037
BUREAU	JOURNAL OF RESEARCH OF THE NATIONAL BUREAU OF STANDARDS: C. ENGINEERING AND INSTRUMENTATION	71-0664
BURIED	DYNAMIC RESPONSE OF A CYLINDER BURIED IN AN EARTH BEAM - RESULTS OF A FINITE ELEMENT ANALYSIS	72-0592
BURIED	EFFECT OF BACKPACKING AND INTERNAL PRESSURIZATION ON STRESSES TRANSMITTED TO BURIED CYLINDERS	73-0897
BURIED	FINITE ELEMENT ANALYSIS OF BURIED CYLINDERS	71-0817
BURIED	FINITE-ELEMENT ANALYSIS OF HEAT FLOW AROUND BURIED PIPES	73-1153
BURIED	FUNDAMENTAL STUDIES OF MEDIUM-STRUCTURE INTERACTION. REPORT 1. FINITE ELEMENT ANALYSIS OF BURIED CYL	72-0854
BURIED	NUMERICAL SOLUTIONS FOR DETERMINING WAVE-INDUCED PRESSURE DISTRIBUTIONS AROUND BURIED PIPELINES	74-1334
BURIED	REDUCTION OF STRESSES IN BURIED STRUCTURES	71-0698
BURIED	RESPONSE OF BURIED CAPSULES IN THE HIGH-OVERPRESSURE REGION	70-0093
BURIED	SCATTERING BY BURIED OBSTACLES	74-0739
BURIED	STRESS DISTRIBUTIONS AROUND SHALLOW BURIED RIGID PIPES	74-0142
BUTTRESS	BUTTRESS DAMS ON COMPLEX ROCK FOUNDATIONS	64-0013
BUTTRESS	STRESSES IN BUTTRESS DAMS	65-0035
BUTTRESSES	A FINITE ELEMENT STRESS ANALYSIS OF THE VERTICAL BUTTRESSES OF A NUCLEAR CONTAINMENT VESSEL	69-0280
BWR	APPLICATION OF FINITE ELEMENT METHOD TO THE THREE DIMENSIONAL STRESS AND VIBRATION ANALYSIS OF BWR P	71-0180
BWR	STRESS ANALYSIS OF BWR PRESSURE VESSEL WITH BERSAFE AND FLHE	73-0406
C-0	NUMERICAL SOLUTION OF THE BIHARMONIC PROBLEM BY MIXED FINITE OF CLASS C-0	73-0803
C-5A	SPECTRUM TRUNCATION AND DAMAGE TOLERANCE STUDY ASSOCIATED WITH THE C-5A OUTBOARD PYLON AFT TRUSS LUG	74-0535
CAB	A SURVEY OF ANALYTICAL METHODS FOR DYNAMIC SIMULATION OF CAB	73-0936
CABLE	ANALYSIS OF CABLE NETS IN MIXED FORMULATION	73-0311
CABLE	ANALYSIS OF PNEUMATIC SHELLS WITH OR WITHOUT CABLE FORMULATION	74-0591
CABLE	APPLICATION OF THE FINITE-ELEMENT METHOD TO TOWED CABLE DYNAMICS	73-1141
CABLE	APPLICATION OF THE FINITE ELEMENT METHOD TO TOWED CABLE DYNAMICS	75-0809
CABLE	EXPERIMENTAL AND FINITE ELEMENT ANALYSIS OF CABLE ROOF STRUCTURES INCLUDING PRECAST PANELS	74-1139
CABLE	FINITE ELEMENT ANALYSIS OF CABLE STAYED BRIDGES	72-0494
CABLE	FINITE ELEMENT ANALYSIS OF NONLINEAR CABLE REINFORCED MEMBRANES	74-1073
CABLE	FINITE ELEMENT ANALYSIS OF THREE-DIMENSIONAL CABLE STRUCTURES	74-1275
CABLE	FINITE ELEMENT STRESS ANALYSIS OF A MULTICONDUCTOR ARMOR CABLE	73-0095
CABLE	MECHANICS OF CABLE MOORING SYSTEMS. VOLUME IV. A COMPUTER PROGRAM FOR ANALYZING THE STEADY STATE RES	72-0653
CABLE	MECHANICS OF CABLE MOORING SYSTEMS. VOLUME IV. A COMPUTER PROGRAM FOR ANALYZING THE STEADY STATE RES	72-0653
CABLE	MECHANICS OF CABLE MOORING SYSTEMS. VOLUME 1. THREE DIMENSIONAL RESPONSE OF DEEP WATER MOORING LINES	72-0654
CABLE	MECHANICS OF CABLE MOORING SYSTEMS. VOLUME V11. THE STEADY-STATE BEHAVIOR OF A PYRAMID ARRAY SYSTEM	73-0724
CABLE	MECHANICS OF CABLE MOORING SYSTEMS. VOLUME VI. A COMPUTER PROGRAM FOR ANALYZING THE STEADY STATE CON	73-0725
CABLE	MEMBRANE ANALOGY FOR ANISOTROPIC CABLE NETWORKS	74-0233
CABLE	NONLINEAR STATIC AND DYNAMIC RESPONSE OF UNDERWATER CABLE STRUCTURES USING THE FINITE ELEMENT METHOD	75-0460
CABLE	WHIRLING CABLE SUBJECTED TO VISCOUS DRAG	74-0457
CABLES	CURVED FINITE ELEMENT APPROXIMATION TO NONLINEAR CABLES	72-0379
CABLES	MODELLING HEAT FLOW FROM UNDERGROUND POWER CABLES	74-1158
CABLES	NONLINEAR DYNAMICS OF CABLES WITH LOW INITIAL TENSION	72-0740
CABLES	TRANSIENT HEAT CONDUCTION PROBLEMS IN POWER CABLES SOLVED BY FINITE ELEMENT METHOD	73-0007
CAISSON	ANALYSE TRIDIMENSIONALE DU COMPORTEMENT NON-LINEAIRE D'UN CAISSON DE REACTEUR NUCLEAIRE EN BETON PRE	71-0144
CAISSON	DETERMINATION DU CHAMP DES CONTRAINTES DANS LE CAISSON DU REACTEUR DE BUGEY L'	67-0121
CAISSON	NONLINEAR ANALYSIS OF PRESTRESSED CONCRETE FAST REACTOR PRESSURE VESSEL. (ANALYSE NON LINEAIRE DES C	70-0145
CAISSONS	STRESS ANALYSIS OF MULTI-CELLULAR CAISSONS	74-0338
CALAC	EVALUATION OF CALAC SEMI-MONOCOQUE MEMBRANE ELEMENT	70-0278
CALCESTRUZZO	DISTRIBUZIONI OTTIMALI DI ARMATURE DI PRECOMPRESSIONE IN TRAVI-PARETE DI CALCESTRUZZO. (OPTIMAL DIST	75-0600
CALCOLO	CALCULATION METHOD FOR THE STUDY OF CRACKED RIGID STRUCTURES (METODO DI CALCOLO PER LO STUDIO DI STR	71-0637
CALCOLO	CALCULATION OF AXISYMMETRICAL OSCILLATION OF FLEXIBLE TURBINE DISCS. (CALCOLO DELLE OSCILLAZIONI ASS	74-1328
CALCOLO	CHARACTERISTICS OF A HYBRID CALCULATION SYSTEM BASED ON THE FINITE ELEMENT METHOD. (CARATTERISTICHE	72-0715
CALUCLATION	APPLICATION OF FINITE ELEMENT METHODS FOR CALCULATION OF TRANSVERSE OSCILLATIONS IN ONE DIMENSIONAL	71-0599
CALVARIUM	FINITE ELEMENT MODEL SYSTEM OF A RABBIT CALVARIUM: AN INVESTIGATION OF OPTIMAL COMPRESSIVE LOADING	73-0206
CALVARIUM	MEASUREMENTS OF HEALING AT AN OSTEOTOMY IN A RABBIT CALVARIUM: THE INFLUENCE OF APPLIED COMPRESSIVE	74-0975
CAMAS	A FINITE-ELEMENT, PLANAR-FLOW MODEL OF CAMAS PRAIRIE, IDAHO	72-0980
CAMBER	THE EFFECT OF CAMBER ON THE NATURAL FREQUENCIES OF LOW ASPECT RATIO TURBOMACHINERY BLADES	73-1092
CAMP	FINITE ELEMENT ANALYSES OF QUICK CAMP MODULE	75-0017
CAMPI	STUDIO DI CAMPI ELETTRICI E MAGNETICI STANZIONARI CON IL METODO DEGLI ELEMENTI FINITI. APPLICAZIONE	75-0207
CAMPI	SU LA DETERMINAZIONE DEI CAMPI MAGNETICI LINEARI E NON LINEARI MEDIANTE L APPLICAZIONE DEI METODI DE	74-1061
CANDIDATE	STRUCTURAL EVALUATION OF CANDIDATE DESIGNS FOR THE LARGE SPACE TELESCOPE PRIMARY MIRROR	75-0708
CANDU	FULMOD - AN INELASTIC ANALYSIS PROGRAM TO PREDICT THE OPERATING BEHAVIOUR OF CANDU FUEL ELEMENTS	75-0574
CANNING	FINITE ELEMENT STRESS ANALYSIS OF INTERACTING FUEL PELLET AND CANNING UNDER FAST REACTOR CONDITION	73-0814
CANOPY	NASTRAN DIFFERENTIAL STIFFNESS ANALYSIS OF AN AIRCRAFT CANOPY	71-0037
CANTILEVER	EFFECTS OF SHEAR DEFORMATION, ROTARY INERTIA, AND ELASTICITY OF THE SUPPORT ON THE RESONANCE FREQUEN	75-0369
CANTILEVER	ELASTOPLASTIC ANALYSIS OF A CANTILEVER BEAM	73-0984
CANTILEVER	FINITE ELEMENT ANALYSIS OF VIBRATION AND FLUTTER OF CANTILEVER ANISOTROPIC PLATES	74-0484
CANTILEVER	FINITE ELEMENT CALCULATIONS OF THE COMPLIANCE OF A TAPERED DOUBLE CANTILEVER BEAM SPECIMEN FOR DIFFE	72-0665
CANTILEVER	FINITE-ELEMENT ANALYSIS OF VIBRATION AND FLUTTER OF CANTILEVER ANISOTROPIC PLATES	74-0933
CANTILEVER	GALERKIN FINITE ELEMENT SOLUTION FOR THE STABILITY OF CANTILEVER COLUMNS SUBJECTED TO TANGENTIAL LOA	75-0534
CANTILEVER	LARGE DEFLECTION OF CANTILEVER BEAMS	71-0649
CANTILEVER	ON VIBRATION ANALYSIS OF RECTANGULAR CANTILEVER PLATES WITH A STEPPED CROSS-SECTION BY THE FINITE EL	71-0608
CANTILEVER	PARALLELOGRAM ELEMENT IN THE SOLUTION OF RHOMBIC CANTILEVER PLATE BENDING	66-0022
CANTILEVER	RANDOM VIBRATION RESPONSE OF CANTILEVER PLATES USING THE FINITE ELEMENT METHOD	70-0181
CANTILEVER	TRANSVERSE VIBRATIONS OF CANTILEVER BARS OF VARIABLE CROSS SECTIONS	66-0001
CANTILEVERED	ANALYTICAL DISPLACEMENTS AND VIBRATIONS OF CANTILEVERED UNSYMMETRIC FIBER COMPOSITE LAMINATES	75-0736
CANTILEVERED	THE VIBRATION OF STEPPED CANTILEVERED PLATES	67-0129
CANTILEVERED	VIBRATION ANALYSIS OF A CANTILEVERED SQUARE PLATE BY THE STIFFNESS MATRIX METHOD	66-0012
CANTILEVERED	VIBRATION ANALYSIS OF CANTILEVERED CURVED PLATES USING A NEW CYLINDRICAL SHELL FINITE ELEMENT	68-0130
CANTILEVERED	VIBRATORY BEHAVIOR OF TWISTED CANTILEVERED PLATES	75-0507
CANTILIVER	GALERKIN FINITE-ELEMENT SOLUTION FOR STABILITY OF CANTILIVER COLUMNS SUBJECTED TO TANGENTIAL LOADS	75-0618
CANTILVER	FINITE ELEMENT MODELING OF A CANTILVER BEAM	67-0132
CANYON	FINITE-ELEMENT MODEL FOR POSSIBLE ISOSTATIC REBOUND IN GRAND CANYON	75-0037
CAPACITANCE	DIRECT METHOD OF OBTAINING CAPACITANCE FROM FINITE-ELEMENT MATRICES	72-0289
CAPACITANCE	NUMERICAL ANALYSIS OF THE LINE CAPACITANCE AND CROSSTALK FACTOR FOR INSULATED WIRE PAIRS	74-0685
CAPACITE	ETUDE DE LA CAPACITE PORTANTE DES PIEUX PAR LA METHODE DES ELEMENTS FINIS. INFLUENCE DE LA DILATANCE	75-0624
CAPACITE	LA CAPACITE PORTANTE DES PIEUX. (BEARING CAPACITY OF PILES)	75-0463
CAPACITY	AN EQUATION SOLVER OF VERY LARGE CAPACITY	71-0677
CAPACITY	EINFACHES NAEHERUNGSVERFAHREN FUR DIE GRENZTRAGFAEHIGKEITSANLYSE VON SPANNBETON-REAKTORDRUCKBEHAELTE	73-0854
CAPACITY	ETUDE DE LA CAPACITE PORTANTE DES PIEUX PAR LA METHODE DES ELEMENTS FINIS. INFLUENCE DE LA DILATANCE	75-0624
CAPACITY	LA CAPACITE PORTANTE DES PIEUX. (BEARING CAPACITY OF PILES)	75-0463
CAPACITY	LARGE CAPACITY EQUATION SOLVER FOR STRUCTURAL ANALYSIS	74-0983
CAPACITY	STRUCTURAL CAPACITY OF REINFORCED CONCRETE COLUMNS SUBJECTED TO FIRE INDUCED THERMAL GRADIENTS	73-0913
CAPACITY	ULTIMATE CAPACITY OF CIRCULAR FOOTINGS ON LAYERED SOILS	70-0280
CAPACITY	ULTIMATE STRAIN CAPACITY AND TEMPERATURE RISE STUDIES. TRUMBULL POND DAM	72-0705
CAPE-2D	CAPE-2D, A COMPUTER PROGRAM FOR THE STRESS ANALYSIS OF PLANE AND AXISYMMETRIC COMPOSITE STRUCTURES.	69-0083
CAPS	STRESS DETERMINATION IN 2 CONDENSER HEAT EXCHANGER BAFFLE CAPS BY METHOD OF FINITE ELEMENTS AND A ST	74-0661
CAPS	TWO-DIMENSIONAL TIME-DEPENDENT ANALYSIS OF PERFORATED END CAPS FOR NUCLEAR REACTOR PRESSURE VESSELS	71-0239
CAPS	VIERENDEEL BENT CAPS	72-0758
CAPSULES	RESPONSE OF BURIED CAPSULES IN THE HIGH-OVERPRESSURE REGION	70-0093
CAR	APPLICATION OF FINITE ELEMENT METHOD TO THE STUDY OF STRENGTH OF CAR BODY STRUCTURE AND WHEEL	74-0281
CAR	BOX CAR BODY BOLSTER PROBED IN STRESS STUDY	74-0654

CAR	COMPUTER ANALYSIS OF A RAILROAD FREIGHT CAR BOLSTER UTILIZING THE FINITE ELEMENT METHOD	71-0283
CAR	COMPUTING SYSTEM FOR STRUCTURAL ANALYSIS OF CAR BODIES	72-0324
CAR	DYNAMIC ANALYSIS OF A CAR CHASSIS FRAME USING FINITE ELEMENT METHOD	72-0265
CAR	FREE VIBRATION OF MODEL OF CAR BODY	71-0312
CAR	IS THE COMPUTATION OF NOISE LEVEL INSIDE A CAR FEASIBLE	71-0631
CARATTERISTICHE	CHARACTERISTICS OF A HYBRID CALCULATION SYSTEM BASED ON THE FINITE ELEMENT METHOD. (CARATTERISTICHE	72-0715
CARBIDE	PROBABILISTIC EVALUATION OF THE RELATIVE BEHAVIOUR OF TUNGSTEN CARBIDE COMPONENTS IN STRUCTURAL APPL	75-0579
CARBON	BASELINE SOLUTIONS FOR THE SMOOTH WALL THERMOCHEMICAL ABLATION RESPONSE OF GRAPHITE AND CARBON PHENO	71-0728
CARBON	NOTCH EFFECT IN LOW CYCLE FATIGUE - REPORT 1: LOW-CYCLE FATIGUE STRENGTH OF 0.48 PERCENT CARBON STEE	73-1090
CARDIAC	FINITE ELEMENT ANALYSIS OF HUMAN CARDIAC STRUCTURES	74-0438
CARDS	AUTOMATIC PREPARATION OF DATA CARDS FOR THE FINITE ELEMENT METHOD	66-0086
CARGO	SIMULATION OF FINITE ELEMENT METHOD FOR LOADING PROCESS OF GRAIN CARGO AND INTERACTION WITH SHIP'S S	75-0161
CARGO	TRANSVERSE STRENGTH AT FORE AND AFT SECTION OF TANKER CARGO HOLD	74-0618
CARGOES	PREDICTION OF TEMPERATURE DISTRIBUTIONS IN CARGOES WITH THE AID OF A COMPUTER PROGRAM USING THE METH	74-1215
CARLO	LOAD LIMIT ANALYSIS, COMBINATION OF STATISTICAL METHODS (MONTE CARLO METHOD) AND FINITE ELEMENT METH	71-0756
CARLO	PRESTRESSED CONCRETE REACTOR PRESSURE VESSEL FAILURE ANALYSIS, THE INCORPORATION OF STATISTICAL METH	71-0120
CARRIER	SIMPLIFIED CALCULATION OF THE INTERACTION FORCES BETWEEN HULL AND TANK SYSTEM FOR A MOSS ROSENBERG T	75-0561
CARRYING	EINFACHES NAEHERUNGSVERFAHREN FUR DIE GRENZTPAGFAEHIGKEITSANLYSE VON SPANNBETON-REAKTORDRUCKBEHAELTE	73-0854
CARRYING	GALERKIN FINITE ELEMENT ANALYSIS OF A UNIFORM BEAM CARRYING A CONCENTRATED MASS AND ROTARY INERTIA W	74-0552
CARRYING	NOTCH STRESSES AT THE LOAD CARRYING ELASTIC CORE	73-0266
CARTESIAN	EXTRAPOLATED FINITE ELEMENT PROCEDURES OVER CARTESIAN PRODUCT DOMAINS	75-0044
CARTESIAN	THIN SHELL THEORY IN CARTESIAN COORDINATES - METHODS FOR FINITE ELEMENTS	70-0359
CARTRIDGE	ANALYTICAL PERFORMANCE OF STEEL, ALUMINUM AND PLASTIC FOR A NEW DESIGN HIGH PRESSURE THIN-WALLED CAR	74-0756
CARTRIDGE	FINITE ELEMENT DESIGN STUDY FOR 30-MM CARTRIDGE CASE	74-0376
CASCADE	FINITE ELEMENT ANALYSIS OF THE FLOW THROUGH A CASCADE OF AEROFOILS	73-0761
CASCADE	FLOW THROUGH A CASCADE OF AEROFOILS	74-0077
CASCADE	THE FINITE ELEMENT METHOD APPLIED TO COMPRESSIBLE CASCADE FLOW	75-0179
CASCADES	INFLUENCE OF COOLING, FREE-STREAM TURBULENCE AND SURFACE ROUGHNESS ON THE AERODYNAMIC BEHAVIOR OF CA	74-0970
CASING	CORRELATION BETWEEN PRESSURE AND THERMAL TRANSIENT STRESS LEVELS DERIVED BY THEORY AND EXPERIMENT ON	73-0469
CASING	FINITE ELEMENT ANALYSIS OF CASING THREADS	75-0330
CASING	FINITE-ELEMENT ANALYSIS OF CASING THREADS	75-0417
CASING	TWO-DIMENSIONAL PHOTOELASTIC STUDY OF STEAM-TURBINE CASING FLANGES WITH INTERFACE RELIEF	74-0603
CASINGS	FESTIGKEITSUNTERSUCHUNGEN UND-BERECHNUNGEN AN PUMPENSPIRALGEHAEUSEN. (INVESTIGATIONS AND CALCULATION	75-0238
CAST	SIMULATION OF ELASTIC MODULUS AND POISSON'S RATIO OF SPHEROIDAL GRAPHITE CAST IRON	74-0955
CASTELLATED	ANALYSIS OF CASTELLATED BEAMS BY THE FINITE ELEMENT METHOD	72-0620
CASTINGS	STUDY OF THE FLEXURAL BEHAVIOUR OF CASTINGS	74-0377
CATAMARAN	STRUCTURAL ANALYSIS AND DESIGN OF A CATAMARAN CROSS-STRUCTURE BY FINITE ELEMENT METHOD	73-0052
CATCHMENT	A FINITE ELEMENT DETERMINISTIC CATCHMENT MODEL	74-1361
CAUSEWAY	STUDY OF FLUID MOVEMENTS THROUGH CAUSEWAY	75-0083
CAVING	PREDICTING THE BLOCK CAVING BEHAVIOUR OF OREBODIES	69-0044
CAVING	STRUCTURAL DESIGN DATA FOR CONCRETE DRIFT LININGS IN BLOCK CAVING	73-0773
CAVITIES	A FINITE ELEMENT SOLUTION OF GENERAL FLUID DYNAMICS PROBLEMS - NATURAL CONVECTION IN RECTANGULAR CAV	70-0008
CAVITIES	ACOUSTIC ANALYSIS OF SOLID ROCKET MOTOR CAVITIES BY A FINITE ELEMENT METHOD	71-0119
CAVITIES	ANALYTICAL METHODS FOR PREDICTING SUBSIDENCE ABOVE SOLUTION-MINED CAVITIES	74-1145
CAVITIES	FINITE ELEMENT STUDY OF PRESSURIZED CAVITIES IN GEOLOGIC MEDIA	71-0805
CAVITIES	TIME-DEPENDENT ANALYSIS TO PREDICT CLOSURE IN SALT CAVITIES	74-1149
CAVITY	A NUMERICAL INVESTIGATION OF STABLY STRATIFIED, WIND DRIVEN CAVITY FLOW BY THE FINITE ELEMENT METHOD	74-1372
CAVITY	CALCULATION OF STRESS CONCENTRATION DUE TO A HYPERBOLOID CAVITY IN A THIN PLATE USING NASTRAN	71-0260
CAVITY	CAVITY STABILITY: FINITE ELEMENT ANALYSIS FOR STEADY AND TRANSIENT CREEP	74-1082
CAVITY	DISTRIBUTION OF STRESS AROUND A TUNNEL CAVITY, USING THE FINITE ELEMENT METHOD	71-0843
CAVITY	LINEAR ACCELERATOR CAVITY FIELD CALCULATION BY FINITE ELEMENT	73-0008
CAVITY	LONG TERM CREEP CLOSURE OF SOLUTION CAVITY SYSTEM	74-1150
CAVITY	TENSILE FIELD CALCULATION ABOUT A PRESSURIZED CAVITY: MONERO	72-0684
CBL	THEORETICAL BASIS FOR CBL/NONSAP, A NONLINEAR STRUCTURAL ANALYSIS PROGRAM	74-0740
CDC	EVALUATION OF THE COMPUTER PROGRAM CREEP-BLAST (CDC VERSION)	74-0684
CDC	HYPERMATRIX SCHEME FOR FINITE-ELEMENT ANALYSIS ON CDC STAR-100 COMPUTER	75-0818
CDC	IMPACT OF CDC STAR-100 COMPUTER ON FINITE ELEMENT SYSTEMS	75-0039
CDC	SESAM 69 A GENERAL PURPOSE FINITE ELEMENT METHOD PROGRAM CDC/NASTRAN APPLICATIONS MANUAL	72-0552
CDC	USER'S GUIDE FOR ANALYSIS OF FINITE ELASTOPLASTIC DEFORMATION: THE FIPDEF AND FIPAX PROGRAMS FOR THE	74-0799
CEL	USERS MANUAL FOR CEL/NONSAP, A NONLINEAR STRUCTURAL ANALYSIS PROGRAM	74-0706
CELL	EVALUATION OF A THREE-DIMENSIONAL STRESS CELL FOR GRANULAR SOILS	72-0882
CELL	EVALUATION OF STRESS CELL PERFORMANCE	74-0664
CELL	FINITE ELEMENT ANALYSIS OF LOAD CELL RESPONSE IN HIGH SPEED TENSILE TESTING	73-1071
CELL	STRESS ANALYSIS AND DESIGN OF SILICON SOLAR CELL ARRAYS AND RELATED MATERIAL PROPERTIES	72-0840
CELL	THREE-DIMENSIONAL BAR CELL FOR ELASTIC STRESS ANALYSIS	70-0219
CELL	USE OF FINITE ELEMENT METHOD IN A MARKER AND CELL PROGRAM FOR SOLVING TIME-DEPENDENT FLUID FLOW PROB	75-0043
CELLS	DEVELOPMENT OF ADVANCED INTERCONNECTORS FOR SOLAR CELLS	74-0371
CELLULAR	CELLULAR STRUCTURES OF ARBITRARY PLAN GEOMETRY	72-0201
CELLULAR	FINITE ELEMENT ANALYSIS OF CELLULAR STRUCTURES	69-0350
CELLULAR	SIMPLE FINITE ELEMENT SOLUTION FOR PLATES OF HOMOGENEOUS, SANDWICH AND CELLULAR CONSTRUCTION	75-0140
CEMENT	FRACTURE IN SOIL CEMENT USING THE FINITE ELEMENT METHOD	74-0434
CEMENT	STRESS - DEFORMATION PREDICTION IN CEMENT - TREATED SOIL PAVEMENTS	71-0286
CENTER	EINE FINITE-ELEMENTE-LOESUNG FUR DIE TORSIONSSTEIFIGKEIT UND DEN SCHUBMITTELPUNKT BELIEBIGER QUERSCH	74-0256
CENTER	FINITE ELEMENTS SOLUTION FOR TORSION RIGIDITY AND SHEAR CENTER OF ARBITRARY CROSS-SECTIONS	74-0168
CENTER	PROCEEDINGS OF THE NAVY-NASTRAN COLLOQUIUM (4TH) HELD AT NAVY SHIP RESEARCH AND DEVELOPMENT CENTER,	73-0888
CENTER	PROCEEDINGS OF THE 5TH NAVY-NASTRAN COLLOQUIUM, HELD AT NAVAL SHIP RESEARCH AND DEVELOPMENT CENTER,	74-0709
CENTER-CRACKED	EVALUATION OF THE USE OF A SINGULARITY ELEMENT IN FINITE ELEMENT ANALYSIS OF CENTER-CRACKED PLATES	74-0870
CENTRAL	ELASTO-PLASTIC ANALYSIS OF METAL SHEETS WITH A CENTRAL CRACK	67-0137
CENTRAL	NON-LINEAR EFFECTS OF HIGH TEMPERATURE ON THE VIBRATION OF A SIMPLY SUPPORTED BEAM WITH A CENTRAL MA	73-0838
CENTRAL	OPTIMUM POSITION OF THE CENTRAL CLAY CORE OF A ROCKFILL DAM IN RESPECT TO ARCHING AND HYDRAULIC FRAC	73-0447
CENTRAL	STRESSES IN A ROTATING DISK WITH CENTRAL AND NONCENTRAL HOLES (STRESS ANALYSIS BY THE FINITE ELEMENT	74-1337
CENTRAL	THEORETICAL AND EXPERIMENTAL ANALYSIS OF ELASTOPLASTIC DEFORMATION IN A UNIAXIALLY LOADED STRIP OF 2	73-0720
CENTRAL	THREE-DIMENSIONAL BEHAVIOR OF A CENTRAL CORE DAM	72-0714
CENTRAL	THREE-DIMENSIONAL STRESS ANALYSIS OF A FINITE SLAB CONTAINING A TRANSVERSE CENTRAL CRACK	75-0732
CENTRALIZED	COMPUTER PROGRAM CONSTRUCTION AND MAINTENANCE - FUTURE OF CENTRALIZED FINITE ELEMENT ACTIVITY	75-0014
CENTRALLY	ARBITRARY SYMMETRIC LOADING PROBLEMS OF CENTRALLY CRACKED RECTANGULAR PLATES	75-0318
CENTRIFUGAL	APPLICATION OF THE FINITE ELEMENT METHOD TO PREDICT STATIC AND DYNAMIC RESPONSE OF AN UNSHROUDED CEN	71-0574
CENTRIFUGAL	CENTRIFUGAL MODEL STUDY AND FINITE ELEMENT ANALYSIS OF STRESS NEAR A FAULT	67-0159
CENTRIFUGAL	STRESS ANALYSIS OF A CERAMIC BLADE DESIGN SUBJECTED TO CENTRIFUGAL AND THERMAL LOADS	73-0790
CENTRIFUGED	FINITE ELEMENT ANALYSIS OF CENTRIFUGED AND BUILT-UP SLOPES	74-0415
CENTRIFUGED	FINITE ELEMENT ANALYSIS OF CENTRIFUGED AND BUILT UP SLOPES	74-1376
CERAMIC	ANALYSIS OF STRUCTURAL ACOUSTIC INTERACTIONS IN METAL CERAMIC TRANSDUCERS	73-0401
CERAMIC	ANALYSIS OF THERMAL STRESSES IN CORED CERAMIC BRICKS USED IN HYPERSONIC WIND TUNNEL HEATERS	72-0655
CERAMIC	APPLICATION OF FINITE ELEMENT ANALYSIS TO CERAMIC COMPONENTS	75-0040
CERAMIC	CALCULATION OF THERMAL STRESSES IN TUBULAR CERAMIC COMPONENTS	73-0670
CERAMIC	CERAMIC ROTATING BLADES: SOME CRITICAL DESIGN PARAMETERS FOR GAS TURBINE APPLICATIONS	74-0146
CERAMIC	DESIGN AND ANALYSIS OF A CERAMIC STATOR VANE	75-0366
CERAMIC	FINITE-ELEMENT MODELING OF PIEZOELECTRIC CERAMIC HYDROPHONES	75-0595
CERAMIC	STRESS ANALYSIS OF A CERAMIC BLADE DESIGN SUBJECTED TO CENTRIFUGAL AND THERMAL LOADS	73-0790
CERAMIC	TWO-DIMENSIONAL FINITE ELEMENT MODEL OF A CERAMIC BODY	73-0010
CERAMICS	APPLICATION OF THE SINGLE-EDGE NOTCHED BEAM TO FRACTURE TOUGHNESS TESTING OF CERAMICS	74-0623
CERAMICS	HIGH STRENGTH CERAMICS	72-0878

CERTIFICATION	PRESSURE VESSEL CERTIFICATION BASED ON FRACTURE MECHANICS TECHNOLOGY	74-0948
CFRP	STRESS ANALYSIS OF LAMINATED CFRP PLATES CONTAINING A HOLE	74-1175
CFRP-AL	HYBRID EFFECT ON BENDING BEHAVIORS OF CFRP-AL LAMINATE	74-1213
CH-53A	CH-53A FLEXIBLE FRAME VIBRATION ANALYSIS/TEST CORRELATION	73-1093
CHAIN	AN ANALYSIS OF GENERAL CHAIN SYSTEMS	72-0787
CHAMBER	CYCLIC FATIGUE ANALYSIS OF ROCKET THRUST CHAMBERS. VOLUME 1: OFHC COPPER CHAMBER LOW CYCLE FATIGUE	74-0694
CHAMBERS	CYCLIC FATIGUE ANALYSIS OF ROCKET THRUST CHAMBERS. VOLUME 2: ATTITUDE CONTROL THRUSTER HIGH CYCLE FA	74-0693
CHAMBERS	CYCLIC FATIGUE ANALYSIS OF ROCKET THRUST CHAMBERS. VOLUME 1: OFHC COPPER CHAMBER LOW CYCLE FATIGUE	74-0694
CHAMBERS	STRENGTH OF REINFORCED CONCRETE CHAMBERS UNDER EXTERNAL PRESSURE	75-0293
CHAMBERS	STRENGTH OF REINFORCED CONCRETE CHAMBERS UNDER EXTERNAL PRESSURE	75-0512
CHAMPLAIN	EVALUATION OF THE STABILITY OF NATURAL SLOPES IN PLASTIC CHAMPLAIN CLAYS	74-0264
CHAMPLAIN	IMMEDIATE SETTLEMENTS OF THREE TEST EMBANKMENTS ON CHAMPLAIN CLAY	74-0275
CHAMPS	CALCUL DES CHAMPS ELECTROMAGNETIQUES PAR LA METHODE DES ELEMENTS FINIS (COMPUTATION OF ELECTROMAGNET	73-0399
CHANNEL	A FINITE ELEMENT METHOD FOR CALCULATING OPEN CHANNEL FLOW	69-0510
CHANNEL	FINITE ELEMENT ANALYSIS OF SLOW NON-NEWTONIAN CHANNEL FLOW	72-0308
CHANNEL	FINITE ELEMENT ANALYSIS OF LONG WAVES IN OPEN CHANNEL SYSTEMS	74-0033
CHANNEL	VARIATIONAL PRINCIPLE FOR MAGNETOHYDRODYNAMIC CHANNEL FLOW	70-0192
CHANNEL-HEAD-CO	THREE-DIMENSIONAL FINITE ELEMENT ANALYSIS OF A STEAM GENERATOR CHANNEL-HEAD-COMPLEX - COMPARISON OF	75-0653
CHANNELS	A FINITE ELEMENT MODEL FOR TWO-DIMENSIONAL STEADY FLOW THROUGH CONTRACTIONS IN NATURAL CHANNELS	71-0810
CHANNELS	FINITE ELEMENT ANALYSIS OF WAVE PROPAGATION IN OPEN CHANNELS	75-0570
CHANNELS	SLOSHING OF LIQUID IN CIRCULAR CHANNELS AND SPHERICAL TANKS	60-0005
CHANNELS	SLOW PARTICULATE VISCOUS FLOW IN CHANNELS AND TUBES. APPLICATION TO BIOMECHANICS	71-0248
CHARACTERISATIO	CHARACTERISATION ELASTIQUE D'UN STRATIFIE RESINE-VERRE PAR LA METHODE DES DEPLACEMENTS (DETERMINATIO	75-0272
CHARACTERISTIC	CALCULATION OF THE CHARACTERISTIC FREQUENCIES OF FLUID-FILLED SPHERICAL SHELLS	73-0511
CHARACTERISTIC	COMPARISON OF THE RESULTS OF THE DYNAMIC RESPONSE OF CYLINDRICAL SHELLS BY CHARACTERISTIC AND BY FIN	74-0781
CHARACTERISTIC	SCALAR ALGORITHM OF DETERMINATION OF STIFFNESS MATRIX AND OF ANOTHER FINITE ELEMENT CHARACTERISTIC	74-1013
CHARACTERISTICS	A COMPUTER BASED PROCEDURE FOR PREDICTING THE TRANSIENT RESPONSE AND FAILURE OF A TWO-DIMENSIONAL CO	71-0630
CHARACTERISTICS	A STUDY OF THE CONVERGENCE CHARACTERISTICS AND ACCURACY OF CONFORMING FINITE ELEMENT SOLUTIONS	72-0967
CHARACTERISTICS	ACCURACY OF FINITE-ELEMENT SOLUTIONS FOR MODAL CHARACTERISTICS OF SHELLS OF REVOLUTION	70-0231
CHARACTERISTICS	ACCURACY, STABILITY AND OSCILLATION CHARACTERISTICS OF FINITE ELEMENT METHOD FOR SOLVING HEAT CONDUC	69-0007
CHARACTERISTICS	ACCURACY, STABILITY, AND OSCILLATION CHARACTERISTICS OF TRANSIENT TWO-DIMENSIONAL HEAT CONDUCTION	75-0849
CHARACTERISTICS	AN INVESTIGATION OF THE DYNAMIC RESPONSE CHARACTERISTICS OF THE BON TEMPE DAM	66-0047
CHARACTERISTICS	APPLICATION OF A FINITE ELEMENT METHOD TO EVALUATION OF OIL WHIRL CHARACTERISTICS	74-0590
CHARACTERISTICS	CHARACTERISTICS OF A HYBRID CALCULATION SYSTEM BASED ON THE FINITE ELEMENT METHOD. (CARATTERISTICHE	72-0715
CHARACTERISTICS	CHARACTERISTICS OF FINITE-ELEMENTS FOR ANALYSIS OF BEAMS SUBJECTED TO IMPACT LOAD	73-1144
CHARACTERISTICS	CHARACTERISTICS OF MANIPULATION ERRORS IN SOLVING LOAD DEFLECTION EQUATIONS	70-0355
CHARACTERISTICS	CONVERGENCE CHARACTERISTICS OF CYLINDRICAL SHELL FINITE ELEMENTS	72-0943
CHARACTERISTICS	CORRELATION OF FINITE ELEMENT STRUCTURAL DYNAMIC ANALYSIS WITH MEASURED FREE VIBRATION CHARACTERISTI	74-0486
CHARACTERISTICS	DETERMINING SEEPAGE CHARACTERISTICS OF MILL-TAILING DAMS BY THE FINITE ELEMENT METHOD	71-0477
CHARACTERISTICS	EFFECT OF DENSIFICATION ON THE ENGINEERING CHARACTERISTICS OF ORGANIC SOILS - VOL. 1 AND 2.	73-0805
CHARACTERISTICS	FINITE ELEMENT - ANALOGUE METHOD FOR DETERMINING THE DYNAMIC CHARACTERISTICS OF AN ARCH DAM-RESERVOI	73-0678
CHARACTERISTICS	FINITE ELEMENT BASED PROCEDURE FOR SIMULATING THE TRANSIENT RESPONSE AND FAILURE OF A TWO-DIMENSIONA	72-0123
CHARACTERISTICS	GEOMETRIC STIFFNESS CHARACTERISTICS OF A ROTATING ELASTIC APPENDAGE	74-1296
CHARACTERISTICS	INFLUENCE CHARACTERISTICS FOR SLAB BRIDGES	71-0188
CHARACTERISTICS	MATERIAL BEHAVIOUR CHARACTERISTICS FOR REINFORCED CONCRETE SHELLS STRESSED BEYOND THE ELASTIC RANGE	70-0166
CHARACTERISTICS	METHODS FOR DETERMINING CHARACTERISTICS OF ACOUSTIC WAVES IN ROCKET MOTORS	75-0075
CHARACTERISTICS	METHODS FOR PREDICTING THE AERODYNAMIC AND STABILITY AND CONTROL CHARACTERISTICS OF STOL AIRCRAFT. V	73-0718
CHARACTERISTICS	METHODS FOR PREDICTING THE AERODYNAMIC AND STABILITY AND CONTROL CHARACTERISTICS OF STOL AIRCRAFT. V	73-0719
CHARACTERISTICS	METHODS OF PREDICTING THE AERODYNAMIC AND STABILITY AND CONTROL CHARACTERISTICS OF STOL AIRCRAFT. VO	73-0717
CHARACTERISTICS	RADIAL AND AXIAL TRANSMISSIBILITY CHARACTERISTICS OF TYPICAL ROCKET VEHICLE STRUCTURE TECHNICAL SUMM	69-0475
CHARACTERISTICS	RASCHET METODOM KONECHNYKH ELEMENTOV DINAMICHESKIKH KHARAKTERISTIK ZEMLYANOI PLOTINY SOVMESTNO S OS	73-0437
CHARACTERISTICS	SENSITIVITY ANALYSIS OF TORSIONAL VIBRATION CHARACTERISTICS OF HELICOPTER ROTOR BLADES. PART 1: STRU	74-0846
CHARACTERISTICS	SENSITIVITY OF ROTOR BLADE VIBRATION CHARACTERISTICS TO TORSIONAL OSCILLATIONS	74-0611
CHARACTERISTICS	STABILITY AND OSCILLATION CHARACTERISTICS OF FINITE ELEMENT, FINITE DIFFERENCE AND WEIGHTED RESIDUAL	73-0139
CHARACTERISTICS	STABILITY AND OSCILLATION CHARACTERISTICS OF FINITE ELEMENT, FINITE-DIFFERENCE, AND WEIGHTED-RESIDUA	75-0615
CHARACTERISTICS	THE CALCULATION OF STATIC AND DYNAMIC CHARACTERISTICS OF FLIGHT VEHICLE STRUCTURES BY THE FINITE ELE	73-1018
CHARACTERISTICS	THE EFFECT OF IN PLANE STATIC STRESSES ON THE VIBRATION CHARACTERISTICS OF CURVED SANDWICH PLATES	71-0588
CHARACTERISTICS	THE USE OF EIGEN-DERIVATIVES TO PREDICT THE EFFECT OF DESIGN CHANGES UPON THE DYNAMIC CHARACTERISTIC	74-0737
CHARACTERISTICS	THE VIBRATION CHARACTERISTICS OF A TENSIONED PLATE CONTAINING A FATIGUE CRACK	68-0134
CHARACTERISTICS	VIBRATION CHARACTERISTICS OF THIN, PRESSURIZED SHELLS OF REVOLUTION PARTIALLY FILLED WITH LIQUID	72-0604
CHARACTERISTICS	VIBRATION CHARACTERISTICS OF FREE THIN CYLINDRICAL SHELLS	73-0396
CHARACTERISTICS	VIBRATION CHARACTERISTICS OF THE NORTH FORK DAM MODEL	74-0845
CHARACTERISTICS	VIBRATIONAL CHARACTERISTICS OF THREE-DIMENSIONAL SOLIDS, WITH APPLICATIONS TO EARTH DAMS	69-0432
CHARACTERISTICS	VIBRATIONAL CHARACTERISTICS OF SPACE FRAMES	69-0434
CHARACTERIZATIO	A METHOD FOR MATERIAL CHARACTERIZATION EMPLOYING THE FINITE-ELEMENT TECHNIQUE WITH APPLICATION TO HO	72-0941
CHARACTERIZATIO	ANALYSIS OF THE BIAXIAL STRIP AND SHEAR LAP TESTS FOR SOLID PROPELLANT CHARACTERIZATION	67-0019
CHARACTERIZATIO	CHARACTERIZATION OF ELASTIC SOLIDS USING FINITE ELEMENT METHODS	72-0712
CHARACTERIZATIO	CONTINUUM CHARACTERIZATION OF JOINTED ROCK MASSES - 1, 2	73-0196
CHARACTERIZATIO	FINITE ELEMENT APPLICATIONS IN THE CHARACTERIZATION OF ELASTIC SOLIDS	71-0138
CHARACTERIZATIO	MECHANICAL CHARACTERIZATION AND FINITE-ELEMENT ANALYSIS OF ELASTIC-PLASTIC, WORK-HARDENING SOILS	72-0974
CHARACTERIZATIO	NUMERICAL BIAXIAL CHARACTERIZATION FOR CONCRETE	74-0641
CHARACTERIZATIO	SILAGE - SILO INTERACTION USING MATERIAL CHARACTERIZATION AND FINITE ELEMENT ANALYSIS	75-0792
CHARACTERIZING	EXTENSION OF CLASSICAL EXPERIMENTAL TECHNIQUES FOR CHARACTERIZING COMPOSITE - MATERIAL BEHAVIOUR	72-0088
CHARGE	ETUDE THEORIQUE DU COMPORTEMENT DES PIEUX SOUS CHARGE VERTICALE. INTRODUCTION DE LA DILATANCE. (THEO	75-0309
CHARLESTOWN	MODELING AN OCEAN POND. A TWO-DIMENSIONAL, FINITE ELEMENT HYDRODYNAMIC MODEL OF NINIGRET POND, CHARL	75-0686
CHARPY	LOAD POINT COMPLIANCE OF THE CHARPY IMPACT SPECIMEN	73-0775
CHASSIS	COMPUTER AIDED DESIGN APPLIED TO A MODEL OF A CHASSIS TYPE STRUCTURE USING FINITE ELEMENT TECHNIQUES	69-0421
CHASSIS	DYNAMIC ANALYSIS OF AN AUTOMOBILE CHASSIS FRAME	70-0135
CHASSIS	DYNAMIC ANALYSIS OF A CAR CHASSIS FRAME USING FINITE ELEMENT METHOD	72-0265
CHASSIS	STRESS ANALYSIS OF AN AUTOMOBILE CHASSIS FRAME	70-0137
CHATTER	PARAMETRIC ANALYSIS OF CONTACT CHATTER IN RELAYS	74-0487
CHAUSSEES	APPLICATION DES CALCULS DE FONDATIONS A L'EVALUATION DU COMPORTEMENT DES CHAUSSEES EM DASH 1, 2. (A	71-0429
CHECKING	A SPECIAL NASTRAN PROGRAM FOR INPUT CHECKING AND UNDEFORMED STRUCTURE PLOTTING	71-0234
CHECKING	PROCEDURE FOR EFFICIENT GENERATING, CHECKING, AND DISPLAYING NASTRAN INPUT AND OUTPUT DATA FOR ANALY	71-0098
CHECKOUT	FINITE ELEMENT METHOD FOR DETERMINATION OF THERMAL STRESSES INTERACTIVE GRAPHICS FOR FINITE ELEMENT	74-0155
CHECKOUT	FINITE ELEMENT MODEL DATA CHECKOUT WITH INTERACTIVE GRAPHICS	74-0195
CHEMICAL	A FINITE ELEMENT ANALYSIS OF DISSOLVED OXYGEN DRAWDOWN AND SULFATE PRODUCTION IN STRIP MINE SPOIL DA	75-0795
CHIEBUNGSMETHOD	SPEKTRALANALYSE VON ROTORBLATTERN NACH DERMATRIZENVERS CHIEBUNGSMETHODE UND MATRIZENKRAFTEMETHODE	66-0123
CHILES	CHILES: A FINITE ELEMENT COMPUTER PROGRAM THAT CALCULATES THE INTENSITIES OF LINEAR ELASTIC SINGULAR	73-0959
CHIP	INCIPIENT CHIP FORMATION IN METAL CUTTING - A THREE- DIMENSION FINITE-ELEMENT ANALYSIS	73-1146
CHISLENNOE	CHISLENNOE RESHENIE PLOSKIKH ZADACH TEPLOPROVODNOSTI DLYA OBLASTEI SLOZHNOI FORMY (NUMERICAL SOLUTIO	74-0659
CHISLENNYI	CHISLENNYI METOD RASCHETA SOBSTVENNYKH I VYNUZHDENNYKH KOLEBANII SOSTAVNYKH OBOLOCHECHNYKH KONSTRUKT	73-0461
CHOLESKY	A COMPARATIVE STUDY OF SEVERAL CORE STORAGE SCHEMES FOR LARGE SPARSE POSITIVE DEFINITE MATRICES WITH	72-0678
CHORD	NATURAL FREQUENCIES OF WIDE CHORD COMPRESSOR BLADES	74-0231
CHORDWISE	ANALYSIS OF THE EFFECTS ON LIFE OF LEADING-EDGE HOLES IN AN AIRFOIL SUBJECTED TO ARBITRARY SPANWISE	75-0700
CHOVANI	STATICKE RESENI FYZIKALNE NELINEARNIHO CHOVANI ZELEZOBETO- NOVYCH DESEK (ANALYSIS OF PHYSICALLY NONL	74-0514
CIAGLOSCI	FUNKCJA KSZTALTU NIESPELNIAPACA WARUNKOW CIAGLOSCI W ANALIZIE PLYT METODA ELEMENTOW SKONCZONYCH. (SH	75-0263
CIAL	OCENA DOKLADNOSCI ROZWIAZAN PROBLEMOW BRZEGOWYCH SPREZYSTYCH CIAL DYSKERTYZOWANYCH. (ASSESSING THE A	74-0921
CIEPLA	METODA ELEMENTOW SKONSZONYCH ROZWIAZYWANIA OSIOWO-SYMETRYCZNEGO PRZEWODNICTWA CIEPLA (FINITE-ELEMENT	73-0434
CILIARY	A FINITE ELEMENT APPROACH TO PHYSIOLOGICAL TRANSPORT PHENOMENA BY CILIARY AND PERISTALTIC PUMPING	72-0353
CINEMATIQUEMENT	FAMILLE D ELEMENTS FINIS DE COQUE PLAN CINEMATIQUEMENT ADMISSIBLES POUR L'ANALYSE DES STRUCTURES	74-0488
CIRCUIT	APPLICABILITY OF APPROXIMATE AND EXACT TRANSIENT HEAT TRANSFER ANALYSES TO HEATING PROCESSES USED TO	75-0244

147

CIRCULAR	A COMPARISON OF FINITE ELEMENT AND FOURIER SERIES SOLUTIONS AS APPLIED TO RADIALLY LOADED CIRCULAR R	68-0024
CIRCULAR	A CURVED CIRCULAR CYLINDRICAL SHELL FINITE ELEMENT FOR HIGHER ORDER SHELL THEORY	69-0528
CIRCULAR	A FINITE ELEMENT SOLUTION OF THE CONFINED FLOW OVER A CIRCULAR CYLINDER	74-0023
CIRCULAR	A NEW METHOD OF ANALYSIS TO OBTAIN EXACT SOLUTIONS FOR STRESSES AND STRAINS IN CIRCULAR INCLUSIONS	74-1063
CIRCULAR	A NONLINEAR FINITE ELEMENT ANALYSIS OF SHALLOW CIRCULAR ARCHES	69-0269
CIRCULAR	ACCURACY OF THE FINITE ELEMENT ANALYSIS FOR THE ELASTIC PLATE WITH A CIRCULAR HOLE	73-0155
CIRCULAR	ANALYSIS AND DESIGN OF AXISYMMETRIC REINFORCEMENT AROUND A CIRCULAR HOLE IN A THIN FLAT PLATE BY THE	74-1113
CIRCULAR	ANALYSIS OF A CIRCULAR RAFT IN ADHESIVE CONTACT WITH A THICK ELASTIC LAYER	74-1152
CIRCULAR	ANALYSIS OF CIRCULAR FOOTINGS ON LAYERED SOILS	70-0281
CIRCULAR	ANALYSIS OF CIRCULAR AND ANGULAR LAMINATED THICK PLATES	74-0426
CIRCULAR	ANALYSIS OF CLAMPED CIRCULAR PLATES OF VARIABLE THICKNESS SUBJECTED TO UNIFORMLY DISTRIBUTED LOAD BY	73-0583
CIRCULAR	ANNULAR AND CIRCULAR SECTOR FINITE ELEMENTS FOR PLATE BENDING	70-0100
CIRCULAR	APPLICATION OF CURVED FINITE ELEMENTS TO CIRCULAR ARCHES	71-0360
CIRCULAR	APPLICATION OF HILL FUNCTIONS TO CIRCULAR PLATE PROBLEMS	75-0572
CIRCULAR	ASYMMETRIC BENDING OF CIRCULAR PLATES USING THE FINITE ELEMENT METHOD	75-0175
CIRCULAR	BENDING OF CIRCULAR PLATES OF HARDENING MATERIAL	67-0112
CIRCULAR	COMMENT ON LIMITATIONS OF CERTAIN FINITE ELEMENTS WHEN APPLIED TO ARCHES - FURTHER STUDIES IN APPLIC	72-0097
CIRCULAR	COMPARISON OF SOME FINITE-ELEMENT MODELS FOR CIRCULAR CYLINDRICAL SHELLS	73-1158
CIRCULAR	CONTINUUM AND FINITE ELEMENT BRANCHING STUDIES OF THE CIRCULAR PLATE.	72-0427
CIRCULAR	DEFLECTION FUNCTION FOR THE ASYMMETRICAL BENDING OF CIRCULAR PLATES	73-0468
CIRCULAR	DEFORMATION AND FAILURE OF BORON-EPOXY PLATE WITH CIRCULAR HOLE	72-0536
CIRCULAR	DYNAMIC ANALYSIS OF CIRCULAR AND SECTOR THICK, LAYERED PLATES	75-0198
CIRCULAR	ELASTIC BENDING OF CIRCULAR PLATES OF VARIABLE THICKNESS. AN ANALYTICAL AND EXPERIMENTAL STUDY	70-0163
CIRCULAR	ELASTIC SETTLEMENT OF A CIRCULAR RAFT IN ADHESIVE CONTACT WITH A TRANSVERSELY ISOTROPIC MEDIUM	75-0828
CIRCULAR	ELASTOPLASTIC INDENTATION OF HALF-SPACE BY AN INFINITELY LONG RIGID CIRCULAR CYLINDER	71-0395
CIRCULAR	EXACT SOLUTION TO THE RADIALLY END CONSTRAINED CIRCULAR CYLINDER UNDER TRIAXIAL LOADING	71-0433
CIRCULAR	FATIGUE CRACK INITIATION IN A BORON EPOXY PLATE WITH A CIRCULAR HOLE	74-0820
CIRCULAR	FINITE ELEMENT ANALYSIS OF CYLINDRICAL SHELLS. SOLUTION USING RECTANGULAR CIRCULAR CYLINDRICAL ELEME	70-0178
CIRCULAR	FINITE ELEMENT MODELING OF A CIRCULAR RING USING HALF AND QUARTER SYMMETRY	66-0119
CIRCULAR	FINITE ELEMENT SIMULATION OF ACOUSTIC FILTERS WITH VARIABLE CIRCULAR CROSS SECTION	75-0871
CIRCULAR	FINITE-ELEMENT METHOD WITH ELEMENTS WITH CIRCULAR RING SECTOR FORM	73-0105
CIRCULAR	FORCED VIBRATIONS OF CIRCULAR FOUNDATIONS ON LAYERED MEDIA	74-0762
CIRCULAR	FREE VIBRATIONS OF CIRCULAR MEMBRANES UNDER ARBITRARY TENSION BY THE FINITE ELEMENT METHOD	69-0190
CIRCULAR	FREQUENCY ANALYSIS OF ORTHOTROPIC CIRCULAR CYLINDRICAL PANELS BY THE FINITE STRIP METHOD	73-0274
CIRCULAR	FURTHER STUDIES IN THE APPLICATION OF CURVED FINITE ELEMENTS TO CIRCULAR ARCHES	71-0644
CIRCULAR	LARGE AMPLITUDE, FREE VIBRATION GEOMETRICALLY NONLINEAR CIRCULAR ARCHES	72-0606
CIRCULAR	LARGE DEFLECTION GEOMETRICALLY NON-LINEAR FINITE ELEMENT ANALYSIS OF CIRCULAR ARCHES	73-0021
CIRCULAR	NATURAL FREQUENCIES OF CIRCULAR ARCHES	71-0302
CIRCULAR	NUMERICAL STUDIES USING CIRCULAR ARCH FINITE ELEMENTS	74-0600
CIRCULAR	NUMERICAL STUDY OF UNIAXIAL COMPRESSION IN CIRCULAR ELASTIC-PLASTIC COLUMNS	73-0238
CIRCULAR	ON THE DISTRIBUTION OF STRESSES AROUND CIRCULAR TUNNELS IN NON-HOMOGENEOUS ROCKS	66-0029
CIRCULAR	ON THE STATES OF STRESS AND DEFORMATION AROUND THE BOTTOM OF A CIRCULAR VERTICAL SHAFT	69-0323
CIRCULAR	ON THE VALIDITY OF THE FINITE ELEMENT ANALYSIS OF CIRCULAR ARCHES REPRESENTED BY AN ASSEMBLAGE OF BE	75-0479
CIRCULAR	PARAMETRISCHE KERBSPANNUNGSUNTERSUCHUNGEN AN DER LOCHSCHEIBE MIT DER METHODE DER FINITEN ELEMENTE. (	75-0603
CIRCULAR	PROPAGATION OF FAILURE IN A CIRCULAR CYLINDER OF ROCK SUBJECTED TO A COMPRESSIVE FORCE	73-0793
CIRCULAR	RADIAL INERTIA EFFECTS ON AN IDEALLY PLASTIC CIRCULAR PLATE UNDER IMPULSIVE AXIAL COMPRESSION	69-0324
CIRCULAR	ROCK FAILURE AROUND A CIRCULAR OPENING IN A GRAVITY FIELD WITH TECTONIC FORCES	70-0071
CIRCULAR	SLOSHING OF LIQUID IN CIRCULAR CHANNELS AND SPHERICAL TANKS	60-0005
CIRCULAR	STATIC, VIBRATION AND BUCKLING ANALYSIS OF AXISYMMETRIC CIRCULAR PLATES USING FINITE ELEMENTS	73-0157
CIRCULAR	STRESS AND FAILURE ANALYSIS OF A GLASS-EPOXY COMPOSITE PLATE WITH A CIRCULAR HOLE	73-0276
CIRCULAR	STRESS CONCENTRATION FACTORS FOR CIRCULAR NOTCHES	74-0791
CIRCULAR	STRESS CONCENTRATION FACTORS FOR CIRCULAR, REINFORCED PENETRATIONS IN PRESSURIZED CYLINDRICAL SHELLS	75-0726
CIRCULAR	STRESS DISTRIBUTION IN THE LIGAMENT ZONE OF PERFORATED CIRCULAR DISCS WITH SPECIAL REFERENCE TO THE	73-0827
CIRCULAR	STRESSES AND DISPLACEMENTS AROUND A CIRCULAR TUNNEL IN A THREE LAYER MEDIUM	69-0038
CIRCULAR	STRESSES AND DISPLACEMENTS AROUND TWO ADJACENT CIRCULAR OPENINGS NEAR TO THE GROUND SURFACE	74-0993
CIRCULAR	STRUCTURAL DESIGN OF CIRCULAR DIVERSION, POWER AND PRESSURE TUNNELS	71-0315
CIRCULAR	SYMMETRICAL BENDING OF CIRCULAR PLATES USING FINITE ELEMENTS	72-0428
CIRCULAR	THE BEHAVIOUR OF SQUARE SHEAR WEBS HAVING A CIRCULAR HOLE	67-0108
CIRCULAR	THE FINITE ELEMENT METHOD APPLIED TO THE STATIC AND DYNAMIC BUCKLING OF CIRCULAR ARCHES	70-0497
CIRCULAR	THREE DIMENSIONAL FINITE ELEMENT ANALYSIS OF THICK LAMINATED COMPOSITES - INCLUDING INTERLAMINAR AND	73-0882
CIRCULAR	THREE-DIMENSIONAL ANALYSIS FOR THE STRESS DISTRIBUTION NEAR CIRCULAR HOLES IN LAMINATED COMPOSITES	74-0793
CIRCULAR	THREE-DIMENSIONAL FINITE ELEMENT ANALYSIS OF THICK LAMINATED COMPOSITES - INCLUDING INTERLAMINAR AND	74-1151
CIRCULAR	TRANSIENT NATURAL CONVECTION COOLING OF A VERTICAL CIRCULAR CYLINDER	74-1169
CIRCULAR	TREUGOL'NY I ELEMENT DLYA KRUGOVOI TSILINDRICHESKOI OBOLOCHKI (TRIANGULAR FINITE ELEMENT FOR CIRCULA	73-0493
CIRCULAR	ULTIMATE CAPACITY OF CIRCULAR FOOTINGS ON LAYERED SOILS	70-0280
CIRCULAR	VIBRATION AND BUCKLING ANALYSIS OF AXISYMMETRIC POLAR ORTHOTROPIC CIRCULAR PLATES	74-0574
CIRCULAR	VIBRATION OF CIRCULAR AND ANNULAR PLATES USING FINITE ELEMENTS	72-0194
CIRCULAR	VYPOCET KRUHOVYCH LOMENICOVYCH KONSTRUKCII METODOU KONECNYCH PASOV. (CALCULATION OF CIRCULAR FOLDED-	74-0612
CIRCULAR-CYLIND	ERRORS IN SHELL FINITE-ELEMENT MODELS FOR VIBRATION OF CIRCULAR-CYLINDERS	75-0203
CIRCULATION	FINITE ELEMENT APPLICATION IN LAKE CIRCULATION	75-0493
CIRCULATION	FINITE ELEMENT CIRCULATION ANALYSIS OF VARIABLE-DEPTH SHALLOW LAKES	73-0727
CIRCULATION	FINITE ELEMENT LAKE CIRCULATION AND THERMAL ANALYSIS	74-0043
CIRCULATION	FINITE ELEMENT MODEL OF TWO LAYER COASTAL CIRCULATION	74-1192
CIRCULATION	FINITE ELEMENT MODELLING OF HYDRODYNAMIC CIRCULATION	74-1370
CIRCULATION	FINITE ELEMENT MODELS FOR OCEAN CIRCULATION PROBLEMS	75-0206
CIRCULATION	FINITE-DIFFERENCE AND FINITE-ELEMENT METHODS IN GLOBAL WIND-DRIVEN OCEAN CIRCULATION	75-0345
CIRCULATION	HIGHER-ORDER FINITE ELEMENT ANALYSIS OF LAKE CIRCULATION	73-1002
CIRCULATION	MATHEMATICAL MODELING OF NEAR COASTAL CIRCULATION	74-0720
CIRCULATION	MATHEMATICAL MODELS OF THE MASSACHUSETTS BAY. PART 1. FINITE ELEMENT MODELING OF TWO-DIMENSIONAL HYD	73-0800
CIRCULATION	NUMERICAL INVESTIGATIONS OF LAKE CIRCULATION AROUND ISLANDS BY THE FINITE ELEMENT METHOD	72-0233
CIRCULATION	NUMERICAL INVESTIGATION OF LAKE CIRCULATION AROUND ISLANDS BY THE FINITE ELEMENT METHOD	72-0337
CIRCULATION	WIND DRIVEN LAKE CIRCULATION BY THE FINITE ELEMENT METHOD	70-0041
CIRCULATION	3-D FINITE ELEMENT MODEL FOR LAKE CIRCULATION	71-0157
CIRCUMFERENTIAL	CIRCUMFERENTIAL PRESTRESSING LOAD ANALYSIS DUE TO WIRE/STRAND WINDING SYSTEMS ON PRESTRESSED CONCRET	74-1202
CIRCUMFERENTIAL	ASAAS: ASYMMETRIC STRESS ANALYSIS OF AXISYMMETRIC SOLIDS WITH ORTHOTROPIC, TEMPERATURE-DEPENDENT MAT	71-0696
CIRCUMFERENTIAL	MATRIX DISPLACEMENT ANALYSIS OF SHELLS WITH AN AXISYMMETRIC MIDSURFACE BUT HAVING CIRCUMFERENTIALLY	68-0031
CIRENE	INTERNAL SUPPORTING STRUCTURES OF THE (CIRENE) REACTOR CORE - RESULTS OF STATIC STRESS ANALYSIS FOR	73-0818
CITATION	A FINITE ELEMENT BIBLIOGRAPHY - PART III: CITATION LISTING	74-0286
CIVIL	APPLICATION OF FINITE ELEMENT METHODS IN CIVIL ENGINEERING	69-0031
CIVIL	APPLICATION OF THE FAST FOURIER TRANSFORM TO LINEAR SYSTEMS IN CIVIL ENGINEERING	73-1119
CIVIL	DYNAMIC ANALYSIS OF CIVIL ENGINEERING STRUCTURES	71-0264
CIVIL	FINITE ELEMENT METHODS IN CIVIL ENGINEERING	72-0336
CIVIL	FINITE ELEMENT METHOD - A VERSATILE TOOL FOR CIVIL ENGINEERING	72-0627
CIVIL	POROUS MEDIA - SOME PROBLEMS IN CIVIL AND PETROLEUM ENGINEERING	74-0349
CIVIL	PORTABILITY AND ADAPTABILITY OF CIVIL ENGINEERING SOFTWARE	73-0343
CLAD	DIGITAL CALCULATION OF ELECTRIC FIELDS IN METAL CLAD INSTALLATIONS	73-0212
CLAD	FINITE ELEMENT ANALYSIS OF CLAD STRUCTURES	73-0353
CLAD	FINITE-ELEMENT ELASTIC-PLASTIC ANALYSIS OF RESIDUAL STRESSES DUE TO CLAD WELDING IN REACTOR VESSELS	75-0779
CLADDING	FINITE ELEMENT FRACTURE MECHANICS ANALYSIS OF CREEP RUPTURE OF FUEL ELEMENT CLADDING	74-0929
CLADDING	INFLUENCE OF DEFECTS ON FUEL ELEMENT CLADDING STRESSES	73-0863
CLAMPED	A FORTRAN PROGRAM FOR CALCULATING THE FREQUENCIES AND MODE SHAPES OF A GUIDED CLAMPED, GUIDED PINNED	72-0890
CLAMPED	A FORTRAN PROGRAM FOR CALCULATING THE FREQUENCIES AND MODE SHAPES OF A GUIDED CLAMPED, GUIDED PINNED	72-0890

CLAMPED	ANALYSIS OF CLAMPED CIRCULAR PLATES OF VARIABLE THICKNESS SUBJECTED TO UNIFORMLY DISTRIBUTED LOAD BY	73-0583
CLAMPED	CONFORMING FINITE ELEMENT METHODS FOR THE CLAMPED PLATE PROBLEM	75-0174
CLAMPED	DYNAMIC PLASTIC LARGE DEFLECTIONS OF CLAMPED RECTANGULAR PLATE	74-0392
CLAMPED	FREQUENCY COEFFICIENTS OF A SPHERICAL SHELL FOR VARYING THICKNESS AND OPENING ANGLE WITH GUIDED PINN	72-0595
CLAMPED	THERMAL MECHANICAL ANALYSIS PROCEDURE FOR A SEMICIRCULAR PLATE WITH CLAMPED STRAIGHT EDGE	75-0651
CLAMPING	DISTRIBUTION OF THE CLAMPING PRESSURE IN FRICTION-GRIP BOLTED JOINTS	74-0250
CLASSE	SUR L APPROXIMATION, PAR ELEMENTS FINIS D ORDRE UN, ET LA RESOLUTION, PAR PENALISATION-DUALITE, D UN	75-0845
CLASSIFICATION	CLASSIFICATION OF FINITE ELEMENT STRESSES ACCORDING TO ASME SECTION 111 STRESS CATEGORIES	74-0610
CLAY	BEHAVIOR OF EMBANKMENTS ON NEW LISKHEARD VARVED CLAY	73-0721
CLAY	ELASTIC-PLASTIC LARGE DEFORMATION RESPONSE OF CLAY TO FOOTING LOADS	74-0699
CLAY	FINITE ELEMENT ANALYSIS OF STRAIN-SOFTENING CLAY	72-0028
CLAY	FINITE ELEMENT PROGRAM FEECON FOR UNDRAINED DEFORMATION ANALYSES OF GRANULAR EMBANKMENTS ON SOFT CLA	72-0875
CLAY	IMMEDIATE SETTLEMENTS OF THREE TEST EMBANKMENTS ON CHAMPLAIN CLAY	74-0275
CLAY	INITIAL SETTLEMENT OF STRUCTURES ON CLAY	71-0304
CLAY	INITIATION OF FAILURE IN SLOPES IN OVERCONSOLIDATED CLAYS AND CLAY SHALES.	70-0186
CLAY	INTERACTIVE ANALYSIS OF A RAFTED MULTISTOREY SPACE FRAME RESSTING ON AN INHOMOGENEOUS CLAY STRATUM	74-0447
CLAY	LONG-TERM DEFORMATION IN CLAY	73-0713
CLAY	OBSERVATIONS ON THE BEHAVIOUR OF A PILED-RAFT FOUNDATION ON LONDON CLAY.	73-0421
CLAY	OPTIMUM POSITION OF THE CENTRAL CLAY CORE OF A ROCKFILL DAM IN RESPECT TO ARCHING AND HYDRAULIC FRAC	73-0447
CLAY	PERFORMANCE OF AN EMBANKMENT CONSTRUCTED ON VARVED CLAY	73-0743
CLAY	PERFORMANCE OF TIED-BACK WALLS IN CLAY	74-0665
CLAY	SETTLEMENTS AND STRENGTHENING OF SOFT CLAY ACCELERATED BY SAND DRAINS	73-0438
CLAY	STRESS-PATH DEPENDENT DEFORMATION MODULI OF CLAY	75-0050
CLAY	STUDIES OF SEISMIC RESPONSE OF CLAY BANKS	71-0833
CLAY	TEMPORARY EXCAVATION IN VARVED CLAY	75-0049
CLAY	TEST OF VALIDITY OF DARCY'S LAW FOR A CLAY SOIL	74-0070
CLAYS	ANALYSIS OF PLANE STRAIN CONSOLIDATION IN SATURATED CLAYS BY FINITE ELEMENT METHOD	71-0815
CLAYS	EVALUATION OF THE STABILITY OF NATURAL SLOPES IN PLASTIC CHAMPLAIN CLAYS.	74-0264
CLAYS	INITIATION OF FAILURE IN SLOPES IN OVERCONSOLIDATED CLAYS AND CLAY SHALES.	70-0186
CLAYS	NONLINEAR STRESS DEFORMATION ANALYSES OF LAKE AGASSIZ CLAYS USING FINITE ELEMENT METHOD	74-1072
CLAYS	SLOPES IN STIFF FISSURED CLAYS AND SHALES	69-0094
CLAYS	SOIL RESISTANCE MOVEMENT RELATIONSHIPS FOR LATERALLY LOADED PILES IN CLAYS AT GREAT DEPTHS	70-0162
CLAYS	STRESS-STRAIN PROPERTIES OF COMPACTED CLAYS	74-0650
CLEAN	STRESS ANALYSIS FOR CLEAN LEAN DESIGN	71-0386
CLEARANCE	THE PIECEWISE LINEAR ANALYSIS OF TWO CONNECTED STRUCTURES INCLUDING THE EFFECT OF CLEARANCE AT THE C	71-0379
CLEARANCES	MULTIDEGREE-OF-FREEDOM ELASTIC SYSTEMS HAVING MULTIPLE CLEARANCES	73-0653
CLIMATIC	SHELL STRUCTURES AND CLIMATIC INFLUENCES.	72-0389
CLIMATIC	SHELL STRUCTURES AND CLIMATIC INFLUENCES	72-0399
CLOSED	ANALYSIS OF LARGE CLOSED SHIP STRUCTURES: APPLICATIONS TO TANKER STRUCTURAL DESIGN	70-0289
CLOSED	CLOSED FORM, FINITE ELEMENT SOLUTIONS FOR PLATE VIBRATIONS	69-0168
CLOSED	MATHEMATICAL MODEL FOR CLOSED HEAD IMPACT	74-0934
CLOSED-BOUNDARY	ANALYSIS OF OPEN STRUCTURES BY FINITE ELEMENT RESOLUTION OF EQUIVALENT CLOSED-BOUNDARY PROBLEM	74-0111
CLOSING	ANALYSIS OF SPACE CLOSING SPRINGS IN ORTHODONTICS	74-0236
CLOSURE	A FINITE ELEMENT ANALYSIS OF FATIGUE CRACK CLOSURE	74-0759
CLOSURE	APPLICATION OF A SUBSTRUCTURING TECHNIQUE TO THE PROBLEM OF CRACK EXTENSION AND CLOSURE	74-0772
CLOSURE	CLOSURE ANALYSIS USING SELF ENERGIZING, DOUBLE CONICAL GASKETS	74-0675
CLOSURE	DESIGN OF A COMPOSITE STEEL AND CONCRETE CLOSURE FOR A PRESTRESSED CONCRETE REACTOR VESSEL.	72-0187
CLOSURE	FINITE ELEMENT ANALYSIS OF FATIGUE CRACK PROPAGATION - INCLUDING THE EFFECTS OF CRACK CLOSURE	74-1112
CLOSURE	FINITE ELEMENT FOR REINFORCED CONCRETE FRAME STUDY - CLOSURE	75-0087
CLOSURE	INFLUENCE OF CRACK CLOSURE AND ELASTO-PLASTIC FLOW ON THE BENDING OF A CRACKED PLATE	75-0735
CLOSURE	LONG TERM CREEP CLOSURE OF SOLUTION CAVITY SYSTEM	74-1150
CLOSURE	STRESS, DISPLACEMENT, LINE INTEGRAL AND CLOSURE ENERGY DETERMINATIONS OF CRACK TIP STRESS INTENSITY	74-0521
CLOSURE	TIME-DEPENDENT ANALYSIS TO PREDICT CLOSURE IN SALT CAVITIES	74-1149
CLOSURES	FREE VIBRATIONS OF CYLINDRICAL SHELLS WITH VARIOUS END CLOSURES	74-0571
CLOSURES	ULTIMATE LOAD BEHAVIOUR OF PCRV TOP CLOSURES. THEORETICAL AND EXPERIMENTAL INVESTIGATIONS	73-0835
CLOUGH	SUR L ELEMENT DE CLOUGH ET TOCHER (CLOUGH-TOCHER ELEMENT)	74-1022
CLOUGH-TOCHER	CLOUGH-TOCHER TRIANGULAR PLATE BENDING ELEMENT IN VIBRATION	69-0093
CLOUGH-TOCHER	SUR L ELEMENT DE CLOUGH ET TOCHER (CLOUGH-TOCHER ELEMENT)	74-1022
CLYINDER	METODY RASCHETNOGO OPREDELENIYA TEMPERATURNYKH NAPRYAZHENII V KRYSHKAKH TSILINDROV DVIGATELEI VNUTRE	74-0179
CO-ORDINATES	BODY ORIENTED (NATURAL) CO-ORDINATES FOR GENERATING THREE-DIMENSIONAL MESHES	74-0908
CO-ORDINATES	FINITE ELEMENT ANALYSIS IN POLAR CO-ORDINATES OF THE SAINT VENANT TORSION PROBLEM	74-0500
CO-ORDINATES	FINITE ELEMENTS FOR FIELD PROBLEMS IN CYLINDRICAL CO-ORDINATES	73-0171
CO-ORDINATES	NON-LINEAR TRANSIENT FINITE ELEMENT ANALYSIS WITH CONVECTED CO-ORDINATES	73-0378
CO-ORDINATES	RESTRICTIONS IN THE APPLICATION OF AUTOMATIC GENERATION SCHEMES BY ISOPARAMETRIC CO-ORDINATES	74-1377
COAL	DESIGN OPTIMIZATION IN UNDERGROUND COAL SYSTEMS	74-1095
COAL	FORECASTING THE ROCK PRESSURE AND ITS EFFECT ON THE LONGWALL FACES AND ROADWAYS IN THE BITUMINOUS CO	74-0374
COAL	STRESS DISTRIBUTION AND ROCK FRACTURE ZONES IN THE ROOF OF LONGWALL FACE IN COAL MINE	73-0247
COAL	STRUCTURAL ANALYSIS OF A COAL MINE OPENING IN ELASTIC, MULTILAYERED MATERIAL	74-0235
COAL	STRUCTURAL ANALYSIS OF A COAL MINE OPENING IN ELASTIC, MULTILAYERED MATERIAL	74-0840
COAL	VORAUSSAGE DES GEBIRGSDRUCKES UND SEINER AUSGIRKUNGEN AUF STREBEN UND STRECKEN IM STEINKOHLENBERGBAU	73-0679
COALESCENCE	DEFORMATION MODE OF VOID-GROWTH AND COALESCENCE IN THE PROCESS OF DUCTILE FRACTURE	75-0683
COASTAL	FINITE ELEMENT MODEL OF TWO LAYER COASTAL CIRCULATION	74-1192
COASTAL	HYDRODYNAMICAL-NUMERICAL MODELS FOR COASTAL WATERS AND OPEN OCEAN AREAS	75-0571
COASTAL	MANAGEMENT OF RETARDATION OF SALT WATER INTRUSION IN COASTAL AQUIFERS	75-0692
COASTAL	MATHEMATICAL MODEL OF HEATED EFFLUENTS IN COASTAL WATER	73-0921
COASTAL	MATHEMATICAL MODELING OF NEAR COASTAL CIRCULATION	75-0720
COASTAL	NUMERICAL SOLUTION OF THE STEADY INTERFACE IN A CONFINED COASTAL AQUIFER	74-1007
COASTAL	ON SEAWATER ENCROACHMENT IN COASTAL AQUIFERS	74-0577
COASTAL	THE NUMERICAL SOLUTION OF THE STEADY INTERFACE IN A COASTAL CONFINED AQUIFER	73-0711
CODE	A FINITE ELEMENT COMPUTER CODE, AXICAP FOR CREEP ANALYSIS	69-0248
CODE	A NONLINEAR FINITE ELEMENT CODE FOR ANALYZING THE BLAST RESPONSE OF UNDERGROUND STRUCTURES	70-0054
CODE	A THREE-DIMENSIONAL FINITE ELEMENT COMPUTER CODE FOR THE ANALYSIS OF COMPLEX STRUCTURES	71-0275
CODE	AFPAV COMPUTER CODE FOR STRUCTURAL ANALYSIS OF AIRFIELD PAVEMENTS	75-0684
CODE	AIRCRAFT FUEL TANK VULNERABILITY TO HYDRAULIC RAM: MODIFICATTION OF THE NORTHROP FINITE ELEMENT COMP	74-0724
CODE	AN ANALYSIS OF FINITE ELEMENT CODE TORT 11	75-0752
CODE	BOOLEAN MATRICES AS A STANDARD FACILITY IN A MATRIX CODE	65-0057
CODE	CODE FOR STRESS ANALYSIS BY THE THREE DIMENSIONAL FINITE ELEMENT METHOD (JAPANESE)	69-0509
CODE	CODE TAFEST: NUMERICAL SOLUTION TO TRANSIENT HEAT-CONDUCTION PROBLEMS USING FINITE ELEMENTS IN SPACE	74-0901
CODE	COMPUTER CODE TO GENERATE FINITE ELEMENT MESH FOR PROBLEMS OF COMPLEX AXISYMMETRIC GEOMETRY	74-0360
CODE	COUPLE - A PSEUDO FINITE ELEMENT CONDUCTION CODE FOR THE TRANSIENT THERMAL RESPONE OF AXISYMMETRIC O	72-0778
CODE	DYNAMIC ELASTIC-PLASTIC ANALYSIS OF AXISYMMETRIC STRUCTURES BY FINITE ELEMENT METHOD COMPUTER CODE	73-0844
CODE	ELASTIC-PLASTIC STRESS-STRAIN RELATIONS USED IN CODE EPSOLA	73-0784
CODE	ELFED A COMPUTER CODE TO GENERATE A FINITE ELEMENT MESH FOR PROBLEMS OF COMPLEX AXISYMMETRIC GEOMETR	74-1270
CODE	ELFED: A COMPUTER CODE TO GENERATE FINITE ELEMENT MESH FOR PROBLEMS OF COMPLEX AXISYMMETRIC GEOMETRY	73-0943
CODE	EXTENSIONS TO THE STAGS COMPUTER CODE	73-0364
CODE	EXTENSIONS TO THE STAGS COMPUTER CODE	73-0917
CODE	FEMESH: A FINITE ELEMENT CODE PREPROCESSOR	73-0949
CODE	FIESTA - FINITE ELEMENT STRESS AND TEMPERATURE ANALYSIS. A DESCRIPTION OF THE THEORY USED FOR THE CO	71-0718
CODE	FINITE ELEMENT ANALYSIS OF PAVEMENT STRUCTURES USING AFPAV CODE LINEAR ELASTIC ANALYSIS	73-0658
CODE	FINITE-ELEMENT ANALYSIS OF PAVEMENT STRUCTURES USING AFPAV CODE (NONLINEAR ELASTIC ANALYSIS)	75-0748
CODE	FINITE-ELEMENT CODE TO SIMULATE ANISOTROPIC-PLASMA DIFFUSION IN LINEAR MULTIPOLES	75-0409
CODE	FINITE-ELEMENT PLASMA-DIFFUSION CODE FOR ARBITRARY TOROIDAL GEOMETRY	75-0500

149

CODE	GRAPHICS APPLICATIONS FOR FINITE ELEMENT CODE PROCESSING	74-0893
CODE	HEATPAN, A FINITE ELEMENT CODE FOR HEAT TRANSFER PROBLEMS	69-0076
CODE	INPUT SOIL PROPERTIES FOR THE FINITE ELEMENT CODE NTEDYNS	71-0694
CODE	ISOFINEL: ISOPARAMETRIC FINITE ELEMENT CODE FOR ELASTIC ANALYSIS OF TWO-DIMENSIONAL BODIES	75-0676
CODE	MESH GENERATION CODE AND A CONTOUR PLOTTING ROUTINE	72-0802
CODE	MESHGEN, A COMPUTER CODE FOR AUTOMATIC FINITE ELEMENT MESH GENERATION	69-0096
CODE	PARAMETRIC STUDIES OF LAYERED PAVEMENT SYSTEMS USING AFPAV CODE WITH FAST EQUATION SOLVER	75-0747
CODE	REBO: COMPUTER CODE FOR THE GENERATION OF A THREE-DIMENSIONAL FINITE-ELEMENT GRID OF THE NOZZLE-CYLI	74-1310
CODE	SASL - A FINITE ELEMENT CODE FOR THE STATIC ANALYSIS OF AXISYMMETRIC AND PLANE SOLIDS SUBJECTED TO A	72-0779
CODE	SINGER: A COMPUTER CODE FOR GENERAL ANALYSIS OF TWO-DIMENSIONAL CONCRETE STRUCTURES. VOL. 11. PROGRA	75-0723
CODE	SINGER: A COMPUTER CODE FOR GENERAL ANALYSIS OF TWO-DIMENSIONAL REINFORCED CONCRETE STRUCTURES. VOLU	75-0724
CODE	SINGER: A COMPUTER CODE FOR GENERAL ANALYSIS TO TWO-DIMENSIONAL CONCRETE STRUCTURES. VOLUME 4. DEMON	75-0730
CODE	SINGER: A COMPUTER CODE FOR GENERAL ANALYSIS OF TWO-DIMENSIONAL REINFORCED CONCRETE STRUCTURES. VOLU	75-0731
CODE	STARS - A COMPUTER CODE FOR STRUCTURAL DYNAMIC ANALYSIS OF REACTOR SYSTEM	71-0204
CODE	STRAP - A COMPUTER CODE FOR THE STATIC AND DYNAMIC ANALYSIS OF REACTOR STRUCTURAL SYSTEMS	71-0078
CODE	STRAP: COMPUTER CODE FOR STATIC AND DYNAMIC STRUCTURAL ANALYSIS AND STUDIES MADE USING THE CODE	70-0409
CODE	STRAP: COMPUTER CODE FOR STATIC AND DYNAMIC STRUCTURAL ANALYSIS AND STUDIES MADE USING THE CODE	70-0409
CODE	STRUCTURE-MEDIUM INTERACTION AND DESIGN PROCEDURES STUDY. VOLUME IV. FEAT CODE USER'S GUIDE	69-0265
CODE	THE TRIDIMENSIONAL THERMOELASTIC COMPUTER CODE TITUS	71-0156
CODE	TRANAL: A 3-D FINITE ELEMENT CODE FOR TRANSIENT NONLINEAR ANALYSIS	74-0707
CODE	TRESS - A FINITE ELEMENT CODE FOR STRESS ANALYSIS PROBLEMS	71-0715
CODE	TRIPLET: A TWO-DIMENSIONAL, MULTIGROUP, TRIANGULAR MESH, PLANAR GEOMETRY, EXPLICIT TRANSPORT CODE	73-0955
CODE	USE OF THE ' TITUS ' FINITE ELEMENT CODE FOR THE NUCLEAR STRUCTURES AND INTERNAL STRESS-ANALYSIS	73-0840
CODE	USER'S MANUAL FOR NTEDYNS - A NONLINEAR FINITE ELEMENT CODE FOR ANALYZING THE BLAST RESPONSE OF UNDE	71-0695
CODE	VERIFICATION OF THE FINITE ELEMENT COMPUTER CODE SAFE-2D	74-0873
CODE	VERIFICATION OF THE PEC CONTAINMENT VESSEL FOR SOME OPERATING CONDITIONS BY USE OF THE FINITE ELEMEN	74-0883
CODE	3-D SEISMIC CODE FOR ILLIAC-1V	73-0868
CODE	3-D STRESS WAVE CODE FOR THE ILLTAC 1V	74-0780
CODES	ASSESSMENT OF COMMERCIALLY AVAILABLE FINITE ELEMENT CODES FOR THE DYNAMIC ANALYSIS OF STRUCTURES	73-0954
CODES	NEW TRANSPORT METHODS AND CODES	74-0878
CODES	ON THE AUTOMATIC SELECTION OF INDEPENDENT DISPLACEMENT VARIABLES IN GENERAL PURPOSE FINITE ELEMENT C	73-0676
CODES	THE DEVELOPMENT OF COMPUTER GRAPHICS FOR LARGE SCALE FINITE ELEMENT CODES	71-0042
CODES	THE DEVELOPMENT OF LARGE SCALE DIGITAL COMPUTER CODES FOR PRODUCTION STRUCTURAL ANALYSIS	71-0721
CODES	THREE-DIMENSIONAL MESH GENERATOR FOR FINITE ELEMENT COMPUTER CODES	74-0864
CODES	TWO-DIMENSIONAL FINITE ELEMENT COMPUTER CODES FOR STRESS ANALYSIS AND FIELD PROBLEMS	73-0055
COEFFICIENT	CONSISTENT MATRIX FORMULATIONS FOR STRUCTURAL ANALYSIS USING INFLUENCE COEFFICIENT TECHNIQUES	64-0017
COEFFICIENT	PROCEDURE FOR SOLVING LINEAR EQUATION SYSTEMS WITH SPARSELY OCCUPIED COEFFICIENT MATRICES ASSOCIATED	70-0352
COEFFICIENT	THE CONDITION OF SOME FINITE ELEMENT COEFFICIENT MATRICES	70-0393
COEFFICIENTS	A FINITE ELEMENT METHOD FOR CALCULATING AERODYNAMIC COEFFICIENTS OF A SUBSONIC AIRPLANE	73-0655
COEFFICIENTS	BOUNDS TO THE INFLUENCE COEFFICIENTS BY THE ASSUMED STRESS METHOD	70-0165
COEFFICIENTS	COMPUTATION OF INFLUENCE COEFFICIENTS FOR AIRCRAFT STRUCTURES WITH DISCONTINUITIES AND SWEEPBACK	47-0002
COEFFICIENTS	DUALITY BETWEEN DISPLACEMENT AND EQUILIBRIUM METHOD WITH A VIEW TO OBTAINING UPPER AND LOWER BOUNDS	62-0013
COEFFICIENTS	FINITE ELEMENT - GALERKIN METHODS FOR HYPERBOLIC EQUATIONS WITH DISCONTINUOUS COEFFICIENTS	73-0726
COEFFICIENTS	FINITE ELEMENT METHOD FOR ELLIPTIC EQUATIONS WITH DISCONTINUOUS COEFFICIENTS	70-0193
COEFFICIENTS	FINITE ELEMENT METHOD FOR NON SMOOTH DOMAINS AND COEFFICIENTS	72-0504
COEFFICIENTS	FREQUENCY COEFFICIENTS OF A SPHERICAL SHELL FOR VARYING THICKNESS AND OPENING ANGLE WITH GUIDED PINN	72-0595
COEFFICIENTS	FUNCTIONAL COEFFICIENTS IN THE ANALYSIS OF GROUNDWATER FLOW USING FINITE ELEMENTS	73-0077
COEFFICIENTS	FURTHER DEVELOPMENTS IN CONSISTENT UNSTEADY SUPERSONIC AERODYNAMIC COEFFICIENTS	72-0559
COEFFICIENTS	KINEMATICALLY CONSISTENT UNSTEADY AERODYNAMIC COEFFICIENTS IN SUPERSONIC FLOW	70-0014
COEFFICIENTS	KINEMATICALLY CONSISTENT UNSTEADY AERODYNAMIC COEFFICIENTS IN SUPERSONIC FLOW	70-0316
COEFFICIENTS	STIFFNESS COEFFICIENTS FOR EMBEDDED FOOTINGS	75-0540
COEUR	TENUE DES STRUCTURES DE GRAPHITE VIS-A-VIS DES SOLLICITATIONS RENCONTREES DANS UN COEUR DE REACTEUR	73-0817
COHESIVE	BEHAVIOUR OF STRIP FOOTINGS ON LAYERED COHESIVE SOILS	69-0223
COHESIVE	DYNAMIC RESPONSE OF COHESIVE EARTH DAMS TO EARTHQUAKES	67-0131
COHESIVE	PREDICTION OF LONG TERM DEFORMATION OF A COMPACTED COHESIVE SOIL EMBANKMENT OVER A SOFT FOUNDATION	74-0727
COHESIVE	RIGID PLATE ON NONLINEARLY DEFORMABLE COHESIVE BASE (PLANE PROBLEM)	75-0264
COIL	SCHWINGUNGEN ZYLINDRISCHER SCHRAUBENFEDERN (VIBRATIONS OF CYLINDRICAL COIL SPRINGS)	74-0331
COLD	STRESS ANALYSIS IN SOME COLD FORGING TOOL COMPONENTS AND THEIR DESIGN GUIDES	74-0589
COLDWORKED	STRESS ANALYSIS OF COLDWORKED FASTENER HOLES	74-0783
COLLAGEN	MEASUREMENTS OF HEALING AT AN OSTEOTOMY IN A RABBIT CALVARIUM: THE INFLUENCE OF APPLIED COMPRESSIVE	74-0975
COLLAPSE	CALCULATIONAL PROCEDURE FOR DETERMINING CREEP COLLAPSE OF LWR FUEL RODS	74-1001
COLLAPSE	COLLAPSE ANALYSIS OF BOX-GIRDER COMPONENTS USING FINITE ELEMENTS	74-0951
COLLAPSE	COLLAPSE LOAD OF REINFORCED CONCRETE PLATES.	72-0406
COLLAPSE	PLASTIC COLLAPSE LOADS FOR PIPE ELBOWS USING INELASTIC ANALYSIS	73-0941
COLLAPSE	PLASTIC COLLAPSE LOADS FOR PIPE ELBOWS USING INELASTIC ANALYSIS	74-0507
COLLAPSE	SAFE-COLAPS: A FINITE ELEMENT PROGRAM FOR THE CREEP COLLAPSE ANALYSIS OF NUCLEAR FUEL RODS	75-0573
COLLECTION	A COLLECTION OF PROBLEMS SOLVED WITH ASKA	70-0401
COLLINEAR	EIGENVALUES OF COLLINEAR BEAM STRUCTURES USING FINITE ELEMENT TECHNIQUES AND VARIOUS DYNAMIC REPRESE	67-0124
COLLISEUM	RECORD CONCRETE SHELL ROOF FEATURED IN COLLISEUM	72-0218
COLLISION-IMPAR	APPLICATION OF THE COLLISION-IMPARTED VELOCITY METHOD FOR ANALYZING THE RESPONSES OF CONTAINMENT AND	73-0770
COLLOCATION	A FINITE ELEMENT COLLOCATION SOLUTION OF DIFFERENTIAL EQUATIONS	72-0050
COLLOCATION	A FINITE ELEMENT COLLOCATION METHOD FOR THE HEAT EQUATION	72-0488
COLLOCATION	A FINITE ELEMENT COLLOCATION METHOD FOR QUASILINEAR PARABOLIC EQUATIONS	73-1000
COLLOCATION	A SUPERCONVERGENCE RESULT FOR THE APPROXIMATE SOLUTION OF THE HEAT EQUATION BY A COLLOCATION METHOD	72-0510
COLLOCATION	COLLOCATION AND FINITE ELEMENTS - A COMBINED METHOD	73-0734
COLLOCATION	FINITE ELEMENT COLLOCATION METHOD FOR QUASILINEAR PARABOLIC EQUATIONS	73-0734
COLLOCATION	ON THE APPLICATIONS OF COLLOCATION OF ANALYTIC SOLUTION TO FINITE ELEMENT METHOD	69-0384
COLLOCATION	ORTHOGONAL COLLOCATION ON FINITE ELEMENTS	75-0364
COLLOCATION	THE RELATION BETWEEN THE GALERKIN AND COLLOCATION METHODS USING SMOOTH SPLINES	74-0912
COLLOCATION	UNSTEADY SUBSONIC COLLOCATION METHOD FOR WINGS WITH AND WITHOUT CONTROL SURFACES	72-0012
COLLOQUIM	NAVY-NASTRAN COLLOQUIM AT WASHINGTON, D. C.	72-0669
COLLOQUIUM	NAVY-NASTRAN COLLOQUIUM, WASHINGTON, D.C.	70-0404
COLLOQUIUM	NSRDC: NASTRAN COLLOQUIUM, WASHINGTON, D.C.	70-0404
COLLOQUIUM	PROCEEDINGS OF THE NAVY-NASTRAN COLLOQUIUM (4TH) HELD AT NAVY SHIP RESEARCH AND DEVELOPMENT CENTER,	73-0888
COLLOQUIUM	PROCEEDINGS OF THE 5TH NAVY-NASTRAN COLLOQUIUM, HELD AT NAVAL SHIP RESEARCH AND DEVELOPMENT CENTER,	74-0709
COLORADO	EARTHQUAKE TRIGGERING BY FLUID INJECTION AT RANGELY, COLORADO	72-0634
COLUMBIA	MATHEMATICAL MODEL OF THE COLUMBIA RIVER FROM THE PACIFIC OCEAN TO BONNEVILLE DAM. PART 1. THEORY, P	69-0464
COLUMBIA	MATHEMATICAL MODEL OF THE COLUMBIA RIVER FROM THE PACIFIC OCEAN TO BONNEVILLE DAM. PART II: INPUT-OU	70-0417
COLUMN	ACCURACY OF SOME NUMERICAL METHODS FOR COLUMN BUCKLING	75-0583
COLUMN	DUAL ANALYSIS OF COLUMN SUPPORTED SLABS	74-0399
COLUMN	LOWER BOUNDS TO COLUMN BUCKLING LOADS	73-0485
COLUMN	RC COLUMN EARTHQUAKE RESPONSE IN TWO DIMENSIONS	74-0651
COLUMN	STRONGEST COLUMN BY FINITE ELEMENT DISPLACEMENT METHOD	73-0136
COLUMN-SUPPORTE	DYNAMIC ANALYSIS OF COLUMN-SUPPORTED HYPERBOLOIDAL SHELLS	74-0188
COLUMNS	A FINITE ELEMENT ANALYSIS FOR THE STRENGTH OF BEAM COLUMNS	66-0037
COLUMNS	A FINITE ELEMENT ANALYSIS OF THE ELASTIC BEHAVIOR OF BEAM COLUMNS	66-0122
COLUMNS	A FINITE ELEMENT APPROACH TO THE STRUCTURAL INSTABILITY OF BEAM COLUMNS, FRAMES, AND ARCHES	70-0434
COLUMNS	A UNIFIED FINITE ELEMENT APPROACH TO COLUMNS SUBJECTED TO NONCONSERVATIVE LOADS, WITH DAMPING EFFECT	74-0795
COLUMNS	ANALYSIS OF ELASTIC-PLASTIC BEHAVIOR OF BEAMS AND COLUMNS BY PLANE FINITE ELEMENT MODEL	73-1073
COLUMNS	DISCUSSION OF PROCEEDINGS, PAPER 6612, ELASTIC BUCKLING OF TWO-COMPONENT COLUMNS	70-0321
COLUMNS	ETUDES DES POUTRES ET POTEAUX PAR LA METHODE DES ELEMENTS FINIS. (STABILITY ANALYSIS OF BEAMS AND C	74-1008
COLUMNS	FINITE DEFORMATION OF BIAXIALLY LOADED COLUMNS	72-0021

COLUMNS	FINITE ELEMENT ANALYSIS OF NONLINEAR VIBRATION OF BEAM COLUMNS	73-0028
COLUMNS	FINITE ELEMENT SOLUTIONS FOR BUCKLING OF COLUMNS AND BEAMS	71-0361
COLUMNS	GALERKIN FINITE ELEMENT SOLUTION FOR THE STABILITY OF CANTILEVER COLUMNS SUBJECTED TO TANGENTIAL LOA	75-0534
COLUMNS	GALERKIN FINITE-ELEMENT SOLUTION FOR STABILITY OF CANTILIVER COLUMNS SUBJECTED TO TANGENTIAL LOADS	75-0618
COLUMNS	MATRIX ANALYSIS OF BEAM COLUMNS BY THE METHOD OF FINITE ELEMENTS	64-0012
COLUMNS	MATRIX ANALYSIS OF LOCAL INSTABILITY IN PLATES, STIFFENED PANELS AND COLUMNS	72-0542
COLUMNS	NUMERICAL STUDY OF UNIAXIAL COMPRESSION IN CIRCULAR ELASTIC-PLASTIC COLUMNS	73-0238
COLUMNS	ORTHOTROPIC FLAT PLATES SUPPORTED ON COLUMNS ELONGATED IN PLANE	72-0492
COLUMNS	SAVINGS IN NASTRAN DECOMPOSITION TIME BY SEQUENCING TO REDUCE ACTIVE COLUMNS	71-0161
COLUMNS	STABILITY ANALYSIS OF COLUMNS USING FINITE ELEMENTS	69-0478
COLUMNS	STATIC AND DYNAMIC BUCKLING OF THIN-WALLED COLUMNS USING FINITE ELEMENTS	70-0505
COLUMNS	STRUCTURAL CAPACITY OF REINFORCED CONCRETE COLUMNS SUBJECTED TO FIRE INDUCED THERMAL GRADIENTS	73-0913
COMBAT	CONCEPT FEASIBILITY COMBAT TRACKED VEHICLE SIGNATURE DUPLICATOR	64-0028
COMBINATIONS	BERECHNUNG DES REIBUNGSLOSEN STROEMUNGSFELDES VON DREIDIMENSIONALEN AUFTRIEBSBEHAFTETENTRAGFLUEGELN	73-0533
COMBUSTION	ECONOMICAL DESIGN ANALYSIS OF PISTONS FOR INTERNAL COMBUSTION ENGINES	73-0188
COMBUSTION	METODY RASCHETNOGO OPREDELENIYA TEMPERATURNYKH NAPRYAZHENII V KRYSHKAKH TSILINDROV DVIGATELEI VNUTRE	74-0179
COMBUSTION	OPREDELENIE STATSIONARNYKH TEMPERATURNYKH POLEI V DETALYAKH DVIGATELEI VNUTRENNEGO SGORANIYA METODOM	73-0454
COMBUSTION	RASCHET POLEI DEFORMATSII I NAPRYAZHENII V DETALYAKH DVIGATELEI VNUTRENNEGO SGORANIYA (CALCULATION O	73-0465
COMMERCIAL	ASAS - A COMMERCIAL FINITE ELEMENT SYSTEM WITH DISPLACEMENT AND FORCE CONTINUITY OPTIONS	73-0947
COMMERCIALLY	ASSESSMENT OF COMMERCIALLY AVAILABLE FINITE ELEMENT CODES FOR THE DYNAMIC ANALYSIS OF STRUCTURES	73-0954
COMMUNICATION	DISCUSSION OF A SHORT COMMUNICATION BY S. GOPALACHARYULU	74-0404
COMMUNICATION	SOLUTION OF NONLINEAR EQUATIONS BY EXPLICIT TIME INTEGRATION - A SHORT COMMUNICATION	73-0606
COMMUNIST	MECHANICS OF DEFORMABLE SOLIDS - EURASIAN COMMUNIST COUNTRIES	73-0794
COMOC	COMOC THERMAL ANALYSIS VARIANT, USER'S MANUAL	72-0701
COMPACT	A COMPACT FINITE-ELEMENT METHOD FOR THE ANALYSIS OF ANISOTROPIC LAMINATED PLATES	73-1166
COMPACT	A COMPACT STORAGE SCHEME FOR THE SOLUTION OF SYMMETRICAL LINEAR SIMULTANEOUS EQUATIONS	66-0116
COMPACT	ANALYTIC INVESTIGATION OF COMPACT REINFORCEMENT FOR RADIAL NOZZLES IN SPHERICAL SHELLS	71-0107
COMPACT	VIBRATION AND STRESS ANALYSIS OF SOFT-BONDED SHUTTLE INSULATION TILES. MODAL ANALYSIS WITH COMPACT W	74-0708
COMPACTED	PREDICTION OF LONG TERM DEFORMATION OF A COMPACTED COHESIVE SOIL EMBANKMENT OVER A SOFT FOUNDATION	74-0727
COMPACTED	STRESS-STRAIN PROPERTIES OF COMPACTED CLAYS	74-0650
COMPACTION	FINITE ELEMENT ANALYSIS OF SOIL COMPACTION	74-0931
COMPARAISON	COMPARAISON DES DIVERSES MODELISATIONS NUMERIQUES D'UN SYSTEME MAGNETIQUE SATURABLE. (COMPARISON OF	75-0295
COMPARATIVE	A COMPARATIVE STUDY OF NUMERICAL METHODS OF ELASTIC-PLASTIC ANALYSIS	67-0047
COMPARATIVE	A COMPARATIVE STUDY OF THE CONSISTENT AND SIMPLIFIED FINITE ELEMENTS ANALYSIS OF EIGENVALUE PROBLEMS	70-0034
COMPARATIVE	A COMPARATIVE STUDY OF THE CONSISTENT AND SIMPLIFIED FINITE ELEMENT ANALYSES OF EIGENVALUE PROBLEMS	70-0373
COMPARATIVE	A COMPARATIVE STUDY OF SEVERAL CORE STORAGE SCHEMES FOR LARGE SPARSE POSITIVE DEFINITE MATRICES WITH	72-0678
COMPARATIVE	A COMPARATIVE STUDY OF DIFFERENT NUMERICAL SOLUTION TECHNIQUES AS APPLIED TO A NONLINEAR STRUCTURAL	73-1117
COMPARATIVE	A COMPARATIVE STUDY OF VARIABLE AND CONSTANT THICKNESS HIGH PRECISION TRIANGULAR PLATE BENDING ELEME	74-0857
COMPARATIVE	COMPARATIVE ANALYSIS OF THE CUTHILL-MCKEE AND THE REVERSE CUTHILL-MCKEE ORDERING ALGORITHMS FOR SPAR	73-1099
COMPARATIVE	COMPARATIVE CONSIDERATION OF PARAMETRIC STRESS CONCENTRATION INVESTIGATIONS WITH FINITE ELEMENTS AND	72-0423
COMPARATIVE	COMPARATIVE EFFICIENCY OF CERTAIN FINITE ELEMENT AND FINITE DIFFERENCE METHODS FOR A HYPERBOLIC PROB	73-0965
COMPARATIVE	COMPARATIVE OBSERVATION ON PARAMETRIC STRESS STUDIES WITH FINITE ELEMENTS WITH COMPLEX STRESS FUNCTI	72-0250
COMPARATIVE	COMPARATIVE PERTURBATION STUDIES OF THE ELASTICA	69-0255
COMPARATIVE	COMPARATIVE STUDY OF NUMERICAL METHODS OF ELASTIC-PLASTIC ANALYSIS	67-0153
COMPARATIVE	COMPARATIVE STUDY OF FRESH-SALT WATER INTERFACES USING FINITE ELEMENT AND SIMPLE APPROACHES	75-0468
COMPARATIVE	FINITE ELEMENT METHOD AND LAPLACE TRANSFORM COMPARATIVE SOLUTIONS OF TRANSIENT HEAT CONDUCTION PROBL	74-0575
COMPARATIVE	THE COMPARATIVE EFFICIENCY OF CERTAIN FINITE ELEMENT AND FINITE DIFFERENCE SCHEMES FOR A HYPERBOLIC	74-0910
COMPARIOSN	COMPARIOSN OF RESULTS OBTAINED FOR THE PROBLEM OF SYMMETRIC FLUID OSCILLATIONS IN TANKS BY THE FINIT	71-0761
COMPARISON	A COMPARISON AND EVALUATION OF INCREMENTAL FINITE ELEMENT FORMULATIONS	75-0023
COMPARISON	A COMPARISON BETWEEN TWO-HIGH-PRECISION FINITE ELEMENTS	75-0033
COMPARISON	A COMPARISON OF CURRENT WORK-HARDENING MODELS USED IN THE ANALYSIS OF PLASTIC DEFORMATIONS	73-0769
COMPARISON	A COMPARISON OF CURVED BEAM FINITE ELEMENTS WHEN USED IN VIBRATION PROBLEMS	71-0227
COMPARISON	A COMPARISON OF FINITE ELEMENT AND FOURIER SERIES SOLUTIONS AS APPLIED TO RADIALLY LOADED CIRCULAR R	68-0024
COMPARISON	A COMPARISON OF FINITE ELEMENT AND FINITE DIFFERENCE METHODS IN ELASTOSTATIC PROBLEMS	69-0273
COMPARISON	A COMPARISON OF FIRST AND SECOND ORDER AXIALLY SYMMETRIC FINITE ELEMENTS	73-0997
COMPARISON	A COMPARISON OF METHODS USED IN FLUTTER RESEARCH	72-0600
COMPARISON	A COMPARISON OF RECENT SHALLOW SHELL FINITE-ELEMENT ANALYSES	70-0232
COMPARISON	A COMPARISON OF SEVERAL COMPUTER SOLUTIONS TO THREE STRUCTURAL SHELL ANALYSIS PROBLEMS	73-0884
COMPARISON	A COMPARISON OF SEVERAL COMPUTER SOLUTIONS TO THREE STRUCTURAL SHELL ANALYSIS PROBLEMS	73-0907
COMPARISON	A COMPARISON OF SOME FINITE ELEMENT AND FINITE-DIFFERENCE METHODS FOR A SIMPLE SLOSHING PROBLEM	71-0055
COMPARISON	A FINITE ELEMENT ANALYSIS OF BODIES OF REVOLUTION. A COMPARISON BETWEEN CURRENTLY USED TYPES OF ELEM	74-1319
COMPARISON	A THEORETICAL AND NUMERICAL COMPARISON OF ELASTIC NONLINEAR FINITE ELEMENT METHODS	72-0709
COMPARISON	ANALYSIS OF TUBULAR T-CONNECTIONS BY THE FINITE ELEMENT METHOD: COMPARISON WITH EXPERIMENTS	72-0380
COMPARISON	APPLICATION OF SMALL AND LARGE-DISPLACEMENT THEORY TO A LARGE FLEXIBLE SOLAR ARRAY AND COMPARISON WI	71-0231
COMPARISON	CALCULATING STRESSES IN A PRESSURE VESSEL WITH FILL TUBE - COMPARISON OF THE FINITE ELEMENT AND PHOT	72-0782
COMPARISON	CALCULATION OF AXISYMMETRIC LONGITUDINAL MODES FOR FLUID-ELASTIC TANK-ULLAGE GAS SYSTEMS AND COMPARI	64-0019
COMPARISON	COMPARAISON DES DIVERSES MODELISATIONS NUMERIQUES D'UN SYSTEME MAGNETIQUE SATURABLE. (COMPARISON OF	75-0295
COMPARISON	COMPARISON AND EVALUATION OF STIFFNESS MATRICES	66-0045
COMPARISON	COMPARISON BETWEEN SPARSE STIFFNESS MATRIX AND SUB-STRUCTURE METHODS	73-1068
COMPARISON	COMPARISON BETWEEN THEORETICAL AND MEASURED EARTH PRESSURES ACTING ON A LARGE MOTORWAY RETAINING WAL	74-0917
COMPARISON	COMPARISON OF ANALYTICAL AND EXPERIMENTAL RESULTS IN THE DESIGN OF A HOLLOW GRAVITY DAM	67-0109
COMPARISON	COMPARISON OF ANALYTICAL AND NUMERICAL SOLUTIONS IN HEAT CONDUCTION PROBLEMS	74-1170
COMPARISON	COMPARISON OF CALCULATION BY THE FINITE ELEMENT METHOD WITH EXPERIMENT AND AN EXAMPLE OF A DYNAMIC S	72-0597
COMPARISON	COMPARISON OF EQUIVALENCE AND FINITE ELEMENT METHODS	75-0483
COMPARISON	COMPARISON OF EXTRAPOLATION TECHNIQUES WITH HIGH ORDER FINITE ELEMENT METHODS FOR DIFFUSION EQUATION	73-0586
COMPARISON	COMPARISON OF FINITE ELEMENT AND UNIQUE LIMIT ANALYSIS SOLUTIONS FOR CERTAIN REINFORCED CONCRETE SLA	69-0184
COMPARISON	COMPARISON OF FINITE ELEMENT AND FINITE DIFFERENCE METHODS	71-0143
COMPARISON	COMPARISON OF FINITE-ELEMENT AND EXPERIMENTAL STUDIES ON DEFORMATIONS OF ZIRCONIUM NOTCHED BEND SPEC	71-0359
COMPARISON	COMPARISON OF FINITE ELEMENTS FOR PLATE BENDING	72-0625
COMPARISON	COMPARISON OF FINITE ELEMENT METHODS FOR A MODEL PROBLEM	74-0134
COMPARISON	COMPARISON OF FINITE-DIFFERENCE AND FINITE-ELEMENT SOLUTION TECHNIQUE IN TRANSIENT HEAT-CONDUCTION	75-0348
COMPARISON	COMPARISON OF MEMBRANE FINITE ELEMENT FORMULATIONS	70-0287
COMPARISON	COMPARISON OF NUMERICAL METHODS FOR DETERMINING STRESS INTENSITY FACTORS	72-0176
COMPARISON	COMPARISON OF PLATE BENDING FINITE ELEMENT FORMULATIONS	70-0288
COMPARISON	COMPARISON OF PRACTICAL METHODS OF ANALYSING ELASTIC LATERALLY LOADED PLATES	71-0425
COMPARISON	COMPARISON OF RECENT SHALLOW SHELL FINITE ELEMENT ANALYSES	71-0362
COMPARISON	COMPARISON OF SOME CURVED TWO DIMENSIONAL FINITE ELEMENTS	73-0383
COMPARISON	COMPARISON OF SOME FINITE-ELEMENT MODELS FOR CIRCULAR CYLINDRICAL SHELLS	73-1158
COMPARISON	COMPARISON OF SOME FORMULATIONS IN THERMAL ANALYSIS OF STRUCTURES	75-0888
COMPARISON	COMPARISON OF SOME PLATE BENDING FINITE ELEMENTS	72-0413
COMPARISON	COMPARISON OF STRUCTURAL LOADS; STATIC VERSUS DYNAMIC	70-0464
COMPARISON	COMPARISON OF STRUCTURAL TEST RESULTS WITH PREDICTIONS OF FINITE ELEMENT ANALYSIS	73-0158
COMPARISON	COMPARISON OF THE CAPABILITY OF OUR HARDENING RULES TO PREDICT A MATERIAL'S PLASTIC BEHAVIOR	75-0523
COMPARISON	COMPARISON OF THE FINITE ELEMENT AND FINITE DIFFERENCE METHODS FOR THE ANALYSIS OF STEADY TWO DIMENS	75-0223
COMPARISON	COMPARISON OF THE NUMERICAL INTEGRATION TECHNIQUE AND THE FINITE ELEMENT METHOD IN THE ANALYSIS OF T	73-0848
COMPARISON	COMPARISON OF THE RESULTS OF THE DYNAMIC RESPONSE OF CYLINDRICAL SHELLS BY CHARACTERISTIC AND BY FIN	74-0781
COMPARISON	COMPARISON OF THE THREE-DIMENSIONAL ANALYSIS OF A SHEAR WALL BUILDING BY TWO DIFFERENT METHODS	75-0705
COMPARISON	COMPARISON OF THEORETICAL AND EXPERIMENTAL SHEAR STRESS IN THE ADHESIVE LAYER OF A LAP JOINT MODEL	74-1176
COMPARISON	COMPARISON OF THREE DIMENSIONAL FINITE ELEMENTS	69-0073
COMPARISON	COMPARISON OF TIME MARCHING SCHEMES FOR THE TRANSIENT HEAT CONDUCTION EQUATION	75-0477
COMPARISON	COMPARISON OF TWO HIGH-PRECISION TRIANGULAR FINITE ELEMENTS FOR ARBITRARY DEEP SHELLS	71-0071
COMPARISON	COMPARISON OF TYPICAL INELASTIC ANALYSIS PREDICTIONS WITH BENCHMARK PROBLEM EXPERIMENTAL RESULTS	75-0696

COMPARISON	CRACK TIP STRAIN - A COMPARISON OF FEM CALCULATIONS AND MEASUREMENTS	75-0721
COMPARISON	CRACKING OF EARTH AND ROCKFILL DAMS: COMPARISON OF OBSERVED AND THEORETICAL TENSILE STRAINS IN THE C	71-0699
COMPARISON	CREATION AND COMPARISON OF FINITE DIFFERENCE ANALOGS OF SOME FINITE ELEMENT SCHEMES	74-0888
COMPARISON	DEFORMATION OF THE END DIASTOLIC LEFT VENTRICLE, COMPARISON OF THEORY WITH EXPERIMENT	71-0131
COMPARISON	DETERMINATION OF STRESS CONCENTRATIONS AT NOZZLE TO CYLINDER INTERSECTIONS AND COMPARISON WITH EXPER	73-0491
COMPARISON	DIGITAL COMPUTATION OF STRESSES AND DEFLECTIONS IN A BOX BEAM - A PERFORMANCE COMPARISON BETWEEN FIN	69-0355
COMPARISON	ELASTIC WAVE PROPAGATION. A COMPARISON BETWEEN FINITE ELEMENT PREDICTIONS AND EXACT SOLUTION	67-0076
COMPARISON	EXAMPLE OF LARGE-SCALE STRUCTURAL ANALYSIS. COMPARISON BETWEEN FINITE ELEMENT CALCULATION AND FULL S	74-0177
COMPARISON	FINITE DIFFERENCE SOLUTIONS FOR LARGE DEFORMATIONS OF CYLINDRICAL SHELLS - A COMPARISON WITH FINITE	72-0270
COMPARISON	FINITE ELEMENT METHOD OF ANALYSIS FOR ARCH DAMS SHELLS AND COMPARISON WITH FINITE DIFFERENCE PROCEDU	65-0036
COMPARISON	FINITE ELEMENTS VERSUS FINITE DIFFERENCES, A COMPARISON OF THE TWO METHODS FOR THE SOLUTION OF THE D	73-1055
COMPARISON	IN VIVO STRESSES IN THE HUMAN LEFT VENTRICULAR WALL: ANALYSIS ACCOUNTING FOR THE IRREGULAR THREE DIM	72-0136
COMPARISON	NASTRAN ANALYSIS OF A TURBINE BLADE AND COMPARISON WITH TEST AND FIELD DATA	75-0367
COMPARISON	PHOTOELASTIC INVESTIGATION OF STRESS CONCENTRATIONS IN SPHERE-CYLINDER TRANSITION REGIONS: INCLUDING	68-0258
COMPARISON	STATIC, STABILITY, AND DYNAMIC ANALYSIS OF SHELLS OF REVOLUTION BY NUMERICAL INTEGRATION - A COMPARI	74-0389
COMPARISON	THE COMPARISON OF ITERATIVE AND DIRECT SOLUTION TECHNIQUES IN THE ANALYSIS OF TIME-DEPENDENT STRESS	72-0080
COMPARISON	THE COMPARISON OF THE FINITE ELEMENT METHOD WITH THE VARIATIONAL-DIFFERENCE METHOD IN THE THEORY OF	67-0096
COMPARISON	THE COMPARISON OF THE RESPONSE OF A HIGHWAY BRIDGE TO UNIFORM GROUND SHOCK AND MOVING GROUND EXCITAT	72-0737
COMPARISON	THEORETICAL AND NUMERICAL COMPARISON OF ELASTIC NONLINEAR FINITE ELEMENT METHODS.	72-0452
COMPARISON	THEORETICAL PREDICTION OF DISC BRAKE TEMPERATURES AND A COMPARISON WITH EXPERIMENTAL DATA	72-0306
COMPARISON	THREE-DIMENSIONAL FINITE ELEMENT ANALYSIS OF A STEAM GENERATOR CHANNEL-HEAD-COMPLEX - COMPARISON OF	75-0653
COMPARISON	TIMING COMPARISON STUDY FOR SOME HIGH ORDER FINITE ELEMENT PROCEDURES AND A LOW ORDER FINITE DIFFRE	72-0767
COMPARISON	VERGLEICH NUMERISCHER LOESUNGEN NACH DEM DIFFERENZENVERFAHREN UND NACH DER FINITE-ELEMENT-METHODE AM	71-0016
COMPARISONS	COMPARISONS OF CONSISTENT MASS MATRIX SCHEMES	69-0437
COMPARISONS	ELASTIC-PLASTIC-CREEP ANALYSIS OF THERMAL RATCHETTING IN STRAIGHT PIPE AND COMPARISONS WITH TEST RES	73-0393
COMPARISONS	EXPERIMENTAL ELASTIC STRESS ANALYSES OF CYLINDER-TO-CYLINDER SHELL MODELS AND COMPARISONS WITH THEOR	71-0065
COMPATIBILITY	COMPATIBILITY	75-0781
COMPATIBILITY	ELEMENT STIFFNESS MATRICES FOR BOUNDARY COMPATIBILITY AND FOR PRESCRIBED BOUNDARY STRESSES	66-0061
COMPATIBILITY	TRIANGULAR PLATE BENDING ELEMENT WITH ENFORCED COMPATIBILITY	71-0593
COMPATIBLE	A COMPATIBLE ELEMENT FOR BUCKLING ANALYSIS OF SKEW PLATES WITH PARALLEL EDGES	73-0328
COMPATIBLE	A COMPATIBLE MEMBRANE QUADRILATERAL ELEMENT	72-0483
COMPATIBLE	A COMPATIBLE PLATE BENDING ANNULAR SECTOR FINITE ELEMENT OF NONUNIFORM THICKNESS	74-0430
COMPATIBLE	A COMPATIBLE QUADRILATERAL PLATE BENDING ELEMENT	72-0482
COMPATIBLE	A COMPATIBLE TRIANGULAR PLATE BENDING FINITE ELEMENT	68-0209
COMPATIBLE	A COMPATIBLE TRIANGULAR PLATE BENDING FINITE ELEMENT	70-0385
COMPATIBLE	A FULLY COMPATIBLE ANNULAR SEGMENT FINITE ELEMENT	72-0582
COMPATIBLE	A FULLY COMPATIBLE ANNULAR SEGMENT FINITE ELEMENT	72-0563
COMPATIBLE	A FULLY COMPATIBLE PLATE BENDING ELEMENT	67-0142
COMPATIBLE	A NEW, FULLY COMPATIBLE FINITE ELEMENT FOR PLATE BENDING EIN NEUES VOLLVERTRAGLICHES ENDLICHES ELEME	68-0032
COMPATIBLE	COMPATIBLE TRIANGULAR FINITE ELEMENTS	74-0330
COMPATIBLE	DERIVATION OF COMPATIBLE STIFFNESS MATRICES FOR LARGE DISCRETE ELEMENTS	67-0080
COMPATIBLE	GENERAL LINEAR GEOMETRIC MATRIX FOR A FULLY COMPATIBLE FINITE ELEMENT	72-0294
COMPATIBLE	HIGHER ORDER COMPATIBLE TRIANGULAR FINITE ELEMENTS	74-0152
COMPATIBLE	HORIZONTALLY CURVED HIGHWAY BRIDGES. A FULLY COMPATIBLE ANNULAR SEGMENT FINITE ELEMENT	72-0834
COMPATIBLE	MESH GRADING TECHNIQUES FOR COMPATIBLE AND EQUILIBRIUM ELEMENTS	74-0397
COMPATIBLE	OPTIM 11: (MAGIC) COMPATIBLE LARGE SCALE AUTOMATED MINIMUM WEIGHT DESIGN PROGRAM - V 2. PROGRAMMER'S	74-1036
COMPATIBLE	OPTIM 11: A (MAGIC) COMPATIBLE LARGE SCALE AUTOMATED MINIMUM WEIGHT DESIGN PROGRAM - 1. ENGINEERS AN	74-1048
COMPATIBLE	OPTIM 11: A MAGIC COMPATIBLE LARGE-SCALE AUTOMATED MINIMUM WEIGHT DESIGN PROGRAM. VOLUME 1. ENGINEER	74-1096
COMPATIBLE	OPTIM 11: MAGIC COMPATIBLE LARGE SCALE AUTOMATED MINIMUM WEIGHT DESIGN PROGRAM. VOLUME 11. PROGRAMME	74-1093
COMPATIBLE	PLATE BUCKLING ANALYSIS USING A FULLY COMPATIBLE FINITE ELEMENT	69-0065
COMPATIBLE	STRUCTURALLY COMPATIBLE FLUID FINITE-ELEMENT FOR SOLID FLUID INTERACTION STUDIES	75-0344
COMPATIBLE	THE GENERATION OF INTERELEMENT, COMPATIBLE STIFFNESS AND MASS MATRICES BY THE USE OF INTERPOLATION F	66-0014
COMPLEMENTAIRES	FORMULATIONS DUALES DES ENERGIES POTENTIELLES ET COMPLEMENTAIRES. APPLICATION A LA METHODE DES ELEME	71-0752
COMPLEMENTARY	AN INCREMENTAL COMPLEMENTARY ENERGY METHOD OF NONLINEAR STRESS ANALYSIS	69-0281
COMPLEMENTARY	DUAL FORMULATION FOR POTENTIAL ENERGY AND COMPLEMENTARY ENERGY APPLICATION TO THE FINITE ELEMENT MET	72-0062
COMPLEMENTARY	GENERAL POTENTIAL ENERGY AND COMPLEMENTARY ENERGY MODELS BASED ON STRESS PARAMETERS	74-1240
COMPLEMENTARY-D	A MATHEMATICAL THEORY OF COMPLEMENTARY-DUAL VARIATIONAL PRINCIPLES AND MIXED FINITE ELEMENT APPROXIM	73-0804
COMPLEMENTARY-D	ON COMPLEMENTARY-DUAL VARIATIONAL PRINCIPLES IN MATHEMATICAL PHYSICS	74-0342
COMPLETE	APPLICATION OF FINITE ELEMENT METHODS TO COMPLETE AUTOMOBILE STRUCTURAL DESIGN EVALUATION	74-0312
COMPLETE	AVERAGE AND COMPLETE INCOMPRESSIBILITY IN THE FINITE ELEMENT METHOD	75-0342
COMPLETE	COMPLETE POLYNOMIAL DISPLACEMENT FIELDS FOR FINITE ELEMENT METHODS	68-0062
COMPLETE	COMPLETE POLYNOMIAL DISPLACEMENT FIELDS FOR FINITE ELEMENT METHODS	68-0092
COMPLETE	CONFORMING FINITE ELEMENTS BASED ON COMPLETE POLYNOMIALS	74-0302
COMPLETE	THE EFFECT OF COMPOSITE ACTION ON THE ELASTOPLASTIC ANALYSIS OF COMPLETE BUILDING STRUCTURES	74-0441
COMPLETE	VIBRATION STUDY OF COMPLETE TREE SYSTEMS USING FINITE ELEMENT METHOD	75-0790
COMPLETE	VOLLSTAENDIGE SPANNUNGS-DEHNUNGS-TEMPERATURAENDERUNGSBEZIEHUNG FUER DIE SCHWEISSEIGENSPANNUNGSBERECH	75-0807
COMPLETELY	HIGH ORDER FINITE ELEMENT FOR COMPLETELY INCOMPRESSIBLE CREEPING FLOW	73-0144
COMPLETENESS	COMPLETENESS AND CONVERGENCE PROPERTIES OF FINITE ELEMENT DISPLACEMENT FUNCTIONS.. A GENERAL TREATME	67-0053
COMPLETENESS	COMPLETENESS AND CONVERGENCE IN THE FINITE ELEMENT METHOD	68-0056
COMPLETENESS	COMPLETENESS AND CONVERGENCE IN THE FINITE ELEMENT METHOD	68-0279
COMPLETENESS	ON COMPLETENESS OF SHAPE FUNCTIONS FOR FINITE ELEMENT ANALYSIS	72-0105
COMPLETENESS	ON THE COMPLETENESS OF SHAPE FUNCTIONS FOR FINITE ELEMENT ANALYSIS	72-0885
COMPLEX	A FINITE ELEMENT ANALYSIS FOR STEADY AND OSCILLATORY SUBSONIC FLOW AROUND COMPLEX CONFIGURATIONS	74-1317
COMPLEX	A FINITE ELEMENT APPROACH FOR THE ANALYSIS OF RANDOMLY EXCITED COMPLEX ELASTIC STRUCTURES	67-0057
COMPLEX	A FINITE-ELEMENT ANALYSIS FOR STEADY-AND OSCILLATORY SUPERSONIC FLOWS AROUND COMPLEX CONFIGURATIONS	74-1316
COMPLEX	A FINITE-ELEMENT FORMULATION FOR SUBSONIC FLOWS AROUND COMPLEX CONFIGURATIONS	73-0742
COMPLEX	A METHOD FOR REDUCING THE ANALYSIS OF COMPLEX REDUNDANT STRUCTURES TO A ROUTINE PROCEDURE	52-0002
COMPLEX	A TECHNIQUE FOR THE FINITE ELEMENT ANALYSIS OF LARGE COMPLEX STRUCTURAL SYSTEMS	69-0514
COMPLEX	A THREE-DIMENSIONAL FINITE ELEMENT COMPUTER CODE FOR THE ANALYSIS OF COMPLEX STRUCTURES	71-0275
COMPLEX	ACCURACY OF COMPLEX FINITE ELEMENTS	72-0291
COMPLEX	AN INTERACTIVE HYBRID TECHNIQUE FOR CRASHWORTHY DESIGN OF COMPLEX VEHICULAR STRUCTURAL SYSTEMS	74-0317
COMPLEX	AN INVESTIGATION INTO THE ECONOMICS OF USING COMPLEX ELEMENTS IN PLANE STRESS ANALYSIS	74-0427
COMPLEX	ANALYSIS OF REAL STRUCTURES FOR CREEP, PLASTICITY AND OTHER COMPLEX CONSTITUTIVE LAWS.	69-0287
COMPLEX	ANALYTICAL PREDICTION OF DEFECTS OCCURRENCE IN SIMPLE AND COMPLEX	69-0513
COMPLEX	APPLICATION OF FRACTURE MECHANICS TO GRAPHITE UNDER COMPLEX STRESS CONDITIONS	72-0771
COMPLEX	BUCKLING UNDER COMBINED LOADING OF THIN FLAT-WALLED STRUCTURES BY A COMPLEX FINITE STRIP METHOD	74-0788
COMPLEX	BUTTRESS DAMS ON COMPLEX ROCK FOUNDATIONS	64-0013
COMPLEX	CALCULATION OF GEOMETRIC STIFFNESS MATRICES FOR COMPLEX STRUCTURES	66-0058
COMPLEX	CHISLENNOE RESHENIE PLOSKIKH ZADACH TEPLOPROVODNOSTI DLYA OBLASTEI SLOZHNOI FORMY (NUMERICAL SOLUTIO	74-0659
COMPLEX	COMPARATIVE CONSIDERATION OF PARAMETRIC STRESS CONCENTRATION INVESTIGATIONS WITH FINITE ELEMENTS AND	72-0423
COMPLEX	COMPARATIVE OBSERVATION ON PARAMETRIC STRESS STUDIES WITH FINITE ELEMENTS WITH COMPLEX STRESS FUNCTI	72-0250
COMPLEX	COMPLEX EIGENVALUE SOLUTION TO A SPINNING SKYLAB PROBLEM	71-0207
COMPLEX	COMPUTER CODE TO GENERATE FINITE ELEMENT MESH FOR PROBLEMS OF COMPLEX AXISYMMETRIC GEOMETRY	74-0360
COMPLEX	CONTRIBUTION A L'ETUDE DU COMPLEXE BARRAGE-FONDATION ET EXPLICATION DES SEISMES DUS AU REMPLISSAGE D	72-0554
COMPLEX	CURRENT COMPUTATION ON COMPLEX STRUCTURES BY FINITE-ELEMENT METHOD	75-0456
COMPLEX	DERIVATION OF CRACK ELEMENT-STIFFNESS MATRIX BY THE COMPLEX VARIABLE APPROACH	73-0870
COMPLEX	DISCRETE-ELEMENT METHODS FOR STABILITY ANALYSIS OF COMPLEX STRUCTURES	68-0140
COMPLEX	DYNAMIC ANALYSIS OF A COMPLEX SHELL STRUCTURE UNDER IMPULSIVE ASYMMETRIC LOADING	68-0226
COMPLEX	DYNAMIC SIMULATION OF COMPLEX SYSTEMS UTILIZING EXPERIMENTAL AND ANALYTICAL TECHNIQUES	75-0864
COMPLEX	ELASTODYNAMICS OF COMPLEX STRUCTURAL SYSTEMS	71-0563
COMPLEX	ELFED A COMPUTER CODE TO GENERATE A FINITE ELEMENT MESH FOR PROBLEMS OF COMPLEX AXISYMMETRIC GEOMETR	74-1270
COMPLEX	ELFED: A COMPUTER CODE TO GENERATE FINITE ELEMENT MESH FOR PROBLEMS OF COMPLEX AXISYMMETRIC GEOMETRY	73-0943

COMPLEX	FINITE-ELEMENT ANALYSIS OF COMPLEX PANEL RESPONSE TO RANDOM LOADS	68-0093
COMPLEX	FREE VIBRATIONS OF FINITE ELEMENT PLATES SUBJECTED TO COMPLEX MIDDLE-PLANE FORCE SYSTEMS	72-0280
COMPLEX	FREE VIBRATIONS OF FINITE ELEMENT PLATES SUBJECTED TO COMPLEX MIDDLE-PLANE FORCE SYSTEMS	73-0097
COMPLEX	FREE VIBRATIONS OF FINITE ELEMENT PLATES SUBJECTED TO COMPLEX MIDDLE-PLANE FORCE SYSTEMS - REPLY	73-0122
COMPLEX	IMPACT ON COMPLEX MECHANICAL STRUCTURES	71-0598
COMPLEX	INTERPRETATION OF COMPLEX STRESS ANALYSIS BY MULTIPLE REGRESSION	74-1203
COMPLEX	LUSH: A COMPUTER PROGRAM FOR COMPLEX RESPONSE ANALYSIS OF SOIL-STRUCTURE SYSTEMS	74-0744
COMPLEX	MINIMUM WEIGHT DESIGN OF COMPLEX STRUCTURES SUBJECT TO A FREQUENCY CONSTRAINT	70-0374
COMPLEX	NATURAL FREQUENCIES OF COMPLEX FREE OR SUBMERGED STRUCTURES BY THE FINITE ELEMENT METHOD	65-0026
COMPLEX	NONLINEAR DYNAMIC ANALYSIS OF COMPLEX STRUCTURES	73-0193
COMPLEX	NONSAP - A GENERAL FINITE ELEMENT PROGRAM FOR NONLINEAR DYNAMIC ANALYSIS OF COMPLEX STRUCTURES	73-0740
COMPLEX	OPTIMIZATION OF COMPLEX STRUCTURES TO SATISFY STATIC, DYNAMIC AND AEROELASTIC REQUIREMENTS	74-0272
COMPLEX	PHOTOELASTIC TESTS ON COMPLEX INTERSECTIONS OF REINFORCED THIN-SHELL CYLINDERS	75-0363
COMPLEX	RANDOM VIBRATION ANALYSIS SYSTEM FOR COMPLEX STRUCTURES, PART 2, COMPUTER PROGRAM DESCRIPTION	69-0262
COMPLEX	RANDOM-VIBRATION ANALYSIS SYSTEM FOR COMPLEX STRUCTURES, PART 1, ENGINEERING USER'S GUIDE	68-0106
COMPLEX	RASCHET KILEBANII PLASTIN SLOZHNDI FORMY S NEODNORODNYMI GRANICHNYMI USLOVIYAMI. (CALCULATION OF VIB	74-0988
COMPLEX	SIMPLIFIED INELASTIC HIGH TEMPERATURE STRUCTURAL ANALYSIS OF MODERATELY COMPLEX SPATIALLY THREE DIME	74-0445
COMPLEX	SOME DEVELOPMENTS IN THE ANALYSIS OF COMPLEX SHIP STRUCTURES	72-0332
COMPLEX	STABILITY ANALYSIS OF COMPLEX STRUCTURES USING DISCRETE ELEMENT TECHNIQUE	67-0068
COMPLEX	STATIC ANALYSIS OF A COMPLEX SHELL STRUCTURE UNDER ASYMMETRIC LOADING	67-0120
COMPLEX	STIFFNESS AND DEFLECTION ANALYSIS OF COMPLEX STRUCTURES	56-0001
COMPLEX	STRESS ANALYSIS IN HEATED, COMPLEX SHAPES	62-0002
COMPLEX	STRESS ANALYSIS OF COMPLEX SHIP COMPONENTS BY A NUMERICAL PROCEDURE USING CURVED FINITE ELEMENTS	73-0879
COMPLEX	STRESS, STABILITY AND VIBRATION OF COMPLEX SHELLS OF REVOLUTION: ANALYSIS AND USERS MANUAL FOR BOSOR	69-0441
COMPLEX	STRUCTURAL ANALYSIS AND EXPERIMENTAL CORRELATION OF A COMPLEX SHELL UNDER ASYMMETRIC LOADS	69-0155
COMPLEX	SUBSONIC AND SUPERSONIC INDICIAL AERODYNAMICS AND AERODYNAMIC TRANSFER FUNCTION FOR COMPLEX CONFIGUR	74-1320
COMPLEX	THE ANALYSIS OF COMPLEX SHIP STRUCTURES BY THE FINITE ELEMENT TECHNIQUE	64-0006
COMPLEX	THE TRANSIENT LINEAR ELASTIC RESPONSE ANALYSIS COMPLEX THIN SHELLS OF REVOLUTION SUBJECTED TO ARBIT	70-0080
COMPLEX	THE USE OF FINITE ELEMENT METHODS FOR STATIC AND DYNAMIC ANALYSES OF COMPLEX SHELLS	68-0247
COMPLEX	THE USE OF NASTRAN IN THE ANALYSIS OF LARGE COMPLEX AIRFRAME STRUCTURES, MODELING TECHNIQUES AND ORG	71-0110
COMPLEX	USE OF SIMPLE THREE-DIMENSIONAL ACOUSTIC FINITE ELEMENTS FOR DETERMINING NATURAL MODES AND FREQUENCI	72-0010
COMPLEXE	CONTRIBUTION A L'ETUDE DU COMPLEXE BARRAGE-FONDATION ET EXPLICATION DES SEISMES DUS AU REMPLISSAGE D	72-0554
COMPLEXES	ANALYSE DES CONTRAINTES DAN DES STRUCTURES COMPLEXES EN PRESENCE DE FLUAGE, DE VARIATIONS DE DIMENSI	71-0081
COMPLEXITY	COMPLEXITY BOUNDS FOR REGULAR FINITE DIFFERENCE AND FINITE ELEMENT GRIDS	72-0872
COMPLEXITY	COMPLEXITY BOUNDS FOR REGULAR FINITE DIFFERENCE AND FINITE ELEMENT GRIDS	73-0043
COMPLIANCE	COMPLIANCE OF A TAPERED DCB SPECIMEN - CONFIGURATIONS BY A FINITE ELEMENT METHOD	74-0545
COMPLIANCE	FINITE ELEMENT CALCULATIONS OF THE COMPLIANCE OF A TAPERED DOUBLE CANTILEVER BEAM SPECIMEN FOR DIFFE	72-0665
COMPLIANCE	LOAD POINT COMPLIANCE OF THE CHARPY IMPACT SPECIMEN	73-0775
COMPLIANCE	LOCAL COMPLIANCE PROBLEM OF LUNG ELASTICITY	75-0660
COMPLIANCE	ON THE USE OF FINITE ELEMENTS AND COMPLIANCE CALCULATIONS FOR DETERMINING CRACK TIP STRESS INTENSITY	69-0448
COMPLIANCES	COMPLIANCES OF LAYERED ELASTIC SYSTEMS	74-1204
COMPONENT	A HYBRID METHOD OF COMPONENT MODE SYNTHESIS	71-0681
COMPONENT	ANALYTICAL METHODS IN COMPONENT MODAL SYNTHESIS	75-0824
COMPONENT	CREEP DATA ACQUISITION AND APPLICATION TO REACTOR COMPONENT DESIGN	72-0113
COMPONENT	DESIGN, MANUFACTURE, DEVELOPMENT, TEST, AND EVALUATION OF BORON/ALUMINUM STRUCTURAL COMPONENTS FOR S	74-0798
COMPONENT	DIESEL ENGINE COMPONENT DESIGN USING THE FINITE ELEMENT METHOD AND INTERACTIVE GRAPHICS	74-0237
COMPONENT	DYNAMIC ANALYSIS OF STRUCTURAL SYSTEMS USING COMPONENT MODES	64-0024
COMPONENT	INELASTIC ANALYSIS AND SATISFACTION OF DESIGN CRITERIA OF A HIGH TEMPERATURE COMPONENT	75-0265
COMPONENT	THE IMPORTANCE OF THE VERTICAL COMPONENT OF EARTHQUAKE MOTIONS	66-0019
COMPONENT	VIBRATION ANALYSIS OF STRUCTURES BY COMPONENT MODE SUBSTITUTION	71-0303
COMPONENTS	A FRACTURE MECHANICS APPROACH TO THE PROPAGATION OF FATIGUE CRACKS IN REACTOR COMPONENTS BY ACOUSTIC	71-0132
COMPONENTS	A HIGH PRECISION AXISYMMETRIC TRIANGULAR ELEMENT USED IN THE ANALYSIS OF HYDRAULIC TURBINE COMPONENT	70-0038
COMPONENTS	AN ENGINEERING APPROACH TO FINITE ELEMENT ANALYSIS OF NUCLEAR COMPONENTS	73-1054
COMPONENTS	APPLICATION OF FINITE ELEMENT ANALYSIS TO CERAMIC COMPONENTS	75-0040
COMPONENTS	APPLICATION OF THE FINITE ELEMENT METHOD FOR THE SAFETY EVALUATION OF REACTOR COMPONENTS	73-0853
COMPONENTS	BERECHNUNG VON SPANNUNGEN UND TEMPERATURFELDERN IN FLOSSENWAENDEN UND AUFGESCHWEISSIEN BAUTEILEN. (C	74-0544
COMPONENTS	CALCULATION OF THERMAL STRESSES IN TUBULAR CERAMIC COMPONENTS	73-0670
COMPONENTS	COLLAPSE ANALYSIS OF BOX-GIRDER COMPONENTS USING FINITE ELEMENTS	74-0951
COMPONENTS	COMPUTER AIDED DESIGN OF THERMALLY LOADED AXISYMMETRIC DIESEL ENGINE COMPONENTS	74-0596
COMPONENTS	COMPUTER CALCULATIONS OF STRESSES IN AXISYMMETRIC THERMALLY LOADED COMPONENTS	68-0069
COMPONENTS	DESIGN ASPECTS OF HIGH DUTY PRESSURE COMPONENTS	74-0199
COMPONENTS	DESIGN, MANUFACTURE, DEVELOPMENT, TEST, AND EVALUATION OF BORON/ALUMINUM STRUCTURAL COMPONENTS FOR S	74-0798
COMPONENTS	ELASTIC PLASTIC ANALYSIS OF PRESSURE VESSEL COMPONENTS	69-0380
COMPONENTS	ENERGY FOR CRACK GROWTH IN MODEL RUBBER COMPONENTS	72-0568
COMPONENTS	ENGINEERING APPROACH TO FINITE ELEMENT ANALYSIS OF NUCLEAR COMPONENTS	73-0839
COMPONENTS	FINITE ELEMENT COMPUTER PROGRAM FOR PREDICTING THE NONLINEAR STATIC AND DYNAMIC BEHAVIOR OF VISCOELA	75-0191
COMPONENTS	LOW CYCLE FATIGUE STRENGTH OF PIPING COMPONENTS	73-0366
COMPONENTS	ON THE USE OF FINITE ELEMENTS IN FRACTURE ANALYSIS OF PRESSURE VESSEL COMPONENTS	75-0514
COMPONENTS	PRESSURE DISTRIBUTION AND DEFORMATIONS OF MACHINED COMPONENTS IN CONTACT	75-0372
COMPONENTS	PROBABILISTIC EVALUATION OF THE RELATIVE BEHAVIOUR OF TUNGSTEN CARBIDE COMPONENTS IN STRUCTURAL APPL	75-0579
COMPONENTS	RASCHET POLEI DEFORMATSII I NAPRYAZHENII V DETALYAKH DVIGATELEI VNUTRENNEGO SGORANIYA (CALCULATION O	73-0465
COMPONENTS	STRESS ANALYSIS IN SOME COLD FORGING TOOL COMPONENTS AND THEIR DESIGN GUIDES	74-0589
COMPONENTS	STRESS ANALYSIS OF COMPLEX SHIP COMPONENTS BY A NUMERICAL PROCEDURE USING CURVED FINITE ELEMENTS	73-0879
COMPONENTS	STRESS ANALYSIS OF INDUSTRIAL COMPONENTS WITH PLASTIC AND FINITE ELEMENT MODELS	74-0512
COMPONENTS	STRESS DETERMINATION IN 2 CONDENSER HEAT EXCHANGER BAFFLE CAPS BY METHOD OF FINITE ELEMENTS AND A ST	74-0661
COMPONENTS	THE APPLICATION OF THE SUPER ELEMENT METHOD IN ANALYSIS AND DESIGN OF SHIP STRUCTURES AND MACHINERY	72-0657
COMPORTAMENTO	ANALISI NUMERICA DEL COMPORTAMENTO PLASTICO DI UNA STRUTTURA BIDIMENSIONALE PIANA. (NUMERICAL ANALYS	74-1029
COMPORTAREA	COMPORTAREA ELASTICA A DIGURILOR DE PAMINT SOLICIATE SEISMIC (ELASTIC BEHAVIOR OF THE EARTH DAMS SEI	74-1185
COMPORTEMENT	ANALYSE TRIDIMENSIONALE DU COMPORTEMENT NON-LINEAIRE D'UN CAISSON DE REACTEUR NUCLEAIRE EN BETON PRE	71-0144
COMPORTEMENT	APPLICATION DES CALCULUS DE FONDATIONS A L'EVALUATION DU COMPORTEMENT DES CHAUSSEES EM DASH 1, 2. (A	71-0429
COMPORTEMENT	ETUDE THEORIQUE DU COMPORTEMENT D'UNE DALLE DE BETON. (THEORETICAL STUDY OF THE BEHAVIOR OF A CONCRE	75-0273
COMPORTEMENT	ETUDE THEORIQUE DU COMPORTEMENT DES PIEUX SOUS CHARGE VERTICALE. INTRODUCTION DE LA DILATANCE. (THEO	75-0309
COMPOSITE	A FINITE ELEMENT SIMULATION OF MECHANICAL PROPERTIES OF COMPOSITE MATERIALS	74-1294
COMPOSITE	A FINITE-ELEMENT METHOD OF ANALYSIS FOR COMPOSITE BEAMS	68-0263
COMPOSITE	A STUDY ON COMPOSITE STRUCTURES EMPLOYING DIFFERENT TYPES OF FINITE ELEMENTS	69-0231
COMPOSITE	AN ANALYTIC METHOD FOR PREDICTING THE FLOW AND FRACTURE IN NOTCHED FIBER COMPOSITE MATERIALS	75-0689
COMPOSITE	ANALIZA ORGAN WYMUSZONYCH ZLOZONYCH UKLADOW LINIOWYCH METODA SZTYWNYCH ELEMENTOW SKONCZONYCH (ANALYS	72-0246
COMPOSITE	ANALYSIS OF AXISYMMETRIC COMPOSITE STRUCTURES BY THE FINITE ELEMENT METHOD	66-0066
COMPOSITE	ANALYSIS OF AXISYMMETRIC COMPOSITE BODIES BY THE FINITE ELEMENT METHOD	68-0242
COMPOSITE	ANALYSIS OF COMPOSITE REINFORCED CUTOUTS AND CRACKS	74-0824
COMPOSITE	ANALYTICAL DISPLACEMENTS AND VIBRATIONS OF CANTILEVERED UNSYMMETRIC FIBER COMPOSITE LAMINATES	75-0736
COMPOSITE	APPLICATION OF METHOD OF FINITE ELEMENTS TO ANALYSIS OF COMPOSITE FLOOR SYSTEMS	75-0418
COMPOSITE	APPLICATION OF THREE-DIMENSIONAL FINITE ELEMENT ANALYSES TO THE MICROMECHANICS OF FIBROUS COMPOSITE	74-1079
COMPOSITE	AUTOMATED STRUCTURAL DESIGN AND ANALYSIS OF ADVANCED COMPOSITE WING MODELS	73-0221
COMPOSITE	AVERAGE ELASTIC PROPERTIES AND STRESS FIELDS FOR COMPOSITE CONTINUA	69-0106
COMPOSITE	AVERAGE ELASTIC PROPERTIES AND STRESS FIELDS FOR COMPOSITE MATERIALS	69-0435
COMPOSITE	CALCULATION AND DESIGN OF JOINTS MADE FROM COMPOSITE MATERIALS	70-0156
COMPOSITE	CAPE-2D, A COMPUTER PROGRAM FOR THE STRESS ANALYSIS OF PLANE AND AXISYMMETRIC COMPOSITE STRUCTURES.	69-0083
COMPOSITE	COMPOSITE WING FOR TRANSONIC IMPROVEMENT. VOLUME 11. ADVANCED ANALYSIS EVALUATION	71-0657
COMPOSITE	COMPUTER SIMULATION OF THE DEFORMATION OF COMPOSITE AND POLYCRYSTALLINE MATERIALS BY FINITE ELEMENT	72-0937
COMPOSITE	CREEP OF A UNIAXIAL METAL MATRIX COMPOSITE SUBJECTED TO AXIAL AND NORMAL LATERAL LOADS	71-0041
COMPOSITE	CREEP OF A UNIAXIAL METAL MATRIX COMPOSITE SUBJECTED TO AXIAL AND NORMAL LATERAL LOADS	71-0727

153

COMPOSITE	DEFORMATION AND FRACTURE OF PARTICULATE COMPOSITE	72-0221
COMPOSITE	DESIGN OF A COMPOSITE STEEL AND CONCRETE CLOSURE FOR A PRESTRESSED CONCRETE REACTOR VESSEL.	72-0187
COMPOSITE	DEVELOPMENT OF AN UNDERSTANDING OF THE FATIGUE PHENOMENA OF BONDED AND BOLTED JOINTS IN ADVANCED FIL	72-0798
COMPOSITE	DISCRETE ELEMENT ANALYSIS OF COMPOSITE STRUCTURES	75-0201
COMPOSITE	DYNAMIC RESPONSE OF THE HUMAN ARM AS A COMPOSITE STRUCTURE	72-0611
COMPOSITE	EFFECT OF FIBRE DISTRIBUTION ON STRESS AND STRAIN CONCENTRATION AT HOLES IN COMPOSITE PLATES	72-0616
COMPOSITE	EFFECT OF PLASTICITY AND CRACK BLUNTING ON THE STRESS DISTRIBUTION IN ORTHOTROPIC COMPOSITE MATERIAL	73-0145
COMPOSITE	ELASTOPLASTIC CRACK PROPAGATION IN A TRANSVERSELY LOADED UNIDIRECTIONAL COMPOSITE	74-0217
COMPOSITE	EVALUATION OF COMPOSITE SHEAR PANELS WITH CUTOUTS	72-0615
COMPOSITE	EXTENSION OF CLASSICAL EXPERIMENTAL TECHNIQUES FOR CHARACTERIZING COMPOSITE - MATERIAL BEHAVIOUR	72-0088
COMPOSITE	FAILURE MECHANISMS IN COMPOSITE SYSTEMS	75-0688
COMPOSITE	FINITE ELEMENT ANALYSIS OF AXISYMMETRIC COMPOSITE STRUCTURES	65-0030
COMPOSITE	FINITE ELEMENT ANALYSIS OF SKEW COMPOSITE GIRDER BRIDGES	67-0150
COMPOSITE	FINITE ELEMENT ANALYSIS OF SKEW COMPOSITE PLATES	67-0160
COMPOSITE	FINITE ELEMENT ANALYSIS OF AXISYMMETRICAL SHELLS OF COMPOSITE MATERIAL	70-0483
COMPOSITE	FINITE ELEMENT ANALYSIS OF A COMPOSITE MATERIAL INTERFACE	73-0777
COMPOSITE	FINITE ELEMENT ANALYSIS SYSTEM FOR THE MECHANICAL BEHAVIOR OF ORIENTED FIBER COMPOSITE MATERIALS UND	73-0869
COMPOSITE	FINITE ELEMENT ANALYSIS OF NON-LINEAR COMPOSITE MATERIALS BY USE OF OVERLAY SYSTEMS	74-0928
COMPOSITE	FINITE ELEMENT ANALYSIS OF COMPOSITE MATERIAL BEHAVIOURS FROM MECHANICAL PROPERTIES OF CONSTITUENT M	75-0181
COMPOSITE	FINITE ELEMENT ANALYSIS OF COMPOSITE MATERIAL BEHAVIORS FROM MECHANICAL PROPERTIES OF CONSTITUTENT M	75-0874
COMPOSITE	FINITE ELEMENT STUDY OF COMPOSITE ACTION IN WALLS	69-0291
COMPOSITE	FINITE-ELEMENT ANALYSIS OF EDGE EFFECTS IN ANGLE-PLY COMPOSITE LAMINATES	74-0843
COMPOSITE	HIGH-PERFORMANCE COMPOSITE MATERIALS FOR VEHICLE CONSTRUCTION: A FINITE ELEMENT COMPUTER PROGRAM FOR	73-0767
COMPOSITE	HIGH-PERFORMANCE COMPOSITE MATERIALS FOR VEHICLE CONSTRUCTION: A FINITE ELEMENT COMPUTER PROGRAM FOR	73-0767
COMPOSITE	INELASTIC ANALYSIS OF A UNIDIRECTIONAL COMPOSITE SUBJECTED TO TRANSVERSE NORMAL LOADING	70-0015
COMPOSITE	INITIAL YIELD SURFACE OF A UNIDIRECTIONALLY REINFORCED COMPOSITE	71-0280
COMPOSITE	INTERACTION STRESSES IN COMPOSITE SYSTEMS	70-0215
COMPOSITE	INTERACTIVE PROGRAM IN DESIGN AND ANALYSIS OF COMPOSITE MATERIALS	74-0854
COMPOSITE	LOAD DISTRIBUTION IN A COMPOSITE STEEL BOX GIRDER BRIDGE	73-0765
COMPOSITE	MATRIX ANALYSIS OF A COMPOSITE STRUCTURE OF PLATE AND LINE ELEMENTS	71-0250
COMPOSITE	MATRIX AND INTERFACE STRESSES IN A DISCONTINUOUS FIBER COMPOSITE MODEL	68-0040
COMPOSITE	MICROMECHANICAL ANALYSIS OF CRACK PROPAGATION IN AN ELASTOPLASTIC COMPOSITE MATERIAL	74-0565
COMPOSITE	MICROMECHANICS ANALYSIS OF COMPOSITE MATERIALS USING FINITE ELEMENT METHODS	72-0770
COMPOSITE	MIXED ISOPARAMETRIC LAMINATED COMPOSITE SHELL ELEMENTS	75-0450
COMPOSITE	NONLINEAR BUCKLING ANALYSIS OF COMPOSITE SHELLS	74-0973
COMPOSITE	NONLINEAR BUCKLING ANALYSIS OF COMPOSITE SHELLS	75-0214
COMPOSITE	NONLINEAR FINITE ELEMENT ANALYSIS OF LAMINATED COMPOSITE SHELLS	74-0751
COMPOSITE	ON THE DETERMINATION OF EFFECTIVE MODULI OF COMPOSITE MATERIALS BY A THREE-DIMENSIONAL FINITE ELEMEN	74-1123
COMPOSITE	OPTIMUM DESIGN OF ADVANCED COMPOSITE STRUCTURES FOR STATIC LOADS	72-0613
COMPOSITE	OPTIMUM DESIGN OF COMPOSITE SHELLS SUBJECT TO NATURAL FREQUENCY CONSTRAINTS	73-0154
COMPOSITE	PLASTIC FLOW AND FAILURE OF DISCONTINUOUS FIBRE COMPOSITE MATERIALS	70-0157
COMPOSITE	PRACTICAL FINITE ELEMENT METHOD OF FAILURE PREDICTION FOR COMPOSITE MATERIAL STRUCTURES	75-0644
COMPOSITE	PRACTICAL PROBLEMS ASSOCIATED WITH THE APPLICATION OF THE FINITE ELEMENT METHOD TO COMPOSITE MATERIA	74-0483
COMPOSITE	PREDICTION OF STATIC AND FATIGUE DAMAGE AND CRACK PROPAGATION IN COMPOSITE MATERIALS	75-0643
COMPOSITE	SAFE-AXISYM, A COMPUTER PROGRAM FOR THE DESIGN OF COMPOSITE BODIES OF REVOLUTION. INPUT INSTRUCTIONS	65-0014
COMPOSITE	SAFE-PLANE, A COMPUTER PROGRAM FOR THE STRESS ANALYSIS AND DESIGN OF TWO-DIMENSIONAL COMPOSITE BODIE	67-0017
COMPOSITE	SANDWICH AND COMPOSITE ELEMENTS	75-0136
COMPOSITE	STATIC AND DYNAMIC ANALYSIS OF VISCOELASTOPLASTIC FIBER-REINFORCED COMPOSITE SHELLS IN MISSILE STRUC	73-0651
COMPOSITE	STOCHASTIC FINITE ELEMENT SIMULATION OF THE NONLINEAR STRUCTURAL RESPONSE OF FIBROUS COMPOSITE MATER	75-0762
COMPOSITE	STRESS ANALYSIS OF COMPOSITE MATERIALS BY THE HYBRID FINITE ELEMENT METHOD	75-0799
COMPOSITE	STRESS AND FAILURE ANALYSIS OF A GLASS-EPOXY COMPOSITE PLATE WITH A CIRCULAR HOLE	73-0276
COMPOSITE	STRESSES AROUND HOLES IN STIFFENED COMPOSITE PANELS USING LAURENT-SERIES AND FINITE ELEMENT METHODS	74-1252
COMPOSITE	STRESSES AROUND HOLES IN STIFFENED COMPOSITE PANELS USING LAURENT-SERIES AND FINITE ELEMENT METHODS	75-0733
COMPOSITE	STUDIES ON TWO PHASE COMPOSITE MATERIALS THROUGH COMPUTER SIMULATION	73-0315
COMPOSITE	THE BEHAVIOR OF ADVANCED FILAMENTARY COMPOSITE PLATES WITH CUTOUTS	73-0891
COMPOSITE	THE BEHAVIOUR OF ADVANCED FILAMENTARY COMPOSITE PLATES WITH CUTOUTS	70-0025
COMPOSITE	THE EFFECT OF COMPOSITE ACTION ON THE ELASTOPLASTIC ANALYSIS OF COMPLETE BUILDING STRUCTURES	74-0441
COMPOSITE	THE STOCHASTIC FINITE ELEMENT SIMULATION OF THE NONLINEAR STRUCTURAL RESPONSE OF FIBROUS COMPOSITE M	75-0798
COMPOSITE	THE THERMO-ELASTO-PLASTIC ANALYSIS OF A RE-ENTRY BODY TYPE COMPOSITE SOLID STRUCTURE USING A NEW FIN	66-0093
COMPOSITE	THEORETICAL POST YIELDING BEHAVIOUR OF COMPOSITE LAMINATES - 1. INELASTIC MICROMECHANICS	73-0168
COMPOSITE	THEORETICAL POST YIELDING BEHAVIOUR OF COMPOSITE LAMINATES -2.	73-0408
COMPOSITE	THEORETICAL STUDY OF THE EFFECT OF THE INTERFACE ON COMPOSITE TOUGHNESS	73-0386
COMPOSITE	THERMAL STRESSES IN STEEL-CONCRETE BRIDGES	75-0154
COMPOSITE	TRANSIENT TEMPERATURE RESPONSE OF COMPOSITE SLABS	74-1038
COMPOSITE	TRANSIENT THERMAL STRESS ANALYSIS OF COMPOSITE STRUCTURES INCLUDING CONTINUOUSLY VARYING PROPERTIES	75-0681
COMPOSITE	VARIATIONAL PRINCIPLE FOR THE LAPLACE'S OPERATOR WITH APPLICATION IN THE TORSION OF COMPOSITE RODS	74-1044
COMPOSITE-OVERL	COMPOSITE-OVERLAY REINFORCEMENT OF CUTOUTS AND CRACKS IN METAL SHEET	73-0878
COMPOSITES	AN INVESTIGATION OF INITIAL YIELD SURFACES FOR UNIDIRECTIONAL REINFORCED COMPOSITES	72-0837
COMPOSITES	APPLICATION OF OPTIMALLY CRITERION TO FIBER-REINFORCED COMPOSITES	73-0431
COMPOSITES	CREEP OF B/AL COMPOSITES AS INFLUENCED BY RESIDUAL STRESSES, BOND STRENGTH, AND FIBER PACKING GEOMET	74-1208
COMPOSITES	DESIGNING WITH THREE DIRECTIONAL COMPOSITES	74-0204
COMPOSITES	ELASTIC-PLASTIC FINITE ELEMENT ANALYSIS OF SHORT FIBRE COMPOSITES	74-0138
COMPOSITES	FATIGUE OF PRENOTCHED GRAPHITE FIBER COMPOSITES IN COMPRESSION	72-0215
COMPOSITES	FINITE ELEMENT ANALYSIS OF BENDING-EXTENSIONAL COUPLING IN LAMINATED COMPOSITES	70-0292
COMPOSITES	FINITE ELEMENT ANALYSIS OF INTERLAMINAR SHEAR FIBROUS COMPOSITES	71-0129
COMPOSITES	FINITE ELEMENT METHOD FOR CALCULATING STRESS INTENSITY FACTORS AND ITS APPLICATION TO COMPOSITES	74-1000
COMPOSITES	FRACTURE OF GRAPHITE FIBER REINFORCED COMPOSITES	74-0695
COMPOSITES	INELASTIC MICROMECHANICS OF CURING STRESSES IN COMPOSITES	75-0284
COMPOSITES	INTERFACE FAILURES IN COMPOSITES	74-1209
COMPOSITES	MECHANICAL PROPERTIES OF TAPE COMPOSITES	69-0317
COMPOSITES	PLASTICITY THEORY OF FIBROUS COMPOSITES - PART 11	73-0750
COMPOSITES	PLASTICITY THEORY OF FIBROUS COMPOSITES - PART 1	73-0751
COMPOSITES	SAFE PCPS, A COMPUTER PROGRAM FOR THE STRESS ANALYSIS OF COMPOSITES BODIES OF REVOLUTION. INPUT INST	65-0013
COMPOSITES	SPECIAL FINITE ELEMENTS FOR FLAW AND DISCONTINUITY STRESSES IN COMPOSITES	73-0846
COMPOSITES	STRESS ANALYSIS OF RIBBON REINFORCED COMPOSITES	70-0221
COMPOSITES	STRESS ANALYSIS OF TAPE REINFORCED COMPOSITES	69-0068
COMPOSITES	STRESS CONCENTRATIONS NEAR A DISCONTINUITY IN FIBROUS COMPOSITES	71-0316
COMPOSITES	STRESS DISTRIBUTION NEAR A BROKEN POINT OF FILAMENT IN UNIDIRECTIONALLY FIBER REINFORCED COMPOSITES	73-0811
COMPOSITES	THE FINITE ELEMENT ANALYSIS OF LAMINATED COMPOSITES	71-0731
COMPOSITES	THE FINITE ELEMENT ANALYSIS OF LAMINATED COMPOSITES	72-0955
COMPOSITES	THERMOMECHANICAL RESPONSE OF INELASTIC FIBRE COMPOSITES	75-0235
COMPOSITES	THREE DIMENSIONAL ANALYSIS OF STRESS CONCENTRATIONS NEAR HOLES IN LAMINATED COMPOSITES	72-0713
COMPOSITES	THREE DIMENSIONAL FINITE ELEMENT ANALYSIS OF THICK LAMINATED COMPOSITES - INCLUDING INTERLAMINAR AND	73-0882
COMPOSITES	THREE DIMENSIONAL SCATTERED-LIGHT STRESS ANALYSIS OF DISCONTINUOUS FIBER-REINFORCED COMPOSITES	73-0218
COMPOSITES	THREE-DIMENSIONAL ANALYSIS FOR THE STRESS DISTRIBUTION NEAR CIRCULAR HOLES IN LAMINATED COMPOSITES	74-0793
COMPOSITES	THREE-DIMENSIONAL FINITE ELEMENT ANALYSIS OF LAMINATED COMPOSITES	72-0345
COMPOSITES	THREE-DIMENSIONAL FINITE ELEMENT ANALYSIS OF SPHERICAL PARTICLE COMPOSITES	74-0137
COMPOSITES	THREE-DIMENSIONAL FINITE ELEMENT ANALYSIS OF THICK LAMINATED COMPOSITES - INCLUDING INTERLAMINAR AND	74-1151
COMPOSITES	TRANSVERSE LOADING OF UNIDIRECTIONAL FIBER COMPOSITES	69-0119
COMPOSITES	TRANSVERSE PROPERTIES OF FIBROUS COMPOSITES	69-0069
COMPOSITES	TRANSVERSE PROPERTIES OF UNIDIRECTIONAL ALUMINUM MATRIX FIBROUS COMPOSITES	69-0167
COMPOSITES	YIELDING IN UNIDIRECTIONAL COMPOSITES UNDER EXTERNAL LOADS AND TEMPERATURE CHANGES.	73-0169

COMPOSITIES	ELASTIC AND PLASTIC INTERLAMINAR SHEAR DEFORMATION IN LAMINATED COMPOSITIES UNDER GENERALIZED PLANE	71-0158
COMPOUND	CHISLENNYI METOD RASCHETA SOBSTVENNYKH I VYNUZHDENNYKH KOLEBANII SOSTAVNYKH OBOLOCHECHNYKH KONSTRUKT	73-0461
COMPOUND	INERTIA LOADING IN FINITE-ELEMENT ANALYSIS OF STRUCTURES SUBJECT TO COMPOUND MOTION	75-0351
COMPOUND	PARAMETRIC STUDY OF A BEAM WITH A COMPOUND SIDE-BRANCH RESONATOR AS A DEVICE TO EVALUATE PRELIMINARY	70-0465
COMPOUND	STATISTICAL ASPECTS OF THE STABILITY OF COMPOUND SYSTEMS WITH BILINEAR ELEMENTS	73-0986
COMPOUND	STRESSES IN SHORT THICK COMPOUND CYLINDERS	69-0385
COMPRESSED	NUMERICAL ANALYSIS ON STRESS DISTRIBUTION IN PARTIAL COMPRESSED SPECIMEN (HARDBOARD) BY FINITE ELEME	72-0920
COMPRESSIBLE	A COMPUTER PROGRAMME FOR ELASTIC PLASTIC CALCULATIONS WITH DUAL PERTURBATION EXPANSION AND VARIATION	75-0882
COMPRESSIBLE	A FINITE ELEMENT COMPUTATIONAL THEORY FOR THE MECHANICS AND THERMODYNAMICS OF A VISCOUS, COMPRESSIBL	71-0787
COMPRESSIBLE	A VARIATIONAL FORMULATION OF THE COMPRESSIBLE THROUGHFLOW PROBLEM	74-1343
COMPRESSIBLE	A VARIATIONAL FORMULATION OF THE COMPRESSIBLE THROUGHFLOW PROBLEM	75-0021
COMPRESSIBLE	AN APPLICATION OF LEAST SQUARES FINITE ELEMENT METHODS TO TWO DIMENSIONAL STEADY INVISCID COMPRESSIB	75-0114
COMPRESSIBLE	ANALYSIS OF COMPRESSIBLE HYDRODYNAMICS BY THE FINITE ELEMENT METHOD	71-0196
COMPRESSIBLE	APPLICATION OF THE FINITE ELEMENT TECHNIQUE TO COMPRESSIBLE FLOW PROBLEMS	70-0311
COMPRESSIBLE	APPLICATION OF THE FINITE ELEMENT TECHNIQUE TO COMPRESSIBLE FLOW PROBLEMS	73-0003
COMPRESSIBLE	APPLICATION OF THE FINITE ELEMENT TECHNIQUE TO COMPRESSIBLE FLOW PROBLEMS	74-0090
COMPRESSIBLE	APPLICATION OF THE PSEUDO-FUNCTIONAL FINITE ELEMENT METHOD TO COMPRESSIBLE FLOW PROBLEMS	74-1360
COMPRESSIBLE	COMPRESSIBLE FLOW BY THE FINITE ELEMENT METHOD	69-0417
COMPRESSIBLE	COMPRESSIBLE SEAL FLOW ANALYSIS USING THE FINITE ELEMENT METHOD WITH GALERKIN SOLUTION TECHNIQUE	74-0688
COMPRESSIBLE	COUPLED VIBRATIONS OF A STRUCTURE SUBMERGED IN A COMPRESSIBLE FLUID	69-0507
COMPRESSIBLE	DEVELOPMENT OF A FINITE ELEMENT APPROACH FOR APPROXIMATE ANALYSIS OF UNSTEADY COMPRESSIBLE FLUID FLO	72-0828
COMPRESSIBLE	DYNAMIC STRESS ANALYSIS OF POROUS ELASTIC SOLIDS SATURATED WITH COMPRESSIBLE FLUIDS	71-0651
COMPRESSIBLE	ELASTIC-PLASTIC PROBLEMS OF COMPRESSIBLE STRAIGHT BAR UNDER COMBINED LOADING CONDITION	75-0313
COMPRESSIBLE	FINITE ELEMENT ANALYSIS OF PERTURBED COMPRESSIBLE FLOW	72-0001
COMPRESSIBLE	FINITE ELEMENT ANALYSIS OF SEVERAL COMPRESSIBLE AND INCOMPRESSIBLE VISCOUS FLOW PROBLEMS	74-1119
COMPRESSIBLE	FINITE ELEMENT CALCULATION OF AXISYMMETRIC SUBCRITICAL COMPRESSIBLE FLOW	74-1064
COMPRESSIBLE	FINITE ELEMENT METHODS FOR UNSTEADY COMPRESSIBLE GAS DYNAMICS	75-0045
COMPRESSIBLE	FINITE ELEMENT SOLUTION OF THE STEADY STATE COMPRESSIBLE LUBRICATION PROBLEM	70-0007
COMPRESSIBLE	FINITE ELEMENT THEORY FOR THE MECHANICS AND THERMODYNAMICS OF A VISCOUS, COMPRESSIBLE MULTI-SPECIE F	71-0005
COMPRESSIBLE	FINITE ELEMENTS FOR COMPRESSIBLE AND INCOMPRESSIBLE CONTINUA	69-0133
COMPRESSIBLE	FINITE-ELEMENT ANALYSIS OF COMPRESSIBLE AND INCOMPRESSIBLE VISCOUS-FLOW AND HEAT-TRANSFER PROBLEMS	75-0347
COMPRESSIBLE	FLOW OF COMPRESSIBLE FLUID IN POROUS ELASTIC MEDIA	73-1004
COMPRESSIBLE	HIGHER ORDER FINITE ELEMENT METHODS FOR THE SOLUTION OF COMPRESSIBLE POROUS BEARING PROBLEMS	75-0341
COMPRESSIBLE	IMPORTANCE OF VARIATIONAL CALCULUS IN FLUID MECHANICS AND ITS APPLICATION TO COMPRESSIBLE POTENTIAL	73-1103
COMPRESSIBLE	LINEARIZED COMPRESSIBLE FLOW BY THE FINITE ELEMENT METHOD	69-0025
COMPRESSIBLE	MOVEMENTS OF ARTICULATED CONDUITS UNDER EARTH DAMS ON COMPRESSIBLE FOUNDATIONS	73-0477
COMPRESSIBLE	NUMERICAL-SOLUTION OF EULERIAN EQUATIONS OF COMPRESSIBLE FLOW BY A FINITE-ELEMENT METHOD WHICH FOLLO	75-0535
COMPRESSIBLE	ON SOME GENERALIZATIONS OF THE INCREMENTAL STIFFNESS RELATIONS FOR FINITE DEFORMATIONS OF COMPRESSIB	71-0189
COMPRESSIBLE	PERTURBATION VARIATIONAL SOLUTION OF COMPRESSIBLE FLOWS	74-0025
COMPRESSIBLE	PERTURBATION-VARIATIONAL SOLUTION OF COMPRESSIBLE FLOWS	74-1088
COMPRESSIBLE	SOLUTION OF THE COMPRESSIBLE FLOW EQUATIONS	71-0003
COMPRESSIBLE	THE FINITE ELEMENT METHOD APPLIED TO COMPRESSIBLE CASCADE FLOW	75-0179
COMPRESSIBLE	THE MBB-UFE SUBSONIC PANEL METHOD. PART 2 - THE LIFT-CONNECTED DISPLACEMENT PROBLEM IN COMPRESSIBLE	72-0865
COMPRESSIBLE	THREE DIMENSIONAL ANALYSIS OF COMPRESSIBLE POTENTIAL FLOWS WITH THE FINITE ELEMENT METHOD	73-0058
COMPRESSIBLE	THREE DIMENSIONAL ANALYSIS OF COMPRESSIBLE POTENTIAL FLOWS	75-0625
COMPRESSIBLE	TWO AND THREE DIMENSIONAL ANALYSIS OF SUBSONIC COMPRESSIBLE POTENTIAL FLOWS IN DUCTS AND AROUND LIFT	74-0026
COMPRESSIBLLE	COUPLED VIBRATIONS OF A STRUCTURE SUBMERGED IN A COMPRESSIBLE FLUID	69-0001
COMPRESSION	APPLICATION OF THE FINITE ELEMENT METHOD TO THE ANALYSIS OF ROCK SAMPLES UNDER UNIAXIAL COMPRESSION	70-0182
COMPRESSION	CONTROLLED FAILURE OF HOLLOW ROCK CYLINDERS IN UNIAXIAL COMPRESSION	72-0196
COMPRESSION	ELASTIC-PLASTIC STRESS-STRAIN RELATIONS FOR MATERIALS WITH DIFFERENT YIELD STRENGTHS IN TENSION AND	72-0764
COMPRESSION	FATIGUE OF GRAPHITE FIBRE-REINFORCED EPOXY IN COMPRESSION.	72-0383
COMPRESSION	FATIGUE OF PRENOTCHED GRAPHITE FIBER COMPOSITES IN COMPRESSION	72-0215
COMPRESSION	FULL-RANGE ANALYSIS OF STEEL PLATES AND STIFFENED PLATING UNDER UNIAXIAL COMPRESSION	75-0803
COMPRESSION	NON-UNIFORM DEFORMATION OF A BLOCK IN PLANE-STRAIN COMPRESSION CAUSED BY FRICTION. PART 4, ANALYSIS	71-0786
COMPRESSION	NON-UNIFORM DEFORMATION OF MATERIAL IN AXIALLY SYMMETRIC COMPRESSION CAUSED BY FRICTION. PART 3, THE	72-0923
COMPRESSION	NON-UNIFORM DEFORMATION OF MATERIAL IN AXIALLY SYMMETRIC COMPRESSION CAUSED BY FRICTION. PART 3, THE	72-0923
COMPRESSION	NUMERICAL STUDY OF UNIAXIAL COMPRESSION IN CIRCULAR ELASTIC-PLASTIC COLUMNS	73-0238
COMPRESSION	NUMERICAL STUDY OF UNIAXIAL AND TRIAXIAL ROCK COMPRESSION TESTS	73-0896
COMPRESSION	ON THE NON-UNIFORM DEFORMATION OF A BLOCK IN PLANE-STRAIN COMPRESSION CAUSED BY FRICTION -3	71-0328
COMPRESSION	ON THE NONUNIFORM DEFORMATION OF MATERIAL IN AXIALLY SYMMETRIC COMPRESSION CAUSED BY FRICTION - 2	71-0321
COMPRESSION	ON THE STATES OF STRESS AND DEFORMATION OF CYLINDRICAL SPECIMENS OF BRITTLE MATERIAL UNDER UNIAXIAL	69-0150
COMPRESSION	RADIAL INERTIA EFFECTS ON AN IDEALLY PLASTIC CIRCULAR PLATE UNDER IMPULSIVE AXIAL COMPRESSION	69-0324
COMPRESSION	SAAS 111: FINITE ELEMENT STRESS ANALYSIS OF AXISYMMETRIC AND PLANE SOLIDS WITH DIFFERENT ORTHOTROPIC	71-0741
COMPRESSION	STRESS DISTRIBUTION UNDER UNIAXIAL ELASTIC PLANE STRAIN COMPRESSION WITH FRICTION	74-0526
COMPRESSION	STRESS STRAIN RELATIONSHIP FROM COMPRESSION TESTS ON NONLINEAR MATERIALS	69-0189
COMPRESSION	STUDY ON STATE OF STRESS AND DEFORMATION OF CYLINDRICAL SPECIMENS OF BRITTLE MATERIAL UNDER UNIAXIAL	69-0152
COMPRESSION	TEMPERATURE RISE IN THE HIGH SPEED COMPRESSION OF RIGHT CYLINDRICAL BILLETS	74-1153
COMPRESSION	THE EFFECT OF END RESTRAINT CONDITION ON THE FRACTURE MECHANISM AND STRENGTH OF ROCK-LIKE MATERIALS	72-0134
COMPRESSIVE	EFFECT OF SHAPE AND HETEROGENEITY OF ROCK SAMPLES ON THEIR COMPRESSIVE STRENGTH	69-0315
COMPRESSIVE	FINITE ELEMENT ANALYSIS ON UNIAXIAL COMPRESSIVE STRENGTH OF CYLINDRICAL BRITTLE SPECIMEN	70-0155
COMPRESSIVE	FINITE ELEMENT MODEL SYSTEM OF A RABBIT CALVARIUM: AN INVESTIGATION OF OPTIMAL COMPRESSIVE LOADING	73-0206
COMPRESSIVE	MEASUREMENTS OF HEALING AT AN OSTEOTOMY IN A RABBIT CALVARIUM: THE INFLUENCE OF APPLIED COMPRESSIVE	74-0975
COMPRESSIVE	PROPAGATION OF FAILURE IN A CIRCULAR CYLINDER OF ROCK SUBJECTED TO A COMPRESSIVE FORCE	73-0793
COMPRESSOR	APPLICATION OF THE FINITE ELEMENT METHOD TO PREDICT STATIC AND DYNAMIC RESPONSE OF AN UNSHROUDED CEN	71-0574
COMPRESSOR	FREQUENCY TUNING OF COMPRESSOR ROTOR BLADES	66-0112
COMPRESSOR	MODELING OF GAS TURBINE ENGINE COMPRESSOR BLADES FOR VIBRATION ANALYSIS	75-0508
COMPRESSOR	MODIFICATIONS TO ARL COMPUTER PROGRAMS USED FOR DESIGN OF AXIAL COMPRESSOR AIRFOILS	74-0508
COMPRESSOR	NATURAL FREQUENCIES OF WIDE CHORD COMPRESSOR BLADES	74-0231
COMPRESSOR	VIBRATION ANALYSIS OF ROTATING COMPRESSOR BLADES	73-0417
COMPUTABLE	COMPUTABLE ERROR BOUNDS FOR THE FINITE ELEMENT METHOD FOR ELLIPTIC BOUNDARY VALUE PROBLEMS	74-1219
COMPUTATION	A METHOD FOR COMPUTATION OF VIBRATION MODES AND FREQUENCIES OF ORTHOTROPIC THIN SHELLS OF REVOLUTION	69-0037
COMPUTATION	A METHOD OF COMPUTATION FOR STRUCTURAL DYNAMICS	59-0002
COMPUTATION	A NOTE ON NUMERICAL COMPUTATION OF ELASTIC CONTACT PROBLEMS	75-0621
COMPUTATION	ALGEBRAIC-GEOMETRY FOUNDATIONS FOR FINITE-ELEMENT COMPUTATION	73-0974
COMPUTATION	APPROXIMATION METHOD FOR COMPUTATION OF NATURAL FREQUENCIES OF SPIRAL SECTIONS	71-0775
COMPUTATION	CALCUL DES CHAMPS ELECTROMAGNETIQUES PAR LA METHODE DES ELEMENTS FINIS (COMPUTATION OF ELECTROMAGNET	73-0399
COMPUTATION	COMPUTATION AND SOLUTION PROCEDURE FOR NONLINEAR ANALYSIS BY COMBINED FINITE ELEMENT-FINITE DIFFEREN	72-0571
COMPUTATION	COMPUTATION OF DERIVATIVES IN THE FINITE ELEMENT METHOD	70-0304
COMPUTATION	COMPUTATION OF FLOWS IN TURBOMACHINES	74-0076
COMPUTATION	COMPUTATION OF IMPELLERS BY METHOD OF FINITE ELEMENTS	74-0646
COMPUTATION	COMPUTATION OF INFLUENCE COEFFICIENTS FOR AIRCRAFT STRUCTURES WITH DISCONTINUITIES AND SWEEPBACK	47-0002
COMPUTATION	COMPUTATION OF POST-YIELD BEHAVIOUR IN NOTCH-BEND AND TENSION TESTPIECES.	70-0244
COMPUTATION	COMPUTATION OF SHELLS WITH FINITE DISPLACEMENTS: MIXED FINITE ELEMENTS (BERECHNUNG VON SCHALEN MIT E	73-0788
COMPUTATION	COMPUTATION OF STRESS RESULTANTS IN FINITE ELEMENT ANALYSIS	66-0056
COMPUTATION	COMPUTATION OF STRESSES IN TRIANGULAR FINITE ELEMENTS	66-0072
COMPUTATION	COMPUTATION OF STRESS RESULTANTS FROM ELEMENT STIFFNESS MATRICES	66-0125
COMPUTATION	COMPUTATION OF STRESS INTENSITY FACTORS BY A SPECIAL FINITE ELEMENT TECHNIQUE	71-0287
COMPUTATION	COMPUTATION OF THE RESPONSE OF COUPLED PLATE ACOUSTIC SYSTEMS USING PLATE FINITE ELEMENTS AND ACOUST	71-0072
COMPUTATION	COMPUTATION OF VIBRATION MODAL SHAPES AND NATURAL FREQUENCIES OF A GENERAL ORTHOTROPIC LAMINATED, TH	71-0309
COMPUTATION	COMPUTATION OF WAVE PROPAGATION BY MEANS OF THE FINITE ELEMENT METHOD	74-0032
COMPUTATION	CURRENT COMPUTATION ON COMPLEX STRUCTURES BY FINITE-ELEMENT METHOD	75-0456
COMPUTATION	DIGITAL COMPUTATION OF STRESSES AND DEFLECTIONS IN A BOX BEAM	67-0073

155

COMPUTATION	DIGITAL COMPUTATION OF STRESSES AND DEFLECTIONS IN A BOX BEAM - A PERFORMANCE COMPARISON BETWEEN FIN	69-0355
COMPUTATION	DIVISION OF COMPUTATION FIELD FOR THE FINITE ELEMENT METHOD	74-0083
COMPUTATION	EVALUATION OF FINITE ELEMENT METHODS FOR THE COMPUTATION OF ELASTIC STRESS INTENSITY FACTORS	72-0249
COMPUTATION	EVALUATION OF USE OF THE FINITE-ELEMENT METHOD IN COMPUTATION OF TEMPERATURE	71-0467
COMPUTATION	FINITE ELEMENT MODEL FOR BACKWATER COMPUTATION	75-0851
COMPUTATION	FINITE-ELEMENT METHOD FOR AXISYMMETRIC FLOW COMPUTATION IN A TURBOMACHINE	75-0359
COMPUTATION	FORMULATION, COMPUTATION, AND SOLUTION PROCEDURES FOR MATERIAL AND/OR GEOMETRIC NONLINEAR STRUCTURAL	72-0346
COMPUTATION	IS THE COMPUTATION OF NOISE LEVEL INSIDE A CAR FEASIBLE	71-0631
COMPUTATION	METHOD FOR THE COMPUTATION OF CYCLIC STRESSES IN A GAS-TURBINE WHEEL	74-0972
COMPUTATION	METHOD SZTYWNYCH ELEMENTOW SKONCZONYCH W OBLICZENIACH KONSTRUCJI OKRETOWYCH. (STIFF FINITE ELEMENT M	74-0923
COMPUTATION	METHODS AND POSSIBILITIES FOR ELECTRONIC COMPUTATION OF PLATES IN FLEXURE	69-0329
COMPUTATION	NUMERICAL COMPUTATION OF BOUNDARY VALUE PROBLEMS IN SOIL MECHANICS	73-0269
COMPUTATION	ON A SPECTRAL PROBLEM IN VIBRATION MECHANICS: COMPUTATION OF ELASTIC TANKS PARTIALLY FILLED WITH LIQ	75-0267
COMPUTATION	ON THE STRESS COMPUTATION IN FINITE ELEMENT MODELS BASED UPON DISPLACEMENT APPROXIMATIONS	74-0598
COMPUTATION	PRAKTISCHE BEISPIELE DER DIGITALEN BERECHNUNG VON GRUNDWASSERSTROMUNGS PROBLEMEN NACH DER METHODE EN	73-0081
COMPUTATION	THE AUTOMATIC COMPUTATION OF THE NATURAL FREQUENCIES OF STRUCTURAL FRAMES USING AN EXACT MATRIX TECH	73-0318
COMPUTATION	THE COMPUTATION OF TEMPERATURE DISTRIBUTIONS AND THERMAL STRESSES USING FINITE ELEMENT TECHNIQUES	71-0092
COMPUTATION	THE COMPUTATION OF TEARING ENERGY OF NICKED RUBBER STRIPS IN EXTENSION	74-0315
COMPUTATION	THE ECONOMICAL COMPUTATION OF STIFFNESS OF LARGE STRUCTURAL ELEMENTS	66-0090
COMPUTATION	USE OF HYBRID COMPUTATION FOR SOLVING TRANSIENT FLOW PROBLEM IN POROUS MEDIA BY FINITE ELEMENT METHO	74-0068
COMPUTATIONAL	A COMPUTATIONAL PROCEDURE FOR GRADIENT ITERATIVE TECHNIQUES IN THE FINITE ELEMENT METHOD	68-0229
COMPUTATIONAL	A COMPUTATIONAL TECHNIQUE FOR THE EFFICIENT HANDLING OF LARGE MATRICES	73-1083
COMPUTATIONAL	A FINITE ELEMENT COMPUTATIONAL THEORY FOR THE MECHANICS AND THERMODYNAMICS OF A VISCOUS, COMPRESSIBL	71-0787
COMPUTATIONAL	A GRADIENT COMPUTATIONAL PROCEDURE FOR THE SOLUTION OF LARGE PROBLEMS ARISING FROM THE FINITE ELEMEN	70-0059
COMPUTATIONAL	A HIGHLY STABLE EXPLICIT INTEGRATION TECHNIQUE FOR COMPUTATIONAL CONTINUUM MECHANICS	74-1277
COMPUTATIONAL	BASIC COMPUTATIONAL PROBLEMS IN THE FINITE ELEMENT ANALYSIS OF SHELLS	71-0678
COMPUTATIONAL	COMPUTATIONAL ANALYSIS AND EVALUATION OF THE FINITE ELEMENT METHOD FOR A CLASS OF NUCLEAR REACTOR CO	74-0595
COMPUTATIONAL	COMPUTATIONAL ASPECTS OF FINITE ELEMENT METHOD FOR A FRAME STRUCTURE	72-0175
COMPUTATIONAL	COMPUTATIONAL ASPECTS OF THE FINITE ELEMENT METHOD	72-0512
COMPUTATIONAL	COMPUTATIONAL CONSIDERATIONS IN APPLICATION OF THE FINITE ELEMENT METHOD FOR ANALYSIS OF UNSTEADY FL	73-0068
COMPUTATIONAL	COMPUTATIONAL EFFICIENCY ASPECTS OF THE FINITE ELEMENT NONLINEAR ANALYSIS OF SHELLS	75-0031
COMPUTATIONAL	COMPUTATIONAL EXPERIMENTATION ON FINITE ELEMENT METHOD IN BARE SLAB CRITICALITY CALCULATIONS	73-0015
COMPUTATIONAL	COMPUTATIONAL FRACTURE MECHANICS	71-0564
COMPUTATIONAL	COMPUTATIONAL METHODS IN SOLID MECHANICS AND STRESS ANALYSIS	69-0006
COMPUTATIONAL	COMPUTATIONAL METHOD FOR DYNAMIC ANALYSIS OF STRUCTURES WITH LARGE SIZE STIFFNESS AND MASS MATRICES.	73-0830
COMPUTATIONAL	COMPUTATIONAL METHODS IN NUCLEAR REACTOR STRUCTURAL DESIGN FOR HIGH TEMPERATURE APPLICATIONS: AN INT	73-1001
COMPUTATIONAL	COMPUTATIONAL METHODS IN NONLINEAR MECHANICS	74-1232
COMPUTATIONAL	COMPUTATIONAL METHODS FOR INVISCID AND VISCOUS TWO-AND-THREE DIMENSIONAL FLOW FIELDS	75-0716
COMPUTATIONAL	COMPUTATIONAL PROCEDURES FOR INELASTIC FINITE-ELEMENT ANALYSIS	72-0933
COMPUTATIONAL	COMPUTATIONAL PROCEDURES FOR INELASTIC STRESS ANALYSIS	73-0859
COMPUTATIONAL	COMPUTATIONAL STABILITY IN NONLINEAR DISCRETE STRUCTURAL SYSTEMS	69-0145
COMPUTATIONAL	COMPUTATIONAL TECHNIQUES AS AN AID IN PHYSICAL METALLURGY	71-0554
COMPUTATIONAL	COMPUTATIONAL TECHNIQUES IN FINITE ELEMENT ANALYSIS OF STRUCTURAL PROBLEMS	71-0556
COMPUTATIONAL	COMPUTATIONAL TECHNIQUES, NONLINEAR EQUATIONS, ISOPARAMETRIC ELEMENTS, FIVE LECTURES	73-0634
COMPUTATIONAL	COMPUTATIONAL TECHNIQUES FOR FINITE ELEMENT ANALYSIS	73-1036
COMPUTATIONAL	DEVELOPMENT OF FINITE ELEMENT COMPUTATIONAL METHODS FOR THREE DIMENSIONAL ELASTIC PLASTIC CRACK ANAL	70-0356
COMPUTATIONAL	DISCRETIZATION AND COMPUTATIONAL ERRORS IN HIGHER-ORDER FINITE ELEMENTS	71-0089
COMPUTATIONAL	DYNAN LECTURE NOTES WITH COMPUTATIONAL EXAMPLES	71-0581
COMPUTATIONAL	FINITE ELEMENT COMPUTATIONAL THEORY FOR THREE-DIMENSIONAL BOUNDARY LAYER FLOW	72-0632
COMPUTATIONAL	MATHEMATICAL AND COMPUTATIONAL PROBLEMS IN REACTOR CALCULATIONS	73-0806
COMPUTATIONAL	MATHEMATICAL AND COMPUTATIONAL PROBLEMS IN REACTOR CALCULATIONS	73-0973
COMPUTATIONAL	NUMERICAL SOLUTION TO THE DYNAMICS OF VISCOUS FLUID FLOW BY A FINITE ELEMENT ALGORITHM: A FIRST STEP	71-0004
COMPUTATIONAL	NUMERICAL SOLUTION TO THE DYNAMICS OF VISCOUS FLUID FLOW BY A FINITE ELEMENT ALGORITHM: A FIRST STEP	71-0791
COMPUTATIONAL	RECENT ADVANCES IN COMPUTATIONAL METHODS IN STRUCTURAL MECHANICS AND DESIGN,	72-0575
COMPUTATIONAL	REPORT ON POST CONF. MEETING ON COMPUTATIONAL ASPECTS OF ASPECTS OF FINITE ELEMENT METHOD ISD STUTTG	72-0475
COMPUTATIONAL	STATUS REPORT ON COMPUTATIONAL TECHNIQUES FOR FINITE ELEMENT ANALYSES	74-0242
COMPUTATIONAL	THE COMPUTATIONAL EFFICIENCY OF A NEW MINIMIZATION ALGORITHM FOR EIGENVALUE ANALYSIS	71-0595
COMPUTATIONAL	THE GENERALIZED INVERSE - SOME PROPERTIES, A COMPUTATIONAL ALGORITHM AND A STRUCTURAL APPLICATION	68-0248
COMPUTATIONALLY	COMPUTATIONALLY EFFICIENT FINITE ELEMENT ANALYSIS OF VISCOUS FLOW PROBLEMS	74-1050
COMPUTATIONALLY	COMPUTATIONALLY EFFICIENT FINITE ELEMENT ANALYSIS OF VISCOUS FLOW PROBLEMS	74-1258
COMPUTATIONS	A METHOD FOR SIMULATING CONCENTRATED FORCES AND LOCAL REINFORCEMENTS IN STRESS COMPUTATIONS	72-0081
COMPUTATIONS	ADJOINING APPROPRIATE SINGULAR ELEMENTS TO TRANSPORT THEORY COMPUTATIONS	74-0946
COMPUTATIONS	APPLICATION OF FINITE ELEMENT COMPUTATIONS TO SOME SHIP STRUCTURAL PROBLEMS	69-0337
COMPUTATIONS	AUTOMATED MESH REFINEMENT STRATEGIES FOR FINITE ELEMENT COMPUTATIONS	75-0877
COMPUTATIONS	EXPERIENCE WITH FINITE ELEMENT COMPUTATIONS FOR NONLINEAR STRUCTURAL PROBLEMS	74-1239
COMPUTATIONS	FINITE ELEMENTS FOR THE COMPUTATIONS OF HYDRODYNAMIC MASS	69-0346
COMPUTATIONS	ON COMPUTATIONS INVOLVING STIFFNESS MATRICES STORED IN RECTANGULAR FORM	75-0576
COMPUTATIONS	ON THE INTERRELATION BETWEEN STABILITY AND COMPUTATIONS	73-0674
COMPUTATIONS	SINGULAR ELEMENTS IN VARIATIONAL AND FINITE ELEMENT TRANSPORT COMPUTATIONS	73-0902
COMPUTATIONS	THEORETICAL COMPUTATIONS ON RIDGE ACOUSTIC SURFACE WAVES USING THE FINITE-ELEMENT METHOD	71-0011
COMPUTATIONS	TRAGLAST VON T-TRAEGERN. EXPERIMENT UND F.E. BERECHNUNGEN (ULTIMATE STRENGTH OF T-BEAMS. EXPERIMENTS	74-0935
COMPUTATIONS	USE OF SINGULARITY PROGRAMMING IN FINITE-DIFFERENCE AND FINITE ELEMENT COMPUTATIONS OF TEMPERATURE	73-0100
COMPUTE	FINITE-ELEMENT METHOD TO COMPUTE STATIONARY SUBCRITICAL FLOWS	75-0439
COMPUTE	THE EXTENSION AND APPLICATION OF SARD KERNEL THEOREMS TO COMPUTE FINITE ELEMENT ERROR BOUNDS	72-0519
COMPUTED	MONOTON FINITE-ELEMENT MATRICES AND THEIR COMPUTED CONDITION NUMBERS	75-0820
COMPUTED	VIBRATION OF SUBMERGED STRUCTURES AS COMPUTED BY THE FINITE ELEMENT METHOD	75-0227
COMPUTER	A COMPARISON OF SEVERAL COMPUTER SOLUTIONS TO THREE STRUCTURAL SHELL ANALYSIS PROBLEMS	73-0884
COMPUTER	A COMPARISON OF SEVERAL COMPUTER SOLUTIONS TO THREE STRUCTURAL SHELL ANALYSIS PROBLEMS	73-0907
COMPUTER	A COMPUTER BASED PROCEDURE FOR PREDICTING THE TRANSIENT RESPONSE AND FAILURE OF A TWO-DIMENSIONAL CO	71-0630
COMPUTER	A COMPUTER METHOD FOR THE SECOND ORDER ELASTIC ANALYSIS OF PLANE FRAMED STRUCTURES	73-0310
COMPUTER	A COMPUTER MODEL FOR PREDICTING THE LOAD-DEFLECTION RESPONSE OF EXPEDIENT SOIL SURFACING	73-0880
COMPUTER	A COMPUTER PROGRAM FOR THE STATIC AND DYNAMIC FINITE ELEMENT ANALYSIS OF AXISYMMETRIC THIN SHELLS	66-0017
COMPUTER	A COMPUTER PROGRAM FOR THE ANALYSIS OF THIN SHELLS	67-0041
COMPUTER	A COMPUTER PROGRAM FOR THE ANALYSIS OF LINEARLY ELASTIC PLANE-STRESS, PLANE-STRAIN PROBLEMS	68-0110
COMPUTER	A COMPUTER PROGRAM FOR THE ANALYSIS OF TWO-DIMENSIONAL HEAT CONDUCTION USING THE FINITE ELEMENT TECH	69-0066
COMPUTER	A COMPUTER PROGRAM FOR THE FINITE ELEMENT ANALYSIS OF A CLASS OF PLATE BENDING PROBLEMS	69-0191
COMPUTER	A COMPUTER PROGRAM FOR THE CALCULATION OF TORSION FACTORS AND SHEAR STRESS FACTORS FOR BOX SECTIONS	69-0485
COMPUTER	A COMPUTER PROGRAM FOR THE ANALYSIS OF TUBULAR K JOINTS	69-0488
COMPUTER	A COMPUTER PROGRAM FOR EARTHQUAKE ANALYSIS OF DAMS	70-0469
COMPUTER	A COMPUTER PROGRAM FOR ANALYSIS OF STIFFENED PLATES UNDER COMBINED INPLANE AND LATERAL LOADS	71-0703
COMPUTER	A COMPUTER PROGRAM TO ANALYZE BEAM-COLUMN UNDER MOVEABLE LOADS	68-0268
COMPUTER	A COMPUTER PROGRAM TO DISPLAY THE DYNAMIC STRESS ANALYSIS OF STRUCTURES	70-0077
COMPUTER	A COMPUTER PROGRAMME FOR ELASTIC-PLASTIC CALCULATIONS WITH USE OF TRIM-6 FINITE ELEMENTS	72-0993
COMPUTER	A COMPUTER PROGRAMME FOR ELASTIC PLASTIC CALCULATIONS WITH DUAL PERTURBATION EXPANSION AND VARIATION	75-0882
COMPUTER	A DIGITAL COMPUTER PROGRAM FOR THE FINITE ELEMENT ANALYSIS OF SOLIDS WITH NONLINEAR MATERIAL PROPERT	65-0032
COMPUTER	A FINITE ELEMENT COMPUTER CODE, AXICAP FOR CREEP ANALYSIS	69-0248
COMPUTER	A FINITE ELEMENT COMPUTER PROGRAM FOR THE PREDICTION OF THE BEHAVIOR OF REINFORCED AND PRESTRESSED C	72-0799
COMPUTER	A GENERAL DIGITAL COMPUTER ANALYSIS OF STATICALLY INDETERMINATE STRUCTURES	59-0007
COMPUTER	A SURVEY OF HEAT CONDUCTION COMPUTER PROGRAMS	71-0017
COMPUTER	A TECHNICAL EVALUATION OF THE NASTRAN COMPUTER PROGRAM	71-0550
COMPUTER	A THREE-DIMENSIONAL FINITE ELEMENT COMPUTER CODE FOR THE ANALYSIS OF COMPLEX STRUCTURES	71-0275
COMPUTER	ADAPTABILITY OF COMPUTER GRAPHICS TO STUDIES OF ATHEROSCLEROSIS PATHOGENESIS	72-0206

156

COMPUTER	ADDITION OF AN ARBITRARY BODY ANALYSIS CAPABILITY TO THE BOEING TEA 236 FINITE ELEMENT COMPUTER PROG	73-0785
COMPUTER	AFPAV COMPUTER CODE FOR STRUCTURAL ANALYSIS OF AIRFIELD PAVEMENTS	75-0684
COMPUTER	AIRCRAFT FUEL TANK VULNERABILITY TO HYDRAULIC RAM: MODIFICATTION OF THE NORTHROP FINITE ELEMENT COMP	74-0724
COMPUTER	AN ADAPTIVE COMPUTER PROGRAM FOR DIV. (P(X,Y) GRAD U) EQUALS F(X,Y,U) IN A POLYGONAL REGION	75-0133
COMPUTER	AN ASSESSMENT OF CURRENT CAPABILITY FOR COMPUTER ANALYSIS OF SHELL STRUCTURES	71-0706
COMPUTER	AN INTERACTIVE COMPUTER GRAPHICS PROGRAM FOR NASTRAN	71-0074
COMPUTER	ANALYTIC MODELING OF ROCK-STRUCTURE INTERACTION. VOLUME 3. COMPUTER PROGRAM	73-0918
COMPUTER	ANALYTIC MODELING OF ROCK-STRUCTURE INTERACTION. VOLUME 2. USERS GUIDE FOR A COMPUTER PROGRAM	73-0919
COMPUTER	ANTRAS COMPUTER PROGRAM FOR LINEAR STRESS ANALYSIS BY FINITE ELEMENT METHOD	74-0601
COMPUTER	APPLICATION OF COMPUTER GRAPHICS TO FINITE ELEMENT MESH GENERATION	73-0546
COMPUTER	APPLICATION, TESTING, AND EVALUATION OF DOUBLY CURVED SHELL ELEMENTS FOR DYNAMIC ANALYSIS, PART III.	69-0256
COMPUTER	APPLICATIONS OF THE FINITE-ELEMENT METHOD AND COMPUTER GRAPHICS. (A VIBRATIONAL ANALYSIS OF A SONAR	72-0729
COMPUTER	ASKA A COMPUTER SYSTEM FOR STRUCTURAL ENGINEERS	69-0345
COMPUTER	AYER HEAT CONDUCTION COMPUTER PROGRAM	74-0885
COMPUTER	BERSAFE - A COMPUTER SYSTEM FOR STRESS ANALYSIS BY FINITE ELEMENTS	70-0212
COMPUTER	CAPE-2D, A COMPUTER PROGRAM FOR THE STRESS ANALYSIS OF PLANE AND AXISYMMETPIC COMPOSITE STRUCTURES.	69-0083
COMPUTER	CHILES: A FINITE ELEMENT COMPUTER PROGRAM THAT CALCULATES THE INTENSITIES OF LINEAR ELASTIC SINGULAR	73-0959
COMPUTER	COMPANY TRAINING OF ENGINEERING IN STRUCTURAL ANALYSIS COMPUTER PROGRAM USAGE	72-0460
COMPUTER	COMPUTER - AIDED DESIGN - STANDARD OIL OF INDIANA BUILDING	73-0182
COMPUTER	COMPUTER AIDED DESIGN APPLIED TO A MODEL OF A CHASSIS TYPE STRUCTURE USING FINITE ELEMENT TECHNIQUES	69-0421
COMPUTER	COMPUTER AIDED DESIGN AND CONSTRUCTION BY MEANS OF CURVED SURFACE AND SOLID ELEMENTS	73-0165
COMPUTER	COMPUTER AIDED DESIGN OF THERMALLY LOADED AXISYMMETRIC DIESEL ENGINE COMPONENTS	74-0596
COMPUTER	COMPUTER AIDED DESIGN OF MACHINE TOOL STRUCTURE	74-1189
COMPUTER	COMPUTER AIDED STRUCTURAL ANALYSIS: THE MACHINE INDEPENDENT SYSTEM ASKA	70-0397
COMPUTER	COMPUTER AIDED TEACHING OF THE FINITE ELEMENT DISPLACEMENT METHOD	69-0430
COMPUTER	COMPUTER ANALYSIS OF SHELL STRUCTURES	69-0060
COMPUTER	COMPUTER ANALYSIS OF LARGE LINEAR FRAMES	69-0408
COMPUTER	COMPUTER ANALYSIS OF A RAILROAD FREIGHT CAR BOLSTER UTILIZING THE FINITE ELEMENT METHOD	71-0283
COMPUTER	COMPUTER ANALYSIS OF RIGID FRAMES	71-0401
COMPUTER	COMPUTER ANALYSIS OF FRUIT DETACHMENT DURING TREE SHAKING	74-1162
COMPUTER	COMPUTER ANALYSIS OF SEGMENTALLY ERECTED BRIDGES	75-0172
COMPUTER	COMPUTER ANALYSIS OF STEEL FRAME IN FIRE	75-0173
COMPUTER	COMPUTER APPLICATIONS IN ROCK MECHANICS	72-0190
COMPUTER	COMPUTER BASED ANALYSES OF THE RESPONSE OF BOX TYPE STRUCTURES TO RANDOM PRESSURES	73-0227
COMPUTER	COMPUTER CALCULATIONS OF STRESSES IN AXISYMMETRIC THERMALLY LOADED COMPONENTS	68-0069
COMPUTER	COMPUTER CODE TO GENERATE FINITE ELEMENT MESH FOR PROBLEMS OF COMPLEX AXISYMMETRIC GEOMETRY	74-0360
COMPUTER	COMPUTER GENERATED DISPLAYS OF STRUCTURES IN VIBRATION	74-1305
COMPUTER	COMPUTER GRAPHICS FOR 3-D FINITE ELEMENT MODELS	74-0290
COMPUTER	COMPUTER GRAPHICS FOR STRUCTURAL ANALYSIS	74-0832
COMPUTER	COMPUTER IMPLEMENTATION OF THE FINITE ELEMENT METHOD	71-0376
COMPUTER	COMPUTER IMPLEMENTATION OF THE FINITE ELEMENT PROCEDURE	71-0633
COMPUTER	COMPUTER IMPLEMENTATION OF THE FINITE ELEMENT PROCEDURE	71-0650
COMPUTER	COMPUTER IMPLEMENTATION OF NONLINEAR FINITE ELEMENT ANALYSIS	73-0407
COMPUTER	COMPUTER IMPLEMENTATION OF THE CONSTRAINT METHOD	75-0001
COMPUTER	COMPUTER INTEGRATION OF CONGRUENTLY TRANSFORMED MATRICES	71-0580
COMPUTER	COMPUTER INTEGRATION FOR FINITE ELEMENTS	73-0329
COMPUTER	COMPUTER METHOD FOR COMPUTING CRITICAL SPEEDS OF NON-UNIFORM SHAFTS GEARED SYSTEMS AND CURVED ASSEMB	73-0675
COMPUTER	COMPUTER METHODS FOR THE STRUCTURAL ANALYSIS OF MACHINE TOOLS	69-0325
COMPUTER	COMPUTER MODELLING OF PLUG SEEDLING SURVIVAL	75-0176
COMPUTER	COMPUTER ORIENTED ANALYSIS OF SHELL STRUCTURES	71-0559
COMPUTER	COMPUTER ORIENTED ANALYSIS OF SHELL STRUCTURES	71-0762
COMPUTER	COMPUTER PREDICTION OF FATIQUE CRACK PROPAGATION UNDER RANDOM LOADING	73-0851
COMPUTER	COMPUTER PROGRAM CONSTRUCTION AND MAINTENANCE - FUTURE OF CENTRALIZED FINITE ELEMENT ACTIVITY	75-0014
COMPUTER	COMPUTER PROGRAM FOR ELASTIC-PLASTIC ANALYSIS OF AXISYMMETRICALLY LOADED SHELLS OF REVOLUTION	68-0239
COMPUTER	COMPUTER PROGRAM FOR PRISMATIC FOLDED PLATES BY PLATE AND BEAM ELEMENTS	70-0371
COMPUTER	COMPUTER PROGRAM FOR GENERATING FINITE ELEMENT MODELS OF MINE STRUCTURES	71-0493
COMPUTER	COMPUTER PROGRAM FOR PIT SLOPE STABILITY ANALYSIS BY THE FINITE ELEMENT STRESS ANALYSIS AND LIMITING	72-0225
COMPUTER	COMPUTER PROGRAM FOR STATIC AND DYNAMIC ANALYSIS OF LINEAR STRUCTURAL SYSTEMS	72-0683
COMPUTER	COMPUTER PROGRAM FOR SOLUTION OF LARGE, SPARSE, UNSYMMETRIC SYSTEMS OF LINEAR EQUATIONS	72-0719
COMPUTER	COMPUTER PROGRAM FOR AUTOMATIC MESH GENERATION FOR THE 2-D FINITE ELEMENT PROGRAM WILAX	73-0565
COMPUTER	COMPUTER PROGRAM JET 3 TO CALCULATE THE LARGE ELASTIC PLASTIC DYNAMICALLY INDUCED DEFORMATIONS OF FR	72-0609
COMPUTER	COMPUTER PROGRAMS FOR THE PLASTIC ANALYSIS OF STRUCTURES USING DISCRETE ELEMENT METHODS	67-0004
COMPUTER	COMPUTER PROGRAMS FOR PLOTTING FINITE ELEMENT PATTERNS OF TWO-AND THREE-DIMENSIONAL STRUCTURES	70-0416
COMPUTER	COMPUTER PROGRAMS FOR THE SOLUTION OF SYSTEMS OF LINEAR ALGEBRAIC EQUATIONS	73-0932
COMPUTER	COMPUTER PROGRAMS FOR THE SOLUTION OF SYSTEMS OF LINEAR ALGEBRAIC EQUATIONS	73-1040
COMPUTER	COMPUTER PROGRAMS FOR SIMULATING PHYSICAL PHENOMENA IN THREE SPACE DIMENSIONS AND TIME	74-0866
COMPUTER	COMPUTER SIMULATION OF SECOND ORDER FAULTS	71-0010
COMPUTER	COMPUTER SIMULATION OF DYNAMIC STRESS, DEFORMATION, AND FRACTURE OF GEAR TEETH	72-0114
COMPUTER	COMPUTER SIMULATION OF THE DEFORMATION OF COMPOSITE AND POLYCRYSTALLINE MATERIALS BY FINITE ELEMENT	72-0937
COMPUTER	COMPUTER SIMULATION OF LAMINATED ROOF REINFORCED WITH GROUTED BOLTS	73-1045
COMPUTER	COMPUTER SIMULATION OF AN IN SITU SHALE-OIL RETORT	74-1059
COMPUTER	COMPUTER SIMULATION OF VIBRATION OF WHOLE TREE SYSTEMS USING FINITE ELEMENT METHODS	74-1161
COMPUTER	COMPUTER SOLUTIONS OF PLASTICITY PROBLEMS	72-0829
COMPUTER	COMPUTER STRESS ANALYSIS	68-0099
COMPUTER	COMPUTER SYSTEM FOR ANALYSIS OF LARGE FRAMEWORKS	67-0122
COMPUTER	COMPUTER SYSTEM FOR THE ANALYSIS OF LARGE SCALE PROBLEMS IN SOLID, FLUID AND THERMODYNAMICS	73-0554
COMPUTER	COMPUTER TECHNIQUE: THE FRONT SOLUTION	72-0631
COMPUTER	COMPUTER WORD LENGTH AND SOLUTION ACCURACY. EXPERIENCE WITH ELAS AND THE IBM 360	73-0335
COMPUTER	CRASH: A COMPUTER SIMULATION OF NONLINEAR TRANSIENT RESPONSE OF STRUCTURES	72-0569
COMPUTER	DESCRIPTION OF THE IBM-360 VERSION OF THE FINITE ELEMENTS COMPUTER PROGRAM SAFE-3D	72-0775
COMPUTER	DETERMINATION OF STRESSES AND DEFORMATIONS ON PISTONS BY MEANS OF COMPUTER PROGRAMS AND PHOTOELASTIC	75-0553
COMPUTER	DEVELOPMENT AND APPLICATIONS OF A GENERAL COMPUTER PROGRAM FOR EVALUATING STABILITY OF OPENINGS IN R	74-0994
COMPUTER	DEVELOPMENT AND USE OF COMPUTER PROGRAMS FOR FINITE ELEMENT ANALYSIS. (LA MISE AU POINT ET L'UTILIZA	69-0289
COMPUTER	DEVELOPMENT OF METHODS FOR COMPUTER SIMULATION OF BEAM-COLUMNS AND GRID-BEAM AND SLAB SYSTEMS	73-1101
COMPUTER	DIE ANWENDUNG VON INTERAKTIVER COMPUTER-GRAPHIK BEI DER BERECHNUNG VON TRAGWERKEN. (APPLICATION OF I	74-0102
COMPUTER	DIGITAL COMPUTER PROGRAMS FOR STEADY-STATE OR TRANSIENT TEMPERATURE ANALYSIS OF PLANE AND AXISYMMETR	72-0915
COMPUTER	DYNAMIC ELASTIC-PLASTIC ANALYSIS OF AXISYMMETRIC STRUCTURES BY FINITE ELEMENT METHOD COMPUTER CODE	73-0844
COMPUTER	DYNAMIC STRUCTURAL ANALYSIS WITH THE NASTRAN COMPUTER PROGRAM	70-0213
COMPUTER	DYNAMIC TESTING AND COMPUTER ANALYSIS OF AUTOMOTIVE FRAMES	72-0089
COMPUTER	ECONOMICAL COMPUTER TECHNIQUES FOR SOLID FINITE ELEMENTS	68-0086
COMPUTER	ECONOMICAL COMPUTER TECHNIQUES FOR NUMERICALLY INTEGRATED FINITE ELEMENTS	69-0138
COMPUTER	EFFICIENT COMPUTER ANALYSIS OF LARGE THREE-DIMENSIONAL PIPING SYSTEMS BY THE FINITE ELEMENT METHOD	72-0378
COMPUTER	EINSATZ DES COMPUTERS BEI DER ENTWICKLUNG EINER PANZERWANNE (THE USE OF A COMPUTER FOR DEVELOPING A	72-0211
COMPUTER	EINSATZ MODERNER RECHENVERFAHREN ZUR ERMITTLUNG DER SPANNUNGEN UND VERFORMUNGEN BEI RADSAETZEN. (USE	75-0186
COMPUTER	ELAD PROGRAM MANUAL, A COMPUTER PROGRAM FOR THE ELASTIC ANALYSIS OF DOME STRUCTURES AND AXISYMMETRIC	69-0064
COMPUTER	ELAS - A GENERAL PURPOSE COMPUTER PROGRAM FOR THE EQUILIBRIUM PROBLEMS OF LINEAR STRUCTURES	68-0158
COMPUTER	ELAS: A GENERAL-PURPOSE COMPUTER PROGRAM FOR THE EQUILIBRIUM PROBLEMS OF LINEAR STRUCTURES. VOLUME 2	69-0462
COMPUTER	ELFED TO GENERATE A FINITE ELEMENT MESH FOR PROBLEMS OF COMPLEX AXISYMMETRIC GEOMETR	74-1270
COMPUTER	ELFED: A COMPUTER CODE TO GENERATE FINITE ELEMENT MESH FOR PROBLEMS OF COMPLEX AXISYMMETRIC GEOMETRY	73-0943
COMPUTER	ENGINEERING OPERATIONS REPORT: NERVA THIN SHELL FINITE ELEMENT STRESS ANALYSIS METHODOLOGY COMPUTER	72-0916
COMPUTER	ENGINEERING OPERATIONS REPORT: NERVA AXISYMMETRIC AND PLANE STRUCTURES FINITE ELEMENT STRESS ANALYSI	72-0917
COMPUTER	EPAD, A COMPUTER PROGRAM FOR THE ELASTIC-PLASTIC ANALYSIS OF DOME STRUCTURES AND ARBITRARY SOLIDS OF	70-0036

157

COMPUTER	EURCYL: A COMPUTER PROGRAM TO GENERATE FINITE ELEMENT MESHES FOR CYLINDER-CYLINDER INTERSECTIONS	73-0962
COMPUTER	EVALUATION OF COMPUTER AIDED METHODS FOR THE DESIGN OF MACHINE TOOL STRUCTURES	74-0684
COMPUTER	EVALUATION OF THE COMPUTER PROGRAM CREEP-BLAST (CDC VERSION)	73-0364
COMPUTER	EXTENSIONS TO THE STAGS COMPUTER CODE	73-0917
COMPUTER	EXTENSIONS TO THE STAGS COMPUTER CODE	73-0356
COMPUTER	FAST MODAL EXTRACTION IN NASTRAN VIA THE FEER COMPUTER PROGRAM	70-0418
COMPUTER	FATIGUE STRENGTH OF LUGS CONTAINING LINERS. VOLUME II. COMPUTER PROGRAM USED FOR ANALYSIS	68-0152
COMPUTER	FEATS, A COMPUTER PROGRAM FOR THE FINITE ELEMENT THERMAL STRESS ANALYSIS OF PLANE OR AXISYMMETRIC SO	69-0492
COMPUTER	FEDGE - A GENERAL PURPOSE COMPUTER PROGRAM FOR FINITE ELEMENT DATA GENERATION. VOLUME 2 - PROGRAM MA	69-0491
COMPUTER	FEDGE - A GENERAL-PURPOSE COMPUTER PROGRAM FOR FINITE ELEMENT DATA GENERATION. VOLUME 1 - USER'S MAN	71-0811
COMPUTER	FESTRAN: A FINITE ELEMENT COMPUTER PROGRAM	73-0385
COMPUTER	FINITE ELEMENT ANALYSIS AND COMPUTER GRAPHICS VISUALIZATION OF FLOW AROUND PITCHING AND PLUNGING AIR	74-1292
COMPUTER	FINITE ELEMENT ANALYSIS, MODELLING AND COMPUTER GRAPHICS	70-0293
COMPUTER	FINITE ELEMENT COMPUTER ANALYSIS OF KOCHMUS HULL 520.	70-0294
COMPUTER	FINITE ELEMENT COMPUTER ANALYSIS OF MITSUBISHI HULL 1670	72-0364
COMPUTER	FINITE ELEMENT COMPUTER PROGRAMS FOR FLOW IN SATURATED UNSATURATED POROUS MEDIA	73-0525
COMPUTER	FINITE ELEMENT COMPUTER PROGRAMS FOR HEAT CONDUCTION PROBLEMS	74-0800
COMPUTER	FINITE ELEMENT COMPUTER PROGRAM FOR THE ANALYSIS OF LAYERED ORTHOTROPIC STRUCTURES	74-1094
COMPUTER	FINITE ELEMENT COMPUTER PROGRAM FOR THE SOLUTION OF NONLINEAR AXISYMMETRIC CONTACT PROBLEMS WITH INT	75-0191
COMPUTER	FINITE ELEMENT COMPUTER PROGRAM FOR PREDICTING THE NONLINEAR STATIC AND DYNAMIC BEHAVIOR OF VISCOELA	74-0450
COMPUTER	FINITE ELEMENT METHOD AND COMPUTER PROGRAMS	73-0648
COMPUTER	FINITE ELEMENT SOLUTION OF PARTIAL DIFFERENTIAL EQUATIONS FEMAC COMPUTER PROGRAM	74-0327
COMPUTER	FINITE ELEMENT SOLUTIONS OF SELECTED PARTIAL DIFFERENTIAL EQUATIONS "FEMAC COMPUTER PROGRAM"	74-0747
COMPUTER	FIRES-RC, A COMPUTER PROGRAM FOR THE FIRE RESPONSE OF STRUCTURES-REINFORCED CONCRETE FRAMES	74-0850
COMPUTER	FIRES-T, A COMPUTER PROGRAM FOR THE FIRE RESPONSE OF STRUCTURES - THERMAL	69-0013
COMPUTER	FUNCTIONAL APPROXIMATIONS FOR SOLVING BOUNDARY VALUE PROBLEMS BY COMPUTER	73-0260
COMPUTER	GENERAL PURPOSE COMPUTER PROGRAM FOR THE DYNAMIC SIMULATION OF VEHICLE GUIDEWAY INTERACTIONS	74-0194
COMPUTER	GENERATION OF FINITE ELEMENT MODELS VIA COMPUTER GRAPHICS	74-0875
COMPUTER	GNATS: A FINITE ELEMENT COMPUTER PROGRAM FOR THE GENERAL NONLINEAR ANALYSIS OF TWO-DIMENSIONAL STRUC	71-0486
COMPUTER	GRAPHICAL DISPLAY OF COMPUTER SIMULATED UNBALANCED ROTOR RESPONSE	73-0767
COMPUTER	HIGH-PERFORMANCE COMPOSITE MATERIALS FOR VEHICLE CONSTRUCTION: A FINITE ELEMENT COMPUTER PROGRAM FOR	73-0682
COMPUTER	HYBRID MODELS IN NUMERICAL AND COMPUTER METHODS.	75-0818
COMPUTER	HYPERMATRIX SCHEME FOR FINITE-ELEMENT SYSTEMS ON CDC STAR-100 COMPUTER	75-0039
COMPUTER	IMPACT OF CDC STAR-100 COMPUTER ON FINITE ELEMENT SYSTEMS	65-0002
COMPUTER	IMPROVED ANALYTIC LONGITUDINAL RESPONSE ANALYSIS FOR AXISYMMETRIC LAUNCH VEHICLES, COMPUTER PROGRAM	72-0639
COMPUTER	INTERACTIVE COMPUTER GRAPHICS IN STRUCTURAL ANALYSIS: ONLINE 72	70-0211
COMPUTER	LARGE SCALE COMPUTER PROGRAMS FOR STRUCTURAL ANALYSIS	74-0744
COMPUTER	LUSH: A COMPUTER PROGRAM FOR COMPLEX RESPONSE ANALYSIS OF SOIL-STRUCTURE SYSTEMS	71-0367
COMPUTER	MANIPULATION ERRORS IN COMPUTER SOLUTION OF CRITICAL SIZE STRUCTURAL EQUATIONS	69-0143
COMPUTER	MATRIX AND DIGITAL COMPUTER METHODS IN STRUCTURAL ANALYSIS	72-0653
COMPUTER	MECHANICS OF CABLE MOORING SYSTEMS. VOLUME IV. A COMPUTER PROGRAM FOR ANALYZING THE STEADY STATE RES	73-0725
COMPUTER	MECHANICS OF CABLE MOORING SYSTEMS. VOLUME VI. A COMPUTER PROGRAM FOR ANALYZING THE STEADY STATE CON	71-0748
COMPUTER	MESH GENERATING COMPUTER PROGRAM FOR FINITE ELEMENT ANALYSES OF SIMPLE SLOPES IN HORIZONTALLY LAYERE	69-0096
COMPUTER	MESHGEN, A COMPUTER CODE FOR AUTOMATIC FINITE ELEMENT MESH GENERATION	73-0718
COMPUTER	METHODS FOR PREDICTING THE AERODYNAMIC AND STABILITY AND CONTROL CHARACTERISTICS OF STOL AIRCRAFT. V	74-0508
COMPUTER	MODIFICATIONS TO ARL COMPUTER PROGRAMS USED FOR DESIGN OF AXIAL COMPRESSOR AIRFOILS	73-0978
COMPUTER	ON THE SOLUTION OF PLANE STRESS PROBLEMS BY FINITE ELEMENTS COMPUTER PROGRAMS	74-0362
COMPUTER	ORGANIZATION AND FUNCTIONAL PURPOSE OF A FINITE ELEMENT COMPUTER PROGRAM	74-0871
COMPUTER	ORGANIZATION OF GNATS: A GENERAL NONLINEAR ANALYSIS COMPUTER PROGRAM FOR TWO-DIMENSIONAL STRUCTURES	71-0745
COMPUTER	PART 1 - TASK 3.2.9. BENDING AND BUCKLING OF A ONE-DIMENSIONAL MODEL SIMULATING A CONICAL SHELL. PAR	72-0765
COMPUTER	PART 1. THEORY REPORT FOR CREEP-PLAST COMPUTER PROGRAM: ANALYSIS OF TWO-DIMENSIONAL PROBLEMS UNDER S	74-0884
COMPUTER	PLACID: A GENERAL FINITE ELEMENT COMPUTER PROGRAM FOR STRESS ANALYSIS OF PLANE AND AXISYMMETRIC SOLI	72-0803
COMPUTER	PLOTS - THE PLOTTING SUBROUTINE INCORPORATED IN THE BALLISTIC RESEARCH LABORATORIES ELECTRONIC SCIEN	72-0803
COMPUTER	PLOTS - THE PLOTTING SUBROUTINE INCORPORATED IN THE BALLISTIC RESEARCH LABORATORIES ELECTRONIC SCIEN	74-0318
COMPUTER	POWER SPECTRAL DENSITY ANALYSIS OF VEHICLE VIBRATION USING THE NASTRAN COMPUTER PROGRAM	69-0098
COMPUTER	PRACTICAL METHOD FOR COMPUTER CALCULATION OF STRUCTURES	74-1215
COMPUTER	PREDICTION OF TEMPERATURE DISTRIBUTIONS IN CARGOES WITH THE AID OF A COMPUTER PROGRAM USING THE METH	74-0851
COMPUTER	PROBLEM OF USING THE METHOD OF FINITE ELEMENTS TO CONSTRUCT AN ALGORITHM FOR CALCULATING A THIN-WALL	73-0771
COMPUTER	QUAD-4: A COMPUTER PROGRAM FOR EVALUATING THE SEISMIC RESPONSE OF SOIL STRUCTURES BY VARIABLE DAMPIN	69-0262
COMPUTER	RANDOM VIBRATION ANALYSIS SYSTEM FOR COMPLEX STRUCTURES. PART 2. COMPUTER PROGRAM DESCRIPTION	74-0251
COMPUTER	REALIZATSIYA METODA KONECHNYKH ELEMENTOV NA EVM DLYA RASCHETA DVUMERNYKH ELEKTRICHESKIKH I MAGNITNYK	74-1310
COMPUTER	REBO: COMPUTER CODE FOR THE GENERATION OF A THREE-DIMENSIONAL FINITE-ELEMENT GRID OF THE NOZZLE-CYLI	74-0692
COMPUTER	RETSCP: A COMPUTER PROGRAM FOR ANALYSIS OF ROCKET ENGINE THERMAL STRAINS WITH CYCLIC PLASTICITY	65-0013
COMPUTER	SAFE PCOS, A COMPUTER PROGRAM FOR THE STRESS ANALYSIS OF COMPOSITES BODIES OF REVOLUTION. INPUT INST	67-0018
COMPUTER	SAFE SHELL, A COMPUTER PROGRAM FOR THE STRESS ANALYSIS OF THIN SHELL BODIES OF REVOLUTION. A USER'S	65-0014
COMPUTER	SAFE-AXISYM, A COMPUTER PROGRAM FOR THE DESIGN OF COMPOSITE BODIES OF REVOLUTION. INPUT INSTRUCTIONS	67-0148
COMPUTER	SAFE-CREEP: A COMPUTER PROGRAM FOR THE VISCOELASTIC ANALYSIS OF AXISYMMETRIC AND PLANE CONCRETE STRU	67-0017
COMPUTER	SAFE-PLANE, A COMPUTER PROGRAM FOR THE STRESS ANALYSIS AND DESIGN OF TWO-DIMENSIONAL COMPOSITE BODIE	69-0107
COMPUTER	SEAL-SHELL-2, A COMPUTER PROGRAM FOR THE STRESS ANALYSIS OF A THICK SHELL OF REVOLUTION WITH AXISYMM	72-0835
COMPUTER	SHIP (SIMPLIFIED-HELMHOLTZ-INTEGRAL PROGRAM): A FAST COMPUTER PROGRAM FOR CALCULATING THE ACOUSTIC R	75-0723
COMPUTER	SINGER: A COMPUTER CODE FOR GENERAL ANALYSIS OF TWO-DIMENSIONAL CONCRETE STRUCTURES. VOL. 11. PROGRA	75-0724
COMPUTER	SINGER: A COMPUTER CODE FOR GENERAL ANALYSIS OF TWO-DIMENSIONAL REINFORCED CONCRETE STRUCTURES. VOLU	75-0730
COMPUTER	SINGER: A COMPUTER CODE FOR GENERAL ANALYSIS TO TWO-DIMENSIONAL CONCRETE STRUCTURES. VOLUME 4. DEMON	75-0731
COMPUTER	SINGER: A COMPUTER CODE FOR GENERAL ANALYSIS OF TWO-DIMENSIONAL REINFORCED CONCRETE STRUCTURES. VOLU	73-0964
COMPUTER	SLADE D: A COMPUTER PROGRAM FOR THE DYNAMIC ANALYSIS OF THIN SHELLS	70-0079
COMPUTER	SLADE, A COMPUTER PROGRAM FOR THE STATIC ANALYSIS OF THIN SHELLS	71-0341
COMPUTER	SOLUTION OF AUTOMOTIVE STRUCTURAL PROBLEMS USING THE FINITE ELEMENT METHOD AND COMPUTER GRAPHICS	75-0778
COMPUTER	SOME ASPECTS OF NON-LINEAR FINITE ELEMENT COMPUTER PROGRAMS	73-0535
COMPUTER	SOME RESULTS OF FUSELAGE ANALYSIS BY THE FINITE ELEMENT METHOD USING A COMPUTER	70-0154
COMPUTER	SPECIAL PURPOSE COMPUTER PROGRAMS	71-0204
COMPUTER	STARS - A COMPUTER CODE FOR STRUCTURAL DYNAMIC ANALYSIS OF REACTOR SYSTEM	71-0078
COMPUTER	STRAP - A COMPUTER CODE FOR THE STATIC AND DYNAMIC ANALYSIS OF REACTOR STRUCTURAL SYSTEMS	70-0409
COMPUTER	STRAP: COMPUTER CODE FOR STATIC AND DYNAMIC STRUCTURAL ANALYSIS AND STUDIES MADE USING THE CODE	74-0304
COMPUTER	STRUCTURAL FINITE ELEMENT ANALYSIS AIDED BY COMPUTER GRAPHICS	74-0902
COMPUTER	STRUCTURAL MECHANICS COMPUTER PROGRAMS	66-0118
COMPUTER	STRUDL-A COMPUTER SYSTEM FOR STRUCTURAL DESIGN	73-0315
COMPUTER	STUDIES ON TWO PHASE COMPOSITE MATERIALS THROUGH COMPUTER SIMULATION	72-0588
COMPUTER	TACOS: A FINITE ELEMENT COMPUTER PROGRAM FOR THE TRANSIENT ANALYSIS OF CYLINDRICAL OBESE SHELLS	72-0481
COMPUTER	THE BANDIT COMPUTER PROGRAM FOR THE REDUCTION OF MATRIX BANDWIDTH FOR NASTRAN	71-0042
COMPUTER	THE DEVELOPMENT OF COMPUTER GRAPHICS FOR LARGE SCALE FINITE ELEMENT CODES	71-0721
COMPUTER	THE DEVELOPMENT OF LARGE SCALE DIGITAL COMPUTER CODES FOR PRODUCTION STRUCTURAL ANALYSIS	72-0099
COMPUTER	THE EFFICIENT USE OF COMPUTER RESOURCES	69-0447
COMPUTER	THE IMPACT OF THE DIGITAL COMPUTER ON STRUCTURAL ANALYSIS	70-0003
COMPUTER	THE IMPACT OF THE DIGITAL COMPUTER ON THE ENGINEERING SCIENCES - PART I	74-0697
COMPUTER	THE MIDCOHV WEIGHT AND BALANCE COMPUTER PROGRAM (WTBAL)	69-0183
COMPUTER	THE NASTRAN COMPUTER PROGRAM FOR STRUCTURAL ANALYSIS	71-0173
COMPUTER	THE NASTRAN COMPUTER PROGRAM FOR STRUCTURAL ANALYSIS	71-0682
COMPUTER	THE NASTRAN COMPUTER PROGRAM FOR STRUCTURAL ANALYSIS	68-0215
COMPUTER	THE ROLE OF COMPUTER GRAPHICS IN THE STRUCTURAL DESIGN PROCESS	71-0156
COMPUTER	THE TRIDIMENSIONAL THERMOELASTIC COMPUTER CODE TITUS	70-0334
COMPUTER	THE VERIFICATION AND ACCEPTANCE OF COMPUTER PROGRAMS FOR DESIGN ANALYSIS	71-0105
COMPUTER	THERMOVISCOELASTIC AXISYMMETRIC STRESS ANALYSIS BY FINITE ELEMENT AND FINITE DIFFERENCE COMPUTER TEC	

COMPUTER	THREE DIMENSIONAL SUPERTANKER COMPUTER PROGRAM	71-0661
COMPUTER	THREE-DIMENSIONAL FINITE-ELEMENT COMPUTER PROGRAM - USERS GUIDE	74-0755
COMPUTER	THREE-DIMENSIONAL MESH GENERATOR FOR FINITE ELEMENT COMPUTER CODES	74-0864
COMPUTER	TRANSPIRATION AND FILM COOLING BOUNDARY LAYER COMPUTER PROGRAM. VOLUME 2, COMPUTER PROGRAM AND USER'	71-0655
COMPUTER	TRANSPIRATION AND FILM COOLING BOUNDARY LAYER COMPUTER PROGRAM. VOLUME 2, COMPUTER PROGRAM AND USER'	71-0655
COMPUTER	TWO-DIMENSIONAL FINITE ELEMENT COMPUTER CODES FOR STRESS ANALYSIS AND FIELD PROBLEMS	73-0055
COMPUTER	USE OF FINITE ELEMENT COMPUTER PROGRAMS IN FRACTURE MECHANICS	71-0022
COMPUTER	USE OF PAS-1 FINITE ELEMENT COMPUTER SYSTEM FOR STATIC AND DYNAMIC ANALYSIS OF 2-DIMENSIONAL AND 3-D	75-0070
COMPUTER	USER'S GUIDE FOR THE COMPUTER PROGRAM DESCRIBED IN TAM REPORT NO. 680 "DETERMINATION OF VOLUME CHANG	69-0473
COMPUTER	USER'S MANUAL FOR UMR FINITE ELEMENT COMPUTER PROGRAM - STRATA	69-0429
COMPUTER	USERS MANUAL FOR DIGITAL COMPUTER PROGRAM OK5 NUMERICAL ANALYSIS OF FINITE STRAINS AND DISPLACEMENTS	72-0718
COMPUTER	USERS' MANUAL FOR DIGITAL COMPUTER PROGRAM OK1. NUMERICAL ANALYSIS OF FINITE AXISYMMETRIC DEFORMATIO	72-0861
COMPUTER	VERIFICATION AND QUALIFICATION OF A 3-D FINITE ELEMENT COMPUTER ANALYSIS	71-0487
COMPUTER	VERIFICATION OF A HYBRID GENERAL PURPOSE SHELL AND BODY OF REVOLUTION COMPUTER PROGRAM	71-0333
COMPUTER	VERIFICATION OF THE FINITE ELEMENT COMPUTER CODE SAFE-2D	74-0873
COMPUTER	VIEW FACTOR COMPUTER PROGRAM (PROGRAM VIEW) USER'S MANUAL	73-0789
COMPUTER	VISCEL A GENERAL PURPOSE COMPUTER PROGRAM FOR ANALYSIS OF LINEAR VISCOELASTIC STRUCTURES, VOL. 2	72-0789
COMPUTER-AIDED	AN INTERACTIVE COMPUTER-AIDED MODELING SYSTEM FOR THE DISCRETIZATION OF BRIDGE STRUCTURES FOR FINITE	73-1165
COMPUTER-AIDED	APPLICATION OF COMPUTER-AIDED AIRCRAFT DESIGN IN A MULTIDISCIPLINARY ENVIRONMENT	73-0527
COMPUTER-AIDED	COMPUTER-AIDED DESIGN OF SHELL STRUCTURES USING THE FINITE ELEMENT METHOD	72-0998
COMPUTER-AIDED	COMPUTER-AIDED INPUT/OUTPUT FOR USE WITH FINITE ELEMENT METHOD OF STRUCTURAL ANALYSIS	70-0330
COMPUTER-AIDED	COMPUTER-AIDED INPUT/OUTPUT FOR THE USE WITH FINITE ELEMENT STRUCTURAL ANALYSES	73-0923
COMPUTER-AIDED	COMPUTERUNTERSTUETZTES SYSTEM FUER DIE CONSTRUKTION DES WERKZEUGMASCHINENAUFBAUS (COMPUTER-AIDED SYS	73-0482
COMPUTER-AIDED	FORMULATION OF STRESS STRAIN INDUCED DIFFUSION OF HYDROGEN AND ITS SOLUTION BY COMPUTER-AIDED FINITE	75-0326
COMPUTER-AIDED	INEXPENSIVE, INTERACTIVE COMPUTER-AIDED DESIGN APPLIED TO FINITE ELEMENT STRESS ANALYSIS	74-0592
COMPUTER-AIDED	SHELLS, PLATES AND MEMBRANES - A COMPUTER-AIDED DESIGN SYSTEM	74-1159
COMPUTER-DEVELO	COMPUTER-DEVELOPED STATICAL METHODS	69-0374
COMPUTER-GENERA	APPLICATIONS OF CONTINUOUS TONE COMPUTER-GENERATED IMAGES IN STRUCTURAL MECHANICS	74-1035
COMPUTER-GRAPHI	DIE ANWENDUNG VON INTERAKTIVER COMPUTER-GRAPHIK BEI DER BERECHNUNG VON TRAGWERKEN. (APPLICATION OF I	72-0102
COMPUTER-PROGRA	BIDIMENSIONAL COMPUTER-PROGRAM FOR CALCULATION OF GROWTH-RATES OF INCLUDED AXIALLY SYMMETRICAL PLASM	75-0420
COMPUTER-PROGRA	DAS PROGRAMMSYSTEM ANTRAS EDV-SYSTEM ZUR BERECHNUNG BELIEBIGER STATISCHER SYSTEME NACH DER METHODE D	74-0918
COMPUTER-UNTERS	COMPUTER-UNTERSTUTZTES KONSTRUIEREN, PRAKTISCHE BEDEUTUNG DER FINIT-ELEMENT-VERFAHREN	71-0552
COMPUTERIZED	COMPUTERIZED ANALYSIS OF NON-ISOTROPIC STRUCTURES	69-0192
COMPUTERIZED	COMPUTERIZED STRUCTURAL ANALYSIS OF THE WORLD'S LARGEST LIGHT-GAGE STEEL PRIMARY STRUCTURAL SYSTEM	73-0503
COMPUTERIZED	COMPUTERIZED SUBSTRUCTURE ANALYSIS	75-0419
COMPUTERIZED	PRACTICAL APPLICATION OF A COMPUTERIZED STRUCTURAL ANALYSIS SYSTEM WHICH ADOPTS A FINITE ELEMENT TEC	69-0225
COMPUTERIZED	PROGRESS IN COMPUTERIZED STRUCTURAL ANALYSIS OF LARGE VESSELS	74-0949
COMPUTERIZED	USE OF COMPUTERIZED SYMBOLIC INTEGRATION IN FINITE ELEMENT DEVELOPMENT	74-0520
COMPUTERS	A LOOK INTO THE FUTURE: HOW COMPUTERS WILL INFLUENCE ENGINEERING	67-0139
COMPUTERS	CATCHING WAVES WITH COMPUTERS	74-0280
COMPUTERS	COMPUTERS AND MECHANICAL DESIGN	73-0195
COMPUTERS	EINSATZ DES COMPUTERS BEI DER ENTWICKLUNG EINER PANZERWANNE (THE USE OF A COMPUTER FOR DEVELOPING A	72-0211
COMPUTERS	FINITE ELEMENT STRUCTURAL ANALYSIS USING COMPUTERS	70-0216
COMPUTERS	IMPACT OF COMPUTERS ON THE PRACTICE OF STRUCTURAL ENGINEERING IN CONCRETE	72-0414
COMPUTERS	IMPACT OF COMPUTERS ON THE PRACTICE OF STRUCTURAL ENGINEERING IN CONCRETE	72-0415
COMPUTERS	MATRIX METHODS AND USE OF COMPUTERS IN STRUCTURAL ENGINEERING	71-0615
COMPUTERS	PRAKTISCHE RESULTATEN, DOEL EN NUT VAN HET ONTWERPEN VAN MECHANISCHE ELEMENTEN MET BEHULP VAN COMPUT	73-0400
COMPUTERS	PRAKTISCHE RESULTATEN, DOEL EN NUT VAN HET ONTWERPEN VAN MECHANISCHE ELEMENTEN MET BEHULP VAN COMPUT	73-0400
COMPUTERS	SHIFTING ROLES OF COMPUTERS, EXPERIMENTS AND ANALYSIS IN APPLIED MECHANICS	71-0604
COMPUTERS	USE OF COMPUTERS IN WELDING TECHNOLOGY	74-1160
COMPUTERUNTERST	COMPUTERUNTERSTUETZTES SYSTEM FUER DIE CONSTRUKTION DES WERKZEUGMASCHINENAUFBAUS (COMPUTER-AIDED SYS	73-0482
COMPUTING	A METHOD OF COMPUTING NUMERICALLY INTEGRATED STIFFNESS MATRICES	72-0546
COMPUTING	A RELAXATION PROCEDURE FOR COMPUTING STRESSES IN BUILT-UP STRUCTURES	69-0063
COMPUTING	AN AUTOMATIC SYSTEM FOR KINEMATIC ANALYSIS ASKA HIGH SPEED COMPUTING OF ELASTIC STRUCTURES	70-0358
COMPUTING	AXISYMMETRIC ELEMENT ANALYSIS USING ANALYTICAL COMPUTING	73-0333
COMPUTING	COMPUTER METHOD FOR COMPUTING CRITICAL SPEEDS OF NON-UNIFORM SHAFTS GEARED SYSTEMS AND CURVED ASSEMB	73-0675
COMPUTING	COMPUTING SYSTEM FOR STRUCTURAL ANALYSIS OF CAR BODIES	72-0324
COMPUTING	COMPUTING WITH SPARSE MATRICES	73-0926
COMPUTING	EVALUATION OF THE USE OF FINITE ELEMENT METHOD IN COMPUTING TEMPERATURE	69-0017
COMPUTING	FINITE ELEMENTS AND SCIENTIFIC COMPUTING	75-0042
COMPUTING	NOTE ON AN APPROXIMATE METHOD FOR COMPUTING NONCONSERVATIVE GENERALIZED FORCES ON FINITELY DEFORMED	70-0097
COMPUTING	NOTE ON AN APPROXIMATE METHOD FOR COMPUTING CONSISTENT CONJUGATE STRESSES IN ELASTIC FINITE ELEMENTS	73-0050
CONCAVE	FINITE-ELEMENT NONLINEAR BUCKLING ANALYSIS OF A PSEUDO-CYLINDRICAL CONCAVE POLYHEDRAL SHELL UNDER EX	73-1147
CONCENTRATED	A METHOD FOR SIMULATING CONCENTRATED FORCES AND LOCAL REINFORCEMENTS IN STRESS COMPUTATIONS	72-0081
CONCENTRATED	DER GERISSENE ZUSTAND ZWEISEITIG GELAGERTER PLATTEN UNTER EINZELLASTEN - NICHTLINEARE BERECHNUNG MIT	75-0858
CONCENTRATED	GALERKIN FINITE ELEMENT ANALYSIS OF A UNIFORM BEAM CARRYING A CONCENTRATED MASS AND ROTARY INERTIA W	74-0552
CONCENTRATED	INFLUENCE OF THE UNPRECISION OF EXECUTION ON THE BEHAVIOUR OF A SPHERICAL SHELL WITH CONCENTRATED LO	73-0802
CONCENTRATED	NONLINEAR VIBRATIONS OF NONUNIFORM BEAMS WITH CONCENTRATED MASSES	74-1300
CONCENTRATED	STIFFNESS ANALYSIS OF SHEET METAL SHELLS UNDER CONCENTRATED LOADS	74-0325
CONCENTRATED	STRESSES DUE TO CONCENTRATED LOADS IN RESTRAINED BENDING TESTS	67-0097
CONCENTRATION	CALCULATION OF STRESS CONCENTRATION DUE TO A HYPERBOLOID CAVITY IN A THIN PLATE USING NASTRAN	71-0260
CONCENTRATION	COMPARATIVE CONSIDERATION OF PARAMETRIC STRESS CONCENTRATION INVESTIGATIONS WITH FINITE ELEMENTS AND	72-0423
CONCENTRATION	EFFECT OF FIBRE DISTRIBUTION ON STRESS AND STRAIN CONCENTRATION AT HOLES IN COMPOSITE PLATES	72-0616
CONCENTRATION	FACTORS OF STRESS CONCENTRATION AT A HOLE IN FINITE STRIPS UNDER BENDING	72-0744
CONCENTRATION	NUMERICAL EXPERIMENTS ON THE DETERMINATION OF STRESS CONCENTRATION FACTORS	74-0292
CONCENTRATION	NUMERICAL PREDICTION OF TRACER CONCENTRATION IN AQUIFERS WITH TRANSIENT GROUNDWATER FLOW BY FINITE-E	75-0610
CONCENTRATION	PARAMETRISCHE KERBSPANNUNGSUNTERSUCHUNGEN AN DER LOCHSCHEIBE MIT DER METHODE DER FINITEN ELEMENTE. (	75-0603
CONCENTRATION	SIMPLE FINITE ELEMENT METHOD FOR STRESS CONCENTRATION PROBLEMS IN PRESSURE VESSELS	71-0669
CONCENTRATION	STRAIN CONCENTRATION ANALYSIS USING THE FINITE ELEMENT METHOD	71-0090
CONCENTRATION	STRAIN CONCENTRATION BEHAVIOUR IN A NOTCHED ROUND BAR SUBJECTED TO CREEP	73-0440
CONCENTRATION	STRESS CONCENTRATION FACTORS AROUND A PENETRATION IN A STRIP OR SLAB OF FINITE DIMENSIONS	69-0409
CONCENTRATION	STRESS CONCENTRATION FACTORS FOR BONDED LAP JOINTS	71-0842
CONCENTRATION	STRESS CONCENTRATION FACTORS FOR CIRCULAR NOTCHES	74-0791
CONCENTRATION	STRESS CONCENTRATION FACTORS FOR A FATIGUE TEST SPECIMEN WITH A U-TYPE NOTCH IN ONE EDGE BY FINITE E	74-1344
CONCENTRATION	STRESS CONCENTRATION FACTORS FOR CIRCULAR, REINFORCED PENETRATIONS IN PRESSURIZED CYLINDRICAL SHELLS	75-0726
CONCENTRATION	THERMAL-STRESS CONCENTRATION CAUSED BY STRUCTURAL DISCONTINUITIES	69-0308
CONCENTRATIONS	CALCULATION OF STRESS CONCENTRATIONS ON SUPPORT STRUCTURES UNDER THERMOMECHANICAL LOADING (BERECHNUN	73-0796
CONCENTRATIONS	DETERMINATION OF STRESS CONCENTRATIONS AT NOZZLE TO CYLINDER INTERSECTIONS AND COMPARISON WITH EXPER	73-0491
CONCENTRATIONS	EFFECT OF FILLETS ON STRESS CONCENTRATIONS IN CYLINDRICAL SHELLS WITH STEP CHANGES IN OUTSIDE DIAMET	71-0108
CONCENTRATIONS	FINITE ELEMENT ANALYSIS OF STRESS CONCENTRATIONS AND SINGULARITIES	72-0841
CONCENTRATIONS	IMPROVED FINITE ELEMENT METHOD FOR STRESS CONCENTRATIONS	73-0127
CONCENTRATIONS	NOTE ON THE STRESS CONCENTRATIONS AT THE END OF A CYLINDRICAL HOLE	70-0139
CONCENTRATIONS	PHOTOELASTIC INVESTIGATION OF STRESS CONCENTRATIONS IN SPHERE-CYLINDER TRANSITION REGIONS: INCLUDING	68-0258
CONCENTRATIONS	REVIEW OF CONTINUUM, FINITE ELEMENT AND HYBRID TECHNIQUES IN THE ANALYSIS OF STRESS CONCENTRATIONS I	73-1051
CONCENTRATIONS	REVIEW OF CONTINUUM, FINITE ELEMENT AND HYBRID TECHNIQUES IN ANALYSIS OF STRESS CONCENTRATIONS IN ST	74-0914
CONCENTRATIONS	SELF-STRESS CONCENTRATIONS	71-0317
CONCENTRATIONS	SPECIAL FINITE ELEMENTS FOR THE ANALYSIS OF STRESS CONCENTRATIONS AND SINGULARITIES	71-0219
CONCENTRATIONS	STRESS CONCENTRATIONS IN TIMBER BEAMS	67-0038
CONCENTRATIONS	STRESS CONCENTRATIONS IN HULL SHAPES FROM SURFACE DISCONTINUITIES	69-0130
CONCENTRATIONS	STRESS CONCENTRATIONS NEAR A DISCONTINUITY IN FIBROUS COMPOSITES	71-0316
CONCENTRATIONS	STRESS CONCENTRATIONS AND SINGULARITIES AT INTERFACE CORNERS	71-0419
CONCENTRATIONS	STRESS CONCENTRATIONS NEAR HOLES IN LAMINATES	74-0329

CONCENTRATIONS	THREE DIMENSIONAL ANALYSIS OF STRESS CONCENTRATIONS NEAR HOLES IN LAMINATED COMPOSITES	72-0713
CONCENTRIC	CONTRIBUTION TO NUMERICAL-SOLUTION OF DEVELOPING LAMINAR-FLOW IN ENTRANCE REGION OF CONCENTRIC ANNUL	75-0448
CONCEPT	A GENERALIZATION OF THE FINITE ELEMENT CONCEPT AND ITS APPLICATION TO A CLASS OF PROBLEMS IN NONLINE	68-0212
CONCEPT	ANALYSIS FOR CRACKS IN SHIP STRUCTURE USING THE FINITE ELEMENT METHOD WITH THE CONCEPT OF CRACK SUB-	74-0248
CONCEPT	ANALYSIS FOR CRACKS IN SHIP STRUCTURES USING THE FINITE-ELEMENT METHOD WITH THE CONCEPT OF CRACK SUB	74-1367
CONCEPT	APPROXIMATION IN ELASTICITY BASED ON THE CONCEPT OF FUNCTION SPACE	47-0001
CONCEPT	CONCEPT FEASIBILITY COMBAT TRACKED VEHICLE SIGNATURE DUPLICATOR	64-0028
CONCEPT	CONTINUOUS CONCEPT OF A SPACE GRID	73-0755
CONCEPT	ISOPARAMETRIC FINITE ELEMENT SYSTEM: A NEW CONCEPT IN FINITE ELEMENT ANALYSIS	68-0223
CONCEPT	LIMITATIONS ON THE APPLICATION OF EFFECTIVE BREADTH CONCEPT IN BOX GIRDER DESIGN	74-0493
CONCEPT	ON A GENERALIZATION OF THE FINITE ELEMENT CONCEPT AND ITS APPLICATION TO A CLASS OF PROBLEMS IN NONL	69-0206
CONCEPT	PLANETARY EXPLORER STRUCTURAL CONCEPT EVALUATION MODEL USING NASTRAN	71-0742
CONCEPT	THE ELEMENT-CHAIN CONCEPT IN FINITE ELEMENT BASED PROGRAMMING	72-0934
CONCEPT	THE ISOPARAMETRIC FINITE ELEMENT SYSTEM - A NEW CONCEPT IN FINITE ELEMENT ANALYSIS	69-0139
CONCEPT	THE VALIDITY OF THE INITIAL STRAIN CONCEPT	61-0010
CONCEPTION	NOUVELLE CONCEPTION DES DRAINS DANS LES BARRAGES EN TERRE HOMOGENE DE PETITE ET MOYENNE DIMENSIONS (	73-0404
CONCEPTS	BASIC CONCEPTS OF THE MECHANICS OF DISCRETIZED BODIES WITH AN INTRODUCTION TO DISCRETE ELEMENT CALCU	73-0146
CONCEPTS	CLASSICAL AND NATURAL CONDUCTION AND DEFORMATION MODES CONCEPTS FOR AXISYMMETRIC VARIATIONAL METHODS	72-1003
CONCEPTS	CONCEPTS AND APPLICATIONS OF FINITE ELEMENT ANALYSIS	74-1373
CONCEPTS	FINITE ELEMENT METHOD: BASIC CONCEPTS AND APPLICATIONS	73-0646
CONCEPTS	NEW CONCEPTS FOR FINITE ELEMENTS OF SHELLS	68-0195
CONCEPTS	REDUNDANT STRUCTURE CONCEPTS IN FINITE ELEMENT ANALYSES	75-0338
CONCEPTS	REDUNDANT STRUCTURE CONCEPTS IN FINITE ELEMENT ANALYSES	75-0626
CONCEPTS	VERIFICATION OF FRACTURE MECHANICS CONCEPTS TO PREDICT CRACKING OF FLEXIBLE PAVEMENTS	73-0772
CONCRETE	A FINITE ELEMENT COMPUTER PROGRAM FOR THE PREDICTION OF THE BEHAVIOR OF REINFORCED AND PRESTRESSED C	72-0799
CONCRETE	A FINITE-ELEMENT MODEL TO SIMULATE THE NONLINEAR RESPONSE OF REINFORCED CONCRETE FRAMES WITH MASONRY	73-1168
CONCRETE	A FINITE-ELEMENT STUDY ON THE NONLINEAR BEHAVIOR OF REINFORCED CONCRETE FLEXURAL MEMBERS DUE TO SHOR	73-1154
CONCRETE	A STUDY OF BEHAVIOUR OF REINFORCED CONCRETE ELEMENTS USING FINITE ELEMENTS	70-0335
CONCRETE	A THREE-DIMENSIONAL FINITE ELEMENT ANALYSIS OF A PRESTRESSED CONCRETE REACTOR VESSEL MODEL	69-0084
CONCRETE	ADVANCES IN THE ANALYSIS OF PRESTRESSED CONCRETE PRESSURE VESSELS	71-0162
CONCRETE	AIRCRAFT-PAVEMENT INTERACTION STUDIES. PHASE 1: A FINITE ELEMENT MODEL OF A JOINTED CONCRETE PAVEMEN	73-0894
CONCRETE	AN ANALYSIS OF A REINFORCED AND A PRESTRESSED CONCRETE BEAM BY FINITE-ELEMENT METHOD	72-0940
CONCRETE	ANALYSIS OF BRIDGE DECKS USING POLYMER IMPREGNATED CONCRETE	73-0673
CONCRETE	ANALYSIS OF CONCRETE BOX GIRDERS BEFORE AND AFTER CRACKING	72-0247
CONCRETE	ANALYSIS OF NONLINEAR CONCRETE FRAMES USING SMALL GYPSUM MODELS AND THE FINITE ELEMENT METHOD	70-0498
CONCRETE	ANALYSIS OF NONLINEAR REINFORCED CONCRETE FRAME BY FINITE ELEMENT METHOD	74-1104
CONCRETE	ANALYSIS OF STRESSES IN CONCRETE NEAR A REINFORCING BAR DUE TO BOND AND TRANSVERSE CRACKING	70-0351
CONCRETE	ANALYTICAL AND EXPERIMENTAL STUDIES OF MULTI-CELL CONCRETE BOX GIRDER BRIDGES	74-1134
CONCRETE	APPLICATION OF FRACTURE MECHANICS TO PRESTRESSED CONCRETE PRESSURE VESSELS	73-0228
CONCRETE	BERECHNUNG VON STAHLBETONSCHEIBEN IM ZUSTAND II BEI ANNAHME EINES WIRKLICHKEITSNAHEN WERKSTOFFVERHAL	74-0978
CONCRETE	BIAXIAL STRESS-STRAIN RELATIONS FOR CONCRETE	72-0203
CONCRETE	CALCULATIONS FOR PRESTRESSED CONCRETE PRESSURE VESSELS	71-0643
CONCRETE	CIRCUMFERENTIAL PRESTRESSING LOAD ANALYSIS DUE TO WIRE/STRAND WINDING SYSTEMS ON PRESTRESSED CONCRET	74-1202
CONCRETE	COLLAPSE LOAD OF REINFORCED CONCRETE PLATES.	72-0406
CONCRETE	COMPARISON OF FINITE ELEMENT AND UNIQUE LIMIT ANALYSIS SOLUTIONS FOR CERTAIN REINFORCED CONCRETE SLA	69-0184
CONCRETE	CONSTITUTIVE EQUATIONS AND PUNCH-INDENTATION OF CONCRETE	75-0286
CONCRETE	CONSTITUTIVE RELATIONS OF CONCRETE AND PUNCH-INDENTATION PROBLEMS	73-0877
CONCRETE	CONSTITUTIVE RELATIONS FOR CONCRETE	75-0541
CONCRETE	CRACKING ANALYSIS OF REINFORCED CONCRETE PLATES	75-0085
CONCRETE	CREEP AND SHRINKAGE LAW FOR CONCRETE AT VARIABLE HUMIDITY	74-0668
CONCRETE	DESIGN OF A COMPOSITE STEEL AND CONCRETE CLOSURE FOR A PRESTRESSED CONCRETE REACTOR VESSEL.	72-0187
CONCRETE	DESIGN OF A COMPOSITE STEEL AND CONCRETE CLOSURE FOR A PRESTRESSED CONCRETE REACTOR VESSEL.	72-0187
CONCRETE	DESIGN OF CONCRETE TRACK SLAB ON ELASTIC FOUNDATION USING THE FINITE ELEMENT METHOD	73-0479
CONCRETE	DETERMINATION OF VOLUME CHANGE STRESSES IN PLAIN AND REINFORCED CONCRETE USING FINITE ELEMENT ANALYS	69-0477
CONCRETE	DEVELOPMENT OF A DESIGN FOR A PRECAST CONCRETE BRIDGE BEAM OF J-SECTION	73-0392
CONCRETE	DIRECT MEASUREMENT OF BOND SLIP IN REINFORCED CONCRETE	69-0428
CONCRETE	DISCUSSION ON FINITE ELEMENT ANALYSIS OF PRESTRESSED CONCRETE REACTOR VESSELS	72-0648
CONCRETE	DISTRIBUZIONI OTTIMALI DI ARMATURE DI PRECOMPRESSIONE IN TRAVI-PARETE DI CALCESTRUZZO. (OPTIMAL DIST	75-0560
CONCRETE	DYNAMIC RESPONSE OF SOIL/CONCRETE INTERFACES AT HIGH PRESSURE	74-0836
CONCRETE	EARTHQUAKE RESPONSE OF CONCRETE GRAVITY DAMS	70-0042
CONCRETE	EINFACHES NAEHERUNGSVERFAHREN FUR DIE GRENZTRAGFAEHIGKEITSANLYSE VON SPANNBETON-REAKTORDRUCKBEHAELTE	73-0854
CONCRETE	ELASTIC-PLASTIC ANALYSIS OF REINFORCED CONCRETE MEMBERS BY FEM	73-0292
CONCRETE	ENTWURF UND SPANNUNGSBERECHNUNG EINES SPANNBETON DRUCKBEHAELTERS AUS FERTIGTEILEN. (DESIGN AND STRES	73-0856
CONCRETE	ETUDE THEORIQUE DU COMPORTEMENT D'UNE DALLE DE BETON. (THEORETICAL STUDY OF THE BEHAVIOR OF A CONCRE	75-0273
CONCRETE	EXPERIMENTELL UNDERSOKNING AV ELEMENTFOGARS HALLFASTHETSEGENSKAPER (EXPERIMENTAL INVESTIGATION OF ST	72-0192
CONCRETE	FAILURE OF THE CONCRETE LINING IN AN INCLINED SHAFT AT GREAT DEPTH.	74-0996
CONCRETE	FINITE ELEMENT ANALYSIS OF REINFORCED CONCRETE BEAMS	67-0058
CONCRETE	FINITE ELEMENT ANALYSIS OF REINFORCED CONCRETE	67-0161
CONCRETE	FINITE ELEMENT ANALYSIS OF HYPERBOLOIDAL PRESTRESSED CONCRETE CONTAINMENT VESSELS	69-0526
CONCRETE	FINITE ELEMENT ANALYSIS OF REINFORCED CONCRETE SLABS	71-0133
CONCRETE	FINITE ELEMENT ANALYSIS OF REINFORCED CONCRETE STRUCTURES	72-0051
CONCRETE	FINITE ELEMENT ANALYSIS OF REINFORCED CONCRETE MEMBERS	72-0583
CONCRETE	FINITE ELEMENT ANALYSIS OF PRESTRESSED CONCRETE VOIDED BRIDGE DECKS	73-0106
CONCRETE	FINITE ELEMENT ANALYSIS OF A DIAGONAL CRACKING BEHAVIOUR IN REINFORCED CONCRETE	73-0462
CONCRETE	FINITE ELEMENT ANALYSIS OF REINFORCED CONCRETE	73-0494
CONCRETE	FINITE ELEMENT ANALYSIS OF REINFORCED CONCRETE	74-0356
CONCRETE	FINITE ELEMENT ANALYSIS OF CONCRETE BEAMS	74-0560
CONCRETE	FINITE ELEMENT ANALYSIS OF SOME CONCRETE NONLINEARITIES. THEORY AND EXAMPLES	74-1287
CONCRETE	FINITE ELEMENT ANALYSIS OF PRESTRESSED CONCRETE PRESSURE VESSELS	74-1289
CONCRETE	FINITE ELEMENT ANALYSIS OF SHEAR STRENGTH OF REINFORCED CONCRETE BEAMS	74-1306
CONCRETE	FINITE ELEMENT APPROACH TO POLYMER CONCRETE BRIDGE DECK DESIGN ANALYSIS	72-0918
CONCRETE	FINITE ELEMENT FOR REINFORCED CONCRETE FRAME STUDY	73-0204
CONCRETE	FINITE ELEMENT FOR REINFORCED CONCRETE FRAME STUDY	74-0110
CONCRETE	FINITE ELEMENT FOR REINFORCED CONCRETE FRAME STUDY - SPACE DISCUSSION	74-0246
CONCRETE	FINITE ELEMENT FOR REINFORCED CONCRETE FRAME STUDY - CLOSURE	75-0087
CONCRETE	FINITE ELEMENT METHOD IN ANALYSIS OF REINFORCED AND PRESTRESSED CONCRETE STRUCTURES	71-0270
CONCRETE	FINITE ELEMENT STUDY OF REINFORCED CONCRETE BEAMS WITH DIAGONAL TENSION CRACKS	74-1308
CONCRETE	FINITE ELEMENT STUDY OF A REINFORCED CONCRETE CYLINDRICAL SHELL THROUGH ELASTIC, CRACKING, AND ULTIM	75-0317
CONCRETE	FINITE ELEMENT ULTIMATE LOAD ANALYSIS OF THREE DIMENSIONAL CONCRETE STRUCTURES	74-0672
CONCRETE	FINITE ELEMENTS FOR CALCULATION OF PRESTRESSED CONCRETE REACTOR PRESSURE VESSELS (FINITE ELEMENTE ZU	73-0797
CONCRETE	FINITE ELEMENTS METHOD IN DESIGN OF PRESTRESSED REACTOR CONCRETE CONSTRUCTION	73-0550
CONCRETE	FINITE-ELEMENT ANALYSIS OF CONCRETE SLABS AND ITS IMPLICATIONS FOR RIGID PAVEMENT DESIGN	73-0538
CONCRETE	FIRES-RC. A COMPUTER PROGRAM FOR THE FIRE RESPONSE OF STRUCTURES-REINFORCED CONCRETE FRAMES	74-0747
CONCRETE	IMPACT OF COMPUTERS ON THE PRACTICE OF STRUCTURAL ENGINEERING IN CONCRETE	72-0414
CONCRETE	IMPACT OF COMPUTERS ON THE PRACTICE OF STRUCTURAL ENGINEERING IN CONCRETE	72-0415
CONCRETE	INELASTIC ANALYSIS OF REINFORCED CONCRETE FRAMES	74-0587
CONCRETE	INELASTIC FINITE ELEMENT ANALYSIS OF REINFORCED CONCRETE PANELS UNDER IN-PLANE LOADS	74-0482
CONCRETE	INELASTIC MODEL FOR CYCLIC BIAXIAL LOADING OF REINFORCED CONCRETE	74-0503
CONCRETE	INFLATION FORMING OF STEEL FIBER-REINFORCED CONCRETE DOMES	74-0705
CONCRETE	INFLUENCE OF END-CLOSURE STIFFNESS ON BEHAVIOR OF CONCRETE CYLINDRICAL HULLS SUBJECTED TO HYDROSTATI	71-0725
CONCRETE	LAYERED PAVEMENT SYSTEMS. PART 1. LAYERED SYSTEM DESIGN. PART 11. FATIGUE OF PLAIN CONCRETE	72-0867
CONCRETE	LIMIT ANALYSIS OF REINFORCED CONCRETE SLABS BY FINITE ELEMENT METHOD	71-0802

CONCRETE	LIMIT ANALYSIS OF REINFORCED CONCRETE SLABS BY A FINITE ELEMENT METHOD	72-0985
CONCRETE	MATERIAL BEHAVIOUR CHARACTERISTICS FOR REINFORCED CONCRETE SHELLS STRESSED BEYOND THE ELASTIC RANGE	70-0166
CONCRETE	MODIFIED APPROACH FOR ESTIMATING THE CRACKING MOMENT OF REINFORCED CONCRETE BEAMS	75-0485
CONCRETE	NON LINEAR ANALYSIS OF REINFORCED CONCRETE MEMBERS BY FINITE ELEMENT METHOD. PART 1: ANALYSIS, PART	71-0313
CONCRETE	NON-LINEAR ANALYSIS OF REINFORCED CONCRETE SLABS	72-0191
CONCRETE	NON-LINEAR FINITE ELEMENT ANALYSIS OF REINFORCED CONCRETE BY THE FINITE ELEMENT METHOD	68-0241
CONCRETE	NONLINEAR ANALYSIS OF REINFORCED CONCRETE BY THE FINITE ELEMENT METHOD	68-0126
CONCRETE	NONLINEAR ANALYSIS OF PRESTRESSED CONCRETE FAST REACTOR PRESSURE VESSEL. (ANALYSE NON LINEAIRE DES C	70-0145
CONCRETE	NONLINEAR ANALYSIS OF REINFORCED CONCRETE FRAMES AND PANELS	70-0436
CONCRETE	NONLINEAR ANALYSIS OF TWO-DIMENSIONAL PROBLEMS IN CONCRETE CREEP	71-0299
CONCRETE	NONLINEAR ANALYSIS OF REINFORCED CONCRETE BUILDINGS	73-0324
CONCRETE	NONLINEAR ANALYSIS OF PLANAR REINFORCED CONCRETE STRUCTURES	74-0533
CONCRETE	NONLINEAR BEHAVIOR OF REINFORCED CONCRETE BEAMS BY THE FINITE-ELEMENT METHOD	72-0954
CONCRETE	NONLINEAR FINITE ELEMENT ANALYSIS OF REINFORCED CONCRETE PLATES	73-0530
CONCRETE	NONLINEAR FINITE ELEMENT ANALYSIS OF REINFORCED CONCRETE PLATES	73-0845
CONCRETE	NONLINEAR FINITE ELEMENT ANALYSIS OF REINFORCED CONCRETE STRUCTURES	74-0442
CONCRETE	NONLINEAR FINITE ELEMENT ANALYSIS OF REINFORCED CONCRETE FRAMES	74-1124
CONCRETE	NUMERICAL BIAXIAL CHARACTERIZATION FOR CONCRETE	74-0641
CONCRETE	OPTIMUM DESIGN OF CONCRETE STRUCTURES	74-0979
CONCRETE	PATTERN LOADING ON REINFORCED CONCRETE FLAT PLATES	71-0290
CONCRETE	PAVEMENT DESIGN AND CONTINUOUSLY REINFORCED PAVEMENT PERFORMANCE	73-0749
CONCRETE	PLANE STRESS REINFORCED CONCRETE FINITE ELEMENTS	74-0108
CONCRETE	POLYMER-IMPREGNATED CONCRETE TUNNEL SUPPORT AND LINING	73-0736
CONCRETE	POST-TENSIONED CONCRETE NUCLEAR STRUCTURES	70-0040
CONCRETE	PREFABRICATED PRETENSIONED LIGHTWEIGHT CONCRETE FOLDED PLATE	73-0808
CONCRETE	PRESTRESSED CONCRETE REACTOR VESSEL ANALYTICAL METHODS DEVELOPMENTS	68-0141
CONCRETE	PRESTRESSED CONCRETE PRESSURE VESSELS	69-0052
CONCRETE	PRESTRESSED CONCRETE REACTOR PRESSURE VESSEL FAILURE ANALYSIS. THE INCORPORATION OF STATISTICAL METH	71-0120
CONCRETE	PRESTRESSED CONCRETE DEEP SLABS WITH OPENINGS	73-0427
CONCRETE	PRINCIPAL STRESSES AT THE INTERMEDIATE SUPPORT OF PRESTRESSED CONCRETE CONTINUOUS BEAMS	70-0352
CONCRETE	PROBABILISTIC FORMULATION FOR ANALYTICAL MODELING OC CONCRETE STRUCTURES	71-0709
CONCRETE	PROBABILISTIC MODELING OF CONCRETE STRUCTURES	72-0236
CONCRETE	RASCHET NAPRYAZHENIL L PEREMESHCHENII V SVODE PODZEMNOGO MASHINNOGO ZALA GES I V OKRUZHAYUSHCHEM SKA	74-0542
CONCRETE	RECENT DEVELOPMENTS IN FINITE ELEMENT ANALYSIS OF PRESTRESSED CONCRETE REACTOR VESSELS	74-0469
CONCRETE	RECENT DEVELOPMENTS IN THE FINITE ELEMENT ANALYSIS OF PRE-STRESSED CONCRETE REACTOR VESSELS	74-0582
CONCRETE	RECORD CONCRETE SHELL ROOF FEATURED IN COLLISEUM	72-0218
CONCRETE	REFINED CRACKED CONCRETE ANALYSIS OF CONCRETE CONTAINMENT STRUCTURES SUBJECT TO OPERATIONAL AND ENVI	74-0943
CONCRETE	REFINED CRACKED CONCRETE ANALYSIS OF CONCRETE CONTAINMENT STRUCTURES SUBJECT TO OPERATIONAL AND ENVI	74-0943
CONCRETE	REINFORCED CONCRETE - A FINITE-ELEMENT FORMULATION	72-0929
CONCRETE	REINFORCED CONCRETE ON CONSTITUTIVE RELATIONS	75-0753
CONCRETE	REINFORCING REQUIREMENTS FOR CONCRETE BEAMS WITH LARGE WEB OPENINGS	73-0198
CONCRETE	RESENI MAZNIHO STAVU ZELEZOBETONOVYCH DESEK (PLASTIC ANALYSIS OF REINFORCED CONCRETE PLATES)	74-0495
CONCRETE	RIDDLE OF BONDED AND UNBONDED TENDONS IN PRESTRESSED CONCRETE REACTOR VESSELS	74-0519
CONCRETE	ROTATIONAL STIFFNESS OF CONCRETE SLABS	74-0660
CONCRETE	SAFE-CREEP: A COMPUTER PROGRAM FOR THE VISCOELASTIC ANALYSIS OF AXISYMMETRIC AND PLANE CONCRETE STRU	67-0148
CONCRETE	SHORT TERM DEFLECTIONS OF CONCRETE FLAT PLATES	73-0277
CONCRETE	SINGER: A COMPUTER CODE FOR GENERAL ANALYSIS OF TWO-DIMENSIONAL CONCRETE STRUCTURES. VOL. 11. PROGRA	75-0723
CONCRETE	SINGER: A COMPUTER CODE FOR GENERAL ANALYSIS OF TWO-DIMENSIONAL REINFORCED CONCRETE STRUCTURES. VOLU	75-0724
CONCRETE	SINGER: A COMPUTER CODE FOR GENERAL ANALYSIS TO TWO-DIMENSIONAL CONCRETE STRUCTURES. VOLUME 4. DEMON	75-0730
CONCRETE	SINGER: A COMPUTER CODE FOR GENERAL ANALYSIS OF TWO-DIMENSIONAL REINFORCED CONCRETE STRUCTURES. VOLU	75-0731
CONCRETE	SOME CREEP EFFECTS IN STRESS ANALYSIS WITH PARTICULAR REFERENCE TO CONCRETE PRESSURE VESSELS	69-0284
CONCRETE	STATICKE RESENI FYZIKALNE NELINEARNIHO CHOVANI ZELEZOBETO- NOVYCH DESEK (ANALYSIS OF PHYSICALLY NONL	74-0514
CONCRETE	STEP AND LOADS EFFECT ON STRESSES IN PRESTRESSED CONCRETE SHORT BRACKETS	71-0334
CONCRETE	STRENGTH OF REINFORCED CONCRETE CHAMBERS UNDER EXTERNAL PRESSURE	75-0293
CONCRETE	STRENGTH OF REINFORCED CONCRETE CHAMBERS UNDER EXTERNAL PRESSURE	75-0512
CONCRETE	STRESS ANALYSIS OF A PRESTRESSED CONCRETE NUCLEAR PRESSURE VESSEL BY THE FINITE ELEMENT METHOD USING	71-0154
CONCRETE	STRESS ANALYSIS OF CONCRETE TRACK SLABS ON AN ELASTIC FOUNDATION BY THE FINITE ELEMENT METHOD	74-1049
CONCRETE	STRESS STRAIN RESPONSE AND FRACTURE OF A CONCRETE MODEL IN BIAXIAL LOADING	71-0295
CONCRETE	STRESSES IN CONTINUOUS CONCRETE BOX GIRDER BRIDGES	71-0233
CONCRETE	STRUCTURAL BEHAVIOR OF A TWO SPAN REINFORCED CONCRETE BOX GIRDER BRIDGE MODEL. VOLUME 11. REDUCTION,	71-0659
CONCRETE	STRUCTURAL CAPACITY OF REINFORCED CONCRETE COLUMNS SUBJECTED TO FIRE INDUCED THERMAL GRADIENTS	73-0913
CONCRETE	STRUCTURAL DESIGN DATA FOR CONCRETE DRIFT LININGS IN BLOCK CAVING	73-0773
CONCRETE	STUDY OF ELASTIC-PLASTIC MATRIX ANALYSIS FOR FOUNDATION AND REINFORCED CONCRETE IN JAPAN	69-0278
CONCRETE	STUDY OF FORCE DISPLACEMENT RELATIONSHIPS FOR THE FINITE ELEMENT ANALYSIS OF REINFORCED CONCRETE	73-0645
CONCRETE	TENSILE CRACK PROPAGATION IN REINFORCED CONCRETE BEAMS BY THE FINITE ELEMENT TECHNIQUES	69-0264
CONCRETE	TENSILE CREEP ANALYSIS OF CONCRETE STRUCTURES	74-0439
CONCRETE	THE APPLICATION OF THREE-DIMENSIONAL FINITE ELEMENTS TO THE ANALYSIS OF PODDED BOILER TYPE PRESTRESS	71-0047
CONCRETE	THE DETERMINATION OF STRESSES AND DEFORMATIONS OF REINFORCED CONCRETE AFTER CRACKING	64-0004
CONCRETE	THE EARTHQUAKE EXPERIENCE AT KOYNA DAM AND STRESSES IN CONCRETE GRAVITY DAMS	72-0589
CONCRETE	THE INELASTIC ANALYSIS OF REINFORCED AND PRESTRESSED CONCRETE BEAMS	72-0667
CONCRETE	THE INELASTIC ANALYSIS OF REINFORCED CONCRETE SLABS	74-1333
CONCRETE	THERMAL LOADING OF THIN-SHELL CONCRETE COOLING TOWERS	74-0667
CONCRETE	THERMOMECHANICAL CREEP OF AGEING CONCRETE - A UNIFIED APPROACH	74-0607
CONCRETE	THICK WALLED MULTIPLE OPENING REINFORCED CONCRETE CONDUITS	72-0666
CONCRETE	THIN-WALLED PRESTRESSED CONCRETE PRESSURE VESSEL FOR HIGH TEMPERATURE REACTORS: EXPERIMENTAL INVESTI	73-0857
CONCRETE	THREE DIMENSIONAL RUPTURE ANALYSIS OF A PRESTRESSED CONCRETE PRESSURE VESSEL INCLUDING CREEP EFFECTS	73-0826
CONCRETE	THREE DIMENSIONAL RUPTURE ANALYSIS OF A PRESTRESSED CONCRETE PRESSURE VESSEL INCLUDING CREEP EFFECTS	74-0538
CONCRETE	TIME DEPENDENT REINFORCED CONCRETE SLAB DEFLECTIONS	74-0586
CONCRETE	TRANSVERSE REINFORCEMENT OF CONTINUOUS, TROUGH-SHAPED CONCRETE RAILROAD STRUCTURES. AN IMPLEMENTATIO	71-0719
CONCRETE	TWO DIMENSIONAL NON LINEAR ANALYSIS OF REINFORCED CONCRETE MEMBERS BY THE FINITE ELEMENT METHOD, PAR	71-0210
CONCRETE	USER'S GUIDE FOR THE COMPUTER PROGRAM DESCRIBED IN TAM REPORT NO. 680 "DETERMINATION OF VOLUME CHANG	69-0473
CONCRETE	VISCOELASTIC ANALYSIS OF PRESTRESSED CONCRETE SHELL OF REVOLUTION	73-0464
CONCRETE	YIELD CRITERION FOR REINFORCED CONCRETE SLABS	73-0985
CONCRETE	ZUR WIRKLICHKEITSNAHEN BERECHNUNG VON STAHLBETONPLATTEN MIT DER FINITE-ELEMENT-METHODE. (APPROXIMATE	75-0832
CONDENSATION	ANALYTICAL STUDY OF THE PROBLEM OF EVAPORATION AND CONDENSATION USING THE FINITE ELEMENT METHOD	74-0960
CONDENSATION	AUTOMATIC GENERATION AND CONDENSATION OF LARGE MASS AND STIFFNESS MATRICES	72-0466
CONDENSATION	CONDENSATION OF FINITE ELEMENT EIGENPROBLEMS	72-0168
CONDENSATION	CONDUCTION THROUGH DROPLETS DURING DROPWISE CONDENSATION	72-0019
CONDENSATION	MASS CONDENSATION, A SEMI-AUTOMATIC METHOD FOR REDUCING THE SIZE OF VIBRATION PROBLEMS	69-0220
CONDENSATION	THE STATIC CONDENSATION ALGORITHM	74-0328
CONDENSATION	THE USE OF CONDENSATION TECHNIQUES FOR SOLVING DYNAMICS PROBLEMS	74-0320
CONDENSER	STRESS DETERMINATION IN 2 CONDENSER HEAT EXCHANGER BAFFLE CAPS BY METHOD OF FINITE ELEMENTS AND A ST	74-0661
CONDITION	ACCURACY AND CONDITION OF CURVED (ISOPARAMETRIC) FINITE ELEMENTS	73-0057
CONDITION	ANGLE CONDITION IN THE FINITE ELEMENT METHOD	74-1083
CONDITION	BOUNDS ON EXTERNAL EIGENVALUES OF FINITE ELEMENT STIFFNESS AND MASS MATRICES AND THEIR SPECTRAL COND	72-0256
CONDITION	CONDITION OF FINITE ELEMENT MATRICES GENERATED FROM NONUNIFORM MESHES	72-0104
CONDITION	CONDITION OF FINITE ELEMENT MATRICES	74-1078
CONDITION	DER GERISSENE ZUSTAND ZWEISEITIG GELAGERTER PLATTEN UNTER EINZELLASTEN - NICHTLINEARE BERECHNUNG MIT	75-0858
CONDITION	ELASTIC-PLASTIC PROBLEMS OF COMPRESSIBLE STRAIGHT BAR UNDER COMBINED LOADING CONDITION	75-0313
CONDITION	ENERGY CONDITION FOR INITIATION OF INTERFACIAL MICROCRACKS AT INCLUSIONS	74-0546
CONDITION	FINITE ELEMENT STRESS ANALYSIS OF INTERACTING FUEL PELLET AND CANNING UNDER FAST REACTOR CONDITION	73-0814
CONDITION	INFLUENCE OF POISSON'S RATIO ON THE CONDITION OF THE FINITE ELEMENT STIFFNESS MATRIX	71-0519

CONDITION	INTERNAL SUPPORTING STRUCTURES OF THE (CIRENE) REACTOR CORE - RESULTS OF STATIC STRESS ANALYSIS FOR	73-0818
CONDITION	MONOTON FINITE-ELEMENT MATRICES AND THEIR COMPUTED CONDITION NUMBERS	75-0820
CONDITION	ON THE ANGLE CONDITION IN THE FINITE ELEMENT METHOD	74-0907
CONDITION	ON THE RIGID DISPLACEMENT CONDITION.	72-0405
CONDITION	THE CONDITION OF SOME FINITE ELEMENT COEFFICIENT MATRICES	70-0393
CONDITION	THE EFFECT OF END RESTRAINT CONDITION ON THE FRACTURE MECHANISM AND STRENGTH OF ROCK-LIKE MATERIALS	72-0134
CONDITION	THE L2 AND L00 CONDITION NUMBERS OF THE FINITE ELEMENT MATRICES AND THE POINTWISE CONVERGENCE OF THE	72-0061
CONDITIONED	BASIS OF A WELL CONDITIONED FORCE PROGRAM FOR EQUILIBRIUM MODELS VIA THE SOUTHWELL SLAB ANALOGIES	67-0111
CONDITIONING	AUTOMATIC ANALYSIS OF FUSELAGES AND PROBLEMS OF CONDITIONING	64-0027
CONDITIONING	NUMERICAL CONDITIONING OF STIFFNESS MATRIX FORMULATIONS FOR FRAME STRUCTURES	68-0022
CONDITIONS	(L SUP 2) ERROR ESTIMATES FOR FINITE ELEMENTS WITH INTERPOLATED BOUNDARY CONDITIONS	73-0885
CONDITIONS	A UNIFIED FINITE ELEMENT APPROACH TO COLUMNS SUBJECTED TO NONCONSERVATIVE LOADS, WITH DAMPING EFFECT	74-0795
CONDITIONS	AN ELASTIC-PLASTIC FINITE ELEMENT ANALYSIS OF CRACK TIP FIELDS UNDER BIAXIAL LOADING CONDITIONS	73-0760
CONDITIONS	AN EXTENSION OF FINITE-ELEMENT METHOD TO INCLUDE NONLINEAR BOUNDARY CONDITIONS	75-0032
CONDITIONS	APPLICATION OF FRACTURE MECHANICS TO GRAPHITE UNDER COMPLEX STRESS CONDITIONS	72-0771
CONDITIONS	APPLICATION OF THE FINITE-ELEMENT METHOD TO HEAT-TRANSFER PROBLEMS. PART 11. TRANSIENT TWO-DIMENSION	71-0759
CONDITIONS	APPLICATION OF THE FINITE-ELEMENT METHOD TO HEAT-TRANSFER PROBLEMS. PART 1. FINITE-ELEMENT METHOD AP	71-0760
CONDITIONS	APPLICATION OF THE FINITE ELEMENT METHOD FOR SOLVING ORDINARY LINEAR DIFFERENTIAL EQUATION WITH BOUN	74-0084
CONDITIONS	APPROXIMATE BOUNDARY CONDITIONS IN THE FINITE ELEMENT METHOD	72-0995
CONDITIONS	AUTOMATIC EMBANKMENT ANALYSIS. EQUILIBRIUM AND INSTABILITY CONDITIONS	66-0015
CONDITIONS	BERECHNUNGSMETHODEN ZUR ERMITTLUNG VON BELASTUNGSBEDINGUNGEN UND GRENZTRAGFAEHIGKEITSVERHALTEN VON K	73-0858
CONDITIONS	CUBIC SPLINE GALERKIN APPROXIMATIONS TO PARABOLIC SYSTEMS WITH COUPLED NON-LINEAR BOUNDARY CONDITION	75-0232
CONDITIONS	CYCLIC-STRESS ANALYSIS OF NOTCHES FOR SUPERSONIC TRANSPORT CONDITIONS	74-0834
CONDITIONS	DYNAMIC RESPONSE OF STRUCTURES SUBJECTED TO TIME-DEPENDENT BOUNDARY CONDITIONS USING THE FINITE ELEM	72-0129
CONDITIONS	ELASTIC-PLASTIC FINITE ELEMENT ANALYSIS OF CRACK TIP FIELDS UNDER BIAXIAL LOADING CONDITIONS	74-0296
CONDITIONS	FINITE ELEMENT METHOD APPLIED TO TRANSIENT TWO-DIMENSIONAL HEAT TRANSFER WITH CONVECTION AND RADIATI	70-0127
CONDITIONS	FINITE-ELEMENT METHOD APPLIED TO HEAT CONDUCTION IN SOLIDS WITH NONLINEAR BOUNDARY CONDITIONS	73-0005
CONDITIONS	FRACTURE TOUGHNESS PARAMETERS AND ELASTIC-PLASTIC ANALYSIS OF NON-MODERATE FRACTURE CONDITIONS USING	73-0258
CONDITIONS	FRACTURE TOUGHNESS PARAMETERS AND ELASTIC-PLASTIC ANALYSIS OF NON-MODERATE FRACTURE CONDITIONS USING	73-1126
CONDITIONS	FREE VIBRATION OF THICK, LAYERED CYLINDERS HAVING FINITE LENGTH WITH VARIOUS BOUNDARY CONDITIONS	72-0491
CONDITIONS	FREQUENCY COEFFICIENTS OF A SPHERICAL SHELL FOR VARYING THICKNESS AND OPENING ANGLE WITH GUIDED PINN	72-0595
CONDITIONS	GALERKIN METHODS FOR PARABOLIC EQUATIONS WITH NONLINEAR BOUNDARY CONDITIONS	73-1023
CONDITIONS	HEAT TRANSFER WITH NONLINEAR BOUNDARY CONDITIONS VIA A VARIATIONAL PRINCIPLE	71-0443
CONDITIONS	INTERPOLATED BOUNDARY CONDITIONS IN THE FINITE ELEMENT METHOD	75-0388
CONDITIONS	L2 ERROR ESTIMATES FOR FINITE ELEMENTS WITH INTERPOLATED BOUNDARY CONDITIONS	73-0194
CONDITIONS	MATCHING OF ESSENTIAL BOUNDARY CONDITIONS IN THE FINITE ELEMENT METHOD	74-1260
CONDITIONS	NATURAL BOUNDARY CONDITIONS IN FINITE-ELEMENT METHOD	72-0276
CONDITIONS	NATURAL VIBRATIONS OF THIN, FLAT WALLED STRUCTURES WITH DIFFERENT BOUNDARY CONDITIONS	71-0509
CONDITIONS	NON-UNIFORM DEFORMATION OF A BLOCK IN PLANE-STRAIN COMPRESSION CAUSED BY FRICTION. PART 4. ANALYSIS	71-0786
CONDITIONS	NONLINEAR FINITE ELEMENT SOLUTION OF ELECTRICAL MACHINES UNDER NO-LOAD AND FULL-LOAD CONDITIONS	74-0472
CONDITIONS	ON DIRICHLET PROBLEMS USING SUBSPACES WITH NEARLY ZERO BOUNDARY CONDITIONS	72-0514
CONDITIONS	ON THE NUMERICAL CONVERGENCE OF MATRIX EIGENVALUE PROBLEMS DUE TO CONSTRAINT CONDITIONS	74-1057
CONDITIONS	RASCHET KILEBANII PLASTIN SLOZHNOI FORMY S NEODNORODNYMI GRANICHNYMI USLOVIYAMI. (CALCULATION OF VIB	74-0988
CONDITIONS	SLAB BRIDGES WITH ARBITRARY SHAPE AND SUPPORT CONDITIONS - A GENERAL METHOD OF ANALYSIS BASED ON FIN	68-0233
CONDITIONS	SPONTANEOUS IGNITION - FINITE ELEMENT SOLUTIONS FOR STEADY AND TRANSIENT CONDITIONS	74-0353
CONDITIONS	SUBSPACES WITH ACCURATELY INTERPOLATED BOUNDARY CONDITIONS	72-0527
CONDITIONS	SUFFICIENT CONDITIONS FOR CONVERGENCE IN THE FINITE ELEMENT METHOD FOR ANY SOLUTION OF FINITE ENERGY	72-0065
CONDITIONS	SYSTEMATIC ENFORCEMENT OF STRESS BOUNDARY CONDITIONS IN THE ASSUMED STRESS HYBRID MODEL BASED ON THE	71-0771
CONDITIONS	THE EFFECTS OF SOIL PARAMETERS AND BOUNDARY CONDITIONS ON THE CONSOLIDATION OF AN ELASTIC LAYER	70-0422
CONDITIONS	THE INELASTIC BEHAVIOUR OF STRUCTURES SUBJECTED TO CYCLIC, THERMAL AND MECHANICAL STRESSING CONDITIO	60-0004
CONDITIONS	THE TREATMENT OF NATURAL BOUNDARY CONDITIONS IN THE FINITE ELEMENT AND FINITE DIFFERENCE METHODS	73-1098
CONDITIONS	THEORETICAL AND EXPERIMENTAL ANALYSIS OF STIFFENED PANEL UNDER DYNAMIC CONDITIONS	70-0130
CONDITIONS	THEORETICAL TRANSIENT BEHAVIOUR OF SATURATED AND UNSATURATED SOILS UNDER LOAD AND CHANGING MOISTURE	73-0901
CONDITIONS	THERMAL STRESSES IN OPEN CYLINDRICAL SHELLS OF ARBITRARY BOUNDARY CONDITIONS BY A FINITE STRIP METHO	72-0371
CONDITIONS	UEBER DAS VERHALTEN VON SPANNBETONDRUCKBEHAELTRERN IM GRENZTRAGFAEHIGKEITSZUSTAND UND DESSEN EINFLU	73-0855
CONDITIONS	UEBER PAUSCHALE FEHLERSCHRANKEN FUER APPROXIMATIONEN VON SCHALENVERFORMUNGEN. (GLOBAL BOUNDARY CONDI	75-0292
CONDITIONS	VERIFICATION OF THE PEC CONTAINMENT VESSEL FOR SOME OPERATING CONDITIONS BY USE OF THE FINITE ELEMEN	74-0883
CONDITIONS	VIBRATION OF RECTANGULAR PLATES WITH MIXED BOUNDARY CONDITIONS	73-1086
CONDITIONS	VIBRATION OF SQUARE PLATES HAVING VARIOUS EDGE CONDITIONS AND LOADINGS USING A REFINED FINITE ELEMEN	75-0024
CONDUCTANCE	FINITE CONDUCTANCE ELEMENT METHOD OF CONDUCTION HEAT TRANSFER	73-0963
CONDUCTION	A COMPUTER PROGRAM FOR THE ANALYSIS OF TWO-DIMENSIONAL HEAT CONDUCTION USING THE FINITE ELEMENT TECH	69-0066
CONDUCTION	A DERIVATION OF PERTURBED EQUATIONS OF MOTION AND HEAT CONDUCTION OF A THERMOELASTIC SOLID	72-0735
CONDUCTION	A DISCRETE LEAST-SQUARES FINITE ELEMENT METHOD FOR TRANSIENT HEAT CONDUCTION PROBLEMS	74-1191
CONDUCTION	A FINITE ELEMENT FORMULATION OF HEAT CONDUCTION AND HEAT AND MASS TRANSFER PROBLEMS	74-0350
CONDUCTION	A SURVEY OF HEAT CONDUCTION COMPUTER PROGRAMS	71-0017
CONDUCTION	ACCURACY, STABILITY AND OSCILLATION CHARACTERISTICS OF FINITE ELEMENT METHOD FOR SOLVING HEAT CONDUC	69-0007
CONDUCTION	ACCURACY, STABILITY, AND OSCILLATION CHARACTERISTICS OF TRANSIENT TWO-DIMENSIONAL HEAT CONDUCTION	75-0849
CONDUCTION	ANALYSIS OF AXISYMMETRIC LINEAR HEAT CONDUCTION PROBLEMS BY FINITE ELEMENT METHOD	69-0015
CONDUCTION	ANALYTIC METHODS IN CONDUCTION HEAT TRANSFER	71-0186
CONDUCTION	APACHE: A THREE-DIMENSIONAL FINITE ELEMENT PROGRAM FOR STEADY-STATE OR TRANSIENT HEAT CONDUCTION ANA	73-0971
CONDUCTION	APPLICATION OF CONFORMAL MAPPING AND VARIATIONAL METHOD TO THE STUDY OF HEAT CONDUCTION IN POLYGONAL	71-0444
CONDUCTION	APPLICATION OF FINITE ELEMENT METHOD TO THE PROBLEM OF HEAT CONDUCTION AND FLUID MECHANICS	72-0486
CONDUCTION	APPLICATION OF FINITE ELEMENTS TO HEAT CONDUCTION PROBLEMS INVOLVING LATENT HEAT	73-0520
CONDUCTION	APPLICATION OF LAPLACE TRANSFORM AND FINITE ELEMENT METHOD TO SOLUTION OF NONSTATIONARY THERMAL COND	74-0171
CONDUCTION	APPLICATION OF THE FINITE ELEMENT METHOD TO HEAT CONDUCTION ANALYSIS	66-0076
CONDUCTION	APPLICATION OF THE FINITE ELEMENT METHOD TO HEAT CONDUCTION IN SOLIDS	67-0005
CONDUCTION	APPROXIMATIONS IN FINITE-ELEMENT HEAT CONDUCTION ANALYSIS	74-1167
CONDUCTION	AYER HEAT CONDUCTION COMPUTER PROGRAM	74-0885
CONDUCTION	CALCULATION BY FINITE ELEMENT TECHNIQUE OF STEADY-STATE OR TRANSIENT HEAT-FLOW CONDUCTION	74-0637
CONDUCTION	CLASSICAL AND NATURAL CONDUCTION AND DEFORMATION MODES CONCEPTS FOR AXISYMMETRIC VARIATIONAL METHODS	72-1003
CONDUCTION	COMPARISON OF ANALYTICAL AND NUMERICAL SOLUTIONS IN HEAT CONDUCTION PROBLEMS	74-1170
CONDUCTION	COMPARISON OF THE FINITE ELEMENT AND FINITE DIFFERENCE METHODS FOR THE ANALYSIS OF STEADY TWO DIMENS	75-0223
CONDUCTION	COMPARISON OF TIME MARCHING SCHEMES FOR THE TRANSIENT HEAT CONDUCTION EQUATION	75-0477
CONDUCTION	CONDUCTION THROUGH DROPLETS DURING DROPWISE CONDENSATION	72-0019
CONDUCTION	COUPLE - A PSEUDO FINITE ELEMENT CONDUCTION CODE FOR THE TRANSIENT THERMAL RESPONE OF AXISYMMETRIC 3	72-0778
CONDUCTION	DIRECT AND INVERSE HEAT CONDUCTION IN ROCK MATERIALS USING THE FINITE ELEMENT METHOD	74-1098
CONDUCTION	DIRECT ITERATION METHOD FOR THE INCORPORATION OF PHASE CHANGE IN FINITE ELEMENT HEAT CONDUCTION PROG	74-0894
CONDUCTION	DISCRETE VARIABLE METHOD APPLIED TO TRANSIENT HEAT CONDUCTION PROBLEMS	74-1163
CONDUCTION	DUAL ANALYSIS FOR HEAT CONDUCTION PROBLEMS BY FINITE ELEMENTS	72-0401
CONDUCTION	EFFICIENT SOLUTION PROCESSES FOR FINITE ELEMENT ANALYSIS OF TRANSIENT HEAT CONDUCTION	69-0014
CONDUCTION	FINITE CONDUCTANCE ELEMENT METHOD OF CONDUCTION HEAT TRANSFER	73-0963
CONDUCTION	FINITE ELEMENT ANALYSIS OF HEAT CONDUCTION IN ANISOTROPIC SOLIDS	68-0161
CONDUCTION	FINITE ELEMENT ANALYSIS OF TRANSIENT HEAT CONDUCTION. APPLICATION OF THE WEIGHTED RESIDUAL PROCESS	74-0277
CONDUCTION	FINITE ELEMENT COMPUTER PROGRAMS FOR HEAT CONDUCTION PROBLEMS	73-0525
CONDUCTION	FINITE ELEMENT FORMULATION OF THE HEAT CONDUCTION EQUATION IN GENERAL ORTHOGONAL CURVILINEAR COORDIN	75-0848
CONDUCTION	FINITE ELEMENT METHOD APPLIED TO THE SOLUTION OF HEAT CONDUCTION PROBLEMS	69-0009
CONDUCTION	FINITE ELEMENT METHOD AND LAPLACE TRANSFORM COMPARATIVE SOLUTIONS OF TRANSIENT HEAT CONDUCTION PROBL	74-0575
CONDUCTION	FINITE ELEMENT METHODS IN HEAT CONDUCTION PROBLEMS	74-0095
CONDUCTION	FINITE ELEMENT SOLUTION OF NON-LINEAR HEAT CONDUCTION PROBLEMS WITH SPECIAL REFERENCE TO PHASE CHANG	74-0348
CONDUCTION	FINITE ELEMENT TECHNIQUE APPLIED TO HEAT CONDUCTION IN SOLIDS WITH TEMPERATURE DEPENDENT THERMAL CON	69-0004
CONDUCTION	FINITE-ELEMENT METHOD APPLIED TO HEAT CONDUCTION IN SOLIDS WITH NONLINEAR BOUNDARY CONDITIONS	73-0005
CONDUCTION	HEAT CONDUCTION AND STRESS ANALYSIS OF SOLID PROPELLANT ROCKET MOTOR NOZZLES	69-0059

CONDUCTION	HYBRIDE FINITE VERFAHREN FUER LINEARE UND NICHTLINEARE INSTATIONAERE WAERMELEITPROBLEME (HYBRID FINI	73-0484
CONDUCTION	NEW TRIANGULAR ELEMENT FOR FINITE DIFFERENCE SOLUTION OF AXISYMMETRIC CONDUCTION PROBLEMS IN CYLINDR	74-0381
CONDUCTION	ON THE ACCURACY OF FINITE ELEMENT SOLUTIONS TO THE TRANSIENT HEAT CONDUCTION EQUATION	74-0182
CONDUCTION	ON THE CONVERGENCE OF THE FINITE ELEMENT SOLUTION IN A NONLINEAR HEAT CONDUCTION PROBLEM	74-1272
CONDUCTION	ON THE PHYSICAL MEANINGS OF THE MATRICES DERIVED FROM FINITE ELEMENT METHOD OF HEAT CONDUCTION PROBL	73-0297
CONDUCTION	PERIODIC HEAT CONDUCTION IN A TWO PHASE, TWO DIMENSIONAL SOLID DOMAIN	74-1164
CONDUCTION	REFINED HEAT CONDUCTION FINITE ELEMENT FOR PERMAFROST ANALYSIS	73-0594
CONDUCTION	SEMI-ANALYTICAL FINITE ELEMENT PROCEDURE FOR CONDUCTION IN ANISOTROPIC AXISYMMETRIC SOLIDS	74-0270
CONDUCTION	STABILITY AND OSCILLATION CHARACTERISTICS OF FINITE ELEMENT, FINITE-DIFFERENCE, AND WEIGHTED-RESIDUA	75-0615
CONDUCTION	STEADY CONDUCTION OF HEAT IN LINEAR AND NONLINEAR FULLY ANISOTROPIC MEDIA BY FINITE ELEMENTS	74-0359
CONDUCTION	STEADY STATE AND TRANSIENT HEAT CONDUCTION AND STRESS ANALYSES OF TWO AND THREE DIMENSIONAL STRUCTUR	71-0675
CONDUCTION	THE APPLICATION OF HIGHER ORDER FINITE ELEMENTS TO THREE DIMENSIONAL HEAT CONDUCTION PROBLEMS	70-0266
CONDUCTION	THE FINITE ELEMENT METHOD IN DEFORMATION AND HEAT CONDUCTION PROBLEMS	68-0238
CONDUCTION	THE HEAT CONDUCTION AND THERMAL STRESS ANALYSIS BY THE FINITE ELEMENT METHOD	68-0073
CONDUCTION	TRANSIENT HEAT CONDUCTION PROBLEMS IN POWER CABLES SOLVED BY FINITE ELEMENT METHOD	73-0007
CONDUCTION	TRANSIENT TWO DIMENSIONAL HEAT CONDUCTION PROBLEMS SOLVED BY THE FINITE ELEMENT METHOD	74-0364
CONDUCTION	UNSYMMETRICAL TRANSIENT HEAT CONDUCTION: ROTATING DISK APPLICATIONS	70-0005
CONDUCTION	USE OF FINITE ELEMENT METHODS IN SOLUTION OF TRANSIENT HEAT CONDUCTION PROBLEMS	69-0016
CONDUCTIVE	COUPLED CONVECTIVE AND CONDUCTIVE HEAT TRANSFER BY FINITE ELEMENT METHODS	74-0050
CONDUCTIVE	COUPLED CONVECTIVE/CONDUCTIVE HEAT TRANSFER INCLUDING VELOCITY FIELD EVALUATION	75-0116
CONDUCTIVITY	APPLICATION OF CONFORMAL MAPPING AND VARIATIONAL METHOD TO THE STUDY OF HEAT CONDUCTION IN POLYGONAL	71-0444
CONDUCTIVITY	CHISLENNOE RESHENIE PLOSKIKH ZADACH TEPLOPROVODNOSTI DLYA OBLASTEI SLOZHNOI FORMY (NUMERICAL SOLUTIO	74-0659
CONDUCTIVITY	FINITE ELEMENT TECHNIQUE APPLIED TO HEAT CONDUCTION IN SOLIDS WITH TEMPERATURE DEPENDENT THERMAL CON	69-0004
CONDUCTIVITY	FINITE ELEMENTE ZUR BERECHNUNG DER WAERMEUEBERTRAGUNG FUER EBENE UND AXISYMMETRISCHE PROBLEME (FINIT	73-0449
CONDUCTIVITY	SOLVING A NONLINEAR PROBLEM OF NONSTEADY THERMAL CONDUCTIVITY IN SOLID BODIES OF ARBITRARY FORM	68-0019
CONDUITS	FINITE ELEMENT ANALYSIS OF THE STEADY FLOW OF NON NEWTONIAN FLUIDS THROUGH PARALLEL SIDED CONDUITS	74-0056
CONDUITS	FINITE ELEMENT SOLUTION FOR FLOW IN NONCIRCULAR CONDUITS	74-0116
CONDUITS	MOVEMENTS OF ARTICULATED CONDUITS UNDER EARTH DAMS ON COMPRESSIBLE FOUNDATIONS	73-0477
CONDUITS	THICK WALLED MULTIPLE OPENING REINFORCED CONCRETE CONDUITS	72-0666
CONE	BERECHNUNG DES KEGELDRUCKVERSUCHES MIT EINER FINITE-ELEMENT-METHODE. (CALCULATION OF THE CONE INDENT	75-0187
CONE	CALCULATION OF CONE INDENTATION TEST USING A FINITE-ELEMENT METHOD	75-0438
CONE	WAVE PROPAGATION IN A THIN HOLLOW CONE BY A FINITE ELEMENT METHOD	72-0045
CONE-INDENTATIO	FINITE-ELEMENT METHOD FOR CALCULATING THE CONE-INDENTATION TEST	75-0813
CONES	ELASTIC DEFLECTION IN THIN WALLED CONES ROTATING ABOUT THEIR AXES OF SYMMETRY	73-0545
CONES	FINITE ELEMENTS FOR THE VIBRATION OF CONES AND CYLINDERS	75-0340
CONFIGURATION	AN ALGORITHM FOR REDUCING THE BANDWIDTH OF A MATRIX OF SYMMETRICAL CONFIGURATION	65-0067
CONFIGURATION	MECHANICS OF CABLE MOORING SYSTEMS. VOLUME VI. A COMPUTER PROGRAM FOR ANALYZING THE STEADY STATE CON	73-0725
CONFIGURATION	NODAL ANALYSIS OF THE MATED SPACE SHUTTLE CONFIGURATION	71-0097
CONFIGURATION	OPTIMUM CONFIGURATION OF TRUSS STRUCTURES	74-1315
CONFIGURATION	PHOTOELASTIC AND FINITE ELEMENT INVESTIGATION OF A NONSYMMETRICAL PLUG HATCH CONFIGURATION	73-0025
CONFIGURATION	VIBRATION ANALYSIS OF A 1/15-SCALE DYNAMIC MODAL OF A SPACE SHUTTLE CONFIGURATION	71-0609
CONFIGURATIONAL	OPTIMUM CONFIGURATIONAL AND DIMENSIONAL DESIGN OF TRUSS STRUCTURES	74-0588
CONFIGURATIONS	A FINITE ELEMENT ANALYSIS FOR STEADY AND OSCILLATORY SUBSONIC FLOW AROUND COMPLEX CONFIGURATIONS	74-1317
CONFIGURATIONS	A FINITE-ELEMENT ANALYSIS FOR STEADY-AND-OSCILLATORY SUPERSONIC FLOWS AROUND COMPLEX CONFIGURATIONS	74-1316
CONFIGURATIONS	A FINITE-ELEMENT FORMULATION FOR SUBSONIC FLOWS AROUND COMPLEX CONFIGURATIONS	73-0742
CONFIGURATIONS	APPLICATION OF FINITE-ELEMENT THEORY TO AIRPLANE CONFIGURATIONS	71-0458
CONFIGURATIONS	COMPLIANCE OF A TAPERED DCB SPECIMEN - CONFIGURATIONS BY A FINITE ELEMENT METHOD	74-0545
CONFIGURATIONS	COMPUTATIONAL ANALYSIS AND EVALUATION OF THE FINITE ELEMENT METHOD FOR A CLASS OF NUCLEAR REACTOR CO	74-0595
CONFIGURATIONS	ELASTIC ANALYSIS OF CYLINDRICAL CONFIGURATIONS WITH STRESS SINGULARITIES	71-0325
CONFIGURATIONS	FINITE ELEMENT CALCULATIONS OF THE COMPLIANCE OF A TAPERED DOUBLE CANTILEVER BEAM SPECIMEN FOR DIFFE	72-0665
CONFIGURATIONS	OPTIMUM MESH CONFIGURATIONS IN THE FINITE ELEMENT METHOD	74-1346
CONFIGURATIONS	STRESS ANALYSIS OF CONFIGURATIONS INVOLVING SMALL FILLETS	75-0361
CONFIGURATIONS	SUBSONIC AND SUPERSONIC INDICIAL AERODYNAMICS AND AERODYNAMIC TRANSFER FUNCTION FOR COMPLEX CONFIGUR	74-1320
CONFINED	A FINITE ELEMENT SOLUTION OF THE CONFINED FLOW OVER A CIRCULAR CYLINDER	74-0023
CONFINED	NUMERICAL SOLUTION OF THE STEADY INTERFACE IN A CONFINED COASTAL AQUIFER	74-1007
CONFINED	SIGNIFICANCE OF VARIABLY CONFINED TRIAXIAL TESTING ROADBUILDING MATERIALS	74-0554
CONFINED	STRUCTURAL EFFECTS OF CONFINED DYNAMIC LOADS	72-0151
CONFINED	STUDY BY THE FINITE-ELEMENT METHOD OF THE INFLUENCE OF FRACTURES IN CONFINED AQUIFERS.	75-0604
CONFINED	THE NUMERICAL SOLUTION OF THE STEADY INTERFACE IN A COASTAL CONFINED AQUIFER	73-0711
CONFINED	THEORY OF FLOW IN A CONFINED TWO AQUIFER SYSTEM	69-0382
CONFORMAL	APPLICATION OF CONFORMAL MAPPING AND VARIATIONAL METHOD TO THE STUDY OF HEAT CONDUCTION IN POLYGONAL	71-0444
CONFORMING	A CONFORMING FINITE ELEMENT FOR PLATE BENDING	68-0071
CONFORMING	A CONFORMING QUADRILATERAL ELEMENT FOR BUCKLING ANALYSIS OF STIFFENED PLATES	73-0613
CONFORMING	A CONFORMING QUARTIC TRIANGULAR ELEMENT FOR PLATE BENDING	69-0137
CONFORMING	A CONFORMING TRIANGULAR FINITE ELEMENT PLATE BENDING SOLUTION	70-0020
CONFORMING	A CONFORMING TRIANGULAR ELEMENT IN THE ANALYSIS OF THIN AND SEMI-THICK PLATES IN BENDING	70-0308
CONFORMING	A HIGHER ORDER CONFORMING RECTANGULAR ELEMENT	73-0690
CONFORMING	A METHOD TO GENERATE BOTH UPPER AND LOWER BOUNDS TO PLATE EIGENVALUES BY CONFORMING DISPLACEMENT FIN	71-0166
CONFORMING	A STUDY OF THE CONVERGENCE CHARACTERISTICS AND ACCURACY OF CONFORMING FINITE ELEMENT SOLUTIONS	72-0967
CONFORMING	A TECHNIQUE FOR MESH GRADING APPLIED TO CONFORMING PLATE BENDING FINITE ELEMENTS	73-1058
CONFORMING	COMMENT ON A HIGHER ORDER CONFORMING RECTANGULAR PLATE ELEMENT	73-0999
CONFORMING	CONFORMING AND NONCONFORMING FINITE ELEMENT METHODS FOR SOLVING THE STATIONARY STOKES EQUATIONS - 1	73-0390
CONFORMING	CONFORMING AND NONCONFORMING FINITE ELEMENT METHODS FOR SOLVING THE PLATE PROBLEM	73-0621
CONFORMING	CONFORMING FINITE ELEMENT WITH INTERNAL EQUILIBRIUM FOR TWO DIMENSIONAL PROBLEMS	72-0386
CONFORMING	CONFORMING FINITE ELEMENTS BASED ON COMPLETE POLYNOMIALS	74-0302
CONFORMING	CONFORMING FINITE ELEMENT METHOD FOR THE SHELL PROBLEM	75-0096
CONFORMING	CONFORMING FINITE ELEMENT METHODS FOR THE CLAMPED PLATE PROBLEM	75-0174
CONFORMING	CONFORMING RECTANGULAR AND TRIANGULAR PLATE BENDING ELEMENTS	71-0720
CONFORMING	CONVERGENCE OF CONFORMING FINITE-ELEMENT METHODS FOR SHELL PROBLEMS	75-0494
CONFORMING	CONVERGENCE OF EIGENVALUE SOLUTIONS IN CONFORMING PLATE BENDING FINITE ELEMENTS	72-0195
CONFORMING	ERROR ESTIMATES FOR CONFORMING FINITE ELEMENT SOLUTIONS	74-0940
CONFORMING	KINDS OF CONVERGENCE AND IMPROVED CONVERGENCE OF CONFORMING FINITE ELEMENT SOLUTIONS IN PLATE BENDIN	70-0323
CONFORMING	ON THE CONFORMING CUBIC TRIANGULAR ELEMENT FOR PLATE BENDING	72-0439
CONFORMING	SOME CONFORMING, RECTANGULAR, PLATE ELEMENTS	75-0094
CONFORMING	THEORETICAL AND NUMERICAL CONVERGENCE RATES FOR SOME CONFORMING FINITE ELEMENTS	74-0941
CONFORMING	TRIANGULAR ELEMENTS IN PLATE BENDING CONFORMING AND NON-CONFORMING SOLUTIONS	66-0011
CONGRUENTLY	COMPUTER INTEGRATION OF CONGRUENTLY TRANSFORMED MATRICES	71-0580
CONICAL	A CONICAL ELEMENT FOR DISPLACEMENT ANALYSIS OF AXI-SYMMETRIC SHELLS	69-0120
CONICAL	ASYMMETRICAL BENDING AND VIBRATION OF A CONICAL SHELL FINITE ELEMENT	74-0085
CONICAL	AXISYMMETRIC VIBRATION OF CONICAL SHELLS	70-0350
CONICAL	CLOSURE ANALYSIS USING SELF ENERGIZING, DOUBLE CONICAL GASKETS	74-0675
CONICAL	CONICAL FINITE ELEMENT STRESS ANALYSIS OF AN AXISYMMETRIC TANK ON AN ELASTIC FOUNDATION	70-0494
CONICAL	CONICAL SEGMENT METHOD FOR ANALYZING OPEN CROWN SHELLS OF REVOLUTION FOR EDGE LOADING	63-0007
CONICAL	CONICAL SHELL FINITE ELEMENT	74-0593
CONICAL	DYNAMIC STABILITY OF CONICAL SHELLS BY FINITE ELEMENTS	74-0467
CONICAL	ELASTO-PLASTIC ANALYSIS OF A ROTATING MODEL CONICAL TURBINE DISC	75-0281
CONICAL	MODAL SENSITIVITY STUDY OF A CONICAL RE-ENTRY VEHICLE SUBJECTED TO AERODYNAMICALLY INDUCED ACOUSTIC	69-0431
CONICAL	PART 1 - TASK 3.2.9. BENDING AND BUCKLING OF A ONE-DIMENSIONAL MODEL SIMULATING A CONICAL SHELL. PAR	71-0745
CONICAL	STRESS ANALYSIS OF A CONICAL ACRYLIC VIEWPORT	70-0106
CONICAL	THE EFFECT OF AN INITIAL AXISYMMETRIC IMPERFECTION ON THE NATURAL FREQUENCIES OF CONICAL SHELLS	70-0368
CONIDA	CONIDA: A FINITE ELEMENT PROGRAM FOR THE STRESS ANALYSIS OF AXISYMMETRIC THIN SHELLS	74-1087
CONJUGATE	A DISCUSSION ON A MODIFIED CONJUGATE GRADIENT METHOD BY W.L. WOOD	74-0402

CONJUGATE	CONJUGATE APPROXIMATION FUNCTIONS IN FINITE ELEMENT ANALYSIS	71-0043
CONJUGATE	DYNAMIC STABILITY ANALYSIS BY THE CONJUGATE GRADIENT METHOD	72-0749
CONJUGATE	GENERALIZED CONJUGATE FUNCTIONS FOR MIXED FINITE ELEMENT APPROXIMATIONS OF BOUNDARY VALUE PROBLEMS	72-0515
CONJUGATE	MIXED CONJUGATE FINITE ELEMENT APPROXIMATIONS OF LINEAR OPERATORS	72-0085
CONJUGATE	NOTE ON A MODIFIED CONJUGATE GRADIENT METHOD	73-0615
CONJUGATE	NOTE ON AN APPROXIMATE METHOD FOR COMPUTING CONSISTENT CONJUGATE STRESSES IN ELASTIC FINITE ELEMENTS	73-0050
CONJUGATE	THE CONJUGATE NEWTON METHOD	73-0560
CONJUGATE	THE SOLUTION OF STRUCTURAL EQUILIBRIUM EQUATIONS BY THE CONJUGATE GRADIENT METHOD WITH PARTICULAR RE	71-0268
CONJUGATE	THEORY OF CONJUGATE PROJECTIONS IN FINITE ELEMENT ANALYSIS	71-0194
CONJUGATED-GRAD	APPLICATION OF A CONJUGATED-GRADIENT-PROCESS IN FINITE-ELEMENT-CALCULATION OF NON-LINEAR STRUCTURES	74-0214
CONJUNCTION	SETTLEMENT ANALYSIS OF A STRIP FOOTING USING A CRITICAL STATE SOIL MODEL IN CONJUNCTION WITH FINITE	71-0533
CONNECTED	DYNAMIC BEHAVIOR OF IN-LINE SHEAR WALLS CONNECTED BY FLOOR SLABS	72-0689
CONNECTED	INCOMPRESSIBLE FLOW IN MULTIPLY CONNECTED REGIONS	74-0904
CONNECTED	STRESS ANALYSIS OF FRAME STRUCTURE CONNECTED WITH PLATES	74-0265
CONNECTED	THE PIECEWISE LINEAR ANALYSIS OF TWO CONNECTED STRUCTURES INCLUDING THE EFFECT OF CLEARANCE AT THE C	71-0379
CONNECTING	A TECHNIQUE FOR CONNECTING BEAM ELEMENTS TO A PLATE MODEL OF A COMPLICATED BOX SECTION	74-0271
CONNECTING	EFFECT OF BEAM-SHEARWALL JUNCTION DEFORMATIONS ON THE FLEXIBILITY OF THE CONNECTING BEAMS	73-0213
CONNECTING-ROD	FINITE-ELEMENT STRESS ANALYSIS OF A NON-LINEAR PROBLEM. A CONNECTING-ROD EYE LOADED BY MEANS OF A PI	70-0239
CONNECTION	CONNECTION BETWEEN THE FINITE DIFFERENCE LIKE METHODS AND THE METHODS BASED ON INITIAL VALUE PROBLEM	74-0877
CONNECTION	RELATIONSHIP BETWEEN FINITE ELEMENT AND MEASURED JOINT-FLOOR CONNECTION RIGIDITIES	75-0738
CONNECTIONS	EFFECTS OF FLEXIBLE CONNECTIONS ON BODY STRUCTURAL RESPONSE	74-0140
CONNECTIONS	FINITE ELEMENT ANALYSIS OF BONDED CONNECTIONS	73-1096
CONNECTIONS	FINITE ELEMENT ANALYSIS OF THREADED CONNECTIONS	74-1179
CONNECTIONS	INADEQUACY OF NODAL CONNECTIONS IN A STIFFNESS SOLUTION FOR PLATE BENDING	65-0021
CONNECTIONS	ON THE STRESS DISTRIBUTION IN NOZZLE TO CYLINDER CONNECTIONS FOR SMALL DIAMETER RATIOS	72-0426
CONNECTIONS	THE FATIGUE STRENGTH OF FILLET WELDED CONNECTIONS	71-0684
CONNECTIONS	THE PIECEWISE LINEAR ANALYSIS OF TWO CONNECTED STRUCTURES INCLUDING THE EFFECT OF CLEARANCE AT THE C	71-0379
CONPLOT	CONPLOT: A GENERAL PURPOSE CONTOUR PLOTTING PROGRAM	74-0887
CONSERVATION	AN APPLICATION OF LEAST SQUARES FINITE ELEMENT METHODS TO TWO DIMENSIONAL STEADY INVISCID COMPRESSIB	75-0114
CONSERVATION	FINITE ELEMENT SOLUTION FOR ENERGY CONSERVATION USING A HIGHLY STABLE EXPLICIT INTEGRATION ALGORITHM	72-0633
CONSIDERING	ANALYSIS OF STIFFENED PLATES CONSIDERING NONLINEAR MATERIAL AND GEOMETRIC BEHAVIOUR	75-0327
CONSIDERING	FINITE ELEMENT ANALYSIS OF 2-DIMENSIONAL FLOW IN SOILS CONSIDERING WATER UPTAKE BY ROOTS .1. THEORY	75-0137
CONSIDERING	FINITE ELEMENT ANALYSIS OF 2-DIMENSIONAL FLOW IN SOILS CONSIDERING WATER UPTAKE BY ROOTS .2. FIELD A	75-0147
CONSIDERING	NON-LINEAR ANALYSIS OF PLATES CONSIDERING GEOMETRIC AND MATERIAL EFFECTS	71-0702
CONSIDERING	ZUR BERECHNUNG EINFACH-SYMMETRISCHER I-TRAEGER NACH DER THEORIE - 2. ORDNUNG UNTER BERUECKSICHTIGUNG	75-0510
CONSISTENCY	EFFECT OF THE VARIATION OF CONSISTENCY IN RESIDUAL SOILS ON LOAD TRANSFER BY BORED PILES	75-0812
CONSISTENT	A COMPARATIVE STUDY OF THE CONSISTENT AND SIMPLIFIED FINITE ELEMENTS ANALYSIS OF EIGENVALUE PROBLEMS	70-0034
CONSISTENT	A COMPARATIVE STUDY OF THE CONSISTENT AND SIMPLIFIED FINITE ELEMENT ANALYSES OF EIGENVALUE PROBLEMS	70-0373
CONSISTENT	A CONSISTENT DISCRETE ELEMENT TECHNIQUE FOR THIN-WALLED ASSEMBLAGES	68-0104
CONSISTENT	A CONSISTENT FINITE ELEMENT METHOD FOR RANDOM RESPONSE PROBLEMS	71-0202
CONSISTENT	A CONSISTENT FINITE ELEMENT METHOD FOR RANDOM RESPONSE PROBLEMS	72-0896
CONSISTENT	A CONSISTENT FINITE ELEMENT TECHNIQUE FOR WORKSOFTENING BEHAVIOUR	74-1233
CONSISTENT	COMPARISONS OF CONSISTENT MASS MATRIX SCHEMES	69-0437
CONSISTENT	CONSISTENT FINITE ELEMENT MODEL FOR TWO DIMENSIONAL CONTINUUM	70-0361
CONSISTENT	CONSISTENT MASS MATRIX FOR DISTRIBUTED SYSTEMS	63-0001
CONSISTENT	CONSISTENT MATRIX FORMULATIONS FOR STRUCTURAL ANALYSIS USING INFLUENCE COEFFICIENT TECHNIQUES	64-0017
CONSISTENT	CONSISTENT STIFFNESS MATRICES IN THE ANALYSIS OF SHELLS	67-0078
CONSISTENT	CONSISTENT STRESSES FOR THE FINITE ELEMENT STIFFNESS METHOD	72-0698
CONSISTENT	CONTINUUM BASES FOR CONSISTENT NUMERICAL FORMULATIONS OF FINITE STRAINS IN ELASTIC AND INELASTIC STR	75-0299
CONSISTENT	DEVELOPMENT AND APPLICATIONS OF SUPERSONIC UNSTEADY CONSISTENT AERODYNAMICS FOR INTERFERING PARALLEL	72-0716
CONSISTENT	FURTHER DEVELOPMENTS IN CONSISTENT UNSTEADY SUPERSONIC AERODYNAMIC COEFFICIENTS	72-0559
CONSISTENT	KINEMATICALLY CONSISTENT UNSTEADY AERODYNAMIC COEFFICIENTS IN SUPERSONIC FLOW	70-0014
CONSISTENT	KINEMATICALLY CONSISTENT UNSTEADY AERODYNAMIC COEFFICIENTS IN SUPERSONIC FLOW	70-0316
CONSISTENT	MODE SHAPES AND FREQUENCIES BY FINITE ELEMENT METHOD USING CONSISTENT AND LUMPED MASSES	71-0510
CONSISTENT	NOTE ON AN APPROXIMATE METHOD FOR COMPUTING CONSISTENT CONJUGATE STRESSES IN ELASTIC FINITE ELEMENTS	73-0050
CONSISTENT	ON APPLICATION OF THE FINITE ELEMENT METHOD TO LIMIT ANALYSIS CONSISTENT FINITE ELEMENT TECHNIQUE FO	74-1234
CONSISTENT	ON THE CALCULATION OF CONSISTENT STRESS DISTRIBUTIONS IN FINITE ELEMENT APPROXIMATIONS	71-0191
CONSISTENT	PREDICTION OF PLATE VIBRATIONS USING A CONSISTENT MASS MATRIX	66-0043
CONSISTENTLY	CONVERGENCE OF CONSISTENTLY DERIVED TIMOSHENKO BEAM FINITE ELEMENTS	72-0396
CONSOLIDATION	A FINITE-ELEMENT PROGRAM FOR THE ANALYSIS OF EMBANKMENTS OVER SOFT, SATURATED SOILS INCLUDING CONSOL	73-1160
CONSOLIDATION	A NUMERICAL METHOD FOR THE SOLUTION OF BIOT'S CONSOLIDATION THEORY	72-0688
CONSOLIDATION	AN INVESTIGATION OF THE STABILITY OF NUMERICAL SOLUTIONS OF BIOT'S EQUATIONS OF CONSOLIDATION	74-0746
CONSOLIDATION	ANALYSIS OF PLANE STRAIN CONSOLIDATION IN SATURATED CLAYS BY FINITE ELEMENT METHOD	71-0815
CONSOLIDATION	APPLICATION OF THE FINITE ELEMENT METHOD TO CONSOLIDATION PROBLEMS	72-0421
CONSOLIDATION	CONSOLIDATION OF A LAYER UNDER A STRIP LOAD	72-0222
CONSOLIDATION	DIE BERECHNUNG VON KONSOLIDATIONSVORGAENGEN MIT DER FINITE-ELEMENT-METHODE. (CALCULATION OF CONSOLID	71-0012
CONSOLIDATION	EVALUATION OF SOME NUMERICAL SCHEMES FOR CONSOLIDATION	73-0349
CONSOLIDATION	EVALUATION OF TWO FINITE ELEMENT FORMULATIONS FOR ONE-DIMENSIONAL CONSOLIDATION	72-0453
CONSOLIDATION	FINITE ELEMENT ANALYSIS OF CONSOLIDATION IN CREEP	72-0420
CONSOLIDATION	FINITE ELEMENT ANALYSIS OF CONSOLIDATION PROBLEM	74-0081
CONSOLIDATION	FINITE ELEMENT ANALYSIS OF CONSOLIDATION OF UNSATUPATED SOIL	74-0082
CONSOLIDATION	FINITE ELEMENT METHOD APPLIED TO BIOT'S CONSOLIDATION THEORY	71-0518
CONSOLIDATION	METOD KONECHNYKH ELEMENTOV V RASCHETAKH KONSOLIDATSII VODONASYSHCHENNYKH GRUNTOV. (FINITE ELEMENT ME	75-0469
CONSOLIDATION	ON SOLUTION OF PLANE STRAIN CONSOLIDATION PROBLEMS BY FINITE ELEMENT METHOD	71-0532
CONSOLIDATION	PLANE STRAIN CONSOLIDATION BY FINITE ELEMENTS	69-0471
CONSOLIDATION	PLANE STRAIN CONSOLIDATION BY FINITE ELEMENTS	70-0043
CONSOLIDATION	SOME NUMERICAL PROCEDURES FOR ANALYSIS OF ONE DIMENSIONAL CONSOLIDATION	72-0928
CONSOLIDATION	THE EFFECTS OF SOIL PARAMETERS AND BOUNDARY CONDITIONS ON THE CONSOLIDATION OF AN ELASTIC LAYER	70-0422
CONSTANT	A COMPARATIVE STUDY OF VARIABLE AND CONSTANT THICKNESS HIGH PRECISION TRIANGULAR PLATE BENDING ELEME	74-0857
CONSTANT	A SIX NODE TRIANGULAR EQUILIBRIUM (CONSTANT STRESS) MEMBRANE ELEMENT	73-0783
CONSTANT	ANALYSIS OF THICK-WALLED PRESSURE VESSELS USING THE CONSTANT STRAIN AXISYMMETRIC FINITE ELEMENT	71-0653
CONSTANT	CALCULATION OF PLATE OF CONSTANT THICKNESS BY FINITE ELEMENT METHOD	69-0227
CONSTANT	CONSTANT CURVATURE BEAM FINITE ELEMENTS FOR IN-PLANE VIBRATION	72-0254
CONSTANT	ELASTIC-PLASTIC PLATE BENDING WITH CONSTANT CURVATURE ELEMENTS	74-0308
CONSTANT	ELEMENT STIFFNESS MATRICES FOR CONSTANT AND LINEAR STRESS - TRIANGULAR, CONSTANT AND LINEAR STRESS T	67-0006
CONSTANT	ELEMENT STIFFNESS MATRICES FOR CONSTANT AND LINEAR STRESS - TRIANGULAR, CONSTANT AND LINEAR STRESS T	67-0006
CONSTANT	EXPERIENCE IN THE USE OF THE CONSTANT STRESS TRIANGLE FOR FINITE ELEMENT STRESS ANALYSIS	69-0407
CONSTANT	NOTCH EFFECT IN LOW CYCLE FATIGUE - REPORT 1: LOW-CYCLE FATIGUE STRENGTH OF 0.48 PERCENT CARBON STEE	73-1090
CONSTANT	ON SHELLS OF CONSTANT STRENGTH	72-0431
CONSTANT	ON THE UNITY OF THE CONSTANT STRAIN-CONSTANT MOMENT FINITE ELEMENT METHODS	73-0164
CONSTANT	THE CONSTANT BENDING MOMENT PLATE BENDING ELEMENT	71-0182
CONSTANT	TRIANGULAR FINITE ELEMENTS FOR PLATE BENDING WITH CONSTANT AND LINEARLY VARYING BENDING MOMENTS	70-0019
CONSTANT	VARIATIONAL SOLUTIONS BY USE OF STEPWISE CONSTANT FUNCTIONS: ONE LINEAR CASE	72-0286
CONSTANT	ZUR BERECHNUNG, BEMESSUNG UND BEWEHRUNG RECHTECKIGER PLATTEN KONSTNATER DICKE MIT OEFFNUNGEN ODER AU	73-0664
CONSTANT-STIFFN	THE ALPHA CONSTANT-STIFFNESS METHOD FOR THE ANALYSIS OF NONLINEAR PROBLEMS	72-0566
CONSTANTES	DETERMINATION OF THE TECHNICAL CONSTANTS OF LAMINATES IN OBLIQUE DIRECTIONS. THEORY AND EXPERIMENTAL	73-0748
CONSTANTS	DETERMINATION OF THE TECHNICAL CONSTANTS OF LAMINATES IN OBLIQUE DIRECTIONS. THEORY AND EXPERIMENTAL	73-0748
CONSTANTS	IDENTIFICATION OF AQUIFER CONSTANTS: TESTS OF TWO INVERSE METHODS	74-0051
CONSTANTS	PREDICTION OF ELASTIC CONSTANTS OF PARTICLEBOARD	74-0259
CONSTITUENT	FINITE ELEMENT ANALYSIS OF COMPOSITE MATERIAL BEHAVIOURS FROM MECHANICAL PROPERTIES OF CONSTITUENT M	75-0181
CONSTITUENTS	MATHEMATICAL MODELING OF MOVEMENT OF DISSOLVED CONSTITUENTS IN GROUND WATER AQUIFERS BY THE FINITE E	70-0490
CONSTITUTENT	FINITE ELEMENT ANALYSIS OF COMPOSITE MATERIAL BEHAVIORS FROM MECHANICAL PROPERTIES OF CONSTITUENT M	75-0874

CONSTITUTIVE	ANALYSIS OF REAL STRUCTURES FOR CREEP, PLASTICITY AND OTHER COMPLEX CONSTITUTIVE LAWS.	69-0287
CONSTITUTIVE	CONSTITUTIVE EQUATIONS AND PUNCH-INDENTATION OF CONCRETE	75-0286
CONSTITUTIVE	CONSTITUTIVE MATRIX AND FINITE ELEMENT FORMULATION FOR A DEFORMATION THEORY OF PLASTICITY	71-0700
CONSTITUTIVE	CONSTITUTIVE RELATIONS OF CONCRETE AND PUNCH-INDENTATION PROBLEMS	73-0877
CONSTITUTIVE	CONSTITUTIVE RELATIONS FOR CONCRETE	75-0541
CONSTITUTIVE	ELASTO-PLASTIC STRESS ANALYSIS. A GENERALIZATION FOR VARIOUS CONSTITUTIVE RELATIONS INCLUDING STRAIN	72-0547
CONSTITUTIVE	IMPROVED APPROXIMATION OF CONSTITUTIVE ELASTO-PLASTIC STRESS STRAIN RELATIONSHIP FOR FINITE ELEMENT	74-0566
CONSTITUTIVE	NOTE ON INCREMENTAL EQUILIBRIUM EQUATIONS AND APPROXIMATE CONSTITUTIVE RELATIONS IN LARGE INELASTIC	74-0984
CONSTITUTIVE	REINFORCED CONCRETE ON CONSTITUTIVE RELATIONS	75-0753
CONSTITUTIVE	THERMO-PLASTIC CONSTITUTIVE RELATIONS AND A VARIATIONAL SOLUTION TECHNIQUE USING FINITE ELEMENTS	70-0486
CONSTITUTTIVE	DEVELOPMENT OF A SOLID ROCKET PROPELLANT NONLINEAR CONSTITUTTIVE THEORY	75-0722
CONSTRAINED	BENDING PROBLEM OF AXIALLY CONSTRAINED BEAMS	73-0397
CONSTRAINED	CONSTRAINED AND SHEAR MODULI FOR FINITE ELEMENTS	74-0774
CONSTRAINED	CONSTRAINED AND SHEAR MODULI FOR FINITE ELEMENTS	75-0228
CONSTRAINED	CONSTRAINED VARIATIONAL PRINCIPLES AND PENALTY FUNCTION METHODS IN FINITE ELEMENT ANALYSIS	72-0551
CONSTRAINED	DYNAMIC RESPONSE OF A CONSTRAINED FIBROUS SYSTEM SUBJECTED TO TRANSVERSE IMPACT. II. A MECHANICAL MO	70-0369
CONSTRAINED	EXACT SOLUTION TO THE RADIALLY END CONSTRAINED CIRCULAR CYLINDER UNDER TRIAXIAL LOADING	71-0433
CONSTRAINED	FORCE AND TORQUE ANALYSIS OF CONSTRAINED SPACE MECHANISMS AND PLANE MECHANISMS WITH OFFSET LINKS BY	74-0569
CONSTRAINED	MODAL RESPONSE ANALYSIS OF CONSTRAINED FINITE-ELEMENTS ELASTIC CONTINUUM	75-0767
CONSTRAINED	NONLINEAR MECHANICS OF CONSTRAINED MATERIAL CONTINUA: FOUNDATIONS OF THE THEORY	74-0640
CONSTRAINED	TRANSIENT RESPONSE OF INELASTICALLY CONSTRAINED RIGID BODY SYSTEMS	73-0291
CONSTRAINT	COMPUTER IMPLEMENTATION OF THE CONSTRAINT METHOD	75-0001
CONSTRAINT	CONSTRAINT TECHNIQUE IN FINITE ELEMENT ANALYSIS	70-0491
CONSTRAINT	CONVERGENCE OF THE FINITE-ELEMENT METHOD UNDER THE CONSTRAINT OF MATERIAL INCOMPRESSIBILITY	73-1135
CONSTRAINT	CURVED ROTATIONAL SHELL ELEMENTS BY THE CONSTRAINT METHOD	72-0924
CONSTRAINT	FLEXIBILITY AND STIFFNESS MATRICES FOR AN OPEN-TUBE WARPING CONSTRAINT FINITE ELEMENT	72-0076
CONSTRAINT	MINIMUM WEIGHT DESIGN WITH STABILITY CONSTRAINT	70-0274
CONSTRAINT	MINIMUM WEIGHT DESIGN OF COMPLEX STRUCTURES SUBJECT TO A FREQUENCY CONSTRAINT	70-0374
CONSTRAINT	ON CONSTRAINT EQUATIONS - A NEW APPROACH	74-1017
CONSTRAINT	ON THE NUMERICAL CONVERGENCE OF MATRIX EIGENVALUE PROBLEMS DUE TO CONSTRAINT CONDITIONS	74-1057
CONSTRAINT	OPTIMIZATION OF A SUPERSONIC PANEL SUBJECT TO A FLUTTER CONSTRAINT. A FINITE ELEMENT SOLUTION	71-0073
CONSTRAINT	THE CONSTRAINT METHOD - A NEW FINITE ELEMENT TECHNIQUE	73-1027
CONSTRAINT	THE CONSTRAINT METHOD FOR FINITE ELEMENT STRESS ANALYSIS	75-0432
CONSTRAINTS	APPLICATION OF GENERALIZED CONSTRAINTS IN THE STIFFNESS METHOD OF STRUCTURAL ANALYSIS	66-0030
CONSTRAINTS	AUTOMATED DESIGN OPTIMIZATION OF SUPERSONIC AIRPLANE WING STRUCTURES UNDER DYNAMIC CONSTRAINTS	72-0178
CONSTRAINTS	FRAME OPTIMIZATION INCLUDING FREQUENCY CONSTRAINTS	75-0082
CONSTRAINTS	LINEAR EQUALITY CONSTRAINTS IN FINITE ELEMENT APPROXIMATION	75-0475
CONSTRAINTS	OPTIMAL DESIGN OF STRUCTURES WITH BUCKLING CONSTRAINTS	73-0246
CONSTRAINTS	OPTIMUM DESIGN OF COMPOSITE SHELLS SUBJECT TO NATURAL FREQUENCY CONSTRAINTS	73-0154
CONSTRICTED	STRUCTURAL RESPONSE OF RELAXED AND CONSTRICTED ARTERIOLES	68-0200
CONSTRUCT	PROBLEM OF USING THE METHOD OF FINITE ELEMENTS TO CONSTRUCT AN ALGORITHM FOR CALCULATING A THIN-WALL	74-0851
CONSTRUCTED	PERFORMANCE OF AN EMBANKMENT CONSTRUCTED ON VARVED CLAY	73-0743
CONSTRUCTED	PHOTOELASTIC AND FINITE ELEMENT ANALYSIS OF EMBANKMENTS CONSTRUCTED OVER SOFT SOILS	70-0151
CONSTRUCTION	ANALYSIS OF EMBANKMENT CONSTRUCTION BY FINITE-ELEMENTS	72-0972
CONSTRUCTION	ANALYTICAL PROBLEMS IN MODELING SLURRY WALL CONSTRUCTION	73-0764
CONSTRUCTION	APPLIED FINITE ELEMENT DISPLACEMENT ANALYSIS OF A SIMPLE INTEGRAL CONSTRUCTION VEHICLE BODY SHELL	72-0264
CONSTRUCTION	ARCH DAM CONSTRUCTION STRESSES DUE TO HYDROSTATIC AND GRAVITY LOADS	75-0658
CONSTRUCTION	BERECHNUNGSBEISPIELE UND ERFAHRUNGEN MIT DER METHODE DER FINITEN ELFMENTE IN MASCHINENBAU. (EXAMPLES	71-0390
CONSTRUCTION	CALCULATION OF THE REINFORCED THIN-WALLED CONSTRUCTION BY THE FINITE ELEMENT	75-0756
CONSTRUCTION	COMPUTER AIDED DESIGN AND CONSTRUCTION BY MEANS OF CURVED SURFACE AND SOLID ELEMENTS	73-0165
CONSTRUCTION	COMPUTER PROGRAM CONSTRUCTION AND MAINTENANCE - FUTURE OF CENTRALIZED FINITE ELEMENT ACTIVITY	75-0014
CONSTRUCTION	CONSTRUCTION OF A FINITE ELEMENT APPROXIMATION WHICH CROSSES MATERIAL INTERFACES	74-0896
CONSTRUCTION	CONSTRUCTION OF CURVILINEAR COORDINATE SYSTEMS AND APPLICATIONS TO MESH GENERATION	74-1295
CONSTRUCTION	CONSTRUCTION OF NORMAL DISPLACEMENT FUNCTIONS OF TRIANGULAR FINITE ELEMENT OF PLATES AND SHELLS	72-0563
CONSTRUCTION	CONSTRUCTION SELF-WEIGHT STRESSES OF PRECAST SHEAR WALLS	71-0203
CONSTRUCTION	CONSTRUCTION STRESSES IN DWORSHAK DAM	65-0029
CONSTRUCTION	DESIGN APPRAISAL OF OFFSHORE PLATFORMS OF WELDED TUBULAR CONSTRUCTION	74-1165
CONSTRUCTION	DESIGN, ANALYSIS AND CONSTRUCTION OF THE RESEARCH PLATFORM (NORDSEE)	75-0261
CONSTRUCTION	ENVIRONMENTAL CONSIDERATIONS OF DAM CONSTRUCTION AND OPERATION IN SEISMICALLY ACTIVE URBAN AREAS	73-0470
CONSTRUCTION	FINITE ELEMENTS METHOD IN DESIGN OF PRESTRESSED REACTOR CONCRETE CONSTRUCTION	73-0550
CONSTRUCTION	GEOMETRIC ASPECTS OF THE FINITE ELEMENT METHOD: CONSTRUCTION OF CURVILINEAR COORDINATE SYSTEMS AND T	72-0534
CONSTRUCTION	HIGH-PERFORMANCE COMPOSITE MATERIALS FOR VEHICLE CONSTRUCTION: A FINITE ELEMENT COMPUTER PROGRAM FOR	73-0767
CONSTRUCTION	INFLUENCE OF CONSTRUCTION PROCEDURES ON GROUND PRESSURES AROUND A TUNNEL	74-0997
CONSTRUCTION	LARGE STRUCTURES FOR MANNED SPACECRAFT: MATHEMATICAL ANALYSIS, DESIGN, CONSTRUCTION AND TESTS	74-1327
CONSTRUCTION	OBSERVED AND PREDICTED DEFORMATIONS IN A LARGE EMBANKMENT DAM DURING CONSTRUCTION	71-0301
CONSTRUCTION	ON THE CONSTRUCTION AND USE OF FINITE ELEMENTS	70-0265
CONSTRUCTION	ON THE CONSTRUCTION AND CONVERGENCE OF A FINITE ELEMENT SOLUTION OF LAPLACE'S EQUATION	72-0340
CONSTRUCTION	PROBLEMS OF LONGITUDINAL STRENGTH IN THE CONSTRUCTION OF VERY LARGE TANK TRUCKS	73-0251
CONSTRUCTION	SIMPLE FINITE ELEMENT SOLUTION FOR PLATES OF HOMOGENEOUS, SANDWICH AND CELLULAR CONSTRUCTION	75-0140
CONSTRUCTION	STRESS ANALYSIS FOR STRUCTURES UNDER CONSTRUCTION	69-0209
CONSTRUCTION	THE CONSTRUCTION OF BASIC FUNCTIONS FOR CURVED ELEMENTS IN THE FINITE ELEMENT METHOD	72-0558
CONSTRUCTION	TOUR MAINE-MONTPARNASSE QUELQUES PROBLEMES DE STRUCTURE (SPECIFIC PROBLEMS ENCOUNTERED IN THE CONSTR	74-0539
CONSTRUCTION	TUNNEL CONSTRUCTION - A BIBLIOGRAPHY WITH ABSTRACTS	74-0748
CONSTRUCTION	VYUZITI METODY KONECNYCH PRVKU PRO PROJEKTOVANI KOMBINOVANYCH KONSTRUCKCI (FINITE ELEMENT METHOD IDE	73-0667
CONSTRUCTIONAL	CONSTRUCTIONAL DEFORMATIONS IN ROCKFILL DAM	73-0281
CONSTRUCTIONS	METHODS OF FINITE ELEMENTS IN ANALYSES OF CONTINUOUS AND COMBINED CONSTRUCTIONS	74-0789
CONSTRUKTION	COMPUTERUNTERSTUETZTES SYSTEM FUER DIE CONSTRUKTION DES WERKZEUGMASCHINENAUFBAUS (COMPUTER-AIDED SYS	73-0462
CONSUMPTION	NUMERICAL SOLUTION OF A DIFFUSION CONSUMPTION PROBLEM WITH A FREE BOUNDARY	74-0835
CONTACT	A FINITE ELEMENT METHOD OF VISCOELASTIC STRESS ANALYSES WITH APPLICATION TO ROLLING CONTACT PROBLEMS	69-0330
CONTACT	A METHOD FOR DETERMINING THE SURFACE CONTACT STRESSES RESULTING FROM INTERFERENCE FITS	70-0101
CONTACT	A NOTE ON NUMERICAL COMPUTATION OF ELASTIC CONTACT PROBLEMS	75-0621
CONTACT	A TRIANGULAR PLATE BENDING ELEMENT FOR CONTACT PROBLEMS	73-0693
CONTACT	ANALYSIS OF A CIRCULAR RAFT IN ADHESIVE CONTACT WITH A THICK ELASTIC LAYER	74-1152
CONTACT	ANALYSIS OF CONTACT PROBLEMS BY THE MATRIX METHOD	69-0388
CONTACT	ANALYSIS OF NONLINEAR CONTACT PROBLEMS BY THE FINITE-ELEMENT METHOD	72-0952
CONTACT	ANALYSIS OF VISCOELASTIC CONTACT STRESSES IN AGRICULTURAL PRODUCTS USING A FINITE ELEMENT METHOD	74-0634
CONTACT	ANALYSIS OF VISCOELASTIC CONTACT STRESSES IN AGRICULTURAL PRODUCTS USING A FINITE ELEMENT METHOD	74-1199
CONTACT	AREAS OF CONTACT AND PRESSURE DISTRIBUTION IN BOLTED JOINTS	71-0329
CONTACT	CONTACT OF AN INFLATED TOROIDAL MEMBRANE WITH A FLAT SURFACE AS AN APPROACH TO THE TIRE DEFLECTION P	75-0164
CONTACT	CONTRIBUTION TO THE SOLUTION OF CONTACT PROBLEMS WITH THE AID OF BOND ELEMENTS	75-0321
CONTACT	DIRECT FINITE ELEMENT ANALYSIS OF ELASTIC CONTACT PROBLEMS	72-0733
CONTACT	ELASTIC SETTLEMENT OF A CIRCULAR RAFT IN ADHESIVE CONTACT WITH A TRANSVERSELY ISOTROPIC MEDIUM	75-0828
CONTACT	FIELD INSTRUMENTATION FOR LONG TERM MEASUREMENT OF PILE LOAD AND RAFT CONTACT PRESSURE	73-0150
CONTACT	FINITE ELEMENT ANALYSIS OF ELASTIC CONTACT PROBLEMS USING DIFFERENTIAL DISPLACEMENTS	70-0126
CONTACT	FINITE ELEMENT ANALYSIS OF ELASTIC CONTACT PROBLEM	72-0550
CONTACT	FINITE ELEMENT ANALYSIS OF ELASTIC CONTACT PROBLEMS	73-0219
CONTACT	FINITE ELEMENT ANALYSIS OF CONTACT PROBLEMS	73-0312
CONTACT	FINITE ELEMENT APPROACH TO THE CONTACT PROBLEM OF A PLATE ON THE ELASTIC HALF SPACE	72-0830
CONTACT	FINITE ELEMENT COMPUTER PROGRAM FOR THE SOLUTION OF NONLINEAR AXISYMMETRIC CONTACT PROBLEMS WITH INT	74-1094
CONTACT	FINITE ELEMENT METHOD FOR CONTACT PROBLEMS OF SOLID BODIES - 2.	71-0435
CONTACT	MAGNETOTELLURIC RESPONSE OF A 2-DIMENSIONAL SLOPING CONTACT BY FINITE ELEMENT METHOD	73-0130
CONTACT	NONLINEAR ANALYSIS OF A ROLLING CONTACT PROBLEM BY FINITE ELEMENTS	74-1216

165

CONTACT	PARAMETRIC ANALYSIS OF CONTACT CHATTER IN RELAYS	74-0487
CONTACT	PRESSURE DISTRIBUTION AND DEFORMATIONS OF MACHINED COMPONENTS IN CONTACT.	73-0372
CONTACT	THERMAL, THERMOELASTIC, AND WEAR SIMULATION OF A HIGH-ENERGY SLIDING CONTACT PROBLEM	73-0430
CONTACT	UNBONDED CONTACT PROBLEM OF A PLATE ON THE ELASTIC HALF SPACE	74-0208
CONTACT	ZUR LOESUNG DES KONTAKTPROBLEMS ELASTISCHER KOERPER MIT AUSGEDEHNTER KONTAKTFLAECHE DURCH QUADRATISC	74-1135
CONTACT	ZUR LOESUNG DES KONTAKTPROBLEMS ELASTISCHER KOERPER MIT AUSGEDEHNTER KONTAKTFLAECHE DURCH QUADRATISC	74-1135
CONTACT-IMPACT	FINITE ELEMENT FORMULATION AND SOLUTION OF CONTACT-IMPACT PROBLEMS IN CONTINUUM MECHANICS	74-0806
CONTACT-IMPACT	FINITE ELEMENT FORMULATION AND SOLUTION OF CONTACT-IMPACT PROBLEMS IN CONTINUUM MECHANICS-II	75-0740
CONTAINED	DYNAMICS OF A FLEXIBLE BULKHEAD AND CONTAINED FLUID	73-0082
CONTAINER	A TORSIONAL STRENGTH ANALYSIS ON THE CONTAINER SHIP BY MEANS OF THE FINITE ELEMENT PROCEDURE AND FUL	71-0243
CONTAINER	FINITE ELEMENT ANALYSIS OF CONTAINER SHIPS	72-0372
CONTAINER	FINITE ELEMENT ANALYSIS OF FLUID MOTION IN A CONTAINER	74-0079
CONTAINER	LIQUID SLOSHING IN AN ELASTIC CONTAINER	66-0096
CONTAINER	NUMERICAL HULL VIBRATION ANALYSIS OF A FAR EAST CONTAINER SHIP	74-1325
CONTAINERS	CALCULATION OF THE FREE OSCILLATIONS OF A LIQUID IN AXISYMMETRIC MOTIONLESS CONTAINERS OF ARBITRARY	74-0387
CONTAINERSHIP	STRUCTURAL ANALYSIS OF SL-7 CONTAINERSHIP UNDER COMBINED LOADING OF VERTICAL, LATERAL AND TORSIONAL	74-0734
CONTAINING	APPLICATION OF FINITE ELEMENT METHOD TO CYCLIC LOADING OF ELASTIC-PLASTIC STRUCTURES CONTAINING EFFE	73-0103
CONTAINING	BEAM BENDING FINITE ELEMENT CONTAINING INTEGRATION PARAMETERS	71-0179
CONTAINING	FATIGUE CRACK FORMATION AND PROPAGATION IN PAVEMENTS CONTAINING SOIL-CEMENT BASES.	72-0435
CONTAINING	FATIQUE STRENGTH OF LUGS CONTAINING LINERS. VOLUME II. COMPUTER PROGRAM USED FOR ANALYSIS	70-0418
CONTAINING	FINITE ELEMENT ANALYSIS OF CYLINDRICAL SHELLS UTILIZING CURVED ELEMENTS CONTAINING RIGID BODY DEFORM	71-0823
CONTAINING	FINITE ELEMENT ANALYSIS OF PERFORATED PLATES CONTAINING TRIANGULAR PENETRATION PATTERNS OF 5 AND 10	75-0216
CONTAINING	FINITE ELEMENT ANALYSIS OF PERFORATED PLATES CONTAINING TRIANGULAR PENETRATION PATTERNS OF 5 AND 10	75-0520
CONTAINING	FINITE ELEMENT METHODS FOR ELLIPTIC BOUNDARY VALUE PROBLEMS CONTAINING SINGULARITIES	72-0360
CONTAINING	FINITE ELEMENT METHODS FOR ELASTIC BODIES CONTAINING CRACKS	73-1029
CONTAINING	MECHANICAL BEHAVIOR OF A GRAPHITE-EPOXY LAMINATE CONTAINING A HOLE	72-0614
CONTAINING	NUMERICAL SOLUTION OF STEADY STATE DIFFUSION PROBLEMS CONTAINING SINGULARITIES	74-0036
CONTAINING	SIMULATION ON THE TENSILE BEHAVIOUR OF TWO-PHASE ALLOY CONTAINING A FIBER UNDER TENSILE LOAD	73-0542
CONTAINING	STRESS ANALYSIS OF LAMINATED CFRP PLATES CONTAINING A HOLE	74-1175
CONTAINING	STRESS AND DISPLACEMENTS AROUND OPENINGS IN ROCK CONTAINING AN INELASTIC DISCONTINUITY	75-0170
CONTAINING	STRESS INTENSITY FACTORS FOR CRACKS IN A PLATE CONTAINING A HOLE AND IN A SPINNING DISC	73-0467
CONTAINING	STRESSES AND DISPLACEMENTS AROUND OPENINGS IN ROCK CONTAINING AN ELASTIC DISCONTINUITY	75-0169
CONTAINING	THE VIBRATION CHARACTERISTICS OF A TENSIONED PLATE CONTAINING A FATIGUE CRACK	68-0134
CONTAINING	THREE-DIMENSIONAL STRESS ANALYSIS OF A FINITE SLAB CONTAINING A TRANSVERSE CENTRAL CRACK	75-0732
CONTAINMENT	A FINITE ELEMENT STRESS ANALYSIS OF THE VERTICAL BUTTRESSES OF A NUCLEAR CONTAINMENT VESSEL	69-0280
CONTAINMENT	APPLICATION OF THE COLLISION-IMPARTED VELOCITY METHOD FOR ANALYZING THE RESPONSES OF CONTAINMENT AND	73-0770
CONTAINMENT	FINITE ELEMENT ANALYSIS OF HYPERBOLOIDAL PRESTRESSED CONCRETE CONTAINMENT VESSELS	69-0526
CONTAINMENT	FINITE ELEMENT ANALYSIS OF CONTAINMENT VESSELS	71-0080
CONTAINMENT	FINITE ELEMENT ANALYSIS OF LARGE TRANSIENT ELASTIC-PLASTIC DEFORMATIONS OF SIMPLE STRUCTURES, WITH A	72-0849
CONTAINMENT	REFINED CRACKED CONCRETE ANALYSIS OF CONCRETE CONTAINMENT STRUCTURES SUBJECT TO OPERATIONAL AND ENVI	74-0943
CONTAINMENT	STATIC AND HEAT TRANSFER ANALYSIS OF A SPHERE-CONE INTERSECTION IN A NUCLEAR CONTAINMENT VESSEL	75-0166
CONTAINMENT	STRUCTURAL ANALYSIS OF AIRCRAFT IMPACT ON A NUCLEAR CONTAINMENT VESSEL AND ASSOCIATED STRUCTURES	69-0279
CONTAINMENT	VERIFICATION OF THE PEC CONTAINMENT VESSEL FOR SOME OPERATING CONDITIONS BY USE OF THE FINITE ELEMEN	74-0883
CONTAINMENT	VISCOELASTIC ANALYSIS OF P C CONTAINMENT USING FINITE ELEMENT METHOD	73-0034
CONTAMINATION	ELECTRO-OSMOSIS AND THE CONTAMINATION OF GROUND FLUIDS	73-0609
CONTAMINATION	GALERKIN-FINITE ELEMENT SIMULATION OF GROUNDWATER CONTAMINATION ON LONG-ISLAND, NEW-YORK	73-0093
CONTAMINATION	SIMULATION OF GROUND WATER CONTAMINATION USING A GALERKIN FINITE ELEMENT TECHNIQUE	74-1362
CONTENT	THE USE OF THE FINITE ELEMENT METHOD IN OBTAINING SOIL-WATER PRESSURE AND SOIL-WATER CONTENT DISTRIB	75-0112
CONTINENTAL	RAYLEIGH WAVES AT A CONTINENTAL BOUNDARY BY FINITE ELEMENT METHOD	72-0016
CONTINUA	A NEW ASSUMED STRESS HYBRID FINITE ELEMENT MODEL FOR SOLID CONTINUA	71-0371
CONTINUA	A NEW FINITE ELEMENT MODEL FOR SOLID CONTINUA	68-0156
CONTINUA	A STUDY OF THE PLANE STRESS OR STRAIN FINITE ELEMENT ANALYSIS FOR SOLUTION OF STRESS DISTRIBUTION IN	69-0487
CONTINUA	ANALYSIS OF CONTINUA BY THE FINITE ELEMENT AND INTEGRAL EQUATION METHOD	75-0370
CONTINUA	AVERAGE ELASTIC PROPERTIES AND STRESS FIELDS FOR COMPOSITE CONTINUA	69-0106
CONTINUA	BASIS OF FINITE ELEMENT METHODS FOR SOLID CONTINUA	69-0213
CONTINUA	CONTINUA AND DISCONTINUA	66-0008
CONTINUA	ELASTO-PLASTIC MATRIX DISPLACEMENT ANALYSIS OF THREE-DIMENSIONAL CONTINUA	66-0004
CONTINUA	FINITE ELEMENT APPROACH TO WAVE MOTIONS IN ELASTIC-PLASTIC CONTINUA	68-0041
CONTINUA	FINITE ELEMENT FORMULATION FOR PIEZOELECTRIC CONTINUA BY A VARIATIONAL THEOREM	68-0230
CONTINUA	FINITE ELEMENT FORMULATION OF FINITE DEFORMATIONS AND IRREVERSIBLE THERMODYNAMICS OF NONLINEAR CONTI	69-0207
CONTINUA	FINITE ELEMENT FORMULATION OF PROBLEMS OF FINITE DEFORMATION AND IRREVERSIBLE THERMODYNAMICS OF NONL	71-0190
CONTINUA	FINITE ELEMENTS FOR COMPRESSIBLE AND INCOMPRESSIBLE CONTINUA	69-0133
CONTINUA	FINITE ELEMENTS IN NONLINEAR CONTINUA	71-0200
CONTINUA	FORMULATIONS OF FINITE ELEMENT METHODS FOR SOLID CONTINUA	71-0679
CONTINUA	FOUNDATIONS OF FINITE ELEMENT METHODS FOR SOLID CONTINUA	71-0213
CONTINUA	GENERAL MIXED FINITE ELEMENT METHODS OF NONLINEAR CONTINUA	72-0831
CONTINUA	GENERAL MIXED FINITE ELEMENT METHODS OF NONLINEAR CONTINUA	73-0091
CONTINUA	NEW DISPLACEMENT HYBRID FINITE ELEMENT MODELS FOR SOLID CONTINUA	70-0115
CONTINUA	NONLINEAR MECHANICS OF CONSTRAINED MATERIAL CONTINUA: FOUNDATIONS OF THE THEORY	74-0640
CONTINUA	RATE OF CONVERGENCE FOR THE ITERATIVE APPROACH IN ELASTIC-PLASTIC ANALYSIS OF CONTINUA	73-0518
CONTINUA	STIFFNESS MATRIX IN PLASTIC-ELASTIC PROBLEMS OF CONTINUA	67-0115
CONTINUA	THE APPLICATION OF FINITE ELEMENT METHODS TO THE SOLUTION OF STRESS AND DIFFUSION PROBLEMS IN CONTIN	68-0034
CONTINUA	THE ELASTO-PLASTIC CALCULATION OF GENERAL STRUCTURES AND CONTINUA	68-0207
CONTINUA	THE OPTIMUM APPROACH TO ANALYSIS OF ELASTIC CONTINUA	71-0177
CONTINUA	VARIATIONAL FORMULATIONS OF NUMERICAL METHODS IN SOLID CONTINUA	71-0670
CONTINUA	VARIATIONAL PRINCIPLE FOR THE DYNAMIC ANALYSIS OF CONTINUA BY HYBRID FINITE ELEMENT METHOD	71-0494
CONTINUA	VARIATIONAL PRINCIPLES FOR ELASTOPLASTIC CONTINUA	75-0837
CONTINUITIES	EFFECTIVENESS OF EXCESSIVE NODAL CONTINUITIES IN THE FINITE ELEMENT ANALYSIS OF THIN RECTANGULAR AND	70-0346
CONTINUITIES	THE EFFECTIVENESS OF EXCESSIVE NODAL CONTINUITIES IN THE FINITE ELEMENT ANALYSIS OF THIN RECTANGULAR	70-0105
CONTINUITY	ASAS - A COMMERCIAL FINITE ELEMENT SYSTEM WITH DISPLACEMENT AND FORCE CONTINUITY OPTIONS	73-0947
CONTINUITY	FINITE ELEMENT FORMULATION BY VARIATIONAL PRINCIPLES WITH RELAXED CONTINUITY REQUIREMENTS	72-0516
CONTINUITY	FINITE ELEMENT METHODS BY VARIATIONAL PRINCIPLES WITH RELAXED CONTINUITY REQUIREMENTS	72-0342
CONTINUITY	FINITE ELEMENT METHOD WITH CONSIDERATION TO THE CONTINUITY OF NORMAL SLOPE	72-0752
CONTINUITY	FINITE ELEMENT METHOD WITH CONSIDERATION TO THE CONTINUITY OF NORMAL SLOPE	72-0813
CONTINUITY	FINITE ELEMENTS WITH RELAXED CONTINUITY - AN INVESTIGATION OF THE VARIATIONAL BASIS AND NUMERICAL PE	74-1065
CONTINUITY	FUNKCJA KSZTALTU NIESPELNIAPACA WARUNKOW CIAGLOSCI W ANALIZIE PLYT METODA ELEMENTOW SKONCZONYCH. (SH	75-0263
CONTINUITY	VARIATIONAL PRINCIPLES OF ELASTOSTATICS WITH RELAXED CONTINUITY REQUIREMENTS AND THEIR APPLICABILITY	74-0758
CONTINUOUS	A FINITE ELEMENT APPROACH TO THE DYNAMIC ANALYSIS OF CONTINUOUS HIGHWAY BRIDGES	70-0591
CONTINUOUS	A FINITE ELEMENT TECHNIQUE FOR THE IDEALIZATION OF CONTINUOUS SHELL STRUCTURES	68-0290
CONTINUOUS	APPLICATIONS OF CONTINUOUS TONE COMPUTER-GENERATED IMAGES IN STRUCTURAL MECHANICS	74-1035
CONTINUOUS	COMPARISON OF EXTRAPOLATION TECHNIQUES WITH HIGH ORDER FINITE ELEMENT METHODS FOR DIFFUSION EQUATION	73-0586
CONTINUOUS	CONTINUOUS AND DISCONTINUOUS FINITE ELEMENT METHODS FOR SOLVING THE TRANSPORT EQUATION	75-0099
CONTINUOUS	CONTINUOUS CONCEPT OF A SPACE GRID	73-0755
CONTINUOUS	FREE VIBRATION OF ORTHOTROPIC VARIABLE THICKNESS, CONTINUOUS RECTANGULAR PLATES	70-0270
CONTINUOUS	FREQUENCY ANALYSIS OF CERTAIN SINGLE AND CONTINUOUS SPAN BRIDGES	71-0337
CONTINUOUS	METHODS OF FINITE ELEMENTS IN ANALYSES OF CONTINUOUS AND COMBINED CONSTRUCTIONS	74-0789
CONTINUOUS	ON THE ACCURACY OF CERTAIN C DEGREE CONTINUOUS FINITE ELEMENT REPRESENTATIONS	74-0367
CONTINUOUS	PIECE-WISE CONTINUOUS FINITE ELEMENT FOR THE STRESS ANALYSIS OF PLATES AND SHELLS	75-0758
CONTINUOUS	PRINCIPAL STRESSES AT THE INTERMEDIATE SUPPORT OF PRESTRESSED CONCRETE CONTINUOUS BEAMS	70-0352
CONTINUOUS	QUADRATIC MATRIX EQUATIONS FOR DETERMINING VIBRATION MODES AND FREQUENCIES OF CONTINUOUS ELASTIC SYS	66-0064
CONTINUOUS	REVIEW OF BASIC THEORY AND METHODS OF ANALYSIS OF CONTINUOUS STRUCTURES	73-0697
CONTINUOUS	STRESSES IN CONTINUOUS CONCRETE BOX GIRDER BRIDGES	71-0233

CONTINUOUS	TRANSVERSE REINFORCEMENT OF CONTINUOUS, TROUGH-SHAPED CONCRETE RAILROAD STRUCTURES, AN IMPLEMENTATIO	71-0719
CONTINUOUSLY	PAVEMENT DESIGN AND CONTINUOUSLY REINFORCED CONCRETE PAVEMENT PERFORMANCE	73-0749
CONTINUOUSLY	PRODUCTION OF MOTION PICTURES OF A CONTINUOUSLY DEFORMING STRUCTURE USING NASTRAN	71-0223
CONTINUOUSLY	TRANSIENT THERMAL STRESS ANALYSIS OF COMPOSITE STRUCTURES INCLUDING CONTINUOUSLY VARYING PROPERTIES	75-0681
CONTINUUA	FINITE ELEMENT PERTURBATION ANALYSIS OF NONLINEAR DYNAMIC RESPONSE OF ELASTIC CONTINUUA	74-0452
CONTINUUM	A COMPUTER BASED PROCEDURE FOR PREDICTING THE TRANSIENT RESPONSE AND FAILURE OF A TWO-DIMENSIONAL CO	71-0630
CONTINUUM	A HIGHLY STABLE EXPLICIT INTEGRATION TECHNIQUE FOR COMPUTATIONAL CONTINUUM MECHANICS	74-1277
CONTINUUM	A MATHEMATICAL THEORY OF COMPLEMENTARY-DUAL VARIATIONAL PRINCIPLES AND MIXED FINITE ELEMENT APPROXIM	73-0804
CONTINUUM	A SURVEY OF FINITE-ELEMENT METHODS IN CONTINUUM MECHANICS	72-0751
CONTINUUM	ANALYSIS OF CONTINUUM BY THE FINITE ELEMENT METHOD	69-0352
CONTINUUM	APPLICATION OF FINITE ELEMENT METHOD TO CONTINUUM MECHANICS PROBLEMS	71-0051
CONTINUUM	CONSISTENT FINITE ELEMENT MODEL FOR TWO DIMENSIONAL CONTINUUM	70-0361
CONTINUUM	CONTINUUM AND FINITE ELEMENT BRANCHING STUDIES OF THE CIRCULAR PLATE.	72-0427
CONTINUUM	CONTINUUM APPROACH TOWARD DYNAMICS OF GRIDWORKS	73-0174
CONTINUUM	CONTINUUM BASES FOR CONSISTENT NUMERICAL FORMULATIONS OF FINITE STRAINS IN ELASTIC AND INELASTIC STR	75-0299
CONTINUUM	CONTINUUM CALCULATION FROM ELASTIC PLASTIC REINFORCED MATERIAL USING A FINITE ELEMENT METHOD (DIE BE	72-0658
CONTINUUM	CONTINUUM CHARACTERIZATION OF JOINTED ROCK MASSES - 1, 2	73-0196
CONTINUUM	CONTINUUM MECHANICS AS AN APPROACH TO ROCK MASS PROBLEMS	68-0172
CONTINUUM	CONTINUUM MECHANICS - A BRIEF REVIEW	71-0407
CONTINUUM	FINITE ELEMENT ANALYSIS OF CONTINUUM PROBLEMS AND ITS APPLICATION TO THE INCOMPRESSIBLE LUBRICATION	71-0539
CONTINUUM	FINITE ELEMENT BASED PROCEDURE FOR SIMULATING THE TRANSIENT RESPONSE AND FAILURE OF A TWO-DIMENSIONA	72-0123
CONTINUUM	FINITE ELEMENT ELASTIC PLASTIC CREEP ANALYSIS OF TWO DIMENSIONAL CONTINUUM WITH TEMPERATURE DEPENDEN	73-0226
CONTINUUM	FINITE ELEMENT FORMULATION AND SOLUTION OF CONTACT-IMPACT PROBLEMS IN CONTINUUM MECHANICS	74-0806
CONTINUUM	FINITE ELEMENT FORMULATION AND SOLUTION OF CONTACT-IMPACT PROBLEMS IN CONTINUUM MECHANICS-II	75-0740
CONTINUUM	FINITE ELEMENT METHODS IN CONTINUUM MECHANICS.	72-0377
CONTINUUM	FINITE ELEMENTS ON THE BASIS OF CONTINUUM MECHANICS	75-0234
CONTINUUM	LECTURES ON FINITE ELEMENT METHODS IN CONTINUUM MECHANICS	71-0714
CONTINUUM	MODAL RESPONSE ANALYSIS OF CONSTRAINED FINITE-ELEMENTS ELASTIC CONTINUUM	75-0767
CONTINUUM	NUMERICAL SOLUTION TO THE DYNAMICS OF VISCOUS FLUID FLOW BY A FINITE ELEMENT ALGORITHM: A FIRST STEP	71-0004
CONTINUUM	NUMERICAL SOLUTION TO THE DYNAMICS OF VISCOUS FLUID FLOW BY A FINITE ELEMENT ALGORITHM: A FIRST STEP	71-0791
CONTINUUM	ON SOME ASPECTS OF CONTINUUM MECHANICS	73-0626
CONTINUUM	RESPONSE OF STRUCTURE TO THE PSEUDO-SOUND FIELD OF A JET (USING A COMBINED CONTINUUM AND FINITE ELEM	67-0146
CONTINUUM	REVIEW OF CONTINUUM, FINITE ELEMENT AND HYBRID TECHNIQUES IN THE ANALYSIS OF STRESS CONCENTRATIONS I	73-1051
CONTINUUM	REVIEW OF CONTINUUM, FINITE ELEMENT AND HYBRID TECHNIQUES IN ANALYSIS OF STRESS CONCENTRATIONS IN ST	74-0914
CONTINUUM	SOIL-STRUCTURE INTERACTION - CONTINUUM OR FINITE ELEMENT	74-0954
CONTINUUM	TEST RESULTS FROM THE BOUNDARY LAYER FACILITY - RESPONSE OF STRUCTURE TO THE PSEUDO-SOUND FIELD OF A	68-0262
CONTINUUM	THE ACCURACY OF THE FINITE ELEMENT METHODS IN CONTINUUM PROBLEMS	66-0027
CONTINUUM	THE FINITE ELEMENT METHOD IN STRUCTURAL AND CONTINUUM MECHANICS	67-0093
CONTOUR	ALGORITHMS FOR THE PRODUCTION OF CONTOUR MAPS OVER AN IRREGULAR TRIANGULAR MESH	72-0825
CONTOUR	CALCULATION OF STRESS INTENSITY FACTORS IN ANISOTROPIC MATERIALS BY A CONTOUR INTEGRAL METHOD	75-0257
CONTOUR	CONPLOT: A GENERAL PURPOSE CONTOUR PLOTTING PROGRAM	74-0887
CONTOUR	MESH AND CONTOUR PLOT FOR TRIANGLE AND ISOPARAMETRIC ELEMENTS	70-0445
CONTOUR	MESH GENERATION CODE AND A CONTOUR PLOTTING ROUTINE	72-0802
CONTOUR	THE NASTRAN CONTOUR PLOTTER	72-0703
CONTRACTIONS	A FINITE ELEMENT MODEL FOR TWO-DIMENSIONAL STEADY FLOW THROUGH CONTRACTIONS IN NATURAL CHANNELS	71-0810
CONTRACTIVE	STRESS ANALYSIS OF CONTRACTIVE OR DILATIVE SOIL	71-0579
CONTRAINTES	ANALYSE DES CONTRAINTES DAN DES STRUCTURES COMPLEXES EN PRESENCE DE FLUAGE, DE VARIATIONS DE DIMENSI	71-0081
CONTRAINTES	DETERMINATION DU CHAMP DES CONTRAINTES DANS LE CAISSON DU REACTEUR DE BUGEY L'	67-0121
CONTRAINTES	NONLINEAR ANALYSIS OF PRESTRESSED CONCRETE FAST REACTOR PRESSURE VESSEL, (ANALYSE NON LINEAIRE DES C	70-0145
CONTRAINTES	TENDANCES ACTUELLES DES RECHERCHES SUR LES CONTRAINTES ET DEFORMATIONS DUES AU BRIDAGE EN RELATION A	74-0273
CONTROL	ANALYSIS OF FEEDBACK CONTROL STABILIZED VIBRATION BY THE FINITE ELEMENT METHOD	74-1188
CONTROL	BOUNDS AND ERROR CONTROL FOR EIGENVALUES	73-0319
CONTROL	CYCLIC FATIGUE ANALYSIS OF ROCKET THRUST CHAMBERS. VOLUME 2: ATTITUDE CONTROL THRUSTER HIGH CYCLE FA	74-0693
CONTROL	FRACTURE CONTROL: PAST, PRESENT AND FUTURE	73-0516
CONTROL	METHODS FOR PREDICTING THE AERODYNAMIC AND STABILITY AND CONTROL CHARACTERISTICS OF STOL AIRCRAFT. V	73-0718
CONTROL	METHODS FOR PREDICTING THE AERODYNAMIC AND STABILITY AND CONTROL CHARACTERISTICS OF STOL AIRCRAFT. V	73-0719
CONTROL	METHODS OF PREDICTING THE AERODYNAMIC AND STABILITY AND CONTROL CHARACTERISTICS OF STOL AIRCRAFT. VO	73-0717
CONTROL	STABILITY CONTROL ANALYSIS OR ROTATING PLATES BY FINITE ELEMENT. EMPHASIS ON SLOTS AND HOLES	71-0284
CONTROL	UNSTEADY SUBSONIC COLLOCATION METHOD FOR WINGS WITH AND WITHOUT CONTROL SURFACES	72-0012
CONTROLLED	CONTROLLED FAILURE OF HOLLOW ROCK CYLINDERS IN UNIAXIAL COMPRESSION	72-0196
CONTROLLED	PROBLEM CONTROLLED GRID GENERATION FOR THE NUMERICAL SOLUTION OF PARTIAL DIFFERENTIAL EQUATIONS	75-0725
CONTROLLED	THREE-DIMENSIONAL MECHANICAL STRESSES IN TOROIDAL MAGNETS FOR CONTROLLED THERMONUCLEAR REACTORS	73-0669
CONTROLLING	CONTROLLING OPEN-PIT SLOPE FAILURES AT SHIRLEY BASIN	73-0149
CONVAIR	NASTRAN EXPERIENCES OF FORT WORTH OPERATION, CONVAIR AEROSPACE DIVISION OF GENERAL DYNAMICS	71-0019
CONVECTED	CONVECTED COORDINATES FOR TRANSIENT NONLINEAR FINITE-ELEMENT ANALYSIS	73-0850
CONVECTED	NON-LINEAR TRANSIENT FINITE ELEMENT ANALYSIS WITH CONVECTED CO-ORDINATES	73-0378
CONVECTED	NONLINEAR TRANSIENT ANALYSIS OF SHELLS AND SOLIDS OF REVOLUTION BY CONVECTED ELEMENTS	73-0519
CONVECTED	NONLINEAR TRANSIENT ANALYSIS OF SHELLS AND SOLIDS OF REVOLUTION BY CONVECTED ELEMENTS	74-0476
CONVECTION	A FINITE ELEMENT SOLUTION OF GENERAL FLUID DYNAMICS PROBLEMS - NATURAL CONVECTION IN RECTANGULAR CAV	70-0008
CONVECTION	A FINITE ELEMENT SOLUTION OF THE ONE DIMENSIONAL DIFFUSION CONVECTION EQUATION	70-0258
CONVECTION	ACCURACY AND CONVERGENCE OF FINITE ELEMENT GALERKIN APPROXIMATIONS OF TIME DEPENDENT PROBLEMS WITH E	73-0987
CONVECTION	APPLICATION OF THE FINITE ELEMENT METHOD TO CONVECTION HEAT TRANSFER BETWEEN PARALLEL PLANES	71-0244
CONVECTION	APPLICATION OF THE FINITE-ELEMENT METHOD TO HEAT-TRANSFER PROBLEMS. PART 11. TRANSIENT TWO-DIMENSION	71-0759
CONVECTION	FINITE ELEMENT METHOD APPLIED TO TRANSIENT TWO-DIMENSIONAL HEAT TRANSFER WITH CONVECTION AND RADIATI	70-0127
CONVECTION	MATHEMATICAL FEATURES OF FINITE ELEMENT APPROXIMATIONS OF CERTAIN FLOW PROBLEMS WITH EMPHASIS ON CON	74-0002
CONVECTION	NOTE ON THE ACCURACY AND CONVERGENCE OF FINITE ELEMENT APPROXIMATIONS OF THE CONVECTION EQUATION	75-0490
CONVECTION	NUMERICAL TECHNIQUES FOR CONVECTION/DIFFUSION PROBLEMS	75-0123
CONVECTION	SOLUTION OF A TWO DIMENSIONAL DIFFUSION CONVECTION EQUATION DESCRIBING THE MASS TRANSFER OF A NEUTRA	74-0037
CONVECTION	TRANSIENT NATURAL CONVECTION COOLING OF A VERTICAL CIRCULAR CYLINDER	74-1169
CONVECTIVE	A NUMERICAL MODEL FOR THE SIMULATION OF TWO DIMENSIONAL CONVECTIVE DISPERSION IN SHALLOW ESTUARIES	74-0052
CONVECTIVE	CONVECTIVE TRANSPORT FINITE ELEMENT ANALOG	75-0289
CONVECTIVE	COUPLED CONVECTIVE AND CONDUCTIVE HEAT TRANSFER BY FINITE ELEMENT METHODS	74-0050
CONVECTIVE	COUPLED CONVECTIVE/CONDUCTIVE HEAT TRANSFER INCLUDING VELOCITY FIELD EVALUATION	75-0116
CONVECTIVE	FINITE ELEMENT ANALYSIS OF LAMINAR CONVECTIVE HEAT TRANSFER IN VERTICAL DUCTS WITH ARBITRARY CROSS-S	75-0555
CONVECTIVE	ON THE STUDY OF CONVECTIVE DISPERSION EQUATION	74-0005
CONVECTIVE	TEMPERATURVERLAUF IM QUERSCHNITT GESCHWEISSTER ROHRWAENDE BEI UEBERWIEGEND KONVEKTIVEM WAERMEUEBERGA	75-0628
CONVECTIVE-DISP	A STUDY OF CONVECTIVE-DISPERSION EQUATION BY ISOPARAMETRIC FINITE ELEMENTS, TRANSPORT OF POLLUTANTS	75-0773
CONVECTIVE-DISP	STUDY OF CONVECTIVE-DISPERSION EQUATION BY ISOPARAMETRIC FINITE ELEMENTS	75-0074
CONVERGENCE	A CLASS OF FUNCTIONALS GIVING IMPROVED CONVERGENCE WITH FINITE ELEMENTS	72-0553
CONVERGENCE	A CONVERGENCE INVESTIGATION OF THE DIRECT STIFFNESS METHOD	66-0102
CONVERGENCE	A NOTE ON ACCURACY AND CONVERGENCE OF FINITE ELEMENT APPROXIMATIONS	71-0565
CONVERGENCE	A NOTE ON CONVERGENCE OF FINITE ELEMENT SOLUTIONS	71-0252
CONVERGENCE	A STUDY OF THE CONVERGENCE CHARACTERISTICS AND ACCURACY OF CONFORMING FINITE ELEMENT SOLUTIONS	72-0967
CONVERGENCE	A TEST ON CONVERGENCE OF EIGENSOLUTIONS BY SIMULTANEOUS ITERATION METHOD	73-0317
CONVERGENCE	A VARIATIONAL PRINCIPLE AND THE CONVERGENCE OF A FINITE-ELEMENT METHOD BASED ON ASSUMED STRESS DISTR	69-0260
CONVERGENCE	ACCURACY AND CONVERGENCE OF FINITE ELEMENT APPROXIMATION	68-0228
CONVERGENCE	ACCURACY AND CONVERGENCE OF FINITE ELEMENT GALERKIN APPROXIMATIONS OF TIME DEPENDENT PROBLEMS WITH E	73-0987
CONVERGENCE	ACCURACY AND CONVERGENCE OF MIXED FINITE-ELEMENT APPROXIMATIONS OF THIN BARS, MEMBRANES, AND PLATES	73-1120
CONVERGENCE	ACCURACY AND CONVERGENCE OF MIXED FINITE ELEMENT/GALERKIN APPROXIMATIONS OF TIME DEPENDENT PROBLEMS WITH E	75-0006
CONVERGENCE	AN EVALUATION OF FINITE ELEMENTS WITH IMPROVED ASSURED CONVERGENCE FOR PLANE STRESS AND PLANE STRAIN	74-0429
CONVERGENCE	COMMENTS ON CONVERGENCE OF FINITE ELEMENT SOLUTIONS	70-0234
CONVERGENCE	COMPLETENESS AND CONVERGENCE PROPERTIES OF FINITE ELEMENT DISPLACEMENT FUNCTIONS., A GENERAL TREATME	67-0053

167

CONVERGENCE	COMPLETENESS AND CONVERGENCE IN THE FINITE ELEMENT METHOD	68-0056
CONVERGENCE	COMPLETENESS AND CONVERGENCE IN THE FINITE ELEMENT METHOD	68-0279
CONVERGENCE	CONVERGENCE AND ACCURACY IN THE FINITE ELEMENT METHOD	75-0630
CONVERGENCE	CONVERGENCE AND ERROR ESTIMATES FOR FINITE ELEMENT APPROXIMATIONS OF STATIONARY NAVIER-STOKES EQUATI	74-0122
CONVERGENCE	CONVERGENCE AND NUMERICAL ACCURACY WITH SPECIAL REFERENCE TO PLATE BENDING	69-0221
CONVERGENCE	CONVERGENCE AND STABILITY OF NONLINEAR FINITE-ELEMENT EQUATIONS	75-0441
CONVERGENCE	CONVERGENCE CHARACTERISTICS OF CYLINDRICAL SHELL FINITE ELEMENTS	72-0943
CONVERGENCE	CONVERGENCE CRITERIA FOR ITERATIVE PROCESSES	72-0664
CONVERGENCE	CONVERGENCE OF A FINITE ELEMENT PROCEDURE FOR SOLVING BOUNDARY VALUE PROBLEMS OF THE FOURTH ORDER	70-0131
CONVERGENCE	CONVERGENCE OF A MIXED FINITE ELEMENT SCHEME FOR PLATE BENDING	73-0023
CONVERGENCE	CONVERGENCE OF A MIXED FINITE-ELEMENT METHOD FOR PLATE BENDING PROBLEMS	73-0107
CONVERGENCE	CONVERGENCE OF ANOTHER MIXED FINITE ELEMENT METHOD FOR PLATE BENDING PROBLEMS	72-0367
CONVERGENCE	CONVERGENCE OF CONFORMING FINITE-ELEMENT METHODS FOR SHELL PROBLEMS	75-0494
CONVERGENCE	CONVERGENCE OF CONSISTENTLY DERIVED TIMOSHENKO BEAM FINITE ELEMENTS	72-0396
CONVERGENCE	CONVERGENCE OF EIGENVALUE SOLUTIONS IN CONFORMING PLATE BENDING FINITE ELEMENTS	72-0195
CONVERGENCE	CONVERGENCE OF FINITE ELEMENT SOLUTIONS REPRESENTED BY A NON-CONFORMING BASIS	72-0812
CONVERGENCE	CONVERGENCE OF FINITE ELEMENT SOLUTIONS IN VISCOELASTICITY	74-0628
CONVERGENCE	CONVERGENCE OF FINITE ELEMENT SOLUTIONS IN VISCOUS FLOW PROBLEMS	75-0337
CONVERGENCE	CONVERGENCE OF FINITE-ELEMENT METHOD IN NEUTRON-TRANSPORT	75-0408
CONVERGENCE	CONVERGENCE OF LUMPED FINITE ELEMENT SCHEMES FOR SELECTED INITIAL VALUE PROBLEMS	73-1026
CONVERGENCE	CONVERGENCE OF MIXED FINITE ELEMENT APPROXIMATIONS OF A CLASS OF LINEAR-BOUNDARY VALUE PROBLEMS	73-0599
CONVERGENCE	CONVERGENCE OF SEMIANALYTICAL SOLUTION OF PLATES	73-0497
CONVERGENCE	CONVERGENCE OF SIMPLIFIED HYBRID DISPLACEMENT METHOD FOR PLATE BENDING	72-0412
CONVERGENCE	CONVERGENCE OF THE FINITE ELEMENT METHOD IN THE THEORY OF ELASTICITY	68-0095
CONVERGENCE	CONVERGENCE OF THE FINITE ELEMENT METHOD	70-0377
CONVERGENCE	CONVERGENCE OF THE FINITE-ELEMENT METHOD UNDER THE CONSTRAINT OF MATERIAL INCOMPRESSIBILITY	73-1135
CONVERGENCE	CONVERGENCE OF THE NASTRAN PLATE ELEMENTS FOR SHELL STRESS ANALYSIS	71-0261
CONVERGENCE	CONVERGENCE STUDIES OF EIGENVALUE SOLUTIONS USING TWO FINITE PLATE ELEMENTS	70-0085
CONVERGENCE	CONVERGENCE, ACCURACY AND STABILITY OF FINITE ELEMENT APPROXIMATIONS OF A CLASS OF NON-LINEAR HYPERB	73-0159
CONVERGENCE	EXPERIENCE WITH THE PATCH TEST FOR CONVERGENCE OF FINITE ELEMENTS	72-0365
CONVERGENCE	FINITE ELEMENT CONVERGENCE FOR SINGULAR DATA	73-0079
CONVERGENCE	FINITE ELEMENT METHOD OF MIXED TYPE AND ITS CONVERGENCE IN LINEAR SHELL PROBLEMS	73-1041
CONVERGENCE	FINITE ELEMENT SCHEMES: STABILITY AND CONVERGENCE	72-0302
CONVERGENCE	KINDS OF CONVERGENCE AND IMPROVED CONVERGENCE OF CONFORMING FINITE ELEMENT SOLUTIONS IN PLATE BENDIN	70-0323
CONVERGENCE	KINDS OF CONVERGENCE AND IMPROVED CONVERGENCE OF CONFORMING FINITE ELEMENT SOLUTIONS IN PLATE BENDIN	70-0323
CONVERGENCE	MAXIMUM PRINCIPLE AND UNIFORM CONVERGENCE FOR THE FINITE ELEMENT METHOD	73-0852
CONVERGENCE	NOTE ON THE ACCURACY AND CONVERGENCE OF FINITE ELEMENT APPROXIMATIONS OF THE CONVECTION EQUATION	75-0490
CONVERGENCE	NOTE ON THE CONVERGENCE OF FINITE ELEMENT APPROXIMATIONS FOR PROBLEMS FORMULATED IN CURVILINEAR COOR	74-0267
CONVERGENCE	NUMERICAL SUFFICIENCY TEST FOR MONOTONIC CONVERGENCE OF FINITE ELEMENT MODELS	75-0620
CONVERGENCE	ON A NUMERICAL SUFFICIENCY TEST FOR MONOTONIC CONVERGENCE OF FINITE ELEMENT MODELS	75-0533
CONVERGENCE	ON THE CONSTRUCTION AND CONVERGENCE OF A FINITE ELEMENT SOLUTION OF LAPLACE'S EQUATION	72-0340
CONVERGENCE	ON THE CONVERGENCE OF FINITE ELEMENT METHODS	68-0155
CONVERGENCE	ON THE CONVERGENCE OF SOR ITERATIONS FOR FINITE ELEMENT APPROXIMATIONS TO ELLIPTIC BOUNDARY VALUE PR	71-0449
CONVERGENCE	ON THE CONVERGENCE OF THE FINITE ELEMENT METHOD	71-0513
CONVERGENCE	ON THE CONVERGENCE OF THE METHOD OF FINITE ELEMENTS IN THE ANALYSIS OF MEMBRANE DYNAMICS	72-0537
CONVERGENCE	ON THE CONVERGENCE OF THE FINITE ELEMENT METHOD FOR PROBLEMS WITH SINGULARITY	73-0173
CONVERGENCE	ON THE CONVERGENCE OF FINITE ELEMENT SCHEMES BASED ON NON-CONFORMING AND SIMPLIFIED HYBRID DISPLACEM	73-0841
CONVERGENCE	ON THE CONVERGENCE OF THE FINITE ELEMENT SOLUTION IN A NONLINEAR HEAT CONDUCTION PROBLEM	74-1272
CONVERGENCE	ON THE NUMERICAL CONVERGENCE OF MATRIX EIGENVALUE PROBLEMS DUE TO CONSTRAINT CONDITIONS	74-1057
CONVERGENCE	ORDER OF CONVERGENCE OF A FINITE ELEMENT SCHEME	74-0550
CONVERGENCE	ORDER OF CONVERGENCE OF A FINITE-ELEMENT METHOD IN REGIONS WITH CURVED BOUNDARIES	75-0817
CONVERGENCE	ORDERS OF CONVERGENCE IN FINITE ELEMENT METHODS	72-0057
CONVERGENCE	PLATE BENDING WITH NONCONFORMING ELEMENT: FINE-MESH CONVERGENCE TEST	65-0049
CONVERGENCE	POINTWISE CONVERGENCE OF FINITE-ELEMENTS	75-0821
CONVERGENCE	PROBLEMS ASSOCIATED WITH THE CONVERGENCE OF ISOPARAMETRIC AND MIXOPARAMETRIC FINITE ELEMENTS	71-0375
CONVERGENCE	RATE OF CONVERGENCE OF THE FINITE ELEMENT METHOD	72-0502
CONVERGENCE	RATE OF CONVERGENCE FOR THE ITERATIVE APPROACH IN ELASTIC-PLASTIC ANALYSIS OF CONTINUA	73-0518
CONVERGENCE	RESULTS ON THE CONVERGENCE OF THE FINITE ELEMENT METHOD IN STRUCTURAL AND NONSTRUCTURAL CASES	74-0416
CONVERGENCE	SOME CONSIDERATIONS ON THE CONVERGENCE OF HYBRID STRESS METHOD	73-0303
CONVERGENCE	SOME DISTRIBUTIONAL AND CONVERGENCE PROPERTIES OF THE FINITE ELEMENT METHOD, WITH APPLICATIONS IN NO	73-0598
CONVERGENCE	SOME INTERIOR SUPER CONVERGENCE ESTIMATES FOR SPLINE APPROXIMATION	73-0641
CONVERGENCE	SUFFICIENT CONDITIONS FOR CONVERGENCE IN THE FINITE ELEMENT METHOD FOR ANY SOLUTION OF FINITE ENERGY	72-0065
CONVERGENCE	THE CONVERGENCE OF FINITE ELEMENT METHOD IN SOLVING LINEAR ELASTIC PROBLEMS	67-0083
CONVERGENCE	THE L2 AND L00 CONDITION NUMBERS OF THE FINITE ELEMENT MATRICES AND THE POINTWISE CONVERGENCE OF THE	72-0061
CONVERGENCE	THE RATE OF CONVERGENCE FOR THE FINITE ELEMENT METHOD	71-0034
CONVERGENCE	THEORETICAL AND NUMERICAL CONVERGENCE RATES FOR SOME CONFORMING FINITE ELEMENTS	74-0941
CONVERGENCE	UNIFORM CONVERGENCE OF VARIATIONAL FINITE ELEMENT AND RITZ METHODS	74-0420
CONVERGENCE	VARIATIONAL PROCEDURES AND CONVERGENCE OF FINITE ELEMENT METHODS	71-0070
CONVERGING	A RAPIDLY CONVERGING TRIANGULAR PLATE ELEMENT	69-0245
CONVEYING	DYNAMIC INSTABILITY OF DUCTS CONVEYING FLUID	75-0691
COOLANT	REACTOR COOLANT PUMP FLYWHEELS OVERSPEED EVALUATION	74-0506
COOLDOWN	FRACTURE OF ROCKET GRAIN MODELS SUBJECTED TO COOLDOWN: A STUDY OF TEMPERATURE-INITIATED FRACTURES OF	74-0536
COOLED	ANALYSIS OF LOADING AND THERMAL EFFECTS ON FUELLED GRAPHITE BRICKS FOR A HIGH-TEMPERATURE GAS COOLED	71-0085
COOLING	ANALYSIS OF COOLING TOWERS BY THE MATRIX FINITE ELEMENT METHOD- 2	70-0148
COOLING	DOUBLE CURVATURE SHELL FINITE-ELEMENT AND ITS USE IN THE DYNAMIC ANALYSIS OF COOLING TOWERS AND OTHE	70-0312
COOLING	FINITE ELEMENT ANALYSIS OF STRUCTURAL INTEGRITY OF A REACTOR PRESSURE VESSEL DURING EMERGENCY CORE C	70-0026
COOLING	FINITE ELEMENT MODEL FOR TRANSIENT TWO-LAYER COOLING POND BEHAVIOR	75-0465
COOLING	INFLUENCE OF COOLING, FREE-STREAM TURBULENCE AND SURFACE ROUGHNESS ON THE AERODYNAMIC BEHAVIOR OF CA	74-0970
COOLING	KOLBENKUEHLUNG (PISTON COOLING)	75-0266
COOLING	NONLINEAR DYNAMIC ANALYSIS OF COOLING TOWER.	72-0219
COOLING	NUMERICAL ANALYSIS OF COOLING TOWER PLUMES AND FOG INCIDENCE	74-0040
COOLING	STRESSES AND DEFLECTIONS IN COOLING TOWER SHELLS DUE TO WIND LOADING	68-0088
COOLING	THE ANALYSIS OF COOLING TOWER RING BEAMS	68-0063
COOLING	THERMAL LOADING OF THIN-SHELL CONCRETE COOLING TOWERS	74-0667
COOLING	TRANSIENT NATURAL CONVECTION COOLING OF A VERTICAL CIRCULAR CYLINDER	74-1169
COOLING	TRANSPIRATION AND FILM COOLING BOUNDARY LAYER COMPUTER PROGRAM. VOLUME 2, COMPUTER PROGRAM AND USER'	71-0655
COOLING-TOWER	BUCKLING OF COOLING-TOWER SHELLS: BIFURCATION RESULTS	75-0384
COOLING-TOWER	BUCKLING OF COOLING-TOWER SHELLS: STATE-OF-THE-ART	75-0385
COORDINATE	CONSTRUCTION OF CURVILINEAR COORDINATE SYSTEMS AND APPLICATIONS TO MESH GENERATION	74-1295
COORDINATE	FINITE ELEMENT APPENDAGE EQUATIONS FOR HYBRID COORDINATE DYNAMIC ANALYSIS	72-0022
COORDINATE	GEOMETRIC ASPECTS OF THE FINITE ELEMENT METHOD: CONSTRUCTION OF CURVILINEAR COORDINATE SYSTEMS AND T	72-0534
COORDINATE	NATURAL COORDINATE SYSTEMS, AN AUTOMATIC INPUT DATA GENERATION SCHEME FOR A FINITE ELEMENT METHOD	70-0018
COORDINATE	NOTE ON THE CONVERGENCE OF FINITE ELEMENT APPROXIMATIONS FOR PROBLEMS FORMULATED IN CURVILINEAR COOR	74-0267
COORDINATE	SOME ASPECTS OF THE NATURAL COORDINATE SYSTEM IN THE FINITE-ELEMENT METHOD	69-0105
COORDINATES	AN AUTOMATIC MESH GENERATION SCHEME FOR PLANE AND CURVED SURFACES BY ISOPARAMETRIC COORDINATES	71-0271
COORDINATES	APPLICATIONS OF TRILINEAR COORDINATES TO SOME PROBLEMS IN PLANE ELASTICITY	72-0728
COORDINATES	AUTOMATED STRUCTURAL SYNTHESIS USING A RE NUMBER OF DESIGN COORDINATES	73-0584
COORDINATES	AXISYMMETRIC FINITE ELEMENT IN NATURAL COORDINATES	75-0826
COORDINATES	CONVECTED COORDINATES FOR TRANSIENT NONLINEAR FINITE-ELEMENT ANALYSIS	73-0850
COORDINATES	FINITE ELEMENT FORMULATION OF THE HEAT CONDUCTION EQUATION IN GENERAL ORTHOGONAL CURVILINEAR COORDIN	73-0848
COORDINATES	FINITE ELEMENT NEUTRON DIFFUSION IN CURVILINEAR COORDINATES.	72-0039
COORDINATES	FINITE ELEMENTS IN CURVILINEAR COORDINATES - SECOND ORDER SURFACES	73-0753

168

COORDINATES	FINITE PLATE-BENDING ELEMENTS IN POLAR COORDINATES	68-0234
COORDINATES	INTEGRATION OF AREA AND VOLUME COORDINATES IN FINITE-ELEMENT METHOD	69-0370
COORDINATES	INTERACTION OF AREA COORDINATES IN MATRIX STRUCTURAL ANALYSIS	68-0150
COORDINATES	NEW TRIANGULAR ELEMENT FOR FINITE DIFFERENCE SOLUTION OF AXISYMMETRIC CONDUCTION PROBLEMS IN CYLINDR	74-0381
COORDINATES	ON FINITE ELEMENT INTEGRATION IN NATURAL COORDINATES	73-0683
COORDINATES	STABILITY ANALYSIS OF STRUCTURES BY A REDUCED SYSTEM OF GENERALIZED COORDINATES	70-0168
COORDINATES	THIN SHELL THEORY IN CARTESIAN COORDINATES - METHODS FOR FINITE ELEMENTS	70-0359
COORDINATES	3-DIMENSIONAL ISOPARAMETRIC ELEMENT IN CURVILINEAR COORDINATES	72-0098
COORDINATION	INTERACTIVE PROCEDURE FOR CURVILINEAR COORDINATION OF A PLANAR REGION WITH APPLICATIONS TO FINITE-EL	73-0040
COORDINATION	REPORT OF INTERAGENCY CONFERENCE ON COORDINATION OF RESEARCH ACTIVITIES	73-1100
COPLANAR	APPLICATION OF FINITE ELEMENT ANALYSIS TO COPLANAR WAVEGUIDE	73-0924
COPLANAR	FINITE BOUNDARY CORRECTIONS TO THE COPLANAR WAVEGUIDE ANALYSIS	73-0398
COPPER	CYCLIC FATIGUE ANALYSIS OF ROCKET THRUST CHAMBERS. VOLUME 1: OFHC COPPER CHAMBER LOW CYCLE FATIGUE	74-0694
COPPER-ELECTROP	STRESS ANALYSIS OF A BOLT-NUT JOINT BY THE FINITE ELEMENT METHOD AND THE COPPER-ELECTROPLATING METHO	73-0220
COPPER-ELECTROP	STRESS ANALYSIS OF A BOLT-NUT JOINT BY FINITE ELEMENT METHOD AND COPPER-ELECTROPLATING METHOD - 2. S	74-0153
COQUE	FAMILLE D ELEMENTS FINIS DE COQUE PLAN CINEMATIQUEMENT ADMISSIBLES POUR L'ANALYSE DES STRUCTURES	74-0488
COQUES	ANALYSE STATIQUE ET DYNAMIQUE DES COQUES PAR LA METHODE DES ELEMENTS FINIS	74-0509
COQUES	LES PLAQUES ET LES COQUES MINCES OU EPAISSES (THIN AND THICK PLATES AND SHELLS)	71-0432
COQUES	UN NOVEL ELEMENT DE COQUES GENERALES	74-0407
CORE	A COMPARATIVE STUDY OF SEVERAL CORE STORAGE SCHEMES FOR LARGE SPARSE POSITIVE DEFINITE MATRICES WITH	72-0678
CORE	ANALYTICAL PROBLEMS ASSOCIATED WITH CORE SUPPORT STRUCTURES OF PWR	71-0039
CORE	DESIGN ANALYSIS FOR LMFBR CORE RESTRAINT	71-0208
CORE	DYNAMIC RESPONSE OF FAST-REACTOR CORE SUBASSEMBLIES	74-0540
CORE	FINITE ELEMENT ANALYSIS OF STRUCTURAL INTEGRITY OF A REACTOR PRESSURE VESSEL DURING EMERGENCY CORE C	70-0026
CORE	INTERFACE CORE DISCRETE-ELEMENT FOR STIFFENED SHELLS OF REVOLUTION	67-0092
CORE	INTERNAL SUPPORTING STRUCTURES OF PRESSURE TUBES TYPE REACTOR CORE. ANALYTICAL MODEL FOR STATIC ANAL	71-0049
CORE	INTERNAL SUPPORTING STRUCTURES OF THE (CIRENE) REACTOR CORE - RESULTS OF STATIC STRESS ANALYSIS FOR	73-0818
CORE	NOTCH STRESSES AT THE LOAD CARRYING ELASTIC CORE	73-0266
CORE	OPTIMUM POSITION OF THE CENTRAL CLAY CORE OF A ROCKFILL DAM IN RESPECT TO ARCHING AND HYDRAULIC FRAC	73-0447
CORE	ORGANIZATION TECHNIQUES OF THE FINITE ELEMENT METHOD FOR REDUCING CORE STORAGE REQUIREMENTS	74-0074
CORE	ORGANIZATION TECHNIQUES OF THE FINITE ELEMENT METHOD FOR REDUCING CORE STORAGE REQUIREMENTS	74-1120
CORE	PARAMETRISCHE KERBSPANNUNGSUNTERSUCHUNG AM ELASTISCHEN KERN.(PARAMETRICAL INVESTIGATION OF NOTCH STR	72-0208
CORE	RESPONSE OF REACTOR CORE SUBASSEMBLIES TO IMPULSIVE LOADINGS	73-0076
CORE	THREE-DIMENSIONAL BEHAVIOR OF A CENTRAL CORE DAM	72-0714
CORE-STIFFENED	DISCRETE-ELEMENT STATIC ANALYSIS OF CORE-STIFFENED SHELLS OF REVOLUTION FOR ASYMMETRIC MECHANICAL LO	68-0148
CORED	ANALYSIS OF THERMAL STRESSES IN CORED CERAMIC BRICKS USED IN HYPERSONIC WIND TUNNEL HEATERS	72-0655
CORES	NATURAL FREQUENCIES AND MODE SHAPES OF TRANSFORMER CORES	65-0048
CORES	THE EFFECTS OF THIN CORES AND NONHOMOGENEITY ON DYNAMIC RESPONSE OF EARTH DAMS	67-0021
CORNEO-SCLERAL	A MATHEMATICAL MODEL OF THE CORNEO-SCLERAL SHELL	71-0150
CORNEO-SCLERAL	ANALYSIS OF THE CORNEO-SCLERAL SHELL BY THE METHOD OF DIRECT STIFFNESS	71-0151
CORNEO-SCLERAL	MATHEMATICAL MODEL OF THE CORNEO-SCLERAL SHELL AS APPLIED TO INTRAOCULAR PRESSURE VOLUME RELATIONS A	72-0471
CORNEO-SCLERAL	VISCOELASTIC PRESPONSE OF THE CORNEO-SCLERAL SHELL UNDER TONOMETER LOADING	75-0855
CORNER	ANALYSIS OF CORNER SUPPORTED SKEW SLABS	68-0203
CORNER	ANALYSIS OF CORNER SUPPORTED RECTANGULAR SLABS	70-0046
CORNER	APPLICATION OF FINITE ELEMENT METHOD FOR THE DETAILED ANALYSIS OF HATCH CORNER STRESSES	73-0735
CORNER	CORNER SINGULARITIES IN ELLIPTIC PROBLEMS BY FINITE ELEMENT METHODS	71-0456
CORNERS	FINITE ELEMENT METHOD FOR DOMAINS WITH CORNERS	70-0244
CORNERS	FINITE ELEMENT SCHEME FOR DOMAINS WITH CORNERS	72-0261
CORNERS	STRESS CONCENTRATIONS AND SINGULARITIES AT INTERFACE CORNERS	71-0419
CORNERS	THE CHANGES IN THE (TWO-DIMENSIONAL) STRESS ROUNDED CORNERS, CAUSED BY A VARYING INTERNAL PRESSURE A	67-0141
CORRECTED	CORRECTED ASSESSMENT OF CYLINDRICAL SHELL FINITE ELEMENT OF BOGNER, FOX AND SCHMIT WHEN APPLIED TO A	73-0024
CORRECTION	ROTARY INERTIA CORRECTION FOR MASS MATRICES	63-0011
CORRECTIONS	CALCULATION OF AREAL RAINFALL USING FINITE ELEMENT TECHNIQUES WITH ALTITUDINAL CORRECTIONS	72-0344
CORRECTIONS	FINITE BOUNDARY CORRECTIONS TO THE COPLANAR WAVEGUIDE ANALYSIS	73-0398
CORRELATION	A CORRELATION STUDY OF METHODS OF MATRIX STRUCTURAL ANALYSIS	69-0110
CORRELATION	A NASTRAN CORRELATION STUDY FOR VIBRATIONS OF A CROSS-STIFFENED SHIP'S DECK	72-0756
CORRELATION	APPLICATION, TESTING, AND EVALUATION OF DOUBLY CURVED SHELL ELEMENTS FOR DYNAMIC ANALYSIS. PART 1. T	69-0246
CORRELATION	CH-53A FLEXIBLE FRAME VIBRATION ANALYSIS/TEST CORRELATION	73-1093
CORRELATION	CORRELATION BETWEEN MODEL STUDY AND FINITE ELEMENT ANALYSIS IN AN INVESTIGATION OF SLOPE STABILITY	72-0227
CORRELATION	CORRELATION BETWEEN PRESSURE AND THERMAL TRANSIENT STRESS LEVELS DERIVED BY THEORY AND EXPERIMENT ON	73-0469
CORRELATION	CORRELATION OF FINITE ELEMENT STRUCTURAL DYNAMIC ANALYSIS WITH MEASURED FREE VIBRATION CHARACTERISTI	74-0486
CORRELATION	CYCLIC ANALYSIS OF A PROPAGATING CRACK AND ITS CORRELATION WITH FATIGUE CRACK GROWTH	75-0188
CORRELATION	EMBANKMENT ANALYSIS AND FIELD CORRELATION	68-0057
CORRELATION	STRUCTURAL ANALYSIS AND EXPERIMENTAL CORRELATION OF A COMPLEX SHELL UNDER ASYMMETRIC LOADS	69-0155
CORRELATION	STRUCTURAL ANALYSIS OF A THREE-DIMENSIONAL PERFORATED PLATE BY THE FINITE ELEMENT METHOD WITH TEST C	74-0945
CORRESPONDENCE	STRESS ANALYSIS IN VISCOELASTIC BODIES USING FINITE ELEMENTS AND A CORRESPONDENCE RULE WITH ELASTICI	69-0270
CORRUGATED	ANALYSIS OF DOUBLE CORRUGATED SHELL STRUCTURES BY THE FINITE ELEMENT METHOD	74-1116
COST	EXPLICIT INTEGRATION - STABILITY, SOLUTION PROPERTIES, COST	75-0298
COST	FINITE ELEMENT AIRPLANE COST ANALYSIS	74-0301
COSTLY	AVOIDING DANGEROUS AND COSTLY FAN FAILURES	75-0061
COUETTE	A FINITE ELEMENT SOLUTION OF THE TRANSIENT MAGNETOHYDRODYNAMIC COUETTE FLOW	74-1103
COULEE	BUILDING THREE NEW MILLION-HORSEPOWER HYDRAULIC TURBINES FOR GRAND COULEE	74-0981
COUPLAGE	COUPLAGE ENTRE L ECOULEMENT AUTOUR D UN ARRIERE-CORPS ET LE JET PROPULSIF EN THEORIE DE FLUIDE PARFA	74-1137
COUPLE	COUPLE - A PSEUDO FINITE ELEMENT CONDUCTION CODE FOR THE TRANSIENT THERMAL RESPONE OF AXISYMMETRIC 0	72-0778
COUPLED	A METHOD FOR THE TRANSIENT NONLINEAR COUPLED THERMOELASTIC ANALYSIS OF AXISYMMETRIC BODIES USING FIN	74-1140
COUPLED	A NEW APPROACH TO THE FINITE-ELEMENT FORMULATION AND SOLUTION OF A CLASS OF PROBLEMS IN COUPLED THER	72-0732
COUPLED	A NEW APPROACH TO THE FINITE-ELEMENT FORMULATION AND SOLUTION OF A CLASS OF PROBLEMS IN COUPLED THER	73-0600
COUPLED	ANALYSIS OF ELASTICALLY COUPLED SHELLS ON ELASTIC FOUNDATION BY HYBRID METHOD	71-0276
COUPLED	ANALYSIS OF GENERAL COUPLED THERMOELASTICITY PROBLEMS BY THE FINITE ELEMENT METHOD	68-0129
COUPLED	ANALYSIS OF NONLINEAR DYNAMIC, COUPLED THERMOVISCOELASTICITY PROBLEMS BY THE FINITE ELEMENT METHOD	71-0557
COUPLED	APPLICATIONS OF NASTRAN TO COUPLED STRUCTURAL AND HYDRODYNAMIC RESPONSES IN AIRCRAFT HYDRAULIC SYSTE	71-0744
COUPLED	APPROXIMATE SOLUTIONS IN LINEAR COUPLED THERMOELASTICITY	68-0220
COUPLED	COMPUTATION OF THE RESPONSE OF COUPLED PLATE ACOUSTIC SYSTEMS USING PLATE FINITE ELEMENTS AND ACOUST	71-0072
COUPLED	COUPLED CONVECTIVE AND CONDUCTIVE HEAT TRANSFER BY FINITE ELEMENT METHODS	74-0050
COUPLED	COUPLED CONVECTIVE/CONDUCTIVE HEAT TRANSFER INCLUDING VELOCITY FIELD EVALUATION	75-0116
COUPLED	COUPLED HEAT AND MOISTURE TRANSPORT MODEL FOR ARCTIC SOILS	74-0578
COUPLED	COUPLED LOCAL BUCKLING IN WIDE-FLANGE BEAM-COLUMNS	73-0235
COUPLED	COUPLED RESPONSE OF SPATIAL VIBRATORY STRUCTURES MOUNTED TO ISOTROPIC PLATE ELEMENTS	71-0228
COUPLED	COUPLED ROTOR-AIRFRAME VIBRATION PREDICTION METHODS	74-0498
COUPLED	COUPLED SHEAR WALL ANALYSIS BY LAGRANGE MULTIPLIERS	75-0202
COUPLED	COUPLED SHEAR WALLS. IMPROVED METHODS FOR ELASTIC ANALYSIS	74-0818
COUPLED	COUPLED VIBRATIONS OF A STRUCTURE SUBMERGED IN A COMPRESSIBLE FLUID	69-0001
COUPLED	COUPLED VIBRATIONS OF A STRUCTURE SUBMERGED IN A COMPRESSIBLE FLUID	69-0507
COUPLED	COUPLED VIBRATIONS OF THIN-WALLED BEAMS OF OPEN SECTION USING FINITE ELEMENT METHOD	70-0230
COUPLED	COUPLED VIBRATIONS OF THIN-WALLED BEAMS OF OPEN SECTION USING FINITE ELEMENT METHOD	71-0568
COUPLED	COUPLED, NONCONSERVATIVE STABILITY-FINITE ELEMENT	72-0220
COUPLED	CUBIC SPLINE GALERKIN APPROXIMATIONS TO PARABOLIC SYSTEMS WITH COUPLED NON-LINEAR BOUNDARY CONDITION	75-0232
COUPLED	CURVED BEAM FINITE ELEMENTS FOR COUPLED BENDING AND TORSIONAL VIBRATION	72-0720
COUPLED	FINITE ELEMENT ANALYSIS OF NONLINEAR PROBLEMS IN THE DYNAMICAL THEORY OF COUPLED THERMOELASTICITY	69-0202
COUPLED	FINITE ELEMENT ANALYSIS OF COUPLED IRREVERSIBLE VECTOR PROCESSES	71-0701
COUPLED	FINITE ELEMENT ANALYSIS OF COUPLED VIBRATION OF TAPERED TWISTED BLADES	72-0766
COUPLED	FINITE ELEMENT SOLUTION OF COUPLED ELECTROKINETIC AND HYDRODYNAMIC FLOW IN POROUS MEDIA	72-0398

169

COUPLED	FINITE-ELEMENT METHODS FOR COUPLED FIELD-STRESS PROBLEMS	74-0635
COUPLED	NEW APPROACH TO FINITE ELEMENT FORMULATION AND SOLUTION OF A CLASS OF PROBLEMS IN COUPLED THERMOELAS	73-0013
COUPLED	ON THE NUMERICAL SOLUTION OF A CLASS OF NONLINEAR PROBLEMS IN DYNAMIC COUPLED THERMOELASTICITY	71-0441
COUPLED	SEEPAGE IN FRACTURED ROCK: FINITE ELEMENT ANALYSIS OF ROCK MASS BEHAVIOR UNDER COUPLED ACTION OF BOD	71-0825
COUPLED	STEADY STATE IN SOIL SUBJECTED TO SEVERAL COUPLED FORCES: A FINITE ELEMENT FORMULATION	70-0207
COUPLED	TRANSIENT RESPONSE OF A COUPLED PLATE-ACOUSTIC SYSTEM USING PLATE AND ACOUSTIC FINITE ELEMENTS	71-0463
COUPLED	TRANSIENT RESPONSE OF A COUPLED PLATE-ACOUSTIC SYSTEM USING PLATE AND ACOUSTIC FINITE ELEMENTS	72-0275
COUPLED	WEIGHTED RESIDUAL PROCESSES IN FINITE ELEMENT WITH PARTICULAR REFERENCE TO SOME TRANSIENT AND COUPLE	71-0503
COUPLING	COUPLAGE ENTRE L ECOULEMENT AUTOUR D UN ARRIERE-CORPS ET LE JET PROPULSIF EN THEORIE DE FLUIDE PARFA	74-1137
COUPLING	COUPLING MATRICES FOR HIGH ORDER FINITE ELEMENT ANALYSIS OF ACOUSTIC - WAVE PROPAGATION	72-0040
COUPLING	DEVELOPMENT OF A GUN-TURRET MODEL AND OF THE EQUATIONS FOR COUPLING IT WITH A HELICOPTER FINITE-ELEM	73-0752
COUPLING	DYNAMIC COUPLING IN AN UNSYMMETRICALLY TAPERED BEAM	74-1301
COUPLING	DYNAMIC SHELL THEORY COUPLING THICKNESS STRESS WAVE EFFECTS WITH GROSS STRUCTURAL RESPONSE	73-0379
COUPLING	FINITE ELEMENT ANALYSIS OF BENDING-EXTENSIONAL COUPLING IN LAMINATED COMPOSITES	70-0292
COUPLING	FINITE ELEMENT COUPLING MATRICES	69-0300
COUPLING	MODAL COUPLING PROCEDURES ADAPTED TO NASTRAN ANALYSIS OF THE 1/8 - SCALE SHUTTLE STRUCTURAL DYNAMICS	75-0677
COUPLING	MODAL COUPLING PROCEDURES ADAPTED TO NASTRAN ANALYSIS OF THE 1/8-SCALE SHUTTLE STRUCTURAL DYNAMICS M	75-0678
COURBES	ANALYSE DES BOILES MINCES PAR ELEMENTS FINIS COURBES	69-0524
COURSE	FINITE ELEMENT TECHNIQUES - A SHORT COURSE OF FUNDAMENTALS AND APPLICATIONS	73-0612
COURSE	USER EDUCATION: THE FINITE ELEMENT SHORT COURSE	72-0437
COVERS	METODY RASCHETNOGO OPREDELENIYA TEMPERATURNYKH NAPRYAZHENII V KRYSHKAKH TSILINDROV DVIGATELEI VNUTRE	74-0179
CRACK	A FINITE ELEMENT ANALYSIS OF FATIGUE CRACK CLOSURE	74-0759
CRACK	A HIGH-ORDER FINITE ELEMENT FOR TWO-DIMENSIONAL CRACK PROBLEMS	74-1128
CRACK	A HYBRID FINITE ELEMENT FOR DETERMINATION OF STRESS INTENSITY FACTORS AT A CRACK TIP	73-0643
CRACK	A HYBRID-ELEMENT APPROACH TO CRACK PROBLEMS IN PLANE ELASTICITY	73-0569
CRACK	A SUPER-ELEMENT FOR CRACK ANALYSIS	73-0754
CRACK	AN ELASTIC-PLASTIC FINITE ELEMENT ANALYSIS OF CRACK TIP FIELDS UNDER BIAXIAL LOADING CONDITIONS	73-0760
CRACK	ANALYSIS FOR CRACKS IN SHIP STRUCTURE USING THE FINITE ELEMENT METHOD WITH THE CONCEPT OF CRACK SUB-	74-0248
CRACK	ANALYSIS FOR CRACKS IN SHIP STRUCTURES USING THE FINITE-ELEMENT METHOD WITH THE CONCEPT OF CRACK SUB	74-1367
CRACK	ANALYSIS OF CRACK PROPAGATION IN WELDED STRUCTURES	74-1144
CRACK	ANALYSIS OF STRESS AND STRAIN DISTRIBUTIONS AT CRACK TIP BY FINITE ELEMENT METHOD	69-0193
CRACK	ANALYTICAL APPROACH TO THE NON-PROPAGATING CRACK PROBLEM USING THE FINITE ELEMENT METHOD	71-0389
CRACK	APPLICATION OF A SUBSTRUCTURING TECHNIQUE TO THE PROBLEM OF CRACK EXTENSION AND CLOSURE	74-0772
CRACK	CALCULATION OF STRESS INTENSITY FACTORS AT CRACK TIPS USING SPECIAL FINITE ELEMENTS	72-0071
CRACK	CALCULATION OF STRESS INTENSITY FACTORS AND FATIGUE CRACK PROPAGATION OF SEMI-ELLIPTICAL, PART-THROU	73-0360
CRACK	COMBINED MODE CRACK EXTENSION IN ADHESIVE JOINTS	72-0111
CRACK	COMPUTER PREDICTION OF FATIGUE CRACK PROPAGATION UNDER RANDOM LOADING	73-0851
CRACK	CRACK BRANCHING IN HOMALITE-100 SHEETS	72-0843
CRACK	CRACK BRANCHING IN HOMALITE-100 SHEETS	74-0291
CRACK	CRACK ELEMENTS	75-0558
CRACK	CRACK GROWTH MONITORING BY STRAIN SENSING	73-0486
CRACK	CRACK PROPAGATION AND ARREST IN IMPACTED PLATES	72-0599
CRACK	CRACK PROPAGATION IN SEA ICE - FINITE ELEMENT APPROACH	72-0029
CRACK	CRACK TIP FINITE ELEMENTS FOR PLANE ELASTICITY	71-0764
CRACK	CRACK TIP FINITE ELEMENTS ARE UNNECESSARY	73-1016
CRACK	CRACK TIP FINITE ELEMENTS ARE UNNECESSARY	75-0472
CRACK	CRACK TIP STRAIN - A COMPARISON OF FEM CALCULATIONS AND MEASUREMENTS	75-0721
CRACK	CYCLIC ANALYSIS OF A PROPAGATING CRACK AND ITS CORRELATION WITH FATIGUE CRACK GROWTH	75-0188
CRACK	CYCLIC ANALYSIS OF A PROPAGATING CRACK AND ITS CORRELATION WITH FATIGUE CRACK GROWTH	75-0188
CRACK	CYCLIC LOADING AND CRACK PROPAGATION - AN ELASTOPLASTIC FINITE ELEMENT STUDY	74-1214
CRACK	DERIVATION OF CRACK ELEMENT-STIFFNESS MATRIX BY THE COMPLEX VARIABLE APPROACH	73-0870
CRACK	DEVELOPMENT OF FINITE ELEMENT COMPUTATIONAL METHODS FOR THREE DIMENSIONAL ELASTIC PLASTIC CRACK ANAL	70-0356
CRACK	EFFECT OF PLASTICITY AND CRACK BLUNTING ON THE STRESS DISTRIBUTION IN ORTHOTROPIC COMPOSITE MATERIAL	73-0145
CRACK	EFFECTS OF A PEAK OVERLOAD ON FATIGUE CRACK PROPAGATION RATE-PLANE STRESS ANALYSIS OF CRACKED PLATE	74-1365
CRACK	EFFECTS OF LOCAL STRENGTH VARIATIONS ON PLASTIC ENCLAVE DEVELOPMENT AT CRACK TIPS	72-0822
CRACK	ELASTIC CRACK ANALYSIS BY A FINITE ELEMENT HYBRID METHOD	71-0212
CRACK	ELASTIC-PLASTIC ANALYSIS OF A PROPAGATING CRACK UNDER CYCLIC LOADING	75-0857
CRACK	ELASTIC-PLASTIC BEHAVIOR OF A LONGITUDINAL SEMI-ELLIPTIC CRACK IN A THICK PRESSURE VESSEL	73-0952
CRACK	ELASTIC-PLASTIC FINITE ELEMENT ANALYSIS OF CRACK TIP FIELDS UNDER BIAXIAL LOADING CONDITIONS	74-0296
CRACK	ELASTO-PLASTIC ANALYSIS OF METAL SHEETS WITH A CENTRAL CRACK	67-0137
CRACK	ELASTOPLASTIC CRACK PROPAGATION IN A TRANSVERSELY LOADED UNIDIRECTIONAL COMPOSITE	74-0217
CRACK	ENERGY FOR CRACK GROWTH IN MODEL RUBBER COMPONENTS	72-0568
CRACK	EXPLORATORY INVESTIGATION OF RAPID CRACK PROPAGATION AND CRACK ARREST	73-0757
CRACK	EXPLORATORY INVESTIGATION OF RAPID CRACK PROPAGATION AND CRACK ARREST	73-0757
CRACK	FAST FRACTURE RESISTANCE AND CRACK ARREST IN STRUCTURAL STEELS	73-0766
CRACK	FATIGUE CRACK FORMATION AND PROPAGATION IN PAVEMENTS CONTAINING SOIL-CEMENT BASES.	72-0435
CRACK	FATIGUE CRACK INITIATION IN A BORON EPOXY PLATE WITH A CIRCULAR HOLE	74-0820
CRACK	FINITE ELEMENT ANALYSIS OF CRACK INITIATION IN A BLOCK MODEL EXPERIMENT	72-0455
CRACK	FINITE ELEMENT ANALYSIS OF CRACK PROPAGATION IN THREE-DIMENSIONAL SOLIDS UNDER CYCLING LOADING	73-0529
CRACK	FINITE ELEMENT ANALYSIS OF CRACK PROPAGATION IN 3-DIMENSIONAL SOLIDS UNDER CYCLIC LOADING	74-0479
CRACK	FINITE ELEMENT ANALYSIS OF FATIGUE CRACK PROPAGATION - INCLUDING THE EFFECTS OF CRACK CLOSURE	74-1112
CRACK	FINITE ELEMENT ANALYSIS OF FATIGUE CRACK PROPAGATION - INCLUDING THE EFFECTS OF CRACK CLOSURE	74-1112
CRACK	FINITE ELEMENT APPROACH TO THE CRITICAL CYCLIC STRESS REQUIRED TO PROPAGATE A CRACK	73-0463
CRACK	FINITE ELEMENT CALCULATIONS OF THE COMPLIANCE OF A TAPERED DOUBLE CANTILEVER BEAM SPECIMEN FOR DIFFE	72-0665
CRACK	FINITE ELEMENT CRACK ANALYSIS USING THE J-INTEGRAL METHOD	73-1106
CRACK	FINITE ELEMENT CRACK TIP MODELING WHEN EVALUATING J-INTEGRAL	75-0637
CRACK	FINITE ELEMENT METHOD FOR PREDICTION OF CRACK BEHAVIOUR	70-0342
CRACK	FINITE ELEMENT STRESS ANALYSIS OF A CRACK IN A BI-MATERIAL PLATE	72-0165
CRACK	FINITE ELEMENT TREATMENT OF A UNIFORMLY MOVING ELASTIC-PLASTIC CRACK TIP	74-0470
CRACK	FINITE ELEMENTS FOR DETERMINATION OF CRACK TIP ELASTIC STRESS INTENSITY FACTORS	71-0436
CRACK	FINITE ELEMENTS FOR THREE-DIMENSIONAL ELASTIC CRACK ANALYSIS	74-0136
CRACK	FRACTURE MECHANICS FOR AN INTERFACIAL CRACK BETWEEN ADHESIVELY BONDED DISSIMILAR MATERIALS	72-0869
CRACK	HIGH-PERFORMANCE COMPOSITE MATERIALS FOR VEHICLE CONSTRUCTION: A FINITE ELEMENT COMPUTER PROGRAM FOR	73-0767
CRACK	HYBRID-ELEMENT APPROACH TO CRACK PROBLEMS IN PLANE ELASTICITY	73-0508
CRACK	INFLUENCE OF CRACK CLOSURE AND ELASTO-PLASTIC FLOW ON THE BENDING OF A CRACKED PLATE	75-0735
CRACK	INFLUENCE OF NON-SINGULAR STRESS TERMS AND SPECIMEN GEOMETRY ON SMALL SCALE YIELDING AT CRACK TIPS I	73-0203
CRACK	LIMITATIONS TO THE SMALL SCALE YIELDING APPROXIMATION FOR CRACK TIP PLASTICITY	74-0145
CRACK	MICROMECHANICAL ANALYSIS OF CRACK PROPAGATION IN AN ELASTOPLASTIC COMPOSITE MATERIAL	74-0565
CRACK	NONLINEAR ANALYSES OF THE STRESS FIELD NEAR A CRACK OR HOLE IN A PLATE	75-0694
CRACK	NONLINEAR CRACK ANALYSIS WITH FINITE ELEMENTS	73-0369
CRACK	NONLINEAR CRACK ANALYSIS WITH FINITE ELEMENTS	73-0875
CRACK	ON THE FINITE ELEMENT ANALYSIS OF CRACK AND INCLUSION PROBLEMS IN ELASTIC-PLASTIC MATERIALS	74-0227
CRACK	ON THE METHOD OF VIRTUAL CRACK EXTENSIONS	75-0237
CRACK	ON THE ROLE OF DEFECTS IN CRACK INITIATION IN WELDED STRUCTURES	74-1142
CRACK	ON THE USE OF FINITE ELEMENTS AND COMPLIANCE CALCULATIONS FOR DETERMINING CRACK TIP STRESS INTENSITY	69-0448
CRACK	PLANE STRESS CRACK TRAJECTORIES IN LOW MODULUS MATERIALS	75-0646
CRACK	PREDICTION OF SLOW CRACK GROWTH BY A FINITE-ELEMENT METHOD	75-0416
CRACK	PREDICTION OF STATIC AND FATIGUE DAMAGE AND CRACK PROPAGATION IN COMPOSITE MATERIALS	75-0643
CRACK	PROPAGATION AND OPENING OF A THROUGH CRACK IN PIPE SUBJECT TO COMBINED CYCLIC THERMOMECHANICAL LOADI	75-0516
CRACK	SIZE OF PLASTIC ZONE AT TIP OF A CRACK IN PLANE STRAIN SLATE BY FINITE ELEMENT METHOD	73-0006
CRACK	SMALL SCALE YIELDING NEAR A CRACK IN PLANE STRAIN-FINITE ELEMENT ANALYSIS	71-0355
CRACK	SMALL SCALE YIELDING ANALYSIS OF MIXED MODE PLANE STRAIN CRACK PROBLEMS	73-0801

CRACK	STEADILY GROWING ELASTIC-PLASTIC CRACK TIP IN A FINITE ELEMENT TREATMENT	73-0087
CRACK	STIFFNESS DERIVATIVE FINITE ELEMENT TECHNIQUE FOR DETERMINATION OF CRACK TIP STRESS INTENSITY FACTOR	74-0956
CRACK	STIFFNESS DERIVATIVE FINITE ELEMENT TECHNIQUE FOR DETERMINATION OF CRACK TIP STRESS INTENSITY FACTOR	74-1358
CRACK	STRESS CRACK PHENOMENA	75-0744
CRACK	STRESS INTENSITY FACTOR FOR A STRAIGHT CRACK APPROACHING A HOLE	69-0158
CRACK	STRESS INTENSITY FACTOR OF A SURFACE CRACK IN A PRESSURIZED CYLINDER	75-0255
CRACK	STRESS, DISPLACEMENT, LINE INTEGRAL AND CLOSURE ENERGY DETERMINATIONS OF CRACK TIP STRESS INTENSITY	74-0621
CRACK	TENSILE CRACK PROPAGATION IN REINFORCED CONCRETE BEAMS BY THE FINITE ELEMENT TECHNIQUES	69-0264
CRACK	THE ANALYSIS OF STRESS DISTRIBUTION AROUND THE TIP OF A CRACK BY FINITE ELEMENT METHOD WITH THE AID	71-0517
CRACK	THE ELASTIC STRESS ANALYSIS OF A BI-MATERIAL PLATE WITH A CRACK NORMAL TO THE INTERFACES	72-0819
CRACK	THE VIBRATION CHARACTERISTICS OF A TENSIONED PLATE CONTAINING A FATIGUE CRACK	68-0134
CRACK	THREE-DIMENSIONAL CRACK ANALYSIS APPLIED TO AN LWR NOZZLE-CYLINDER INTERSECTION	75-0165
CRACK	THREE-DIMENSIONAL STRESS ANALYSIS OF A FINITE SLAB CONTAINING A TRANSVERSE CENTRAL CRACK	75-0148
CRACK	TWO NEARLY POLYGONAL HOLES, MATHEMATICAL CRACK PROBLEMS	69-0061
CRACK-ARREST	PHOTOELASTIC INVESTIGATION ON THE CRACK-ARREST CAPABILITY OF A HOLE	72-0086
CRACK-ARREST	PHOTOELASTIC INVESTIGATION ON THE CRACK-ARREST CAPABILITY OF A PRETENSIONED STIFFENED PLATE	75-0088
CRACK-GROWTH	FINITE ELEMENT REPRESENTATION OF STABLE CRACK-GROWTH	73-0086
CRACK-TIP	FINITE ELEMENTS FOR DETERMINATION OF CRACK-TIP ELASTIC STRESS INTENSITY FACTORS	71-0785
CRACKED	ARBITRARY SYMMETRIC LOADING PROBLEMS OF CENTRALLY CRACKED RECTANGULAR PLATES	75-0318
CRACKED	CALCULATION METHOD FOR THE STUDY OF CRACKED RIGID STRUCTURES (METODO DI CALCOLO PER LO STUDIO DI STR	71-0637
CRACKED	CALCULATION OF STRESS INTENSITY FACTORS USING FINITE ELEMENT METHOD WITH CRACKED ELEMENTS	70-0224
CRACKED	CALCULATION OF STRESS INTENSITY FACTORS USING THE FINITE ELEMENT METHOD WITH CRACKED ELEMENTS	70-0475
CRACKED	DETERMINATION OF STRESS INTENSITY FACTORS IN CRACKED PLATES BY THE FINITE ELEMENT METHOD	73-0148
CRACKED	DEVELOPMENT OF A CRACKED FINITE ELEMENT	70-0279
CRACKED	DUGDALE MODEL SOLUTIONS FOR A SINGLE EDGE CRACKED PLATE	74-0221
CRACKED	DYNAMICS OF CRACKED STRUCTURES USING FINITE ELEMENTS	74-1115
CRACKED	EFFECTS OF A PEAK OVERLOAD ON FATIGUE CRACK PROPAGATION RATE-PLANE STRESS ANALYSIS OF CRACKED PLATE	74-1365
CRACKED	ELASTIC ANALYSIS OF CRACKED THIN SHELLS BY THE FINITE ELEMENT METHOD	71-0048
CRACKED	ELASTIC PLASTIC CRACKED PLATES IN PLANE STRAIN	69-0321
CRACKED	ELASTIC-PLASTIC ANALYSIS FOR CRACKED MEMBERS	75-0519
CRACKED	ELASTO PLASTIC STRESS IN CRACKED PLATES	65-0042
CRACKED	ELASTODYNAMICS OF CRACKED STRUCTURES USING FINITE ELEMENTS	74-0625
CRACKED	FINITE ELEMENT ANALYSIS OF CRACKED STRUCTURES SUBJECTED TO SHOCK LOADS	75-0258
CRACKED	FINITE ELEMENT FRACTURE MECHANICS ANALYSIS OF TWO-DIMENSIONAL AND AXISYMMETRIC ELASTIC AND ELASTIC-P	74-1092
CRACKED	FINITE ELEMENT METHOD FOR CALCULATING STRESS INTENSITY FACTORS FOR CRACKED PLATES IN BENDING	71-0424
CRACKED	FINITE ELEMENT VIBRATION ANALYSIS OF CRACKED PLATES IN TENSION	68-0133
CRACKED	FINITE-ELEMENT SOLUTION FOR A CRACKED TWO-LAYERED ELASTIC CYLINDER	71-0397
CRACKED	HEAVY SECTION STEEL TECHNOLOGY PROGRAM TECHNICAL OR PROGRAMMATIC MANUSCRIPT NO. 25, VARIABLE THICKNE	73-0948
CRACKED	INFLUENCE OF CRACK CLOSURE AND ELASTO-PLASTIC FLOW ON THE BENDING OF A CRACKED PLATE	75-0735
CRACKED	NONLINEAR ANALYSIS OF CRACKED BODY BY FINITE ELEMENT METHOD	75-0315
CRACKED	PLASTIC INTENSITY FACTORS FOR CRACKED PLATES	71-0164
CRACKED	REFINED CRACKED CONCRETE ANALYSIS OF CONCRETE CONTAINMENT STRUCTURES SUBJECT TO OPERATIONAL AND ENVI	74-0943
CRACKED	THE APPLICATION OF FINITE ELEMENT ANALYSIS TO THE STUDY OF TWO-DIMENSIONAL CRACKED BODIES	72-0695
CRACKING	A FINITE ELEMENT COMPUTER PROGRAM FOR THE PREDICTION OF THE BEHAVIOR OF REINFORCED AND PRESTRESSED C	72-0799
CRACKING	ANALYSIS OF CONCRETE BOX GIRDERS BEFORE AND AFTER CRACKING	72-0247
CRACKING	ANALYSIS OF STRESSES IN CONCRETE NEAR A REINFORCING BAR DUE TO BOND AND TRANSVERSE CRACKING	70-0351
CRACKING	CRACKING ANALYSIS OF REINFORCED CONCRETE PLATES	75-0085
CRACKING	CRACKING OF EARTH AND ROCKFILL DAMS. A THEORETICAL INVESTIGATION BY MEANS OF THE FINITE ELEMENT METH	69-0085
CRACKING	CRACKING OF EARTH AND ROCKFILL DAMS	69-0294
CRACKING	CRACKING OF EARTH AND ROCKFILL DAMS: COMPARISON OF OBSERVED AND THEORETICAL TENSILE STRAINS IN THE C	71-0699
CRACKING	FINITE ELEMENT ANALYSIS OF A DIAGONAL CRACKING BEHAVIOUR IN REINFORCED CONCRETE	73-0462
CRACKING	FINITE ELEMENT ANALYSIS OF TRANSVERSE CRACKING IN LOW-EMBANKMENT DAMS	74-0696
CRACKING	FINITE ELEMENT INVESTIGATION OF SECONDARY CRACKING DURING BRITTLE FRACTURE	71-0804
CRACKING	FINITE ELEMENT STUDY OF A REINFORCED CONCRETE CYLINDRICAL SHELL THROUGH ELASTIC, CRACKING, AND ULTIM	75-0317
CRACKING	MODIFIED APPROACH FOR ESTIMATING THE CRACKING MOMENT OF REINFORCED CONCRETE BEAMS	75-0485
CRACKING	PAVEMENT CRACKING IN WEST TEXAS DUE TO FREEZE-THAW CYCLING	75-0276
CRACKING	RECENT TRENDS OF RESEARCH INTO RESTRAINT STRESSES AND STRAINS IN RELATION TO WELD CRACKING	73-0483
CRACKING	TENDANCES ACTUELLES DES RECHERCHES SUR LES CONTRAINTES ET DEFORMATIONS DUES AU BRIDAGE EN RELATION A	74-0273
CRACKING	THE DETERMINATION OF STRESSES AND DEFORMATIONS OF REINFORCED CONCRETE AFTER CRACKING	64-0004
CRACKING	VERIFICATION OF FRACTURE MECHANICS CONCEPTS TO PREDICT CRACKING OF FLEXIBLE PAVEMENTS	73-0772
CRACKING	ZUSAMMENHANG ZWISCHEN OBERFLAECHENBESCHAFFENHEIT, VERBUND UND SPRENGWIRKUNG VON BEWEHRUNGSSTAEHLEN U	73-0450
CRACKS	A FRACTURE MECHANICS APPROACH TO THE PROPAGATION OF FATIGUE CRACKS IN REACTOR COMPONENTS BY ACOUSTIC	71-0132
CRACKS	ANALYSIS FOR CRACKS IN SHIP STRUCTURE USING THE FINITE-ELEMENT METHOD WITH THE CONCEPT OF CRACK SUB-	74-0248
CRACKS	ANALYSIS FOR CRACKS IN SHIP STRUCTURES USING THE FINITE-ELEMENT METHOD WITH THE CONCEPT OF CRACK SUB	74-1367
CRACKS	ANALYSIS OF COMPOSITE REINFORCED CUTOUTS AND CRACKS	74-0824
CRACKS	CALCULATION OF STRESS INTENSITY FACTORS AND FATIGUE CRACK PROPAGATION OF SEMI-ELLIPTICAL, PART-THROU	73-0360
CRACKS	COMPOSITE-OVERLAY REINFORCEMENT OF CUTOUTS AND CRACKS IN METAL SHEET	73-0878
CRACKS	CRACKS, NOTCHES AND FINITE ELEMENTS	75-0253
CRACKS	DEFORMATION MECHANICS OF DEEP SURFACE FLAW CRACKS	72-0785
CRACKS	FINITE ELEMENT ANALYSIS OF CRACKS IN REGION OF A BI-MATERIAL INTERFACE	74-0164
CRACKS	FINITE ELEMENT METHODS FOR ELASTIC BODIES CONTAINING CRACKS	73-1029
CRACKS	FINITE ELEMENT STUDY OF REINFORCED CONCRETE BEAMS WITH DIAGONAL TENSION CRACKS	74-1308
CRACKS	FINITE ELEMENT TECHNIQUES APPLIED TO CRACKS INTERACTING WITH SELECTED SINGULARITIES	75-0196
CRACKS	FRACTURE MECHANICS OF TINY CRACKS NEAR FASTENERS	74-0735
CRACKS	LEAKAGE THROUGH CRACKS IN LNG TANKAGE	74-0749
CRACKS	MATHEMATICAL ANALYSIS OF STRESS CRACKS. VOLUME 2.	75-0742
CRACKS	MATHEMATICAL ANALYSIS OF STRESS CRACKS. VOLUME 1	75-0743
CRACKS	ON CRACKS IN WELDS AND WELDED STRUCTURES	71-0688
CRACKS	PROPAGATION OF ELLIPTICAL SURFACE CRACKS AND NONLINEAR FRACTURE MECHANICS BY THE FINITE ELEMENT METH	74-0906
CRACKS	QUANTITATIVE STRAIN-AND-STRESS STATE CRITERION FOR FAILURE IN THE VICINITY OF SHARP CRACKS	74-0639
CRACKS	STRAIN ENERGY RELEASE RATES FOR TURNING CRACKS	72-0146
CRACKS	STRESS AND STRAIN DISTRIBUTIONS AT THE ROOT OF CRACKS	69-0290
CRACKS	STRESS INTENSITY FACTORS FOR CRACKS IN A PLATE CONTAINING A HOLE AND IN A SPINNING DISC	73-0467
CRACKS	THEORETICAL STUDY OF THE INFLUENCE OF INCLUSIONS UPON THE INITIATION AND GROWTH OF FATIGUE CRACKS IN	71-0394
CRACKS	THREE-DIMENSIONAL ELASTIC ANALYSIS OF SEMI-ELLIPTICAL SURFACE CRACKS SUBJECT TO THERMAL SHOCK	75-0256
CRASH	A USER-ORIENTED PROGRAM FOR CRASH DYNAMICS	74-0321
CRASH	CRASH: A COMPUTER SIMULATION OF NONLINEAR TRANSIENT RESPONSE OF STRUCTURES	72-0569
CRASH	STUDY ON AUTOMOBILE CRASH WORTHINESS	70-0217
CRASH-IMPACT	MODELING AND ANALYSIS TECHNIQUES FOR PREDICTION OF STRUCTURAL AND BIODYNAMIC CRASH-IMPACT RESPONSE	75-0300
CRASHES	BERECHNUNGSMETHODEN ZUR ERMITTLUNG VON BELASTUNGSBEDINGUNGEN UND GRENZTRAGFAEHIGKEITSVERHALTEN VON K	73-0858
CRASHWORTHINESS	ANALYTICAL FINITE ELEMENT SIMULATION MODEL FOR STRUCTURAL CRASHWORTHINESS PREDICTION	73-0778
CRASHWORTHINESS	MODULAR APPROACH TO STRUCTURAL SIMULATION FOR VEHICLE CRASHWORTHINESS PREDICTION	75-0737
CRASHWORTHY	AN INTERACTIVE HYBRID TECHNIQUE FOR CRASHWORTHY DESIGN OF COMPLEX VEHICULAR STRUCTURAL SYSTEMS	74-0317
CRATER	APPLICATION OF FINITE ELEMENT METHOD IN DETERMINING STABILITY OF CRATER SLOPES, PRELIMINARY REPORT	58-0026
CRATER	APPLICATION OF FINITE ELEMENT METHOD IN DETERMINING STABILITY OF CRATER SLOPES	72-0888
CRATERING	SEEPAGE AND GROUNDWATER EFFECTS ASSOCIATED WITH EXPLOSIVE CRATERING	72-0656
CREATION	CREATION AND COMPARISON OF FINITE DIFFERENCE ANALOGS OF SOME FINITE ELEMENT SCHEMES	74-0888
CREATION	ICON: THE INTERACTIVE CREATION OF NASTRAN DATA. A SYSTEM DESCRIPTION	75-0274
CREEP	A DESIGN ORIENTED APPROACH TO CREEP AND PLASTICITY IN FINITE ELEMENT PROGRAMS	70-0125
CREEP	A FINITE ELEMENT COMPUTER CODE, AXICAP FOR CREEP ANALYSIS	69-0248
CREEP	A FINITE-ELEMENT PROGRAM FOR THE ANALYSIS OF EMBANKMENTS OVER SOFT, SATURATED SOILS INCLUDING CONSOL	73-1160
CREEP	A GENERAL APPROACH TO THE PRACTICAL SOLUTION OF CREEP PROBLEMS	59-0003

CREEP	ANALYSIS OF ELASTIC-PLASTIC CREEP BUCKLING OF AXISYMMETRIC SHELLS BY FINITE-ELEMENT METHOD	75-0355
CREEP	ANALYSIS OF LARGE ELASTIC-PLASTIC CREEP DEFORMATIONS OF AXISYMMETRIC SHELLS BY THE FINITE ELEMENT ME	74-0563
CREEP	ANALYSIS OF MULTI-AXIAL ANISOTROPIC CREEP, ROOF RESIN BOLTS, AND BED SEPARATION IN ROCK STRUCTURES U	74-1122
CREEP	ANALYSIS OF REAL STRUCTURES FOR CREEP, PLASTICITY AND OTHER COMPLEX CONSTITUTIVE LAWS.	69-0287
CREEP	ANALYSIS OF RECTANGULAR PLATES IN TRANSIENT CREEP BENDING	69-0425
CREEP	ANALYSIS OF STRESSES IN PRESSURIZED WELDED PIPE IN THE CREEP RANGE	71-0404
CREEP	APPLICATION OF FINITE ELEMENT METHODS FOR THE ANALYSIS OF THERMAL CREEP, IRRADIATION INDUCED CREEP,	71-0242
CREEP	APPLICATION OF FINITE ELEMENT METHODS FOR THE ANALYSIS OF THERMAL CREEP, IRRADIATION INDUCED CREEP,	71-0242
CREEP	APPLICATION OF FINITE ELEMENT METHOD TO PROBLEMS OF NON-LINEAR BEHAVIOR WITH CREEP	75-0811
CREEP	APPLICATIONS OF THE FINITE ELEMENT METHOD TO THE STRESS ANALYSIS OF MATERIALS SUBJECT TO CREEP	71-0548
CREEP	CALCULATIONAL PROCEDURE FOR DETERMINING CREEP COLLAPSE OF LWR FUEL RODS	74-1001
CREEP	CAVITY STABILITY: FINITE ELEMENT ANALYSIS FOR STEADY AND TRANSIENT CREEP	74-1082
CREEP	COMBINED ELASTIC-PLASTIC CREEP ANALYSIS OF TWO-DIMENSIONAL BODIES	71-0232
CREEP	CREEP ANALYSIS OF AXISYMMETRIC BODIES	68-0081
CREEP	CREEP ANALYSIS OF TRANSVERSELY ISOTROPIC BODIES SUBJECTED TO TIME DEPENDENT LOADING	73-0170
CREEP	CREEP AND SHRINKAGE LAW FOR CONCRETE AT VARIABLE HUMIDITY	74-0668
CREEP	CREEP BENDING ANALYSIS OF PLATES BY THE FINITE ELEMENT METHOD	71-0415
CREEP	CREEP BENDING TESTS OF LARGE WELD OVERLAY AND BASE METAL SPECIMENS AT 1000 DEGREE F. (537.7 DEG. C)	75-0521
CREEP	CREEP DATA ACQUISITION AND APPLICATION TO REACTOR COMPONENT DESIGN	72-0113
CREEP	CREEP DEFORMATION OF A MULTI-STOREY STRUCTURE IN A FIRE	74-1197
CREEP	CREEP OF A UNIAXIAL METAL MATRIX COMPOSITE SUBJECTED TO AXIAL AND NORMAL LATERAL LOADS	71-0041
CREEP	CREEP OF A UNIAXIAL METAL MATRIX COMPOSITE SUBJECTED TO AXIAL AND NORMAL LATERAL LOADS	71-0727
CREEP	CREEP OF B/AL COMPOSITES AS INFLUENCED BY RESIDUAL STRESSES, BOND STRENGTH, AND FIBER PACKING GEOMET	74-1208
CREEP	CREEP OF FABRICATED STRUCTURES	68-0001
CREEP	CREEP STRESS AND STRAIN ANALYSIS OF NOTCHED PLATES AND BARS BY MEANS OF THE FINITE ELEMENT METHOD	70-0141
CREEP	DESIGN ORIENTED APPROACH TO CREEP AND PLASTICITY IN FINITE ELEMENT PROGRAMS	71-0281
CREEP	ELASTIC PLASTIC AND CREEP ANALYSIS OF PIPELINES BY FINITE ELEMENTS	73-0419
CREEP	ELASTIC-PLASTIC AND CREEP ANALYSIS VIA THE MARC 2 FINITE ELEMENT PROGRAM	71-0032
CREEP	ELASTOPLASTIC AND CREEP ANALYSIS WITH THE ASKA PROGRAM SYSTEM	74-0863
CREEP	ESTIMATES OF THE CREEP RUPTURE LIFETIME OF STRUCTURES USING THE FINITE ELEMENT METHOD	75-0833
CREEP	EVALUATION OF A CLASS OF METHODS FOR BOUNDING STEADY-STATE CREEP DEFORMATION	74-0892
CREEP	FINITE ELEMENT ANALYSIS OF CREEP PROBLEMS IN SOIL MECHANICS	71-0809
CREEP	FINITE ELEMENT ANALYSIS OF CONSOLIDATION IN CREEP	72-0420
CREEP	FINITE ELEMENT ANALYSIS OF CREEP IN ROCK-SALT PILLAR MODELS	72-0935
CREEP	FINITE ELEMENT CREEP BUCKLING ANALYSIS OF VISCO-ELASTIC SHALLOW SPHERICAL SHELL	75-0159
CREEP	FINITE ELEMENT ELASTIC PLASTIC CREEP ANALYSIS OF TWO DIMENSIONAL CONTINUUM WITH TEMPERATURE DEPENDEN	73-0226
CREEP	FINITE ELEMENT FRACTURE MECHANICS ANALYSIS OF CREEP RUPTURE OF FUEL ELEMENT CLADDING	74-0929
CREEP	IN SITU CREEP ANALYSIS OF ROOM IN FROZEN SOIL	72-0128
CREEP	LONG TERM CREEP CLOSURE OF SOLUTION CAVITY SYSTEM	74-1150
CREEP	MATRIX ANALYSIS OF CREEP AND PLASTICITY PROBLEMS	65-0001
CREEP	MATRIX DIFFERENTIAL EQUATION AND HIGHER-ORDER NUMERICAL METHODS FOR PROBLEMS OF NON-LINEAR CREEP, VI	72-0198
CREEP	NON-UNIFORM FLOW OF POLYCRYSTALS BY GRAIN-BOUNDARY SLIDING ACCOMMODATED BY POWER-LAW CREEP	75-0562
CREEP	NONLINEAR ANALYSIS OF TWO-DIMENSIONAL PROBLEMS IN CONCRETE CREEP	71-0299
CREEP	NONLINEAR CREEP ANALYSIS AT ELEVATED TEMPERATURE	71-0050
CREEP	NONLINEAR CREEP ANALYSIS BY ASSUMED STRESS FINITE ELEMENT METHODS	74-0631
CREEP	NONLINEAR CREEP ANALYSIS BY ASSUMED STRESS FINITE ELEMENT METHODS	74-0922
CREEP	NONLINEAR THERMO ELASTIC PLASTIC AND CREEP ANALYSIS BY FINITE ELEMENT METHOD	74-0092
CREEP	PART 1. THEORY REPORT FOR CREEP-PLAST COMPUTER PROGRAM: ANALYSIS OF TWO-DIMENSIONAL PROBLEMS UNDER S	72-0765
CREEP	PREDICTION OF THE CREEP BEHAVIOUR OF A FLANGED JOINT	74-0757
CREEP	PRESSURE VESSEL WALL THICKNESS LIMITS BASED ON THERMAL STRESS RACHETTING AND CREEP FATIGUE INTERACTI	74-0968
CREEP	SAFE-COLAPS: A FINITE ELEMENT PROGRAM FOR THE CREEP COLLAPSE ANALYSIS OF NUCLEAR FUEL RODS	75-0573
CREEP	SIMPLIFIED INELASTIC (PLASTIC AND CREEP) ANALYSIS OF PIPE ELBOWS SUBJECTED TO INPLANE AND OUT-OF-PLA	75-0654
CREEP	SOME CREEP EFFECTS IN STRESS ANALYSIS WITH PARTICULAR REFERENCE TO CONCRETE PRESSURE VESSELS	69-0284
CREEP	STRAIN CONCENTRATION BEHAVIOUR IN A NOTCHED ROUND BAR SUBJECTED TO CREEP	73-0440
CREEP	STRESS ANALYSIS OF A GRAPHITE FUEL ELEMENT INCLUDING RADIATION CREEP AND DIMENSIONAL CHANGES	71-0217
CREEP	STRESS ANALYSIS OF TWO-DIMENSIONAL PROBLEMS UNDER SIMULTANEOUS CREEP AND PLASTICITY	71-0220
CREEP	TENSILE CREEP ANALYSIS OF CONCRETE STRUCTURES	74-0439
CREEP	THE COMPARISON OF ITERATIVE AND DIRECT SOLUTION TECHNIQUES IN THE ANALYSIS OF TIME-DEPENDENT STRESS	72-0080
CREEP	THERMOMECHANICAL CREEP OF AGEING CONCRETE - A UNIFIED APPROACH	74-0607
CREEP	THREE DIMENSIONAL RUPTURE ANALYSIS OF A PRESTRESSED CONCRETE PRESSURE VESSEL INCLUDING CREEP EFFECTS	73-0826
CREEP	THREE DIMENSIONAL RUPTURE ANALYSIS OF A PRESTRESSED CONCRETE PRESSURE VESSEL INCLUDING CREEP EFFECTS	74-0538
CREEP	THREE-DIMENSIONAL FINITE ELEMENT METHOD OF THERMOELASTOPLASTICITY WITH CREEP EFFECT	73-0837
CREEP	VISCO-PLASTICITY, PLASTICITY AND CREEP IN ELASTIC SOLIDS. A UNIFIED NUMERICAL SOLUTION APPROACH	74-1279
CREEP	VISCOPLASTICITY, PLASTICITY AND CREEP. A UNIFIED NUMERICAL SOLUTION APPROACH	74-1205
CREEP-BLAST	EVALUATION OF THE COMPUTER PROGRAM CREEP-BLAST (CDC VERSION)	74-0684
CREEP-PLAST	PART 1. THEORY REPORT FOR CREEP-PLAST COMPUTER PROGRAM: ANALYSIS OF TWO-DIMENSIONAL PROBLEMS UNDER S	72-0765
CREEPING	APPLICATION OF THE FINITE ELEMENT METHOD TO CREEPING FLOW PROBLEMS	70-0024
CREEPING	CREEPING FLOW IN RECTANGULAR DUCTS BY FINITE ELEMENT METHOD	75-0569
CREEPING	HIGH ORDER FINITE ELEMENT FOR COMPLETELY INCOMPRESSIBLE CREEPING FLOW	73-0144
CREST	FINITE ELEMENT METHOD APPLIED TO ANALYSIS OF FLOW OVER A SPILLWAY CREST	73-0300
CRESTS	CRACKING OF EARTH AND ROCKFILL DAMS: COMPARISON OF OBSERVED AND THEORETICAL TENSILE STRAINS IN THE C	71-0699
CRIMES	VARIATIONAL CRIMES IN THE FINITE ELEMENT METHOD	72-0517
CRITERIA	ASSESSMENT OF DESIGN TOOLS AND CRITERIA FOR URBAN RAIL TRACK STRUCTURES. VOLUME 11. AT-GRADE SLAB TR	74-0813
CRITERIA	CONVERGENCE CRITERIA FOR ITERATIVE PROCESSES	72-0664
CRITERIA	CRITERIA FOR FINITE ELEMENT DISCRETIZATION OF SHELLS OF REVOLUTION	73-0152
CRITERIA	FINITE ELEMENT MODELING CRITERIA FOR UNDERGROUND OPENINGS IN ROCK	74-0630
CRITERIA	INELASTIC ANALYSIS AND SATISFACTION OF DESIGN CRITERIA OF A HIGH TEMPERATURE COMPONENT	75-0265
CRITERIA	OPTIMALITY CRITERIA IN STRUCTURAL DESIGN	71-0758
CRITERIA	REVIEW OF OPTIMALITY CRITERIA APPROACHES TO STRUCTURAL OPTIMIZATION	74-0652
CRITERIA	ROUNDOFF CRITERIA IN DIRECT STIFFNESS SOLUTIONS	68-0204
CRITERIA	STABILITY CRITERIA FOR TWO FINITE ELEMENT SCHEMES FOR PARABOLIC EQUATION	75-0568
CRITERION	APPLICATION OF OPTIMALLY CRITERION TO FIBER-REINFORCED COMPOSITES	73-0431
CRITERION	CRITERION FOR SELECTING STIFFNESS MATRICES	65-0061
CRITERION	FINITE ELEMENT ANALYSIS OF ELASTIC-PLASTIC PLANE STRESS PROBLEMS BASED UPON TRESCA YIELD CRITERION	70-0179
CRITERION	ON TRESCA YIELD CRITERION IN FINITE ELEMENT ANALYSIS OF PLANE STRESS PROBLEMS	73-0314
CRITERION	QUANTITATIVE STRAIN-AND-STRESS STATE CRITERION FOR FAILURE IN THE VICINITY OF SHARP CRACKS	74-0639
CRITERION	STUDY ON J//I//C FRACTURE CRITERION	74-1034
CRITERION	USE OF THE LEAST SQUARES CRITERION IN THE FINITE ELEMENT FORMULATION	73-0270
CRITERION	YIELD CRITERION FOR REINFORCED CONCRETE SLABS	73-0985
CRITICAL	A CRITICAL REVIEW OF NASTRAN	71-0246
CRITICAL	ADAPTATION OF CRITICAL STATE SOIL MECHANICS FOR USE IN FINITE ELEMENTS	71-0534
CRITICAL	CALCULATION OF THE CRITICAL LOAD OF ELASTIC THIN WALLED BARS BY THE METHOD OF FINITE ELEMENTS	72-0446
CRITICAL	CERAMIC ROTATING BLADES: SOME CRITICAL DESIGN PARAMETERS FOR GAS TURBINE APPLICATIONS	74-0146
CRITICAL	COMPUTER METHOD FOR COMPUTING CRITICAL SPEEDS OF NON-UNIFORM SHAFTS GEARED SYSTEMS AND CURVED ASSEMB	73-0675
CRITICAL	DETERMINATION OF THE LOWEST CRITICAL BUCKLING STRESS OF VARIOUS SHAPED PLATES BY FINITE-ELEMENTS	73-1134
CRITICAL	DISCUSSION ON THE ADAPTION OF CRITICAL STATE SOLID MECHANICS THEORY FOR USE IN FINITE ELEMENTS	71-0007
CRITICAL	FINITE ELEMENT ANALYSIS OF SOME CRITICAL FAST ASSEMBLIES	73-0128
CRITICAL	FINITE ELEMENT APPROACH TO THE CRITICAL CYCLIC STRESS REQUIRED TO PROPAGATE A CRACK	73-0463
CRITICAL	LARGE DEFORMATION AND CRITICAL LOADS ANALYSIS OF FRAMED STRUCTURES	74-1293
CRITICAL	MANIPULATION ERRORS IN COMPUTER SOLUTION OF CRITICAL SIZE STRUCTURAL EQUATIONS	71-0367
CRITICAL	SETTLEMENT ANALYSIS OF A STRIP FOOTING USING A CRITICAL STATE SOIL MODEL IN CONJUNCTION WITH FINITE	71-0533
CRITICALITY	COMPUTATIONAL EXPERIMENTATION ON FINITE ELEMENT METHOD IN BARE SLAB CRITICALITY CALCULATIONS	73-0015

172

CROSS	ANALYSIS OF RAREFIED GAS FLOW THROUGH AN ARBITRARY CROSS SECTION BY THE FINITE ELEMENT METHOD	73-0365
CROSS	CALCUL DES PONTS BIAIS A POUTRES MULTIPLES SANS ENTRETOISES PAR LA METHODE DES ELEMENTS (CALCULATION	72-0184
CROSS	EINE FINITE-ELEMENTE-LOESUNG FUR DIE TORSIONSSTEIFIGKEIT UND DEN SCHUBMITTELPUNKT BELIEBIGER QUERSCH	74-0256
CROSS	FINITE ELEMENT ANALYSIS ON THE TORSION OF A BAR WITH UNIFORM CROSS SECTION	68-0186
CROSS	FINITE ELEMENT SIMULATION OF ACOUSTIC FILTERS WITH VARIABLE CIRCULAR CROSS SECTION	75-0871
CROSS	MODELING TECHNIQUES OF THIN-WALLED BEAMS WITH OPEN CROSS SECTIONS	72-0573
CROSS	TRANSVERSE VIBRATIONS OF CANTILEVER BARS OF VARIABLE CROSS SECTIONS	66-0001
CROSS-ANISOTROP	FINITE ELEMENT ANALYSIS OF A PLATE ON A LAYERED CROSS-ANISOTROPIC FOUNDATION	74-0451
CROSS-JOINT	STRESS ANALYSIS OF A TUBULAR CROSS-JOINT WITHOUT ANY INTERNAL STIFFENING FOR OFFSHORE STRUCTURES	74-0549
CROSS-LINKAGE	PROBLEMS OF CROSS-LINKAGE IN USING METHOD OF FINITE ELEMENTS	74-0165
CROSS-SECTION	BENDING AND TORSION OF THIN-WALLED STRAIGHT AND CURVED BEAMS WITH VARIABLE OPEN CROSS-SECTION	69-0493
CROSS-SECTION	ELASTIC-PLASTIC ANALYSIS OF BEAMS WITH UNIFORM CROSS-SECTION UNDER COMBINED LOADINGS	71-0777
CROSS-SECTION	ELASTIC-PLASTIC ANALYSIS OF A BAR WITH UNIFORM CROSS-SECTION UNDER COMBINED LOADINGS 1 - COMBINATION	71-0778
CROSS-SECTION	ELASTIC-PLASTIC ANALYSIS OF BEAMS WITH UNIFORM CROSS-SECTION UNDER COMBINED LOADINGS (2)	73-1070
CROSS-SECTION	FINITE-ELEMENT ANALYSIS OF FLOW IN BLOOD VESSEL WITH ARBITRARY CROSS-SECTION	72-0936
CROSS-SECTION	FORMULATION AND EVALUATION OF A TRIANGULAR CROSS-SECTION RING DISCRETE ELEMENT	66-0083
CROSS-SECTION	FORMULATION AND EVALUATION OF A TRAPEZOIDAL CROSS-SECTION RING DISCRETE ELEMENT	66-0084
CROSS-SECTION	ON VIBRATION ANALYSIS OF RECTANGULAR CANTILEVER PLATES WITH A STEPPED CROSS-SECTION BY THE FINITE EL	71-0608
CROSS-SECTION	PRISMATIC SHELLS WITH HEXAGONAL CROSS-SECTION	74-1043
CROSS-SECTION	TEMPERATURVERLAUF IM QUERSCHNITT GESCHWEISSTER ROHRWAENDE BEI UEBERWIEGEND KONVEKTIVEM WAERMEUEBERGA	75-0628
CROSS-SECTION	TRANSIENT LAMINAR FLOW IN DUCTS OF ARBITRARY CROSS-SECTION BY FINITE ELEMENT METHODS	72-0013
CROSS-SECTION	ZUR BERECHNUNG EINFACH-SYMMETRISCHER I-TRAEGER NACH DER THEORIE - 2. ORDNUNG UNTER BERUECKSICHTIGUNG	75-0510
CROSS-SECTIONS	FINITE ELEMENT ANALYSIS OF LAMINAR CONVECTIVE HEAT TRANSFER IN VERTICAL DUCTS WITH ARBITRARY CROSS-S	75-0555
CROSS-SECTIONS	FINITE ELEMENTS SOLUTION FOR TORSION RIGIDITY AND SHEAR CENTER OF ARBITRARY CROSS-SECTIONS	74-0168
CROSS-STIFFENED	A NASTRAN CORRELATION STUDY FOR VIBRATIONS OF A CROSS-STIFFENED SHIP'S DECK	72-0756
CROSS-STRUCTURE	STRUCTURAL ANALYSIS AND DESIGN OF A CATAMARAN CROSS-STRUCTURE BY FINITE ELEMENT METHOD	73-0052
CROSS-WALLED	ON THE BENDING STIFFNESS OF SLABS IN CROSS-WALLED STRUCTURES	74-0398
CROSSES	CONSTRUCTION OF A FINITE ELEMENT APPROXIMATION WHICH CROSSES MATERIAL INTERFACES	74-0896
CROSSTALK	NUMERICAL ANALYSIS OF THE LINE CAPACITANCE AND CROSSTALK FACTOR FOR INSULATED WIRE PAIRS	74-0685
CROWN	CONICAL SEGMENT METHOD FOR ANALYZING OPEN CROWN SHELLS OF REVOLUTION FOR EDGE LOADING	63-0007
CROWN	INFLUENCE OF THE UNPRECISION OF EXECUTION ON THE BEHAVIOUR OF A SPHERICAL SHELL WITH CONCENTRATED LO	73-0802
CROWN	STRESS ANALYSIS OF FIRST MOLARS WTH FULL CROWN PREPARATIONS BY THREE-DIMENSIONAL PHOTOELASTICITY AND	72-0939
CRUCIFORM	A CRUCIFORM ELEMENT FOR SURFACES OF INFLATABLE STRUCTURES	75-0118
CRUCIFORM	CRUCIFORM SPECIMENS FOR BIAXIAL FATIGUE TESTS - AN INVESTIGATION USING FINITE-ELEMENT ANALYSIS AND P	71-0795
CRUSH	VEHICLE CRUSH PREDICTION USING FINITE-ELEMENT TECHNIQUES	73-0280
CRUST	INFLUENCE OF FLUID INJECTION ON THE STATE OF STRESS IN THE EARTH'S CRUST	72-0624
CRUSTAL	CRUSTAL STRUCTURE FROM SURFACE LOAD TILTS USING A FINITE ELEMENT MODEL	72-0005
CRYSTAL	FINITE-ELEMENT ANALYSES OF INITIAL ELASTIC DEFORMATION OF VERTICALLY DEFORMED CRYSTAL BLOCKS	72-0970
CRYSTALLINE	A LARGE DEFORMATION ANALYSIS OF CRYSTALLINE ELASTIC-VISCOPLASTIC MATERIALS	73-0602
CRYSTALLINE	A NEW APPROACH TO THE FINITE-ELEMENT FORMULATION AND SOLUTION OF A CLASS OF PROBLEMS IN COUPLED THER	72-0732
CRYSTALLINE	A NEW APPROACH TO THE FINITE-ELEMENT FORMULATION AND SOLUTION OF A CLASS OF PROBLEMS IN COUPLED THER	73-0600
CRYSTALLINE	NEW APPROACH TO FINITE ELEMENT FORMULATION AND SOLUTION OF A CLASS OF PROBLEMS IN COUPLED THERMOELAS	73-0013
CRYSTALS	EFFECT OF GRAIN BOUNDARY FOR THE ELASTIC BEHAVIOR OF AL AND CU CRYSTALS (ANALYSIS BY THE FINITE ELEM	74-0947
CS-2	SABOR CS-2 USER'S MANUAL	70-0408
CUBIC	AN IMPROVED SHELL OF REVOLUTION ELEMENT UTILISING CUBIC DISPLACEMENT FUNCTIONS	70-0451
CUBIC	CUBIC SPLINE GALERKIN APPROXIMATIONS TO PARABOLIC SYSTEMS WITH COUPLED NON-LINEAR BOUNDARY CONDITION	75-0232
CUBIC	CURVED CUBIC TRIANGULAR FINITE ELEMENT FOR POTENTIAL FLOW PROBLEMS	73-0423
CUBIC	ON THE CONFORMING CUBIC TRIANGULAR ELEMENT FOR PLATE BENDING	72-0439
CUBIC	RAYLEIGH-RITZ APPROXIMATION OF PIECEWISE CUBIC POLYNOMIALS	66-0087
CUBIC	SPECIFICATION OF GEOMETRICAL PARAMETERS FOR ELEMENTS WITH CUBIC MAPPING FUNCTIONS	73-1082
CULVERT	GEOMETRIC HIGHWAY AND CULVERT DESIGN	74-1342
CULVERT	INVESTIGATION OF SOIL-STRUCTURE INTERACTION: FLEXIBLE CULVERT PIPES	72-0661
CULVERT	PERFORMANCE OF CULVERT UNDER WINSCAR DAM	75-0638
CULVERT	STRUCTURAL BEHAVIOR OF A FLEXIBLE METAL CULVERT UNDER A DEEP EARTH EMBANKMENT USING METHOD A BACKFIL	73-0739
CULVERTS	BALANCED DESIGN AND FINITE-ELEMENT ANALYSIS OF CULVERTS.	72-0462
CULVERTS	FORCES ON RIGID CULVERTS UNDER HIGH FILLS	67-0008
CUMULATIVE	FINITE ELEMENT AND CUMULATIVE DAMAGE ANALYSIS OF A KEYHOLE TEST SPECIMEN	75-0163
CUMULATIVE	FINITE ELEMENT AND CUMULATIVE DAMAGE ANALYSIS OF A KEYHOLE TEST SPECIMEN	75-0774
CUMULATIVE	THE USE OF ELASTIC-PLASTIC FINITE ELEMENT ANALYSIS IN THE CALCULATION OF CUMULATIVE FATIGUE DAMAGE	74-0314
CURING	INELASTIC MICROMECHANICS OF CURING STRESSES IN COMPOSITES	75-0284
CURRENT-DENSITY	FINITE ELEMENT CURRENT-DENSITY REPRESENTATION IN NUMERICAL-SOLUTION OF MICROSTRIP PROBLEMS	75-0400
CURRENTS	EDDY CURRENTS SCREENING PROBLEMS SOLVED BY MEANS OF FINITE ELEMENT DIFFERENCE AND FINITE ELEMENT MET	75-0552
CURRENTS	FINITE ELEMENT SOLUTION OF THE WIND DRIVEN CURRENTS AND ITS MASS TRANSPORT IN LAKES	74-1368
CURSHL	CURSHL: A HIGH-PRECISION FINITE ELEMENT FOR SHELLS OF ARBITRARY SHAPE	71-0430
CURVATURE	A METHOD FOR COMPUTATION OF VIBRATION MODES AND FREQUENCIES OF ORTHOTROPIC THIN SHELLS OF REVOLUTION	69-0037
CURVATURE	BEHAVIOR WITH DIMINISHING CURVATURE OF STRAIN-BASED ARCH FINITE ELEMENTS	73-0088
CURVATURE	CONSTANT CURVATURE BEAM FINITE ELEMENTS FOR IN-PLANE VIBRATION	72-0254
CURVATURE	DOUBLE CURVATURE SHELL FINITE-ELEMENT AND ITS USE IN THE DYNAMIC ANALYSIS OF COOLING TOWERS AND OTHE	70-0312
CURVATURE	ELASTIC-PLASTIC PLATE BENDING WITH CONSTANT CURVATURE ELEMENTS	74-0308
CURVATURE	FINITE ELEMENT ANALYSIS OF STRUCTURES WITH SURFACES OF ARBITRARY CURVATURE	70-0102
CURVATURE	STIFFNESS MATRICES FOR THIN TRIANGULAR ELEMENTS OF NONZERO GAUSSIAN CURVATURE	66-0073
CURVATURE	THE SEISMIC DESIGN STUDY OF A DOUBLE CURVATURE ARCH DAM	69-0053
CURVED	A COMPARISON OF CURVED BEAM FINITE ELEMENTS WHEN USED IN VIBRATION PROBLEMS	71-0227
CURVED	A CURVED BEAM ELEMENT	68-0202
CURVED	A CURVED CIRCULAR CYLINDRICAL SHELL FINITE ELEMENT FOR HIGHER ORDER SHELL THEORY	69-0528
CURVED	A CURVED CYLINDRICAL SHELL DISCRETE ELEMENT	67-0009
CURVED	A CURVED ELEMENT APPROXIMATION IN THE ANALYSIS OF AXISYMMETRIC THIN SHELLS	70-0063
CURVED	A CURVED FINITE ELEMENT FOR CYLINDRICAL SHELLS	68-0294
CURVED	A CURVED FINITE ELEMENT FOR THIN ELASTIC SHELLS	69-0486
CURVED	A CURVED, CYLINDRICAL SHELL, FINITE ELEMENT	68-0039
CURVED	A DOUBLY CURVED TRIANGULAR SHELL ELEMENT	68-0149
CURVED	A FINITE CYLINDRICAL SHELL ELEMENT AND THE VIBRATIONS OF A CURVED FAN BLADE	68-0131
CURVED	A FINITE ELEMENT ANALYSIS OF A DOUBLY CURVED GOTHIC VAULT	71-0827
CURVED	A FINITE ELEMENT FORMULATION FOR STABILITY ANALYSIS OF DOUBLY CURVED THIN SHELL STRUCTURES	69-0513
CURVED	A FINITE ELEMENT PROCEDURE FOR PLATES WITH CURVED BOUNDARIES	75-0030
CURVED	A FINITE ELEMENT PROCEDURE FOR PLATES WITH CURVED BOUNDARIES	75-0750
CURVED	A HIGHER ORDER IN-PLANE PARALLELOGRAM ELEMENT AND ITS APPLICATION TO SKEWED CURVED BOX GIRDER BRIDGE	71-0002
CURVED	A REFINED CURVED CYLINDRICAL SHELL FINITE ELEMENT	68-0038
CURVED	A REFINED CURVED ELEMENT FOR THIN SHELLS OF REVOLUTION	71-0214
CURVED	A SEQUEL TO TECHNICAL NOTE 13 - THE CURVED TETRAHEDRONAL AND TRIANGULAR ELEMENTS TEC AND TRIC FOR TH	69-0048
CURVED	A STIFFNESS MATRIX FOR A CURVED MEMBRANE SHELL	69-0055
CURVED	A THEORETICAL AND EXPERIMENTAL STUDY ON THE FREE VIBRATION OF CURVED SLAB BRIDGES	71-0496
CURVED	ACCURACY AND CONDITION OF CURVED (ISOPARAMETRIC) FINITE ELEMENTS	73-0057
CURVED	ACCURATE FINITE ELEMENT MODELLING OF FLAT AND CURVED STIFFENED PANELS	72-0538
CURVED	ALGORITHM FOR THE LARGE DEFLECTION GEOMETRICALLY NON-LINEAR BEHAVIOUR OF PLANE AND CURVED STRUCTURES	72-0083
CURVED	AN AUTOMATIC MESH GENERATION SCHEME FOR PLANE AND CURVED SURFACES BY ISOPARAMETRIC COORDINATES	71-0271
CURVED	ANALYSIS OF CURVED BOX GIRDER BRIDGES BY FINITE STRIP METHOD	71-0485
CURVED	ANALYSIS OF CURVED BOX GIRDER BRIDGES	73-0294
CURVED	ANALYSIS OF CURVED FOLDED PLATE STRUCTURES	70-0431
CURVED	ANALYSIS OF CURVED FOLDED PLATE STRUCTURES	71-0139
CURVED	ANALYSIS OF CURVED TARGET-TYPE THRUST REVERSERS	75-0152
CURVED	ANALYSIS OF SHELLS OF REVOLUTION BY A CURVED FINITE ELEMENT	75-0680

173

CURVED	ANALYSIS OF THICK AND THIN SHELL STRUCTURE BY CURVED FINITE ELEMENTS	70-0016
CURVED	ANALYSIS OF THIN SHELLS USING CURVED FINITE ELEMENTS	69-0523
CURVED	ANALYSIS OF THIN SHELLS WITH A DOUBLY CURVED ARBITRARY QUADRILATERAL FINITE ELEMENT.	72-0443
CURVED	ANALYSIS OF THIN-WALLED CURVED BEAMS BY THE TRANSFER MATRIX METHOD	72-0335
CURVED	APPLICATION OF CURVED FINITE ELEMENTS TO CIRCULAR ARCHES	71-0360
CURVED	APPLICATION OF THE HIGH PRECISION TRIANGULAR PLATE BENDING ELEMENT TO PROBLEMS WITH CURVED BOUNDARIE	69-0071
CURVED	APPLICATION, TESTING AND EVALUATION OF DOUBLY CURVED SHELL ELEMENTS FOR DYNAMIC ANALYSIS. PART II. F	69-0117
CURVED	APPLICATION, TESTING, AND EVALUATION OF DOUBLY CURVED SHELL ELEMENTS FOR DYNAMIC ANALYSIS. PART I. T	69-0246
CURVED	APPLICATION, TESTING, AND EVALUATION OF DOUBLY CURVED SHELL ELEMENTS FOR DYNAMIC ANALYSIS, PART III.	69-0256
CURVED	APPLICATIONS OF CURVED FINITE ELEMENT TO SANDWICH BEAMS USING SHEAR STRAIN FORMULATION	74-0523
CURVED	APPLICATIONS OF CURVED FINITE ELEMENTS TO SANDWICH SHELLS USING SHEAR STRAIN FORMULATION	75-0433
CURVED	BASIC FUNCTIONS FOR CURVED ELEMENTS IN THE MATHEMATICAL THEORY OF FINITE ELEMENTS	75-0091
CURVED	BASIS FUNCTIONS FOR CURVED ELEMENTS IN THE FINITE ELEMENT METHOD	72-0807
CURVED	BENDING AND TORSION OF THIN-WALLED STRAIGHT AND CURVED BEAMS WITH VARIABLE OPEN CROSS-SECTION	69-0493
CURVED	BLENDING FUNCTION FINITE ELEMENTS FOR CURVED BOUNDARIES	75-0093
CURVED	COMMENT ON LIMITATIONS OF CERTAIN FINITE ELEMENTS WHEN APPLIED TO ARCHES - FURTHER STUDIES IN APPLIC	72-0097
CURVED	COMMENT ON THE PAPER "POSSIBLE LOSS OF ACCURACY IN CURVED (ISOPARAMETRIC) FINITE ELEMENTS" BY HENSHE	72-0279
CURVED	COMPARISON OF SOME CURVED TWO DIMENSIONAL FINITE ELEMENTS	73-0383
CURVED	COMPUTER AIDED DESIGN AND CONSTRUCTION BY MEANS OF CURVED SURFACE AND SOLID ELEMENTS	73-0165
CURVED	COMPUTER METHOD FOR COMPUTING CRITICAL SPEEDS OF NON-UNIFORM SHAFTS GEARED SYSTEMS AND CURVED ASSEMB	73-0675
CURVED	CONTRIBUTION TO SOLUTION OF BODIES WITH CURVED BOUNDARY BY FINITE ELEMENT METHOD	72-1001
CURVED	CURVED BEAM FINITE ELEMENTS FOR COUPLED BENDING AND TORSIONAL VIBRATION	72-0720
CURVED	CURVED BOUNDARIES IN THE FINITE ELEMENT METHOD	73-0638
CURVED	CURVED BOUNDARY ELEMENTS FOR FINITE ELEMENT METHOD	72-0859
CURVED	CURVED BOUNDARY ELEMENTS, GENERAL FORMS OF POLYNOMIAL MAPPINGS	74-1256
CURVED	CURVED CUBIC TRIANGULAR FINITE ELEMENT FOR POTENTIAL FLOW PROBLEMS	73-0423
CURVED	CURVED ELEMENTS IN FINITE ELEMENT METHOD. 1.	73-0045
CURVED	CURVED ELEMENTS IN FINITE ELEMENT METHOD - 2	74-0172
CURVED	CURVED ELEMENTS IN THE FINITE ELEMENT METHOD	74-1261
CURVED	CURVED ELEMENTS, ERROR BOUNDS, TIME DEPENDENT PROBLEMS FIVE LECTURES	73-0636
CURVED	CURVED FINITE ELEMENTS IN THE ANALYSIS OF SOLID SHELL AND PLATE STRUCTURES	69-0357
CURVED	CURVED FINITE ELEMENT FOR THIN ELASTIC SHELLS	70-0173
CURVED	CURVED FINITE ELEMENTS IN SHELL CALCULATIONS	70-0227
CURVED	CURVED FINITE ELEMENTS IN THE ANALYSIS OF SHELL STRUCTURES	71-0076
CURVED	CURVED FINITE ELEMENT APPROXIMATION TO NONLINEAR CABLES	72-0379
CURVED	CURVED FINITE ELEMENTS BY METHODS OF INITIAL STRAINS	73-0030
CURVED	CURVED FINITE ELEMENTS FOR THE ANALYSIS OF SHALLOW AND DEEP ARCHES	74-0332
CURVED	CURVED ISOPARAMETRIC QUADRILATERAL ELEMENTS FOR FINITE ANALYSIS	68-0008
CURVED	CURVED ISOPARAMETRIC QUADRILATERAL ELEMENTS FOR FINITE ELEMENT ANALYSIS	68-0067
CURVED	CURVED LOW ORDER ANGULAR ELEMENTS IN TWO DIMENSIONAL NEUTRON TRANSPORT PROBLEMS	74-1322
CURVED	CURVED ROTATIONAL SHELL ELEMENTS BY THE CONSTRAINT METHOD	72-0924
CURVED	CURVED THICK SHELL AND MEMBRANE ELEMENTS WITH PARTICULAR REFERENCE TO AXISYMMETRIC PROBLEMS	68-0005
CURVED	CURVED TRIANGULAR ELEMENTS FOR THE ANALYSIS OF SHELLS	68-0030
CURVED	CURVED TRIANGULAR FINITE ELEMENTS FOR THE ANALYSIS OF THICK ELASTIC PLATES AND SHELLS	71-0814
CURVED	CURVED, CYLINDRICAL SHELL, FINITE ELEMENT	72-0096
CURVED	DIRECT STIFFNESS METHOD OF ANALYSIS OF SHELLS OF REVOLUTION UTILIZING CURVED ELEMENTS	66-0042
CURVED	DOUBLY CURVED TRIANGULAR FINITE ELEMENTS FOR SHELLS OF ARBITRARY SHAPE	73-0881
CURVED	DYNAMIC ANALYSIS OF SHELLS USING CURVED FINITE ELEMENTS	69-0296
CURVED	DYNAMIC ANALYSIS OF CURVED BOX-GIRDER BRIDGES	75-0280
CURVED	ELASTIC SNAP THROUGH ANALYSIS OF CURVED PLATES USING DISCRETE ELEMENTS	72-0899
CURVED	ELASTIC SNAP-THROUGH ANALYSIS OF CURVED PLATES USING DISCRETE ELEMENTS	72-0927
CURVED	EXPERIMENTAL AND ANALYTICAL INVESTIGATION OF A HORIZONTALLY CURVED BOX-BEAM HIGHWAY BRIDGE MODEL	71-0024
CURVED	EXPERIMENTAL AND THEORETICAL INVESTIGATION OF THE BEHAVIOR OF BOX GIRDERS CURVED IN PLAN	75-0659
CURVED	EXPLICIT ADDITION OF RIGID-BODY MOTIONS IN CURVED FINITE ELEMENTS	73-0037
CURVED	FINITE ELEMENT ANALYSIS OF SKEW, CURVED BOX-GIRDER BRIDGE	70-0241
CURVED	FINITE ELEMENT ANALYSIS OF BRIDGES CURVED IN PLAN	71-0547
CURVED	FINITE ELEMENT ANALYSIS OF CYLINDRICAL SHELLS UTILIZING CURVED ELEMENTS CONTAINING RIGID BODY DEFORM	71-0823
CURVED	FINITE ELEMENT ANALYSIS OF PLATES WITH CURVED EDGES.	72-0173
CURVED	FINITE ELEMENT ANALYSIS OF JET IMPINGEMENT ON AXISYMMETRIC CURVED DEFLECTORS	75-0396
CURVED	FINITE ELEMENT PROCEDURE FOR POSTBUCKLING ANALYSIS OF INITIALLY CURVED PLATES	71-0421
CURVED	FINITE ELEMENT TECHNIQUES FOR CURVED BOUNDARIES	73-0622
CURVED	FINITE ELEMENTE FUR GEKRUEMMTE FLAECHENTRAGWERKE. (FINITE ELEMENTS SOLUTIONS FOR CURVED SHELL STRUCT	71-0391
CURVED	FINITE-ELEMENT FOR BUCKLING OF CURVED BEAMS AND SHELLS WITH SHEAR	75-0607
CURVED	FINITE-ELEMENT PROCEDURES FOR PROBLEMS WITH CURVED BOUNDARIES	72-0977
CURVED	FINITE-ELEMENT SOLUTIONS WITHIN CURVED BOUNDARIES	72-0038
CURVED	FREE VIBRATION OF A CURVED BEAM	71-0606
CURVED	FREE VIBRATION OF CURVED SANDWICH BEAMS BY METHOD OF FINITE ELEMENTS	71-0351
CURVED	FREE VIBRATION OF CURVED AND STRAIGHT BEAM SLAB OR BOX GIRDER BRIDGES	72-0487
CURVED	FREE VIBRATION OF DOUBLY CURVED RECTANGULAR PLATES INCLUDING THE PRESENCE OF MEMBRANE STRESSES WITH	69-0468
CURVED	FREE VIBRATIONS OF CURVED SANDWICH BEAMS	71-0729
CURVED	FURTHER STUDIES IN THE APPLICATION OF CURVED FINITE ELEMENTS TO CIRCULAR ARCHES	71-0644
CURVED	HIGH-PRECISION CURVED SHELL FINITE ELEMENT	71-0422
CURVED	HORIZONTALLY CURVED HIGHWAY BRIDGES. A FULLY COMPATIBLE ANNULAR SEGMENT FINITE ELEMENT	72-0834
CURVED	IMPLICIT RIGID BODY MOTION IN CURVED FINITE ELEMENTS	71-0174
CURVED	INSTABILITY OF HORIZONTALLY CURVED MEMBERS. FLANGE BUCKLING STUDIES	69-0480
CURVED	INTERPOLATION THEORY OVER CURVED ELEMENTS WITH APPLICATION TO THE FINITE ELEMENT METHOD	72-0356
CURVED	JET IMPINGEMENT ON AXISYMMETRIC CURVED DEFLECTORS	74-0017
CURVED	LAGRANGE INTERPOLATION ON CURVED FINITE ELEMENTS	72-0163
CURVED	LIMITATIONS OF CERTAIN CURVED FINITE ELEMENTS WHEN APPLIED TO ARCHES	71-0279
CURVED	MODEL ANALYSIS OF CURVED BOX-BEAM HIGHWAY BRIDGE	71-0282
CURVED	MODELING EARDRUM AS A DOUBLY CURVED SHELL USING FINITE-ELEMENT METHOD	75-0047
CURVED	NUMERICAL ANALYSIS OF THIN SHELLS BY CURVED TRIANGULAR ELEMENTS BASED ON DISCRETE - KIRCHHOFF HYPOTH	69-0091
CURVED	ON THE ENERGY FUNCTIONS FOR CURVED SANDWICH PLATES	71-0587
CURVED	ON THE EXPLICIT REPRESENTATION OF TRANSFER-MATRICES OF CURVED BEAMS AND THEIR APPLICATIONS	75-0565
CURVED	ORDER OF CONVERGENCE OF A FINITE-ELEMENT METHOD IN REGIONS WITH CURVED BOUNDARIES	75-0817
CURVED	OVERCOMING LOSS OF ACCURACY WHEN USING CURVED FINITE ELEMENTS	75-0092
CURVED	PLANAR AND CURVED SHELL ELEMENTS	69-0186
CURVED	POSSIBLE LOSS OF ACCURACY IN CURVED FINITE ELEMENTS	72-0278
CURVED	POST-BUCKLING ANALYSIS OF SLIGHTLY CURVED PLATES BY THE FINITE ELEMENT METHOD	69-0365
CURVED	RIGID BODY DISPLACEMENTS OF CURVED ELEMENTS IN THE ANALYSIS OF SHELLS BY THE MATRIX-DISPLACEMENT MET	67-0031
CURVED	RIGID BODY MOTIONS IN CURVED FINITE ELEMENTS	70-0035
CURVED	RIGID-BODY MOTIONS AND STRAIN-DISPLACEMENT EQUATIONS OF CURVED SHELL FINITE ELEMENTS	72-0267
CURVED	STACUSS 1: A DISCRETE-ELEMENT PROGRAM FOR THE STATIC ANALYSIS OF SINGLE-LAYER CURVED STIFFENED SHELL	69-0160
CURVED	STIFFNESS MATRICES FOR THE GENERAL DEFORMATION (OUT-OF-PLANE AND INPLANE) OF CURVED BEAM MEMBERS BAS	75-0124
CURVED	STIFFNESS MATRIX FOR CURVED TAPERED AND STRAIGHT TAPERED SHELL STIFFENERS	69-0074
CURVED	STRESS ANALYSIS OF COMPLEX SHIP COMPONENTS BY A NUMERICAL PROCEDURE USING CURVED FINITE ELEMENTS	73-0879
CURVED	STRESS ANALYSIS OF CURVED PIPES WITH END RESTRAINTS	75-0667
CURVED	STRONGLY CURVED FINITE ELEMENT FOR SHELL ANALYSIS	73-0244
CURVED	STUDIES IN DOUBLY CURVED SHELL ELEMENTS	74-0133
CURVED	TECHNIQUES IN TRIANGULAR MESH GENERATION FLAT AND CURVED SURFACES	69-0394
CURVED	THE ANALYSIS OF CURVILINEAR ORTHOTROPIC CURVED BRIDGE DECKS	69-0162
CURVED	THE APPLICATION OF A CURVED MIXED TYPE SHELL ELEMENT	70-0118

CURVED	THE APPLICATION OF A CURVED MIXED-TYPE SHELL ELEMENT	70-0440
CURVED	THE COMBINED EFFECT TO CURVED BOUNDARIES AND NUMERICAL INTEGRATION IN ISOPARAMETRIC FINITE ELEMENT M	72-0357
CURVED	THE CONSTRUCTION OF BASIC FUNCTIONS FOR CURVED ELEMENTS IN THE FINITE ELEMENT METHOD	72-0558
CURVED	THE CURVED TETRAHEDRONAL AND TRIANGULAR ELEMENTS TEC AND TRIC FOR THE MATRIX DISPLACEMENT METHOD	69-0047
CURVED	THE EFFECT OF IN PLANE STATIC STRESSES ON THE VIBRATION CHARACTERISTICS OF CURVED SANDWICH PLATES	71-0588
CURVED	THE FINITE ELEMENT METHOD IN DOMAINS WITH CURVED BOUNDARIES	73-0080
CURVED	THREE-DIMENSIONAL FINITE ELEMENT ANALYSIS OF THICK-WALLED PIPE-NOZZLE JUNCTIONS WITH CURVED TRANSITI	71-0153
CURVED	TRANSFINITE ELEMENT METHODS: BLENDING FUNCTION INTERPOLATION OVER ARBITRARY CURVED ELEMENT DOMAINS	73-1006
CURVED	USE OF PARABOLIC ARCS IN MATCHING CURVED BOUNDARIES IN FINITE-ELEMENT METHOD	75-0589
CURVED	USE OF STRAIGHT BEAM FINITE ELEMENTS FOR ANALYSIS OF VIBRATIONS OF CURVED BEAMS	73-0616
CURVED	VIBRATION ANALYSIS OF CANTILEVERED CURVED PLATES USING A NEW CYLINDRICAL SHELL FINITE ELEMENT	68-0130
CURVED	VIBRATION OF CURVED STRUCTURES USING QUADRILATERAL FINITE ELEMENTS	72-0074
CURVED	VIBRATION OF MULTISUPPORTED CURVED BEAMS	74-1297
CURVED	VIBRATION OF THICK, CURVED, SHELLS WITH PARTICULAR REFERENCE TO TURBINE BLADES	70-0017
CURVED	VIBRATIONS OF TAPERED AND CURVED MEMBERS BY FINITE ELEMENT METHODS	69-0517
CURVED-BEAM	GENERALIZED STIFFNESS MATRIX OF A CURVED-BEAM ELEMENT	69-0165
CURVES	SIMPLE EQUATIONS FOR THE MAGNETIZATION AND RELUCTIVITY CURVES OF STEEL	75-0156
CURVILINEAR	A CURVILINEAR SHELL FINITE ELEMENT	69-0153
CURVILINEAR	A NEW FAMILY OF CURVILINEAR PLATE BENDING ELEMENTS FOR VIBRATION AND STABILITY	72-0645
CURVILINEAR	A NEW FAMILY OF CURVILINEAR PLATE BENDING ELEMENTS FOR VIBRATION AND STABILITY	72-0999
CURVILINEAR	CONSTRUCTION OF CURVILINEAR COORDINATE SYSTEMS AND APPLICATIONS TO MESH GENERATION	74-1295
CURVILINEAR	CURVILINEAR FINITE ELEMENTS FOR 2-DIMENSIONAL SATURABLE MAGNETIC FIELDS	74-0562
CURVILINEAR	FINITE ELEMENT FORMULATION OF THE HEAT CONDUCTION EQUATION IN GENERAL ORTHOGONAL CURVILINEAR COORDIN	75-0848
CURVILINEAR	FINITE ELEMENT NEUTRON DIFFUSION IN CURVILINEAR COORDINATES	72-0039
CURVILINEAR	FINITE ELEMENTS IN CURVILINEAR COORDINATES - SECOND ORDER SURFACES	73-0753
CURVILINEAR	GEOMETRIC ASPECTS OF THE FINITE ELEMENT METHOD: CONSTRUCTION OF CURVILINEAR COORDINATE SYSTEMS AND T	72-0534
CURVILINEAR	INTERACTIVE PROCEDURE FOR CURVILINEAR COORDINATION OF A PLANAR REGION WITH APPLICATIONS TO FINITE-EL	73-0040
CURVILINEAR	NOTE ON THE CONVERGENCE OF FINITE ELEMENT APPROXIMATIONS FOR PROBLEMS FORMULATED IN CURVILINEAR COOR	74-0267
CURVILINEAR	NUMERICAL METHOD FOR GENERATING ORTHOGONAL CURVILINEAR MESHES	70-0443
CURVILINEAR	THE ANALYSIS OF CURVILINEAR ORTHOTROPIC CURVED BRIDGE DECKS	69-0162
CURVILINEAR	3-DIMENSIONAL ISOPARAMETRIC ELEMENT IN CURVILINEAR COORDINATES	72-0098
CUT	GROUNDWATER PREDICTIONS FOR A HIGHWAY CUT USING FINITE ELEMENTS	73-0346
CUT	ROCK MECHANICS STUDIES FOR A HIGH CUT AT A NUCLEAR POWER STATION IN PENNSYLVANIA	70-0122
CUT-GROWTH	FINITE ELEMENT ANALYSIS OF CUT-GROWTH IN SHEETS OF HIGHLY ELASTIC MATERIALS	73-0514
CUT-OUTS	NATURAL FREQUENCIES AND MODE SHAPES OF PLATES WITH INTERIOR CUT-OUTS	70-0402
CUT-OUTS	THE MATRIX ANALYSIS OF STRUCTURES WITH CUT-OUTS AND MODIFICATIONS	56-0003
CUTHILL-MCKEE	COMPARATIVE ANALYSIS OF THE CUTHILL-MCKEE AND THE REVERSE CUTHILL-MCKEE ORDERING ALGORITHMS FOR SPAR	73-1099
CUTHILL-MCKEE	COMPARATIVE ANALYSIS OF THE CUTHILL-MCKEE AND THE REVERSE CUTHILL-MCKEE ORDERING ALGORITHMS FOR SPAR	73-1099
CUTOUT	INVESTIGATION OF THE STRESS-STRAINED STATE OF SPHERICAL SHELLS WITH AN ECCENTRIC CUTOUT ON THE BASIS	73-0222
CUTOUT	NASTRAN BUCKLING ANALYSIS OF A LARGE STIFFENED CYLINDRICAL SHELL WITH A CUTOUT	71-0045
CUTOUTS	AN INVESTIGATION OF THE EFFECTS OF SQUARE CUTOUTS ON THE NATURAL FREQUENCIES AND MODE SHAPES OF RECT	71-0638
CUTOUTS	ANALYSIS OF COMPOSITE REINFORCED CUTOUTS AND CRACKS	74-0824
CUTOUTS	COMPOSITE-OVERLAY REINFORCEMENT OF CUTOUTS AND CRACKS IN METAL SHEET	73-0878
CUTOUTS	EVALUATION OF COMPOSITE SHEAR PANELS WITH CUTOUTS	72-0615
CUTOUTS	ISSLEDOVANIE NAPRYAZHENNOGO SOSTOYANIYA REBRISTYKH TSILINDRICHESKIKH OBOLOCHEK S PRYAMOUGOL'NYMI OTV	74-0944
CUTOUTS	METOD KONECHNYKH ELEMENTOV V PRIMENENII K RASCHETU TSILINDRICHESKIKH OBOLOCHEK S PRYAMOUGOL'NYMI OTV	73-0536
CUTOUTS	THE BEHAVIOUR OF ADVANCED FILAMENTARY COMPOSITE PLATES WITH CUTOUTS	73-0891
CUTOUTS	THE BEHAVIOUR OF ADVANCED FILAMENTARY COMPOSITE PLATES WITH CUTOUTS	70-0025
CUTS	A DESIGN PROCEDURE FOR HIGH CUTS IN JOINTED HARD ROCK. THREE DIMENSIONAL SOLUTIONS. VOLUME 1	71-0656
CUTS	MESH GENERATION PROGRAM FOR HIGHWAY EXCAVATION CUTS	74-1250
CUTTING	DETERMINATION OF THE STRESS-STRAINED STATE WHEN CUTTING METAL WITH A WEDGE	73-0237
CUTTING	INCIPIENT CHIP FORMATION IN METAL CUTTING - A THREE- DIMENSION FINITE-ELEMENT ANALYSIS	73-1146
CUTTING	INVESTIGATION OF VENEER CUTTING BY ELASTIC FINITE ELEMENT MODELS	74-1110
CUTTING	RESIDUAL STRESS PRODUCED BY METAL CUTTING	71-0414
CYCLE	A METHOD FOR DETERMINING LOCAL ELASTOPLASTIC STRESS AND STRAIN IN METALLURGICALLY BONDED NOTCHED LAM	74-0686
CYCLE	CYCLIC FATIGUE ANALYSIS OF ROCKET THRUST CHAMBERS. VOLUME 2: ATTITUDE CONTROL THRUSTER HIGH CYCLE FA	74-0693
CYCLE	CYCLIC FATIGUE ANALYSIS OF ROCKET THRUST CHAMBERS. VOLUME 1: OFHC COPPER CHAMBER LOW CYCLE FATIGUE	74-0694
CYCLE	HOW FINITE ELEMENT METHODS IMPROVE THE DESIGN CYCLE	74-0372
CYCLE	LOW CYCLE FATIGUE STRENGTH OF PIPING COMPONENTS	73-0366
CYCLE	NOTCH EFFECT IN LOW CYCLE FATIGUE - REPORT 1: LOW-CYCLE FATIGUE STRENGTH OF 0.48 PERCENT CARBON STEE	73-1090
CYCLIC	APPLICATION OF FINITE ELEMENT METHOD TO CYCLIC LOADING OF ELASTIC-PLASTIC STRUCTURES CONTAINING EFFE	73-0103
CYCLIC	ASPECTS OF MECHANICAL BEHAVIOR OF ROCK UNDER STATIC AND CYCLIC LOADING. PART A. MECHANICAL BEHAVIOR	72-0761
CYCLIC	ASPECTS OF MECHANICAL BEHAVIOR OF ROCK UNDER STATIC AND CYCLIC LOADING. PART A: MECHANICAL BEHAVIOR	72-0762
CYCLIC	CYCLIC ANALYSIS OF A PROPAGATING CRACK AND ITS CORRELATION WITH FATIGUE CRACK GROWTH	75-0188
CYCLIC	CYCLIC FATIGUE ANALYSIS OF ROCKET THRUST CHAMBERS. VOLUME 2: ATTITUDE CONTROL THRUSTER HIGH CYCLE FA	74-0693
CYCLIC	CYCLIC FATIGUE ANALYSIS OF ROCKET THRUST CHAMBERS. VOLUME 1: OFHC COPPER CHAMBER LOW CYCLE FATIGUE	74-0694
CYCLIC	CYCLIC LOADING AND CRACK PROPAGATION - AN ELASTOPLASTIC FINITE ELEMENT STUDY	74-1214
CYCLIC	CYCLIC MECHANICAL TESTS AND AN APPROPRIATE ANALYTICAL STRESS-STRAIN MODEL FOR A36 STEEL	74-0815
CYCLIC	CYCLIC PLASTIC ANALYSIS OF STRUCTURAL STEEL JOINTS	73-0799
CYCLIC	DISCRETE ELEMENT METHODS FOR THE PLASTIC ANALYSIS OF STRUCTURES SUBJECTED TO CYCLIC LOADING	70-0023
CYCLIC	ELASTIC PLASTIC DEFORMATION OF CYLINDRICAL PRESSURE VESSELS UNDER CYCLIC LOADING	74-0604
CYCLIC	ELASTIC-PLASTIC ANALYSIS OF A PROPAGATING CRACK UNDER CYCLIC LOADING	75-0857
CYCLIC	ELASTO-PLASTIC CYCLIC ANALYSIS OF STRUCTURAL MEMBERS	72-0202
CYCLIC	FINITE ELEMENT ANALYSIS OF CRACK PROPAGATION IN 3-DIMENSIONAL SOLIDS UNDER CYCLIC LOADING	74-0479
CYCLIC	FINITE ELEMENT APPROACH TO THE CRITICAL CYCLIC STRESS REQUIRED TO PROPAGATE A CRACK	73-0463
CYCLIC	FINITE ELEMENT CYCLIC THERMOPLASTICITY ANALYSIS BY THE METHOD OF SUBVOLUMES	75-0787
CYCLIC	FINITE-ELEMENT METHOD IN CYCLIC THERMO-ELASTOPLASTICITY	73-1111
CYCLIC	GAS TURBINE ENGINE DISK CYCLIC LIFE PREDICTION	75-0509
CYCLIC	INELASTIC MODEL FOR CYCLIC BIAXIAL LOADING OF REINFORCED CONCRETE	74-0503
CYCLIC	ISOTROPIC-KINEMATIC HARDENING MODEL FOR ELASTIC-PLASTIC CYCLIC STRUCTURAL ANALYSIS	75-0210
CYCLIC	METHOD FOR THE COMPUTATION OF CYCLIC STRESSES IN A GAS-TURBINE WHEEL	74-0972
CYCLIC	NONLINEAR BEHAVIOR OF SHELLS OF REVOLUTION UNDER CYCLIC LOADING	73-0242
CYCLIC	NOTE ON STRESS AND STRAIN REDISTRIBUTION IN A NOTCHED PLATE SPECIMEN DURING CYCLIC LOADING	69-0194
CYCLIC	PROPAGATION AND OPENING OF A THROUGH CRACK IN PIPE SUBJECT TO COMBINED CYCLIC THERMOMECHANICAL LOADI	75-0516
CYCLIC	RETSCP: A COMPUTER PROGRAM FOR ANALYSIS OF ROCKET ENGINE THERMAL STRAINS WITH CYCLIC PLASTICITY	74-0692
CYCLIC	THE INELASTIC BEHAVIOUR OF STRUCTURES SUBJECTED TO CYCLIC, THERMAL AND MECHANICAL STRESSING CONDITIO	60-0004
CYCLIC	THERMO-ELASTIC-PLASTIC CYCLIC ANALYSIS BY FINITE ELEMENT METHOD	73-1003
CYCLIC-STRESS	CYCLIC-STRESS ANALYSIS OF NOTCHES FOR SUPERSONIC TRANSPORT CONDITIONS	74-0834
CYCLINDRICAL	BENDING OF CYLINDRICAL SHELLS - SOLUTION USING CYCLINDRICAL STRIP ELEMENTS - CONTRIBUTION TO FINITE	70-0363
CYCLINDRICAL	FREE VIBRATION OF RING STIFFENED CYCLINDRICAL SHELLS	70-0172
CYCLING	FINITE ELEMENT ANALYSIS OF CRACK PROPAGATION IN THREE-DIMENSIONAL SOLIDS UNDER CYCLING LOADING	73-0529
CYCLING	PAVEMENT CRACKING IN WEST TEXAS DUE TO FREEZE-THAW CYCLING	75-0276
CYLINDER	A FINITE ELEMENT SOLUTION OF THE CONFINED FLOW OVER A CIRCULAR CYLINDER	74-0023
CYLINDER	A FINITE-ELEMENT APPROACH TO THE ANALYSIS OF TANGENTIALLY POLARIZED PIEZOELECTRIC-CERAMIC FREE-FLOOD	74-0802
CYLINDER	AN EXPERIMENTAL AND THEORETICAL STUDY OF HOLLOW CYLINDER TESTS ON SAND	71-0008
CYLINDER	BEHAVIOR OF A BI-MATERIAL CYLINDER INCLUDING THE EFFECTS OF FRACTURE	69-0466
CYLINDER	DETERMINATION OF STRESS CONCENTRATIONS AT NOZZLE TO CYLINDER INTERSECTIONS AND COMPARISON WITH EXPER	73-0491
CYLINDER	DIFFRACTION OF OBLIQUE WAVES BY AN INFINITE CYLINDER	75-0605
CYLINDER	DYNAMIC RESPONSE OF A CYLINDER BURIED IN AN EARTH BEAM - RESULTS OF A FINITE ELEMENT ANALYSIS	72-0592
CYLINDER	ELASTOPLASTIC INDENTATION OF HALF-SPACE BY AN INFINITELY LONG RIGID CIRCULAR CYLINDER	71-0395
CYLINDER	EXACT SOLUTION TO THE RADIALLY END CONSTRAINED CIRCULAR CYLINDER UNDER TRIAXIAL LOADING	71-0433
CYLINDER	EXPERIMENTAL AND FINITE ELEMENT STRESS ANALYSIS OF A THIN SHELLED CYLINDER TO CYLINDER MODEL	71-0066

CYLINDER	EXPERIMENTAL AND FINITE ELEMENT STRESS ANALYSIS OF A THIN SHELLED CYLINDER TO CYLINDER MODEL	71-0066
CYLINDER	FINITE ELEMENT ANALYSIS OF AN EXPANDED I - SECTION BEAM AND AN AXISYMMETRIC FLANGED CYLINDER	68-0002
CYLINDER	FINITE ELEMENT ANALYSIS OF A THICK WALLED MICROPOLAR CYLINDER LOADED AXISYMMETRICALLY	73-0847
CYLINDER	FINITE-ELEMENT SOLUTION FOR A CRACKED TWO-LAYERED ELASTIC CYLINDER	71-0397
CYLINDER	NON-UNIFORM DEFORMATION OF MATERIAL IN AXIALLY SYMMETRIC COMPRESSION CAUSED BY FRICTION. PART 3, THE	72-0923
CYLINDER	NUMERICAL SOLUTION FOR THE FLOW AROUND A CYLINDER AT REYNOLDS NUMBERS OF 40, 200 AND 500	69-0378
CYLINDER	ON THE STRESS DISTRIBUTION IN NOZZLE TO CYLINDER CONNECTIONS FOR SMALL DIAMETER RATIOS	72-0426
CYLINDER	PROPAGATION OF FAILURE IN A CIRCULAR CYLINDER OF ROCK SUBJECTED TO A COMPRESSIVE FORCE	73-0793
CYLINDER	STRESS INTENSITY FACTOR OF A SURFACE CRACK IN A PRESSURIZED CYLINDER	75-0255
CYLINDER	TRANSIENT NATURAL CONVECTION COOLING OF A VERTICAL CIRCULAR CYLINDER	74-1169
CYLINDER-CONE-C	A STRUCTURAL ANALYSIS OF CYLINDER-CONE-CYLINDER SHELLS BY THE FINITE ELEMENT METHOD	72-0796
CYLINDER-CYLIND	CALCULATION OF THERMAL STRESSES IN A CYLINDER-CYLINDER INTERSECTION BY MEANS OF FINITE ELEMENTS	73-0571
CYLINDER-CYLIND	EURCYL: A COMPUTER PROGRAM TO GENERATE FINITE ELEMENT MESHES FOR CYLINDER-CYLINDER INTERSECTIONS	73-0962
CYLINDER-CYLIND	THE APPLICATION OF THREE-DIMENSIONAL FINITE ELEMENTS TO A CYLINDER-CYLINDER INTERSECTION	70-0068
CYLINDER-TO-CYL	EXPERIMENTAL ELASTIC STRESS ANALYSES OF CYLINDER-TO-CYLINDER SHELL MODELS AND COMPARISONS WITH THEOR	71-0065
CYLINDER-TO-CYL	THEORETICAL AND EXPERIMENTAL STRESS ANALYSES OF ORNL THIN-SHELL CYLINDER-TO-CYLINDER MODEL 3	75-0690
CYLINDER-TO-CYL	THEORETICAL AND EXPERIMENTAL STRESS ANALYSES OF ORNL THIN-SHELL CYLINDER-TO-CYLINDER MODEL 4	75-0695
CYLINDERS	ANALYSES OF AXISYMMETRIC UPSETTING AND PLANE STRAIN SIDE PRESSING OF SOLID CYLINDERS BY THE FINITE E	70-0083
CYLINDERS	ANALYSES OF AXYSYMMETRIC UPSETTING AND PLANE-STRAIN SIDE-PRESSING OF SOLID CYLINDERS BY THE FINITE E	71-0794
CYLINDERS	ANALYSIS OF STRENGTH AND RESIDUAL STRESSES IN FILAMENT REINFORCED ALUMINUM CYLINDERS	75-0517
CYLINDERS	APPLICATION OF FRACTURE MECHANICS TO THERMALLY STRESSED CYLINDERS USING A FINITE ELEMENT METHOD	69-0089
CYLINDERS	APPLICATION OF FRACTURE MECHANICS TO FAILURE OF THICK WALLED CYLINDERS	71-0511
CYLINDERS	CONTROLLED FAILURE OF HOLLOW ROCK CYLINDERS IN UNIAXIAL COMPRESSION	72-0196
CYLINDERS	EFFECT OF BACKPACKING AND INTERNAL PRESSURIZATION ON STRESSES TRANSMITTED TO BURIED CYLINDERS	73-0897
CYLINDERS	ELASTIC PLASTIC DEFORMATION OF THICK WALLED CYLINDERS	69-0427
CYLINDERS	END DISPLACEMENTS OF SEMI-INFINITE CYLINDERS DUE TO ANNULAR LOADINGS	69-0399
CYLINDERS	FINITE ELEMENT ANALYSIS OF BURIED CYLINDERS	71-0817
CYLINDERS	FINITE ELEMENT ANALYSIS OF AXISYMMETRIC VIBRATIONS OF CYLINDERS	72-0255
CYLINDERS	FINITE ELEMENT ANALYSIS OF INTERSECTING CYLINDERS	73-0866
CYLINDERS	FINITE ELEMENT SOLUTION OF NORMALLY INTERSECTING CYLINDERS	73-0723
CYLINDERS	FINITE ELEMENTS FOR THE VIBRATION OF CONES AND CYLINDERS	75-0340
CYLINDERS	FRACTURE TOUGHNESS TESTING WITH SECTIONS OF CYLINDERS	74-0999
CYLINDERS	FREE VIBRATION OF THICK, LAYERED CYLINDERS HAVING FINITE LENGTH WITH VARIOUS BOUNDARY CONDITIONS	72-0491
CYLINDERS	FUNDAMENTAL STUDIES OF MEDIUM-STRUCTURE INTERACTION. REPORT 1. FINITE ELEMENT ANALYSIS OF BURIED CYL	72-0854
CYLINDERS	MEASURED AND CALCULATED VIBRATION PROPERTIES OF RING-STIFFENED HONEY-COMB CYLINDERS	71-0834
CYLINDERS	MENISCI IN ARRAYS OF CYLINDERS - NUMERICAL-SIMULATION BY FINITE-ELEMENTS	75-0449
CYLINDERS	PHOTOELASTIC TESTS ON COMPLEX INTERSECTIONS OF REINFORCED THIN-SHELL CYLINDERS	75-0363
CYLINDERS	SCATTERING BY DIELECTRIC CYLINDERS	74-1025
CYLINDERS	SECOND ORDER THEORY OF OSCILLATING CYLINDERS	70-0430
CYLINDERS	STRESS INTENSITY FACTORS FOR THICK-WALLED CYLINDERS	74-0300
CYLINDERS	STRESSES IN SHORT THICK COMPOUND CYLINDERS	69-0385
CYLINDRIC	FINITE ELEMENT SOLUTION OF THE INTERACTION OF A PLANE ACOUSTIC BLAST WAVE AND A CYLINDRIC STRUCTURE	74-0790
CYLINDRICAL	A CURVED CIRCULAR CYLINDRICAL SHELL FINITE ELEMENT FOR HIGHER ORDER SHELL THEORY	69-0528
CYLINDRICAL	A CURVED CYLINDRICAL SHELL DISCRETE ELEMENT	67-0009
CYLINDRICAL	A CURVED FINITE ELEMENT FOR CYLINDRICAL SHELLS	68-0294
CYLINDRICAL	A CURVED, CYLINDRICAL SHELL, FINITE ELEMENT	68-0039
CYLINDRICAL	A CYLINDRICAL SHELL DISCRETE ELEMENT	67-0007
CYLINDRICAL	A FINITE CYLINDRICAL SHELL ELEMENT AND THE VIBRATIONS OF A CURVED FAN BLADE	68-0131
CYLINDRICAL	A FINITE ELEMENT ANALYSIS FOR CYLINDRICAL SHELLS AND PLATES	66-0034
CYLINDRICAL	A FINITE ELEMENT ANALYSIS OF ECCENTRICALLY STIFFENED-CIRCULAR CYLINDRICAL SHELLS	69-0159
CYLINDRICAL	A FINITE ELEMENT DISPLACEMENT APPROACH FOR THE ELASTO-PLASTIC ANALYSIS OF THIN CYLINDRICAL SHELLS	74-1077
CYLINDRICAL	A FORTRAN PROGRAM FOR CALCULATING THE FREQUENCIES AND MODE SHAPES OF AN ARBITRARY SHAPE CYLINDRICAL	72-0594
CYLINDRICAL	A HIGH-PRECISION TRIANGULAR CYLINDRICAL SHELL FINITE ELEMENT	71-0163
CYLINDRICAL	A NEW HYBRID CYLINDRICAL SHELL FINITE ELEMENT	71-0118
CYLINDRICAL	A REFINED CURVED CYLINDRICAL SHELL FINITE ELEMENT	68-0038
CYLINDRICAL	A STABILITY ANALYSIS OF CYLINDRICAL PANELS USING A FINITE ELEMENT FORMULATION	71-0636
CYLINDRICAL	AN APPLICATION OF THE FINITE ELEMENT METHOD TO THE CALCULATION OF CYLINDRICAL SHELLS WITH RECTANGULA	74-1236
CYLINDRICAL	AN EXTENSION OF THE SHALLOW TO THE NON-SHALLOW STIFFNESS MATRIX FOR A CYLINDRICAL SHELL FINITE ELEME	70-0104
CYLINDRICAL	ANALYSIS OF CYLINDRICAL SHELL ROOFS WITH POST TENSIONED EDGEBEAMS	74-1042
CYLINDRICAL	ANALYSIS OF ECCENTRICALLY STIFFENED CYLINDRICAL SHELLS	72-0217
CYLINDRICAL	ANALYTICAL AND EXPERIMENTAL INVESTIGATION OF A NOZZLE TO CYLINDRICAL SHELL JUNCTION	74-0638
CYLINDRICAL	APPLICATION OF FINITE ELEMENT METHOD - STRESS CALCULATIONS ON 90 DEGREES NOZZLES OF THIN-WALLED CYLI	71-0253
CYLINDRICAL	APPLICATION OF THE FINITE ELEMENT METHOD TO THE ANALYSIS OF CYLINDRICAL SHELLS	73-0828
CYLINDRICAL	BENDING OF CYLINDRICAL SHELLS - SOLUTION USING CYCLINDRICAL STRIP ELEMENTS - CONTRIBUTION TO FINITE	70-0363
CYLINDRICAL	BUCKLING OF CYLINDRICAL SHELLS BY WIND PRESSURE	74-0642
CYLINDRICAL	BUCKLING OF OPEN CYLINDRICAL SHELLS	67-0115
CYLINDRICAL	CASE BOND STRESS CALCULATIONS FOR FLAPPED CYLINDRICAL ANALOGS OF SOLID PROPELLANT ROCKET MOTORS. BON	72-0842
CYLINDRICAL	COMPARISON OF SOME FINITE-ELEMENT MODELS FOR CIRCULAR CYLINDRICAL SHELLS	73-1158
CYLINDRICAL	COMPARISON OF THE RESULTS OF THE DYNAMIC RESPONSE OF CYLINDRICAL SHELLS BY CHARACTERISTIC AND BY FIN	74-0781
CYLINDRICAL	CONVERGENCE CHARACTERISTICS OF CYLINDRICAL SHELL FINITE ELEMENTS	72-0943
CYLINDRICAL	CORRECTED ASSESSMENT OF CYLINDRICAL SHELL FINITE ELEMENT OF BOGNER, FOX AND SCHMIT WHEN APPLIED TO A	73-0024
CYLINDRICAL	CURVED, CYLINDRICAL SHELL, FINITE ELEMENT	72-0096
CYLINDRICAL	DYNAMIC RESPONSES OF CYLINDRICAL SHELLS INCLUDING GEOMETRIC AND MATERIAL NONLINEARITIES	74-0150
CYLINDRICAL	DYNAMIC STABILITY OF CYLINDRICAL SHELLS BY FINITE ELEMENTS	68-0292
CYLINDRICAL	DYNAMICAL ANALYSIS OF AXIALLY NON-UNIFORM, THIN CYLINDRICAL SHELLS. PART 1 - MATRIX FORMULATION	70-0413
CYLINDRICAL	DYNAMICAL ANALYSIS OF AXIALLY NONUNIFORM, THIN CYLINDRICAL SHELLS. PART 2 FREE VIBRATION	71-0716
CYLINDRICAL	EFFECT OF FILLETS ON STRESS CONCENTRATIONS IN CYLINDRICAL SHELLS WITH STEP CHANGES IN OUTSIDE DIAMET	71-0108
CYLINDRICAL	EFFECTS OF INSERTS ON THE ELASTIC BEHAVIOUR OF CYLINDRICAL MATERIALS LOADED BETWEEN ROUGH END-PLATES	71-0318
CYLINDRICAL	EIGENMODES OF CYLINDRICAL PLASMA USING FINITE ELEMENT METHOD	74-0174
CYLINDRICAL	ELASTIC ANALYSIS OF CYLINDRICAL CONFIGURATIONS WITH STRESS SINGULARITIES	71-0325
CYLINDRICAL	ELASTIC PLASTIC DEFORMATION OF CYLINDRICAL PRESSURE VESSELS UNDER CYCLIC LOADING	74-0604
CYLINDRICAL	FIELD ANALYSIS FOR PREFABRICATED RIBBED CYLINDRICAL SHELLS	73-0899
CYLINDRICAL	FINITE DEFLECTION DISCRETE ELEMENT ANALYSIS OF SANDWICH PLATES AND CYLINDRICAL SHELLS WITH LAMINATED	69-0236
CYLINDRICAL	FINITE DEFLECTION STRUCTURAL ANALYSIS USING PLATE AND CYLINDRICAL SHELL DISCRETE ELEMENTS	68-0193
CYLINDRICAL	FINITE DIFFERENCE SOLUTIONS FOR LARGE DEFORMATIONS OF CYLINDRICAL SHELLS - A COMPARISON WITH FINITE	72-0270
CYLINDRICAL	FINITE ELEMENT ANALYSIS OF SANDWICH PLATES AND CYLINDRICAL SHELLS WITH LAMINATED FACES	68-0121
CYLINDRICAL	FINITE ELEMENT ANALYSIS ON UNIAXIAL COMPRESSIVE STRENGTH OF CYLINDRICAL BRITTLE SPECIMEN	70-0155
CYLINDRICAL	FINITE ELEMENT ANALYSIS OF CYLINDRICAL SHELLS. SOLUTION USING RECTANGULAR CIRCULAR CYLINDRICAL ELEME	70-0178
CYLINDRICAL	FINITE ELEMENT ANALYSIS OF CYLINDRICAL SHELLS. SOLUTION USING RECTANGULAR CIRCULAR CYLINDRICAL ELEME	70-0178
CYLINDRICAL	FINITE ELEMENT ANALYSIS OF CYLINDRICAL SHELLS UTILIZING CURVED ELEMENTS CONTAINING RIGID BODY DEFORM	71-0823
CYLINDRICAL	FINITE ELEMENT ANALYSIS OF THE UNSTEADY FLOW OF A VISCOUS PLASTIC FLUID IN A CYLINDRICAL PIPE	74-0055
CYLINDRICAL	FINITE ELEMENT METHOD FOR SOLVING NEUTRON TRANSPORT PROBLEMS IN TWO-DIMENSIONAL CYLINDRICAL GEOMETRY	74-1081
CYLINDRICAL	FINITE ELEMENT STUDY OF A REINFORCED CONCRETE CYLINDRICAL SHELL THROUGH ELASTIC, CRACKING, AND ULTIM	75-0317
CYLINDRICAL	FINITE ELEMENTS FOR FIELD PROBLEMS IN CYLINDRICAL CO-ORDINATES	73-0171
CYLINDRICAL	FINITE-ELEMENT PREDICTIONS OF TRANSIENT ELASTIC-PLASTIC LARGE DEFLECTIONS OF STIFFENED AND/OR UNSTIF	74-0825
CYLINDRICAL	FREE VIBRATION AND RESPONSE TO RANDOM PRESSURE FIELD OF ANISOTROPIC THIN CYLINDRICAL SHELLS	74-0830
CYLINDRICAL	FREE VIBRATION OF CYLINDRICAL SHELLS PARTIALLY FILLED WITH LIQUID	71-0839
CYLINDRICAL	FREE VIBRATION OF THIN CYLINDRICAL SHELLS WITH A DISCONTINUITY IN THE THICKNESS	69-0439
CYLINDRICAL	FREE VIBRATIONS OF CYLINDRICAL SHELLS WITH VARIOUS END CLOSURES	74-0571
CYLINDRICAL	FREQUENCY ANALYSIS OF ORTHOTROPIC CIRCULAR CYLINDRICAL PANELS BY THE FINITE STRIP METHOD	73-0274
CYLINDRICAL	HOLLOW-ENDED ROLLER. A SOLUTION FOR IMPROVING FATIGUE LIFE IN ASYMMETRICALLY LOADED CYLINDRICAL ROLL	71-0393
CYLINDRICAL	INFLUENCE OF END-CLOSURE STIFFNESS ON BEHAVIOR OF CONCRETE CYLINDRICAL HULLS SUBJECTED TO HYDROSTATI	71-0725

CYLINDRICAL   ISSLEDOVANIE NAPRYAZHENNOGO SOSTOYANIYA REBRISTYKH ISILINDRICHESKIKH OBOLOCHEK S PRYAMOUGOL'NYMI OTV   74-0944
CYLINDRICAL   MEASUREMENT OF SLOPE CHANGE ON CYLINDRICAL SHELL SURFACES   73-0439
CYLINDRICAL   METOD KONECHNYKH ELEMENTOV V PRIMENENII K RASCHETU TSILINDRICHESKIKH OBOLOCHEK S PRYAMOUGOL'NYMI OTV   73-0536
CYLINDRICAL   NASTRAN BUCKLING ANALYSIS OF A LARGE STIFFENED CYLINDRICAL SHELL WITH A CUTOUT   71-0045
CYLINDRICAL   NEW CYLINDRICAL SHELL FINITE ELEMENT BASED ON SIMPLE INDEPENDENT STRAIN FUNCTIONS   72-0094
CYLINDRICAL   NEW HYBRID CYLINDRICAL SHELL FINITE ELEMENT   72-0092
CYLINDRICAL   NEW TRIANGULAR ELEMENT FOR FINITE DIFFERENCE SOLUTION OF AXISYMMETRIC CONDUCTION PROBLEMS IN CYLINDR   74-0381
CYLINDRICAL   NOTE ON THE STRESS CONCENTRATIONS AT THE END OF A CYLINDRICAL HOLE   70-0139
CYLINDRICAL   NUMERICAL STUDY OF PROPER MODES OF A HOMOGENEOUS CYLINDRICAL PLASMA BY METHOD OF FINITE ELEMENTS   74-0160
CYLINDRICAL   ON THE STATES OF STRESS AND DEFORMATION OF CYLINDRICAL SPECIMENS OF BRITTLE MATERIAL UNDER UNIAXIAL   69-0150
CYLINDRICAL   SCHWINGUNGEN ZYLINDRISCHER SCHRAUBENFEDERN (VIBRATIONS OF CYLINDRICAL COIL SPRINGS)   74-0331
CYLINDRICAL   STABILITY INVESTIGATION OF A CYLINDRICAL SHELL WITH FINITE ELEMENTS   72-0127
CYLINDRICAL   STRESS ANALYSIS OF PRESSURIZED CYLINDRICAL BORE IN RECTANGULAR BLOCK   69-0312
CYLINDRICAL   STRESS CONCENTRATION FACTORS FOR CIRCULAR, REINFORCED PENETRATIONS IN PRESSURIZED CYLINDRICAL SHELLS   75-0726
CYLINDRICAL   STRESS DISTRIBUTIONS IN THIN-WALLED INTERSECTING CYLINDRICAL SHELLS SUBJECTED TO INTERNAL PRESSURE A   71-0023
CYLINDRICAL   STUDY OF THE MHD INSTABILITIES OF CYLINDRICAL BELT PINCH USING THE FINITE ELEMENT METHOD   74-0700
CYLINDRICAL   STUDY ON STATE OF STRESS AND DEFORMATION OF CYLINDRICAL SPECIMENS OF BRITTLE MATERIAL UNDER UNIAXIAL   69-0152
CYLINDRICAL   TACOS: A FINITE ELEMENT COMPUTER PROGRAM FOR THE TRANSIENT ANALYSIS OF CYLINDRICAL OBESE SHELLS   72-0588
CYLINDRICAL   TEMPERATURE RISE IN THE HIGH SPEED COMPRESSION OF RIGHT CYLINDRICAL BILLETS   74-1153
CYLINDRICAL   THE APPLICATION OF FINITE ELEMENTS TO THE LARGE DEFLECTION GEOMETRICALLY NONLINEAR BEHAVIOR OF CYLIN   72-0815
CYLINDRICAL   THE APPLICATION OF THE FINITE ELEMENT METHOD TO THE ANALYSIS OF FRACTURE OF CYLINDRICAL SHELLS   73-0728
CYLINDRICAL   THERMAL STRESSES IN OPEN CYLINDRICAL SHELLS OF ARBITRARY BOUNDARY CONDITIONS BY A FINITE STRIP METHO   72-0371
CYLINDRICAL   THERMAL STRESSES IN THIN CYLINDRICAL SHELLS   73-0822
CYLINDRICAL   TRANSIENT ANALYSIS OF THICK CYLINDRICAL SHELLS   71-0590
CYLINDRICAL   TREUGOL'NYI I ELEMENT DLYA KRUGOVOI TSILINDRICHESKOI OBOLOCHKI (TRIANGULAR FINITE ELEMENT FOR CIRCULA   73-0493
CYLINDRICAL   TRIANGULAR FINITE ELEMENT FOR CYLINDRICAL SHELLS   70-0488
CYLINDRICAL   VIBRATION ANALYSIS OF CANTILEVERED CURVED PLATES USING A NEW CYLINDRICAL SHELL FINITE ELEMENT   68-0130
CYLINDRICAL   VIBRATION ANALYSIS OF CYLINDRICAL PANELS   72-0485
CYLINDRICAL   VIBRATION CHARACTERISTICS OF FREE THIN CYLINDRICAL SHELLS   73-0396
CYLINDRICAL   VIBRATION OF CYLINDRICAL SHELLS BY HYBRID FINITE ELEMENT METHOD   72-0253
CYLINDRICAL   VIBRATION OF CYLINDRICAL SHELLS BY A HYBRID FINITE ELEMENT METHOD   72-0795
CYLINDRICAL-SHE   ELASTIC-PLASTIC FORMULATION FOR A CYLINDRICAL-SHELL FINITE-ELEMENT   75-0841
D-C   FINITE ELEMENT ANALYSIS OF MAGNETICALLY SATURATED D-C MACHINES   71-0566
D'ELEMENTS   ANALYSE NUMERIQUE D'UN PROBLEME DE STEFAN A DEUX PHASES PAR UNE METHODE D'ELEMENTS FINIS. (NUMERICAL   75-0387
D'ELEMENTS   QUELQUES METHODES D'ELEMENTS FINIS POUR DE PROBLEME D'UNE PLAQUE ENCASTREE   74-0406
D'ERREUR   ESTIMATIONS D'ERREUR POUR DES ELEMENTS FINIS DROITS PRESQUE DEGENERES   74-1247
DAISY   MARK5C/DAISY SERIES OF PROGRAMS   72-0160
DALLE   ETUDE THEORIQUE DU COMPORTEMENT D'UNE DALLE DE BETON. (THEORETICAL STUDY OF THE BEHAVIOR OF A CONCRE   75-0273
DAM   A SEISMIC STRENGTH OF KOLKEWADI DAM   73-0197
DAM   AN INVESTIGATION OF THE DYNAMIC RESPONSE CHARACTERISTICS OF THE BON TEMPE DAM   66-0047
DAM   ANALYSIS OF THE SEEPAGE THROUGH AN EARTH DAM BY FINITE ELEMENT METHOD   72-0585
DAM   ARCH DAM CONSTRUCTION STRESSES DUE TO HYDROSTATIC AND GRAVITY LOADS   75-0658
DAM   BEHAVIOUR OF THE ROCK MASS AT THE SITE OF A ROCKFILL DAM   74-0992
DAM   COMPARISON OF ANALYTICAL AND EXPERIMENTAL RESULTS IN THE DESIGN OF A HOLLOW GRAVITY DAM   67-0109
DAM   CONSTRUCTION STRESSES IN DWORSHAK DAM   65-0029
DAM   CONSTRUCTIONAL DEFORMATIONS IN ROCKFILL DAM   73-0281
DAM   CONTRIBUTION A L'ETUDE DU COMPLEXE BARRAGE-FONDATION ET EXPLICATION DES SEISMES DUS AU REMPLISSAGE D   72-0554
DAM   DYNAMIC ANALYSIS OF THE SLIDE IN THE LOWER SAN FERNANDO DAM DURING THE EARTHQUAKE OF FEBRUARY 9, 197   75-0277
DAM   EMBANKMENT DAM FOR THE MARIMBONDO HYDRO PROJECT   75-0270
DAM   ENVIRONMENTAL CONSIDERATIONS OF DAM CONSTRUCTION AND OPERATION IN SEISMICALLY ACTIVE URBAN AREAS   73-0470
DAM   FINITE ELEMENT ANALYSIS OF SOME MECHANICAL PROBLEMS OF HOLLOW GRAVITY DAM   70-0470
DAM   FINITE ELEMENT ANALYSIS OF A NONSTEADY GROUNDWATER SEEPAGE PROBLEM DAM, DRAINAGE   74-1375
DAM   FINITE ELEMENT STUDY OF THE TRIAXIAL STRESS STATE AROUND AN INSPECTION TUNNEL IN AN ARCH DAM   74-0831
DAM   FOUNDATION TESTING FOR AUBURN DAM   70-0119
DAM   MATHEMATICAL MODEL OF THE COLUMBIA RIVER FROM THE PACIFIC OCEAN TO BONNEVILLE DAM. PART 1. THEORY, P   69-0464
DAM   MATHEMATICAL MODEL OF THE COLUMBIA RIVER FROM THE PACIFIC OCEAN TO BONNEVILLE DAM. PART II: INPUT-OU   70-0417
DAM   MICA DAM EMBANKMENT STRESS ANALYSIS   75-0054
DAM   OBSERVED AND PREDICTED DEFORMATIONS IN A LARGE EMBANKMENT DAM DURING CONSTRUCTION   71-0301
DAM   OPTIMUM POSITION OF THE CENTRAL CLAY CORE OF A ROCKFILL DAM IN RESPECT TO ARCHING AND HYDRAULIC FRAC   73-0447
DAM   PERFORMANCE OF CULVERT UNDER WINSCAR DAM   75-0638
DAM   RASCHET METODOM KONECHNYKH ELEMENTOV DINAMICHESKIKH KHARAKTEERISTIK ZEMLYANOI PLOTINY SOVMESTNO S OS   73-0437
DAM   SEISMIC ANALYSIS OF EARTH DAM RESERVOIR SYSTEMS   73-0412
DAM   SPECIAL STUDY ON THE STABILITY OF A DAM SPILLWAY   72-0469
DAM   STATIC AND DYNAMIC ANALYSIS OF AN ARCH DAM   74-1101
DAM   STATIC AND SEISMIC BEHAVIOUR OF AN EARTH DAM   67-0026
DAM   STRESS ANALYSIS OF A GRAVITY DAM BY THE FINITE ELEMENT METHOD   63-0004
DAM   STRESSES AND MOVEMENTS IN OROVILLE DAM   72-0200
DAM   STUDIES ON THE DETERMINATION OF SEEPAGE-LINE OF FILL DAM BY THE METHOD OF FINITE ELEMENT. WATER FLOW   74-1348
DAM   THE EARTHQUAKE EXPERIENCE AT KOYNA DAM AND STRESSES IN CONCRETE GRAVITY DAMS   72-0589
DAM   THE SEISMIC DESIGN STUDY OF A DOUBLE CURVATURE ARCH DAM   69-0053
DAM   THE STRESS DISTRIBUTION OF NORFOLK DAM   62-0001
DAM   THREE-DIMENSIONAL BEHAVIOR OF A CENTRAL CORE DAM   72-0714
DAM   THREE-DIMENSIONAL STRESS ANALYSIS OF ARCH DAMS BY THE FINITE ELEMENT METHOD: PART I, ARCH DAM NO. 1,   66-0099
DAM   THREE-DIMENSIONAL STRESS ANALYSIS OF ARCH DAMS BY THE FINITE ELEMENT METHOD: PART I, ARCH DAM NO. 1,   66-0099
DAM   ULTIMATE STRAIN CAPACITY AND TEMPERATURE RISE STUDIES, TRUMBULL POND DAM   72-0705
DAM   UTILISATION DE MODELES MATHEMATIQUES MULTIDIMENSIONNELS POUR L'ETUDE DES ECOULEMENTS DANS LES DIGUES   73-0381
DAM   VIBRATION CHARACTERISTICS OF THE NORTH FORK DAM MODEL   74-0845
DAM   VOGT BOUNDARY FOR FINITE ELEMENT ARCH DAM ANALYSIS   74-0156
DAM-RESERVOIR   DYNAMICS OF GRAVITY DAM-RESERVOIR SYSTEMS   73-0225
DAM-RESERVOIR   FINITE ELEMENT - ANALOGUE METHOD FOR DETERMINING THE DYNAMIC CHARACTERISTICS OF AN ARCH DAM-RESERVOI   73-0678
DAMAGE   A STRUCTURAL EVALUATION OF RAPID METHODS OF BACKFILLING FOR BOMB DAMAGE REPAIR   74-0821
DAMAGE   A STRUCTURAL EVALUATION OF RAPID METHODS OF BACKFILLING FOR BOMB DAMAGE REPAIR - PHASE II   75-0682
DAMAGE   EFFECT OF PERFORATION DAMAGE ON WELL PRODUCTIVITY   74-0936
DAMAGE   EVALUATION OF DAMAGE TOLERANCE IN AIRCRAFT STRUCTURES   75-0559
DAMAGE   FINITE ELEMENT AND CUMULATIVE DAMAGE ANALYSIS OF A KEYHOLE TEST SPECIMEN   75-0163
DAMAGE   FINITE ELEMENT AND CUMULATIVE DAMAGE ANALYSIS OF A KEYHOLE TEST SPECIMEN   75-0774
DAMAGE   PREDICTION OF STATIC AND FATIGUE DAMAGE AND CRACK PROPAGATION IN COMPOSITE MATERIALS   75-0643
DAMAGE   SPECTRUM TRUNCATION AND DAMAGE TOLERANCE STUDY ASSOCIATED WITH THE C-5A OUTBOARD PYLON AFT TRUSS LUG   74-0535
DAMAGE   STRUCTURAL RESPONSE TO IMPACT DAMAGE   75-0325
DAMAGE   THE USE OF ELASTIC-PLASTIC FINITE ELEMENT ANALYSIS IN THE CALCULATION OF CUMULATIVE FATIGUE DAMAGE   74-0314
DAMAGES   REDUCTION GEAR DAMAGES RELATED TO EXTERNAL INFLUENCES   75-0307
DAMPED   A VARIATIONAL FORMULATION OF DAMPED ACOUSTO-STRUCTURAL PROBLEMS   66-0133
DAMPED   ANWENDUNG DER METHODE FINITER ELEMENTE BEI DER ANALYSE DES DYNAMISCHEN VERHALTENS GEDAEMPFTER WERKZE   75-0580
DAMPED   METHODS OF MODELING AND ANALYZING VISCOELASTICALLY DAMPED STRUCTURES   71-0473
DAMPED   METHODS OF MODELING AND ANALYZING VISCOELASTICALLY DAMPED STRUCTURES   71-0836
DAMPED   THE DAMPED TRANSIENT BEHAVIOR OF FINITE BEAMS AND PLATES   69-0520
DAMPED   THE MATRIX EIGENVALUE PROBLEM FOR DAMPED STRUCTURAL VIBRATION   66-0121
DAMPED   USE OF THE FINITE ELEMENT DAMPED FORCED RESPONSE STRAIN ENERGY DISTRIBUTION FOR VIBRATION REDUCTION   72-0759
DAMPING   A UNIFIED FINITE ELEMENT APPROACH TO COLUMNS SUBJECTED TO NONCONSERVATIVE LOADS, WITH DAMPING EFFECT   74-0795
DAMPING   ANALYSIS OF LONGITUDINAL WAVE IN A BAR EXHIBITING VISCO-ELASTIC BEHAVIOR OR WITH MATERIAL DAMPING   71-0780
DAMPING   APPLICATION OF THE FINITE ELEMENT METHOD TO NONCONSERVATIVE STABILITY PROBLEMS WITH DAMPING   72-0610
DAMPING   EFFECT OF DAMPING ON EXCITABILITY OF HIGH-ORDER NORMAL MODES   75-0709

DAMPING	EFFECTS OF RISE TIME AND DAMPING ON FINITE ELEMENT ANALYSIS OF RESPONSE OF STRUCTURES	71-0087
DAMPING	FINITE ELEMENT ANALYSIS OF TWO-DIMENSIONAL ADDED MASS AND DAMPING	75-0394
DAMPING	FINITE ELEMENT SOLUTION FOR ADDED MASS AND DAMPING	74-0018
DAMPING	HYSTERETIC DAMPING SYSTEMS APPLIED TO SOIL STRUCTURE INTERACTION BY FINITE ELEMENT METHOD	73-0663
DAMPING	NUMERICAL STUDY OF DAMPING IN VISCOELASTIC SANDWICH BEAMS	73-0443
DAMPING	QJAD-4: A COMPUTER PROGRAM FOR EVALUATING THE SEISMIC RESPONSE OF SOIL STRUCTURES BY VARIABLE DAMPIN	73-0771
DAMPING	SEISMIC RESPONSE BY VARIABLE DAMPING FINITE ELEMENTS	74-0143
DAMPING	SOLUMN INSTABILITY UNDER NONCONSERVATIVE FORCES, WITH INTERNAL AND EXTERNAL DAMPING FINITE ELEMENT U	73-0566
DAMPING	THE FINITE ELEMENT TECHNIQUE APPLIED TO AN ACOUSTIC DAMPING PROBLEM	66-0048
DAMPING	TRANSVERSE VIBRATION OF BEAMS WITH INTERNAL DAMPING, ROTATORY INERTIA AND SHEAR	63-0020
DAMS	A COMPUTER PROGRAM FOR EARTHQUAKE ANALYSIS OF DAMS	70-0469
DAMS	A FINITE ELEMENT ANALYSIS OF DISSOLVED OXYGEN DRAWDOWN AND SULFATE PRODUCTION IN STRIP MINE SPOIL DA	75-0795
DAMS	ANALYSIS OF ARCH DAMS USING A SPACE-FRAME MODEL	73-0176
DAMS	ANALYSIS OF EARTHEN DAMS BY THE METHOD OF FINITE ELEMENTS	70-0454
DAMS	ANALYSIS OF NONELASTIC AND NONLINEAR VIBRATION OF ROCK DAMS USING FINITE ELEMENT METHOD	73-0680
DAMS	ANALYSIS OF SEEPAGE THROUGH DAMS AND IN FOUNDATION BY FINITE ELEMENT METHOD	73-1110
DAMS	ANALYSIS OF THIN ARCH DAMS BY THE FINITE ELEMENT METHOD	64-0001
DAMS	APPROACH TO ANALYSIS OF ARCH DAMS IN WIDE VALLEYS	73-0153
DAMS	ARCH DAMS ANALYSED BY A LINEAR FINITE ELEMENT SHELL THE FINITE STRIP METHOD IN THE ANALYSIS OF ELAST	68-0189
DAMS	BUTTRESS DAMS ON COMPLEX ROCK FOUNDATIONS	64-0013
DAMS	COMPORTAREA ELASTICA A DIGUPILOP DE PAMINT SOLICIATE SEISMIC (ELASTIC BEHAVIOR OF THE EARTH DAMS SEI	74-1185
DAMS	CRACKING OF EARTH AND ROCKFILL DAMS. A THEOPETICAL INVESTIGATION BY MEANS OF THE FINITE ELEMENT METH	69-0085
DAMS	CRACKING OF EARTH AND ROCKFILL DAMS	69-0294
DAMS	CRACKING OF EARTH AND ROCKFILL DAMS: COMPARISON OF OBSERVED AND THEORETICAL TENSILE STRAINS IN THE C	71-0699
DAMS	CRACKING OF EARTH AND ROCKFILL DAMS: COMPARISON OF OBSERVED AND THEORETICAL TENSILE STRAINS IN THE C	71-0699
DAMS	DAMS ON ROCKS OF VARYING ELASTICITY	74-0671
DAMS	DETERMINING SEEPAGE CHARACTERISTICS OF MILL-TAILING DAMS BY THE FINITE ELEMENT METHOD	71-0477
DAMS	DYNAMIC RESPONSE OF EARTH DAMS	67-0024
DAMS	DYNAMIC RESPONSE OF COHESIVE EARTH DAMS TO EARTHQUAKES	67-0131
DAMS	DYNAMICS OF EARTH DAMS WITH FOUNDATION INTERACTION	71-0343
DAMS	EARTHQUAKE ANALYSIS OF EARTH DAMS	69-0072
DAMS	EARTHQUAKE ANALYSIS OF GRAVITY DAMS INCLUDING HYDRODYNAMIC INTERACTION	73-0350
DAMS	EARTHQUAKE RESPONSE OF EARTH DAMS	67-0010
DAMS	EARTHQUAKE RESPONSE OF CONCRETE GRAVITY DAMS	70-0042
DAMS	EARTHQUAKE RESPONSE OF ARCH DAMS	70-0113
DAMS	EARTHQUAKE RESPONSE OF GRAVITY DAMS INCLUDING RESERVOIR INTERACTION EFFECTS	72-0675
DAMS	EARTHQUAKE STRESS ANALYSIS IN EARTH DAMS	65-0043
DAMS	EFFECT OF RESERVOIR FILLING ON STRESSES AND MOVEMENT IN EARTH AND ROCKFILL DAMS. A REPORT OF AN INVE	72-0836
DAMS	FINITE ELEMENT ANALYSIS OF SEEPAGE THROUGH DAMS	67-0152
DAMS	FINITE ELEMENT ANALYSIS FOR SEISMIC RESPONSE OF EARTH DAMS	68-0091
DAMS	FINITE ELEMENT ANALYSIS OF ARCH DAMS AND FOUNDATIONS	71-0668
DAMS	FINITE ELEMENT ANALYSIS OF TRANSVERSE CRACKING IN LOW-EMBANKMENT DAMS	74-0696
DAMS	FINITE ELEMENT METHOD OF ANALYSIS FOR ARCH DAMS SHELLS AND COMPARISON WITH FINITE DIFFERENCE PROCEDU	65-0036
DAMS	FINITE-ELEMENT ANALYSIS OF SEEPAGE THROUGH DAMS	67-0025
DAMS	HYDRAULIC FRACTURING IN ZONED EARTH AND ROCKFILL DAMS: A REPORT OF AN INVESTIGATION	73-0873
DAMS	HYDRODYNAMIC PRESSURES ON HIGH DAMS DUE TO VERTICAL EARTHQUAKE MOTIONS	69-0438
DAMS	INFLUENCE OF RESERVOIR WATER ON EARTHQUAKE RESPONSE OF EARTH DAMS	70-0375
DAMS	INVESTIGATION OF THE EFFECT OF FOUNDATION YIELD ON THE FREQUENCY AND SHAPE OF INHERENT VIBRATIONS OF	72-0419
DAMS	MOVEMENTS OF ARTICULATED CONDUITS UNDER EARTH DAMS ON COMPRESSIBLE FOUNDATIONS	73-0477
DAMS	NOUVELLE CONCEPTION DES DRAINS DANS LES BARRAGES EN TERRE HOMOGENE DE PETITE ET MOYENNE DIMENSIONS (	73-0404
DAMS	RESPCNSE OF EARTH DAMS TO TRAVELLING SEISMIC WAVES	69-0092
DAMS	SOLUTION OF STRESSES AND TRANSFORMATIONS OF EARTH DAMS BY THE METHOD OF FINITE ELEMENTS	69-0426
DAMS	STRESS ANALYSIS AND SPECIAL PROBLEMS OF PRESTRESSED DAMS	61-0006
DAMS	STRESSES IN BUTTRESS DAMS	65-0035
DAMS	TENSION ZONES IN DECKED ROCKFILL DAMS	73-0540
DAMS	THE ANALYSIS OF ARCH DAMS UNDER STATIC LOADING BY THE FINITE ELEMENT METHOD	68-0059
DAMS	THE EARTHQUAKE EXPERIENCE AT KOYNA DAM AND STRESSES IN CONCRETE GRAVITY DAMS	72-0589
DAMS	THE EFFECTS OF THIN CORES AND NONHOMOGENEITY ON DYNAMIC RESPONSE OF EARTH DAMS	67-0021
DAMS	THREE DIMENSIONAL ANALYSIS OF TWO ARCH DAMS BY THE FINITE ELEMENT METHOD	68-0018
DAMS	THREE DIMENSIONAL ANALYSIS OF ARCH DAMS AND THEIR FOUNDATIONS	68-0066
DAMS	THREE-DIMENSIONAL FINITE ELEMENT ANALYSIS OF DAMS	71-0733
DAMS	THREE-DIMENSIONAL FINITE ELEMENT ANALYSES OF DAMS	73-0202
DAMS	THREE-DIMENSIONAL STRESS ANALYSIS OF ARCH DAMS BY THE FINITE ELEMENT METHOD: PART I, ARCH DAM NO. 1,	66-0099
DAMS	VIBRATIONAL CHARACTERISTICS OF THREE-DIMENSIONAL SOLIDS, WITH APPLICATIONS TO EARTH DAMS	69-0432
DANGEROUS	AVOIDING DANGEROUS AND COSTLY FAN FAILURES	75-0061
DANUTA	DANUTA THREE DIMENSIONAL FINITE ELEMENT PROGRAM USED IN ANALYSIS OF TURBOMACHINERY	72-0296
DARCY	DARCY FLOW SOLUTIONS WITH A FREE SURFACE	67-0081
DARCY'S	TEST OF VALIDITY OF DARCY'S LAW FOR A CLAY SOIL	74-0070
DATA	ACCURATE FINITE ELEMENT APPROXIMATION FOR PROBLEMS WITH SINGULAR DATA	75-0098
DATA	ANALYSIS OF BAND PRESSURE IN A GUN TUBE USING FINITE ELEMENTS AND SHELL PUSHING DATA	72-0793
DATA	ANALYTICAL EVALUATION OF STRESS-STRAIN TEST DATA.	74-0442
DATA	AUTOMATED DATA GENERATION AND LARGE MATRIX OPERATION FOR 3 DIMENSIONAL PLATE STRUCTURE ANALYSIS	73-0339
DATA	AUTOMATED INPUT DATA PREPARATION FOR NASTRAN	69-0080
DATA	AUTOMATIC PREPARATION OF DATA CARDS FOR THE FINITE ELEMENT METHOD	66-0086
DATA	CREEP DATA ACQUISITION AND APPLICATION TO REACTOR COMPONENT DESIGN	72-0113
DATA	DATA GENERATION FOR A 3-D FINITE ELEMENT SYSTEM	74-0361
DATA	DATA GENERATION FOR A 3-D FINITE ELEMENT SYSTEM	74-0475
DATA	DATA GENERATION FOR FINITE-ELEMENT STRUCTURAL ANALYSIS OF THREE-DIMENSIONAL NAVAL SHIP STRUCTURES	75-0707
DATA	DATA GENERATOR FOR THE IDEALIZATION FOR FINITE ELEMENT STRUCTURAL ANALYSIS OF NAVAL SHIP FLAT PLATED	72-0860
DATA	EMPIRICAL INVESTIGATION OF REORDERING AND DATA MANAGEMENT FOR FINITE ELEMENT SYSTEMS OF EQUATIONS	73-0944
DATA	FEDGE - A GENERAL PURPOSE COMPUTER PROGRAM FOR FINITE ELEMENT DATA GENERATION. VOLUME 2 - PROGRAM MA	69-0492
DATA	FEDGE - A GENERAL-PURPOSE COMPUTER PROGRAM FOR FINITE ELEMENT DATA GENERATION. VOLUME 1 - USER'S MAN	69-0491
DATA	FINITE ELEMENT ANALYSIS OF TIME-DEPENDENT PUMPING TEST DATA IN AN UNCONFINED AQUIFER	75-0497
DATA	FINITE ELEMENT APPLICATION TO STRAIN GAGE DATA REDUCTION	73-0561
DATA	FINITE ELEMENT CONVERGENCE FOR SINGULAR DATA	73-0079
DATA	FINITE ELEMENT METHOD FOR DETERMINATION OF THERMAL STRESSES INTERACTIVE GRAPHICS FOR FINITE ELEMENT	74-0155
DATA	FINITE ELEMENT MODEL DATA CHECKOUT WITH INTERACTIVE GRAPHICS	74-0195
DATA	ICON: THE INTERACTIVE CREATION OF NASTRAN DATA. A SYSTEM DESCRIPTION	75-0274
DATA	INTERACTIVE FINITE ELEMENT DATA GENERATION FOR THREE-DIMENSIONAL PLATE STRUCTURES	74-0909
DATA	INTERACTIVE FINITE ELEMENT DATA GENERATION USING THE SUB-STRUCTURING TECHNIQUE	74-1291
DATA	LEAST SQUARES SMOOTHING OF EXPERIMENTAL DATA USING FINITE ELEMENTS	68-0224
DATA	MATHEMATICAL MODELING VIA DIRECT USE OF VIBRATION DATA	69-0310
DATA	MODAL COUPLING PROCEDURES ADAPTED TO NASTRAN ANALYSIS OF THE 1/8-SCALE SHUTTLE STRUCTURAL DYNAMICS M	75-0678
DATA	NASTRAN ANALYSIS OF A TURBINE BLADE AND COMPARISON WITH TEST AND FIELD DATA	75-0367
DATA	NATURAL COORDINATE SYSTEMS, AN AUTOMATIC INPUT DATA GENERATION SCHEME FOR A FINITE ELEMENT METHOD	70-0018
DATA	PROCEDURE FOR EFFICIENT GENERATING, CHECKING, AND DISPLAYING NASTRAN INPUT AND OUTPUT DATA FOR ANALY	71-0098
DATA	SKIN-STRINGER PANEL NORMAL MODE RESPONSE EXPERIMENTAL DATA	71-0691
DATA	STRUCTURAL DESIGN DATA FOR CONCRETE DRIFT LININGS IN BLOCK CAVING	73-0773
DATA	TECHNIQUES FOR HANDLING INPUT DATA FOR FINITE ELEMENT ANALYSIS OF REGIONAL GROUNDWATER FLOW	75-0806
DATA	THEORETICAL PREDICTION OF DISC BRAKE TEMPERATURES AND A COMPARISON WITH EXPERIMENTAL DATA	72-0306
DATA	USING INTERACTIVE GRAPHICS FOR THE PREPARATION AND MANAGEMENT OF FINITE ELEMENT DATA	74-0176
DATA	VEHICLE STRUCTURAL ANALYSIS, PRE AND POST PROCESSOR DATA HANDLING	71-0093

DATA-GENERATION	A DATA-GENERATION PROCEDURE FOR THE FINITE ELEMENT ANALYSIS OF PLANE FRAMES AND GRILLAGES	68-0179
DATA-STRUCTURES	DATA-STRUCTURES IN FINITE ELEMENT ANALYSIS	74-0114
DECAY	DISCONTINUITY STRESS DECAY IN OFFSHORE PLATFORM JOINTS	74-0491
DECELERATION	STRUCTURAL ANALYSIS OF AERODYNAMIC DECELERATION SYSTEMS BY THE FINITE ELEMENT METHOD	67-0064
DECK	A NASTRAN CORRELATION STUDY FOR VIBRATIONS OF A CROSS-STIFFENED SHIP'S DECK	72-0756
DECK	FINITE ELEMENT ANALYSIS OF A BROAD ARCH BRIDGE WITH A SLAB DECK	71-0724
DECK	FINITE ELEMENT ANALYSIS OF METAL DECK SHEAR DIAPHRAGMS	74-0128
DECK	FINITE ELEMENT APPROACH TO POLYMER CONCRETE BRIDGE DECK DESIGN ANALYSIS	72-0918
DECKED	TENSION ZONES IN DECKED ROCKFILL DAMS	73-0540
DECKHOUSE	EXPERIMENTAL AND THEORETICAL STUDIES OF HULL-DECKHOUSE INTERACTION - 2-LEVEL DECKHOUSE	71-0672
DECKS	ANALYSIS OF BRIDGE DECKS USING POLYMER IMPREGNATED CONCRETE	73-0673
DECKS	CALCULATION OF THE BRIDGE DECKS BY USING NUMERICAL METHODS. (LE CALCUL DES TABLIERS DE PONTS PAR DES	71-0411
DECKS	FINITE ELEMENT ANALYSIS OF PRESTRESSED CONCRETE VOIDED BRIDGE DECKS	73-0106
DECKS	FINITE-ELEMENT ANALYSIS OF BRIDGE DECKS	72-0672
DECKS	MULTI-CELL BRIDGE DECKS - MODEL TESTS AND THEORETICAL ANALYSIS	72-0381
DECKS	THE ANALYSIS OF CURVILINEAR ORTHOTROPIC CURVED BRIDGE DECKS	69-0162
DECKS	THE ANALYSIS OF SKEW BRIDGE DECKS, A NEW FINITE ELEMENT APPROACH	69-0234
DECOMPOSITION	FINITE ELEMENT METHOD FOR PLATE BENDING ANALYSIS BY DECOMPOSITION OF DIFFERENTIAL OPERATOR	71-0558
DECOMPOSITION	SAVINGS IN NASTRAN DECOMPOSITION TIME BY SEQUENCING TO REDUCE ACTIVE COLUMNS	71-0161
DECONVOLUTION	DECONVOLUTION OF SEISMIC RESPONSE FOR LINEAR SYSTEMS	73-0774
DEDENDUM	MEASUREMENT OF ACTUAL DEDENDUM STRAINS, AND THE INFLUENCE OF DEDENDUM RADIUS ON STRESSES AND DEDENDU	72-0118
DEDENDUM	MEASUREMENT OF ACTUAL DEDENDUM STRAINS, AND THE INFLUENCE OF DEDENDUM RADIUS ON STRESSES AND DEDENDU	72-0118
DEDENDUM	MEASUREMENT OF ACTUAL DEDENDUM STRAINS, AND THE INFLUENCE OF DEDENDUM RADIUS ON STRESSES AND DEDENDU	72-0118
DEEP	A DEEP HOLE STRESS MEASUREMENT DEVICE	74-0804
DEEP	AN INVESTIGATION ON THE PLASTICITY MAGNIFICATION FACTOR IN DEEP SURFACE FLOWS	74-0766
DEEP	ANALYSIS OF SOIL MOVEMENT AROUND A DEEP EXCAVATION	70-0039
DEEP	BUCKLING OF DEEP SHELLS	73-0849
DEEP	COMPARISON OF TWO HIGH-PRECISION TRIANGULAR FINITE ELEMENTS FOR ARBITRARY DEEP SHELLS	71-0071
DEEP	CURVED FINITE ELEMENTS FOR THE ANALYSIS OF SHALLOW AND DEEP ARCHES	74-0332
DEEP	DEEP SEA ENGINEERING	71-0739
DEEP	DEFORMATION MECHANICS OF DEEP SURFACE FLAW CRACKS	72-0785
DEEP	FINITE ELEMENT ANALYSES OF SOIL MOVEMENTS CAUSED BY DEEP EXCAVATION AND DEWATERING	69-0530
DEEP	LOAD DEFORMATION ANALYSIS OF DEEP PILE FOUNDATION	72-0931
DEEP	MECHANICS OF CABLE MOORING SYSTEMS. VOLUME 1. THREE DIMENSIONAL RESPONSE OF DEEP WATER MOORING LINES	72-0654
DEEP	MODIFIED FINITE ELEMENT METHOD APPLIED TO R C DEEP BEAMS	73-0476
DEEP	ON THE STRESS ANALYSIS AT JOINTS OF DEEP BEAMS	67-0055
DEEP	PRESTRESSED CONCRETE DEEP SLABS WITH OPENINGS	73-0427
DEEP	STRUCTURAL BEHAVIOR OF A FLEXIBLE METAL CULVERT UNDER A DEEP EARTH EMBANKMENT USING METHOD A BACKFIL	73-0739
DEEP	THERMAL STRESS DEVELOPED IN S PLATE, STRAIGHT PLATE, AND DEEP DISH WHEELS	73-0172
DEEPENING	THEORETICAL BASIS OF THE BOREHOLE DEEPENING METHOD OF ABSOLUTE STRESS MEASUREMENT	70-0047
DEFECT	FINITE ELEMENT MODELING OF MAGNETIC FIELD DEFECT INTERACTIONS	75-0068
DEFECTS	ANALYTICAL PREDICTION OF DEFECTS OCCURRENCE IN SIMPLE AND COMPLEX	69-0157
DEFECTS	INFLUENCE OF DEFECTS ON FUEL ELEMENT CLADDING STRESSES	73-0863
DEFECTS	ON THE ROLE OF DEFECTS IN CRACK INITIATION IN WELDED STRUCTURES	74-1142
DEFICIENCY	DEFICIENCY IN CURRENT FINITE ELEMENTS FOR THIN SHELL APPLICATIONS	73-0184
DEFINE	DEVELOPMENT OF THE STIFFNESS METHOD TO DEFINE BOUNDS ON ELASTIC BEHAVIOUR OF STRUCTURES	52-0011
DEFINIT	DEFINIT - A NEW ELEMENT DEFINITION CAPABILITY FOR NASTRAN: USER'S MANUAL	73-0738
DEFINITE	A COMPARATIVE STUDY OF SEVERAL CORE STORAGE SCHEMES FOR LARGE SPARSE POSITIVE DEFINITE MATRICES WITH	72-0678
DEFLECTION	A FINITE ELEMENT FORMULATION FOR LARGE DEFLECTION ELASTIC STRUCTURAL ANALYSIS APPLIED TO PLATE BENDI	70-0485
DEFLECTION	A FINITE ELEMENT PROGRAM FOR THE DYNAMIC LARGE DEFLECTION, ELASTIC-PLASTIC ANALYSIS OF STIFFENED SHE	72-0908
DEFLECTION	A METHOD OF FINITE DEFLECTION ELASTIC PLASTIC ANALYSIS OF FRAMED STRUCTURES SUBJECTED TO LARGE AXIAL	69-0239
DEFLECTION	A VARIATIONAL METHOD FOR LARGE DEFLECTION ELASTO-PLASTIC THEORY	69-0115
DEFLECTION	ALGORITHM FOR THE LARGE DEFLECTION GEOMETRICALLY NON-LINEAR BEHAVIOUR OF PLANE AND CURVED STRUCTURES	72-0083
DEFLECTION	ANALYSIS OF LARGE DEFLECTION OF PLATES BY THE FINITE ELEMENT METHOD	69-0148
DEFLECTION	AUTOMATED LARGE DEFLECTION AND STABILITY ANALYSIS OF THREE DIMENSIONAL BAR STRUCTURES	69-0393
DEFLECTION	AUTOMATED METHOD FOR THE LARGE DEFLECTION AND INSTABILITY ANALYSIS OF THREE-DIMENSIONAL TRUSS AND FR	66-0105
DEFLECTION	BEHAVIOR OF TRIANGULAR SHELL ELEMENT STIFFNESS MATRICES ASSOCIATED WITH POLYHEDRAL DEFLECTION DISTRI	68-0159
DEFLECTION	CHARACTERISTICS OF MANIPULATION ERRORS IN SOLVING LARGE DEFLECTION EQUATIONS	70-0355
DEFLECTION	CONTACT OF AN INFLATED TOROIDAL MEMBRANE WITH A FLAT SURFACE AS AN APPROACH TO THE TIRE DEFLECTION P	75-0164
DEFLECTION	DEFLECTION ANALYSIS OF EXPANDED OPEN-WEB STEEL BEAMS	74-0224
DEFLECTION	DEFLECTION AND VIBRATION ANALYSIS OF A FLEXIBLE FAN BLADE	73-0556
DEFLECTION	DEFLECTION FUNCTION FOR THE ASYMMETRICAL BENDING OF CIRCULAR PLATES	73-0468
DEFLECTION	DERIVATION OF STIFFNESS MATRICES FOR THE ANALYSIS OF LARGE DEFLECTION AND STABILITY PROBLEMS	65-0066
DEFLECTION	DEVELOPMENTS IN DISCRETE ELEMENT FINITE DEFLECTION STRUCTURAL ANALYSIS BY FUNCTION MINIMIZATION	68-0144
DEFLECTION	DEVELOPMENTS IN DISCRETE ELEMENT FINITE DEFLECTION STRUCTURAL ANALYSIS BY FUNCTION MINIMIZATION	68-0243
DEFLECTION	DYNAPLAS: A FINITE ELEMENT PROGRAM FOR THE DYNAMIC, LARGE DEFLECTION, ELASTIC-PLASTIC ANALYSIS OF ST	72-0773
DEFLECTION	ECONOMICAL SOLUTION TECHNIQUES FOR LOAD DEFLECTION EQUATIONS	74-0414
DEFLECTION	EFFECT OF INITIAL DISPLACEMENT ON PROBLEM OF LARGE DEFLECTION AND STABILITY	67-0113
DEFLECTION	ELASTIC DEFLECTION IN THIN WALLED CONES ROTATING ABOUT THEIR AXES OF SYMMETRY	73-0545
DEFLECTION	EVALUATION OF STRESSES AND DEFLECTION OF SPUR GEAR TEETH UNDER STRAIN	72-0115
DEFLECTION	FINITE DEFLECTION DISCRETE ELEMENT ANALYSIS OF SANDWICH PLATES AND CYLINDRICAL SHELLS WITH LAMINATED	69-0236
DEFLECTION	FINITE DEFLECTION STRUCTURAL ANALYSIS USING PLATE AND SHELL DISCRETE ELEMENTS	68-0143
DEFLECTION	FINITE DEFLECTION STRUCTURAL ANALYSIS USING PLATE AND CYLINDRICAL SHELL DISCRETE ELEMENTS	68-0193
DEFLECTION	FINITE DEFLECTION, DISCRETE ELEMENT ANALYSIS OF SHELLS	68-0029
DEFLECTION	FINITE ELEMENT ANALYSIS OF LARGE TRANSIENT ELASTIC-PLASTIC DEFORMATIONS OF SIMPLE STRUCTURES, WITH A	72-0849
DEFLECTION	FINITE ELEMENT LARGE DEFLECTION ANALYSIS OF PLATES	69-0208
DEFLECTION	FINITE ELEMENT LARGE DEFLECTION ANALYSIS OF ELASTIC-PLASTIC SHELLS OF REVOLUTION SUBJECTED TO AXISYM	72-0137
DEFLECTION	FINITE ELEMENT LARGE DEFLECTION ANALYSIS OF THIN-WALLED BEAMS	74-0444
DEFLECTION	FINITE ELEMENT LARGE DEFLECTION ANALYSIS OF SHALLOW SHELLS	75-0356
DEFLECTION	FINITE ELEMENT METHOD AND DENDIXEN-OSTENFELD SLOPE DEFLECTION METHOD	74-0557
DEFLECTION	FINITE ELEMENT PROCEDURE FOR LARGE DEFLECTION ANALYSIS OF PLATES WITH INITIAL DEFLECTIONS	71-0327
DEFLECTION	FINITE-ELEMENT LARGE DEFLECTION ANALYSIS OF PLATES	69-0196
DEFLECTION	INCLUSION OF SHEAR DEFLECTION IN THE STIFFNESS MATRIX OF A BEAM ELEMENT	70-0307
DEFLECTION	INFLUENCE OF HIGHER ORDER TERMS IN THE LARGE DEFLECTION ANALYSIS OF FRAMEWORKS	68-0259
DEFLECTION	LARGE DEFLECTION ANALYSIS OF ELASTIC-PLASTIC SHELLS OF REVOLUTION	69-0174
DEFLECTION	LARGE DEFLECTION ANALYSIS OF ELASTIC-PLASTIC PLATES AND SHELLS	69-0178
DEFLECTION	LARGE DEFLECTION ANALYSIS OF ELASTIC-PLASTIC SHELLS OF REVOLUTION	70-0507
DEFLECTION	LARGE DEFLECTION ANALYSIS OF PRESTRESSED NETWORKS	72-0169
DEFLECTION	LARGE DEFLECTION ANALYSIS OF ARBITRARY SHALLOW SHELLS BY THE FINITE ELEMENT METHOD	73-0937
DEFLECTION	LARGE DEFLECTION ANALYSIS OF PLATES AND SHALLOW SHELLS USING THE FINITE ELEMENT METHOD	73-0989
DEFLECTION	LARGE DEFLECTION DYNAMIC ANALYSIS OF THIN SHELLS USING FINITE ELEMENT METHOD	70-0379
DEFLECTION	LARGE DEFLECTION ELASTIC-PLASTIC ANALYSIS OF SHELLS OF REVOLUTION	71-0263
DEFLECTION	LARGE DEFLECTION ELASTIC-PLASTIC DYNAMIC RESPONSE OF STIFFENED SHELLS OF REVOLUTION	72-0608
DEFLECTION	LARGE DEFLECTION ELASTIC-PLASTIC DYNAMIC RESPONSE OF STIFFENED SHELLS OF REVOLUTION	74-0496
DEFLECTION	LARGE DEFLECTION GEOMETRICALLY NON-LINEAR FINITE ELEMENT ANALYSIS OF CIRCULAR ARCHES	73-0021
DEFLECTION	LARGE DEFLECTION OF CANTILEVER BEAMS	71-0649
DEFLECTION	LARGE DEFLECTION OF ELASTIC PLATES UNDER PATCH LOADING	72-0120
DEFLECTION	LARGE DEFLECTION SPATIAL BUCKLING OF THIN WALLED BEAMS AND FRAMES	73-1085
DEFLECTION	LATERAL RESISTANCE AND DEFLECTION OF VERTICAL PILES	73-0912
DEFLECTION	LINEARIZED LARGE DEFLECTION PROBLEM	73-0685
DEFLECTION	MACHINERY SPACE DEFLECTION AND VIBRATION ANALYSES USING FINITE ELEMENT METHODS	72-0474
DEFLECTION	MIXED FINITE ELEMENT ANALYSIS OF PLATES IN BENDING, SMALL DEFLECTION THEORY OF ELASTIC AND ELASTO-PL	71-0783

179

DEFLECTICN	ON THE DERIVATION OF STIFFNESS MATRICES FOR THE ANALYSIS OF LARGE DEFLECTION AND STABILITY PROBLEMS	66-0050
DEFLECTION	ON THE HYBRID STRESS FINITE ELEMENT MODEL FOR INCREMENTAL ANALYSIS OF LARGE DEFLECTION PROBLEMS	73-0371
DEFLECTION	STIFFNESS AND DEFLECTION ANALYSIS OF COMPLEX STRUCTURES	56-0001
DEFLECTION	STRESS AND DEFLECTION ANALYSIS OF MECHANICALLY FASTENED JOINTS	70-0067
DEFLECTION	THE APPLICATION OF FINITE ELEMENTS TO THE LARGE DEFLECTION GEOMETRICALLY NONLINEAR BEHAVIOR OF CYLIN	72-0815
DEFLECTION	THE EFFECT OF INITIAL DISPLACEMENTS OF PROBLEMS OF LARGE DEFLECTION AND STABILITY	67-0048
DEFLECTIONS	AN ANALYSIS OF THE LARGE DEFLECTIONS OF BEAMS USING THE RAYLEIGH-RITZ FINITE ELEMENT METHOD	68-0162
DEFLECTIONS	ANALYSIS OF STRESSES AND DEFLECTIONS IN FRAME SUPPORTED TENTS	74-0689
DEFLECTIONS	DEPENDENCE OF PLATE-BENDING FINITE ELEMENT DEFLECTIONS AND EIGENVALUES ON POISSONS RATIO	74-0567
DEFLECTIONS	DEPENDENCE OF PLATE-BENDING FINITE ELEMENT DEFLECTIONS AND EIGENVALUES ON POISSONS RATIO	74-0583
DEFLECTIONS	DIGITAL COMPUTATION OF STRESSES AND DEFLECTIONS IN A BOX BEAM	67-0073
DEFLECTIONS	DIGITAL COMPUTATION OF STRESSES AND DEFLECTIONS IN A BOX BEAM - A PERFORMANCE COMPARISON BETWEEN FIN	69-0355
DEFLECTIONS	DYNAMIC PLASTIC LARGE DEFLECTIONS OF CLAMPED RECTANGULAR PLATE	74-0392
DEFLECTIONS	EXPERIMENTAL AND THEORETICAL INVESTIGATION OF PAVEMENT DEFLECTIONS	69-0154
DEFLECTIONS	EXPERIMENTAL AND THEORETICAL BENDING DEFLECTIONS OF LAMINATED PLATES	72-0144
DEFLECTIONS	FINITE DEFLECTIONS, BUCKLING AND POSTBUCKLING OF AN ARCH	69-0123
DEFLECTIONS	FINITE ELEMENT PROCEDURE FOR LARGE DEFLECTION ANALYSIS OF PLATES WITH INITIAL DEFLECTIONS	71-0327
DEFLECTIONS	FINITE-ELEMENT PREDICTIONS OF TRANSIENT ELASTIC-PLASTIC LARGE DEFLECTIONS OF STIFFENED AND/OR UNSTIF	74-0825
DEFLECTIONS	INCREMENTAL ANALYSIS OF LARGE DEFLECTIONS OF SHELLS OF REVOLUTION	71-0296
DEFLECTIONS	LARGE DEFLECTIONS OF STRUCTURES SUBJECT TO HEATING AND EXTERNAL LOADS	60-0002
DEFLECTIONS	LARGE DEFLECTIONS OF ELASTIC BEAMS AND PLATES USING THE FINITE ELEMENT METHOD	68-0101
DEFLECTIONS	OPTIMAL SYNTHESIS OF FLEXIBLE LINK MECHANISMS WITH LARGE STATIC DEFLECTIONS	74-0525
DEFLECTIONS	SHORT TERM DEFLECTIONS OF CONCRETE FLAT PLATES	73-0277
DEFLECTIONS	STRESSES AND DEFLECTIONS IN COOLING TOWER SHELLS DUE TO WIND LOADING	68-0088
DEFLECTIONS	THEORETICAL AND EXPERIMENTAL STUDIES OF TRANSIENT ELASTIC-PLASTIC LARGE DEFLECTIONS OF GEOMETRICALLY	75-0830
DEFLECTIONS	THEORETICAL AND EXPERIMENTAL STUDIES OF TRANSIENT ELASTIC-PLASTIC LARGE DEFLECTIONS OF GEOMETRICALLY	75-0831
DEFLECTIONS	TIME DEPENDENT REINFORCED CONCRETE SLAB DEFLECTIONS	74-0586
DEFLECTIONS	VELKE PRIEHYBY DOSIEK METODOU KONECNYCH PRVKOV (LARGE DEFLECTIONS OF PLATES ANALYZED BY FINITE ELEME	73-0373
DEFLECTOR	APPLICATION OF THE COLLISION-IMPARTED VELOCITY METHOD FOR ANALYZING THE RESPONSES OF CONTAINMENT AND	73-0770
DEFLECTORS	FINITE ELEMENT ANALYSIS OF JET IMPINGEMENT ON AXISYMMETRIC CURVED DEFLECTORS	75-0396
DEFLECTORS	JET IMPINGEMENT ON AXISYMMETRIC CURVED DEFLECTORS	74-0017
DEFORMABILITY	DEFORMABILITY OF THE HULL STEEL-WORK AND DEFORMATIONS OF THE ENGINE-ROOM OF LARGE TANKERS	74-0532
DEFORMABILITY	NUMERICAL ANALYSES OF DEFORMABILITY TESTS IN JOINTED ROCK. ' JOINT PERTURBATION ' AND ' NO TENSION '	71-0439
DEFORMABLE	FLOW IN ROCKS WITH DEFORMABLE FRACTURES	74-0065
DEFORMABLE	MECHANICS OF CABLE MOORING SYSTEMS. VOLUME V1. A COMPUTER PROGRAM FOR ANALYZING THE STEADY STATE CON	73-0725
DEFORMABLE	MECHANICS OF DEFORMABLE SOLIDS - EURASIAN COMMUNIST COUNTRIES	73-0764
DEFORMABLE	RIGID PLATE ON NONLINEARLY DEFORMABLE COHESIVE BASE (PLANE PROBLEM)	75-0264
DEFORMABLES	UNE METHODE DE CALCUL PAR ELEMENTS FINIS DES MOUVEMENTS DE LIQUIDES DANS DES RESERVOIRS RIGIDES OU D	74-1171
DEFORMATII	ASUPRA ANALIZEI PRIN METODA ELEMENTULUI FINIT A STARILOR DE EFORTURI SI DEFORMATII DIN MASIVELE DE P	70-0471
DEFORMATION	A CASE STUDY OF IN-SITU ROCK DEFORMATION BEHAVIOR FOR THE DESIGN OF GROUND SUPPORT SYSTEM	72-0668
DEFORMATION	A DISCRETE-ELEMENT METHOD OF ANALYSIS FOR COMBINED BENDING AND SHEAR DEFORMATION OF A BEAM	69-0494
DEFORMATION	A FINITE ELEMENT MODEL FOR LAMINATED PLATES INCLUDING TRANSVERSE SHEAR DEFORMATION	72-0818
DEFORMATION	A GENERAL APPROACH TO PROBLEMS OF LARGE DEFORMATION, AND PLASTICITY USING ISOPARAMETRIC ELEMENT	71-0273
DEFORMATION	A LARGE DEFORMATION ANALYSIS OF CRYSTALLINE ELASTIC-VISCOPLASTIC MATERIALS	73-0602
DEFORMATION	A NUMERICAL SOLUTION OF A CLASS OF PROBLEMS OF FINITE ELASTIC DEFORMATION	66-0097
DEFORMATION	ANALYSIS OF THERMAL DEFORMATION OF MACHINE TOOL BY THE FINITE ELEMENT METHOD.	72-0432
DEFORMATION	ANALYSIS ON THE LARGE DEFORMATION OF PLATE STRUCTURES	74-0435
DEFORMATION	APPLICATION OF FINITE ELASTICITY THEORY TO THE DEFORMATION OF RUBBERS	62-0005
DEFORMATION	APPLICATION OF THE DIRECT STIFFNESS METHOD TO PLANE PROBLEMS INVOLVING LARGE, TIME-DEPENDENT DEFORMA	66-0108
DEFORMATION	APPLICATION OF THE DOUBLE LIGAMENT TENSILE TEST IN DEFORMATION PROCESSING	75-0042
DEFORMATION	APPLICATION OF THE FINITE ELEMENT DISPLACEMENT METHOD TO PROBLEMS OF ELASTOPLASTIC DEFORMATION	72-0199
DEFORMATION	APPLICATIONS OF FINITE ELEMENT STRESS ANALYSIS AND STRESS-STRAIN PROPERTIES IN DETERMINING NOTCH FAT	73-1114
DEFORMATION	CALCULATION OF ELASTIC-PLASTIC DEFORMATION BY THE FINITE ELEMENT METHOD OF NUMERICAL ANALYSIS	71-0793
DEFORMATION	CALCULATION OF RESIDUAL STRESSES IN PLASTIC DEFORMATION PROCESSES	72-0241
DEFORMATION	CLASSICAL AND NATURAL CONDUCTION AND DEFORMATION MODES CONCEPTS FOR AXISYMMETRIC VARIATIONAL METHODS	72-1003
DEFORMATION	COMPUTER SIMULATION OF DYNAMIC STRESS, DEFORMATION, AND FRACTURE OF GEAR TEETH	72-0114
DEFORMATION	COMPUTER SIMULATION OF THE DEFORMATION OF COMPOSITE AND POLYCRYSTALLINE MATERIALS BY FINITE ELEMENT	72-0937
DEFORMATION	CONSTITUTIVE MATRIX AND FINITE ELEMENT FORMULATION FOR A DEFORMATION THEORY OF PLASTICITY	71-0700
DEFORMATION	CREEP DEFORMATION OF A MULTI-STOREY STRUCTURE IN A FIRE	74-1197
DEFORMATION	DEFORMATION AND FAILURE OF BORON-EPOXY PLATE WITH CIRCULAR HOLE	72-0536
DEFORMATION	DEFORMATION AND FRACTURE OF PARTICULATE COMPOSITE	72-0221
DEFORMATION	DEFORMATION MECHANICS OF DEEP SURFACE FLAW CRACKS	72-0785
DEFORMATION	DEFORMATION MODE OF VOID-GROWTH AND COALESCENCE IN THE PROCESS OF DUCTILE FRACTURE	75-0683
DEFORMATION	DEFORMATION OF THE ARTERIAL VASA VASORUM AT NORMAL AND HYPERTENSIVE ARTERIAL PRESSURE	71-0149
DEFORMATION	DEFORMATION OF THE DIASTOLIC LEFT VENTRICLE - 2. NONLINEAR GEOMETRIC EFFECTS	74-0974
DEFORMATION	DEFORMATION OF THE END DIASTOLIC LEFT VENTRICLE. COMPARISON OF THEORY WITH EXPERIMENT	71-0131
DEFORMATION	DYNAMIC ANALYSIS OF THE DEFORMATION OF STRUCTURES AND OF METAL FORMING	74-0823
DEFORMATION	DYNAMIC ANALYSIS OF STRUCTURAL DEFORMATION AND METAL FORMING	75-0058
DEFORMATION	DYPLAS. A FINITE ELEMENT DYNAMIC ELASTIC-PLASTIC LARGE DEFORMATION ANALYSIS PROGRAM	74-0306
DEFORMATION	EFFECT OF TRANSVERSE SHEAR DEFORMATION IN VIBRATIONS OF PLANAR STRUCTURES COMPOSED OF BEAM TYPE ELEM	73-0402
DEFORMATION	EFFECTS OF SHEAR DEFORMATION, ROTARY INERTIA, AND ELASTICITY OF THE SUPPORT ON THE RESONANCE FREQUEN	75-0369
DEFORMATION	ELASTIC AND PLASTIC INTERLAMINAR SHEAR DEFORMATION IN LAMINATED COMPOSITIES UNDER GENERALIZED PLANE	71-0158
DEFORMATION	ELASTIC DEFORMATION OF LIGHTWEIGHT MIRRORS METHOD OF LIMIT POINT CALCULATION IN FINITE ELEMENT STRUC	72-0232
DEFORMATION	ELASTIC DEFORMATION OF POLYCRYSTALLINE METAL - 1. INFLUENCE OF GRAIN SHAPE ON DISTRIBUTIONS OF STRES	72-0463
DEFORMATION	ELASTIC DEFORMATION OF POLYCRYSTALLINE METAL - 2 EFFECT OF SLENDERNESS OF GRAINS ON DEFORMATION MODE	73-0275
DEFORMATION	ELASTIC DEFORMATION OF POLYCRYSTALLINE METAL - 2 EFFECT OF SLENDERNESS OF GRAINS ON DEFORMATION MODE	73-0275
DEFORMATION	ELASTIC DEFORMATION OF POLYCRYSTALLINE METALS - 3	73-0361
DEFORMATION	ELASTIC DEFORMATION OF SQUARE EXTRUSION DIES DUE TO LOAD AND TEMPERATURE	75-0581
DEFORMATION	ELASTIC PLASTIC DEFORMATION OF THICK WALLED CYLINDERS	69-0427
DEFORMATION	ELASTIC PLASTIC DEFORMATION OF CYLINDRICAL PRESSURE VESSELS UNDER CYCLIC LOADING	74-0604
DEFORMATION	ELASTIC-PLASTIC LARGE DEFORMATION RESPONSE OF CLAY TO FOOTING LOADS	74-0699
DEFORMATION	ELASTO-PLASTIC FINITE ELEMENT ANALYSIS OF SOIL DEFORMATION	73-1157
DEFORMATION	ELEMENTMETODEN OG DEFORMATIONSMETODEN (FINITE ELEMENT AND DEFORMATION METHODS IN STRESS ANALYSIS)	74-0580
DEFORMATION	EVALUATION OF A CLASS OF METHODS FOR BOUNDING STEADY-STATE CREEP DEFORMATION	74-0892
DEFORMATION	FINITE DEFORMATION OF ELASTIC MEMBRANES BY THE FINITE ELEMENT METHOD	67-0105
DEFORMATION	FINITE DEFORMATION OF BIAXIALLY LOADED COLUMNS	72-0021
DEFORMATION	FINITE DEFORMATION OF ELASTIC PLATES, SHELLS, AND MEMBRANES BY THE FINITE ELEMENT METHOD	72-0805
DEFORMATION	FINITE ELASTO-PLASTIC DEFORMATION - 1. THEORY AND NUMERICAL EXAMPLES	74-0202
DEFORMATION	FINITE ELEMENT ANALYSIS OF PLATE BENDING WITH TRANSVERSE SHEAR DEFORMATION	70-0109
DEFORMATION	FINITE ELEMENT ANALYSIS OF PLATES ALLOWING FOR TRANSVERSE SHEAR DEFORMATION EFFECTS	73-0996
DEFORMATION	FINITE ELEMENT ANALYSIS OF GENERAL DEFORMATION OF SHEET METALS	74-0501
DEFORMATION	FINITE ELEMENT ANALYSIS OF THE SURFACE DEFORMATION DUE TO A UNIFORM LOADING ON A LAYER OF GIBSON SOI	75-0240
DEFORMATION	FINITE ELEMENT FORMULATION OF PROBLEMS OF FINITE DEFORMATION AND IRREVERSIBLE THERMODYNAMICS OF NONL	71-0190
DEFORMATION	FINITE ELEMENT FORMULATIONS FOR PROBLEMS OF LARGE ELASTIC-PLASTIC DEFORMATION	74-0796
DEFORMATION	FINITE ELEMENT FORMULATIONS FOR LARGE DEFORMATION DYNAMIC ANALYSIS	75-0564
DEFORMATION	FINITE ELEMENT LARGE DEFORMATION ANALYSIS OF A TAPERED AORTA	75-0856
DEFORMATION	FINITE ELEMENT METHOD APPLIED TO PROBLEMS IN STRESSES AND DEFORMATION OF SOIL	67-0085
DEFORMATION	FINITE ELEMENT METHOD OF PREDICTING STRESS DISTRIBUTIONS AND SOIL DEFORMATION UNDER A TRACTIVE DEVIC	69-0389
DEFORMATION	FINITE ELEMENT PROCEDURE FOR THE LARGE DEFORMATION DYNAMIC RESPONSE OF AXISYMMETRIC SOLIDS	74-0527
DEFORMATION	FINITE ELEMENT PROCEDURE FOR THE LARGE DEFORMATION DYNAMIC RESPONSE OF AXISYMMETRIC SOLIDS	74-0967
DEFORMATION	FINITE ELEMENT PROCEDURE FOR LARGE DEFORMATION DYNAMIC RESPONSE OF AXISYMMETRIC SOLIDS	75-0013
DEFORMATION	FINITE ELEMENT PROGRAM FEECON FOR UNDRAINED DEFORMATION ANALYSES OF GRANULAR EMBANKMENTS ON SOFT CLA	72-0875

180

DEFORMATION	FINITE-ELEMENT ANALYSES OF INITIAL ELASTIC DEFORMATION OF VERTICALLY DEFORMED CRYSTAL BLOCKS	72-0970
DEFORMATION	FINITE-ELEMENT MODEL FOR THE MECHANICAL BEHAVIOUR OF THE LEFT VENTRICLE. PREDICTION OF DEFORMATION I	72-0304
DEFORMATION	FINITE-ELEMENT MODELING OF SURFACE DEFORMATION ASSOCIATED WITH VOLCANISM	75-0596
DEFORMATION	IMPROVED DEFORMATION FUNCTIONS FOR THE FINITE ELEMENT ANALYSIS OF BEAM SYSTEMS	69-0075
DEFORMATION	INCLUSION OF TRANSVERSE SHEAR DEFORMATION IN FINITE ELEMENT DISPLACEMENT FORMULATIONS	74-0662
DEFORMATION	INCLUSION OF TRANSVERSE-SHEAR DEFORMATION IN FINITE-ELEMENT DISPLACEMENT FORMULATIONS - COMMENT	75-0406
DEFORMATION	INCORPORATION OF FOUNDATION DEFORMATION IN A FINITE ELEMENT SHELL ANALYSIS	68-0227
DEFORMATION	INCREMENTAL FINITE ELEMENT ANALYSIS OF LARGE ELASTIC DEFORMATION PROBLEMS	71-0753
DEFORMATION	INCREMENTAL NUMERICAL SOLUTION OF A SIMPLE DEFORMATION PROBLEM IN SOIL MECHANICS	70-0380
DEFORMATION	INCREMENTAL SOLUTION OF AXISYMMETRIC PLATE AND SHELL FINITE DEFORMATION	71-0776
DEFORMATION	INTRODUCTION TO SHEARS DEFORMATION INTO A THIN PLATE DISPLACEMENTS FORMULATIONS	73-0617
DEFORMATION	LARGE DEFORMATION AND CRITICAL LOADS ANALYSIS OF FRAMED STRUCTURES	74-1293
DEFORMATION	LOAD DEFORMATION ANALYSIS OF DEEP PILE FOUNDATION	72-0931
DEFORMATION	LONG-TERM DEFORMATION IN CLAY	73-0713
DEFORMATION	NEAR-CRACK DEFORMATION FIELDS IN STRAIN-HARDENING MATERIAL	74-0938
DEFORMATION	NON-UNIFORM DEFORMATION OF A BLOCK IN PLANE-STRAIN COMPRESSION CAUSED BY FRICTION. PART 4. ANALYSIS	71-0786
DEFORMATION	NON-UNIFORM DEFORMATION OF MATERIAL IN AXIALLY SYMMETRIC COMPRESSION CAUSED BY FRICTION. PART 3. THE	72-0923
DEFORMATION	NONLINEAR STRESS DEFORMATION ANALYSES OF LAKE AGASSIZ CLAYS USING FINITE ELEMENT METHOD	74-1072
DEFORMATION	ON ANALYSIS OF LARGE DEFORMATION PROBLEMS OF BEAM	71-0660
DEFORMATION	ON THE NON-UNIFORM DEFORMATION OF A BLOCK IN PLANE-STRAIN COMPRESSION CAUSED BY FRICTION -3	71-0328
DEFORMATION	ON THE NONUNIFORM DEFORMATION OF MATERIAL IN AXIALLY SYMMETRIC COMPRESSION CAUSED BY FRICTION - 2	71-0321
DEFORMATION	ON THE STATES OF STRESS AND DEFORMATION OF CYLINDRICAL SPECIMENS OF BRITTLE MATERIAL UNDER UNIAXIAL	69-0150
DEFORMATION	ON THE STATES OF STRESS AND DEFORMATION AROUND THE BOTTOM OF A CIRCULAR VERTICAL SHAFT	69-0323
DEFORMATION	PLANE PLASTIC DEFORMATION OF SOIL.	71-0006
DEFORMATION	PLASTICITY AND LARGE DEFORMATION PROBLEMS BY FINITE ELEMENT METHOD	71-0529
DEFORMATION	PRACTICAL SOLUTION OF PLASTIC DEFORMATION PROBLEMS IN THE ELASTIC PLASTIC RANGE	59-0004
DEFORMATION	PREDICTION OF LONG TERM DEFORMATION OF A COMPACTED COHESIVE SOIL EMBANKMENT OVER A SOFT FOUNDATION	74-0727
DEFORMATION	PRESSURE DISTRIBUTION AND BELT DEFORMATION IN V-BELT DRIVES	74-1028
DEFORMATION	PSEUDO STATIC DEFORMATION AND FREQUENCIES OF ROTATING TURBOMACHINERY BLADES	72-0239
DEFORMATION	RASCHET POLEI DEFORMATSII I NAPRYAZHENII V DETALYAKH DVIGATELEI VNUTRENNEGO SGORANIYA (CALCULATION O	73-0465
DEFORMATION	SHEAR DEFORMATION ANALYSIS OF BEAMS AND PLATES BY FINITE ELEMENT METHOD	75-0873
DEFORMATION	SHELL DEFORMATION UNDER TOOL LOADS REPRESENTATION OF SINGULARITIES WITH ISOPARAMETRIC FINITE ELEMENT	74-0365
DEFORMATION	SIMPLIFIED MINIMUM INTERNAL ENERGY OF DEFORMATION POSTULATE AND ITS APPLICATION TO METALFORMING PROC	74-1155
DEFORMATION	SOME TWO-FIELD VARIATIONAL PRINCIPLES FOR NONLINEAR DEFORMATION ANALYSIS OF SHELLS	75-0294
DEFORMATION	STIFFNESS MATRICES FOR THE GENERAL DEFORMATION (OUT-OF-PLANE AND INPLANE) OF CURVED BEAM MEMBERS BAS	75-0124
DEFORMATION	STRESS - DEFORMATION PREDICTION IN CEMENT - TREATED SOIL PAVEMENTS	71-0286
DEFORMATION	STRESS AND DEFORMATION ANALYSIS OF IRRADIATION INDUCED SWELLING	74-0305
DEFORMATION	STRESS AND DEFORMATION ANALYSIS OF IRRADIATION INDUCED SWELLING	74-0882
DEFORMATION	STRESS-PATH DEPENDENT DEFORMATION MODULI OF CLAY	75-0050
DEFORMATION	STUDY ON STATE OF STRESS AND DEFORMATION OF CYLINDRICAL SPECIMENS OF BRITTLE MATERIAL UNDER UNIAXIAL	69-0152
DEFORMATION	SYSTEMATIC ENFORCEMENT OF STRESS BOUNDARY CONDITIONS IN THE ASSUMED STRESS HYBRID MODEL BASED ON THE	71-0771
DEFORMATION	TECHNIQUE OF MEASURING INITIAL DEFORMATION AROUND AN OPENING	71-0331
DEFORMATION	THE EFFECT OF DEFORMATION ON THE BEHAVIOUR OF HYDRODYNAMIC JOURNAL BEARINGS	71-0009
DEFORMATION	THE FINITE ELEMENT METHOD IN DEFORMATION AND HEAT CONDUCTION PROBLEMS	68-0238
DEFORMATION	THE FINITE ELEMENT METHOD FOR PREDICTING STRESS DISTRIBUTION AND SOIL DEFORMATION UNDER A TRACTIVE D	69-0522
DEFORMATION	THE FINITE ELEMENT METHOD FOR PREDICTING STRESS DISTRIBUTION AND SOIL DEFORMATION UNDER A TRACTIVE D	71-0801
DEFORMATION	THE MATRIX DEFORMATION METHOD OF STRUCTURAL ANALYSIS	71-0079
DEFORMATION	THE ROLE OF FINITE DEFORMATION ANALYSIS IN PLANE STRESS AND STRAIN FRACTURES	74-0322
DEFORMATION	THE USES OF FINITE ELEMENT TECHNIQUES IN SOIL MECHANICS WITH PARTICULAR REFERENCE TO DEFORMATION PLA	72-0002
DEFORMATION	THEORETICAL AND EXPERIMENTAL ANALYSIS OF ELASTOPLASTIC DEFORMATION IN A UNIAXIALLY LOADED STRIP OF 2	73-0720
DEFORMATION	THERMAL DEFORMATION PREDICTION IN ORBITING OBSERVATORY STRUCTURES	69-0292
DEFORMATION	THERMAL DEFORMATION VECTOR FOR A BILINEAR TEMPERATURE DISTRIBUTION IN AN ANISOTROPIC QUADRILATERAL M	75-0670
DEFORMATION	TRIANGULAR FINITE ELEMENT FOR PLATE BENDING ANALYSIS WITH TRANSVERSE SHEAR DEFORMATION	71-0555
DEFORMATION	UNDRAINED VISCO-ELASTIC ANALYSIS OF SOIL DEFORMATION	72-0424
DEFORMATION	USER'S GUIDE FOR ANALYSIS OF FINITE ELASTOPLASTIC DEFORMATION: THE FIPDEF AND FIPAX PROGRAMS FOR THE	74-0799
DEFORMATION	WECHSELWIRKUNG ZWISCHEN DEFORMATION UND DURCHSTROEMUNG IM KLUEFTIGEN, ANISOTROPEN GEBIRGE. (INTERACT	72-0612
DEFORMATION	WECHSELWIRKUNG ZWISCHEN DEFORMATION UND DURCHSTROEMUNG IM KLUEFTIGEN, ANISOTROPEN GEBIRGE. (INTERACT	72-0612
DEFORMATION	ZUR BERECHNUNG EINFACH-SYMMETRISCHER I-TRAEGER NACH DER THEORIE - 2. ORDNUNG UNTER BERUECKSICHTIGUNG	75-0510
DEFORMATIONEN	EFFECT OF DISCONTINUITIES ON STRESSES AND DEFORMATIONS OF ROCK AROUND A TUNNEL. (EINFLUSS VON DISKON	70-0170
DEFORMATIONS	A COMPARISON OF CURRENT WORK-HARDENING MODELS USED IN THE ANALYSIS OF PLASTIC DEFORMATIONS	73-0769
DEFORMATIONS	A NOTE ON THE FINITE DEFORMATIONS OF A THICK ELASTIC SHELL OF REVOLUTION	71-0193
DEFORMATIONS	ANALYSIS OF EMBANKMENT STRESSES AND DEFORMATIONS	67-0012
DEFORMATIONS	ANALYSIS OF FINITE DEFORMATIONS OF ELASTIC SOLIDS BY THE FINITE ELEMENT METHOD	70-0096
DEFORMATIONS	ANALYSIS OF LARGE DEFORMATIONS OF ELASTIC MEMBRANES BY THE FINITE ELEMENT METHOD	66-0057
DEFORMATIONS	ANALYSIS OF LARGE ELASTIC-PLASTIC CREEP DEFORMATIONS OF AXISYMMETRIC SHELLS BY THE FINITE ELEMENT ME	74-0563
DEFORMATIONS	ANALYSIS OF ROCK MASS DEFORMATIONS DUE TO EXCAVATIONS	73-0395
DEFORMATIONS	APPLICATION OF THE FINITE ELEMENT METHOD ON RESIDUAL STRESSES AND DEFORMATIONS IN WELDS	68-0178
DEFORMATIONS	APPROXIMATIONS AND NUMERICAL ANALYSIS OF FINITE DEFORMATIONS OF ELASTIC SOLIDS	74-0343
DEFORMATIONS	COMPARISON OF FINITE-ELEMENT AND EXPERIMENTAL STUDIES ON DEFORMATIONS OF ZIRCONIUM NOTCHED BEND SPEC	71-0359
DEFORMATIONS	COMPUTER PROGRAM JET 3 TO CALCULATE THE LARGE ELASTIC PLASTIC DYNAMICALLY INDUCED DEFORMATIONS OF FR	72-0609
DEFORMATIONS	CONSTRUCTIONAL DEFORMATIONS IN ROCKFILL DAM	73-0281
DEFORMATIONS	DEFORMABILITY OF THE HULL STEEL-WORK AND DEFORMATIONS OF THE ENGINE-ROOM OF LARGE TANKERS	74-0532
DEFORMATIONS	DETERMINATION OF STRESSES AND DEFORMATIONS ON PISTONS BY MEANS OF COMPUTER PROGRAMS AND PHOTOELASTIC	75-0553
DEFORMATIONS	EFFECT OF BEAM-SHEARWALL JUNCTION DEFORMATIONS ON THE FLEXIBILITY OF THE CONNECTING BEAMS	73-0213
DEFORMATIONS	EFFECT OF DISCONTINUITIES ON STRESSES AND DEFORMATIONS OF ROCK AROUND A TUNNEL. (EINFLUSS VON DISKON	70-0170
DEFORMATIONS	EFFECT OF SEEPAGE ON EMBANKMENT DEFORMATIONS DUE TO WATER LOADING	74-1329
DEFORMATIONS	EFFECT OF SHEAR DEFORMATIONS AND ROTARY INERTIA ON OPTIMUM BEAM FREQUENCIES	75-0231
DEFORMATIONS	EINSATZ MODERNER RECHENVERFAHREN ZUR ERMITTLUNG DER SPANNUNGEN UND VERFORMUNGEN BEI RADSAETZEN. (USE	75-0186
DEFORMATIONS	ELASTIC-PLASTIC DEFORMATIONS OF FLEXURALLY STIFF FRAMEWORKS FROM SECOND ORDER STRESS THEORY	73-1061
DEFORMATIONS	EMBANKMENT DEFORMATIONS DUE TO WATER LOADS	73-0782
DEFORMATIONS	FINITE DEFORMATIONS IN ELASTIC-PLASTIC MEDIA SUBJECTED TO IMPULSIVE LOADINGS BY THE FINITE ELEMENT M	72-0947
DEFORMATIONS	FINITE DIFFERENCE SOLUTIONS FOR LARGE DEFORMATIONS OF CYLINDRICAL SHELLS - A COMPARISON WITH FINITE	72-0270
DEFORMATIONS	FINITE ELEMENT ANALYSIS OF SOIL DEFORMATIONS	69-0134
DEFORMATIONS	FINITE ELEMENT ANALYSIS OF LAMINATED ANISOTROPIC PLATES INCLUDING TRANSVERSE SHEAR DEFORMATIONS	70-0506
DEFORMATIONS	FINITE ELEMENT ANALYSIS OF CYLINDRICAL SHELLS UTILIZING CURVED ELEMENTS CONTAINING RIGID BODY DEFORM	71-0823
DEFORMATIONS	FINITE ELEMENT ANALYSIS OF FINITE DEFORMATIONS OF PLATES, SHELLS, AND MEMBRANES	72-0578
DEFORMATIONS	FINITE ELEMENT ANALYSIS OF LARGE TRANSIENT ELASTIC-PLASTIC DEFORMATIONS OF SIMPLE STRUCTURES, WITH A	72-0849
DEFORMATIONS	FINITE ELEMENT FORMULATION OF FINITE DEFORMATIONS AND IRREVERSIBLE THERMODYNAMICS OF NONLINEAR CONTI	69-0207
DEFORMATIONS	FINITE ELEMENT METHOD FOR SOIL DEFORMATIONS	70-0495
DEFORMATIONS	FINITE-ELEMENT ANALYSIS OF LARGE ELASTIC-PLASTIC TRANSIENT DEFORMATIONS OF SIMPLE STRUCTURES	71-0326
DEFORMATIONS	FORCES AND DEFORMATIONS AT THE LATERAL STIFFENER OF THIN WALLED U-BARS	73-0236
DEFORMATIONS	FORMULATION AND APPLICATION OF CERTAIN PRIMAL AND MIXED FINITE ELEMENT MODELS OF FINITE DEFORMATIONS	74-0344
DEFORMATIONS	FUEL ROD DEFORMATIONS - A TIME DEPENDENT FINITE ELEMENT ANALYSIS ON AXISYMMETRIC FUEL RODS, INCLUDIN	70-0414
DEFORMATIONS	LUMPED PARAMETER APPROACH TO AXISYMMETRIC DYNAMIC SOIL DEFORMATIONS	69-0170
DEFORMATIONS	MARINE PROPELLER BLADES STRESSES AND DEFORMATIONS WITH GIVEN, FROZEN, HYDRODYNAMIC LOADING	73-0474
DEFORMATIONS	NOTE ON INCREMENTAL EQUILIBRIUM EQUATIONS AND APPROXIMATE CONSTITUTIVE RELATIONS IN LARGE INELASTIC	74-0984
DEFORMATIONS	NUMERICAL ANALYSIS FOR FINITE AXISYMMETRIC DEFORMATIONS OF INCOMPRESSIBLE ELASTIC SOLIDS OF REVOLUTI	70-0094
DEFORMATIONS	NUMERICAL DETERMINATION OF STRESSES AND DEFORMATIONS IN ROCK TAKING INTO ACCOUNT DISCONTINUITIES	70-0066
DEFORMATIONS	NUMERICAL DETERMINATION OF STRESSES AND DEFORMATIONS IN ROCK TAKING INTO ACCOUNT DISCONTINUITIES	70-0167
DEFORMATIONS	NUMERISCHE BERECHNUNG INSTATIONAERER TEMPERATURFELDER UND THERMISCH BEDINGTER VERFORMUNGEN VON WERKZ	75-0847
DEFORMATIONS	OBSERVED AND PREDICTED DEFORMATIONS IN A LARGE EMBANKMENT DAM DURING CONSTRUCTION	71-0301
DEFORMATIONS	ON SOME GENERALIZATIONS OF THE INCREMENTAL STIFFNESS RELATIONS FOR FINITE DEFORMATIONS OF COMPRESSIB	71-0189

181

DEFORMATIONS	ON THE ASYMPTOTICALLY SPHERICAL DEFORMATIONS OF ARBITRARY MEMBRANES OF REVOLUTION FIXED ALONG AN EDG	71-0640
DEFORMATIONS	PENALTY FUNCTION APPROACH TO PROBLEMS OF PLASTIC FLOW OF METALS WITH LARGE SURFACE DEFORMATIONS	75-0282
DEFORMATIONS	PRESSURE DISTRIBUTION AND DEFORMATIONS OF MACHINED COMPONENTS IN CONTACT.	73-0372
DEFORMATIONS	SLOW FINITE DEFORMATIONS OF VISCOUS SOLIDS	69-0423
DEFORMATIONS	STRESSES AND DEFORMATIONS IN MELTING PLATES	69-0476
DEFORMATIONS	STRESSES AND DEFORMATIONS OF VERTICAL SLOPES IN ELASTO-PLASTIC ROCKS	70-0114
DEFORMATIONS	STRESSES IN JAIL GULCH EMBANKMENT	73-0391
DEFORMATIONS	TENDANCES ACTUELLES DES RECHERCHES SUR LES CONTRAINTES ET DEFORMATIONS DUES AU BRIDAGE EN RELATION A	74-0273
DEFORMATIONS	THE DETERMINATION OF STRESSES AND DEFORMATIONS OF REINFORCED CONCRETE AFTER CRACKING	64-0004
DEFORMATIONS	UEBER PAUSCHALE FEHLERSCHRANKEN FUER APPROXIMATIONEN VON SCHALENVERFORMUNGEN. (GLOBAL BOUNDARY CONDI	75-0292
DEFORMATIONS	UPPER AND LOWER BOUNDS TO STRUCTURAL DEFORMATIONS BY DUAL ANALYSIS IN FINITE ELEMENTS	67-0074
DEFORMATIONS	USERS' MANUAL FOR DIGITAL COMPUTER PROGRAM OK1. NUMERICAL ANALYSIS OF FINITE AXISYMMETRIC DEFORMATIO	72-0861
DEFORMATIONS	ZUR BERECHNUNG VON PLATTEN NACH DER THEORIE ZWEITER. ORDNUNG MIT HILFE EINES HYBRIDEN DEFORMATIONS M	73-0558
DEFORMATIONSMET	ELEMENTMETODEN OG DEFORMATIONSMETODEN (FINITE ELEMENT AND DEFORMATION METHODS IN STRESS ANALYSIS)	74-0580
DEFORMATSII	RASCHET POLEI DEFORMATSII I NAPRYAZHENII V DETALYAKH DVIGATELEI VNUTRENNEGO SGORANIYA (CALCULATION O	73-0465
DEFORMED	A METHOD OF ELASTIC-PLASTIC ANALYSIS OF LARGELY DEFORMED PLATE PROBLEMS	72-0319
DEFORMED	A STUDY OF INTERNAL STRESSES IN STATICALLY DEFORMED PNEUMATIC TIRES. A MATHEMATICAL MODEL FOR THE PN	70-0421
DEFORMED	FINITE-ELEMENT ANALYSES OF INITIAL ELASTIC DEFORMATION OF VERTICALLY DEFORMED CRYSTAL BLOCKS	72-0970
DEFORMED	NOTE ON AN APPROXIMATE METHOD FOR COMPUTING NONCONSERVATIVE GENERALIZED FORCES ON FINITELY DEFORMED	70-0097
DEFORMED	STABILITY ANALYSIS OF AN EXPERIMENTALLY DEFORMED SINGLE LAYER OF INDIANA LIMESTONE USING FINITE ELEM	74-0212
DEFORMING	PRODUCTION OF MOTION PICTURES OF A CONTINUOUSLY DEFORMING STRUCTURE USING NASTRAN	71-0223
DEGENERATE	ELEMENTS FINIS ET PROBLEMES ELLIPTIQUES DEGENERES (FINITE ELEMENTS AND DEGENERATE ELLIPTIC PROBLEMS)	73-0582
DEGENERATE	ERROR ESTIMATE FOR QUADRILATERAL ISOPARAMETRIC FINITE-ELEMENTS OF TYPE Q1 WHICH CAN DEGENERATE INTO	75-0401
DEGENERATE	FINITE ELEMENT SOLUTION OF DEGENERATE INTERFACE PROBLEMS	75-0101
DEGENERATED	FINITE ELEMENTS AND DEGENERATED ELLIPTIC PROBLEMS	73-0357
DEGENERATED	NUMERICAL SOLUTION BY METHODS OF FINITE ELEMENTS OF DEGENERATED ELLIPTICAL PROBLEMS	72-0284
DEGENERATING	A TECHNIQUE FOR DEGENERATING BRICK-TYPE ISOPARAMETRIC ELEMENTS USING HIERARCHICAL MIDSIDE NODES	74-0403
DEGENERATION	DEGENERATION OF BRICK-TYPE ISOPARAMETRIC ELEMENTS	73-1031
DEGENERES	ELEMENTS FINIS ET PROBLEMES ELLIPTIQUES DEGENERES (FINITE ELEMENTS AND DEGENERATE ELLIPTIC PROBLEMS)	73-0582
DEGENERES	ESTIMATIONS D'ERREUR POUR DES ELEMENTS FINIS DROITS PRESQUE DEGENERES	74-1247
DEGENERES	RESOLUTION NUMERIQUE PAR DES METHODES D ELEMENTS FINIS DE PROBLEMES ELLIPTIQUES DEGENERES	72-0702
DEGREE	CREEP BENDING TESTS OF LARGE WELD OVERLAY AND BASE METAL SPECIMENS AT 1000 DEGREE F. (537.7 DEG. C)	73-0521
DEGREE	FINITE ELEMENT WITH A HIGH DEGREE OF REGULARITY	70-0052
DEGREE	ON THE ACCURACY OF CERTAIN C DEGREE CONTINUOUS FINITE ELEMENT REPRESENTATIONS	74-0367
DEGREE-OF-FREED	ANALYSIS OF ASYMMETRIC STIFFENED SHELL TYPE STRUCTURES BY FINITE ELEMENT METHOD II. SIX DEGREE-OF-FR	69-0373
DEGREES	APPLICATION OF FINITE ELEMENT METHOD - STRESS CALCULATIONS ON 90 DEGREES NOZZLES OF THIN-WALLED CYLI	71-0253
DEGREES	APPLICATION OF THE FINITE ELEMENT METHOD TO CASES REQUIRING THE COMBINATION OF ELEMENTS POSSESSING D	72-0580
DEGREES	FURTHER LIMITATIONS ON GENERAL HEXAHEDRON FINITE ELEMENTS WITH DERIVATIVE DEGREES OF FREEDOM.	72-0395
DEGREES	REDUCTION OF THE DEGREES OF FREEDOM IN SOLVING DYNAMIC PROBLEMS BY THE FINITE ELEMENT METHOD	73-0647
DEGREES	THREE-DIMENSIONAL ELASTIC FINITE ELEMENT ANALYSIS USING GRADIENT DEGREES OF FREEDOM	73-0843
DELINQUENT	DELINQUENT FINITE ELEMENTS FOR SHELL IDEALIZATION	75-0478
DELTA	METHOD FOR EXTERNAL APPROXIMATION USING FINITE ELEMENTS FOR DELTA - 2 AND ITERATIVE METHOD FOR SOLVI	72-0282
DELTA	METHOD FOR EXTERNAL APPROXIMATION USING FINITE ELEMENTS FOR DELTA -2 AND ITERATIVE METHOD FOR SOLVIN	72-0283
DEMONSTRATION	PROGRAMMER'S MANUAL ADDITIONS AND DEMONSTRATION PROBLEMS FOR A THERMOSTRUCTURAL CAPABILITY FOR NASTR	75-0734
DEMONSTRATION	SINGER: A COMPUTER CODE FOR GENERAL ANALYSIS TO TWO-DIMENSIONAL CONCRETE STRUCTURES. VOLUME 4. DEMON	75-0470
DENDIXEN-OSTENF	FINITE ELEMENT METHOD AND DENDIXEN-OSTENFELD SLOPE DEFLECTION METHOD	74-0557
DENSIFICATION	EFFECT OF DENSIFICATION ON THE ENGINEERING CHARACTERISTICS OF ORGANIC SOILS - VOL. 1 AND 2.	73-0805
DENSITY	A DISCUSSION - ON THE DENSITY OF FINITE ELEMENT MATRICES	73-1066
DENSITY	A FINITE ELEMENT SOLUTION FOR TWO-DIMENSIONAL DENSITY STRATIFIED FLOW	73-0911
DENSITY	FINITE ELEMENT MODEL FOR TWO-DIMENSIONAL DENSITY STRATIFIED FLOW	73-0064
DENSITY	ON THE DENSITY OF FINITE ELEMENT MATRICES	72-0545
DENSITY	POWER SPECTRAL DENSITY ANALYSIS OF VEHICLE VIBRATION USING THE NASTRAN COMPUTER PROGRAM	74-0318
DENTAL	FINITE ELEMENT ANALYSIS OF DENTAL STRUCTURES - AXISYMMETRIC AND PLANE STRESS IDEALIZATIONS	75-0525
DENTAL	SOME FACTORS AFFECTING DENTAL IMPLANT DESIGN	74-1146
DENTAL	STRESS DISTRIBUTION IN BONE ARISING FROM LOADING ON ENDOSTEAL DENTAL IMPLANTS	73-0501
DEPENDENCE	DEPENDENCE OF PLATE-BENDING FINITE ELEMENT DEFLECTIONS AND EIGENVALUES ON POISSONS RATIO	74-0452
DEPENDENCE	DEPENDENCE OF PLATE-BENDING FINITE ELEMENT DEFLECTIONS AND EIGENVALUES ON POISSONS RATIO	74-0583
DEPLACEMENTS	CHARACTERISATION ELASTIQUE D'UN STRATIFIE RESINE-VERRE PAR LA METHODE DES DEPLACEMENTS (DETERMINATIO	75-0272
DEPLACEMENTS	DEPLACEMENTS DANS LES STRUCTURES MINCES A FAIBLE RIGIDITE GEOMETRIQUE-DETERMINATION NUMERIQUE (NUMER	72-0213
DEPOSITS	BERECHNUNG UND BEURTEILUNG DER STANDSICHERHEIT VON ABBAUHOHLRAEUMEN IN STEILEN GANGERZLAGERSTAETTEN	74-0597
DEPOSITS	SEISMIC RESPONSE OF SOIL DEPOSITS UNDERLAIN BY SLOPING ROCK BOUNDARIES	69-0484
DEPTH	FAILURE OF THE CONCRETE LINING IN AN INCLINED SHAFT AT GREAT DEPTH.	74-0996
DEPTH	SHEAR STRESSES IN BEAMS WITH DISCONTINUITY IN DEPTH	75-0487
DEPTHS	SOIL RESISTANCE MOVEMENT RELATIONSHIPS FOR LATERALLY LOADED PILES IN CLAYS AT GREAT DEPTHS	70-0162
DERIVATION	A DERIVATION OF PERTURBED EQUATIONS OF MOTION AND HEAT CONDUCTION OF A THERMOELASTIC SOLID	72-0735
DERIVATION	A DERIVATION PROCEDURE FOR THE DYNAMIC FLEXIBILITY MATRIX OF A TRIANGULAR BENDING ELEMENT	70-0103
DERIVATION	BASIS OF DERIVATION OF MATRICES FOR THE DIRECT STIFFNESS METHOD	63-0008
DERIVATION	COMMENTS ON DERIVATION OF ELEMENT STIFFNESS MATRICES BY T. H. H. PIAN	65-0060
DERIVATION	DERIVATION OF A STIFFNESS MATRIX FOR A RIGHT TRIANGULAR PLATE IN BENDING AND SUBJECTED TO INITIAL ST	62-0007
DERIVATION	DERIVATION OF COMPATIBLE STIFFNESS MATRICES FOR LARGE DISCRETE ELEMENTS	67-0080
DERIVATION	DERIVATION OF CRACK ELEMENT-STIFFNESS MATRIX BY THE COMPLEX VARIABLE APPROACH	73-0870
DERIVATION	DERIVATION OF ELEMENT STIFFNESS MATRICES BY ASSUMED STRESS DISTRIBUTIONS	64-0007
DERIVATION	DERIVATION OF ELEMENT STIFFNESS MATRICES	64-0008
DERIVATION	DERIVATION OF GEOMETRIC STIFFNESS AND MASS MATRICES FOR FINITE ELEMENT HYBRID MODELS	74-0844
DERIVATION	DERIVATION OF STIFFNESS MATRICES FOR THE ANALYSIS OF LARGE DEFLECTION AND STABILITY PROBLEMS	65-0066
DERIVATION	DERIVATION OF STIFFNESS MATRICES FOR PROBLEMS IN PLANE ELASTICITY BY GALERKIN'S METHOD	69-0250
DERIVATION	DERIVATION OF THE STIFFNESS MATRIX FOR PLATES BY GALERKIN METHOD	69-0402
DERIVATION	FUNCTIONAL ANALYSIS DERIVATION AND GENERALIZATION OF HYBRID VARIATIONAL METHODS	73-1104
DERIVATION	ON DERIVATION OF STIFFNESS MATRICES WITH CO ROTATION FIELDS FOR PLATES AND SHELLS	71-0254
DERIVATION	ON THE DERIVATION OF STIFFNESS MATRICES FOR THE ANALYSIS OF LARGE DEFLECTION AND STABILITY PROBLEMS	66-0050
DERIVATION	TRAPEZOIDAL FINITE ELEMENTS: THEIR DERIVATION AND USE FOR AXISYMMETRIC ROTATING BODIES	73-0192
DERIVATIONS	APPLICATION, TESTING AND EVALUATION OF DOUBLY CURVED SHELL ELEMENTS FOR DYNAMIC ANALYSIS. PART II. F	69-0117
DERIVATIVE	FURTHER LIMITATIONS ON GENERAL HEXAHEDRON FINITE ELEMENTS WITH DERIVATIVE DEGREES OF FREEDOM.	72-0395
DERIVATIVE	GALERKIN APPROXIMATION OF THE TIME DERIVATIVE IN THE FINITE ELEMENT ANALYSIS	73-0573
DERIVATIVE	GALERKIN APPROXIMATION OF THE TIME DERIVATIVE IN THE FINITE ELEMENT ANALYSIS OF GROUNDWATER FLOW	74-0658
DERIVATIVE	STIFFNESS DERIVATIVE FINITE ELEMENT TECHNIQUE FOR DETERMINATION OF CRACK TIP STRESS INTENSITY FACTOR	74-0956
DERIVATIVE	STIFFNESS DERIVATIVE FINITE ELEMENT TECHNIQUE FOR DETERMINATION OF CRACK TIP STRESS INTENSITY FACTOR	74-1358
DERIVATIVES	COMPUTATION OF DERIVATIVES IN THE FINITE ELEMENT METHOD	70-0304
DERIVATIVES	FINESSE - A FINITE ELEMENT SYSTEM. PART V FINITE ELEMENT DESIGN - DERIVATIVES FOR STRUCTURAL SHAPE D	72-0963
DERIVATIVES	FINITE ELEMENT METHOD FOR HYDRODYNAMIC DISPERSION EQUATION WITH MIXED PARTIAL DERIVATIVES	72-0030
DERIVATIVES	PROGRAM FOR TRIANGULAR BENDING ELEMENTS WITH DERIVATIVES SMOOTHING	72-0473
DERIVATIVES	STEADY STATE EQUATIONS OF MOTION, EQUILIBRIUM SHAPE AND STABILITY DERIVATIVES OF ELASTIC AIRPLANES E	72-0541
DERIVING	A NEW APPROACH FOR DERIVING GOOD ELEMENT STIFFNESS MATRICES	75-0128
DERIVING	RATIONALIZATION IN DERIVING ELEMENT STIFFNESS MATRIX BY ASSUMED STRESS APPROACH	68-0138
DERMATRIZENVERS	SPEKTRALANALYSE VON ROTORBLAETTERN NACH DERMATRIZENVERS CHIEBUNGSMETHODE UND MATRIZENKRAFTEMETHODE	66-0123
DESCOID	FINITE ELEMENT SOLUTION OF THE FLOW AROUND A TWO-DIMENSIONAL DESCOID	72-0298
DESCRIBED	USER'S GUIDE FOR THE COMPUTER PROGRAM DESCRIBED IN TAM REPORT NO. 680 "DETERMINATION OF VOLUME CHANG	69-0473
DESEK	RESENI MAZNIHO STAVU ZELEZOBETONOVYCH DESEK (PLASTIC ANALYSIS OF REINFORCED CONCRETE PLATES)	74-0495
DESEK	STATICKE RESENI FYZIKALNE NELINEARNIHO CHOVANI ZELEZOBETO- NOVYCH DESEK (ANALYSIS OF PHYSICALLY NONL	74-0514
DESH	DESH, EIN EINFACHES FLACHES SCHALENELEMENT UEBER DREIECKIGEM GRUNDRISS. (DESH, A SIMPLE TRIANGULAR S	75-0304
DESH	DESH, EIN EINFACHES FLACHES SCHALENELEMENT UEBER DREIECKIGEM GRUNDRISS. (DESH, A SIMPLE TRIANGULAR S	75-0304
DESIGN	A CASE STUDY OF IN-SITU ROCK DEFORMATION BEHAVIOR FOR THE DESIGN OF GROUND SUPPORT SYSTEM	72-0668

DESIGN	A DESIGN ORIENTED APPROACH TO CREEP AND PLASTICITY IN FINITE ELEMENT PROGRAMS	70-0125
DESIGN	A DESIGN PROCEDURE FOR HIGH CUTS IN JOINTED HARD ROCK. THREE DIMENSIONAL SOLUTIONS. VOLUME 1	71-0656
DESIGN	A DESIGN STUDY FOR THE INCORPORATION OF AEROELASTIC CAPABILITY INTO NASTRAN	71-0112
DESIGN	A FINITE ELEMENT METHOD FOR OPTIMUM BEAM DESIGN	71-0685
DESIGN	A FINITE ELEMENT PROGRAM FOR DESIGN AND ANALYSIS OF THREE-DIMENSIONAL FLAT SPRINGS	74-0738
DESIGN	A FINITE ELEMENT PROCEDURE FOR NUMERICAL OPTIMIZATION IN MECHANICAL DESIGN	74-1218
DESIGN	A FINITE-ELEMENT APPROACH TO OPTIMAL STRUCTURAL DESIGN	73-1162
DESIGN	A RATIONAL ANALYSIS AND DESIGN PROCEDURE FOR WOOD JOIST FLOOR SYSTEMS	74-0730
DESIGN	A SEISMIC DESIGN OF NUCLEAR REACTOR BUILDING. STRESS ANALYSIS AND STIFFNESS EVALUATION OF THE ENTIRE	71-0251
DESIGN	AN APPLICATION OF FINITE ELEMENT TECHNIQUE IN BRIDGE DESIGN	70-0225
DESIGN	AN INTERACTIVE GRAPHICS SHIP DESIGN ORIENTED PROGRAM PACKAGE FOR TIME SHARING SYSTEMS	73-0341
DESIGN	AN INTERACTIVE HYBRID TECHNIQUE FOR CRASHWORTHY DESIGN OF COMPLEX VEHICULAR STRUCTURAL SYSTEMS	74-0317
DESIGN	ANALYSIS AND DESIGN OF MULTIPLE INPUT FLEXIBLE LINK MECHANISMS	70-0354
DESIGN	ANALYSIS AND DESIGN CAPABILITY OF THE STRUDL PROGRAM	71-0054
DESIGN	ANALYSIS AND DESIGN OF SKEW BOX GIRDER BRIDGES	72-0659
DESIGN	ANALYSIS AND DESIGN OF AXISYMMETRIC REINFORCEMENT AROUND A CIRCULAR HOLE IN A THIN FLAT PLATE BY THE	74-1113
DESIGN	ANALYSIS FOR THE DESIGN OF BEAM TESTING APPARATUS	75-0527
DESIGN	ANALYSIS OF LARGE CLOSED SHIP STRUCTURES: APPLICATIONS TO TANKER STRUCTURAL DESIGN	70-0289
DESIGN	ANALYTICAL PERFORMANCE OF STEEL, ALUMINUM AND PLASTIC FOR A NEW DESIGN HIGH PRESSURE THIN-WALLED CAR	74-0756
DESIGN	ANCHOR-PILE DESIGN FOR OCEAN-FLOOR ENVIRONMENTS USING FINITE-ELEMENT ANALYSIS	75-0459
DESIGN	APPLICATION DES CALCULUS DE FONDATIONS A L'EVALUATION DU COMPORTEMENT DES CHAUSSEES EM DASH 1, 2. (A	71-0429
DESIGN	APPLICATION OF COMPUTER-AIDED AIRCRAFT DESIGN IN A MULTIDISCIPLINARY ENVIRONMENT	73-0527
DESIGN	APPLICATION OF ELASTIC PLASTIC ANALYSIS TO THE DESIGN OF PRESSURE VESSELS	68-0157
DESIGN	APPLICATION OF FINITE ELEMENT TECHNIQUE TO DESIGN AND MANUFACTURE OF ASTRONOMICAL MIRRORS	69-0302
DESIGN	APPLICATION OF FINITE ELEMENT METHODS FOR THE ANALYSIS OF THERMAL CREEP, IRRADIATION INDUCED CREEP,	71-0242
DESIGN	APPLICATION OF FINITE ELEMENT METHODS TO COMPLETE AUTOMOBILE STRUCTURAL DESIGN EVALUATION	74-0312
DESIGN	APPLICATION OF THE FINITE ELEMENT METHOD TO PRESSURE VESSEL DESIGN (JAPANESE)	70-0446
DESIGN	APPLICATION OF THE FINITE ELEMENT METHOD TO THE DESIGN OF UNDERGROUND BULKHEADS	72-0387
DESIGN	ASKA 105. A PROBLEM ORIENTED LANGUAGE FOR STRUCTURAL ENGINEERS. VOLUME 1, DESIGN	69-0460
DESIGN	ASSESSMENT OF DESIGN TOOLS AND CRITERIA FOR URBAN RAIL TRACK STRUCTURES. VOLUME 11. AT-GRADE SLAB TR	74-0813
DESIGN	AUTOMATED DESIGN OPTIMIZATION OF SUPERSONIC AIRPLANE WING STRUCTURES UNDER DYNAMIC CONSTRAINTS	72-0178
DESIGN	AUTOMATED OPTIMUM DESIGN OF AIRCRAFT WINGS TO SATISFY STRENGTH, STABILITY, FREQUENCY AND FLUTTER REQ	72-0605
DESIGN	AUTOMATED PROCEDURE FOR DESIGN OF WING STRUCTURES TO SATISFY STRENGTH AND FLUTTER REQUIREMENTS	73-0867
DESIGN	AUTOMATED STRUCTURAL DESIGN AND ANALYSIS OF ADVANCED COMPOSITE WING MODELS	73-0221
DESIGN	AUTOMATED STRUCTURAL SYNTHESIS USING A RE NUMBER OF DESIGN COORDINATES	73-0584
DESIGN	AUTOMATIC DESIGN OF SHELL STRUCTURES	71-0336
DESIGN	BALANCED DESIGN AND FINITE-ELEMENT ANALYSIS OF CULVERTS.	72-0462
DESIGN	BERECHNUNG VON LAUFRAEDERN MIT DER METHODE DER FINITEN ELEMENTE (DESIGN OF ROTATING WHEELS BY THE FI	74-0547
DESIGN	BRITTLE MATERIALS: A DESIGN CHALLENGE	75-0769
DESIGN	BUCKLING PROBLEMS IN THE DESIGN OF FIBREGLASS-REINFORCED PLASTIC SHIPS	72-0164
DESIGN	CALCULATION AND DESIGN OF JOINTS MADE FROM COMPOSITE MATERIALS	70-0156
DESIGN	CERAMIC ROTATING BLADES: SOME CRITICAL DESIGN PARAMETERS FOR GAS TURBINE APPLICATIONS	74-0146
DESIGN	COMPARISON OF ANALYTICAL AND EXPERIMENTAL RESULTS IN THE DESIGN OF A HOLLOW GRAVITY DAM	67-0109
DESIGN	COMPUTATIONAL METHODS IN NUCLEAR REACTOR STRUCTURAL DESIGN FOR HIGH TEMPERATURE APPLICATIONS: AN INT	73-1001
DESIGN	COMPUTER – AIDED DESIGN – STANDARD OIL OF INDIANA BUILDING	73-0182
DESIGN	COMPUTER AIDED DESIGN APPLIED TO A MODEL OF A CHASSIS TYPE STRUCTURE USING FINITE ELEMENT TECHNIQUES	69-0421
DESIGN	COMPUTER AIDED DESIGN AND CONSTRUCTION BY MEANS OF CURVED SURFACE AND SOLID ELEMENTS	73-0165
DESIGN	COMPUTER AIDED DESIGN OF THERMALLY LOADED AXISYMMETRIC DIESEL ENGINE COMPONENTS	74-0596
DESIGN	COMPUTER AIDED DESIGN OF MACHINE TOOL STRUCTURE	74-1189
DESIGN	COMPUTER-AIDED DESIGN OF SHELL STRUCTURES USING THE FINITE ELEMENT METHOD	72-0998
DESIGN	COMPUTERS AND MECHANICAL DESIGN	73-0195
DESIGN	COMPUTERUNTERSTUETZTES SYSTEM FUER DIE CONSTRUKTION DES WERKZEUGMASCHINENAUFBAUS (COMPUTER-AIDED SYS	73-0482
DESIGN	CREEP DATA ACQUISITION AND APPLICATION TO REACTOR COMPONENT DESIGN	72-0113
DESIGN	DESIGN ANALYSIS FOR LMFBR CORE RESTRAINT	71-0208
DESIGN	DESIGN AND ANALYSIS EXPERIENCE WITH A LIQUID METAL HEAT EXCHANGER FOR PFTF SERVICE	72-0772
DESIGN	DESIGN AND ANALYSIS OF A CERAMIC STATOR VANE	75-0366
DESIGN	DESIGN AND CALCULATION WITH FINITE ELEMENTS	74-0161
DESIGN	DESIGN APPRAISAL OF OFFSHORE PLATFORMS OF WELDED TUBULAR CONSTRUCTION	74-1165
DESIGN	DESIGN ASPECTS OF HIGH DUTY PRESSURE COMPONENTS	74-0199
DESIGN	DESIGN METHOD FOR FLEXIBLE AIRFIELD PAVEMENTS	74-1174
DESIGN	DESIGN OF A COMPOSITE STEEL AND CONCRETE CLOSURE FOR A PRESTRESSED CONCRETE REACTOR VESSEL.	72-0187
DESIGN	DESIGN OF CONCRETE TRACK SLAB ON ELASTIC FOUNDATION USING THE FINITE ELEMENT METHOD	73-0479
DESIGN	DESIGN OF INCLINED TIED ARCH RAILWAY BRIDGES OVER THE M56	71-0297
DESIGN	DESIGN OF LOW-THERMAL-DISTORTION LST METERING STRUCTURE	75-0840
DESIGN	DESIGN OF NEW SHIP TYPES BY MEANS OF FINITE ELEMENT METHOD	72-0562
DESIGN	DESIGN OF OPTICAL MIRROR STRUCTURES	71-0658
DESIGN	DESIGN OF OPTIMUM STRUCTURES FOR DYNAMIC LOADS	71-0255
DESIGN	DESIGN OF PRISMATIC SHELLS	72-0404
DESIGN	DESIGN OF PROPELLANT FINS SUBJECTED TO AXIAL ACCELERATION	71-0734
DESIGN	DESIGN OF ROOF BOLT PATTERNS FOR JOINTED ROCK	74-1318
DESIGN	DESIGN OF TWO AND THREE DIMENSIONAL STRUCTURES BY THE FINITE ELEMENT METHOD (IN GERMAN)	74-1364
DESIGN	DESIGN OPTIMIZATION IN UNDERGROUND COAL SYSTEMS	74-1095
DESIGN	DESIGN ORIENTED APPROACH TO CREEP AND PLASTICITY IN FINITE ELEMENT PROGRAMS	71-0281
DESIGN	DESIGN, ANALYSIS AND CONSTRUCTION OF THE RESEARCH PLATFORM (NORDSEE)	75-0261
DESIGN	DESIGN, DEVELOPMENT, DOCUMENTATION AND SUPPORT OF A MAJOR FINITE ELEMENT SYSTEM	73-0205
DESIGN	DESIGN, MANUFACTURE, DEVELOPMENT, TEST, AND EVALUATION OF BORON/ALUMINUM STRUCTURAL COMPONENTS FOR S	74-0798
DESIGN	DEVELOPMENT OF A DESIGN FOR A PRECAST CONCRETE BRIDGE BEAM OF U-SECTION	73-0392
DESIGN	DIESEL ENGINE COMPONENT DESIGN USING THE FINITE ELEMENT METHOD AND INTERACTIVE GRAPHICS	74-0237
DESIGN	EARTHQUAKE-RESISTANT DESIGN OF ENGINEERING STRUCTURES	73-0284
DESIGN	ECONOMICAL DESIGN ANALYSIS OF PISTONS FOR INTERNAL COMBUSTION ENGINES	73-0188
DESIGN	ENTWURF UND SPANNUNGSBERECHNUNG EINES SPANNBETON DRUCKBEHAELTERS AUS FERTIGTEILEN. (DESIGN AND STRES	73-0856
DESIGN	ESTIMATIONS OF THE EFFECT OF DESIGN CHANGES ON THE MODES AND FREQUENCIES OF VIBRATING STRUCTURAL ELE	74-0819
DESIGN	EVALUATION OF COMPUTER AIDED METHODS FOR THE DESIGN OF MACHINE TOOL STRUCTURES	73-0577
DESIGN	EXPERIENCES FROM STRESS ANALYSIS OF AXI-SYMMETRIC PROBLEMS IN MACHINE DESIGN	69-0247
DESIGN	FINESSE – A FINITE ELEMENT SYSTEM, PART V FINITE ELEMENT DESIGN – DERIVATIVES FOR STRUCTURAL SHAPE D	72-0963
DESIGN	FINESSE – A FINITE ELEMENT SYSTEM, PART V FINITE ELEMENT DESIGN – DERIVATIVES FOR STRUCTURAL SHAPE D	72-0963
DESIGN	FINITE ELEMENT ANALYSIS IN BRITTLE MATERIAL DESIGN	70-0360
DESIGN	FINITE ELEMENT APPROACH TO OPTIMAL DESIGN OF PLASTIC STRUCTURES IN PLANE STRESS.	72-0397
DESIGN	FINITE ELEMENT APPROACH TO POLYMER CONCRETE BRIDGE DECK DESIGN ANALYSIS	72-0918
DESIGN	FINITE ELEMENT DESIGN STUDY FOR 30-MM CARTRIDGE CASE	74-0376
DESIGN	FINITE ELEMENT METHODS IN REACTOR DESIGN CALCULATIONS	71-0571
DESIGN	FINITE ELEMENT METHOD FOR OPTIMAL DESIGN OF VARIABLE THICKNESS SHEETS	73-0131
DESIGN	FINITE ELEMENT METHOD IN AUTOMOBILE DESIGN	73-0326
DESIGN	FINITE ELEMENT METHODS IN PRODUCT DESIGN	74-0117
DESIGN	FINITE ELEMENT PROGRAMS FOR SLAB BRIDGE DESIGN	71-0082
DESIGN	FINITE ELEMENT TECHNIQUES IN SHIP STRUCTURE DESIGN IN FINITE ELEMENT TECHNIQUES	69-0335
DESIGN	FINITE ELEMENTS METHOD IN DESIGN OF PRESTRESSED REACTOR CONCRETE CONSTRUCTION	73-0550
DESIGN	FINITE-ELEMENT ANALYSIS OF CONCRETE SLABS AND ITS IMPLICATIONS FOR RIGID PAVEMENT DESIGN	73-0538
DESIGN	GEOMETRIC HIGHWAY AND CULVERT DESIGN	74-1342
DESIGN	GRAVITY FLOW BIN DESIGN FOR LUNAR SOIL	72-0240
DESIGN	GRIP: AN INFORMATION SYSTEM FOR DESIGN AUTOMATION	73-0210
DESIGN	HOW FINITE ELEMENT METHODS IMPROVE THE DESIGN CYCLE	74-0372

DESIGN	HULL AND SUPERSTRUCTURE VIBRATIONS DESIGN CALCULATIONS BY FINITE ELEMENTS	75-0556
DESIGN	HYBRID FINITE ELEMENT OPTIMIZATION PROGRAMS FOR DISCS USING A DYNAMICS DESIGN TECHNOLOGY	74-0691
DESIGN	IMPACT OF FINITE ELEMENT TECHNIQUES ON PRACTICAL DESIGN OF SHIP STRUCTURES	69-0344
DESIGN	IMPROVEMENT IN PLATE MODELING AND PLATE DESIGN BY THE FINITE ELEMENT METHOD	75-0791
DESIGN	INDETS - INTEGRATED DESIGN OF TANKER STRUCTURES	75-0524
DESIGN	INELASTIC ANALYSIS AND SATISFACTION OF DESIGN CRITERIA OF A HIGH TEMPERATURE COMPONENT	75-0265
DESIGN	INEXPENSIVE, INTERACTIVE COMPUTER-AIDED DESIGN APPLIED TO FINITE ELEMENT STRESS ANALYSIS	74-0592
DESIGN	INTERACTIVE PROGRAM IN DESIGN AND ANALYSIS OF COMPOSITE MATERIALS	74-0854
DESIGN	INTERFACING FINITE ELEMENT METHODS WITH PRODUCT DESIGN ENGINEERING	73-0189
DESIGN	INTERNATIONAL MACHINE TOOL DESIGN AND RESEARCH CONFERENCE, 12TH PROCEEDINGS, 1971.	72-0224
DESIGN	INTERNATIONAL MACHINE TOOL DESIGN AND RESEARCH CONFERENCE, 13TH PROCEEDINGS, 1972	73-0504
DESIGN	LARGE STRUCTURES FOR MANNED SPACECRAFT: MATHEMATICAL ANALYSIS, DESIGN, CONSTRUCTION AND TESTS	74-1327
DESIGN	LARGE USER-ORIENTATED SYSTEMS OF PROGRAMS FOR STRUCTURAL ANALYSIS AND DESIGN	74-0482
DESIGN	LAYERED PAVEMENT SYSTEMS. PART 1. LAYERED SYSTEM DESIGN. PART 11. FATIQUE OF PLAIN CONCRETE	72-0867
DESIGN	LIMITATIONS ON THE APPLICATION OF EFFECTIVE BREADTH CONCEPT IN BOX GIRDER DESIGN	74-0493
DESIGN	MAN MACHINE INTERACTIVE STRUCTURAL ANALYSIS AS A PRELIMINARY DESIGN AID	69-0293
DESIGN	METHOD OF MINIMUM WEIGHT DESIGN WITH REQUIREMENTS IMPOSED ON STRESSES AND NATURAL FREQUENCIES	70-0149
DESIGN	MINIMUM WEIGHT DESIGN WITH STABILITY CONSTRAINT	70-0274
DESIGN	MINIMUM WEIGHT DESIGN OF COMPLEX STRUCTURES SUBJECT TO A FREQUENCY CONSTRAINT	70-0374
DESIGN	MINIMUM WEIGHT DESIGN OF FINITE ELEMENT STRUCTURES	74-0148
DESIGN	MODIFICATION OF LAYERED THEORY FOR PAVEMENT DESIGN ANALYSIS	73-0574
DESIGN	MODIFICATIONS TO APL COMPUTER PROGRAMS USED FOR DESIGN OF AXIAL COMPRESSOR AIRFOILS	74-0508
DESIGN	NASTRAN AS ANALYSIS TOOL IN A STRUCTURAL DESIGN OPTIMIZATION PROCESS	71-0235
DESIGN	NEW APPROACH TO ANALYSIS AND DESIGN OF ELECTROMECHANICAL FILTERS BY FINITE-ELEMENT TECHNIQUE	71-0476
DESIGN	NEW DESIGN FACTOR FOR A SHRINK FITTED ASSEMBLY	75-0167
DESIGN	NOUVELLE CONCEPTION DES DRAINS DANS LES BARRAGES EN TERRE HOMOGENE DE PETITE ET MOYENNE DIMENSIONS (	73-0404
DESIGN	ON THE DESIGN OF DIE-SETS	75-0582
DESIGN	ON THE STRUCTURAL DESIGN OF TUBULAR JOINTS	74-0492
DESIGN	OPTIM 11: (MAGIC) COMPATIBLE LARGE SCALE AUTOMATED MINIMUM WEIGHT DESIGN PROGRAM - V 2. PROGRAMMER'S	74-1036
DESIGN	OPTIM 11: A (MAGIC) COMPATIBLE LARGE SCALE AUTOMATED MINIMUM WEIGHT DESIGN PROGRAM - 1. ENGINEERS AN	74-1048
DESIGN	OPTIM 11: A MAGIC COMPATIBLE LARGE-SCALE AUTOMATED MINIMUM WEIGHT DESIGN PROGRAM. VOLUME 1. ENGINEER	74-1096
DESIGN	OPTIM 11: MAGIC COMPATIBLE LARGE SCALE AUTOMATED MINIMUM WEIGHT DESIGN PROGRAM. VOLUME 11. PROGRAMME	74-1093
DESIGN	OPTIMAL DESIGN FOR PRESCRIBED BUCKLING LOADS	74-0568
DESIGN	OPTIMAL DESIGN OF STRUCTURES WITH BUCKLING CONSTRAINTS	73-0246
DESIGN	OPTIMAL DESIGN OF WATER WELLS IN UNCONSOLIDATED SEDIMENTS	74-1182
DESIGN	OPTIMALITY CRITERIA IN STRUCTURAL DESIGN	71-0758
DESIGN	OPTIMIZATION OF STRUCTURAL DESIGN 1	73-0904
DESIGN	OPTIMUM CONFIGURATIONAL AND DIMENSIONAL DESIGN OF TRUSS STRUCTURES	74-0588
DESIGN	OPTIMUM DESIGN OF ADVANCED COMPOSITE STRUCTURES FOR STATIC LOADS	72-0613
DESIGN	OPTIMUM DESIGN OF COMPOSITE SHELLS SUBJECT TO NATURAL FREQUENCY CONSTRAINTS	73-0154
DESIGN	OPTIMUM DESIGN OF CONCRETE STRUCTURES	74-0979
DESIGN	OPTIMUM DESIGN OF FRAMED STRUCTURES	69-0251
DESIGN	OPTIMUM DESIGN OF STRESSED SKIN STRUCTURES	73-0472
DESIGN	OPTIMUM PLASTIC DESIGN FOR MULTIPLE SETS OF LOADS	74-0980
DESIGN	PARAMETRIC STUDY OF A BEAM WITH A COMPOUND SIDE-BRANCH RESONATOR AS A DEVICE TO EVALUATE PRELIMINARY	70-0465
DESIGN	PAVEMENT DESIGN AND CONTINUOUSLY REINFORCED CONCRETE PAVEMENT PERFORMANCE	73-0749
DESIGN	PRACTICAL APPLICATION OF FINITE ELEMENT METHODS IN THE DESIGN OF STRUCTURES	69-0420
DESIGN	PROGRAM DESIGN BY A MULTIDISCIPLINARY TEAM	75-0854
DESIGN	RAPID CASE DESIGN USING FINITE ELEMENT STRESS ANALYSIS AND NASTRAN	73-0730
DESIGN	RATIONAL DESIGN OF TUNNEL SUPPORTS: TUNNEL SUPPORT LOADING CAUSED BY ROCK FAILURE	75-0687
DESIGN	REANALYSIS FOR LIMITED STRUCTURAL DESIGN MODIFICATIONS	72-0650
DESIGN	RECENT ADVANCES IN COMPUTATIONAL METHODS IN STUCTURAL MECHANICS AND DESIGN,	72-0575
DESIGN	RECENT ADVANCES IN MATRIX METHODS IN STRUCTUAL ANALYSIS AND DESIGN	70-0290
DESIGN	SAFE-AXISYM, A COMPUTER PROGRAM FOR THE DESIGN OF COMPOSITE BODIES OF REVOLUTION. INPUT INSTRUCTIONS	65-0014
DESIGN	SAFE-PLANE, A COMPUTER PROGRAM FOR THE STRESS ANALYSIS AND DESIGN OF TWO-DIMENSIONAL COMPOSITE BODIE	67-0017
DESIGN	SCHWINGUNGSMAESSIGE AUSLEGUNG DER BESCHAUFELUNG AXIALER TURBOMASCHINEN (DESIGN OF DISPLACEMENT-FREE	74-0253
DESIGN	SHELLS, PLATES AND MEMBRANES - A COMPUTER-AIDED DESIGN SYSTEM	74-1159
DESIGN	SLAB BRIDGE DESIGN AND DRAWING. AN AUTOMATED PROCESS	70-0347
DESIGN	SOME APPLICATIONS OF FINITE ELEMENT ANALYSIS IN RUBBER SPRING DESIGN	72-0539
DESIGN	SOME DESIGN ASPECTS OF IC ENGINE PISTON HEADS	73-0578
DESIGN	SOME FACTORS AFFECTING DENTAL IMPLANT DESIGN	74-1146
DESIGN	STRESS ANALYSIS AND DESIGN OF SILICON SOLAR CELL ARRAYS AND RELATED MATERIAL PROPERTIES	72-0840
DESIGN	STRESS ANALYSIS FOR CLEAN LEAN DESIGN	71-0386
DESIGN	STRESS ANALYSIS IN SOME COLD FORGING TOOL COMPONENTS AND THEIR DESIGN GUIDES	74-0589
DESIGN	STRESS ANALYSIS OF A CERAMIC BLADE DESIGN SUBJECTED TO CENTRIFUGAL AND THERMAL LOADS	73-0790
DESIGN	STRUCTURAL ANALYSIS AND DESIGN OPTIMIZATION OF SPACECRAFT USING NASTRAN	71-0068
DESIGN	STRUCTURAL ANALYSIS AND DESIGN OF A CATAMARAN CROSS-STRUCTURE BY FINITE ELEMENT METHOD	73-0052
DESIGN	STRUCTURAL DESIGN AND ANALYSIS OF THE SPACE SHUTTLE	70-0050
DESIGN	STRUCTURAL DESIGN DATA FOR CONCRETE DRIFT LININGS IN BLOCK CAVING	73-0773
DESIGN	STRUCTURAL DESIGN OF SUPERTANKERS	70-0152
DESIGN	STRUCTURAL DESIGN OF CIRCULAR DIVERSION, POWER AND PRESSURE TUNNELS	71-0315
DESIGN	STRUCTURAL DESIGN OF NUCLEAR PLANT FACILITIES	73-0812
DESIGN	STRUCTURE-MEDIUM INTERACTION AND DESIGN PROCEDURES STUDY. VOLUME I. ANALYSIS METHOD, THEORY, VERIFIC	69-0166
DESIGN	STRUCTURE-MEDIUM INTERACTION AND DESIGN PROCEDURES STUDY. VOLUME IV. FEAT CODE USER'S GUIDE	69-0265
DESIGN	STRUDL-A COMPUTER SYSTEM FOR STRUCTURAL DESIGN	66-0118
DESIGN	STUDY OF EFFECTS OF DESIGN DETAILS ON STRUCTURAL RESPONSE TO ACOUSTIC EXCITATION	72-0864
DESIGN	SURVEY OF ADVANCED STRUCTURAL DESIGN ANALYSIS TECHNIQUES	69-0306
DESIGN	THE APPLICATION OF THE BERSAFE FINITE ELEMENT SYSTEM TO NUCLEAR DESIGN PROBLEMS	71-0117
DESIGN	THE APPLICATION OF THE SUPER ELEMENT METHOD IN ANALYSIS AND DESIGN OF SHIP STRUCTURES AND MACHINERY	72-0657
DESIGN	THE KANSAS TEST TRACK. NON-CONVENTIONAL TRACK STRUCTURES. DESIGN REPORT	72-0797
DESIGN	THE ROLE OF COMPUTER GRAPHICS IN THE STRUCTURAL DESIGN PROCESS	68-0215
DESIGN	THE ROLE OF MECHANICS IN DESIGN - STRUCTURAL JOINTS	74-0770
DESIGN	THE SEISMIC DESIGN STUDY OF A DOUBLE CURVATURE ARCH DAM	69-0053
DESIGN	THE USE OF EIGEN-DERIVATIVES TO PREDICT THE EFFECT OF DESIGN CHANGES UPON THE DYNAMIC CHARACTERISTIC	74-0737
DESIGN	THE VERIFICATION AND ACCEPTANCE OF COMPUTER PROGRAMS FOR DESIGN ANALYSIS	70-0334
DESIGN	THEORETICAL INVESTIGATION OF ROCK AND SUPPORT INTERACTION - DEVELOP MORE RATIONAL DESIGN METHODS AND	74-0782
DESIGN	THEORY OF EQUIVALENCE AND ITS APPLICATION TO VARIOUS PROBLEMS IN ELASTICITY AND DESIGN OF SKEW BRIDG	72-0470
DESIGN	THERMAL SLEEVE DESIGN FOR PRESSURE VESSELS	74-1129
DESIGN	THERMOSTRUCTURAL DESIGN CONSIDERATIONS TO ACHIEVE THE LARGE SPACE TELESCOPE LINE-OF-SIGHT REQUIREMEN	74-0839
DESIGN	THIN-SHELL STRUCTURES: THEORY, EXPERIMENT, AND DESIGN	74-0828
DESIGN	THIN-WALLED PRESTRESSED CONCRETE PRESSURE VESSEL FOR HIGH TEMPERATURE REACTORS: EXPERIMENTAL INVESTI	73-0857
DESIGN	TUNNEL DESIGN CONSIDERATIONS: ANALYSIS OF MEDIUM-SUPPORT INTERACTION	74-1062
DESIGN	UEBER DAS VERHALTEN VON SPANNEBETONDRUCKBEHAELTRERN IM GRENZTRAGFAEHIGKEITSZUSTAND UND DESSEN EINFUL	73-0855
DESIGN	USE OF SMALL MODELS IN DESIGN AND ANALYSIS OF PRESTRESSED-CONCRETE REACTOR VESSELS	70-0433
DESIGN	USE OF SMALL MODELS IN DESIGN AND ANALYSIS OF PRESTRESSED-CONCRETE REACTOR VESSELS	71-0426
DESIGN	USING FINITE ELEMENTS IN AUTOMOTIVE DESIGN	74-0104
DESIGN	USING FINITE ELEMENTS IN AUTOMOTIVE DESIGN	74-1299
DESIGN	VYUZITI METODY KONECNYCH PRVKU PRO PROJEKTOVANI KOMBINOVANYCH KONSTRUCKCI (FINITE ELEMENT METHOD IDE	73-0667
DESIGN	ZUR BERECHNUNG, BEMESSUNG UND BEWEHRUNG RECHTECKIGER PLATTEN KONSTNATER DICKE MIT OEFFNUNGEN ODER AU	73-0664
DESIGN-ANALYSIS	NUMERICAL DESIGN-ANALYSIS FOR PILES IN SANDS	74-0333
DESIGN-STRUCTUR	THE ROLE OF MECHANICS IN DESIGN-STRUCTURAL JOINTS	74-0767
DESIGNERS	ON BUCKLING OF UNIDIRECTIONAL BORON-ALUMINUM STIFFENERS - A CAUTION TO DESIGNERS	75-0634

DESIGNING	DER EINSATZ DER FINIT-ELEMENT-METHODE IN PRESSENBAU (APPLICATION OF THE FINITE ELEMENT METHOD FOR TH	74-0615
DESIGNING	DESIGNING BY FINITE ELEMENTS CATCHING ON IN AUTOMOTIVE ENGINEERING	74-0113
DESIGNING	DESIGNING WITH THREE DIRECTIONAL COMPOSITES	74-0204
DESIGNING	PRAKTISCHE RESULTATEN, DOEL EN NUT VAN HET ONTWERPEN VAN MECHANISCHE ELEMENTEN MET BEHULP VAN COMPUT	73-0400
DESIGNING	STRIP PROGRAM FOR DESIGNING LOAD BEARING SURFACE STRUCTURES (PROGRAMME STRIP POUR LE CALCUL DES STRU	71-0408
DESIGNS	AUTOMATED EVALUATION OF INTERCEPTOR ROCKET MOTOR DESIGNS UNDER COMBINED OPERATIONAL AND NUCLEAR EFFE	74-0698
DESIGNS	FINITE ELEMENT ANALYSIS OF SEVERAL SWAGE MANDREL DESIGNS	74-0794
DESIGNS	STRUCTURAL EVALUATION OF CANDIDATE DESIGNS FOR THE LARGE SPACE TELESCOPE PRIMARY MIRROR	75-0708
DESSEN	UEBER DAS VERHALTEN VON SPANNEBETONDRUCKBEHAELTRERN IM GRENZTRAGFAEHIGKEITSZUSTAND UND DESSEN EINFUL	73-0855
DESTRESSING	DESTRESSING TEST AT THE GALENA MINE, WALLACE, IDAHO	72-0126
DETACHMENT	COMPUTER ANALYSIS OF FRUIT DETACHMENT DURING TREE SHAKING	74-1162
DETAILED	APPLICATION OF FINITE ELEMENT METHOD FOR THE DETAILED ANALYSIS OF HATCH CORNER STRESSES	73-0735
DETAILED	DETAILED STRUCTURAL ANALYSIS OF THE HUMAN TIBIA	75-0379
DETAILS	STUDY OF EFFECTS OF DESIGN DETAILS ON STRUCTURAL RESPONSE TO ACOUSTIC EXCITATION	72-0864
DETALYAKH	OPREDELENIE STATSIONARNYKH TEMPERATURNYKH POLEI V DETALYAKH DVIGATELEI VNUTRENNEGO SGORANIYA METODOM	73-0454
DETALYAKH	RASCHET POLEI DEFORMATSII I NAPRYAZHENII V DETALYAKH DVIGATELEI VNUTRENNEGO SGORANIYA (CALCULATION O	73-0465
DETECTING	AN INVESTIGATION INTO THE METHODS OF DETECTING DEWETTED ZONES IN SOLID ROCKET PROPELLANTS USING HOLO	73-0890
DETERMINATE	MATRIX-DISPLACEMENT DIRECT-ELEMENT METHOD FOR FORCE AND TORQUE ANALYSIS OF DETERMINATE AND INDETERMI	74-0531
DETERMINATIONS	NX BOREHOLE JACK MODULUS DETERMINATIONS IN HOMOGENEOUS, ISOTROPIC, ELASTIC MATERIALS	74-0295
DETERMINATIONS	STRESS, DISPLACEMENT, LINE INTEGRAL AND CLOSURE ENERGY DETERMINATIONS OF CRACK TIP STRESS INTENSITY	74-0621
DETERMINAZIONE	SU LA DETERMINAZIONE DEI CAMPI MAGNETICI LINEARI E NON LINEARI MEDIANTE L APPLICAZIONE DEI METODI DE	74-1061
DETERMINAZIONE	SULLA DETERMINAZIONE DELLE FREQUENZE PROPRIE DI ALBERI SU PIU SUPPORTI ELASTICI. (DETERMINATION OF F	74-0916
DETERMINISTIC	A FINITE ELEMENT DETERMINISTIC CATCHMENT MODEL	74-1361
DETERMINISTIC	RESPONSE OF NUCLEAR STRUCTURAL SYSTEMS TO TRANSIENT AND RANDOM EXCITATIONS, USING BOTH DETERMINISTIC	74-1047
DETONATION	FINITE ELEMENT SOLUTION FOR WAVE PROPAGATION IN LAYERED MEDIA CAUSED BY A NUCLEAR DETONATION	67-0016
DETONATIONS	APPROXIMATE METHOD FOR CALCULATING THE RESPONSE OF EMPLACEMENT STRUCTURES SUBJECTED TO GROUND SHOCK	73-0654
DEVELOP	THEORETICAL INVESTIGATION OF ROCK AND SUPPORT INTERACTION - DEVELOP MORE RATIONAL DESIGN METHODS AND	74-0782
DEVELOP	THEORETICAL INVESTIGATION OF ROCK AND SUPPORT INTERACTION - DEVELOP MORE RATIONAL DESIGN METHODS AND	74-0782
DEVICE	A DEEP HOLE STRESS MEASUREMENT DEVICE	74-0804
DEVICE	FINITE ELEMENT METHOD OF PREDICTING STRESS DISTRIBUTIONS AND SOIL DEFORMATION UNDER A TRACTIVE DEVIC	69-0389
DEVICE	PARAMETRIC STUDY OF A BEAM WITH A COMPOUND SIDE-BRANCH RESONATOR AS A DEVICE TO EVALUATE PRELIMINARY	70-0465
DEVICE	THE FINITE ELEMENT METHOD FOR PREDICTING STRESS DISTRIBUTION AND SOIL DEFORMATION UNDER A TRACTIVE D	69-0522
DEVICE	THE FINITE ELEMENT METHOD FOR PREDICTING STRESS DISTRIBUTION AND SOIL DEFORMATION UNDER A TRACTIVE D	71-0801
DEVICES	FEASIBILITY OF USING FINITE ELEMENTS IN ANALYSIS OF SECOND BREAKDOWN IN SEMICONDUCTOR DEVICES	73-0874
DEVICES	FINITE-ELEMENT SOLUTIONS FOR DEVICES WITH PERMANENT MAGNETS	75-0464
DEVICES	STUDY ON FOIL BEARINGS - AN APPLICATION TO TAPE MEMORY DEVICES	75-0311
DEVICES	TWO-DIMENSIONAL FINITE ELEMENT SIMULATION OF SEMICONDUCTOR DEVICES	74-0260
DEWATERING	FINITE ELEMENT ANALYSES OF SOIL MOVEMENTS CAUSED BY DEEP EXCAVATION AND DEWATERING	69-0530
DEWATERING	FINITE ELEMENT ANALYSIS OF ELECTRO-OSMOTIC DEWATERING	73-0592
DEWETTED	AN INVESTIGATION INTO THE METHODS OF DETECTING DEWETTED ZONES IN SOLID ROCKET PROPELLANTS USING HOLO	73-0890
DIAGONAL	DIAGONAL TENSILE STRENGTH OF BRICKWORK	70-0158
DIAGONAL	DIE BEULLASTEN DIAGONAL VERSTEIFTER RECHTECKPLATTEN (BUCKLING LOADS IN DIAGONALLY STIFFENED RECTANGU	73-0405
DIAGONAL	FINITE ELEMENT ANALYSIS OF A DIAGONAL CRACKING BEHAVIOUR IN REINFORCED CONCRETE	73-0462
DIAGONAL	FINITE ELEMENT STUDY OF REINFORCED CONCRETE BEAMS WITH DIAGONAL TENSION CRACKS	74-1308
DIAGONALLY	DIAGONALLY DOMINANT MATRICES FOR THE FINITE ELEMENT METHOD IN HYDROLOGY	73-0410
DIAGONALLY	DIE BEULLASTEN DIAGONAL VERSTEIFTER RECHTECKPLATTEN (BUCKLING LOADS IN DIAGONALLY STIFFENED RECTANGU	73-0405
DIAMETER	EFFECT OF FILLETS ON STRESS CONCENTRATIONS IN CYLINDRICAL SHELLS WITH STEP CHANGES IN OUTSIDE DIAMET	71-0108
DIAMETER	ON THE STRESS DISTRIBUTION IN NOZZLE TO CYLINDER CONNECTIONS FOR SMALL DIAMETER RATIOS	72-0426
DIAMETER	SHORT COMMUNICATIONS: TORSION OF AXI-SYMMETRIC SOLIDS OF VARIABLE DIAMETER - INCLUDING ACCELERATION	74-0203
DIAMETRAL	FINITE ELEMENT ANALYSIS OF DIAMETRAL TEST OF POLYMER MOLDINGS	73-0129
DIAPHRAGM-WEB-F	STRESS SINGULARITIES AT DIAPHRAGM-WEB-FLANGE JUNCTIONS IN BOX GIRDERS	75-0242
DIAPHRAGMS	A NOTE ON THE ELASTIC ANALYSIS OF BOX-GIRDER DIAPHRAGMS USING FINITE ELEMENTS	74-1230
DIAPHRAGMS	ANALYSIS OF LIGHT GATE STEEL SHEAR DIAPHRAGMS	73-0733
DIAPHRAGMS	ANALYSIS OF MULTI-CELL BRIDGES WITHOUT TRANSVERSE DIAPHRAGMS, A FINITE ELEMENT APPROACH	69-0235
DIAPHRAGMS	DIAPHRAGMS IN SINGLE- AND DOUBLE-CELL BOX GIRDER BRIDGES WITH VARYING ANGLE OF SKEW	72-0130
DIAPHRAGMS	FINITE ELEMENT ANALYSIS OF METAL DECK SHEAR DIAPHRAGMS	74-0128
DIASTOLIC	DEFORMATION OF THE DIASTOLIC LEFT VENTRICLE - 2. NONLINEAR GEOMETRIC EFFECTS	74-0974
DIASTOLIC	DEFORMATION OF THE END DIASTOLIC LEFT VENTRICLE. COMPARISON OF THEORY WITH EXPERIMENT	71-0131
DICKE	ZUR BERECHNUNG, BEMESSUNG UND BEWEHRUNG RECHTECKIGER PLATTEN KONSTNATER DICKE MIT OEFFNUNGEN ODER AU	73-0664
DIE-LESS	PRELIMINARY INVESTIGATION OF THE DIE-LESS DRAWING OF TITANIUM AND SOME STEELS	75-0770
DIE-SETS	ON THE DESIGN OF DIE-SETS	75-0582
DIE-SWELL	DIE-SWELL RECONSIDERED: SOME NUMERICAL SOLUTIONS USING A FINITE ELEMENT PROGRAM	73-0498
DIELECTRIC	APPLICATION OF A UNIMOMENT TECHNIQUE TO A BICONICAL ANTENNA WITH INHOMOGENEOUS DIELECTRIC LOADING	75-0375
DIELECTRIC	COMMENT ON "SOLUTION OF LAPLACE'S EQUATION IN A REGION WITH DIELECTRIC INTERFACES OF ARBITRARY SHAPE	70-0238
DIELECTRIC	FINITE ELEMENT SOLUTION OF DIELECTRIC LOADED WAVEGUIDES	70-0206
DIELECTRIC	NUMERICAL SOLUTION OF DIELECTRIC LOADED WAVEGUIDES: I. FINITE-ELEMENT ANALYSIS	70-0199
DIELECTRIC	SCATTERING BY DIELECTRIC CYLINDERS	74-1025
DIES	ELASTIC DEFORMATION OF SQUARE EXTRUSION DIES DUE TO LOAD AND TEMPERATURE	75-0581
DIESEL	COMPUTER AIDED DESIGN OF THERMALLY LOADED AXISYMMETRIC DIESEL ENGINE COMPONENTS	74-0596
DIESEL	DIESEL ENGINE COMPONENT DESIGN USING THE FINITE ELEMENT METHOD AND INTERACTIVE GRAPHICS	74-0237
DIFFERENCE	A COMPARISON OF FINITE ELEMENT AND FINITE DIFFERENCE METHODS IN ELASTOSTATIC PROBLEMS	69-0273
DIFFERENCE	A FINITE DIFFERENCE METHOD FOR UNSTEADY FLOW IN VARIABLY SATURATED POROUS MEDIA	71-0586
DIFFERENCE	A GENERAL FINITE DIFFERENCE METHOD FOR ARBITRARY MESHES	72-0883
DIFFERENCE	A RECTANGULAR PLATE BENDING ELEMENTS THE USE OF WHICH IS EQUIVALENT TO THE USE OF THE FINITE DIFFERE	71-0398
DIFFERENCE	AN EVALUATION OF FINITE DIFFERENCE AND FINITE ELEMENT TECHNIQUES FOR ANALYSIS OF GENERAL SHELLS	70-0254
DIFFERENCE	CALCULATION OF HARMONIC FIELDS IN WHICH SINGULARITIES ARE PRESENT USING FINITE ELEMENT AND FINITE DI	73-0979
DIFFERENCE	COMPARATIVE EFFICIENCY OF CERTAIN FINITE ELEMENT AND FINITE DIFFERENCE METHODS FOR A HYPERBOLIC PROB	73-0965
DIFFERENCE	COMPARISON OF FINITE ELEMENT AND FINITE DIFFERENCE METHODS	71-0143
DIFFERENCE	COMPARISON OF THE FINITE ELEMENT AND FINITE DIFFERENCE METHODS FOR THE ANALYSIS OF STEADY TWO DIMENS	75-0223
DIFFERENCE	COMPLEXITY BOUNDS FOR REGULAR FINITE DIFFERENCE AND FINITE ELEMENT GRIDS	72-0872
DIFFERENCE	COMPLEXITY BOUNDS FOR REGULAR FINITE DIFFERENCE AND FINITE ELEMENT GRIDS	73-0043
DIFFERENCE	COMPUTATION AND SOLUTION PROCEDURE FOR NONLINEAR ANALYSIS BY COMBINED FINITE ELEMENT-FINITE DIFFEREN	72-0571
DIFFERENCE	CONNECTION BETWEEN THE FINITE DIFFERENCE LIKE METHODS AND THE METHODS BASED ON INITIAL VALUE PROBLEM	74-0877
DIFFERENCE	CREATION AND COMPARISON OF FINITE DIFFERENCE ANALOGS OF SOME FINITE ELEMENT SCHEMES	74-0888
DIFFERENCE	DIFFERENCE AND ELEMENT METHODS FOR POTENTIAL FLOW	74-0019
DIFFERENCE	EDDY CURRENTS SCREENING PROBLEMS SOLVED BY MEANS OF FINITE ELEMENT DIFFERENCE AND FINITE ELEMENT MET	75-0552
DIFFERENCE	EQUIVALENCY BETWEEN FINITE ELEMENT METHOD AND FINITE DIFFERENCE METHOD AND A NEW DISCRETIZATION APPR	73-0480
DIFFERENCE	EVALUATION AND ANALYSES OF SOME FINITE ELEMENT AND FINITE DIFFERENCE PROCEDURES FOR TIME-DEPENDENT P	74-0745
DIFFERENCE	FINITE DIFFERENCE AND FINITE ELEMENT METHODS OF APPROXIMATION	71-0457
DIFFERENCE	FINITE DIFFERENCE AND FINITE ELEMENT METHODS FOR THE SEMICONDUCTOR FLOW EQUATIONS	74-0028
DIFFERENCE	FINITE DIFFERENCE SOLUTIONS FOR LARGE DEFORMATIONS OF CYLINDRICAL SHELLS - A COMPARISON WITH FINITE	74-0270
DIFFERENCE	FINITE ELEMENT AND FINITE DIFFERENCE ENERGY TECHNIQUES FOR THE NUMERICAL SOLUTION OF PARTIAL DIFFERE	73-0570
DIFFERENCE	FINITE ELEMENT METHOD OF ANALYSIS FOR ARCH DAMS SHELLS AND COMPARISON WITH FINITE DIFFERENCE PROCEDU	65-0036
DIFFERENCE	LAGRANGIAN FINITE ELEMENT AND FINITE DIFFERENCE METHODS FOR POISSON PROBLEMS	74-1276
DIFFERENCE	MATRIX DIFFERENCE EQUATION ANALYSIS OF VIBRATING PERIODIC STRUCTURES	75-0151
DIFFERENCE	NEW DIFFERENCE BASED FINITE-ELEMENT METHOD	75-0444
DIFFERENCE	NEW TRIANGULAR ELEMENT FOR FINITE DIFFERENCE SOLUTION OF AXISYMMETRIC CONDUCTION PROBLEMS IN CYLINDR	74-0381
DIFFERENCE	RELATIVE EFFICIENCY OF FINITE DIFFERENCE AND FINITE ELEMENT METHODS - 1. HYPERBOLIC PROBLEMS AND SPL	74-0986
DIFFERENCE	SEEPAGE IN MISSISSIPPI RIVER BANKS. REPORT 1. ANALYSIS OF TRANSIENT SEEPAGE USING A VISCOUS-FLOW MOD	73-0909
DIFFERENCE	SOME DIFFERENCE SCHEMES FOR THE BIHARMONIC EQUATION	75-0204
DIFFERENCE	SPLINE APPROXIMATION AND DIFFERENCE SCHEMES FOR THE HEAT EQUATION	72-0518
DIFFERENCE	STABILITY AND OSCILLATION CHARACTERISTICS OF FINITE ELEMENT, FINITE DIFFERENCE AND WEIGHTED RESIDUAL	73-0139

DIFFERENCE	SU LA DETERMINAZIONE DEI CAMPI MAGNETICI LINEARI E NON LINEARI MEDIANTE L APPLICAZIONE DEI METODI DE	74-1061
DIFFERENCE	THE COMPARATIVE EFFICIENCY OF CERTAIN FINITE ELEMENT AND FINITE DIFFERENCE SCHEMES FOR A HYPERBOLIC	74-0910
DIFFERENCE	THE FINITE DIFFERENCE AND LOCALIZED RITZ METHODS	71-0447
DIFFERENCE	THE RELATIVE EFFICIENCY OF FINITE DIFFERENCE AND FINITE ELEMENT METHODS. 1. HYPERBOLIC PROBLEMS AND	74-0911
DIFFERENCE	THE TREATMENT OF NATURAL BOUNDARY CONDITIONS IN THE FINITE ELEMENT AND FINITE DIFFERENCE METHODS	73-1098
DIFFERENCE	THEORY OF A FINITE DIFFERENCE METHOD ON IRREGULAR NETWORKS	74-0375
DIFFERENCE	THERMOVISCOELASTIC AXISYMMETRIC STRESS ANALYSIS BY FINITE ELEMENT AND FINITE DIFFERENCE COMPUTER TEC	71-0105
DIFFERENCE	TIMING COMPARISON STUDY FOR SOME HIGH ORDER FINITE ELEMENT PROCEDURES AND A LOW ORDER FINITE DIFFERE	72-0767
DIFFERENCE-ELEM	VARIABLE GRID FINITE DIFFERENCE-ELEMENT SOLUTION OF ELLIPTIC PARTIAL DIFFERENTIAL EQUATIONS	74-0862
DIFFERENCES	APPLICATION OF FINITE DIFFERENCES FOR SOLVING THE TWO-DIMENSIONAL ELASTICITY PROBLEM BY MEANS OF THE	75-0482
DIFFERENCES	COMPARIOSN OF RESULTS OBTAINED FOR THE PROBLEM OF SYMMETRIC FLUID OSCILLATIONS IN TANKS BY THE FINIT	71-0761
DIFFERENCES	FINITE ELEMENTS VERSUS FINITE DIFFERENCES, A COMPARISON OF THE TWO METHODS FOR THE SOLUTION OF THE D	73-1055
DIFFERENCES	SIGNIFICANT APPLICATIONS OF FINITE ELEMENTS AND FINITE DIFFERENCES METHODS TO SOLVE PROBLEMS OF THER	73-0864
DIFFERENCES	VERGLEICH NUMERISCHER LOESUNGEN NACH DEM DIFFERENZENVERFAHREN UND NACH DER FINITE-ELEMENT-METHODE AM	71-0016
DIFFERENT	A COMPARATIVE STUDY OF DIFFERENT NUMERICAL SOLUTION TECHNIQUES AS APPLIED TO A NONLINEAR STRUCTURAL	73-1117
DIFFERENT	A STUDY ON COMPOSITE STRUCTURES EMPLOYING DIFFERENT TYPES OF FINITE ELEMENTS	69-0231
DIFFERENT	APPLICATION OF THE FINITE ELEMENT METHOD TO CASES REQUIRING THE COMBINATION OF ELEMENTS POSSESSING D	72-0580
DIFFERENT	COMPARAISON DES DIVERSES MODELISATIONS NUMERIQUES D'UN SYSTEME MAGNETIQUE SATURABLE. (COMPARISON OF	75-0295
DIFFERENT	COMPARISON OF THE THREE-DIMENSIONAL ANALYSIS OF A SHEAR WALL BUILDING BY TWO DIFFERENT METHODS	75-0705
DIFFERENT	ELASTIC-PLASTIC STRESS-STRAIN RELATIONS FOR MATERIALS WITH DIFFERENT YIELD STRENGTHS IN TENSION AND	72-0764
DIFFERENT	EQUIVALENT FINITE ELEMENTS ON DIFFERENT BASES	72-0287
DIFFERENT	FINITE ELEMENT CALCULATIONS OF THE COMPLIANCE OF A TAPERED DOUBLE CANTILEVER BEAM SPECIMEN FOR DIFFE	72-0665
DIFFERENT	FINITE ELEMENT METHODS BY DIFFERENT VARIATIONAL PRINCIPALS IN ELASTICITY	70-0407
DIFFERENT	FINITE ELEMENT STIFFNESS METHODS BY DIFFERENT VARIATIONAL PRINCIPLES IN ELASTICITY	70-0455
DIFFERENT	NATURAL VIBRATIONS OF THIN, FLAT WALLED STRUCTURES WITH DIFFERENT BOUNDARY CONDITIONS	71-0509
DIFFERENT	SAAS 111: FINITE ELEMENT STRESS ANALYSIS OF AXISYMMETRIC AND PLANE SOLIDS WITH DIFFERENT ORTHOTROPIC	71-0741
DIFFERENTIAL	A FINITE ELEMENT COLLOCATION SOLUTION OF DIFFERENTIAL EQUATIONS	72-0050
DIFFERENTIAL	APPLICATION OF THE FINITE ELEMENT METHOD FOR SOLVING ORDINARY LINEAR DIFFERENTIAL EQUATION WITH BOUN	74-0084
DIFFERENTIAL	APPLICATIONS OF VARIATIONAL EQUATIONS TO ORDINARY AND PARTIAL DIFFERENTIAL EQUATIONS - MULTIPLE SOLU	75-0010
DIFFERENTIAL	DIFFERENTIAL EXPANSION IN ELASTIC LAMINATES	73-0190
DIFFERENTIAL	DIFFERENTIAL OF VARIATIONAL FORMS	71-0445
DIFFERENTIAL	FINITE ELEMENT ANALYSIS OF ELASTIC CONTACT PROBLEMS USING DIFFERENTIAL DISPLACEMENTS	70-0126
DIFFERENTIAL	FINITE ELEMENT AND FINITE DIFFERENCE ENERGY TECHNIQUES FOR THE NUMERICAL SOLUTION OF PARTIAL DIFFERE	73-0570
DIFFERENTIAL	FINITE ELEMENT GALERKIN METHOD SOLUTIONS TO SELECTED ELLIPTIC AND PARABOLIC DIFFERENTIAL EQUATIONS	71-0027
DIFFERENTIAL	FINITE ELEMENT METHOD FOR PLATE BENDING ANALYSIS BY DECOMPOSITION OF DIFFERENTIAL OPERATOR	71-0558
DIFFERENTIAL	FINITE ELEMENT METHOD FOR THE GENERAL SOLUTION OF ORDINARY DIFFERENTIAL EQUATIONS	75-0473
DIFFERENTIAL	FINITE ELEMENT SOLUTIONS TO DIFFERENTIAL EQUATIONS	70-0084
DIFFERENTIAL	FINITE ELEMENT SOLUTION OF PARTIAL DIFFERENTIAL EQUATIONS FEMAC COMPUTER PROGRAM	73-0648
DIFFERENTIAL	FINITE ELEMENT SOLUTIONS OF SELECTED PARTIAL DIFFERENTIAL EQUATIONS "FEMAC COMPUTER PROGRAM"	74-0327
DIFFERENTIAL	FINITE ELEMENT SOLUTION OF A NON-LINEAR DIFFERENTIAL EQUATION USING KACHANOV'S AND SUCCESSIVE APPROX	75-0108
DIFFERENTIAL	MATRIX DIFFERENTIAL EQUATION AND HIGHER-ORDER NUMERICAL METHODS FOR PROBLEMS OF NON-LINEAR CREEP, VI	72-0198
DIFFERENTIAL	METHODS FOR SOLUTION OF PARTIAL DIFFERENTIAL EQUATIONS	73-0521
DIFFERENTIAL	NASTRAN DIFFERENTIAL STIFFNESS ANALYSIS OF AN AIRCRAFT CANOPY	71-0037
DIFFERENTIAL	NUMERICAL SOLUTION OF PARTIAL DIFFERENTIAL EQUATIONS, 11	70-0072
DIFFERENTIAL	NUMERICAL SOLUTION OF SOME NONSTATIONARY PARTIAL DIFFERENTIAL EQUATIONS: THE FINITE ELEMENT METHOD A	72-0930
DIFFERENTIAL	NUMERICAL SOLUTION OF PARTIAL DIFFERENTIAL EQUATIONS	73-0968
DIFFERENTIAL	ON THE DISCRETIZATION OF THE TIME VARIABLE FOR PARABOLIC PARTIAL DIFFERENTIAL EQUATIONS	75-0103
DIFFERENTIAL	PIECEWISE HERMITE INTERPOLATION IN ONE AND TWO VARIABLES WITH APPLICATIONS TO PARTIAL DIFFERENTIAL E	68-0201
DIFFERENTIAL	PIECEWISE POLYNOMIAL INTERPOLATIONS AND THEIR APPLICATIONS TO PARTIAL DIFFERENTIAL EQUATIONS	70-0458
DIFFERENTIAL	PROBLEM CONTROLLED GRID GENERATION FOR THE NUMERICAL SOLUTION OF PARTIAL DIFFERENTIAL EQUATIONS	75-0725
DIFFERENTIAL	RAYLEIGH RITZ GALERKIN METHODS FOR SYMMETRIC POSITIVE DIFFERENTIAL EQUATIONS APPLICATION TO FINITE E	73-1030
DIFFERENTIAL	SELF ADJOINT DIFFERENTIAL EQUATIONS AND VARIATIONAL PRINCIPLE - 1.	71-0388
DIFFERENTIAL	SELF ADJOINT DIFFERENTIAL EQUATIONS AND VARIATIONAL PRINCIPLE - 2.	72-0106
DIFFERENTIAL	THE FINITE ELEMENT METHOD FOR ELLIPTIC DIFFERENTIAL EQUATIONS	70-0027
DIFFERENTIAL	THE FINITE ELEMENT METHOD IN PARTIAL DIFFERENTIAL EQUATIONS	70-0456
DIFFERENTIAL	THE MATHEMATICAL FOUNDATIONS OF THE FINITE ELEMENT METHOD - WITH APPLICATIONS TO PARTIAL DIFFERENTIA	73-0286
DIFFERENTIAL	THE SECOND VARIAION TEST WITH ALGEBRAIC AND DIFFERENTIAL CONTRASTS	69-0342
DIFFERENTIAL	VARIABLE GRID FINITE DIFFERENCE-ELEMENT SOLUTION OF ELLIPTIC PARTIAL DIFFERENTIAL EQUATIONS	74-0862
DIFFERENTIAL	VARIATIONAL PRINCIPLES AND FINITE ELEMENT METHOD IN PARTIAL DIFFERENTIAL EQUATIONS	71-0452
DIFFERENTIAL	VARIATIONAL PRINCIPLE OF LINEAR DIFFERENTIAL EQUATIONS	72-0107
DIFFERENZE	SU LA DETERMINAZIONE DEI CAMPI MAGNETICI LINEARI E NON LINEARI MEDIANTE L APPLICAZIONE DEI METODI DE	74-1061
DIFFERENZENVERF	VERGLEICH NUMERISCHER LOESUNGEN NACH DEM DIFFERENZENVERFAHREN UND NACH DER FINITE-ELEMENT-METHODE AM	71-0016
DIFFICULT	APPLICATION OF NEW AUSTRIAN TUNNELLING METHOD IN DIFFICULT BUILTOVER AREAS IN FRANKFURT/MAIN METRO	73-1025
DIFFICULTIES	FINITE ELEMENT DIFFICULTIES IN NONLINEAR AND INCOMPRESSIBLE MATERIAL	74-1242
DIFFICULTIES	LARGE NATURAL STRAINS AND SOME SPECIAL DIFFICULTIES DUE TO NONLINEARITY AND INCOMPRESSIBILITY IN FIN	74-0528
DIFFICULTIES	LARGE NATURAL STRAINS AND SOME SPECIAL DIFFICULTIES DUE TO NONLINEARITY AND INCOMPRESSIBILITY IN FIN	74-1027
DIFFRACTION	DIFFRACTION OF OBLIQUE WAVES BY AN INFINITE CYLINDER	75-0605
DIFFRACTION	STRAIN DISTRIBUTION IN AN ALUMINA BICRYSTAL STUDIES BY THE FINITE ELEMENT METHOD AND X-RAY DIFFRACTI	74-1070
DIFFUSE	A METHOD FOR THE THERMAL ANALYSIS OF SPACECRAFT INCLUDING ALL MULTIPLE REFLECTIONS AND SHADING AMONG	70-0424
DIFFUSION	A DIFFUSION ANALYSIS OF ION IN ELECTROLYSIS BY FINITE ELEMENT METHOD	73-0296
DIFFUSION	A FINITE ELEMENT SOLUTION OF THE ONE DIMENSIONAL DIFFUSION CONVECTION EQUATION	70-0258
DIFFUSION	A FINITE ELEMENT TREATMENT OF NEUTRON DIFFUSION	71-0373
DIFFUSION	A MODEL STUDY OF GAS DIFFUSION IN ALVEOLAR SACS	73-0029
DIFFUSION	ACCURACY AND CONVERGENCE OF FINITE ELEMENT GALERKIN APPROXIMATIONS OF TIME DEPENDENT PROBLEMS WITH E	73-0987
DIFFUSION	ACCURACY AND CONVERGENCE OF FINITE ELEMENT/GALERKIN APPROXIMATIONS OF TIME DEPENDENT PROBLEMS WITH E	75-0006
DIFFUSION	AN OPTIMUM IDEALIZATION IN FINITE ELEMENT METHOD FOR A TWO-GROUP NEUTRON DIFFUSION PROBLEM	74-1143
DIFFUSION	APLIKACE MKP NA RESENI PROBLEMU DIFUZE V KOVECH. (APPLICATION OF THE FINITE ELEMENT METHOD TO THE SO	74-0609
DIFFUSION	APPLICATION OF FINITE ELEMENT TECHNIQUES FOR THE NUMERICAL SOLUTION OF THE NEUTRON TRANSPORT AND DIF	71-0013
DIFFUSION	APPLICATION OF FINITE ELEMENT SOLUTION TECHNIQUE TO NEUTRON DIFFUSION AND TRANSPORT EQUATIONS	71-0015
DIFFUSION	APPLICATION OF FINITE ELEMENT METHOD TO MULTIGROUP NEUTRON DIFFUSION EQUATION	72-0310
DIFFUSION	APPLICATION OF FINITE ELEMENT METHOD TO 2-DIMENSIONAL DIFFUSION PROBLEMS	73-0098
DIFFUSION	APPLICATION OF THE FINITE ELEMENT METHOD TO TWO DIMENSIONAL MULTIGROUP NEUTRON DIFFUSION	72-0971
DIFFUSION	APPLICATION OF THE FINITE ELEMENT METHOD IN METAL DIFFUSION	74-1366
DIFFUSION	APPLICATIONS OF FINITE ELEMENT METHODS IN REACTOR MATHEMATICS - NUMERICAL SOLUTION OF THE NEUTRON DI	72-0535
DIFFUSION	COMPARISON OF EXTRAPOLATION TECHNIQUES WITH HIGH ORDER FINITE ELEMENT METHODS FOR DIFFUSION EQUATION	73-0586
DIFFUSION	DIFFUSION OF HYDROGEN IN FILLET WELDS	75-0373
DIFFUSION	DIFFUSION ON SOLID SPHERICAL PARTICLE IN VISCOUS FLUID FLOW AT FINITE VALUES OF PE NUMBERS	69-0419
DIFFUSION	DUAL FINITE ELEMENT METHODS FOR NEUTRON DIFFUSION	71-0489
DIFFUSION	ENERGY-DEPENDENT FINITE-ELEMENT METHOD FOR FEW-GROUP DIFFUSION EQUATIONS	75-0499
DIFFUSION	FEM2D: PROGRAM FOR SOLVING THE TWO-DIMENSIONAL DIFFUSION EQUATION BY THE METHOD OF FINITE ELEMENTS	74-1089
DIFFUSION	FINITE ELEMENT METHODS FOR THE NEUTRON DIFFUSION EQUATIONS	71-0372
DIFFUSION	FINITE ELEMENT METHOD APPLIED TO NEUTRON DIFFUSION PROBLEMS	73-0977
DIFFUSION	FINITE ELEMENT NEUTRON DIFFUSION IN CURVILINEAR COORDINATES	73-0039
DIFFUSION	FINITE ELEMENTS VERSUS FINITE DIFFERENCES, A COMPARISON OF THE TWO METHODS FOR THE SOLUTION OF THE D	73-1055
DIFFUSION	FINITE-ELEMENT CODE TO SIMULATE ANISOTROPIC-PLASMA DIFFUSION IN LINEAR MULTIPOLES	75-0409
DIFFUSION	FORMULATION OF STRESS STRAIN INDUCED DIFFUSION OF HYDROGEN AND ITS SOLUTION BY COMPUTER-AIDED FINITE	75-0326
DIFFUSION	MATHEMATICAL FEATURES OF FINITE ELEMENT APPROXIMATIONS OF CERTAIN FLOW PROBLEMS WITH EMPHASIS ON CON	74-0002
DIFFUSION	NUMERICAL SOLUTION OF STEADY STATE DIFFUSION PROBLEMS CONTAINING SINGULARITIES	74-0036
DIFFUSION	NUMERICAL SOLUTION OF A DIFFUSION CONSUMPTION PROBLEM WITH A FREE BOUNDARY	74-0835
DIFFUSION	NUMERICAL TECHNIQUES FOR CONVECTION/DIFFUSION PROBLEMS	75-0123
DIFFUSION	ON THE FINITE ELEMENT ANALYSIS OF DIFFUSION PROBLEMS	73-0331
DIFFUSION	SINGULARITIES IN THE FINITE ELEMENT APPROXIMATION OF TWO DIMENSIONAL DIFFUSION PROBLEMS	73-1015

186

DIFFUSION   SOLUTION OF A TWO DIMENSIONAL DIFFUSION CONVECTION EQUATION DESCRIBING THE MASS TRANSFER OF A NEUTRA   74-0037
DIFFUSION   SOLUTION OF DIFFUSION AND P1 FINITE-ELEMENT EQUATIONS BY ITERATION   75-0404
DIFFUSION   SONIC FATIGUE OF DIFFUSION BOUNDED TITANIUM SANDWICH STRUCTURES   72-0290
DIFFUSION   STRESS DIFFUSION IN THICK STIFFENED PANELS   74-0620
DIFFUSION   THE APPLICATION OF FINITE ELEMENT METHODS TO THE SOLUTION OF STRESS AND DIFFUSION PROBLEMS IN CONTIN   68-0034
DIFFUSION   THE USE OF INTERPOLATORY POLYNOMIALS FOR A FINITE ELEMENT SOLUTION OF THE MULTIGROUP DIFFUSION EQUAT   72-0524
DIFFUSION   TIMING COMPARISON STUDY FOR SOME HIGH ORDER FINITE ELEMENT PROCEDURES AND A LOW ORDER FINITE DIFFERE   72-0767
DIFFUSION-CONVE   A GENERAL NUMERICAL SOLUTION OF THE TWO-DIMENSIONAL DIFFUSION-CONVECTION EQUATION BY THE FINITE ELEM   70-0004
DIFFUSION-CONVE   APPLICATION OF FINITE ELEMENT METHOD TO GENERAL NUMERICAL SOLUTION OF 2-DIMENSIONAL DIFFUSION-CONVEC   70-0269
DIFFUSION-CONVE   FINITE ELEMENT SOLUTION OF DIFFUSION-CONVECTION EQUATION   72-0018
DIFFUSION-CONVE   RAYLEIGH-RITZ AND GALERKIN FINITE ELEMENTS FOR DIFFUSION-CONVECTION PROBLEMS   73-0017
DIFFUSION-CONVE   USE OF FINITE-ELEMENT METHOD IN THE SOLUTION OF DIFFUSION-CONVECTION EQUATIONS   71-0490
DIFFUSION-KINET   A LEAST SQUARES FINITE ELEMENT ANALYSIS OF NONLINEAR BIOPHYSICAL DIFFUSION-KINETICS EQUATIONS   74-1257
DIFFUSIONAL   GENERAL VARIATIONAL ANALYSIS OF HYDRODYNAMIC, THERMAL AND DIFFUSIONAL BOUNDARY LAYERS   70-0191
DIFFUSIVE   DIFFUSIVE EQUILIBRIUM MODELS   73-1112
DIFFUSIVITY   MEASUREMENT OF THE WATER DIFFUSIVITY OF POROUS MATERIALS   74-1138
DIFUZE   APLIKACE MKP NA RESENI PROBLEMU DIFUZE V KOVECH. (APPLICATION OF THE FINITE ELEMENT METHOD TO THE SO   74-0609
DIGITAL   A DIGITAL COMPUTER PROGRAM FOR THE FINITE ELEMENT ANALYSIS OF SOLIDS WITH NONLINEAR MATERIAL PROPERT   65-0032
DIGITAL   A GENERAL DIGITAL COMPUTER ANALYSIS OF STATICALLY INDETERMINATE STRUCTURES   59-0007
DIGITAL   DIGITAL ANALYSIS OF NON-LINEAR STRUCTURES BY THE FORCE METHOD   64-0023
DIGITAL   DIGITAL CALCULATION OF ELECTRIC FIELDS IN METAL CLAD INSTALLATIONS   73-0212
DIGITAL   DIGITAL COMPUTATION OF STRESSES AND DEFLECTIONS IN A BOX BEAM   67-0073
DIGITAL   DIGITAL COMPUTATION OF STRESSES AND DEFLECTIONS IN A BOX BEAM - A PERFORMANCE COMPARISON BETWEEN FIN   69-0355
DIGITAL   DIGITAL COMPUTER PROGRAMS FOR STEADY-STATE OR TRANSIENT TEMPERATURE ANALYSIS OF PLANE AND AXISYMMETR   72-0915
DIGITAL   MATRIX AND DIGITAL COMPUTER METHODS IN STRUCTURAL ANALYSIS   69-0143
DIGITAL   PRAKTISCHE BEISPIELE DER DIGITALEN BERECHNUNG VON GRUNDWASSERSTROMUNGS PROBLEMEN NACH DER METHODE EN   73-0081
DIGITAL   PROBLEM OF USING THE METHOD OF FINITE ELEMENTS TO CONSTRUCT AN ALGORITHM FOR CALCULATING A THIN-WALL   74-0851
DIGITAL   THE DEVELOPMENT OF LARGE SCALE DIGITAL COMPUTER CODES FOR PRODUCTION STRUCTURAL ANALYSIS   71-0721
DIGITAL   THE IMPACT OF THE DIGITAL COMPUTER ON STRUCTURAL ANALYSIS   69-0447
DIGITAL   THE IMPACT OF THE DIGITAL COMPUTER ON THE ENGINEERING SCIENCES - PART I   70-0003
DIGITAL   USERS MANUAL FOR DIGITAL COMPUTER PROGRAM OK5 NUMERICAL ANALYSIS OF FINITE STRAINS AND DISPLACEMENTS   72-0718
DIGITAL   USERS' MANUAL FOR DIGITAL COMPUTER PROGRAM OK1. NUMERICAL ANALYSIS OF FINITE AXISYMMETRIC DEFORMATIO   72-0861
DIGITALE   AUFBEREITUNG MATHEMATISCHER MODELLE FUR DIE DIGITALE BERECHNUNG VON GRUNDWASSERSTROMUNGSPROBLEMEN NA   71-0543
DIGITALEN   GRUNDLAGEN DER DIGITALEN LOSUNG VON GRUNDWASSERSTROMUNGSPROBLEMEN NACH DER METHODE ENDLICHER ELEMENT   71-0542
DIGITALEN   PRAKTISCHE BEISPIELE DER DIGITALEN BERECHNUNG VON GRUNDWASSERSTROMUNGS PROBLEMEN NACH DER METHODE EN   73-0081
DIGUES   UTILISATION DE MODELES MATHEMATIQUES MULTIDIMENSIONNELS POUR L'ETUDE DES ECOULEMENTS DANS LES DIGUES   73-0031
DIGURILOR   COMPORTAREA ELASTICA A DIGURILOR DE PAMINT SOLICIATE SEISMIC (ELASTIC BEHAVIOR OF THE EARTH DAMS SEI   74-1185
DILATANCE   ETUDE DE LA CAPACITE PORTANTE DES PIEUX PAR LA METHODE DES ELEMENTS FINIS. INFLUENCE DE LA DILATANCE   75-0624
DILATANCE   ETUDE THEORIQUE DU COMPORTEMENT DES PIEUX SOUS CHARGE VERTICALE. INTRODUCTION DE LA DILATANCE. (THEO   75-0309
DILATANCY   ETUDE DE LA CAPACITE PORTANTE DES PIEUX PAR LA METHODE DES ELEMENTS FINIS. INFLUENCE DE LA DILATANCE   75-0624
DILATANCY   ETUDE THEORIQUE DU COMPORTEMENT DES PIEUX SOUS CHARGE VERTICALE. INTRODUCTION DE LA DILATANCE. (THEO   75-0309
DILATANT   DUPLICATION OF DILATANT BEHAVIOR IN THE ANALYSIS OF JOINTED ROCKS   72-0757
DILATIVE   STRESS ANALYSIS OF CONTRACTIVE OR DILATIVE SOIL   71-0579
DIMENSION   FREE BOUNDARIES AND FINITE-ELEMENTS IN ONE DIMENSION   75-0447
DIMENSION   FREE BOUNDARIES AND FINITE-ELEMENTS IN ONE DIMENSION   75-0587
DIMENSION   INCIPIENT CHIP FORMATION IN METAL CUTTING - A THREE- DIMENSION FINITE-ELEMENT ANALYSIS   73-1146
DIMENSION   TWO DIMENSION, PLATE, SHELL AND FINITE PRISM ISOPARAMETRIC ELEMENTS AND THEIR APPLICATIONS   71-0772
DIMENSIONAL   A DESIGN PROCEDURE FOR HIGH CUTS IN JOINTED HARD ROCK. THREE DIMENSIONAL SOLUTIONS. VOLUME 1   71-0656
DIMENSIONAL   A FINITE ELEMENT METHOD FOR THREE DIMENSIONAL FUNCTION APPROXIMATION   71-0654
DIMENSIONAL   A FINITE ELEMENT METHOD FOR THREE DIMENSIONAL ANALYSIS OF METAL FORMING   74-1109
DIMENSIONAL   A FINITE ELEMENT MODEL FOR TWO DIMENSIONAL FLOW   74-0015
DIMENSIONAL   A FINITE ELEMENT SOLUTION OF THE ONE DIMENSIONAL DIFFUSION CONVECTION EQUATION   70-0258
DIMENSIONAL   A FINITE ELEMENT SOLUTION FOR A CLASS OF TWO DIMENSIONAL VISCOUS FLUID FLOW DYNAMICS PROBLEMS   71-0544
DIMENSIONAL   A GENERAL PURPOSE THREE DIMENSIONAL STRESS PROGRAM   71-0686
DIMENSIONAL   A NUMERICAL MODEL FOR THE SIMULATION OF TWO DIMENSIONAL CONVECTIVE DISPERSION IN SHALLOW ESTUARIES   74-0052
DIMENSIONAL   A NUMERICAL PROCEDURE FOR TWO DIMENSIONAL HEATING AND MELTING CALCULATIONS, WITH APPLICATIONS TO LAS   72-0708
DIMENSIONAL   A QUARTIC, TWO DIMENSIONAL ISOPARAMETRIC ELEMENT   68-0187
DIMENSIONAL   A THREE DIMENSIONAL SOLUTION OF THE TRANSIENT FIELD PROBLEM USING ISOPARAMETRIC FINITE ELEMENTS   72-0601
DIMENSIONAL   A THREE DIMENSIONAL "VARIABLE NODE" ISOPARAMETRIC SOLID ELEMENT   74-0779
DIMENSIONAL   A VARIATIONAL FINITE ELEMENT METHOD FOR TWO DIMENSIONAL STEADY VISCOUS FLOWS   72-0355
DIMENSIONAL   AN APPLICATION OF LEAST SQUARES FINITE ELEMENT METHODS TO TWO DIMENSIONAL STEADY INVISCID COMPRESSIB   75-0114
DIMENSIONAL   AN AUTOMATIC MESH GENERATOR USING TWO AND THREE DIMENSIONAL ISOPARAMETRIC ELEMENTS   73-0066
DIMENSIONAL   AN INCREMENTAL FINITE ELEMENT ANALYSIS OF TWO DIMENSIONAL LARGE STRAIN AND LARGE DISPLACEMENT PROBLE   72-0338
DIMENSIONAL   ANALYSE TRIDIMENSIONELLE DE LA STABILITE D'UN TUNNEL AU VOISINAGE DU FRONT DE TAILLE DANS UNE ROCHE   74-0995
DIMENSIONAL   APPLICATION OF FINITE ELEMENT METHOD TO THE THREE DIMENSIONAL STRESS AND VIBRATION ANALYSIS OF BWR P   74-0180
DIMENSIONAL   APPLICATION OF FINITE ELEMENT METHODS FOR CALUCLATION OF TRANSVERSE OSCILLATIONS IN ONE DIMENSIONAL   71-0599
DIMENSIONAL   APPLICATION OF PHASE-SPACE FINITE ELEMENTS TO ONE DIMENSIONAL NEUTRON TRANSPORT EQUATION   73-0123
DIMENSIONAL   APPLICATION OF PHASE SPACE FINITE ELEMENTS TO THE TWO DIMENSIONAL NEUTRON TRANSPORT EQUATION IN X-Y   73-0453
DIMENSIONAL   APPLICATION OF THE BOUNDARY-INTEGRAL EQUATION METHOD TO THREE DIMENSIONAL STRESS ANALYSIS   73-0252
DIMENSIONAL   APPLICATION OF THE FINITE ELEMENT METHOD TO 2 AND 3 DIMENSIONAL ELASTIC-PLASTIC PROBLEMS OF FRACTURE   70-0501
DIMENSIONAL   APPLICATION OF THE FINITE ELEMENT METHOD TO TWO DIMENSIONAL MULTIGROUP NEUTRON DIFFUSION   72-0971
DIMENSIONAL   APPLICATION OF THE FINITE ELEMENT METHOD TO TWO DIMENSIONAL LAGRANGIAN HYDRODYNAMICS   74-0016
DIMENSIONAL   AUTOMATED DATA GENERATION AND LARGE MATRIX OPERATION FOR 3 DIMENSIONAL PLATE STRUCTURE ANALYSIS   73-0339
DIMENSIONAL   AUTOMATED LARGE DEFLECTION AND STABILITY ANALYSIS OF THREE DIMENSIONAL BAR STRUCTURES   69-0393
DIMENSIONAL   AUTOMATIC MESH GENERATION IN TWO AND THREE DIMENSIONAL INTERCONNECTED DOMAINS   70-0251
DIMENSIONAL   AUTOMATIC MESH GENERATION FOR THREE DIMENSIONAL STRUCTURES BASED ON THEIR THREE VIEWS   73-0340
DIMENSIONAL   AUTOMATIC TRIANGULATION OF TWO DIMENSIONAL REGIONS   72-0697
DIMENSIONAL   BERECHNUNG DES REIBUNGSLOSEN STROEMUNGSFELDES VON DREIDIMENSIONALEN AUFTRIEBSBEHAFTETENTRAGFLUEGELN   73-0533
DIMENSIONAL   CALCULATION OF TWO AND THREE DIMENSIONAL FLOW PROBLEMS ACCORDING TO SUB-STRUCTURE THEORY WITH ISOPAR   74-0215
DIMENSIONAL   CODE FOR STRESS ANALYSIS BY THE THREE DIMENSIONAL FINITE ELEMENT METHOD (JAPANESE)   69-0509
DIMENSIONAL   COMPARISON OF SOME CURVED TWO DIMENSIONAL FINITE ELEMENTS   73-0333
DIMENSIONAL   COMPARISON OF THE FINITE ELEMENT AND FINITE DIFFERENCE METHODS FOR THE ANALYSIS OF STEADY TWO DIMENS   75-0223
DIMENSIONAL   COMPARISON OF THREE DIMENSIONAL FINITE ELEMENTS   69-0073
DIMENSIONAL   COMPUTATIONAL METHODS FOR INVISCID AND VISCOUS TWO-AND-THREE DIMENSIONAL FLOW FIELDS   75-0716
DIMENSIONAL   CONFORMING FINITE ELEMENT WITH INTERNAL EQUILIBRIUM FOR TWO DIMENSIONAL PROBLEMS   72-0386
DIMENSIONAL   CONSISTENT FINITE ELEMENT MODEL FOR TWO DIMENSIONAL CONTINUUM   70-0361
DIMENSIONAL   CURVED LOW ORDER ANGULAR ELEMENTS IN TWO DIMENSIONAL NEUTRON TRANSPORT PROBLEMS   74-1322
DIMENSIONAL   DANUTA THREE DIMENSIONAL FINITE ELEMENT PROGRAM USED IN ANALYSIS OF TURBOMACHINERY   72-0296
DIMENSIONAL   DESIGN OF TWO AND THREE DIMENSIONAL STRUCTURES BY THE FINITE ELEMENT METHOD (IN GERMAN)   74-1364
DIMENSIONAL   DEVELOPMENT OF FINITE ELEMENT COMPUTATIONAL METHODS FOR THREE DIMENSIONAL ELASTIC PLASTIC CRACK ANAL   70-0356
DIMENSIONAL   DYNAMIC ANALYSIS OF THE GROUND USING THE THREE DIMENSIONAL FINITE ELEMENT   73-0293
DIMENSIONAL   DYNAMIC FINITE ELEMENT METHOD CALCULATION OF ELASTIC WAVE PROPAGATION IN TWO DIMENSIONAL STRUCTURES   74-0166
DIMENSIONAL   ELASTIC PLASTIC ANALYSIS OF THREE DIMENSIONAL PROBLEMS BY THE FINITE ELEMENT METHOD   69-0387
DIMENSIONAL   EXTENSION OF THREE DIMENSIONAL ANALYSIS TO SHELL STRUCTURES USING THE FINITE ELEMENT IDEALIZATION   74-0849
DIMENSIONAL   FINITE ELEMENT ANALYSIS OF TWO DIMENSIONAL STRUCTURES   63-0015
DIMENSIONAL   FINITE ELEMENT ELASTIC PLASTIC CREEP ANALYSIS OF TWO DIMENSIONAL CONTINUUM WITH TEMPERATURE DEPENDEN   73-0226
DIMENSIONAL   FINITE ELEMENT SOLUTION OF TWO DIMENSIONAL EXTERIOR FIELD PROBLEMS   71-0491
DIMENSIONAL   FINITE ELEMENT SOLUTION OF THE THREE DIMENSIONAL FLOW PROBLEMS AND OF REYNOLD'S EQUATION FOR INCOMPR   72-0011
DIMENSIONAL   FINITE ELEMENT SOLUTION OF MULTIGROUP TRANSPORT EQUATION IN 2 DIMENSIONAL GEOMETRIES   74-0648
DIMENSIONAL   FINITE ELEMENT SOLUTION OF TWO DIMENSIONAL UNSTEADY AND UNSATURATED FLOW IN POROUS MEDIA   75-0113
DIMENSIONAL   FINITE ELEMENT ULTIMATE LOAD ANALYSIS OF THREE DIMENSIONAL CONCRETE STRUCTURES   74-0672
DIMENSIONAL   FINITE ELEMENT WEIGHTED RESIDUAL SOLUTION TO ONE DIMENSIONAL FIELD PROBLEMS   73-0388

187

DIMENSIONAL	IN VIVO STRESSES IN THE HUMAN LEFT VENTRICULAR WALL: ANALYSIS ACCOUNTING FOR THE IRREGULAR THREE DIM	72-0136
DIMENSIONAL	INCREMENTAL THEORY OF THREE DIMENSIONAL TRANSIENT THERMOELASTOPLASTICITY - FORMULATION AND SOLUTION	73-0555
DIMENSIONAL	ISO-PARAMETRIC AND ASSOCIATED ELEMENT FAMILIES FOR TWO AND THREE DIMENSIONAL ANALYSIS	69-0285
DIMENSIONAL	MECHANICS OF CABLE MOORING SYSTEMS. VOLUME 1. THREE DIMENSIONAL RESPONSE OF DEEP WATER MOORING LINES	72-0654
DIMENSIONAL	NON-LINEAR STRESS ANALYSIS OF TWO DIMENSIONAL PROBLEMS WITH SPECIAL REFERENCE TO ROCK AND SOIL MECHA	68-0183
DIMENSIONAL	NONLINEAR FINITE-ELEMENT FORMULATION FOR TRANSIENT ANALYSIS OF THREE DIMENSIONAL THIN STRUCTURES	74-0874
DIMENSIONAL	NUMERICAL INTEGRATION CONSIDERATIONS IN TWO AND THREE DIMENSIONAL ISOPARAMETRIC FINITE ELEMENTS	75-0130
DIMENSIONAL	ON THE SOLUTION OF THE NAVIER-STOKES EQUATION IN TWO DIMENSIONAL AND AXIAL SYMMETRIC PROBLEMS	74-0007
DIMENSIONAL	OPTIMUM CONFIGURATIONAL AND DIMENSIONAL DESIGN OF TRUSS STRUCTURES	74-0588
DIMENSIONAL	PERIODIC HEAT CONDUCTION IN A TWO PHASE, TWO DIMENSIONAL SOLID DOMAIN	74-1164
DIMENSIONAL	SIMPLIFIED INELASTIC HIGH TEMPERATURE STRUCTURAL ANALYSIS OF MODERATELY COMPLEX SPATIALLY THREE DIME	74-0445
DIMENSIONAL	SINGULARITIES IN THE FINITE ELEMENT APPROXIMATION OF TWO DIMENSIONAL DIFFUSION PROBLEMS	73-1015
DIMENSIONAL	SOLID SAP. A STATIC ANALYSIS PROGRAM FOR THREE DIMENSIONAL SOLID STRUCTURES	72-0855
DIMENSIONAL	SOLUTION OF A THREE DIMENSIONAL AXISYMMETRICAL PROBLEM OF THERMOPLASTICITY AS APPLIED TO THICK TURBI	74-0384
DIMENSIONAL	SOLUTION OF A TWO DIMENSIONAL DIFFUSION CONVECTION EQUATION DESCRIBING THE MASS TRANSFER OF A NEUTRA	74-0037
DIMENSIONAL	SOME NUMERICAL PROCEDURES FOR ANALYSIS OF ONE DIMENSIONAL CONSOLIDATION	72-0928
DIMENSIONAL	STATIC THREE DIMENSIONAL ANALYSIS OF ELEVATED STEEL TRANSPORTATION STRUCTURES	74-1187
DIMENSIONAL	STEADY STATE AND TRANSIENT HEAT CONDUCTION AND STRESS ANALYSES OF TWO AND THREE DIMENSIONAL STRUCTUR	71-0675
DIMENSIONAL	STRESS ANALYSIS OF A GRAPHITE FUEL ELEMENT INCLUDING RADIATION CREEP AND DIMENSIONAL CHANGES	71-0217
DIMENSIONAL	STRESS ANALYSIS OF TWO AND THREE DIMENSIONAL STRUCTURES BY THE FINITE ELEMENT METHOD	72-0794
DIMENSIONAL	THE APPLICATION OF HIGHER ORDER FINITE ELEMENTS TO THREE DIMENSIONAL HEAT CONDUCTION PROBLEMS	70-0266
DIMENSIONAL	THREE DIMENSIONAL ANISOTROPIC AND INHOMOGENEOUS ELASTIC MEDIA MATRIX ANALYSIS FOR SMALL AND LARGE DI	65-0008
DIMENSIONAL	THREE DIMENSIONAL ANALYSIS OF TWO ARCH DAMS BY THE FINITE ELEMENT METHOD	68-0018
DIMENSIONAL	THREE DIMENSIONAL ANALYSIS OF ARCH DAMS AND THEIR FOUNDATIONS	68-0066
DIMENSIONAL	THREE DIMENSIONAL ANALYSIS OF BUILDINGS COMPOSED OF FLOOR AND WALL PANELS	71-0387
DIMENSIONAL	THREE DIMENSIONAL ANALYSIS OF STRESS CONCENTRATIONS NEAR HOLES IN LAMINATED COMPOSITES	72-0713
DIMENSIONAL	THREE DIMENSIONAL ANALYSIS OF COMPRESSIBLE POTENTIAL FLOWS WITH THE FINITE ELEMENT METHOD	73-0058
DIMENSIONAL	THREE DIMENSIONAL ANALYSIS OF COMPRESSIBLE POTENTIAL FLOWS	75-0625
DIMENSIONAL	THREE DIMENSIONAL ELASTIC AND ELASTO-PLASTIC ANALYSIS OF REACTOR PRESSURE VESSELS	69-0453
DIMENSIONAL	THREE DIMENSIONAL ELASTIC PLASTIC ANALYSIS FOR FRACTURE MECHANICS - 2.	71-0358
DIMENSIONAL	THREE DIMENSIONAL FINITE ELEMENT ANALYSIS OF JOINTED ROCK SLOPES	70-0502
DIMENSIONAL	THREE DIMENSIONAL FINITE ELEMENT STRESS ANALYSIS APPLIED TO TWO PROBLEMS IN ROCK MECHANICS	72-0189
DIMENSIONAL	THREE DIMENSIONAL FINITE ELEMENT STUDIES. PART ONE: SERVICE ROUTINES	72-0721
DIMENSIONAL	THREE DIMENSIONAL FINITE ELEMENT ANALYSIS	72-0965
DIMENSIONAL	THREE DIMENSIONAL FINITE ELEMENT ANALYSIS OF THICK LAMINATED COMPOSITES - INCLUDING INTERLAMINAR AND	73-0882
DIMENSIONAL	THREE DIMENSIONAL FINITE ELEMENT METHOD FOR PLATE BENDING	74-0243
DIMENSIONAL	THREE DIMENSIONAL FINITE ELEMENT APPROXIMATION IN ELECTROMAGNETIC EXPLORATION	74-0358
DIMENSIONAL	THREE DIMENSIONAL INELASTIC ANALYSIS OF A SHELL APPURTENANCE PER SECTION 111 - APPENDIX F.	74-0716
DIMENSIONAL	THREE DIMENSIONAL NONLINEAR MAGNETIC FIELD BOUNDARY VALUE PROBLEM AND ITS NUMERICAL SOLUTION	70-0060
DIMENSIONAL	THREE DIMENSIONAL RUPTURE ANALYSIS OF A PRESTRESSED CONCRETE PRESSURE VESSEL INCLUDING CREEP EFFECTS	73-0826
DIMENSIONAL	THREE DIMENSIONAL RUPTURE ANALYSIS OF A PRESTRESSED CONCRETE PRESSURE VESSEL INCLUDING CREEP EFFECTS	74-0538
DIMENSIONAL	THREE DIMENSIONAL SCATTERED-LIGHT STRESS ANALYSIS OF DISCONTINUOUS FIBER-REINFORCED COMPOSITES	73-0218
DIMENSIONAL	THREE DIMENSIONAL SEISMIC ANALYSIS OF STRUCTURE-GROUND SYSTEMS BY THE FINITE ELEMENT METHOD	74-0268
DIMENSIONAL	THREE DIMENSIONAL STRESS ANALYSIS	70-0086
DIMENSIONAL	THREE DIMENSIONAL STRUCTURES/MEDIA INTERACTION, DYNAX DEVELOPMENT AND APPLICATIONS.	70-0378
DIMENSIONAL	THREE DIMENSIONAL STRESS ANALYSIS OF PENETRATIONS IN THICK WALLED PRESSURE VESSELS	74-0422
DIMENSIONAL	THREE DIMENSIONAL SUPERTANKER COMPUTER PROGRAM	71-0661
DIMENSIONAL	THREE DIMENSIONAL THEORY OF ELASTICITY	69-0099
DIMENSIONAL	TRANSIENT TWO DIMENSIONAL UNCONFINED AQUIFER PROBLEMS	74-0069
DIMENSIONAL	TRANSIENT TWO DIMENSIONAL HEAT CONDUCTION PROBLEMS SOLVED BY THE FINITE ELEMENT METHOD	74-0364
DIMENSIONAL	TWO AND THREE DIMENSIONAL POTENTIAL FLOW BY THE METHOD OF SINGULARITIES	69-0449
DIMENSIONAL	TWO AND THREE DIMENSIONAL POTENTIAL FLOW ANALYSIS BY THE METHOD OF SINGULARITIES	70-0357
DIMENSIONAL	TWO AND THREE DIMENSIONAL ANALYSIS OF SUBSONIC COMPRESSIBLE POTENTIAL FLOWS IN DUCTS AND AROUND LIFT	74-0026
DIMENSIONAL	TWO AND THREE DIMENSIONAL GROUNDWATER SEEPAGE BY FINITE ELEMENTS	74-0463
DIMENSIONAL	TWO DIMENSIONAL NON LINEAR ANALYSIS OF REINFORCED CONCRETE MEMBERS BY THE FINITE ELEMENT METHOD, PAR	71-0210
DIMENSIONAL	TWO DIMENSIONAL STOCHASTIC MODEL OF A HETEROGENEOUS GEOLOGIC SYSTEM	73-0715
DIMENSIONAL	VARIATIONAL FINITE ELEMENT METHODS FOR TWO DIMENSIONAL AND AXISYMMETRIC NAVIER STOKES EQUATIONS	74-0012
DIMENSIONAL	VARIATIONAL SOLUTION OF TWO DIMENSIONAL FLUID FLOW PROBLEMS	73-1014
DIMENSIONAL	YIELDING OF A TWO DIMENSIONAL VOID ASSEMBLY IN AN ORGANIC GLASS	73-1016
DIMENSIONIERUNG	MALEN DIMENSIONIERUNG VON TRAGWERKEN	74-0683
DIMENSIONIEUNGS	UNTERSUCHUNGEN ZUM FESTIGKEITSVERHALTEN VON METALLKLEBUNGEN IM HIMBLICK AUF DIE ERSTELLUNG EINES PRA	74-0289
DIMENSIONING	ON THE DIMENSIONING OF UNDERGROUND STRUCTURES	69-0147
DIMENSIONING	UNTERSUCHUNGEN ZUM FESTIGKEITSVERHALTEN VON METALLKLEBUNGEN IM HIMBLICK AUF DIE ERSTELLUNG EINES PRA	74-0289
DIMENSIONS	AN ASSUMED STRESS APPROACH TO REFINED ISOPARAMETRIC ELEMENTS IN THREE DIMENSIONS	74-0421
DIMENSIONS	ANALYSE DES CONTRAINTES DAN DES STRUCTURES COMPLEXES EN PRESENCE DE FLUAGE, DE VARIATIONS DE DIMENSI	71-0081
DIMENSIONS	BERECHNUNG INSTATIONAERER TEMPERATUREFELDER MIT FINITEN ELEMENTEN IN RAUM-UND ZEITDIMENSION. (CALCUL	74-0987
DIMENSIONS	CALCULATION OF STRESS INTENSITY FACTORS IN TWO AND THREE DIMENSIONS USING FINITE ELEMENTS	75-0254
DIMENSIONS	CALCULATION OF TRANSIENT TEMPERATURE FIELDS WITH FINITE ELEMENTS IN SPACE AND TIME DIMENSIONS	74-0379
DIMENSIONS	COMPUTER PROGRAMS FOR SIMULATING PHYSICAL PHENOMENA IN THREE SPACE DIMENSIONS AND TIME	74-0866
DIMENSIONS	FINITE ELEMENT MATRIX DISPLACEMENT METHOD FOR TIME DEPENDENT STRESS ANALYSIS OF REACTOR MATERIALS IN	71-0674
DIMENSIONS	FORTRAN SUBROUTINES FOR THE SOLUTION OF LAPLACE'S EQUATION OVER A GENERAL REGION IN TWO DIMENSIONS	70-0425
DIMENSIONS	FUEL ELEMENT STRESSES AT DISCONTINUITIES AND INTERACTIONS BY FINITE ELEMENTS IN TWO DIMENSIONS	73-0083
DIMENSIONS	HELPFUL FORMULAS FOR INTEGRATING POLYNOMIALS IN THREE DIMENSIONS	64-0031
DIMENSIONS	NOUVELLE CONCEPTION DES DRAINS DANS LES BARRAGES EN TERRE HOMOGENE DE PETITE ET MOYENNE DIMENSIONS (	73-0404
DIMENSIONS	RC COLUMN EARTHQUAKE RESPONSE IN TWO DIMENSIONS	74-0651
DIMENSIONS	STRESS CONCENTRATION FACTORS AROUND A PENETRATION IN A STRIP OR SLAB OF FINITE DIMENSIONS	69-0409
DIMENSIONS	UNE METHODE VARIATIONNELLE ELEMENTS FINIS POUR LA RESOLUTION NUMERIQUE UN PROBLEME EXTERIEUR DANS R*	73-0576
DIMINISHING	BEHAVIOR WITH DIMINISHING CURVATURE OF STRAIN-BASED ARCH FINITE ELEMENTS	73-0088
DINAMICHESKIKH	RASCHET METODOM KONECHNYKH ELEMENTOV DINAMICHESKIKH KHARAKTEERISTIK ZEMLYANOI PLOTINY SOVMESTNO S OS	73-0437
DIPOLE	FINITE ELEMENT ANALYSIS OF ELASTO-PLASTIC MATERIALS BY USE OF DISLOCATION DIPOLE SYSTEMS	74-0269
DIPPING	UNDERGROUND MEASUREMENTS IN A STEEPLY DIPPING OREBODY	68-0085
DIRECT	A CONVERGENCE INVESTIGATION OF THE DIRECT STIFFNESS METHOD	66-0102
DIRECT	ANALYSIS OF AXISYMMETRIC SHELLS BY THE DIRECT STIFFNESS METHOD	63-0006
DIRECT	ANALYSIS OF AXISYMMETRIC SOLIDS BY THE DIRECT STIFFNESS METHOD	66-0049
DIRECT	ANALYSIS OF THE CORNEO-SCLERAL SHELL BY THE METHOD OF DIRECT STIFFNESS	71-0151
DIRECT	APPLICATION OF THE DIRECT STIFFNESS METHOD TO PLANE PROBLEMS INVOLVING LARGE, TIME-DEPENDENT DEFORMA	66-0108
DIRECT	ASKA 105 - A TECHNICAL MANUAL FOR THE DIRECT SOLUTION AND INVERSION PACKAGE	68-0255
DIRECT	BASIS OF DERIVATION OF MATRICES FOR THE DIRECT STIFFNESS METHOD	63-0008
DIRECT	COMMENT ON MATRICES FOR THE DIRECT STIFFNESS METHOD BY R. J. MELOSH	64-0022
DIRECT	DEVELOPMENT AND APPLICATION OF THE DIRECT STIFFNESS METHOD FOR OUT-OF-PLANE BENDING USING A TRIANGUL	62-0009
DIRECT	DEVELOPMENTS IN STRUCTURAL ANALYSIS BY DIRECT ENERGY MINIMIZATION	68-0070
DIRECT	DIRECT AND INVERSE HEAT CONDUCTION IN ROCK MATERIALS USING THE FINITE ELEMENT METHOD	74-1098
DIRECT	DIRECT BEAM STIFFNESS MATRIX CALCULATIONS INCLUDING SHEAR EFFECTS	62-0003
DIRECT	DIRECT FINITE ELEMENT ANALYSIS OF ELASTIC CONTACT PROBLEMS	72-0733
DIRECT	DIRECT FLEXIBILITY FINITE ELEMENT ELASTOPLASTIC ANALYSIS	71-0095
DIRECT	DIRECT ITERATION METHOD FOR THE INCORPORATION OF PHASE CHANGE IN FINITE ELEMENT HEAT CONDUCTION PROG	74-0894
DIRECT	DIRECT MEASUREMENT OF BOND SLIP IN REINFORCED CONCRETE	69-0428
DIRECT	DIRECT METHOD OF OBTAINING CAPACITANCE FROM FINITE-ELEMENT MATRICES	72-0289
DIRECT	DIRECT STIFFNESS METHOD OF ANALYSIS OF SHELLS OF REVOLUTION UTILIZING CURVED ELEMENTS	66-0042
DIRECT	DIRECT TENSILE TESTING OF ANISOTROPIC ROCKS	74-0989
DIRECT	FORCE AND TORQUE ANALYSIS OF CONSTRAINED SPACE MECHANISMS AND PLANE MECHANISMS WITH OFFSET LINKS BY	74-0569
DIRECT	GEOMETRICALLY NONLINEAR ANALYSIS BY THE DIRECT STIFFNESS METHOD	70-0331

DIRECT	GEOMETRICALLY NONLINEAR STRUCTURAL ANALYSIS BY THE DIRECT STIFFNESS METHOD	71-0560
DIRECT	MATHEMATICAL MODELING VIA DIRECT USE OF VIBRATION DATA	69-0310
DIRECT	ON THE SOLUTION OF TWO-DIMENSIONAL PROBLEMS OF ELASTOPLASTICITY BY THE FINITE ELEMENT AND DIRECT STI	67-0002
DIRECT	PLANE ELASTICITY PROBLEMS AND THE DIRECT STIFFNESS METHOD	61-0009
DIRECT	PRACTICAL APPLICATION OF A COMPUTERIZED STRUCTURAL ANALYSIS SYSTEM WHICH ADOPTS A FINITE ELEMENT TEC	69-0225
DIRECT	ROUNDOFF CRITERIA IN DIRECT STIFFNESS SOLUTIONS	68-0204
DIRECT	SPARSE MATRIX TECHNIQUES IN DIRECT SOLUTION OF FINITE ELEMENT EQUATIONS	74-0115
DIRECT	STRUCTURAL ANALYSIS OF AN ARTERIOLE BY THE DIRECT STIFFNESS METHOD	66-0109
DIRECT	SURVEY OF SPARSE MATRIX METHODS IN THE DIRECT SOLUTION OF FINITE ELEMENT EQUATIONS	73-0572
DIRECT	THE COMPARISON OF ITERATIVE AND DIRECT SOLUTION TECHNIQUES IN THE ANALYSIS OF TIME-DEPENDENT STRESS	72-0080
DIRECT-ELEMENT	MATRIX-DISPLACEMENT DIRECT-ELEMENT METHOD FOR FORCE AND TORQUE ANALYSIS OF DETERMINATE AND INDETERMI	74-0531
DIRECT-STIFFNES	A GENERALIZATION OF THE DIRECT-STIFFNESS METHOD OF STRUCTURAL ANALYSIS	64-0002
DIRECTIONAL	DESIGNING WITH THREE DIRECTIONAL COMPOSITES	74-0204
DIRECTIONAL	FINITE-ELEMENT MODELING OF DIRECTIONAL SOLIDIFICATION OF SUPER ALLOY EUTECTICS	75-0350
DIRECTIONS	DETERMINATION OF THE TECHNICAL CONSTANTS OF LAMINATES IN OBLIQUE DIRECTIONS. THEORY AND EXPERIMENTAL	73-0748
DIRECTIONS	DETERMINATION OF THE TECHNICAL CONSTANTS OF LAMINATES IN OBLIQUE DIRECTIONS. THEORY AND EXPERIMENTAL	73-0748
DIRECTIONS	TRENDS AND DIRECTIONS IN THE APPLICATIONS OF NUMERICAL ANALYSIS.	71-0667
DIRICHLET	APPROXIMATION BY FIRST ORDER FINITE ELEMENTS AND SOLUTION BY PENALTY AND DUALITY OF NONLINEAR DIRICH	74-0245
DIRICHLET	APPROXIMATIONS EXTERNES, PAR ELEMENTS FINIS DE LAGRANGE D ORDRE, UN ET DEUX, DU PROBLEME DE DIRICHLE	72-0495
DIRICHLET	ON DIRICHLET PROBLEMS USING SUBSPACES WITH NEARLY ZERO BOUNDARY CONDITIONS	72-0514
DIRICHLET	SUR L APPROXIMATION, PAR ELEMENTS FINIS D ORDRE UN, ET LA RESOLUTION, PAR PENALISATION-DUALITE, D UN	75-0845
DIRICHLET	SUR L APPROXIMATION, PAR ELEMENTS FINIS D ORDRE UN, ET LA RESOLUTION, PAR PENALISATION-DUALITE, D UN	75-0845
DIRICHLET	SURVEY OF SOME FINITE ELEMENT METHOD PROPOSED FOR TREATING DIRICHLET PROBLEM	75-0612
DIRICHTLET	NICHTLINEARE DIRICHTLET PROBLEME UND EINFACHE FINITE ELEMENT VERFAHREN	74-1263
DISC	A STUDY OF INTERVERTEBRAL DISC MECHANICS BY THE FINITE ELEMENT METHOD	74-1121
DISC	ELASTO-PLASTIC ANALYSIS OF A ROTATING MODEL CONICAL TURBINE DISC	75-0281
DISC	FINITE ELEMENT STRESS ANALYSIS OF AN INTERVERTEBRAL DISC	74-0244
DISC	NONLINEAR BEHAVIOR OF THE HUMAN INTERVERTEBRAL DISC UNDER AXIAL LOADING	74-0919
DISC	STRESS INTENSITY FACTORS FOR CRACKS IN A PLATE CONTAINING A HOLE AND IN A SPINNING DISC	73-0467
DISC	THEORETICAL PREDICTION OF DISC BRAKE TEMPERATURES AND A COMPARISON WITH EXPERIMENTAL DATA	72-0306
DISCHI	CALCULATION OF AXISYMMETRICAL OSCILLATION OF FLEXIBLE TURBINE DISCS. (CALCOLO DELLE OSCILLAZIONI ASS	74-1328
DISCONNECTED	FINITE AND DISCONNECTED SUBGROUPS OF SU(3) AND THEIR APPLICATION TO THE ELEMENTARY PARTICLE SPECTRUM	66-0129
DISCONTINUA	CONTINUA AND DISCONTINUA	66-0008
DISCONTINUA	DEVELOPMENT OF A NUMERICAL MODEL FOR DISCONTINUA	74-0961
DISCONTINUITIES	COMPUTATION OF INFLUENCE COEFFICIENTS FOR AIRCRAFT STRUCTURES WITH DISCONTINUITIES AND SWEEPBACK	47-0002
DISCONTINUITIES	DETERMINATION OF STRESSES DUE TO DISCONTINUITIES IN FINITE PLATES OF ISOTROPIC AND ORTHOTROPIC MATER	73-1052
DISCONTINUITIES	DOUBLE STRESS-WAVE DISCONTINUITIES IN FINITE, SHEAR-CORRECTED BEAMS	68-0210
DISCONTINUITIES	EFFECT OF DISCONTINUITIES ON STRESSES AND DEFORMATIONS OF ROCK AROUND A TUNNEL, (EINFLUSS VON DISKON	70-0170
DISCONTINUITIES	ELASTIC-PLASTIC ANALYSIS OF THICK WALLED PRESSURE VESSELS WITH SHARP DISCONTINUITIES	71-0155
DISCONTINUITIES	FINITE ELEMENT ANALYSIS OF ELASTIC WAVE SCATTERING FROM DISCONTINUITIES	73-0547
DISCONTINUITIES	FUEL ELEMENT STRESSES AT DISCONTINUITIES AND INTERACTIONS BY FINITE ELEMENTS IN TWO DIMENSIONS	73-0083
DISCONTINUITIES	MULTIPLE STRESS WAVE DISCONTINUITIES IN FINITE BEAMS AND PLATES	67-0114
DISCONTINUITIES	NUMERICAL DETERMINATION OF STRESSES AND DEFORMATIONS IN ROCK TAKING INTO ACCOUNT DISCONTINUITIES	70-0066
DISCONTINUITIES	NUMERICAL DETERMINATION OF STRESSES AND DEFORMATIONS IN ROCK TAKING INTO ACCOUNT DISCONTINUITIES	70-0167
DISCONTINUITIES	STRESS CONCENTRATIONS IN HULL SHAPES FROM SURFACE DISCONTINUITIES	69-0130
DISCONTINUITIES	THERMAL-STRESS CONCENTRATION CAUSED BY STRUCTURAL DISCONTINUITIES	69-0308
DISCONTINUITY	DISCONTINUITY STRESS DECAY IN OFFSHORE PLATFORM JOINTS	74-0491
DISCONTINUITY	DISCONTINUITY STRESS ANALYSIS OF PRESSURE-VESSELS USING FINITE-ELEMENT METHOD	75-0346
DISCONTINUITY	FREE VIBRATION OF THIN CYLINDRICAL SHELLS WITH A DISCONTINUITY IN THE THICKNESS	69-0439
DISCONTINUITY	SHEAR STRESSES IN BEAMS WITH DISCONTINUITY IN DEPTH	75-0487
DISCONTINUITY	SPECIAL FINITE ELEMENTS FOR FLAW AND DISCONTINUITY STRESSES IN COMPOSITES	73-0846
DISCONTINUITY	STRESS AND DISPLACEMENTS AROUND OPENINGS IN ROCK CONTAINING AN INELASTIC DISCONTINUITY	75-0170
DISCONTINUITY	STRESS CONCENTRATIONS NEAR A DISCONTINUITY IN FIBROUS COMPOSITES	71-0316
DISCONTINUITY	STRESSES AND DISPLACEMENTS AROUND OPENINGS IN ROCK CONTAINING AN ELASTIC DISCONTINUITY	75-0169
DISCONTINUOUS	A VARIATIONAL FORMULATION FOR DISCONTINUOUS FIELDS IN LINEARLY ELASTIC FINITE ELEMENT APPLICATIONS	72-0962
DISCONTINUOUS	CONTINUOUS AND DISCONTINUOUS FINITE ELEMENT METHODS FOR SOLVING THE TRANSPORT EQUATION	75-0099
DISCONTINUOUS	DISCONTINUOUS FINITE ELEMENTS IN THERMAL ANALYSIS	70-0256
DISCONTINUOUS	DISCONTINUOUS FINITE-ELEMENT APPROXIMATIONS FOR ANALYSIS OF SHOCK-WAVES IN NONLINEARLY ELASTIC-MATER	75-0451
DISCONTINUOUS	FINITE ELEMENT - GALERKIN METHODS FOR HYPERBOLIC EQUATIONS WITH DISCONTINUOUS COEFFICIENTS	73-0726
DISCONTINUOUS	FINITE ELEMENT ANALYSIS OF DISCONTINUOUS SHELLS	68-0301
DISCONTINUOUS	FINITE ELEMENT FORMULATIONS FOR ELASTIC PLATES BY GENERAL VARIATIONAL STATEMENTS WITH DISCONTINUOUS	73-0559
DISCONTINUOUS	FINITE ELEMENT FORMULATIONS FOR ELASTIC PLATES BY GENERAL VARIATIONAL STATEMENTS WITH DISCONTINUOUS	73-1124
DISCONTINUOUS	FINITE ELEMENT METHOD FOR ELLIPTIC EQUATIONS WITH DISCONTINUOUS COEFFICIENTS	70-0193
DISCONTINUOUS	LINEAR DISCONTINUOUS FINITE-ELEMENT SCHEME IN TIME-DEPENDENT TRANSPORT CALCULATIONS	75-0498
DISCONTINUOUS	LOCAL AND GLOBAL SMOOTHING OF DISCONTINUOUS FINITE ELEMENT FUNCTIONS USING A LEAST SQUARES METHOD	74-0378
DISCONTINUOUS	MATRIX AND INTERFACE STRESSES IN A DISCONTINUOUS FIBER COMPOSITE MODEL	68-0040
DISCONTINUOUS	PLASTIC FLOW AND FAILURE OF DISCONTINUOUS FIBRE COMPOSITE MATERIALS	70-0157
DISCONTINUOUS	THREE DIMENSIONAL SCATTERED-LIGHT STRESS ANALYSIS OF DISCONTINUOUS FIBER-REINFORCED COMPOSITES	73-0218
DISCRETE	A CONSISTENT DISCRETE ELEMENT TECHNIQUE FOR THIN-WALLED ASSEMBLAGES	68-0104
DISCRETE	A CURVED CYLINDRICAL SHELL DISCRETE ELEMENT	67-0009
DISCRETE	A CYLINDRICAL SHELL DISCRETE ELEMENT	67-0007
DISCRETE	A DISCRETE ELEMENT METHOD FOR ANALYSIS OF PLANE ELASTIC-PLASTIC STRESS PROBLEMS	65-0063
DISCRETE	A DISCRETE ELEMENT METHOD FOR THE ANALYSIS OF PLANE ELASTO-PLASTIC STRESS PROBLEMS	66-0062
DISCRETE	A DISCRETE ELEMENT PROCEDURE FOR THIN-SHELL INSTABILITY ANALYSIS	67-0029
DISCRETE	A DISCRETE ELEMENT STRESS AND DISPLACEMENT ANALYSIS OF ELASTOPLASTIC PLATES	70-0108
DISCRETE	A DISCRETE LEAST SQUARES AND FINITE ELEMENT METHOD	69-0322
DISCRETE	A DISCRETE LEAST-SQUARES FINITE ELEMENT METHOD FOR TRANSIENT HEAT CONDUCTION PROBLEMS	74-1191
DISCRETE	A FINITE-DEFLECTION ANALYSIS OF SHALLOW ARCHES BY THE DISCRETE ELEMENT METHOD	71-0077
DISCRETE	A NUMERICAL SOLUTION TECHNIQUE FOR A CLASS OF TWO-DIMENSIONAL PROBLEMS IN FLUID DYNAMICS FORMULATED	69-0019
DISCRETE	AN EVALUATION OF CURRENT NASTRAN DISCRETE ELEMENT MODELS FOR MONOCOQUE AND SEMI-MONOCOQUE STRUCTURES	71-0236
DISCRETE	ANALYSIS OF SHALLOW SHELL STRUCTURES BY A DISCRETE ELEMENT SYSTEM	66-0113
DISCRETE	ANISOTROPIC PARAMETRIC PLATE DISCRETE ELEMENTS	73-1047
DISCRETE	APPLICATION OF THE DISCRETE ELEMENT METHOD TO THE BUCKLING ANALYSIS OF RECTANGULAR PLATES UNDER ARBI	68-0054
DISCRETE	BASIC CONCEPTS OF THE MECHANICS OF DISCRETIZED BODIES WITH AN INTRODUCTION TO DISCRETE ELEMENT CALCU	73-0146
DISCRETE	COMMENT ON DISCRETE ELEMENT IDEALIZATION OF AN INCOMPRESSIBLE LIQUID FOR VIBRATION ANALYSIS	73-0694
DISCRETE	COMPUTATIONAL STABILITY IN NONLINEAR DISCRETE STRUCTURAL SYSTEMS	69-0145
DISCRETE	COMPUTER PROGRAMS FOR THE PLASTIC ANALYSIS OF STRUCTURES USING DISCRETE ELEMENT METHODS	67-0004
DISCRETE	DERIVATION OF COMPATIBLE STIFFNESS MATRICES FOR LARGE DISCRETE ELEMENTS	67-0080
DISCRETE	DEVELOPMENTS IN DISCRETE ELEMENT FINITE DEFLECTION STRUCTURAL ANALYSIS BY FUNCTION MINIMIZATION	68-0144
DISCRETE	DEVELOPMENTS IN DISCRETE ELEMENT FINITE DEFLECTION STRUCTURAL ANALYSIS BY FUNCTION MINIMIZATION	68-0243
DISCRETE	DISCRETE APPROXIMATIONS RELATED TO NONLINEAR THEORIES OF SOLIDS	71-0259
DISCRETE	DISCRETE APPROXIMATIONS OF ELASTIC-PLASTIC BODIES BY VARIATIONAL METHODS	72-0710
DISCRETE	DISCRETE ELEMENT ANALYSIS OF COMPOSITE STRUCTURES	75-0201
DISCRETE	DISCRETE ELEMENT APPROACH TO STRUCTURAL INSTABILITY ANALYSIS	63-0005
DISCRETE	DISCRETE ELEMENT FINITE DISPLACEMENT ANALYSIS OF ANISOTROPIC SANDWICH SHELLS	70-0504
DISCRETE	DISCRETE ELEMENT IDEALIZATION OF AN INCOMPRESSIBLE LIQUID FOR VIBRATION ANALYSIS	70-0391
DISCRETE	DISCRETE ELEMENT METHODS FOR THE PLASTIC ANALYSIS OF STRUCTURES SUBJECTED TO CYCLIC LOADING	70-0023
DISCRETE	DISCRETE ELEMENT PROCEDURE FOR THIN SHELL INSTABILITY ANALYSIS	67-0028
DISCRETE	DISCRETE ELEMENT STRUCTURAL THEORY OF FLUIDS	70-0009
DISCRETE	DISCRETE METHODS IN THE ANALYSIS OF PLATE SYSTEMS	71-0835
DISCRETE	DISCRETE MODEL ANALYSIS OF ELASTIC-PLASTIC PLATES	68-0009
DISCRETE	DISCRETE TRIANGULAR APPROXIMATION OF MOMENT AND DISPLACEMENT SURFACES FOR PLATE BENDING	71-0269

DISCRETE	DISCRETE VARIABLE METHOD APPLIED TO TRANSIENT HEAT CONDUCTION PROBLEMS	74-1163
DISCRETE	ELASTIC SNAP THROUGH ANALYSIS OF CURVED PLATES USING DISCRETE ELEMENTS	72-0899
DISCRETE	ELASTIC SNAP-THROUGH ANALYSIS OF CURVED PLATES USING DISCRETE ELEMENTS	72-0927
DISCRETE	ELASTIC STABILITY OF SHELLS OF REVOLUTION BY THE VARIATIONAL APPROACH USING DISCRETE ELEMENTS	66-0106
DISCRETE	ELASTIC-PLASTIC DISCRETE ELEMENT ANALYSIS OF AXISYMMETRICALLY LOADED SHELLS OF REVOLUTION	67-0099
DISCRETE	EVALUATION OF DISCRETE METHODS FOR THE LINEAR DYNAMIC RESPONSE OF ELASTIC AND VISCOELASTIC SOLIDS	70-0439
DISCRETE	FINITE DEFLECTION DISCRETE ELEMENT ANALYSIS OF SANDWICH PLATES AND CYLINDRICAL SHELLS WITH LAMINATED	69-0236
DISCRETE	FINITE DEFLECTION STRUCTURAL ANALYSIS USING PLATE AND SHELL DISCRETE ELEMENTS	68-0143
DISCRETE	FINITE DEFLECTION STRUCTURAL ANALYSIS USING PLATE AND CYLINDRICAL SHELL DISCRETE ELEMENTS	68-0193
DISCRETE	FINITE DEFLECTION, DISCRETE ELEMENT ANALYSIS OF SHELLS	68-0029
DISCRETE	FINITE ELEMENT MODEL FOR SHELLS BASED ON THE DISCRETE KIRCHHOFF HYPOTHESIS.	72-0388
DISCRETE	FINITE ELEMENTS FORMULATED BY THE WEIGHTED DISCRETE LEAST SQUARES METHOD	74-0200
DISCRETE	FINITE-ELEMENT SCHEME BASED ON DISCRETE KIRCHHOFF ASSUMPTION	75-0544
DISCRETE	FORMULATION AND EVALUATION OF A TOROIDAL RING DISCRETE ELEMENT	66-0082
DISCRETE	FORMULATION AND EVALUATION OF A TRIANGULAR CROSS-SECTION RING DISCRETE ELEMENT	66-0083
DISCRETE	FORMULATION AND EVALUATION OF A TRAPEZOIDAL CROSS-SECTION RING DISCRETE ELEMENT	66-0084
DISCRETE	FORMULATION OF DISCRETE ELEMENT MODELS FOR STRESS AND VIBRATION ANALYSIS OF PLATES	70-0328
DISCRETE	FORMULATION OF GENERAL DISCRETE MODELS OF THERMOMECHANICAL BEHAVIOR OF MATERIALS WITH MEMORY	69-0381
DISCRETE	NUMERICAL ANALYSIS OF THIN SHELLS BY CURVED TRIANGULAR ELEMENTS BASED ON DISCRETE – KIRCHHOFF HYPOTH	69-0091
DISCRETE	OCENA DOKLADNOSCI ROZWIAZAN PROBLEMOW BRZEGOWYCH SPREZYSTYCH CIAL DYSKERTYZOWANYCH. (ASSESSING THE A	74-0921
DISCRETE	RAMIFICATIONS OF OPTIMUM IDEALIZATION GEOMETRY IN DISCRETE ELEMENT ANALYSIS	75-0182
DISCRETE	RAY-EFFECT MITIGATION IN DISCRETE ORDINATE-LIKE ANGULAR FINITE-ELEMENT APPROXIMATIONS IN NEUTRON-TRA	75-0496
DISCRETE	SOME PROBLEMS IN THE DISCRETE ELEMENT REPRESENTATION OF AIRCRAFT STRUCTURES	64-0021
DISCRETE	STABILITY ANALYSIS OF COMPLEX STRUCTURES USING DISCRETE ELEMENT TECHNIQUE	67-0068
DISCRETE	THE APPLICATION OF GRADIENT MINIMIZATION METHODS AND HIGHER ORDER DISCRETE ELEMENTS TO SHELL BUCKLIN	71-0237
DISCRETE-ELEMEN	A DISCRETE-ELEMENT METHOD OF ANALYSIS FOR COMBINED BENDING AND SHEAR DEFORMATION OF A BEAM	69-0494
DISCRETE-ELEMEN	A DISCRETE-ELEMENT METHOD OF MULTIPLE-LOADING ANALYSIS FOR TWO-WAY BRIDGE FLOOR SLABS	70-0435
DISCRETE-ELEMEN	A DISCRETE-ELEMENT METHOD OF ANALYSIS FOR ORTHOGONAL SLAB AND GRID BRIDGE FLOOR SYSTEMS	72-0690
DISCRETE-ELEMEN	A FEASIBILITY STUDY OF THE ANALYSIS OF ASYMMETRIC SHELL BY THE DISCRETE-ELEMENT METHOD	68-0115
DISCRETE-ELEMEN	A SUMMARY OF DISCRETE-ELEMENT METHODS OF ANALYSIS FOR PAVEMENT SLABS	72-0674
DISCRETE-ELEMEN	DISCRETE-ELEMENT ANALYSIS OF SHELLS OF REVOLUTION	68-0168
DISCRETE-ELEMEN	DISCRETE-ELEMENT METHODS FOR THE PLASTIC ANALYSIS OF STRUCTURES	67-0040
DISCRETE-ELEMEN	DISCRETE-ELEMENT METHODS FOR STABILITY ANALYSIS OF COMPLEX STRUCTURES	68-0140
DISCRETE-ELEMEN	DISCRETE-ELEMENT MODELLING AND SENSITIVITY STUDIES OF STATICALLY LOADED SINGLE-LAYER AND BOUNDED DOU	70-0129
DISCRETE-ELEMEN	DISCRETE-ELEMENT PLASTIC ANALYSIS OF STRUCTURES IN A STATE OF MODIFIED PLANE STRAIN	69-0140
DISCRETE-ELEMEN	DISCRETE-ELEMENT STATIC ANALYSIS OF BONDED, DOUBLE-LAYER, BRANCHED, THIN SHELLS OF REVOLUTION, PART	68-0102
DISCRETE-ELEMEN	DISCRETE-ELEMENT STATIC ANALYSIS OF CORE-STIFFENED SHELLS OF REVOLUTION FOR ASYMMETRIC MECHANICAL LO	68-0148
DISCRETE-ELEMEN	DISCRETE-ELEMENT STATIC ANALYSIS OF BONDED, DOUBLE-LAYER, BRANCHED, THIN SHELLS OF REVOLUTION, PART	70-0081
DISCRETE-ELEMEN	EXPERIMENTAL VERIFICATION OF DISCRETE-ELEMENT SOLUTIONS FOR PLATES AND PAVEMENT SLABS	70-0411
DISCRETE-ELEMEN	INTERFACE CORE DISCRETE-ELEMENT FOR STIFFENED SHELLS OF REVOLUTION	67-0092
DISCRETE-ELEMEN	PROGRESS REPORT ON DISCRETE-ELEMENT ELASTIC AND ELASTIC-PLASTIC ANALYSIS OF SHELLS OF REVOLUTION SUB	68-0170
DISCRETE-ELEMEN	STACUSS 1: A DISCRETE-ELEMENT PROGRAM FOR THE STATIC ANALYSIS OF SINGLE-LAYER CURVED STIFFENED SHELL	69-0160
DISCRETIZATION	A GRADIENT COMPUTATIONAL PROCEDURE FOR THE SOLUTION OF LARGE PROBLEMS ARISING FROM THE FINITE ELEMEN	70-0059
DISCRETIZATION	AN INTERACTIVE COMPUTER-AIDED MODELING SYSTEM FOR THE DISCRETIZATION OF BRIDGE STRUCTURES FOR FINITE	73-1165
DISCRETIZATION	CRITERIA FOR FINITE ELEMENT DISCRETIZATION OF SHELLS OF REVOLUTION	73-0152
DISCRETIZATION	DISCRETIZATION AND COMPUTATIONAL ERRORS IN HIGHER-ORDER FINITE ELEMENTS	71-0089
DISCRETIZATION	DISCRETIZATION AND ROUND-OFF ERROR IN THE FINITE ELEMENT ANALYSIS OF ELLIPTIC BOUNDARY VALUE PROBLEM	71-0514
DISCRETIZATION	DISCRETIZATION ERROR BOUNDS FOR TRANSFINITE ELEMENTS	72-0533
DISCRETIZATION	EQUIVALENCY BETWEEN FINITE ELEMENT METHOD AND FINITE DIFFERENCE METHOD AND A NEW DISCRETIZATION APPR	73-0480
DISCRETIZATION	LOCAL ERROR BOUNDS IN BOUNDARY VALUE PROBLEMS USING FINITE ELEMENT DISCRETIZATION	72-0092
DISCRETIZATION	ON THE DEVELOPMENT OF A REFINED FINITE-ELEMENT FREE FROM DISCRETIZATION – DISPERSION FOR ELASTIC WAV	73-1148
DISCRETIZATION	ON THE DISCRETIZATION OF THE TIME VARIABLE FOR PARABOLIC PARTIAL DIFFERENTIAL EQUATIONS	75-0103
DISCRETIZATION-	DISCRETIZATION-DISPERSION IN FINITE ELEMENT MODELLING OF WAVE PROPAGATION IN SOLIDS	74-1253
DISCRETIZATIONS	FINITE-ELEMENT MULTISTEP DISCRETIZATIONS OF PARABOLIC BOUNDARY-VALUE PROBLEMS	75-0501
DISCRETIZED	BASIC CONCEPTS OF THE MECHANICS OF DISCRETIZED BODIES WITH AN INTRODUCTION TO DISCRETE ELEMENT CALCU	73-0146
DISCS	CALCULATION OF AXISYMMETRICAL OSCILLATION OF FLEXIBLE TURBINE DISCS. (CALCOLO DELLE OSCILLAZIONI ASS	74-1328
DISCS	HYBRID FINITE ELEMENT OPTIMIZATION PROGRAMS FOR DISCS USING A DYNAMICS DESIGN TECHNOLOGY	74-0691
DISCS	STRESS DISTRIBUTION IN THE LIGAMENT ZONE OF PERFORATED CIRCULAR DISCS WITH SPECIAL REFERENCE TO THE	73-0827
DISCS	VIBRATION ANALYSIS OF PLATES, SHELLS AND BLADED RIMMED DISCS USING THE TRANSFER-MATRIX-FINITE-ELEMEN	73-1151
DISEASE	FINITE ELEMENT CALCULATION OF LEFT VENTRICULAR WALL STRESS IN NORMAL AND DISEASE STATES	73-0588
DISH	THERMAL STRESS DEVELOPED IN S PLATE, STRAIGHT PLATE, AND DEEP DISH WHEELS	73-0172
DISK	A THERMAL, THERMOELASTIC, AND WEAR ANALYSIS OF HIGH-ENERGY DISK BRAKES	74-0839
DISK	GAS TURBINE ENGINE DISK CYCLIC LIFE PREDICTION	75-0509
DISK	NUMERICAL STRESS ANALYSIS OF INTERVERTEBRAL DISK	72-0125
DISK	STRESSES IN A ROTATING DISK WITH CENTRAL AND NONCENTRAL HOLES (STRESS ANALYSIS BY THE FINITE ELEMENT	74-1337
DISK	UNSYMMETRICAL TRANSIENT HEAT CONDUCTION: ROTATING DISK APPLICATIONS	70-0005
DISKONTINUITAET	EFFECT OF DISCONTINUITIES ON STRESSES AND DEFORMATIONS OF ROCK AROUND A TUNNEL. (EINFLUSS VON DISKON	70-0170
DISKOV	ISSLEDOVANIE NAPRYAZHENNO-DEFORMIROVANNOGO SOSTOYANIYA SIMMETRICHNYKH DISKOV SO STUPITSEI. (INVESTIG	74-0965
DISKS	ACCURATE CALCULATION OF STRESSES IN BEAMS, DISKS AND PLATES BY MEANS OF FINITE ELEMENT METHODS AND I	70-0133
DISKS	FINITE ELEMENT APPROACH TO THE VIBRATION ANALYSIS OF ELASTIC DISKS ON A FLEXIBLE SHAFT	75-0368
DISKS	ISSLEDOVANIE NAPRYAZHENNO-DEFORMIROVANNOGO SOSTOYANIYA SIMMETRICHNYKH DISKOV SO STUPITSEI. (INVESTIG	74-0965
DISKS	SOLUTION OF A THREE DIMENSIONAL AXISYMMETRICAL PROBLEM OF THERMOPLASTICITY AS APPLIED TO THICK TURBI	74-0384
DISKS	STUDY OF UMBRELLA OSCILLATIONS OF TURBINE DISKS OF ARBITRARY PROFILE BY THE FINITE ELEMENT METHOD	75-0332
DISKS	VIBRATION ANALYSIS OF AXIAL FLOW TURBINE DISKS USING FINITE ELEMENTS	75-0195
DISKS	VIBRATION ANALYSIS OF AXIAL-FLOW TURBINE DISKS USING FINITE-ELEMENTS	75-0415
DISKS	2-DIMENSIONAL FINITE-ELEMENT ANALYSIS OF ELECTROELASTIC NORMAL MODES OF FERROELECTRIC DISKS AND SHEL	75-0594
DISLOCATION	ANALYSIS OF DISLOCATION STRESS FIELD BY FINITE ELEMENT METHOD	75-0862
DISLOCATION	FINITE ELEMENT ANALYSIS OF ELASTO-PLASTIC MATERIALS BY USE OF DISLOCATION DIPOLE SYSTEMS	74-0269
DISPERISON	FINITE ELEMENT METHOD FOR HYDRODYNAMIC DISPERSION EQUATION WITH MIXED PARTIAL DERIVATIVES	72-0030
DISPERSION	A NUMERICAL MODEL FOR THE SIMULATION OF TWO DIMENSIONAL CONVECTIVE DISPERSION IN SHALLOW ESTUARIES	74-0052
DISPERSION	A NUMERICAL MODEL OF DISPERSION IN ESTUARIES	74-0045
DISPERSION	DISPERSION IN STEADY UNSTEADY SEEPAGE FLOWFIELD	74-0072
DISPERSION	FINITE ELEMENT NUMERICAL MODELLING OF FLOW AND DISPERSION IN ESTUARIES	73-0070
DISPERSION	FINITE ELEMENT SOLUTION FOR EFFLUENT DISPERSION	74-1371
DISPERSION	LOVE WAVE DISPERSION THROUGH IRREGULAR STRUCTURES USING FINITE-ELEMENT ALGORITHMS	75-0431
DISPERSION	NUMERICAL MODEL OF DISPERSION IN ESTUARIES	74-0905
DISPERSION	ON THE DEVELOPMENT OF A REFINED FINITE-ELEMENT FREE FROM DISCRETIZATION – DISPERSION FOR ELASTIC WAV	73-1148
DISPERSION	ON THE STUDY OF CONVECTIVE DISPERSION EQUATION	74-0005
DISPERSION	TIDAL PROPAGATION AND DISPERSION IN ESTUARIES	75-0390
DISPERSION-FREE	FINITE ELEMENT ANALYSIS OF DISPERSION-FREE GUIDES FOR ELASTIC SURFACE WAVES	73-0009
DISPLACEMENT	A CONICAL ELEMENT FOR DISPLACEMENT ANALYSIS OF AXI-SYMMETRIC SHELLS	69-0120
DISPLACEMENT	A DISCRETE ELEMENT STRESS AND DISPLACEMENT ANALYSIS OF ELASTOPLASTIC PLATES	70-0108
DISPLACEMENT	A FINITE ELEMENT DISPLACEMENT APPROACH FOR THE ELASTO-PLASTIC ANALYSIS OF THIN CYLINDRICAL SHELLS	74-1077
DISPLACEMENT	A FINITE ELEMENT FORMULATION FOR PROBLEMS OF LARGE STRAIN AND LARGE DISPLACEMENT	70-0069
DISPLACEMENT	A FINITE ELEMENT METHOD FOR GEOMETRICALLY NONLINEAR LARGE DISPLACEMENT PROBLEMS IN THIN, ELASTIC PLA	69-0005
DISPLACEMENT	A FINITE ELEMENT MODEL FOR MISCIBLE DISPLACEMENT IN GROUNDWATER AQUIFERS	75-0783
DISPLACEMENT	A FORMULATION OF MATRIX DISPLACEMENT METHOD	66-0103
DISPLACEMENT	A MATRIX DISPLACEMENT METHOD ON PRE AND POST BUCKLING ANALYSIS OF LINERS FOR REACTOR VESSELS	71-0267
DISPLACEMENT	A METHOD TO GENERATE BOTH UPPER AND LOWER BOUNDS TO PLATE EIGENVALUES BY CONFORMING DISPLACEMENT FIN	71-0166
DISPLACEMENT	A SEQUEL TO TECHNICAL NOTE 13 – THE CURVED TETRAHEDRONAL AND TRIANGULAR ELEMENTS TEC AND TRIC FOR TH	69-0048
DISPLACEMENT	A SEQUEL TO TECHNICAL NOTE 15: THE SHEBA FAMILY OF SHELL ELEMENTS FOR THE MATRIX DISPLACEMENT METHOD	69-0444
DISPLACEMENT	A STUDY OF THE MATRIX DISPLACEMENT METHOD AS APPLIED TO SHELLS OF REVOLUTION	66-0111

DISPLACEMENT	A SURVEY OF THE ANALYSIS OF SHELLS BY THE DISPLACEMENT METHOD	66-0041
DISPLACEMENT	A TAPERED THICKNESS TRIM 6 ELEMENT FOR THE MATRIX DISPLACEMENT METHOD	66-0007
DISPLACEMENT	AN IMPROVED MATRIX DISPLACEMENT METHOD FOR VIBRATION OF NONUNIFORM BEAMS	71-0841
DISPLACEMENT	AN IMPROVED SHELL OF REVOLUTION ELEMENT UTILISING CUBIC DISPLACEMENT FUNCTIONS	70-0451
DISPLACEMENT	AN INCREMENTAL FINITE ELEMENT ANALYSIS OF TWO DIMENSIONAL LARGE STRAIN AND LARGE DISPLACEMENT PROBLE	72-0338
DISPLACEMENT	ANALYSIS OF GEOMETRICALLY NONLINEAR PLATES AND SHELLS BY THE FINITE ELEMENT DISPLACEMENT METHOD	68-0050
DISPLACEMENT	ANALYSIS OF SHELLS BY THE FINITE ELEMENT DISPLACEMENT METHOD	68-0185
DISPLACEMENT	ANALYSIS OF STRESS DISTRIBUTION IN SHEAR WALLS BY THE FINITE ELEMENT DISPLACEMENT METHOD	67-0033
DISPLACEMENT	ANALYSIS OF THE ELASTIC PLASTIC PROBLEMS BY THE MATRIX DISPLACEMENT METHOD	68-0173
DISPLACEMENT	ANISOTROPIC TAPERED ELEMENTS USING DISPLACEMENT MODELS	72-0391
DISPLACEMENT	APPLICATION OF AN ASSUMED DISPLACEMENT HYBRID FINITE ELEMENT PROCEDURE TO TWO-DIMENSIONAL PROBLEMS I	74-0729
DISPLACEMENT	APPLICATION OF MATRIX DISPLACEMENT METHOD TO LINEAR ELASTIC ANALYSIS OF SHELLS OF REVOLUTION	65-0025
DISPLACEMENT	APPLICATION OF MATRIX DISPLACEMENT METHODS IN THE STUDY OF PANEL FLUTTER	69-0406
DISPLACEMENT	APPLICATION OF SIMPLIFIED HYBRID DISPLACEMENT METHOD TO PLATE AND SHELL PROBLEMS	73-1037
DISPLACEMENT	APPLICATION OF THE FINITE ELEMENT DISPLACEMENT METHOD TO PROBLEMS OF ELASTOPLASTIC DEFORMATION	72-0199
DISPLACEMENT	APPLICATION OF THE MATRIX DISPLACEMENT METHOD TO THE ANALYSIS OF PRESSURE VESSELS	69-0051
DISPLACEMENT	APPLIED FINITE ELEMENT DISPLACEMENT ANALYSIS OF A SIMPLE INTEGRAL CONSTRUCTION VEHICLE BODY SHELL	72-0264
DISPLACEMENT	ARBITRARY QUADRILATERAL SPAR WEBS FOR THE MATRIX DISPLACEMENT METHOD	66-0003
DISPLACEMENT	ASAS - A COMMERCIAL FINITE ELEMENT SYSTEM WITH DISPLACEMENT AND FORCE CONTINUITY OPTIONS	73-0947
DISPLACEMENT	ASSUMED DISPLACEMENT HYBRID FINITE ELEMENT MODEL FOR LINEAR FRACTURE MECHANICS	75-0146
DISPLACEMENT	BENDING OF THIN PLATES - DISPLACEMENT MODELS	68-0049
DISPLACEMENT	CHARACTERISATION ELASTIQUE D'UN STRATIFIE RESINE-VERRE PAR LA METHODE DES DEPLACEMENTS (DETERMINATIO	75-0272
DISPLACEMENT	COMPLETE POLYNOMIAL DISPLACEMENT FIELDS FOR FINITE ELEMENT METHODS	68-0062
DISPLACEMENT	COMPLETE POLYNOMIAL DISPLACEMENT FIELDS FOR FINITE ELEMENT METHODS	68-0092
DISPLACEMENT	COMPLETENESS AND CONVERGENCE PROPERTIES OF FINITE ELEMENT DISPLACEMENT FUNCTIONS.. A GENERAL TREATME	67-0053
DISPLACEMENT	COMPUTER AIDED TEACHING OF THE FINITE ELEMENT DISPLACEMENT METHOD	69-0430
DISPLACEMENT	CONSTRUCTION OF NORMAL DISPLACEMENT FUNCTIONS OF TRIANGULAR FINITE ELEMENT OF PLATES AND SHELLS	72-0563
DISPLACEMENT	CONVERGENCE OF SIMPLIFIED HYBRID DISPLACEMENT METHOD FOR PLATE BENDING	72-0412
DISPLACEMENT	DETERMINATION OF THE INITIAL BUCKLING STRESS OF FLAT PLATES USING THE DISPLACEMENT METHOD	69-0465
DISPLACEMENT	DEVELOPMENT OF AN ANISOTROPIC SINGULARITY FINITE ELEMENT UTILIZING THE HYBRID DISPLACEMENT METHOD	74-1099
DISPLACEMENT	DISCRETE ELEMENT FINITE DISPLACEMENT ANALYSIS OF ANISOTROPIC SANDWICH SHELLS	70-0504
DISPLACEMENT	DISCRETE TRIANGULAR APPROXIMATION OF MOMENT AND DISPLACEMENT SURFACES FOR PLATE BENDING	71-0269
DISPLACEMENT	DISPLACEMENT AND EQUILIBRIUM MODELS IN THE FINITE ELEMENT METHOD	65-0016
DISPLACEMENT	DISPLACEMENT AND STRAIN FIELDS FROM UNDERGROUND EXPLOSIONS USING ANALYTICAL AND FINITE ELEMENT METHO	72-0587
DISPLACEMENT	DISPLACEMENT METHOD IN STRUCTURAL ANALYSIS AS APPLIED TO RECTANGULAR SHELL BEAMS	72-0411
DISPLACEMENT	DUALITY BETWEEN DISPLACEMENT AND EQUILIBRIUM METHOD WITH A VIEW TO OBTAINING UPPER AND LOWER BOUNDS	62-0013
DISPLACEMENT	EFFECT OF INITIAL DISPLACEMENT ON PROBLEM OF LARGE DEFLECTION AND STABILITY	67-0113
DISPLACEMENT	ELASTO PLASTIC ANALYSIS OF PLANE STRUCTURES BY THE FINITE ELEMENT DISPLACEMENT METHOD	70-0299
DISPLACEMENT	ELASTO-PLASTIC MATRIX DISPLACEMENT ANALYSIS OF THREE-DIMENSIONAL CONTINUA	66-0004
DISPLACEMENT	ESTIMATING MATRIX DISPLACEMENT SOLUTIONS OF TWO-DIMENSIONAL PROBLEMS BY LARGE ELEMENT TECHNIQUE	70-0160
DISPLACEMENT	FINITE DISPLACEMENT PLATE FLEXURE BY THE USE OF MATRIX INCREMENTAL APPROACH	72-0384
DISPLACEMENT	FINITE DISPLACEMENT BEAM THEORY	75-0529
DISPLACEMENT	FINITE ELEMENT ANALYSIS OF VIBRATION OF THIN ELASTIC PLATES BY SIMPLIFIED HYBRID DISPLACEMENT METHOD	72-0565
DISPLACEMENT	FINITE ELEMENT DISPLACEMENT ANALYSIS OF PLATE BENDING BASED ON RECTANGULAR ELEMENTS	66-0032
DISPLACEMENT	FINITE ELEMENT DISPLACEMENT ANALYSIS OF A LUNG	72-0026
DISPLACEMENT	FINITE ELEMENT DISPLACEMENT METHOD FOR LARGE AMPLITUDE FREE FLEXURAL VIBRATIONS OF BEAMS AND PLATES	73-0185
DISPLACEMENT	FINITE ELEMENT FORMULATIONS FOR LARGE DISPLACEMENT AND LARGE STRAIN ANALYSIS	73-0457
DISPLACEMENT	FINITE ELEMENT MATRIX DISPLACEMENT METHOD FOR TIME DEPENDENT STRESS ANALYSIS OF REACTOR MATERIALS IN	71-0674
DISPLACEMENT	FINITE ELEMENT SOLUTIONS FOR PLATE BENDING PROBLEMS BY SIMPLIFIED HYBRID DISPLACEMENT METHOD	72-0780
DISPLACEMENT	FLUTTER OF SKEW PANELS BY THE MATRIX DISPLACEMENT APPROACH	70-0201
DISPLACEMENT	FORCE AND TORQUE ANALYSIS OF CONSTRAINED SPACE MECHANISMS AND PLANE MECHANISMS WITH OFFSET LINKS BY	74-0569
DISPLACEMENT	FRACTURE MECHANICS APPLICATION OF AN ASSUMED DISPLACEMENT HYBRID FINITE ELEMENT PROCEDURE	75-0212
DISPLACEMENT	FRONT DISPLACEMENT MODEL IN A FRACTURED RESERVOIR	74-0071
DISPLACEMENT	FURTHER DEVELOPMENTS IN THE USE OF LINEAR DISPLACEMENT FUNCTIONS IN THE FINITE ELEMENT ANALYSIS OF O	71-0344
DISPLACEMENT	GENERALISED MATRIX FORCE AND DISPLACEMENT METHODS FOR LINEAR STRUCTURAL ANALYSIS	65-0064
DISPLACEMENT	GEOMETRIC NONLINEARITY AND THE FINITE ELEMENT DISPLACEMENT METHOD	74-0679
DISPLACEMENT	HYPAR DISPLACEMENT FIELDS FOR RECTANGULAR FINITE ELEMENTS	73-0240
DISPLACEMENT	IMPROVEMENTS ON THE ANALYSIS OF SHELLS OF REVOLUTION BY THE MATRIX DISPLACEMENT METHOD	66-0070
DISPLACEMENT	INCLUSION OF TRANSVERSE SHEAR DEFORMATION IN FINITE ELEMENT DISPLACEMENT FORMULATIONS	74-0662
DISPLACEMENT	INCLUSION OF TRANSVERSE-SHEAR DEFORMATION IN FINITE-ELEMENT DISPLACEMENT FORMULATIONS - COMMENT	75-0406
DISPLACEMENT	INCOMPATIBLE DISPLACEMENT MODELS	71-0262
DISPLACEMENT	INCREMENTAL THEORY OF LARGE STRAIN AND LARGE DISPLACEMENT PROBLEMS AND ITS FINITE ELEMENT FORMULATIO	72-0274
DISPLACEMENT	INCREMENTAL VARIATIONAL METHOD FOR THE LARGE DISPLACEMENT ANALYSIS OF SHELLS WITH GEOMETRIC IMPERFEC	72-0894
DISPLACEMENT	INTERFERENCE-FIT FASTENER DISPLACEMENT MEASUREMENT BY SPECKLE PHOTOGRAPHY	74-1178
DISPLACEMENT	LARGE DISPLACEMENT ANALYSIS OF ASIXYMMETRIC SHELLS	69-0276
DISPLACEMENT	LARGE DISPLACEMENT ANALYSIS OF THIN PLATES	73-0261
DISPLACEMENT	LARGE DISPLACEMENT ANALYSIS OF VISCOELASTIC BEAMS AND FRAMES BY FINITE-ELEMENT METHOD	74-0644
DISPLACEMENT	LARGE DISPLACEMENT FORMULATION FOR ELASTIC BODIES	70-0429
DISPLACEMENT	LARGE DISPLACEMENT, NONLINEAR TRANSIENT ANALYSIS BY FINITE ELEMENTS	74-0324
DISPLACEMENT	LARGE STRAIN, LARGE DISPLACEMENT ANALYSIS	71-0508
DISPLACEMENT	MATRIX DISPLACEMENT ANALYSIS OF ANISOTROPIC SHELLS BY TRIANGULAR ELEMENTS	65-0005
DISPLACEMENT	MATRIX DISPLACEMENT ANALYSIS OF PLATES AND SHELLS	66-0010
DISPLACEMENT	MATRIX DISPLACEMENT ANALYSIS OF SHELLS WITH AN AXISYMMETRIC MIDSURFACE BUT HAVING CIRCUMFERENTIALLY	68-0031
DISPLACEMENT	MATRIX DISPLACEMENT METHODS IN FRACTURE MECHANICS ANALYSIS OF REACTOR VESSELS	71-0226
DISPLACEMENT	MATRIX DISPLACEMENT ANALYSIS OF PLATES AND SHELLS, PART 11, GENERAL FORMULATION OF THE LINEAR THEORY	71-0621
DISPLACEMENT	MATRIX DISPLACEMENT ANALYSIS OF SHELLS AND PLATES INCLUDING TRANSVERSE SHEAR STRAIN EFFECTS	72-0643
DISPLACEMENT	MEMBRANE PARALLELOGRAM ELEMENT WITH LINEARLY VARYING EDGE STRAIN FOR THE MATRIX DISPLACEMENT METHOD	66-0006
DISPLACEMENT	MIXED DISPLACEMENT 3D ISOPARAMETRIC FINITE ELEMENTS AND A STRAIN ENERGY BASIS FOR STUDIES OF ELEMENT	70-0286
DISPLACEMENT	NATURAL FACTOR FORMULATION OF THE STIFFNESS FOR THE MATRIX DISPLACEMENT METHOD	75-0057
DISPLACEMENT	NATURAL GEOMETRY OF SURFACE WITH SPECIFIC REFERENCE TO THE MATRIX DISPLACEMENT ANALYSIS OF SHELLS. 1	73-0709
DISPLACEMENT	NEW DISPLACEMENT HYBRID FINITE ELEMENT MODELS FOR SOLID CONTINUA	70-0115
DISPLACEMENT	NONLINEAR ANALYSIS OF SHELLS OF REVOLUTION BY THE MATRIX DISPLACEMENT METHOD	67-0079
DISPLACEMENT	NONLINEAR DYNAMIC ANALYSIS OF SHELLS OF REVOLUTION BY MATRIX DISPLACEMENT METHOD	70-0110
DISPLACEMENT	ON THE AUTOMATIC SELECTION OF INDEPENDENT DISPLACEMENT VARIABLES IN GENERAL PURPOSE FINITE ELEMENT C	73-0676
DISPLACEMENT	ON THE CONVERGENCE OF FINITE ELEMENT SCHEMES BASED ON NON-CONFORMING AND SIMPLIFIED HYBRID DISPLACEM	73-0841
DISPLACEMENT	ON THE EXTREMAL PROPERTIES OF THE SOLUTION IN DYNAMICS OF RIGID VISCOPLASTIC BODIES ALLOWING FOR LAR	73-1080
DISPLACEMENT	ON THE REDUCTION OF NUMERICAL ERROR IN THE MATRIX DISPLACEMENT METHOD	74-0417
DISPLACEMENT	ON THE RIGID DISPLACEMENT CONDITION.	72-0405
DISPLACEMENT	ON THE STRESS COMPUTATION IN FINITE ELEMENT MODELS BASED UPON DISPLACEMENT APPROXIMATIONS	74-0598
DISPLACEMENT	OPTIMAL BEAM FREQUENCIES BY THE FINITE ELEMENT DISPLACEMENT METHOD	73-0187
DISPLACEMENT	PLATE AND SHELL PROBLEMS, FINITE ELEMENT DISPLACEMENT APPROACH	66-0005
DISPLACEMENT	RECENT DEVELOPMENTS IN MATRIX DISPLACEMENT METHOD FOR ELASTIC-PLASTIC PROBLEMS IN JAPAN	71-0265
DISPLACEMENT	SABOR 1, A FORTRAN PROGRAM FOR THE LINEAR ELASTIC AXISYMMETRIC LOADING BY USING THE MATRIX DISPLACEM	65-0027
DISPLACEMENT	SECTOR ELEMENTS FOR MATRIX DISPLACEMENT ANALYSIS	73-1049
DISPLACEMENT	SHAPE FUNCTION FORMULATIONS FOR ELEMENTS OTHER THAN DISPLACEMENT MODELS	72-0472
DISPLACEMENT	SIMPLIFIED HYBRID DISPLACEMENT METHOD APPLIED TO PLATE BUCKLING PROBLEMS	72-0781
DISPLACEMENT	SIMPLIFIED HYBRID DISPLACEMENT METHOD FOR LINEAR FINITE ELEMENT ANALYSIS OF GENERAL SHELLS	74-0614
DISPLACEMENT	SOME FINITE ELEMENT SOLUTIONS FOR PLATE BENDING PROBLEMS BY SIMPLIFIED HYBRID DISPLACEMENT METHOD	72-0268
DISPLACEMENT	SOME NEW ELEMENTS FOR THE MATRIX DISPLACEMENT METHOD	68-0017
DISPLACEMENT	STRESS ANALYSIS BY THE FINITE ELEMENT DISPLACEMENT METHOD	68-0036
DISPLACEMENT	STRESS, DISPLACEMENT, LINE INTEGRAL AND CLOSURE ENERGY DETERMINATIONS OF CRACK TIP STRESS INTENSITY	74-0621
DISPLACEMENT	STRONGEST COLUMN BY FINITE ELEMENT DISPLACEMENT METHOD	73-0136

191

DISPLACEMENT	STRUCTURAL ANALYSIS BY THE MATRIX DISPLACEMENT METHOD	61-0003
DISPLACEMENT	STUDY OF FORCE DISPLACEMENT RELATIONSHIPS FOR THE FINITE ELEMENT ANALYSIS OF REINFORCED CONCRETE	73-0645
DISPLACEMENT	TETRAHEDRON ELEMENTS WITH LINEARLY VARYING STRAIN FOR THE MATRIX DISPLACEMENT METHOD	65-0006
DISPLACEMENT	THE APPLICATION OF THE MATRIX DISPLACEMENT METHOD IN PLANE ELASTO-PLASTIC STRESS PROBLEMS	64-0009
DISPLACEMENT	THE CURVED TETRAHEDRONAL AND TRIANGULAR ELEMENTS TEC AND TRIC FOR THE MATRIX DISPLACEMENT METHOD	69-0047
DISPLACEMENT	THE HERMES 8 ELEMENT FOR THE MATRIX DISPLACEMENT METHOD	68-0011
DISPLACEMENT	THE HYBRID DISPLACEMENT METHOD APPLIED TO PLATE AND SHELL PROBLEMS	75-0089
DISPLACEMENT	THE LINEAR ELASTIC DYNAMIC ANALYSIS OF SHELLS OF REVOLUTION BY THE MATRIX DISPLACEMENT METHOD	66-0046
DISPLACEMENT	THE LUMINA ELEMENT FOR THE MATRIX DISPLACEMENT METHOD	68-0010
DISPLACEMENT	THE MBB-UFE SUBSONIC PANEL METHOD. PART 2 - THE LIFT-CONNECTED DISPLACEMENT PROBLEM IN COMPRESSIBLE	72-0865
DISPLACEMENT	THE NATURAL FACTOR APPROACH IN THE DISPLACEMENT METHOD - FOUNDATION AND FURTHER DEVELOPMENTS	75-0111
DISPLACEMENT	THE SHEBA FAMILY OF SHELL ELEMENTS FOR THE MATRIX DISPLACEMENT METHOD	68-0014
DISPLACEMENT	THE SHEBA FAMILY OF SHELL ELEMENTS FOR THE MATRIX DISPLACEMENT METHOD. 111, LARGE DISPLACEMENTS	69-0499
DISPLACEMENT	THE TET 20 AND TEA 8 ELEMENTS FOR THE MATRIX DISPLACEMENT METHOD	68-0012
DISPLACEMENT	THE TRIAX 6 ELEMENT FOR AXISYMMETRIC ANALYSIS BY THE MATRIX DISPLACEMENT METHOD: PART 1, FOUNDATIONS	66-0124
DISPLACEMENT	THE TUBA FAMILY OF PLATE ELEMENTS FOR THE MATRIX DISPLACEMENT METHOD	68-0013
DISPLACEMENT	THERMO-ELASTIC ANALYSIS OF RECTANGULAR PLATES IN PLANE STRESS BY THE FINITE-ELEMENT DISPLACEMENT MET	67-0087
DISPLACEMENT	TRIANGULAR ELEMENTS WITH LINEARLY VARYING STRAIN FOR THE MATRIX DISPLACEMENT METHOD	65-0053
DISPLACEMENT	TRIAX 6 ELEMENT FOR AXISYMMETRIC ANALYSIS BY THE MATRIX DISPLACEMENT METHOD	66-0009
DISPLACEMENT	USE OF HIGHER ORDER DISPLACEMENT FUNCTIONS IN THE FREE VIBRATION ANALYSIS OF SHELLS OF REVOLUTION HA	74-1055
DISPLACEMENT	VARIATIONAL FORMULATION OF FINITE DISPLACEMENT ANALYSIS	70-0284
DISPLACEMENT	ZUR BERECHNUNG VON PLATTEN NACH DER THEORIE ZWEITER. ORDNUNG MIT HILFE EINES HYBRIDEN DEFORMATIONS M	73-0558
DISPLACEMENT-FR	SCHWINGUNGSMAESSIGE AUSLEGUNG DER BESCHAUFELUNG AXIALER TURBOMASCHINEN (DESIGN OF DISPLACEMENT-FREE	74-0253
DISPLACEMENT-HY	THREE-DIMENSIONAL LINEAR FRACTURE MECHANICS ANALYSIS BY A DISPLACEMENT-HYBRID FINITE-ELEMENT MODEL	75-0884
DISPLACEMENTS	A SEQUEL TO TECHNICAL NOTE 15: THE SHEBA FAMILY OF SHELL ELEMENTS FOR THE MATRIX DISPLACEMENT METHOD	69-0444
DISPLACEMENTS	ANALYTICAL DISPLACEMENTS AND VIBRATIONS OF CANTILEVERED UNSYMMETRIC FIBER COMPOSITE LAMINATES	75-0736
DISPLACEMENTS	APPLICATION OF NATURAL KINEMATIC MODES TO MATRIX ANALYSIS: SMALL AND LARGE DISPLACEMENTS AND STRAINS	67-0138
DISPLACEMENTS	CALCULATION OF DISPLACEMENTS AND STRESSES IN ANISOTROPIC ROCKS BY USING METHOD OF FINITE ELEMENTS	73-0265
DISPLACEMENTS	COMPUTATION OF SHELLS WITH FINITE DISPLACEMENTS: MIXED FINITE ELEMENTS (BERECHNUNG VON SCHALEN MIT E	73-0788
DISPLACEMENTS	DEPLACEMENTS DANS LES STRUCTURES MINCES A FAIBLE RIGIDITE GEOMETRIQUE-DETERMINATION NUMERIQUE (NUMER	72-0213
DISPLACEMENTS	DIE BERECHNUNG DER BEWEGUNGEN UND SPANNUNGEN DES GEBIRGES VOM ABBAU BIS ZUR TAGESOBERFLAECHE NACH DE	73-0456
DISPLACEMENTS	END DISPLACEMENTS OF SEMI-INFINITE CYLINDERS DUE TO ANNULAR LOADINGS	69-0399
DISPLACEMENTS	FINITE ELEMENT ANALYSIS OF ELASTIC CONTACT PROBLEMS USING DIFFERENTIAL DISPLACEMENTS	70-0126
DISPLACEMENTS	FINITE STRAINS AND DISPLACEMENTS OF ELASTIC MEMBRANES BY THE FINITE ELEMENT METHOD	67-0062
DISPLACEMENTS	FINITE-ELEMENT SOLUTION OF STRESSES AND DISPLACEMENTS IN A SOIL-CULVERT SYSTEM	73-1156
DISPLACEMENTS	INTRODUCTION TO SHEARS DEFORMATION INTO A THIN PLATE DISPLACEMENTS FORMULATIONS	73-0617
DISPLACEMENTS	MATRIX ANALYSIS OF SHELLS - SMALL AND LARGE DISPLACEMENTS	68-0016
DISPLACEMENTS	MATRIX ANALYSIS OF THREE-DIMENSIONAL ELASTIC MEDIA: SMALL AND LARGE DISPLACEMENTS	65-0003
DISPLACEMENTS	NEW VARIATIONAL PRINCIPLE FOR FINITE ELASTIC DISPLACEMENTS	72-0454
DISPLACEMENTS	ON ASSUMED DISPLACEMENTS FOR THE RECTANGULAR PLATE BENDING ELEMENT	67-0020
DISPLACEMENTS	RADIAL DISPLACEMENTS OF A HOLLOW SHAFT SUBJECTED TO A UNIFORM BAND OF EXTERNAL PRESSURE	72-0242
DISPLACEMENTS	RASCHET NAPRYAZHENIL L PEREMESHCHENII V SVODE PODZEMNOGO MASHINNOGO ZALA GES I V OKRUZHAYUSHCHEM SKA	74-0542
DISPLACEMENTS	RIGID BODY DISPLACEMENTS OF CURVED ELEMENTS IN THE ANALYSIS OF SHELLS BY THE MATRIX-DISPLACEMENT MET	67-0031
DISPLACEMENTS	SOME GENERAL CONSIDERATIONS ON THE NATURAL MODE TECHNIQUE, PART II: LARGE DISPLACEMENTS	69-0445
DISPLACEMENTS	STRESS AND DISPLACEMENTS AROUND OPENINGS IN ROCK CONTAINING AN INELASTIC DISCONTINUITY	75-0170
DISPLACEMENTS	STRESSES AND DISPLACEMENTS AROUND A CIRCULAR TUNNEL IN A THREE LAYER MEDIUM	69-0038
DISPLACEMENTS	STRESSES AND DISPLACEMENTS AROUND TWO ADJACENT CIRCULAR OPENINGS NEAR TO THE GROUND SURFACE	74-0993
DISPLACEMENTS	STRESSES AND DISPLACEMENTS AROUND OPENINGS IN HOMOGENEOUS ROCK	75-0168
DISPLACEMENTS	STRESSES AND DISPLACEMENTS AROUND OPENINGS IN ROCK CONTAINING AN ELASTIC DISCONTINUITY	75-0169
DISPLACEMENTS	THE EFFECT OF INITIAL DISPLACEMENTS OF PROBLEMS OF LARGE DEFLECTION AND STABILITY	67-0048
DISPLACEMENTS	THE SHEBA FAMILY OF SHELL ELEMENTS FOR THE MATRIX DISPLACEMENT METHOD. 111, LARGE DISPLACEMENTS	69-0499
DISPLACEMENTS	THREE DIMENSIONAL ANISOTROPIC AND INHOMOGENEOUS ELASTIC MEDIA MATRIX ANALYSIS FOR SMALL AND LARGE DI	65-0008
DISPLACEMENTS	USERS MANUAL FOR DIGITAL COMPUTER PROGRAM OK5 NUMERICAL ANALYSIS OF FINITE STRAINS AND DISPLACEMENTS	72-0718
DISPLAY	A COMPUTER PROGRAM TO DISPLAY THE DYNAMIC STRESS ANALYSIS OF STRUCTURES	70-0077
DISPLAY	GRAPHICAL DISPLAY OF COMPUTER SIMULATED UNBALANCED ROTOR RESPONSE	71-0486
DISPLAYING	PROCEDURE FOR EFFICIENT GENERATING, CHECKING, AND DISPLAYING NASTRAN INPUT AND OUTPUT DATA FOR ANALY	71-0098
DISPLAYS	COMPUTER GENERATED DISPLAYS OF STRUCTURES IN VIBRATION	74-1305
DISPLAYS	DISPLAYS OF KINEMATIC AND ELASTIC SYSTEMS	71-0053
DISSECTION	AUTOMATIC NESTED DISSECTION	74-0860
DISSECTION	NESTED DISSECTION OF A REGULAR FINITE ELEMENT MESH	73-0042
DISSIMILAR	FRACTURE MECHANICS FOR AN INTERFACIAL CRACK BETWEEN ADHESIVELY BONDED DISSIMILAR MATERIALS	72-0869
DISSIPATION	A NEW LOOK AT STRUCTURAL ENERGY DISSIPATION	74-0816
DISSIPATION	ANALYTICAL MODEL FOR DETERMINING ENERGY DISSIPATION IN DYNAMICALLY LOADED STRUCTURES	75-0328
DISSIPATION	THERMOMECHANICAL DISSIPATION ANALYSIS OF THERMOVISCOELASTIC SOLIDS BY FINITE ELEMENTS	72-0403
DISSIPATIVE	NUMERICAL METHOD FOR INVESTIGATING THE PROPAGATION OF SURFACE WAVES ON DISSIPATIVE GUIDES	74-0613
DISSOLVED	A FINITE ELEMENT ANALYSIS OF DISSOLVED OXYGEN DRAWDOWN AND SULFATE PRODUCTION IN STRIP MINE SPOIL DA	75-0795
DISSOLVED	MATHEMATICAL MODELING OF MOVEMENT OF DISSOLVED CONSTITUENTS IN GROUND WATER AQUIFERS BY THE FINITE E	70-0400
DISTORTED	NASTRAN-GAP, AN INTEGRATED APPROACH TO THE ANALYSIS OF RADIATION PATTERNS OF DISTORTED REFLECTORS	71-0062
DISTORTION	PREDICTION OF MARINE PROPELLER DISTORTION AND STRESSES USING A SUPERPARAMETRIC THICK SHELL FINITE EL	73-0370
DISTORTION	THEORY FOR DISTORTION STUDIES OF LUNG PARENCHYMA BASED ON ALVEOLAR MEMBRANE PROPERTIES	74-0222
DISTORTION	WARPING AND DISTORTION AT I-SECTION JOINTS	73-0722
DISTORTIONS	BERECHNUNG DER SCHWEISSEIGENSPANNUNGEN UND SCHWEISSFORMAENDERUNGEN MIT ELASTISCH-PLASTISCHEN FINITEN	75-0271
DISTRIBUTED	ANALYSIS OF CLAMPED CIRCULAR PLATES OF VARIABLE THICKNESS SUBJECTED TO UNIFORMLY DISTRIBUTED LOAD BY	73-0583
DISTRIBUTED	CONSISTENT MASS MATRIX FOR DISTRIBUTED SYSTEMS	63-0001
DISTRIBUTED	DISTRIBUTED MASS MATRIX FOR PLATE ELEMENTS IN BENDING	65-0018
DISTRIBUTED	DISTRIBUTED MASS MATRIX FOR PLATE ELEMENT BENDING (DISCUSSION)	66-0038
DISTRIBUTED	DYNAMICS OF DISTRIBUTED PARAMETER ROTOR SYSTEMS: TRANSFER MATRIX AND FINITE ELEMENT TECHNIQUES	70-0461
DISTRIBUTED	FINITE ELEMENT MODEL FOR DISTRIBUTED PARAMETER TURBOMOTOR SYSTEMS	71-0332
DISTRIBUTED	STATIC FLEXIBILITY MATRICES FOR A RECTANGULAR PLATE ELEMENT IN BENDING, TWISTING, AND SHEAR. UNIFORM	68-0217
DISTRIBUTED	THE RESPONSE OF SHELLS TO DISTRIBUTED RANDOM LOADS USING NASTRAN	71-0134
DISTRIBUTION	A STUDY OF THE PLANE STRESS OR STRAIN FINITE ELEMENT ANALYSIS FOR SOLUTION OF STRESS DISTRIBUTION IN	69-0487
DISTRIBUTION	A VARIATIONAL PRINCIPLE AND THE CONVERGENCE OF A FINITE-ELEMENT METHOD BASED ON ASSUMED STRESS DISTR	69-0260
DISTRIBUTION	ANALYSIS OF STRESS DISTRIBUTION IN SHEAR WALLS BY THE FINITE ELEMENT DISPLACEMENT METHOD	67-0033
DISTRIBUTION	APPLICATION OF AN ' ELEMENT-METHOD ' PROGRAM TO SOLVING TEMPERATURE DISTRIBUTION PROBLEMS (ANWENDUNG	71-0464
DISTRIBUTION	APPLICATION OF THE FINITE ELEMENT METHOD TO STRESS DISTRIBUTION IN SOIL MEDIA	67-0075
DISTRIBUTION	APPLICATION OF THE FINITE ELEMENT METHOD TO PROBLEMS OF TEMPERATURE DISTRIBUTION ANALYSIS FOR BODIES	74-1117
DISTRIBUTION	AREAS OF CONTACT AND PRESSURE DISTRIBUTION IN BOLTED JOINTS	71-0329
DISTRIBUTION	BEST FINITE ELEMENTS DISTRIBUTION AROUND A SINGULARITY	72-0252
DISTRIBUTION	CALCULATION OF TEMPERATURE DISTRIBUTION WITHIN A MELTING-FREEZING MATERIAL USING FINITE ELEMENT TECH	72-0043
DISTRIBUTION	DISTRIBUTION OF STRESS AROUND A TUNNEL CAVITY, USING THE FINITE ELEMENT METHOD	71-0843
DISTRIBUTION	DISTRIBUTION OF THE CLAMPING PRESSURE IN FRICTION-GRIP BOLTED JOINTS	74-0250
DISTRIBUTION	EFFECT OF FIBRE DISTRIBUTION ON STRESS AND STRAIN CONCENTRATION AT HOLES IN COMPOSITE PLATES	72-0616
DISTRIBUTION	EFFECT OF PLASTICITY AND CRACK BLUNTING ON THE STRESS DISTRIBUTION IN ORTHOTROPIC COMPOSITE MATERIAL	73-0145
DISTRIBUTION	ELASTIC AND PLASTIC STRESS DISTRIBUTION ON BASE PLATES IN THE NEIGHBOURHOOD OF FILLET WELDS IN LAP J	70-0175
DISTRIBUTION	FINITE ELEMENT ANALYSIS OF THE MAGNETIC FIELD DISTRIBUTION INSIDE A ROTATING FERROMAGNETIC BAR	74-1037
DISTRIBUTION	FINITE ELEMENT EQUILIBRIUM MODEL FOR GENERAL BODY FORCE DISTRIBUTION	75-0268
DISTRIBUTION	LOAD DISTRIBUTION IN A COMPOSITE STEEL BOX GIRDER BRIDGE	73-0765
DISTRIBUTION	NUMERICAL ANALYSIS ON STRESS DISTRIBUTION IN PARTIAL COMPRESSED SPECIMEN (HARDBOARD) BY FINITE ELEME	72-0920
DISTRIBUTION	NUMERICAL CALCULATIONS OF VELOCITY AND PRESSURE DISTRIBUTION AROUND OSCILLATING AIRFOILS	74-0642
DISTRIBUTION	OIL-FILM TEMPERATURE DISTRIBUTION IN AN INFINITELY WIDE SLIDER BEARING: AN APPLICATION OF THE FINITE	73-1128
DISTRIBUTION	ON THE APPLICATION OF FINITE-ELEMENT TECHNIQUES TO PROBLEMS CONCERNING POTENTIAL DISTRIBUTION AND ST	69-0268
DISTRIBUTION	ON THE DISTRIBUTION OF STRESSES AROUND CIRCULAR TUNNELS IN NON-HOMOGENEOUS ROCKS	66-0029

192

DISTRIBUTION	ON THE LOCAL DISTRIBUTION OF ERRORS BY FINITE ELEMENT APPROXIMATIONS	73-0304
DISTRIBUTION	ON THE STRESS DISTRIBUTION IN NOZZLE TO CYLINDER CONNECTIONS FOR SMALL DIAMETER RATIOS	72-0426
DISTRIBUTION	PLATE VIBRATION PROBLEMS USING THE FINITE ELEMENT METHOD WITH ASSUMED STRESS DISTRIBUTION AND TRIANG	66-0068
DISTRIBUTION	PRESSURE DISTRIBUTION AND DEFORMATIONS OF MACHINED COMPONENTS IN CONTACT.	73-0372
DISTRIBUTION	PRESSURE DISTRIBUTION AND BELT DEFORMATION IN V-BELT DRIVES	74-1028
DISTRIBUTION	RESEARCH NOTE: A THREE-DIMENSIONAL OIL FILM TEMPERATURE DISTRIBUTION IN TILTING THRUST BEARINGS	74-0608
DISTRIBUTION	STRAIN DISTRIBUTION AROUND UNDERGROUND OPENINGS	69-0212
DISTRIBUTION	STRAIN DISTRIBUTION AROUND UNDERGROUND OPENINGS. STATISTICAL RELATIONSHIPS FOR CERTAIN ROCK PROPERTI	71-0710
DISTRIBUTION	STRAIN DISTRIBUTION AROUND UNDERGROUND OPENINGS. FINITE ELEMENT ANALYSIS OF JOINTED SYSTEMS	72-0838
DISTRIBUTION	STRAIN DISTRIBUTION IN AN ALUMINA BICRYSTAL STUDIES BY THE FINITE ELEMENT METHOD AND X-RAY DIFFRACTI	74-1070
DISTRIBUTION	STRESS AND STRAIN DISTRIBUTION IN THE VICINITY OF INTERFERENCE FIT FASTENERS	73-0363
DISTRIBUTION	STRESS DISTRIBUTION DUE TO GRAVITY IN A VERTICAL ROCK BANK	68-0090
DISTRIBUTION	STRESS DISTRIBUTION AND ROCK FRACTURE ZONES IN THE ROOF OF LONGWALL FACE IN COAL MINE	73-0247
DISTRIBUTION	STRESS DISTRIBUTION IN BONE ARISING FROM LOADING ON ENDOSTEAL DENTAL IMPLANTS	73-0501
DISTRIBUTION	STRESS DISTRIBUTION NEAR A BROKEN POINT OF FILAMENT IN UNIDIRECTIONALLY FIBER REINFORCED COMPOSITES	73-0811
DISTRIBUTION	STRESS DISTRIBUTION IN THE LIGAMENT ZONE OF PERFORATED CIRCULAR DISCS WITH SPECIAL REFERENCE TO THE	73-0827
DISTRIBUTION	STRESS DISTRIBUTION UNDER UNIAXIAL ELASTIC PLANE STRAIN COMPRESSION WITH FRICTION	74-0526
DISTRIBUTION	THE ANALYSIS OF STRESS DISTRIBUTION AROUND THE TIP OF A CRACK BY FINITE ELEMENT METHOD WITH THE AID	71-0517
DISTRIBUTION	THE FINITE ELEMENT METHOD FOR PREDICTING STRESS DISTRIBUTION AND SOIL DEFORMATION UNDER A TRACTIVE D	69-0522
DISTRIBUTION	THE FINITE ELEMENT METHOD FOR PREDICTING STRESS DISTRIBUTION AND SOIL DEFORMATION UNDER A TRACTIVE D	71-0801
DISTRIBUTION	THE STRESS DISTRIBUTION OF NORFOLK DAM	62-0001
DISTRIBUTION	THERMAL DEFORMATION VECTOR FOR A BILINEAR TEMPERATURE DISTRIBUTION IN AN ANISOTROPIC QUADRILATERAL M	75-0670
DISTRIBUTION	THREE-DIMENSIONAL ANALYSIS FOR THE STRESS DISTRIBUTION NEAR CIRCULAR HOLES IN LAMINATED COMPOSITES	74-0793
DISTRIBUTION	UNDRAINED STRESS DISTRIBUTION BY NUMERICAL METHODS	68-0045
DISTRIBUTION	USE OF THE FINITE ELEMENT DAMPED FORCED RESPONSE STRAIN ENERGY DISTRIBUTION FOR VIBRATION REDUCTION	72-0759
DISTRIBUTION	VIBRATION REDUCTION BY USING BOTH THE FINITE ELEMENT STRAIN ENERGY DISTRIBUTION AND MOBILITY TECHNIQ	74-1303
DISTRIBUTIONAL	SOME DISTRIBUTIONAL AND CONVERGENCE PROPERTIES OF THE FINITE ELEMENT METHOD, WITH APPLICATIONS IN NO	73-0598
DISTRIBUTIONS	A FINITE ELEMENT METHOD FOR THE DETERMINATION OF NON-STATIONARY TEMPERATURE DISTRIBUTIONS AND THERMA	66-0074
DISTRIBUTIONS	A FINITE ELEMENT STUDY OF ELASTIC-PLASTIC STRESS DISTRIBUTIONS IN NOTCHED SPECIMENS UNDER TENSION	68-0177
DISTRIBUTIONS	AN INVESTIGATION INTO THE USE OF FINITE ELEMENT METHODS FOR ANALYSING STRESS DISTRIBUTIONS IN BLOCK	71-0530
DISTRIBUTIONS	ANALYSIS OF STRESS AND STRAIN DISTRIBUTIONS AT CRACK TIP BY FINITE ELEMENT METHOD	69-0193
DISTRIBUTIONS	ANALYSIS OF STRESS AND STRAIN DISTRIBUTIONS IN TRIAXIAL TESTS USING THE METHOD OF FINITE ELEMENTS	73-0903
DISTRIBUTIONS	ANALYSIS OF THE EFFECTS ON LIFE OF LEADING-EDGE HOLES IN AN AIRFOIL SUBJECTED TO ARBITRARY SPANWISE	75-0700
DISTRIBUTIONS	BEHAVIOR OF TRIANGULAR SHELL ELEMENT STIFFNESS MATRICES ASSOCIATED WITH POLYHEDRAL DEFLECTION DISTRI	68-0159
DISTRIBUTIONS	CALCULATION OF TEMPERATURE DISTRIBUTIONS IN THIN SHELLS OF REVOLUTION BY THE FINITE ELEMENT METHOD	71-0018
DISTRIBUTIONS	DERIVATION OF ELEMENT STIFFNESS MATRICES BY ASSUMED STRESS DISTRIBUTIONS	64-0007
DISTRIBUTIONS	DISTRIBUZIONI OTTIMALI DI ARMATURE DI PRECOMPRESSIONE IN TRAVI-PARETE DI CALCESTRUZZO. (OPTIMAL DIST	75-0560
DISTRIBUTIONS	ELASTIC DEFORMATION OF POLYCRYSTALLINE METAL - 1. INFLUENCE OF GRAIN SHAPE ON DISTRIBUTIONS OF STRES	72-0463
DISTRIBUTIONS	FINITE ELEMENT CALCULATION OF STRESS DISTRIBUTIONS IN ARCHES	74-0094
DISTRIBUTIONS	FINITE ELEMENT METHOD OF PREDICTING STRESS DISTRIBUTIONS AND SOIL DEFORMATION UNDER A TRACTIVE DEVIC	69-0389
DISTRIBUTIONS	INFLUENCE OF INTERLAMINATE FLAWS ON TRANSIENT TEMPERATURE DISTRIBUTIONS	73-0223
DISTRIBUTIONS	JET-WING LIFTING - SURFACE THEORY USING ELEMENTARY VORTEX DISTRIBUTIONS	75-0245
DISTRIBUTIONS	NUMERICAL SOLUTIONS FOR DETERMINING WAVE-INDUCED PRESSURE DISTRIBUTIONS AROUND BURIED PIPELINES	74-1334
DISTRIBUTIONS	ON THE CALCULATION OF CONSISTENT STRESS DISTRIBUTIONS IN FINITE ELEMENT APPROXIMATIONS	71-0191
DISTRIBUTIONS	PREDICTION OF TEMPERATURE DISTRIBUTIONS IN CARGOES WITH THE AID OF A COMPUTER PROGRAM USING THE METH	74-1215
DISTRIBUTIONS	STRENGTH DISTRIBUTIONS OF SINGLE FILAMENTS	75-0305
DISTRIBUTIONS	STRESS AND STRAIN DISTRIBUTIONS AT THE ROOT OF CRACKS	69-0290
DISTRIBUTIONS	STRESS DISTRIBUTIONS AROUND PLUG MATCHES WITH ANCHOR BLOCKS	69-0135
DISTRIBUTIONS	STRESS DISTRIBUTIONS IN THIN-WALLED INTERSECTING CYLINDRICAL SHELLS SUBJECTED TO INTERNAL PRESSURE A	71-0023
DISTRIBUTIONS	STRESS DISTRIBUTIONS AROUND SHALLOW BURIED RIGID PIPES	74-0142
DISTRIBUTIONS	THE CALCULATION OF THREE-DIMENSIONAL TEMPERATURE DISTRIBUTIONS AND THERMAL STRESSES USING FINITE ELE	71-0187
DISTRIBUTIONS	THE COMPUTATION OF TEMPERATURE DISTRIBUTIONS AND THERMAL STRESSES USING FINITE ELEMENT TECHNIQUES	71-0092
DISTRIBUTIONS	THE FINITE ELEMENT METHOD FOR FLEXURE OF SLABS WHEN STRESS DISTRIBUTIONS ARE ASSUMED	66-0069
DISTRIBUTIONS	THE USE OF THE FINITE ELEMENT METHOD IN OBTAINING SOIL-WATER PRESSURE AND SOIL-WATER CONTENT DISTRIB	75-0112
DISTRIBUTIONS	USING THE FINITE ELEMENT METHOD TO DETERMINE TEMPERATURE DISTRIBUTIONS IN ORTHOGONAL MACHINING	74-1212
DISTRIBUZIONI	DISTRIBUZIONI OTTIMALI DI ARMATURE DI PRECOMPRESSIONE IN TRAVI-PARETE DI CALCESTRUZZO. (OPTIMAL DIST	75-0560
DIV	AN ADAPTIVE COMPUTER PROGRAM FOR DIV. (P(X,Y) GRAD U) EQUALS F(X,Y,U) IN A POLYGONAL REGION	75-0133
DIVERSES	COMPARAISON DES DIVERSES MODELISATIONS NUMERIQUES D'UN SYSTEME MAGNETIQUE SATURABLE. (COMPARISON OF	75-0295
DIVERSION	STRUCTURAL DESIGN OF CIRCULAR DIVERSION, POWER AND PRESSURE TUNNELS	71-0315
DIVIDER	NONLINEAR SEISMIC ANALYSIS OF CURRENT DIVIDER SUPPORT SYSTEM FOR PACIFIC DC INTERTIE	73-0200
DIVISION	DIVISION OF COMPUTATION FIELD FOR THE FINITE ELEMENT METHOD	74-0083
DIVISION	NASTRAN EXPERIENCES OF FORT WORTH OPERATION. CONVAIR AEROSPACE DIVISION OF GENERAL DYNAMICS	71-0019
DIVISION	THE INFLUENCE OF DIVISION ON THE RESULTS IN THE FINITE ELEMENT METHOD	70-0447
DOASK	RIESENIE VAZKOPRUZNYCH ANIZOTROPICKYCH DOASK METODOU KONECNYCH PRVKOV (SOLUTION OF VISCOELASTIC ANIS	74-0477
DOCUMENT	SAMECS STRUCTURAL ANALYSIS SYSTEM - USER'S DOCUMENT	69-0495
DOCUMENT	SINGER: A COMPUTER CODE FOR GENERAL ANALYSIS OF TWO-DIMENSIONAL CONCRETE STRUCTURES. VOL. 11. PROGRA	75-0723
DOCUMENTATION	DESIGN, DEVELOPMENT, DOCUMENTATION AND SUPPORT OF A MAJOR FINITE ELEMENT SYSTEM	73-0205
DOCUMENTATION	ELAS: A GENERAL-PURPOSE COMPUTER PROGRAM FOR THE EQUILIBRIUM PROBLEMS OF LINEAR STRUCTURES. VOLUME 2	69-0462
DOER	VARIATIONAL PRINCIPLES OF ELASTOSTATICS WITH RELAXED CONTINUITY REQUIREMENTS AND THEIR APPLICABILITY	74-0758
DOKLADNOSCI	OCENA DOKLADNOSCI ROZWIAZAN PROBLEMOW BRZEGOWYCH SPREZYSTYCH CIAL DYSKERTYZOWANYCH. (ASSESSING THE A	74-0921
DOMAIN	APPLICATION OF FINITE-ELEMENT METHOD TO APPROXIMATION OF OPTIMAL DOMAIN PROBLEM - SOLUTION METHODS F	75-0349
DOMAIN	ERROR ESTIMATES FOR FINITE ELEMENT METHOD FOR HEAT TRANSFER PROBLEMS IN THE SPACE-TIME DOMAIN	75-0104
DOMAIN	FINITE ELEMENT METHOD IN NONLINEAR DOMAIN. (METHODE DES ELEMENTS FINIS DANS LE DOMAINE NON LINEAIRE)	71-0288
DOMAIN	PERIODIC HEAT CONDUCTION IN A TWO PHASE, TWO DIMENSIONAL SOLID DOMAIN	74-1164
DOMAINE	APPLICATION DE LA T METHODE DES ELEMENTS FINIS A LA RESOLUTION D'UN PROBLEME DE DOMAINE OPTIMAL	74-0412
DOMAINE	FINITE ELEMENT METHOD IN NONLINEAR DOMAIN. (METHODE DES ELEMENTS FINIS DANS LE DOMAINE NON LINEAIRE)	71-0288
DOMAINES	ANALYSE DES VOILES MINCES DANS LES DOMAINES LINEAIRES ET GEOMETRIQUEMENT NONLINEAIRES PAR LA METHODE	71-0122
DOMAINS	APPROXIMATION AND BOUNDS FOR EIGENVALUES OF VARIOUS DOMAINS USING THE INTERMEDIATE PROBLEM TECHNIQUE	72-0711
DOMAINS	AUTOMATIC MESH GENERATION IN TWO AND THREE DIMENSIONAL INTERCONNECTED DOMAINS	70-0251
DOMAINS	AUTOMATIC TRIANGULATION OF ARBITRARY PLANAR DOMAINS FOR THE FINITE ELEMENT METHOD	74-0480
DOMAINS	EXTRAPOLATED FINITE ELEMENT PROCEDURES OVER CARTESIAN PRODUCT DOMAINS	75-0044
DOMAINS	FINITE ELEMENT METHOD FOR DOMAINS WITH CORNERS	70-0244
DOMAINS	FINITE ELEMENT METHOD FOR INFINITE DOMAINS	72-0288
DOMAINS	FINITE ELEMENT METHOD FOR NON SMOOTH DOMAINS AND COEFFICIENTS	72-0504
DOMAINS	FINITE ELEMENT SCHEME FOR DOMAINS WITH CORNERS	72-0261
DOMAINS	SARD KERNEL THEOREMS ON TRIANGULAR AND RECTANGULAR DOMAINS WITH EXTENSIONS AND APPLICATIONS TO FINIT	72-0361
DOMAINS	SOLUTION OF MIXED BOUNDARY VALUE PROBLEMS IN FINITE DOMAINS BY FOURIERS METHOD	72-0790
DOMAINS	THE FINITE ELEMENT METHOD IN DOMAINS WITH CURVED BOUNDARIES	73-0080
DOMAINS	TRANSFINITE ELEMENT METHODS: BLENDING FUNCTION INTERPOLATION OVER ARBITRARY CURVED ELEMENT DOMAINS	73-1006
DOME	FLAD PROGRAM MANUAL, A COMPUTER PROGRAM FOR THE ELASTIC ANALYSIS OF DOME STRUCTURES AND AXISYMMETRIC	69-0064
DOME	EPAD, A COMPUTER PROGRAM FOR THE ELASTIC-PLASTIC ANALYSIS OF DOME STRUCTURES AND ARBITRARY SOLIDS OF	70-0036
DOME	THEORETICAL ANALYSIS OF THICK DOME STRUCTURES AND ARBITRARY SOLIDS OF REVOLUTION SUBJECTED TO UNSYMM	70-0037
DOMES	FINITE ELEMENT ANALYSES OF SALT DOMES WITH STORED HOT WASTES	74-1148
DOMES	INEXTENSIONAL VIBRATIONS OF PARABOLOIDAL SHELLS OF REVOLUTION. DOMES AND SHELLS	71-0427
DOMES	INFLATION FORMING OF STEEL FIBER-REINFORCED CONCRETE DOMES	74-0705
DOMINANT	DIAGONALLY DOMINANT MATRICES FOR THE FINITE ELEMENT METHOD IN HYDROLOGY	73-0410
DOOR	FINAL REPORT ON NASTRAN/GRAPHICS, 1971 FORD DOOR PROJECT	73-0595
DOOR	FINITE ELEMENT STRUCTURAL ANALYSIS AS APPLIED TO AN AUTOMOTIVE DOOR STRUCTURE	74-0310
DOSIEK	VELKE PRIEHYBY DOSIEK METODOU KONECNYCH PRVKOV (LARGE DEFLECTIONS OF PLATES ANALYZED BY FINITE ELEME	73-0373
DOSTIGNUCA	DOSTIGNUCA I TENDENCIJE U ISPITIVANJIMA ALATNIH MASINA (ACHIEVEMENTS AND TENDENCIES IN MACHINE TOOLS	73-0426
DOUBLE	ANALYSIS OF DOUBLE CORRUGATED SHELL STRUCTURES BY THE FINITE ELEMENT METHOD	74-1116
DOUBLE	APPLICATION OF THE DOUBLE LIGAMENT TENSILE TEST IN DEFORMATION PROCESSING	75-0827

193

DOUBLE	CLOSURE ANALYSIS USING SELF ENERGIZING, DOUBLE CONICAL GASKETS	74-0675
DOUBLE	DOUBLE CURVATURE SHELL FINITE-ELEMENT AND ITS USE IN THE DYNAMIC ANALYSIS OF COOLING TOWERS AND OTHE	70-0312
DOUBLE	DOUBLE STRESS-WAVE DISCONTINUITIES IN FINITE, SHEAR-CORRECTED BEAMS AND PLATES	68-0210
DOUBLE	FINITE ELEMENT CALCULATIONS OF THE COMPLIANCE OF A TAPERED DOUBLE CANTILEVER BEAM SPECIMEN FOR DIFFE	72-0665
DOUBLE	THE SEISMIC DESIGN STUDY OF A DOUBLE CURVATURE ARCH DAM	69-0053
DOUBLE-CELL	DIAPHRAGMS IN SINGLE- AND DOUBLE-CELL BOX GIRDER BRIDGES WITH VARYING ANGLE OF SKEW	72-0130
DOUBLE-LAYER	DISCRETE-ELEMENT MODELLING AND SENSITIVITY STUDIES OF STATICALLY LOADED SINGLE-LAYER AND BOUNDED DOU	70-0129
DOUBLE-LAYER	DISCRETE-ELEMENT STATIC ANALYSIS OF BONDED, DOUBLE-LAYER, BRANCHED, THIN SHELLS OF REVOLUTION, PART	68-0102
DOUBLE-LAYER	DISCRETE-ELEMENT STATIC ANALYSIS OF BONDED, DOUBLE-LAYER, BRANCHED, THIN SHELLS OF REVOLUTION, PART	70-0081
DOUBLET	USE OF FINITE ELEMENTS IN THE ANALYSIS OF A HYDROTHERMAL DOUBLET	74-0049
DOUBLY	A DOUBLY CURVED TRIANGULAR SHELL ELEMENT	68-0149
DOUBLY	A FINITE ELEMENT ANALYSIS OF A DOUBLY CURVED GOTHIC VAULT	71-0827
DOUBLY	A FINITE ELEMENT FORMULATION FOR STABILITY ANALYSIS OF DOUBLY CURVED THIN SHELL STRUCTURES	69-0513
DOUBLY	ANALYSIS OF THIN SHELLS WITH A DOUBLY CURVED ARBITRARY QUADRILATERAL FINITE ELEMENT.	72-0443
DOUBLY	APPLICATION, TESTING AND EVALUATION OF DOUBLY CURVED SHELL ELEMENTS FOR DYNAMIC ANALYSIS, PART II. F	69-0117
DOUBLY	APPLICATION, TESTING, AND EVALUATION OF DOUBLY CURVED SHELL ELEMENTS FOR DYNAMIC ANALYSIS, PART 1. T	69-0246
DOUBLY	APPLICATION, TESTING, AND EVALUATION OF DOUBLY CURVED SHELL ELEMENTS FOR DYNAMIC ANALYSIS, PART III.	69-0256
DOUBLY	DOUBLY CURVED TRIANGULAR FINITE ELEMENTS FOR SHELLS OF ARBITRARY SHAPE	73-0881
DOUBLY	FREE VIBRATION OF DOUBLY CURVED RECTANGULAR PLATES INCLUDING THE PRESENCE OF MEMBRANE STRESSES WITH	69-0468
DOUBLY	MODELING EARDRUM AS A DOUBLY CURVED SHELL USING FINITE-ELEMENT METHOD	75-0047
DOUBLY	STUDIES IN DOUBLY CURVED SHELL ELEMENTS	74-0133
DOUBLY-CURVED	DYNAMIC ANALYSIS OF SHELLS USING DOUBLY-CURVED FINITE ELEMENTS	68-0082
DOUBLY-CURVED	DYNAMIC ANALYSIS OF SHALLOW SHELLS WITH A DOUBLY-CURVED TRIANGULAR FINITE ELEMENT	71-0353
DOUBLY-CURVED	ELASTIC INSTABILITY PREDICTIONS FOR DOUBLY-CURVED SHELLS	68-0075
DOUBLY-CURVED	STATIC ANALYSIS OF SHELLS OF REVOLUTION USING DOUBLY-CURVED QUADRILATERAL ELEMENTS DERIVED FROM ALTE	69-0368
DOZVUKOVOM	RASCHIOT USTOICHIVOSTI IZGIBNO-KRUTILNYKH KOLEBANIJ KRULA V DOZVUKOVOM POTOKE METODOM KONECHNYKH ELE	72-0903
DRAG	WHIRLING CABLE SUBJECTED TO VISCOUS DRAG	74-0457
DRAIN	NOUVELLE CONCEPTION DES DRAINS DANS LES BARRAGES EN TERRE HOMOGENE DE PETITE ET MOYENNE DIMENSIONS (	73-0404
DRAINAGE	FINITE ELEMENT ANALYSIS OF A NONSTEADY GROUNDWATER SEEPAGE PROBLEM DAM, DRAINAGE	74-1375
DRAINAGE	GRAVITY FLOW TO EXCAVATIONS AND DRAINAGE TRENCHES IN LAYERED AQUIFERS	71-0014
DRAINED	CALCULATION OF STEADY-STATE WATER-TABLE HEIGHTS IN DRAINED SOILS BY MEANS OF FINITE-ELEMENT METHOD	75-0457
DRAINED	THE USE OF THE FINITE ELEMENT METHOD IN OBTAINING SOIL-WATER PRESSURE AND SOIL-WATER CONTENT DISTRIB	75-0112
DRAINS	NOUVELLE CONCEPTION DES DRAINS DANS LES BARRAGES EN TERRE HOMOGENE DE PETITE ET MOYENNE DIMENSIONS (	73-0404
DRAINS	SETTLEMENTS AND STRENGTHENING OF SOFT CLAY ACCELERATED BY SAND DRAINS	73-0438
DRAWDOWN	A FINITE ELEMENT ANALYSIS OF DISSOLVED OXYGEN DRAWDOWN AND SULFATE PRODUCTION IN STRIP MINE SPOIL DA	75-0795
DRAWDOWN	SEEPAGE ANALYSIS OF EARTH BANKS UNDER DRAWDOWN	72-0014
DRAWING	A FINITE ELEMENT ANALYSIS OF THE PLANE STRAIN SHEET DRAWING PROBLEM	71-0812
DRAWING	FINITE ELEMENT ANALYSIS OF THE PLANE STRAIN DRAWING PROBLEM	74-1154
DRAWING	PRELIMINARY INVESTIGATION OF THE DIE-LESS DRAWING OF TITANIUM AND SOME STEELS	75-0770
DRAWING	SLAB BRIDGE DESIGN AND DRAWING, AN AUTOMATED PROCESS	70-0347
DREIACHSIGER	MESSVERFAHREN RECHENVERFAHREN ZUR ERMITTLUNG DREIACHSIGER EIGENSPANNUNGSZUSTAENDE	73-0507
DREIDIMENSIONAL	BERECHNUNG DES REIBUNGSLOSEN STROEMUNGSFELDES VON DREIDIMENSIONALEN AUFTRIEBSBEHAFTETENTRAGFLUEGELN	73-0533
DREIECKFOERMIGE	BERECHNUNG VON PLATTEN NACH DEM RITZ' SCHEN VERFAHREN MIT HILFE DREIECKFOERMIGER MASCHENNETZE	66-0094
DREIECKIGEM	DESH, EIN EINFACHES FLACHES SCHALENELEMENT UEBER DREIECKIGEM GRUNDRISS. (DESH, A SIMPLE TRIANGULAR S	75-0304
DREIECKSFORMIGE	BEITRAG ZUR SIMULATION NICHTSTATIONAERER HORIZONTAL-EVENER GRUNDWASSERSTROMUNGSFELDER AUF DER GRUNDLA	74-0088
DRIFT	STRUCTURAL DESIGN DATA FOR CONCRETE DRIFT LININGS IN BLOCK CAVING	73-0773
DRIVEN	A NUMERICAL INVESTIGATION OF STABLY STRATIFIED, WIND DRIVEN CAVITY FLOW BY THE FINITE ELEMENT METHOD	74-1372
DRIVEN	FINITE ELEMENT ANALYSIS OF DRIVEN PILES	74-1105
DRIVEN	FINITE ELEMENT SOLUTION OF THE WIND DRIVEN CURRENTS AND ITS MASS TRANSPORT IN LAKES	74-1368
DRIVEN	SETTLEMENT ANALYSIS OF DRIVEN PILES	75-0319
DRIVEN	WIND DRIVEN LAKE CIRCULATION BY THE FINITE ELEMENT METHOD	70-0041
DRIVER	FINITE ELEMENT STUDIES OF PILE DRIVER TUNNELS INCLUDING CONSIDERATIONS OF SUPPORT REQUIREMENTS	72-0679
DRIVES	PRESSURE DISTRIBUTION AND BELT DEFORMATION IN V-BELT DRIVES	74-1028
DROITS	ESTIMATIONS D'ERREUR POUR DES ELEMENTS FINIS DROITS PRESQUE DEGENERES	74-1247
DROPLETS	CONDUCTION THROUGH DROPLETS DURING DROPWISE CONDENSATION	72-0019
DROPWISE	CONDUCTION THROUGH DROPLETS DURING DROPWISE CONDENSATION	72-0019
DRUCKBEHAELTERS	ENTWURF UND SPANNUNGSBERECHNUNG EINES SPANNBETON DRUCKBEHAELTERS AUS FERTIGTEILEN. (DESIGN AND STRES	73-0856
DRUMS	AN AXISYMMETRIC FINITE ELEMENT ANALYSIS OF THE MECHANICAL AND THERMAL STRESSES IN BRAKE DRUMS	74-0311
DUAL	A COMPUTER PROGRAMME FOR ELASTIC PLASTIC CALCULATIONS WITH DUAL PERTURBATION EXPANSION AND VARIATION	75-0882
DUAL	ANALYSIS OF PLATE STRUCTURES BY A DUAL FINITE ELEMENT METHOD	69-0482
DUAL	APPLICATION OF DUAL MINIMUM THEOREMS TO THE FINITE ELEMENT SOLUTION OF POTENTIAL PROBLEMS WITH SPECI	72-0036
DUAL	APPLICATION OF THE DUAL ANALYSIS PRINCIPLE	70-0306
DUAL	DUAL ANALYSIS BY FINITE ELEMENTS: LINEAR AND NONLINEAR APPLICATIONS	72-0476
DUAL	DUAL ANALYSIS FOR HEAT CONDUCTION PROBLEMS BY FINITE ELEMENTS	72-0401
DUAL	DUAL ANALYSIS OF COLUMN SUPPORTED SLABS	74-0399
DUAL	DUAL FINITE ELEMENT METHODS FOR NEUTRON DIFFUSION	71-0489
DUAL	DUAL FORMULATION FOR POTENTIAL ENERGY AND COMPLEMENTARY ENERGY APPLICATION TO THE FINITE ELEMENT MET	72-0062
DUAL	DUAL ITERATIVE TECHNIQUES FOR SOLVING A FINITE ELEMENT APPROXIMATION OF THE BIHARMONIC EQUATION	75-0480
DUAL	DUALITY IN STRUCTURAL ANALYSIS BY FINITE ELEMENTS, STATIC GEOMETRIC ANALOGIES, THE DUAL PRINCIPLES O	71-0583
DUAL	THE DUAL PRINCIPLE OF ELASTODYNAMICS FINITE ELEMENT APPLICATIONS	71-0501
DUAL	TWO DUAL FINITE ELEMENT METHODS DERIVED FROM EXTENDED VARIATIONAL PRINCIPLE	73-0109
DUAL	UPPER AND LOWER BOUNDS TO STRUCTURAL DEFORMATIONS BY DUAL ANALYSIS IN FINITE ELEMENTS	67-0074
DUALE	ANALYSE DYNAMIQUE DUALE DES STRUCTURES PAR LA METHODE DES ELEMENTS FINIS	73-0623
DUALES	FORMULATIONS DUALES DES ENERGIES POTENTIELLES ET COMPLEMENTAIRES. APPLICATION A LA METHODE DES ELEME	71-0752
DUALISTIC	VARIATIONAL PRINCIPLES AND DUALISTIC SCHEME FOR INTERSECTION PROBLEMS IN ELASTICITY	69-0305
DUALITY	APPROXIMATION BY FIRST ORDER FINITE ELEMENTS AND SOLUTION BY PENALTY AND DUALITY OF NONLINEAR DIRICH	74-0245
DUALITY	DUALITY BETWEEN DISPLACEMENT AND EQUILIBRIUM METHOD WITH A VIEW TO OBTAINING UPPER AND LOWER BOUNDS	62-0013
DUALITY	DUALITY IN FINITE ELEMENT METHODS	67-0143
DUALITY	DUALITY IN FINITE ELEMENTS METHODS	68-0065
DUALITY	DUALITY IN STRUCTURAL ANALYSIS BY FINITE ELEMENTS, STATIC GEOMETRIC ANALOGIES, THE DUAL PRINCIPLES O	71-0583
DUALITY	DUALITY IN THE ANALYSIS OF SHELLS BY THE FINITE ELEMENT METHOD	71-0124
DUALITY	ELASTIC EQUILIBRIUM MODELS, STRESS FUNCTIONS, AND DUALITY NON LINEARITIES AND STABILITY, FIVE LECTUR	70-0631
DUCT	UNSTEADY MHD DUCT FLOW BY THE FINITE ELEMENT METHOD	73-0263
DUCTED	LOADS AND RESPONSE OF LARGE DUCTED PROPELLER SYSTEMS	73-0810
DUCTILE	DEFORMATION MODE OF VOID-GROWTH AND COALESCENCE IN THE PROCESS OF DUCTILE FRACTURE	75-0683
DUCTILE	SIMULATION ON THE INITIATION AND GROWTH OF DUCTILE FRACTURE VOIDS	75-0836
DUCTS	CREEPING FLOW IN RECTANGULAR DUCTS BY FINITE ELEMENT METHOD	75-0569
DUCTS	DYNAMIC INSTABILITY OF DUCTS CONVEYING FLUID	75-0691
DUCTS	FINITE ELEMENT ANALYSIS OF LAMINAR CONVECTIVE HEAT TRANSFER IN VERTICAL DUCTS WITH ARBITRARY CROSS-S	75-0555
DUCTS	THE FINITE ELEMENT SOLUTION OF FLOW AND HEAT TRANSFER IN ELLIPTICAL DUCTS	74-0039
DUCTS	THE PREDICTION OF FULLY DEVELOPED TURBULENT FLOW IN DUCTS BY THE FINITE ELEMENT METHOD	75-0117
DUCTS	TRANSIENT LAMINAR FLOW IN DUCTS OF ARBITRARY CROSS-SECTION BY FINITE ELEMENT METHODS	72-0013
DUCTS	TWO AND THREE DIMENSIONAL ANALYSIS OF SUBSONIC COMPRESSIBLE POTENTIAL FLOWS IN DUCTS AND AROUND LIFT	74-0026
DUGDALE	DUGDALE MODEL SOLUTIONS FOR A SINGLE EDGE CRACKED PLATE	74-0221
DUPLICATION	DUPLICATION OF DILATANT BEHAVIOR IN THE ANALYSIS OF JOINTED ROCKS	72-0757
DUPLICATOR	CONCEPT FEASIBILITY COMBAT TRACKED VEHICLE SIGNATURE DUPLICATOR	64-0028
DURBAN	BENTONITE PILES IN DURBAN	75-0797
DURCH	SPANNUNGEN IN MEMBRANWAENDEN DURCH WAERMEBEANSPRUCHUNG UND MECHANISCHE BELASTUNG. (STRESSES IN MEMBR	75-0066
DURCH	ZUR LOESUNG DES KONTAKTPROBLEMS ELASTISCHER KOERPER MIT AUSGEDEHNTER KONTAKTFLAECHE DURCH QUADRATISC	74-1135
DURCHSTROIMUNG	WECHSELWIRKUNG ZWISCHEN DEFORMATION UND DURCHSTROIMUNG IN KLUEFTIGEN, ANISOTROPEN GEBIRGE. (INTERACT	72-0612
DUTCH	PROPOSALS FOR THE APPLICATION OF THE FINITE ELEMENT TECHNIQUE TO SOME AIRCRAFT AERODYNAMICAL PROBLEM	72-0628
DUTCH	THE PREDICTION OF REGULAR WAVE FORCES ON FLOATING BODIES WITH THE FINITE ELEMENT METHOD (IN DUTCH)	71-0743

194

DUTY	DESIGN ASPECTS OF HIGH DUTY PRESSURE COMPONENTS	74-0199
DUTY	EFFECTS OF SHOE FORCE GEOMETRY ON HEAVY DUTY INTERNAL SHOE BRAKE PERFORMANCE	68-0219
DVIGATELEI	METODY RASCHETNOGO OPREDELENIYA TEMPERATURNYKH NAPRYAZHENII V KRYSHKAKH TSILINDROV DVIGATELEI VNUTRE	74-0179
DVIGATELEI	OPREDELENIE STATSIONARNYKH TEMPERATURNYKH POLEI V DETALYAKH DVIGATELEI VNUTRENNEGO SGORANIYA METODOM	73-0454
DVIGATELEI	RASCHET POLEI DEFORMATSII I NAPRYAZHENII V DETALYAKH DVIGATELEI VNUTRENNEGO SGORANIYA (CALCULATION O	73-0465
DVUMERNYKH	REALIZATSIYA METODA KONECHNYKH ELEMENTOV NA EVM DLYA RASCHETA DVUMERNYKH ELEKTRICHESKIKH I MAGNITNYK	74-0251
DWORSHAK	CONSTRUCTION STRESSES IN DWORSHAK DAM	65-0029
DYNAMIC	A COMPUTER PROGRAM FOR THE STATIC AND DYNAMIC FINITE ELEMENT ANALYSIS OF AXISYMMETRIC THIN SHELLS	66-0017
DYNAMIC	A COMPUTER PROGRAM TO DISPLAY THE DYNAMIC STRESS ANALYSIS OF STRUCTURES	70-0077
DYNAMIC	A DERIVATION PROCEDURE FOR THE DYNAMIC FLEXIBILITY MATRIX OF A TRIANGULAR BENDING ELEMENT	70-0103
DYNAMIC	A DYNAMIC ELASTOPLASTIC ANALYSIS OF THIN SHELLS OF REVOLUTION UNDER ASYMMETRIC MECHANICAL AND THERMA	72-0068
DYNAMIC	A DYNAMIC FINITE ELEMENT MODEL OF THE HUMAN BRAIN	74-1076
DYNAMIC	A DYNAMIC PHOTOELASTIC ANALYSIS OF DYNAMIC TEAR TEST SPECIMEN	74-0736
DYNAMIC	A DYNAMIC PHOTOELASTIC ANALYSIS OF DYNAMIC TEAR TEST SPECIMEN	74-0736
DYNAMIC	A FINITE ELEMENT ANALYSIS FOR DYNAMIC BEHAVIOR OF STIFFENED PLATES	69-0392
DYNAMIC	A FINITE ELEMENT APPROACH TO THE DYNAMIC ANALYSIS OF CONTINUOUS HIGHWAY BRIDGES	72-0591
DYNAMIC	A FINITE ELEMENT APPROACH TO THE DETERMINATION OF THE DYNAMIC RESPONSE OF BEAMS SUPPORTING MOVING MA	74-1114
DYNAMIC	A FINITE ELEMENT PROGRAM FOR THE DYNAMIC LARGE DEFLECTION, ELASTIC-PLASTIC ANALYSIS OF STIFFENED SHE	72-0908
DYNAMIC	A GENERALIZATION OF FINITE ELEMENT METHOD TO DYNAMIC VISCOELASTIC ANALYSIS	75-0106
DYNAMIC	A MODAL SYNTHESIS TECHNIQUE FOR DETERMINING DYNAMIC PROPERTIES FOR A STRUCTURE FOR MASS AND STIFFNES	74-0319
DYNAMIC	A REFINED FINITE ELEMENT ANALYSIS OF THIN SHELL STRUCTURES INCLUDING DYNAMIC LOADINGS	67-0157
DYNAMIC	A STATIC AND DYNAMIC FINITE ELEMENT SHELL ANALYSIS WITH EXPERIMENTAL VERIFICATION	69-0156
DYNAMIC	A STATIC AND DYNAMIC FINITE ELEMENT SHELL ANALYSIS WITH EXPERIMENTAL VERIFICATION	71-0769
DYNAMIC	A SURVEY OF ANALYTICAL METHODS FOR DYNAMIC SIMULATION OF CAB	73-0936
DYNAMIC	A TWO-DIMENSIONAL AXISYMMETRIC SHELL FLUID MODEL FOR LAUNCH VEHICLE LONGITUDINAL DYNAMIC ANALYSIS	69-0046
DYNAMIC	AIRCRAFT-PAVEMENT INTERACTION STUDIES, PHASE 1: A FINITE ELEMENT MODEL OF A JOINTED CONCRETE PAVEMEN	73-0894
DYNAMIC	AN INVESTIGATION OF THE DYNAMIC RESPONSE CHARACTERISTICS OF THE BON TEMPE DAM	66-0047
DYNAMIC	ANALYSIS OF BUILDINGS ON ELASTIC MEDIUM UNDER DYNAMIC OR SEISMIC LOADS	74-0456
DYNAMIC	ANALYSIS OF DYNAMIC RESPONSE OF SAFEGUARD STRUCTURES	70-0462
DYNAMIC	ANALYSIS OF FOUNDATIONS ON A SEMI-INFINITE ELASTIC MASS UNDER DYNAMIC LOADS	72-0763
DYNAMIC	ANALYSIS OF NONLINEAR DYNAMIC, COUPLED THERMOVISCOELASTICITY PROBLEMS BY THE FINITE ELEMENT METHOD	71-0527
DYNAMIC	ANALYSIS OF STRUCTURAL VIBRATIONS AND DYNAMIC RESPONSE	71-0059
DYNAMIC	ANALYTICAL AND EXPERIMENTAL INVESTIGATION OF A 1/8 SCALE DYNAMIC MODEL OF THE SHUTTLE ORBITER. VOLUM	74-1186
DYNAMIC	ANWENDUNG DER METHODE FINITER ELEMENTE BEI DER ANALYSE DES DYNAMISCHEN VERHALTENS GEDAEMPFTER WERKZE	75-0580
DYNAMIC	APPLICATION OF FINITE ELEMENT METHOD TO THE DYNAMIC ANALYSIS OF TALL STRUCTURES	71-0111
DYNAMIC	APPLICATION OF FINITE ELEMENT DYNAMIC ANALYSIS TO FRACTURE MECHANICS	75-0861
DYNAMIC	APPLICATION OF THE FINITE ELEMENT METHOD TO PREDICT STATIC AND DYNAMIC RESPONSE OF AN UNSHROUDED CEN	71-0574
DYNAMIC	APPLICATION, TESTING AND EVALUATION OF DOUBLY CURVED SHELL ELEMENTS FOR DYNAMIC ANALYSIS, PART II. F	69-0117
DYNAMIC	APPLICATION, TESTING, AND EVALUATION OF DOUBLY CURVED SHELL ELEMENTS FOR DYNAMIC ANALYSIS, PART I. T	69-0246
DYNAMIC	APPLICATION, TESTING, AND EVALUATION OF DOUBLY CURVED SHELL ELEMENTS FOR DYNAMIC ANALYSIS, PART III.	69-0256
DYNAMIC	ASSESSMENT OF COMMERCIALLY AVAILABLE FINITE ELEMENT CODES FOR THE DYNAMIC ANALYSIS OF STRUCTURES	73-0954
DYNAMIC	AUTOMATED DESIGN OPTIMIZATION OF SUPERSONIC AIRPLANE WING STRUCTURES UNDER DYNAMIC CONSTRAINTS	72-0178
DYNAMIC	COMBINING EXPERIMENTAL AND ANALYTICAL TECHNIQUES FOR DYNAMIC SYSTEM ANALYSIS	73-0321
DYNAMIC	COMPARISON OF CALCULATION BY THE FINITE ELEMENT METHOD WITH EXPERIMENT AND AN EXAMPLE OF A DYNAMIC S	72-0597
DYNAMIC	COMPARISON OF STRUCTURAL LOADS: STATIC VERSUS DYNAMIC	70-0464
DYNAMIC	COMPARISON OF THE RESULTS OF THE DYNAMIC RESPONSE OF CYLINDRICAL SHELLS BY CHARACTERISTIC AND BY FIN	74-0781
DYNAMIC	COMPUTATIONAL METHOD FOR DYNAMIC ANALYSIS OF STRUCTURES WITH LARGE SIZE STIFFNESS AND MASS MATRICES.	73-0830
DYNAMIC	COMPUTER PROGRAM FOR STATIC AND DYNAMIC ANALYSIS OF LINEAR STRUCTURAL SYSTEMS	72-0683
DYNAMIC	COMPUTER SIMULATION OF DYNAMIC STRESS, DEFORMATION, AND FRACTURE OF GEAR TEETH	72-0114
DYNAMIC	CORRELATION OF FINITE ELEMENT STRUCTURAL DYNAMIC ANALYSIS WITH MEASURED FREE VIBRATION CHARACTERISTI	74-0486
DYNAMIC	DESIGN OF OPTIMUM STRUCTURES FOR DYNAMIC LOADS	71-0255
DYNAMIC	DEVELOPMENT OF TECHNOLOGY FOR MODELING OF A 1/8-SCALE DYNAMIC MODEL OF THE SHUTTLE SOLID ROCKET BOOS	74-0703
DYNAMIC	DOUBLE CURVATURE SHELL FINITE-ELEMENT AND ITS USE IN THE DYNAMIC ANALYSIS OF COOLING TOWERS AND OTHE	70-0312
DYNAMIC	DYNAMIC ANALYSIS	75-0171
DYNAMIC	DYNAMIC ANALYSIS OF STRUCTURAL SYSTEMS USING COMPONENT MODES	64-0024
DYNAMIC	DYNAMIC ANALYSIS OF SOIL - STRUCTURE SYSTEM USING FINITE ELEMENT METHOD	67-0001
DYNAMIC	DYNAMIC ANALYSIS OF SHELLS USING DOUBLY-CURVED FINITE ELEMENTS	68-0082
DYNAMIC	DYNAMIC ANALYSIS OF A COMPLEX SHELL STRUCTURE UNDER IMPULSIVE ASYMMETRIC LOADING	68-0226
DYNAMIC	DYNAMIC ANALYSIS OF SHELLS USING CURVED FINITE ELEMENTS	69-0296
DYNAMIC	DYNAMIC ANALYSIS OF AN AUTOMOBILE CHASSIS FRAME	70-0135
DYNAMIC	DYNAMIC ANALYSIS OF AUTOMOTIVE STRUCTURAL SYSTEMS	70-0340
DYNAMIC	DYNAMIC ANALYSIS OF CIVIL ENGINEERING STRUCTURES	71-0264
DYNAMIC	DYNAMIC ANALYSIS OF SHALLOW SHELLS WITH A DOUBLY-CURVED TRIANGULAR FINITE ELEMENT	71-0353
DYNAMIC	DYNAMIC ANALYSIS OF MECHANICAL STRUCTURES	71-0381
DYNAMIC	DYNAMIC ANALYSIS OF MACHINE TOOL STRUCTURES BY THE FINITE ELEMENT METHOD	71-0416
DYNAMIC	DYNAMIC ANALYSIS OF STIFFENED PANEL STRUCTURES	71-0522
DYNAMIC	DYNAMIC ANALYSIS OF STRUCTURES USING THE RANK FORCE METHOD	71-0553
DYNAMIC	DYNAMIC ANALYSIS OF A FLEXIBLE VEHICLE MOVING ALONG A FLEXIBLE SUPPORT	71-0597
DYNAMIC	DYNAMIC ANALYSIS OF A CAR CHASSIS FRAME USING FINITE ELEMENT METHOD	72-0265
DYNAMIC	DYNAMIC ANALYSIS OF STRUCTURES	72-0298
DYNAMIC	DYNAMIC ANALYSIS OF FRAME STRUCTURES BY THE FORCE METHOD	72-0313
DYNAMIC	DYNAMIC ANALYSIS OF AXISYMMETRIC SHELLS UNDER ARBITRARY TRANSIENT PRESSURES.	72-0467
DYNAMIC	DYNAMIC ANALYSIS OF SANDWICH BEAMS	72-0586
DYNAMIC	DYNAMIC ANALYSIS OF BEAM-SLAB HIGHWAY BRIDGES	73-0199
DYNAMIC	DYNAMIC ANALYSIS OF VISCOELASTOPLASTIC ANISOTROPIC SHELLS	73-0253
DYNAMIC	DYNAMIC ANALYSIS OF THE GROUND USING THE THREE DIMENSIONAL FINITE ELEMENT	73-0293
DYNAMIC	DYNAMIC ANALYSIS OF FRAME AND PLATE STRUCTURES WITH GEOMETRIC AND MATERIAL NONLINEARITIES	73-0649
DYNAMIC	DYNAMIC ANALYSIS OF HYDROELASTIC SYSTEMS USING THE FINITE-ELEMENT METHOD	73-1137
DYNAMIC	DYNAMIC ANALYSIS OF COLUMN-SUPPORTED HYPERBOLOIDAL SHELLS	74-0188
DYNAMIC	DYNAMIC ANALYSIS OF THE DEFORMATION OF STRUCTURES AND OF METAL FORMING	74-0823
DYNAMIC	DYNAMIC ANALYSIS OF A SYSTEM OF HINGE-CONNECTED RIGID BODIES WITH NONRIGID APPENDAGES	74-0852
DYNAMIC	DYNAMIC ANALYSIS OF STRUCTURAL DEFORMATION AND METAL FORMING	75-0058
DYNAMIC	DYNAMIC ANALYSIS OF CIRCULAR AND SECTOR THICK, LAYERED PLATES	75-0198
DYNAMIC	DYNAMIC ANALYSIS OF THE SLIDE IN THE LOWER SAN FERNANDO DAM DURING THE EARTHQUAKE OF FEBRUARY 9, 197	75-0277
DYNAMIC	DYNAMIC ANALYSIS OF CURVED BOX-GIRDER BRIDGES	75-0280
DYNAMIC	DYNAMIC ANALYSIS OF FOOTINGS ON LAYERED MEDIA	75-0586
DYNAMIC	DYNAMIC AND STOCHASTIC ASPECTS IN THE STRUCTURAL ANALYSIS OF VLCCS AND ULCCS	74-0298
DYNAMIC	DYNAMIC BEHAVIOR OF LIQUIDS IN ELASTIC TANKS	71-0141
DYNAMIC	DYNAMIC BEHAVIOR OF ECCENTRICALLY STIFFENED PLATES	71-0796
DYNAMIC	DYNAMIC BEHAVIOR OF STEEL FOUNDATIONS FOR TURBO-ALTERNATORS	71-0837
DYNAMIC	DYNAMIC BEHAVIOR OF IN-LINE SHEAR WALLS CONNECTED BY FLOOR SLABS	72-0689
DYNAMIC	DYNAMIC BEHAVIOR OF HYDROSTATIC THRUST BEARING	75-0189
DYNAMIC	DYNAMIC BEHAVIOUR OF SHELL ROOFS	69-0228
DYNAMIC	DYNAMIC BEHAVIOUR OF ECCENTRICALLY STIFFENED PLATES	71-0605
DYNAMIC	DYNAMIC COUPLING IN AN UNSYMMETRICALLY TAPERED BEAM	74-1301
DYNAMIC	DYNAMIC ELASTIC-PLASTIC ANALYSIS OF AXISYMMETRIC STRUCTURES BY FINITE ELEMENT METHOD COMPUTER CODE	73-0844
DYNAMIC	DYNAMIC ELASTOPLASTIC RESPONSE OF GEOMETRICALLY NONLINEAR ARBITRARY SHELLS OF REVOLUTION UNDER IMPUL	70-0098
DYNAMIC	DYNAMIC FINITE ELEMENT ANALYSIS OF ARBITRARY THIN SHELLS	71-0061
DYNAMIC	DYNAMIC FINITE ELEMENT ANALYSIS OF AXI-SYMMETRIC STRESS WAVE PROPAGATION IN SOIL-FILLED BIN	72-0622
DYNAMIC	DYNAMIC FINITE ELEMENT METHOD CALCULATION OF ELASTIC WAVE PROPAGATION IN TWO DIMENSIONAL STRUCTURES	74-0166
DYNAMIC	DYNAMIC FINITE ELEMENT ANALYSIS WITH APPLICATIONS TO SEISMOLOGICAL PROBLEMS	75-0785
DYNAMIC	DYNAMIC INSTABILITY OF DUCTS CONVEYING FLUID	75-0691

195

DYNAMIC	DYNAMIC LOADING OF A FLUID-FILLED SPHERICAL SHELL.	72-0444
DYNAMIC	DYNAMIC LOADS ON GROUTED EMPLACEMENT SYSTEMS	70-0406
DYNAMIC	DYNAMIC MATRIX ANALYSIS OF FRAMED STRUCTURES	62-0012
DYNAMIC	DYNAMIC MODELING OF HALF-PLANE BY FINITE ELEMENTS	74-0953
DYNAMIC	DYNAMIC PLASTIC LARGE DEFLECTIONS OF CLAMPED RECTANGULAR PLATE	74-0392
DYNAMIC	DYNAMIC PROGRAMMING APPROACH TO THE FORMULATION AND SOLUTION OF FINITE ELEMENT EQUATIONS	75-0059
DYNAMIC	DYNAMIC RESPONSE ANALYSIS OF GEOMETRICALLY NON-LINEAR STRUCTURES SUBJECTED TO HIGH IMPACT	72-0193
DYNAMIC	DYNAMIC RESPONSE ANALYSIS OF GEOMETRICALLY NON-LINEAR STRUCTURES SUBJECTED TO HIGH IMPACT	72-0555
DYNAMIC	DYNAMIC RESPONSE ANALYSIS OF THE IRREGULAR GROUND BY MEANS OF THE FINITE ELEMENT METHOD	75-0860
DYNAMIC	DYNAMIC RESPONSE BY A STEP BY STEP ANALYSIS	62-0014
DYNAMIC	DYNAMIC RESPONSE BY LARGE STEP INTEGRATION	73-0567
DYNAMIC	DYNAMIC RESPONSE OF EARTH DAMS	67-0024
DYNAMIC	DYNAMIC RESPONSE OF COHESIVE EARTH DAMS TO EARTHQUAKES	67-0131
DYNAMIC	DYNAMIC RESPONSE OF ARBITRARY SHELLS OF REVOLUTION UNDER COMBINED IMPULSIVE NONISOTHERMAL AND IRREGU	70-0365
DYNAMIC	DYNAMIC RESPONSE OF A CONSTRAINED FIBROUS SYSTEM SUBJECTED TO TRANSVERSE IMPACT. II. A MECHANICAL MO	70-0369
DYNAMIC	DYNAMIC RESPONSE OF REINFORCED OPENINGS IN ROCK	71-0311
DYNAMIC	DYNAMIC RESPONSE OF NONLINEAR SPACE FRAMES BY FINITE ELEMENT METHODS	71-0584
DYNAMIC	DYNAMIC RESPONSE OF STRUCTURES SUBJECTED TO TIME-DEPENDENT BOUNDARY CONDITIONS USING THE FINITE ELEM	72-0129
DYNAMIC	DYNAMIC RESPONSE OF MODEL PAVEMENT STRUCTURE	72-0226
DYNAMIC	DYNAMIC RESPONSE OF A CYLINDER BURIED IN AN EARTH BEAM - RESULTS OF A FINITE ELEMENT ANALYSIS	72-0592
DYNAMIC	DYNAMIC RESPONSE OF THE HUMAN ARM AS A COMPOSITE STRUCTURE	72-0611
DYNAMIC	DYNAMIC RESPONSE OF VISCOELASTIC MATERIALS TO SINUSOIDALLY VARYING LOADINGS	73-1072
DYNAMIC	DYNAMIC RESPONSE OF VISCOELASTIC MATERIALS	73-1075
DYNAMIC	DYNAMIC RESPONSE OF FAST-REACTOR CORE SUBASSEMBLIES	74-0540
DYNAMIC	DYNAMIC RESPONSE OF STRUCTURES IN LAYERED SOILS	74-0761
DYNAMIC	DYNAMIC RESPONSE OF SOIL/CONCRETE INTERFACES AT HIGH PRESSURE	74-0836
DYNAMIC	DYNAMIC RESPONSE OF ELASTIC HEAD MODELS	75-0378
DYNAMIC	DYNAMIC RESPONSE ON NONLINEAR MEDIA AT LARGE STRAINS	72-0237
DYNAMIC	DYNAMIC RESPONSES OF CYLINDRICAL SHELLS INCLUDING GEOMETRIC AND MATERIAL NONLINEARITIES	74-0150
DYNAMIC	DYNAMIC SHELL THEORY COUPLING THICKNESS STRESS WAVE EFFECTS WITH GROSS STRUCTURAL RESPONSE	73-0379
DYNAMIC	DYNAMIC SIMULATION OF COMPLEX SYSTEMS UTILIZING EXPERIMENTAL AND ANALYTICAL TECHNIQUES	75-0864
DYNAMIC	DYNAMIC STABILITY ANALYSIS BY THE CONJUGATE GRADIENT METHOD	72-0749
DYNAMIC	DYNAMIC STABILITY OF CYLINDRICAL SHELLS BY FINITE ELEMENTS	68-0292
DYNAMIC	DYNAMIC STABILITY OF CONICAL SHELLS BY FINITE ELEMENTS	70-0467
DYNAMIC	DYNAMIC STABILITY OF PLATES BY FINITE ELEMENTS	71-0573
DYNAMIC	DYNAMIC STABILITY OF A NONLINEAR BEAM SUBJECTED TO BOTH LONGITUDINAL AND TRANSVERSE EXCITATION	73-0553
DYNAMIC	DYNAMIC STABILITY OF TIMOSHENKO BEAMS BY FINITE ELEMENT METHOD	75-0194
DYNAMIC	DYNAMIC STABILITY OF TIMOSHENKO BEAMS BY FINITE ELEMENT METHOD	75-0422
DYNAMIC	DYNAMIC STIFFNESS MATRIX FORMULATION BY MEANS OF HERMITIAN POLYNOMIALS	66-0059
DYNAMIC	DYNAMIC STRAINS IN A FOUR BAR MECHANISM	73-0677
DYNAMIC	DYNAMIC STRESS ANALYSIS OF AXISYMMETRIC STRUCTURES UNDER ARBITRARY LOADING	69-0433
DYNAMIC	DYNAMIC STRESS ANALYSIS OF POROUS ELASTIC SOLIDS SATURATED WITH COMPRESSIBLE FLUIDS	71-0651
DYNAMIC	DYNAMIC STRUCTURAL ANALYSIS WITH THE NASTRAN COMPUTER PROGRAM	70-0213
DYNAMIC	DYNAMIC SUBSTRUCTURES METHOD FOR SHOCK ANALYSIS	68-0285
DYNAMIC	DYNAMIC TESTING AND COMPUTER ANALYSIS OF AUTOMOTIVE FRAMES	72-0089
DYNAMIC	DYNAMIC TORSIONAL RESPONSE OF THE HUMAN LEG RELATIVE TO SKIING INJURIES	73-0471
DYNAMIC	DYNAMIC TRANSIENT NON-LINEAR BEHAVIOUR OF THICK AND THIN PLATES	75-0125
DYNAMIC	DYNAPLAS: A FINITE ELEMENT PROGRAM FOR THE DYNAMIC, LARGE DEFLECTION, ELASTIC-PLASTIC ANALYSIS OF ST	72-0773
DYNAMIC	DYNASOR 11 - A FINITE ELEMENT PROGRAM FOR THE DYNAMIC NONLINEAR ANALYSIS OF SHELLS OF REVOLUTION (US	71-0666
DYNAMIC	DYPLAS, A FINITE ELEMENT DYNAMIC ELASTIC-PLASTIC LARGE DEFORMATION ANALYSIS PROGRAM	74-0306
DYNAMIC	EFFECT OF TIME-DEPENDENT MATERIAL PROPERTIES ON DYNAMIC RESPONSE	74-0900
DYNAMIC	EFFECTS OF SHEAR WALLS ON DYNAMIC RESPONSE OF FRAMES	69-0230
DYNAMIC	EIGENVALUES OF COLLINEAR BEAM STRUCTURES USING FINITE ELEMENT TECHNIQUES AND VARIOUS DYNAMIC REPRESE	67-0124
DYNAMIC	ELASTIC DYNAMIC RESPONSE OF AXISYMMETRIC STRUCTURES	69-0275
DYNAMIC	ELASTIC PLASTIC DYNAMIC ANALYSIS OF AXISYMMETRIC SOLIDS	73-0657
DYNAMIC	EVALUATION OF DISCRETE METHODS FOR THE LINEAR DYNAMIC RESPONSE OF ELASTIC AND VISCOELASTIC SOLIDS	70-0439
DYNAMIC	EVALUATION OF NUMERICAL TIME INTEGRATION METHODS AS APPLIED TO ELASTIC-PLASTIC DYNAMIC PROBLEMS INVO	74-0858
DYNAMIC	FINITE BEAM ELEMENTS FOR DYNAMIC ANALYSIS	70-0372
DYNAMIC	FINITE DYNAMIC MODEL FOR INFINITE MEDIA	69-0171
DYNAMIC	FINITE ELEMENT - ANALOGUE METHOD FOR DETERMINING THE DYNAMIC CHARACTERISTICS OF AN ARCH DAM-RESERVOI	73-0678
DYNAMIC	FINITE ELEMENT ALGORITHM FOR THE DETERMINATION OF DYNAMIC BUCKLING	75-0302
DYNAMIC	FINITE ELEMENT ANALYSIS OF DYNAMIC SHELL BEHAVIOUR	67-0013
DYNAMIC	FINITE ELEMENT ANALYSIS OF DYNAMIC RESPONSE	72-0681
DYNAMIC	FINITE ELEMENT ANALYSIS OF STATIC AND DYNAMIC STRESSES AROUND SUDDENLY PUNCHED HOLES IN PLATES AND S	73-1088
DYNAMIC	FINITE ELEMENT APPENDAGE EQUATIONS FOR HYBRID COORDINATE DYNAMIC ANALYSIS	72-0022
DYNAMIC	FINITE ELEMENT COMPUTER PROGRAM FOR PREDICTING THE NONLINEAR STATIC AND DYNAMIC BEHAVIOR OF VISCOELA	75-0191
DYNAMIC	FINITE ELEMENT DYNAMIC ANALYSIS OF SHALLOW SHELL STRUCTURES	70-0164
DYNAMIC	FINITE ELEMENT DYNAMIC ANALYSIS OF AN AUTOMOTIVE FRAME	73-0256
DYNAMIC	FINITE ELEMENT FORMULATIONS FOR LARGE DEFORMATION DYNAMIC ANALYSIS	75-0564
DYNAMIC	FINITE ELEMENT MODEL FOR DYNAMIC ANALYSIS OF TIMOSHENKO BEAM	75-0275
DYNAMIC	FINITE ELEMENT PERTURBATION ANALYSIS OF NONLINEAR DYNAMIC RESPONSE OF ELASTIC CONTINUUA	74-0452
DYNAMIC	FINITE ELEMENT PROCEDURE FOR THE LARGE DEFORMATION DYNAMIC RESPONSE OF AXISYMMETRIC SOLIDS	74-0527
DYNAMIC	FINITE ELEMENT PROCEDURE FOR THE LARGE DEFORMATION DYNAMIC RESPONSE OF AXISYMMETRIC SOLIDS	74-0967
DYNAMIC	FINITE ELEMENT PROCEDURE FOR LARGE DEFORMATION DYNAMIC RESPONSE OF AXISYMMETRIC SOLIDS	75-0013
DYNAMIC	FINITE ELEMENT SOLUTION OF AXISYMMETRICAL DYNAMIC PROBLEMS OF SHELLS OF REVOLUTION	66-0020
DYNAMIC	FINITE ELEMENT SOLUTION TO DYNAMIC STABILITY OF BARS	68-0035
DYNAMIC	FINITE ELEMENT STRESS FORMULATION FOR DYNAMIC ELASTIC-PLASTIC ANALYSIS	75-0484
DYNAMIC	FINITE ELEMENTS AND DYNAMIC VISCOELASTICITY	71-0479
DYNAMIC	FINITE-ELEMENT FORMULATION OF MIXED AND STRESS VARIATIONAL PRINCIPLES FOR STRUCTURAL DYNAMIC ANALYSI	72-0964
DYNAMIC	FINITE-ELEMENT TECHNIQUES FOR THE DYNAMIC ANALYSIS OF UNSYMMETRICALLY LAMINATED PLATES	73-1131
DYNAMIC	GENERAL PURPOSE COMPUTER PROGRAM FOR THE DYNAMIC SIMULATION OF VEHICLE GUIDEWAY INTERACTIONS	73-0260
DYNAMIC	HIGHER ORDER FINITE ELEMENT METHOD FOR STATIC AND DYNAMIC ANALYSIS OF SHEAR WALLS	71-0368
DYNAMIC	IMPROVED TECHNIQUES FOR DYNAMIC ANALYSIS OF EARTHMOVING EQUIPMENT	74-0294
DYNAMIC	IN-VACUO MODAL DYNAMIC RESPONSE OF THE HUMAN SKULL	73-0460
DYNAMIC	INCREMENTAL STIFFNESS METHOD FOR FINITE ELEMENT ANALYSIS OF THE NONLINEAR DYNAMIC PROBLEM	71-0573
DYNAMIC	INELASTIC RESPONCE OF FRAMES TO DYNAMIC LOADS	71-0611
DYNAMIC	INTEGRATED PROGRAM FOR THE DYNAMIC STRUCTURAL CALCULATION OF MOBIL OFFSHORE UNITS	74-0490
DYNAMIC	LARGE DEFLECTION DYNAMIC ANALYSIS OF THIN SHELLS USING FINITE ELEMENT METHOD	70-0379
DYNAMIC	LARGE DEFLECTION ELASTIC-PLASTIC DYNAMIC RESPONSE OF STIFFENED SHELLS OF REVOLUTION	72-0608
DYNAMIC	LARGE DEFLECTION ELASTIC-PLASTIC DYNAMIC RESPONSE OF STIFFENED SHELLS OF REVOLUTION	74-0496
DYNAMIC	LINEAR DYNAMIC ANALYSES OF LAMINATED PLATES AND SHELLS BY THE HYBRID-STRESS FINITE-ELEMENT METHOD	73-0786
DYNAMIC	LUMPED PARAMETER APPROACH TO AXISYMMETRIC DYNAMIC SOIL DEFORMATIONS	69-0170
DYNAMIC	MATHEMATICAL PREDICTION OF DYNAMIC TIRE BEHAVIOR	73-0579
DYNAMIC	MATRIX DYNAMIC AND INSTABILITY ANALYSIS WITH NON-UNIFORM ELEMENTS	70-0062
DYNAMIC	MULTIDEGREE DYNAMIC ANALYSIS OF TALL BUILDINGS SUBJECTED TO WIND AS A STOCHASTIC PROCESS	74-1196
DYNAMIC	NON-LINEAR DYNAMIC ANALYSIS OF AXISYMMETRIC SHELLS	75-0474
DYNAMIC	NONLINEAR DYNAMIC ANALYSIS OF SHELLS OF REVOLUTION BY MATRIX DISPLACEMENT METHOD	70-0110
DYNAMIC	NONLINEAR DYNAMIC ANALYSIS OF COOLING TOWER.	72-0219
DYNAMIC	NONLINEAR DYNAMIC ANALYSIS OF COMPLEX STRUCTURES	73-0193
DYNAMIC	NONLINEAR DYNAMIC ANALYSIS OF AXISYMMETRIC SOLIDS BY THE FINITE ELEMENT METHOD	73-0942
DYNAMIC	NONLINEAR DYNAMIC ANALYSIS OF FRAMES WITH FILLER PANELS	74-0226
DYNAMIC	NONLINEAR DYNAMIC ANALYSIS OF AXISYMMETRIC SOLIDS BY THE FINITE ELEMENT METHOD	74-0510

DYNAMIC	NONLINEAR DYNAMIC ANALYSIS OF SHELLS OF REVOLUTION WITH ASYMMETRIC PROPERTIES BY THE FINITE ELEMENT	75-0217
DYNAMIC	NONLINEAR DYNAMIC ANALYSIS OF SHELLS OF REVOLUTION WITH ASYMMETRIC PROPERTIES BY THE FINITE ELEMENT	75-0657
DYNAMIC	NONLINEAR DYNAMIC ANALYSIS OF TORSIONAL SHELLS BY COMBINED USE OF FINITE ELEMENT AND MODE SUPERPOSIT	75-0859
DYNAMIC	NONLINEAR FINITE ELEMENT DYNAMIC ANALYSIS OF AXISYMMETRIC SOLIDS	74-0775
DYNAMIC	NONLINEAR FINITE-ELEMENT METHOD FOR PLATES AND ITS APPLICATION TO DYNAMIC RESPONSE OF REACTOR FUEL S	74-0670
DYNAMIC	NONLINEAR STATIC AND DYNAMIC RESPONSE OF UNDERWATER CABLE STRUCTURES USING THE FINITE ELEMENT METHOD	75-0460
DYNAMIC	NONSAP - A GENERAL FINITE ELEMENT PROGRAM FOR NONLINEAR DYNAMIC ANALYSIS OF COMPLEX STRUCTURES	73-0740
DYNAMIC	NONSAP: A STRUCTURAL ANALYSIS PROGRAM FOR STATIC AND DYNAMIC RESPONSE OF NONLINEAR SYSTEMS	74-0848
DYNAMIC	ON THE DYNAMIC BUCKLING OF SHELLS OF REVOLUTION	73-1102
DYNAMIC	ON THE GENERALIZATION OF STRESS FUNCTION PROCEDURE FOR DYNAMIC ANALYSIS OF PLATES	73-0661
DYNAMIC	ON THE NUMERICAL SOLUTION OF A CLASS OF NONLINEAR PROBLEMS IN DYNAMIC COUPLED THERMOELASTICITY	71-0441
DYNAMIC	OPTIMIZATION OF COMPLEX STRUCTURES TO SATISFY STATIC, DYNAMIC AND AEROELASTIC REQUIREMENTS	74-0272
DYNAMIC	PREDICTION OF LONG-TERM STRESS RANGES: USER'S MANUAL - BRIDGE DYNAMIC STRESS ANALYSIS	74-0807
DYNAMIC	RASCHET METODOM KONECHNYKH ELEMENTOV DINAMICHESKIKH KHARAKTEERISTIK ZEMLYANOI PLOTINY SOVMESTNO S OS	73-0437
DYNAMIC	RATE OF CHANGE OF DYNAMIC RESPONSE	72-1004
DYNAMIC	REDUCTION OF THE DEGREES OF FREEDOM IN SOLVING DYNAMIC PROBLEMS BY THE FINITE ELEMENT METHOD	73-0647
DYNAMIC	RELATIONSHIP OF SUBSTRUCTURING TO IMPROVEMENT OF DYNAMIC MODELING	71-0323
DYNAMIC	SAP IV. A STRUCTURAL ANALYSIS PROGRAM FOR STATIC AND DYNAMIC RESPONSE OF LINEAR SYSTEMS	73-0650
DYNAMIC	SLADE D: A COMPUTER PROGRAM FOR THE DYNAMIC ANALYSIS OF THIN SHELLS	73-0964
DYNAMIC	SOLUTION OF FLUID DYNAMIC PROBLEMS BY FINITE ELEMENTS	74-1184
DYNAMIC	SOME CONSIDERATIONS ON THE DYNAMIC STRESS ANALYSIS OF A STRIP BY FINITE ELEMENT METHOD	71-0591
DYNAMIC	SPACE TIME FINITE ELEMENTS IN DYNAMIC PROBLEMS	72-0402
DYNAMIC	STARS - A COMPUTER CODE FOR STRUCTURAL DYNAMIC ANALYSIS OF REACTOR SYSTEM	71-0204
DYNAMIC	STATIC AND DYNAMIC APPLICATIONS OF A HIGH-PRECISION TRIANGULAR PLATE BENDING ELEMENT	69-0086
DYNAMIC	STATIC AND DYNAMIC ANALYSIS OF F-14A BORON HORIZONTAL STABILIZER	71-0123
DYNAMIC	STATIC AND DYNAMIC ANALYSIS OF SANDWICH STRUCTURES BY METHOD OF FINITE ELEMENTS	71-0352
DYNAMIC	STATIC AND DYNAMIC ANALYSIS OF MULTILAYERED SANDWICH PLATES	72-0493
DYNAMIC	STATIC AND DYNAMIC ANALYSIS OF VISCOELASTOPLASTIC FIBER-REINFORCED COMPOSITE SHELLS IN MISSILE STRUC	73-0651
DYNAMIC	STATIC AND DYNAMIC ANALYSIS OF WOOD-JOIST FLOORS BY THE FINITE-ELEMENT METHOD	73-1161
DYNAMIC	STATIC AND DYNAMIC ANALYSIS OF AN ARCH DAM	74-1101
DYNAMIC	STATIC AND DYNAMIC ANALYSIS OF NONLINEAR STRUCTURES	75-0718
DYNAMIC	STATIC AND DYNAMIC BEHAVIOUR OF GUYED MASTS	68-0271
DYNAMIC	STATIC AND DYNAMIC BUCKLING OF THIN-WALLED COLUMNS USING FINITE ELEMENTS	70-0505
DYNAMIC	STATIC AND DYNAMIC BEHAVIOUR OF RECTANGULAR PLATES USING HIGHER ORDER FINITE STRIPS	72-0458
DYNAMIC	STATIC AND DYNAMIC ELASTO-PLASTIC ANALYSIS BY THE METHOD OF FINITE ELEMENTS IN SPACE AND TIME	72-0642
DYNAMIC	STATIC AND DYNAMIC FINITE ELEMENT ANALYSIS OF SANDWICH STRUCTURES	68-0003
DYNAMIC	STATIC AND DYNAMIC GEOMETRIC AND MATERIAL NONLINEAR ANALYSIS	74-1302
DYNAMIC	STATIC AND DYNAMIC NONLINEAR STRUCTURAL PROBLEMS	72-0848
DYNAMIC	STATIC AND DYNAMIC STRESSES IN SLOPES	66-0025
DYNAMIC	STATIC, STABILITY, AND DYNAMIC ANALYSIS OF SHELLS OF REVOLUTION BY NUMERICAL INTEGRATION - A COMPARI	74-0389
DYNAMIC	STOCHASTIC DYNAMIC RESPONSE ANALYSIS OF OFFSHORE PLATFORMS, WITH PARTICULAR REFERENCE TO GRAVITY-TYP	75-0220
DYNAMIC	STRAP - A COMPUTER CODE FOR THE STATIC AND DYNAMIC ANALYSIS OF REACTOR STRUCTURAL SYSTEMS	71-0078
DYNAMIC	STRAP: COMPUTER CODE FOR STATIC AND DYNAMIC STRUCTURAL ANALYSIS AND STUDIES MADE USING THE CODE	70-0409
DYNAMIC	STRESS-WAVE PROPAGATION AND DYNAMIC EFFECTS IN SHELLS (SPADES) MICROMECHANICAL STRESS-WAVE RESPONSE	72-0851
DYNAMIC	STRUCTURAL DYNAMIC ANALYSIS OF ELECTRONIC ASSEMBLIES USING NASTRAN RESTART/FORMAT CHANGE CAPABILITY	71-0230
DYNAMIC	STRUCTURAL DYNAMIC PROPERTIES OF TACTICAL MISSILE JOINTS - PHASE 1	70-0423
DYNAMIC	STRUCTURAL DYNAMIC PROPERTIES OF TACTICAL MISSILE JOINTS - PHASE 3	74-0811
DYNAMIC	STRUCTURAL EFFECTS OF CONFINED DYNAMIC LOADS	72-0151
DYNAMIC	SURVEY OF SOLUTION PROCEDURES FOR NONLINEAR STATIC AND DYNAMIC ANALYSES	74-0307
DYNAMIC	THE APPLICATION OF DYNAMIC RELAXATION TO THE FINITE ELEMENT METHOD OF STRUCTURAL ANALYSIS	68-0108
DYNAMIC	THE CALCULATION OF STATIC AND DYNAMIC CHARACTERISTICS OF FLIGHT VEHICLE STRUCTURES BY THE FINITE ELE	73-1018
DYNAMIC	THE EFFECTS OF THIN CORES AND NONHOMOGENEITY ON DYNAMIC RESPONSE OF EARTH DAMS	67-0021
DYNAMIC	THE FINITE ELEMENT DYNAMIC STABILITY ANALYSIS OF THIN SHELLS SUBMERGED IN FLUIDS	71-0057
DYNAMIC	THE FINITE ELEMENT METHOD AND ITS APPLICATION TO SOIL DYNAMIC PROBLEMS	68-0288
DYNAMIC	THE FINITE ELEMENT METHOD APPLIED TO THE STATIC AND DYNAMIC BUCKLING OF CIRCULAR ARCHES	70-0497
DYNAMIC	THE LINEAR ELASTIC DYNAMIC ANALYSIS OF SHELLS OF REVOLUTION BY THE MATRIX DISPLACEMENT METHOD	66-0046
DYNAMIC	THE NONLINEAR DYNAMIC ANALYSIS OF SHELLS OF REVOLUTION WITH ASYMMETRIC PROPERTIES BY THE FINITE ELEM	71-0818
DYNAMIC	THE TRANSIENT DYNAMIC ANALYSIS OF THIN SHELLS BY THE FINITE ELEMENT METHODS	71-0142
DYNAMIC	THE USE OF EIGEN-DERIVATIVES TO PREDICT THE EFFECT OF DESIGN CHANGES UPON THE DYNAMIC CHARACTERISTIC	74-0737
DYNAMIC	THE USE OF FINITE ELEMENT METHODS FOR STATIC AND DYNAMIC ANALYSES OF COMPLEX SHELLS	68-0247
DYNAMIC	THEORETICAL AND EXPERIMENTAL ANALYSIS OF STIFFENED PANEL UNDER DYNAMIC CONDITIONS	70-0130
DYNAMIC	THEORETICAL BASIS FOR HYBRID FINITE ELEMENTS IN DYNAMIC PROBLEMS	75-0489
DYNAMIC	TRANSIENT DYNAMIC RESPONSE OF VISCOPLASTIC STRUCTURES	72-0234
DYNAMIC	USE OF PAS-1 FINITE ELEMENT COMPUTER SYSTEM FOR STATIC AND DYNAMIC ANALYSIS OF 2-DIMENSIONAL AND 3-D	75-0070
DYNAMIC	USING NASTRAN FOR DYNAMIC ANALYSIS OF VEHICLE SYSTEMS	74-0316
DYNAMIC	VARIATIONAL PRINCIPLE FOR THE DYNAMIC ANALYSIS OF CONTINUA BY HYBRID FINITE ELEMENT METHOD	71-0492
DYNAMIC	VERTICAL RESPONSE OF A RIGID BASE TO DYNAMIC LOADING	69-0316
DYNAMIC	VIBRATION ANALYSIS OF A 1/15-SCALE DYNAMIC MODAL OF A SPACE SHUTTLE CONFIGURATION	71-0609
DYNAMICAL	A STUDY OF A DYNAMICAL FINITE ELEMENT ANALYSIS FOR APPLICATION TO AXIAL WAVE PROPAGATION PROBLEMS IN	71-0589
DYNAMICAL	DYNAMICAL ANALYSIS OF AXIALLY NON-UNIFORM, THIN CYLINDRICAL SHELLS. PART 1 - MATRIX FORMULATION	70-0413
DYNAMICAL	DYNAMICAL ANALYSIS OF AXIALLY NONUNIFORM, THIN CYLINDRICAL SHELLS. PART 2 FREE VIBRATION	71-0716
DYNAMICAL	DYNAMICAL FINITE ELEMENT ANALYSIS FOR ELASTIC WAVES IN BEAMS AND PLATES	68-0211
DYNAMICAL	FINITE ELEMENT ANALYSIS OF NONLINEAR PROBLEMS IN THE DYNAMICAL THEORY OF COUPLED THERMOELASTICITY	69-0202
DYNAMICALLY	ANALYTICAL MODEL FOR DETERMINING ENERGY DISSIPATION IN DYNAMICALLY LOADED STRUCTURES	75-0328
DYNAMICALLY	COMPUTER PROGRAM JET 3 TO CALCULATE THE LARGE ELASTIC PLASTIC DYNAMICALLY INDUCED DEFORMATIONS OF FR	72-0609
DYNAMICALLY	FINITE ELEMENT METHOD FOR THE CALCULATION OF LOCUS PATHS IN DYNAMICALLY LOADED BEARINGS	73-0487
DYNAMICS	A FINITE ELEMENT SOLUTION OF GENERAL FLUID DYNAMICS PROBLEMS - NATURAL CONVECTION IN RECTANGULAR CAV	70-0008
DYNAMICS	A FINITE ELEMENT SOLUTION FOR A CLASS OF TWO DIMENSIONAL VISCOUS FLUID FLOW DYNAMICS PROBLEMS	71-0544
DYNAMICS	A METHOD OF COMPUTATION FOR STRUCTURAL DYNAMICS	59-0002
DYNAMICS	A NOTE ON THE ANALYSIS OF NONLINEAR DYNAMICS OF ELASTIC MEMBRANES BY THE FINITE ELEMENT METHOD	74-0341
DYNAMICS	A NUMERICAL SOLUTION TECHNIQUE FOR A CLASS OF TWO-DIMENSIONAL PROBLEMS IN FLUID DYNAMICS FORMULATED	69-0019
DYNAMICS	A USER-ORIENTED PROGRAM FOR CRASH DYNAMICS	74-0321
DYNAMICS	ADVANCES IN NUMERICAL FLUID DYNAMICS	73-0813
DYNAMICS	ANALYSIS OF THE DYNAMICS OF ELASTIC STRUCTURES BY THE FINITE ELEMENT METHOD	72-0893
DYNAMICS	ANALYSIS OF THE DYNAMICS OF THIN PRIMARY MIRRORS FOR LARGE ASTRONOMICAL TELESCOPES	73-1084
DYNAMICS	ANALYTICAL DYNAMICS AND NONRIGID SPACECRAFT SIMULATION	74-0771
DYNAMICS	APPLICATION OF FINITE ELEMENT METHODS IN FLUID DYNAMICS	71-0369
DYNAMICS	APPLICATION OF NASTRAN TO A SPACE SHUTTLE DYNAMICS MODEL	71-0247
DYNAMICS	APPLICATION OF THE FINITE-ELEMENT METHOD TO TOWED CABLE DYNAMICS	73-1141
DYNAMICS	APPLICATION OF THE FINITE ELEMENT METHOD TO TOWED CABLE DYNAMICS	75-0809
DYNAMICS	APPLICATIONS OF FINITE ELEMENT METHODS IN FLUID DYNAMICS	75-0184
DYNAMICS	BASIC PRINCIPLES OF STRUCTURAL DYNAMICS	71-0505
DYNAMICS	CONTINUUM APPROACH TOWARD DYNAMICS OF GRIDWORKS	73-0174
DYNAMICS	DYNAMICS OF A FLEXIBLE BULKHEAD AND CONTAINED FLUID	73-0082
DYNAMICS	DYNAMICS OF BUILDING - SOIL INTERACTION	72-0856
DYNAMICS	DYNAMICS OF CRACKED STRUCTURES USING FINITE ELEMENTS	74-1115
DYNAMICS	DYNAMICS OF DISTRIBUTED PARAMETER ROTOR SYSTEMS: TRANSFER MATRIX AND FINITE ELEMENT TECHNIQUES	70-0642
DYNAMICS	DYNAMICS OF EARTH DAMS WITH FOUNDATION INTERACTION	71-0343
DYNAMICS	DYNAMICS OF GRAVITY DAM-RESERVOIR SYSTEMS	73-0225
DYNAMICS	DYNAMICS OF PRISMATIC AND OPEN SECTION MEMBER GRIDS	69-0070
DYNAMICS	DYNAMICS OF SUBMARINE STRUCTURES, PART 2	68-0025
DYNAMICS	DYNAMICS OF SUBMARINE STRUCTURES, VOLUME 1	68-0251

DYNAMICS	DYNAMICS OF SUBMARINE STRUCTURES, PART 1.	69-0054
DYNAMICS	FINITE ELEMENT ANALYSIS METHODS IN DYNAMICS OF PULSATILE FLOW	69-0519
DYNAMICS	FINITE ELEMENT APPLICATIONS IN FLUID DYNAMICS	69-0003
DYNAMICS	FINITE ELEMENT FORMULATION WITH REFERENCE TO FLUID DYNAMICS	74-1245
DYNAMICS	FINITE ELEMENT METHOD IN FLUID DYNAMICS AND HEAT TRANSFER	67-0151
DYNAMICS	FINITE ELEMENT METHODS FOR UNSTEADY COMPRESSIBLE GAS DYNAMICS	75-0045
DYNAMICS	FINITE ELEMENT SOLUTION ALGORITHM FOR VISCOUS INCOMPRESSIBLE FLUID DYNAMICS	73-0063
DYNAMICS	FINITE ELEMENT SOLUTION ALGORITHM FOR INCOMPRESSIBLE FLUID DYNAMICS	74-0006
DYNAMICS	FINITE ELEMENT SOLUTION TO SOME PLANE PROBLEMS IN SOIL DYNAMICS	74-0448
DYNAMICS	FINITE ELEMENT SOLUTIONS OF A NONLINEAR REACTOR DYNAMICS PROBLEM	74-1268
DYNAMICS	FINITE ELEMENTS FOR FLUID DYNAMICS. VARIATIONAL FORMULATIONS FOR NONLINEAR, INITIAL-VALUE PROBLEMS	74-0710
DYNAMICS	FRACTURE DYNAMICS OF HOMALITE-100 SHEETS	72-0809
DYNAMICS	HABEAS- A STRUCTURAL DYNAMICS ANALYSIS SYSTEM	70-0121
DYNAMICS	HYBRID FINITE ELEMENT OPTIMIZATION PROGRAMS FOR DISCS USING A DYNAMICS DESIGN TECHNOLOGY	74-0691
DYNAMICS	ILLUSTRATIONS OF AUTOMOTIVE FINITE ELEMENT MODELS - DYNAMICS	74-0391
DYNAMICS	INCREMENTAL SOLUTION PROCEDURE FOR NONLINEAR STRUCTURAL DYNAMICS PROBLEMS	75-0656
DYNAMICS	INFLUENCE OF ASCENT HEATING ON THE SEPARATION DYNAMICS OF A SPACECRAFT FAIRING	72-0906
DYNAMICS	MATHEMATICAL OBSERVATIONS IN STRUCTURAL DYNAMICS	73-0156
DYNAMICS	METHOD OF FINITE SPACE-TIME ELEMENTS IN DYNAMICS OF STRUCTURES	75-0823
DYNAMICS	MODAL COUPLING PROCEDURES ADAPTED TO NASTRAN ANALYSIS OF THE 1/8 - SCALE SHUTTLE STRUCTURAL DYNAMICS	75-0677
DYNAMICS	MODAL COUPLING PROCEDURES ADAPTED TO NASTRAN ANALYSIS OF THE 1/8-SCALE SHUTTLE STRUCTURAL DYNAMICS M	75-0678
DYNAMICS	NASTRAN EXPERIENCES OF FORT WORTH OPERATION, CONVAIR AEROSPACE DIVISION OF GENERAL DYNAMICS	71-0019
DYNAMICS	NONLINEAR DYNAMICS OF SOLIDS BY FINITE ELEMENT METHODS	69-0490
DYNAMICS	NONLINEAR DYNAMICS OF SOLIDS BY THE FINITE ELEMENT METHOD	71-0116
DYNAMICS	NONLINEAR DYNAMICS OF CABLES WITH LOW INITIAL TENSION	72-0740
DYNAMICS	NOTE ON THE ANALYSIS OF NONLINEAR DYNAMICS OF ELASTIC MEMBRANES BY THE FINITE ELEMENT METHOD	74-0430
DYNAMICS	NUMERICAL METHODS IN FLUID DYNAMICS	72-0844
DYNAMICS	NUMERICAL METHODS IN FLUID DYNAMICS	73-0689
DYNAMICS	NUMERICAL SOLUTION TO THE DYNAMICS OF VISCOUS FLUID FLOW BY A FINITE ELEMENT ALGORITHM: A FIRST STEP	71-0004
DYNAMICS	NUMERICAL SOLUTION TO THE DYNAMICS OF VISCOUS FLUID FLOW BY A FINITE ELEMENT ALGORITHM: A FIRST STEP	71-0791
DYNAMICS	ON THE APPLICATION OF THE FINITE ELEMENT METHOD TO A CLASS OF PROBLEMS IN FLUID DYNAMICS	69-0400
DYNAMICS	ON THE CONVERGENCE OF THE METHOD OF FINITE ELEMENTS IN THE ANALYSIS OF MEMBRANE DYNAMICS	72-0537
DYNAMICS	ON THE EXTREMAL PROPERTIES OF THE SOLUTION IN DYNAMICS OF RIGID VISCOPLASTIC BODIES ALLOWING FOR LAR	73-1080
DYNAMICS	ON THE UNCONDITIONAL STABILITY OF AN IMPLICIT ALGORITHM FOR NONLINEAR STRUCTURAL DYNAMICS	75-0825
DYNAMICS	SENSITIVITY ANALYSIS OF TORSIONAL VIBRATION CHARACTERISTICS OF HELICOPTER ROTOR BLADES. PART 1: STRU	74-0846
DYNAMICS	SNAP DYNAMICS	72-0911
DYNAMICS	SOLUTION SCHEMES FOR PROBLEMS OF NONLINEAR STRUCTURAL DYNAMICS	74-0505
DYNAMICS	SPACE-TIME SOLUTIONS OF NONLINEAR REACTOR DYNAMICS BY FINITE-ELEMENT METHOD	75-0503
DYNAMICS	STATICS AND DYNAMICS OF HYDROSTATICALLY LOADED SHELLS BY FINITE ELEMENT METHOD	71-0044
DYNAMICS	STRUCTURAL DYNAMICS PROGRAMS	70-0412
DYNAMICS	STRUCTURAL DYNAMICS BY FINITE ELEMENTS: MODAL AND FOURIER ANALYSIS	73-0836
DYNAMICS	STUDY OF PROPELLANT DYNAMICS IN A SHUTTLE TYPE LAUNCH VEHICLE	72-0690
DYNAMICS	THE USE OF CONDENSATION TECHNIQUES FOR SOLVING DYNAMICS PROBLEMS	74-0320
DYNAMICS	USE OF FINITE ELEMENT METHOD IN THE DYNAMICS OF PLATES AND SHELLS	70-0313
DYNAMICS	VARIATIONAL FORMULATION OF DYNAMICS OF FLUID-SATURATED POROUS ELASTIC SOLIDS	72-0017
DYNAMICS	VIKING DYNAMICS - AN OVERVIEW	74-1304
DYNAMICS	ZUR SYSTEMATIK NUMERISCHER NAEHERUNGSVERFAHREN IN DER STATIK UND DYNAMIK DER KONSTRUKTIONEN (METHODS	72-0557
DYNAMIK	ZUR SYSTEMATIK NUMERISCHER NAEHERUNGSVERFAHREN IN DER STATIK UND DYNAMIK DER KONSTRUKTIONEN (METHODS	72-0557
DYNAMIQUE	ANALYSE DYNAMIQUE DUALE DES STRUCTURES PAR LA METHODE DES ELEMENTS FINIS	73-0623
DYNAMIQUE	ANALYSE STATIQUE ET DYNAMIQUE DES COQUES PAR LA METHODE DES ELEMENTS FINIS	74-0509
DYNAMISCHEN	ANWENDUNG DER METHODEN DER FINITEN ELEMENTE MIT DEM PROGRAMMSYSTEM ASKA UNTER BESONDERER BERUCKSICHT	71-0624
DYNAMISCHEN	ANWENDUNG DER METHODE FINITER ELEMENTE BEI DER ANALYSE DES DYNAMISCHEN VERHALTENS GEDAEMPFTER WERKZE	75-0800
DYNAMISCHEN	ASKA - EINE PROGRAMMIERSPRACHE ZUR STATISCHEN UND DYNAMISCHEN ANALYSE	69-0457
DYNAMISCHEN	TURBAN VERSION 1.0. EIN SPEZIELLES PROGRAMM ZUR LINEAREN UND NICHTLINEAREN STATISCHEN UND DYNAMISCHE	71-0622
DYNAN	DYNAN LECTURE NOTES WITH COMPUTATIONAL EXAMPLES	71-0581
DYNAN	DYNAN USERS MANUEL	71-0625
DYNAPLAS	DYNAPLAS: A FINITE ELEMENT PROGRAM FOR THE DYNAMIC, LARGE DEFLECTION, ELASTIC-PLASTIC ANALYSIS OF ST	72-0773
DYNASOR	DYNASOR 11 - A FINITE ELEMENT PROGRAM FOR THE DYNAMIC NONLINEAR ANALYSIS OF SHELLS OF REVOLUTION (US	71-0666
DYNAX	THREE DIMENSIONAL STRUCTURES/MEDIA INTERACTION, DYNAX DEVELOPMENT AND APPLICATIONS,	70-0378
DYPLAS	DYPLAS, A FINITE ELEMENT DYNAMIC ELASTIC-PLASTIC LARGE DEFORMATION ANALYSIS PROGRAM	74-0600
DYSKERTYZOWANYC	OCENA DOKLADNOSCI ROZWIAZAN PROBLEMOW BRZEGOWYCH SPREZYSTYCH CIAL DYSKERTYZOWANYCH. (ASSESSING THE A	74-0921
EAC	EAC/EASE USER INFORMATION MANUAL	70-0320
EARDRUM	MODELING EARDRUM AS A DOUBLY CURVED SHELL USING FINITE-ELEMENT METHOD	75-0047
EARDRUM	SIMULATING BEHAVIOR OF EARDRUM BY FINITE-ELEMENT METHOD	74-0511
EARTH	ANALYSIS OF THE SEEPAGE THROUGH AN EARTH DAM BY FINITE ELEMENT METHOD	72-0585
EARTH	COMPARISON BETWEEN THEORETICAL AND MEASURED EARTH PRESSURES ACTING ON A LARGE MOTORWAY RETAINING WAL	74-0917
EARTH	COMPORTAREA ELASTICA A DIGURILOR DE PAMINT SOLICIATE SEISMIC (ELASTIC BEHAVIOR OF THE EARTH DAMS SEI	74-1185
EARTH	CRACKING OF EARTH AND ROCKFILL DAMS. A THEORETICAL INVESTIGATION BY MEANS OF THE FINITE ELEMENT METH	69-0085
EARTH	CRACKING OF EARTH AND ROCKFILL DAMS	69-0294
EARTH	CRACKING OF EARTH AND ROCKFILL DAMS: COMPARISON OF OBSERVED AND THEORETICAL TENSILE STRAINS IN THE C	71-0699
EARTH	CRACKING OF EARTH AND ROCKFILL DAMS: COMPARISON OF OBSERVED AND THEORETICAL TENSILE STRAINS IN THE C	71-0699
EARTH	DYNAMIC RESPONSE OF EARTH DAMS	67-0024
EARTH	DYNAMIC RESPONSE OF COHESIVE EARTH DAMS TO EARTHQUAKES	67-0131
EARTH	DYNAMIC RESPONSE OF A CYLINDER BURIED IN AN EARTH BEAM - RESULTS OF A FINITE ELEMENT ANALYSIS	72-0592
EARTH	DYNAMICS OF EARTH DAMS WITH FOUNDATION INTERACTION	71-0343
EARTH	EARTH PRESSURES AGAINST THE ABUTMENTS AND WING WALLS OF STANDARD MOTORWAY BRIDGES	75-0629
EARTH	EARTH PRESSURES ON STRUCTURES DUE TO FAULT MOVEMENT	73-0403
EARTH	EARTH VIBRATION EFFECTS AND ABATEMENT FOR MILITARY FACILITIES. REPORT 3. ANALYSIS METHOD FOR FOOTING	72-0806
EARTH	EARTHQUAKE ANALYSIS OF EARTH DAMS	69-0072
EARTH	EARTHQUAKE RESPONSE OF EARTH DAMS	67-0010
EARTH	EARTHQUAKE STRESS ANALYSIS IN EARTH DAMS	65-0043
EARTH	EFFECT OF RESERVOIR FILLING ON STRESSES AND MOVEMENT IN EARTH AND ROCKFILL DAMS. A REPORT OF AN INVE	72-0836
EARTH	FINITE ELEMENT ANALYSIS FOR SEISMIC RESPONSE OF EARTH DAMS	68-0091
EARTH	FINITE ELEMENT ANALYSIS OF ELASTO-PLASTIC SOILS. REPORT NO. 4. FINITE ELEMENT ANALYSIS OF ELASTO-PLA	72-0788
EARTH	FINITE ELEMENT ANALYSIS FOR THE SEISMIC STABILITY OF EARTH STRUCTURES	74-1106
EARTH	FINITE-ELEMENT TREATMENT OF QUADRUPOLE GROUND MOTIONS IN AXISYMMETRIC EARTH STRUCTURES	75-0434
EARTH	HYDRAULIC FRACTURING IN ZONED EARTH AND ROCKFILL DAMS: A REPORT OF AN INVESTIGATION	73-0873
EARTH	INFLUENCE OF RESERVOIR WATER ON EARTHQUAKE RESPONSE OF EARTH DAMS	70-0375
EARTH	MOVEMENTS OF ARTICULATED CONDUITS UNDER EARTH DAMS ON COMPRESSIBLE FOUNDATIONS	70-0477
EARTH	NOUVELLE CONCEPTION DES DRAINS DANS LES BARRAGES EN TERRE HOMOGENE DE PETITE ET MOYENNE DIMENSIONS (	73-0404
EARTH	ON THE APPLICATION OF FINITE-ELEMENT TECHNIQUES TO PROBLEMS CONCERNING POTENTIAL DISTRIBUTION AND ST	69-0268
EARTH	PERFORMANCE OF EARTH AND EARTH-SUPPORTED STRUCTURES	72-0449
EARTH	RESPONSE OF EARTH BANKS DURING EARTHQUAKES	67-0039
EARTH	RESPONSE OF EARTH DAMS TO TRAVELLING SEISMIC WAVES	69-0092
EARTH	SEEPAGE ANALYSIS OF EARTH BANKS UNDER DRAWDOWN	72-0014
EARTH	SEISMIC ANALYSIS OF EARTH DAM RESERVOIR SYSTEMS	73-0412
EARTH	SOLUTION OF STRESSES AND TRANSFORMATIONS OF EARTH DAMS BY THE METHOD OF FINITE ELEMENTS	69-0426
EARTH	STATIC AND SEISMIC BEHAVIOUR OF AN EARTH DAM	67-0026
EARTH	STRUCTURAL BEHAVIOR OF A FLEXIBLE METAL CULVERT UNDER A DEEP EARTH EMBANKMENT USING METHOD A BACKFIL	73-0739
EARTH	THE BEHAVIOR OF STATISTICALLY HETEROGENEOUS EXCAVATED EARTH SLOPES	74-1330
EARTH	THE EFFECTS OF THIN CORES AND NONHOMOGENEITY ON DYNAMIC RESPONSE OF EARTH DAMS	67-0021
EARTH	UTILISATION DE MODELES MATHEMATIQUES MULTIDIMENSIONNELS POUR L'ETUDE DES ECOULEMENTS DANS LES DIGUES	73-0381

EARTH	VIBRATIONAL CHARACTERISTICS OF THREE-DIMENSIONAL SOLIDS, WITH APPLICATIONS TO EARTH DAMS	69-0432
EARTH-SUPPORTED	PERFORMANCE OF EARTH AND EARTH-SUPPORTED STRUCTURES	72-0449
EARTH'S	INFLUENCE OF FLUID INJECTION ON THE STATE OF STRESS IN THE EARTH'S CRUST	72-0624
EARTHEN	ANALYSIS OF EARTHEN DAMS BY THE METHOD OF FINITE ELEMENTS	70-0454
EARTHFILL	RASCHET METODOM KONECHNYKH ELEMENTOV DINAMICHESKIKH KHARAKTEERISTIK ZEMLYANOI PLOTINY SOVMESTNO S OS	73-0437
EARTHMOVING	IMPROVED TECHNIQUES FOR DYNAMIC ANALYSIS OF EARTHMOVING EQUIPMENT	74-0294
EARTHQUAKE	A COMPUTER PROGRAM FOR EARTHQUAKE ANALYSIS OF DAMS	70-0469
EARTHQUAKE	DYNAMIC ANALYSIS OF THE SLIDE IN THE LOWER SAN FERNANDO DAM DURING THE EARTHQUAKE OF FEBRUARY 9, 197	75-0277
EARTHQUAKE	EARTHQUAKE ANALYSIS OF EARTH DAMS	69-0072
EARTHQUAKE	EARTHQUAKE ANALYSIS OF GRAVITY DAMS INCLUDING HYDRODYNAMIC INTERACTION	73-0350
EARTHQUAKE	EARTHQUAKE ANALYSIS OF STRUCTURE-FOUNDATION SYSTEMS	73-0876
EARTHQUAKE	EARTHQUAKE ANALYSIS OF AXISYMMETRIC TOWERS PARTIALLY SUBMERGED IN WATER	75-0076
EARTHQUAKE	EARTHQUAKE BEHAVIOUR OF RESERVOIR-DAM SYSTEMS	68-0043
EARTHQUAKE	EARTHQUAKE BEHAVIOUR OF RESERVOIR-DAM SYSTEMS (DISCUSSION)	69-0283
EARTHQUAKE	EARTHQUAKE FINITE ELEMENT ANALYSIS OF STRUCTURE-FOUNDATION SYSTEMS	74-0558
EARTHQUAKE	EARTHQUAKE FINITE-ELEMENT ANALYSIS OF STRUCTURE-FOUNDATION SYSTEMS	75-0410
EARTHQUAKE	EARTHQUAKE RESPONSE OF EARTH DAMS	67-0010
EARTHQUAKE	EARTHQUAKE RESPONSE OF CONCRETE GRAVITY DAMS	70-0042
EARTHQUAKE	EARTHQUAKE RESPONSE OF ARCH DAMS	70-0113
EARTHQUAKE	EARTHQUAKE RESPONSE OF GRAVITY DAMS INCLUDING RESERVOIR INTERACTION EFFECTS	72-0675
EARTHQUAKE	EARTHQUAKE STRESS ANALYSIS IN EARTH DAMS	65-0043
EARTHQUAKE	EARTHQUAKE TRIGGERING BY FLUID INJECTION AT RANGELY, COLORADO	72-0634
EARTHQUAKE	ELASTIC-PLASTIC EARTHQUAKE RESPONSE OF SOIL-BUILDING SYSTEMS	72-0717
EARTHQUAKE	HYDRODYNAMIC PRESSURES ON HIGH DAMS DUE TO VERTICAL EARTHQUAKE MOTIONS	69-0438
EARTHQUAKE	INFLUENCE OF RESERVOIR WATER ON EARTHQUAKE RESPONSE OF EARTH DAMS	70-0375
EARTHQUAKE	RC COLUMN EARTHQUAKE RESPONSE IN TWO DIMENSIONS	74-0651
EARTHQUAKE	STUDY OF 1964 ALASKAN EARTHQUAKE USING FINITE ELEMENT AND GENERALIZED INVERSION TECHNIQUES	72-0003
EARTHQUAKE	THE EARTHQUAKE EXPERIENCE AT KOYNA DAM AND STRESSES IN CONCRETE GRAVITY DAMS	72-0589
EARTHQUAKE	THE IMPORTANCE OF THE VERTICAL COMPONENT OF EARTHQUAKE MOTIONS	66-0019
EARTHQUAKE-RESI	EARTHQUAKE-RESISTANT DESIGN OF ENGINEERING STRUCTURES	73-0284
EARTHQUAKES	CALCULATION OF VIBRATIONS OF STRUCTURES CAUSED BY EARTHQUAKES USING THE FINITE ELEMENT METHOD	71-0690
EARTHQUAKES	CONTRIBUTION A L'ETUDE DU COMPLEXE BARRAGE-FONDATION ET EXPLICATION DES SEISMES DUS AU REMPLISSAGE D	72-0554
EARTHQUAKES	DYNAMIC RESPONSE OF COHESIVE EARTH DAMS TO EARTHQUAKES	67-0131
EARTHQUAKES	EFFECTS OF SOURCE STRUCTURE AND SITE RESPONSE ON GROUND MOTION IN THE NEAR FIELD OF EARTHQUAKES	73-0732
EARTHQUAKES	RESPONSE OF EARTH BANKS DURING EARTHQUAKES	67-0039
EARTHS	FINITE ELEMENT MODEL OF EARTHS GRAVITY FIELD	74-0516
EASE	EAC/EASE USER INFORMATION MANUAL	70-0320
EAST	NUMERICAL HULL VIBRATION ANALYSIS OF A FAR EAST CONTAINER SHIP	74-1325
EBENE	FINITE ELEMENTE ZUR BERECHNUNG DER WAERMEUEBERTRAGUNG FUER EBENE UND AXISYMMETRISCHE PROBLEME (FINIT	73-0449
EBENER	KOMBINATION EINES INTEGRALGLEICHUNGS VERFAHRENS MIT METHODE DER FINITEN ELEMENTEN ZUR BERECHNUNG EBE	74-0856
ECCENTRIC	ANALYSIS OF ORTHOTROPIC FOLDED PLATES WITH ECCENTRIC STIFFENERS	70-0437
ECCENTRIC	FINITE-ELEMENT ANALYSIS OF ORTHOTROPIC PLATES WITH ECCENTRIC STIFFENERS	72-0958
ECCENTRIC	INVESTIGATION OF THE STRESS-STRAINED STATE OF SPHERICAL SHELLS WITH AN ECCENTRIC CUTOUT ON THE BASIS	73-0222
ECCENTRICALLY	A FINITE ELEMENT ANALYSIS OF ECCENTRICALLY STIFFENED-CIRCULAR CYLINDRICAL SHELLS	69-0159
ECCENTRICALLY	AN ELASTIC FINITE-ELEMENT ANALYSIS OF THE BUCKLING OF ECCENTRICALLY STIFFENED PLATES AND SHELLS	73-1136
ECCENTRICALLY	ANALYSIS OF ECCENTRICALLY STIFFENED CYLINDRICAL SHELLS	72-0217
ECCENTRICALLY	BIFURCATION OF ECCENTRICALLY LOADED MEMBERS AND OF STIFFENED PLATES	70-0057
ECCENTRICALLY	DYNAMIC BEHAVIOR OF ECCENTRICALLY STIFFENED PLATES	71-0796
ECCENTRICALLY	DYNAMIC BEHAVIOUR OF ECCENTRICALLY STIFFENED PLATES	71-0605
ECCENTRICALLY	FINITE ELEMENT ANALYSIS OF ELASTIC-PLASTIC PLATES AND ECCENTRICALLY STIFFENED PLATES	71-0585
ECCENTRICALLY	FINITE ELEMENT ANALYSIS OF PLATES AND ECCENTRICALLY STIFFENED PLATES	73-0691
ECCENTRICALLY	FULL RANGE ANALYSIS OF ECCENTRICALLY STIFFENED PLATES	74-0149
ECCENTRICALLY	INELASTIC RESPONSE OF PLATES AND ECCENTRICALLY STIFFENED PLATES	74-0458
ECONOMICAL	AN ECONOMICAL METHOD FOR DETERMINING THE SMALLEST EIGENVALUES OF LARGE LINEAR SYSTEMS	71-0551
ECONOMICAL	ECONOMICAL COMPUTER TECHNIQUES FOR SOLID FINITE ELEMENTS	68-0086
ECONOMICAL	ECONOMICAL COMPUTER TECHNIQUES FOR NUMERICALLY INTEGRATED FINITE ELEMENTS	69-0138
ECONOMICAL	ECONOMICAL DESIGN ANALYSIS OF PISTONS FOR INTERNAL COMBUSTION ENGINES	73-0188
ECONOMICAL	ECONOMICAL SOLUTION TECHNIQUES FOR LOAD DEFLECTION EQUATIONS	74-0414
ECONOMICAL	THE ECONOMICAL COMPUTATION OF STIFFNESS OF LARGE STRUCTURAL ELEMENTS	66-0090
ECONOMICAL	THE ECONOMICAL SOLUTION OF ELASTIC PROBLEMS FOR A RANGE OF POISSONS RATIO	74-0786
ECONOMICS	AN INVESTIGATION INTO THE ECONOMICS OF USING COMPLEX ELEMENTS IN PLANE STRESS ANALYSIS	74-0427
ECONOMISERS	EIGENVALUE ECONOMISERS IN VIBRATION PROBLEMS	63-0013
ECONOMIZATION	AUTOMATIC MASTERS FOR EIGENVALUE ECONOMIZATION	75-0557
ECOULEMENT	COUPLAGE ENTRE L ECOULEMENT AUTOUR D UN ARRIERE-CORPS ET LE JET PROPULSIF EN THEORIE DE FLUIDE PARFA	74-1137
ECOULEMENTS	CALCUL NUMERIQUE DES ECOULEMENTS DES FLUIDES DE BINGHAM ET DES FLUIDES NEWTONIENS INCOMPRESSIBLES PA	72-0349
ECOULEMENTS	UTILISATION DE MODELES MATHEMATIQUES MULTIDIMENSIONNELS POUR L'ETUDE DES ECOULEMENTS DANS LES DIGUES	73-0381
EDDY	APPLICATION OF THE FINITE ELEMENT METHOD TO EDDY CURRENT PROBLEMS	74-1156
EDDY	EDDY CURRENTS SCREENING PROBLEMS SOLVED BY MEANS OF FINITE ELEMENT DIFFERENCE AND FINITE ELEMENT MET	75-0552
EDDY-CURRENT	FINITE-ELEMENT NETWORK MODELS AND THEIR APPLICATION TO EDDY-CURRENT PROBLEMS	75-0143
EDDY-CURRENT	FINITE-ELEMENT SOLUTION OF EDDY-CURRENT PROBLEM IN MAGNETIC STRUCTURES	74-0106
EDGE	AN ELASTIC-PLASTIC FINITE ELEMENT ANALYSIS FOR PLATES WITH EDGE BEAMS	69-0185
EDGE	ANALYSIS OF SLABS WITH EDGE BEAMS	71-0075
EDGE	APPLICATION OF THE FINITE ELEMENT METHOD OF ANALYSIS TO HYPERBOLIC PARABOLOIDS WITH EDGE BEAMS	75-0250
EDGE	BUCKLING OF PLATE GIRDER WEBS UNDER PARTIAL EDGE LOADINGS	70-0283
EDGE	CONICAL SEGMENT METHOD FOR ANALYZING OPEN CROWN SHELLS OF REVOLUTION FOR EDGE LOADING	63-0007
EDGE	DUGDALE MODEL SOLUTIONS FOR A SINGLE EDGE CRACKED PLATE	74-0221
EDGE	FINITE-ELEMENT ANALYSIS OF EDGE EFFECTS IN ANGLE-PLY COMPOSITE LAMINATES	74-0843
EDGE	HEAVY SECTION STEEL TECHNOLOGY PROGRAM TECHNICAL OR PROGRAMMATIC MANUSCRIPT NO. 25. VARIABLE THICKNE	73-0948
EDGE	MEMBRANE PARALLELOGRAM ELEMENT WITH LINEARLY VARYING EDGE STRAIN FOR THE MATRIX DISPLACEMENT METHOD	66-0006
EDGE	ON THE ASYMPTOTICALLY SPHERICAL DEFORMATIONS OF ARBITRARY MEMBRANES OF REVOLUTION FIXED ALONG AN EDG	71-0640
EDGE	ON THE FICTION OF LEADING EDGE SUCTION	71-0712
EDGE	STRESS CONCENTRATION FACTORS FOR A FATIGUE TEST SPECIMEN WITH A U-TYPE NOTCH IN ONE EDGE BY FINITE E	74-1344
EDGE	THERMAL MECHANICAL ANALYSIS PROCEDURE FOR A SEMICIRCULAR PLATE WITH CLAMPED STRAIGHT EDGE	75-0651
EDGE	VIBRATION OF SQUARE PLATES HAVING VARIOUS EDGE CONDITIONS AND LOADINGS USING A REFINED FINITE ELEMEN	75-0024
EDGE-CRACKED	EXPERIMENTAL AND ANALYTICAL STRAINS IN A EDGE-CRACKED SHEET	69-0483
EDGEBEAMS	ANALYSIS OF CYLINDRICAL SHELL ROOFS WITH POST TENSIONED EDGEBEAMS	74-1042
EDGES	A COMPATIBLE ELEMENT FOR BUCKLING ANALYSIS OF SKEW PLATES WITH PARALLEL EDGES	73-0328
EDGES	ANALYSIS OF AN ORTHOTROPIC PLATE WITH TWO OPPOSITE EDGES SIMPLY SUPPORTED AND THE OTHER TWO FREE	73-0741
EDGES	FINITE ELEMENT ANALYSIS OF PLATES WITH CURVED EDGES.	72-0173
EDUCATION	SOFTWARE USER: EDUCATION AND QUALIFICATION	72-0441
EDUCATION	USER EDUCATION: THE FINITE ELEMENT SHORT COURSE	72-0437
EDV-SYSTEM	DAS PROGRAMMSYSTEM ANTRAS EDV-SYSTEM ZUR BERECHNUNG BELIEBIGER STATISCHER SYSTEME NACH DER METHODE D	74-0918
EDWARD	ANALYSIS OF BEHAVIOR OF EDWARD HYATT POWER PLANT	75-0055
EFEKTOW	ZASTOSOWANIE METODY SZTYWNYCH ELEMENTOW SKONCZONYCH DO OBLICZEN ORGAN WALOW OKRETOYCH Z UWZGLEDNIENI	74-0962
EFFECT	APPROACH TO INVESTIGATE NOTCH EFFECT ON LOW-CYCLE FATIGUE WITH FINITE ELEMENT	73-0509
EFFECT	EFFECT OF BACKPACKING AND INTERNAL PRESSURIZATION ON STRESSES TRANSMITTED TO BURIED CYLINDERS	73-0897
EFFECT	EFFECT OF BEAM-SHEARWALL JUNCTION DEFORMATIONS ON THE FLEXIBILITY OF THE CONNECTING BEAMS	73-0213
EFFECT	EFFECT OF DAMPING ON EXCITABILITY OF HIGH-ORDER NORMAL MODES	75-0709
EFFECT	EFFECT OF DENSIFICATION ON THE ENGINEERING CHARACTERISTICS OF ORGANIC SOILS - VOL. 1 AND 2.	73-0805
EFFECT	EFFECT OF DISCONTINUITIES ON STRESSES AND DEFORMATIONS OF ROCK AROUND A TUNNEL. (EINFLUSS VON DISKON	70-0170
EFFECT	EFFECT OF ELASTIC MODULAR RATIO AND ANISOTROPY ON THE STRESS FIELD IN A TWO-DIMENSIONAL MATRIX INTO	71-0647
EFFECT	EFFECT OF FIBRE DISTRIBUTION ON STRESS AND STRAIN CONCENTRATION AT HOLES IN COMPOSITE PLATES	72-0616

EFFECT	EFFECT OF FILLETS ON STRESS CONCENTRATIONS IN CYLINDRICAL SHELLS WITH STEP CHANGES IN OUTSIDE DIAMET	71-0108
EFFECT	EFFECT OF GEOLOGIC IRREGULARITIES OF SEISMIC RESPONSE	74-0898
EFFECT	EFFECT OF GRAIN BOUNDARY FOR THE ELASTIC BEHAVIOR OF AL AND CU CRYSTALS (ANALYSIS BY THE FINITE ELEM	74-0947
EFFECT	EFFECT OF HIGHER ORDER TERMS IN CERTAIN NONLINEAR FINITE ELEMENT MODELS	73-0125
EFFECT	EFFECT OF HIGHER ORDER TERMS IN CERTAIN NONLINEAR FINITE ELEMENT MODELS - COMMENT	74-0473
EFFECT	EFFECT OF INITIAL DISPLACEMENT ON PROBLEM OF LARGE DEFLECTION AND STABILITY	67-0113
EFFECT	EFFECT OF INITIAL STATE FORCES ON THE HYDROELASTIC VIBRATION OF TUBES	73-0589
EFFECT	EFFECT OF LATERAL PRESSURE ON NATURAL FREQUENCIES OF HYPERBOLOIDAL SHELLS OF REVOLUTION	71-0278
EFFECT	EFFECT OF NONLINEAR MATERIAL RESPONSE ON THE BEHAVIOUR OF PAVEMENTS UNDER TRAFFIC	70-0171
EFFECT	EFFECT OF OPENINGS ON THE LATERAL STIFFNESS OF INFILLED FRAMES	71-0300
EFFECT	EFFECT OF OPENINGS ON STRESSES IN RIGID PAVEMENTS	73-0344
EFFECT	EFFECT OF OUT-OF-PLANENESS OF MEMBRANE QUADRILATERAL FINITE ELEMENTS	73-0031
EFFECT	EFFECT OF PERFORATION DAMAGE ON WELL PRODUCTIVITY	74-0936
EFFECT	EFFECT OF PLASTICITY AND CRACK BLUNTING ON THE STRESS DISTRIBUTION IN ORTHOTROPIC COMPOSITE MATERIAL	73-0145
EFFECT	EFFECT OF RESERVOIR FILLING ON STRESSES AND MOVEMENT IN EARTH AND ROCKFILL DAMS. A REPORT OF AN INVE	72-0836
EFFECT	EFFECT OF SEEPAGE ON EMBANKMENT DEFORMATIONS DUE TO WATER LOADING	74-1329
EFFECT	EFFECT OF SHAPE AND HETEROGENEITY OF ROCK SAMPLES ON THEIR COMPRESSIVE STRENGTH	69-0315
EFFECT	EFFECT OF SHEAR DEFORMATIONS AND ROTARY INERTIA ON OPTIMUM BEAM FREQUENCIES	75-0231
EFFECT	EFFECT OF SYSTEM PARAMETER VARIATIONS ON NATURAL FREQUENCIES	75-0661
EFFECT	EFFECT OF THE VARIATION OF CONSISTENCY IN RESIDUAL SOILS ON LOAD TRANSFER BY BORED PILES	75-0812
EFFECT	EFFECT OF TIME-DEPENDENT MATERIAL PROPERTIES ON DYNAMIC RESPONSE	74-0900
EFFECT	EFFECT OF TRANSVERSE SHEAR DEFORMATION IN VIBRATIONS OF PLANAR STRUCTURES COMPOSED OF BEAM TYPE ELEM	73-0402
EFFECT	EFFECT OF 2450-MHZ RADIATION ON THE RABBIT EYE	75-0365
EFFECT	ELASTIC DEFORMATION OF POLYCRYSTALLINE METAL - 2 EFFECT OF SLENDERNESS OF GRAINS ON DEFORMATION MODE	73-0275
EFFECT	ESTIMATIONS OF THE EFFECT OF DESIGN CHANGES ON THE MODES AND FREQUENCIES OF VIBRATING STRUCTURAL ELE	74-0819
EFFECT	FORECASTING THE ROCK PRESSURE AND ITS EFFECT ON THE LONGWALL FACES AND ROADWAYS IN THE BITUMINOUS CO	74-0374
EFFECT	HYBRID EFFECT ON BENDING BEHAVIORS OF CFRP-AL LAMINATE	74-1213
EFFECT	INVESTIGATION OF THE EFFECT OF FOUNDATION YIELD ON THE FREQUENCY AND SHAPE OF INHERENT VIBRATIONS OF	72-0419
EFFECT	KERBSPANNUNGSANALYSE NACH DER FINITE-ELEMENT-METHODE AN KEHLNAEHTEN. (NOTCH EFFECT ANALYSIS USING TH	75-0673
EFFECT	NOTCH EFFECT IN LOW CYCLE FATIGUE - REPORT 1: LOW-CYCLE FATIGUE STRENGTH OF 0.48 PERCENT CARBON STEE	73-1090
EFFECT	NOTCH EFFECT IN LOW-CYCLE FATIGUE. CONSIDERATIONS ON NOTCH EFFECT IN LOW-CYCLE FATIGUE WITH FINITE-E	75-0343
EFFECT	NOTCH EFFECT IN LOW-CYCLE FATIGUE. CONSIDERATIONS ON NOTCH EFFECT IN LOW-CYCLE FATIGUE WITH FINITE-E	75-0343
EFFECT	NOTCH EFFECT IN LOW-CYCLE FATIGUE - 2. EFFECT OF STRESS CHANGE AND MEAN STRESS	75-0537
EFFECT	NOTCH EFFECT IN LOW-CYCLE FATIGUE - 2. EFFECT OF STRESS CHANGE AND MEAN STRESS	75-0537
EFFECT	ON THE EFFECT OF APPROXIMATING HIGHER ORDER TERMS IN CERTAIN FINITE ELEMENT MODELS OF NONLINEAR STRU	73-0603
EFFECT	ON THE EFFECT OF THE FORM OF THE STRAIN ENERGY FUNCTION ON THE SOLUTION OF A BOUNDARY-VALUE PROBLEM	72-0429
EFFECT	PROBABILISTIC MODEL FOR MATERIAL STRENGTH VARIATION AND SIZE EFFECT	72-0876
EFFECT	REINFORCED FIELDS OF TRIANGULAR ELEMENTS WITH LINEARLY VARYING STRAIN: EFFECT OF INITIAL STRAINS	65-0004
EFFECT	STEP AND LOADS EFFECT ON STRESSES IN PRESTRESSED CONCRETE SHORT BRACKETS	71-0334
EFFECT	THE COMBINED EFFECT TO CURVED BOUNDARIES AND NUMERICAL INTEGRATION IN ISOPARAMETRIC FINITE ELEMENT M	72-0357
EFFECT	THE EFFECT OF AN INITIAL AXISYMMETRIC IMPERFECTION ON THE NATURAL FREQUENCIES OF CONICAL SHELLS	70-0368
EFFECT	THE EFFECT OF CAMBER ON THE NATURAL FREQUENCIES OF LOW ASPECT RATIO TURBOMACHINERY BLADES	73-1092
EFFECT	THE EFFECT OF COMPOSITE ACTION ON THE ELASTOPLASTIC ANALYSIS OF COMPLETE BUILDING STRUCTURES	74-0441
EFFECT	THE EFFECT OF DEFORMATION ON THE BEHAVIOUR OF HYDRODYNAMIC JOURNAL BEARINGS	71-0009
EFFECT	THE EFFECT OF END RESTRAINT CONDITION ON THE FRACTURE MECHANISM AND STRENGTH OF ROCK-LIKE MATERIALS	72-0134
EFFECT	THE EFFECT OF GRAVITY UPON THE MELT-THROUGH TIME OF A SOLID SUBJECTED TO A HIGH INTENSITY LASER	72-0693
EFFECT	THE EFFECT OF IN PLANE STATIC STRESSES ON THE VIBRATION CHARACTERISTICS OF CURVED SANDWICH PLATES	71-0588
EFFECT	THE EFFECT OF INITIAL DISPLACEMENTS OF PROBLEMS OF LARGE DEFLECTION AND STABILITY	67-0048
EFFECT	THE EFFECT OF LUMPED PARAMETERS ON BEAM FREQUENCIES	63-0012
EFFECT	THE EFFECT OF STRESS HISTORY ON THE RESILIENT RESPONSE OF SOILS	73-0914
EFFECT	THE PIECEWISE LINEAR ANALYSIS OF TWO CONNECTED STRUCTURES INCLUDING THE EFFECT OF CLEARANCE AT THE C	71-0379
EFFECT	THE USE OF EIGEN-DERIVATIVES TO PREDICT THE EFFECT OF DESIGN CHANGES UPON THE DYNAMIC CHARACTERISTIC	74-0737
EFFECT	THEORETICAL STUDY OF THE EFFECT OF THE INTERFACE ON COMPOSITE TOUGHNESS	73-0386
EFFECT	THREE-DIMENSIONAL FINITE ELEMENT METHOD OF THERMOELASTOPLASTICITY WITH CREEP EFFECT	73-0837
EFFECTIVE	EFFECTIVE STRESS-STRAIN-STRENGTH MODEL FOR SOILS	75-0056
EFFECTIVE	LIMITATIONS ON THE APPLICATION OF EFFECTIVE BREADTH CONCEPT IN BOX GIRDER DESIGN	74-0493
EFFECTIVE	ON THE DETERMINATION OF EFFECTIVE ELASTIC-PLASTIC PROPERTIES FOR THE EQUIVALENT SOLID PLATE ANALYSI	74-0494
EFFECTIVE	ON THE DETERMINATION OF EFFECTIVE MODULI OF COMPOSITE MATERIALS BY A THREE-DIMENSIONAL FINITE ELEMEN	74-1123
EFFECTIVENESS	EFFECTIVENESS OF EXCESSIVE NODAL CONTINUITIES IN THE FINITE ELEMENT ANALYSIS OF THIN RECTANGULAR AND	70-0346
EFFECTIVENESS	FINITE ELEMENT PREDICTION OF FIREPROOFING EFFECTIVENESS	72-0047
EFFECTIVENESS	THE EFFECTIVENESS OF EXCESSIVE NODAL CONTINUITIES IN THE FINITE ELEMENT ANALYSIS OF THIN RECTANGULAR	70-0105
EFFECTS	A FINITE ELEMENT ANALYSIS INCLUDING TRANSVERSE SHEAR EFFECTS FOR APPLICATIONS TO LAMINATED PLATES	71-0216
EFFECTS	A FINITE-ELEMENT PROGRAM FOR THE ANALYSIS OF EMBANKMENTS OVER SOFT, SATURATED SOILS INCLUDING CONSOL	73-1160
EFFECTS	A NUMERICAL PROCEDURE FOR TWO DIMENSIONAL HEATING AND MELTING CALCULATIONS, WITH APPLICATIONS TO LAS	72-0708
EFFECTS	A UNIFIED FINITE ELEMENT APPROACH TO COLUMNS SUBJECTED TO NONCONSERVATIVE LOADS, WITH DAMPING EFFECT	74-0795
EFFECTS	AN INVESTIGATION OF THE EFFECTS OF SQUARE CUTOUTS ON THE NATURAL FREQUENCIES AND MODE SHAPES OF RECT	71-0638
EFFECTS	ANALYSIS OF LOADING AND THERMAL EFFECTS ON FUELLED GRAPHITE BRICKS FOR A HIGH-TEMPERATURE GAS COOLED	71-0085
EFFECTS	ANALYSIS OF SOIL-STRUCTURE INTERACTION EFFECTS UNDER SEISMIC EXCITATION	71-0064
EFFECTS	ANALYSIS OF THE EFFECTS ON LIFE OF LEADING-EDGE HOLES IN AN AIRFOIL SUBJECTED TO ARBITRARY SPANWISE	75-0700
EFFECTS	ANALYSIS OF VISCOPLASTIC EFFECTS IN PRESSURE VESSELS BY THE FINITE ELEMENT METHOD	74-0573
EFFECTS	APPLICATION OF FINITE ELEMENT METHOD TO CYCLIC LOADING OF ELASTIC-PLASTIC STRUCTURES CONTAINING EFFE	73-0103
EFFECTS	AUTOMATED EVALUATION OF INTERCEPTOR ROCKET MOTOR DESIGNS UNDER COMBINED OPERATIONAL AND NUCLEAR EFFE	74-0698
EFFECTS	BEHAVIOR OF A BI-MATERIAL CYLINDER INCLUDING THE EFFECTS OF FRACTURE	69-0466
EFFECTS	BLOWDOWN EFFECTS ON A FUEL ROD	75-0575
EFFECTS	DEFORMATION OF THE DIASTOLIC LEFT VENTRICLE - 2. NONLINEAR GEOMETRIC EFFECTS	74-0974
EFFECTS	DIRECT BEAM STIFFNESS MATRIX CALCULATIONS INCLUDING SHEAR EFFECTS	62-0003
EFFECTS	DYNAMIC SHELL THEORY COUPLING THICKNESS STRESS WAVE EFFECTS WITH GROSS STRUCTURAL RESPONSE	73-0379
EFFECTS	EARTH VIBRATION EFFECTS AND ABATEMENT FOR MILITARY FACILITIES. REPORT 3. ANALYSIS METHOD FOR FOOTING	72-0806
EFFECTS	EARTHQUAKE RESPONSE OF GRAVITY DAMS INCLUDING RESERVOIR INTERACTION EFFECTS	72-0675
EFFECTS	EFFECTS OF A PEAK OVERLOAD ON FATIGUE CRACK PROPAGATION RATE-PLANE STRESS ANALYSIS OF CRACKED PLATE	74-1365
EFFECTS	EFFECTS OF FLEXIBLE CONNECTIONS ON BODY STRUCTURAL RESPONSE	74-0410
EFFECTS	EFFECTS OF FOUNDATION ROTATION ON SEISMIC INERTIA FORCES OF NUCLEAR POWER PLANT STRUCTURES	73-0177
EFFECTS	EFFECTS OF GEOMETRY CHANGE ON ESSENTIAL FEATURES OF INELASTIC BEHAVIOUR	73-0448
EFFECTS	EFFECTS OF INSERTS ON THE INELASTIC BEHAVIOUR OF CYLINDRICAL MATERIALS LOADED BETWEEN ROUGH END-PLATES	73-0318
EFFECTS	EFFECTS OF LOCAL STRENGTH VARIATIONS ON PLASTIC ENCLAVE DEVELOPMENT AT CRACK TIPS	72-0822
EFFECTS	EFFECTS OF MATERIAL NONLINEARITIES ON VIBRATION	73-0323
EFFECTS	EFFECTS OF QUADRATURE ERRORS IN FINITE ELEMENT APPROXIMATION OF STEADY STATE, EIGENVALUE, AND PARABO	72-0343
EFFECTS	EFFECTS OF RISE TIME AND DAMPING ON FINITE ELEMENT ANALYSIS OF RESPONSE OF STRUCTURES	71-0087
EFFECTS	EFFECTS OF SHEAR DEFORMATION, ROTARY INERTIA, AND ELASTICITY OF THE SUPPORT ON THE RESONANCE FREQUEN	75-0369
EFFECTS	EFFECTS OF SHEAR WALLS ON DYNAMIC RESPONSE OF FRAMES	69-0230
EFFECTS	EFFECTS OF SHOE FORCE GEOMETRY ON HEAVY DUTY INTERNAL SHOE BRAKE PERFORMANCE	68-0219
EFFECTS	EFFECTS OF SOURCE STRUCTURE AND SITE RESPONSE ON GROUND MOTION IN THE NEAR FIELD OF EARTHQUAKES	73-0732
EFFECTS	EFFECTS OF SPECIMEN RADIUS ON THE STRESS STATE NEAR A ROOF BOLT ANCHOR: A FINITE ELEMENT DETERMINATI	73-0229
EFFECTS	EFFECTS OF TURBULENCE ON FLOOD FLOW EXPLORED BY FINITE ELEMENT ANALYSIS.	72-0410
EFFECTS	EVALUATION OF TORSIONAL EFFECTS IN THE SEISMIC ANALYSIS OF SYMMETRIC AND UNSYMMETRIC NUCLEAR POWER P	73-0175
EFFECTS	EXPERIMENTAL STUDY OF FLEXIBILITY FACTORS AND STRESSES OF WELDING ELBOWS WITH END EFFECTS	73-0422
EFFECTS	FINITE ELEMENT ANALYSIS OF PLATES ALLOWING FOR TRANSVERSE SHEAR DEFORMATION EFFECTS	73-0996
EFFECTS	FINITE ELEMENT ANALYSIS OF THE EFFECTS OF INSTALLATION ON PILE LOAD-SETTLEMENT BEHAVIOUR	74-0763
EFFECTS	FINITE ELEMENT ANALYSIS OF FATIGUE CRACK PROPAGATION - INCLUDING THE EFFECTS OF CRACK CLOSURE	74-1112
EFFECTS	FINITE ELEMENT THEORY OF PLATES AND SHELLS INCLUDING TRANSVERSE SHEAR STRAIN EFFECTS	70-0395
EFFECTS	FINITE-ELEMENT ANALYSIS OF EDGE EFFECTS IN ANGLE-PLY COMPOSITE LAMINATES	74-0843
EFFECTS	FUEL ROD DEFORMATIONS - A TIME DEPENDENT FINITE ELEMENT ANALYSIS ON AXISYMMETRIC FUEL RODS, INCLUDIN	70-0414

EFFECTS	INTERNAL PRESSURE EFFECTS ON THE VIBRATION OF PARTIALLY FILLED ELASTIC TANKS	75-0241
EFFECTS	MASS LOADING EFFECTS ON VIBRATION OF RING AND SHELL STRUCTURES	68-0245
EFFECTS	MATRIX DISPLACEMENT ANALYSIS OF SHELLS AND PLATES INCLUDING TRANSVERSE SHEAR STRAIN EFFECTS	72-0647
EFFECTS	NEUARTIGE ANWENDUNG GEOMECHANISCHER UNTERSUCHUNGSMETHODEN ZUR BEURTEILUNG DER ABBAUWIRKUNGEN IM GANG	75-0530
EFFECTS	NON-LINEAR ANALYSIS OF PLATES CONSIDERING GEOMETRIC AND MATERIAL EFFECTS	71-0702
EFFECTS	NON-LINEAR EFFECTS OF HIGH TEMPERATURE ON THE VIBRATION OF A SIMPLY SUPPORTED BEAM WITH A CENTRAL MA	73-0838
EFFECTS	ON SHELL EFFECTS ON FERRO-CEMENT VESSELS	73-0455
EFFECTS	ON THE EFFECTS OF QUADRATURE ERRORS IN THE FINITE ELEMENT METHOD	72-0258
EFFECTS	ON THE EXTREMAL PROPERTIES OF THE SOLUTION IN DYNAMICS OF RIGID VISCOPLASTIC BODIES ALLOWING FOR LAR	73-1080
EFFECTS	RADIAL INERTIA EFFECTS ON AN IDEALLY PLASTIC CIRCULAR PLATE UNDER IMPULSIVE AXIAL COMPRESSION	69-0324
EFFECTS	RESPONSE OF STRUCTURES TO COMBINED BLAST EFFECTS	73-0424
EFFECTS	SEEPAGE AND GROUNDWATER EFFECTS ASSOCIATED WITH EXPLOSIVE CRATERING	72-0656
EFFECTS	SEISMIC EFFECTS ON A PROPOSED UNDERGROUND REACTOR FACILITY	73-0953
EFFECTS	SEISMIC RESPONSE EFFECTS ON EMBEDDED STRUCTURES	72-0902
EFFECTS	SHORT COMMUNICATIONS: TORSION OF AXI-SYMMETRIC SOLIDS OF VARIABLE DIAMETER - INCLUDING ACCELERATION	74-0203
EFFECTS	SOME CREEP EFFECTS IN STRESS ANALYSIS WITH PARTICULAR REFERENCE TO CONCRETE PRESSURE VESSELS	69-0284
EFFECTS	SOME EFFECTS OF SYSTEM IDEALIZATIONS, SINGULARITIES AND MESH PATTERNS ON FINITE ELEMENT SOLUTIONS	74-0939
EFFECTS	STRESS ANALYSIS BY THE FINITE ELEMENT METHOD - THERMAL EFFECTS	67-0022
EFFECTS	STRESS-WAVE PROPAGATION AND DYNAMIC EFFECTS IN SHELLS (SPADES) MICROMECHANICAL STRESS-WAVE RESPONSE	72-0851
EFFECTS	STRUCTURAL EFFECTS OF CONFINED DYNAMIC LOADS	72-0151
EFFECTS	STUDY OF EFFECTS OF DESIGN DETAILS ON STRUCTURAL RESPONSE TO ACOUSTIC EXCITATION	72-0864
EFFECTS	THE EFFECTS OF SOIL PARAMETERS AND BOUNDARY CONDITIONS ON THE CONSOLIDATION OF AN ELASTIC LAYER	70-0422
EFFECTS	THE EFFECTS OF THIN CORES AND NONHOMOGENEITY ON DYNAMIC RESPONSE OF EARTH DAMS	67-0021
EFFECTS	THREE DIMENSIONAL FINITE ELEMENT ANALYSIS OF THICK LAMINATED COMPOSITES - INCLUDING INTERLAMINAR AND	73-0882
EFFECTS	THREE DIMENSIONAL RUPTURE ANALYSIS OF A PRESTRESSED CONCRETE PRESSURE VESSEL INCLUDING CREEP EFFECTS	73-0826
EFFECTS	THREE DIMENSIONAL RUPTURE ANALYSIS OF A PRESTRESSED CONCRETE PRESSURE VESSEL INCLUDING CREEP EFFECTS	74-0538
EFFECTS	THREE-DIMENSIONAL FINITE ELEMENT ANALYSIS OF THICK LAMINATED COMPOSITES - INCLUDING INTERLAMINAR AND	74-1151
EFFECTS	TRANSIENT THERMAL STRESS TECHNIQUES IN NUCLEAR REACTORS. APPLIED MECHANICS ASPECTS OF NUCLEAR EFFECT	71-0474
EFFECTS	VORAUSSAGE DES GEBIRGSDRUCKES UND SEINER AUSWIRKUNGEN AUF STREBEN UND STRECKEN IM STEINKOHLENBERGBAU	73-0679
EFFECTS	ZASTOSOWANIE METODY SZTYWNYCH ELEMENTOW SKONCZONYCH DO OBLICZEN ORGAN WALOW OKRETOWYCH Z UWZGLEDNIENI	74-0962
EFFICIENCY	COMPARATIVE EFFICIENCY OF CERTAIN FINITE ELEMENT AND FINITE DIFFERENCE METHODS FOR A HYPERBOLIC PROB	73-0965
EFFICIENCY	COMPUTATIONAL EFFICIENCY ASPECTS OF THE FINITE ELEMENT NONLINEAR ANALYSIS OF SHELLS	75-0031
EFFICIENCY	EFFICIENCY EVALUATION OF A TWO-DIMENSIONAL INCOMPATIBLE FINITE ELEMENT	73-0988
EFFICIENCY	FINITE ELEMENT ANALYSIS OF PERFORATED PLATES CONTAINING TRIANGULAR PENETRATION PATTERNS OF 5 AND 10	75-0216
EFFICIENCY	FINITE ELEMENT ANALYSIS OF PERFORATED PLATES CONTAINING TRIANGULAR PENETRATION PATTERNS OF 5 AND 10	75-0520
EFFICIENCY	ON IMPROVING EFFICIENCY OF AN ALGORITHM FOR STRUCTURAL OPTIMIZATION AND A USERS MANUAL FOR PROGRAM T	74-0778
EFFICIENCY	PRAKTISCHE RESULTATEN, DOEL EN NUT VAN HET ONTWERPEN VAN MECHANISCHE ELEMENTEN MET BEHULP VAN COMPUT	73-0400
EFFICIENCY	RELATIVE EFFICIENCY OF FINITE DIFFERENCE AND FINITE ELEMENT METHODS - 1. HYPERBOLIC PROBLEMS AND SPL	74-0986
EFFICIENCY	REVIEW OF NASTRAN DEVELOPMENT RELATIVE TO EFFICIENCY OF EXECUTION	73-0938
EFFICIENCY	THE COMPARATIVE EFFICIENCY OF CERTAIN FINITE ELEMENT AND FINITE DIFFERENCE SCHEMES FOR A HYPERBOLIC	74-0910
EFFICIENCY	THE COMPUTATIONAL EFFICIENCY OF A NEW MINIMIZATION ALGORITHM FOR EIGENVALUE ANALYSIS	71-0595
EFFICIENCY	THE RELATIVE EFFICIENCY OF FINITE DIFFERENCE AND FINITE ELEMENT METHODS. 1. HYPERBOLIC PROBLEMS AND	74-0911
EFFICIENT	A COMPUTATIONAL TECHNIQUE FOR THE EFFICIENT HANDLING OF LARGE MATRICES	73-1083
EFFICIENT	A MORE EFFICIENT FRONT SOLUTION: ALLOCATING ASSEMBLY LOCATIONS BY LONGEVITY CONSIDERATIONS	73-1069
EFFICIENT	AN EFFICIENT ITERATIVE PROCEDURE FOR USE WITH THE FINITE-ELEMENT METHOD	73-1145
EFFICIENT	AN EFFICIENT TRIANGULAR SHELL ELEMENT	70-0382
EFFICIENT	AN EFFICIENT TRIANGULAR SHELL ELEMENT	70-0441
EFFICIENT	AUTOMATIC GENERATION OF FINITE ELEMENT MESH WITHIN A BANDWIDTH AND AN EFFICIENT SOLUTION OF THE BAND	73-0338
EFFICIENT	COMPUTATIONALLY EFFICIENT FINITE ELEMENT ANALYSIS OF VISCOUS FLOW PROBLEMS	74-1050
EFFICIENT	COMPUTATIONALLY EFFICIENT FINITE ELEMENT ANALYSIS OF VISCOUS FLOW PROBLEMS	74-1258
EFFICIENT	EFFICIENT COMPUTER ANALYSIS OF LARGE THREE-DIMENSIONAL PIPING SYSTEMS BY THE FINITE ELEMENT METHOD	72-0378
EFFICIENT	EFFICIENT FINITE ELEMENT ANALYSIS OF RECTANGULAR AND SKEW LAMINATED PLATES	73-0436
EFFICIENT	EFFICIENT FINITE ELEMENT METHODS FOR STRESS INTENSITY FACTOR	75-0148
EFFICIENT	EFFICIENT FINITE-ELEMENT METHODS FOR STRESS INTENSITY FACTORS USING WEIGHT FUNCTIONS	75-0411
EFFICIENT	EFFICIENT FINITE-ELEMENT METHODS FOR STRESS INTENSITY FACTORS USING WEIGHT FUNCTIONS - REPLY	75-0421
EFFICIENT	EFFICIENT FORMULATIONS OF THE FINITE ELEMENT METHOD IN LINEAR AND NONLINEAR FRACTURE MECHANICS	75-0403
EFFICIENT	EFFICIENT METHOD FOR SOLUTION OF VISCOELASTIC PROBLEMS	73-0232
EFFICIENT	EFFICIENT SOLUTION OF LOAD-DEFLECTING EQUATIONS	69-0396
EFFICIENT	EFFICIENT SOLUTION PROCESSES FOR FINITE ELEMENT ANALYSIS OF TRANSIENT HEAT CONDUCTION	69-0014
EFFICIENT	EFFICIENT TECHNIQUES FOR FINITE ELEMENT ANALYSIS OF ELECTRIC MACHINES	73-0134
EFFICIENT	ELASTO-PLASTIC ANALYSIS USING AN EFFICIENT FORMULATION OF THE FINITE ELEMENT METHOD	75-0012
EFFICIENT	PROCEDURE FOR EFFICIENT GENERATING, CHECKING, AND DISPLAYING NASTRAN INPUT AND OUTPUT DATA FOR ANALY	71-0098
EFFICIENT	THE EFFICIENT USE OF COMPUTER RESOURCES	72-0090
EFFLUENT	FINITE ELEMENT SOLUTION FOR EFFLUENT DISPERSION	74-1371
EFFLUENTS	MATHEMATICAL MODEL OF HEATED EFFLUENTS IN COASTAL WATER	73-0921
EFORTURI	ASUPRA ANALIZEI PRIN METODA ELEMENTULUI FINIT A STARILOR DE EFORTURI SI DEFORMATII DIN MASIVELE DE R	70-0471
EIGEN	NUMERICAL STUDY OF EIGEN OSCILLATIONS OF A SCREW PINCH WITH FINITE ELEMENTS	74-0159
EIGEN-DERIVATIV	THE USE OF EIGEN-DERIVATIVES TO PREDICT THE EFFECT OF DESIGN CHANGES UPON THE DYNAMIC CHARACTERISTIC	74-0737
EIGENMODES	DETERMINATION OF EIGENMODES OF HOMOGENEOUS PLASMA WITH ELLIPTICAL SECTION USING BI-DIMENSIONAL FINIT	74-0553
EIGENMODES	EIGENMODES OF CYLINDRICAL PLASMA USING FINITE ELEMENT METHOD	74-0174
EIGENPROBLEM	ON A COMBINED STURM SEQUENCE AND INVERSE ITERATION TECHNIQUE FOR EIGENPROBLEM SOLUTION OF SPINNING S	73-0413
EIGENPROBLEMS	ACCURACY OF FINITE ELEMENT EIGENPROBLEMS	71-0088
EIGENPROBLEMS	CONDENSATION OF FINITE ELEMENT EIGENPROBLEMS	72-0168
EIGENPROBLEMS	GRADIENT METHODS FOR FINITE ELEMENT EIGENPROBLEMS	69-0103
EIGENPROBLEMS	OPTIMAL GRADIENT MINIMIZATION SCHEME FOR FINITE ELEMENT EIGENPROBLEMS	72-0292
EIGENPROBLEMS	THE APPLICATION OF GRADIENT MINIMIZATION METHODS AND HIGHER ORDER DISCRETE ELEMENTS TO SHELL BUCKLIN	71-0237
EIGENSCHAFTEN	FINEL, UNIVERSELLES PROGRAMMSYSTEM ZUR BERECHNUNG DER ELASTISCHEN EIGENSCHAFTEN VON MASCHINENBAUTEIL	72-0132
EIGENSOLUTIONS	A TEST ON CONVERGENCE OF EIGENSOLUTIONS BY SIMULTANEOUS ITERATION METHOD	73-0317
EIGENSPANNUNGSZ	MESSVERFAHREN RECHENVERFAHREN ZUR ERMITTLUNG DREIACHSIGER EIGENSPANNUNGSZUSTAENDE	73-0507
EIGENVALUE	A COMPARATIVE STUDY OF THE CONSISTENT AND SIMPLIFIED FINITE ELEMENTS ANALYSIS OF EIGENVALUE PROBLEMS	70-0034
EIGENVALUE	A COMPARATIVE STUDY OF THE CONSISTENT AND SIMPLIFIED FINITE ELEMENT ANALYSES OF EIGENVALUE PROBLEMS	70-0373
EIGENVALUE	A FINITE ELEMENT EIGENVALUE SYSTEM	68-0192
EIGENVALUE	AUTOMATIC MASTERS FOR EIGENVALUE ECONOMIZATION	75-0557
EIGENVALUE	COMPLEX EIGENVALUE SOLUTION TO A SPINNING SKYLAB PROBLEM	71-0207
EIGENVALUE	CONVERGENCE OF EIGENVALUE SOLUTIONS IN CONFORMING PLATE BENDING FINITE ELEMENTS	72-0195
EIGENVALUE	CONVERGENCE STUDIES OF EIGENVALUE SOLUTIONS USING TWO FINITE PLATE ELEMENTS	70-0085
EIGENVALUE	DISCRETIZATION AND ROUND-OFF ERROR IN THE FINITE ELEMENT ANALYSIS OF ELLIPTIC BOUNDARY VALUE PROBLEM	71-0514
EIGENVALUE	EFFECTS OF QUADRATURE ERRORS IN FINITE ELEMENT APPROXIMATION OF STEADY STATE, EIGENVALUE, AND PARABO	72-0343
EIGENVALUE	EIGENVALUE ANALYSIS BY FINITE ELEMENT METHODS	69-0377
EIGENVALUE	EIGENVALUE APPROXIMATION BY FINITE ELEMENT METHOD	73-0039
EIGENVALUE	EIGENVALUE BOUNDS FOR FINITE ELEMENT MODELS	72-0570
EIGENVALUE	EIGENVALUE BOUNDS FOR FINITE ELEMENT MODELS	72-0932
EIGENVALUE	EIGENVALUE ECONOMISERS IN VIBRATION PROBLEMS	63-0013
EIGENVALUE	EIGENVALUE PROBLEMS WITH A ' MIXED ' PLATE ELEMENT	69-0081
EIGENVALUE	EIGENVALUE PROBLEMS (A - LAMBDA B) X EQUALS 0 FOR SYMMETRIC MATRICES OF HIGH ORDER	74-0205
EIGENVALUE	ERROR BOUNDS FOR EIGENVALUE ANALYSIS BY ELIMINATION OF VARIABLES	71-0594
EIGENVALUE	KRON'S METHOD: AN ALGORITHM FOR THE EIGENVALUE ANALYSIS OF LARGE-SCALE STRUCTURAL SYSTEMS	73-0758
EIGENVALUE	ON THE NUMERICAL CONVERGENCE OF MATRIX EIGENVALUE PROBLEMS DUE TO CONSTRAINT CONDITIONS	74-1057
EIGENVALUE	ON THE NUMERICAL SOLUTION OF LARGE EIGENVALUE PROBLEMS ARISING IN PANEL FLUTTER ANALYSIS BY THE FINI	74-0942
EIGENVALUE	RECENT ADVANCES IN NUMERICAL ANALYSIS OF STRUCTURAL EIGENVALUE PROBLEMS	73-0316
EIGENVALUE	SOLUTION METHODS FOR EIGENVALUE PROBLEMS IN STRUCTURAL MECHANICS	73-1095
EIGENVALUE	SOLUTION OF EIGENVALUE PROBLEMS BY STURM SEQUENCE METHOD.	72-0400

EIGENVALUE	STRUCTURAL EIGENVALUE PROBLEMS: ELIMINATION OF UNWANTED VARIABLES	65-0050
EIGENVALUE	TESTING AND ASSESSING FINITE ELEMENTS BY AN EIGENVALUE TECHNIQUE	68-0188
EIGENVALUE	THE COMPUTATIONAL EFFICIENCY OF A NEW MINIMIZATION ALGORITHM FOR EIGENVALUE ANALYSIS	71-0595
EIGENVALUE	THE EIGENVALUE PROBLEM	72-0506
EIGENVALUE	THE MATRIX EIGENVALUE PROBLEM FOR DAMPED STRUCTURAL VIBRATION	66-0121
EIGENVALUES	A BOUND THEOREM IN EIGENVALUES AND ITS PRACTICAL APPLICATIONS	71-0128
EIGENVALUES	A METHOD TO GENERATE BOTH UPPER AND LOWER BOUNDS TO PLATE EIGENVALUES BY CONFORMING DISPLACEMENT FIN	71-0166
EIGENVALUES	AN ECONOMICAL METHOD FOR DETERMINING THE SMALLEST EIGENVALUES OF LARGE LINEAR SYSTEMS	71-0551
EIGENVALUES	APPROXIMATION AND BOUNDS FOR EIGENVALUES OF VARIOUS DOMAINS USING THE INTERMEDIATE PROBLEM TECHNIQUE	72-0711
EIGENVALUES	APPROXIMATION OF STEKLOV EIGENVALUES OF NON-SELFADJOINT SECOND ORDER ELLIPTIC OPERATORS	72-0509
EIGENVALUES	ASSESSMENT OF ACCURACIES OF FINITE ELEMENT EIGENVALUES	72-0075
EIGENVALUES	ASSESSMENT OF ACCURACIES OF FINITE ELEMENT EIGENVALUES	72-0747
EIGENVALUES	BOUNDS AND ERROR CONTROL FOR EIGENVALUES	73-0319
EIGENVALUES	BOUNDS FOR EIGENVALUES IN SOME VIBRATION AND STABILITY PROBLEMS.	72-0390
EIGENVALUES	BOUNDS ON EXTERNAL EIGENVALUES OF FINITE ELEMENT STIFFNESS AND MASS MATRICES AND THEIR SPECTRAL COND	72-0256
EIGENVALUES	DEPENDENCE OF PLATE-BENDING FINITE ELEMENT DEFLECTIONS AND EIGENVALUES ON POISSONS RATIO	74-0567
EIGENVALUES	DEPENDENCE OF PLATE-BENDING FINITE ELEMENT DEFLECTIONS AND EIGENVALUES ON POISSONS RATIO	74-0583
EIGENVALUES	EIGENVALUES OF COLLINEAR BEAM STRUCTURES USING FINITE ELEMENT TECHNIQUES AND VARIOUS DYNAMIC REPRESE	67-0124
EIGENVALUES	EIGENVALUES OF LARGE MATRICES	69-0341
EIGENVECTOR	SIMULTANEOUS EIGENVECTOR ITERATION APPLIED TO LARGE SYSTEMS	74-0680
EIGENVEKTOREN	EINIGE VERFAHREN ZUR BERECHNUNG VON EIGENWERTEN UND EIGENVEKTOREN VON NICHT HERMITESCHEN MATRIZEN U	72-0646
EIGENWERTEN	EINIGE VERFAHREN ZUR BERECHNUNG VON EIGENWERTEN UND EIGENVEKTOREN VON NICHT HERMITESCHEN MATRIZEN U	72-0646
EIGHT-NODE	MESH 3D: A THREE-DIMENSIONAL FINITE ELEMENT MESH GENERATOR PROGRAM FOR EIGHT-NODE ISOPARAMETRIC ELEM	73-0972
EINACHSIGEN	BERECHNUNG EINES EINACHSIGEN STAHLDRAHT-KUNSTSTOFFVERBUNDES UNTER EINSATZ DER FINIT-ELEMENT METHODE	72-0644
EINEM	THEORIE UND EXPERIMENT ZU EINEM NEUEN INJEKTIONSVERFAHREN. (THEORY AND EXPERIMENT OF A NEW GROUTING	74-0991
EINFACH-SYMMETR	ZUR BERECHNUNG EINFACH-SYMMETRISCHER I-TRAEGER NACH DER THEORIE - 2. ORDNUNG UNTER BERUECKSICHTIGUNG	75-0510
EINFACHE	EINE EINFACHE KONSTRUKTION VON KOORDINATENFUNKTIONEN FUR DIE NUMERISCHE LOSUNG ZWEIDIMENSIONALER RAN	66-0134
EINFACHE	NICHTLINEARE DIRICHLET PROBLEME UND EINFACHE FINITE ELEMENT VERFAHREN	74-1263
EINFACHES	DESH, EIN EINFACHES FLACHES SCHALENELEMENT UEBER DREIECKIGEM GRUNDRISS. (DESH, A SIMPLE TRIANGULAR S	75-0304
EINFACHES	EINFACHES NAEHERUNGSVERFAHREN FUR DIE GRENZTRAGFAEHIGKEITSANLYSE VON SPANNBETON-REAKTORDRUCKBEHAELTE	73-0854
EINFLUSS	EFFECT OF DISCONTINUITIES ON STRESSES AND DEFORMATIONS OF ROCK AROUND A TUNNEL. (EINFLUSS VON DISKON	70-0170
EINFULUSS	UEBER DAS VERHALTEN VON SPANNBETONDRUCKBEHAELTRERN IM GRENZTRAGFAEHIGKEITSZUSTAND UND DESSEN EINFUL	73-0855
EINIGE	EINIGE ALLGEMEINE PROGRAMMSYSTEME FUR FINITE ELEMENTE	73-0581
EINIGE	EINIGE ENTWICKLUNGEN IN DER MATRIZEN-VERSCHIEBUNGSMETHODE	69-0341
EINIGE	EINIGE VERFAHREN ZUR BERECHNUNG VON EIGENWERTEN UND EIGENVEKTOREN VON NICHT HERMITESCHEN MATRIZEN U	72-0646
EINWIRKUNGEN	BERECHNUNGSMETHODEN ZUR ERMITTLUNG VON BELASTUNGSBEDINGUNGEN UND GRENZTRAGFAEHIGKEITSVERHALTEN VON K	73-0858
EINZELLASTEN	DER GERISSENE ZUSTAND ZWEISEITIG GELAGERTER PLATTEN UNTER EINZELLASTEN - NICHTLINEARE BERECHNUNG MIT	75-0858
EINZELNEN	ZUSAMMENWIRKEN VON EINZELNEN FERTIGTEILEN ALS GROSSFLAECHIGE SCHEIBE (INTERACTION OF INDIVIDUAL PREF	73-0552
ELAD	ELAD PROGRAM MANUAL, A COMPUTER PROGRAM FOR THE ELASTIC ANALYSIS OF DOME STRUCTURES AND AXISYMMETRIC	69-0064
ELAS	COMPUTER WORD LENGTH AND SOLUTION ACCURACY. EXPERIENCE WITH ELAS AND THE IBM 360	73-0335
ELAS	ELAS - A GENERAL PURPOSE COMPUTER PROGRAM FOR THE EQUILIBRIUM PROBLEMS OF LINEAR STRUCTURES	68-0158
ELAS	ELAS: A GENERAL-PURPOSE COMPUTER PROGRAM FOR THE EQUILIBRIUM PROBLEMS OF LINEAR STRUCTURES. VOLUME 2	69-0462
ELASTIC	A COMPUTER METHOD FOR THE SECOND ORDER ELASTIC ANALYSIS OF PLANE FRAMED STRUCTURES	73-0310
ELASTIC	A COMPUTER PROGRAM FOR THE ANALYSIS OF LINEARLY ELASTIC PLANE-STRESS, PLANE-STRAIN PROBLEMS	68-0110
ELASTIC	A COMPUTER PROGRAMME FOR ELASTIC PLASTIC CALCULATIONS WITH DUAL PERTURBATION EXPANSION AND VARIATION	75-0882
ELASTIC	A CURVED FINITE ELEMENT FOR THIN ELASTIC SHELLS	69-0486
ELASTIC	A FINITE ELEMENT ANALYSIS OF THE ELASTIC BEHAVIOR OF BEAM COLUMNS	66-0122
ELASTIC	A FINITE ELEMENT ANALYSIS OF AN AIRCRAFT LANDING MAT ON AN ELASTIC FOUNDATION	68-0280
ELASTIC	A FINITE ELEMENT APPROACH FOR THE ANALYSIS OF RANDOMLY EXCITED COMPLEX ELASTIC STRUCTURES	67-0057
ELASTIC	A FINITE ELEMENT APPROACH TO ELASTIC SOIL-STRUCTURE INTERACTION	70-0013
ELASTIC	A FINITE ELEMENT APPROACH TO SCATTERING FROM ELASTIC SPHERICAL SHELLS	74-0702
ELASTIC	A FINITE ELEMENT FORMULATION FOR LARGE DEFLECTION ELASTIC STRUCTURAL ANALYSIS APPLIED TO PLATE BENDI	70-0485
ELASTIC	A FINITE ELEMENT METHOD OF SOLUTION FOR LINEARLY ELASTIC BEAM-COLUMNS	66-0117
ELASTIC	A FINITE ELEMENT METHOD FOR GEOMETRICALLY NONLINEAR LARGE DISPLACEMENT PROBLEMS IN THIN, ELASTIC PLA	69-0005
ELASTIC	A METHOD OF FINITE DEFLECTION ELASTIC PLASTIC ANALYSIS OF FRAMED STRUCTURES SUBJECTED TO LARGE AXIAL	69-0239
ELASTIC	A METHOD OF SOLUTION FOR THE COMPLICATED BUCKLING PROBLEMS OF ELASTIC PLATES WITH COMBINED USE OF RA	68-0097
ELASTIC	A NOTE ON NUMERICAL COMPUTATION OF ELASTIC CONTACT PROBLEMS	75-0621
ELASTIC	A NOTE ON THE ANALYSIS OF NONLINEAR DYNAMICS OF ELASTIC MEMBRANES BY THE FINITE ELEMENT METHOD	74-0341
ELASTIC	A NOTE ON THE ELASTIC ANALYSIS OF BOX-GIRDER DIAPHRAGMS USING FINITE ELEMENTS	74-1230
ELASTIC	A NOTE ON THE FINITE DEFORMATIONS OF A THICK ELASTIC SHELL OF REVOLUTION	71-0193
ELASTIC	A NUMERICAL METHOD FOR THE ANALYSIS OF LONGITUDINAL ELASTIC PLASTIC STRESS WAVE PROPAGATION	73-0526
ELASTIC	A NUMERICAL SOLUTION OF A CLASS OF PROBLEMS OF FINITE ELASTIC DEFORMATION	66-0097
ELASTIC	A STUDY OF FREE-VIBRATIONS OF THIN ELASTIC SHELLS SUBJECTED TO INITIAL LOADS USING FINITE-ELEMENTS	73-1143
ELASTIC	A STUDY OF THE PLANE STRESS OR STRAIN FINITE ELEMENT ANALYSIS FOR SOLUTION OF STRESS DISTRIBUTION IN	69-0487
ELASTIC	A THEORETICAL AND NUMERICAL COMPARISON OF ELASTIC NONLINEAR FINITE ELEMENT METHODS	72-0709
ELASTIC	A VARIATIONAL FORMULATION FOR DISCONTINUOUS FIELDS IN LINEARLY ELASTIC FINITE ELEMENT APPLICATIONS	72-0962
ELASTIC	ACCURACY AND CONVERGENCE OF MIXED FINITE-ELEMENT APPROXIMATIONS OF THIN BARS, MEMBRANES, AND PLATES	73-1120
ELASTIC	ACCURACY OF THE FINITE ELEMENT ANALYSIS OF THE ELASTIC PLATE WITH A CIRCULAR HOLE	73-0155
ELASTIC	ACCURATE CALCULATION OF STRESSES IN BEAMS, DISKS AND PLATES BY MEANS OF FINITE ELEMENT METHODS AND I	70-0133
ELASTIC	AN AUTOMATIC SYSTEM FOR KINEMATIC ANALYSIS ASKA HIGH SPEED COMPUTING OF ELASTIC STRUCTURES	70-0358
ELASTIC	AN ELASTIC FINITE-ELEMENT ANALYSIS OF THE BUCKLING OF ECCENTRICALLY STIFFENED PLATES AND SHELLS	73-1136
ELASTIC	AN EQUILIBRIUM STRESS FIELD MODEL FOR FINITE ELEMENT SOLUTIONS OF TWO-DIMENSIONAL ELASTIC PROBLEMS	68-0163
ELASTIC	ANALISI NUMERICA DEL COMPORTAMENTO ELASTICO DI UNA STRUTTURA BIDIMENSIONALE PIANA. (NUMERICAL ANALYS	74-1029
ELASTIC	ANALYSIS OF A CIRCULAR RAFT IN ADHESIVE CONTACT WITH A THICK ELASTIC LAYER	74-1152
ELASTIC	ANALYSIS OF BUILDINGS ON ELASTIC MEDIUM UNDER DYNAMIC OR SEISMIC LOADS	74-0456
ELASTIC	ANALYSIS OF ELASTIC PLATES IN FLEXURE BY A SIMPLIFIED FINITE ELEMENT METHOD	67-0035
ELASTIC	ANALYSIS OF ELASTIC STABILITY OF SHELLS OF REVOLUTION BY THE FINITE ELEMENT METHOD	67-0056
ELASTIC	ANALYSIS OF ELASTIC PLATES BY A SIMPLIFIED FINITE ELEMENT METHOD	67-0156
ELASTIC	ANALYSIS OF ELASTIC SHELLS OF REVOLUTION WITH MEMBRANE AND FLEXURE STRESSES UNDER ARBITRARY LOADING	68-0289
ELASTIC	ANALYSIS OF ELASTICALLY COUPLED SHELLS ON ELASTIC FOUNDATION BY HYBRID METHOD	71-0276
ELASTIC	ANALYSIS OF FINITE DEFORMATIONS OF ELASTIC SOLIDS BY THE FINITE ELEMENT METHOD	70-0096
ELASTIC	ANALYSIS OF FOUNDATIONS ON A SEMI-INFINITE ELASTIC MASS UNDER DYNAMIC LOADS	72-0763
ELASTIC	ANALYSIS OF LARGE DEFORMATIONS OF ELASTIC MEMBRANES BY THE FINITE ELEMENT METHOD	66-0057
ELASTIC	ANALYSIS OF PLATES ON ELASTIC FOUNDATIONS	73-0684
ELASTIC	ANALYSIS OF THE DYNAMICS OF ELASTIC STRUCTURES BY THE FINITE ELEMENT METHOD	72-0893
ELASTIC	ANALYSIS OF THE ELASTIC PLASTIC PROBLEMS BY THE MATRIX DISPLACEMENT METHOD	68-0173
ELASTIC	ANALYSIS OF THERMAL ELASTIC PLASTIC STRESS AND STRAIN DURING WELDING BY FINITE ELEMENT METHOD	71-0382
ELASTIC	APPLICATION OF ELASTIC PLASTIC ANALYSIS TO THE DESIGN OF PRESSURE VESSELS	68-0157
ELASTIC	APPLICATION OF FINITE ELEMENT METHODS TO LARGE SCALE ELASTIC PLASTIC PROBLEMS OF FRACTURE MECHANICS	69-0372
ELASTIC	APPLICATION OF MATRIX DISPLACEMENT METHOD TO LINEAR ELASTIC ANALYSIS OF SHELLS OF REVOLUTION	65-0025
ELASTIC	APPLICATION OF RITZ'S METHOD TO THIN ELASTIC SHELL ANALYSIS	71-0083
ELASTIC	APPLICATION OF THE FINITE ELEMENT METHOD TO THE BUCKLING OF RECTANGULAR PLATES AND PLATES ON ELASTIC	73-0478
ELASTIC	APPLICATION TO PLANE STRESS AND PLANE STRAIN--LINEAR ELASTIC MATERIAL	68-0048
ELASTIC	APPROXIMATIONS AND NUMERICAL ANALYSIS OF FINITE DEFORMATIONS OF ELASTIC SOLIDS	74-0343
ELASTIC	ARCH DAMS ANALYSED BY A LINEAR FINITE ELEMENT SHELL THE FINITE STRIP METHOD IN THE ANALYSIS OF ELAST	68-0189
ELASTIC	AVERAGE ELASTIC PROPERTIES AND STRESS FIELDS FOR COMPOSITE CONTINUA	69-0106
ELASTIC	AVERAGE ELASTIC PROPERTIES AND STRESS FIELDS FOR COMPOSITE MATERIALS	69-0435
ELASTIC	CALCULATION OF THE CRITICAL LOAD OF ELASTIC THIN WALLED BARS BY THE METHOD OF FINITE ELEMENTS	70-0446
ELASTIC	CALCULATION OF THE ELASTIC BUCKLING LOADS OF THIN FLAT REINFORCED PLATES USING TRIANGULAR FINITE ELE	73-1024
ELASTIC	CALCULATION OF THE ELASTIC BUCKLING LOADS OF THIN FLAT REINFORCED PLATES USING TRIANGULAR FINITE ELE	75-0177
ELASTIC	CHARACTERISATION ELASTIQUE D'UN STRATIFIE RESINE-VERRE PAR LA METHODE DES DEPLACEMENTS (DETERMINATIO	75-0272
ELASTIC	CHARACTERIZATION OF ELASTIC SOLIDS USING FINITE ELEMENT METHODS	72-0712

202

ELASTIC	CHILES: A FINITE ELEMENT COMPUTER PROGRAM THAT CALCULATES THE INTENSITIES OF LINEAR ELASTIC SINGULAR	73-0959
ELASTIC	COMPARISON OF PRACTICAL METHODS OF ANALYSING ELASTIC LATERALLY LOADED PLATES	71-0425
ELASTIC	COMPLIANCES OF LAYERED ELASTIC SYSTEMS	74-1204
ELASTIC	COMPORTAREA ELASTICA A DIGURILOR DE PAMINT SOLICIATE SEISMIC (ELASTIC BEHAVIOR OF THE EARTH DAMS SEI	74-1185
ELASTIC	COMPUTER PROGRAM JET 3 TO CALCULATE THE LARGE ELASTIC PLASTIC DYNAMICALLY INDUCED DEFORMATIONS OF FR	72-0609
ELASTIC	CONCERNING THE FINITE ELEMENT METHOD FOR THE SOLUTION OF PROBLEMS OF ELASTIC EQUILIBRIUM	71-0516
ELASTIC	CONICAL FINITE ELEMENT STRESS ANALYSIS OF AN AXISYMMETRIC TANK ON AN ELASTIC FOUNDATION	70-0494
ELASTIC	CONTINUUM BASES FOR CONSISTENT NUMERICAL FORMULATIONS OF FINITE STRAINS IN ELASTIC AND INELASTIC STR	75-0299
ELASTIC	CONTINUUM CALCULATION FROM ELASTIC PLASTIC REINFORCED MATERIAL USING A FINITE ELEMENT METHOD (DIE BE	72-0658
ELASTIC	COUPLED SHEAR WALLS. IMPROVED METHODS FOR ELASTIC ANALYSIS	74-0818
ELASTIC	CURVED FINITE ELEMENT FOR THIN ELASTIC SHELLS	70-0173
ELASTIC	CURVED TRIANGULAR FINITE ELEMENTS FOR THE ANALYSIS OF THICK ELASTIC PLATES AND SHELLS	71-0814
ELASTIC	DESIGN OF CONCRETE TRACK SLAB ON ELASTIC FOUNDATION USING THE FINITE ELEMENT METHOD	73-0479
ELASTIC	DETERMINATION OF STRESSES AT NODAL POINTS FROM FINITE ELEMENT SOLUTIONS TO ELASTIC PROBLEMS	71-0307
ELASTIC	DEVELOPMENT OF FINITE ELEMENT COMPUTATIONAL METHODS FOR THREE DIMENSIONAL ELASTIC PLASTIC CRACK ANAL	70-0356
ELASTIC	DEVELOPMENT OF THE STIFFNESS METHOD TO DEFINE BOUNDS ON ELASTIC BEHAVIOUR OF STRUCTURES	62-0011
ELASTIC	DIFFERENTIAL EXPANSION IN ELASTIC LAMINATES	73-0190
ELASTIC	DIRECT FINITE ELEMENT ANALYSIS OF ELASTIC CONTACT PROBLEMS	72-0733
ELASTIC	DISCUSSION OF PROCEEDINGS, PAPER 6612, ELASTIC BUCKLING OF TWO-COMPONENT COLUMNS	70-0321
ELASTIC	DISPLAYS OF KINEMATIC AND ELASTIC SYSTEMS	71-0053
ELASTIC	DYNAMIC BEHAVIOR OF LIQUIDS IN ELASTIC TANKS	71-0141
ELASTIC	DYNAMIC FINITE ELEMENT METHOD CALCULATION OF ELASTIC WAVE PROPAGATION IN TWO DIMENSIONAL STRUCTURES	74-0166
ELASTIC	DYNAMIC RESPONSE OF ELASTIC HEAD MODELS	75-0378
ELASTIC	DYNAMIC STRESS ANALYSIS OF POROUS ELASTIC SOLIDS SATURATED WITH COMPRESSIBLE FLUIDS	71-0651
ELASTIC	DYNAMICAL FINITE ELEMENT ANALYSIS FOR ELASTIC WAVES IN BEAMS AND PLATES	68-0211
ELASTIC	EFFECT OF ELASTIC MODULAR RATIO AND ANISOTROPY ON THE STRESS FIELD IN A TWO-DIMENSIONAL MATRIX INTO	71-0647
ELASTIC	EFFECT OF GRAIN BOUNDARY FOR THE ELASTIC BEHAVIOR OF AL AND CU CRYSTALS (ANALYSIS BY THE FINITE ELEM	74-0947
ELASTIC	EFFECTS OF INSERTS ON THE ELASTIC BEHAVIOUR OF CYLINDRICAL MATERIALS LOADED BETWEEN ROUGH END-PLATES	71-0318
ELASTIC	ELAD PROGRAM MANUAL, A COMPUTER PROGRAM FOR THE ELASTIC ANALYSIS OF DOME STRUCTURES AND AXISYMMETRIC	69-0064
ELASTIC	ELASTIC ANALYSIS FOR BEHAVIOR OF ROCKFILL	70-0032
ELASTIC	ELASTIC ANALYSIS OF CRACKED THIN SHELLS BY THE FINITE ELEMENT METHOD	71-0048
ELASTIC	ELASTIC ANALYSIS OF CYLINDRICAL CONFIGURATIONS WITH STRESS SINGULARITIES	71-0325
ELASTIC	ELASTIC ANALYSIS OF LARGE SPACEBOUND MIRRORS	72-0619
ELASTIC	ELASTIC ANALYSIS OF A SKULL	73-0415
ELASTIC	ELASTIC ANALYSIS OF A REFRACTORY SPRUNG ARCH	74-0386
ELASTIC	ELASTIC ANALYSIS OF SPATIAL SYSTEMS OF INTERCONNECTED SHEAR WALLS AND FRAMES	74-1195
ELASTIC	ELASTIC AND EQUILIBRIUM MATRICES OF GENERAL SEMI-MONOCOGUE MEMBRANE ELEMENTS	70-0159
ELASTIC	ELASTIC AND PLASTIC ANALYSIS BY FINITE ELEMENT METHOD	68-0119
ELASTIC	ELASTIC AND PLASTIC STRESS DISTRIBUTION ON BASE PLATES IN THE NEIGHBOURHOOD OF FILLET WELDS IN LAP J	70-0175
ELASTIC	ELASTIC AND PLASTIC INTERLAMINAR SHEAR DEFORMATION IN LAMINATED COMPOSITIES UNDER GENERALIZED PLANE	71-0158
ELASTIC	ELASTIC BENDING OF CIRCULAR PLATES OF VARIABLE THICKNESS. AN ANALYTICAL AND EXPERIMENTAL STUDY	70-0163
ELASTIC	ELASTIC CRACK ANALYSIS BY A FINITE ELEMENT HYBRID METHOD	71-0212
ELASTIC	ELASTIC DEFLECTION IN THIN WALLED CONES ROTATING ABOUT THEIR AXES OF SYMMETRY	73-0545
ELASTIC	ELASTIC DEFORMATION OF LIGHTWEIGHT MIRRORS METHOD OF LIMIT POINT CALCULATION IN FINITE ELEMENT STRUC	72-0232
ELASTIC	ELASTIC DEFORMATION OF POLYCRYSTALLINE METAL - 1. INFLUENCE OF GRAIN SHAPE ON DISTRIBUTIONS OF STRES	74-0463
ELASTIC	ELASTIC DEFORMATION OF POLYCRYSTALLINE METAL - 2 EFFECT OF SLENDERNESS OF GRAINS ON DEFORMATION MODE	73-0275
ELASTIC	ELASTIC DEFORMATION OF POLYCRYSTALLINE METALS - 3	73-0361
ELASTIC	ELASTIC DEFORMATION OF SQUARE EXTRUSION DIES DUE TO LOAD AND TEMPERATURE	75-0581
ELASTIC	ELASTIC DYNAMIC RESPONSE OF AXISYMMETRIC STRUCTURES	69-0275
ELASTIC	ELASTIC EQUILIBRIUM MODELS, STRESS FUNCTIONS, AND DUALITY NON LINEARITIES AND STABILITY. FIVE LECTUR	73-0631
ELASTIC	ELASTIC INSTABILITY PREDICTIONS FOR DOUBLY-CURVED SHELLS	68-0075
ELASTIC	ELASTIC LATERAL BUCKLING OF STEEL BEAMS	70-0364
ELASTIC	ELASTIC NONLINEAR FINITE ELEMENT ANALYSIS OF PLANAR STRUCTURES	71-0830
ELASTIC	ELASTIC PLASTIC ANALYSIS OF PRESSURE VESSEL COMPONENTS	69-0380
ELASTIC	ELASTIC PLASTIC ANALYSIS OF THREE DIMENSIONAL PROBLEMS BY THE FINITE ELEMENT METHOD	69-0387
ELASTIC	ELASTIC PLASTIC ANALYSIS OF NOTCHED SPECIMENS USING THE FINITE ELEMENT METHOD	70-0140
ELASTIC	ELASTIC PLASTIC ANALYSIS OF SAINT-VENANT TORSION PROBLEM BY A HYBRID STRESS MODEL	72-0385
ELASTIC	ELASTIC PLASTIC ANALYSIS BY ASSUMED STRESS HYBRID MODEL	74-0443
ELASTIC	ELASTIC PLASTIC ANALYSIS OF SHALLOW FOUNDATIONS	74-0446
ELASTIC	ELASTIC PLASTIC AND CREEP ANALYSIS OF PIPELINES BY FINITE ELEMENTS	73-0419
ELASTIC	ELASTIC PLASTIC BUCKLING OF STIFFENED RECTANGULAR PLATES	69-0122
ELASTIC	ELASTIC PLASTIC CRACKED PLATES IN PLANE STRAIN	69-0321
ELASTIC	ELASTIC PLASTIC DEFORMATION OF THICK WALLED CYLINDERS	69-0427
ELASTIC	ELASTIC PLASTIC DEFORMATION OF CYLINDRICAL PRESSURE VESSELS UNDER CYCLIC LOADING	74-0604
ELASTIC	ELASTIC PLASTIC DYNAMIC ANALYSIS OF AXISYMMETRIC SOLIDS	73-0657
ELASTIC	ELASTIC PLASTIC FINITE ELEMENT ANALYSIS OF PLATES	74-0093
ELASTIC	ELASTIC PLATE SPALLATION	72-0682
ELASTIC	ELASTIC SETTLEMENT OF A CIRCULAR RAFT IN ADHESIVE CONTACT WITH A TRANSVERSELY ISOTROPIC MEDIUM	75-0828
ELASTIC	ELASTIC SHEAR ANALYSIS OF GENERAL PRISMATIC BEAMS	68-0114
ELASTIC	ELASTIC SNAP THROUGH ANALYSIS OF CURVED PLATES USING DISCRETE ELEMENTS	72-0899
ELASTIC	ELASTIC SNAP-TRHOUGH ANALYSIS OF CURVED PLATES USING DISCRETE ELEMENTS	72-0927
ELASTIC	ELASTIC STABILITY OF SHELLS OF REVOLUTION BY THE VARIATIONAL APPROACH USING DISCRETE ELEMENTS	66-0106
ELASTIC	ELASTIC STRENGTH OF PROPELLERS: AN ANALYSIS BY MATRIX METHODS	70-0468
ELASTIC	ELASTIC STRESSES IN LAYERED SNOW PACKS	72-0925
ELASTIC	ELASTIC TORSIONAL ANALYSIS OF IRREGULAR SHAPES	65-0020
ELASTIC	ELASTIC TORSIONAL ANALYSIS OF IRREGULAR SHAPES	66-0078
ELASTIC	ELASTIC WAVE PROPAGATION, A COMPARISON BETWEEN FINITE ELEMENT PREDICTIONS AND EXACT SOLUTION	67-0076
ELASTIC	ELASTIC, PLASTIC BEHAVIOUR OF ROTO - 1	73-0458
ELASTIC	EVALUATION OF DISCRETE METHODS FOR THE LINEAR DYNAMIC RESPONSE OF ELASTIC AND VISCOELASTIC SOLIDS	70-0439
ELASTIC	EVALUATION OF FINITE ELEMENT METHODS FOR THE COMPUTATION OF ELASTIC STRESS INTENSITY FACTORS	72-0249
ELASTIC	EXPERIMENTAL ELASTIC STRESS ANALYSES OF CYLINDER-TO-CYLINDER SHELL MODELS AND COMPARISONS WITH THEOR	71-0065
ELASTIC	FEATHER: FINITE ELEMENT ANALYSIS FOR THREE-DIMENSIONAL ELASTIC RESPONSE	72-0607
ELASTIC	FINEL, UNIVERSELLES PROGRAMMSYSTEM ZUR BERECHNUNG DER ELASTISCHEN EIGENSCHAFTEN VON MASCHINENBAUTEIL	72-0132
ELASTIC	FINITE DEFORMATION OF ELASTIC MEMBRANES BY THE FINITE ELEMENT METHOD	67-0105
ELASTIC	FINITE DEFORMATION OF ELASTIC PLATES, SHELLS, AND MEMBRANES BY THE FINITE ELEMENT METHOD	72-0805
ELASTIC	FINITE ELEMENT ANALYSIS OF SEEPAGE IN ELASTIC MEDIA	69-0024
ELASTIC	FINITE ELEMENT ANALYSIS OF CONTACT PROBLEMS USING DIFFERENTIAL DISPLACEMENTS	70-0126
ELASTIC	FINITE ELEMENT ANALYSIS OF ALTERNATIVE AXIAL LOADING OF AN ELASTIC PLATE PRESSED BETWEEN TWO ELASTIC	71-0357
ELASTIC	FINITE ELEMENT ANALYSIS OF ALTERNATIVE AXIAL LOADING OF AN ELASTIC PLATE PRESSED BETWEEN TWO ELASTIC	71-0357
ELASTIC	FINITE ELEMENT ANALYSIS OF ELASTIC CONTACT PROBLEM	72-0550
ELASTIC	FINITE ELEMENT ANALYSIS OF VIBRATION OF THIN ELASTIC PLATES BY SIMPLIFIED HYBRID DISPLACEMENT METHOD	72-0565
ELASTIC	FINITE ELEMENT ANALYSIS OF FINITE SIZED PLATES BOUNDED TO AN ELASTIC HALF SPACE	72-0880
ELASTIC	FINITE ELEMENT ANALYSIS OF PROBLEMS IN NONCONSERVATIVE ELASTIC STABILITY	72-0957
ELASTIC	FINITE ELEMENT ANALYSIS OF DISPERSION-FREE GUIDES FOR ELASTIC SURFACE WAVES	73-0009
ELASTIC	FINITE ELEMENT ANALYSIS OF PIEZOELECTRIC ELASTIC WAVEGUIDES	73-0115
ELASTIC	FINITE ELEMENT ANALYSIS OF ELASTIC CONTACT PROBLEMS	73-0219
ELASTIC	FINITE ELEMENT ANALYSIS OF CUT-GROWTH IN SHEETS OF HIGHLY ELASTIC MATERIALS	73-0514
ELASTIC	FINITE ELEMENT ANALYSIS OF ELASTIC WAVE SCATTERING FROM DISCONTINUITIES	73-0547
ELASTIC	FINITE ELEMENT ANALYSIS OF PAVEMENT STRUCTURES USING AFPAV CODE LINEAR ELASTIC ANALYSIS	73-0658
ELASTIC	FINITE ELEMENT ANALYSIS OF SLABS ON ELASTIC SOLIDS	74-0120
ELASTIC	FINITE ELEMENT ANALYSIS OF ELASTIC PLASTIC THICK-WALLED TUBES	74-0368
ELASTIC	FINITE ELEMENT ANALYSIS OF THIN ELASTIC SHELLS WITH RESIDUAL ENERGY BALANCING AND THE ROLE OF THE RI	75-0034

ELASTIC	FINITE ELEMENT ANALYSIS OF FREQUENCY SPECTRA FOR ELASTIC WAVEGUIDES	75-0247
ELASTIC	FINITE ELEMENT APPLICATIONS IN THE CHARACTERIZATION OF ELASTIC SOLIDS	71-0138
ELASTIC	FINITE ELEMENT APPROACH TO THE CONTACT PROBLEM OF A PLATE ON THE ELASTIC HALF SPACE	72-0830
ELASTIC	FINITE ELEMENT APPROACH TO ACOUSTIC RADIATION FROM ELASTIC STRUCTURES	74-0121
ELASTIC	FINITE ELEMENT APPROACH TO THE VIBRATION ANALYSIS OF ELASTIC DISKS ON A FLEXIBLE SHAFT	75-0368
ELASTIC	FINITE ELEMENT ELASTIC BUCKLING ANALYSIS	67-0107
ELASTIC	FINITE ELEMENT ELASTIC THIN SHELL PRE-BUCKLING AND POST-BUCKLING ANALYSIS	71-0821
ELASTIC	FINITE ELEMENT ELASTIC PLASTIC CREEP ANALYSIS OF TWO DIMENSIONAL CONTINUUM WITH TEMPERATURE DEPENDEN	73-0226
ELASTIC	FINITE ELEMENT FORMULATION FOR LINEAR THERMOVISCO ELASTIC MATERIALS	69-0351
ELASTIC	FINITE ELEMENT FORMULATIONS FOR ELASTIC PLATES BY GENERAL VARIATIONAL STATEMENTS WITH DISCONTINUOUS	73-0559
ELASTIC	FINITE ELEMENT FORMULATIONS FOR ELASTIC PLATES BY GENERAL VARIATIONAL STATEMENTS WITH DISCONTINUOUS	73-1124
ELASTIC	FINITE ELEMENT FRACTURE MECHANICS ANALYSIS OF TWO-DIMENSIONAL AND AXISYMMETRIC ELASTIC AND ELASTIC-P	74-1092
ELASTIC	FINITE ELEMENT METHOD IN ELASTIC INSTABILITY ANALYSIS	69-0108
ELASTIC	FINITE ELEMENT METHOD FOR ELASTIC PLASTIC PLATES	72-0167
ELASTIC	FINITE ELEMENT METHODS FOR ELASTIC BODIES CONTAINING CRACKS	73-1029
ELASTIC	FINITE ELEMENT PERTURBATION ANALYSIS OF NONLINEAR DYNAMIC RESPONSE OF ELASTIC CONTINUUA	74-0452
ELASTIC	FINITE ELEMENT POST BUCKLING ANALYSIS OF THIN ELASTIC PLATES	68-0123
ELASTIC	FINITE ELEMENT STUDY OF THE ELASTIC BEHAVIOR OF PLANE FRAMES WITH FILLER WALLS	71-0828
ELASTIC	FINITE ELEMENT STUDY OF A REINFORCED CONCRETE CYLINDRICAL SHELL THROUGH ELASTIC, CRACKING, AND ULTIM	75-0317
ELASTIC	FINITE ELEMENTS FOR DETERMINATION OF CRACK TIP ELASTIC STRESS INTENSITY FACTORS	71-0436
ELASTIC	FINITE ELEMENTS FOR DETERMINATION OF CRACK-TIP ELASTIC STRESS INTENSITY FACTORS	71-0785
ELASTIC	FINITE ELEMENTS FOR THREE-DIMENSIONAL ELASTIC CRACK ANALYSIS	74-0136
ELASTIC	FINITE PLANE STRAIN OF INCOMPRESSIBLE ELASTIC SOLIDS BY THE FINITE ELEMENT METHOD	68-0128
ELASTIC	FINITE STRAINS AND DISPLACEMENTS OF ELASTIC MEMBRANES BY THE FINITE ELEMENT METHOD	67-0062
ELASTIC	FINITE STRIP METHOD OF ANALYSIS OF ELASTIC SLABS	68-0190
ELASTIC	FINITE-ELEMENT ANALYSIS OF FRACTURE PROPAGATION IN TWO-DIMENSIONAL ELASTIC BRITTLE SOLIDS	72-0948
ELASTIC	FINITE-ELEMENT ANALYSES OF INITIAL ELASTIC DEFORMATION OF VERTICALLY DEFORMED CRYSTAL BLOCKS	72-0970
ELASTIC	FINITE-ELEMENT ANALYSIS OF THIN ELASTIC SHELLS WITH RESIDUAL ENERGY BALANCING AND ROLE OF RIGID BODY	75-0048
ELASTIC	FINITE-ELEMENT ANALYSIS OF PAVEMENT STRUCTURES USING AFPAV CODE (NONLINEAR ELASTIC ANALYSIS)	75-0748
ELASTIC	FINITE-ELEMENT APPROACH TO ACOUSTIC SCATTERING FROM ELASTIC STRUCTURES	75-0069
ELASTIC	FINITE-ELEMENT SOLUTION FOR A CRACKED TWO-LAYERED ELASTIC CYLINDER	71-0397
ELASTIC	FLEXIBLE PLATE FINITE ELEMENT ON ELASTIC FOUNDATION	70-0128
ELASTIC	FLEXURE OF A BEAM SUPPORTED ON AN ELASTIC FOUNDATION	72-0871
ELASTIC	FLOW OF COMPRESSIBLE FLUID IN POROUS ELASTIC MEDIA	73-1004
ELASTIC	FORMULATION AND APPLICATION OF CERTAIN PRIMAL AND MIXED FINITE ELEMENT MODELS OF FINITE DEFORMATIONS	74-0344
ELASTIC	GEOMETRIC STIFFNESS CHARACTERISTICS OF A ROTATING ELASTIC APPENDAGE	74-1296
ELASTIC	HIGH PRECISION FINITE ELEMENTS FOR PLANE ELASTIC PROBLEMS	72-0862
ELASTIC	HIGHER-ORDER FINITE-ELEMENT ANALYSIS OF TOPOGRAPHIC GUIDES SUPPORTING ELASTIC SURFACE WAVES	73-0114
ELASTIC	IDENTIFICATION OF NONLINEAR ELASTIC SOLIDS BY A FINITE ELEMENT METHOD	74-0530
ELASTIC	INCREMENTAL FINITE ELEMENT ANALYSIS OF LARGE ELASTIC DEFORMATION PROBLEMS	71-0753
ELASTIC	INFLUENCE OF NON-SINGULAR STRESS TERMS AND SPECIMEN GEOMETRY ON SMALL SCALE YIELDING AT CRACK TIPS I	73-0203
ELASTIC	INTERNAL PRESSURE EFFECTS ON THE VIBRATION OF PARTIALLY FILLED ELASTIC TANKS	75-0241
ELASTIC	INVESTIGATION OF ELASTIC PLATE SPALLATION BY FINITE ELEMENTS	74-0382
ELASTIC	INVESTIGATION OF VENEER CUTTING BY ELASTIC FINITE ELEMENT MODELS	74-1110
ELASTIC	ISOFINEL: ISOPARAMETRIC FINITE ELEMENT CODE FOR ELASTIC ANALYSIS OF TWO-DIMENSIONAL BODIES	75-0676
ELASTIC	LARGE DEFLECTION OF ELASTIC PLATES UNDER PATCH LOADING	72-0120
ELASTIC	LARGE DEFLECTIONS OF ELASTIC BEAMS AND PLATES USING THE FINITE ELEMENT METHOD	68-0101
ELASTIC	LARGE DISPLACEMENT FORMULATION FOR ELASTIC BODIES	70-0429
ELASTIC	LINEAR ELASTIC FINITE ELEMENT ANALYSIS OF MASONRY WALLS ON BUILDINGS	74-1015
ELASTIC	LIQUID SLOSHING IN AN ELASTIC CONTAINER	66-0096
ELASTIC	LUMPED PARAMETER METHODS APPLIED TO ELASTIC VIBRATIONS	63-0016
ELASTIC	MATERIAL BEHAVIOUR CHARACTERISTICS FOR REINFORCED CONCRETE SHELLS STRESSED BEYOND THE ELASTIC RANGE	70-0166
ELASTIC	MATHEMATICAL MODELING OF SPINNING ELASTIC BODIES FOR MODAL ANALYSIS	73-0441
ELASTIC	MATRIX ANALYSIS OF THREE-DIMENSIONAL ELASTIC MEDIA: SMALL AND LARGE DISPLACEMENTS	65-0003
ELASTIC	MATUS: A THREE-DIMENSIONAL FINITE ELEMENT PROGRAM FOR SMALL-STRAIN ELASTIC ANALYSIS	74-0970
ELASTIC	MIXED FINITE ELEMENT ANALYSIS OF PLATES IN BENDING, SMALL DEFLECTION THEORY OF ELASTIC AND ELASTO-PL	71-0783
ELASTIC	MODAL RESPONSE ANALYSIS OF CONSTRAINED FINITE-ELEMENTS ELASTIC CONTINUUM	75-0767
ELASTIC	MODEL OF ELASTIC PLATE FROM FINITE ELEMENTS IN A SUPERSONIC FLOW	75-0038
ELASTIC	MODERN FUSELAGE ANALYSIS AND THE ELASTIC AIRCRAFT	63-0023
ELASTIC	MULTIDEGREE-OF-FREEDOM ELASTIC SYSTEMS HAVING MULTIPLE CLEARANCES	73-0653
ELASTIC	NEW VARIATIONAL PRINCIPLE FOR FINITE ELASTIC DISPLACEMENTS	72-0454
ELASTIC	NON-LINEAR ELASTIC ANALYSIS OF UNGUYED TOWERS AND STACKS	74-1157
ELASTIC	NONLINEAR ANALYSIS OF ELASTIC FRAMED STRUCTURES	68-0222
ELASTIC	NONLINEAR ANALYSIS OF ELASTIC SHELLS OF REVOLUTION UNDER AXISYMMETRIC LOADING	70-0161
ELASTIC	NONLINEAR THERMO ELASTIC PLASTIC AND CREEP ANALYSIS BY FINITE ELEMENT METHOD	74-0092
ELASTIC	NONLINEAR THERMOELASTICITY, IRREVERSIBLE THERMODYNAMICS AND ELASTIC INSTABILITY	73-0762
ELASTIC	NOTCH STRESSES AT THE LOAD CARRYING ELASTIC CORE	73-0266
ELASTIC	NOTE ON AN APPROXIMATE METHOD FOR COMPUTING CONSISTENT CONJUGATE STRESSES IN ELASTIC FINITE ELEMENTS	75-0050
ELASTIC	NOTE ON THE ANALYSIS OF NONLINEAR DYNAMICS OF ELASTIC MEMBRANES BY THE FINITE ELEMENT METHOD	74-0230
ELASTIC	NOTES ON THE FINITE ELEMENT ANALYSIS OF THE AXISYMMETRIC ELASTIC SOLID	74-0185
ELASTIC	NUMERICAL ANALYSIS FOR FINITE AXISYMMETRIC DEFORMATIONS OF INCOMPRESSIBLE ELASTIC SOLIDS OF REVOLUTI	70-0094
ELASTIC	NX BOREHOLE JACK MODULUS DETERMINATIONS IN HOMOGENEOUS, ISOTROPIC, ELASTIC MATERIALS	74-0295
ELASTIC	OCENA DOKLADNOSCI ROZWIAZAN PROBLEMOW BRZEGOWYCH SPREZYSTYCH CIAL DYSKERTYZOWANYCH. (ASSESSING THE A	74-0921
ELASTIC	ON A SPECTRAL PROBLEM IN VIBRATION MECHANICS: COMPUTATION OF ELASTIC TANKS PARTIALLY FILLED WITH LIQ	75-0267
ELASTIC	ON FURTHER APPLICATION OF THE FINITE ELEMENT METHOD TO THREE-DIMENSIONAL ELASTIC ANALYSIS	70-0252
ELASTIC	ON THE DEVELOPMENT OF A REFINED FINITE-ELEMENT FREE FROM DISCRETIZATION - DISPERSION FOR ELASTIC WAV	73-1148
ELASTIC	ON THE RESOLUTION OF NONLINEAR ELASTIC AND ELASTOPLASTIC PROBLEMS BY THE METHOD OF FINITE ELEMENTS.	74-1326
ELASTIC	PARAMETRISCHE KERBSPANNUNGSUNTERSUCHUNG AM ELASTISCHEN KERN.(PARAMETRICAL INVESTIGATION OF NOTCH STR	72-0208
ELASTIC	PERIODIC SHOCK EXCITATION OF ELASTIC STRUCTURES	69-0163
ELASTIC	PLATES AND BEAMS ON ELASTIC FOUNDATIONS: LINEAR AND NONLINEAR BEHAVIOUR	68-0042
ELASTIC	PLATES AND TANKS ON ELASTIC FOUNDATIONS, AN APPLICATION OF FINITE ELEMENT METHOD	65-0010
ELASTIC	PRACTICAL SOLUTION OF PLASTIC DEFORMATION PROBLEMS IN THE ELASTIC PLASTIC RANGE	59-0004
ELASTIC	PREDICTION OF ELASTIC CONSTANTS OF PARTICLEBOARD	74-0259
ELASTIC	PROGRESS REPORT ON DISCRETE-ELEMENT ELASTIC AND ELASTIC-PLASTIC ANALYSIS OF SHELLS OF REVOLUTION SUB	68-0170
ELASTIC	QUADRATIC MATRIX EQUATIONS FOR DETERMINING VIBRATION MODES AND FREQUENCIES OF CONTINUOUS ELASTIC SYS	66-0064
ELASTIC	SABOR 1. A FORTRAN PROGRAM FOR THE LINEAR ELASTIC AXISYMMETRIC LOADING BY USING THE MATRIX DISPLACEM	65-0027
ELASTIC	SECOND GENERATION BOUNDARY INTEGRAL EQUATION PROGRAM FOR THREE-DIMENSIONAL ELASTIC ANALYSIS	75-0376
ELASTIC	SIMULATION OF ELASTIC MODULUS AND POISSON'S RATIO OF SPHEROIDAL GRAPHITE CAST IRON	74-0955
ELASTIC	SOME NONCONFORMING VARIANTS ON THE TRILINEAR ISOPARAMETRIC BRICK FOR LINEAR ELASTIC PROBLEMS	72-0531
ELASTIC	STEADY STATE EQUATIONS OF MOTION, EQUILIBRIUM SHAPE AND STABILITY DERIVATIVES OF ELASTIC AIRPLANES E	72-0541
ELASTIC	STIFFNESS ANALYSIS OF ELASTIC PLASTIC PLATES	65-0040
ELASTIC	STRESS ANALYSIS OF CONCRETE TRACK SLABS ON AN ELASTIC FOUNDATION BY THE FINITE ELEMENT METHOD	74-1049
ELASTIC	STRESS DISTRIBUTION UNDER UNIAXIAL ELASTIC PLANE STRAIN COMPRESSION WITH FRICTION	74-0526
ELASTIC	STRESS INTENSITY FACTORS FOR A NOTCHED BEAM ON AN ELASTIC FOUNDATION	75-0566
ELASTIC	STRESSES AND DISPLACEMENTS AROUND OPENINGS IN ROCK CONTAINING AN ELASTIC DISCONTINUITY	75-0169
ELASTIC	STRUCTURAL ANALYSIS OF A COAL MINE OPENING IN ELASTIC, MULTILAYERED MATERIAL	74-0235
ELASTIC	STRUCTURAL ANALYSIS OF A COAL MINE OPENING IN ELASTIC, MULTILAYERED MATERIAL	74-0840
ELASTIC	SULLA DETERMINAZIONE DELLE FREQUENZE PROPRIE DI ALBERI SU PIU SUPPORTI ELASTICI. (DETERMINATION OF F	74-0916
ELASTIC	THE CONVERGENCE OF FINITE ELEMENT METHOD IN SOLVING LINEAR ELASTIC PROBLEMS	67-0083
ELASTIC	THE ECONOMICAL SOLUTION OF ELASTIC PROBLEMS FOR A RANGE OF POISSONS RATIO	74-0786
ELASTIC	THE EFFECTS OF SOIL PARAMETERS AND BOUNDARY CONDITIONS ON THE CONSOLIDATION OF AN ELASTIC LAYER	70-0422
ELASTIC	THE ELASTIC STRESS ANALYSIS OF A BI-MATERIAL PLATE WITH A CRACK NORMAL TO THE INTERFACES	72-0819

ELASTIC	THE FINITE ELEMENT ANALYSIS OF PHYSICALLY AND KINEMATICALLY NONLINEAR ELASTIC SOLIDS	69-0146
ELASTIC	THE FINITE ELEMENT ANALYSIS OF THIN ELASTIC SHELLS	73-1076
ELASTIC	THE FINITE ELEMENT METHOD FOR ANALYSIS OF ELASTIC ISOTROPIC AND ORTHOTROPIC SLABS	64-0014
ELASTIC	THE FINITE ELEMENT METHOD IN ELASTIC INSTABILITY ANALYSIS	69-0339
ELASTIC	THE FINITE ELEMENT METHOD AND LOCAL BOUNDS FOR BOUNDARY VALUE PROBLEMS OF ELASTIC STRUCTURES	73-1056
ELASTIC	THE FINITE STRIP METHOD IN THE ANALYSIS OF ELASTIC PLATES WITH TWO SIMPLY SUPPORTED ENDS	68-0068
ELASTIC	THE LINEAR ELASTIC DYNAMIC ANALYSIS OF SHELLS OF REVOLUTION BY THE MATRIX DISPLACEMENT METHOD	66-0046
ELASTIC	THE OPTIMUM APPROACH TO ANALYSIS OF ELASTIC CONTINUA	71-0177
ELASTIC	THE TRANSIENT LINEAR ELASTIC RESPONSE ANALYSIS COMPLEX THIN SHELLS OF REVOLUTION SUBJECTED TO ARBITR	70-0080
ELASTIC	THEORETICAL AND NUMERICAL COMPARISON OF ELASTIC NONLINEAR FINITE ELEMENT METHODS.	72-0452
ELASTIC	THEORETICAL FORMULATION OF FINITE ELEMENT METHODS IN LINEAR ELASTIC ANALYSIS OF GENERAL SHELLS	72-0496
ELASTIC	THERMAL ELASTIC PLASTIC STRESS ANALYSIS BY THE FINITE ELEMENT METHOD	69-0391
ELASTIC	THREE DIMENSIONAL ANISOTROPIC AND INHOMOGENEOUS ELASTIC MEDIA MATRIX ANALYSIS FOR SMALL AND LARGE DI	65-0008
ELASTIC	THREE DIMENSIONAL ELASTIC AND ELASTO-PLASTIC ANALYSIS OF REACTOR PRESSURE VESSELS	69-0453
ELASTIC	THREE DIMENSIONAL ELASTIC PLASTIC ANALYSIS FOR FRACTURE MECHANICS - 2.	71-0358
ELASTIC	THREE-DIMENSIONAL ANALYSIS OF ELASTIC SOLIDS I. ANALYSIS PROCEDURE	69-0222
ELASTIC	THREE-DIMENSIONAL BAR CELL FOR ELASTIC STRESS ANALYSIS	70-0219
ELASTIC	THREE-DIMENSIONAL ELASTIC FINITE ELEMENT ANALYSIS USING GRADIENT DEGREES OF FREEDOM	73-0843
ELASTIC	THREE-DIMENSIONAL ELASTIC ANALYSIS OF SEMI-ELLIPTICAL SURFACE CRACKS SUBJECT TO THERMAL SHOCK	75-0256
ELASTIC	THREE-DIMENSIONAL FINITE-ELEMENT ELASTIC ANALYSIS OF PRISMATIC SOLIDS	73-1138
ELASTIC	TRANSIENT RESPONSE OF LINEAR ELASTIC STRUCTURES DETERMINED BY THE MATRIX EXPONENTIAL METHOD	71-0240
ELASTIC	TRANSIENT RESPONSE OF ELASTIC, VISCO-PLASTIC BEAMS	73-0780
ELASTIC	TRIANGULAR FINITE ELEMENTS FOR THE BENDING ANALYSIS OF THICK ELASTIC PLATES	70-0466
ELASTIC	UNBONDED CONTACT PROBLEM OF A PLATE ON THE ELASTIC HALF SPACE	74-0208
ELASTIC	USE OF SINGULARITY PROGRAMMING IN FINITE ELEMENT CALCULATIONS OF ELASTIC STRESS INTENSITY FACTORS. P	75-0252
ELASTIC	USE OF THE FINITE ELEMENT METHOD IN EXAMINING ELASTIC BUCKLING OF THIN PLATES OF ANY SHAPE	75-0153
ELASTIC	USERS MANUAL FOR DIGITAL COMPUTER PROGRAM OK5 NUMERICAL ANALYSIS OF FINITE STRAINS AND DISPLACEMENTS	72-0718
ELASTIC	USEPS' MANUAL FOR DIGITAL COMPUTER PROGRAM OK1. NUMERICAL ANALYSIS OF FINITE AXISYMMETRIC DEFORMATIO	72-0861
ELASTIC	VARIATIONAL FORMULATIONS AND FINITE ELEMENT METHOD FOR THIN ELASTIC SHELLS AND PLATES	68-0064
ELASTIC	VARIATIONAL FORMULATION OF DYNAMICS OF FLUID-SATURATED POROUS ELASTIC SOLIDS	72-0017
ELASTIC	VIBRATION OF AN ELASTIC BODY IMMERSED IN A FLUID	72-0312
ELASTIC	VISCO-PLASTICITY, PLASTICITY AND CREEP IN ELASTIC SOLIDS. A UNIFIED NUMERICAL SOLUTION APPROACH	74-1279
ELASTIC	ZUR LOESUNG DES KONTAKTPROBLEMS ELASTISCHER KOERPER MIT AUSGEDEHNTER KONTAKTFLAECHE DURCH QUADRATISC	74-1135
ELASTIC-MATERIA	DISCONTINUOUS FINITE-ELEMENT APPROXIMATIONS FOR ANALYSIS OF SHOCK-WAVES IN NONLINEARLY ELASTIC-MATER	75-0451
ELASTIC-PLASTIC	A COMPARATIVE STUDY OF NUMERICAL METHODS OF ELASTIC-PLASTIC ANALYSIS	67-0047
ELASTIC-PLASTIC	A COMPUTER PROGRAMME FOR ELASTIC-PLASTIC CALCULATIONS WITH USE OF TRIM-6 FINITE ELEMENTS	72-0993
ELASTIC-PLASTIC	A DISCRETE ELEMENT METHOD FOR ANALYSIS OF PLANE ELASTIC-PLASTIC STRESS PROBLEMS	65-0063
ELASTIC-PLASTIC	A FINITE ELEMENT PROGRAM FOR THE DYNAMIC LARGE DEFLECTION, ELASTIC-PLASTIC ANALYSIS OF STIFFENED SHE	72-0908
ELASTIC-PLASTIC	A FINITE ELEMENT STUDY OF ELASTIC-PLASTIC STRESS DISTRIBUTIONS IN NOTCHED SPECIMENS UNDER TENSION	68-0177
ELASTIC-PLASTIC	A METHOD OF ELASTIC-PLASTIC PLANE STRESS AND PLANE STRAIN ANALYSIS	66-0114
ELASTIC-PLASTIC	A METHOD OF ELASTIC-PLASTIC ANALYSIS OF LARGELY DEFORMED PLATE PROBLEMS	72-0319
ELASTIC-PLASTIC	A NEW MODEL FOR ELASTIC-PLASTIC TRUSSES	74-0805
ELASTIC-PLASTIC	A SIMPLE FINITE ELEMENT MODEL FOR ELASTIC-PLASTIC PLATE BENDING	72-0889
ELASTIC-PLASTIC	A STUDY OF ELASTIC-PLASTIC ANALYSIS BY THE ASSUMED-STRESS HYBRID FINITE-ELEMENT MODEL, WITH APPLICAT	74-0718
ELASTIC-PLASTIC	AN APPLICATION OF THE FINITE ELEMENT METHOD TO ELASTIC-PLASTIC PROBLEMS OF PLANE STRESS	70-0031
ELASTIC-PLASTIC	AN ELASTIC-PLASTIC FINITE ELEMENT ANALYSIS FOR PLATES WITH EDGE BEAMS	69-0185
ELASTIC-PLASTIC	AN ELASTIC-PLASTIC FINITE ELEMENT ANALYSIS OF CRACK TIP FIELDS UNDER BIAXIAL LOADING CONDITIONS	73-0760
ELASTIC-PLASTIC	AN INCREMENTAL FINITE ELEMENT ANALYSIS OF TWO DIMENSIONAL LARGE STRAIN AND LARGE DISPLACEMENT PROBLE	72-0338
ELASTIC-PLASTIC	ANALYSIS OF ELASTIC-PLASTIC SHELLS OF REVOLUTION UNDER AXI-SYMMETRIC LOADING BY THE FINITE ELEMENT M	67-0102
ELASTIC-PLASTIC	ANALYSIS OF ELASTIC-PLASTIC BUCKLING OF PLATES BY THE FINITE ELEMENT METHOD	69-0263
ELASTIC-PLASTIC	ANALYSIS OF ELASTIC-PLASTIC SHELLS OF REVOLUTION	70-0220
ELASTIC-PLASTIC	ANALYSIS OF ELASTIC-PLASTIC BEHAVIOR OF BEAMS AND COLUMNS BY PLANE FINITE ELEMENT MODEL	73-1073
ELASTIC-PLASTIC	ANALYSIS OF ELASTIC-PLASTIC CREEP BUCKLING OF AXISYMMETRIC SHELLS BY FINITE-ELEMENT METHOD	75-0355
ELASTIC-PLASTIC	ANALYSIS OF LARGE ELASTIC-PLASTIC CREEP DEFORMATIONS OF AXISYMMETRIC SHELLS BY THE FINITE ELEMENT ME	74-0563
ELASTIC-PLASTIC	APPLICATION OF FINITE ELEMENT ELASTIC-PLASTIC STRESS ANALYSIS TO NOTCHED FATIGUE SPECIMEN BEHAVIOUR	71-0185
ELASTIC-PLASTIC	APPLICATION OF FINITE ELEMENT METHOD TO CYCLIC LOADING OF ELASTIC-PLASTIC STRUCTURES CONTAINING EFFE	73-0103
ELASTIC-PLASTIC	APPLICATION OF THE FINITE ELEMENT METHOD TO 2 AND 3 DIMENSIONAL ELASTIC-PLASTIC PROBLEMS OF FRACTURE	70-0501
ELASTIC-PLASTIC	ASSUMED STRESS HYBRID FINITE ELEMENT METHOD FOR FRACTURE MECHANICS AND ELASTIC-PLASTIC ANALYSIS	72-0584
ELASTIC-PLASTIC	CALCULATION OF ELASTIC-PLASTIC DEFORMATION BY THE FINITE ELEMENT METHOD OF NUMERICAL ANALYSIS	71-0793
ELASTIC-PLASTIC	COMBINED ANALYTICAL - EXPERIMENTAL FRACTURE STUDY OF THE TWO LEADING THEORIES OF ELASTIC-PLASTIC FRA	73-0951
ELASTIC-PLASTIC	COMBINED ELASTIC-PLASTIC CREEP ANALYSIS OF TWO-DIMENSIONAL BODIES	71-0232
ELASTIC-PLASTIC	COMPARATIVE STUDY OF NUMERICAL METHODS OF ELASTIC-PLASTIC ANALYSIS	67-0153
ELASTIC-PLASTIC	COMPUTER PROGRAM FOR ELASTIC-PLASTIC ANALYSIS OF AXISYMMETRICALLY LOADED SHELLS OF REVOLUTION	68-0239
ELASTIC-PLASTIC	DISCRETE APPROXIMATIONS OF ELASTIC-PLASTIC BODIES BY VARIATIONAL METHODS	72-0710
ELASTIC-PLASTIC	DISCRETE MODEL ANALYSIS OF ELASTIC-PLASTIC PLATES	68-0009
ELASTIC-PLASTIC	DYNAMIC ELASTIC-PLASTIC ANALYSIS OF AXISYMMETRIC STRUCTURES BY FINITE ELEMENT METHOD COMPUTER CODE	73-0844
ELASTIC-PLASTIC	DYNAPLAS: A FINITE ELEMENT PROGRAM FOR THE DYNAMIC, LARGE DEFLECTION, ELASTIC-PLASTIC ANALYSIS OF ST	72-0773
ELASTIC-PLASTIC	DYPLAS, A FINITE ELEMENT DYNAMIC ELASTIC-PLASTIC LARGE DEFORMATION ANALYSIS PROGRAM	74-0706
ELASTIC-PLASTIC	ELASTIC-PLASTIC ANALYSIS OF THREE-DIMENSIONAL MEDIA	66-0002
ELASTIC-PLASTIC	ELASTIC-PLASTIC ANALYSIS OF TWO-DIMENSIONAL STRESS SYSTEMS BY THE FINITE ELEMENT METHOD	67-0049
ELASTIC-PLASTIC	ELASTIC-PLASTIC ANALYSIS OF FLAT PLATES BY THE FINITE ELEMENT METHOD	68-0111
ELASTIC-PLASTIC	ELASTIC-PLASTIC ANALYSIS OF FRAMED STRUCTURES	69-0197
ELASTIC-PLASTIC	ELASTIC-PLASTIC ANALYSIS OF SOME PRESSURE VESSEL HEADS	69-0379
ELASTIC-PLASTIC	ELASTIC-PLASTIC ANALYSIS OF THICK WALLED PRESSURE VESSELS WITH SHARP DISCONTINUITIES	71-0155
ELASTIC-PLASTIC	ELASTIC-PLASTIC ANALYSIS OF BEAMS WITH UNIFORM CROSS-SECTION UNDER COMBINED LOADINGS	71-0777
ELASTIC-PLASTIC	ELASTIC-PLASTIC ANALYSIS OF A BAR WITH UNIFORM CROSS-SECTION UNDER COMBINED LOADINGS 1 - COMBINATION	71-0778
ELASTIC-PLASTIC	ELASTIC-PLASTIC ANALYSIS OF AXISYMMETRIC PLATES AND SHELLS	71-0779
ELASTIC-PLASTIC	ELASTIC-PLASTIC ANALYSIS BY QUADRATIC PROGRAMMING	72-0238
ELASTIC-PLASTIC	ELASTIC-PLASTIC ANALYSIS OF REINFORCED CONCRETE MEMBERS BY FEM	73-0292
ELASTIC-PLASTIC	ELASTIC-PLASTIC ANALYSIS OF BEAMS WITH UNIFORM CROSS-SECTION UNDER COMBINED LOADINGS (2)	73-1070
ELASTIC-PLASTIC	ELASTIC-PLASTIC ANALYSIS OF A WHEEL ROLLING ON A RIGID TRACK	74-1040
ELASTIC-PLASTIC	ELASTIC-PLASTIC ANALYSIS OF INTERFERENCE FIT FASTENERS	74-1177
ELASTIC-PLASTIC	ELASTIC-PLASTIC ANALYSIS FOR CRACKED MEMBERS	75-0519
ELASTIC-PLASTIC	ELASTIC-PLASTIC ANALYSIS OF A PROPAGATING CRACK UNDER CYCLIC LOADING	75-0857
ELASTIC-PLASTIC	ELASTIC-PLASTIC AND THERMAL STRESS ANALYSIS	68-0169
ELASTIC-PLASTIC	ELASTIC-PLASTIC AND CREEP ANALYSIS VIA THE MARC 2 FINITE ELEMENT PROGRAM	71-0032
ELASTIC-PLASTIC	ELASTIC-PLASTIC BEHAVIOR OF A LONGITUDINAL SEMI-ELLIPTIC CRACK IN A THICK PRESSURE VESSEL	73-0952
ELASTIC-PLASTIC	ELASTIC-PLASTIC BUCKLING OF PLATES USING THE FINITE ELEMENT METHOD	67-0094
ELASTIC-PLASTIC	ELASTIC-PLASTIC BUCKLING OF PLATES BY FINITE ELEMENT METHOD	69-0253
ELASTIC-PLASTIC	ELASTIC-PLASTIC DEFORMATIONS OF FLEXURALLY STIFF FRAMEWORKS FROM SECOND ORDER STRESS THEORY	73-1061
ELASTIC-PLASTIC	ELASTIC-PLASTIC DISCRETE ELEMENT ANALYSIS OF AXISYMMETRICALLY LOADED SHELLS OF REVOLUTION	67-0099
ELASTIC-PLASTIC	ELASTIC-PLASTIC EARTHQUAKE RESPONSE OF SOIL-BUILDING SYSTEMS	72-0717
ELASTIC-PLASTIC	ELASTIC-PLASTIC FINITE ELEMENT ANALYSIS OF SHORT FIBRE COMPOSITES	74-0138
ELASTIC-PLASTIC	ELASTIC-PLASTIC FINITE ELEMENT ANALYSIS OF CRACK TIP FIELDS UNDER BIAXIAL LOADING CONDITIONS	74-0296
ELASTIC-PLASTIC	ELASTIC-PLASTIC FINITE ELEMENT ANALYSIS OF PLATES - DISCUSSION	75-0139
ELASTIC-PLASTIC	ELASTIC-PLASTIC FINITE ELEMENT ANALYSIS USING SUPERPOSITION	75-0314
ELASTIC-PLASTIC	ELASTIC-PLASTIC FINITE-ELEMENT ANALYSIS USING SUPERPOSITION	75-0429
ELASTIC-PLASTIC	ELASTIC-PLASTIC FINITE ELEMENT ANALYSIS OF PLATES	75-0599
ELASTIC-PLASTIC	ELASTIC-PLASTIC FORMULATION FOR A CYLINDRICAL-SHELL FINITE-ELEMENT	75-0841
ELASTIC-PLASTIC	ELASTIC-PLASTIC LARGE DEFORMATION RESPONSE OF CLAY TO FOOTING LOADS	74-0699
ELASTIC-PLASTIC	ELASTIC-PLASTIC PLATE BENDING WITH CONSTANT CURVATURE ELEMENTS	74-0308
ELASTIC-PLASTIC	ELASTIC-PLASTIC PROBLEMS OF COMPRESSIBLE STRAIGHT BAR UNDER COMBINED LOADING CONDITION	75-0313

205

ELASTIC-PLASTIC	ELASTIC-PLASTIC SOLUTION OF A TWO-DIMENSIONAL TUBE PROBLEM BY THE FINITE ELEMENT METHOD	74-0769
ELASTIC-PLASTIC	ELASTIC-PLASTIC STABILITY ANALYSIS OF MINE-WASTE EMBANKMENTS	75-0835
ELASTIC-PLASTIC	ELASTIC-PLASTIC STRESS ANALYSIS FOR A NOTCHED BAR IN PLANE STRAIN BENDING	71-0410
ELASTIC-PLASTIC	ELASTIC-PLASTIC STRESS-STRAIN RELATIONS FOR MATERIALS WITH DIFFERENT YIELD STRENGTHS IN TENSION AND	72-0764
ELASTIC-PLASTIC	ELASTIC-PLASTIC STRESS-STRAIN RELATIONS USED IN CODE EPSOLA	73-0784
ELASTIC-PLASTIC	EPAD, A COMPUTER PROGRAM FOR THE ELASTIC-PLASTIC ANALYSIS OF DOME STRUCTURES AND ARBITRARY SOLIDS OF	70-0036
ELASTIC-PLASTIC	EVALUATION OF NUMERICAL TIME INTEGRATION METHODS AS APPLIED TO ELASTIC-PLASTIC DYNAMIC PROBLEMS INVO	74-0858
ELASTIC-PLASTIC	FINITE DEFORMATIONS IN ELASTIC-PLASTIC MEDIA SUBJECTED TO IMPULSIVE LOADINGS BY THE FINITE ELEMENT M	72-0947
ELASTIC-PLASTIC	FINITE ELEMENT ANALYSIS OF ELASTIC-PLASTIC PLANE STRESS PROBLEMS BASED UPON TRESCA YIELD CRITERION	70-0179
ELASTIC-PLASTIC	FINITE ELEMENT ANALYSES OF ELASTIC-PLASTIC PROBLEMS IN THE MECHANICS OF GEOLOGIC MEDIA: AN OVERVIEW	70-0203
ELASTIC-PLASTIC	FINITE ELEMENT ANALYSIS OF ELASTIC-PLASTIC PLATES AND ECCENTRICALLY STIFFENED PLATES	71-0585
ELASTIC-PLASTIC	FINITE ELEMENT ANALYSIS OF LARGE TRANSIENT ELASTIC-PLASTIC DEFORMATIONS OF SIMPLE STRUCTURES, WITH A	72-0849
ELASTIC-PLASTIC	FINITE ELEMENT ANALYSIS FOR WAVE PROPAGATION IN ELASTIC-PLASTIC SOLIDS	73-0990
ELASTIC-PLASTIC	FINITE ELEMENT APPROACH TO WAVE MOTIONS IN ELASTIC-PLASTIC CONTINUA	68-0041
ELASTIC-PLASTIC	FINITE ELEMENT FORMULATIONS FOR PROBLEMS OF LARGE ELASTIC-PLASTIC DEFORMATION	74-0796
ELASTIC-PLASTIC	FINITE ELEMENT FRACTURE MECHANICS ANALYSIS OF TWO-DIMENSIONAL AND AXISYMMETRIC ELASTIC AND ELASTIC-P	74-1092
ELASTIC-PLASTIC	FINITE ELEMENT INCREMENTAL ELASTIC-PLASTIC ANALYSIS OF PRESSURE VESSELS	70-0253
ELASTIC-PLASTIC	FINITE ELEMENT LARGE DEFLECTION ANALYSIS OF ELASTIC-PLASTIC SHELLS OF REVOLUTION SUBJECTED TO AXISYM	72-0137
ELASTIC-PLASTIC	FINITE ELEMENT STRESS FORMULATION FOR DYNAMIC ELASTIC-PLASTIC ANALYSIS	75-0484
ELASTIC-PLASTIC	FINITE ELEMENT TREATMENT OF A UNIFORMLY MOVING ELASTIC-PLASTIC CRACK TIP	74-0470
ELASTIC-PLASTIC	FINITE-ELEMENT ANALYSIS OF LARGE ELASTIC-PLASTIC TRANSIENT DEFORMATIONS OF SIMPLE STRUCTURES	71-0326
ELASTIC-PLASTIC	FINITE-ELEMENT ELASTIC-PLASTIC ANALYSIS OF RESIDUAL STRESSES DUE TO CLAD WELDING IN REACTOR VESSELS	75-0779
ELASTIC-PLASTIC	FINITE-ELEMENT PREDICTIONS OF TRANSIENT ELASTIC-PLASTIC LARGE DEFLECTIONS OF STIFFENED AND/OR UNSTIF	74-0825
ELASTIC-PLASTIC	FRACTURE TOUGHNESS PARAMETERS AND ELASTIC-PLASTIC ANALYSIS OF NON-MODERATE FRACTURE CONDITIONS USING	73-0258
ELASTIC-PLASTIC	FRACTURE TOUGHNESS PARAMETERS AND ELASTIC-PLASTIC ANALYSIS OF NON-MODERATE FRACTURE CONDITIONS USING	73-1126
ELASTIC-PLASTIC	HEAVY SECTION STEEL TECHNOLOGY PROGRAM. TECHNICAL REPORT NO. 4, OCTOBER 1969. A TWO-DIMENSIONAL ELAS	69-0474
ELASTIC-PLASTIC	ISOTROPIC-KINEMATIC HARDENING MODEL FOR ELASTIC-PLASTIC CYCLIC STRUCTURAL ANALYSIS	75-0210
ELASTIC-PLASTIC	LARGE DEFLECTION ANALYSIS OF ELASTIC-PLASTIC SHELLS OF REVOLUTION	69-0174
ELASTIC-PLASTIC	LARGE DEFLECTION ANALYSIS OF ELASTIC-PLASTIC PLATES AND SHELLS	69-0178
ELASTIC-PLASTIC	LARGE DEFLECTION ANALYSIS OF ELASTIC-PLASTIC SHELLS OF REVOLUTION	70-0507
ELASTIC-PLASTIC	LARGE DEFLECTION ELASTIC-PLASTIC ANALYSIS OF SHELLS OF REVOLUTION	71-0263
ELASTIC-PLASTIC	LARGE DEFLECTION ELASTIC-PLASTIC DYNAMIC RESPONSE OF STIFFENED SHELLS OF REVOLUTION	72-0608
ELASTIC-PLASTIC	LARGE DEFLECTION ELASTIC-PLASTIC DYNAMIC RESPONSE OF STIFFENED SHELLS OF REVOLUTION	74-0496
ELASTIC-PLASTIC	MECHANICAL CHARACTERIZATION AND FINITE-ELEMENT ANALYSIS OF ELASTIC-PLASTIC, WORK-HARDENING SOILS	72-0974
ELASTIC-PLASTIC	NUMERICAL STUDY OF UNIAXIAL COMPRESSION IN CIRCULAR ELASTIC-PLASTIC COLUMNS	73-0238
ELASTIC-PLASTIC	ON THE DETERMINATION OF EFFECTIVE ELASTIC-PLASTIC PROPERTIES FOR THE EQUIVALENT SOLID PLATE ANALYSI	74-0494
ELASTIC-PLASTIC	ON THE FINITE ELEMENT ANALYSIS OF CRACK AND INCLUSION PROBLEMS IN ELASTIC-PLASTIC MATERIALS	74-0227
ELASTIC-PLASTIC	ON THE FRACTURE MECHANICS ANALYSIS OF ELASTIC-PLASTIC MATERIALS USING THE FINITE ELEMENT METHOD	73-1067
ELASTIC-PLASTIC	PARAMETER EINES ELASTO-PLASTISCHEN STOFFANSATZES FUER KOERNIGE ERDSTOFFE (DETERMINATION OF PARAMETER	74-0657
ELASTIC-PLASTIC	PLASTIC STRESS-STRAIN MATRIX AND ITS APPLICATION FOR THE SOLUTION OF ELASTIC-PLASTIC PROBLEMS BY THE	68-0171
ELASTIC-PLASTIC	PROGRESS IN THREE-DIMENSIONAL ELASTIC-PLASTIC STRESS ANALYSIS FOR FRACTURE MECHANICS	71-0314
ELASTIC-PLASTIC	PROGRESS REPORT ON DISCRETE-ELEMENT ELASTIC AND ELASTIC-PLASTIC ANALYSIS OF SHELLS OF REVOLUTION SUB	68-0170
ELASTIC-PLASTIC	QUADRILATERAL FINITE ELEMENT IN ELASTIC-PLASTIC PLANE STRESS ANALYSIS	67-0067
ELASTIC-PLASTIC	RATE OF CONVERGENCE FOR THE ITERATIVE APPROACH IN ELASTIC-PLASTIC ANALYSIS OF CONTINUA	73-0518
ELASTIC-PLASTIC	RECENT DEVELOPMENTS IN MATRIX DISPLACEMENT METHOD FOR ELASTIC-PLASTIC PROBLEMS IN JAPAN	71-0265
ELASTIC-PLASTIC	REFINED FINITE ELEMENT ANALYSIS OF ELASTIC-PLASTIC THIN SHELLS OF REVOLUTION	69-0497
ELASTIC-PLASTIC	RESULT OF CALCULATIONS WITH TRIM 6 AND TRIAX 6 ELASTIC-PLASTIC ELEMENTS	73-1122
ELASTIC-PLASTIC	SIMPLE FINITE ELEMENT MODEL FOR ELASTIC-PLASTIC PLATE BENDING.	72-0447
ELASTIC-PLASTIC	STEADILY GROWING ELASTIC-PLASTIC CRACK TIP IN A FINITE ELEMENT TREATMENT	73-0087
ELASTIC-PLASTIC	STRESS STRAIN MATRIX AND ITS APPLICATIONS FOR THE SOLUTION OF ELASTIC-PLASTIC PROBLEMS BY THE FINITE	68-0282
ELASTIC-PLASTIC	STRUCTURAL INELASTICITY. V11. ELASTIC-PLASTIC ANALYSIS OF A WHEEL ROLLING ON A RIGID TRACK	73-0886
ELASTIC-PLASTIC	STUDY OF ELASTIC-PLASTIC MATRIX ANALYSIS FOR FOUNDATION AND REINFORCED CONCRETE IN JAPAN	69-0278
ELASTIC-PLASTIC	THE FINITE ELEMENT ANALYSIS OF ELASTIC-PLASTIC THICK-WALLED TUBES	74-0768
ELASTIC-PLASTIC	THE USE OF ELASTIC-PLASTIC FINITE ELEMENT ANALYSIS IN THE CALCULATION OF CUMULATIVE FATIGUE DAMAGE	74-0314
ELASTIC-PLASTIC	THEORETICAL AND EXPERIMENTAL STUDIES OF TRANSIENT ELASTIC-PLASTIC LARGE DEFLECTIONS OF GEOMETRICALLY	75-0830
ELASTIC-PLASTIC	THEORETICAL AND EXPERIMENTAL STUDIES OF TRANSIENT ELASTIC-PLASTIC LARGE DEFLECTIONS OF GEOMETRICALLY	75-0831
ELASTIC-PLASTIC	THREE-DIMENSIONAL ELASTIC-PLASTIC STRESS AND STRAIN ANALYSIS FOR FRACTURE MECHANICS. PHASE 1. SIMPLE	70-0188
ELASTIC-PLASTIC	THREE-DIMENSIONAL FINITE ELEMENT ANALYSIS INCLUDING ELASTIC-PLASTIC MATERIAL RESPONSE	74-1126
ELASTIC-PLASTIC	ELASTIC-PLASTIC-CREEP ANALYSIS OF THERMAL RATCHETTING IN STRAIGHT PIPE AND COMPARISONS WITH TEST RES	73-0393
ELASTIC-VISCOPL	A LARGE DEFORMATION ANALYSIS OF CRYSTALLINE ELASTIC-VISCOPLASTIC MATERIALS	73-0602
ELASTIC-VISCOPL	NONLINEAR ELASTIC-VISCOPLASTIC ANALYSIS USING THE FINITE ELEMENT METHOD	74-0750
ELASTICA	COMPARATIVE PERTURBATION STUDIES OF THE ELASTICA	69-0255
ELASTICA	COMPORTAREA ELASTICA A DIGURILOR DE PAMINT SOLICIATE SEISMIC (ELASTIC BEHAVIOR OF THE EARTH DAMS SEI	74-1185
ELASTICA	FINITE ELEMENT SOLUTION TO AN ELASTICA PROBLEM OF BEAMS	70-0112
ELASTICALLY	ANALYSIS OF ELASTICALLY COUPLED SHELLS ON ELASTIC FOUNDATION BY HYBRID METHOD	71-0276
ELASTICALLY	FINITE ELEMENT FORMULATION OF STRESS STRAIN MATRIX FOR AN ELASTICALLY PERFECTLY PLASTIC TRESCA MATER	70-0275
ELASTICI	SULLA DETERMINAZIONE DELLE FREQUENZE PROPRIE DI ALBERI SU PIU SUPPORTI ELASTICI. (DETERMINATION OF F	74-0916
ELASTICITA	ON THE RESOLUTION OF NONLINEAR ELASTIC AND ELASTOPLASTIC PROBLEMS BY THE METHOD OF FINITE ELEMENTS.	74-1326
ELASTICITE	ELEMENTS FINIS RAFFINES EN ELASTICITE BIDIMENSIONNELLE	69-0411
ELASTICITY	A HYBRID-ELEMENT APPROACH TO CRACK PROBLEMS IN PLANE ELASTICITY	73-0569
ELASTICITY	APPLICATION OF FINITE ELASTICITY THEORY TO THE DEFORMATION OF RUBBERS	62-0005
ELASTICITY	APPLICATION OF FINITE DIFFERENCES FOR SOLVING THE TWO-DIMENSIONAL ELASTICITY PROBLEM BY MEANS OF THE	75-0482
ELASTICITY	APPLICATIONS DE LA METHODE DE GAUSS-SEIDEL A LA RESOLUTION DES PROBLEMS D'ELASTICITE. (APPLICATION O	72-0145
ELASTICITY	APPLICATIONS OF TRILINEAR COORDINATES TO SOME PROBLEMS IN PLANE ELASTICITY	72-0728
ELASTICITY	APPROXIMATION IN ELASTICITY BASED ON THE CONCEPT OF FUNCTION SPACE	47-0001
ELASTICITY	BOUSSINESQ PROBLEM OF PLANE MICROPOLAR ELASTICITY	74-0197
ELASTICITY	CONVERGENCE OF THE FINITE ELEMENT METHOD IN THE THEORY OF ELASTICITY	68-0095
ELASTICITY	CRACK TIP FINITE ELEMENTS FOR PLANE ELASTICITY	71-0764
ELASTICITY	DAMS ON ROCKS OF VARYING ELASTICITY	74-0671
ELASTICITY	DERIVATION OF STIFFNESS MATRICES FOR PROBLEMS IN PLANE ELASTICITY BY GALERKIN'S METHOD	69-0250
ELASTICITY	EFFECTS OF SHEAR DEFORMATION, ROTARY INERTIA, AND ELASTICITY OF THE SUPPORT ON THE RESONANCE FREQUEN	75-0369
ELASTICITY	ELASTICITY EQUATIONS FOR INCOMPRESSIBLE AND NEARLY INCOMPRESSIBLE MATERIALS BY A VARATIONAL THEOREM	65-0019
ELASTICITY	FINITE ELEMENT APPROXIMATIONS IN NONLINEAR ELASTICITY	71-0470
ELASTICITY	FINITE ELEMENT FORMULATION OF GEOMETRICALLY NONLINEAR PROBLEMS OF ELASTICITY	71-0114
ELASTICITY	FINITE ELEMENT GALERKIN METHOD FOR MIXED INITIAL-BOUNDARY VALUE PROBLEMS IN ELASTICITY THEORY	71-0520
ELASTICITY	FINITE ELEMENT METHODS BY DIFFERENT VARIATIONAL PRINCIPALS IN ELASTICITY	70-0407
ELASTICITY	FINITE ELEMENT METHODS IN ELASTICITY	74-1118
ELASTICITY	FINITE ELEMENT STIFFNESS METHODS BY DIFFERENT VARIATIONAL PRINCIPLES IN ELASTICITY	70-0455
ELASTICITY	GENERAL VARIATIONAL PRINCIPLES IN NONLINEAR AND LINEAR ELASTICITY WITH APPLICATIONS	72-0914
ELASTICITY	GOVERNING EQUATIONS AND EXTREMUM PRINCIPLES OF ELASTICITY AND PLASTICITY GENERATED FROM A SINGLE FUN	73-0666
ELASTICITY	HYBRID-ELEMENT APPROACH TO CRACK PROBLEMS IN PLANE ELASTICITY	73-0508
ELASTICITY	INVESTIGATION OF THE STRESS-STRAINED STATE OF SPHERICAL SHELLS WITH AN ECCENTRIC CUTOUT ON THE BASIS	73-0222
ELASTICITY	LOCAL COMPLIANCE PROBLEM OF LUNG ELASTICITY	75-0660
ELASTICITY	MIXED VARIATIONAL PRINCIPLES, FINITE ELEMENTS AND FINITE ELASTICITY	74-1254
ELASTICITY	NUMERICAL FORMULATION OF NONLINEAR ELASTICITY PROBLEMS	67-0063
ELASTICITY	ON A VARIATIONAL THEOREM FOR INCOMPRESSIBLE AND NEARLY INCOMPRESSIBLE ORTHOTROPIC ELASTICITY	68-0249
ELASTICITY	ON THE EFFECT OF THE FORM OF THE STRAIN ENERGY FUNCTION ON THE SOLUTION OF A BOUNDARY-VALUE PROBLEM	72-0429
ELASTICITY	ON THE NUMERICAL SOLUTION OF A CLASS OF PROBLEMS IN A LINEAR FIRST STRAIN-GRADIENT THEORY OF ELASTIC	70-0318
ELASTICITY	ON THE NUMERICAL SOLUTION OF CERTAIN PROBLEMS IN FINITE ELASTICITY BY THE FINITE ELEMENT METHOD	75-0542
ELASTICITY	PLANE ELASTICITY PROBLEMS AND THE DIRECT STIFFNESS METHOD	61-0009
ELASTICITY	PROJECTIVE METHOD APPLIED TO THREE-DIMENSIONAL ELASTICITY EQUATIONS	74-0617
ELASTICITY	REFINED FINITE ELEMENTS IN 2-DIMENSIONAL ELASTICITY	69-0320

ELASTICITY	SECTOR FINITE ELEMENTS IN THE THEORY OF PLANE ELASTICITY	71-0766
ELASTICITY	SOLUTION OF PROBLEMS IN ELASTICITY BY THE FRAMEWORK METHOD	41-0001
ELASTICITY	SOLUTION OF THE MIXED PROBLEM OF THE THEORY OF ELASTICITY AND THE THEORY OF PLASTICITY OF SOILS	74-1136
ELASTICITY	SOME MATHEMATICAL PROPERTIES OF CERTAIN MIXED GALERKIN APPROXIMATIONS IN NONLINEAR ELASTICITY	74-1248
ELASTICITY	SOME RESULTS OF FINITE ELEMENT APPLICATIONS IN FINITE ELASTICITY	73-0183
ELASTICITY	SOME RESULTS OF FINITE ELEMENT APPLICATIONS ON NONLINEAR ELASTICITY	73-0605
ELASTICITY	STABILITY OF FINITE ELEMENT SCHEMES FOR VIBRATION PROBLEMS OF THE EQUATION OF ELASTICITY THEORY	71-0521
ELASTICITY	STRAIN ENERGY FUNCTIONS AND MATHEMATICAL MODELS OF LUNG ELASTICITY	73-0591
ELASTICITY	STRESS ANALYSIS IN VISCOELASTIC BODIES USING FINITE ELEMENTS AND A CORRESPONDENCE RULE WITH ELASTICI	69-0270
ELASTICITY	THE COMPARISON OF THE FINITE ELEMENT METHOD WITH THE VARIATIONAL-DIFFERENCE METHOD IN THE THEORY OF	67-0096
ELASTICITY	THE ENERGY METHOD OF NETWORKS OF ARBITRARY SHAPE IN PROBLEMS OF THE THEORY OF ELASTICITY	59-0001
ELASTICITY	THE METHOD OF INTERFACE STRESS ADJUSTMENT AND ITS USES IN SOME PLANE ELASTICITY PROBLEMS	51-0005
ELASTICITY	THEORY OF EQUIVALENCE AND ITS APPLICATION TO VARIOUS PROBLEMS IN ELASTICITY AND DESIGN OF SKEW BRIDG	72-0470
ELASTICITY	THREE DIMENSIONAL THEORY OF ELASTICITY	69-0099
ELASTICITY	VARIATIONAL PRINCIPLES AND DUALISTIC SCHEME FOR INTERSECTION PROBLEMS IN ELASTICITY	69-0305
ELASTICITY	VARIATIONAL PRINCIPLE FOR INCOMPRESSIBLE AND NEARLY-INCOMPRESSIBLE ANISOTROPIC ELASTICITY	69-0319
ELASTICO	ANALISI NUMERICA DEL COMPORTAMENTO ELASTICO DI UNA STRUTTURA BIDIMENSIONALE PIANA. (NUMERICAL ANALYS	74-1029
ELASTIQUE	CHARACTERISATION ELASTIQUE D'UN STRATIFIE RESINE-VERRE PAR LA METHODE DES DEPLACEMENTS	75-0272
ELASTISCH-PLAST	CONTINUUM CALCULATION FROM ELASTIC PLASTIC REINFORCED MATERIAL USING A FINITE ELEMENT METHOD (DIE BE	72-0658
ELASTISCH-PLAST	BERECHNUNG DER SCHWEISSEIGENSPANNUNGEN UND SCHWEISSFORMAENDERUNGEN MIT ELASTISCH-PLASTISCHEN FINITEN	75-0271
ELASTISCHE	ACCURATE CALCULATION OF STRESSES IN BEAMS, DISKS AND PLATES BY MEANS OF FINITE ELEMENT METHODS AND I	70-0133
ELASTISCHEN	FINEL, UNIVERSELLES PROGRAMMSYSTEM ZUR BERECHNUNG DER ELASTISCHEN EIGENSCHAFTEN VON MASCHINENBAUTEIL	72-0132
ELASTISCHEN	PARAMETRISCHE KERBSPANNUNGSUNTERSUCHUNG AM ELASTISCHEN KERN.(PARAMETRICAL INVESTIGATION OF NOTCH STR	72-0208
ELASTISCHER	ANWENDUNG DER MATRIZENVERSCHIEBUNGSMETHODE AUF ERZWUNGENE SCHWINGUNGEN (SCHWINGUNGSPROBLEME) (ARBEIT	71-0627
ELASTISCHER	NUMERISCHE BERECHNUNG NICHTLINEARER ELASTISCHER KOERPER	63-0017
ELASTISCHER	ZUR LOESUNG DES KONTAKTPROBLEMS ELASTISCHER KOERPER MIT AUSGEDEHNTER KONTAKTFLAECHE DURCH QUADRATISC	74-1135
ELASTIZITATSFIG	BERECHNUNG DER ELASTIZITATSEIGENSCHAFTEN VON MASCHINENBAUTEILEN NACH DER METHODE FINITER ELEMENTE	73-1046
ELASTO	AXISYMMETRIC SOLUTION OF ELASTO PLASTIC PROBLEMS BY FINITE ELEMENT METHODS	69-0361
ELASTO	ELASTO PLASTIC ANALYSIS OF PLANE STRUCTURES BY THE FINITE ELEMENT DISPLACEMENT METHOD	70-0299
ELASTO	ELASTO PLASTIC STRESS IN CRACKED PLATES	65-0042
ELASTO	LARGE STRAIN, VISCOELASTIC AND ELASTO VISCOPLASTIC NUMERICAL ANALYSIS BY MEANS OF THE FINITE ELEMENT	75-0563
ELASTO-PLASTIC	A DISCRETE ELEMENT METHOD FOR THE ANALYSIS OF PLANE ELASTO-PLASTIC STRESS PROBLEMS	66-0062
ELASTO-PLASTIC	A FINITE ELEMENT DISPLACEMENT APPROACH FOR THE ELASTO-PLASTIC ANALYSIS OF THIN CYLINDRICAL SHELLS	74-1077
ELASTO-PLASTIC	A VARIATIONAL METHOD FOR LARGE DEFLECTION ELASTO-PLASTIC ANALYSIS	69-0115
ELASTO-PLASTIC	ANALYSE TRIDIMENSIONELLE DE LA STABILITE D'UN TUNNEL AU VOISINAGE DU FRONT DE TAILLE DANS UNE ROCHE	74-0995
ELASTO-PLASTIC	ANALYSIS OF ELASTO-PLASTIC THERMAL STRESSES BY THE FINITE ELEMENT METHOD	70-0142
ELASTO-PLASTIC	APPLICATION OF ELASTO-PLASTIC ANALYSIS IN ROCK MECHANICS BY FINITE ELEMENT METHOD	71-0349
ELASTO-PLASTIC	BERECHNUNG DER SCHWEISSEIGENSPANNUNGEN UND SCHWEISSFORMAENDERUNGEN MIT ELASTISCH-PLASTISCHEN FINITEN	75-0271
ELASTO-PLASTIC	ELASTO-PLASTIC ANALYSIS OF UNDERGROUND OPENINGS BY THE FINITE ELEMENT METHOD	66-0067
ELASTO-PLASTIC	ELASTO-PLASTIC ANALYSIS OF METAL SHEETS WITH A CENTRAL CRACK	67-0137
ELASTO-PLASTIC	ELASTO-PLASTIC ANALYSIS OF ORTHOTROPIC PLATES AND SHELLS	68-0166
ELASTO-PLASTIC	ELASTO-PLASTIC ANALYSIS OF PLATES USING THE FINITE ELEMENT METHOD	71-0038
ELASTO-PLASTIC	ELASTO-PLASTIC ANALYSIS OF THREE-DIMENSIONAL STRUCTURES USING THE ISOPARAMETRIC ELEMENT	72-0248
ELASTO-PLASTIC	ELASTO-PLASTIC ANALYSIS OF AUTOMOBILE BODY STRUCTURE BY THE FINITE ELEMENT METHOD	74-0196
ELASTO-PLASTIC	ELASTO-PLASTIC ANALYSIS OF STRESS IN A GAS-TURBINE WHEEL	74-0323
ELASTO-PLASTIC	ELASTO-PLASTIC ANALYSIS USING AN EFFICIENT FORMULATION OF THE FINITE ELEMENT METHOD	75-0012
ELASTO-PLASTIC	ELASTO-PLASTIC ANALYSIS OF A ROTATING MODEL CONICAL TURBINE DISC	75-0281
ELASTO-PLASTIC	ELASTO-PLASTIC CYCLIC ANALYSIS OF STRUCTURAL MEMBERS	72-0202
ELASTO-PLASTIC	ELASTO-PLASTIC FINITE ELEMENT ANALYSIS OF SOIL DEFORMATION	73-1157
ELASTO-PLASTIC	ELASTO-PLASTIC INDENTATION OF A LAYERED MEDIUM	74-0225
ELASTO-PLASTIC	ELASTO-PLASTIC MATRIX DISPLACEMENT ANALYSIS OF THREE-DIMENSIONAL CONTINUA	66-0004
ELASTO-PLASTIC	ELASTO-PLASTIC ORTHOTROPIC PLATES AND SHELLS	69-0467
ELASTO-PLASTIC	ELASTO-PLASTIC SEISMIC RESPONSE ANALYSIS OF FRAMED STRUCTURES BY MODE SUPERPOSITION	73-0325
ELASTO-PLASTIC	ELASTO-PLASTIC SOLUTIONS OF ENGINEERING PROBLEMS ' INITIAL STRESS ' FINITE ELEMENT APPROACH	69-0282
ELASTO-PLASTIC	ELASTO-PLASTIC STRESS ANALYSIS. A GENERALIZATION FOR VARIOUS CONSTITUTIVE RELATIONS INCLUDING STRAIN	72-0547
ELASTO-PLASTIC	FINITE ELASTO-PLASTIC DEFORMATION - 1. THEORY AND NUMERICAL EXAMPLES	74-0202
ELASTO-PLASTIC	FINITE ELEMENT ANALYSIS OF ELASTO-PLASTIC SOILS. REPORT NO. 4. FINITE ELEMENT ANALYSIS OF ELASTO-PLA	72-0788
ELASTO-PLASTIC	FINITE ELEMENT ANALYSIS OF ELASTO-PLASTIC SOILS. REPORT NO. 4. FINITE ELEMENT ANALYSIS OF ELASTO-PLA	72-0788
ELASTO-PLASTIC	FINITE ELEMENT ANALYSIS OF ELASTO-PLASTIC SHELLS	74-0249
ELASTO-PLASTIC	FINITE ELEMENT ANALYSIS OF ELASTO-PLASTIC MATERIALS BY USE OF DISLOCATION DIPOLE SYSTEMS	74-0269
ELASTO-PLASTIC	FINITE ELEMENT ANALYSIS OF ELASTO-PLASTIC TEE JOINTS	74-1313
ELASTO-PLASTIC	FINITE ELEMENT APPROXIMATION, AND SOLUTION BY A PENALTY-DUALITY ALGORITHM OF AN ELASTO-PLASTIC PROBL	75-0041
ELASTO-PLASTIC	FINITE ELEMENT SOLUTION OF EXTRUSION OF ELASTO-PLASTIC WORK HARDENING MATERIALS	73-1127
ELASTO-PLASTIC	IMPROVED APPROXIMATION OF CONSTITUTIVE ELASTO-PLASTIC STRESS STRAIN RELATIONSHIP FOR FINITE ELEMENT	74-0566
ELASTO-PLASTIC	IMPROVED CALCULATION OF ELASTO-PLASTIC SYSTEMS BY THE FINITE ELEMENT METHOD	70-0076
ELASTO-PLASTIC	INFLUENCE OF CRACK CLOSURE AND ELASTO-PLASTIC FLOW ON THE BENDING OF A CRACKED PLATE	75-0735
ELASTO-PLASTIC	LARGE STRAIN ELASTO-PLASTIC FINITE ELEMENT ANALYSIS	71-0121
ELASTO-PLASTIC	LARGE STRAIN, ELASTO-PLASTIC NUMERICAL ANALYSIS BY MEANS OF FINITE ELEMENT METHOD	71-0790
ELASTO-PLASTIC	LARGE-DEFLECTION ELASTO-PLASTIC BUCKLING ANALYSIS OF PLATES USING FINITE ELEMENTS	73-0394
ELASTO-PLASTIC	METHOD OF FINITE ELEMENTS FOR CALCULATION OF HETEROGENEOUS BODIES OF REVOLUTION IN ELASTO-PLASTIC ME	75-0144
ELASTO-PLASTIC	METHODS OF ELASTO-PLASTIC ANALYSIS	71-0617
ELASTO-PLASTIC	MIXED FINITE ELEMENT ANALYSIS OF PLATES IN BENDING. SMALL DEFLECTION THEORY OF ELASTIC AND ELASTO-PL	71-0783
ELASTO-PLASTIC	MIXED FINITE ELEMENT ANALYSIS OF ELASTO-PLASTIC PLATES IN BENDING	72-0135
ELASTO-PLASTIC	PROCEDURE FOR SOLVING PROBLEMS OF ELASTO-PLASTIC FLOW	73-0217
ELASTO-PLASTIC	STATIC AND DYNAMIC ELASTO-PLASTIC ANALYSIS BY THE METHOD OF FINITE ELEMENTS IN SPACE AND TIME	72-0642
ELASTO-PLASTIC	STRESSES AND DEFORMATIONS OF VERTICAL SLOPES IN ELASTO-PLASTIC ROCKS	70-0114
ELASTO-PLASTIC	STUDY ON FRACTURE TOUGHNESS EVALUATION BY ELASTO-PLASTIC ANALYSIS AROUND NOTCHES	74-0619
ELASTO-PLASTIC	THE APPLICATION OF THE MATRIX DISPLACEMENT METHOD IN PLANE ELASTO-PLASTIC STRESS PROBLEMS	64-0009
ELASTO-PLASTIC	THE ELASTO-PLASTIC AND LARGE-DISPLACEMENT RESPONSE OF PLATES TO BLAST LOADING	73-0731
ELASTO-PLASTIC	THE ELASTO-PLASTIC CALCULATION OF GENERAL STRUCTURES AND CONTINUA	68-0207
ELASTO-PLASTIC	THE ELASTO-PLASTIC INDENTATION OF A HALF-SPACE BY A RIGID SPHERE	71-0113
ELASTO-PLASTIC	THE SIGNIFICANCE OF THE RATE OF PLASTIC WORK IN ELASTO-PLASTIC ANALYSIS	74-0437
ELASTO-PLASTIC	THE SIGNIFICANCE OF THE RATE OF PLASTIC WORK IN ELASTO-PLASTIC ANALYSIS	74-0764
ELASTO-PLASTIC	THREE DIMENSIONAL ELASTIC AND ELASTO-PLASTIC ANALYSIS OF REACTOR PRESSURE VESSELS	69-0453
ELASTO-PLASTIC	THREE-DIMENSIONAL ELASTO-PLASTIC FINITE ELEMENT ANALYSIS	75-0236
ELASTO-PLASTICI	ON THE RESOLUTION OF NONLINEAR ELASTIC AND ELASTOPLASTIC PROBLEMS BY THE METHOD OF FINITE ELEMENTS.	74-1326
ELASTO-PLASTICI	LAGRANGIAN AND EULERIAN FINITE ELEMENT FORMULATION FOR LARGE STRAIN ELASTO-PLASTICITY	75-0209
ELASTO-PLASTICI	MATRIX DIFFERENTIAL EQUATION AND HIGHER-ORDER NUMERICAL METHODS FOR PROBLEMS OF NON-LINEAR CREEP, VI	72-0198
ELASTO-PLASTIQU	ANALYSE TRIDIMENSIONELLE DE LA STABILITE D'UN TUNNEL AU VOISINAGE DU FRONT DE TAILLE DANS UNE ROCHE	74-0995
ELASTO-PLASTISC	DIE ELASTO-PLASTISCHE BERECHNUNG VON ALLGEMEINEN TRAGWERKEN UND KONTINUA	69-0443
ELASTO-PLASTISC	PARAMETER EINES ELASTO-PLASTISCHEN STOFFANSATZES FUER KOERNIGE ERDSTOFFE (DETERMINATION OF PARAMETER	74-0657
ELASTO-STATIC	LEAST SQUARE-FINITE ELEMENT FOR ELASTO-STATIC PROBLEMS	74-0279
ELASTO-STATIC	SOLUTION OF ELASTO-STATIC BOUNDARY VALUE PROBLEMS BY THE FINITE ELEMENT METHOD	64-0025
ELASTO-STATIC	SOLUTION OF ELASTO-STATIC BOUNDARY VALUE PROBLEMS BY THE FINITE ELEMENT METHOD	64-0030
ELASTO-STATIC	THREE-DIMENSIONAL ELASTO-STATIC PROBLEMS USING ISOPARAMETRIC FINITE ELEMENTS	71-0751
ELASTO-VISCOPLA	ELASTO-VISCOPLASTIC FINITE ELEMENT ANALYSIS BY PERTURBATION METHOD	74-1271
ELASTO-VISCOPLA	LARGE STRAIN, VISCOELASTIC AND ELASTO-VISCOPLASTIC NUMERICAL-ANALYSIS BY MEANS OF FINITE-ELEMENT MET	75-0435
ELASTODYNAMIC	ASSUMED STRESS HYBRID FINITE ELEMENT MODEL FOR LINEAR ELASTODYNAMIC ANALYSIS	73-0089
ELASTODYNAMIC	ELASTODYNAMIC BEHAVIOR OF LAMINATED ORTHOTROPIC PLATES UNDER INITIAL STRESS	75-0249
ELASTODYNAMICS	DUALITY IN STRUCTURAL ANALYSIS BY FINITE ELEMENTS. STATIC GEOMETRIC ANALOGIES. THE DUAL PRINCIPLES O	71-0583
ELASTODYNAMICS	ELASTODYNAMICS OF COMPLEX STRUCTURAL SYSTEMS	71-0563
ELASTODYNAMICS	ELASTODYNAMICS OF CRACKED STRUCTURES USING FINITE ELEMENTS	74-0625

207

ELASTODYNAMICS	MEAN-SQUARE ERRORS OF FINITE ELEMENT APPROXIMATIONS ON LINEAR ELASTOSTATICS, ELASTODYNAMICS, AND THE	74-0107
ELASTODYNAMICS	ON THE SEMI-DISCRETE GALERKIN METHOD FOR HYPERBOLIC PROBLEMS AND ITS APPLICATION TO PROBLEMS IN ELAS	72-0457
ELASTODYNAMICS	SOME DISTRIBUTIONAL AND CONVERGENCE PROPERTIES OF THE FINITE ELEMENT METHOD, WITH APPLICATIONS IN NO	73-0598
ELASTODYNAMICS	THE DUAL PRINCIPLE OF ELASTODYNAMICS FINITE ELEMENT APPLICATIONS	71-0501
ELASTODYNAMICS	THEMIS FINAL REPORT, VOL. IV, RECENT DEVELOPMENTS IN THE NUMERICAL SOLUTION OF CERTAIN PROBLEMS IN N	74-0721
ELASTOHYDRODYNA	A UNIFIED TREATMENT OF THICK AND THIN FILM ELASTOHYDRODYNAMIC PROBLEMS USING HIGHER ORDER ELEMENT PR	75-0890
ELASTOHYDRODYNA	APPLICATION OF A GRADIENT METHOD IN FINITE ELEMENT SOLUTION OF ELASTOHYDRODYNAMIC PROBLEMS	74-0129
ELASTOHYDRODYNA	NUMERICAL SOLUTION OF THE ELASTOHYDRODYNAMIC LUBRICATION PROBLEM USING FINITE ELEMENTS	72-0042
ELASTOHYDRODYNA	SOLUTION OF THE ELASTOHYDRODYNAMIC FINITE JOURNAL BEARING PROBLEM	72-0033
ELASTOHYDRODYNA	TRANSIENT ELASTOHYDRODYNAMIC LUBRICATION BY FINITE ELEMENT METHODS	74-0087
ELASTOPLASTIC	A DISCRETE ELEMENT STRESS AND DISPLACEMENT ANALYSIS OF ELASTOPLASTIC PLATES	70-0108
ELASTOPLASTIC	A DYNAMIC ELASTOPLASTIC ANALYSIS OF THIN SHELLS OF REVOLUTION UNDER ASYMMETRIC MECHANICAL AND THERMA	72-0068
ELASTOPLASTIC	A METHOD FOR DETERMINING LOCAL ELASTOPLASTIC STRESS AND STRAIN IN METALLURGICALLY BONDED NOTCHED LAM	74-0686
ELASTOPLASTIC	A MODIFIED GRADIENT METHOD FOR FINITE ELEMENT ELASTOPLASTIC ANALYSIS BY QUADRATIC PROGRAMMING	73-0998
ELASTOPLASTIC	AN ELASTOPLASTIC ANALYSIS OF A UNIAXIALLY LOADED SHEET WITH AN INTERFERENCE - FIT BOLT	74-0687
ELASTOPLASTIC	APPLICATION OF THE FINITE ELEMENT DISPLACEMENT METHOD TO PROBLEMS OF ELASTOPLASTIC DEFORMATION	72-0199
ELASTOPLASTIC	CYCLIC LOADING AND CRACK PROPAGATION - AN ELASTOPLASTIC FINITE ELEMENT STUDY	74-1214
ELASTOPLASTIC	DIRECT FLEXIBILITY FINITE ELEMENT ELASTOPLASTIC ANALYSIS	71-0095
ELASTOPLASTIC	DYNAMIC ELASTOPLASTIC RESPONSE OF GEOMETRICALLY NONLINEAR ARBITRARY SHELLS OF REVOLUTION UNDER IMPUL	70-0098
ELASTOPLASTIC	ELASTOPLASTIC ANALYSIS OF PLANE STRAIN AND AXISYMMETRIC FLAT PUNCH INDENTATION BY THE FINITE ELEMENT	70-0343
ELASTOPLASTIC	ELASTOPLASTIC ANALYSIS OF A CANTILEVER BEAM	73-0984
ELASTOPLASTIC	ELASTOPLASTIC AND CREEP ANALYSIS WITH THE ASKA PROGRAM SYSTEM	74-0863
ELASTOPLASTIC	ELASTOPLASTIC CRACK PROPAGATION IN A TRANSVERSELY LOADED UNIDIRECTIONAL COMPOSITE	74-0217
ELASTOPLASTIC	ELASTOPLASTIC INDENTATION OF HALF-SPACE BY AN INFINITELY LONG RIGID CIRCULAR CYLINDER	71-0395
ELASTOPLASTIC	FINITE ELEMENT ELASTOPLASTIC ANALYSIS BY QUADRATIC PROGRAMMING: THE MULTISTAGE METHOD	73-0842
ELASTOPLASTIC	FINITE ELEMENT ELASTOPLASTIC ANALYSIS OF UNDERGROUND OPENINGS BY QUADRATIC PROGRAMMING	74-1288
ELASTOPLASTIC	FINITE ELEMENT INCREMENTAL ANALYSIS OF ELASTOPLASTIC PLATE BENDING	73-0557
ELASTOPLASTIC	HIGH-PERFORMANCE COMPOSITE MATERIALS FOR VEHICLE CONSTRUCTION: A FINITE ELEMENT COMPUTER PROGRAM FOR	73-0767
ELASTOPLASTIC	METHODS OF ELASTOPLASTIC ANALYSIS	69-0343
ELASTOPLASTIC	MICROMECHANICAL ANALYSIS OF CRACK PROPAGATION IN AN ELASTOPLASTIC COMPOSITE MATERIAL	74-0565
ELASTOPLASTIC	ON THE RESOLUTION OF NONLINEAR ELASTIC AND ELASTOPLASTIC PROBLEMS BY THE METHOD OF FINITE ELEMENTS.	74-1326
ELASTOPLASTIC	THE EFFECT OF COMPOSITE ACTION ON THE ELASTOPLASTIC ANALYSIS OF COMPLETE BUILDING STRUCTURES	74-0441
ELASTOPLASTIC	THEORETICAL AND EXPERIMENTAL ANALYSIS OF ELASTOPLASTIC DEFORMATION IN A UNIAXIALLY LOADED STRIP OF 2	73-0720
ELASTOPLASTIC	USER'S GUIDE FOR ANALYSIS OF FINITE ELASTOPLASTIC DEFORMATION: THE FIPDEF AND FIPAX PROGRAMS FOR THE	74-0799
ELASTOPLASTIC	VARIATIONAL PRINCIPLES FOR ELASTOPLASTIC CONTINUA	75-0837
ELASTOPLASTICIT	CALCUL DE MASSIFS EN ELASTOPLASTICITE APPLICATIONS A LA MECANIQUE DES ROCHES ET DES SOLS. (APPLICATI	74-0223
ELASTOPLASTICIT	CALCUL DE MASSIFS EN ELASTOPLASTICITE APPLICATIONS A LA MECANIQUE DES ROCHES ET DES SOLS. (APPLICATI	74-0223
ELASTOPLASTICIT	ON THE SOLUTION OF TWO-DIMENSIONAL PROBLEMS OF ELASTOPLASTICITY BY THE FINITE ELEMENT AND DIRECT STI	67-0002
ELASTOPLASTICIT	SHAKEDOWN THEORY IN PERFECT ELASTOPLASTICITY WITH ASSOCIATED AND NON ASSOCIATED FLOW LAWS - A FINITE	69-0172
ELASTOPLASTICIT	SOLUTION OF NONLINEAR PROBLEMS OF ELASTOPLASTICITY BY THE FINITE ELEMENT METHOD	68-0007
ELASTOSTATIC	A COMPARISON OF FINITE ELEMENT AND FINITE DIFFERENCE METHODS IN ELASTOSTATIC PROBLEMS	69-0273
ELASTOSTATIC	PSEUDO 3-DIMENSIONAL FINITE-ELEMENT FORMULATION FOR ELASTOSTATIC PROBLEMS AND ITS GEOPHYSICAL APPLIC	75-0597
ELASTOSTATIC	PSEUDO-THREE-DIMENSIONAL FINITE ELEMENT FORMULATION FOR ELASTOSTATIC PROBLEMS	74-0167
ELASTOSTATICS	MEAN-SQUARE ERRORS OF FINITE ELEMENT APPROXIMATIONS ON LINEAR ELASTOSTATICS, ELASTODYNAMICS, AND THE	74-0107
ELASTOSTATICS	VARIATIONAL PRINCIPLES OF ELASTOSTATICS WITH RELAXED CONTINUITY REQUIREMENTS AND THEIR APPLICABILITY	74-0758
ELASTOSTATIK	METHODE DER ELASTOSTATIK PROGRAMMIERUNG UND ANWENDUNG	70-0381
ELASTOSTATIK	VARIATIONAL PRINCIPLES OF ELASTOSTATICS WITH RELAXED CONTINUITY REQUIREMENTS AND THEIR APPLICABILITY	74-0758
ELASTOSTATIK-EL	ELASTOSTATIK-ELEMENTMETHODE (ESEM)	69-0415
ELBOW	INFLUENCE OF END-EFFECTS ON STRESSES AND FLEXIBILITY OF A PIPING ELBOW WITH IN-PLANE MOMENT	74-0107
ELBOWS	EXPERIMENTAL STUDY OF FLEXIBILITY FACTORS AND STRESSES OF WELDING ELBOWS WITH END EFFECTS	73-0422
ELBOWS	PLASTIC COLLAPSE LOADS FOR PIPE ELBOWS USING INELASTIC ANALYSIS	73-0941
ELBOWS	PLASTIC COLLAPSE LOADS FOR PIPE ELBOWS USING INELASTIC ANALYSIS	74-0507
ELBOWS	SIMPLIFIED INELASTIC (PLASTIC AND CREEP) ANALYSIS OF PIPE ELBOWS SUBJECTED TO INPLANE AND OUT-OF-PLA	75-0654
ELBOWS	THREE-DIMENSIONAL STRESS ANALYSES OF MITERED ELBOWS	75-0650
ELECTRIC	DIGITAL CALCULATION OF ELECTRIC FIELDS IN METAL CLAD INSTALLATIONS	73-0212
ELECTRIC	EFFICIENT TECHNIQUES FOR FINITE ELEMENT ANALYSIS OF ELECTRIC MACHINES	73-0134
ELECTRIC	FINITE ELEMENT ANALYSIS OF MERCURY SLOSH IN THE SOLAR ELECTRIC PROPULSION STAGE	75-0757
ELECTRIC	FINITE-ELEMENT ANALYSIS OF NONLINEAR MAGNETIC FIELDS IN ELECTRIC MACHINES	70-0484
ELECTRIC	REALIZATSIYA METODA KONECHNYKH ELEMENTOV NA EVM DLYA RASCHETA DVUMERNYKH ELEKTRICHESKIKH I MAGNITNYK	74-0251
ELECTRIC	SATURATED MAGNETIC ENERGY FUNCTIONAL FOR FINITE-ELEMENT ANALYSIS OF ELECTRIC MACHINES	75-0636
ELECTRICAL	ANALYSIS OF STEADY ELECTRICAL AND MAGNETIC-FIELDS BY FINITE-ELEMENT METHOD - APPLICATION TO TRANSFOR	75-0550
ELECTRICAL	ELECTROMAGNETIC AND ELECTRICAL MODELING BY THE FINITE ELEMENT METHOD	71-0462
ELECTRICAL	FINITE ELEMENT VIBRATION ANALYSIS OF MULTICONDUCTOR ELECTRICAL TRANSMISSION LINES	73-0590
ELECTRICAL	NONLINEAR FINITE ELEMENT SOLUTION OF ELECTRICAL MACHINES UNDER NO-LOAD AND FULL-LOAD CONDITIONS	74-0472
ELECTRICAL	STUDIO DI CAMPI ELETTRICI E MAGNETICI STANZIONAPI CON IL METODO DEGLI ELEMENTI FINITI. APPLICAZIONE	75-0207
ELECTRO-OSMOSIS	APPLICATIONS OF ELECTRO-OSMOSIS TO GROUND-WATER FLOW PROBLEMS	75-0336
ELECTRO-OSMOSIS	ELECTRO-OSMOSIS AND THE CONTAMINATION OF UNDERGROUND FLUIDS	73-0609
ELECTRO-OSMOTIC	FINITE ELEMENT ANALYSIS OF ELECTRO-OSMOTIC DEWATERING	73-0592
ELECTRO-OSMOTIC	NUMERICAL ANALYSIS OF ELECTRO-OSMOTIC FLOW IN SOILS	73-0444
ELECTRODE-NIPPL	FINITE-ELEMENT ANALYSIS OF AN ELECTRODE-NIPPLE JOINT	73-0843
ELECTRODES	FINITE ELEMENT SIMULATION OF ELECTROMECHANICAL RESONATORS WITH ARBITRARY ELECTRODES	75-0492
ELECTROELASTIC	2-DIMENSIONAL FINITE-ELEMENT ANALYSIS OF ELECTROELASTIC NORMAL MODES OF FERROELECTRIC DISKS AND SHEL	75-0594
ELECTROELASTIC	3-DIMENSIONAL FINITE-ELEMENT ANALYSIS OF ELECTROELASTIC NORMAL MODES OF A TANGENTIALLY POLARIZED, SE	75-0591
ELECTROKINETIC	FINITE ELEMENT SOLUTION OF COUPLED ELECTROKINETIC AND HYDRODYNAMIC FLOW IN POROUS MEDIA	72-0398
ELECTROLYSIS	A DIFFUSION ANALYSIS OF ION IN ELECTROLYSIS BY FINITE ELEMENT METHOD	73-0296
ELECTROMAGNETIC	CALCUL DES CHAMPS ELECTROMAGNETIQUES PAR LA METHODE DES ELEMENTS FINIS (COMPUTATION OF ELECTROMAGNET	73-0399
ELECTROMAGNETIC	ELECTROMAGNETIC AND ELECTRICAL MODELING BY THE FINITE ELEMENT METHOD	71-0462
ELECTROMAGNETIC	FINITE ELEMENT ANALYSIS OF ELECTROMAGNETIC FIELD PROBLEMS	74-0080
ELECTROMAGNETIC	FINITE ELEMENT METHOD TO ELECTROMAGNETIC INDUCTION PROBLEMS IN GEOPHYSICS	72-0309
ELECTROMAGNETIC	FINITE ELEMENTS IN THE SOLUTION OF ELECTROMAGNETIC INDUCTION PROBLEMS	74-0252
ELECTROMAGNETIC	SINGULARITIES IN ELECTROMAGNETIC TRANSMISSION LINES	72-0072
ELECTROMAGNETIC	THREE DIMENSIONAL FINITE ELEMENT APPROXIMATION IN ELECTROMAGNETIC EXPLORATION	74-0358
ELECTROMAGNETIQ	CALCUL DES CHAMPS ELECTROMAGNETIQUES PAR LA METHODE DES ELEMENTS FINIS (COMPUTATION OF ELECTROMAGNET	73-0399
ELECTROMECHANIC	FINITE ELEMENT SIMULATION OF TWO-DIMENSIONAL ELECTROMECHANICAL RESONATORS	74-1014
ELECTROMECHANIC	FINITE ELEMENT SIMULATION OF ENERGY-TRAPPED ELECTROMECHANICAL RESONATORS	75-0142
ELECTROMECHANIC	FINITE ELEMENT SIMULATION OF ELECTROMECHANICAL RESONATORS WITH ARBITRARY ELECTRODES	75-0492
ELECTROMECHANIC	NEW APPROACH TO ANALYSIS AND DESIGN OF ELECTROMECHANICAL FILTERS BY FINITE-ELEMENT TECHNIQUE	71-0476
ELECTRON	FINITE ELEMENT METHOD IN ELECTRON OPTICS	75-0081
ELECTRONIC	METHODS AND POSSIBILITIES FOR ELECTRONIC COMPUTATION OF PLATES IN FLEXURE	69-0329
ELECTRONIC	PLOTS - THE PLOTTING SUBROUTINE INCORPORATED IN THE BALLISTIC RESEARCH LABORATORIES ELECTRONIC SCIEN	72-0803
ELECTRONIC	PROBLEM OF USING THE METHOD OF FINITE ELEMENTS TO CONSTRUCT AN ALGORITHM FOR CALCULATING A THIN-WALL	74-0851
ELECTRONIC	STRUCTURAL DYNAMIC ANALYSIS OF ELECTRONIC ASSEMBLIES USING NASTRAN RESTART/FORMAT CHANGE CAPABILITY	71-0230
ELECTROSTATIC	LAPLACIAN ELECTROSTATIC FIELD CALCULATIONS BY FINITE ELEMENTS WITH AUTOMATIC GRID GENERATION	73-0084
ELECTROSTRICTIV	FINITE ELEMENT ANALYSIS OF FLEXURE - TYPE VIBRATORS WITH ELECTROSTRICTIVE TRANSDUCERS	70-0338
ELECTROTECHNICA	IL METODO DEGLI ELEMENTI FINITI: POSSIBILITA DI APPLICAZIONE A PROBLEMI DI INTERESSE DEGLI ELETTROTE	75-0200
ELEKTRICHESKIKH	REALIZATSIYA METODA KONECHNYKH ELEMENTOV NA EVM DLYA RASCHETA DVUMERNYKH ELEKTRICHESKIKH I MAGNITNYK	74-0251
ELEMENT-CHAIN	THE ELEMENT-CHAIN CONCEPT IN FINITE ELEMENT BASED PROGRAMMING	72-0934
ELEMENT-FINITE	COMPUTATION AND SOLUTION PROCEDURE FOR NONLINEAR ANALYSIS BY COMBINED FINITE ELEMENT-FINITE DIFFEREN	72-0571
ELEMENT-METHOD	APPLICATION OF AN * ELEMENT-METHOD * PROGRAM TO SOLVING TEMPERATURE DISTRIBUTION PROBLEMS (ANWENDUNG	71-0464
ELEMENT-STIFFNE	DERIVATION OF CRACK ELEMENT-STIFFNESS MATRIX BY THE COMPLEX VARIABLE APPROACH	73-0870
ELEMENTA	CALCULATION OF THE REINFORCED THIN-WALLED STRUCTURES BY THE METHOD OF FINITE ELEMENTS (RASCHET PODKR	75-0685
ELEMENTA	OPREDELENIE STATSIONARNYKH TEMPERATURNYKH POLEI V DETALYAKH DVIGATELEI VNUTRENNEGO SGORANIYA METODOM	73-0454

ELEMENTARY	FINITE AND DISCONNECTED SUBGROUPS OF SU(3) AND THEIR APPLICATION TO THE ELEMENTARY PARTICLE SPECTRUM	66-0129
ELEMENTARY	JET-WING LIFTING - SURFACE THEORY USING ELEMENTARY VORTEX DISTRIBUTIONS	75-0245
ELEMENTE	ANWENDUNG DER METHODEN DER FINITEN ELEMENTE MIT DEM PROGRAMMSYSTEM ASKA UNTER BESONDERER BERUCKSICHT	71-0624
ELEMENTE	ANWENDUNG DER METHODE FINITER FLEMENTE BEI DER ANALYSE DES DYNAMISCHEN VERHALTENS GEDAEMPFTER WERKZE	75-0580
ELEMENTE	AUFBEREITUNG MATHEMATISCHER MODELLE FUR DIE DIGITALE BERECHNUNG VON GRUNDWASSERSTPOMUNGSPROBLEMEN NA	75-0543
ELEMENTE	BERECHNUNG DER ELASTIZITATSFIGENSCHAFTEN VON MASCHINENBAUTEILEN NACH DER METHODE FINITER ELEMENTE	73-1046
ELEMENTE	BERECHNUNG INSTATIONARER GRUND - UND SICKERWASSERSTROMUNGEN MIT FREIER OBERFLACHE NACH DER METHODE F	71-0541
ELEMENTE	BERECHNUNG VON LAUFRAEDERN MIT DER METHODE DER FINITEN ELEMENTE (DESIGN OF ROTATING WHEELS BY THE FI	74-0547
ELEMENTE	BERECHNUNGSBEISPIELE UND ERFAHRUNGEN MIT DER METHODE DER FINITEN ELEMENTE IN MASCHINENBAU. (EXAMPLES	71-0390
ELEMENTE	COMPUTATION OF SHELLS WITH FINITE DISPLACEMENTS: MIXED FINITE ELEMENTS (BERECHNUNG VON SCHALEN MIT E	73-0788
ELEMENTE	DAS PROGRAMMSYSTEM ANTRAS EDV-SYSTEM ZUR BERECHNUNG BELIEBIGER STATISCHER SYSTEME NACH DER METHODE D	74-0918
ELEMENTE	DIE BERECHNUNG DER BEWEGUNGEN UND SPANNUNGEN DES GEBIRGES VOM ABBAU BIS ZUR TAGESOBERFLAECHE NACH DE	73-0456
ELEMENTE	DIE BERECHNUNG VON FLACHWASSERWELLEN NACH DER METHODE DER FINITEN ELEMENTE	72-0359
ELEMENTE	DIE LOSUNG NICHT-LINEARER PROBLEME NACH DER METHODE DER FINITEN ELEMENTE	74-0673
ELEMENTE	DIE METHODE DER ENDLICHEN ELEMENTE UND IHRE ANWENDUNG IM MASCHINENBAU	71-0628
ELEMENTE	EIN NEUES VARIATIONSPRINZIP MIT ANWENDUNG AUF SCHWINGUNGEN FLACHER SCHALEN NACH DER METHODE DER FINI	73-0563
ELEMENTE	EINIGE ALLGEMEINE PROGRAMMSYSTEME FUR FINITE ELEMENTE	73-0581
ELEMENTE	FEHLERBETRACHTUNGEN AUFGRUND DER METHODE DER GEWICHTETEN RESIDUEN FUR DAS VERFAHREN DER FINITEN ELEM	72-0368
ELEMENTE	FESTIGKEITSUNTERSUCHUNG MIT DER METHODE DER FINITN ELEMENTE. (STRENGTH OF THE MATERIALS INVESTIGATE	74-1009
ELEMENTE	FINEL. UNIVERSELLES PROGRAMMSYSTEM ZUR BERECHNUNG DER ELASTISCHEN EIGENSCHAFTEN VON MASCHINENBAUTEIL	72-0132
ELEMENTE	FINITE ELEMENTE FUR GEKRUEMMTE FLAECHENTRAGWERKE. (FINITE ELEMENTS SOLUTIONS FOR CURVED SHELL STRUCT	71-0391
ELEMENTE	FINITE ELEMENTE ZUR BERECHNUNG DER WAERMEUEBERTRAGUNG FUER EBENE UND AXISYMMETRISCHE PROBLEME (FINIT	73-0449
ELEMENTE	FINITE ELEMENTE ZUR BERECHNUNG VON SPANNBETONREAKTORDRUCKBEHALTERN	73-0702
ELEMENTE	FINITE ELEMENTS FOR CALCULATION OF PRESTRESSED CONCRETE REACTOR PRESSURE VESSELS (FINITE ELEMENTE ZU	73-0797
ELEMENTE	GRUNDLAGEN DER DIGITALEN LOSUNG VON GRUNDWASSERSTROMUNGSPROBLEMEN NACH DER METHODE ENDLICHER ELEMENT	71-0542
ELEMENTE	KONSTRUKTION UND BERECHNUNG MIT DER METHODE FINITER ELEMENTE	73-0707
ELEMENTE	PARAMETRISCHE KERBSPANNUNGSUNTERSUCHUNGEN AN DER LOCHSCHEIBE MIT DER METHODE DER FINITEN ELEMENTE. (	75-0603
ELEMENTE	PRAKTISCHE BEISPIELE DER DIGITALEN BERECHNUNG VON GRUNDWASSERSTROMUNGS PROBLEMEN NACH DER METHODE EN	73-0081
ELEMENTE	STABILITAT BEI DER METHODE DER FINITEN ELEMENTE FUR QUASILINEARE ELLIPTISCHE RANDWERTPROBLEME	74-1283
ELEMENTE	STABILITY INVESTIGATIONS OF PLATES USING FINITE ELEMENT METHOD BASED ON AN EXPANDED VARIATION PRINCI	71-0663
ELEMENTE	STARR-PLASTISCHE TRAGLASTBERECHNUNGEN MITTELS DER METHODE DER FINITEN ELEMENTE	71-0635
ELEMENTE	VARIATIONAL PRINCIPLES OF ELASTOSTATICS WITH RELAXED CONTINUITY REQUIREMENTS AND THEIR APPLICABILITY	74-0758
ELEMENTE	ZUR BERECHNUNG EINFACH-SYMMETRISCHER I-TRAEGER NACH DER THEORIE - 2. ORDNUNG UNTER BERUECKSICHTIGUNG	75-0510
ELEMENTEN	ACCURATE CALCULATION OF STRESSES IN BEAMS, DISKS AND PLATES BY MEANS OF FINITE ELEMENT METHODS AND I	70-0133
ELEMENTEN	BERECHNUNG DER SCHWEISSEIGENSPANNUNGEN UND SCHWEISSFORMAENDERUNGEN MIT ELASTISCH-PLASTISCHEN FINITEN	75-0271
ELEMENTEN	BERECHNUNG INSTATIONAERER TEMPERATUREFELDER MIT FINITEN ELEMENTEN IN RAUM-UND ZEITDIMENSION. (CALCUL	74-0987
ELEMENTEN	DER GERISSENE ZUSTAND ZWEISEITIG GELAGERTER PLATTEN UNTER EINZELLASTEN - NICHTLINEARE BERECHNUNG MIT	75-0858
ELEMENTEN	EINE FINITE-ELEMENT-METHODE MIT KREISRINGSEKTORFORMIGEN ELEMENTEN	73-1115
ELEMENTEN	KOMBINATION EINES INTEGRALGLEICHUNGS VERFAHRENS MIT METHODE DER FINITEN ELEMENTEN ZUR BERECHNUNG EBE	74-0856
ELEMENTEN	KONSTRUKTION UND BERECHNUNG MIT FINITEN ELEMENTEN	74-0678
ELEMENTEN	PRAKTISCHE RESULTATEN, DOEL EN NUT VAN HET ONTWERPEN VAN MECHANISCHE ELEMENTEN MET BEHULP VAN COMPUT	73-0400
ELEMENTEN	UEBER DIE PRAKTISCHE ANWENDUNG VON FINITEN ELEMENTEN IN DER PLATTENSTATIK (APPLICATION OF THE FINITE	75-0805
ELEMENTEN	VOLLSTAENDIGE SPANNUNGS-DEHNUNGS-TEMPERATURAENDERUNGSBEZIEHUNG FUER DIE SCHWEISSEIGENSPANNUNGSBERECH	75-0807
ELEMENTEN	ZUR BERECHNUNG DER BIEGESTEIFIGKEIT ABGESETZTER STAEBE UND WELLEN UNTER ANWENDUNG VON FINITEN ELEMEN	75-0481
ELEMENTENMETHOD	APPLICATION OF AN ' ELEMENT-METHOD ' PROGRAM TO SOLVING TEMPERATURE DISTRIBUTION PROBLEMS (ANWENDUNG	71-0464
ELEMENTENMETHOD	BERECHNUNG VON SPANNUNGSUMLAGERUNGEN IN FELS UND BODEN MIT HILFE DER ELEMENTENMETHODE	69-0364
ELEMENTER	SPAENDINGSHYBRIDE FINITE ELEMENTER TIL SVINGNINGSPROBLEMER. (STRESS HYBRID FINITE ELEMENT METHOD FOR	75-0208
ELEMENTFOGARS	EXPEIMENTELL UNDERSOKNING AV ELEMENTFOGARS HALLFASTHETSEGENSKAPER (EXPERIMENTAL INVESTIGATION OF T	72-0192
ELEMENTI	CHARACTERISTICS OF A HYBRID CALCULATION SYSTEM BASED ON THE FINITE ELEMENT METHOD. (CARATTERISTICHE	72-0715
ELEMENTI	IL METODO DEGLI ELEMENTI FINITI: POSSIBILITA DI APPLICAZIONE A PROBLEMI DI INTERESSE DEGLI ELETTROTE	75-0200
ELEMENTI	LA TECNICA DEGLI ELEMENTI FINITI PER PROBLEMI DI LUBRIFICAZIONE INCOMPRESSIBILE. (FINITE ELEMENT MET	75-0471
ELEMENTI	ON THE RESOLUTION OF NONLINEAR ELASTIC AND ELASTOPLASTIC PROBLEMS BY THE METHOD OF FINITE ELEMENTS.	74-1326
ELEMENTI	OPTIMIZZAZIONE DI RETICOLI TRIANGOLARI PER IL METODO DEGLI ELEMENTI FINITI IN IDROLOGIA	72-0987
ELEMENTI	PROCESS FOR THE AUTOMATIC GENERATION OF FLAT TRIANGULAR LATTICES FOR APPLICATION TO THE FINITE ELEME	71-0366
ELEMENTI	STUDIO DI CAMPI ELETTRICI E MAGNETICI STANZIONARI CON IL METODO DEGLI ELEMENTI FINITI. APPLICAZIONE	75-0207
ELEMENTMETHODE	CONTINUUM CALCULATION FROM ELASTIC PLASTIC REINFORCED MATERIAL USING A FINITE ELEMENT METHOD (DIE BE	72-0658
ELEMENTMETODEN	ELEMENTMETODEN OG DEFORMATIONSMETODEN (FINITE ELEMENT AND DEFORMATION METHODS IN STRESS ANALYSIS)	74-0580
ELEMENTOV	APPLICATION OF THE METHOD OF FINITE ELEMENTS TO THE CALCULATION OF SHIP STRUCTURES (ISPOLZOVANIE MET	73-0933
ELEMENTOV	ISSLEDOVANIE NAPRYAZHENNOGO SOSTOYANIYA REBRISTYKH ISILINDRICHESKIKH OBOLOCHEK S PRYAMOUGOL'NYMI OTV	74-0944
ELEMENTOV	ISSLEDOVANIE PROCHNOSTI I ZHESTKOSTI PLASTIN. NAGRUZHENNYKH V SVOEI PLOSKOSTI, METHODOM KONECHNYKH E	73-0382
ELEMENTOV	K VOPROSU OB AVTOMATIZATSII ZADANIYA INFORMATSII V METODE KONECHNYKH ELEMENTOV. (PROBLEM OF AUTOMATI	73-0382
ELEMENTOV	METOD KONECHNYKH ELEMENTOV V PRIMENENII K RASCHETU TSILINDRICHESKIKH OBOLOCHEK S PRYAMOUGOL'NYMI OTV	73-0536
ELEMENTOV	METOD KONECHNYKH ELEMENTOV V RASCHETAKH KONSOLIDATSII VODONASYSHCHENNYKH GRUNTOV. (FINITE ELEMENT ME	75-0469
ELEMENTOV	RASCHET METODOM KONECHNYKH ELEMENTOV DINAMICHESKIKH KHARAKTEERISTIK ZEMLYANOI PLOTINY SOVMESTNO S OS	74-0437
ELEMENTOV	RASCHIOT USTOICHIVOSTI IZGIBNO-KRUTILNYKH KOLEBANIJ KRULA V DOZVUKOVOM POTOKE METODOM KONECHNYKH FLE	72-0903
ELEMENTOV	RASHET TREKHMERNYKH MAGNITNYKE POLEI METODOM KONECHNYKH ELEMENTOV. (CALCULATION OF THREE-DIMENSIONAL	75-0766
ELEMENTOV	REALIZATSIYA METODA KONECHNYKH ELEMENTOV NA EVM DLYA RASCHETA DVUMERNYKH FLEKTRICHESKIKH I MAGNITNYK	74-0251
ELEMENTOW	ALGORYTM OBLICZEN ORGAN WYMUSZONYCH METODA SZTYWNYCH ELEMENTOW SKONCZONYCH (ALGORITHM FOR CALCULATIN	74-0581
ELEMENTOW	ANALIZA ORGAN WYMUSZONYCH ZLOZONYCH UKLADOW LINIOWYCH METODA SZTYWNYCH ELEMENTOW SKONCZONYCH (ANALYS	74-0246
ELEMENTOW	APPLICATION OF HERMITE POLYNOMIALS TO THE DETERMINATION OF THE STIFFNESS MATRIX OF PLATE BY FINITE E	71-0365
ELEMENTOW	FUNKCJA KSZTALTU NIESPELNIAPACA WARUNKOW CIAGLOSCI W ANALIZIE PLYT METODA ELEMENTOW SKONCZONYCH. (SH	75-0263
ELEMENTOW	HYBRYDOWA METODA ELEMENTOW SKONCZONYCH W ZASTOSOWANIU DO OBLICZEN UPZADZEN OKRETOWYCH. (HYBRID	74-0963
ELEMENTOW	METHOD SZTYWNYCH ELEMENTOW SKONCZONYCH W OBLICZENIACH KONSTRUCJI OKRETOWYCH. (STIFF FINITE ELEMENT M	74-0923
ELEMENTOW	METODA ELEMENTOW SKONSZONYCH ROZWIAZYWANIA OSIOWO-SYMETRYCZNEGO PRZEWODNICTWA CIEPLA (FINITE-ELEMENT	75-0434
ELEMENTOW	METODA ELEMENTOW SKONSZONYCH W MECHANICE GRUNTOW I MECHANICE GOROTWORU (FINITE ELEMENT METHOD IN SOI	73-0500
ELEMENTOW	OBLICZANIE ORGAN METODA SZTYWNYCH ELEMENTOW SKONCZONYCH. (CALCULATING VIBRATIONS BY THE RIGID-FINITE	75-0649
ELEMENTOW	ZASTOSOWANIE ITERACJI SEIDLA W METODZIE ELEMENTOW SKONCZONYCCH NA PRZYKLADZIE OBLICZEN STATYCZNYCH P	74-1031
ELEMENTOW	ZASTOSOWANIE METODY SZTYWNYCH ELEMENTOW SKONCZONYCH DO OBLICZEN ORGAN WALOW OKRETOWYCH Z UWZGLEDNIENI	74-0962
ELEMENTU	APPLICATION OF HERMITE POLYNOMIALS TO THE DETERMINATION OF THE STIFFNESS MATRIX OF PLATE BY FINITE E	71-0365
ELEMENTULUI	ASUPRA ANALIZEI PRIN METODA ELEMENTULUI FINIT A STARILOR DE EFORTURI SI DEFORMATII IN MASIVELE DE P	70-0471
ELEMETE	PROGRAMME ZUR METHODE DER FINITEN ELEMETE	69-0416
ELETTRICI	STUDIO DI CAMPI ELETTRICI E MAGNETICI STANZIONARI CON IL METODO DEGLI ELEMENTI FINITI. APPLICAZIONE	75-0207
ELETTROTECNICI	IL METODO DEGLI ELEMENTI FINITI: POSSIBILITA DI APPLICAZIONE A PROBLEMI DI INTERESSE DEGLI ELETTROTE	75-0200
ELEVATED	NONLINEAR CREEP ANALYSIS AT ELEVATED TEMPERATURE	71-0050
ELEVATED	STATIC THREE DIMENSIONAL ANALYSIS OF ELEVATED STEEL TRANSPORTATION STRUCTURES	74-1187
ELFED	ELFED A COMPUTER CODE TO GENERATE A FINITE ELEMENT MESH FOR PROBLEMS OF COMPLEX AXISYMMETRIC GEOMETR	74-1270
ELFED	ELFED: A COMPUTER CODE TO GENERATE FINITE ELEMENT MESH FOR PROBLEMS OF COMPLEX AXISYMMETRIC GEOMETRY	73-0943
ELIMINATION	BLOCK ELIMINATION ON FINITE ELEMENT SYSTEMS OF EQUATIONS	72-0530
ELIMINATION	ERROR BOUNDS FOR EIGENVALUE ANALYSIS BY ELIMINATION OF VARIABLES	71-0594
ELIMINATION	STRUCTURAL EIGENVALUE PROBLEMS: ELIMINATION OF UNWANTED VARIABLES	65-0050
ELIMINATION-TYP	USE OF SPARSITY IN THE SOLUTION OF FINITE ELEMENT SYSTEMS OF EQUATIONS BY GAUSSIAN ELIMINATION-TYPE	74-0868
ELLIPTIC	APPLICATION OF THE FINITE ELEMENT METHOD TO ELLIPTIC PROBLEMS	74-0570
ELLIPTIC	APPROXIMATION OF STEKLOV EIGENVALUES OF NON-SELFADJOINT SECOND ORDER ELLIPTIC OPERATORS	72-0509
ELLIPTIC	COMPUTABLE ERROR BOUNDS FOR THE FINITE ELEMENT METHOD FOR ELLIPTIC BOUNDARY VALUE PROBLEMS	74-1219
ELLIPTIC	CORNER SINGULARITIES IN ELLIPTIC PROBLEMS BY FINITE ELEMENT METHODS	71-0456
ELLIPTIC	DISCRETIZATION AND ROUND-OFF ERROR IN THE FINITE ELEMENT ANALYSIS OF ELLIPTIC BOUNDARY VALUE PROBLEM	70-0514
ELLIPTIC	ELEMENTS FINIS ET PROBLEMES ELLIPTIQUES DEGENERES (FINITE ELEMENTS AND DEGENERATE ELLIPTIC PROBLEMS)	73-0582
ELLIPTIC	ELLIPTIC BOUNDARY VALUE PROBLEMS WITH NONVARIATIONAL PERTURBATION AND THE FINITE ELEMENT METHOD	73-1022
ELLIPTIC	ERROR ANALYSIS OF FINITE ELEMENT METHODS WITH TRIANGLES FOR ELLIPTIC BOUNDARY VALUE PROBLEMS	73-0072
ELLIPTIC	FINITE ELEMENT GALERKIN METHOD SOLUTIONS TO SELECTED ELLIPTIC AND PARABOLIC DIFFERENTIAL EQUATIONS	71-0027
ELLIPTIC	FINITE ELEMENT METHOD FOR ELLIPTIC EQUATIONS WITH DISCONTINUOUS COEFFICIENTS	70-0193
ELLIPTIC	FINITE ELEMENT METHODS FOR ELLIPTIC BOUNDARY VALUE PROBLEMS CONTAINING SINGULARITIES	72-0360

ELLIPTIC	FINITE ELEMENTS AND DEGENERATED ELLIPTIC PROBLEMS	73-0357
ELLIPTIC	FUNCTIONAL ANALYSIS AND THE FINITE ELEMENT METHOD APPLIED TO ELLIPTIC BOUNDARY VALUE PROBLEMS	72-0056
ELLIPTIC	HYBRID FINITE ELEMENT METHODS FOR SOLVING 2ND ORDER ELLIPTIC EQUATIONS	74-1267
ELLIPTIC	LOCAL MESH REFINEMENT WITH FINITE ELEMENTS FOR ELLIPTIC PROBLEMS	73-0073
ELLIPTIC	ON THE CONVERGENCE OF SOR ITERATIONS FOR FINITE ELEMENT APPROXIMATIONS TO ELLIPTIC BOUNDARY VALUE PR	71-0449
ELLIPTIC	PROPERTIES OF SOLUTIONS OF ELLIPTIC BOUNDARY VALUE PROBLEMS	72-0499
ELLIPTIC	QUASIOPTIMAL FINITE-ELEMENT METHOD FOR ELLIPTIC INTERFACE PROBLEMS	75-0427
ELLIPTIC	STRESSES AROUND FLEXIBLE ELLIPTIC PIPES	73-0215
ELLIPTIC	THE FINITE ELEMENT METHOD FOR ELLIPTIC DIFFERENTIAL EQUATIONS	70-0027
ELLIPTIC	VARIABLE GRID FINITE DIFFERENCE-ELEMENT SOLUTION OF ELLIPTIC PARTIAL DIFFERENTIAL EQUATIONS	74-0862
ELLIPTICAL	DETERMINATION OF EIGENMODES OF HOMOGENEOUS PLASMA WITH ELLIPTICAL SECTION USING BI-DIMENSIONAL FINIT	74-0553
ELLIPTICAL	NUMERICAL SOLUTION BY METHODS OF FINITE ELEMENTS OF DEGENERATED ELLIPTICAL PROBLEMS	72-0284
ELLIPTICAL	PROPAGATION IN ELLIPTICAL AND PARABOLIC WAVEGUIDES	74-1173
ELLIPTICAL	PROPAGATION OF ELLIPTICAL SURFACE CRACKS AND NONLINEAR FRACTURE MECHANICS BY THE FINITE ELEMENT METH	74-0906
ELLIPTICAL	THE FINITE ELEMENT SOLUTION OF FLOW AND HEAT TRANSFER IN ELLIPTICAL DUCTS	74-0039
ELLIPTICAL	TRANSFORMATIONS FOR ELLIPTICAL HOLES APPLIED TO FINITE ELEMENT ANALYSIS	74-0132
ELLIPTIQUES	ELEMENTS FINIS ET PROBLEMES ELLIPTIQUES DEGENERES (FINITE ELEMENTS AND DEGENERATE ELLIPTIC PROBLEMS)	73-0582
ELLIPTIQUES	LA METHODE DES ELEMENTS FINIS POUR LES PROBLEMES AUX LIMITES ELLIPTIQUES	74-1228
ELLIPTIQUES	RESOLUTION NUMERIQUE PAR DES METHODES D ELEMENTS FINIS DE PROBLEMES ELLIPTIQUES DEGENERES	72-0702
ELLIPTISCHE	STABILITAT BEI DER METHODE DER FINITEN ELEMENTE FUR QUASILINEARE ELLIPTISCHE RANDWERTPROBLEME	74-1283
ELLIPTISCHEN	FINITE ELEMENT VERFAHREN BEI QUASILINEAREN ELLIPTISCHEN RANDWERTPROBLEM	74-1262
ELONGATED	ORTHOTROPIC FLAT PLATES SUPPORTED ON COLUMNS ELONGATED IN PLANE	72-0492
EMBANKMENT	ANALYSIS OF EMBANKMENT STRESSES AND DEFORMATIONS	67-0012
EMBANKMENT	ANALYSIS OF EMBANKMENT CONSTRUCTION BY FINITE-ELEMENTS	72-0972
EMBANKMENT	AUTOMATIC EMBANKMENT ANALYSIS, EQUILIBRIUM AND INSTABILITY CONDITIONS	66-0015
EMBANKMENT	EFFECT OF SEEPAGE ON EMBANKMENT DEFORMATIONS DUE TO WATER LOADING	74-1329
EMBANKMENT	EMBANKMENT ANALYSIS AND FIELD CORRELATION	68-0057
EMBANKMENT	EMBANKMENT DAM FOR THE MARIMBONDO HYDRO PROJECT	75-0270
EMBANKMENT	EMBANKMENT DEFORMATIONS DUE TO WATER LOADS	73-0782
EMBANKMENT	FOUNDATION FAILURE OF NEW LISKEARD EMBANKMENT	73-0549
EMBANKMENT	MICA DAM EMBANKMENT STRESS ANALYSIS	75-0054
EMBANKMENT	OBSERVED AND PREDICTED DEFORMATIONS IN A LARGE EMBANKMENT DAM DURING CONSTRUCTION	71-0301
EMBANKMENT	PERFORMANCE OF AN EMBANKMENT CONSTRUCTED ON VARVED CLAY	73-0743
EMBANKMENT	PREDICTION OF LONG TERM DEFORMATION OF A COMPACTED COHESIVE SOIL EMBANKMENT OVER A SOFT FOUNDATION	74-0277
EMBANKMENT	STRESSES AND DEFORMATIONS IN JAIL GULCH EMBANKMENT	73-0391
EMBANKMENT	STRUCTURAL BEHAVIOR OF A FLEXIBLE METAL CULVERT UNDER A DEEP EARTH EMBANKMENT USING METHOD A BACKFIL	73-0739
EMBANKMENTS	A FINITE ELEMENT MODEL FOR STABILITY ANALYSIS OF MINE WASTE EMBANKMENTS SUBJECTED TO SEEPAGE	74-1166
EMBANKMENTS	A FINITE-ELEMENT PROGRAM FOR THE ANALYSIS OF EMBANKMENTS OVER SOFT, SATURATED SOILS INCLUDING CONSOL	73-1160
EMBANKMENTS	BEHAVIOR OF EMBANKMENTS ON NEW LISKHEARD VARVED CLAY	73-0721
EMBANKMENTS	ELASTIC-PLASTIC STABILITY ANALYSIS OF MINE-WASTE EMBANKMENTS	75-0835
EMBANKMENTS	FINITE ELEMENT ANALYSIS OF THE BEHAVIOR OF EMBANKMENTS	70-0499
EMBANKMENTS	FINITE ELEMENT PROGRAM FEECON FOR UNDRAINED DEFORMATION ANALYSES OF GRANULAR EMBANKMENTS ON SOFT CLA	72-0875
EMBANKMENTS	IMMEDIATE SETTLEMENTS OF THREE TEST EMBANKMENTS ON CHAMPLAIN CLAY	74-0275
EMBANKMENTS	NUMERICAL ANALYSIS OF EMBANKMENTS OVER SOFT SOILS	75-0719
EMBANKMENTS	PHOTOELASTIC AND FINITE ELEMENT ANALYSIS OF EMBANKMENTS CONSTRUCTED OVER SOFT SOILS	70-0151
EMBANKMENTS	PREDICTION OF UNDRAINED MOVEMENTS CAUSED BY EMBANKMENTS ON MUSKEG	70-0070
EMBEDDED	EFFECT OF ELASTIC MODULAR RATIO AND ANISOTROPY ON THE STRESS FIELD IN A TWO-DIMENSIONAL MATRIX INTO	71-0647
EMBEDDED	SEISMIC RESPONSE EFFECTS ON EMBEDDED STRUCTURES	72-0902
EMBEDDED	STIFFNESS COEFFICIENTS FOR EMBEDDED FOOTINGS	75-0540
EMBEDDED	VERTICAL VIBRATION OF EMBEDDED FOOTINGS	72-0231
EMBEDDED	VERTICAL VIBRATIONS OF FOOTINGS EMBEDDED IN LAYERED MEDIA	69-0436
EMERGENCY	FINITE ELEMENT ANALYSIS OF STRUCTURAL INTEGRITY OF A REACTOR PRESSURE VESSEL DURING EMERGENCY CORE C	70-0026
EMPIRICAL	EMPIRICAL INVESTIGATION OF REORDERING AND DATA MANAGEMENT FOR FINITE ELEMENT SYSTEMS OF EQUATIONS	73-0944
EMPLACEMENT	APPROXIMATE METHOD FOR CALCULATING THE RESPONSE OF EMPLACEMENT STRUCTURES SUBJECTED TO GROUND SHOCK	73-0654
EMPLACEMENT	DYNAMIC LOADS ON GROUTED EMPLACEMENT SYSTEMS	70-0406
ENCASTREE	QUELQUES METHODES D'ELEMENTS FINIS POUR DE PROBLEME D'UNE PLAQUE ENCASTREE	74-0406
ENCLAVE	EFFECTS OF LOCAL STRENGTH VARIATIONS ON PLASTIC ENCLAVE DEVELOPMENT AT CRACK TIPS	72-0822
ENCLOSING	FINITE ELEMENT METHOD FOR A SOLID ENCLOSING AN INVISCID INCOMPRESSIBLE FLUID	73-0090
ENCLOSURES	ACOUSTIC FINITE ELEMENT APPROACH FOR STUDYING BOUNDARY FLEXIBILITY AND SOUND TRANSMISSION BETWEEN IR	73-0096
ENCLOSURES	AN ACOUSTIC FINITE ELEMENT APPROACH FOR STUDYING BOUNDARY FLEXIBILITY AND SOUND TRANSMISSION BETWEEN	73-1087
ENCLOSURES	USE OF SIMPLE THREE-DIMENSIONAL ACOUSTIC FINITE ELEMENTS FOR DETERMINING NATURAL MODES AND FREQUENCI	72-0010
ENCOUNTERED	TOUR MAINE-MONTPARNASSE QUELQUES PROBLEMES DE STRUCTURE (SPECIFIC PROBLEMS ENCOUNTERED IN THE CONSTR	74-0539
ENCROACHMENT	ON SEAWATER ENCROACHMENT IN COASTAL AQUIFERS	74-0577
END	A UNIFIED FINITE ELEMENT APPROACH TO COLUMNS SUBJECTED TO NONCONSERVATIVE LOADS, WITH DAMPING EFFECT	74-0795
END	ANALYSIS OF A STUB END BY FINITE ELEMENT METHOD	72-0210
END	ANCHORAGE ZONE STRESSES IN AXIALLY POST-TENSIONED I-SECTION MEMBERS WITH END BLOCKS	71-0305
END	DEFORMATION OF THE END DIASTOLIC LEFT VENTRICLE, COMPARISON OF THEORY WITH EXPERIMENT	71-0131
END	END DISPLACEMENTS OF SEMI-INFINITE CYLINDERS DUE TO ANNULAR LOADINGS	69-0399
END	EXACT SOLUTION TO THE RADIALLY END CONSTRAINED CIRCULAR CYLINDER UNDER TRIAXIAL LOADING	71-0433
END	EXPERIMENTAL STUDY OF FLEXIBILITY FACTORS AND STRESSES OF WELDING ELBOWS WITH END EFFECTS	73-0422
END	FREE VIBRATIONS OF CYLINDRICAL SHELLS WITH VARIOUS END CLOSURES	74-0571
END	FUEL ROD DEFORMATIONS - A TIME DEPENDENT FINITE ELEMENT ANALYSIS ON AXISYMMETRIC FUEL RODS, INCLUDIN	70-0414
END	HIGH STRENGTH END FITTINGS FOR FRP ROD AND ROPE	74-0605
END	NOTE ON THE STRESS CONCENTRATIONS AT THE END OF A CYLINDRICAL HOLE	70-0139
END	PLANE STRESS ELEMENT OF VARYING THICKNESS APPLIED TO END BLOCK ANALYSIS	74-0396
END	STRESS ANALYSIS OF CURVED PIPES WITH END RESTRAINTS	75-0667
END	THE EFFECT OF END RESTRAINT CONDITION ON THE FRACTURE MECHANISM AND STRENGTH OF ROCK-LIKE MATERIALS	72-0134
END	TWO-DIMENSIONAL TIME-DEPENDENT ANALYSIS OF PERFORATED END CAPS FOR NUCLEAR REACTOR PRESSURE VESSELS	71-0239
END-CLOSURE	INFLUENCE OF END-CLOSURE STIFFNESS ON BEHAVIOR OF CONCRETE CYLINDRICAL HULLS SUBJECTED TO HYDROSTATI	71-0725
END-EFFECTS	INFLUENCE OF END-EFFECTS ON STRESSES AND FLEXIBILITY OF A PIPING ELBOW WITH IN-PLANE MOMENT	74-0534
END-PLATES	EFFECTS OF INSERTS ON THE ELASTIC BEHAVIOUR OF CYLINDRICAL MATERIALS LOADED BETWEEN ROUGH END-PLATES	71-0318
ENDLICHEN	COMPUTATION OF SHELLS WITH FINITE DISPLACEMENTS: MIXED FINITE ELEMENTS (BERECHNUNG VON SCHALEN MIT E	73-0788
ENDLICHEN	DIE METHODE DER ENDLICHEN ELEMENTE UND IHRE ANWENDUNG IM MASCHINENBAU	71-0628
ENDLICHER	AUFBEREITUNG MATHEMATISCHER MODELLE FUR DIE BERECHNUNG VON GRUNDWASSERSTROMUNGSPROBLEMEN NA	71-0543
ENDLICHER	GRUNDLAGEN DER DIGITALEN LOSUNG VON GRUNDWASSERSTROMUNGSPROBLEMEN NACH DER METHODE ENDLICHER ELEMENT	71-0542
ENDLICHER	PRAKTISCHE BEISPIELE DER DIGITALEN BERECHNUNG VON GRUNDWASSERSTROMUNGS PROBLEMEN NACH DER METHODE EN	73-0081
ENDLICHES	A NEW, FULLY COMPATIBLE FINITE ELEMENT FOR PLATE BENDING EIN NEUES VOLLVERTRAGLICHES ENDLICHES ELEME	68-0032
ENDOPROSTHESES	FINITE ELEMENT STUDIES OF FEMORAL ENDOPROSTHESES FOR HIP RECONSTRUCTION	75-0381
ENDOSTEAL	STRESS DISTRIBUTION IN BONE ARISING FROM LOADING ON ENDOSTEAL DENTAL IMPLANTS	73-0501
ENDS	ARCH DAMS ANALYSED BY A LINEAR FINITE ELEMENT SHELL THE FINITE STRIP METHOD IN THE ANALYSIS OF ELAST	68-0189
ENDS	THE FINITE STRIP METHOD IN THE ANALYSIS OF ELASTIC PLATES WITH TWO SIMPLY SUPPORTED ENDS	68-0068
ENERGETISCHEN	UBER DIE ANWENDUNG DER ENERGETISCHEN METHODE AUF STABILITATSPROBLEME	38-0001
ENERGIES	FORMULATIONS DUALES DES ENERGIES POTENTIELLES ET COMPLEMENTAIRES. APPLICATION A LA METHODE DES ELEME	74-0752
ENERGIES	FRACTURE SURFACE ENERGIES BY FINITE ELEMENT STRESS ANALYSIS	73-0915
ENERGIZING	CLOSURE ANALYSIS USING SELF ENERGIZING, DOUBLE CONICAL GASKETS	74-0675
ENERGY	A NEW LOOK AT STRUCTURAL ENERGY DISSIPATION	74-0816
ENERGY	A STRAIN ENERGY BASIS FOR STUDIES OF ELEMENT STIFFNESS MATRICES	72-0621
ENERGY	AN INCREMENTAL COMPLEMENTARY ENERGY METHOD OF NONLINEAR STRESS ANALYSIS	69-0281
ENERGY	ANALYTICAL MODEL FOR DETERMINING ENERGY DISSIPATION IN DYNAMICALLY LOADED STRUCTURES	75-0328
ENERGY	COMBINED ANALYTICAL - EXPERIMENTAL FRACTURE STUDY OF THE TWO LEADING THEORIES OF ELASTIC-PLASTIC FRA	73-0951
ENERGY	DETERMINATION OF ENERGY RELEASE RATES AND STRESS - INTENSITY FACTORS BY THE FINITE-ELEMENT METHOD	72-0183
ENERGY	DEVELOPMENTS IN STRUCTURAL ANALYSIS BY DIRECT ENERGY MINIMIZATION	68-0070

ENERGY	DUAL FORMULATION FOR POTENTIAL ENERGY AND COMPLEMENTARY ENERGY APPLICATION TO THE FINITE ELEMENT MET	72-0062
ENERGY	DUAL FORMULATION FOR POTENTIAL ENERGY AND COMPLEMENTARY ENERGY APPLICATION TO THE FINITE ELEMENT MET	72-0062
ENERGY	ENERGY APPROACHES TO FINITE-DIFFERENCE AND FINITE ELEMENT METHODS	71-0046
ENERGY	ENERGY CONDITION FOR INITIATION OF INTERFACIAL MICROCRACKS AT INCLUSIONS	74-0546
ENERGY	ENERGY FOR CRACK GROWTH IN MODEL RUBBER COMPONENTS	72-0568
ENERGY	ENERGY FORMS IN THE FINITE ELEMENT TECHNIQUES	72-0680
ENERGY	ENERGY THEOREMS AND STRUCTURAL ANALYSIS	54-0001
ENERGY	ENERGY THEOREMS AND STRUCTURAL ANALYSIS	67-0136
ENERGY	FINITE ELEMENT ANALYSIS OF SHELLS OF REVOLUTION BY MINIMIZATION OF THE POTENTIAL ENERGY FUNCTIONAL	69-0114
ENERGY	FINITE ELEMENT ANALYSIS OF SHELLS OF REVOLUTION BY MINIMIZATION OF THE POTENTIAL ENERGY FUNCTIONAL	69-0354
ENERGY	FINITE ELEMENT ANALYSIS OF SHELLS OF REVOLUTION BY MINIMIZATION OF THE POTENTIAL ENERGY FUNCTIONAL	70-0481
ENERGY	FINITE ELEMENT ANALYSIS OF INCOMPRESSIBLE MATERIAL BY RESIDUAL ENERGY BALANCING	74-1041
ENERGY	FINITE ELEMENT ANALYSIS OF A MULTI-COMPONENT KINETIC ENERGY PROJECTILE	74-1180
ENERGY	FINITE ELEMENT ANALYSIS OF THIN ELASTIC SHELLS WITH RESIDUAL ENERGY BALANCING AND THE ROLE OF THE RI	75-0034
ENERGY	FINITE ELEMENT ANALYSIS OF A MULTI-COMPONENT KINETIC ENERGY PROJECTILE	75-0754
ENERGY	FINITE ELEMENT AND FINITE DIFFERENCE ENERGY TECHNIQUES FOR THE NUMERICAL SOLUTION OF PARTIAL DIFFERE	73-0570
ENERGY	FINITE ELEMENT CALCULATION OF STRESS INTENSITY FACTORS USING ENERGY TECHNIQUES	73-0825
ENERGY	FINITE ELEMENT SOLUTION FOR ENERGY CONSERVATION USING A HIGHLY STABLE EXPLICIT INTEGRATION ALGORITHM	72-0633
ENERGY	FINITE-DIFFERENCE ENERGY METHOD FOR NONLINEAR SHELL ANALYSIS	71-0399
ENERGY	FINITE-ELEMENT ANALYSIS OF THIN ELASTIC SHELLS WITH RESIDUAL ENERGY BALANCING AND ROLE OF RIGID BODY	75-0048
ENERGY	GENERAL POTENTIAL ENERGY AND COMPLEMENTARY ENERGY MODELS BASED ON STRESS PARAMETERS	74-1240
ENERGY	GENERAL POTENTIAL ENERGY AND COMPLEMENTARY ENERGY MODELS BASED ON STRESS PARAMETERS	74-1240
ENERGY	MIXED DISPLACEMENT 3D ISOPARAMETRIC FINITE ELEMENTS AND A STRAIN ENERGY BASIS FOR STUDIES OF ELEMENT	70-0286
ENERGY	MODIFIED POTENTIAL ENERGY MASS REPRESENTATION FOR FREQUENCY PREDICTION	66-0053
ENERGY	NONLINEAR FINITE ELEMENT ANALYSIS INCLUDING HIGHER ORDER STRAIN ENERGY TERMS	70-0065
ENERGY	ON THE EFFECT OF THE FORM OF THE STRAIN ENERGY FUNCTION ON THE SOLUTION OF A BOUNDARY-VALUE PROBLEM	72-0429
ENERGY	ON THE ENERGY FUNCTIONS FOR CURVED SANDWICH PLATES	71-0587
ENERGY	OPTIMIZATION OF FINITE ELEMENT GRIDS BASED ON MINIMUM POTENTIAL ENERGY	71-0747
ENERGY	OPTIMIZATION OF FINITE ELEMENT GRIDS BASED ON MINIMUM POTENTIAL ENERGY	72-0109
ENERGY	RESIDUAL ENERGY BALANCING TECHNIQUE IN THE GENERATION OF PLATE BENDING FINITE ELEMENTS	74-0594
ENERGY	SATURATED MAGNETIC ENERGY FUNCTIONAL FOR FINITE-ELEMENT ANALYSIS OF ELECTRIC MACHINES	75-0636
ENERGY	SIMPLIFIED MINIMUM INTERNAL ENERGY OF DEFORMATION POSTULATE AND ITS APPLICATION TO METALFORMING PROC	74-1155
ENERGY	STRAIN ENERGY FUNCTIONS AND MATHEMATICAL MODELS OF LUNG ELASTICITY	73-0591
ENERGY	STRAIN ENERGY RELEASE RATES FOR TURNING CRACKS	72-0146
ENERGY	STRESS, DISPLACEMENT, LINE INTEGRAL AND CLOSURE ENERGY DETERMINATIONS OF CRACK TIP STRESS INTENSITY	74-0621
ENERGY	SUFFICIENT CONDITIONS FOR CONVERGENCE IN THE FINITE ELEMENT METHOD FOR ANY SOLUTION OF FINITE ENERGY	72-0065
ENERGY	THE COMPUTATION OF TEARING ENERGY OF NICKED RUBBER STRIPS IN EXTENSION	74-0315
ENERGY	THE ENERGY METHOD OF NETWORKS OF ARBITRARY SHAPE IN PROBLEMS OF THE THEORY OF ELASTICITY	59-0001
ENERGY	USE OF THE FINITE ELEMENT DAMPED FORCED RESPONSE STRAIN ENERGY DISTRIBUTION FOR VIBRATION REDUCTION	72-0759
ENERGY	VIBRATION REDUCTION BY USING BOTH THE FINITE ELEMENT STRAIN ENERGY DISTRIBUTION AND MOBILITY TECHNIQ	74-1303
ENERGY-DEPENDEN	ENERGY-DEPENDENT FINITE-ELEMENT METHOD FOR FEW-GROUP DIFFUSION EQUATIONS	75-0499
ENERGY-TRAPPED	FINITE ELEMENT SIMULATION OF ENERGY-TRAPPED ELECTROMECHANICAL RESONATORS	75-0142
ENFORCED	TRIANGULAR PLATE BENDING ELEMENT WITH ENFORCED COMPATIBILITY	71-0593
ENFORCEMENT	SYSTEMATIC ENFORCEMENT OF STRESS BOUNDARY CONDITIONS IN THE ASSUMED STRESS HYBRID MODEL BASED ON THE	71-0771
ENGINE	APPLICATION OF THE COLLISION-IMPARTED VELOCITY METHOD FOR ANALYZING THE RESPONSES OF CONTAINMENT AND	73-0770
ENGINE	COMPUTER AIDED DESIGN OF THERMALLY LOADED AXISYMMETRIC DIESEL ENGINE COMPONENTS	74-0596
ENGINE	DIESEL ENGINE COMPONENT DESIGN USING THE FINITE ELEMENT METHOD AND INTERACTIVE GRAPHICS	74-0237
ENGINE	FINITE ELEMENT ANALYSIS OF LARGE TRANSIENT ELASTIC-PLASTIC DEFORMATIONS OF SIMPLE STRUCTURES, WITH A	72-0849
ENGINE	FLEXURAL RIGIDITY OF A THIN WALLED BUILD-UP ROTOR FOR JET ENGINE. MEASUREMENT BY STATIC LOAD TEST AND	73-0756
ENGINE	GAS TURBINE ENGINE DISK CYCLIC LIFE PREDICTION	75-0509
ENGINE	MODELING OF GAS TURBINE ENGINE COMPRESSOR BLADES FOR VIBRATION ANALYSIS	75-0508
ENGINE	OPREDELENIE STATSIONARNYKH TEMPERATURNYKH POLEI V DETALYAKH DVIGATELEI VNUTRENNEGO SGORANIYA METODOM	73-0454
ENGINE	RETSCP: A COMPUTER PROGRAM FOR ANALYSIS OF ROCKET ENGINE THERMAL STRAINS WITH CYCLIC PLASTICITY	74-0692
ENGINE	SIGNIFICANT APPLICATIONS OF FINITE ELEMENTS AND FINITE DIFFERENCES METHODS TO SOLVE PROBLEMS OF THER	73-0864
ENGINE	SOME DESIGN ASPECTS OF IC ENGINE PISTON HEADS	73-0578
ENGINE-ROOM	DEFORMABILITY OF THE HULL STEEL-WORK AND DEFORMATIONS OF THE ENGINE-ROOM OF LARGE TANKERS	74-0532
ENGINEER	FINITE ELEMENTS AND THE AG ENGINEER	75-0647
ENGINEER'S	FINITE ELEMENT ANALYSIS, AN AUTOMOBILE ENGINEER'S TOOL	74-0238
ENGINEER'S	MAGIC 11: AN AUTOMATED GENERAL PURPOSE SYSTEM FOR STRUCTURAL ANALYSIS. VOLUME 1. ENGINEER'S MANUAL (	71-0757
ENGINEER'S	MAGIC 111: AN AUTOMATED GENERAL PURPOSE SYSTEM FOR STRUCTURAL ANALYSIS VOLUME 1. ENGINEER'S MANUAL	72-0727
ENGINEERING	A LOOK INTO THE FUTURE: HOW COMPUTERS WILL INFLUENCE ENGINEERING	67-0139
ENGINEERING	A SURVEY OF THE APPLICATION OF FINITE ELEMENT METHODS TO STRESS ANALYSIS PROBLEMS WITH PARTICULAR EM	71-0632
ENGINEERING	AN ENGINEERING APPROACH TO FINITE ELEMENT ANALYSIS OF NUCLEAR COMPONENTS	73-1054
ENGINEERING	ANSYS (ENGINEERING ANALYSIS SYSTEM) PROGRAM	72-0156
ENGINEERING	APPLICATION OF FINITE ELEMENT METHODS IN CIVIL ENGINEERING	69-0031
ENGINEERING	APPLICATION OF FINITE ELEMENT METHODS TO LUBRICATION - ENGINEERING APPROACH	72-0006
ENGINEERING	APPLICATION OF THE FAST FOURIER TRANSFORM TO LINEAR SYSTEMS IN CIVIL ENGINEERING	73-1119
ENGINEERING	APPLICATION OF THE TWO-DIMENSIONAL INTEGRAL EQUATION METHOD TO ENGINEERING PROBLEMS	75-0377
ENGINEERING	APPLICATIONS OF THE FINITE ELEMENT METHOD IN GEOTECHNICAL ENGINEERING, PROCEEDINGS OF THE SYMPOSIUM	73-0273
ENGINEERING	ASKA, SOFTWARE ENGINEERING FOR THE SEVENTIES IN STRUCTURAL ANALYSIS	73-0704
ENGINEERING	COMPANY TRAINING OF ENGINEERING IN STRUCTURAL ANALYSIS COMPUTER PROGRAM USAGE	72-0460
ENGINEERING	CONFERENCE ON APPLICATION OF FINITE ELEMENT METHODS TO STRESS ANALYSIS PROBLEMS IN NUCLEAR ENGINEERI	71-0774
ENGINEERING	DEEP SEA ENGINEERING	71-0739
ENGINEERING	DESIGNING BY FINITE ELEMENTS CATCHING ON IN AUTOMOTIVE ENGINEERING	74-0113
ENGINEERING	DIGITAL COMPUTER PROGRAMS FOR STEADY-STATE OR TRANSIENT TEMPERATURE ANALYSIS OF PLANE AND AXISYMMETR	72-0915
ENGINEERING	DYNAMIC ANALYSIS OF CIVIL ENGINEERING STRUCTURES	71-0264
ENGINEERING	EARTHQUAKE-RESISTANT DESIGN OF ENGINEERING STRUCTURES	73-0284
ENGINEERING	EFFECT OF DENSIFICATION ON THE ENGINEERING CHARACTERISTICS OF ORGANIC SOILS - VOL. 1 AND 2.	73-0805
ENGINEERING	ELASTO-PLASTIC SOLUTIONS OF ENGINEERING PROBLEMS ' INITIAL STRESS ' FINITE ELEMENT APPROACH	69-0282
ENGINEERING	ENGINEERING APPLICATIONS OF NUMERICAL INTEGRATION IN STIFFNESS METHOD	66-0040
ENGINEERING	ENGINEERING APPROACH TO FINITE ELEMENT ANALYSIS OF NUCLEAR COMPONENTS	73-0839
ENGINEERING	ENGINEERING OPERATIONS REPORT: NERVA THIN SHELL FINITE ELEMENT STRESS ANALYSIS METHODOLOGY COMPUTER	72-0916
ENGINEERING	ENGINEERING OPERATIONS REPORT: NERVA AXISYMMETRIC AND PLANE STRUCTURES FINITE ELEMENT STRESS ANALYSI	72-0917
ENGINEERING	FINITE ELEMENT ANALYSIS OF THIN-WALLED STRUCTURES BASED ON THE MODERN ENGINEERING THEORY OF BEAMS	71-0140
ENGINEERING	FINITE ELEMENT ANALYSIS OF ENGINEERING SYSTEMS GOVERNED BY A NON-LINEAR QUASI-HARMONIC EQUATION	75-0664
ENGINEERING	FINITE ELEMENT METHODS IN CIVIL ENGINEERING	72-0336
ENGINEERING	FINITE ELEMENT METHOD - A VERSATILE TOOL FOR CIVIL ENGINEERING	72-0627
ENGINEERING	FINITE ELEMENT METHOD IN ROLLING BEARING ENGINEERING	73-0054
ENGINEERING	FINITE ELEMENT METHODS IN ENGINEERING	74-0400
ENGINEERING	FINITE ELEMENT METHOD IN FOOD ENGINEERING	74-1200
ENGINEERING	FINITE ELEMENTS METHOD IN AEROSPACE ENGINEERING PROBLEMS	74-0409
ENGINEERING	FINITE-ELEMENT METHOD AND ITS APPLICATIONS IN AN ENGINEERING LABORATORY	70-0202
ENGINEERING	HANDBOOK FOR THE ENGINEERING STRUCTURAL ANALYSIS OF SOLID PROPELLANTS	71-0642
ENGINEERING	HYDROLOGIC ENGINEERING METHODS FOR WATER RESOURCES DEVELOPMENT. VOLUME 10. PRINCIPLES OF GROUND-WATE	72-0696
ENGINEERING	ICE ENGINEERING: VISCOELASTIC FINITE ELEMENT FORMULATION	74-0861
ENGINEERING	ICES STRUDL-II ENGINEERING USERS MANUAL. VOLS. 1, 11, 111	69-0398
ENGINEERING	IMPACT OF COMPUTERS ON THE PRACTICE OF STRUCTURAL ENGINEERING IN CONCRETE	72-0414
ENGINEERING	IMPACT OF COMPUTERS ON THE PRACTICE OF STRUCTURAL ENGINEERING IN CONCRETE	72-0415
ENGINEERING	IMPACT OF FINITE ELEMENT METHOD ON STRUCTURAL ENGINEERING	75-0007
ENGINEERING	INTERFACING FINITE ELEMENT METHODS WITH PRODUCT DESIGN ENGINEERING	73-0189
ENGINEERING	JOURNAL OF RESEARCH OF THE NATIONAL BUREAU OF STANDARDS: C. ENGINEERING AND INSTRUMENTATION	71-0664
ENGINEERING	MATRIX METHODS AND USE OF COMPUTERS IN STRUCTURAL ENGINEERING	71-0615

211

ENGINEERING	METHODS OF PREDICTING THE AERODYNAMIC AND STABILITY AND CONTROL CHARACTERISTICS OF STOL AIRCRAFT. VO	73-0717
ENGINEERING	OVERVIEW, TRENDS AND PROJECTIONS: THEORY AND APPLICATIONS OF THE FINITE ELEMENT METHOD IN GEOTECHNIC	72-0988
ENGINEERING	POROUS MEDIA - SOME PROBLEMS IN CIVIL AND PETROLEUM ENGINEERING	74-0349
ENGINEERING	PORTABILITY AND ADAPTABILITY OF CIVIL ENGINEERING SOFTWARE	73-0343
ENGINEERING	RANDOM-VIBRATION ANALYSIS SYSTEM FOR COMPLEX STRUCTURES. PART 1, ENGINEERING USER'S GUIDE	68-0106
ENGINEERING	RECENT TRENDS IN STUDIES AT INTERNATIONAL CONFERENCE ON STRUCTURAL ENGINEERING	71-0347
ENGINEERING	RUBBER IN ENGINEERING	73-0367
ENGINEERING	SOME APPLICATIONS OF ENGINEERING MECHANICS IN PRESSURE VESSEL ANALYSIS	72-0892
ENGINEERING	THE FINITE ELEMENT METHOD IN ENGINEERING SCIENCE	71-0274
ENGINEERING	THE GROWTH OF INTERACTIVE GRAPHICS AS AN ENGINEERING TOOL.	69-0347
ENGINEERING	THE IMPACT OF THE DIGITAL COMPUTER ON THE ENGINEERING SCIENCES - PART I	70-0003
ENGINEERING	THEORY AND APPLICATION OF THE FINITE ELEMENT METHOD IN GEOTECHNICAL ENGINEERING	72-0704
ENGINEERING	USE OF THE FINITE ELEMENT IN FOUNDATION ENGINEERING	72-0422
ENGINEERS	ASKA A COMPUTER SYSTEM FOR STRUCTURAL ENGINEERS	69-0345
ENGINEERS	ASKA 105 - A PROBLEM ORIENTED LANGUAGE FOR STRUCTURAL ENGINEERS, VOLUME 11, APPLICATION	69-0461
ENGINEERS	ASKA 105, A PROBLEM ORIENTED LANGUAGE FOR STRUCTURAL ENGINEERS, VOLUME 1, DESIGN	69-0460
ENGINEERS	FUNDAMENTALS OF FINITE ELEMENT TECHNIQUES FOR STRUCTURAL ENGINEERS	73-1050
ENGINEERS	FUNDAMENTALS OF FINITE ELEMENT METHODS FOR ENGINEERS	74-0719
ENGINEERS	OPTIM 11: A (MAGIC) COMPATIBLE LARGE SCALE AUTOMATED MINIMUM WEIGHT DESIGN PROGRAM - 1. ENGINEERS AN	74-1048
ENGINEERS	OPTIM 11: A MAGIC COMPATIBLE LARGE-SCALE AUTOMATED MINIMUM WEIGHT DESIGN PROGRAM. VOLUME 1. ENGINEER	74-1096
ENGINEERS	THE FINITE ELEMENT METHOD FOR ENGINEERS	75-0887
ENGINES	ECONOMICAL DESIGN ANALYSIS OF PISTONS FOR INTERNAL COMBUSTION ENGINES	73-0188
ENGINES	METODY RASCHETNOGO OPREDELENIYA TEMPERATURNYKH NAPRYAZHENII V KRYSHKAKH TSILINDROV DVIGATELEI VNUTRE	74-0179
ENGINES	RASCHET POLEI DEFORMATSII I NAPRYAZHENII V DETALYAKH DVIGATELEI VNUTRENNEGO SGORANIYA (CALCULATION O	73-0465
ENGLISCHEN	DIE BERECHNUNG DER BEWEGUNGEN UND SPANNUNGEN DES GEBIRGES VOM ABBAU BIS ZUR TAGESOBERFLAECHE NACH DE	73-0456
ENHANCEMENTS	ENHANCEMENTS OF THE FINITE ELEMENT METHOD THROUGH MULTIDISCIPLINARY APPLICATIONS	75-0842
ENTRANCE	CONTRIBUTION TO NUMERICAL-SOLUTION OF DEVELOPING LAMINAR-FLOW IN ENTRANCE REGION OF CONCENTRIC ANNUL	75-0448
ENTRANCE	NUMERICAL SOLUTION OF THE VISCOUS FLOW IN THE ENTRANCE REGION OF PARALLEL PLATES	73-0022
ENTRE	COUPLAGE ENTRE L ECOULEMENT AUTOUR D UN ARRIERE-CORPS ET LE JET PROPULSIF EN THEORIE DE FLUIDE PARFA	74-1137
ENTRETOISES	CALCUL DES PONTS BIAIS A POUTRES MULTIPLES SANS ENTRETOISES PAR LA METHODE DES ELEMENTS (CALCULATION	72-0184
ENTRY	ANALYSIS OF BOTTOM ENTRY NOZZLES FOR ATMOSPHERIC STORAGE TANKS	75-0331
ENTWICKLUNG	EINSATZ DES COMPUTERS BEI DER ENTWICKLUNG EINER PANZERWANNE (THE USE OF A COMPUTER FOR DEVELOPING A	72-0211
ENTWICKLUNGEN	EINIGE ENTWICKLUNGEN IN DER MATRIZEN-VERSCHIEBUNGSMETHODE	69-0453
ENTWURF	ENTWURF UND SPANNUNGSBERECHNUNG EINES SPANNBETON DRUCKBEHAELTERS AUS FERTIGTEILEN. (DESIGN AND STRES	73-0856
ENVIRONMENT	APPLICATION OF COMPUTER-AIDED AIRCRAFT DESIGN IN A MULTIDISCIPLINARY ENVIRONMENT	73-0527
ENVIRONMENT	GRAVITY FLOW OF POWDER IN A LUNAR ENVIRONMENT. PT. 2-ANALYSIS OF FLOW INITIATION	74-0484
ENVIRONMENT	THE FATE OF RADIONUCLIDES IN A GROUNDWATER ENVIRONMENT	74-0061
ENVIRONMENTAL	ENVIRONMENTAL CONSIDERATIONS OF DAM CONSTRUCTION AND OPERATION IN SEISMICALLY ACTIVE URBAN AREAS	73-0470
ENVIRONMENTAL	PREDICTIONS IN ENVIRONMENTAL HYDRODYNAMICS USING FINITE ELEMENT METHOD 1, THEORETICAL DEVELOPMENT	75-0003
ENVIRONMENTAL	PREDICTIONS IN ENVIRONMENTAL HYDRODYNAMICS USING THE FINITE ELEMENT METHOD 2. APPLICATIONS	75-0064
ENVIRONMENTAL	REFINED CRACKED CONCRETE ANALYSIS OF CONCRETE CONTAINMENT STRUCTURES SUBJECT TO OPERATIONAL AND ENVI	74-0943
ENVIRONMENTAL	SINGLE-TRACK SUBWAY ENVIRONMENTAL SIMULATION MODEL. PHASE 1	71-0693
ENVIRONMENTS	ANCHOR-PILE DESIGN FOR OCEAN-FLOOR ENVIRONMENTS USING FINITE-ELEMENT ANALYSIS	75-0459
ENVIRONMENTS	FORMING OF 7075-T6 ALUMINUM IN HIGH PRESSURE ENVIRONMENTS	71-0322
ENVIRONMENTS	VISCOELASTIC RESPONSE OF GRAPHITE MATERIALS IN IRRADIATION ENVIRONMENTS	70-0314
EPAD	EPAD, A COMPUTER PROGRAM FOR THE ELASTIC-PLASTIC ANALYSIS OF DOME STRUCTURES AND ARBITRARY SOLIDS OF	70-0036
EPAISSES	LES PLAQUES ET LES COQUES MINCES OU EPAISSES (THIN AND THICK PLATES AND SHELLS)	71-0432
EPITAXIAL	APPLICATION OF FINITE ELEMENT METHOD TO MASS TRANSPORT LIMITED EPITAXIAL GROWTH PROCESSES	71-0488
EPOXY	FATIGUE CRACK INITIATION IN A BORON EPOXY PLATE WITH A CIRCULAR HOLE	74-0820
EPOXY	FATIGUE OF GRAPHITE FIBRE-REINFORCED EPOXY IN COMPRESSION.	72-0383
EPSOLA	ELASTIC-PLASTIC STRESS-STRAIN RELATIONS USED IN CODE EPSOLA	73-0784
EQUAL	OSCILLATION STABILITY OF INTERNALLY AND EXTERNALLY NOTCHED PLANE BARS OF EQUAL FORM FACTOR FROM 3943	73-0745
EQUALITY	LINEAR EQUALITY CONSTRAINTS IN FINITE ELEMENT APPROXIMATION	75-0475
EQUALS	AN ADAPTIVE COMPUTER PROGRAM FOR DIV. (P(X,Y) GRAD U) EQUALS F(X,Y,U) IN A POLYGONAL REGION	75-0133
EQUALS	EIGENVALUE PROBLEMS (A - LAMBDA B) X EQUALS 0 FOR SYMMETRIC MATRICES OF HIGH ORDER	74-0205
EQUATION	A FINITE ELEMENT ANALOGUE OF THE NAVIER-STOKES EQUATION	70-0099
EQUATION	A FINITE ELEMENT COLLOCATION METHOD FOR THE HEAT EQUATION	72-0488
EQUATION	A FINITE ELEMENT SOLUTION TECHNIQUE FOR THE BOLTZMANN EQUATION	69-0026
EQUATION	A FINITE ELEMENT SOLUTION OF THE ONE DIMENSIONAL DIFFUSION CONVECTION EQUATION	70-0258
EQUATION	A FINITE SOLUTION OF PUCHER'S EQUATION	75-0782
EQUATION	A GENERAL NUMERICAL SOLUTION OF THE TWO-DIMENSIONAL DIFFUSION-CONVECTION EQUATION BY THE FINITE ELEM	70-0004
EQUATION	A MIXED FINITE ELEMENT METHOD FOR THE BIHARMONIC EQUATION	74-1227
EQUATION	A NUMERICAL SOLUTION OF THE BOLTZMANN EQUATION BY THE FINITE ELEMENT TECHNIQUE	70-0095
EQUATION	A STUDY OF CONVECTIVE-DISPERSION EQUATION BY ISOPARAMETRIC FINITE ELEMENTS. TRANSPORT OF POLLUTANTS	75-0773
EQUATION	A SUPERCONVERGENCE RESULT FOR THE APPROXIMATE SOLUTION OF THE HEAT EQUATION BY A COLLOCATION METHOD	72-0510
EQUATION	ACCURACY, STABILITY AND OSCILLATION CHARACTERISTICS OF FINITE ELEMENT METHOD FOR SOLVING HEAT CONDUC	69-0007
EQUATION	AN EQUATION SOLVER OF VERY LARGE CAPACITY	71-0677
EQUATION	ANALYSIS OF CONTINUA BY THE FINITE ELEMENT AND INTEGRAL EQUATION METHOD	75-0370
EQUATION	APPLICATION OF FINITE ELEMENTS TO THE SOLUTION OF HELMHOLT'S EQUATION	68-0020
EQUATION	APPLICATION OF FINITE ELEMENT METHOD TO GENERAL NUMERICAL SOLUTION OF 2-DIMENSIONAL DIFFUSION-CONVEC	70-0269
EQUATION	APPLICATION OF FINITE ELEMENT METHOD TO MULTIGROUP NEUTRON DIFFUSION EQUATION	72-0310
EQUATION	APPLICATION OF LAPLACE TRANSFORM AND FINITE ELEMENT METHOD TO SOLUTION OF NONSTATIONARY THERMAL COND	74-0171
EQUATION	APPLICATION OF PHASE-SPACE FINITE ELEMENTS TO ONE DIMENSIONAL NEUTRON TRANSPORT EQUATION	74-0123
EQUATION	APPLICATION OF PHASE SPACE FINITE ELEMENTS TO THE TWO DIMENSIONAL NEUTRON TRANSPORT EQUATION IN X-Y	73-0453
EQUATION	APPLICATION OF THE BOUNDARY-INTEGRAL EQUATION METHOD TO THREE DIMENSIONAL STRESS ANALYSIS	73-0252
EQUATION	APPLICATION OF THE FINITE ELEMENT METHOD FOR SOLVING ORDINARY LINEAR DIFFERENTIAL EQUATION WITH BOUN	74-0084
EQUATION	APPLICATION OF THE TWO-DIMENSIONAL INTEGRAL EQUATION METHOD TO ENGINEERING PROBLEMS	75-0377
EQUATION	APPLICATIONS OF FINITE ELEMENT METHODS IN REACTOR MATHEMATICS - NUMERICAL SOLUTION OF THE NEUTRON DI	72-0535
EQUATION	APPLICATIONS OF FINITE ELEMENT METHODS IN REACTOR MATHEMATICS. NUMERICAL SOLUTION OF THE NEUTRON TRA	74-0853
EQUATION	AXISYMETRIC TRIANGULAR ELEMENT FOR THE SCALAR HELMHOLTZ EQUATION	73-0688
EQUATION	COMMENT ON "SOLUTION OF LAPLACE'S EQUATION IN A REGION WITH DIELECTRIC INTERFACES OF ARBITRARY SHAPE	70-0238
EQUATION	COMPARISON OF TIME MARCHING SCHEMES FOR THE TRANSIENT HEAT CONDUCTION EQUATION	75-0477
EQUATION	CONTINUOUS AND DISCONTINUOUS FINITE ELEMENT METHODS FOR SOLVING THE TRANSPORT EQUATION	75-0099
EQUATION	DUAL ITERATIVE TECHNIQUES FOR SOLVING A FINITE ELEMENT APPROXIMATION OF THE BIHARMONIC EQUATION	75-0480
EQUATION	EQUATION SOLVING ALGORITHMS FOR THE FINITE ELEMENT METHOD	71-0127
EQUATION	FEM2D: PROGRAM FOR SOLVING THE TWO-DIMENSIONAL DIFFUSION EQUATION BY THE METHOD OF FINITE ELEMENTS	74-1089
EQUATION	FINITE ELEMENT ANALYSIS OF PROBLEMS FORMULATED BY AN INTEGRAL EQUATION; APPLICATION TO POTENTIAL FLO	68-0191
EQUATION	FINITE ELEMENT ANALYSIS OF THE UNSTEADY, LINEARIZED BOLTZMANN EQUATION	70-0033
EQUATION	FINITE ELEMENT ANALYSIS OF THE LINEARIZED BGK BOLTZMANN EQUATION.	72-0408
EQUATION	FINITE ELEMENT ANALYSIS OF ENGINEERING SYSTEMS GOVERNED BY A NON-LINEAR QUASI-HARMONIC EQUATION	75-0664
EQUATION	FINITE ELEMENT FORMULATION OF THE HEAT CONDUCTION EQUATION IN GENERAL ORTHOGONAL CURVILINEAR COORDIN	75-0848
EQUATION	FINITE ELEMENT METHOD FOR HYDRODYNAMIC DISPERSION EQUATION WITH MIXED PARTIAL DERIVATIVES	72-0030
EQUATION	FINITE ELEMENT METHODS FOR THE TRANSPORT EQUATION	74-0925
EQUATION	FINITE ELEMENT SOLUTION OF THE THREE DIMENSIONAL FLOW PROBLEMS AND OF REYNOLD'S EQUATION FOR INCOMPR	72-0011
EQUATION	FINITE ELEMENT SOLUTION OF DIFFUSION-CONVECTION EQUATION	72-0018
EQUATION	FINITE ELEMENT SOLUTION TECHNIQUE FOR NEUTRON TRANSPORT EQUATION	72-0730
EQUATION	FINITE ELEMENT SOLUTIONS OF THE WAVE EQUATION	74-0216
EQUATION	FINITE ELEMENT SOLUTION OF BOUSSINESQ'S EQUATION FOR UNSTEADY GROUNDWATER FLOW	74-0459
EQUATION	FINITE ELEMENT SOLUTION OF MULTIGROUP TRANSPORT EQUATION IN 2 DIMENSIONAL GEOMETRIES	74-0648
EQUATION	FINITE ELEMENT SOLUTION OF A NON-LINEAR DIFFERENTIAL EQUATION USING KACHANOV'S AND SUCCESSIVE APPROX	75-0108
EQUATION	FINITE ELEMENT TECHNIQUES FOR SOLUTION OF POISSON'S EQUATION	71-0494
EQUATION	FINITE ELEMENTS VERSUS FINITE DIFFERENCES, A COMPARISON OF THE TWO METHODS FOR THE SOLUTION OF THE D	73-1055

EQUATION	FORTRAN SUBROUTINES FOR THE SOLUTION OF LAPLACE'S EQUATION OVER A GENERAL REGION IN TWO DIMENSIONS	70-0425
EQUATION	GENERAL SOLUTION ROUTINES FOR SYMMETRIC EQUATION SYSTEMS	75-0129
EQUATION	INTEGRAL EQUATION METHOD FOR SUBSONIC FLOW PAST AIRFOILS IN VENTILATED WIND TUNNELS	75-0065
EQUATION	INTEGRAL FINITE-ELEMENT APPROACH FOR SOLVING LAPLACE EQUATION	73-0121
EQUATION	LARGE CAPACITY EQUATION SOLVER FOR STRUCTURAL ANALYSIS	74-0983
EQUATION	LEAST SQUARES FINITE ELEMENT ANALYSIS OF TWO-DIMENSIONAL LAMINAR BOUNDARY-LAYER EQUATION WITH SEPARA	75-0796
EQUATION	MATRIX DIFFERENCE EQUATION ANALYSIS OF VIBRATING PERIODIC STRUCTURES	75-0151
EQUATION	MATRIX DIFFERENTIAL EQUATION AND HIGHER-ORDER NUMERICAL METHODS FOR PROBLEMS OF NON-LINEAR CREEP, VI	72-0198
EQUATION	NON-SELF-ADJOINT VARIATIONAL PROCEDURE FOR FINITE-ELEMENT APPROXIMATION OF TRANSPORT EQUATION	75-0072
EQUATION	NONLINEAR EQUATION OF UNSTEADY GROUND-WATER FLOW	73-0262
EQUATION	NOTE ON THE ACCURACY AND CONVERGENCE OF FINITE ELEMENT APPROXIMATIONS OF THE CONVECTION EQUATION	75-0490
EQUATION	NUMERICAL INTEGRATION OF THE EQUATION OF MOTION	71-0507
EQUATION	NUMERICAL SOLUTION OF THE QUASI LINEAR POISSON EQUATION IN A NON-UNIFORM TRIANGLE MESH	67-0090
EQUATION	ON A FINITE ELEMENT METHOD FOR SOLVING THE NEUTRON TRANSPORT EQUATION	74-1255
EQUATION	ON A FINITE ELEMENT METHOD FOR SOLVING THE NEUTRON TRANSPORT EQUATION	74-1355
EQUATION	ON THE ACCURACY OF FINITE ELEMENT SOLUTIONS TO THE TRANSIENT HEAT CONDUCTION EQUATION	74-0182
EQUATION	ON THE CONSTRUCTION AND CONVERGENCE OF A FINITE ELEMENT SOLUTION OF LAPLACE'S EQUATION	72-0340
EQUATION	ON THE SOLUTION OF THE NAVIER-STOKES EQUATION IN TWO DIMENSIONAL AND AXIAL SYMMETRIC PROBLEMS	74-0007
EQUATION	ON THE STUDY OF CONVECTIVE DISPERSION EQUATION	74-0005
EQUATION	PARAMETRIC STUDIES OF LAYERED PAVEMENT SYSTEMS USING AFPAV CODE WITH FAST EQUATION SOLVER	75-0747
EQUATION	PROCEDURE FOR SOLVING LINEAR EQUATION SYSTEMS WITH SPARSELY OCCUPIED COEFFICIENT MATRICES ASSOCIATED	74-0352
EQUATION	RECTANGULAR AND TRIANGULAR FINITE ELEMENTS FOR POISSON'S EQUATION	75-0802
EQUATION	SECOND GENERATION BOUNDARY INTEGRAL EQUATION PROGRAM FOR THREE-DIMENSIONAL ELASTIC ANALYSIS	75-0376
EQUATION	SOLUTION OF A TWO DIMENSIONAL DIFFUSION CONVECTION EQUATION DESCRIBING THE MASS TRANSFER OF A NEUTRA	74-0037
EQUATION	SOLUTION OF MULTI-DIMENSIONAL NEUTRON TRANSPORT EQUATION BY FINITE ELEMENT METHOD	72-0564
EQUATION	SOLVING THE LINEAR BALANCE EQUATION GLOBALLY	74-0810
EQUATION	SOME DIFFERENCE SCHEMES FOR THE BIHARMONIC EQUATION	75-0204
EQUATION	SOME SUPERCONVERGENCE RESULTS FOR AN H-GALERKIN PROCEDURE FOR THE HEAT EQUATION	74-0411
EQUATION	SPLINE APPROXIMATION AND DIFFERENCE SCHEMES FOR THE HEAT EQUATION	72-0518
EQUATION	STABILITY CRITERIA FOR TWO FINITE ELEMENT SCHEMES FOR PARABOLIC EQUATION	75-0568
EQUATION	STABILITY OF FINITE ELEMENT SCHEMES FOR VIBRATION PROBLEMS OF THE EQUATION OF ELASTICITY THEORY	71-0521
EQUATION	STUDY OF CONVECTIVE-DISPERSION EQUATION BY ISOPARAMETRIC FINITE ELEMENTS	75-0074
EQUATION	TETRAHEDRAL POLYNOMIAL FINITE ELEMENTS FOR THE HELMHOLTZ EQUATION.	72-0392
EQUATION	THE USE OF INTERPOLATORY POLYNOMIALS FOR A FINITE ELEMENT SOLUTION OF THE MULTIGROUP DIFFUSION EQUAT	72-0524
EQUATION	TIMING COMPARISON STUDY FOR SOME HIGH ORDER FINITE ELEMENT PROCEDURES AND A LOW ORDER FINITE DIFFERE	72-0767
EQUATION	TRIANGULAR FINITE ELEMENTS FOR THE GENERALIZED BESSEL EQUATION OF ORDER M	73-1094
EQUATION	TRIANGULAR MESH METHODS FOR THE NEUTRON TRANSPORT EQUATION	73-1062
EQUATION	TWO TYPES OF PIECEWISE QUADRATIC SPACES AND THEIR ORDER OF ACCURACY FOR POISSON'S EQUATION	72-0520
EQUATION	USE OF VARIATIONAL PRINCIPLE IN STUDY ON RELATIVISTIC BOLTZMANN EQUATION	71-0442
EQUATIONS	A COMPACT STORAGE SCHEME FOR THE SOLUTION OF SYMMETRICAL LINEAR SIMULTANEOUS EQUATIONS	66-0116
EQUATIONS	A DERIVATION OF PERTURBED EQUATIONS OF MOTION AND HEAT CONDUCTION OF A THERMOELASTIC SOLID	72-0735
EQUATIONS	A FINITE ELEMENT COLLOCATION SOLUTION OF DIFFERENTIAL EQUATIONS	72-0050
EQUATIONS	A FINITE ELEMENT COLLOCATION METHOD FOR QUASILINEAR PARABOLIC EQUATIONS	73-1000
EQUATIONS	A FINITE ELEMENT SOLUTION ALGORITHM FOR THE NAVIER-STOKES EQUATIONS	74-0784
EQUATIONS	A GENERAL SOLUTION OF THE VON KARMAN PLATE EQUATIONS BY THE FINITE ELEMENT METHOD	70-0487
EQUATIONS	A HYBRID FINITE ELEMENT METHOD FOR TWO-DIMENSIONAL NAVIER STOKES EQUATIONS	75-0027
EQUATIONS	A LEAST SQUARES FINITE ELEMENT ANALYSIS OF NONLINEAR BIOPHYSICAL DIFFUSION-KINETICS EQUATIONS	74-1257
EQUATIONS	A METHOD FOR THE NUMERICAL INTEGRATION OF THE EQUATIONS OF MOTION ARISING FROM A FINITE ELEMENT ANAL	70-0061
EQUATIONS	A METHOD TO GALERKIN TYPE ACHIEVING OPTIMUM L(2) ACCURACY FOR FIRST ORDER HYPERBOLICS AND EQUATIONS	72-0521
EQUATIONS	A NON VARIATIONAL FINITE ELEMENT ANALYSIS FOR THE NAVIER STOKES EQUATIONS	74-0010
EQUATIONS	A NUMERICAL SOLUTION OF THE NAVIER-STOKES EQUATIONS USING THE FINITE ELEMENT TECHNIQUE	73-0061
EQUATIONS	AN APPROXIMATE NUMERICAL SOLUTION OF THE NAVIER STOKES EQUATIONS BY GALERKIN METHOD	74-0011
EQUATIONS	AN AUTOMATIC RECORDING SCHEME FOR SIMULTANEOUS EQUATIONS DERIVED FROM NETWORK SYSTEMS	70-0285
EQUATIONS	AN INVESTIGATION OF THE STABILITY OF NUMERICAL SOLUTIONS OF BIOT'S EQUATIONS OF CONSOLIDATION	74-0746
EQUATIONS	AN ITERATIVE METHOD FOR LARGE SYSTEMS OF LINEAR STRUCTURAL EQUATIONS	73-1020
EQUATIONS	APPLICATION OF FINITE ELEMENT TECHNIQUES FOR THE NUMERICAL SOLUTION OF THE NEUTRON TRANSPORT AND DIF	71-0013
EQUATIONS	APPLICATION OF FINITE ELEMENT SOLUTION TECHNIQUE TO NEUTRON DIFFUSION AND TRANSPORT EQUATIONS	71-0015
EQUATIONS	APPLICATION OF INTEGRAL EQUATIONS TO FLUID FLOWS IN UNBOUNDED REGIONS	74-0029
EQUATIONS	APPLICATIONS OF VARIATIONAL EQUATIONS TO ORDINARY AND PARTIAL DIFFERENTIAL EQUATIONS - MULTIPLE SOLU	75-0010
EQUATIONS	APPLICATIONS OF VARIATIONAL EQUATIONS TO ORDINARY AND PARTIAL DIFFERENTIAL EQUATIONS - MULTIPLE SOLU	75-0010
EQUATIONS	BANDED MATRIX TECHNIQUES AND SOLUTIONS OF LINEAR SIMULTANEOUS EQUATIONS OF FINITE ELEMENT ANALYSES	70-0302
EQUATIONS	BLOCK ELIMINATION ON FINITE ELEMENT SYSTEMS OF EQUATIONS	72-0530
EQUATIONS	CHARACTERISTICS OF MANIPULATION ERRORS IN SOLVING LOAD DEFLECTION EQUATIONS	70-0355
EQUATIONS	COMPARISON OF EXTRAPOLATION TECHNIQUES WITH HIGH ORDER FINITE ELEMENT METHODS FOR DIFFUSION EQUATION	73-0586
EQUATIONS	COMPUTATIONAL TECHNIQUES, NONLINEAR EQUATIONS, ISOPARAMETRIC ELEMENTS, FIVE LECTURES	73-0634
EQUATIONS	COMPUTER PROGRAM FOR SOLUTION OF LARGE, SPARSE, UNSYMMETRIC SYSTEMS OF LINEAR EQUATIONS	72-0719
EQUATIONS	COMPUTER PROGRAMS FOR THE SOLUTION OF SYSTEMS OF LINEAR ALGEBRAIC EQUATIONS	73-0932
EQUATIONS	COMPUTER PROGRAMS FOR THE SOLUTION OF SYSTEMS OF LINEAR ALGEBRAIC EQUATIONS	73-1040
EQUATIONS	CONFORMING AND NONCONFORMING FINITE ELEMENT METHODS FOR SOLVING THE STATIONARY STOKES EQUATIONS - 1	73-0390
EQUATIONS	CONSTITUTIVE EQUATIONS AND PUNCH-INDENTATION OF CONCRETE	75-0286
EQUATIONS	CONVERGENCE AND ERROR ESTIMATES FOR FINITE ELEMENT APPROXIMATIONS OF STATIONARY NAVIER-STOKES EQUATI	74-0122
EQUATIONS	CONVERGENCE AND STABILITY OF NONLINEAR FINITE-ELEMENT EQUATIONS	75-0441
EQUATIONS	CONVERGENCE, ACCURACY AND STABILITY OF FINITE ELEMENT APPROXIMATIONS OF A CLASS OF NON-LINEAR HYPERB	73-0159
EQUATIONS	DEVELOPMENT OF A GUN-TURRET MODEL AND OF THE EQUATIONS FOR COUPLING IT WITH A HELICOPTER FINITE-ELEM	73-0752
EQUATIONS	DYNAMIC PROGRAMMING APPROACH TO THE FORMULATION AND SOLUTION OF FINITE ELEMENT EQUATIONS	75-0059
EQUATIONS	ECONOMICAL SOLUTION TECHNIQUES FOR LOAD DEFLECTION EQUATIONS	74-0414
EQUATIONS	EFFICIENT SOLUTION OF LOAD-DEFLECTING EQUATIONS	69-0396
EQUATIONS	ELASTICITY EQUATIONS FOR INCOMPRESSIBLE AND NEARLY INCOMPRESSIBLE MATERIALS BY A VARIATIONAL THEOREM	65-0019
EQUATIONS	EMPIRICAL INVESTIGATION OF REORDERING AND DATA MANAGEMENT FOR FINITE ELEMENT SYSTEMS OF EQUATIONS	73-0944
EQUATIONS	ENERGY-DEPENDENT FINITE-ELEMENT METHOD FOR FEW-GROUP DIFFUSION EQUATIONS	75-0499
EQUATIONS	EXACT FINITE ELEMENT SOLUTIONS TO THE HELMHOLTZ EQUATIONS	73-0652
EQUATIONS	FINITE DIFFERENCE AND FINITE ELEMENT METHODS FOR THE SEMICONDUCTOR FLOW EQUATIONS	74-0028
EQUATIONS	FINITE ELEMENT - GALERKIN METHODS FOR HYPERBOLIC EQUATIONS WITH DISCONTINUOUS COEFFICIENTS	73-0726
EQUATIONS	FINITE ELEMENT ANALOGUE OF NAVIER-STOKES EQUATIONS	69-0002
EQUATIONS	FINITE ELEMENT ANALYSIS OF SURFACE WAVE PROBLEMS USING INTEGRAL EQUATIONS	74-0624
EQUATIONS	FINITE ELEMENT AND FINITE DIFFERENCE ENERGY TECHNIQUES FOR THE NUMERICAL SOLUTION OF PARTIAL DIFFERE	73-0570
EQUATIONS	FINITE ELEMENT APPENDAGE EQUATIONS FOR HYBRID COORDINATE DYNAMIC ANALYSIS	72-0022
EQUATIONS	FINITE ELEMENT COLLOCATION METHOD FOR QUASILINEAR PARABOLIC EQUATIONS	73-0047
EQUATIONS	FINITE ELEMENT GALERKIN METHOD SOLUTIONS TO SELECTED ELLIPTIC AND PARABOLIC DIFFERENTIAL EQUATIONS	71-0027
EQUATIONS	FINITE ELEMENT METHOD FOR ELLIPTIC EQUATIONS WITH DISCONTINUOUS COEFFICIENTS	70-0193
EQUATIONS	FINITE ELEMENT METHODS FOR THE NEUTRON DIFFUSION EQUATIONS	71-0372
EQUATIONS	FINITE ELEMENT METHODS FOR SYMMETRIC HYPERBOLIC EQUATIONS	73-0116
EQUATIONS	FINITE ELEMENT METHODS FOR PARABOLIC SYSTEMS	74-0173
EQUATIONS	FINITE ELEMENT METHODS FOR NONLINEAR HYPERBOLIC EQUATIONS	74-1238
EQUATIONS	FINITE ELEMENT METHODS FOR PARABOLIC EQUATIONS	74-1280
EQUATIONS	FINITE ELEMENT METHOD FOR THE GENERAL SOLUTION OF ORDINARY DIFFERENTIAL EQUATIONS	75-0473
EQUATIONS	FINITE ELEMENT MODELS OF NONLINEAR OPERATOR EQUATIONS	71-0199
EQUATIONS	FINITE ELEMENT SOLUTIONS TO DIFFERENTIAL EQUATIONS	70-0084
EQUATIONS	FINITE ELEMENT SOLUTION OF INELASTIC BEAM EQUATIONS	73-0234
EQUATIONS	FINITE ELEMENT SOLUTION OF PARTIAL DIFFERENTIAL EQUATIONS FEMAC COMPUTER PROGRAM	73-0648
EQUATIONS	FINITE ELEMENT SOLUTIONS OF THE SHALLOW WATER EQUATIONS SURFACE RUN OFF	74-0046
EQUATIONS	FINITE ELEMENT SOLUTIONS OF SELECTED PARTIAL DIFFERENTIAL EQUATIONS "FEMAC COMPUTER PROGRAM"	74-0327

213

EQUATIONS	FINITE ELEMENT SOLUTIONS FOR THE EQUATIONS OF GROUND-WATER FLOW	74-1341
EQUATIONS	FINITE-ELEMENT METHOD FOR 1ST ORDER HYPERBOLIC EQUATIONS	75-0590
EQUATIONS	FINITE-ELEMENT SOLUTION OF NAVIER-STOKES EQUATIONS FOR TRANSIENT 2-DIMENSIONAL INCOMPRESSIBLE FLOW	75-0134
EQUATIONS	FINITE-ELEMENT-TYPE SOLUTION OF INTEGRAL EQUATIONS	73-0995
EQUATIONS	GALERKIN METHODS FOR PARABOLIC EQUATIONS	70-0448
EQUATIONS	GALERKIN METHODS FOR PARABOLIC EQUATIONS WITH NONLINEAR BOUNDARY CONDITIONS	73-1023
EQUATIONS	GOVERNING EQUATIONS AND EXTREMUM PRINCIPLES OF ELASTICITY AND PLASTICITY GENERATED FROM A SINGLE FUN	73-0666
EQUATIONS	HYBRID FINITE ELEMENT METHODS FOR SOLVING 2ND ORDER ELLIPTIC EQUATIONS	74-1267
EQUATIONS	INTEGRATED SEQUENTIAL SOLVER FOR LARGE MATRIX EQUATIONS	74-0276
EQUATIONS	INTEGRATIONS OF PRIMITIVE EQUATIONS ON A SPHERE USING FINITE ELEMENT METHOD	74-0649
EQUATIONS	INTEGRATIONS OF THE PRIMITIVE EQUATIONS ON A SPHERE USING THE FINITE ELEMENT METHOD	74-1347
EQUATIONS	INTERACTIVE SOLUTION OF LINEAR EQUATIONS IN FINITE ELEMENT ANALYSIS	73-1163
EQUATIONS	INVARIANT IMBEDDING AND SOLUTION OF FINITE ELEMENT EQUATIONS	74-0158
EQUATIONS	MANIPULATION ERRORS IN COMPUTER SOLUTION OF CRITICAL SIZE STRUCTURAL EQUATIONS	71-0367
EQUATIONS	METHODS FOR SOLUTION OF PARTIAL DIFFERENTIAL EQUATIONS	73-0521
EQUATIONS	NAVIER STOKES EQUATIONS USING MIXED INTERPOLATION	74-0014
EQUATIONS	NOTE ON INCREMENTAL EQUILIBRIUM EQUATIONS AND APPROXIMATE CONSTITUTIVE RELATIONS IN LARGE INELASTIC	74-0984
EQUATIONS	NUMERICAL SOLUTION OF PARTIAL DIFFERENTIAL EQUATIONS, 11	70-0072
EQUATIONS	NUMERICAL SOLUTION OF NAVIER-STOKES EQUATIONS BY FINITE ELEMENT METHOD	72-0008
EQUATIONS	NUMERICAL SOLUTION OF SOME NONSTATIONARY PARTIAL DIFFERENTIAL EQUATIONS: THE FINITE ELEMENT METHOD A	72-0930
EQUATIONS	NUMERICAL SOLUTION OF PARTIAL DIFFERENTIAL EQUATIONS	73-0968
EQUATIONS	NUMERICAL SOLUTION OF THE STATIONARY NAVIER-STOKES EQUATIONS BY FINITE ELEMENT METHODS	74-0408
EQUATIONS	NUMERICAL SOLUTIONS OF NONLINEAR PLASMA EQUATIONS BY THE FINITE-ELEMENT METHOD	75-0671
EQUATIONS	NUMERICAL TREATMENT OF ALGEBRAIC INTEGRAL EQUATIONS BY VARIATIONAL METHODS	75-0374
EQUATIONS	NUMERICAL-SOLUTION OF EULERIAN EQUATIONS OF COMPRESSIBLE FLOW BY A FINITE-ELEMENT METHOD WHICH FOLLO	75-0535
EQUATIONS	ON AN APPROACH TO TECHNIQUES FOR THE ANALYSIS OF THE STRUCTURES OF LARGE SYSTEMS OF EQUATIONS	62-0015
EQUATIONS	ON CONSTRAINT EQUATIONS - A NEW APPROACH	74-1017
EQUATIONS	ON LOCAL APPROXIMATIONS OF OPERATOR EQUATIONS	72-0063
EQUATIONS	ON NUMERICAL SOLUTIONS OF EQUATIONS OF HYPERBOLIC-ELLIPTIC TYPE	74-0741
EQUATIONS	ON THE DISCRETIZATION OF THE TIME VARIABLE FOR PARABOLIC PARTIAL DIFFERENTIAL EQUATIONS	75-0103
EQUATIONS	ON THE FINITE ELEMENT FORMULATION OF NAVIER-STOKES EQUATIONS	74-0720
EQUATIONS	ON THE NUMERICAL TREATMENT OF THE NAVIER-STOKES EQUATIONS FOR AN INCOMPRESSIBLE FLUID	73-0809
EQUATIONS	ON THE SOLUTION OF SETS OF NONLINEAR EQUATIONS ARISING IN THE APPLICATION OF FINITE ELEMENT METHODS	75-0127
EQUATIONS	ON THE STABILITY OF EXPLICIT METHODS FOR THE NUMERICAL INTEGRATION OF THE EQUATIONS OF MOTION IN FIN	72-0172
EQUATIONS	PIECEWISE HERMITE INTERPOLATION IN ONE AND TWO VARIABLES WITH APPLICATIONS TO PARTIAL DIFFERENTIAL E	68-0201
EQUATIONS	PIECEWISE POLYNOMIAL INTERPOLATIONS AND THEIR APPLICATIONS TO PARTIAL DIFFERENTIAL EQUATIONS	70-0458
EQUATIONS	PROBLEM CONTROLLED GRID GENERATION FOR THE NUMERICAL SOLUTION OF PARTIAL DIFFERENTIAL EQUATIONS	75-0725
EQUATIONS	PROJECTIVE METHOD APPLIED TO THREE-DIMENSIONAL ELASTICITY EQUATIONS	74-0617
EQUATIONS	QUADRATIC MATRIX EQUATIONS FOR DETERMINING VIBRATION MODES AND FREQUENCIES OF CONTINUOUS ELASTIC SYS	66-0064
EQUATIONS	RAYLEIGH RITZ GALERKIN METHODS FOR SYMMETRIC POSITIVE DIFFERENTIAL EQUATIONS APPLICATION TO FINITE E	73-1030
EQUATIONS	RIGID-BODY MOTIONS AND STRAIN-DISPLACEMENT EQUATIONS OF CURVED SHELL FINITE ELEMENTS	72-0267
EQUATIONS	SELF ADJOINT DIFFERENTIAL EQUATIONS AND VARIATIONAL PRINCIPLE - 1.	71-0388
EQUATIONS	SELF ADJOINT DIFFERENTIAL EQUATIONS AND VARIATIONAL PRINCIPLE - 2.	72-0106
EQUATIONS	SIMPLE EQUATIONS FOR THE MAGNETIZATION AND RELUCTIVITY CURVES OF STEEL	75-0156
EQUATIONS	SOLUTION OF DIFFUSION AND P1 FINITE-ELEMENT EQUATIONS BY ITERATION	75-0404
EQUATIONS	SOLUTION OF LARGE SYSTEMS OF EQUATIONS	75-0608
EQUATIONS	SOLUTION OF LINEAR EQUATIONS WITH SKYLINE-STORED SYMMETRIC MATRIX	75-0662
EQUATIONS	SOLUTION OF NONLINEAR EQUATIONS BY EXPLICIT TIME INTEGRATION - A SHORT COMMUNICATION	73-0606
EQUATIONS	SOLUTION OF THE COMPRESSIBLE FLOW EQUATIONS	71-0003
EQUATIONS	SOLUTIONS OF LAPLACE'S EQUATIONS	61-0002
EQUATIONS	SOLVING A CLASS OF LARGE SPARSE LINEAR SYSTEMS OF EQUATIONS BY PARTITIONING	73-0580
EQUATIONS	SPARSE MATRIX TECHNIQUES IN DIRECT SOLUTION OF FINITE ELEMENT EQUATIONS	74-0115
EQUATIONS	STEADY STATE EQUATIONS OF MOTION, EQUILIBRIUM SHAPE AND STABILITY DERIVATIVES OF ELASTIC AIRPLANES E	72-0541
EQUATIONS	SURVEY OF SPARSE MATRIX METHODS IN THE DIRECT SOLUTION OF FINITE ELEMENT EQUATIONS	73-0572
EQUATIONS	TECHNIQUE FOR FORMULATING BEAM EQUATIONS	75-0584
EQUATIONS	THE FINITE ELEMENT METHOD FOR ELLIPTIC DIFFERENTIAL EQUATIONS	70-0027
EQUATIONS	THE FINITE ELEMENT METHOD IN PARTIAL DIFFERENTIAL EQUATIONS	70-0456
EQUATIONS	THE MATHEMATICAL FOUNDATIONS OF THE FINITE ELEMENT METHOD - WITH APPLICATIONS TO PARTIAL DIFFERENTIA	73-0286
EQUATIONS	THE SOLUTION OF STRUCTURAL EQUILIBRIUM EQUATIONS BY THE CONJUGATE GRADIENT METHOD WITH PARTICULAR RE	71-0268
EQUATIONS	THE SOLUTION OF THE NAVIER STOKES EQUATIONS	74-1220
EQUATIONS	THE USE OF NUMERICAL INTEGRATION IN FINITE ELEMENT METHODS FOR SOLVING PARABOLIC EQUATIONS	72-0351
EQUATIONS	THE USE OF THE FINITE ELEMENT METHOD IN THE SOLUTION OF THE FLOW EQUATIONS IN SOILS	74-0449
EQUATIONS	TWO-DIMENSIONAL FINITE ELEMENT ANALYSIS OF SEMI-CONDUCTOR STEADY STATE TRANSPORT EQUATIONS	74-1224
EQUATIONS	UNCONDITIONALLY STABLE FINITE ELEMENT SCHEMES FOR PARABOLIC EQUATIONS	74-1286
EQUATIONS	USE OF FINITE-ELEMENT METHOD IN THE SOLUTION OF DIFFUSION-CONVECTION EQUATIONS	71-0490
EQUATIONS	USE OF SPARSITY IN THE SOLUTION OF FINITE ELEMENT SYSTEMS OF EQUATIONS BY GAUSSIAN ELIMINATION-TYPE	74-0868
EQUATIONS	VARIABLE GRID FINITE DIFFERENCE-ELEMENT SOLUTION OF ELLIPTIC PARTIAL DIFFERENTIAL EQUATIONS	74-0862
EQUATIONS	VARIATIONAL FINITE ELEMENT METHODS FOR TWO DIMENSIONAL AND AXISYMMETRIC NAVIER STOKES EQUATIONS	74-0012
EQUATIONS	VARIATIONAL PRINCIPLES AND FINITE ELEMENT METHOD IN PARTIAL DIFFERENTIAL EQUATIONS	71-0452
EQUATIONS	VARIATIONAL PRINCIPLE OF LINEAR DIFFERENTIAL EQUATIONS	72-0107
EQUILATERAL	FINITE ELEMENT SOLUTIONS FOR AN EQUILATERAL TRIANGLE	74-1281
EQUILIBRE	ANALYSE DE LA STABILITY DES TALUS ROCHEUX PAR UNE METHODE GENERALISEE DE L EQUILIBRE LIMITE (ROCK SL	74-1019
EQUILIBRIUM	A FAMILY OF EQUILIBRIUM PLATE BENDING ELEMENTS	74-0431
EQUILIBRIUM	A GENERAL EQUILIBRIUM BEAM ELEMENT WITH OFFSETS	73-0351
EQUILIBRIUM	A GENERAL EQUILIBRIUM BENDING ELEMENT	73-0699
EQUILIBRIUM	A GENERAL EQUILIBRIUM AXIAL ELEMENT	73-0729
EQUILIBRIUM	A GENERAL EQUILIBRIUM TORQUE ELEMENT	74-0187
EQUILIBRIUM	A MODE AMPLITUDE TECHNIQUE FOR GENERAL FINITE ELEMENT EQUILIBRIUM	74-0283
EQUILIBRIUM	A SIX NODE TRIANGULAR EQUILIBRIUM (CONSTANT STRESS) MEMBRANE ELEMENT	73-0783
EQUILIBRIUM	A SIX NODE TRIANGULAR SHEAR PANEL FOR EQUILIBRIUM MODELS	73-1011
EQUILIBRIUM	A STUDY OF THE EQUILIBRIUM MODEL FOR USE IN FINITE ELEMENT ANALYSIS	66-0132
EQUILIBRIUM	A SYNTHESIZED WARPED QUADRILATERAL ELEMENT GIVING SIMPLIFIED INPUT FOR EQUILIBRIUM MODELS	73-1013
EQUILIBRIUM	A TRIANGULAR EQUILIBRIUM ELEMENT WITH LINEARLY VARYING BENDING MOMENTS FOR PLATE BENDING PROBLEMS	67-0054
EQUILIBRIUM	A TRIANGULAR EQUILIBRIUM (LINEAR STRESS) MEMBRANE ELEMENT	75-0883
EQUILIBRIUM	A WARPED QUADRILATERAL EQUILIBRIUM MEMBRANE ELEMENT	75-0885
EQUILIBRIUM	ACCURACY OF EQUILIBRIUM SLOPE STABILITY ANALYSIS	73-0517
EQUILIBRIUM	AN EQUILIBRIUM MEMBRANE QUADRILATERAL ELEMENT	72-0484
EQUILIBRIUM	AN EQUILIBRIUM MODEL FOR PLATE BENDING	68-0072
EQUILIBRIUM	AN EQUILIBRIUM STRESS FIELD MODEL FOR FINITE ELEMENT SOLUTIONS OF TWO-DIMENSIONAL ELASTIC PROBLEMS	68-0163
EQUILIBRIUM	ANALYSE DE LA STABILITY DES TALUS ROCHEUX PAR UNE METHODE GENERALISEE DE L EQUILIBRE LIMITE (ROCK SL	74-1019
EQUILIBRIUM	APPROXIMATE THREE-DIMENSIONAL STRESS SOLUTIONS FOR SOME LAMINATED PLATE PROBLEMS USING A HIGHER ORDE	70-0419
EQUILIBRIUM	AUTOMATIC EMBANKMENT ANALYSIS, EQUILIBRIUM AND INSTABILITY CONDITIONS	66-0015
EQUILIBRIUM	BASIS OF A WELL CONDITIONED FORCE PROGRAM FOR EQUILIBRIUM MODELS VIA THE SOUTHWELL SLAB ANALOGIES	67-0111
EQUILIBRIUM	COMPUTER PROGRAM FOR PIT SLOPE STABILITY ANALYSIS BY THE FINITE ELEMENT STRESS ANALYSIS AND LIMITING	72-0225
EQUILIBRIUM	CONCERNING THE FINITE ELEMENT METHOD FOR THE SOLUTION OF PROBLEMS OF ELASTIC EQUILIBRIUM	71-0516
EQUILIBRIUM	CONFORMING FINITE ELEMENT WITH INTERNAL EQUILIBRIUM FOR TWO DIMENSIONAL PROBLEMS	72-0386
EQUILIBRIUM	DIFFUSIVE EQUILIBRIUM MODELS	73-1112
EQUILIBRIUM	DISPLACEMENT AND EQUILIBRIUM MODELS IN THE FINITE ELEMENT METHOD	65-0016
EQUILIBRIUM	DUALITY BETWEEN DISPLACEMENT AND EQUILIBRIUM METHOD WITH A VIEW TO OBTAINING UPPER AND LOWER BOUNDS	62-0013
EQUILIBRIUM	ELAS - A GENERAL PURPOSE COMPUTER PROGRAM FOR THE EQUILIBRIUM PROBLEMS OF LINEAR STRUCTURES	68-0158
EQUILIBRIUM	ELAS: A GENERAL-PURPOSE COMPUTER PROGRAM FOR THE EQUILIBRIUM PROBLEMS OF LINEAR STRUCTURES. VOLUME 2	69-0462
EQUILIBRIUM	ELASTIC AND EQUILIBRIUM MATRICES OF GENERAL SEMI-MONOCOGUE MEMBRANE ELEMENTS	70-0159

214

EQUILIBRIUM	ELASTIC EQUILIBRIUM MODELS, STRESS FUNCTIONS, AND DUALITY NON LINEARITIES AND STABILITY, FIVE LECTUR	73-0631
EQUILIBRIUM	EQUILIBRIUM MODELS	75-0008
EQUILIBRIUM	EQUILIBRIUM STRESS FIELD MODELS IN FINITE ELEMENTS	67-0011
EQUILIBRIUM	FINITE ELEMENT EQUILIBRIUM MODEL FOR GENERAL BODY FORCE DISTRIBUTION	75-0268
EQUILIBRIUM	FINITE ELEMENT PLATE BENDING EQUILIBRIUM ANALYSIS	69-0045
EQUILIBRIUM	FINITE-ELEMENT SOLUTIONS OF PLASMA EQUILIBRIUM	75-0455
EQUILIBRIUM	FLAT QUADRILATERAL EQUILIBRIUM MEMBRANE ELEMENTS	73-1116
EQUILIBRIUM	MESH GRADING TECHNIQUES FOR COMPATIBLE AND EQUILIBRIUM ELEMENTS	74-0397
EQUILIBRIUM	NOTE ON INCREMENTAL EQUILIBRIUM EQUATIONS AND APPROXIMATE CONSTITUTIVE RELATIONS IN LARGE INELASTIC	74-0984
EQUILIBRIUM	RIGID BODY MOTIONS AND EQUILIBRIUM IN FINITE ELEMENTS	74-0102
EQUILIBRIUM	SLOPE STABILITY ANALYSIS BY THE FINITE ELEMENT STRESS ANALYSIS AND LIMITING EQUILIBRIUM METHOD	70-0222
EQUILIBRIUM	STEADY STATE EQUATIONS OF MOTION, EQUILIBRIUM SHAPE AND STABILITY DERIVATIVES OF ELASTIC AIRPLANES E	72-0541
EQUILIBRIUM	THE ANALYSIS OF CLASSICAL PLATE BENDING PROBLEMS USING AN EQUILIBRIUM FINITE ELEMENT MODEL	68-0302
EQUILIBRIUM	THE SOLUTION OF STRUCTURAL EQUILIBRIUM EQUATIONS BY THE CONJUGATE GRADIENT METHOD WITH PARTICULAR RE	71-0268
EQUILIBRIUM	THE TRIANGULAR EQUILIBRIUM ELEMENT IN THE SOLUTION OF PLATE BENDING PROBLEMS	68-0122
EQUILIBRIUM	VARIATIONAL METHODS FOR THE SOLUTION OF PROBLEMS OF EQUILIBRIUM AND VIBRATION	43-0001
EQUIPMENT	IMPROVED TECHNIQUES FOR DYNAMIC ANALYSIS OF EARTHMOVING EQUIPMENT	74-0294
EQUIPMENT	STRUCTURAL ANALYSIS OF PETROCHEMICAL EQUIPMENT BY THE FINITE ELEMENT METHOD	73-0377
EQUIPOTENTIAL	EQUIPOTENTIAL ZONING OF TWO-DIMENSIONAL MESHES	63-0025
EQUIPOTENTIAL	EQUIPOTENTIAL ZONING OF TWO-DIMENSIONAL MESHES	64-0033
EQUIVALENCE	COMPARISON OF EQUIVALENCE AND FINITE ELEMENT METHODS	75-0483
EQUIVALENCE	THEORY OF EQUIVALENCE AND ITS APPLICATION TO VARIOUS PROBLEMS IN ELASTICITY AND DESIGN OF SKEW BRIDG	72-0470
EQUIVALENCY	EQUIVALENCY BETWEEN FINITE ELEMENT METHOD AND FINITE DIFFERENCE METHOD AND A NEW DISCRETIZATION APPR	73-0480
EQUIVALENT	A RECTANGULAR PLATE BENDING ELEMENTS THE USE OF WHICH IS EQUIVALENT TO THE USE OF THE FINITE DIFFERE	71-0398
EQUIVALENT	ANALYSIS OF OPEN STRUCTURES BY FINITE ELEMENT RESOLUTION OF EQUIVALENT CLOSED-BOUNDARY PROBLEM	74-0111
EQUIVALENT	COMBINED ANALYTICAL - EXPERIMENTAL FRACTURE STUDY OF THE TWO LEADING THEORIES OF ELASTIC-PLASTIC FRA	73-0951
EQUIVALENT	EQUIVALENT FINITE ELEMENTS ON DIFFERENT BASES	72-0287
EQUIVALENT	EQUIVALENT MASS MATRICES FOR RECTANGULAR PLATES IN BENDING	66-0063
EQUIVALENT	ON THE DETERMINATION OF EFFECTIVE ELASTIC-PLASTIC PROPER TIES FOR THE EQUIVALENT SOLID PLATE ANALYSI	74-0494
ERDEBENSCHWINGU	BERECHNUNG DER ERDEBENSCHWINGUNGEN VON STRUKTUREN MIT DER FINITE ELEMENT METHODE, MECHANISCHE MODELL	71-0170
ERDSTOFFE	PARAMETER EINES ELASTO-PLASTISCHEN STOFFANSATZES FUER KOERNIGE ERDSTOFFE (DETERMINATION OF PARAMETER	74-0657
ERECTED	COMPUTER ANALYSIS OF SEGMENTALLY ERECTED BRIDGES	75-0172
ERFAHRUNG	THEORIE, PROGRAMMENTWICKLUNG UND ERFAHRUNG ANVORGESPANNTEN NETZWERKKONSTRUKTIONEN	72-0640
ERMITTLUNG	BERECHNUNGSMETHODEN ZUR ERMITTLUNG VON BELASTUNGSBEDINGUNGEN UND GRENZTRAGFAEHIGKEITSVERHALTEN VON K	73-0858
ERMITTLUNG	EINSATZ MODERNER RECHENVERFAHREN ZUR ERMITTLUNG DER SPANNUNGEN UND VERFORMUNGEN BEI RADSAETZEN. (USE	75-0186
ERMITTLUNG	MESSVERFAHREN RECHENVERFAHREN ZUR ERMITTLUNG DREIACHSIGER EIGENSPANNUNGSZUSTAENDE	73-0507
ERROR	(L SUP 2) ERROR ESTIMATES FOR FINITE ELEMENTS WITH INTERPOLATED BOUNDARY CONDITIONS	73-0885
ERROR	A STUDY OF NUMERICAL ERROR IN STRUCTURAL SYSTEMS	70-0394
ERROR	AN ERROR EXPANSION FOR A FINITE-ELEMENT SOLUTION OF A TWO-POINT BOUNDARY-VALUE PROBLEM	75-0801
ERROR	BOUNDS AND ERROR CONTROL FOR EIGENVALUES	73-0319
ERROR	COMPUTABLE ERROR BOUNDS FOR THE FINITE ELEMENT METHOD FOR ELLIPTIC BOUNDARY VALUE PROBLEMS	74-1219
ERROR	CONVERGENCE AND ERROR ESTIMATES FOR FINITE ELEMENT APPROXIMATIONS OF STATIONARY NAVIER-STOKES EQUATI	74-0122
ERROR	CURVED ELEMENTS, ERROR BOUNDS, TIME DEPENDENT PROBLEMS FIVE LECTURES	73-0636
ERROR	DISCRETIZATION AND ROUND-OFF ERROR IN THE FINITE ELEMENT ANALYSIS OF ELLIPTIC BOUNDARY VALUE PROBLEM	71-0514
ERROR	DISCRETIZATION ERROR BOUNDS FOR TRANSFINITE ELEMENTS	72-0533
ERROR	ERROR ANALYSIS OF A REMOTE TRANSIENT HEAT FLUX SENSOR	74-0753
ERROR	ERROR ANALYSIS OF FINITE ELEMENT METHODS WITH TRIANGLES FOR ELLIPTIC BOUNDARY VALUE PROBLEMS	73-0072
ERROR	ERROR ANALYSIS OF VARIOUS FINITE ELEMENT PATTERNS	73-0510
ERROR	ERROR BOUNDS FOR EIGENVALUE ANALYSIS BY ELIMINATION OF VARIABLES	71-0594
ERROR	ERROR BOUNDS FOR FINITE ELEMENT METHOD	71-0033
ERROR	ERROR BOUNDS FOR SEMIDISCRETE GALERKIN APPROXIMATIONS OF PARABOLIC PROBLEMS WITH APPLICATIONS TO PET	70-0291
ERROR	ERROR BOUNDS IN AN OPTIMIZATION PROBLEM USING FINITE-ELEMENT METHOD	73-0049
ERROR	ERROR ESTIMATE FOR QUADRILATERAL ISOPARAMETRIC FINITE-ELEMENTS OF TYPE Q1 WHICH CAN DEGENERATE INTO	75-0401
ERROR	ERROR ESTIMATES FOR THE FINITE ELEMENT METHOD	72-0686
ERROR	ERROR ESTIMATES FOR CONFORMING FINITE ELEMENT SOLUTIONS	74-0940
ERROR	ERROR ESTIMATES FOR FINITE ELEMENT METHOD FOR HEAT TRANSFER PROBLEMS IN THE SPACE-TIME DOMAIN	75-0104
ERROR	ERROR ESTIMATES FOR A FINITE-ELEMENT APPROXIMATION OF A MINIMAL SURFACE	75-0491
ERROR	EVALUATION OF THE ERROR BOUNDS IN AN OPTIMIZATION PROBLEM USING THE FINITE-ELEMENT METHOD	73-0211
ERROR	FINITE ELEMENT ERROR BOUNDS	75-0100
ERROR	INHERITED ERROR IN FINITE ELEMENT ANALYSES OF STRUCTURES	73-0452
ERROR	INTERPRETATION OF FINITE ELEMENT PROCEDURE AS STRESS ERROR MINIMIZATION PROCEDURE	72-0984
ERROR	LOCAL ERROR BOUNDS IN BOUNDARY VALUE PROBLEMS USING FINITE ELEMENT DISCRETIZATION	72-0982
ERROR	L2 ERROR ESTIMATES FOR FINITE ELEMENTS WITH INTERPOLATED BOUNDARY CONDITIONS	73-0194
ERROR	NONUNIFORM ERROR ESTIMATES FOR THE FINITE ELEMENT METHOD	74-0890
ERROR	ON THE REDUCTION OF NUMERICAL ERROR IN THE MATRIX DISPLACEMENT METHOD	74-0417
ERROR	REMOVAL OF TRUNCATION ERROR IN FINITE ELEMENT ANALYSIS	75-0131
ERROR	SARD KERNEL THEOREMS ON TRIANGULAR AND RECTANGULAR DOMAINS WITH EXTENSIONS AND APPLICATIONS TO FINIT	72-0361
ERROR	SOLUTION OF MIXED BOUNDARY VALUE PROBLEMS WITH LOCAL ERROR BOUND BY THE FINITE ELEMENT METHOD	73-0795
ERROR	SOME L(2) ERROR ESTIMATES FOR PARABOLIC GALERKIN METHODS	72-0511
ERROR	THE EXTENSION AND APPLICATION OF SARD KERNEL THEOREMS TO COMPUTE FINITE ELEMENT ERROR BOUNDS	72-0519
ERRORS	BOUNDARY AND INTERIOR APPROXIMATION ERRORS IN THE FINITE ELEMENT METHOD	73-0209
ERRORS	CHARACTERISTICS OF MANIPULATION ERRORS IN SOLVING LOAD DEFLECTION EQUATIONS	70-0355
ERRORS	DISCRETIZATION AND COMPUTATIONAL ERRORS IN HIGHER-ORDER FINITE ELEMENTS	71-0089
ERRORS	EFFECTS OF QUADRATURE ERRORS IN FINITE ELEMENT APPROXIMATION OF STEADY STATE, EIGENVALUE, AND PARABO	72-0343
ERRORS	ERRORS IN SHELL FINITE-ELEMENT MODELS FOR VIBRATION OF CIRCULAR-CYLINDERS	75-0203
ERRORS	MANIPULATION ERRORS IN FINITE ELEMENT ANALYSIS	68-0253
ERRORS	MANIPULATION ERRORS IN FINITE ELEMENT ANALYSIS OF STRUCTURES	69-0349
ERRORS	MANIPULATION ERRORS IN THE FINITE ELEMENT ANALYSES	71-0176
ERRORS	MANIPULATION ERRORS IN COMPUTER SOLUTION OF CRITICAL SIZE STRUCTURAL EQUATIONS	71-0367
ERRORS	MEAN-SQUARE ERRORS OF FINITE ELEMENT APPROXIMATIONS ON LINEAR ELASTOSTATICS, ELASTODYNAMICS, AND THE	74-0107
ERRORS	NUMERICAL ACCURACY AND ROUNDING-OFF ERRORS	69-0348
ERRORS	ON THE EFFECTS OF QUADRATURE ERRORS IN THE FINITE ELEMENT METHOD	72-0258
ERRORS	ON THE LOCAL DISTRIBUTION OF ERRORS BY FINITE ELEMENT APPROXIMATIONS	73-0304
ERRORS	PERTURBATION ERRORS IN FINITE ELEMENT ANALYSIS	72-0277
ERRORS	SOME CONSIDERATIONS ON ROUND-OFF ERRORS OF THE FINITE ELEMENT METHOD	72-0260
ERRORS	UNBOUNDED ERRORS IN NUMERICAL ANALYSIS OF STRUCTURES	74-1210
ERSTELLUNG	UNTERSUCHUNGEN ZUM FESTIGKEITSVERHALTEN VON METALLKLEBUNGEN IM HIMBLICK AUF DIE ERSTELLUNG EINES PRA	74-0289
ERWEITERTEN	STABILITY INVESTIGATIONS OF PLATES USING FINITE ELEMENT METHOD BASED ON AN EXPANDED VARIATION PRINCI	71-0663
ERZWUNGENE	ANWENDUNG DER MATRIZENVERSCHIEBUNGSMETHODE AUF ERZWUNGENE SCHWINGUNGEN (SCHWINGUNGSPROBLEME) (ARBEIT	71-0627
ESEM	ELASTOSTATIK-ELEMENTMETHODE (ESEM)	69-0415
ESPACES	METHODE DES ELEMENTS FINIS; UNE FAMILE D'APPROXIMATION DES ESPACES DE SOBOLEV PAR LES TRANSLATES DE	70-0301
ESPIADAS	SOBRE TORRES ESPIADAS - METODO DE CALCULO E UM EXEMPLO DE APLICACAO. (METHOD AND AN EXAMPLE OF CALCU	74-0915
ESSENTIAL	EFFECTS OF GEOMETRY CHANGE ON ESSENTIAL FEATURES OF INELASTIC BEHAVIOUR	73-0448
ESSENTIAL	MATCHING OF ESSENTIAL BOUNDARY CONDITIONS IN THE FINITE ELEMENT METHOD	74-1260
ESSO	EXAMPLE OF LARGE-SCALE STRUCTURAL ANALYSIS. COMPARISON BETWEEN FINITE ELEMENT CALCULATION AND FULL S	74-0177
ESTABLISHMENT	ESTABLISHMENT OF A VALID FRACTURE TOUGHNESS SPECIMEN GEOMETRY USING A FINITE ELEMENT ANALYSIS	74-0135
ESTABLISHMENT	UNTERSUCHUNGEN ZUM FESTIGKEITSVERHALTEN VON METALLKLEBUNGEN IM HIMBLICK AUF DIE ERSTELLUNG EINES PRA	74-0289
ESTIMATE	ERROR ESTIMATE FOR QUADRILATERAL ISOPARAMETRIC FINITE-ELEMENTS OF TYPE Q1 WHICH CAN DEGENERATE INTO	75-0401
ESTIMATE	FINITE ELEMENT PARAMETRIC STUDY OF VERTICAL STRAIN INFLUENCE FACTORS AND THE PRESSUREMETER TEST TO E	74-1127
ESTIMATED	A NEW MEASURING METHOD OF RESIDUAL STRESSES WITH THE AID OF FINITE ELEMENT METHOD AND RELIABILITY OF	75-0875
ESTIMATION	ESTIMATION OF THE MECHANICAL PROPERTIES OF ROCK MASSES	73-0935
ESTIMATIONS	ESTIMATIONS D'ERREUR POUR DES ELEMENTS FINIS DROITS PRESQUE DEGENERES	74-1247

ESTIMATIONS   ESTIMATIONS OF THE EFFECT OF DESIGN CHANGES ON THE MODES AND FREQUENCIES OF VIBRATING STRUCTURAL ELE   74-0819
ESTUARIES   A FINITE ELEMENT MODEL OF TIDES IN ESTUARIES   74-0044
ESTUARIES   A FINITE ELEMENT MODEL FOR SEDIMENT TRANSPORT IN ESTUARIES   74-1102
ESTUARIES   A NUMERICAL MODEL FOR THE SIMULATION OF TWO DIMENSIONAL CONVECTIVE DISPERSION IN SHALLOW ESTUARIES   74-0052
ESTUARIES   A NUMERICAL MODEL OF DISPERSION IN ESTUARIES   74-0045
ESTUARIES   FINITE ELEMENT NUMERICAL MODELLING OF FLOW AND DISPERSION IN ESTUARIES   73-0070
ESTUARIES   GALERKIN FINITE ELEMENT SOLUTIONS FOR POLLUTION PROBLEMS IN PARTIALLY MIXED ESTUARIES   74-0047
ESTUARIES   NUMERICAL MODEL OF DISPERSION IN ESTUARIES   74-0905
ESTUARIES   TIDAL PROPAGATION AND DISPERSION IN ESTUARIES   75-0390
ESTUARY   A NUMERICAL MODEL OF TRANSIENT WATER QUALITY IN A ONE-DIMENSIONAL ESTUARY BASED ON THE FINITE ELEMEN   74-1369
ESTUARY   NUMERICAL MODEL FOR THE PREDICTION OF TRANSIENT WATER QUALITY IN ESTUARY NETWORKS   72-0635
EUEGETECHNIK   EXPERIMENTS IN FORMING AND JOINTING OF TITANIUM AND TITANIUM ALLOYS FOR AIRCRAFT AND SPACE VEHICLES   70-0134
EULER'S   GENERATION OF FINITE ELEMENTS BY EULER'S VARIATIONAL METHOD   72-0596
EULERIAN   LAGRANGIAN AND EULERIAN FINITE ELEMENT FORMULATION FOR LARGE STRAIN ELASTO-PLASTICITY   75-0209
EULERIAN   NUMERICAL-SOLUTION OF EULERIAN EQUATIONS OF COMPRESSIBLE FLOW BY A FINITE-ELEMENT METHOD WHICH FOLLO   75-0535
EURASIAN   MECHANICS OF DEFORMABLE SOLIDS - EURASIAN COMMUNIST COUNTRIES   73-0794
EURCYL   EURCYL: A COMPUTER PROGRAM TO GENERATE FINITE ELEMENT MESHES FOR CYLINDER-CYLINDER INTERSECTIONS   73-0962
EURCYL   EURCYL: A PROGRAM TO GENERATE FINITE ELEMENT MESHES FOR PRESSURE VESSEL NOZZLES   74-1086
EUTECTICS   FINITE-ELEMENT MODELING OF DIRECTIONAL SOLIDIFICATION OF SUPER ALLOY EUTECTICS   75-0350
EVALUATE   PARAMETRIC STUDY OF A BEAM WITH A COMPOUND SIDE-BRANCH RESONATOR AS A DEVICE TO EVALUATE PRELIMINARY   70-0465
EVALUATING   A THEORETICAL METHOD FOR EVALUATING STABILITY OF OPENINGS IN ROCK   72-0877
EVALUATING   DEVELOPMENT AND APPLICATIONS OF THEORETICAL METHODS FOR EVALUATING STABILITY OF OPENINGS IN ROCK   73-0791
EVALUATING   DEVELOPMENT AND APPLICATIONS OF A GENERAL COMPUTER PROGRAM FOR EVALUATING STABILITY OF OPENINGS IN R   74-0994
EVALUATING   FACTOR OF SAFETY APPROACH FOR EVALUATING SEISMIC STABILITY OF SLOPES   75-0645
EVALUATING   FINITE ELEMENT CRACK TIP MODELING WHEN EVALUATING J-INTEGRAL   75-0637
EVALUATING   NEUARTIGE ANWENDUNG GEOMECHANISCHER UNTERSUCHUNGSMETHODEN ZUR BEURTEILUNG DER ABBAUWIRKUNGEN IM GANG   75-0530
EVALUATING   QUAD-4: A COMPUTER PROGRAM FOR EVALUATING THE SEISMIC RESPONSE OF SOIL STRUCTURES BY VARIABLE DAMPIN   73-0771
EVALUATING   SOIL-STRUCTURE INTERACTION ANALYSES FOR EVALUATING SEISMIC RESPONSE   74-0745
EVAPORATION   ANALYTICAL STUDY OF THE PROBLEM OF EVAPORATION AND CONDENSATION USING THE FINITE ELEMENT METHOD   74-0960
EVENT   BERECHNUNGSMETHODEN ZUR ERMITTLUNG VON BELASTUNGSBEDINGUNGEN UND GRENZTRAGFAEHIGKEITSVERHALTEN VON K   73-0858
EVOLUTION   THE EVOLUTION OF SHELL STRUCTURES IN OCEAN TECHNOLOGY   69-0042
EVOLUTION   THE EVOLUTION OF THE ISOPARAMETRIC ELEMENTS   75-0425
EXACT   A NEW METHOD OF ANALYSIS TO OBTAIN EXACT SOLUTIONS FOR STRESSES AND STRAINS IN CIRCULAR INCLUSIONS   74-1063
EXACT   AN EXACT BOUNDARY TECHNIQUE FOR IMPROVED ACCURACY IN THE FINITE ELEMENT METHOD   73-1035
EXACT   APPLICABILITY OF APPROXIMATE AND EXACT TRANSIENT HEAT TRANSFER ANALYSES TO HEATING PROCESSES USED TO   75-0244
EXACT   COMMENT ON EXACT SOLUTION OF CERTAIN PROBLEMS BY FINITE ELEMENT METHOD   69-0371
EXACT   ELASTIC WAVE PROPAGATION. A COMPARISON BETWEEN FINITE ELEMENT PREDICTIONS AND EXACT SOLUTION   67-0076
EXACT   EXACT FINITE ELEMENT SOLUTIONS TO THE HELMHOLTZ EQUATIONS   73-0652
EXACT   EXACT SOLUTION OF CERTAIN PROBLEMS BY THE FINITE-ELEMENT METHOD   69-0258
EXACT   EXACT SOLUTION TO THE RADIALLY END CONSTRAINED CIRCULAR CYLINDER UNDER TRIAXIAL LOADING   71-0433
EXACT   EXACT THEORY OF SPATIAL STRUCTURES   73-0781
EXACT   FLEXIBILITY MATRIX FOR AN EXACT TORSION-BENDING FINITE ELEMENT   72-0214
EXACT   THE AUTOMATIC COMPUTATION OF THE NATURAL FREQUENCIES OF STRUCTURAL FRAMES USING AN EXACT MATRIX TECH   73-0318
EXACT   ZUR GENAUEN BERECHNUNG KONSTRUKTIV ORTHOTROPER PLATTEN (EXACT CALCULATION OF STRUCTURALLY ORTHOTROPI   70-0051
EXCAVATED   DEVELOPMENT OF FAILURE AROUND EXCAVATED SLOPES   74-1330
EXCAVATED   THE BEHAVIOR OF STATISTICALLY HETEROGENEOUS EXCAVATED EARTH SLOPES   70-0039
EXCAVATION   ANALYSIS OF SOIL MOVEMENT AROUND A DEEP EXCAVATION   73-0313
EXCAVATION   EXCAVATION IN ROCK, AN APPRECIATION OF THE FINITE ELEMENT METHOD OF ANALYSIS   69-0530
EXCAVATION   FINITE ELEMENT ANALYSES OF SOIL MOVEMENTS CAUSED BY DEEP EXCAVATION AND DEWATERING   74-1250
EXCAVATION   MESH GENERATION PROGRAM FOR HIGHWAY EXCAVATION CUTS   74-0542
EXCAVATION   RASCHET NAPRYAZHENIL L PEREMESHCHENII V SVODE PODZEMNOGO MASHINNOGO ZALA GES I V OKRUZHAYUSHCHEM SKA   72-0629
EXCAVATION   SEEPAGE INTO AN EXCAVATION IN A MEDIUM POSSESSING STRESS-DEPENDENT PERMEABILITY   75-0049
EXCAVATION   TEMPORARY EXCAVATION IN VARVED CLAY   72-0800
EXCAVATION   THE USE OF A LASER FOR ARPA MILITARY GEOPHYSICS PROGRAM (ROCK MECHANICS AND RAPID EXCAVATION)   73-0395
EXCAVATIONS   ANALYSIS OF ROCK MASS DEFORMATIONS DUE TO EXCAVATIONS   74-0597
EXCAVATIONS   BERECHNUNG UND BEURTEILUNG DER STANDSICHERHEIT VON ABBAUHOHLRAEUMEN IN STEILEN GANGERZLAGERSTAETTEN   71-0014
EXCAVATIONS   GRAVITY FLOW TO EXCAVATIONS AND DRAINAGE TRENCHES IN LAYERED AQUIFERS   70-0146
EXCAVATIONS   SIMULATION OF PROGRESSIVE FAILURE IN ROCK AROUND MINING EXCAVATIONS   73-0858
EXCEPTIONAL   BERECHNUNGSMETHODEN ZUR ERMITTLUNG VON BELASTUNGSBEDINGUNGEN UND GRENZTRAGFAEHIGKEITSVERHALTEN VON K   74-0383
EXCESS   STRESSES IN NEARLY INCOMPRESSIBLE MATERIALS BY FINITE ELEMENTS WITH APPLICATION TO THE CALCULATION O   70-0346
EXCESSIVE   EFFECTIVENESS OF EXCESSIVE NODAL CONTINUITIES IN THE FINITE ELEMENT ANALYSIS OF THIN RECTANGULAR AND   70-0371
EXCESSIVE   THE EFFECTIVENESS OF EXCESSIVE NODAL CONTINUITIES IN THE FINITE ELEMENT ANALYSIS OF THIN RECTANGULAR   73-0665
EXCHANGER   CALCULATION OF NATURAL FREQUENCIES OF HEAT EXCHANGER TUBES   72-0772
EXCHANGER   DESIGN AND ANALYSIS EXPERIENCE WITH A LIQUID METAL HEAT EXCHANGER FOR PFTF SERVICE   74-0661
EXCHANGER   STRESS DETERMINATION IN 2 CONDENSER HEAT EXCHANGER BAFFLE CAPS BY METHOD OF FINITE ELEMENTS AND A ST   72-0451
EXCHANGERS   NONSTEADY PYROMETRIC CALCULATIONS AS THE BASIS FOR STRESS ANALYSIS OF STEAM GENERATORS AND HEAT EXCH   75-0709
EXCITABILITY   EFFECT OF DAMPING ON EXCITABILITY OF HIGH-ORDER NORMAL MODES   71-0094
EXCITATION   ANALYSIS OF SOIL-STRUCTURE INTERACTION EFFECTS UNDER SEISMIC EXCITATION   73-0553
EXCITATION   DYNAMIC STABILITY OF A NONLINEAR BEAM SUBJECTED TO BOTH LONGITUDINAL AND TRANSVERSE EXCITATION   74-0803
EXCITATION   EXCITATION OF PLANE THERMOELASTIC WAVES IN A LAYER WITH A FINITE RATE OF HEAT PROPAGATION   73-0834
EXCITATION   FINITE ELEMENT TREATMENT OF SOIL-STRUCTURE INTERACTION PROBLEM FOR NUCLEAR POWER PLANT UNDER SEISMIC   71-0546
EXCITATION   JET NOISE EXCITATION OF AN INTEGRALLY STIFFENED PANEL   68-0232
EXCITATION   METHOD FOR ESTIMATING RESPONSE OF PAYLOAD SECONDARY STRUCTURES TO RANDOM EXCITATION   75-0711
EXCITATION   OPTIMIZATION OF STRUCTURES UNDERGOING HARMONIC OR STOCHASTIC EXCITATION   69-0163
EXCITATION   PERIODIC SHOCK EXCITATION OF ELASTIC STRUCTURES   73-0285
EXCITATION   RESPONSE OF NONLINEAR BEAM TO RANDOM EXCITATION   72-0864
EXCITATION   STUDY OF EFFECTS OF DESIGN DETAILS ON STRUCTURAL RESPONSE TO ACOUSTIC EXCITATION   74-1332
EXCITATION   STUDY OF THE ACOUSTIC EXCITATION OF STRUCTURES   72-0737
EXCITATION   THE COMPARISON OF THE RESPONSE OF A HIGHWAY BRIDGE TO UNIFORM GROUND SHOCK AND MOVING GROUND EXCITAT   70-0348
EXCITATIONS   FINITE ELEMENT ANALYSES OF SHELL RESPONSE FOR ARBITRARY EXCITATIONS   74-0453
EXCITATIONS   FINITE ELEMENT ANALYSIS OF STRUCTURAL INSTABILITY BY ASSOCIATION OF PULSATING EXCITATIONS   75-0230
EXCITATIONS   FINITE ELEMENT METHODS FOR THE RESPONSE OF STRUCTURES TO RANDOM EXCITATIONS   73-0278
EXCITATIONS   ON RESPONSE OF INITIALLY STRESSED STRUCTURES TO RANDOM EXCITATIONS   74-1047
EXCITATIONS   RESPONSE OF NUCLEAR STRUCTURAL SYSTEMS TO TRANSIENT AND RANDOM EXCITATIONS. USING BOTH DETERMINISTIC   73-0496
EXCITATIONS   RESPONSE OF RECTANGULAR ORTHOTROPIC PLATES TO RANDOM EXCITATIONS   67-0057
EXCITED   A FINITE ELEMENT APPROACH FOR THE ANALYSIS OF RANDOMLY EXCITED COMPLEX ELASTIC STRUCTURES   74-1309
EXCITED   VIBROACOUSTIC RESPONSE OF TURBULENCE EXCITED THIN RECTANGULAR FINITE PLATES IN HEAVY AND FLUID MEDIA   73-0802
EXECUTION   INFLUENCE OF THE UNPRECISION OF EXECUTION ON THE BEHAVIOUR OF A SPHERICAL SHELL WITH CONCENTRATED LO   74-0938
EXECUTION   REVIEW OF NASTRAN DEVELOPMENT RELATIVE TO EFFICIENCY OF EXECUTION   74-0915
EXEMPLO   SOBRE TORRES ESPIADAS - METODO DE CALCULO E UM EXEMPLO DE APLICACAO. (METHOD AND AN EXAMPLE OF CALCU   72-0660
EXHIBITING   A MATHEMATICAL THEORY OF PLASTICITY FOR STRAIN HARDENING MATERIALS EXHIBITING INITIAL ANISOTROPY   71-0780
EXHIBITING   ANALYSIS OF LONGITUDINAL WAVE IN A BAR EXHIBITING VISCO-ELASTIC BEHAVIOR OR WITH MATERIAL DAMPING   71-0130
EXLPORER   NORMAL MODE ANALYSIS OF THE RADIO ASTRONOMY EXLPORER BOOMS AND SPACECRAFT   74-0224
EXPANDED   DEFLECTION ANALYSIS OF EXPANDED OPEN-WEB STEEL BEAMS   68-0002
EXPANDED   FINITE ELEMENT ANALYSIS OF AN EXPANDED I - SECTION BEAM AND AN AXISYMMETRIC FLANGED CYLINDER   71-0663
EXPANDED   STABILITY INVESTIGATIONS OF PLATES USING FINITE ELEMENT METHOD BASED ON AN EXPANDED VARIATION PRINCI   75-0882
EXPANSION   A COMPUTER PROGRAMME FOR ELASTIC PLASTIC CALCULATIONS WITH DUAL PERTURBATION EXPANSION AND VARIATIO   75-0801
EXPANSION   AN ERROR EXPANSION FOR A FINITE-ELEMENT SOLUTION OF A TWO-POINT BOUNDARY-VALUE PROBLEM   73-0190
EXPANSION   DIFFERENTIAL EXPANSION IN ELASTIC LAMINATES   73-0916
EXPANSION   MEASUREMENT OF THERMAL EXPANSION IN SOLID PROPELLANTS   75-0543
EXPANSION   POWER SERIES EXPANSION OF THE GENERAL STIFFNESS MATRIX FOR BEAM ELEMENTS   73-0880
EXPEDIENT   A COMPUTER MODEL FOR PREDICTING THE LOAD-DEFLECTION RESPONSE OF EXPEDIENT SOIL SURFACING

EXPERIMENT	COMPARISON OF CALCULATION BY THE FINITE ELEMENT METHOD WITH EXPERIMENT AND AN EXAMPLE OF A DYNAMIC S	72-0597
EXPERIMENT	CORRELATION BETWEEN PRESSURE AND THERMAL TRANSIENT STRESS LEVELS DERIVED BY THEORY AND EXPERIMENT ON	73-0469
EXPERIMENT	DEFORMATION OF THE END DIASTOLIC LEFT VENTRICLE. COMPARISON OF THEORY WITH EXPERIMENT	71-0131
EXPERIMENT	FINITE ELEMENT ANALYSIS OF CRACK INITIATION IN A BLOCK MODEL EXPERIMENT	72-0455
EXPERIMENT	NUMERICAL ANALYSIS OF SHIP STRUCTURAL ELEMENTS BY MEANS OF FINITE ELEMENT METHOD AND PHOTO-ELASTIC E	69-0125
EXPERIMENT	NUMERICAL EXPERIMENT WITH FINITE ELEMENT I	70-0303
EXPERIMENT	THEORIE UND EXPERIMENT ZU EINEM NEUEN INJEKTIONSVERFAHREN. (THEORY AND EXPERIMENT OF A NEW GROUTING	74-0991
EXPERIMENT	THEORIE UND EXPERIMENT ZU EINEM NEUEN INJEKTIONSVERFAHREN. (THEORY AND EXPERIMENT OF A NEW GROUTING	74-0991
EXPERIMENT	THERMAL ANALYSIS OF M552 EXPERIMENT FOR MATERIALS PROCESSING IN SPACE	74-0704
EXPERIMENT	THERMAL ANALYSIS OF M551 EXPERIMENT FOR MATERIALS PROCESSING IN SPACE	74-0725
EXPERIMENT	THIN-SHELL STRUCTURES: THEORY, EXPERIMENT, AND DESIGN	74-0828
EXPERIMENT	THREE-DIMENSIONAL FINITE ELEMENT ANALYSIS OF A STEAM GENERATOR CHANNEL-HEAD-COMPLEX - COMPARISON OF	75-0653
EXPERIMENT	TRAGLAST VON T-TRAEGERN. EXPERIMENT UND F.E. BERECHNUNGEN (ULTIMATE STRENGTH OF T-BEAMS. EXPERIMENTS	74-0935
EXPERIMENTAL	A NUMERICAL AND EXPERIMENTAL INVESTIGATION OF THE USE OF J-INTEGRALS	72-0858
EXPERIMENTAL	A STATIC AND DYNAMIC FINITE ELEMENT SHELL ANALYSIS WITH EXPERIMENTAL VERIFICATION	69-0156
EXPERIMENTAL	A STATIC AND DYNAMIC FINITE ELEMENT SHELL ANALYSIS WITH EXPERIMENTAL VERIFICATION	71-0769
EXPERIMENTAL	A THEORETICAL AND EXPERIMENTAL STUDY ON THE FREE VIBRATION OF CURVED SLAB BRIDGES	71-0496
EXPERIMENTAL	AN EXPERIMENTAL AND THEORETICAL STUDY OF HOLLOW CYLINDER TESTS ON SAND	71-0008
EXPERIMENTAL	ANALYTICAL AND EXPERIMENTAL INVESTIGATION OF A NOZZLE TO CYLINDRICAL SHELL JUNCTION	74-0638
EXPERIMENTAL	ANALYTICAL AND EXPERIMENTAL STUDIES OF MULTI-CELL CONCRETE BOX GIRDER BRIDGES	74-1134
EXPERIMENTAL	ANALYTICAL AND EXPERIMENTAL EVALUATION OF MODULUS OF SOIL REACTION	74-1168
EXPERIMENTAL	ANALYTICAL AND EXPERIMENTAL INVESTIGATION OF A 1/8 SCALE DYNAMIC MODEL OF THE SHUTTLE ORBITER. VOLUM	74-1186
EXPERIMENTAL	ANALYTICAL AND EXPERIMENTAL INVESTIGATION OF FATIGUE IN A SHEET SPECIMEN WITH AN INTERFERENCE-FIT BO	75-0701
EXPERIMENTAL	APPLICATION, TESTING, AND EVALUATION OF DOUBLY CURVED SHELL ELEMENTS FOR DYNAMIC ANALYSIS. PART 1, T	69-0246
EXPERIMENTAL	COMBINED ANALYTICAL - EXPERIMENTAL FRACTURE STUDY OF THE TWO LEADING THEORIES OF ELASTIC-PLASTIC FRA	73-0951
EXPERIMENTAL	COMBINING EXPERIMENTAL AND ANALYTICAL TECHNIQUES FOR DYNAMIC SYSTEM ANALYSIS	73-0321
EXPERIMENTAL	COMPARISON OF ANALYTICAL AND EXPERIMENTAL RESULTS IN THE DESIGN OF A HOLLOW GRAVITY DAM	67-0109
EXPERIMENTAL	COMPARISON OF FINITE-ELEMENT AND EXPERIMENTAL STUDIES ON DEFORMATIONS OF ZIRCONIUM NOTCHED BEND SPEC	71-0359
EXPERIMENTAL	COMPARISON OF THEORETICAL AND EXPERIMENTAL SHEAR STRESS IN THE ADHESIVE LAYER OF A LAP JOINT MODEL	74-1176
EXPERIMENTAL	COMPARISON OF TYPICAL INELASTIC ANALYSIS PREDICTIONS WITH BENCHMARK PROBLEM EXPERIMENTAL RESULTS	75-0696
EXPERIMENTAL	DETERMINATION OF STRESS CONCENTRATIONS AT NOZZLE TO CYLINDER INTERSECTIONS AND COMPARISON WITH EXPER	73-0491
EXPERIMENTAL	DETERMINATION OF THE TECHNICAL CONSTANTS OF LAMINATES IN OBLIQUE DIRECTIONS. THEORY AND EXPERIMENTAL	73-0748
EXPERIMENTAL	DYNAMIC SIMULATION OF COMPLEX SYSTEMS UTILIZING EXPERIMENTAL AND ANALYTICAL TECHNIQUES	75-0864
EXPERIMENTAL	ELASTIC BENDING OF CIRCULAR PLATES OF VARIABLE THICKNESS. AN ANALYTICAL AND EXPERIMENTAL STUDY	70-0163
EXPERIMENTAL	EXPERIMENTAL AND ANALYTICAL STRAINS IN A EDGE-CRACKED SHEET	69-0483
EXPERIMENTAL	EXPERIMENTAL AND ANALYTICAL INVESTIGATION OF A HORIZONTALLY CURVED BOX-BEAM HIGHWAY BRIDGE MODEL	71-0024
EXPERIMENTAL	EXPERIMENTAL AND FINITE ELEMENT STRESS ANALYSIS OF A THIN SHELLED CYLINDER TO CYLINDER MODEL	71-0066
EXPERIMENTAL	EXPERIMENTAL AND FINITE ELEMENT ANALYSIS OF CABLE ROOF STRUCTURES INCLUDING PRECAST PANELS	74-1139
EXPERIMENTAL	EXPERIMENTAL AND THEORETICAL INVESTIGATION OF PAVEMENT DEFLECTIONS	69-0154
EXPERIMENTAL	EXPERIMENTAL AND THEORETICAL INVESTIGATION OF PLANE STRESS FRACTURE OF 2024 - T351 ALUMINUM ALLOY	70-0147
EXPERIMENTAL	EXPERIMENTAL AND THEORETICAL STUDIES OF HULL-DECKHOUSE INTERACTION - 2-LEVEL DECKHOUSE	71-0672
EXPERIMENTAL	EXPERIMENTAL AND THEORETICAL BENDING DEFLECTIONS OF LAMINATED PLATES	72-0144
EXPERIMENTAL	EXPERIMENTAL AND THEORETICAL STUDY ON A 1/10TH SCALE MODEL PCRV	73-0829
EXPERIMENTAL	EXPERIMENTAL AND THEORETICAL FATIGUE-CRACK PROPAGATION LIVES OF VARIOUSLY NOTCHED PLATES	74-0602
EXPERIMENTAL	EXPERIMENTAL AND THEORETICAL INVESTIGATION OF THE BEHAVIOR OF BOX GIRDERS CURVED IN PLAN	75-0659
EXPERIMENTAL	EXPERIMENTAL ELASTIC STRESS ANALYSES OF CYLINDER-TO-CYLINDER SHELL MODELS AND COMPARISONS WITH THEOR	71-0065
EXPERIMENTAL	EXPERIMENTAL STUDY OF FLEXIBILITY FACTORS AND STRESSES OF WELDING ELBOWS WITH END EFFECTS	73-0422
EXPERIMENTAL	EXPERIMENTAL VERIFICATION OF DISCRETE-ELEMENT SOLUTIONS FOR PLATES AND PAVEMENT SLABS	70-0411
EXPERIMENTAL	EXPERIMENTELL UNDERSOKNING AV ELEMENTFOGARS HALLFASTHETSEGENSKAPER (EXPERIMENTAL INVESTIGATION OF ST	72-0192
EXPERIMENTAL	EXTENSION OF CLASSICAL EXPERIMENTAL TECHNIQUES FOR CHARACTERIZING COMPOSITE - MATERIAL BEHAVIOUR	72-0088
EXPERIMENTAL	FINITE ELEMENT ANALYSIS AND EXPERIMENTAL INVESTIGATION OF SHEET METAL STRETCHING	72-0229
EXPERIMENTAL	FINITE ELEMENT AND EXPERIMENTAL ANALYSIS OF AN INTERNAL FIXATION PLATE MODEL	75-0380
EXPERIMENTAL	FORMULATION AND EXPERIMENTAL VERIFICATION OF AN AXISYMMETRIC FINITE-ELEMENT STRUCTURAL ANALYSIS	71-0354
EXPERIMENTAL	LEAST SQUARES SMOOTHING OF EXPERIMENTAL DATA USING FINITE ELEMENTS	68-0224
EXPERIMENTAL	MATHEMATICAL AND EXPERIMENTAL STUDIES ON THE MECHANICS OF PLATED TRANSVERSE FRACTURES	74-0656
EXPERIMENTAL	NUMERICAL AND EXPERIMENTAL INVESTIGATION ON THE USE OF J-INTEGRAL	73-0245
EXPERIMENTAL	NUMERICAL AND EXPERIMENTAL ANALYSIS OF A PSEUDOCYLINDRICAL SHELL	74-1133
EXPERIMENTAL	SKIN-STRINGER PANEL NORMAL MODE RESPONSE EXPERIMENTAL DATA	71-0691
EXPERIMENTAL	STATIC ANALYSIS VIA SUBSTRUCTURING OF AN EXPERIMENTAL VEHICLE FRONT-END BODY STRUCTURE	74-0313
EXPERIMENTAL	STRUCTURAL ANALYSIS AND EXPERIMENTAL CORRELATION OF A COMPLEX SHELL UNDER ASYMMETRIC LOADS	69-0155
EXPERIMENTAL	THE SCANDINAVIAN PCRV MODEL PROJECT: STRESS CALCULATION AND EXPERIMENTAL VERIFICATION	71-0021
EXPERIMENTAL	THEORETICAL AND EXPERIMENTAL INVESTIGATION ON A STEAM GENERATOR TUBE PLATE	69-0424
EXPERIMENTAL	THEORETICAL AND EXPERIMENTAL ANALYSIS OF STIFFENED PANEL UNDER DYNAMIC CONDITIONS	70-0130
EXPERIMENTAL	THEORETICAL AND EXPERIMENTAL CONSIDERATIONS ON FLOW OF VISCOUS LIQUIDS PAST PLATE AT SMALL AND AVERA	70-0144
EXPERIMENTAL	THEORETICAL AND EXPERIMENTAL VIBRATION ANALYSIS OF AN OBLIQUE SPACE FRAME	70-0460
EXPERIMENTAL	THEORETICAL AND EXPERIMENTAL ANALYSIS OF ELASTOPLASTIC DEFORMATION IN A UNIAXIALLY LOADED STRIP OF 2	73-0720
EXPERIMENTAL	THEORETICAL AND EXPERIMENTAL STRESS ANALYSES OF ORNL THIN-SHELL CYLINDER-TO-CYLINDER MODEL 3	75-0690
EXPERIMENTAL	THEORETICAL AND EXPERIMENTAL STRESS ANALYSES OF ORNL THIN-SHELL CYLINDER-TO-CYLINDER MODEL 4	75-0695
EXPERIMENTAL	THEORETICAL AND EXPERIMENTAL STUDIES OF TRANSIENT ELASTIC-PLASTIC LARGE DEFLECTIONS OF GEOMETRICALLY	75-0830
EXPERIMENTAL	THEORETICAL AND EXPERIMENTAL STUDIES OF TRANSIENT ELASTIC-PLASTIC LARGE DEFLECTIONS OF GEOMETRICALLY	75-0831
EXPERIMENTAL	THEORETICAL PREDICTION OF DISC BRAKE TEMPERATURES AND A COMPARISON WITH EXPERIMENTAL DATA	72-0306
EXPERIMENTAL	THIN-WALLED PRESTRESSED CONCRETE PRESSURE VESSEL FOR HIGH TEMPERATURE REACTORS: EXPERIMENTAL INVESTI	73-0857
EXPERIMENTAL	TRANSVERSE VIBRATIONS OF SHIP'S PROPULSION SYSTEMS. PART 2: EXPERIMENTAL ANALYSIS	74-1324
EXPERIMENTAL	ULTIMATE LOAD BEHAVIOUR OF PCRV TOP CLOSURES. THEORETICAL AND EXPERIMENTAL INVESTIGATIONS	73-0835
EXPERIMENTALLY	STABILITY ANALYSIS OF AN EXPERIMENTALLY DEFORMED SINGLE LAYER OF INDIANA LIMESTONE USING FINITE ELEM	74-0212
EXPERIMENTATION	COMPUTATIONAL EXPERIMENTATION ON FINITE ELEMENT METHOD IN BARE SLAB CRITICALITY CALCULATIONS	73-0015
EXPERIMENTELL	EXPERIMENTELL UNDERSOKNING AV ELEMENTFOGARS HALLFASTHETSEGENSKAPER (EXPERIMENTAL INVESTIGATION OF ST	72-0192
EXPERIMENTELLE	THEORETICAL AND EXPERIMENTAL CONSIDERATIONS ON FLOW OF VISCOUS LIQUIDS PAST PLATE AT SMALL AND AVERA	70-0144
EXPERIMENTS	ANALYSIS OF TUBULAR T-CONNECTIONS BY THE FINITE ELEMENT METHOD: COMPARISON WITH EXPERIMENTS	72-0380
EXPERIMENTS	EXPERIMENTS IN FORMING AND JOINTING OF TITANIUM AND TITANIUM ALLOYS FOR AIRCRAFT AND SPACE VEHICLES	70-0134
EXPERIMENTS	MODEL EXPERIMENTS AND FINITE ELEMENT ANALYSIS OF STRESS IN AN OPEN SHIP	66-0054
EXPERIMENTS	NUMERICAL EXPERIMENTS ON THE DETERMINATION OF STRESS CONCENTRATION FACTORS	74-0292
EXPERIMENTS	SHIFTING ROLES OF COMPUTERS, EXPERIMENTS AND ANALYSIS IN APPLIED MECHANICS	71-0604
EXPERIMENTS	STRESS ANALYSIS IN HULL STRUCTURAL MEMBERS BY PHOTOELASTIC EXPERIMENTS AND FINITE ELEMENT METHODS -	70-0341
EXPERIMENTS	TRAGLAST VON T-TRAEGERN. EXPERIMENT UND F.E. BERECHNUNGEN (ULTIMATE STRENGTH OF T-BEAMS. EXPERIMENTS	74-0935
EXPLICATION	CONTRIBUTION A L'ETUDE DU COMPLEXE BARRAGE-FONDATION ET EXPLICATION DES SEISMES DUS AU REMPLISSAGE D	72-0554
EXPLICATION	CONTRIBUTION A L'ETUDE DU COMPLEXE BARRAGE-FONDATION ET EXPLICATION DES SEISMES DUS AU REMPLISSAGE D	72-0554
EXPLICIT	A HIGHLY STABLE EXPLICIT INTEGRATION TECHNIQUE FOR COMPUTATIONAL CONTINUUM MECHANICS	74-1277
EXPLICIT	DEVELOPMENT OF EXPLICIT STIFFNESS AND MASS MATRICES FOR A TRIANGULAR PLATE ELEMENT	69-0242
EXPLICIT	EXPLICIT ADDITION OF RIGID-BODY MOTIONS IN CURVED FINITE ELEMENTS	73-0037
EXPLICIT	EXPLICIT EXPRESSIONS FOR TRIANGULAR TOWA ELEMENT STIFFNESS MATRIX	68-0160
EXPLICIT	EXPLICIT INTEGRATION - STABILITY, SOLUTION PROPERTIES, COST	75-0298
EXPLICIT	EXPLICIT TRIANGULAR BENDING ELEMENT MATRIX	74-0288
EXPLICIT	FINITE ELEMENT SOLUTION FOR ENERGY CONSERVATION USING A HIGHLY STABLE EXPLICIT INTEGRATION ALGORITHM	72-0633
EXPLICIT	ON THE EXPLICIT REPRESENTATION OF TRANSFER-MATRICES OF CURVED BEAMS AND THEIR APPLICATIONS	75-0565
EXPLICIT	ON THE STABILITY OF EXPLICIT METHODS FOR THE NUMERICAL INTEGRATION OF THE EQUATIONS OF MOTION IN FIN	72-0172
EXPLICIT	SOLUTION OF NONLINEAR EQUATIONS BY EXPLICIT TIME INTEGRATION - A SHORT COMMUNICATION	73-0606
EXPLICIT	TRIPLET: A TWO-DIMENSIONAL, MULTIGROUP, TRIANGULAR MESH, PLANAR GEOMETRY, EXPLICIT TRANSPORT CODE	73-0955
EXPLOITATION	DIE BERECHNUNG DER BEWEGUNGEN UND SPANNUNGEN DES GEBIRGES VOM ABBAU BIS ZUR TAGESOBERFLAECHE NACH DE	73-0456
EXPLORATION	THREE DIMENSIONAL FINITE ELEMENT APPROXIMATION IN ELECTROMAGNETIC EXPLORATION	74-0358
EXPLORATORY	EXPLORATORY INVESTIGATION OF RAPID CRACK PROPAGATION AND CRACK ARREST	73-0757
EXPLORER	PLANETARY EXPLORER STRUCTURAL CONCEPT EVALUATION MODEL USING NASTRAN	71-0742

EXPLOSIONS	DISPLACEMENT AND STRAIN FIELDS FROM UNDERGROUND EXPLOSIONS USING ANALYTICAL AND FINITE ELEMENT METHO	72-0587
EXPLOSIONS	RESPONSE OF BOX-LIKE STRUCTURES TO WEAK EXPLOSIONS	75-0197
EXPLOSIVE	MAGAZINE HEADWALL RESPONSE TO EXPLOSIVE BLAST	74-0754
EXPLOSIVE	SEEPAGE AND GROUNDWATER EFFECTS ASSOCIATED WITH EXPLOSIVE CRATERING	72-0656
EXPONENTIAL	TRANSIENT RESPONSE OF LINEAR ELASTIC STRUCTURES DETERMINED BY THE MATRIX EXPONENTAL METHOD	71-0240
EXPOSED	PERFORMANCE OF STRUCTURAL WOOD MEMBERS EXPOSED TO FIRE	75-0060
EXPOSITORY	NONLINEAR ANALYSIS BY THE FINITE ELEMENT METHOD AND SOME EXPOSITORY EXAMPLES	73-0309
EXPRESSIONS	EXPLICIT EXPRESSIONS FOR TRIANGULAR TOWA ELEMENT STIFFNESS MATRIX	68-0160
EXTENDED	A NUMERICAL STUDY OF NONLINEAR INSTABILITY BY MEANS OF AN EXTENDED FINITE ELEMENT METHOD	74-0632
EXTENDED	AN EXTENDED THEORY FOR THE FINITE ELEMENT METHOD	72-0350
EXTENDED	APPLICATION OF EXTENDED VARIATIONAL PRINCIPLES TO FINITE ELEMENT ANALYSIS	72-0529
EXTENDED	EXTENDED INTERPOLATION IN FINITE ELEMENT ANALYSIS	71-0183
EXTENDED	STRESS WAVE ANALYSIS IN LAYERED THERMOVISCOELASTIC MATERIALS BY THE EXTENDED RITZ METHOD, VOLUME II	68-0125
EXTENDED	TWO DUAL FINITE ELEMENT METHODS DERIVED FROM EXTENDED VARIATIONAL PRINCIPLE	73-0109
EXTENDED	ZUR LOESUNG DES KONTAKTPROBLEMS ELASTISCHER KOERPER MIT AUSGEDEHNTER KONTAKTFLAECHE DURCH QUADRATISC	74-1135
EXTENSION	AN EXTENSION OF FINITE-ELEMENT METHOD TO INCLUDE NONLINEAR BOUNDARY CONDITIONS	75-0032
EXTENSION	AN EXTENSION OF THE SHALLOW TO THE NON-SHALLOW STIFFNESS MATRIX FOR A CYLINDRICAL SHELL FINITE ELEME	70-0104
EXTENSION	APPLICATION OF A SUBSTRUCTURING TECHNIQUE TO THE PROBLEM OF CRACK EXTENSION AND CLOSURE	74-0772
EXTENSION	COMBINED MODE CRACK EXTENSION IN ADHESIVE JOINTS	72-0111
EXTENSION	EXTENSION OF CLASSICAL EXPERIMENTAL TECHNIQUES FOR CHARACTERIZING COMPOSITE - MATERIAL BEHAVIOUR	72-0088
EXTENSION	EXTENSION OF THREE DIMENSIONAL ANALYSIS TO SHELL STRUCTURES USING THE FINITE ELEMENT IDEALIZATION	74-0849
EXTENSION	FINITE ELEMENT FORMULATION OF PROBLEMS OF FINITE DEFORMATION AND IRREVERSIBLE THERMODYNAMICS OF NONL	71-0190
EXTENSION	M.I.T. TEST SECTION INSTRUMENTATION, MASSACHUSETTS BAY TRANSPORTATION AUTHORITY, HAYMARKET-NORTH EXT	72-0673
EXTENSION	PLANE FRAMEWORK METHODS FOR PLATES IN EXTENSION	66-0126
EXTENSION	THE COMPUTATION OF TEARING ENERGY OF NICKED RUBBER STRIPS IN EXTENSION	74-0315
EXTENSION	THE EXTENSION AND APPLICATION OF SARD KERNEL THEOREMS TO COMPUTE FINITE ELEMENT ERROR BOUNDS	72-0519
EXTENSIONAL	APPLICATION OF THE FINITE ELEMENT METHOD TO THE EXTENSIONAL VIBRATIONS OF PIEZOELECTRIC PLATES	72-0073
EXTENSIONS	EXTENSIONS OF THE SUCCESSIVE OVER-RELAXATION THEORY WITH APPLICATIONS TO FINITE ELEMENT APPROXIMATIO	73-0980
EXTENSIONS	EXTENSIONS TO THE STAGS COMPUTER CODE	73-0364
EXTENSIONS	EXTENSIONS TO THE STAGS COMPUTER CODE	73-0917
EXTENSIONS	FINITE ELEMENT FORMULATION OF FINITE DEFORMATIONS AND IRREVERSIBLE THERMODYNAMICS OF NONLINEAR CONTI	69-0207
EXTENSIONS	ON THE METHOD OF VIRTUAL CRACK EXTENSIONS	75-0237
EXTENSIONS	SARD KERNEL THEOREMS ON TRIANGULAR AND RECTANGULAR DOMAINS WITH EXTENSIONS AND APPLICATIONS TO FINIT	72-0361
EXTERIEUR	UNE METHODE VARIATIONNELLE ELEMENTS FINIS POUR LA RESOLUTION NUMERIQUE UN PROBLEME EXTERIEUR DANS R*	73-0576
EXTERIOR	FINITE ELEMENT SOLUTION OF TWO DIMENSIONAL EXTERIOR FIELD PROBLEMS	71-0491
EXTERIOR	UNE METHODE VARIATIONNELLE ELEMENTS FINIS POUR LA RESOLUTION NUMERIQUE UN PROBLEME EXTERIEUR DANS R*	73-0576
EXTERNAL	APPLICATION OF FINITE ELEMENT METHOD TO AXISYMMETRIC BUCKLING OF SHALLOW SPHERICAL SHELLS UNDER EXTE	73-0111
EXTERNAL	BERECHNUNGSMETHODEN ZUR ERMITTLUNG VON BELASTUNGSBEDINGUNGEN UND GRENZTRAGFAEHIGKEITSVERHALTEN VON K	73-0858
EXTERNAL	BOUNDS ON EXTERNAL EIGENVALUES OF FINITE ELEMENT STIFFNESS AND MASS MATRICES AND THEIR SPECTRAL COND	72-0256
EXTERNAL	FINITE-ELEMENT NONLINEAR BUCKLING ANALYSIS OF A PSEUDO-CYLINDRICAL CONCAVE POLYHEDRAL SHELL UNDER EX	73-1147
EXTERNAL	LARGE DEFLECTIONS OF STRUCTURES SUBJECT TO HEATING AND EXTERNAL LOADS	60-0002
EXTERNAL	METHOD FOR EXTERNAL APPROXIMATION USING FINITE ELEMENTS FOR DELTA - 2 AND ITERATIVE METHOD FOR SOLVI	72-0282
EXTERNAL	METHOD FOR EXTERNAL APPROXIMATION USING FINITE ELEMENTS FOR DELTA -2 AND ITERATIVE METHOD FOR SOLVIN	72-0283
EXTERNAL	PENETRATION IN SHELLS UNDER EXTERNAL PRESSURE	70-0237
EXTERNAL	RADIAL DISPLACEMENTS OF A HOLLOW SHAFT SUBJECTED TO A UNIFORM BAND OF EXTERNAL PRESSURE	72-0242
EXTERNAL	REDUCTION GEAR DAMAGES RELATED TO EXTERNAL INFLUENCES	75-0307
EXTERNAL	SEEPAGE IN FRACTURED ROCK: FINITE ELEMENT ANALYSIS OF ROCK MASS BEHAVIOR UNDER COUPLED ACTION OF BOD	71-0825
EXTERNAL	SOLUMN INSTABILITY UNDER NONCONSERVATIVE FORCES, WITH INTERNAL AND EXTERNAL DAMPING FINITE ELEMENT U	73-0566
EXTERNAL	STRENGTH OF REINFORCED CONCRETE CHAMBERS UNDER EXTERNAL PRESSURE	75-0293
EXTERNAL	STRENGTH OF REINFORCED CONCRETE CHAMBERS UNDER EXTERNAL PRESSURE	75-0512
EXTERNAL	THE TRANSIENT LINEAR ELASTIC RESPONSE ANALYSIS COMPLEX THIN SHELLS OF REVOLUTION SUBJECTED TO ARBITR	70-0080
EXTERNAL	VARIATION METHOD ON FINITE ELEMENTS FOR NUMERICAL RESOLUTION TO PROBLEMS WITH EXTERNAL R3	73-0359
EXTERNAL	YIELDING IN UNIDIRECTIONAL COMPOSITES UNDER EXTERNAL LOADS AND TEMPERATURE CHANGES.	73-0109
EXTERNALLY	APPLICATION OF FINITE ELEMENT ANALYSIS TO HYDRODYNAMIC AND EXTERNALLY PRESSURIZED POCKET BEARINGS	72-0295
EXTERNALLY	OSCILLATION STABILITY OF INTERNALLY AND EXTERNALLY NOTCHED PLANE BARS OF EQUAL FORM FACTOR FROM 3943	73-0745
EXTERNE	SURE UNE METHODE D'APPROXIMATION EXTERNE PAR ELEMENTS FINIS D ORDRE DEUX	72-0742
EXTERNES	APPROXIMATIONS EXTERNES, PAR ELEMENTS FINIS DE LAGRANGE D ORDRE, UN ET DEUX, DU PROBLEME DE DIRICHLE	72-0495
EXTRACTION	AUTOMATIC EXTRACTION OF RIGID BODY MODES FROM STRESS AND STRAIN ELEMENTS	72-0913
EXTRACTION	FAST MODAL EXTRACTION IN NASTRAN VIA THE FEER COMPUTER PROGRAM	73-0356
EXTRAPOLATED	EXTRAPOLATED FINITE ELEMENT PROCEDURES OVER CARTESIAN PRODUCT DOMAINS	75-0044
EXTRAPOLATION	COMBINED APPLICATION OF FINITE ELEMENT METHODS AND RICHARDSON EXTRAPOLATION TO THE TORSION PROBLEM	72-0066
EXTRAPOLATION	COMPARISON OF EXTRAPOLATION TECHNIQUES WITH HIGH ORDER FINITE ELEMENT METHODS FOR DIFFUSION EQUATION	73-0586
EXTRAPOLATION	COMPILED APPLICATION OF FINITE ELEMENT METHODS AND RICHARDSON EXTRAPOLATION TO THE TORSION PROBLEM	73-1012
EXTRAPOLATION	RICHARDSON EXTRAPOLATION FOR PARABOLIC GALERKIN METHODS	72-0522
EXTREMAL	ON THE EXTREMAL PROPERTIES OF THE SOLUTION IN DYNAMICS OF RIGID VISCOPLASTIC BODIES ALLOWING FOR LAR	73-1080
EXTREMELY	PROBLEM OF USING THE METHOD OF FINITE ELEMENTS TO CONSTRUCT AN ALGORITHM FOR CALCULATING A THIN-WALL	74-0851
EXTREMUM	FINITE ELEMENT APPLICATIONS OF A PRECISION TRIANGULAR PLATE BENDING ELEMENT TO LINEAR BOUNDARY VALUE	75-0788
EXTREMUM	GOVERNING EQUATIONS AND EXTREMUM PRINCIPLES OF ELASTICITY AND PLASTICITY GENERATED FROM A SINGLE FUN	73-0666
EXTRUSION	ANALYSIS OF HYDROSTATIC EXTRUSION BY THE FINITE ELEMENT METHOD	72-0020
EXTRUSION	APPLICATION OF FINITE ELEMENT METHOD TO PLANE STRAIN EXTRUSION PROCESSES	74-1243
EXTRUSION	ELASTIC DEFORMATION OF SQUARE EXTRUSION DIES DUE TO LOAD AND TEMPERATURE	75-0581
EXTRUSION	FINITE ELEMENT SOLUTION OF EXTRUSION OF ELASTO-PLASTIC WORK HARDENING MATERIALS	73-1127
EXTRUSION	FLOW OF PLASTIC AND VISCO-PLASTIC SOLIDS WITH SPECIAL REFERENCE TO EXTRUSION AND FORMING PROCESSES	74-0210
EYE	EFFECT OF 2450-MHZ RADIATION ON THE RABBIT EYE	75-0365
EYE	FINITE-ELEMENT STRESS ANALYSIS OF A NON-LINEAR PROBLEM. A CONNECTING-ROD EYE LOADED BY MEANS OF A PI	70-0239
E11101	ENGINEERING OPERATIONS REPORT: NERVA THIN SHELL FINITE ELEMENT STRESS ANALYSIS METHODOLOGY COMPUTER	72-0916
E11401	ENGINEERING OPERATIONS REPORT: NERVA AXISYMMETRIC AND PLANE STRUCTURES FINITE ELEMENT STRESS ANALYSI	72-0917
E11405	ENGINEERING OPERATIONS REPORT: NERVA AXISYMMETRIC AND PLANE STRUCTURES FINITE ELEMENT STRESS ANALYSI	72-0917
E12202	DIGITAL COMPUTER PROGRAMS FOR STEADY-STATE OR TRANSIENT TEMPERATURE ANALYSIS OF PLANE AND AXISYMMETR	72-0915
E12205	DIGITAL COMPUTER PROGRAMS FOR STEADY-STATE OR TRANSIENT TEMPERATURE ANALYSIS OF PLANE AND AXISYMMETR	72-0915
E12206	DIGITAL COMPUTER PROGRAMS FOR STEADY-STATE OR TRANSIENT TEMPERATURE ANALYSIS OF PLANE AND AXISYMMETR	72-0915
E12207	DIGITAL COMPUTER PROGRAMS FOR STEADY-STATE OR TRANSIENT TEMPERATURE ANALYSIS OF PLANE AND AXISYMMETR	72-0915
F-14A	STATIC AND DYNAMIC ANALYSIS OF F-14A BORON HORIZONTAL STABILIZER	71-0123
FABRICATED	A NUMERICAL THERMO-MECHANICAL MODEL FOR THE WELDING AND SUBSEQUENT LOADING OF A FABRICATED STRUCTURE	72-0847
FABRICATED	CREEP OF FABRICATED STRUCTURES	68-0001
FABRICATED	NUMERICAL, THERMO-MECHANICAL MODEL FOR THE WELDING AND SUBSEQUENT LOADING OF A FABRICATED STRUCTURE	73-0418
FACE	ANALYSE TRIDIMENSIONELLE DE LA STABILITE D'UN TUNNEL AU VOISINAGE DU FRONT DE TAILLE DANS UNE ROCHE	74-0995
FACE	STRESS DISTRIBUTION AND ROCK FRACTURE ZONES IN THE ROOF OF LONGWALL FACE IN COAL MINE	73-0247
FACES	FINITE DEFLECTION DISCRETE ELEMENT ANALYSIS OF SANDWICH PLATES AND CYLINDRICAL SHELLS WITH LAMINATED	69-0236
FACES	FINITE ELEMENT ANALYSIS OF SANDWICH PLATES AND CYLINDRICAL SHELLS WITH LAMINATED FACES	68-0121
FACES	FORECASTING THE ROCK PRESSURE AND ITS EFFECT ON THE LONGWALL FACES AND ROADWAYS IN THE BITUMINOUS CO	74-0374
FACES	VORAUSSAGE DES GEBIRGSDRUCKES UND SEINER AUSQIRKUNGEN AUF STREBEN UND STRECKEN IM STEINKOHLENBERGBAU	73-0679
FACET	SHELL ANALYSIS USING A SIMPLE FACET ELEMENT	72-0986
FACILITIES	EARTH VIBRATION EFFECTS AND ABATEMENT FOR MILITARY FACILITIES. REPORT 3. ANALYSIS METHOD FOR FOOTING	72-0806
FACILITIES	STRUCTURAL DESIGN OF NUCLEAR PLANT FACILITIES	73-0812
FACILITY	BOOLEAN MATRICES AS A STANDARD FACILITY IN A MATRIX CODE	65-0057
FACILITY	SEISMIC EFFECTS ON A PROPOSED UNDERGROUND REACTOR FACILITY	73-0953
FACILITY	TEST RESULTS FROM THE BOUNDARY LAYER FACILITY - RESPONSE OF STRUCTURE TO THE PSEUDO-SOUND FIELD OF A	68-0262
FACT	THERMAL STRESSES IN FACT REACTOR BOILER TUBES AT THE SODIUM/GAS INTERFACE	73-0820
FACTOR	A CONSIDERATION ON THE FINITE ELEMENT METHOD FOR CALCULATING STRESS INTENSITY FACTOR KI USING THE J	75-0863
FACTOR	AN INVESTIGATION ON THE PLASTICITY MAGNIFICATION FACTOR IN DEEP SURFACE FLOWS	74-0766
FACTOR	EFFICIENT FINITE ELEMENT METHODS FOR STRESS INTENSITY FACTOR	75-0148

FACTOR	FACTOR OF SAFETY APPROACH FOR EVALUATING SEISMIC STABILITY OF SLOPES	75-0645
FACTOR	METHOD OF EVALUATION OF THREE-DIMENSIONAL STRESS INTENSITY FACTOR USING THE FINITE ELEMENT METHOD	75-0316
FACTOR	NATURAL FACTOR FORMULATION OF THE STIFFNESS FOR THE MATRIX DISPLACEMENT METHOD	75-0057
FACTOR	NEW DESIGN FACTOR FOR A SHRINK FITTED ASSEMBLY	75-0167
FACTOR	NUMERICAL ANALYSIS OF THE LINE CAPACITANCE AND CROSSTALK FACTOR FOR INSULATED WIRE PAIRS	74-0685
FACTOR	OSCILLATION STABILITY OF INTERNALLY AND EXTERNALLY NOTCHED PLANE BARS OF EQUAL FORM FACTOR FROM 3943	73-0745
FACTOR	STRESS INTENSITY FACTOR FOR A STRAIGHT CRACK APPROACHING A HOLE	69-0158
FACTOR	STRESS INTENSITY FACTOR OF A SURFACE CRACK IN A PRESSURIZED CYLINDER	75-0255
FACTOR	THE NATURAL FACTOR APPROACH IN THE DISPLACEMENT METHOD - FOUNDATION AND FURTHER DEVELOPMENTS	75-0111
FACTOR	THE SIMPLE METHOD OF EVALUATION OF STRESS INTENSITY FACTOR USING THE FINITE ELEMENT METHOD	74-0829
FACTOR	VIEW FACTOR COMPUTER PROGRAM (PROGRAM VIEW) USER'S MANUAL	73-0789
FACTORS	A COMPUTER PROGRAM FOR THE CALCULATION OF TORSION FACTORS AND SHEAR STRESS FACTORS FOR BOX SECTIONS	69-0485
FACTORS	A COMPUTER PROGRAM FOR THE CALCULATION OF TORSION FACTORS AND SHEAR STRESS FACTORS FOR BOX SECTIONS	69-0485
FACTORS	A HYBRID FINITE ELEMENT FOR DETERMINATION OF STRESS INTENSITY FACTORS AT A CRACK TIP	73-0643
FACTORS	AN ANALYSIS OF SOME FACTORS THAT AFFECT HYDROPHONE SENSITIVITY	72-0739
FACTORS	APPLICATIONS OF THE FINITE ELEMENT METHOD TO THE CALCULATIONS OF STRESS INTENSITY FACTORS	72-0900
FACTORS	CALCULATION OF STRESS INTENSITY FACTORS USING FINITE ELEMENT METHOD WITH CRACKED ELEMENTS	70-0224
FACTORS	CALCULATION OF STRESS INTENSITY FACTORS USING THE FINITE ELEMENT METHOD WITH CRACKED ELEMENTS	70-0475
FACTORS	CALCULATION OF STRESS INTENSITY FACTORS AT CRACK TIPS USING SPECIAL FINITE ELEMENTS	72-0071
FACTORS	CALCULATION OF STRESS INTENSITY FACTORS AND FATIGUE CRACK PROPAGATION OF SEMI-ELLIPTICAL, PART-THROU	73-0360
FACTORS	CALCULATION OF STRESS INTENSITY FACTORS FOR COMBINED TENSILE AND SHEAR LOADING	75-0205
FACTORS	CALCULATION OF STRESS INTENSITY FACTORS IN TWO AND THREE DIMENSIONS USING FINITE ELEMENTS	75-0254
FACTORS	CALCULATION OF STRESS INTENSITY FACTORS IN ANISOTROPIC MATERIALS BY A CONTOUR INTEGRAL METHOD	75-0257
FACTORS	COMPARISON OF NUMERICAL METHODS FOR DETERMINING STRESS INTENSITY FACTORS	72-0176
FACTORS	COMPUTATION OF STRESS INTENSITY FACTORS BY A SPECIAL FINITE ELEMENT TECHNIQUE	71-0287
FACTORS	DETERMINATION OF ENERGY RELEASE RATES AND STRESS - INTENSITY FACTORS BY THE FINITE-ELEMENT METHOD	72-0183
FACTORS	DETERMINATION OF STRESS INTENSITY FACTORS IN CRACKED PLATES BY THE FINITE ELEMENT METHOD	73-0148
FACTORS	EFFICIENT FINITE-ELEMENT METHODS FOR STRESS INTENSITY FACTORS USING WEIGHT FUNCTIONS	75-0411
FACTORS	EFFICIENT FINITE-ELEMENT METHODS FOR STRESS INTENSITY FACTORS USING WEIGHT FUNCTIONS - REPLY	75-0421
FACTORS	EVALUATION OF FINITE ELEMENT METHODS FOR THE COMPUTATION OF ELASTIC STRESS INTENSITY FACTORS	72-0249
FACTORS	EXPERIMENTAL STUDY OF FLEXIBILITY FACTORS AND STRESSES OF WELDING ELBOWS WITH END EFFECTS	73-0422
FACTORS	FACTORS OF STRESS CONCENTRATION AT A HOLE IN FINITE STRIPS UNDER BENDING	72-0744
FACTORS	FINITE ELEMENT CALCULATION OF STRESS INTENSITY FACTORS USING ENERGY TECHNIQUES	73-0825
FACTORS	FINITE ELEMENT CALCULATION OF STRESS INTENSITY FACTORS USING SUPERPOSITION	75-0251
FACTORS	FINITE ELEMENT METHOD FOR CALCULATING STRESS INTENSITY FACTORS FOR CRACKED PLATES IN BENDING	71-0424
FACTORS	FINITE ELEMENT METHOD FOR CALCULATING STRESS INTENSITY FACTORS AND ITS APPLICATION TO COMPOSITES	74-1000
FACTORS	FINITE ELEMENT PARAMETRIC STUDY OF VERTICAL STRAIN INFLUENCE FACTORS AND THE PRESSUREMETER TEST TO E	74-1127
FACTORS	FINITE ELEMENT TREATMENT OF SINGULARITIES OF BOUNDARY VALUE PROBLEMS AND ITS APPLICATION TO ANALYSIS	73-0306
FACTORS	FINITE ELEMENTS FOR DETERMINATION OF CRACK TIP ELASTIC STRESS INTENSITY FACTORS	71-0436
FACTORS	FINITE ELEMENTS FOR DETERMINATION OF CRACK-TIP ELASTIC STRESS INTENSITY FACTORS	71-0785
FACTORS	FORM FACTORS OF FLAT BARS FOR FATIGUE STRENGTH TESTS. (FORMZAHLEN VON FLACHSTAEBEN FUER SCHWINGFESTI	71-0641
FACTORS	NUMERICAL EXPERIMENTS ON THE DETERMINATION OF STRESS CONCENTRATION FACTORS	74-0239
FACTORS	ON THE USE OF FINITE ELEMENTS AND COMPLIANCE CALCULATIONS FOR DETERMINING CRACK TIP STRESS INTENSITY	69-0448
FACTORS	PLASTIC INTENSITY FACTORS FOR CRACKED PLATES	71-0164
FACTORS	SOME FACTORS AFFECTING DENTAL IMPLANT DESIGN	74-1146
FACTORS	STIFFNESS DERIVATIVE FINITE ELEMENT TECHNIQUE FOR DETERMINATION OF CRACK TIP STRESS INTENSITY FACTOR	74-0956
FACTORS	STIFFNESS DERIVATIVE FINITE ELEMENT TECHNIQUE FOR DETERMINATION OF CRACK TIP STRESS INTENSITY FACTOR	74-1358
FACTORS	STRESS CONCENTRATION FACTORS AROUND A PENETRATION IN A STRIP OR SLAB OF FINITE DIMENSIONS	69-0409
FACTORS	STRESS CONCENTRATION FACTORS FOR BONDED LAP JOINTS	71-0842
FACTORS	STRESS CONCENTRATION FACTORS FOR CIRCULAR NOTCHES	74-0791
FACTORS	STRESS CONCENTRATION FACTORS FOR A FATIGUE TEST SPECIMEN WITH A U-TYPE NOTCH IN ONE EDGE BY FINITE E	74-1344
FACTORS	STRESS CONCENTRATION FACTORS FOR CIRCULAR, REINFORCED PENETRATIONS IN PRESSURIZED CYLINDRICAL SHELLS	75-0726
FACTORS	STRESS INTENSITY FACTORS CALCULATED GENERALLY BY FINITE ELEMENT TECHNIQUE	69-0369
FACTORS	STRESS INTENSITY FACTORS IN PLATE STRUCTURES CALCULATED BY THE FINITE ELEMENT METHOD (IN JAPANESE)	71-0788
FACTORS	STRESS INTENSITY FACTORS FOR HEXAGONAL FRACTURE TOUGHNESS SPECIMENS	71-0799
FACTORS	STRESS INTENSITY FACTORS FOR CRACKS IN A PLATE CONTAINING A HOLE AND IN A SPINNING DISC	73-0467
FACTORS	STRESS INTENSITY FACTORS FOR ANISOTROPIC FRACTURE TEST SPECIMENS OF SEVERAL GEOMETRIES	74-0293
FACTORS	STRESS INTENSITY FACTORS FOR THICK-WALLED CYLINDERS	74-0300
FACTORS	STRESS INTENSITY FACTORS FOR A NOTCHED BEAM ON AN ELASTIC FOUNDATION	75-0566
FACTORS	STRESS, DISPLACEMENT, LINE INTEGRAL AND CLOSURE ENERGY DETERMINATIONS OF CRACK TIP STRESS INTENSITY	74-0621
FACTORS	THE FINITE ELEMENT CALCULATION OF STRESS INTENSITY FACTORS USING SUPERPOSITION	75-0886
FACTORS	USE OF SINGULARITY PROGRAMMING IN FINITE ELEMENT CALCULATIONS OF ELASTIC STRESS INTENSITY FACTORS. P	75-0252
FAIBLE	DEPLACEMENTS DANS LES STRUCTURES MINCES A FAIBLE RIGIDITE GEOMETRIQUE-DETERMINATION NUMERIQUE (NUMER	72-0213
FAILURE	A COMPUTER BASED PROCEDURE FOR PREDICTING THE TRANSIENT RESPONSE AND FAILURE OF A TWO-DIMENSIONAL CO	71-0630
FAILURE	APPLICATION OF FRACTURE MECHANICS TO FAILURE OF THICK WALLED CYLINDERS	71-0511
FAILURE	CONTROLLED FAILURE OF HOLLOW ROCK CYLINDERS IN UNIAXIAL COMPRESSION	72-0196
FAILURE	DEFORMATION AND FAILURE OF BORON-EPOXY PLATE WITH CIRCULAR HOLE	72-0536
FAILURE	DEVELOPMENT OF FAILURE AROUND EXCAVATED SLOPES	70-0051
FAILURE	FAILURE MECHANISMS IN COMPOSITE SYSTEMS	75-0688
FAILURE	FAILURE OF THE CONCRETE LINING IN AN INCLINED SHAFT AT GREAT DEPTH.	74-0996
FAILURE	FAILURE OF THIN-WALLED MEMBERS UNDER PATCH LOADINGS	72-0230
FAILURE	FINITE ELEMENT BASED PROCEDURE FOR SIMULATING THE TRANSIENT RESPONSE AND FAILURE OF A TWO-DIMENSIONA	72-0123
FAILURE	FOUNDATION FAILURE OF NEW LISKEARD EMBANKMENT	73-0549
FAILURE	INITIATION OF FAILURE IN SLOPES IN OVERCONSOLIDATED CLAYS AND CLAY SHALES.	70-0186
FAILURE	INVESTIGATION OF BOND FAILURE IN FIBRE-REINFORCED MATERIALS BY THE FINITE ELEMENT METHOD	72-0034
FAILURE	PLASTIC FLOW AND FAILURE OF DISCONTINUOUS FIBRE COMPOSITE MATERIALS	70-0157
FAILURE	PRACTICAL FINITE ELEMENT METHOD OF FAILURE PREDICTION FOR COMPOSITE MATERIAL STRUCTURES	75-0644
FAILURE	PRESTRESSED CONCRETE REACTOR PRESSURE VESSEL FAILURE ANALYSIS, THE INCORPORATION OF STATISTICAL METH	71-0120
FAILURE	PROPAGATION OF FAILURE IN A CIRCULAR CYLINDER OF ROCK SUBJECTED TO A COMPRESSIVE FORCE	73-0793
FAILURE	QUANTITATIVE STRAIN-AND-STRESS STATE CRITERION FOR FAILURE IN THE VICINITY OF SHARP CRACKS	74-0639
FAILURE	RATIONAL DESIGN OF TUNNEL SUPPORTS: TUNNEL SUPPORT LOADING CAUSED BY ROCK FAILURE	75-0687
FAILURE	ROCK FAILURE AROUND A CIRCULAR OPENING IN A GRAVITY FIELD WITH TECTONIC FORCES	70-0071
FAILURE	SIMULATION OF PROGRESSIVE FAILURE IN ROCK AROUND MINING EXCAVATIONS	70-0146
FAILURE	STRESS AND FAILURE ANALYSIS OF A GLASS-EPOXY COMPOSITE PLATE WITH A CIRCULAR HOLE	73-0276
FAILURE	THEORETICAL STUDY OF THE SEQUENCE OF FAILURE IN PRECAST PANEL SHEAR WALLS	73-0490
FAILURES	AVOIDING DANGEROUS AND COSTLY FAN FAILURES	75-0061
FAILURES	CONTROLLING OPEN-PIT SLOPE FAILURES AT SHIRLEY BASIN	73-0149
FAILURES	INTERFACE FAILURES IN COMPOSITES	74-1209
FAIRING	INFLUENCE OF ASCENT HEATING ON THE SEPARATION DYNAMICS OF A SPACECRAFT FAIRING	72-0906
FALL	VARIATIONAL PROPERTIES OF STEADY FALL IN STOKES FLOW	72-0285
FALTWER	A FINITE ELEMENT METHOD FOR RANDOM FOLDED PLATE STRUCTURES (EINE FINITE ELEMENT METHODE ZUR BERECHNU	73-0787
FAMILE	METHODE DES ELEMENTS FINIS: UNE FAMILE D'APPROXIMATION DES ESPACES DE SOBOLEV PAR LES TRANSLATES DE	70-0301
FAMILIES	ISO-PARAMETRIC AND ASSOCIATED ELEMENT FAMILIES FOR TWO AND THREE DIMENSIONAL ANALYSIS	69-0285
FAMILIES	ONE PARAMETER FAMILIES OF VARIATIONAL PRINCIPLES	72-0503
FAMILLE	FAMILLE D ELEMENTS FINIS DE COQUE PLAN CINEMATIQUEMENT ADMISSIBLES POUR L'ANALYSE DES STRUCTURES	74-0488
FAMILY	A FAMILY OF EQUILIBRIUM PLATE BENDING ELEMENTS	74-0431
FAMILY	A NEW FAMILY OF CURVILINEAR PLATE BENDING ELEMENTS FOR VIBRATION AND STABILITY	72-0645
FAMILY	A NEW FAMILY OF CURVILINEAR PLATE BENDING ELEMENTS FOR VIBRATION AND STABILITY	72-0999
FAMILY	A REMARK ON THE SERENDIPITY FAMILY	73-1081
FAMILY	A SEQUEL TO TECHNICAL NOTE 14 ON THE TUBA FAMILY OF PLATE ELEMENTS	68-0015
FAMILY	A SEQUEL TO TECHNICAL NOTE 15: THE SHEBA FAMILY OF SHELL ELEMENTS FOR THE MATRIX DISPLACEMENT METHOD	69-0444
FAMILY	HERMITEAN RECTANGLE FINITE ELEMENT FAMILY	73-1009

FAMILY	THE FUGA FAMILY FOR FOLDED PLATE STRUCTURES	69-0333
FAMILY	THE FUGA FAMILY OF ELEMENTS FOR FOLDED PLATE STRUCTURES A FURTHER SEQUEL TO TECHNICAL NOTE 14	69-0446
FAMILY	THE SHEBA FAMILY OF SHELL ELEMENTS FOR THE MATRIX DISPLACEMENT METHOD	68-0014
FAMILY	THE SHEBA FAMILY OF SHELL ELEMENTS FOR THE MATRIX DISPLACEMENT METHOD. 111. LARGE DISPLACEMENTS	69-0499
FAMILY	THE TUBA FAMILY OF PLATE ELEMENTS FOR THE MATRIX DISPLACEMENT METHOD	68-0013
FAMSOR	FAMSOR, A FINITE ELEMENT PROGRAM FOR THE FREQUENCIES AND MODE SHAPES OF SHELLS OF REVOLUTION	70-0090
FAN	A FINITE CYLINDRICAL SHELL ELEMENT AND THE VIBRATIONS OF A CURVED FAN BLADE	68-0131
FAN	AVOIDING DANGEROUS AND COSTLY FAN FAILURES	75-0061
FAN	DEFLECTION AND VIBRATION ANALYSIS OF A FLEXIBLE FAN BLADE	73-0556
FAR	NUMERICAL HULL VIBRATION ANALYSIS OF A FAR EAST CONTAINER SHIP	74-1325
FAST	APPLICATION OF THE FAST FOURIER TRANSFORM TO LINEAR SYSTEMS IN CIVIL ENGINEERING	73-1119
FAST	FAST FRACTURE RESISTANCE AND CRACK ARREST IN STRUCTURAL STEELS	73-0766
FAST	FAST MODAL EXTRACTION IN NASTRAN VIA THE FEER COMPUTER PROGRAM	73-0356
FAST	FAST REACTOR PROGRAMME. FINITE ELEMENT TEMPERATURE ANALYSIS IN FUEL PINS	73-0939
FAST	FINITE ELEMENT ANALYSIS OF SOME CRITICAL FAST ASSEMBLIES	73-0128
FAST	FINITE ELEMENT STRESS ANALYSIS OF INTERACTING FUEL PELLET AND CANNING UNDER FAST REACTOR CONDITION	73-0814
FAST	NONLINEAR ANALYSIS OF PRESTRESSED CONCRETE FAST REACTOR PRESSURE VESSEL. (ANALYSE NON LINEAIRE DES C	70-0145
FAST	PARAMETRIC STUDIES OF LAYERED PAVEMENT SYSTEMS USING AFPAV CODE WITH FAST EQUATION SOLVER	75-0747
FAST	SHIP (SIMPLIFIED-HELMHOLTZ-INTEGRAL PROGRAM): A FAST COMPUTER PROGRAM FOR CALCULATING THE ACOUSTIC R	72-0835
FAST-REACTOR	DYNAMIC RESPONSE OF FAST-REACTOR CORE SUBASSEMBLIES	74-0540
FASTENED	STRESS AND DEFLECTION ANALYSIS OF MECHANICALLY FASTENED JOINTS	70-0067
FASTENER	INTERFERENCE-FIT FASTENER DISPLACEMENT MEASUREMENT BY SPECKLE PHOTOGRAPHY	74-1178
FASTENER	STRESS ANALYSIS OF COLDWORKED FASTENER HOLES	74-0783
FASTENERS	ELASTIC-PLASTIC ANALYSIS OF INTERFERENCE FIT FASTENERS	74-1177
FASTENERS	FRACTURE MECHANICS OF TINY CRACKS NEAR FASTENERS	74-0735
FASTENERS	STRESS AND STRAIN DISTRIBUTION IN THE VICINITY OF INTERFERENCE FIT FASTENERS	73-0363
FATE	THE FATE OF RADIONUCLIDES IN A GROUNDWATER ENVIRONMENT	74-0061
FATIGUE	A FINITE ELEMENT ANALYSIS OF FATIGUE CRACK CLOSURE	74-0759
FATIGUE	A FRACTURE MECHANICS APPROACH TO THE PROPAGATION OF FATIGUE CRACKS IN REACTOR COMPONENTS BY ACOUSTIC	71-0132
FATIGUE	ANALYSIS OF FATIGUE OF PAVING MIXTURES FROM THE FRACTURE MECHANICS VIEWPOINT	72-0540
FATIGUE	ANALYTICAL AND EXPERIMENTAL INVESTIGATION OF FATIGUE IN A SHEET SPECIMEN WITH AN INTERFERENCE-FIT 30	75-0701
FATIGUE	APPLICATIONS OF FINITE ELEMENT STRESS ANALYSIS AND STRESS-STRAIN PROPERTIES IN DETERMINING NOTCH FAT	73-1114
FATIGUE	APPROACH TO INVESTIGATE NOTCH EFFECT ON LOW-CYCLE FATIGUE WITH FINITE ELEMENT	73-0509
FATIGUE	CALCULATION OF STRESS INTENSITY FACTORS AND FATIGUE CRACK PROPAGATION OF SEMI-ELLIPTICAL, PART-THROU	73-0360
FATIGUE	CRUCIFORM SPECIMENS FOR BIAXIAL FATIGUE TESTS - AN INVESTIGATION USING FINITE-ELEMENT ANALYSIS AND P	71-0795
FATIGUE	CYCLIC ANALYSIS OF A PROPAGATING CRACK AND ITS CORRELATION WITH FATIGUE CRACK GROWTH	75-0188
FATIGUE	CYCLIC FATIGUE ANALYSIS OF ROCKET THRUST CHAMBERS. VOLUME 2: ATTITUDE CONTROL THRUSTER HIGH CYCLE FA	74-0693
FATIGUE	CYCLIC FATIGUE ANALYSIS OF ROCKET THRUST CHAMBERS. VOLUME 2: ATTITUDE CONTROL THRUSTER HIGH CYCLE FA	74-0693
FATIGUE	CYCLIC FATIGUE ANALYSIS OF ROCKET THRUST CHAMBERS. VOLUME 1: OFHC COPPER CHAMBER LOW CYCLE FATIGUE	74-0694
FATIGUE	CYCLIC FATIGUE ANALYSIS OF ROCKET THRUST CHAMBERS. VOLUME 1: OFHC COPPER CHAMBER LOW CYCLE FATIGUE	74-0694
FATIGUE	EFFECTS OF A PEAK OVERLOAD ON FATIGUE CRACK PROPAGATION RATE-PLANE STRESS ANALYSIS OF CRACKED PLATE	74-1365
FATIGUE	FATIGUE CRACK FORMATION AND PROPAGATION IN PAVEMENTS CONTAINING SOIL-CEMENT BASES.	72-0435
FATIGUE	FATIGUE CRACK INITIATION IN A BORON EPOXY PLATE WITH A CIRCULAR HOLE	74-0820
FATIGUE	FATIGUE OF GRAPHITE FIBRE-REINFORCED EPOXY IN COMPRESSION.	72-0383
FATIGUE	FATIGUE OF PRENOTCHED GRAPHITE FIBER COMPOSITES IN COMPRESSION	72-0215
FATIGUE	FINITE ELEMENT ANALYSIS OF FATIGUE CRACK PROPAGATION - INCLUDING THE EFFECTS OF CRACK CLOSURE	74-1112
FATIGUE	FORM FACTORS OF FLAT BARS FOR FATIGUE STRENGTH TESTS. (FORMZAHLEN VON FLACHSTAEBEN FUER SCHWINGFESTI	71-0641
FATIGUE	HOLLOW-ENDED ROLLER. A SOLUTION FOR IMPROVING FATIGUE LIFE IN ASYMMETRICALLY LOADED CYLINDRICAL ROL	71-0393
FATIGUE	LOW CYCLE FATIGUE STRENGTH OF PIPING COMPONENTS	73-0366
FATIGUE	NOTCH EFFECT IN LOW CYCLE FATIGUE - REPORT 1: LOW-CYCLE FATIGUE STRENGTH OF 0.48 PERCENT CARBON STEE	73-1090
FATIGUE	NOTCH EFFECT IN LOW CYCLE FATIGUE - REPORT 1: LOW-CYCLE FATIGUE STRENGTH OF 0.48 PERCENT CARBON STEE	73-1090
FATIGUE	NOTCH EFFECT IN LOW-CYCLE FATIGUE. CONSIDERATIONS ON NOTCH EFFECT IN LOW-CYCLE FATIGUE WITH FINITE-E	75-0343
FATIGUE	NOTCH EFFECT IN LOW-CYCLE FATIGUE. CONSIDERATIONS ON NOTCH EFFECT IN LOW-CYCLE FATIGUE WITH FINITE-E	75-0343
FATIGUE	NOTCH EFFECT IN LOW-CYCLE FATIGUE - 2. EFFECT OF STRESS CHANGE AND MEAN STRESS	75-0537
FATIGUE	PREDICTION OF STATIC AND FATIGUE DAMAGE AND CRACK PROPAGATION IN COMPOSITE MATERIALS	75-0643
FATIGUE	PRESSURE VESSEL WALL THICKNESS LIMITS BASED ON THERMAL STRESS RACHETTING AND CREEP FATIGUE INTERACTI	74-0968
FATIGUE	SONIC FATIGUE OF DIFFUSION BOUNDED TITANIUM SANDWICH STRUCTURES	72-0290
FATIGUE	STRESS CONCENTRATION FACTORS FOR A FATIGUE TEST SPECIMEN WITH A U-TYPE NOTCH IN ONE EDGE BY FINITE E	74-1344
FATIGUE	SYMPOSIUM ON ACOUSTIC FATIGUE	73-0898
FATIGUE	THE FATIGUE STRENGTH OF FILLET WELDED CONNECTIONS	71-0684
FATIGUE	THE USE OF ELASTIC-PLASTIC FINITE ELEMENT ANALYSIS IN THE CALCULATION OF CUMULATIVE FATIGUE DAMAGE	74-0314
FATIGUE	THE VIBRATION CHARACTERISTICS OF A TENSIONED PLATE CONTAINING A FATIGUE CRACK	68-0134
FATIGUE	THEORETICAL STUDY OF THE INFLUENCE OF INCLUSIONS UPON THE INITIATION AND GROWTH OF FATIGUE CRACKS IN	71-0394
FATIGUE-CRACK	EXPERIMENTAL AND THEORETICAL FATIGUE-CRACK PROPAGATION LIVES OF VARIOUSLY NOTCHED PLATES	74-0602
FATIGUE-CRACK	FINITE-ELEMENT METHOD APPLIED TO PREDICTING FATIGUE-CRACK GROWTH	73-0513
FATIGUE	APPLICATION OF FINITE ELEMENT ELASTIC-PLASTIC STRESS ANALYSIS TO NOTCHED FATIGUE SPECIMEN BEHAVIOUR	71-0185
FATIGUE	APPLICATION OF PROGRAM FATIQUE TEST TO MEMBER JOINTS OF HULLS	72-0814
FATIGUE	COMPUTER PREDICTION OF FATIGUE CRACK PROPAGATION UNDER RANDOM LOADING	73-0851
FATIGUE	DEVELOPMENT OF AN UNDERSTANDING OF THE FATIGUE PHENOMENA OF BONDED AND BOLTED JOINTS IN ADVANCED FIL	72-0798
FATIGUE	FATIGUE STRENGTH OF LUGS CONTAINING LINERS. VOLUME II. COMPUTER PROGRAM USED FOR ANALYSIS	70-0418
FATIGUE	LAYERED PAVEMENT SYSTEMS. PART 1. LAYERED SYSTEM DESIGN. PART 11. FATIGUE OF PLAIN CONCRETE	72-0867
FAULT	CENTRIFUGAL MODEL STUDY AND FINITE ELEMENT ANALYSIS OF STRESS NEAR A FAULT	67-0159
FAULT	EARTH PRESSURES ON STRUCTURES DUE TO FAULT MOVEMENT	73-0403
FAULTS	COMPUTER SIMULATION OF SECOND ORDER FAULTS	71-0010
FEABL	FEABL (FINITE ELEMENT ANALYSIS BASIC LIBRARY) USER'S GUIDE	72-0748
FEASIBILITY	A FEASIBILITY STUDY OF THE ANALYSIS OF ASYMMETRIC SHELL BY THE DISCRETE-ELEMENT METHOD	68-0115
FEASIBILITY	CONCEPT FEASIBILITY COMBAT TRACKED VEHICLE SIGNATURE DUPLICATOR	64-0028
FEASIBILITY	FEASIBILITY OF USING FINITE ELEMENTS IN ANALYSIS OF SECOND BREAKDOWN IN SEMICONDUCTOR DEVICES	73-0874
FEASIBILITY	FINITE ELEMENT ANALYSIS OF TUBULAR JOINTS: A REPORT ON A FEASIBILITY STUDY	67-0147
FEASIBLE	IS THE COMPUTATION OF NOISE LEVEL INSIDE A CAR FEASIBLE	71-0631
FEAT	STRUCTURE-MEDIUM INTERACTION AND DESIGN PROCEDURES STUDY. VOLUME IV. FEAT CODE USER'S GUIDE	69-0265
FEATHER	FEATHER: FINITE ELEMENT ANALYSIS FOR THREE-DIMENSIONAL ELASTIC RESPONSE	72-0607
FEATS	FEATS, A COMPUTER PROGRAM FOR THE FINITE ELEMENT THERMAL STRESS ANALYSIS OF PLANE OR AXISYMMETRIC SO	68-0152
FEATURED	RECORD CONCRETE SHELL ROOF FEATURED IN COLLISEUM	72-0218
FEATURES	EFFECTS OF GEOMETRY CHANGE ON ESSENTIAL FEATURES OF INELASTIC BEHAVIOUR	73-0448
FEATURES	MATHEMATICAL FEATURES OF FINITE ELEMENT APPROXIMATIONS OF CERTAIN FLOW PROBLEMS WITH EMPHASIS ON CON	74-0002
FEBRUARY	DYNAMIC ANALYSIS OF THE SLIDE IN THE LOWER SAN FERNANDO DAM DURING THE EARTHQUAKE OF FEBRUARY 9, 197	75-0277
FEDGE	FEDGE - A GENERAL PURPOSE COMPUTER PROGRAM FOR FINITE ELEMENT DATA GENERATION. VOLUME 2 - PROGRAM MA	69-0492
FEDGE	FEDGE - A GENERAL-PURPOSE COMPUTER PROGRAM FOR FINITE ELEMENT DATA GENERATION. VOLUME 1 - USER'S MAN	69-0491
FEECON	FINITE ELEMENT PROGRAM FEECON FOR UNDRAINED DEFORMATION ANALYSES OF GRANULAR EMBANKMENTS ON SOFT CLA	72-0875
FEEDBACK	ANALYSIS OF FEEDBACK CONTROL STABILIZED VIBRATION BY THE FINITE ELEMENT METHOD	74-1188
FEER	FAST MODAL EXTRACTION IN NASTRAN VIA THE FEER COMPUTER PROGRAM	73-0356
FEHLERBETRACHTU	FEHLERBETRACHTUNGEN AUFGRUND DER METHODE DER GEWICHTETEN RESIDUEN FUR DAS VERFAHREN DER FINITEN ELEM	72-0368
FEHLERSCHRANKEN	UEBER PAUSCHALE FEHLERSCHRANKEN FUER APPROXIMATIONEN VON SCHALENVERFORMUNGEN. (GLOBAL BOUNDARY CONDI	75-0292
FELDELEMENTE	BEITRAG ZUR SIMULATION NICHTSTATIONARER HORIZONTAL-EVENER GRUNDWASSERSTROMUNGSFELDER AUF DER GRUNDLA	74-0088
FELS	BERECHNUNG VON SPANNUNGSUMLAGERUNGEN IN FELS UND BODEN MIT HILFE DER ELEMENTENMETHODE	69-0364
FEM	CRACK TIP STRAIN - A COMPARISON OF FEM CALCULATIONS AND MEASUREMENTS	75-0721
FEM	ELASTIC-PLASTIC ANALYSIS OF REINFORCED CONCRETE MEMBERS BY FEM	73-0292
FEM	FUNDAMENTAL VIBRATIONAL FREQUENCY BY A 3D - FEM PROCEDURE	75-0710
FEMAC	FINITE ELEMENT SOLUTION OF PARTIAL DIFFERENTIAL EQUATIONS FEMAC COMPUTER PROGRAM	73-0648
FEMAC	FINITE ELEMENT SOLUTIONS OF SELECTED PARTIAL DIFFERENTIAL EQUATIONS "FEMAC COMPUTER PROGRAM"	74-0327
FEMESH	FEMESH: A FINITE ELEMENT CODE PREPROCESSOR	73-0949

FEMORAL	FINITE ELEMENT STUDIES OF FEMORAL ENDOPROSTHESES FOR HIP RECONSTRUCTION	75-0381
FEMUR	ON THE MATHEMATICAL ANALYSIS OF STRESS IN THE HUMAN FEMUR	72-0207
FEMUR	STRESS ANALYSIS OF HUMAN FEMUR	73-0327
FEM2D	FEM2D: PROGRAM FOR SOLVING THE TWO-DIMENSIONAL DIFFUSION EQUATION BY THE METHOD OF FINITE ELEMENTS	74-1089
FERNANDO	DYNAMIC ANALYSIS OF THE SLIDE IN THE LOWER SAN FERNANDO DAM DURING THE EARTHQUAKE OF FEBRUARY 9, 197	75-0277
FERRO-CEMENT	ON SHELL EFFECTS ON FERRO-CEMENT VESSELS	73-0455
FERROELECTRIC	2-DIMENSIONAL FINITE-ELEMENT ANALYSIS OF ELECTROELASTIC NORMAL MODES OF FERROELECTRIC DISKS AND SHEL	75-0594
FERROELECTRIC	3-DIMENSIONAL FINITE-ELEMENT ANALYSIS OF ELECTROELASTIC NORMAL MODES OF A TANGENTIALLY POLARIZED, SE	75-0591
FERROMAGNETIC	FINITE ELEMENT ANALYSIS OF THE MAGNETIC FIELD DISTRIBUTION INSIDE A ROTATING FERROMAGNETIC BAR	74-1037
FERTIGTEILEN	ENTWURF UND SPANNUNGSBERECHNUNG EINES SPANNBETON DRUCKBEHAELTERS AUS FERTIGTEILEN. (DESIGN AND STRES	73-0856
FERTIGTEILEN	ZUSAMMENWIRKEN VON EINZELNEN FERTIGTEILEN ALS GROSSFLAECHIGE SCHEIBE (INTERACTION OF INDIVIDUAL PREF	73-0552
FERTIGTEILTAFEL	ZUM TRAGVERHALTEN VON WAENDEN AUS FERTIGTEILTAFELN (BEHAVIOR OF LARGE PANEL SHEAR WALLS)	74-0299
FESS	FESS AND FINESSE PROGRAMS	72-0158
FESS	FINITE ELEMENT SOLUTION SYSTEMS (FESS).	67-0098
FESSURATE	CALCULATION METHOD FOR THE STUDY OF CRACKED RIGID STRUCTURES (METODO DI CALCOLO PER LO STUDIO DI STR	71-0637
FESTIGKEITSNACH	FESTIGKEITSNACHWEISE	74-0258
FESTIGKEITSUNTE	FESTIGKEITSUNTERSUCHUNG MIT DER METHODE DER FINITEN ELEMENTE. (STRENGTH OF THE MATERIALS INVESTIGATE	74-1009
FESTIGKEITSUNTE	FESTIGKEITSUNTERSUCHUNGEN UND-BERECHNUNGEN AN PUMPENSPIRALGEHAEUSEN. (INVESTIGATIONS AND CALCULATION	75-0238
FESTIGKEITSVERH	UNTERSUCHUNGEN ZUM FESTIGKEITSVERHALTEN VON METALLKLEBUNGEN IM HINBLICK AUF DIE ERSTELLUNG EINES PRA	74-0289
FESTIGKEITSVERH	MODERNE VERFAHREN ZUR UNTERSUCHUNG DER STEIFIGKEITS-UND FESTIGKEITSVERHALTNISSE VON RAHMENKORPERN. I	69-0401
FESTIGKERIT	ANWENDUNG DES FINITE ELEMENT VERFAHRENS SUR BESTIMMUNG DER FESTIGKEIT VON SCHWEISSKONSTRUKTIONEN. (	72-0185
FESTRAN	FESTRAN: A FINITE ELEMENT COMPUTER PROGRAM	71-0811
FETE	FETE PROGRAM PACKAGE	72-0161
FEW-GROUP	ENERGY-DEPENDENT FINITE-ELEMENT METHOD FOR FEW-GROUP DIFFUSION EQUATIONS	75-0499
FIBER	AN ANALYTIC METHOD FOR PREDICTING THE FLOW AND FRACTURE IN NOTCHED FIBER COMPOSITE MATERIALS	75-0689
FIBER	ANALYTICAL DISPLACEMENTS AND VIBRATIONS OF CANTILEVERED UNSYMMETRIC FIBER COMPOSITE LAMINATES	75-0736
FIBER	ARBITRARILY SHAPED INHOMOGENEOUS OPTICAL FIBER OR INTEGRATED OPTICAL WAVEGUIDES	75-0372
FIBER	CREEP OF B/AL COMPOSITES AS INFLUENCED BY RESIDUAL STRESSES, BOND STRENGTH, AND FIBER PACKING GEOMET	74-1208
FIBER	FATIGUE OF PRENOTCHED GRAPHITE FIBER COMPOSITES IN COMPRESSION	72-0215
FIBER	FINITE ELEMENT ANALYSIS SYSTEM FOR THE MECHANICAL BEHAVIOR OF ORIENTED FIBER COMPOSITE MATERIALS UND	73-0869
FIBER	FRACTURE OF GRAPHITE FIBER REINFORCED COMPOSITES	74-0695
FIBER	MATRIX AND INTERFACE STRESSES IN A DISCONTINUOUS FIBER COMPOSITE MODEL	68-0040
FIBER	MECHANICS OF LOAD TRANSFER AT THE FIBER/MATRIX INTERFACE	72-0879
FIBER	SIMULATION ON THE TENSILE BEHAVIOUR OF TWO-PHASE ALLOY CONTAINING A FIBER UNDER TENSILE LOAD	73-0542
FIBER	STRESS DISTRIBUTION NEAR A BROKEN POINT OF FILAMENT IN UNIDIRECTIONALLY FIBER REINFORCED COMPOSITES	73-0811
FIBER	TRANSVERSE LOADING OF UNIDIRECTIONAL FIBER COMPOSITES	69-0119
FIBER	WAVE PROPAGATION IN A VISCOELASTIC FIBER SUBJECTED TO TRANSVERSE IMPACT	71-0639
FIBER-REINFORCE	APPLICATION OF OPTIMALLY CRITERION TO FIBER-REINFORCED COMPOSITES	73-0431
FIBER-REINFORCE	INFLATION FORMING OF STEEL FIBER-REINFORCED CONCRETE DOMES	74-0705
FIBER-REINFORCE	STATIC AND DYNAMIC ANALYSIS OF VISCOELASTOPLASTIC FIBER-REINFORCED COMPOSITE SHELLS IN MISSILE STRUC	73-0651
FIBER-REINFORCE	THREE DIMENSIONAL SCATTERED-LIGHT STRESS ANALYSIS OF DISCONTINUOUS FIBER-REINFORCED COMPOSITES	73-0218
FIBRE	EFFECT OF FIBRE DISTRIBUTION ON STRESS AND STRAIN CONCENTRATION AT HOLES IN COMPOSITE PLATES	72-0616
FIBRE	ELASTIC-PLASTIC FINITE ELEMENT ANALYSIS OF SHORT FIBRE COMPOSITES	74-0138
FIBRE	PLASTIC FLOW AND FAILURE OF DISCONTINUOUS FIBRE COMPOSITE MATERIALS	70-0157
FIBRE	THERMOMECHANICAL RESPONSE OF INELASTIC FIBRE COMPOSITES	75-0235
FIBRE-REINFORCE	FATIGUE OF GRAPHITE FIBRE-REINFORCED EPOXY IN COMPRESSION.	72-0383
FIBRE-REINFORCE	INVESTIGATION OF BOND FAILURE IN FIBRE-REINFORCED MATERIALS BY THE FINITE ELEMENT METHOD	72-0034
FIBRE-REINFORCE	QUASI-LINEAR PROGRAMMING ALGORITHM FOR OPTIMIZING FIBRE-REINFORCED STRUCTURES OF FIXED STIFFNESS	75-0225
FIBREGLASS-REIN	BUCKLING PROBLEMS IN THE DESIGN OF FIBREGLASS-REINFORCED PLASTIC SHIPS	72-0164
FIBROUS	AN ANALYSIS OF THE PHYSICAL BEHAVIOR OF NONWOVEN FIBROUS MATERIAL BY THE FINITE-ELEMENT METHOD	73-1149
FIBROUS	APPLICATION OF THREE-DIMENSIONAL FINITE ELEMENT ANALYSES TO THE MICROMECHANICS OF FIBROUS COMPOSITE	74-1079
FIBROUS	DYNAMIC RESPONSE OF A CONSTRAINED FIBROUS SYSTEM SUBJECTED TO TRANSVERSE IMPACT. II. A MECHANICAL MO	70-0369
FIBROUS	FINITE ELEMENT ANALYSIS OF INTERLAMINAR SHEAR FIBROUS COMPOSITES	71-0129
FIBROUS	PLASTICITY THEORY OF FIBROUS COMPOSITES - PART 11	73-0750
FIBROUS	PLASTICITY THEORY OF FIBROUS COMPOSITES - PART 1	73-0751
FIBROUS	STOCHASTIC FINITE ELEMENT SIMULATION OF THE NONLINEAR STRUCTURAL RESPONSE OF FIBROUS COMPOSITE MATER	75-0762
FIBROUS	STRESS CONCENTRATIONS NEAR A DISCONTINUITY IN FIBROUS COMPOSITES	71-0316
FIBROUS	THE STOCHASTIC FINITE ELEMENT SIMULATION OF THE NONLINEAR STRUCTURAL RESPONSE OF FIBROUS COMPOSITE M	75-0798
FIBROUS	TRANSVERSE PROPERTIES OF FIBROUS COMPOSITES	69-0069
FIBROUS	TRANSVERSE PROPERTIES OF UNIDIRECTIONAL ALUMINUM MATRIX FIBROUS COMPOSITES	69-0167
FICTION	ON THE FICTION OF LEADING EDGE SUCTION	71-0712
FIELD	A FINITE ELEMENT SOLUTION OF TIME DEPENDENT FIELD PROBLEMS	67-0155
FIELD	A FINITE ELEMENT SOLUTION OF TIME DEPENDENT FIELD PROBLEMS USING A REDUCED VARIATIONAL FORMULATION	69-0504
FIELD	A HIGH ORDER TRIANGULAR FIELD ELEMENTS	71-0535
FIELD	A HIGHER ORDER TRIANGULAR FINITE ELEMENT FOR THE SOLUTION OF FIELD PROBLEMS IN ORTHOTROPIC MEDIA	72-0370
FIELD	A THREE DIMENSIONAL SOLUTION OF THE TRANSIENT FIELD PROBLEM USING ISOPARAMETRIC FINITE ELEMENTS	72-0601
FIELD	AN EQUILIBRIUM STRESS FIELD MODEL FOR FINITE ELEMENT SOLUTIONS OF TWO-DIMENSIONAL ELASTIC PROBLEMS	68-0163
FIELD	ANALYSIS OF ACOUSTIC FIELD IN IRREGULARLY SHAPED ROOMS BY FINITE ELEMENT METHOD	73-0133
FIELD	ANALYSIS OF DISLOCATION STRESS FIELD BY FINITE ELEMENT METHOD	75-0862
FIELD	BERECHNUNG DES REIBUNGSLOSEN STROEMUNGSFELDES VON DREIDIMENSIONALEN AUFTRIEBSBEHAFTETENTRAGFLUEGELN	73-0533
FIELD	COUPLED CONVECTIVE/CONDUCTIVE HEAT TRANSFER INCLUDING VELOCITY FIELD EVALUATION	75-0116
FIELD	DIVISION OF COMPUTATION FIELD IN THE FINITE ELEMENT METHOD	74-0083
FIELD	EFFECT OF ELASTIC MODULAR RATIO AND ANISOTROPY ON THE STRESS FIELD IN A TWO-DIMENSIONAL MATRIX INTO	71-0647
FIELD	EFFECTS OF SOURCE STRUCTURE AND SITE RESPONSE ON GROUND MOTION IN THE NEAR FIELD OF EARTHQUAKES	73-0732
FIELD	EMBANKMENT ANALYSIS AND FIELD CORRELATION	68-0057
FIELD	EQUILIBRIUM STRESS FIELD MODELS IN FINITE ELEMENTS	67-0011
FIELD	FIELD ANALYSIS FOR PREFABRICATED RIBBED CYLINDRICAL SHELLS	73-0899
FIELD	FIELD INSTRUMENTATION FOR LONG TERM MEASUREMENT OF PILE LOAD AND RAFT CONTACT PRESSURE	73-0150
FIELD	FIELD PROBLEMS SOLUTIONS BY FINITE METHODS	68-0118
FIELD	FIELD SOLUTIONS FOR TURBULENT SEEPAGE FLOW	71-0206
FIELD	FIELD TESTS OF A STEEL-COMPOSITE BOX-GIRDER BRIDGE	75-0633
FIELD	FINITE ELEMENT ANALYSIS OF IRREGULARLY SHAPED ACOUSTIC FIELD	71-0711
FIELD	FINITE ELEMENT ANALYSIS OF ELECTROMAGNETIC FIELD PROBLEMS	74-0080
FIELD	FINITE ELEMENT ANALYSIS OF THE MAGNETIC FIELD DISTRIBUTION INSIDE A ROTATING FERROMAGNETIC BAR	74-1037
FIELD	FINITE ELEMENT ANALYSIS OF 2-DIMENSIONAL FLOW IN SOILS CONSIDERING WATER UPTAKE BY ROOTS .2. FIELD A	75-0147
FIELD	FINITE ELEMENT MODEL OF EARTHS GRAVITY FIELD	74-0516
FIELD	FINITE ELEMENT MODELING OF MAGNETIC FIELD DEFECT INTERACTIONS	75-0068
FIELD	FINITE ELEMENT PROGRAM PACKAGE FOR AXISYMMETRIC SCALAR FIELD PROBLEMS	73-0113
FIELD	FINITE ELEMENT PROGRAM PACKAGE FOR AXISYMMETRIC VECTOR FIELD PROBLEMS	75-0138
FIELD	FINITE ELEMENT SOLUTION OF SATURABLE MAGNETIC FIELD PROBLEMS	70-0205
FIELD	FINITE ELEMENT SOLUTION OF TWO DIMENSIONAL EXTERIOR FIELD PROBLEMS	71-0491
FIELD	FINITE ELEMENT SOLUTION OF SATURATED TRAVELLING MAGNETIC FIELD PROBLEMS	75-0600
FIELD	FINITE ELEMENT STRESS FIELD SOLUTION OF THE PROBLEM OF SAINT VENANT TORSION	73-1042
FIELD	FINITE ELEMENT WEIGHTED RESIDUAL SOLUTION TO ONE DIMENSIONAL FIELD PROBLEMS	73-0388
FIELD	FINITE ELEMENTS FOR FIELD PROBLEMS IN CYLINDRICAL CO-ORDINATES	73-0171
FIELD	FINITE ELEMENTS IN THE SOLUTION OF FIELD PROBLEMS	65-0034
FIELD	FINITE-ELEMENT CALCULATION OF PRESSURE HISTORY IN A GAS FIELD	75-0546
FIELD	FINITE-ELEMENT FIELD ANALYSIS OF AN INHOMOGENEOUS ANISOTROPIC RELUCTANCE MACHINE ROTOR	73-0020
FIELD	FINITE-ELEMENT SOLUTION OF UNBOUNDED FIELD PROBLEMS	72-0025
FIELD	FREE VIBRATION AND RESPONSE TO RANDOM PRESSURE FIELD OF ANISOTROPIC THIN CYLINDRICAL SHELLS	74-0830
FIELD	GENERALIZED STEADY STATE FIELD PROBLEM AND ITS SOLUTION	74-0059
FIELD	IMPROVED METHOD FOR OBTAINING THE GENERAL-DISPLACEMENT FIELD FROM A HOLOGRAPHIC INTERFEROGRAM	72-0015

FIELD	INTEGRAL REPRESENTATION OF FIELD VARIABLES FOR THE FINITE ELEMENT SOLUTION OF VISCOUS FLOW PROBLEMS	74-0468
FIELD	ITERATIVE APPROACH TO FINITE-ELEMENT METHOD IN FIELD PROBLEMS	74-0124
FIELD	LAPLACIAN ELECTROSTATIC FIELD CALCULATIONS BY FINITE ELEMENTS WITH AUTOMATIC GRID GENERATION	73-0084
FIELD	LINEAR ACCELERATOR CAVITY FIELD CALCULATION BY FINITE ELEMENT	73-0008
FIELD	MAGNET FIELD AND FORCE BY FINITE ELEMENT TECHNIQUES	69-0035
FIELD	NASTRAN ANALYSIS OF A TURBINE BLADE AND COMPARISON WITH TEST AND FIELD DATA	75-0367
FIELD	NONLINEAR ANALYSES OF THE STRESS FIELD NEAR A CRACK OR HOLE IN A PLATE	75-0694
FIELD	NUMERICAL CALCULATION OF MERIDIONAL FLOW FIELD IN TURBOMACHINES USING FINITE ELEMENTS METHOD	74-0481
FIELD	RESPONSE OF STRUCTURE TO THE PSEUDO-SOUND FIELD OF A JET (USING A COMBINED CONTINUUM AND FINITE ELEM	67-0146
FIELD	ROCK FAILURE AROUND A CIRCULAR OPENING IN A GRAVITY FIELD WITH TECTONIC FORCES	70-0071
FIELD	SCALAR FINITE-ELEMENT PROGRAM PACKAGE FOR TWO-DIMENSIONAL FIELD PROBLEMS	71-0478
FIELD	SOLUTION OF THREE-DIMENSIONAL FIELD PROBLEMS BY THE FINITE ELEMENT METHOD	67-0091
FIELD	SOME REMARKS ON FINITE ELEMENT ANALYSIS OF TIME DEPENDENT FIELD PROBLEMS	73-0307
FIELD	TENSILE FIELD CALCULATION ABOUT A PRESSURIZED CAVITY: MONERO	72-0684
FIELD	TEST RESULTS FROM THE BOUNDARY LAYER FACILITY - RESPONSE OF STRUCTURE TO THE PSEUDO-SOUND FIELD OF A	68-0262
FIELD	THE FINITE ELEMENT METHOD AND BOUNDARY SOLUTION PROCEDURES AS GENERAL APPROXIMATION METHODS FOR FIEL	75-0262
FIELD	THE HARMONIC FINITE ELEMENT MODEL FOR USE IN FIELD PROBLEMS	70-0389
FIELD	THERMAL-MECHANICAL FRAGMENTATION OF HARD ROCK - FIELD TESTS	74-0998
FIELD	THREE DIMENSIONAL NONLINEAR MAGNETIC FIELD BOUNDARY VALUE PROBLEM AND ITS NUMERICAL SOLUTION	70-0060
FIELD	THREE-DIMENSIONAL FIELD PROBLEMS BY HIGHER ORDER FINITE ELEMENTS	70-0257
FIELD	THREE-DIMENSIONAL STRESS ANALYSIS AND FIELD PROBLEMS BY THE FINITE ELEMENT METHOD	70-0183
FIELD	TRANSIENT AND NONLINEAR-ANALYSIS OF FIELD PROBLEMS BY FINITE ELEMENTS	72-0975
FIELD	TRANSIENT FIELD PROBLEMS: TWO-DIMENSIONAL AND THREE-DIMENSIONAL ANALYSIS BY ISO-PARAMETRIC FINITE EL	70-0001
FIELD	TWO-DIMENSIONAL FINITE ELEMENT COMPUTER CODES FOR STRESS ANALYSIS AND FIELD PROBLEMS	73-0055
FIELD-PLOTTING	FINPLT; A FINITE-ELEMENT FIELD-PLOTTING PROGRAM	72-0299
FIELD-STRESS	FINITE-ELEMENT METHODS FOR COUPLED FIELD-STRESS PROBLEMS	74-0635
FIELDS	A VARIATIONAL FORMULATION FOR DISCONTINUOUS FIELDS IN LINEARLY ELASTIC FINITE ELEMENT APPLICATIONS	72-0962
FIELDS	AN ELASTIC-PLASTIC FINITE ELEMENT ANALYSIS OF CRACK TIP FIELDS UNDER BIAXIAL LOADING CONDITIONS	73-0760
FIELDS	ANALYSIS OF MAGNETIC FIELDS BY FINITE ELEMENT METHOD	73-0126
FIELDS	ANALYSIS OF TURBOALTERNATOR MAGNETIC FIELDS BY FINITE ELEMENTS	71-0460
FIELDS	AVERAGE ELASTIC PROPERTIES AND STRESS FIELDS FOR COMPOSITE CONTINUA	69-0106
FIELDS	AVERAGE ELASTIC PROPERTIES AND STRESS FIELDS FOR COMPOSITE MATERIALS	69-0435
FIELDS	BERECHNUNG INSTATIONAERER TEMPERATUREFELDER MIT FINITEN ELEMENTEN IN RAUM-UND ZEITDIMENSION. (CALCUL	74-0987
FIELDS	BERECHNUNG VON SPANNUNGEN UND TEMPERATUREFELDERN IN FLOSSENWAENDEN UND AUFGESCHWEISSTEN BAUTEILEN. (C	74-0544
FIELDS	CALCUL DES CHAMPS ELECTROMAGNETIQUES PAR LA METHODE DES ELEMENTS FINIS (COMPUTATION OF ELECTROMAGNET	73-0399
FIELDS	CALCULATION OF HARMONIC FIELDS IN WHICH SINGULARITIES ARE PRESENT USING FINITE ELEMENT AND FINITE DI	73-0979
FIELDS	CALCULATION OF TRANSIENT TEMPERATURE FIELDS WITH FINITE ELEMENTS IN SPACE AND TIME DIMENSIONS	74-0379
FIELDS	COMPLETE POLYNOMIAL DISPLACEMENT FIELDS FOR FINITE ELEMENT METHODS	68-0062
FIELDS	COMPLETE POLYNOMIAL DISPLACEMENT FIELDS FOR FINITE ELEMENT METHODS	68-0092
FIELDS	COMPUTATIONAL METHODS FOR INVISCID AND VISCOUS TWO-AND-THREE DIMENSIONAL FLOW FIELDS	75-0716
FIELDS	CURVILINEAR FINITE ELEMENTS FOR 2-DIMENSIONAL SATURABLE MAGNETIC FIELDS	74-0562
FIELDS	DIGITAL CALCULATION OF ELECTRIC FIELDS IN METAL CLAD INSTALLATIONS	73-0212
FIELDS	DISPLACEMENT AND STRAIN FIELDS FROM UNDERGROUND EXPLOSIONS USING ANALYTICAL AND FINITE ELEMENT METHO	72-0587
FIELDS	ELASTIC-PLASTIC FINITE ELEMENT ANALYSIS OF CRACK TIP FIELDS UNDER BIAXIAL LOADING CONDITIONS	74-0296
FIELDS	FINITE ELEMENT FORMULATIONS FOR ELASTIC PLATES BY GENERAL VARIATIONAL STATEMENTS WITH DISCONTINUOUS	73-0559
FIELDS	FINITE ELEMENT FORMULATIONS FOR ELASTIC PLATES BY GENERAL VARIATIONAL STATEMENTS WITH DISCONTINUOUS	73-1124
FIELDS	FINITE ELEMENT SIMULATION OF THERMALLY INDUCED FLOW FIELDS	74-1235
FIELDS	FINITE-ELEMENT ANALYSIS OF NONLINEAR MAGNETIC FIELDS IN ELECTRIC MACHINES	70-0484
FIELDS	HYPAR DISPLACEMENT FIELDS FOR RECTANGULAR FINITE ELEMENTS	73-0240
FIELDS	NEAR-CRACK DEFORMATION FIELDS IN STRAIN-HARDENING MATERIAL	74-0938
FIELDS	NUMERISCHE BERECHNUNG INSTATIONAERER TEMPERATUPFELDER UND THERMISCH BEDINGTER VERFORMUNGEN VON WERK?	75-0847
FIELDS	ON DERIVATION OF STIFFNESS MATRICES WITH CO ROTATION FIELDS FOR PLATES AND SHELLS	71-0254
FIELDS	OPREDELENIE STATSIONARNYKH TEMPERATURNYKH POLEI V DETALYAKH DVIGATELEI VNUTRENNEGO SGORANIYA METODOM	73-0454
FIELDS	PREDICTING MAGNETIC LEAKAGE FIELDS BY FINITE ELEMENT TECHNIQUES	75-0046
FIELDS	RASCHET POLEI DEFORMATSII I NAPRYAZHENII V DETALYAKH DVIGATELEI VNUTRENNEGO SGORANIYA (CALCULATION O	73-0465
FIELDS	RASHET TREKHMERNYKH MAGNITNYKE POLEI METODOM KONECHNYKH ELEMENTOV. (CALCULATION OF THREE-DIMENSIONAL	75-0766
FIELDS	REALIZATSIYA METODA KONECHNYKH ELEMENTOV NA EVM DLYA RASCHETA DVUMERNYKH ELEKTRICHESKIKH I MAGNITNYE	74-0251
FIELDS	REINFORCED FIELDS OF TRIANGULAR ELEMENTS WITH LINEARLY VARYING STRAIN: EFFECT OF INITIAL STRAINS	65-0004
FIELDS	STRESS FIELDS AROUND MULTIPLE INCLUSIONS	69-0067
FIELDS	STUDIO DI CAMPI ELETTRICI E MAGNETICI STANZIONARI CON IL METODO DEGLI ELEMENTI FINITI. APPLICAZIONE	75-0207
FIELDS	THERMAL STRESS AND TEMPERATURE FIELDS - A FINITE-ELEMENT APPROACH	71-0782
FIESTA	FIESTA - FINITE ELEMENT STRESS AND TEMPERATURE ANALYSIS. A DESCRIPTION OF THE THEORY USED FOR THE CC	71-0718
FILAMENT	ANALYSIS OF STRENGTH AND RESIDUAL STRESSES IN FILAMENT REINFORCED ALUMINUM CYLINDERS	75-0517
FILAMENT	STRESS DISTRIBUTION NEAR A BROKEN POINT OF FILAMENT IN UNIDIRECTIONALLY FIBER REINFORCED COMPOSITES	73-0811
FILAMENTARY	DEVELOPMENT OF AN UNDERSTANDING OF THE FATIGUE PHENOMENA OF BONDED AND BOLTED JOINTS IN ADVANCED FIL	72-0798
FILAMENTARY	THE BEHAVIOR OF ADVANCED FILAMENTARY COMPOSITE PLATES WITH CUTOUTS	73-0891
FILAMENTARY	THE BEHAVIOUR OF ADVANCED FILAMENTARY COMPOSITE PLATES WITH CUTOUTS	70-0025
FILAMENTS	STRENGTH DISTRIBUTIONS OF SINGLE FILAMENTS	75-0305
FILL	CALCULATING STRESSES IN A PRESSURE VESSEL WITH FILL TUBE - COMPARISON OF THE FINITE ELEMENT AND PHOT	72-0782
FILL	STUDIES ON THE DETERMINATION OF SEEPAGE-LINE OF FILL DAM BY THE METHOD OF FINITE ELEMENT. WATER FLOW	74-1348
FILLED	EVALUATION OF ATTENUATION FROM LOSSLESS TRIANGULAR-FINITE-ELEMENT SOLUTIONS FOR INHOMOGENEOUSLY FILL	73-0059
FILLED	FREE VIBRATION OF CYLINDRICAL SHELLS PARTIALLY FILLED WITH LIQUID	71-0839
FILLED	INTERNAL PRESSURE EFFECTS ON THE VIBRATION OF PARTIALLY FILLED ELASTIC TANKS	75-0241
FILLED	ON A SPECTRAL PROBLEM IN VIBRATION MECHANICS: COMPUTATION OF ELASTIC TANKS PARTIALLY FILLED WITH LIQ	75-0267
FILLED	VIBRATION CHARACTERISTICS OF THIN, PRESSURIZED SHELLS OF REVOLUTION PARTIALLY FILLED WITH LIQUID	72-0604
FILLED	VIBRATIONS OF TANKS PARTIALLY FILLED WITH LIQUIDS	74-0089
FILLER	A FINITE-ELEMENT MODEL TO SIMULATE THE NONLINEAR RESPONSE OF REINFORCED CONCRETE FRAMES WITH MASONRY	73-1168
FILLER	FINITE ELEMENT STUDY OF THE ELASTIC BEHAVIOR OF PLANE FRAMES WITH FILLER WALLS	71-0828
FILLER	NONLINEAR DYNAMIC ANALYSIS OF FRAMES WITH FILLER PANELS	74-0226
FILLET	DIFFUSION OF HYDROGEN IN FILLET WELDS	75-0373
FILLET	ELASTIC AND PLASTIC STRESS DISTRIBUTION ON BASE PLATES IN THE NEIGHBOURHOOD OF FILLET WELDS IN LAP J	70-0175
FILLET	KERBSPANNUNGSANALYSE NACH DER FINITE-ELEMENT-METHODE AN KEHLNAEHTEN. (NOTCH EFFECT ANALYSIS USING TH	75-0673
FILLET	ON MAXIMUM FILLET STRESSES IN BREECH RING	72-0736
FILLET	STRENGTH OF TRANSVERSE FILLET WELDED JOINTS	74-0144
FILLET	THE FATIGUE STRENGTH OF FILLET WELDED CONNECTIONS	71-0684
FILLETS	EFFECT OF FILLETS ON STRESS CONCENTRATIONS IN CYLINDRICAL SHELLS WITH STEP CHANGES IN OUTSIDE DIAMET	71-0108
FILLETS	STRESS ANALYSIS OF CONFIGURATIONS INVOLVING SMALL FILLETS	75-0361
FILLING	CONTRIBUTION A L'ETUDE DU COMPLEXE BARRAGE-FONDATION ET EXPLICATION DES SEISMES DUS AU REMPLISSAGE D	72-0554
FILLING	EFFECT OF RESERVOIR FILLING ON STRESSES AND MOVEMENT IN EARTH AND ROCKFILL DAMS. A REPORT OF AN INVE	72-0836
FILLS	ANALYSIS OF LOAD-BEARING FILLS OVER SOFT SUBSOILS	71-0330
FILLS	FORCES ON RIGID CULVERTS UNDER HIGH FILLS	67-0008
FILM	A UNIFIED TREATMENT OF THICK AND THIN FILM ELASTOHYDRODYNAMIC PROBLEMS USING HIGHER ORDER ELEMENT PR	75-0890
FILM	FINITE ELEMENT ANALYSIS OF FLUID FILM LUBRICATION - A SURVEY	74-0041
FILM	FINITE ELEMENT ANALYSIS OF FLUID FILM LUBRICATION - A SURVEY	74-1100
FILM	FINITE ELEMENT ANALYSIS OF FLUID FILM LUBRICATION SURVEY	75-0880
FILM	ON THERMO-ELASTO-HYDRODYNAMIC FLUID FILM BEARINGS	75-0532
FILM	RESEARCH NOTE: A THREE-DIMENSIONAL OIL FILM TEMPERATURE DISTRIBUTION IN TILTING THRUST BEARINGS	74-0608
FILM	TRANSPIRATION AND FILM COOLING BOUNDARY LAYER COMPUTER PROGRAM. VOLUME 2, COMPUTER PROGRAM AND USER'	71-0655
FILMS	APPLICATION OF FINITE ELEMENT METHODS TO LUBRICATION: SQUEEZE FILMS BETWEEN POROUS SURFACES	75-0764
FILTERS	FINITE ELEMENT SIMULATION OF ACOUSTIC FILTERS WITH VARIABLE CIRCULAR CROSS SECTION	75-0871
FILTERS	NEW APPROACH TO ANALYSIS AND DESIGN OF ELECTROMECHANICAL FILTERS BY FINITE-ELEMENT TECHNIQUE	71-0476
FINAL	F 104 G FINAL MOTIVE: GROUND VIBRATION CALCULATIONS FOR THE STARFIGHTER F 104 G	69-0456

222

FINAL	FINAL REPORT ON NASTRAN/GRAPHICS, 1971 FORD DOOR PROJECT	73-0595
FINAL	M.I.T. TEST SECTION INSTRUMENTATION, MASSACHUSETTS BAY TRANSPORTATION AUTHORITY, HAYMARKET-NORTH EXT	72-0673
FINAL	THEMIS FINAL REPORT, VOL. IV, RECENT DEVELOPMENTS IN THE NUMERICAL SOLUTION OF CERTAIN PROBLEMS IN N	74-0721
FINAL	THEMIS FINAL REPORT, VOL. II, NONLINEAR MECHANICS IN SOLIDS AND STRUCTURES	74-0722
FINAL	THEMIS FINAL REPORT, VOL. 1, BASIC RESEARCH IN METHODS OF APPROXIMATION IN THE NONLINEAR MECHANICS O	74-0723
FINE-MESH	PLATE BENDING WITH NONCONFORMING ELEMENT: FINE-MESH CONVERGENCE TEST	65-0049
FINEL	FINEL, UNIVERSELLES PROGRAMMSYSTEM ZUR BERECHNUNG DER ELASTISCHEN EIGENSCHAFTEN VON MASCHINENBAUTEIL	72-0132
FINEL	FINEL, UNIVERSELLES PROGRAMMSYSTEM ZUR BERECHNUNG DER ELASTISCHEN EIGENSCHAFTEN VON MASCHINENBAUTEIL	72-0132
FINESSE	FESS AND FINESSE PROGRAMS	72-0158
FINESSE	FINESSE - A FINITE ELEMENT SYSTEM, PART V FINITE ELEMENT DESIGN - DERIVATIVES FOR STRUCTURAL SHAPE D	72-0963
FINIT	ASUPRA ANALIZEI PRIN METODA ELEMENTULUI FINIT A STARILOR DE EFORTURI SI DEFORMATII DIN MASIVELE DE P	70-0471
FINIT-ELEMENT	BERECHNUNG EINES EINACHSIGEN STAHLDRAHT-KUNSTSTOFFVERBUNDES UNTER EINSATZ DER FINIT-ELEMENT METHODE	72-0644
FINIT-ELEMENT-M	DER EINSATZ DER FINIT-ELEMENT-METHODE IN PRESSENBAU (APPLICATION OF THE FINITE ELEMENT METHOD FOR TH	74-0615
FINIT-ELEMENT-V	COMPUTER-UNTERSTUTZTES KONSTRUIEREN, PRAKTISCHE BEDEUTUNG DER FINIT-ELEMENT-VERFAHREN	71-0552
FINITE-AMPLITUD	A FINITE ELEMENT ANALYSIS OF SHOCK AND FINITE-AMPLITUDE WAVES IN ONE-DIMENSIONAL HYPERELASTIC BODIES	72-0731
FINITE-CIRCULAR	SHIP (SIMPLIFIED-HELMHOLTZ-INTEGRAL PROGRAM): A FAST COMPUTER PROGRAM FOR CALCULATING THE ACOUSTIC R	72-0835
FINITE-DEFLECTI	A FINITE-DEFLECTION ANALYSIS OF SHALLOW ARCHES BY THE DISCRETE ELEMENT METHOD	71-0077
FINITE-DIFFERENCE	A COMPARISON OF SOME FINITE ELEMENT AND FINITE-DIFFERENCE METHODS FOR A SIMPLE SLOSHING PROBLEM	71-0055
FINITE-DIFFEREN	COMPARISON OF FINITE-DIFFERENCE AND FINITE-ELEMENT SOLUTION TECHNIQUE IN TRANSIENT HEAT-CONDUCTION	75-0348
FINITE-DIFFEREN	ENERGY APPROACHES TO FINITE-DIFFERENCE AND FINITE ELEMENT METHODS	71-0046
FINITE-DIFFEREN	FINITE-DIFFERENCE AND FINITE-ELEMENT METHODS IN GLOBAL WIND-DRIVEN OCEAN CIRCULATION	75-0345
FINITE-DIFFEREN	FINITE-DIFFERENCE ENERGY METHOD FOR NONLINEAR SHELL ANALYSIS	71-0399
FINITE-DIFFEREN	STABILITY AND OSCILLATION CHARACTERISTICS OF FINITE ELEMENT, FINITE-DIFFERENCE, AND WEIGHTED-RESIDUA	75-0615
FINITE-DIFFEREN	USE OF SINGULARITY PROGRAMMING IN FINITE-DIFFERENCE AND FINITE ELEMENT COMPUTATIONS OF TEMPERATURE	73-0100
FINITE-DIFFEREN	FINITE-DIFFERENCES VIA FINITE-ELEMENTS	71-0665
FINITE-ELEMENT-	APPLICATION OF A CONJUGATED-GRADIENT-PROCESS IN FINITE-ELEMENT-CALCULATION OF NON-LINEAR STRUCTURES	74-0214
FINITE-ELEMENT-	WATER FLOW THROUGH SATURATED-UNSATURATED POROUS MEDIA USING FINITE-ELEMENT-GALERKIN METHODS	74-0879
FINITE-ELEMENT-	ZUR WIRKLICHKEITSNAHEN BERECHNUNG VON STAHLBETONPLATTEN MIT DER FINITE-ELEMENT-METHODE. (APPROXIMATE	75-0832
FINITE-ELEMENT-	ANWENDUNG DER FINITE-ELEMENT-METHODE BEI DER BERECHNUNG VON REAKTORDRUCKBEHAELTERN	71-0036
FINITE-ELEMENT-	ANWENDUNG DER FINITE-ELEMENT-METHODE ZUR BERECHNUNG VON SPANNUNGEN IN WASSERGESAETTIGTEN BOEDEN (FIN	73-0502
FINITE-ELEMENT-	BERECHNUNG DES KEGELDRUCKVERSUCHES MIT EINER FINITE-ELEMENT-METHODE. (CALCULATION OF THE CONE INDENT	75-0187
FINITE-ELEMENT-	DIE BERECHNUNG VON KONSOLIDATIONSVORGAENGEN MIT DER FINITE-ELEMENT-METHODE. (CALCULATION OF CONSOLID	71-0012
FINITE-ELEMENT-	EINE FINITE-ELEMENT-METHODE MIT KREISRINGSEKTORFORMIGEN ELEMENTEN	73-1115
FINITE-ELEMENT-	KERBSPANNUNGSANALYSE NACH DER FINITE-ELEMENT-METHODE AN KEHLNAEHTEN. (NOTCH EFFECT ANALYSIS USING TH	75-0673
FINITE-ELEMENT-	VERGLEICH NUMERISCHER LOESUNGEN NACH DEM DIFFERENZENVERFAHREN UND NACH DER FINITE-ELEMENT-METHODE AM	71-0016
FINITE-ELEMENT-	ZUR WIRKLICHKEITSNAHEN BERECHNUNG VON STAHLBETONPLATTEN MIT DER FINITE-ELEMENT-METHODE. (APPROXIMATE	75-0832
FINITE-ELEMENT-	FINITE-ELEMENT-TYPE SOLUTION OF INTEGRAL EQUATIONS	73-0995
FINITE-ELEMENTE	EINE FINITE-ELEMENTE-LOESUNG FUR DIE TORSIONSSTEIFIGKEIT UND DEN SCHUBMITTELPUNKT BELIEBIGER QUERSCH	74-0256
FINITE-WIDTH	NUMERICAL SIMULATION OF FINITE-WIDTH THRUST BEARINGS, TAKING INTO ACCOUNT VISCOSITY VARIATION WITH T	75-0623
FINITELY	NOTE ON AN APPROXIMATE METHOD FOR COMPUTING NONCONSERVATIVE GENERALIZED FORCES ON FINITELY DEFORMED	70-0097
FINITER	ANWENDUNG DER METHODE FINITER ELEMENTE BEI DER ANALYSE DES DYNAMISCHEN VERHALTENS GEDAEMPFTER WERKZE	75-0580
FINITER	BERECHNUNG DER ELASTIZITATSEIGENSCHAFTEN VON MASCHINENBAUTEILEN NACH DER METHODE FINITER ELEMENTE	73-1046
FINITER	BERECHNUNG INSTATIONARER GRUND - UND SICKERWASSERSTROMUNGEN MIT FREIER OBERFLACHE NACH DER METHODE F	71-0541
FINITER	FINEL, UNIVERSELLES PROGRAMMSYSTEM ZUR BERECHNUNG DER ELASTISCHEN EIGENSCHAFTEN VON MASCHINENBAUTEIL	72-0132
FINITER	ZUR BERECHNUNG EINFACH-SYMMETRISCHER I-TRAEGER NACH DER THEORIE - 2. ORDNUNG UNTER BERUECKSICHTIGUNG	75-0510
FINITI	CHARACTERISTICS OF A HYBRID CALCULATION SYSTEM BASED ON THE FINITE ELEMENT METHOD. (CARATTERISTICHE	72-0715
FINITI	IL METODO DEGLI ELEMENTI FINITI: POSSIBILITA DI APPLICAZIONE A PROBLEMI DI INTERESSE DEGLI FLETTROTE	75-0200
FINITI	LA TECNICA DEGLI ELEMENTI FINITI PER PROBLEMI DI LUBRIFICAZIONE INCOMPRESSIBILE. (FINITE ELEMENT MET	75-0471
FINITI	ON THE RESOLUTION OF NONLINEAR ELASTIC AND ELASTOPLASTIC PROBLEMS BY THE METHOD OF FINITE ELEMENTS.	74-1326
FINITI	OPTIMIZZAZIONE DI RETICOLI TRIANGOLARI PER IL METODO DEGLI ELEMENTI FINITI IN IDROLOGIA	72-0987
FINITI	PROCESS FOR THE AUTOMATIC GENERATION OF FLAT TRIANGULAR LATTICES FOR APPLICATION TO THE FINITE ELEME	71-0366
FINITI	STUDIO DI CAMPI ELETTRICI E MAGNETICI STANZIONARI CON IL METODO DEGLI ELEMENTI FINITI. APPLICAZIONE	75-0207
FINNED	FULLY DEVELOPED VISCOUS FLOW IN INTERNALLY FINNED TUBES	75-0648
FINNED-TUBE	BERECHNUNG VON SPANNUNGEN UND TEMPERATURFELDERN IN FLOSSENWAENDEN UND AUFGESCHWEISSTEN BAUTEILEN. (C	74-0544
FINPLA	TRANSVERSE REINFORCEMENT OF CONTINUOUS, TROUGH-SHAPED CONCRETE RAILROAD STRUCTURES, AN IMPLEMENTATIO	71-0719
FINPLT	FINPLT; A FINITE-ELEMENT FIELD-PLOTTING PROGRAM	72-0299
FINS	DESIGN OF PROPELLANT FINS SUBJECTED TO AXIAL ACCELERATION	71-0734
FINTE	SOLUTION OF FLUID DYNAMIC PROBLEMS BY FINTE ELEMENTS	74-1184
FINTIE	ON GENERAL PURPOSE PROGRAMS FOR FINTIE ELEMENT ANALYSIS WITH SPECIAL REFERENCE TO GEOMETRIC AND MATE	70-0087
FIPAX	USER'S GUIDE FOR ANALYSIS OF FINITE ELASTOPLASTIC DEFORMATION: THE FIPDEF AND FIPAX PROGRAMS FOR THE	74-0799
FIPDEF	USER'S GUIDE FOR ANALYSIS OF FINITE ELASTOPLASTIC DEFORMATION: THE FIPDEF AND FIPAX PROGRAMS FOR THE	74-0799
FIRE	COMPUTER ANALYSIS OF STEEL FRAME IN FIRE	75-0173
FIRE	CREEP DEFORMATION OF A MULTI-STOREY STRUCTURE IN A FIRE	74-1197
FIRE	FIRES-RC, A COMPUTER PROGRAM FOR THE FIRE RESPONSE OF STRUCTURES-REINFORCED CONCRETE FRAMES	74-0747
FIRE	FIRES-T, A COMPUTER PROGRAM FOR THE FIRE RESPONSE OF STRUCTURES - THERMAL	74-0850
FIRE	PERFORMANCE OF STRUCTURAL WOOD MEMBERS EXPOSED TO FIRE	75-0060
FIRE	STRUCTURAL CAPACITY OF REINFORCED CONCRETE COLUMNS SUBJECTED TO FIRE INDUCED THERMAL GRADIENTS	73-0913
FIRE	USE OF THE FINITE ELEMENT TECHNIQUE TO DETERMINE THE FIRE RESISTANCE CAPABILITIES OF INSULATED STRUC	71-0147
FIREPROOFING	FINITE ELEMENT PREDICTION OF FIREPROOFING EFFECTIVENESS	72-0047
FIRES-PC	FIRES-RC, A COMPUTER PROGRAM FOR THE FIRE RESPONSE OF STRUCTURES-REINFORCED CONCRETE FRAMES	74-0747
FIRES-T	FIRES-T, A COMPUTER PROGRAM FOR THE FIRE RESPONSE OF STRUCTURES - THERMAL	74-0850
FISSURATION	TENDANCES ACTUELLES DES RECHERCHES SUR LES CONTRAINTES ET DEFORMATIONS DUES AU BRIDAGE EN RELATION A	74-0273
FISSURED	METHODS OF CALCULATION OF THREE-DIMENSIONAL PROBLEMS OF PERCOLATION OF FISSURED ROCK BY FINITE ELEME	70-0472
FISSURED	SLOPES IN STIFF FISSURED CLAYS AND SHALES	69-0094
FISSURED	WECHSELWIRKUNG ZWISCHEN DEFORMATION UND DURCHSTROEMUNG IM KLUEFTIGEN, ANISOTROPEN GEBIRGE. (INTERACT	72-0612
FIT	AN ELASTOPLASTIC ANALYSIS OF A UNIAXIALLY LOADED SHEET WITH AN INTERFERENCE - FIT BOLT	74-0687
FIT	ELASTIC-PLASTIC ANALYSIS OF INTERFERENCE FIT FASTENERS	74-1177
FIT	STRESS AND STRAIN DISTRIBUTION IN THE VICINITY OF INTERFERENCE FIT FASTENERS	73-0363
FITS	A METHOD FOR DETERMINING THE SURFACE CONTACT STRESSES RESULTING FROM INTERFERENCE FITS	70-0101
FITS	FINITE ELEMENT COMPUTER PROGRAM FOR THE SOLUTION OF NONLINEAR AXISYMMETRIC CONTACT PROBLEMS WITH INT	74-1094
FITTED	NEW DESIGN FACTOR FOR A SHRINK FITTED ASSEMBLY	75-0167
FITTING	LEAST SQUARES SURFACE FITTING BY FINITE ELEMENTS AND AN APPLICATION TO STRESS SMOOTHING	67-0154
FITTINGS	HIGH STRENGTH END FITTINGS FOR FRP ROD AND ROPE	74-0605
FIVE	COMPUTATIONAL TECHNIQUES, NONLINEAR EQUATIONS, ISOPARAMETRIC ELEMENTS, FIVE LECTURES	73-0634
FIVE	CURVED ELEMENTS, ERROR BOUNDS, TIME DEPENDENT PROBLEMS FIVE LECTURES	73-0636
FIVE	ELASTIC EQUILIBRIUM MODELS, STRESS FUNCTIONS, AND DUALITY NON LINEARITIES AND STABILITY, FIVE LECTUR	73-0631
FIVE	PLATES AND SHELLS, INELASTIC ANALYSIS, TIME DEPENDENT INELASTICITY, FIVE LECTURES	73-0633
FIVE	SURVEY OF UNDERLYING THEORY, APPROXIMATION BY PIECEWISE POLYNOMIALS, ILLEGAL ELEMENTS AND THE PATCH	73-0635
FIXATION	A FINITE ELEMENT MODEL FOR PREDICTING LONG BONE FRACTURE FIXATION	71-0819
FIXATION	FINITE ELEMENT AND EXPERIMENTAL ANALYSIS OF AN INTERNAL FIXATION PLATE MODEL	75-0380
FIXED	ON THE ASYMPTOTICALLY SPHERICAL DEFORMATIONS OF ARBITRARY MEMBRANES OF REVOLUTION FIXED ALONG AN EDG	71-0640
FIXED	QUASI-LINEAR PROGRAMMING ALGORITHM FOR OPTIMIZING FIBRE-REINFORCED STRUCTURES OF FIXED STIFFNESS	75-0225
FIXED	ZUR BERECHNUNG DER BIEGESTEIFIGKEIT ABGESETZTER STAEBE UND WELLEN UNTER ANWENDUNG VON FINITEN ELEMEN	75-0481
FLA	ASKA 105 A TECHNICAL MANUAL AND PROGRAMMING THEORY FOR THE ELEMENTS, FLA, TRIM, SPARTA AND TET	68-0254
FLACHER	EIN NEUES VARIATIONSPRINZIP MIT ANWENDUNG AUF SCHWINGUNGEN FLACHER SCHALEN NACH DER METHODE DER FINI	73-0563
FLACHES	DESH, EIN EINFACHES FLACHES SCHALENELEMENT UEBER DREIECKIGEM GRUNDRISS. (DESH, A SIMPLE TRIANGULAR S	73-0304
FLACHSTAEBE	OSCILLATION STABILITY OF INTERNALLY AND EXTERNALLY NOTCHED PLANE BARS OF EQUAL FORM FACTOR FROM 3943	73-0745
FLACHSTAEBEN	FORM FACTORS OF FLAT BARS FOR FATIGUE STRENGTH TESTS. (FORMZAHLEN VON FLACHSTAEBEN FUER SCHWINGFESTI	71-0641
FLACHWASSERWELL	DIE BERECHNUNG VON FLACHWASSERWELLEN NACH DER METHODE DER FINITEN ELEMENTE	72-0359
FLAECHENTRAGWER	BERECHNUNG PRISMATISCHER FLAECHENTRAGWERKE MIT HILFE VON FINITEN STREIFENELEMENTEN. (CALCULATION OF	72-0133
FLAECHENTRAGWER	FINITE ELEMENTE FUR GEKRUEMMTE FLAECHENTRAGWERKE. (FINITE ELEMENTS SOLUTIONS FOR CURVED SHELL STRUCT	71-0391
FLAECHENTRAGWER	DAS FLAECHENTRAGWERKSPROGRAMM VON STRIP. (STRIP PROGRAM FOR PLATE STRUCTURES)	72-0150

FLANGE	INSTABILITY OF HORIZONTALLY CURVED MEMBERS. FLANGE BUCKLING STUDIES	69-0480
FLANGED	ANALYSIS OF MODERATELY THICK FLANGED REACTOR PRESSURE VESSELS	73-0821
FLANGED	ANALYSIS OF MODERATELY THICK FLANGED REACTOR VESSELS	75-0320
FLANGED	FINITE ELEMENT ANALYSIS OF AN EXPANDED I - SECTION BEAM AND AN AXISYMMETRIC FLANGED CYLINDER	68-0002
FLANGED	PREDICTION OF THE CREEP BEHAVIOUR OF A FLANGED JOINT	74-0757
FLANGES	TWO-DIMENSIONAL PHOTOELASTIC STUDY OF STEAM-TURBINE CASING FLANGES WITH INTERFACE RELIEF	74-0603
FLAPPED	CASE BOND STRESS CALCULATIONS FOR FLAPPED CYLINDRICAL ANALOGS OF SOLID PROPELLANT ROCKET MOTORS. BON	72-0842
FLAT	A FINITE ELEMENT PROGRAM FOR DESIGN AND ANALYSIS OF THREE-DIMENSIONAL FLAT SPRINGS	74-0738
FLAT	A FLAT TRIANGULAR SHELL ELEMENT STIFFNESS MATRIX	66-0052
FLAT	A STUDY OF STIFFNESS MATRICES FOR THE ANALYSIS OF FLAT PLATES	68-0105
FLAT	A TRIANGULAR FLAT PLATE BENDING ELEMENT	68-0252
FLAT	ACCURATE FINITE ELEMENT MODELLING OF FLAT AND CURVED STIFFENED PANELS	72-0538
FLAT	ANALYSIS AND DESIGN OF AXISYMMETRIC REINFORCEMENT AROUND A CIRCULAR HOLE IN A THIN FLAT PLATE BY THE	74-1113
FLAT	ANALYSIS OF ASYMMETRIC STIFFENED SHELL TYPE STRUCTURES BY THE FINITE ELEMENT METHOD: I FLAT RECTANGU	68-0283
FLAT	CALCULATION OF THE ELASTIC BUCKLING LOADS OF THIN FLAT REINFORCED PLATES USING TRIANGULAR FINITE ELE	73-1024
FLAT.	CALCULATION OF THE ELASTIC BUCKLING LOADS OF THIN FLAT REINFORCED PLATES USING TRIANGULAR FINITE ELE	75-0177
FLAT	CONTACT OF AN INFLATED TOROIDAL MEMBRANE WITH A FLAT SURFACE AS AN APPROACH TO THE TIRE DEFLECTION P	75-0164
FLAT	DATA GENERATOR FOR THE IDEALIZATION FOR FINITE ELEMENT STRUCTURAL ANALYSIS OF NAVAL SHIP FLAT PLATED	72-0860
FLAT	DETERMINATION OF THE INITIAL BUCKLING STRESS OF FLAT PLATES USING THE DISPLACEMENT METHOD	69-0465
FLAT	ELASTIC-PLASTIC ANALYSIS OF FLAT PLATES BY THE FINITE ELEMENT METHOD	68-0111
FLAT	ELASTOPLASTIC ANALYSIS OF PLANE STRAIN AND AXISYMMETRIC FLAT PUNCH INDENTATION BY THE FINITE ELEMENT	70-0343
FLAT	FLAT FINITE ELEMENT FOR THIN SHELL ANALYSIS DERIVED BY ASSUMED STRESS APPROACH	73-0147
FLAT	FLAT QUADRILATERAL EQUILIBRIUM MEMBRANE ELEMENTS	73-1116
FLAT	FLUTTER OF FLAT FINITE-ELEMENT PANELS IN A SUPERSONIC POTENTIAL FLOW	75-0814
FLAT	FORM FACTORS OF FLAT BARS FOR FATIGUE STRENGTH TESTS. (FORMZAHLEN VON FLACHSTAEBEN FUER SCHWINGFEST-I	71-0641
FLAT	NATURAL VIBRATIONS OF THIN, FLAT WALLED STRUCTURES WITH DIFFERENT BOUNDARY CONDITIONS	71-0509
FLAT	ORTHOTROPIC FLAT PLATES SUPPORTED ON COLUMNS ELONGATED IN PLANE	72-0492
FLAT	PATTERN LOADING ON REINFORCED CONCRETE FLAT PLATES	71-0290
FLAT	PROCESS FOR THE AUTOMATIC GENERATION OF FLAT TRIANGULAR LATTICES FOR APPLICATION TO THE FINITE ELEME	71-0366
FLAT	SHORT TERM DEFLECTIONS OF CONCRETE FLAT PLATES	73-0277
FLAT	SINGLE ELEMENT PANEL FOR FLAT PLATE STRUCTURES	75-0279
FLAT	SUBSTRUCTURE ANALYSIS OF MULTISTOREY FLAT SLAB BUILDINGS	74-0394
FLAT	TECHNIQUES IN TRIANGULAR MESH GENERATION FLAT AND CURVED SURFACES	69-0394
FLAT	VIBRATION STUDIES OF A FLAT PLATE AND BUILT-UP WING	71-0238
FLAT-WALLED	BUCKLING UNDER COMBINED LOADING OF THIN FLAT-WALLED STRUCTURES BY A COMPLEX FINITE STRIP METHOD	74-0788
FLAT-WALLED	GENERAL SINUSOIDAL STIFFNESS MATRICES FOR BUCKLING AND VIBRATION ANALYSES OF THIN FLAT-WALLED STRUCT	68-0278
FLAW	DEFORMATION MECHANICS OF DEEP SURFACE FLAW CRACKS	72-0785
FLAW	SPECIAL FINITE ELEMENTS FOR FLAW AND DISCONTINUITY STRESSES IN COMPOSITES	73-0846
FLAWED	DER GERISSENE ZUSTAND ZWEISEITIG GELAGERTER PLATTEN UNTER EINZELLASTEN - NICHTLINEARE BERECHNUNG MIT	75-0858
FLAWED	THREE-DIMENSIONAL ELASTIC-PLASTIC STRESS AND STRAIN ANALYSIS FOR FRACTURE MECHANICS. PHASE 1. SIMPLE	70-0188
FLAWS	INFLUENCE OF INTERLAMINATE FLAWS ON TRANSIENT TEMPERATURE DISTRIBUTIONS	73-0223
FLECHIES	APPLICATION DE LA METHODE DES ELEMENTS FINIS AU CALCUL DES PLAQUES FLECHIES (APPLICATION OF THE FINI	73-0019
FLESSIONALI	CALCULATION OF AXISYMMETRICAL OSCILLATION OF FLEXIBLE TURBINE DISCS. (CALCOLO DELLE OSCILLAZIONI ASS	74-1328
FLEXIBILITY	A DERIVATION PROCEDURE FOR THE DYNAMIC FLEXIBILITY MATRIX OF A TRIANGULAR BENDING ELEMENT	70-0103
FLEXIBILITY	ACOUSTIC FINITE ELEMENT APPROACH FOR STUDYING BOUNDARY FLEXIBILITY AND SOUND TRANSMISSION BETWEEN IR	73-0096
FLEXIBILITY	AN ACOUSTIC FINITE ELEMENT APPROACH FOR STUDYING BOUNDARY FLEXIBILITY AND SOUND TRANSMISSION BETWEEN	73-1087
FLEXIBILITY	BOUNDS ON THE SPECTRAL AND MAXIMUM NORMS OF THE FINITE ELEMENT STIFFNESS, FLEXIBILITY AND MASS MATRI	73-0409
FLEXIBILITY	DIRECT FLEXIBILITY FINITE ELEMENT ELASTOPLASTIC ANALYSIS	71-0095
FLEXIBILITY	EFFECT OF BEAM-SHEARWALL JUNCTION DEFORMATIONS ON THE FLEXIBILITY OF THE CONNECTING BEAMS	73-0213
FLEXIBILITY	ELEMENT STATIC FLEXIBILITY AND STIFFNESS MATRICES FOR THE FINITE ELEMENT ANALYSIS OF BEAM, PLATE, AN	67-0070
FLEXIBILITY	EXPERIMENTAL STUDY OF FLEXIBILITY FACTORS AND STRESSES OF WELDING ELBOWS WITH END EFFECTS	73-0422
FLEXIBILITY	FLEXIBILITY AND STIFFNESS MATRICES FOR AN OPEN-TUBE WARPING CONSTRAINT FINITE ELEMENT	72-0076
FLEXIBILITY	FLEXIBILITY MATRIX FOR AN EXACT TORSION-BENDING FINITE ELEMENT	72-0214
FLEXIBILITY	INFLUENCE OF END-EFFECTS ON STRESSES AND FLEXIBILITY OF A PIPING ELBOW WITH IN-PLANE MOMENT	74-0534
FLEXIBILITY	NEW TRIANGULAR PLATE-BENDING FINITE ELEMENT WITH TRANSVERSE SHEAR FLEXIBILITY	74-0627
FLEXIBILITY	NEW TRIANGULAR PLATE-BENDING FINITE ELEMENT WITH TRANSVERSE SHEAR FLEXIBILITY	74-0964
FLEXIBILITY	STATIC FLEXIBILITY MATRICES FOR A RECTANGULAR PLATE ELEMENT IN BENDING, TWISTING, AND SHEAR. UNIFORM	68-0217
FLEXIBILITY	THE FLEXIBILITY OF A TUBULAR WELDED JOINT IN A VEHICLE FRAME	74-0209
FLEXIBLE	ANALYSIS AND DESIGN OF MULTIPLE INPUT FLEXIBLE LINK MECHANISMS	70-0354
FLEXIBLE	APPLICATION OF SMALL AND LARGE-DISPLACEMENT THEORY TO A LARGE FLEXIBLE SOLAR ARRAY AND COMPARISON WI	71-0231
FLEXIBLE	CALCULATION OF AXISYMMETRICAL OSCILLATION OF FLEXIBLE TURBINE DISCS. (CALCOLO DELLE OSCILLAZIONI ASS	74-1328
FLEXIBLE	CH-53A FLEXIBLE FRAME VIBRATION ANALYSIS/TEST CORRELATION	73-1093
FLEXIBLE	DEFLECTION AND VIBRATION ANALYSIS OF A FLEXIBLE FAN BLADE	73-0556
FLEXIBLE	DESIGN METHOD FOR FLEXIBLE AIRFIELD PAVEMENTS	74-1174
FLEXIBLE	DYNAMIC ANALYSIS OF A FLEXIBLE VEHICLE MOVING ALONG A FLEXIBLE SUPPORT	71-0597
FLEXIBLE	DYNAMIC ANALYSIS OF A FLEXIBLE VEHICLE MOVING ALONG A FLEXIBLE SUPPORT	71-0597
FLEXIBLE	DYNAMICS OF A FLEXIBLE BULKHEAD AND CONTAINED FLUID	73-0082
FLEXIBLE	EFFECTS OF FLEXIBLE CONNECTIONS ON BODY STRUCTURAL RESPONSE	74-0140
FLEXIBLE	FINITE ELEMENT APPROACH TO THE VIBRATION ANALYSIS OF ELASTIC DISKS ON A FLEXIBLE SHAFT	75-0368
FLEXIBLE	FINITE ELEMENTS, FINITE ROTATIONS AND SMALL STRAINS OF FLEXIBLE SHELLS	69-0516
FLEXIBLE	FLEXIBLE PLATE FINITE ELEMENT ON ELASTIC FOUNDATION	70-0128
FLEXIBLE	INVESTIGATION OF SOIL-STRUCTURE INTERACTION: FLEXIBLE CULVERT PIPES	72-0661
FLEXIBLE	NONLINEAR THEORY FOR PREDICTING THE PERFORMANCE OF FLEXIBLE HIGHWAY PAVEMENTS	70-0143
FLEXIBLE	ON SMALL VIBRATIONS AND PERTURBATIONS OF FLEXIBLE BODIES UNDERGOING ARBITRARY NOMINAL MOTION	70-0376
FLEXIBLE	OPTIMAL SYNTHESIS OF FLEXIBLE LINK MECHANISMS WITH LARGE STATIC DEFLECTIONS	74-0525
FLEXIBLE	STRESSES AROUND FLEXIBLE ELLIPTIC PIPES	73-0215
FLEXIBLE	STRUCTURAL BEHAVIOR OF A FLEXIBLE METAL CULVERT UNDER A DEEP EARTH EMBANKMENT USING METHOD A BACKFIL	73-0739
FLEXIBLE	THE ANALYSIS OF FLEXIBLE BULKHEADS IN SANDS	74-0369
FLEXIBLE	THE MOTION OF INTERCONNECTED FLEXIBLE BODIES	75-0727
FLEXIBLE	VERIFICATION OF FRACTURE MECHANICS CONCEPTS TO PREDICT CRACKING OF FLEXIBLE PAVEMENTS	73-0772
FLEXIBLE	WOOD SHEAR PANELS BONDED WITH FLEXIBLE ADHESIVES	75-0086
FLEXING	VIBRATIONAL ANALYSIS OF THE FLEXING HEAD OF THE BOS-6 TRANSDUCER	71-0708
FLEXION	APPLICATIONS DE LA METHODE DES ELEMENTS FINIS A LA FLEXION DES PLAQUES	69-0511
FLEXION	BORNES SUPERIEURES ET INFERIEURES DANS L'ANALYSE MATRICIELLE DES PLAQUES EN FLEXION TORSION	64-0015
FLEXTENSIONAL	DEVELOPMENT OF A FINITE ELEMENT MODEL FOR THE CLASS V FLEXTENSIONAL UNDERWATER TRANSDUCER SHELL	71-0754
FLEXTENSIONAL	DEVELOPMENT OF A FINITE ELEMENT MODEL FOR THE CLASS 5 FLEXTENSIONAL UNDERWATER TRANSDUCER SHELL	71-0840
FLEXTENSIONAL	DEVELOPMENT OF A FINITE ELEMENT MODEL FOR THE CLASS IV FLEXTENSIONAL UNDERWATER TRANSDUCER SHELL	72-0593
FLEXUAL	FLEXUAL RIGIDITY OF A THIN WALLED BUILD-UP ROTOR FOR JET ENGINE. MEASUREMENT BY STATIC LOAD TEST AND	73-0756
FLEXURAL	A FINITE-ELEMENT STUDY ON THE NONLINEAR BEHAVIOR OF REINFORCED CONCRETE FLEXURAL MEMBERS DUE TO SHOR	73-1154
FLEXURAL	FINITE ELEMENT DISPLACEMENT METHOD FOR LARGE AMPLITUDE FREE FLEXURAL VIBRATIONS OF BEAMS AND PLATES	73-0185
FLEXURAL	FLEXURAL VIBRATIONS OF RECTANGULAR AND OTHER POLYGONAL PLATES	71-0417
FLEXURAL	INVESTIGATION OF FLEXURAL STIFFNESS OF RECTANGULAR PLATES BY THE FINITE ELEMENT METHOD	73-0991
FLEXURAL	PUNCHING STRENGTH OF SLABS, A FLEXURAL APPROACH USING FINITE ELEMENTS	74-1307
FLEXURAL	STUDY OF THE FLEXURAL BEHAVIOUR OF CASTINGS	74-0377
FLEXURALLY	ELASTIC-PLASTIC DEFORMATIONS OF FLEXURALLY STIFF FRAMEWORKS FROM SECOND ORDER STRESS THEORY	73-1061
FLEXURE	A STUDY OF FINITE ELEMENTS APPLIED TO PLATE FLEXURE	66-0016
FLEXURE	ANALYSIS OF ELASTIC PLATES IN FLEXURE BY A SIMPLIFIED FINITE ELEMENT METHOD	67-0035
FLEXURE	ANALYSIS OF ELASTIC SHELLS OF REVOLUTION WITH MEMBRANE AND FLEXURE STRESSES UNDER ARBITRARY LOADING	68-0289
FLEXURE	FINITE DISPLACEMENT PLATE FLEXURE BY THE USE OF MATRIX INCREMENTAL APPROACH	72-0384
FLEXURE	FINITE ELEMENT ANALYSIS OF FLEXURE - TYPE VIBRATORS WITH ELECTROSTRICTIVE TRANSDUCERS	70-0338
FLEXURE	FLEXURE ANALYSIS OF AXIALLY-LOADED BEAM BY FINITE-ELEMENT METHOD WITH INTEGRAL PARAMETERS	75-0462
FLEXURE	FLEXURE OF A BEAM SUPPORTED ON AN ELASTIC FOUNDATION	72-0871

224

FLEXURE	INELASTIC BUCKLING ANALYSIS OF PIPES SUBJECTED TO INTERNAL PRESSURE, FLEXURE AND AXIAL LOADING	74-0515
FLEXURE	METHODS AND POSSIBILITIES FOR ELECTRONIC COMPUTATION OF PLATES IN FLEXURE	69-0329
FLEXURE	SECTOR ELEMENT FOR THIN PLATE FLEXURE	72-0174
FLEXURE	THE FINITE ELEMENT METHOD FOR FLEXURE OF SLABS WHEN STRESS DISTRIBUTIONS ARE ASSUMED	66-0069
FLEXURE	THE FINITE ELEMENT METHOD APPLIED TO PLATE FLEXURE	66-0095
FLHE	STRESS ANALYSIS OF BWR PRESSURE VESSEL WITH BERSAFE AND FLHE	73-0406
FLIGHT	CONTRIBUTIONS TO FLUID MECHANICS AND FLIGHT MECHANICS	74-1311
FLIGHT	FUEL ROD FINITE ELEMENT STRESS ANALYSIS: SEGMENTED FLIGHT REACTOR FUEL VERSUS SBDR FUEL	71-0798
FLIGHT	THE CALCULATION OF STATIC AND DYNAMIC CHARACTERISTICS OF FLIGHT VEHICLE STRUCTURES BY THE FINITE ELE	73-1018
FLIGHT	USING THE NASTRAN THERMAL ANALYZER TO SIMULATE A FLIGHT SCIENTIFIC INSTRUMENT PACKAGE	74-1011
FLOATING	STUDY OF MOTION AND STRENGTH OF FLOATING MARINE STRUCTURES IN WAVES	74-0489
FLOATING	THE PREDICTION OF REGULAR WAVE FORCES ON FLOATING BODIES WITH THE FINITE ELEMENT METHOD (IN DUTCH)	71-0743
FLOOD	EFFECTS OF TURBULENCE ON FLOOD FLOW EXPLORED BY FINITE ELEMENT ANALYSIS.	72-0410
FLOOD	FINITE-ELEMENT SIMULATION OF FLOOD HYDROGRAPHS	75-0632
FLOOR	A DISCRETE-ELEMENT METHOD OF MULTIPLE-LOADING ANALYSIS FOR TWO-WAY BRIDGE FLOOR SLABS	70-0435
FLOOR	A DISCRETE-ELEMENT METHOD OF ANALYSIS FOR ORTHOGONAL SLAB AND GRID BRIDGE FLOOR SYSTEMS	72-0690
FLOOR	A RATIONAL ANALYSIS AND DESIGN PROCEDURE FOR WOOD JOIST FLOOR SYSTEMS	74-0730
FLOOR	APPLICATION OF METHOD OF FINITE ELEMENTS TO ANALYSIS OF COMPOSITE FLOOR SYSTEMS	75-0418
FLOOR	DYNAMIC BEHAVIOR OF IN-LINE SHEAR WALLS CONNECTED BY FLOOR SLABS	72-0689
FLOOR	INFLUENCE OF SOIL/STRUCTURE INTERACTION PARAMETERS ON FLOOR RESPONSE SPECTRA	73-0833
FLOOR	THREE DIMENSIONAL ANALYSIS OF BUILDINGS COMPOSED OF FLOOR AND WALL PANELS	71-0387
FLOORS	STATIC AND DYNAMIC ANALYSIS OF WOOD-JOIST FLOORS BY THE FINITE-ELEMENT METHOD	73-1161
FLOSSENWAENDEN	BERECHNUNG VON SPANNUNGEN UND TEMPERATURFELDERN IN FLOSSENWAENDEN UND AUFGESCHWEISSTEN BAUTEILEN. (C	74-0544
FLOW	A FINITE DIFFERENCE METHOD FOR UNSTEADY FLOW IN VARIABLY SATURATED POROUS MEDIA	71-0586
FLOW	A FINITE ELEMENT ANALYSIS OF TRANSIENT RAREFIED GAS FLOW	73-0604
FLOW	A FINITE ELEMENT ANALYSIS FOR STEADY AND OSCILLATORY SUBSONIC FLOW AROUND COMPLEX CONFIGURATIONS	74-1317
FLOW	A FINITE ELEMENT APPROACH TO TWO-PHASE FLOW IN POROUS MEDIA	75-0392
FLOW	A FINITE ELEMENT METHOD FOR LOW REYNOLDS NUMBER FLOW AND ITS APPLICATION TO BIOMECHANICS	69-0304
FLOW	A FINITE ELEMENT METHOD FOR CALCULATING OPEN CHANNEL FLOW	69-0510
FLOW	A FINITE ELEMENT METHOD FOR FLOW CALCULATIONS IN TURBOMACHINES	74-0726
FLOW	A FINITE ELEMENT MODEL FOR TWO-DIMENSIONAL STEADY FLOW THROUGH CONTRACTIONS IN NATURAL CHANNELS	71-0810
FLOW	A FINITE ELEMENT MODEL FOR TWO DIMENSIONAL FLOW	74-0015
FLOW	A FINITE ELEMENT SOLUTION FOR A CLASS OF TWO DIMENSIONAL VISCOUS FLUID FLOW DYNAMICS PROBLEMS	71-0544
FLOW	A FINITE ELEMENT SOLUTION FOR TWO-DIMENSIONAL DENSITY STRATIFIED FLOW	73-0911
FLOW	A FINITE ELEMENT SOLUTION OF THE CONFINED FLOW OVER A CIRCULAR CYLINDER	74-0023
FLOW	A FINITE ELEMENT SOLUTION OF THE TRANSIENT MAGNETOHYDRODYNAMIC COUETTE FLOW	74-1103
FLOW	A FINITE ELEMENT SOLUTION FOR TWO-DIMENSIONAL STRATIFIED FLOW PROBLEMS	75-0391
FLOW	A FINITE-ELEMENT ANALYSIS OF STEADY, TWO-DIMENSIONAL, INCOMPRESSIBLE, LAMINAR FLOW	72-0981
FLOW	A FINITE-ELEMENT METHOD FOR LIFTING SURFACES IN STEADY INCOMPRESSIBLE SUBSONIC FLOW	74-1321
FLOW	A LAGRANGIAN FINITE ELEMENT SOLUTION TO UNSTEADY FLOW	74-1264
FLOW	A METHOD OF ANALYZING TRANSIENT FLUID FLOW IN MULTILAYERED AQUIFERS	69-0142
FLOW	A MIXED FINITE ELEMENT MODEL FOR USE IN POTENTIAL FLOW PROBLEMS	73-0078
FLOW	A NOTE ON SOME VARIATIONAL STATEMENTS FOR THE SLOW FLOW OF A NAVIER-POISSON FLUID	74-1054
FLOW	A NUMERICAL INVESTIGATION OF STABLY STRATIFIED, WIND DRIVEN CAVITY FLOW BY THE FINITE ELEMENT METHOD	74-1372
FLOW	AN ANALYTIC METHOD FOR PREDICTING THE FLOW AND FRACTURE IN NOTCHED FIBER COMPOSITE MATERIALS	75-0689
FLOW	AN APPLICATION OF LEAST SQUARES FINITE ELEMENT METHODS TO TWO DIMENSIONAL STEADY INVISCID COMPRESSIB	75-0114
FLOW	AN APPLICATION OF THE FINITE ELEMENT TECHNIQUE TO BOUNDARY VALUE PROBLEMS OF POTENTIAL FLOW	70-0002
FLOW	AN ITERATIVE FINITE ELEMENT METHOD OF VISCOUS FLOW	74-1353
FLOW	ANALYSIS OF FLOW OF VISCOUS FLUIDS BY FINITE-ELEMENT METHOD	72-0032
FLOW	ANALYSIS OF INTERFACE PROBLEM IN GROUNDWATER FLOW BY FINITE ELEMENT METHOD	74-0467
FLOW	ANALYSIS OF NONSTEADY FLOW WITH A FREE SURFACE USING FINITE ELEMENT METHOD	71-0482
FLOW	ANALYSIS OF RADIAL FLOW IMPELLERS BY THE MATRIX FINITE ELEMENT METHOD	71-0412
FLOW	ANALYSIS OF RAREFIED GAS FLOW THROUGH AN ARBITRARY CROSS SECTION BY THE FINITE ELEMENT METHOD	73-0365
FLOW	ANALYSIS OF STEADY INCOMPRESSIBLE VISCOUS FLOW	74-0013
FLOW	ANALYSIS OF THREE-DIMENSIONAL UNSTEADY VISCOUS FLOW AROUND OSCILLATING WINGS	74-0982
FLOW	ANALYSIS OF UNSTEADY FLOW TOWARD AN ARTESIAN WELL BY 3-DIMENSIONAL FINITE ELEMENTS	74-0351
FLOW	ANALYSIS OF UNSTEADY FLOW TOWARD ARTESIAN WELLS BY THREE-DIMENSIONAL FINITE ELEMENTS	74-0742
FLOW	ANALYSIS OF UNSTEADY FLOW TOWARD AN ARTESIAN WELL BY THREE-DIMENSIONAL FINITE ELEMENTS	74-1351
FLOW	APPLICATION OF FINITE ELEMENT METHOD WITH SECTIONAL LINEARIZATION TO FLOW PROBLEMS	75-0611
FLOW	APPLICATION OF THE FINITE ELEMENT METHOD TO TRANSIENT FLOW IN POROUS MEDIA	68-0094
FLOW	APPLICATION OF THE FINITE ELEMENT METHOD TO CREEPING FLOW PROBLEMS	70-0024
FLOW	APPLICATION OF THE FINITE ELEMENT METHOD TO UNSTEADY FLOW PROBLEMS	70-0310
FLOW	APPLICATION OF THE FINITE ELEMENT TECHNIQUE TO COMPRESSIBLE FLOW PROBLEMS	70-0311
FLOW	APPLICATION OF THE FINITE ELEMENT TECHNIQUE TO COMPRESSIBLE FLOW PROBLEMS	73-0003
FLOW	APPLICATION OF THE FINITE ELEMENT METHOD TO THE VIBRATION ANALYSIS OF AXIAL FLOW TURBINES	73-1125
FLOW	APPLICATION OF THE FINITE ELEMENT TECHNIQUE TO COMPRESSIBLE FLOW PROBLEMS	74-0090
FLOW	APPLICATION OF THE PSEUDO-FUNCTIONAL FINITE ELEMENT METHOD TO VISCOUS FLOW PROBLEMS	74-1183
FLOW	APPLICATION OF THE PSEUDO-FUNCTIONAL FINITE ELEMENT METHOD TO COMPRESSIBLE FLOW PROBLEMS	74-1360
FLOW	APPLICATION, TESTING, AND EVALUATION OF DOUBLY CURVED SHELL ELEMENTS FOR DYNAMIC ANALYSIS, PART III.	69-0256
FLOW	APPLICATIONS OF ELECTRO-OSMOSIS TO GROUND-WATER FLOW PROBLEMS	75-0336
FLOW	APPLICATIONS OF FINITE ELEMENT METHODS TO UNSTEADY VISCOUS FLOW AROUND A BOX-GIRDER BRIDGE OSCILLATI	74-0827
FLOW	APPLICATIONS OF FINITE ELEMENT METHODS TO UNSTEADY VISCOUS FLOW AROUND A BOX-GIRDER BRIDGE OSCILLATI	74-0827
FLOW	BERECHNUNG DES REIBUNGSLOSEN STROEMUNGSFELDES VON DREIDIMENSIONALEN AUFTRIEBSBEHAFTETENTRAGFLUEGELN	73-0533
FLOW	CALCULATION OF POTENTIAL FLOW ABOUT ARBITRARY THREE-DIMENSIONAL LIFTING BODIES	72-0722
FLOW	CALCULATION OF TWO AND THREE DIMENSIONAL FLOW PROBLEMS ACCORDING TO SUB-STRUCTURE THEORY WITH ISOPAR	74-0215
FLOW	COMPRESSIBLE FLOW BY THE FINITE ELEMENT METHOD	69-0417
FLOW	COMPRESSIBLE SEAL FLOW ANALYSIS USING THE FINITE ELEMENT METHOD WITH GALERKIN SOLUTION TECHNIQUE	74-0688
FLOW	COMPUTATIONAL CONSIDERATIONS IN APPLICATION OF THE FINITE ELEMENT METHOD FOR ANALYSIS OF UNSTEADY FL	73-0068
FLOW	COMPUTATIONAL METHODS FOR INVISCID AND VISCOUS TWO-AND-THREE DIMENSIONAL FLOW FIELDS	75-0716
FLOW	COMPUTATIONALLY EFFICIENT FINITE ELEMENT ANALYSIS OF VISCOUS FLOW PROBLEMS	74-1050
FLOW	COMPUTATIONALLY EFFICIENT FINITE ELEMENT ANALYSIS OF VISCOUS FLOW PROBLEMS	74-1258
FLOW	CONVERGENCE OF FINITE ELEMENT SOLUTIONS IN VISCOUS FLOW PROBLEMS	75-0337
FLOW	COUPLAGE ENTRE L ECOULEMENT AUTOUR D UN ARRIERE-CORPS ET LE JET PROPULSIF EN THEORIE DE FLUIDE PARFA	74-1137
FLOW	COUPLAGE ENTRE L ECOULEMENT AUTOUR D UN ARRIERE-CORPS ET LE JET PROPULSIF EN THEORIE DE FLUIDE PARFA	74-1137
FLOW	CREEPING FLOW IN RECTANGULAR DUCTS BY FINITE ELEMENT METHOD	75-0569
FLOW	CURVED CUBIC TRIANGULAR FINITE ELEMENT FOR POTENTIAL FLOW PROBLEMS	73-0423
FLOW	DARCY FLOW SOLUTIONS WITH A FREE SURFACE	67-0081
FLOW	DEVELOPMENT OF A FINITE ELEMENT APPROACH FOR APPROXIMATE ANALYSIS OF UNSTEADY COMPRESSIBLE FLUID FLO	72-0828
FLOW	DIFFERENCE AND ELEMENT METHODS FOR POTENTIAL FLOW	74-0019
FLOW	DIFFUSION ON SOLID SPHERICAL PARTICLE IN VISCOUS FLUID FLOW AT FINITE VALUES OF PE NUMBERS	69-0419
FLOW	EFFECTS OF TURBULENCE ON FLOOD FLOW EXPLORED BY FINITE ELEMENT ANALYSIS.	72-0410
FLOW	FIELD SOLUTIONS FOR TURBULENT SEEPAGE FLOW	71-0206
FLOW	FINITE DIFFERENCE AND FINITE ELEMENT METHODS FOR THE SEMICONDUCTOR FLOW EQUATIONS	74-0028
FLOW	FINITE ELEMENT ANALYSIS OF PROBLEMS FORMULATED BY AN INTEGRAL EQUATION; APPLICATION TO POTENTIAL FLO	68-0191
FLOW	FINITE ELEMENT ANALYSIS METHODS IN DYNAMICS OF PULSATILE FLOW	69-0519
FLOW	FINITE ELEMENT ANALYSIS OF NON-DARCY FLOW	70-0268
FLOW	FINITE ELEMENT ANALYSIS OF GENERAL FLUID FLOW PROBLEMS	71-0536
FLOW	FINITE ELEMENT ANALYSIS OF SLUICE GATE FLOW	71-0540
FLOW	FINITE ELEMENT ANALYSIS OF PERTURBED COMPRESSIBLE FLOW	72-0001
FLOW	FINITE ELEMENT ANALYSIS OF SLOW NON-NEWTONIAN CHANNEL FLOW	72-0308
FLOW	FINITE ELEMENT ANALYSIS OF UNSTEADY INCOMPRESSIBLE FLOW AROUND AN OSCILLATING OBSTACLE OF ARBITRARY	73-0067
FLOW	FINITE ELEMENT ANALYSIS AND COMPUTER GRAPHICS VISUALIZATION OF FLOW AROUND PITCHING AND PLUNGING AIR	73-0385

225

FLOW	FINITE ELEMENT ANALYSIS OF THE FLOW THROUGH A CASCADE OF AEROFOILS	73-0761
FLOW	FINITE ELEMENT ANALYSIS OF STEADY FLUID AND METAL FLOW	74-0054
FLOW	FINITE ELEMENT ANALYSIS OF THE UNSTEADY FLOW OF A VISCOUS PLASTIC FLUID IN A CYLINDRICAL PIPE	74-0055
FLOW	FINITE ELEMENT ANALYSIS OF THE STEADY FLOW OF NON NEWTONIAN FLUIDS THROUGH PARALLEL SIDED CONDUITS	74-0056
FLOW	FINITE ELEMENT ANALYSIS OF THREE-DIMENSIONAL GROUNDWATER FLOW PROBLEMS	74-0254
FLOW	FINITE ELEMENT ANALYSIS OF STEADY FLOW OF NON-NEWTONIAN FLUIDS	74-0465
FLOW	FINITE ELEMENT ANALYSIS OF INTERFACE PROBLEM IN GROUND-WATER FLOW	74-0497
FLOW	FINITE ELEMENT ANALYSIS OF UNCONFINED FLOW WITH A VARIABLE TIME-STEP PROCEDURE	74-0776
FLOW	FINITE ELEMENT ANALYSIS OF SEVERAL COMPRESSIBLE AND INCOMPRESSIBLE VISCOUS FLOW PROBLEMS	74-1119
FLOW	FINITE ELEMENT ANALYSIS OF TRANSONIC FLOW	74-1223
FLOW	FINITE ELEMENT ANALYSIS OF 2-DIMENSIONAL FLOW IN SOILS CONSIDERING WATER UPTAKE BY ROOTS .1. THEORY	75-0137
FLOW	FINITE ELEMENT ANALYSIS OF 2-DIMENSIONAL FLOW IN SOILS CONSIDERING WATER UPTAKE BY ROOTS .2. FIELD A	75-0147
FLOW	FINITE ELEMENT ANALYSIS OF STEADY FLUID AND METAL FLOW	75-0178
FLOW	FINITE ELEMENT ANALYSIS OF VISCOUS FLOW BY STREAM FUNCTION	75-0526
FLOW	FINITE ELEMENT ANALYSIS OF VISCOUS, INCOMPRESSIBLE FLUID FLOW	75-0674
FLOW	FINITE ELEMENT ANALYSIS OF VISCOUS, INCOMPRESSIBLE FLUID FLOW	75-0760
FLOW	FINITE ELEMENT ANALYSIS OF INCOMPRESSIBLE MICROPOLAR VISCOUS FLUID FLOW	75-0866
FLOW	FINITE ELEMENT ANALYSIS OF POTENTIAL FLOW PROBLEMS	75-0869
FLOW	FINITE ELEMENT CALCULATION OF AXISYMMETRIC SUBCRITICAL COMPRESSIBLE FLOW	74-1064
FLOW	FINITE ELEMENT COMPUTER PROGRAMS FOR FLOW IN SATURATED UNSATURATED POROUS MEDIA	72-0364
FLOW	FINITE ELEMENT COMPUTATIONAL THEORY FOR THREE-DIMENSIONAL BOUNDARY LAYER FLOW	72-0632
FLOW	FINITE ELEMENT FORMULATION FOR SOLUTION OF STEADY FLOW OF INCOMPRESSIBLE VISCOUS FLUID	73-0299
FLOW	FINITE ELEMENT GALERKIN METHOD FOR ANALYSIS OF FLOW IN FRACTURED POROUS MEDIA	74-0066
FLOW	FINITE ELEMENT METHOD FOR INCOMPRESSIBLE SLOW VISCOUS FLOW WITH A FREE SURFACE	69-0303
FLOW	FINITE ELEMENT METHOD APPLIED TO UNSTEADY NON-DARCY FLOW	70-0088
FLOW	FINITE ELEMENT METHOD APPLIED TO ANALYSIS OF FLOW OVER A SPILLWAY CREST	73-0300
FLOW	FINITE ELEMENT METHODS IN FLOW PROBLEMS	74-0001
FLOW	FINITE ELEMENT METHODS FOR FLOW IN POROUS MEDIA	74-0057
FLOW	FINITE ELEMENT METHOD FOR TRANSIENT LINEAR VISCOUS FLOW PROBLEMS	74-0474
FLOW	FINITE ELEMENT METHODS IN FLOW PROBLEMS	74-0643
FLOW	FINITE ELEMENT METHODS IN FLOW PROBLEMS	74-0752
FLOW	FINITE ELEMENT METHODS IN FLUID FLOW	75-0729
FLOW	FINITE ELEMENT MODEL FOR TWO-DIMENSIONAL DENSITY STRATIFIED FLOW	73-0064
FLOW	FINITE ELEMENT MODELING OF FLOW THROUGH POROUS MEDIA	75-0706
FLOW	FINITE ELEMENT NUMERICAL MODELLING OF FLOW AND DISPERSION IN ESTUARIES	73-0070
FLOW	FINITE ELEMENT SIMULATION OF THERMALLY INDUCED FLOW FIELDS	74-1235
FLOW	FINITE ELEMENT SOLUTION OF THE FLOW AROUND A TWO-DIMENSIONAL DESCOID	69-0298
FLOW	FINITE ELEMENT SOLUTION OF STEADY STATE POTENTIAL FLOW PROBLEMS	70-0480
FLOW	FINITE ELEMENT SOLUTION OF VISCOUS FLOW IN AN ARBITRARY INTERNAL PASSAGE	71-0461
FLOW	FINITE ELEMENT SOLUTION OF THE THREE DIMENSIONAL FLOW PROBLEMS AND OF REYNOLD'S EQUATION FOR INCOMPR	72-0011
FLOW	FINITE ELEMENT SOLUTION OF UNSTEADY AND UNSATURATED FLOW IN POROUS MEDIA	72-0064
FLOW	FINITE ELEMENT SOLUTION OF COUPLED ELECTROKINETIC AND HYDRODYNAMIC FLOW IN POROUS MEDIA	72-0398
FLOW	FINITE ELEMENT SOLUTION FOR FLOW IN NONCIRCULAR CONDUITS	74-0116
FLOW	FINITE ELEMENT SOLUTION OF THE INCOMPRESSIBLE FLOW OVER AN AIRFOIL IN A NON UNIFORM STREAM	74-0432
FLOW	FINITE ELEMENT SOLUTION OF BOUSSINESQ'S EQUATION FOR UNSTEADY GROUNDWATER FLOW	74-0459
FLOW	FINITE ELEMENT SOLUTIONS OF TRANSIENT, TWO-REGIME FLOW TOWARDS WELLS	74-0466
FLOW	FINITE ELEMENT SOLUTIONS FOR THE EQUATIONS OF GROUND-WATER FLOW	74-1341
FLOW	FINITE ELEMENT SOLUTION OF TWO-REGIME FLOW TOWARDS WELLS	74-1352
FLOW	FINITE ELEMENT SOLUTION OF TWO DIMENSIONAL UNSTEADY AND UNSATURATED FLOW IN POROUS MEDIA	75-0113
FLOW	FINITE ELEMENTS FOR TWO PHASE FLOW IN POROUS MEDIA	74-0060
FLOW	FINITE ELEMENTS IN THE SOLUTION OF THE INVERSE PROBLEM IN GROUNDWATER FLOW	74-0063
FLOW	FINITE-ELEMENT ANALYSIS OF FLOW IN BLOOD VESSEL WITH ARBITRARY CROSS-SECTION	72-0936
FLOW	FINITE-ELEMENT ANALYSIS OF TAPERED VISCOUS FLOW	72-0961
FLOW	FINITE-ELEMENT ANALYSIS OF HEAT FLOW AROUND BURIED PIPES	73-1153
FLOW	FINITE-ELEMENT METHOD FOR AXISYMMETRIC FLOW COMPUTATION IN A TURBOMACHINE	75-0359
FLOW	FINITE-ELEMENT SOLUTION OF NAVIER-STOKES EQUATIONS FOR TRANSIENT 2-DIMENSIONAL INCOMPRESSIBLE FLOW	75-0134
FLOW	FLOW BETWEEN LINES OF A GIVEN PRESSURE	74-0020
FLOW	FLOW IN ROCKS WITH DEFORMABLE FRACTURES	74-0065
FLOW	FLOW OF COMPRESSIBLE FLUID IN POROUS ELASTIC MEDIA	73-1004
FLOW	FLOW OF PLASTIC AND VISCO-PLASTIC SOLIDS WITH SPECIAL REFERENCE TO EXTRUSION AND FORMING PROCESSES	74-0210
FLOW	FLOW THROUGH A CASCADE OF AEROFOILS	74-0077
FLOW	FLUTTER OF FLAT FINITE-ELEMENT PANELS IN A SUPERSONIC POTENTIAL FLOW	75-0814
FLOW	FORMULATION OF A VARIATIONAL PRINCIPLE - FINITE ELEMENT METHOD FOR VISCOUS FLOW	72-0354
FLOW	FREE BOUNDARY POTENTIAL FLOW USING FINITE ELEMENTS	71-0829
FLOW	FULLY DEVELOPED VISCOUS FLOW IN INTERNALLY FINNED TUBES	75-0648
FLOW	FUNCTIONAL COEFFICIENTS IN THE ANALYSIS OF GROUNDWATER FLOW USING FINITE ELEMENTS	73-0077
FLOW	GALERKIN APPROACH TO UNSATURATED FLOW IN SOILS	74-0058
FLOW	GALERKIN APPROACH TO SATURATED-UNSATURATED FLOW IN POROUS MEDIA	75-0393
FLOW	GALERKIN APPROXIMATION OF THE TIME DERIVATIVE IN THE FINITE ELEMENT ANALYSIS OF GROUNDWATER FLOW	74-0658
FLOW	GALERKIN FINITE ELEMENT FORMULATION FOR INCOMPRESSIBLE FLOW	70-0006
FLOW	GENERALIZED VARIATIONAL PRINCIPLES FOR USE IN FLOW PROBLEMS THROUGH POROUS MEDIA	74-0067
FLOW	GRAVITY FLOW BIN DESIGN FOR LUNAR SOIL	72-0240
FLOW	GRAVITY FLOW OF POWDER IN A LUNAR ENVIRONMENT. PT. 2-ANALYSIS OF FLOW INITIATION	71-0484
FLOW	GRAVITY FLOW OF POWDER IN A LUNAR ENVIRONMENT. PT. 2-ANALYSIS OF FLOW INITIATION	71-0484
FLOW	GRAVITY FLOW TO EXCAVATIONS AND DRAINAGE TRENCHES IN LAYERED AQUIFERS	71-0014
FLOW	GROUNDWATER FLOW - THE FINITE ELEMENT METHOD	68-0260
FLOW	GROUNDWATER FLOW IN A SANDY TIDAL BEACH: II. 2-DIMENSIONAL FINITE ELEMENT ANALYSIS	72-0300
FLOW	GROUNDWATER FLOW IN PARTIALLY SATURATED SOILS	72-0745
FLOW	HIGH ORDER FINITE ELEMENT FOR COMPLETELY INCOMPRESSIBLE CREEPING FLOW	73-0144
FLOW	INCOMPRESSIBLE FLOW IN MULTIPLY CONNECTED REGIONS	74-0904
FLOW	INFLUENCE OF CRACK CLOSURE AND ELASTO-PLASTIC FLOW ON THE BENDING OF A CRACKED PLATE	75-0735
FLOW	INTEGRAL EQUATION METHOD FOR SUBSONIC FLOW PAST AIRFOILS IN VENTILATED WIND TUNNELS	75-0065
FLOW	INTEGRAL REPRESENTATION OF FIELD VARIABLES FOR THE FINITE ELEMENT SOLUTION OF VISCOUS FLOW PROBLEMS	74-0468
FLOW	INVESTIGATION OF THE STATIONARY FLOW OF A RIGID PLASTIC MATERIAL BY THE NUMERICAL FINITE ELEMENT MET	73-0231
FLOW	KERNEL FUNCTION FOR NONPLANAR OSCILLATING SURFACES IN SUPERSONIC FLOW	71-0472
FLOW	KINEMATICALLY CONSISTENT UNSTEADY AERODYNAMIC COEFFICIENTS IN SUPERSONIC FLOW	70-0014
FLOW	KINEMATICALLY CONSISTENT UNSTEADY AERODYNAMIC COEFFICIENTS IN SUPERSONIC FLOW	70-0316
FLOW	LINEARIZED COMPRESSIBLE FLOW BY THE FINITE ELEMENT METHOD	69-0025
FLOW	LOCAL POTENTIAL APPROACH TO FINITE ELEMENT METHOD IN UNSTEADY VISCOUS INCOMPRESSIBLE FLUID FLOW	73-0298
FLOW	LOW REYNOLDS NUMBER DEVELOPING FLOW	69-0020
FLOW	MATHEMATICAL FEATURES OF FINITE ELEMENT APPROXIMATIONS OF CERTAIN FLOW PROBLEMS WITH EMPHASIS ON CON	74-0002
FLOW	METHOD OF ANALYZING NON-DARCY FLOW PROBLEMS BY FINITE ELEMENTS	75-0158
FLOW	METODA ELEMENTOW SKONSZONYCH ROZWIAZYWANIA OSIOWO-SYMETRYCZNEGO PRZEWODNICTWA CIEPLA (FINITE-ELEMENT	73-0434
FLOW	MODEL OF ELASTIC PLATE FROM FINITE ELEMENTS IN A SUPERSONIC FLOW	75-0038
FLOW	MODELLING GROUNDWATER FLOW BY THE FINITE ELEMENT METHOD	72-0637
FLOW	MODELLING HEAT FLOW FROM UNDERGROUND POWER CABLES	74-1158
FLOW	NEWTONIAN AND NON-NEWTONIAN VISCOUS FLOW	75-0109
FLOW	NON-DARCY FLOW SOLVED BY FINITE ELEMENT ANALYSIS	69-0366
FLOW	NON-LINEAR FLOW IN POROUS MEDIA BY FINITE ELEMENTS	69-0023
FLOW	NON-UNIFORM FLOW OF POLYCRYSTALS BY GRAIN-BOUNDARY SLIDING ACCOMMODATED BY POWER-LAW CREEP	75-0562
FLOW	NONLINEAR EQUATION OF UNSTEADY GROUND-WATER FLOW	73-0262
FLOW	NUMERICAL ANALYSIS OF ELECTRO-OSMOTIC FLOW IN SOILS	73-0444

FLOW	NUMERICAL CALCULATION OF MERIDIONAL FLOW FIELD IN TURBOMACHINES USING FINITE ELEMENTS METHOD	74-0481
FLOW	NUMERICAL METHODS FOR UNSTEADY NON-DARCY FLOW	74-0062
FLOW	NUMERICAL METHODS IN FLUID FLOW PROBLEMS. VOL. 2.	75-0714
FLOW	NUMERICAL METHODS IN FLUID FLOW PROBLEMS. VOL. 1	75-0715
FLOW	NUMERICAL PREDICTION OF TRACER CONCENTRATION IN AQUIFERS WITH TRANSIENT GROUNDWATER FLOW BY FINITE-E	75-0610
FLOW	NUMERICAL SOLUTION FOR THE FLOW AROUND A CYLINDER AT REYNOLDS NUMBERS OF 40, 200 AND 500	69-0378
FLOW	NUMERICAL SOLUTION OF THE VISCOUS FLOW IN THE ENTRANCE REGION OF PARALLEL PLATES	73-0022
FLOW	NUMERICAL SOLUTION TO THE DYNAMICS OF VISCOUS FLUID FLOW BY A FINITE ELEMENT ALGORITHM: A FIRST STEP	71-0004
FLOW	NUMERICAL SOLUTION TO THE DYNAMICS OF VISCOUS FLUID FLOW BY A FINITE ELEMENT ALGORITHM: A FIRST STEP	71-0791
FLOW	NUMERICAL-SOLUTION OF EULERIAN EQUATIONS OF COMPRESSIBLE FLOW BY A FINITE-ELEMENT METHOD WHICH FOLLO	75-0535
FLOW	ON FLOW NEAR SINGULAR POINTS OF A WALL BOUNDARY	74-0008
FLOW	ON SOME FREE SURFACE ON TRANSIENT FLOW PROBLEMS OF SEEPAGE AND IRROTATIONAL FLOW	72-0070
FLOW	ON SOME FREE SURFACE ON TRANSIENT FLOW PROBLEMS OF SEEPAGE AND IRROTATIONAL FLOW	72-0070
FLOW	ON THE SOLUTION OF TRANSIENT FREE-SURFACE FLOW PROBLEMS IN POROUS MEDIA BY THE FINITE ELEMENT METHOD	73-1019
FLOW	PENALTY FUNCTION APPROACH TO PROBLEMS OF PLASTIC FLOW OF METALS WITH LARGE SURFACE DEFORMATIONS	75-0282
FLOW	PLANE POISEUILLE FLOW OF A RAREFIED GAS BASED ON FINITE ELEMENT METHOD	71-0575
FLOW	PLASTIC FLOW AND FAILURE OF DISCONTINUOUS FIBRE COMPOSITE MATERIALS	70-0157
FLOW	POTENTIAL FLOW ANALYSIS BY FINITE ELEMENTS	72-0004
FLOW	PRAKTISCHE BEISPIELE DER DIGITALEN BERECHNUNG VON GRUNDWASSERSTRÖMUNGS PROBLEMEN NACH DER METHODE EN	73-0081
FLOW	PROCEDURE FOR SOLVING PROBLEMS OF ELASTO-PLASTIC FLOW	73-0217
FLOW	SEEPAGE IN FRACTURED ROCK: FINITE ELEMENT ANALYSIS OF ROCK MASS BEHAVIOR UNDER COUPLED ACTION OF BOD	71-0825
FLOW	SEEPAGE IN MISSISSIPPI RIVER BANKS. REPORT 1, ANALYSIS OF TRANSIENT SEEPAGE USING VISCOUS FLOW MODEL	70-0405
FLOW	SENSITIVITY ANALYSIS OF INPUT PARAMETERS IN NUMERICAL MODELING OF STEADY STATE REGIONAL GROUNDWATER	74-0926
FLOW	SHAKEDOWN THEORY IN PERFECT ELASTOPLASTICITY WITH ASSOCIATED AND NON ASSOCIATED FLOW LAWS - A FINITE	69-0172
FLOW	SLOW PARTICULATE VISCOUS FLOW IN CHANNELS AND TUBES. APPLICATION TO BIOMECHANICS	71-0248
FLOW	SLOW VISCOUS FLOW AND ITS APPLICATION TO BIOMECHANICS	71-0619
FLOW	SOLUTION OF THE COMPRESSIBLE FLOW EQUATIONS	71-0003
FLOW	SOLUTION OF TRANSIENT GROUNDWATER FLOW PROBLEMS BY FINITE ELEMENT METHOD	72-0046
FLOW	SOME APPLICATIONS OF FINITE ELEMENTS FOR FLOW PROBLEMS	73-0698
FLOW	STABILITY ANALYSIS OF THE TORSIONAL BENDING VIBRATIONS OF WING IN A SUBSONIC FLOW BY METHOD OF THE F	75-0755
FLOW	STEADY FLOW ANALYSIS OF INCOMPRESSIBLE VISCOUS FLOW BY THE FINITE ELEMENT METHOD	73-0332
FLOW	STEADY FLOW ANALYSIS OF INCOMPRESSIBLE VISCOUS FLOW BY THE FINITE ELEMENT METHOD	73-0332
FLOW	STEADY FLOW AND TRANSIENT GROUNDWATER FLOW IN SLOPES	70-0273
FLOW	STEADY STATE FLOW IN RIGID NETWORKS OF FRACTURES	74-0278
FLOW	STRESS ANALYSIS PROGRAM BY FINITE ELEMENT METHOD FOR RADIAL FLOW IMPELLERS	74-1016
FLOW	STUDIES ON THE DETERMINATION OF SEEPAGE-LINE OF FILL DAM BY THE METHOD OF FINITE ELEMENT. WATER FLOW	74-1348
FLOW	SUITABILITY OF THE FINITE ELEMENT METHOD FOR ANALYSIS OF UNSTEADY FLOW AROUND OSCILLATING AIRFOILS	73-0069
FLOW	TECHNIQUES FOR HANDLING INPUT DATA FOR FINITE ELEMENT ANALYSIS OF REGIONAL GROUNDWATER FLOW	75-0806
FLOW	THE APPLICATION OF THE FINITE ELEMENT TECHNIQUE TO POTENTIAL FLOW PROBLEMS: PART I	69-0087
FLOW	THE APPLICATION OF THE FINITE ELEMENT TECHNIQUE TO POTENTIAL FLOW PROBLEMS: PART II	69-0395
FLOW	THE APPLICATION OF THE FINITE ELEMENT TECHNIQUE TO POTENTIAL FLOW PROBLEMS	71-0370
FLOW	THE APPLICATION OF THE PSEUDO-FUNCTIONAL FINITE ELEMENT METHOD TO VISCOUS FLOW PROBLEMS	74-0287
FLOW	THE FINITE ELEMENT ANALYSIS OF TRANSIENT RAREFIED GAS FLOW	74-0027
FLOW	THE FINITE ELEMENT METHOD FOR FLUID FLOW	69-0018
FLOW	THE FINITE ELEMENT METHOD IN FLUID FLOW ANALYSIS	71-0249
FLOW	THE FINITE ELEMENT METHOD APPLIED TO COMPRESSIBLE CASCADE FLOW	75-0179
FLOW	THE FINITE ELEMENT SOLUTION OF FLOW AND HEAT TRANSFER IN ELLIPTICAL DUCTS	74-0039
FLOW	THE MBB-UFE SUBSONIC PANEL METHOD. PART 2 - THE LIFT-CONNECTED DISPLACEMENT PROBLEM IN COMPRESSIBLE	72-0865
FLOW	THE MERIDIONAL THROUGH-FLOW CALCULATION IN AN AXIAL FLOW MACHINE BY THE FINITE ELEMENT METHOD	75-0115
FLOW	THE PREDICTION OF FULLY DEVELOPED TURBULENT FLOW IN DUCTS BY THE FINITE ELEMENT METHOD	75-0117
FLOW	THE USE OF FINITE ELEMENT METHODS IN HEAT FLOW ANALYSIS	68-0004
FLOW	THE USE OF THE FINITE ELEMENT METHOD IN THE SOLUTION OF THE FLOW EQUATIONS IN SOILS	74-0449
FLOW	THEORETICAL AND EXPERIMENTAL CONSIDERATIONS ON FLOW OF VISCOUS LIQUIDS PAST PLATE AT SMALL AND AVERA	70-0144
FLOW	THEORY OF FLOW IN A CONFINED TWO AQUIFER SYSTEM	69-0382
FLOW	THEORY OF GROUNDWATER FLOW	70-0117
FLOW	TRANSIENT FLOW OF GROUNDWATER TO WELLS IN MULTIPLE-AQUIFER SYSTEMS	69-0501
FLOW	TRANSIENT FREE-SURFACE FLOW IN POROUS MEDIA BY FINITE ELEMENT METHOD	72-0297
FLOW	TRANSIENT HEAT FLOW IN SOLIDS	74-0461
FLOW	TRANSIENT LAMINAR FLOW IN DUCTS OF ARBITRARY CROSS-SECTION BY FINITE ELEMENT METHODS	72-0013
FLOW	TWO AND THREE DIMENSIONAL POTENTIAL FLOW BY THE METHOD OF SINGULARITIES	69-0449
FLOW	TWO AND THREE DIMENSIONAL POTENTIAL FLOW ANALYSIS BY THE METHOD OF SINGULARITIES	70-0357
FLOW	TWO AND THREE-DIMENSIONAL FLOW USING FINITE ELEMENTS	69-0008
FLOW	TWO-AND THREE-DIMENSIONAL FLOW ANALYSIS USING FINITE ELEMENTS	69-0021
FLOW	TWO-DIMENSIONAL FINITE ELEMENT MODELING OF TRANSIENT FLOW IN REGIONAL AQUIFER SYSTEMS. ANALYSIS OF G	74-0743
FLOW	TWO-DIMENSIONAL FINITE ELEMENT MODELING OF TRANSIENT FLOW IN REGIONAL AQUIFER SYSTEMS. ANALYSIS OF G	74-0743
FLOW	UNSTEADY MHD DUCT FLOW BY THE FINITE ELEMENT METHOD	73-0263
FLOW	UNSTEADY RADIAL FLOW IN GAS RESERVOIRS BY FINITE ELEMENTS	74-0462
FLOW	USE OF FINITE ELEMENT METHOD IN A MARKER AND CELL PROGRAM FOR SOLVING TIME-DEPENDENT FLUID FLOW PROB	75-0043
FLOW	USE OF HYBRID COMPUTATION FOR SOLVING TRANSIENT FLOW PROBLEM IN POROUS MEDIA BY FINITE ELEMENT METHO	74-0068
FLOW	USE OF THE FINITE ELEMENT METHOD IN SOLVING TRANSIENT FLOW PROBLEMS IN AQUIFER SYSTEMS	68-0167
FLOW	VARIATIONAL APPROACH TO NON-DARCY FLOW	70-0089
FLOW	VARIATIONAL PRINCIPLES FOR A RAREFIED GAS FLOW PAST AND ARBITRARY THREE-DIMENSIONAL BODY	70-0189
FLOW	VARIATIONAL PRINCIPLE FOR MAGNETOHYDRODYNAMIC CHANNEL FLOW	70-0192
FLOW	VARIATIONAL PRINCIPLES FOR FLUID FLOW IN POROUS MEDIA	71-0844
FLOW	VARIATIONAL PROPERTIES OF STEADY FALL IN STOKES FLOW	72-0285
FLOW	VARIATIONAL SOLUTION OF TWO DIMENSIONAL FLUID FLOW PROBLEMS	73-1014
FLOW	VIBRATION ANALYSIS OF AXIAL FLOW TURBINE DISKS USING FINITE ELEMENTS	75-0195
FLOW	VISCOPLASTICITY, PLASTICITY AND PLASTIC FLOW	74-1278
FLOW	VISCOUS, INCOMPRESSIBLE FLOW WITH SPECIAL REFERENCE TO NON-NEWTONIAN (PLASTIC) FLUIDS	74-0053
FLOW	WATER FLOW THROUGH SATURATED-UNSATURATED POROUS MEDIA USING FINITE-ELEMENT-GALERKIN METHODS	74-0879
FLOW	WEIGHTED RESIDUAL METHODS AND THEIR RELATION TO FINITE ELEMENT METHODS IN FLOW PROBLEMS	74-0003
FLOWFIELD	DISPERSION IN STEADY UNSTEADY SEEPAGE FLOWFIELD	74-0072
FLOWGRAPHS	APPLICATION OF FLOWGRAPHS TO FREE VIBRATION OF STRUCTURES	68-0196
FLOWN	GROUNDWATER FLOWN IN A SANDY TIDAL BEACH .1. ONE-DIMENSIONAL FINITE ELEMENT ANALYSIS	71-0567
FLOWS	A COMPUTER PROGRAMME FOR ELASTIC PLASTIC CALCULATIONS WITH DUAL PERTURBATION EXPANSION AND VARIATION	75-0882
FLOWS	A FINITE-ELEMENT ANALYSIS FOR STEADY-AND OSCILLATORY SUPERSONIC FLOWS AROUND COMPLEX CONFIGURATIONS	74-1316
FLOWS	A FINITE-ELEMENT FORMULATION FOR SUBSONIC FLOWS AROUND COMPLEX CONFIGURATIONS	73-0742
FLOWS	A VARIATIONAL FINITE ELEMENT METHOD FOR TWO DIMENSIONAL STEADY VISCOUS FLOWS	72-0355
FLOWS	A VARIATIONAL METHOD IN POTENTIAL FLOWS WITH A FREE SURFACE	72-0723
FLOWS	AN INVESTIGATION ON THE PLASTICITY MAGNIFICATION FACTOR IN DEEP SURFACE FLOWS	74-0766
FLOWS	APPLICATION OF INTEGRAL EQUATIONS TO FLUID FLOWS IN UNBOUNDED REGIONS	74-0029
FLOWS	CALCULATION OF FLOWS IN POROUS MEDIA BY THE FINITE ELEMENT METHOD	70-0473
FLOWS	COMPUTATION OF FLOWS IN TURBOMACHINES	74-0076
FLOWS	FINITE ELEMENT ANALYSIS OF FLUID FLOWS	68-0112
FLOWS	FINITE ELEMENT ANALYSIS OF IRROTATIONAL FLOWS OF AN IDEAL FLUID	71-0806
FLOWS	FINITE ELEMENT ANALYSIS OF TWO-DIMENSIONAL SLOW NON-NEWTONIAN FLOWS	72-0035
FLOWS	FINITE ELEMENT SOLUTION THEORY FOR THREE-DIMENSIONAL BOUNDARY FLOWS	74-0579
FLOWS	FINITE ELEMENT WEIGHTED RESIDUAL SOLUTION OF ONE-DIMENSIONAL, UNSTEADY AND UNSATURATED FLOWS IN PORO	73-0712
FLOWS	FINITE-ELEMENT METHOD TO COMPUTE STATIONARY SUBCRITICAL FLOWS	75-0439
FLOWS	FREE-SURFACE IDEAL FLUID FLOWS BY FINITE ELEMENTS	73-0254
FLOWS	FREE-SURFACE IDEAL FLUID FLOWS BY FINITE ELEMENTS	74-0125
FLOWS	GALERKIN METHOD FOR A CLASS OF STEADY, 2-DIMENSIONAL, INCOMPRESSIBLE, LAMINAR BOUNDARY-LAYER FLOWS	75-0437

227

FLOWS	IMPORTANCE OF VARIATIONAL CALCULUS IN FLUID MECHANICS AND ITS APPLICATION TO COMPRESSIBLE POTENTIAL	73-1103
FLOWS	LEAST SQUARES FINITE ELEMENT ANALYSIS OF LAMINAR BOUNDARY LAYER FLOWS	74-0616
FLOWS	MECHANICS OF CABLE MOORING SYSTEMS. VOLUME 1. THREE DIMENSIONAL RESPONSE OF DEEP WATER MOORING LINES	72-0654
FLOWS	NUMERICAL INVESTIGATION OF AXI-SYMMETRIC VISCOPLASTIC FLOWS BY FINITE-ELEMENT PROCEDURE	73-0132
FLOWS	PERTURBATION VARIATIONAL SOLUTION OF COMPRESSIBLE FLOWS	74-0025
FLOWS	PERTURBATION-VARIATIONAL SOLUTION OF COMPRESSIBLE FLOWS	74-1088
FLOWS	SOLUTION OF VISCOUS INCOMPRESSIBLE JET AND FREE-SURFACE FLOWS USING FINITE ELEMENT METHODS	74-0357
FLOWS	STABILITY AND ACCURACY OF NUMERICAL APPROXIMATIONS TO TIME DEPENDENT FLOWS	75-0183
FLOWS	THE PROGRAM " ASSEN " FOR STEADY FREE SURFACE FLOWS	74-0075
FLOWS	THREE DIMENSIONAL ANALYSIS OF COMPRESSIBLE POTENTIAL FLOWS WITH THE FINITE ELEMENT METHOD	73-0058
FLOWS	THREE DIMENSIONAL ANALYSIS OF COMPRESSIBLE POTENTIAL FLOWS	75-0625
FLOWS	TWO AND THREE DIMENSIONAL ANALYSIS OF SUBSONIC COMPRESSIBLE POTENTIAL FLOWS IN DUCTS AND AROUND LIFT	74-0026
FLUAGE	ANALYSE DES CONTRAINTES DAN DES STRUCTURES COMPLEXES EN PRESENCE DE FLUAGE, DE VARIATIONS DE DIMENSI	71-0081
FLUE	TEMPERATURVERLAUF IM QUERSCHNITT GESCHWEISSTER ROHRWAENDE BEI UEBERWIEGEND KONVEKTIVEM WAERMEUEBERGA	75-0628
FLUEGEL-RUMPF-K	BERECHNUNG DES REIBUNGSLOSEN STROEMUNGSFELDES VON DREIDIMENSIONALEN AUFTRIEBSBEHAFTETENTRAGFLUEGELN	73-0533
FLUESSIGKEITEN	THEORETICAL AND EXPERIMENTAL CONSIDERATIONS ON FLOW OF VISCOUS LIQUIDS PAST PLATE AT SMALL AND AVERA	70-0144
FLUESSIGKEITSSC	VERGLEICH NUMERISCHER LOESUNGEN NACH DEM DIFFERENZENVERFAHREN UND NACH DER FINITE-ELEMENT-METHODE AM	71-0016
FLUGZENGASTURZ	BERECHNUNGSMETHODEN ZUR ERMITTLUNG VON BELASTUNGSBEDINGUNGEN UND GRENZTRAGFAEHIGKEITSVERHALTEN VON K	73-0858
FLUID	A FINITE ELEMENT COMPUTATIONAL THEORY FOR THE MECHANICS AND THERMODYNAMICS OF A VISCOUS, COMPRESSIBL	71-0787
FLUID	A FINITE ELEMENT SOLUTION OF GENERAL FLUID DYNAMICS PROBLEMS - NATURAL CONVECTION IN RECTANGULAR CAV	70-0008
FLUID	A FINITE ELEMENT SOLUTION FOR A CLASS OF TWO DIMENSIONAL VISCOUS FLUID FLOW DYNAMICS PROBLEMS	71-0544
FLUID	A METHOD OF ANALYZING TRANSIENT FLUID FLOW IN MULTILAYERED AQUIFERS	69-0142
FLUID	A NOTE ON SOME VARIATIONAL STATEMENTS FOR THE SLOW FLOW OF A NAVIER-POISSON FLUID	74-1054
FLUID	A NUMERICAL SOLUTION TECHNIQUE FOR A CLASS OF TWO-DIMENSIONAL PROBLEMS IN FLUID DYNAMICS FORMULATED	69-0019
FLUID	A TWO-DIMENSIONAL AXISYMMETRIC SHELL FLUID MODEL FOR LAUNCH VEHICLE LONGITUDINAL DYNAMIC ANALYSIS	69-0046
FLUID	ACOUSITC FINITE ELEMENT ANALYSIS OF AXISYMMETRIC FLUID REGIONS	73-0763
FLUID	ADVANCES IN NUMERICAL FLUID DYNAMICS	73-0813
FLUID	ANALYSIS OF VISCOELASTIC FLUID BY FINITE ELEMENT METHOD	75-0865
FLUID	APPLICATION OF FINITE ELEMENT ANALYSIS IN FLUID MECHANICS	71-0026
FLUID	APPLICATION OF FINITE ELEMENT METHODS IN FLUID DYNAMICS	71-0369
FLUID	APPLICATION OF FINITE ELEMENT METHOD TO THE PROBLEM OF HEAT CONDUCTION AND FLUID MECHANICS	72-0486
FLUID	APPLICATION OF INTEGRAL EQUATIONS TO FLUID FLOWS IN UNBOUNDED REGIONS	74-0029
FLUID	APPLICATIONS OF FINITE ELEMENT METHODS IN FLUID DYNAMICS	75-0184
FLUID	COMPARIOSN OF RESULTS OBTAINED FOR THE PROBLEM OF SYMMETRIC FLUID OSCILLATIONS IN TANKS BY THE FINIT	71-0761
FLUID	COMPUTER SYSTEM FOR THE ANALYSIS OF LARGE SCALE PROBLEMS IN SOLID, FLUID AND THERMODYNAMICS	73-0554
FLUID	CONTRIBUTIONS TO FLUID MECHANICS AND FLIGHT MECHANICS	74-1311
FLUID	COUPLED VIBRATIONS OF A STRUCTURE SUBMERGED IN A COMPRESSIBLLE FLUID	69-0001
FLUID	COUPLED VIBRATIONS OF A STRUCTURE SUBMERGED IN A COMPRESSIBLE FLUID	69-0507
FLUID	DEVELOPMENT OF A FINITE ELEMENT APPROACH FOR APPROXIMATE ANALYSIS OF UNSTEADY COMPRESSIBLE FLUID FLO	72-0828
FLUID	DIFFUSION ON SOLID SPHERICAL PARTICLE IN VISCOUS FLUID FLOW AT FINITE VALUES OF RE NUMBERS	69-0419
FLUID	DYNAMIC INSTABILITY OF DUCTS CONVEYING FLUID	75-0691
FLUID	DYNAMICS OF A FLEXIBLE BULKHEAD AND CONTAINED FLUID	73-0082
FLUID	EARTHQUAKE TRIGGERING BY FLUID INJECTION AT RANGELY, COLORADO	72-0634
FLUID	FINITE ELEMENT ANALYSIS OF FLUID FLOWS	68-0112
FLUID	FINITE ELEMENT ANALYSIS OF GENERAL FLUID FLOW PROBLEMS	71-0536
FLUID	FINITE ELEMENT ANALYSIS OF IRROTATIONAL FLOWS OF AN IDEAL FLUID	71-0806
FLUID	FINITE ELEMENT ANALYSIS OF FLUID FILM LUBRICATION - A SURVEY	74-0041
FLUID	FINITE ELEMENT ANALYSIS OF STEADY FLUID AND METAL FLOW	74-0054
FLUID	FINITE ELEMENT ANALYSIS OF THE UNSTEADY FLOW OF A VISCOUS PLASTIC FLUID IN A CYLINDRICAL PIPE	74-0055
FLUID	FINITE ELEMENT ANALYSIS OF FLUID MOTION IN A CONTAINER	74-0079
FLUID	FINITE ELEMENT ANALYSIS OF FLUID FILM LUBRICATION - A SURVEY	74-1100
FLUID	FINITE ELEMENT ANALYSIS OF STEADY FLUID AND METAL FLOW	75-0178
FLUID	FINITE ELEMENT ANALYSIS OF VISCOUS, INCOMPRESSIBLE FLUID FLOW	75-0674
FLUID	FINITE ELEMENT ANALYSIS OF VISCOUS, INCOMPRESSIBLE FLUID FLOW	75-0760
FLUID	FINITE ELEMENT ANALYSIS OF INCOMPRESSIBLE MICROPOLAR VISCOUS FLUID FLOW	75-0866
FLUID	FINITE ELEMENT ANALYSIS OF FLUID FILM LUBRICATION SURVEY	75-0880
FLUID	FINITE ELEMENT APPLICATIONS IN FLUID DYNAMICS	69-0003
FLUID	FINITE ELEMENT FORMULATION FOR SOLUTION OF STEADY FLOW OF INCOMPRESSIBLE VISCOUS FLUID	73-0299
FLUID	FINITE ELEMENT FORMULATION WITH REFERENCE TO FLUID DYNAMICS	74-1245
FLUID	FINITE ELEMENT METHOD IN FLUID DYNAMICS AND HEAT TRANSFER	67-0151
FLUID	FINITE ELEMENT METHODS FOR FLUID PROBLEMS	72-0352
FLUID	FINITE ELEMENT METHOD FOR A SOLID ENCLOSING AN INVISCID INCOMPRESSIBLE FLUID	73-0090
FLUID	FINITE ELEMENT METHODS FOR THE SOLUTION OF SOME INCOMPRESSIBLE NON-NEWTONIAN FLUID MECHANICS PROBLEM	75-0224
FLUID	FINITE ELEMENT METHOD FOR AXISYMMETRIC JETS OF FLUID WITH GRAVITY FORCE	75-0428
FLUID	FINITE ELEMENT METHODS IN FLUID FLOW	75-0729
FLUID	FINITE ELEMENT SOLUTION ALGORITHM FOR VISCOUS INCOMPRESSIBLE FLUID DYNAMICS	73-0063
FLUID	FINITE ELEMENT SOLUTION FOR GENERAL FLUID MOTION	73-0249
FLUID	FINITE ELEMENT SOLUTION ALGORITHM FOR INCOMPRESSIBLE FLUID DYNAMICS	74-0006
FLUID	FINITE ELEMENT THEORY FOR THE MECHANICS AND THERMODYNAMICS OF A VISCOUS, COMPRESSIBLE MULTI-SPECIE F	71-0005
FLUID	FINITE ELEMENTS FOR FLUID DYNAMICS. VARIATIONAL FORMULATIONS FOR NONLINEAR, INITIAL-VALUE PROBLEMS	74-0710
FLUID	FINITE-ELEMENT ANALYSIS OF SLOW INCOMPRESSIBLE VISCOUS FLUID MOTION	74-0154
FLUID	FLOW OF COMPRESSIBLE FLUID IN POROUS ELASTIC MEDIA	73-1004
FLUID	FLUID FINITE ELEMENTS FOR ADDED-MASS CALCULATIONS	72-0009
FLUID	FLUID MECHANICS AND FINITE ELEMENTS	72-0707
FLUID	FREE-SURFACE IDEAL FLUID FLOWS BY FINITE ELEMENTS	73-0254
FLUID	FREE-SURFACE IDEAL FLUID FLOWS BY FINITE ELEMENTS	74-0125
FLUID	IMPORTANCE OF VARIATIONAL CALCULUS IN FLUID MECHANICS AND ITS APPLICATION TO COMPRESSIBLE POTENTIAL	73-1103
FLUID	IMPROVED LINEAR AXISYMMETRIC SHELL FLUID MODEL FOR LAUNCH VEHICLE LONGITUDINAL RESPONSE ANALYSIS	65-0044
FLUID	INFLUENCE OF FLUID INJECTION ON THE STATE OF STRESS IN THE EARTH'S CRUST	72-0624
FLUID	LOCAL POTENTIAL APPROACH TO FINITE ELEMENT METHOD IN UNSTEADY VISCOUS INCOMPRESSIBLE FLUID FLOW	73-0298
FLUID	NUMERICAL METHODS IN FLUID DYNAMICS	72-0844
FLUID	NUMERICAL METHODS IN FLUID DYNAMICS	73-0689
FLUID	NUMERICAL METHODS IN FLUID FLOW PROBLEMS. VOL. 2.	75-0714
FLUID	NUMERICAL METHODS IN FLUID FLOW PROBLEMS. VOL. 1	75-0715
FLUID	NUMERICAL SOLUTION TO THE DYNAMICS OF VISCOUS FLUID FLOW BY A FINITE ELEMENT ALGORITHM: A FIRST STEP	71-0004
FLUID	NUMERICAL SOLUTION TO THE DYNAMICS OF VISCOUS FLUID FLOW BY A FINITE ELEMENT ALGORITHM: A FIRST STEP	71-0791
FLUID	ON THE APPLICATION OF THE FINITE ELEMENT METHOD TO A CLASS OF PROBLEMS IN FLUID DYNAMICS	69-0400
FLUID	ON THE NUMERICAL TREATMENT OF THE NAVIER-STOKES EQUATIONS FOR AN INCOMPRESSIBLE FLUID	73-0809
FLUID	ON THERMO-ELASTO-HYDRODYNAMIC FLUID FILM BEARINGS	75-0532
FLUID	SOLUTION OF FLUID DYNAMIC PROBLEMS BY FINTE ELEMENTS	74-1184
FLUID	SOME APPLICATIONS OF THE FINITE ELEMENT METHOD TO FLUID MECHANICS	74-1274
FLUID	SOME LINEAR AND NON-LINEAR PROBLEMS IN FLUID MECHANICS. FINITE ELEMENT METHOD FORMULATION	73-0523
FLUID	STRUCTURALLY COMPATIBLE FLUID FINITE-ELEMENT FOR SOLID FLUID INTERACTION STUDIES	75-0344
FLUID	STRUCTURALLY COMPATIBLE FLUID FINITE-ELEMENT FOR SOLID FLUID INTERACTION STUDIES	75-0344
FLUID	STUDY OF FLUID MOVEMENTS THROUGH CAUSEWAY	75-0083
FLUID	THE FINITE ELEMENT METHOD FOR FLUID FLOW	69-0018
FLUID	THE FINITE ELEMENT METHOD IN FLUID MECHANICS	71-0197
FLUID	THE FINITE ELEMENT METHOD IN FLUID FLOW ANALYSIS	71-0249
FLUID	THE FINITE ELEMENT METHOD APPLIED TO FLUID MECHANICS	71-0676
FLUID	USE OF FINITE ELEMENT METHOD IN A MARKER AND CELL PROGRAM FOR SOLVING TIME-DEPENDENT FLUID FLOW PROB	75-0043
FLUID	VARIATIONAL PRINCIPLES FOR FLUID FLOW IN POROUS MEDIA	71-0844
FLUID	VARIATIONAL SOLUTION OF TWO DIMENSIONAL FLUID FLOW PROBLEMS	73-1014

228

FLUID	VERGLEICH NUMERISCHER LOESUNGEN NACH DEM DIFFERENZENVERFAHREN UND NACH DER FINITE-ELEMENT-METHODE AM	71-0016
FLUID	VIBRATION OF AN ELASTIC BODY IMMERSED IN A FLUID	72-0312
FLUID-BORNE	VIBROACOUSTIC RESPONSE OF TURBULENCE EXCITED THIN RECTANGULAR FINITE PLATES IN HEAVY AND FLUID MEDIA	74-1309
FLUID-DYNAMICS	FREE VIBRATIONS OF FLUID-BORNE STRUCTURES: INVESTIGATIONS ON A SIMPLE MODEL	74-0976
FLUID-ELASTIC	FINITE ELEMENT METHODS IN FLUID-DYNAMICS	73-0060
FLUID-FILLED	CALCULATION OF AXISYMMETRIC LONGITUDINAL MODES FOR FLUID-ELASTIC TANK-ULLAGE GAS SYSTEMS AND COMPARI	64-0019
FLUID-FILLED	CALCULATION OF THE CHARACTERISTIC FREQUENCIES OF FLUID-FILLED SPHERICAL SHELLS	73-0511
FLUID-SATURATED	DYNAMIC LOADING OF A FLUID-FILLED SPHERICAL SHELL.	72-0444
FLUID-STRUCTURE	VARIATIONAL FORMULATION OF DYNAMICS OF FLUID-SATURATED POROUS ELASTIC SOLIDS	72-0017
FLUID-STRUCTURE	AIRCRAFT FUEL TANK VULNERABILITY TO HYDRAULIC RAM: MODIFICATTION OF THE NORTHROP FINITE ELEMENT COMP	74-0724
FLUIDE	ROLE OF PART-INVERSION IN FLUID-STRUCTURE PROBLEMS WITH MIXED VARIABLES	70-0209
FLUIDES	COUPLAGE ENTRE L ECOULEMENT AUTOUR D UN ARRIERE-CORPS ET LE JET PROPULSIF EN THEORIE DE FLUIDE PARFA	74-1137
FLUIDES	CALCUL NUMERIQUE DES ECOULEMENTS DES FLUIDES DE BINGHAM ET DES FLUIDES NEWTONIENS INCOMPRESSIBLES PA	72-0349
FLUIDS	CALCUL NUMERIQUE DES ECOULEMENTS DES FLUIDES DE BINGHAM ET DES FLUIDES NEWTONIENS INCOMPRESSIBLES PA	72-0349
FLUIDS	A FINITE ELEMENT SOLUTION OF THE STABILITY OF SUPERPOSED FLUIDS	74-1226
FLUIDS	ANALYSIS OF FLOW OF VISCOUS FLUIDS BY FINITE-ELEMENT METHOD	72-0032
FLUIDS	DISCRETE ELEMENT STRUCTURAL THEORY OF FLUIDS	70-0009
FLUIDS	DYNAMIC STRESS ANALYSIS OF POROUS ELASTIC SOLIDS SATURATED WITH COMPRESSIBLE FLUIDS	71-0651
FLUIDS	ELECTRO-OSMOSIS AND THE CONTAMINATION OF UNDERGROUND FLUIDS	73-0609
FLUIDS	FINITE ELEMENT ANALYSIS OF THE STEADY FLOW OF NON NEWTONIAN FLUIDS THROUGH PARALLEL SIDED CONDUITS	74-0056
FLUIDS	FINITE ELEMENT ANALYSIS OF STEADY FLOW OF NON-NEWTONIAN FLUIDS	74-0465
FLUIDS	FINITE ELEMENT APPLICATIONS TO NEWTONIAN FLUIDS	68-0197
FLUIDS	FINITE ELEMENT SOLUTION OF THE THREE DIMENSIONAL FLOW PROBLEMS AND OF REYNOLD'S EQUATION FOR INCOMPR	72-0011
FLUIDS	THE FINITE ELEMENT DYNAMIC STABILITY ANALYSIS OF THIN SHELLS SUBMERGED IN FLUIDS	71-0057
FLUIDS	VISCOUS, INCOMPRESSIBLE FLOW WITH SPECIAL REFERENCE TO NON-NEWTONIAN (PLASTIC) FLUIDS	74-0053
FLUIDS	WAVE PROPAGATION IN FLUIDS	74-0034
FLUTTER	A COMPARISON OF METHODS USED IN FLUTTER RESEARCH	72-0600
FLUTTER	AN APPLICATION OF FINITE ELEMENT METHODS TO PANEL FLUTTER OPTIMIZATION	71-0258
FLUTTER	APPLICATION OF MATRIX DISPLACEMENT METHODS IN THE STUDY OF PANEL FLUTTER	69-0406
FLUTTER	AUTOMATED OPTIMUM DESIGN OF AIRCRAFT WINGS TO SATISFY STRENGTH, STABILITY, FREQUENCY AND FLUTTER REQ	72-0605
FLUTTER	AUTOMATED PROCEDURE FOR DESIGN OF WING STRUCTURES TO SATISFY STRENGTH AND FLUTTER REQUIREMENTS	73-0867
FLUTTER	FINITE ELEMENT ANALYSIS OF SUPERSONIC PANEL FLUTTER	73-0481
FLUTTER	FINITE ELEMENT ANALYSIS OF VIBRATION AND FLUTTER OF CANTILEVER ANISOTROPIC PLATES	74-0484
FLUTTER	FINITE ELEMENT FLUTTER ANALYSIS OF MULTI-WEB WING STRUCTURES	75-0004
FLUTTER	FINITE ELEMENTS APPLIED TO PANEL FLUTTER	67-0066
FLUTTER	FINITE-ELEMENT ANALYSIS OF VIBRATION AND FLUTTER OF CANTILEVER ANISOTROPIC PLATES	74-0933
FLUTTER	FLUTTER OF FLAT FINITE-ELEMENT PANELS IN A SUPERSONIC POTENTIAL FLOW	75-0814
FLUTTER	FLUTTER OF SKEW PANELS BY THE MATRIX DISPLACEMENT APPROACH	70-0201
FLUTTER	ON APPLYING FINITE ELEMENTS TO PANEL FLUTTER	67-0135
FLUTTER	ON THE NUMERICAL SOLUTION OF LARGE EIGENVALUE PROBLEMS ARISING IN PANEL FLUTTER ANALYSIS BY THE FINI	74-0942
FLUTTER	OPTIMIZATION OF A SUPERSONIC PANEL SUBJECT TO A FLUTTER CONSTRAINT. A FINITE ELEMENT SOLUTION	71-0073
FLUTTER	OPTIMIZATION OF STRUCTURES TO SATISFY FLUTTER REQUIREMENTS	69-0311
FLUTTER	PANEL FLUTTER OPTIMIZATION - REFINED FINITE-ELEMENT APPROACH	75-0358
FLUTTER	SOME CONSIDERATIONS ON METHODS FOR FLUTTER ANALYSIS	69-0136
FLUTTER	SOME FLUTTER SOLUTIONS USING FINITE ELEMENTS	70-0011
FLUX	ERROR ANALYSIS OF A REMOTE TRANSIENT HEAT FLUX SENSOR	74-0753
FLUX	TRANSFORMER LEAKAGE FLUX PROGRAM BASED ON FINITE ELEMENT METHOD	73-0004
FLYWHEELS	REACTOR COOLANT PUMP FLYWHEELS OVERSPEED EVALUATION	74-0506
FOG	NUMERICAL ANALYSIS OF COOLING TOWER PLUMES AND FOG INCIDENCE	74-0040
FOIL	STUDY ON FOIL BEARINGS - AN APPLICATION TO TAPE MEMORY DEVICES	75-0311
FOLD	NONLINEAR FINITE ELEMENT FOLD MODEL	73-0014
FOLDED	A FINITE ELEMENT METHOD FOR RANDOM FOLDED PLATE STRUCTURES (EINE FINITE ELEMENT METHODE ZUR BERECHNU	73-0787
FOLDED	A FINITE ELEMENT SOLUTION FOR FOLDED PLATE STRUCTURES	67-0072
FOLDED	A FINITE ELEMENT TECHNIQUE FOR ANALYZING FOLDED PLATE TYPE STRUCTURES WITH SKEWED SUPPORTS	69-0525
FOLDED	A HIGH-ORDER FINITE ELEMENT FOR PRISMATIC AND NONPRISMATIC FOLDED PLATE STRUCTURES	75-0889
FOLDED	A REFINED FINITE ELEMENT ANALYSIS OF FOLDED PLATE STRUCTURES	71-0671
FOLDED	ANALYSIS OF BOX GIRDERS BY FOLDED PLATE AND STRIP METHODS	73-0982
FOLDED	ANALYSIS OF CURVED FOLDED PLATE STRUCTURES	70-0431
FOLDED	ANALYSIS OF CURVED FOLDED PLATE STRUCTURES	71-0139
FOLDED	ANALYSIS OF ORTHOTROPIC FOLDED PLATES	70-0272
FOLDED	ANALYSIS OF ORTHOTROPIC FOLDED PLATES WITH ECCENTRIC STIFFENERS	70-0437
FOLDED	BERECHNUNG PRISMATISCHER FLAECHENTRAGWERKE MIT HILFE VON FINITEN STREIFFNELEMENTEN. (CALCULATION OF	72-0133
FOLDED	COMPUTER PROGRAM FOR PRISMATIC FOLDED PLATES BY PLATE AND BEAM ELEMENTS	70-0371
FOLDED	FINITE ELEMENT ANALYSIS OF PRESTRESSED FOLDED PLATE STRUCTURES	74-1066
FOLDED	FINITE SEGMENT ANALYSIS OF FOLDED PLATES	69-0169
FOLDED	FOLDED PLATE STRUCTURES BY THE FINITE STRIP METHOD	69-0359
FOLDED	GENERAL ANALYSIS OF FOLDED PLATES BY THE FINITE ELEMENT METHOD	69-0200
FOLDED	HIGH-ORDER FINITE ELEMENT TECHNIQUE FOR PRISMATIC AND NONPRISMATIC FOLDED PLATE STRUCTURES	74-1130
FOLDED	MODEL TEST AND ANALYSIS OF OBLIQUE FOLDED PLATE STRUCTURE	73-0230
FOLDED	PREFABRICATED PRETENSIONED LIGHTWEIGHT CONCRETE FOLDED PLATE	73-0808
FOLDED	REFINED FINITE ELEMENTS FOR FOLDED PLATE STRUCTURES	72-0574
FOLDED	THE FUGA FAMILY FOR FOLDED PLATE STRUCTURES	69-0333
FOLDED	THE FUGA FAMILY OF ELEMENTS FOR FOLDED PLATE STRUCTURES A FURTHER SEQUEL TO TECHNICAL NOTE 14	69-0446
FOLDED	TRANSVERSE REINFORCEMENT OF CONTINUOUS, TROUGH-SHAPED CONCRETE RAILROAD STRUCTURES. AN IMPLEMENTATIO	71-0719
FOLDED-PLATE	VYPOCET KRUHOVYCH LOMENICOVYCH KONSTRUKCII METODOU KONECNYCH PASOV. (CALCULATION OF CIRCULAR FOLDED-	74-0612
FOLDS	THE DEVELOPMENT OF APPLIED METHODS IN PROBLEMS OF STATIC CALCULATION OF THIN-WALLED THREE-DIMENSIONA	73-0934
FONCTIONS	FONCTIONS SPLINES ET METHODE D ELEMENTS FINIS (SPLINE FUNCTIONS AND FINITE ELEMENT METHOD)	75-0846
FONCTIONS	LES FONCTIONS DE TENSION DANS LA METHODE DES ELEMENTS FINIS	73-1113
FONDATIONS	APPLICATION DES CALCULS DE FONDATIONS A L'EVALUATION DU COMPORTEMENT DES CHAUSSEES EN DASH 1, 2. (A	74-0429
FOOD	FINITE ELEMENT METHOD IN FOOD ENGINEERING	74-1200
FOOTING	EARTH VIBRATION EFFECTS AND ABATEMENT FOR MILITARY FACILITIES. REPORT 3. ANALYSIS METHOD FOR FOOTING	72-0806
FOOTING	ELASTIC-PLASTIC LARGE DEFORMATION RESPONSE OF CLAY TO FOOTING LOADS	74-0699
FOOTING	SETTLEMENT ANALYSIS OF A STRIP FOOTING USING A CRITICAL STATE SOIL MODEL IN CONJUNCTION WITH FINITE	71-0533
FOOTING	STRIP FOOTING ANALYSIS	72-0755
FOOTINGS	ANALYSIS OF CIRCULAR FOOTINGS ON LAYERED SOILS	70-0281
FOOTINGS	BEHAVIOUR OF STRIP FOOTINGS ON LAYERED COHESIVE SOILS	69-0223
FOOTINGS	DYNAMIC ANALYSIS OF FOOTINGS ON LAYERED MEDIA	75-0586
FOOTINGS	FINITE ELEMENT PARAMETRIC STUDY OF VERTICAL STRAIN INFLUENCE FACTORS AND THE PRESSUREMETER TEST TO E	74-1127
FOOTINGS	MATS AND COMBINED FOOTINGS. ANALYSIS BY THE FINITE ELEMENT METHOD	71-0289
FOOTINGS	STIFFNESS COEFFICIENTS FOR EMBEDDED FOOTINGS	75-0540
FOOTINGS	SURFACE FOOTINGS ON SOFT ROCKS	74-0990
FOOTINGS	ULTIMATE CAPACITY OF CIRCULAR FOOTINGS ON LAYERED SOILS	70-0280
FOOTINGS	VERTICAL VIBRATION OF EMBEDDED FOOTINGS	72-0231
FOOTINGS	VERTICAL VIBRATIONS OF FOOTINGS EMBEDDED IN LAYERED MEDIA	69-0436
FORBIDDEN	FORBIDDEN SHAPES IN THE FINITE ELEMENT METHOD	71-0705
FORCE	ASAS - A COMMERCIAL FINITE ELEMENT SYSTEM WITH DISPLACEMENT AND FORCE CONTINUITY OPTIONS	73-0947
FORCE	BASIS OF A WELL CONDITIONED FORCE PROGRAM FOR EQUILIBRIUM MODELS VIA THE SOUTHWELL SLAB ANALOGIES	67-0111
FORCE	DIGITAL ANALYSIS OF NON-LINEAR STRUCTURES BY THE FORCE METHOD	64-0023
FORCE	DYNAMIC ANALYSIS OF STRUCTURES USING THE RANK FORCE METHOD	71-0553
FORCE	DYNAMIC ANALYSIS OF FRAME STRUCTURES BY THE FORCE METHOD	72-0313
FORCE	EFFECTS OF SHOE FORCE GEOMETRY ON HEAVY DUTY INTERNAL SHOE BRAKE PERFORMANCE	68-0219
FORCE	FINITE ELEMENT EQUILIBRIUM MODEL FOR GENERAL BODY FORCE DISTRIBUTION	75-0268

FORCE	FINITE ELEMENT METHOD FOR AXISYMMETRIC JETS OF FLUID WITH GRAVITY FORCE	75-0428
FORCE	FORCE AND MOMENTS ON ASYMMETRIC AND YAWED BODIES IN A FREE SURFACE	73-0908
FORCE	FORCE AND TORQUE ANALYSIS OF CONSTRAINED SPACE MECHANISMS AND PLANE MECHANISMS WITH OFFSET LINKS BY	74-0569
FORCE	FREE VIBRATIONS OF FINITE ELEMENT PLATES SUBJECTED TO COMPLEX MIDDLE-PLANE FORCE SYSTEMS	72-0280
FORCE	FREE VIBRATIONS OF FINITE ELEMENT PLATES SUBJECTED TO COMPLEX MIDDLE-PLANE FORCE SYSTEMS	73-0097
FORCE	FREE VIBRATIONS OF FINITE ELEMENT PLATES SUBJECTED TO COMPLEX MIDDLE-PLANE FORCE SYSTEMS - REPLY	73-0122
FORCE	GENERALISED MATRIX FORCE AND DISPLACEMENT METHODS FOR LINEAR STRUCTURAL ANALYSIS	65-0064
FORCE	INITIAL STRAINS IN THE MATRIX FORCE METHOD OF STRUCTURAL ANALYSIS	60-0008
FORCE	MAGNET FIELD AND FORCE BY FINITE ELEMENT TECHNIQUES	69-0035
FORCE	MATRIX-DISPLACEMENT DIRECT-ELEMENT METHOD FOR FORCE AND TORQUE ANALYSIS OF DETERMINATE AND INDETERMI	74-0531
FORCE	OPTIMIZATION OF REDUNDANCY SELECTION IN FINITE-ELEMENT FORCE METHOD	70-0228
FORCE	PROPAGATION OF FAILURE IN A CIRCULAR CYLINDER OF ROCK SUBJECTED TO A COMPRESSIVE FORCE	73-0793
FORCE	SOME APPLICATIONS OF MATRIX FORCE METHOD	66-0092
FORCE	SOME NEW DEVELOPMENTS IN MATRIX FORCE ANALYSIS	71-0225
FORCE	STRESS DISTRIBUTIONS IN THIN-WALLED INTERSECTING CYLINDRICAL SHELLS SUBJECTED TO INTERNAL PRESSURE A	71-0023
FORCE	STRUCTURAL ANALYSIS BY THE MATRIX FORCE METHOD WITH APPLICATION TO AIRCRAFT WINGS	56-0004
FORCE	STRUCTURAL OPTIMIZATION USING THE FORCE METHOD	75-0440
FORCE	STUDY OF FORCE DISPLACEMENT RELATIONSHIPS FOR THE FINITE ELEMENT ANALYSIS OF REINFORCED CONCRETE	73-0645
FORCE	TETRAHEDRON ELEMENTS IN THE MATRIX FORCE METHOD OF STRUCTURAL ANALYSIS	64-0011
FORCE	THE MATRIX FORCE METHOD OF STRUCTURAL ANALYSIS AND SOME NEW APPLICATIONS	56-0005
FORCE	TRIANGULAR PLATE ELEMENTS IN THE MATRIX FORCE METHOD OF STRUCTURAL ANALYSIS	63-0010
FORCED	ALGORYTM OBLICZEN ORGAN WYMUSZONYCH METODA SZTYWNYCH ELEMENTOW SKONCZONYCH (ALGORITHM FOR CALCULATIN	74-0581
FORCED	ANALIZA ORGAN WYMUSZONYCH ZLOZONYCH UKLADOW LINIOWYCH METODA SZTYWNYCH ELEMENTOW SKONCZONYCH (ANALYS	72-0246
FORCED	CHISLENNYI METOD RASCHETA SOBSTVENNYKH I VYNUZHDENNYKH KOLEBANII SOSTAVNYKH OBOLOCHECHNYKH KONSTRUKT	73-0461
FORCED	FINITE ELEMENT ANALYSIS OF THE FORCED OSCILLATION OF SHIP HULL FORMS	74-0814
FORCED	FORCED VIBRATIONS OF CIRCULAR FOUNDATIONS ON LAYERED MEDIA	74-0762
FORCED	FREQUENCY-DEPENDENT SHAPE FUNCTIONS FOR FREE AND FORCED STRUCTURAL VIBRATION CALCULATION	73-1091
FORCED	INVESTIGATION ON FREE AND FORCED VIBRATIONS OF A LONG TANKER WITH OVERLAPPING PROPELLER ARRANGEMENT	73-0981
FORCED	USE OF THE FINITE ELEMENT DAMPED FORCED RESPONSE STRAIN ENERGY DISTRIBUTION FOR VIBRATION REDUCTION	72-0759
FORCES	A METHOD FOR SIMULATING CONCENTRATED FORCES AND LOCAL REINFORCEMENTS IN STRESS COMPUTATIONS	72-0081
FORCES	A METHOD OF FINITE DEFLECTION ELASTIC PLASTIC ANALYSIS OF FRAMED STRUCTURES SUBJECTED TO LARGE AXIAL	69-0239
FORCES	ANALYSIS OF PLATES SUBJECTED TO IN-PLANE FORCES USING LARGE FINITE ELEMENTS	70-0271
FORCES	DISTRIBUZIONI OTTIMALI DI ARMATURE DI PRECOMPRESSIONE IN TRAVI-PARETE DI CALCESTRUZZO. (OPTIMAL DIST	75-0560
FORCES	EFFECT OF INITIAL STATE FORCES ON THE HYDROELASTIC VIBRATION OF TUBES	73-0589
FORCES	EFFECTS OF FOUNDATION ROTATION ON SEISMIC INERTIA FORCES OF NUCLEAR POWER PLANT STRUCTURES	73-0177
FORCES	FORCES AND DEFORMATIONS AT THE LATERAL STIFFENER OF THIN WALLED U-BARS	73-0236
FORCES	FORCES ON RIGID CULVERTS UNDER HIGH FILLS	67-0008
FORCES	INELASTIC RESPONSE OF NUCLEAR PIPING SUBJECTED TO RUPTURE FORCES	75-0515
FORCES	NOTE ON AN APPROXIMATE METHOD FOR COMPUTING NONCONSERVATIVE GENERALIZED FORCES ON FINITELY DEFORMED	70-0097
FORCES	NOTE ON RECTANGULAR FINITE ELEMENTS WITH IN-PLANE FORCES	75-0080
FORCES	OSCILLATIONS AND WAVE FORCES IN A MAN-MADE HARBOUR IN THE OPEN SEA	74-0151
FORCES	OSCILLATIONS AND WAVE FORCES IN AN OFFSHORE HARBOR - APPLICATIONS OF HYBRID FINITE ELEMENT METHOD TO	74-1024
FORCES	ROCK FAILURE AROUND A CIRCULAR OPENING IN A GRAVITY FIELD WITH TECTONIC FORCES	70-0071
FORCES	SEEPAGE IN FRACTURED ROCK: FINITE ELEMENT ANALYSIS OF ROCK MASS BEHAVIOR UNDER COUPLED ACTION OF BOD	71-0825
FORCES	SEEPAGE IN FRACTURED ROCK: FINITE ELEMENT ANALYSIS OF ROCK MASS BEHAVIOR UNDER COUPLED ACTION OF BOD	71-0825
FORCES	SIMPLIFIED CALCULATION OF THE INTERACTION FORCES BETWEEN HULL AND TANK SYSTEM FOR A MOSS ROSENBERG T	75-0561
FORCES	SOLUMN INSTABILITY UNDER NONCONSERVATIVE FORCES, WITH INTERNAL AND EXTERNAL DAMPING FINITE ELEMENT U	73-0566
FORCES	STEADY STATE IN SOIL SUBJECTED TO SEVERAL COUPLED FORCES: A FINITE ELEMENT FORMULATION	70-0207
FORCES	THE PREDICTION OF REGULAR WAVE FORCES ON FLOATING BODIES WITH THE FINITE ELEMENT METHOD (IN DUTCH)	71-0743
FORD	FINAL REPORT ON NASTRAN/GRAPHICS, 1971 FORD DOOR PROJECT	73-0595
FORE	TRANSVERSE STRENGTH AT FORE AND AFT SECTION OF TANKER CARGO HOLD	74-0618
FORE-CASTING	VORAUSSAGE DES GEBIRGSDRUCKES UND SEINER AUSWIRKUNGEN AUF STREBEN UND STRECKEN IM STEINKOHLENBERGBAU	73-0679
FORECASTING	FORECASTING THE ROCK PRESSURE AND ITS EFFECT ON THE LONGWALL FACES AND ROADWAYS IN THE BITUMINOUS CO	74-0374
FORECASTING	THE APPLICATION OF NODE-ELEMENT RULES FOR FORECASTING PROBLEMS IN THE GENERATION OF FINITE ELEMENT M	75-0614
FORGING	STRESS ANALYSIS IN SOME COLD FORGING TOOL COMPONENTS AND THEIR DESIGN GUIDES	74-0589
FORK	VIBRATION CHARACTERISTICS OF THE NORTH FORK DAM MODEL	74-0845
FORMAT	FORMAT, A GENERAL PURPOSE MATRIX ANALYSIS SYSTEM IN FORTRAN	68-0266
FORMAT	STRUCTURAL DYNAMIC ANALYSIS OF ELECTRONIC ASSEMBLIES USING NASTRAN RESTART/FORMAT CHANGE CAPABILITY	71-0230
FORMAT-FORTRAN	FORMAT-FORTRAN MATRIX ABSTRACTION TECHNIQUE	68-0214
FORMATION	FATIGUE CRACK FORMATION AND PROPAGATION IN PAVEMENTS CONTAINING SOIL-CEMENT BASES.	72-0435
FORMATION	INCIPIENT CHIP FORMATION IN METAL CUTTING - A THREE- DIMENSION FINITE-ELEMENT ANALYSIS	73-1146
FORMFINDUNG	EIN VERFAHREN DUR DIE FORMFINDUNG VEON BELIEBIGEN VORGESPANNTEN NETZWERKKONSTRUKTIONEN	72-0641
FORMING	A FINITE ELEMENT METHOD FOR THREE DIMENSIONAL ANALYSIS OF METAL FORMING	74-1109
FORMING	APPLICATION OF FINITE-ELEMENT METHOD AND MOIRE TECHNIQUE TO SHEET-METAL FORMING PROBLEMS	71-0784
FORMING	DYNAMIC ANALYSIS OF THE DEFORMATION OF STRUCTURES AND OF METAL FORMING	74-0823
FORMING	DYNAMIC ANALYSIS OF STRUCTURAL DEFORMATION AND METAL FORMING	75-0058
FORMING	EXPERIMENTS IN FORMING AND JOINTING OF TITANIUM AND TITANIUM ALLOYS FOR AIRCRAFT AND SPACE VEHICLES	70-0134
FORMING	FINITE ELEMENT PROCEDURE FOR ANALYSIS OF METAL FORMING PROCESSES	74-0380
FORMING	FLOW OF PLASTIC AND VISCO-PLASTIC SOLIDS WITH SPECIAL REFERENCE TO EXTRUSION AND FORMING PROCESSES	74-0210
FORMING	FORMING OF 7075-T6 ALUMINUM IN HIGH PRESSURE ENVIRONMENTS	71-0322
FORMING	INFLATION FORMING OF STEEL FIBER-REINFORCED CONCRETE DOMES	74-0705
FORMING	USE OF METHOD OF FINITE ELEMENTS FOR STUDY OF STATIONARY METAL FORMING TECHNIQUES	72-0024
FORMULA	A FORMULA FOR CERTAIN TYPES OF STIFFNESS MATRICES OF STRUCTURAL ELEMENTS	63-0003
FORMULA	A GENERAL FORMULA FOR STIFFNESS MATRICES OF STRUCTURAL ELEMENTS	63-0002
FORMULA	AN INVERSE FORM OF RAMBERG - OSGOOD FORMULA	69-0479
FORMULATED	A NUMERICAL SOLUTION TECHNIQUE FOR A CLASS OF TWO-DIMENSIONAL PROBLEMS IN FLUID DYNAMICS FORMULATED	69-0019
FORMULATED	FINITE ELEMENT ANALYSIS OF PROBLEMS FORMULATED BY AN INTEGRAL EQUATION: APPLICATION TO POTENTIAL FLO	68-0191
FORMULATED	FINITE ELEMENTS FORMULATED BY THE WEIGHTED DISCRETE LEAST SQUARES METHOD	74-0200
FORMULATED	NOTE ON THE CONVERGENCE OF FINITE ELEMENT APPROXIMATIONS FOR PROBLEMS FORMULATED IN CURVILINEAR COOR	74-0267
FORMULATING	TECHNIQUE FOR FORMULATING BEAM EQUATIONS	75-0584
FORMULATION	A FINITE ELEMENT FORMULATION FOR MATERIALS WITH SPECIFIED CHANGE OF VOLUME	68-0044
FORMULATION	A FINITE ELEMENT FORMULATION FOR STABILITY ANALYSIS OF DOUBLY CURVED THIN SHELL STRUCTURES	69-0513
FORMULATION	A FINITE ELEMENT FORMULATION FOR PROBLEMS OF LARGE STRAIN AND LARGE DISPLACEMENT	70-0069
FORMULATION	A FINITE ELEMENT FORMULATION FOR THE GENERAL STABILITY ANALYSIS OF THIN WALLED MEMBERS	70-0337
FORMULATION	A FINITE ELEMENT FORMULATION FOR LARGE DEFLECTION ELASTIC STRUCTURAL ANALYSIS APPLIED TO PLATE BENDI	70-0485
FORMULATION	A FINITE ELEMENT FORMULATION OF HEAT CONDUCTION AND HEAT AND MASS TRANSFER PROBLEMS	74-0350
FORMULATION	A FINITE ELEMENT SOLUTION OF TIME DEPENDENT FIELD PROBLEMS USING A REDUCED VARIATIONAL FORMULATION	69-0504
FORMULATION	A FINITE-ELEMENT FORMULATION FOR SUBSONIC FLOWS AROUND COMPLEX CONFIGURATIONS	73-0742
FORMULATION	A FORMULATION OF MATRIX DISPLACEMENT METHOD	66-0103
FORMULATION	A MIXED FINITE ELEMENT SHELL FORMULATION	71-0063
FORMULATION	A MIXED FINITE ELEMENT FORMULATION FOR PLATE PROBLEMS	72-0556
FORMULATION	A MIXED FINITE-ELEMENT SHALLOW SHELL FORMULATION	69-0078
FORMULATION	A NEW APPROACH TO THE FINITE-ELEMENT FORMULATION AND SOLUTION OF A CLASS OF PROBLEMS IN COUPLED THER	70-0732
FORMULATION	A NEW APPROACH TO THE FINITE-ELEMENT FORMULATION AND SOLUTION OF A CLASS OF PROBLEMS IN COUPLED THER	73-0600
FORMULATION	A STABILITY ANALYSIS OF CYLINDRICAL PANELS USING A FINITE ELEMENT FORMULATION	71-0636
FORMULATION	A VARIATIONAL FORMULATION OF DAMPED ACOUSTO-STRUCTURAL PROBLEMS	66-0133
FORMULATION	A VARIATIONAL FORMULATION FOR DISCONTINUOUS FIELDS IN LINEARLY ELASTIC FINITE ELEMENT APPLICATIONS	72-0962
FORMULATION	A VARIATIONAL FORMULATION OF THE COMPRESSIBLE THROUGHFLOW PROBLEM	74-1343
FORMULATION	A VARIATIONAL FORMULATION OF THE COMPRESSIBLE THROUGHFLOW PROBLEM	75-0021
FORMULATION	ANALYSIS OF CABLE NETS IN MIXED FORMULATION	73-0311
FORMULATION	ANALYSIS OF PNEUMATIC SHELLS WITH OR WITHOUT CABLE FORMULATION	74-0591
FORMULATION	APPLICATIONS OF CURVED FINITE ELEMENT TO SANDWICH BEAMS USING SHEAR STRAIN FORMULATION	74-0523

FORMULATION	APPLICATIONS OF CURVED FINITE ELEMENTS TO SANDWICH SHELLS USING SHEAR STRAIN FORMULATION	75-0433
FORMULATION	BASIS FOR FORMULATION OF FINITE ELEMENT MODELS	72-0273
FORMULATION	CONSTITUTIVE MATRIX AND FINITE ELEMENT FORMULATION FOR A DEFORMATION THEORY OF PLASTICITY	71-0700
FORMULATION	DUAL FORMULATION FOR POTENTIAL ENERGY AND COMPLEMENTARY ENERGY APPLICATION TO THE FINITE ELEMENT MET	72-0062
FORMULATION	DYNAMIC PROGRAMMING APPROACH TO THE FORMULATION AND SOLUTION OF FINITE ELEMENT EQUATIONS	75-0059
FORMULATION	DYNAMIC STIFFNESS MATRIX FORMULATION BY MEANS OF HERMITIAN POLYNOMIALS	66-0059
FORMULATION	DYNAMICAL ANALYSIS OF AXIALLY NON-UNIFORM, THIN CYLINDRICAL SHELLS. PART 1 - MATRIX FORMULATION	70-0413
FORMULATION	ELASTIC-PLASTIC FORMULATION FOR A CYLINDRICAL-SHELL FINITE-ELEMENT	75-0841
FORMULATION	ELASTO-PLASTIC ANALYSIS USING AN EFFICIENT FORMULATION OF THE FINITE ELEMENT METHOD	75-0012
FORMULATION	FINITE ELEMENT APPROACH TO THE INTEGRATED POTENTIAL FORMULATION OF GENERAL UNSTEADY SUPERSONIC AEROD	73-0910
FORMULATION	FINITE ELEMENT FORMULATION FOR PIEZOELECTRIC CONTINUA BY A VARIATIONAL THEOREM	68-0230
FORMULATION	FINITE ELEMENT FORMULATION OF FINITE DEFORMATIONS AND IRREVERSIBLE THERMODYNAMICS OF NONLINEAR CONTI	69-0207
FORMULATION	FINITE ELEMENT FORMULATION FOR LINEAR THERMOVISCO ELASTIC MATERIALS	69-0351
FORMULATION	FINITE ELEMENT FORMULATION OF THE INCOMPRESSIBLE LUBRICATION PROBLEM	70-0022
FORMULATION	FINITE ELEMENT FORMULATION OF STRESS STRAIN MATRIX FOR AN ELASTICALLY PERFECTLY PLASTIC TRESCA MATER	70-0275
FORMULATION	FINITE ELEMENT FORMULATION OF GENERAL THERMO-ELASTICITY PROBLEMS	71-0001
FORMULATION	FINITE ELEMENT FORMULATION OF GEOMETRICALLY NONLINEAR PROBLEMS OF ELASTICITY	71-0114
FORMULATION	FINITE ELEMENT FORMULATION OF PROBLEMS OF FINITE DEFORMATION AND IRREVERSIBLE THERMODYNAMICS OF NONL	71-0190
FORMULATION	FINITE ELEMENT FORMULATION BY VARIATIONAL PRINCIPLES WITH RELAXED CONTINUITY REQUIREMENTS	72-0516
FORMULATION	FINITE ELEMENT FORMULATION FOR SOLUTION OF STEADY FLOW OF INCOMPRESSIBLE VISCOUS FLUID	73-0299
FORMULATION	FINITE ELEMENT FORMULATION AND SOLUTION OF CONTACT-IMPACT PROBLEMS IN CONTINUUM MECHANICS	74-0806
FORMULATION	FINITE ELEMENT FORMULATION WITH REFERENCE TO FLUID DYNAMICS	74-1245
FORMULATION	FINITE ELEMENT FORMULATION AND SOLUTION OF CONTACT-IMPACT PROBLEMS IN CONTINUUM MECHANICS-II	75-0740
FORMULATION	FINITE ELEMENT FORMULATION OF THE HEAT CONDUCTION EQUATION IN GENERAL ORTHOGONAL CURVILINEAR COORDIN	75-0848
FORMULATION	FINITE ELEMENT MATRIX FORMULATION OF POST-BUCKLING STABILITY AND IMPERFECTION SENSITIVITY	70-0246
FORMULATION	FINITE ELEMENT METHOD. BASIC PRINCIPLES AND FORMULATION	69-0328
FORMULATION	FINITE ELEMENT STRESS FORMULATION FOR WAVE PROPAGATION	71-0831
FORMULATION	FINITE ELEMENT STRESS FORMULATION FOR WAVE PROPAGATION	73-1043
FORMULATION	FINITE ELEMENT STRESS FORMULATION FOR DYNAMIC ELASTIC-PLASTIC ANALYSIS	75-0484
FORMULATION	FINITE-ELEMENT FORMULATION OF MIXED AND STRESS VARIATIONAL PRINCIPLES FOR STRUCTURAL DYNAMIC ANALYSI	72-0964
FORMULATION	FORMULATION AND APPLICATION OF CERTAIN PRIMAL AND MIXED FINITE ELEMENT MODELS OF FINITE DEFORMATIONS	74-0344
FORMULATION	FORMULATION AND EVALUATION OF A TOROIDAL RING DISCRETE ELEMENT	66-0082
FORMULATION	FORMULATION AND EVALUATION OF A TRIANGULAR CROSS-SECTION RING DISCRETE ELEMENT	66-0083
FORMULATION	FORMULATION AND EVALUATION OF A TRAPEZOIDAL CROSS-SECTION RING DISCRETE ELEMENT	66-0084
FORMULATION	FORMULATION AND EXPERIMENTAL VERIFICATION OF AN AXISYMMETRIC FINITE-ELEMENT STRUCTURAL ANALYSIS	71-0354
FORMULATION	FORMULATION OF A NEW TRIANGULAR PLATE BENDING ELEMENT	68-0051
FORMULATION	FORMULATION OF A VARIATIONAL PRINCIPLE - FINITE ELEMENT METHOD FOR VISCOUS FLOW	72-0354
FORMULATION	FORMULATION OF DISCRETE ELEMENT MODELS FOR STRESS AND VIBRATION ANALYSIS OF PLATES	70-0328
FORMULATION	FORMULATION OF GENERAL DISCRETE MODELS OF THERMOMECHANICAL BEHAVIOR OF MATERIALS WITH MEMORY	69-0381
FORMULATION	FORMULATION OF STRESS STRAIN INDUCED DIFFUSION OF HYDROGEN AND ITS SOLUTION BY COMPUTER-AIDED FINITE	75-0326
FORMULATION	FORMULATION, COMPUTATION, AND SOLUTION PROCEDURES FOR MATERIAL AND/OR GEOMETRIC NONLINEAR STRUCTURAL	72-0346
FORMULATION	GALERKIN FINITE ELEMENT FORMULATION FOR INCOMPRESSIBLE FLOW	70-0006
FORMULATION	GLOBAL-LOCAL FINITE ELEMENT STRUCTURAL ANALYSIS, ANALYTICAL CONSIDERATIONS/ELEMENT FORMULATION	74-1090
FORMULATION	ICE ENGINEERING: VISCOELASTIC FINITE ELEMENT FORMULATION	74-0861
FORMULATION	INCREMENTAL THEORY OF LARGE STRAIN AND LARGE DISPLACEMENT PROBLEMS AND ITS FINITE ELEMENT FORMULATIO	72-0274
FORMULATION	INCREMENTAL THEORY OF THREE DIMENSIONAL TRANSIENT THERMOELASTOPLASTICITY - FORMULATION AND SOLUTION	73-0555
FORMULATION	LAGRANGIAN AND EULERIAN FINITE ELEMENT FORMULATION FOR LARGE STRAIN ELASTO-PLASTICITY	75-0209
FORMULATION	LARGE DISPLACEMENT FORMULATION FOR ELASTIC BODIES	70-0429
FORMULATION	MATRIX DISPLACEMENT ANALYSIS OF PLATES AND SHELLS, PART 11, GENERAL FORMULATION OF THE LINEAR THEORY	71-0621
FORMULATION	MATRIX FORMULATION OF STRUCTURAL STABILITY PROBLEMS	65-0039
FORMULATION	NATURAL FACTOR FORMULATION OF THE STIFFNESS FOR THE MATRIX DISPLACEMENT METHOD	75-0057
FORMULATION	NEW APPROACH TO FINITE ELEMENT FORMULATION AND SOLUTION OF A CLASS OF PROBLEMS IN COUPLED THERMOELAS	73-0013
FORMULATION	NONLINEAR FINITE-ELEMENT FORMULATION FOR TRANSIENT ANALYSIS OF THREE DIMENSIONAL THIN STRUCTURES	74-0874
FORMULATION	NONLINEAR STRUCTURAL ANALYSIS; TENSOR FORMULATION	65-0046
FORMULATION	NUMERICAL FORMULATION OF A CLASS OF PROBLEMS IN NONLINEAR VISCOELASTICITY	67-0061
FORMULATION	NUMERICAL FORMULATION OF NONLINEAR ELASTICITY PROBLEMS	67-0063
FORMULATION	ON THE FINITE ELEMENT FORMULATION OF NAVIER-STOKES EQUATIONS	74-0720
FORMULATION	PROBABILISTIC FORMULATION FOR ANALYTICAL MODELING OC CONCRETE STRUCTURES	71-0709
FORMULATION	PSEUDO 3-DIMENSIONAL FINITE-ELEMENT FORMULATION FOR ELASTOSTATIC PROBLEMS AND ITS GEOPHYSICAL APPLIC	75-0597
FORMULATION	PSEUDO-THREE-DIMENSIONAL FINITE ELEMENT FORMULATION FOR ELASTOSTATIC PROBLEMS	74-0167
FORMULATION	REINFORCED CONCRETE - A FINITE-ELEMENT FORMULATION	72-0929
FORMULATION	SHOCK TUNNEL TESTS OF PRELOADED AND ARCHED WALL PANELS FINITE ELEMENT STRESS FORMULATION FOR WAVE PR	73-0892
FORMULATION	SOLUTION OF A FREE-SURFACE BOUNDARY VALUE PROBLEM USING AN INVERSE FORMULATION AND THE FINITE ELEMEN	75-0466
FORMULATION	SOME LINEAR AND NON-LINEAR PROBLEMS IN FLUID MECHANICS. FINITE ELEMENT METHOD FORMULATION	73-0523
FORMULATION	STABILITY OF PLATES USING A MIXED FINITE ELEMENT FORMULATION	74-0218
FORMULATION	STEADY STATE IN SOIL SUBJECTED TO SEVERAL COUPLED FORCES: A FINITE ELEMENT FORMULATION	70-0207
FORMULATION	THEORETICAL FORMULATION OF FINITE ELEMENT METHODS IN LINEAR ELASTIC ANALYSIS OF GENERAL SHELLS	72-0496
FORMULATION	USE OF THE LEAST SQUARES CRITERION IN THE FINITE ELEMENT FORMULATION	73-0270
FORMULATION	VARIATIONAL FORMULATION OF FINITE DISPLACEMENT ANALYSIS	70-0284
FORMULATION	VARIATIONAL FORMULATION OF DYNAMICS OF FLUID-SATURATED POROUS ELASTIC SOLIDS	72-0017
FORMULATION	VARIATIONAL FORMULATION FOR A CLASS OF FREE BOUNDARY PROBLEMS ARISING IN HYDRODYNAMIC LUBRICATION	75-0531
FORMULATIONS	A COMPARISON AND EVALUATION OF INCREMENTAL FINITE ELEMENT FORMULATIONS	75-0023
FORMULATIONS	A SURVEY OF FINITE ELEMENT FORMULATIONS OF GEOMETRICALLY NONLINEAR PROBLEMS	71-0171
FORMULATIONS	COMPARISON OF MEMBRANE FINITE ELEMENT FORMULATIONS	70-0287
FORMULATIONS	COMPARISON OF PLATE BENDING FINITE ELEMENT FORMULATIONS	70-0288
FORMULATIONS	COMPARISON OF SOME FORMULATIONS IN THERMAL ANALYSIS OF STRUCTURES	75-0888
FORMULATIONS	CONSISTENT MATRIX FORMULATIONS FOR STRUCTURAL ANALYSIS USING INFLUENCE COEFFICIENT TECHNIQUES	64-0017
FORMULATIONS	CONTINUUM BASES FOR CONSISTENT NUMERICAL FORMULATIONS OF FINITE STRAINS IN ELASTIC AND INELASTIC STR	75-0299
FORMULATIONS	EFFICIENT FORMULATIONS OF THE FINITE ELEMENT METHOD IN LINEAR AND NONLINEAR FRACTURE MECHANICS	75-0403
FORMULATIONS	EVALUATION OF TWO FINITE ELEMENT FORMULATIONS FOR ONE-DIMENSIONAL CONSOLIDATION	72-0453
FORMULATIONS	FINITE ELEMENT FORMULATIONS OF GEOMETRICALLY NONLINEAR PROBLEMS	69-0180
FORMULATIONS	FINITE ELEMENT FORMULATIONS OF NONLINEAR BOUNDARY VALUE PROBLEMS	71-0497
FORMULATIONS	FINITE ELEMENT FORMULATIONS FOR LARGE DISPLACEMENT AND LARGE STRAIN ANALYSIS	73-0457
FORMULATIONS	FINITE ELEMENT FORMULATIONS FOR ELASTIC PLATES BY GENERAL VARIATIONAL STATEMENTS WITH DISCONTINUOUS	73-0559
FORMULATIONS	FINITE ELEMENT FORMULATIONS FOR ELASTIC PLATES BY GENERAL VARIATIONAL STATEMENTS WITH DISCONTINUOUS	73-1124
FORMULATIONS	FINITE ELEMENT FORMULATIONS FOR PROBLEMS OF LARGE ELASTIC-PLASTIC DEFORMATION	74-0796
FORMULATIONS	FINITE ELEMENT FORMULATIONS FOR LARGE DEFORMATION DYNAMIC ANALYSIS	75-0564
FORMULATIONS	FINITE ELEMENTS FOR FLUID DYNAMICS. VARIATIONAL FORMULATIONS FOR NONLINEAR, INITIAL-VALUE PROBLEMS	74-0710
FORMULATIONS	FORMULATIONS DUALES DES ENERGIES POTENTIELLES ET COMPLEMENTAIRES. APPLICATION A LA METHODE DES ELEME	71-0752
FORMULATIONS	FORMULATIONS OF FINITE ELEMENT METHODS FOR SOLID CONTINUA	71-0679
FORMULATIONS	INCLUSION OF TRANSVERSE SHEAR DEFORMATION IN FINITE ELEMENT DISPLACEMENT FORMULATIONS	74-0662
FORMULATIONS	INCLUSION OF TRANSVERSE-SHEAR DEFORMATION IN FINITE-ELEMENT DISPLACEMENT FORMULATIONS - COMMENT	75-0406
FORMULATIONS	INCREMENTAL FORMULATIONS FOR PROBLEMS WITH GEOMETRIC AND MATERIAL NONLINEARITIES	72-0315
FORMULATIONS	INTRODUCTION TO SHEARS DEFORMATION INTO A THIN PLATE DISPLACEMENTS FORMULATIONS	73-0617
FORMULATIONS	MIXED FORMULATIONS FOR FINITE ELEMENT SHELL ANALYSIS	71-0222
FORMULATIONS	NUMERICAL CONDITIONING OF STIFFNESS MATRIX FORMULATIONS FOR FRAME STRUCTURES	68-0022
FORMULATIONS	SHAPE FUNCTION FORMULATIONS FOR ELEMENTS OTHER THAN DISPLACEMENT MODELS	72-0472
FORMULATIONS	VARIATIONAL FORMULATIONS AND FINITE ELEMENT METHOD FOR THIN ELASTIC SHELLS AND PLATES	68-0064
FORMULATIONS	VARIATIONAL FORMULATIONS OF NUMERICAL METHODS	70-0392
FORMULATIONS	VARIATIONAL FORMULATIONS OF NUMERICAL METHODS IN SOLID CONTINUA	71-0670
FORMUNGS	EXPERIMENTS IN FORMING AND JOINTING OF TITANIUM AND TITANIUM ALLOYS FOR AIRCRAFT AND SPACE VEHICLES	70-0134
FORMY	CHISLENNOE RESHENIE PLOSKIKH ZADACH TEPLOPROVODNOSTI DLYA OBLASTEI SLOZHNOI FORMY (NUMERICAL SOLUTIO	74-0659

FORMY	RASCHET KILEBANII PLASTIN SLOZHNOI FORMY S NEODNORODNYMI GRANICHNYMI USLOVIYAMI. (CALCULATION OF VIB	74-0988
FORMZAHL	OSCILLATION STABILITY OF INTERNALLY AND EXTERNALLY NOTCHED PLANE BARS OF EQUAL FORM FACTOR FROM 3943	73-0745
FORMZAHLEN	FORM FACTORS OF FLAT BARS FOR FATIGUE STRENGTH TESTS. (FORMZAHLEN VON FLACHSTAEBEN FUER SCHWINGFESTI	71-0641
FORSCHUNG	GROPRECHENANLAGEN IN DER FORSCHUNG	68-0264
FORT	NASTRAN EXPERIENCES OF FORT WORTH OPERATION, CONVAIR AEROSPACE DIVISION OF GENERAL DYNAMICS	71-0019
FORTRAN	A FORTRAN PROGRAM FOR CALCULATING THE FREQUENCIES AND MODE SHAPES OF AN ARBITRARY SHAPE CYLINDRICAL	72-0594
FORTRAN	A FORTRAN PROGRAM FOR CALCULATING THE FREQUENCIES AND MODE SHAPES OF A GUIDED CLAMPED, GUIDED PINNED	72-0890
FORTRAN	FINITE ELEMENT ANALYSIS OF ANISOTROPIC PLATES USING 0-19 ELEMENT: PART 11. INSTRUCTIONS FOR USERS AN	74-0731
FORTRAN	FORMAT, A GENERAL PURPOSE MATRIX ANALYSIS SYSTEM IN FORTRAN	68-0266
FORTRAN	FORTRAN SUBROUTINES FOR THE SOLUTION OF LAPLACE'S EQUATION OVER A GENERAL REGION IN TWO DIMENSIONS	70-0425
FORTRAN	FORTRAN SUBROUTINES FOR FINITE ELEMENT ANALYSIS	72-0694
FORTRAN	SABOR 1, A FORTRAN PROGRAM FOR THE LINEAR ELASTIC AXISYMMETRIC LOADING BY USING THE MATRIX DISPLACEM	65-0027
FORTRAN	STRING EIN PROGRAMMSYSTEM FUR DIE MASCHINENUNABHANGIGE BEARBEITUNG VON ZEICHENKETTEN IN FORTRAN	68-0265
FOUNDATION	A FINITE ELEMENT ANALYSIS OF AN AIRCRAFT LANDING MAT ON AN ELASTIC FOUNDATION	68-0280
FOUNDATION	ANALYSIS OF ELASTICALLY COUPLED SHELLS ON ELASTIC FOUNDATION BY HYBRID METHOD	71-0276
FOUNDATION	ANALYSIS OF SEEPAGE THROUGH DAMS AND IN FOUNDATION BY FINITE ELEMENT METHOD	73-1110
FOUNDATION	APPLICATION DES CALCULUS DE FONDATIONS A L'EVALUATION DU COMPORTEMENT DES CHAUSSEES FM DASH 1, 2. (A	71-0429
FOUNDATION	APPLICATION OF THE FINITE ELEMENT METHOD TO THE BUCKLING OF RECTANGULAR PLATES AND PLATES ON ELASTIC	73-0478
FOUNDATION	CONICAL FINITE ELEMENT STRESS ANALYSIS OF AN AXISYMMETRIC TANK ON AN ELASTIC FOUNDATION	70-0494
FOUNDATION	CONTRIBUTION A L'ETUDE DU COMPLEXE BARRAGE-FONDATION ET EXPLICATION DES SEISMES DUS AU REMPLISSAGE D	72-0554
FOUNDATION	DESIGN OF CONCRETE TRACK SLAB ON ELASTIC FOUNDATION USING THE FINITE ELEMENT METHOD	73-0479
FOUNDATION	DYNAMICS OF EARTH DAMS WITH FOUNDATION INTERACTION	71-0343
FOUNDATION	EFFECTS OF FOUNDATION ROTATION ON SEISMIC INERTIA FORCES OF NUCLEAR POWER PLANT STRUCTURES	73-0177
FOUNDATION	FINITE ELEMENT ANALYSIS OF PLATES ON A TWO PARAMETER FOUNDATION MODEL.	72-0425
FOUNDATION	FINITE ELEMENT ANALYSIS OF A PLATE ON A LAYERED CROSS-ANISOTROPIC FOUNDATION	74-0451
FOUNDATION	FLEXIBLE PLATE FINITE ELEMENT ON ELASTIC FOUNDATION	70-0128
FOUNDATION	FLEXURE OF A BEAM SUPPORTED ON AN ELASTIC FOUNDATION	72-0871
FOUNDATION	FOUNDATION FAILURE OF NEW LISKEARD EMBANKMENT	73-0549
FOUNDATION	FOUNDATION INSTRUMENTATION FOR THE NATIONAL WESTMINISTER BANK TOWER	74-1147
FOUNDATION	FOUNDATION TESTING FOR AUBURN DAM	70-0119
FOUNDATION	INCORPORATION OF FOUNDATION DEFORMATION IN A FINITE ELEMENT SHELL ANALYSIS	68-0227
FOUNDATION	INVESTIGATION OF THE EFFECT OF FOUNDATION YIELD ON THE FREQUENCY AND SHAPE OF INHERENT VIBRATIONS OF	72-0419
FOUNDATION	LOAD DEFORMATION ANALYSIS OF DEEP PILE FOUNDATION	72-0931
FOUNDATION	MATHEMATICAL FOUNDATION OF THE FINITE ELEMENT METHOD	67-0126
FOUNDATION	NON-LINEAR ANALYSIS OF THE ROTATIONAL STIFFNESS OF A NUCLEAR STEAM GENERATOR FOUNDATION	69-0286
FOUNDATION	OBSERVATIONS ON THE BEHAVIOUR OF A PILED-RAFT FOUNDATION ON LONDON CLAY.	73-0421
FOUNDATION	PHOTOELASTIC DETERMINATION OF THE STRESSES IN THE FOUNDATION OF THE NUCLEAR-POWER STATION BUGEY 2	75-0834
FOUNDATION	PREDICTION OF LONG TERM DEFORMATION OF A COMPACTED COHESIVE SOIL EMBANKMENT OVER A SOFT FOUNDATION	74-0277
FOUNDATION	RASCHET METODOM KONECHNYKH ELEMENTOV DINAMICHESKIKH KHARAKTEERISTIK ZEMLYANOI PLOTINY SOVMESTNO S OS	73-0437
FOUNDATION	STRESS ANALYSIS OF CONCRETE TRACK SLABS ON AN ELASTIC FOUNDATION BY THE FINITE ELEMENT METHOD	74-1049
FOUNDATION	STRESS AND VIBRATION ANALYSIS ON THE SYSTEM OF FRAME AND FOUNDATION	69-0218
FOUNDATION	STRESS INTENSITY FACTORS FOR A NOTCHED BEAM ON AN ELASTIC FOUNDATION	75-0566
FOUNDATION	STUDY OF ELASTIC-PLASTIC MATRIX ANALYSIS FOR FOUNDATION AND REINFORCED CONCRETE IN JAPAN	69-0278
FOUNDATION	THE NATURAL FACTOR APPROACH IN THE DISPLACEMENT METHOD - FOUNDATION AND FURTHER DEVELOPMENTS	75-0111
FOUNDATION	USE OF THE FINITE ELEMENT IN FOUNDATION ENGINEERING	72-0422
FOUNDATIONS	A TECHNIQUE FOR THE SETTLEMENT AND STRESS ANALYSIS OF MAT FOUNDATIONS USING THE FINITE ELEMENT METHO	72-0950
FOUNDATIONS	ACCURACY AND CONVERGENCE OF MIXED FINITE-ELEMENT APPROXIMATIONS OF THIN BARS, MEMBRANES, AND PLATES	73-1120
FOUNDATIONS	ALGEBRAIC-GEOMETRY FOUNDATIONS FOR FINITE-ELEMENT COMPUTATION	73-0974
FOUNDATIONS	ANALYSIS OF FOUNDATIONS ON A SEMI-INFINITE ELASTIC MASS UNDER DYNAMIC LOADS	72-0763
FOUNDATIONS	ANALYSIS OF PLATES ON ELASTIC FOUNDATIONS	73-0684
FOUNDATIONS	BUTTRESS DAMS ON COMPLEX ROCK FOUNDATIONS	64-0013
FOUNDATIONS	DYNAMIC BEHAVIOR OF STEEL FOUNDATIONS FOR TURBO-ALTERNATORS	71-0837
FOUNDATIONS	ELASTIC PLASTIC ANALYSIS OF SHALLOW FOUNDATIONS	74-0446
FOUNDATIONS	FINITE ELEMENT ANALYSIS OF ARCH DAMS AND FOUNDATIONS	71-0668
FOUNDATIONS	FINITE ELEMENT PROGRAM FEECON FOR UNDRAINED DEFORMATION ANALYSES OF GRANULAR EMBANKMENTS ON SOFT CLA	72-0875
FOUNDATIONS	FORCED VIBRATIONS OF CIRCULAR FOUNDATIONS ON LAYERED MEDIA	74-0762
FOUNDATIONS	FOUNDATIONS OF FINITE ELEMENT METHODS FOR SOLID CONTINUA	71-0213
FOUNDATIONS	LECTURES ON MATHEMATICAL FOUNDATIONS OF THE FINITE ELEMENT METHOD	72-0777
FOUNDATIONS	MOVEMENTS OF ARTICULATED CONDUITS UNDER EARTH DAMS ON COMPRESSIBLE FOUNDATIONS	73-0477
FOUNDATIONS	NONLINEAR MECHANICS OF CONSTRAINED MATERIAL CONTINUA: FOUNDATIONS OF THE THEORY	74-0640
FOUNDATIONS	NUMERICAL PROCEDURE FOR THE ANALYSIS OF STRUCTURES SUPPORTED BY VISCOELASTIC FOUNDATIONS	73-0564
FOUNDATIONS	PLATES AND BEAMS ON ELASTIC FOUNDATIONS: LINEAR AND NONLINEAR BEHAVIOUR	68-0042
FOUNDATIONS	PLATES AND TANKS ON ELASTIC FOUNDATIONS, AN APPLICATION OF FINITE ELEMENT METHOD	65-0010
FOUNDATIONS	POSSIBILITY OF USING FINITE ELEMENT METHOD IN THE ANALYSIS OF JOINT PERFORMANCE OF THE HYDRAULIC STR	72-0418
FOUNDATIONS	SURVEY LECTURES ON THE MATHEMATICAL FOUNDATIONS OF THE FINITE ELEMENT METHOD	72-0670
FOUNDATIONS	THE FINITE-ELEMENT METHOD AS APPLIED TO SHALLOW FOUNDATIONS	74-1074
FOUNDATIONS	THE MATHEMATICAL FOUNDATIONS OF THE FINITE ELEMENT METHOD - WITH APPLICATIONS TO PARTIAL DIFFERENTIA	73-0286
FOUNDATIONS	THE TRIAX 6 ELEMENT FOR AXISYMMETRIC ANALYSIS BY THE MATRIX DISPLACEMENT METHOD: PART 1, FOUNDATIONS	66-0124
FOUNDATIONS	THEORETICAL FOUNDATIONS ON THE FINITE ELEMENT METHOD	68-0055
FOUNDATIONS	THEORETICAL FOUNDATIONS OF THE FINITE ELEMENT METHOD (DISCUSSION)	70-0074
FOUNDATIONS	THEORETICAL FOUNDATIONS OF THE FINITE ELEMENT METHOD	70-0453
FOUNDATIONS	THREE DIMENSIONAL ANALYSIS OF ARCH DAMS AND THEIR FOUNDATIONS	68-0066
FOUNDATIONS	UNUSUAL FOUNDATIONS OVER THE MERSEY TUNNEL	70-0197
FOUR	DYNAMIC STRAINS IN A FOUR BAR MECHANISM	73-0677
FOUR	NASTRAN: USER EXPERIENCE WITH FOUR EXAMPLE PROBLEMS	72-0804
FOUR	VIBRATION OF FOUR POINT-SUPPORTED PLATES BY A FINITE ELEMENT METHOD	71-0592
FOURIER	A COMPARISON OF FINITE ELEMENT AND FOURIER SERIES SOLUTIONS AS APPLIED TO RADIALLY LOADED CIRCULAR R	68-0024
FOURIER	APPLICATION OF THE FAST FOURIER TRANSFORM TO LINEAR SYSTEMS IN CIVIL ENGINEERING	73-1119
FOURIER	FOURIER ANALYSIS OF THE FINITE ELEMENT METHOD IN RITZ-GALERKIN THEORY	69-0297
FOURIER	FOURIER ANALYSIS OF TRANSIENT PROBLEMS BY THE FINITE ELEMENT METHOD	74-1331
FOURIER	STRUCTURAL DYNAMICS BY FINITE ELEMENTS: MODAL AND FOURIER ANALYSIS	73-0836
FOURIER	USE OF FOURIER SERIES IN THE FINITE ELEMENT METHOD	68-0103
FOURIERS	SOLUTION OF MIXED BOUNDARY VALUE PROBLEMS IN FINITE DOMAINS BY FOURIERS METHOD	72-0790
FOURTH	ANALYSIS OF OUT-OF-PLANE STRESS BY THE SIX NODES, FOURTH ORDER ACCURACY TRIANGULAR ELEMENTS	69-0375
FOURTH	CONVERGENCE OF A FINITE ELEMENT PROCEDURE FOR SOLVING BOUNDARY VALUE PROBLEMS OF THE FOURTH ORDER	70-0131
FOX	CORRECTED ASSESSMENT OF CYLINDRICAL SHELL FINITE ELEMENT OF BOGNER, FOX AND SCHMIT WHEN APPLIED TO A	73-0024
FRACTURE	A FINITE ELEMENT CONTRIBUTION TO FRACTURE MECHANICS	72-0320
FRACTURE	A FINITE ELEMENT MODEL FOR PREDICTING LONG BONE FRACTURE FIXATION	71-0819
FRACTURE	A FRACTURE MECHANICS APPROACH TO THE PROPAGATION OF FATIGUE CRACKS IN REACTOR COMPONENTS BY ACOUSTIC	71-0132
FRACTURE	A NOTE ON THE FINITE ELEMENT METHOD IN LINEAR FRACTURE MECHANICS	70-0489
FRACTURE	A STATISTICAL THEORY FOR THE FRACTURE OF BRITTLE STRUCTURES SUBJECTED TO NONUNIFORM POLYAXIAL STRESS	73-0928
FRACTURE	AN ANALYTIC METHOD FOR PREDICTING THE FLOW AND FRACTURE IN NOTCHED FIBER COMPOSITE MATERIALS	75-0689
FRACTURE	ANALYSIS OF FATIGUE OF PAVING MIXTURES FROM THE FRACTURE MECHANICS VIEWPOINT	72-0540
FRACTURE	APPLICATION OF AN ASSUMED DISPLACEMENT HYBRID FINITE ELEMENT PROCEDURE TO TWO-DIMENSIONAL PROBLEMS I	74-0729
FRACTURE	APPLICATION OF FINITE ELEMENT ANALYSIS METHOD TO TWO-DIMENSIONAL PROBLEMS IN FRACTURE MECHANICS	69-0033
FRACTURE	APPLICATION OF FINITE ELEMENT METHODS TO LARGE SCALE ELASTIC PLASTIC PROBLEMS OF FRACTURE MECHANICS	69-0372
FRACTURE	APPLICATION OF FINITE ELEMENT METHOD TO FRACTURE MECHANICS	71-0178
FRACTURE	APPLICATION OF FINITE ELEMENT METHOD TO MIXED-MODE FRACTURE	73-1123
FRACTURE	APPLICATION OF FINITE ELEMENT TECHNIQUES TO POST-YIELD ANALYSIS OF PROPOSED STANDARD 3 POINT BEND FR	74-0118
FRACTURE	APPLICATION OF FINITE ELEMENT DYNAMIC ANALYSIS TO FRACTURE MECHANICS	75-0861
FRACTURE	APPLICATION OF FRACTURE MECHANICS TO THERMALLY STRESSED CYLINDERS USING A FINITE ELEMENT METHOD	69-0089
FRACTURE	APPLICATION OF FRACTURE MECHANICS TO FAILURE OF THICK WALLED CYLINDERS	71-0511

FRACTURE	APPLICATION OF FRACTURE MECHANICS TO GRAPHITE UNDER COMPLEX STRESS CONDITIONS	72-0771
FRACTURE	APPLICATION OF FRACTURE MECHANICS TO PRESTRESSED CONCRETE PRESSURE VESSELS	73-0228
FRACTURE	APPLICATION OF GRIFFITH'S THEORY TO ANALYSIS OF PROGRESSIVE FRACTURE	75-0150
FRACTURE	APPLICATION OF QUADRATIC ISOPARAMETRIC FINITE ELEMENTS IN LINEAR FRACTURE MECHANICS	74-0521
FRACTURE	APPLICATION OF THE FINITE ELEMENT METHOD TO 2 AND 3 DIMENSIONAL ELASTIC-PLASTIC PROBLEMS OF FRACTURE	70-0501
FRACTURE	APPLICATION OF THE FINITE ELEMENT METHOD TO PROBLEMS IN LINEAR FRACTURE MECHANICS	74-0903
FRACTURE	APPLICATION OF THE METHOD OF FINITE ELEMENT ANALYSIS TO TWO-DIMENSIONAL PROBLEMS IN FRACTURE MECHANI	68-0100
FRACTURE	APPLICATION OF THE SINGLE-EDGE NOTCHED BEAM TO FRACTURE TOUGHNESS TESTING OF CERAMICS	74-0623
FRACTURE	APPLICATIONS OF THE ASSUMED-STRESS HYBRID METHOD TO FINITE ELEMENT ANALYSIS IN FRACTURE MECHANICS	75-0495
FRACTURE	ASSUMED DISPLACEMENT HYBRID FINITE ELEMENT MODEL FOR LINEAR FRACTURE MECHANICS	75-0146
FRACTURE	ASSUMED STRESS HYBRID FINITE ELEMENT METHOD FOR FRACTURE MECHANICS AND ELASTIC-PLASTIC ANALYSIS	72-0584
FRACTURE	BEHAVIOR OF A BI-MATERIAL CYLINDER INCLUDING THE EFFECTS OF FRACTURE	69-0466
FRACTURE	COMBINED ANALYTICAL - EXPERIMENTAL FRACTURE STUDY OF THE TWO LEADING THEORIES OF ELASTIC-PLASTIC FRA	73-0951
FRACTURE	COMBINED ANALYTICAL - EXPERIMENTAL FRACTURE STUDY OF THE TWO LEADING THEORIES OF ELASTIC-PLASTIC FRA	73-0951
FRACTURE	COMBINED MACROSCOPIC AND MICROSCOPIC APPROACH TO THE FRACTURE OF METALS	74-0897
FRACTURE	COMPUTATIONAL FRACTURE MECHANICS	71-0564
FRACTURE	COMPUTER SIMULATION OF DYNAMIC STRESS, DEFORMATION, AND FRACTURE OF GEAR TEETH	72-0114
FRACTURE	DEFORMATION AND FRACTURE OF PARTICULATE COMPOSITE	72-0221
FRACTURE	DEFORMATION MODE OF VOID-GROWTH AND COALESCENCE IN THE PROCESS OF DUCTILE FRACTURE	75-0683
FRACTURE	EFFICIENT FORMULATIONS OF THE FINITE ELEMENT METHOD IN LINEAR AND NONLINEAR FRACTURE MECHANICS	75-0403
FRACTURE	ESTABLISHMENT OF A VALID FRACTURE TOUGHNESS SPECIMEN GEOMETRY USING A FINITE ELEMENT ANALYSIS	74-0135
FRACTURE	EXPERIMENTAL AND THEORETICAL INVESTIGATION OF PLANE STRESS FRACTURE OF 2024 - T351 ALUMINUM ALLOY	70-0147
FRACTURE	FAST FRACTURE RESISTANCE AND CRACK ARREST IN STRUCTURAL STEELS	73-0766
FRACTURE	FINITE ELEMENT FRACTURE MECHANICS ANALYSIS OF CREEP RUPTURE OF FUEL ELEMENT CLADDING	74-0929
FRACTURE	FINITE ELEMENT FRACTURE MECHANICS ANALYSIS OF TWO-DIMENSIONAL AND AXISYMMETRIC ELASTIC AND ELASTIC-P	74-1092
FRACTURE	FINITE ELEMENT IN NUMERICAL ANALYSIS OF ADHESIVE FRACTURE	73-0085
FRACTURE	FINITE ELEMENT INVESTIGATION OF SECONDARY CRACKING DURING BRITTLE FRACTURE	71-0804
FRACTURE	FINITE ELEMENT METHOD IN LINEAR FRACTURE MECHANICS	70-0476
FRACTURE	FINITE ELEMENT METHOD FOR LINEAR FRACTURE MECHANICS	74-1359
FRACTURE	FINITE ELEMENT MODELS OF FRACTURE	73-0012
FRACTURE	FINITE ELEMENT TECHNIQUES IN FRACTURE MECHANICS	72-0746
FRACTURE	FINITE-ELEMENT ANALYSIS OF A FRACTURE TOUGHNESS TEST SPECIMEN IN NON-LINEAR RANGE	72-0093
FRACTURE	FINITE-ELEMENT ANALYSIS OF FRACTURE PROPAGATION IN TWO-DIMENSIONAL ELASTIC BRITTLE SOLIDS	72-0948
FRACTURE	FINITE-ELEMENT TECHNIQUES IN FRACTURE MECHANICS	75-0780
FRACTURE	FRACTURE (ULTIMATE STRENGTH) ANALYSES OF ASPHALT PAVEMENT LAYERS RESULTING FROM TRAFFIC LOADING	73-0548
FRACTURE	FRACTURE ANALYSIS IN AREAS OF HIGH NOMINAL STRAIN	73-0495
FRACTURE	FRACTURE AND SAFETY ANALYSIS OF NUCLEAR PRESSURE VESSELS	73-0257
FRACTURE	FRACTURE CONTROL: PAST, PRESENT AND FUTURE	73-0516
FRACTURE	FRACTURE DYNAMICS OF HOMALITE-100 SHEETS	72-0809
FRACTURE	FRACTURE IN SOIL CEMENT USING THE FINITE ELEMENT METHOD	74-0434
FRACTURE	FRACTURE MECHANICS APPROACH TO ADHESIVE JOINTS	71-0726
FRACTURE	FRACTURE MECHANICS APPROACH TO ADHESIVE JOINTS	72-0181
FRACTURE	FRACTURE MECHANICS APPLICATION OF AN ASSUMED DISPLACEMENT HYBRID FINITE ELEMENT PROCEDURE	75-0212
FRACTURE	FRACTURE MECHANICS FOR AN INTERFACIAL CRACK BETWEEN ADHESIVELY BONDED DISSIMILAR MATERIALS	72-0869
FRACTURE	FRACTURE MECHANICS OF TINY CRACKS NEAR FASTENERS	74-0735
FRACTURE	FRACTURE OF GRAPHITE FIBER REINFORCED COMPOSITES	74-0695
FRACTURE	FRACTURE OF ROCKET GRAIN MODELS SUBJECTED TO COOLDOWN: A STUDY OF TEMPERATURE-INITIATED FRACTURES OF	74-0536
FRACTURE	FRACTURE SURFACE ENERGIES BY FINITE ELEMENT STRESS ANALYSIS	73-0915
FRACTURE	FRACTURE TOUGHNESS PARAMETERS AND ELASTIC-PLASTIC ANALYSIS OF NON-MODERATE FRACTURE CONDITIONS USING	73-0258
FRACTURE	FRACTURE TOUGHNESS PARAMETERS AND ELASTIC-PLASTIC ANALYSIS OF NON-MODERATE FRACTURE CONDITIONS USING	73-0258
FRACTURE	FRACTURE TOUGHNESS PARAMETERS AND ELASTIC-PLASTIC ANALYSIS OF NON-MODERATE FRACTURE CONDITIONS USING	73-1126
FRACTURE	FRACTURE TOUGHNESS PARAMETERS AND ELASTIC-PLASTIC ANALYSIS OF NON-MODERATE FRACTURE CONDITIONS USING	73-1126
FRACTURE	FRACTURE TOUGHNESS TESTING WITH SECTIONS OF CYLINDERS	74-0999
FRACTURE	FURTHER APPLICATION OF QUADRATIC ISOPARAMETRIC FINITE-ELEMENTS TO LINEAR FRACTURE MECHANICS OF PLATE	75-0639
FRACTURE	HEAVY SECTION STEEL TECHNOLOGY PROGRAM. TECHNICAL REPORT NO. 4, OCTOBER 1969. A TWO-DIMENSIONAL ELAS	69-0474
FRACTURE	LOAD LIMIT ANALYSIS. COMBINATION OF STATISTICAL METHODS (MONTE CARLO METHOD) AND FINITE ELEMENT METH	71-0768
FRACTURE	MATRIX DISPLACEMENT METHODS IN FRACTURE MECHANICS ANALYSIS OF REACTOR VESSELS	71-0226
FRACTURE	MECHANICS OF FRACTURE IN ADHESIVE JOINTS	72-0691
FRACTURE	NOTE ON THE FINITE ELEMENT METHOD IN LINEAR FRACTURE MECHANICS	70-0474
FRACTURE	NUMERICAL ANALYSIS IN ORTHOTROPIC LINEAR FRACTURE MECHANICS	73-0512
FRACTURE	NUMERICAL TECHNIQUES IN LINEAR AND NONLINEAR FRACTURE MECHANICS	75-0260
FRACTURE	ON THE FINITE ELEMENT METHOD IN LINEAR FRACTURE MECHANICS	69-0032
FRACTURE	ON THE FRACTURE MECHANICS ANALYSIS OF ELASTIC-PLASTIC MATERIALS USING THE FINITE ELEMENT METHOD	73-1067
FRACTURE	ON THE USE OF FINITE ELEMENTS IN FRACTURE ANALYSIS OF PRESSURE VESSEL COMPONENTS	75-0514
FRACTURE	OPTIMUM POSITION OF THE CENTRAL CLAY CORE OF A ROCKFILL DAM IN RESPECT TO ARCHING AND HYDRAULIC FRAC	73-0447
FRACTURE	PRESSURE VESSEL CERTIFICATION BASED ON FRACTURE MECHANICS TECHNOLOGY	74-0948
FRACTURE	PROGRESS IN THREE-DIMENSIONAL ELASTIC-PLASTIC STRESS ANALYSIS FOR FRACTURE MECHANICS	71-0314
FRACTURE	PROPAGATION OF ELLIPTICAL SURFACE CRACKS AND NONLINEAR FRACTURE MECHANICS BY THE FINITE ELEMENT METH	74-0906
FRACTURE	SIGNIFICANCE OF SYMMETRY IN MULTIPLE FRACTURE ANALYSIS	74-0869
FRACTURE	SIMULATION ON THE INITIATION AND GROWTH OF DUCTILE FRACTURE VOIDS	75-0836
FRACTURE	SOME EXAMINATIONS ON AVAILABILITY OF FINITE ELEMENT METHODS IN LINEAR FRACTURE MECHANICS	73-0112
FRACTURE	STRESS DISTRIBUTION AND ROCK FRACTURE ZONES IN THE ROOF OF LONGWALL FACE IN COAL MINE	73-0247
FRACTURE	STRESS INTENSITY FACTORS FOR HEXAGONAL FRACTURE TOUGHNESS SPECIMENS	71-0799
FRACTURE	STRESS INTENSITY FACTORS FOR ANISOTROPIC FRACTURE TEST SPECIMENS OF SEVERAL GEOMETRIES	74-0293
FRACTURE	STRESS STRAIN RESPONSE AND FRACTURE OF A CONCRETE MODEL IN BIAXIAL LOADING	71-0295
FRACTURE	STUDY ON FRACTURE TOUGHNESS EVALUATION BY ELASTO-PLASTIC ANALYSIS AROUND NOTCHES	74-0619
FRACTURE	STUDY ON J//I//C FRACTURE CRITERION	74-1034
FRACTURE	SURVEY AND EVALUATION OF THE FINITE ELEMENT METHOD IN LINEAR FRACTURE MECHANICS ANALYSIS	71-0094
FRACTURE	THE APPLICATION OF THE FINITE ELEMENT METHOD TO FRACTURE MECHANICS	72-0331
FRACTURE	THE APPLICATION OF THE FINITE ELEMENT METHOD TO THE ANALYSIS OF FRACTURE OF CYLINDRICAL SHELLS	73-0728
FRACTURE	THE EFFECT OF END RESTRAINT CONDITION ON THE FRACTURE MECHANISM AND STRENGTH OF ROCK-LIKE MATERIALS	72-0134
FRACTURE	THE USE OF FINITE ELEMENT TECHNIQUES IN FRACTURE MECHANICS	72-0926
FRACTURE	THREE DIMENSIONAL ELASTIC PLASTIC ANALYSIS FOR FRACTURE MECHANICS - 2.	71-0358
FRACTURE	THREE-DIMENSIONAL ELASTIC-PLASTIC STRESS AND STRAIN ANALYSIS FOR FRACTURE MECHANICS. PHASE 1. SIMPLE	70-0188
FRACTURE	THREE-DIMENSIONAL FINITE ELEMENT ANALYSIS FOR FRACTURE MECHANICS	72-0162
FRACTURE	THREE-DIMENSIONAL FINITE ELEMENT ANALYSIS FOR FRACTURE MECHANICS	72-0921
FRACTURE	THREE-DIMENSIONAL LINEAR FRACTURE MECHANICS ANALYSIS BY A DISPLACEMENT-HYBRID FINITE-ELEMENT MODEL	75-0884
FRACTURE	UNIMOD - APPLICATIONS ORIENTED FINITE ELEMENT SCHEME FOR THE ANALYSIS OF FRACTURE MECHANICS PROBLEMS	74-1356
FRACTURE	USE OF FINITE ELEMENT COMPUTER PROGRAMS IN FRACTURE MECHANICS	71-0022
FRACTURE	USE OF FINITE ELEMENT ANALYSIS IN FRACTURE MECHANICS	74-0837
FRACTURE	USE OF FRACTURE MECHANICS AND FINITE ELEMENT TECHNIQUES FOR DETERMINING THE SAFETY OF OFFSHORE STRUC	74-1211
FRACTURE	USE OF ISOPARAMETRIC FINITE-ELEMENTS IN LINEAR FRACTURE MECHANICS	75-0352
FRACTURE	USE OF SINGULARITY PROGRAMMING IN FINITE ELEMENT CALCULATIONS OF ELASTIC STRESS INTENSITY FACTORS. P	75-0252
FRACTURE	VERIFICATION OF FRACTURE MECHANICS CONCEPTS TO PREDICT CRACKING OF FLEXIBLE PAVEMENTS	73-0772
FRACTURED	FINITE ELEMENT GALERKIN METHOD FOR ANALYSIS OF FLOW IN FRACTURED POROUS MEDIA	74-0066
FRACTURED	FRONT DISPLACEMENT MODEL IN A FRACTURED RESERVOIR	74-0071
FRACTURED	SEEPAGE IN FRACTURED ROCK: FINITE ELEMENT ANALYSIS OF ROCK MASS BEHAVIOR UNDER COUPLED ACTION OF BOD	71-0825
FRACTURED	THE USE OF ROCK BOLTS OR WIRE ROPE TO INCREASE THE STRENGTH OF FRACTURED MODEL PILLARS	71-0722
FRACTURES	BOUNDARIES AND FRACTURES IN FINITE ELEMENT MODELS OF GEOLOGICAL STRUCTURES	73-0352
FRACTURES	FLOW IN ROCKS WITH DEFORMABLE FRACTURES	74-0065
FRACTURES	FRACTURE OF ROCKET GRAIN MODELS SUBJECTED TO COOLDOWN: A STUDY OF TEMPERATURE-INITIATED FRACTURES OF	74-0536
FRACTURES	MATHEMATICAL AND EXPERIMENTAL STUDIES ON THE MECHANICS OF PLATED TRANSVERSE FRACTURES	74-0656

FRACTURES	STEADY STATE FLOW IN RIGID NETWORKS OF FRACTURES	74-0278
FRACTURES	STUDY BY THE FINITE-ELEMENT METHOD OF THE INFLUENCE OF FRACTURES IN CONFINED AQUIFERS.	75-0604
FRACTURES	THE ROLE OF FINITE DEFORMATION ANALYSIS IN PLANE STRESS AND STRAIN FRACTURES	74-0322
FRACTURING	HYDRAULIC FRACTURING IN ZONED EARTH AND ROCKFILL DAMS: A REPORT OF AN INVESTIGATION	73-0873
FRAEIJS	SUR L ELEMENT DE FRAEIJS DE VEUBEKE ET SANDER (FRAEIJS DE VEUBEKE-SANDER ELEMENT)	74-0924
FRAEIJS	SUR L ELEMENT DE FRAEIJS DE VEUBEKE ET SANDER (FRAEIJS DE VEUBEKE-SANDER ELEMENT)	74-0924
FRAGILITY	BRAIN TISSUE FRAGILITY - A FINITE STRAIN ANALYSIS BY A HYBRID FINITE-ELEMENT METHOD	75-0162
FRAGILITY	BRAIN TISSUE FRAGILITY - A FINITE STRAIN ANALYSIS BY A HYBRID FINITE-ELEMENT METHOD	75-0243
FRAGILITY	BRAIN-TISSUE FRAGILITY - FINITE STRAIN ANALYSIS BY A HYBRID FINITE-ELEMENT METHOD	75-0631
FRAGILITY	MECHANICS OF BRAIN TISSUE FRAGILITY	75-0214
FRAGMENT	APPLICATION OF THE COLLISION-IMPARTED VELOCITY METHOD FOR ANALYZING THE RESPONSES OF CONTAINMENT AND	73-0770
FRAGMENT	FINITE ELEMENT ANALYSIS OF LARGE TRANSIENT ELASTIC-PLASTIC DEFORMATIONS OF SIMPLE STRUCTURES, WITH A	72-0849
FRAGMENTATION	INFLUENCE OF TEMPERATURE DEPENDENT PROPERTIES ON THERMAL ROCK FRAGMENTATION	75-0222
FRAGMENTATION	ROCK FRAGMENTATION BY SUBSURFACE THERMAL INCLUSIONS - A FINITE ELEMENT STUDY	73-1159
FRAGMENTATION	THERMAL-MECHANICAL FRAGMENTATION OF HARD ROCK - FIELD TESTS	74-0998
FRAME	AN ITERATION METHOD FOR THE NONLINEAR BUCKLING OF FRAME STRUCTURES	66-0107
FRAME	ANALYSIS OF NONLINEAR REINFORCED CONCRETE FRAME BY FINITE ELEMENT METHOD	74-1104
FRAME	ANALYSIS OF STRESSES AND DEFLECTIONS IN FRAME SUPPORTED TENTS	74-0689
FRAME	AUTOMATED METHOD FOR THE LARGE DEFLECTION AND INSTABILITY ANALYSIS OF THREE-DIMENSIONAL TRUSS AND FR	66-0105
FRAME	CH-53A FLEXIBLE FRAME VIBRATION ANALYSIS/TEST CORRELATION	73-1093
FRAME	COMPUTATIONAL ASPECTS OF FINITE ELEMENT METHOD FOR A FRAME STRUCTURE	72-0175
FRAME	COMPUTER ANALYSIS OF STEEL FRAME IN FIRE	75-0173
FRAME	DYNAMIC ANALYSIS OF AN AUTOMOBILE CHASSIS FRAME	70-0135
FRAME	DYNAMIC ANALYSIS OF A CAR CHASSIS FRAME USING FINITE ELEMENT METHOD	72-0265
FRAME	DYNAMIC ANALYSIS OF FRAME STRUCTURES BY THE FORCE METHOD	72-0313
FRAME	DYNAMIC ANALYSIS OF FRAME AND PLATE STRUCTURES WITH GEOMETRIC AND MATERIAL NONLINEARITIES	73-0649
FRAME	FINITE ELEMENT DYNAMIC ANALYSIS OF AN AUTOMOTIVE FRAME	73-0256
FRAME	FINITE ELEMENT FOR REINFORCED CONCRETE FRAME STUDY	73-0204
FRAME	FINITE ELEMENT FOR REINFORCED CONCRETE FRAME STUDY	74-0110
FRAME	FINITE ELEMENT FOR REINFORCED CONCRETE FRAME STUDY - SPACE DISCUSSION	74-0246
FRAME	FINITE ELEMENT FOR REINFORCED CONCRETE FRAME STUDY - CLOSURE	75-0087
FRAME	FINITE ELEMENTS IN TRANSVERSE FRAME ANALYSIS	72-0182
FRAME	FRAME OPTIMIZATION INCLUDING FREQUENCY CONSTRAINTS	75-0082
FRAME	INTERACTION OF FRAME AND SHEAR WALL WITH OPENINGS	74-1190
FRAME	INTERACTIVE ANALYSIS OF A RAFTED MULTISTOREY SPACE FRAME RESTING ON AN INHOMOGENEOUS CLAY STRATUM	74-0447
FRAME	LIMIT ANALYSIS OF FRAME SYSTEMS STIFFENED BY PANELS	74-0584
FRAME	NUMERICAL CONDITIONING OF STIFFNESS MATRIX FORMULATIONS FOR FRAME STRUCTURES	68-0022
FRAME	NUMERICAL PROCEDURES FOR THE DETERMINATION OF THE BEHAVIOUR OF A SHEAR WALL FRAME SYSTEM	73-0216
FRAME	STRESS ANALYSIS OF AN AUTOMOBILE CHASSIS FRAME	70-0137
FRAME	STRESS ANALYSIS OF FRAME STRUCTURE CONNECTED WITH PLATES	74-0265
FRAME	STRESS AND VIBRATION ANALYSIS ON THE SYSTEM OF FRAME AND FOUNDATION	69-0218
FRAME	TANGENT STIFFNESS MATRIX FOR SPACE FRAME MEMBERS	69-0397
FRAME	THE FLEXIBILITY OF A TUBULAR WELDED JOINT IN A VEHICLE FRAME	74-0209
FRAME	THE STRESS ANALYSIS OF AXISYMMETRIC SHELLS USING A PLANE FRAME ANALYSIS PROGRAM	69-0238
FRAME	THEORETICAL AND EXPERIMENTAL VIBRATION ANALYSIS OF AN OBLIQUE SPACE FRAME	70-0460
FRAME-PLATE	FINITE ELEMENT ANALYSIS OF ORTHOGONAL FRAME-PLATE INTERACTION	70-0508
FRAMED	A COMPUTER METHOD FOR THE SECOND ORDER ELASTIC ANALYSIS OF PLANE FRAMED STRUCTURES	73-0310
FRAMED	A METHOD OF FINITE DEFLECTION ELASTIC PLASTIC ANALYSIS OF FRAMED STRUCTURES SUBJECTED TO LARGE AXIAL	69-0239
FRAMED	DYNAMIC MATRIX ANALYSIS OF FRAMED STRUCTURES	62-0012
FRAMED	ELASTIC-PLASTIC ANALYSIS OF FRAMED STRUCTURES	69-0197
FRAMED	ELASTO-PLASTIC SEISMIC RESPONSE ANALYSIS OF FRAMED STRUCTURES BY MODE SUPERPOSITION	73-0325
FRAMED	FINITE ELEMENTS FOR VIBRATION OF FRAMED SHEAR WALLS	73-0036
FRAMED	LARGE DEFORMATION AND CRITICAL LOADS ANALYSIS OF FRAMED STRUCTURES	74-1293
FRAMED	NONLINEAR ANALYSIS OF ELASTIC FRAMED STRUCTURES	68-0222
FRAMED	OPTIMUM DESIGN OF FRAMED STRUCTURES	69-0251
FRAMED	SEISMIC BEHAVIOUR OF FRAMED TUBES	75-0306
FRAMED	STRUCTURAL ANALYSIS OF LONGITUDINALLY FRAMED SHIPS	72-0436
FRAMES	A DATA-GENERATION PROCEDURE FOR THE FINITE ELEMENT ANALYSIS OF PLANE FRAMES AND GRILLAGES	68-0179
FRAMES	A FINITE ELEMENT APPROACH TO THE STRUCTURAL INSTABILITY OF BEAM COLUMNS, FRAMES, AND ARCHES	70-0434
FRAMES	A FINITE ELEMENT APPROACH TO STABILITY ANALYSIS IN FRAMES	74-0455
FRAMES	A FINITE-ELEMENT METHOD OF SOLUTION FOR STRUCTURAL FRAMES	67-0133
FRAMES	A FINITE-ELEMENT MODEL TO SIMULATE THE NONLINEAR RESPONSE OF REINFORCED CONCRETE FRAMES WITH MASONRY	73-1168
FRAMES	ANALYSIS OF FRAMES WITH SHEAR WALLS BY FINITE ELEMENTS	69-0199
FRAMES	ANALYSIS OF FRAMES WITH SHEAR WALLS WITH FINITE ELEMENTS	69-0500
FRAMES	ANALYSIS OF NONLINEAR CONCRETE FRAMES USING SMALL GYPSUM MODELS AND THE FINITE ELEMENT METHOD	70-0498
FRAMES	BIFURCATION OF SPACE FRAMES	70-0055
FRAMES	COMPUTER ANALYSIS OF LARGE LINEAR FRAMES	69-0408
FRAMES	COMPUTER ANALYSIS OF RIGID FRAMES	71-0401
FRAMES	DYNAMIC RESPONSE OF NONLINEAR SPACE FRAMES BY FINITE ELEMENT METHODS	71-0584
FRAMES	DYNAMIC TESTING AND COMPUTER ANALYSIS OF AUTOMOTIVE FRAMES	72-0089
FRAMES	EFFECT OF OPENINGS ON THE LATERAL STIFFNESS OF INFILLED FRAMES	71-0300
FRAMES	EFFECTS OF SHEAR WALLS ON DYNAMIC RESPONSE OF FRAMES	69-0230
FRAMES	ELASTIC ANALYSIS OF SPATIAL SYSTEMS OF INTERCONNECTED SHEAR WALLS AND FRAMES	74-1195
FRAMES	FINITE ELEMENT ANALYSIS OF STRUCTURAL FRAMES	67-0032
FRAMES	FINITE ELEMENT ANALYSIS FOR BUCKLING OF SEMI-RIGID SPACE FRAMES	68-0270
FRAMES	FINITE ELEMENT FOR THE INTERACTION OF FRAMES AND SHEAR WALLS	73-0681
FRAMES	FINITE ELEMENT STUDY OF THE ELASTIC BEHAVIOR OF PLANE FRAMES WITH FILLER WALLS	71-0828
FRAMES	FIRES-RC. A COMPUTER PROGRAM FOR THE FIRE RESPONSE OF STRUCTURES-REINFORCED CONCRETE FRAMES	74-0747
FRAMES	INELASTIC ANALYSIS OF REINFORCED CONCRETE FRAMES	74-0587
FRAMES	INELASTIC RESPONSE OF FRAMES TO DYNAMIC LOADS	71-0611
FRAMES	LARGE DEFLECTION SPATIAL BUCKLING OF THIN WALLED BEAMS AND FRAMES	73-1085
FRAMES	LARGE DISPLACEMENT ANALYSIS OF VISCOELASTIC BEAMS AND FRAMES BY FINITE-ELEMENT METHOD	74-0644
FRAMES	LIGHT GAGE STEEL INFILL PANELS IN MULTISTORY STEEL FRAMES	74-0229
FRAMES	NATURAL PERIODS OF INFILLED FRAMES	72-0122
FRAMES	NONLINEAR ANALYSIS OF REINFORCED CONCRETE FRAMES AND PANELS	70-0436
FRAMES	NONLINEAR DYNAMIC ANALYSIS OF FRAMES WITH FILLER PANELS	74-0226
FRAMES	NONLINEAR FINITE ELEMENT ANALYSIS OF REINFORCED CONCRETE FRAMES	74-1124
FRAMES	STRUCTURAL MODELING OF A WELDED JOINT TYPICAL FOR VEHICLE FRAMES	73-0539
FRAMES	THE AUTOMATIC COMPUTATION OF THE NATURAL FREQUENCIES OF STRUCTURAL FRAMES USING AN EXACT MATRIX TECH	73-0318
FRAMES	THE NATURAL FREQUENCIES OF PORTAL FRAMES II	65-0047
FRAMES	VIBRATION OF FRAMES AND OTHER STRUCTURES WITH BANDED STIFFNESS MATRIX	70-0322
FRAMES	VIBRATIONAL CHARACTERISTICS OF SPACE FRAMES	69-0434
FRAMEWORK	A FRAMEWORK METHOD FOR STRETCHED PLATES	71-0697
FRAMEWORK	A SURVEY OF FINITE ELEMENT FRAMEWORK STABILITY ANALYSIS	69-0418
FRAMEWORK	PLANE FRAMEWORK METHODS FOR PLATES IN EXTENSION	66-0126
FRAMEWORK	SOLUTION OF PROBLEMS IN ELASTICITY BY THE FRAMEWORK METHOD	41-0001
FRAMEWORKS	AXIAL-FLEXURAL VIBRATION OF FRAMEWORKS USING FINITE-ELEMENT APPROACH	73-0051
FRAMEWORKS	COMPUTER SYSTEM FOR ANALYSIS OF LARGE FRAMEWORKS	67-0122
FRAMEWORKS	ELASTIC-PLASTIC DEFORMATIONS OF FLEXURALLY STIFF FRAMEWORKS FROM SECOND ORDER STRESS THEORY	73-1061
FRAMEWORKS	INFLUENCE OF HIGHER ORDER TERMS IN THE LARGE DEFLECTION ANALYSIS OF FRAMEWORKS	68-0259
FRAMEWORKS	STATICAL CALCULATIONS OF PLANE AND SPACE BAR STRUCTURES AS WELL AS STUMPY FRAMEWORKS USING A MODIFIE	73-0747
FRANKFURT	APPLICATION OF NEW AUSTRIAN TUNNELLING METHOD IN DIFFICULT BUILTOVER AREAS IN FRANKFURT/MAIN METRO	73-1025

234

FREE	A FINITE ELEMENT PROGRAM FOR FREE FORM SHELLS	75-0786
FREE	A THEORETICAL AND EXPERIMENTAL STUDY ON THE FREE VIBRATION OF CURVED SLAB BRIDGES	71-0496
FREE	A VARIATIONAL METHOD IN POTENTIAL FLOWS WITH A FREE SURFACE	72-0723
FREE	A VARIATIONAL METHOD FOR FREE BOUNDARY PROBLEMS	75-0105
FREE	ANALYSIS OF AN ORTHOTROPIC PLATE WITH TWO OPPOSITE EDGES SIMPLY SUPPORTED AND THE OTHER TWO FREE	73-0741
FREE	ANALYSIS OF NONSTEADY FLOW WITH A FREE SURFACE USING FINITE ELEMENT METHOD	71-0482
FREE	APPLICATION OF FLOWGRAPHS TO FREE VIBRATION OF STRUCTURES	68-0196
FREE	CALCULATION OF THE FREE OSCILLATIONS OF A LIQUID IN AXISYMMETRIC MOTIONLESS CONTAINERS OF ARBITRARY	74-0387
FREE	COMPUTER PROGRAM JET 3 TO CALCULATE THE LARGE ELASTIC PLASTIC DYNAMICALLY INDUCED DEFORMATIONS OF FR	72-0609
FREE	CONTRIBUTION A L'ANALYSE DES VIBRATIONS LIBRES D'UNE STRUCTURE. APPLICATION AUX AUBAGES DE TURBOMACH	75-0804
FREE	CORRELATION OF FINITE ELEMENT STRUCTURAL DYNAMIC ANALYSIS WITH MEASURED FREE VIBRATION CHARACTERISTI	74-0486
FREE	DARCY FLOW SOLUTIONS WITH A FREE SURFACE	67-0081
FREE	DYNAMICAL ANALYSIS OF AXIALLY NONUNIFORM, THIN CYLINDRICAL SHELLS. PART 2 FREE VIBRATION	71-0716
FREE	FINITE ELEMENT ANALYSIS OF FREE VIBRATION AND STABILITY OF STRUCTURES SUBJECTED TO INITIAL STRESS	68-0300
FREE	FINITE ELEMENT DISPLACEMENT METHOD FOR LARGE AMPLITUDE FREE FLEXURAL VIBRATIONS OF BEAMS AND PLATES	73-0185
FREE	FINITE ELEMENT METHOD FOR INCOMPRESSIBLE SLOW VISCOUS FLOW WITH A FREE SURFACE	69-0303
FREE	FINITE ELEMENT METHOD OF ANALYZING STEADY SEEPAGE WITH A FREE SURFACE	70-0092
FREE	FINITE ELEMENT METHODS: TWO-DIMENSIONAL SEEPAGE WITH A FREE SURFACE	71-0475
FREE	FINITE ELEMENT METHODS FOR THE SOLUTION OF SOME INCOMPRESSIBLE NON-NEWTONIAN FLUID MECHANICS PROBLEM	75-0224
FREE	FINITE-ELEMENT ANALYSIS FOR FREE VIBRATION OF GENERAL ANISOTROPIC LAMINATED THIN SHELLS	72-0618
FREE	FORCE AND MOMENTS ON ASYMMETRIC AND YAWED BODIES IN A FREE SURFACE	73-0908
FREE	FREE BOUNDARIES AND FINITE-ELEMENTS IN ONE DIMENSION	75-0447
FREE	FREE BOUNDARIES AND FINITE-ELEMENTS IN ONE DIMENSION	75-0587
FREE	FREE BOUNDARY POTENTIAL FLOW USING FINITE ELEMENTS	71-0829
FREE	FREE VIBRATION ANALYSIS OF STIFFENED PLATES	72-0303
FREE	FREE VIBRATION ANALYSIS OF SPINNING STRUCTURAL SYSTEMS	73-1077
FREE	FREE VIBRATION AND RESPONSE TO RANDOM PRESSURE FIELD OF ANISOTROPIC THIN CYLINDRICAL SHELLS	74-0830
FREE	FREE VIBRATION AND TRANSIENT RESPONSE OF THICK AND THIN PLATES USING THE FINITE ELEMENT METHOD	74-0517
FREE	FREE VIBRATION OF A CURVED BEAM	71-0606
FREE	FREE VIBRATION OF A SIMPLY SUPPORTED BEAM WITH NONLINEAR MATERIAL PROPERTIES	71-0689
FREE	FREE VIBRATION OF CURVED SANDWICH BEAMS BY METHOD OF FINITE ELEMENTS	71-0351
FREE	FREE VIBRATION OF CURVED AND STRAIGHT BEAM SLAB OR BOX GIRDER BRIDGES	72-0487
FREE	FREE VIBRATION OF CYLINDRICAL SHELLS PARTIALLY FILLED WITH LIQUID	71-0839
FREE	FREE VIBRATION OF DOUBLY CURVED RECTANGULAR PLATES INCLUDING THE PRESENCE OF MEMBRANE STRESSES WITH	69-0468
FREE	FREE VIBRATION OF MODEL OF CAR BODY	71-0312
FREE	FREE VIBRATION OF ORTHOTROPIC VARIABLE THICKNESS, CONTINUOUS RECTANGULAR PLATES	70-0270
FREE	FREE VIBRATION OF RING STIFFENED CYCLINDRICAL SHELLS	70-0172
FREE	FREE VIBRATION OF SHELLS OF REVOLUTION USING RING ELEMENTS	67-0088
FREE	FREE VIBRATION OF SHELLS OF REVOLUTION USING FINITE ELEMENT METHOD	74-0232
FREE	FREE VIBRATION OF THIN CYLINDRICAL SHELLS WITH A DISCONTINUITY IN THE THICKNESS	69-0439
FREE	FREE VIBRATION OF THICK, LAYERED RECTANGULAR PLATES BY FINITE LAYER METHOD	72-0440
FREE	FREE VIBRATION OF THICK, LAYERED CYLINDERS HAVING FINITE LENGTH WITH VARIOUS BOUNDARY CONDITIONS	72-0491
FREE	FREE VIBRATIONS AND RANDOM RESPONSE OF AN INTEGRALLY-STIFFENED PANEL	70-0138
FREE	FREE VIBRATIONS OF CIRCULAR MEMBRANES UNDER ARBITRARY TENSION BY THE FINITE ELEMENT METHOD	69-0190
FREE	FREE VIBRATIONS OF CURVED SANDWICH BEAMS	71-0729
FREE	FREE VIBRATIONS OF CYLINDRICAL SHELLS WITH VARIOUS END CLOSURES	74-0571
FREE	FREE VIBRATIONS OF FINITE ELEMENT PLATES SUBJECTED TO COMPLEX MIDDLE-PLANE FORCE SYSTEMS	72-0280
FREE	FREE VIBRATIONS OF FINITE ELEMENT PLATES SUBJECTED TO COMPLEX MIDDLE-PLANE FORCE SYSTEMS	73-0097
FREE	FREE VIBRATIONS OF FINITE ELEMENT PLATES SUBJECTED TO COMPLEX MIDDLE-PLANE FORCE SYSTEMS - REPLY	73-0122
FREE	FREE VIBRATIONS OF FLUID-BORNE STRUCTURES: INVESTIGATIONS ON A SIMPLE MODEL	74-0976
FREE	FREE VIBRATIONS OF PRESSURE LOADED PARABOLOIDAL SHELLS	71-0601
FREE	FREE VIBRATIONS OF PLATE-BEAM SYSTEMS	73-0322
FREE	FREQUENCY-DEPENDENT SHAPE FUNCTIONS FOR FREE AND FORCED STRUCTURAL VIBRATION CALCULATION	73-1091
FREE	INVESTIGATION ON FREE AND FORCED VIBRATIONS OF A LONG TANKER WITH OVERLAPPING PROPELLER ARRANGEMENT	73-0981
FREE	LARGE AMPLITUDE, FREE VIBRATION GEOMETRICALLY NONLINEAR CIRCULAR ARCHES	72-0606
FREE	MODAL RESPONSE OF FREE ROCKETS UNDER THRUST LOADING	74-0855
FREE	NATURAL FREQUENCIES OF COMPLEX FREE OR SUBMERGED STRUCTURES BY THE FINITE ELEMENT METHOD	65-0026
FREE	NATURAL VIBRATION OF A FREE SURFACE	62-0008
FREE	NUMERICAL ANALYSIS OF FREE SURFACE SEEPAGE PROBLEMS	71-0469
FREE	NUMERICAL SOLUTION OF A DIFFUSION CONSUMPTION PROBLEM WITH A FREE BOUNDARY	74-0835
FREE	NUMERICAL-SOLUTION OF EULERIAN EQUATIONS OF COMPRESSIBLE FLOW BY A FINITE-ELEMENT METHOD WHICH FOLLO	75-0535
FREE	NUMERYCZNE ROZWIAZANIE PROBLEMU ORGAN WLASNYCH NA PRZYKLADZIE PLYT. (NUMERICAL SOLUTION OF THE PROBL	75-0296
FREE	ON SOME FREE SURFACE ON TRANSIENT FLOW PROBLEMS OF SEEPAGE AND IRROTATIONAL FLOW	72-0070
FREE	ON THE DEVELOPMENT OF A REFINED FINITE-ELEMENT FREE FROM DISCRETIZATION - DISPERSION FOR ELASTIC WAV	73-1148
FREE	ON THE NUMERICAL SOLUTION OF FREE BOUNDARY PROBLEMS	75-0577
FREE	THE PROGRAM " ASSEN " FOR STEADY FREE SURFACE FLOWS	74-0075
FREE	USE OF FINITE ELEMENT METHOD IN A MARKER AND CELL PROGRAM FOR SOLVING TIME-DEPENDENT FLUID FLOW PROB	75-0043
FREE	USE OF HIGHER ORDER DISPLACEMENT FUNCTIONS IN THE FREE VIBRATION ANALYSIS OF SHELLS OF REVOLUTION HA	74-1055
FREE	VARIATIONAL FORMULATION FOR A CLASS OF FREE BOUNDARY PROBLEMS ARISING IN HYDRODYNAMIC LUBRICATION	75-0531
FREE	VIBRATION CHARACTERISTICS OF FREE THIN CYLINDRICAL SHELLS	73-0396
FREE-FLOODED	A FINITE-ELEMENT APPROACH TO THE ANALYSIS OF TANGENTIALLY POLARIZED PIEZOELECTRIC-CERAMIC FREE-FLOOD	74-0802
FREE-FLOODED-RI	SHIP (SIMPLIFIED-HELMHOLTZ-INTEGRAL PROGRAM): A FAST COMPUTER PROGRAM FOR CALCULATING THE ACOUSTIC R	72-0835
FREE-FREE	FREE-FREE STRUCTURES IN FINITE ELEMENT ANALYSIS	70-0442
FREE-FREE	SOME RESULTS ON THE FREE-FREE OSCILLATIONS OF AIRCRAFT TYPE STRUCTURES	65-0007
FREE-STREAM	INFLUENCE OF COOLING, FREE-STREAM TURBULENCE AND SURFACE ROUGHNESS ON THE AERODYNAMIC BEHAVIOR OF CA	74-0970
FREE-SURFACE	FREE-SURFACE IDEAL FLUID FLOWS BY FINITE ELEMENTS	73-0254
FREE-SURFACE	FREE-SURFACE IDEAL FLUID FLOWS BY FINITE ELEMENTS	74-0125
FREE-SURFACE	ON THE SOLUTION OF TRANSIENT FREE-SURFACE FLOW PROBLEMS IN POROUS MEDIA BY THE FINITE ELEMENT METHOD	73-1019
FREE-SURFACE	SOLUTION OF A FREE-SURFACE BOUNDARY VALUE PROBLEM USING AN INVERSE FORMULATION AND THE FINITE ELEMEN	75-0466
FREE-SURFACE	SOLUTION OF VISCOUS INCOMPRESSIBLE JET AND FREE-SURFACE FLOWS USING FINITE ELEMENT METHODS	74-0357
FREE-SURFACE	TRANSIENT FREE-SURFACE FLOW IN POROUS MEDIA BY FINITE ELEMENT METHOD	72-0297
FREE-VIBRATIONS	A STUDY OF FREE-VIBRATIONS OF THIN ELASTIC SHELLS SUBJECTED TO INITIAL LOADS USING FINITE-ELEMENTS	73-1143
FREEDOM	APPLICATION OF THE FINITE ELEMENT METHOD TO CASES REQUIRING THE COMBINATION OF ELEMENTS POSSESSING D	72-0580
FREEDOM	FURTHER LIMITATIONS ON GENERAL HEXAHEDRON FINITE ELEMENTS WITH DERIVATIVE DEGREES OF FREEDOM.	72-0395
FREEDOM	REDUCTION OF THE DEGREES OF FREEDOM IN SOLVING DYNAMIC PROBLEMS BY THE FINITE ELEMENT METHOD	73-0647
FREEDOM	THREE-DIMENSIONAL ELASTIC FINITE ELEMENT ANALYSIS USING GRADIENT DEGREES OF FREEDOM	73-0843
FREEZE-THAW	PAVEMENT CRACKING IN WEST TEXAS DUE TO FREEZE-THAW CYCLING	75-0276
FREIER	BERECHNUNG INSTATIONARER GRUND - UND SICKERWASSERSTROMUNGEN MIT FREIER OBERFLACHE NACH DER METHODE F	71-0541
FREIGHT	COMPUTER ANALYSIS OF A RAILROAD FREIGHT CAR BOLSTER UTILIZING THE FINITE ELEMENT METHOD	71-0283
FREQUENCIES	A FORTRAN PROGRAM FOR CALCULATING THE FREQUENCIES AND MODE SHAPES OF AN ARBITRARY SHAPE CYLINDRICAL	72-0594
FREQUENCIES	A FORTRAN PROGRAM FOR CALCULATING THE FREQUENCIES AND MODE SHAPES OF A GUIDED CLAMPED, GUIDED PINNED	72-0890
FREQUENCIES	A METHOD FOR COMPUTATION OF VIBRATION MODES AND FREQUENCIES OF ORTHOTROPIC THIN SHELLS OF REVOLUTION	69-0037
FREQUENCIES	AN INVESTIGATION OF THE EFFECTS OF SQUARE CUTOUTS ON THE NATURAL FREQUENCIES AND MODE SHAPES OF RECT	71-0638
FREQUENCIES	APPROXIMATION METHOD FOR COMPUTATION OF NATURAL FREQUENCIES OF SPIRAL SECTIONS	71-0775
FREQUENCIES	CALCULATION OF NATURAL FREQUENCIES OF HEAT EXCHANGER TUBES	73-0665
FREQUENCIES	CALCULATION OF THE CHARACTERISTIC FREQUENCIES OF FLUID-FILLED SPHERICAL SHELLS	73-0511
FREQUENCIES	CALCULATION OF THE NATURAL FREQUENCIES AND THE PRINCIPAL MODES OF HELICOPTER BLADES	72-0907
FREQUENCIES	COMPUTATION OF VIBRATION MODAL SHAPES AND NATURAL FREQUENCIES OF A GENERAL ORTHOTROPIC LAMINATED, TH	71-0309
FREQUENCIES	EFFECT OF LATERAL PRESSURE ON NATURAL FREQUENCIES OF HYPERBOLOIDAL SHELLS OF REVOLUTION	71-0278
FREQUENCIES	EFFECT OF SHEAR DEFORMATIONS AND ROTARY INERTIA ON OPTIMUM BEAM FREQUENCIES	75-0231
FREQUENCIES	EFFECT OF SYSTEM PARAMETER VARIATIONS ON NATURAL FREQUENCIES	75-0661
FREQUENCIES	EFFECTS OF SHEAR DEFORMATION, ROTARY INERTIA, AND ELASTICITY OF THE SUPPORT ON THE RESONANCE FREQUEN	75-0369

FREQUENCIES	ESTIMATIONS OF THE EFFECT OF DESIGN CHANGES ON THE MODES AND FREQUENCIES OF VIBRATING STRUCTURAL ELE	74-0819
FREQUENCIES	FAMSOR, A FINITE ELEMENT PROGRAM FOR THE FREQUENCIES AND MODE SHAPES OF SHELLS OF REVOLUTION	70-0090
FREQUENCIES	METHOD OF MINIMUM WEIGHT DESIGN WITH REQUIREMENTS IMPOSED ON STRESSES AND NATURAL FREQUENCIES	70-0149
FREQUENCIES	MODE SHAPES AND FREQUENCIES BY FINITE ELEMENT METHOD USING CONSISTENT AND LUMPED MASSES	71-0610
FREQUENCIES	NATURAL FREQUENCIES OF COMPLEX FREE OR SUBMERGED STRUCTURES BY THE FINITE ELEMENT METHOD	65-0026
FREQUENCIES	NATURAL FREQUENCIES AND MODE SHAPES OF TRANSFORMER CORES	65-0048
FREQUENCIES	NATURAL FREQUENCIES AND MODE SHAPES OF PLATES WITH INTERIOR CUT-OUTS	70-0402
FREQUENCIES	NATURAL FREQUENCIES OF CIRCULAR ARCHES	71-0302
FREQUENCIES	NATURAL FREQUENCIES OF BOX-TYPE STRUCTURES BY A FINITE-ELEMENT METHOD	74-0193
FREQUENCIES	NATURAL FREQUENCIES OF WIDE CHORD COMPRESSOR BLADES	74-0231
FREQUENCIES	OPTIMAL BEAM FREQUENCIES BY THE FINITE ELEMENT DISPLACEMENT METHOD	73-0187
FREQUENCIES	PSEUDO STATIC DEFORMATION AND FREQUENCIES OF ROTATING TURBOMACHINERY BLADES	72-0239
FREQUENCIES	QUADRATIC MATRIX EQUATIONS FOR DETERMINING VIBRATION MODES AND FREQUENCIES OF CONTINUOUS ELASTIC SYS	66-0064
FREQUENCIES	SULLA DETERMINAZIONE DELLE FREQUENZE PROPRIE DI ALBERI SU PIU SUPPORTI ELASTICI. (DETERMINATION OF F	74-0916
FREQUENCIES	THE AUTOMATIC COMPUTATION OF THE NATURAL FREQUENCIES OF STRUCTURAL FRAMES USING AN EXACT MATRIX TECH	73-0318
FREQUENCIES	THE EFFECT OF AN INITIAL AXISYMMETRIC IMPERFECTION ON THE NATURAL FREQUENCIES OF CONICAL SHELLS	70-0368
FREQUENCIES	THE EFFECT OF CAMBER ON THE NATURAL FREQUENCIES OF LOW ASPECT RATIO TURBOMACHINERY BLADES	73-1092
FREQUENCIES	THE EFFECT OF LUMPED PARAMETERS ON BEAM FREQUENCIES	63-0012
FREQUENCIES	THE NATURAL FREQUENCIES OF PORTAL FRAMES II	65-0047
FREQUENCIES	USE OF SIMPLE THREE-DIMENSIONAL ACOUSTIC FINITE ELEMENTS FOR DETERMINING NATURAL MODES AND FREQUENCI	72-0010
FREQUENCIES	VARIATION WITH TEMPERATURE OF RESONANT FREQUENCIES OF ANISOTROPIC PLATES	74-1046
FREQUENCY	AUTOMATED OPTIMUM DESIGN OF AIRCRAFT WINGS TO SATISFY STRENGTH, STABILITY, FREQUENCY AND FLUTTER REQ	72-0605
FREQUENCY	FINITE ELEMENT ANALYSIS OF FREQUENCY SPECTRA FOR ELASTIC WAVEGUIDES	75-0247
FREQUENCY	FRAME OPTIMIZATION INCLUDING FREQUENCY CONSTRAINTS	75-0082
FREQUENCY	FREQUENCY ANALYSIS OF CERTAIN SINGLE AND CONTINUOUS SPAN BRIDGES	71-0337
FREQUENCY	FREQUENCY ANALYSIS OF SKEW ORTHOTROPIC PLATES BY THE FINITE STRIP METHOD	71-0413
FREQUENCY	FREQUENCY ANALYSIS OF ORTHOTROPIC CIRCULAR CYLINDRICAL PANELS BY THE FINITE STRIP METHOD	73-0274
FREQUENCY	FREQUENCY COEFFICIENTS OF A SPHERICAL SHELL FOR VARYING THICKNESS AND OPENING ANGLE WITH GUIDED PINN	72-0595
FREQUENCY	FREQUENCY TUNING OF COMPRESSOR ROTOR BLADES	66-0112
FREQUENCY	FUNDAMENTAL VIBRATIONAL FREQUENCY BY A 3D - FEM PROCEDURE	75-0710
FREQUENCY	INVESTIGATION OF THE EFFECT OF FOUNDATION YIELD ON THE FREQUENCY AND SHAPE OF INHERENT VIBRATIONS OF	72-0419
FREQUENCY	MINIMUM WEIGHT DESIGN OF COMPLEX STRUCTURES SUBJECT TO A FREQUENCY CONSTRAINT	70-0374
FREQUENCY	MODIFIED POTENTIAL ENERGY MASS REPRESENTATION FOR FREQUENCY PREDICTION	66-0053
FREQUENCY	OPTIMUM DESIGN OF COMPOSITE SHELLS SUBJECT TO NATURAL FREQUENCY CONSTRAINTS	73-0154
FREQUENCY-DEPEN	FREQUENCY-DEPENDENT SHAPE FUNCTIONS FOR FREE AND FORCED STRUCTURAL VIBRATION CALCULATION	73-1091
FREQUENZE	SULLA DETERMINAZIONE DELLE FREQUENZE PROPRIE DI ALBERI SU PIU SUPPORTI ELASTICI. (DETERMINATION OF F	74-0916
FRESH-SALT	COMPARATIVE STUDY OF FRESH-SALT WATER INTERFACES USING FINITE ELEMENT AND SIMPLE APPROACHES	75-0468
FRICTION	FINITE ELEMENT ANALYSIS OF ALTERNATIVE AXIAL LOADING OF AN ELASTIC PLATE PRESSED BETWEEN TWO ELASTIC	71-0357
FRICTION	NON-UNIFORM DEFORMATION OF A BLOCK IN PLANE-STRAIN COMPRESSION CAUSED BY FRICTION. PART 4, ANALYSIS	71-0786
FRICTION	NON-UNIFORM DEFORMATION OF MATERIAL IN AXIALLY SYMMETRIC COMPRESSION CAUSED BY FRICTION. PART 3, THE	72-0923
FRICTION	ON THE NON-UNIFORM DEFORMATION OF A BLOCK IN PLANE-STRAIN COMPRESSION CAUSED BY FRICTION -3	71-0328
FRICTION	ON THE NONUNIFORM DEFORMATION OF MATERIAL IN AXIALLY SYMMETRIC COMPRESSION CAUSED BY FRICTION - 2	71-0321
FRICTION	STRESS DISTRIBUTION UNDER UNIAXIAL ELASTIC PLANE STRAIN COMPRESSION WITH FRICTION	74-0526
FRICTION-GRIP	DISTRIBUTION OF THE CLAMPING PRESSURE IN FRICTION-GRIP BOLTED JOINTS	74-0250
FRICTIONAL	FINITE ELEMENT ANALYSIS OF ELASTO-PLASTIC SOILS. REPORT NO. 4, FINITE ELEMENT ANALYSIS OF ELASTO-PLA	72-0788
FRIEDRICH'S	A FINITE ELEMENT METHOD FOR FRIEDRICH'S SYMMETRIC POSITIVE SYSTEMS	71-0146
FRONT	A FRONT SOLUTION FOR FINITE ELEMENT TECHNIQUES	69-0412
FRONT	A MORE EFFICIENT FRONT SOLUTION: ALLOCATING ASSEMBLY LOCATIONS BY LONGEVITY CONSIDERATIONS	73-1069
FRONT	ANALYSE TRIDIMENSIONELLE DE LA STABILITE D'UN TUNNEL AU VOISINAGE DU FRONT DE TAILLE DANS UNE ROCHE	74-0995
FRONT	APPLICATION OF SAMIS TO THE ANALYSIS OF A FRONT SUSPENSION UPPER ARM	70-0260
FRONT	COMPUTER TECHNIQUE: THE FRONT SOLUTION	72-0631
FRONT	FRONT DISPLACEMENT MODEL IN A FRACTURED RESERVOIR	74-0071
FRONT	GALERKIN-FINITE ELEMENT TECHNIQUE FOR CALCULATING TRANSIENT POSITION OF SALTWATER FRONT	75-0135
FRONT-END	STATIC ANALYSIS VIA SUBSTRUCTURING OF AN EXPERIMENTAL VEHICLE FRONT-END BODY STRUCTURE	74-0313
FRONTAL	A FRONTAL SOLUTION PROGRAM FOR FINITE ELEMENT ANALYSIS	74-0073
FRONTAL	ELEMENT RESEQUENCING FOR FRONTAL SOLUTIONS	75-0132
FRONTAL	LA METHODE FRONTALE POUR LA RESOLUTION DES SYSTEMES LINEARESS. (FRONTAL METHOD FOR SOLVING LINEAR SY	73-1097
FRONTALE	LA METHODE FRONTALE POUR LA RESOLUTION DES SYSTEMES LINEARESS. (FRONTAL METHOD FOR SOLVING LINEAR SY	73-1097
FROZEN	IN SITU CREEP ANALYSIS OF ROOM IN FROZEN SOIL	72-0128
FROZEN	MARINE PROPELLER BLADES STRESSES AND DEFORMATIONS WITH GIVEN, FROZEN, HYDRODYNAMIC LOADING	73-0474
FRP	HIGH STRENGTH END FITTINGS FOR FRP ROD AND ROPE	74-0605
FRUIT	COMPUTER ANALYSIS OF FRUIT DETACHMENT DURING TREE SHAKING	74-1162
FUEL	A RADIATIVE IGNITION MODEL OF A SOLID FUEL	74-1312
FUEL	AIRCRAFT FUEL TANK VULNERABILITY TO HYDRAULIC RAM: MODIFICATTION OF THE NORTHROP FINITE ELEMENT COMP	74-0724
FUEL	ANALYSIS OF NUCLEAR FUEL ROD LOCALIZED STRESSES	73-0708
FUEL	APPLICATION OF THE FINITE ELEMENT METHOD TO THE NONLINEAR ANALYSIS OF NUCLEAR REACTOR FUEL BEHAVIOR	74-1246
FUEL	APPLICATION OF THREE-DIMENSIONAL FINITE ELEMENTS FOR ANALYSIS OF IRRADIATION INDUCED STRESSES IN NUC	72-0078
FUEL	BLOWDOWN EFFECTS ON A FUEL ROD	75-0575
FUEL	CALCULATIONAL PROCEDURE FOR DETERMINING CREEP COLLAPSE OF LWR FUEL RODS	74-1001
FUEL	FAST REACTOR PROGRAMME. FINITE ELEMENT TEMPERATURE ANALYSIS IN FUEL PINS	73-0939
FUEL	FINITE ELEMENT ANALYSIS OF STRESSES IN FUEL ELEMENTS OF A HIGH-TEMPERATURE GAS-COOLED REACTOR	71-0115
FUEL	FINITE ELEMENT APPROACH TO PRESSURE WAVE ATTENUATION BY REACTOR FUEL SUBASSEMBLIES	75-0522
FUEL	FINITE ELEMENT FRACTURE MECHANICS ANALYSIS OF CREEP RUPTURE OF FUEL ELEMENT CLADDING	74-0929
FUEL	FINITE ELEMENT STRESS ANALYSIS OF INTERACTING FUEL PELLET AND CANNING UNDER FAST REACTOR CONDITION	73-0814
FUEL	FINITE ELEMENT STUDY OF PRESSURE WAVE ATTENUATION BY REACTOR FUEL SUBASSEMBLIES	75-0215
FUEL	FINITE-ELEMENT ANALYSIS OF RADIATION-DAMAGE STRESSES IN THE GRAPHITE OF MATRIX FUEL ELEMENTS	70-0048
FUEL	FUEL ELEMENT STRESSES AT DISCONTINUITIES AND INTERACTIONS BY FINITE ELEMENTS IN TWO DIMENSIONS	73-0083
FUEL	FUEL ROD DEFORMATIONS - A TIME DEPENDENT FINITE ELEMENT ANALYSIS ON AXISYMMETRIC FUEL RODS, INCLUDIN	70-0414
FUEL	FUEL ROD DEFORMATIONS - A TIME DEPENDENT FINITE ELEMENT ANALYSIS ON AXISYMMETRIC FUEL RODS, INCLUDIN	70-0414
FUEL	FUEL ROD FINITE ELEMENT STRESS ANALYSIS: SEGMENTED FLIGHT REACTOR FUEL VERSUS S8DR FUEL	71-0798
FUEL	FUEL ROD FINITE ELEMENT STRESS ANALYSIS: SEGMENTED FLIGHT REACTOR FUEL VERSUS S8DR FUEL	71-0798
FUEL	FUEL ROD FINITE ELEMENT STRESS ANALYSIS: SEGMENTED FLIGHT REACTOR FUEL VERSUS S8DR FUEL	71-0798
FUEL	FULMOD - AN INELASTIC ANALYSIS PROGRAM TO PREDICT THE OPERATING BEHAVIOUR OF CANDU FUEL ELEMENTS	75-0574
FUEL	INFLUENCE OF DEFECTS ON FUEL ELEMENT CLADDING STRESSES	73-0863
FUEL	MATHEMATICAL MODELING AND ANALYSIS OF FUEL RODS	74-0930
FUEL	NONLINEAR FINITE-ELEMENT METHOD FOR PLATES AND ITS APPLICATION TO DYNAMIC RESPONSE OF REACTOR FUEL S	74-0670
FUEL	SAFE-COLAPS: A FINITE ELEMENT PROGRAM FOR THE CREEP COLLAPSE ANALYSIS OF NUCLEAR FUEL RODS	75-0573
FUEL	SOME MODELS FOR THE ANALYSIS OF STRESSES IN A TUBULAR FUEL PIN FOR HIGH TEMPERATURE REACTORS BY FINI	73-0816
FUEL	STRESS ANALYSIS OF A GRAPHITE FUEL ELEMENT INCLUDING RADIATION CREEP AND DIMENSIONAL CHANGES	71-0217
FUEL	THERMAL AND STRESS ANALYSIS ON PRISMATIC NUCLEAR FUEL ELEMENTS	75-0120
FUEL	THERMAL AND STRESS ANALYSIS OF A FUEL ROD	75-0713
FUEL	THREE-DIMENSIONAL STUDY OF NUCLEAR FUEL ROD BEHAVIOUR DURING STARTUP	73-0815
FUEL	THREE-DIMENSIONAL STUDY OF NUCLEAR FUEL ROD BEHAVIOR DURING STARTUP	74-1003
FUEL	VIBRATORY WEAR OF FUEL RODS	75-0771
FUELLED	ANALYSIS OF LOADING AND THERMAL EFFECTS ON FUELLED GRAPHITE BRICKS FOR A HIGH-TEMPERATURE GAS COOLED	71-0085
FUGA	THE FUGA FAMILY FOR FOLDED PLATE STRUCTURES	69-0333
FUGA	THE FUGA FAMILY OF ELEMENTS FOR FOLDED PLATE STRUCTURES A FURTHER SEQUEL TO TECHNICAL NOTE 14	69-0446
FULL	A TORSIONAL STRENGTH ANALYSIS ON THE CONTAINER SHIP BY MEANS OF THE FINITE ELEMENT PROCEDURE AND FUL	71-0243
FULL	EXAMPLE OF LARGE-SCALE STRUCTURAL ANALYSIS. COMPARISON BETWEEN FINITE ELEMENT CALCULATION AND FULL S	74-0177
FULL	FULL RANGE ANALYSIS OF ECCENTRICALLY STIFFENED PLATES	74-0149
FULL	STRESS ANALYSIS OF FIRST MOLARS WITH FULL CROWN PREPARATIONS BY THREE-DIMENSIONAL PHOTOELASTICITY AND	72-0939
FULL-LOAD	NONLINEAR FINITE ELEMENT SOLUTION OF ELECTRICAL MACHINES UNDER NO-LOAD AND FULL-LOAD CONDITIONS	74-0477

FULL-RANGE	FULL-RANGE ANALYSIS OF STEEL PLATES AND STIFFENED PLATING UNDER UNIAXIAL COMPRESSION	75-0803
FULL-SCALE	CORRELATION OF FINITE ELEMENT STRUCTURAL DYNAMIC ANALYSIS WITH MEASURED FREE VIBRATION CHARACTERISTI	74-0486
FULL-SCALE	PROPELLER-EXCITED VIBRATION WITH PARTICULAR REFERENCE TO FULL-SCALE MEASUREMENTS	75-0018
FULL-WIDTH	SHEAR LAG ANALYSIS OF RECTANGULAR FULL-WIDTH TUBE JUNCTIONS	72-0087
FULMOD	FULMOD - AN INELASTIC ANALYSIS PROGRAM TO PREDICT THE OPERATING BEHAVIOUR OF CANDU FUEL ELEMENTS	75-0574
FUNCTION	A FINITE ELEMENT METHOD FOR THREE DIMENSIONAL FUNCTION APPROXIMATION	71-0654
FUNCTION	A PENALTY FUNCTION APPROACH TO THE MINIMIZATION OF QUADRATIC FUNCTIONALS IN FINITE ELEMENT ANALYSIS	74-0418
FUNCTION	APPROXIMATION IN ELASTICITY BASED ON THE CONCEPT OF FUNCTION SPACE	47-0001
FUNCTION	BLENDING FUNCTION FINITE ELEMENTS FOR CURVED BOUNDARIES	75-0093
FUNCTION	CONSTRAINED VARIATIONAL PRINCIPLES AND PENALTY FUNCTION METHODS IN FINITE ELEMENT ANALYSIS	72-0551
FUNCTION	DEFLECTION FUNCTION FOR THE ASYMMETRICAL BENDING OF CIRCULAR PLATES	73-0468
FUNCTION	DEVELOPMENTS IN DISCRETE ELEMENT FINITE DEFLECTION STRUCTURAL ANALYSIS BY FUNCTION MINIMIZATION	68-0144
FUNCTION	DEVELOPMENTS IN DISCRETE ELEMENT FINITE DEFLECTION STRUCTURAL ANALYSIS BY FUNCTION MINIMIZATION	68-0243
FUNCTION	DEVELOPMENTS IN STRUCTURAL ANALYSIS AND SYNTHESIS BY FUNCTION MINIMIZATION	71-0384
FUNCTION	FINITE ELEMENT ANALYSIS OF VISCOUS FLOW BY STREAM FUNCTION	75-0526
FUNCTION	FUNKCJA KSZTALTU NIESPELNIAPACA WARUNKOW CIAGLOSCI W ANALIZIE PLYT METODA ELEMENTOW SKONCZONYCH. (SH	75-0263
FUNCTION	KERNEL FUNCTION FOR NONPLANAR OSCILLATING SURFACES IN SUPERSONIC FLOW	71-0472
FUNCTION	ON THE EFFECT OF THE FORM OF THE STRAIN ENERGY FUNCTION ON THE SOLUTION OF A BOUNDARY-VALUE PROBLEM	72-0429
FUNCTION	ON THE GENERALIZATION OF STRESS FUNCTION PROCEDURE FOR DYNAMIC ANALYSIS OF PLATES	73-0661
FUNCTION	PENALTY FUNCTION APPROACH TO PROBLEMS OF PLASTIC FLOW OF METALS WITH LARGE SURFACE DEFORMATIONS	75-0282
FUNCTION	SHAPE FUNCTION FORMULATIONS FOR ELEMENTS OTHER THAN DISPLACEMENT MODELS	72-0472
FUNCTION	SHAPE FUNCTION SUBROUTINE FOR AN ISOPARAMETRIC THIN PLATE ELEMENT	73-0618
FUNCTION	SOLUTION OF TRANSPORT PROBLEMS BY USE OF TRIAL FUNCTION TECHNIQUES	72-0991
FUNCTION	STABILITY IN LQ OF L2-PROJECTION INTO FINITE ELEMENT FUNCTION SPACES	75-0071
FUNCTION	STRESS FUNCTION APPROACH	75-0016
FUNCTION	SUBSONIC AND SUPERSONIC INDICIAL AERODYNAMICS AND AERODYNAMIC TRANSFER FUNCTION FOR COMPLEX CONFIGUR	74-1320
FUNCTION	TRANSFINITE ELEMENT METHODS: BLENDING FUNCTION INTERPOLATION OVER ARBITRARY CURVED ELEMENT DOMAINS	73-1006
FUNCTION	TRIGONOMETRIC FUNCTION REPRESENTATIONS FOR RECTANGULAR PLATE BENDING ELEMENTS	71-0765
FUNCTIONAL	FINITE ELEMENT ANALYSIS OF SHELLS OF REVOLUTION BY MINIMIZATION OF THE POTENTIAL ENERGY FUNCTIONAL	69-0114
FUNCTIONAL	FINITE ELEMENT ANALYSIS OF SHELLS OF REVOLUTION BY MINIMIZATION OF THE POTENTIAL ENERGY FUNCTIONAL	69-0354
FUNCTIONAL	FINITE ELEMENT ANALYSIS OF SHELLS OF REVOLUTION BY MINIMIZATION OF THE POTENTIAL ENERGY FUNCTIONAL	70-0481
FUNCTIONAL	FUNCTIONAL ANALYSIS AND THE FINITE ELEMENT METHOD APPLIED TO ELLIPTIC BOUNDARY VALUE PROBLEMS	72-0056
FUNCTIONAL	FUNCTIONAL ANALYSIS DERIVATION AND GENERALIZATION OF HYBRID VARIATIONAL METHODS	73-1104
FUNCTIONAL	FUNCTIONAL APPROXIMATIONS FOR SOLVING BOUNDARY VALUE PROBLEMS BY COMPUTER	69-0013
FUNCTIONAL	FUNCTIONAL COEFFICIENTS IN THE ANALYSIS OF GROUNDWATER FLOW USING FINITE ELEMENTS	73-0077
FUNCTIONAL	FUNCTIONAL MINIMIZATION IN NONLINEAR SOLID MECHANICS	74-1251
FUNCTIONAL	GOVERNING EQUATIONS AND EXTREMUM PRINCIPLES OF ELASTICITY AND PLASTICITY GENERATED FROM A SINGLE FUN	73-0666
FUNCTIONAL	NEW VARIATIONAL FUNCTIONAL FOR THE FINITE-ELEMENT METHOD AND ITS APPLICATION TO PLATE SHELL PROBLEMS	72-0108
FUNCTIONAL	ORGANIZATION AND FUNCTIONAL PURPOSE OF A FINITE ELEMENT COMPUTER PROGRAM	74-0362
FUNCTIONAL	SATURATED MAGNETIC ENERGY FUNCTIONAL FOR FINITE-ELEMENT ANALYSIS OF ELECTRIC MACHINES	75-0636
FUNCTIONAL	THE PSEUDO FUNCTIONAL FINITE ELEMENT METHOD	74-0004
FUNCTIONALS	A CLASS OF FUNCTIONALS GIVING IMPROVED CONVERGENCE WITH FINITE ELEMENTS	72-0553
FUNCTIONALS	A PENALTY FUNCTION APPROACH TO THE MINIMIZATION OF QUADRATIC FUNCTIONALS IN FINITE ELEMENT ANALYSIS	74-0418
FUNCTIONALS	BOUNDS FOR A CLASS OF LINEAR FUNCTIONALS WITH APPLICATIONS TO HERMITE INTERPOLATION	71-0582
FUNCTIONS	AN IMPROVED SHELL OF REVOLUTION ELEMENT UTILISING CUBIC DISPLACEMENT FUNCTIONS	70-0451
FUNCTIONS	APPLICATION OF HILL FUNCTIONS TO CIRCULAR PLATE PROBLEMS	75-0572
FUNCTIONS	APPLICATIONS OF HILL FUNCTIONS (APPLIED MATHEMATIC'S FINITE ELEMENT) TO APPLIED MECHANICS PROBLEMS	72-0884
FUNCTIONS	APPROXIMATION BY FINITE ELEMENT FUNCTIONS USING REGULARIZATION	74-1229
FUNCTIONS	APPROXIMATION BY FINITE-ELEMENT FUNCTIONS USING LOCAL REGULARIZATION	75-0436
FUNCTIONS	BASIC FUNCTIONS FOR CURVED ELEMENTS IN THE MATHEMATICAL THEORY OF FINITE ELEMENTS	75-0091
FUNCTIONS	BASIS FUNCTIONS FOR CURVED ELEMENTS IN THE FINITE ELEMENT METHOD	72-0807
FUNCTIONS	BASIS FUNCTIONS IN FINITE ELEMENT THEORY AND APPLICATION	74-0419
FUNCTIONS	BLENDING FUNCTIONS AND FINITE ELEMENTS	74-0889
FUNCTIONS	COMPARATIVE CONSIDERATION OF PARAMETRIC STRESS CONCENTRATION INVESTIGATIONS WITH FINITE ELEMENTS AND	72-0423
FUNCTIONS	COMPARATIVE OBSERVATION ON PARAMETRIC STRESS STUDIES WITH FINITE ELEMENTS WITH COMPLEX STRESS FUNCTI	72-0250
FUNCTIONS	COMPLETENESS AND CONVERGENCE PROPERTIES OF FINITE ELEMENT DISPLACEMENT FUNCTIONS., A GENERAL TREATME	67-0053
FUNCTIONS	CONJUGATE APPROXIMATION FUNCTIONS IN FINITE ELEMENT ANALYSIS	71-0043
FUNCTIONS	CONSTRUCTION OF NORMAL DISPLACEMENT FUNCTIONS OF TRIANGULAR FINITE ELEMENT OF PLATES AND SHELLS	72-0563
FUNCTIONS	EFFICIENT FINITE-ELEMENT METHODS FOR STRESS INTENSITY FACTORS USING WEIGHT FUNCTIONS	75-0411
FUNCTIONS	EFFICIENT FINITE-ELEMENT METHODS FOR STRESS INTENSITY FACTORS USING WEIGHT FUNCTIONS - REPLY	75-0421
FUNCTIONS	ELASTIC EQUILIBRIUM MODELS, STRESS FUNCTIONS, AND DUALITY NON LINEARITIES AND STABILITY, FIVE LECTUR	73-0631
FUNCTIONS	ELEMENT TYPES AND BASE FUNCTIONS	73-0522
FUNCTIONS	FINITE ELEMENTS WITH HARMONIC INTERPOLATION FUNCTIONS	72-0058
FUNCTIONS	FINITE ELEMENTS WITH HARMONIC INTERPOLATION FUNCTIONS	73-1065
FUNCTIONS	FONCTIONS SPLINES ET METHODE D ELEMENTS FINIS (SPLINE FUNCTIONS AND FINITE ELEMENT METHOD)	75-0846
FUNCTIONS	FREQUENCY-DEPENDENT SHAPE FUNCTIONS FOR FREE AND FORCED STRUCTURAL VIBRATION CALCULATION	73-1091
FUNCTIONS	FURTHER DEVELOPMENTS IN THE USE OF LINEAR DISPLACEMENT FUNCTIONS IN THE FINITE ELEMENT ANALYSIS OF O	71-0344
FUNCTIONS	GENERALIZED CONJUGATE FUNCTIONS FOR MIXED FINITE ELEMENT APPROXIMATIONS OF BOUNDARY VALUE PROBLEMS	72-0515
FUNCTIONS	GENERATION OF SERENDIPEAN SHAPE FUNCTIONS IN FINITE ELEMENT METHOD	74-1012
FUNCTIONS	IMPROVED DEFORMATION FUNCTIONS FOR THE FINITE ELEMENT ANALYSIS OF BEAM SYSTEMS	69-0075
FUNCTIONS	INFLUENCE FUNCTIONS IN FINITE ELEMENT METHOD	70-0226
FUNCTIONS	INVESTIGATION OF THE FINITE ELEMENT METHOD FOR DETERMINING GREEN'S FUNCTIONS	75-0751
FUNCTIONS	LOCAL AND GLOBAL SMOOTHING OF DISCONTINUOUS FINITE ELEMENT FUNCTIONS USING A LEAST SQUARES METHOD	74-0378
FUNCTIONS	METHODE DES ELEMENTS FINIS; UNE FAMILE D'APPROXIMATION DES ESPACES DE SOBOLEV PAR LES TRANSLATES DE	70-0301
FUNCTIONS	NEW CYLINDRICAL SHELL FINITE ELEMENT BASED ON SIMPLE INDEPENDENT STRAIN FUNCTIONS	72-0094
FUNCTIONS	NONLINEAR ANALYSES USING SPLINE FUNCTIONS	71-0466
FUNCTIONS	ON COMPLETENESS OF SHAPE FUNCTIONS FOR FINITE ELEMENT ANALYSIS	72-0105
FUNCTIONS	ON THE COMPLETENESS OF SHAPE FUNCTIONS FOR FINITE ELEMENT ANALYSIS	72-0885
FUNCTIONS	ON THE ENERGY FUNCTIONS FOR CURVED SANDWICH PLATES	71-0587
FUNCTIONS	SHAPE FUNCTIONS AND ACCURACY OF ARCH FINITE ELEMENTS	73-0038
FUNCTIONS	SOME ELEMENT TYPES AND TRIAL FUNCTIONS IN THE FINITE ELEMENT METHOD	70-0242
FUNCTIONS	SPECIFICATION OF GEOMETRICAL PARAMETERS FOR ELEMENTS WITH CUBIC MAPPING FUNCTIONS	73-1082
FUNCTIONS	SPLINE FUNCTIONS AND FINITE-ELEMENT METHOD	75-0430
FUNCTIONS	STIFFNESS MATRICES FOR THE GENERAL DEFORMATION (OUT-OF-PLANE AND INPLANE) OF CURVED BEAM MEMBERS BAS	75-0124
FUNCTIONS	STRAIN ENERGY FUNCTIONS AND MATHEMATICAL MODELS OF LUNG ELASTICITY	73-0591
FUNCTIONS	THE CONSTRUCTION OF BASIC FUNCTIONS FOR CURVED ELEMENTS IN THE FINITE ELEMENT METHOD	72-0558
FUNCTIONS	THREE-DIMENSIONAL HARMONIC FUNCTIONS NEAR TERMINATION OR INTERSECTION OF GRADIENT SINGULARITY LINES:	74-0178
FUNCTIONS	USE OF AUXILIARY FUNCTIONS IN THE FINITE ELEMENT METHOD	72-0938
FUNCTIONS	USE OF HIGHER ORDER DISPLACEMENT FUNCTIONS IN THE FREE VIBRATION ANALYSIS OF SHELLS OF REVOLUTION HA	74-1055
FUNCTIONS	USE OF SINGULAR FUNCTIONS WITH FINITE ELEMENT APPROXIMATIONS	73-0102
FUNCTIONS	VARIATIONAL SOLUTIONS BY USE OF STEPWISE CONSTANT FUNCTIONS: ONE LINEAR CASE	72-0286
FUNDAMENTAL	FUNDAMENTAL STUDIES OF MEDIUM-STRUCTURE INTERACTION. REPORT 1. FINITE ELEMENT ANALYSIS OF BURIED CYL	72-0854
FUNDAMENTAL	FUNDAMENTAL STUDIES ON STEADY SHIP WAVE PROBLEMS BY THE FINITE ELEMENT METHOD	74-0564
FUNDAMENTAL	FUNDAMENTAL STUDY OF UNDERWATER WELDING	72-0382
FUNDAMENTAL	FUNDAMENTAL VIBRATIONAL FREQUENCY BY A 3D - FEM PROCEDURE	75-0710
FUNDAMENTAL	THE FUNDAMENTAL NOTIONS	72-0498
FUNDAMENTALS	FINITE ELEMENT ANALYSIS: FUNDAMENTALS	75-0772
FUNDAMENTALS	FINITE ELEMENT TECHNIQUES - A SHORT COURSE OF FUNDAMENTALS AND APPLICATIONS	73-0612
FUNDAMENTALS	FUNDAMENTALS OF FINITE ELEMENT TECHNIQUES FOR STRUCTURAL ENGINEERS	73-1050
FUNDAMENTALS	FUNDAMENTALS OF FINITE ELEMENT METHODS FOR ENGINEERS	74-0719
FUNDAMENTALS	FUNDAMENTALS OF THE FINITE ELEMENT METHOD	74-0191
FUNDAMENTALS	THE FINITE ELEMENT METHOD - FUNDAMENTALS AND APPLICATIONS	73-0001

237

FUNKCJA	FUNKCJA KSZTALTU NIESPELNIAPACA WARUNKOW CIAGLOSCI W ANALIZIE PLYT METODA ELEMENTOW SKONCZONYCH. (SH	75-0263
FURNACE	STRUCTURAL ANALYSIS OF WELDED WALL FOR BOILER FURNACE	74-0653
FUSELAGE	CORRELATION OF FINITE ELEMENT STRUCTURAL DYNAMIC ANALYSIS WITH MEASURED FREE VIBRATION CHARACTERISTI	74-0486
FUSELAGE	MODERN FUSELAGE ANALYSIS AND THE ELASTIC AIRCRAFT	63-0023
FUSELAGE	ON THE THEORY OF CALCULATION OF THE OVERALL STRENGTH OF A FUSELAGE BY THE METHOD OF FINITE ELEMENTS	72-0692
FUSELAGE	POSTBUCKLING BEHAVIOUR OF A SECTION REPRESENTATIVE OF THE B-1 AFT INTERMEDIATE FUSELAGE	73-0489
FUSELAGE	SOME RESULTS OF FUSELAGE ANALYSIS BY THE FINITE ELEMENT METHOD USING A COMPUTER	73-0535
FUSELAGE	STRUCTURAL ANALYSIS OF ARAVA FUSELAGE USING FINITE ELEMENT METHOD	71-0578
FUSELAGES	AUTOMATIC ANALYSIS OF FUSELAGES AND PROBLEMS OF CONDITIONING	64-0027
FUSELAGES	BERECHNUNG DES REIBUNGSLOSEN STROEMUNGSFELDES VON DREIDIMENSIONALEN AUFTRIEBSBEHAFTETENTRAGFLUEGELN	70-0533
FUTURE	A LOOK INTO THE FUTURE: HOW COMPUTERS WILL INFLUENCE ENGINEERING	67-0139
FUTURE	COMPUTER PROGRAM CONSTRUCTION AND MAINTENANCE - FUTURE OF CENTRALIZED FINITE ELEMENT ACTIVITY	75-0014
FUTURE	FRACTURE CONTROL; PAST, PRESENT AND FUTURE	73-0516
FUTURE	FUTURE DEVELOPMENTS IN STRUCTURAL ANALYSIS	74-0390
FUZE	NORMAL MODE ANALYSIS OF A FUZE SUPPORT STRUCTURE USING NASTRAN. PART 1	72-0853
FYZIKALNE	STATICKE RESENI FYZIKALNE NELINEARNIHO CHOVANI ZELEZOBETO- NOVYCH DESEK (ANALYSIS OF PHYSICALLY NONL	74-0514
GABLE	STRESSES IN GABLE ROOF H. P. SHELLS.	72-0394
GAGE	FINITE ELEMENT APPLICATION TO STRAIN GAGE DATA REDUCTION	73-0561
GAGE	LIGHT GAGE STEEL INFILL PANELS IN MULTISTORY STEEL FRAMES	74-0229
GAGE	UNTERSUCHUNGEN ZUR SPURMASSTABILITAET BEI RADSAETZEN MIT KLOTZGEBREMSTEN VOLLRAEDERN (INVESTIGATIONS	75-0810
GALENA	DESTRESSING TEST AT THE GALENA MINE, WALLACE, IDAHO	72-0126
GALERKIN	A METHOD TO GALERKIN TYPE ACHIEVING OPTIMUM L(2) ACCURACY FOR FIRST ORDER HYPERBOLICS AND EQUATIONS	72-0521
GALERKIN	ACCURACY AND CONVERGENCE OF FINITE ELEMENT GALERKIN APPROXIMATIONS OF TIME DEPENDENT PROBLEMS WITH E	73-0987
GALERKIN	ACCURACY AND CONVERGENCE OF FINITE ELEMENT/GALERKIN APPROXIMATIONS OF TIME DEPENDENT PROBLEMS WITH E	75-0006
GALERKIN	AN APPPOXIMATE NUMERICAL SOLUTION OF THE NAVIER STOKES EQUATIONS BY GALERKIN METHOD	74-0011
GALERKIN	COMPRESSIBLE SEAL FLOW ANALYSIS USING THE FINITE ELEMENT METHOD WITH GALERKIN SOLUTION TECHNIQUE	74-0688
GALERKIN	CUBIC SPLINE GALERKIN APPROXIMATIONS TO PARABOLIC SYSTEMS WITH COUPLED NON-LINEAR BOUNDARY CONDITION	75-0232
GALERKIN	DERIVATION OF THE STIFFNESS MATRIX FOR PLATES BY GALERKIN METHOD	69-0402
GALERKIN	ERROR BOUNDS FOR SEMIDISCRETE GALERKIN APPROXIMATIONS OF PARABOLIC PROBLEMS WITH APPLICATIONS TO PET	70-0291
GALERKIN	FINITE ELEMENT - GALERKIN METHODS FOR HYPERBOLIC EQUATIONS WITH DISCONTINUOUS COEFFICIENTS	73-0726
GALERKIN	FINITE ELEMENT GALERKIN METHOD SOLUTIONS TO SELECTED ELLIPTIC AND PARABOLIC DIFFERENTIAL EQUATIONS	71-0027
GALERKIN	FINITE ELEMENT GALERKIN METHOD FOR MIXED INITIAL-BOUNDARY VALUE PROBLEMS IN ELASTICITY THEORY	71-0520
GALERKIN	FINITE ELEMENT GALERKIN METHOD FOR ANALYSIS OF FLOW IN FRACTURED POROUS MEDIA	74-0066
GALERKIN	FINITE ELEMENT METHOD: A GALERKIN APPROACH	71-0451
GALERKIN	GALERKIN APPROACH TO UNSATURATED FLOW IN SOILS	74-0058
GALERKIN	GALERKIN APPROACH TO SATURATED-UNSATURATED FLOW IN POROUS MEDIA	75-0393
GALERKIN	GALERKIN APPROXIMATION OF THE TIME DERIVATIVE IN THE FINITE ELEMENT ANALYSIS	73-0573
GALERKIN	GALERKIN APPROXIMATION OF THE TIME DERIVATIVE IN THE FINITE ELEMENT ANALYSIS OF GROUNDWATER FLOW	74-0658
GALERKIN	GALERKIN FINITE ELEMENT FORMULATION FOR INCOMPRESSIBLE FLOW	70-0000
GALERKIN	GALERKIN FINITE ELEMENT METHOD FOR VIBRATION PROBLEMS	73-0026
GALERKIN	GALERKIN FINITE ELEMENT METHOD FOR VIBRATION PROBLEMS	73-0659
GALERKIN	GALERKIN FINITE ELEMENT SOLUTIONS FOR POLLUTION PROBLEMS IN PARTIALLY MIXED ESTUARIES	74-0047
GALERKIN	GALERKIN FINITE ELEMENT ANALYSIS OF A UNIFORM BEAM CARRYING A CONCENTRATED MASS AND ROTARY INERTIA W	74-0552
GALERKIN	GALERKIN FINITE ELEMENT SOLUTION FOR THE STABILITY OF CANTILEVER COLUMNS SUBJECTED TO TANGENTIAL LOA	75-0534
GALERKIN	GALERKIN FINITE-ELEMENT SIMULATION OF A GEOTHERMAL RESERVOIR	73-0900
GALERKIN	GALERKIN FINITE-ELEMENT SOLUTION FOR STABILITY OF CANTILIVER COLUMNS SUBJECTED TO TANGENTIAL LOADS	75-0618
GALERKIN	GALERKIN METHOD AS APPLIED TO PROBLEMS IN VISCOELASTICITY	74-1039
GALERKIN	GALERKIN METHOD FOR A CLASS OF STEADY, 2-DIMENSIONAL, INCOMPRESSIBLE, LAMINAR BOUNDARY-LAYER FLOWS	75-0437
GALERKIN	GALERKIN METHODS FOR PARABOLIC EQUATIONS	70-0448
GALERKIN	GALERKIN METHODS FOR PARABOLIC EQUATIONS WITH NONLINEAR BOUNDARY CONDITIONS	73-1023
GALERKIN	NUMERICAL PERFORMANCE OF A FINITE ELEMENT GALERKIN PROCEDURE	75-0022
GALERKIN	ON THE SEMI-DISCRETE GALERKIN METHOD FOR HYPERBOLIC PROBLEMS AND ITS APPLICATION TO PROBLEMS IN ELAS	72-0457
GALERKIN	RAYLEIGH RITZ GALERKIN METHODS FOR SYMMETRIC POSITIVE DIFFERENTIAL EQUATIONS APPLICATION TO FINITE E	73-1030
GALERKIN	RAYLEIGH-RITZ AND GALERKIN FINITE ELEMENTS FOR DIFFUSION-CONVECTION PROBLEMS	73-0017
GALERKIN	RICHARDSON EXTRAPOLATION FOR PARABOLIC GALERKIN METHODS	72-0522
GALERKIN	ROTOR BLADE VIBRATIONS BY GALERKIN FINITE-ELEMENT METHOD	75-0389
GALERKIN	SEMIDISCRETE GALERKIN TECHNIQUES WITH TIME INTERPOLATION AND SPLITTING UP FOR PLASMA SIMULATION	75-0121
GALERKIN	SIMULATION OF GROUND WATER CONTAMINATION USING A GALERKIN FINITE ELEMENT TECHNIQUE	74-1362
GALERKIN	SOME L(2) ERROR ESTIMATES FOR PARABOLIC GALERKIN METHODS	72-0511
GALERKIN	SOME MATHEMATICAL PROPERTIES OF CERTAIN MIXED GALERKIN APPROXIMATIONS IN NONLINEAR ELASTICITY	74-1248
GALERKIN	THE RELATION BETWEEN THE GALERKIN AND COLLOCATION METHODS USING SMOOTH SPLINES	74-0912
GALERKIN	WATER MOVEMENT THROUGH SATURATED-UNSATURATED POROUS MEDIA: A FINITE-ELEMENT GALERKIN MODEL	75-0761
GALERKIN-FINITE	GALERKIN-FINITE ELEMENT SIMULATION OF GROUNDWATER CONTAMINATION ON LONG-ISLAND, NEW-YORK	73-0093
GALERKIN-FINITE	GALERKIN-FINITE ELEMENT TECHNIQUE FOR CALCULATING TRANSIENT POSITION OF SALTWATER FRONT	75-0135
GALERKIN-FINITE	GALERKIN-FINITE ELEMENT ANALYSIS OF HYDROTHERMAL SYSTEM AT WAIRAKEI, NEW-ZEALAND	75-0601
GALERKIN'S	APPLICATION OF GALERKIN'S PROCEDURE TO AQUIFER ANALYSIS	72-0636
GALERKIN'S	DERIVATION OF STIFFNESS MATRICES FOR PROBLEMS IN PLANE ELASTICITY BY GALERKIN'S METHOD	69-0250
GALERKIN'S	RESERVOIR SIMULATION BY GALERKIN'S METHOD	73-0662
GALERKIN'S	STIFFNESS METHODS FOR PLATES BY GALERKIN'S METHOD	69-0249
GANGERZBERGEAU	NEUARTIGE ANWENDUNG GEOMECHANISCHER UNTERSUCHUNGSMETHODEN ZUR BEURTEILUNG DER ABBAUWIRKUNGEN IM GANG	75-0530
GANGERZLAGERSTA	BERECHNUNG UND BEURTEILUNG DER STANDSICHERHEIT VON ABBAUHOHLRAEUMEN IN STEILEN GANGERZLAGERSTAETTEN	74-0597
GAPS	IMPEDANCE OF INTERSTITIAL GAPS IN TRANSDUCER ARRAYS	70-0410
GAS	A FINITE ELEMENT ANALYSIS OF TRANSIENT RAREFIED GAS FLOW	73-0604
GAS	A MODEL STUDY OF GAS DIFFUSION IN ALVEOLAR SACS	73-0029
GAS	ANALYSIS OF LOADING AND THERMAL EFFECTS ON FUELLED GRAPHITE BRICKS FOR A HIGH-TEMPERATURE GAS COOLED	71-0085
GAS	ANALYSIS OF RAREFIED GAS FLOW THROUGH AN ARBITRARY CROSS SECTION BY THE FINITE ELEMENT METHOD	73-0365
GAS	APPLICATION OF FINITE ELEMENT ANALYSIS TO UNDERGROUND STORAGE OF NATURAL GAS	70-0198
GAS	BERECHNUNGSMETHODEN ZUR ERMITTLUNG VON BELASTUNGSBEDINGUNGEN UND GRENZTRAGFAEHIGKEITSVERHALTEN VON K	73-0858
GAS	CALCULATION OF AXISYMMETRIC LONGITUDINAL MODES FOR FLUID-ELASTIC TANK-ULLAGE GAS SYSTEMS AND COMPARI	64-0019
GAS	CERAMIC ROTATING BLADES: SOME CRITICAL DESIGN PARAMETERS FOR GAS TURBINE APPLICATIONS	74-0146
GAS	FINITE ELEMENT ANALYSIS OF A GAS TURBINE BLADE	74-0971
GAS	FINITE ELEMENT METHODS FOR UNSTEADY COMPRESSIBLE GAS DYNAMICS	75-0045
GAS	FINITE-ELEMENT CALCULATION OF PRESSURE HISTORY IN A GAS FIELD	75-0546
GAS	GAS TURBINE ENGINE DISK CYCLIC LIFE PREDICTION	75-0509
GAS	MODELING OF GAS TURBINE ENGINE COMPRESSOR BLADES FOR VIBRATION ANALYSIS	75-0508
GAS	PLANE POISEUILLE FLOW OF A RAREFIED GAS BASED ON FINITE ELEMENT METHOD	71-0575
GAS	STUDY ON THE VIBRATION OF THIN PLATES IN A GAS STREAM BY THE FINITE ELEMENT METHOD	74-0454
GAS	TEMPERATURVERLAUF IM QUERSCHNITT GESCHWEISSTER ROHRWAENDE BEI UEBERWIEGEND KONVEKTIVEM WAERMEUEBERGA	75-0628
GAS	THE DETERMINATION OF A WATER/GAS INTERFACE USING FINITE ELEMENTS	74-0346
GAS	THE FINITE ELEMENT ANALYSIS OF TRANSIENT RAREFIED GAS FLOW	74-0027
GAS	THERMAL STRESSES IN FACT REACTOR BOILER TUBES AT THE SODIUM/GAS INTERFACE	73-0820
GAS	UNSTEADY RADIAL FLOW IN GAS RESERVOIRS BY FINITE ELEMENTS	74-0462
GAS	VARIATIONAL PRINCIPLES FOR A RAREFIED GAS FLOW PAST AND ARBITRARY THREE-DIMENSIONAL BODY	70-0189
GAS-COOLED	FINITE ELEMENT ANALYSIS OF STRESSES IN FUEL ELEMENTS OF A HIGH-TEMPERATURE GAS-COOLED REACTOR	71-0115
GAS-COOLED	STRESS DISTRIBUTION IN THE LIGAMENT ZONE OF PERFORATED CIRCULAR DISCS WITH SPECIAL REFERENCE TO THE	73-0827
GAS-LIQUID	AN INTRODUCTION TO THE FINITE ELEMENT ANALYSIS OF THE MOTION OF A GAS-LIQUID INTERFACE IN A POROUS M	72-0904
GAS-LIQUID	FINITE ELEMENT ANALYSIS OF THE MOTION OF A GAS-LIQUID INTERFACE IN A POROUS MEDIUM	75-0672
GAS-TURBINE	ELASTO-PLASTIC ANALYSIS OF STRESS IN A GAS-TURBINE WHEEL	74-0323
GAS-TURBINE	METHOD FOR THE COMPUTATION OF CYCLIC STRESSES IN A GAS-TURBINE WHEEL	74-0972
GASDRUCKWELLE	BERECHNUNGSMETHODEN ZUR ERMITTLUNG VON BELASTUNGSBEDINGUNGEN UND GRENZTRAGFAEHIGKEITSVERHALTEN VON K	73-0858
GASEOUS	FINITE ELEMENT METHOD APPLIED TO IDEAL GASEOUS MIXTURES.	72-0393
GASEOUS	THIN WINDOWS FOR GASEOUS AND LIQUID TARGETS: AN OPTIMIZATION PROCEDURE	75-0741

238

GASKETS	CLOSURE ANALYSIS USING SELF ENERGIZING, DOUBLE CONICAL GASKETS	74-0675
GATE	ANALYSIS OF LIGHT GATE STEEL SHEAR DIAPHRAGMS	73-0733
GATE	FINITE ELEMENT ANALYSIS OF SLUICE GATE FLOW	71-0540
GATES	FINITE-ELEMENT ANALYSIS OF HIGH-PRESSURE, LARGE VALVE GATES	75-0329
GATES	FINITE-ELEMENT ANALYSIS OF HIGH-PRESSURE LARGE VALVE GATES	75-0413
GAUSS-SEIDEL	APPLICATIONS DE LA METHODE DE GAUSS-SEIDEL A LA RESOLUTION DES PROBLEMS D'ELASTICITE. (APPLICATION O	72-0145
GAUSS-SEIDEL	APPLICATIONS DE LA METHODE DE GAUSS-SEIDEL A LA RESOLUTION DES PROBLEMS D'ELASTICITE. (APPLICATION O	72-0145
GAUSSIAN	GAUSSIAN QUADRATURE FORMULAS FOR TRIANGLES	73-0906
GAUSSIAN	STIFFNESS MATRICES FOR THIN TRIANGULAR ELEMENTS OF NONZERO GAUSSIAN CURVATURE	66-0073
GAUSSIAN	USE OF SPARSITY IN THE SOLUTION OF FINITE ELEMENT SYSTEMS OF EQUATIONS BY GAUSSIAN ELIMINATION-TYPE	74-0868
GEAR	APPLICATION OF FINITE ELEMENTS TO THE ANALYSIS OF GEAR TOOTH STRESSES	72-0116
GEAR	COMPUTER SIMULATION OF DYNAMIC STRESS, DEFORMATION, AND FRACTURE OF GEAR TEETH	72-0114
GEAR	EVALUATION OF STRESSES AND DEFLECTION OF SPUR GEAR TEETH UNDER STRAIN	72-0115
GEAR	NEW APPROACH TO GEAR TOOTH ROOT STRESSES	72-0117
GEAR	REDUCTION GEAR DAMAGES RELATED TO EXTERNAL INFLUENCES	75-0307
GEARED	COMPUTER METHOD FOR COMPUTING CRITICAL SPEEDS OF NON-UNIFORM SHAFTS GEARED SYSTEMS AND CURVED ASSEMB	73-0675
GEBIRGE	WECHSELWIRKUNG ZWISCHEN DEFORMATION UND DURCHSTROIMUNG IM KLUEFTIGEN, ANISOTROPEN GEBIRGE. (INTERACT	72-0612
GEBIRGES	DIE BERECHNUNG DER BEWEGUNGEN UND SPANNUNGEN DES GEBIRGES VOM ABBAU BIS ZUR TAGESOBERFLAECHE NACH DE	73-0456
GEBIRGSDRUCKES	VORAUSSAGE DES GEBIRGSDRUCKES UND SEINER AUSOIRKUNGEN AUF STREBEN UND STRECKEN IM STEINKOHLENBERGBAU	73-0679
GEDAEMPFTER	ANWENDUNG DER METHODE FINITER ELEMENTE BEI DER ANALYSE DES DYNAMISCHEN VERHALTENS GEDAEMPFTER WERKZE	75-0580
GEDAMPFTER	ANWENDUNG DER MATRIZENVERSCHIEBUNGSMETHODE AUF ERZWUNGENE SCHWINGUNGEN (SCHWINGUNGSPROBLEME) (ARBEIT	71-0627
GEKRUEMMTE	FINITE ELEMENTE FUR GEKRUEMMTE FLAECHENTRAGWERKE. (FINITE ELEMENTS SOLUTIONS FOR CURVED SHELL STRUCT	71-0391
GELAGERTER	DER GEPISSENE ZUSTAND ZWEISEITIG GELAGERTER PLATTEN UNTER EINZELLASTEN - NICHTLINEARE BERECHNUNG MIT	75-0858
GELOCKERTEN	VARIATIONAL PRINCIPLES OF ELASTOSTATICS WITH RELAXED CONTINUITY REQUIREMENTS AND THEIR APPLICABILITY	74-0758
GEMISCHTE	COMPUTATION OF SHELLS WITH FINITE DISPLACEMENTS: MIXED FINITE ELEMENTS (BERECHNUNG VON SCHALEN MIT E	73-0788
GENAUE	ACCURATE CALCULATION OF STRESSES IN BEAMS, DISKS AND PLATES BY MEANS OF FINITE ELEMENT METHODS AND I	70-0133
GENAUEN	ZUR GENAUEN BERECHNUNG KONSTRUKTIV ORTHOTROPER PLATTEN (EXACT CALCULATION OF STRUCTURALLY ORTHOTROPI	72-0188
GENERAL	A FINITE ELEMENT FORMULATION FOR THE GENERAL STABILITY ANALYSIS OF THIN WALLED MEMBERS	70-0337
GENERAL	A FINITE ELEMENT SOLUTION OF GENERAL FLUID DYNAMICS PROBLEMS - NATURAL CONVECTION IN RECTANGULAR CAV	70-0008
GENERAL	A GENERAL APPROACH TO THE PRACTICAL SOLUTION OF CREEP PROBLEMS	59-0003
GENERAL	A GENERAL APPROACH TO PROBLEMS OF LARGE DEFORMATION, AND PLASTICITY USING ISOPARAMETRIC ELEMENT	71-0273
GENERAL	A GENERAL CALCULATION OF THE STRUCTURAL LOADS ON A SHIP IN A SEAWAY	71-0345
GENERAL	A GENERAL DIGITAL COMPUTER ANALYSIS OF STATICALLY INDETERMINATE STRUCTURES	59-0007
GENERAL	A GENERAL EQUILIBRIUM BEAM ELEMENT WITH OFFSETS	73-0351
GENERAL	A GENERAL EQUILIBRIUM BENDING ELEMENT	73-0699
GENERAL	A GENERAL EQUILIBRIUM AXIAL ELEMENT	73-0729
GENERAL	A GENERAL EQUILIBRIUM TORQUE ELEMENT	74-0187
GENERAL	A GENERAL FINITE DIFFERENCE METHOD FOR ARBITRARY MESHES	72-0883
GENERAL	A GENERAL FORMULA FOR STIFFNESS MATRICES OF STRUCTURAL ELEMENTS	63-0002
GENERAL	A GENERAL HIGH-ORDER FINITE-ELEMENT WAVEGUIDE ANALYSIS PROGRAM	69-0512
GENERAL	A GENERAL METHOD FOR THE SHAPE FINDING OF LIGHTWEIGHT TENSION STRUCTURES	74-1217
GENERAL	A GENERAL METHOD OF ANALYSIS FOR TWO-DIMENSIONAL PILE GROUPS	71-0298
GENERAL	A GENERAL NUMERICAL SOLUTION OF THE TWO-DIMENSIONAL DIFFUSION-CONVECTION EQUATION BY THE FINITE ELEM	70-0004
GENERAL	A GENERAL PURPOSE THREE DIMENSIONAL STRESS ANALYSIS PROGRAM	71-0686
GENERAL	A GENERAL QUADRILATERAL PLATE ELEMENT	73-0807
GENERAL	A GENERAL SOLUTION OF THE VON KARMAN PLATE EQUATIONS BY THE FINITE ELEMENT METHOD	70-0487
GENERAL	A GENERAL SYSTEM FOR STRUCTURAL FINITE ELEMENT ANALYSIS	68-0107
GENERAL	A GENERAL TECHNIQUE FOR BUCKLING ANALYSIS WITH A FINITE ELEMENT SOLUTION APPLICATION	66-0036
GENERAL	A GENERAL THEORY OF FINITE ELEMENTS	69-0034
GENERAL	A GENERAL THEORY OF VARIATIONAL METHODS WITH APPLICATION TO FINITE ELEMENTS	72-0897
GENERAL	A METHOD FOR COMPUTATION OF VIBRATION MODES AND FREQUENCIES OF ORTHOTROPIC THIN SHELLS OF REVOLUTION	69-0037
GENERAL	A MODE AMPLITUDE TECHNIQUE FOR GENERAL FINITE ELEMENT EQUILIBRIUM	74-0283
GENERAL	AN ANALYSIS OF GENERAL CHAIN SYSTEMS	72-0787
GENERAL	AN EVALUATION OF FINITE DIFFERENCE AND FINITE ELEMENT TECHNIQUES FOR ANALYSIS OF GENERAL SHELLS	70-0254
GENERAL	ANALYSIS OF GENERAL COUPLED THERMOELASTICITY PROBLEMS BY THE FINITE ELEMENT METHOD	68-0129
GENERAL	APPLICATION OF FINITE ELEMENT METHOD TO GENERAL NUMERICAL SOLUTION OF 2-DIMENSIONAL DIFFUSION-CONVEC	70-0269
GENERAL	COMPLETENESS AND CONVERGENCE PROPERTIES OF FINITE ELEMENT DISPLACEMENT FUNCTIONS., A GENERAL TREATME	67-0053
GENERAL	COMPUTATION OF VIBRATION MODAL SHAPES AND NATURAL FREQUENCIES OF A GENERAL ORTHOTROPIC LAMINATED, TH	71-0309
GENERAL	CONPLOT: A GENERAL PURPOSE CONTOUR PLOTTING PROGRAM	74-0887
GENERAL	CURVED BOUNDARY ELEMENTS, GENERAL FORMS OF POLYNOMIAL MAPPINGS	74-1256
GENERAL	DEVELOPMENT AND APPLICATIONS OF A GENERAL COMPUTER PROGRAM FOR EVALUATING STABILITY OF OPENINGS IN R	74-0994
GENERAL	ELAS - A GENERAL PURPOSE COMPUTER PROGRAM FOR THE EQUILIBRIUM PROBLEMS OF LINEAR STRUCTURES	68-0158
GENERAL	ELASTIC AND EQUILIBRIUM MATRICES OF GENERAL SEMI-MONOCOGUE MEMBRANE ELEMENTS	70-0159
GENERAL	ELASTIC SHEAR ANALYSIS OF GENERAL PRISMATIC BEAMS	68-0114
GENERAL	FEDGE - A GENERAL PURPOSE COMPUTER PROGRAM FOR FINITE ELEMENT DATA GENERATION. VOLUME 2 - PROGRAM MA	69-0492
GENERAL	FINITE ELEMENT ANALYSIS OF GENERAL FLUID FLOW PROBLEMS	71-0536
GENERAL	FINITE ELEMENT ANALYSIS OF GENERAL DEFORMATION OF SHEET METALS	74-0501
GENERAL	FINITE ELEMENT APPROACH TO THE INTEGRATED POTENTIAL FORMULATION OF GENERAL UNSTEADY SUPERSONIC AEROD	73-0910
GENERAL	FINITE ELEMENT EQUILIBRIUM MODEL FOR GENERAL BODY FORCE DISTRIBUTION	75-0268
GENERAL	FINITE ELEMENT FORMULATION OF GENERAL THERMO-ELASTICITY PROBLEMS	71-0001
GENERAL	FINITE ELEMENT FORMULATIONS FOR ELASTIC PLATES BY GENERAL VARIATIONAL STATEMENTS WITH DISCONTINUOUS	73-0559
GENERAL	FINITE ELEMENT FORMULATIONS FOR ELASTIC PLATES BY GENERAL VARIATIONAL STATEMENTS WITH DISCONTINUOUS	73-1124
GENERAL	FINITE ELEMENT FORMULATION OF THE HEAT CONDUCTION EQUATION IN GENERAL ORTHOGONAL CURVILINEAR COORDIN	75-0848
GENERAL	FINITE ELEMENT METHOD FOR THE GENERAL SOLUTION OF ORDINARY DIFFERENTIAL EQUATIONS	75-0473
GENERAL	FINITE ELEMENT SOLUTION FOR GENERAL FLUID MOTION	73-0249
GENERAL	FINITE-ELEMENT ANALYSIS FOR FREE VIBRATION OF GENERAL ANISOTROPIC LAMINATED THIN SHELLS	72-0618
GENERAL	FORMAT, A GENERAL PURPOSE MATRIX ANALYSIS SYSTEM IN FORTRAN	68-0266
GENERAL	FORMULATION OF GENERAL DISCRETE MODELS OF THERMOMECHANICAL BEHAVIOR OF MATERIALS WITH MEMORY	69-0381
GENERAL	FORTRAN SUBROUTINES FOR THE SOLUTION OF LAPLACE'S EQUATION OVER A GENERAL REGION IN TWO DIMENSIONS	70-0425
GENERAL	FURTHER APPLICATION OF QUADRATIC ISOPARAMETRIC FINITE-ELEMENTS TO LINEAR FRACTURE MECHANICS OF PLATE	75-0639
GENERAL	FURTHER LIMITATIONS ON GENERAL HEXAHEDRON FINITE ELEMENTS WITH DERIVATIVE DEGREES OF FREEDOM.	72-0395
GENERAL	GENERAL ANALYSIS OF FOLDED PLATES BY THE FINITE ELEMENT METHOD	69-0200
GENERAL	GENERAL BEAM ELEMENT FOR ASKA	69-0458
GENERAL	GENERAL DESCRIPTION OF SESAM-69, SUPER ELEMENT STRUCTURAL ANALYSIS (PROGRAM) MODULES	71-0612
GENERAL	GENERAL FINITE ELEMENT VIBRATIONAL ANALYSIS OF PLANAR MECHANISMS	74-1107
GENERAL	GENERAL LAGRANGE AND HERMITE INTERPOLATION IN R**N WITH APPLICATIONS TO FINITE ELEMENT METHODS	72-0110
GENERAL	GENERAL LINEAR GEOMETRIC MATRIX FOR A FULLY COMPATIBLE FINITE ELEMENT	72-0294
GENERAL	GENERAL MEMBRANE ISOPARAMETRIC ELEMENTS	75-0442
GENERAL	GENERAL METHOD FOR STABILITY ANALYSIS OF ROTATING SHAFTS	75-0528
GENERAL	GENERAL METHOD OF ANALYSIS BASED ON FINITE ELEMENTS	68-0277
GENERAL	GENERAL MIXED FINITE ELEMENT METHODS OF NONLINEAR CONTINUA	72-0831
GENERAL	GENERAL MIXED FINITE ELEMENT METHODS OF NONLINEAR CONTINUA	73-0091
GENERAL	GENERAL POTENTIAL ENERGY AND COMPLEMENTARY ENERGY MODELS BASED ON STRESS PARAMETERS	74-1240
GENERAL	GENERAL PURPOSE COMPUTER PROGRAM FOR THE DYNAMIC SIMULATION OF VEHICLE GUIDEWAY INTERACTIONS	73-0260
GENERAL	GENERAL PURPOSE PROGRAMS FOR NONLINEAR FINITE ELEMENT ANALYSIS	70-0247
GENERAL	GENERAL PURPOSE PROGRAM OF PLANE STRESS ANALYSIS BY FINITE ELEMENT METHOD, AND ITS APPLICATIONS	72-0186
GENERAL	GENERAL PURPOSE PROGRAM FOR FINITE ELEMENT ANALYSIS WITH REFERENCE TO INTERACTION BETWEEN SOFTWARE A	73-0334
GENERAL	GENERAL PURPOSE PROGRAM FOR SHELL STRUCTURES: ISTRAN-S	75-0511
GENERAL	GENERAL QUADRILATERAL PLATE ELEMENT	75-0339
GENERAL	GENERAL SINUSOIDAL STIFFNESS MATRICES FOR BUCKLING AND VIBRATION ANALYSES OF THIN FLAT-WALLED STRUCT	68-0278
GENERAL	GENERAL SOLUTION ROUTINES FOR SYMMETRIC EQUATION SYSTEMS	75-0129
GENERAL	GENERAL TREATMENT OF STRUCTURAL MODIFICATIONS	72-0166

GENERAL   GENERAL VARIATIONAL ANALYSIS OF HYDRODYNAMIC, THERMAL AND DIFFUSIONAL BOUNDARY LAYERS   70-0191
GENERAL   GENERAL VARIATIONAL PRINCIPLES IN NONLINEAR AND LINEAR ELASTICITY WITH APPLICATIONS   72-0914
GENERAL   GNATS: A FINITE ELEMENT COMPUTER PROGRAM FOR THE GENERAL NONLINEAR ANALYSIS OF TWO-DIMENSIONAL STRUC   74-0875
GENERAL   HARDWARE AND SOFTWARE CONSIDERATIONS IN GENERAL PURPOSE PROGRAMS FOR FINITE ELEMENT ANALYSIS   70-0261
GENERAL   IMPROVED GENERAL QUADRILATERAL PLATE ELEMENTS   75-0036
GENERAL   MAGIC II. AN AUTOMATED GENERAL PURPOSE SYSTEM FOR STRUCTURAL ANALYSIS   71-0495
GENERAL   MAGIC 11: AN AUTOMATED GENERAL PURPOSE SYSTEM FOR STRUCTURAL ANALYSIS. VOLUME 111. PROGRAMMER'S MANU   71-0756
GENERAL   MAGIC 11: AN AUTOMATED GENERAL PURPOSE SYSTEM FOR STRUCTURAL ANALYSIS. VOLUME 1. ENGINEER'S MANUAL (71-0757
GENERAL   MAGIC 111: AN AUTOMATED GENERAL PURPOSE SYSTEM FOR STRUCTURAL ANALYSIS. VOLUME 111. PROGRAMMER'S MAN   72-0726
GENERAL   MAGIC 111: AN AUTOMATED GENERAL PURPOSE SYSTEM FOR STRUCTURAL ANALYSIS VOLUME 1. ENGINEER'S MANUAL   72-0727
GENERAL   MAGIC, AN AUTOMATED GENERAL PURPOSE SYSTEM FOR STRUCTURAL ANALYSIS. VOLUME III, PROGRAMMER'S MANUAL   69-0090
GENERAL   MAGIC, AN AUTOMATED GENERAL PURPOSE SYSTEM FOR STRUCTURAL ANALYSIS. VOLUME II, USER'S MANUAL   69-0144
GENERAL   MAGIC, AN AUTOMATED GENERAL PURPOSE SYSTEM FOR STRUCTURAL ANALYSIS   69-0173
GENERAL   MATRIX ANALYSIS OF AXISYMMETRIC SHELLS UNDER GENERAL LOADING   66-0075
GENERAL   MATRIX DISPLACEMENT ANALYSIS OF PLATES AND SHELLS, PART 11, GENERAL FORMULATION OF THE LINEAR THEORY   71-0621
GENERAL   NASTRAN EXPERIENCES OF FORT WORTH OPERATION, CONVAIR AEROSPACE DIVISION OF GENERAL DYNAMICS   71-0019
GENERAL   NASTRAN, NASA'S GENERAL PURPOSE STRUCTURAL ANALYSIS   70-0370
GENERAL   NON-UNIFORM DEFORMATION OF A BLOCK IN PLANE-STRAIN COMPRESSION CAUSED BY FRICTION. PART 4, ANALYSIS   71-0786
GENERAL   NONLINEAR ANALYSIS OF RC SHELLS OF GENERAL FORM   75-0053
GENERAL   NONSAP - A GENERAL FINITE ELEMENT PROGRAM FOR NONLINEAR DYNAMIC ANALYSIS OF COMPLEX STRUCTURES   73-0740
GENERAL   ON GENERAL PURPOSE PROGRAMS FOR FINTIE ELEMENT ANALYSIS WITH SPECIAL REFERENCE TO GEOMETRIC AND MATE   70-0087
GENERAL   ON THE AUTOMATIC SELECTION OF INDEPENDENT DISPLACEMENT VARIABLES IN GENERAL PURPOSE FINITE ELEMENT C   73-0676
GENERAL   ORGANIZATION OF GNATS: A GENERAL NONLINEAR ANALYSIS COMPUTER PROGRAM FOR TWO-DIMENSIONAL STRUCTURES   74-0871
GENERAL   PLACID: A GENERAL FINITE ELEMENT COMPUTER PROGRAM FOR STRESS ANALYSIS OF PLANE AND AXISYMMETRIC SOLI   74-0884
GENERAL   PLASTICITY OF POLYCRYSTALS AND GENERAL DISCUSSIONS ON PLASTICITY THEORY WITH IT   75-0312
GENERAL   POWER SERIES EXPANSION OF THE GENERAL STIFFNESS MATRIX FOR BEAM ELEMENTS   75-0543
GENERAL   PREPROCESSORS FOR GENERAL PURPOSE FINITE ELEMENT PROGRAMS   74-1033
GENERAL   PROGRAM SPECIFICATION NV 300. GENERAL FINITE ELEMENT PROGRAM FOR 2- AND 3-DIMENSIONAL MEMBRANE STRUC   68-0184
GENERAL   REDUCED INTEGRATION TECHNIQUE IN GENERAL ANALYSIS OF PLATES AND SHELLS   71-0374
GENERAL   SAP-A GENERAL STRUCTURAL ANALYSIS PROGRAM FOR LINEAR SYSTEMS   72-0329
GENERAL   SESAM 69 A GENERAL PURPOSE FINITE ELEMENT METHOD PROGRAM CDC/NASTRAN APPLICATIONS MANUAL   72-0552
GENERAL   SESAM-69 - A GENERAL PURPOSE FINITE ELEMENT METHOD PROGRAM   74-0183
GENERAL   SHELL ANALYSIS WITH LARGE GENERAL PURPOSE PROGRAMS   71-0402
GENERAL   SIMPLIFIED HYBRID DISPLACEMENT METHOD FOR LINEAR FINITE ELEMENT ANALYSIS OF GENERAL SHELLS   74-0614
GENERAL   SINGER: A COMPUTER CODE FOR GENERAL ANALYSIS OF TWO-DIMENSIONAL CONCRETE STRUCTURES. VOL. 11. PROGRA   75-0723
GENERAL   SINGER: A COMPUTER CODE FOR GENERAL ANALYSIS OF TWO-DIMENSIONAL REINFORCED CONCRETE STRUCTURES. VOLU   75-0724
GENERAL   SINGER: A COMPUTER CODE FOR GENERAL ANALYSIS TO TWO-DIMENSIONAL CONCRETE STRUCTURES. VOLUME 4. DEMON   75-0730
GENERAL   SINGER: A COMPUTER CODE FOR GENERAL ANALYSIS OF TWO-DIMENSIONAL REINFORCED CONCRETE STRUCTURES. VOLU   75-0731
GENERAL   SLAB BRIDGES WITH ARBITRARY SHAPE AND SUPPORT CONDITIONS - A GENERAL METHOD OF ANALYSIS BASED ON FIN   68-0233
GENERAL   SOME GENERAL CONSIDERATIONS ON THE NATURAL MODE TECHNIQUE   69-0049
GENERAL   SOME GENERAL CONSIDERATIONS ON THE NATURAL MODE TECHNIQUE, PART II: LARGE DISPLACEMENTS   69-0445
GENERAL   STIFFNESS MATRICES FOR THE GENERAL DEFORMATION (OUT-OF-PLANE AND INPLANE) OF CURVED BEAM MEMBERS BAS   75-0124
GENERAL   SURVEY OF GENERAL PURPOSE PROGRAMS FOR FINITE ELEMENT ANALYSIS   72-0323
GENERAL   THE ELASTO-PLASTIC CALCULATION OF GENERAL STRUCTURES AND CONTINUA   68-0207
GENERAL   THE FINITE ELEMENT METHOD AND BOUNDARY SOLUTION PROCEDURES AS GENERAL APPROXIMATION METHODS FOR FIEL   75-0262
GENERAL   THEORETICAL FORMULATION OF FINITE ELEMENT METHODS IN LINEAR ELASTIC ANALYSIS OF GENERAL SHELLS   72-0496
GENERAL   THREE-DIMENSIONAL HARMONIC FUNCTIONS NEAR TERMINATION OR INTERSECTION OF GRADIENT SINGULARITY LINES: 74-0178
GENERAL   VERIFICATION OF A HYBRID GENERAL PURPOSE SHELL AND BODY OF REVOLUTION COMPUTER PROGRAM   71-0333
GENERAL   VIBRATIONAL ANALYSIS OF GENERAL SHELLS UTILIZING FINITE-ELEMENT TECHNIQUES   73-1133
GENERAL   VISCEL A GENERAL PURPOSE COMPUTER PROGRAM FOR ANALYSIS OF LINEAR VISCOELASTIC STRUCTURES, VOL. 2   72-0789
GENERAL-DISPLAC   IMPROVED METHOD FOR OBTAINING THE GENERAL-DISPLACEMENT FIELD FROM A HOLOGRAPHIC INTERFEROGRAM   72-0015
GENERAL-PURPOSE   ELAS: A GENERAL-PURPOSE COMPUTER PROGRAM FOR THE EQUILIBRIUM PROBLEMS OF LINEAR STRUCTURES. VOLUME 2   69-0462
GENERAL-PURPOSE   FEDGE - A GENERAL-PURPOSE COMPUTER PROGRAM FOR FINITE ELEMENT DATA GENERATION. VOLUME 1 - USER'S MAN   69-0491
GENERALES   UN NOVEL ELEMENT DE COQUES GENERALES   74-0407
GENERALISED   GENERALISED MATRIX FORCE AND DISPLACEMENT METHODS FOR LINEAR STRUCTURAL ANALYSIS   65-0064
GENERALISEE   ANALYSE DE LA STABILITY DES TALUS ROCHEUX PAR UNE METHODE GENERALISEE DE L EQUILIBRE LIMITE (ROCK SL   74-1019
GENERALITY   THE FINITE ELEMENT METHOD: FROM INTUITION TO GENERALITY   70-0010
GENERALIZATION   A GENERALIZATION OF THE DIRECT-STIFFNESS METHOD OF STRUCTURAL ANALYSIS   64-0002
GENERALIZATION   A GENERALIZATION OF THE FINITE ELEMENT CONCEPT AND ITS APPLICATION TO A CLASS OF PROBLEMS IN NONLINE   68-0212
GENERALIZATION   A GENERALIZATION OF FINITE ELEMENT METHOD TO DYNAMIC VISCOELASTIC ANALYSIS   75-0106
GENERALIZATION   ACCURATE CALCULATION OF STRESSES IN BEAMS, DISKS AND PLATES BY MEANS OF FINITE ELEMENT METHODS AND I   70-0133
GENERALIZATION   ELASTO-PLASTIC STRESS ANALYSIS. A GENERALIZATION FOR VARIOUS CONSTITUTIVE RELATIONS INCLUDING STRAIN   72-0547
GENERALIZATION   FUNCTIONAL ANALYSIS DERIVATION AND GENERALIZATION OF HYBRID VARIATIONAL METHODS   73-1104
GENERALIZATION   GENERALIZATION OF PLATE FINITE ELEMENTS TO SHELLS   72-0171
GENERALIZATION   GENERALIZATION OF THE ELEMENT METHOD   71-0646
GENERALIZATION   ON A GENERALIZATION OF THE FINITE ELEMENT CONCEPT AND ITS APPLICATION TO A CLASS OF PROBLEMS IN NONL   69-0206
GENERALIZATION   ON THE GENERALIZATION OF STRESS FUNCTION PROCEDURE FOR DYNAMIC ANALYSIS OF PLATES   73-0661
GENERALIZATIONS   ON SOME GENERALIZATIONS OF THE INCREMENTAL STIFFNESS RELATIONS FOR FINITE DEFORMATIONS OF COMPRESSIB   71-0189
GENERALIZED   A " LOCAL " BASIS OF GENERALIZED SPLINES OVER RIGHT TRIANGLES DETERMINED FROM A NONUNIFORM PARTITION   72-0525
GENERALIZED   ANALYSE DE LA STABILITY DES TALUS ROCHEUX PAR UNE METHODE GENERALISEE DE L EQUILIBRE LIMITE (ROCK SL   74-1019
GENERALIZED   APPLICATION OF GENERALIZED CONSTRAINTS IN THE STIFFNESS METHOD OF STRUCTURAL ANALYSIS   66-0030
GENERALIZED   ELASTIC AND PLASTIC INTERLAMINAR SHEAR DEFORMATION IN LAMINATED COMPOSITES UNDER GENERALIZED PLANE   71-0158
GENERALIZED   FINITE ELEMENT VISCOELASTIC STRESS ANALYSIS STUDY USING THE GENERALIZED KELVIN SOLID MODEL   73-0957
GENERALIZED   GENERALIZED CONJUGATE FUNCTIONS FOR MIXED FINITE ELEMENT APPROXIMATIONS OF BOUNDARY VALUE PROBLEMS   72-0515
GENERALIZED   GENERALIZED HYBRID STRESS FINITE-ELEMENT MODELS   73-0027
GENERALIZED   GENERALIZED STEADY STATE FIELD PROBLEM AND ITS SOLUTION   74-0059
GENERALIZED   GENERALIZED STIFFNESS MATRIX OF A CURVED-BEAM ELEMENT   69-0165
GENERALIZED   GENERALIZED STRESS MODELS FOR FINITE ELEMENT ANALYSIS   74-1067
GENERALIZED   GENERALIZED VARIATIONAL PRINCIPLES IN THE FINITE ELEMENT METHOD   69-0116
GENERALIZED   GENERALIZED VARIATIONAL PRINCIPLES FOR USE IN FLOW PROBLEMS THROUGH POROUS MEDIA   74-0067
GENERALIZED   NOTE ON AN APPROXIMATE METHOD FOR COMPUTING NONCONSERVATIVE GENERALIZED FORCES ON FINITELY DEFORMED   70-0097
GENERALIZED   ON THE APPLICATION OF GENERALIZED VARIATIONAL PRINCIPLES IN THE FINITE ELEMENT METHOD   68-0083
GENERALIZED   STABILITY ANALYSIS OF STRUCTURES BY A REDUCED SYSTEM OF GENERALIZED COORDINATES   70-0168
GENERALIZED   STUDY OF 1964 ALASKAN EARTHQUAKE USING FINITE ELEMENT AND GENERALIZED INVERSION TECHNIQUES   72-0003
GENERALIZED   THE GENERALIZED INVERSE - SOME PROPERTIES, A COMPUTATIONAL ALGORITHM AND A STRUCTURAL APPLICATION   68-0248
GENERALIZED   TRIANGULAR FINITE ELEMENTS FOR THE GENERALIZED BESSEL EQUATION OF ORDER M   73-1094
GENERATE   A METHOD TO GENERATE BOTH UPPER AND LOWER BOUNDS TO PLATE EIGENVALUES BY CONFORMING DISPLACEMENT FIN   71-0166
GENERATE   COMPUTER CODE TO GENERATE FINITE ELEMENT MESH FOR PROBLEMS OF COMPLEX AXISYMMETRIC GEOMETRY   74-0360
GENERATE   ELFED A COMPUTER CODE TO GENERATE A FINITE ELEMENT MESH FOR PROBLEMS OF COMPLEX AXISYMMETRIC GEOMETR   74-1270
GENERATE   ELFED: A COMPUTER CODE TO GENERATE FINITE ELEMENT MESH FOR PROBLEMS OF COMPLEX AXISYMMETRIC GEOMETRY   73-0943
GENERATE   EURCYL: A COMPUTER PROGRAM TO GENERATE FINITE ELEMENT MESHES FOR CYLINDER-CYLINDER INTERSECTIONS   73-0962
GENERATE   EURCYL: A PROGRAM TO GENERATE FINITE ELEMENT MESHES FOR PRESSURE VESSEL NOZZLES   74-1086
GENERATED   CLASS OF FINITE ELEMENTS GENERATED BY LAGRANGE INTERPOLATION - 2.   73-0044
GENERATED   COMPUTER GENERATED DISPLAYS OF STRUCTURES IN VIBRATION   74-1305
GENERATED   CONDITION OF FINITE ELEMENT MATRICES GENERATED FROM NONUNIFORM MESHES   72-0104
GENERATED   GOVERNING EQUATIONS AND EXTREMUM PRINCIPLES OF ELASTICITY AND PLASTICITY GENERATED FROM A SINGLE FUN   73-0666
GENERATED   ON A CLASS OF FINITE ELEMENTS GENERATED BY LAGRANGE INTERPOLATION.   72-0101
GENERATED   UNBALANCE GENERATED VIBRATIONS AND STABILITY OF TURBINE ROTORS   73-0224
GENERATING   BODY ORIENTED (NATURAL) CO-ORDINATES FOR GENERATING THREE-DIMENSIONAL MESHES   74-0908
GENERATING   COMPUTER PROGRAM FOR GENERATING FINITE ELEMENT MODELS OF MINE STRUCTURES   71-0493
GENERATING   MESH GENERATING COMPUTER PROGRAM FOR FINITE ELEMENT ANALYSES OF SIMPLE SLOPES IN HORIZONTALLY LAYERE   71-0748
GENERATING   NUMERICAL METHOD FOR GENERATING ORTHOGONAL CURVILINEAR MESHES   70-0443

GENERATING	PROCEDURE FOR EFFICIENT GENERATING, CHECKING, AND DISPLAYING NASTRAN INPUT AND OUTPUT DATA FOR ANALY	71-0098
GENERATION	A SELF-ORGANIZING MESH GENERATION PROGRAM	74-1051
GENERATION	A SELF-ORGANIZING MESH GENERATION PROGRAM	74-1249
GENERATION	AN AUTOMATIC MESH GENERATION SCHEME FOR PLANE AND CURVED SURFACES BY ISOPARAMETRIC COORDINATES	71-0271
GENERATION	AN IMBEDDING TECHNIQUE FOR THE GENERATION OF WEAK-WEAK FINITE ELEMENT APPROXIMATIONS OF LINEAR AND N	72-0577
GENERATION	APPLICATION OF COMPUTER GRAPHICS TO FINITE ELEMENT MESH GENERATION	73-0546
GENERATION	AUTOMATED DATA GENERATION AND LARGE MATRIX OPERATION FOR 3 DIMENSIONAL PLATE STRUCTURE ANALYSIS	73-0339
GENERATION	AUTOMATIC GENERATION OF FINITE ELEMENT MATRICES	71-0165
GENERATION	AUTOMATIC GENERATION AND CONDENSATION OF LARGE MASS AND STIFFNESS MATRICES	72-0466
GENERATION	AUTOMATIC GENERATION OF FINITE ELEMENT MESH WITHIN A BANDWIDTH AND AN EFFICIENT SOLUTION OF THE BAND	73-0338
GENERATION	AUTOMATIC GENERATION OF TRIANGULATIONS FOR PIECEWISE POLYNOMIAL APPROXIMATION	73-1118
GENERATION	AUTOMATIC GENERATION OF FINITE ELEMENTS	74-1335
GENERATION	AUTOMATIC MESH GENERATION IN TWO AND THREE DIMENSIONAL INTERCONNECTED DOMAINS	70-0251
GENERATION	AUTOMATIC MESH GENERATION FOR FINITE ELEMENT ANALYSIS	72-0328
GENERATION	AUTOMATIC MESH GENERATION FOR THREE DIMENSIONAL STRUCTURES BASED ON THEIR THREE VIEWS	73-0340
GENERATION	BERGEN, AN AUTOMATIC FINITE ELEMENT MESH GENERATION PROGRAM FOR ARBITRARY STRUCTURES	72-0776
GENERATION	COMPUTER PROGRAM FOR AUTOMATIC MESH GENERATION FOR THE 2-D FINITE ELEMENT PROGRAM WILAX	73-0565
GENERATION	CONSTRUCTION OF CURVILINEAR COORDINATE SYSTEMS AND APPLICATIONS TO MESH GENERATION	74-1295
GENERATION	DATA GENERATION FOR A 3-D FINITE ELEMENT SYSTEM	74-0361
GENERATION	DATA GENERATION FOR A 3-D FINITE ELEMENT SYSTEM	74-0475
GENERATION	DATA GENERATION FOR FINITE-ELEMENT STRUCTURAL ANALYSIS OF THREE-DIMENSIONAL NAVAL SHIP STRUCTURES	75-0707
GENERATION	FEDGE - A GENERAL PURPOSE COMPUTER PROGRAM FOR FINITE ELEMENT DATA GENERATION. VOLUME 2 - PROGRAM MA	69-0492
GENERATION	FEDGE - A GENERAL-PURPOSE COMPUTER PROGRAM FOR FINITE ELEMENT DATA GENERATION. VOLUME 1 - USER'S MAN	69-0491
GENERATION	FINITE ELEMENT MESH GENERATION PROGRAM FOR ARBITRARY TWO-AND THREE-DIMENSIONAL STRUCTURES	73-0966
GENERATION	GENERATION OF FINITE ELEMENTS BY EULER'S VARIATIONAL METHOD	72-0596
GENERATION	GENERATION OF FINITE ELEMENT MODELS VIA COMPUTER GRAPHICS	74-0194
GENERATION	GENERATION OF SERENDIPEAN SHAPE FUNCTIONS IN FINITE ELEMENT METHOD	74-1012
GENERATION	GENERATION OF TWO- AND THREE-DIMENSIONAL FINITE ELEMENTS	71-0040
GENERATION	GEOMETRIC ASPECTS OF THE FINITE ELEMENT METHOD: CONSTRUCTION OF CURVILINEAR COORDINATE SYSTEMS AND T	72-0534
GENERATION	INTERACTIVE FINITE ELEMENT DATA GENERATION FOR THREE-DIMENSIONAL PLATE STRUCTURES	74-0909
GENERATION	INTERACTIVE FINITE ELEMENT DATA GENERATION USING THE SUB-STRUCTURING TECHNIQUE	74-1291
GENERATION	LAPLACIAN ELECTROSTATIC FIELD CALCULATIONS BY FINITE ELEMENTS WITH AUTOMATIC GRID GENERATION	73-0084
GENERATION	MESH GENERATION - SURVEY	73-0048
GENERATION	MESH GENERATION CODE AND A CONTOUR PLOTTING ROUTINE	72-0802
GENERATION	MESH GENERATION PROGRAM FOR HIGHWAY EXCAVATION CUTS	74-1250
GENERATION	MESHGEN, A COMPUTER CODE FOR AUTOMATIC FINITE ELEMENT MESH GENERATION	69-0096
GENERATION	NATURAL COORDINATE SYSTEMS, AN AUTOMATIC INPUT DATA GENERATION SCHEME FOR A FINITE ELEMENT METHOD	70-0018
GENERATION	PROBLEM CONTROLLED GRID GENERATION FOR THE NUMERICAL SOLUTION OF PARTIAL DIFFERENTIAL EQUATIONS	75-0725
GENERATION	PROCESS FOR THE AUTOMATIC GENERATION OF FLAT TRIANGULAR LATTICES FOR APPLICATION TO THE FINITE ELEME	71-0366
GENERATION	QMESH: A SELF-ORGANIZING MESH GENERATION PROGRAM	74-0872
GENERATION	QMESH: A SELF-ORGANIZING MESH GENERATION PROGRAM	74-0899
GENERATION	REBO: COMPUTER CODE FOR THE GENERATION OF A THREE-DIMENSIONAL FINITE-ELEMENT GRID OF THE NOZZLE-CYLI	74-1310
GENERATION	RESIDUAL ENERGY BALANCING TECHNIQUE IN THE GENERATION OF PLATE BENDING FINITE ELEMENTS	74-0594
GENERATION	RESTRICTIONS IN THE APPLICATION OF AUTOMATIC GENERATION SCHEMES BY ISOPARAMETRIC CO-ORDINATES	74-1377
GENERATION	SECOND GENERATION BOUNDARY INTEGRAL EQUATION PROGRAM FOR THREE-DIMENSIONAL ELASTIC ANALYSIS	75-0376
GENERATION	TECHNIQUES IN TRIANGULAR MESH GENERATION FLAT AND CURVED SURFACES	69-0394
GENERATION	THE APPLICATION OF NODE-ELEMENT RULES FOR FORECASTING PROBLEMS IN THE GENERATION OF FINITE ELEMENT M	75-0614
GENERATION	THE GENERATION OF INTERELEMENT, COMPATIBLE STIFFNESS AND MASS MATRICES BY THE USE OF INTERPOLATION F	66-0014
GENERATION	TRIANGULAR MESH GENERATION PROGRAM	67-0037
GENERATION	TRIANGULAR MESH GENERATION PROGRAM	67-0127
GENERATION	TWO-DIMENSIONAL AUTOMATIC MESH GENERATION FOR STRUCTURAL ANALYSIS	70-0058
GENERATION	USERS MANUAL FOR QMESH, A SELF-ORGANIZING MESH GENERATION PROGRAM	74-0880
GENERATOR	AN AUTOMATIC MESH GENERATOR USING TWO AND THREE DIMENSIONAL ISOPARAMETRIC ELEMENTS	73-0066
GENERATOR	AN IRREGULAR TRIANGLE MESH GENERATOR	64-0032
GENERATOR	DATA GENERATOR FOR THE IDEALIZATION FOR FINITE ELEMENT STRUCTURAL ANALYSIS OF NAVAL SHIP FLAT PLATED	72-0860
GENERATOR	ELEMENT STIFFNESS MATRIX GENERATOR	71-0106
GENERATOR	MESH 3D: A THREE-DIMENSIONAL FINITE ELEMENT MESH GENERATOR PROGRAM FOR EIGHT-NODE ISOPARAMETRIC ELEM	73-0972
GENERATOR	MISSILE BODY INPUT GENERATOR (BING), A NASTRAN PRE-PROCESSOR; THEORETICAL DEVELOPMENT, USER'S MANUAL	75-0717
GENERATOR	NON-LINEAR ANALYSIS OF THE ROTATIONAL STIFFNESS OF A NUCLEAR STEAM GENERATOR FOUNDATION	69-0286
GENERATOR	PLANFORM INPUT GENERATOR (PING), A NASTRAN PRE-PROCESSOR FOR LIFTING SURFACES-THEORETICAL DEVELOPMEN	73-0776
GENERATOR	PLANFORM INPUT GENERATOR (PING), A NASTRAN PROCESSOR FOR LIFTING SURFACES-THEORETICAL DEVELOPMENT, U	73-0780
GENERATOR	PREDICTION OF LONG-TERM STRESS RANGES: USER'S MANUAL - BRIDGE LOAD GENERATOR	73-0737
GENERATOR	THEORETICAL AND EXPERIMENTAL INVESTIGATION ON A STEAM GENERATOR TUBE PLATE	69-0424
GENERATOR	THREE-DIMENSIONAL FINITE ELEMENT ANALYSIS OF A STEAM GENERATOR CHANNEL-HEAD-COMPLEX - COMPARISON OF	75-0653
GENERATOR	THREE-DIMENSIONAL MESH GENERATOR FOR FINITE ELEMENT COMPUTER CODES	74-0864
GENERATORS	NONSTEADY PYROMETRIC CALCULATIONS AS THE BASIS FOR STRESS ANALYSIS OF STEAM GENERATORS AND HEAT EXCH	72-0451
GENERAZION	PROCESS FOR THE AUTOMATIC GENERATION OF FLAT TRIANGULAR LATTICES FOR APPLICATION TO THE FINITE ELEME	71-0366
GEOHYDRAULICS	APPLICATION OF FINITE ELEMENT METHODS IN GEOHYDRAULICS	74-0078
GEOKINETIC	MODELING AN INPUT-OUTPUT GEOKINETIC SYSTEM UTILIZING A FINITE ELEMENT APPROACH	75-0746
GEOLOGIC	EFFECT OF GEOLOGIC IRREGULARITIES OF SEISMIC RESPONSE	74-0898
GEOLOGIC	FINITE ELEMENT ANALYSES OF ELASTIC-PLASTIC PROBLEMS IN THE MECHANICS OF GEOLOGIC MEDIA: AN OVERVIEW	70-0203
GEOLOGIC	FINITE ELEMENT STUDY OF PRESSURIZED CAVITIES IN GEOLOGIC MEDIA	71-0805
GEOLOGIC	TWO DIMENSIONAL STOCHASTIC MODEL OF A HETEROGENEOUS GEOLOGIC SYSTEM	73-0715
GEOLOGICAL	BOUNDARIES AND FRACTURES IN FINITE ELEMENT MODELS OF GEOLOGICAL STRUCTURES	73-0352
GEOLOGICAL	FINITE ELEMENTS FOR GEOLOGICAL MODELLING	70-0049
GEOLOGICAL	MATHEMATICAL MODELS FOR GEOLOGICAL MATERIALS FOR WAVE PROPAGATION STUDIES	71-0740
GEOLOGICAL	SOME APPLICATIONS OF FINITE ELEMENTS TO MODEL GEOLOGICAL STRUCTURES	73-0143
GEOMECHANICAL	NEUARTIGE ANWENDUNG GEOMECHANISCHER UNTERSUCHUNGSMETHODEN ZUR BEURTEILUNG DER ABBAUWIRKUNGEN IM GANG	75-0530
GEOMECHANISCHER	NEUARTIGE ANWENDUNG GEOMECHANISCHER UNTERSUCHUNGSMETHODEN ZUR BEURTEILUNG DER ABBAUWIRKUNGEN IM GANG	75-0530
GEOMETRIC	ANALYSIS OF STIFFENED PLATES CONSIDERING NONLINEAR MATERIAL AND GEOMETRIC BEHAVIOUR	75-0327
GEOMETRIC	CALCULATION OF GEOMETRIC STIFFNESS MATRICES FOR COMPLEX STRUCTURES	66-0058
GEOMETRIC	COMBINED MATERIAL AND GEOMETRIC NONLINEARITY FOR THIN STEEL PLATES	75-0397
GEOMETRIC	DEFORMATION OF THE DIASTOLIC LEFT VENTRICLE - 2. NONLINEAR GEOMETRIC EFFECTS	74-0974
GEOMETRIC	DERIVATION OF GEOMETRIC STIFFNESS AND MASS MATRICES FOR FINITE ELEMENT HYBRID MODELS	74-0844
GEOMETRIC	DUALITY IN STRUCTURAL ANALYSIS BY FINITE ELEMENTS, STATIC GEOMETRIC ANALOGIES, THE DUAL PRINCIPLES O	71-0583
GEOMETRIC	DYNAMIC ANALYSIS OF FRAME AND PLATE STRUCTURES WITH GEOMETRIC AND MATERIAL NONLINEARITIES	73-0649
GEOMETRIC	DYNAMIC RESPONSES OF CYLINDRICAL SHELLS INCLUDING GEOMETRIC AND MATERIAL NONLINEARITIES	74-0150
GEOMETRIC	FINITE ELEMENT ANALYSIS OF COMBINED PROBLEMS ON NONLINEAR MATERIAL AND GEOMETRIC BEHAVIOUR	69-0175
GEOMETRIC	FINITE ELEMENT ANALYSIS OF COMBINED PROBLEMS OF MATERIAL AND GEOMETRIC BEHAVIOUR	69-0508
GEOMETRIC	FINITE ELEMENT ANALYSIS OF BUCKLING AND POST-BUCKLING BEHAVIORS OF ARCHES WITH GEOMETRIC IMPERFECTIO	73-0186
GEOMETRIC	FORMULATION, COMPUTATION, AND SOLUTION PROCEDURES FOR MATERIAL AND/OR GEOMETRIC NONLINEAR STRUCTURAL	72-0346
GEOMETRIC	GENERAL LINEAR GEOMETRIC MATRIX FOR A FULLY COMPATIBLE FINITE ELEMENT	72-0294
GEOMETRIC	GEOMETRIC ASPECTS OF THE FINITE ELEMENT METHOD	72-0523
GEOMETRIC	GEOMETRIC ASPECTS OF THE FINITE ELEMENT METHOD: CONSTRUCTION OF CURVILINEAR COORDINATE SYSTEMS AND T	72-0534
GEOMETRIC	GEOMETRIC HIGHWAY AND CULVERT DESIGN	74-1342
GEOMETRIC	GEOMETRIC METHOD FOR THE PREDICTION OF STRESSES IN INCLUSIONS, OREBODIES, AND MINING SYSTEMS	74-0219
GEOMETRIC	GEOMETRIC NONLINEAR ANALYSIS OF THIN PLATES BY FINITE ELEMENTS	70-0492
GEOMETRIC	GEOMETRIC NONLINEARITY AND THE FINITE ELEMENT DISPLACEMENT METHOD	74-0679
GEOMETRIC	GEOMETRIC STIFFNESS CHARACTERISTICS OF A ROTATING ELASTIC APPENDAGE	74-1296
GEOMETRIC	INCREMENTAL FORMULATIONS FOR PROBLEMS WITH GEOMETRIC AND MATERIAL NONLINEARITIES	72-0315
GEOMETRIC	INCREMENTAL VARIATIONAL METHOD FOR THE LARGE DISPLACEMENT ANALYSIS OF SHELLS WITH GEOMETRIC IMPERFEC	72-0894
GEOMETRIC	NON-LINEAR ANALYSIS OF PLATES CONSIDERING GEOMETRIC AND MATERIAL EFFECTS	71-0702

241

GEOMETRIC	NONLINEAR MATERIAL AND GEOMETRIC BEHAVIOUR OF SHELL STRUCTURES	71-0084
GEOMETRIC	ON GENERAL PURPOSE PROGRAMS FOR FINITE ELEMENT ANALYSIS WITH SPECIAL REFERENCE TO GEOMETRIC AND MATE	70-0087
GEOMETRIC	PROBABILISTIC BUCKLING BEHAVIOR OF STRUCTURES WITH RANDOM LACK-OF-FIT AND GEOMETRIC IMPERFECTIONS	73-0862
GEOMETRIC	STATIC AND DYNAMIC GEOMETRIC AND MATERIAL NONLINEAR ANALYSIS	74-1302
GEOMETRIC	STATIC GEOMETRIC AND MATERIAL NONLINEAR ANALYSIS	72-0314
GEOMETRIC	STATIC STRESS ANALYSIS OF AXISYMMETRIC SOLIDS WITH MATERIAL AND GEOMETRIC NONLINEARITIES BY THE FINI	73-0958
GEOMETRIC	SURVEY OF STATIC GEOMETRIC AND MATERIAL NONLINEAR ANALYSIS BY THE FINITE ELEMENT METHOD	72-0783
GEOMETRIC	UNIFIED APPROACH TO THREE FINITE ELEMENT THEORIES FOR GEOMETRIC NONLINEARITY	74-0599
GEOMETRICAL	DEPLACEMENTS DANS LES STRUCTURES MINCES A FAIBLE RIGIDITE GEOMETRIQUE-DETERMINATION NUMERIQUE (NUMER	72-0213
GEOMETRICAL	SPECIFICATION OF GEOMETRICAL PARAMETERS FOR ELEMENTS WITH CUBIC MAPPING FUNCTIONS	73-1082
GEOMETRICALLY	A FINITE ELEMENT METHOD FOR GEOMETRICALLY NONLINEAR LARGE DISPLACEMENT PROBLEMS IN THIN, ELASTIC PLA	69-0005
GEOMETRICALLY	A SURVEY OF FINITE ELEMENT FORMULATIONS OF GEOMETRICALLY NONLINEAR PROBLEMS	71-0171
GEOMETRICALLY	ALGORITHM FOR THE LARGE DEFLECTION GEOMETRICALLY NON-LINEAR BEHAVIOUR OF PLANE AND CURVED STRUCTURES	72-0083
GEOMETRICALLY	ANALYSIS OF GEOMETRICALLY NONLINEAR PLATES AND SHELLS BY THE FINITE ELEMENT DISPLACEMENT METHOD	68-0050
GEOMETRICALLY	APPLICATION OF A REFINED SHALLOW SHELL FINITE ELEMENT TO GEOMETRICALLY NONLINEAR STRUCTURES	73-0358
GEOMETRICALLY	DYNAMIC ELASTOPLASTIC RESPONSE OF GEOMETRICALLY NONLINEAR ARBITRARY SHELLS OF REVOLUTION UNDER IMPUL	70-0098
GEOMETRICALLY	DYNAMIC RESPONSE ANALYSIS OF GEOMETRICALLY NON-LINEAR STRUCTURES SUBJECTED TO HIGH IMPACT	72-0193
GEOMETRICALLY	DYNAMIC RESPONSE ANALYSIS OF GEOMETRICALLY NON-LINEAR STRUCTURES SUBJECTED TO HIGH IMPACT	72-0555
GEOMETRICALLY	EVALUATION OF SOLUTION PROCEDURES FOR MATERIAL AND/OR GEOMETRICALLY NONLINEAR STRUCTURAL ANALYSIS	73-0362
GEOMETRICALLY	FINITE ELEMENT ANALYSIS OF THE GEOMETRICALLY NONLINEAR PROBLEMS	69-0149
GEOMETRICALLY	FINITE ELEMENT ANALYSIS OF GEOMETRICALLY NONLINEAR PROBLEMS	73-0308
GEOMETRICALLY	FINITE ELEMENT FORMULATIONS OF GEOMETRICALLY NONLINEAR PROBLEMS	69-0180
GEOMETRICALLY	FINITE ELEMENT FORMULATION OF GEOMETRICALLY NONLINEAR PROBLEMS OF ELASTICITY	71-0114
GEOMETRICALLY	FINITE ELEMENTS AND THE ANALYSIS OF GEOMETRICALLY NONLINEAR PROBLEMS	69-0179
GEOMETRICALLY	GEOMETRICALLY NONLINEAR FINITE-ELEMENT ANALYSIS	69-0332
GEOMETRICALLY	GEOMETRICALLY NONLINEAR ANALYSIS BY THE DIRECT STIFFNESS METHOD	70-0331
GEOMETRICALLY	GEOMETRICALLY NONLINEAR STRUCTURAL ANALYSIS BY THE DIRECT STIFFNESS METHOD	71-0560
GEOMETRICALLY	GEOMETRICALLY NONLINEAR FINITE ELEMENT ANALYSIS	72-0347
GEOMETRICALLY	GEOMETRICALLY NONLINEAR BEHAVIOUR OF THIN-WALLED MEMBERS USING FINITE-ELEMENTS	73-1142
GEOMETRICALLY	GEOMETRICKY NELINEARNE ULOHY PLOCHYCH SKRUPIN RIESENE PRIRASTKOVOU METODOU KONECNYCH PRVKOV. (GEOMET	74-1032
GEOMETRICALLY	LARGE AMPLITUDE, FREE VIBRATION GEOMETRICALLY NONLINEAR CIRCULAR ARCHES	72-0606
GEOMETRICALLY	LARGE DEFLECTION GEOMETRICALLY NON-LINEAR FINITE ELEMENT ANALYSIS OF CIRCULAR ARCHES	73-0021
GEOMETRICALLY	LINEAR BUCKLING AND GEOMETRICALLY NONLINEAR ANALYSIS OF PLANAR PLATE-STIFFENER TYPE STRUCTURES BY TH	74-1097
GEOMETRICALLY	PERTURBATION METHOD IN THE ANALYSIS OF GEOMETRICALLY NON LINEAR AND STABILITY PROBLEMS	72-0321
GEOMETRICALLY	THE APPLICATION OF FINITE ELEMENTS TO THE LARGE DEFLECTION GEOMETRICALLY NONLINEAR BEHAVIOR OF CYLIN	72-0815
GEOMETRICALLY	THEORETICAL AND EXPERIMENTAL STUDIES OF TRANSIENT ELASTIC-PLASTIC LARGE DEFLECTIONS OF GEOMETRICALLY	75-0830
GEOMETRICALLY	THEORETICAL AND EXPERIMENTAL STUDIES OF TRANSIENT ELASTIC-PLASTIC LARGE DEFLECTIONS OF GEOMETRICALLY	75-0831
GEOMETRICKY	GEOMETRICKY NELINEARNE ULOHY PLOCHYCH SKRUPIN RIESENE PRIRASTKOVOU METODOU KONECNYCH PRVKOV. (GEOMET	74-1032
GEOMETRIES	FINITE ELEMENT SOLUTION OF MULTIGROUP TRANSPORT EQUATION IN 2 DIMENSIONAL GEOMETRIES	74-0648
GEOMETRIES	STRESS INTENSITY FACTORS FOR ANISOTROPIC FRACTURE TEST SPECIMENS OF SEVERAL GEOMETRIES	70-0293
GEOMETRIQUE-DET	DEPLACEMENTS DANS LES STRUCTURES MINCES A FAIBLE RIGIDITE GEOMETRIQUE-DETERMINATION NUMERIQUE (NUMER	72-0213
GEOMETRIQUEMENT	ANALYSE DES VOILES MINCES DANS LES DOMAINES LINEAIRES ET GEOMETRIQUEMENT NONLINEAIRES PAR LA METHODE	71-0122
GEOMETRY	APPLICATION OF PHASE SPACE FINITE ELEMENTS TO THE TWO DIMENSIONAL NEUTRON TRANSPORT EQUATION IN X-Y	73-0453
GEOMETRY	CELLULAR STRUCTURES OF ARBITRARY PLAN GEOMETRY	72-0201
GEOMETRY	COMPUTER CODE TO GENERATE FINITE ELEMENT MESH FOR PROBLEMS OF COMPLEX AXISYMMETRIC GEOMETRY	74-0360
GEOMETRY	CREEP OF B/AL COMPOSITES AS INFLUENCED BY RESIDUAL STRESSES, BOND STRENGTH, AND FIBER PACKING GEOMET	74-1208
GEOMETRY	EFFECTS OF GEOMETRY CHANGE ON ESSENTIAL FEATURES OF INELASTIC BEHAVIOUR	73-0448
GEOMETRY	EFFECTS OF SHOE FORCE GEOMETRY ON HEAVY DUTY INTERNAL SHOE BRAKE PERFORMANCE	68-0219
GEOMETRY	ELFED A COMPUTER CODE TO GENERATE A FINITE ELEMENT MESH FOR PROBLEMS OF COMPLEX AXISYMMETRIC GEOMETR	74-1270
GEOMETRY	ELFED: A COMPUTER CODE TO GENERATE FINITE ELEMENT MESH FOR PROBLEMS OF COMPLEX AXISYMMETRIC GEOMETRY	73-0943
GEOMETRY	ESTABLISHMENT OF A VALID FRACTURE TOUGHNESS SPECIMEN GEOMETRY USING A FINITE ELEMENT ANALYSIS	74-0135
GEOMETRY	FINITE ELEMENT METHOD FOR SOLVING NEUTRON TRANSPORT PROBLEMS IN TWO-DIMENSIONAL CYLINDRICAL GEOMETRY	74-1081
GEOMETRY	FINITE-ELEMENT PLASMA-DIFFUSION CODE FOR ARBITRARY TOROIDAL GEOMETRY	75-0500
GEOMETRY	IN VIVO STRESSES IN THE HUMAN LEFT VENTRICULAR WALL: ANALYSIS ACCOUNTING FOR THE IRREGULAR THREE DIM	72-0136
GEOMETRY	IN VIVO STRESSES IN THE HUMAN LEFT VENTRICULAR WALL: ANALYSIS ACCOUNTING FOR THE IRREGULAR THREE DIM	72-0136
GEOMETRY	INFLUENCE OF NON-SINGULAR STRESS TERMS AND SPECIMEN GEOMETRY ON SMALL SCALE YIELDING AT CRACK TIPS I	73-0203
GEOMETRY	MIXED DISPLACEMENT 3D ISOPARAMETRIC FINITE ELEMENTS AND A STRAIN ENERGY BASIS FOR STUDIES OF ELEMENT	70-0286
GEOMETRY	NATURAL GEOMETRY OF SURFACE WITH SPECIFIC REFERENCE TO THE MATRIX DISPLACEMENT ANALYSIS OF SHELLS. 1	73-0709
GEOMETRY	POLAR GEOMETRY WAVEGUIDES BY FINITE-ELEMENT METHOD	74-0109
GEOMETRY	RAMIFICATIONS OF OPTIMUM IDEALIZATION GEOMETRY IN DISCRETE ELEMENT ANALYSIS	75-0182
GEOMETRY	TRIPLET: A TWO-DIMENSIONAL, MULTIGROUP, TRIANGULAR MESH, PLANAR GEOMETRY, EXPLICIT TRANSPORT CODE	73-0955
GEOPHYSICAL	PSEUDO 3-DIMENSIONAL FINITE-ELEMENT FORMULATION FOR ELASTOSTATIC PROBLEMS AND ITS GEOPHYSICAL APPLIC	75-0597
GEOPHYSICS	FINITE ELEMENT METHOD TO ELECTROMAGNETIC INDUCTION PROBLEMS IN GEOPHYSICS	72-0309
GEOPHYSICS	THE USE OF A LASER FOR ARPA MILITARY GEOPHYSICS PROGRAM (ROCK MECHANICS AND RAPID EXCAVATION)	72-0800
GEOPOTENTIAL	INVESTIGATION OF FINITE ELEMENT REPRESENTATIONS OF THE GEOPOTENTIAL	75-0290
GEOTECHNICAL	ANALYSIS OF GEOTECHNICAL PROBLEMS BY MEANS OF THE FINITE ELEMENTS METHOD	72-0376
GEOTECHNICAL	APPLICATIONS OF THE FINITE ELEMENT METHOD IN GEOTECHNICAL ENGINEERING. PROCEEDINGS OF THE SYMPOSIUM	73-0273
GEOTECHNICAL	OVERVIEW, TRENDS AND PROJECTIONS: THEORY AND APPLICATIONS OF THE FINITE ELEMENT METHOD IN GEOTECHNIC	72-0988
GEOTECHNICAL	THEORY AND APPLICATION OF THE FINITE ELEMENT METHOD IN GEOTECHNICAL ENGINEERING	72-0704
GEOTHERMAL	GALERKIN FINITE-ELEMENT SIMULATION OF A GEOTHERMAL RESERVOIR	73-0900
GERISSENE	DER GERISSENE ZUSTAND ZWEISEITIG GELAGERTER PLATTEN UNTER EINZELLASTEN - NICHTLINEARE BERECHNUNG MIT	75-0858
GERMAN	DESIGN OF TWO AND THREE DIMENSIONAL STRUCTURES BY THE FINITE ELEMENT METHOD (IN GERMAN)	74-1364
GESCHWEISSTER	TEMPERATURVERLAUF IM QUERSCHNITT GESCHWEISSTER ROHRWAENDE BEI UEBERWIEGEND KONVEKTIVEM WAERMEUEBERGA	75-0628
GEWICHTETEN	FEHLERBETRACHTUNGEN AUFGRUND DER METHODE DER GEWICHTETEN RESIDUEN FUR DAS VERFAHREN DER FINITEN ELEM	72-0368
GIBSON	FINITE ELEMENT ANALYSIS OF THE SURFACE DEFORMATION DUE TO A UNIFORM LOADING ON A LAYER OF GIBSON SOI	75-0240
GIFTS	GIFTS SYSTEM	74-1060
GIFTS	GIFTS SYSTEM FOR THE PDP-15 AND OTHER CONSIDERATIONS IN DEVELOPING GRAPHICS AND CAPABILITIES FOR FIN	73-0671
GIRDER	A HIGHER ORDER IN-PLANE PARALLELOGRAM ELEMENT AND ITS APPLICATION TO SKEWED CURVED BOX GIRDER BRIDGE	71-0002
GIRDER	ANALYSIS AND DESIGN OF SKEW BOX GIRDER BRIDGES	72-0659
GIRDER	ANALYSIS OF BOX GIRDER BRIDGES BY FINITE STRIP METHOD	69-0050
GIRDER	ANALYSIS OF CURVED BOX GIRDER BRIDGES BY FINITE STRIP METHOD	71-0485
GIRDER	ANALYSIS OF CURVED BOX GIRDER BRIDGES	73-0294
GIRDER	ANALYSIS OF SLAB AND GIRDER BRIDGES BY THE FINITE STRIP METHOD	70-0210
GIRDER	ANALYTICAL AND EXPERIMENTAL STUDIES OF MULTI-CELL CONCRETE BOX GIRDER BRIDGES	74-1134
GIRDER	BUCKLING OF PLATE GIRDER WEBS UNDER PARTIAL EDGE LOADINGS	70-0283
GIRDER	DIAPHRAGMS IN SINGLE- AND DOUBLE-CELL BOX GIRDER BRIDGES WITH VARYING ANGLE OF SKEW	72-0130
GIRDER	FINITE ELEMENT ANALYSIS OF SKEW COMPOSITE GIRDER BRIDGES	67-0150
GIRDER	FINITE ELEMENT ANALYSIS OF SKEW BOX GIRDER BRIDGES	72-0209
GIRDER	FREE VIBRATION OF CURVED AND STRAIGHT BEAM SLAB OR BOX GIRDER BRIDGES	72-0487
GIRDER	LIMITATIONS ON THE APPLICATION OF EFFECTIVE BREADTH CONCEPT IN BOX GIRDER DESIGN	74-0493
GIRDER	LOAD DISTRIBUTION IN A COMPOSITE STEEL BOX GIRDER BRIDGE	73-0765
GIRDER	NEW FINITE ELEMENTS WITH APPLICATION TO BOX GIRDER BRIDGES.	72-0465
GIRDER	SEMI-ANALYTIC SOLUTION OF SKEW BOX GIRDER BRIDGES	75-0578
GIRDER	SHEAR LAG IN STEEL BOX GIRDER BRIDGES	75-0310
GIRDER	STRESSES IN CONTINUOUS CONCRETE BOX GIRDER BRIDGES	71-0233
GIRDER	STRUCTURAL BEHAVIOR OF A TWO SPAN REINFORCED CONCRETE BOX GIRDER BRIDGE MODEL. VOLUME 11. REDUCTION,	71-0659
GIRDERS	ANALYSIS OF BOX GIRDERS BY FOLDED PLATE AND STRIP METHODS	73-0982
GIRDERS	ANALYSIS OF CONCRETE BOX GIRDERS BEFORE AND AFTER CRACKING	72-0247
GIRDERS	BOX SHAPED GIRDERS. (NOSNIKY UZAVRETEHO PRIEREZU)	70-0169
GIRDERS	EXPERIMENTAL AND THEORETICAL INVESTIGATION OF THE BEHAVIOR OF BOX GIRDERS CURVED IN PLAN	75-0659
GIRDERS	LAMELLAR TEARING IN WELDED GIRDERS	70-0218
GIRDERS	SHEAR LAG IN STIFFENED BOX GIRDERS	70-0243

242

GIRDERS	STRESS SINGULARITIES AT DIAPHRAGM-WEB-FLANGE JUNCTIONS IN BOX GIRDERS	75-0242
GIRDERS	ULTIMATE STRENGTH OF BOX GIRDERS BY FINITE ELEMENT METHOD	75-0794
GIROSKOPOWYCH	ZASTOSOWANIE METODY SZTYWNYCH ELEMENTOW SKONCZONYCH DO OBLICZEN ORGAN WALOW OKRETOYCH Z UWZGLEDNIENI	74-0962
GLASS	YIELDING OF A TWO DIMENSIONAL VOID ASSEMBLY IN AN ORGANIC GLASS	73-0416
GLASS-EPOXY	STRESS AND FAILURE ANALYSIS OF A GLASS-EPOXY COMPOSITE PLATE WITH A CIRCULAR HOLE	73-0276
GLEICHER	OSCILLATION STABILITY OF INTERNALLY AND EXTERNALLY NOTCHED PLANE BARS OF EQUAL FORM FACTOR FROM 3943	73-0745
GLOBAL	FINITE-DIFFERENCE AND FINITE-ELEMENT METHODS IN GLOBAL WIND-DRIVEN OCEAN CIRCULATION	75-0345
GLOBAL	LOCAL AND GLOBAL SMOOTHING OF DISCONTINUOUS FINITE ELEMENT FUNCTIONS USING A LEAST SQUARES METHOD	74-0378
GLOBAL	UEBER PAUSCHALE FEHLERSCHRANKEN FUER APPROXIMATIONEN VON SCHALENVERFORMUNGEN. (GLOBAL BOUNDARY CONDI	75-0292
GLOBAL-LOCAL	GLOBAL-LOCAL FINITE ELEMENTS	71-0184
GLOBAL-LOCAL	GLOBAL-LOCAL FINITE ELEMENT STRUCTURAL ANALYSIS. ANALYTICAL CONSIDERATIONS/ELEMENT FORMULATION	74-1090
GLOBALLY	SOLVING THE LINEAR BALANCE EQUATION GLOBALLY	74-0810
GNATS	GNATS: A FINITE ELEMENT COMPUTER PROGRAM FOR THE GENERAL NONLINEAR ANALYSIS OF TWO-DIMENSIONAL STRUC	74-0875
GNATS	ORGANIZATION OF GNATS: A GENERAL NONLINEAR ANALYSIS COMPUTER PROGRAM FOR TWO-DIMENSIONAL STRUCTURES	74-0871
GOOD	A NEW APPROACH FOR DERIVING GOOD ELEMENT STIFFNESS MATRICES	75-0128
GOPALACHARYULU	DISCUSSION OF A SHORT COMMUNICATION BY S. GOPALACHARYULU	74-0404
GOROTWORU	METODA ELEMENTOW SKONSZONYCH W MECHANICE GRUNTOW I MECHANICE GOROTWORU (FINITE ELEMENT METHOD IN SOI	73-0500
GOTHIC	A FINITE ELEMENT ANALYSIS OF A DOUBLY CURVED GOTHIC VAULT	71-0827
GOVERNED	FINITE ELEMENT ANALYSIS OF ENGINEERING SYSTEMS GOVERNED BY A NON-LINEAR QUASI-HARMONIC EQUATION	75-0664
GOVERNED	FINITE ELEMENT APPLICATIONS OF A PRECISION TRIANGULAR PLATE BENDING ELEMENT TO LINEAR BOUNDARY VALUE	75-0788
GRAD	AN ADAPTIVE COMPUTER PROGRAM FOR DIV. (P(X,Y) GRAD U) EQUALS F(X,Y,U) IN A POLYGONAL REGION	75-0133
GRADIENT	A COMPUTATIONAL PROCEDURE FOR GRADIENT ITERATIVE TECHNIQUES IN THE FINITE ELEMENT METHOD	68-0229
GRADIENT	A DISCUSSION ON A MODIFIED CONJUGATE GRADIENT METHOD BY W.L. WOOD	74-0402
GRADIENT	A GRADIENT COMPUTATIONAL PROCEDURE FOR THE SOLUTION OF LARGE PROBLEMS ARISING FROM THE FINITE ELEMEN	70-0059
GRADIENT	A MODIFIED GRADIENT METHOD FOR FINITE ELEMENT ELASTOPLASTIC ANALYSIS BY QUADRATIC PROGRAMMING	73-0998
GRADIENT	APPLICATION OF A GRADIENT METHOD IN FINITE ELEMENT SOLUTION OF ELASTOHYDRODYNAMIC PROBLEMS	74-0129
GRADIENT	BUCKLING OF INELASTIC I-BEAMS UNDER MOMENT GRADIENT	74-0765
GRADIENT	DYNAMIC STABILITY ANALYSIS BY THE CONJUGATE GRADIENT METHOD	72-0749
GRADIENT	GRADIENT METHODS FOR FINITE ELEMENT EIGENPROBLEMS	69-0103
GRADIENT	MORE ON GRADIENT ITERATIVE METHODS IN FINITE-ELEMENT ANALYSIS	69-0102
GRADIENT	NOTE ON A MODIFIED CONJUGATE GRADIENT METHOD	73-0615
GRADIENT	NUMERICAL RESULTS FROM THE APPLICATION OF GRADIENT ITERATIVE TECHNIQUES TO THE FINITE ELEMENT VIBRAT	72-0138
GRADIENT	OPTIMAL GRADIENT MINIMIZATION SCHEME FOR FINITE ELEMENT EIGENPROBLEMS	72-0292
GRADIENT	THE APPLICATION OF GRADIENT MINIMIZATION METHODS AND HIGHER ORDER DISCRETE ELEMENTS TO SHELL BUCKLIN	71-0237
GRADIENT	THE SOLUTION OF STRUCTURAL EQUILIBRIUM EQUATIONS BY THE CONJUGATE GRADIENT METHOD WITH PARTICULAR RE	71-0268
GRADIENT	THREE-DIMENSIONAL ELASTIC FINITE ELEMENT ANALYSIS USING GRADIENT DEGREES OF FREEDOM	73-0843
GRADIENT	THREE-DIMENSIONAL HARMONIC FUNCTIONS NEAR TERMINATION OR INTERSECTION OF GRADIENT SINGULARITY LINES:	74-0178
GRADIENTS	APPLICATION OF THE FINITE ELEMENT TECHNIQUE TO THE STUDY OF AREAS OF HIGH STRESS GRADIENTS WITHIN A	73-0259
GRADIENTS	STRUCTURAL CAPACITY OF REINFORCED CONCRETE COLUMNS SUBJECTED TO FIRE INDUCED THERMAL GRADIENTS	73-0913
GRADING	A TECHNIQUE FOR MESH GRADING APPLIED TO CONFORMING PLATE BENDING FINITE ELEMENTS	73-1058
GRADING	MESH GRADING TECHNIQUES FOR COMPATIBLE AND EQUILIBRIUM ELEMENTS	74-0397
GRAIN	EFFECT OF GRAIN BOUNDARY FOR THE ELASTIC BEHAVIOR OF AL AND CU CRYSTALS (ANALYSIS BY THE FINITE ELEM	74-0947
GRAIN	ELASTIC DEFORMATION OF POLYCRYSTALLINE METAL - 1. INFLUENCE OF GRAIN SHAPE ON DISTRIBUTIONS OF STRES	72-0463
GRAIN	FRACTURE OF ROCKET GRAIN MODELS SUBJECTED TO COOLDOWN: A STUDY OF TEMPERATURE-INITIATED FRACTURES OF	74-0536
GRAIN	FRACTURE OF ROCKET GRAIN MODELS SUBJECTED TO COOLDOWN: A STUDY OF TEMPERATURE-INITIATED FRACTURES OF	74-0536
GRAIN	SIMULATION OF FINITE ELEMENT METHOD FOR LOADING PROCESS OF GRAIN CARGO AND INTERACTION WITH SHIP'S S	75-0161
GRAIN	THREE-DIMENSIONAL GRAIN STRESS ANALYSIS USING THE FINITE ELEMENT METHOD	71-0732
GRAIN	THREE-DIMENSIONAL PHOTOELASTIC AND FINITE-ELEMENT ANALYSIS OF A PROPELLANT GRAIN	72-0031
GRAIN-BOUNDARY	NON-UNIFORM FLOW OF POLYCRYSTALS BY GRAIN-BOUNDARY SLIDING ACCOMMODATED BY POWER-LAW CREEP	75-0562
GRAINS	APPLICATION OF THE FINITE ELEMENT METHOD TO STRESS ANALYSIS OF SOLID PROPELLANT ROCKET GRAINS	65-0009
GRAINS	ELASTIC DEFORMATION OF POLYCRYSTALLINE METAL - 1. INFLUENCE OF GRAIN SHAPE ON DISTRIBUTIONS OF STRES	72-0463
GRAINS	ELASTIC DEFORMATION OF POLYCRYSTALLINE METAL - 2 EFFECT OF SLENDERNESS OF GRAINS ON DEFORMATION MODE	73-0275
GRAND	BUILDING THREE NEW MILLION-HORSEPOWER HYDRAULIC TURBINES FOR GRAND COULEE	74-0981
GRAND	FINITE-ELEMENT MODEL FOR POSSIBLE ISOSTATIC REBOUND IN GRAND CANYON	75-0037
GRANICHNYMI	RASCHET KILEBANII PLASTIN SLOZHNOI FORMY S NEODNORODNYMI GRANICHNYMI USLOVIYAMI. (CALCULATION OF VIB	74-0988
GRANITE	FINITE ELEMENT MODEL FOR LOWER GRANITE RESERVOIR	73-0065
GRANULAR	EVALUATION OF A THREE-DIMENSIONAL STRESS CELL FOR GRANULAR SOILS	72-0882
GRANULAR	FINITE ELEMENT PROGRAM FEECON FOR UNDRAINED DEFORMATION ANALYSES OF GRANULAR EMBANKMENTS ON SOFT CLA	72-0875
GRANULAR	PARAMETER EINES ELASTO-PLASTISCHEN STOFFANSATZES FUER KOERNIGE ERDSTOFFE (DETERMINATION OF PARAMETER	74-0657
GRAPHIC	INGA - INTERACTIVE GRAPHIC ANALYSIS	74-0681
GRAPHICAL	GRAPHICAL DISPLAY OF COMPUTER SIMULATED UNBALANCED ROTOR RESPONSE	71-0486
GRAPHICAL	INTERACTIVE TECHNIQUES WITH GRAPHICAL OUTPUT FOR BRIDGE ANALYSES	75-0622
GRAPHICS	ADAPTABILITY OF COMPUTER GRAPHICS TO STUDIES OF ATHEROSCLEROSIS PATHOGENESIS	72-0206
GRAPHICS	AN INTERACTIVE COMPUTER GRAPHICS PROGRAM FOR NASTRAN	71-0074
GRAPHICS	AN INTERACTIVE GRAPHICS SHIP DESIGN ORIENTED PROGRAM PACKAGE FOR TIME SHARING SYSTEMS	73-0341
GRAPHICS	APPLICATION OF COMPUTER GRAPHICS TO FINITE ELEMENT MESH GENERATION	73-0546
GRAPHICS	APPLICATION OF INTERACTIVE GRAPHICS TO THE NUMERICAL METHODS USED IN STRUCTURAL ANALYSIS	73-0384
GRAPHICS	APPLICATIONS OF THE FINITE-ELEMENT METHOD AND COMPUTER GRAPHICS. (A VIBRATIONAL ANALYSIS OF A SONAR	72-0729
GRAPHICS	COMPUTER GRAPHICS FOR 3-D FINITE ELEMENT MODELS	74-0290
GRAPHICS	COMPUTER GRAPHICS FOR STRUCTURAL ANALYSIS	74-0832
GRAPHICS	DIE ANWENDUNG VON INTERAKTIVER COMPUTER-GRAPHIK BEI DER BERECHNUNG VON TRAGWERKEN. (APPLICATION OF I	72-0102
GRAPHICS	DIESEL ENGINE COMPONENT DESIGN USING THE FINITE ELEMENT METHOD AND INTERACTIVE GRAPHICS	74-0237
GRAPHICS	FINAL REPORT ON NASTRAN/GRAPHICS. 1971 FORD DOOR PROJECT	73-0595
GRAPHICS	FINITE ELEMENT ANALYSIS AND COMPUTER GRAPHICS VISUALIZATION OF FLOW AROUND PITCHING AND PLUNGING AIR	73-0385
GRAPHICS	FINITE ELEMENT ANALYSIS, MODELLING AND COMPUTER GRAPHICS	74-1292
GRAPHICS	FINITE ELEMENT METHOD FOR DETERMINATION OF THERMAL STRESSES INTERACTIVE GRAPHICS FOR FINITE ELEMENT	74-0155
GRAPHICS	FINITE ELEMENT MODEL DATA CHECKOUT WITH INTERACTIVE GRAPHICS	74-0195
GRAPHICS	GENERATION OF FINITE ELEMENT MODELS VIA COMPUTER GRAPHICS	74-0194
GRAPHICS	GIFTS SYSTEM FOR THE PDP-15 AND OTHER CONSIDERATIONS IN DEVELOPING GRAPHICS AND CAPABILITIES FOR FIN	73-0671
GRAPHICS	GRAPHICS APPLICATIONS FOR FINITE ELEMENT CODE PROCESSING	74-0893
GRAPHICS	INTERACTIVE COMPUTER GRAPHICS IN STRUCTURAL ANALYSIS: ONLINE 72	72-0639
GRAPHICS	INTERACTIVE GRAPHICS - AN AID FOR DEVELOPING FINITE ELEMENT MODELS	74-0139
GRAPHICS	INTERACTIVE GRAPHICS AIDED STRUCTURAL ANALYSIS	74-0867
GRAPHICS	SOLUTION OF AUTOMOTIVE STRUCTURAL PROBLEMS USING THE FINITE ELEMENT METHOD AND COMPUTER GRAPHICS	71-0091
GRAPHICS	STRUCTURAL FINITE ELEMENT ANALYSIS AIDED BY COMPUTER GRAPHICS	74-0304
GRAPHICS	THE DEVELOPMENT OF COMPUTER GRAPHICS FOR LARGE SCALE FINITE ELEMENT CODES	71-0042
GRAPHICS	THE GROWTH OF INTERACTIVE GRAPHICS AS AN ENGINEERING TOOL.	69-0347
GRAPHICS	THE ROLE OF COMPUTER GRAPHICS IN THE STRUCTURAL DESIGN PROCESS	68-0215
GRAPHICS	USING INTERACTIVE GRAPHICS FOR THE PREPARATION AND MANAGEMENT OF FINITE ELEMENT DATA	74-0176
GRAPHISCHE	INGA - INTERAKTIVE GRAPHISCHE ANALYSE BENUTZERHAND	73-0701
GRAPHITE	ANALYSIS OF LOADING AND THERMAL EFFECTS ON FUELLED GRAPHITE BRICKS FOR A HIGH-TEMPERATURE GAS COOLED	71-0085
GRAPHITE	APPLICATION OF FRACTURE MECHANICS TO GRAPHITE UNDER COMPLEX STRESS CONDITIONS	72-0771
GRAPHITE	BASELINE SOLUTIONS FOR THE SMOOTH WALL THERMOCHEMICAL ABLATION RESPONSE OF GRAPHITE AND CARBON PHENO	71-0728
GRAPHITE	DEVELOPMENT OF A GRAPHITE HORIZONTAL STABILIZER	73-0931
GRAPHITE	FATIGUE OF GRAPHITE FIBRE-REINFORCED EPOXY IN COMPRESSION.	72-0383
GRAPHITE	FATIGUE OF PRENOTCHED GRAPHITE FIBER COMPOSITES IN COMPRESSION	72-0215
GRAPHITE	FINITE-ELEMENT ANALYSIS OF RADIATION-DAMAGE STRESSES IN THE GRAPHITE OF MATRIX FUEL ELEMENTS	70-0048
GRAPHITE	FRACTURE OF GRAPHITE FIBER REINFORCED COMPOSITES	74-0695
GRAPHITE	NEW MATERIAL MODEL FOR THE NONLINEAR BIAXIAL BEHAVIOR OF ATJ-S GRAPHITE	75-0063
GRAPHITE	SIMULATION OF ELASTIC MODULUS AND POISSON'S RATIO OF SPHEROIDAL GRAPHITE CAST IRON	74-0955
GRAPHITE	STRESS ANALYSIS OF A GRAPHITE FUEL ELEMENT INCLUDING RADIATION CREEP AND DIMENSIONAL CHANGES	71-0217
GRAPHITE	TENUE DES STRUCTURES DE GRAPHITE VIS-A-VIS DES SOLLICITATIONS RENCONTREES DANS UN COEUR DE REACTEUR	73-0817

GRAPHITE	TENUE DES STRUCTURES DE GRAPHITE VIS-A-VIS DES SOLLICITATIONS RENCONTREES DANS UN COEUR DE REACTEUR	73-0817
GRAPHITE	VISCOELASTIC RESPONSE OF GRAPHITE MATERIALS IN IRRADIATION ENVIRONMENTS	70-0314
GRAPHITE-EPOXY	MECHANICAL BEHAVIOR OF A GRAPHITE-EPOXY LAMINATE CONTAINING A HOLE	72-0614
GRAVITY	A FINITE ELEMENT MODEL FOR ANISOTROPIC YIELDING IN GRAVITY LOADED ROCK	69-0012
GRAVITY	ANALYSIS OF PILE-SUPPORTED GRAVITY LOCK	74-0585
GRAVITY	ARCH DAM CONSTRUCTION STRESSES DUE TO HYDROSTATIC AND GRAVITY LOADS	75-0658
GRAVITY	COMPARISON OF ANALYTICAL AND EXPERIMENTAL RESULTS IN THE DESIGN OF A HOLLOW GRAVITY DAM	67-0109
GRAVITY	CONTRIBUTION A L'ETUDE DU COMPLEXE BARRAGE-FONDATION ET EXPLICATION DES SEISMES DUS AU REMPLISSAGE D	72-0554
GRAVITY	DYNAMICS OF GRAVITY DAM-RESERVOIR SYSTEMS	73-0225
GRAVITY	EARTHQUAKE ANALYSIS OF GRAVITY DAMS INCLUDING HYDRODYNAMIC INTERACTION	73-0350
GRAVITY	EARTHQUAKE RESPONSE OF CONCRETE GRAVITY DAMS	70-0042
GRAVITY	EARTHQUAKE RESPONSE OF GRAVITY DAMS INCLUDING RESERVOIR INTERACTION EFFECTS	72-0675
GRAVITY	FINITE ELEMENT ANALYSIS OF SOME MECHANICAL PROBLEMS OF HOLLOW GRAVITY DAM	70-0470
GRAVITY	FINITE ELEMENT METHOD FOR AXISYMMETRIC JETS OF FLUID WITH GRAVITY FORCE	75-0428
GRAVITY	FINITE ELEMENT MODEL OF EARTHS GRAVITY FIELD	74-0516
GRAVITY	GRAVITY FLOW BIN DESIGN FOR LUNAR SOIL	72-0240
GRAVITY	GRAVITY FLOW OF POWDER IN A LUNAR ENVIRONMENT. PT. 2-ANALYSIS OF FLOW INITIATION	71-0484
GRAVITY	GRAVITY FLOW TO EXCAVATIONS AND DRAINAGE TRENCHES IN LAYERED AQUIFERS	71-0014
GRAVITY	NONLINEAR CONSIDERATION OF GRAVITY IN A STIFFNESS TEST OF A WEAK STRUCTURE AT SMALL STRAINS	71-0035
GRAVITY	OFFSHORE GRAVITY BASED OIL PRODUCTION PLATFORM INTERACTION WITH THE SEA BED	75-0221
GRAVITY	ROCK FAILURE AROUND A CIRCULAR OPENING IN A GRAVITY FIELD WITH TECTONIC FORCES	70-0071
GRAVITY	STRESS ANALYSIS OF A GRAVITY DAM BY THE FINITE ELEMENT METHOD	63-0004
GRAVITY	STRESS DISTRIBUTION DUE TO GRAVITY IN A VERTICAL ROCK BANK	68-0090
GRAVITY	THE EARTHQUAKE EXPERIENCE AT KOYNA DAM AND STRESSES IN CONCRETE GRAVITY DAMS	72-0589
GRAVITY	THE EFFECT OF GRAVITY UPON THE MELT-THROUGH TIME OF A SOLID SUBJECTED TO A HIGH INTENSITY LASER	72-0693
GRAVITY-TYPE	STOCHASTIC DYNAMIC RESPONSE ANALYSIS OF OFFSHORE PLATFORMS, WITH PARTICULAR REFERENCE TO GRAVITY-TYP	75-0220
GRAY	A METHOD FOR THE THERMAL ANALYSIS OF SPACECRAFT INCLUDING ALL MULTIPLE REFLECTIONS AND SHADING AMONG	70-0424
GREEN'S	INVESTIGATION OF THE FINITE ELEMENT METHOD FOR DETERMINING GREEN'S FUNCTIONS	75-0751
GRENZEN	MODERN TECHNIQUES FOR CALCULATING STRESSES IN TANKS, AND THEIR LIMITATIONS, (MODERNE VERFAHREN ZUR S	70-0153
GRENZTRAGFAEHIG	EINFACHES NAEHERUNGSVERFAHREN FUR DIE GRENZTRAGFAEHIGKEITSANLYSE VON SPANNBETON-REAKTORDRUCKBEHAELTE	73-0854
GRENZTRAGFAEHIG	BERECHNUNGSMETHODEN ZUR ERMITTLUNG VON BELASTUNGSBEDINGUNGEN UND GRENZTRAGFAEHIGKEITSVERHALTEN VON K	73-0858
GRENZTRAGFAEHIG	UEBER DAS VERHALTEN VON SPANNBETONDRUCKBEHAELTRERN IM GRENZTRAGFAEHIGKEITSZUSTAND UND DESSEN EINFUL	73-0855
GRID	A DISCRETE-ELEMENT METHOD OF ANALYSIS FOR ORTHOGONAL SLAB AND GRID BRIDGE FLOOR SYSTEMS	72-0690
GRID	A VARIATIONAL APPROACH TO GRID OPTIMIZATION IN THE FINITE ELEMENT METHOD	72-0572
GRID	APPLICATION OF GRID SELECTION PROCEDURES FOR IMPROVED FINITE ELEMENT STRESS ANALYSIS	74-0326
GRID	CONTINUOUS CONCEPT OF A SPACE GRID	73-0755
GRID	FURTHER DEVELOPMENTS IN GRID SELECTION PROCEDURES IN THE FINITE ELEMENT METHOD	75-0028
GRID	LAPLACIAN ELECTROSTATIC FIELD CALCULATIONS BY FINITE ELEMENTS WITH AUTOMATIC GRID GENERATION	73-0084
GRID	MACRO-APPROACH FOR RIBBED AND GRID PLATE SYSTEMS	75-0009
GRID	PROBLEM CONTROLLED GRID GENERATION FOR THE NUMERICAL SOLUTION OF PARTIAL DIFFERENTIAL EQUATIONS	75-0725
GRID	REBO: COMPUTER CODE FOR THE GENERATION OF A THREE-DIMENSIONAL FINITE-ELEMENT GRID OF THE NOZZLE-CYLI	74-1310
GRID	STATISTICAL STATE OF STRESS STUDIED BY GRID ANALYSIS: "NUMERICAL METHODS OF ANALYSIS"	49-0001
GRID	VARIABLE GRID FINITE DIFFERENCE-ELEMENT SOLUTION OF ELLIPTIC PARTIAL DIFFERENTIAL EQUATIONS	74-0862
GRID-BEAM	DEVELOPMENT OF METHODS FOR COMPUTER SIMULATION OF BEAM-COLUMNS AND GRID-BEAM AND SLAB SYSTEMS	73-1101
GRIDS	COMPLEXITY BOUNDS FOR REGULAR FINITE DIFFERENCE AND FINITE ELEMENT GRIDS	72-0872
GRIDS	COMPLEXITY BOUNDS FOR REGULAR FINITE DIFFERENCE AND FINITE ELEMENT GRIDS	73-0043
GRIDS	DYNAMICS OF PRISMATIC AND OPEN SECTION MEMBER GRIDS	69-0070
GRIDS	GUIDELINES FOR SELECTING FINITE ELEMENT GRIDS BASED ON AN OPTIMIZATION STUDY	74-0303
GRIDS	OPTIMIZATION OF FINITE ELEMENT GRIDS BASED ON MINIMUM POTENTIAL ENERGY	71-0747
GRIDS	OPTIMIZATION OF FINITE ELEMENT GRIDS BASED ON MINIMUM POTENTIAL ENERGY	72-0109
GRIDS	PROCEDURES FOR SELECTING NEAR OPTIMUM FINITE ELEMENT GRIDS	73-0537
GRIDS	VIBRATIONS OF GRIDS BY THE FINITE ELEMENT METHOD	73-1089
GRIDWORKS	CONTINUUM APPROACH TOWARD DYNAMICS OF GRIDWORKS	73-0174
GRIFFITH'S	APPLICATION OF GRIFFITH'S THEORY TO ANALYSIS OF PROGRESSIVE FRACTURE	75-0150
GRILLAGES	A DATA-GENERATION PROCEDURE FOR THE FINITE ELEMENT ANALYSIS OF PLANE FRAMES AND GRILLAGES	68-0179
GRILLAGES	DATA GENERATOR FOR THE IDEALIZATION FOR FINITE ELEMENT STRUCTURAL ANALYSIS OF NAVAL SHIP FLAT PLATED	72-0860
GRIP	GRIP: AN INFORMATION SYSTEM FOR DESIGN AUTOMATION	73-0210
GROPRECHENANLAG	GROPRECHENANLAGEN IN DER FORSCHUNG	68-0264
GROSS	DYNAMIC SHELL THEORY COUPLING THICKNESS STRESS WAVE EFFECTS WITH GROSS STRUCTURAL RESPONSE	73-0379
GROSSFLAECHIGE	ZUSAMMENWIRKEN VON EINZELNEN FERTIGTEILEN ALS GROSSFLAECHIGE SCHEIBE (INTERACTION OF INDIVIDUAL PREF	73-0552
GROUND	A CASE STUDY OF IN-SITU ROCK DEFORMATION BEHAVIOR FOR THE DESIGN OF GROUND SUPPORT SYSTEM	72-0668
GROUND	APPROXIMATE METHOD FOR CALCULATING THE RESPONSE OF EMPLACEMENT STRUCTURES SUBJECTED TO GROUND SHOCK	73-0654
GROUND	DYNAMIC ANALYSIS OF THE GROUND USING THE THREE DIMENSIONAL FINITE ELEMENT	73-0293
GROUND	DYNAMIC RESPONSE ANALYSIS OF THE IRREGULAR GROUND BY MEANS OF THE FINITE ELEMENT METHOD	75-0860
GROUND	EFFECTS OF SOURCE STRUCTURE AND SITE RESPONSE ON GROUND MOTION IN THE NEAR FIELD OF EARTHQUAKES	73-0732
GROUND	F 104 G FINAL MOTIVE: GROUND VIBRATION CALCULATIONS FOR THE STARFIGHTER F 104 G	69-0456
GROUND	FINITE-ELEMENT TREATMENT OF QUADRUPOLE GROUND MOTIONS IN AXISYMMETRIC EARTH STRUCTURES	75-0434
GROUND	INFLUENCE OF CONSTRUCTION PROCEDURES ON GROUND PRESSURES AROUND A TUNNEL	74-0997
GROUND	MATHEMATICAL MODELING OF MOVEMENT OF DISSOLVED CONSTITUENTS IN GROUND WATER AQUIFERS BY THE FINITE E	70-0490
GROUND	MESH GENERATING COMPUTER PROGRAM FOR FINITE ELEMENT ANALYSES OF SIMPLE SLOPES IN HORIZONTALLY LAYERE	71-0748
GROUND	SIMULATION OF GROUND WATER CONTAMINATION USING A GALERKIN FINITE ELEMENT TECHNIQUE	74-1362
GROUND	STRESSES AND DISPLACEMENTS AROUND TWO ADJACENT CIRCULAR OPENINGS NEAR TO THE GROUND SURFACE	74-0993
GROUND	THE COMPARISON OF THE RESPONSE OF A HIGHWAY BRIDGE TO UNIFORM GROUND SHOCK AND MOVING GROUND EXCITAT	72-0737
GROUND	THE COMPARISON OF THE RESPONSE OF A HIGHWAY BRIDGE TO UNIFORM GROUND SHOCK AND MOVING GROUND EXCITAT	72-0737
GROUND-WATER	A STUDY OF CONVECTIVE-DISPERSION EQUATION BY ISOPARAMETRIC FINITE ELEMENTS. TRANSPORT OF POLLUTANTS	75-0773
GROUND-WATER	APPLICATIONS OF ELECTRO-OSMOSIS TO GROUND-WATER FLOW PROBLEMS	75-0336
GROUND-WATER	FINITE ELEMENT ANALYSIS OF INTERFACE PROBLEM IN GROUND-WATER FLOW	74-0497
GROUND-WATER	FINITE ELEMENT SOLUTIONS FOR THE EQUATIONS OF GROUND-WATER FLOW	74-1341
GROUND-WATER	HYDROLOGIC ENGINEERING METHODS FOR WATER RESOURCES DEVELOPMENT. VOLUME 10. PRINCIPLES OF GROUND-WATE	72-0696
GROUND-WATER	NONLINEAR EQUATION OF UNSTEADY GROUND-WATER FLOW	73-0262
GROUND-WATER	PRAKTISCHE BEISPIELE DER DIGITALEN BERECHNUNG VON GRUNDWASSERSTROMUNGS PROBLEMEN NACH DER METHODE EN	73-0081
GROUNDWATER	A FINITE ELEMENT MODEL FOR MISCIBLE DISPLACEMENT IN GROUNDWATER AQUIFERS	75-0783
GROUNDWATER	ANALYSIS OF INTERFACE PROBLEM IN GROUNDWATER FLOW BY FINITE ELEMENT METHOD	74-0467
GROUNDWATER	FINITE ELEMENT ANALYSIS OF THREE-DIMENSIONAL GROUNDWATER FLOW PROBLEMS	74-0254
GROUNDWATER	FINITE ELEMENT ANALYSIS OF A NONSTEADY GROUNDWATER SEEPAGE PROBLEM DAM, DRAINAGE	74-1375
GROUNDWATER	FINITE ELEMENT SOLUTION OF BOUSSINESQ'S EQUATION FOR UNSTEADY GROUNDWATER FLOW	74-0459
GROUNDWATER	FINITE ELEMENTS IN THE SOLUTION OF THE INVERSE PROBLEM IN GROUNDWATER FLOW	74-0063
GROUNDWATER	FUNCTIONAL COEFFICIENTS IN THE ANALYSIS OF GROUNDWATER FLOW USING FINITE ELEMENTS	73-0077
GROUNDWATER	GALERKIN APPROXIMATION OF THE TIME DERIVATIVE IN THE FINITE ELEMENT ANALYSIS OF GROUNDWATER FLOW	74-0658
GROUNDWATER	GALERKIN-FINITE ELEMENT SIMULATION OF GROUNDWATER CONTAMINATION ON LONG-ISLAND, NEW-YORK	73-0093
GROUNDWATER	GROUNDWATER FLOW - THE FINITE ELEMENT METHOD	68-0260
GROUNDWATER	GROUNDWATER FLOW IN A SANDY TIDAL BEACH: II. 2-DIMENSIONAL FINITE ELEMENT ANALYSIS	72-0300
GROUNDWATER	GROUNDWATER FLOW IN PARTIALLY SATURATED SOILS	72-0745
GROUNDWATER	GROUNDWATER FLOWN IN A SANDY TIDAL BEACH .1. ONE-DIMENSIONAL FINITE ELEMENT ANALYSIS	71-0567
GROUNDWATER	GROUNDWATER PREDICTIONS FOR A HIGHWAY CUT USING FINITE ELEMENTS	73-0346
GROUNDWATER	MODELLING GROUNDWATER FLOW BY THE FINITE ELEMENT METHOD	72-0637
GROUNDWATER	NUMERICAL PREDICTION OF TRACER CONCENTRATION IN AQUIFERS WITH TRANSIENT GROUNDWATER FLOW BY FINITE-E	75-0610
GROUNDWATER	SEEPAGE AND GROUNDWATER EFFECTS ASSOCIATED WITH EXPLOSIVE CRATERING	72-0656
GROUNDWATER	SENSITIVITY ANALYSIS OF INPUT PARAMETERS IN NUMERICAL MODELING OF STEADY STATE REGIONAL GROUNDWATER	74-0926
GROUNDWATER	SOLUTION OF TRANSIENT GROUNDWATER FLOW PROBLEMS BY FINITE ELEMENT METHOD	72-0046
GROUNDWATER	STEADY STATE AND TRANSIENT GROUNDWATER FLOW IN SLOPES	70-0273
GROUNDWATER	TECHNIQUES FOR HANDLING INPUT DATA FOR FINITE ELEMENT ANALYSIS OF REGIONAL GROUNDWATER FLOW	75-0806

244

GROUNDWATER	THE FATE OF RADIONUCLIDES IN A GROUNDWATER ENVIRONMENT	74-0061
GROUNDWATER	THEORY OF GROUNDWATER FLOW	70-0117
GROUNDWATER	TRANSIENT FLOW OF GROUNDWATER TO WELLS IN MULTIPLE-AQUIFER SYSTEMS	74-0463
GROUNDWATER	TWO AND THREE DIMENSIONAL GROUNDWATER SEEPAGE BY FINITE ELEMENTS	74-0501
GROUNDWATER	TWO-DIMENSIONAL FINITE ELEMENT MODELING OF TRANSIENT FLOW IN REGIONAL AQUIFER SYSTEMS. ANALYSIS OF G	74-0743
GROUNDWATER	USE OF THE OVER-RELAXATION TECHNIQUE IN THE SIMULATION OF LARGE GROUNDWATER BASINS BY THE FINITE ELE	75-0606
GROUP-THEORETIC	USE OF GROUP-THEORETIC METHODS IN THE DEVELOPMENT OF NONLINEAR SHELL FINITE ELEMENTS	71-0704
GROUPS	A GENERAL METHOD OF ANALYSIS FOR TWO-DIMENSIONAL PILE GROUPS	71-0298
GROUPS	THREE-DIMENSIONAL FINITE ELEMENT ANALYSIS OF VERTICALLY LOADED PILE GROUPS	75-0239
GROUTED	COMPUTER SIMULATION OF LAMINATED ROOF REINFORCED WITH GROUTED BOLTS	73-1045
GROUTED	DYNAMIC LOADS ON GROUTED EMPLACEMENT SYSTEMS	70-0406
GROUTING	THEORIE UND EXPERIMENT ZU EINEM NEUEN INJEKTIONSVERFAHREN. (THEORY AND EXPERIMENT OF A NEW GROUTING	74-0991
GROWING	STEADILY GROWING ELASTIC-PLASTIC CRACK TIP IN A FINITE ELEMENT TREATMENT	73-0087
GROWTH	APPLICATION OF FINITE ELEMENT METHOD TO MASS TRANSPORT LIMITED EPITAXIAL GROWTH PROCESSES	71-0488
GROWTH	CRACK GROWTH MONITORING BY STRAIN SENSING	73-0486
GROWTH	CYCLIC ANALYSIS OF A PROPAGATING CRACK AND ITS CORRELATION WITH FATIGUE CRACK GROWTH	75-0188
GROWTH	ENERGY FOR CRACK GROWTH IN MODEL RUBBER COMPONENTS	72-0568
GROWTH	FINITE-ELEMENT METHOD APPLIED TO PREDICTING FATIGUE-CRACK GROWTH	73-0513
GROWTH	PREDICTION OF SLOW CRACK GROWTH BY A FINITE-ELEMENT METHOD	75-0416
GROWTH	SIMULATION ON THE INITIATION AND GROWTH OF DUCTILE FRACTURE VOIDS	75-0836
GROWTH	THE GROWTH OF INTERACTIVE GRAPHICS AS AN ENGINEERING TOOL.	69-0347
GROWTH	THEORETICAL STUDY OF THE INFLUENCE OF INCLUSIONS UPON THE INITIATION AND GROWTH OF FATIGUE CRACKS IN	71-0394
GROWTH-RATES	BIDIMENSIONAL COMPUTER-PROGRAM FOR CALCULATION OF GROWTH-RATES OF INCLUDED AXIALLY SYMMETRICAL PLASM	75-0420
GRUND	BERECHNUNG INSTATIONARER GRUND - UND SICKERWASSERSTROMUNGEN MIT FREIER OBERFLACHE NACH DER METHODE F	71-0541
GRUND	FEHLERBETRACHTUNGEN AUFGRUND DER METHODE DER GEWICHTETEN RESIDUEN FUR DAS VERFAHREN DER FINITEN ELEM	72-0368
GRUNDLAGE	BEITRAG ZUR SIMULATION NICHTSTATIONARER HORIZONTAL-EVENER GRUNDWASSERSTROMUNGSFELDER AUF DER GRUNDLA	74-0088
GRUNDLAGE	STABILITY INVESTIGATIONS OF PLATES USING FINITE ELEMENT METHOD BASED ON AN EXPANDED VARIATION PRINCI	71-0663
GRUNDLAGEN	GRUNDLAGEN DER DIGITALEN LOSUNG VON GRUNDWASSERSTROMUNGSPROBLEMEN NACH DER METHODE ENDLICHER ELEMENT	71-0542
GRUNDRISS	DESH. EIN EINFACHES FLACHES SCHALENELEMENT UEBER DREIECKIGEM GRUNDRISS. (DESH, A SIMPLE TRIANGULAR S	75-0304
GRUNDWASSERSTRO	PRAKTISCHE BEISPIELE DER DIGITALEN BERECHNUNG VON GRUNDWASSERSTROMUNGS PROBLEMEN NACH DER METHODE EN	73-0081
GRUNDWASSERSTRO	BEITRAG ZUR SIMULATION NICHTSTATIONARER HORIZONTAL-EVENER GRUNDWASSERSTROMUNGSFELDER AUF DER GRUNDLA	74-0088
GRUNDWASSERSTRO	AUFBEREITUNG MATHEMATISCHER MODELLE FUR DIE DIGITALE BERECHNUNG VON GRUNDWASSERSTROMUNGSPROBLEMEN NA	71-0543
GRUNDWASSERSTRO	GRUNDLAGEN DER DIGITALEN LOSUNG VON GRUNDWASSERSTROMUNGSPROBLEMEN NACH DER METHODE ENDLICHER ELEMENT	71-0542
GRUNTOV	METOD KONECHNYKH ELEMENTOV V RASCHETAKH KONSOLIDATSII VODONASYSHCHENNYKH GRUNTOV. (FINITE ELEMENT ME	75-0469
GRUNTOW	METODA ELEMENTOW SKONSZONYCH W MECHANICE GRUNTOW I MECHANICE GOROTWORU (FINITE ELEMENT METHOD IN SOI	73-0500
GUIDE	ANALYTIC MODELING OF ROCK-STRUCTURE INTERACTION. VOLUME 2. USERS GUIDE FOR A COMPUTER PROGRAM	73-0919
GUIDE	AXISYMMETRIC RING AND PLATE STRESS ANALYSIS PROGRAM - USERS GUIDE	70-0333
GUIDE	FEABL (FINITE ELEMENT ANALYSIS BASIC LIBRARY) USER'S GUIDE	72-0748
GUIDE	FINITE ELEMENT METHOD FOR WAVE GUIDE PROBLEMS	75-0872
GUIDE	RANDOM-VIBRATION ANALYSIS SYSTEM FOR COMPLEX STRUCTURES. PART 1, ENGINEERING USER'S GUIDE	68-0106
GUIDE	SINGER: A COMPUTER CODE FOR GENERAL ANALYSIS OF TWO-DIMENSIONAL REINFORCED CONCRETE STRUCTURES. VOLU	75-0731
GUIDE	STRUCTURE-MEDIUM INTERACTION AND DESIGN PROCEDURES STUDY. VOLUME IV. FEAT CODE USER'S GUIDE	69-0265
GUIDE	THREE-DIMENSIONAL FINITE-ELEMENT COMPUTER PROGRAM - USERS GUIDE	74-0755
GUIDE	USER'S GUIDE FOR ANALYSIS OF FINITE ELASTOPLASTIC DEFORMATION: THE FIPDEF AND FIPAX PROGRAMS FOR THE	74-0799
GUIDE	USER'S GUIDE FOR THE COMPUTER PROGRAM DESCRIBED IN TAM REPORT NO. 680 "DETERMINATION OF VOLUME CHANG	69-0473
GUIDED	A FORTRAN PROGRAM FOR CALCULATING THE FREQUENCIES AND MODE SHAPES OF A GUIDED CLAMPED, GUIDED PINNED	72-0890
GUIDED	A FORTRAN PROGRAM FOR CALCULATING THE FREQUENCIES AND MODE SHAPES OF A GUIDED CLAMPED, GUIDED PINNED	72-0890
GUIDED	FREQUENCY COEFFICIENTS OF A SPHERICAL SHELL FOR VARYING THICKNESS AND OPENING ANGLE WITH GUIDED PINN	72-0595
GUIDELINES	GUIDELINES FOR SELECTING FINITE ELEMENT GRIDS BASED ON AN OPTIMIZATION STUDY	74-0303
GUIDES	FINITE ELEMENT ANALYSIS OF DISPERSION-FREE GUIDES FOR ELASTIC SURFACE WAVES	73-0009
GUIDES	HIGHER-ORDER FINITE-ELEMENT ANALYSIS OF TOPOGRAPHIC GUIDES SUPPORTING ELASTIC SURFACE WAVES	73-0114
GUIDES	NUMERICAL METHOD FOR INVESTIGATING THE PROPAGATION OF SURFACE WAVES ON DISSIPATIVE GUIDES	74-0613
GUIDES	STRESS ANALYSIS IN SOME COLD FORGING TOOL COMPONENTS AND THEIR DESIGN GUIDES	74-0589
GUIDEWAY	GENERAL PURPOSE COMPUTER PROGRAM FOR THE DYNAMIC SIMULATION OF VEHICLE GUIDEWAY INTERACTIONS	73-0260
GUIDEWAY	SUBTERRANEAN GUIDEWAY RESPONSES TO LEVITATION VEHICLE LOADING	74-0927
GUIDING	EVALUATION OF ATTENUATION FROM LOSSLESS TRIANGULAR-FINITE-ELEMENT SOLUTIONS FOR INHOMOGENEOUSLY FILL	73-0059
GUIDING	HIGH-ORDER FINITE ELEMENTS FOR INHOMOGENEOUS ACOUSTIC GUIDING STRUCTURES	73-0137
GULCH	STRESSES AND DEFORMATIONS IN JAIL GULCH EMBANKMENT	73-0391
GUN	ANALYSIS OF BAND PRESSURE IN A GUN TUBE USING FINITE ELEMENTS AND SHELL PUSHING DATA	72-0793
GUN-TURRET	DEVELOPMENT OF A GUN-TURRET MODEL AND OF THE EQUATIONS FOR COUPLING IT WITH A HELICOPTER FINITE-ELEM	73-0752
GURTIN'S	A SIMPLIFIED VERSION OF GURTIN'S VARIATIONAL PRINCIPLE	74-1244
GUSSET	TESTS OF GUSSET PLATE MODELS	71-0338
GUYAN	GUYAN REDUCTION SOLUTIONS RECYCLED FOR IMPROVED ACCURACY	71-0159
GUYED	SOBRE TORRES ESPIADAS - METODO DE CALCULO E UM EXEMPLO DE APLICACAO. (METHOD AND AN EXAMPLE OF CALCU	74-0915
GUYED	STATIC AND DYNAMIC BEHAVIOUR OF GUYED MASTS	68-0271
GYPSUM	ANALYSIS OF NONLINEAR CONCRETE FRAMES USING SMALL GYPSUM MODELS AND THE FINITE ELEMENT METHOD	70-0498
GYROSCOPIC	ZASTOSOWANIE METODY SZTYWNYCH ELEMENTOW SKONCZONYCH DO OBLICZEN ORGAN WALOW OKRETOYCH Z UWZGLEDNIENI	74-0962
H-GALERKIN	SOME SUPERCONVERGENCE RESULTS FOR AN H-GALERKIN PROCEDURE FOR THE HEAT EQUATION	74-0411
HABEAS-	HABEAS: A STRUCTURAL DYNAMICS ANALYSIS SYSTEM	70-0121
HALF	FINITE ELEMENT ANALYSIS OF FINITE SIZED PLATES BOUNDED TO AN ELASTIC HALF SPACE	72-0880
HALF	FINITE ELEMENT APPROACH TO THE CONTACT PROBLEM OF A PLATE ON THE ELASTIC HALF SPACE	72-0830
HALF	FINITE ELEMENT MODELING OF A CIRCULAR RING USING HALF AND QUARTER SYMMETRY	66-0119
HALF	UNBONDED CONTACT PROBLEM OF A PLATE ON THE ELASTIC HALF SPACE	74-0208
HALF-PLANE	DYNAMIC MODELING OF HALF-PLANE BY FINITE ELEMENTS	74-0953
HALF-SPACE	A FINITE ELEMENT STUDY OF THE PENETRATION OF A SOIL HALF-SPACE BY A RIGID PROJECTILE	69-0518
HALF-SPACE	ELASTOPLASTIC INDENTATION OF HALF-SPACE BY AN INFINITELY LONG RIGID CIRCULAR CYLINDER	71-0395
HALF-SPACE	IMPACT ON A TRANSVERSELY ANISOTROPIC HALF-SPACE	74-0950
HALF-SPACE	THE ELASTO-PLASTIC INDENTATION OF A HALF-SPACE BY A RIGID SPHERE	71-0113
HALL	FINITE-ELEMENT METHOD FOR POTENTIAL CALCULATIONS IN A HALL PLATE	75-0454
HALLFASTHETSEGE	EXPERIMENTELL UNDERSOKNING AV ELEMENTFOGARS HALLFASTHETSEGENSKAPER (EXPERIMENTAL INVESTIGATION OF ST	72-0192
HANDBOOK	HANDBOOK FOR THE ENGINEERING STRUCTURAL ANALYSIS OF SOLID PROPELLANTS	71-0642
HANDLING	A COMPUTATIONAL TECHNIQUE FOR THE EFFICIENT HANDLING OF LARGE MATRICES	73-1083
HANDLING	A SIMPLE MATRIX-VECTOR HANDLING SCHEME FOR THREE-DIMENSIONAL AND SHELL ANALYSIS	70-0317
HANDLING	PRACTICAL HANDLING OF FINITE-ELEMENT METHOD 2	75-0547
HANDLING	PRACTICAL HANDLING OF FINITE-ELEMENT METHOD 1	75-0551
HANDLING	TECHNIQUES FOR HANDLING INPUT DATA FOR FINITE ELEMENT ANALYSIS OF REGIONAL GROUNDWATER FLOW	75-0806
HANDLING	VEHICLE STRUCTURAL ANALYSIS, PRE AND POST PROCESSOR DATA HANDLING	71-0093
HARBOR	OSCILLATIONS AND WAVE FORCES IN AN OFFSHORE HARBOR - APPLICATIONS OF HYBRID FINITE ELEMENT METHOD TO	74-1024
HARBOUR	HARBOUR OSCILLATION - A NUMERICAL TREATMENT FOR UNDAMPED NATURAL MODES	70-0208
HARBOUR	OSCILLATIONS AND WAVE FORCES IN A MAN-MADE HARBOUR IN THE OPEN SEA	74-0151
HARD	A DESIGN PROCEDURE FOR HIGH CUTS IN JOINTED HARD ROCK. THREE DIMENSIONAL SOLUTIONS. VOLUME 1	71-0656
HARD	LOAD TRANSFER MECHANISM FOR PIERS SOCKETTED IN HARD SOILS OR ROCK	74-0969
HARD	THERMAL-MECHANICAL FRAGMENTATION OF HARD ROCK - FIELD TESTS	74-0998
HARDBOARD	NUMERICAL ANALYSIS ON STRESS DISTRIBUTION IN PARTIAL COMPRESSED SPECIMEN (HARDBOARD) BY FINITE ELEME	72-0920
HARDENING	A MATHEMATICAL THEORY OF PLASTICITY FOR STRAIN HARDENING MATERIALS EXHIBITING INITIAL ANISOTROPY	72-0660
HARDENING	ANISOTROPIC WORK HARDENING IN METALS. FINITE ELEMENT APPLICATION TO PLANE STRESS AND PLANE STRAIN	73-0950
HARDENING	BENDING OF CIRCULAR PLATES OF HARDENING MATERIAL	67-0112
HARDENING	COMPARISON OF THE CAPABILITY OF OUR HARDENING RULES TO PREDICT A MATERIAL'S PLASTIC BEHAVIOR	75-0523
HARDENING	FINITE ELEMENT SOLUTION OF EXTRUSION OF ELASTO-PLASTIC WORK HARDENING MATERIALS	73-1127
HARDENING	ISOTROPIC-KINEMATIC HARDENING MODEL FOR ELASTIC-PLASTIC CYCLIC STRUCTURAL ANALYSIS	75-0210
HARDENING	J INTEGRAL ESTIMATES FOR STRAIN HARDENING MATERIALS USING FULLY PLASTIC SOLUTIONS	74-0690
HARDENING	STRAIN HARDENING TORSION OF SOLID BARS AS A MINIMIZATION PROBLEM	70-0174

HARDWARE	GENERAL PURPOSE PROGRAM FOR FINITE ELEMENT ANALYSIS WITH REFERENCE TO INTERACTION BETWEEN SOFTWARE A	73-0334
HARDWARE	HARDWARE AND SOFTWARE CONSIDERATIONS IN GENERAL PURPOSE PROGRAMS FOR FINITE ELEMENT ANALYSIS	70-0261
HARMONIC	CALCULATION OF HARMONIC FIELDS IN WHICH SINGULARITIES ARE PRESENT USING FINITE ELEMENT AND FINITE DI	73-0979
HARMONIC	FINITE ELEMENT STUDY OF HARMONIC WAVE PROPAGATION IN PERIODIC STRUCTURES	74-0130
HARMONIC	FINITE ELEMENTS WITH HARMONIC INTERPOLATION FUNCTIONS	72-0058
HARMONIC	FINITE ELEMENTS WITH HARMONIC INTERPOLATION FUNCTIONS	73-1065
HARMONIC	OPTIMIZATION OF STRUCTURES UNDERGOING HARMONIC OR STOCHASTIC EXCITATION	75-0711
HARMONIC	THE HARMONIC FINITE ELEMENT MODEL FOR USE IN FIELD PROBLEMS	70-0389
HARMONIC	THREE-DIMENSIONAL HARMONIC FUNCTIONS NEAR TERMINATION OR INTERSECTION OF GRADIENT SINGULARITY LINES:	74-0178
HATCH	APPLICATION OF FINITE ELEMENT METHOD FOR THE DETAILED ANALYSIS OF HATCH CORNER STRESSES	73-0735
HATCH	PHOTOELASTIC AND FINITE ELEMENT INVESTIGATION OF A NONSYMMETRICAL PLUG HATCH CONFIGURATION	73-0025
HAUTE	TENUE DES STRUCTURES DE GRAPHITE VIS-A-VIS DES SOLLICITATIONS RENCONTREES DANS UN COEUR DE REACTEUR	73-0817
HAYMARKET-NORTH	M.I.T. TEST SECTION INSTRUMENTATION, MASSACHUSETTS BAY TRANSPORTATION AUTHORITY, HAYMARKET-NORTH EXT	72-0673
HAZARD	PROBABILISTIC ASSESSMENT OF AIRCRAFT HAZARD FOR NUCLEAR POWER PLANTS	72-0235
HCDA	FINITE-ELEMENT ANALYSIS OF STRUCTURAL RESPONSE IN HCDA	75-0407
HEAD	ACTUAL STRESS OF HIGH HEAD PUMP-TURBINE RUNNER	75-0567
HEAD	DEVELOPMENT OF FINITE ELEMENT HEAD INJURY MODEL	75-0506
HEAD	DYNAMIC RESPONSE OF ELASTIC HEAD MODELS	75-0378
HEAD	MATHEMATICAL MODEL FOR CLOSED HEAD IMPACT	74-0934
HEAD	STRESS ANALYSIS OF SELF-SEALING TYPE PISTON HEAD BY FINITE-ELEMENT METHOD	72-0259
HEAD	STRESS ANALYSIS OF THE "SELF-SEALING" TYPE PISTON HEAD BY THE FINITE ELEMENT METHOD	72-0617
HEAD	VIBRATIONAL ANALYSIS OF THE FLEXING HEAD OF THE BOS-6 TRANSDUCER	71-0708
HEADS	ELASTIC-PLASTIC ANALYSIS OF SOME PRESSURE VESSEL HEADS	69-0379
HEADS	SOME DESIGN ASPECTS OF IC ENGINE PISTON HEADS	73-0578
HEADWALL	MAGAZINE HEADWALL RESPONSE TO EXPLOSIVE BLAST	74-0754
HEALING	MEASUREMENTS OF HEALING AT AN OSTEOTOMY IN A RABBIT CALVARIUM: THE INFLUENCE OF APPLIED COMPRESSIVE	74-0975
HEART	FINITE-ELEMENT MODEL FOR THE MECHANICAL BEHAVIOUR OF THE LEFT VENTRICLE, PREDICTION OF DEFORMATION I	72-0304
HEAT	A COMPUTER PROGRAM FOR THE ANALYSIS OF TWO-DIMENSIONAL HEAT CONDUCTION USING THE FINITE ELEMENT TECH	69-0066
HEAT	A DERIVATION OF PERTURBED EQUATIONS OF MOTION AND HEAT CONDUCTION OF A THERMOELASTIC SOLID	72-0735
HEAT	A DISCRETE LEAST-SQUARES FINITE ELEMENT METHOD FOR TRANSIENT HEAT CONDUCTION PROBLEMS	74-1191
HEAT	A FINITE ELEMENT COLLOCATION METHOD FOR THE HEAT EQUATION	72-0488
HEAT	A FINITE ELEMENT FORMULATION OF HEAT CONDUCTION AND HEAT AND MASS TRANSFER PROBLEMS	74-0350
HEAT	A FINITE ELEMENT FORMULATION OF HEAT CONDUCTION AND HEAT AND MASS TRANSFER PROBLEMS	74-0350
HEAT	A SUPERCONVERGENCE RESULT FOR THE APPROXIMATE SOLUTION OF THE HEAT EQUATION BY A COLLOCATION METHOD	72-0510
HEAT	A SURVEY OF HEAT CONDUCTION COMPUTER PROGRAMS	71-0017
HEAT	ACCURACY, STABILITY AND OSCILLATION CHARACTERISTICS OF FINITE ELEMENT METHOD FOR SOLVING HEAT CONDUC	69-0007
HEAT	ACCURACY, STABILITY, AND OSCILLATION CHARACTERISTICS OF TRANSIENT TWO-DIMENSIONAL HEAT CONDUCTION	75-0849
HEAT	ANALYSIS OF AXISYMMETRIC LINEAR HEAT CONDUCTION PROBLEMS BY FINITE ELEMENT METHOD	69-0015
HEAT	ANALYTIC METHODS IN CONDUCTION HEAT TRANSFER	71-0186
HEAT	APACHE: A THREE-DIMENSIONAL FINITE ELEMENT PROGRAM FOR STEADY-STATE OR TRANSIENT HEAT CONDUCTION ANA	73-0971
HEAT	APPLICABILITY OF APPROXIMATE AND EXACT TRANSIENT HEAT TRANSFER ANALYSES TO HEATING PROCESSES USED TO	75-0244
HEAT	APPLICATION OF CONFORMAL MAPPING AND VARIATIONAL METHOD TO THE STUDY OF HEAT CONDUCTION IN POLYGONAL	74-0044
HEAT	APPLICATION OF FINITE ELEMENT METHOD TO THE PROBLEM OF HEAT CONDUCTION AND FLUID MECHANICS	72-0486
HEAT	APPLICATION OF FINITE ELEMENTS TO HEAT CONDUCTION PROBLEMS INVOLVING LATENT HEAT	73-0520
HEAT	APPLICATION OF FINITE ELEMENTS TO HEAT CONDUCTION PROBLEMS INVOLVING LATENT HEAT	73-0520
HEAT	APPLICATION OF THE FINITE ELEMENT METHOD TO HEAT CONDUCTION ANALYSIS	66-0076
HEAT	APPLICATION OF THE FINITE ELEMENT METHOD TO HEAT CONDUCTION IN SOLIDS	67-0005
HEAT	APPLICATION OF THE FINITE ELEMENT METHOD TO CONVECTION HEAT TRANSFER BETWEEN PARALLEL PLANES	70-0244
HEAT	APPLICATION OF THE FINITE-ELEMENT METHOD TO HEAT-TRANSFER PROBLEMS. PART 11. TRANSIENT TWO-DIMENSION	71-0759
HEAT	APPROXIMATIONS IN FINITE-ELEMENT HEAT CONDUCTION ANALYSIS	74-1167
HEAT	AYER HEAT CONDUCTION COMPUTER PROGRAM	74-0885
HEAT	CALCULATION OF NATURAL FREQUENCIES OF HEAT EXCHANGER TUBES	73-0665
HEAT	COMPARISON OF ANALYTICAL AND NUMERICAL SOLUTIONS IN HEAT CONDUCTION PROBLEMS	74-1170
HEAT	COMPARISON OF THE FINITE ELEMENT AND FINITE DIFFERENCE METHODS FOR THE ANALYSIS OF STEADY TWO DIMENS	75-0223
HEAT	COMPARISON OF TIME MARCHING SCHEMES FOR THE TRANSIENT HEAT CONDUCTION EQUATION	75-0477
HEAT	COUPLED CONVECTIVE AND CONDUCTIVE HEAT TRANSFER BY FINITE ELEMENT METHODS	74-0050
HEAT	COUPLED CONVECTIVE/CONDUCTIVE HEAT TRANSFER INCLUDING VELOCITY FIELD EVALUATION	75-0116
HEAT	COUPLED HEAT AND MOISTURE TRANSPORT MODEL FOR ARCTIC SOILS	74-0520
HEAT	DESIGN AND ANALYSIS EXPERIENCE WITH A LIQUID METAL HEAT EXCHANGER FOR PFTF SERVICE	72-0772
HEAT	DIRECT AND INVERSE HEAT CONDUCTION IN ROCK MATERIALS USING THE FINITE ELEMENT METHOD	74-1098
HEAT	DIRECT ITERATION METHOD FOR THE INCORPORATION OF PHASE CHANGE IN FINITE ELEMENT HEAT CONDUCTION PROG	74-0894
HEAT	DISCRETE VARIABLE METHOD APPLIED TO TRANSIENT HEAT CONDUCTION PROBLEMS	74-1163
HEAT	DUAL ANALYSIS FOR HEAT CONDUCTION PROBLEMS BY FINITE ELEMENTS	72-0401
HEAT	EFFICIENT SOLUTION PROCESSES FOR FINITE ELEMENT ANALYSIS OF TRANSIENT HEAT CONDUCTION	69-0014
HEAT	ERROR ANALYSIS OF A REMOTE TRANSIENT HEAT FLUX SENSOR	74-0753
HEAT	ERROR ESTIMATES FOR FINITE ELEMENT METHOD FOR HEAT TRANSFER PROBLEMS IN THE SPACE-TIME DOMAIN	75-0104
HEAT	EXCITATION OF PLANE THERMOELASTIC WAVES IN A LAYER WITH A FINITE RATE OF HEAT PROPAGATION	74-0803
HEAT	FINITE CONDUCTANCE ELEMENT METHOD OF CONDUCTION HEAT TRANSFER	73-0963
HEAT	FINITE ELEMENT ANALYSIS OF HEAT CONDUCTION IN ANISOTROPIC SOLIDS	68-0161
HEAT	FINITE ELEMENT ANALYSIS OF TRANSIENT HEAT CONDUCTION. APPLICATION OF THE WEIGHTED RESIDUAL PROCESS	74-0277
HEAT	FINITE ELEMENT ANALYSIS OF STEADY-STATE NONLINEAR HEAT TRANSFER PROBLEMS	74-0572
HEAT	FINITE ELEMENT ANALYSIS OF LINEAR AND NONLINEAR HEAT TRANSFER	74-0626
HEAT	FINITE ELEMENT ANALYSIS OF LAMINAR CONVECTIVE HEAT TRANSFER IN VERTICAL DUCTS WITH ARBITRARY CROSS-S	75-0555
HEAT	FINITE ELEMENT COMPUTER PROGRAMS FOR HEAT CONDUCTION PROBLEMS	75-0525
HEAT	FINITE ELEMENT FORMULATION OF THE HEAT CONDUCTION EQUATION IN GENERAL ORTHOGONAL CURVILINEAR COORDIN	75-0848
HEAT	FINITE ELEMENT METHOD IN FLUID DYNAMICS AND HEAT TRANSFER	67-0151
HEAT	FINITE ELEMENT METHOD APPLIED TO THE SOLUTION OF HEAT CONDUCTION PROBLEMS	69-0009
HEAT	FINITE ELEMENT METHOD APPLIED TO TRANSIENT TWO-DIMENSIONAL HEAT TRANSFER WITH CONVECTION AND RADIATI	70-0127
HEAT	FINITE ELEMENT METHOD AND LAPLACE TRANSFORM COMPARATIVE SOLUTIONS OF TRANSIENT HEAT CONDUCTION PROBL	74-0575
HEAT	FINITE ELEMENT METHODS IN HEAT CONDUCTION PROBLEMS	75-0095
HEAT	FINITE ELEMENT SOLUTION OF HEAT TRANSMISSION IN STEEL INGOTS	72-0311
HEAT	FINITE ELEMENT SOLUTION OF NON-LINEAR HEAT CONDUCTION PROBLEMS WITH SPECIAL REFERENCE TO PHASE CHANG	74-0348
HEAT	FINITE ELEMENT TECHNIQUE APPLIED TO HEAT CONDUCTION IN SOLIDS WITH TEMPERATURE DEPENDENT THERMAL CON	69-0004
HEAT	FINITE-ELEMENT ANALYSIS OF HEAT FLOW AROUND BURIED PIPES	73-1153
HEAT	FINITE-ELEMENT METHOD APPLIED TO HEAT CONDUCTION IN SOLIDS WITH NONLINEAR BOUNDARY CONDITIONS	73-0005
HEAT	HEAT CONDUCTION AND STRESS ANALYSIS OF SOLID PROPELLANT ROCKET MOTOR NOZZLES	69-0059
HEAT	HEAT TRANSFER PROBLEMS AND TEMPERATURE STRESSES	69-0326
HEAT	HEAT TRANSFER WITH NONLINEAR BOUNDARY CONDITIONS VIA A VARIATIONAL PRINCIPLE	71-0443
HEAT	HEATRAN, A FINITE ELEMENT CODE FOR HEAT TRANSFER PROBLEMS	69-0076
HEAT	HETRAP: A HEAT TRANSFER ANALYSIS PROGRAM	74-0870
HEAT	HYBRIDE FINITE VERFAHREN FUER LINEARE UND NICHTLINEARE INSTATIONAERE WAERMELEITPROBLEME (HYBRID FINI	73-0484
HEAT	METODA ELEMENTOW SKONSZONYCH ROZWIAZYWANIA OSIOWO-SYMETRYCZNEGO PRZEWODNICTWA CIEPLA (FINITE-ELEMENT	73-0434
HEAT	MODELLING HEAT FLOW FROM UNDERGROUND POWER CABLES	74-1158
HEAT	NONSTEADY PYROMETRIC CALCULATIONS AS THE BASIS FOR STRESS ANALYSIS OF STEAM GENERATORS AND HEAT EXCH	72-0451
HEAT	ON THE ACCURACY OF FINITE ELEMENT SOLUTIONS TO THE TRANSIENT HEAT CONDUCTION EQUATION	74-0182
HEAT	ON THE CONVERGENCE OF THE FINITE ELEMENT SOLUTION IN A NONLINEAR HEAT CONDUCTION PROBLEM	74-1272
HEAT	ON THE PHYSICAL MEANINGS OF THE MATRICES DERIVED FROM FINITE ELEMENT METHOD OF HEAT CONDUCTION PROBL	73-0297
HEAT	PERIODIC HEAT CONDUCTION IN A TWO PHASE, TWO DIMENSIONAL SOLID DOMAIN	74-1164
HEAT	REFINED HEAT CONDUCTION FINITE ELEMENT FOR PERMAFROST ANALYSIS	73-0594
HEAT	SOME SUPERCONVERGENCE RESULTS FOR AN H-GALERKIN PROCEDURE FOR THE HEAT EQUATION	74-0411
HEAT	SPLINE APPROXIMATION AND DIFFERENCE SCHEMES FOR THE HEAT EQUATION	72-0518
HEAT	STABILITY AND OSCILLATION CHARACTERISTICS OF FINITE ELEMENT, FINITE-DIFFERENCE, AND WEIGHTED-RESIDUA	75-0615

246

HEAT	STATIC AND HEAT TRANSFER ANALYSIS OF A SPHERE-CONE INTERSECTION IN A NUCLEAR CONTAINMENT VESSEL	75-0166
HEAT	STEADY CONDUCTION OF HEAT IN LINEAR AND NONLINEAR FULLY ANISOTROPIC MEDIA BY FINITE ELEMENTS	74-0359
HEAT	STEADY STATE AND TRANSIENT HEAT CONDUCTION AND STRESS ANALYSES OF TWO AND THREE DIMENSIONAL STRUCTUR	71-0675
HEAT	STRESS DETERMINATION IN 2 CONDENSER HEAT EXCHANGER BAFFLE CAPS BY METHOD OF FINITE ELEMENTS AND A ST	74-0661
HEAT	TEMPERATURVERLAUF IM QUERSCHNITT GESCHWEISSTER ROHRWAENDE BEI UEBERWIEGEND KONVEKTIVEM WAERMEUEBERGA	75-0628
HEAT	THE APPLICATION OF HIGHER ORDER FINITE ELEMENTS TO THREE DIMENSIONAL HEAT CONDUCTION PROBLEMS	70-0266
HEAT	THE FINITE ELEMENT METHOD IN DEFORMATION AND HEAT CONDUCTION PROBLEMS	68-0238
HEAT	THE FINITE ELEMENT SOLUTION OF FLOW AND HEAT TRANSFER IN ELLIPTICAL DUCTS	74-0039
HEAT	THE HEAT CONDUCTION AND THERMAL STRESS ANALYSIS BY THE FINITE ELEMENT METHOD	68-0073
HEAT	THE USE OF FINITE ELEMENT METHODS IN HEAT FLOW ANALYSIS	68-0004
HEAT	TRANSIENT HEAT CONDUCTION PROBLEMS IN POWER CABLES SOLVED BY FINITE ELEMENT METHOD	73-0007
HEAT	TRANSIENT HEAT FLOW IN SOLIDS	74-0461
HEAT	TRANSIENT TWO DIMENSIONAL HEAT CONDUCTION PROBLEMS SOLVED BY THE FINITE ELEMENT METHOD	74-0364
HEAT	UNSYMMETRICAL TRANSIENT HEAT CONDUCTION: ROTATING DISK APPLICATIONS	70-0005
HEAT	USE OF FINITE ELEMENT METHODS IN SOLUTION OF TRANSIENT HEAT CONDUCTION PROBLEMS	69-0016
HEAT-CONDUCTION	APPLICATION OF THE FINITE-ELEMENT METHOD TO HEAT-TRANSFER PROBLEMS. PART 1. FINITE-ELEMENT METHOD AP	71-0760
HEAT-CONDUCTION	CODE TAFEST: NUMERICAL SOLUTION TO TRANSIENT HEAT-CONDUCTION PROBLEMS USING FINITE ELEMENTS IN SPACE	74-0901
HEAT-CONDUCTION	COMPARISON OF FINITE-DIFFERENCE AND FINITE-ELEMENT SOLUTION TECHNIQUE IN TRANSIENT HEAT-CONDUCTION	75-0348
HEAT-FLOW	CALCULATION BY FINITE ELEMENT TECHNIQUE OF STEADY-STATE OR TRANSIENT HEAT-FLOW CONDUCTION	74-0637
HEAT-TRANSFER	APPLICATION OF THE FINITE-ELEMENT METHOD TO HEAT-TRANSFER PROBLEMS. PART 11. TRANSIENT TWO-DIMENSION	71-0759
HEAT-TRANSFER	APPLICATION OF THE FINITE-ELEMENT METHOD TO HEAT-TRANSFER PROBLEMS. PART 1. FINITE-ELEMENT METHOD AP	71-0760
HEAT-TRANSFER	FINITE-ELEMENT ANALYSIS OF COMPRESSIBLE AND INCOMPRESSIBLE VISCOUS-FLOW AND HEAT-TRANSFER PROBLEMS	75-0347
HEATED	MATHEMATICAL MODEL OF HEATED EFFLUENTS IN COASTAL WATER	73-0921
HEATED	STRESS ANALYSIS IN HEATED, COMPLEX SHAPES	62-0002
HEATED	TWO-DIMENSIONAL HEATED PIPELINE IN PERMAFROST PIPELINES	74-0541
HEATERS	ANALYSIS OF THERMAL STRESSES IN CORED CERAMIC BRICKS USED IN HYPERSONIC WIND TUNNEL HEATERS	72-0655
HEATING	A NUMERICAL PROCEDURE FOR TWO DIMENSIONAL HEATING AND MELTING CALCULATIONS, WITH APPLICATIONS TO LAS	72-0708
HEATING	APPLICABILITY OF APPROXIMATE AND EXACT TRANSIENT HEAT TRANSFER ANALYSES TO HEATING PROCESSES USED TO	75-0244
HEATING	INFLUENCE OF ASCENT HEATING ON THE SEPARATION DYNAMICS OF A SPACECRAFT FAIRING	72-0906
HEATING	LARGE DEFLECTIONS OF STRUCTURES SUBJECT TO HEATING AND EXTERNAL LOADS	60-0002
HEATRAN	HEATRAN, A FINITE ELEMENT CODE FOR HEAT TRANSFER PROBLEMS	69-0076
HEAVY	EFFECTS OF SHOE FORCE GEOMETRY ON HEAVY DUTY INTERNAL SHOE BRAKE PERFORMANCE	68-0219
HEAVY	HEAVY SECTION STEEL TECHNOLOGY PROGRAM. TECHNICAL REPORT NO. 4, OCTOBER 1969. A TWO-DIMENSIONAL ELAS	69-0474
HEAVY	HEAVY SECTION STEEL TECHNOLOGY PROGRAM TECHNICAL OR PROGRAMMATIC MANUSCRIPT NO. 25. VARIABLE THICKNE	73-0948
HEAVY	VIBROACOUSTIC RESPONSE OF TURBULENCE EXCITED THIN RECTANGULAR FINITE PLATES IN HEAVY AND FLUID MEDIA	74-1309
HEIGHTS	CALCULATION OF STEADY-STATE WATER-TABLE HEIGHTS IN DRAINED SOILS BY MEANS OF FINITE-ELEMENT METHOD	75-0457
HELICAL	HYBRIDE SCHALENELEMENTE MIT EINER ANWENDUNG AUF WENDELSCHALEN. (HYBRID SHELL ELEMENTS IN APPLICATION	75-0609
HELICOPTER	A NASTRAN VIBRATION MODEL OF THE AH-IG HELICOPTER AIRFRAME. VOLUME 1	74-1338
HELICOPTER	A NASTRAN VIBRATION MODEL OF THE AH-IG HELICOPTER AIRFRAME. VOLUME 11	74-1339
HELICOPTER	CALCULATION OF THE NATURAL FREQUENCIES AND THE PRINCIPAL MODES OF HELICOPTER BLADES	72-0907
HELICOPTER	CORRELATION OF FINITE ELEMENT STRUCTURAL DYNAMIC ANALYSIS WITH MEASURED FREE VIBRATION CHARACTERISTI	74-0486
HELICOPTER	DEVELOPMENT OF A GUN-TURRET MODEL AND OF THE EQUATIONS FOR COUPLING IT WITH A HELICOPTER FINITE-ELEM	74-0752
HELICOPTER	SENSITIVITY ANALYSIS OF TORSIONAL VIBRATION CHARACTERISTICS OF HELICOPTER ROTOR BLADES. PART 1: STRU	74-0846
HELLINGER-REISS	A FINITE ELEMENT APPLICATION OF THE HELLINGER-REISSNER VARIATIONAL THEOREM	68-0060
HELLINGER-REISS	RECTANGULAR FINITE ELEMENT FOR PLATE BENDING ANALYSIS BASED ON HELLINGER-REISSNERS VARIATIONAL PRINC	71-0545
HELMHOLT'S	APPLICATION OF FINITE ELEMENTS TO THE SOLUTION OF HELMHOLT'S EQUATION	68-0020
HELMHOLTZ	AXISYMETRIC TRIANGULAR ELEMENT FOR THE SCALAR HELMHOLTZ EQUATION	73-0688
HELMHOLTZ	EXACT FINITE ELEMENT SOLUTIONS TO THE HELMHOLTZ EQUATIONS	73-0652
HELMHOLTZ	TETRAHEDRAL POLYNOMIAL FINITE ELEMENTS FOR THE HELMHOLTZ EQUATION.	72-0392
HERMES	THE HERMES 8 ELEMENT FOR THE MATRIX DISPLACEMENT METHOD	68-0011
HERMITE	APPLICATION OF HERMITE POLYNOMIALS TO THE DETERMINATION OF THE STIFFNESS MATRIX OF PLATE BY FINITE E	71-0365
HERMITE	BOUNDS FOR A CLASS OF LINEAR FUNCTIONALS WITH APPLICATIONS TO HERMITE INTERPOLATION	71-0582
HERMITE	GENERAL LAGRANGE AND HERMITE INTERPOLATION IN R**N WITH APPLICATIONS TO FINITE ELEMENT METHODS	72-0110
HERMITE	HERMITE INTERPOLATION TYPE RITZ METHODS FOR TWO POINT BOUNDARY VALUE PROBLEMS	66-0089
HERMITE	PIECEWISE HERMITE INTERPOLATION IN ONE AND TWO VARIABLES WITH APPLICATIONS TO PARTIAL DIFFERENTIAL E	68-0201
HERMITE'A	APPLICATION OF HERMITE POLYNOMIALS TO THE DETERMINATION OF THE STIFFNESS MATRIX OF PLATE BY FINITE E	71-0365
HERMITEAN	HERMITEAN RECTANGLE FINITE ELEMENT FAMILY	73-1009
HERMITESCHEN	DAS VERFAHREN VON RAYLEIGH-RITZ MIT HERMITESCHEN INTERPOLATIONSPOLYNOMEN	63-0018
HERMITESCHEN	EINIGE VERFAHREN ZUR BERECHNUNG VON EIGENWERTEN UND EIGENVEKTOREN VON NICHT HERMITESCHEN MATRIZEN U	72-0646
HERMITIAN	DYNAMIC STIFFNESS MATRIX FORMULATION BY MEANS OF HERMITIAN POLYNOMIALS	66-0059
HETEROGENEITY	EFFECT OF SHAPE AND HETEROGENEITY OF ROCK SAMPLES ON THEIR COMPRESSIVE STRENGTH	69-0315
HETEROGENEOUS	METHOD OF FINITE ELEMENTS FOR CALCULATION OF HETEROGENEOUS BODIES OF REVOLUTION IN ELASTO-PLASTIC ME	75-0144
HETEROGENEOUS	SOME REMARKS CONCERNING HETEROGENEOUS ANISOTROPIC PLATES.	72-0430
HETEROGENEOUS	THE BEHAVIOR OF STATISTICALLY HETEROGENEOUS EXCAVATED EARTH SLOPES	74-1330
HETEROGENEOUS	TWO DIMENSIONAL STOCHASTIC MODEL OF A HETEROGENEOUS GEOLOGIC SYSTEM	73-0715
HETRAP	HETRAP: A HEAT TRANSFER ANALYSIS PROGRAM	74-0870
HEXAGONAL	PRISMATIC SHELLS WITH HEXAGONAL CROSS-SECTION	74-1043
HEXAGONAL	STRESS INTENSITY FACTORS FOR HEXAGONAL FRACTURE TOUGHNESS SPECIMENS	71-0799
HEXAHEDRON	FURTHER LIMITATIONS ON GENERAL HEXAHEDRON FINITE ELEMENTS WITH DERIVATIVE DEGREES OF FREEDOM.	72-0395
HEXE	PART 11: LUMINA AND HEXE ELEMENTS	68-0261
HIERARCHICAL	A TECHNIQUE FOR DEGENERATING BRICK-TYPE ISOPARAMETRIC ELEMENTS USING HIERARCHICAL MIDSIDE NODES	74-0403
HIGH-ENERGY	A THERMAL, THERMOELASTIC, AND WEAR ANALYSIS OF HIGH-ENERGY DISK BRAKES	74-0839
HIGH-ENERGY	THERMAL, THERMOELASTIC, AND WEAR SIMULATION OF A HIGH-ENERGY SLIDING CONTACT PROBLEM	73-0430
HIGH-ORDER	A GENERAL HIGH-ORDER FINITE-ELEMENT WAVEGUIDE ANALYSIS PROGRAM	69-0512
HIGH-ORDER	A HIGH-ORDER FINITE ELEMENT FOR TWO-DIMENSIONAL CRACK PROBLEMS	74-1128
HIGH-ORDER	A HIGH-ORDER FINITE ELEMENT FOR PRISMATIC AND NONPRISMATIC FOLDED PLATE STRUCTURES	75-0889
HIGH-ORDER	ANALYSIS OF TRANSFORMER LEAKAGE PHENOMENA BY HIGH-ORDER FINITE ELEMENTS	73-0135
HIGH-ORDER	EFFECT OF DAMPING ON EXCITABILITY OF HIGH-ORDER NORMAL MODES	75-0709
HIGH-ORDER	HIGH-ORDER FINITE ELEMENT ANALYSIS FOR TRANSIENT TEMPERATURE ANALYSIS OF INHOMOGENEOUS MATERIALS	69-0301
HIGH-ORDER	HIGH-ORDER FINITE ELEMENTS FOR INHOMOGENEOUS ACOUSTIC GUIDING STRUCTURES	73-0137
HIGH-ORDER	HIGH-ORDER FINITE ELEMENT TECHNIQUE FOR PRISMATIC AND NONPRISMATIC FOLDED PLATE STRUCTURES	74-1130
HIGH-ORDER	NUMERICAL METHODS OF HIGH-ORDER ACCURACY FOR NONLINEAR BOUNDARY VALUE PROBLEMS	68-0194
HIGH-OVERPRESSU	RESPONSE OF BURIED CAPSULES IN THE HIGH-OVERPRESSURE REGION	70-0093
HIGH-PERFORMANC	HIGH-PERFORMANCE COMPOSITE MATERIALS FOR VEHICLE CONSTRUCTION: A FINITE ELEMENT COMPUTER PROGRAM FOR	73-0767
HIGH-PRECISION	A HIGH-PRECISION TRIANGULAR CYLINDRICAL SHELL FINITE ELEMENT	71-0163
HIGH-PRECISION	COMPARISON OF TWO HIGH-PRECISION TRIANGULAR FINITE ELEMENTS FOR ARBITRARY DEEP SHELLS	71-0071
HIGH-PRECISION	CURSHL: A HIGH-PRECISION FINITE ELEMENT FOR SHELLS OF ARBITRARY SHAPE	71-0430
HIGH-PRECISION	HIGH-PRECISION CURVED SHELL FINITE ELEMENT	71-0422
HIGH-PRECISION	HIGH-PRECISION FINITE-ELEMENT ANALYSIS OF SHELLS OF REVOLUTION	73-1164
HIGH-PRECISION	HIGH-PRECISION PLATE AND SHELL FINITE ELEMENTS	71-0437
HIGH-PRECISION	STATIC AND DYNAMIC APPLICATIONS OF A HIGH-PRECISION TRIANGULAR PLATE BENDING ELEMENT	69-0086
HIGH-PRESSURE	FINITE-ELEMENT ANALYSIS OF HIGH-PRESSURE, LARGE VALVE GATES	75-0329
HIGH-PRESSURE	FINITE-ELEMENT ANALYSIS OF HIGH-PRESSURE LARGE VALVE GATES	75-0413
HIGH-RISE	TOUR MAINE-MONTPARNASSE QUELQUES PROBLEMES DE STRUCTURE (SPECIFIC PROBLEMS ENCOUNTERED IN THE CONSTR	74-0539
HIGH-TEMPERATUR	ANALYSIS OF LOADING AND THERMAL EFFECTS ON FUELLED GRAPHITE BRICKS FOR A HIGH-TEMPERATURE GAS COOLED	71-0085
HIGH-TEMPERATUR	FINITE ELEMENT ANALYSIS OF STRESSES IN FUEL ELEMENTS OF A HIGH-TEMPERATURE GAS-COOLED REACTOR	71-0115
HIGHER-ORDER	DISCRETIZATION AND COMPUTATIONAL ERRORS IN HIGHER-ORDER FINITE ELEMENTS	71-0089
HIGHER-ORDER	HIGHER-ORDER FINITE ELEMENT FOR PLANE STRESS	67-0082
HIGHER-ORDER	HIGHER-ORDER FINITE ELEMENT FOR PLANE STRESS	68-0089
HIGHER-ORDER	HIGHER-ORDER FINITE ELEMENT ANALYSIS OF LAKE CIRCULATION	73-1002
HIGHER-ORDER	HIGHER-ORDER FINITE-ELEMENT ANALYSIS OF TOPOGRAPHIC GUIDES SUPPORTING ELASTIC SURFACE WAVES	73-0114
HIGHER-ORDER	HIGHER-ORDER LINEAR FINITE ELEMENT METHODS	74-0826

HIGHER-ORDER	MATRIX DIFFERENTIAL EQUATION AND HIGHER-ORDER NUMERICAL METHODS FOR PROBLEMS OF NON-LINEAR CREEP, VI	72-0198
HIGHER-ORDER	PRACTICAL ANALYSIS OF STEEL STRUCTURES USING HIGHER-ORDER ELEMENTS	72-0326
HIGHER-ORDER	TRANSVERSE IMPACT PROBLEMS BY HIGHER-ORDER BEAM FINITE-ELEMENTS	75-0819
HIGHWAY	A FINITE ELEMENT APPROACH TO THE DYNAMIC ANALYSIS OF CONTINUOUS HIGHWAY BRIDGES	72-0591
HIGHWAY	DYNAMIC ANALYSIS OF BEAM-SLAB HIGHWAY BRIDGES	73-0199
HIGHWAY	EXPERIMENTAL AND ANALYTICAL INVESTIGATION OF A HORIZONTALLY CURVED BOX-BEAM HIGHWAY BRIDGE MODEL	71-0024
HIGHWAY	GEOMETRIC HIGHWAY AND CULVERT DESIGN	74-1342
HIGHWAY	GROUNDWATER PREDICTIONS FOR A HIGHWAY CUT USING FINITE ELEMENTS	73-0346
HIGHWAY	HORIZONTALLY CURVED HIGHWAY BRIDGES. A FULLY COMPATIBLE ANNULAR SEGMENT FINITE ELEMENT	72-0834
HIGHWAY	MESH GENERATION PROGRAM FOR HIGHWAY EXCAVATION CUTS	74-1250
HIGHWAY	MODEL ANALYSIS OF CURVED BOX-BEAM HIGHWAY BRIDGE	71-0282
HIGHWAY	NONLINEAR THEORY FOR PREDICTING THE PERFORMANCE OF FLEXIBLE HIGHWAY PAVEMENTS	70-0143
HIGHWAY	SEISMIC RESPONSE OF HIGHWAY BRIDGES	75-0879
HIGHWAY	THE COMPARISON OF THE RESPONSE OF A HIGHWAY BRIDGE TO UNIFORM GROUND SHOCK AND MOVING GROUND EXCITAT	72-0737
HILL	APPLICATION OF HILL FUNCTIONS TO CIRCULAR PLATE PROBLEMS	75-0572
HILL	APPLICATIONS OF HILL FUNCTIONS (APPLIED MATHEMATIC'S FINITE ELEMENT) TO APPLIED MECHANICS PROBLEMS	72-0884
HIMBLICK	UNTERSUCHUNGEN ZUM FESTIGKEITSVERHALTEN VON METALLKLEBUNGEN IM HIMBLICK AUF DIE ERSTELLUNG EINES PRA	74-0289
HINGE	GALERKIN FINITE ELEMENT ANALYSIS OF A UNIFORM BEAM CARRYING A CONCENTRATED MASS AND ROTARY INERTIA W	74-0552
HINGE	HINGE	71-0634
HINGE-CONNECTED	DYNAMIC ANALYSIS OF A SYSTEM OF HINGE-CONNECTED RIGID BODIES WITH NONRIGID APPENDAGES	74-0852
HIP	FINITE ELEMENT STUDIES OF FEMORAL ENDOPROSTHESES FOR HIP RECONSTRUCTION	75-0381
HIPPED	ANALYSIS OF HIPPED ROOF HYPERBOLIC PARABOLOID STRUCTURES	72-0216
HISTORY	A HISTORY OF THE FINITE ELEMENT METHOD	75-0398
HISTORY	FINITE-ELEMENT CALCULATION OF PRESSURE HISTORY IN A GAS FIELD	75-0546
HISTORY	THE EFFECT OF STRESS HISTORY ON THE RESILIENT RESPONSE OF SOILS	73-0914
HOLD	TRANSVERSE STRENGTH AT FORE AND AFT SECTION OF TANKER CARGO HOLD	74-0618
HOLE	A DEEP HOLE STRESS MEASUREMENT DEVICE	74-0804
HOLE	ACCURACY OF THE FINITE ELEMENT ANALYSIS FOR THE ELASTIC PLATE WITH A CIRCULAR HOLE	73-0155
HOLE	ANALYSIS AND DESIGN OF AXISYMMETRIC REINFORCEMENT AROUND A CIRCULAR HOLE IN A THIN FLAT PLATE BY THE	74-1113
HOLE	DEFORMATION AND FAILURE OF BORON-EPOXY PLATE WITH CIRCULAR HOLE	72-0536
HOLE	FACTORS OF STRESS CONCENTRATION AT A HOLE IN FINITE STRIPS UNDER BENDING	72-0744
HOLE	FATIGUE CRACK INITIATION IN A BORON EPOXY PLATE WITH A CIRCULAR HOLE	74-0820
HOLE	MECHANICAL BEHAVIOR OF A GRAPHITE-EPOXY LAMINATE CONTAINING A HOLE	72-0614
HOLE	NONLINEAR ANALYSES OF THE STRESS FIELD NEAR A CRACK OR HOLE IN A PLATE	75-0694
HOLE	NOTE ON THE STRESS CONCENTRATIONS AT THE END OF A CYLINDRICAL HOLE	70-0139
HOLE	PARAMETRISCHE KERBSPANNUNGSUNTERSUCHUNGEN AN DER LOCHSCHEIBE MIT DER METHODE DER FINITEN ELEMENTE. (	75-0603
HOLE	PHOTOELASTIC INVESTIGATION ON THE CRACK-ARREST CAPABILITY OF A HOLE	72-0086
HOLE	STRESS ANALYSIS OF LAMINATED CFRP PLATES CONTAINING A HOLE	74-1175
HOLE	STRESS AND FAILURE ANALYSIS OF A GLASS-EPOXY COMPOSITE PLATE WITH A CIRCULAR HOLE	73-0276
HOLE	STRESS INTENSITY FACTOR FOR A STRAIGHT CRACK APPROACHING A HOLE	69-0158
HOLE	STRESS INTENSITY FACTORS FOR CRACKS IN A PLATE CONTAINING A HOLE AND IN A SPINNING DISC	73-0467
HOLE	THE BEHAVIOUR OF SQUARE SHEAR WEBS HAVING A CIRCULAR HOLE	67-0108
HOLE	THEORETICAL AND EXPERIMENTAL ANALYSIS OF ELASTOPLASTIC DEFORMATION IN A UNIAXIALLY LOADED STRIP OF 2	73-0720
HOLES	ANALYSIS OF THE EFFECTS ON LIFE OF LEADING-EDGE HOLES IN AN AIRFOIL SUBJECTED TO ARBITRARY SPANWISE	75-0700
HOLES	EFFECT OF FIBRE DISTRIBUTION ON STRESS AND STRAIN CONCENTRATION AT HOLES IN COMPOSITE PLATES	72-0616
HOLES	FINITE ELEMENT ANALYSIS OF STATIC AND DYNAMIC STRESSES AROUND SUDDENLY PUNCHED HOLES IN PLATES AND S	73-1088
HOLES	STABILITY CONTROL ANALYSIS OR ROTATING PLATES BY FINITE ELEMENT. EMPHASIS ON SLOTS AND HOLES	71-0284
HOLES	STRESS ANALYSIS OF COLDWORKED FASTENER HOLES	74-0783
HOLES	STRESS CONCENTRATIONS NEAR HOLES IN LAMINATES	74-0329
HOLES	STRESSES AROUND HOLES IN STIFFENED COMPOSITE PANELS USING LAURENT-SERIES AND FINITE ELEMENT METHODS	74-1252
HOLES	STRESSES AROUND HOLES IN STIFFENED COMPOSITE PANELS USING LAURENT-SERIES AND FINITE ELEMENT METHODS	75-0733
HOLES	STRESSES IN A ROTATING DISK WITH CENTRAL AND NONCENTRAL HOLES (STRESS ANALYSIS BY THE FINITE ELEMENT	74-1337
HOLES	THREE DIMENSIONAL ANALYSIS OF STRESS CONCENTRATIONS NEAR HOLES IN LAMINATED COMPOSITES	72-0713
HOLES	THREE DIMENSIONAL FINITE ELEMENT ANALYSIS OF THICK LAMINATED COMPOSITES - INCLUDING INTERLAMINAR AND	73-0882
HOLES	THREE-DIMENSIONAL ANALYSIS FOR THE STRESS DISTRIBUTION NEAR CIRCULAR HOLES IN LAMINATED COMPOSITES	74-0793
HOLES	THREE-DIMENSIONAL FINITE ELEMENT ANALYSIS OF THICK LAMINATED COMPOSITES - INCLUDING INTERLAMINAR AND	74-1151
HOLES	TRANSFORMATIONS FOR ELLIPTICAL HOLES APPLIED TO FINITE ELEMENT ANALYSIS	74-0132
HOLES	TWO NEARLY POLYGONAL HOLES. MATHEMATICAL CRACK PROBLEMS	69-0061
HOLES	ZUR BERECHNUNG, BEMESSUNG UND BEWEHRUNG RECHTECKIGER PLATTEN KONSTNATER DICKE MIT OEFFNUNGEN ODER AU	73-0664
HOLLOW	AN EXPERIMENTAL AND THEORETICAL STUDY OF HOLLOW CYLINDER TESTS ON SAND	71-0008
HOLLOW	COMPARISON OF ANALYTICAL AND EXPERIMENTAL RESULTS IN THE DESIGN OF A HOLLOW GRAVITY DAM	67-0109
HOLLOW	CONTROLLED FAILURE OF HOLLOW ROCK CYLINDERS IN UNIAXIAL COMPRESSION	72-0196
HOLLOW	FINITE ELEMENT ANALYSIS OF SOME MECHANICAL PROBLEMS OF HOLLOW GRAVITY DAM	70-0470
HOLLOW	NON-UNIFORM DEFORMATION OF MATERIAL IN AXIALLY SYMMETRIC COMPRESSION CAUSED BY FRICTION. PART 3. THE	72-0923
HOLLOW	RADIAL DISPLACEMENTS OF A HOLLOW SHAFT SUBJECTED TO A UNIFORM BAND OF EXTERNAL PRESSURE	72-0242
HOLLOW	WAVE PROPAGATION IN A THIN HOLLOW CONE BY A FINITE ELEMENT METHOD	72-0045
HOLLOW-ENDED	HOLLOW-ENDED ROLLER. A SOLUTION FOR IMPROVING FATIGUE LIFE IN ASYMMETRICALLY LOADED CYLINDRICAL ROLL	71-0393
HOLOGRAPHIC	AN INVESTIGATION INTO THE METHODS OF DETECTING DEWETTED ZONES IN SOLID ROCKET PROPELLANTS USING HOLO	73-0890
HOLOGRAPHIC	HOLOGRAPHIC INSTRUMENTATION OF TURBINE BLADES	73-0466
HOLOGRAPHIC	IMPROVED METHOD FOR OBTAINING THE GENERAL-DISPLACEMENT FIELD FROM A HOLOGRAPHIC INTERFEROGRAM	72-0015
HOLOGRAPHY	A METHOD FOR MATERIAL CHARACTERIZATION EMPLOYING THE FINITE-ELEMENT TECHNIQUE WITH APPLICATION TO HO	72-0941
HOMALITE-100	CRACK BRANCHING IN HOMALITE-100 SHEETS	72-0843
HOMALITE-100	CRACK BRANCHING IN HOMALITE-100 SHEETS	74-0291
HOMALITE-100	FRACTURE DYNAMICS OF HOMALITE-100 SHEETS	72-0809
HOMOGENE	NOUVELLE CONCEPTION DES DRAINS DANS LES BARRAGES EN TERRE HOMOGENE DE PETITE ET MOYENNE DIMENSIONS (	73-0404
HOMOGENEOUS	DETERMINATION OF EIGENMODES OF HOMOGENEOUS PLASMA WITH ELLIPTICAL SECTION USING BI-DIMENSIONAL FINIT	74-0553
HOMOGENEOUS	FINITE ELEMENT BUCKLING ANALYSIS OF HOMOGENEOUS AND SANDWICH PLATES	75-0613
HOMOGENEOUS	FINITE ELEMENT SOLUTION OF THE HOMOGENEOUS WAVEGUIDE PROBLEMS	69-0010
HOMOGENEOUS	NOUVELLE CONCEPTION DES DRAINS DANS LES BARRAGES EN TERRE HOMOGENE DE PETITE ET MOYENNE DIMENSIONS (	73-0404
HOMOGENEOUS	NUMERICAL STUDY OF PROPER MODES OF A HOMOGENEOUS CYLINDRICAL PLASMA BY METHOD OF FINITE ELEMENTS	74-0160
HOMOGENEOUS	NX BOREHOLE JACK MODULUS DETERMINATIONS IN HOMOGENEOUS, ISOTROPIC, ELASTIC MATERIALS	74-0295
HOMOGENEOUS	SIMPLE FINITE ELEMENT SOLUTION FOR PLATES OF HOMOGENEOUS, SANDWICH AND CELLULAR CONSTRUCTION	75-0140
HOMOGENEOUS	STRESSES AND DISPLACEMENTS AROUND OPENINGS IN HOMOGENEOUS ROCK	75-0168
HONEY-COMB	MEASURED AND CALCULATED VIBRATION PROPERTIES OF RING-STIFFENED HONEY-COMB CYLINDERS	71-0834
HOOKS	FINITE ELEMENT STRESS ANALYSIS OF TURBINE BLADE HOOKS	70-0479
HORIZONS	HORIZONS IN THE STRUCTURAL USE OF ADVANCED MATERIALS - THRAUSTICS	70-0415
HORIZONS	NEW HORIZONS IN ROCK MECHANICS	73-0428
HORIZONS	NEW HORIZONS IN ROCK MECHANICS	73-0446
HORIZONTAL	DEVELOPMENT OF A GRAPHITE HORIZONTAL STABILIZER	73-0931
HORIZONTAL	IN-SITU HORIZONTAL STRESSES IN LATERITE BY FINITE ELEMENT ANALYSIS	73-1048
HORIZONTAL	STATIC AND DYNAMIC ANALYSIS OF F-14A BORON HORIZONTAL STABILIZER	71-0123
HORIZONTAL-EVEN	BEITRAG ZUR SIMULATION NICHTSTATIONARER HORIZONTAL-EVENER GRUNDWASSERSTROMUNGSFELDER AUF DER GRUNDLA	74-0088
HORIZONTALLY	EXPERIMENTAL AND ANALYTICAL INVESTIGATION OF A HORIZONTALLY CURVED BOX-BEAM HIGHWAY BRIDGE MODEL	71-0024
HORIZONTALLY	HORIZONTALLY CURVED HIGHWAY BRIDGES. A FULLY COMPATIBLE ANNULAR SEGMENT FINITE ELEMENT	72-0834
HORIZONTALLY	INSTABILITY OF HORIZONTALLY CURVED MEMBERS. FLANGE BUCKLING STUDIES	69-0480
HORIZONTALLY	MESH GENERATING COMPUTER PROGRAM FOR FINITE ELEMENT ANALYSES OF SIMPLE SLOPES IN HORIZONTALLY LAYERE	71-0748
HOT	FINITE ELEMENT ANALYSES OF SALT DOMES WITH STORED HOT WASTES	74-1148
HUB	ISSLEDOVANIE NAPRYAZHENNO-DEFORMIROVANNOGO SOSTOYANIYA SIMMETRICHNYKH DISKOV SO STUPITSEI. (INVESTIG	74-0965
HUB	SHAKEDOWN OF ROLLING WHEEL UNDER HUB LOADING	74-0663
HUBS	FINITE ELEMENT ANALYSIS OF STRESSES IN SHAFTS DUE TO INTERFERENCE-FIT HUBS	69-0353
HULL	APPLICATION OF FINITE ELEMENT METHOD TO ADDED VIRTUAL MASS OF SHIP HULL VIBRATION	70-0463
HULL	DEFORMABILITY OF THE HULL STEEL-WORK AND DEFORMATIONS OF THE ENGINE-ROOM OF LARGE TANKERS	74-0532

248

HULL	EINSATZ DES COMPUTERS BEI DER ENTWICKLUNG EINER PANZERWANNE (THE USE OF A COMPUTER FOR DEVELOPING A	72-0211
HULL	FINITE ELEMENT ANALYSIS OF THE FORCED OSCILLATION OF SHIP HULL FORMS	74-0814
HULL	FINITE ELEMENT COMPUTER ANALYSIS OF KOCHMUS HULL 520.	70-0293
HULL	FINITE ELEMENT COMPUTER ANALYSIS OF MITSUBISHI HULL 1670	70-0294
HULL	HULL AND SUPERSTRUCTURE VIBRATIONS DESIGN CALCULATIONS BY FINITE ELEMENTS	75-0556
HULL	NUMERICAL HULL VIBRATION ANALYSIS OF A FAR EAST CONTAINER SHIP	74-1325
HULL	RECENT DEVELOPMENTS IN HULL AND SHAFT VIBRATION ANALYSIS	70-0176
HULL	SIMPLIFIED CALCULATION OF THE INTERACTION FORCES BETWEEN HULL AND TANK SYSTEM FOR A MOSS ROSENBERG T	75-0561
HULL	STRESS ANALYSIS IN HULL STRUCTURAL MEMBERS BY PHOTOELASTIC EXPERIMENTS AND FINITE ELEMENT METHODS -	70-0341
HULL	STRESS CONCENTRATIONS IN HULL SHAPES FROM SURFACE DISCONTINUITIES	69-0130
HULL	3-DIMENSIONAL ANALYSIS OF HULL STRUCTURE BY FINITE ELEMENT METHOD	73-0160
HULL-DECKHOUSE	EXPERIMENTAL AND THEORETICAL STUDIES OF HULL-DECKHOUSE INTERACTION - 2-LEVEL DECKHOUSE	71-0672
HULL-DECKHOUSE	HULL-DECKHOUSE INTERACTION BY FINITE ELEMENT CALCULATIONS	68-0132
HULLS	APPLICATION OF PROGRAM FATIQUE TEST TO MEMBER JOINTS OF HULLS	72-0814
HULLS	APPLICATION OF THE FINITE ELEMENT METHOD TO THE CALCULATION OF SHIP HULLS	71-0431
HULLS	INFLUENCE OF END-CLOSURE STIFFNESS ON BEHAVIOR OF CONCRETE CYLINDRICAL HULLS SUBJECTED TO HYDROSTATI	71-0725
HUMAN	A DYNAMIC FINITE ELEMENT MODEL OF THE HUMAN BRAIN	74-1076
HUMAN	A FINITE ELEMENT ANALYSIS OF WAVE PROPAGATION IN HUMAN SPINE	73-0792
HUMAN	DETAILED STRUCTURAL ANALYSIS OF THE HUMAN TIBIA	75-0379
HUMAN	DYNAMIC RESPONSE OF THE HUMAN ARM AS A COMPOSITE STRUCTURE	72-0611
HUMAN	DYNAMIC TORSIONAL RESPONSE OF THE HUMAN LEG RELATIVE TO SKIING INJURIES	73-0471
HUMAN	FINITE ELEMENT ANALYSIS OF HUMAN CARDIAC STRUCTURES	74-0438
HUMAN	IN VIVO STRESSES IN THE HUMAN LEFT VENTRICULAR WALL: ANALYSIS ACCOUNTING FOR THE IRREGULAR THREE DIM	72-0136
HUMAN	IN-VACUO MODAL DYNAMIC RESPONSE OF THE HUMAN SKULL	73-0460
HUMAN	NONLINEAR BEHAVIOR OF THE HUMAN INTERVERTEBRAL DISC UNDER AXIAL LOADING	74-0919
HUMAN	ON THE MATHEMATICAL ANALYSIS OF STRESS IN THE HUMAN FEMUR	72-0207
HUMAN	STRESS ANALYSIS OF HUMAN FEMUR	73-0327
HUMAN	STRESS ANALYSIS OF HUMAN TEETH	73-0505
HUMAN	STRESS ANALYSIS OF THE HUMAN AORTIC VALVE	73-0166
HUMAN	THE FINITE ELEMENT STRESS ANALYSIS OF THE HUMAN AORTIC VALVE TOWARD THE DEVELOPMENT OF A PROSTHETIC	74-1201
HUMIDITY	CREEP AND SHRINKAGE LAW FOR CONCRETE AT VARIABLE HUMIDITY	74-0668
HYATT	ANALYSIS OF BEHAVIOR OF EDWARD HYATT POWER PLANT	75-0055
HYBOS	THERMAL AND STRESS ANALYSIS PROGRAM - HYBOS	72-0159
HYBRID	A HYBRID FINITE ELEMENT FOR DETERMINATION OF STRESS INTENSITY FACTORS AT A CRACK TIP	73-0643
HYBRID	A HYBRID FINITE ELEMENT METHOD FOR TWO-DIMENSIONAL NAVIER STOKES EQUATIONS	75-0027
HYBRID	A HYBRID METHOD OF COMPONENT MODE SYNTHESIS	71-0681
HYBRID	A HYBRID STRESS ELEMENT FOR THIN SHELL ANALYSIS	74-0433
HYBRID	A NEW ASSUMED STRESS HYBRID FINITE ELEMENT MODEL FOR SOLID CONTINUA	71-0371
HYBRID	A NEW HYBRID CYLINDRICAL SHELL FINITE ELEMENT	71-0118
HYBRID	A POWERFUL HYBRID METHOD IN FINITE ELEMENT ANALYSIS	71-0218
HYBRID	A STUDY OF ELASTIC-PLASTIC ANALYSIS BY THE ASSUMED-STRESS HYBRID FINITE-ELEMENT MODEL, WITH APPLICAT	74-0718
HYBRID	ALTERNATE HYBRID STRESS FINITE ELEMENT MODELS	75-0476
HYBRID	AN ASSUMED STRESS HYBRID FINITE ELEMENT METHOD FOR AN INCOMPRESSIBLE AND NEAR INCOMPRESSIBLE MATERIA	69-0259
HYBRID	AN ASSUMED STRESS HYBRID FINITE ELEMENT MODEL FOR THE ANALYSIS OF AN AXISYMMETRIC THICK WALLED PRESS	71-0025
HYBRID	AN ASSUMED STRESS HYBRID FINITE ELEMENT FOR THREE-DIMENSIONAL PLASTIC STRUCTURAL ANALYSIS	74-0717
HYBRID	AN INTERACTIVE HYBRID TECHNIQUE FOR CRASHWORTHY DESIGN OF COMPLEX VEHICULAR STRUCTURAL SYSTEMS	74-0317
HYBRID	ANALYSIS OF ELASTICALLY COUPLED SHELLS ON ELASTIC FOUNDATION BY HYBRID METHOD	71-0276
HYBRID	ANALYSIS OF SAINT-VENANT TORSION PROBLEM BY A HYBRID STRESS MODEL	69-0277
HYBRID	APPLICATION OF AN ASSUMED DISPLACEMENT HYBRID FINITE ELEMENT PROCEDURE TO TWO-DIMENSIONAL PROBLEMS !	74-0729
HYBRID	APPLICATION OF SIMPLIFIED HYBRID DISPLACEMENT METHOD TO PLATE AND SHELL PROBLEMS	73-1037
HYBRID	APPLICATIONS OF THE ASSUMED-STRESS HYBRID METHOD TO FINITE ELEMENT ANALYSIS IN FRACTURE MECHANICS	75-0495
HYBRID	ASSUMED DISPLACEMENT HYBRID FINITE ELEMENT MODEL FOR LINEAR FRACTURE MECHANICS	75-0146
HYBRID	ASSUMED STRESS HYBRID FINITE ELEMENT METHOD FOR FRACTURE MECHANICS AND ELASTIC-PLASTIC ANALYSIS	72-0584
HYBRID	ASSUMED STRESS HYBRID FINITE ELEMENT MODEL FOR LINEAR ELASTODYNAMIC ANALYSIS	73-0089
HYBRID	BRAIN TISSUE FRAGILITY - A FINITE STRAIN ANALYSIS BY A HYBRID FINITE-ELEMENT METHOD	75-0162
HYBRID	BRAIN TISSUE FRAGILITY - A FINITE STRAIN ANALYSIS BY A HYBRID FINITE-ELEMENT METHOD	75-0243
HYBRID	BRAIN-TISSUE FRAGILITY - FINITE STRAIN ANALYSIS BY A HYBRID FINITE-ELEMENT METHOD	75-0631
HYBRID	CALCULATION OF MECHANICAL IMPEDANCE BY FINITE ELEMENT HYBRID MODEL	74-0126
HYBRID	CHARACTERISTICS OF A HYBRID CALCULATION SYSTEM BASED ON THE FINITE ELEMENT METHOD. (CARATTERISTICHE	72-0715
HYBRID	CONVERGENCE OF SIMPLIFIED HYBRID DISPLACEMENT METHOD FOR PLATE BENDING	72-0412
HYBRID	DERIVATION OF GEOMETRIC STIFFNESS AND MASS MATRICES FOR FINITE ELEMENT HYBRID MODELS	74-0844
HYBRID	DEVELOPMENT AND APPLICATION OF HYBRID STRESS FINITE ELEMENTS TO AN AUTOMOTIVE STRUCTURE	74-1221
HYBRID	DEVELOPMENT AND EVALUATION OF A TRIANGULAR THIN SHELL ELEMENT BASED UPON THE HYBRID ASSUMED STRESS M	69-0252
HYBRID	DEVELOPMENT OF AN ANISOTROPIC SINGULARITY FINITE ELEMENT UTILIZING THE HYBRID DISPLACEMENT METHOD	74-1099
HYBRID	ELASTIC CRACK ANALYSIS BY A FINITE ELEMENT HYBRID METHOD	71-0212
HYBRID	ELASTIC PLASTIC ANALYSIS OF SAINT-VENANT TORSION PROBLEM BY A HYBRID STRESS MODEL	72-0385
HYBRID	ELASTIC PLASTIC ANALYSIS BY ASSUMED STRESS HYBRID MODEL	74-0443
HYBRID	FINITE ELEMENT ANALYSIS OF VIBRATION OF THIN ELASTIC PLATES BY SIMPLIFIED HYBRID DISPLACEMENT METHOD	72-0565
HYBRID	FINITE ELEMENT APPENDAGE EQUATIONS FOR HYBRID COORDINATE DYNAMIC ANALYSIS	72-0022
HYBRID	FINITE ELEMENT SOLUTIONS FOR PLATE BENDING PROBLEMS BY SIMPLIFIED HYBRID DISPLACEMENT METHOD	72-0780
HYBRID	FRACTURE MECHANICS APPLICATION OF AN ASSUMED DISPLACEMENT HYBRID FINITE ELEMENT PROCEDURE	75-0212
HYBRID	FUNCTIONAL ANALYSIS DERIVATION AND GENERALIZATION OF HYBRID VARIATIONAL METHODS	73-1104
HYBRID	GENERALIZED HYBRID STRESS FINITE-ELEMENT MODELS	73-0027
HYBRID	HYBRID EFFECT ON BENDING BEHAVIORS OF CFRP-AL LAMINATE	74-1213
HYBRID	HYBRID FINITE ELEMENT ANALYSIS WITH PARTICULAR REFERENCE TO AXISYMMETRIC STRUCTURES	70-0248
HYBRID	HYBRID FINITE ELEMENT OPTIMIZATION PROGRAMS FOR DISCS USING A DYNAMICS DESIGN TECHNOLOGY	74-0691
HYBRID	HYBRID FINITE ELEMENTS IN THE PLANE STRESS PROBLEM	74-1071
HYBRID	HYBRID FINITE ELEMENT METHODS FOR SOLVING 2ND ORDER ELLIPTIC EQUATIONS	74-1267
HYBRID	HYBRID FINITE-ELEMENT METHODS	75-0445
HYBRID	HYBRID MODE ANALYSIS OF MICROSTRIP BY FINITE ELEMENT METHODS	71-0465
HYBRID	HYBRID MODELS	71-0211
HYBRID	HYBRID MODELS IN NUMERICAL AND COMPUTER METHODS	73-0682
HYBRID	HYBRID PLATE BENDING ELEMENTS	72-0792
HYBRID	HYBRIDE FINITE VERFAHREN FUER LINEARE UND NICHTLINEARE INSTATIONAERE WAERMELEITPROBLEME (HYBRID FINI	73-0484
HYBRID	HYBRIDE SCHALENELEMENTE MIT EINER ANWENDUNG AUF WENDELSCHALEN. (HYBRID SHELL ELEMENTS IN APPLICATION	75-0609
HYBRID	HYBRYDOWA METODA ELEMENTOW SKONCZONYCH W ZASTOSOWANIU DO OBLICZEN ORGAN URZADZEN OKRETOWYCH. (HYBRID	74-0963
HYBRID	NEW DISPLACEMENT HYBRID FINITE ELEMENT MODELS FOR SOLID CONTINUA	70-0115
HYBRID	NEW HYBRID CYLINDRICAL SHELL FINITE ELEMENT	72-0092
HYBRID	NONLINEAR ANALYSIS BY ASSUMED STRESS HYBRID MODELS	74-1004
HYBRID	NONLINEAR HYBRID STRESS FINITE ELEMENT ANALYSIS OF LAMINATED SHELLS	74-1350
HYBRID	NONLINEAR HYBRID STRESS FINITE ELEMENT ANALYSIS OF PNEUMATIC TIRES	75-0029
HYBRID	NONLINEAR RESPONSE OF PLATES TO IMPULSIVE LOADING USING HYBRID FINITE ELEMENTS	74-1018
HYBRID	OBSERVATIONS REGARDING ASSUMED STRESS HYBRID PLATE ELEMENTS	74-0370
HYBRID	ON HYBRID FINITE ELEMENTS	72-0069
HYBRID	ON THE CONVERGENCE OF FINITE ELEMENT SCHEMES BASED ON NON-CONFORMING AND SIMPLIFIED HYBRID DISPLACEM	73-0841
HYBRID	ON THE HYBRID STRESS FINITE ELEMENT MODEL FOR INCREMENTAL ANALYSIS OF LARGE DEFLECTION PROBLEMS	73-0371
HYBRID	OSCILLATIONS AND WAVE FORCES IN AN OFFSHORE HARBOR - APPLICATIONS OF HYBRID FINITE ELEMENT METHOD TO	74-1024
HYBRID	REVIEW OF CONTINUUM, FINITE ELEMENT AND HYBRID TECHNIQUES IN THE ANALYSIS OF STRESS CONCENTRATIONS I	73-1051
HYBRID	REVIEW OF CONTINUUM, FINITE ELEMENT AND HYBRID TECHNIQUES IN ANALYSIS OF STRESS CONCENTRATIONS IN ST	74-0914
HYBRID	SIMPLIFIED HYBRID DISPLACEMENT METHOD APPLIED TO PLATE BUCKLING PROBLEMS	72-0781
HYBRID	SIMPLIFIED HYBRID DISPLACEMENT METHOD FOR LINEAR FINITE ELEMENT ANALYSIS OF GENERAL SHELLS	74-0614
HYBRID	SOME CONSIDERATIONS ON THE CONVERGENCE OF HYBRID STRESS METHOD	73-0303

HYBRID	SOME FINITE ELEMENT SOLUTIONS FOR PLATE BENDING PROBLEMS BY SIMPLIFIED HYBRID DISPLACEMENT METHOD	72-0268
HYBRID	SOME PLANE QUADRILATERAL HYBRID FINITE ELEMENTS	69-0082
HYBRID	SOME RECENT STUDIES IN ASSUMED STRESS HYBRID MODELS	72-0262
HYBRID	SPAENDINGSHYBRIDE FINITE ELEMENTER TIL SVINGNINGSPROBLEMER. (STRESS HYBRID FINITE ELEMENT METHOD FOR	75-0208
HYBRID	STRESS ANALYSIS OF COMPOSITE MATERIALS BY THE HYBRID FINITE ELEMENT METHOD	75-0799
HYBRID	SYSTEMATIC ENFORCEMENT OF STRESS BOUNDARY CONDITIONS IN THE ASSUMED STRESS HYBRID MODEL BASED ON THE	71-0771
HYBRID	THE HYBRID DISPLACEMENT METHOD APPLIED TO PLATE AND SHELL PROBLEMS	75-0089
HYBRID	THEOPETICAL BASIS FOR HYBRID FINITE ELEMENTS IN DYNAMIC PROBLEMS	75-0489
HYBRID	THERMAL STRESS ANALYSIS OF PLATES AND SHALLOW SHELLS BY HYBRID FINITE-ELEMENT METHOD	74-0485
HYBRID	TWO HYBRID ELEMENTS FOR ANALYSIS OF THICK, THIN AND SANDWICH PLATES	72-0544
HYBRID	USE OF HYBRID COMPUTATION FOR SOLVING TRANSIENT FLOW PROBLEM IN POROUS MEDIA BY FINITE ELEMENT METHO	74-0068
HYBRID	VARIATIONAL PRINCIPLE FOR THE DYNAMIC ANALYSIS OF CONTINUA BY HYBRID FINITE ELEMENT METHOD	71-0492
HYBRID	VERIFICATION OF A HYBRID GENERAL PURPOSE SHELL AND BODY OF REVOLUTION COMPUTER PROGRAM	71-0333
HYBRID	VIBRATION ANALYSIS OF LAMINATED PLATES AND SHELLS BY A HYBRID STRESS ELEMENT	73-0120
HYBRID	VIBRATION OF CYLINDRICAL SHELLS BY HYBRID FINITE ELEMENT METHOD	72-0253
HYBRID	VIBRATION OF CYLINDRICAL SHELLS BY A HYBRID FINITE ELEMENT METHOD	72-0795
HYBRID	ZUR BERECHNUNG EINFACH-SYMMETRISCHER I-TRAEGER NACH DER THEORIE - 2. ORDNUNG UNTER BERUECKSICHTIGUNG	75-0510
HYBRID	ZUR BERECHNUNG VON PLATTEN NACH DER THEORIE ZWEITER. ORDNUNG MIT HILFE EINES HYBRIDEN DEFORMATIONS M	73-0558
HYBRID-ELEMENT	A HYBRID-ELEMENT APPROACH TO CRACK PROBLEMS IN PLANE ELASTICITY	73-0569
HYBRID-ELEMENT	HYBRID-ELEMENT APPROACH TO CRACK PROBLEMS IN PLANE ELASTICITY	73-0508
HYBRID-ELEMENT	HYBRID-ELEMENT METHOD FOR WATER WAVES	75-0808
HYBRID-STRESS	LINEAR DYNAMIC ANALYSES OF LAMINATED PLATES AND SHELLS BY THE HYBRID-STRESS FINITE-ELEMENT METHOD	73-0786
HYBRID-STRESS	STATIC, VIBRATION, AND THERMAL STRESS ANALYSES OF LAMINATED PLATE AND SHELLS BY THE HYBRID-STRESS FI	72-0602
HYBRID-STRESS	THE DEVELOPMENT OF A SERIES OF HYBRID-STRESS FINITE ELEMENTS	75-0335
HYBRIDE	HYBRIDE FINITE VERFAHREN FUER LINEARE UND NICHTLINEARE INSTATIONAERE WAERMELEITPROBLEME (HYBRID FINI	73-0484
HYBRIDE	HYBRIDE SCHALENELEMENTE MIT EINER ANWENDUNG AUF WENDELSCHALEN. (HYBRID SHELL ELEMENTS IN APPLICATION	75-0609
HYBRIDEN	ZUR BERECHNUNG VON PLATTEN NACH DER THEORIE ZWEITER. ORDNUNG MIT HILFE EINES HYBRIDEN DEFORMATIONS M	73-0558
HYBRIDER	ZUR BERECHNUNG EINFACH-SYMMETRISCHER I-TRAEGER NACH DER THEORIE - 2. ORDNUNG UNTER BERUECKSICHTIGUNG	75-0510
HYBRYDOWA	HYBRYDOWA METODA ELEMENTOW SKONCZONYCH W ZASTOSOWANIU DO OBLICZEN ORGAN URZADZEN OKRETOWYCH. (HYBRID	74-0963
HYDRAULIC	A HIGH PRECISION AXISYMMETRIC TRIANGULAR ELEMENT USED IN THE ANALYSIS OF HYDRAULIC TURBINE COMPONENT	70-0038
HYDRAULIC	AIRCRAFT FUEL TANK VULNERABILITY TO HYDRAULIC RAM: MODIFICATTION OF THE NORTHROP FINITE ELEMENT COMP	74-0724
HYDRAULIC	APPLICATIONS OF NASTRAN TO COUPLED STRUCTURAL AND HYDRODYNAMIC RESPONSES IN AIRCRAFT HYDRAULIC SYSTE	71-0744
HYDRAULIC	BUILDING THREE NEW MILLION-HORSEPOWER HYDRAULIC TURBINES FOR GRAND COULEE	74-0981
HYDRAULIC	HYDRAULIC FRACTURING IN ZONED EARTH AND ROCKFILL DAMS: A REPORT OF AN INVESTIGATION	73-0873
HYDRAULIC	OPTIMUM POSITION OF THE CENTRAL CLAY CORE OF A ROCKFILL DAM IN RESPECT TO ARCHING AND HYDRAULIC FRAC	73-0447
HYDRAULIC	POSSIBILITY OF USING FINITE ELEMENT METHOD IN THE ANALYSIS OF JOINT PERFORMANCE OF THE HYDRAULIC STR	72-0418
HYDRAULIC	PROCEEDINGS OF INTERNATIONAL ASSOCIATION FOR HYDRAULIC RESEARCH AND PROCEEDINGS OF INTERNATIONAL SYM	75-0534
HYDRAULIC	SUPPORT PERFORMANCE OF HYDRAULIC BACKFILL	74-0257
HYDRO	EMBANKMENT DAM FOR THE MAPIMBONDO HYDRO PROJECT	75-0270
HYDRO-DYNAMICS	APPLICATION OF FINITE ELEMENT METHOD TO HYDRO-DYNAMICS	73-0301
HYDRODYNAMIC	ANALYSIS OF HYDRODYNAMIC PROBLEMS BY THE FINITE ELEMENT METHOD	69-0027
HYDRODYNAMIC	ANALYSIS OF HYDRODYNAMIC AND PLATE STRUCTURES PROBLEMS BY THE FINITE ELEMENT METHOD	71-0091
HYDRODYNAMIC	APPLICATION OF FINITE ELEMENT METHOD TO HYDRODYNAMIC LUBRICATION PROBLEMS (PART 1, "INFINITE-WIDTH B	71-0537
HYDRODYNAMIC	APPLICATION OF FINITE ELEMENT METHOD TO HYDRODYNAMIC LUBRICATION PROBLEMS (PART 2, " FINITE WIDTH BE	71-0538
HYDRODYNAMIC	APPLICATION OF FINITE ELEMENT ANALYSIS TO HYDRODYNAMIC AND EXTERNALLY PRESSURIZED POCKET BEARINGS	72-0295
HYDRODYNAMIC	APPLICATIONS OF NASTRAN TO COUPLED STRUCTURAL AND HYDRODYNAMIC RESPONSES IN AIRCRAFT HYDRAULIC SYSTE	71-0744
HYDRODYNAMIC	EARTHQUAKE ANALYSIS OF GRAVITY DAMS INCLUDING HYDRODYNAMIC INTERACTION	73-0350
HYDRODYNAMIC	FINITE ELEMENT ANALYSIS OF A HYDRODYNAMIC BEARING	74-0773
HYDRODYNAMIC	FINITE ELEMENT METHOD FOR HYDRODYNAMIC DISPERISON EQUATION WITH MIXED PARTIAL DERIVATIVES	72-0030
HYDRODYNAMIC	FINITE ELEMENT METHOD IN HYDRODYNAMIC LUBRICATION	72-0301
HYDRODYNAMIC	FINITE ELEMENT MODELLING OF HYDRODYNAMIC CIRCULATION	74-1370
HYDRODYNAMIC	FINITE ELEMENT SOLUTION OF COUPLED ELECTROKINETIC AND HYDRODYNAMIC FLOW IN POROUS MEDIA	72-0398
HYDRODYNAMIC	FINITE ELEMENTS FOR THE COMPUTATIONS OF HYDRODYNAMIC MASS	69-0346
HYDRODYNAMIC	GENERAL VARIATIONAL ANALYSIS OF HYDRODYNAMIC, THERMAL AND DIFFUSIONAL BOUNDARY LAYERS	70-0191
HYDRODYNAMIC	HYDRODYNAMIC PRESSURES ON HIGH DAMS DUE TO VERTICAL EARTHQUAKE MOTIONS	69-0438
HYDRODYNAMIC	LA TECNICA DEGLI ELEMENTI FINITI PER PROBLEMI DI LUBRIFICAZIONE INCOMPRESSIBILE. (FINITE ELEMENT MET	75-0471
HYDRODYNAMIC	MARINE PROPELLER BLADES STRESSES AND DEFORMATIONS WITH GIVEN, FROZEN, HYDRODYNAMIC LOADING	73-0474
HYDRODYNAMIC	MATHEMATICAL MODELS OF THE MASSACHUSETTS BAY. PART 1. FINITE ELEMENT MODELING OF TWO-DIMENSIONAL HYD	73-0800
HYDRODYNAMIC	MODELING AN OCEAN POND. A TWO-DIMENSIONAL, FINITE ELEMENT HYDRODYNAMIC MODEL OF NINIGRET POND, CHARL	75-0686
HYDRODYNAMIC	THE EFFECT OF DEFORMATION ON THE BEHAVIOUR OF HYDRODYNAMIC JOURNAL BEARINGS	71-0009
HYDRODYNAMIC	VARIATIONAL FORMULATION FOR A CLASS OF FREE BOUNDARY PROBLEMS ARISING IN HYDRODYNAMIC LUBRICATION	75-0531
HYDRODYNAMICAL	FINITE ELEMENT METHOD AS AN ASPECT OF THE PRINCIPLE OF MAXIMUM UNIFORMITY: NEW HYDRODYNAMICAL RAMIFA	74-0009
HYDRODYNAMICAL-	HYDRODYNAMICAL-NUMERICAL MODELS FOR COASTAL WATERS AND OPEN OCEAN AREAS	75-0571
HYDRODYNAMICS	ANALYSIS OF COMPRESSIBLE HYDRODYNAMICS BY THE FINITE ELEMENT METHOD	71-0196
HYDRODYNAMICS	APPLICATION OF THE FINITE ELEMENTS METHOD IN TWO-DIMENSIONAL HYDRODYNAMICS USING THE LAGRANGE VARIAB	73-0976
HYDRODYNAMICS	APPLICATION OF THE FINITE ELEMENT METHOD TO TWO DIMENSIONAL LAGRANGIAN HYDRODYNAMICS	74-0016
HYDRODYNAMICS	FINITE ELEMENT APPROACH TO HYDRODYNAMICS AND MESH STABILIZATION	74-1298
HYDRODYNAMICS	PREDICTIONS IN ENVIRONMENTAL HYDRODYNAMICS USING FINITE ELEMENT METHOD 1. THEORETICAL DEVELOPMENT	75-0003
HYDRODYNAMICS	PREDICTIONS IN ENVIRONMENTAL HYDRODYNAMICS USING THE FINITE ELEMENT METHOD 2. APPLICATIONS	75-0064
HYDROELASTIC	DYNAMIC ANALYSIS OF HYDROELASTIC SYSTEMS USING THE FINITE-ELEMENT METHOD	73-1137
HYDROELASTIC	EFFECT OF INITIAL STATE FORCES ON THE HYDROELASTIC VIBRATION OF TUBES	73-0589
HYDROELASTIC	HYDROELASTIC ANALYSIS OF AXISYMMETRIC SYSTEMS BY A FINITE ELEMENT METHOD	68-0084
HYDROELASTIC	THE NASTRAN HYDROELASTIC ANALYZER	71-0738
HYDROGEN	DIFFUSION OF HYDROGEN IN FILLET WELDS	75-0373
HYDROGEN	FORMULATION OF STRESS STRAIN INDUCED DIFFUSION OF HYDROGEN AND ITS SOLUTION BY COMPUTER-AIDED FINITE	75-0326
HYDROGMAGNETIC	NUMERICAL STUDY OF HYDROGMAGNETIC STABILITY USING FINITE ELEMENT METHOD	73-0092
HYDROGRAPHS	FINITE-ELEMENT SIMULATION OF FLOOD HYDROGRAPHS	73-0632
HYDROGRAPHS	SIMULATION OF RUNOFF HYDROGRAPHS FROM NATURAL WATERSHEDS BY FINITE-ELEMENT METHOD	73-1140
HYDROLOGIC	HYDROLOGIC ENGINEERING METHODS FOR WATER RESOURCES DEVELOPMENT. VOLUME 10. PRINCIPLES OF GROUND-WATE	72-0696
HYDROLOGY	DIAGONALLY DOMINANT MATRICES FOR THE FINITE ELEMENT METHOD IN HYDROLOGY	73-0410
HYDROLOGY	HYDROLOGIC ENGINEERING METHODS FOR WATER RESOURCES DEVELOPMENT. VOLUME 10. PRINCIPLES OF GROUND-WATE	72-0696
HYDROLOGY	NUMERICAL METHODS IN SUBSURFACE HYDOLOGY: WITH AN INTRODUCTION TO THE FINITE ELEMENT METHOD	71-0800
HYDROMAGNETIC	ANALYSIS OF HYDROMAGNETIC PLASMA STABILITY BY FINITE-ELEMENT METHOD	72-0307
HYDROMAGNETIC	HYDROMAGNETIC STABILITY STUDIES USING THE FINITE ELEMENT METHOD	75-0878
HYDROMAGNETIC	STUDIES OF HYDROMAGNETIC STABILITY	74-0035
HYDROPHONE	AN ANALYSIS OF SOME FACTORS THAT AFFECT HYDROPHONE SENSITIVITY	72-0739
HYDROPHONES	FINITE-ELEMENT MODELING OF PIEZOELECTRIC CERAMIC HYDROPHONES	75-0595
HYDROSTATIC	ANALYSIS OF HYDROSTATIC EXTRUSION BY THE FINITE ELEMENT METHOD	72-0020
HYDROSTATIC	ARCH DAM CONSTRUCTION STRESSES DUE TO HYDROSTATIC AND GRAVITY LOADS	75-0658
HYDROSTATIC	DYNAMIC BEHAVIOR OF HYDROSTATIC THRUST BEARING	75-0189
HYDROSTATIC	FINITE ELEMENT METHOD OF ANALYSIS OF THE HYDROSTATIC BULGING OF A SHEET METAL - 1.	74-0958
HYDROSTATIC	INFLUENCE OF END-CLOSURE STIFFNESS ON BEHAVIOR OF CONCRETE CYLINDRICAL HULLS SUBJECTED TO HYDROSTATI	71-0725
HYDROSTATICALLY	STATICS AND DYNAMICS OF HYDROSTATICALLY LOADED SHELLS BY FINITE ELEMENT METHOD	71-0044
HYDROTHERMAL	FINITE ELEMENT ANALYSIS OF HYDROTHERMAL SYSTEMS	74-0048
HYDROTHERMAL	FINITE ELEMENT APPROACH TO HYDROTHERMAL ANALYSIS OF SMALL LAKES	72-0363
HYDROTHERMAL	FINITE ELEMENT APPROACH TO THE MODELING OF HYDROTHERMAL SYSTEMS	73-0011
HYDROTHERMAL	GALERKIN-FINITE ELEMENT ANALYSIS OF HYDROTHERMAL SYSTEM AT WAIRAKEI, NEW-ZEALAND	75-0601
HYDROTHERMAL	HYDROTHERMAL ANALYSIS BY FINITE ELEMENT METHOD	72-0023
HYDROTHERMAL	TRANSIENT HYDROTHERMAL ANALYSIS OF SMALL LAKES	73-0445
HYDROTHERMAL	USE OF FINITE ELEMENTS IN THE ANALYSIS OF A HYDROTHERMAL DOUBLET	74-0049
HYPAR	HYPAR DISPLACEMENT FIELDS FOR RECTANGULAR FINITE ELEMENTS	73-0240
HYPERBOLIC	ANALYSIS OF HIPPED ROOF HYPERBOLIC PARABOLOID STRUCTURES	72-0216

HYPERBOLIC   ANALYSIS OF HYPERBOLIC PARABOLOID SHELLS   70-0324
HYPERBOLIC   ANALYSIS OF THIN-SHELL HYPERBOLIC PARABOLOID SHELLS   72-0121
HYPERBOLIC   APPLICATION OF THE FINITE ELEMENT METHOD OF ANALYSIS TO HYPERBOLIC PARABOLOIDS WITH EDGE BEAMS   75-0250
HYPERBOLIC   COMPARATIVE EFFICIENCY OF CERTAIN FINITE ELEMENT AND FINITE DIFFERENCE METHODS FOR A HYPERBOLIC PROB   73-0965
HYPERBOLIC   CONVERGENCE, ACCURACY AND STABILITY OF FINITE ELEMENT APPROXIMATIONS OF A CLASS OF NON-LINEAR HYPERB   73-0159
HYPERBOLIC   FINITE ELEMENT - GALERKIN METHODS FOR HYPERBOLIC EQUATIONS WITH DISCONTINUOUS COEFFICIENTS   73-0726
HYPERBOLIC   FINITE ELEMENT ANALYSIS OF HYPERBOLIC PARABOLOID SHELLS   71-0245
HYPERBOLIC   FINITE ELEMENT ANALYSIS OF HYPERBOLIC PARABOLOID SHELLS   72-0887
HYPERBOLIC   FINITE ELEMENT METHODS FOR SYMMETRIC HYPERBOLIC EQUATIONS   73-0116
HYPERBOLIC   FINITE ELEMENT METHOD FOR NONLINEAR HYPERBOLIC EQUATIONS   74-1238
HYPERBOLIC   FINITE-ELEMENT METHOD FOR 1ST ORDER HYPERBOLIC EQUATIONS   75-0590
HYPERBOLIC   LUMPED PARAMETER ELEMENT FOR THE ANALYSIS OF HYPERBOLIC PARABOLOID SHELLS   72-0148
HYPERBOLIC   ON THE ACCURACY AND NUMERICAL STABILITY OF THE FINITE ELEMENT APPROXIMATION FOR PARABOLIC AND HYPERB   75-0102
HYPERBOLIC   ON THE SEMI-DISCRETE GALERKIN METHOD FOR HYPERBOLIC PROBLEMS AND ITS APPLICATION TO PROBLEMS IN ELAS   72-0457
HYPERBOLIC   RELATIVE EFFICIENCY OF FINITE DIFFERENCE AND FINITE ELEMENT METHODS - 1. HYPERBOLIC PROBLEMS AND SPL   74-0986
HYPERBOLIC   THE COMPARATIVE EFFICIENCY OF CERTAIN FINITE ELEMENT AND FINITE DIFFERENCE SCHEMES FOR A HYPERBOLIC   74-0910
HYPERBOLIC   THE RELATIVE EFFICIENCY OF FINITE DIFFERENCE AND FINITE ELEMENT METHODS. 1. HYPERBOLIC PROBLEMS AND   74-0911
HYPERBOLIC-ELLI   ON NUMERICAL SOLUTIONS OF EQUATIONS OF HYPERBOLIC-ELLIPTIC TYPE   74-0741
HYPERBOLICS   A METHOD TO GALERKIN TYPE ACHIEVING OPTIMUM L(2) ACCURACY FOR FIRST ORDER HYPERBOLICS AND EQUATIONS   72-0521
HYPERBOLOID   CALCULATION OF STRESS CONCENTRATION DUE TO A HYPERBOLOID CAVITY IN A THIN PLATE USING NASTRAN   71-0260
HYPERBOLOID   SOLUTION OF A SINGLE HYPERBOLOID SHELL BY FINITE ELEMENT METHOD   73-0151
HYPERBOLOIDAL   DYNAMIC ANALYSIS OF COLUMN-SUPPORTED HYPERBOLOIDAL SHELLS   74-0188
HYPERBOLOIDAL   EFFECT OF LATERAL PRESSURE ON NATURAL FREQUENCIES OF HYPERBOLOIDAL SHELLS OF REVOLUTION   71-0278
HYPERBOLOIDAL   FINITE ELEMENT ANALYSIS OF HYPERBOLOIDAL PRESTRESSED CONCRETE CONTAINMENT VESSELS   69-0526
HYPERBOLOIDAL   STABILITY OF HYPERBOLOIDAL SHELLS   75-0488
HYPERBOLOIDAL   STABILITY OF PRESSURIZED HYPERBOLOIDAL SHELLS   75-0585
HYPERBOLOIDAL   VIBRATIONS OF PRESSURE LOADED HYPERBOLOIDAL SHELLS   72-0874
HYPERCIRCLE   SOLUTION BOUNDS IN SOME STRESS PROBLEMS BY THE HYPERCIRCLE AND FINITE ELEMENT METHODS   72-0706
HYPERCIRCLE   THE HYPERCIRCLE IN MATHEMATICAL PHYSICS   57-0001
HYPERCIRCLE   THE HYPERCIRCLE METHOD   74-1284
HYPERCIRCLE   TRIANGULATION IN THE HYPERCIRCLE METHOD FOR PLANE PROBLEMS   52-0001
HYPERELASTIC   A FINITE ELEMENT ANALYSIS OF SHOCK AND FINITE-AMPLITUDE WAVES IN ONE-DIMENSIONAL HYPERELASTIC BODIES   72-0731
HYPERELASTIC   NONLINEAR ANALYSIS OF ARBITRARY, HYPERELASTIC MEMBRANE SHELLS BY THE FINITE-ELEMENT METHOD   73-1132
HYPERELASTIC   STABILITY AND POSTBUCKLING BEHAVIOUR OF HYPERELASTIC BODIES AT FINITE STRAIN BY THE FINITE ELEMENT M   73-0562
HYPERELASTIC   STABILITY AND POSTBUCKLING BEHAVIOR OF HYPERELASTIC BODIES AT FINITE STRAIN BY THE FINITE ELEMENT ME   74-0838
HYPERMATRIX   HYPERMATRIX SCHEME FOR FINITE-ELEMENT SYSTEMS ON CDC STAR-100 COMPUTER   75-0818
HYPERSONIC   ANALYSIS OF THERMAL STRESSES IN CORED CERAMIC BRICKS USED IN HYPERSONIC WIND TUNNEL HEATERS   72-0655
HYPERTENSIVE   DEFORMATION OF THE ARTERIAL VASA VASORUM AT NORMAL AND HYPERTENSIVE ARTERIAL PRESSURE   71-0149
HYPERVELOCITY   STUDY OF HYPERVELOCITY METEOROID IMPACT ON ORBITAL SPACE STATIONS   73-0905
HYPOTHESIS   FINITE ELEMENT MODEL FOR SHELLS BASED ON THE DISCRETE KIRCHHOFF HYPOTHESIS.   72-0388
HYPOTHESIS   NUMERICAL ANALYSIS OF THIN SHELLS BY CURVED TRIANGULAR ELEMENTS BASED ON DISCRETE - KIRCHHOFF HYPOTH   69-0091
HYSTERETIC   HYSTERETIC DAMPING SYSTEMS APPLIED TO SOIL STRUCTURE INTERACTION BY FINITE ELEMENT METHOD   73-0663
I-BEAMS   BUCKLING OF INELASTIC I-BEAMS UNDER MOMENT GRADIENT   74-0765
I-SECTION   ANCHORAGE ZONE STRESSES IN AXIALLY POST-TENSIONED I-SECTION MEMBERS WITH END BLOCKS   71-0305
I-SECTION   WARPING AND DISTORTION AT I-SECTION JOINTS   73-0722
I-SHAPED   ZUR BERECHNUNG EINFACH-SYMMETRISCHER I-TRAEGER NACH DER THEORIE - 2. ORDNUNG UNTER BERUECKSICHTIGUNG   75-0510
I-TRAEGER   ZUR BERECHNUNG EINFACH-SYMMETRISCHER I-TRAEGER NACH DER THEORIE - 2. ORDNUNG UNTER BERUECKSICHTIGUNG   75-0510
IBM   COMPUTER WORD LENGTH AND SOLUTION ACCURACY. EXPERIENCE WITH ELAS AND THE IBM 360   73-0335
IBM   DESCRIPTION AND INSTRUCTION FOR USE OF THE SAFE-SHELL. 8 FINITE ELEMENT PROGRAM IN THE IBM/360 VERSI   73-0945
IBM-360   DESCRIPTION OF THE IBM-360 VERSION OF THE FINITE ELEMENTS COMPUTER PROGRAM SAFE-3D   72-0775
IBRIDO   CHARACTERISTICS OF A HYBRID CALCULATION SYSTEM BASED ON THE FINITE ELEMENT METHOD. (CARATTERISTICHE   72-0715
ICE   AN ANALYSIS OF THE SMALL-SCALE STRENGTH TESTING OF ICE   72-0881
ICE   CRACK PROPAGATION IN SEA ICE - FINITE ELEMENT APPROACH   72-0029
ICE   ICE ENGINEERING: VISCOELASTIC FINITE ELEMENT FORMULATION   74-0861
ICES   ICES STRUDL-II ENGINEERING USERS MANUAL. VOLS. 1, 11, 111   69-0398
ICES   THE DEVELOPMENT OF ICES STRUDL   72-0327
ICES-STRUDL-11   ICES-STRUDL-11   68-0257
ICON   ICON; THE INTERACTIVE CREATION OF NASTRAN DATA. A SYSTEM DESCRIPTION   75-0274
IDAHO   A FINITE-ELEMENT, PLANAR-FLOW MODEL OF CAMAS PRAIRIE, IDAHO   72-0980
IDAHO   DESTRESSING TEST AT THE GALENA MINE, WALLACE, IDAHO   72-0126
IDEAL   A USEFUL FORM OF THE VARIATIONAL PRINCIPLE OF IDEAL MHD FOR ONE-DIMENSIONAL NUMERICAL STABILITY CALC   73-0746
IDEAL   FINITE ELEMENT ANALYSIS OF IRROTATIONAL FLOWS OF AN IDEAL FLUID   71-0806
IDEAL   FINITE ELEMENT METHOD APPLIED TO IDEAL GASEOUS MIXTURES.   72-0393
IDEAL   FREE-SURFACE IDEAL FLUID FLOWS BY FINITE ELEMENTS   73-0254
IDEAL   FREE-SURFACE IDEAL FLUID FLOWS BY FINITE ELEMENTS   74-0125
IDEAL   STRONGLY PLANAR AC-LATTICES IN WHICH IDEAL OF FINITE-ELEMENTS IS STANDARD   75-0816
IDEAL   USEFUL FORM OF THE VARIATIONAL PRINCIPLE OF IDEAL MHD FOR ONE-DIMENSIONAL NUMERICAL STABILITY CALCUL   73-0946
IDEALIZATION   A FINITE ELEMENT TECHNIQUE FOR THE IDEALIZATION OF CONTINUOUS SHELL STRUCTURES   68-0290
IDEALIZATION   AN OPTIMUM IDEALIZATION IN FINITE ELEMENT METHOD FOR A TWO-GROUP NEUTRON DIFFUSION PROBLEM   74-1143
IDEALIZATION   COMMENT ON DISCRETE ELEMENT IDEALIZATION OF AN INCOMPRESSIBLE LIQUID FOR VIBRATION ANALYSIS   73-0694
IDEALIZATION   DATA GENERATOR FOR THE IDEALIZATION FOR FINITE ELEMENT STRUCTURAL ANALYSIS OF NAVAL SHIP FLAT PLATED   72-0860
IDEALIZATION   DELINQUENT FINITE ELEMENTS FOR SHELL IDEALIZATION   75-0478
IDEALIZATION   DIGITAL COMPUTATION OF STRESSES AND DEFLECTIONS IN A BOX BEAM - A PERFORMANCE COMPARISON BETWEEN FIN   69-0355
IDEALIZATION   DISCRETE ELEMENT IDEALIZATION OF AN INCOMPRESSIBLE LIQUID FOR VIBRATION ANALYSIS   70-0391
IDEALIZATION   EXTENSION OF THREE DIMENSIONAL ANALYSIS TO SHELL STRUCTURES USING THE FINITE ELEMENT IDEALIZATION   74-0849
IDEALIZATION   RAMIFICATIONS OF OPTIMUM IDEALIZATION GEOMETRY IN DISCRETE ELEMENT ANALYSIS   75-0182
IDEALIZATIONS   A STUDY OF VARIOUS IDEALIZATIONS FOR WING STRUCTURES AND NUMERICAL PROCEDURES INVOLVED USING MATRIX   69-0257
IDEALIZATIONS   A THEOREM FOR OPTIMUM IDEALIZATIONS IN FINITE-ELEMENT ANALYSIS   72-0791
IDEALIZATIONS   FINITE ELEMENT ANALYSIS OF DENTAL STRUCTURES - AXISYMMETRIC AND PLANE STRESS IDEALIZATIONS   75-0525
IDEALIZATIONS   SOME EFFECTS OF SYSTEM IDEALIZATIONS, SINGULARITIES AND MESH PATTERNS ON FINITE ELEMENT SOLUTIONS   74-0939
IDEALIZATIONS   THEOREM FOR OPTIMUM FINITE-ELEMENT IDEALIZATIONS   73-0255
IDEALIZED   IN VIVO STRESSES IN THE HUMAN LEFT VENTRICULAR WALL: ANALYSIS ACCOUNTING FOR THE IRREGULAR THREE DIM   72-0136
IDEALLY   RADIAL INERTIA EFFECTS ON AN IDEALLY PLASTIC CIRCULAR PLATE UNDER IMPULSIVE AXIAL COMPRESSION   69-0324
IDEAS   VYUZITI METODY KONECNYCH PRVKU PRO PROJEKTOVANI KOMBINOVANYCH KONSTRUCKCI (FINITE ELEMENT METHOD IDE   73-0667
IDENTIFICATION   IDENTIFICATION OF AQUIFER CONSTANTS: TESTS OF TWO INVERSE METHODS   74-0051
IDENTIFICATION   IDENTIFICATION OF NONLINEAR MATERIALS BY FINITE ELEMENT METHODS   73-0927
IDENTIFICATION   IDENTIFICATION OF NONLINEAR ELASTIC SOLIDS BY A FINITE ELEMENT METHOD   74-0530
IDENTIFICATION   NONLINEAR STRUCTURES FOR SYSTEM IDENTIFICATION   74-1030
IDENTIFICATION   STATISTICAL IDENTIFICATION OF STRUCTURES   74-0141
IDROLOGIA   OPTIMIZZAZIONE DI RETICOLI TRIANGOLARI PER IL METODO DEGLI ELEMENTI FINITI IN IDROLOGIA   72-0987
IEEE-G-MTT   IEEE-G-MTT INTERNATIONAL MICROWAVE SYMPOSIUM, 1973: APPLICATIONS IN THE 70'S   73-0374
IGNITION   A RADIATIVE IGNITION MODEL OF A SOLID FUEL   74-1312
IGNITION   SPONTANEOUS IGNITION - FINITE ELEMENT SOLUTIONS FOR STEADY AND TRANSIENT CONDITIONS   74-0353
ILLEGAL   SURVEY OF UNDERLYING THEORY, APPROXIMATION BY PIECEWISE POLYNOMIALS, ILLEGAL ELEMENTS AND THE PATCH   73-0635
ILLIAC-1V   3-D SEISMIC CODE FOR ILLIAC-1V   73-0868
ILLINOIS   ANALYSIS OF LIQUID-WASTE INJECTION WELLS IN ILLINOIS BY MATHEMATICAL MODELS   74-0518
ILLTAC   3-D STRESS WAVE CODE FOR THE ILLTAC 1V   74-0780
ILLUSTRATIONS   ILLUSTRATIONS OF AUTOMOTIVE FINITE ELEMENT MODELS - STATICS   74-0373
ILLUSTRATIONS   ILLUSTRATIONS OF AUTOMOTIVE FINITE ELEMENT MODELS - DYNAMICS   74-0391
IMAGES   APPLICATIONS OF CONTINUOUS TONE COMPUTER-GENERATED IMAGES IN STRUCTURAL MECHANICS   74-1035
IMBEDDING   AN IMBEDDING TECHNIQUE FOR THE GENERATION OF WEAK-WEAK FINITE ELEMENT APPROXIMATIONS OF LINEAR AND N   72-0577
IMBEDDING   INVARIANT IMBEDDING AND SOLUTION OF FINITE ELEMENT EQUATIONS   74-0158

IMMERSED	VIBRATION OF AN ELASTIC BODY IMMERSED IN A FLUID	72-0312
IMPACT	APPLICATION OF THE COLLISION-IMPARTED VELOCITY METHOD FOR ANALYZING THE RESPONSES OF CONTAINMENT AND	73-0770
IMPACT	BALLISTIC IMPACT OF TEXTILE STRUCTURES.	72-0433
IMPACT	CHARACTERISTICS OF FINITE-ELEMENTS FOR ANALYSIS OF BEAMS SUBJECTED TO IMPACT LOAD	73-1144
IMPACT	DYNAMIC RESPONSE ANALYSIS OF GEOMETRICALLY NON-LINEAR STRUCTURES SUBJECTED TO HIGH IMPACT	72-0193
IMPACT	DYNAMIC RESPONSE ANALYSIS OF GEOMETRICALLY NON-LINEAR STRUCTURES SUBJECTED TO HIGH IMPACT	72-0555
IMPACT	DYNAMIC RESPONSE OF A CONSTRAINED FIBROUS SYSTEM SUBJECTED TO TRANSVERSE IMPACT. II. A MECHANICAL MO	70-0369
IMPACT	IMPACT OF CDC STAR-100 COMPUTER ON FINITE ELEMENT SYSTEMS	75-0039
IMPACT	IMPACT OF COMPUTERS ON THE PRACTICE OF STRUCTURAL ENGINEERING IN CONCRETE	72-0414
IMPACT	IMPACT OF COMPUTERS ON THE PRACTICE OF STRUCTURAL ENGINEERING IN CONCRETE	72-0415
IMPACT	IMPACT OF FINITE ELEMENT TECHNIQUES ON PRACTICAL DESIGN OF SHIP STRUCTURES	69-0344
IMPACT	IMPACT OF FINITE ELEMENT METHOD ON STRUCTURAL ENGINEERING	75-0007
IMPACT	IMPACT ON A TRANSVERSELY ANISOTROPIC HALF-SPACE	74-0950
IMPACT	IMPACT ON COMPLEX MECHANICAL STRUCTURES	71-0598
IMPACT	LOAD POINT COMPLIANCE OF THE CHARPY IMPACT SPECIMEN	73-0775
IMPACT	MATHEMATICAL MODEL FOR CLOSED HEAD IMPACT	74-0934
IMPACT	SOME RECENT FINITE ELEMENT APPROACHES IN ANALYSIS OF PLASTIC IMPACT	69-0299
IMPACT	STRUCTURAL ANALYSIS OF AIRCRAFT IMPACT ON A NUCLEAR CONTAINMENT VESSEL AND ASSOCIATED STRUCTURES	69-0279
IMPACT	STRUCTURAL RESPONSE TO IMPACT DAMAGE	75-0325
IMPACT	STUDY OF HYPERVELOCITY METEOROID IMPACT ON ORBITAL SPACE STATIONS	73-0905
IMPACT	THE IMPACT OF THE DIGITAL COMPUTER ON STRUCTURAL ANALYSIS	69-0447
IMPACT	THE IMPACT OF THE DIGITAL COMPUTER ON THE ENGINEERING SCIENCES - PART I	70-0003
IMPACT	TRANSVERSE IMPACT PROBLEMS BY HIGHER-ORDER BEAM FINITE-ELEMENTS	75-0819
IMPACT	WATER IMPACT ANALYSIS OF SPACE SHUTTLE SOLID ROCKET BY THE FINITE ELEMENT METHOD	74-0760
IMPACT	WAVE PROPAGATION IN A VISCOELASTIC FIBER SUBJECTED TO TRANSVERSE IMPACT	71-0639
IMPACTED	CRACK PROPAGATION AND ARREST IN IMPACTED PLATES	72-0599
IMPEDANCE	CALCULATION OF MECHANICAL IMPEDANCE BY FINITE ELEMENT HYBRID MODEL	74-0126
IMPEDANCE	FINITE ELEMENT FOR TIMOSHENKO BEAM BASED ON MECHANICAL IMPEDANCE	71-0607
IMPEDANCE	IMPEDANCE OF INTERSTITIAL GAPS IN TRANSDUCER ARRAYS	70-0410
IMPEDANCE	SHIP (SIMPLIFIED-HELMHOLTZ-INTEGRAL PROGRAM): A FAST COMPUTER PROGRAM FOR CALCULATING THE ACOUSTIC R	72-0835
IMPELLER	SUBSTRUCTURES ANALYSIS OF IMPELLER VIBRATION MODES	75-0283
IMPELLERS	ANALYSIS OF RADIAL FLOW IMPELLERS BY THE MATRIX FINITE ELEMENT METHOD	71-0412
IMPELLERS	COMPUTATION OF IMPELLERS BY METHOD OF FINITE ELEMENTS	74-0646
IMPELLERS	STRESS ANALYSIS PROGRAM BY FINITE ELEMENT METHOD FOR RADIAL FLOW IMPELLERS	74-1016
IMPERFECT	THE USE OF FINITE ELEMENT MID-INCREMENT STIFFNESS MATRICES IN THE POST BUCKLING ANALYSIS OF IMPERFEC	71-0383
IMPERFECTION	FINITE ELEMENT MATRIX FORMULATION OF POST-BUCKLING STABILITY AND IMPERFECTION SENSITIVITY	70-0246
IMPERFECTION	THE EFFECT OF AN INITIAL AXISYMMETRIC IMPERFECTION ON THE NATURAL FREQUENCIES OF CONICAL SHELLS	70-0368
IMPERFECTIONS	FINITE ELEMENT ANALYSIS OF BUCKLING AND POST-BUCKLING BEHAVIORS OF ARCHES WITH GEOMETRIC IMPERFECTIO	73-0186
IMPERFECTIONS	INCREMENTAL VARIATIONAL METHOD FOR THE LARGE DISPLACEMENT ANALYSIS OF SHELLS WITH GEOMETRIC IMPERFEC	72-0894
IMPERFECTIONS	PROBABILISTIC BUCKLING BEHAVIOR OF STRUCTURES WITH RANDOM LACK-OF-FIT AND GEOMETRIC IMPERFECTIONS	73-0862
IMPINGEMENT	FINITE ELEMENT ANALYSIS OF JET IMPINGEMENT ON AXISYMMETRIC CURVED DEFLECTORS	75-0396
IMPINGEMENT	JET IMPINGEMENT ON AXISYMMETRIC CURVED DEFLECTORS	74-0017
IMPLANT	SOME FACTORS AFFECTING DENTAL IMPLANT DESIGN	74-1146
IMPLANTS	FINITE ELEMENT STRESS ANALYSIS OF PROSTHETIC TOOTH IMPLANTS	75-0073
IMPLANTS	STRESS DISTRIBUTION IN BONE ARISING FROM LOADING ON ENDOSTEAL DENTAL IMPLANTS	73-0501
IMPLICIT	IMPLICIT RIGID BODY MOTION IN CURVED FINITE ELEMENTS	71-0174
IMPLICIT	ON THE UNCONDITIONAL STABILITY OF AN IMPLICIT ALGORITHM FOR NONLINEAR STRUCTURAL DYNAMICS	75-0825
IMPREGNATED	ANALYSIS OF BRIDGE DECKS USING POLYMER IMPREGNATED CONCRETE	73-0673
IMPROVE	HOW FINITE ELEMENT METHODS IMPROVE THE DESIGN CYCLE	74-0372
IMPROVE	RESEARCH TO IMPROVE TUNNEL SUPPORT SYSTEMS	74-0777
IMPROVE	STUDIES TO IMPROVE TUNNEL SUPPORT SYSTEMS	74-0957
IMPULSIVE	DYNAMIC ANALYSIS OF A COMPLEX SHELL STRUCTURE UNDER IMPULSIVE ASYMMETRIC LOADING	68-0226
IMPULSIVE	DYNAMIC ELASTOPLASTIC RESPONSE OF GEOMETRICALLY NONLINEAR ARBITRARY SHELLS OF REVOLUTION UNDER IMPUL	70-0098
IMPULSIVE	DYNAMIC RESPONSE OF ARBITRARY SHELLS OF REVOLUTION UNDER COMBINED IMPULSIVE NONISOTHERMAL AND IRREGU	70-0365
IMPULSIVE	FINITE DEFORMATIONS IN ELASTIC-PLASTIC MEDIA SUBJECTED TO IMPULSIVE LOADINGS BY THE FINITE ELEMENT M	70-0947
IMPULSIVE	NONLINEAR RESPONSE OF PLATES TO IMPULSIVE LOADING USING HYBRID FINITE ELEMENTS	74-1018
IMPULSIVE	RADIAL INERTIA EFFECTS ON AN IDEALLY PLASTIC CIRCULAR PLATE UNDER IMPULSIVE AXIAL COMPRESSION	69-0324
IMPULSIVE	RESPONSE OF REACTOR CORE SUBASSEMBLIES TO IMPULSIVE LOADINGS	73-0076
IN-LINE	DYNAMIC BEHAVIOR OF IN-LINE SHEAR WALLS CONNECTED BY FLOOR SLABS	72-0689
IN-PILE	ANALYSIS OF AN IN-PILE REACTOR TUBE	75-0193
IN-PLANE	A HIGHER ORDER IN-PLANE PARALLELOGRAM ELEMENT AND ITS APPLICATION TO SKEWED CURVED BOX GIRDER BRIDGE	71-0002
IN-PLANE	ANALYSIS OF IN-PLANE VIBRATION OF SHEAR WALLS BY A FINITE ELEMENT METHOD	72-0090
IN-PLANE	ANALYSIS OF PLATES SUBJECTED TO IN-PLANE FORCES USING LARGE FINITE ELEMENTS	70-0271
IN-PLANE	CONSTANT CURVATURE BEAM FINITE ELEMENTS FOR IN-PLANE VIBRATION	72-0254
IN-PLANE	INELASTIC FINITE ELEMENT ANALYSIS OF REINFORCED CONCRETE PANELS UNDER IN-PLANE LOADS	70-0482
IN-PLANE	INFLUENCE OF END-EFFECTS ON STRESSES AND FLEXIBILITY OF A PIPING ELBOW WITH IN-PLANE MOMENT	74-0534
IN-PLANE	NONLINEAR IN-PLANE BENDING OF A PLANE STRESS-STRAIN FINITE ELEMENT MODEL	74-0211
IN-PLANE	NOTE ON RECTANGULAR FINITE ELEMENTS WITH IN-PLANE FORCES	75-0080
IN-PLANE	STRESS DISTRIBUTIONS IN THIN-WALLED INTERSECTING CYLINDRICAL SHELLS SUBJECTED TO INTERNAL PRESSURE A	71-0023
IN-SITU	A CASE STUDY OF IN-SITU ROCK DEFORMATION BEHAVIOR FOR THE DESIGN OF GROUND SUPPORT SYSTEM	72-0686
IN-SITU	IN-SITU HORIZONTAL STRESSES IN LATERITE BY FINITE ELEMENT ANALYSIS	73-1048
IN-VACUO	IN-VACUO MODAL DYNAMIC RESPONSE OF THE HUMAN SKULL	73-0460
INCIDENCE	NUMERICAL ANALYSIS OF COOLING TOWER PLUMES AND FOG INCIDENCE	74-0040
INCIDENT	THE USE OF THE FINITE ELEMENT METHOD IN OBTAINING SOIL-WATER PRESSURE AND SOIL-WATER CONTENT DISTRIB	75-0112
INCIPIENT	INCIPIENT CHIP FORMATION IN METAL CUTTING - A THREE- DIMENSION FINITE-ELEMENT ANALYSIS	73-1146
INCLINED	DESIGN OF INCLINED TIED ARCH RAILWAY BRIDGES OVER THE M56	71-0297
INCLINED	FAILURE OF THE CONCRETE LINING IN AN INCLINED SHAFT AT GREAT DEPTH.	74-0996
INCLUSION	A METHOD FOR THE PREDICTION OF STRESSES IN AN ISOTROPIC INCLUSION OR OREBODY OF IRREGULAR SHAPE	72-0827
INCLUSION	EFFECT OF ELASTIC MODULAR RATIO AND ANISOTROPY ON THE STRESS FIELD IN A TWO-DIMENSIONAL MATRIX INTO	71-0647
INCLUSION	INCLUSION OF SHEAR DEFLECTION IN THE STIFFNESS MATRIX OF A BEAM ELEMENT	70-0307
INCLUSION	INCLUSION OF TRANSVERSE SHEAR DEFORMATION IN FINITE ELEMENT DISPLACEMENT FORMULATIONS	74-0662
INCLUSION	INCLUSION OF TRANSVERSE-SHEAR DEFORMATION IN FINITE-ELEMENT DISPLACEMENT FORMULATIONS - COMMENT	75-0406
INCLUSION	ON THE FINITE ELEMENT ANALYSIS OF CRACK AND INCLUSION PROBLEMS IN ELASTIC-PLASTIC MATERIALS	74-0227
INCLUSIONS	A NEW METHOD OF ANALYSIS TO OBTAIN EXACT SOLUTIONS FOR STRESSES AND STRAINS IN CIRCULAR INCLUSIONS	74-1063
INCLUSIONS	ENERGY CONDITION FOR INITIATION OF INTERFACIAL MICROCRACKS AT INCLUSIONS	74-0546
INCLUSIONS	GEOMETRIC METHOD FOR THE PREDICTION OF STRESSES IN INCLUSIONS, OREBODIES, AND MINING SYSTEMS	74-0219
INCLUSIONS	ROCK FRAGMENTATION BY SUBSURFACE THERMAL INCLUSIONS - A FINITE ELEMENT STUDY	73-1159
INCLUSIONS	STRESS FIELDS AROUND MULTIPLE INCLUSIONS	69-0067
INCLUSIONS	THEORETICAL STUDY OF THE INFLUENCE OF INCLUSIONS UPON THE INITIATION AND GROWTH OF FATIGUE CRACKS IN	71-0394
INCOMPATIBLE	A NOTE ON CERTAIN INCOMPATIBLE ELEMENTS	73-0994
INCOMPATIBLE	EFFICIENCY EVALUATION OF A TWO-DIMENSIONAL INCOMPATIBLE FINITE ELEMENT	73-0988
INCOMPATIBLE	INCOMPATIBLE DISPLACEMENT MODELS	71-0262
INCOMPRESSIBILE	LA TECNICA DEGLI ELEMENTI FINITI PER PROBLEMI DI LUBRIFICAZIONE INCOMPRESSIBILE. (FINITE ELEMENT MET	75-0471
INCOMPRESSIBILITY	AVERAGE AND COMPLETE INCOMPRESSIBILITY IN THE FINITE ELEMENT METHOD	75-0342
INCOMPRESSIBILITY	CONVERGENCE OF THE FINITE-ELEMENT METHOD UNDER THE CONSTRAINT OF MATERIAL INCOMPRESSIBILITY	73-1135
INCOMPRESSIBILI	LARGE NATURAL STRAINS AND SOME SPECIAL DIFFICULTIES DUE TO NONLINEARITY AND INCOMPRESSIBILITY IN FIN	74-0528
INCOMPRESSIBILI	LARGE NATURAL STRAINS AND SOME SPECIAL DIFFICULTIES DUE TO NONLINEARITY AND INCOMPRESSIBILITY IN FIN	74-1027
INCOMPRESSIBLE	A FINITE-ELEMENT ANALYSIS OF STEADY, TWO-DIMENSIONAL, INCOMPRESSIBLE, LAMINAR FLOW	72-0981
INCOMPRESSIBLE	A FINITE-ELEMENT METHOD FOR LIFTING SURFACES IN STEADY INCOMPRESSIBLE SUBSONIC FLOW	74-1321
INCOMPRESSIBLE	A REFINED AXISYMMETRIC FINITE ELEMENT FOR THE ANALYSIS OF NEARLY INCOMPRESSIBLE SOLIDS	70-0309
INCOMPRESSIBLE	AN ASSUMED STRESS HYBRID FINITE ELEMENT METHOD FOR AN INCOMPRESSIBLE AND NEAR INCOMPRESSIBLE MATERIA	69-0259
INCOMPRESSIBLE	AN ASSUMED STRESS HYBRID FINITE ELEMENT METHOD FOR AN INCOMPRESSIBLE AND NEAR INCOMPRESSIBLE MATERIA	69-0259

252

INCOMPRESSIBLE	ANALYSIS OF STEADY INCOMPRESSIBLE VISCOUS FLOW	
INCOMPRESSIBLE	COMMENT ON DISCRETE ELEMENT IDEALIZATION OF AN INCOMPRESSIBLE LIQUID FOR VIBRATION ANALYSIS	74-0013
INCOMPRESSIBLE	DISCRETE ELEMENT IDEALIZATION OF AN INCOMPRESSIBLE LIQUID FOR VIBRATION ANALYSIS	73-0694
INCOMPRESSIBLE	ELASTICITY EQUATIONS FOR INCOMPRESSIBLE AND NEARLY INCOMPRESSIBLE MATERIALS BY A VARATIONAL THEOREM	70-0391
INCOMPRESSIBLE	ELASTICITY EQUATIONS FOR INCOMPRESSIBLE AND NEARLY INCOMPRESSIBLE MATERIALS BY A VARATIONAL THEOREM	65-0019
INCOMPRESSIBLE	FINITE ELEMENT ANALYSIS OF CONTINUUM PROBLEMS AND ITS APPLICATION TO THE INCOMPRESSIBLE LUBRICATION	65-0019
INCOMPRESSIBLE	FINITE ELEMENT ANALYSIS OF UNSTEADY INCOMPRESSIBLE FLOW AROUND AN OSCILLATING OBSTACLE OF ARBITRARY	71-0539
INCOMPRESSIBLE	FINITE ELEMENT ANALYSIS OF INCOMPRESSIBLE MATERIAL BY RESIDUAL ENERGY BALANCING	73-0067
INCOMPRESSIBLE	FINITE ELEMENT ANALYSIS OF SEVERAL COMPRESSIBLE AND INCOMPRESSIBLE VISCOUS FLOW PROBLEMS	74-1041
INCOMPRESSIBLE	FINITE ELEMENT ANALYSIS OF VISCOUS, INCOMPRESSIBLE FLUID FLOW	74-1119
INCOMPRESSIBLE	FINITE ELEMENT ANALYSIS OF VISCOUS, INCOMPRESSIBLE FLUID FLOW	75-0674
INCOMPRESSIBLE	FINITE ELEMENT ANALYSIS OF INCOMPRESSIBLE MICROPOLAR VISCOUS FLUID FLOW	75-0760
INCOMPRESSIBLE	FINITE ELEMENT DIFFICULTIES IN NONLINEAR AND INCOMPRESSIBLE MATERIAL	75-0866
INCOMPRESSIBLE	FINITE ELEMENT FORMULATION OF THE INCOMPRESSIBLE LUBRICATION PROBLEM	74-1242
INCOMPRESSIBLE	FINITE ELEMENT FORMULATION FOR SOLUTION OF STEADY FLOW OF INCOMPRESSIBLE VISCOUS FLUID	70-0022
INCOMPRESSIBLE	FINITE ELEMENT METHOD FOR INCOMPRESSIBLE SLOW VISCOUS FLOW WITH A FREE SURFACE	73-0299
INCOMPRESSIBLE	FINITE ELEMENT METHOD FOR A SOLID ENCLOSING AN INVISCID INCOMPRESSIBLE FLUID	69-0303
INCOMPRESSIBLE	FINITE ELEMENT METHODS FOR THE SOLUTION OF SOME INCOMPRESSIBLE NON-NEWTONIAN FLUID MECHANICS PROBLEM	73-0090
INCOMPRESSIBLE	FINITE ELEMENT SOLUTION OF THE INCOMPRESSIBLE LUBRICATION PROBLEM	75-0224
INCOMPRESSIBLE	FINITE ELEMENT SOLUTION OF THE THREE DIMENSIONAL FLOW PROBLEMS AND OF REYNOLD'S EQUATION FOR INCOMPR	69-0021
INCOMPRESSIBLE	FINITE ELEMENT SOLUTION ALGORITHM FOR VISCOUS INCOMPRESSIBLE FLUID DYNAMICS	72-0011
INCOMPRESSIBLE	FINITE ELEMENT SOLUTION ALGORITHM FOR INCOMPRESSIBLE FLUID DYNAMICS	73-0063
INCOMPRESSIBLE	FINITE ELEMENT SOLUTION OF THE INCOMPRESSIBLE FLOW OVER AN AIRFOIL IN A NON UNIFORM STREAM	74-0006
INCOMPRESSIBLE	FINITE ELEMENTS FOR COMPRESSIBLE AND INCOMPRESSIBLE CONTINUA	74-0432
INCOMPRESSIBLE	FINITE PLANE STRAIN OF INCOMPRESSIBLE ELASTIC SOLIDS BY THE FINITE ELEMENT METHOD	69-0133
INCOMPRESSIBLE	FINITE-ELEMENT ANALYSIS OF SLOW INCOMPRESSIBLE VISCOUS FLUID MOTION	68-0128
INCOMPRESSIBLE	FINITE-ELEMENT ANALYSIS OF COMPRESSIBLE AND INCOMPRESSIBLE VISCOUS-FLOW AND HEAT-TRANSFER PROBLEMS	74-0154
INCOMPRESSIBLE	FINITE-ELEMENT SOLUTION OF NAVIER-STOKES EQUATIONS FOR TRANSIENT 2-DIMENSIONAL INCOMPRESSIBLE FLOW	75-0347
INCOMPRESSIBLE	GALERKIN FINITE ELEMENT FORMULATION FOR INCOMPRESSIBLE FLOW	75-0134
INCOMPRESSIBLE	GALERKIN METHOD FOR A CLASS OF STEADY, 2-DIMENSIONAL, INCOMPRESSIBLE, LAMINAR BOUNDARY-LAYER FLOWS	70-0006
INCOMPRESSIBLE	HIGH ORDER FINITE ELEMENT FOR COMPLETELY INCOMPRESSIBLE CREEPING FLOW	75-0437
INCOMPRESSIBLE	INCOMPRESSIBLE FLOW IN MULTIPLY CONNECTED REGIONS	73-0144
INCOMPRESSIBLE	LOCAL POTENTIAL APPROACH TO FINITE ELEMENT METHOD IN UNSTEADY VISCOUS INCOMPRESSIBLE FLUID FLOW	74-0904
INCOMPRESSIBLE	NUMERICAL ANALYSIS FOR FINITE AXISYMMETRIC DEFORMATIONS OF INCOMPRESSIBLE ELASTIC SOLIDS OF REVOLUTI	73-0298
INCOMPRESSIBLE	ON A VARIATIONAL THEOREM FOR INCOMPRESSIBLE AND NEARLY INCOMPRESSIBLE ORTHOTROPIC ELASTICITY	70-0094
INCOMPRESSIBLE	ON A VARIATIONAL THEOREM FOR INCOMPRESSIBLE AND NEARLY INCOMPRESSIBLE ORTHOTROPIC ELASTICITY	68-0249
INCOMPRESSIBLE	ON SOME GENERALIZATIONS OF THE INCREMENTAL STIFFNESS RELATIONS FOR FINITE DEFORMATIONS OF COMPRESSIB	68-0249
INCOMPRESSIBLE	ON THE NUMERICAL TREATMENT OF THE NAVIER-STOKES EQUATIONS FOR AN INCOMPRESSIBLE FLUID	71-0189
INCOMPRESSIBLE	SOLUTION OF VISCOUS INCOMPRESSIBLE JET AND FREE-SURFACE FLOWS USING FINITE ELEMENT METHODS	73-0809
INCOMPRESSIBLE	STEADY FLOW ANALYSIS OF INCOMPRESSIBLE VISCOUS FLOW BY THE FINITE ELEMENT METHOD	74-0357
INCOMPRESSIBLE	STRESSES IN NEARLY INCOMPRESSIBLE MATERIALS BY FINITE ELEMENTS WITH APPLICATION TO THE CALCULATION O	73-0332
INCOMPRESSIBLE	THE INCOMPRESSIBLE LUBRICATION PROBLEM	74-0383
INCOMPRESSIBLE	USERS' MANUAL FOR DIGITAL COMPUTER PROGRAM QK1. NUMERICAL ANALYSIS OF FINITE AXISYMMETRIC DEFORMATIO	69-0022
INCOMPRESSIBLE	VARIATIONAL PRINCIPLE FOR INCOMPRESSIBLE AND NEARLY-INCOMPRESSIBLE ANISOTROPIC ELASTICITY	72-0861
INCOMPRESSIBLE	VISCOUS, INCOMPRESSIBLE FLOW WITH SPECIAL REFERENCE TO NON-NEWTONIAN (PLASTIC) FLUIDS	69-0319
INCOMPRESSIBLES	CALCUL NUMERIQUE DES ECOULEMENTS DES FLUIDES DE BINGHAM ET DES FLUIDES NEWTONIENS INCOMPRESSIBLES PA	74-0053
INCORE	AN INCORE FINITE ELEMENT PROGRAM FOR RESEARCH AND TEACHING PURPOSE	72-0349
INCORPORATED	PLOTS - THE PLOTTING SUBROUTINE INCORPORATED IN THE BALLISTIC RESEARCH LABORATORIES ELECTRONIC SCIEN	74-0666
INCORPORATING	CALCULATION OF HARMONIC FIELDS IN WHICH SINGULARITIES ARE PRESENT USING FINITE ELEMENT AND FINITE DI	72-0803
INCREASE	THE USE OF ROCK BOLTS OR WIRE ROPE TO INCREASE THE STRENGTH OF FRACTURED MODEL PILLARS	73-0979
INCREMENTAL	A COMPARISON AND EVALUATION OF INCREMENTAL FINITE ELEMENT FORMULATIONS	71-0722
INCREMENTAL	AN INCREMENTAL COMPLEMENTARY ENERGY METHOD OF NONLINEAR STRESS ANALYSIS	75-0023
INCREMENTAL	AN INCREMENTAL FINITE ELEMENT ANALYSIS OF TWO DIMENSIONAL LARGE STRAIN AND LARGE DISPLACEMENT PROBLE	69-0281
INCREMENTAL	FINITE DISPLACEMENT PLATE FLEXURE BY THE USE OF MATRIX INCREMENTAL APPROACH	72-0338
INCREMENTAL	FINITE ELEMENT INCREMENTAL ELASTIC-PLASTIC ANALYSIS OF PRESSURE VESSELS	72-0384
INCREMENTAL	FINITE ELEMENT INCREMENTAL ANALYSIS OF ELASTOPLASTIC PLATE BENDING	70-0253
INCREMENTAL	GEOMETRICKY NELINEARNE ULOHY PLOCHYCH SKRUPIN RIESENE PRIRASTKOVOU METODOU KONECNYCH PRVKOV. (GEOMET	73-0557
INCREMENTAL	INCREMENTAL ANALYSIS OF LARGE DEFLECTIONS OF SHELLS OF REVOLUTION	74-1032
INCREMENTAL	INCREMENTAL FINITE ELEMENT ANALYSIS OF LARGE ELASTIC DEFORMATION PROBLEMS	71-0296
INCREMENTAL	INCREMENTAL FINITE ELEMENT MATRICES	71-0753
INCREMENTAL	INCREMENTAL FINITE ELEMENT MATRICES	73-0475
INCREMENTAL	INCREMENTAL FINITE-ELEMENT MATRICES	74-0354
INCREMENTAL	INCREMENTAL FORMULATIONS FOR PROBLEMS WITH GEOMETRIC AND MATERIAL NONLINEARITIES	74-0556
INCREMENTAL	INCREMENTAL NUMERICAL SOLUTION OF A SIMPLE DEFORMATION PROBLEM IN SOIL MECHANICS	75-0635
INCREMENTAL	INCREMENTAL SOLUTION OF AXISYMMETRIC PLATE AND SHELL FINITE DEFORMATION	72-0315
INCREMENTAL	INCREMENTAL SOLUTION PROCEDURE FOR NONLINEAR STRUCTURAL DYNAMICS PROBLEMS	70-0380
INCREMENTAL	INCREMENTAL STIFFNESS METHOD FOR FINITE ELEMENT ANALYSIS OF THE NONLINEAR DYNAMIC PROBLEM	71-0776
INCREMENTAL	INCREMENTAL THEORY OF LARGE STRAIN AND LARGE DISPLACEMENT PROBLEMS AND ITS FINITE ELEMENT FORMULATIO	75-0656
INCREMENTAL	INCREMENTAL THEORY OF THREE DIMENSIONAL TRANSIENT THERMOELASTOPLASTICITY - FORMULATION AND SOLUTION	71-0525
INCREMENTAL	INCREMENTAL THERMOMECHANICAL THEORY OF VISCOELASTOPLASTIC SOLIDS AND SOLUTION BY FINITE ELEMENTS	72-0274
INCREMENTAL	INCREMENTAL VARIATIONAL METHOD FOR THE LARGE DISPLACEMENT ANALYSIS OF SHELLS WITH GEOMETRIC IMPERFEC	73-0555
INCREMENTAL	NON-LINEAR MATERIAL ANALYSIS WITH FINITE ELEMENTS AND INCREMENTAL METHODS	73-0861
INCREMENTAL	NOTE ON INCREMENTAL EQUILIBRIUM EQUATIONS AND APPROXIMATE CONSTITUTIVE RELATIONS IN LARGE INELASTIC	72-0894
INCREMENTAL	ON SOME GENERALIZATIONS OF THE INCREMENTAL STIFFNESS RELATIONS FOR FINITE DEFORMATIONS OF COMPRESSIB	68-0199
INCREMENTAL	ON THE HYBRID STRESS FINITE ELEMENT MODEL FOR INCREMENTAL ANALYSIS OF LARGE DEFLECTION PROBLEMS	74-0984
INCREMENTAL	SNAP-THROUGH BUCKLING ANALYSIS OF SHELLS OF REVOLUTION USING INCREMENTAL STIFFNESS MATRICES	71-0189
INDENTATION	BERECHNUNG DES KEGELDRUCKVERSUCHES MIT EINER FINITE-ELEMENT-METHODE. (CALCULATION OF THE CONE INDENT	73-0371
INDENTATION	CALCULATION OF CONE INDENTATION TEST USING A FINITE-ELEMENT METHOD	74-1051
INDENTATION	ELASTO-PLASTIC INDENTATION OF A LAYERED MEDIUM	75-0187
INDENTATION	ELASTOPLASTIC ANALYSIS OF PLANE STRAIN AND AXISYMMETRIC FLAT PUNCH INDENTATION BY THE FINITE ELEMENT	75-0438
INDENTATION	ELASTOPLASTIC INDENTATION OF HALF-SPACE BY AN INFINITELY LONG RIGID CIRCULAR CYLINDER	74-0225
INDENTATION	THE ELASTO-PLASTIC INDENTATION OF A HALF-SPACE BY A RIGID SPHERE	70-0343
INDEPENDENT	COMPUTER AIDED STRUCTURAL ANALYSIS: THE MACHINE INDEPENDENT SYSTEM ASKA	71-0395
INDEPENDENT	NEW CYLINDRICAL SHELL FINITE ELEMENT BASED ON SIMPLE INDEPENDENT STRAIN FUNCTIONS	71-0113
INDEPENDENT	ON THE AUTOMATIC SELECTION OF INDEPENDENT DISPLACEMENT VARIABLES IN GENERAL PURPOSE FINITE ELEMENT C	70-0397
INDEPENDENT	STIFFNESS MATRICES FOR THE GENERAL DEFORMATION (OUT-OF-PLANE AND INPLANE) OF CURVED BEAM MEMBERS BAS	72-0094
INDETERMINATE	A GENERAL DIGITAL COMPUTER ANALYSIS OF STATICALLY INDETERMINATE STRUCTURES	73-0676
INDETERMINATE	MATRIX-DISPLACEMENT DIRECT-ELEMENT METHOD FOR FORCE AND TORQUE ANALYSIS OF DETERMINATE AND INDETERMI	75-0124
INDETERMINATE	NONLINEAR STRESS ANALYSIS OF INDETERMINATE STRUCTURAL ELEMENTS SUBJECTED TO BIAXIAL STRESS STATES US	59-0007
INDETS	INDETS - INTEGRATED DESIGN OF TANKER STRUCTURES	74-0531
INDIANA	COMPUTER - AIDED DESIGN - STANDARD OIL OF INDIANA BUILDING	73-1152
INDIANA	SOME STRUCTURAL PROBLEMS - STANDARD OIL OF INDIANA BUILDING	75-0524
INDIANA	STABILITY ANALYSIS OF AN EXPERIMENTALLY DEFORMED SINGLE LAYER OF INDIANA LIMESTONE USING FINITE ELEM	73-0182
INDICIAL	SUBSONIC AND SUPERSONIC INDICIAL AERODYNAMICS AND AERODYNAMIC TRANSFER FUNCTION FOR COMPLEX CONFIGUR	73-0180
INDIRECT	INDIRECT TENSILE TESTING OF ANISOTROPIC ROCKS	74-0212
INDIVIDUAL	ZUSAMMENWIRKEN VON EINZELNEN FERTIGTEILEN ALS GROSSFLAECHIGE SCHEIBE (INTERACTION OF INDIVIDUAL PREF	74-1320
INDUCED	A FRACTURE MECHANICS APPROACH TO THE PROPAGATION OF FATIGUE CRACKS IN REACTOR COMPONENTS BY ACOUSTIC	73-0376
INDUCED	ANALYSIS OF THERMALLY INDUCED STRUCTURAL VIBRATIONS OF FINITE ELEMENT TECHNIQUES	73-0552
INDUCED	APPLICATION OF FINITE ELEMENT METHODS FOR THE ANALYSIS OF THERMAL CREEP, IRRADIATION INDUCED CREEP,	71-0132
INDUCED	APPLICATION OF THREE-DIMENSIONAL FINITE ELEMENTS FOR ANALYSIS OF IRRADIATION INDUCED STRESSES IN NUC	68-0272
INDUCED	APPLICATIONS OF PULSED LASER INDUCED STRESS WAVES	71-0242
		72-0078
		73-0956

253

INDUCED	BOTTOM STABILITY UNDER WATER INDUCED LOADING	72-0373
INDUCED	CALCULATION OF THERMALLY INDUCED STRAINS IN MACHINE TOOL ELEMENTS	72-0464
INDUCED	COMPUTER PROGRAM JET 3 TO CALCULATE THE LARGE ELASTIC PLASTIC DYNAMICALLY INDUCED DEFORMATIONS OF FR	72-0609
INDUCED	FINITE ELEMENT SIMULATION OF THERMALLY INDUCED FLOW FIELDS	74-1235
INDUCED	FORMULATION OF STRESS STRAIN INDUCED DIFFUSION OF HYDROGEN AND ITS SOLUTION BY COMPUTER-AIDED FINITE	75-0326
INDUCED	MODAL SENSITIVITY STUDY OF A CONICAL RE-ENTRY VEHICLE SUBJECTED TO AERODYNAMICALLY INDUCED ACOUSTIC	69-0431
INDUCED	STRESS AND DEFORMATION ANALYSIS OF IRRADIATION INDUCED SWELLING	74-0305
INDUCED	STRESS AND DEFORMATION ANALYSIS OF IRRADIATION INDUCED SWELLING	74-0882
INDUCED	STRUCTURAL CAPACITY OF REINFORCED CONCRETE COLUMNS SUBJECTED TO FIRE INDUCED THERMAL GRADIENTS	73-0913
INDUCER	STRESS AND VIBRATION ANALYSIS OF TURBOPUMP INDUCER BLADES BY FINITE ELEMENTS	71-0363
INDUCER	STRESSES AND VIBRATION ANALYSIS OF INDUCER BLADES USING FINITE ELEMENT TECHNIQUE	70-0305
INDUCTANCE	ANALYSIS OF THE MUTUAL INDUCTANCE PARTICLE VELOCIMETER	74-0712
INDUCTION	FINITE ELEMENT METHOD TO ELECTROMAGNETIC INDUCTION PROBLEMS IN GEOPHYSICS	72-0309
INDUCTION	FINITE ELEMENTS IN THE SOLUTION OF ELECTROMAGNETIC INDUCTION PROBLEMS	74-0252
INDUSTRIAL	STRESS ANALYSIS OF INDUSTRIAL COMPONENTS WITH PLASTIC AND FINITE ELEMENT MODELS	74-0512
INELASTIC	A STUDY OF MATRIX ANALYSIS METHODS FOR INELASTIC STRUCTURES	63-0022
INELASTIC	BUCKLING OF INELASTIC I-BEAMS UNDER MOMENT GRADIENT	74-0765
INELASTIC	COMPARISON OF TYPICAL INELASTIC ANALYSIS PREDICTIONS WITH BENCHMARK PROBLEM EXPERIMENTAL RESULTS	75-0696
INELASTIC	COMPUTATIONAL PROCEDURES FOR INELASTIC FINITE-ELEMENT ANALYSIS	72-0933
INELASTIC	COMPUTATIONAL PROCEDURES FOR INELASTIC STRESS ANALYSIS	73-0859
INELASTIC	CONTINUUM BASES FOR CONSISTENT NUMERICAL FORMULATIONS OF FINITE STRAINS IN ELASTIC AND INELASTIC STR	75-0299
INELASTIC	EFFECTS OF GEOMETRY CHANGE ON ESSENTIAL FEATURES OF INELASTIC BEHAVIOUR	73-0448
INELASTIC	FINITE ELEMENT ANALYSIS OF INELASTIC STRUCTURES	69-0224
INELASTIC	FINITE ELEMENT ANALYSIS FOR THE TIME-DEPENDENT INELASTIC MATERIAL BEHAVIOR	75-0749
INELASTIC	FINITE ELEMENT SOLUTION OF INELASTIC BEAM EQUATIONS	73-0234
INELASTIC	FINITE-ELEMENT APPROACH FOR THE PREDICTION OF INELASTIC BEHAVIOR OF SHEAR WALL-FRAME SYSTEMS	72-0983
INELASTIC	FULMOD - AN INELASTIC ANALYSIS PROGRAM TO PREDICT THE OPERATING BEHAVIOUR OF CANDU FUEL ELEMENTS	75-0574
INELASTIC	INELASTIC ANALYSIS AND SATISFACTION OF DESIGN CRITERIA OF A HIGH TEMPERATURE COMPONENT	75-0265
INELASTIC	INELASTIC ANALYSIS OF SLOPES BY THE FINITE ELEMENT METHOD	68-0298
INELASTIC	INELASTIC ANALYSIS OF A UNIDIRECTIONAL COMPOSITE SUBJECTED TO TRANSVERSE NORMAL LOADING	70-0015
INELASTIC	INELASTIC ANALYSIS OF REINFORCED CONCRETE FRAMES	74-0587
INELASTIC	INELASTIC ANALYSIS OF A BRANCH SHELL JUNCTION	74-0682
INELASTIC	INELASTIC ANALYSIS OF A STRAIGHT TUBE UNDER COMBINED BENDING, PRESSURE AND THERMAL LOADS	75-0218
INELASTIC	INELASTIC ANALYSIS OF A STRAIGHT TUBE UNDER COMBINED BENDING, PRESSURE AND THERMAL LOADS	75-0513
INELASTIC	INELASTIC ANALYSIS OF TUBESHEETS BY THE FINITE ELEMENT METHOD	75-0652
INELASTIC	INELASTIC BUCKLING ANALYSIS OF PIPES SUBJECTED TO INTERNAL PRESSURE, FLEXURE AND AXIAL LOADING	74-0515
INELASTIC	INELASTIC BUCKLING OF STEEL BEAMS UNDER NON UNIFORM MOMENT	75-0149
INELASTIC	INELASTIC FINITE ELEMENT ANALYSIS OF REINFORCED CONCRETE PANELS UNDER IN-PLANE LOADS	70-0482
INELASTIC	INELASTIC MICROMECHANICS OF CURING STRESSES IN COMPOSITES	75-0284
INELASTIC	INELASTIC MODEL FOR CYCLIC BIAXIAL LOADING OF REINFORCED CONCRETE	74-0507
INELASTIC	INELASTIC RESPONSE OF FRAMES TO DYNAMIC LOADS	71-0611
INELASTIC	INELASTIC RESPONSE OF PLATES AND ECCENTRICALLY STIFFENED PLATES	74-0458
INELASTIC	INELASTIC RESPONSE OF NUCLEAR PIPING SUBJECTED TO RUPTURE FORCES	75-0515
INELASTIC	MATRIX ANALYSIS METHODS FOR INELASTIC STRUCTURES	65-0051
INELASTIC	MATRIX ANALYSIS METHODS FOR ANISOTROPIC INELASTIC STRUCTURES	66-0104
INELASTIC	NOTE ON INCREMENTAL EQUILIBRIUM EQUATIONS AND APPROXIMATE CONSTITUTIVE RELATIONS IN LARGE INELASTIC	74-0984
INELASTIC	PLASTIC COLLAPSE LOADS FOR PIPE ELBOWS USING INELASTIC ANALYSIS	73-0941
INELASTIC	PLASTIC COLLAPSE LOADS FOR PIPE ELBOWS USING INELASTIC ANALYSIS	73-0507
INELASTIC	PLATES AND SHELLS, INELASTIC ANALYSIS, TIME DEPENDENT INELASTICITY, FIVE LECTURES	73-0633
INELASTIC	SIMPLIFIED INELASTIC HIGH TEMPERATURE STRUCTURAL ANALYSIS OF MODERATELY COMPLEX SPATIALLY THREE DIME	74-0445
INELASTIC	SIMPLIFIED INELASTIC (PLASTIC AND CREEP) ANALYSIS OF PIPE ELBOWS SUBJECTED TO INPLANE AND OUT-OF-PLA	75-0654
INELASTIC	SPHERICAL WAVES IN INELASTIC MATERIAL	70-0075
INELASTIC	STRESS AND DISPLACEMENTS AROUND OPENINGS IN ROCK CONTAINING AN INELASTIC DISCONTINUITY	75-0170
INELASTIC	STUDY AND DEVELOPMENT OF SIMPLE MATRIX METHODS OF INELASTIC STRUCTURES	66-0115
INELASTIC	THE INELASTIC ANALYSIS OF REINFORCED AND PRESTRESSED CONCRETE BEAMS	72-0667
INELASTIC	THE INELASTIC ANALYSIS OF REINFORCED CONCRETE SLABS	74-1333
INELASTIC	THE INELASTIC BEHAVIOUR OF STRUCTURES SUBJECTED TO CYCLIC, THERMAL AND MECHANICAL STRESSING CONDITIO	60-0004
INELASTIC	THEORETICAL POST YIELDING BEHAVIOUR OF COMPOSITE LAMINATES - 1. INELASTIC MICROMECHANICS	73-0168
INELASTIC	THERMOMECHANICAL RESPONSE OF INELASTIC FIBRE COMPOSITES	75-0235
INELASTIC	THREE DIMENSIONAL INELASTIC ANALYSIS OF A SHELL APPURTENANCE PER SECTION 111 - APPENDIX F.	74-0716
INELASTICALLY	TRANSIENT RESPONSE OF INELASTICALLY CONSTRAINED RIGID BODY SYSTEMS	73-0291
INELASTICITY	PLATES AND SHELLS, INELASTIC ANALYSIS, TIME DEPENDENT INELASTICITY, FIVE LECTURES	73-0633
INELASTICITY	STRUCTURAL INELASTICITY. V11. ELASTIC-PLASTIC ANALYSIS OF A WHEEL ROLLING ON A RIGID TRACK	73-0886
INELASTICITY	STRUCTURAL INELASTICITY. V1. A FINITE ELEMENT METHOD FOR PLASTICITY PROBLEMS	73-0887
INEQUALITIES	ONE-SIDED APPROXIMATION AND VARIATIONAL INEQUALITIES	74-0335
INEQUALITY	ON A VARIATIONAL INEQUALITY AND ITS APPROXIMATION, IN THE THEORY OF SEMICONDUCTORS	75-0829
INERTIA	CALCULATION OF THE VIRTUAL MASS, MOVEMENT AND INERTIA BY THE FINITE ELEMENT METHOD	68-0221
INERTIA	EFFECT OF SHEAR DEFORMATIONS AND ROTARY INERTIA ON OPTIMUM BEAM FREQUENCIES	75-0231
INERTIA	EFFECTS OF FOUNDATION ROTATION ON SEISMIC INERTIA FORCES OF NUCLEAR POWER PLANT STRUCTURES	73-0177
INERTIA	EFFECTS OF SHEAR DEFORMATION, ROTARY INERTIA, AND ELASTICITY OF THE SUPPORT ON THE RESONANCE FREQUEN	70-0369
INERTIA	GALERKIN FINITE ELEMENT ANALYSIS OF A UNIFORM BEAM CARRYING A CONCENTRATED MASS AND ROTARY INERTIA W	74-0552
INERTIA	INERTIA LOADING IN FINITE-ELEMENT ANALYSIS OF STRUCTURES SUBJECT TO COMPOUND MOTION	75-0351
INERTIA	RADIAL INERTIA EFFECTS ON AN IDEALLY PLASTIC CIRCULAR PLATE UNDER IMPULSIVE AXIAL COMPRESSION	69-0324
INERTIA	ROTARY INERTIA CORRECTION FOR MASS MATRICES	63-0011
INERTIA	TRANSVERSE VIBRATION OF BEAMS WITH INTERNAL DAMPING, ROTATORY INERTIA AND SHEAR	63-0020
INEXPENSIVE	INEXPENSIVE, INTERACTIVE COMPUTER-AIDED DESIGN APPLIED TO FINITE ELEMENT STRESS ANALYSIS	74-0592
INEXTENSIONAL	INEXTENSIONAL VIBRATIONS OF PARABOLOIDAL SHELLS OF REVOLUTION. DOMES AND SHELLS	71-0427
INFERIEURES	BORNES SUPERIEURES ET INFERIEURES DANS L'ANALYSE MATRICIELLE DES PLAQUES EN FLEXION TORSION	64-0015
INFILL	LIGHT GAGE STEEL INFILL PANELS IN MULTISTORY STEEL FRAMES	74-0229
INFILLED	EFFECT OF OPENINGS ON THE LATERAL STIFFNESS OF INFILLED FRAMES	71-0300
INFILLED	NATURAL PERIODS OF INFILLED FRAMES	72-0122
INFINITE	DIFFRACTION OF OBLIQUE WAVES BY AN INFINITE CYLINDER	75-0605
INFINITE	FINITE DYNAMIC MODEL FOR INFINITE MEDIA	69-0171
INFINITE	FINITE ELEMENT METHOD FOR INFINITE DOMAINS	72-0288
INFINITE	SOME REMARKS ON INFINITE VENEZIANO REPRESENTATIONS	69-0472
INFINITE-WIDTH	APPLICATION OF FINITE ELEMENT METHOD TO HYDRODYNAMIC LUBRICATION PROBLEMS (PART 1, "INFINITE-WIDTH B	71-0537
INFINITELY	ELASTOPLASTIC INDENTATION OF HALF-SPACE BY AN INFINITELY LONG RIGID CIRCULAR CYLINDER	71-0395
INFINITELY	OIL-FILM TEMPERATURE DISTRIBUTION IN AN INFINITELY WIDE SLIDER BEARING: AN APPLICATION OF THE FINITE	73-1128
INFLATABLE	A CRUCIFORM ELEMENT FOR SURFACES OF INFLATABLE STRUCTURES	75-0118
INFLATABLE	ANALYSIS OF INFLATABLE SHELLS BY THE FINITE ELEMENT METHOD	71-0822
INFLATABLE	FINITE ELEMENT ANALYSIS OF INFLATABLE SHELLS	73-0241
INFLATABLE	FINITE ELEMENT ANALYSIS OF INFLATABLE SHELLS	74-0559
INFLATED	CONTACT OF AN INFLATED TOROIDAL MEMBRANE WITH A FLAT SURFACE AS AN APPROACH TO THE TIRE DEFLECTION P	75-0164
INFLATED	ON THE ASYMPTOTICALLY SPHERICAL DEFORMATIONS OF ARBITRARY MEMBRANES OF REVOLUTION FIXED ALONG AN EDG	71-0640
INFLATION	INFLATION FORMING OF STEEL FIBER-REINFORCED CONCRETE DOMES	74-0705
INFLUENCE	A LOOK INTO THE FUTURE: HOW COMPUTERS WILL INFLUENCE ENGINEERING	67-0139
INFLUENCE	BOUNDS TO THE INFLUENCE COEFFICIENTS BY THE ASSUMED STRESS METHOD	70-0165
INFLUENCE	COMPUTATION OF INFLUENCE COEFFICIENTS FOR AIRCRAFT STRUCTURES WITH DISCONTINUITIES AND SWEEPBACK	47-0002
INFLUENCE	CONSISTENT MATRIX FORMULATIONS FOR STRUCTURAL ANALYSIS USING INFLUENCE COEFFICIENT TECHNIQUES	64-0017
INFLUENCE	DUALITY BETWEEN DISPLACEMENT AND EQUILIBRIUM METHOD WITH A VIEW TO OBTAINING UPPER AND LOWER BOUNDS	62-0013
INFLUENCE	ELASTIC DEFORMATION OF POLYCRYSTALLINE METAL - 1. INFLUENCE OF GRAIN SHAPE ON DISTRIBUTIONS OF STRES	72-0463
INFLUENCE	ETUDE DE LA CAPACITE PORTANTE DES PIEUX PAR LA METHODE DES ELEMENTS FINIS. INFLUENCE DE LA DILATANCE	75-0624

254

INFLUENCE	ETUDE DE LA CAPACITE PORTANTE DES PIEUX PAR LA METHODE DES ELEMENTS FINIS. INFLUENCE DE LA DILATANCE	75-0624
INFLUENCE	FINITE ELEMENT PARAMETRIC STUDY OF VERTICAL STRAIN INFLUENCE FACTORS AND THE PRESSUREMETER TEST TO E	74-1127
INFLUENCE	INFLUENCE CHARACTERISTICS FOR SLAB BRIDGES	71-0188
INFLUENCE	INFLUENCE FUNCTIONS IN FINITE ELEMENT METHOD	70-0226
INFLUENCE	INFLUENCE OF ASCENT HEATING ON THE SEPARATION DYNAMICS OF A SPACECRAFT FAIRING	72-0906
INFLUENCE	INFLUENCE OF CONSTRUCTION PROCEDURES ON GROUND PRESSURES AROUND A TUNNEL	74-0997
INFLUENCE	INFLUENCE OF COOLING, FREE-STREAM TURBULENCE AND SURFACE ROUGHNESS ON THE AERODYNAMIC BEHAVIOR OF CA	74-0970
INFLUENCE	INFLUENCE OF CRACK CLOSURE AND ELASTO-PLASTIC FLOW ON THE BENDING OF A CRACKED PLATE	75-0735
INFLUENCE	INFLUENCE OF DEFECTS ON FUEL ELEMENT CLADDING STRESSES	73-0863
INFLUENCE	INFLUENCE OF END-CLOSURE STIFFNESS ON BEHAVIOR OF CONCRETE CYLINDRICAL HULLS SUBJECTED TO HYDROSTATI	71-0725
INFLUENCE	INFLUENCE OF END-EFFECTS ON STRESSES AND FLEXIBILITY OF A PIPING ELBOW WITH IN-PLANE MOMENT	74-0534
INFLUENCE	INFLUENCE OF FLUID INJECTION ON THE STATE OF STRESS IN THE EARTH'S CRUST	72-0624
INFLUENCE	INFLUENCE OF HIGHER ORDER TERMS IN THE LARGE DEFLECTION ANALYSIS OF FRAMEWORKS	68-0259
INFLUENCE	INFLUENCE OF INTERLAMINAR FLAWS ON TRANSIENT TEMPERATURE DISTRIBUTIONS	73-0223
INFLUENCE	INFLUENCE OF JOINTS IN PANELIZED STRUCTURAL SYSTEMS	72-0204
INFLUENCE	INFLUENCE OF NON-SINGULAR STRESS TERMS AND SPECIMEN GEOMETRY ON SMALL SCALE YIELDING AT CRACK TIPS I	73-0203
INFLUENCE	INFLUENCE OF POISSON'S RATIO ON THE CONDITION OF THE FINITE ELEMENT STIFFNESS MATRIX	71-0519
INFLUENCE	INFLUENCE OF RESERVOIR WATER ON EARTHQUAKE RESPONSE OF EARTH DAMS	70-0375
INFLUENCE	INFLUENCE OF ROCK PROPERTIES VARIABILITY ON MINE OPENING STABILITY ANALYSIS	74-1021
INFLUENCE	INFLUENCE OF SOIL/STRUCTURE INTERACTION PARAMETERS ON FLOOR RESPONSE SPECTRA	73-0833
INFLUENCE	INFLUENCE OF TEMPERATURE DEPENDENT PROPERTIES ON THERMAL ROCK FRAGMENTATION	75-0222
INFLUENCE	INFLUENCE OF THE UNPRECISION OF EXECUTION ON THE BEHAVIOUR OF A SPHERICAL SHELL WITH CONCENTRATED LO	73-0802
INFLUENCE	MEASUREMENT OF ACTUAL DEDENDUM STRAINS, AND THE INFLUENCE OF DEDENDUM RADIUS ON STRESSES AND DEDENDU	72-0118
INFLUENCE	MEASUREMENTS OF HEALING AT AN OSTEOTOMY IN A RABBIT CALVARIUM: THE INFLUENCE OF APPLIED COMPRESSIVE	74-0975
INFLUENCE	STUDY BY THE FINITE-ELEMENT METHOD OF THE INFLUENCE OF FRACTURES IN CONFINED AQUIFERS.	75-0604
INFLUENCE	THE INFLUENCE OF DIVISION ON THE RESULTS IN THE FINITE ELEMENT METHOD	70-0447
INFLUENCE	THEORETICAL STUDY OF THE INFLUENCE OF INCLUSIONS UPON THE INITIATION AND GROWTH OF FATIGUE CRACKS IN	71-0394
INFLUENCE	UEBER DAS VERHALTEN VON SPANNBETONDRUCKBEHAELTRERN IM GRENZTRAGFAEHIGKEITSZUSTAND UND DESSEN EINFUL	73-0855
INFLUENCED	CREEP OF B/AL COMPOSITES AS INFLUENCED BY RESIDUAL STRESSES, BOND STRENGTH, AND FIBER PACKING GEOMET	74-1208
INFLUENCES	BERECHNUNGSMETHODEN ZUR ERMITTLUNG VON BELASTUNGSBEDINGUNGEN UND GRENZTRAGFAEHIGKEITSVERHALTEN VON K	73-0858
INFLUENCES	REDUCTION GEAR DAMAGES RELATED TO EXTERNAL INFLUENCES	75-0307
INFLUENCES	SHELL STRUCTURES AND CLIMATIC INFLUENCES.	72-0389
INFLUENCES	SHELL STRUCTURES AND CLIMATIC INFLUENCES	72-0399
INFORMATSII	K VOPROSU OB AVTOMATIZATSII ZADANIYA INFORMATSII V METODE KONECHNYKH ELEMENTOV. (PROBLEM OF AUTOMATI	75-0382
INGA	INGA - INTERACTIVE GRAPHIC ANALYSIS	74-0681
INGA	INGA - INTERAKTIVE GRAPHISCHE ANALYSE BENUTZERHAND	73-0701
INGOTS	FINITE ELEMENT SOLUTION OF HEAT TRANSMISSION IN STEEL INGOTS	72-0311
INH	STUDY ON FINITE ELEMENT METHOD FOR STRUCTURAL ANALYSIS - ISTRAN/PL (INH STRUCTURE ANALYSIS PLATE STR	73-0167
INHERENT	INVESTIGATION OF THE EFFECT OF FOUNDATION YIELD ON THE FREQUENCY AND SHAPE OF INHERENT VIBRATIONS OF	72-0419
INHERITED	INHERITED ERROR IN FINITE ELEMENT ANALYSES OF STRUCTURES	73-0452
INHOMOGENEOUS	AN INHOMOGENEOUS, TAPERED FINITE ELEMENT	73-1038
INHOMOGENEOUS	APPLICATION OF A UNIMOMENT TECHNIQUE TO A BICONICAL ANTENNA WITH INHOMOGENEOUS DIELECTRIC LOADING	75-0375
INHOMOGENEOUS	APPLICATION OF THE METHOD OF FINITE ELEMENTS TO CALCULATIONS OF THE STRESSED AND STRAINED STATE OF I	75-0775
INHOMOGENEOUS	ARBITRARILY SHAPED INHOMOGENEOUS OPTICAL FIBER OR INTEGRATED OPTICAL WAVEGUIDES	75-0372
INHOMOGENEOUS	FINITE ELEMENT METHODS FOR INHOMOGENEOUS WAVEGUIDES	69-0498
INHOMOGENEOUS	FINITE-ELEMENT FIELD ANALYSIS OF AN INHOMOGENEOUS ANISOTROPIC RELUCTANCE MACHINE ROTOR	73-0020
INHOMOGENEOUS	FINITE-ELEMENT METHODS FOR INHOMOGENEOUS WAVEGUIDES	69-0039
INHOMOGENEOUS	HIGH-ORDER FINITE ELEMENT ANALYSIS FOR TRANSIENT TEMPERATURE ANALYSIS OF INHOMOGENEOUS MATERIALS	69-0301
INHOMOGENEOUS	HIGH-ORDER FINITE ELEMENTS FOR INHOMOGENEOUS ACOUSTIC GUIDING STRUCTURES	73-0137
INHOMOGENEOUS	INTERACTIVE ANALYSIS OF A RAFTED MULTISTOREY SPACE FRAME RESSTING ON AN INHOMOGENEOUS CLAY STRATUM	74-0447
INHOMOGENEOUS	MAGNETOTELLURIC RESPONSE OF LATERALLY INHOMOGENEOUS AND ANISOTROPIC MEDIA	75-0850
INHOMOGENEOUS	THREE DIMENSIONAL ANISOTROPIC AND INHOMOGENEOUS ELASTIC MEDIA MATRIX ANALYSIS FOR SMALL AND LARGE DI	65-0008
INHOMOGENEOUSLY	EVALUATION OF ATTENUATION FROM LOSSLESS TRIANGULAR-FINITE-ELEMENT SOLUTIONS FOR INHOMOGENEOUSLY FILL	73-0059
INITIAL	A FINITE ELEMENT METHOD FOR INITIAL VALUE PROBLEMS	71-0145
INITIAL	A MATHEMATICAL THEORY OF PLASTICITY FOR STRAIN HARDENING MATERIALS EXHIBITING INITIAL ANISOTROPY	72-0660
INITIAL	A STUDY OF FREE-VIBRATIONS OF THIN ELASTIC SHELLS SUBJECTED TO INITIAL LOADS USING FINITE-ELEMENTS	73-1143
INITIAL	AN ANALYSIS OF VARIOUS TIME STEPPING SCHEMES FOR INITIAL VALUE PROBLEMS	73-0607
INITIAL	AN INVESTIGATION OF INITIAL YIELD SURFACES FOR UNIDIRECTIONAL REINFORCED COMPOSITES	72-0837
INITIAL	CONNECTION BETWEEN THE FINITE DIFFERENCE LIKE METHODS AND THE METHODS BASED ON INITIAL VALUE PROBLEM	74-0877
INITIAL	CONVERGENCE OF LUMPED FINITE ELEMENT SCHEMES FOR SELECTED INITIAL VALUE PROBLEMS	73-1026
INITIAL	CURVED FINITE ELEMENTS BY METHODS OF INITIAL STRAINS	73-0070
INITIAL	DERIVATION OF A STIFFNESS MATRIX FOR A RIGHT TRIANGULAR PLATE IN BENDING AND SUBJECTED TO INITIAL ST	62-0007
INITIAL	DETERMINATION OF THE INITIAL BUCKLING STRESS OF FLAT PLATES USING THE DISPLACEMENT METHOD	69-0465
INITIAL	EFFECT OF INITIAL DISPLACEMENT ON PROBLEM OF LARGE DEFLECTION AND STABILITY	67-0113
INITIAL	EFFECT OF INITIAL STATE FORCES ON THE HYDROELASTIC VIBRATION OF TUBES	73-0589
INITIAL	ELASTO-PLASTIC SOLUTIONS OF ENGINEERING PROBLEMS ' INITIAL STRESS ' FINITE ELEMENT APPROACH	69-0282
INITIAL	ELASTODYNAMIC BEHAVIOR OF LAMINATED ORTHOTROPIC PLATES UNDER INITIAL STRESS	75-0249
INITIAL	FINITE ELEMENT ANALYSIS OF FREE VIBRATION AND STABILITY OF STRUCTURES SUBJECTED TO INITIAL STRESS	68-0300
INITIAL	FINITE ELEMENT METHODS FOR INITIAL VALUE PROBLEMS	72-0059
INITIAL	FINITE ELEMENT METHODS FOR SOLVING INITIAL BOUNDARY VALUE PROBLEMS	74-1125
INITIAL	FINITE ELEMENT PROCEDURE FOR LARGE DEFLECTION ANALYSIS OF PLATES WITH INITIAL DEFLECTIONS	71-0327
INITIAL	FINITE ELEMENT PROCEDURE FOR NONLINEAR PREBUCKLING AND INITIAL POSTBUCKLING ANALYSIS	72-0170
INITIAL	FINITE-ELEMENT ANALYSES OF INITIAL ELASTIC DEFORMATION OF VERTICALLY DEFORMED CRYSTAL BLOCKS	72-0970
INITIAL	INITIAL SETTLEMENT OF STRUCTURES ON CLAY	71-0304
INITIAL	INITIAL STRAINS IN THE MATRIX FORCE METHOD OF STRUCTURAL ANALYSIS	60-0008
INITIAL	INITIAL YIELD SURFACE OF A UNIDIRECTIONALLY REINFORCED COMPOSITE	71-0280
INITIAL	MATHEMATICAL MODEL OF THE COLUMBIA RIVER FROM THE PACIFIC OCEAN TO BONNEVILLE DAM. PART II: INPUT-OU	70-0417
INITIAL	NONLINEAR DYNAMICS OF CABLES WITH LOW INITIAL TENSION	72-0740
INITIAL	REINFORCED FIELDS OF TRIANGULAR ELEMENTS WITH LINEARLY VARYING STRAIN: EFFECT OF INITIAL STRAINS	65-0004
INITIAL	STIFFNESS MATRIX FOR BENDING OF A RECTANGULAR PLATE ELEMENT WITH INITIAL MEMBRANE STRESSES	62-0006
INITIAL	TECHNIQUE OF MEASURING INITIAL DEFORMATION AROUND AN OPENING	71-0331
INITIAL	THE EFFECT OF AN INITIAL AXISYMMETRIC IMPERFECTION ON THE NATURAL FREQUENCIES OF CONICAL SHELLS	70-0368
INITIAL	THE EFFECT OF INITIAL DISPLACEMENTS OF PROBLEMS OF LARGE DEFLECTION AND STABILITY	67-0048
INITIAL	THE VALIDITY OF THE INITIAL STRAIN CONCEPT	61-0010
INITIAL	VARIATIONAL PRINCIPLES FOR LINEAR INITIAL VALUE PROBLEMS	64-0029
INITIAL-BOUNDAR	FINITE ELEMENT GALERKIN METHOD FOR MIXED INITIAL-BOUNDARY VALUE PROBLEMS IN ELASTICITY THEORY	71-0520
INITIAL-BOUNDAR	NONLINEAR INITIAL-BOUNDARY VALUE SOLUTIONS BY THE FINITE ELEMENT METHOD	74-0812
INITIAL-VALUE	FINITE ELEMENTS FOR FLUID DYNAMICS. VARIATIONAL FORMULATIONS FOR NONLINEAR, INITIAL-VALUE PROBLEMS	74-0710
INITIALLY	FINITE ELEMENT ANALYSIS OF VIBRATIONS OF INITIALLY STRESSED THIN SHELLS OF REVOLUTION	74-1056
INITIALLY	FINITE ELEMENT PROCEDURE FOR POSTBUCKLING ANALYSIS OF INITIALLY CURVED PLATES	71-0421
INITIALLY	ON RESPONSE OF INITIALLY STRESSED STRUCTURES TO RANDOM EXCITATIONS	73-0278
INITIATION	ENERGY CONDITION FOR INITIATION OF INTERFACIAL MICROCRACKS AT INCLUSIONS	74-0546
INITIATION	FATIGUE CRACK INITIATION IN A BORON EPOXY PLATE WITH A CIRCULAR HOLE	74-0820
INITIATION	FINITE ELEMENT ANALYSIS OF CRACK INITIATION IN A BLOCK MODEL EXPERIMENT	72-0455
INITIATION	GRAVITY FLOW OF POWDER IN A LUNAR ENVIRONMENT. PT. 2-ANALYSIS OF FLOW INITIATION	71-0484
INITIATION	INITIATION OF FAILURE IN SLOPES IN OVERCONSOLIDATED CLAYS AND CLAY SHALES.	70-0186
INITIATION	ON THE ROLE OF DEFECTS IN CRACK INITIATION IN WELDED STRUCTURES	74-1142
INITIATION	SIMULATION ON THE INITIATION AND GROWTH OF DUCTILE FRACTURE VOIDS	75-0836
INITIATION	THEORETICAL STUDY OF THE INFLUENCE OF INCLUSIONS UPON THE INITIATION AND GROWTH OF FATIGUE CRACKS IN	71-0394
INJECTION	ANALYSIS OF LIQUID-WASTE INJECTION WELLS IN ILLINOIS BY MATHEMATICAL MODELS	74-0518
INJECTION	EARTHQUAKE TRIGGERING BY FLUID INJECTION AT RANGELY, COLORADO	72-0634
INJECTION	INFLUENCE OF FLUID INJECTION ON THE STATE OF STRESS IN THE EARTH'S CRUST	72-0624

INJEKTIONSVERFA	THEORIE UND EXPERIMENT ZU EINEM NEUEN INJEKTIONSVEPFAHREN. (THEORY AND EXPERIMENT OF A NEW GROUTING	74-0991
INJURIES	DYNAMIC TORSIONAL RESPONSE OF THE HUMAN LEG RELATIVE TO SKIING INJURIES	73-0471
INJURY	DEVELOPMENT OF FINITE ELEMENT HEAD INJURY MODEL	75-0506
INNENUND	OSCILLATION STABILITY OF INTERNALLY AND EXTERNALLY NOTCHED PLANE BARS OF EQUAL FORM FACTOR FROM 3943	73-0745
INNER	CONTRIBUTION TO NUMERICAL-SOLUTION OF DEVELOPING LAMINAR-FLOW IN ENTRANCE REGION OF CONCENTRIC ANNUL	75-0448
INPLANE	A COMPUTER PROGRAM FOR ANALYSIS OF STIFFENED PLATES UNDER COMBINED INPLANE AND LATERAL LOADS	71-0703
INPLANE	ANALYSIS OF INPLANE VIBRATION OF BOX-TYPE STRUCTURES BY A FINITE ELEMENT METHOD	72-0091
INPLANE	APPROXIMATE THREE-DIMENSIONAL SOLUTIONS FOR SYMMETRIC LAMINATES UNDER INPLANE LOADING	71-0392
INPLANE	SIMPLIFIED INELASTIC (PLASTIC AND CREEP) ANALYSIS OF PIPE ELBOWS SUBJECTED TO INPLANE AND OUT-OF-PLA	75-0654
INPLANE	STIFFNESS MATRICES FOR THE GENERAL DEFORMATION (OUT-OF-PLANE AND INPLANE) OF CURVED BEAM MEMBERS BAS	75-0124
INPUT	A SPECIAL NASTRAN PROGRAM FOR INPUT CHECKING AND UNDEFORMED STRUCTURE PLOTTING	71-0234
INPUT	A SYNTHESIZED WARPED QUADRILATERAL ELEMENT GIVING SIMPLIFIED INPUT FOR EQUILIBRIUM MODELS	73-1013
INPUT	ANALYSIS AND DESIGN OF MULTIPLE INPUT FLEXIBLE LINK MECHANISMS	70-0354
INPUT	AUTOMATED INPUT DATA PREPARATION FOR NASTRAN	69-0080
INPUT	COMPUTER-AIDED INPUT/OUTPUT FOR USE WITH FINITE ELEMENT METHOD OF STRUCTURAL ANALYSIS	70-0330
INPUT	COMPUTER-AIDED INPUT/OUTPUT FOR THE USE WITH FINITE ELEMENT STRUCTURAL ANALYSES	73-0923
INPUT	INPUT SOIL PROPERTIES FOR THE FINITE ELEMENT CODE NTEDYNS	71-0694
INPUT	MISSILE BODY INPUT GENERATOR (BING). A NASTRAN PRE-PROCESSOR; THEORETICAL DEVELOPMENT. USER'S MANUAL	75-0717
INPUT	NASTRAN USER INTERFACES, AUTOMATED INPUT INNOVATIONS	71-0126
INPUT	NATURAL COORDINATE SYSTEMS, AN AUTOMATIC INPUT DATA GENERATION SCHEME FOR A FINITE ELEMENT METHOD	70-0018
INPUT	PLANFORM INPUT GENERATOR (PING). A NASTRAN PRE-PROCESSOR FOR LIFTING SURFACES-THEORETICAL DEVFLOPMEN	73-0776
INPUT	PLANFORM INPUT GENERATOR (PING). A NASTRAN PROCESSOR FOR LIFTING SURFACES-THEORETICAL DEVELOPMENT. U	73-0780
INPUT	PROCEDURE FOR EFFICIENT GENERATING, CHECKING, AND DISPLAYING NASTRAN INPUT AND OUTPUT DATA FOR ANALY	71-0098
INPUT	SAFE PCRS, A COMPUTER PROGRAM FOR THE STRESS ANALYSIS OF COMPOSITES BODIES OF REVOLUTION. INPUT INST	65-0013
INPUT	SAFE-AXISYM, A COMPUTER PROGRAM FOR THE DESIGN OF COMPOSITE BODIES OF REVOLUTION. INPUT INSTRUCTIONS	65-0014
INPUT	SENSITIVITY ANALYSIS OF INPUT PARAMETERS IN NUMERICAL MODELING OF STEADY STATE REGIONAL GROUNDWATER	74-0926
INPUT	TECHNIQUES FOR HANDLING INPUT DATA FOR FINITE ELEMENT ANALYSIS OF REGIONAL GROUNDWATER FLOW	75-0806
INPUT-OUTPUT	MATHEMATICAL MODEL OF THE COLUMBIA RIVER FROM THE PACIFIC OCEAN TO BONNEVILLE DAM. PART II: INPUT-OU	70-0417
INPUT-OUTPUT	MODELING AN INPUT-OUTPUT GEOKINETIC SYSTEM UTILIZING A FINITE ELEMENT APPROACH	75-0746
INSBESONDERE	MODERNE VERFAHREN ZUR UNTERSUCHUNG DER STEIFIGKEITS-UND FESTIGKEITSVERHALTNISSE VON RAHMENKORPERN, I	69-0401
INSERTS	EFFECTS OF INSERTS ON THE ELASTIC BEHAVIOUR OF CYLINDRICAL MATERIALS LOADED BETWEEN ROUGH END-PLATES	71-0318
INSPECTION	FINITE ELEMENT STUDY OF THE TRIAXIAL STRESS STATE AROUND AN INSPECTION TUNNEL IN AN ARCH DAM	74-0831
INSTABILITIES	NUMERICAL ANALYSIS OF MAGNETOHYDRODYNAMIC INSTABILITIES BY FINITE ELEMENT METHOD	72-0041
INSTABILITIES	STUDY OF THE MHD INSTABILITIES OF CYLINDRICAL BELT PINCH USING THE FINITE ELEMENT METHOD	74-0700
INSTABILITY	A DISCRETE ELEMENT PROCEDURE FOR THIN-SHELL INSTABILITY ANALYSIS	67-0029
INSTABILITY	A FINITE ELEMENT APPROACH TO THE STRUCTURAL INSTABILITY OF BEAM COLUMNS, FRAMES, AND ARCHES	70-0434
INSTABILITY	A NUMERICAL STUDY OF NONLINEAR INSTABILITY BY MEANS OF AN EXTENDED FINITE ELEMENT METHOD	74-0632
INSTABILITY	AUTOMATED METHOD FOR THE LARGE DEFLECTION AND INSTABILITY ANALYSIS OF THREE-DIMENSIONAL TRUSS AND FR	66-0105
INSTABILITY	AUTOMATIC EMBANKMENT ANALYSIS. EQUILIBRIUM AND INSTABILITY CONDITIONS	66-0015
INSTABILITY	DISCRETE ELEMENT APPROACH TO STRUCTURAL INSTABILITY ANALYSIS	63-0005
INSTABILITY	DISCRETE ELEMENT PROCEDURE FOR THIN SHELL INSTABILITY ANALYSIS	67-0028
INSTABILITY	DYNAMIC INSTABILITY OF DUCTS CONVEYING FLUID	75-0691
INSTABILITY	ELASTIC INSTABILITY PREDICTIONS FOR DOUBLY-CURVED SHELLS	68-0075
INSTABILITY	FINITE ELEMENT ANALYSIS OF STRUCTURAL INSTABILITY BY ASSOCIATION OF PULSATING EXCITATIONS	74-0453
INSTABILITY	FINITE ELEMENT METHOD IN ELASTIC INSTABILITY ANALYSIS	69-0108
INSTABILITY	FINITE ELEMENT METHOD IN PLATE AND SHELL INSTABILITY ANALYSIS	73-0983
INSTABILITY	FINITE ELEMENT STRUCTURAL ANALYSIS OF LOCAL INSTABILITY	72-0179
INSTABILITY	FINITE ELEMENT STRUCTURAL ANALYSIS OF LOCAL INSTABILITY	74-0098
INSTABILITY	FINITE ELEMENT STRUCTURAL ANALYSIS OF LOCAL INSTABILITY - REPLY	74-0099
INSTABILITY	INSTABILITY ANALYSIS BY THE FINITE ELEMENT METHOD	75-0180
INSTABILITY	INSTABILITY OF HORIZONTALLY CURVED MEMBERS. FLANGE BUCKLING STUDIES	69-0480
INSTABILITY	INSTABILITY OF THIN SHELLS BY THE FINITE ELEMENT METHOD	70-0187
INSTABILITY	INSTABILITY OF THIN SHELLS BY THE FINITE ELEMENT METHOD	70-0477
INSTABILITY	MATRIX ANALYSIS OF LOCAL INSTABILITY IN PLATES, STIFFENED PANELS AND COLUMNS	72-0542
INSTABILITY	MATRIX DYNAMIC AND INSTABILITY ANALYSIS WITH NON-UNIFORM ELEMENTS	70-0062
INSTABILITY	NEW FINITE-ELEMENT SCHEME FOR INSTABILITY ANALYSIS OF THIN SHELLS	75-0357
INSTABILITY	NONLINEAR THERMOELASTICITY, IRREVERSIBLE THERMODYNAMICS AND ELASTIC INSTABILITY	73-0762
INSTABILITY	SHELL INSTABILITY ANALYSIS APPLIED TO A RADOME	74-0261
INSTABILITY	SOLUMN INSTABILITY UNDER NONCONSERVATIVE FORCES, WITH INTERNAL AND EXTERNAL DAMPING FINITE ELEMENT U	73-0566
INSTABILITY	THE FINITE ELEMENT METHOD IN ELASTIC INSTABILITY ANALYSIS	69-0339
INSTALLATION	FINITE ELEMENT ANALYSIS OF THE EFFECTS OF INSTALLATION ON PILE LOAD-SETTLEMENT BEHAVIOUR	74-0763
INSTALLATION	OPTIMIZATION OF TENSION LEVEL AND STINGER LENGTH FOR OFFSHORE PIPELINE INSTALLATION	74-0504
INSTALLATION	SEISMIC ANALYSIS OF A THREE-PHASE BREAKER INSTALLATION	74-0622
INSTALLATIONS	DIGITAL CALCULATION OF ELECTRIC FIELDS IN METAL CLAD INSTALLATIONS	73-0212
INSTATIONAERE	HYBRIDE FINITE VERFAHREN FUER LINEARE UND NICHTLINEARE INSTATIONAERE WAERMELEITPROBLEME (HYBRID FINI	73-0484
INSTATIONAERER	BERECHNUNG INSTATIONAERER TEMPERATUREFELDER MIT FINITEN ELEMENTEN IN RAUM-UND ZEITDIMENSION. (CALCUL	74-0987
INSTATIONAERER	NUMERISCHE BERECHNUNG INSTATIONAERER TEMPERATURFELDER UND THERMISCH BEDINGTER VERFORMUNGEN VON WERKZ	75-0847
INSTATIONAREP	BERECHNUNG INSTATIONARER GRUND - UND SICKERWASSERSTROMUNGEN MIT FREIER OBEPFLACHE NACH DER METHODE F	71-0541
INSTRUCTION	DESCRIPTION AND INSTRUCTION FOR USE OF THE SAFE-SHELL. B FINITE ELEMENT PROGRAM IN THE IBM/360 VERSI	73-0945
INSTRUMENT	USING THE NASTRAN THERMAL ANALYZER TO SIMULATE A FLIGHT SCIENTIFIC INSTRUMENT PACKAGE	74-1011
INSTRUMENTATION	FIELD INSTRUMENTATION FOR LONG TERM MEASUREMENT OF PILE LOAD AND RAFT CONTACT PRESSURE	74-0150
INSTRUMENTATION	FOUNDATION INSTRUMENTATION FOR THE NATIONAL WESTMINISTER BANK TOWER	74-1147
INSTRUMENTATION	HOLOGRAPHIC INSTRUMENTATION OF TURBINE BLADES	73-0466
INSTRUMENTATION	JOURNAL OF RESEARCH OF THE NATIONAL BUREAU OF STANDARDS: C. ENGINEERING AND INSTRUMENTATION	71-0664
INSTRUMENTATION	M.I.T. TEST SECTION INSTRUMENTATION, MASSACHUSETTS BAY TRANSPORTATION AUTHORITY. HAYMARKET-NORTH EXT	72-0673
INSULATED	NUMERICAL ANALYSIS OF THE LINE CAPACITANCE AND CROSSTALK FACTOR FOR INSULATED WIRE PAIRS	74-0685
INSULATED	USE OF THE FINITE ELEMENT TECHNIQUE TO DETERMINE THE FIRE RESISTANCE CAPABILITIES OF INSULATED STRUC	71-0147
INSULATION	VIBRATION AND STRESS ANALYSIS OF SOFT-BONDED SHUTTLE INSULATION TILES. MODAL ANALYSIS WITH COMPACT W	74-0708
INTEGRAL	A CONSIDERATION ON THE FINITE ELEMENT METHOD FOR CALCULATING STRESS INTENSITY FACTOR KI USING THE J	75-0863
INTEGRAL	ANALYSIS OF CONTINUA BY THE FINITE ELEMENT AND INTEGRAL EQUATION METHOD	75-0370
INTEGRAL	APPLICATION OF INTEGRAL EQUATIONS TO FLUID FLOWS IN UNBOUNDED REGIONS	74-0029
INTEGRAL	APPLICATION OF THE TWO-DIMENSIONAL INTEGRAL EQUATION METHOD TO ENGINEERING PROBLEMS	75-0377
INTEGRAL	APPLIED FINITE ELEMENT DISPLACEMENT ANALYSIS OF A SIMPLE INTEGRAL CONSTRUCTION VEHICLE BODY SHELL	72-0264
INTEGRAL	CALCULATION OF STRESS INTENSITY FACTORS IN ANISOTROPIC MATERIALS BY A CONTOUR INTEGRAL METHOD	75-0257
INTEGRAL	FINITE ELEMENT ANALYSIS OF PROBLEMS FORMULATED BY AN INTEGRAL EQUATION; APPLICATION TO POTENTIAL FLO	68-0191
INTEGRAL	FINITE ELEMENT ANALYSIS OF SURFACE WAVE PROBLEMS USING INTEGRAL EQUATIONS	74-0624
INTEGRAL	FINITE ELEMENT INTEGRAL NEUTRON TRANSPORT	73-1032
INTEGRAL	FINITE-ELEMENT-TYPE SOLUTION OF INTEGRAL EQUATIONS	73-0995
INTEGRAL	FLEXURE ANALYSIS OF AXIALLY-LOADED BEAM BY FINITE-ELEMENT METHOD WITH INTEGRAL PARAMETERS	75-0462
INTEGRAL	INTEGRAL EQUATION METHOD FOR SUBSONIC FLOW PAST AIRFOILS IN VENTILATED WIND TUNNELS	75-0065
INTEGRAL	INTEGRAL FINITE-ELEMENT APPROACH FOR SOLVING LAPLACE EQUATION	73-0121
INTEGRAL	INTEGRAL REPRESENTATION OF FIELD VARIABLES FOR THE FINITE ELEMENT SOLUTION OF VISCOUS FLOW PROBLEMS	74-0468
INTEGRAL	J INTEGRAL ESTIMATES FOR STRAIN HARDENING MATERIALS USING FULLY PLASTIC SOLUTIONS	74-0690
INTEGRAL	NUMERICAL TREATMENT OF ALGEBRAIC INTEGRAL EQUATIONS BY VARIATIONAL METHODS	75-0374
INTEGRAL	NUMERICAL USE OF INTEGRAL PROCEDURES	65-0041
INTEGRAL	SECOND GENERATION BOUNDARY INTEGRAL EQUATION PROGRAM FOR THREE-DIMENSIONAL ELASTIC ANALYSIS	75-0376
INTEGRAL	STRESS, DISPLACEMENT, LINE INTEGRAL AND CLOSURE ENERGY DETERMINATIONS OF CRACK TIP STRESS INTENSITY	74-0621
INTEGRAL	USE OF INTEGRAL PARAMETERS IN ROD PROBLEMS	74-1194
INTEGRAL	2-DIMENSIONAL FINITE-ELEMENT METHOD FOR INTEGRAL NEUTRON-TRANSPORT CALCULATIONS	75-0443
INTEGRALGLEICHU	KOMBINATION EINES INTEGRALGLEICHUNGS VERFAHRENS MIT METHODE DER FINITE ELEMENTEN ZUR BERECHNUNG FRE	74-0856
INTEGRALLY	JET NOISE EXCITATION OF AN INTEGRALLY STIFFENED PANEL	71-0546
INTEGRALLY-STIF	FREE VIBRATIONS AND RANDOM RESPONSE OF AN INTEGRALLY-STIFFENED PANEL	70-0138

INTEGRATED	A METHOD OF COMPUTING NUMERICALLY INTEGRATED STIFFNESS MATRICES	72-0546
INTEGRATED	ARBITRARILY SHAPED INHOMOGENEOUS OPTICAL FIBER OR INTEGRATED OPTICAL WAVEGUIDES	75-0372
INTEGRATED	ECONOMICAL COMPUTER TECHNIQUES FOR NUMERICALLY INTEGRATED FINITE ELEMENTS	69-0138
INTEGRATED	FINITE ELEMENT APPROACH TO THE INTEGRATED POTENTIAL FORMULATION OF GENERAL UNSTEADY SUPERSONIC AEROD	73-0910
INTEGRATED	INDETS - INTEGRATED DESIGN OF TANKER STRUCTURES	75-0524
INTEGRATED	INTEGRATED PROGRAM FOR THE DYNAMIC STRUCTURAL CALCULATION OF MOBIL OFFSHORE UNITS	74-0490
INTEGRATED	INTEGRATED SEQUENTIAL SOLVER FOR LARGE MATRIX EQUATIONS	74-0276
INTEGRATED	INTEGRATED THEORY OF FINITE ELEMENT METHODS	73-0531
INTEGRATED	ISOPARAMETRIC AND ALLIED NUMERICALLY INTEGRATED ELEMENTS - A REVIEW	71-0763
INTEGRATED	ISOPARAMETRIC AND OTHER NUMERICALLY INTEGRATED ELEMENTS	71-0272
INTEGRATED	NASTRAN-GAP, AN INTEGRATED APPROACH TO THE ANALYSIS OF RADIATION PATTERNS OF DISTORTED REFLECTORS	71-0062
INTEGRATING	HELPFUL FORMULAS FOR INTEGRATING POLYNOMIALS IN THREE DIMENSIONS	64-0031
INTEGRATION	A HIGHLY STABLE EXPLICIT INTEGRATION TECHNIQUE FOR COMPUTATIONAL CONTINUUM MECHANICS	74-1277
INTEGRATION	A METHOD FOR THE NUMERICAL INTEGRATION OF THE EQUATIONS OF MOTION ARISING FROM A FINITE ELEMENT ANAL	70-0061
INTEGRATION	BEAM BENDING FINITE ELEMENT CONTAINING INTEGRATION PARAMETERS	71-0179
INTEGRATION	COMPARISON OF THE NUMERICAL INTEGRATION TECHNIQUE AND THE FINITE ELEMENT METHOD IN THE ANALYSIS OF T	73-0848
INTEGRATION	COMPUTER INTEGRATION OF CONGRUENTLY TRANSFORMED MATRICES	71-0580
INTEGRATION	COMPUTER INTEGRATION FOR FINITE ELEMENTS	73-0329
INTEGRATION	DYNAMIC RESPONSE BY LARGE STEP INTEGRATION	73-0567
INTEGRATION	ENGINEERING APPLICATIONS OF NUMERICAL INTEGRATION IN STIFFNESS METHOD	66-0040
INTEGRATION	EVALUATION OF NUMERICAL TIME INTEGRATION METHODS AS APPLIED TO ELASTIC-PLASTIC DYNAMIC PROBLEMS INVO	74-0858
INTEGRATION	EXPERIENCES WITH ORTHOGONAL POLYNOMIALS AND "BEST" NUMERICAL INTEGRATION FORMULAS ON A TRIANGLE; WIT	74-0977
INTEGRATION	EXPLICIT INTEGRATION - STABILITY, SOLUTION PROPERTIES, COST	75-0298
INTEGRATION	FINITE ELEMENT SOLUTION FOR ENERGY CONSERVATION USING A HIGHLY STABLE EXPLICIT INTEGRATION ALGORITHM	72-0633
INTEGRATION	IMPROVED NUMERICAL INTEGRATION OF THICK SHELL FINITE ELEMENTS	71-0396
INTEGRATION	INTEGRATION FOR THE FINITE ELEMENTS - AN ALGORITHM IN ORTHOGONAL AXES	73-0295
INTEGRATION	INTEGRATION OF AREA AND VOLUME COORDINATES IN FINITE-ELEMENT METHOD	69-0370
INTEGRATION	MORE ON REDUCED INTEGRATION AND ISOPARAMETRIC ELEMENTS	72-0548
INTEGRATION	NUMERICAL INTEGRATION APPLIED TO FINITE ELEMENT METHODS	66-0039
INTEGRATION	NUMERICAL INTEGRATION APPLIED TO FINITE ELEMENT METHODS	70-0390
INTEGRATION	NUMERICAL INTEGRATION OF THE EQUATION OF MOTION	71-0507
INTEGRATION	NUMERICAL INTEGRATION IN THE FINITE ELEMENT METHOD	74-0576
INTEGRATION	NUMERICAL INTEGRATION METHODS IN FINITE ELEMENTS	74-0841
INTEGRATION	NUMERICAL INTEGRATION CONSIDERATIONS IN TWO AND THREE DIMENSIONAL ISOPARAMETRIC FINITE ELEMENTS	75-0130
INTEGRATION	ON FINITE ELEMENT INTEGRATION IN NATURAL COORDINATES	73-0683
INTEGRATION	ON THE STABILITY OF EXPLICIT METHODS FOR THE NUMERICAL INTEGRATION OF THE EQUATIONS OF MOTION IN FIN	72-0172
INTEGRATION	REDUCED INTEGRATION TECHNIQUE IN GENERAL ANALYSIS OF PLATES AND SHELLS	71-0374
INTEGRATION	SOLUTION OF NONLINEAR EQUATIONS BY EXPLICIT TIME INTEGRATION - A SHORT COMMUNICATION	73-0606
INTEGRATION	STATIC, STABILITY, AND DYNAMIC ANALYSIS OF SHELLS OF REVOLUTION BY NUMERICAL INTEGRATION - A COMPARI	74-0389
INTEGRATION	THE COMBINED EFFECT TO CURVED BOUNDARIES AND NUMERICAL INTEGRATION IN ISOPARAMETRIC FINITE ELEMENT M	72-0357
INTEGRATION	THE USE OF NUMERICAL INTEGRATION IN FINITE ELEMENT METHODS FOR SOLVING PARABOLIC EQUATIONS	72-0351
INTEGRATION	TRANSIENT SHELL RESPONSE BY NUMERICAL TIME INTEGRATION	72-0305
INTEGRATION	USE OF COMPUTERIZED SYMBOLIC INTEGRATION IN FINITE ELEMENT DEVELOPMENT	74-0520
INTEGRATIONS	INTEGRATIONS OF PRIMITIVE EQUATIONS ON A SPHERE USING FINITE ELEMENT METHOD	74-0649
INTEGRATIONS	INTEGRATIONS OF THE PRIMITIVE EQUATIONS ON A SPHERE USING THE FINITE ELEMENT METHOD	74-1347
INTEGRITY	FINITE ELEMENT ANALYSIS OF STRUCTURAL INTEGRITY OF A REACTOR PRESSURE VESSEL DURING EMERGENCY CORE C	70-0026
INTENSITIES	CHILES: A FINITE ELEMENT COMPUTER PROGRAM THAT CALCULATES THE INTENSITIES OF LINEAR ELASTIC SINGULAR	73-0959
INTENSITY	A CONSIDERATION ON THE FINITE ELEMENT METHOD FOR CALCULATING STRESS INTENSITY FACTOR KI USING THE J	75-0863
INTENSITY	A HYBRID FINITE ELEMENT FOR DETERMINATION OF STRESS INTENSITY FACTORS AT A CRACK TIP	73-0643
INTENSITY	APPLICATIONS OF THE FINITE ELEMENT METHOD TO THE CALCULATIONS OF STRESS INTENSITY FACTORS	72-0900
INTENSITY	CALCULATION OF STRESS INTENSITY FACTORS USING FINITE ELEMENT METHOD WITH CRACKED ELEMENTS	70-0224
INTENSITY	CALCULATION OF STRESS INTENSITY FACTORS USING THE FINITE ELEMENT METHOD WITH CRACKED ELEMENTS	70-0475
INTENSITY	CALCULATION OF STRESS INTENSITY FACTORS AT CRACK TIPS USING SPECIAL FINITE ELEMENTS	72-0071
INTENSITY	CALCULATION OF STRESS INTENSITY FACTORS AND FATIGUE CRACK PROPAGATION OF SEMI-ELLIPTICAL, PART-THROU	73-0360
INTENSITY	CALCULATION OF STRESS INTENSITY FACTORS FOR COMBINED TENSILE AND SHEAR LOADING	75-0205
INTENSITY	CALCULATION OF STRESS INTENSITY FACTORS IN TWO AND THREE DIMENSIONS USING FINITE ELEMENTS	75-0254
INTENSITY	CALCULATION OF STRESS INTENSITY FACTORS IN ANISOTROPIC MATERIALS BY A CONTOUR INTEGRAL METHOD	75-0257
INTENSITY	COMPARISON OF NUMERICAL METHODS FOR DETERMINING STRESS INTENSITY FACTORS	72-0176
INTENSITY	COMPUTATION OF STRESS INTENSITY FACTORS BY A SPECIAL FINITE ELEMENT TECHNIQUE	71-0287
INTENSITY	DETERMINATION OF ENERGY RELEASE RATES AND STRESS - INTENSITY FACTORS BY THE FINITE-ELEMENT METHOD	72-0183
INTENSITY	DETERMINATION OF STRESS INTENSITY FACTORS IN CRACKED PLATES BY THE FINITE ELEMENT METHOD	73-0148
INTENSITY	EFFICIENT FINITE ELEMENT METHODS FOR STRESS INTENSITY FACTOR	75-0148
INTENSITY	EFFICIENT FINITE-ELEMENT METHODS FOR STRESS INTENSITY FACTORS USING WEIGHT FUNCTIONS	75-0411
INTENSITY	EFFICIENT FINITE-ELEMENT METHODS FOR STRESS INTENSITY FACTORS USING WEIGHT FUNCTIONS - REPLY	75-0421
INTENSITY	EVALUATION OF FINITE ELEMENT METHODS FOR THE COMPUTATION OF ELASTIC STRESS INTENSITY FACTORS	72-0249
INTENSITY	FINITE ELEMENT CALCULATION OF STRESS INTENSITY FACTORS USING ENERGY TECHNIQUES	73-0825
INTENSITY	FINITE ELEMENT CALCULATION OF STRESS INTENSITY FACTORS USING SUPERPOSITION	75-0251
INTENSITY	FINITE ELEMENT METHOD FOR CALCULATING STRESS INTENSITY FACTORS FOR CRACKED PLATES IN BENDING	71-0424
INTENSITY	FINITE ELEMENT METHOD FOR CALCULATING STRESS INTENSITY FACTORS AND ITS APPLICATION TO COMPOSITES	74-1000
INTENSITY	FINITE ELEMENT TREATMENT OF SINGULARITIES OF BOUNDARY VALUE PROBLEMS AND ITS APPLICATION TO ANALYSIS	73-0306
INTENSITY	FINITE ELEMENTS FOR DETERMINATION OF CRACK TIP ELASTIC STRESS INTENSITY FACTORS	71-0436
INTENSITY	FINITE ELEMENTS FOR DETERMINATION OF CRACK-TIP ELASTIC STRESS INTENSITY FACTORS	71-0785
INTENSITY	METHOD OF EVALUATION OF THREE-DIMENSIONAL STRESS INTENSITY FACTOR USING THE FINITE ELEMENT METHOD	75-0316
INTENSITY	ON THE USE OF FINITE ELEMENTS AND COMPLIANCE CALCULATIONS FOR DETERMINING CRACK TIP STRESS INTENSITY	69-0448
INTENSITY	PLASTIC INTENSITY FACTORS FOR CRACKED PLATES	71-0164
INTENSITY	SOME STRESS INTENSITY CALCULATIONS USING FINITE ELEMENTS	74-0423
INTENSITY	STIFFNESS DERIVATIVE FINITE ELEMENT TECHNIQUE FOR DETERMINATION OF CRACK TIP STRESS INTENSITY FACTOR	74-0956
INTENSITY	STIFFNESS DERIVATIVE FINITE ELEMENT TECHNIQUE FOR DETERMINATION OF CRACK TIP STRESS INTENSITY FACTOR	74-1358
INTENSITY	STRESS INTENSITY FACTOR FOR A STRAIGHT CRACK APPROACHING A HOLE	69-0158
INTENSITY	STRESS INTENSITY FACTORS CALCULATED GENERALLY BY FINITE ELEMENT TECHNIQUE	69-0369
INTENSITY	STRESS INTENSITY FACTORS IN PLATE STRUCTURES CALCULATED BY THE FINITE ELEMENT METHOD (IN JAPANESE)	71-0788
INTENSITY	STRESS INTENSITY FACTORS FOR HEXAGONAL FRACTURE TOUGHNESS SPECIMENS	71-0799
INTENSITY	STRESS INTENSITY FACTORS FOR CRACKS IN A PLATE CONTAINING A HOLE AND IN A SPINNING DISC	73-0467
INTENSITY	STRESS INTENSITY FACTORS FOR ANISOTROPIC FRACTURE TEST SPECIMENS OF SEVERAL GEOMETRIES	74-0293
INTENSITY	STRESS INTENSITY FACTORS FOR THICK-WALLED CYLINDERS	74-0300
INTENSITY	STRESS INTENSITY FACTOR OF A SURFACE CRACK IN A PRESSURIZED CYLINDER	75-0255
INTENSITY	STRESS INTENSITY FACTORS FOR A NOTCHED BEAM ON AN ELASTIC FOUNDATION	75-0566
INTENSITY	STRESS, DISPLACEMENT, LINE INTEGRAL AND CLOSURE ENERGY DETERMINATIONS OF CRACK TIP STRESS INTENSITY	74-0621
INTENSITY	THE EFFECT OF GRAVITY UPON THE MELT-THROUGH TIME OF A SOLID SUBJECTED TO A HIGH INTENSITY LASER	72-0693
INTENSITY	THE FINITE ELEMENT CALCULATION OF STRESS INTENSITY FACTORS USING SUPERPOSITION	75-0886
INTENSITY	THE SIMPLE METHOD OF EVALUATION OF STRESS INTENSITY FACTOR USING THE FINITE ELEMENT METHOD	74-0829
INTENSITY	USE OF SINGULARITY PROGRAMMING IN FINITE ELEMENT CALCULATIONS OF ELASTIC STRESS INTENSITY FACTORS. P	75-0252
INTERACTING	FINITE ELEMENT STRESS ANALYSIS OF INTERACTING FUEL PELLET AND CANNING UNDER FAST REACTOR CONDITION	73-0814
INTERACTING	FINITE ELEMENT TECHNIQUES APPLIED TO CRACKS INTERACTING WITH SELECTED SINGULARITIES	75-0196
INTERACTION	A FINITE ELEMENT APPROACH TO ELASTIC SOIL-STRUCTURE INTERACTION	70-0013
INTERACTION	AIRCRAFT-PAVEMENT INTERACTION STUDIES, PHASE 1: A FINITE ELEMENT MODEL OF A JOINTED CONCRETE PAVEMEN	73-0894
INTERACTION	AIRCRAFT-PAVEMENT INTERACTION STUDIES, PHASE 1: A FINITE ELEMENT MODEL OF A JOINTED CONCRETE PAVEMEN	73-0894
INTERACTION	ANALYSIS OF SOIL-STRUCTURE INTERACTION EFFECTS UNDER SEISMIC EXCITATION	71-0064
INTERACTION	ANALYTIC MODELING OF ROCK-STRUCTURE INTERACTION	72-0808
INTERACTION	ANALYTIC MODELING OF ROCK-STRUCTURE INTERACTION. VOLUME 3. COMPUTER PROGRAM	73-0918
INTERACTION	ANALYTIC MODELING OF ROCK-STRUCTURE INTERACTION. VOLUME 2. USERS GUIDE FOR A COMPUTER PROGRAM	73-0919
INTERACTION	ANALYTIC MODELING OF ROCK-STRUCTURE INTERACTION. VOLUME 1	73-0920

257

INTERACTION	APPLICATION OF FINITE ELEMENT METHOD TO SOIL-FOUNDATION INTERACTION ANALYSES	74-1080
INTERACTION	DYNAMICS OF BUILDING - SOIL INTERACTION	72-0856
INTERACTION	DYNAMICS OF EARTH DAMS WITH FOUNDATION INTERACTICN	71-0343
INTERACTION	EARTHQUAKE ANALYSIS OF GRAVITY DAMS INCLUDING HYDRODYNAMIC INTERACTION	73-0350
INTERACTION	EARTHQUAKE RESPONSE OF GRAVITY DAMS INCLUDING RESERVOIR INTERACTION EFFECTS	72-0675
INTERACTION	EXPERIMENTAL AND THEORETICAL STUDIES OF HULL-DECKHOUSE INTERACTION - 2-LEVEL DECKHOUSE	71-0672
INTERACTION	FINITE ELEMENT ANALYSES OF SOIL-STRUCTURE INTERACTION IN U-FRAME LOCKS	69-0531
INTERACTION	FINITE ELEMENT ANALYSIS OF ORTHOGONAL FRAME-PLATE INTERACTION	70-0508
INTERACTION	FINITE ELEMENT FOR THE INTERACTION OF FRAMES AND SHEAR WALLS	73-0681
INTERACTION	FINITE ELEMENT SOLUTION OF THE INTERACTION OF A PLANE ACOUSTIC BLAST WAVE AND A CYLINDRIC STRUCTURE	74-0790
INTERACTION	FINITE ELEMENT TREATMENT OF SOIL-STRUCTURE INTERACTION PROBLEM FOR NUCLEAR POWER PLANT UNDER SEISMIC	73-0834
INTERACTION	FUNDAMENTAL STUDIES OF MEDIUM-STRUCTURE INTERACTION. REPORT 1. FINITE ELEMENT ANALYSIS OF BURIED CYL	72-0854
INTERACTION	GENERAL PURPOSE PROGRAM FOR FINITE ELEMENT ANALYSIS WITH REFERENCE TO INTERACTION BETWEEN SOFTWARE A	68-0132
INTERACTION	HULL-DECKHOUSE INTERACTION BY FINITE ELEMENT CALCULATIONS	73-0663
INTERACTION	HYSTERETIC DAMPING SYSTEMS APPLIED TO SOIL STRUCTURE INTERACTION BY FINITE ELEMENT METHOD	73-0833
INTERACTION	INFLUENCE OF SOIL/STRUCTUPE INTERACTION PARAMFTERS ON FLOOR RESPONSE SPECTRA	68-0150
INTERACTION	INTERACTION OF AREA COORDINATES IN MATRIX STRUCTURAL ANALYSIS	74-1190
INTERACTION	INTERACTION OF FRAME AND SHEAR WALL WITH OPENINGS	73-1057
INTERACTION	INTERACTION OF ROCK AND TUNNEL LINING	72-0450
INTERACTION	INTERACTION OF SOLID SHEAR WALLS AND THEIR SUPPORTING STRUCTURES.	70-0215
INTERACTION	INTERACTION STRESSES IN COMPOSITE SYSTEMS	72-0661
INTERACTION	INVESTIGATION OF SOIL-STRUCTURE INTERACTION: FLEXIBLE CULVERT PIPES	75-0052
INTERACTION	LINER-MEDIUM INTERACTION IN TUNNELS	75-0221
INTERACTION	OFFSHORE GRAVITY BASED OIL PRODUCTION PLATFORM INTERACTION WITH THE SEA BED	74-0968
INTERACTION	PRESSURE VESSEL WALL THICKNESS LIMITS BASED ON THERMAL STRESS RACHETTING AND CREEP FATIGUE INTERACTI	75-0792
INTERACTION	SILAGE - SILO INTERACTION USING MATERIAL CHARACTERIZATION AND FINITE ELEMENT ANALYSIS	75-0561
INTERACTION	SIMPLIFIED CALCULATION OF THE INTERACTION FORCES BETWEEN HULL AND TANK SYSTEM FOR A MOSS ROSENBERG T	74-1052
INTERACTION	SIMPLIFIED SOIL-STRUCTURE INTERACTION ANALYSIS WITH STRAIN DEPENDENT SOIL PROPERTIES	75-0161
INTERACTION	SIMULATION OF FINITE ELEMENT METHOD FOR LOADING PROCESS OF GRAIN CARGO AND INTERACTION WITH SHIP'S S	74-0745
INTERACTION	SOIL-STRUCTURE INTERACTION ANALYSES FOR EVALUATING SEISMIC RESPONSE	74-0954
INTERACTION	SOIL-STRUCTURE INTERACTION - CONTINUUM OR FINITE ELEMENT	75-0344
INTERACTION	STRUCTURALLY COMPATIBLE FLUID FINITE-ELEMENT FOR SOLID FLUID INTERACTION STUDIES	72-0863
INTERACTION	STRUCTURE- MEDIUM INTERACTION PHENOMENOLOGY	69-0166
INTERACTION	STRUCTURE-MEDIUM INTERACTION AND DESIGN PROCEDURES STUDY. VOLUME I. ANALYSIS METHOD, THEORY, VERIFIC	69-0265
INTERACTION	STRUCTURE-MEDIUM INTERACTION AND DESIGN PROCEDURES STUDY. VOLUME IV. FEAT CODE USER'S GUIDE	72-0817
INTERACTION	SUMMARY OF SOIL-STRUCTURE INTERACTION	70-0378
INTERACTION	THEORETICAL INVESTIGATION OF ROCK AND SUPPORT INTERACTION - DEVELOP MORE RATIONAL DESIGN METHODS AND	74-1062
INTERACTION	THREE DIMENSIONAL STRUCTURES/MEDIA INTERACTION. DYNAX DEVELOPMENT AND APPLICATIONS,	72-0612
INTERACTION	TUNNEL DESIGN CONSIDERATIONS: ANALYSIS OF MEDIUM-SUPPORT INTERACTION	73-0552
INTERACTION	WECHSELWIRKUNG ZWISCHEN DEFORMATION UND DURCHSTROIMUNG IM KLUEFTIGEN, ANISOTROPEN GEBIRGE. (INTERACT	74-0724
INTERACTION	ZUSAMMENWIRKEN VON EINZELNEN FERTIGTEILEN ALS GROSSFLAECHIGE SCHEIBE (INTERACTION OF INDIVIDUAL PREF	72-0850
INTERACTION-THE	AIRCRAFT FUEL TANK VULNERABILITY TO HYDRAULIC RAM: MODIFICATTION OF THE NORTHROP FINITE ELEMENT COMP	73-0401
INTERACTIONS	ANALYSIS OF STRUCTURAL-ACOUSTIC INTERACTIONS IN METAL-CERAMIC TRANSDUCERS	75-0068
INTERACTIONS	ANALYSIS OF STRUCTURAL ACOUSTIC INTERACTIONS IN METAL CERAMIC TRANSDUCERS	73-0083
INTERACTIONS	FINITE ELEMENT MODELING OF MAGNETIC FIELD DEFECT INTERACTIONS	73-0260
INTERACTIONS	FUEL ELEMENT STRESSES AT DISCONTINUITIES AND INTERACTIONS BY FINITE ELEMENTS IN TWO DIMENSIONS	71-0074
INTERACTIONS	GENERAL PURPOSE COMPUTER PROGRAM FOR THE DYNAMIC SIMULATION OF VEHICLE GUIDEWAY INTERACTIONS	73-1165
INTERACTIVE	AN INTERACTIVE COMPUTER GRAPHICS PROGRAM FOR NASTRAN	73-0341
INTERACTIVE	AN INTERACTIVE COMPUTER-AIDED MODELING SYSTEM FOR THE DISCRETIZATION OF BRIDGE STRUCTURES FOR FINITE	74-0317
INTERACTIVE	AN INTERACTIVE GRAPHICS SHIP DESIGN ORIENTED PROGRAM PACKAGE FOR TIME SHARING SYSTEMS	73-0384
INTERACTIVE	AN INTERACTIVE HYBRID TECHNIQUE FOR CRASHWORTHY DESIGN OF COMPLEX VEHICULAR STRUCTURAL SYSTEMS	72-0102
INTERACTIVE	APPLICATION OF INTERACTIVE GRAPHICS TO THE NUMERICAL METHODS USED IN STRUCTURAL ANALYSIS	74-0237
INTERACTIVE	DIE ANWENDUNG VON INTERAKTIVER COMPUTER-GRAPHIK BEI DER BERECHNUNG VON TRAGWERKEN. (APPLICATION OF I	74-0155
INTERACTIVE	DIESEL ENGINE COMPONENT DESIGN USING THE FINITE ELEMENT METHOD AND INTERACTIVE GRAPHICS	74-0195
INTERACTIVE	FINITE ELEMENT METHOD FOR DETERMINATION OF THERMAL STRESSES INTERACTIVE GRAPHICS FOR FINITE ELEMENT	75-0274
INTERACTIVE	FINITE ELEMENT MODEL DATA CHECKOUT WITH INTERACTIVE GRAPHICS	74-0592
INTERACTIVE	ICON: THE INTERACTIVE CREATION OF NASTRAN DATA. A SYSTEM DESCRIPTION	74-0681
INTERACTIVE	INEXPENSIVE, INTERACTIVE COMPUTER-AIDED DESIGN APPLIED TO FINITE ELEMENT STRESS ANALYSIS	74-0447
INTERACTIVE	INGA - INTERACTIVE GRAPHIC ANALYSIS	72-0639
INTERACTIVE	INTERACTIVE ANALYSIS OF A RAFTED MULTISTOREY SPACE FRAME RESSTING ON AN INHOMOGENEOUS CLAY STRATUM	74-0909
INTERACTIVE	INTERACTIVE COMPUTER GRAPHICS IN STRUCTURAL ANALYSIS: ONLINE 72	74-1291
INTERACTIVE	INTERACTIVE FINITE ELEMENT DATA GENERATION FOR THREE-DIMENSIONAL PLATE STRUCTURES	74-0139
INTERACTIVE	INTERACTIVE FINITE ELEMENT DATA GENERATION USING THE SUB-STRUCTURING TECHNIQUE	74-0867
INTERACTIVE	INTERACTIVE GRAPHICS - AN AID FOR DEVELOPING FINITE ELEMENT MODELS	73-0040
INTERACTIVE	INTERACTIVE GRAPHICS AIDED STRUCTURAL ANALYSIS	74-0854
INTERACTIVE	INTERACTIVE PROCEDURE FOR CURVILINEAR COORDINATION OF A PLANAR REGION WITH APPLICATIONS TO FINITE-EL	73-1163
INTERACTIVE	INTERACTIVE PROGRAM IN DESIGN AND ANALYSIS OF COMPOSITE MATERIALS	75-0622
INTERACTIVE	INTERACTIVE SOLUTION OF LINEAR EQUATIONS IN FINITE ELEMENT ANALYSIS	69-0293
INTERACTIVE	INTERACTIVE TECHNIQUES WITH GRAPHICAL OUTPUT FOR BRIDGE ANALYSES	69-0347
INTERACTIVE	MAN MACHINE INTERACTIVE STRUCTURAL ANALYSIS AS A PRELIMINARY DESIGN AID	74-0176
INTERACTIVE	THE GROWTH OF INTERACTIVE GRAPHICS AS AN ENGINEERING TOOL.	73-1100
INTERACTIVE	USING INTERACTIVE GRAPHICS FOR THE PREPARATION AND MANAGEMENT OF FINITE ELEMENT DATA	73-0701
INTERAGENCY	REPORT OF INTERAGENCY CONFERENCE ON COORDINATION OF RESEARCH ACTIVITIES	72-0102
INTERAKTIVE	INGA - INTERAKTIVE GRAPHISCHE ANALYSE BENUTZERHAND	74-0698
INTERAKTIVER	DIE ANWENDUNG VON INTERAKTIVER COMPUTER-GRAPHIK BEI DER BERECHNUNG VON TRAGWERKEN. (APPLICATION OF I	57-0042
INTERCEPTOR	AUTOMATED EVALUATION OF INTERCEPTOR ROCKET MOTOR DESIGNS UNDER COMBINED OPERATIONAL AND NUCLEAR EFFE	71-0224
INTERCHANGE	PROCEEDINGS OF THE NOSETIP STRESS ANALYSIS TECHNICAL INTERCHANGE MEETING	73-0696
INTERCHANGEABIL	BASIS FOR ELEMENT INTERCHANGEABILITY IN FINITE ELEMENT PROGRAMS	70-0251
INTERCONNECT	ADVANCED INTERCONNECT SYSTEMS FOR LIGHTWEIGHT SOLAR ARRAYS	74-1195
INTERCONNECTED	AUTOMATIC MESH GENERATION IN TWO AND THREE DIMENSIONAL INTERCONNECTED DOMAINS	75-0727
INTERCONNECTED	ELASTIC ANALYSIS OF SPATIAL SYSTEMS OF INTERCONNECTED SHEAR WALLS AND FRAMES	74-0371
INTERCONNECTED	THE MOTION OF INTERCONNECTED FLEXIBLE BODIES	66-0014
INTERCONNECTORS	DEVELOPMENT OF ADVANCED INTERCONNECTORS FOR SOLAR CELLS	75-0200
INTERELEMENT	THE GENERATION OF INTERELEMENT, COMPATIBLE STIFFNESS AND MASS MATRICES BY THE USE OF INTERPOLATION F	73-0608
INTERESSE	IL METODO DEGLI ELEMENTI FINITI: POSSIBILITA DI APPLICAZIONE A PROBLEMI DI INTERESSE DEGLI ELETTROTE	74-0904
INTERFACE	AN APPLICATION OF THE FINITE ELEMENT METHOD TO TRANSIENT INTERFACE PROBLEMS IN POROUS MEDIA	74-0467
INTERFACE	AN INTRODUCTION TO THE FINITE ELEMENT ANALYSIS OF THE MOTION OF A GAS-LIQUID INTERFACE IN A POROUS M	73-0777
INTERFACE	ANALYSIS OF INTERFACE PROBLEM IN GROUNDWATER FLOW BY FINITE ELEMENT METHOD	74-0164
INTERFACE	FINITE ELEMENT ANALYSIS OF A COMPOSITE MATERIAL INTERFACE	74-0497
INTERFACE	FINITE ELEMENT ANALYSIS OF CRACKS IN REGION OF A BI-MATERIAL INTERFACE	75-0672
INTERFACE	FINITE ELEMENT ANALYSIS OF INTERFACE PROBLEM IN GROUND-WATER FLOW	75-0101
INTERFACE	FINITE ELEMENT ANALYSIS OF THE MOTION OF A GAS-LIQUID INTERFACE IN A POROUS MEDIUM	72-0513
INTERFACE	FINITE ELEMENT SOLUTION OF DEGENERATE INTERFACE PROBLEMS	67-0092
INTERFACE	HIGHER ORDER SINGULARITIES FOR INTERFACE PROBLEMS	74-1209
INTERFACE	INTERFACE CORE DISCRETE-ELEMENT FOR STIFFENED SHELLS OF REVOLUTION	68-0040
INTERFACE	INTERFACE FAILURES IN COMPOSITES	72-0879
INTERFACE	MATRIX AND INTERFACE STRESSES IN A DISCONTINUOUS FIBER COMPOSITE MODEL	74-1007
INTERFACE	MECHANICS OF LOAD TRANSFER AT THE FIBER/MATRIX INTERFACE	75-0427
INTERFACE	NUMERICAL SOLUTION OF THE STEADY INTERFACE IN A CONFINED COASTAL AQUIFER	71-0419
INTERFACE	QUASIOPTIMAL FINITE-ELEMENT METHOD FOR ELLIPTIC INTERFACE PROBLEMS	74-0346
INTERFACE	STRESS CONCENTRATIONS AND SINGULARITIES AT INTERFACE CORNERS	
INTERFACE	THE DETERMINATION OF A WATER/GAS INTERFACE USING FINITE ELEMENTS	

INTERFACE   THE METHOD OF INTERFACE STRESS ADJUSTMENT AND ITS USES IN SOME PLANE ELASTICITY PROBLEMS   61-0005
INTERFACE   THE NUMERICAL SOLUTION OF THE STEADY INTERFACE IN A COASTAL CONFINED AQUIFER   73-0711
INTERFACE   THEORETICAL STUDY OF THE EFFECT OF THE INTERFACE ON COMPOSITE TOUGHNESS   73-0386
INTERFACE   THERMAL STRESSES IN FACT REACTOR BOILER TUBES AT THE SODIUM/GAS INTERFACE   73-0820
INTERFACE   TWO-DIMENSIONAL PHOTOELASTIC STUDY OF STEAM-TURBINE CASING FLANGES WITH INTERFACE RELIEF   74-0603
INTERFACE   WEAPON-MOUNT INTERFACE STUDY - TRIPOD MOUNT M3   72-0845
INTERFACES   COMMENT ON "SOLUTION OF LAPLACE'S EQUATION IN A REGION WITH DIELECTRIC INTERFACES OF ARBITRARY SHAPE   70-0238
INTERFACES   COMPARATIVE STUDY OF FRESH-SALT WATER INTERFACES USING FINITE ELEMENT AND SIMPLE APPROACHES   75-0868
INTERFACES   CONSTRUCTION OF A FINITE ELEMENT APPROXIMATION WHICH CROSSES MATERIAL INTERFACES   74-0896
INTERFACES   DYNAMIC RESPONSE OF SOIL/CONCRETE INTERFACES AT HIGH PRESSURE   74-0836
INTERFACES   FINITE ELEMENT FOR ROCK JOINTS AND INTERFACES   73-0411
INTERFACES   NASTRAN USER INTERFACES, AUTOMATED INPUT INNOVATIONS   71-0126
INTERFACES   NUMERICAL-SOLUTION OF EULERIAN EQUATIONS OF COMPRESSIBLE FLOW BY A FINITE-ELEMENT METHOD WHICH FOLLO   75-0535
INTERFACES   SOLUTION OF PROBLEMS WITH INTERFACES AND SINGULARITIES   74-0822
INTERFACES   SOLUTION OF PROBLEMS WITH INTERFACES AND SINGULARITIES   74-0891
INTERFACES   THE ELASTIC STRESS ANALYSIS OF A BI-MATERIAL PLATE WITH A CRACK NORMAL TO THE INTERFACES   72-0819
INTERFACIAL   ENERGY CONDITION FOR INITIATION OF INTERFACIAL MICROCRACKS AT INCLUSIONS   74-0546
INTERFACIAL   FRACTURE MECHANICS FOR AN INTERFACIAL CRACK BETWEEN ADHESIVELY BONDED DISSIMILAR MATERIALS   72-0869
INTERFACING   INTERFACING FINITE ELEMENT METHODS WITH PRODUCT DESIGN ENGINEERING   73-0189
INTERFERENCE   A METHOD FOR DETERMINING THE SURFACE CONTACT STRESSES RESULTING FROM INTERFERENCE FITS   70-0101
INTERFERENCE   AN ELASTOPLASTIC ANALYSIS OF A UNIAXIALLY LOADED SHEET WITH AN INTERFERENCE - FIT BOLT   74-0687
INTERFERENCE   ELASTIC-PLASTIC ANALYSIS OF INTERFERENCE FIT FASTENERS   74-1177
INTERFERENCE   FINITE ELEMENT COMPUTER PROGRAM FOR THE SOLUTION OF NONLINEAR AXISYMMETRIC CONTACT PROBLEMS WITH INT   74-1094
INTERFERENCE   RECENT EXPERIENCE IN USING FINITE ELEMENT METHODS FOR THE SOLUTION OF PROBLEMS IN AERODYNAMIC INTERF   70-0339
INTERFERENCE   STRESS AND STRAIN DISTRIBUTION IN THE VICINITY OF INTERFERENCE FIT FASTENERS   73-0363
INTERFERENCE-FI   ANALYTICAL AND EXPERIMENTAL INVESTIGATION OF FATIGUE IN A SHEET SPECIMEN WITH AN INTERFERENCE-FIT BO   75-0701
INTERFERENCE-FI   FINITE ELEMENT ANALYSIS OF STRESSES IN SHAFTS DUE TO INTERFERENCE-FIT HUBS   69-0353
INTERFERENCE-FI   INTERFERENCE-FIT FASTENER DISPLACEMENT MEASUREMENT BY SPECKLE PHOTOGRAPHY   74-1178
INTERFERING   DEVELOPMENT AND APPLICATIONS OF SUPERSONIC UNSTEADY CONSISTENT AERODYNAMICS FOR INTERFERING PARALLEL   72-0716
INTERFEROGRAM   IMPROVED METHOD FOR OBTAINING THE GENERAL-DISPLACEMENT FIELD FROM A HOLOGRAPHIC INTERFEROGRAM   72-0015
INTERFEROMETRY   AN INVESTIGATION INTO THE METHODS OF DETECTING DEWETTED ZONES IN SOLID ROCKET PROPELLANTS USING HOLO   73-0890
INTERIOR   BOUNDARY AND INTERIOR APPROXIMATION ERRORS IN THE FINITE ELEMENT METHOD   73-0209
INTERIOR   INTERIOR MAXIMUM NORM ESTIMATES FOR SOME SIMPLE FINITE ELEMENT METHODS   74-1023
INTERIOR   NATURAL FREQUENCIES AND MODE SHAPES OF PLATES WITH INTERIOR CUT-OUTS   70-0402
INTERIOR   SOME INTERIOR SUPER CONVERGENCE ESTIMATES FOR SPLINE APPROXIMATION   73-0641
INTERLAMINAR   ELASTIC AND PLASTIC INTERLAMINAR SHEAR DEFORMATION IN LAMINATED COMPOSITES UNDER GENERALIZED PLANE   71-0158
INTERLAMINAR   FINITE ELEMENT ANALYSIS OF INTERLAMINAR SHEAR FIBROUS COMPOSITES   71-0129
INTERLAMINAR   THREE DIMENSIONAL FINITE ELEMENT ANALYSIS OF THICK LAMINATED COMPOSITES - INCLUDING INTERLAMINAR AND   73-0882
INTERLAMINAR   THREE-DIMENSIONAL FINITE ELEMENT ANALYSIS OF THICK LAMINATED COMPOSITES - INCLUDING INTERLAMINAR AND   74-1151
INTERLAMINATE   INFLUENCE OF INTERLAMINATE FLAWS ON TRANSIENT TEMPERATURE DISTRIBUTIONS   73-0223
INTERMEDIATE   APPROXIMATION AND BOUNDS FOR EIGENVALUES OF VARIOUS DOMAINS USING THE INTERMEDIATE PROBLEM TECHNIQUE   72-0711
INTERMEDIATE   POSTBUCKLING BEHAVIOUR OF A SECTION REPRESENTATIVE OF THE B-1 AFT INTERMEDIATE FUSELAGE   73-0489
INTERMEDIATE   PRINCIPAL STRESSES AT THE INTERMEDIATE SUPPORT OF PRESTRESSED CONCRETE CONTINUOUS BEAMS   70-0352
INTERNAL   A STUDY OF INTERNAL STRESSES IN STATICALLY DEFORMED PNEUMATIC TIRES. A MATHEMATICAL MODEL FOR THE PN   70-0421
INTERNAL   BERECHNUNG DER SCHWEISSEIGENSPANNUNGEN UND SCHWEISSFORMAENDERUNGEN MIT ELASTISCH-PLASTISCHEN FINITEN   75-0271
INTERNAL   CONFORMING FINITE ELEMENT WITH INTERNAL EQUILIBRIUM FOR TWO DIMENSIONAL PROBLEMS   72-0386
INTERNAL   ECONOMICAL DESIGN ANALYSIS OF PISTONS FOR INTERNAL COMBUSTION ENGINES   73-0188
INTERNAL   EFFECT OF BACKPACKING AND INTERNAL PRESSURIZATION ON STRESSES TRANSMITTED TO BURIED CYLINDERS   73-0897
INTERNAL   EFFECTS OF SHOE FORCE GEOMETRY ON HEAVY DUTY INTERNAL SHOE BRAKE PERFORMANCE   68-0219
INTERNAL   FINITE ELEMENT AND EXPERIMENTAL ANALYSIS OF AN INTERNAL FIXATION PLATE MODEL   75-0380
INTERNAL   FINITE ELEMENT SOLUTION OF VISCOUS FLOW IN AN ARBITRARY INTERNAL PASSAGE   71-0461
INTERNAL   INELASTIC BUCKLING ANALYSIS OF PIPES SUBJECTED TO INTERNAL PRESSURE, FLEXURE AND AXIAL LOADING   74-0515
INTERNAL   INTERNAL PRESSURE EFFECTS ON THE VIBRATION OF PARTIALLY FILLED ELASTIC TANKS   75-0241
INTERNAL   INTERNAL SUPPORTING STRUCTURES OF PRESSURE TUBES TYPE REACTOR CORE, ANALYTICAL MODEL FOR STATIC ANAL   71-0049
INTERNAL   INTERNAL SUPPORTING STRUCTURES OF THE (CIRENE) REACTOR CORE - RESULTS OF STATIC STRESS ANALYSIS FOR   73-0818
INTERNAL   METODY RASCHETNOGO OPREDELENIYA TEMPERATURNYKH NAPRYAZHENII V KRYSHKAKH TSILINDROV DVIGATELEI VNUTRE   74-0179
INTERNAL   OPREDELENIE STATSIONARNYKH TEMPERATURNYKH POLEI V DETALYAKH DVIGATELEI VNUTRENNEGO SGORANIYA METODOM   73-0454
INTERNAL   RASCHET POLEI DEFORMATSII I NAPRYAZHENII V DETALYAKH DVIGATELEI VNUTRENNEGO SGORANIYA (CALCULATION O   73-0465
INTERNAL   SIMPLIFIED MINIMUM INTERNAL ENERGY OF DEFORMATION POSTULATE AND ITS APPLICATION TO METALFORMING PROC   74-1155
INTERNAL   SOLUMN INSTABILITY UNDER NONCONSERVATIVE FORCES, WITH INTERNAL AND EXTERNAL DAMPING FINITE ELEMENT U   73-0566
INTERNAL   STRESS ANALYSIS OF A TUBULAR CROSS-JOINT WITHOUT ANY INTERNAL STIFFENING FOR OFFSHORE STRUCTURES   74-0549
INTERNAL   STRESS DISTRIBUTIONS IN THIN-WALLED INTERSECTING CYLINDRICAL SHELLS SUBJECTED TO INTERNAL PRESSURE A   71-0023
INTERNAL   THE BOEING SST PROTYPE INTERNAL LOADS ANALYSIS SYSTEM AND PROCEDURES   71-0103
INTERNAL   THE CHANGES IN THE (TWO-DIMENSIONAL) STRESS ROUNDED CORNERS, CAUSED BY A VARYING INTERNAL PRESSURE A   67-0141
INTERNAL   TRANSVERSE VIBRATION OF BEAMS WITH INTERNAL DAMPING, ROTATORY INERTIA AND SHEAR   63-0020
INTERNAL   USE OF THE ' TITUS ' FINITE ELEMENT CODE FOR THE NUCLEAR STRUCTURES AND INTERNAL STRESS-ANALYSIS   73-0840
INTERNAL   VIBRATION ANALYSIS OF AN INTERNAL REFERENCE UNIT   69-0176
INTERNALLY   FULLY DEVELOPED VISCOUS FLOW IN INTERNALLY FINNED TUBES   75-0648
INTERNALLY   OSCILLATION STABILITY OF INTERNALLY AND EXTERNALLY NOTCHED PLANE BARS OF EQUAL FORM FACTOR FROM 3943   73-0745
INTERPOLATED   (L SUP 2) ERROR ESTIMATES FOR FINITE ELEMENTS WITH INTERPOLATED BOUNDARY CONDITIONS   73-0885
INTERPOLATED   INTERPOLATED BOUNDARY CONDITIONS IN THE FINITE ELEMENT METHOD   75-0388
INTERPOLATED   L2 ERROR ESTIMATES FOR FINITE ELEMENTS WITH INTERPOLATED BOUNDARY CONDITIONS   73-0194
INTERPOLATED   SUBSPACES WITH ACCURATELY INTERPOLATED BOUNDARY CONDITIONS   72-0527
INTERPOLATION   APPLICATION OF THE SMOOTH-SURFACE INTERPOLATION TO THE FINITE ELEMENT ANALYSIS   69-0088
INTERPOLATION   BICUBIC INTERPOLATION OVER TRIANGLES   69-0367
INTERPOLATION   BOUNDS FOR A CLASS OF LINEAR FUNCTIONALS WITH APPLICATIONS TO HERMITE INTERPOLATION   71-0582
INTERPOLATION   CLASS OF FINITE ELEMENTS GENERATED BY LAGRANGE INTERPOLATION - 2.   73-0044
INTERPOLATION   EXTENDED INTERPOLATION IN FINITE ELEMENT ANALYSIS   71-0183
INTERPOLATION   FINITE ELEMENTS WITH HARMONIC INTERPOLATION FUNCTIONS   72-0058
INTERPOLATION   FINITE ELEMENTS WITH HARMONIC INTERPOLATION FUNCTIONS   73-1065
INTERPOLATION   GENERAL LAGRANGE AND HERMITE INTERPOLATION IN R**N WITH APPLICATIONS TO FINITE ELEMENT METHODS   72-0110
INTERPOLATION   HERMITE INTERPOLATION TYPE RITZ METHODS FOR TWO POINT BOUNDARY VALUE PROBLEMS   66-0089
INTERPOLATION   INTERPOLATION POLYNOMIALS ON THE TRIANGLE   70-0235
INTERPOLATION   INTERPOLATION POLYNOMIALS ON THE TRIANGLE ELEMENT   70-0326
INTERPOLATION   INTERPOLATION THEORY OVER CURVED ELEMENTS WITH APPLICATION TO THE FINITE ELEMENT METHOD   72-0356
INTERPOLATION   LAGRANGE INTERPOLATION ON CURVED FINITE ELEMENTS   72-0163
INTERPOLATION   NAVIER STOKES EQUATIONS USING MIXED INTERPOLATION   74-0014
INTERPOLATION   ON A CLASS OF FINITE ELEMENTS GENERATED BY LAGRANGE INTERPOLATION.   72-0101
INTERPOLATION   ON LAGRANGIAN INTERPOLATION IN N VARIABLES   70-0452
INTERPOLATION   PIECEWISE ANALYTIC INTERPOLATION AND APPROXIMATION IN TRIANGULATED POLYGONS   72-0508
INTERPOLATION   PIECEWISE HERMITE INTERPOLATION IN ONE AND TWO VARIABLES WITH APPLICATIONS TO PARTIAL DIFFERENTIAL E   68-0201
INTERPOLATION   PIECEWISE POLYNOMIAL INTERPOLATION AND APPROXIMATION   65-0058
INTERPOLATION   SEMIDISCRETE GALERKIN TECHNIQUES WITH TIME INTERPOLATION AND SPLITTING UP FOR PLASMA SIMULATION   75-0121
INTERPOLATION   SPLINE INTERPOLATION TECHNIQUES FOR VARIATIONAL METHODS   73-1079
INTERPOLATION   THE APPLICATION OF THE SMOOTH-SURFACE INTERPOLATION TECHNIQUE TO THE FINITE-ELEMENT ANALYSIS OF THIN   66-0120
INTERPOLATION   THE GENERATION OF INTERELEMENT, COMPATIBLE STIFFNESS AND MASS MATRICES BY THE USE OF INTERPOLATION F   66-0014
INTERPOLATION   TRANSFINITE ELEMENT METHODS: BLENDING FUNCTION INTERPOLATION OVER ARBITRARY CURVED ELEMENT DOMAINS   73-1006
INTERPOLATION   TRICUBIC POLYNOMIAL INTERPOLATION   71-0377
INTERPOLATIONS   PIECEWISE POLYNOMIAL INTERPOLATIONS AND THEIR APPLICATIONS TO PARTIAL DIFFERENTIAL EQUATIONS   70-0458
INTERPOLATIONS   PIECEWISE POLYNOMIAL INTERPOLATIONS IN THE FINITE ELEMENT METHOD   73-1028
INTERPOLATIONSP   DAS VERFAHREN VON RAYLEIGH-RITZ MIT HERMITESCHEN INTERPOLATIONSPOLYNOMEN   63-0018
INTERPOLATORY   THE USE OF INTERPOLATORY POLYNOMIALS FOR A FINITE ELEMENT SOLUTION OF THE MULTIGROUP DIFFUSION EQUAT   72-0524

259

INTERPRET	PRACTICAL HINTS ON HOW TO UNDERSTAND, MODEL, USE AND INTERPRET FINITE ELEMENTS	73-0141
INTERPRETATION	INTERPRETATION OF COMPLEX STRESS ANALYSIS BY MULTIPLE REGRESSION	74-1203
INTERPRETATION	INTERPRETATION OF FINITE ELEMENT PROCEDURE AS STRESS ERROR MINIMIZATION PROCEDURE	72-0984
INTERPRETATION	INTERPRETATION OF FINITE ELEMENT STRESSES ACCORDING TO ASME SECTION 111	75-0655
INTERPRETATION	MODELLING AND INTERPRETATION OF RESULTS	75-0399
INTERPRETATION	STRUCTURAL BEHAVIOR OF A TWO SPAN REINFORCED CONCRETE BOX GIRDER BRIDGE MODEL. VOLUME 11. REDUCTION,	71-0659
INTERPRETIVE	COMPUTATIONAL METHODS IN NUCLEAR REACTOR STRUCTURAL DESIGN FOR HIGH TEMPERATURE APPLICATIONS: AN INT	73-1001
INTERPRETIVE	STRUCTURAL ANALYSIS AND MATRIX INTERPRETIVE SYSTEM (SAMIS)	66-0098
INTERPRETIVE	STRUCTURAL ANALYSIS AND MATRIX INTERPRETIVE SYSTEM (SAMIS) USER REPORT	67-0125
INTERRELATION	ON THE INTERRELATION BETWEEN STABILITY AND COMPUTATIONS	73-0674
INTERSECTING	FINITE ELEMENT ANALYSIS OF INTERSECTING CYLINDERS	73-0866
INTERSECTING	FINITE ELEMENT SOLUTION OF NORMALLY INTERSECTING CYLINDERS	73-0723
INTERSECTING	STRESS DISTRIBUTIONS IN THIN-WALLED INTERSECTING CYLINDRICAL SHELLS SUBJECTED TO INTERNAL PRESSURE A	71-0023
INTERSECTION	AN ANALYSIS OF AN INTERSECTION PROBLEM USING THE SEMILOOP SHELL ELEMENT	74-0401
INTERSECTION	ANALYSIS OF THE 747 AIRCRAFT WING BODY INTERSECTION	68-0275
INTERSECTION	CALCULATION OF THERMAL STRESSES IN A CYLINDER-CYLINDER INTERSECTION BY MEANS OF FINITE ELEMENTS	73-0571
INTERSECTION	STATIC AND HEAT TRANSFER ANALYSIS OF A SPHERE-CONE INTERSECTION IN A NUCLEAR CONTAINMENT VESSEL	75-0166
INTERSECTION	THE APPLICATION OF THREE-DIMENSIONAL FINITE ELEMENTS TO A CYLINDER-CYLINDER INTERSECTION	70-0068
INTERSECTION	THERMAL STRESSES IN A NOZZLE-DRUM INTERSECTION	73-0824
INTERSECTION	THREE-DIMENSIONAL CRACK ANALYSIS APPLIED TO AN LWR NOZZLE-CYLINDER INTERSECTION	75-0165
INTERSECTION	THREE-DIMENSIONAL HARMONIC FUNCTIONS NEAR TERMINATION OR INTERSECTION OF GRADIENT SINGULARITY LINES:	74-0178
INTERSECTION	VARIATIONAL PRINCIPLES AND DUALISTIC SCHEME FOR INTERSECTION PROBLEMS IN ELASTICITY	69-0305
INTERSECTIONS	DETERMINATION OF STRESS CONCENTRATIONS AT NOZZLE TO CYLINDER INTERSECTIONS AND COMPARISON WITH EXPER	73-0491
INTERSECTIONS	EURCYL: A COMPUTER PROGRAM TO GENERATE FINITE ELEMENT MESHES FOR CYLINDER-CYLINDER INTERSECTIONS	73-0962
INTERSECTIONS	PHOTOELASTIC TESTS ON COMPLEX INTERSECTIONS OF REINFORCED THIN-SHELL CYLINDERS	75-0363
INTERSECTIONS	STRUCTURAL ANALYSIS OF SHELL INTERSECTIONS	69-0217
INTERSTITIAL	IMPEDANCE OF INTERSTITIAL GAPS IN TRANSDUCER ARRAYS	70-0410
INTERTIE	NONLINEAR SEISMIC ANALYSIS OF CURRENT DIVIDER SUPPORT SYSTEM FOR PACIFIC DC INTERTIE	73-0200
INTERVERTEBRAL	A STUDY OF INTERVERTEBRAL DISC MECHANICS BY THE FINITE ELEMENT METHOD	74-1121
INTERVERTEBRAL	FINITE ELEMENT STRESS ANALYSIS OF AN INTERVERTEBRAL DISC	74-0244
INTERVERTEBRAL	NONLINEAR BEHAVIOR OF THE HUMAN INTERVERTEBRAL DISC UNDER AXIAL LOADING	74-0919
INTERVERTEBRAL	NUMERICAL STRESS ANALYSIS OF INTERVERTEBRAL DISK	72-0125
INTRAOCULAR	MATHEMATICAL MODEL OF THE CORNEO-SCLERAL SHELL AS APPLIED TO INTRAOCULAR PRESSURE VOLUME RELATIONS A	72-0471
INTRUSION	MANAGEMENT OF RETARDATION OF SALT WATER INTRUSION IN COASTAL AQUIFERS	75-0692
INTRUSION	NUMERICAL DETERMINATION OF SEA WATER INTRUSION IN AQUIFERS	74-1181
INVARIANT	INVARIANT IMBEDDING AND SOLUTION OF FINITE ELEMENT EQUATIONS	74-0158
INVERSE	AN INVERSE FORM OF RAMBERG - OSGOOD FORMULA	69-0479
INVERSE	DIRECT AND INVERSE HEAT CONDUCTION IN ROCK MATERIALS USING THE FINITE ELEMENT METHOD	74-1098
INVERSE	FINITE ELEMENTS IN THE SOLUTION OF THE INVERSE PROBLEM IN GROUNDWATER FLOW	74-0063
INVERSE	IDENTIFICATION OF AQUIFER CONSTANTS: TESTS OF TWO INVERSE METHODS	74-0051
INVERSE	ON A COMBINED STURM SEQUENCE AND INVERSE ITERATION TECHNIQUE FOR EIGENPROBLEM SOLUTION OF SPINNING S	73-0413
INVERSE	SOLUTION OF A FREE-SURFACE BOUNDARY VALUE PROBLEM USING AN INVERSE FORMULATION AND THE FINITE ELEMEN	75-0466
INVERSE	THE GENERALIZED INVERSE - SOME PROPERTIES, A COMPUTATIONAL ALGORITHM AND A STRUCTURAL APPLICATION	68-0248
INVERSION	ASKA 105 - A TECHNICAL MANUAL FOR THE DIRECT SOLUTION AND INVERSION PACKAGE	68-0255
INVERSION	INVERSION TECHNIQUE FOR STATIC FINITE ELEMENT METHOD	74-0170
INVERSION	STUDY OF 1964 ALASKAN EARTHQUAKE USING FINITE ELEMENT AND GENERALIZED INVERSION TECHNIQUES	72-0003
INVISCID	AN APPLICATION OF LEAST SQUARES FINITE ELEMENT METHODS TO TWO DIMENSIONAL STEADY INVISCID COMPRESSIB	75-0114
INVISCID	BERECHNUNG DES REIBUNGSLOSEN STROEMUNGSFELDES VON DREIDIMENSIONALEN AUFTRIEBSBEHAFTETENTRAGFLUEGELN	73-0533
INVISCID	COMPUTATIONAL METHODS FOR INVISCID AND VISCOUS TWO-AND-THREE DIMENSIONAL FLOW FIELDS	75-0116
INVISCID	COUPLAGE ENTRE L ECOULEMENT AUTOUR D UN ARRIERE-CORPS ET LE JET PROPULSIF EN THEORIE DE FLUIDE PARFA	74-1137
INVISCID	FINITE ELEMENT METHOD FOR A SOLID ENCLOSING AN INVISCID INCOMPRESSIBLE FLUID	73-0090
ION	A DIFFUSION ANALYSIS OF ION IN ELECTROLYSIS BY FINITE ELEMENT METHOD	73-0906
IRON	SIMULATION OF ELASTIC MODULUS AND POISSON'S RATIO OF SPHEROIDAL GRAPHITE CAST IRON	74-0955
IRRADIATION	APPLICATION OF FINITE ELEMENT METHODS FOR THE ANALYSIS OF THERMAL CREEP, IRRADIATION INDUCED CREEP,	71-0242
IRRADIATION	APPLICATION OF THREE-DIMENSIONAL FINITE ELEMENTS FOR ANALYSIS OF IRRADIATION INDUCED STRESSES IN NUC	72-0078
IRRADIATION	STRESS AND DEFORMATION ANALYSIS OF IRRADIATION INDUCED SWELLING	74-0305
IRRADIATION	STRESS AND DEFORMATION ANALYSIS OF IRRADIATION INDUCED SWELLING	74-0882
IRRADIATION	VISCOELASTIC RESPONSE OF GRAPHITE MATERIALS IN IRRADIATION ENVIRONMENTS	70-0314
IRREGULAR	A METHOD FOR THE PREDICTION OF STRESSES IN AN ISOTROPIC INCLUSION OR OREBODY OF IRREGULAR SHAPE	72-0827
IRREGULAR	A NUMERICAL APPROACH TO THE STUDY OF IRREGULAR SHIP MOTIONS	74-0030
IRREGULAR	ACOUSTIC FINITE ELEMENT APPROACH FOR STUDYING BOUNDARY FLEXIBILITY AND SOUND TRANSMISSION BETWEEN IR	73-0096
IRREGULAR	ALGORITHMS FOR THE PRODUCTION OF CONTOUR MAPS OVER AN IRREGULAR TRIANGULAR MESH	72-0825
IRREGULAR	AN ACOUSTIC FINITE ELEMENT APPROACH FOR STUDYING BOUNDARY FLEXIBILITY AND SOUND TRANSMISSION BETWEEN	73-1087
IRREGULAR	AN IRREGULAR TRIANGLE MESH GENERATOR	64-0032
IRREGULAR	DYNAMIC RESPONSE ANALYSIS OF THE IRREGULAR GROUND BY MEANS OF THE FINITE ELEMENT METHOD	74-0860
IRREGULAR	DYNAMIC RESPONSE OF ARBITRARY SHELLS OF REVOLUTION UNDER COMBINED IMPULSIVE NONISOTHERMAL AND IRREGU	70-0365
IRREGULAR	ELASTIC TORSIONAL ANALYSIS OF IRREGULAR SHAPES	65-0020
IRREGULAR	ELASTIC TORSIONAL ANALYSIS OF IRREGULAR SHAPES	66-0078
IRREGULAR	IN VIVO STRESSES IN THE HUMAN LEFT VENTRICULAR WALL: ANALYSIS ACCOUNTING FOR THE IRREGULAR THREE DIM	72-0136
IRREGULAR	LOVE WAVE DISPERSION THROUGH IRREGULAR STRUCTURES USING FINITE-ELEMENT ALGORITHMS	75-0431
IRREGULAR	PROBLEM OF USING THE METHOD OF FINITE ELEMENTS TO CONSTRUCT AN ALGORITHM FOR CALCULATING A THIN-WALL	74-0851
IRREGULAR	THEORY OF A FINITE DIFFERENCE METHOD ON IRREGULAR NETWORKS	74-0375
IRREGULARITIES	EFFECT OF GEOLOGIC IRREGULARITIES OF SEISMIC RESPONSE	74-0898
IRREGULARLY	ANALYSIS OF ACOUSTIC FIELD IN IRREGULARLY SHAPED ROOMS BY FINITE ELEMENT METHOD	73-0133
IRREGULARLY	FINITE ELEMENT ANALYSIS OF IRREGULARLY SHAPED ACOUSTIC FIELD	71-0711
IRREGULARLY	STRUCTURAL ANALYSIS OF TALL BUILDINGS HAVING IRREGULARLY POSITIONED SHEAR WALLS	73-0282
IRREVERSIBLE	FINITE ELEMENT ANALYSIS OF COUPLED IRREVERSIBLE VECTOR PROCESSES	71-0701
IRREVERSIBLE	FINITE ELEMENT FORMULATION OF FINITE DEFORMATIONS AND IRREVERSIBLE THERMODYNAMICS OF NONLINEAR CONTI	69-0207
IRREVERSIBLE	FINITE ELEMENT FORMULATION OF PROBLEMS OF FINITE DEFORMATION AND IRREVERSIBLE THERMODYNAMICS OF NONL	71-0190
IRREVERSIBLE	NONLINEAR THERMOELASTICITY, IRREVERSIBLE THERMODYNAMICS AND ELASTIC INSTABILITY	73-0762
IRRIGIDITE	CALCULATION METHOD FOR THE STUDY OF CRACKED RIGID STRUCTURES (METODO DI CALCOLO PER LO STUDIO DI STR	71-0637
IRROTATIONAL	FINITE ELEMENT ANALYSIS OF IRROTATIONAL FLOWS OF AN IDEAL FLUID	71-0806
IRROTATIONAL	ON SOME FREE SURFACE ON TRANSIENT FLOW PROBLEMS OF SEEPAGE AND IRROTATIONAL FLOW	72-0070
ISD	REPORT ON POST CONF. MEETING ON COMPUTATIONAL ASPECTS OF ASPECTS OF FINITE ELEMENT METHOD ISD STUTTG	72-0475
ISILINDRICHESKI	ISSLEDOVANIE NAPRYAZHENNOGO SOSTOYANIYA REBRISTYKH ISILINDRICHESKIKH OBOLOCHEK S PRYAMOUGOL'NYMI OTV	74-0944
ISLAND	MODELING AN OCEAN POND, A TWO-DIMENSIONAL, FINITE ELEMENT HYDRODYNAMIC MODEL OF NINIGRET POND, CHARL	75-0686
ISLANDS	NUMERICAL INVESTIGATIONS OF LAKE CIRCULATION AROUND ISLANDS BY THE FINITE ELEMENT METHOD	72-0233
ISLANDS	NUMERICAL INVESTIGATION OF LAKE CIRCULATION AROUND ISLANDS BY THE FINITE ELEMENT METHOD	72-0337
ISO-PARAMETRIC	ISO-PARAMETRIC AND ASSOCIATED ELEMENT FAMILIES FOR TWO AND THREE DIMENSIONAL ANALYSIS	69-0285
ISO-PARAMETRIC	TRANSIENT FIELD PROBLEMS: TWO-DIMENSIONAL AND THREE-DIMENSIONAL ANALYSIS BY ISO-PARAMETRIC FINITE EL	70-0001
ISOFINEL	ISOFINEL: ISOPARAMETRIC FINITE ELEMENT CODE FOR ELASTIC ANALYSIS OF TWO-DIMENSIONAL BODIES	75-0676
ISOLATED	BEHAVIOR OF ISOLATED SHEAR WALLS SUBJECTED TO LATERAL LOADS	75-0704
ISOLATED	POST-TENSIONED SLAB BRIDGES ON ISOLATED SUPPORTS	75-0084
ISOPARAMETRIC	A GENERAL APPROACH TO PROBLEMS OF LARGE DEFORMATION, AND PLASTICITY USING ISOPARAMETRIC ELEMENT	71-0273
ISOPARAMETRIC	A QUARTIC, TWO DIMENSIONAL ISOPARAMETRIC ELEMENT	68-0187
ISOPARAMETRIC	A STUDY OF CONVECTIVE-DISPERSION EQUATION BY ISOPARAMETRIC FINITE ELEMENTS. TRANSPORT OF POLLUTANTS	75-0773
ISOPARAMETRIC	A TECHNIQUE FOR DEGENERATING BRICK-TYPE ISOPARAMETRIC ELEMENTS USING HIERARCHICAL MIDSIDE NODES	74-0403
ISOPARAMETRIC	A THREE DIMENSIONAL SOLUTION OF THE TRANSIENT FIELD PROBLEM USING ISOPARAMETRIC FINITE ELEMENTS	72-0601
ISOPARAMETRIC	A THREE DIMENSIONAL "VARIABLE NODE" ISOPARAMETRIC SOLID ELEMENT	74-0779
ISOPARAMETRIC	ACCURACY AND CONDITION OF CURVED (ISOPARAMETRIC) FINITE ELEMENTS	73-0057
ISOPARAMETRIC	AN ASSUMED STRESS APPROACH TO REFINED ISOPARAMETRIC ELEMENTS IN THREE DIMENSIONS	74-0421
ISOPARAMETRIC	AN ASSUMED STRESS VERSION OF THE WILSON 8-NODE ISOPARAMETRIC BRICK	72-0532

ISOPARAMETRIC	AN AUTOMATIC MESH GENERATION SCHEME FOR PLANE AND CURVED SURFACES BY ISOPARAMETRIC COORDINATES	71-0271
ISOPARAMETRIC	AN AUTOMATIC MESH GENERATOR USING TWO AND THREE DIMENSIONAL ISOPARAMETRIC ELEMENTS	73-0066
ISOPARAMETRIC	APPLICATION OF QUADRATIC ISOPARAMETRIC FINITE ELEMENTS IN LINEAR FRACTURE MECHANICS	74-0521
ISOPARAMETRIC	BASIS FOR ISOPARAMETRIC STRESS ELEMENTS	73-0532
ISOPARAMETRIC	BOPACE 3-D (THE BOEING PLASTIC ANALYSIS CAPABILITY FOR 3-DIMENSIONAL SOLIDS USING ISOPARAMETRIC FINI	75-0699
ISOPARAMETRIC	CALCULATION OF TWO AND THREE DIMENSIONAL FLOW PROBLEMS ACCORDING TO SUB-STRUCTURE THEORY WITH ISOPAR	74-0215
ISOPARAMETRIC	COMMENT ON THE PAPER "POSSIBLE LOSS OF ACCURACY IN CURVED (ISOPARAMETRIC) FINITE ELEMENTS" BY HENSHE	72-0279
ISOPARAMETRIC	COMPUTATIONAL TECHNIQUES, NONLINEAR EQUATIONS, ISOPARAMETRIC ELEMENTS, FIVE LECTURES	73-0634
ISOPARAMETRIC	CURVED ISOPARAMETRIC QUADRILATERAL ELEMENTS FOR FINITE ANALYSIS	68-0008
ISOPARAMETRIC	CURVED ISOPARAMETRIC QUADRILATERAL ELEMENTS FOR FINITE ELEMENT ANALYSIS	68-0067
ISOPARAMETRIC	DEGENERATION OF BRICK-TYPE ISOPARAMETRIC ELEMENTS	73-1031
ISOPARAMETRIC	ELASTO-PLASTIC ANALYSIS OF THREE-DIMENSIONAL STRUCTURES USING THE ISOPARAMETRIC ELEMENT	72-0248
ISOPARAMETRIC	ERROR ESTIMATE FOR QUADRILATERAL ISOPARAMETRIC FINITE-ELEMENTS OF TYPE Q1 WHICH CAN DEGENERATE INTO	75-0401
ISOPARAMETRIC	FINITE ELEMENT PROCEDURES FOR SEEPAGE ANALYSIS USING AN ISOPARAMETRIC ELEMENT	72-0366
ISOPARAMETRIC	FURTHER APPLICATION OF QUADRATIC ISOPARAMETRIC FINITE-ELEMENTS TO LINEAR FRACTURE MECHANICS OF PLATE	75-0639
ISOPARAMETRIC	GENERAL MEMBRANE ISOPARAMETRIC ELEMENTS	75-0442
ISOPARAMETRIC	ISOFINEL: ISOPARAMETRIC FINITE ELEMENT CODE FOR ELASTIC ANALYSIS OF TWO-DIMENSIONAL BODIES	75-0676
ISOPARAMETRIC	ISOPARAMETRIC AND ALLIED NUMERICALLY INTEGRATED ELEMENTS - A REVIEW	71-0763
ISOPARAMETRIC	ISOPARAMETRIC AND OTHER NUMERICALLY INTEGRATED ELEMENTS	71-0272
ISOPARAMETRIC	ISOPARAMETRIC ELEMENT FORMS IN FINITE ELEMENT ANALYSIS	71-0502
ISOPARAMETRIC	ISOPARAMETRIC FINITE ELEMENT SYSTEM: A NEW CONCEPT IN FINITE ELEMENT ANALYSIS	68-0223
ISOPARAMETRIC	ISOPARAMETRIC FINITE ELEMENTS IN TWO AND THREE-DIMENSIONAL ANALYSIS	68-0237
ISOPARAMETRIC	ISOPARAMETRIC FINITE-ELEMENT METHODS FOR 2-DIMENSIONAL TRANSPORT CALCULATIONS	75-0354
ISOPARAMETRIC	LOCAL LEAST SQUARES STRESS SMOOTHING FOR PARABOLIC ISOPARAMETRIC ELEMENTS	75-0593
ISOPARAMETRIC	MESH AND CONTOUR PLOT FOR TRIANGLE AND ISOPARAMETRIC ELEMENTS	70-0445
ISOPARAMETRIC	MESH 3D: A THREE-DIMENSIONAL FINITE ELEMENT MESH GENERATOR PROGRAM FOR EIGHT-NODE ISOPARAMETRIC ELEM	73-0972
ISOPARAMETRIC	MIXED DISPLACEMENT 3D ISOPARAMETRIC FINITE ELEMENTS AND A STRAIN ENERGY BASIS FOR STUDIES OF ELEMENT	70-0286
ISOPARAMETRIC	MIXED ISOPARAMETRIC LAMINATED COMPOSITE SHELL ELEMENTS	75-0450
ISOPARAMETRIC	MORE ON REDUCED INTEGRATION AND ISOPARAMETRIC ELEMENTS	72-0548
ISOPARAMETRIC	NUMERICAL INTEGRATION CONSIDERATIONS IN TWO AND THREE DIMENSIONAL ISOPARAMETRIC FINITE ELEMENTS	75-0130
ISOPARAMETRIC	PROBLEMS ASSOCIATED WITH THE CONVERGENCE OF ISOPARAMETRIC AND MIXOPARAMETRIC FINITE ELEMENTS	71-0375
ISOPARAMETRIC	PROGRAMMER'S MANUAL ADDITIONS AND DEMONSTRATION PROBLEMS FOR A THERMOSTRUCTURAL CAPABILITY FOR NASTR	75-0734
ISOPARAMETRIC	PSEUDO ISOPARAMETRIC FINITE ELEMENTS FOR SHELL AND PLATE ANALYSIS	69-0040
ISOPARAMETRIC	REPRESENTATION OF SINGULARITIES WITH ISOPARAMETRIC FINITE ELEMENTS	74-1075
ISOPARAMETRIC	RESTRICTIONS IN THE APPLICATION OF AUTOMATIC GENERATION SCHEMES BY ISOPARAMETRIC CO-ORDINATES	74-1377
ISOPARAMETRIC	SHAPE FUNCTION SUBROUTINE FOR AN ISOPARAMETRIC THIN PLATE ELEMENT	73-0618
ISOPARAMETRIC	SHELL DEFORMATION UNDER TOOL LOADS REPRESENTATION OF SINGULARITIES WITH ISOPARAMETRIC FINITE ELEMENT	74-0365
ISOPARAMETRIC	SOME MODIFICATIONS OF AN ISOPARAMETRIC SHELL ELEMENT	73-1010
ISOPARAMETRIC	SOME NONCONFORMING VARIANTS ON THE TRILINEAR ISOPARAMETRIC BRICK FOR LINEAR ELASTIC PROBLEMS	72-0531
ISOPARAMETRIC	STEADY STATE AND TRANSIENT HEAT CONDUCTION AND STRESS ANALYSES OF TWO AND THREE DIMENSIONAL STRUCTUR	71-0675
ISOPARAMETRIC	STUDY OF CONVECTIVE-DISPERSION EQUATION BY ISOPARAMETRIC FINITE ELEMENTS	75-0074
ISOPARAMETRIC	THE COMBINED EFFECT TO CURVED BOUNDARIES AND NUMERICAL INTEGRATION IN ISOPARAMETRIC FINITE ELEMENT M	72-0357
ISOPARAMETRIC	THE EVOLUTION OF THE ISOPARAMETRIC ELEMENTS	75-0425
ISOPARAMETRIC	THE ISOPARAMETRIC FINITE ELEMENT SYSTEM - A NEW CONCEPT IN FINITE ELEMENT ANALYSIS	69-0139
ISOPARAMETRIC	THE PLANE ISOPARAMETRIC STRUCTURAL ELEMENT	70-0262
ISOPARAMETRIC	THREE NODE ISOPARAMETRIC STRESS AXIAL ELEMENT	72-0992
ISOPARAMETRIC	THREE-DIMENSIONAL ELASTO-STATIC PROBLEMS USING ISOPARAMETRIC FINITE ELEMENTS	71-0751
ISOPARAMETRIC	THREE-DIMENSIONAL ISOPARAMETRIC ELEMENTS AND THEIR USE FOR STATICAL ANALYSIS OF MASSIVE WATER-POWER	71-0364
ISOPARAMETRIC	TWO DIMENSION, PLATE, SHELL AND FINITE PRISM ISOPARAMETRIC ELEMENTS AND THEIR APPLICATIONS	71-0772
ISOPARAMETRIC	USE OF ISOPARAMETRIC FINITE-ELEMENTS IN LINEAR FRACTURE MECHANICS	75-0352
ISOPARAMETRIC	VIBRATION PROBLEMS USING ISOPARAMETRIC SHELL ELEMENTS	72-0549
ISOPARAMETRIC	3-DIMENSIONAL ISOPARAMETRIC ELEMENT IN CURVILINEAR COORDINATES	72-0098
ISOSTATIC	FINITE-ELEMENT MODEL FOR POSSIBLE ISOSTATIC REBOUND IN GRAND CANYON	75-0037
ISOTROPIC	A METHOD FOR THE PREDICTION OF STRESSES IN AN ISOTROPIC INCLUSION OR OREBODY OF IRREGULAR SHAPE	72-0827
ISOTROPIC	COUPLED RESPONSE OF SPATIAL VIBRATORY STRUCTURES MOUNTED TO ISOTROPIC PLATE ELEMENTS	71-0228
ISOTROPIC	CREEP ANALYSIS OF TRANSVERSELY ISOTROPIC BODIES SUBJECTED TO TIME DEPENDENT LOADING	73-0170
ISOTROPIC	DETERMINATION OF STRESSES DUE TO DISCONTINUITIES IN FINITE PLATES OF ISOTROPIC AND ORTHOTROPIC MATER	73-1052
ISOTROPIC	ELASTIC SETTLEMENT OF A CIRCULAR RAFT IN ADHESIVE CONTACT WITH A TRANSVERSELY ISOTROPIC MEDIUM	75-0828
ISOTROPIC	NX BOREHOLE JACK MODULUS DETERMINATIONS IN HOMOGENEOUS, ISOTROPIC, ELASTIC MATERIALS	74-0295
ISOTROPIC	THE FINITE ELEMENT METHOD FOR ANALYSIS OF ELASTIC ISOTROPIC AND ORTHOTROPIC SLABS	64-0014
ISOTROPIC-KINEM	ISOTROPIC-KINEMATIC HARDENING MODEL FOR ELASTIC-PLASTIC CYCLIC STRUCTURAL ANALYSIS	75-0210
ISPITIVANJIMA	DOSTIGNUCA I TENDENCIJE U ISPITIVANJIMA ALATNIH MASINA (ACHIEVEMENTS AND TENDENCIES IN MACHINE TOOLS	73-0426
ISPOLZOVANIE	APPLICATION OF THE METHOD OF FINITE ELEMENTS TO THE CALCULATION OF SHIP STRUCTURES (ISPOLZOVANIE MET	73-0933
ISSLEDOVANIE	ISSLEDOVANIE NAPRYAZHENNOGO SOSTOYANIYA REBRISTYKH ISILINDRICHESKIKH OBOLOCHEK S PRYAMOUGOL'NYMI OTV	74-0944
ISSLEDOVANIE	ISSLEDOVANIE NAPRYAZHENNO-DEFORMIROVANNOGO SOSTOYANIYA SIMMETRICHNYKH DISKOV SO STUPITSEI. (INVESTIG	74-0965
ISSLEDOVANIE	ISSLEDOVANIE PROCHNOSTI I ZHESTKOSTI PLASTIN, NAGRUZHENNYKH V SVOEI PLOSKOSTI, METHODOM KONECHNYKH E	73-0382
ISTRAN	STUDY ON FINITE ELEMENT METHOD FOR STRUCTURAL ANALYSIS - ISTRAN/PL (INH STRUCTURE ANALYSIS PLATE STR	73-0167
ISTRAN-S	GENERAL PURPOSE PROGRAM FOR SHELL STRUCTURES: ISTRAN-S	75-0511
ITERACJI	ZASTOSOWANIE ITERACJI SEIDLA W METODZIE ELEMENTOW SKONCZONYCCH NA PRZYKLADZIE OBLICZEN STATYCZNYCH P	74-1031
ITERATION	A TEST ON CONVERGENCE OF EIGENSOLUTIONS BY SIMULTANEOUS ITERATION METHOD	73-0317
ITERATION	AN ITERATION METHOD FOR THE NONLINEAR BUCKLING OF FRAME STRUCTURES	66-0107
ITERATION	APPLICATION OF SIMULTANEOUS ITERATION METHOD TO TORSIONAL VIBRATION PROBLEMS	73-0660
ITERATION	DIRECT ITERATION METHOD FOR THE INCORPORATION OF PHASE CHANGE IN FINITE ELEMENT HEAT CONDUCTION PROG	74-0894
ITERATION	MATRIX ITERATION AND ACCELERATION PROCESSES IN FINITE ELEMENT PROBLEMS OF STRUCTURAL MECHANICS	70-0132
ITERATION	ON A COMBINED STURM SEQUENCE AND INVERSE ITERATION TECHNIQUE FOR EIGENPROBLEM SOLUTION OF SPINNING S	73-0413
ITERATION	SIMULTANEOUS EIGENVECTOR ITERATION APPLIED TO LARGE SYSTEMS	74-0680
ITERATION	SOLUTION OF DIFFUSION AND P1 FINITE-ELEMENT EQUATIONS BY ITERATION	75-0404
ITERATIONS	ON THE CONVERGENCE OF SOR ITERATIONS FOR FINITE ELEMENT APPROXIMATIONS TO ELLIPTIC BOUNDARY VALUE PR	71-0449
ITERATIONS	ZASTOSOWANIE ITERACJI SEIDLA W METODZIE ELEMENTOW SKONCZONYCCH NA PRZYKLADZIE OBLICZEN STATYCZNYCH P	74-1031
ITERATIVE	A COMPUTATIONAL PROCEDURE FOR GRADIENT ITERATIVE TECHNIQUES IN THE FINITE ELEMENT METHOD	68-0229
ITERATIVE	AN EFFICIENT ITERATIVE PROCEDURE FOR USE WITH THE FINITE-ELEMENT METHOD	73-1145
ITERATIVE	AN ITERATIVE FINITE ELEMENT METHOD OF VISCOUS FLOW	74-1353
ITERATIVE	AN ITERATIVE IMPROVEMENT FOR FINITE ELEMENT ANALYSIS	74-0428
ITERATIVE	AN ITERATIVE IMPROVEMENT FOR NON-CONFORMING BENDING ELEMENTS	75-0844
ITERATIVE	AN ITERATIVE METHOD FOR LARGE SYSTEMS OF LINEAR EQUATIONS	73-1020
ITERATIVE	APPROXIMATIONS EXTERNES, PAR ELEMENTS FINIS DE LAGRANGE D ORDRE, UN ET DEUX, DU PROBLEME DE DIRICHLE	72-0495
ITERATIVE	CONVERGENCE CRITERIA FOR ITERATIVE PROCESSES	72-0664
ITERATIVE	DUAL ITERATIVE TECHNIQUES FOR SOLVING A FINITE ELEMENT APPROXIMATION OF THE BIHARMONIC EQUATION	75-0480
ITERATIVE	ITERATIVE APPROACH TO FINITE-ELEMENT METHOD IN FIELD PROBLEMS	74-0124
ITERATIVE	ITERATIVE METHOD FOR THE ANALYSIS OF LARGE STRUCTURAL SYSTEMS	68-0236
ITERATIVE	ITERATIVE METHODS FOR FINITE ELEMENT SYSTEMS	70-0196
ITERATIVE	ITERATIVE SOLUTION OF LINEAR SYSTEMS ARISING FROM FINITE ELEMENT TECHNIQUES	75-0126
ITERATIVE	ITERATIVE SOLUTIONS FOR THE FINITE ELEMENT METHOD	72-0820
ITERATIVE	ITERATIVE TECHNIQUE FOR THE STRESS ANALYSIS OF PCR STRUCTURES USING THE FINITE ELEMENT METHOD	64-0026
ITERATIVE	METHOD FOR EXTERNAL APPROXIMATION USING FINITE ELEMENTS FOR DELTA - 2 AND ITERATIVE METHOD FOR SOLVI	72-0282
ITERATIVE	METHOD FOR EXTERNAL APPROXIMATION USING FINITE ELEMENTS FOR DELTA -2 AND ITERATIVE METHOD FOR SOLVIN	72-0283
ITERATIVE	MORE ON GRADIENT ITERATIVE METHODS IN FINITE-ELEMENT ANALYSIS	69-0102
ITERATIVE	NUMERICAL RESULTS FROM THE APPLICATION OF GRADIENT ITERATIVE TECHNIQUES TO THE FINITE ELEMENT VIBRAT	72-0138
ITERATIVE	RATE OF CONVERGENCE FOR THE ITERATIVE APPROACH IN ELASTIC-PLASTIC ANALYSIS OF CONTINUA	73-0518
ITERATIVE	STUDY OF ITERATIVE SOLUTIONS FOR FINITE ELEMENT METHOD	71-0453
ITERATIVE	THE COMPARISON OF ITERATIVE AND DIRECT SOLUTION TECHNIQUES IN THE ANALYSIS OF TIME-DEPENDENT STRESS	72-0080
IZGIBNO-KRUTILN	RASCHIOT USTOICHIVOSTI IZGIBNO-KRUTILNYKH KOLEBANIJ KRULA V DOZVUKOVOM POTOKE METODOM KONECHNYKH ELF	72-0903

J-INTEGRAL	COMBINED ANALYTICAL - EXPERIMENTAL FRACTURE STUDY OF THE TWO LEADING THEORIES OF ELASTIC-PLASTIC FRA	73-0951
J-INTEGRAL	FINITE ELEMENT CRACK ANALYSIS USING THE J-INTEGRAL METHOD	73-1106
J-INTEGRAL	FINITE ELEMENT CRACK TIP MODELING WHEN EVALUATING J-INTEGRAL	75-0637
J-INTEGRAL	NUMERICAL AND EXPERIMENTAL INVESTIGATION ON THE USE OF J-INTEGRAL	73-0245
J-INTEGRALS	A NUMERICAL AND EXPERIMENTAL INVESTIGATION OF THE USE OF J-INTEGRALS	72-0858
JACK	NX BOREHOLE JACK MODULUS DETERMINATIONS IN HOMOGENEOUS, ISOTROPIC, ELASTIC MATERIALS	74-0295
JAIL	STRESSES AND DEFORMATIONS IN JAIL GULCH EMBANKMENT	73-0391
JAPAN	FINITE ELEMENT ANALYSIS OF SHELL STRUCTURES IN JAPAN	69-0151
JAPAN	RECENT DEVELOPMENTS IN MATRIX DISPLACEMENT METHOD FOR ELASTIC-PLASTIC PROBLEMS IN JAPAN	71-0265
JAPAN	RECENT DEVELOPMENTS OF FINITE ELEMENT METHODS IN JAPAN	75-0226
JAPAN	SOME PROGRAM DEVELOPMENTS OF MATRIX STRUCTURAL ANALYSIS IN JAPAN	72-0322
JAPAN	STUDY OF ELASTIC-PLASTIC MATRIX ANALYSIS FOR FOUNDATION AND REINFORCED CONCRETE IN JAPAN	69-0278
JAPANESE	APPLICATION OF FINITE ELEMENT METHOD TO BOILING WATER REACTOR POWER PLANT (JAPANESE)	72-0816
JAPANESE	APPLICATION OF THE FINITE ELEMENT METHOD TO PRESSURE VESSEL DESIGN (JAPANESE)	70-0446
JAPANESE	CODE FOR STRESS ANALYSIS BY THE THREE DIMENSIONAL FINITE ELEMENT METHOD, (JAPANESE)	69-0509
JAPANESE	EXAMPLES OF STRESS ANALYSIS OF STRUCTURES BY THE FINITE ELEMENT METHOD, (JAPANESE)	70-0450
JAPANESE	FINITE ELEMENT ANALYSIS IN NUCLEAR POWER PLANTS (JAPANESE)	72-0685
JAPANESE	STRESS INTENSITY FACTORS IN PLATE STRUCTURES CALCULATED BY THE FINITE ELEMENT METHOD (IN JAPANESE)	71-0788
JAPANESE	THE ANALYSIS OF STRESS DISTRIBUTION AROUND THE TIP OF A CRACK BY FINITE ELEMENT METHOD WITH THE AID	71-0517
JET	COMPUTER PROGRAM JET 3 TO CALCULATE THE LARGE ELASTIC PLASTIC DYNAMICALLY INDUCED DEFORMATIONS OF FR	72-0609
JET	COUPLAGE ENTRE L ECOULEMENT AUTOUR D UN ARRIERE-CORPS ET LE JET PROPULSIF EN THEORIE DE FLUIDE PARFA	74-1137
JET	FINITE ELEMENT ANALYSIS OF JET IMPINGEMENT ON AXISYMMETRIC CURVED DEFLECTORS	75-0396
JET	FLEXUAL RIGIDITY OF A THIN WALLED BUILD-UP ROTOR FOR JET ENGINE. MEASUREMENT BY STATIC LOAD TEST AND	73-0756
JET	JET IMPINGEMENT ON AXISYMMETRIC CURVED DEFLECTORS	74-0017
JET	JET NOISE EXCITATION OF AN INTEGRALLY STIFFENED PANEL	71-0546
JET	RESPONSE OF STRUCTURE TO THE PSEUDO-SOUND FIELD OF A JET (USING A COMBINED CONTINUUM AND FINITE ELEM	67-0044
JET	SOLUTION OF VISCOUS INCOMPRESSIBLE JET AND FREE-SURFACE FLOWS USING FINITE ELEMENT METHODS	74-0357
JET	TEST RESULTS FROM THE BOUNDARY LAYER FACILITY - RESPONSE OF STRUCTURE TO THE PSEUDO-SOUND FIELD OF A	68-0262
JET-WING	JET-WING LIFTING - SURFACE THEORY USING ELEMENTARY VORTEX DISTRIBUTIONS	75-0245
JETS	COUPLAGE ENTRE L ECOULEMENT AUTOUR D UN ARRIERE-CORPS ET LE JET PROPULSIF EN THEORIE DE FLUIDE PARFA	74-1137
JETS	FINITE ELEMENT METHOD FOR AXISYMMETRIC JETS OF FLUID WITH GRAVITY FORCE	75-0428
JOINT	COMPARISON OF THEORETICAL AND EXPERIMENTAL SHEAR STRESS IN THE ADHESIVE LAYER OF A LAP JOINT MODEL	74-1176
JOINT	FINITE-ELEMENT ANALYSIS OF AN ELECTRODE-NIPPLE JOINT	75-0843
JOINT	NUMERICAL ANALYSES OF DEFORMABILITY TESTS IN JOINTED ROCK. ' JOINT PERTURBATION ' AND ' NO TENSION	71-0439
JOINT	POSSIBILITY OF USING FINITE ELEMENT METHOD IN THE ANALYSIS OF JOINT PERFORMANCE OF THE HYDRAULIC STR	72-0418
JOINT	PREDICTION OF THE CREEP BEHAVIOUR OF A FLANGED JOINT	74-0757
JOINT	STRESS ANALYSIS OF A BOLT-NUT JOINT BY THE FINITE ELEMENT METHOD AND THE COPPER-ELECTROPLATING METHO	73-0220
JOINT	STRESS ANALYSIS OF A BOLT-NUT JOINT BY FINITE ELEMENT METHOD AND COPPER-ELECTROPLATING METHOD - 2. S	74-0153
JOINT	STRUCTURAL MODELING OF A WELDED JOINT TYPICAL FOR VEHICLE FRAMES	73-0539
JOINT	THE FLEXIBILITY OF A TUBULAR WELDED JOINT IN A VEHICLE FRAME	74-0209
JOINT	WELDED JOINT STIFFNESS OBTAINED USING SOLID FINITE ELEMENTS	75-0190
JOINT-FLOOR	RELATIONSHIP BETWEEN FINITE ELEMENT AND MEASURED JOINT-FLOOR CONNECTION RIGIDITIES	75-0738
JOINTED	A DESIGN PROCEDURE FOR HIGH CUTS IN JOINTED HARD ROCK. THREE DIMENSIONAL SOLUTIONS. VOLUME 1	71-0656
JOINTED	A MODEL FOR THE MECHANICS OF JOINTED ROCK	68-0079
JOINTED	AIRCRAFT-PAVEMENT INTERACTION STUDIES. PHASE 1: A FINITE ELEMENT MODEL OF A JOINTED CONCRETE PAVEMEN	73-0894
JOINTED	AN INVESTIGATION INTO THE USE OF FINITE ELEMENT METHODS FOR ANALYSING STRESS DISTRIBUTIONS IN BLOCK	71-0530
JOINTED	ANALYSIS AND MODEL STUDIES OF UNDERGROUND OPENINGS IN JOINTED ROCK. RESEARCH ON ROCK BOLT REINFORCEM	71-0438
JOINTED	ANALYSIS AND MODEL STUDIES OF UNDERGROUND OPENINGS IN JOINTED ROCK. RESEARCH ON ROCK BOLT REINFORCEM	71-0750
JOINTED	ANALYSIS OF NONLINEAR PROBLEMS IN ROCK MECHANICS WITH PARTICULAR REFERENCE TO JOINTED ROCK SYSTEMS	70-0282
JOINTED	CONTINUUM CHARACTERIZATION OF JOINTED ROCK MASSES - 1, 2	73-0196
JOINTED	DESIGN OF ROOF BOLT PATTERNS FOR JOINTED ROCK	74-1318
JOINTED	DUPLICATION OF DILATANT BEHAVIOR IN THE ANALYSIS OF JOINTED ROCKS	72-0757
JOINTED	FINITE ELEMENT ANALYSES OF SLOPES IN JOINTED ROCK	68-0269
JOINTED	NONLINEAR FINITE ELEMENT ANALYSIS OF MULTI JOINTED ROCK MASSES	72-0956
JOINTED	NUMERICAL ANALYSES OF DEFORMABILITY TESTS IN JOINTED ROCK. ' JOINT PERTURBATION ' AND ' NO TENSION '	71-0439
JOINTED	NUMERICAL AND PHYSICAL MODELLING OF REINFORCEMENT SYSTEMS FOR TUNNELS IN JOINTED ROCK	73-0872
JOINTED	STRAIN DISTRIBUTION AROUND UNDERGROUND OPENINGS. FINITE ELEMENT ANALYSIS OF JOINTED SYSTEMS	72-0838
JOINTED	THREE DIMENSIONAL FINITE ELEMENT ANALYSIS OF JOINTED ROCK SLOPES	70-0502
JOINTING	EXPERIMENTS IN FORMING AND JOINTING OF TITANIUM AND TITANIUM ALLOYS FOR AIRCRAFT AND SPACE VEHICLES	70-0134
JOINTLY	RASCHET METODOM KONECHNYKH ELEMENTOV DINAMICHESKIKH KHARAKTEERISTIK ZEMLYANOI PLOTINY SOVMESTNO S OS	73-0437
JOINTS	A COMPUTER PROGRAM FOR THE ANALYSIS OF TUBULAR K JOINTS	69-0488
JOINTS	ANALYSIS OF BONDED JOINTS IN VEHICULAR STRUCTURES	73-0375
JOINTS	APPLICATION OF PROGRAM FATIQUE TEST TO MEMBER JOINTS OF HULLS	72-0814
JOINTS	AREAS OF CONTACT AND PRESSURE DISTRIBUTION IN BOLTED JOINTS	71-0329
JOINTS	BALANCED STRESSES IN POST-YIELDED MULTI-MATERIAL STRUCTURAL JOINTS	69-0271
JOINTS	CALCULATION AND DESIGN OF JOINTS MADE FROM COMPOSITE MATERIALS	70-0156
JOINTS	COMBINED MODE CRACK EXTENSION IN ADHESIVE JOINTS	72-0111
JOINTS	CYCLIC PLASTIC ANALYSIS OF STRUCTURAL STEEL JOINTS	73-0799
JOINTS	DEVELOPMENT OF AN UNDERSTANDING OF THE FATIQUE PHENOMENA OF BONDED AND BOLTED JOINTS IN ADVANCED FIL	72-0798
JOINTS	DISCONTINUITY STRESS DECAY IN OFFSHORE PLATFORM JOINTS	74-0491
JOINTS	DISTRIBUTION OF THE CLAMPING PRESSURE IN FRICTION-GRIP BOLTED JOINTS	74-0250
JOINTS	ELASTIC AND PLASTIC STRESS DISTRIBUTION ON BASE PLATES IN THE NEIGHBOURHOOD OF FILLET WELDS IN LAP J	70-0175
JOINTS	EXPERIMENTELL UNDERSOKNING AV ELEMENTFOGARS HALLFASTHETSEGENSKAPER (EXPERIMENTAL INVESTIGATION OF ST	72-0192
JOINTS	FINITE ELEMENT ANALYSIS OF TUBULAR JOINTS: A REPORT ON A FEASIBILITY STUDY	67-0147
JOINTS	FINITE ELEMENT ANALYSIS OF TUBULAR K JOINTS	70-0427
JOINTS	FINITE ELEMENT ANALYSIS OF LUBRICATION IN NATURAL JOINTS	74-0086
JOINTS	FINITE ELEMENT ANALYSIS OF ELASTO-PLASTIC TEE JOINTS	74-1313
JOINTS	FINITE ELEMENT FOR ROCK JOINTS AND INTERFACES	73-0411
JOINTS	FRACTURE MECHANICS APPROACH TO ADHESIVE JOINTS	71-0726
JOINTS	FRACTURE MECHANICS APPROACH TO ADHESIVE JOINTS	72-0181
JOINTS	INFLUENCE OF JOINTS IN PANELIZED STRUCTURAL SYSTEMS	72-0204
JOINTS	MECHANICS OF FRACTURE IN ADHESIVE JOINTS	72-0691
JOINTS	ON STRENGTH OF TRANSVERSE WEBS AND THEIR JOINTS OF LARGE TANKER	69-0327
JOINTS	ON THE STRESS ANALYSIS AT JOINTS OF DEEP BEAMS	67-0055
JOINTS	ON THE STRUCTURAL DESIGN OF TUBULAR JOINTS	74-0492
JOINTS	STRENGTH OF TRANSVERSE FILLET WELDED JOINTS	74-0144
JOINTS	STRESS ANALYSIS OF ADHESIVE-BONDED LAP JOINTS	74-0548
JOINTS	STRESS ANALYSIS OF TWO JOINTS BY THE FINITE-ELEMENT METHOD	72-0942
JOINTS	STRESS ANALYSIS OF TUBULAR JOINTS FOR OFFSHORE STRUCTURES	73-0207
JOINTS	STRESS AND DEFLECTION ANALYSIS OF MECHANICALLY FASTENED JOINTS	70-0067
JOINTS	STRESS CONCENTRATION FACTORS FOR BONDED LAP JOINTS	71-0842
JOINTS	STRUCTURAL DYNAMIC PROPERTIES OF TACTICAL MISSILE JOINTS - PHASE 1	70-0423
JOINTS	STRUCTURAL DYNAMIC PROPERTIES OF TACTICAL MISSILE JOINTS - PHASE 3	74-0811
JOINTS	THE ROLE OF MECHANICS IN DESIGN-STRUCTURAL JOINTS	74-0767
JOINTS	THE ROLE OF MECHANICS IN DESIGN - STRUCTURAL JOINTS	74-0770
JOINTS	UNTERSUCHUNGEN ZUM FESTIGKEITSVERHALTEN VON METALLKLEBUNGEN IM HINBLICK AUF DIE ERSTELLUNG EINES PRA	74-0289
JOINTS	WARPING AND DISTORTION AT I-SECTION JOINTS	73-0722
JOIST	A RATIONAL ANALYSIS AND DESIGN PROCEDURE FOR WOOD JOIST FLOOR SYSTEMS	74-0730
JOURNAL	JOURNAL OF RESEARCH OF THE NATIONAL BUREAU OF STANDARDS: C. ENGINEERING AND INSTRUMENTATION	71-0664
JOURNAL	SOLUTION OF THE ELASTOHYDRODYNAMIC FINITE JOURNAL BEARING PROBLEM	72-0033
JOURNAL	STABILITY OF TILTING-PAD JOURNAL BEARINGS	71-0687
JOURNAL	THE EFFECT OF DEFORMATION ON THE BEHAVIOUR OF HYDRODYNAMIC JOURNAL BEARINGS	71-0009

JUNCTION	ANALYSIS OF A PRESSURE-VESSEL JUNCTION BY THE FINITE-ELEMENT METHOD	72-0949
JUNCTION	ANALYTICAL AND EXPERIMENTAL INVESTIGATION OF A NOZZLE TO CYLINDRICAL SHELL JUNCTION	74-0638
JUNCTION	EFFECT OF BEAM-SHEARWALL JUNCTION DEFORMATIONS ON THE FLEXIBILITY OF THE CONNECTING BEAMS	73-0213
JUNCTION	INELASTIC ANALYSIS OF A BRANCH SHELL JUNCTION	74-0682
JUNCTION	REBO: COMPUTER CODE FOR THE GENERATION OF A THREE-DIMENSIONAL FINITE-ELEMENT GRID OF THE NOZZLE-CYLI	74-1310
JUNCTIONS	SHEAR LAG ANALYSIS OF RECTANGULAR FULL-WIDTH TUBE JUNCTIONS	72-0087
JUNCTIONS	STRESS SINGULARITIES AT DIAPHRAGM-WEB-FLANGE JUNCTIONS IN BOX GIRDERS	75-0242
JUNCTIONS	THREE-DIMENSIONAL FINITE ELEMENT ANALYSIS OF THICK-WALLED PIPE-NOZZLE JUNCTIONS WITH CURVED TRANSITI	71-0153
K-31	FINITE ELEMENT MODEL WITH APPLICATIONS TO BUILDINGS K-33 AND K-31	70-0420
K-33	FINITE ELEMENT MODEL WITH APPLICATIONS TO BUILDINGS K-33 AND K-31	70-0420
KACHANOV'S	FINITE ELEMENT SOLUTION OF A NON-LINEAR DIFFERENTIAL EQUATION USING KACHANOV'S AND SUCCESSIVE APPROX	75-0108
KANSAS	THE KANSAS TEST TRACK. NON-CONVENTIONAL TRACK STRUCTURES. DESIGN REPORT	72-0797
KARMAN	A GENERAL SOLUTION OF THE VON KARMAN PLATE EQUATIONS BY THE FINITE ELEMENT METHOD	70-0487
KAROSSERIEN	MODERNE VERFAHREN ZUR UNTERSUCHUNG DER STEIFIGKEITS-UND FESTIGKEITSVERHALTNISSE VON RAHMENKORPERN, I	69-0401
KEGELDRUCKVERSU	BERECHNUNG DES KEGELDRUCKVERSUCHES MIT EINER FINITE-ELEMENT-METHODE. (CALCULATION OF THE CONE INDENT	75-0187
KEHLNAEHTEN	KERBSPANNUNGSANALYSE NACH DER FINITE-ELEMENT-METHODE AN KEHLNAEHTEN. (NOTCH EFFECT ANALYSIS USING TH	75-0673
KELVIN	FINITE ELEMENT VISCOELASTIC STRESS ANALYSIS STUDY USING THE GENERALIZED KELVIN SOLID MODEL	73-0957
KERBEN	PARAMETRISCHE UNTERSUCHUNGEN VON KERBEN MIT DER MATRIZENVERSCHIEBUNGSMETHODE UND BILDSCHIRMUNTERSTUT	72-0638
KERBSPANNUNGSAN	KERBSPANNUNGSANALYSE NACH DER FINITE-ELEMENT-METHODE AN KEHLNAEHTEN. (NOTCH EFFECT ANALYSIS USING TH	75-0673
KERBSPANNUNGSUN	PARAMETRISCHE KERBSPANNUNGSUNTERSUCHUNG AM ELASTISCHEN KERN.(PARAMETRICAL INVESTIGATION OF NOTCH STR	72-0208
KERBSPANNUNGSUN	PARAMETRISCHE KERBSPANNUNGSUNTERSUCHUNGEN AN DER LOCHSCHEIBE MIT DER METHODE DER FINITEN ELEMENTE. (	75-0503
KERN	PARAMETRISCHE KERBSPANNUNGSUNTERSUCHUNG AM ELASTISCHEN KERN.(PARAMETRICAL INVESTIGATION OF NOTCH STR	72-0208
KERNEL	KERNEL FUNCTION FOR NONPLANAR OSCILLATING SURFACES IN SUPERSONIC FLOW	71-0472
KERNEL	SARD KERNEL THEOREMS ON TRIANGULAR AND RECTANGULAR DOMAINS WITH EXTENSIONS AND APPLICATIONS TO FINIT	72-0361
KERNEL	THE EXTENSION AND APPLICATION OF SARD KERNEL THEOREMS TO COMPUTE FINITE ELEMENT ERROR BOUNDS	72-0519
KERNKRAFTWERKEN	BERECHNUNG DER ERDBEBENSCHWINGUNGEN VON STRUKTUREN MIT DER FINITE ELEMENT METHODE, MECHANISCHE MODELL	74-0170
KERNKRAFTWERKS	BERECHNUNGSMETHODEN ZUR ERMITTLUNG VON BELASTUNGSBEDINGUNGEN UND GRENZTRAGFAEHIGKEITSVERHALTEN VON K	73-0858
KEYHOLE	FINITE ELEMENT AND CUMULATIVE DAMAGE ANALYSIS OF A KEYHOLE TEST SPECIMEN	75-0163
KEYHOLE	FINITE ELEMENT AND CUMULATIVE DAMAGE ANALYSIS OF A KEYHOLE TEST SPECIMEN	75-0774
KEYWORD	A FINITE ELEMENT BIBLIOGRAPHY - PART II: KEYWORD LISTING	74-0285
KHARAKTEERISTIK	RASCHET METODOM KONECHNYKH ELEMENTOV DINAMICHESKIKH KHARAKTERISTIK ZEMLYANOI PLOTINY SOVMESTNO S OS	73-0437
KILEBANII	RASCHET KILEBANII PLASTIN SLOZHNOI FORMY S NEODNORODNYMI GRANICHNYMI USLOVIYAMI. (CALCULATION OF VIB	74-0988
KINEMATIC	AN AUTOMATIC SYSTEM FOR KINEMATIC ANALYSIS ASKA HIGH SPEED COMPUTING OF ELASTIC STRUCTURES	70-0358
KINEMATIC	AN AUTOMATIC SYSTEM FOR KINEMATIC ANALYSIS, ASKA PART 1	70-0396
KINEMATIC	AN AUTOMATIC SYSTEM FOR KINEMATIC ANALYSIS ASKA PART 1	71-0626
KINEMATIC	APPLICATION OF NATURAL KINEMATIC MODES TO MATRIX ANALYSIS: SMALL AND LARGE DISPLACEMENTS AND STRAINS	67-0138
KINEMATIC	ASKA - AUTOMATIC SYSTEM FOR KINEMATIC ANALYSIS	65-0059
KINEMATIC	ASKA AUTOMATIC SYSTEM FOR KINEMATIC ANALYSIS	69-0307
KINEMATIC	AUTOMATIC SYSTEM FOR KINEMATIC ANALYSIS, NASKA	65-0054
KINEMATIC	DISPLAYS OF KINEMATIC AND ELASTIC SYSTEMS	71-0053
KINEMATICALLY	KINEMATICALLY CONSISTENT UNSTEADY AERODYNAMIC COEFFICIENTS IN SUPERSONIC FLOW	70-0014
KINEMATICALLY	KINEMATICALLY CONSISTENT UNSTEADY AERODYNAMIC COEFFICIENTS IN SUPERSONIC FLOW	70-0316
KINEMATICALLY	THE FINITE ELEMENT ANALYSIS OF PHYSICALLY AND KINEMATICALLY NONLINEAR ELASTIC SOLIDS	69-0146
KINETIC	FINITE ELEMENT ANALYSIS OF A MULTI-COMPONENT KINETIC ENERGY PROJECTILE	74-1180
KINETIC	FINITE ELEMENT ANALYSIS OF A MULTI-COMPONENT KINETIC ENERGY PROJECTILE	75-0754
KIRCHHOFF	FINITE ELEMENT MODEL FOR SHELLS BASED ON THE DISCRETE KIRCHHOFF HYPOTHESIS.	72-0388
KIRCHHOFF	FINITE-ELEMENT SCHEME BASED ON DISCRETE KIRCHHOFF ASSUMPTION	75-0544
KIRCHHOFF	NUMERICAL ANALYSIS OF THIN SHELLS BY CURVED TRIANGULAR ELEMENTS BASED ON DISCRETE - KIRCHHOFF HYPOTH	69-0091
KLEINEN	THEORETICAL AND EXPERIMENTAL CONSIDERATIONS ON FLOW OF VISCOUS LIQUIDS PAST PLATE AT SMALL AND AVERA	70-0144
KLOTZGEBREMSTEN	UNTERSUCHUNGEN ZUR SPURMASSTABILITAET BEI RADSAETZEN MIT KLOTZGEBREMSTEN VOLLRAEDERN (INVESTIGATIONS	75-0810
KLUEFTIGEN	WECHSELWIRKUNG ZWISCHEN DEFORMATION UND DURCHSTROEMUNG IM KLUEFTIGEN, ANISOTROPEN GEBIRGE. (INTERACT	72-0612
KOCHMUS	FINITE ELEMENT COMPUTER ANALYSIS OF KOCHMUS HULL 520.	70-0293
KOERNIGE	PARAMETER EINES ELASTO-PLASTISCHEN STOFFANSATZES FUER KOERNIGE ERDSTOFFE (DETERMINATION OF PARAMETER	74-0657
KOERPER	ACCURATE CALCULATION OF STRESSES IN BEAMS, DISKS AND PLATES BY MEANS OF FINITE ELEMENT METHODS AND I	70-0133
KOERPER	NUMERISCHE BERECHNUNG NICHTLINEARER ELASTISCHER KOERPER	63-0017
KOERPER	ZUR LOESUNG DES KONTAKTPROBLEMS ELASTISCHER KOERPER MIT AUSGEDEHNTER KONTAKTFLAECHE DURCH QUADRATISC	74-1135
KOITER	A KOITER TYPE METHOD FOR FINITE ELEMENT ANALYSIS OF NONLINEAR STRUCTURAL BEHAVIOUR	70-0064
KOITER-TYPE	A KOITER-TYPE METHOD FOR FINITE ELEMENT ANALYSIS OF NONLINEAR STRUCTURAL BEHAVIOR	70-0493
KOITER'S	ADAPTION OF KOITER'S METHOD TO FINITE ELEMENT ANALYSIS OF SNAP-THROUGH BUCKLING BEHAVIOUR	71-0109
KOLBENKUEHLUNG	KOLBENKUEHLUNG (PISTON COOLING)	75-0266
KOLEBANII	CHISLENNYI METOD RASCHETA SOBSTVENNYKH I VYNUZHDENNYKH KOLEBANII SOSTAVNYKH OBOLOCHECHNYKH KONSTRUKT	73-0461
KOLEBANIJ	MODIFICIROVANNYJ KONECHNYJ ELEMENT DIJA RASCHIOTA KOLEBANIJ TONKIKH PLASTIN	72-0891
KOLEBANIJ	RASCHIOT USTOICHIVOSTI IZGIBNO-KRUTILNYKH KOLEBANIJ KRULA V DOZVUKOVOM POTOKE METODOM KONECHNYKH ELE	72-0903
KOLKEWADI	A SEISMIC STRENGTH OF KOLKEWADI DAM	73-0197
KOMBINATION	KOMBINATION EINES INTEGRALGLEICHUNGS VERFAHRENS MIT METHODE DER FINITEN ELEMENTEN ZUR BERECHNUNG EBE	74-0856
KOMBINOVANYCH	VYUZITI METODY KONECNYCH PRVKU PRO PROJEKTOVANI KOMBINOVANYCH KONSTRUCCI (FINITE ELEMENT METHOD IDE	73-0667
KOMPLIZIERTER	NEUESTE ERFAHRUNGEN BEI DER RECHNERISCHEN UND KONSTRUKTIVEN AUSLEGUNG KOMPLIZIERTER SCHALENSYSTEME	71-0052
KOMPRESSIBLEM	IMPORTANCE OF VARIATIONAL CALCULUS IN FLUID MECHANICS AND ITS APPLICATION TO COMPRESSIBLE POTENTIAL	71-1103
KONECHNOGO	CALCULATION OF THE REINFORCED THIN-WALLED STRUCTURES BY THE METHOD OF FINITE ELEMENTS (RASCHET PODKR	75-0685
KONECHNYJ	MODIFICIROVANNYJ KONECHNYJ ELEMENT DIJA RASCHIOTA KOLEBANIJ TONKIKH PLASTIN	72-0891
KONECHNYKH	APPLICATION OF THE METHOD OF FINITE ELEMENTS TO THE CALCULATION OF SHIP STRUCTURES (ISPOLZOVANIE MET	73-0933
KONECHNYKH	ISSLEDOVANIE NAPRYAZHENNOGO SOSTOYANIYA REBRISTYKH ISILINDRICHESKIKH OBOLOCHEK S PRYAMOUGOL'NYMI OTV	74-0944
KONECHNYKH	ISSLEDOVANIE PROCHNOSTI I ZHESTKOSTI PLASTIN. NAGRUZHENNYKH V SVOEI PLOSKOSTI. METHODOM KONECHNYKH E	73-0382
KONECHNYKH	K VOPROSU OB AVTOMATIZATSII ZADANIYA INFORMATSII V METODE KONECHNYKH ELEMENTOV. (PROBLEM OF AUTOMATI	73-0382
KONECHNYKH	METOD KONECHNYKH ELEMENTOV V PRIMENENII K RASCHETU TSILINDRICHESKIKH OBOLOCHEK S PRYAMOUGOL'NYMI OTV	73-0536
KONECHNYKH	METOD KONECHNYKH ELEMENTOV V RASCHETAKH KONSOLIDATSII VODONASYSHCHENNYKH GRUNTOV. (FINITE ELEMENT ME	75-0469
KONECHNYKH	RASCHET METODOM KONECHNYKH ELEMENTOV DINAMICHESKIKH KHARAKTERISTIK ZEMLYANOI PLOTINY SOVMESTNO S OS	73-0437
KONECHNYKH	RASCHIOT USTOICHIVOSTI IZGIBNO-KRUTILNYKH KOLEBANIJ KRULA V DOZVUKOVOM POTOKE METODOM KONECHNYKH ELE	72-0903
KONECHNYKH	RASHET TREKHMERNYKH MAGNITNYC POLEI METODOM KONECHNYKH ELEMENTOV. (CALCULATION OF THREE-DIMENSIONAL	74-0766
KONECHNYKH	REALIZATSIYA METODA KONECHNYKH ELEMENTOV NA EVM DLYA RASCHETA DVUMERNYKH ELEKTRICHESKIKH I MAGNITNYK	74-0251
KONECNYCH	FINITE ELEMENTS METHOD IN SOIL MECHANICS (METODA KONECNYCH PRVKU V MECHANICE SEMIN).	71-0468
KONECNYCH	GEOMETRICKY NELINEARNE ULOHY PLOCHYCH SKRUPIN RIESENE PRIRASTKOVOU METODOU KONECNYCH PRVKOV. (GEOMET	74-1032
KONECNYCH	RIESENIE VAZKOPRUZNYCH ANIZOTROPICKYCH DOSAK METODOU KONECNYCH PRVKOV (SOLUTION OF VISCOELASTIC ANIS	74-0477
KONECNYCH	VELKE PRIEHYBY DOSIEK METODOU KONECNYCH PRVKOV (LARGE DEFLECTIONS OF PLATES ANALYZED BY FINITE ELEME	74-0373
KONECNYCH	VYPOCET KRUHOVYCH LOMENICOVYCH KONSTRUKCII METODOU KONECNYCH PASOV. (CALCULATION OF CIRCULAR FOLDED-	74-0612
KONECNYCH	VYUZITI METODY KONECNYCH PRVKU PRO PROJEKTOVANI KOMBINOVANYCH KONSTRUCCI (FINITE ELEMENT METHOD IDE	73-0667
KONSOLIDATIONSV	DIE BERECHNUNG VON KONSOLIDATIONSVORGAENGEN MIT DER FINITE-ELEMENTE-METHODE. (CALCULATION OF CONSOLID	71-0012
KONSOLIDATSII	METOD KONECHNYKH ELEMENTOV V RASCHETAKH KONSOLIDATSII VODONASYSHCHENNYKH GRUNTOV. (FINITE ELEMENT ME	75-0469
KONSTNATER	ZUR BERECHNUNG, BEMESSUNG UND BEWEHRUNG RECHTECKIGER PLATTEN KONSTNATER DICKE MIT OEFFNUNGEN ODER AU	73-0664
KONSTRUCJI	METHOD SZTYWNYCH ELEMENTOW SKONCZONYCH W OBLICZENIACH KONSTRUCJI OKRETOWYCH. (STIFF FINITE ELEMENT M	74-0923
KONSTRUCKCI	VYUZITI METODY KONECNYCH PRVKU PRO PROJEKTOVANI KOMBINOVANYCH KONSTRUCKCI (FINITE ELEMENT METHOD IDE	73-0667
KONSTRUIEREN	COMPUTER-UNTERSTUTZTES KONSTRUIEREN, PRAKTISCHE BEDEUTUNG DER FINIT-ELEMENT-VERFAHREN	71-0552
KONSTRUKCII	VYPOCET KRUHOVYCH LOMENICOVYCH KONSTRUKCII METODOU KONECNYCH PASOV. (CALCULATION OF CIRCULAR FOLDED-	74-0612
KONSTRUKCJI	OBLICZANIE KONSTRUKCJI TOROIDALNEGO ZBIORNIKA NA WODE. (STATICAL ANALYSIS OF A TOROIDAL WATER RESERV	75-0211
KONSTRUKTION	EINE EINFACHE KONSTRUKTION VON KOORDINATENFUNKTIONEN FUR DIE NUMERISCHE LOSUNG ZWEIDIMENSIONALER RAN	66-0134
KONSTRUKTION	KONSTRUKTION UND BERECHNUNG MIT DER METHODE FINITE ELEMENTE	73-0707
KONSTRUKTION	KONSTRUKTION UND BERECHNUNG MIT FINITEN ELEMENTEN	74-0678
KONSTRUKTION	UEBER DAS VERHALTEN VON SPANNBETONDRUCKBEHAELTERN IM GRENZTRAGFAEHIGKEITSZUSTAND UND DESSEN EINFLU	73-0855
KONSTRUKTIONEN	ZUR SYSTEMATIK NUMERISCHER NAEHERUNGSVERFAHREN IN DER STATIK UND DYNAMIK DER KONSTRUKTIONEN (METHODS	72-0557
KONSTRUKTIV	ZUR GENAUEN BERECHNUNG KONSTRUKTIV ORTHOTROPER PLATTEN (EXACT CALCULATION OF STRUCTURALLY ORTHOTROPI	74-0188
KONSTRUKTIVEN	NEUESTE ERFAHRUNGEN BEI DER RECHNERISCHEN UND KONSTRUKTIVEN AUSLEGUNG KOMPLIZIERTER SCHALENSYSTEME	71-0052
KONSTRUKTSII	CALCULATION OF THE REINFORCED THIN-WALLED STRUCTURES BY THE METHOD OF FINITE ELEMENTS (RASCHET PODKR	75-0685

KONSTRUKTSII	CHISLENNYI METOD RASCHETA SOBSTVENNYKH I VYNUZHDENNYKH KOLEBANII SOSTAVNYKH OBOLOCHECHNYKH KONSTRUKT	73-0461
KONTAKTFLAECHE	ZUR LOESUNG DES KONTAKTPROBLEMS ELASTISCHER KOERPER MIT AUSGEDEHNTER KONTAKTFLAECHE DURCH QUADRATISC	74-1135
KONTAKTPROBLEMS	ZUR LOESUNG DES KONTAKTPROBLEMS ELASTISCHER KOERPER MIT AUSGEDEHNTER KONTAKTFLAECHE DURCH QUADRATISC	74-1135
KONTINUA	CONTINUUM CALCULATION FROM ELASTIC PLASTIC REINFORCED MATERIAL USING A FINITE ELEMENT METHOD (DIE BE	72-0658
KONTINUA	DIE ELASTO-PLASTISCHE BERECHNUNG VON ALLGEMEINEN TRAGWERKEN UND KONTINUA	69-0443
KONVEKTIVEM	TEMPERATURVERLAUF IM QUERSCHNITT GESCHWEISSTER ROHRWAENDE BEI UEBERWIEGEND KONVEKTIVEM WAERMEUEBERGA	75-0628
KONZENTRATIONS	KOMBINATION EINES INTEGRALGLEICHUNGS VERFAHRENS MIT METHODE DER FINITEN ELEMENTEN ZUR BERECHNUNG EBE	74-0856
KOORDINATENFUNK	EINE EINFACHE KONSTRUKTION VON KOORDINATENFUNKTIONEN FUR DIE NUMERISCHE LOSUNG ZWEIDIMENSIONALER RAN	66-0134
KOVECH	APLIKACE MKP NA RESENI PROBLEMU DIFUZE V KOVECH. (APPLICATION OF THE FINITE ELEMENT METHOD TO THE SO	74-0609
KOYNA	THE EARTHQUAKE EXPERIENCE AT KOYNA DAM AND STRESSES IN CONCRETE GRAVITY DAMS	72-0589
KREISRINGSEKTOR	EINE FINITE-ELEMENT-METHODE MIT KREISRINGSEKTORFORMIGEN ELEMENTEN	73-1115
KRON'S	KRON'S METHOD: AN ALGORITHM FOR THE EIGENVALUE ANALYSIS OF LARGE-SCALE STRUCTURAL SYSTEMS	73-0758
KRUGOVOI	TREUGOL'NY I ELEMENT DLYA KRUGOVOI TSILINDRICHESKOI OBOLOCHKI (TRIANGULAR FINITE ELEMENT FOR CIRCULA	73-0493
KRUHOVYCH	VYPOCET KRUHOVYCH LOMENICOVYCH KONSTRUKCII METODOU KONECNYCH PASOV. (CALCULATION OF CIRCULAR FOLDED-	74-0612
KRULA	RASCHIOT USTOICHIVOSTI IZGIBNO-KRUTILNYKH KOLEBANIJ KRULA V DOZVUKOVOM POTOKE METODOM KONECHNYKH ELE	72-0903
KRYSHKAKH	METODY RASCHETNOGO OPREDELENIYA TEMPERATURNYKH NAPRYAZHENII V KRYSHKAKH TSILINDROV DVIGATELEI VNUTRE	74-0179
KSZTALTU	FUNKCJA KSZTALTU NIESPELNIAPACA WARUNKOW CIAGLOSCI W ANALIZIE PLYT METODA ELEMENTOW SKONCZONYCH. (SH	75-0263
KURZZEITBELASTU	ZUSAMMENHANG ZWISCHEN OBERFLAECHENBESCHAFFENHEIT, VERBUND UND SPRENGWIRKUNG VON BEWEHRUNGSSTAEHLEN U	73-0450
LABORATORIES	PLOTS - THE PLOTTING SUBROUTINE INCORPORATED IN THE BALLISTIC RESEARCH LABORATORIES ELECTRONIC SCIEN	72-0803
LACK-OF-FIT	PROBABILISTIC BUCKLING BEHAVIOR OF STRUCTURES WITH RANDOM LACK-OF-FIT AND GEOMETRIC IMPERFECTIONS	73-0862
LAG	SHEAR LAG ANALYSIS OF RECTANGULAR FULL-WIDTH TUBE JUNCTIONS	72-0087
LAG	SHEAR LAG IN STEEL BOX GIRDER BRIDGES	75-0310
LAG	SHEAR LAG IN STIFFENED BOX GIRDERS	70-0243
LAGRANGE	APPLICATION OF THE FINITE ELEMENTS METHOD IN TWO-DIMENSIONAL HYDRODYNAMICS USING THE LAGRANGE VARIAB	73-0976
LAGRANGE	APPROXIMATIONS EXTERNES, PAR ELEMENTS FINIS DE LAGRANGE D ORDRE, UN ET DEUX, DU PROBLEME DE DIRICHLE	72-0445
LAGRANGE	CLASS OF FINITE ELEMENTS GENERATED BY LAGRANGE INTERPOLATION - 2.	73-0044
LAGRANGE	COUPLED SHEAR WALL ANALYSIS BY LAGRANGE MULTIPLIERS	75-0202
LAGRANGE	GENERAL LAGRANGE AND HERMITE INTERPOLATION IN R**N WITH APPLICATIONS TO FINITE ELEMENT METHODS	72-0110
LAGRANGE	LAGRANGE INTERPOLATION ON CURVED FINITE ELEMENTS	72-0163
LAGRANGE	LAGRANGE MULTIPLIER TECHNIQUES IN STRUCTURAL ANALYSIS	65-0045
LAGRANGE	ON A CLASS OF FINITE ELEMENTS GENERATED BY LAGRANGE INTERPOLATION.	72-0101
LAGRANGIAN	A LAGRANGIAN FINITE ELEMENT SOLUTION TO UNSTEADY FLOW	74-1264
LAGRANGIAN	APPLICATION OF THE FINITE ELEMENT METHOD TO TWO DIMENSIONAL LAGRANGIAN HYDRODYNAMICS	74-0016
LAGRANGIAN	FINITE ELEMENT METHOD WITH LAGRANGIAN MULTIPLIERS	73-0046
LAGRANGIAN	LAGRANGIAN AND EULERIAN FINITE ELEMENT FORMULATION FOR LARGE STRAIN ELASTO-PLASTICITY	75-0209
LAGRANGIAN	LAGRANGIAN FINITE ELEMENT AND FINITE DIFFERENCE METHODS FOR POISSON PROBLEMS	74-1276
LAGRANGIAN	ON LAGRANGIAN INTERPOLATION IN N VARIABLES	70-0452
LAGRANGIAN	THE FINITE ELEMENT METHOD WITH LAGRANGIAN MULTIPLIERS	72-0901
LAKE	FINITE ELEMENT APPLICATION IN LAKE CIRCULATION	75-0493
LAKE	FINITE ELEMENT LAKE CIRCULATION AND THERMAL ANALYSIS	74-0043
LAKE	HIGHER-ORDER FINITE ELEMENT ANALYSIS OF LAKE CIRCULATION	73-1002
LAKE	NONLINEAR STRESS DEFORMATION ANALYSES OF LAKE AGASSIZ CLAYS USING FINITE ELEMENT METHOD	74-1072
LAKE	NUMERICAL INVESTIGATIONS OF LAKE CIRCULATION AROUND ISLANDS BY THE FINITE ELEMENT METHOD	72-0233
LAKE	NUMERICAL INVESTIGATION OF LAKE CIRCULATION AND ISLANDS BY THE FINITE ELEMENT METHOD	72-0337
LAKE	WIND DRIVEN LAKE CIRCULATION BY THE FINITE ELEMENT METHOD	70-0041
LAKE	3-D FINITE ELEMENT MODEL FOR LAKE CIRCULATION	71-0157
LAKES	FINITE ELEMENT APPROACH TO HYDROTHERMAL ANALYSIS OF SMALL LAKES	72-0363
LAKES	FINITE ELEMENT CIRCULATION ANALYSIS OF VARIABLE-DEPTH SHALLOW LAKES	73-0727
LAKES	FINITE ELEMENT SOLUTION OF THE WIND DRIVEN CURRENTS AND ITS MASS TRANSPORT IN LAKES	74-1368
LAKES	TRANSIENT HYDROTHERMAL ANALYSIS OF SMALL LAKES	73-0445
LAMBDA	EIGENVALUE PROBLEMS (A - LAMBDA B) X EQUALS 0 FOR SYMMETRIC MATRICES OF HIGH ORDER	74-0205
LAMELLAR	LAMELLAR TEARING IN WELDED GIRDERS	70-0218
LAMINAR	A FINITE-ELEMENT ANALYSIS OF STEADY, TWO-DIMENSIONAL, INCOMPRESSIBLE, LAMINAR FLOW	72-0981
LAMINAR	FINITE ELEMENT ANALYSIS OF LAMINAR CONVECTIVE HEAT TRANSFER IN VERTICAL DUCTS WITH ARBITRARY CROSS-S	75-0555
LAMINAR	GALERKIN METHOD FOR A CLASS OF STEADY, 2-DIMENSIONAL, INCOMPRESSIBLE, LAMINAR BOUNDARY-LAYER FLOWS	75-0437
LAMINAR	LEAST SQUARES FINITE ELEMENT ANALYSIS OF LAMINAR BOUNDARY LAYER FLOWS	74-0616
LAMINAR	LEAST SQUARES FINITE ELEMENT ANALYSIS OF TWO-DIMENSIONAL LAMINAR BOUNDARY-LAYER EQUATION WITH SEPARA	75-0796
LAMINAR	THE NUMERICAL CALCULATION OF LAMINAR BOUNDARY-LAYER SEPARATION	74-0785
LAMINAR	TRANSIENT LAMINAR FLOW IN DUCTS OF ARBITRARY CROSS-SECTION BY FINITE ELEMENT METHODS	72-0013
LAMINAR-FLOW	CONTRIBUTION TO NUMERICAL-SOLUTION OF DEVELOPING LAMINAR-FLOW IN ENTRANCE REGION OF CONCENTRIC ANNUL	75-0448
LAMINATE	HYBRID EFFECT ON BENDING BEHAVIORS OF CFRP-AL LAMINATE	74-1213
LAMINATE	MECHANICAL BEHAVIOR OF A GRAPHITE-EPOXY LAMINATE CONTAINING A HOLE	72-0614
LAMINATED	A COMPACT FINITE-ELEMENT METHOD FOR THE ANALYSIS OF ANISOTROPIC LAMINATED PLATES	73-1166
LAMINATED	A FINITE ELEMENT ANALYSIS OF LAMINATED ORTHOTROPIC PLATES AND SHALLOW SHELLS	70-0123
LAMINATED	A FINITE ELEMENT ANALYSIS OF LAMINATED ISOTROPIC TUBES	70-0319
LAMINATED	A FINITE ELEMENT ANALYSIS INCLUDING TRANSVERSE SHEAR EFFECTS FOR APPLICATIONS TO LAMINATED PLATES	71-0216
LAMINATED	A FINITE ELEMENT MODEL FOR LAMINATED PLATES INCLUDING TRANSVERSE SHEAR DEFORMATION	72-0818
LAMINATED	ANALYSIS OF CIRCULAR AND ANGULAR LAMINATED THICK PLATES	74-0426
LAMINATED	ANALYSIS OF LAMINATED SHELL FOR REVOLUTION	66-0023
LAMINATED	APPROXIMATE THREE-DIMENSIONAL STRESS SOLUTIONS FOR SOME LAMINATED PLATE PROBLEMS USING A HIGHER ORDE	70-0419
LAMINATED	CHARACTERISATION ELASTIQUE D'UN STRATIFIE RESINE-VERRE PAR LA METHODE DES DEPLACEMENTS (DETERMINATIO	75-0272
LAMINATED	COMPUTATION OF VIBRATION MODAL SHAPES AND NATURAL FREQUENCIES OF A GENERAL ORTHOTROPIC LAMINATED, TH	71-0309
LAMINATED	COMPUTER SIMULATION OF LAMINATED ROOF REINFORCED WITH GROUTED BOLTS	73-1045
LAMINATED	EFFICIENT FINITE ELEMENT ANALYSIS OF RECTANGULAR AND SKEW LAMINATED PLATES	73-0436
LAMINATED	ELASTIC AND PLASTIC INTERLAMINAR SHEAR DEFORMATION IN LAMINATED COMPOSITES UNDER GENERALIZED PLANE	71-0158
LAMINATED	ELASTODYNAMIC BEHAVIOR OF LAMINATED ORTHOTROPIC PLATES UNDER INITIAL STRESS	75-0249
LAMINATED	EXPERIMENTAL AND THEORETICAL BENDING DEFLECTIONS OF LAMINATED PLATES	72-0144
LAMINATED	FINITE DEFLECTION DISCRETE ELEMENT ANALYSIS OF SANDWICH PLATES AND CYLINDRICAL SHELLS WITH LAMINATED	69-0236
LAMINATED	FINITE ELEMENT ANALYSIS OF SANDWICH PLATES AND CYLINDRICAL SHELLS WITH LAMINATED FACES	68-0121
LAMINATED	FINITE ELEMENT ANALYSIS OF BENDING-EXTENSIONAL COUPLING IN LAMINATED COMPOSITES	70-0292
LAMINATED	FINITE ELEMENT ANALYSIS OF LAMINATED ANISOTROPIC PLATES INCLUDING TRANSVERSE SHEAR DEFORMATIONS	70-0506
LAMINATED	FINITE ELEMENT SOLUTIONS FOR LAMINATED THICK PLATES	72-0143
LAMINATED	FINITE-ELEMENT ANALYSIS FOR FREE VIBRATION OF GENERAL ANISOTROPIC LAMINATED THIN SHELLS	72-0618
LAMINATED	FINITE-ELEMENT TECHNIQUES FOR THE DYNAMIC ANALYSIS OF UNSYMMETRICALLY LAMINATED PLATES	73-1131
LAMINATED	LAMINATED ORTHOTROPIC PLATES AND SHALLOW SHELLS	71-0400
LAMINATED	LINEAR DYNAMIC ANALYSES OF LAMINATED PLATES AND SHELLS BY THE HYBRID-STRESS FINITE-ELEMENT METHOD	73-0786
LAMINATED	MIXED ISOPARAMETRIC LAMINATED COMPOSITE SHELL ELEMENTS	75-0450
LAMINATED	NATURAL VIBRATIONS OF LAMINATED ORTHOTROPIC SHELLS OF REVOLUTION	70-0367
LAMINATED	NONLINEAR FINITE ELEMENT ANALYSIS OF LAMINATED COMPOSITE SHELLS	74-0751
LAMINATED	NONLINEAR HYBRID STRESS FINITE ELEMENT ANALYSIS OF LAMINATED SHELLS	74-1350
LAMINATED	STATIC, VIBRATION, AND THERMAL STRESS ANALYSES OF LAMINATED PLATE AND SHELLS BY THE HYBRID-STRESS FI	72-0602
LAMINATED	STRESS ANALYSIS OF LAMINATED CFRP PLATES CONTAINING A HOLE	74-1175
LAMINATED	THE FINITE ELEMENT ANALYSIS OF LAMINATED COMPOSITES	71-0731
LAMINATED	THE FINITE ELEMENT ANALYSIS OF LAMINATED COMPOSITES	72-0955
LAMINATED	THREE DIMENSIONAL ANALYSIS OF STRESS CONCENTRATIONS NEAR HOLES IN LAMINATED COMPOSITES	72-0713
LAMINATED	THREE DIMENSIONAL FINITE ELEMENT ANALYSIS OF THICK LAMINATED COMPOSITES - INCLUDING INTERLAMINAR AND	73-0882
LAMINATED	THREE-DIMENSIONAL ANALYSIS FOR THE STRESS DISTRIBUTION NEAR CIRCULAR HOLES IN LAMINATED COMPOSITES	74-0793
LAMINATED	THREE-DIMENSIONAL FINITE ELEMENT ANALYSIS OF LAMINATED COMPOSITES	72-0345
LAMINATED	THREE-DIMENSIONAL FINITE ELEMENT ANALYSIS OF THICK LAMINATED COMPOSITES - INCLUDING INTERLAMINAR AND	74-1151
LAMINATED	VIBRATION ANALYSIS OF LAMINATED PLATES AND SHELLS BY A HYBRID STRESS ELEMENT	73-0120
LAMINATES	A METHOD FOR DETERMINING LOCAL ELASTOPLASTIC STRESS AND STRAIN IN METALLURGICALLY BONDED NOTCHED LAM	74-0686
LAMINATES	ANALYTICAL DISPLACEMENTS AND VIBRATIONS OF CANTILEVERED UNSYMMETRIC FIBER COMPOSITE LAMINATES	75-0736

264

LAMINATES	APPROXIMATE THREE-DIMENSIONAL SOLUTIONS FOR SYMMETRIC LAMINATES UNDER INPLANE LOADING	71-0392
LAMINATES	DETERMINATION OF THE TECHNICAL CONSTANTS OF LAMINATES IN OBLIQUE DIRECTIONS. THEORY AND EXPERIMENTAL	73-0748
LAMINATES	DIFFERENTIAL EXPANSION IN ELASTIC LAMINATES	73-0190
LAMINATES	FINITE-ELEMENT ANALYSIS OF EDGE EFFECTS IN ANGLE-PLY COMPOSITE LAMINATES	74-0843
LAMINATES	STRESS CONCENTRATIONS NEAR HOLES IN LAMINATES	74-0329
LAMINATES	THEORETICAL POST YIELDING BEHAVIOUR OF COMPOSITE LAMINATES - 1. INELASTIC MICROMECHANICS	73-0168
LAMINATES	THEORETICAL POST YIELDING BEHAVIOUR OF COMPOSITE LAMINATES -2.	73-0408
LAND	FINITE ELEMENT ANALYSIS OF LAND SUBSIDENCE	69-0233
LAND	THE USE OF THE FINITE ELEMENT METHOD IN OBTAINING SOIL-WATER PRESSURE AND SOIL-WATER CONTENT DISTRIB	75-0112
LANDING	A FINITE ELEMENT ANALYSIS OF AN AIRCRAFT LANDING MAT ON AN ELASTIC FOUNDATION	68-0280
LANGUAGE	ASKA 105 - A PROBLEM ORIENTED LANGUAGE FOR STRUCTURAL ENGINEERS, VOLUME 11, APPLICATION	69-0461
LANGUAGE	ASKA 105, A PROBLEM ORIENTED LANGUAGE FOR STRUCTURAL ENGINEERS, VOLUME 1, DESIGN	69-0460
LANGUAGES	STRUCTURAL ANALYSIS BY PROBLEM ORIENTED LANGUAGES	71-0623
LAP	ANALYSIS OF THE BIAXIAL STRIP AND SHEAR LAP TESTS FOR SOLID PROPELLANT CHARACTERIZATION	67-0019
LAP	COMPARISON OF THEORETICAL AND EXPERIMENTAL SHEAR STRESS IN THE ADHESIVE LAYER OF A LAP JOINT MODEL	74-1176
LAP	ELASTIC AND PLASTIC STRESS DISTRIBUTION ON BASE PLATES IN THE NEIGHBOURHOOD OF FILLET WELDS IN LAP J	70-0175
LAP	STRESS ANALYSIS OF ADHESIVE-BONDED LAP JOINTS	74-0548
LAP	STRESS CONCENTRATION FACTORS FOR BONDED LAP JOINTS	71-0842
LAPLACE	APPLICATION OF LAPLACE TRANSFORM AND FINITE ELEMENT METHOD TO SOLUTION OF NONSTATIONARY THERMAL COND	74-0171
LAPLACE	FINITE ELEMENT METHOD AND LAPLACE TRANSFORM COMPARATIVE SOLUTIONS OF TRANSIENT HEAT CONDUCTION PROBL	74-0575
LAPLACE	INTEGRAL FINITE-ELEMENT APPROACH FOR SOLVING LAPLACE EQUATION	73-0121
LAPLACE	THE NUMERICAL SOLUTION OF VISCO-ELASTIC PROBLEMS USING LAPLACE TRANSFORMS	74-0436
LAPLACE'S	COMMENT ON "SOLUTION OF LAPLACE'S EQUATION IN A REGION WITH DIELECTRIC INTERFACES OF ARBITRARY SHAPE	70-0238
LAPLACE'S	FORTRAN SUBROUTINES FOR THE SOLUTION OF LAPLACE'S EQUATION OVER A GENERAL REGION IN TWO DIMENSIONS	70-0425
LAPLACE'S	ON THE CONSTRUCTION AND CONVERGENCE OF A FINITE ELEMENT SOLUTION OF LAPLACE'S EQUATION	72-0340
LAPLACE'S	SOLUTIONS OF LAPLACE'S EQUATIONS	61-0002
LAPLACE'S	VARIATIONAL PRINCIPLE FOR THE LAPLACE'S OPERATOR WITH APPLICATION IN THE TORSION OF COMPOSITE RODS	74-1044
LAPLACIAN	LAPLACIAN ELECTROSTATIC FIELD CALCULATIONS BY FINITE ELEMENTS WITH AUTOMATIC GRID GENERATION	73-0084
LARGE	A COMPARATIVE STUDY OF SEVERAL CORE STORAGE SCHEMES FOR LARGE SPARSE POSITIVE DEFINITE MATRICES WITH	72-0678
LARGE	A COMPUTATIONAL TECHNIQUE FOR THE EFFICIENT HANDLING OF LARGE MATRICES	73-1083
LARGE	A FINITE ELEMENT FORMULATION FOR PROBLEMS OF LARGE STRAIN AND LARGE DISPLACEMENT	70-0069
LARGE	A FINITE ELEMENT FORMULATION FOR PROBLEMS OF LARGE STRAIN AND LARGE DISPLACEMENT	70-0069
LARGE	A FINITE ELEMENT FORMULATION FOR LARGE DEFLECTION ELASTIC STRUCTURAL ANALYSIS APPLIED TO PLATE BENDI	70-0485
LARGE	A FINITE ELEMENT METHOD FOR GEOMETRICALLY NONLINEAR LARGE DISPLACEMENT PROBLEMS IN THIN, ELASTIC PLA	69-0005
LARGE	A FINITE ELEMENT PROGRAM FOR THE DYNAMIC LARGE DEFLECTION, ELASTIC-PLASTIC ANALYSIS OF STIFFENED SHE	72-0908
LARGE	A GENERAL APPROACH TO PROBLEMS OF LARGE DEFORMATION, AND PLASTICITY USING ISOPARAMETRIC ELEMENT	71-0273
LARGE	A GRADIENT COMPUTATIONAL PROCEDURE FOR THE SOLUTION OF LARGE PROBLEMS ARISING FROM THE FINITE ELEMEN	70-0059
LARGE	A LARGE DEFORMATION ANALYSIS OF CRYSTALLINE ELASTIC-VISCOPLASTIC MATERIALS	73-0602
LARGE	A METHOD OF FINITE DEFLECTION ELASTIC PLASTIC ANALYSIS OF FRAMED STRUCTURES SUBJECTED TO LARGE AXIAL	69-0239
LARGE	A SEQUEL TO TECHNICAL NOTE 15: THE SHEBA FAMILY OF SHELL ELEMENTS FOR THE MATRIX DISPLACEMENT METHOD	69-0444
LARGE	A TECHNIQUE FOR THE FINITE ELEMENT ANALYSIS OF LARGE COMPLEX STRUCTURAL SYSTEMS	69-0514
LARGE	A VARIATIONAL METHOD FOR LARGE DEFLECTION ELASTO-PLASTIC THEORY	69-0115
LARGE	ALGORITHM FOR THE LARGE DEFLECTION GEOMETRICALLY NON-LINEAR BEHAVIOUR OF PLANE AND CURVED STRUCTURES	72-0083
LARGE	AN ANALYSIS OF THE LARGE DEFLECTIONS OF BEAMS USING THE RAYLEIGH-RITZ FINITE ELEMENT METHOD	68-0162
LARGE	AN ECONOMICAL METHOD FOR DETERMINING THE SMALLEST EIGENVALUES OF LARGE LINEAR SYSTEMS	71-0551
LARGE	AN EQUATION SOLVER OF VERY LARGE CAPACITY	71-0677
LARGE	AN INCREMENTAL FINITE ELEMENT ANALYSIS OF TWO DIMENSIONAL LARGE STRAIN AND LARGE DISPLACEMENT PROBLE	72-0338
LARGE	AN INCREMENTAL FINITE ELEMENT ANALYSIS OF TWO DIMENSIONAL LARGE STRAIN AND LARGE DISPLACEMENT PROBLE	72-0338
LARGE	AN ITERATIVE METHOD FOR LARGE SYSTEMS OF LINEAR STRUCTURAL EQUATIONS	73-1020
LARGE	ANALYSIS OF LARGE CLOSED SHIP STRUCTURES: APPLICATIONS TO TANKER STRUCTURAL DESIGN	70-0289
LARGE	ANALYSIS OF LARGE DEFORMATIONS OF ELASTIC MEMBRANES BY THE FINITE ELEMENT METHOD	66-0057
LARGE	ANALYSIS OF LARGE DEFLECTION OF PLATES BY THE FINITE ELEMENT METHOD	69-0148
LARGE	ANALYSIS OF LARGE ELASTIC-PLASTIC CREEP DEFORMATIONS OF AXISYMMETRIC SHELLS BY THE FINITE ELEMENT ME	74-0563
LARGE	ANALYSIS OF PLATES IN BENDING USING LARGE FINITE ELEMENTS	70-0185
LARGE	ANALYSIS OF PLATES SUBJECTED TO IN-PLANE FORCES USING LARGE FINITE ELEMENTS	70-0271
LARGE	ANALYSIS OF THE DYNAMICS OF THIN PRIMARY MIRRORS FOR LARGE ASTRONOMICAL TELESCOPES	73-1084
LARGE	ANALYSIS ON THE LARGE DEFORMATION OF PLATE STRUCTURES	74-0435
LARGE	APPLICATION OF FINITE ELEMENT METHODS TO LARGE SCALE ELASTIC PLASTIC PROBLEMS OF FRACTURE MECHANICS	69-0372
LARGE	APPLICATION OF NATURAL KINEMATIC MODES TO MATRIX ANALYSIS: SMALL AND LARGE DISPLACEMENTS AND STRAINS	67-0138
LARGE	APPLICATION OF SMALL AND LARGE-DISPLACEMENT THEORY TO A LARGE FLEXIBLE SOLAR ARRAY AND COMPARISON WI	71-0231
LARGE	APPLICATION OF THE DIRECT STIFFNESS METHOD TO PLANE PROBLEMS INVOLVING LARGE, TIME-DEPENDENT DEFORMA	66-0108
LARGE	ASKA, A LARGE SCALE SOFTWARE SYSTEM FOR FINITE ELEMENT ANALYSIS	69-0331
LARGE	AUTOMATED DATA GENERATION AND LARGE MATRIX OPERATION FOR 3 DIMENSIONAL PLATE STRUCTURE ANALYSIS	73-0339
LARGE	AUTOMATED LARGE DEFLECTION AND STABILITY ANALYSIS OF THREE DIMENSIONAL BAR STRUCTURES	69-0393
LARGE	AUTOMATED METHOD FOR THE LARGE DEFLECTION AND INSTABILITY ANALYSIS OF THREE-DIMENSIONAL TRUSS AND FR	66-0105
LARGE	AUTOMATIC GENERATION AND CONDENSATION OF LARGE MASS AND STIFFNESS MATRICES	72-0466
LARGE	BLOCKING TECHNIQUE FOR LARGE SCALE STRUCTURAL ANALYSIS	73-0243
LARGE	COMPARISON BETWEEN THEORETICAL AND MEASURED EARTH PRESSURES ACTING ON A LARGE MOTORWAY RETAINING WAL	74-0917
LARGE	COMPUTATIONAL METHOD FOR DYNAMIC ANALYSIS OF STRUCTURES WITH LARGE SIZE STIFFNESS AND MASS MATRICES.	73-0830
LARGE	COMPUTER ANALYSIS OF LARGE LINEAR FRAMES	69-0408
LARGE	COMPUTER PROGRAM FOR SOLUTION OF LARGE, SPARSE, UNSYMMETRIC SYSTEMS OF LINEAR EQUATIONS	72-0719
LARGE	COMPUTER PROGRAM JET 3 TO CALCULATE THE LARGE ELASTIC PLASTIC DYNAMICALLY INDUCED DEFORMATIONS OF FR	72-0609
LARGE	COMPUTER SYSTEM FOR ANALYSIS OF LARGE FRAMEWORKS	67-0122
LARGE	COMPUTER SYSTEM FOR THE ANALYSIS OF LARGE SCALE PROBLEMS IN SOLID, FLUID AND THERMODYNAMICS	73-0554
LARGE	CREEP BENDING TESTS OF LARGE WELD OVERLAY AND BASE METAL SPECIMENS AT 1000 DEGREE F. (537.7 DEG. C)	75-0521
LARGE	DEFORMABILITY OF THE HULL STEEL-WORK AND DEFORMATIONS OF THE ENGINE-ROOM OF LARGE TANKERS	74-0532
LARGE	DERIVATION OF COMPATIBLE STIFFNESS MATRICES FOR LARGE DISCRETE ELEMENTS	67-0080
LARGE	DERIVATION OF STIFFNESS MATRICES FOR THE ANALYSIS OF LARGE DEFLECTION AND STABILITY PROBLEMS	65-0066
LARGE	DEVELOPMENT AND MAINTENANCE OF LARGE FINITE ELEMENT SOFTWARE SYSTEMS	74-0865
LARGE	DEVELOPMENT OF ADVANCED STRUCTURAL OPTIMIZATION PROGRAMS AND THEIR APPLICATION TO LARGE ORDER STRAIN	66-0028
LARGE	DYNAMIC PLASTIC LARGE DEFLECTIONS OF CLAMPED RECTANGULAR PLATE	74-0392
LARGE	DYNAMIC RESPONSE BY LARGE STEP INTEGRATION	73-0567
LARGE	DYNAMIC RESPONSE ON NONLINEAR MEDIA AT LARGE STRAINS	72-0237
LARGE	DYNAPLAS: A FINITE ELEMENT PROGRAM FOR THE DYNAMIC, LARGE DEFLECTION, ELASTIC-PLASTIC ANALYSIS OF ST	72-0773
LARGE	DYPLAS, A FINITE ELEMENT DYNAMIC ELASTIC-PLASTIC LARGE DEFORMATION ANALYSIS PROGRAM	74-0306
LARGE	EFFECT OF INITIAL DISPLACEMENT ON PROBLEM OF LARGE DEFLECTION AND STABILITY	67-0113
LARGE	EFFICIENT COMPUTER ANALYSIS OF LARGE THREE-DIMENSIONAL PIPING SYSTEMS BY THE FINITE ELEMENT METHOD	72-0378
LARGE	EIGENVALUES OF LARGE MATRICES	69-0341
LARGE	ELASTIC ANALYSIS OF LARGE SPACEBOUND MIRRORS	72-0619
LARGE	ELASTIC-PLASTIC LARGE DEFORMATION RESPONSE OF CLAY TO FOOTING LOADS	74-0699
LARGE	ESTIMATING MATRIX DISPLACEMENT SOLUTIONS OF TWO-DIMENSIONAL PROBLEMS BY LARGE ELEMENT TECHNIQUE	70-0160
LARGE	FINITE DIFFERENCE SOLUTIONS FOR LARGE DEFORMATIONS OF CYLINDRICAL SHELLS - A COMPARISON WITH FINITE	72-0270
LARGE	FINITE ELEMENT ANALYSIS OF LARGE TRANSIENT ELASTIC-PLASTIC DEFORMATIONS OF SIMPLE STRUCTURES, WITH A	72-0849
LARGE	FINITE ELEMENT DISPLACEMENT METHOD FOR LARGE AMPLITUDE FREE FLEXURAL VIBRATIONS OF BEAMS AND PLATES	73-0185
LARGE	FINITE ELEMENT FORMULATIONS FOR LARGE DISPLACEMENT AND LARGE STRAIN ANALYSIS	73-0457
LARGE	FINITE ELEMENT FORMULATIONS FOR LARGE DISPLACEMENT AND LARGE STRAIN ANALYSIS	73-0457
LARGE	FINITE ELEMENT FORMULATIONS FOR PROBLEMS OF LARGE ELASTIC-PLASTIC DEFORMATION	74-0796
LARGE	FINITE ELEMENT FORMULATIONS FOR LARGE DEFORMATION DYNAMIC ANALYSIS	75-0564
LARGE	FINITE ELEMENT LARGE DEFLECTION ANALYSIS OF PLATES	69-0208
LARGE	FINITE ELEMENT LARGE DEFLECTION ANALYSIS OF ELASTIC-PLASTIC SHELLS OF REVOLUTION SUBJECTED TO AXISYM	72-0137
LARGE	FINITE ELEMENT LARGE DEFLECTION ANALYSIS OF THIN-WALLED BEAMS	74-0444
LARGE	FINITE ELEMENT LARGE DEFLECTION ANALYSIS OF SHALLOW SHELLS	75-0356

LARGE	FINITE ELEMENT LARGE DEFORMATION ANALYSIS OF A TAPERED AORTA	75-0856
LARGE	FINITE ELEMENT PROCEDURE FOR LARGE DEFLECTION ANALYSIS OF PLATES WITH INITIAL DEFLECTIONS	71-0327
LARGE	FINITE ELEMENT PROCEDURE FOR THE LARGE DEFORMATION DYNAMIC RESPONSE OF AXISYMMETRIC SOLIDS	74-0527
LARGE	FINITE ELEMENT PROCEDURE FOR THE LARGE DEFORMATION DYNAMIC RESPONSE OF AXISYMMETRIC SOLIDS	74-0967
LARGE	FINITE ELEMENT PROCEDURE FOR LARGE DEFORMATION DYNAMIC RESPONSE OF AXISYMMETRIC SOLIDS	75-0013
LARGE	FINITE-ELEMENT ANALYSIS OF LARGE ELASTIC-PLASTIC TRANSIENT DEFORMATIONS OF SIMPLE STRUCTURES	71-0326
LARGE	FINITE-ELEMENT ANALYSIS OF HIGH-PRESSURE, LARGE VALVE GATES	75-0329
LARGE	FINITE-ELEMENT ANALYSIS OF HIGH-PRESSURE LARGE VALVE GATES	75-0413
LARGE	FINITE-ELEMENT LARGE DEFLECTION ANALYSIS OF PLATES	69-0196
LARGE	FINITE-ELEMENT PREDICTIONS OF TRANSIENT ELASTIC-PLASTIC LARGE DEFLECTIONS OF STIFFENED AND/OR UNSTIF	74-0825
LARGE	HOW FINITE ELEMENT METHODS ARE INTRODUCED IN LARGE AND SMALL ORGANIZATIONS	74-0366
LARGE	INCREMENTAL ANALYSIS OF LARGE DEFLECTIONS OF SHELLS OF REVOLUTION	71-0296
LARGE	INCREMENTAL FINITE ELEMENT ANALYSIS OF LARGE ELASTIC DEFORMATION PROBLEMS	71-0753
LARGE	INCREMENTAL THEORY OF LARGE STRAIN AND LARGE DISPLACEMENT PROBLEMS AND ITS FINITE ELEMENT FORMULATIO	72-0274
LARGE	INCREMENTAL THEORY OF LARGE STRAIN AND LARGE DISPLACEMENT PROBLEMS AND ITS FINITE ELEMENT FORMULATIO	72-0274
LARGE	INCREMENTAL VARIATIONAL METHOD FOR THE LARGE DISPLACEMENT ANALYSIS OF SHELLS WITH GEOMETRIC IMPERFEC	72-0894
LARGE	INFLUENCE OF HIGHER ORDER TERMS IN THE LARGE DEFLECTION ANALYSIS OF FRAMEWORKS	68-0259
LARGE	INTEGRATED SEQUENTIAL SOLVER FOR LARGE MATRIX EQUATIONS	74-0276
LARGE	ITERATIVE METHOD FOR THE ANALYSIS OF LARGE STRUCTURAL SYSTEMS	68-0236
LARGE	LAGRANGIAN AND EULERIAN FINITE ELEMENT FORMULATION FOR LARGE STRAIN ELASTO-PLASTICITY	75-0209
LARGE	LARGE AMPLITUDE, FREE VIBRATION GEOMETRICALLY NONLINEAR CIRCULAR ARCHES	72-0606
LARGE	LARGE CAPACITY EQUATION SOLVER FOR STRUCTURAL ANALYSIS	74-0983
LARGE	LARGE DEFLECTION ANALYSIS OF ELASTIC-PLASTIC SHELLS OF REVOLUTION	69-0174
LARGE	LARGE DEFLECTION ANALYSIS OF ELASTIC-PLASTIC PLATES AND SHELLS	69-0178
LARGE	LARGE DEFLECTION ANALYSIS OF ELASTIC-PLASTIC SHELLS OF REVOLUTION	70-0507
LARGE	LARGE DEFLECTION ANALYSIS OF PRESTRESSED NETWORKS	72-0169
LARGE	LARGE DEFLECTION ANALYSIS OF ARBITRARY SHALLOW SHELLS BY THE FINITE ELEMENT METHOD	73-0937
LARGE	LARGE DEFLECTION ANALYSIS OF PLATES AND SHALLOW SHELLS USING THE FINITE ELEMENT METHOD	73-0989
LARGE	LARGE DEFLECTION DYNAMIC ANALYSIS OF THIN SHELLS USING FINITE ELEMENT METHOD	70-0379
LARGE	LARGE DEFLECTION ELASTIC-PLASTIC ANALYSIS OF SHELLS OF REVOLUTION	71-0263
LARGE	LARGE DEFLECTION ELASTIC-PLASTIC DYNAMIC RESPONSE OF STIFFENED SHELLS OF REVOLUTION	72-0608
LARGE	LARGE DEFLECTION ELASTIC-PLASTIC DYNAMIC RESPONSE OF STIFFENED SHELLS OF REVOLUTION	74-0496
LARGE	LARGE DEFLECTION GEOMETRICALLY NON-LINEAR FINITE ELEMENT ANALYSIS OF CIRCULAR ARCHES	73-0021
LARGE	LARGE DEFLECTION OF CANTILEVER BEAMS	71-0649
LARGE	LARGE DEFLECTION OF ELASTIC PLATES UNDER PATCH LOADING	72-0120
LARGE	LARGE DEFLECTION SPATIAL BUCKLING OF THIN WALLED BEAMS AND FRAMES	73-1085
LARGE	LARGE DEFLECTIONS OF STRUCTURES SUBJECT TO HEATING AND EXTERNAL LOADS	60-0002
LARGE	LARGE DEFLECTIONS OF ELASTIC BEAMS AND PLATES USING THE FINITE ELEMENT METHOD	68-0101
LARGE	LARGE DEFORMATION AND CRITICAL LOADS ANALYSIS OF FRAMED STRUCTURES	74-1293
LARGE	LARGE DISPLACEMENT ANALYSIS OF ASIXYMMETRIC SHELLS	69-0276
LARGE	LARGE DISPLACEMENT ANALYSIS OF THIN PLATES	73-0261
LARGE	LARGE DISPLACEMENT ANALYSIS OF VISCOELASTIC BEAMS AND FRAMES BY FINITE-ELEMENT METHOD	74-0644
LARGE	LARGE DISPLACEMENT FORMULATION FOR ELASTIC BODIES	70-0429
LARGE	LARGE DISPLACEMENT, NONLINEAR TRANSIENT ANALYSIS BY FINITE ELEMENTS	74-0324
LARGE	LARGE NATURAL STRAINS AND SOME SPECIAL DIFFICULTIES DUE TO NONLINEARITY AND INCOMPRESSIBILITY IN FIN	74-0528
LARGE	LARGE NATURAL STRAINS AND SOME SPECIAL DIFFICULTIES DUE TO NONLINEARITY AND INCOMPRESSIBILITY IN FIN	74-1027
LARGE	LARGE SCALE ANALYSIS OF CURRENT AIRCRAFT	70-0383
LARGE	LARGE SCALE COMPUTER PROGRAMS FOR STRUCTURAL ANALYSIS	70-0211
LARGE	LARGE STRAIN ELASTO-PLASTIC FINITE ELEMENT ANALYSIS	71-0121
LARGE	LARGE STRAIN, ELASTO-PLASTIC NUMERICAL ANALYSIS BY MEANS OF FINITE ELEMENT METHOD	71-0790
LARGE	LARGE STRAIN, LARGE DISPLACEMENT ANALYSIS	71-0508
LARGE	LARGE STRAIN, LARGE DISPLACEMENT ANALYSIS	71-0508
LARGE	LARGE STRAIN, VISCOELASTIC NUMERICAL ANALYSIS BY MEANS OF FINITE ELEMENT METHOD	72-0768
LARGE	LARGE STRAIN, VISCOELASTIC AND ELASTO-VISCOPLASTIC NUMERICAL-ANALYSIS BY MEANS OF FINITE-ELEMENT MET	75-0435
LARGE	LARGE STRAIN, VISCOELASTIC AND ELASTO VISCOPLASTIC NUMERICAL ANALYSIS BY MEANS OF THE FINITE ELEMENT	75-0563
LARGE	LARGE STRUCTURES FOR MANNED SPACECRAFT: MATHEMATICAL ANALYSIS, DESIGN, CONSTRUCTION AND TESTS	74-1327
LARGE	LARGE USER-ORIENTATED SYSTEMS OF PROGRAMS FOR STRUCTURAL ANALYSIS AND DESIGN	74-0482
LARGE	LINEARIZED LARGE DEFLECTION PROBLEM	73-0685
LARGE	LOADS AND RESPONSE OF LARGE DUCTED PROPELLER SYSTEMS	73-0810
LARGE	MATRIX ANALYSIS OF SHELLS - SMALL AND LARGE DISPLACEMENTS	68-0016
LARGE	MATRIX ANALYSIS OF THREE-DIMENSIONAL ELASTIC MEDIA: SMALL AND LARGE DISPLACEMENTS	65-0003
LARGE	NASTRAN BUCKLING ANALYSIS OF A LARGE STIFFENED CYLINDRICAL SHELL WITH A CUTOUT	71-0045
LARGE	NONLINEAR FINITE ELEMENT AND ITS APPLICATION TO LARGE SCALE STRUCTURAL PROBLEMS	75-0386
LARGE	NOTE ON INCREMENTAL EQUILIBRIUM EQUATIONS AND APPROXIMATE CONSTITUTIVE RELATIONS IN LARGE INELASTIC	74-0984
LARGE	OBSERVED AND PREDICTED DEFORMATIONS IN A LARGE EMBANKMENT DAM DURING CONSTRUCTION	71-0301
LARGE	ON AN APPROACH TO TECHNIQUES FOR THE ANALYSIS OF THE STRUCTURES OF LARGE SYSTEMS OF EQUATIONS	62-0015
LARGE	ON ANALYSIS OF LARGE DEFORMATION PROBLEMS OF BEAM	71-0660
LARGE	ON STRENGTH OF TRANSVERSE WEBS AND THEIR JOINTS OF LARGE TANKER	69-0327
LARGE	ON THE ASYMPTOTICALLY SPHERICAL DEFORMATIONS OF ARBITRARY MEMBRANES OF REVOLUTION FIXED ALONG AN EDG	71-0640
LARGE	ON THE DERIVATION OF STIFFNESS MATRICES FOR THE ANALYSIS OF LARGE DEFLECTION AND STABILITY PROBLEMS	66-0050
LARGE	ON THE EXTREMAL PROPERTIES OF THE SOLUTION IN DYNAMICS OF RIGID VISCOPLASTIC BODIES ALLOWING FOR LAR	73-1080
LARGE	ON THE HYBRID STRESS FINITE ELEMENT MODEL FOR INCREMENTAL ANALYSIS OF LARGE DEFLECTION PROBLEMS	73-0371
LARGE	ON THE NUMERICAL SOLUTION OF LARGE EIGENVALUE PROBLEMS ARISING IN PANEL FLUTTER ANALYSIS BY THE FINI	74-0942
LARGE	OPTIM 11: (MAGIC) COMPATIBLE LARGE SCALE AUTOMATED MINIMUM WEIGHT DESIGN PROGRAM - V 2. PROGRAMMER'S	74-1016
LARGE	OPTIM 11: A (MAGIC) COMPATIBLE LARGE SCALE AUTOMATED MINIMUM WEIGHT DESIGN PROGRAM - 1. ENGINEERS AN	74-1048
LARGE	OPTIM 11: MAGIC COMPATIBLE LARGE SCALE AUTOMATED MINIMUM WEIGHT DESIGN PROGRAM. VOLUME 11. PROGRAMME	74-1093
LARGE	OPTIMAL SYNTHESIS OF FLEXIBLE LINK MECHANISMS WITH LARGE STATIC DEFLECTIONS	74-0525
LARGE	PENALTY FUNCTION APPROACH TO PROBLEMS OF PLASTIC FLOW OF METALS WITH LARGE SURFACE DEFORMATIONS	75-0282
LARGE	PLASTICITY AND LARGE DEFORMATION PROBLEMS BY FINITE ELEMENT METHOD	71-0529
LARGE	PROBLEMS IN BUILDING AND MANIPULATING LARGE STIFFNESS MATRICES	71-0596
LARGE	PROBLEMS OF LONGITUDINAL STRENGTH IN THE CONSTRUCTION OF VERY LARGE TANK TRUCKS	73-0251
LARGE	PROGRESS IN COMPUTERIZED STRUCTURAL ANALYSIS OF LARGE VESSELS	74-0949
LARGE	REINFORCING REQUIREMENTS FOR CONCRETE BEAMS WITH LARGE WEB OPENINGS	73-0198
LARGE	SHELL ANALYSIS WITH LARGE GENERAL PURPOSE PROGRAMS	71-0402
LARGE	SIMULTANEOUS EIGENVECTOR ITERATION APPLIED TO LARGE SYSTEMS	74-0680
LARGE	SOLUTION OF LARGE SYSTEMS OF EQUATIONS	75-0608
LARGE	SOLVING A CLASS OF LARGE SPARSE LINEAR SYSTEMS OF EQUATIONS BY PARTITIONING	73-0580
LARGE	SOME GENERAL CONSIDERATIONS ON THE NATURAL MODE TECHNIQUE, PART II: LARGE DISPLACEMENTS	69-0445
LARGE	STRUCTURAL EVALUATION OF CANDIDATE DESIGNS FOR THE LARGE SPACE TELESCOPE PRIMARY MIRROR	75-0708
LARGE	STRUCTURAL MODEL STUDIES OF LARGE PIPE BIFURCATION	73-0283
LARGE	THE APPLICATION OF FINITE ELEMENTS TO THE LARGE DEFLECTION GEOMETRICALLY NONLINEAR BEHAVIOR OF CYLIN	72-0815
LARGE	THE DEVELOPMENT OF COMPUTER GRAPHICS FOR LARGE SCALE FINITE ELEMENT CODES	71-0042
LARGE	THE DEVELOPMENT OF LARGE SCALE DIGITAL COMPUTER CODES FOR PRODUCTION STRUCTURAL ANALYSIS	71-0721
LARGE	THE ECONOMICAL COMPUTATION OF STIFFNESS OF LARGE STRUCTURAL ELEMENTS	66-0090
LARGE	THE EFFECT OF INITIAL DISPLACEMENTS OF PROBLEMS OF LARGE DEFLECTION AND STABILITY	67-0048
LARGE	THE SHEBA FAMILY OF SHELL ELEMENTS FOR THE MATRIX DISPLACEMENT METHOD. 111, LARGE DISPLACEMENTS	69-0499
LARGE	THE USE OF NASTRAN IN THE ANALYSIS OF LARGE COMPLEX AIRFRAME STRUCTURES, MODELING TECHNIQUES AND ORG	71-0110
LARGE	THEORETICAL AND EXPERIMENTAL STUDIES OF TRANSIENT ELASTIC-PLASTIC LARGE DEFLECTIONS OF GEOMETRICALLY	75-0830
LARGE	THEORETICAL AND EXPERIMENTAL STUDIES OF TRANSIENT ELASTIC-PLASTIC LARGE DEFLECTIONS OF GEOMETRICALLY	75-0831
LARGE	THERMOSTRUCTURAL DESIGN CONSIDERATIONS TO ACHIEVE THE LARGE SPACE TELESCOPE LINE-OF-SIGHT REQUIREMEN	75-0839
LARGE	THREE DIMENSIONAL ANISOTROPIC AND INHOMOGENEOUS ELASTIC MEDIA MATRIX ANALYSIS FOR SMALL AND LARGE DI	65-0008
LARGE	USE OF THE OVER-RELAXATION TECHNIQUE IN THE SIMULATION OF LARGE GROUNDWATER BASINS BY THE FINITE ELE	75-0606

LARGE	VELKE PRIEHYBY DOSIEK METODOU KONECNYCH PRVKOV (LARGE DEFLECTIONS OF PLATES ANALYZED BY FINITE ELEME	73-0373
LARGE	ZUM TRAGVERHALTEN VON WAENDEN AUS FERTIGTEILTAFELN (BEHAVIOR OF LARGE PANEL SHEAR WALLS)	74-0299
LARGE	ZUSAMMENWIRKEN VON EINZELNEN FERTIGTEILEN ALS GROSSFLAECHIGE SCHEIBE (INTERACTION OF INDIVIDUAL PREF	73-0552
LARGE-DEFLECTIO	LARGE-DEFLECTION ELASTO-PLASTIC BUCKLING ANALYSIS OF PLATES USING FINITE ELEMENTS	73-0394
LARGE-DISPLACEM	APPLICATION OF SMALL AND LARGE-DISPLACEMENT THEORY TO A LARGE FLEXIBLE SOLAR ARRAY AND COMPARISON WI	71-0231
LARGE-DISPLACEM	THE ELASTO-PLASTIC AND LARGE-DISPLACEMENT RESPONSE OF PLATES TO BLAST LOADING	73-0731
LARGE-SCALE	EXAMPLE OF LARGE-SCALE STRUCTURAL ANALYSIS. COMPARISON BETWEEN FINITE ELEMENT CALCULATION AND FULL S	74-0177
LARGE-SCALE	KRON'S METHOD: AN ALGORITHM FOR THE EIGENVALUE ANALYSIS OF LARGE-SCALE STRUCTURAL SYSTEMS	73-0758
LARGE-SCALE	OPTIM 11: A MAGIC COMPATIBLE LARGE-SCALE AUTOMATED MINIMUM WEIGHT DESIGN PROGRAM. VOLUME 1. ENGINEER	74-1096
LARGELY	A METHOD OF ELASTIC-PLASTIC ANALYSIS OF LARGELY DEFORMED PLATE PROBLEMS	72-0319
LARGEST	COMPUTERIZED STRUCTURAL ANALYSIS OF THE WORLD'S LARGEST LIGHT-GAGE STEEL PRIMARY STRUCTURAL SYSTEM	73-0503
LASER	A NUMERICAL PROCEDURE FOR TWO DIMENSIONAL HEATING AND MELTING CALCULATIONS, WITH APPLICATIONS TO LAS	72-0708
LASER	APPLICATIONS OF PULSED LASER INDUCED STRESS WAVES	73-0956
LASER	THE EFFECT OF GRAVITY UPON THE MELT-THROUGH TIME OF A SOLID SUBJECTED TO A HIGH INTENSITY LASER	72-0693
LASER	THE USE OF A LASER FOR ARPA MILITARY GEOPHYSICS PROGRAM (ROCK MECHANICS AND RAPID EXCAVATION)	72-0800
LATENT	APPLICATION OF FINITE ELEMENTS TO HEAT CONDUCTION PROBLEMS INVOLVING LATENT HEAT	73-0520
LATERAL	A COMPUTER PROGRAM FOR ANALYSIS OF STIFFENED PLATES UNDER COMBINED INPLANE AND LATERAL LOADS	71-0703
LATERAL	BEHAVIOR OF ISOLATED SHEAR WALLS SUBJECTED TO LATERAL LOADS	75-0704
LATERAL	CREEP OF A UNIAXIAL METAL MATRIX COMPOSITE SUBJECTED TO AXIAL AND NORMAL LATERAL LOADS	71-0041
LATERAL	CREEP OF A UNIAXIAL METAL MATRIX COMPOSITE SUBJECTED TO AXIAL AND NORMAL LATERAL LOADS	71-0727
LATERAL	EFFECT OF LATERAL PRESSURE ON NATURAL FREQUENCIES OF HYPERBOLOIDAL SHELLS OF REVOLUTION	71-0278
LATERAL	EFFECT OF OPENINGS ON THE LATERAL STIFFNESS OF INFILLED FRAMES	71-0300
LATERAL	ELASTIC LATERAL BUCKLING OF STEEL BEAMS	70-0364
LATERAL	FORCES AND DEFORMATIONS AT THE LATERAL STIFFENER OF THIN WALLED U-BARS	73-0236
LATERAL	LATERAL RESISTANCE AND DEFLECTION OF VERTICAL PILES	73-0912
LATERAL	STRUCTURAL ANALYSIS OF SL-7 CONTAINERSHIP UNDER COMBINED LOADING OF VERTICAL, LATERAL AND TORSIONAL	74-0734
LATERAL	THREE-DIMENSIONAL ANALYSIS OF SHEAR WALL BUILDINGS TO LATERAL LOAD	74-0220
LATERALLY	COMPARISON OF PRACTICAL METHODS OF ANALYSING ELASTIC LATERALLY LOADED PLATES	71-0425
LATERALLY	MAGNETOTELLURIC RESPONSE OF LATERALLY INHOMOGENEOUS AND ANISOTROPIC MEDIA	75-0850
LATERALLY	SOIL RESISTANCE MOVEMENT RELATIONSHIPS FOR LATERALLY LOADED PILES IN CLAYS AT GREAT DEPTHS	70-0162
LATERITE	IN-SITU HORIZONTAL STRESSES IN LATERITE BY FINITE ELEMENT ANALYSIS	73-1048
LATTICE	A LATTICE ANALOGY FOR THE SOLUTION OF PLANE STRESS	43-0001
LATTICES	PROCESS FOR THE AUTOMATIC GENERATION OF FLAT TRIANGULAR LATTICES FOR APPLICATION TO THE FINITE ELEME	71-0366
LAUFRAEDERN	BERECHNUNG VON LAUFRAEDERN MIT DER METHODE DER FINITEN ELEMENTE (DESIGN OF ROTATING WHEELS BY THE FI	74-0547
LAUNCH	A TWO-DIMENSIONAL AXISYMMETRIC SHELL FLUID MODEL FOR LAUNCH VEHICLE LONGITUDINAL DYNAMIC ANALYSIS	69-0046
LAUNCH	EVALUATION OF A FINITE ELEMENT ANALYSIS FOR THE LONGITUDINAL VIBRATIONS OF LIQUID-PROPELLANT LAUNCH	70-0204
LAUNCH	IMPROVED ANALYTIC LONGITUDINAL RESPONSE ANALYSIS FOR AXISYMMETRIC LAUNCH VEHICLES, COMPUTER PROGRAM	65-0002
LAUNCH	IMPROVED LINEAR AXISYMMETRIC SHELL FLUID MODEL FOR LAUNCH VEHICLE LONGITUDINAL RESPONSE ANALYSIS	65-0044
LAUNCH	MEASURED AND PREDICTED LONGITUDINAL VIBRATIONS OF A LIQUID PROPELLANT TWO-STAGE LAUNCH VEHICLE	70-0366
LAUNCH	STUDY OF PROPELLANT DYNAMICS IN A SHUTTLE TYPE LAUNCH VEHICLE	72-0699
LAURENT-SERIES	STRESSES AROUND HOLES IN STIFFENED COMPOSITE PANELS USING LAURENT-SERIES AND FINITE ELEMENT METHODS	74-1252
LAURENT-SERIES	STRESSES AROUND HOLES IN STIFFENED COMPOSITE PANELS USING LAURENT-SERIES AND FINITE ELEMENT METHODS	75-0733
LAYER	ANALYSIS OF A CIRCULAR RAFT IN ADHESIVE CONTACT WITH A THICK ELASTIC LAYER	74-1152
LAYER	COMPARISON OF THEORETICAL AND EXPERIMENTAL SHEAR STRESS IN THE ADHESIVE LAYER OF A LAP JOINT MODEL	74-1176
LAYER	CONSOLIDATION OF A LAYER UNDER A STRIP LOAD	72-0222
LAYER	EXCITATION OF PLANE THERMOELASTIC WAVES IN A LAYER WITH A FINITE RATE OF HEAT PROPAGATION	74-0803
LAYER	FINITE ELEMENT ANALYSIS OF THE SURFACE DEFORMATION DUE TO A UNIFORM LOADING ON A LAYER OF GIBSON SOI	75-0240
LAYER	FINITE ELEMENT COMPUTATIONAL THEORY FOR THREE-DIMENSIONAL BOUNDARY LAYER FLOW	72-0632
LAYER	FINITE ELEMENT MODEL OF TWO LAYER COASTAL CIRCULATION	74-1192
LAYER	FREE VIBRATION OF THICK, LAYERED RECTANGULAR PLATES BY FINITE LAYER METHOD	72-0440
LAYER	LEAST SQUARES FINITE ELEMENT ANALYSIS OF LAMINAR BOUNDARY LAYER FLOWS	74-0616
LAYER	MATHEMATICAL MODELS OF THE MASSACHUSETTS BAY. PART 1. FINITE ELEMENT MODELING OF TWO-DIMENSIONAL HYD	73-0800
LAYER	ON THE ASYMPTOTICALLY SPHERICAL DEFORMATIONS OF ARBITRARY MEMBRANES OF REVOLUTION FIXED ALONG AN EDG	71-0640
LAYER	STABILITY ANALYSIS OF AN EXPERIMENTALLY DEFORMED SINGLE LAYER OF INDIANA LIMESTONE USING FINITE ELEM	74-0212
LAYER	STRESSES AND DISPLACEMENTS AROUND A CIRCULAR TUNNEL IN A THREE LAYER MEDIUM	69-0038
LAYER	TEST RESULTS FROM THE BOUNDARY LAYER FACILITY - RESPONSE OF STRUCTURE TO THE PSEUDO-SOUND FIELD OF A	68-0262
LAYER	THE EFFECTS OF SOIL PARAMETERS AND BOUNDARY CONDITIONS ON THE CONSOLIDATION OF AN ELASTIC LAYER	70-0422
LAYER	THE STRESSES IN AN ADHESIVE LAYER	72-0910
LAYER	TRANSPIRATION AND FILM COOLING BOUNDARY LAYER COMPUTER PROGRAM. VOLUME 2, COMPUTER PROGRAM AND USER'	71-0655
LAYERED	A FINITE ELEMENT METHOD FOR BENDING ANALYSIS OF LAYERED STRUCTURAL SYSTEMS	67-0134
LAYERED	ANALYSIS OF CIRCULAR FOOTINGS ON LAYERED SOILS	70-0281
LAYERED	ANALYSIS OF LAYERED SYSTEMS SUBJECTED TO WHEEL LOADS	72-0868
LAYERED	ANALYSIS OF SIMPLY SUPPORTED THICK, LAYERED PLATES	71-0428
LAYERED	BEHAVIOUR OF STRIP FOOTINGS ON LAYERED COHESIVE SOILS	69-0223
LAYERED	COMPLIANCES OF LAYERED ELASTIC SYSTEMS	74-1204
LAYERED	DYNAMIC ANALYSIS OF CIRCULAR AND SECTOR THICK, LAYERED PLATES	75-0198
LAYERED	DYNAMIC ANALYSIS OF FOOTINGS ON LAYERED MEDIA	75-0586
LAYERED	DYNAMIC RESPONSE OF STRUCTURES IN LAYERED SOILS	74-0761
LAYERED	EARTH VIBRATION EFFECTS AND ABATEMENT FOR MILITARY FACILITIES. REPORT 3. ANALYSIS METHOD FOR FOOTING	72-0806
LAYERED	ELASTIC STRESSES IN LAYERED SNOW PACKS	72-0925
LAYERED	ELASTO-PLASTIC INDENTATION OF A LAYERED MEDIUM	74-0225
LAYERED	FINITE ELEMENT ANALYSIS OF A PLATE ON A LAYERED CROSS-ANISOTROPIC FOUNDATION	74-0451
LAYERED	FINITE ELEMENT ANALYSIS OF LAYERED WOOD SYSTEMS	75-0288
LAYERED	FINITE ELEMENT COMPUTER PROGRAM FOR THE ANALYSIS OF LAYERED ORTHOTROPIC STRUCTURES	74-0800
LAYERED	FINITE ELEMENT SOLUTION FOR WAVE PROPAGATION IN LAYERED MEDIA CAUSED BY A NUCLEAR DETONATION	67-0016
LAYERED	FORCED VIBRATIONS OF CIRCULAR FOUNDATIONS ON LAYERED MEDIA	74-0762
LAYERED	FREE VIBRATION OF THICK, LAYERED RECTANGULAR PLATES BY FINITE LAYER METHOD	72-0440
LAYERED	FREE VIBRATION OF THICK, LAYERED CYLINDERS HAVING FINITE LENGTH WITH VARIOUS BOUNDARY CONDITIONS	72-0491
LAYERED	GRAVITY FLOW TO EXCAVATIONS AND DRAINAGE TRENCHES IN LAYERED AQUIFERS	71-0014
LAYERED	LAYERED PAVEMENT SYSTEMS. PART 1. LAYERED SYSTEM DESIGN. PART 11. FATIGUE OF PLAIN CONCRETE	72-0867
LAYERED	LAYERED PAVEMENT SYSTEMS. PART 1. LAYERED SYSTEM DESIGN. PART 11. FATIGUE OF PLAIN CONCRETE	72-0867
LAYERED	MESH GENERATING COMPUTER PROGRAM FOR FINITE ELEMENT ANALYSES OF SIMPLE SLOPES IN HORIZONTALLY LAYERE	71-0748
LAYERED	MODIFICATION OF LAYERED THEORY FOR PAVEMENT DESIGN ANALYSIS	73-0574
LAYERED	NONLINEAR LAYERED ANALYSIS OF RC PLATES AND SHELLS	73-0208
LAYERED	NONLINEAR LAYERED FINITE-ELEMENT ANALYSIS OF REINFORCED-CONCRETE PLATES AND SHELLS	72-0946
LAYERED	PARAMETRIC STUDIES OF LAYERED PAVEMENT SYSTEMS USING AFPAV CODE WITH FAST EQUATION SOLVER	75-0747
LAYERED	STRESS WAVE ANALYSIS IN LAYERED THERMOVISCOELASTIC MATERIALS BY THE EXTENDED RITZ METHOD, VOLUME II	68-0125
LAYERED	THE PROPAGATION OF LOVE WAVES ACROSS NONHORIZONTALLY LAYERED STRUCTURES	71-0600
LAYERED	ULTIMATE CAPACITY OF CIRCULAR FOOTINGS ON LAYERED SOILS	70-0280
LAYERED	VERTICAL VIBRATIONS OF FOOTINGS EMBEDDED IN LAYERED MEDIA	69-0436
LAYERS	FRACTURE (ULTIMATE STRENGTH) ANALYSES OF ASPHALT PAVEMENT LAYERS RESULTING FROM TRAFFIC LOADING	73-0548
LAYERS	GENERAL VARIATIONAL ANALYSIS OF HYDRODYNAMIC, THERMAL AND DIFFUSIONAL BOUNDARY LAYERS	70-0191
LEADING	COMBINED ANALYTICAL - EXPERIMENTAL FRACTURE STUDY OF THE TWO LEADING THEORIES OF ELASTIC-PLASTIC FRA	73-0951
LEADING	ON THE FICTION OF LEADING EDGE SUCTION	71-0712
LEADING-EDGE	ANALYSIS OF THE EFFECTS ON LIFE OF LEADING-EDGE HOLES IN AN AIRFOIL SUBJECTED TO ARBITRARY SPANWISE	75-0700
LEAKAGE	ANALYSIS OF TRANSFORMER LEAKAGE PHENOMENA BY HIGH-ORDER FINITE ELEMENTS	73-0135
LEAKAGE	LEAKAGE THROUGH CRACKS IN LNG TANKAGE	74-0749
LEAKAGE	PREDICTING MAGNETIC LEAKAGE FIELDS BY FINITE ELEMENT TECHNIQUES	75-0004
LEAKAGE	TRANSFORMER LEAKAGE FLUX PROGRAM BASED ON FINITE ELEMENT METHOD	73-0004
LEAN	STRESS ANALYSIS FOR CLEAN LEAN DESIGN	71-0386
LEAST	A DISCRETE LEAST SQUARES AND FINITE ELEMENT METHOD	69-0322
LEAST	A LEAST SQUARES FINITE ELEMENT SOLUTION OF NONLINEAR OPERATORS	72-0060

LEAST	A LEAST SQUARES FINITE ELEMENT ANALYSIS OF NONLINEAR BIOPHYSICAL DIFFUSION-KINETICS EQUATIONS	74-1257
LEAST	A TWO-DIMENSIONAL FINITE ELEMENT - LEAST SQUARES SOLUTION	75-0025
LEAST	AN APPLICATION OF LEAST SQUARES TO ONE-DIMENSIONAL TRANSIENT PROBLEMS	74-0347
LEAST	AN APPLICATION OF LEAST SQUARES FINITE ELEMENT METHODS TO TWO DIMENSIONAL STEADY INVISCID COMPRESSIB	75-0114
LEAST	FINITE ELEMENT APPLICATIONS OF THE LEAST SQUARES METHOD	72-0052
LEAST	FINITE ELEMENTS FORMULATED BY THE WEIGHTED DISCRETE LEAST SQUARES METHOD	74-0200
LEAST	LEAST SQUARE OPTIMAL APPROXIMATION SIMULATION TECHNIQUES	73-0871
LEAST	LEAST SQUARE POLYNOMIAL SPLINE APPROXIMATION	72-0526
LEAST	LEAST SQUARE-FINITE ELEMENT FOR ELASTO-STATIC PROBLEMS	74-0279
LEAST	LEAST SQUARES ANALYSIS USING FINITE ELEMENTS	68-0287
LEAST	LEAST SQUARES FINITE ELEMENT ANALYSIS OF LAMINAR BOUNDARY LAYER FLOWS	74-0616
LEAST	LEAST SQUARES FINITE ELEMENT ANALYSIS OF TWO-DIMENSIONAL LAMINAR BOUNDARY-LAYER EQUATION WITH SEPARA	75-0796
LEAST	LEAST SQUARES SMOOTHING OF EXPERIMENTAL DATA USING FINITE ELEMENTS	68-0224
LEAST	LEAST SQUARES SURFACE FITTING BY FINITE ELEMENTS AND AN APPLICATION TO STRESS SMOOTHING	67-0154
LEAST	LOCAL AND GLOBAL SMOOTHING OF DISCONTINUOUS FINITE ELEMENT FUNCTIONS USING A LEAST SQUARES METHOD	74-0378
LEAST	LOCAL LEAST SQUARES STRESS SMOOTHING FOR PARABOLIC ISOPARAMETRIC ELEMENTS	75-0593
LEAST	TRANSIENT TWO-DIMENSIONAL PROBLEMS UTILIZING THE LEAST SQUARES ALGORITHM	75-0219
LEAST	USE OF THE LEAST SQUARES CRITERION IN THE FINITE ELEMENT FORMULATION	73-0270
LEAST-SQUARES	A DISCRETE LEAST-SQUARES FINITE ELEMENT METHOD FOR TRANSIENT HEAT CONDUCTION PROBLEMS	74-1191
LEAST-SQUARES	LEAST-SQUARES VARIATIONAL PRINCIPLE FOR FINITE-ELEMENT APPLICATIONS	75-0412
LECTURE	DYNAN LECTURE NOTES WITH COMPUTATIONAL EXAMPLES	71-0581
LECTURES	COMPUTATIONAL TECHNIQUES, NONLINEAR EQUATIONS, ISOPARAMETRIC ELEMENTS, FIVE LECTURES	73-0634
LECTURES	CURVED ELEMENTS, ERROR BOUNDS, TIME DEPENDENT PROBLEMS FIVE LECTURES	73-0636
LECTURES	ELASTIC EQUILIBRIUM MODELS, STRESS FUNCTIONS, AND DUALITY NON LINEARITIES AND STABILITY, FIVE LECTUR	73-0631
LECTURES	LECTURES ON FINITE ELEMENT METHODS IN CONTINUUM MECHANICS	71-0714
LECTURES	LECTURES ON MATHEMATICAL FOUNDATIONS OF THE FINITE ELEMENT METHOD	72-0777
LECTURES	PLATES AND SHELLS, INELASTIC ANALYSIS, TIME DEPENDENT INELASTICITY, FIVE LECTURES	73-0633
LECTURES	SURVEY LECTURES ON THE MATHEMATICAL FOUNDATIONS OF THE FINITE ELEMENT METHOD	72-0670
LECTURES	SURVEY OF UNDERLYING THEORY, APPROXIMATION BY PIECEWISE POLYNOMIALS, ILLEGAL ELEMENTS AND THE PATCH	73-0635
LEFT	DEFORMATION OF THE DIASTOLIC LEFT VENTRICLE - 2. NONLINEAR GEOMETRIC EFFECTS	74-0974
LEFT	DEFORMATION OF THE END DIASTOLIC LEFT VENTRICLE, COMPARISON OF THEORY WITH EXPERIMENT	71-0131
LEFT	FINITE ELEMENT ANALYSIS OF LEFT VENTRICULAR MYOCARDIAL STRESSES	74-0669
LEFT	FINITE ELEMENT CALCULATION OF LEFT VENTRICULAR WALL STRESS IN NORMAL AND DISEASE STATES	73-0588
LEFT	FINITE-ELEMENT MODEL FOR THE MECHANICAL BEHAVIOUR OF THE LEFT VENTRICLE, PREDICTION OF DEFORMATION I	72-0304
LEFT	IN VIVO STRESSES IN THE HUMAN LEFT VENTRICULAR WALL: ANALYSIS ACCOUNTING FOR THE IRREGULAR THREE DIM	72-0136
LEG	DYNAMIC TORSIONAL RESPONSE OF THE HUMAN LEG RELATIVE TO SKIING INJURIES	73-0471
LENGTH	COMPUTER WORD LENGTH AND SOLUTION ACCURACY. EXPERIENCE WITH ELAS AND THE IBM 360	73-0335
LENGTH	FREE VIBRATION OF THICK, LAYERED CYLINDERS HAVING FINITE LENGTH WITH VARIOUS BOUNDARY CONDITIONS	72-0491
LENGTH	OPTIMIZATION OF TENSION LEVEL AND STINGER LENGTH FOR OFFSHORE PIPELINE INSTALLATION	74-0504
LEVEES	PRELIMINARY FINITE ELEMENT ANALYSIS, ATCHAFALAYA BASIN PROTECTION LEVEES, TEST SECTION 3	69-0463
LEVEL	IS THE COMPUTATION OF NOISE LEVEL INSIDE A CAR FEASIBLE	71-0631
LEVEL	OPTIMIZATION OF TENSION LEVEL AND STINGER LENGTH FOR OFFSHORE PIPELINE INSTALLATION	74-0504
LEVELS	CORRELATION BETWEEN PRESSURE AND THERMAL TRANSIENT STRESS LEVELS DERIVED BY THEORY AND EXPERIMENT ON	73-0469
LEVITATION	SUBTERRANEAN GUIDEWAY RESPONSES TO LEVITATION VEHICLE LOADING	74-0927
LIBRARY	FEABL (FINITE ELEMENT ANALYSIS BASIC LIBRARY) USER'S GUIDE	72-0748
LIBRES	CONTRIBUTION A L'ANALYSE DES VIBRATIONS LIBRES D'UNE STRUCTURE. APPLICATION AUX AUBAGES DE TURBOMACH	75-0804
LIFE	ANALYSIS OF THE EFFECTS ON LIFE OF LEADING-EDGE HOLES IN AN AIRFOIL SUBJECTED TO ARBITRARY SPANWISE	75-0700
LIFE	APPLICATIONS OF FINITE ELEMENT STRESS ANALYSIS AND STRESS-STRAIN PROPERTIES IN DETERMINING NOTCH FAT	75-1114
LIFE	GAS TURBINE ENGINE DISK CYCLIC LIFE PREDICTION	75-0509
LIFE	HOLLOW-ENDED ROLLER. A SOLUTION FOR IMPROVING FATIGUE LIFE IN ASYMMETRICALLY LOADED CYLINDRICAL ROLL	71-0393
LIFE	V-BELT LIFE PREDICTION AND POWER RATING	75-0838
LIFETIME	ESTIMATES OF THE CREEP RUPTURE LIFETIME OF STRUCTURES USING THE FINITE ELEMENT METHOD	75-0833
LIFT-CONNECTED	THE MBB-UFE SUBSONIC PANEL METHOD. PART 2 - THE LIFT-CONNECTED DISPLACEMENT PROBLEM IN COMPRESSIBLE	72-0865
LIFTING	A FINITE-ELEMENT METHOD FOR LIFTING SURFACES IN STEADY INCOMPRESSIBLE SUBSONIC FLOW	74-1321
LIFTING	BERECHNUNG DES REIBUNGSLOSEN STROEMUNGSFELDES VON DREIDIMENSIONALEN AUFTRIEBSBEHAFTETENTRAGFLUEGELN	73-0533
LIFTING	CALCULATION OF POTENTIAL FLOW ABOUT ARBITRARY THREE-DIMENSIONAL LIFTING BODIES	72-0722
LIFTING	FINITE ELEMENT TECHNIQUE IN LIFTING SURFACE PROBLEMS	74-0022
LIFTING	JET-WING LIFTING - SURFACE THEORY USING ELEMENTARY VORTEX DISTRIBUTIONS	75-0245
LIFTING	LIFTING SURFACE PROBLEMS ANALYSIS	73-0333
LIFTING	PLANFORM INPUT GENERATOR (PING), A NASTRAN PRE-PROCESSOR FOR LIFTING SURFACES-THEORETICAL DEVELOPMEN	73-0776
LIFTING	PLANFORM INPUT GENERATOR (PING), A NASTRAN PROCESSOR FOR LIFTING SURFACES-THEORETICAL DEVELOPMENT. U	73-0780
LIFTING	THE APPLICATION OF FINITE ELEMENT METHOD TO THE LIFTING SURFACE PROBLEM	75-0868
LIFTING	TWO AND THREE DIMENSIONAL ANALYSIS OF SUBSONIC COMPRESSIBLE POTENTIAL FLOWS IN DUCTS AND AROUND LIFT	74-0026
LIFTING	VARIATIONAL APPROACH TO THE LIFTING SURFACE PROBLEM	73-0779
LIGAMENT	APPLICATION OF THE DOUBLE LIGAMENT TENSILE TEST IN DEFORMATION PROCESSING	75-0827
LIGAMENT	FINITE ELEMENT ANALYSIS OF PERFORATED PLATES CONTAINING TRIANGULAR PENETRATION PATTERNS OF 5 AND 10	75-0216
LIGAMENT	FINITE ELEMENT ANALYSIS OF PERFORATED PLATES CONTAINING TRIANGULAR PENETRATION PATTERNS OF 5 AND 10	75-0520
LIGAMENT	STRESS DISTRIBUTION IN THE LIGAMENT ZONE OF PERFORATED CIRCULAR DISCS WITH SPECIAL REFERENCE TO THE	73-0827
LIGHT	ANALYSIS OF LIGHT GATE STEEL SHEAR DIAPHRAGMS	73-0733
LIGHT	LIGHT GAGE STEEL INFILL PANELS IN MULTISTORY STEEL FRAMES	74-0229
LIGHT	NEW LIGHT ON THE NAYAK ALPHA TECHNIQUE	73-0614
LIGHT-GAGE	COMPUTERIZED STRUCTURAL ANALYSIS OF THE WORLD'S LARGEST LIGHT-GAGE STEEL PRIMARY STRUCTURAL SYSTEM	73-0503
LIGHTWEIGHT	A GENERAL METHOD FOR THE SHAPE FINDING OF LIGHTWEIGHT TENSION STRUCTURES	74-1217
LIGHTWEIGHT	ADVANCED INTERCONNECT SYSTEMS FOR LIGHTWEIGHT SOLAR ARRAYS	73-0696
LIGHTWEIGHT	ELASTIC DEFORMATION OF LIGHTWEIGHT MIRRORS METHOD OF LIMIT POINT CALCULATION IN FINITE ELEMENT STRUC	72-0232
LIGHTWEIGHT	PREFABRICATED PRETENSIONED LIGHTWEIGHT CONCRETE FOLDED PLATE	73-0808
LIKE	CONNECTION BETWEEN THE FINITE DIFFERENCE LIKE METHODS AND THE METHODS BASED ON INITIAL VALUE PROBLEM	74-0877
LIMESTONE	STABILITY ANALYSIS OF AN EXPERIMENTALLY DEFORMED SINGLE LAYER OF INDIANA LIMESTONE USING FINITE ELEM	74-0212
LIMIT	A FINITE ELEMENT LINEAR PROGRAMMING METHOD FOR THE LIMIT ANALYSIS OF THIN PLATES	75-1064
LIMIT	A METHOD OF LIMIT POINT CALCULATION IN FINITE ELEMENT STRUCTURAL ANALYSIS	72-0786
LIMIT	ANALYSE DE LA STABILITY DES TALUS ROCHEUX PAR UNE METHODE GENERALISEE DE L EQUILIBRE LIMITE (ROCK SL	74-1019
LIMIT	BERECHNUNGSMETHODEN ZUR ERMITTLUNG VON BELASTUNGSBEDINGUNGEN UND GRENZTRAGFAEHIGKEITSVERHALTEN VON K	73-0858
LIMIT	COMPARISON OF FINITE ELEMENT AND UNIQUE LIMIT ANALYSIS SOLUTIONS FOR CERTAIN REINFORCED CONCRETE SLA	69-0184
LIMIT	DISCUSSION ON "PLANE STRESS LIMIT ANALYSIS BY FINITE ELEMENTS"	71-0792
LIMIT	ELASTIC DEFORMATION OF LIGHTWEIGHT MIRRORS METHOD OF LIMIT POINT CALCULATION IN FINITE ELEMENT STRUC	72-0232
LIMIT	FINITE ELEMENT LIMIT ANALYSIS USING LINEAR PROGRAMMING.	72-0461
LIMIT	FINITE ELEMENT TECHNIQUES IN PLANE LIMIT PROBLEMS	74-1206
LIMIT	LIMIT ANALYSIS OF FRAME SYSTEMS STIFFENED BY PANELS	74-0584
LIMIT	LIMIT ANALYSIS OF REINFORCED CONCRETE SLABS BY FINITE ELEMENT METHOD	71-0802
LIMIT	LIMIT ANALYSIS OF REINFORCED CONCRETE SLABS BY A FINITE ELEMENT METHOD	72-0985
LIMIT	LOAD LIMIT ANALYSIS, COMBINATION OF STATISTICAL METHODS (MONTE CARLO METHOD) AND FINITE ELEMENT METH	71-0768
LIMIT	LOWER BOUND LIMIT ANALYSIS OF PLANE STRESS PROBLEMS BY A FINITE ELEMENT METHOD	70-0428
LIMIT	MIXED FINITE ELEMENT MODELS IN LIMIT ANALYSIS	74-1314
LIMIT	NONPROPORTIONAL LOADING LIMIT FOR STRUCTURES	74-0228
LIMIT	NUMERICAL METHODS FOR THE LIMIT ANALYSIS OF PLATES	68-0181
LIMIT	ON APPLICATION OF THE FINITE ELEMENT METHOD TO LIMIT ANALYSIS CONSISTENT FINITE ELEMENT TECHNIQUE FO	74-1234
LIMIT	ON APPLICATION OF THE FINITE ELEMENT METHOD TO LIMIT ANALYSIS	74-1237
LIMIT	PLANE STRESS LIMIT ANALYSIS BY FINITE ELEMENTS	70-0255
LIMITATIONS	COMMENT ON LIMITATIONS OF CERTAIN FINITE ELEMENTS WHEN APPLIED TO ARCHES - FURTHER STUDIES IN APPLIC	72-0097
LIMITATIONS	FURTHER LIMITATIONS ON GENERAL HEXAHEDRON FINITE ELEMENTS WITH DERIVATIVE DEGREES OF FREEDOM.	72-0395
LIMITATIONS	LIMITATIONS OF CERTAIN CURVED FINITE ELEMENTS WHEN APPLIED TO ARCHES	71-0279
LIMITATIONS	LIMITATIONS OF THE FINITE ELEMENT METHOD IN STRUCTURAL ANALYSIS	73-0420

LIMITATIONS	LIMITATIONS ON THE APPLICATION OF EFFECTIVE BREADTH CONCEPT IN BOX GIRDER DESIGN	74-0493
LIMITATIONS	LIMITATIONS TO THE SMALL SCALE YIELDING APPROXIMATION FOR CRACK TIP PLASTICITY	74-0145
LIMITATIONS	MODERN TECHNIQUES FOR CALCULATING STRESSES IN TANKS, AND THEIR LIMITATIONS, (MODERNE VERFAHREN ZUR S	70-0153
LIMITE	ANALYSE DE LA STABILITY DES TALUS ROCHEUX PAR UNE METHODE GENERALISEE DE L EQUILIBRE LIMITE (ROCK SL	74-1019
LIMITED	APPLICATION OF FINITE ELEMENT METHOD TO MASS TRANSPORT LIMITED EPITAXIAL GROWTH PROCESSES	71-0488
LIMITED	REANALYSIS FOR LIMITED STRUCTURAL DESIGN MODIFICATIONS	72-0650
LIMITES	LA METHODE DES ELEMENTS FINIS POUR LES PROBLEMES AUX LIMITES ELLIPTIQUES	74-1228
LIMITING	APPLICATION OF RESPONSE LIMITING TO FINITE-ELEMENT STRUCTURAL-ANALYSIS PROGRAMS	75-0424
LIMITING	COMPUTER PROGRAM FOR PIT SLOPE STABILITY ANALYSIS BY THE FINITE ELEMENT STRESS ANALYSIS AND LIMITING	72-0225
LIMITING	ON THE APPLICATION OF RESPONSE LIMITING TO FINITE ELEMENT STRUCTURAL ANALYSIS PROGRAMS	75-0192
LIMITING	ON THE APPLICATION OF RESPONSE LIMITING TO FINITE ELEMENT STRUCTURAL ANALYSIS PROGRAMS	75-0822
LIMITING	SLOPE STABILITY ANALYSIS BY THE FINITE ELEMENT STRESS ANALYSIS AND LIMITING EQUILIBRIUM METHOD	70-0222
LIMITS	PRESSURE VESSEL WALL THICKNESS LIMITS BASED ON THERMAL STRESS RACHETTING AND CREEP FATIGUE INTERACTI	74-0968
LINE	A METHOD OF ANALYSIS OF LINE STRUCTURES BY TRANSFER MATRICES DERIVED FROM FINITE ELEMENTS	74-0808
LINE	MATRIX ANALYSIS OF A COMPOSITE STRUCTURE OF PLATE AND LINE ELEMENTS	71-0250
LINE	NUMERICAL ANALYSIS OF THE LINE CAPACITANCE AND CROSSTALK FACTOR FOR INSULATED WIRE PAIRS	74-0685
LINE	NUMERICAL SOLUTION OF SOME NONSTATIONARY PARTIAL DIFFERENTIAL EQUATIONS: THE FINITE ELEMENT METHOD A	72-0930
LINE	STRESS, DISPLACEMENT, LINE INTEGRAL AND CLOSURE ENERGY DETERMINATIONS OF CRACK TIP STRESS INTENSITY	74-0621
LINE-OF-SIGHT	THERMOSTRUCTURAL DESIGN CONSIDERATIONS TO ACHIEVE THE LARGE SPACE TELESCOPE LINE-OF-SIGHT REQUIREMEN	75-0839
LINEAIRE	ANALYSE NUMERIQUE DU CHAMP MAGNETIQUE D'UN ALTERNATEUR PAR ELEMENTS FINIS ET SURRELAXATION PONCTUELL	73-0625
LINEAIRE	FINITE ELEMENT METHOD IN NONLINEAR DOMAIN, (METHODE DES ELEMENTS FINIS DANS LE DOMAINE NON LINEAIRE)	71-0288
LINEAIRE	NONLINEAR ANALYSIS OF PRESTRESSED CONCRETE FAST REACTOR PRESSURE VESSEL, (ANALYSE NON LINEAIRE DES C	70-0145
LINEAIRES	ANALYSE DES VOILES MINCES DANS LES DOMAINES LINEAIRES ET GEOMETRIQUEMENT NONLINEAIRES PAR LA METHODE	71-0122
LINEAIRES	SUR L APPROXIMATION, PAR ELEMENTS FINIS D ORDRE UN, ET LA RESOLUTION, PAR PENALISATION-DUALITE, D UN	75-0845
LINEAR	A COMPACT STORAGE SCHEME FOR THE SOLUTION OF SYMMETRICAL LINEAR SIMULTANEOUS EQUATIONS	66-0116
LINEAR	A FINITE ELEMENT LINEAR PROGRAMMING METHOD FOR THE LIMIT ANALYSIS OF THIN PLATES	73-1064
LINEAR	A MATHEMATICAL THEORY OF COMPLEMENTARY-DUAL VARIATIONAL PRINCIPLES AND MIXED FINITE ELEMENT APPROXIM	73-0804
LINEAR	A NOTE ON THE FINITE ELEMENT METHOD IN LINEAR FRACTURE MECHANICS	70-0489
LINEAR	A TRIANGULAR EQUILIBRIUM (LINEAR STRESS) MEMBRANE ELEMENT	75-0883
LINEAR	A TRIANGULAR THIN SHELL FINITE ELEMENT: LINEAR ANALYSIS	75-0698
LINEAR	AN ECONOMICAL METHOD FOR DETERMINING THE SMALLEST EIGENVALUES OF LARGE LINEAR SYSTEMS	71-0551
LINEAR	AN IMBEDDING TECHNIQUE FOR THE GENERATION OF WEAK-WEAK FINITE ELEMENT APPROXIMATIONS OF LINEAR AND N	72-0577
LINEAR	AN ITERATIVE METHOD FOR LARGE SYSTEMS OF LINEAR STRUCTURAL EQUATIONS	73-1020
LINEAR	ANALIZA ORGAN WYMUSZONYCH ZLOZONYCH UKLADOW LINIOWYCH METODA SZTYWNYCH ELEMENTOW SKONCZONYCH (ANALYS	72-0246
LINEAR	ANALYSIS OF AXISYMMETRIC LINEAR HEAT CONDUCTION PROBLEMS BY FINITE ELEMENT METHOD	69-0015
LINEAR	ANTRAS COMPUTER PROGRAM FOR LINEAR STRESS ANALYSIS BY FINITE ELEMENT METHOD	74-0601
LINEAR	APPLICATION OF MATRIX DISPLACEMENT METHOD TO LINEAR ELASTIC ANALYSIS OF SHELLS OF REVOLUTION	65-0025
LINEAR	APPLICATION OF QUADRATIC ISOPARAMETRIC FINITE ELEMENTS IN LINEAR FRACTURE MECHANICS	74-0521
LINEAR	APPLICATION OF THE FAST FOURIER TRANSFORM TO LINEAR SYSTEMS IN CIVIL ENGINEERING	73-1119
LINEAR	APPLICATION OF THE FINITE ELEMENT METHOD FOR SOLVING ORDINARY LINEAR DIFFERENTIAL EQUATION WITH BOUN	74-0084
LINEAR	APPLICATION OF THE FINITE ELEMENT METHOD TO PROBLEMS IN LINEAR FRACTURE MECHANICS	74-0903
LINEAR	APPROXIMATE SOLUTION IN LINEAR VISCOELASTICITY	66-0101
LINEAR	APPROXIMATE SOLUTIONS IN LINEAR COUPLED THERMOELASTICITY	68-0220
LINEAR	ARCH DAMS ANALYSED BY A LINEAR FINITE ELEMENT SHELL THE FINITE STRIP METHOD IN THE ANALYSIS OF ELAST	68-0189
LINEAR	ASSUMED DISPLACEMENT HYBRID FINITE ELEMENT MODEL FOR LINEAR FRACTURE MECHANICS	75-0146
LINEAR	ASSUMED STRESS HYBRID FINITE ELEMENT MODEL FOR LINEAR ELASTODYNAMIC ANALYSIS	73-0089
LINEAR	BANDED MATRIX TECHNIQUES AND SOLUTIONS OF LINEAR SIMULTANEOUS EQUATIONS OF FINITE ELEMENT ANALYSES	70-0302
LINEAR	BOUNDS FOR A CLASS OF LINEAR FUNCTIONALS WITH APPLICATIONS TO HERMITE INTERPOLATION	71-0582
LINEAR	CHILES: A FINITE ELEMENT COMPUTER PROGRAM THAT CALCULATES THE INTENSITIES OF LINEAR ELASTIC SINGULAR	73-0959
LINEAR	COMPUTER ANALYSIS OF LARGE LINEAR FRAMES	69-0408
LINEAR	COMPUTER PROGRAM FOR STATIC AND DYNAMIC ANALYSIS OF LINEAR STRUCTURAL SYSTEMS	72-0683
LINEAR	COMPUTER PROGRAM FOR SOLUTION OF LARGE, SPARSE, UNSYMMETRIC SYSTEMS OF LINEAR EQUATIONS	72-0719
LINEAR	COMPUTER PROGRAMS FOR THE SOLUTION OF SYSTEMS OF LINEAR ALGEBRAIC EQUATIONS	73-0932
LINEAR	COMPUTER PROGRAMS FOR THE SOLUTION OF SYSTEMS OF LINEAR ALGEBRAIC EQUATIONS	73-1040
LINEAR	DAS PROGRAMMSYSTEM ANTRAS EDV-SYSTEM ZUR BERECHNUNG BELIEBIGER STATISCHER SYSTEME NACH DER METHODE D	74-0918
LINEAR	DECONVOLUTION OF SEISMIC RESPONSE FOR LINEAR SYSTEMS	73-0774
LINEAR	DUAL ANALYSIS BY FINITE ELEMENTS: LINEAR AND NONLINEAR APPLICATIONS	72-0476
LINEAR	EFFICIENT FORMULATIONS OF THE FINITE ELEMENT METHOD IN LINEAR AND NONLINEAR FRACTURE MECHANICS	75-0403
LINEAR	ELAS - A GENERAL PURPOSE COMPUTER PROGRAM FOR THE EQUILIBRIUM PROBLEMS OF LINEAR STRUCTURES	68-0158
LINEAR	ELAS: A GENERAL-PURPOSE COMPUTER PROGRAM FOR THE EQUILIBRIUM PROBLEMS OF LINEAR STRUCTURES, VOLUME 2	69-0462
LINEAR	ELEMENT STIFFNESS MATRICES FOR CONSTANT AND LINEAR STRESS - TRIANGULAR, CONSTANT AND LINEAR STRESS T	67-0006
LINEAR	ELEMENT STIFFNESS MATRICES FOR CONSTANT AND LINEAR STRESS - TRIANGULAR, CONSTANT AND LINEAR STRESS T	67-0006
LINEAR	EVALUATION OF DISCRETE METHODS FOR THE LINEAR DYNAMIC RESPONSE OF ELASTIC AND VISCOELASTIC SOLIDS	70-0439
LINEAR	FINITE ELEMENT ANALYSIS OF PAVEMENT STRUCTURES USING AFPAV CODE LINEAR ELASTIC ANALYSIS	73-0658
LINEAR	FINITE ELEMENT ANALYSIS OF LINEAR AND NONLINEAR HEAT TRANSFER	74-0626
LINEAR	FINITE ELEMENT APPLICATIONS IN LINEAR AND NONLINEAR THERMOVISCOELASTICITY	67-0065
LINEAR	FINITE ELEMENT APPLICATIONS OF A PRECISION TRIANGULAR PLATE BENDING ELEMENT TO LINEAR BOUNDARY VALUE	75-0788
LINEAR	FINITE ELEMENT FORMULATION FOR LINEAR THERMOVISCO ELASTIC MATERIALS	69-0351
LINEAR	FINITE ELEMENT LIMIT ANALYSIS USING LINEAR PROGRAMMING.	72-0461
LINEAR	FINITE ELEMENT METHOD IN LINEAR FRACTURE MECHANICS	70-0476
LINEAR	FINITE ELEMENT METHOD OF MIXED TYPE AND ITS CONVERGENCE IN LINEAR SHELL PROBLEMS	73-1041
LINEAR	FINITE ELEMENT METHOD IN LINEAR VISCOELASTICITY	74-0213
LINEAR	FINITE ELEMENT METHOD FOR TRANSIENT LINEAR VISCOUS FLOW PROBLEMS	74-0474
LINEAR	FINITE ELEMENT METHOD FOR LINEAR FRACTURE MECHANICS	74-1359
LINEAR	FINITE ELEMENTS IN LINEAR VISCOELASTICITY	68-0281
LINEAR	FINITE-ELEMENT CODE TO SIMULATE ANISOTROPIC-PLASMA DIFFUSION IN LINEAR MULTIPOLES	75-0409
LINEAR	FURTHER APPLICATION OF QUADRATIC ISOPARAMETRIC FINITE-ELEMENTS TO LINEAR FRACTURE MECHANICS OF PLATE	75-0639
LINEAR	FURTHER DEVELOPMENTS IN THE USE OF LINEAR DISPLACEMENT FUNCTIONS IN THE FINITE ELEMENT ANALYSIS OF O	71-0344
LINEAR	GENERAL LINEAR GEOMETRIC MATRIX FOR A FULLY COMPATIBLE FINITE ELEMENT	72-0294
LINEAR	GENERAL VARIATIONAL PRINCIPLES IN NONLINEAR AND LINEAR ELASTICITY WITH APPLICATIONS	72-0914
LINEAR	GENERALISED MATRIX FORCE AND DISPLACEMENT METHODS FOR LINEAR STRUCTURAL ANALYSIS	65-0064
LINEAR	HIGHER-ORDER LINEAR FINITE ELEMENT METHODS	74-0826
LINEAR	HYBRIDE FINITE VERFAHREN FUER LINEARE UND NICHTLINEARE INSTATIONAERE WAERMELEITPROBLEME (HYBRID FINI	73-0484
LINEAR	IMPROVED LINEAR AXISYMMETRIC SHELL FLUID MODEL FOR LAUNCH VEHICLE LONGITUDINAL RESPONSE ANALYSIS	65-0044
LINEAR	INTERACTIVE SOLUTION OF LINEAR EQUATIONS IN FINITE ELEMENT ANALYSIS	73-1163
LINEAR	ITERATIVE SOLUTION OF LINEAR SYSTEMS ARISING FROM FINITE ELEMENT TECHNIQUES	75-0126
LINEAR	LA METHODE FRONTALE POUR LA RESOLUTION DES SYSTEMES LINEAIRES. (FRONTAL METHOD FOR SOLVING LINEAR SY	73-1097
LINEAR	LINEAR ACCELERATOR CAVITY FIELD CALCULATION BY FINITE ELEMENT	73-0008
LINEAR	LINEAR AND NON LINEAR ANALYSIS OF SHELLS OF REVOLUTION WITH ASYMMETRICAL STIFFNESS PROPERTIES	68-0151
LINEAR	LINEAR AND NON LINEAR ANALYSIS OF SHELLS OF REVOLUTION WITH ASYMMETRICAL STIFFNESS PROPERTIES	68-0151
LINEAR	LINEAR AND NON-LINEAR ANALYSIS OF THIN SHALLOW SHELLS BY MIXED FINITE ELEMENTS	75-0665
LINEAR	LINEAR AND NONLINEAR STATIC ANALYSIS OF AXISYMMETRICALLY LOADED THIN SHELLS OF REVOLUTION	69-0215
LINEAR	LINEAR AND NONLINEAR ANALYSIS OF SHELLS	72-0832
LINEAR	LINEAR AND NONLINEAR FINITE ELEMENT ANALYSIS OF WOOD STRUCTURAL MEMBERS	73-0117
LINEAR	LINEAR AND NONLINEAR OSCILLATIONS IN MECHANICS	74-0677
LINEAR	LINEAR BUCKLING AND GEOMETRICALLY NONLINEAR ANALYSIS OF PLANAR PLATE-STIFFENER TYPE STRUCTURES BY TH	74-1097
LINEAR	LINEAR DISCONTINUOUS FINITE-ELEMENT SCHEME IN TIME-DEPENDENT TRANSPORT CALCULATIONS	75-0498
LINEAR	LINEAR DYNAMIC ANALYSES OF LAMINATED PLATES AND SHELLS BY THE HYBRID-STRESS FINITE-ELEMENT METHOD	73-0786
LINEAR	LINEAR ELASTIC FINITE ELEMENT ANALYSIS OF MASONRY WALLS ON BUILDINGS	74-1015
LINEAR	LINEAR EQUALITY CONSTRAINTS IN FINITE ELEMENT APPROXIMATION	75-0475
LINEAR	LINEAR METHODS OF STRUCTURAL ANALYSIS	71-0028
LINEAR	LINEAR PROGRAMMING APPROACH TO SHAKEDOWN ANALYSIS OF STRUCTURES	74-0181

269

LINEAR	LINEAR SHELL THEORY FOR RAPID VARIATION SOLUTIONS	71-0406
LINEAR	LINEAR STABILITY ANALYSIS OF BEAM-COLUMNS	73-0499
LINEAR	LINEAR WAVE PROPAGATION PROBLEMS AND THE FINITE ELEMENT METHOD	74-1340
LINEAR	LINEAR WAVE PROPAGATION PROBLEMS AND THE FINITE ELEMENT METHOD	75-0395
LINEAR	MATHEMATICAL THEORY OF LINEAR STRUCTURES	71-0498
LINEAR	MATRIX DISPLACEMENT ANALYSIS OF PLATES AND SHELLS, PART 11, GENERAL FORMULATION OF THE LINEAR THEORY	71-0621
LINEAR	MEAN-SQUARE ERRORS OF FINITE ELEMENT APPROXIMATIONS ON LINEAR ELASTOSTATICS, ELASTODYNAMICS, AND THE	74-0107
LINEAR	MIXED CONJUGATE FINITE ELEMENT APPROXIMATIONS OF LINEAR OPERATORS	72-0085
LINEAR	MIXED FINITE-ELEMENT APPROXIMATIONS OF LINEAR BOUNDARY-VALUE PROBLEMS	75-0453
LINEAR	NON LINEAR ANALYSIS OF REINFORCED CONCRETE MEMBERS BY FINITE ELEMENT METHOD, PART 1: ANALYSIS, PART	71-0313
LINEAR	NOTE ON THE FINITE ELEMENT METHOD IN LINEAR FRACTURE MECHANICS	70-0474
LINEAR	NUMERICAL ANALYSIS IN ORTHOTROPIC LINEAR FRACTURE MECHANICS	73-0512
LINEAR	NUMERICAL SOLUTION OF THE QUASI LINEAR POISSON EQUATION IN A NON-UNIFORM TRIANGLE MESH	67-0090
LINEAR	NUMERICAL TECHNIQUES IN LINEAR AND NONLINEAR FRACTURE MECHANICS	75-0260
LINEAR	ON THE FINITE ELEMENT METHOD IN LINEAR FRACTURE MECHANICS	69-0032
LINEAR	ON THE NUMERICAL SOLUTION OF A CLASS OF PROBLEMS IN A LINEAR FIRST STRAIN-GRADIENT THEORY OF ELASTIC	70-0318
LINEAR	PERTURBATION METHOD IN THE ANALYSIS OF GEOMETRICALLY NON LINEAR AND STABILITY PROBLEMS	72-0321
LINEAR	PLATES AND BEAMS ON ELASTIC FOUNDATIONS: LINEAR AND NONLINEAR BEHAVIOUR	68-0042
LINEAR	POST-BUCKLING AND NON-LINEAR ANALYSIS BY FINITE ELEMENT METHOD AS A SUPPLEMENT TO A LINEAR ANALYSIS	75-0505
LINEAR	PROCEDURE FOR SOLVING LINEAR EQUATION SYSTEMS WITH SPARSELY OCCUPIED COEFFICIENT MATRICES ASSOCIATED	74-0352
LINEAR	REFINED FINITE ELEMENT ANALYSIS OF LINEAR AND NONLINEAR TWO-DIMENSIONAL STRUCTURES	66-0024
LINEAR	RELIABILITY STUDY OF LINEAR VARIATION TRIANGULAR FINITE ELEMENT	72-0281
LINEAR	SABOR 1, A FORTRAN PROGRAM FOR THE LINEAR ELASTIC AXISYMMETRIC LOADING BY USING THE MATRIX DISPLACEM	65-0027
LINEAR	SAP IV, A STRUCTURAL ANALYSIS PROGRAM FOR STATIC AND DYNAMIC RESPONSE OF LINEAR SYSTEMS	73-0650
LINEAR	SAP-A GENERAL STRUCTURAL ANALYSIS PROGRAM FOR LINEAR SYSTEMS	72-0329
LINEAR	SHAKEDOWN THEORY IN PERFECT ELASTOPLASTICITY WITH ASSOCIATED AND NON ASSOCIATED FLOW LAWS - A FINITE	69-0172
LINEAR	SIMPLIFIED HYBRID DISPLACEMENT METHOD FOR LINEAR FINITE ELEMENT ANALYSIS OF GENERAL SHELLS	74-0614
LINEAR	SOLUTION OF LINEAR EQUATIONS WITH SKYLINE-STORED SYMMETRIC MATRIX	75-0662
LINEAR	SOLVING A CLASS OF LARGE SPARSE LINEAR SYSTEMS OF EQUATIONS BY PARTITIONING	73-0650
LINEAR	SOLVING THE LINEAR BALANCE EQUATION GLOBALLY	74-0810
LINEAR	SOME EXAMINATIONS ON AVAILABILITY OF FINITE ELEMENT METHODS IN LINEAR FRACTURE MECHANICS	73-0112
LINEAR	SOME LINEAR AND NON-LINEAR PROBLEMS IN FLUID MECHANICS, FINITE ELEMENT METHOD FORMULATION	73-0523
LINEAR	SOME NON LINEAR PROBLEMS IN SOIL AND ROCK MECHANICS - FINITE ELEMENT SOLUTION	69-0363
LINEAR	SOME NONCONFORMING VARIANTS ON THE TRILINEAR ISOPARAMETRIC BRICK FOR LINEAR ELASTIC PROBLEMS	72-0531
LINEAR	SOME PROPERTIES OF LINEAR STRAIN TRIANGLES AND OPTIMAL FINITE ELEMENT MODELS	73-1044
LINEAR	STATIC STRESSES BY LINEAR AND NONLINEAR METHODS	75-0278
LINEAR	STEADY CONDUCTION OF HEAT IN LINEAR AND NONLINEAR FULLY ANISOTROPIC MEDIA BY FINITE ELEMENTS	74-0359
LINEAR	STUDY ON FINITE ELEMENT METHOD FOR STRUCTURAL ANALYSIS - ISTRAN/PL (INH STRUCTURE ANALYSIS PLATE STR	73-0167
LINEAR	SU LA DETERMINAZIONE DEI CAMPI MAGNETICI LINEARI E NON LINEARI MEDIANTE L APPLICAZIONE DEI METODI DE	74-1061
LINEAR	SURVEY AND EVALUATION OF THE FINITE ELEMENT METHOD IN LINEAR FRACTURE MECHANICS ANALYSIS	71-0094
LINEAR	THE CONVERGENCE OF FINITE ELEMENT METHOD IN SOLVING LINEAR ELASTIC PROBLEMS	67-0083
LINEAR	THE LINEAR ELASTIC DYNAMIC ANALYSIS OF SHELLS OF REVOLUTION BY THE MATRIX DISPLACEMENT METHOD	66-0046
LINEAR	THE PIECEWISE LINEAR ANALYSIS OF TWO CONNECTED STRUCTURES INCLUDING THE EFFECT OF CLEARANCE AT THE C	71-0379
LINEAR	THE TRANSIENT LINEAR ELASTIC RESPONSE ANALYSIS COMPLEX THIN SHELLS OF REVOLUTION SUBJECTED TO ARBITR	70-0080
LINEAR	THEORETICAL FORMULATION OF FINITE ELEMENT METHODS IN LINEAR ELASTIC ANALYSIS OF GENERAL SHELLS	72-0496
LINEAR	THREE-DIMENSIONAL LINEAR FRACTURE MECHANICS ANALYSIS BY A DISPLACEMENT-HYBRID FINITE-ELEMENT MODEL	75-0884
LINEAR	TRANSIENT RESPONSE OF LINEAR ELASTIC STRUCTURES DETERMINED BY THE MATRIX EXPONENTIAL METHOD	71-0240
LINEAR	TRANSIENT VIBRATION ANALYSIS OF LINEAR SYSTEMS USING TRANSITION MATRICES	68-0053
LINEAR	TRIANGULAR THIN SHELL FINITE ELEMENT: LINEAR ANALYSIS	75-0602
LINEAR	TWO DIMENSIONAL NON LINEAR ANALYSIS OF REINFORCED CONCRETE MEMBERS BY THE FINITE ELEMENT METHOD, PAR	71-0210
LINEAR	USE OF ISOPARAMETRIC FINITE-ELEMENTS IN LINEAR FRACTURE MECHANICS	75-0352
LINEAR	VARIATIONAL PRINCIPLES FOR LINEAR INITIAL VALUE PROBLEMS	64-0029
LINEAR	VARIATIONAL PRINCIPLE OF LINEAR DIFFERENTIAL EQUATIONS	72-0107
LINEAR	VARIATIONAL SOLUTIONS BY USE OF STEPWISE CONSTANT FUNCTIONS: ONE LINEAR CASE	72-0286
LINEAR	VISCEL A GENERAL PURPOSE COMPUTER PROGRAM FOR ANALYSIS OF LINEAR VISCOELASTIC STRUCTURES, VOL. 2	72-0789
LINEAR-BOUNDARY	CONVERGENCE OF MIXED FINITE ELEMENT APPROXIMATIONS OF A CLASS OF LINEAR-BOUNDARY VALUE PROBLEMS	73-0599
LINEARE	HYBRIDE FINITE VERFAHREN FUER LINEARE UND NICHTLINEARE INSTATIONAERE WAERMELEITPROBLEME (HYBRID FINI	73-0484
LINEARE	ON THE RESOLUTION OF NONLINEAR ELASTIC AND ELASTOPLASTIC PROBLEMS BY THE METHOD OF FINITE ELEMENTS.	74-1326
LINEAREN	TURBAN VERSION 1.0. EIN SPEZIELLES PROGRAMM ZUR LINEAREN UND NICHTLINEAREN STATISCHEN UND DYNAMISCHE	71-0622
LINEARESS	LA METHODE FRONTALE POUR LA RESOLUTION DES SYSTEMES LINEARESS. (FRONTAL METHOD FOR SOLVING LINEAR SY	73-1097
LINEARI	SU LA DETERMINAZIONE DEI CAMPI MAGNETICI LINEARI E NON LINEARI MEDIANTE L APPLICAZIONE DEI METODI DE	74-1061
LINEARI	SU LA DETERMINAZIONE DEI CAMPI MAGNETICI LINEARI E NON LINEARI MEDIANTE L APPLICAZIONE DEI METODI DE	74-1061
LINEARITIES	ELASTIC EQUILIBRIUM MODELS, STRESS FUNCTIONS, AND DUALITY NON LINEARITIES AND STABILITY, FIVE LECTUR	73-0631
LINEARIZATION	APPLICATION OF FINITE ELEMENT METHOD WITH SECTIONAL LINEARIZATION TO FLOW PROBLEMS	75-0611
LINEARIZED	FINITE ELEMENT ANALYSIS OF THE UNSTEADY, LINEARIZED BOLTZMANN EQUATION	70-0033
LINEARIZED	FINITE ELEMENT ANALYSIS OF THE LINEARIZED BGK BOLTZMANN EQUATION.	72-0408
LINEARIZED	LINEARIZED COMPRESSIBLE FLOW BY THE FINITE ELEMENT METHOD	69-0025
LINEARIZED	LINEARIZED LARGE DEFLECTION PROBLEM	73-0685
LINEARLY	A COMPUTER PROGRAM FOR THE ANALYSIS OF LINEARLY ELASTIC PLANE-STRESS, PLANE-STRAIN PROBLEMS	68-0110
LINEARLY	A FINITE ELEMENT METHOD OF SOLUTION FOR LINEARLY ELASTIC BEAM-COLUMNS	66-0117
LINEARLY	A TRIANGULAR EQUILIBRIUM ELEMENT WITH LINEARLY VARYING BENDING MOMENTS FOR PLATE BENDING PROBLEMS	67-0054
LINEARLY	A VARIATIONAL FORMULATION FOR DISCONTINUOUS FIELDS IN LINEARLY ELASTIC FINITE ELEMENT APPLICATIONS	72-0962
LINEARLY	AXISYMMETRIC VIBRATIONS OF LINEARLY TAPERED ANNULAR PLATES	74-1266
LINEARLY	MEMBRANE PARALLELOGRAM ELEMENT WITH LINEARLY VARYING EDGE STRAIN FOR THE MATRIX DISPLACEMENT METHOD	66-0006
LINEARLY	REINFORCED FIELDS OF TRIANGULAR ELEMENTS WITH LINEARLY VARYING STRAIN: EFFECT OF INITIAL STRAINS	65-0004
LINEARLY	TETRAHEDRON ELEMENTS WITH LINEARLY VARYING STRAIN FOR THE MATRIX DISPLACEMENT METHOD	65-0006
LINEARLY	TRIANGULAR ELEMENTS WITH LINEARLY VARYING STRAIN FOR THE MATRIX DISPLACEMENT METHOD	65-0053
LINEARLY	TRIANGULAR FINITE ELEMENTS FOR PLATE BENDING WITH CONSTANT AND LINEARLY VARYING BENDING MOMENTS	70-0019
LINER	CASE LINER BOND ANALYSIS	74-0817
LINER-MEDIUM	LINER-MEDIUM INTERACTION IN TUNNELS	75-0052
LINERS	A MATRIX DISPLACEMENT METHOD ON PRE AND POST BUCKLING ANALYSIS OF LINERS FOR REACTOR VESSELS	71-0267
LINERS	FATIGUE STRENGTH OF LUGS CONTAINING LINERS. VOLUME II. COMPUTER PROGRAM USED FOR ANALYSIS	70-0418
LINES	FINITE ELEMENT VIBRATION ANALYSIS OF MULTICONDUCTOR ELECTRICAL TRANSMISSION LINES	73-0590
LINES	FLOW BETWEEN LINES OF A GIVEN PRESSURE	74-0020
LINES	MECHANICS OF CABLE MOORING SYSTEMS. VOLUME 1. THREE DIMENSIONAL RESPONSE OF DEEP WATER MOORING LINES	72-0654
LINES	SINGULARITIES IN ELECTROMAGNETIC TRANSMISSION LINES	72-0072
LINES	THREE-DIMENSIONAL HARMONIC FUNCTIONS NEAR TERMINATION OR INTERSECTION OF GRADIENT SINGULARITY LINES:	74-0178
LINING	FAILURE OF THE CONCRETE LINING IN AN INCLINED SHAFT AT GREAT DEPTH.	74-0996
LINING	INTERACTION OF ROCK AND TUNNEL LINING	73-1057
LINING	POLYMER-IMPREGNATED CONCRETE TUNNEL SUPPORT AND LINING	73-0736
LININGS	STRUCTURAL DESIGN DATA FOR CONCRETE DRIFT LININGS IN BLOCK CAVING	73-0773
LINIOWYCH	ANALIZA DRGAN WYMUSZONYCH ZLOZONYCH UKLADOW LINIOWYCH METODA SZTYWNYCH ELEMENTOW SKONCZONYCH (ANALYS	72-0246
LINK	ANALYSIS AND DESIGN OF MULTIPLE INPUT FLEXIBLE LINK MECHANISMS	70-0354
LINK	OPTIMAL SYNTHESIS OF FLEXIBLE LINK MECHANISMS WITH LARGE STATIC DEFLECTIONS	74-0525
LINKS	FORCE AND TORQUE ANALYSIS OF CONSTRAINED SPACE MECHANISMS AND PLANE MECHANISMS WITH OFFSET LINKS BY	74-0569
LIQUID	CALCULATION OF THE FREE OSCILLATIONS OF A LIQUID IN AXISYMMETRIC MOTIONLESS CONTAINERS OF ARBITRARY	74-0387
LIQUID	COMMENT ON DISCRETE ELEMENT IDEALIZATION OF AN INCOMPRESSIBLE LIQUID FOR VIBRATION ANALYSIS	73-0694
LIQUID	DESIGN AND ANALYSIS EXPERIENCE WITH A LIQUID METAL HEAT EXCHANGER FOR PFTF SERVICE	72-0772
LIQUID	DISCRETE ELEMENT IDEALIZATION OF AN INCOMPRESSIBLE LIQUID FOR VIBRATION ANALYSIS	70-0391
LIQUID	FINITE ELEMENT ANALYSIS FOR LIQUID SLOSHING PROBLEMS	69-0029
LIQUID	FREE VIBRATION OF CYLINDRICAL SHELLS PARTIALLY FILLED WITH LIQUID	71-0839
LIQUID	LIQUID SLOSHING IN AN ELASTIC CONTAINER	66-0096

270

LIQUID	MEASURED AND PREDICTED LONGITUDINAL VIBRATIONS OF A LIQUID PROPELLANT TWO-STAGE LAUNCH VEHICLE	70-0366
LIQUID	SLOSHING OF LIQUID IN CIRCULAR CHANNELS AND SPHERICAL TANKS	60-0005
LIQUID	THIN WINDOWS FOR GASEOUS AND LIQUID TARGETS: AN OPTIMIZATION PROCEDURE	75-0741
LIQUID	VIBRATION CHARACTERISTICS OF THIN, PRESSURIZED SHELLS OF REVOLUTION PARTIALLY FILLED WITH LIQUID	72-0604
LIQUID-PROPELLA	EVALUATION OF A FINITE ELEMENT ANALYSIS FOR THE LONGITUDINAL VIBRATIONS OF LIQUID-PROPELLANT LAUNCH	70-0204
LIQUID-WASTE	ANALYSIS OF LIQUID-WASTE INJECTION WELLS IN ILLINOIS BY MATHEMATICAL MODELS	74-0518
LIQUIDES	UNE METHODE DE CALCUL PAR ELEMENTS FINIS DES MOUVEMENTS DE LIQUIDES DANS DES RESERVOIRS RIGIDES OU D	74-1171
LIQUIDS	DYNAMIC BEHAVIOR OF LIQUIDS IN ELASTIC TANKS	71-0141
LIQUIDS	ON A SPECTRAL PROBLEM IN VIBRATION MECHANICS: COMPUTATION OF ELASTIC TANKS PARTIALLY FILLED WITH LIQ	75-0267
LIQUIDS	THEORETICAL AND EXPERIMENTAL CONSIDERATIONS ON FLOW OF VISCOUS LIQUIDS PAST PLATE AT SMALL AND AVERA	70-0144
LIQUIDS	VIBRATIONS OF TANKS PARTIALLY FILLED WITH LIQUIDS	74-0089
LISKEARD	FOUNDATION FAILURE OF NEW LISKEARD EMBANKMENT	73-0549
LISKHEARD	BEHAVIOR OF EMBANKMENTS ON NEW LISKHEARD VARVED CLAY	73-0721
LISTING	A FINITE ELEMENT BIBLIOGRAPHY - PART I: AUTHOR LISTING	74-0284
LISTING	A FINITE ELEMENT BIBLIOGRAPHY - PART II: KEYWORD LISTING	74-0285
LISTING	A FINITE ELEMENT BIBLIOGRAPHY - PART III: CITATION LISTING	74-0286
LISTING	FINITE ELEMENT ANALYSIS OF ANISOTROPIC PLATES USING 0-19 ELEMENT: PART 11. INSTRUCTIONS FOR USERS AN	74-0731
LISTING	MISSILE BODY INPUT GENERATOR (BING), A NASTRAN PRE-PROCESSOR: THEORETICAL DEVELOPMENT, USER'S MANUAL	75-0717
LISTING	PLANFORM INPUT GENERATOR (PING), A NASTRAN PROCESSOR FOR LIFTING SURFACES-THEORETICAL DEVELOPMENT. U	74-0780
LIVES	EXPERIMENTAL AND THEORETICAL FATIGUE-CRACK PROPAGATION LIVES OF VARIOUSLY NOTCHED PLATES	74-0602
LLOYD'S	SHIP STRUCTURAL ANALYSIS IN LLOYD'S REGISTER OF SHIPPING	75-0285
LMFBR	APPLICATION OF FINITE ELEMENT METHODS FOR THE ANALYSIS OF THERMAL CREEP, IRRADIATION INDUCED CREEP,	71-0242
LMFBR	DESIGN ANALYSIS FOR LMFBR CORE RESTRAINT	71-0208
LNG	LEAKAGE THROUGH CRACKS IN LNG TANKAGE	74-0749
LNG	SIMPLIFIED CALCULATION OF THE INTERACTION FORCES BETWEEN HULL AND TANK SYSTEM FOR A MOSS ROSENBERG T	75-0561
LOAD	ANALYSIS OF CLAMPED CIRCULAR PLATES OF VARIABLE THICKNESS SUBJECTED TO UNIFORMLY DISTRIBUTED LOAD BY	73-0583
LOAD	BERECHNUNGSMETHODEN ZUR ERMITTLUNG VON BELASTUNGSBEDINGUNGEN UND GRENZTRAGFAEHIGKEITSVERHALTEN VON K	73-0858
LOAD	CALCULATION OF THE CRITICAL LOAD OF ELASTIC THIN WALLED BARS BY THE METHOD OF FINITE ELEMENTS	72-0446
LOAD	CHARACTERISTICS OF FINITE-ELEMENTS FOR ANALYSIS OF BEAMS SUBJECTED TO IMPACT LOAD	73-1144
LOAD	CHARACTERISTICS OF MANIPULATION ERRORS IN SOLVING LOAD DEFLECTION EQUATIONS	70-0355
LOAD	CIRCUMFERENTIAL PRESTRESSING LOAD ANALYSIS DUE TO WIRE/STRAND WINDING SYSTEMS ON PRESTRESSED CONCRET	74-1202
LOAD	COLLAPSE LOAD OF REINFORCED CONCRETE PLATES.	72-0406
LOAD	CONSOLIDATION OF A LAYER UNDER A STRIP LOAD	72-0222
LOAD	CRUSTAL STRUCTURE FROM SURFACE LOAD TILTS USING A FINITE ELEMENT MODEL	72-0005
LOAD	DER GERISSENE ZUSTAND ZWEISEITIG GELAGERTER PLATTEN UNTER EINZELLASTEN - NICHTLINEARE BERECHNUNG MIT	75-0858
LOAD	ECONOMICAL SOLUTION TECHNIQUES FOR LOAD DEFLECTION EQUATIONS	74-0414
LOAD	EFFECT OF THE VARIATION OF CONSISTENCY IN RESIDUAL SOILS ON LOAD TRANSFER BY BORED PILES	75-0812
LOAD	ELASTIC DEFORMATION OF SQUARE EXTRUSION DIES DUE TO LOAD AND TEMPERATURE	75-0581
LOAD	ELASTIC-PLASTIC ANALYSIS OF A BAR WITH UNIFORM CROSS-SECTION UNDER COMBINED LOADINGS 1 - COMBINATION	71-0778
LOAD	FIELD INSTRUMENTATION FOR LONG TERM MEASUREMENT OF PILE LOAD AND RAFT CONTACT PRESSURE	73-0150
LOAD	FINITE ELEMENT ANALYSIS OF LOAD CELL RESPONSE IN HIGH SPEED TENSILE TESTING	73-1071
LOAD	FINITE ELEMENT ULTIMATE LOAD ANALYSIS OF THREE DIMENSIONAL CONCRETE STRUCTURES	74-0672
LOAD	FLEXUAL RIGIDITY OF A THIN WALLED BUILD-UP ROTOR FOR JET ENGINE. MEASUREMENT BY STATIC LOAD TEST AND	73-0756
LOAD	INFLUENCE OF THE UNPRECISION OF EXECUTION ON THE BEHAVIOUR OF A SPHERICAL SHELL WITH CONCENTRATED LO	73-0802
LOAD	LOAD DEFORMATION ANALYSIS OF DEEP PILE FOUNDATION	72-0931
LOAD	LOAD DISTRIBUTION IN A COMPOSITE STEEL BOX GIRDER BRIDGE	73-0765
LOAD	LOAD LIMIT ANALYSIS. COMBINATION OF STATISTICAL METHODS (MONTE CARLO METHOD) AND FINITE ELEMENT METH	71-0768
LOAD	LOAD POINT COMPLIANCE OF THE CHARPY IMPACT SPECIMEN	73-0775
LOAD	LOAD TRANSFER MECHANISM FOR PIERS SOCKETTED IN HARD SOILS OR ROCK	74-0969
LOAD	MECHANICS OF LOAD TRANSFER AT THE FIBER/MATRIX INTERFACE	72-0879
LOAD	NOTCH EFFECT IN LOW CYCLE FATIGUE - REPORT 1: LOW-CYCLE FATIGUE STRENGTH OF 0.48 PERCENT CARBON STEE	73-1090
LOAD	NOTCH STRESSES AT THE LOAD CARRYING ELASTIC CORE	73-0266
LOAD	PREDICTION OF LONG-TERM STRESS RANGES: USER'S MANUAL - BRIDGE LOAD GENERATOR	73-0737
LOAD	SIMULATION ON THE TENSILE BEHAVIOUR OF TWO-PHASE ALLOY CONTAINING A FIBER UNDER TENSILE LOAD	73-0542
LOAD	STRESS ANALYSIS OF A BOLT-NUT JOINT BY THE FINITE ELEMENT METHOD AND THE COPPER-ELECTROPLATING METHO	73-0220
LOAD	STRESS ANALYSIS OF A BOLT-NUT JOINT BY FINITE ELEMENT METHOD AND COPPER-ELECTROPLATING METHOD - 2. S	74-0153
LOAD	STRIP PROGRAM FOR DESIGNING LOAD BEARING SURFACE STRUCTURES (PROGRAMME STRIP POUR LE CALCUL DES STRU	71-0408
LOAD	THEORETICAL TRANSIENT BEHAVIOUR OF SATURATED AND UNSATURATED SOILS UNDER LOAD AND CHANGING MOISTURE	73-0901
LOAD	THREE-DIMENSIONAL ANALYSIS OF SHEAR WALL BUILDINGS TO LATERAL LOAD	74-0220
LOAD	UEBER DAS VERHALTEN VON SPANNEBETONDRUCKBEHAELTRERN IM GRENZTRAGFAEHIGKEITSZUSTAND UND DESSEN EINFUL	73-0855
LOAD	ULTIMATE LOAD BEHAVIOUR OF PCRV TOP CLOSURES. THEORETICAL AND EXPERIMENTAL INVESTIGATIONS	73-0835
LOAD	ZUSAMMENHANG ZWISCHEN OBERFLAECHENBESCHAFFENHEIT, VERBUND UND SPRENGWIRKUNG VON BEWEHRUNGSSTAEHLEN U	73-0450
LOAD-BEARING	ANALYSIS OF LOAD-BEARING FILLS OVER SOFT SUBSOILS	71-0330
LOAD-DEFLECTING	EFFICIENT SOLUTION OF LOAD-DEFLECTING EQUATIONS	69-0396
LOAD-DEFLECTION	A COMPUTER MODEL FOR PREDICTING THE LOAD-DEFLECTION RESPONSE OF EXPEDIENT SOIL SURFACING	73-0880
LOAD-SETTLEMENT	FINITE ELEMENT ANALYSIS OF THE EFFECTS OF INSTALLATION ON PILE LOAD-SETTLEMENT BEHAVIOUR	74-0763
LOADED	A COMPARISON OF FINITE ELEMENT AND FOURIER SERIES SOLUTIONS AS APPLIED TO RADIALLY LOADED CIRCULAR R	68-0024
LOADED	A FINITE ELEMENT MODEL FOR ANISOTROPIC YIELDING IN GRAVITY LOADED ROCK	69-0012
LOADED	AN ELASTOPLASTIC ANALYSIS OF A UNIAXIALLY LOADED SHEET WITH AN INTERFERENCE - FIT BOLT	74-0687
LOADED	ANALYTICAL MODEL FOR DETERMINING ENERGY DISSIPATION IN DYNAMICALLY LOADED STRUCTURES	75-0328
LOADED	BIFURCATION OF ECCENTRICALLY LOADED MEMBERS AND OF STIFFENED PLATES	70-0057
LOADED	COMPARISON OF PRACTICAL METHODS OF ANALYSING ELASTIC LATERALLY LOADED PLATES	71-0425
LOADED	COMPUTER AIDED DESIGN OF THERMALLY LOADED AXISYMMETRIC DIESEL ENGINE COMPONENTS	74-0596
LOADED	COMPUTER CALCULATIONS OF STRESSES IN AXISYMMETRIC THERMALLY LOADED COMPONENTS	68-0069
LOADED	COMPUTER PROGRAM FOR ELASTIC-PLASTIC ANALYSIS OF AXISYMMETRICALLY LOADED SHELLS OF REVOLUTION	68-0239
LOADED	DISCRETE-ELEMENT MODELLING AND SENSITIVITY STUDIES OF STATICALLY LOADED SINGLE-LAYER AND BOUNDED DOU	70-0129
LOADED	EFFECTS OF INSERTS ON THE ELASTIC BEHAVIOUR OF CYLINDRICAL MATERIALS LOADED BETWEEN ROUGH END-PLATES	71-0318
LOADED	ELASTIC-PLASTIC DISCRETE ELEMENT ANALYSIS OF AXISYMMETRICALLY LOADED SHELLS OF REVOLUTION	67-0099
LOADED	ELASTOPLASTIC CRACK PROPAGATION IN A TRANSVERSELY LOADED UNIDIRECTIONAL COMPOSITE	74-0217
LOADED	FINITE DEFORMATION OF BIAXIALLY LOADED COLUMNS	72-0021
LOADED	FINITE ELEMENT ANALYSIS OF AN AXISYMMETRICALLY LOADED ORTHOTROPIC SHELL OF REVOLUTION.	72-0438
LOADED	FINITE ELEMENT ANALYSIS OF A THICK WALLED MICROPOLAR CYLINDER LOADED AXISYMMETRICALLY	73-0847
LOADED	FINITE ELEMENT METHOD FOR THE CALCULATION OF LOCUS PATHS IN DYNAMICALLY LOADED BEARINGS	73-0487
LOADED	FINITE ELEMENT SOLUTION OF DIELECTRIC LOADED WAVEGUIDES	70-0206
LOADED	FINITE-ELEMENT STRESS ANALYSIS OF A NON-LINEAR PROBLEM. A CONNECTING-ROD EYE LOADED BY MEANS OF A PI	70-0239
LOADED	FREE VIBRATIONS OF PRESSURE LOADED PARABOLOIDAL SHELLS	71-0601
LOADED	HOLLOW-ENDED ROLLER. A SOLUTION FOR IMPROVING FATIGUE LIFE IN ASYMMETRICALLY LOADED CYLINDRICAL ROLL	71-0393
LOADED	ISSLEDOVANIE PROCHNOSTI I ZHESTKOSTI PLASTIN, NAGRUZHENNYKH V SVOEI PLOSKOSTI, METHODOM KONECHNYKH E	73-0382
LOADED	LINEAR AND NONLINEAR STATIC ANALYSIS OF AXISYMMETRICALLY LOADED THIN SHELLS OF REVOLUTION	69-0215
LOADED	NEW FINITE ELEMENT METHOD FOR ANALYZING SYMMETRICALLY LOADED THIN SHELLS OF REVOLUTION	73-0389
LOADED	NUMERICAL SOLUTION OF DIELECTRIC LOADED WAVEGUIDES: I. FINITE-ELEMENT ANALYSIS	70-0199
LOADED	SOIL RESISTANCE MOVEMENT RELATIONSHIPS FOR LATERALLY LOADED PILES IN CLAYS AT GREAT DEPTHS	70-0162
LOADED	STATICS AND DYNAMICS OF HYDROSTATICALLY LOADED SHELLS BY FINITE ELEMENT METHOD	71-0044
LOADED	THEORETICAL AND EXPERIMENTAL ANALYSIS OF ELASTOPLASTIC DEFORMATION IN A UNIAXIALLY LOADED STRIP OF 2	73-0720
LOADED	THREE-DIMENSIONAL FINITE ELEMENT ANALYSIS OF VERTICALLY LOADED PILE GROUPS	75-0239
LOADED	VIBRATIONS OF PRESSURE LOADED HYPERBOLOIDAL SHELLS	72-0874
LOADING	A FINITE-ELEMENT STUDY ON THE NONLINEAR BEHAVIOR OF REINFORCED CONCRETE FLEXURAL MEMBERS DUE TO SHOR	73-1154
LOADING	A METHOD FOR DETERMINING LOCAL ELASTOPLASTIC STRESS AND STRAIN IN METALLURGICALLY BONDED NOTCHED LAM	74-0686
LOADING	A NUMERICAL THERMO-MECHANICAL MODEL FOR THE WELDING AND SUBSEQUENT LOADING OF A FABRICATED STRUCTURE	72-0847
LOADING	AN ELASTIC-PLASTIC FINITE ELEMENT ANALYSIS OF CRACK TIP FIELDS UNDER BIAXIAL LOADING CONDITIONS	73-0760
LOADING	AN INVESTIGATION OF THE BENDING OF SYMMETRIC PLATES UNDER SYMMETRIC TRANSVERSE LOADING	72-0662
LOADING	ANALYSIS OF ELASTIC SHELLS OF REVOLUTION WITH MEMBRANE AND FLEXURE STRESSES UNDER ARBITRARY LOADING	68-0289

271

LOADING	ANALYSIS OF ELASTIC-PLASTIC SHELLS OF REVOLUTION UNDER AXI-SYMMETRIC LOADING BY THE FINITE ELEMENT M	67-0102
LOADING	ANALYSIS OF LOADING AND THERMAL EFFECTS ON FUELLED GRAPHITE BRICKS FOR A HIGH-TEMPERATURE GAS COOLED	71-0085
LOADING	ANALYSIS OF RING-STIFFENED SHELLS OF REVOLUTION UNDER COMBINED THERMAL AND MECHANICAL LOADING	71-0342
LOADING	APPLICATION OF A UNIMOMENT TECHNIQUE TO A BICONICAL ANTENNA WITH INHOMOGENEOUS DIELECTRIC LOADING	75-0375
LOADING	APPLICATION OF FINITE ELEMENT METHOD TO CYCLIC LOADING OF ELASTIC-PLASTIC STRUCTURES CONTAINING EFFE	73-0103
LOADING	APPLICATION OF THE DISCRETE ELEMENT METHOD TO THE BUCKLING ANALYSIS OF RECTANGULAR PLATES UNDER ARBI	68-0054
LOADING	APPROXIMATE THREE-DIMENSIONAL SOLUTIONS FOR SYMMETRIC LAMINATES UNDER INPLANE LOADING	71-0392
LOADING	ARBITRARY SYMMETRIC LOADING PROBLEMS OF CENTRALLY CRACKED RECTANGULAR PLATES	75-0318
LOADING	ASPECTS OF MECHANICAL BEHAVIOR OF ROCK UNDER STATIC AND CYCLIC LOADING. PART A. MECHANICAL BEHAVIOR	72-0761
LOADING	ASPECTS OF MECHANICAL BEHAVIOR OF ROCK UNDER STATIC AND CYCLIC LOADING. PART A. MECHANICAL BEHAVIOR	72-0761
LOADING	ASPECTS OF MECHANICAL BEHAVIOR OF ROCK UNDER STATIC AND CYCLIC LOADING. PART A: MECHANICAL BEHAVIOR	72-0762
LOADING	ASPECTS OF MECHANICAL BEHAVIOR OF ROCK UNDER STATIC AND CYCLIC LOADING. PART A: MECHANICAL BEHAVIOR	72-0762
LOADING	BERECHNUNGSMETHODEN ZUR ERMITTLUNG VON BELASTUNGSBEDINGUNGEN UND GRENZTRAGFAEHIGKEITSVERHALTEN VON K	73-0858
LOADING	BOTTOM STABILITY UNDER WATER INDUCED LOADING	72-0373
LOADING	BUCKLING UNDER COMBINED LOADING OF THIN FLAT-WALLED STRUCTURES BY A COMPLEX FINITE STRIP METHOD	74-0788
LOADING	CALCULATION OF STRESS CONCENTRATIONS ON SUPPORT STRUCTURES UNDER THERMOMECHANICAL LOADING (BERECHNUN	73-0796
LOADING	CALCULATION OF STRESS INTENSITY FACTORS FOR COMBINED TENSILE AND SHEAR LOADING	75-0205
LOADING	COMPUTER PREDICTION OF FATIGUE CRACK PROPAGATION UNDER RANDOM LOADING	73-0851
LOADING	CONICAL SEGMENT METHOD FOR ANALYZING OPEN CROWN SHELLS OF REVOLUTION FOR EDGE LOADING	63-0007
LOADING	CREEP ANALYSIS OF TRANSVERSELY ISOTROPIC BODIES SUBJECTED TO TIME DEPENDENT LOADING	73-0170
LOADING	CYCLIC LOADING AND CRACK PROPAGATION - AN ELASTOPLASTIC FINITE ELEMENT STUDY	74-1214
LOADING	DISCRETE ELEMENT METHODS FOR THE PLASTIC ANALYSIS OF STRUCTURES SUBJECTED TO CYCLIC LOADING	70-0023
LOADING	DISCRETE-ELEMENT STATIC ANALYSIS OF CORE-STIFFENED SHELLS OF REVOLUTION FOR ASYMMETRIC MECHANICAL LO	68-0148
LOADING	DYNAMIC ANALYSIS OF A COMPLEX SHELL STRUCTURE UNDER IMPULSIVE ASYMMETRIC LOADING	68-0226
LOADING	DYNAMIC LOADING OF A FLUID-FILLED SPHERICAL SHELL.	72-0444
LOADING	DYNAMIC STRESS ANALYSIS OF AXISYMMETRIC STRUCTURES UNDER ARBITRARY LOADING	69-0433
LOADING	EFFECT OF SEEPAGE ON EMBANKMENT DEFORMATIONS DUE TO WATER LOADING	74-1329
LOADING	ELASTIC PLASTIC DEFORMATION OF CYLINDRICAL PRESSURE VESSELS UNDER CYCLIC LOADING	74-0604
LOADING	ELASTIC-PLASTIC ANALYSIS OF A PROPAGATING CRACK UNDER CYCLIC LOADING	75-0857
LOADING	ELASTIC-PLASTIC FINITE ELEMENT ANALYSIS OF CRACK TIP FIELDS UNDER BIAXIAL LOADING CONDITIONS	74-0296
LOADING	ELASTIC-PLASTIC PROBLEMS OF COMPRESSIBLE STRAIGHT BAR UNDER COMBINED LOADING CONDITION	75-0313
LOADING	ETUDE THEORIQUE DU COMPORTEMENT DES PIEUX SOUS CHARGE VERTICALE. INTRODUCTION DE LA DILATANCE. (THEO	75-0309
LOADING	EXACT SOLUTION TO THE RADIALLY END CONSTRAINED CIRCULAR CYLINDER UNDER TRIAXIAL LOADING	71-0433
LOADING	FINITE ELEMENT ANALYSIS OF OPEN SHELLS OF REVOLUTION UNDER NONSYMMETRICAL LOADING	68-0295
LOADING	FINITE ELEMENT ANALYSIS OF ALTERNATIVE AXIAL LOADING OF AN ELASTIC PLATE PRESSED BETWEEN TWO ELASTIC	71-0357
LOADING	FINITE ELEMENT ANALYSIS OF CRACK PROPAGATION IN THREE-DIMENSIONAL SOLIDS UNDER CYCLING LOADING	73-0529
LOADING	FINITE ELEMENT ANALYSIS OF CRACK PROPAGATION IN 3-DIMENSIONAL SOLIDS UNDER CYCLIC LOADING	74-0479
LOADING	FINITE ELEMENT ANALYSIS OF THE SURFACE DEFORMATION DUE TO A UNIFORM LOADING ON A LAYER OF GIBSON SOI	75-0240
LOADING	FINITE ELEMENT LARGE DEFLECTION ANALYSIS OF ELASTIC-PLASTIC SHELLS OF REVOLUTION SUBJECTED TO AXISYM	72-0137
LOADING	FINITE ELEMENT METHOD FOR STRESS ANALYSIS OF BODIES OF REVOLUTION SUBJECTED TO AXISYMMETRIC LOADING	71-0409
LOADING	FINITE ELEMENT MODEL SYSTEM OF A RABBIT CALVARIUM: AN INVESTIGATION OF OPTIMAL COMPRESSIVE LOADING	73-0206
LOADING	FRACTURE (ULTIMATE STRENGTH) ANALYSES OF ASPHALT PAVEMENT LAYERS RESULTING FROM TRAFFIC LOADING	73-0548
LOADING	INELASTIC ANALYSIS OF A UNIDIRECTIONAL COMPOSITE SUBJECTED TO TRANSVERSE NORMAL LOADING	70-0015
LOADING	INELASTIC BUCKLING ANALYSIS OF PIPES SUBJECTED TO INTERNAL PRESSURE, FLEXURE AND AXIAL LOADING	74-0515
LOADING	INELASTIC MODEL FOR CYCLIC BIAXIAL LOADING OF REINFORCED CONCRETE	74-0503
LOADING	INERTIA LOADING IN FINITE-ELEMENT ANALYSIS OF STRUCTURES SUBJECT TO COMPOUND MOTION	75-0351
LOADING	INFLUENCE OF END-CLOSURE STIFFNESS ON BEHAVIOR OF CONCRETE CYLINDRICAL HULLS SUBJECTED TO HYDROSTATI	71-0725
LOADING	LARGE DEFLECTION OF ELASTIC PLATES UNDER PATCH LOADING	72-0120
LOADING	MARINE PROPELLER BLADES STRESSES AND DEFORMATIONS WITH GIVEN, FROZEN, HYDRODYNAMIC LOADING	73-0474
LOADING	MASS LOADING EFFECTS ON VIBRATION OF RING AND SHELL STRUCTURES	68-0245
LOADING	MATRIX ANALYSIS OF AXISYMMETRIC SHELLS UNDER GENERAL LOADING	66-0075
LOADING	MODAL RESPONSE OF FREE ROCKETS UNDER THRUST LOADING	74-0855
LOADING	NON-LINEAR RESPONSE OF A BEAM TO PERIODIC LOADING.	72-0407
LOADING	NONLINEAR ANALYSIS OF ELASTIC SHELLS OF REVOLUTION UNDER AXISYMMETRIC LOADING	70-0161
LOADING	NONLINEAR BEHAVIOR OF SHELLS OF REVOLUTION UNDER CYCLIC LOADING	73-0242
LOADING	NONLINEAR BEHAVIOR OF THE HUMAN INTERVERTEBRAL DISC UNDER AXIAL LOADING	74-0919
LOADING	NONLINEAR RESPONSE OF PLATES TO IMPULSIVE LOADING USING HYBRID FINITE ELEMENTS	74-1018
LOADING	NONPROPORTIONAL LOADING LIMIT FOR STRUCTURES	74-0228
LOADING	NOTE ON STRESS AND STRAIN REDISTRIBUTION IN A NOTCHED PLATE SPECIMEN DURING CYCLIC LOADING	69-0194
LOADING	NUMERICAL, THERMO-MECHANICAL MODEL FOR THE WELDING AND SUBSEQUENT LOADING OF A FABRICATED STRUCTURE	73-0418
LOADING	PATTERN LOADING ON REINFORCED CONCRETE FLAT PLATES	71-0290
LOADING	PROGRESS REPORT ON DISCRETE-ELEMENT ELASTIC AND ELASTIC-PLASTIC ANALYSIS OF SHELLS OF REVOLUTION SUB	68-0170
LOADING	PROPAGATION AND OPENING OF A THROUGH CRACK IN PIPE SUBJECT TO COMBINED CYCLIC THERMOMECHANICAL LOADI	75-0516
LOADING	PURE MOMENT LOADING OF AXISYMMETRIC FINITE ELEMENT MODELS	73-1007
LOADING	RATIONAL DESIGN OF TUNNEL SUPPORTS: TUNNEL SUPPORT LOADING CAUSED BY ROCK FAILURE	75-0687
LOADING	RIGID PAVEMENT STRESSES UNDER AIRCRAFT LOADING	71-0423
LOADING	SABOR 1, A FORTRAN PROGRAM FOR THE LINEAR ELASTIC AXISYMMETRIC LOADING BY USING THE MATRIX DISPLACEM	65-0027
LOADING	SHAKEDOWN OF ROLLING WHEEL UNDER HUB LOADING	75-0516
LOADING	SIMULATION OF FINITE ELEMENT METHOD FOR LOADING PROCESS OF GRAIN CARGO AND INTERACTION WITH SHIP'S S	75-0161
LOADING	SPANNUNGEN IN MEMBRANWAENDEN DURCH WAERMEBEANSPRUCHUNG UND MECHANISCHE BELASTUNG. (STRESSES IN MEMBR	75-0066
LOADING	STATIC ANALYSIS OF A COMPLEX SHELL STRUCTURE UNDER ASYMMETRIC LOADING	67-0120
LOADING	STATIC FLEXIBILITY MATRICES FOR A RECTANGULAR PLATE ELEMENT IN BENDING, TWISTING, AND SHEAR. UNIFORM	68-0217
LOADING	STRESS DISTRIBUTION IN BONE ARISING FROM LOADING ON ENDOSTEAL DENTAL IMPLANTS	73-0501
LOADING	STRESS STRAIN RESPONSE AND FRACTURE OF A CONCRETE MODEL IN BIAXIAL LOADING	71-0295
LOADING	STRESSES AND DEFLECTIONS IN COOLING TOWER SHELLS DUE TO WIND LOADING	68-0088
LOADING	STRUCTURAL ANALYSIS OF SL-7 CONTAINERSHIP UNDER COMBINED LOADING OF VERTICAL, LATERAL AND TORSIONAL	74-0734
LOADING	STRUCTURAL RESPONSE AND LOADING OF WALL PANELS	71-0735
LOADING	SUBTERRANEAN GUIDEWAY RESPONSES TO LEVITATION VEHICLE LOADING	74-0927
LOADING	THE ANALYSIS OF ARCH DAMS UNDER STATIC LOADING BY THE FINITE ELEMENT METHOD	68-0059
LOADING	THE ELASTO-PLASTIC AND LARGE-DISPLACEMENT RESPONSE OF PLATES TO BLAST LOADING	73-0731
LOADING	THE TRANSIENT LINEAR ELASTIC RESPONSE ANALYSIS COMPLEX THIN SHELLS OF REVOLUTION SUBJECTED TO ARBITR	70-0080
LOADING	THERMAL LOADING OF THIN-SHELL CONCRETE COOLING TOWERS	74-0667
LOADING	TRANSVERSE LOADING OF UNIDIRECTIONAL FIBER COMPOSITES	69-0119
LOADING	VERTICAL RESPONSE OF A RIGID BASE TO DYNAMIC LOADING	69-0316
LOADING	VISCOELASTIC RESPONSE OF THE CORNEO-SCLERAL SHELL UNDER TONOMETER LOADING	75-0855
LOADINGS	A DYNAMIC ELASTOPLASTIC ANALYSIS OF THIN SHELLS OF REVOLUTION UNDER ASYMMETRIC MECHANICAL AND THERMA	72-0068
LOADINGS	A REFINED FINITE ELEMENT ANALYSIS OF THIN SHELL STRUCTURES INCLUDING DYNAMIC LOADINGS	67-0157
LOADINGS	BUCKLING OF PLATE GIRDER WEBS UNDER PARTIAL EDGE LOADINGS	70-0283
LOADINGS	DYNAMIC ELASTOPLASTIC RESPONSE OF GEOMETRICALLY NONLINEAR ARBITRARY SHELLS OF REVOLUTION UNDER IMPUL	70-0098
LOADINGS	DYNAMIC RESPONSE OF ARBITRARY SHELLS OF REVOLUTION UNDER COMBINED IMPULSIVE NONISOTHERMAL AND IRREGU	70-0365
LOADINGS	DYNAMIC RESPONSE OF VISCOELASTIC MATERIALS TO SINUSOIDALLY VARYING LOADINGS	73-1072
LOADINGS	ELAD PROGRAM MANUAL, A COMPUTER PROGRAM FOR THE ELASTIC ANALYSIS OF DOME STRUCTURES AND AXISYMMETRIC	69-0064
LOADINGS	ELASTIC-PLASTIC ANALYSIS OF BEAMS WITH UNIFORM CROSS-SECTION UNDER COMBINED LOADINGS	71-0777
LOADINGS	ELASTIC-PLASTIC ANALYSIS OF A BAR WITH UNIFORM CROSS-SECTION UNDER COMBINED LOADINGS 1 - COMBINATION	71-0778
LOADINGS	ELASTIC-PLASTIC ANALYSIS OF BEAMS WITH UNIFORM CROSS-SECTION UNDER COMBINED LOADINGS (2)	73-1070
LOADINGS	END DISPLACEMENTS OF SEMI-INFINITE CYLINDERS DUE TO ANNULAR LOADINGS	69-0399
LOADINGS	EPAD, A COMPUTER PROGRAM FOR THE ELASTIC-PLASTIC ANALYSIS OF DOME STRUCTURES AND ARBITRARY SOLIDS OF	70-0036
LOADINGS	FAILURE OF THIN-WALLED MEMBERS UNDER PATCH LOADINGS	72-0230
LOADINGS	FINITE DEFORMATIONS IN ELASTIC-PLASTIC MEDIA SUBJECTED TO IMPULSIVE LOADINGS BY THE FINITE ELEMENT M	72-0947
LOADINGS	FINITE ELEMENT ANALYSIS OF AXISYMMETRIC SOLIDS WITH ARBITRARY LOADINGS	67-0145
LOADINGS	FINITE ELEMENTS FOR AXISYMMETRIC SOLIDS UNDER ARBITRARY LOADINGS WITH NODES ON ORIGIN.	72-0251
LOADINGS	FINITE ELEMENTS FOR AXISYMMETRIC SOLIDS UNDER ARBITRARY LOADINGS WITH NODES ON ORIGIN	73-0094

272

LOADINGS	MODAL SENSITIVITY STUDY OF A CONICAL RE-ENTRY VEHICLE SUBJECTED TO AERODYNAMICALLY INDUCED ACOUSTIC	69-0431
LOADINGS	REFINED CRACKED CONCRETE ANALYSIS OF CONCRETE CONTAINMENT STRUCTURES SUBJECT TO OPERATIONAL AND ENVI	74-0943
LOADINGS	RESPONSE OF REACTOR CORE SUBASSEMBLIES TO IMPULSIVE LOADINGS	74-0076
LOADINGS	SASL - A FINITE ELEMENT CODE FOR THE STATIC ANALYSIS OF AXISYMMETRIC AND PLANE SOLIDS SUBJECTED TO A	72-0779
LOADINGS	THEORETICAL ANALYSIS OF THICK DOME STRUCTURES AND ARBITRARY SOLIDS OF REVOLUTION SUBJECTED TO UNSYMM	70-0037
LOADINGS	VIBRATION OF SQUARE PLATES HAVING VARIOUS EDGE CONDITIONS AND LOADINGS USING A REFINED FINITE ELEMEN	75-0024
LOADS	A COMPUTER PROGRAM FOR ANALYSIS OF STIFFENED PLATES UNDER COMBINED INPLANE AND LATERAL LOADS	71-0703
LOADS	A COMPUTER PROGRAM TO ANALYZE BEAM-COLUMN UNDER MOVEABLE LOADS	68-0268
LOADS	A GENERAL CALCULATION OF THE STRUCTURAL LOADS ON A SHIP IN A SEAWAY	71-0345
LOADS	A STUDY OF FREE-VIBRATIONS OF THIN ELASTIC SHELLS SUBJECTED TO INITIAL LOADS USING FINITE-ELEMENTS	73-1143
LOADS	A UNIFIED FINITE ELEMENT APPROACH TO COLUMNS SUBJECTED TO NONCONSERVATIVE LOADS, WITH DAMPING EFFECT	74-0795
LOADS	AEROELASTIC LOADS PREDICTIONS USING FINITE-ELEMENT AERODYNAMICS	75-0815
LOADS	ANALYSIS OF BUILDINGS ON ELASTIC MEDIUM UNDER DYNAMIC OR SEISMIC LOADS	74-0456
LOADS	ANALYSIS OF FOUNDATIONS ON A SEMI-INFINITE ELASTIC MASS UNDER DYNAMIC LOADS	72-0763
LOADS	ANALYSIS OF LAYERED SYSTEMS SUBJECTED TO WHEEL LOADS	72-0868
LOADS	ARCH DAM CONSTRUCTION STRESSES DUE TO HYDROSTATIC AND GRAVITY LOADS	74-0658
LOADS	AUTOMATED EVALUATION OF INTERCEPTOR ROCKET MOTOR DESIGNS UNDER COMBINED OPERATIONAL AND NUCLEAR EFFE	74-0698
LOADS	BEHAVIOR OF ISOLATED SHEAR WALLS SUBJECTED TO LATERAL LOADS	75-0704
LOADS	CALCULATION OF THE ELASTIC BUCKLING LOADS OF THIN FLAT REINFORCED PLATES USING TRIANGULAR FINITE ELE	73-1024
LOADS	CALCULATION OF THE ELASTIC BUCKLING LOADS OF THIN FLAT REINFORCED PLATES USING TRIANGULAR FINITE ELE	75-0177
LOADS	COMPARISON OF STRUCTURAL LOADS; STATIC VERSUS DYNAMIC	70-0464
LOADS	CREEP OF A UNIAXIAL METAL MATRIX COMPOSITE SUBJECTED TO AXIAL AND NORMAL LATERAL LOADS	71-0041
LOADS	CREEP OF A UNIAXIAL METAL MATRIX COMPOSITE SUBJECTED TO AXIAL AND NORMAL LATERAL LOADS	71-0727
LOADS	DESIGN OF OPTIMUM STRUCTURES FOR DYNAMIC LOADS	71-0255
LOADS	DIE BEULLASTEN DIAGONAL VERSTEIFTER RECHTECKPLATTEN (BUCKLING LOADS IN DIAGONALLY STIFFENED RECTANGU	73-0405
LOADS	DYNAMIC LOADS ON GROUTED EMPLACEMENT SYSTEMS	70-0406
LOADS	ELASTIC-PLASTIC LARGE DEFORMATION RESPONSE OF CLAY TO FOOTING LOADS	74-0699
LOADS	EMBANKMENT DEFORMATIONS DUE TO WATER LOADS	73-0782
LOADS	FINITE ELEMENT ANALYSIS OF SHELLS UNDER NONUNIFORM SNOW LOADS	69-0113
LOADS	FINITE ELEMENT ANALYSIS OF CRACKED STRUCTURES SUBJECTED TO SHOCK LOADS	75-0258
LOADS	FINITE-ELEMENT ANALYSIS OF COMPLEX PANEL RESPONSE TO RANDOM LOADS	68-0093
LOADS	GALERKIN FINITE ELEMENT SOLUTION FOR THE STABILITY OF CANTILEVER COLUMNS SUBJECTED TO TANGENTIAL LOA	75-0534
LOADS	GALERKIN FINITE-ELEMENT SOLUTION FOR STABILITY OF CANTILIVER COLUMNS SUBJECTED TO TANGENTIAL LOADS	75-0618
LOADS	INELASTIC ANALYSIS OF A STRAIGHT TUBE UNDER COMBINED BENDING, PRESSURE AND THERMAL LOADS	75-0218
LOADS	INELASTIC ANALYSIS OF A STRAIGHT TUBE UNDER COMBINED BENDING, PRESSURE AND THERMAL LOADS	75-0513
LOADS	INELASTIC FINITE ELEMENT ANALYSIS OF REINFORCED CONCRETE PANELS UNDER IN-PLANE LOADS	70-0482
LOADS	INELASTIC RESPONCE OF FRAMES TO DYNAMIC LOADS	71-0611
LOADS	LARGE DEFLECTIONS OF STRUCTURES SUBJECT TO HEATING AND EXTERNAL LOADS	60-0002
LOADS	LARGE DEFORMATION AND CRITICAL LOADS ANALYSIS OF FRAMED STRUCTURES	74-1293
LOADS	LOADS AND RESPONSE OF LARGE DUCTED PROPELLER SYSTEMS	73-0810
LOADS	LOWER BOUNDS TO COLUMN BUCKLING LOADS	73-0485
LOADS	OPTIMAL DESIGN FOR PRESCRIBED BUCKLING LOADS	74-0568
LOADS	OPTIMUM DESIGN OF ADVANCED COMPOSITE STRUCTURES FOR STATIC LOADS	72-0613
LOADS	OPTIMUM PLASTIC DESIGN FOR MULTIPLE SETS OF LOADS	74-0980
LOADS	PARAMETRIC STUDY OF A BEAM WITH A COMPOUND SIDE-BRANCH RESONATOR AS A DEVICE TO EVALUATE PRELIMINARY	70-0465
LOADS	PLASTIC COLLAPSE LOADS FOR PIPE ELBOWS USING INELASTIC ANALYSIS	73-0941
LOADS	PLASTIC COLLAPSE LOADS FOR PIPE ELBOWS USING INELASTIC ANALYSIS	74-0507
LOADS	RESPONSE OF BEAMS AND PLATES TO MOVING LOADS	71-0277
LOADS	RESPONSE OF RECTANGULAR PLATES TO MOVING LOADS BY A FINITE ELEMENT PROCEDURE	69-0515
LOADS	SEEPAGE IN FRACTURED ROCK: FINITE ELEMENT ANALYSIS OF ROCK MASS BEHAVIOR UNDER COUPLED ACTION OF BOD	71-0825
LOADS	SHELL DEFORMATION UNDER TOOL LOADS REPRESENTATION OF SINGULARITIES WITH ISOPARAMETRIC FINITE ELEMENT	74-0365
LOADS	SHOCK LOADS ON PIPING SYSTEMS	72-0760
LOADS	SIGNIFICANT APPLICATIONS OF FINITE ELEMENTS AND FINITE DIFFERENCES METHODS TO SOLVE PROBLEMS OF THER	73-0864
LOADS	STACUSS 1: A DISCRETE-ELEMENT PROGRAM FOR THE STATIC ANALYSIS OF SINGLE-LAYER CURVED STIFFENED SHELL	69-0160
LOADS	STEP AND LOADS EFFECT ON STRESSES IN PRESTRESSED CONCRETE SHORT BRACKETS	71-0334
LOADS	STIFFNESS ANALYSIS OF SHEET METAL SHELLS UNDER CONCENTRATED LOADS	74-0325
LOADS	STRESS ANALYSIS OF A CERAMIC BLADE DESIGN SUBJECTED TO CENTRIFUGAL AND THERMAL LOADS	73-0790
LOADS	STRESSES DUE TO CONCENTRATED LOADS IN RESTRAINED BENDING TESTS	67-0097
LOADS	STRUCTURAL ANALYSIS AND EXPERIMENTAL CORRELATION OF A COMPLEX SHELL UNDER ASYMMETRIC LOADS	69-0155
LOADS	STRUCTURAL EFFECTS OF CONFINED DYNAMIC LOADS	72-0151
LOADS	THE BOEING SST PROTYPE INTERNAL LOADS ANALYSIS SYSTEM AND PROCEDURES	71-0103
LOADS	THE RESPONSE OF SHELLS TO DISTRIBUTED RANDOM LOADS USING NASTRAN	71-0134
LOADS	YIELDING IN UNIDIRECTIONAL COMPOSITES UNDER EXTERNAL LOADS AND TEMPERATURE CHANGES.	73-0169
LOCAL	A " LOCAL " BASIS OF GENERALIZED SPLINES OVER RIGHT TRIANGLES DETERMINED FROM A NONUNIFORM PARTITION	72-0525
LOCAL	A METHOD FOR DETERMINING LOCAL ELASTOPLASTIC STRESS AND STRAIN IN METALLURGICALLY BONDED NOTCHED LAM	74-0686
LOCAL	A METHOD FOR SIMULATING CONCENTRATED FORCES AND LOCAL REINFORCEMENTS IN STRESS COMPUTATIONS	72-0081
LOCAL	APPROXIMATION BY FINITE-ELEMENT FUNCTIONS USING LOCAL REGULARIZATION	75-0436
LOCAL	COUPLED LOCAL BUCKLING IN WIDE-FLANGE BEAM-COLUMNS	73-0235
LOCAL	EFFECTS OF LOCAL STRENGTH VARIATIONS ON PLASTIC ENCLAVE DEVELOPMENT AT CRACK TIPS	72-0822
LOCAL	FINITE ELEMENT METHOD VIA LOCAL POTENTIAL	74-0038
LOCAL	FINITE ELEMENT STRUCTURAL ANALYSIS OF LOCAL INSTABILITY	72-0179
LOCAL	FINITE ELEMENT STRUCTURAL ANALYSIS OF LOCAL INSTABILITY	74-0098
LOCAL	FINITE ELEMENT STRUCTURAL ANALYSIS OF LOCAL INSTABILITY - REPLY	74-0099
LOCAL	LOCAL AND GLOBAL SMOOTHING OF DISCONTINUOUS FINITE ELEMENT FUNCTIONS USING A LEAST SQUARES METHOD	74-0378
LOCAL	LOCAL COMPLIANCE PROBLEM OF LUNG ELASTICITY	75-0660
LOCAL	LOCAL ERROR BOUNDS IN BOUNDARY VALUE PROBLEMS USING FINITE ELEMENT DISCRETIZATION	72-0982
LOCAL	LOCAL LEAST SQUARES STRESS SMOOTHING FOR PARABOLIC ISOPARAMETRIC ELEMENTS	75-0593
LOCAL	LOCAL MESH REFINEMENT WITH FINITE ELEMENTS FOR ELLIPTIC PROBLEMS	73-0073
LOCAL	LOCAL MESH REFINEMENT USING RECTANGULAR BLENDED FINITE-ELEMENTS	75-0452
LOCAL	LOCAL POTENTIAL APPROACH TO FINITE ELEMENT METHOD IN UNSTEADY VISCOUS INCOMPRESSIBLE FLUID FLOW	73-0298
LOCAL	MATRIX ANALYSIS OF LOCAL INSTABILITY IN PLATES, STIFFENED PANELS AND COLUMNS	72-0542
LOCAL	ON LOCAL APPROXIMATIONS OF OPERATOR EQUATIONS	72-0063
LOCAL	ON THE LOCAL DISTRIBUTION OF ERRORS BY FINITE ELEMENT APPROXIMATIONS	73-0304
LOCAL	SOLUTION OF MIXED BOUNDARY VALUE PROBLEMS WITH LOCAL ERROR BOUND BY THE FINITE ELEMENT METHOD	73-0795
LOCAL	THE FINITE ELEMENT METHOD AND LOCAL BOUNDS FOR BOUNDARY VALUE PROBLEMS OF ELASTIC STRUCTURES	73-1056
LOCALIZED	ANALYSIS OF NUCLEAR FUEL ROD LOCALIZED STRESSES	73-0708
LOCALIZED	THE FINITE DIFFERENCE AND LOCALIZED RITZ METHODS	71-0447
LOCATION	LOCATION OF POINTS AT WHICH FINITE ELEMENT SOLUTIONS ARE VERY ACCURATE	74-0112
LOCATIONS	A MORE EFFICIENT FRONT SOLUTION: ALLOCATING ASSEMBLY LOCATIONS BY LONGEVITY CONSIDERATIONS	73-1069
LOCATIONS	SELECTION OF LOCATIONS FOR LONG TERM STORAGE OF WHEAT	75-0323
LOCHSCHEIBE	PARAMETRISCHE KERBSPANNUNGSUNTERSUCHUNGEN AN DER LOCHSCHEIBE MIT DER METHODE DER FINITEN ELEMENTE. (	75-0603
LOCK	ANALYSIS OF PILE-SUPPORTED GRAVITY LOCK	74-0585
LOCK	FINITE ELEMENT ANALYSIS OF PORT ALLEN LOCK	71-0293
LOCK	NAVIGATION LOCK STRUCTURE ANALYSIS BY FINITE ELEMENT METHOD	71-0816
LOCKS	FINITE ELEMENT ANALYSES OF PORT ALLEN AND OLD RIVER LOCKS	69-0203
LOCKS	FINITE ELEMENT ANALYSES OF SOIL-STRUCTURE INTERACTION IN U-FRAME LOCKS	69-0531
LOCUS	FINITE ELEMENT METHOD FOR THE CALCULATION OF LOCUS PATHS IN DYNAMICALLY LOADED BEARINGS	73-0487
LOESUNG	ZUR LOESUNG DES KONTAKTPROBLEMS ELASTISCHER KOERPER MIT AUSGEDEHNTER KONTAKTFLAECHE DURCH QUADRATISC	74-1135
LOESUNGEN	VERGLEICH NUMERISCHER LOESUNGEN NACH DEM DIFFERENZENVERFAHREN UND NACH DER FINITE-ELEMENT-METHODE AM	71-0016
LOMENICOVYCH	VYPOCET KRUHOVYCH LOMENICOVYCH KONSTRUKCII METODOU KONECNYCH PASOV. (CALCULATION OF CIRCULAR FOLDED-	74-0612
LONDON	OBSERVATIONS ON THE BEHAVIOUR OF A PILED-RAFT FOUNDATION ON LONDON CLAY.	73-0421
LONG	A FINITE ELEMENT MODEL FOR PREDICTING LONG BONE FRACTURE FIXATION	71-0819

LONG	ANALYSIS OF LONG RECTANGULAR TANKS	70-0276
LONG	BENDING MOMENTS IN LONG WALL TANKS	67-0110
LONG	ELASTOPLASTIC INDENTATION OF HALF-SPACE BY AN INFINITELY LONG RIGID CIRCULAR CYLINDER	71-0395
LONG	FIELD INSTRUMENTATION FOR LONG TERM MEASUREMENT OF PILE LOAD AND RAFT CONTACT PRESSURE	73-0150
LONG	FINITE ELEMENT ANALYSIS OF LONG WAVES IN OPEN CHANNEL SYSTEMS	74-0033
LONG	INVESTIGATION ON FREE AND FORCED VIBRATIONS OF A LONG TANKER WITH OVERLAPPING PROPELLER ARRANGEMENT	73-0981
LONG	LONG TERM CREEP CLOSURE OF SOLUTION CAVITY SYSTEM	74-1150
LONG	PREDICTION OF LONG TERM DEFORMATION OF A COMPACTED COHESIVE SOIL EMBANKMENT OVER A SOFT FOUNDATION	74-0727
LONG	SELECTION OF LOCATIONS FOR LONG TERM STORAGE OF WHEAT	75-0323
LONG	TIDAL AND LONG WAVE PROPAGATION - A FINITE ELEMENT APPROACH	73-1059
LONG-ISLAND	GALERKIN-FINITE ELEMENT SIMULATION OF GROUNDWATER CONTAMINATION ON LONG-ISLAND, NEW-YORK	73-0093
LONG-PERIOD	FINITE ELEMENT ANALYSIS OF LONG-PERIOD WATER WAVES	73-1008
LONG-TERM	LONG-TERM DEFORMATION IN CLAY	73-0713
LONG-TERM	PREDICTION OF LONG-TERM STRESS RANGES: USER'S MANUAL - BRIDGE LOAD GENERATOR	73-0737
LONG-TERM	PREDICTION OF LONG-TERM STRESS RANGES: USER'S MANUAL - BRIDGE DYNAMIC STRESS ANALYSIS	74-0807
LONGEVITY	A MORE EFFICIENT FRONT SOLUTION: ALLOCATING ASSEMBLY LOCATIONS BY LONGEVITY CONSIDERATIONS	73-1069
LONGITUDINAL	A NUMERICAL METHOD FOR THE ANALYSIS OF LONGITUDINAL ELASTIC PLASTIC STRESS WAVE PROPAGATION	73-0526
LONGITUDINAL	A TWO-DIMENSIONAL AXISYMMETRIC SHELL FLUID MODEL FOR LAUNCH VEHICLE LONGITUDINAL DYNAMIC ANALYSIS	69-0046
LONGITUDINAL	ANALYSIS OF LONGITUDINAL WAVE IN A BAR EXHIBITING VISCO-ELASTIC BEHAVIOR OR WITH MATERIAL DAMPING	71-0780
LONGITUDINAL	CALCULATION OF AXISYMMETRIC LONGITUDINAL MODES FOR FLUID-ELASTIC TANK-ULLAGE GAS SYSTEMS AND COMPARI	64-0019
LONGITUDINAL	DYNAMIC STABILITY OF A NONLINEAR BEAM SUBJECTED TO BOTH LONGITUDINAL AND TRANSVERSE EXCITATION	73-0553
LONGITUDINAL	ELASTIC-PLASTIC BEHAVIOR OF A LONGITUDINAL SEMI-ELLIPTIC CRACK IN A THICK PRESSURE VESSEL	73-0952
LONGITUDINAL	EVALUATION OF A FINITE ELEMENT ANALYSIS FOR THE LONGITUDINAL VIBRATIONS OF LIQUID-PROPELLANT LAUNCH	70-0204
LONGITUDINAL	IMPROVED ANALYTIC LONGITUDINAL RESPONSE ANALYSIS FOR AXISYMMETRIC LAUNCH VEHICLES, COMPUTER PROGRAM	65-0002
LONGITUDINAL	IMPROVED LINEAR AXISYMMETRIC SHELL FLUID MODEL FOR LAUNCH VEHICLE LONGITUDINAL RESPONSE ANALYSIS	65-0044
LONGITUDINAL	MEASURED AND PREDICTED LONGITUDINAL VIBRATIONS OF A LIQUID PROPELLANT TWO-STAGE LAUNCH VEHICLE	70-0366
LONGITUDINAL	PROBLEMS OF LONGITUDINAL STRENGTH IN THE CONSTRUCTION OF VERY LARGE TANK TRUCKS	73-0251
LONGITUDINALLY	STRUCTURAL ANALYSIS OF LONGITUDINALLY FRAMED SHIPS	72-0436
LONGWALL	FORECASTING THE ROCK PRESSURE AND ITS EFFECT ON THE LONGWALL FACES AND ROADWAYS IN THE BITUMINOUS CO	74-0374
LONGWALL	STRESS DISTRIBUTION AND ROCK FRACTURE ZONES IN THE ROOF OF LONGWALL FACE IN COAL MINE	73-0247
LONGWALL	VORAUSSAGE DES GEBIRGSDRUCKES UND SEINER AUSWIRKUNGEN AUF STREBEN UND STRECKEN IM STEINKOHLENBERGBAU	73-0679
LOSS	COMMENT ON THE PAPER "POSSIBLE LOSS OF ACCURACY IN CURVED (ISOPARAMETRIC) FINITE ELEMENTS" BY HENSHE	75-0279
LOSS	OVERCOMING LOSS OF ACCURACY WHEN USING CURVED FINITE ELEMENTS	75-0092
LOSS	POSSIBLE LOSS OF ACCURACY IN CURVED FINITE ELEMENTS	72-0278
LOSS	PREDICTION OF TRANSMISSION LOSS IN MUFFLERS BY THE FINITE ELEMENT METHOD	75-0160
LOSSLESS	EVALUATION OF ATTENUATION FROM LOSSLESS TRIANGULAR-FINITE-ELEMENT SOLUTIONS FOR INHOMOGENEOUSLY FILL	73-0059
LOSUNG	APPLICATION OF AN ' ELEMENT-METHOD ' PROGRAM TO SOLVING TEMPERATURE DISTRIBUTION PROBLEMS (ANWENDUNG	71-0464
LOSUNG	DIE LOSUNG NICHT-LINEARER PROBLEME NACH DER METHODE DER FINITEN ELEMENTE	74-0673
LOSUNG	EINE EINFACHE KONSTRUKTION VON KOORDINATENFUNKTIONEN FUR DIE NUMERISCHE LOSUNG ZWEIDIMENSIONALER RAN	66-0134
LOSUNG	GRUNDLAGEN DER DIGITALEN LOSUNG VON GRUNDWASSERSTROMUNGSPROBLEMEN NACH DER METHODE ENDLICHER ELEMENT	71-0542
LOVE	LOVE WAVE DISPERSION THROUGH IRREGULAR STRUCTURES USING FINITE-ELEMENT ALGORITHMS	75-0431
LOVE	THE PROPAGATION OF LOVE WAVES ACROSS NONHORIZONTALLY LAYERED STRUCTURES	71-0600
LOW	A FINITE ELEMENT METHOD FOR LOW REYNOLDS NUMBER FLOW AND ITS APPLICATION TO BIOMECHANICS	69-0304
LOW	AEROELASTIC PROBLEMS OF LOW ASPECT RATIO WINGS (PART 1)	56-0006
LOW	CURVED LOW ORDER ANGULAR ELEMENTS IN TWO DIMENSIONAL NEUTRON TRANSPORT PROBLEMS	74-1322
LOW	CYCLIC FATIGUE ANALYSIS OF ROCKET THRUST CHAMBERS. VOLUME 1: OFHC COPPER CHAMBER LOW CYCLE FATIGUE	74-0694
LOW	FINITE ELEMENT SOLUTION OF LOW BOND NUMBER SLOSHING	75-0728
LOW	LOW CYCLE FATIGUE STRENGTH OF PIPING COMPONENTS	73-0366
LOW	LOW REYNOLDS NUMBER DEVELOPING FLOW	69-0020
LOW	NONLINEAR DYNAMICS OF CABLES WITH LOW INITIAL TENSION	72-0740
LOW	NOTCH EFFECT IN LOW CYCLE FATIGUE - REPORT 1: LOW-CYCLE FATIGUE STRENGTH OF 0.48 PERCENT CARBON STEE	73-1090
LOW	PLANE STRESS CRACK TRAJECTORIES IN LOW MODULUS MATERIALS	75-0646
LOW	THE EFFECT OF CAMBER ON THE NATURAL FREQUENCIES OF LOW ASPECT RATIO TURBOMACHINERY BLADES	73-1092
LOW	THE FINITE ELEMENT METHOD IN LOW SPEED AERODYNAMICS	75-0702
LOW	TIMING COMPARISON STUDY FOR SOME HIGH ORDER FINITE ELEMENT PROCEDURES AND A LOW ORDER FINITE DIFFERE	72-0767
LOW-CYCLE	APPROACH TO INVESTIGATE NOTCH EFFECT ON LOW-CYCLE FATIGUE WITH FINITE ELEMENT	73-0509
LOW-CYCLE	NOTCH EFFECT IN LOW CYCLE FATIGUE - REPORT 1: LOW-CYCLE FATIGUE STRENGTH OF 0.48 PERCENT CARBON STEE	73-1090
LOW-CYCLE	NOTCH EFFECT IN LOW-CYCLE FATIGUE. CONSIDERATIONS ON NOTCH EFFECT IN LOW-CYCLE FATIGUE WITH FINITE-E	75-0343
LOW-CYCLE	NOTCH EFFECT IN LOW-CYCLE FATIGUE. CONSIDERATIONS ON NOTCH EFFECT IN LOW-CYCLE FATIGUE WITH FINITE-E	75-0343
LOW-CYCLE	NOTCH EFFECT IN LOW-CYCLE FATIGUE - 2. EFFECT OF STRESS CHANGE AND MEAN STRESS	75-0537
LOW-EMBANKMENT	FINITE ELEMENT ANALYSIS OF TRANSVERSE CRACKING IN LOW-EMBANKMENT DAMS	74-0696
LOW-THERMAL-DIS	DESIGN OF LOW-THERMAL-DISTORTION LST METERING STRUCTURE	75-0840
LOWER	A METHOD TO GENERATE BOTH UPPER AND LOWER BOUNDS TO PLATE EIGENVALUES BY CONFORMING DISPLACEMENT FIN	71-0166
LOWER	BENDING AND STRETCHING OF PLATES, SPECIAL MODELS FOR UPPER AND LOWER BOUNDS	66-0026
LOWER	DUALITY BETWEEN DISPLACEMENT AND EQUILIBRIUM METHOD WITH A VIEW TO OBTAINING UPPER AND LOWER BOUNDS	62-0013
LOWER	DYNAMIC ANALYSIS OF THE SLIDE IN THE LOWER SAN FERNANDO DAM DURING THE EARTHQUAKE OF FEBRUARY 9, 197	75-0277
LOWER	FINITE ELEMENT MODEL FOR LOWER GRANITE RESERVOIR	73-0065
LOWER	LOWER BOUND LIMIT ANALYSIS OF PLANE STRESS PROBLEMS BY A FINITE ELEMENT METHOD	70-0428
LOWER	LOWER BOUNDS TO COLUMN BUCKLING LOADS	73-0485
LOWER	UPPER AND LOWER BOUNDS IN MATRIX STRUCTURAL ANALYSIS	64-0018
LOWER	UPPER AND LOWER BOUNDS TO STRUCTURAL DEFORMATIONS BY DUAL ANALYSIS IN FINITE ELEMENTS	67-0074
LOWEST	DETERMINATION OF THE LOWEST CRITICAL BUCKLING STRESS OF VARIOUS SHAPED PLATES BY FINITE-ELEMENTS	73-1134
LST	DESIGN OF LOW-THERMAL-DISTORTION LST METERING STRUCTURE	75-0840
LUBRICATION	APPLICATION OF FINITE ELEMENT METHOD TO HYDRODYNAMIC LUBRICATION PROBLEMS (PART 1, "INFINITE-WIDTH B	71-0537
LUBRICATION	APPLICATION OF FINITE ELEMENT METHOD TO HYDRODYNAMIC LUBRICATION PROBLEMS (PART 2, " FINITE WIDTH BE	71-0538
LUBRICATION	APPLICATION OF FINITE ELEMENT METHODS TO LUBRICATION - ENGINEERING APPROACH	72-0006
LUBRICATION	APPLICATION OF FINITE ELEMENT METHODS TO THERMOHYDRODYNAMIC LUBRICATION	73-0074
LUBRICATION	APPLICATION OF FINITE ELEMENT METHODS TO LUBRICATION: SQUEEZE FILMS BETWEEN POROUS SURFACES	75-0764
LUBRICATION	FINITE ELEMENT ANALYSIS OF CONTINUUM PROBLEMS AND ITS APPLICATION TO THE INCOMPRESSIBLE LUBRICATION	71-0539
LUBRICATION	FINITE ELEMENT ANALYSIS OF FLUID FILM LUBRICATION - A SURVEY	74-0041
LUBRICATION	FINITE ELEMENT ANALYSIS OF LUBRICATION IN NATURAL JOINTS	74-0086
LUBRICATION	FINITE ELEMENT ANALYSIS OF FLUID FILM LUBRICATION - A SURVEY	74-1100
LUBRICATION	FINITE ELEMENT ANALYSIS OF FLUID FILM LUBRICATION SURVEY	75-0880
LUBRICATION	FINITE ELEMENT FORMULATION OF THE INCOMPRESSIBLE LUBRICATION PROBLEM	70-0022
LUBRICATION	FINITE ELEMENT METHOD IN HYDRODYNAMIC LUBRICATION	72-0301
LUBRICATION	FINITE ELEMENT SOLUTION OF THE INCOMPRESSIBLE LUBRICATION PROBLEM	69-0021
LUBRICATION	FINITE ELEMENT SOLUTION OF THE STEADY STATE COMPRESSIBLE LUBRICATION PROBLEM	70-0007
LUBRICATION	FINITE ELEMENT SOLUTION OF LUBRICATION PROBLEMS	71-0683
LUBRICATION	LA TECNICA DEGLI ELEMENTI FINITI PER PROBLEMI DI LUBRIFICAZIONE INCOMPRESSIBILE. (FINITE ELEMENT MET	75-0471
LUBRICATION	LUBRICATION PROBLEMS - THE SELECTION OF MATHEMATICAL MODELS	75-0881
LUBRICATION	NUMERICAL SOLUTION OF THE ELASTOHYDRODYNAMIC LUBRICATION PROBLEM USING FINITE ELEMENTS	72-0042
LUBRICATION	THE INCOMPRESSIBLE LUBRICATION PROBLEM	69-0022
LUBRICATION	THE ROLE OF NUMERICAL METHODS IN THE STUDY OF LUBRICATION	74-0042
LUBRICATION	TRANSIENT ELASTOHYDRODYNAMIC LUBRICATION BY FINITE ELEMENT METHODS	74-0087
LUBRICATION	VARIATIONAL FORMULATION FOR A CLASS OF FREE BOUNDARY PROBLEMS ARISING IN HYDRODYNAMIC LUBRICATION	75-0531
LUBRIFICAZIONE	LA TECNICA DEGLI ELEMENTI FINITI PER PROBLEMI DI LUBRIFICAZIONE INCOMPRESSIBILE. (FINITE ELEMENT MET	75-0471
LUFT	EXPERIMENTS IN FORMING AND JOINTING OF TITANIUM AND TITANIUM ALLOYS FOR AIRCRAFT AND SPACE VEHICLES	70-0134
LUGS	FATIGUE STRENGTH OF LUGS CONTAINING LINERS. VOLUME II. COMPUTER PROGRAM USED FOR ANALYSIS	70-0418
LUGS	SPECTRUM TRUNCATION AND DAMAGE TOLERANCE STUDY ASSOCIATED WITH THE C-5A OUTBOARD PYLON AFT TRUSS LUG	74-0535
LUMINA	PART 11: LUMINA AND HEXE ELEMENTS	68-0261
LUMINA	THE LUMINA ELEMENT FOR THE MATRIX DISPLACEMENT METHOD	68-0010
LUMPED	CONVERGENCE OF LUMPED FINITE ELEMENT SCHEMES FOR SELECTED INITIAL VALUE PROBLEMS	73-1026

LUMPED	LUMPED FINITE ELEMENT BASES FOR BEAMS AND PLATES	72-0263
LUMPED	LUMPED PARAMETER APPROACH TO AXISYMMETRIC DYNAMIC SOIL DEFORMATIONS	69-0170
LUMPED	LUMPED PARAMETER ELEMENT FOR THE ANALYSIS OF HYPERBOLIC PARABOLOID SHELLS	72-0148
LUMPED	LUMPED PARAMETER METHODS APPLIED TO ELASTIC VIBRATIONS	63-0016
LUMPED	MODE SHAPES AND FREQUENCIES BY FINITE ELEMENT METHOD USING CONSISTENT AND LUMPED MASSES	71-0610
LUMPED	THE EFFECT OF LUMPED PARAMETERS ON BEAM FREQUENCIES	63-0012
LUMPING	LUMPING OF FINITE ELEMENT MATRICES	74-0645
LUNAR	FINITE ELEMENT ANALYSIS OF ELASTO-PLASTIC SOILS. REPORT NO. 4, FINITE ELEMENT ANALYSIS OF ELASTO-PLA	72-0788
LUNAR	GRAVITY FLOW BIN DESIGN FOR LUNAR SOIL	72-0240
LUNAR	GRAVITY FLOW OF POWDER IN A LUNAR ENVIRONMENT. PT. 2-ANALYSIS OF FLOW INITIATION	71-0484
LUNG	FINITE ELEMENT DISPLACEMENT ANALYSIS OF A LUNG	72-0026
LUNG	LOCAL COMPLIANCE PROBLEM OF LUNG ELASTICITY	75-0660
LUNG	STRAIN ENERGY FUNCTIONS AND MATHEMATICAL MODELS OF LUNG ELASTICITY	73-0591
LUNG	THEORY FOR DISTORTION STUDIES OF LUNG PARENCHYMA BASED ON ALVEOLAR MEMBRANE PROPERTIES	74-0222
LUSH	LUSH: A COMPUTER PROGRAM FOR COMPLEX RESPONSE ANALYSIS OF SOIL-STRUCTURE SYSTEMS	74-0744
LWBR	FUEL ROD DEFORMATIONS - A TIME DEPENDENT FINITE ELEMENT ANALYSIS ON AXISYMMETRIC FUEL RODS, INCLUDIN	70-0414
LWR	CALCULATIONAL PROCEDURE FOR DETERMINING CREEP COLLAPSE OF LWR FUEL RODS	74-1001
LWR	THREE-DIMENSIONAL CRACK ANALYSIS APPLIED TO AN LWR NOZZLE-CYLINDER INTERSECTION	75-0165
LYNX	THE ASAS ANALYSIS OF THE LYNX	74-0932
L00	THE L2 AND L00 CONDITION NUMBERS OF THE FINITE ELEMENT MATRICES AND THE POINTWISE CONVERGENCE OF THE	72-0061
L2-PROJECTION	STABILITY IN L0 OF L2-PROJECTION INTO FINITE ELEMENT FUNCTION SPACES	75-0071
MACHINE	ANALYSIS OF THERMAL DEFORMATION OF MACHINE TOOL BY THE FINITE ELEMENT METHOD.	72-0432
MACHINE	ANWENDUNG DER METHODE FINITER ELEMENTE BEI DER ANALYSE DES DYNAMISCHEN VERHALTENS GEDAEMPFTER WERKZE	75-0580
MACHINE	BERECHNUNGSBEISPIELE UND ERFAHRUNGEN MIT DER METHODE DER FINITEN ELEMENTE IN MASCHINENBAU. (EXAMPLES	71-0390
MACHINE	CALCULATION OF THERMALLY INDUCED STRAINS IN MACHINE TOOL ELEMENTS	72-0464
MACHINE	COMPUTER AIDED DESIGN OF MACHINE TOOL STRUCTURE	74-1189
MACHINE	COMPUTER AIDED STRUCTURAL ANALYSIS: THE MACHINE INDEPENDENT SYSTEM ASKA	70-0397
MACHINE	COMPUTER METHODS FOR THE STRUCTURAL ANALYSIS OF MACHINE TOOLS	69-0325
MACHINE	COMPUTERUNTERSTUETZTES SYSTEM FUER DIE CONSTRUKTION DES WERKZEJGMASCHINENAUFBAUS (COMPUTER-AIDED SYS	73-0482
MACHINE	DOSTIGNUCA I TENDENCIJE U ISPITIVANJIMA ALATNIH MASINA (ACHIEVEMENTS AND TENDENCIES IN MACHINE TOOLS	73-0426
MACHINE	DYNAMIC ANALYSIS OF MACHINE TOOL STRUCTURES BY THE FINITE ELEMENT METHOD	71-0416
MACHINE	EVALUATION OF COMPUTER AIDED METHODS FOR THE DESIGN OF MACHINE TOOL STRUCTURES	73-0577
MACHINE	EXPERIENCES FROM STRESS ANALYSIS OF AXI-SYMMETRIC PROBLEMS IN MACHINE DESIGN	69-0247
MACHINE	FINEL, UNIVERSELLES PROGRAMMSYSTEM ZUR BERECHNUNG DER ELASTISCHEN EIGENSCHAFTEN VON MASCHINENBAUTEIL	72-0132
MACHINE	FINITE ELEMENT METHOD FOR MACHINE TOOL STRUCTURAL ANALYSIS	71-0136
MACHINE	FINITE ELEMENT METHOD FOR DETERMINATION OF MACHINE TOOL STRUCTURES	72-0095
MACHINE	FINITE ELEMENT METHOD FOR THE ANALYSIS OF MACHINE TOOL STRUCTURES.	72-0448
MACHINE	FINITE-ELEMENT FIELD ANALYSIS OF AN INHOMOGENEOUS ANISOTROPIC RELUCTANCE MACHINE ROTOR	73-0020
MACHINE	INTERNATIONAL MACHINE TOOL DESIGN AND RESEARCH CONFERENCE, 12TH PROCFEDINGS, 1971.	72-0224
MACHINE	INTERNATIONAL MACHINE TOOL DESIGN AND RESEARCH CONFERENCE, 13TH PROCEEDINGS, 1972	73-0504
MACHINE	INTRODUCTION TO FINITE ELEMENT METHOD AND ITS APPLICATION TO MACHINE TOOL STRUCTURES	72-0112
MACHINE	MAN MACHINE INTERACTIVE STRUCTURAL ANALYSIS AS A PRELIMINARY DESIGN AID	69-0293
MACHINE	NUMERISCHE BERECHNUNG INSTATIONAERER TEMPERATURFELDER UND THERMISCH BEDINGTER VERFORMUNGEN VON WERKZ	75-0847
MACHINE	THE MERIDIONAL THROUGH-FLOW CALCULATION IN AN AXIAL FLOW MACHINE BY THE FINITE ELEMENT METHOD	75-0115
MACHINED	PRESSURE DISTRIBUTION AND DEFORMATIONS OF MACHINED COMPONENTS IN CONTACT.	73-0372
MACHINERY	APPLICATION OF THE FINITE ELEMENT METHOD TO MACHINERY	74-0198
MACHINERY	MACHINERY SPACE DEFLECTION AND VIBRATION ANALYSES USING FINITE ELEMENT METHODS	72-0474
MACHINERY	THE APPLICATION OF THE SUPER ELEMENT METHOD IN ANALYSIS AND DESIGN OF SHIP STRUCTURES AND MACHINERY	72-0657
MACHINES	EFFICIENT TECHNIQUES FOR FINITE ELEMENT ANALYSIS OF ELECTRIC MACHINES	73-0134
MACHINES	FINITE ELEMENT ANALYSIS OF MAGNETICALLY SATURATED D-C MACHINES	71-0566
MACHINES	FINITE-ELEMENT ANALYSIS OF NONLINEAR MAGNETIC FIELDS IN ELECTRIC MACHINES	70-0484
MACHINES	NONLINEAR FINITE ELEMENT SOLUTION OF ELECTRICAL MACHINES UNDER NO-LOAD AND FULL-LOAD CONDITIONS	74-0472
MACHINES	SATURATED MAGNETIC ENERGY FUNCTIONAL FOR FINITE-ELEMENT ANALYSIS OF ELECTRIC MACHINES	75-0636
MACHINING	USING THE FINITE ELEMENT METHOD TO DETERMINE TEMPERATURE DISTRIBUTIONS IN ORTHOGONAL MACHINING	74-1212
MACIERZY	APPLICATION OF HERMITE POLYNOMIALS TO THE DETEPMINATION OF THE STIFFNESS MATRIX OF PLATE BY FINITE E	71-0365
MACIERZY	WIELOSTOPNIOWA SYNTEZA MACIERZY SZTYWNOSCI (MULTI-STAGE SYNTHESIS OF THE STIFFNESS MATRICES)	73-0668
MACKAY-MARG	MECHANICS OF MACKAY-MARG TONOMETER	73-0433
MACRO-APPROACH	MACRO-APPROACH FOR RIBBED AND GRID PLATE SYSTEMS	75-0079
MACROSCOPIC	COMBINED MACROSCOPIC AND MICROSCOPIC APPROACH TO THE FRACTURE OF METALS	74-0897
MAGAZINE	MAGAZINE HEADWALL RESPONSE TO EXPLOSIVE BLAST	74-0754
MAGIC	MAGIC II. AN AUTOMATED GENERAL PURPOSE SYSTEM FOR STRUCTURAL ANALYSIS	71-0495
MAGIC	MAGIC 11: AN AUTOMATED GENERAL PURPOSE SYSTEM FOR STRUCTURAL ANALYSIS. VOLUME 111. PROGRAMMER'S MANU	71-0756
MAGIC	MAGIC 11: AN AUTOMATED GENERAL PURPOSE SYSTEM FOR STRUCTURAL ANALYSIS. VOLUME 1. ENGINEER'S MANUAL (	71-0757
MAGIC	MAGIC 111: AN AUTOMATED GENERAL PURPOSE SYSTEM FOR STRUCTURAL ANALYSIS. VOLUME 111. PROGRAMMER'S MAN	72-0726
MAGIC	MAGIC 111: AN AUTOMATED GENERAL PURPOSE SYSTEM FOR STRUCTURAL ANALYSIS VOLUME 1. ENGINEER'S MANUAL	72-0727
MAGIC	MAGIC, AN AUTOMATED GENERAL PURPOSE SYSTEM FOR STRUCTURAL ANALYSIS. VOLUME III. PROGRAMMER'S MANUAL	69-0090
MAGIC	MAGIC, AN AUTOMATED GENERAL PURPOSE SYSTEM FOR STRUCTURAL ANALYSIS, VOLUME II, USER'S MANUAL	69-0144
MAGIC	MAGIC, AN AUTOMATED GENERAL PURPOSE SYSTEM FOR STRUCTURAL ANALYSIS.	69-0173
MAGIC	OPTIM 11: (MAGIC) COMPATIBLE LARGE SCALE AUTOMATED MINIMUM WEIGHT DESIGN PROGRAM - V 2. PROGRAMMER'S	74-1036
MAGIC	OPTIM 11: A (MAGIC) COMPATIBLE LARGE SCALE AUTOMATED MINIMUM WEIGHT DESIGN PROGRAM - 1. ENGINEERS AN	74-1048
MAGIC	OPTIM 11: A MAGIC COMPATIBLE LARGE-SCALE AUTOMATED MINIMUM WEIGHT DESIGN PROGRAM. VOLUME 1. ENGINEER	74-1096
MAGIC	OPTIM 11: MAGIC COMPATIBLE LARGE SCALE AUTOMATED MINIMUM WEIGHT DESIGN PROGRAM. VOLUME 11. PROGRAMME	74-1093
MAGNET	MAGNET FIELD AND FORCE BY FINITE ELEMENT TECHNIQUES	69-0035
MAGNETIC	ANALYSIS OF MAGNETIC FIELDS BY FINITE ELEMENT METHOD	73-0126
MAGNETIC	ANALYSIS OF TURBOALTERNATOR MAGNETIC FIELDS BY FINITE ELEMENTS	71-0460
MAGNETIC	COMPARAISON DES DIVERSES MODELISATIONS NUMERIQUES D'UN SYSTEME MAGNETIQUE SATURABLE. (COMPARISON OF	75-0295
MAGNETIC	CURVILINEAR FINITE ELEMENTS FOR 2-DIMENSIONAL SATURABLE MAGNETIC FIELDS	74-0562
MAGNETIC	FINITE ELEMENT ANALYSIS OF THE MAGNETIC FIELD DISTRIBUTION INSIDE A ROTATING FERROMAGNETIC BAR	74-1037
MAGNETIC	FINITE ELEMENT MODELING OF MAGNETIC FIELD DEFECT INTERACTIONS	75-0068
MAGNETIC	FINITE ELEMENT SOLUTION OF SATURABLE MAGNETIC FIELD PROBLEMS	70-0205
MAGNETIC	FINITE ELEMENT SOLUTION OF SATURATED TRAVELLING MAGNETIC FIELD PROBLEMS	75-0600
MAGNETIC	FINITE-ELEMENT ANALYSIS OF NONLINEAR MAGNETIC FIELDS IN ELECTRIC MACHINES	70-0484
MAGNETIC	FINITE-ELEMENT SOLUTION OF EDDY-CURRENT PROBLEM IN MAGNETIC STRUCTURES	74-0106
MAGNETIC	PREDICTING MAGNETIC LEAKAGE FIELDS BY FINITE ELEMENT TECHNIQUES	75-0406
MAGNETIC	RASHET TREKHMERNYKH MAGNITNYKE POLEI METODOM KONECHNYKH ELEMENTOV. (CALCULATION OF THREE-DIMENSIONAL	75-0766
MAGNETIC	REALIZATSIYA METODA KONECHNYKH ELEMENTOV NA EVM DLYA RASCHETA DVUMERNYKH ELEKTRICHESKIKH I MAGNITNYK	74-0251
MAGNETIC	SATURATED MAGNETIC ENERGY FUNCTIONAL FOR FINITE-ELEMENT ANALYSIS OF ELECTRIC MACHINES	75-0636
MAGNETIC	STUDIO DI CAMPI ELETTRICI E MAGNETICI STANZIONARI CON IL METODO DEGLI ELEMENTI FINITI. APPLICAZIONE	75-0207
MAGNETIC	SU LA DETERMINAZIONE DEI CAMPI MAGNETICI LINEARI E NON LINEARI MEDIANTE L APPLICAZIONE DEI METODI DE	74-1061
MAGNETIC	THREE DIMENSIONAL NONLINEAR MAGNETIC FIELD BOUNDARY VALUE PROBLEM AND ITS NUMERICAL SOLUTION	70-0060
MAGNETIC-FIELDS	ANALYSIS OF STEADY ELECTRICAL AND MAGNETIC-FIELDS BY FINITE-ELEMENT METHOD - APPLICATION TO TRANSFOR	75-0550
MAGNETICALLY	FINITE ELEMENT ANALYSIS OF MAGNETICALLY SATURATED D-C MACHINES	71-0566
MAGNETICI	STUDIO DI CAMPI ELETTRICI E MAGNETICI STANZIONARI CON IL METODO DEGLI ELEMENTI FINITI. APPLICAZIONE	75-0207
MAGNETICI	SU LA DETERMINAZIONE DEI CAMPI MAGNETICI LINEARI E NON LINEARI MEDIANTE L APPLICAZIONE DEI METODI DE	74-1061
MAGNETIQUE	ANALYSE NUMERIQUE DU CHAMP MAGNETIQUE D'UN ALTERNATEUR PAR ELEMENTS FINIS ET SURRELAXATION PONCTUELL	73-0625
MAGNETIQUE	COMPARAISON DES DIVERSES MODELISATIONS NUMERIQUES D'UN SYSTEME MAGNETIQUE SATURABLE. (COMPARISON OF	75-0295
MAGNETIQUE	ETUDE NUMERIQUE DU CHAMP MAGNETIQUE DANS UN ALTERNATEUR TETRAPOLAIRE PAR LA METHODE DES ELEMENTS FIN	74-0413
MAGNETIZATION	SIMPLE EQUATIONS FOR THE MAGNETIZATION AND RELUCTIVITY CURVES OF STEEL	75-0156
MAGNETOHYDRODYN	A FINITE ELEMENT SOLUTION OF THE TRANSIENT MAGNETOHYDRODYNAMIC COUETTE FLOW	74-1103
MAGNETOHYDRODYN	NUMERICAL ANALYSIS OF MAGNETOHYDRODYNAMIC INSTABILITIES BY FINITE ELEMENT METHOD	72-0041
MAGNETOHYDRODYN	THALIA - ONE-DIMENSIONAL MAGNETOHYDRODYNAMIC STABILITY PROGRAM USING METHOD OF FINITE-ELEMENTS	75-0458
MAGNETOHYDRODYN	THALIA: A ONE-DIMENSIONAL MAGNETOHYDRODYNAMIC STABILITY PROGRAM USING THE METHOD OF FINITE ELEMENTS	74-1085

MAGNETOHYDRODYN	VARIATIONAL PRINCIPLE FOR MAGNETOHYDRODYNAMIC CHANNEL FLOW	70-0192
MAGNETOHYDRODYN	NEW FINITE ELEMENT APPROACH TO THE NORMAL MODE ANALYSIS IN MAGNETOHYDRODYNAMICS	74-0876
MAGNETOTELLURIC	MAGNETOTELLURIC RESPONSE OF A 2-DIMENSIONAL SLOPING CONTACT BY FINITE ELEMENT METHOD	73-0130
MAGNETOTELLURIC	MAGNETOTELLURIC RESPONSE OF LATERALLY INHOMOGENEOUS AND ANISOTROPIC MEDIA	75-0850
MAGNETS	FINITE-ELEMENT SOLUTIONS FOR DEVICES WITH PERMANENT MAGNETS	75-0464
MAGNETS	THREE-DIMENSIONAL MECHANICAL STRESSES IN TOROIDAL MAGNETS FOR CONTROLLED THERMONUCLEAR REACTORS	73-0669
MAGNIFICATION	AN INVESTIGATION ON THE PLASTICITY MAGNIFICATION FACTOR IN DEEP SURFACE FLOWS	74-0766
MAGNITNYKE	RASHET TREKHMERNYKH MAGNITNYKE POLEI METODOM KONECHNYKH ELEMENTOV. (CALCULATION OF THREE-DIMENSIONAL	75-0766
MAGNITNYKH	REALIZATSIYA METODA KONECHNYKH ELEMENTOV NA EVM DLYA RASCHETA DVUMERNYKH ELEKTRICHESKIKH I MAGNITNYK	75-0251
MAIN	APPLICATION OF NEW AUSTRIAN TUNNELLING METHOD IN DIFFICULT BUILTOVER AREAS IN FRANKFURT/MAIN METRO	73-1025
MAINE-MONTPARNA	TOUR MAINE-MONTPARNASSE QUELQUES PROBLEMES DE STRUCTURE (SPECIFIC PROBLEMS ENCOUNTERED IN THE CONSTR	74-0539
MAINE-MONTPARNA	TOUR MAINE-MONTPARNASSE QUELQUES PROBLEMES DE STRUCTURE (SPECIFIC PROBLEMS ENCOUNTERED IN THE CONSTR	74-0539
MAINTENANCE	COMPUTER PROGRAM CONSTRUCTION AND MAINTENANCE - FUTURE OF CENTRALIZED FINITE ELEMENT ACTIVITY	75-0014
MAINTENANCE	DEVELOPMENT AND MAINTENANCE OF LARGE FINITE ELEMENT SOFTWARE SYSTEMS	74-0865
MALEN	MALEN DIMENSIONIERUNG VON TRAGWERKEN	74-0683
MAMBRANES	FINITE ELEMENT ANALYSIS OF NONLINEAR CABLE REINFORCED MAMBRANES	74-1073
MAN	MAN MACHINE INTERACTIVE STRUCTURAL ANALYSIS AS A PRELIMINARY DESIGN AID	69-0293
MANAGEMENT	EMPIRICAL INVESTIGATION OF REORDERING AND DATA MANAGEMENT FOR FINITE ELEMENT SYSTEMS OF EQUATIONS	73-0944
MANAGEMENT	MANAGEMENT OF RETARDATION OF SALT WATER INTRUSION IN COASTAL AQUIFERS	75-0692
MANAGEMENT	USING INTERACTIVE GRAPHICS FOR THE PREPARATION AND MANAGEMENT OF FINITE ELEMENT DATA	74-0176
MANDIBLE	MATHEMATICAL MODELLING AND STUCTURAL ANALYSIS OF THE MANDIBLE	72-0124
MANDREL	FINITE ELEMENT ANALYSIS OF SEVERAL SWAGE MANDREL DESIGNS	74-0794
MANIPULATING	PROBLEMS IN BUILDING AND MANIPULATING LARGE STIFFNESS MATRICES	71-0596
MANIPULATION	CHARACTERISTICS OF MANIPULATION ERRORS IN SOLVING LOAD DEFLECTION EQUATIONS	70-0355
MANIPULATION	MANIPULATION ERRORS IN FINITE ELEMENT ANALYSIS	68-0253
MANIPULATION	MANIPULATION ERRORS IN FINITE ELEMENT ANALYSIS OF STRUCTURES	69-0349
MANIPULATION	MANIPULATION ERRORS IN THE FINITE ELEMENT ANALYSES	71-0176
MANIPULATION	MANIPULATION ERRORS IN COMPUTER SOLUTION OF CRITICAL SIZE STRUCTURAL EQUATIONS	71-0367
MANIPULATION	USE OF SYMBOLIC MANIPULATION IN DEVELOPMENT OF TWO-DIMENSIONAL FINITE ELEMENTS	74-0103
MANNED	LARGE STRUCTURES FOR MANNED SPACECRAFT: MATHEMATICAL ANALYSIS, DESIGN, CONSTRUCTION AND TESTS	74-1327
MANUAL	A TECHNICAL MANUAL FOR THE BEAM ELEMENTS	69-0455
MANUAL	AIRCRAFT FUEL TANK VULNERABILITY TO HYDRAULIC RAM: MODIFICATTION OF THE NORTHROP FINITE ELEMENT COMP	74-0724
MANUAL	ANSYS USER'S MANUAL	71-0561
MANUAL	ASKA 105 - A TECHNICAL MANUAL FOR THE DIRECT SOLUTION AND INVERSION PACKAGE	68-0255
MANUAL	ASKA 105 A TECHNICAL MANUAL AND PROGRAMMING THEORY FOR THE ELEMENTS, FLA, TRIM, SPARTA AND TET	68-0254
MANUAL	CAPE-2D, A COMPUTER PROGRAM FOR THE STRESS ANALYSIS OF PLANE AND AXISYMMETRIC COMPOSITE STRUCTURES.	69-0083
MANUAL	COMOC THERMAL ANALYSIS VARIANT, USER'S MANUAL	72-0701
MANUAL	DEFINIT - A NEW ELEMENT DEFINITION CAPABILITY FOR NASTRAN: USER'S MANUAL	73-0738
MANUAL	DEVELOPMENT AND APPLICATIONS OF SUPERSONIC UNSTEADY CONSISTENT AERODYNAMICS FOR INTERFERING PARALLEL	72-0716
MANUAL	DYNASOR 11 - A FINITE ELEMENT PROGRAM FOR THE DYNAMIC NONLINEAR ANALYSIS OF SHELLS OF REVOLUTION (US	71-0666
MANUAL	EAC/EASE USER INFORMATION MANUAL	70-0320
MANUAL	ELAD PROGRAM MANUAL, A COMPUTEP PROGRAM FOR THE ELASTIC ANALYSIS OF DOME STRUCTURES AND AXISYMMETRIC	69-0064
MANUAL	FEDGE - A GENERAL PURPOSE COMPUTER PROGRAM FOR FINITE ELEMENT DATA GENERATION. VOLUME 2 - PROGRAM MA	69-0492
MANUAL	FEDGE - A GENERAL-PURPOSE COMPUTER PROGRAM FOR FINITE ELEMENT DATA GENERATION. VOLUME 1 - USER'S MAN	69-0491
MANUAL	ICES STRUDL-II ENGINEERING USERS MANUAL. VOLS. 1, 11, 111	69-0398
MANUAL	MAGIC 11: AN AUTOMATED GENERAL PURPOSE SYSTEM FOR STRUCTURAL ANALYSIS. VOLUME 111. PROGRAMMER'S MANU	71-0756
MANUAL	MAGIC 11: AN AUTOMATED GENERAL PURPOSE SYSTEM FOR STRUCTURAL ANALYSIS. VOLUME 1. ENGINEER'S MANUAL (	71-0757
MANUAL	MAGIC 111: AN AUTOMATED GENERAL PURPOSE SYSTEM FOR STRUCTURAL ANALYSIS. VOLUME 111. PROGRAMMER'S MAN	72-0726
MANUAL	MAGIC 111: AN AUTOMATED GENERAL PURPOSE SYSTEM FOR STRUCTURAL ANALYSIS VOLUME 1. ENGINEER'S MANUAL	72-0727
MANUAL	MAGIC, AN AUTOMATED GENERAL PURPOSE SYSTEM FOR STRUCTURAL ANALYSIS. VOLUME III, PROGRAMMER'S MANUAL	69-0090
MANUAL	MAGIC, AN AUTOMATED GENERAL PURPOSE SYSTEM FOR STRUCTURAL ANALYSIS, VOLUME II, USER'S MANUAL	69-0144
MANUAL	MISSILE BODY INPUT GENERATOR (BING), A NASTRAN PRE-PROCESSOR; THEORETICAL DEVELOPMENT, USER'S MANUAL	75-0717
MANUAL	MRI/STARDYNE USER INFORMATION MANUAL	73-0541
MANUAL	MSC/NASTRAN APPLICATION MANUAL	72-0567
MANUAL	ON IMPROVING EFFICIENCY OF AN ALGORITHM FOR STRUCTURAL OPTIMIZATION AND A USERS MANUAL FOR PROGRAM T	74-0778
MANUAL	OPTIM 11: (MAGIC) COMPATIBLE LARGE SCALE AUTOMATED MINIMUM WEIGHT DESIGN PROGRAM - V 2. PROGRAMMER'S	74-1036
MANUAL	OPTIM 11: A (MAGIC) COMPATIBLE LARGE SCALE AUTOMATED MINIMUM WEIGHT DESIGN PROGRAM - 1. ENGINEERS AN	74-1048
MANUAL	OPTIM 11: A MAGIC COMPATIBLE LARGE-SCALE AUTOMATED MINIMUM WEIGHT DESIGN PROGRAM. VOLUME 1. ENGINEER	74-1096
MANUAL	OPTIM 11: MAGIC COMPATIBLE LARGE SCALE AUTOMATED MINIMUM WEIGHT DESIGN PROGRAM. VOLUME 11. PROGRAMME	74-1093
MANUAL	PLANFORM INPUT GENERATOR (PING), A NASTRAN PROCESSOR FOR LIFTING SURFACES-THEORETICAL DEVELOPMENT, U	73-0780
MANUAL	PREDICTION OF LONG-TERM STRESS RANGES: USER'S MANUAL - BRIDGE LOAD GENERATOR	73-0737
MANUAL	PREDICTION OF LONG-TERM STRESS RANGES: USER'S MANUAL - BRIDGE DYNAMIC STRESS ANALYSIS	74-0807
MANUAL	PROGRAMMER'S MANUAL ADDITIONS AND DEMONSTRATION PROBLEMS FOR A THERMOSTRUCTURAL CAPABILITY FOR NASTR	70-0734
MANUAL	SABOR CS-2 USER'S MANUAL	70-0408
MANUAL	SAFE SHELL, A COMPUTER PROGRAM FOR THE STRESS ANALYSIS OF THIN SHELL BODIES OF REVOLUTION. A USER'S	67-0018
MANUAL	SAFE-CREEP: A COMPUTER PROGRAM FOR THE VISCOELASTIC ANALYSIS OF AXISYMMETRIC AND PLANE CONCRETE STRU	67-0148
MANUAL	SESAM 69 A GENERAL PURPOSE FINITE ELEMENT METHOD PROGRAM CDC/NASTRAN APPLICATIONS MANUAL	72-0552
MANUAL	SP-222 NASTRAN THEORETICAL MANUAL	70-0387
MANUAL	SP-222 NASTRAN USERS MANUAL	70-0386
MANUAL	STARDYNE STRUCTURAL ANALYSIS SYSTEM, ANALYTICAL CAPABILITY MANUAL	68-0235
MANUAL	STRESS, STABILITY AND VIBRATION OF COMPLEX SHELLS OF REVOLUTION: ANALYSIS AND USERS MANUAL FOR BOSOR	69-0441
MANUAL	TANKER TRANSVERSE STRENGTH ANALYSIS: USER'S MANUAL	72-0750
MANUAL	THE NASTRAN PROGRAMMER'S MANUAL	72-0480
MANUAL	THE NASTRAN THEORETICAL MANUAL	70-0298
MANUAL	THE NASTRAN USER'S MANUAL	72-0479
MANUAL	TRANSPIRATION AND FILM COOLING BOUNDARY LAYER COMPUTER PROGRAM. VOLUME 2, COMPUTER PROGRAM AND USER'	71-0655
MANUAL	USER INFORMATION MANUAL	73-0543
MANUAL	USER'S MANUAL FOR NTEDYNS - A NONLINEAR FINITE ELEMENT CODE FOR ANALYZING THE BLAST RESPONSE OF UNDE	71-0695
MANUAL	USER'S MANUAL FOR PROGRAM BEAM	73-1105
MANUAL	USER'S MANUAL FOR UMR FINITE ELEMENT COMPUTER PROGRAM - STRATA	69-0429
MANUAL	USERS MANUAL FOR CEL/NONSAP, A NONLINEAR STRUCTURAL ANALYSIS PROGRAM	74-0706
MANUAL	USERS MANUAL FOR DIGITAL COMPUTER PROGRAM OK5 NUMERICAL ANALYSIS OF FINITE STRAINS AND DISPLACEMENTS	72-0718
MANUAL	USERS MANUAL FOR PROGRAM BEAM	73-0714
MANUAL	USERS MANUAL FOR QMESH, A SELF-ORGANIZING MESH GENERATION PROGRAM	73-0880
MANUAL	USERS' MANUAL FOR DIGITAL COMPUTER PROGRAM OK1. NUMERICAL ANALYSIS OF FINITE AXISYMMETRIC DEFORMATIO	72-0861
MANUAL	VIEW FACTOR COMPUTER PROGRAM (PROGRAM VIEW) USER'S MANUAL	73-0789
MANUALS	MARC-CDC USER INFORMATION MANUALS	71-0562
MANUEL	DYNAN USERS MANUEL	71-0625
MANUFACTURE	APPLICATION OF FINITE ELEMENT TECHNIQUE TO DESIGN AND MANUFACTURE OF ASTRONOMICAL MIRRORS	69-0302
MANUFACTURE	DESIGN, MANUFACTURE, DEVELOPMENT, TEST, AND EVALUATION OF BORON/ALUMINUM STRUCTURAL COMPONENTS FOR S	74-0798
MANUSCRIPT	HEAVY SECTION STEEL TECHNOLOGY PROGRAM TECHNICAL OR PROGRAMMATIC MANUSCRIPT NO. 25. VARIABLE THICKNE	73-0948
MAPPING	APPLICATION OF CONFORMAL MAPPING AND VARIATIONAL METHOD TO THE STUDY OF HEAT CONDUCTION IN POLYGONAL	73-0444
MAPPING	SPECIFICATION OF GEOMETRICAL PARAMETERS FOR ELEMENTS WITH CUBIC MAPPING FUNCTIONS	73-1082
MAPPINGS	CURVED BOUNDARY ELEMENTS, GENERAL FORMS OF POLYNOMIAL MAPPINGS	74-1256
MAPS	ALGORITHMS FOR THE PRODUCTION OF CONTOUR MAPS OVER AN IRREGULAR TRIANGULAR MESH	72-0825
MARC	ELASTIC-PLASTIC AND CREEP ANALYSIS VIA THE MARC 2 FINITE ELEMENT PROGRAM	71-0032
MARC	MARC SERIES OF PROGRAMS	72-0154
MARC-CDC	MARC-CDC USER INFORMATION MANUALS	71-0562
MARCH	PROCEEDINGS OF THE NAVY-NASTRAN COLLOQUIUM (4TH) HELD AT NAVY SHIP RESEARCH AND DEVELOPMENT CENTER,	73-0888
MARCHING	COMPARISON OF TIME MARCHING SCHEMES FOR THE TRANSIENT HEAT CONDUCTION EQUATION	75-0477
MARIMBONDO	EMBANKMENT DAM FOR THE MARIMBONDO HYDRO PROJECT	75-0270
MARINE	MARINE PROPELLER BLADES STRESSES AND DEFORMATIONS WITH GIVEN, FROZEN, HYDRODYNAMIC LOADING	73-0474

MARINE	PREDICTION OF MARINE PROPELLER DISTORTION AND STRESSES USING A SUPERPARAMETRIC THICK SHELL FINITE EL	73-0370
MARINE	STUDY OF MOTION AND STRENGTH OF FLOATING MARINE STRUCTURES IN WAVES	74-0489
MARITIME	BULLETIN DE L ASSOCIATION TECHNIQUE MARITIME ET AERONAUTIQUE (BULLETIN OF THE TECHNICAL MARITIME AND	74-1108
MARITIME	BULLETIN DE L ASSOCIATION TECHNIQUE MARITIME ET AERONAUTIQUE (BULLETIN OF THE TECHNICAL MARITIME AND	74-1108
MARKER	USE OF FINITE ELEMENT METHOD IN A MARKER AND CELL PROGRAM FOR SOLVING TIME-DEPENDENT FLUID FLOW PROB	75-0043
MARK5C	MARK5C/DAISY SERIES OF PROGRAMS	72-0160
MASCHENNETZE	BERECHNUNG VON PLATTEN NACH DEM RITZ' SCHEN VERFAHREN MIT HILFE DREIECKFOERMIGER MASCHENNETZE	66-0094
MASCHINENBAU	BERECHNUNGSBEISPIELE UND ERFAHRUNGEN MIT DER METHODE DER FINITEN ELEMENTE IN MASCHINENBAU. (EXAMPLES	71-0390
MASCHINENBAU	DIE METHODE DER ENDLICHEN ELEMENTE UND IHRE ANWENDUNG IM MASCHINENBAU	71-0628
MASCHINENBAUTEI	BERECHNUNG DER ELASTIZITATSFIGENSCHAFTEN VON MASCHINENBAUTEILEN NACH DER METHODE FINITER ELFMENTE	73-1046
MASCHINENBAUTEI	FINEL, UNIVERSELLES PROGRAMMSYSTEM ZUR BERECHNUNG DER ELASTISCHEN EIGENSCHAFTEN VON MASCHINENBAUTEIL	72-0132
MASCHINENUNABHA	STRING EIN PROGRAMMSYSTEM FUR DIE MASCHINENUNABHANGIGE BEARBEITUNG VON ZEICHENKETTEN IN FORTRAN	68-0265
MASHINNOGO	RASCHET NAPRYAZHENIL L PEREMESHCHENII V SVODE PODZEMNOGO MASHINNOGO ZALA GES I V OKRUZHAYUSHCHEM SKA	74-0542
MASINA	DOSTIGNUCA I TENDENCIJE U ISPITIVANJIMA ALATNIH MASINA (ACHIEVEMENTS AND TENDENCIES IN MACHINE TOOLS	73-0426
MASIVELE	ASUPRA ANALIZEI PRIN METODA ELEMENTULUI FINIT A STARILOR DE EFORTURI SI DEFORMATII DIN MASIVELE DE P	70-0471
MASONRY	A FINITE-ELEMENT MODEL TO SIMULATE THE NONLINEAR RESPONSE OF REINFORCED CONCRETE FRAMES WITH MASONRY	73-1168
MASONRY	LINEAR ELASTIC FINITE ELEMENT ANALYSIS OF MASONRY WALLS ON BUILDINGS	74-1015
MASONRY	STABILITY OF MASONRY PIERS	75-0248
MASS	A FINITE ELEMENT FORMULATION OF HEAT CONDUCTION AND HEAT AND MASS TRANSFER PROBLEMS	74-0350
MASS	A MODAL SYNTHESIS TECHNIQUE FOR DETERMINING DYNAMIC PROPERTIES FOR A STRUCTURE FOR MASS AND STIFFNES	74-0319
MASS	AN APPLICATION OF LEAST SQUARES FINITE ELEMENT METHODS TO TWO DIMENSIONAL STEADY INVISCID COMPRESSIB	75-0114
MASS	ANALYSIS OF FOUNDATIONS ON A SEMI-INFINITE ELASTIC MASS UNDER DYNAMIC LOADS	72-0763
MASS	ANALYSIS OF ROCK MASS DEFORMATIONS DUE TO EXCAVATIONS	73-0395
MASS	APPLICATION OF FINITE ELEMENT METHOD TO ADDED VIRTUAL MASS OF SHIP HULL VIBRATION	70-0463
MASS	APPLICATION OF FINITE ELEMENT METHOD TO MASS TRANSPORT LIMITED EPITAXIAL GROWTH PROCESSES	71-0488
MASS	AUTOMATIC GENERATION AND CONDENSATION OF LARGE MASS AND STIFFNESS MATRICES	72-0466
MASS	BEHAVIOUR OF THE ROCK MASS AT THE SITE OF A ROCKFILL DAM	74-0992
MASS	BOUNDS ON EXTERNAL EIGENVALUES OF FINITE ELEMENT STIFFNESS AND MASS MATRICES AND THEIR SPECTRAL COND	72-0256
MASS	BOUNDS ON THE SPECTRAL AND MAXIMUM NORMS OF THE FINITE ELEMENT STIFFNESS, FLEXIBILITY AND MASS MATRI	73-0409
MASS	CALCULATION OF THE VIRTUAL MASS, MOVEMENT AND INERTIA BY THE FINITE ELEMENT METHOD	68-0221
MASS	COMPARISONS OF CONSISTENT MASS MATRIX SCHEMES	69-0437
MASS	COMPUTATIONAL METHOD FOR DYNAMIC ANALYSIS OF STRUCTURES WITH LARGE SIZE STIFFNESS AND MASS MATRICES.	73-0830
MASS	CONSISTENT MASS MATRIX FOR DISTRIBUTED SYSTEMS	63-0001
MASS	CONTINUUM MECHANICS AS AN APPROACH TO ROCK MASS PROBLEMS	68-0172
MASS	DERIVATION OF GEOMETRIC STIFFNESS AND MASS MATRICES FOR FINITE ELEMENT HYBRID MODELS	74-0844
MASS	DEVELOPMENT OF EXPLICIT STIFFNESS AND MASS MATRICES FOR A TRIANGULAR PLATE ELEMENT	69-0242
MASS	DISTRIBUTED MASS MATRIX FOR PLATE ELEMENTS IN BENDING	65-0018
MASS	DISTRIBUTED MASS MATRIX FOR PLATE ELEMENT BENDING (DISCUSSION)	66-0038
MASS	EQUIVALENT MASS MATRICES FOR RECTANGULAR PLATES IN BENDING	66-0063
MASS	FINITE ELEMENT ANALYSIS OF TWO-DIMENSIONAL ADDED MASS AND DAMPING	75-0394
MASS	FINITE ELEMENT SOLUTION FOR ADDED MASS AND DAMPING	74-0018
MASS	FINITE ELEMENT SOLUTION OF THE WIND DRIVEN CURRENTS AND ITS MASS TRANSPORT IN LAKES	74-1368
MASS	FINITE ELEMENTS FOR THE COMPUTATIONS OF HYDRODYNAMIC MASS	69-0346
MASS	GALERKIN FINITE ELEMENT ANALYSIS OF A UNIFORM BEAM CARRYING A CONCENTRATED MASS AND ROTARY INERTIA W	74-0552
MASS	MASS CONDENSATION, A SEMI-AUTOMATIC METHOD FOR REDUCING THE SIZE OF VIBRATION PROBLEMS	69-0220
MASS	MASS LOADING EFFECTS ON VIBRATION OF RING AND SHELL STRUCTURES	68-0245
MASS	MASS MATRIX FOR A PRISMATIC BEAM SEGMENT	68-0116
MASS	MODIFIED POTENTIAL ENERGY MASS REPRESENTATION FOR FREQUENCY PREDICTION	66-0053
MASS	NON-LINEAR EFFECTS OF HIGH TEMPERATURE ON THE VIBRATION OF A SIMPLY SUPPORTED BEAM WITH A CENTRAL MA	73-0838
MASS	PREDICTION OF PLATE VIBRATIONS USING A CONSISTENT MASS MATRIX	66-0043
MASS	REDUCTION OF STIFFNESS AND MASS MATRICES	65-0037
MASS	ROTARY INERTIA CORRECTION FOR MASS MATRICES	63-0011
MASS	SAMMSOR 111: A FINITE ELEMENT PROGRAM TO DETERMINE STIFFNESS AND MASS MATRICES OF RING-STIFFENED SHE	72-0774
MASS	SEEPAGE IN FRACTURED ROCK: FINITE ELEMENT ANALYSIS OF ROCK MASS BEHAVIOR UNDER COUPLED ACTION OF BOD	71-0825
MASS	SOLUTION OF A TWO DIMENSIONAL DIFFUSION CONVECTION EQUATION DESCRIBING THE MASS TRANSFER OF A NEUTRA	74-0037
MASS	STRUCTURAL RESPONSE TO MOVING PROJECTILE MASS BY THE FINITE ELEMENT METHOD	75-0693
MASS	THE GENERATION OF INTERELEMENT, COMPATIBLE STIFFNESS AND MASS MATRICES BY THE USE OF INTERPOLATION F	66-0014
MASSACHUSETTS	M.I.T. TEST SECTION INSTRUMENTATION, MASSACHUSETTS BAY TRANSPORTATION AUTHORITY, HAYMARKET-NORTH EXT	72-0673
MASSACHUSETTS	MATHEMATICAL MODELS OF THE MASSACHUSETTS BAY. PART 1. FINITE ELEMENT MODELING OF TWO-DIMENSIONAL HYD	73-0800
MASSES	A FINITE ELEMENT APPROACH TO THE DETERMINATION OF THE DYNAMIC RESPONSE OF BEAMS SUPPORTING MOVING MA	74-1114
MASSES	AN INVESTIGATION INTO THE USE OF FINITE ELEMENT METHODS FOR ANALYSING STRESS DISTRIBUTIONS IN BLOCK	71-0530
MASSES	ASUPRA ANALIZEI PRIN METODA ELEMENTULUI FINIT A STARILOR DE EFORTURI SI DEFORMATII DIN MASIVELE DE P	70-0471
MASSES	CONTINUUM CHARACTERIZATION OF JOINTED ROCK MASSES - 1, 2	73-0196
MASSES	ESTIMATION OF THE MECHANICAL PROPERTIES OF ROCK MASSES	73-0935
MASSES	MATHEMATICAL MODELS OF THE MASSACHUSETTS BAY. PART 1. FINITE ELEMENT MODELING OF TWO-DIMENSIONAL HYD	73-0800
MASSES	MODE SHAPES AND FREQUENCIES BY FINITE ELEMENT METHOD USING CONSISTENT AND LUMPED MASSES	71-0610
MASSES	NONLINEAR FINITE ELEMENT ANALYSIS OF MULTI JOINTED ROCK MASSES	71-0956
MASSES	NONLINEAR VIBRATIONS OF NONUNIFORM BEAMS WITH CONCENTRATED MASSES	74-1300
MASSIF	RASCHET NAPRYAZHENIL L PEREMESHCHENII V SVODE PODZEMNOGO MASHINNOGO ZALA GES I V OKRUZHAYUSHCHEM SKA	74-0542
MASSIFS	CALCUL DE MASSIFS EN ELASTOPLASTICITE APPLICATIONS A LA MECANIQUE DES ROCHES ET DES SOLS. (APPLICATI	74-0223
MASSIVE	RASCHET NAPRYAZHENIL L PEREMESHCHENII V SVODE PODZEMNOGO MASHINNOGO ZALA GES I V OKRUZHAYUSHCHEM SKA	74-0542
MASSIVE	THREE-DIMENSIONAL ISOPARAMETRIC ELEMENTS AND THEIR USE FOR STATICAL ANALYSIS OF MASSIVE WATER-POWER	71-0364
MAST	STRUCTURAL ANALYSIS OF MAST BY FINITE ELEMENT TECHNIQUE	74-0385
MASTERS	AUTOMATIC MASTERS FOR EIGENVALUE ECONOMIZATION	75-0557
MASTS	STATIC AND DYNAMIC BEHAVIOUR OF GUYED MASTS	68-0271
MAT	A FINITE ELEMENT ANALYSIS OF AN AIRCRAFT LANDING MAT ON AN ELASTIC FOUNDATION	68-0280
MAT	A TECHNIQUE FOR THE SETTLEMENT AND STRESS ANALYSIS OF MAT FOUNDATIONS USING THE FINITE ELEMENT METHO	72-0950
MATCHES	STRESS DISTRIBUTIONS AROUND PLUG MATCHES WITH ANCHOR BLOCKS	69-0135
MATCHING	MATCHING OF ESSENTIAL BOUNDARY CONDITIONS IN THE FINITE ELEMENT METHOD	74-1260
MATCHING	USE OF PARABOLIC ARCS IN MATCHING CURVED BOUNDARIES IN FINITE-ELEMENT METHOD	75-0589
MATED	NODAL ANALYSIS OF THE MATED SPACE SHUTTLE CONFIGURATION	71-0097
MATERIAL	A COMPUTER BASED PROCEDURE FOR PREDICTING THE TRANSIENT RESPONSE AND FAILURE OF A TWO-DIMENSIONAL CO	71-0630
MATERIAL	A DIGITAL COMPUTER PROGRAM FOR THE FINITE ELEMENT ANALYSIS OF SOLIDS WITH NONLINEAR MATERIAL PROPERT	65-0032
MATERIAL	A METHOD FOR MATERIAL CHARACTERIZATION EMPLOYING THE FINITE-ELEMENT TECHNIQUE WITH APPLICATION TO HO	72-0941
MATERIAL	AN ANALYSIS OF THE PHYSICAL BEHAVIOR OF NONWOVEN FIBROUS MATERIAL BY THE FINITE-ELEMENT METHOD	73-1149
MATERIAL	AN ASSUMED STRESS HYBRID FINITE ELEMENT METHOD FOR AN INCOMPRESSIBLE AND NEAR INCOMPRESSIBLE MATERIA	69-0259
MATERIAL	AN INCREMENTAL FINITE ELEMENT ANALYSIS OF TWO DIMENSIONAL LARGE STRAIN AND LARGE DISPLACEMENT PROBLE	72-0338
MATERIAL	ANALYSIS OF LONGITUDINAL WAVE IN A BAR EXHIBITING VISCO-ELASTIC BEHAVIOR OR WITH MATERIAL DAMPING	71-0780
MATERIAL	ANALYSIS OF STIFFENED PLATES CONSIDERING NONLINEAR MATERIAL AND GEOMETRIC BEHAVIOUR	75-0327
MATERIAL	APPLICATION TO PLANE STRESS AND PLANE STRAIN--LINEAR ELASTIC MATERIAL	68-0048
MATERIAL	ASAAS: ASYMMETRIC STRESS ANALYSIS OF AXISYMMETRIC SOLIDS WITH ORTHOTROPIC, TEMPERATURE-DEPENDENT MAT	71-0696
MATERIAL	BENDING OF CIRCULAR PLATES OF HARDENING MATERIAL	67-0112
MATERIAL	BERECHNUNG VON STAHLBETONSCHEIBEN IM ZUSTAND II BEI ANNAHME EINES WIRKLICHKEITSNAHEN WERKSTOFFVERHAL	74-0978
MATERIAL	BRLESC FINITE ELEMENT PROGRAM FOR AXISYMMETRIC, PLANE STRAIN, AND PLANE STRESS, ORTHOTROPIC SOLIDS W	71-0749
MATERIAL	CALCULATION OF TEMPERATURE DISTRIBUTION WITHIN A MELTING-FREEZING MATERIAL USING FINITE ELEMENT TECH	72-0043
MATERIAL	COMBINED MATERIAL AND GEOMETRIC NONLINEARITY FOR THIN STEEL PLATES	75-0397
MATERIAL	COMPARISON OF EXTRAPOLATION TECHNIQUES WITH HIGH ORDER FINITE ELEMENT METHODS FOR DIFFUSION EQUATION	73-0586
MATERIAL	CONSTRUCTION OF A FINITE ELEMENT APPROXIMATION WHICH CROSSES MATERIAL INTERFACES	74-0896
MATERIAL	CONTINUUM CALCULATION FROM ELASTIC PLASTIC REINFORCED MATERIAL USING A FINITE ELEMENT METHOD (DIE BE	72-0658
MATERIAL	CONTINUUM CALCULATION FROM ELASTIC PLASTIC REINFORCED MATERIAL USING A FINITE ELEMENT METHOD (DIE BE	72-0658
MATERIAL	CONVERGENCE OF THE FINITE-ELEMENT METHOD UNDER THE CONSTRAINT OF MATERIAL INCOMPRESSIBILITY	73-1135
MATERIAL	DER SPANNUNGS UND BEANSPRUCHUNGSZUSTAND DES BODENMATERIALS IN DER NACHBARSCHAFT EINER BODENBOESCHUNG	75-0371

277

MATERIAL	DYNAMIC ANALYSIS OF FRAME AND PLATE STRUCTURES WITH GEOMETRIC AND MATERIAL NONLINEARITIES	73-0649
MATERIAL	DYNAMIC RESPONSES OF CYLINDRICAL SHELLS INCLUDING GEOMETRIC AND MATERIAL NONLINEARITIES	74-0150
MATERIAL	EFFECT OF NONLINEAR MATERIAL RESPONSE ON THE BEHAVIOUR OF PAVEMENTS UNDER TRAFFIC	70-0171
MATERIAL	EFFECT OF TIME-DEPENDENT MATERIAL PROPERTIES ON DYNAMIC RESPONSE	74-0900
MATERIAL	EFFECTS OF MATERIAL NONLINEARITIES ON VIBRATION	73-0323
MATERIAL	EVALUATION OF SOLUTION PROCEDURES FOR MATERIAL AND/OR GEOMETRICALLY NONLINEAR STRUCTURAL ANALYSIS	73-0362
MATERIAL	EXTENSION OF CLASSICAL EXPERIMENTAL TECHNIQUES FOR CHARACTERIZING COMPOSITE - MATERIAL BEHAVIOUR	72-0088
MATERIAL	FINITE ELEMENT ANALYSIS FOR ARBITRARY AXISYMMETRIC STRUCTURES ORTHOTROPIC, TEMPERATURE-DEPENDENT MAT	68-0076
MATERIAL	FINITE ELEMENT ANALYSIS OF COMBINED PROBLEMS ON NONLINEAR MATERIAL AND GEOMETRIC BEHAVIOUR	69-0175
MATERIAL	FINITE ELEMENT ANALYSIS WITH MATERIAL NONLINEARTIES: THEORY AND PRACTICE	69-0459
MATERIAL	FINITE ELEMENT ANALYSIS OF COMBINED PROBLEMS OF MATERIAL AND GEOMETRIC BEHAVIOUR	69-0508
MATERIAL	FINITE ELEMENT ANALYSIS IN BRITTLE MATERIAL DESIGN	70-0360
MATERIAL	FINITE ELEMENT ANALYSIS OF AXISYMMETRICAL SHELLS OF COMPOSITE MATERIAL	70-0483
MATERIAL	FINITE ELEMENT ANALYSIS WITH MATERIAL NONLINEARITIES, THEORY AND PRACTICE	71-0168
MATERIAL	FINITE ELEMENT ANALYSIS OF A COMPOSITE MATERIAL INTERFACE	73-0777
MATERIAL	FINITE ELEMENT ANALYSIS OF INCOMPRESSIBLE MATERIAL BY RESIDUAL ENERGY BALANCING	74-1041
MATERIAL	FINITE ELEMENT ANALYSIS OF COMPOSITE MATERIAL BEHAVIOURS FROM MECHANICAL PROPERTIES OF CONSTITUENT M	75-0181
MATERIAL	FINITE ELEMENT ANALYSIS FOR THE TIME-DEPENDENT INELASTIC MATERIAL BEHAVIOR	75-0749
MATERIAL	FINITE ELEMENT ANALYSIS OF COMPOSITE MATERIAL BEHAVIORS FROM MECHANICAL PROPERTIES OF CONSTITUTENT M	75-0874
MATERIAL	FINITE ELEMENT BASED PROCEDURE FOR SIMULATING THE TRANSIENT RESPONSE AND FAILURE OF A TWO-DIMENSIONA	72-0123
MATERIAL	FINITE ELEMENT DIFFICULTIES IN NONLINEAR AND INCOMPRESSIBLE MATERIAL	74-1242
MATERIAL	FINITE ELEMENT ELASTIC PLASTIC CREEP ANALYSIS OF TWO DIMENSIONAL CONTINUUM WITH TEMPERATURE DEPENDEN	73-0226
MATERIAL	FINITE ELEMENT FORMULATION OF STRESS STRAIN MATRIX FOR AN ELASTICALLY PERFECTLY PLASTIC TRESCA MATER	70-0275
MATERIAL	FINITE ELEMENT OF UNDERGROUND STRESSES UTILIZING STOCHASTICALLY SIMULATED MATERIAL PROPERTIES	70-0111
MATERIAL	FINITE ELEMENT STRESS ANALYSIS OF AXISYMMETRIC SOLIDS WITH ORTHOTROPIC TEMPERATURE DEPENDENT MATERIA	67-0089
MATERIAL	FINITE-ELEMENT ANALYSIS WITH MATERIAL NONLINEARITIES	69-0459
MATERIAL	FORMULATION, COMPUTATION, AND SOLUTION PROCEDURES FOR MATERIAL AND/OR GEOMETRIC NONLINEAR STRUCTURAL	72-0346
MATERIAL	FREE VIBRATION OF A SIMPLY SUPPORTED BEAM WITH NONLINEAR MATERIAL PROPERTIES	71-0689
MATERIAL	INCREMENTAL FORMULATIONS FOR PROBLEMS WITH GEOMETRIC AND MATERIAL NONLINEARITIES	72-0315
MATERIAL	INVESTIGATION OF THE STATIONARY FLOW OF A RIGID PLASTIC MATERIAL BY THE NUMERICAL FINITE ELEMENT MET	73-0231
MATERIAL	MATERIAL BEHAVIOUR CHARACTERISTICS FOR REINFORCED CONCRETE SHELLS STRESSED BEYOND THE ELASTIC RANGE	70-0166
MATERIAL	MATERIAL NONLINEARITIES	72-0647
MATERIAL	MATERIAL NONLINEARITIES IN THE FINITE ELEMENT ANALYSIS	74-0647
MATERIAL	MICROMECHANICAL ANALYSIS OF CRACK PROPAGATION IN AN ELASTOPLASTIC COMPOSITE MATERIAL	74-0565
MATERIAL	NEAR-CRACK DEFORMATION FIELDS IN STRAIN-HARDENING MATERIAL	74-0938
MATERIAL	NEW MATERIAL MODEL FOR THE NONLINEAR BIAXIAL BEHAVIOR OF ATJ-S GRAPHITE	75-0063
MATERIAL	NON-LINEAR ANALYSIS OF PLATES CONSIDERING GEOMETRIC AND MATERIAL EFFECTS	71-0702
MATERIAL	NON-LINEAR MATERIAL ANALYSIS WITH FINITE ELEMENTS AND INCREMENTAL METHODS	68-0199
MATERIAL	NON-UNIFORM DEFORMATION OF MATERIAL IN AXIALLY SYMMETRIC COMPRESSION CAUSED BY FRICTION. PART 3, THE	72-0923
MATERIAL	NONLINEAR MATERIAL AND GEOMETRIC BEHAVIOUR OF SHELL STRUCTURES	71-0084
MATERIAL	NONLINEAR MECHANICS OF CONSTRAINED MATERIAL CONTINUA: FOUNDATIONS OF THE THEORY	74-0640
MATERIAL	ON GENERAL PURPOSE PROGRAMS FOR FINTIE ELEMENT ANALYSIS WITH SPECIAL REFERENCE TO GEOMETRIC AND MATE	70-0087
MATERIAL	ON THE NONUNIFORM DEFORMATION OF MATERIAL IN AXIALLY SYMMETRIC COMPRESSION CAUSED BY FRICTION - 2	71-0321
MATERIAL	ON THE STATES OF STRESS AND DEFORMATION OF CYLINDRICAL SPECIMENS OF BRITTLE MATERIAL UNDER UNIAXIAL	69-0150
MATERIAL	PRACTICAL FINITE ELEMENT METHOD OF FAILURE PREDICTION FOR COMPOSITE MATERIAL STRUCTURES	74-0644
MATERIAL	PRACTICAL PROBLEMS ASSOCIATED WITH THE APPLICATION OF THE FINITE ELEMENT METHOD TO COMPOSITE MATERIA	74-0483
MATERIAL	PROBALILISTIC MODEL FOR MATERIAL STRENGTH VARIATION AND SIZE EFFECT	72-0876
MATERIAL	SAAS II: FINITE ELEMENT STRESS ANALYSIS OF AXISYMMETRIC SOLIDS WITH ORTHOTROPIC, TEMPERATURE DEPENDE	68-0096
MATERIAL	SAAS 111: FINITE ELEMENT STRESS ANALYSIS OF AXISYMMETRIC AND PLANE SOLIDS WITH DIFFERENT ORTHOTROPIC	71-0741
MATERIAL	SILAGE - SILO INTERACTION USING MATERIAL CHARACTERIZATION AND FINITE ELEMENT ANALYSIS	75-0792
MATERIAL	SOME CURRENT FINITE-ELEMENT MODELS FOR MATERIAL BEHAVIOR	75-0549
MATERIAL	SPHERICAL WAVES IN INELASTIC MATERIAL	70-0075
MATERIAL	STATIC AND DYNAMIC GEOMETRIC AND MATERIAL NONLINEAR ANALYSIS	74-1302
MATERIAL	STATIC GEOMETRIC AND MATERIAL NONLINEAR ANALYSIS	72-0314
MATERIAL	STATIC STRESS ANALYSIS OF AXISYMMETRIC SOLIDS WITH MATERIAL AND GEOMETRIC NONLINEARITIES BY THE FINI	73-0958
MATERIAL	STRESS ANALYSIS AND DESIGN OF SILICON SOLAR CELL ARRAYS AND RELATED MATERIAL PROPERTIES	72-0840
MATERIAL	STRESS ANALYSIS OF ROCK AS A NO-TENSION MATERIAL	68-0174
MATERIAL	STRUCTURAL ANALYSIS OF A COAL MINE OPENING IN ELASTIC, MULTILAYERED MATERIAL	74-0235
MATERIAL	STRUCTURAL ANALYSIS OF A COAL MINE OPENING IN ELASTIC, MULTILAYERED MATERIAL	74-0840
MATERIAL	STUDY ON STATE OF STRESS AND DEFORMATION OF CYLINDRICAL SPECIMENS OF BRITTLE MATERIAL UNDER UNIAXIAL	69-0152
MATERIAL	SURVEY OF STATIC GEOMETRIC AND MATERIAL NONLINEAR ANALYSIS BY THE FINITE ELEMENT METHOD	72-0783
MATERIAL	THEORETICAL AND EXPERIMENTAL ANALYSIS OF ELASTOPLASTIC DEFORMATION IN A UNIAXIALLY LOADED STRIP OF 2	73-0720
MATERIAL	THREE-DIMENSIONAL FINITE ELEMENT ANALYSIS INCLUDING ELASTIC-PLASTIC MATERIAL RESPONSE	74-1126
MATERIAL	VISCO-PLASTICITY AND PLASTICITY-AN ALTERNATIVE FOR FINITE ELEMENT SOLUTION OF MATERIAL NONLINEARITIE	74-0410
MATERIAL'S	COMPARISON OF THE CAPABILITY OF OUR HARDENING RULES TO PREDICT A MATERIAL'S PLASTIC BEHAVIOR	75-0523
MATERIALS	A FINITE ELEMENT FORMULATION FOR MATERIALS WITH SPECIFIED CHANGE OF VOLUME	68-0044
MATERIALS	A FINITE ELEMENT SIMULATION OF MECHANICAL PROPERTIES OF COMPOSITE MATERIALS	74-1294
MATERIALS	A LARGE DEFORMATION ANALYSIS OF CRYSTALLINE ELASTIC-VISCOPLASTIC MATERIALS	73-0602
MATERIALS	A MATHEMATICAL THEORY OF PLASTICITY FOR STRAIN HARDENING MATERIALS EXHIBITING INITIAL ANISOTROPY	72-0660
MATERIALS	AN ANALYTIC METHOD FOR PREDICTING THE FLOW AND FRACTURE IN NOTCHED FIBER COMPOSITE MATERIALS	75-0689
MATERIALS	APPLICATION OF THREE-DIMENSIONAL FINITE ELEMENT ANALYSES TO THE MICROMECHANICS OF FIBROUS COMPOSITE	74-1079
MATERIALS	APPLICATIONS OF THE FINITE ELEMENT METHOD TO THE STRESS ANALYSIS OF MATERIALS SUBJECT TO CREEP	71-0548
MATERIALS	AVERAGE ELASTIC PROPERTIES AND STRESS FIELDS FOR COMPOSITE MATERIALS	69-0435
MATERIALS	BRITTLE MATERIALS: A DESIGN CHALLENGE	75-0769
MATERIALS	CALCULATION AND DESIGN OF JOINTS MADE FROM COMPOSITE MATERIALS	70-0156
MATERIALS	CALCULATION OF STRESS INTENSITY FACTORS IN ANISOTROPIC MATERIALS BY A CONTOUR INTEGRAL METHOD	75-0257
MATERIALS	CHARACTERISATION ELASTIQUE D'UN STRATIFIE RESINE-VERRE PAR LA METHODE DES DEPLACEMENTS (DETERMINATIO	75-0272
MATERIALS	COMPUTER SIMULATION OF THE DEFORMATION OF COMPOSITE AND POLYCRYSTALLINE MATERIALS BY FINITE ELEMENT	72-0937
MATERIALS	COUPLE - A PSEUDO FINITE ELEMENT CONDUCTION CODE FOR THE TRANSIENT THERMAL RESPONE OF AXISYMMETRIC O	72-0778
MATERIALS	DETERMINATION OF STRESSES DUE TO DISCONTINUITIES IN FINITE PLATES OF ISOTROPIC AND ORTHOTROPIC MATER	73-1052
MATERIALS	DEVELOPMENT OF AN UNDERSTANDING OF THE FATIGUE PHENOMENA OF BONDED AND BOLTED JOINTS IN ADVANCED FIL	72-0798
MATERIALS	DIRECT AND INVERSE HEAT CONDUCTION IN ROCK MATERIALS USING THE FINITE ELEMENT METHOD	74-1098
MATERIALS	DYNAMIC RESPONSE OF VISCOELASTIC MATERIALS TO SINUSOIDALLY VARYING LOADINGS	73-1072
MATERIALS	DYNAMIC RESPONSE OF VISCOELASTIC MATERIALS	73-1075
MATERIALS	EFFECT OF PLASTICITY AND CRACK BLUNTING ON THE STRESS DISTRIBUTION IN ORTHOTROPIC COMPOSITE MATERIAL	73-0145
MATERIALS	EFFECTS OF INSERTS ON THE ELASTIC BEHAVIOUR OF CYLINDRICAL MATERIALS LOADED BETWEEN ROUGH END-PLATES	71-0318
MATERIALS	ELASTIC-PLASTIC STRESS-STRAIN RELATIONS FOR MATERIALS WITH DIFFERENT YIELD STRENGTHS IN TENSION AND	72-0764
MATERIALS	ELASTICITY EQUATIONS FOR INCOMPRESSIBLE AND NEARLY INCOMPRESSIBLE MATERIALS BY A VARATIONAL THEOREM	65-0019
MATERIALS	FESTIGKEITSUNTERSUCHUNG MIT DER METHODE DER FINITEN ELEMENTE. (STRENGTH OF THE MATERIALS INVESTIGATE	74-1009
MATERIALS	FINITE ELEMENT ANALYSIS OF ELASTO-PLASTIC SOILS. REPORT NO. 4, FINITE ELEMENT ANALYSIS OF ELASTO-PLA	72-0788
MATERIALS	FINITE ELEMENT ANALYSIS OF CUT-GROWTH IN SHEETS OF HIGHLY ELASTIC MATERIALS	73-0514
MATERIALS	FINITE ELEMENT ANALYSIS SYSTEM FOR THE MECHANICAL BEHAVIOR OF ORIENTED FIBER COMPOSITE MATERIALS UND	73-0869
MATERIALS	FINITE ELEMENT ANALYSIS OF ELASTO-PLASTIC MATERIALS BY USE OF DISLOCATION DIPOLE SYSTEMS	74-0269
MATERIALS	FINITE ELEMENT ANALYSIS OF NON-LINEAR COMPOSITE MATERIALS BY USE OF OVERLAY SYSTEMS	74-0928
MATERIALS	FINITE ELEMENT ANALYSIS OF WAVES IN NONLINEAR MATERIALS	75-0110
MATERIALS	FINITE ELEMENT ANALYSIS OF COMPOSITE MATERIAL BEHAVIOURS FROM MECHANICAL PROPERTIES OF CONSTITUENT M	75-0181
MATERIALS	FINITE ELEMENT ANALYSIS OF COMPOSITE MATERIAL BEHAVIORS FROM MECHANICAL PROPERTIES OF CONSTITUTENT M	75-0874
MATERIALS	FINITE ELEMENT FORMULATION FOR LINEAR THERMOVISCO ELASTIC MATERIALS	69-0351
MATERIALS	FINITE ELEMENT MATRIX DISPLACEMENT METHOD FOR TIME DEPENDENT STRESS ANALYSIS OF REACTOR MATERIALS IN	71-0674
MATERIALS	FINITE ELEMENT SOLUTION OF EXTRUSION OF ELASTO-PLASTIC WORK HARDENING MATERIALS	73-1127
MATERIALS	FORMULATION OF GENERAL DISCRETE MODELS OF THERMOMECHANICAL BEHAVIOR OF MATERIALS WITH MEMORY	69-0381
MATERIALS	FRACTURE MECHANICS FOR AN INTERFACIAL CRACK BETWEEN ADHESIVELY BONDED DISSIMILAR MATERIALS	72-0869

278

MATERIALS	HIGH STRENGTH MATERIALS AND VEHICLE WEIGHT REDUCTION ANALYSIS	75-0157
MATERIALS	HIGH-ORDER FINITE ELEMENT ANALYSIS FOR TRANSIENT TEMPERATURE ANALYSIS OF INHOMOGENEOUS MATERIALS	69-0301
MATERIALS	HIGH-PERFORMANCE COMPOSITE MATERIALS FOR VEHICLE CONSTRUCTION: A FINITE ELEMENT COMPUTER PROGRAM FOR	73-0767
MATERIALS	HORIZONS IN THE STRUCTURAL USE OF ADVANCED MATERIALS - THRAUSTICS	70-0415
MATERIALS	IDENTIFICATION OF NONLINEAR MATERIALS BY FINITE ELEMENT METHODS	73-0927
MATERIALS	INFLUENCE OF NON-SINGULAR STRESS TERMS AND SPECIMEN GEOMETRY ON SMALL SCALE YIELDING AT CRACK TIPS I	73-0203
MATERIALS	INTERACTIVE PROGRAM IN DESIGN AND ANALYSIS OF COMPOSITE MATERIALS	74-0854
MATERIALS	INVESTIGATION OF BOND FAILURE IN FIBRE-REINFORCED MATERIALS BY THE FINITE ELEMENT METHOD	72-0034
MATERIALS	J INTEGRAL ESTIMATES FOR STRAIN HARDENING MATERIALS USING FULLY PLASTIC SOLUTIONS	74-0690
MATERIALS	MATHEMATICAL MODELS FOR GEOLOGICAL MATERIALS FOR WAVE PROPAGATION STUDIES	71-0740
MATERIALS	MATRIX METHODS IN MECHANICS OF MATERIALS	70-0459
MATERIALS	MEASUREMENT OF THE WATER DIFFUSIVITY OF POROUS MATERIALS	74-1138
MATERIALS	MICROMECHANICS ANALYSIS OF COMPOSITE MATERIALS USING FINITE ELEMENT METHODS	72-0770
MATERIALS	NON-METALLIC ANTENNA-SUPPORT MATERIALS	73-0716
MATERIALS	NX BOREHOLE JACK MODULUS DETERMINATIONS IN HOMOGENEOUS, ISOTROPIC, ELASTIC MATERIALS	74-0295
MATERIALS	ON THE DETERMINATION OF EFFECTIVE MODULI OF COMPOSITE MATERIALS BY A THREE-DIMENSIONAL FINITE ELEMEN	74-1123
MATERIALS	ON THE FINITE ELEMENT ANALYSIS OF CRACK AND INCLUSION PROBLEMS IN ELASTIC-PLASTIC MATERIALS	74-0227
MATERIALS	ON THE FRACTURE MECHANICS ANALYSIS OF ELASTIC-PLASTIC MATERIALS USING THE FINITE ELEMENT METHOD	73-1067
MATERIALS	PARAMETER EINES ELASTO-PLASTISCHEN STOFFANSATZES FUER KOERNIGE ERDSTOFFE (DETERMINATION OF PARAMETER	74-0657
MATERIALS	PLANE STRESS CRACK TRAJECTORIES IN LOW MODULUS MATERIALS	75-0646
MATERIALS	PLASTIC FLOW AND FAILURE OF DISCONTINUOUS FIBRE COMPOSITE MATERIALS	70-0157
MATERIALS	PLASTIC STRESS STRAIN MATRIX OF ANISOTROPIC MATERIALS AND ITS APPLICATION	69-0383
MATERIALS	PREDICTION OF STATIC AND FATIGUE DAMAGE AND CRACK PROPAGATION IN COMPOSITE MATERIALS	75-0643
MATERIALS	RHEOLOGICAL THEORY OF STRENGTH OF MATERIALS	69-0313
MATERIALS	SIGNIFICANCE OF VARIABLY CONFINED TRIAXIAL TESTING ROADBUILDING MATERIALS	74-0554
MATERIALS	SOME RECENT DEVELOPMENTS IN THE THEORY AND APPLICATION OF FINITE ELEMENTS TO MIXED PROBLEMS AND SHOC	73-0637
MATERIALS	STOCHASTIC FINITE ELEMENT SIMULATION OF THE NONLINEAR STRUCTURAL RESPONSE OF FIBROUS COMPOSITE MATER	75-0762
MATERIALS	STRESS ANALYSIS OF COMPOSITE MATERIALS BY THE HYBRID FINITE ELEMENT METHOD	75-0799
MATERIALS	STRESS STRAIN RELATIONSHIP FROM COMPRESSION TESTS ON NONLINEAR MATERIALS	69-0189
MATERIALS	STRESS WAVE ANALYSIS IN LAYERED THERMOVISCOELASTIC MATERIALS BY THE EXTENDED RITZ METHOD. VOLUME II	68-0125
MATERIALS	STRESSES IN NEARLY INCOMPRESSIBLE MATERIALS BY FINITE ELEMENTS WITH APPLICATION TO THE CALCULATION O	74-0383
MATERIALS	STUDIES ON TWO PHASE COMPOSITE MATERIALS THROUGH COMPUTER SIMULATION	73-0315
MATERIALS	THE EFFECT OF END RESTRAINT CONDITION ON THE FRACTURE MECHANISM AND STRENGTH OF ROCK-LIKE MATERIALS	72-0134
MATERIALS	THE STOCHASTIC FINITE ELEMENT SIMULATION OF THE NONLINEAR STRUCTURAL RESPONSE OF FIBROUS COMPOSITE M	75-0798
MATERIALS	THERMAL ANALYSIS OF M552 EXPERIMENT FOR MATERIALS PROCESSING IN SPACE	74-0704
MATERIALS	THERMAL ANALYSIS OF M551 EXPERIMENT FOR MATERIALS PROCESSING IN SPACE	74-0725
MATERIALS	TRANSIENT THERMAL STRESS TECHNIQUES IN NUCLEAR REACTORS. APPLIED MECHANICS ASPECTS OF NUCLEAR EFFECT	71-0474
MATERIALS	TSAAS: FINITE ELEMENT THERMAL AND STRESS ANALYSIS OF AXISYMMETRIC SOLIDS WITH ORTHOTROPIC TEMPERATUR	74-0886
MATERIALS	VISCOELASTIC RESPONSE OF GRAPHITE MATERIALS IN IRRADIATION ENVIRONMENTS	70-0314
MATHEMATIC'S	APPLICATIONS OF HILL FUNCTIONS (APPLIED MATHEMATIC'S FINITE ELEMENT) TO APPLIED MECHANICS PROBLEMS	72-0884
MATHEMATICAL	A MATHEMATICAL MODEL OF THE CORNEO-SCLERAL SHELL	71-0150
MATHEMATICAL	A MATHEMATICAL THEORY OF PLASTICITY FOR STRAIN HARDENING MATERIALS EXHIBITING INITIAL ANISOTROPY	72-0660
MATHEMATICAL	A MATHEMATICAL THEORY OF COMPLEMENTARY-DUAL VARIATIONAL PRINCIPLES AND MIXED FINITE ELEMENT APPROXIM	73-0804
MATHEMATICAL	A STUDY OF INTERNAL STRESSES IN STATICALLY DEFORMED PNEUMATIC TIRES. A MATHEMATICAL MODEL FOR THE PN	70-0421
MATHEMATICAL	ANALYSIS OF LIQUID-WASTE INJECTION WELLS IN ILLINOIS BY MATHEMATICAL MODELS	74-0518
MATHEMATICAL	BASIC FUNCTIONS FOR CURVED ELEMENTS IN THE MATHEMATICAL THEORY OF FINITE ELEMENTS	75-0091
MATHEMATICAL	FINITE ELEMENT APPLICATIONS IN MATHEMATICAL PHYSICS	72-0067
MATHEMATICAL	FINITE ELEMENT METHOD - MATHEMATICAL REVIVAL	74-0169
MATHEMATICAL	LARGE STRUCTURES FOR MANNED SPACECRAFT: MATHEMATICAL ANALYSIS, DESIGN, CONSTRUCTION AND TESTS	74-1327
MATHEMATICAL	LECTURES ON MATHEMATICAL FOUNDATIONS OF THE FINITE ELEMENT METHOD	72-0777
MATHEMATICAL	LUBRICATION PROBLEMS - THE SELECTION OF MATHEMATICAL MODELS	75-0881
MATHEMATICAL	MATHEMATICAL ANALYSIS OF STRESS CRACKS. VOLUME 2.	75-0742
MATHEMATICAL	MATHEMATICAL ANALYSIS OF STRESS CRACKS. VOLUME 1	75-0743
MATHEMATICAL	MATHEMATICAL AND COMPUTATIONAL PROBLEMS IN REACTOR CALCULATIONS	73-0806
MATHEMATICAL	MATHEMATICAL AND COMPUTATIONAL PROBLEMS IN REACTOR CALCULATIONS	73-0973
MATHEMATICAL	MATHEMATICAL AND EXPERIMENTAL STUDIES ON THE MECHANICS OF PLATED TRANSVERSE FRACTURES	74-0656
MATHEMATICAL	MATHEMATICAL ASPECTS OF THE FINITE ELEMENT METHOD	71-0723
MATHEMATICAL	MATHEMATICAL ASPECTS OF FINITE ELEMENT APPROXIMATIONS IN THEORETICAL MECHANICS	75-0852
MATHEMATICAL	MATHEMATICAL FEATURES OF FINITE ELEMENT APPROXIMATIONS OF CERTAIN FLOW PROBLEMS WITH EMPHASIS ON CON	74-0002
MATHEMATICAL	MATHEMATICAL FOUNDATION OF THE FINITE ELEMENT METHOD	67-0126
MATHEMATICAL	MATHEMATICAL MODEL FOR CLOSED HEAD IMPACT	74-0934
MATHEMATICAL	MATHEMATICAL MODEL OF THE COLUMBIA RIVER FROM THE PACIFIC OCEAN TO BONNEVILLE DAM. PART 1. THEORY, P	69-0464
MATHEMATICAL	MATHEMATICAL MODEL OF THE COLUMBIA RIVER FROM THE PACIFIC OCEAN TO BONNEVILLE DAM. PART II: INPUT-OU	70-0417
MATHEMATICAL	MATHEMATICAL MODEL OF THE CORNEO-SCLERAL SHELL AS APPLIED TO INTRAOCULAR PRESSURE VOLUME RELATIONS A	72-0471
MATHEMATICAL	MATHEMATICAL MODEL OF HEATED EFFLUENTS IN COASTAL WATER	73-0921
MATHEMATICAL	MATHEMATICAL MODELING VIA DIRECT USE OF VIBRATION DATA	69-0310
MATHEMATICAL	MATHEMATICAL MODELING OF MOVEMENT OF DISSOLVED CONSTITUENTS IN GROUND WATER AQUIFERS BY THE FINITE E	70-0490
MATHEMATICAL	MATHEMATICAL MODELING OF SPINNING ELASTIC BODIES FOR MODAL ANALYSIS	73-0441
MATHEMATICAL	MATHEMATICAL MODELING AND ANALYSIS OF FUEL RODS	74-0930
MATHEMATICAL	MATHEMATICAL MODELING OF NEAR COASTAL CIRCULATION	75-0720
MATHEMATICAL	MATHEMATICAL MODELLING AND STUCTURAL ANALYSIS OF THE MANDIBLE	72-0124
MATHEMATICAL	MATHEMATICAL MODELS FOR GEOLOGICAL MATERIALS FOR WAVE PROPAGATION STUDIES	71-0740
MATHEMATICAL	MATHEMATICAL MODELS OF THE MASSACHUSETTS BAY. PART 1. FINITE ELEMENT MODELING OF TWO-DIMENSIONAL HYD	73-0800
MATHEMATICAL	MATHEMATICAL OBSERVATIONS IN STRUCTURAL DYNAMICS	73-0156
MATHEMATICAL	MATHEMATICAL PREDICTION OF DYNAMIC TIRE BEHAVIOR	73-0579
MATHEMATICAL	MATHEMATICAL PROGRAMMING	72-0100
MATHEMATICAL	MATHEMATICAL SIMULATION OF THE SUBSIDENCE OF VENICE 1 THEORY	73-0992
MATHEMATICAL	MATHEMATICAL THEORY OF LINEAR STRUCTURES	71-0498
MATHEMATICAL	NON-LINEAR EFFECTS OF HIGH TEMPERATURE ON THE VIBRATION OF A SIMPLY SUPPORTED BEAM WITH A CENTRAL MA	73-0838
MATHEMATICAL	ON COMPLEMENTARY-DUAL VARIATIONAL PRINCIPLES IN MATHEMATICAL PHYSICS	74-0342
MATHEMATICAL	ON THE MATHEMATICAL ANALYSIS OF STRESS IN THE HUMAN FEMUR	72-0207
MATHEMATICAL	SOME ASPECTS OF RECENT CONTRIBUTIONS TO THE MATHEMATICAL THEORY OF FINITE ELEMENTS	72-0811
MATHEMATICAL	SOME ASPECTS OF RECENT CONTRIBUTIONS TO THE MATHEMATICAL THEORY OF FINITE ELEMENTS	73-0601
MATHEMATICAL	SOME ASPECTS OF THE MATHEMATICAL THEORY OF FINITE ELEMENTS	72-0243
MATHEMATICAL	SOME CONTRIBUTIONS TO THE MATHEMATICAL THEORY OF MIXED FINITE ELEMENT APPROXIMATIONS	73-0302
MATHEMATICAL	SOME MATHEMATICAL ASPECTS OF FINITE ELEMENT CALCULATIONS	73-0110
MATHEMATICAL	SOME MATHEMATICAL PROPERTIES OF CERTAIN MIXED GALERKIN APPROXIMATIONS IN NONLINEAR ELASTICITY	74-1248
MATHEMATICAL	STRAIN ENERGY FUNCTIONS AND MATHEMATICAL MODELS OF LUNG ELASTICITY	73-0591
MATHEMATICAL	SURVEY LECTURES ON THE MATHEMATICAL FOUNDATIONS OF THE FINITE ELEMENT METHOD	72-0670
MATHEMATICAL	TECHNICAL, PHYSICAL AND MATHEMATICAL PRINCIPLES OF THE FINITE ELEMENT METHOD	71-0676
MATHEMATICAL	THE FINITE ELEMENT METHOD IN MATHEMATICAL PHYSICS	73-0289
MATHEMATICAL	THE HYPERCIRCLE IN MATHEMATICAL PHYSICS	57-0001
MATHEMATICAL	THE MATHEMATICAL FOUNDATIONS OF THE FINITE ELEMENT METHOD - WITH APPLICATIONS TO PARTIAL DIFFERENTIA	73-0286
MATHEMATICAL	TWO NEARLY POLYGONAL HOLES, MATHEMATICAL CRACK PROBLEMS	69-0061
MATHEMATICAL	UTILISATION DE MODELES MATHEMATIQUES MULTIDIMENSIONNELS POUR L'ETUDE DES ECOULEMENTS DANS LES DIGUES	73-0781
MATHEMATICIANS	TRANSACTIONS OF ARMY MATHEMATICIANS RESEARCH CONFERENCE	73-0925
MATHEMATICS	APPLICATIONS OF FINITE ELEMENT METHODS IN REACTOR MATHEMATICS - NUMERICAL SOLUTION OF THE NEUTRON DI	72-0535
MATHEMATICS	APPLICATIONS OF FINITE ELEMENT METHODS IN REACTOR MATHEMATICS. NUMERICAL SOLUTION OF THE NEUTRON TRA	74-0853
MATHEMATICS	INTRODUCTION TO THE MATHEMATICS OF THE FINITE ELEMENT METHOD	72-0054
MATHEMATICS	SOME ASPECTS OF THE MATHEMATICS OF FINITE ELEMENTS	75-0090
MATHEMATICS	SOME RECENT ADVANCES IN THE MATHEMATICS OF FINITE ELEMENTS	73-0287
MATHEMATICS	THE MATHEMATICS OF FINITE ELEMENTS AND APPLICATIONS	73-0288

MATHEMATIQUES	UTILISATION DE MODELES MATHEMATIQUES MULTIDIMENSIONNELS POUR L'ETUDE DES ECOULEMENTS DANS LES DIGUES	73-0381
MATHEMATISCHER	AUFBEREITUNG MATHEMATISCHER MODELLE FUR DIE DIGITALE BERECHNUNG VON GRUNDWASSERSTROMUNGSPROBLEMEN NA	71-0543
MATRIC	THE MATRIC SOLUTION OF CERTAIN NONLINEAR PROBLEMS IN STRUCTURAL ANALYSIS	56-0002
MATRICE	MATRICE DE RIGIDITE D'UN ELEMENT DE POUTRE SOLLICITE A LA TORSION-FLEXION (STIFFNESS MATRIX OF A BEA	73-0442
MATRICES	A COMPARATIVE STUDY OF SEVERAL CORE STORAGE SCHEMES FOR LARGE SPARSE POSITIVE DEFINITE MATRICES WITH	72-0678
MATRICES	A COMPUTATIONAL TECHNIQUE FOR THE EFFICIENT HANDLING OF LARGE MATRICES	73-1083
MATRICES	A DISCUSSION - ON THE DENSITY OF FINITE ELEMENT MATRICES	73-1066
MATRICES	A FORMULA FOR CERTAIN TYPES OF STIFFNESS MATRICES OF STRUCTURAL ELEMENTS	63-0003
MATRICES	A GENERAL FORMULA FOR STIFFNESS MATRICES OF STRUCTURAL ELEMENTS	63-0002
MATRICES	A METHOD OF ANALYSIS OF LINE STRUCTURES BY TRANSFER MATRICES DERIVED FROM FINITE ELEMENTS	74-0808
MATRICES	A METHOD OF COMPUTING NUMERICALLY INTEGRATED STIFFNESS MATRICES	72-0546
MATRICES	A NEW APPROACH FOR DERIVING GOOD ELEMENT STIFFNESS MATRICES	75-0128
MATRICES	A STRAIN ENERGY BASIS FOR STUDIES OF ELEMENT STIFFNESS MATRICES	72-0621
MATRICES	A STUDY OF STIFFNESS MATRICES FOR THE ANALYSIS OF FLAT PLATES	68-0105
MATRICES	AN AUTOMATIC NODE RELABELLING SCHEME FOR BANDWIDTH MINIMIZATION OF STIFFNESS MATRICES	68-0256
MATRICES	AN IMPROVED METHOD FOR REDUCING THE BANDWIDTH OF SPARSE SYMMETRIC MATRICES	72-0990
MATRICES	AUTOMATIC GENERATION OF FINITE ELEMENT MATRICES	71-0165
MATRICES	AUTOMATIC GENERATION AND CONDENSATION OF LARGE MASS AND STIFFNESS MATRICES	72-0466
MATRICES	BANDWIDTH MINIMIZATION OF STIFFNESS MATRICES	70-0045
MATRICES	BASIS OF DERIVATION OF MATRICES FOR THE DIRECT STIFFNESS METHOD	63-0008
MATRICES	BEHAVIOR OF TRIANGULAR SHELL ELEMENT STIFFNESS MATRICES ASSOCIATED WITH POLYHEDRAL DEFLECTION DISTRI	68-0159
MATRICES	BOOLEAN MATRICES AS A STANDARD FACILITY IN A MATRIX CODE	65-0057
MATRICES	BOUNDS ON EXTERNAL EIGENVALUES OF FINITE ELEMENT STIFFNESS AND MASS MATRICES AND THEIR SPECTRAL COND	72-0256
MATRICES	BOUNDS ON THE SPECTRAL AND MAXIMUM NORMS OF THE FINITE ELEMENT STIFFNESS, FLEXIBILITY AND MASS MATRI	73-0409
MATRICES	CALCULATION OF GEOMETRIC STIFFNESS MATRICES FOR COMPLEX STRUCTURES	66-0058
MATRICES	CALCULATION OF STIFFNESS MATRICES FOR FINITE ELEMENTS OF THIN SHELLS OF ARBITRARY SHAPE	68-0127
MATRICES	COMMENT ON MATRICES FOR THE DIRECT STIFFNESS METHOD BY R. J. MELOSH	64-0022
MATRICES	COMMENT ON STIFFNESS MATRICES FOR SECTOR ELEMENTS	70-0233
MATRICES	COMMENTS ON DERIVATION OF ELEMENT STIFFNESS MATRICES BY T. H. H. PIAN	65-0060
MATRICES	COMPARATIVE ANALYSIS OF THE CUTHILL-MCKEE AND THE REVERSE CUTHILL-MCKEE ORDERING ALGORITHMS FOR SPAR	73-1099
MATRICES	COMPARISON AND EVALUATION OF STIFFNESS MATRICES	66-0045
MATRICES	COMPUTATION OF STRESS RESULTANTS FROM ELEMENT STIFFNESS MATRICES	66-0125
MATRICES	COMPUTATIONAL METHOD FOR DYNAMIC ANALYSIS OF STRUCTURES WITH LARGE SIZE STIFFNESS AND MASS MATRICES.	73-0830
MATRICES	COMPUTER INTEGRATION OF CONGRUENTLY TRANSFORMED MATRICES	71-0580
MATRICES	COMPUTING WITH SPARSE MATRICES	73-0926
MATRICES	CONDITION OF FINITE ELEMENT MATRICES GENERATED FROM NONUNIFORM MESHES	72-0104
MATRICES	CONDITION OF FINITE ELEMENT MATRICES	74-1078
MATRICES	CONSISTENT STIFFNESS MATRICES IN THE ANALYSIS OF SHELLS	67-0078
MATRICES	COUPLING MATRICES FOR HIGH ORDER FINITE ELEMENT ANALYSIS OF ACOUSTIC - WAVE PROPAGATION	72-0040
MATRICES	CRITERION FOR SELECTING STIFFNESS MATRICES	65-0061
MATRICES	DERIVATION OF COMPATIBLE STIFFNESS MATRICES FOR LARGE DISCRETE ELEMENTS	67-0080
MATRICES	DERIVATION OF ELEMENT STIFFNESS MATRICES BY ASSUMED STRESS DISTRIBUTIONS	64-0007
MATRICES	DERIVATION OF ELEMENT STIFFNESS MATRICES	64-0008
MATRICES	DERIVATION OF GEOMETRIC STIFFNESS AND MASS MATRICES FOR FINITE ELEMENT HYBRID MODELS	74-0844
MATRICES	DERIVATION OF STIFFNESS MATRICES FOR THE ANALYSIS OF LARGE DEFLECTION AND STABILITY PROBLEMS	65-0066
MATRICES	DERIVATION OF STIFFNESS MATRICES FOR PROBLEMS IN PLANE ELASTICITY BY GALERKIN'S METHOD	69-0250
MATRICES	DEVELOPMENT OF EXPLICIT STIFFNESS AND MASS MATRICES FOR A TRIANGULAR PLATE ELEMENT	69-0242
MATRICES	DIAGONALLY DOMINANT MATRICES FOR THE FINITE ELEMENT METHOD IN HYDROLOGY	73-0410
MATRICES	DIRECT METHOD OF OBTAINING CAPACITANCE FROM FINITE-ELEMENT MATRICES	72-0289
MATRICES	EIGENVALUE PROBLEMS (A - LAMBDA B) X EQUALS 0 FOR SYMMETRIC MATRICES OF HIGH ORDER	74-0205
MATRICES	EIGENVALUES OF LARGE MATRICES	69-0341
MATRICES	ELASTIC AND EQUILIBRIUM MATRICES OF GENERAL SEMI-MONOCOGUE MEMBRANE ELEMENTS	70-0159
MATRICES	ELEMENT STATIC FLEXIBILITY AND STIFFNESS MATRICES FOR THE FINITE ELEMENT ANALYSIS OF BEAM, PLATE, AN	67-0070
MATRICES	ELEMENT STIFFNESS MATRICES FOR BOUNDARY COMPATIBILITY AND FOR PRESCRIBED BOUNDARY STRESSES	66-0061
MATRICES	ELEMENT STIFFNESS MATRICES FOR CONSTANT AND LINEAR STRESS - TRIANGULAR, CONSTANT AND LINEAR STRESS T	67-0006
MATRICES	ELEMENT STIFFNESS MATRICES OF THICK-WALLED ORTHOTROPIC SHELLS WITH APPLICATIONS	69-0261
MATRICES	EQUIVALENT MASS MATRICES FOR RECTANGULAR PLATES IN BENDING	66-0063
MATRICES	FINITE ELEMENT COUPLING MATRICES	69-0300
MATRICES	FINITE ELEMENT STIFFNESS MATRICES FOR ANALYSIS OF PLATE BENDING	66-0021
MATRICES	FLEXIBILITY AND STIFFNESS MATRICES FOR AN OPEN-TUBE WARPING CONSTRAINT FINITE ELEMENT	72-0076
MATRICES	GENERAL SINUSOIDAL STIFFNESS MATRICES FOR BUCKLING AND VIBRATION ANALYSES OF THIN FLAT-WALLED STRUCT	68-0278
MATRICES	INCREMENTAL FINITE ELEMENT MATRICES	73-0475
MATRICES	INCREMENTAL FINITE ELEMENT MATRICES	74-0354
MATRICES	INCREMENTAL FINITE ELEMENT MATRICES	74-0556
MATRICES	INCREMENTAL FINITE-ELEMENT MATRICES	75-0635
MATRICES	LUMPING OF FINITE ELEMENT MATRICES	74-0645
MATRICES	MONOTON FINITE-ELEMENT MATRICES AND THEIR COMPUTED CONDITION NUMBERS	75-0820
MATRICES	NOTE ON MINIMIZING THE BANDWIDTH OF SPARSE, SYMMETRIC MATRICES	73-1107
MATRICES	ON COMPUTATIONS INVOLVING STIFFNESS MATRICES STORED IN RECTANGULAR FORM	75-0576
MATRICES	ON DERIVATION OF STIFFNESS MATRICES WITH C0 ROTATION FIELDS FOR PLATES AND SHELLS	71-0254
MATRICES	ON FINITE ELEMENT MATRICES	72-0103
MATRICES	ON SPARSE MATRICES IN FINITE ELEMENT WAVE PROPAGATION	73-0290
MATRICES	ON THE DENSITY OF FINITE ELEMENT MATRICES	72-0545
MATRICES	ON THE DERIVATION OF STIFFNESS MATRICES FOR THE ANALYSIS OF LARGE DEFLECTION AND STABILITY PROBLEMS	66-0050
MATRICES	ON THE PHYSICAL MEANINGS OF THE MATRICES DERIVED FROM FINITE ELEMENT METHOD OF HEAT CONDUCTION PROBL	73-0297
MATRICES	PROBLEMS IN BUILDING AND MANIPULATING LARGE STIFFNESS MATRICES	71-0596
MATRICES	PROCEDURE FOR SOLVING LINEAR EQUATION SYSTEMS WITH SPARSELY OCCUPIED COEFFICIENT MATRICES ASSOCIATED	74-0352
MATRICES	PROGRAM TO REDUCE THE SIZE OF STRUCTURAL MATRICES	72-0912
MATRICES	REDUCING THE BANDWIDTH OF SPARSE SYMMETRIC MATRICES	69-0469
MATRICES	REDUCING THE BANDWIDTH OF SPARSE SYMMETRIC MATRICES	69-0502
MATRICES	REDUCTION OF STIFFNESS AND MASS MATRICES	65-0037
MATRICES	ROTARY INERTIA CORRECTION FOR MASS MATRICES	63-0011
MATRICES	SAMMSOR 111: A FINITE ELEMENT PROGRAM TO DETERMINE STIFFNESS AND MASS MATRICES OF RING-STIFFENED SHE	72-0774
MATRICES	SEVERAL STRATEGIES FOR REDUCING THE BANDWIDTH OF MATRICES	72-0996
MATRICES	SNAP-THROUGH BUCKLING ANALYSIS OF SHELLS OF REVOLUTION USING INCREMENTAL STIFFNESS MATRICES	74-1349
MATRICES	SPARSE MATRICES IN FINITE ELEMENT PROGRAM DEVELOPMENT	73-0337
MATRICES	STATIC FLEXIBILITY MATRICES FOR A RECTANGULAR PLATE ELEMENT IN BENDING, TWISTING, AND SHEAR. UNIFORM	68-0217
MATRICES	STIFFNESS MATRICES FOR THIN TRIANGULAR ELEMENTS OF NONZERO GAUSSIAN CURVATURE	66-0073
MATRICES	STIFFNESS MATRICES FOR PLATE BENDING ELEMENTS	69-0129
MATRICES	STIFFNESS MATRICES FOR SECTOR ELEMENTS	69-0219
MATRICES	STIFFNESS MATRICES FOR THE GENERAL DEFORMATION (OUT-OF-PLANE AND INPLANE) OF CURVED BEAM MEMBERS BAS	75-0124
MATRICES	THE CONDITION OF SOME FINITE ELEMENT COEFFICIENT MATRICES	70-0393
MATRICES	THE GENERATION OF INTERELEMENT, COMPATIBLE STIFFNESS AND MASS MATRICES BY THE USE OF INTERPOLATION F	66-0014
MATRICES	THE L2 AND L00 CONDITION NUMBERS OF THE FINITE ELEMENT MATRICES AND THE POINTWISE CONVERGENCE OF THE	72-0061
MATRICES	THE USE OF FINITE ELEMENT MID-INCREMENT STIFFNESS MATRICES IN THE POST BUCKLING ANALYSIS OF IMPERFEC	71-0383
MATRICES	TRANSIENT VIBRATION ANALYSIS OF LINEAR SYSTEMS USING TRANSITION MATRICES	68-0053
MATRICES	TWISTED BEAM ELEMENT MATRICES FOR BENDING	70-0082
MATRICES	WIELOSTOPNIOWA SYNTEZA MACIERZY SZTYWNOSCI (MULTI-STAGE SYNTHESIS OF THE STIFFNESS MATRICES)	73-0668
MATRICIELLE	BORNES SUPERIEURES ET INFERIEURES DANS L'ANALYSE MATRICIELLE DES PLAQUES EN FLEXION TORSION	64-0015
MATRIX	A CORRELATION STUDY OF METHODS OF MATRIX STRUCTURAL ANALYSIS	69-0110
MATRIX	A DERIVATION PROCEDURE FOR THE DYNAMIC FLEXIBILITY MATRIX OF A TRIANGULAR BENDING ELEMENT	70-0103
MATRIX	A FLAT TRIANGULAR SHELL ELEMENT STIFFNESS MATRIX	66-0052

MATRIX	A FORMULATION OF MATRIX DISPLACEMENT METHOD	66-0103
MATRIX	A MATRIX DISPLACEMENT METHOD ON PRE AND POST BUCKLING ANALYSIS OF LINERS FOR REACTOR VESSELS	71-0267
MATRIX	A MATRIX METHOD FOR VIBRATION AND STABILITY PROBLEMS	66-0091
MATRIX	A SEQUEL TO TECHNICAL NOTE 13 - THE CURVED TETRAHEDRONAL AND TRIANGULAR ELEMENTS TEC AND TRIC FOR TH	69-0048
MATRIX	A SEQUEL TO TECHNICAL NOTE 15: THE SHEBA FAMILY OF SHELL ELEMENTS FOR THE MATRIX DISPLACEMENT METHOD	69-0444
MATRIX	A SIMPLE ALGORITHM FOR THE STIFFNESS MATRIX OF TRIANGULAR PLATE BENDING ELEMENTS	71-0152
MATRIX	A STIFFNESS MATRIX FOR THE ANALYSIS OF THIN PLATES IN BENDING	61-0001
MATRIX	A STIFFNESS MATRIX FOR A CURVED MEMBRANE SHELL	69-0055
MATRIX	A STIFFNESS MATRIX FOR SHALLOW SHELL FINITE ELEMENTS	69-0229
MATRIX	A STIFFNESS MATRIX METHOD OF NATURAL MODE ANALYSIS	58-0002
MATRIX	A STUDY OF MATRIX ANALYSIS METHODS FOR INELASTIC STRUCTURES	63-0022
MATRIX	A STUDY OF THE MATRIX DISPLACEMENT METHOD AS APPLIED TO SHELLS OF REVOLUTION	66-0111
MATRIX	A STUDY OF VARIOUS IDEALIZATIONS FOR WING STRUCTURES AND NUMERICAL PROCEDURES INVOLVED USING MATRIX	69-0257
MATRIX	A SURVEY OF SPARSE MATRIX TECHNOLOGY	72-0989
MATRIX	A TAPERED THICKNESS TRIM 6 ELEMENT FOR THE MATRIX DISPLACEMENT METHOD	66-0007
MATRIX	ALGORITHM FOR MATRIX BANDWIDTH REDUCTION	72-0895
MATRIX	AN ALGORITHM FOR REDUCING THE BANDWIDTH OF A MATRIX OF SYMMETRICAL CONFIGURATION	65-0067
MATRIX	AN ALGORITHM FOR REDUCING THE BANDWITH AND PROFILE OF A SPARSE MATRIX	74-0792
MATRIX	AN EXTENSION OF THE SHALLOW TO THE NON-SHALLOW STIFFNESS MATRIX FOR A CYLINDRICAL SHELL FINITE ELEME	70-0104
MATRIX	AN IMPROVED MATRIX DISPLACEMENT METHOD FOR VIBRATION OF NONUNIFORM BEAMS	71-0841
MATRIX	ANALYSIS OF CONTACT PROBLEMS BY THE MATRIX METHOD	69-0388
MATRIX	ANALYSIS OF COOLING TOWERS BY THE MATRIX FINITE ELEMENT METHOD- 2	70-0148
MATRIX	ANALYSIS OF RADIAL FLOW IMPELLERS BY THE MATRIX FINITE ELEMENT METHOD	71-0412
MATRIX	ANALYSIS OF TENSION STRUCTURES BY MATRIX METHODS	67-0119
MATRIX	ANALYSIS OF THE ELASTIC PLASTIC PROBLEMS BY THE MATRIX DISPLACEMENT METHOD	68-0173
MATRIX	ANALYSIS OF THIN-WALLED CURVED BEAMS BY THE TRANSFER MATRIX METHOD	72-0335
MATRIX	APPLICATION OF HERMITE POLYNOMIALS TO THE DETERMINATION OF THE STIFFNESS MATRIX OF PLATE BY FINITE E	71-0365
MATRIX	APPLICATION OF MATRIX DISPLACEMENT METHOD TO LINEAR ELASTIC ANALYSIS OF SHELLS OF REVOLUTION	65-0025
MATRIX	APPLICATION OF MATRIX DISPLACEMENT METHODS IN THE STUDY OF PANEL FLUTTER	69-0406
MATRIX	APPLICATION OF MATRIX METHODS IN STRUCTURAL THEORY	69-0454
MATRIX	APPLICATION OF NATURAL KINEMATIC MODES TO MATRIX ANALYSIS: SMALL AND LARGE DISPLACEMENTS AND STRAINS	67-0138
MATRIX	APPLICATION OF THE MATRIX DISPLACEMENT METHOD TO THE ANALYSIS OF PRESSURE VESSELS	69-0051
MATRIX	ARBITRARY QUADRILATERAL SPAR WEBS FOR THE MATRIX DISPLACEMENT METHOD	66-0003
MATRIX	AUTOMATED DATA GENERATION AND LARGE MATRIX OPERATION FOR 3 DIMENSIONAL PLATE STRUCTURE ANALYSIS	73-0339
MATRIX	BANDED MATRIX TECHNIQUES AND SOLUTIONS OF LINEAR SIMULTANEOUS EQUATIONS OF FINITE ELEMENT ANALYSES	70-0302
MATRIX	BOOLEAN MATRICES AS A STANDARD FACILITY IN A MATRIX CODE	65-0057
MATRIX	COMPARISON BETWEEN SPARSE STIFFNESS MATRIX AND SUB-STRUCTURE METHODS	73-1068
MATRIX	COMPARISONS OF CONSISTENT MASS MATRIX SCHEMES	69-0437
MATRIX	CONF. ON MATRIX METHODS IN STRUCTURAL MECHANICS	68-0284
MATRIX	CONFERENCE ON MATRIX METHODS IN STRUCTURAL MECHANICS	65-0065
MATRIX	CONFERENCE ON MATRIX METHODS IN STRUCTURAL MECHANICS	69-0470
MATRIX	CONFERENCE ON MATRIX METHODS IN STRUCTURAL MECHANICS	71-0773
MATRIX	CONSISTENT MASS MATRIX FOR DISTRIBUTED SYSTEMS	63-0001
MATRIX	CONSISTENT MATRIX FORMULATIONS FOR STRUCTURAL ANALYSIS USING INFLUENCE COEFFICIENT TECHNIQUES	64-0017
MATRIX	CONSTITUTIVE MATRIX AND FINITE ELEMENT FORMULATION FOR A DEFORMATION THEORY OF PLASTICITY	71-0700
MATRIX	CREEP OF A UNIAXIAL METAL MATRIX COMPOSITE SUBJECTED TO AXIAL AND NORMAL LATERAL LOADS	71-0041
MATRIX	CREEP OF A UNIAXIAL METAL MATRIX COMPOSITE SUBJECTED TO AXIAL AND NORMAL LATERAL LOADS	71-0727
MATRIX	DERIVATION OF A STIFFNESS MATRIX FOR A RIGHT TRIANGULAR PLATE IN BENDING AND SUBJECTED TO INITIAL ST	62-0007
MATRIX	DERIVATION OF CRACK ELEMENT-STIFFNESS MATRIX BY THE COMPLEX VARIABLE APPROACH	73-0870
MATRIX	DERIVATION OF THE STIFFNESS MATRIX FOR PLATES BY GALERKIN METHOD	69-0402
MATRIX	DIRECT BEAM STIFFNESS MATRIX CALCULATIONS INCLUDING SHEAR EFFECTS	62-0003
MATRIX	DISTRIBUTED MASS MATRIX FOR PLATE ELEMENTS IN BENDING	65-0018
MATRIX	DISTRIBUTED MASS MATRIX FOR PLATE ELEMENT BENDING (DISCUSSION)	66-0038
MATRIX	DYNAMIC MATRIX ANALYSIS OF FRAMED STRUCTURES	62-0012
MATRIX	DYNAMIC STIFFNESS MATRIX FORMULATION BY MEANS OF HERMITIAN POLYNOMIALS	66-0059
MATRIX	DYNAMICAL ANALYSIS OF AXIALLY NON-UNIFORM, THIN CYLINDRICAL SHELLS. PART 1 - MATRIX FORMULATION	70-0413
MATRIX	DYNAMICS OF DISTRIBUTED PARAMETER ROTOR SYSTEMS: TRANSFER MATRIX AND FINITE ELEMENT TECHNIQUES	70-0461
MATRIX	EFFECT OF ELASTIC MODULAR RATIO AND ANISOTROPY ON THE STRESS FIELD IN A TWO-DIMENSIONAL MATRIX INTO	71-0647
MATRIX	ELASTIC STRENGTH OF PROPELLERS: AN ANALYSIS BY MATRIX METHODS	70-0468
MATRIX	ELASTO-PLASTIC MATRIX DISPLACEMENT ANALYSIS OF THREE-DIMENSIONAL CONTINUA	66-0004
MATRIX	ELEMENT STIFFNESS MATRIX GENERATOR	71-0106
MATRIX	ESTIMATING MATRIX DISPLACEMENT SOLUTIONS OF TWO-DIMENSIONAL PROBLEMS BY LARGE ELEMENT TECHNIQUE	70-0160
MATRIX	EXPLICIT EXPRESSIONS FOR TRIANGULAR TOWA ELEMENT STIFFNESS MATRIX	68-0160
MATRIX	EXPLICIT TRIANGULAR BENDING ELEMENT MATRIX	74-0288
MATRIX	FINITE DISPLACEMENT PLATE FLEXURE BY THE USE OF MATRIX INCREMENTAL APPROACH	72-0384
MATRIX	FINITE ELEMENT FORMULATION OF STRESS STRAIN MATRIX FOR AN ELASTICALLY PERFECTLY PLASTIC TRESCA MATER	70-0275
MATRIX	FINITE ELEMENT MATRIX FORMULATION OF POST-BUCKLING STABILITY AND IMPERFECTION SENSITIVITY	70-0246
MATRIX	FINITE ELEMENT MATRIX DISPLACEMENT METHOD FOR TIME DEPENDENT STRESS ANALYSIS OF REACTOR MATERIALS IN	71-0674
MATRIX	FINITE-ELEMENT ANALYSIS OF RADIATION-DAMAGE STRESSES IN THE GRAPHITE OF MATRIX FUEL ELEMENTS	70-0048
MATRIX	FLEXIBILITY MATRIX FOR AN EXACT TORSION-BENDING FINITE ELEMENT	72-0214
MATRIX	FLUTTER OF SKEW PANELS BY THE MATRIX DISPLACEMENT APPROACH	70-0201
MATRIX	FORCE AND TORQUE ANALYSIS OF CONSTRAINED SPACE MECHANISMS AND PLANE MECHANISMS WITH OFFSET LINKS BY	74-0569
MATRIX	FORMAT-FORTRAN MATRIX ABSTRACTION TECHNIQUE	68-0214
MATRIX	FORMAT, A GENERAL PURPOSE MATRIX ANALYSIS SYSTEM IN FORTRAN	68-0266
MATRIX	GENERAL LINEAR GEOMETRIC MATRIX FOR A FULLY COMPATIBLE FINITE ELEMENT	72-0294
MATRIX	GENERALISED MATRIX FORCE AND DISPLACEMENT METHODS FOR LINEAR STRUCTURAL ANALYSIS	65-0064
MATRIX	GENERALIZED STIFFNESS MATRIX OF A CURVED-BEAM ELEMENT	69-0165
MATRIX	IMPROVEMENTS ON THE ANALYSIS OF SHELLS OF REVOLUTION BY THE MATRIX DISPLACEMENT METHOD	66-0070
MATRIX	INCLUSION OF SHEAR DEFLECTION IN THE STIFFNESS MATRIX OF A BEAM ELEMENT	70-0307
MATRIX	INFLUENCE OF POISSON'S RATIO ON THE CONDITION OF THE FINITE ELEMENT STIFFNESS MATRIX	71-0519
MATRIX	INITIAL STRAINS IN THE MATRIX FORCE METHOD OF STRUCTURAL ANALYSIS	60-0008
MATRIX	INTEGRATED SEQUENTIAL SOLVER FOR LARGE MATRIX EQUATIONS	74-0276
MATRIX	INTERACTION OF AREA COORDINATES IN MATRIX STRUCTURAL ANALYSIS	68-0150
MATRIX	MASS MATRIX FOR A PRISMATIC BEAM SEGMENT	68-0116
MATRIX	MATRICE DE RIGIDITE D'UN ELEMENT DE POUTRE SOLLICITE A LA TORSION-FLEXION (STIFFNESS MATRIX OF A BEA	73-0442
MATRIX	MATRIX ANALYSIS METHODS FOR INELASTIC STRUCTURES	65-0051
MATRIX	MATRIX ANALYSIS METHODS FOR ANISOTROPIC INELASTIC STRUCTURES	66-0104
MATRIX	MATRIX ANALYSIS OF AXISYMMETRIC SHELLS UNDER GENERAL LOADING	66-0075
MATRIX	MATRIX ANALYSIS OF A COMPOSITE STRUCTURE OF PLATE AND LINE ELEMENTS	71-0250
MATRIX	MATRIX ANALYSIS OF BEAM COLUMNS BY THE METHOD OF FINITE ELEMENTS	64-0012
MATRIX	MATRIX ANALYSIS OF CREEP AND PLASTICITY PROBLEMS	65-0001
MATRIX	MATRIX ANALYSIS OF LOCAL INSTABILITY IN PLATES, STIFFENED PANELS AND COLUMNS	72-0542
MATRIX	MATRIX ANALYSIS OF MULTILAYERED AND SANDWICH SHELLS BY FINITE ELEMENT METHOD	73-0101
MATRIX	MATRIX ANALYSIS OF SHELL STRUCTURES	64-0003
MATRIX	MATRIX ANALYSIS OF SHELLS - SMALL AND LARGE DISPLACEMENTS	68-0016
MATRIX	MATRIX ANALYSIS OF THREE-DIMENSIONAL ELASTIC MEDIA: SMALL AND LARGE DISPLACEMENTS	65-0003
MATRIX	MATRIX ANALYSIS OF THIN SHELLS USING FINITE ELEMENTS	69-0527
MATRIX	MATRIX AND DIGITAL COMPUTER METHODS IN STRUCTURAL ANALYSIS	69-0143
MATRIX	MATRIX AND INTERFACE STRESSES IN A DISCONTINUOUS FIBER COMPOSITE MODEL	68-0040
MATRIX	MATRIX DIFFERENCE EQUATION ANALYSIS OF VIBRATING PERIODIC STRUCTURES	75-0151
MATRIX	MATRIX DIFFERENTIAL EQUATION AND HIGHER-ORDER NUMERICAL METHODS FOR PROBLEMS OF NON-LINEAR CREEP, VI	72-0198
MATRIX	MATRIX DISPLACEMENT ANALYSIS OF ANISOTROPIC SHELLS BY TRIANGULAR ELEMENTS	65-0005

281

MATRIX	MATRIX DISPLACEMENT ANALYSIS OF PLATES AND SHELLS	66-0010
MATRIX	MATRIX DISPLACEMENT ANALYSIS OF SHELLS WITH AN AXISYMMETRIC MIDSURFACE BUT HAVING CIRCUMFERENTIALLY	68-0031
MATRIX	MATRIX DISPLACEMENT METHODS IN FRACTURE MECHANICS ANALYSIS OF REACTOR VESSELS	71-0226
MATRIX	MATRIX DISPLACEMENT ANALYSIS OF PLATES AND SHELLS, PART 11, GENERAL FORMULATION OF THE LINEAR THEORY	71-0621
MATRIX	MATRIX DISPLACEMENT ANALYSIS OF SHELLS AND PLATES INCLUDING TRANSVERSE SHEAR STRAIN EFFECTS	72-0643
MATRIX	MATRIX DYNAMIC AND INSTABILITY ANALYSIS WITH NON-UNIFORM ELEMENTS	70-0062
MATRIX	MATRIX FORMULATION OF STRUCTURAL STABILITY PROBLEMS	65-0039
MATRIX	MATRIX ITERATION AND ACCELERATION PROCESSES IN FINITE ELEMENT PROBLEMS OF STRUCTURAL MECHANICS	70-0132
MATRIX	MATRIX METHOD OF ANALYSIS OF SHIP STRUCTURES 3RD REPORT	68-0225
MATRIX	MATRIX METHODS AND USE OF COMPUTERS IN STRUCTURAL ENGINEERING	71-0615
MATRIX	MATRIX METHODS IN MECHANICS OF MATERIALS	70-0459
MATRIX	MATRIX METHODS IN STRUCTURAL MECHANICS	66-0065
MATRIX	MATRIX METHODS OF AEROSPACE STRUCTURAL ANALYSIS	68-0198
MATRIX	MATRIX METHODS OF STRUCTURAL ANALYSIS	62-0004
MATRIX	MATRIX METHODS OF STRUCTURAL ANALYSIS	63-0014
MATRIX	MATRIX METHODS OF STRUCTURAL ANALYSIS. A PRECIS OF RECENT DEVELOPMENTS	63-0019
MATRIX	MATRIX METHODS OF STRUCTURAL ANALYSIS	71-0620
MATRIX	MATRIX STRUCTURAL ANALYSIS	71-0614
MATRIX	MECHANICS OF LOAD TRANSFER AT THE FIBER/MATRIX INTERFACE	72-0879
MATRIX	MEMBRANE PARALLELOGRAM ELEMENT WITH LINEARLY VARYING EDGE STRAIN FOR THE MATRIX DISPLACEMENT METHOD	66-0006
MATRIX	NATURAL FACTOR FORMULATION OF THE STIFFNESS FOR THE MATRIX DISPLACEMENT METHOD	75-0057
MATRIX	NATURAL GEOMETRY OF SURFACE WITH SPECIFIC REFERENCE TO THE MATRIX DISPLACEMENT ANALYSIS OF SHELLS, 1	73-0709
MATRIX	NEW APPROACH FOR PLATE VIBRATIONS. COMBINATION OF TRANSFER MATRIX AND FINITE ELEMENT TECHNIQUE	71-0434
MATRIX	NONLINEAR ANALYSIS OF SHELLS OF REVOLUTION BY THE MATRIX DISPLACEMENT METHOD	67-0079
MATRIX	NONLINEAR DYNAMIC ANALYSIS OF SHELLS OF REVOLUTION BY MATRIX DISPLACEMENT METHOD	70-0110
MATRIX	NUMERICAL CONDITIONING OF STIFFNESS MATRIX FORMULATIONS FOR FRAME STRUCTURES	68-0022
MATRIX	ON THE MATRIX THEORY OF STRUCTURES	60-0007
MATRIX	ON THE NUMERICAL CONVERGENCE OF MATRIX EIGENVALUE PROBLEMS DUE TO CONSTRAINT CONDITIONS	74-1057
MATRIX	ON THE REDUCTION OF NUMERICAL ERROR IN THE MATRIX DISPLACEMENT METHOD	74-0417
MATRIX	PLASTIC STRESS STRAIN MATRIX OF ANISOTROPIC MATERIALS AND ITS APPLICATION	69-0383
MATRIX	PLASTIC STRESS-STRAIN MATRIX AND ITS APPLICATION FOR THE SOLUTION OF ELASTIC-PLASTIC PROBLEMS BY THE	68-0171
MATRIX	POWER SERIES EXPANSION OF THE GENERAL STIFFNESS MATRIX FOR BEAM ELEMENTS	75-0543
MATRIX	PREDICTION OF PLATE VIBRATIONS USING A CONSISTENT MASS MATRIX	66-0043
MATRIX	QUADRATIC MATRIX EQUATIONS FOR DETERMINING VIBRATION MODES AND FREQUENCIES OF CONTINUOUS ELASTIC SYS	66-0064
MATRIX	RATIONALIZATION IN DERIVING ELEMENT STIFFNESS MATRIX BY ASSUMED STRESS APPROACH	68-0138
MATRIX	RECENT ADVANCES IN MATRIX METHODS OF STRUCTURAL ANALYSIS	64-0016
MATRIX	RECENT ADVANCES IN MATRIX METHODS IN STRUCTURAL ANALYSIS AND DESIGN	70-0290
MATRIX	RECENT DEVELOPMENTS OF MATRIX THEORY OF STRUCTURES	59-0006
MATRIX	RECENT DEVELOPMENTS IN MATRIX DISPLACEMENT METHOD FOR ELASTIC-PLASTIC PROBLEMS IN JAPAN	71-0265
MATRIX	RELATIONSHIPS BETWEEN MATRIX METHODS IN FINITE ELEMENTS	67-0086
MATRIX	SABOR 1, A FORTRAN PROGRAM FOR THE LINEAR ELASTIC AXISYMMETRIC LOADING BY USING THE MATRIX DISPLACEM	65-0027
MATRIX	SCALAR ALGORITHM OF DETERMINATION OF STIFFNESS MATRIX AND OF ANOTHER FINITE ELEMENT CHARACTERISTIC	74-1013
MATRIX	SECTOR ELEMENTS FOR MATRIX DISPLACEMENT ANALYSIS	73-1049
MATRIX	SHOCK ANALYSIS BY MATRIX METHODS	67-0051
MATRIX	SOLUTION OF LINEAR EQUATIONS WITH SKYLINE-STORED SYMMETRIC MATRIX	75-0662
MATRIX	SOME APPLICATIONS OF MATRIX FORCE METHOD	66-0092
MATRIX	SOME APPLICATIONS OF TOPOLOGICAL CONSIDERATION AND WEIGHT MATRIX METHOD TO FINITE ELEMENT ANALYSIS	73-0305
MATRIX	SOME NEW DEVELOPMENTS IN MATRIX FORCE ANALYSIS	71-0225
MATRIX	SOME NEW ELEMENTS FOR THE MATRIX DISPLACEMENT METHOD	68-0017
MATRIX	SOME PROGRAM DEVELOPMENTS OF MATRIX STRUCTURAL ANALYSIS IN JAPAN	72-0322
MATRIX	SPARSE MATRIX OPERATIONS IN NASTRAN	73-0336
MATRIX	SPARSE MATRIX TECHNIQUES IN DIRECT SOLUTION OF FINITE ELEMENT EQUATIONS	74-0115
MATRIX	STIFFNESS MATRIX FOR BENDING OF A RECTANGULAR PLATE ELEMENT WITH INITIAL MEMBRANE STRESSES	62-0006
MATRIX	STIFFNESS MATRIX FOR SHALLOW RECTANGULAR SHELL ELEMENT	67-0014
MATRIX	STIFFNESS MATRIX FOR A TRIANGULAR SANDWICH ELEMENT IN BENDING	67-0050
MATRIX	STIFFNESS MATRIX FOR SHALLOW RECTANGULAR SHELL ELEMENT	68-0074
MATRIX	STIFFNESS MATRIX FOR CURVED TAPERED AND STRAIGHT TAPERED SHELL STIFFENERS	69-0074
MATRIX	STIFFNESS MATRIX IN PLASTIC-ELASTIC PROBLEMS OF CONTINUA	67-0116
MATRIX	STRESS STRAIN MATRIX AND ITS APPLICATIONS FOR THE SOLUTION OF ELASTIC-PLASTIC PROBLEMS BY THE FINITE	68-0282
MATRIX	STRUCTURAL ANALYSIS BY THE MATRIX FORCE METHOD WITH APPLICATION TO AIRCRAFT WINGS	56-0004
MATRIX	STRUCTURAL ANALYSIS BY THE MATRIX DISPLACEMENT METHOD	61-0003
MATRIX	STRUCTURAL ANALYSIS AND MATRIX INTERPRETIVE SYSTEM (SAMIS)	66-0098
MATRIX	STRUCTURAL ANALYSIS AND MATRIX INTERPRETIVE SYSTEM (SAMIS) USER REPORT	67-0125
MATRIX	STRUCTURAL ANALYSIS: A UNIFIED CLASSICAL AND MATRIX APPROACH	72-0055
MATRIX	STRUCTURAL VIBRATION - A MATRIX APPROACH	71-0385
MATRIX	STUDY AND DEVELOPMENT OF SIMPLE MATRIX METHODS OF INELASTIC STRUCTURES	66-0115
MATRIX	STUDY OF ELASTIC-PLASTIC MATRIX ANALYSIS FOR FOUNDATION AND REINFORCED CONCRETE IN JAPAN	69-0278
MATRIX	STUDY ON WEIGHT MATRIX IN STRUCTURAL ANALYSIS	75-0062
MATRIX	SURVEY OF SPARSE MATRIX METHODS IN THE DIRECT SOLUTION OF FINITE ELEMENT EQUATIONS	73-0572
MATRIX	TANGENT STIFFNESS MATRIX FOR SPACE FRAME MEMBERS	69-0397
MATRIX	TETRAHEDRON ELEMENTS IN THE MATRIX FORCE METHOD OF STRUCTURAL ANALYSIS	64-0011
MATRIX	TETRAHEDRON ELEMENTS WITH LINEARLY VARYING STRAIN FOR THE MATRIX DISPLACEMENT METHOD	65-0006
MATRIX	THE ANALYSIS AND APPLICATION OF SPARSE MATRIX ALGORITHMS IN THE FINITE ELEMENT METHOD	72-0079
MATRIX	THE APPLICATION OF THE MATRIX DISPLACEMENT METHOD IN PLANE ELASTO-PLASTIC STRESS PROBLEMS	64-0009
MATRIX	THE AUTOMATIC COMPUTATION OF THE NATURAL FREQUENCIES OF STRUCTURAL FRAMES USING AN EXACT MATRIX TECH	73-0318
MATRIX	THE BANDIT COMPUTER PROGRAM FOR THE REDUCTION OF MATRIX BANDWIDTH FOR NASTRAN	72-0481
MATRIX	THE CURVED TETRAHEDRONAL AND TRIANGULAR ELEMENTS TEC AND TRIC FOR THE MATRIX DISPLACEMENT METHOD	69-0047
MATRIX	THE HERMES 8 ELEMENT FOR THE MATRIX DISPLACEMENT METHOD	68-0011
MATRIX	THE LINEAR ELASTIC DYNAMIC ANALYSIS OF SHELLS OF REVOLUTION BY THE MATRIX DISPLACEMENT METHOD	66-0046
MATRIX	THE LUMINA ELEMENT FOR THE MATRIX DISPLACEMENT METHOD	68-0010
MATRIX	THE MATRIX ANALYSIS OF STRUCTURES WITH CUT-OUTS AND MODIFICATIONS	56-0003
MATRIX	THE MATRIX DEFORMATION METHOD OF STRUCTURAL ANALYSIS	71-0079
MATRIX	THE MATRIX EIGENVALUE PROBLEM FOR DAMPED STRUCTURAL VIBRATION	66-0121
MATRIX	THE MATRIX FORCE METHOD OF STRUCTURAL ANALYSIS AND SOME NEW APPLICATIONS	56-0005
MATRIX	THE MATRIX THEORY OF STATICS	57-0002
MATRIX	THE SHEBA FAMILY OF SHELL ELEMENTS FOR THE MATRIX DISPLACEMENT METHOD	68-0014
MATRIX	THE SHEBA FAMILY OF SHELL ELEMENTS FOR THE MATRIX DISPLACEMENT METHOD. 111, LARGE DISPLACEMENTS	69-0499
MATRIX	THE TET 20 AND TEA 8 ELEMENTS FOR THE MATRIX DISPLACEMENT METHOD	68-0012
MATRIX	THE TRIAX 6 ELEMENT FOR AXISYMMETRIC ANALYSIS BY THE MATRIX DISPLACEMENT METHOD: PART 1, FOUNDATIONS	66-0124
MATRIX	THE TUBA FAMILY OF PLATE ELEMENTS FOR THE MATRIX DISPLACEMENT METHOD	68-0013
MATRIX	THEORY OF MATRIX STRUCTURAL ANALYSIS	68-0139
MATRIX	THREE DIMENSIONAL ANISOTROPIC AND INHOMOGENEOUS ELASTIC MEDIA MATRIX ANALYSIS FOR SMALL AND LARGE DI	65-0008
MATRIX	TRANSIENT RESPONSE OF LINEAR ELASTIC STRUCTURES DETERMINED BY THE MATRIX EXPONENTIAL METHOD	71-0240
MATRIX	TRANSVERSE PROPERTIES OF UNIDIRECTIONAL ALUMINUM MATRIX FIBROUS COMPOSITES	69-0167
MATRIX	TRIANGULAR ELEMENTS WITH LINEARLY VARYING STRAIN FOR THE MATRIX DISPLACEMENT METHOD	65-0053
MATRIX	TRIANGULAR PLATE ELEMENTS IN THE MATRIX FORCE METHOD OF STRUCTURAL ANALYSIS	63-0010
MATRIX	TRIAX 6 ELEMENT FOR AXISYMMETRIC ANALYSIS BY THE MATRIX DISPLACEMENT METHOD	66-0009
MATRIX	UPPER AND LOWER BOUNDS IN MATRIX STRUCTURAL ANALYSIS	64-0018
MATRIX	VIBRATION ANALYSIS OF A CANTILEVERED SQUARE PLATE BY THE STIFFNESS MATRIX METHOD	66-0012
MATRIX	VIBRATION OF FRAMES AND OTHER STRUCTURES WITH BANDED STIFFNESS MATRIX	70-0322
MATRIX	775 SELECTED REFERENCES ON THE FINITE ELEMENT METHOD AND MATRIX METHODS OF STRUCTURAL ANALYSIS	69-0241
MATRIX-DISPLACE	MATRIX-DISPLACEMENT DIRECT-ELEMENT METHOD FOR FORCE AND TORQUE ANALYSIS OF DETERMINATE AND INDETERMI	74-0531

MATRIX-DISPLACE	RIGID BODY DISPLACEMENTS OF CURVED ELEMENTS IN THE ANALYSIS OF SHELLS BY THE MATRIX-DISPLACEMENT MET	67-0031
MATRIX-TRANSFOR	STRUCTURAL ANALYSIS OF SWEPT-BACK WINGS BY MATRIX-TRANSFORMATION	51-0001
MATRIX-VECTOR	A SIMPLE MATRIX-VECTOR HANDLING SCHEME FOR THREE-DIMENSIONAL AND SHELL ANALYSIS	70-0317
MATRIZEN	EINIGE VERFAHREN ZUR BERECHNUNG VON EIGENWERTEN UND EIGENVEKTOREN VON NICHT HERMITESCHEN MATRIZEN U	72-0646
MATRIZEN-VERSCH	EINIGE ENTWICKLUNGEN IN DER MATRIZEN-VERSCHIEBUNGSMETHODE	69-0451
MATRIZENKRAFTEM	SPEKTRALANALYSE VON ROTORBLÄTTERN NACH DERMATRIZENVERS CHIEBUNGSMETHODE UND MATRIZENKRAFTEMETHODE	66-0123
MATRIZENTHEORIE	DIE MATRIZENTHEORIE DER STATIK	57-0003
MATRIZENTHEORIE	MATRIZENTHEORIE DER STATIK	60-0009
MATRIZENVERSCHI	ANWENDUNG DER MATRIZENVERSCHIEBUNGSMETHODE AUF ERZWUNGENE SCHWINGUNGEN (SCHWINGUNGSPROBLEME) (ARBEIT	71-0627
MATRIZENVERSCHI	BERECHNUNG RÄUMLICHER SPANNUNGSVERTEILUNGEN IN STAUDÄMMEN NACH DER MATRIZENVERSCHIEBUNGSMETHODE	67-0003
MATRIZENVERSCHI	DIE ANWENDUNG DER MATRIZENVERSCHIEBUNGSMETHODE ZUM BERECHNEN VON SPANNUNGSKONZENTRATIONEN	68-0208
MATRIZENVERSCHI	DIE VERWENDUNG DER MATRIZENVERSCHIEBUNGSMETHODE ZUR BERECHNUNG VON SPANNUNGSKOMPONENTEN	67-0140
MATRIZENVERSCHI	DIE VERWENDUNG DER MATRIZENVERSCHIEBUNGSMETHODE ZUR BERECHNUNG VON SPANNUNGSKONZENTRATIONEN	70-0400
MATRIZENVERSCHI	PARAMETRISCHE UNTERSUCHUNGEN VON KERBEN MIT DER MATRIZENVERSCHIEBUNGSMETHODE UND BILDSCHIRMUNTERSTUT	72-0678
MATS	MATS AND COMBINED FOOTINGS. ANALYSIS BY THE FINITE ELEMENT METHOD	71-0289
MATUS	MATUS: A THREE-DIMENSIONAL FINITE ELEMENT PROGRAM FOR SMALL-STRAIN ELASTIC ANALYSIS	73-0970
MAXIMUM	BOUNDS ON THE SPECTRAL AND MAXIMUM NORMS OF THE FINITE ELEMENT STIFFNESS, FLEXIBILITY AND MASS MATRI	73-0409
MAXIMUM	FINITE ELEMENT METHOD AS AN ASPECT OF THE PRINCIPLE OF MAXIMUM UNIFORMITY: NEW HYDRODYNAMICAL RAMIFA	74-0009
MAXIMUM	INTERIOR MAXIMUM NORM ESTIMATES FOR SOME SIMPLE FINITE ELEMENT METHODS	74-1023
MAXIMUM	MAXIMUM PRINCIPLE AND UNIFORM CONVERGENCE FOR THE FINITE ELEMENT METHOD	73-0852
MAXIMUM	ON MAXIMUM FILLET STRESSES IN BREECH RING	72-0736
MAZNIHO	RESENI MAZNIHO STAVU ZELEZOBETONOVYCH DESEK (PLASTIC ANALYSIS OF REINFORCED CONCRETE PLATES)	74-0495
MBB-UFE	THE MBB-UFE SUBSONIC PANEL METHOD. PART 2 - THE LIFT-CONNECTED DISPLACEMENT PROBLEM IN COMPRESSIBLE	72-0865
MEAN	NOTCH EFFECT IN LOW-CYCLE FATIGUE - 2. EFFECT OF STRESS CHANGE AND MEAN STRESS	75-0537
MEAN-SQUARE	MEAN-SQUARE ERRORS OF FINITE ELEMENT APPROXIMATIONS ON LINEAR ELASTOSTATICS, ELASTODYNAMICS, AND THE	74-0107
MEASURED	COMPARISON BETWEEN THEORETICAL AND MEASURED EARTH PRESSURES ACTING ON A LARGE MOTORWAY RETAINING WAL	74-0917
MEASURED	CORRELATION OF FINITE ELEMENT STRUCTURAL DYNAMIC ANALYSIS WITH MEASURED FREE VIBRATION CHARACTERISTI	74-0486
MEASURED	MEASURED AND CALCULATED VIBRATION PROPERTIES OF RING-STIFFENED HONEY-COMB CYLINDERS	71-0834
MEASURED	MEASURED AND PREDICTED LONGITUDINAL VIBRATIONS OF A LIQUID PROPELLANT TWO-STAGE LAUNCH VEHICLE	70-0366
MEASURED	RELATIONSHIP BETWEEN FINITE ELEMENT AND MEASURED JOINT-FLOOR CONNECTION RIGIDITIES	75-0738
MEASUREMENT	A DEEP HOLE STRESS MEASUREMENT DEVICE	74-C804
MEASUREMENT	DIRECT MEASUREMENT OF BOND SLIP IN REINFORCED CONCRETE	69-0428
MEASUREMENT	FIELD INSTRUMENTATION FOR LONG TERM MEASUREMENT OF PILE LOAD AND RAFT CONTACT PRESSURE	73-0150
MEASUREMENT	FLEXUAL RIGIDITY OF A THIN WALLED BUILD-UP ROTOR FOR JET ENGINE. MEASUREMENT BY STATIC LOAD TEST AND	73-0756
MEASUREMENT	INTERFERENCE-FIT FASTENER DISPLACEMENT MEASUREMENT BY SPECKLE PHOTOGRAPHY	74-1178
MEASUREMENT	MEASUREMENT OF ACTUAL DEDENDUM STRAINS, AND THE INFLUENCE OF DEDENDUM RADIUS ON STRESSES AND DEDENDU	72-0118
MEASUREMENT	MEASUREMENT OF SLOPE CHANGE ON CYLINDRICAL SHELL SURFACES	73-0439
MEASUREMENT	MEASUREMENT OF THE WATER DIFFUSIVITY OF POROUS MATERIALS	74-1138
MEASUREMENT	MEASUREMENT OF THERMAL EXPANSION IN SOLID PROPELLANTS	73-0916
MEASUREMENT	STRESS SEPARATION IN PHOTOELASTICIMETRIC MEASUREMENT BY THE FINITE ELEMENT METHOD	71-0807
MEASUREMENT	THEORETICAL BASIS OF THE BOREHOLE DEEPENING METHOD OF ABSOLUTE STRESS MEASUREMENT	70-0047
MEASUREMENTS	CRACK TIP STRAIN - A COMPARISON OF FEM CALCULATIONS AND MEASUREMENTS	75-0721
MEASUREMENTS	EXAMPLE OF LARGE-SCALE STRUCTURAL ANALYSIS. COMPARISON BETWEEN FINITE ELEMENT CALCULATION AND FULL S	74-0177
MEASUREMENTS	MEASUREMENTS OF HEALING AT AN OSTEOTOMY IN A RABBIT CALVARIUM: THE INFLUENCE OF APPLIED COMPRESSIVE	74-0975
MEASUREMENTS	PLATE MEASUREMENTS USING FINITE ELEMENTS	70-0229
MEASUREMENTS	PROPELLER-EXCITED VIBRATION WITH PARTICULAR REFERENCE TO FULL-SCALE MEASUREMENTS	75-0018
MEASUREMENTS	UNDERGROUND MEASUREMENTS IN A STEEPLY DIPPING OREBODY	68-0079
MEASURING	A NEW MEASURING METHOD OF RESIDUAL STRESSES WITH THE AID OF FINITE ELEMENT METHOD AND RELIABILITY OF	75-0875
MEASURING	TECHNIQUE OF MEASURING INITIAL DEFORMATION AROUND AN OPENING	71-0331
MECANIQUE	CALCUL DE MASSIFS EN ELASTOPLASTICITE APPLICATIONS A LA MECANIQUE DES ROCHES ET DES SOLS. (APPLICATI	74-0223
MECHANICAL	A DYNAMIC ELASTOPLASTIC ANALYSIS OF THIN SHELLS OF REVOLUTION UNDER ASYMMETRIC MECHANICAL AND THERMA	72-0068
MECHANICAL	A FINITE ELEMENT PROCEDURE FOR NUMERICAL OPTIMIZATION IN MECHANICAL DESIGN	74-1218
MECHANICAL	A FINITE ELEMENT SIMULATION OF MECHANICAL PROPERTIES OF COMPOSITE MATERIALS	74-1294
MECHANICAL	AN AXISYMMETRIC FINITE ELEMENT ANALYSIS OF THE MECHANICAL AND THERMAL STRESSES IN BRAKE DRUMS	74-0311
MECHANICAL	ANALYSIS OF RING-STIFFENED SHELLS OF REVOLUTION UNDER COMBINED THERMAL AND MECHANICAL LOADING	71-0342
MECHANICAL	ASPECTS OF MECHANICAL BEHAVIOR OF ROCK UNDER STATIC AND CYCLIC LOADING. PART A. MECHANICAL BEHAVIOR	72-0761
MECHANICAL	ASPECTS OF MECHANICAL BEHAVIOR OF ROCK UNDER STATIC AND CYCLIC LOADING. PART A. MECHANICAL BEHAVIOR	72-0761
MECHANICAL	ASPECTS OF MECHANICAL BEHAVIOR OF ROCK UNDER STATIC AND CYCLIC LOADING. PART A: MECHANICAL BEHAVIOR	72-0762
MECHANICAL	ASPECTS OF MECHANICAL BEHAVIOR OF ROCK UNDER STATIC AND CYCLIC LOADING. PART A: MECHANICAL BEHAVIOR	72-0762
MECHANICAL	CALCULATION OF MECHANICAL IMPEDANCE BY FINITE ELEMENT HYBRID MODEL	74-0126
MECHANICAL	COMPUTERS AND MECHANICAL DESIGN	73-0195
MECHANICAL	CYCLIC MECHANICAL TESTS AND AN APPROPRIATE ANALYTICAL STRESS-STRAIN MODEL FOR A36 STEEL	74-0815
MECHANICAL	DISCRETE-ELEMENT STATIC ANALYSIS OF CORE-STIFFENED SHELLS OF REVOLUTION FOR ASYMMETRIC MECHANICAL LO	68-0148
MECHANICAL	DYNAMIC ANALYSIS OF MECHANICAL STRUCTURES	71-0381
MECHANICAL	DYNAMIC ELASTOPLASTIC RESPONSE OF GEOMETRICALLY NONLINEAR ARBITRARY SHELLS OF REVOLUTION UNDER IMPUL	70-0098
MECHANICAL	DYNAMIC RESPONSE OF ARBITRARY SHELLS OF REVOLUTION UNDER COMBINED IMPULSIVE NONISOTHERMAL AND IRREGU	70-0365
MECHANICAL	DYNAMIC RESPONSE OF A CONSTRAINED FIBROUS SYSTEM SUBJECTED TO TRANSVERSE IMPACT. II. A MECHANICAL MO	70-0369
MECHANICAL	ESTIMATION OF THE MECHANICAL PROPERTIES OF ROCK MASSES	73-0935
MECHANICAL	FINITE ELEMENT ANALYSIS OF SOME MECHANICAL PROBLEMS OF HOLLOW GRAVITY DAM	70-0470
MECHANICAL	FINITE ELEMENT ANALYSIS SYSTEM FOR THE MECHANICAL BEHAVIOR OF ORIENTED FIBER COMPOSITE MATERIALS UND	73-0869
MECHANICAL	FINITE ELEMENT ANALYSIS OF COMPOSITE MATERIAL BEHAVIOURS FROM MECHANICAL PROPERTIES OF CONSTITUENT M	75-0181
MECHANICAL	FINITE ELEMENT ANALYSIS OF COMPOSITE MATERIAL BEHAVIORS FROM MECHANICAL PROPERTIES OF CONSTITUTENT M	75-0874
MECHANICAL	FINITE ELEMENT FOR TIMOSHENKO BEAM BASED ON MECHANICAL IMPEDANCE	71-0507
MECHANICAL	FINITE-ELEMENT MODEL FOR THE MECHANICAL BEHAVIOUR OF THE LEFT VENTRICLE. PREDICTION OF DEFORMATION I	72-0304
MECHANICAL	IMPACT ON COMPLEX MECHANICAL STRUCTURES	71-0598
MECHANICAL	MECHANICAL BEHAVIOR OF A GRAPHITE-EPOXY LAMINATE CONTAINING A HOLE	72-0614
MECHANICAL	MECHANICAL CHARACTERIZATION AND FINITE-ELEMENT ANALYSIS OF ELASTIC-PLASTIC, WORK-HARDENING SOILS	72-0974
MECHANICAL	MECHANICAL PROPERTIES OF TAPE COMPOSITES	69-0317
MECHANICAL	PRAKTISCHE RESULTATEN. DOEL EN NUT VAN HET ONTWERPEN VAN MECHANISCHE ELEMENTEN MET BEHULP VAN COMPUT	73-0400
MECHANICAL	SIMPLIFICATION OF HIGH ORDER MECHANICAL SYSTEMS USING THE ROUTH APPROXIMATION	75-0853
MECHANICAL	SPANNUNGEN IN MEMBRANWÄNDEN DURCH WÄRMEBEANSPRUCHUNG UND MECHANISCHE BELASTUNG. (STRESSES IN MEMBR	75-0066
MECHANICAL	STACUSS 1: A DISCRETE-ELEMENT PROGRAM FOR THE STATIC ANALYSIS OF SINGLE-LAYER CURVED STIFFENED SHELL	69-0160
MECHANICAL	STRUCTURAL MODEL FOR THE MECHANICAL BEHAVIOUR OF TRABECULAR BONE	73-0473
MECHANICAL	THE INELASTIC BEHAVIOUR OF STRUCTURES SUBJECTED TO CYCLIC, THERMAL AND MECHANICAL STRESSING CONDITIO	60-0004
MECHANICAL	THERMAL AND MECHANICAL ANALYSIS OF WELDED STRUCTURES	73-0819
MECHANICAL	THERMAL MECHANICAL ANALYSIS PROCEDURE FOR A SEMICIRCULAR PLATE WITH CLAMPED STRAIGHT EDGE	75-0651
MECHANICAL	THREE-DIMENSIONAL MECHANICAL STRESSES IN TOROIDAL MAGNETS FOR CONTROLLED THERMONUCLEAR REACTORS	73-0669
MECHANICALLY	STRESS AND DEFLECTION ANALYSIS OF MECHANICALLY FASTENED JOINTS	70-0067
MECHANICE	FINITE ELEMENTS METHOD IN SOIL MECHANICS (METODA KONECNYCH PRVKU V MECHANICE SEMIN),	74-0468
MECHANICE	METODA ELEMENTOW SKONSZONYCH W MECHANICE GRUNTOW I MECHANICE GOROTWORU (FINITE ELEMENT METHOD IN SOI	73-0500
MECHANICE	METODA ELEMENTOW SKONSZONYCH W MECHANICE GRUNTOW I MECHANICE GOROTWORU (FINITE ELEMENT METHOD IN SOI	73-0500
MECHANICS	A FINITE ELEMENT COMPUTATIONAL THEORY FOR THE MECHANICS AND THERMODYNAMICS OF A VISCOUS, COMPRESSIBL	71-0787
MECHANICS	A FINITE ELEMENT CONTRIBUTION TO FRACTURE MECHANICS	72-0320
MECHANICS	A FRACTURE MECHANICS APPROACH TO THE PROPAGATION OF FATIGUE CRACKS IN REACTOR COMPONENTS BY ACOUSTIC	71-0132
MECHANICS	A HIGHLY STABLE EXPLICIT INTEGRATION TECHNIQUE FOR COMPUTATIONAL CONTINUUM MECHANICS	74-1277
MECHANICS	A MATHEMATICAL THEORY OF COMPLEMENTARY-DUAL VARIATIONAL PRINCIPLES AND MIXED FINITE ELEMENT APPROXIM	73-0804
MECHANICS	A MODEL FOR THE MECHANICS OF JOINTED ROCK	68-0079
MECHANICS	A NOTE ON THE FINITE ELEMENT METHOD IN LINEAR FRACTURE MECHANICS	70-0489
MECHANICS	A STUDY OF INTERVERTEBRAL DISC MECHANICS BY THE FINITE ELEMENT METHOD	74-1121
MECHANICS	A SURVEY OF FINITE-ELEMENT METHODS IN CONTINUUM MECHANICS	72-0751
MECHANICS	A UNIFIED THEORY OF VARIATIONAL PRINCIPLES IN MECHANICS	75-0155
MECHANICS	ADAPTATION OF CRITICAL STATE SOIL MECHANICS FOR USE IN FINITE ELEMENTS	71-0534

283

MECHANICS	ADVANCES IN APPLIED MECHANICS, VOLUME 12.	72-0375
MECHANICS	ANALYSIS OF FATIGUE OF PAVING MIXTURES FROM THE FRACTURE MECHANICS VIEWPOINT	72-0540
MECHANICS	ANALYSIS OF NONLINEAR PROBLEMS IN ROCK MECHANICS WITH PARTICULAR REFERENCE TO JOINTED ROCK SYSTEMS	70-0282
MECHANICS	ANALYSIS OF SLOPE STABILITY BY FINITE ELEMENT METHOD AND ITS APPLICATION PROGRAM. SOIL MECHANICS	72-0919
MECHANICS	APPLICATION OF AN ASSUMED DISPLACEMENT HYBRID FINITE ELEMENT PROCEDURE TO TWO-DIMENSIONAL PROBLEMS I	74-0729
MECHANICS	APPLICATION OF ELASTO-PLASTIC ANALYSIS IN ROCK MECHANICS BY FINITE ELEMENT METHOD	71-0349
MECHANICS	APPLICATION OF FINITE ELEMENT ANALYSIS METHOD TO TWO-DIMENSIONAL PROBLEMS IN FRACTURE MECHANICS	69-0033
MECHANICS	APPLICATION OF FINITE ELEMENT METHODS TO LARGE SCALE ELASTIC PLASTIC PROBLEMS OF FRACTURE MECHANICS	69-0372
MECHANICS	APPLICATION OF FINITE ELEMENT ANALYSIS IN FLUID MECHANICS	71-0026
MECHANICS	APPLICATION OF FINITE ELEMENT METHOD TO CONTINUUM MECHANICS PROBLEMS	71-0051
MECHANICS	APPLICATION OF FINITE ELEMENT METHOD TO FRACTURE MECHANICS	71-0178
MECHANICS	APPLICATION OF FINITE ELEMENT METHOD TO THE PROBLEM OF HEAT CONDUCTION AND FLUID MECHANICS	72-0486
MECHANICS	APPLICATION OF FINITE ELEMENT DYNAMIC ANALYSIS TO FRACTURE MECHANICS	75-0861
MECHANICS	APPLICATION OF FRACTURE MECHANICS TO THERMALLY STRESSED CYLINDERS USING A FINITE ELEMENT METHOD	69-0089
MECHANICS	APPLICATION OF FRACTURE MECHANICS TO FAILURE OF THICK WALLED CYLINDERS	71-0511
MECHANICS	APPLICATION OF FRACTURE MECHANICS TO GRAPHITE UNDER COMPLEX STRESS CONDITIONS	72-0771
MECHANICS	APPLICATION OF FRACTURE MECHANICS TO PRESTRESSED CONCRETE PRESSURE VESSELS	73-0228
MECHANICS	APPLICATION OF QUADRATIC ISOPARAMETRIC FINITE ELEMENTS IN LINEAR FRACTURE MECHANICS	74-0521
MECHANICS	APPLICATION OF THE FINITE ELEMENT METHOD OF ANALYSIS IN SOLVING BOUNDARY VALUE PROBLEMS IN ROCK MECH	66-0017
MECHANICS	APPLICATION OF THE FINITE ELEMENT METHOD TO PROBLEMS OF ROCK MECHANICS	66-0088
MECHANICS	APPLICATION OF THE FINITE ELEMENT METHOD TO PROBLEMS IN SOIL AND ROCK MECHANICS	67-0118
MECHANICS	APPLICATION OF THE FINITE ELEMENT METHOD TO 2 AND 3 DIMENSIONAL ELASTIC-PLASTIC PROBLEMS OF FRACTURE	70-0501
MECHANICS	APPLICATION OF THE FINITE ELEMENT METHOD TO PROBLEMS IN LINEAR FRACTURE MECHANICS	74-0903
MECHANICS	APPLICATION OF THE METHOD OF FINITE ELEMENT ANALYSIS TO TWO-DIMENSIONAL PROBLEMS IN FRACTURE MECHANI	68-0100
MECHANICS	APPLICATION OF THE THREE-DIMENSIONAL FINITE ELEMENT METHOD TO SOME PROBLEMS IN SOIL MECHANICS	70-0510
MECHANICS	APPLICATIONS OF CONTINUOUS TONE COMPUTER-GENERATED IMAGES IN STRUCTURAL MECHANICS	74-1035
MECHANICS	APPLICATIONS OF HILL FUNCTIONS (APPLIED MATHEMATIC'S FINITE ELEMENT) TO APPLIED MECHANICS PROBLEMS	72-0884
MECHANICS	APPLICATIONS OF THE ASSUMED-STRESS HYBRID METHOD TO FINITE ELEMENT ANALYSIS IN FRACTURE MECHANICS	75-0495
MECHANICS	ASSUMED DISPLACEMENT HYBRID FINITE ELEMENT MODEL FOR LINEAR FRACTURE MECHANICS	75-0146
MECHANICS	ASSUMED STRESS HYBRID FINITE ELEMENT METHOD FOR FRACTURE MECHANICS AND ELASTIC-PLASTIC ANALYSIS	72-0584
MECHANICS	BASIC CONCEPTS OF THE MECHANICS OF DISCRETIZED BODIES WITH AN INTRODUCTION TO DISCRETE ELEMENT CALCU	73-0146
MECHANICS	BASIC RESEARCH IN METHODS OF APPROXIMATION IN THE NONLINEAR MECHANICS OF SOLIDS AND STRUCTURES	72-0857
MECHANICS	BASIC RESEARCH IN METHODS OF APPROXIMATION IN THE NONLINEAR MECHANICS OF SOLIDS AND STRUCTURES, VOLU	74-1091
MECHANICS	CALCUL DE MASSIFS EN ELASTOPLASTICITE APPLICATIONS A LA MECANIQUE DES ROCHES ET DES SOLS. (APPLICATI	74-0223
MECHANICS	COMPUTATIONAL FRACTURE MECHANICS	71-0564
MECHANICS	COMPUTATIONAL METHODS IN SOLID MECHANICS AND STRESS ANALYSIS	69-0006
MECHANICS	COMPUTATIONAL METHODS IN NONLINEAR MECHANICS	74-1232
MECHANICS	COMPUTER APPLICATIONS IN ROCK MECHANICS	72-0190
MECHANICS	CONF. ON MATRIX METHODS IN STRUCTURAL MECHANICS	68-0284
MECHANICS	CONFERENCE ON MATRIX METHODS IN STRUCTURAL MECHANICS	65-0065
MECHANICS	CONFERENCE ON MATRIX METHODS IN STRUCTURAL MECHANICS	69-0470
MECHANICS	CONFERENCE ON MATRIX METHODS IN STRUCTURAL MECHANICS	71-0773
MECHANICS	CONTINUUM MECHANICS AS AN APPROACH TO ROCK MASS PROBLEMS	68-0172
MECHANICS	CONTINUUM MECHANICS - A BRIEF REVIEW	71-0407
MECHANICS	CONTRIBUTIONS TO FLUID MECHANICS AND FLIGHT MECHANICS	74-1311
MECHANICS	CONTRIBUTIONS TO FLUID MECHANICS AND FLIGHT MECHANICS	74-1311
MECHANICS	DEFORMATION MECHANICS OF DEEP SURFACE FLAW CRACKS	72-0785
MECHANICS	DISCUSSION OF FINITE ELEMENT TECHNIQUES IN STRUCTURAL MECHANICS	70-0263
MECHANICS	DISCUSSION ON THE ADAPTION OF CRITICAL STATE SOLID MECHANICS THEORY FOR USE IN FINITE ELEMENTS	71-0007
MECHANICS	EFFICIENT FORMULATIONS OF THE FINITE ELEMENT METHOD IN LINEAR AND NONLINEAR FRACTURE MECHANICS	75-0403
MECHANICS	ERROR BOUNDS FOR SEMIDISCRETE GALERKIN APPROXIMATIONS OF PARABOLIC PROBLEMS WITH APPLICATIONS TO PET	70-0291
MECHANICS	FINITE ELEMENT ANALYSES OF ELASTIC-PLASTIC PROBLEMS IN THE MECHANICS OF GEOLOGIC MEDIA: AN OVERVIEW	70-0203
MECHANICS	FINITE ELEMENT ANALYSIS OF CREEP PROBLEMS IN SOIL MECHANICS	71-0809
MECHANICS	FINITE ELEMENT FORMULATION AND SOLUTION OF CONTACT-IMPACT PROBLEMS IN CONTINUUM MECHANICS	74-0806
MECHANICS	FINITE ELEMENT FRACTURE MECHANICS ANALYSIS OF CREEP RUPTURE OF FUEL ELEMENT CLADDING	74-0929
MECHANICS	FINITE ELEMENT FRACTURE MECHANICS ANALYSIS OF TWO-DIMENSIONAL AND AXISYMMETRIC ELASTIC AND ELASTIC-P	74-1092
MECHANICS	FINITE ELEMENT METHOD FOR PROBLEMS IN SOIL MECHANICS	68-0078
MECHANICS	FINITE ELEMENT METHOD IN LINEAR FRACTURE MECHANICS	70-0476
MECHANICS	FINITE ELEMENT METHOD FOR WOOD MECHANICS	71-0294
MECHANICS	FINITE ELEMENT METHOD FOR WOOD MECHANICS	71-0613
MECHANICS	FINITE ELEMENT METHODS IN CONTINUUM MECHANICS.	72-0377
MECHANICS	FINITE ELEMENT METHOD AND ROCK MECHANICS	73-0551
MECHANICS	FINITE ELEMENT METHOD FOR LINEAR FRACTURE MECHANICS	74-1359
MECHANICS	FINITE ELEMENT METHOD FOR BOUND STATE CALCULATIONS IN QUANTUM MECHANICS	75-0005
MECHANICS	FINITE ELEMENT METHODS FOR THE SOLUTION OF SOME INCOMPRESSIBLE NON-NEWTONIAN FLUID MECHANICS PROBLEM	75-0224
MECHANICS	FINITE ELEMENT TECHNIQUES IN ROCK MECHANICS	68-0028
MECHANICS	FINITE ELEMENT TECHNIQUES IN STRUCTURAL MECHANICS	70-0116
MECHANICS	FINITE ELEMENT TECHNIQUES IN FRACTURE MECHANICS	72-0746
MECHANICS	FINITE ELEMENT THEORY FOR THE MECHANICS AND THERMODYNAMICS OF A VISCOUS, COMPRESSIBLE MULTI-SPECIE F	71-0005
MECHANICS	FINITE ELEMENTS METHOD IN SOIL MECHANICS (METODA KONECNYCH PRVKU V MECHANICE SEMIN).	71-0468
MECHANICS	FINITE ELEMENTS ON THE BASIS OF CONTINUUM MECHANICS	75-0234
MECHANICS	FINITE-ELEMENT METHOD APPLIED TO SOME PROBLEMS IN SOIL MECHANICS	68-0077
MECHANICS	FINITE-ELEMENT TECHNIQUES IN FRACTURE MECHANICS	75-0780
MECHANICS	FLUID MECHANICS AND FINITE ELEMENTS	72-0707
MECHANICS	FRACTURE MECHANICS APPROACH TO ADHESIVE JOINTS	71-0726
MECHANICS	FRACTURE MECHANICS APPROACH TO ADHESIVE JOINTS	72-0181
MECHANICS	FRACTURE MECHANICS APPLICATION OF AN ASSUMED DISPLACEMENT HYBRID FINITE ELEMENT PROCEDURE	75-0212
MECHANICS	FRACTURE MECHANICS FOR AN INTERFACIAL CRACK BETWEEN ADHESIVELY BONDED DISSIMILAR MATERIALS	72-0869
MECHANICS	FRACTURE MECHANICS OF TINY CRACKS NEAR FASTENERS	74-0735
MECHANICS	FUNCTIONAL MINIMIZATION IN NONLINEAR SOLID MECHANICS	74-1251
MECHANICS	FURTHER APPLICATION OF QUADRATIC ISOPARAMETRIC FINITE-ELEMENTS TO LINEAR FRACTURE MECHANICS OF PLATE	75-0639
MECHANICS	IMPORTANCE OF VARIATIONAL CALCULUS IN FLUID MECHANICS AND ITS APPLICATION TO COMPRESSIBLE POTENTIAL	73-1103
MECHANICS	INCREMENTAL NUMERICAL SOLUTION OF A SIMPLE DEFORMATION PROBLEM IN SOIL MECHANICS	70-0380
MECHANICS	LECTURES ON FINITE ELEMENT METHODS IN CONTINUUM MECHANICS	71-0714
MECHANICS	LINEAR AND NONLINEAR OSCILLATIONS IN MECHANICS	74-0677
MECHANICS	MATHEMATICAL AND EXPERIMENTAL STUDIES ON THE MECHANICS OF PLATED TRANSVERSE FRACTURES	74-0656
MECHANICS	MATHEMATICAL ASPECTS OF FINITE ELEMENT APPROXIMATIONS IN THEORETICAL MECHANICS	75-0852
MECHANICS	MATRIX DISPLACEMENT METHODS IN FRACTURE MECHANICS ANALYSIS OF REACTOR VESSELS	71-0226
MECHANICS	MATRIX ITERATION AND ACCELERATION PROCESSES IN FINITE ELEMENT PROBLEMS OF STRUCTURAL MECHANICS	70-0132
MECHANICS	MATRIX METHODS IN MECHANICS OF MATERIALS	70-0459
MECHANICS	MATRIX METHODS IN STRUCTURAL MECHANICS	66-0065
MECHANICS	MECHANICS AND THERMOMECHANICS	73-0929
MECHANICS	MECHANICS OF BRAIN TISSUE FRAGILITY	73-0214
MECHANICS	MECHANICS OF CABLE MOORING SYSTEMS. VOLUME IV. A COMPUTER PROGRAM FOR ANALYZING THE STEADY STATE RES	72-0653
MECHANICS	MECHANICS OF CABLE MOORING SYSTEMS. VOLUME 1. THREE DIMENSIONAL RESPONSE OF DEEP WATER MOORING LINES	72-0654
MECHANICS	MECHANICS OF CABLE MOORING SYSTEMS. VOLUME V11. THE STEADY-STATE BEHAVIOR OF A PYRAMID ARRAY SYSTEM	73-0724
MECHANICS	MECHANICS OF CABLE MOORING SYSTEMS. VOLUME V1. A COMPUTER PROGRAM FOR ANALYZING THE STEADY STATE CON	73-0725
MECHANICS	MECHANICS OF DEFORMABLE SOLIDS - EURASIAN COMMUNIST COUNTRIES	73-0794
MECHANICS	MECHANICS OF FRACTURE IN ADHESIVE JOINTS	72-0691
MECHANICS	MECHANICS OF LOAD TRANSFER AT THE FIBER/MATRIX INTERFACE	72-0879
MECHANICS	MECHANICS OF MACKAY-MARG TONOMETER	73-0433
MECHANICS	METODA ELEMENTOW SKONSZONYCH W MECHANICE GRUNTOW I MECHANICE GOROTWORU (FINITE ELEMENT METHOD IN SOI	73-0500

284

MECHANICS	NEW HORIZONS IN ROCK MECHANICS	73-0428
MECHANICS	NEW HORIZONS IN ROCK MECHANICS	73-0446
MECHANICS	NON-LINEAR STRESS ANALYSIS OF TWO DIMENSIONAL PROBLEMS WITH SPECIAL REFERENCE TO ROCK AND SOIL MECHA	68-0183
MECHANICS	NONDETERMINISTIC PROBLEMS IN STRUCTURAL MECHANICS	71-0746
MECHANICS	NONLINEAR MECHANICS OF CONSTRAINED MATERIAL CONTINUA: FOUNDATIONS OF THE THEORY	74-0640
MECHANICS	NOTE ON THE FINITE ELEMENT METHOD IN LINEAR FRACTURE MECHANICS	70-0474
MECHANICS	NUMERICAL ANALYSIS IN ORTHOTROPIC LINEAR FRACTURE MECHANICS	73-0512
MECHANICS	NUMERICAL COMPUTATION OF BOUNDARY VALUE PROBLEMS IN SOIL MECHANICS	73-0269
MECHANICS	NUMERICAL SOLUTION TO THE DYNAMICS OF VISCOUS FLUID FLOW BY A FINITE ELEMENT ALGORITHM: A FIRST STEP	71-0004
MECHANICS	NUMERICAL SOLUTION TO THE DYNAMICS OF VISCOUS FLUID FLOW BY A FINITE ELEMENT ALGORITHM: A FIRST STEP	71-0791
MECHANICS	NUMERICAL TECHNIQUES IN LINEAR AND NONLINEAR FRACTURE MECHANICS	75-0260
MECHANICS	ON A SPECTRAL PROBLEM IN VIBRATION MECHANICS: COMPUTATION OF ELASTIC TANKS PARTIALLY FILLED WITH LIQ	75-0267
MECHANICS	ON SOME ASPECTS OF CONTINUUM MECHANICS	73-0626
MECHANICS	ON THE FINITE ELEMENT METHOD IN LINEAR FRACTURE MECHANICS	69-0032
MECHANICS	ON THE FRACTURE MECHANICS ANALYSIS OF ELASTIC-PLASTIC MATERIALS USING THE FINITE ELEMENT METHOD	73-1067
MECHANICS	PRESSURE VESSEL CERTIFICATION BASED ON FRACTURE MECHANICS TECHNOLOGY	74-0948
MECHANICS	PROBLEMS IN ROCK MECHANICS	65-0033
MECHANICS	PROCEEDINGS OF INTERNATIONAL ASSOCIATION FOR HYDRAULIC RESEARCH AND PROCEEDINGS OF INTERNATIONAL SYM	73-0534
MECHANICS	PROGRESS IN THREE-DIMENSIONAL ELASTIC-PLASTIC STRESS ANALYSIS FOR FRACTURE MECHANICS	71-0314
MECHANICS	PROPAGATION OF ELLIPTICAL SURFACE CRACKS AND NONLINEAR FRACTURE MECHANICS BY THE FINITE ELEMENT METH	74-0906
MECHANICS	RECENT ADVANCES IN COMPUTATIONAL METHODS IN STRUCTURAL MECHANICS AND DESIGN.	72-0575
MECHANICS	ROCK MECHANICS - THEORY AND PRACTICE	70-0107
MECHANICS	ROCK MECHANICS AND RISK IN OPEN PIT MINING	73-0544
MECHANICS	ROCK MECHANICS STUDIES FOR A HIGH CUT AT A NUCLEAR POWER STATION IN PENNSYLVANIA	70-0122
MECHANICS	SHIFTING ROLES OF COMPUTERS, EXPERIMENTS AND ANALYSIS IN APPLIED MECHANICS	71-0604
MECHANICS	SOLUTION METHODS FOR EIGENVALUE PROBLEMS IN STRUCTURAL MECHANICS	73-1095
MECHANICS	SOLUTION OF SOME PLANE STRAIN PROBLEMS IN SOIL MECHANICS USING THE METHOD OF FINITE ELEMENTS	69-0521
MECHANICS	SOLUTION OF STRESS-DEFORMATION PROBLEMS IN SOIL AND ROCK MECHANICS USING FINITE ELEMENT METHODS	69-0272
MECHANICS	SOME APPLICATIONS OF ENGINEERING MECHANICS IN PRESSURE VESSEL ANALYSIS	72-0892
MECHANICS	SOME APPLICATIONS OF THE FINITE ELEMENT METHOD TO FLUID MECHANICS	74-1274
MECHANICS	SOME CONTRIBUTIONS TO NONLINEAR SOLID MECHANICS	73-0705
MECHANICS	SOME CONTRIBUTIONS TO NONLINEAR SOLID MECHANICS	74-0674
MECHANICS	SOME EXAMINATIONS ON AVAILABILITY OF FINITE ELEMENT METHODS IN LINEAR FRACTURE MECHANICS	73-0112
MECHANICS	SOME LINEAR AND NON-LINEAR PROBLEMS IN FLUID MECHANICS. FINITE ELEMENT METHOD FORMULATION	73-0523
MECHANICS	SOME NON LINEAR PROBLEMS IN SOIL AND ROCK MECHANICS - FINITE ELEMENT SOLUTION	69-0363
MECHANICS	STRESSES IN ANISOTROPIC MEDIA WITH PARTICULAR REFERENCE TO PROBLEMS OF ROCK MECHANICS	66-0079
MECHANICS	STRESSES IN ANISOTROPIC MEDIA OF ROCK MECHANICS	66-0127
MECHANICS	STRUCTURAL MECHANICS SOFTWARE. A BIBLIOGRAPHY WITH ABSTRACTS	74-0842
MECHANICS	STRUCTURAL MECHANICS COMPUTER PROGRAMS	74-0902
MECHANICS	SURVEY AND EVALUATION OF THE FINITE ELEMENT METHOD IN LINEAR FRACTURE MECHANICS ANALYSIS	71-0094
MECHANICS	THE APPLICATION OF THE FINITE ELEMENT METHOD TO SEEPAGE AND STRESS-DEFORMATION PROBLEMS IN SOIL MECH	70-0267
MECHANICS	THE APPLICATION OF THE FINITE ELEMENT METHOD TO FRACTURE MECHANICS	72-0331
MECHANICS	THE FINITE ELEMENT METHOD IN STRUCTURAL MECHANICS	65-0012
MECHANICS	THE FINITE ELEMENT METHOD IN STRUCTURAL MECHANICS	66-0051
MECHANICS	THE FINITE ELEMENT METHOD IN STRUCTURAL AND CONTINUUM MECHANICS	67-0093
MECHANICS	THE FINITE ELEMENT METHOD IN SOLID MECHANICS	70-0295
MECHANICS	THE FINITE ELEMENT METHOD IN FLUID MECHANICS	71-0197
MECHANICS	THE FINITE ELEMENT METHOD APPLIED TO FLUID MECHANICS	74-0676
MECHANICS	THE ROLE OF MECHANICS IN DESIGN-STRUCTURAL JOINTS	74-0767
MECHANICS	THE ROLE OF MECHANICS IN DESIGN - STRUCTURAL JOINTS	74-0770
MECHANICS	THE USE OF A LASER FOR ARPA MILITARY GEOPHYSICS PROGRAM (ROCK MECHANICS AND RAPID EXCAVATION)	72-0800
MECHANICS	THE USE OF FINITE ELEMENT TECHNIQUES IN FRACTURE MECHANICS	72-0926
MECHANICS	THE USES OF FINITE ELEMENT TECHNIQUES IN SOIL MECHANICS WITH PARTICULAR REFERENCE TO DEFORMATION PLA	72-0000
MECHANICS	THEMIS FINAL REPORT, VOL. II, NONLINEAR MECHANICS IN SOLIDS AND STRUCTURES	74-0722
MECHANICS	THEMIS FINAL REPORT, VOL. 1, BASIC RESEARCH IN METHODS OF APPROXIMATION IN THE NONLINEAR MECHANICS O	74-0723
MECHANICS	THREE DIMENSIONAL ELASTIC PLASTIC ANALYSIS FOR FRACTURE MECHANICS - 2.	71-0358
MECHANICS	THREE DIMENSIONAL FINITE ELEMENT STRESS ANALYSIS APPLIED TO TWO PROBLEMS IN ROCK MECHANICS	72-0189
MECHANICS	THREE-DIMENSIONAL ELASTIC-PLASTIC STRESS AND STRAIN ANALYSIS FOR FRACTURE MECHANICS. PHASE 1. SIMPLE	70-0188
MECHANICS	THREE-DIMENSIONAL FINITE ELEMENT ANALYSIS FOR FRACTURE MECHANICS	72-0162
MECHANICS	THREE-DIMENSIONAL FINITE ELEMENT ANALYSIS FOR FRACTURE MECHANICS	72-0921
MECHANICS	THREE-DIMENSIONAL LINEAR FRACTURE MECHANICS ANALYSIS BY A DISPLACEMENT-HYBRID FINITE-ELEMENT MODEL	75-0884
MECHANICS	TRANSIENT THERMAL STRESS TECHNIQUES IN NUCLEAR REACTORS. APPLIED MECHANICS ASPECTS OF NUCLEAR EFFECT	71-0474
MECHANICS	UNIMOD - APPLICATIONS ORIENTED FINITE ELEMENT SCHEME FOR THE ANALYSIS OF FRACTURE MECHANICS PROBLEMS	74-1356
MECHANICS	USE OF FINITE ELEMENT COMPUTER PROGRAMS IN FRACTURE MECHANICS	71-0022
MECHANICS	USE OF FINITE ELEMENT ANALYSIS IN FRACTURE MECHANICS	74-0837
MECHANICS	USE OF FRACTURE MECHANICS AND FINITE ELEMENT TECHNIQUES FOR DETERMINING THE SAFETY OF OFFSHORE STRUC	74-1211
MECHANICS	USE OF ISOPARAMETRIC FINITE-ELEMENTS IN LINEAR FRACTURE MECHANICS	75-0352
MECHANICS	VERIFICATION OF FRACTURE MECHANICS CONCEPTS TO PREDICT CRACKING OF FLEXIBLE PAVEMENTS	73-0772
MECHANICS-II	FINITE ELEMENT FORMULATION AND SOLUTION OF CONTACT-IMPACT PROBLEMS IN CONTINUUM MECHANICS-II	75-0740
MECHANISCHE	BERECHNUNG DER ERDEBENSCHWINGUNGEN VON STRUKTUREN MIT DER FINITE ELEMENT METHODE. MECHANISCHE MODELL	71-0170
MECHANISCHE	PRAKTISCHE RESULTATEN, DOEL EN NUT VAN HET ONTWERPEN VAN MECHANISCHE ELEMENTEN MET BEHULP VAN COMPUT	73-0400
MECHANISCHE	SPANNUNGEN IN MEMBRANWAENDEN DURCH WAERMEBEANSPRUCHUNG UND MECHANISCHE BELASTUNG. (STRESSES IN MEMBR	75-0066
MECHANISM	DYNAMIC STRAINS IN A FOUR BAR MECHANISM	73-0677
MECHANISM	LOAD TRANSFER MECHANISM FOR PIERS SOCKETTED IN HARD SOILS OR ROCK	74-0969
MECHANISM	MECHANISM OF PARALLEL SURFACE THRUST BEARING	75-0322
MECHANISM	THE EFFECT OF END RESTRAINT CONDITION ON THE FRACTURE MECHANISM AND STRENGTH OF ROCK-LIKE MATERIALS	72-0134
MECHANISMS	ANALYSIS AND DESIGN OF MULTIPLE INPUT FLEXIBLE LINK MECHANISMS	70-0354
MECHANISMS	FAILURE MECHANISMS IN COMPOSITE SYSTEMS	75-0688
MECHANISMS	FORCE AND TORQUE ANALYSIS OF CONSTRAINED SPACE MECHANISMS AND PLANE MECHANISMS WITH OFFSET LINKS BY	74-0569
MECHANISMS	FORCE AND TORQUE ANALYSIS OF CONSTRAINED SPACE MECHANISMS AND PLANE MECHANISMS WITH OFFSET LINKS BY	74-0569
MECHANISMS	GENERAL FINITE ELEMENT VIBRATIONAL ANALYSIS OF PLANAR MECHANISMS	74-1107
MECHANISMS	MATRIX-DISPLACEMENT DIRECT-ELEMENT METHOD FOR FORCE AND TORQUE ANALYSIS OF DETERMINATE AND INDETERMI	74-0531
MECHANISMS	OPTIMAL SYNTHESIS OF FLEXIBLE LINK MECHANISMS WITH LARGE STATIC DEFLECTIONS	74-0525
MEDIA	A FINITE DIFFERENCE METHOD FOR UNSTEADY FLOW IN VARIABLY SATURATED POROUS MEDIA	71-0586
MEDIA	A FINITE ELEMENT APPROACH TO TWO-PHASE FLOW IN POROUS MEDIA	75-0392
MEDIA	A HIGHER ORDER TRIANGULAR FINITE ELEMENT FOR THE SOLUTION OF FIELD PROBLEMS IN ORTHOTROPIC MEDIA	72-0370
MEDIA	AN APPLICATION OF THE FINITE ELEMENT METHOD TO TRANSIENT INTERFACE PROBLEMS IN POROUS MEDIA	73-0608
MEDIA	APPLICATION OF THE FINITE ELEMENT METHOD TO STRESS DISTRIBUTION IN SOIL MEDIA	67-0075
MEDIA	APPLICATION OF THE FINITE ELEMENT METHOD TO TRANSIENT FLOW IN POROUS MEDIA	68-0094
MEDIA	CALCULATION OF FLOWS IN POROUS MEDIA BY THE FINITE ELEMENT METHOD	70-0473
MEDIA	DYNAMIC ANALYSIS OF FOOTINGS ON LAYERED MEDIA	75-0586
MEDIA	DYNAMIC RESPONSE ON NONLINEAR MEDIA AT LARGE STRAINS	72-0237
MEDIA	EARTH VIBRATION EFFECTS AND ABATEMENT FOR MILITARY FACILITIES. REPORT 3. ANALYSIS METHOD FOR FOOTING	72-0806
MEDIA	ELASTIC-PLASTIC ANALYSIS OF THREE-DIMENSIONAL MEDIA	66-0002
MEDIA	FINITE DEFORMATIONS IN ELASTIC-PLASTIC MEDIA SUBJECTED TO IMPULSIVE LOADINGS BY THE FINITE ELEMENT M	72-0947
MEDIA	FINITE DYNAMIC MODEL FOR INFINITE MEDIA	69-0171
MEDIA	FINITE ELEMENT ANALYSIS OF SEEPAGE IN ELASTIC MEDIA	59-0024
MEDIA	FINITE ELEMENT ANALYSIS OF NONLINEAR SOIL MEDIA	69-0132
MEDIA	FINITE ELEMENT ANALYSES OF ELASTIC-PLASTIC PROBLEMS IN THE MECHANICS OF GEOLOGIC MEDIA: AN OVERVIEW	70-0203
MEDIA	FINITE ELEMENT COMPUTER PROGRAMS FOR FLOW IN SATURATED UNSATURATED POROUS MEDIA	72-0364
MEDIA	FINITE ELEMENT GALERKIN METHOD FOR ANALYSIS OF FLOW IN FRACTURED POROUS MEDIA	74-0066
MEDIA	FINITE ELEMENT METHODS FOR FLOW IN POROUS MEDIA	74-0057

MEDIA	FINITE ELEMENT MODELING OF FLOW THROUGH POROUS MEDIA	75-0706
MEDIA	FINITE ELEMENT SOLUTION FOR WAVE PROPAGATION IN LAYERED MEDIA CAUSED BY A NUCLEAR DETONATION	67-0016
MEDIA	FINITE ELEMENT SOLUTION OF UNSTEADY AND UNSATURATED FLOW IN POROUS MEDIA	72-0064
MEDIA	FINITE ELEMENT SOLUTION OF COUPLED ELECTROKINETIC AND HYDRODYNAMIC FLOW IN POROUS MEDIA	72-0398
MEDIA	FINITE ELEMENT SOLUTION OF TWO DIMENSIONAL UNSTEADY AND UNSATURATED FLOW IN POROUS MEDIA	75-0113
MEDIA	FINITE ELEMENT STRESS ANALYSIS OF POROUS MEDIA	69-0440
MEDIA	FINITE ELEMENT STUDIES OF SOILS AND POROUS MEDIA	71-0504
MEDIA	FINITE ELEMENT STUDY OF PRESSURIZED CAVITIES IN GEOLOGIC MEDIA	71-0805
MEDIA	FINITE ELEMENT WEIGHTED RESIDUAL SOLUTION OF ONE-DIMENSIONAL, UNSTEADY AND UNSATURATED FLOWS IN PORO	73-0712
MEDIA	FINITE ELEMENTS FOR TWO PHASE FLOW IN POROUS MEDIA	74-0762
MEDIA	FLOW OF COMPRESSIBLE FLUID IN POROUS ELASTIC MEDIA	73-1004
MEDIA	FORCED VIBRATIONS OF CIRCULAR FOUNDATIONS ON LAYERED MEDIA	74-0762
MEDIA	GALERKIN APPROACH TO SATURATED-UNSATURATED FLOW IN POROUS MEDIA	75-0393
MEDIA	GENERALIZED VARIATIONAL PRINCIPLES FOR USE IN FLOW PROBLEMS THROUGH POROUS MEDIA	74-0067
MEDIA	MAGNETOTELLURIC RESPONSE OF LATERALLY INHOMOGENEOUS AND ANISOTROPIC MEDIA	75-0850
MEDIA	MATRIX ANALYSIS OF THREE-DIMENSIONAL ELASTIC MEDIA: SMALL AND LARGE DISPLACEMENTS	65-0003
MEDIA	NON-LINEAR FLOW IN POROUS MEDIA BY FINITE ELEMENTS	69-0023
MEDIA	ON THE SOLUTION OF TRANSIENT FREE-SUPFACE FLOW PROBLEMS IN POROUS MEDIA BY THE FINITE ELEMENT METHOD	73-1019
MEDIA	POROUS MEDIA - SOME PROBLEMS IN CIVIL AND PETROLEUM ENGINEERING	74-0349
MEDIA	STEADY CONDUCTION OF HEAT IN LINEAR AND NONLINEAR FULLY ANISOTROPIC MEDIA BY FINITE ELEMENTS	74-0359
MEDIA	STRESSES IN ANISOTROPIC MEDIA WITH PARTICULAR REFERENCE TO PROBLEMS OF ROCK MECHANICS	66-0079
MEDIA	STRESSES IN ANISOTROPIC MEDIA OF ROCK MECHANICS	66-0127
MEDIA	THREE DIMENSIONAL ANISOTROPIC AND INHOMOGENEOUS ELASTIC MEDIA MATRIX ANALYSIS FOR SMALL AND LARGE DI	65-0008
MEDIA	THREE DIMENSIONAL STRUCTURES/MEDIA INTERACTION, DYNAX DEVELOPMENT AND APPLICATIONS,	70-0378
MEDIA	TRANSIENT FREE-SURFACE FLOW IN POROUS MEDIA BY FINITE ELEMENT METHOD	72-0297
MEDIA	USE OF HYBRID COMPUTATION FOR SOLVING TRANSIENT FLOW PROBLEM IN POROUS MEDIA BY FINITE ELEMENT METHO	74-0068
MEDIA	VARIATIONAL PRINCIPLES FOR FLUID FLOW IN POROUS MEDIA	71-0844
MEDIA	VERTICAL VIBRATIONS OF FOOTINGS EMBEDDED IN LAYERED MEDIA	69-0436
MEDIA	VIBROACOUSTIC RESPONSE OF TURBULENCE EXCITED THIN RECTANGULAR FINITE PLATES IN HEAVY AND FLUID MEDIA	74-1309
MEDIA	WATER FLOW THROUGH SATURATED-UNSATURATED POROUS MEDIA USING FINITE-ELEMENT-GALERKIN METHODS	74-0879
MEDIA	WATER MOVEMENT THROUGH SATURATED-UNSATURATED POROUS MEDIA: A FINITE-ELEMENT GALERKIN MODEL	75-0761
MEDIANTE	SU LA DETERMINAZIONE DEI CAMPI MAGNETICI LINEARI E NON LINEARI MEDIANTE L APPLICAZIONE DEI METODI DE	74-1061
MEDIUM	AN INTRODUCTION TO THE FINITE ELEMENT ANALYSIS OF THE MOTION OF A GAS-LIQUID INTERFACE IN A POROUS M	72-0904
MEDIUM	ANALYSIS OF BUILDINGS ON ELASTIC MEDIUM UNDER DYNAMIC OR SEISMIC LOADS	74-0456
MEDIUM	ELASTIC SETTLEMENT OF A CIRCULAR RAFT IN ADHESIVE CONTACT WITH A TRANSVERSELY ISOTROPIC MEDIUM	75-0828
MEDIUM	ELASTO-PLASTIC INDENTATION OF A LAYERED MEDIUM	74-0225
MEDIUM	ETUDE DE LA CAPACITE PORTANTE DES PIEUX PAR LA METHODE DES ELEMENTS FINIS. INFLUENCE DE LA DILATANCE	75-0624
MEDIUM	FINITE ELEMENT ANALYSIS OF THE MOTION OF A GAS-LIQUID INTERFACE IN A POROUS MEDIUM	75-0672
MEDIUM	METHOD OF FINITE ELEMENTS FOR CALCULATION OF HETEROGENEOUS BODIES OF REVOLUTION IN ELASTO-PLASTIC ME	75-0144
MEDIUM	NOUVELLE CONCEPTION DES DRAINS DANS LES BARRAGES EN TERRE HOMOGENE DE PETITE ET MOYENNE DIMENSIONS (	73-0404
MEDIUM	SEEPAGE INTO AN EXCAVATION IN A MEDIUM POSSESSING STRESS-DEPENDENT PERMEABILITY	72-0629
MEDIUM	SOLUTION OF WAVE PROPAGATION PROBLEMS IN SATURATED MEDIUM	72-0784
MEDIUM	STRESSES AND DISPLACEMENTS AROUND A CIRCULAR TUNNEL IN A THREE LAYER MEDIUM	69-0038
MEDIUM	STRUCTURE- MEDIUM INTERACTION PHENOMENOLOGY	72-0863
MEDIUM-SPEED	SIGNIFICANT APPLICATIONS OF FINITE ELEMENTS AND FINITE DIFFERENCES METHODS TO SOLVE PROBLEMS OF THER	73-0864
MEDIUM-STRUCTUR	FUNDAMENTAL STUDIES OF MEDIUM-STRUCTURE INTERACTION. REPORT 1. FINITE ELEMENT ANALYSIS OF BURIED CYL	72-0854
MEDIUM-SUPPORT	TUNNEL DESIGN CONSIDERATIONS: ANALYSIS OF MEDIUM-SUPPORT INTERACTION	74-1062
MEETING	PROCEEDINGS OF THE NOSETIP STRESS ANALYSIS TECHNICAL INTERCHANGE MEETING	67-0042
MEETING	REPORT ON POST CONF. MEETING ON COMPUTATIONAL ASPECTS OF ASPECTS OF FINITE ELEMENT METHOD ISD STUTTG	72-0475
MELOSH	COMMENT ON MATRICES FOR THE DIRECT STIFFNESS METHOD BY R. J. MELOSH	64-0022
MELT-THROUGH	THE EFFECT OF GRAVITY UPON THE MELT-THROUGH TIME OF A SOLID SUBJECTED TO A HIGH INTENSITY LASER	72-0693
MELTING	A NUMERICAL PROCEDURE FOR TWO DIMENSIONAL HEATING AND MELTING CALCULATIONS, WITH APPLICATIONS TO LAS	72-0708
MELTING	STRESSES AND DEFORMATIONS IN MELTING PLATES	69-0440
MELTING-FREEZIN	CALCULATION OF TEMPERATURE DISTRIBUTION WITHIN A MELTING-FREEZING MATERIAL USING FINITE ELEMENT TECH	72-0043
MEMBER	APPLICATION OF PROGRAM FATIGUE TEST TO MEMBER JOINTS OF HULLS	72-0814
MEMBER	DYNAMICS OF PRISMATIC AND OPEN SECTION MEMBER GRIDS	69-0070
MEMBERS	A FINITE ELEMENT FORMULATION FOR THE GENERAL STABILITY ANALYSIS OF THIN WALLED MEMBERS	70-0337
MEMBERS	A FINITE-ELEMENT STUDY ON THE NONLINEAR BEHAVIOR OF REINFORCED CONCRETE FLEXURAL MEMBERS DUE TO SHOR	73-1154
MEMBERS	ANCHORAGE ZONE STRESSES IN AXIALLY POST-TENSIONED I-SECTION MEMBERS WITH END BLOCKS	71-0305
MEMBERS	BIFURCATION OF ECCENTRICALLY LOADED MEMBERS AND OF STIFFENED PLATES	70-0057
MEMBERS	ELASTIC-PLASTIC ANALYSIS OF REINFORCED CONCRETE MEMBERS BY FEM	73-0292
MEMBERS	ELASTIC-PLASTIC ANALYSIS FOR CRACKED MEMBERS	75-0519
MEMBERS	ELASTO-PLASTIC CYCLIC ANALYSIS OF STRUCTURAL MEMBERS	72-0202
MEMBERS	FAILURE OF THIN-WALLED MEMBERS UNDER PATCH LOADINGS	72-0230
MEMBERS	FINITE ELEMENT ANALYSIS OF WOOD STRUCTURAL MEMBERS	70-0180
MEMBERS	FINITE ELEMENT ANALYSIS OF REINFORCED CONCRETE MEMBERS	72-0583
MEMBERS	FINITE ELEMENT ANALYSIS OF THIN-WALLED MEMBERS OF OPEN SECTION	72-0966
MEMBERS	GEOMETRICALLY NONLINEAR BEHAVIOUR OF THIN-WALLED MEMBERS USING FINITE-ELEMENTS	73-1142
MEMBERS	INSTABILITY OF HORIZONTALLY CURVED MEMBERS. FLANGE BUCKLING STUDIES	69-0480
MEMBERS	LINEAR AND NONLINEAR FINITE ELEMENT ANALYSIS OF WOOD STRUCTURAL MEMBERS	73-0117
MEMBERS	MECHANICS OF CABLE MOORING SYSTEMS. VOLUME VI. A COMPUTER PROGRAM FOR ANALYZING THE STEADY STATE CON	73-0725
MEMBERS	NON LINEAR ANALYSIS OF REINFORCED CONCRETE MEMBERS BY FINITE ELEMENT METHOD. PART 1: ANALYSIS, PART	71-0313
MEMBERS	PERFORMANCE OF STRUCTURAL WOOD MEMBERS EXPOSED TO FIRE	75-0060
MEMBERS	REFINED FINITE ELEMENT STABILITY ANALYSIS OF THIN-WALLED MEMBERS	75-0784
MEMBERS	STIFFNESS MATRICES FOR THE GENERAL DEFORMATION (OUT-OF-PLANE AND INPLANE) OF CURVED BEAM MEMBERS BAS	75-0124
MEMBERS	STRESS ANALYSIS IN HULL STRUCTURAL MEMBERS BY PHOTOELASTIC EXPERIMENTS AND FINITE ELEMENT METHODS -	70-0341
MEMBERS	TANGENT STIFFNESS MATRIX FOR SPACE FRAME MEMBERS	69-0397
MEMBERS	TWO DIMENSIONAL NON LINEAR ANALYSIS OF REINFORCED CONCRETE MEMBERS BY THE FINITE ELEMENT METHOD, PAR	71-0210
MEMBERS	USE OF THE FINITE ELEMENT TECHNIQUE TO DETERMINE THE FIRE RESISTANCE CAPABILITIES OF INSULATED STRUC	71-1071
MEMBERS	VIBRATIONS OF TAPERED AND CURVED MEMBERS BY FINITE ELEMENT METHODS	69-0517
MEMBRANE	A COMPATIBLE MEMBRANE QUADRILATERAL ELEMENT	72-0483
MEMBRANE	A SIX NODE TRIANGULAR EQUILIBRIUM (CONSTANT STRESS) MEMBRANE ELEMENT	73-0783
MEMBRANE	A STIFFNESS MATRIX FOR A CURVED MEMBRANE SHELL	69-0055
MEMBRANE	A TRIANGULAR EQUILIBRIUM (LINEAR STRESS) MEMBRANE ELEMENT	75-0883
MEMBRANE	A WARPED QUADRILATERAL STRAIN MEMBRANE ELEMENTS	75-0141
MEMBRANE	A WARPED QUADRILATERAL EQUILIBRIUM MEMBRANE ELEMENT	75-0885
MEMBRANE	AN EQUILIBRIUM MEMBRANE QUADRILATERAL ELEMENT	72-0484
MEMBRANE	ANALYSIS MODELS FOR PLATE ELEMENTS IN THE MEMBRANE STATE OF STRESS	67-0046
MEMBRANE	ANALYSIS OF ELASTIC SHELLS OF REVOLUTION WITH MEMBRANE AND FLEXURE STRESSES UNDER ARBITRARY LOADING	68-0289
MEMBRANE	APPLICATION OF THE DISCRETE ELEMENT METHOD TO THE BUCKLING ANALYSIS OF RECTANGULAR PLATES UNDER ARBI	68-0054
MEMBRANE	COMPARISON OF MEMBRANE FINITE ELEMENT FORMULATIONS	70-0287
MEMBRANE	CONTACT OF AN INFLATED TOROIDAL MEMBRANE WITH A FLAT SURFACE AS AN APPROACH TO THE TIRE DEFLECTION P	75-0164
MEMBRANE	CURVED THICK SHELL AND MEMBRANE ELEMENTS WITH PARTICULAR REFERENCE TO AXISYMMETRIC PROBLEMS	68-0054
MEMBRANE	EFFECT OF OUT-OF-PLANENESS OF MEMBRANE QUADRILATERAL FINITE ELEMENTS	73-0031
MEMBRANE	ELASTIC AND EQUILIBRIUM MATRICES OF GENERAL SEMI-MONOCOGUE MEMBRANE ELEMENTS	70-0159
MEMBRANE	EVALUATION OF CALAC SEMI-MONOCOQUE MEMBRANE ELEMENT	70-0278
MEMBRANE	FINITE ELEMENT ANALYSIS OF NONLINEAR MEMBRANE STRUCTURES	72-0909
MEMBRANE	FINITE ELEMENT ANALYSIS OF NONLINEAR MEMBRANE STRUCTURES	72-0944
MEMBRANE	FLAT QUADRILATERAL EQUILIBRIUM MEMBRANE ELEMENTS	73-1116
MEMBRANE	FREE VIBRATION OF DOUBLY CURVED RECTANGULAR PLATES INCLUDING THE PRESENCE OF MEMBRANE STRESSES WITH	69-0468
MEMBRANE	GENERAL MEMBRANE ISOPARAMETRIC ELEMENTS	75-0442
MEMBRANE	MEMBRANE ANALOGY FOR ANISOTROPIC CABLE NETWORKS	74-0233

286

MEMBRANE	MEMBRANE ANALYSIS OF UNFRAMED HIGH RISE BUILDINGS	73-0451
MEMBRANE	MEMBRANE AND PLATE BENDING ELEMENTS	75-0405
MEMBRANE	MEMBRANE PARALLELOGRAM ELEMENT WITH LINEAR VARYING EDGE STRAIN FOR THE MATRIX DISPLACEMENT METHOD	66-0006
MEMBRANE	NONLINEAR ANALYSIS OF ARBITRARY, HYPERELASTIC MEMBRANE SHELLS BY THE FINITE-ELEMENT METHOD	73-1132
MEMBRANE	ON THE CONVERGENCE OF THE METHOD OF FINITE ELEMENTS IN THE ANALYSIS OF MEMBRANE DYNAMICS	72-0537
MEMBRANE	PROGRAM SPECIFICATION NV 300. GENERAL FINITE ELEMENT PROGRAM FOR 2- AND 3-DIMENSIONAL MEMBRANE STRUC	68-0184
MEMBRANE	SIMPLE TRIANGULAR MEMBRANE FINITE ELEMENT	74-0881
MEMBRANE	SPANNUNGEN IN MEMBRANWAENDEN DURCH WAERMEBEANSPRUCHUNG UND MECHANISCHE BELASTUNG. (STRESSES IN MEMBR	75-0066
MEMBRANE	STIFFNESS MATRIX FOR BENDING OF A RECTANGULAR PLATE ELEMENT WITH INITIAL MEMBRANE STRESSES	62-0006
MEMBRANE	TEMPERATURES AND STRESSES IN A BOILER MEMBRANE WALL TUBE	74-1053
MEMBRANE	THEORY FOR DISTORTION STUDIES OF LUNG PARENCHYMA BASED ON ALVEOLAR MEMBRANE PROPERTIES	74-0222
MEMBRANE	THERMAL DEFORMATION VECTOR FOR A BILINEAR TEMPERATURE DISTRIBUTION IN AN ANISOTROPIC QUADRILATERAL M	75-0670
MEMBRANE	WARPED SEMI-MONOCOQUE QUADRILATERAL MEMBRANE ELEMENT FOR THE FINITE ELEMENT METHODS	70-0195
MEMBRANES	A NOTE ON THE ANALYSIS OF NONLINEAR DYNAMICS OF ELASTIC MEMBRANES BY THE FINITE ELEMENT METHOD	74-0341
MEMBRANES	A STUDY OF A DYNAMICAL FINITE ELEMENT ANALYSIS FOR APPLICATION TO AXIAL WAVE PROPAGATION PROBLEMS IN	71-0589
MEMBRANES	ACCURACY AND CONVERGENCE OF MIXED FINITE-ELEMENT APPROXIMATIONS OF THIN BARS, MEMBRANES, AND PLATES	73-1120
MEMBRANES	ANALYSIS OF LARGE DEFORMATIONS OF ELASTIC MEMBRANES BY THE FINITE ELEMENT METHOD	66-0057
MEMBRANES	FINITE DEFORMATION OF ELASTIC MEMBRANES BY THE FINITE ELEMENT METHOD	67-0105
MEMBRANES	FINITE DEFORMATION OF ELASTIC PLATES, SHELLS, AND MEMBRANES BY THE FINITE ELEMENT METHOD	72-0805
MEMBRANES	FINITE ELEMENT ANALYSIS OF FINITE DEFORMATIONS OF PLATES, SHELLS, AND MEMBRANES	72-0578
MEMBRANES	FINITE STRAINS AND DISPLACEMENTS OF ELASTIC MEMBRANES BY THE FINITE ELEMENT METHOD	67-0062
MEMBRANES	FREE VIBRATIONS OF CIRCULAR MEMBRANES UNDER ARBITRARY TENSION BY THE FINITE ELEMENT METHOD	69-0190
MEMBRANES	NOTE ON THE ANALYSIS OF NONLINEAR DYNAMICS OF ELASTIC MEMBRANES BY THE FINITE ELEMENT METHOD	74-0230
MEMBRANES	ON THE ASYMPTOTICALLY SPHERICAL DEFORMATIONS OF ARBITRARY MEMBRANES OF REVOLUTION FIXED ALONG AN EDG	71-0640
MEMBRANES	SHELLS, PLATES AND MEMBRANES - A COMPUTER-AIDED DESIGN SYSTEM	74-1159
MEMBRANES	USE OF TRIGONOMETRIC TERMS IN FINITE ELEMENT METHOD WITH APPLICATION TO VIBRATING MEMBRANES	74-0127
MEMBRANES	USE OF TRIGONOMETRIC TERMS IN FINITE ELEMENT METHOD WITH APPLICATION TO VIBRATING MEMBRANES	74-0555
MEMBRANES	USERS MANUAL FOR DIGITAL COMPUTER PROGRAM OK5 NUMERICAL ANALYSIS OF FINITE STRAINS AND DISPLACEMENTS	72-0718
MEMBRANWAENDEN	SPANNUNGEN IN MEMBRANWAENDEN DURCH WAERMEBEANSPRUCHUNG UND MECHANISCHE BELASTUNG. (STRESSES IN MEMBR	75-0066
MEMORY	FORMULATION OF GENERAL DISCRETE MODELS OF THERMOMECHANICAL BEHAVIOR OF MATERIALS WITH MEMORY	69-0381
MEMORY	STUDY ON FOIL BEARINGS - AN APPLICATION TO TAPE MEMORY DEVICES	75-0311
MENISCI	MENISCI IN ARRAYS OF CYLINDERS - NUMERICAL-SIMULATION BY FINITE-ELEMENTS	75-0449
MERCURY	FINITE ELEMENT ANALYSIS OF MERCURY SLOSH IN THE SOLAR ELECTRIC PROPULSION STAGE	75-0757
MERIDIONAL	A METHOD FOR COMPUTATION OF VIBRATION MODES AND FREQUENCIES OF ORTHOTROPIC THIN SHELLS OF REVOLUTION	69-0037
MERIDIONAL	NUMERICAL CALCULATION OF MERIDIONAL FLOW FIELD IN TURBOMACHINES USING FINITE ELEMENTS METHOD	74-0481
MERIDIONAL	THE MERIDIONAL THROUGH-FLOW CALCULATION IN AN AXIAL FLOW MACHINE BY THE FINITE ELEMENT METHOD	75-0115
MERIDIONAL	USE OF HIGHER ORDER DISPLACEMENT FUNCTIONS IN THE FREE VIBRATION ANALYSIS OF SHELLS OF REVOLUTION HA	74-1055
MERSEY	UNUSUAL FOUNDATIONS OVER THE MERSEY TUNNEL	70-0197
MESH	A SELF-ORGANIZING MESH GENERATION PROGRAM	74-1051
MESH	A SELF-ORGANIZING MESH GENERATION PROGRAM	74-1249
MESH	A TECHNIQUE FOR MESH GRADING APPLIED TO CONFORMING PLATE BENDING FINITE ELEMENTS	73-1058
MESH	ALGORITHMS FOR THE PRODUCTION OF CONTOUR MAPS OVER AN IRREGULAR TRIANGULAR MESH	72-0825
MESH	AN ALPHANUMERIC FINITE ELEMENT MESH PLOTTER	72-0543
MESH	AN AUTOMATIC MESH GENERATION SCHEME FOR PLANE AND CURVED SURFACES BY ISOPARAMETRIC COORDINATES	71-0271
MESH	AN AUTOMATIC MESH GENERATOR USING TWO AND THREE DIMENSIONAL ISOPARAMETRIC ELEMENTS	73-0066
MESH	AN IRREGULAR TRIANGLE MESH GENERATOR	64-0032
MESH	APPLICATION OF COMPUTER GRAPHICS TO FINITE ELEMENT MESH GENERATION	73-0546
MESH	AUTOMATED MESH REFINEMENT STRATEGIES FOR FINITE ELEMENT COMPUTATIONS	75-0877
MESH	AUTOMATIC GENERATION OF FINITE ELEMENT MESH WITHIN A BANDWIDTH AND AN EFFICIENT SOLUTION OF THE BAND	73-0338
MESH	AUTOMATIC MESH GENERATION IN TWO AND THREE DIMENSIONAL INTERCONNECTED DOMAINS	70-0251
MESH	AUTOMATIC MESH GENERATION FOR FINITE ELEMENT ANALYSIS	72-0328
MESH	AUTOMATIC MESH GENERATION FOR THREE DIMENSIONAL STRUCTURES BASED ON THEIR THREE VIEWS	73-0340
MESH	BERGEN, AN AUTOMATIC FINITE ELEMENT MESH GENERATION PROGRAM FOR ARBITRARY STRUCTURES	72-0776
MESH	COMPUTER CODE TO GENERATE FINITE ELEMENT MESH FOR PROBLEMS OF COMPLEX AXISYMMETRIC GEOMETRY	74-0360
MESH	COMPUTER PROGRAM FOR AUTOMATIC MESH GENERATION FOR THE 2-D FINITE ELEMENT PROGRAM WILAX	73-0565
MESH	CONSTRUCTION OF CURVILINEAR COORDINATE SYSTEMS AND APPLICATIONS TO MESH GENERATION	74-1295
MESH	ELFED A COMPUTER CODE TO GENERATE A FINITE ELEMENT MESH FOR PROBLEMS OF COMPLEX AXISYMMETRIC GEOMETR	74-1270
MESH	ELFED: A COMPUTER CODE TO GENERATE FINITE ELEMENT MESH FOR PROBLEMS OF COMPLEX AXISYMMETRIC GEOMETRY	73-0943
MESH	FINITE ELEMENT APPROACH TO HYDRODYNAMICS AND MESH STABILIZATION	74-1298
MESH	FINITE ELEMENT MESH GENERATION PROGRAM FOR ARBITRARY TWO-AND THREE-DIMENSIONAL STRUCTURES	74-0966
MESH	GEOMETRIC ASPECTS OF THE FINITE ELEMENT METHOD: CONSTRUCTION OF CURVILINEAR COORDINATE SYSTEMS AND T	72-0534
MESH	LOCAL MESH REFINEMENT WITH FINITE ELEMENTS FOR ELLIPTIC PROBLEMS	73-0073
MESH	LOCAL MESH REFINEMENT USING RECTANGULAR BLENDED FINITE-ELEMENTS	75-0452
MESH	MESH AND CONTOUR PLOT FOR TRIANGLE AND ISOPARAMETRIC ELEMENTS	70-0445
MESH	MESH GENERATING COMPUTER PROGRAM FOR FINITE ELEMENT ANALYSES OF SIMPLE SLOPES IN HORIZONTALLY LAYERE	71-0748
MESH	MESH GENERATION - SURVEY	73-0048
MESH	MESH GENERATION CODE AND A CONTOUR PLOTTING ROUTINE	72-0802
MESH	MESH GENERATION PROGRAM FOR HIGHWAY EXCAVATION CUTS	74-1250
MESH	MESH GRADING TECHNIQUES FOR COMPATIBLE AND EQUILIBRIUM ELEMENTS	74-0397
MESH	MESH 3D: A THREE-DIMENSIONAL FINITE ELEMENT MESH GENERATOR PROGRAM FOR EIGHT-NODE ISOPARAMETRIC ELEM	73-0972
MESH	MESH 3D: A THREE-DIMENSIONAL FINITE ELEMENT MESH GENERATOR PROGRAM FOR EIGHT-NODE ISOPARAMETRIC ELEM	73-0972
MESH	MESHGEN, A COMPUTER CODE FOR AUTOMATIC FINITE ELEMENT MESH GENERATION	69-0096
MESH	NESTED DISSECTION OF A REGULAR FINITE ELEMENT MESH	73-0042
MESH	NUMERICAL SOLUTION OF THE QUASI LINEAR POISSON EQUATION IN A NON-UNIFORM TRIANGLE MESH	67-0090
MESH	OPTIMUM MESH CONFIGURATIONS IN THE FINITE ELEMENT METHOD	74-1346
MESH	QMESH: A SELF-ORGANIZING MESH GENERATION PROGRAM	74-0872
MESH	QMESH: A SELF-ORGANIZING MESH GENERATION PROGRAM	74-0899
MESH	RULES GOVERNING THE NUMBER OF NODES AND ELEMENTS IN A FINITE ELEMENT MESH	70-0053
MESH	SOME EFFECTS OF SYSTEM IDEALIZATIONS, SINGULARITIES AND MESH PATTERNS ON FINITE ELEMENT SOLUTIONS	74-0939
MESH	TECHNIQUES IN TRIANGULAR MESH GENERATION FLAT AND CURVED SURFACES	69-0394
MESH	THREE-DIMENSIONAL MESH GENERATOR FOR FINITE ELEMENT COMPUTER CODES	74-0864
MESH	TRIANGULAR MESH GENERATION PROGRAM	67-0037
MESH	TRIANGULAR MESH GENERATION PROGRAM	67-0127
MESH	TRIANGULAR MESH METHODS FOR THE NEUTRON TRANSPORT EQUATION	73-1062
MESH	TRIPLET: A TWO-DIMENSIONAL, MULTIGROUP, TRIANGULAR MESH, PLANAR GEOMETRY, EXPLICIT TRANSPORT CODE	73-0955
MESH	TWO-DIMENSIONAL AUTOMATIC MESH GENERATION FOR STRUCTURAL ANALYSIS	70-0058
MESH	USERS MANUAL FOR QMESH, A SELF-ORGANIZING MESH GENERATION PROGRAM	74-0880
MESHES	A GENERAL FINITE DIFFERENCE METHOD FOR ARBITRARY MESHES	72-0883
MESHES	BODY ORIENTED (NATURAL) CO-ORDINATES FOR GENERATING THREE-DIMENSIONAL MESHES	74-0908
MESHES	CONDITION OF FINITE ELEMENT MATRICES GENERATED FROM NONUNIFORM MESHES	72-0104
MESHES	EQUIPOTENTIAL ZONING OF TWO-DIMENSIONAL MESHES	63-0025
MESHES	EQUIPOTENTIAL ZONING OF TWO-DIMENSIONAL MESHES	64-0033
MESHES	EURCYL: A COMPUTER PROGRAM TO GENERATE FINITE ELEMENT MESHES FOR CYLINDER-CYLINDER INTERSECTIONS	73-0962
MESHES	EURCYL: A PROGRAM TO GENERATE FINITE ELEMENT MESHES FOR PRESSURE VESSEL NOZZLES	74-1086
MESHES	NUMERICAL METHOD FOR GENERATING ORTHOGONAL CURVILINEAR MESHES	70-0443
MESHES	RELABELING OF FINITE ELEMENT MESHES USING A RANDOM PROCESS	72-0754
MESHES	THE APPLICATION OF NODE-ELEMENT RULES FOR FORECASTING PROBLEMS IN THE GENERATION OF FINITE ELEMENT M	75-0614
MESHGEN	MESHGEN, A COMPUTER CODE FOR AUTOMATIC FINITE ELEMENT MESH GENERATION	69-0096
MESSUERFAHREN	MESSUERFAHREN RECHENVERFAHREN ZUR ERMITTLUNG DREIACHSIGER EIGENSPANNUNGSZUSTAENDE	73-0507
MESY	MESY - EIN PROGRAMMSYSTEM ZUR UNTERSUCHUNG VON TRAGWERKEN. (STRUCTURAL ANALYSIS BY MESY PROGRAMMING	75-0360
MESY	MESY - EIN PROGRAMMSYSTEM ZUR UNTERSUCHUNG VON TRAGWERKEN. (STRUCTURAL ANALYSIS BY MESY PROGRAMMING	75-0360
METAL	A FINITE ELEMENT METHOD FOR THREE DIMENSIONAL ANALYSIS OF METAL FORMING	74-1109

287

METAL	ANALYSIS OF STRUCTURAL ACOUSTIC INTERACTIONS IN METAL CERAMIC TRANSDUCERS	73-0401
METAL	ANALYSIS OF THERMAL STRESSES AND METAL MOVEMENT DURING WELDING 1. ANALYTICAL STUDY	74-1026
METAL	APPLICATION OF THE FINITE ELEMENT METHOD IN METAL DIFFUSION	74-1366
METAL	CALCULATION OF SOLIDIFICATION OF MOLTEN METAL BY FINITE-ELEMENT METHOD	74-1357
METAL	COMPOSITE-OVERLAY REINFORCEMENT OF CUTOUTS AND CRACKS IN METAL SHEET	73-0878
METAL	CREEP BENDING TESTS OF LARGE WELD OVERLAY AND BASE METAL SPECIMENS AT 1000 DEGREE F. (537.7 DEG. C)	75-0521
METAL	CREEP OF A UNIAXIAL METAL MATRIX COMPOSITE SUBJECTED TO AXIAL AND NORMAL LATERAL LOADS	71-0041
METAL	CREEP OF A UNIAXIAL METAL MATRIX COMPOSITE SUBJECTED TO AXIAL AND NORMAL LATERAL LOADS	71-0727
METAL	DESIGN AND ANALYSIS EXPERIENCE WITH A LIQUID METAL HEAT EXCHANGER FOR PFTF SERVICE	72-0772
METAL	DETERMINATION OF THE STRESS-STRAINED STATE WHEN CUTTING METAL WITH A WEDGE	73-0237
METAL	DIGITAL CALCULATION OF ELECTRIC FIELDS IN METAL CLAD INSTALLATIONS	73-0212
METAL	DYNAMIC ANALYSIS OF THE DEFORMATION OF STRUCTURES AND OF METAL FORMING	74-0823
METAL	DYNAMIC ANALYSIS OF STRUCTURAL DEFORMATION AND METAL FORMING	75-0058
METAL	ELASTIC DEFORMATION OF POLYCRYSTALLINE METAL - 1. INFLUENCE OF GRAIN SHAPE ON DISTRIBUTIONS OF STRES	72-0463
METAL	ELASTIC DEFORMATION OF POLYCRYSTALLINE METAL - 2 EFFECT OF SLENDERNESS OF GRAINS ON DEFORMATION MODE	73-0275
METAL	ELASTO-PLASTIC ANALYSIS OF METAL SHEETS WITH A CENTRAL CRACK	67-0137
METAL	FINITE ELEMENT ANALYSIS AND EXPERIMENTAL INVESTIGATION OF SHEET METAL STRETCHING	72-0229
METAL	FINITE ELEMENT ANALYSIS OF STEADY FLUID AND METAL FLOW	74-0054
METAL	FINITE ELEMENT ANALYSIS OF METAL DECK SHEAR DIAPHRAGMS	74-0128
METAL	FINITE ELEMENT ANALYSIS OF STEADY FLUID AND METAL FLOW	75-0178
METAL	FINITE ELEMENT METHOD OF ANALYSIS OF THE HYDROSTATIC BULGING OF A SHEET METAL - 1.	74-0958
METAL	FINITE ELEMENT PROCEDURE FOR ANALYSIS OF METAL FORMING PROCESSES	74-0380
METAL	INCIPIENT CHIP FORMATION IN METAL CUTTING - A THREE- DIMENSION FINITE-ELEMENT ANALYSIS	73-1146
METAL	RESIDUAL STRESS PRODUCED BY METAL CUTTING	71-0414
METAL	STIFFNESS ANALYSIS OF SHEET METAL SHELLS UNDER CONCENTRATED LOADS	74-0325
METAL	STRUCTURAL BEHAVIOR OF A FLEXIBLE METAL CULVERT UNDER A DEEP EARTH EMBANKMENT USING METHOD A BACKFIL	73-0739
METAL	UNTERSUCHUNGEN ZUM FESTIGKEITSVERHALTEN VON METALLKLEBUNGEN IM HIMBLICK AUF DIE ERSTELLUNG EINES PRA	74-0289
METAL	USE OF METHOD OF FINITE ELEMENTS FOR STUDY OF STATIONARY METAL FORMING TECHNIQUES	72-0024
METAL-CERAMIC	ANALYSIS OF STRUCTURAL-ACOUSTIC INTERACTIONS IN METAL-CERAMIC TRANSDUCERS	72-0850
METALFORMING	SIMPLIFIED MINIMUM INTERNAL ENERGY OF DEFORMATION POSTULATE AND ITS APPLICATION TO METALFORMING PROC	74-1155
METALLKLEBUNGEN	UNTERSUCHUNGEN ZUM FESTIGKEITSVERHALTEN VON METALLKLEBUNGEN IM HIMBLICK AUF DIE ERSTELLUNG EINES PRA	74-0289
METALLURGICALLY	A METHOD FOR DETERMINING LOCAL ELASTOPLASTIC STRESS AND STRAIN IN METALLURGICALLY BONDED NOTCHED LAM	74-0686
METALLURGY	COMPUTATIONAL TECHNIQUES AS AN AID IN PHYSICAL METALLURGY	71-0554
METALS	ANISOTROPIC WORK HARDENING IN METALS. FINITE ELEMENT APPLICATION TO PLANE STRESS AND PLANE STRAIN	73-0950
METALS	APLIKACE MKP NA RESENI PROBLEMU DIFUZE V KOVECH. (APPLICATION OF THE FINITE ELEMENT METHOD TO THE SO	74-0609
METALS	COMBINED MACROSCOPIC AND MICROSCOPIC APPROACH TO THE FRACTURE OF METALS	74-0689
METALS	ELASTIC DEFORMATION OF POLYCRYSTALLINE METALS - 3	73-0361
METALS	FINITE ELEMENT ANALYSIS OF GENERAL DEFORMATION OF SHEET METALS	74-0501
METALS	PENALTY FUNCTION APPROACH TO PROBLEMS OF PLASTIC FLOW OF METALS WITH LARGE SURFACE DEFORMATIONS	75-0282
METALS	SOLUTION OF THE PROBLEM OF PROCESSING METALS BY PRESSURE USING THE METHOD OF FINITE ELEMENTS	74-1363
METALS	SOLVING THE PROBLEM OF PROCESSING METALS BY PRESSURE THROUGH THE METHOD OF FINITE ELEMENTS	74-1361
METEOROID	STUDY OF HYPERVELOCITY METEOROID IMPACT ON ORBITAL SPACE STATIONS	73-0905
METEOROLOGICAL	A SIMPLE FINITE ELEMENT METHOD FOR METEOROLOGICAL PROBLEMS	73-1121
METERING	DESIGN OF LOW-THERMAL-DISTORTION LST METERING STRUCTURE	75-0840
METHOD-	ANALYSIS OF COOLING TOWERS BY THE MATRIX FINITE ELEMENT METHOD- 2	70-0148
METHOD-LINEAR	THE FINITE ELEMENT METHOD-LINEAR AND NONLINEAR APPLICATIONS	74-0337
METHODE	A FINITE ELEMENT METHOD FOR RANDOM FOLDED PLATE STRUCTURES (EINE FINITE ELEMENT METHODE ZUR BERECHNU	73-0787
METHODE	ANALYSE DE LA STABILITY DES TALUS ROCHEUX PAR UNE METHODE GENERALISEE DE L EQUILIBRE LIMITE (ROCK SL	74-1019
METHODE	ANALYSE DES VOILES MINCES DANS LES DOMAINES LINEAIRES ET GEOMETRIQUEMENT NONLINEAIRES PAR LA METHODE	71-0122
METHODE	ANALYSE DYNAMIQUE DUALE DES STRUCTURES PAR LA METHODE DES ELEMENTS FINIS	73-0623
METHODE	ANALYSE NUMERIQUE D'UN PROBLEME DE STEFAN A DEUX PHASES PAR UNE METHODE D'ELEMENTS FINIS. (NUMERICAL	75-0387
METHODE	ANALYSE STATIQUE ET DYNAMIQUE DES COQUES PAR LA METHODE DES ELEMENTS FINIS	74-0509
METHODE	ANWENDUNG DER METHODE FINITER ELEMENTE BEI DER ANALYSE DES DYNAMISCHEN VERHALTENS GEDAEMPFTER WERKZE	75-0580
METHODE	APPLICATION DE LA METHODE DES ELEMENTS FINIS AU CALCUL DES PLAQUES FLECHIES (APPLICATION OF THE FINI	73-0019
METHODE	APPLICATION DE LA METHODE DES ELEMENTS FINIS - UN PROCEDE DE SOUS-ASSEMBLAGE	74-0405
METHODE	APPLICATION DE LAT METHODE DES ELEMENTS FINIS A LA RESOLUTION D'UN PROBLEME DE DOMAINE OPTIMAL	74-0412
METHODE	APPLICATIONS DE LA METHODE DES ELEMENTS FINIS A LA FLEXION DES PLAQUES	69-0511
METHODE	APPLICATIONS DE LA METHODE DE GAUSS-SEIDEL A LA RESOLUTION DES PROBLEMS D'ELASTICITE. (APPLICATION O	72-0145
METHODE	APPROXIMATIONS EXTERNES, PAR ELEMENTS FINIS DE LAGRANGE D ORDRE, UN ET DEUX, DU PROBLEME DE DIRICHLE	72-0495
METHODE	AUFBEREITUNG MATHEMATISCHER MODELLE FUR DIE DIGITALE BERECHNUNG VON GRUNDWASSERSTROMUNGSPROBLEMEN NA	71-0543
METHODE	BERECHNUNG DER ELASTIZITATSEIGENSCHAFTEN VON MASCHINENBAUTEILEN NACH DER METHODE FINITER ELFMENTE	73-1046
METHODE	BERECHNUNG DER ERDEBENSCHWINGUNGEN VON STRUKTUREN MIT DER FINITE ELEMENT METHODE. MECHANISCHE MODELL	71-0170
METHODE	BERECHNUNG EINES EINACHSIGEN STAHLDRAHT-KUNSTSTOFFVERBUNDES UNTER EINSATZ DER FINIT-ELEMENT METHODE	70-0644
METHODE	BERECHNUNG INSTATIONARER GRUND - UND SICKERWASSERSTROMUNGEN MIT FREIER OBERFLACHE NACH DER METHODE F	71-0541
METHODE	BERECHNUNG VON LAUFRAEDERN MIT DER METHODE DER FINITEN ELEMENTE (DESIGN OF ROTATING WHEELS BY THE FI	74-0547
METHODE	BERECHNUNGSBEISPIELE UND ERFAHRUNGEN MIT DER METHODE DER FINITEN ELEMENTE IN MASCHINENBAU. (EXAMPLES	71-0390
METHODE	CALCUL DES CHAMPS ELECTROMAGNETIQUES PAR LA METHODE DES ELEMENTS FINIS (COMPUTATION OF ELECTROMAGNET	73-0399
METHODE	CALCUL DES PONTS BIAIS A POUTRES MULTIPLES SANS ENTRETOISES PAR LA METHODE DES ELEMENTS (CALCULATION	72-0184
METHODE	CALCUL NUMERIQUE DES ECOULEMENTS DES FLUIDES DE BINGHAM ET DES FLUIDES NEWTONIENS INCOMPRESSIBLES PA	72-0349
METHODE	CHARACTERISATION ELASTIQUE D'UN STRATIFIE RESINE-VERRE PAR LA METHODE DES DEPLACEMENTS (DETERMINATIO	75-0272
METHODE	DAS PROGRAMMSYSTEM ANTRAS EDV-SYSTEM ZUR BERECHNUNG BELIEBIGER STATISCHER SYSTEME NACH DER METHODE D	74-0918
METHODE	DIE BERECHNUNG DER BEWEGUNGEN UND SPANNUNGEN DES GEBIRGES VOM ABBAU BIS ZUR TAGESOBERFLAECHE NACH DE	73-0456
METHODE	DIE BERECHNUNG VON FLACHWASSERWELLEN NACH DER METHODE DER FINITEN ELEMENTE	72-0359
METHODE	DIE LOSUNG NICHT-LINEARER PROBLEME NACH DER METHODE DER FINITEN ELEMENTE	74-0673
METHODE	DIE METHODE DER ENDLICHEN ELEMENTE UND IHRE ANWENDUNG IM MASCHINENBAU	71-0628
METHODE	EIN NEUES VARIATIONSPRINZIP MIT ANWENDUNG AUF SCHWINGUNGEN FLACHER SCHALEN NACH DER METHODE DER FINI	70-0563
METHODE	ETUDE DE LA CAPACITE PORTANTE DES PIEUX PAR LA METHODE DES ELEMENTS FINIS. INFLUENCE DE LA DILATANCE	75-0624
METHODE	ETUDE NUMERIQUE DU CHAMP MAGNETIQUE DANS UN ALTERNATEUR TETRAPOLAIRE PAR LA METHODE DES ELEMENTS FIN	74-0413
METHODE	ETUDES DES POUTRES ET POTEAUX PAR LA METHODE DES ELEMENTS FINIS. (STABILITY ANALYSIS OF BEAMS AND C	74-1008
METHODE	FEHLERBETRACHTUNGEN AUFGRUND DER METHODE DER GEWICHTETEN RESIDUEN FUR DAS VERFAHREN DER FINITEN ELEM	72-0368
METHODE	FESTIGKEITSUNTERSUCHUNG MIT DER METHODE DER FINITEN ELEMENTE. (STRENGTH OF THE MATERIALS INVESTIGATE	74-1009
METHODE	FINEL. UNIVERSELLES PROGRAMMSYSTEM ZUR BERECHNUNG DER ELASTISCHEN EIGENSCHAFTEN VON MASCHINENBAUTEIL	72-0132
METHODE	FINITE ELEMENT METHOD IN NONLINEAR DOMAIN. (METHODE DES ELEMENTS FINIS DANS LE DOMAINE NON LINEAIRE)	71-0288
METHODE	FONCTIONS SPLINES ET METHODE D ELEMENTS FINIS (SPLINE FUNCTIONS AND FINITE ELEMENT METHOD)	75-0846
METHODE	FORMULATIONS DUALES DES ENERGIES POTENTIELLES ET COMPLEMENTAIRES. APPLICATION A LA METHODE DES ELEME	71-0752
METHODE	GRUNDLAGEN DER DIGITALEN LOSUNG VON GRUNDWASSERSTROMUNGSPROBLEMEN NACH DER METHODE ENDLICHER ELEMENT	71-0542
METHODE	KOMBINATION EINES INTEGRALGLEICHUNGS VERFAHRENS MIT METHODE DER FINITEN ELEMENTEN ZUR BERECHNUNG EBE	74-0856
METHODE	KONSTRUKTION UND BERECHNUNG MIT DER METHODE FINITE ELEMENTE	73-0707
METHODE	LA METHODE DES ELEMENTS FINIS	72-0724
METHODE	LA METHODE DES ELEMENTS FINIS POUR LES PROBLEMES AUX LIMITES ELLIPTIQUES	74-1228
METHODE	LA METHODE FRONTALE POUR LA RESOLUTION DES SYSTEMES LINEARESS. (FRONTAL METHOD FOR SOLVING LINEAR SY	73-1097
METHODE	LES FONCTIONS DE TENSION DANS LA METHODE DES ELEMENTS FINIS	73-1113
METHODE	METHODE DER ELASTOSTATIK PROGRAMMIERUNG UND ANWENDUNG	70-0381
METHODE	METHODE DES ELEMENTS FINIS; UNE FAMILE D'APPROXIMATION DES ESPACES DE SOBOLEV PAR LES TRANSLATES DE	70-0301
METHODE	METHODE DES ELEMENTS FINIS	73-1053
METHODE	PARAMETRISCHE KERBSPANNUNGSUNTERSUCHUNGEN AN DER LOCHSCHEIBE MIT DER METHODE DER FINITEN ELEMENTE. (	75-0603
METHODE	PRAKTISCHE BEISPIELE DER DIGITALEN BERECHNUNG VON GRUNDWASSERSTROMUNGS PROBLEMEN NACH DER METHODE EN	73-0081
METHODE	PROGRAMME ZUR METHODE DER FINITEN ELEMETE	69-0416
METHODE	STABILITAT BEI DER METHODE DER FINITEN ELEMENTE FUR QUASILINEARE ELLIPTISCHE RANDWERTPROBLEME	74-1283
METHODE	STARR-PLASTISCHE TRAGLASTBERCHNUNGEN MITTELS DER METHODE DER FINITEN ELEMENTE	71-0635
METHODE	SURE UNE METHODE D'APPROXIMATION EXTERNE PAR ELEMENTS FINIS D ORDRE DEUX	72-0742
METHODE	UBER DIE ANWENDUNG DER ENERGETISCHEN METHODE AUF STABILITATSPROBLEME	38-0001

288

METHODE	UNE METHODE DE CALCUL PAR ELEMENTS FINIS DES MOUVEMENTS DE LIQUIDES DANS DES RESERVOIRS RIGIDES OU D	74-1171
METHODE	UNE METHODE VARIATIONNELLE ELEMENTS FINIS POUR LA RESOLUTION NUMERIQUE UN PROBLEME EXTERIEUR DANS R*	73-0576
METHODE	VARIATIONAL PRINCIPLES OF ELASTOSTATICS WITH RELAXED CONTINUITY REQUIREMENTS AND THEIR APPLICABILITY	74-0758
METHODEN	ANWENDUNG DER METHODEN DER FINITEN ELEMENTE MIT DEM PROGRAMMSYSTEM ASKA UNTER BESONDERER BERUCKSICHT	71-0624
METHODES	LE CALCUL DES TABLIERS DE PONTS PAR DES METHODES NUMERIQUES	72-0197
METHODES	QUELQUES METHODES D'ELEMENTS FINIS POUR DE PROBLEME D'UNE PLAQUE ENCASTREE	74-0406
METHODES	RESOLUTION NUMERIQUE PAR DES METHODES D ELEMENTS FINIS DE PROBLEMES ELLIPTIQUES DEGENERES	72-0702
METHODOLOGY	ENGINEERING OPERATIONS REPORT: NERVA THIN SHELL FINITE ELEMENT STRESS ANALYSIS METHODOLOGY COMPUTER	72-0916
METHODOLOGY	ENGINEERING OPERATIONS REPORT: NERVA AXISYMMETRIC AND PLANE STRUCTURES FINITE ELEMENT STRESS ANALYSI	72-0917
METHODOM	ISSLEDOVANIE PROCHNOSTI I ZHESTKOSTI PLASTIN, NAGRUZHENNYKH V SVOEI PLOSKOSTI, METHODOM KONECHNYKH E	73-0382
METOD	CHISLENNYI METOD RASCHETA SOBSTVENNYKH I VYNUZHDENNYKH KOLEBANII SOSTAVNYKH OBOLOCHECHNYKH KONSTRUKT	73-0461
METOD	METOD KONECHNYKH ELEMENTOV V PRIMENENII K RASCHETU TSILINDRICHESKIKH OBOLOCHEK S PRYAMOUGOL'NYMI OTV	73-0536
METOD	METOD KONECHNYKH ELEMENTOV V RASCHETAKH KONSOLIDATSII VODONASYSHCHENNYKH GRUNTOV. (FINITE ELEMENT ME	75-0469
METODE	K VOPROSU OB AVTOMATIZATSII ZADANIYA INFORMATSII V METODE KONECHNYKH ELEMENTOV. (PROBLEM OF AUTOMATI	75-0382
METODI	SU LA DETERMINAZIONE DEI CAMPI MAGNETICI LINEARI E NON LINEARI MEDIANTE L APPLICAZIONE DEI METODI DE	74-1061
METODOM	CALCULATION OF THE REINFORCED THIN-WALLED STRUCTURES BY THE METHOD OF FINITE ELEMENTS (RASCHET PODKR	75-0685
METODOM	ISSLEDOVANIE NAPRYAZHENNOGO SOSTOYANIYA REBRISTYKH ISILINDRICHESKIKH OBOLOCHEK S PRYAMOUGOL'NYMI OTV	74-0944
METODOM	OPREDELENIE STATSIONARNYKH TEMPERATURNYKH POLEI V DETALYAKH DVIGATELEI VNUTRENNEGO SGORANIYA METODOM	73-0454
METODOM	RASCHET METODOM KONECHNYKH ELEMENTOV DINAMICHESKIKH KHARAKTEERISTIK ZEMLYANOI PLOTINY SOVMESTNO S OS	73-0437
METODOM	RASCHET USTOICHIVOSTI IZGIBNO-KRUTILNYKH KOLEBANIJ KRULA V DOZVUKOVOM POTOKE METODOM KONECHNYKH ELF	72-0903
METODOM	RASHET TREKHMERNYKH MAGNITNYKE POLEI METODOM KONECHNYKH ELEMENTOV. (CALCULATION OF THREE-DIMENSIONAL	75-0766
METODOU	GEOMETRICKY NELINEARNE ULOHY PLOCHYCH SKRUPIN RIESENE PRIRASTKOVOU METODOU KONECNYCH PRVKOV. (GEOMET	74-1032
METODOU	RIESENIE VAZKOPRUZNYCH ANIZOTROPICKYCH DOASK METODOU KONECNYCH PRVKOV (SOLUTION OF VISCOELASTIC ANIS	74-0477
METODOU	VELKE PRIEHYBY DOSIEK METODOU KONECNYCH PRVKOV (LARGE DEFLECTIONS OF PLATES ANALYZED BY FINITE ELEME	73-0373
METODOV	VYPOCET KRUHOVYCH LOMENICOVYCH KONSTRUKCII METODOU KONECNYCH PASOV. (CALCULATION OF CIRCULAR FOLDED-	74-0612
METODY	THE DEVELOPMENT OF APPLIED METHODS IN PROBLEMS OF STATIC CALCULATION OF THIN-WALLED THREE-DIMENSIONA	73-0934
METODY	METODY RASCHETNOGO OPREDELENIYA TEMPERATURNYKH NAPRYAZHENII V KRYSHKAKH TSILINDROV DVIGATELEI VNUTRE	74-0179
METODY	VYUZITI METODY KONECNYCH PRVKU PRO PROJEKTOVANI KOMBINOVANYCH KONSTRUCKCI (FINITE ELEMENT METHOD IDE	73-0667
METODY	ZASTOSOWANIE METODY SZTYWNYCH ELEMENTOW SKONCZONYCH DO OBLICZEN ORGAN WALOW OKRETOVYCH Z UWZGLEDNIENI	74-0962
METRISCHER	DIE BERECHNUNG ACHSENSY METRISCHER TRAGWERKE UNTER ACHSENSYMETRISCHER UND NIGHT-SYMMETRISHCER BELAST	69-0452
METRO	APPLICATION OF NEW AUSTRIAN TUNNELLING METHOD IN DIFFICULT BUILTOVER AREAS IN FRANKFURT/MAIN METRO	73-1025
MHD	A USEFUL FORM OF THE VARIATIONAL PRINCIPLE OF IDEAL MHD FOR ONE-DIMENSIONAL NUMERICAL STABILITY CALC	73-0746
MHD	STUDY OF THE MHD INSTABILITIES OF CYLINDRICAL BELT PINCH USING THE FINITE ELEMENT METHOD	74-0700
MHD	UNSTEADY MHD DUCT FLOW BY THE FINITE ELEMENT METHOD	73-0263
MHD	USEFUL FORM OF THE VARIATIONAL PRINCIPLE OF IDEAL MHD FOR ONE-DIMENSIONAL NUMERICAL STABILITY CALCUL	73-0946
MICA	MICA DAM EMBANKMENT STRESS ANALYSIS	75-0054
MICRO-BUCKLING	AN INVESTIGATION OF MICRO-BUCKLING USING PHOTOELASTIC AND FINITE-ELEMENT METHODS	72-0960
MICROCRACKS	ENERGY CONDITION FOR INITIATION OF INTERFACIAL MICROCRACKS AT INCLUSIONS	74-0546
MICROELASTICITY	FINITE ELEMENT APPROACH TO PLANE MICROELASTICITY	72-0205
MICROMECHANICAL	MICROMECHANICAL ANALYSIS OF CRACK PROPAGATION IN AN ELASTOPLASTIC COMPOSITE MATERIAL	74-0565
MICROMECHANICAL	PRACTICAL PROBLEMS ASSOCIATED WITH THE APPLICATION OF THE FINITE ELEMENT METHOD TO COMPOSITE MATERIA	74-0483
MICROMECHANICAL	STRESS-WAVE PROPAGATION AND DYNAMIC EFFECTS IN SHELLS (SPADES) MICROMECHANICAL STRESS-WAVE RESPONSE	72-0851
MICROMECHANICS	APPLICATION OF THREE-DIMENSIONAL FINITE ELEMENT ANALYSES TO THE MICROMECHANICS OF FIBROUS COMPOSITE	74-1079
MICROMECHANICS	INELASTIC MICROMECHANICS OF CURING STRESSES IN COMPOSITES	75-0284
MICROMECHANICS	MICROMECHANICS ANALYSIS OF COMPOSITE MATERIALS USING FINITE ELEMENT METHODS	72-0770
MICROMECHANICS	THEORETICAL POST YIELDING BEHAVIOUR OF COMPOSITE LAMINATES - 1. INELASTIC MICROMECHANICS	73-0168
MICROPOLAR	BENDING OF MICROPOLAR PLATES	74-0263
MICROPOLAR	BOUSSINESQ PROBLEM OF PLANE MICROPOLAR ELASTICITY	74-0197
MICROPOLAR	FINITE ELEMENT ANALYSIS OF A THICK WALLED MICROPOLAR CYLINDER LOADED AXISYMMETRICALLY	73-0847
MICROPOLAR	FINITE ELEMENT ANALYSIS OF INCOMPRESSIBLE MICROPOLAR VISCOUS FLUID FLOW	75-0866
MICROPOLAR	FINITE ELEMENT APPROACH TO BENDING OF MICROPOLAR PLATES	74-0255
MICROSCOPIC	COMBINED MACROSCOPIC AND MICROSCOPIC APPROACH TO THE FRACTURE OF METALS	74-0897
MICROSTRIP	FINITE-ELEMENT CURRENT-DENSITY REPRESENTATION IN NUMERICAL-SOLUTION OF MICROSTRIP PROBLEMS	75-0400
MICROSTRIP	HYBRID MODE ANALYSIS OF MICROSTRIP BY FINITE ELEMENT METHODS	71-0465
MICROWAVE	FINITE ELEMENT ANALYSIS OF PLANAR MICROWAVE NETWORKS	73-0016
MICROWAVE	IEEE-G-MTT INTERNATIONAL MICROWAVE SYMPOSIUM, 1973: APPLICATIONS IN THE 70'S	73-0374
MICROZONATION	APPLICATION OF FINITE ELEMENT METHOD TO A MICROZONATION STUDY OF SALONICA REGION	73-0610
MID-INCREMENT	THE USE OF FINITE ELEMENT MID-INCREMENT STIFFNESS MATRICES IN THE POST BUCKLING ANALYSIS OF IMPERFEC	71-0383
MIDCOHV	THE MIDCOHV WEIGHT AND BALANCE COMPUTER PROGRAM (WTBAL)	74-0697
MIDDLE-PLANE	FREE VIBRATIONS OF FINITE ELEMENT PLATES SUBJECTED TO COMPLEX MIDDLE-PLANE FORCE SYSTEMS	72-0280
MIDDLE-PLANE	FREE VIBRATIONS OF FINITE ELEMENT PLATES SUBJECTED TO COMPLEX MIDDLE-PLANE FORCE SYSTEMS	73-0097
MIDDLE-PLANE	FREE VIBRATIONS OF FINITE ELEMENT PLATES SUBJECTED TO COMPLEX MIDDLE-PLANE FORCE SYSTEMS - REPLY	73-0122
MIDSIDE	A TECHNIQUE FOR DEGENERATING BRICK-TYPE ISOPARAMETRIC ELEMENTS USING HIERARCHICAL MIDSIDE NODES	74-0403
MIDSURFACE	MATRIX DISPLACEMENT ANALYSIS OF SHELLS WITH AN AXISYMMETRIC MIDSURFACE BUT HAVING CIRCUMFERENTIALLY	68-0031
MILIEU	ETUDE DE LA CAPACITE PORTANTE DES PIEUX PAR LA METHODE DES ELEMENTS FINIS. INFLUENCE DE LA DILATANCE	75-0624
MILITARY	EARTH VIBRATION EFFECTS AND ABATEMENT FOR MILITARY FACILITIES. REPORT 3. ANALYSIS METHOD FOR FOOTING	72-0806
MILITARY	THE USE OF A LASER FOR ARPA MILITARY GEOPHYSICS PROGRAM (ROCK MECHANICS AND RAPID EXCAVATION)	72-0800
MILL-TAILING	DETERMINING SEEPAGE CHARACTERISTICS OF MILL-TAILING DAMS BY THE FINITE ELEMENT METHOD	71-0477
MILLION-HORSEPO	BUILDING THREE NEW MILLION-HORSEPOWER HYDRAULIC TURBINES FOR GRAND COULEE	74-0981
MIMIC	MIMIC SOURCE APPROACH TO VISCOELASTIC ANALYSIS	69-0164
MINCES	ANALYSE DES BOILES MINCES PAR ELEMENTS FINIS COURBES	69-0524
MINCES	ANALYSE DES VOILES MINCES DANS LES DOMAINES LINEAIRES ET GEOMETRIQUEMENT NONLINEAIRES PAR LA METHODE	71-0122
MINCES	DEPLACEMENTS DANS LES STRUCTURES MINCES A FAIBLE RIGIDITE GEOMETRIQUE-DETERMINATION NUMERIQUE (NUMER	72-0213
MINCES	LES PLAQUES ET LES COQUES MINCES OU EPAISSES (THIN AND THICK PLATES AND SHELLS)	71-0432
MINE	A FINITE ELEMENT ANALYSIS OF DISSOLVED OXYGEN DRAWDOWN AND SULFATE PRODUCTION IN STRIP MINE SPOIL DA	75-0795
MINE	A FINITE ELEMENT MODEL FOR STABILITY ANALYSIS OF MINE WASTE EMBANKMENTS SUBJECTED TO SEEPAGE	74-1166
MINE	A MULTI-LAYERED FINITE ELEMENT MODEL FOR PREDICTING MINE SUBSIDENCE	68-0293
MINE	COMPUTER PROGRAM FOR GENERATING FINITE ELEMENT MODELS OF MINE STRUCTURES	71-0493
MINE	DESTRESSING TEST AT THE GALENA MINE, WALLACE, IDAHO	72-0126
MINE	FINITE ELEMENT ANALYSIS OF MINE STRUCTURES	72-0687
MINE	INFLUENCE OF ROCK PROPERTIES VARIABILITY ON MINE OPENING STABILITY ANALYSIS	74-1021
MINE	STRESS DISTRIBUTION AND ROCK FRACTURE ZONES IN THE ROOF OF LONGWALL FACE IN COAL MINE	73-0247
MINE	STRUCTURAL ANALYSIS OF A COAL MINE OPENING IN ELASTIC, MULTILAYERED MATERIAL	74-0235
MINE	STRUCTURAL ANALYSIS OF A COAL MINE OPENING IN ELASTIC, MULTILAYERED MATERIAL	74-0840
MINE	THREE-DIMENSIONAL CONSIDERATION OF THE STRESSES SURROUNDING OPEN-PIT MINE SLOPES	73-0402
MINE-WASTE	ELASTIC-PLASTIC STABILITY ANALYSIS OF MINE-WASTE EMBANKMENTS	75-0835
MINES	VORAUSSAGE DES GEBIRGSDRUCKES UND SEINER AUSQIRKUNGEN AUF STREBEN UND STRECKEN IM STEINKOHLENBERGBAU	73-0679
MINIMAL	ERROR ESTIMATES FOR A FINITE-ELEMENT APPROXIMATION OF A MINIMAL SURFACE	75-0491
MINIMIZATION	A PENALTY FUNCTION APPROACH TO THE MINIMIZATION OF QUADRATIC FUNCTIONALS IN FINITE ELEMENT ANALYSIS	74-0418
MINIMIZATION	AN AUTOMATIC NODE RELABELLING SCHEME FOR BANDWIDTH MINIMIZATION OF STIFFNESS MATRICES	68-0256
MINIMIZATION	AN AUTOMATIC RELABELING ALGORITHM FOR BANDWIDTH MINIMIZATION	75-0763
MINIMIZATION	BANDWIDTH MINIMIZATION OF STIFFNESS MATRICES	70-0045
MINIMIZATION	DEVELOPMENTS IN DISCRETE ELEMENT FINITE DEFLECTION STRUCTURAL ANALYSIS BY FUNCTION MINIMIZATION	68-0144
MINIMIZATION	DEVELOPMENTS IN DISCRETE ELEMENT FINITE DEFLECTION STRUCTURAL ANALYSIS BY FUNCTION MINIMIZATION	68-0243
MINIMIZATION	DEVELOPMENTS IN STRUCTURAL ANALYSIS BY DIRECT ENERGY MINIMIZATION	68-0070
MINIMIZATION	DEVELOPMENTS IN STRUCTURAL ANALYSIS AND SYNTHESIS BY FUNCTION MINIMIZATION	71-0384
MINIMIZATION	FINITE ELEMENT ANALYSIS OF SHELLS OF REVOLUTION BY MINIMIZATION OF THE POTENTIAL ENERGY FUNCTIONAL	69-0114
MINIMIZATION	FINITE ELEMENT ANALYSIS OF SHELLS OF REVOLUTION BY MINIMIZATION OF THE POTENTIAL ENERGY FUNCTIONAL	69-0354
MINIMIZATION	FINITE ELEMENT ANALYSIS OF SHELLS OF REVOLUTION BY MINIMIZATION OF THE POTENTIAL ENERGY FUNCTIONAL	70-0481
MINIMIZATION	FUNCTIONAL MINIMIZATION IN NONLINEAR SOLID MECHANICS	74-1251
MINIMIZATION	INTERPRETATION OF FINITE ELEMENT PROCEDURE AS STRESS ERROR MINIMIZATION PROCEDURE	72-0984
MINIMIZATION	OPTIMAL GRADIENT MINIMIZATION SCHEME FOR FINITE ELEMENT EIGENPROBLEMS	72-0292

289

MINIMIZATION	STRAIN HARDENING TORSION OF SOLID BARS AS A MINIMIZATION PROBLEM	70-0174
MINIMIZATION	THE APPLICATION OF GRADIENT MINIMIZATION METHODS AND HIGHER ORDER DISCRETE ELEMENTS TO SHELL BUCKLIN	71-0237
MINIMIZATION	THE COMPUTATIONAL EFFICIENCY OF A NEW MINIMIZATION ALGORITHM FOR EIGENVALUE ANALYSIS	71-0595
MINIMIZING	NOTE ON MINIMIZING THE BANDWIDTH OF SPARSE, SYMMETRIC MATRICES	73-1107
MINIMUM	APPLICATION OF DUAL MINIMUM THEOREMS TO THE FINITE ELEMENT SOLUTION OF POTENTIAL PROBLEMS WITH SPECI	72-0036
MINIMUM	METHOD OF MINIMUM WEIGHT DESIGN WITH REQUIREMENTS IMPOSED ON STRESSES AND NATURAL FREQUENCIES	70-0149
MINIMUM	MINIMUM WEIGHT DESIGN WITH STABILITY CONSTRAINT	70-0274
MINIMUM	MINIMUM WEIGHT DESIGN OF COMPLEX STRUCTURES SUBJECT TO A FREQUENCY CONSTRAINT	70-0374
MINIMUM	MINIMUM WEIGHT DESIGN OF FINITE ELEMENT STRUCTURES	74-0148
MINIMUM	OPTIM 11: (MAGIC) COMPATIBLE LARGE SCALE AUTOMATED MINIMUM WEIGHT DESIGN PROGRAM - V 2. PROGRAMMER'S	74-1036
MINIMUM	OPTIM 11: A (MAGIC) COMPATIBLE LARGE SCALE AUTOMATED MINIMUM WEIGHT DESIGN PROGRAM - 1. ENGINEERS AN	74-1048
MINIMUM	OPTIM 11: A MAGIC COMPATIBLE LARGE-SCALE AUTOMATED MINIMUM WEIGHT DESIGN PROGRAM. VOLUME 1. ENGINEER	74-1096
MINIMUM	OPTIM 11: MAGIC COMPATIBLE LARGE SCALE AUTOMATED MINIMUM WEIGHT DESIGN PROGRAM. VOLUME 11. PROGRAMME	74-1093
MINIMUM	OPTIMIZATION OF FINITE ELEMENT GRIDS BASED ON MINIMUM POTENTIAL ENERGY	71-0747
MINIMUM	OPTIMIZATION OF FINITE ELEMENT GRIDS BASED ON MINIMUM POTENTIAL ENERGY	72-0109
MINIMUM	SIMPLIFIED MINIMUM INTERNAL ENERGY OF DEFORMATION POSTULATE AND ITS APPLICATION TO METALFORMING PROC	74-1155
MINING	BERECHNUNG UND BEURTEILUNG DER STANDSICHERHEIT VON ABBAUHOHLRAEUMEN IN STEILEN GANGERZLAGERSTAETTEN	74-0597
MINING	DIE BERECHNUNG DER BEWEGUNGEN UND SPANNUNGEN DES GEBIRGES VOM ABBAU BIS ZUR TAGESOBERFLAECHE NACH DE	73-0456
MINING	FORECASTING THE ROCK PRESSURE AND ITS EFFECT ON THE LONGWALL FACES AND ROADWAYS IN THE BITUMINOUS CO	74-0374
MINING	GEOMETRIC METHOD FOR THE PREDICTION OF STRESSES IN INCLUSIONS, OREBODIES, AND MINING SYSTEMS	74-0219
MINING	NEUARTIGE ANWENDUNG GEOMECHANISCHER UNTERSUCHUNGSMETHODEN ZUR BEURTEILUNG DER ABBAUWIRKUNGEN IM GANG	73-0530
MINING	ROCK MECHANICS AND RISK IN OPEN PIT MINING	73-0544
MINING	SIMULATION OF PROGRESSIVE FAILURE IN ROCK AROUND MINING EXCAVATIONS	70-0146
MINING	THEORETICAL INVESTIGATION OF ROCK AND SUPPORT INTERACTION - DEVELOP MORE RATIONAL DESIGN METHODS AND	74-0782
MINUTEMAN	MINUTEMAN STV STRESS ANALYSIS AND TESTING	74-0809
MIRROR	DESIGN OF OPTICAL MIRROR STRUCTURES	71-0658
MIRROR	STRUCTURAL EVALUATION OF CANDIDATE DESIGNS FOR THE LARGE SPACE TELESCOPE PRIMARY MIRROR	75-0708
MIRRORS	ANALYSIS OF THE DYNAMICS OF THIN PRIMARY MIRRORS FOR LARGE ASTRONOMICAL TELESCOPES	73-1084
MIRRORS	APPLICATION OF FINITE ELEMENT TECHNIQUE TO DESIGN AND MANUFACTURE OF ASTRONOMICAL MIRRORS	69-0302
MIRRORS	ELASTIC ANALYSIS OF LARGE SPACEBOUND MIRRORS	72-0619
MIRRORS	ELASTIC DEFORMATION OF LIGHTWEIGHT MIRRORS METHOD OF LIMIT POINT CALCULATION IN FINITE ELEMENT STRUC	72-0232
MISCIBLE	A FINITE ELEMENT MODEL FOR MISCIBLE DISPLACEMENT IN GROUNDWATER AQUIFERS	75-0783
MISE	DEVELOPMENT AND USE OF COMPUTER PROGRAMS FOR FINITE ELEMENT ANALYSIS. (LA MISE AU POINT ET L'UTILIZA	69-0289
MISSILE	MISSILE BODY INPUT GENERATOR (BING), A NASTRAN PRE-PROCESSOR; THEORETICAL DEVELOPMENT, USER'S MANUAL	75-0717
MISSILE	STATIC AND DYNAMIC ANALYSIS OF VISCOELASTOPLASTIC FIBER-REINFORCED COMPOSITE SHELLS IN MISSILE STRUC	73-0651
MISSILE	STRUCTURAL DYNAMIC PROPERTIES OF TACTICAL MISSILE JOINTS - PHASE 1	70-0423
MISSILE	STRUCTURAL DYNAMIC PROPERTIES OF TACTICAL MISSILE JOINTS - PHASE 3	74-0811
MISSILES	VIBRATION OF MISSILES	61-0007
MISSISSIPPI	SEEPAGE IN MISSISSIPPI RIVER BANKS. REPORT 1. ANALYSIS OF TRANSIENT SEEPAGE USING VISCOUS FLOW MODEL	70-0405
MISSISSIPPI	SEEPAGE IN MISSISSIPPI RIVER BANKS. REPORT 1. ANALYSIS OF TRANSIENT SEEPAGE USING A VISCOUS-FLOW MOD	73-0909
MITERED	THREE-DIMENSIONAL STRESS ANALYSES OF MITERED ELBOWS	75-0650
MITIGATION	RAY-EFFECT MITIGATION IN DISCRETE ORDINATE-LIKE ANGULAR FINITE-ELEMENT APPROXIMATIONS IN NEUTRON-TRA	75-0496
MITSUBISHI	FINITE ELEMENT COMPUTER ANALYSIS OF MITSUBISHI HULL 1670	70-0294
MITTELS	STARR-PLASTISCHE TRAGLASTBERECHNUNGEN MITTELS DER METHODE DER FINITEN ELEMENTE	71-0635
MITTLEREN	THEORETICAL AND EXPERIMENTAL CONSIDERATIONS ON FLOW OF VISCOUS LIQUIDS PAST PLATE AT SMALL AND AVERA	70-0144
MIXED	A MATHEMATICAL THEORY OF COMPLEMENTARY-DUAL VARIATIONAL PRINCIPLES AND MIXED FINITE ELEMENT APPROXIM	73-0804
MIXED	A MIXED FINITE ELEMENT METHOD FOR THIN SHELLS ANALYSIS	68-0182
MIXED	A MIXED FINITE ELEMENT SHELL FORMULATION	71-0063
MIXED	A MIXED FINITE ELEMENT FORMULATION FOR PLATE PROBLEMS	72-0556
MIXED	A MIXED FINITE ELEMENT MODEL FOR USE IN POTENTIAL FLOW PROBLEMS	73-0078
MIXED	A MIXED FINITE ELEMENT METHOD FOR THE BIHARMONIC EQUATION	74-1227
MIXED	A MIXED FINITE-ELEMENT SHALLOW SHELL FORMULATION	69-0078
MIXED	A SIMPLE MIXED METHOD FOR PLATE AND SHELL PROBLEMS	73-1005
MIXED	A THEORY OF MIXED FINITE ELEMENT APPROXIMATIONS OF NON-SELF-ADJOINT BOUNDARY-VALUE PROBLEMS	74-0340
MIXED	A THEORY OF MIXED FINITE ELEMENT APPROXIMATIONS OF NON-SELF-ADJOINT BOUNDARY VALUE PROBLEMS	74-0715
MIXED	ACCURACY AND CONVERGENCE OF MIXED FINITE-ELEMENT APPROXIMATIONS OF THIN BARS, MEMBRANES, AND PLATES	73-1120
MIXED	ANALYSIS OF CABLE NETS IN MIXED FORMULATION	73-0311
MIXED	COMPUTATION OF SHELLS WITH FINITE DISPLACEMENTS: MIXED FINITE ELEMENTS (BERECHNUNG VON SCHALEN MIT E	73-0788
MIXED	CONVERGENCE OF A MIXED FINITE ELEMENT SCHEME FOR PLATE BENDING	73-0023
MIXED	CONVERGENCE OF A MIXED FINITE-ELEMENT METHOD FOR PLATE BENDING PROBLEMS	73-0107
MIXED	CONVERGENCE OF ANOTHER MIXED FINITE ELEMENT METHOD FOR PLATE BENDING PROBLEMS	72-0367
MIXED	CONVERGENCE OF MIXED FINITE ELEMENT APPROXIMATIONS OF A CLASS OF LINEAR-BOUNDARY VALUE PROBLEMS	73-0599
MIXED	EIGENVALUE PROBLEMS WITH A ' MIXED ' PLATE ELEMENT	69-0081
MIXED	FINITE ELEMENT ANALYSIS OF PLATE BUCKLING USING A MIXED VARIATIONAL PRINCIPLE	71-0020
MIXED	FINITE ELEMENT GALERKIN METHOD FOR MIXED INITIAL-BOUNDARY VALUE PROBLEMS IN ELASTICITY THEORY	71-0520
MIXED	FINITE ELEMENT METHOD FOR HYDRODYNAMIC DISPERSION EQUATION WITH MIXED PARTIAL DERIVATIVES	72-0030
MIXED	FINITE ELEMENT METHOD OF MIXED TYPE AND ITS CONVERGENCE IN LINEAR SHELL PROBLEMS	73-1041
MIXED	FINITE-ELEMENT FORMULATION OF MIXED AND STRESS VARIATIONAL PRINCIPLES FOR STRUCTURAL DYNAMIC ANALYSI	72-0964
MIXED	FORMULATION AND APPLICATION OF CERTAIN PRIMAL AND MIXED FINITE ELEMENT MODELS OF FINITE DEFORMATIONS	74-0344
MIXED	GALERKIN FINITE ELEMENT SOLUTIONS FOR POLLUTION PROBLEMS IN PARTIALLY MIXED ESTUARIES	74-0047
MIXED	GENERAL MIXED FINITE ELEMENT METHODS OF NONLINEAR CONTINUA	72-0831
MIXED	GENERAL MIXED FINITE ELEMENT METHODS OF NONLINEAR CONTINUA	73-0091
MIXED	GENERALIZED CONJUGATE FUNCTIONS FOR MIXED FINITE ELEMENT APPROXIMATIONS OF BOUNDARY VALUE PROBLEMS	72-0515
MIXED	LINEAR AND NON-LINEAR ANALYSIS OF THIN SHALLOW SHELLS BY MIXED FINITE ELEMENTS	75-0665
MIXED	MIXED CONJUGATE FINITE ELEMENT APPROXIMATIONS OF LINEAR OPERATORS	72-0085
MIXED	MIXED DISPLACEMENT 3D ISOPARAMETRIC FINITE ELEMENTS AND A STRAIN ENERGY BASIS FOR STUDIES OF ELEMENT	70-0286
MIXED	MIXED FINITE ELEMENT ANALYSIS OF PLATES IN BENDING, SMALL DEFLECTION THEORY OF ELASTIC AND ELASTO-PL	71-0783
MIXED	MIXED FINITE ELEMENT ANALYSIS OF ELASTO-PLASTIC PLATES IN BENDING	72-0135
MIXED	MIXED FINITE ELEMENT METHOD FOR AXISYMMETRIC SHELLS	72-0149
MIXED	MIXED FINITE ELEMENT METHOD FOR VIBRATION ANALYSIS OF SHALLOW SHELLS	73-0593
MIXED	MIXED FINITE ELEMENT METHODS IN THEORY AND APPLICATIONS	73-1108
MIXED	MIXED FINITE ELEMENT MODELS IN LIMIT ANALYSIS	74-1314
MIXED	MIXED FINITE ELEMENT METHOD FOR BENDING OF PLATES	75-0229
MIXED	MIXED FINITE-ELEMENT APPROXIMATIONS OF LINEAR BOUNDARY-VALUE PROBLEMS	75-0453
MIXED	MIXED FORMULATIONS FOR FINITE ELEMENT SHELL ANALYSIS	71-0222
MIXED	MIXED ISOPARAMETRIC LAMINATED COMPOSITE SHELL ELEMENTS	75-0450
MIXED	MIXED MODELS FOR PLATES	70-0384
MIXED	MIXED QUADRILATERAL ELEMENTS FOR BENDING	72-0623
MIXED	MIXED VARIATIONAL PRINCIPLES, FINITE ELEMENTS AND FINITE ELASTICITY	74-1254
MIXED	NAVIER STOKES EQUATIONS USING MIXED INTERPOLATION	74-0014
MIXED	NUMERICAL SOLUTION OF THE BIHARMONIC PROBLEM BY MIXED FINITE OF CLASS C-0	73-0803
MIXED	ON MIXED FINITE ELEMENT APPROXIMATIONS	74-0701
MIXED	REFINED MIXED METHOD FINITE ELEMENTS FOR SHELLS OF REVOLUTION	71-0101
MIXED	ROLE OF PART-INVERSION IN FLUID-STRUCTURE PROBLEMS WITH MIXED VARIABLES	70-0209
MIXED	SIMPLE "MIXED" METHOD FOR PLATE AND SHELL PROBLEMS	74-0952
MIXED	SMALL SCALE YIELDING ANALYSIS OF MIXED MODE PLANE STRAIN CRACK PROBLEMS	73-0801
MIXED	SOLUTION OF MIXED BOUNDARY VALUE PROBLEMS IN FINITE DOMAINS BY FOURIERS METHOD	72-0790
MIXED	SOLUTION OF MIXED BOUNDARY VALUE PROBLEMS WITH LOCAL ERROR BOUND BY THE FINITE ELEMENT METHOD	73-0795
MIXED	SOLUTION OF THE MIXED PROBLEM OF THE THEORY OF ELASTICITY AND THE THEORY OF PLASTICITY OF SOILS	74-1136
MIXED	SOME CONTRIBUTIONS TO THE MATHEMATICAL THEORY OF MIXED FINITE ELEMENT APPROXIMATIONS	73-0302
MIXED	SOME MATHEMATICAL PROPERTIES OF CERTAIN MIXED GALERKIN APPROXIMATIONS IN NONLINEAR ELASTICITY	74-1248
MIXED	SOME OBSERVATIONS ON PROPERTIES OF CERTAIN MIXED FINITE ELEMENT APPROXIMATIONS	75-0617

MIXED	SOME RECENT DEVELOPMENTS IN THE THEORY AND APPLICATION OF FINITE ELEMENTS TO MIXED PROBLEMS AND SHOC	73-0637
MIXED	STABILITY OF PLATES USING A MIXED FINITE ELEMENT FORMULATION	74-0218
MIXED	THE APPLICATION OF A CURVED MIXED TYPE SHELL ELEMENT	70-0118
MIXED	THEMIS FINAL REPORT. VOL. IV. RECENT DEVELOPMENTS IN THE NUMERICAL SOLUTION OF CERTAIN PROBLEMS IN N	74-0721
MIXED	VIBRATION OF RECTANGULAR PLATES WITH MIXED BOUNDARY CONDITIONS	73-1086
MIXED-DISPLACEM	MIXED-DISPLACEMENT FINITE-ELEMENT ANALYSIS WITH PARTICULAR APPLICATION USING PLANE-STRESS TRIANGLES	72-0245
MIXED-MODE	APPLICATION OF FINITE ELEMENT METHOD TO MIXED-MODE FRACTURE	73-1123
MIXED-TYPE	A REFINED MIXED-TYPE PLATE BENDING ELEMENT	69-0267
MIXED-TYPE	THE APPLICATION OF A CURVED MIXED-TYPE SHELL ELEMENT	70-0440
MIXOPARAMETRIC	PROBLEMS ASSOCIATED WITH THE CONVERGENCE OF ISOPARAMETRIC AND MIXOPARAMETRIC FINITE ELEMENTS	71-0375
MIXTES	ANALYSE DES VOILES MINCES DANS LES DOMAINES LINEAIRES ET GEOMETRIQUEMENT NONLINEAIRES PAR LA METHODE	71-0122
MIXTURES	ANALYSIS OF FATIGUE OF PAVING MIXTURES FROM THE FRACTURE MECHANICS VIEWPOINT	72-0540
MIXTURES	FINITE ELEMENT METHOD APPLIED TO IDEAL GASEOUS MIXTURES.	72-0393
MKP	APLIKACE MKP NA RESENI PROBLEMU DIFUZE V KOVECH. (APPLICATION OF THE FINITE ELEMENT METHOD TO THE SO	74-0609
MOBIL	INTEGRATED PROGRAM FOR THE DYNAMIC STRUCTURAL CALCULATION OF MOBIL OFFSHORE UNITS	74-0490
MOBILITY	VIBRATION REDUCTION BY USING BOTH THE FINITE ELEMENT STRAIN ENERGY DISTRIBUTION AND MOBILITY TECHNIQ	74-1303
MODAL	A MODAL SYNTHESIS TECHNIQUE FOR DETERMINING DYNAMIC PROPERTIES FOR A STRUCTURE FOR MASS AND STIFFNES	74-0319
MODAL	ACCURACY OF FINITE-ELEMENT SOLUTIONS FOR MODAL CHARACTERISTICS OF SHELLS OF REVOLUTION	70-0231
MODAL	ACCURACY OF MODAL STRESS CALCULATIONS BY FINITE ELEMENT METHOD	70-0223
MODAL	ANALYTICAL METHODS IN COMPONENT MODAL SYNTHESIS	75-0824
MODAL	CALCULATION OF AXISYMMETRIC LONGITUDINAL MODES FOR FLUID-ELASTIC TANK-ULLAGE GAS SYSTEMS AND COMPARI	64-0019
MODAL	COMPUTATION OF VIBRATION MODAL SHAPES AND NATURAL FREQUENCIES OF A GENERAL ORTHOTROPIC LAMINATED, TH	71-0309
MODAL	FAST MODAL EXTRACTION IN NASTRAN VIA THE FEER COMPUTER PROGRAM	73-0356
MODAL	IN-VACUO MODAL DYNAMIC RESPONSE OF THE HUMAN SKULL	73-0460
MODAL	MATHEMATICAL MODELING OF SPINNING ELASTIC BODIES FOR MODAL ANALYSIS	73-0441
MODAL	MODAL ANALYSIS OF A SATELLITE PRIMARY STRUCTURE USING A FINITE ELEMENT PROCEDURE	73-0624
MODAL	MODAL COUPLING PROCEDURES ADAPTED TO NASTRAN ANALYSIS OF THE 1/8 - SCALE SHUTTLE STRUCTURAL DYNAMICS	75-0677
MODAL	MODAL COUPLING PROCEDURES ADAPTED TO NASTRAN ANALYSIS OF THE 1/8-SCALE SHUTTLE STRUCTURAL DYNAMICS M	75-0678
MODAL	MODAL RESPONSE ANALYSIS OF CONSTRAINED FINITE-ELEMENTS ELASTIC CONTINUUM	75-0767
MODAL	MODAL RESPONSE OF FREE ROCKETS UNDER THRUST LOADING	74-0855
MODAL	MODAL SENSITIVITY STUDY OF A CONICAL RE-ENTRY VEHICLE SUBJECTED TO AERODYNAMICALLY INDUCED ACOUSTIC	69-0431
MODAL	STRUCTURAL DYNAMICS BY FINITE ELEMENTS: MODAL AND FOURIER ANALYSIS	73-0836
MODAL	VIBRATION ANALYSIS OF A 1/15-SCALE DYNAMIC MODAL OF A SPACE SHUTTLE CONFIGURATION	71-0609
MODAL	VIBRATION AND STRESS ANALYSIS OF SOFT-BONDED SHUTTLE INSULATION TILES. MODAL ANALYSIS WITH COMPACT W	74-0708
MODE	A FORTRAN PROGRAM FOR CALCULATING THE FREQUENCIES AND MODE SHAPES OF AN ARBITRARY SHAPE CYLINDRICAL	72-0594
MODE	A FORTRAN PROGRAM FOR CALCULATING THE FREQUENCIES AND MODE SHAPES OF A GUIDED CLAMPED, GUIDED PINNED	72-0890
MODE	A HYBRID METHOD OF COMPONENT MODE SYNTHESIS	71-0681
MODE	A MODE AMPLITUDE TECHNIQUE FOR GENERAL FINITE ELEMENT EQUILIBRIUM	74-0283
MODE	A STIFFNESS MATRIX METHOD OF NATURAL MODE ANALYSIS	58-0002
MODE	ACOUSTIC RESPONSE OF A SPACECRAFT SOLAR PANEL USING NORMAL MODE METHOD	69-0422
MODE	AN INVESTIGATION OF THE EFFECTS OF SQUARE CUTOUTS ON THE NATURAL FREQUENCIES AND MODE SHAPES OF RECT	71-0638
MODE	COMBINED MODE CRACK EXTENSION IN ADHESIVE JOINTS	72-0111
MODE	DEFORMATION MODE OF VOID-GROWTH AND COALESCENCE IN THE PROCESS OF DUCTILE FRACTURE	75-0683
MODE	ELASTIC DEFORMATION OF POLYCRYSTALLINE METAL - 2 EFFECT OF SLENDERNESS OF GRAINS ON DEFORMATION MODE	73-0275
MODE	ELASTO-PLASTIC SEISMIC RESPONSE ANALYSIS OF FRAMED STRUCTURES BY MODE SUPERPOSITION	73-0325
MODE	FAMSOR, A FINITE ELEMENT PROGRAM FOR THE FREQUENCIES AND MODE SHAPES OF SHELLS OF REVOLUTION	70-0090
MODE	HYBRID MODE ANALYSIS OF MICROSTRIP BY FINITE ELEMENT METHODS	71-0465
MODE	MODE SHAPES AND FREQUENCIES BY FINITE ELEMENT METHOD USING CONSISTENT AND LUMPED MASSES	71-0610
MODE	NATURAL FREQUENCIES AND MODE SHAPES OF TRANSFORMER CORES	65-0048
MODE	NATURAL FREQUENCIES AND MODE SHAPES OF PLATES WITH INTERIOR CUT-OUTS	70-0402
MODE	NEW FINITE ELEMENT APPROACH TO THE NORMAL MODE ANALYSIS IN MAGNETOHYDRODYNAMICS	74-0876
MODE	NONLINEAR DYNAMIC ANALYSIS OF TORSIONAL SHELLS BY COMBINED USE OF FINITE ELEMENT AND MODE SUPERPOSIT	75-0859
MODE	NORMAL MODE ANALYSIS OF THE RADIO ASTRONOMY EXLPORER BOOMS AND SPACECRAFT	71-0130
MODE	NORMAL MODE ANALYSIS OF A FUZE SUPPORT STRUCTURE USING NASTRAN. PART 1	72-0853
MODE	SKIN-STRINGER PANEL NORMAL MODE RESPONSE EXPERIMENTAL DATA	71-0691
MODE	SMALL SCALE YIELDING ANALYSIS OF MIXED MODE PLANE STRAIN CRACK PROBLEMS	73-0801
MODE	SOME GENERAL CONSIDERATIONS ON THE NATURAL MODE TECHNIQUE	69-0049
MODE	SOME GENERAL CONSIDERATIONS ON THE NATURAL MODE TECHNIQUE, PART II: LARGE DISPLACEMENTS	69-0445
MODE	VIBRATION ANALYSIS OF STRUCTURES BY COMPONENT MODE SUBSTITUTION	71-0303
MODEL	A COMPUTER MODEL FOR PREDICTING THE LOAD-DEFLECTION RESPONSE OF EXPEDIENT SOIL SURFACING	73-0880
MODEL	A DYNAMIC FINITE ELEMENT MODEL OF THE HUMAN BRAIN	74-1076
MODEL	A FINITE ELEMENT DETERMINISTIC CATCHMENT MODEL	74-1361
MODEL	A FINITE ELEMENT MODEL FOR ANISOTROPIC YIELDING IN GRAVITY LOADED ROCK	69-0012
MODEL	A FINITE ELEMENT MODEL FOR TWO-DIMENSIONAL STEADY FLOW THROUGH CONTRACTIONS IN NATURAL CHANNELS	71-0810
MODEL	A FINITE ELEMENT MODEL FOR PREDICTING LONG BONE FRACTURE FIXATION	71-0819
MODEL	A FINITE ELEMENT MODEL FOR LAMINATED PLATES INCLUDING TRANSVERSE SHEAR DEFORMATION	72-0818
MODEL	A FINITE ELEMENT MODEL FOR TWO DIMENSIONAL FLOW	74-0015
MODEL	A FINITE ELEMENT MODEL OF TIDES IN ESTUARIES	74-0044
MODEL	A FINITE ELEMENT MODEL FOR SEDIMENT TRANSPORT IN ESTUARIES	74-1102
MODEL	A FINITE ELEMENT MODEL FOR STABILITY ANALYSIS OF MINE WASTE EMBANKMENTS SUBJECTED TO SEEPAGE	74-1166
MODEL	A FINITE ELEMENT MODEL FOR PLASTICITY PROBLEMS	74-1225
MODEL	A FINITE ELEMENT MODEL FOR MISCIBLE DISPLACEMENT IN GROUNDWATER AQUIFERS	75-0783
MODEL	A FINITE-ELEMENT MODEL TO SIMULATE THE NONLINEAR RESPONSE OF REINFORCED CONCRETE FRAMES WITH MASONRY	73-1168
MODEL	A FINITE-ELEMENT, PLANAR-FLOW MODEL OF CAMAS PRAIRIE, IDAHO	72-0980
MODEL	A MATHEMATICAL MODEL OF THE CORNEO-SCLERAL SHELL	71-0150
MODEL	A MIXED FINITE ELEMENT MODEL FOR USE IN POTENTIAL FLOW PROBLEMS	73-0078
MODEL	A MODEL FOR THE MECHANICS OF JOINTED ROCK	68-0079
MODEL	A MODEL STUDY OF GAS DIFFUSION IN ALVEOLAR SACS	73-0029
MODEL	A MULTI-LAYERED FINITE ELEMENT MODEL FOR PREDICTING MINE SUBSIDENCE	68-0293
MODEL	A NASTRAN VIBRATION MODEL OF THE AH-1G HELICOPTER AIRFRAME. VOLUME 1	74-1338
MODEL	A NASTRAN VIBRATION MODEL OF THE AH-1G HELICOPTER AIRFRAME. VOLUME 11	74-1339
MODEL	A NEW ASSUMED STRESS HYBRID FINITE ELEMENT MODEL FOR SOLID CONTINUA	71-0371
MODEL	A NEW FINITE ELEMENT MODEL FOR SOLID CONTINUA	68-0156
MODEL	A NEW MODEL FOR ELASTIC-PLASTIC TRUSSES	74-0805
MODEL	A NUMERICAL MODEL FOR THE SIMULATION OF TWO DIMENSIONAL CONVECTIVE DISPERSION IN SHALLOW ESTUARIES	74-0052
MODEL	A NUMERICAL MODEL OF DISPERSION IN ESTUARIES	74-0045
MODEL	A NUMERICAL MODEL OF TRANSIENT WATER QUALITY IN A ONE-DIMENSIONAL ESTUARY BASED ON THE FINITE ELEMEN	74-1369
MODEL	A NUMERICAL THERMO-MECHANICAL MODEL FOR THE WELDING AND SUBSEQUENT LOADING OF A FABRICATED STRUCTURE	72-0847
MODEL	A RADIATIVE IGNITION MODEL OF A SOLID FUEL	74-1312
MODEL	A SIMPLE FINITE ELEMENT MODEL FOR ELASTIC-PLASTIC PLATE BENDING	72-0889
MODEL	A STUDY OF ELASTIC-PLASTIC ANALYSIS BY THE ASSUMED-STRESS HYBRID FINITE-ELEMENT MODEL, WITH APPLICAT	74-0718
MODEL	A STUDY OF INTERNAL STRESSES IN STATICALLY DEFORMED PNEUMATIC TIRES. A MATHEMATICAL MODEL FOR THE PN	70-0421
MODEL	A STUDY OF THE EQUILIBRIUM MODEL FOR USE IN FINITE ELEMENT ANALYSIS	66-0132
MODEL	A TECHNIQUE FOR CONNECTING BEAM ELEMENTS TO A PLATE MODEL OF A COMPLICATED BOX SECTION	74-0271
MODEL	A TETRA-CORE STRESS ANALYSIS MODEL	72-0845
MODEL	A THREE-DIMENSIONAL FINITE ELEMENT ANALYSIS OF A PRESTRESSED CONCRETE REACTOR VESSEL MODEL	69-0084
MODEL	A TWO-DIMENSIONAL AXISYMMETRIC SHELL FLUID MODEL FOR LAUNCH VEHICLE LONGITUDINAL DYNAMIC ANALYSIS	69-0046
MODEL	AIRCRAFT-PAVEMENT INTERACTION STUDIES. PHASE 1: A FINITE ELEMENT MODEL OF A JOINTED CONCRETE PAVEMEN	73-0894
MODEL	AN ASSUMED STRESS HYBRID FINITE ELEMENT MODEL FOR THE ANALYSIS OF AN AXISYMMETRIC THICK WALLED PRESS	71-0025
MODEL	AN EQUILIBRIUM MODEL FOR PLATE BENDING	68-0072
MODEL	AN EQUILIBRIUM STRESS FIELD MODEL FOR FINITE ELEMENT SOLUTIONS OF TWO-DIMENSIONAL ELASTIC PROBLEMS	68-0163
MODEL	ANALYSIS AND MODEL STUDIES OF UNDERGROUND OPENINGS IN JOINTED ROCK. RESEARCH ON ROCK BOLT REINFORCEM	71-0438

291

MODEL	SETTLEMENT ANALYSIS OF A STRIP FOOTING USING A CRITICAL STATE SOIL MODEL IN CONJUNCTION WITH FINITE	71-0533
MODEL	SIMPLE FINITE ELEMENT MODEL FOR ELASTIC-PLASTIC PLATE BENDING.	72-0447
MODEL	SINGLE-TRACK SUBWAY ENVIRONMENTAL SIMULATION MODEL. PHASE 1	71-0693
MODEL	SOME APPLICATIONS OF FINITE ELEMENTS TO MODEL GEOLOGICAL STRUCTURES	73-0143
MODEL	STRESS STRAIN RESPONSE AND FRACTURE OF A CONCRETE MODEL IN BIAXIAL LOADING	71-0295
MODEL	STRUCTURAL BEHAVIOR OF A TWO SPAN REINFORCED CONCRETE BOX GIRDER BRIDGE MODEL. VOLUME 11. REDUCTION,	71-0659
MODEL	STRUCTURAL MODEL FOR THE MECHANICAL BEHAVIOUR OF TRABECULAR BONE	73-0473
MODEL	STRUCTURAL MODEL STUDIES OF LARGE PIPE BIFURCATION	73-0283
MODEL	SYSTEMATIC ENFORCEMENT OF STRESS BOUNDARY CONDITIONS IN THE ASSUMED STRESS HYBRID MODEL BASED ON THE	71-0771
MODEL	THE ANALYSIS OF CLASSICAL PLATE BENDING PROBLEMS USING AN EQUILIBRIUM FINITE ELEMENT MODEL	68-0302
MODEL	THE HARMONIC FINITE ELEMENT MODEL FOR USE IN FIELD PROBLEMS	70-0389
MODEL	THE RING OF FINITE ELEMENTS IN A NON-STANDARD MODEL OF THE REALS	71-0450
MODEL	THE SCANDINAVIAN PORV MODEL PROJECT: STRESS CALCULATION AND EXPERIMENTAL VERIFICATION	71-0021
MODEL	THE USE OF ROCK BOLTS OR WIRE ROPE TO INCREASE THE STRENGTH OF FRACTURED MODEL PILLARS	71-0722
MODEL	THEORETICAL AND EXPERIMENTAL STRESS ANALYSES OF ORNL THIN-SHELL CYLINDER-TO-CYLINDER MODEL 3	75-0690
MODEL	THEORETICAL AND EXPERIMENTAL STRESS ANALYSES OF ORNL THIN-SHELL CYLINDER-TO-CYLINDER MODEL 4	75-0695
MODEL	THREE-DIMENSIONAL LINEAR FRACTURE MECHANICS ANALYSIS BY A DISPLACEMENT-HYBRID FINITE-ELEMENT MODEL	75-0884
MODEL	TWO DIMENSIONAL STOCHASTIC MODEL OF A HETEROGENEOUS GEOLOGIC SYSTEM	73-0715
MODEL	TWO-DIMENSIONAL FINITE ELEMENT MODEL OF A CERAMIC BODY	73-0010
MODEL	VIBRATION CHARACTERISTICS OF THE NORTH FORK DAM MODEL	74-0845
MODEL	WATER MOVEMENT THROUGH SATURATED-UNSATURATED POROUS MEDIA: A FINITE-ELEMENT GALERKIN MODEL	75-0761
MODEL	ZUR BERECHNUNG VON PLATTEN NACH DER THEORIE ZWEITER. ORDNUNG MIT HILFE EINES HYBRIDEN DEFORMATIONS M	73-0558
MODEL	3-D FINITE ELEMENT MODEL FOR LAKE CIRCULATION	71-0157
MODELES	UTILISATION DE MODELES MATHEMATIQUES MULTIDIMENSIONNELS POUR L'ETUDE DES ECOULEMENTS DANS LES DIGUES	73-0381
MODELING	AN INTERACTIVE COMPUTER-AIDED MODELING SYSTEM FOR THE DISCRETIZATION OF BRIDGE STRUCTURES FOR FINITE	73-1165
MODELING	ANALYTIC MODELING OF ROCK-STRUCTURE INTERACTION	72-0808
MODELING	ANALYTIC MODELING OF ROCK-STRUCTURE INTERACTION. VOLUME 3. COMPUTER PROGRAM	73-0918
MODELING	ANALYTIC MODELING OF ROCK-STRUCTURE INTERACTION. VOLUME 2. USERS GUIDE FOR A COMPUTER PROGRAM	73-0919
MODELING	ANALYTIC MODELING OF ROCK-STRUCTURE INTERACTION. VOLUME 1	73-0920
MODELING	ANALYTICAL MODELING OF BEAM-SLAB BRIDGES	74-1131
MODELING	ANALYTICAL PROBLEMS IN MODELING SLURRY WALL CONSTRUCTION	73-0764
MODELING	COMPARAISON DES DIVERSES MODELISATIONS NUMERIQUES D'UN SYSTEME MAGNETIQUE SATURABLE. (COMPARISON OF	75-0295
MODELING	DEVELOPMENT OF TECHNOLOGY FOR MODELING OF A 1/8-SCALE DYNAMIC MODEL OF THE SHUTTLE SOLID ROCKET BOOS	74-0703
MODELING	DYNAMIC MODELING OF HALF-PLANE BY FINITE ELEMENTS	74-0953
MODELING	ELECTROMAGNETIC AND ELECTRICAL MODELING BY THE FINITE ELEMENT METHOD	71-0462
MODELING	FINITE ELEMENT APPROACH TO THE MODELING OF HYDROTHERMAL SYSTEMS	73-0011
MODELING	FINITE ELEMENT CRACK TIP MODELING WHEN EVALUATING J-INTEGRAL	75-0637
MODELING	FINITE ELEMENT MODELING OF A CIRCULAR RING USING HALF AND QUARTER SYMMETRY	66-0119
MODELING	FINITE ELEMENT MODELING OF A CANTILVER BEAM	67-0132
MODELING	FINITE ELEMENT MODELING STUDIES IN THE NORMAL-MODE METHOD AND NORMAL-MODE SYNTHESIS	71-0067
MODELING	FINITE ELEMENT MODELING AND OPTIMIZATION OF AEROSPACE STRUCTURES	72-0416
MODELING	FINITE ELEMENT MODELING AND OPTIMIZATION OF AEROSPACE STRUCTURES	72-0734
MODELING	FINITE ELEMENT MODELING	73-0018
MODELING	FINITE ELEMENT MODELING CRITERIA FOR UNDERGROUND OPENINGS IN ROCK	74-0630
MODELING	FINITE ELEMENT MODELING OF MAGNETIC FIELD DEFECT INTERACTIONS	75-0068
MODELING	FINITE ELEMENT MODELING OF FLOW THROUGH POROUS MEDIA	75-0706
MODELING	FINITE-ELEMENT MODELING OF DIRECTIONAL SOLIDIFICATION OF SUPER ALLOY EUTECTICS	75-0350
MODELING	FINITE-ELEMENT MODELING OF PIEZOELECTRIC CERAMIC HYDROPHONES	75-0595
MODELING	FINITE-ELEMENT MODELING OF SURFACE DEFORMATION ASSOCIATED WITH VOLCANISM	75-0596
MODELING	IMPROVEMENT IN PLATE MODELING AND PLATE DESIGN BY THE FINITE ELEMENT METHOD	75-0791
MODELING	MATHEMATICAL MODELING VIA DIRECT USE OF VIBRATION DATA	69-0310
MODELING	MATHEMATICAL MODELING OF MOVEMENT OF DISSOLVED CONSTITUENTS IN GROUND WATER AQUIFERS BY THE FINITE E	70-0490
MODELING	MATHEMATICAL MODELING OF SPINNING ELASTIC BODIES FOR MODAL ANALYSIS	73-0441
MODELING	MATHEMATICAL MODELING AND ANALYSIS OF FUEL RODS	74-0930
MODELING	MATHEMATICAL MODELING OF NEAR COASTAL CIRCULATION	75-0720
MODELING	MATHEMATICAL MODELS OF THE MASSACHUSETTS BAY. PART 1. FINITE ELEMENT MODELING OF TWO-DIMENSIONAL HYD	73-0800
MODELING	METHODS OF MODELING AND ANALYZING VISCOELASTICALLY DAMPED STRUCTURES	71-0473
MODELING	METHODS OF MODELING AND ANALYZING VISCOELASTICALLY DAMPED STRUCTURES	71-0836
MODELING	MODELING AN INPUT-OUTPUT GEOKINETIC SYSTEM UTILIZING A FINITE ELEMENT APPROACH	75-0746
MODELING	MODELING AN OCEAN POND. A TWO-DIMENSIONAL, FINITE ELEMENT HYDRODYNAMIC MODEL OF NINIGRET POND, CHARL	75-0686
MODELING	MODELING AND ANALYSIS TECHNIQUES FOR PREDICTION OF STRUCTURAL AND BIODYNAMIC CRASH-IMPACT RESPONSE	75-0300
MODELING	MODELING EARDRUM AS A DOUBLY CURVED SHELL USING FINITE-ELEMENT METHOD	75-0047
MODELING	MODELING OF GAS TURBINE ENGINE COMPRESSOR BLADES FOR VIBRATION ANALYSIS	75-0508
MODELING	MODELING TECHNIQUES OF THIN-WALLED BEAMS WITH OPEN CROSS SECTIONS	72-0573
MODELING	MODELING WAVE PROPAGATION WITH FINITE ELEMENT METHODS	75-0015
MODELING	NASTRAN MODELING STUDIES IN THE NORMAL-MODE AND NORMAL-MODE SYNTHESIS	71-0069
MODELING	PROBABILISTIC FORMULATION FOR ANALYTICAL MODELING OC CONCRETE STRUCTURES	71-0709
MODELING	PROBABILISTIC MODELING OF CONCRETE STRUCTURES	72-0236
MODELING	RELATIONSHIP OF SUBSTRUCTURING TO IMPROVEMENT OF DYNAMIC MODELING	71-0323
MODELING	SENSITIVITY ANALYSIS OF INPUT PARAMETERS IN NUMERICAL MODELING OF STEADY STATE REGIONAL GROUNDWATER	74-0926
MODELING	SOME RESULTS IN THE FINITE ELEMENT MODELING OF THIN PLATES	66-0018
MODELING	STRUCTURAL MODELING OF A WELDED JOINT TYPICAL FOR VEHICLE FRAMES	73-0539
MODELING	THE USE OF NASTRAN IN THE ANALYSIS OF LARGE COMPLEX AIRFRAME STRUCTURES. MODELING TECHNIQUES AND ORG	71-0110
MODELING	TREATMENT OF RANDOMNESS IN FINITE ELEMENT MODELING	70-0353
MODELING	TWO-DIMENSIONAL FINITE ELEMENT MODELING OF TRANSIENT FLOW IN REGIONAL AQUIFER SYSTEMS. ANALYSIS OF G	74-0743
MODELISATIONS	COMPARAISON DES DIVERSES MODELISATIONS NUMERIQUES D'UN SYSTEME MAGNETIQUE SATURABLE. (COMPARISON OF	75-0295
MODELLE	AUFBEREITUNG MATHEMATISCHER MODELLE FUR DIE DIGITALE BERECHNUNG VON GRUNDWASSERSTROMUNGSPROBLEMEN NA	71-0543
MODELLE	BERECHNUNG DER ERDEBENSCHWINGUNGEN VON STRUKTUREN MIT DER FINITE ELEMENT METHODE. MECHANISCHE MODELL	71-0170
MODELLING	ACCURATE FINITE ELEMENT MODELLING OF FLAT AND CURVED STIFFENED PANELS	72-0538
MODELLING	COMPUTER MODELLING OF PLUG SEEDLING SURVIVAL	75-0176
MODELLING	DISCRETE-ELEMENT MODELLING AND SENSITIVITY STUDIES OF STATICALLY LOADED SINGLE-LAYER AND BOUNDED DOU	70-0129
MODELLING	DISCRETIZATION-DISPERSION IN FINITE ELEMENT MODELLING OF WAVE PROPAGATION IN SOLIDS	74-1253
MODELLING	FINITE ELEMENT ANALYSIS, MODELLING AND COMPUTER GRAPHICS	74-1292
MODELLING	FINITE ELEMENT MODELLING OF THREE-DIMENSIONAL STRUCTURES	68-0267
MODELLING	FINITE ELEMENT MODELLING TECHNIQUES FOR THE ANALYSIS OF BUILDING VIBRATIONS	72-0839
MODELLING	FINITE ELEMENT MODELLING OF NATURAL VIBRATION PROBLEMS	73-0320
MODELLING	FINITE ELEMENT MODELLING OF HYDRODYNAMIC CIRCULATION	74-1370
MODELLING	FINITE ELEMENT NUMERICAL MODELLING OF FLOW AND DISPERSION IN ESTUARIES	73-0070
MODELLING	FINITE ELEMENTS FOR GEOLOGICAL MODELLING	70-0049
MODELLING	MATHEMATICAL MODELLING AND STUCTURAL ANALYSIS OF THE MANDIBLE	72-0124
MODELLING	MODELLING AND INTERPRETATION OF RESULTS	75-0399
MODELLING	MODELLING GROUNDWATER FLOW BY THE FINITE ELEMENT METHOD	72-0637
MODELLING	MODELLING HEAT FLOW FROM UNDERGROUND POWER CABLES	74-1158
MODELLING	NUMERICAL AND PHYSICAL MODELLING OF REINFORCEMENT SYSTEMS FOR TUNNELS IN JOINTED ROCK	73-0872
MODELLING	NUMERICAL AND PHYSICAL MODELLING	74-0393
MODELLING	ON THE DEVELOPMENT OF A REFINED FINITE-ELEMENT FREE FROM DISCRETIZATION - DISPERSION FOR ELASTIC WAV	73-1148
MODELLS	ZUR BERECHNUNG VON PLATTEN NACH DER THEORIE ZWEITER. ORDNUNG MIT HILFE EINES HYBRIDEN DEFORMATIONS M	73-0558
MODELS	A COMPARISON OF CURRENT WORK-HARDENING MODELS USED IN THE ANALYSIS OF PLASTIC DEFORMATIONS	73-0769
MODELS	A SIX NODE TRIANGULAR SHEAR PANEL FOR EQUILIBRIUM MODELS	73-1011
MODELS	A SYNTHESIZED WARPED QUADRILATERAL ELEMENT GIVING SIMPLIFIED INPUT FOR EQUILIBRIUM MODELS	73-1013
MODELS	ALTERNATE HYBRID STRESS FINITE ELEMENT MODELS	75-0476
MODELS	AN EVALUATION OF CURRENT NASTRAN DISCRETE ELEMENT MODELS FOR MONOCOQUE AND SEMI-MONOCOQUE STRUCTURES	71-0236

MODELS	ANALYSIS MODELS FOR PLATE ELEMENTS IN THE MEMBRANE STATE OF STRESS	67-0046
MODELS	ANALYSIS OF LIQUID-WASTE INJECTION WELLS IN ILLINOIS BY MATHEMATICAL MODELS	74-0518
MODELS	ANALYSIS OF NONLINEAR CONCRETE FRAMES USING SMALL GYPSUM MODELS AND THE FINITE ELEMENT METHOD	70-0498
MODELS	ANISOTROPIC TAPERED ELEMENTS USING DISPLACEMENT MODELS	72-0391
MODELS	AUTOMATED STRUCTURAL DESIGN AND ANALYSIS OF ADVANCED COMPOSITE WING MODELS	73-0221
MODELS	BASIS FOR FORMULATION OF FINITE ELEMENT MODELS	72-0273
MODELS	BASIS OF A WELL CONDITIONED FORCE PROGRAM FOR EQUILIBRIUM MODELS VIA THE SOUTHWELL SLAB ANALOGIES	67-0111
MODELS	BENDING AND STRETCHING OF PLATES, SPECIAL MODELS FOR UPPER AND LOWER BOUNDS	66-0026
MODELS	BENDING OF THIN PLATES - DISPLACEMENT MODELS	68-0049
MODELS	BOUNDARIES AND FRACTURES IN FINITE ELEMENT MODELS OF GEOLOGICAL STRUCTURES	73-0352
MODELS	COMPARISON OF SOME FINITE-ELEMENT MODELS FOR CIRCULAR CYLINDRICAL SHELLS	73-1158
MODELS	COMPUTER GRAPHICS FOR 3-D FINITE ELEMENT MODELS	74-0290
MODELS	COMPUTER PROGRAM FOR GENERATING FINITE ELEMENT MODELS OF MINE STRUCTURES	71-0493
MODELS	DERIVATION OF GEOMETRIC STIFFNESS AND MASS MATRICES FOR FINITE ELEMENT HYBRID MODELS	74-0844
MODELS	DIFFUSIVE EQUILIBRIUM MODELS	73-1112
MODELS	DIGITAL COMPUTATION OF STRESSES AND DEFLECTIONS IN A BOX BEAM - A PERFORMANCE COMPARISON BETWEEN FIN	69-0355
MODELS	DISPLACEMENT AND EQUILIBRIUM MODELS IN THE FINITE ELEMENT METHOD	65-0016
MODELS	DYNAMIC RESPONSE OF ELASTIC HEAD MODELS	75-0378
MODELS	EFFECT OF HIGHER ORDER TERMS IN CERTAIN NONLINEAR FINITE ELEMENT MODELS	73-0125
MODELS	EFFECT OF HIGHER ORDER TERMS IN CERTAIN NONLINEAR FINITE ELEMENT MODELS - COMMENT	74-0473
MODELS	EIGENVALUE BOUNDS FOR FINITE ELEMENT MODELS	72-0570
MODELS	EIGENVALUE BOUNDS FOR FINITE ELEMENT MODELS	72-0932
MODELS	ELASTIC EQUILIBRIUM MODELS, STRESS FUNCTIONS, AND DUALITY NON LINEARITIES AND STABILITY, FIVE LECTUR	73-0631
MODELS	EQUILIBRIUM MODELS	75-0008
MODELS	EQUILIBRIUM STRESS FIELD MODELS IN FINITE ELEMENTS	67-0011
MODELS	ERRORS IN SHELL FINITE-ELEMENT MODELS FOR VIBRATION OF CIRCULAR-CYLINDERS	75-0203
MODELS	EXPERIMENTAL ELASTIC STRESS ANALYSES OF CYLINDER-TO-CYLINDER SHELL MODELS AND COMPARISONS WITH THEOR	71-0065
MODELS	FINITE ELEMENT ANALYSIS OF SALT PILLAR MODELS	68-0291
MODELS	FINITE ELEMENT ANALYSIS OF CREEP IN ROCK-SALT PILLAR MODELS	72-0935
MODELS	FINITE ELEMENT MODELS FOR SHELLS OF REVOLUTION	68-0027
MODELS	FINITE ELEMENT MODELS OF NONLINEAR OPERATOR EQUATIONS	71-0199
MODELS	FINITE ELEMENT MODELS OF FRACTURE	73-0012
MODELS	FINITE ELEMENT MODELS FOR OCEAN CIRCULATION PROBLEMS	75-0206
MODELS	FINITE-ELEMENT NETWORK MODELS AND THEIR APPLICATION TO EDDY-CURRENT PROBLEMS	75-0143
MODELS	FORMULATION AND APPLICATION OF CERTAIN PRIMAL AND MIXED FINITE ELEMENT MODELS OF FINITE DEFORMATIONS	74-0344
MODELS	FORMULATION OF DISCRETE ELEMENT MODELS FOR STRESS AND VIBRATION ANALYSIS OF PLATES	70-0328
MODELS	FORMULATION OF GENERAL DISCRETE MODELS OF THERMOMECHANICAL BEHAVIOR OF MATERIALS WITH MEMORY	69-0381
MODELS	FRACTURE OF ROCKET GRAIN MODELS SUBJECTED TO COOLDOWN: A STUDY OF TEMPERATURE-INITIATED FRACTURES OF	74-0536
MODELS	FRACTURE OF ROCKET GRAIN MODELS SUBJECTED TO COOLDOWN: A STUDY OF TEMPERATURE-INITIATED FRACTURES OF	74-0536
MODELS	GENERAL POTENTIAL ENERGY AND COMPLEMENTARY ENERGY MODELS BASED ON STRESS PARAMETERS	74-1240
MODELS	GENERALIZED HYBRID STRESS FINITE-ELEMENT MODELS	73-0027
MODELS	GENERALIZED STRESS MODELS FOR FINITE ELEMENT ANALYSIS	74-1067
MODELS	GENERATION OF FINITE ELEMENT MODELS VIA COMPUTER GRAPHICS	74-0194
MODELS	HYBRID MODELS	71-0211
MODELS	HYBRID MODELS IN NUMERICAL AND COMPUTER METHODS	73-0682
MODELS	HYDRODYNAMICAL-NUMERICAL MODELS FOR COASTAL WATERS AND OPEN OCEAN AREAS	75-0571
MODELS	ILLUSTRATIONS OF AUTOMOTIVE FINITE ELEMENT MODELS - STATICS	74-0373
MODELS	ILLUSTRATIONS OF AUTOMOTIVE FINITE ELEMENT MODELS - DYNAMICS	74-0391
MODELS	INCOMPATIBLE DISPLACEMENT MODELS	71-0262
MODELS	INTERACTIVE GRAPHICS - AN AID FOR DEVELOPING FINITE ELEMENT MODELS	74-0139
MODELS	INVESTIGATION OF VENEER CUTTING BY ELASTIC FINITE ELEMENT MODELS	74-1110
MODELS	LUBRICATION PROBLEMS - THE SELECTION OF MATHEMATICAL MODELS	75-0881
MODELS	MATHEMATICAL MODELS FOR GEOLOGICAL MATERIALS FOR WAVE PROPAGATION STUDIES	71-0740
MODELS	MATHEMATICAL MODELS OF THE MASSACHUSETTS BAY. PART 1. FINITE ELEMENT MODELING OF TWO-DIMENSIONAL HYD	73-0800
MODELS	MATHEMATICAL MODELS OF THE MASSACHUSETTS BAY. PART 1. FINITE ELEMENT MODELING OF TWO-DIMENSIONAL HYD	73-0800
MODELS	MIXED FINITE ELEMENT MODELS IN LIMIT ANALYSIS	74-1314
MODELS	MIXED MODELS FOR PLATES	70-0384
MODELS	NEW DISPLACEMENT HYBRID FINITE ELEMENT MODELS FOR SOLID CONTINUA	70-0115
MODELS	NONLINEAR ANALYSIS BY ASSUMED STRESS HYBRID MODELS	74-1004
MODELS	NUMERICAL SUFFICIENCY TEST FOR MONOTONIC CONVERGENCE OF FINITE ELEMENT MODELS	75-0620
MODELS	ON A NUMERICAL SUFFICIENCY TEST FOR MONOTONIC CONVERGENCE OF FINITE ELEMENT MODELS	75-0533
MODELS	ON THE EFFECT OF APPROXIMATING HIGHER ORDER TERMS IN CERTAIN FINITE ELEMENT MODELS OF NONLINEAR STRU	73-0603
MODELS	ON THE STRESS COMPUTATION IN FINITE ELEMENT MODELS BASED UPON DISPLACEMENT APPROXIMATIONS	74-0598
MODELS	PURE MOMENT LOADING OF AXISYMMETRIC FINITE ELEMENT MODELS	73-1007
MODELS	SHAPE FUNCTION FORMULATIONS FOR ELEMENTS OTHER THAN DISPLACEMENT MODELS	72-0472
MODELS	SOME CONSIDERATION FOR THE EVALUATION OF THE FINITE ELEMENT MODELS	73-0710
MODELS	SOME CONSIDERATIONS FOR THE EVALUATION OF FINITE ELEMENT MODELS	73-0706
MODELS	SOME CONSIDERATIONS FOR EVALUATION OF FINITE ELEMENT MODELS	74-0471
MODELS	SOME CURRENT FINITE-ELEMENT MODELS FOR MATERIAL BEHAVIOR	75-0549
MODELS	SOME MODELS FOR THE ANALYSIS OF STRESSES IN A TUBULAR FUEL PIN FOR HIGH TEMPERATURE REACTORS BY FINI	73-0816
MODELS	SOME PROPERTIES OF LINEAR STRAIN TRIANGLES AND OPTIMAL FINITE ELEMENT MODELS	73-1044
MODELS	SOME RECENT STUDIES IN ASSUMED STRESS HYBRID MODELS	72-0262
MODELS	STATIC ANALYSIS OF SHELLS OF REVOLUTION USING DOUBLY-CURVED QUADRILATERAL ELEMENTS DERIVED FROM ALTE	69-0368
MODELS	STATIC-GEOMETRIC ANALOGIES AND FINITE ELEMENT MODELS	71-0499
MODELS	STRAIN ENERGY FUNCTIONS AND MATHEMATICAL MODELS OF LUNG ELASTICITY	73-0591
MODELS	STRESS ANALYSIS OF INDUSTRIAL COMPONENTS WITH PLASTIC AND FINITE ELEMENT MODELS	74-0512
MODELS	STRUCTURAL AND SYSTEM MODELS	75-0669
MODELS	TESTS OF GUSSET PLATE MODELS	71-0338
MODELS	THIN-WALLED PRESTRESSED CONCRETE PRESSURE VESSEL FOR HIGH TEMPERATURE REACTORS: EXPERIMENTAL INVESTI	73-0857
MODELS	USE OF SMALL MODELS IN DESIGN AND ANALYSIS OF PRESTRESSED-CONCRETE REACTOR VESSELS	70-0433
MODELS	USE OF SMALL MODELS IN DESIGN AND ANALYSIS OF PRESTRESSED-CONCRETE REACTOR VESSELS	71-0426
MODELS	UTILISATION DE MODELES MATHEMATIQUES MULTIDIMENSIONNELS POUR L'ETUDE DES ECOULEMENTS DANS LES DIGUES	73-0381
MODERATELY	A FINITE ELEMENT ANALYSIS FOR ' MODERATELY THICK ' RECTANGULAR PLATES IN BENDING	68-0146
MODERATELY	ANALYSIS OF MODERATELY THICK FLANGED REACTOR PRESSURE VESSELS	73-0821
MODERATELY	ANALYSIS OF MODERATELY THICK FLANGED REACTOR VESSELS	75-0320
MODERATELY	FINITE-ELEMENT ANALYSIS OF MODERATELY THICK SHELLS	75-0588
MODERATELY	SIMPLIFIED INELASTIC HIGH TEMPERATURE STRUCTURAL ANALYSIS OF MODERATELY COMPLEX SPATIALLY THREE DIME	74-0445
MODERATELY	STABILITY OF MODERATELY THICK RECTANGULAR PLATES USING A HIGH PRECISION TRIANGULAR FINITE ELEMENT	75-0334
MODERATELY	THE ANALYSIS OF MODERATELY THICK TO THIN SHELLS BY THE FINITE ELEMENT METHOD	70-0315
MODERNER	EINSATZ MODERNER RECHENVERFAHREN ZUR ERMITTLUNG DER SPANNUNGEN UND VERFORMUNGEN BEI RADSAETZEN. (USE	75-0186
MODES	A METHOD FOR COMPUTATION OF VIBRATION MODES AND FREQUENCIES OF ORTHOTROPIC THIN SHELLS OF REVOLUTION	69-0037
MODES	APPLICATION OF NATURAL KINEMATIC MODES TO MATRIX ANALYSIS: SMALL AND LARGE DISPLACEMENTS AND STRAINS	67-0138
MODES	AUTOMATIC EXTRACTION OF RIGID BODY MODES FROM STRESS AND STRAIN ELEMENTS	72-0913
MODES	CALCULATION OF AXISYMMETRIC LONGITUDINAL MODES FOR FLUID-ELASTIC TANK-ULLAGE GAS SYSTEMS AND COMPARI	64-0019
MODES	CALCULATION OF THE NATURAL FREQUENCIES AND THE PRINCIPAL MODES OF HELICOPTER BLADES	72-0907
MODES	CLASSICAL AND NATURAL CONDUCTION AND DEFORMATION MODES CONCEPTS FOR AXISYMMETRIC VARIATIONAL METHODS	72-1003
MODES	DYNAMIC ANALYSIS OF STRUCTURAL SYSTEMS USING COMPONENT MODES	64-0024
MODES	EFFECT OF DAMPING ON EXCITABILITY OF HIGH-ORDER NORMAL MODES	75-0709
MODES	ESTIMATIONS OF THE EFFECT OF DESIGN CHANGES ON THE MODES AND FREQUENCIES OF VIBRATING STRUCTURAL ELE	74-0819
MODES	FINITE ELEMENT ANALYSIS OF THIN ELASTIC SHELLS WITH RESIDUAL ENERGY BALANCING AND THE ROLE OF THE RI	75-0034
MODES	FINITE-ELEMENT ANALYSIS OF THIN ELASTIC SHELLS WITH RESIDUAL ENERGY BALANCING AND ROLE OF RIGID BODY	75-0048
MODES	HARBOUR OSCILLATION - A NUMERICAL TREATMENT FOR UNDAMPED NATURAL MODES	70-0208

MODES	MODES OF MOTION OF A FINITE-ELEMENT BRAIN MODEL	75-0592
MODES	NUMERICAL STUDY OF PROPER MODES OF A HOMOGENEOUS CYLINDRICAL PLASMA BY METHOD OF FINITE ELEMENTS	74-0160
MODES	QUADRATIC MATRIX EQUATIONS FOR DETERMINING VIBRATION MODES AND FREQUENCIES OF CONTINUOUS ELASTIC SYS	66-0064
MODES	SUBSTRUCTURES ANALYSIS OF IMPELLER VIBRATION MODES	75-0283
MODES	USE OF NONCONFORMING MODES IN FINITE ELEMENT ANALYSIS OF PLATES AND SHELLS	73-0865
MODES	USE OF SIMPLE THREE-DIMENSIONAL ACOUSTIC FINITE ELEMENTS FOR DETERMINING NATURAL MODES AND FREQUENCI	72-0010
MODES	VIBRATION MODES AND RANDOM RESPONSE OF A MULTI-BAY PANEL SYSTEM USING FINITE ELEMENTS	67-0128
MODES	2-DIMENSIONAL FINITE-ELEMENT ANALYSIS OF ELECTROELASTIC NORMAL MODES OF FERROELECTRIC DISKS AND SHEL	75-0594
MODES	3-DIMENSIONAL FINITE-ELEMENT ANALYSIS OF ELECTROELASTIC NORMAL MODES OF A TANGENTIALLY POLARIZED, SE	75-0591
MODIFICATION	A FURTHER MODIFICATION TO AHMAD'S SHELL ELEMENT	73-0619
MODIFICATION	A NASTRAN POSTPROCESSOR FOR STRUCTURAL MODIFICATION REANALYSIS	71-0160
MODIFICATION	EVALUATION OF ALGORITHMS FOR STRUCTURAL MODIFICATION	69-0243
MODIFICATION	MODIFICATION OF LAYERED THEORY FOR PAVEMENT DESIGN ANALYSIS	73-0574
MODIFICATIONS	GENERAL TREATMENT OF STRUCTURAL MODIFICATIONS	72-0166
MODIFICATIONS	MODIFICATIONS TO ARL COMPUTER PROGRAMS USED FOR DESIGN OF AXIAL COMPRESSOR AIRFOILS	74-0508
MODIFICATIONS	REANALYSIS FOR LIMITED STRUCTURAL DESIGN MODIFICATIONS	72-0650
MODIFICATIONS	SOME MODIFICATIONS OF AN ISOPARAMETRIC SHELL ELEMENT	73-1010
MODIFICATIONS	THE MATRIX ANALYSIS OF STRUCTURES WITH CUT-OUTS AND MODIFICATIONS	56-0003
MODIFICATTION	AIRCRAFT FUEL TANK VULNERABILITY TO HYDRAULIC RAM: MODIFICATTION OF THE NORTHROP FINITE ELEMENT COMP	74-0724
MODIFICIROVANNYJ	MODIFICIROVANNYJ KONECHNYJ ELEMENT DIJA RASCHIOTA KOLEBANIJ TONKIKH PLASTIN	72-0891
MODIFIED	A DISCUSSION ON A MODIFIED CONJUGATE GRADIENT METHOD BY W.L. WOOD	74-0402
MODIFIED	A FINITE ELEMENT ANALOGUE OF THE MODIFIED RAYLEIGH-RITZ METHOD FOR VIBRATION PROBLEMS	72-0478
MODIFIED	A FINITE ELEMENT APPLICATION OF THE MODIFIED RAYLEIGH-RITZ METHOD	70-0091
MODIFIED	A MODIFIED FINITE ELEMENT METHOD FOR CALCULATION OF VIBRATION OF THIN PLATES	72-0598
MODIFIED	A MODIFIED GRADIENT METHOD FOR FINITE ELEMENT ELASTOPLASTIC ANALYSIS BY QUADRATIC PROGRAMMING	73-0998
MODIFIED	DISCRETE-ELEMENT PLASTIC ANALYSIS OF STRUCTURES IN A STATE OF MODIFIED PLANE STRAIN	69-0140
MODIFIED	FINITE ELEMENT ANALOGUE OF THE MODIFIED RAYLEIGH RITZ METHOD FOR VIBRATION PROBLEMS	73-0181
MODIFIED	MODIFIED APPROACH FOR ESTIMATING THE CRACKING MOMENT OF REINFORCED CONCRETE BEAMS	75-0485
MODIFIED	MODIFIED FINITE ELEMENT METHOD APPLIED TO R C DEEP BEAMS	73-0476
MODIFIED	MODIFIED POTENTIAL ENERGY MASS REPRESENTATION FOR FREQUENCY PREDICTION	66-0053
MODIFIED	NOTE ON A MODIFIED CONJUGATE GRADIENT METHOD	73-0615
MODIFIED	STATICAL CALCULATIONS OF PLANE AND SPACE BAR STRUCTURES AS WELL AS STUMPY FRAMEWORKS USING A MODIFIE	73-0747
MODULAR	EFFECT OF ELASTIC MODULAR RATIO AND ANISOTROPY ON THE STRESS FIELD IN A TWO-DIMENSIONAL MATRIX INTO	71-0647
MODULAR	MODULAR APPROACH TO STRUCTURAL SIMULATION FOR VEHICLE CRASHWORTHINESS PREDICTION	75-0737
MODULE	FINITE ELEMENT ANALYSES OF QUICK CAMP MODULE	75-0017
MODULES	GENERAL DESCRIPTION OF SESAM-69, SUPER ELEMENT STRUCTURAL ANALYSIS (PROGRAM) MODULES	71-0612
MODULI	CONSTRAINED AND SHEAR MODULI FOR FINITE ELEMENTS	74-0774
MODULI	CONSTRAINED AND SHEAR MODULI FOR FINITE ELEMENTS	75-0228
MODULI	ON THE DETERMINATION OF EFFECTIVE MODULI OF COMPOSITE MATERIALS BY A THREE-DIMENSIONAL FINITE ELEMEN	74-1123
MODULI	STRESS-PATH DEPENDENT DEFORMATION MODULI OF CLAY	75-0050
MODULUS	ANALYTICAL AND EXPERIMENTAL EVALUATION OF MODULUS OF SOIL REACTION	74-1168
MODULUS	DETERMINATION OF YOUNG'S MODULUS AND POISSON'S RATIO FOR WC-CO ALLOYS BY THE FINITE ELEMENT METHOD	72-0142
MODULUS	NX BOREHOLE JACK MODULUS DETERMINATIONS IN HOMOGENEOUS, ISOTROPIC, ELASTIC MATERIALS	74-0295
MODULUS	PLANE STRESS CRACK TRAJECTORIES IN LOW MODULUS MATERIALS	75-0646
MODULUS	SIMULATION OF ELASTIC MODULUS AND POISSON'S RATIO OF SPHEROIDAL GRAPHITE CAST IRON	74-0955
MOIRE	APPLICATION OF FINITE-ELEMENT METHOD AND MOIRE TECHNIQUE TO SHEET-METAL FORMING PROBLEMS	71-0784
MOISTURE	COUPLED HEAT AND MOISTURE TRANSPORT MODEL FOR ARCTIC SOILS	74-0578
MOISTURE	THEORETICAL TRANSIENT BEHAVIOUR OF SATURATED AND UNSATURATED SOILS UNDER LOAD AND CHANGING MOISTURE	73-0901
MOLAR	FINITE ELEMENT STRESS ANALYSIS OF A RESTORED AXISYMMETRIC FIRST MOLAR	73-0053
MOLAR	PHOTOELASTIC AND FINITE ELEMENT STRESS ANALYSIS OF A RESTORED AXISYMMETRIC FIRST MOLAR	73-0142
MOLARS	STRESS ANALYSIS OF FIRST MOLARS WTH FULL CROWN PREPARATIONS BY THREE-DIMENSIONAL PHOTOELASTICITY AND	72-0939
MOLDINGS	FINITE ELEMENT ANALYSIS OF DIAMETRAL TEST OF POLYMER MOLDINGS	73-0129
MOLTEN	CALCULATION OF SOLIDIFICATION OF MOLTEN METAL BY FINITE-ELEMENT METHOD	74-1357
MOMENT	BUCKLING OF INELASTIC I-BEAMS UNDER MOMENT GRADIENT	74-0765
MOMENT	DISCRETE TRIANGULAR APPROXIMATION OF MOMENT AND DISPLACEMENT SURFACES FOR PLATE BENDING	71-0269
MOMENT	INELASTIC BUCKLING OF STEEL BEAMS UNDER NON UNIFORM MOMENT	75-0149
MOMENT	INFLUENCE OF END-EFFECTS ON STRESSES AND FLEXIBILITY OF A PIPING ELBOW WITH IN-PLANE MOMENT	74-0534
MOMENT	MODIFIED APPROACH FOR ESTIMATING THE CRACKING MOMENT OF REINFORCED CONCRETE BEAMS	75-0485
MOMENT	ON THE UNITY OF THE CONSTANT STRAIN-CONSTANT MOMENT FINITE ELEMENT METHODS	73-0164
MOMENT	PURE MOMENT LOADING OF AXISYMMETRIC FINITE ELEMENT MODELS	73-1007
MOMENT	THE CONSTANT BENDING MOMENT PLATE BENDING ELEMENT	71-0182
MOMENTS	A TRIANGULAR EQUILIBRIUM ELEMENT WITH LINEARLY VARYING BENDING MOMENTS FOR PLATE BENDING PROBLEMS	67-0054
MOMENTS	BENDING MOMENTS IN LONG WALL TANKS	67-0110
MOMENTS	FORCE AND MOMENTS ON ASYMMETRIC AND YAWED BODIES IN A FREE SURFACE	73-0908
MOMENTS	STRUCTURAL ANALYSIS OF SL-7 CONTAINERSHIP UNDER COMBINED LOADING OF VERTICAL, LATERAL AND TORSIONAL	74-0734
MOMENTS	TRIANGULAR FINITE ELEMENTS FOR PLATE BENDING WITH CONSTANT AND LINEARLY VARYING BENDING MOMENTS	70-0019
MONERO	TENSILE FIELD CALCULATION ABOUT A PRESSURIZED CAVITY: MONERO	72-0684
MONITORING	CRACK GROWTH MONITORING BY STRAIN SENSING	73-0486
MONOCOQUE	AN EVALUATION OF CURRENT NASTRAN DISCRETE ELEMENT MODELS FOR MONOCOQUE AND SEMI-MONOCOQUE STRUCTURES	71-0236
MONOTON	MONOTON FINITE-ELEMENT MATRICES AND THEIR COMPUTED CONDITION NUMBERS	75-0820
MONOTONIC	NUMERICAL SUFFICIENCY TEST FOR MONOTONIC CONVERGENCE OF FINITE ELEMENT MODELS	75-0620
MONOTONIC	ON A NUMERICAL SUFFICIENCY TEST FOR MONOTONIC CONVERGENCE OF FINITE ELEMENT MODELS	75-0533
MONOTONY	PROPERTIES OF STABILITY AND MONOTONY WITH THE FINITE ELEMENT METHOD	69-0266
MONTE	LOAD LIMIT ANALYSIS, COMBINATION OF STATISTICAL METHODS (MONTE CARLO METHOD) AND FINITE ELEMENT METH	71-0768
MONTE	PRESTRESSED CONCRETE REACTOR PRESSURE VESSEL FAILURE ANALYSIS, THE INCORPORATION OF STATISTICAL METH	71-0120
MOORING	MECHANICS OF CABLE MOORING SYSTEMS. VOLUME IV. A COMPUTER PROGRAM FOR ANALYZING THE STEADY STATE RES	72-0653
MOORING	MECHANICS OF CABLE MOORING SYSTEMS. VOLUME 1. THREE DIMENSIONAL RESPONSE OF DEEP WATER MOORING LINES	72-0654
MOORING	MECHANICS OF CABLE MOORING SYSTEMS. VOLUME 1. THREE DIMENSIONAL RESPONSE OF DEEP WATER MOORING LINES	72-0654
MOORING	MECHANICS OF CABLE MOORING SYSTEMS. VOLUME V11. THE STEADY-STATE BEHAVIOR OF A PYRAMID ARRAY SYSTEM	73-0724
MOORING	MECHANICS OF CABLE MOORING SYSTEMS. VOLUME V1. A COMPUTER PROGRAM FOR ANALYZING THE STEADY STATE CON	73-0725
MOSS	SIMPLIFIED CALCULATION OF THE INTERACTION FORCES BETWEEN HULL AND TANK SYSTEM FOR A MOSS ROSENBERG T	75-0561
MOTION	A DERIVATION OF PERTURBED EQUATIONS OF MOTION AND HEAT CONDUCTION OF A THERMOELASTIC SOLID	72-0735
MOTION	A METHOD FOR THE NUMERICAL INTEGRATION OF THE EQUATIONS OF MOTION ARISING FROM A FINITE ELEMENT ANAL	70-0061
MOTION	AN INTRODUCTION TO THE FINITE ELEMENT ANALYSIS OF THE MOTION OF A GAS-LIQUID INTERFACE IN A POROUS M	72-0904
MOTION	EFFECTS OF SOURCE STRUCTURE AND SITE RESPONSE ON GROUND MOTION IN THE NEAR FIELD OF EARTHQUAKES	73-0732
MOTION	FINITE ELEMENT ANALYSIS OF FLUID MOTION IN A CONTAINER	74-0079
MOTION	FINITE ELEMENT ANALYSIS OF THE MOTION OF A GAS-LIQUID INTERFACE IN A POROUS MEDIUM	75-0672
MOTION	FINITE ELEMENT SOLUTION FOR GENERAL FLUID MOTION	73-0249
MOTION	FINITE-ELEMENT ANALYSIS OF SLOW INCOMPRESSIBLE VISCOUS FLUID MOTION	74-0154
MOTION	IMPLICIT RIGID BODY MOTION IN CURVED FINITE ELEMENTS	71-0174
MOTION	INERTIA LOADING IN FINITE-ELEMENT ANALYSIS OF STRUCTURES SUBJECT TO COMPOUND MOTION	75-0351
MOTION	MODES OF MOTION OF A FINITE-ELEMENT BRAIN MODEL	75-0592
MOTION	NUMERICAL INTEGRATION OF THE EQUATION OF MOTION	71-0507
MOTION	ON SMALL VIBRATIONS AND PERTURBATIONS OF FLEXIBLE BODIES UNDERGOING ARBITRARY NOMINAL MOTION	70-0376
MOTION	ON THE STABILITY OF EXPLICIT METHODS FOR THE NUMERICAL INTEGRATION OF THE EQUATIONS OF MOTION IN FIN	72-0172
MOTION	PRODUCTION OF MOTION PICTURES OF A CONTINUOUSLY DEFORMING STRUCTURE USING NASTRAN	71-0223
MOTION	STEADY STATE EQUATIONS OF MOTION, EQUILIBRIUM SHAPE AND STABILITY DERIVATIVES OF ELASTIC AIRPLANES E	72-0541
MOTION	STUDY OF MOTION AND STRENGTH OF FLOATING MARINE STRUCTURES IN WAVES	74-0489
MOTION	THE MOTION OF INTERCONNECTED FLEXIBLE BODIES	75-0727
MOTIONLESS	CALCULATION OF THE FREE OSCILLATIONS OF A LIQUID IN AXISYMMETRIC MOTIONLESS CONTAINERS OF ARBITRARY	74-0387
MOTIONS	A NUMERICAL APPROACH TO THE STUDY OF IRREGULAR SHIP MOTIONS	74-0030
MOTIONS	EXPLICIT ADDITION OF RIGID-BODY MOTIONS IN CURVED FINITE ELEMENTS	73-0037

MOTIONS	FINITE ELEMENT APPROACH TO WAVE MOTIONS IN ELASTIC-PLASTIC CONTINUA	68-0041
MOTIONS	FINITE-ELEMENT TREATMENT OF QUADRUPOLE GROUND MOTIONS IN AXISYMMETRIC EARTH STRUCTURES	75-0434
MOTIONS	HYDRODYNAMIC PRESSURES ON HIGH DAMS DUE TO VERTICAL EARTHQUAKE MOTIONS	69-0438
MOTIONS	RIGID BODY MOTIONS AND EQUILIBRIUM IN FINITE ELEMENTS	74-0102
MOTIONS	RIGID BODY MOTIONS IN CURVED FINITE ELEMENTS	70-0035
MOTIONS	RIGID-BODY MOTIONS AND STRAIN-DISPLACEMENT EQUATIONS OF CURVED SHELL FINITE ELEMENTS	72-0267
MOTIONS	THE IMPORTANCE OF THE VERTICAL COMPONENT OF EARTHQUAKE MOTIONS	66-0019
MOTIVE	F 104 G FINAL MOTIVE: GROUND VIBRATION CALCULATIONS FOR THE STARFIGHTER F 104 G	69-0456
MOTOR	ACOUSTIC ANALYSIS OF SOLID ROCKET MOTOR CAVITIES BY A FINITE ELEMENT METHOD	71-0119
MOTOR	AUTOMATED EVALUATION OF INTERCEPTOR ROCKET MOTOR DESIGNS UNDER COMBINED OPERATIONAL AND NUCLEAR EFFE	74-0698
MOTOR	HEAT CONDUCTION AND STRESS ANALYSIS OF SOLID PROPELLANT ROCKET MOTOR NOZZLES	69-0059
MOTORS	CASE BOND STRESS CALCULATIONS FOR FLAPPED CYLINDRICAL ANALOGS OF SOLID PROPELLANT ROCKET MOTORS. BON	72-0842
MOTORS	METHODS FOR DETERMINING CHARACTERISTICS OF ACOUSTIC WAVES IN ROCKET MOTORS	75-0075
MOTORWAY	COMPARISON BETWEEN THEORETICAL AND MEASURED EARTH PRESSURES ACTING ON A LARGE MOTORWAY RETAINING WAL	74-0917
MOTORWAY	EARTH PRESSURES AGAINST THE ABUTMENTS AND WING WALLS OF STANDARD MOTORWAY BRIDGES	75-0629
MOUNT	WEAPON-MOUNT INTERFACE STUDY - TRIPOD MOUNT M3	72-0846
MOUNTED	COUPLED RESPONSE OF SPATIAL VIBRATORY STRUCTURES MOUNTED TO ISOTROPIC PLATE ELEMENTS	71-0228
MOUVEMENTS	UNE METHODE DE CALCUL PAR ELEMENTS FINIS DES MOUVEMENTS DE LIQUIDES DANS DES RESERVOIRS RIGIDES OU D	74-1171
MOVEABLE	A COMPUTER PROGRAM TO ANALYZE BEAM-COLUMN UNDER MOVEABLE LOADS	68-0268
MOVEMENT	ANALYSIS OF SOIL MOVEMENT AROUND A DEEP EXCAVATION	70-0039
MOVEMENT	ANALYSIS OF THERMAL STRESSES AND METAL MOVEMENT DURING WELDING 1. ANALYTICAL STUDY	74-1026
MOVEMENT	CALCULATION OF THE VIRTUAL MASS, MOVEMENT AND INERTIA BY THE FINITE ELEMENT METHOD	68-0221
MOVEMENT	EARTH PRESSURES ON STRUCTURES DUE TO FAULT MOVEMENT	73-0403
MOVEMENT	EFFECT OF RESERVOIR FILLING ON STRESSES AND MOVEMENT IN EARTH AND ROCKFILL DAMS. A REPORT OF AN INVE	72-0836
MOVEMENT	MATHEMATICAL MODELING OF MOVEMENT OF DISSOLVED CONSTITUENTS IN GROUND WATER AQUIFERS BY THE FINITE E	70-0490
MOVEMENT	SOIL RESISTANCE MOVEMENT RELATIONSHIPS FOR LATERALLY LOADED PILES IN CLAYS AT GREAT DEPTHS	70-0162
MOVEMENT	WATER MOVEMENT THROUGH SATURATED-UNSATURATED POROUS MEDIA: A FINITE-ELEMENT GALERKIN MODEL	75-0761
MOVEMENTS	FINITE ELEMENT ANALYSES OF SOIL MOVEMENTS CAUSED BY DEEP EXCAVATION AND DEWATERING	69-0530
MOVEMENTS	MOVEMENTS OF ARTICULATED CONDUITS UNDER EARTH DAMS ON COMPRESSIBLE FOUNDATIONS	73-0477
MOVEMENTS	PREDICTION OF UNDRAINED MOVEMENTS CAUSED BY EMBANKMENTS ON MUSKEG	70-0070
MOVEMENTS	STRESSES AND MOVEMENTS IN OROVILLE DAM	72-0200
MOVEMENTS	STUDY OF FLUID MOVEMENTS THROUGH CAUSEWAY	75-0083
MOVING	A FINITE ELEMENT APPROACH TO THE DETERMINATION OF THE DYNAMIC RESPONSE OF BEAMS SUPPORTING MOVING MA	74-1114
MOVING	ANALYSIS OF MOVING BODY PROBLEMS IN AERODYNAMICS	74-0024
MOVING	DYNAMIC ANALYSIS OF A FLEXIBLE VEHICLE MOVING ALONG A FLEXIBLE SUPPORT	71-0597
MOVING	FINITE ELEMENT SOLUTION OF PROBLEMS WITH UNKNOWN AND MOVING BOUNDARIES	74-0460
MOVING	FINITE ELEMENT TREATMENT OF A UNIFORMLY MOVING ELASTIC-PLASTIC CRACK TIP	74-0470
MOVING	RESPONSE OF BEAMS AND PLATES TO MOVING LOADS	71-0277
MOVING	RESPONSE OF RECTANGULAR PLATES TO MOVING LOADS BY A FINITE ELEMENT PROCEDURE	69-0515
MOVING	STRUCTURAL RESPONSE TO MOVING PROJECTILE MASS BY THE FINITE ELEMENT METHOD	75-0693
MOVING	THE COMPARISON OF THE RESPONSE OF A HIGHWAY BRIDGE TO UNIFORM GROUND SHOCK AND MOVING GROUND EXCITAT	72-0737
MOYENNE	NOUVELLE CONCEPTION DES DRAINS DANS LES BARRAGES EN TERRE HOMOGENE DE PETITE ET MOYENNE DIMENSIONS (	73-0404
MRI	MRI/STARDYNE USER INFORMATION MANUAL	73-0541
MSC	MSC/NASTRAN APPLICATION MANUAL	72-0567
MUFFLERS	PREDICTION OF TRANSMISSION LOSS IN MUFFLERS BY THE FINITE ELEMENT METHOD	75-0160
MULTI	NONLINEAR FINITE ELEMENT ANALYSIS OF MULTI JOINTED ROCK MASSES	72-0956
MULTI-AXIAL	ANALYSIS OF MULTI-AXIAL ANISOTROPIC CREEP, ROOF RESIN BOLTS, AND BED SEPARATION IN ROCK STRUCTURES U	74-1122
MULTI-BAY	VIBRATION MODES AND RANDOM RESPONSE OF A MULTI-BAY PANEL SYSTEM USING FINITE ELEMENTS	67-0128
MULTI-CELL	ANALYSIS OF MULTI-CELL BRIDGES WITHOUT TRANSVERSE DIAPHRAGMS, A FINITE ELEMENT APPROACH	69-0235
MULTI-CELL	ANALYTICAL AND EXPERIMENTAL STUDIES OF MULTI-CELL CONCRETE BOX GIRDER BRIDGES	74-1134
MULTI-CELL	MULTI-CELL BRIDGE DECKS - MODEL TESTS AND THEORETICAL ANALYSIS	72-0381
MULTI-CELLULAR	STRESS ANALYSIS OF MULTI-CELLULAR CAISSONS	74-0338
MULTI-COMPONENT	FINITE ELEMENT ANALYSIS OF A MULTI-COMPONENT KINETIC ENERGY PROJECTILE	74-1180
MULTI-COMPONENT	FINITE ELEMENT ANALYSIS OF A MULTI-COMPONENT KINETIC ENERGY PROJECTILE	75-0754
MULTI-DIMENSION	SOLUTION OF MULTI-DIMENSIONAL NEUTRON TRANSPORT EQUATION BY FINITE ELEMENT METHOD	72-0564
MULTI-ELEMENT	MULTI-ELEMENT SPATIAL STRUCTURES	73-0768
MULTI-LAYERED	A MULTI-LAYERED FINITE ELEMENT MODEL FOR PREDICTING MINE SUBSIDENCE	68-0293
MULTI-MATERIAL	BALANCED STRESSES IN POST-YIELDED MULTI-MATERIAL STRUCTURAL JOINTS	69-0271
MULTI-SPECIE	FINITE ELEMENT THEORY FOR THE MECHANICS AND THERMODYNAMICS OF A VISCOUS, COMPRESSIBLE MULTI-SPECIE F	71-0005
MULTI-SPECIES	A FINITE ELEMENT COMPUTATIONAL THEORY FOR THE MECHANICS AND THERMODYNAMICS OF A VISCOUS, COMPRESSIBL	71-0787
MULTI-STAGE	WIELOSTOPNIOWA SYNTEZA MACIERZY SZTYWNOSCI (MULTI-STAGE SYNTHESIS OF THE STIFFNESS MATRICES)	73-0668
MULTI-STOREY	CREEP DEFORMATION OF A MULTI-STOREY STRUCTURE IN A FIRE	74-1197
MULTI-WEB	FINITE ELEMENT FLUTTER ANALYSIS OF MULTI-WEB WING STRUCTURES	75-0004
MULTIANGULAR	MULTIANGULAR ELEMENTS IN PLATE BENDING	73-0435
MULTIBEAM	CALCUL DES PONTS BIAIS A POUTRES MULTIPLES SANS ENTRETOISES PAR LA METHODE DES ELEMENTS (CALCULATION	72-0184
MULTICELLULAR	FINITE ELEMENT METHODS FOR ANALYSIS OF MULTICELLULAR STRUCTURES	72-0561
MULTICONDUCTOR	FINITE ELEMENT STRESS ANALYSIS OF A MULTICONDUCTOR ARMOR CABLE	73-0095
MULTICONDUCTOR	FINITE ELEMENT VIBRATION ANALYSIS OF MULTICONDUCTOR ELECTRICAL TRANSMISSION LINES	73-0590
MULTIDEGREE	MULTIDEGREE DYNAMIC ANALYSIS OF TALL BUILDINGS SUBJECTED TO WIND AS A STOCHASTIC PROCESS	74-1196
MULTIDEGREE-OF-	MULTIDEGREE-OF-FREEDOM ELASTIC SYSTEMS HAVING MULTIPLE CLEARANCES	73-0653
MULTIDIMENSIONAL	REVIEW OF CURRENT PROBLEMS FOR MULTIDIMENSIONAL REACTOR STATICS CALCULATIONS	73-0960
MULTIDIMENSIONA	THE MULTIDIMENSIONAL STEFAN PROBLEM: A FINITE ELEMENT APPROACH	74-0464
MULTIDIMENSIONA	UTILISATION DE MODELES MATHEMATIQUES MULTIDIMENSIONNELS POUR L'ETUDE DES ECOULEMENTS DANS LES DIGUES	73-0381
MULTIDIMENSIONN	UTILISATION DE MODELES MATHEMATIQUES MULTIDIMENSIONNELS POUR L'ETUDE DES ECOULEMENTS DANS LES DIGUES	73-0381
MULTIDISCIPLINA	APPLICATION OF COMPUTER-AIDED AIRCRAFT DESIGN IN A MULTIDISCIPLINARY ENVIRONMENT	73-0527
MULTIDISCIPLINA	ENHANCEMENTS OF THE FINITE ELEMENT METHOD THROUGH MULTIDISCIPLINARY APPLICATIONS	75-0842
MULTIDISCIPLINA	PROGRAM DESIGN BY A MULTIDISCIPLINARY TEAM	75-0854
MULTIGROUP	APPLICATION OF FINITE ELEMENT METHOD TO MULTIGROUP NEUTRON DIFFUSION EQUATION	72-0310
MULTIGROUP	APPLICATION OF THE FINITE ELEMENT METHOD TO TWO DIMENSIONAL MULTIGROUP NEUTRON DIFFUSION	72-0971
MULTIGROUP	FINITE ELEMENT SOLUTION OF MULTIGROUP TRANSPORT EQUATION IN 2 DIMENSIONAL GEOMETRIES	74-0648
MULTIGROUP	SPACE FINITE-ELEMENTS 2-DIMENSIONAL MULTIGROUP TRANSPORT CALCULATIONS USING PHASE-SPACE FINITE-ELEME	75-0504
MULTIGROUP	THE USE OF INTERPOLATORY POLYNOMIALS FOR A FINITE ELEMENT SOLUTION OF THE MULTIGROUP DIFFUSION EQUAT	72-0524
MULTIGROUP	TIMING COMPARISON STUDY FOR SOME HIGH ORDER FINITE ELEMENT PROCEDURES AND A LOW ORDER FINITE DIFFERE	72-0767
MULTIGROUP	TRIPLET: A TWO-DIMENSIONAL, MULTIGROUP, TRIANGULAR MESH, PLANAR GEOMETRY, EXPLICIT TRANSPORT CODE	73-0955
MULTIHULLED	THE APPLICATION OF THE FINITE ELEMENT METHOD TO MULTIHULLED SHIP STRUCTURAL ANALYSIS	72-0886
MULTILAYER	A TRIANGULAR ELEMENT FOR BENDING AND VIBRATION OF MULTILAYER SANDWICH PLATES	72-0498
MULTILAYER	APPLICABILITY OF APPROXIMATE AND EXACT TRANSIENT HEAT TRANSFER ANALYSES TO HEATING PROCESSES USED TO	75-0244
MULTILAYER	BENDING AND VIBRATION OF MULTILAYER SANDWICH BEAMS AND PLATES	73-0272
MULTILAYER	BUCKLING OF MULTILAYER PLATES BY FINITE ELEMENTS	67-0103
MULTILAYER	FINITE ELEMENT ANALYSIS OF AXISYMMETRIC MULTILAYER SANDWICH PLATES AND SHELLS	74-0101
MULTILAYER	FINITE ELEMENT ANALYSIS OF MULTILAYER SANDWICH PLATES AND SHELLS	74-0425
MULTILAYER	FINITE ELEMENT BENDING ANALYSIS OF MULTILAYER PLATES	74-0266
MULTILAYER	FINITE-ELEMENT ANALYSIS OF MULTILAYER SANDWICH STRUCTURES	72-0953
MULTILAYER	STABILITY ANALYSIS OF MULTILAYER SANDWICH STRUCTRES	73-0345
MULTILAYER	TRIANGULAR ELEMENT FOR MULTILAYER SANDWICH PLATES	72-0581
MULTILAYERED	A METHOD OF ANALYZING TRANSIENT FLUID FLOW IN MULTILAYERED AQUIFERS	69-0142
MULTILAYERED	MATRIX ANALYSIS OF MULTILAYERED AND SANDWICH SHELLS BY FINITE ELEMENT METHOD	73-0101
MULTILAYERED	STATIC AND DYNAMIC ANALYSIS OF MULTILAYERED SANDWICH PLATES	72-0493
MULTILAYERED	STRUCTURAL ANALYSIS OF A COAL MINE OPENING IN ELASTIC, MULTILAYERED MATERIAL	74-0235
MULTILAYERED	STRUCTURAL ANALYSIS OF A COAL MINE OPENING IN ELASTIC, MULTILAYERED MATERIAL	74-0840
MULTIPIECE	FINITE-ELEMENT STRESS ANALYSIS OF A MULTIPIECE PISTON	69-0274
MULTIPLE	A METHOD FOR THE THERMAL ANALYSIS OF SPACECRAFT INCLUDING ALL MULTIPLE REFLECTIONS AND SHADING AMONG	70-0424

MULTIPLE	ANALYSIS AND DESIGN OF MULTIPLE INPUT FLEXIBLE LINK MECHANISMS	70-0354
MULTIPLE	APPLICATIONS OF VARIATIONAL EQUATIONS TO ORDINARY AND PARTIAL DIFFERENTIAL EQUATIONS - MULTIPLE SOLU	75-0010
MULTIPLE	DATA GENERATOR FOR THE IDEALIZATION FOR FINITE ELEMENT STRUCTURAL ANALYSIS OF NAVAL SHIP FLAT PLATED	72-0860
MULTIPLE	INTERPRETATION OF COMPLEX STRESS ANALYSIS BY MULTIPLE REGRESSION	74-1203
MULTIPLE	MULTIDEGREE-OF-FREEDOM ELASTIC SYSTEMS HAVING MULTIPLE CLEARANCES	73-0653
MULTIPLE	MULTIPLE STRESS WAVE DISCONTINUITIES IN FINITE BEAMS AND PLATES	67-0114
MULTIPLE	OPTIMUM PLASTIC DESIGN FOR MULTIPLE SETS OF LOADS	74-0980
MULTIPLE	SIGNIFICANCE OF SYMMETRY IN MULTIPLE FRACTURE ANALYSIS	74-0869
MULTIPLE	STRESS FIELDS AROUND MULTIPLE INCLUSIONS	69-0067
MULTIPLE	THICK WALLED MULTIPLE OPENING REINFORCED CONCRETE CONDUITS	72-0666
MULTIPLE-AQUIFE	TRANSIENT FLOW OF GROUNDWATER TO WELLS IN MULTIPLE-AQUIFER SYSTEMS	69-0501
MULTIPLE-LOADIN	A DISCRETE-ELEMENT METHOD OF MULTIPLE-LOADING ANALYSIS FOR TWO-WAY BRIDGE FLOOR SLABS	70-0435
MULTIPLES	CALCUL DES PONTS BIAIS A POUTRES MULTIPLES SANS ENTRETOISES PAR LA METHODE DES ELEMENTS (CALCULATION	72-0184
MULTIPLIER	LAGRANGE MULTIPLIER TECHNIQUES IN STRUCTURAL ANALYSIS	65-0045
MULTIPLIERS	COUPLED SHEAR WALL ANALYSIS BY LAGRANGE MULTIPLIERS	75-0202
MULTIPLIERS	FINITE ELEMENT METHOD WITH LAGRANGIAN MULTIPLIERS	73-0046
MULTIPLIERS	THE FINITE ELEMENT METHOD WITH LAGRANGIAN MULTIPLIERS	72-0901
MULTIPLY	INCOMPRESSIBLE FLOW IN MULTIPLY CONNECTED REGIONS	74-0904
MULTIPOINT	MULTIPOINT TAYLOR FORMULAS AND APPLICATIONS TO THE FINITE ELEMENT METHOD	71-0446
MULTIPOLES	FINITE-ELEMENT CODE TO SIMULATE ANISOTROPIC-PLASMA DIFFUSION IN LINEAR MULTIPOLES	75-0409
MULTIREGION	FINITE ELEMENT SOLUTIONS FOR MULTIREGION PROBLEMS	73-1021
MULTISTAGE	FINITE ELEMENT ELASTOPLASTIC ANALYSIS BY QUADRATIC PROGRAMMING: THE MULTISTAGE METHOD	73-0842
MULTISTEP	FINITE-ELEMENT MULTISTEP DISCRETIZATIONS OF PARABOLIC BOUNDARY-VALUE PROBLEMS	75-0501
MULTISTOREY	INTERACTIVE ANALYSIS OF A RAFTED MULTISTOREY SPACE FRAME RESSTING ON AN INHOMOGENEOUS CLAY STRATUM	74-0447
MULTISTOREY	SUBSTRUCTURE ANALYSIS OF MULTISTOREY FLAT SLAB BUILDINGS	74-0394
MULTISTORY	LIGHT GAGE STEEL INFILL PANELS IN MULTISTORY STEEL FRAMES	74-0229
MULTISUPPORTED	VIBRATION OF MULTISUPPORTED CURVED BEAMS	74-1297
MUPDI	TRANSVERSE REINFORCEMENT OF CONTINUOUS, TROUGH-SHAPED CONCRETE RAILROAD STRUCTURES. AN IMPLEMENTATIO	71-0719
MUSKEG	PREDICTION OF UNDRAINED MOVEMENTS CAUSED BY EMBANKMENTS ON MUSKEG	70-0070
MUTUAL	ANALYSIS OF THE MUTUAL INDUCTANCE PARTICLE VELOCIMETER	74-0712
MYOCARDIAL	FINITE ELEMENT ANALYSIS OF LEFT VENTRICULAR MYOCARDIAL STRESSES	74-0669
M437	STRESS ANALYSIS OF 175 MM PROJECTILE, HE, M437	71-0736
M551	THERMAL ANALYSIS OF M551 EXPERIMENT FOR MATERIALS PROCESSING IN SPACE	74-0725
M552	THERMAL ANALYSIS OF M552 EXPERIMENT FOR MATERIALS PROCESSING IN SPACE	74-0704
M56	DESIGN OF INCLINED TIED ARCH RAILWAY BRIDGES OVER THE M56	71-0297
NACHBARSCHAFT	DER SPANNUNGS UND BEANSPRUCHUNGSZUSTAND DES BODENMATERIALS IN DER NACHBARSCHAFT EINER BODENBOESCHUNG	75-0371
NAEHERUNGSVERFA	EINFACHES NAEHERUNGSVERFAHREN FUR DIE GRENZTRAGFAEHIGKEITSANLYSE VON SPANNBETON-REAKTORDRUCKBEHAELTE	73-0854
NAEHERUNGSVERFA	ZUR SYSTEMATIK NUMERISCHER NAEHERUNGSVERFAHREN IN DER STATIK UND DYNAMIK DER KONSTRUKTIONEN (METHODS	72-0557
NAGRUZHENNYKH	ISSLEDOVANIE PROCHNOSTI I ZHESTKOSTI PLASTIN, NAGRUZHENNYKH V SVOEI PLOSKOSTI, METHODOM KONECHNYKH E	73-0382
NAPRYAZHENII	METODY RASCHETNOGO OPREDELENIYA TEMPERATURNYKH NAPRYAZHENII V KRYSHKAKH TSILINDROV DVIGATELEI VNUTRE	74-0179
NAPRYAZHENII	RASCHET POLEI DEFORMATSII I NAPRYAZHENII V DETALYAKH DVIGATELEI VNUTRENNEGO SGORANIYA (CALCULATION O	73-0465
NAPRYAZHENIL	RASCHET NAPRYAZHENIL L PEREMESHCHENII V SVODE PODZEMNOGO MASHINNOGO ZALA GES I V OKRUZHAYUSHCHEM SKA	74-0542
NAPRYAZHENNO-DE	ISSLEDOVANIE NAPRYAZHENNO-DEFORMIROVANNOGO SOSTOYANIYA SIMMETRICHNYKH DISKOV SO STUPITSEI. (INVESTIG	74-0965
NAPRYAZHENNOGO	ISSLEDOVANIE NAPRYAZHENNOGO SOSTOYANIYA REBRISTYKH ISILINDRICHESKIKH OBOLOCHEK S PRYAMOUGOL'NYMI OTV	74-0944
NASA'S	NASTRAN, NASA'S GENERAL PURPOSE STRUCTURAL ANALYSIS PROGRAM	70-0370
NASKA	AUTOMATIC SYSTEM FOR KINEMATIC ANALYSIS, NASKA	65-0054
NASTRAN	A CRITICAL REVIEW OF NASTRAN	71-0246
NASTRAN	A DESIGN STUDY FOR THE INCORPORATION OF AEROELASTIC CAPABILITY INTO NASTRAN	71-0112
NASTRAN	A NASTRAN CORRELATION STUDY FOR VIBRATIONS OF A CROSS-STIFFENED SHIP'S DECK	72-0756
NASTRAN	A NASTRAN POSTPROCESSOR FOR STRUCTURAL MODIFICATION REANALYSIS	71-0160
NASTRAN	A NASTRAN VIBRATION MODEL OF THE AH-IG HELICOPTER AIRFRAME. VOLUME 1	74-1338
NASTRAN	A NASTRAN VIBRATION MODEL OF THE AH-IG HELICOPTER AIRFRAME. VOLUME 11	74-1339
NASTRAN	A SPECIAL NASTRAN PROGRAM FOR INPUT CHECKING AND UNDEFORMED STRUCTURE PLOTTING	71-0234
NASTRAN	A SURVEY OF PRE AND POSTPROCESSORS FOR NASTRAN	74-0714
NASTRAN	A TECHNICAL EVALUATION OF THE NASTRAN COMPUTER PROGRAM	71-0550
NASTRAN	ADAPTATION OF NASTRAN TO THE ANALYSIS OF THE VIKING SPACE VEHICLE	71-0135
NASTRAN	AN EVALUATION OF CURRENT NASTRAN DISCRETE ELEMENT MODELS FOR MONOCOQUE AND SEMI-MONOCOQUE STRUCTURES	71-0236
NASTRAN	AN INTERACTIVE COMPUTER GRAPHICS PROGRAM FOR NASTRAN	71-0074
NASTRAN	APPLICATION OF NASTRAN TO THE ANALYSIS OF A SPACE SHUTTLE ORBITER STRUCTURE	71-0099
NASTRAN	APPLICATION OF NASTRAN TO A SPACE SHUTTLE DYNAMICS MODEL	71-0247
NASTRAN	APPLICATIONS OF NASTRAN TO COUPLED STRUCTURAL AND HYDRODYNAMIC RESPONSES IN AIRCRAFT HYDRAULIC SYSTE	74-0744
NASTRAN	AUTOMATED INPUT DATA PREPARATION FOR NASTRAN	69-0080
NASTRAN	CALCULATION OF STRESS CONCENTRATION DUE TO A HYPERBOLOID CAVITY IN A THIN PLATE USING NASTRAN	71-0260
NASTRAN	CONVERGENCE OF THE NASTRAN PLATE ELEMENTS FOR SHELL STRESS ANALYSIS	71-0261
NASTRAN	DEFINIT - A NEW ELEMENT DEFINITION CAPABILITY FOR NASTRAN: USER'S MANUAL	73-0738
NASTRAN	DYNAMIC STRUCTURAL ANALYSIS WITH THE NASTRAN COMPUTER PROGRAM	70-0213
NASTRAN	EVALUATION OF NASTRAN FROM THE THE USER'S POINT OF VIEW	73-0596
NASTRAN	FAST MODAL EXTRACTION IN NASTRAN VIA THE FEER COMPUTER PROGRAM	73-0356
NASTRAN	FINAL REPORT ON NASTRAN/GRAPHICS, 1971 FORD DOOR PROJECT	73-0595
NASTRAN	ICON: THE INTERACTIVE CREATION OF NASTRAN DATA. A SYSTEM DESCRIPTION	75-0274
NASTRAN	MISSILE BODY INPUT GENERATOR (BING), A NASTRAN PRE-PROCESSOR: THEORETICAL DEVELOPMENT, USER'S MANUAL	75-0717
NASTRAN	MODAL COUPLING PROCEDURES ADAPTED TO NASTRAN ANALYSIS OF THE 1/8 - SCALE SHUTTLE STRUCTURAL DYNAMICS	75-0677
NASTRAN	MODAL COUPLING PROCEDURES ADAPTED TO NASTRAN ANALYSIS OF THE 1/8-SCALE SHUTTLE STRUCTURAL DYNAMICS M	75-0678
NASTRAN	MSC/NASTRAN APPLICATION MANUAL	72-0567
NASTRAN	NASTRAN - A FINITE ELEMENT PROGRAM FOR STRUCTURAL ANALYSIS	72-0131
NASTRAN	NASTRAN ANALYSIS OF A TURBINE BLADE AND COMPARISON WITH TEST AND FIELD DATA	75-0367
NASTRAN	NASTRAN AS ANALYSIS TOOL IN A STRUCTURAL DESIGN OPTIMIZATION PROCESS	71-0235
NASTRAN	NASTRAN BUCKLING ANALYSIS OF A LARGE STIFFENED CYLINDRICAL SHELL WITH A CUTOUT	71-0045
NASTRAN	NASTRAN DIFFERENTIAL STIFFNESS ANALYSIS OF AN AIRCRAFT CANOPY	71-0037
NASTRAN	NASTRAN EXPERIENCES OF FORT WORTH OPERATION, CONVAIR AEROSPACE DIVISION OF GENERAL DYNAMICS	71-0019
NASTRAN	NASTRAN MODELING STUDIES IN THE NORMAL-MODE AND NORMAL-MODE SYNTHESIS	71-0069
NASTRAN	NASTRAN PLOTTING AT A REMOTE TERMINAL	74-0186
NASTRAN	NASTRAN PLOTTING CAPABILITIES	71-0172
NASTRAN	NASTRAN STRUCTURAL/THERMAL ANALYZER	72-0152
NASTRAN	NASTRAN USER INTERFACES, AUTOMATED INPUT INNOVATIONS	71-0126
NASTRAN	NASTRAN USERS EXPERIENCES	71-0713
NASTRAN	NASTRAN USERS EXPERIENCES	72-0753
NASTRAN	NASTRAN, NASA'S GENERAL PURPOSE STRUCTURAL ANALYSIS PROGRAM	70-0370
NASTRAN	NASTRAN: USER EXPERIENCE WITH FOUR EXAMPLE PROBLEMS	72-0804
NASTRAN	NASTRAN: USER'S EXPERIENCES	75-0675
NASTRAN	NEW ELEMENT DEFINITION CAPABILITY FOR NASTRAN	71-0125
NASTRAN	NORMAL MODE ANALYSIS OF A FUZE SUPPORT STRUCTURE USING NASTRAN. PART 1	72-0853
NASTRAN	NSRDC: NASTRAN COLLOQUIUM, WASHINGTON, D.C.	70-0404
NASTRAN	PLANETARY EXPLORER STRUCTURAL CONCEPT EVALUATION MODEL USING NASTRAN	71-0742
NASTRAN	PLANFORM INPUT GENERATOR (PING), A NASTRAN PRE-PROCESSOR FOR LIFTING SURFACES-THEORETICAL DEVELOPMEN	73-0776
NASTRAN	PLANFORM INPUT GENERATOR (PING), A NASTRAN PRE-PROCESSOR FOR LIFTING SURFACES-THEORETICAL DEVELOPMENT, U	73-0780
NASTRAN	POWER SPECTRAL DENSITY ANALYSIS OF VEHICLE VIBRATION USING THE NASTRAN COMPUTER PROGRAM	74-0318
NASTRAN	PRACTICAL ANALYSIS OF PLATE VIBRATIONS USING NASTRAN	71-0058
NASTRAN	PROCEDURE FOR EFFICIENT GENERATING, CHECKING, AND DISPLAYING NASTRAN INPUT AND OUTPUT DATA FOR ANALY	71-0098
NASTRAN	PRODUCTION OF MOTION PICTURES OF A CONTINUOUSLY DEFORMING STRUCTURE USING NASTRAN	71-0223
NASTRAN	PROGRAMMER'S MANUAL ADDITIONS AND DEMONSTRATION PROBLEMS FOR A THERMOSTRUCTURAL CAPABILITY FOR NASTR	75-0734
NASTRAN	RAPID CASE DESIGN USING FINITE ELEMENT STRESS ANALYSIS AND NASTRAN	73-0730

297

NASTRAN	REVIEW OF NASTRAN DEVELOPMENT RELATIVE TO EFFICIENCY OF EXECUTION	73-0938
NASTRAN	SAVINGS IN NASTRAN DECOMPOSITION TIME BY SEQUENCING TO REDUCE ACTIVE COLUMNS	71-0161
NASTRAN	SESAM 69 A GENERAL PURPOSE FINITE ELEMENT METHOD PROGRAM CDC/NASTRAN APPLICATIONS MANUAL	72-0552
NASTRAN	SOLID ELEMENT REQUIREMENT FOR NASTRAN	71-0256
NASTRAN	SOME ORGANIZATIONAL ASPECTS OF NASTRAN	73-1033
NASTRAN	SOME ORGANIZATIONAL ASPECTS OF NASTRAN	74-1005
NASTRAN	SP-222 NASTRAN THEORETICAL MANUAL	70-0387
NASTRAN	SP-222 NASTRAN USERS MANUAL	70-0386
NASTRAN	SPARSE MATRIX OPERATIONS IN NASTRAN	73-0336
NASTRAN	STRUCTURAL ANALYSIS AND DESIGN OPTIMIZATION OF SPACECRAFT USING NASTRAN	71-0068
NASTRAN	STRUCTURAL DYNAMIC ANALYSIS OF ELECTRONIC ASSEMBLIES USING NASTRAN RESTART/FORMAT CHANGE CAPABILITY	71-0230
NASTRAN	TEMPERATURE CALCULATIONS USING THE NASTRAN FINITE ELEMENT PROGRAMME	75-0642
NASTRAN	THE BANDIT COMPUTER PROGRAM FOR THE REDUCTION OF MATRIX BANDWIDTH FOR NASTRAN	72-0481
NASTRAN	THE NASTRAN COMPUTER PROGRAM FOR STRUCTURAL ANALYSIS	69-0183
NASTRAN	THE NASTRAN COMPUTER PROGRAM FOR STRUCTURAL ANALYSIS	71-0173
NASTRAN	THE NASTRAN COMPUTER PROGRAM FOR STRUCTURAL ANALYSIS	71-0682
NASTRAN	THE NASTRAN CONTOUR PLOTTER	72-0703
NASTRAN	THE NASTRAN HYDROELASTIC ANALYZER	71-0738
NASTRAN	THE NASTRAN PROGRAM FOR STRUCTURAL ANALYSIS	72-0325
NASTRAN	THE NASTRAN PROGRAMMER'S MANUAL	72-0480
NASTRAN	THE NASTRAN THEORETICAL MANUAL	70-0298
NASTRAN	THE NASTRAN USER'S MANUAL	72-0479
NASTRAN	THE RESPONSE OF SHELLS TO DISTRIBUTED RANDOM LOADS USING NASTRAN	71-0134
NASTRAN	THE USE OF NASTRAN IN THE ANALYSIS OF LARGE COMPLEX AIRFRAME STRUCTURES, MODELING TECHNIQUES AND ORG	71-0110
NASTRAN	THERMAL AND STRESS ANALYSIS BY NASTRAN	72-0153
NASTRAN	USING NASTRAN FOR DYNAMIC ANALYSIS OF VEHICLE SYSTEMS	74-0316
NASTRAN	USING THE NASTRAN THERMAL ANALYZER TO SIMULATE A FLIGHT SCIENTIFIC INSTRUMENT PACKAGE	74-1011
NASTRAN-GAP	NASTRAN-GAP, AN INTEGRATED APPROACH TO THE ANALYSIS OF RADIATION PATTERNS OF DISTORTED REFLECTORS	71-0062
NATIONAL	FOUNDATION INSTRUMENTATION FOR THE NATIONAL WESTMINISTER BANK TOWER	74-1147
NATIONAL	JOURNAL OF RESEARCH OF THE NATIONAL BUREAU OF STANDARDS: C. ENGINEERING AND INSTRUMENTATION	71-0664
NATRUAL	NATRUAL STRAINS AND STRESSES FOR TRAPEZOIDAL STRUCTURES ANALYSIS	69-0101
NATURAL	A FINITE ELEMENT MODEL FOR TWO-DIMENSIONAL STEADY FLOW THROUGH CONTRACTIONS IN NATURAL CHANNELS	71-0810
NATURAL	A FINITE ELEMENT SOLUTION OF GENERAL FLUID DYNAMICS PROBLEMS - NATURAL CONVECTION IN RECTANGULAR CAV	70-0008
NATURAL	A STIFFNESS MATRIX METHOD OF NATURAL MODE ANALYSIS	58-0002
NATURAL	AN INVESTIGATION OF THE EFFECTS OF SQUARE CUTOUTS ON THE NATURAL FREQUENCIES AND MODE SHAPES OF RECT	71-0638
NATURAL	APPLICATION OF FINITE ELEMENT ANALYSIS TO UNDERGROUND STORAGE OF NATURAL GAS	70-0198
NATURAL	APPLICATION OF NATURAL KINEMATIC MODES TO MATRIX ANALYSIS: SMALL AND LARGE DISPLACEMENTS AND STRAINS	67-0138
NATURAL	APPROXIMATION METHOD FOR COMPUTATION OF NATURAL FREQUENCIES OF SPIRAL SECTIONS	71-0775
NATURAL	AXISYMMETRIC FINITE ELEMENT IN NATURAL COORDINATES	75-0826
NATURAL	BODY ORIENTED (NATURAL) CO-ORDINATES FOR GENERATING THREE-DIMENSIONAL MESHES	74-0908
NATURAL	CALCULATION OF NATURAL FREQUENCIES OF HEAT EXCHANGER TUBES	73-0665
NATURAL	CALCULATION OF THE NATURAL FREQUENCIES AND THE PRINCIPAL MODES OF HELICOPTER BLADES	72-0907
NATURAL	CHISLENNYI METOD RASCHETA SOBSTVENNYKH I VYNUZHDENNYKH KOLEBANII SOSTAVNYKH OBOLOCHECHNYKH KONSTRUKT	73-0461
NATURAL	CLASSICAL AND NATURAL CONDUCTION AND DEFORMATION MODES CONCEPTS FOR AXISYMMETRIC VARIATIONAL METHODS	72-1003
NATURAL	COMPUTATION OF VIBRATION MODAL SHAPES AND NATURAL FREQUENCIES OF A GENERAL ORTHOTROPIC LAMINATED, TH	71-0309
NATURAL	EFFECT OF LATERAL PRESSURE ON NATURAL FREQUENCIES OF HYPERBOLOIDAL SHELLS OF REVOLUTION	71-0278
NATURAL	EFFECT OF SYSTEM PARAMETER VARIATIONS ON NATURAL FREQUENCIES	75-0661
NATURAL	EVALUATION OF THE STABILITY OF NATURAL SLOPES IN PLASTIC CHAMPLAIN CLAYS	74-0264
NATURAL	FINITE ELEMENT ANALYSIS OF LUBRICATION IN NATURAL JOINTS	74-0086
NATURAL	FINITE ELEMENT MODELLING OF NATURAL VIBRATION PROBLEMS	73-0320
NATURAL	HARBOUR OSCILLATION - A NUMERICAL TREATMENT FOR UNDAMPED NATURAL MODES	70-0208
NATURAL	LARGE NATURAL STRAINS AND SOME SPECIAL DIFFICULTIES DUE TO NONLINEARITY AND INCOMPRESSIBILITY IN FIN	74-0528
NATURAL	LARGE NATURAL STRAINS AND SOME SPECIAL DIFFICULTIES DUE TO NONLINEARITY AND INCOMPRESSIBILITY IN FIN	74-1027
NATURAL	METHOD OF MINIMUM WEIGHT DESIGN WITH REQUIREMENTS IMPOSED ON STRESSES AND NATURAL FREQUENCIES	70-0149
NATURAL	NATURAL BOUNDARY CONDITIONS IN FINITE-ELEMENT METHOD	72-0276
NATURAL	NATURAL COORDINATE SYSTEMS, AN AUTOMATIC INPUT DATA GENERATION SCHEME FOR A FINITE ELEMENT METHOD	70-0018
NATURAL	NATURAL FACTOR FORMULATION OF THE STIFFNESS FOR THE MATRIX DISPLACEMENT METHOD	75-0057
NATURAL	NATURAL FREQUENCIES OF COMPLEX FREE OR SUBMERGED STRUCTURES BY THE FINITE ELEMENT METHOD	65-0026
NATURAL	NATURAL FREQUENCIES AND MODE SHAPES OF TRANSFORMER CORES	65-0048
NATURAL	NATURAL FREQUENCIES AND MODE SHAPES OF PLATES WITH INTERIOR CUT-OUTS	70-0402
NATURAL	NATURAL FREQUENCIES OF CIRCULAR ARCHES	71-0302
NATURAL	NATURAL FREQUENCIES OF BOX-TYPE STRUCTURES BY A FINITE-ELEMENT METHOD	74-0193
NATURAL	NATURAL FREQUENCIES OF WIDE CHORD COMPRESSOR BLADES	74-0231
NATURAL	NATURAL GEOMETRY OF SURFACE WITH SPECIFIC REFERENCE TO THE MATRIX DISPLACEMENT ANALYSIS OF SHELLS. 1	73-0709
NATURAL	NATURAL PERIODS OF INFILLED FRAMES	72-0122
NATURAL	NATUPAL VIBRATION OF A FREE SURFACE	62-0008
NATURAL	NATURAL VIBRATIONS OF LAMINATED ORTHOTROPIC SHELLS OF REVOLUTION	70-0367
NATURAL	NATURAL VIBRATIONS OF THIN, FLAT WALLED STRUCTURES WITH DIFFERENT BOUNDARY CONDITIONS	71-0509
NATURAL	ON FINITE ELEMENT INTEGRATION IN NATURAL COORDINATES	73-0683
NATURAL	OPTIMUM DESIGN OF COMPOSITE SHELLS SUBJECT TO NATURAL FREQUENCY CONSTRAINTS	73-0154
NATURAL	SIMULATION OF RUNOFF HYDROGRAPHS FROM NATURAL WATERSHEDS BY FINITE-ELEMENT METHOD	73-1140
NATURAL	SOME ASPECTS OF THE NATURAL COORDINATE SYSTEM IN THE FINITE-ELEMENT METHOD	69-0105
NATURAL	SOME GENERAL CONSIDERATIONS ON THE NATURAL MODE TECHNIQUE	69-0049
NATURAL	SOME GENERAL CONSIDERATIONS ON THE NATURAL MODE TECHNIQUE, PART II: LARGE DISPLACEMENTS	69-0445
NATURAL	THE AUTOMATIC COMPUTATION OF THE NATURAL FREQUENCIES OF STRUCTURAL FRAMES USING AN EXACT MATRIX TECH	73-0318
NATURAL	THE EFFECT OF AN INITIAL AXISYMMETRIC IMPERFECTION ON THE NATURAL FREQUENCIES OF CONICAL SHELLS	70-0368
NATURAL	THE EFFECT OF CAMBER ON THE NATURAL FREQUENCIES OF LOW ASPECT RATIO TURBOMACHINERY BLADES	73-1092
NATURAL	THE NATURAL FACTOR APPROACH IN THE DISPLACEMENT METHOD - FOUNDATION AND FURTHER DEVELOPMENTS	75-0111
NATURAL	THE NATURAL FREQUENCIES OF PORTAL FRAMES II	65-0047
NATURAL	THE TREATMENT OF NATURAL BOUNDARY CONDITIONS IN THE FINITE ELEMENT AND FINITE DIFFERENCE METHODS	73-1098
NATURAL	TRANSIENT NATURAL CONVECTION COOLING OF A VERTICAL CIRCULAR CYLINDER	74-1169
NATURAL	USE OF SIMPLE THREE-DIMENSIONAL ACOUSTIC FINITE ELEMENTS FOR DETERMINING NATURAL MODES AND FREQUENCI	72-0010
NATURE	PROBLEM OF USING THE METHOD OF FINITE ELEMENTS TO CONSTRUCT AN ALGORITHM FOR CALCULATING A THIN-WALL	74-0851
NAVAL	DATA GENERATION FOR FINITE-ELEMENT STRUCTURAL ANALYSIS OF THREE-DIMENSIONAL NAVAL SHIP STRUCTURES	75-0707
NAVAL	DATA GENERATOR FOR THE IDEALIZATION FOR FINITE ELEMENT STRUCTURAL ANALYSIS OF NAVAL SHIP FLAT PLATED	72-0860
NAVAL	PROCEEDINGS OF THE 5TH NAVY-NASTRAN COLLOQUIUM, HELD AT NAVAL SHIP RESEARCH AND DEVELOPMENT CENTER,	74-0709
NAVIER	A HYBRID FINITE ELEMENT METHOD FOR TWO-DIMENSIONAL NAVIER STOKES EQUATIONS	75-0027
NAVIER	A NON VARIATIONAL FINITE ELEMENT ANALYSIS FOR THE NAVIER STOKES EQUATIONS	74-0010
NAVIER	AN APPROXIMATE NUMERICAL SOLUTION OF THE NAVIER STOKES EQUATIONS BY GALERKIN METHOD	74-0011
NAVIER	NAVIER STOKES EQUATIONS USING MIXED INTERPOLATION	74-0014
NAVIER	THE SOLUTION OF THE NAVIER STOKES EQUATIONS	74-1220
NAVIER	VARIATIONAL FINITE ELEMENT METHODS FOR TWO DIMENSIONAL AND AXISYMMETRIC NAVIER STOKES EQUATIONS	74-0012
NAVIER-POISSON	A NOTE ON SOME VARIATIONAL STATEMENTS FOR THE SLOW FLOW OF A NAVIER-POISSON FLUID	74-1054
NAVIER-STOKES	A FINITE ELEMENT ANALOGUE OF THE NAVIER-STOKES EQUATION	70-0099
NAVIER-STOKES	A FINITE ELEMENT SOLUTION ALGORITHM FOR THE NAVIER-STOKES EQUATIONS	74-0784
NAVIER-STOKES	A NUMERICAL SOLUTION OF THE NAVIER-STOKES EQUATIONS USING THE FINITE ELEMENT TECHNIQUE	73-0061
NAVIER-STOKES	CONVERGENCE AND ERROR ESTIMATES FOR FINITE ELEMENT APPROXIMATIONS OF STATIONARY NAVIER-STOKES EQUATI	74-0122
NAVIER-STOKES	FINITE ELEMENT ANALOGUE OF NAVIER-STOKES EQUATIONS	69-0002
NAVIER-STOKES	FINITE-ELEMENT SOLUTION OF NAVIER-STOKES EQUATIONS FOR TRANSIENT 2-DIMENSIONAL INCOMPRESSIBLE FLOW	75-0134
NAVIER-STOKES	NUMERICAL SOLUTION OF NAVIER-STOKES EQUATIONS BY FINITE ELEMENT METHOD	72-0008
NAVIER-STOKES	NUMERICAL SOLUTION OF THE STATIONARY NAVIER-STOKES EQUATIONS BY FINITE ELEMENT METHODS	74-0408
NAVIER-STOKES	ON THE FINITE ELEMENT FORMULATION OF NAVIER-STOKES EQUATIONS	74-0720

NAVIER-STOKES	ON THE NUMERICAL TREATMENT OF THE NAVIER-STOKES EQUATIONS FOR AN INCOMPRESSIBLE FLUID	73-0809
NAVIER-STOKES	ON THE SOLUTION OF THE NAVIER-STOKES EQUATION IN TWO DIMENSIONAL AND AXIAL SYMMETRIC PROBLEMS	74-0007
NAVIGATION	NAVIGATION LOCK STRUCTURE ANALYSIS BY FINITE ELEMENT METHOD	71-0816
NAVY	PROCEEDINGS OF THE NAVY-NASTRAN COLLOQUIUM (4TH) HELD AT NAVY SHIP RESEARCH AND DEVELOPMENT CENTER,	73-0888
NAVY-NASTRAN	NAVY-NASTRAN COLLOQUIUM, WASHINGTON, D.C.	70-0403
NAVY-NASTRAN	NAVY-NASTRAN COLLOQUIM AT WASHINGTON, D. C.	72-0669
NAVY-NASTRAN	PROCEEDINGS OF THE NAVY-NASTRAN COLLOQUIUM (4TH) HELD AT NAVY SHIP RESEARCH AND DEVELOPMENT CENTER,	73-0888
NAVY-NASTRAN	PROCEEDINGS OF THE 5TH NAVY-NASTRAN COLLOQUIUM, HELD AT NAVAL SHIP RESEARCH AND DEVELOPMENT CENTER,	74-0709
NAYAK	NEW LIGHT ON THE NAYAK ALPHA TECHNIQUE	73-0614
NEAR-CRACK	NEAR-CRACK DEFORMATION FIELDS IN STRAIN-HARDENING MATERIAL	74-0938
NEARLY-INCOMPRE	VARIATIONAL PRINCIPLE FOR INCOMPRESSIBLE AND NEARLY-INCOMPRESSIBLE ANISOTROPIC ELASTICITY	69-0319
NEIGHBOURHOOD	ELASTIC AND PLASTIC STRESS DISTRIBUTION ON BASE PLATES IN THE NEIGHBOURHOOD OF FILLET WELDS IN LAP J	70-0175
NELINEARNE	GEOMETRICKY NELINEARNE ULOHY PLOCHYCH SKRUPIN RIESENE PRIRASTKOVOU METODOU KONECNYCH PRVKOV. (GEOMET	74-1032
NELINEARNIHO	STATICKE RESENI FYZIKALNE NELINEARNIHO CHOVANI ZELEZOBETO- NOVYCH DESEK (ANALYSIS OF PHYSICALLY NONL	74-0514
NEODNORODNYMI	RASCHET KILEBANII PLASTIN SLOZHNOI FORMY S NEODNORODNYMI GRANICHNYMI USLOVIYAMI. (CALCULATION OF VIB	74-0988
NERVA	ENGINEERING OPERATIONS REPORT: NERVA THIN SHELL FINITE ELEMENT STRESS ANALYSIS METHODOLOGY COMPUTER	72-0916
NERVA	ENGINEERING OPERATIONS REPORT: NERVA AXISYMMETRIC AND PLANE STRUCTURES FINITE ELEMENT STRESS ANALYSI	72-0917
NESTED	AUTOMATIC NESTED DISSECTION	74-0860
NESTED	NESTED DISSECTION OF A REGULAR FINITE ELEMENT MESH	73-0042
NETS	ANALYSIS OF CABLE NETS IN MIXED FORMULATION	73-0311
NETWORK	AN AUTOMATIC RECORDING SCHEME FOR SIMULTANEOUS EQUATIONS DERIVED FROM NETWORK SYSTEMS	70-0285
NETWORK	FINITE-ELEMENT NETWORK MODELS AND THEIR APPLICATION TO EDDY-CURRENT PROBLEMS	75-0143
NETWORKS	BENDING ELEMENTS FOR PLATE AND SHELL NETWORKS	67-0043
NETWORKS	FINITE ELEMENT ANALYSIS OF PLANAR MICROWAVE NETWORKS	73-0016
NETWORKS	FINITE-ELEMENT METHOD FOR WATER-DISTRIBUTION NETWORKS	75-0269
NETWORKS	LARGE DEFLECTION ANALYSIS OF PRESTRESSED NETWORKS	72-0169
NETWORKS	MEMBRANE ANALOGY FOR ANISOTROPIC CABLE NETWORKS	74-0233
NETWORKS	METHODS OF CALCULATION OF THREE-DIMENSIONAL PROBLEMS OF PERCOLATION OF FISSURED ROCK BY FINITE ELEME	70-0472
NETWORKS	NUMERICAL MODEL FOR THE PREDICTION OF TRANSIENT WATER QUALITY IN ESTUARY NETWORKS	72-0635
NETWORKS	SHAPE DETERMINATION OF PREFABRICATED PRESTRESSED NETWORKS	73-0889
NETWORKS	STEADY STATE FLOW IN RIGID NETWORKS OF FRACTURES	74-0278
NETWORKS	THE ENERGY METHOD OF NETWORKS OF ARBITRARY SHAPE IN PROBLEMS OF THE THEORY OF ELASTICITY	59-0001
NETWORKS	THEORY OF A FINITE DIFFERENCE METHOD ON IRREGULAR NETWORKS	74-0375
NETZWERKE	BERECHNUNG VORGESPANNTER NETZWERKE	70-0399
NETZWERKKONSTRU	EIN VERFAHREN DUR DIE FORMFINDUNG VEON BELIEBIGEN VORGESPANNTEN NETZWERKKONSTRUKTIONEN	72-0641
NETZWERKKONSTRU	THEORIE, PROGRAMMENTWICKLUNG UND ERFAHRUNG ANVORGESPANNTEN NETZWERKKONSTRUKTIONEN	72-0640
NEUARTIGE	NEUARTIGE ANWENDUNG GEOMECHANISCHER UNTERSUCHUNGSMETHODEN ZUR BEURTEILUNG DER ABBAUWIRKUNGEN IM GANG	75-0530
NEUTRALLY	SOLUTION OF A TWO DIMENSIONAL DIFFUSION CONVECTION EQUATION DESCRIBING THE MASS TRANSFER OF A NEUTRA	74-0037
NEUTRON	A FINITE ELEMENT TREATMENT OF NEUTRON DIFFUSION	71-0373
NEUTRON	AN OPTIMUM IDEALIZATION IN FINITE ELEMENT METHOD FOR A TWO-GROUP NEUTRON DIFFUSION PROBLEM	74-1143
NEUTRON	APPLICATION OF FINITE ELEMENT TECHNIQUES FOR THE NUMERICAL SOLUTION OF THE NEUTRON TRANSPORT AND DIF	71-0013
NEUTRON	APPLICATION OF FINITE ELEMENT SOLUTION TECHNIQUE TO NEUTRON DIFFUSION AND TRANSPORT EQUATIONS	71-0015
NEUTRON	APPLICATION OF FINITE ELEMENT METHOD TO MULTIGROUP NEUTRON DIFFUSION EQUATION	72-0310
NEUTRON	APPLICATION OF PHASE-SPACE FINITE ELEMENTS TO ONE DIMENSIONAL NEUTRON TRANSPORT EQUATION	73-0123
NEUTRON	APPLICATION OF PHASE SPACE FINITE ELEMENTS TO THE TWO DIMENSIONAL NEUTRON TRANSPORT EQUATION IN X-Y	73-0453
NEUTRON	APPLICATION OF THE FINITE ELEMENT METHOD TO TWO DIMENSIONAL MULTIGROUP NEUTRON DIFFUSION	72-0971
NEUTRON	APPLICATIONS OF FINITE ELEMENT METHODS IN REACTOR MATHEMATICS - NUMERICAL SOLUTION OF THE NEUTRON DI	72-0535
NEUTRON	APPLICATIONS OF FINITE ELEMENT METHODS IN REACTOR MATHEMATICS. NUMERICAL SOLUTION OF THE NEUTRON TRA	74-0853
NEUTRON	CURVED LOW ORDER ANGULAR ELEMENTS IN TWO DIMENSIONAL NEUTRON TRANSPORT PROBLEMS	74-1322
NEUTRON	DUAL FINITE ELEMENT METHODS FOR NEUTRON DIFFUSION	71-0489
NEUTRON	FINITE ELEMENT INTEGRAL NEUTRON TRANSPORT	73-1032
NEUTRON	FINITE ELEMENT METHODS FOR THE NEUTRON DIFFUSION EQUATIONS	71-0372
NEUTRON	FINITE ELEMENT METHOD APPLIED TO NEUTRON DIFFUSION PROBLEMS	73-0977
NEUTRON	FINITE ELEMENT METHOD FOR SOLVING NEUTRON TRANSPORT PROBLEMS IN TWO-DIMENSIONAL CYLINDRICAL GEOMETRY	74-1081
NEUTRON	FINITE ELEMENT NEUTRON DIFFUSION IN CURVILINEAR COORDINATES	72-0039
NEUTRON	FINITE ELEMENT SOLUTION TECHNIQUE FOR NEUTRON TRANSPORT EQUATION	72-0730
NEUTRON	FINITE ELEMENTS IN NEUTRON TRANSPORT THEORY	73-0961
NEUTRON	ON A FINITE ELEMENT METHOD FOR SOLVING THE NEUTRON TRANSPORT EQUATION	74-1255
NEUTRON	ON A FINITE ELEMENT METHOD FOR SOLVING THE NEUTRON TRANSPORT EQUATION	74-1355
NEUTRON	QUADRATIC FINITE ELEMENT IN NEUTRON TRANSPORT	73-1039
NEUTRON	SOLUTION OF MULTI-DIMENSIONAL NEUTRON TRANSPORT EQUATION BY FINITE ELEMENT METHOD	72-0564
NEUTRON	TIMING COMPARISON STUDY FOR SOME HIGH ORDER FINITE ELEMENT PROCEDURES AND A LOW ORDER FINITE DIFFERE	72-0767
NEUTRON	TRIANGULAR MESH METHODS FOR THE NEUTRON TRANSPORT EQUATION	73-1062
NEUTRON-TRANSPO	APPLICATION OF FINITE-ELEMENTS TO ONE-DIMENSIONAL AND TWO-DIMENSIONAL NEUTRON-TRANSPORT	73-1150
NEUTRON-TRANSPO	CONVERGENCE OF FINITE-ELEMENT METHOD IN NEUTRON-TRANSPORT	75-0408
NEUTRON-TRANSPO	RAY-EFFECT MITIGATION IN DISCRETE ORDINATE-LIKE ANGULAR FINITE-ELEMENT APPROXIMATIONS IN NEUTRON-TRA	75-0496
NEUTRON-TRANSPO	2-DIMENSIONAL FINITE-ELEMENT METHOD FOR INTEGRAL NEUTRON-TRANSPORT CALCULATIONS	75-0443
NEW-YORK	GALERKIN-FINITE ELEMENT SIMULATION OF GROUNDWATER CONTAMINATION ON LONG-ISLAND, NEW-YORK	73-0093
NEW-ZEALAND	GALERKIN-FINITE ELEMENT ANALYSIS OF HYDROTHERMAL SYSTEM AT WAIRAKEI, NEW-ZEALAND	75-0601
NEWTON	THE CONJUGATE NEWTON METHOD	73-0560
NEWTONIAN	FINITE ELEMENT ANALYSIS OF THE STEADY FLOW OF NON NEWTONIAN FLUIDS THROUGH PARALLEL SIDED CONDUITS	74-0056
NEWTONIAN	FINITE ELEMENT APPLICATIONS TO NEWTONIAN FLUIDS	68-0197
NEWTONIAN	NEWTONIAN AND NON-NEWTONIAN VISCOUS FLOW	75-0109
NEWTONIENS	CALCUL NUMERIQUE DES ECOULEMENTS DES FLUIDES DE BINGHAM ET DES FLUIDES NEWTONIENS INCOMPRESSIBLES PA	72-0349
NICHT	EINIGE VERFAHREN ZUR BERECHNUNG VON EIGENWERTEN UND EIGENVEKTOREN VON NICHT HERMITESCHEN MATRIZEN U	72-0646
NICHT-LINEARER	DIE LOSUNG NICHT-LINEARER PROBLEME MIT DER METHODE DER FINITEN ELEMENTE	74-0673
NICHTLINEARE	DER GERISSENE ZUSTAND ZWEISEITIG GELAGERTER PLATTEN UNTER EINZELLASTEN - NICHTLINEARE BERECHNUNG MIT	75-0858
NICHTLINEARE	HYBRIDE FINITE VERFAHREN FUER LINEARE UND NICHTLINEARE INSTATIONAERE WAERMELEITPROBLEME (HYBRID FINI	73-0484
NICHTLINEARE	NICHTLINEARE DIRICHTLET PROBLEME UND EINFACHE FINITE ELEMENT VERFAHREN	74-1263
NICHTLINEAREN	TURBAN VERSION 1.0. EIN SPEZIELLES PROGRAMM ZUR LINEAREN UND NICHTLINEAREN STATISCHEN UND DYNAMISCHE	71-0622
NICHTLINEARER	NUMERISCHE BERECHNUNG NICHTLINEARER ELASTISCHER KOERPER	63-0017
NICHTSTATIONARE	BEITRAG ZUR SIMULATION NICHTSTATIONARER HORIZONTAL-EVENER GRUNDWASSERSTROMUNGSFELDER AUF DER GRUNDLA	74-0088
NICKED	THE COMPUTATION OF TEARING ENERGY OF NICKED RUBBER STRIPS IN EXTENSION	74-0315
NIESPELNIAPACA	FUNKCJA KSZTALTU NIESPELNIAPACA WARUNKOW CIAGLOSCI W ANALIZIE PLYT METODA ELEMENTOW SKONCZONYCH. (SH	75-0263
NIGHT-SYMMETRIS	DIE BERECHNUNG ACHSENSY METRISCHER TRAGWERKE UNTER ACHSENSYMETRISCHER UND NIGHT-SYMMETRISHCER BELAST	69-0452
NINIGRET	MODELING AN OCEAN POND. A TWO-DIMENSIONAL, FINITE ELEMENT HYDRODYNAMIC MODEL OF NINIGRET POND, CHARL	75-0686
NO-LOAD	NONLINEAR FINITE ELEMENT SOLUTION OF ELECTRICAL MACHINES UNDER NO-LOAD AND FULL-LOAD CONDITIONS	74-0472
NO-TENSION	STRESS ANALYSIS OF ROCK AS A NO-TENSION MATERIAL	68-0174
NODAL	AUTOMATIC NODAL TRIANGULATION FOR FINITE ELEMENTS	73-0178
NODAL	DETERMINATION OF STRESSES AT NODAL POINTS FROM FINITE ELEMENT SOLUTIONS TO ELASTIC PROBLEMS	71-0307
NODAL	EFFECTIVENESS OF EXCESSIVE NODAL CONTINUITIES IN THE FINITE ELEMENT ANALYSIS OF THIN RECTANGULAR AND	70-0346
NODAL	INADEQUACY OF NODAL CONNECTIONS IN A STIFFNESS SOLUTION FOR PLATE BENDING	65-0021
NODAL	NODAL ANALYSIS OF A NINE BAY SKIN-STRINGER PANEL	71-0102
NODAL	NODAL ANALYSIS OF THE MATED SPACE SHUTTLE CONFIGURATION	71-0097
NODAL	NODAL VARIABLES FOR ARBITRARY ORDER FINITE ELEMENTS	75-0002
NODAL	THE EFFECTIVENESS OF EXCESSIVE NODAL CONTINUITIES IN THE FINITE ELEMENT ANALYSIS OF THIN RECTANGULAR	70-0105
NODE	A SIX NODE TRIANGULAR EQUILIBRIUM (CONSTANT STRESS) MEMBRANE ELEMENT	73-0783
NODE	A SIX NODE TRIANGULAR SHEAR PANEL FOR EQUILIBRIUM MODELS	73-1011
NODE	A THREE DIMENSIONAL "VARIABLE NODE" ISOPARAMETRIC SOLID ELEMENT	74-0779
NODE	AN AUTOMATIC NODE RELABELLING SCHEME FOR BANDWIDTH MINIMIZATION OF STIFFNESS MATRICES	68-0256
NODE	NODE NUMBERING OPTIMIZATION IN STRUCTURAL ANALYSIS	75-0077
NODE	PLANE STRESS ANALYSIS USING THE FINITE ELEMENT METHOD. TRIANGULAR FINITE ELEMENT WITH SIX PARAMETERS	67-0106

299

NODE	THREE NODE ISOPARAMETRIC STRESS AXIAL ELEMENT	72-0992
NODE-ELEMENT	THE APPLICATION OF NODE-ELEMENT RULES FOR FORECASTING PROBLEMS IN THE GENERATION OF FINITE ELEMENT M	75-0614
NODES	A TECHNIQUE FOR DEGENERATING BRICK-TYPE ISOPARAMETRIC ELEMENTS USING HIERARCHICAL MIDSIDE NODES	74-0403
NODES	ANALYSIS OF OUT-OF-PLANE STRESS BY THE SIX NODES. FOURTH ORDER ACCURACY TRIANGULAR ELEMENTS	69-0375
NODES	FINITE ELEMENTS FOR AXISYMMETRIC SOLIDS UNDER ARBITRARY LOADINGS WITH NODES ON ORIGIN.	72-0251
NODES	FINITE ELEMENTS FOR AXISYMMETRIC SOLIDS UNDER ARBITRARY LOADINGS WITH NODES ON ORIGIN	73-0094
NODES	RULES GOVERNING THE NUMBER OF NODES AND ELEMENTS IN A FINITE ELEMENT MESH	70-0053
NOISE	IS THE COMPUTATION OF NOISE LEVEL INSIDE A CAR FEASIBLE	71-0631
NOISE	JET NOISE EXCITATION OF AN INTEGRALLY STIFFENED PANEL	71-0546
NOMERIQUE	RESOLUTION NOMERIQUE PAR DES METHODES D ELEMENTS FINIS DE PROBLEMES ELLIPTIQUES DEGENERES	72-0702
NOMINAL	FRACTURE ANALYSIS IN AREAS OF HIGH NOMINAL STRAIN	73-0495
NOMINAL	ON SMALL VIBRATIONS AND PERTURBATIONS OF FLEXIBLE BODIES UNDERGOING ARBITRARY NOMINAL MOTION	70-0376
NON-AXISYMMETRI	RING FINITE ELEMENTS FOR AXISYMMETRIC AND NON-AXISYMMETRIC THIN SHELL ANALYSIS	72-0922
NON-CONFORMING	AN ITERATIVE IMPROVEMENT FOR NON-CONFORMING BENDING ELEMENTS	75-0844
NON-CONFORMING	CONVERGENCE OF FINITE ELEMENT SOLUTIONS REPRESENTED BY A NON-CONFORMING BASIS	72-0812
NON-CONFORMING	NON-CONFORMING FINITE ELEMENT METHODS AND THE PATCH TEST	73-0640
NON-CONFORMING	ON THE CONVERGENCE OF FINITE ELEMENT SCHEMES BASED ON NON-CONFORMING AND SIMPLIFIED HYBRID DISPLACEM	73-0841
NON-CONFORMING	TRIANGULAR ELEMENTS IN PLATE BENDING CONFORMING AND NON-CONFORMING SOLUTIONS	66-0011
NON-CONSERVATIV	FINITE ELEMENT METHOD APPLIED TO THE PROBLEM OF STABILITY OF A NON-CONSERVATIVE SYSTEM	71-0291
NON-CONVENTIONA	THE KANSAS TEST TRACK. NON-CONVENTIONAL TRACK STRUCTURES. DESIGN REPORT	72-0797
NON-COPLANAR	FINITE ELEMENT METHOD APPLIED TO CALCULATIONS OF NON-COPLANAR PLATE SYSTEMS	74-0241
NON-DARCY	FINITE ELEMENT ANALYSIS OF NON-DARCY FLOW	70-0268
NON-DARCY	FINITE ELEMENT METHOD APPLIED TO UNSTEADY NON-DARCY FLOW	70-0088
NON-DARCY	METHOD OF ANALYZING NON-DARCY FLOW PROBLEMS BY FINITE ELEMENTS	75-0158
NON-DARCY	NON-DARCY FLOW SOLVED BY FINITE ELEMENT ANALYSIS	69-0366
NON-DARCY	NUMERICAL METHODS FOR UNSTEADY NON-DARCY FLOW	74-0062
NON-DARCY	VARIATIONAL APPROACH TO NON-DARCY FLOW	70-0089
NON-HOMOGENEOUS	ON THE DISTRIBUTION OF STRESSES AROUND CIRCULAR TUNNELS IN NON-HOMOGENEOUS ROCKS	66-0029
NON-HOMOGENEOUS	TRANSIENT RESPONSE OF NON-UNIFORM, NON-HOMOGENEOUS BEAMS	73-0162
NON-ISOTROPIC	COMPUTERIZED ANALYSIS OF NON-ISOTROPIC STRUCTURES	69-0192
NON-LINEAIRE	ANALYSE TRIDIMENSIONALE DU COMPORTEMENT NON-LINEAIRE D'UN CAISSON DE REACTEUR NUCLEAIRE EN BETON PRE	71-0144
NON-LINEAR	A FINITE ELEMENT METHOD FOR THE RESOLUTION OF NON-LINEAR BOUNDARY-VALUE PROBLEMS	75-0107
NON-LINEAR	AIRCRAFT-PAVEMENT INTERACTION STUDIES. PHASE 1: A FINITE ELEMENT MODEL OF A JOINTED CONCRETE PAVEMEN	73-0894
NON-LINEAR	ALGORITHM FOR THE LARGE DEFLECTION GEOMETRICALLY NON-LINEAR BEHAVIOUR OF PLANE AND CURVED STRUCTURES	72-0083
NON-LINEAR	AN APPROXIMATE NON-LINEAR ANALYSIS OF THIN PLATES	68-0286
NON-LINEAR	APPLICATION OF A CONJUGATED-GRADIENT-PROCESS IN FINITE-ELEMENT-CALCULATION OF NON-LINEAR STRUCTURES	74-0214
NON-LINEAR	APPLICATION OF FINITE ELEMENT METHOD TO PROBLEMS OF NON-LINEAR BEHAVIOR WITH CREEP	75-0811
NON-LINEAR	APPLICATION OF THE PSEUDO-FUNCTIONAL FINITE ELEMENT METHOD TO NON-LINEAR PROBLEMS	74-0282
NON-LINEAR	CALCULATION OF NON-LINEAR AXISYMMETRIC VIBRATIONS OF THIN SHELLS OF REVOLUTION BY A FINITE ELEMENT M	75-0067
NON-LINEAR	CONVERGENCE, ACCURACY AND STABILITY OF FINITE ELEMENT APPROXIMATIONS OF A CLASS OF NON-LINEAR HYPERB	73-0159
NON-LINEAR	CUBIC SPLINE GALERKIN APPROXIMATIONS TO PARABOLIC SYSTEMS WITH COUPLED NON-LINEAR BOUNDARY CONDITION	75-0232
NON-LINEAR	DIGITAL ANALYSIS OF NON-LINEAR STRUCTURES BY THE FORCE METHOD	64-0023
NON-LINEAR	DYNAMIC RESPONSE ANALYSIS OF GEOMETRICALLY NON-LINEAR STRUCTURES SUBJECTED TO HIGH IMPACT	72-0193
NON-LINEAR	DYNAMIC RESPONSE ANALYSIS OF GEOMETRICALLY NON-LINEAR STRUCTURES SUBJECTED TO HIGH IMPACT	75-0555
NON-LINEAR	DYNAMIC TRANSIENT NON-LINEAR BEHAVIOUR OF THICK AND THIN PLATES	75-0125
NON-LINEAR	FINITE ELEMENT ANALYSIS OF NON-LINEAR COMPOSITE MATERIALS BY USE OF OVERLAY SYSTEMS	74-0928
NON-LINEAR	FINITE ELEMENT ANALYSIS OF ENGINEERING SYSTEMS GOVERNED BY A NON-LINEAR QUASI-HARMONIC EQUATION	75-0664
NON-LINEAR	FINITE ELEMENT SOLUTION OF NON-LINEAR HEAT CONDUCTION PROBLEMS WITH SPECIAL REFERENCE TO PHASE CHANG	74-0348
NON-LINEAR	FINITE ELEMENT SOLUTION OF A NON-LINEAR DIFFERENTIAL EQUATION USING KACHANOV'S AND SUCCESSIVE APPROX	75-0108
NON-LINEAR	FINITE-ELEMENT ANALYSIS OF A FRACTURE TOUGHNESS TEST SPECIMEN IN NON-LINEAR RANGE	72-0093
NON-LINEAR	FINITE-ELEMENT STRESS ANALYSIS OF A NON-LINEAR PROBLEM. A CONNECTING-ROD EYE LOADED BY MEANS OF A PI	70-0239
NON-LINEAR	LARGE DEFLECTION GEOMETRICALLY NON-LINEAR FINITE ELEMENT ANALYSIS OF CIRCULAR ARCHES	73-0021
NON-LINEAR	LINEAR AND NON-LINEAR ANALYSIS OF THIN SHALLOW SHELLS BY MIXED FINITE ELEMENTS	75-0665
NON-LINEAR	MATRIX DIFFERENTIAL EQUATION AND HIGHER-ORDER NUMERICAL METHODS FOR PROBLEMS OF NON-LINEAR CREEP. VI	72-0198
NON-LINEAR	NON-LINEAR ANALYSIS OF THE ROTATIONAL STIFFNESS OF A NUCLEAR STEAM GENERATOR FOUNDATION	69-0286
NON-LINEAR	NON-LINEAR ANALYSIS OF PLATES CONSIDERING GEOMETRIC AND MATERIAL EFFECTS	71-0702
NON-LINEAR	NON-LINEAR ANALYSIS OF REINFORCED CONCRETE SLABS	72-0191
NON-LINEAR	NON-LINEAR DYNAMIC ANALYSIS OF AXISYMMETRIC SHELLS	75-0474
NON-LINEAR	NON-LINEAR EFFECTS OF HIGH TEMPERATURE ON THE VIBRATION OF A SIMPLY SUPPORTED BEAM WITH A CENTRAL MA	73-0838
NON-LINEAR	NON-LINEAR ELASTIC ANALYSIS OF UNGUYED TOWERS AND STACKS	74-1157
NON-LINEAR	NON-LINEAR FINITE ELEMENT ANALYSIS OF REINFORCED CONCRETE BY THE FINITE ELEMENT METHOD	68-0241
NON-LINEAR	NON-LINEAR FLOW IN POROUS MEDIA BY FINITE ELEMENTS	69-0023
NON-LINEAR	NON-LINEAR MATERIAL ANALYSIS WITH FINITE ELEMENTS AND INCREMENTAL METHODS	68-0199
NON-LINEAR	NON-LINEAR RESPONSE OF A BEAM TO PERIODIC LOADING.	72-0407
NON-LINEAR	NON-LINEAR STRESS ANALYSIS OF TWO DIMENSIONAL PROBLEMS WITH SPECIAL REFERENCE TO ROCK AND SOIL MECHA	68-0183
NON-LINEAR	NON-LINEAR THERMAL STRESS ANALYSIS FOR NUCLEAR POWER PLANT BY FINITE ELEMENT METHOD	71-0181
NON-LINEAR	NON-LINEAR TRANSIENT FINITE ELEMENT ANALYSIS WITH CONVECTED CO-ORDINATES	73-0378
NON-LINEAR	POST-BUCKLING AND NON-LINEAR ANALYSIS BY FINITE ELEMENT METHOD AS A SUPPLEMENT TO A LINEAR ANALYSIS	75-0505
NON-LINEAR	SOME ASPECTS OF NON-LINEAR FINITE ELEMENT COMPUTER PROGRAMS	75-0778
NON-LINEAR	SOME LINEAR AND NON-LINEAR PROBLEMS IN FLUID MECHANICS. FINITE ELEMENT METHOD FORMULATION	73-0523
NON-LINEAR	SOME METHODS OF SOLVING PROBLEMS OF NON-LINEAR THERMOVISCOELASTICITY	74-1207
NON-METALLIC	NON-METALLIC ANTENNA-SUPPORT MATERIALS	73-0716
NON-MODERATE	FRACTURE TOUGHNESS PARAMETERS AND ELASTIC-PLASTIC ANALYSIS OF NON-MODERATE FRACTURE CONDITIONS USING	73-0258
NON-MODERATE	FRACTURE TOUGHNESS PARAMETERS AND ELASTIC-PLASTIC ANALYSIS OF NON-MODERATE FRACTURE CONDITIONS USING	73-1126
NON-NEWTONIAN	FINITE ELEMENT ANALYSIS OF TWO-DIMENSIONAL SLOW NON-NEWTONIAN FLOWS	72-0035
NON-NEWTONIAN	FINITE ELEMENT ANALYSIS OF SLOW NON-NEWTONIAN CHANNEL FLOW	72-0308
NON-NEWTONIAN	FINITE ELEMENT ANALYSIS OF STEADY FLOW OF NON-NEWTONIAN FLUIDS	74-0465
NON-NEWTONIAN	FINITE ELEMENT METHODS FOR THE SOLUTION OF SOME INCOMPRESSIBLE NON-NEWTONIAN FLUID MECHANICS PROBLEM	75-0224
NON-NEWTONIAN	NEWTONIAN AND NON-NEWTONIAN VISCOUS FLOW	75-0109
NON-NEWTONIAN	VISCOUS, INCOMPRESSIBLE FLOW WITH SPECIAL REFERENCE TO NON-NEWTONIAN (PLASTIC) FLUIDS	74-0053
NON-PROPAGATING	ANALYTICAL APPROACH TO THE NON-PROPAGATING CRACK PROBLEM USING THE FINITE ELEMENT METHOD	71-0389
NON-REMOVABLE	FINITE ELEMENT SOLUTION OF BOUNDARY-VALUE PROBLEMS WITH NON-REMOVABLE SINGULARITIES	73-0124
NON-SELF-ADJOIN	A THEORY OF MIXED FINITE ELEMENT APPROXIMATIONS OF NON-SELF-ADJOINT BOUNDARY-VALUE PROBLEMS	74-0340
NON-SELF-ADJOIN	A THEORY OF MIXED FINITE ELEMENT APPROXIMATIONS OF NON-SELF-ADJOINT BOUNDARY VALUE PROBLEMS	74-0715
NON-SELF-ADJOIN	FINITE ELEMENT METHOD FOR NON-SELF-ADJOINT PROBLEMS	73-0271
NON-SELF-ADJOIN	NON-SELF-ADJOINT VARIATIONAL PROCEDURE FOR FINITE-ELEMENT APPROXIMATION OF TRANSPORT EQUATION	75-0072
NON-SELFADJOINT	APPROXIMATION OF STEKLOV EIGENVALUES OF NON-SELFADJOINT SECOND ORDER ELLIPTIC OPERATORS	72-0509
NON-SHALLOW	AN EXTENSION OF THE SHALLOW TO THE NON-SHALLOW STIFFNESS MATRIX FOR A CYLINDRICAL SHELL FINITE ELEME	70-0104
NON-SINGULAR	INFLUENCE OF NON-SINGULAR STRESS TERMS AND SPECIMEN GEOMETRY ON SMALL SCALE YIELDING AT CRACK TIPS I	73-0203
NON-STANDARD	THE RING OF FINITE ELEMENTS IN A NON-STANDARD MODEL OF THE REALS	71-0450
NON-STATIONARY	A FINITE ELEMENT METHOD FOR THE DETERMINATION OF NON-STATIONARY TEMPERATURE DISTRIBUTIONS AND THERMA	66-0074
NON-STATIONARY	HYBRIDE FINITE VERFAHREN FUER LINEARE UND NICHTLINEARE INSTATIONAERE WAERMELEITPROBLEME (HYBRID FINI	70-0484
NON-STRUCTURAL	IMPROVEMENTS OF FINITE ELEMENT SOLUTIONS FOR STRUCTURAL AND NON-STRUCTURAL APPLICATIONS	71-0229
NON-STRUCTURAL	NON-STRUCTURAL APPLICATIONS OF THE FINITE ELEMENT METHOD	73-0330
NON-UNIFORM	COMPUTER METHOD FOR COMPUTING CRITICAL SPEEDS OF NON-UNIFORM SHAFTS GEARED SYSTEMS AND CURVED ASSEMB	73-0675
NON-UNIFORM	DYNAMICAL ANALYSIS OF AXIALLY NON-UNIFORM, THIN CYLINDRICAL SHELLS. PART 1 - MATRIX FORMULATION	70-0413
NON-UNIFORM	MATRIX DYNAMIC AND INSTABILITY ANALYSIS WITH NON-UNIFORM ELEMENTS	70-0062
NON-UNIFORM	NON-UNIFORM DEFORMATION OF A BLOCK IN PLANE-STRAIN COMPRESSION CAUSED BY FRICTION. PART 4, ANALYSIS	71-0786
NON-UNIFORM	NON-UNIFORM DEFORMATION OF MATERIAL IN AXIALLY SYMMETRIC COMPRESSION CAUSED BY FRICTION. PART 3, THE	72-0923
NON-UNIFORM	NON-UNIFORM FLOW OF POLYCRYSTALS BY GRAIN-BOUNDARY SLIDING ACCOMMODATED BY POWER-LAW CREEP	75-0562
NON-UNIFORM	NUMERICAL SOLUTION OF THE QUASI LINEAR POISSON EQUATION IN A NON-UNIFORM TRIANGLE MESH	67-0090
NON-UNIFORM	ON THE NON-UNIFORM DEFORMATION OF A BLOCK IN PLANE-STRAIN COMPRESSION CAUSED BY FRICTION -3	71-0328

NON-UNIFORM	STATIC FLEXIBILITY MATRICES FOR A RECTANGULAR PLATE ELEMENT IN BENDING, TWISTING, AND SHEAR. UNIFORM	68-0217
NON-UNIFORM	TRANSIENT RESPONSE OF NON-UNIFORM, NON-HOMOGENEOUS BEAMS	73-0162
NON-UNIFORM	TRANSVERSE VIBRATIONS OF NON-UNIFORM BEAMS	74-1058
NON-UNIFORM	VIBRATION OF NON-UNIFORM BEAMS	63-0021
NON-UNIFORM	VIBRATIONS OF A NON-UNIFORM BEAM	68-0246
NONCENTRAL	STRESSES IN A ROTATING DISK WITH CENTRAL AND NONCENTRAL HOLES (STRESS ANALYSIS BY THE FINITE ELEMENT	74-1337
NONCIRCULAR	FINITE ELEMENT SOLUTION FOR FLOW IN NONCIRCULAR CONDUITS	74-0116
NONCIRCULAR	PLASTIC STRAINS ASSOCIATED WITH TORSION OF NONCIRCULAR BARS	72-0177
NONCONFORMING	CONFORMING AND NONCONFORMING FINITE ELEMENT METHODS FOR SOLVING THE STATIONARY STOKES EQUATIONS - 1	73-0390
NONCONFORMING	CONFORMING AND NONCONFORMING FINITE ELEMENT METHODS FOR SOLVING THE PLATE PROBLEM	73-0621
NONCONFORMING	NONCONFORMING ELEMENTS IN THE FINITE ELEMENT METHOD	72-0341
NONCONFORMING	NONCONFORMING ELEMENTS IN FINITE ELEMENT METHOD WITH PENALTY	73-0056
NONCONFORMING	NONCONFORMING FINITE ELEMENT ANALYSIS OF SHELLS	75-0539
NONCONFORMING	PLATE BENDING WITH NONCONFORMING ELEMENT: FINE-MESH CONVERGENCE TEST	65-0049
NONCONFORMING	SOME NONCONFORMING FINITE ELEMENTS FOR THE PLATE BENDING PROBLEM	75-0324
NONCONFORMING	SOME NONCONFORMING VARIANTS ON THE TRILINEAR ISOPARAMETRIC BRICK FOR LINEAR ELASTIC PROBLEMS	72-0531
NONCONFORMING	USE OF NONCONFORMING MODES IN FINITE ELEMENT ANALYSIS OF PLATES AND SHELLS	73-0865
NONCONSERVATIVE	A UNIFIED FINITE ELEMENT APPROACH TO COLUMNS SUBJECTED TO NONCONSERVATIVE LOADS, WITH DAMPING EFFECT	74-0795
NONCONSERVATIVE	APPLICATION OF THE FINITE ELEMENT STIFFNESS METHOD TO NONCONSERVATIVE STABILITY PROBLEMS	67-0158
NONCONSERVATIVE	APPLICATION OF THE FINITE ELEMENT METHOD TO NONCONSERVATIVE STABILITY PROBLEMS WITH DAMPING	72-0610
NONCONSERVATIVE	COUPLED, NONCONSERVATIVE STABILITY-FINITE ELEMENT	72-0220
NONCONSERVATIVE	FINITE ELEMENT ANALYSIS OF PROBLEMS IN NONCONSERVATIVE ELASTIC STABILITY	72-0957
NONCONSERVATIVE	NONCONSERVATIVE STABILITY BY FINITE ELEMENT	71-0572
NONCONSERVATIVE	NONCONSERVATIVE STABILITY OF PLATES BY FINITE ELEMENT METHOD	75-0789
NONCONSERVATIVE	NOTE ON AN APPROXIMATE METHOD FOR COMPUTING NONCONSERVATIVE GENERALIZED FORCES ON FINITELY DEFORMED	70-0097
NONCONSERVATIVE	SOLUMN INSTABILITY UNDER NONCONSERVATIVE FORCES, WITH INTERNAL AND EXTERNAL DAMPING FINITE ELEMENT U	73-0566
NONDESTRUCTIVE	EVALUATION OF RIGID PAVEMENTS BY NONDESTRUCTIVE TESTS.	72-0459
NONDETERMINISTI	NONDETERMINISTIC PROBLEMS IN STRUCTURAL MECHANICS	71-0746
NONELASTIC	ANALYSIS OF NONELASTIC AND NONLINEAR VIBRATION OF ROCK DAMS USING FINITE ELEMENT METHOD	73-0680
NONHOMOGENEITY	THE EFFECTS OF THIN CORES AND NONHOMOGENEITY ON DYNAMIC RESPONSE OF EARTH DAMS	67-0021
NONHORIZONTALLY	THE PROPAGATION OF LOVE WAVES ACROSS NONHORIZONTALLY LAYERED STRUCTURES	71-0600
NONISOTHERMAL	DYNAMIC RESPONSE OF ARBITRARY SHELLS OF REVOLUTION UNDER COMBINED IMPULSIVE NONISOTHERMAL AND IRREGU	70-0365
NONLINEAIRES	ANALYSE DES VOILES MINCES DANS LES DOMAINES LINEAIRES ET GEOMETRIQUEMENT NONLINEAIRES PAR LA METHODE	71-0122
NONLINEAR	A COMPARATIVE STUDY OF DIFFERENT NUMERICAL SOLUTION TECHNIQUES AS APPLIED TO A NONLINEAR STRUCTURAL	73-1117
NONLINEAR	A COMPUTER BASED PROCEDURE FOR PREDICTING THE TRANSIENT RESPONSE AND FAILURE OF A TWO-DIMENSIONAL	71-0630
NONLINEAR	A DIGITAL COMPUTER PROGRAM FOR THE FINITE ELEMENT ANALYSIS OF SOLIDS WITH NONLINEAR MATERIAL PROPERT	65-0032
NONLINEAR	A FINITE ELEMENT METHOD FOR GEOMETRICALLY NONLINEAR LARGE DISPLACEMENT PROBLEMS IN THIN, ELASTIC PLA	69-0005
NONLINEAR	A FINITE-ELEMENT MODEL TO SIMULATE THE NONLINEAR RESPONSE OF REINFORCED CONCRETE FRAMES WITH MASONRY	73-1168
NONLINEAR	A FINITE-ELEMENT STUDY ON THE NONLINEAR BEHAVIOR OF REINFORCED CONCRETE FLEXURAL MEMBERS DUE TO SHOR	73-1154
NONLINEAR	A GENERALIZATION OF THE FINITE ELEMENT CONCEPT AND ITS APPLICATION TO A CLASS OF PROBLEMS IN NONLINE	68-0212
NONLINEAR	A KOITER TYPE METHOD FOR FINITE ELEMENT ANALYSIS OF NONLINEAR STRUCTURAL BEHAVIOUR	70-0064
NONLINEAR	A KOITER-TYPE METHOD FOR FINITE ELEMENT ANALYSIS OF NONLINEAR STRUCTURAL BEHAVIOR	70-0493
NONLINEAR	A LEAST SQUARES FINITE ELEMENT SOLUTION OF NONLINEAR PROBLEMS	72-0060
NONLINEAR	A LEAST SQUARES FINITE ELEMENT ANALYSIS OF NONLINEAR BIOPHYSICAL DIFFUSION-KINETICS EQUATIONS	74-1257
NONLINEAR	A METHOD FOR THE TRANSIENT NONLINEAR COUPLED THERMOELASTIC ANALYSIS OF AXISYMMETRIC BODIES USING FIN	74-1140
NONLINEAR	A NONLINEAR FINITE ELEMENT ANALYSIS OF SHALLOW CIRCULAR ARCHES	69-0269
NONLINEAR	A NONLINEAR FINITE ELEMENT CODE FOR ANALYZING THE BLAST RESPONSE OF UNDERGROUND STRUCTURES	70-0054
NONLINEAR	A NOTE ON THE ANALYSIS OF NONLINEAR DYNAMICS OF ELASTIC MEMBRANES BY THE FINITE ELEMENT METHOD	74-0341
NONLINEAR	A NUMERICAL STUDY OF NONLINEAR INSTABILITY BY MEANS OF AN EXTENDED FINITE ELEMENT METHOD	74-0632
NONLINEAR	A SURVEY OF FINITE ELEMENT FORMULATIONS OF GEOMETRICALLY NONLINEAR PROBLEMS	71-0171
NONLINEAR	A THEORETICAL AND NUMERICAL COMPARISON OF ELASTIC NONLINEAR FINITE ELEMENT METHODS	72-0709
NONLINEAR	A TREATISE OF NONLINEAR FINITE ELEMENT ANALYSIS	73-1063
NONLINEAR	A TRIANGULAR THIN SHELL FINITE ELEMENT: NONLINEAR ANALYSIS	75-0697
NONLINEAR	AN EXTENSION OF FINITE-ELEMENT METHOD TO INCLUDE NONLINEAR BOUNDARY CONDITIONS	75-0032
NONLINEAR	AN IMBEDDING TECHNIQUE FOR THE GENERATION OF WEAK-WEAK FINITE ELEMENT APPROXIMATIONS OF LINEAR AND N	72-0577
NONLINEAR	AN INCREMENTAL COMPLEMENTARY ENERGY METHOD OF NONLINEAR STRESS ANALYSIS	69-0281
NONLINEAR	AN ITERATION METHOD FOR THE NONLINEAR BUCKLING OF FRAME STRUCTURES	66-0107
NONLINEAR	ANALYSIS OF GEOMETRICALLY NONLINEAR PLATES AND SHELLS BY THE FINITE ELEMENT DISPLACEMENT METHOD	68-0050
NONLINEAR	ANALYSIS OF NONELASTIC AND NONLINEAR VIBRATION OF ROCK DAMS USING FINITE ELEMENT METHOD	73-0680
NONLINEAR	ANALYSIS OF NONLINEAR PROBLEMS IN ROCK MECHANICS WITH PARTICULAR REFERENCE TO JOINTED ROCK SYSTEMS	70-0282
NONLINEAR	ANALYSIS OF NONLINEAR CONCRETE FRAMES USING SMALL GYPSUM MODELS AND THE FINITE ELEMENT METHOD	70-0498
NONLINEAR	ANALYSIS OF NONLINEAR THERMOELASTIC AND THERMOPLASTIC BEHAVIOUR OF SOLIDS OF REVOLUTION BY THE FINIT	71-0192
NONLINEAR	ANALYSIS OF NONLINEAR DYNAMIC, COUPLED THERMOVISCOELASTICITY PROBLEMS BY THE FINITE ELEMENT METHOD	71-0527
NONLINEAR	ANALYSIS OF NONLINEAR THERMOVISCOELASTICITY PROBLEMS BY THE FINITE ELEMENT METHOD	71-0528
NONLINEAR	ANALYSIS OF NONLINEAR CONTACT PROBLEMS BY THE FINITE-ELEMENT METHOD	72-0952
NONLINEAR	ANALYSIS OF NONLINEAR REINFORCED CONCRETE FRAME BY FINITE ELEMENT METHOD	74-1104
NONLINEAR	ANALYSIS OF STIFFENED PLATES CONSIDERING NONLINEAR MATERIAL AND GEOMETRIC BEHAVIOUR	75-0327
NONLINEAR	APPLICATION OF A REFINED SHALLOW SHELL FINITE ELEMENT TO GEOMETRICALLY NONLINEAR STRUCTURES	73-0358
NONLINEAR	APPLICATION OF NONLINEAR ANALYSIS TO SOLID STRUCTURES	72-0212
NONLINEAR	APPLICATION OF THE FINITE-ELEMENT METHOD TO HEAT-TRANSFER PROBLEMS. PART 1. FINITE-ELEMENT METHOD AP	71-0760
NONLINEAR	APPLICATION OF THE FINITE ELEMENT METHOD TO THE NONLINEAR ANALYSIS OF NUCLEAR REACTOR FUEL BEHAVIOR	74-1246
NONLINEAR	APPLICATIONS OF FINITE-ELEMENT ANALYSIS TO NONLINEAR PLANE STRESS PROBLEMS	72-0976
NONLINEAR	APPROXIMATION BY FIRST ORDER FINITE ELEMENTS AND SOLUTION BY PENALTY AND DUALITY OF NONLINEAR DIRICH	74-0245
NONLINEAR	BASIC RESEARCH IN METHODS OF APPROXIMATION IN THE NONLINEAR MECHANICS OF SOLIDS AND STRUCTURES	72-0857
NONLINEAR	BASIC RESEARCH IN METHODS OF APPROXIMATION IN THE NONLINEAR MECHANICS OF SOLIDS AND STRUCTURES, VOLU	74-1091
NONLINEAR	COMPUTATION AND SOLUTION PROCEDURE FOR NONLINEAR ANALYSIS BY COMBINED FINITE ELEMENT-FINITE DIFFEREN	72-0571
NONLINEAR	COMPUTATIONAL EFFICIENCY ASPECTS OF THE FINITE ELEMENT NONLINEAR ANALYSIS OF SHELLS	75-0031
NONLINEAR	COMPUTATIONAL METHODS IN NONLINEAR MECHANICS	74-1232
NONLINEAR	COMPUTATIONAL STABILITY IN NONLINEAR DISCRETE STRUCTURAL SYSTEMS	69-0145
NONLINEAR	COMPUTATIONAL TECHNIQUES, NONLINEAR EQUATIONS, ISOPARAMETRIC ELEMENTS, FIVE LECTURES	73-0634
NONLINEAR	COMPUTER IMPLEMENTATION OF NONLINEAR FINITE ELEMENT ANALYSIS	73-0407
NONLINEAR	CONVECTED COORDINATES FOR TRANSIENT NONLINEAR FINITE-ELEMENT ANALYSIS	73-0850
NONLINEAR	CONVERGENCE AND STABILITY OF NONLINEAR FINITE-ELEMENT EQUATIONS	75-0441
NONLINEAR	CRASH: A COMPUTER SIMULATION OF NONLINEAR TRANSIENT RESPONSE OF STRUCTURES	72-0569
NONLINEAR	CURVED FINITE ELEMENT APPROXIMATION TO NONLINEAR CABLES	72-0379
NONLINEAR	DEFORMATION OF THE DIASTOLIC LEFT VENTRICLE - 2. NONLINEAR GEOMETRIC EFFECTS	74-0974
NONLINEAR	DER GERISSENE ZUSTAND ZWEISFITIG GELAGERTER PLATTEN UNTER EINZELLASTEN - NICHTLINEARE BERECHNUNG MIT	75-0858
NONLINEAR	DEVELOPMENT OF A SOLID ROCKET PROPELLANT NONLINEAR CONSTITUTTIVE THEORY	75-0722
NONLINEAR	DISCRETE APPROXIMATIONS RELATED TO NONLINEAR THEORIES OF SOLIDS	71-0259
NONLINEAR	DUAL ANALYSIS BY FINITE ELEMENTS: LINEAR AND NONLINEAR APPLICATIONS	72-0476
NONLINEAR	DYNAMIC ELASTOPLASTIC RESPONSE OF GEOMETRICALLY NONLINEAR ARBITRARY SHELLS OF REVOLUTION UNDER IMPUL	70-0098
NONLINEAR	DYNAMIC RESPONSE OF NONLINEAR SPACE FRAMES BY FINITE ELEMENT METHODS	71-0584
NONLINEAR	DYNAMIC RESPONSE ON NONLINEAR MEDIA AT LARGE STRAINS	72-0237
NONLINEAR	DYNAMIC STABILITY OF A NONLINEAR BEAM SUBJECTED TO BOTH LONGITUDINAL AND TRANSVERSE EXCITATION	73-0553
NONLINEAR	DYNASOR 11 - A FINITE ELEMENT PROGRAM FOR THE DYNAMIC NONLINEAR ANALYSIS OF SHELLS OF REVOLUTION (US	71-0666
NONLINEAR	EFFECT OF HIGHER ORDER TERMS IN CERTAIN NONLINEAR FINITE ELEMENT MODELS	73-0125
NONLINEAR	EFFECT OF HIGHER ORDER TERMS IN CERTAIN NONLINEAR FINITE ELEMENT MODELS - COMMENT	74-0473
NONLINEAR	EFFECT OF NONLINEAR MATERIAL RESPONSE ON THE BEHAVIOUR OF PAVEMENTS UNDER TRAFFIC	70-0171
NONLINEAR	EFFICIENT FORMULATIONS OF THE FINITE ELEMENT METHOD IN LINEAR AND NONLINEAR FRACTURE MECHANICS	75-0403
NONLINEAR	ELASTIC NONLINEAR FINITE ELEMENT ANALYSIS OF PLANAR STRUCTURES	71-0830
NONLINEAR	EVALUATION OF SOLUTION PROCEDURES FOR MATERIAL AND/OR GEOMETRICALLY NONLINEAR STRUCTURAL ANALYSIS	73-0362
NONLINEAR	EXPERIENCE WITH FINITE ELEMENT COMPUTATIONS FOR NONLINEAR STRUCTURAL PROBLEMS	74-1239

NONLINEAR	FINITE ELEMENT ANALYSIS OF NONLINEAR STRUCTURES	68-0109
NONLINEAR	FINITE ELEMENT ANALYSIS OF NONLINEAR SOIL MEDIA	69-0132
NONLINEAR	FINITE ELEMENT ANALYSIS OF THE GEOMETRICALLY NONLINEAR PROBLEMS	69-0149
NONLINEAR	FINITE ELEMENT ANALYSIS OF COMBINED PROBLEMS ON NONLINEAR MATERIAL AND GEOMETRIC BEHAVIOUR	69-0175
NONLINEAR	FINITE ELEMENT ANALYSIS OF NONLINEAR PROBLEMS IN THE DYNAMICAL THEORY OF COUPLED THERMOELASTICITY	69-0202
NONLINEAR	FINITE ELEMENT ANALYSIS OF NONLINEAR STRUCTURES	69-0205
NONLINEAR	FINITE ELEMENT ANALYSIS OF NONLINEAR MEMBRANE STRUCTURES	72-0909
NONLINEAR	FINITE ELEMENT ANALYSIS OF NONLINEAR MEMBRANE STRUCTURES	72-0944
NONLINEAR	FINITE ELEMENT ANALYSIS OF NONLINEAR VIBRATION OF BEAM COLUMNS	73-0028
NONLINEAR	FINITE ELEMENT ANALYSIS OF GEOMETRICALLY NONLINEAR PROBLEMS	73-0308
NONLINEAR	FINITE ELEMENT ANALYSIS OF NONLINEAR STRUCTURES	73-0975
NONLINEAR	FINITE ELEMENT ANALYSIS OF STEADY-STATE NONLINEAR HEAT TRANSFER PROBLEMS	74-0572
NONLINEAR	FINITE ELEMENT ANALYSIS OF LINEAR AND NONLINEAR HEAT TRANSFER	74-0626
NONLINEAR	FINITE ELEMENT ANALYSIS OF NONLINEAR CABLE REINFORCED MEMBRANES	74-1073
NONLINEAR	FINITE ELEMENT ANALYSIS OF WAVES IN NONLINEAR MATERIALS	75-0110
NONLINEAR	FINITE ELEMENT ANALYSIS OF TRANSIENT NONLINEAR STRUCTURAL BEHAVIOR	75-0297
NONLINEAR	FINITE ELEMENT APPLICATIONS IN LINEAR AND NONLINEAR THERMOVISCOELASTICITY	67-0065
NONLINEAR	FINITE ELEMENT APPLICATIONS IN NONLINEAR STRUCTURAL ANALYSIS	69-0204
NONLINEAR	FINITE ELEMENT APPROXIMATION OF A CLASS OF NONLINEAR OPERATORS	71-0454
NONLINEAR	FINITE ELEMENT APPROXIMATIONS IN NONLINEAR THERMOVISCOELASTICITY	71-0459
NONLINEAR	FINITE ELEMENT APPROXIMATIONS IN NONLINEAR ELASTICITY	71-0470
NONLINEAR	FINITE ELEMENT BASED PROCEDURE FOR SIMULATING THE TRANSIENT RESPONSE AND FAILURE OF A TWO-DIMENSIONA	72-0123
NONLINEAR	FINITE ELEMENT COMPUTER PROGRAM FOR THE SOLUTION OF NONLINEAR AXISYMMETRIC CONTACT PROBLEMS WITH INT	74-1094
NONLINEAR	FINITE ELEMENT COMPUTER PROGRAM FOR PREDICTING THE NONLINEAR STATIC AND DYNAMIC BEHAVIOR OF VISCOELA	75-0191
NONLINEAR	FINITE ELEMENT DIFFICULTIES IN NONLINEAR AND INCOMPRESSIBLE MATERIAL	74-1242
NONLINEAR	FINITE ELEMENT FORMULATIONS OF GEOMETRICALLY NONLINEAR PROBLEMS	69-0180
NONLINEAR	FINITE ELEMENT FORMULATION OF FINITE DEFORMATIONS AND IRREVERSIBLE THERMODYNAMICS OF NONLINEAR CONTI	69-0207
NONLINEAR	FINITE ELEMENT FORMULATION OF GEOMETRICALLY NONLINEAR PROBLEMS OF ELASTICITY	71-0114
NONLINEAR	FINITE ELEMENT FORMULATIONS OF PROBLEMS OF FINITE DEFORMATION AND IRREVERSIBLE THERMODYNAMICS OF NONL	71-0190
NONLINEAR	FINITE ELEMENT FORMULATIONS OF NONLINEAR BOUNDARY VALUE PROBLEMS	71-0497
NONLINEAR	FINITE ELEMENT METHOD IN NONLINEAR DOMAIN, (METHODE DES ELEMENTS FINIS DANS LE DOMAINE NON LINEAIRE)	71-0288
NONLINEAR	FINITE ELEMENT METHOD FOR NONLINEAR HYPERBOLIC EQUATIONS	74-1238
NONLINEAR	FINITE ELEMENT MODELS OF NONLINEAR OPERATOR EQUATIONS	71-0199
NONLINEAR	FINITE ELEMENT NONLINEAR ANALYSIS FOR PLATES AND SHALLOW SHELLS	68-0116
NONLINEAR	FINITE ELEMENT PERTURBATION ANALYSIS OF NONLINEAR DYNAMIC RESPONSE OF ELASTIC CONTINUUA	74-0452
NONLINEAR	FINITE ELEMENT PROCEDURE FOR NONLINEAR PREBUCKLING AND INITIAL POSTBUCKLING ANALYSIS	72-0170
NONLINEAR	FINITE ELEMENT SOLUTIONS OF A NONLINEAR REACTOR DYNAMICS PROBLEM	74-1268
NONLINEAR	FINITE ELEMENT SOLUTION OF NONLINEAR STRUCTURES BY PERTURBATION TECHNIQUE	74-1273
NONLINEAR	FINITE ELEMENTS AND THE ANALYSIS OF GEOMETRICALLY NONLINEAR PROBLEMS	69-0179
NONLINEAR	FINITE ELEMENTS FOR FLUID DYNAMICS. VARIATIONAL FORMULATIONS FOR NONLINEAR, INITIAL-VALUE PROBLEMS	74-0710
NONLINEAR	FINITE ELEMENTS IN NONLINEAR CONTINUA	71-0200
NONLINEAR	FINITE TIME-SPACE ELEMENTS APPLIED TO NONLINEAR PROBLEMS	70-0398
NONLINEAR	FINITE-DIFFERENCE ENERGY METHOD FOR NONLINEAR SHELL ANALYSIS	71-0399
NONLINEAR	FINITE-ELEMENT ANALYSIS OF NONLINEAR MAGNETIC FIELDS IN ELECTRIC MACHINES	70-0484
NONLINEAR	FINITE-ELEMENT ANALYSIS OF PAVEMENT STRUCTURES USING AFPAV CODE (NONLINEAR ELASTIC ANALYSIS)	75-0748
NONLINEAR	FINITE-ELEMENT METHOD APPLIED TO HEAT CONDUCTION IN SOLIDS WITH NONLINEAR BOUNDARY CONDITIONS	73-0005
NONLINEAR	FINITE-ELEMENT NONLINEAR BUCKLING ANALYSIS OF A PSEUDO-CYLINDRICAL CONCAVE POLYHEDRAL SHELL UNDER EX	73-1147
NONLINEAR	FORMULATION, COMPUTATION, AND SOLUTION PROCEDURES FOR MATERIAL AND/OR GEOMETRIC NONLINEAR STRUCTURAL	72-0346
NONLINEAR	FREE VIBRATION OF A SIMPLY SUPPORTED BEAM WITH NONLINEAR MATERIAL PROPERTIES	71-0689
NONLINEAR	FUNCTIONAL MINIMIZATION IN NONLINEAR SOLID MECHANICS	74-1251
NONLINEAR	GALERKIN METHODS FOR PARABOLIC EQUATIONS WITH NONLINEAR BOUNDARY CONDITIONS	73-1023
NONLINEAR	GENERAL MIXED FINITE ELEMENT METHODS OF NONLINEAR CONTINUA	72-0831
NONLINEAR	GENERAL MIXED FINITE ELEMENT METHODS OF NONLINEAR CONTINUA	73-0091
NONLINEAR	GENERAL PURPOSE PROGRAMS FOR NONLINEAR FINITE ELEMENT ANALYSIS	70-0247
NONLINEAR	GENERAL VARIATIONAL PRINCIPLES IN NONLINEAR AND LINEAR ELASTICITY WITH APPLICATIONS	72-0914
NONLINEAR	GEOMETRIC NONLINEAR ANALYSIS OF THIN PLATES BY FINITE ELEMENTS	70-0492
NONLINEAR	GEOMETRICALLY NONLINEAR FINITE-ELEMENT ANALYSIS	69-0332
NONLINEAR	GEOMETRICALLY NONLINEAR ANALYSIS BY THE DIRECT STIFFNESS METHOD	70-0331
NONLINEAR	GEOMETRICALLY NONLINEAR STRUCTURAL ANALYSIS BY THE DIRECT STIFFNESS METHOD	71-0560
NONLINEAR	GEOMETRICALLY NONLINEAR FINITE ELEMENT ANALYSIS	72-0347
NONLINEAR	GEOMETRICALLY NONLINEAR ANALYSIS OF THIN-WALLED MEMBERS USING FINITE-ELEMENTS	73-1142
NONLINEAR	GEOMETRICKY NELINEARNE ULOHY PLOCHYCH SKRUPIN RIESENE PRIRASTKOVOU METODOU KONECNYCH PRVKOV. (GEOMET	74-1032
NONLINEAR	GNATS: A FINITE ELEMENT COMPUTER PROGRAM FOR THE GENERAL NONLINEAR ANALYSIS OF TWO-DIMENSIONAL STRUC	74-0875
NONLINEAR	HEAT TRANSFER WITH NONLINEAR BOUNDARY CONDITIONS VIA A VARIATIONAL PRINCIPLE	71-0443
NONLINEAR	HYBRIDE FINITE VERFAHREN FUER LINEARE UND NICHTLINEARE INSTATIONAERE WAERMELEITPROBLEME (HYBRID FINI	73-0484
NONLINEAR	IDENTIFICATION OF NONLINEAR MATERIALS BY FINITE ELEMENT METHODS	73-0927
NONLINEAR	IDENTIFICATION OF NONLINEAR ELASTIC SOLIDS BY A FINITE ELEMENT METHOD	74-0530
NONLINEAR	INCREMENTAL SOLUTION PROCEDURE FOR NONLINEAR STRUCTURAL DYNAMICS PROBLEMS	75-0656
NONLINEAR	INCREMENTAL STIFFNESS METHOD FOR FINITE ELEMENT ANALYSIS OF THE NONLINEAR DYNAMIC PROBLEM	71-0525
NONLINEAR	LARGE AMPLITUDE, FREE VIBRATION GEOMETRICALLY NONLINEAR CIRCULAR ARCHES	72-0606
NONLINEAR	LARGE DISPLACEMENT, NONLINEAR TRANSIENT ANALYSIS BY FINITE ELEMENTS	74-0324
NONLINEAR	LINEAR AND NONLINEAR STATIC ANALYSIS OF AXISYMMETRICALLY LOADED THIN SHELLS OF REVOLUTION	69-0215
NONLINEAR	LINEAR AND NONLINEAR ANALYSIS OF SHELLS	72-0832
NONLINEAR	LINEAR AND NONLINEAR FINITE ELEMENT ANALYSIS OF WOOD STRUCTURAL MEMBERS	73-0117
NONLINEAR	LINEAR AND NONLINEAR OSCILLATIONS IN MECHANICS	74-0677
NONLINEAR	LINEAR BUCKLING AND GEOMETRICALLY NONLINEAR ANALYSIS OF PLANAR PLATE-STIFFENER TYPE STRUCTURES BY TH	74-1097
NONLINEAR	NEW MATERIAL MODEL FOR THE NONLINEAR BIAXIAL BEHAVIOR OF ATJ-S GRAPHITE	75-0063
NONLINEAR	NONLINEAR ACOUSTIC RESPONSE ANALYSIS OF PLATES USING THE FINITE ELEMENT METHOD	72-0223
NONLINEAR	NONLINEAR ANALYSES OF THE STRESS FIELD NEAR A CRACK OR HOLE IN A PLATE	75-0694
NONLINEAR	NONLINEAR ANALYSES USING SPLINE FUNCTIONS	71-0466
NONLINEAR	NONLINEAR ANALYSIS	71-0169
NONLINEAR	NONLINEAR ANALYSIS	75-0627
NONLINEAR	NONLINEAR ANALYSIS BY THE FINITE ELEMENT METHOD AND SOME EXPOSITORY EXAMPLES	73-0309
NONLINEAR	NONLINEAR ANALYSIS BY ASSUMED STRESS HYBRID MODELS	74-1004
NONLINEAR	NONLINEAR ANALYSIS OF SHELLS OF REVOLUTION BY THE MATRIX DISPLACEMENT METHOD	67-0079
NONLINEAR	NONLINEAR ANALYSIS OF REINFORCED CONCRETE BY THE FINITE ELEMENT METHOD	68-0126
NONLINEAR	NONLINEAR ANALYSIS OF ELASTIC FRAMED STRUCTURES	68-0222
NONLINEAR	NONLINEAR ANALYSIS OF STRESS AND STRAIN IN SOILS	70-0050
NONLINEAR	NONLINEAR ANALYSIS OF PRESTRESSED CONCRETE FAST REACTOR PRESSURE VESSEL. (ANALYSE NON LINEAIRE DES C	70-0145
NONLINEAR	NONLINEAR ANALYSIS OF ELASTIC SHELLS OF REVOLUTION UNDER AXISYMMETRIC LOADING	70-0161
NONLINEAR	NONLINEAR ANALYSIS OF REINFORCED CONCRETE FRAMES AND PANELS	70-0436
NONLINEAR	NONLINEAR ANALYSIS OF TWO-DIMENSIONAL PROBLEMS IN CONCRETE CREEP	71-0299
NONLINEAR	NONLINEAR ANALYSIS OF REINFORCED CONCRETE BUILDINGS	73-0324
NONLINEAR	NONLINEAR ANALYSIS OF BEAMS AND SHELLS	73-1074
NONLINEAR	NONLINEAR ANALYSIS OF ARBITRARY, HYPERELASTIC MEMBRANE SHELLS BY THE FINITE-ELEMENT METHOD	73-1132
NONLINEAR	NONLINEAR ANALYSIS OF STRUCTURES	74-0363
NONLINEAR	NONLINEAR ANALYSIS OF PLANAR REINFORCED CONCRETE STRUCTURES	74-0533
NONLINEAR	NONLINEAR ANALYSIS OF A ROLLING CONTACT PROBLEM BY FINITE ELEMENTS	74-1216
NONLINEAR	NONLINEAR ANALYSIS OF RC SHELLS OF GENERAL FORM	75-0053
NONLINEAR	NONLINEAR ANALYSIS OF RETICULATED SPACE TRUSSES	75-0287
NONLINEAR	NONLINEAR ANALYSIS OF CRACKED BODY BY FINITE ELEMENT METHOD	75-0315
NONLINEAR	NONLINEAR BEAM AND PLATE ELEMENTS	72-0866

NONLINEAR	NONLINEAR BEHAVIOR OF REINFORCED CONCRETE BEAMS BY THE FINITE-ELEMENT METHOD	72-0954
NONLINEAR	NONLINEAR BEHAVIOR OF SHELLS OF REVOLUTION UNDER CYCLIC LOADING	73-0242
NONLINEAR	NONLINEAR BEHAVIOR OF THE HUMAN INTERVERTEBRAL DISC UNDER AXIAL LOADING	74-0919
NONLINEAR	NONLINEAR BUCKLING ANALYSIS OF SANDWICH ARCHES	71-0292
NONLINEAR	NONLINEAR BUCKLING ANALYSIS OF COMPOSITE SHELLS	74-0973
NONLINEAR	NONLINEAR BUCKLING ANALYSIS OF COMPOSITE SHELLS	75-0214
NONLINEAR	NONLINEAR CONSIDERATION OF GRAVITY IN A STIFFNESS TEST OF A WEAK STRUCTURE AT SMALL STRAINS	71-0035
NONLINEAR	NONLINEAR CRACK ANALYSIS WITH FINITE ELEMENTS	73-0369
NONLINEAR	NONLINEAR CRACK ANALYSIS WITH FINITE ELEMENTS	73-0875
NONLINEAR	NONLINEAR CREEP ANALYSIS AT ELEVATED TEMPERATURE	71-0050
NONLINEAR	NONLINEAR CREEP ANALYSIS BY ASSUMED STRESS FINITE ELEMENT METHODS	74-0631
NONLINEAR	NONLINEAR CREEP ANALYSIS BY ASSUMED STRESS FINITE ELEMENT METHODS	74-0922
NONLINEAR	NONLINEAR DYNAMIC ANALYSIS OF SHELLS OF REVOLUTION BY MATRIX DISPLACEMENT METHOD	70-0110
NONLINEAR	NONLINEAR DYNAMIC ANALYSIS OF COOLING TOWER.	72-0219
NONLINEAR	NONLINEAR DYNAMIC ANALYSIS OF COMPLEX STRUCTURES	73-0193
NONLINEAR	NONLINEAR DYNAMIC ANALYSIS OF AXISYMMETRIC SOLIDS BY THE FINITE ELEMENT METHOD	73-0942
NONLINEAR	NONLINEAR DYNAMIC ANALYSIS OF FRAMES WITH FILLER PANELS	74-0226
NONLINEAR	NONLINEAR DYNAMIC ANALYSIS OF AXISYMMETRIC SOLIDS BY THE FINITE ELEMENT METHOD	74-0510
NONLINEAR	NONLINEAR DYNAMIC ANALYSIS OF SHELLS OF REVOLUTION WITH ASYMMETRIC PROPERTIES BY THE FINITE ELEMENT	75-0217
NONLINEAR	NONLINEAR DYNAMIC ANALYSIS OF SHELLS OF REVOLUTION WITH ASYMMETRIC PROPERTIES BY THE FINITE ELEMENT	75-0657
NONLINEAR	NONLINEAR DYNAMIC ANALYSIS OF TORSIONAL SHELLS BY COMBINED USE OF FINITE ELEMENT AND MODE SUPERPOSIT	75-0859
NONLINEAR	NONLINEAR DYNAMICS OF SOLIDS BY FINITE ELEMENT METHODS	69-0490
NONLINEAR	NONLINEAR DYNAMICS OF SOLIDS BY THE FINITE ELEMENT METHOD	71-0116
NONLINEAR	NONLINEAR DYNAMICS OF CABLES WITH LOW INITIAL TENSION	72-0740
NONLINEAR	NONLINEAR ELASTIC-VISCOPLASTIC ANALYSIS USING THE FINITE ELEMENT METHOD	74-0750
NONLINEAR	NONLINEAR EQUATION OF UNSTEADY GROUND-WATER FLOW	73-0262
NONLINEAR	NONLINEAR FINITE ELEMENT ANALYSIS INCLUDING HIGHER ORDER STRAIN ENERGY TERMS	70-0065
NONLINEAR	NONLINEAR FINITE ELEMENT ANALYSIS OF SANDWICH SHELLS OF REVOLUTION	72-0180
NONLINEAR	NONLINEAR FINITE ELEMENT ANALYSIS OF MULTI JOINTED ROCK MASSES	72-0956
NONLINEAR	NONLINEAR FINITE ELEMENT FOLD MODEL	73-0014
NONLINEAR	NONLINEAR FINITE ELEMENT ANALYSIS OF REINFORCED CONCRETE PLATES	73-0530
NONLINEAR	NONLINEAR FINITE ELEMENT ANALYSIS OF REINFORCED CONCRETE PLATES	73-0845
NONLINEAR	NONLINEAR FINITE ELEMENT ANALYSIS OF REINFORCED CONCRETE STRUCTURES	74-0442
NONLINEAR	NONLINEAR FINITE ELEMENT SOLUTION OF ELECTRICAL MACHINES UNDER NO-LOAD AND FULL-LOAD CONDITIONS	74-0472
NONLINEAR	NONLINEAR FINITE ELEMENT ANALYSIS OF LAMINATED COMPOSITE SHELLS	74-0751
NONLINEAR	NONLINEAR FINITE ELEMENT DYNAMIC ANALYSIS OF AXISYMMETRIC SOLIDS	74-0775
NONLINEAR	NONLINEAR FINITE ELEMENT ANALYSIS OF REINFORCED CONCRETE FRAMES	74-1124
NONLINEAR	NONLINEAR FINITE ELEMENT AND ITS APPLICATION TO LARGE SCALE STRUCTURAL PROBLEMS	75-0386
NONLINEAR	NONLINEAR FINITE-ELEMENT ANALYSIS OF THIN SHELLS	73-1169
NONLINEAR	NONLINEAR FINITE-ELEMENT METHOD FOR PLATES AND ITS APPLICATION TO DYNAMIC RESPONSE OF REACTOR FUEL S	74-0670
NONLINEAR	NONLINEAR FINITE-ELEMENT FORMULATION FOR TRANSIENT ANALYSIS OF THREE DIMENSIONAL THIN STRUCTURES	74-0874
NONLINEAR	NONLINEAR HYBRID STRESS FINITE ELEMENT ANALYSIS OF LAMINATED SHELLS	74-1350
NONLINEAR	NONLINEAR HYBRID STRESS FINITE ELEMENT ANALYSIS OF PNEUMATIC TIRES	75-0029
NONLINEAR	NONLINEAR IN-PLANE BENDING OF A PLANE STRESS-STRAIN FINITE ELEMENT MODEL	74-0211
NONLINEAR	NONLINEAR INITIAL-BOUNDARY VALUE SOLUTIONS BY THE FINITE ELEMENT METHOD	74-0812
NONLINEAR	NONLINEAR LAYERED ANALYSIS OF RC PLATES AND SHELLS	73-0208
NONLINEAR	NONLINEAR LAYERED FINITE-ELEMENT ANALYSIS OF REINFORCED-CONCETE PLATES AND SHELLS	72-0946
NONLINEAR	NONLINEAR MATERIAL AND GEOMETRIC BEHAVIOUR OF SHELL STRUCTURES	71-0084
NONLINEAR	NONLINEAR MECHANICS OF CONSTRAINED MATERIAL CONTINUA: FOUNDATIONS OF THE THEORY	74-0640
NONLINEAR	NONLINEAR METHODS OF STRUCTURAL ANALYSIS	71-0029
NONLINEAR	NONLINEAR OSCILLATIONS USING THE FINITE ELEMENT TECHNIQUE	73-0568
NONLINEAR	NONLINEAR RESPONSE OF PLATES TO IMPULSIVE LOADING USING HYBRID FINITE ELEMENTS	74-1018
NONLINEAR	NONLINEAR SEISMIC ANALYSIS OF CURRENT DIVIDER SUPPORT SYSTEM FOR PACIFIC DC INTERTIE	73-0200
NONLINEAR	NONLINEAR SETTLEMENT ANALYSIS BY FINITE ELEMENT	75-0486
NONLINEAR	NONLINEAR STATIC AND DYNAMIC RESPONSE OF UNDERWATER CABLE STRUCTURES USING THE FINITE ELEMENT METHOD	75-0460
NONLINEAR	NONLINEAR STRESS ANALYSIS OF INDETERMINATE STRUCTURAL ELEMENTS SUBJECTED TO BIAXIAL STRESS STATES US	73-1152
NONLINEAR	NONLINEAR STRESS DEFORMATION ANALYSES OF LAKE AGASSIZ CLAYS USING FINITE ELEMENT METHOD	74-1072
NONLINEAR	NONLINEAR STRUCTURAL ANALYSIS: TENSOR FORMULATION	65-0046
NONLINEAR	NONLINEAR STRUCTURES FOR SYSTEM IDENTIFICATION	74-1030
NONLINEAR	NONLINEAR THEORY FOR PREDICTING THE PERFORMANCE OF FLEXIBLE HIGHWAY PAVEMENTS	70-0143
NONLINEAR	NONLINEAR THERMO ELASTIC PLASTIC AND CREEP ANALYSIS BY FINITE ELEMENT METHOD	74-0092
NONLINEAR	NONLINEAR THERMOELASTICITY, IRREVERSIBLE THERMODYNAMICS AND ELASTIC INSTABILITY	73-0762
NONLINEAR	NONLINEAR TRANSIENT RESPONSES OF STRUCTURES BY SPATIAL FINITE-ELEMENT METHOD	73-0138
NONLINEAR	NONLINEAR TRANSIENT ANALYSIS OF SHELLS AND SOLIDS OF REVOLUTION BY CONVECTED ELEMENTS	73-0519
NONLINEAR	NONLINEAR TRANSIENT ANALYSIS OF SHELLS AND SOLIDS OF REVOLUTTION BY CONVECTED ELEMENTS	74-0476
NONLINEAR	NONLINEAR TRANSIENT ANALYSIS OF AUTOMOTIVE STRUCTURES BY FINITE ELEMENTS	75-0301
NONLINEAR	NONLINEAR VIBRATIONS OF NONUNIFORM BEAMS WITH CONCENTRATED MASSES	74-1300
NONLINEAR	NONSAP – A GENERAL FINITE ELEMENT PROGRAM FOR NONLINEAR DYNAMIC ANALYSIS OF COMPLEX STRUCTURES	73-0740
NONLINEAR	NONSAP – A NONLINEAR STRUCTURAL ANALYSIS PROGRAM	74-1006
NONLINEAR	NONSAP: A STRUCTURAL ANALYSIS PROGRAM FOR STATIC AND DYNAMIC RESPONSE OF NONLINEAR SYSTEMS	74-0848
NONLINEAR	NOTE ON THE ANALYSIS OF NONLINEAR DYNAMICS OF ELASTIC MEMBRANES BY THE FINITE ELEMENT METHOD	74-0230
NONLINEAR	NUMERICAL ANALYSIS OF NONLINEAR PNEUMATIC STRUCTURES	67-0060
NONLINEAR	NUMERICAL FORMULATION OF A CLASS OF PROBLEMS IN NONLINEAR VISCOELASTICITY	67-0061
NONLINEAR	NUMERICAL FORMULATION OF NONLINEAR ELASTICITY PROBLEMS	67-0063
NONLINEAR	NUMERICAL METHODS FOR TIME DEPENDENT NONLINEAR BOUNDARY VALUE PROBLEMS	70-0296
NONLINEAR	NUMERICAL METHODS FOR THE SOLUTION OF NONLINEAR PROBLEMS IN STRUCTURAL ANALYSIS	73-0506
NONLINEAR	NUMERICAL METHODS OF HIGH-ORDER ACCURACY FOR NONLINEAR BOUNDARY VALUE PROBLEMS	68-0194
NONLINEAR	NUMERICAL SOLUTION SCHEMES FOR HIGHLY NONLINEAR STATIC STRUCTURAL BEHAVIOR	75-0679
NONLINEAR	NUMERICAL SOLUTIONS OF NONLINEAR PLASMA EQUATIONS BY THE FINITE-ELEMENT METHOD	75-0671
NONLINEAR	NUMERICAL STUDY OF TRANSITORY OR NONLINEAR WAVES	74-0031
NONLINEAR	NUMERICAL TECHNIQUES IN LINEAR AND NONLINEAR FRACTURE MECHANICS	75-0260
NONLINEAR	NUMERICAL-PERTURBATION METHOD FOR THE NONLINEAR ANALYSIS OF STRUCTURAL VIBRATIONS	74-0499
NONLINEAR	ON A GENERALIZATION OF THE FINITE ELEMENT CONCEPT AND ITS APPLICATION TO A CLASS OF PROBLEMS IN NONL	69-0206
NONLINEAR	ON THE ASYMPTOTICALLY SPHERICAL DEFORMATIONS OF ARBITRARY MEMBRANES OF REVOLUTION FIXED ALONG AN EDG	71-0640
NONLINEAR	ON THE CONVERGENCE OF THE FINITE ELEMENT SOLUTION IN A NONLINEAR HEAT CONDUCTION PROBLEM	74-1272
NONLINEAR	ON THE EFFECT OF APPROXIMATING HIGHER ORDER TERMS IN CERTAIN FINITE ELEMENT MODELS OF NONLINEAR STRU	73-0603
NONLINEAR	ON THE NUMERICAL SOLUTION OF A CLASS OF NONLINEAR PROBLEMS IN DYNAMIC COUPLED THERMOELASTICITY	74-0441
NONLINEAR	ON THE RESOLUTION OF NONLINEAR ELASTIC AND ELASTOPLASTIC PROBLEMS BY THE METHOD OF FINITE ELEMENTS.	74-1326
NONLINEAR	ON THE SOLUTION OF SETS OF NONLINEAR EQUATIONS ARISING IN THE APPLICATION OF FINITE ELEMENT METHODS	75-0127
NONLINEAR	ON THE UNCONDITIONAL STABILITY OF AN IMPLICIT ALGORITHM FOR NONLINEAR STRUCTURAL DYNAMICS	75-0825
NONLINEAR	ORGANIZATION OF GNATS: A GENERAL NONLINEAR ANALYSIS COMPUTER PROGRAM FOR TWO-DIMENSIONAL STRUCTURES	74-0871
NONLINEAR	PART 1 – TASK 3.2.9. BENDING AND BUCKLING OF A ONE-DIMENSIONAL MODEL SIMULATING A CONICAL SHELL. PAR	71-0745
NONLINEAR	PERTURBATION PROCEDURES IN NONLINEAR FINITE ELEMENT ANALYSIS	74-1241
NONLINEAR	PLANE STRESS FINITE ELEMENTS FOR NONLINEAR ANALYSIS	69-0529
NONLINEAR	PLANS – A FINITE ELEMENT PROGRAM FOR NONLINEAR ANALYSIS OF STRUCTURES	74-0966
NONLINEAR	PLANS – A FINITE-ELEMENT PROGRAM FOR NONLINEAR ANALYSIS OF STRUCTURES	75-0011
NONLINEAR	PLATES AND BEAMS ON ELASTIC FOUNDATIONS: LINEAR AND NONLINEAR BEHAVIOUR	68-0042
NONLINEAR	PROGRESS IN NONLINEAR FINITE ELEMENT ANALYSIS USING ASYMPTOTIC SOLUTION TECHNIQUES	72-0316
NONLINEAR	PROPAGATION OF ELLIPTICAL SURFACE CRACKS AND NONLINEAR FRACTURE MECHANICS BY THE FINITE ELEMENT METH	74-0906
NONLINEAR	REFINED FINITE ELEMENT ANALYSIS OF LINEAR AND NONLINEAR TWO-DIMENSIONAL STRUCTURES	66-0024
NONLINEAR	RESPONSE OF NONLINEAR BEAM TO RANDOM EXCITATION	73-0285
NONLINEAR	SOLUTION OF NONLINEAR PROBLEMS OF ELASTOPLASTICITY BY THE FINITE ELEMENT METHOD	68-0007

NONLINEAR	SOLUTION OF NONLINEAR EQUATIONS BY EXPLICIT TIME INTEGRATION - A SHORT COMMUNICATION	73-0606
NONLINEAR	SOLUTION PROCEDURE FOR NONLINEAR STRUCTURAL PROBLEMS	73-0425
NONLINEAR	SOLUTION SCHEMES FOR PROBLEMS OF NONLINEAR STRUCTURAL DYNAMICS	74-0505
NONLINEAR	SOLVING A NONLINEAR PROBLEM OF NONSTEADY THERMAL CONDUCTIVITY IN SOLID BODIES OF ARBITRARY FORM	68-0019
NONLINEAR	SOME APPROXIMATIONS IN THE NONLINEAR ANALYSIS OF RECTANGULAR PLATES USING FINITE ELEMENTS	74-1231
NONLINEAR	SOME CONTRIBUTIONS TO NONLINEAR SOLID MECHANICS	73-0705
NONLINEAR	SOME CONTRIBUTIONS TO NONLINEAR SOLID MECHANICS	74-0674
NONLINEAR	SOME DISTRIBUTIONAL AND CONVERGENCE PROPERTIES OF THE FINITE ELEMENT METHOD, WITH APPLICATIONS IN NO	73-0598
NONLINEAR	SOME MATHEMATICAL PROPERTIES OF CERTAIN MIXED GALERKIN APPROXIMATIONS IN NONLINEAR ELASTICITY	74-1248
NONLINEAR	SOME RECENT DEVELOPMENTS IN THE THEORY AND APPLICATION OF FINITE ELEMENTS TO MIXED PROBLEMS AND SHOC	73-0637
NONLINEAR	SOME RESULTS OF FINITE ELEMENT APPLICATIONS ON NONLINEAR ELASTICITY	73-0605
NONLINEAR	SOME TWO-FIELD VARIATIONAL PRINCIPLES FOR NONLINEAR DEFORMATION ANALYSIS OF SHELLS	75-0294
NONLINEAR	SPACE-TIME SOLUTIONS OF NONLINEAR REACTOR DYNAMICS BY FINITE-ELEMENT METHOD	75-0503
NONLINEAR	STATIC AND DYNAMIC ANALYSIS OF NONLINEAR STRUCTURES	75-0718
NONLINEAR	STATIC AND DYNAMIC GEOMETRIC AND MATERIAL NONLINEAR ANALYSIS	74-1302
NONLINEAR	STATIC AND DYNAMIC NONLINEAR STRUCTURAL PROBLEMS	72-0848
NONLINEAR	STATIC GEOMETRIC AND MATERIAL NONLINEAR ANALYSIS	72-0314
NONLINEAR	STATIC STRESSES BY LINEAR AND NONLINEAR METHODS	75-0278
NONLINEAR	STATICKE RESENI FYZIKALNE NELINEARNIHO CHOVANI ZELEZOBETO- NOVYCH DESEK (ANALYSIS OF PHYSICALLY NONL	74-0514
NONLINEAR	STEADY CONDUCTION OF HEAT IN LINEAR AND NONLINEAR FULLY ANISOTROPIC MEDIA BY FINITE ELEMENTS	74-0359
NONLINEAR	STOCHASTIC FINITE ELEMENT SIMULATION OF THE NONLINEAR STRUCTURAL RESPONSE OF FIBROUS COMPOSITE MATER	75-0762
NONLINEAR	STRESS STRAIN RELATIONSHIP FROM COMPRESSION TESTS ON NONLINEAR MATERIALS	69-0189
NONLINEAR	SU LA DETERMINAZIONE DEI CAMPI MAGNETICI LINEARI E NON LINEARI MEDIANTE L APPLICAZIONE DEI METODI DE	74-1061
NONLINEAR	SUR L APPROXIMATION, PAR ELEMENTS FINIS D ORDRE UN, ET LA RESOLUTION, PAR PENALISATION-DUALITE, D UN	75-0845
NONLINEAR	SURVEY OF SOLUTION PROCEDURES FOR NONLINEAR STATIC AND DYNAMIC ANALYSES	74-0307
NONLINEAR	SURVEY OF STATIC GEOMETRIC AND MATERIAL NONLINEAR ANALYSIS BY THE FINITE ELEMENT METHOD	72-0783
NONLINEAR	THE ALPHA CONSTANT-STIFFNESS METHOD FOR THE ANALYSIS OF NONLINEAR PROBLEMS	72-0566
NONLINEAR	THE APPLICATION OF FINITE ELEMENTS TO THE LARGE DEFLECTION GEOMETRICALLY NONLINEAR BEHAVIOR OF CYLIN	72-0815
NONLINEAR	THE FINITE ELEMENT ANALYSIS OF PHYSICALLY AND KINEMATICALLY NONLINEAR ELASTIC SOLIDS	69-0146
NONLINEAR	THE FINITE ELEMENT METHOD FOR NONLINEAR PROBLEMS	70-0245
NONLINEAR	THE FINITE ELEMENT METHOD-LINEAR AND NONLINEAR APPLICATIONS	74-0337
NONLINEAR	THE MATRIC SOLUTION OF CERTAIN NONLINEAR PROBLEMS IN STRUCTURAL ANALYSIS	56-0002
NONLINEAR	THE NONLINEAR DYNAMIC ANALYSIS OF SHELLS OF REVOLUTION WITH ASYMMETRIC PROPERTIES BY THE FINITE ELEM	71-0818
NONLINEAR	THE STATUS AND OUTLOOK FOR THE SOLUTION OF NONLINEAR PROBLEMS BY THE FINITE ELEMENT METHOD	70-0259
NONLINEAR	THE STOCHASTIC FINITE ELEMENT SIMULATION OF THE NONLINEAR STRUCTURAL RESPONSE OF FIBROUS COMPOSITE M	75-0798
NONLINEAR	THEMIS FINAL REPORT, VOL. IV, RECENT DEVELOPMENTS IN THE NUMERICAL SOLUTION OF CERTAIN PROBLEMS IN N	74-0721
NONLINEAR	THEMIS FINAL REPORT, VOL. II, NONLINEAR MECHANICS IN SOLIDS AND STRUCTURES	74-0722
NONLINEAR	THEMIS FINAL REPORT, VOL. 1, BASIC RESEARCH IN METHODS OF APPROXIMATION IN THE NONLINEAR MECHANICS O	74-0723
NONLINEAR	THEORETICAL AND NUMERICAL COMPARISON OF ELASTIC NONLINEAR FINITE ELEMENT METHODS.	72-0452
NONLINEAR	THEORETICAL BASIS FOR CBL/NONSAP, A NONLINEAR STRUCTURAL ANALYSIS PROGRAM	74-0740
NONLINEAR	THERMAL NONLINEAR BEHAVIOUR OF STRUCTURES	72-0317
NONLINEAR	THREE DIMENSIONAL NONLINEAR MAGNETIC FIELD BOUNDARY VALUE PROBLEM AND ITS NUMERICAL SOLUTION	70-0060
NONLINEAR	TRANAL: A 3-D FINITE ELEMENT CODE FOR TRANSIENT NONLINEAR ANALYSIS	74-0707
NONLINEAR	TRIANGULAR THIN SHELL FINITE ELEMENT: NONLINEAR ANALYSIS	75-0668
NONLINEAR	USE OF GROUP-THEORETIC METHODS IN THE DEVELOPMENT OF NONLINEAR SHELL FINITE ELEMENTS	71-0704
NONLINEAR	USER'S MANUAL FOR NTEDYNS - A NONLINEAR FINITE ELEMENT CODE FOR ANALYZING THE BLAST RESPONSE OF UNDE	71-0695
NONLINEAR	USERS MANUAL FOR CEL/NONSAP, A NONLINEAR STRUCTURAL ANALYSIS PROGRAM	74-0706
NONLINEAR	VARIATIONAL PRINCIPLES, NONLINEAR OPERATORS, AND FINITE ELEMENT APPLICATIONS	72-0576
NONLINEAR-ANALY	TRANSIENT AND NONLINEAR-ANALYSIS OF FIELD PROBLEMS BY FINITE ELEMENTS	72-0975
NONLINEARITIES	DYNAMIC ANALYSIS OF FRAME AND PLATE STRUCTURES WITH GEOMETRIC AND MATERIAL NONLINEARITIES	73-0649
NONLINEARITIES	DYNAMIC RESPONSES OF CYLINDRICAL SHELLS INCLUDING GEOMETRIC AND MATERIAL NONLINEARITIES	74-0150
NONLINEARITIES	EFFECTS OF MATERIAL NONLINEARITIES ON VIBRATION	73-0323
NONLINEARITIES	FINITE ELEMENT ANALYSIS WITH MATERIAL NONLINEARITIES, THEORY AND PRACTICE	71-0168
NONLINEARITIES	FINITE ELEMENT ANALYSIS OF SOME CONCRETE NONLINEARITIES. THEORY AND EXAMPLES	74-1287
NONLINEARITIES	FINITE-ELEMENT ANALYSIS WITH MATERIAL NONLINEARITIES	69-0177
NONLINEARITIES	INCREMENTAL FORMULATIONS FOR PROBLEMS WITH GEOMETRIC AND MATERIAL NONLINEARITIES	72-0315
NONLINEARITIES	MATERIAL NONLINEARITIES	72-0647
NONLINEARITIES	MATERIAL NONLINEARITIES IN THE FINITE ELEMENT ANALYSIS	74-0647
NONLINEARITIES	ON GENERAL PURPOSE PROGRAMS FOR FINTIE ELEMENT ANALYSIS WITH SPECIAL REFERENCE TO GEOMETRIC AND MATE	70-0087
NONLINEARITIES	STATIC STRESS ANALYSIS OF AXISYMMETRIC SOLIDS WITH MATERIAL AND GEOMETRIC NONLINEARITIES BY THE FINI	73-0958
NONLINEARITIES	VISCO-PLASTICITY AND PLASTICITY-AN ALTERNATIVE FOR FINITE ELEMENT SOLUTION OF MATERIAL NONLINEARITIE	74-0410
NONLINEARITY	COMBINED MATERIAL AND GEOMETRIC NONLINEARITY FOR THIN STEEL PLATES	75-0397
NONLINEARITY	GEOMETRIC NONLINEARITY AND THE FINITE ELEMENT DISPLACEMENT METHOD	74-0679
NONLINEARITY	LARGE NATURAL STRAINS AND SOME SPECIAL DIFFICULTIES DUE TO NONLINEARITY AND INCOMPRESSIBILITY IN FIN	74-0528
NONLINEARITY	LARGE NATURAL STRAINS AND SOME SPECIAL DIFFICULTIES DUE TO NONLINEARITY AND INCOMPRESSIBILITY IN FIN	74-1027
NONLINEARITY	UNIFIED APPROACH TO THREE FINITE ELEMENT THEORIES FOR GEOMETRIC NONLINEARITY	74-0599
NONLINEARLY	DISCONTINUOUS FINITE-ELEMENT APPROXIMATIONS FOR ANALYSIS OF SHOCK-WAVES IN NONLINEARLY ELASTIC-MATER	75-0451
NONLINEARLY	RIGID PLATE ON NONLINEARLY DEFORMABLE COHESIVE BASE (PLANE PROBLEM)	75-0264
NONLINEARTIES	FINITE ELEMENT ANALYSIS WITH MATERIAL NONLINEARTIES: THEORY AND PRACTICE	69-0459
NONPLANAR	KERNEL FUNCTION FOR NONPLANAR OSCILLATING SURFACES IN SUPERSONIC FLOW	71-0472
NONPRISMATIC	A HIGH-ORDER FINITE ELEMENT FOR PRISMATIC AND NONPRISMATIC FOLDED PLATE STRUCTURES	75-0889
NONPRISMATIC	HIGH-ORDER FINITE ELEMENT TECHNIQUE FOR PRISMATIC AND NONPRISMATIC FOLDED PLATE STRUCTURES	74-1130
NONPROPORTIONAL	NONPROPORTIONAL LOADING LIMIT FOR STRUCTURES	74-0228
NONRIGID	ANALYTICAL DYNAMICS AND NONRIGID SPACECRAFT SIMULATION	74-0771
NONRIGID	DYNAMIC ANALYSIS OF A SYSTEM OF HINGE-CONNECTED RIGID BODIES WITH NONRIGID APPENDAGES	74-0852
NONSAP	NONSAP - A GENERAL FINITE ELEMENT PROGRAM FOR NONLINEAR DYNAMIC ANALYSIS OF COMPLEX STRUCTURES	73-0740
NONSAP	NONSAP - A NONLINEAR STRUCTURAL ANALYSIS PROGRAM	74-1006
NONSAP	NONSAP: A STRUCTURAL ANALYSIS PROGRAM FOR STATIC AND DYNAMIC RESPONSE OF NONLINEAR SYSTEMS	74-0848
NONSAP	THEORETICAL BASIS FOR CBL/NONSAP, A NONLINEAR STRUCTURAL ANALYSIS PROGRAM	74-0740
NONSAP	USERS MANUAL FOR CEL/NONSAP, A NONLINEAR STRUCTURAL ANALYSIS PROGRAM	74-0706
NONSTATIONARY	APPLICATION OF LAPLACE TRANSFORM AND FINITE ELEMENT METHOD TO SOLUTION OF NONSTATIONARY THERMAL COND	74-0171
NONSTATIONARY	NUMERICAL SOLUTION OF SOME NONSTATIONARY PARTIAL DIFFERENTIAL EQUATIONS: THE FINITE ELEMENT METHOD A	72-0930
NONSTEADY	ANALYSIS OF NONSTEADY FLOW WITH A FREE SURFACE USING FINITE ELEMENT METHOD	71-0482
NONSTEADY	FINITE ELEMENT ANALYSIS OF A NONSTEADY GROUNDWATER SEEPAGE PROBLEM DAM, DRAINAGE	74-1375
NONSTEADY	NONSTEADY PYROMETRIC CALCULATIONS AS THE BASIS FOR STRESS ANALYSIS OF STEAM GENERATORS AND HEAT EXCH	72-0451
NONSTEADY	SOLVING A NONLINEAR PROBLEM OF NONSTEADY THERMAL CONDUCTIVITY IN SOLID BODIES OF ARBITRARY FORM	68-0019
NONSTRUCTURAL	IMPROVEMENTS OF FINITE ELEMENT SOLUTIONS FOR STRUCTURAL AND NONSTRUCTURAL APPLICATIONS	72-0671
NONSTRUCTURAL	RESULTS ON THE CONVERGENCE OF THE FINITE ELEMENT METHOD IN STRUCTURAL AND NONSTRUCTURAL CASES	74-0416
NONSYMMETRICAL	FINITE ELEMENT ANALYSIS OF OPEN SHELLS OF REVOLUTION UNDER NONSYMMETRICAL LOADING	68-0295
NONSYMMETRICAL	PHOTOELASTIC AND FINITE ELEMENT INVESTIGATION OF A NONSYMMETRICAL PLUG HATCH CONFIGURATION	73-0025
NONUNIFORM	A " LOCAL " BASIS OF GENERALIZED SPLINES OVER RIGHT TRIANGLES DETERMINED FROM A NONUNIFORM PARTITION	72-0525
NONUNIFORM	A COMPATIBLE PLATE BENDING ANNULAR SECTOR FINITE ELEMENT OF NONUNIFORM THICKNESS	74-0430
NONUNIFORM	A STATISTICAL THEORY FOR THE FRACTURE OF BRITTLE STRUCTURES SUBJECTED TO NONUNIFORM POLYAXIAL STRESS	73-0928
NONUNIFORM	AN IMPROVED MATRIX DISPLACEMENT METHOD FOR VIBRATION OF NONUNIFORM BEAMS	71-0841
NONUNIFORM	CONDITION OF FINITE ELEMENT MATRICES GENERATED FROM NONUNIFORM MESHES	72-0104
NONUNIFORM	DYNAMICAL ANALYSIS OF AXIALLY NONUNIFORM, THIN CYLINDRICAL SHELLS. PART 2 FREE VIBRATION	71-0716
NONUNIFORM	FINITE ELEMENT ANALYSIS OF SHELLS UNDER NONUNIFORM SNOW LOADS	69-0113
NONUNIFORM	NONLINEAR VIBRATIONS OF NONUNIFORM BEAMS WITH CONCENTRATED MASSES	74-1300
NONUNIFORM	NONUNIFORM ERROR ESTIMATES FOR THE FINITE ELEMENT METHOD	74-0890
NONUNIFORM	ON THE NONUNIFORM DEFORMATION OF MATERIAL IN AXIALLY SYMMETRIC COMPRESSION CAUSED BY FRICTION - 2	74-0321
NONUNIFORM	RASCHET KILEBANII PLASTIN SLOZHNDI FORMY S NEODNORODNYMI GRANICHNYMI USLOVIYAMI. (CALCULATION OF VIB	74-0988
NONVARIATIONAL	ELLIPTIC BOUNDARY VALUE PROBLEMS WITH NONVARIATIONAL PERTURBATION AND THE FINITE ELEMENT METHOD	73-1022

NONWOVEN	AN ANALYSIS OF THE PHYSICAL BEHAVIOR OF NONWOVEN FIBROUS MATERIAL BY THE FINITE-ELEMENT METHOD	73-1149
NONZERO	STIFFNESS MATRICES FOR THIN TRIANGULAR ELEMENTS OF NONZERO GAUSSIAN CURVATURE	66-0073
NORDSEE	DESIGN, ANALYSIS AND CONSTRUCTION OF THE RESEARCH PLATFORM (NORDSEE)	75-0261
NORFOLK	THE STRESS DISTRIBUTION OF NORFOLK DAM	62-0001
NORM	INTERIOR MAXIMUM NORM ESTIMATES FOR SOME SIMPLE FINITE ELEMENT METHODS	74-1023
NORMAL	ACOUSTIC RESPONSE OF A SPACECRAFT SOLAR PANEL USING NORMAL MODE METHOD	69-0422
NORMAL	CONSTRUCTION OF NORMAL DISPLACEMENT FUNCTIONS OF TRIANGULAR FINITE ELEMENT OF PLATES AND SHELLS	72-0563
NORMAL	CREEP OF A UNIAXIAL METAL MATRIX COMPOSITE SUBJECTED TO AXIAL AND NORMAL LATERAL LOADS	71-0041
NORMAL	CREEP OF A UNIAXIAL METAL MATRIX COMPOSITE SUBJECTED TO AXIAL AND NORMAL LATERAL LOADS	71-0727
NORMAL	DEFORMATION OF THE ARTERIAL VASA VASORUM AT NORMAL AND HYPERTENSIVE ARTERIAL PRESSURE	71-0149
NORMAL	EFFECT OF DAMPING ON EXCITABILITY OF HIGH-ORDER NORMAL MODES	75-0709
NORMAL	FINITE ELEMENT CALCULATION OF LEFT VENTRICULAR WALL STRESS IN NORMAL AND DISEASE STATES	73-0588
NORMAL	FINITE ELEMENT METHOD WITH CONSIDERATION TO THE CONTINUITY OF NORMAL SLOPE	72-0752
NORMAL	FINITE ELEMENT METHOD WITH CONSIDERATION TO THE CONTINUITY OF NORMAL SLOPE	72-0813
NORMAL	INELASTIC ANALYSIS OF A UNIDIRECTIONAL COMPOSITE SUBJECTED TO TRANSVERSE NORMAL LOADING	70-0015
NORMAL	INTERNAL SUPPORTING STRUCTURES OF THE (CIRENE) REACTOR CORE - RESULTS OF STATIC STRESS ANALYSIS FOR	73-0818
NORMAL	NEW FINITE ELEMENT APPROACH TO THE NORMAL MODE ANALYSIS IN MAGNETOHYDRODYNAMICS	74-0876
NORMAL	NORMAL MODE ANALYSIS OF THE RADIO ASTRONOMY EXLPORER BOOMS AND SPACECRAFT	71-0130
NORMAL	NORMAL MODE ANALYSIS OF A FUZE SUPPORT STRUCTURE USING NASTRAN, PART 1	72-0353
NORMAL	SKIN-STRINGER PANEL NORMAL MODE RESPONSE EXPERIMENTAL DATA	71-0691
NORMAL	THE ELASTIC STRESS ANALYSIS OF A BI-MATERIAL PLATE WITH A CRACK NORMAL TO THE INTERFACES	72-0819
NORMAL	2-DIMENSIONAL FINITE-ELEMENT ANALYSIS OF ELECTROELASTIC NORMAL MODES OF FERROELECTRIC DISKS AND SHEL	75-0594
NORMAL	3-DIMENSIONAL FINITE-ELEMENT ANALYSIS OF ELECTROELASTIC NORMAL MODES OF A TANGENTIALLY POLARIZED, SE	75-0591
NORMAL-MODE	FINITE ELEMENT MODELING STUDIES IN THE NORMAL-MODE METHOD AND NORMAL-MODE SYNTHESIS	71-0067
NORMAL-MODE	FINITE ELEMENT MODELING STUDIES IN THE NORMAL-MODE METHOD AND NORMAL-MODE SYNTHESIS	71-0067
NORMAL-MODE	NASTRAN MODELING STUDIES IN THE NORMAL-MODE AND NORMAL-MODE SYNTHESIS	71-0069
NORMAL-MODE	NASTRAN MODELING STUDIES IN THE NORMAL-MODE AND NORMAL-MODE SYNTHESIS	71-0069
NORMALLY	FINITE ELEMENT SOLUTION OF NORMALLY INTERSECTING CYLINDERS	73-0723
NORMS	BOUNDS ON THE SPECTRAL AND MAXIMUM NORMS OF THE FINITE ELEMENT STIFFNESS, FLEXIBILITY AND MASS MATRI	73-0409
NORTH	VIBRATION CHARACTERISTICS OF THE NORTH FORK DAM MODEL	74-0845
NORTHROP	AIRCRAFT FUEL TANK VULNERABILITY TO HYDRAULIC RAM: MODIFICATTION OF THE NORTHROP FINITE ELEMENT COMP	74-0724
NORWAY	EXAMPLE OF LARGE-SCALE STRUCTURAL ANALYSIS. COMPARISON BETWEEN FINITE ELEMENT CALCULATION AND FULL S	74-0177
NOSETIP	PROCEEDINGS OF THE NOSETIP STRESS ANALYSIS TECHNICAL INTERCHANGE MEETING	67-0042
NOSETIPS	THERMAL STRESS ANALYSIS OF REENTRY VEHICLE NOSETIPS AT ANGLE OF ATTACK	72-0456
NOSETIPS	THERMAL STRESS ANALYSIS OF RE-ENTRY VEHICLE NOSETIPS AT ANGLE OF ATTACK	72-0852
NOSNIKY	BOX SHAPED GIRDERS, (NOSNIKY UZAVRETEHO PRIEREZU)	70-0169
NOTCH	APPLICATIONS OF FINITE ELEMENT STRESS ANALYSIS AND STRESS-STRAIN PROPERTIES IN DETERMINING NOTCH FAT	73-1114
NOTCH	APPROACH TO INVESTIGATE NOTCH EFFECT ON LOW-CYCLE FATIGUE WITH FINITE ELEMENT	73-0509
NOTCH	KERBSPANNUNGSANALYSE NACH DER FINITE-ELEMENTE-METHODE AN KEHLNAEHTEN. (NOTCH EFFECT ANALYSIS USING TH	75-0673
NOTCH	NOTCH EFFECT IN LOW CYCLE FATIGUE - REPORT 1: LOW-CYCLE FATIGUE STRENGTH OF 0.48 PERCENT CARBON STEE	73-1090
NOTCH	NOTCH EFFECT IN LOW-CYCLE FATIGUE. CONSIDERATIONS ON NOTCH EFFECT IN LOW-CYCLE FATIGUE WITH FINITE-E	75-0343
NOTCH	NOTCH EFFECT IN LOW-CYCLE FATIGUE. CONSIDERATIONS ON NOTCH EFFECT IN LOW-CYCLE FATIGUE WITH FINITE-E	75-0343
NOTCH	NOTCH EFFECT IN LOW-CYCLE FATIGUE - 2. EFFECT OF STRESS CHANGE AND MEAN STRESS	75-0537
NOTCH	NOTCH STRESSES AT THE LOAD CARRYING ELASTIC CORE	73-0266
NOTCH	PARAMETRICAL KERBSPANNUNGSUNTERSUCHUNG AM ELASTISCHEN KERN.(PARAMETRICAL INVESTIGATION OF NOTCH STR	72-0208
NOTCH	STRESS CONCENTRATION FACTORS FOR A FATIGUE TEST SPECIMEN WITH A U-TYPE NOTCH IN ONE EDGE BY FINITE E	74-1344
NOTCH-BEND	COMPUTATION OF POST-YIELD BEHAVIOUR IN NOTCH-BEND AND TENSION TESTPIECES.	72-0244
NOTCHED	A FINITE ELEMENT STUDY OF ELASTIC-PLASTIC STRESS DISTRIBUTIONS IN NOTCHED SPECIMENS UNDER TENSION	68-0177
NOTCHED	A METHOD FOR DETERMINING LOCAL ELASTOPLASTIC STRESS AND STRAIN IN METALLURGICALLY BONDED NOTCHED LAM	74-0686
NOTCHED	AN ANALYTIC METHOD FOR PREDICTING THE FLOW AND FRACTURE IN NOTCHED FIBER COMPOSITE MATERIALS	75-0689
NOTCHED	APPLICATION OF FINITE ELEMENT ELASTIC-PLASTIC STRESS ANALYSIS TO NOTCHED FATIGUE SPECIMEN BEHAVIOUR	71-0185
NOTCHED	APPLICATION OF THE SINGLE-EDGE NOTCHED BEAM TO FRACTURE TOUGHNESS TESTING OF CERAMICS	74-0623
NOTCHED	COMPARISON OF FINITE-ELEMENT AND EXPERIMENTAL STUDIES ON DEFORMATIONS OF ZIRCONIUM NOTCHED BEND SPEC	71-0359
NOTCHED	CREEP STRESS AND STRAIN ANALYSIS OF NOTCHED PLATES AND BARS BY MEANS OF THE FINITE ELEMENT METHOD	70-0141
NOTCHED	ELASTIC PLASTIC ANALYSIS OF NOTCHED SPECIMENS USING THE FINITE ELEMENT METHOD	70-0140
NOTCHED	ELASTIC-PLASTIC STRESS ANALYSIS FOR A NOTCHED BAR IN PLANE STRAIN BENDING	71-0410
NOTCHED	EXPERIMENTAL AND THEORETICAL FATIGUE-CRACK PROPAGATION LIVES OF VARIOUSLY NOTCHED PLATES	74-0602
NOTCHED	FINITE ELEMENT ANALYSIS OF NOTCHED TENSILE SPECIMENS IN PLANE STRESS	71-0557
NOTCHED	NOTE ON STRESS AND STRAIN REDISTRIBUTION IN A NOTCHED PLATE SPECIMEN DURING CYCLIC LOADING	69-0194
NOTCHED	OSCILLATION STABILITY OF INTERNALLY AND EXTERNALLY NOTCHED PLANE BARS OF EQUAL FORM FACTOR FROM 3943	73-0745
NOTCHED	STRAIN CONCENTRATION BEHAVIOUR IN A NOTCHED ROUND BAR SUBJECTED TO CREEP	73-0440
NOTCHED	STRESS INTENSITY FACTORS FOR A NOTCHED BEAM ON AN ELASTIC FOUNDATION	75-0566
NOTCHED	STRESSES IN A PARTLY YIELDED NOTCHED BAR - AN ASSESSMENT OF THREE ALTERNATIVE PROGRAMS	73-0267
NOTCHED	STRESSES IN A PARTLY YIELDED NOTCHED BAR EM DASH AN ASSESSMENT OF THREE ALTERNATIVE PROGRAMS	73-0268
NOTCHES	CRACKS, NOTCHES AND FINITE ELEMENTS	75-0253
NOTCHES	CYCLIC-STRESS ANALYSIS OF NOTCHES FOR SUPERSONIC TRANSPORT CONDITIONS	74-0834
NOTCHES	STRESS CONCENTRATION FACTORS FOR CIRCULAR NOTCHES	74-0791
NOTCHES	STUDY ON FRACTURE TOUGHNESS EVALUATION BY ELASTO-PLASTIC ANALYSIS AROUND NOTCHES	74-0619
NOVYCH	STATICKE RESENI FYZIKALNE NELINEARNIHO CHOVANI ZELEZOBETO- NOVYCH DESEK (ANALYSIS OF PHYSICALLY NONL	74-0514
NOZZLE	ANALYTICAL AND EXPERIMENTAL INVESTIGATION OF A NOZZLE TO CYLINDRICAL SHELL JUNCTION	74-0638
NOZZLE	DETERMINATION OF STRESS CONCENTRATIONS AT NOZZLE TO CYLINDER INTERSECTIONS AND COMPARISON WITH EXPER	73-0491
NOZZLE	FINITE ELEMENT METHOD FOR THE ANALYSIS OF NOZZLE OPENINGS IN SHELLS OF REVOLUTION	69-0254
NOZZLE	ON THE STRESS DISTRIBUTION IN NOZZLE TO CYLINDER CONNECTIONS FOR SMALL DIAMETER RATIOS	72-0426
NOZZLE-CYLINDER	THREE-DIMENSIONAL CRACK ANALYSIS APPLIED TO AN LWR NOZZLE-CYLINDER INTERSECTION	75-0165
NOZZLE-CYLINDRI	REBO: COMPUTER CODE FOR THE GENERATION OF A THREE-DIMENSIONAL FINITE-ELEMENT GRID OF THE NOZZLE-CYLI	74-1310
NOZZLE-DRUM	THERMAL STRESSES IN A NOZZLE-DRUM INTERSECTION	73-0824
NOZZLES	ADVANCED WIRE-WOUND TUNGSTEN NOZZLES	66-0031
NOZZLES	ANALYSIS OF BOTTOM ENTRY NOZZLES FOR ATMOSPHERIC STORAGE TANKS	75-0331
NOZZLES	ANALYTIC INVESTIGATION OF COMPACT REINFORCEMENT FOR RADIAL NOZZLES IN SPHERICAL SHELLS	71-0107
NOZZLES	APPLICATION OF FINITE ELEMENT METHOD - STRESS CALCULATIONS ON 90 DEGREES NOZZLES OF THIN-WALLED CYLI	71-0253
NOZZLES	EURCYL: A PROGRAM TO GENERATE FINITE ELEMENT MESHES FOR PRESSURE VESSEL NOZZLES	74-1086
NOZZLES	HEAT CONDUCTION AND STRESS ANALYSIS OF SOLID PROPELLANT ROCKET MOTOR NOZZLES	69-0059
NOZZLES	STRESS ANALYSIS OF PRESSURE VESSEL NOZZLES	73-0623
NOZZLES	THREE-DIMENSIONAL ANALYSIS OF REACTOR PRESSURE VESSEL NOZZLES	71-0221
NSRDC	NSRDC: NASTRAN COLLOQUIUM, WASHINGTON, D.C.	70-0404
NTEDYNS	INPUT SOIL PROPERTIES FOR THE FINITE ELEMENT CODE NTEDYNS	71-0694
NTEDYNS	USER'S MANUAL FOR NTEDYNS - A NONLINEAR FINITE ELEMENT CODE FOR ANALYZING THE BLAST RESPONSE OF UNDE	71-0695
NUCLEAIRE	ANALYSE TRIDIMENSIONALE DU COMPORTEMENT NON-LINEAIRE D'UN CAISSON DE REACTEUR NUCLEAIRE EN BETON PRE	71-0144
NUCLEAR	A FINITE ELEMENT STRESS ANALYSIS OF THE VERTICAL BUTTRESSES OF A NUCLEAR CONTAINMENT VESSEL	69-0280
NUCLEAR	A SEISMIC DESIGN OF NUCLEAR REACTOR BUILDING, STRESS ANALYSIS AND STIFFNESS EVALUATION OF THE ENTIRE	71-0251
NUCLEAR	A SURVEY OF THE APPLICATION OF FINITE ELEMENT METHODS TO STRESS ANALYSIS PROBLEMS WITH PARTICULAR EM	71-0632
NUCLEAR	AN ENGINEERING APPROACH TO FINITE ELEMENT ANALYSIS OF NUCLEAR COMPONENTS	73-1054
NUCLEAR	ANALYSIS OF NUCLEAR FUEL ROD LOCALIZED STRESSES	73-0708
NUCLEAR	APPLICATION OF THE FINITE ELEMENT METHOD TO THE NONLINEAR ANALYSIS OF NUCLEAR REACTOR FUEL BEHAVIOR	74-1246
NUCLEAR	APPLICATION OF THREE-DIMENSIONAL FINITE ELEMENTS FOR ANALYSIS OF IRRADIATION INDUCED STRESSES IN NUC	72-0078
NUCLEAR	APPROXIMATE METHOD FOR CALCULATING THE RESPONSE OF EMPLACEMENT STRUCTURES SUBJECTED TO GROUND SHOCK	74-0654
NUCLEAR	AUTOMATED EVALUATION OF INTERCEPTOR ROCKET MOTOR DESIGNS UNDER COMBINED OPERATIONAL AND NUCLEAR EFFE	74-0698
NUCLEAR	BERECHNUNGSMETHODEN ZUR ERMITTLUNG VON BELASTUNGSBEDINGUNGEN UND GRENZTRAGFAEHIGKEITSVERHALTEN VON K	73-0858
NUCLEAR	COMPUTATIONAL ANALYSIS AND EVALUATION OF THE FINITE ELEMENT METHOD FOR A CLASS OF NUCLEAR REACTOR CO	74-0595
NUCLEAR	COMPUTATIONAL METHODS IN NUCLEAR REACTOR STRUCTURAL DESIGN FOR HIGH TEMPERATURE APPLICATIONS: AN INT	73-1001
NUCLEAR	CONFERENCE ON APPLICATION OF FINITE ELEMENT METHODS TO STRESS ANALYSIS PROBLEMS IN NUCLEAR ENGINEERI	71-0774
NUCLEAR	EFFECTS OF FOUNDATION ROTATION ON SEISMIC INERTIA FORCES OF NUCLEAR POWER PLANT STRUCTURES	73-0177

NUCLEAR	ENGINEERING APPROACH TO FINITE ELEMENT ANALYSIS OF NUCLEAR COMPONENTS	73-0839
NUCLEAR	EVALUATION OF TORSIONAL EFFECTS IN THE SEISMIC ANALYSIS OF SYMMETRIC AND UNSYMMETRIC NUCLEAR POWER P	73-0175
NUCLEAR	FINITE ELEMENT ANALYSIS IN NUCLEAR POWER PLANTS (JAPANESE)	72-0685
NUCLEAR	FINITE ELEMENT SOLUTION FOR WAVE PROPAGATION IN LAYERED MEDIA CAUSED BY A NUCLEAR DETONATION	67-0016
NUCLEAR	FINITE ELEMENT TREATMENT OF SOIL-STRUCTURE INTERACTION PROBLEM FOR NUCLEAR POWER PLANT UNDER SEISMIC	73-0834
NUCLEAR	FRACTURE AND SAFETY ANALYSIS OF NUCLEAR PRESSURE VESSELS	73-0257
NUCLEAR	INELASTIC RESPONSE OF NUCLEAR PIPING SUBJECTED TO RUPTURE FORCES	75-0515
NUCLEAR	NON-LINEAR ANALYSIS OF THE ROTATIONAL STIFFNESS OF A NUCLEAR STEAM GENERATOR FOUNDATION	69-0286
NUCLEAR	NON-LINEAR THERMAL STRESS ANALYSIS FOR NUCLEAR POWER PLANT BY FINITE ELEMENT METHOD	71-0181
NUCLEAR	POST-TENSIONED CONCRETE NUCLEAR STRUCTURES	70-0040
NUCLEAR	PROBABILISTIC ASSESSMENT OF AIRCRAFT HAZARD FOR NUCLEAR POWER PLANTS	72-0235
NUCLEAR	PWR DEVELOPMENT FOR NUCLEAR PROPULSION: VIBRATION ANALYSIS OF NUCLEAR POWER PLANT	71-0797
NUCLEAR	PWR DEVELOPMENT FOR NUCLEAR PROPULSION: VIBRATION ANALYSIS OF NUCLEAR POWER PLANT	71-0797
NUCLEAR	RESPONSE OF NUCLEAR STRUCTURAL SYSTEMS TO TRANSIENT AND RANDOM EXCITATIONS, USING BOTH DETERMINISTIC	74-1047
NUCLEAR	ROCK MECHANICS STUDIES FOR A HIGH CUT AT A NUCLEAR POWER STATION IN PENNSYLVANIA	70-0122
NUCLEAR	SAFE-COLAPS: A FINITE ELEMENT PROGRAM FOR THE CREEP COLLAPSE ANALYSIS OF NUCLEAR FUEL RODS	75-0573
NUCLEAR	STATIC AND HEAT TRANSFER ANALYSIS OF A SPHERE-CONE INTERSECTION IN A NUCLEAR CONTAINMENT VESSEL	75-0166
NUCLEAR	STRESS ANALYSIS OF A PRESTRESSED CONCRETE NUCLEAR PRESSURE VESSEL BY THE FINITE ELEMENT METHOD USING	71-0154
NUCLEAR	STRUCTURAL ANALYSIS OF AIRCRAFT IMPACT ON A NUCLEAR CONTAINMENT VESSEL AND ASSOCIATED STRUCTURES	69-0279
NUCLEAR	STRUCTURAL DESIGN OF NUCLEAR PLANT FACILITIES	73-0812
NUCLEAR	THE APPLICATION OF THE BERSAFE FINITE ELEMENT SYSTEM TO NUCLEAR DESIGN PROBLEMS	71-0117
NUCLEAR	THERMAL AND STRESS ANALYSIS ON PRISMATIC NUCLEAR FUEL ELEMENTS	75-0120
NUCLEAR	THREE-DIMENSIONAL STUDY OF NUCLEAR FUEL ROD BEHAVIOUR DURING STARTUP	73-0815
NUCLEAR	THREE-DIMENSIONAL STUDY OF NUCLEAR FUEL ROD BEHAVIOR DURING STARTUP	74-1003
NUCLEAR	TRANSIENT THERMAL STRESS TECHNIQUES IN NUCLEAR REACTORS. APPLIED MECHANICS ASPECTS OF NUCLEAR EFFECT	71-0474
NUCLEAR	TRANSIENT THERMAL STRESS TECHNIQUES IN NUCLEAR REACTORS. APPLIED MECHANICS ASPECTS OF NUCLEAR EFFECT	71-0474
NUCLEAR	TWO-DIMENSIONAL TIME-DEPENDENT ANALYSIS OF PERFORATED END CAPS FOR NUCLEAR REACTOR PRESSURE VESSELS	71-0239
NUCLEAR	USE OF THE ' TITUS ' FINITE ELEMENT CODE FOR THE NUCLEAR STRUCTURES AND INTERNAL STRESS-ANALYSIS	73-0840
NUCLEAR-CONTAIN	IMPROVED FINITE-ELEMENT ANALYSIS OF NUCLEAR-CONTAINMENT SHELL STRUCTURES	73-0239
NUCLEAR-POWER	PHOTOELASTIC DETERMINATION OF THE STRESSES IN THE FOUNDATION OF THE NUCLEAR-POWER STATION BUGEY 2	75-0834
NUMBER	A FINITE ELEMENT METHOD FOR LOW REYNOLDS NUMBER FLOW AND ITS APPLICATION TO BIOMECHANICS	69-0304
NUMBER	AUTOMATED STRUCTURAL SYNTHESIS USING A RE NUMBER OF DESIGN COORDINATES	73-0584
NUMBER	BOUNDS ON EXTERNAL EIGENVALUES OF FINITE ELEMENT STIFFNESS AND MASS MATRICES AND THEIR SPECTRAL COND	72-0256
NUMBER	FINITE ELEMENT SOLUTION OF LOW BOND NUMBER SLOSHING	75-0728
NUMBER	LOW REYNOLDS NUMBER DEVELOPING FLOW	69-0020
NUMBER	RULES GOVERNING THE NUMBER OF NODES AND ELEMENTS IN A FINITE ELEMENT MESH	70-0053
NUMBERING	NODE NUMBERING OPTIMIZATION IN STRUCTURAL ANALYSIS	75-0077
NUMBERS	APPLICATION OF THE FINITE ELEMENT METHOD TO CASES REQUIRING THE COMBINATION OF ELEMENTS POSSESSING D	72-0580
NUMBERS	DIFFUSION ON SOLID SPHERICAL PARTICLE IN VISCOUS FLUID FLOW AT FINITE VALUES OF RE NUMBERS	69-0419
NUMBERS	MONOTON FINITE-ELEMENT MATRICES AND THEIR COMPUTED CONDITION NUMBERS	75-0820
NUMBERS	NUMERICAL SOLUTION FOR THE FLOW AROUND A CYLINDER AT REYNOLDS NUMBERS OF 40, 200 AND 500	69-0378
NUMBERS	THE L2 AND L00 CONDITION NUMBERS OF THE FINITE ELEMENT MATRICES AND THE POINTWISE CONVERGENCE OF THE	72-0061
NUMBERS	THEORETICAL AND EXPERIMENTAL CONSIDERATIONS ON FLOW OF VISCOUS LIQUIDS PAST PLATE AT SMALL AND AVERA	70-0144
NUMERICA	ANALISI NUMERICA DEL COMPORTAMENTO ELASTICO DI UNA STRUTTURA BIDIMENSIONALE PIANA. (NUMERICAL ANALYS	74-1029
NUMERICAL	A COMPARATIVE STUDY OF NUMERICAL METHODS OF ELASTIC-PLASTIC ANALYSIS	67-0047
NUMERICAL	A COMPARATIVE STUDY OF DIFFERENT NUMERICAL SOLUTION TECHNIQUES AS APPLIED TO A NONLINEAR STRUCTURAL	73-1117
NUMERICAL	A FINITE ELEMENT PROCEDURE FOR NUMERICAL OPTIMIZATION IN MECHANICAL DESIGN	74-1218
NUMERICAL	A GENERAL NUMERICAL SOLUTION OF THE TWO-DIMENSIONAL DIFFUSION-CONVECTION EQUATION BY THE FINITE ELEM	70-0004
NUMERICAL	A METHOD FOR THE NUMERICAL INTEGRATION OF THE EQUATIONS OF MOTION ARISING FROM A FINITE ELEMENT ANAL	70-0061
NUMERICAL	A NOTE ON NUMERICAL COMPUTATION OF ELASTIC CONTACT PROBLEMS	75-0621
NUMERICAL	A NUMERICAL AND EXPERIMENTAL INVESTIGATION OF THE USE OF J-INTEGRALS	72-0858
NUMERICAL	A NUMERICAL APPROACH TO THE STUDY OF IRREGULAR SHIP MOTIONS	74-0030
NUMERICAL	A NUMERICAL INVESTIGATION OF STABLY STRATIFIED, WIND DRIVEN CAVITY FLOW BY THE FINITE ELEMENT METHOD	74-1372
NUMERICAL	A NUMERICAL METHOD FOR THE SOLUTION OF BIOT'S CONSOLIDATION THEORY	72-0688
NUMERICAL	A NUMERICAL METHOD FOR THE ANALYSIS OF LONGITUDINAL ELASTIC PLASTIC STRESS WAVE PROPAGATION	73-0526
NUMERICAL	A NUMERICAL METHOD OF VISCO-ELASTIC STRESS ANALYSIS	68-0175
NUMERICAL	A NUMERICAL MODEL FOR THE SIMULATION OF TWO DIMENSIONAL CONVECTIVE DISPERSION IN SHALLOW ESTUARIES	74-0052
NUMERICAL	A NUMERICAL MODEL OF DISPERSION IN ESTUARIES	74-0045
NUMERICAL	A NUMERICAL MODEL OF TRANSIENT WATER QUALITY IN A ONE-DIMENSIONAL ESTUARY BASED ON THE FINITE ELEMEN	74-1369
NUMERICAL	A NUMERICAL PROCEDURE FOR TWO DIMENSIONAL HEATING AND MELTING CALCULATIONS, WITH APPLICATIONS TO LAS	72-0708
NUMERICAL	A NUMERICAL SOLUTION OF A CLASS OF PROBLEMS OF FINITE ELASTIC DEFORMATION	66-0097
NUMERICAL	A NUMERICAL SOLUTION TECHNIQUE FOR A CLASS OF TWO-DIMENSIONAL PROBLEMS IN FLUID DYNAMICS FORMULATED	69-0019
NUMERICAL	A NUMERICAL SOLUTION OF THE BOLTZMANN EQUATION BY THE FINITE ELEMENT TECHNIQUE	70-0095
NUMERICAL	A NUMERICAL SOLUTION OF THE NAVIER-STOKES EQUATIONS USING THE FINITE ELEMENT TECHNIQUE	73-0061
NUMERICAL	A NUMERICAL STUDY OF NONLINEAR INSTABILITY BY MEANS OF AN EXTENDED FINITE ELEMENT METHOD	74-0632
NUMERICAL	A NUMERICAL THERMO-MECHANICAL MODEL FOR THE WELDING AND SUBSEQUENT LOADING OF A FABRICATED STRUCTURE	72-0847
NUMERICAL	A STUDY OF NUMERICAL ERROR IN STRUCTURAL SYSTEMS	70-0394
NUMERICAL	A STUDY OF VARIOUS IDEALIZATIONS FOR WING STRUCTURES AND NUMERICAL PROCEDURES INVOLVED USING MATRIX	69-0257
NUMERICAL	A THEORETICAL AND NUMERICAL COMPARISON OF ELASTIC NONLINEAR FINITE ELEMENT METHODS	72-0709
NUMERICAL	A USEFUL FORM OF THE VARIATIONAL PRINCIPLE OF IDEAL MHD FOR ONE-DIMENSIONAL NUMERICAL STABILITY CALC	73-0746
NUMERICAL	ACCURACY OF SOME NUMERICAL METHODS FOR COLUMN BUCKLING	75-0583
NUMERICAL	ADVANCES IN NUMERICAL FLUID DYNAMICS	73-0813
NUMERICAL	AN APPROXIMATE NUMERICAL SOLUTION OF THE NAVIER STOKES EQUATIONS BY GALERKIN METHOD	74-0011
NUMERICAL	AN INVESTIGATION OF THE STABILITY OF NUMERICAL SOLUTIONS OF BIOT'S EQUATIONS OF CONSOLIDATION	74-0746
NUMERICAL	ANALISI NUMERICA DEL COMPORTAMENTO ELASTICO DI UNA STRUTTURA BIDIMENSIONALE PIANA. (NUMERICAL ANALYS	74-1029
NUMERICAL	ANALYSE NUMERIQUE D'UN PROBLEME DE STEFAN A DEUX PHASES PAR UNE METHODE D'ELEMENTS FINIS. (NUMERICAL	75-0387
NUMERICAL	APPLICATION OF FINITE ELEMENT METHOD TO GENERAL NUMERICAL SOLUTION OF 2-DIMENSIONAL DIFFUSION-CONVEC	70-0269
NUMERICAL	APPLICATION OF FINITE ELEMENT TECHNIQUES FOR THE NUMERICAL SOLUTION OF THE NEUTRON TRANSPORT AND DIF	71-0013
NUMERICAL	APPLICATION OF INTERACTIVE GRAPHICS TO THE NUMERICAL METHODS USED IN STRUCTURAL ANALYSIS	73-0384
NUMERICAL	APPLICATIONS OF FINITE ELEMENT METHODS IN REACTOR MATHEMATICS - NUMERICAL SOLUTION OF THE NEUTRON DI	72-0535
NUMERICAL	APPLICATIONS OF FINITE ELEMENT METHODS IN REACTOR MATHEMATICS. NUMERICAL SOLUTION OF THE NEUTRON TRA	74-0853
NUMERICAL	APPLICATIONS OF NUMERICAL ANALYSIS	71-0755
NUMERICAL	APPROXIMATIONS AND NUMERICAL ANALYSIS OF FINITE DEFORMATIONS OF ELASTIC SOLIDS	74-0343
NUMERICAL	ARMY NUMERICAL ANALYSIS CONFERENCE	71-0707
NUMERICAL	CALCULATION OF ELASTIC-PLASTIC DEFORMATION BY THE FINITE ELEMENT METHOD OF NUMERICAL ANALYSIS	71-0793
NUMERICAL	CALCULATION OF THE BRIDGE DECKS BY USING NUMERICAL METHODS, (LE CALCUL DES TABLIERS DE PONTS PAR DES	71-0411
NUMERICAL	CHISLENNOE RESHENIE PLOSKIKH ZADACH TEPLOPROVODNOSTI DLYA OBLASTEI SLOZHNOI FORMY (NUMERICAL SOLUTIO	74-0659
NUMERICAL	CHISLENNYI METOD RASCHETA SOBSTVENNYKH I VYNUZHDENNYKH KOLEBANII SOSTAVNYKH OBOLOCHECHNYKH KONSTRUKT	73-0461
NUMERICAL	CODE TAFEST: NUMERICAL SOLUTION TO TRANSIENT HEAT-CONDUCTION PROBLEMS USING FINITE ELEMENTS IN SPACE	74-0901
NUMERICAL	COMPARATIVE STUDY OF NUMERICAL METHODS OF ELASTIC-PLASTIC ANALYSIS	67-0153
NUMERICAL	COMPARISON OF ANALYTICAL AND NUMERICAL SOLUTIONS IN HEAT CONDUCTION PROBLEMS	74-1170
NUMERICAL	COMPARISON OF NUMERICAL METHODS FOR DETERMINING STRESS INTENSITY FACTORS	72-0176
NUMERICAL	COMPARISON OF THE NUMERICAL INTEGRATION TECHNIQUE AND THE FINITE ELEMENT METHOD IN THE ANALYSIS OF T	73-0848
NUMERICAL	CONTINUUM BASES FOR CONSISTENT NUMERICAL FORMULATIONS OF FINITE STRAINS IN ELASTIC AND INELASTIC STR	75-0299
NUMERICAL	CONVERGENCE AND NUMERICAL ACCURACY WITH SPECIAL REFERENCE TO PLATE BENDING	69-0221
NUMERICAL	DEPLACEMENTS DANS LES STRUCTURES MINCES A FAIBLE RIGIDITE GEOMETRIQUE-DETERMINATION NUMERIQUE (NUMER	74-0213
NUMERICAL	DEVELOPMENT OF A NUMERICAL MODEL FOR DISCONTINUA	74-0961
NUMERICAL	DIE-SWELL RECONSIDERED: SOME NUMERICAL SOLUTIONS USING A FINITE ELEMENT PROGRAM	73-0498
NUMERICAL	ENGINEERING APPLICATIONS OF NUMERICAL INTEGRATION IN STIFFNESS METHOD	66-0040
NUMERICAL	EVALUATION OF NUMERICAL TIME INTEGRATION METHODS AS APPLIED TO ELASTIC-PLASTIC DYNAMIC PROBLEMS INVO	74-0858
NUMERICAL	EVALUATION OF SOME NUMERICAL SCHEMES FOR CONSOLIDATION	73-0349
NUMERICAL	EVALUATION OF TECHNIQUES FOR NUMERICAL CALCULATION OF STORM SURGES	75-0554

NUMERICAL	EXPERIENCES WITH ORTHOGONAL POLYNOMIALS AND "BEST" NUMERICAL INTEGRATION FORMULAS ON A TRIANGLE; WIT	74-0977
NUMERICAL	FINITE ELASTO-PLASTIC DEFORMATION - 1. THEORY AND NUMERICAL EXAMPLES	74-0202
NUMERICAL	FINITE ELEMENT AND FINITE DIFFERENCE ENERGY TECHNIQUES FOR THE NUMERICAL SOLUTION OF PARTIAL DIFFERE	73-0570
NUMERICAL	FINITE ELEMENT IN NUMERICAL ANALYSIS OF ADHESIVE FRACTURE	73-0085
NUMERICAL	FINITE ELEMENT NUMERICAL MODELLING OF FLOW AND DISPERSION IN ESTUARIES	73-0070
NUMERICAL	FINITE ELEMENTS WITH RELAXED CONTINUITY - AN INVESTIGATION OF THE VARIATIONAL BASIS AND NUMERICAL PE	74-1065
NUMERICAL	HARBOUR OSCILLATION - A NUMERICAL TREATMENT FOR UNDAMPED NATURAL MODES	70-0208
NUMERICAL	HYBRID MODELS IN NUMERICAL AND COMPUTER METHODS	73-0682
NUMERICAL	IMPROVED NUMERICAL INTEGRATION OF THICK SHELL FINITE ELEMENTS	71-0396
NUMERICAL	INCREMENTAL NUMERICAL SOLUTION OF A SIMPLE DEFORMATION PROBLEM IN SOIL MECHANICS	70-0380
NUMERICAL	INVESTIGATION OF THE STATIONARY FLOW OF A RIGID PLASTIC MATERIAL BY THE NUMERICAL FINITE ELEMENT MET	73-0231
NUMERICAL	LARGE STRAIN, ELASTO-PLASTIC NUMERICAL ANALYSIS BY MEANS OF FINITE ELEMENT METHOD	71-0790
NUMERICAL	LARGE STRAIN, VISCOELASTIC NUMERICAL ANALYSIS BY MEANS OF FINITE ELEMENT METHOD	72-0768
NUMERICAL	LARGE STRAIN, VISCOELASTIC AND ELASTO VISCOPLASTIC NUMERICAL ANALYSIS BY MEANS OF THE FINITE ELEMENT	75-0563
NUMERICAL	MATRIX DIFFERENTIAL EQUATION AND HIGHER-ORDER NUMERICAL METHODS FOR PROBLEMS OF NON-LINEAR CREEP, VI	72-0198
NUMERICAL	NUMERICAL ACCURACY AND ROUNDING-OFF ERRORS	69-0348
NUMERICAL	NUMERICAL ANALYSES OF DEFORMABILITY TESTS IN JOINTED ROCK. ' JOINT PERTURBATION ' AND ' NO TENSION '	71-0439
NUMERICAL	NUMERICAL ANALYSIS	70-0432
NUMERICAL	NUMERICAL ANALYSIS	73-0628
NUMERICAL	NUMERICAL ANALYSIS	73-1060
NUMERICAL	NUMERICAL ANALYSIS FOR FINITE AXISYMMETRIC DEFORMATIONS OF INCOMPRESSIBLE ELASTIC SOLIDS OF REVOLUTI	70-0094
NUMERICAL	NUMERICAL ANALYSIS IN ORTHOTROPIC LINEAR FRACTURE MECHANICS	73-0512
NUMERICAL	NUMERICAL ANALYSIS OF NONLINEAR PNEUMATIC STRUCTURES	67-0060
NUMERICAL	NUMERICAL ANALYSIS OF SHELL STRUCTURES	67-0077
NUMERICAL	NUMERICAL ANALYSIS OF THIN SHELLS BY CURVED TRIANGULAR ELEMENTS BASED ON DISCRETE - KIRCHHOFF HYPOTH	69-0091
NUMERICAL	NUMERICAL ANALYSIS OF SHIP STRUCTURAL ELEMENTS BY MEANS OF FINITE ELEMENT METHOD AND PHOTO-ELASTIC E	69-0125
NUMERICAL	NUMERICAL ANALYSIS OF VISCO-ELASTIC STRUCTURES BY THE FINITE ELEMENT METHOD	70-0177
NUMERICAL	NUMERICAL ANALYSIS OF FREE SURFACE SEEPAGE PROBLEMS	71-0469
NUMERICAL	NUMERICAL ANALYSIS OF MAGNETOHYDRODYNAMIC INSTABILITIES BY FINITE ELEMENT METHOD	72-0041
NUMERICAL	NUMERICAL ANALYSIS OF AUTOMOBILE STRUCTURES	72-0334
NUMERICAL	NUMERICAL ANALYSIS ON STRESS DISTRIBUTION IN PARTIAL COMPRESSED SPECIMEN (HARDBOARD) BY FINITE ELEME	72-0920
NUMERICAL	NUMERICAL ANALYSIS OF ELECTRO-OSMOTIC FLOW IN SOILS	73-0444
NUMERICAL	NUMERICAL ANALYSIS OF PLASTICITY IN SOILS	73-0611
NUMERICAL	NUMERICAL ANALYSIS OF COOLING TOWER PLUMES AND FOG INCIDENCE	74-0040
NUMERICAL	NUMERICAL ANALYSIS OF THE LINE CAPACITANCE AND CROSSTALK FACTOR FOR INSULATED WIRE PAIRS	74-0685
NUMERICAL	NUMERICAL ANALYSIS OF 2-PHASE STEFAN PROBLEM BY METHOD OF FINITE ELEMENTS	75-0616
NUMERICAL	NUMERICAL ANALYSIS OF EMBANKMENTS OVER SOFT SOILS	75-0719
NUMERICAL	NUMERICAL AND EXPERIMENTAL INVESTIGATION ON THE USE OF J-INTEGRAL	73-0245
NUMERICAL	NUMERICAL AND EXPERIMENTAL ANALYSIS OF A PSEUDOCYLINDRICAL SHELL	74-1133
NUMERICAL	NUMERICAL AND PHYSICAL MODELLING OF REINFORCEMENT SYSTEMS FOR TUNNELS IN JOINTED ROCK	73-0872
NUMERICAL	NUMERICAL AND PHYSICAL MODELLING	74-0393
NUMERICAL	NUMERICAL BIAXIAL CHARACTERIZATION FOR CONCRETE	74-0641
NUMERICAL	NUMERICAL CALCULATION OF MERIDIONAL FLOW FIELD IN TURBOMACHINES USING FINITE ELEMENTS METHOD	74-0481
NUMERICAL	NUMERICAL CALCULATIONS OF VELOCITY AND PRESSURE DISTRIBUTION AROUND OSCILLATING AIRFOILS	74-0847
NUMERICAL	NUMERICAL CALCULATION OF STORM SURGES: AN EVALUATION OF TECHNIQUES	75-0461
NUMERICAL	NUMERICAL COMPUTATION OF BOUNDARY VALUE PROBLEMS IN SOIL MECHANICS	73-0269
NUMERICAL	NUMERICAL CONDITIONING OF STIFFNESS MATRIX FORMULATIONS FOR FRAME STRUCTURES	68-0022
NUMERICAL	NUMERICAL DESIGN-ANALYSIS FOR PILES IN SANDS	74-0333
NUMERICAL	NUMERICAL DETERMINATION OF STRESSES AND DEFORMATIONS IN ROCK TAKING INTO ACCOUNT DISCONTINUITIES	70-0066
NUMERICAL	NUMERICAL DETERMINATION OF STRESSES AND DEFORMATIONS IN ROCK TAKING INTO ACCOUNT DISCONTINUITIES	70-0167
NUMERICAL	NUMERICAL DETERMINATION OF SEA WATER INTRUSION IN AQUIFERS	74-1181
NUMERICAL	NUMERICAL EXPERIMENT WITH FINITE ELEMENT I	70-0303
NUMERICAL	NUMERICAL EXPERIMENTS ON THE DETERMINATION OF STRESS CONCENTRATION FACTORS	74-0292
NUMERICAL	NUMERICAL FORMULATION OF A CLASS OF PROBLEMS IN NONLINEAR VISCOELASTICITY	67-0061
NUMERICAL	NUMERICAL FORMULATION OF NONLINEAR ELASTICITY PROBLEMS	67-0063
NUMERICAL	NUMERICAL HULL VIBRATION ANALYSIS OF A FAR EAST CONTAINER SHIP	74-1325
NUMERICAL	NUMERICAL INTEGRATION APPLIED TO FINITE ELEMENT METHODS	66-0039
NUMERICAL	NUMERICAL INTEGRATION APPLIED TO FINITE ELEMENT METHODS	70-0390
NUMERICAL	NUMERICAL INTEGRATION OF THE EQUATION OF MOTION	71-0507
NUMERICAL	NUMERICAL INTEGRATION IN THE FINITE ELEMENT METHOD	74-0576
NUMERICAL	NUMERICAL INTEGRATION METHODS IN FINITE ELEMENTS	74-0841
NUMERICAL	NUMERICAL INTEGRATION CONSIDERATIONS IN TWO AND THREE DIMENSIONAL ISOPARAMETRIC FINITE ELEMENTS	75-0130
NUMERICAL	NUMERICAL INVESTIGATIONS OF LAKE CIRCULATION AROUND ISLANDS BY THE FINITE ELEMENT METHOD	72-0233
NUMERICAL	NUMERICAL INVESTIGATION OF LAKE CIRCULATION AROUND ISLANDS BY THE FINITE ELEMENT METHOD	72-0337
NUMERICAL	NUMERICAL INVESTIGATION OF AXI-SYMMETRIC VISCOPLASTIC FLOWS BY FINITE-ELEMENT PROCEDURE	73-0132
NUMERICAL	NUMERICAL METHOD FOR GENERATING ORTHOGONAL CURVILINEAR MESHES	70-0443
NUMERICAL	NUMERICAL METHOD FOR INVESTIGATING THE PROPAGATION OF SURFACE WAVES ON DISSIPATIVE GUIDES	74-0613
NUMERICAL	NUMERICAL METHODS FOR THE LIMIT ANALYSIS OF PLATES	68-0181
NUMERICAL	NUMERICAL METHODS FOR TIME DEPENDENT NONLINEAR BOUNDARY VALUE PROBLEMS	70-0296
NUMERICAL	NUMERICAL METHODS FOR THE SOLUTION OF NONLINEAR PROBLEMS IN STRUCTURAL ANALYSIS	73-0506
NUMERICAL	NUMERICAL METHODS FOR UNSTEADY NON-DARCY FLOW	74-0062
NUMERICAL	NUMERICAL METHODS IN SUBSURFACE HYDROLOGY: WITH AN INTRODUCTION TO THE FINITE ELEMENT METHOD	71-0800
NUMERICAL	NUMERICAL METHODS IN FLUID DYNAMICS	72-0844
NUMERICAL	NUMERICAL METHODS IN FLUID DYNAMICS	73-0689
NUMERICAL	NUMERICAL METHODS IN FLUID FLOW PROBLEMS. VOL. 2.	75-0714
NUMERICAL	NUMERICAL METHODS IN FLUID FLOW PROBLEMS. VOL. 1	75-0715
NUMERICAL	NUMERICAL METHODS OF HIGH-ORDER ACCURACY FOR NONLINEAR BOUNDARY VALUE PROBLEMS	68-0194
NUMERICAL	NUMERICAL MODEL FOR TSUNAMI RUN-UP	70-0200
NUMERICAL	NUMERICAL MODEL FOR THE PREDICTION OF TRANSIENT WATER QUALITY IN ESTUARY NETWORKS	72-0635
NUMERICAL	NUMERICAL MODEL OF DISPERSION IN ESTUARIES	74-0905
NUMERICAL	NUMERICAL PERFORMANCE OF A FINITE ELEMENT GALERKIN PROCEDURE	75-0022
NUMERICAL	NUMERICAL PREDICTION OF TRACER CONCENTRATION IN AQUIFERS WITH TRANSIENT GROUNDWATER FLOW BY FINITE-E	75-0610
NUMERICAL	NUMERICAL PROCEDURE FOR THE ANALYSIS OF STRUCTURES SUPPORTED BY VISCOELASTIC FOUNDATIONS	73-0564
NUMERICAL	NUMERICAL PROCEDURES FOR THE DETERMINATION OF THE BEHAVIOUR OF A SHEAR WALL FRAME SYSTEM	73-0216
NUMERICAL	NUMERICAL RESULTS FROM THE APPLICATION OF GRADIENT ITERATIVE TECHNIQUES TO THE FINITE ELEMENT VIBRAT	72-0138
NUMERICAL	NUMERICAL SIMULATION OF FINITE-WIDTH THRUST BEARINGS, TAKING INTO ACCOUNT VISCOSITY VARIATION WITH T	75-0623
NUMERICAL	NUMERICAL SOLUTION BY METHODS OF FINITE ELEMENTS OF DEGENERATED ELLIPTICAL PROBLEMS	72-0284
NUMERICAL	NUMERICAL SOLUTION FOR THE FLOW AROUND A CYLINDER AT REYNOLDS NUMBERS OF 40, 200 AND 500	69-0378
NUMERICAL	NUMERICAL SOLUTION OF THE QUASI LINEAR POISSON EQUATION IN A NON-UNIFORM TRIANGLE MESH	67-0090
NUMERICAL	NUMERICAL SOLUTION OF PARTIAL DIFFERENTIAL EQUATIONS. 11	70-0072
NUMERICAL	NUMERICAL SOLUTION OF DIELECTRIC LOADED WAVEGUIDES: I. FINITE-ELEMENT ANALYSIS	70-0199
NUMERICAL	NUMERICAL SOLUTION OF NAVIER-STOKES EQUATIONS BY FINITE ELEMENT METHOD	72-0008
NUMERICAL	NUMERICAL SOLUTION OF THE ELASTOHYDRODYNAMIC LUBRICATION PROBLEM USING FINITE ELEMENTS	72-0042
NUMERICAL	NUMERICAL SOLUTION OF SOME NONSTATIONARY PARTIAL DIFFERENTIAL EQUATIONS: THE FINITE ELEMENT METHOD A	72-0930
NUMERICAL	NUMERICAL SOLUTION OF THE VISCOUS FLOW IN THE ENTRANCE REGION OF PARALLEL PLATES	73-0022
NUMERICAL	NUMERICAL SOLUTION OF THE BIHARMONIC PROBLEM BY MIXED FINITE OF CLASS C-0	73-0803
NUMERICAL	NUMERICAL SOLUTION OF PARTIAL DIFFERENTIAL EQUATIONS	73-0968
NUMERICAL	NUMERICAL SOLUTION OF STEADY STATE DIFFUSION PROBLEMS CONTAINING SINGULARITIES	74-0036
NUMERICAL	NUMERICAL SOLUTION OF PLATEAUS PROBLEM BY A FINITE ELEMENT METHOD	74-0119
NUMERICAL	NUMERICAL SOLUTION OF THE STATIONARY NAVIER-STOKES EQUATIONS BY FINITE ELEMENT METHODS	74-0408
NUMERICAL	NUMERICAL SOLUTION OF A DIFFUSION CONSUMPTION PROBLEM WITH A FREE BOUNDARY	74-0835
NUMERICAL	NUMERICAL SOLUTION OF THE STEADY INTERFACE IN A CONFINED COASTAL AQUIFER	74-1007

NUMERICAL	NUMERICAL SOLUTION OF 2-POINT BOUNDARY VALUE PROBLEMS	74-1141
NUMERICAL	NUMERICAL SOLUTION SCHEMES FOR HIGHLY NONLINEAR STATIC STRUCTURAL BEHAVIOR	75-0679
NUMERICAL	NUMERICAL SOLUTION TO THE DYNAMICS OF VISCOUS FLUID FLOW BY A FINITE ELEMENT ALGORITHM: A FIRST STEP	71-0004
NUMERICAL	NUMERICAL SOLUTION TO THE DYNAMICS OF VISCOUS FLUID FLOW BY A FINITE ELEMENT ALGORITHM: A FIRST STEP	71-0791
NUMERICAL	NUMERICAL SOLUTIONS FOR DETERMINING WAVE-INDUCED PRESSURE DISTRIBUTIONS AROUND BURIED PIPELINES	74-1334
NUMERICAL	NUMERICAL SOLUTIONS OF NONLINEAR PLASMA EQUATIONS BY THE FINITE-ELEMENT METHOD	75-0671
NUMERICAL	NUMERICAL STRESS ANALYSIS OF INTERVERTEBRAL DISK	72-0125
NUMERICAL	NUMERICAL STUDIES USING CIRCULAR ARCH FINITE ELEMENTS	74-0600
NUMERICAL	NUMERICAL STUDY OF DAMPING IN VISCOELASTIC SANDWICH BEAMS	73-0443
NUMERICAL	NUMERICAL STUDY OF EIGEN OSCILLATIONS OF A SCREW PINCH WITH FINITE ELEMENTS	74-0159
NUMERICAL	NUMERICAL STUDY OF HYDROGMAGNETIC STABILITY USING FINITE ELEMENT METHOD	73-0092
NUMERICAL	NUMERICAL STUDY OF PROPER MODES OF A HOMOGENEOUS CYLINDRICAL PLASMA BY METHOD OF FINITE ELEMENTS	74-0160
NUMERICAL	NUMERICAL STUDY OF TRANSITORY OR NONLINEAR WAVES	74-0031
NUMERICAL	NUMERICAL STUDY OF UNIAXIAL COMPRESSION IN CIRCULAR ELASTIC-PLASTIC COLUMNS	73-0238
NUMERICAL	NUMERICAL STUDY OF UNIAXIAL AND TRIAXIAL ROCK COMPRESSION TESTS	73-0896
NUMERICAL	NUMERICAL SUFFICIENCY TEST FOR MONOTONIC CONVERGENCE OF FINITE ELEMENT MODELS	75-0620
NUMERICAL	NUMERICAL TECHNIQUES FOR CONVECTION/DIFFUSION PROBLEMS	75-0123
NUMERICAL	NUMERICAL TECHNIQUES IN LINEAR AND NONLINEAR FRACTURE MECHANICS	75-0260
NUMERICAL	NUMERICAL TREATMENT OF ALGEBRAIC INTEGRAL EQUATIONS BY VARIATIONAL METHODS	75-0374
NUMERICAL	NUMERICAL USE OF INTEGRAL PROCEDURES	65-0041
NUMERICAL	NUMERICAL, THERMO-MECHANICAL MODEL FOR THE WELDING AND SUBSEQUENT LOADING OF A FABRICATED STRUCTURE	73-0418
NUMERICAL	NUMERISCHE BERECHNUNG INSTATIONAERER TEMPERATURFELDER UND THERMISCH BEDINGTER VERFORMUNGEN VON WERKZ	75-0847
NUMERICAL	NUMERYCZNE ROZWIAZANIE PROBLEMU ORGAN WLASNYCH NA PRZYKLADZIE PLYT. (NUMERICAL SOLUTION OF THE PROBL	75-0296
NUMERICAL	ON A NUMERICAL SUFFICIENCY TEST FOR MONOTONIC CONVERGENCE OF FINITE ELEMENT MODELS	75-0533
NUMERICAL	ON NUMERICAL SOLUTIONS OF EQUATIONS OF HYPERBOLIC-ELLIPTIC TYPE	74-0741
NUMERICAL	ON THE ACCURACY AND NUMERICAL STABILITY OF THE FINITE ELEMENT APPROXIMATION FOR PARABOLIC AND HYPERB	75-0102
NUMERICAL	ON THE NUMERICAL CONVERGENCE OF MATRIX EIGENVALUE PROBLEMS DUE TO CONSTRAINT CONDITIONS	74-1057
NUMERICAL	ON THE NUMERICAL PROBLEMS BY THE FINITE ELEMENT METHODS	70-0264
NUMERICAL	ON THE NUMERICAL SOLUTION OF A CLASS OF PROBLEMS IN A LINEAR FIRST STRAIN-GRADIENT THEORY OF ELASTIC	70-0318
NUMERICAL	ON THE NUMERICAL SOLUTION OF A CLASS OF NONLINEAR PROBLEMS IN DYNAMIC COUPLED THERMOELASTICITY	71-0441
NUMERICAL	ON THE NUMERICAL SOLUTION OF LARGE EIGENVALUE PROBLEMS ARISING IN PANEL FLUTTER ANALYSIS BY THE FINI	74-0942
NUMERICAL	ON THE NUMERICAL SOLUTION OF CERTAIN PROBLEMS IN FINITE ELASTICITY BY THE FINITE ELEMENT METHOD	75-0542
NUMERICAL	ON THE NUMERICAL SOLUTION OF FREE BOUNDARY PROBLEMS	75-0577
NUMERICAL	ON THE NUMERICAL TREATMENT OF THE NAVIER-STOKES EQUATIONS FOR AN INCOMPRESSIBLE FLUID	73-0809
NUMERICAL	ON THE REDUCTION OF NUMERICAL ERROR IN THE MATRIX DISPLACEMENT METHOD	74-0417
NUMERICAL	ON THE STABILITY OF EXPLICIT METHODS FOR THE NUMERICAL INTEGRATION OF THE EQUATIONS OF MOTION IN FIN	72-0172
NUMERICAL	PRACTICAL APPLICATION OF THE FINITE ELEMENT METHOD AND SOME NUMERICAL ASPECTS	69-0309
NUMERICAL	PROBLEM CONTROLLED GRID GENERATION FOR THE NUMERICAL SOLUTION OF PARTIAL DIFFERENTIAL EQUATIONS	75-0725
NUMERICAL	RECENT ADVANCES IN NUMERICAL ANALYSIS OF STRUCTURAL EIGENVALUE PROBLEMS	73-0316
NUMERICAL	RECENT ADVANCES IN NUMERICAL WEATHER PREDICTION	74-0787
NUMERICAL	SEEPAGE IN MISSISSIPPI RIVER BANKS. REPORT 1. ANALYSIS OF TRANSIENT SEEPAGE USING VISCOUS FLOW MODEL	70-0405
NUMERICAL	SENSITIVITY ANALYSIS OF INPUT PARAMETERS IN NUMERICAL MODELING OF STEADY STATE REGIONAL GROUNDWATER	74-0926
NUMERICAL	SOME NUMERICAL PROCEDURES FOR ANALYSIS OF ONE DIMENSIONAL CONSOLIDATION	72-0928
NUMERICAL	STABILITY AND ACCURACY OF NUMERICAL APPROXIMATIONS TO TIME DEPENDENT FLOWS	75-0183
NUMERICAL	STATIC, STABILITY, AND DYNAMIC ANALYSIS OF SHELLS OF REVOLUTION BY NUMERICAL INTEGRATION - A COMPARI	74-0399
NUMERICAL	STATISTICAL STATE OF STRESS STUDIED BY GRID ANALYSIS: "NUMERICAL METHODS OF ANALYSIS"	49-0001
NUMERICAL	STRESS ANALYSIS OF COMPLEX SHIP COMPONENTS BY A NUMERICAL PROCEDURE USING CURVED FINITE ELEMENTS	73-0879
NUMERICAL	STUDIES IN NUMERICAL ANALYSIS	74-1269
NUMERICAL	STUDY OF NUMERICAL METHODS IN SEISMIC PROBLEMS	72-0738
NUMERICAL	THE COMBINED EFFECT TO CURVED BOUNDARIES AND NUMERICAL INTEGRATION IN ISOPARAMETRIC FINITE ELEMENT M	72-0357
NUMERICAL	THE NUMERICAL CALCULATION OF LAMINAR BOUNDARY-LAYER SEPARATION	74-0785
NUMERICAL	THE NUMERICAL SOLUTION OF THE STEADY INTERFACE IN A COASTAL CONFINED AQUIFER	73-0711
NUMERICAL	THE NUMERICAL SOLUTION OF VISCO-ELASTIC PROBLEMS USING LAPLACE TRANSFORMS	74-0436
NUMERICAL	THE ROLE OF NUMERICAL METHODS IN THE STUDY OF LUBRICATION	74-0042
NUMERICAL	THE USE OF NUMERICAL INTEGRATION IN FINITE ELEMENT METHODS FOR SOLVING PARABOLIC EQUATIONS	72-0351
NUMERICAL	THEMIS FINAL REPORT, VOL. IV. RECENT DEVELOPMENTS IN THE NUMERICAL SOLUTION OF CERTAIN PROBLEMS IN N	74-0721
NUMERICAL	THEORETICAL AND NUMERICAL COMPARISON OF ELASTIC NONLINEAR FINITE ELEMENT METHODS.	72-0452
NUMERICAL	THEORETICAL AND NUMERICAL CONVERGENCE RATES FOR SOME CONFORMING FINITE ELEMENTS	74-0941
NUMERICAL	THREE DIMENSIONAL NONLINEAR MAGNETIC FIELD BOUNDARY VALUE PROBLEM AND ITS NUMERICAL SOLUTION	70-0060
NUMERICAL	THREE-DIMENSIONAL HARMONIC FUNCTIONS NEAR TERMINATION OR INTERSECTION OF GRADIENT SINGULARITY LINES:	74-0178
NUMERICAL	TIMING COMPARISON STUDY FOR SOME HIGH ORDER FINITE ELEMENT PROCEDURES AND A LOW ORDER FINITE DIFFERE	72-0767
NUMERICAL	TRANSIENT SHELL RESPONSE BY NUMERICAL TIME INTEGRATION	72-0305
NUMERICAL	TRANSPORT THEORY NUMERICAL METHODS	72-0769
NUMERICAL	TRENDS AND DIRECTIONS IN THE APPLICATIONS OF NUMERICAL ANALYSIS.	71-0667
NUMERICAL	UNBOUNDED ERRORS IN NUMERICAL ANALYSIS OF STRUCTURES	74-1210
NUMERICAL	UNDRAINED STRESS DISTRIBUTION BY NUMERICAL METHODS	68-0045
NUMERICAL	UNE METHODE VARIATIONNELLE ELEMENTS FINIS POUR LA RESOLUTION NUMERIQUE UN PROBLEME EXTERIEUR DANS R*	73-0576
NUMERICAL	USEFUL FORM OF THE VARIATIONAL PRINCIPLE OF IDEAL MHD FOR ONE-DIMENSIONAL NUMERICAL STABILITY CALCUL	73-0946
NUMERICAL	USERS MANUAL FOR DIGITAL COMPUTER PROGRAM OK5 NUMERICAL ANALYSIS OF FINITE STRAINS AND DISPLACEMENTS	72-0718
NUMERICAL	USERS' MANUAL FOR DIGITAL COMPUTER PROGRAM OK1. NUMERICAL ANALYSIS OF FINITE AXISYMMETRIC DEFORMATIO	72-0861
NUMERICAL	VARIATION METHOD ON FINITE ELEMENTS FOR NUMERICAL RESOLUTION TO PROBLEMS WITH EXTERNAL R3	73-0359
NUMERICAL	VARIATIONAL FORMULATIONS OF NUMERICAL METHODS	70-0392
NUMERICAL	VARIATIONAL FORMULATIONS OF NUMERICAL METHODS IN SOLID CONTINUA	71-0670
NUMERICAL	VISCO-PLASTICITY, PLASTICITY AND CREEP IN ELASTIC SOLIDS. A UNIFIED NUMERICAL SOLUTION APPROACH	74-1279
NUMERICAL	VISCOPLASTICITY, PLASTICITY AND CREEP. A UNIFIED NUMERICAL SOLUTION APPROACH	74-1205
NUMERICAL	ZUR SYSTEMATIK NUMERISCHER NAEHERUNGSVERFAHREN IN DER STATIK UND DYNAMIK DER KONSTRUKTIONEN (METHODS	72-0557
NUMERICAL-ANALY	LARGE STRAIN, VISCOELASTIC AND ELASTO-VISCOPLASTIC NUMERICAL-ANALYSIS BY MEANS OF FINITE-ELEMENT MET	75-0435
NUMERICAL-PERTU	NUMERICAL-PERTURBATION METHOD FOR THE NONLINEAR ANALYSIS OF STRUCTURAL VIBRATIONS	74-0499
NUMERICAL-SIMUL	MENISCI IN ARRAYS OF CYLINDERS - NUMERICAL-SIMULATION BY FINITE-ELEMENTS	75-0449
NUMERICAL-SOLUT	CONTRIBUTION TO NUMERICAL-SOLUTION OF DEVELOPING LAMINAR-FLOW IN ENTRANCE REGION OF CONCENTRIC ANNUL	75-0448
NUMERICAL-SOLUT	FINITE-ELEMENT CURRENT-DENSITY REPRESENTATION IN NUMERICAL-SOLUTION OF MICROSTRIP PROBLEMS	75-0400
NUMERICAL-SOLUT	NUMERICAL-SOLUTION OF EULERIAN EQUATIONS OF COMPRESSIBLE FLOW BY A FINITE-ELEMENT METHOD WHICH FOLLO	75-0535
NUMERICALLY	A METHOD OF COMPUTING NUMERICALLY INTEGRATED STIFFNESS MATRICES	72-0546
NUMERICALLY	ECONOMICAL COMPUTER TECHNIQUES FOR NUMERICALLY INTEGRATED FINITE ELEMENTS	69-0138
NUMERICALLY	ISOPARAMETRIC AND ALLIED NUMERICALLY INTEGRATED ELEMENTS - A REVIEW	71-0763
NUMERICALLY	ISOPARAMETRIC AND OTHER NUMERICALLY INTEGRATED ELEMENTS	71-0272
NUMERICALLY	ON NUMERICALLY ACCURATE FINITE ELEMENT SOLUTIONS IN THE FULLY PLASTIC RANGE	74-0529
NUMERICALLY	ON NUMERICALLY ACCURATE FINITE ELEMENTS	74-0797
NUMERIQUE	ANALYSE NUMERIQUE D'UN PROBLEME DE STEFAN A DEUX PHASES PAR UNE METHODE D'ELEMENTS FINIS. (NUMERICAL	74-0387
NUMERIQUE	ANALYSE NUMERIQUE DU CHAMP MAGNETIQUE D'UN ALTERNATEUR PAR ELEMENTS FINIS ET SURRELAXATION PONCTUELL	73-0625
NUMERIQUE	CALCUL NUMERIQUE DES ECOULEMENTS DES FLUIDES DE BINGHAM ET DES FLUIDES NEWTONIENS INCOMPPESSIBLES PA	72-0349
NUMERIQUE	DEPLACEMENTS DANS LES STRUCTURES MINCES A FAIBLE RIGIDITE GEOMETRIQUE-DETERMINATION NUMERIQUE (NUMER	72-0213
NUMERIQUE	ETUDE NUMERIQUE DU CHAMP MAGNETIQUE DANS UN ALTERNATEUR TETRAPOLAIRE PAR LA METHODE DES ELEMENTS FIN	74-0413
NUMERIQUE	UNE METHODE VARIATIONNELLE ELEMENTS FINIS POUR LA RESOLUTION NUMERIQUE UN PROBLEME EXTERIEUR DANS R*	73-0576
NUMERIQUES	CALCULATION OF THE BRIDGE DECKS BY USING NUMERICAL METHODS. (LE CALCUL DES TABLIERS DE PONTS PAR DES	71-0411
NUMERIQUES	COMPARAISON DES DIVERSES MODELISATIONS NUMERIQUES D'UN SYSTEME MAGNETIQUE SATURABLE. (COMPARISON OF	75-0295
NUMERIQUES	LE CALCUL DES TABLIERS DE PONTS PAR DES METHODES NUMERIQUES	72-0197
NUMERISCHE	EINE EINFACHE KONSTRUKTION VON KOORDINATENFUNKTIONEN FUR DIE NUMERISCHE LOSUNG ZWEIDIMENSIONALER RAN	66-0134
NUMERISCHE	NUMERISCHE BERECHNUNG NICHTLINEARER ELASTISCHER KOERPER	63-0017
NUMERISCHE	NUMERISCHE BERECHNUNG INSTATIONAERER TEMPERATURFELDER UND THERMISCH BEDINGTER VERFORMUNGEN VON WERKZ	75-0847
NUMERISCHER	VERGLEICH NUMERISCHER LOESUNGEN NACH DEM DIFFERENZENVERFAHREN UND NACH DER FINITE-ELEMENT-METHODE AM	71-0016
NUMERISCHER	ZUR SYSTEMATIK NUMERISCHER NAEHERUNGSVERFAHREN IN DER STATIK UND DYNAMIK DER KONSTRUKTIONEN (METHODS	72-0557

NUMERYCZNE	NUMERYCZNE ROZWIAZANIE PROBLEMU DRGAN WLASNYCH NA PRZYKLADZIE PLYT. (NUMERICAL SOLUTION OF THE PROBL	75-0296
NUT	PRAKTISCHE RESULTATEN. DOEL EN NUT VAN HET ONTWERPEN VAN MECHANISCHE ELEMENTEN MET BEHULP VAN COMPUT	73-0400
OBERFLACHE	BERECHNUNG INSTATIONARER GRUND – UND SICKERWASSERSTROMUNGEN MIT FREIER OBERFLACHE NACH DER METHODE F	71-0541
OBERFLAECHENBES	ZUSAMMENHANG ZWISCHEN OBERFLAECHENBESCHAFFENHEIT. VERBUND UND SPRENGWIRKUNG VON BEWEHRUNGSSTAEHLEN U	73-0450
OBESF	TACOS: A FINITE ELEMENT COMPUTER PROGRAM FOR THE TRANSIENT ANALYSIS OF CYLINDRICAL OBESE SHELLS	72-0588
OBLASTEI	CHISLENNOE RESHENIE PLOSKIKH ZADACH TEPLOPROVODNOSTI DLYA OBLASTEI SLOZHNOI FORMY (NUMERICAL SOLUTIO	74-0659
OBLICZANIE	OBLICZANIE DRGAN METODA SZTYWNYCH ELEMENTOW SKONCZONYCH. (CALCJLATING VIBRATIONS BY THE RIGID-FINITE	75-0649
OBLICZANIE	OBLICZANIE KONSTRUKCJI TOROIDALNEGO ZBIORNIKA NA WODE. (STATICAL ANALYSIS OF A TOROIDAL WATER RESERV	75-0211
OBLICZEN	ALGORYTM OBLICZEN DRGAN WYMUSZONYCH METODA SZTYWNYCH ELEMENTOW SKONCZONYCH (ALGORITHM FOR CALCULATIN	74-0581
OBLICZEN	HYBRYDOWA METODA ELEMENTOW SKONCZONYCH W ZASTOSOWANIU DO OBLICZEN URZADZEN OKRETOWYCH. (HYBRID	74-0963
OBLICZEN	ZASTOSOWANIE ITERACJI SEIDLA W METODZIE ELEMENTOW SKONCZONYCH NA PRZYKLADZIE OBLICZEN STATYCZNYCH P	74-1031
OBLICZEN	ZASTOSOWANIE METODY SZTYWNYCH ELEMENTOW SKONCZONYCH DO OBLICZEN DRGAN WALOW OKRETOYCH Z UWGLEDNIENI	74-0962
OBLICZENIACH	METHOD SZTYWNYCH ELEMENTOW SKONCZONYCH W OBLICZENIACH KONSTRUCJI OKRETOWYCH. (STIFF FINITE ELEMENT M	74-0923
OBLIQUE	DETERMINATION OF THE TECHNICAL CONSTANTS OF LAMINATES IN OBLIQUE DIRECTIONS. THEORY AND EXPERIMENTAL	73-0748
OBLIQUE	DIFFRACTION OF OBLIQUE WAVES BY AN INFINITE CYLINDER	75-0605
OBLIQUE	MODEL TEST AND ANALYSIS OF OBLIQUE FOLDED PLATE STRUCTURE	73-0230
OBLIQUE	THEORETICAL AND EXPERIMENTAL VIBRATION ANALYSIS OF AN OBLIQUE SPACE FRAME	70-0460
OBLIQUES	DETERMINATION OF THE TECHNICAL CONSTANTS OF LAMINATES IN OBLIQUE DIRECTIONS. THEORY AND EXPERIMENTAL	73-0748
OBLOCHKI	THE DEVELOPMENT OF APPLIED METHODS IN PROBLEMS OF STATIC CALCULATION OF THIN-WALLED THREE-DIMENSIONA	73-0934
OBOLOCHECHNYKH	CHISLENNYI METOD RASCHETA SOBSTVENNYKH I VYNUZHDENNYKH KOLEBANII SOSTAVNYKH OBOLOCHECHNYKH KONSTRUKT	73-0461
OBOLOCHEK	ISSLEDOVANIE NAPRYAZHENNOGO SOSTOYANIYA REBRISTYKH TSILINDRICHESKIKH OBOLOCHEK S PRYAMOUGOL'NYMI OTV	74-0944
OBOLOCHEK	METOD KONECHNYKH ELEMENTOV V PRIMENENII K RASCHETU TSILINDRICHESKIKH OBOLOCHEK S PRYAMOUGOL'NYMI OTV	73-0536
OBOLOCHKI	TREUGOL'NY I ELEMENT DLYA KRUGOVOI TSILINDRICHESKOI OBOLOCHKI (TRIANGULAR FINITE ELEMENT FOR CIRCULA	73-0493
OBORU	STABILITA STEN V PLASTICKEM OBORU (STABILITY IN WEBPLATES IN PLASTIC RANGE)	73-0387
OBSERVATIONS	MATHEMATICAL OBSERVATIONS IN STRUCTURAL DYNAMICS	73-0156
OBSERVATIONS	OBSERVATIONS ON THE BEHAVIOUR OF A PILED-RAFT FOUNDATION ON LONDON CLAY.	73-0421
OBSERVATIONS	OBSERVATIONS REGARDING ASSUMED STRESS HYBRID PLATE ELEMENTS	74-0370
OBSERVATIONS	SOME OBSERVATIONS ON PROPERTIES OF CERTAIN MIXED FINITE ELEMENT APPROXIMATIONS	75-0617
OBSERVATORY	THERMAL DEFORMATION PREDICTION IN ORBITING OBSERVATORY STRUCTURES	69-0292
OBSTACLE	FINITE ELEMENT ANALYSIS OF UNSTEADY INCOMPRESSIBLE FLOW AROUND AN OSCILLATING OBSTACLE OF ARBITRARY	73-0067
OBSTACLES	SCATTERING BY BURIED OBSTACLES	74-0739
OCB	COMPLIANCE OF A TAPERED OCB SPECIMEN - CONFIGURATIONS BY A FINITE ELEMENT METHOD	74-0545
OCEAN	FINITE ELEMENT MODELS FOR OCEAN CIRCULATION PROBLEMS	75-0206
OCEAN	FINITE-DIFFERENCE AND FINITE-ELEMENT METHODS IN GLOBAL WIND-DRIVEN OCEAN CIRCULATION	75-0345
OCEAN	HYDRODYNAMICAL-NUMERICAL MODELS FOR COASTAL WATERS AND OPEN OCEAN AREAS	75-0571
OCEAN	MATHEMATICAL MODEL OF THE COLUMBIA RIVER FROM THE PACIFIC OCEAN TO BONNEVILLE DAM. PART 1. THEORY, P	69-0464
OCEAN	MATHEMATICAL MODEL OF THE COLUMBIA RIVER FROM THE PACIFIC OCEAN TO BONNEVILLE DAM. PART II: INPUT-OU	70-0417
OCEAN	MODELING AN OCEAN POND. A TWO-DIMENSIONAL, FINITE ELEMENT HYDRODYNAMIC MODEL OF NINIGRET POND, CHARL	75-0686
OCEAN	THE EVOLUTION OF SHELL STRUCTURES IN OCEAN TECHNOLOGY	69-0042
OCEAN-FLOOR	ANCHOR-PILE DESIGN FOR OCEAN-FLOOR ENVIRONMENTS USING FINITE-ELEMENT ANALYSIS	75-0459
OCENA	OCENA DOKLADNOSCI ROZWIAZAN PROBLEMOW BRZEGOWYCH SPREZYSTYCH CIAL DYSKERTYZOWANYCH. (ASSESSING THE A	74-0921
ODE	CONNECTION BETWEEN THE FINITE DIFFERENCE LIKE METHODS AND THE METHODS BASED ON INITIAL VALUE PROBLEM	74-0877
ODER	ZUR BERECHNUNG, BEMESSUNG UND BEWEHRUNG RECHTECKIGER PLATTEN KONSTNATER DICKE MIT OEFFNUNGEN ODER AU	73-0664
OEFFNUNGEN	ZUR BERECHNUNG, BEMESSUNG UND BEWEHRUNG RECHTECKIGER PLATTEN KONSTNATER DICKE MIT OEFFNUNGEN ODER AU	73-0664
OFF	FINITE ELEMENT SOLUTIONS OF THE SHALLOW WATER EQUATIONS SURFACE RUN OFF	74-0046
OFF-SHORE	APPLICATION OF THE FINITE ELEMENT METHOD TO OFF-SHORE STRUCTURES	74-0206
OFFSET	FORCE AND TORQUE ANALYSIS OF CONSTRAINED SPACE MECHANISMS AND PLANE MECHANISMS WITH OFFSET LINKS BY	74-0569
OFFSETS	A GENERAL EQUILIBRIUM BEAM ELEMENT WITH OFFSETS	73-0351
OFFSHORE	DESIGN APPRAISAL OF OFFSHORE PLATFORMS OF WELDED TUBULAR CONSTRUCTION	74-1165
OFFSHORE	DISCONTINUITY STRESS DECAY IN OFFSHORE PLATFORM JOINTS	74-0491
OFFSHORE	INTEGRATED PROGRAM FOR THE DYNAMIC STRUCTURAL CALCULATION OF MOBIL OFFSHORE UNITS	74-0490
OFFSHORE	OFFSHORE GRAVITY BASED OIL PRODUCTION PLATFORM INTERACTION WITH THE SEA BED	75-0221
OFFSHORE	OPTIMIZATION OF TENSION LEVEL AND STINGER LENGTH FCR OFFSHORE PIPELINE INSTALLATION	74-0504
OFFSHORE	OSCILLATIONS AND WAVE FORCES IN AN OFFSHORE HARBOR - APPLICATIONS OF HYBRID FINITE ELEMENT METHOD TO	74-1024
OFFSHORE	STOCHASTIC DYNAMIC RESPONSE ANALYSIS OF OFFSHORE PLATFORMS, WITH PARTICULAR REFERENCE TO GRAVITY-TYP	75-0220
OFFSHORE	STRESS ANALYSIS OF A TUBULAR CROSS-JOINT WITHOUT ANY INTERNAL STIFFENING FOR OFFSHORE STRUCTURES	74-0549
OFFSHORE	STRESS ANALYSIS OF TUBULAR JOINTS FOR OFFSHORE STRUCTURES	73-0207
OFFSHORE	USE OF FRACTURE MECHANICS AND FINITE ELEMENT TECHNIQUES FOR DETERMINING THE SAFETY OF OFFSHORE STRUC	74-1211
OFHC	CYCLIC FATIGUE ANALYSIS OF ROCKET THRUST CHAMBERS. VOLUME 1: OFHC COPPER CHAMBER LOW CYCLE FATIGUE	74-0694
OIL	APPLICATION OF A FINITE ELEMENT METHOD TO EVALUATION OF OIL WHIRL CHARACTERISTICS	74-0590
OIL	COMPUTER - AIDED DESIGN - STANDARD OIL OF INDIANA BUILDING	73-0182
OIL	EXAMPLE OF LARGE-SCALE STRUCTURAL ANALYSIS. COMPARISON BETWEEN FINITE ELEMENT CALCULATION AND FULL S	74-0177
OIL	OFFSHORE GRAVITY BASED OIL PRODUCTION PLATFORM INTERACTION WITH THE SEA BED	75-0221
OIL	ON THE FINITE ELEMENT STRESS ANALYSIS OF OIL TANKER STRUCTURES	74-0606
OIL	RESEARCH NOTE: A THREE-DIMENSIONAL OIL FILM TEMPERATURE DISTRIBUTION IN TILTING THRUST BEARINGS	74-0608
OIL	SOME STRUCTURAL PROBLEMS - STANDARD OIL OF INDIANA BUILDING	73-0180
OIL-FILM	OIL-FILM TEMPERATURE DISTRIBUTION IN AN INFINITELY WIDE SLIDER BEARING: AN APPLICATION OF THE FINITE	73-1128
OKRETOWYCH	HYBRYDOWA METODA ELEMENTOW SKONCZONYCH W ZASTOSOWANIU DO OBLICZEN DRGAN URZADZEN OKRETOWYCH. (HYBRID	74-0963
OKRETOWYCH	METHOD SZTYWNYCH ELEMENTOW SKONCZONYCH W OBLICZENIACH KCNSTRUCJI OKRETOWYCH. (STIFF FINITE ELEMENT M	74-0923
OKRETOYCH	ZASTOSOWANIE METODY SZTYWNYCH ELEMENTOW SKONCZONYCH DO OBLICZEN WALOW OKRETOYCH Z UWGLEDNIENI	74-0962
OKRUZHAYUSHCHEM	RASCHET NAPRYAZHENIL L PEREMESHCHENII V SVODE PODZEMNOGO MASHINNOGO ZALA GES I V OKRUZHAYUSHCHEM SKA	74-0542
OK1	USERS' MANUAL FOR DIGITAL COMPUTER PROGRAM OK1. NUMERICAL ANALYSIS OF FINITE AXISYMMETRIC DEFORMATIO	72-0861
OK5	USERS MANUAL FOR DIGITAL COMPUTER PROGRAM OK5 NUMERICAL ANALYSIS OF FINITE STRAINS AND DISPLACEMENTS	72-0718
ONE-AND	MATHEMATICAL MODELS OF THE MASSACHUSETTS BAY. PART 1. FINITE ELEMENT MODELING OF TWO-DIMENSIONAL HYD	73-0800
ONE-DIMENSIONAL	A FINITE ELEMENT ANALYSIS OF SHOCK AND FINITE-AMPLITUDE WAVES IN ONE-DIMENSIONAL HYPERELASTIC BODIES	72-0731
ONE-DIMENSIONAL	A NUMERICAL MODEL OF TRANSIENT WATER QUALITY IN A ONE-DIMENSIONAL ESTUARY BASED ON THE FINITE ELEMEN	74-1369
ONE-DIMENSIONAL	A SECOND ORDER FINITE ELEMENT METHOD FOR THE ONE-DIMENSIONAL STEFAN PROBLEM	74-1222
ONE-DIMENSIONAL	A USEFUL FORM OF THE VARIATIONAL PRINCIPLE OF IDEAL MHD FOR ONE-DIMENSIONAL NUMERICAL STABILITY CALC	73-0746
ONE-DIMENSIONAL	AN APPLICATION OF LEAST SQUARES TO ONE-DIMENSIONAL TRANSIENT PROBLEMS	74-0347
ONE-DIMENSIONAL	ANALYSIS OF ONE-DIMENSIONAL STRESS WAVE BY THE FINITE ELEMENT METHOD	71-0523
ONE-DIMENSIONAL	APPLICATION OF FINITE-ELEMENTS TO ONE-DIMENSIONAL AND TWO-DIMENSIONAL NEUTRON-TRANSPORT	73-1150
ONE-DIMENSIONAL	EVALUATION OF TWO FINITE ELEMENT FORMULATIONS FOR ONE-DIMENSIONAL CONSOLIDATION	74-0453
ONE-DIMENSIONAL	FINITE ELEMENT WEIGHTED RESIDUAL SOLUTION OF ONE-DIMENSIONAL, UNSTEADY AND UNSATURATED FLOWS IN PORO	73-0712
ONE-DIMENSIONAL	GROUNDWATER FLOWN IN A SANDY TIDAL BEACH .1. ONE-DIMENSIONAL FINITE ELEMENT ANALYSIS	71-0567
ONE-DIMENSIONAL	PART 1 - TASK 3.2.9. BENDING AND BUCKLING OF A ONE-DIMENSIONAL MODEL SIMULATING A CONICAL SHELL. PAR	71-0745
ONE-DIMENSIONAL	THALIA - ONE-DIMENSIONAL MAGNETOHYDRODYNAMIC STABILITY PROGRAM USING METHOD OF FINITE-ELEMENTS	75-0458
ONE-DIMENSIONAL	THALIA: A ONE-DIMENSIONAL MAGNETOHYDRODYNAMIC STABILITY PROGRAM USING THE METHOD OF FINITE ELEMENTS	74-1085
ONE-DIMENSIONAL	USEFUL FORM OF THE VARIATIONAL PRINCIPLE OF IDEAL MHD FOR ONE-DIMENSIONAL NUMERICAL STABILITY CALCUL	73-0946
ONE-SIDED	ONE-SIDED APPROXIMATION AND VARIATIONAL INEQUALITIES	74-0335
ONE-SIDED	ONE-SIDED APPROXIMATION AND PLATE BENDING	74-0336
ONLINE	INTERACTIVE COMPUTER GRAPHICS IN STRUCTURAL ANALYSIS: ONLINE 72	72-0639
ONTWERPEN	PRAKTISCHE RESULTATEN. DOEL EN NUT VAN HET ONTWERPEN VAN MECHANISCHE ELEMENTEN MET BEHULP VAN COMPUT	73-0400
OPEN	A FINITE ELEMENT METHOD FOR CALCULATING OPEN CHANNEL FLOW	69-0510
OPEN	ANALYSIS OF OPEN STRUCTURES BY FINITE ELEMENT RESOLUTION OF EQUIVALENT CLOSED-BOUNDARY PROBLEM	74-0111
OPEN	BENDING AND TORSION OF THIN-WALLED STRAIGHT AND CURVED BEAMS WITH VARIABLE OPEN CROSS-SECTION	69-0493
OPEN	BUCKLING OF OPEN CYLINDRICAL SHELLS	67-0115
OPEN	CONICAL SEGMENT METHOD FOR ANALYZING OPEN CROWN SHELLS OF REVOLUTION FOR EDGE LOADING	63-0007
OPEN	COUPLED VIBRATIONS OF THIN-WALLED BEAMS OF OPEN SECTION USING FINITE ELEMENT METHOD	70-0230
OPEN	COUPLED VIBRATIONS OF THIN-WALLED BEAMS OF OPEN SECTION USING FINITE ELEMENT METHOD	71-0568
OPEN	DYNAMICS OF PRISMATIC AND OPEN SECTION MEMBER GRIDS	69-0070
OPEN	FINITE ELEMENT ANALYSIS OF OPEN SHELLS OF REVOLUTION UNDER NONSYMMETRICAL LOADING	68-0295
OPEN	FINITE ELEMENT ANALYSIS OF THIN-WALLED MEMBERS OF OPEN SECTION	72-0966

OPEN	FINITE ELEMENT ANALYSIS OF LONG WAVES IN OPEN CHANNEL SYSTEMS	74-0033
OPEN	FINITE ELEMENT ANALYSIS OF WAVE PROPAGATION IN OPEN CHANNELS	75-0570
OPEN	FINITE ELEMENT FOR TORSION OF THIN WALLED OPEN TUBES	71-0308
OPEN	HYDRODYNAMICAL-NUMERICAL MODELS FOR COASTAL WATERS AND OPEN OCEAN AREAS	75-0571
OPEN	MODEL EXPERIMENTS AND FINITE ELEMENT ANALYSIS OF STRESS IN AN OPEN SHIP	66-0054
OPEN	MODELING TECHNIQUES OF THIN-WALLED BEAMS WITH OPEN CROSS SECTIONS	72-0573
OPEN	OSCILLATIONS AND WAVE FORCES IN A MAN-MADE HARBOUR IN THE OPEN SEA	74-0151
OPEN	ROCK MECHANICS AND RISK IN OPEN PIT MINING	73-0544
OPEN	THERMAL STRESSES IN OPEN CYLINDRICAL SHELLS OF ARBITRARY BOUNDARY CONDITIONS BY A FINITE STRIP METHO	72-0371
OPEN-PIT	CONTROLLING OPEN-PIT SLOPE FAILURES AT SHIPLEY BASIN	73-0100
OPEN-PIT	THREE-DIMENSIONAL CONSIDERATION OF THE STRESSES SURROUNDING OPEN-PIT MINE SLOPES	73-0492
OPEN-TUBE	FLEXIBILITY AND STIFFNESS MATRICES FOR AN OPEN-TUBE WARPING CONSTRAINT FINITE ELEMENT	72-0076
OPEN-WEB	DEFLECTION ANALYSIS OF EXPANDED OPEN-WEB STEEL BEAMS	74-0224
OPENING	FREQUENCY COEFFICIENTS OF A SPHERICAL SHELL FOR VARYING THICKNESS AND OPENING ANGLE WITH GUIDED PINN	72-0595
OPENING	INFLUENCE OF ROCK PROPERTIES VARIABILITY ON MINE OPENING STABILITY ANALYSIS	74-1021
OPENING	PROPAGATION AND OPENING OF A THROUGH CRACK IN PIPE SUBJECT TO COMBINED CYCLIC THERMOMECHANICAL LOADI	75-0516
OPENING	ROCK FAILURE AROUND A CIRCULAR OPENING IN A GRAVITY FIELD WITH TECTONIC FORCES	70-0071
OPENING	STRUCTURAL ANALYSIS OF A COAL MINE OPENING IN ELASTIC, MULTILAYERED MATERIAL	74-0235
OPENING	STRUCTURAL ANALYSIS OF A COAL MINE OPENING IN ELASTIC, MULTILAYERED MATERIAL	74-0840
OPENING	TECHNIQUE OF MEASURING INITIAL DEFORMATION AROUND AN OPENING	71-0331
OPENING	THICK WALLED MULTIPLE OPENING REINFORCED CONCRETE CONDUITS	72-0666
OPENINGS	A THEORETICAL METHOD FOR EVALUATING STABILITY OF OPENINGS IN ROCK	72-0877
OPENINGS	AN APPLICATION OF THE FINITE ELEMENT METHOD TO THE CALCULATION OF CYLINDRICAL SHELLS WITH RECTANGULA	74-1236
OPENINGS	ANALYSIS AND MODEL STUDIES OF UNDERGROUND OPENINGS IN JOINTED ROCK. RESEARCH ON ROCK BOLT REINFORCEM	71-0438
OPENINGS	ANALYSIS AND MODEL STUDIES OF UNDERGROUND OPENINGS IN JOINTED ROCK. RESEARCH ON ROCK BOLT REINFORCEM	71-0750
OPENINGS	ANALYSIS OF SHEAR WALLS WITH OPENINGS	69-0111
OPENINGS	ANALYSIS OF UNDERGROUND OPENINGS IN ROCK BY FINITE ELEMENT METHODS	73-0895
OPENINGS	DATA GENERATOR FOR THE IDEALIZATION FOR FINITE ELEMENT STRUCTURAL ANALYSIS OF NAVAL SHIP FLAT PLATED	72-0860
OPENINGS	DEVELOPMENT AND APPLICATIONS OF THEORETICAL METHODS FOR EVALUATING STABILITY OF OPENINGS IN ROCK	73-0791
OPENINGS	DEVELOPMENT AND APPLICATIONS OF A GENERAL COMPUTER PROGRAM FOR EVALUATING STABILITY OF OPENINGS IN R	74-0994
OPENINGS	DYNAMIC RESPONSE OF REINFORCED OPENINGS IN ROCK	74-0311
OPENINGS	EFFECT OF OPENINGS ON THE LATERAL STIFFNESS OF INFILLED FRAMES	71-0300
OPENINGS	EFFECT OF OPENINGS ON STRESSES IN RIGID PAVEMENTS	73-0344
OPENINGS	ELASTO-PLASTIC ANALYSIS OF UNDERGROUND OPENINGS BY THE FINITE ELEMENT METHOD	66-0067
OPENINGS	FINITE ELEMENT ELASTOPLASTIC ANALYSIS OF UNDERGROUND OPENINGS BY QUADRATIC PROGRAMMING	74-1288
OPENINGS	FINITE ELEMENT METHOD FOR THE ANALYSIS OF NOZZLE OPENINGS IN SHELLS OF REVOLUTION	69-0254
OPENINGS	FINITE ELEMENT MODELING CRITERIA FOR UNDERGROUND OPENINGS IN ROCK	74-0630
OPENINGS	INTERACTION OF FRAME AND SHEAR WALL WITH OPENINGS	74-1190
OPENINGS	PRESTRESSED CONCRETE DEEP SLABS WITH OPENINGS	73-0427
OPENINGS	REINFORCING REQUIREMENTS FOR CONCRETE BEAMS WITH LARGE WEB OPENINGS	73-0198
OPENINGS	STRAIN DISTRIBUTION AROUND UNDERGROUND OPENINGS	69-0212
OPENINGS	STRAIN DISTRIBUTION AROUND UNDERGROUND OPENINGS. STATISTICAL RELATIONSHIPS FOR CERTAIN ROCK PROPERTI	71-0710
OPENINGS	STRAIN DISTRIBUTION AROUND UNDERGROUND OPENINGS. FINITE ELEMENT ANALYSIS OF JOINTED SYSTEMS	72-0838
OPENINGS	STRESS AND DISPLACEMENTS AROUND OPENINGS IN ROCK CONTAINING AN INELASTIC DISCONTINUITY	75-0170
OPENINGS	STRESSES AND DISPLACEMENTS AROUND TWO ADJACENT CIRCULAR OPENINGS NEAR TO THE GROUND SURFACE	74-0993
OPENINGS	STRESSES AND DISPLACEMENTS AROUND OPENINGS IN HOMOGENEOUS ROCK	75-0168
OPENINGS	STRESSES AND DISPLACEMENTS AROUND OPENINGS IN ROCK CONTAINING AN ELASTIC DISCONTINUITY	75-0169
OPERATEUR	APPROXIMATIONS EXTERNES, PAR ELEMENTS FINIS DE LAGRANGE D ORDRE, UN ET DEUX, DU PROBLEME DE DIRICHLE	72-0495
OPERATING	FULMOD - AN INELASTIC ANALYSIS PROGRAM TO PREDICT THE OPERATING BEHAVIOUR OF CANDU FUEL ELEMENTS	75-0574
OPERATING	VERIFICATION OF THE PEC CONTAINMENT VESSEL FOR SOME OPERATING CONDITIONS BY USE OF THE FINITE ELEMEN	74-0883
OPERATIONAL	AUTOMATED EVALUATION OF INTERCEPTOR ROCKET MOTOR DESIGNS UNDER COMBINED OPERATIONAL AND NUCLEAR EFFE	74-0698
OPERATIONAL	REFINED CRACKED CONCRETE ANALYSIS OF CONCRETE CONTAINMENT STRUCTURES SUBJECT TO OPERATIONAL AND ENVI	74-0943
OPERATOR	FINITE ELEMENT METHOD FOR PLATE BENDING ANALYSIS BY DECOMPOSITION OF DIFFERENTIAL OPERATOR	71-0558
OPERATOR	FINITE ELEMENT MODELS OF NONLINEAR OPERATOR EQUATIONS	71-0199
OPERATOR	ON LOCAL APPROXIMATIONS OF OPERATOR EQUATIONS	72-0063
OPERATOR	VARIATIONAL PRINCIPLE FOR THE LAPLACE'S OPERATOR WITH APPLICATION IN THE TORSION OF COMPOSITE RODS	74-1044
OPERATORS	A LEAST SQUARE FINITE ELEMENT SOLUTION OF NONLINEAR OPERATORS	72-0060
OPERATORS	AN IMBEDDING TECHNIQUE FOR THE GENERATION OF WEAK-WEAK FINITE ELEMENT APPROXIMATIONS OF LINEAR AND N	72-0577
OPERATORS	APPROXIMATION OF STEKLOV EIGENVALUES OF NON-SELFADJOINT SECOND ORDER ELLIPTIC OPERATORS	72-0509
OPERATORS	FINITE ELEMENT APPROXIMATION OF A CLASS OF NONLINEAR OPERATORS	71-0454
OPERATORS	MIXED CONJUGATE FINITE ELEMENT APPROXIMATIONS OF LINEAR OPERATORS	72-0085
OPERATORS	ON THE ACCURACY AND NUMERICAL STABILITY OF THE FINITE ELEMENT APPROXIMATION FOR PARABOLIC AND HYPERB	75-0102
OPERATORS	VARIATIONAL PRINCIPLES, NONLINEAR OPERATORS, AND FINITE ELEMENT APPLICATIONS	72-0576
OPPOSITE	ANALYSIS OF AN ORTHOTROPIC PLATE WITH TWO OPPOSITE EDGES SIMPLY SUPPORTED AND THE OTHER TWO FREE	73-0741
OPPOSITE	ARCH DAMS ANALYSED BY A LINEAR FINITE ELEMENT SHELL THE FINITE STRIP METHOD IN THE ANALYSIS OF ELAST	68-0189
OPREDELENIE	OPREDELENIE STATSIONARNYKH TEMPERATURNYKH POLEI V DETALYAKH DVIGATELEI VNUTRENNEGO SGORANIYA METODOM	73-0454
OPREDELENIYA	METODY RASCHETNOGO OPREDELENIYA TEMPERATURNYKH NAPRYAZHENII V KRYSHKAKH TSILINDROV DVIGATELEI VNUTRE	74-0179
OPTICAL	ARBITRARILY SHAPED INHOMOGENEOUS OPTICAL FIBER OR INTEGRATED OPTICAL WAVEGUIDES	75-0372
OPTICAL	ARBITRARILY SHAPED INHOMOGENEOUS OPTICAL FIBER OR INTEGRATED OPTICAL WAVEGUIDES	75-0372
OPTICAL	DESIGN OF OPTICAL MIRROR STRUCTURES	71-0658
OPTICS	FINITE ELEMENT METHOD IN ELECTRON OPTICS	75-0081
OPTIM	OPTIM 11: (MAGIC) COMPATIBLE LARGE SCALE AUTOMATED MINIMUM WEIGHT DESIGN PROGRAM - V 2. PROGRAMMER'S	74-1036
OPTIM	OPTIM 11: A (MAGIC) COMPATIBLE LARGE SCALE AUTOMATED MINIMUM WEIGHT DESIGN PROGRAM - 1. ENGINEERS AN	74-1048
OPTIM	OPTIM 11: A MAGIC COMPATIBLE LARGE-SCALE AUTOMATED MINIMUM WEIGHT DESIGN PROGRAM. VOLUME 1. ENGINEER	74-1096
OPTIM	OPTIM 11: MAGIC COMPATIBLE LARGE SCALE AUTOMATED MINIMUM WEIGHT DESIGN PROGRAM. VOLUME 11. PROGRAMME	74-1093
OPTIMAL	A FINITE-ELEMENT APPROACH TO OPTIMAL STRUCTURAL DESIGN	73-1162
OPTIMAL	APPLICATION DE LA METHODE DES ELEMENTS FINIS A LA RESOLUTION D'UN PROBLEME DE DOMAINE OPTIMAL	75-0412
OPTIMAL	APPLICATION OF FINITE-ELEMENT METHOD TO APPROXIMATION OF OPTIMAL DOMAIN PROBLEM - SOLUTION METHODS F	75-0349
OPTIMAL	DISTRIBUZIONI OTTIMALI DI ARMATURE DI PRECOMPRESSIONE IN TRAVI-PARETE DI CALCESTRUZZO. (OPTIMAL DIST	75-0560
OPTIMAL	FINITE ELEMENT APPROACH TO OPTIMAL DESIGN OF PLASTIC STRUCTURES IN PLANE STRESS.	72-0397
OPTIMAL	FINITE ELEMENT METHOD FOR OPTIMAL DESIGN OF VARIABLE THICKNESS SHEETS	73-0131
OPTIMAL	FINITE ELEMENT MODEL SYSTEM OF A RABBIT CALVARIUM: AN INVESTIGATION OF OPTIMAL COMPRESSIVE LOADING	73-0206
OPTIMAL	LEAST SQUARE OPTIMAL APPROXIMATION SIMULATION TECHNIQUES	73-0871
OPTIMAL	OPTIMAL BEAM FREQUENCIES BY THE FINITE ELEMENT DISPLACEMENT METHOD	73-0187
OPTIMAL	OPTIMAL DESIGN FOR PRESCRIBED BUCKLING LOADS	74-0568
OPTIMAL	OPTIMAL DESIGN OF STRUCTURES WITH BUCKLING CONSTRAINTS	73-0246
OPTIMAL	OPTIMAL DESIGN OF WATER WELLS IN UNCONSOLIDATED SEDIMENTS	74-1182
OPTIMAL	OPTIMAL GRADIENT MINIMIZATION SCHEME FOR FINITE ELEMENT EIGENPROBLEMS	72-0292
OPTIMAL	OPTIMAL SYNTHESIS OF FLEXIBLE LINK MECHANISMS WITH LARGE STATIC DEFLECTIONS	74-0525
OPTIMAL	SOME PROPERTIES OF LINEAR STRAIN TRIANGLES AND OPTIMAL FINITE ELEMENT MODELS	73-1044
OPTIMALITY	OPTIMALITY CRITERIA IN STRUCTURAL DESIGN	71-0758
OPTIMALITY	REVIEW OF OPTIMALITY CRITERIA APPROACHES TO STRUCTURAL OPTIMIZATION	74-0652
OPTIMALLY	APPLICATION OF OPTIMALLY CRITERION TO FIBER-REINFORCED COMPOSITES	73-0431
OPTIMIZATION	A FINITE ELEMENT PROCEDURE FOR NUMERICAL OPTIMIZATION IN MECHANICAL DESIGN	74-1218
OPTIMIZATION	A REVIEW OF SOME ASPECTS OF STRUCTURAL OPTIMIZATION	75-0467
OPTIMIZATION	A VARIATIONAL APPROACH TO GRID OPTIMIZATION IN THE FINITE ELEMENT METHOD	72-0572
OPTIMIZATION	ACCURACY AND OPTIMIZATION IN THE ANALYSIS OF SHIP STRUCTURES. USING FINITE ELEMENT METHODS	71-0418
OPTIMIZATION	AN APPLICATION OF FINITE ELEMENT METHODS TO PANEL FLUTTER OPTIMIZATION	71-0258
OPTIMIZATION	AN IMPROVED AUTOMATED STRUCTURAL OPTIMIZATION PROGRAM	74-0732
OPTIMIZATION	APPLICATION OF THE FINITE ELEMENT METHOD FOR ANALYSIS AND OPTIMIZATION OF SUBMARINE STRUCTURES	71-0350
OPTIMIZATION	AUTOMATED DESIGN OPTIMIZATION OF SUPERSONIC AIRPLANE WING STRUCTURES UNDER DYNAMIC CONSTRAINTS	72-0178
OPTIMIZATION	DESIGN OPTIMIZATION IN UNDERGROUND COAL SYSTEMS	74-1095

OPTIMIZATION	DEVELOPMENT OF ADVANCED STRUCTURAL OPTIMIZATION PROGRAMS AND THEIR APPLICATION TO LARGE ORDER STRAIN	66-0028
OPTIMIZATION	ERROR BOUNDS IN AN OPTIMIZATION PROBLEM USING FINITE-ELEMENT METHOD	73-0049
OPTIMIZATION	EVALUATION OF THE ERROR BOUNDS IN AN OPTIMIZATION PROBLEM USING THE FINITE-ELEMENT METHOD	73-0211
OPTIMIZATION	FINITE ELEMENT MODELING AND OPTIMIZATION OF AEROSPACE STRUCTURES	72-0416
OPTIMIZATION	FINITE ELEMENT MODELING AND OPTIMIZATION OF AEROSPACE STRUCTURES	72-0734
OPTIMIZATION	FINITE ELEMENT OPTIMIZATION OF FINITE STEPPED SLIDER BEARING PROFILES	74-0131
OPTIMIZATION	FRAME OPTIMIZATION INCLUDING FREQUENCY CONSTRAINTS	75-0082
OPTIMIZATION	GUIDELINES FOR SELECTING FINITE ELEMENT GRIDS BASED ON AN OPTIMIZATION STUDY	74-0303
OPTIMIZATION	HYBRID FINITE ELEMENT OPTIMIZATION PROGRAMS FOR DISCS USING A DYNAMICS DESIGN TECHNOLOGY	74-0691
OPTIMIZATION	NASTRAN AS ANALYSIS TOOL IN A STRUCTURAL DESIGN OPTIMIZATION PROCESS	71-0235
OPTIMIZATION	NODE NUMBERING OPTIMIZATION IN STRUCTURAL ANALYSIS	75-0077
OPTIMIZATION	ON IMPROVING EFFICIENCY OF AN ALGORITHM FOR STRUCTURAL OPTIMIZATION AND A USERS MANUAL FOR PROGRAM T	74-0778
OPTIMIZATION	OPTIMIZATION OF A SUPERSONIC PANEL SUBJECT TO A FLUTTER CONSTRAINT, A FINITE ELEMENT SOLUTION	71-0073
OPTIMIZATION	OPTIMIZATION OF COMPLEX STRUCTURES TO SATISFY STATIC, DYNAMIC AND AEROELASTIC REQUIREMENTS	74-0272
OPTIMIZATION	OPTIMIZATION OF FINITE ELEMENT GRIDS BASED ON MINIMUM POTENTIAL ENERGY	71-0747
OPTIMIZATION	OPTIMIZATION OF FINITE ELEMENT GRIDS BASED ON MINIMUM POTENTIAL ENERGY	72-0109
OPTIMIZATION	OPTIMIZATION OF REDUNDANCY SELECTION IN FINITE-ELEMENT FORCE METHOD	70-0228
OPTIMIZATION	OPTIMIZATION OF STRUCTURES TO SATISFY FLUTTER REQUIREMENTS	69-0311
OPTIMIZATION	OPTIMIZATION OF STRUCTURES FOR STRENGTH AND STABILITY REQUIREMENTS	73-0575
OPTIMIZATION	OPTIMIZATION OF STRUCTURAL DESIGN 1	73-0904
OPTIMIZATION	OPTIMIZATION OF STRUCTURES UNDERGOING HARMONIC OR STOCHASTIC EXCITATION	75-0711
OPTIMIZATION	OPTIMIZATION OF TENSION LEVEL AND STINGER LENGTH FOR OFFSHORE PIPELINE INSTALLATION	74-0504
OPTIMIZATION	PANEL FLUTTER OPTIMIZATION - REFINED FINITE-ELEMENT APPROACH	75-0358
OPTIMIZATION	REVIEW OF OPTIMALITY CRITERIA APPROACHES TO STRUCTURAL OPTIMIZATION	74-0652
OPTIMIZATION	SOME CONSIDERATIONS REGARDING STRUCTURAL OPTIMIZATION AND FINITE ELEMENT ANALYSIS	74-0424
OPTIMIZATION	STRUCTURAL ANALYSIS AND DESIGN OPTIMIZATION OF SPACECRAFT USING NASTRAN	71-0068
OPTIMIZATION	STRUCTURAL OPTIMIZATION USING THE FINITE ELEMENT METHOD APPLIED TO A BEAM	75-0233
OPTIMIZATION	STRUCTURAL OPTIMIZATION USING THE FORCE METHOD	75-0440
OPTIMIZATION	THIN WINDOWS FOR GASEOUS AND LIQUID TARGETS: AN OPTIMIZATION PROCEDURE	75-0741
OPTIMIZING	QUASI-LINEAR PROGRAMMING ALGORITHM FOR OPTIMIZING FIBRE-REINFORCED STRUCTURES OF FIXED STIFFNESS	75-0225
OPTIMIZZAZIONE	OTTIMIZZAZIONE DI RETICOLI TRIANGOLARI PER IL METODO DEGLI ELEMENTI FINITI IN IDROLOGIA	72-0987
OPTIMUM	A FINITE ELEMENT METHOD FOR OPTIMUM BEAM DESIGN	71-0685
OPTIMUM	A METHOD TO GALERKIN TYPE ACHIEVING OPTIMUM L(2) ACCURACY FOR FIRST ORDER HYPERBOLICS AND EQUATIONS	72-0521
OPTIMUM	A THEOREM FOR OPTIMUM IDEALIZATIONS IN FINITE-ELEMENT ANALYSIS	72-0791
OPTIMUM	AN OPTIMUM IDEALIZATION IN FINITE ELEMENT METHOD FOR A TWO-GROUP NEUTRON DIFFUSION PROBLEM	74-1143
OPTIMUM	AUTOMATED OPTIMUM DESIGN OF AIRCRAFT WINGS TO SATISFY STRENGTH, STABILITY, FREQUENCY AND FLUTTER REQ	72-0605
OPTIMUM	DESIGN OF OPTIMUM STRUCTURES FOR DYNAMIC LOADS	71-0255
OPTIMUM	EFFECT OF SHEAR DEFORMATIONS AND ROTARY INERTIA ON OPTIMUM BEAM FREQUENCIES	75-0231
OPTIMUM	OPTIMUM CONFIGURATIONAL AND DIMENSIONAL DESIGN OF TRUSS STRUCTURES	74-0588
OPTIMUM	OPTIMUM CONFIGURATION OF TRUSS STRUCTURES	74-1315
OPTIMUM	OPTIMUM DESIGN OF ADVANCED COMPOSITE STRUCTURES FOR STATIC LOADS	72-0613
OPTIMUM	OPTIMUM DESIGN OF COMPOSITE SHELLS SUBJECT TO NATURAL FREQUENCY CONSTRAINTS	73-0154
OPTIMUM	OPTIMUM DESIGN OF CONCRETE STRUCTURES	74-0979
OPTIMUM	OPTIMUM DESIGN OF FRAMED STRUCTURES	69-0251
OPTIMUM	OPTIMUM DESIGN OF STRESSED SKIN STRUCTURES	73-0472
OPTIMUM	OPTIMUM MESH CONFIGURATIONS IN THE FINITE ELEMENT METHOD	74-1346
OPTIMUM	OPTIMUM PLASTIC DESIGN FOR MULTIPLE SETS OF LOADS	74-0980
OPTIMUM	OPTIMUM POSITION OF THE CENTRAL CLAY CORE OF A ROCKFILL DAM IN RESPECT TO ARCHING AND HYDRAULIC FRAC	73-0447
OPTIMUM	PROCEDURES FOR SELECTING NEAR OPTIMUM FINITE ELEMENT GRIDS	73-0537
OPTIMUM	RAMIFICATIONS OF OPTIMUM IDEALIZATION GEOMETRY IN DISCRETE ELEMENT ANALYSIS	75-0182
OPTIMUM	THE OPTIMUM APPROACH TO ANALYSIS OF ELASTIC CONTINUA	71-0177
OPTIMUM	THEOREM FOR OPTIMUM FINITE-ELEMENT IDEALIZATIONS	73-0255
OPTIONS	ASAS - A COMMERCIAL FINITE ELEMENT SYSTEM WITH DISPLACEMENT AND FORCE CONTINUITY OPTIONS	73-0947
ORBITAL	STUDY OF HYPERVELOCITY METEOROID IMPACT ON ORBITAL SPACE STATIONS	73-0905
ORBITER	ANALYTICAL AND EXPERIMENTAL INVESTIGATION OF A 1/8 SCALE DYNAMIC MODEL OF THE SHUTTLE ORBITER, VOLUM	74-1186
ORBITER	APPLICATION OF NASTRAN TO THE ANALYSIS OF A SPACE SHUTTLE ORBITER STRUCTURE	71-0099
ORBITING	THERMAL DEFORMATION PREDICTION IN ORBITING OBSERVATORY STRUCTURES	69-0292
ORDER	A COMPARISON OF FIRST AND SECOND ORDER AXIALLY SYMMETRIC FINITE ELEMENTS	73-0997
ORDER	A COMPUTER METHOD FOR THE SECOND ORDER ELASTIC ANALYSIS OF PLANE FRAMED STRUCTURES	73-0310
ORDER	A CURVED CIRCULAR CYLINDRICAL SHELL FINITE ELEMENT FOR HIGHER ORDER SHELL THEORY	69-0528
ORDER	A FINITE ELEMENT PROCEDURE OF THE SECOND ORDER ACCURACY	70-0277
ORDER	A HIGH ORDER TRIANGULAR FIELD ELEMENTS	71-0535
ORDER	A HIGHER ORDER CONFORMING RECTANGULAR ELEMENT	73-0690
ORDER	A HIGHER ORDER IN-PLANE PARALLELOGRAM ELEMENT AND ITS APPLICATION TO SKEWED CURVED BOX GIRDER BRIDGE	71-0002
ORDER	A HIGHER ORDER TRIANGULAR FINITE ELEMENT FOR THE SOLUTION OF FIELD PROBLEMS IN ORTHOTROPIC MEDIA	72-0370
ORDER	A METHOD TO GALERKIN TYPE ACHIEVING OPTIMUM L(2) ACCURACY FOR FIRST ORDER HYPERBOLICS AND EQUATIONS	72-0521
ORDER	A SECOND ORDER FINITE ELEMENT METHOD FOR THE ONE-DIMENSIONAL STEFAN PROBLEM	74-1222
ORDER	A UNIFIED TREATMENT OF THICK AND THIN FILM ELASTOHYDRODYNAMIC PROBLEMS USING HIGHER ORDER ELEMENT PR	75-0890
ORDER	ANALYSIS OF OUT-OF-PLANE STRESS BY THE SIX NODES, FOURTH ORDER ACCURACY TRIANGULAR ELEMENTS	69-0375
ORDER	APPROXIMATE THREE-DIMENSIONAL STRESS SOLUTIONS FOR SOME LAMINATED PLATE PROBLEMS USING A HIGHER ORDE	70-0419
ORDER	APPROXIMATION BY FIRST ORDER FINITE ELEMENTS AND SOLUTION BY PENALTY AND DUALITY OF NONLINEAR DIRICH	74-0245
ORDER	APPROXIMATION OF STEKLOV EIGENVALUES OF NON-SELFADJOINT SECOND ORDER ELLIPTIC OPERATORS	72-0509
ORDER	COMMENT ON A HIGHER ORDER CONFORMING RECTANGULAR PLATE ELEMENT	73-0999
ORDER	COMPARISON OF EXTRAPOLATION TECHNIQUES WITH HIGH ORDER FINITE ELEMENT METHODS FOR DIFFUSION EQUATION	73-0586
ORDER	COMPUTER SIMULATION OF SECOND ORDER FAULTS	71-0010
ORDER	CONVERGENCE OF A FINITE ELEMENT PROCEDURE FOR SOLVING BOUNDARY VALUE PROBLEMS OF THE FOURTH ORDER	70-0131
ORDER	COUPLING MATRICES FOR HIGH ORDER FINITE ELEMENT ANALYSIS OF ACOUSTIC - WAVE PROPAGATION	72-0040
ORDER	CURVED LOW ORDER ANGULAR ELEMENTS IN TWO DIMENSIONAL NEUTRON TRANSPORT PROBLEMS	74-1322
ORDER	DEVELOPMENT OF ADVANCED STRUCTURAL OPTIMIZATION PROGRAMS AND THEIR APPLICATION TO LARGE ORDER STRAIN	66-0028
ORDER	EFFECT OF HIGHER ORDER TERMS IN CERTAIN NONLINEAR FINITE ELEMENT MODELS	73-0125
ORDER	EFFECT OF HIGHER ORDER TERMS IN CERTAIN NONLINEAR FINITE ELEMENT MODELS - COMMENT	74-0473
ORDER	EIGENVALUE PROBLEMS (A - LAMBDA B) X EQUALS 0 FOR SYMMETRIC MATRICES OF HIGH ORDER	74-0205
ORDER	ELASTIC-PLASTIC DEFORMATIONS OF FLEXURALLY STIFF FRAMEWORKS FROM SECOND ORDER STRESS THEORY	73-1061
ORDER	FINITE ELEMENTS IN CURVILINEAR COORDINATES - SECOND ORDER SURFACES	73-0753
ORDER	FINITE-ELEMENT METHOD FOR 1ST ORDER HYPERBOLIC EQUATIONS	75-0590
ORDER	HIGH ORDER FINITE ELEMENT FOR COMPLETELY INCOMPRESSIBLE CREEPING FLOW	73-0144
ORDER	HIGH ORDER POLYNOMIAL TRIANGULAR FINITE ELEMENTS FOR POTENTIAL PROBLEMS	69-0240
ORDER	HIGH ORDER RECTANGULAR SHALLOW SHELL FINITE ELEMENT	73-0279
ORDER	HIGHER ORDER COMPATIBLE TRIANGULAR FINITE ELEMENTS	74-0152
ORDER	HIGHER ORDER FINITE ELEMENT METHOD FOR STATIC AND DYNAMIC ANALYSIS OF SHEAR WALLS	71-0368
ORDER	HIGHER ORDER FINITE ELEMENT METHODS FOR THE SOLUTION OF COMPRESSIBLE POROUS BEARING PROBLEMS	75-0341
ORDER	HIGHER ORDER SINGULARITIES FOR INTERFACE PROBLEMS	72-0513
ORDER	HYBRID FINITE ELEMENT METHODS FOR SOLVING 2ND ORDER ELLIPTIC EQUATIONS	74-1267
ORDER	INFLUENCE OF HIGHER ORDER TERMS IN THE LARGE DEFLECTION ANALYSIS OF FRAMEWORKS	68-0259
ORDER	METHOD FOR EXTERNAL APPROXIMATION USING FINITE ELEMENTS FOR DELTA - 2 AND ITERATIVE METHOD FOR SOLVI	72-0282
ORDER	METHOD FOR EXTERNAL APPROXIMATION USING FINITE ELEMENTS FOR DELTA -2 AND ITERATIVE METHOD FOR SOLVIN	72-0283
ORDER	NODAL VARIABLES FOR ARBITRARY ORDER FINITE ELEMENTS	75-0002
ORDER	NONLINEAR FINITE ELEMENT ANALYSIS INCLUDING HIGHER ORDER STRAIN ENERGY TERMS	70-0065
ORDER	ON SOME FINITE ELEMENT PROCEDURES FOR SOLVING SECOND ORDER BOUNDARY VALUE PROBLEMS	69-0288
ORDER	ON THE EFFECT OF APPROXIMATING HIGHER ORDER TERMS IN CERTAIN FINITE ELEMENT MODELS OF NONLINEAR STRU	73-0603
ORDER	ORDER OF CONVERGENCE OF A FINITE ELEMENT SCHEME	74-0550
ORDER	ORDER OF CONVERGENCE OF A FINITE-ELEMENT METHOD IN REGIONS WITH CURVED BOUNDARIES	75-0817

311

ORDER	SECOND ORDER THEORY OF OSCILLATING CYLINDERS	70-0430
ORDER	SIMPLIFICATION OF HIGH ORDER MECHANICAL SYSTEMS USING THE ROUTH APPROXIMATION	75-0853
ORDER	STATIC AND DYNAMIC BEHAVIOUR OF RECTANGULAR PLATES USING HIGHER ORDER FINITE STRIPS	72-0458
ORDER	STRESS ANALYSIS OF AXISYMMETRIC SOLIDS UTILIZING HIGHER ORDER QUADRILATERAL FINITE ELEMENTS	69-0356
ORDER	SUR L APPROXIMATION, PAR ELEMENTS FINIS D ORDRE UN, ET LA RESOLUTION, PAR PENALISATION-DUALITE, D UN	75-0845
ORDER	THE APPLICATION OF GRADIENT MINIMIZATION METHODS AND HIGHER ORDER DISCRETE ELEMENTS TO SHELL BUCKLIN	71-0237
ORDER	THE APPLICATION OF HIGHER ORDER FINITE ELEMENTS TO THREE DIMENSIONAL HEAT CONDUCTION PROBLEMS	70-0266
ORDER	THREE-DIMENSIONAL FIELD PROBLEMS BY HIGHER ORDER FINITE ELEMENTS	70-0257
ORDER	TIMING COMPARISON STUDY FOR SOME HIGH ORDER FINITE ELEMENT PROCEDURES AND A LOW ORDER FINITE DIFFERE	72-0767
ORDER	TIMING COMPARISON STUDY FOR SOME HIGH ORDER FINITE ELEMENT PROCEDURES AND A LOW ORDER FINITE DIFFERE	72-0767
ORDER	TRIANGULAR FINITE ELEMENTS FOR THE GENERALIZED BESSEL EQUATION OF ORDER M	73-1094
ORDER	TWO TYPES OF PIECEWISE QUADRATIC SPACES AND THEIR ORDER OF ACCURACY FOR POISSON'S EQUATION	72-0520
ORDER	USE OF HIGHER ORDER DISPLACEMENT FUNCTIONS IN THE FREE VIBRATION ANALYSIS OF SHELLS OF REVOLUTION HA	74-1055
ORDER	ZUR BERECHNUNG VON PLATTEN NACH DER THEORIE ZWEITER. ORDNUNG MIT HILFE EINES HYBRIDEN DEFORMATIONS M	73-0558
ORDERING	COMPARATIVE ANALYSIS OF THE CUTHILL-MCKEE AND THE REVERSE CUTHILL-MCKEE ORDERING ALGORITHMS FOR SPAR	73-1099
ORDERS	ORDERS OF CONVERGENCE IN FINITE ELEMENT METHODS	72-0057
ORDINARY	APPLICATION OF THE FINITE ELEMENT METHOD FOR SOLVING ORDINARY LINEAR DIFFERENTIAL EQUATION WITH BOUN	74-0084
ORDINARY	APPLICATIONS OF VARIATIONAL EQUATIONS TO ORDINARY AND PARTIAL DIFFERENTIAL EQUATIONS - MULTIPLE SOLU	75-0010
ORDINARY	FINITE ELEMENT METHOD FOR THE GENERAL SOLUTION OF ORDINARY DIFFERENTIAL EQUATIONS	75-0473
ORDINATE-LIKE	RAY-EFFECT MITIGATION IN DISCRETE ORDINATE-LIKE ANGULAR FINITE-ELEMENT APPROXIMATIONS IN NEUTRON-TRA	75-0496
ORDINATEUR	DEVELOPMENT AND USE OF COMPUTER PROGRAMS FOR FINITE ELEMENT ANALYSIS. (LA MISE AU POINT ET L'UTILIZA	69-0289
ORDNUNG	ZUR BERECHNUNG EINFACH-SYMMETRISCHER I-TRAEGER NACH DER THEORIE - 2. ORDNUNG UNTER BERUECKSICHTIGUNG	75-0510
ORDNUNG	ZUR BERECHNUNG VON PLATTEN NACH DER THEORIE ZWEITER. ORDNUNG MIT HILFE EINES HYBRIDEN DEFORMATIONS M	73-0558
ORDRE	APPROXIMATIONS EXTERNES, PAR ELEMENTS FINIS DE LAGRANGE D ORDRE, UN ET DEUX, DU PROBLEME DE DIRICHLE	72-0495
ORDRE	SUR L APPROXIMATION, PAR ELEMENTS FINIS D ORDRE UN, ET LA RESOLUTION, PAR PENALISATION-DUALITE, D UN	75-0845
ORDRE	SURE UNE METHODE D'APPROXIMATION EXTERNE PAR ELEMENTS FINIS D ORDRE DEUX	72-0742
ORE	BERECHNUNG UND BEURTEILUNG DER STANDSICHERHEIT VON ABBAUHOHLRAEUMEN IN STEILEN GANGERZLAGERSTAETTEN	74-0597
ORE	NEUARTIGE ANWENDUNG GEOMECHANISCHER UNTERSUCHUNGSMETHODEN ZUR BEURTEILUNG DER ABBAUWIRKUNGEN IM GANG	74-0530
OREBODIES	GEOMETRIC METHOD FOR THE PREDICTION OF STRESSES IN INCLUSIONS, OREBODIES, AND MINING SYSTEMS	74-0219
OREBODIES	PREDICTING THE BLOCK CAVING BEHAVIOUR OF OREBODIES	69-0044
OREBODY	A METHOD FOR THE PREDICTION OF STRESSES IN AN ISOTROPIC INCLUSION OR OREBODY OF IRREGULAR SHAPE	72-0827
OREBODY	UNDERGROUND MEASUREMENTS IN A STEEPLY DIPPING OREBODY	58-0085
ORGANIC	EFFECT OF DENSIFICATION ON THE ENGINEERING CHARACTERISTICS OF ORGANIC SOILS - VOL. 1 AND 2.	73-0805
ORGANIC	YIELDING OF A TWO DIMENSIONAL VOID ASSEMBLY IN AN ORGANIC GLASS	73-0416
ORGANIZATION	APPLICATION, TESTING, AND EVALUATION OF DOUBLY CURVED SHELL ELEMENTS FOR DYNAMIC ANALYSIS, PART III.	69-0256
ORGANIZATION	ORGANIZATION AND FUNCTIONAL PURPOSE OF A FINITE ELEMENT COMPUTER PROGRAM	74-0362
ORGANIZATION	ORGANIZATION OF GNATS: A GENERAL NONLINEAR ANALYSIS COMPUTER PROGRAM FOR TWO-DIMENSIONAL STRUCTURES	74-0871
ORGANIZATION	ORGANIZATION TECHNIQUES OF THE FINITE ELEMENT METHOD FOR REDUCING CORE STORAGE REQUIREMENTS	74-0074
ORGANIZATION	ORGANIZATION TECHNIQUES OF THE FINITE ELEMENT METHOD FOR REDUCING CORE STORAGE REQUIREMENTS	74-1120
ORGANIZATION	THE USE OF NASTRAN IN THE ANALYSIS OF LARGE COMPLEX AIRFRAME STRUCTURES, MODELING TECHNIQUES AND ORG	71-0110
ORGANIZATIONAL	SOME ORGANIZATIONAL ASPECTS OF NASTRAN	73-1033
ORGANIZATIONAL	SOME ORGANIZATIONAL ASPECTS OF NASTRAN	74-1005
ORGANIZATIONS	HOW FINITE ELEMENT METHODS ARE INTRODUCED IN LARGE AND SMALL ORGANIZATIONS	74-0366
ORIENTED	A DESIGN ORIENTED APPROACH TO CREEP AND PLASTICITY IN FINITE ELEMENT PROGRAMS	70-0125
ORIENTED	AN INTERACTIVE GRAPHICS SHIP DESIGN ORIENTED PROGRAM PACKAGE FOR TIME SHARING SYSTEMS	73-0341
ORIENTED	ASKA 105 - A PROBLEM ORIENTED LANGUAGE FOR STRUCTURAL ENGINEERS, VOLUME 11, APPLICATION	69-0461
ORIENTED	ASKA 105, A PROBLEM ORIENTED LANGUAGE FOR STRUCTURAL ENGINEERS, VOLUME 1, DESIGN	69-0460
ORIENTED	BODY ORIENTED (NATURAL) CO-ORDINATES FOR GENERATING THREE-DIMENSIONAL MESHES	74-0908
ORIENTED	COMPUTER ORIENTED ANALYSIS OF SHELL STRUCTURES	71-0559
ORIENTED	COMPUTER ORIENTED ANALYSIS OF SHELL STRUCTURES	71-0762
ORIENTED	DESIGN ORIENTED APPROACH TO CREEP AND PLASTICITY IN FINITE ELEMENT PROGRAMS	71-0281
ORIENTED	FINITE ELEMENT ANALYSIS SYSTEM FOR THE MECHANICAL BEHAVIOR OF ORIENTED FIBER COMPOSITE MATERIALS UND	73-0869
ORIENTED	STRUCTURAL ANALYSIS BY PROBLEM ORIENTED LANGUAGES	71-0623
ORIENTED	UNIMOD - APPLICATIONS ORIENTED FINITE ELEMENT SCHEME FOR THE ANALYSIS OF FRACTURE MECHANICS PROBLEMS	74-1356
ORIGIN	FINITE ELEMENTS FOR AXISYMMETRIC SOLIDS UNDER ARBITRARY LOADINGS WITH NODES ON ORIGIN.	72-0251
ORIGIN	FINITE ELEMENTS FOR AXISYMMETRIC SOLIDS UNDER ARBITRARY LOADINGS WITH NODES ON ORIGIN	73-0094
ORNL	THEORETICAL AND EXPERIMENTAL STRESS ANALYSES OF ORNL THIN-SHELL CYLINDER-TO-CYLINDER MODEL 3	75-0690
ORNL	THEORETICAL AND EXPERIMENTAL STRESS ANALYSES OF ORNL THIN-SHELL CYLINDER-TO-CYLINDER MODEL 4	75-0695
OROVILLE	STRESSES AND MOVEMENTS IN OROVILLE DAM	72-0200
ORSAM	ORSAM - A PROGRAMMING SYSTEM FOR THE FINITE ELEMENT METHOD	73-0528
ORTHODONTICS	ANALYSIS OF SPACE CLOSING SPRINGS IN ORTHODONTICS	74-0236
ORTHOGONAL	A DISCRETE-ELEMENT METHOD OF ANALYSIS FOR ORTHOGONAL SLAB AND GRID BRIDGE FLOOR SYSTEMS	72-0690
ORTHOGONAL	EXPERIENCES WITH ORTHOGONAL POLYNOMIALS AND "BEST" NUMERICAL INTEGRATION FORMULAS ON A TRIANGLE: WIT	74-0977
ORTHOGONAL	FINITE ELEMENT ANALYSIS OF ORTHOGONAL FRAME-PLATE INTERACTION	70-0508
ORTHOGONAL	FINITE ELEMENT FORMULATION OF THE HEAT CONDUCTION EQUATION IN GENERAL ORTHOGONAL CURVILINEAR COORDIN	75-0848
ORTHOGONAL	INTEGRATION FOR THE FINITE ELEMENTS - AN ALGORITHM IN ORTHOGONAL AXES	73-0295
ORTHOGONAL	NUMERICAL METHOD FOR GENERATING ORTHOGONAL CURVILINEAR MESHES	70-0443
ORTHOGONAL	ORTHOGONAL COLLOCATION ON FINITE ELEMENTS	75-0364
ORTHOGONAL	USING THE FINITE ELEMENT METHOD TO DETERMINE TEMPERATURE DISTRIBUTIONS IN ORTHOGONAL MACHINING	74-1212
ORTHOTROPER	ZUR GENAUEN BERECHNUNG KONSTRUKTIV ORTHOTROPER PLATTEN (EXACT CALCULATION OF STRUCTURALLY ORTHOTROPI	72-0188
ORTHOTROPIC	A FINITE ELEMENT ANALYSIS OF LAMINATED ORTHOTROPIC PLATES AND SHALLOW SHELLS	70-0123
ORTHOTROPIC	A HIGHER ORDER TRIANGULAR FINITE ELEMENT FOR THE SOLUTION OF FIELD PROBLEMS IN ORTHOTROPIC MEDIA	72-0370
ORTHOTROPIC	A METHOD FOR COMPUTATION OF VIBRATION MODES AND FREQUENCIES OF ORTHOTROPIC THIN SHELLS OF REVOLUTION	69-0272
ORTHOTROPIC	ANALYSIS OF AN ORTHOTROPIC PLATE WITH TWO OPPOSITE EDGES SIMPLY SUPPORTED AND THE OTHER TWO FREE	73-0741
ORTHOTROPIC	ANALYSIS OF ORTHOTROPIC FOLDED PLATES	70-0272
ORTHOTROPIC	ANALYSIS OF ORTHOTROPIC FOLDED PLATES WITH ECCENTRIC STIFFENERS	70-0437
ORTHOTROPIC	ANALYSIS OF THREE-DIMENSIONAL ORTHOTROPIC SANDWICH PLATE STRUCTURES BY FINITE-ELEMENT METHOD	72-0945
ORTHOTROPIC	ASAAS: ASYMMETRIC STRESS ANALYSIS OF AXISYMMETRIC SOLIDS WITH ORTHOTROPIC, TEMPERATURE-DEPENDENT MAT	71-0696
ORTHOTROPIC	BRLESC FINITE ELEMENT PROGRAM FOR AXISYMMETRIC, PLANE STRAIN, AND PLANE STRESS, ORTHOTROPIC SOLIDS W	71-0749
ORTHOTROPIC	COMPUTATION OF VIBRATION MODAL SHAPES AND NATURAL FREQUENCIES OF A GENERAL ORTHOTROPIC LAMINATED, TH	71-0309
ORTHOTROPIC	DETERMINATION OF STRESSES DUE TO DISCONTINUITIES IN FINITE PLATES OF ISOTROPIC AND ORTHOTROPIC MATER	73-1052
ORTHOTROPIC	EFFECT OF PLASTICITY AND CRACK BLUNTING ON THE STRESS DISTRIBUTION IN ORTHOTROPIC COMPOSITE MATERIAL	73-0145
ORTHOTROPIC	ELASTO-PLASTIC ANALYSIS OF ORTHOTROPIC PLATES AND SHELLS	68-0166
ORTHOTROPIC	ELASTO-PLASTIC ORTHOTROPIC PLATES AND SHELLS	69-0467
ORTHOTROPIC	ELASTODYNAMIC BEHAVIOR OF LAMINATED ORTHOTROPIC PLATES UNDER INITIAL STRESS	75-0249
ORTHOTROPIC	ELEMENT STIFFNESS MATRICES OF THICK-WALLED ORTHOTROPIC SHELLS WITH APPLICATIONS	69-0261
ORTHOTROPIC	FINITE ELEMENT ANALYSIS FOR ARBITRARY AXISYMMETRIC STRUCTURES ORTHOTROPIC, TEMPERATURE-DEPENDENT MAT	68-0076
ORTHOTROPIC	FINITE ELEMENT ANALYSIS OF AN AXISYMMETRICALLY LOADED ORTHOTROPIC SHELL OF REVOLUTION.	72-0438
ORTHOTROPIC	FINITE ELEMENT COMPUTER PROGRAM FOR THE ANALYSIS OF LAYERED ORTHOTROPIC STRUCTURES	74-0800
ORTHOTROPIC	FINITE ELEMENT STRESS ANALYSIS OF AXISYMMETRIC SOLIDS WITH ORTHOTROPIC TEMPERATURE DEPENDENT MATERIA	67-0089
ORTHOTROPIC	FINITE ELEMENT TECHNIQUES FOR ORTHOTROPIC PLANE STRESS AND ORTHOTROPIC PLATE ANALYSIS	68-0274
ORTHOTROPIC	FINITE ELEMENT TECHNIQUES FOR ORTHOTROPIC PLANE STRESS AND ORTHOTROPIC PLATE ANALYSIS	68-0274
ORTHOTROPIC	FINITE-ELEMENT ANALYSIS OF ORTHOTROPIC PLATES WITH ECCENTRIC STIFFENERS	72-0958
ORTHOTROPIC	FREE VIBRATION OF ORTHOTROPIC VARIABLE THICKNESS, CONTINUOUS RECTANGULAR PLATES	70-0270
ORTHOTROPIC	FREQUENCY ANALYSIS OF SKEW ORTHOTROPIC PLATES BY THE FINITE STRIP METHOD	71-0413
ORTHOTROPIC	FREQUENCY ANALYSIS OF ORTHOTROPIC CIRCULAR CYLINDRICAL PANELS BY THE FINITE STRIP METHOD	73-0274
ORTHOTROPIC	FURTHER DEVELOPMENTS IN THE USE OF LINEAR DISPLACEMENT FUNCTIONS IN THE FINITE ELEMENT ANALYSIS OF O	71-0344
ORTHOTROPIC	LAMINATED ORTHOTROPIC PLATES AND SHALLOW SHELLS	71-0400
ORTHOTROPIC	NATURAL VIBRATIONS OF LAMINATED ORTHOTROPIC SHELLS OF REVOLUTION	70-0367
ORTHOTROPIC	NUMERICAL ANALYSIS IN ORTHOTROPIC LINEAR FRACTURE MECHANICS	73-0512
ORTHOTROPIC	ON A VARIATIONAL THEOREM FOR INCOMPRESSIBLE AND NEARLY INCOMPRESSIBLE ORTHOTROPIC ELASTICITY	68-0249
ORTHOTROPIC	ORTHOTROPIC FLAT PLATES SUPPORTED ON COLUMNS ELONGATED IN PLANE	72-0492

312

ORTHOTROPIC	ORTHOTROPIC RIGHT BRIDGES BY THE FINITE STRIP METHOD	69-0028
ORTHOTROPIC	RESPONSE OF RECTANGULAR ORTHOTROPIC PLATES TO RANDOM EXCITATIONS	73-0496
ORTHOTROPIC	SAAS II: FINITE ELEMENT STRESS ANALYSIS OF AXISYMMETRIC SOLIDS WITH ORTHOTROPIC, TEMPERATURE DEPENDE	68-0096
ORTHOTROPIC	SAAS III: FINITE ELEMENT STRESS ANALYSIS OF AXISYMMETRIC AND PLANE SOLIDS WITH DIFFERENT ORTHOTROPIC	71-0741
ORTHOTROPIC	THE ANALYSIS OF CURVILINEAR ORTHOTROPIC CURVED BRIDGE DECKS	69-0162
ORTHOTROPIC	THE FINITE ELEMENT METHOD FOR ANALYSIS OF ELASTIC ISOTROPIC AND ORTHOTROPIC SLABS	64-0014
ORTHOTROPIC	TSAAS: FINITE ELEMENT THERMAL AND STRESS ANALYSIS OF AXISYMMETRIC SOLIDS WITH ORTHOTROPIC TEMPERATUR	74-0886
ORTHOTROPIC	VIBRATION AND BUCKLING ANALYSIS OF AXISYMMETRIC POLAR ORTHOTROPIC CIRCULAR PLATES	74-0574
ORTHOTROPIC	ZUR GENAUEN BERECHNUNG KONSTRUKTIV ORTHOTROPER PLATTEN (EXACT CALCULATION OF STRUCTURALLY ORTHOTROPI	72-0188
OSCILLATING	ANALYSIS OF THREE-DIMENSIONAL UNSTEADY VISCOUS FLOW AROUND OSCILLATING WINGS	74-0982
OSCILLATING	APPLICATIONS OF FINITE ELEMENT METHODS TO UNSTEADY VISCOUS FLOW AROUND A BOX-GIRDER BRIDGE OSCILLATI	74-0827
OSCILLATING	FINITE ELEMENT ANALYSIS OF UNSTEADY INCOMPRESSIBLE FLOW AROUND AN OSCILLATING OBSTACLE OF ARBITRARY	73-0067
OSCILLATING	KERNEL FUNCTION FOR NONPLANAR OSCILLATING SURFACES IN SUPERSONIC FLOW	71-0472
OSCILLATING	NUMERICAL CALCULATIONS OF VELOCITY AND PRESSURE DISTRIBUTION AROUND OSCILLATING AIRFOILS	74-0847
OSCILLATING	SECOND ORDER THEORY OF OSCILLATING CYLINDERS	70-0430
OSCILLATING	SUITABILITY OF THE FINITE ELEMENT METHOD FOR ANALYSIS OF UNSTEADY FLOW AROUND OSCILLATING AIRFOILS	73-0069
OSCILLATION	ACCURACY, STABILITY AND OSCILLATION CHARACTERISTICS OF FINITE ELEMENT METHOD FOR SOLVING HEAT CONDUC	69-0007
OSCILLATION	ACCURACY, STABILITY, AND OSCILLATION CHARACTERISTICS OF TRANSIENT TWO-DIMENSIONAL HEAT CONDUCTION	75-0849
OSCILLATION	CALCULATION OF AXISYMMETRICAL OSCILLATION OF FLEXIBLE TURBINE DISCS. (CALCOLO DELLE OSCILLAZIONI ASS	74-1328
OSCILLATION	FINITE ELEMENT ANALYSIS OF THE FORCED OSCILLATION OF SHIP HULL FORMS	74-0814
OSCILLATION	HARBOUR OSCILLATION - A NUMERICAL TREATMENT FOR UNDAMPED NATURAL MODES	70-0208
OSCILLATION	OSCILLATION STABILITY OF INTERNALLY AND EXTERNALLY NOTCHED PLANE BARS OF EQUAL FORM FACTOR FROM 3943	73-0745
OSCILLATION	STABILITY AND OSCILLATION CHARACTERISTICS OF FINITE ELEMENT, FINITE DIFFERENCE AND WEIGHTED RESIDUAL	73-0139
OSCILLATION	STABILITY AND OSCILLATION CHARACTERISTICS OF FINITE ELEMENT, FINITE-DIFFERENCE, AND WEIGHTED-RESIDUA	75-0615
OSCILLATIONS	APPLICATION OF FINITE ELEMENT METHODS FOR CALCULATION OF TRANSVERSE OSCILLATIONS IN ONE DIMENSIONAL	71-0599
OSCILLATIONS	CALCULATION OF THE FREE OSCILLATIONS OF A LIQUID IN AXISYMMETRIC MOTIONLESS CONTAINERS OF ARBITRARY	74-0387
OSCILLATIONS	COMPARIOSN OF RESULTS OBTAINED FOR THE PROBLEM OF SYMMETRIC FLUID OSCILLATIONS IN TANKS BY THE FINIT	71-0761
OSCILLATIONS	LINEAR AND NONLINEAR OSCILLATIONS IN MECHANICS	74-0677
OSCILLATIONS	NONLINEAR OSCILLATIONS USING THE FINITE ELEMENT TECHNIQUE	73-0568
OSCILLATIONS	NUMERICAL STUDY OF EIGEN OSCILLATIONS OF A SCREW PINCH WITH FINITE ELEMENTS	74-0159
OSCILLATIONS	OSCILLATIONS AND WAVE FORCES IN A MAN-MADE HARBOUR IN THE OPEN SEA	74-0151
OSCILLATIONS	OSCILLATIONS AND WAVE FORCES IN AN OFFSHORE HARBOR - APPLICATIONS OF HYBRID FINITE ELEMENT METHOD TO	74-1024
OSCILLATIONS	SENSITIVITY OF ROTOR BLADE VIBRATION CHARACTERISTICS TO TORSIONAL OSCILLATIONS	74-0611
OSCILLATIONS	SOME RESULTS ON THE FREE-FREE OSCILLATIONS OF AIRCRAFT TYPE STRUCTURES	65-0007
OSCILLATIONS	STUDY OF UMBRELLA OSCILLATIONS OF TURBINE DISKS OF ARBITRARY PROFILE BY THE FINITE ELEMENT METHOD	75-0332
OSCILLATIONS	VERGLEICH NUMERISCHER LOESUNGEN NACH DEM DIFFERENZENVERFAHREN UND NACH DER FINITE-ELEMENT-METHODE AM	71-0016
OSCILLATORY	A FINITE ELEMENT ANALYSIS FOR STEADY AND OSCILLATORY SUBSONIC FLOW AROUND COMPLEX CONFIGURATIONS	74-1317
OSCILLATORY	A FINITE-ELEMENT ANALYSIS FOR STEADY-AND OSCILLATORY SUPERSONIC FLOWS AROUND COMPLEX CONFIGURATIONS	74-1316
OSCILLAZIONI	CALCULATION OF AXISYMMETRICAL OSCILLATION OF FLEXIBLE TURBINE DISCS. (CALCOLO DELLE OSCILLAZIONI ASS	74-1328
OSGOOD	AN INVERSE FORM OF RAMBERG - OSGOOD FORMULA	69-0479
OSIOWO-SYMETRYC	METODA ELEMENTOW SKONSZONYCH ROZWIAZYWANIA OSIOWO-SYMETRYCZNEGO PRZEWODNICTWA CIEPLA (FINITE-ELEMENT	73-0434
OSNOVANIEM	RASCHET METODOM KONECHNYKH ELEMENTOV DINAMICHESKIKH KHARAKTEERISTIK ZEMLYANOI PLOTINY SOVMESTNO S OS	73-0437
OSTEOTOMY	MEASUREMENTS OF HEALING IN AN OSTEOTOMY IN A RABBIT CALVARIUM: THE INFLUENCE OF APPLIED COMPRESSIVE	74-0975
OTTIMALI	DISTRIBUZIONI OTTIMALI DI ARMATURE DI PRECOMPRESSIONE IN TRAVI-PARETE DI CALCESTRUZZO. (OPTIMAL DIST	75-0560
OTVERSTIYAMI	ISSLEDOVANIE NAPRYAZHENNOGO SOSTOYANIYA REBRISTYKH ISILINDRICHESKIKH OBOLOCHEK S PRYAMOUGOL'NYMI OTV	74-0944
OTVERSTIYAMI	METOD KONECHNYKH ELEMENTOV V PRIMENENII K RASCHETU TSILINDRICHESKIKH OBOLOCHEK S PRYAMOUGOL'NYMI OTV	73-0536
OUT-OF-PLANE	ANALYSIS OF OUT-OF-PLANE STRESS BY THE SIX NODES, FOURTH ORDER ACCURACY TRIANGULAR ELEMENTS	69-0375
OUT-OF-PLANE	DEVELOPMENT AND APPLICATION OF THE DIRECT STIFFNESS METHOD FOR OUT-OF-PLANE BENDING USING A TRIANGUL	62-0009
OUT-OF-PLANE	SIMPLIFIED INELASTIC (PLASTIC AND CREEP) ANALYSIS OF PIPE ELBOWS SUBJECTED TO INPLANE AND OUT-OF-PLA	75-0654
OUT-OF-PLANE	STIFFNESS MATRICES FOR THE GENERAL DEFORMATION (OUT-OF-PLANE AND INPLANE) OF CURVED BEAM MEMBERS BAS	75-0124
OUT-OF-PLANENES	EFFECT OF OUT-OF-PLANENESS OF MEMBRANE QUADRILATERAL FINITE ELEMENTS	73-0031
OUTBOARD	SPECTRUM TRUNCATION AND DAMAGE TOLERANCE STUDY ASSOCIATED WITH THE C-5A OUTBOARD PYLON AFT TRUSS LUG	74-0535
OUTLOOK	THE STATUS AND OUTLOOK FOR THE SOLUTION OF NONLINEAR PROBLEMS BY THE FINITE ELEMENT METHOD	70-0259
OUTPUT	COMPUTER-AIDED INPUT/OUTPUT FOR USE WITH FINITE ELEMENT METHOD OF STRUCTURAL ANALYSIS	70-0330
OUTPUT	COMPUTER-AIDED INPUT/OUTPUT FOR THE USE WITH FINITE ELEMENT STRUCTURAL ANALYSES	73-0923
OUTPUT	INTERACTIVE TECHNIQUES WITH GRAPHICAL OUTPUT FOR BRIDGE ANALYSES	75-0622
OUTPUT	PROCEDURE FOR EFFICIENT GENERATING, CHECKING, AND DISPLAYING NASTRAN INPUT AND OUTPUT DATA FOR ANALY	71-0098
OVER-RELAXATION	EXTENSIONS OF THE SUCCESSIVE OVER-RELAXATION THEORY WITH APPLICATIONS TO FINITE ELEMENT APPROXIMATIO	73-0980
OVER-RELAXATION	USE OF THE OVER-RELAXATION TECHNIQUE IN THE SIMULATION OF LARGE GROUNDWATER BASINS BY THE FINITE ELE	75-0606
OVERALL	ON THE THEORY OF CALCULATION OF THE OVERALL STRENGTH OF A FUSELAGE BY THE METHOD OF FINITE ELEMENTS	72-0692
OVERCOMING	OVERCOMING LOSS OF ACCURACY WHEN USING CURVED FINITE ELEMENTS	75-0092
OVERCONSOLIDATE	INITIATION OF FAILURE IN SLOPES IN OVERCONSOLIDATED CLAYS AND CLAY SHALES.	70-0186
OVERHANGS	VIBRATION OF BEAMS WITH OVERHANGS	73-0488
OVERLAPPING	INVESTIGATION ON FREE AND FORCED VIBRATIONS OF A LONG TANKER WITH OVERLAPPING PROPELLER ARRANGEMENT	73-0981
OVERLAY	CREEP BENDING TESTS OF LARGE WELD OVERLAY AND BASE METAL SPECIMENS AT 1000 DEGREE F. (537.7 DEG. C)	75-0521
OVERLAY	FINITE ELEMENT ANALYSIS OF NON-LINEAR COMPOSITE MATERIALS BY USE OF OVERLAY SYSTEMS	74-0928
OVERLOAD	EFFECTS OF A PEAK OVERLOAD ON FATIGUE CRACK PROPAGATION RATE-PLANE STRESS ANALYSIS OF CRACKED PLATE	74-1365
OVERSPEED	REACTOR COOLANT PUMP FLYWHEELS OVERSPEED EVALUATION	74-0506
OXYGEN	A FINITE ELEMENT ANALYSIS OF DISSOLVED OXYGEN DRAWDOWN AND SULFATE PRODUCTION IN STRIP MINE SPOIL DA	75-0795
PACIFIC	MATHEMATICAL MODEL OF THE COLUMBIA RIVER FROM THE PACIFIC OCEAN TO BONNEVILLE DAM. PART 1. THEORY, P	69-0464
PACIFIC	MATHEMATICAL MODEL OF THE COLUMBIA RIVER FROM THE PACIFIC OCEAN TO BONNEVILLE DAM. PART II: INPUT-OU	70-0417
PACIFIC	NONLINEAR SEISMIC ANALYSIS OF CURRENT DIVIDER SUPPORT SYSTEM FOR PACIFIC DC INTERTIE	73-0200
PACKAGE	AN INTERACTIVE GRAPHICS SHIP DESIGN ORIENTED PROGRAM PACKAGE FOR TIME SHARING SYSTEMS	73-0341
PACKAGE	ASKA 105 - A TECHNICAL MANUAL FOR THE DIRECT SOLUTION AND INVERSION PACKAGE	68-0255
PACKAGE	FETE PROGRAM PACKAGE	72-0161
PACKAGE	FINITE ELEMENT PROGRAM PACKAGE FOR AXISYMMETRIC SCALAR FIELD PROBLEMS	73-0113
PACKAGE	FINITE ELEMENT PROGRAM PACKAGE FOR AXISYMMETRIC VECTOR FIELD PROBLEMS	75-0138
PACKAGE	SCALAR FINITE-ELEMENT PROGRAM PACKAGE FOR TWO-DIMENSIONAL FIELD PROBLEMS	71-0478
PACKAGE	USING THE NASTRAN THERMAL ANALYZER TO SIMULATE A FLIGHT SCIENTIFIC INSTRUMENT PACKAGE	74-1011
PACKING	CREEP OF B/AL COMPOSITES AS INFLUENCED BY RESIDUAL STRESSES, BOND STRENGTH, AND FIBER PACKING GEOMET	74-1208
PACKS	ELASTIC STRESSES IN LAYERED SNOW PACKS	72-0925
PAIRS	NUMERICAL ANALYSIS OF THE LINE CAPACITANCE AND CROSSTALK FACTOR FOR INSULATED WIRE PAIRS	74-0685
PAMINT	ASUPRA ANALIZEI PRIN METODA ELEMENTULUI FINIT A STARILOR DE EFORTURI SI DEFORMATII DIN MASIVELE DE P	70-0471
PAMINT	COMPORTAREA ELASTICA A DIGURILOR DE PAMINT SOLICITATE SEISMIC (ELASTIC BEHAVIOR OF THE EARTH DAMS SEI	74-1185
PANEL	A SIX NODE TRIANGULAR SHEAR PANEL FOR EQUILIBRIUM MODELS	73-1011
PANEL	ACOUSTIC RESPONSE OF A SPACECRAFT SOLAR PANEL USING NORMAL MODE METHOD	69-0422
PANEL	AN APPLICATION OF FINITE ELEMENT METHODS TO PANEL FLUTTER OPTIMIZATION	71-0258
PANEL	APPLICATION OF MATRIX DISPLACEMENT METHODS IN THE STUDY OF PANEL FLUTTER	69-0406
PANEL	AUTOMOBILE PANEL SWEEP STIFFNESS ANALYSIS	74-0175
PANEL	DYNAMIC ANALYSIS OF STIFFENED PANEL STRUCTURES	71-0522
PANEL	FINITE ELEMENT ANALYSIS OF SUPERSONIC PANEL FLUTTER	73-0481
PANEL	FINITE ELEMENTS APPLIED TO PANEL FLUTTER	67-0066
PANEL	FINITE-ELEMENT ANALYSIS OF COMPLEX PANEL RESPONSE TO RANDOM LOADS	68-0093
PANEL	FREE VIBRATIONS AND RANDOM RESPONSE OF AN INTEGRALLY-STIFFENED PANEL	70-0138
PANEL	JET NOISE EXCITATION OF AN INTEGRALLY STIFFENED PANEL	71-0546
PANEL	NODAL ANALYSIS OF A NINE BAY SKIN-STRINGER PANEL	71-0102
PANEL	ON APPLYING FINITE ELEMENTS TO PANEL FLUTTER	67-0135
PANEL	ON THE NUMERICAL SOLUTION OF LARGE EIGENVALUE PROBLEMS ARISING IN PANEL FLUTTER ANALYSIS BY THE FINI	74-0942
PANEL	OPTIMIZATION OF A SUPERSONIC PANEL SUBJECT TO A FLUTTER CONSTRAINT, A FINITE ELEMENT SOLUTION	71-0073
PANEL	PANEL FLUTTER OPTIMIZATION - REFINED FINITE-ELEMENT APPROACH	75-0358
PANEL	SINGLE ELEMENT PANEL FOR FLAT PLATE STRUCTURES	75-0279
PANEL	SKIN-STRINGER PANEL NORMAL MODE RESPONSE EXPERIMENTAL DATA	71-0691

PANEL	THE MBB-UFE SUBSONIC PANEL METHOD. PART 2 - THE LIFT-CONNECTED DISPLACEMENT PROBLEM IN COMPRESSIBLE	72-0865
PANEL	THE QUADRILATERAL SHEAR PANEL	51-0002
PANEL	THEORETICAL AND EXPERIMENTAL ANALYSIS OF STIFFENED PANEL UNDER DYNAMIC CONDITIONS	70-0130
PANEL	THEORETICAL STUDY OF THE SEQUENCE OF FAILURE IN PRECAST PANEL SHEAR WALLS	73-0490
PANEL	VARIATIONAL FINITE ELEMENT CALCULATION OF THE ACOUSTIC RESPONSE OF A RECTANGULAR PANEL	71-0100
PANEL	VIBRATION MODES AND RANDOM RESPONSE OF A MULTI-BAY PANEL SYSTEM USING FINITE ELEMENTS	67-0128
PANEL	ZUM TRAGVERHALTEN VON WAENDEN AUS FERTIGTEILTAFELN (BEHAVIOR OF LARGE PANEL SHEAR WALLS)	74-0299
PANEL	ZUSAMMENWIRKEN VON EINZELNEN FERTIGTEILEN ALS GROSSFLAECHIGE SCHEIBE (INTERACTION OF INDIVIDUAL PREF	73-0552
PANEL-VERFAHREN	BERECHNUNG DES REIBUNGSLOSEN STROEMUNGSFELDES VON DREIDIMENSIONALEN AUFTRIEBSBEHAFTETENTRAGFLUEGELN	73-0533
PANELIZED	INFLUENCE OF JOINTS IN PANELIZED STRUCTURAL SYSTEMS	72-0204
PANELS	A STABILITY ANALYSIS OF CYLINDRICAL PANELS USING A FINITE ELEMENT FORMULATION	71-0636
PANELS	ACCURATE FINITE ELEMENT MODELLING OF FLAT AND CURVED STIFFENED PANELS	72-0538
PANELS	BERECHNUNG VON STAHLBETONSCHEIBEN IM ZUSTAND II BEI ANNAHME EINES WIRKLICHKEITSNAHEN WERKSTOFFVERHAL	74-0978
PANELS	DIE BEULLASTEN DIAGONAL VERSTEIFTER RECHTECKPLATTEN (BUCKLING LOADS IN DIAGONALLY STIFFENED RECTANGU	73-0405
PANELS	EVALUATION OF COMPOSITE SHEAR PANELS WITH CUTOUTS	72-0615
PANELS	EXPERIMENTAL AND FINITE ELEMENT ANALYSIS OF CABLE ROOF STRUCTURES INCLUDING PRECAST PANELS	74-1139
PANELS	FINITE ELEMENT ANALYSIS OF SANDWICH PANELS	70-0449
PANELS	FINITE ELEMENT ANALYSIS OF SANDWICH PANELS	71-0662
PANELS	FLUTTER OF FLAT FINITE-ELEMENT PANELS IN A SUPERSONIC POTENTIAL FLOW	75-0814
PANELS	FLUTTER OF SKEW PANELS BY THE MATRIX DISPLACEMENT APPROACH	70-0201
PANELS	FREQUENCY ANALYSIS OF ORTHOTROPIC CIRCULAR CYLINDRICAL PANELS BY THE FINITE STRIP METHOD	73-0274
PANELS	INELASTIC FINITE ELEMENT ANALYSIS OF REINFORCED CONCRETE PANELS UNDER IN-PLANE LOADS	70-0482
PANELS	LIGHT GAGE STEEL INFILL PANELS IN MULTISTORY STEEL FRAMES	74-0229
PANELS	LIMIT ANALYSIS OF FRAME SYSTEMS STIFFENED BY PANELS	74-0584
PANELS	MATRIX ANALYSIS OF LOCAL INSTABILITY IN PLATES, STIFFENED PANELS AND COLUMNS	72-0542
PANELS	NONLINEAR ANALYSIS OF REINFORCED CONCRETE FRAMES AND PANELS	70-0436
PANELS	NONLINEAR DYNAMIC ANALYSIS OF FRAMES WITH FILLER PANELS	74-0226
PANELS	SHOCK TUNNEL TESTS OF PRELOADED AND ARCHED WALL PANELS FINITE ELEMENT STRESS FORMULATION FOR WAVE PR	73-0892
PANELS	STRESS DIFFUSION IN THICK STIFFENED PANELS	74-0620
PANELS	STRESSES AROUND HOLES IN STIFFENED COMPOSITE PANELS USING LAURENT-SERIES AND FINITE ELEMENT METHODS	74-1252
PANELS	STRESSES AROUND HOLES IN STIFFENED COMPOSITE PANELS USING LAURENT-SERIES AND FINITE ELEMENT METHODS	75-0733
PANELS	STRUCTURAL RESPONSE AND LOADING OF WALL PANELS	71-0735
PANELS	THREE DIMENSIONAL ANALYSIS OF BUILDINGS COMPOSED OF FLOOR AND WALL PANELS	71-0387
PANELS	VIBRATION ANALYSIS OF CYLINDRICAL PANELS	72-0485
PANELS	WOOD SHEAR PANELS BONDED WITH FLEXIBLE ADHESIVES	75-0086
PANZERWANNE	EINSATZ DES COMPUTERS BEI DER ENTWICKLUNG EINER PANZERWANNE (THE USE OF A COMPUTER FOR DEVELOPING A	72-0211
PAPA	PAPA, STRUCTURAL ANALYSIS OF PLATES AND SHELLS USING TRAPEZOIDAL AND TRIANGULAR PLATE ELEMENTS	67-0059
PAPER	COMMENT ON THE PAPER "POSSIBLE LOSS OF ACCURACY IN CURVED (ISOPARAMETRIC) FINITE ELEMENTS" BY HENSHE	72-0279
PAPER	DISCUSSION OF A PAPER BY F. W. WILLIAMS	73-1078
PAPER	DISCUSSION OF PROCEEDINGS, PAPER 6612, ELASTIC BUCKLING OF TWO-COMPONENT COLUMNS	70-0321
PARABOLIC	A FINITE ELEMENT COLLOCATION METHOD FOR QUASILINEAR PARABOLIC EQUATIONS	73-1000
PARABOLIC	CUBIC SPLINE GALERKIN APPROXIMATIONS TO PARABOLIC SYSTEMS WITH COUPLED NON-LINEAR BOUNDARY CONDITION	75-0232
PARABOLIC	EFFECTS OF QUADRATURE ERRORS IN FINITE ELEMENT APPROXIMATION OF STEADY STATE, EIGENVALUE, AND PARABO	72-0343
PARABOLIC	ERROR BOUNDS FOR SEMIDISCRETE GALERKIN APPROXIMATIONS OF PARABOLIC PROBLEMS WITH APPLICATIONS TO PET	70-0291
PARABOLIC	FINITE ELEMENT APPROXIMATIONS TO PARABOLIC PROBLEMS	71-0448
PARABOLIC	FINITE ELEMENT COLLOCATION METHOD FOR QUASILINEAR PARABOLIC EQUATIONS	73-0047
PARABOLIC	FINITE ELEMENT GALERKIN METHOD SOLUTIONS TO SELECTED ELLIPTIC AND PARABOLIC DIFFERENTIAL EQUATIONS	71-0027
PARABOLIC	FINITE ELEMENT METHODS FOR PARABOLIC EQUATIONS	74-0173
PARABOLIC	FINITE ELEMENT METHODS FOR PARABOLIC EQUATIONS	74-1280
PARABOLIC	FINITE-ELEMENT MULTISTEP DISCRETIZATIONS OF PARABOLIC BOUNDARY-VALUE PROBLEMS	75-0501
PARABOLIC	GALERKIN METHODS FOR PARABOLIC EQUATIONS	70-0448
PARABOLIC	GALERKIN METHODS FOR PARABOLIC EQUATIONS WITH NONLINEAR BOUNDARY CONDITIONS	73-1023
PARABOLIC	LOCAL LEAST SQUARES STRESS SMOOTHING FOR PARABOLIC ISOPARAMETRIC ELEMENTS	75-0593
PARABOLIC	ON THE ACCURACY AND NUMERICAL STABILITY OF THE FINITE ELEMENT APPROXIMATION FOR PARABOLIC AND HYPERB	75-0102
PARABOLIC	ON THE DISCRETIZATION OF THE TIME VARIABLE FOR PARABOLIC PARTIAL DIFFERENTIAL EQUATIONS	75-0103
PARABOLIC	ON THE PARABOLIC VELAROIDAL SHELL	69-0314
PARABOLIC	PROPAGATION IN ELLIPTICAL AND PARABOLIC WAVEGUIDES	74-1173
PARABOLIC	RICHARDSON EXTRAPOLATION FOR PARABOLIC GALERKIN METHODS	72-0522
PARABOLIC	SOME L(2) ERROR ESTIMATES FOR PARABOLIC GALERKIN METHODS	72-0511
PARABOLIC	STABILITY CRITERIA FOR TWO FINITE ELEMENT SCHEMES FOR PARABOLIC EQUATION	75-0568
PARABOLIC	THE USE OF NUMERICAL INTEGRATION IN FINITE ELEMENT METHODS FOR SOLVING PARABOLIC EQUATIONS	72-0351
PARABOLIC	UNCONDITIONALLY STABLE FINITE ELEMENT SCHEMES FOR PARABOLIC EQUATIONS	74-1286
PARABOLIC	USE OF PARABOLIC ARCS IN MATCHING CURVED BOUNDARIES IN FINITE-ELEMENT METHOD	75-0589
PARABOLOID	ANALYSIS OF HIPPED ROOF HYPERBOLIC PARABOLOID STRUCTURES	72-0216
PARABOLOID	ANALYSIS OF HYPERBOLIC PARABOLOID SHELLS	70-0324
PARABOLOID	ANALYSIS OF THIN-SHELL HYPERBOLIC PARABOLOID SHELLS	72-0121
PARABOLOID	FINITE ELEMENT ANALYSIS OF HYPERBOLIC PARABOLOID SHELLS	71-0245
PARABOLOID	FINITE ELEMENT ANALYSIS OF HYPERBOLIC PARABOLOID SHELLS	72-0887
PARABOLOID	LUMPED PARAMETER ELEMENT FOR THE ANALYSIS OF HYPERBOLIC PARABOLOID SHELLS	72-0148
PARABOLOIDAL	FREE VIBRATIONS OF PRESSURE LOADED PARABOLOIDAL SHELLS	71-0601
PARABOLOIDAL	INEXTENSIONAL VIBRATIONS OF PARABOLOIDAL SHELLS OF REVOLUTION, DOMES AND SHELLS	71-0427
PARABOLOIDS	APPLICATION OF THE FINITE ELEMENT METHOD OF ANALYSIS TO HYPERBOLIC PARABOLOIDS WITH EDGE BEAMS	75-0250
PARACHUTES	STRESS ANALYSIS OF PARACHUTES USING FINITE ELEMENTS	71-0481
PARALLEL	A COMPATIBLE ELEMENT FOR BUCKLING ANALYSIS OF SKEW PLATES WITH PARALLEL EDGES	73-0328
PARALLEL	A THERMO-ELASTIC ANALYSIS OF THE PARALLEL SURFACE THRUST BEARING	74-1069
PARALLEL	APPLICATION OF THE FINITE ELEMENT METHOD TO CONVECTION HEAT TRANSFER BETWEEN PARALLEL PLANES	71-0244
PARALLEL	DEVELOPMENT AND APPLICATIONS OF SUPERSONIC UNSTEADY CONSISTENT AERODYNAMICS FOR INTERFERING PARALLEL	72-0716
PARALLEL	FINITE ELEMENT ANALYSIS OF THE STEADY FLOW OF NON NEWTONIAN FLUIDS THROUGH PARALLEL SIDED CONDUITS	74-0056
PARALLEL	MECHANISM OF PARALLEL SURFACE THRUST BEARING	75-0322
PARALLEL	NUMERICAL SOLUTION OF THE VISCOUS FLOW IN THE ENTRANCE REGION OF PARALLEL PLATES	73-0022
PARALLEL	THERMO-ELASTIC ANALYSIS OF THE PARALLEL SURFACE THRUST BEARING	74-1172
PARALLELOGRAM	A HIGHER ORDER IN-PLANE PARALLELOGRAM ELEMENT AND ITS APPLICATION TO SKEWED CURVED BOX GIRDER BRIDGE	71-0002
PARALLELOGRAM	MEMBRANE PARALLELOGRAM ELEMENT WITH LINEARLY VARYING EDGE STRAIN FOR THE MATRIX DISPLACEMENT METHOD	66-0006
PARALLELOGRAM	PARALLELOGRAM ELEMENT IN THE SOLUTION OF RHOMBIC CANTILEVER PLATE BENDING	66-0022
PARAMETER	DYNAMICS OF DISTRIBUTED PARAMETER ROTOR SYSTEMS: TRANSFER MATRIX AND FINITE ELEMENT TECHNIQUES	70-0461
PARAMETER	EFFECT OF SYSTEM PARAMETER VARIATIONS ON NATURAL FREQUENCIES	75-0661
PARAMETER	FINITE ELEMENT ANALYSIS OF PLATES ON A TWO PARAMETER FOUNDATION MODEL.	72-0425
PARAMETER	FINITE ELEMENT MODEL FOR DISTRIBUTED PARAMETER TURBOMOTOR SYSTEMS	71-0332
PARAMETER	LUMPED PARAMETER APPROACH TO AXISYMMETRIC DYNAMIC SOIL DEFORMATIONS	69-0170
PARAMETER	LUMPED PARAMETER ELEMENT FOR THE ANALYSIS OF HYPERBOLIC PARABOLOID SHELLS	72-0148
PARAMETER	LUMPED PARAMETER METHODS APPLIED TO ELASTIC VIBRATIONS	63-0016
PARAMETER	ONE PARAMETER FAMILIES OF VARIATIONAL PRINCIPLES	72-0503
PARAMETER	PARAMETER EINES ELASTO-PLASTISCHEN STOFFANSATZES FUER KOERNIGE ERDSTOFFE (DETERMINATION OF PARAMETER	74-0657
PARAMETER	PARAMETER EINES ELASTO-PLASTISCHEN STOFFANSATZES FUER KOERNIGE ERDSTOFFE (DETERMINATION OF PARAMETER	74-0657
PARAMETERS	BEAM BENDING FINITE ELEMENT CONTAINING INTEGRATION PARAMETERS	71-0179
PARAMETERS	CERAMIC ROTATING BLADES: SOME CRITICAL DESIGN PARAMETERS FOR GAS TURBINE APPLICATIONS	74-0146
PARAMETERS	FLEXURE ANALYSIS OF AXIALLY-LOADED BEAM BY FINITE-ELEMENT METHOD WITH INTEGRAL PARAMETERS	75-0462
PARAMETERS	FRACTURE TOUGHNESS PARAMETERS AND ELASTIC-PLASTIC ANALYSIS OF NON-MODERATE FRACTURE CONDITIONS USING	73-0258
PARAMETERS	FRACTURE TOUGHNESS PARAMETERS AND ELASTIC-PLASTIC ANALYSIS OF NON-MODERATE FRACTURE CONDITIONS USING	73-1126
PARAMETERS	GENERAL POTENTIAL ENERGY AND COMPLEMENTARY ENERGY MODELS BASED ON STRESS PARAMETERS	74-1240
PARAMETERS	INFLUENCE OF SOIL/STRUCTURE INTERACTION PARAMETERS ON FLOOR RESPONSE SPECTRA	73-0833
PARAMETERS	PLANE STRESS ANALYSIS USING THE FINITE ELEMENT METHOD. TRIANGULAR FINITE ELEMENT WITH SIX PARAMETERS	67-0106

PARAMETERS	SENSITIVITY ANALYSIS OF INPUT PARAMETERS IN NUMERICAL MODELING OF STEADY STATE REGIONAL GROUNDWATER	74-0926
PARAMETERS	SPECIFICATION OF GEOMETRICAL PARAMETERS FOR ELEMENTS WITH CUBIC MAPPING FUNCTIONS	73-1082
PARAMETERS	THE EFFECT OF LUMPED PARAMETERS ON BEAM FREQUENCIES	63-0012
PARAMETERS	THE EFFECTS OF SOIL PARAMETERS AND BOUNDARY CONDITIONS ON THE CONSOLIDATION OF AN ELASTIC LAYER	70-0422
PARAMETERS	USE OF INTEGRAL PARAMETERS IN ROD PROBLEMS	74-1194
PARAMETRIC	ANISOTROPIC PARAMETRIC PLATE DISCRETE ELEMENTS	73-1047
PARAMETRIC	CASE BOND STRESS CALCULATIONS FOR FLAPPED CYLINDRICAL ANALOGS OF SOLID PROPELLANT ROCKET MOTORS. BON	72-0842
PARAMETRIC	COMPARATIVE CONSIDERATION OF PARAMETRIC STRESS CONCENTRATION INVESTIGATIONS WITH FINITE ELEMENTS AND	72-0423
PARAMETRIC	COMPARATIVE OBSERVATION ON PARAMETRIC STRESS STUDIES WITH FINITE ELEMENTS WITH COMPLEX STRESS FUNCTI	72-0250
PARAMETRIC	FINITE ELEMENT PARAMETRIC STUDY OF VERTICAL STRAIN INFLUENCE FACTORS AND THE PRESSUREMETER TEST TO E	74-1127
PARAMETRIC	PARAMETRIC ANALYSIS OF CONTACT CHATTER IN RELAYS	74-0487
PARAMETRIC	PARAMETRIC STUDIES OF LAYERED PAVEMENT SYSTEMS USING AFPAV CODE WITH FAST EQUATION SOLVER	75-0747
PARAMETRIC	PARAMETRIC STUDY OF A BEAM WITH A COMPOUND SIDE-BRANCH RESONATOR AS A DEVICE TO EVALUATE PRELIMINARY	70-0465
PARAMETRICAL	PARAMETRISCHE KERBSPANNUNGSUNTERSUCHUNGEN AN DER LOCHSCHEIBE MIT DER METHODE DER FINITEN ELEMENTE. (	75-0603
PARAMETRISCHE	PARAMETRISCHE KERBSPANNUNGSUNTERSUCHUNG AM ELASTISCHEN KERN. (PARAMETRICAL INVESTIGATION OF NOTCH STR	72-0208
PARAMETRISCHE	PARAMETRISCHE KERBSPANNUNGSUNTERSUCHUNG AM ELASTISCHEN KERN. (PARAMETRICAL INVESTIGATION OF NOTCH STR	72-0208
PARAMETRISCHE	PARAMETRISCHE KERBSPANNUNGSUNTERSUCHUNGEN AN DER LOCHSCHEIBE MIT DER METHODE DER FINITEN ELEMENTE. (	75-0603
PARAMETRISCHE	PARAMETRISCHE UNTERSUCHUNGEN VON KERBEN MIT DER MATRIZENVERSCHIEBUNGSMETHODE UND BILDSCHIRMUNTERSTUT	72-0638
PARASITIC	AVOIDANCE OF PARASITIC SHEAR IN PLANE ELEMENT	75-0383
PARAWING	STRUCTURAL ANALYSIS OF A PARAWING DURING DEPLOYMENT	72-0560
PARENCHYMA	THEORY FOR DISTORTION STUDIES OF LUNG PARENCHYMA BASED ON ALVEOLAR MEMBRANE PROPERTIES	74-0222
PARFAIT	COUPLAGE ENTRE L ECOULEMENT AUTOUR D UN ARRIERE-CORPS ET LE JET PROPULSIF EN THEORIE DE FLUIDE PARFA	74-1137
PART-INVERSION	ROLE OF PART-INVERSION IN FLUID-STRUCTURE PROBLEMS WITH MIXED VARIABLES	70-0209
PARTIAL	APPLICATIONS OF VARIATIONAL EQUATIONS TO ORDINARY AND PARTIAL DIFFERENTIAL EQUATIONS - MULTIPLE SOLU	75-0010
PARTIAL	BUCKLING OF PLATE GIRDER WEBS UNDER PARTIAL EDGE LOADINGS	70-0283
PARTIAL	FINITE ELEMENT AND FINITE DIFFERENCE ENERGY TECHNIQUES FOR THE NUMERICAL SOLUTION OF PARTIAL DIFFERE	73-0570
PARTIAL	FINITE ELEMENT METHOD FOR HYDRODYNAMIC DISPERSION EQUATION WITH MIXED PARTIAL DERIVATIVES	72-0030
PARTIAL	FINITE ELEMENT SOLUTION OF PARTIAL DIFFERENTIAL EQUATIONS FEMAC COMPUTER PROGRAM	73-0648
PARTIAL	FINITE ELEMENT SOLUTIONS OF SELECTED PARTIAL DIFFERENTIAL EQUATIONS "FEMAC COMPUTER PROGRAM"	74-0327
PARTIAL	METHODS FOR SOLUTION OF PARTIAL DIFFERENTIAL EQUATIONS	73-0521
PARTIAL	NUMERICAL ANALYSIS ON STRESS DISTRIBUTION IN PARTIAL COMPRESSED SPECIMEN (HARDBOARD) BY FINITE ELEME	72-0920
PARTIAL	NUMERICAL SOLUTION OF PARTIAL DIFFERENTIAL EQUATIONS. 11	70-0072
PARTIAL	NUMERICAL SOLUTION OF SOME NONSTATIONARY PARTIAL DIFFERENTIAL EQUATIONS: THE FINITE ELEMENT METHOD A	72-0930
PARTIAL	NUMERICAL SOLUTION OF PARTIAL DIFFERENTIAL EQUATIONS	73-0968
PARTIAL	ON THE DISCRETIZATION OF THE TIME VARIABLE FOR PARABOLIC PARTIAL DIFFERENTIAL EQUATIONS	75-0103
PARTIAL	PIECEWISE HERMITE INTERPOLATION IN ONE AND TWO VARIABLES WITH APPLICATIONS TO PARTIAL DIFFERENTIAL E	68-0201
PARTIAL	PIECEWISE POLYNOMIAL INTERPOLATIONS AND THEIR APPLICATIONS TO PARTIAL DIFFERENTIAL EQUATIONS	70-0458
PARTIAL	PROBLEM CONTROLLED GRID GENERATION FOR THE NUMERICAL SOLUTION OF PARTIAL DIFFERENTIAL EQUATIONS	75-0725
PARTIAL	THE FINITE ELEMENT METHOD IN PARTIAL DIFFERENTIAL EQUATIONS	74-0456
PARTIAL	THE MATHEMATICAL FOUNDATIONS OF THE FINITE ELEMENT METHOD - WITH APPLICATIONS TO PARTIAL DIFFERENTIA	73-0286
PARTIAL	THEORY AND EXAMPLES OF PARTIAL APPROXIMATION IN FINITE-ELEMENT METHOD	75-0353
PARTIAL	VARIABLE GRID FINITE DIFFERENCE-ELEMENT SOLUTION OF ELLIPTIC PARTIAL DIFFERENTIAL EQUATIONS	74-0862
PARTIAL	VARIATIONAL PRINCIPLES AND FINITE ELEMENT METHOD IN PARTIAL DIFFERENTIAL EQUATIONS	71-0452
PARTIALLY	EARTHQUAKE ANALYSIS OF AXISYMMETRIC TOWERS PARTIALLY SUBMERGED IN WATER	75-0076
PARTIALLY	FREE VIBRATION OF CYLINDRICAL SHELLS PARTIALLY FILLED WITH LIQUID	71-0839
PARTIALLY	GALERKIN FINITE ELEMENT SOLUTIONS FOR POLLUTION PROBLEMS IN PARTIALLY MIXED ESTUARIES	74-0047
PARTIALLY	GROUNDWATER FLOW IN PARTIALLY SATURATED SOILS	72-0745
PARTIALLY	INTERNAL PRESSURE EFFECTS ON THE VIBRATION OF PARTIALLY FILLED ELASTIC TANKS	75-0241
PARTIALLY	ON A SPECTRAL PROBLEM IN VIBRATION MECHANICS: COMPUT. ION OF ELASTIC TANKS PARTIALLY FILLED WITH LIQ	75-0267
PARTIALLY	VIBRATION CHARACTERISTICS OF THIN, PRESSURIZED SHELLS OF REVOLUTION PARTIALLY FILLED WITH LIQUID	72-0604
PARTIALLY	VIBRATIONS OF TANKS PARTIALLY FILLED WITH LIQUIDS	74-0089
PARTICLE	ANALYSIS OF THE MUTUAL INDUCTANCE PARTICLE VELOCIMETER	74-0712
PARTICLE	DIFFUSION ON SOLID SPHERICAL PARTICLE IN VISCOUS FLUID FLOW AT FINITE VALUES OF RE NUMBERS	69-0419
PARTICLE	FINITE AND DISCONNECTED SUBGROUPS OF SU(3) AND THEIR APPLICATION TO THE ELEMENTARY PARTICLE SPECTRUM	66-0129
PARTICLE	THREE-DIMENSIONAL FINITE ELEMENT ANALYSIS OF SPHERICAL PARTICLE COMPOSITES	74-0137
PARTICLEBOARD	PREDICTION OF ELASTIC CONSTANTS OF PARTICLEBOARD	74-0259
PARTICULAR	A SURVEY OF THE APPLICATION OF FINITE ELEMENT METHODS TO STRESS ANALYSIS PROBLEMS WITH PARTICULAR EM	71-0632
PARTICULAR	ANALYSIS OF NONLINEAR PROBLEMS IN ROCK MECHANICS WITH PARTICULAR REFERENCE TO JOINTED ROCK SYSTEMS	70-0282
PARTICULAR	CURVED THICK SHELL AND MEMBRANE ELEMENTS WITH PARTICULAR REFERENCE TO AXISYMMETRIC PROBLEMS	68-0005
PARTICULAR	EXPERIENCES WITH ORTHOGONAL POLYNOMIALS AND "BEST" NUMERICAL INTEGRATION FORMULAS ON A TRIANGLE: WIT	74-0977
PARTICULAR	HYBRID FINITE ELEMENT ANALYSIS WITH PARTICULAR REFERENCE TO AXISYMMETRIC STRUCTURES	70-0248
PARTICULAR	MIXED-DISPLACEMENT FINITE-ELEMENT ANALYSIS WITH PARTICULAR APPLICATION USING PLANE-STRESS TRIANGLES	72-0245
PARTICULAR	PROPELLER-EXCITED VIBRATION WITH PARTICULAR REFERENCE TO FULL-SCALE MEASUREMENTS	75-0018
PARTICULAR	SOME CREEP EFFECTS IN STRESS ANALYSIS WITH PARTICULAR REFERENCE TO CONCRETE PRESSURE VESSELS	69-0284
PARTICULAR	STOCHASTIC DYNAMIC RESPONSE ANALYSIS OF OFFSHORE PLATFORMS, WITH PARTICULAR REFERENCE TO GRAVITY-TYP	75-0220
PARTICULAR	STRESSES IN ANISOTROPIC MEDIA WITH PARTICULAR REFERENCE TO PROBLEMS OF ROCK MECHANICS	66-0079
PARTICULAR	THE SOLUTION OF STRUCTURAL EQUILIBRIUM EQUATIONS BY THE CONJUGATE GRADIENT METHOD WITH PARTICULAR RE	71-0268
PARTICULAR	THE USES OF FINITE ELEMENT TECHNIQUES IN SOIL MECHANICS WITH PARTICULAR REFERENCE TO DEFORMATION PLA	72-0002
PARTICULAR	VIBRATION OF THICK, CURVED, SHELLS WITH PARTICULAR REFERENCE TO TURBINE BLADES	70-0017
PARTICULAR	WEIGHTED RESIDUAL PROCESSES IN FINITE ELEMENT WITH PARTICULAR REFERENCE TO SOME TRANSIENT AND COUPLE	71-0503
PARTICULATE	DEFORMATION AND FRACTURE OF PARTICULATE COMPOSITE	72-0221
PARTICULATE	SLOW PARTICULATE VISCOUS FLOW IN CHANNELS AND TUBES. APPLICATION TO BIOMECHANICS	71-0248
PARTITIONING	A " LOCAL " BASIS OF GENERALIZED SPLINES OVER RIGHT TRIANGLES DETERMINED FROM A NONUNIFORM PARTITION	72-0525
PARTITIONING	SOLVING A CLASS OF LARGE SPARSE LINEAR SYSTEMS OF EQUATIONS BY PARTITIONING	73-0580
PARTS	FINEL. UNIVERSELLES PROGRAMMSYSTEM ZUR BERECHNUNG DER ELASTISCHEN EIGENSCHAFTEN VON MASCHINENBAUTEIL	72-0132
PARTS	NUMERISCHE BERECHNUNG INSTATIONAERER TEMPERATURFELDER UND THERMISCH BEDINGTER VERFORMUNGEN VON WERKZ	75-0847
PARTS	OPREDELENIE STATSIONARNYKH TEMPERATURNYKH POLEI V DETALYAKH DVIGATELEI VNUTRENNEGO SGORANIYA METODOM	73-0454
PAS-1	USE OF PAS-1 FINITE ELEMENT COMPUTER SYSTEM FOR STATIC AND DYNAMIC ANALYSIS OF 2-DIMENSIONAL AND 3-D	75-0070
PASOV	VYPOCET KRUHOVYCH LOMENICOVYCH KONSTRUKCII METODOU KONECNYCH PASOV. (CALCULATION OF CIRCULAR FOLDED-	74-0612
PASSAGE	FINITE ELEMENT SOLUTION OF VISCOUS FLOW IN AN ARBITRARY INTERNAL PASSAGE	71-0461
PASSAGE	ON THE DEVELOPMENT OF THE "PASSAGE" PROGRAM	73-0342
PATCH	EXPERIENCE WITH THE PATCH TEST FOR CONVERGENCE OF FINITE ELEMENTS	72-0365
PATCH	FAILURE OF THIN-WALLED MEMBERS UNDER PATCH LOADINGS	72-0230
PATCH	LARGE DEFLECTION OF ELASTIC PLATES UNDER PATCH LOADING	72-0120
PATCH	NON-CONFORMING FINITE ELEMENT METHODS AND THE PATCH TEST	73-0640
PATCH	SURVEY OF UNDERLYING THEORY, APPROXIMATION BY PIECEWISE POLYNOMIALS, ILLEGAL ELEMENTS AND THE PATCH	73-0635
PATCH	THE PATCH TEST	74-0339
PATCH	THE SUPERPATCH THEOREM AND OTHER PROPOSITIONS RELATING TO THE PATCH TEST	75-0026
PATCH	VARIATIONAL PRINCIPLES AND THE PATCH TEST	74-1282
PATHOGENESIS	ADAPTABILITY OF COMPUTER GRAPHICS TO STUDIES OF ATHEROSCLEROSIS PATHOGENESIS	72-0206
PATHS	FINITE ELEMENT METHOD FOR THE CALCULATION OF LOCUS PATHS IN DYNAMICALLY LOADED BEARINGS	73-0487
PATTERN	PATTERN LOADING ON REINFORCED CONCRETE FLAT PLATES	71-0290
PATTERNS	COMPUTER PROGRAMS FOR PLOTTING FINITE ELEMENT PATTERNS OF TWO-AND THREE-DIMENSIONAL STRUCTURES	70-0416
PATTERNS	DESIGN OF ROOF BOLT PATTERNS FOR JOINTED ROCK	74-1318
PATTERNS	DIGITAL COMPUTATION OF STRESSES AND DEFLECTIONS IN A BOX BEAM - A PERFORMANCE COMPARISON BETWEEN FIN	69-0355
PATTERNS	ERROR ANALYSIS OF VARIOUS FINITE ELEMENT PATTERNS	73-0510
PATTERNS	FINITE ELEMENT ANALYSIS OF PERFORATED PLATES CONTAINING TRIANGULAR PENETRATION PATTERNS OF 5 AND 10	75-0216
PATTERNS	FINITE ELEMENT ANALYSIS OF PERFORATED PLATES CONTAINING TRIANGULAR PENETRATION PATTERNS OF 5 AND 10	75-0520
PATTERNS	NASTRAN-GAP, AN INTEGRATED APPROACH TO THE ANALYSIS OF RADIATION PATTERNS OF DISTORTED REFLECTORS	71-0062
PATTERNS	SOME EFFECTS OF SYSTEM IDEALIZATIONS, SINGULARITIES AND MESH PATTERNS ON FINITE ELEMENT SOLUTIONS	74-0939
PAUSCHALE	UEBER PAUSCHALE FEHLERSCHRANKEN FUER APPROXIMATIONEN VON SCHALENVERFORMUNGEN. (GLOBAL BOUNDARY CONDI	75-0292
PAVEMENT	A SUMMARY OF DISCRETE-ELEMENT METHODS OF ANALYSIS FOR PAVEMENT SLABS	72-0674

PAVEMENT	AIRCRAFT-PAVEMENT INTERACTION STUDIES, PHASE 1: A FINITE ELEMENT MODEL OF A JOINTED CONCRETE PAVEMEN	73-0894
PAVEMENT	DYNAMIC RESPONSE OF MODEL PAVEMENT STRUCTURE	72-0226
PAVEMENT	EXPERIMENTAL AND THEORETICAL INVESTIGATION OF PAVEMENT DEFLECTIONS	69-0154
PAVEMENT	EXPERIMENTAL VERIFICATION OF DISCRETE-ELEMENT SOLUTIONS FOR PLATES AND PAVEMENT SLABS	70-0411
PAVEMENT	FINITE ELEMENT ANALYSIS OF PAVEMENT STRUCTURES USING AFPAV CODE LINEAR ELASTIC ANALYSIS	73-0658
PAVEMENT	FINITE ELEMENT ANALYSIS OF PILE-SUPPORTED PAVEMENT STRUCTURES	74-0157
PAVEMENT	FINITE ELEMENT ANALYSIS OF PILE-SUPPORTED PAVEMENT STRUCTURES	75-0145
PAVEMENT	FINITE-ELEMENT ANALYSIS OF CONCRETE SLABS AND ITS IMPLICATIONS FOR RIGID PAVEMENT DESIGN	73-0538
PAVEMENT	FINITE-ELEMENT ANALYSIS OF PAVEMENT STRUCTURES USING AFPAV CODE (NONLINEAR ELASTIC ANALYSIS)	75-0748
PAVEMENT	FRACTURE (ULTIMATE STRENGTH) ANALYSES OF ASPHALT PAVEMENT LAYERS RESULTING FROM TRAFFIC LOADING	73-0548
PAVEMENT	LAYERED PAVEMENT SYSTEMS. PART 1. LAYERED SYSTEM DESIGN. PART 11. FATIQUE OF PLAIN CONCRETE	72-0867
PAVEMENT	MODIFICATION OF LAYERED THEORY FOR PAVEMENT DESIGN ANALYSIS	73-0574
PAVEMENT	PARAMETRIC STUDIES OF LAYERED PAVEMENT SYSTEMS USING AFPAV CODE WITH FAST EQUATION SOLVER	75-0747
PAVEMENT	PAVEMENT CRACKING IN WEST TEXAS DUE TO FREEZE-THAW CYCLING	75-0276
PAVEMENT	PAVEMENT DESIGN AND CONTINUOUSLY REINFORCED CONCRETE PAVEMENT PERFORMANCE	73-0749
PAVEMENT	PAVEMENT DESIGN AND CONTINUOUSLY REINFORCED CONCRETE PAVEMENT PERFORMANCE	73-0749
PAVEMENT	RIGID PAVEMENT STRESSES UNDER AIRCRAFT LOADING	71-0423
PAVEMENTS	AFPAV COMPUTER CODE FOR STRUCTURAL ANALYSIS OF AIRFIELD PAVEMENTS	75-0684
PAVEMENTS	AN ADVANCED ANALYSIS OF RIGID PAVEMENTS	72-0417
PAVEMENTS	DESIGN METHOD FOR FLEXIBLE AIRFIELD PAVEMENTS	74-1174
PAVEMENTS	EFFECT OF NONLINEAR MATERIAL RESPONSE ON THE BEHAVIOUR OF PAVEMENTS UNDER TRAFFIC	70-0171
PAVEMENTS	EFFECT OF OPENINGS ON STRESSES IN RIGID PAVEMENTS	73-0344
PAVEMENTS	EVALUATION OF RIGID PAVEMENTS BY NONDESTRUCTIVE TESTS.	72-0459
PAVEMENTS	FATIGUE CRACK FORMATION AND PROPAGATION IN PAVEMENTS CONTAINING SOIL-CEMENT BASES.	72-0435
PAVEMENTS	FINITE ELEMENT ANALYSIS OF PAVEMENTS	68-0058
PAVEMENTS	NONLINEAR THEORY FOR PREDICTING THE PERFORMANCE OF FLEXIBLE HIGHWAY PAVEMENTS	70-0143
PAVEMENTS	STRESS - DEFORMATION PREDICTION IN CEMENT - TREATED SOIL PAVEMENTS	71-0286
PAVEMENTS	VERIFICATION OF FRACTURE MECHANICS CONCEPTS TO PREDICT CRACKING OF FLEXIBLE PAVEMENTS	73-0772
PAVING	ANALYSIS OF FATIGUE OF PAVING MIXTURES FROM THE FRACTURE MECHANICS VIEWPOINT	72-0540
PAYLOAD	METHOD FOR ESTIMATING RESPONSE OF PAYLOAD SECONDARY STRUCTURES TO RANDOM EXCITATION	68-0232
PCPS	SAFE PCPS, A COMPUTER PROGRAM FOR THE STRESS ANALYSIS OF COMPOSITES BODIES OF REVOLUTION. INPUT INST	65-0013
PCPV	EXPERIMENTAL AND THEORETICAL STUDY ON A 1/10TH SCALE MODEL PCPV	73-0829
PCPV	STRESS DISTRIBUTION IN THE LIGAMENT ZONE OF PERFORATED CIRCULAR DISCS WITH SPECIAL REFERENCE TO THE	73-0827
PCPV	THIN-WALLED PRESTRESSED CONCRETE PRESSURE VESSEL FOR HIGH TEMPERATURE REACTORS: EXPERIMENTAL INVESTI	73-0857
PCPV'S	UEBER DAS VERHALTEN VON SPANNEBETONDRUCKBEHAELTRERN IM GRENZTRAGFAEHIGKEITSZUSTAND UND DESSEN EINFUL	73-0855
PCPVS	THE CALCULATION OF TIME-DEPENDENT STRESSES IN PCPVS BY AN APPROXIMATE METHOD	71-0086
PCR	ITERATIVE TECHNIQUE FOR THE STRESS ANALYSIS OF PCR STRUCTURES USING THE FINITE ELEMENT METHOD	64-0026
PCRV	ULTIMATE LOAD BEHAVIOUR OF PCRV TOP CLOSURES. THEORETICAL AND EXPERIMENTAL INVESTIGATIONS	73-0835
PDP-15	GIFTS SYSTEM FOR THE PDP-15 AND OTHER CONSIDERATIONS IN DEVELOPING GRAPHICS AND CAPABILITIES FOR FIN	73-0671
PEAK	EFFECTS OF A PEAK OVERLOAD ON FATIGUE CRACK PROPAGATION RATE-PLANE STRESS ANALYSIS OF CRACKED PLATE	74-1365
PEC	VERIFICATION OF THE PEC CONTAINMENT VESSEL FOR SOME OPERATING CONDITIONS BY USE OF THE FINITE ELEMEN	74-0883
PELLET	FINITE ELEMENT STRESS ANALYSIS OF INTERACTING FUEL PELLET AND CANNING UNDER FAST REACTOR CONDITION	73-0814
PELLET	FUEL ROD DEFORMATIONS - A TIME DEPENDENT FINITE ELEMENT ANALYSIS ON AXISYMMETRIC FUEL RODS, INCLUDIN	70-0414
PENALISATION-DU	SUR L APPROXIMATION, PAR ELEMENTS FINIS D ORDRE UN, ET LA RESOLUTION, PAR PENALISATION-DUALITE, D UN	75-0845
PENALIZATION-DU	SUR L APPROXIMATION, PAR ELEMENTS FINIS D ORDRE UN, ET LA RESOLUTION, PAR PENALISATION-DUALITE, D UN	75-0845
PENALTY	A PENALTY FUNCTION APPROACH TO THE MINIMIZATION OF QUADRATIC FUNCTIONALS IN FINITE ELEMENT ANALYSIS	74-0418
PENALTY	APPROXIMATION BY FIRST ORDER FINITE ELEMENTS AND SOLUTION BY PENALTY AND DUALITY OF NONLINEAR DIRICH	74-0245
PENALTY	CONSTRAINED VARIATIONAL PRINCIPLES AND PENALTY FUNCTION METHODS IN FINITE ELEMENT ANALYSIS	72-0551
PENALTY	FINITE ELEMENT METHOD WITH PENALTY	71-0512
PENALTY	NONCONFORMING ELEMENTS IN FINITE ELEMENT METHOD WITH PENALTY	73-0056
PENALTY	PENALTY FUNCTION APPROACH TO PROBLEMS OF PLASTIC FLOW OF METALS WITH LARGE SURFACE DEFORMATIONS	75-0282
PENALTY	THE FINITE ELEMENT METHOD WITH PENALTY	75-0967
PENALTY-DUALITY	FINITE ELEMENT APPROXIMATION, AND SOLUTION BY A PENALTY-DUALITY ALGORITHM OF AN ELASTO-PLASTIC PROBL	75-0041
PENETRATION	A FINITE ELEMENT STUDY OF THE PENETRATION OF A SOIL HALF-SPACE BY A RIGID PROJECTILE	69-0518
PENETRATION	BIT PENETRATION INTO ROCK - A FINITE ELEMENT STUDY	75-0800
PENETRATION	FINITE ELEMENT ANALYSIS OF PERFORATED PLATES CONTAINING TRIANGULAR PENETRATION PATTERNS OF 5 AND 10	75-0216
PENETRATION	FINITE ELEMENT ANALYSIS OF PERFORATED PLATES CONTAINING TRIANGULAR PENETRATION PATTERNS OF 5 AND 10	75-0520
PENETRATION	PENETRATION IN SHELLS UNDER EXTERNAL PRESSURE	70-0237
PENETRATION	STRESS CONCENTRATION FACTORS AROUND A PENETRATION IN A STRIP OR SLAB OF FINITE DIMENSIONS	69-0409
PENETRATIONS	STRESS CONCENTRATION FACTORS FOR CIRCULAR, REINFORCED PENETRATIONS IN PRESSURIZED CYLINDRICAL SHELLS	74-0726
PENETRATIONS	THREE DIMENSIONAL STRESS ANALYSIS OF PENETRATIONS IN THICK WALLED PRESSURE VESSELS	74-0422
PENNSYLVANIA	ROCK MECHANICS STUDIES FOR A HIGH CUT AT A NUCLEAR POWER STATION IN PENNSYLVANIA	70-0122
PENSTOCK	STRUCTURAL ANALYSIS OF SYMMETRICAL PENSTOCK BIFURCATIONS BY THE FINITE ELEMENT METHOD	71-0310
PERCOLATION	METHODS OF CALCULATION OF THREE-DIMENSIONAL PROBLEMS OF PERCOLATION OF FISSURED ROCK BY FINITE ELEME	70-0472
PERCOLATION	WECHSELWIRKUNG ZWISCHEN DEFORMATION UND DURCHSTROMUNG IM KLUEFTIGEN, ANISOTROPEN GEBIRGE. (INTERACT	72-0612
PEREKTYTII	APPLICATION OF THE METHOD OF FINITE ELEMENTS TO THE CALCULATION OF SHIP STRUCTURES (ISPOLZOVANIE MET	73-0933
PEREMESHCHENII	RASCHET NAPRYAZHENIL L PEREMESHCHENII V SVODE PODZEMNOGO MASHINNOGO ZALA GES I V OKRUZHAYUSHCHEM SKA	74-0542
PERFECT	SHAKEDOWN THEORY IN PERFECT ELASTOPLASTICITY WITH ASSOCIATED AND NON ASSOCIATED FLOW LAWS - A FINITE	69-0172
PERFECTLY	FINITE ELEMENT FORMULATION OF STRESS STRAIN MATRIX FOR AN ELASTICALLY PERFECTLY PLASTIC TRESCA MATER	70-0275
PERFORATED	FINITE ELEMENT ANALYSIS OF PERFORATED SHELLS	72-0228
PERFORATED	FINITE ELEMENT ANALYSIS OF PERFORATED PLATES CONTAINING TRIANGULAR PENETRATION PATTERNS OF 5 AND 10	75-0216
PERFORATED	FINITE ELEMENT ANALYSIS OF PERFORATED PLATES CONTAINING TRIANGULAR PENETRATION PATTERNS OF 5 AND 10	75-0520
PERFORATED	SHEAR BUCKLING OF SQUARE PERFORATED PLATES	74-0713
PERFORATED	STRESS DISTRIBUTION IN THE LIGAMENT ZONE OF PERFORATED CIRCULAR DISCS WITH SPECIAL REFERENCE TO THE	73-0827
PERFORATED	STRUCTURAL ANALYSIS OF A THREE-DIMENSIONAL PERFORATED PLATE BY THE FINITE ELEMENT METHOD WITH TEST C	74-0945
PERFORATED	TWO-DIMENSIONAL TIME-DEPENDENT ANALYSIS OF PERFORATED END CAPS FOR NUCLEAR REACTOR PRESSURE VESSELS	71-0239
PERFORATION	EFFECT OF PERFORATION DAMAGE ON WELL PRODUCTIVITY	74-0936
PERFORMANCE	ANALYTICAL PERFORMANCE OF STEEL, ALUMINUM AND PLASTIC FOR A NEW DESIGN HIGH PRESSURE THIN-WALLED CAR	74-0756
PERFORMANCE	DIGITAL COMPUTATION OF STRESSES AND DEFLECTIONS IN A BOX BEAM - A PERFORMANCE COMPARISON BETWEEN FIN	69-0355
PERFORMANCE	EFFECTS OF SHOE FORCE GEOMETRY ON HEAVY DUTY INTERNAL SHOE BRAKE PERFORMANCE	68-0219
PERFORMANCE	EVALUATION OF STRESS CELL PERFORMANCE	74-0664
PERFORMANCE	FINITE ELEMENTS WITH RELAXED CONTINUITY - AN INVESTIGATION OF THE VARIATIONAL BASIS AND NUMERICAL PE	74-1065
PERFORMANCE	NONLINEAR THEORY FOR PREDICTING THE PERFORMANCE OF FLEXIBLE HIGHWAY PAVEMENTS	70-0143
PERFORMANCE	NUMERICAL PERFORMANCE OF A FINITE ELEMENT GALERKIN PROCEDURE	75-0022
PERFORMANCE	PAVEMENT DESIGN AND CONTINUOUSLY REINFORCED CONCRETE PAVEMENT PERFORMANCE	73-0749
PERFORMANCE	PERFORMANCE OF AN EMBANKMENT CONSTRUCTED ON VARVED CLAY	73-0743
PERFORMANCE	PERFORMANCE OF CULVERT UNDER WINSCAR DAM	75-0638
PERFORMANCE	PERFORMANCE OF EARTH AND EARTH-SUPPORTED STRUCTURES	72-0449
PERFORMANCE	PERFORMANCE OF STRUCTURAL WOOD MEMBERS EXPOSED TO FIRE	75-0060
PERFORMANCE	PERFORMANCE OF TIED-BACK WALLS IN CLAY	74-0665
PERFORMANCE	POSSIBILITY OF USING FINITE ELEMENT METHOD IN THE ANALYSIS OF JOINT PERFORMANCE OF THE HYDRAULIC STR	72-0418
PERFORMANCE	SUPPORT PERFORMANCE OF HYDRAULIC BACKFILL	74-0257
PERFORMING	BESTRAN: A TECHNIQUE FOR PERFORMING STRUCTURAL ANALYSES	75-0759
PERIODIC	FINITE ELEMENT STUDY OF HARMONIC WAVE PROPAGATION IN PERIODIC STRUCTURES	74-0130
PERIODIC	MATRIX DIFFERENCE EQUATION ANALYSIS OF VIBRATING PERIODIC STRUCTURES	75-0151
PERIODIC	NON-LINEAR RESPONSE OF A BEAM TO PERIODIC LOADING.	72-0407
PERIODIC	PERIODIC HEAT CONDUCTION IN A TWO PHASE, TWO DIMENSIONAL SOLID DOMAIN	74-1164
PERIODIC	PERIODIC SHOCK EXCITATION OF ELASTIC STRUCTURES	69-0163
PERIODIC	RANDOM RESPONSE OF PERIODIC STRUCTURES BY A FINITE-ELEMENT TECHNIQUE	75-0426
PERIODIC-STRUCT	FINITE-ELEMENT ANALYSIS OF PERIODIC-STRUCTURES	73-1129
PERIODS	NATURAL PERIODS OF INFILLED FRAMES	72-0122
PERISTALTIC	A FINITE ELEMENT APPROACH TO PHYSIOLOGICAL TRANSPORT PHENOMENA BY CILIARY AND PERISTALTIC PUMPING	72-0353

PERISTALTIC	ANALYSIS OF PERISTALTIC PUMPING	72-0044
PERMAFROST	FINITE ELEMENT PERMAFROST THAW SETTLEMENT MODEL	74-0543
PERMAFROST	REFINED HEAT CONDUCTION FINITE ELEMENT FOR PERMAFROST ANALYSIS	73-0594
PERMAFROST	TWO-DIMENSIONAL HEATED PIPELINE IN PERMAFROST PIPELINES	74-0541
PERMANENT	FINITE-ELEMENT SOLUTIONS FOR DEVICES WITH PERMANENT MAGNETS	75-0464
PERMEABILITY	SEEPAGE INTO AN EXCAVATION IN A MEDIUM POSSESSING STRESS-DEPENDENT PERMEABILITY	72-0629
PERTURBATION	A COMPUTER PROGRAMME FOR ELASTIC PLASTIC CALCULATIONS WITH DUAL PERTURBATION EXPANSION AND VARIATION	75-0882
PERTURBATION	COMPARATIVE PERTURBATION STUDIES OF THE ELASTICA	69-0255
PERTURBATION	ELASTO-VISCOPLASTIC FINITE ELEMENT ANALYSIS BY PERTURBATION METHOD	74-1271
PERTURBATION	ELLIPTIC BOUNDARY VALUE PROBLEMS WITH NONVARIATIONAL PERTURBATION AND THE FINITE ELEMENT METHOD	73-1022
PERTURBATION	FINITE ELEMENT PERTURBATION ANALYSIS OF NONLINEAR DYNAMIC RESPONSE OF ELASTIC CONTINUUA	74-0452
PERTURBATION	FINITE ELEMENT SOLUTION OF NONLINEAR STRUCTURES BY PERTURBATION TECHNIQUE	74-1273
PERTURBATION	NUMERICAL ANALYSES OF DEFORMABILITY TESTS IN JOINTED ROCK. ' JOINT PERTURBATION ' AND ' NO TENSION '	71-0439
PERTURBATION	PERTURBATION ERRORS IN FINITE ELEMENT METHOD	72-0277
PERTURBATION	PERTURBATION METHOD IN THE ANALYSIS OF GEOMETRICALLY NON LINEAR AND STABILITY PROBLEMS	72-0321
PERTURBATION	PERTURBATION PROCEDURES IN NONLINEAR FINITE ELEMENT ANALYSIS	74-1241
PERTURBATION	PERTURBATION VARIATIONAL SOLUTION OF COMPRESSIBLE FLOWS	74-0025
PERTURBATION-VA	PERTURBATION-VARIATIONAL SOLUTION OF COMPRESSIBLE FLOWS	74-1088
PERTURBATIONS	ON SMALL VIBRATIONS AND PERTURBATIONS OF FLEXIBLE BODIES UNDERGOING ARBITRARY NOMINAL MOTION	70-0376
PERTURBATIONS	THE PROBLEMS OF PERTURBATIONS IN THE FINITE ELEMENT METHOD	72-0505
PERTURBED	A DERIVATION OF PERTURBED EQUATIONS OF MOTION AND HEAT CONDUCTION OF A THERMOELASTIC SOLID	72-0735
PERTURBED	FINITE ELEMENT ANALYSIS OF PERTURBED COMPRESSIBLE FLOW	72-0001
PETROCHEMICAL	STRUCTURAL ANALYSIS OF PETROCHEMICAL EQUIPMENT BY THE FINITE ELEMENT METHOD	73-0377
PETROLEUM	ERROR BOUNDS FOR SEMIDISCRETE GALERKIN APPROXIMATIONS OF PARABOLIC PROBLEMS WITH APPLICATIONS TO PET	70-0291
PETROLEUM	FINITE ELEMENT METHOD IN PETROLEUM RESERVOIR ANALYSIS	72-0049
PETROLEUM	POROUS MEDIA - SOME PROBLEMS IN CIVIL AND PETROLEUM ENGINEERING	74-0349
PETF	DESIGN AND ANALYSIS EXPERIENCE WITH A LIQUID METAL HEAT EXCHANGER FOR PETF SERVICE	72-0772
PHASE	A STRUCTURAL EVALUATION OF RAPID METHODS OF BACKFILLING FOR BOMB DAMAGE REPAIR - PHASE 11	75-0682
PHASE	AIRCRAFT-PAVEMENT INTERACTION STUDIES, PHASE 1: A FINITE ELEMENT MODEL OF A JOINTED CONCRETE PAVEMEN	73-0894
PHASE	APPLICATION OF PHASE SPACE FINITE ELEMENTS TO THE TWO DIMENSIONAL NEUTRON TRANSPORT EQUATION IN X-Y	73-0453
PHASE	DIRECT ITERATION METHOD FOR THE INCORPORATION OF PHASE CHANGE IN FINITE ELEMENT HEAT CONDUCTION PROG	74-0894
PHASE	FINITE ELEMENT SOLUTION OF NON-LINEAR HEAT CONDUCTION PROBLEMS WITH SPECIAL REFERENCE TO PHASE CHANG	74-0348
PHASE	FINITE ELEMENTS FOR TWO PHASE FLOW IN POROUS MEDIA	74-0060
PHASE	PERIODIC HEAT CONDUCTION IN A TWO PHASE, TWO DIMENSIONAL SOLID DOMAIN	74-1164
PHASE	SINGLE-TRACK SUBWAY ENVIRONMENTAL SIMULATION MODEL. PHASE 1	71-0693
PHASE	STRUCTURAL DYNAMIC PROPERTIES OF TACTICAL MISSILE JOINTS - PHASE 1	70-0423
PHASE	STRUCTURAL DYNAMIC PROPERTIES OF TACTICAL MISSILE JOINTS - PHASE 3	74-0811
PHASE	STUDIES ON TWO PHASE COMPOSITE MATERIALS THROUGH COMPUTER SIMULATION	73-0315
PHASE	THREE-DIMENSIONAL ELASTIC-PLASTIC STRESS AND STRAIN ANALYSIS FOR FRACTURE MECHANICS. PHASE 1. SIMPLE	70-0188
PHASE	TRANSPORT SOLUTIONS USING FINITE ELEMENTS IN SPACE-ANGLE PHASE SPACE	71-0480
PHASE-SPACE	APPLICATION OF PHASE-SPACE FINITE ELEMENTS TO ONE DIMENSIONAL NEUTRON TRANSPORT EQUATION	73-0123
PHASE-SPACE	SPACE FINITE-ELEMENTS 2-DIMENSIONAL MULTIGROUP TRANSPORT CALCULATIONS USING PHASE-SPACE FINITE-ELEME	75-0504
PHASE-SPACE	TWO-DIMENSIONAL TRANSPORT CALCULATIONS USING PHASE-SPACE FINITE ELEMENTS	72-0027
PHASES	ANALYSE NUMERIQUE D'UN PROBLEME DE STEFAN A DEUX PHASES PAR UNE METHODE D'ELEMENTS FINIS. (NUMERICAL	75-0387
PHENOLIC	BASELINE SOLUTIONS FOR THE SMOOTH WALL THERMOCHEMICAL ABLATION RESPONSE OF GRAPHITE AND CARBON PHENO	71-0728
PHENOLIC	STRESS-WAVE PROPAGATION AND DYNAMIC EFFECTS IN SHELLS (SPADES) MICROMECHANICAL STRESS-WAVE RESPONSE	72-0851
PHENOMENOLOGY	STRUCTURE- MEDIUM INTERACTION PHENOMENOLOGY	72-0863
PHENOMENON	ANALYSIS OF QUICK SAND PHENOMENON BY FINITE ELEMENT METHOD	73-0191
PHENOMENON	ON THE ASYMPTOTICALLY SPHERICAL DEFORMATIONS OF ARBITRARY MEMBRANES OF REVOLUTION FIXED ALONG AN EDG	71-0640
PHILOSPHY	THIN-WALLED PRESTRESSED CONCRETE PRESSURE VESSEL FOR HIGH TEMPERATURE REACTORS: EXPERIMENTAL INVESTI	73-0857
PHOTO-ELASTIC	NUMERICAL ANALYSIS OF SHIP STRUCTURAL ELEMENTS BY MEANS OF FINITE ELEMENT METHOD AND PHOTO-ELASTIC E	69-0125
PHOTO-ELASTIC-C	CRUCIFORM SPECIMENS FOR BIAXIAL FATIGUE TESTS - AN INVESTIGATION USING FINITE-ELEMENT ANALYSIS AND P	71-0795
PHOTOELASTIC	A DYNAMIC PHOTOELASTIC ANALYSIS OF DYNAMIC TEAR TEST SPECIMEN	74-0736
PHOTOELASTIC	AN INVESTIGATION OF MICRO-BUCKLING USING PHOTOELASTIC AND FINITE-ELEMENT METHODS	72-0960
PHOTOELASTIC	CALCULATING STRESSES IN A PRESSURE VESSEL WITH FILL TUBE - COMPARISON OF THE FINITE ELEMENT AND PHOT	72-0782
PHOTOELASTIC	PHOTOELASTIC AND FINITE ELEMENT ANALYSIS OF EMBANKMENTS CONSTRUCTED OVER SOFT SOILS	70-0151
PHOTOELASTIC	PHOTOELASTIC AND FINITE ELEMENT INVESTIGATION OF A NONSYMMETRICAL PLUG HATCH CONFIGURATION	73-0025
PHOTOELASTIC	PHOTOELASTIC AND FINITE-ELEMENT ANALYSIS OF A QUADRIPARTITE VAULT	73-0119
PHOTOELASTIC	PHOTOELASTIC AND FINITE ELEMENT STRESS ANALYSIS OF A RESTORED AXISYMMETRIC FIRST MOLAR	73-0142
PHOTOELASTIC	PHOTOELASTIC DETERMINATION OF THE STRESSES IN THE FOUNDATION OF THE NUCLEAR-POWER STATION BUGEY 2	75-0834
PHOTOELASTIC	PHOTOELASTIC INVESTIGATION OF STRESS CONCENTRATIONS IN SPHERE-CYLINDER TRANSITION REGIONS: INCLUDING	68-0258
PHOTOELASTIC	PHOTOELASTIC INVESTIGATION OF STRESS CONCENTRATIONS IN SPHERE-CYLINDER TRANSITION REGIONS: INCLUDING	68-0258
PHOTOELASTIC	PHOTOELASTIC INVESTIGATION ON THE CRACK-ARREST CAPABILITY OF A HOLE	72-0086
PHOTOELASTIC	PHOTOELASTIC INVESTIGATION ON THE CRACK-ARREST CAPABILITY OF A PRETENSIONED STIFFENED PLATE	75-0088
PHOTOELASTIC	PHOTOELASTIC TESTS ON COMPLEX INTERSECTIONS OF REINFORCED THIN-SHELL CYLINDERS	75-0363
PHOTOELASTIC	STRESS ANALYSIS IN HULL STRUCTURAL MEMBERS BY PHOTOELASTIC EXPERIMENTS AND FINITE ELEMENT METHODS -	70-0341
PHOTOELASTIC	THREE-DIMENSIONAL PHOTOELASTIC AND FINITE-ELEMENT ANALYSIS OF A PROPELLANT GRAIN	72-0031
PHOTOELASTIC	TWO-DIMENSIONAL PHOTOELASTIC STUDY OF STEAM-TURBINE CASING FLANGES WITH INTERFACE RELIEF	74-0603
PHOTOELASTICIME	STRESS SEPARATION IN PHOTOELASTICIMETRIC MEASUREMENT BY THE FINITE ELEMENT METHOD	71-0807
PHOTOELASTICITY	DETERMINATION OF STRESSES AND DEFORMATIONS ON PISTONS BY MEANS OF COMPUTER PROGRAMS AND PHOTOELASTIC	75-0553
PHOTOELASTICITY	STRESS ANALYSIS OF FIRST MOLARS WITH FULL CROWN PREPARATIONS BY THREE-DIMENSIONAL PHOTOELASTICITY AND	72-0939
PHOTOGRAMMETRY	FINITE ELEMENT METHOD FOR BLOCK ADJUSTMENT PROBLEMS OF PHOTOGRAMMETRY	73-0264
PHOTOGRAPHY	INTERFERENCE-FIT FASTENER DISPLACEMENT MEASUREMENT BY SPECKLE PHOTOGRAPHY	74-1178
PHYSICAL	AN ANALYSIS OF THE PHYSICAL BEHAVIOR OF NONWOVEN FIBROUS MATERIAL BY THE FINITE-ELEMENT METHOD	73-1149
PHYSICAL	COMPUTATIONAL TECHNIQUES AS AN AID IN PHYSICAL METALLURGY	71-0554
PHYSICAL	COMPUTER PROGRAMS FOR SIMULATING PHYSICAL PHENOMENA IN THREE SPACE DIMENSIONS AND TIME	74-0866
PHYSICAL	NUMERICAL AND PHYSICAL MODELLING OF REINFORCEMENT SYSTEMS FOR TUNNELS IN JOINTED ROCK	73-0872
PHYSICAL	NUMERICAL AND PHYSICAL MODELLING	74-0393
PHYSICAL	ON THE PHYSICAL MEANINGS OF THE MATRICES DERIVED FROM FINITE ELEMENT METHOD OF HEAT CONDUCTION PROBL	73-0297
PHYSICAL	TECHNICAL, PHYSICAL AND MATHEMATICAL PRINCIPLES OF THE FINITE ELEMENT METHOD	71-0676
PHYSICALLY	STATICKE RESENI FYZIKALNE NELINEARNIHO CHOVANI ZELEZOBETO- NOVYCH DESEK (ANALYSIS OF PHYSICALLY NONL	74-0514
PHYSICALLY	THE FINITE ELEMENT ANALYSIS OF PHYSICALLY AND KINEMATICALLY NONLINEAR ELASTIC SOLIDS	69-0146
PHYSICS	FINITE ELEMENT APPLICATIONS IN MATHEMATICAL PHYSICS	72-0067
PHYSICS	FINITE ELEMENT METHOD APPLIED TO REACTOR PHYSICS PROBLEMS	71-0483
PHYSICS	FINITE ELEMENT METHODS IN REACTOR PHYSICS ANALYSIS	75-0020
PHYSICS	ON COMPLEMENTARY-DUAL VARIATIONAL PRINCIPLES IN MATHEMATICAL PHYSICS	74-0342
PHYSICS	ON THE APPLICATION OF THE FINITE ELEMENT METHOD IN REACTOR PHYSICS	75-0122
PHYSICS	THE FINITE ELEMENT METHOD IN MATHEMATICAL PHYSICS	73-0289
PHYSICS	THE HYPERCIRCLE IN MATHEMATICAL PHYSICS	57-0001
PHYSIOLOGICAL	A FINITE ELEMENT APPROACH TO PHYSIOLOGICAL TRANSPORT PHENOMENA BY CILIARY AND PERISTALTIC PUMPING	72-0353
PIAN	COMMENTS ON DERIVATION OF ELEMENT STIFFNESS MATRICES BY T. H. H. PIAN	65-0060
PIAN	THE RECTANGULAR PLANE STRESS ELEMENT BY TURNER, PIAN AND WILSON	74-1265
PIANA	ANALISI NUMERICA DEL COMPORTAMENTO ELASTICO DI UNA STRUTTURA BIDIMENSIONALE PIANA. (NUMERICAL ANALYS	74-1029
PIANI	PROCESS FOR THE AUTOMATIC GENERATION OF FLAT TRIANGULAR LATTICES FOR APPLICATION TO THE FINITE ELEME	71-0366
PICTURES	PRODUCTION OF MOTION PICTURES OF A CONTINUOUSLY DEFORMING STRUCTURE USING NASTRAN	71-0223
PIECE-WISE	PIECE-WISE CONTINUOUS FINITE ELEMENT FOR THE STRESS ANALYSIS OF PLATES AND SHELLS	75-0758
PIECES	APPLICATION OF FINITE ELEMENT TECHNIQUES TO POST-YIELD ANALYSIS OF PROPOSED STANDARD 3 POINT BEND FR	74-0118
PIECEWISE	AUTOMATIC GENERATION OF TRIANGULATIONS FOR PIECEWISE POLYNOMIAL APPPOXIMATION	73-1118
PIECEWISE	COMPARISON OF EXTRAPOLATION TECHNIQUES WITH HIGH ORDER FINITE ELEMENT METHODS FOR DIFFUSION EQUATION	73-0586
PIECEWISE	PIECEWISE ANALYTIC INTERPOLATION AND APPROXIMATION IN TRIANGULATED POLYGONS	72-0508
PIECEWISE	PIECEWISE HERMITE INTERPOLATION IN ONE AND TWO VARIABLES WITH APPLICATIONS TO PARTIAL DIFFERENTIAL E	68-0201
PIECEWISE	PIECEWISE POLYNOMIAL INTERPOLATION AND APPROXIMATION	65-0058

317

PIECEWISE	PIECEWISE POLYNOMIAL INTERPOLATIONS AND THEIR APPLICATIONS TO PARTIAL DIFFERENTIAL EQUATIONS	70-0458
PIECEWISE	PIECEWISE POLYNOMIALS AND FINITE ELEMENT METHOD	73-0041
PIECEWISE	PIECEWISE POLYNOMIAL INTERPOLATIONS IN THE FINITE ELEMENT METHOD	73-1028
PIECEWISE	RAYLEIGH-RITZ APPROXIMATION OF PIECEWISE CUBIC POLYNOMIALS	66-0087
PIECEWISE	SURVEY OF UNDERLYING THEORY. APPROXIMATION BY PIECEWISE POLYNOMIALS, ILLEGAL ELEMENTS AND THE PATCH	73-0635
PIECEWISE	THE PIECEWISE LINEAR ANALYSIS OF TWO CONNECTED STRUCTURES INCLUDING THE EFFECT OF CLEARANCE AT THE C	71-0379
PIECEWISE	TWO TYPES OF PIECEWISE QUADRATIC SPACES AND THEIR ORDER OF ACCURACY FOR POISSON'S EQUATION	72-0520
PIERS	LOAD TRANSFER MECHANISM FOR PIERS SOCKETTED IN HARD SOILS OR ROCK	74-0969
PIERS	STABILITY OF MASONRY PIERS	75-0248
PIEUX	ETUDE DE LA CAPACITE PORTANTE DES PIEUX PAR LA METHODE DES ELEMENTS FINIS. INFLUENCE DE LA DILATANCE	75-0624
PIEUX	ETUDE THEORIQUE DU COMPORTEMENT DES PIEUX SOUS CHARGE VERTICALE. INTRODUCTION DE LA DILATANCE. (THEO	75-0309
PIEUX	LA CAPACITE PORTANTE DES PIEUX. (BEARING CAPACITY OF PILES)	75-0463
PIEZO-ELECTRIC	VIBRATIONAL RESPONSE OF SONAR TRANSDUCERS USING PIEZO-ELECTRIC FINITE ELEMENTS	74-0522
PIEZOELECTRIC	ANALYSIS AND REDESIGN OF SONAR PIEZOELECTRIC RESONATORS USING THE FINITE ELEMENT METHOD.	72-0374
PIEZOELECTRIC	APPLICATION OF THE FINITE ELEMENT METHOD TO THE EXTENSIONAL VIBRATIONS OF PIEZOELECTRIC PLATES	72-0073
PIEZOELECTRIC	FINITE ELEMENT ANALYSIS OF PIEZOELECTRIC ELASTIC WAVEGUIDES	73-0115
PIEZOELECTRIC	FINITE ELEMENT ANALYSIS OF THE PIEZOELECTRIC TOPOGRAPHIC WAVEGUIDE	74-1111
PIEZOELECTRIC	FINITE ELEMENT FORMULATION FOR PIEZOELECTRIC CONTINUA BY A VARIATIONAL THEOREM	68-0230
PIEZOELECTRIC	FINITE ELEMENT METHOD FOR PIEZOELECTRIC VIBRATION	70-0012
PIEZOELECTRIC	FINITE-ELEMENT MODELING OF PIEZOELECTRIC CERAMIC HYDROPHONES	75-0595
PIEZOELECTRIC-C	A FINITE-ELEMENT APPROACH TO THE ANALYSIS OF TANGENTIALLY POLARIZED PIEZOELECTRIC-CERAMIC FREE-FLOOD	74-0802
PILE	A GENERAL METHOD OF ANALYSIS FOR TWO-DIMENSIONAL PILE GROUPS	71-0298
PILE	FIELD INSTRUMENTATION FOR LONG TERM MEASUREMENT OF PILE LOAD AND RAFT CONTACT PRESSURE	73-0150
PILE	FINITE ELEMENT ANALYSIS OF THE EFFECTS OF INSTALLATION ON PILE LOAD-SETTLEMENT BEHAVIOUR	74-0763
PILE	FINITE ELEMENT STUDIES OF PILE DRIVER TUNNELS INCLUDING CONSIDERATIONS OF SUPPORT REQUIREMENTS	72-0679
PILE	LOAD DEFORMATION ANALYSIS OF DEEP PILE FOUNDATION	72-0931
PILE	THREE-DIMENSIONAL FINITE ELEMENT ANALYSIS OF VERTICALLY LOADED PILE GROUPS	75-0239
PILE-SUPPORTED	ANALYSIS OF PILE-SUPPORTED GRAVITY LOCK	74-0585
PILE-SUPPORTED	FINITE ELEMENT ANALYSIS OF PILE-SUPPORTED PAVEMENT STRUCTURES	74-0157
PILE-SUPPORTED	FINITE ELEMENT ANALYSIS OF PILE-SUPPORTED PAVEMENT STRUCTURES	75-0145
PILED-RAFT	OBSERVATIONS ON THE BEHAVIOUR OF A PILED-RAFT FOUNDATION ON LONDON CLAY.	73-0421
PILES	BENTONITE PILES IN DURBAN	75-0797
PILES	EFFECT OF THE VARIATION OF CONSISTENCY IN RESIDUAL SOILS ON LOAD TRANSFER BY BORED PILES	75-0812
PILES	ETUDE DE LA CAPACITE PORTANTE DES PIEUX PAR LA METHODE DES ELEMENTS FINIS. INFLUENCE DE LA DILATANCE	75-0624
PILES	ETUDE THEORIQUE DU COMPORTEMENT DES PIEUX SOUS CHARGE VERTICALE. INTRODUCTION DE LA DILATANCE. (THEO	75-0309
PILES	FINITE ELEMENT ANALYSIS OF DRIVEN PILES	74-1105
PILES	LA CAPACITE PORTANTE DES PIEUX. (BEARING CAPACITY OF PILES)	75-0463
PILES	LATERAL RESISTANCE AND DEFLECTION OF VERTICAL PILES	73-0912
PILES	NUMERICAL DESIGN-ANALYSIS FOR PILES IN SANDS	74-0333
PILES	SETTLEMENT ANALYSIS OF DRIVEN PILES	75-0319
PILES	SOIL RESISTANCE MOVEMENT RELATIONSHIPS FOR LATERALLY LOADED PILES IN CLAYS AT GREAT DEPTHS	70-0162
PILLAR	FINITE ELEMENT ANALYSIS OF SALT PILLAR MODELS	68-0291
PILLAR	FINITE ELEMENT ANALYSIS OF CREEP IN ROCK-SALT PILLAR MODELS	72-0935
PILLARS	THE USE OF ROCK BOLTS OR WIRE ROPE TO INCREASE THE STRENGTH OF FRACTURED MODEL PILLARS	71-0722
PIN	FINITE-ELEMENT STRESS ANALYSIS OF A NON-LINEAR PROBLEM. A CONNECTING-ROD EYE LOADED BY MEANS OF A PI	70-0239
PIN	SOME MODELS FOR THE ANALYSIS OF STRESSES IN A TUBULAR FUEL PIN FOR HIGH TEMPERATURE REACTORS BY FINI	73-0816
PINCH	NUMERICAL STUDY OF EIGEN OSCILLATIONS OF A SCREW PINCH WITH FINITE ELEMENTS	74-0159
PINCH	STUDY OF THE MHD INSTABILITIES OF CYLINDRICAL BELT PINCH USING THE FINITE ELEMENT METHOD	74-0700
PING	PLANFORM INPUT GENERATOR (PING), A NASTRAN PRE-PROCESSOR FOR LIFTING SURFACES-THEORETICAL DEVELOPMEN	73-0776
PING	PLANFORM INPUT GENERATOR (PING), A NASTRAN PROCESSOR FOR LIFTING SURFACES-THEORETICAL DEVELOPMENT. U	73-0780
PINNED	A FORTRAN PROGRAM FOR CALCULATING THE FREQUENCIES AND MODE SHAPES OF A GUIDED CLAMPED, GUIDED PINNED	72-0890
PINNED	A FORTRAN PROGRAM FOR CALCULATING THE FREQUENCIES AND MODE SHAPES OF A GUIDED CLAMPED, GUIDED PINNED	72-0890
PINNED	FREQUENCY COEFFICIENTS OF A SPHERICAL SHELL FOR VARYING THICKNESS AND OPENING ANGLE WITH GUIDED PINN	72-0595
PINNED	FREQUENCY COEFFICIENTS OF A SPHERICAL SHELL FOR VARYING THICKNESS AND OPENING ANGLE WITH GUIDED PINN	72-0595
PINS	FAST REACTOR PROGRAMME. FINITE ELEMENT TEMPERATURE ANALYSIS IN FUEL PINS	73-0939
PIPE	ANALYSIS OF STRESSES IN PRESSURIZED WELDED PIPE IN THE CREEP RANGE	71-0404
PIPE	ELASTIC-PLASTIC-CREEP ANALYSIS OF THERMAL RATCHETTING IN STRAIGHT PIPE AND COMPARISONS WITH TEST RES	73-0393
PIPE	FINITE ELEMENT ANALYSIS OF THE UNSTEADY FLOW OF A VISCOUS PLASTIC FLUID IN A CYLINDRICAL PIPE	74-0055
PIPE	PLASTIC COLLAPSE LOADS FOR PIPE ELBOWS USING INELASTIC ANALYSIS	73-0941
PIPE	PLASTIC COLLAPSE LOADS FOR PIPE ELBOWS USING INELASTIC ANALYSIS	74-0507
PIPE	PROPAGATION AND OPENING OF A THROUGH CRACK IN PIPE SUBJECT TO COMBINED CYCLIC THERMOMECHANICAL LOADI	75-0516
PIPE	SIMPLIFIED INELASTIC (PLASTIC AND CREEP) ANALYSIS OF PIPE ELBOWS SUBJECTED TO INPLANE AND OUT-OF-PLA	75-0654
PIPE	STRUCTURAL MODEL STUDIES OF LARGE PIPE BIFURCATION	73-0283
PIPE-NOZZLE	THREE-DIMENSIONAL FINITE ELEMENT ANALYSIS OF THICK-WALLED PIPE-NOZZLE JUNCTIONS WITH CURVED TRANSITI	71-0153
PIPELINE	OPTIMIZATION OF TENSION LEVEL AND STINGER LENGTH FOR OFFSHORE PIPELINE INSTALLATION	74-0504
PIPELINE	PIPELINE STRESS ANALYSIS	75-0078
PIPELINE	TWO-DIMENSIONAL HEATED PIPELINE IN PERMAFROST PIPELINES	74-0541
PIPELINES	ELASTIC PLASTIC AND CREEP ANALYSIS OF PIPELINES BY FINITE ELEMENTS	73-0419
PIPELINES	NUMERICAL SOLUTIONS FOR DETERMINING WAVE-INDUCED PRESSURE DISTRIBUTIONS AROUND BURIED PIPELINES	74-1334
PIPELINES	TWO-DIMENSIONAL HEATED PIPELINE IN PERMAFROST PIPELINES	74-0541
PIPES	FINITE-ELEMENT ANALYSIS OF HEAT FLOW AROUND BURIED PIPES	73-1153
PIPES	INELASTIC BUCKLING ANALYSIS OF PIPES SUBJECTED TO INTERNAL PRESSURE, FLEXURE AND AXIAL LOADING	74-0515
PIPES	INVESTIGATION OF SOIL-STRUCTURE INTERACTION: FLEXIBLE CULVERT PIPES	72-0661
PIPES	STRESS ANALYSIS OF CURVED PIPES WITH END RESTRAINTS	75-0667
PIPES	STRESS DISTRIBUTIONS AROUND SHALLOW BURIED RIGID PIPES	74-0142
PIPES	STRESSES AROUND FLEXIBLE ELLIPTIC PIPES	73-0215
PIPING	EFFICIENT COMPUTER ANALYSIS OF LARGE THREE-DIMENSIONAL PIPING SYSTEMS BY THE FINITE ELEMENT METHOD	72-0378
PIPING	INELASTIC RESPONSE OF NUCLEAR PIPING SUBJECTED TO RUPTURE FORCES	75-0515
PIPING	INFLUENCE OF END-EFFECTS ON STRESSES AND FLEXIBILITY OF A PIPING ELBOW WITH IN-PLANE MOMENT	74-0534
PIPING	LOW CYCLE FATIGUE STRENGTH OF PIPING COMPONENTS	73-0366
PIPING	SHOCK LOADS ON PIPING SYSTEMS	72-0760
PIPING	SIMPLIFIED INELASTIC HIGH TEMPERATURE STRUCTURAL ANALYSIS OF MODERATELY COMPLEX SPATIALLY THREE DIME	74-0445
PIPING	SPECIAL STRUCTURAL ELEMENTS FOR PIPING ANALYSIS	74-0502
PIPING	THREE-DIMENSIONAL THERMAL AND STRESS ANALYSIS OF A PIPING TEE	75-0703
PISTON	FINITE-ELEMENT STRESS ANALYSIS OF A MULTIPIECE PISTON	69-0274
PISTON	KOLBENKUEHLUNG (PISTON COOLING)	75-0266
PISTON	SOME DESIGN ASPECTS OF IC ENGINE PISTON HEADS	73-0578
PISTON	STRESS ANALYSIS OF SELF-SEALING TYPE PISTON HEAD BY FINITE-ELEMENT METHOD	72-0259
PISTON	STRESS ANALYSIS OF THE "SELF-SEALING" TYPE PISTON HEAD BY THE FINITE ELEMENT METHOD	72-0617
PISTONS	DETERMINATION OF STRESSES AND DEFORMATIONS ON PISTONS BY MEANS OF COMPUTER PROGRAMS AND PHOTOELASTIC	75-0553
PISTONS	ECONOMICAL DESIGN ANALYSIS OF PISTONS FOR INTERNAL COMBUSTION ENGINES	73-0188
PIT	COMPUTER PROGRAM FOR PIT SLOPE STABILITY ANALYSIS BY THE FINITE ELEMENT STRESS ANALYSIS AND LIMITING	72-0225
PIT	ROCK MECHANICS AND RISK IN OPEN PIT MINING	75-0544
PITCHING	FINITE ELEMENT ANALYSIS AND COMPUTER GRAPHICS VISUALIZATION OF FLOW AROUND PITCHING AND PLUNGING AIR	73-0385
PLACID	PLACID: A GENERAL FINITE ELEMENT COMPUTER PROGRAM FOR STRESS ANALYSIS OF PLANE AND AXISYMMETRIC SOLI	74-0884
PLAIN	DETERMINATION OF VOLUME CHANGE STRESSES IN PLAIN AND REINFORCED CONCRETE USING FINITE ELEMENT ANALYS	69-0477
PLAIN	LAYERED PAVEMENT SYSTEMS. PART 1. LAYERED SYSTEM DESIGN. PART 11. FATIQUE OF PLAIN CONCRETE	72-0867
PLAIN	THE USES OF FINITE ELEMENT TECHNIQUES IN SOIL MECHANICS WITH PARTICULAR REFERENCE TO DEFORMATION PLA	75-0002
PLAIN	USER'S GUIDE FOR THE COMPUTER PROGRAM DESCRIBED IN TAM REPORT NO. 680 "DETERMINATION OF VOLUME CHANG	69-0473
PLAN	CELLULAR STRUCTURES OF ARBITRARY PLAN GEOMETRY	72-0201
PLAN	EXPERIMENTAL AND THEORETICAL INVESTIGATION OF THE BEHAVIOR OF BOX GIRDERS CURVED IN PLAN	75-0659
PLAN	FAMILLE D ELEMENTS FINIS DE COQUE PLAN CINEMATIQUEMENT ADMISSIBLES POUR L'ANALYSE DES STRUCTURES	74-0488

PLAN	FINITE ELEMENT ANALYSIS OF BRIDGES CURVED IN PLAN	71-0547
PLANAR	AUTOMATIC TRIANGULATION OF ARBITRARY PLANAR DOMAINS FOR THE FINITE ELEMENT METHOD	74-0480
PLANAR	EFFECT OF TRANSVERSE SHEAR DEFORMATION IN VIBRATIONS OF PLANAR STRUCTURES COMPOSED OF BEAM TYPE ELEM	73-0402
PLANAR	ELASTIC NONLINEAR FINITE ELEMENT ANALYSIS OF PLANAR STRUCTURES	71-0830
PLANAR	FINITE ELEMENT ANALYSIS OF PLANAR MICROWAVE NETWORKS	73-0016
PLANAR	GENERAL FINITE ELEMENT VIBRATIONAL ANALYSIS OF PLANAR MECHANISMS	74-1107
PLANAR	INTERACTIVE PROCEDURE FOR CURVILINEAR COORDINATION OF A PLANAR REGION WITH APPLICATIONS TO FINITE-EL	73-0040
PLANAR	LINEAR BUCKLING AND GEOMETRICALLY NONLINEAR ANALYSIS OF PLANAR PLATE-STIFFENER TYPE STRUCTURES BY TH	74-1097
PLANAR	NONLINEAR ANALYSIS OF PLANAR REINFORCED CONCRETE STRUCTURES	74-0533
PLANAR	PLANAR AND CURVED SHELL ELEMENTS	69-0186
PLANAR	PLANAR FINITE ELEMENT AT RANDOM TEMPERATURE	73-0140
PLANAR	SHELL ANALYSIS USING PLANAR TRIANGULAR ELEMENTS	69-0041
PLANAR	STRONGLY PLANAR AC-LATTICES IN WHICH IDEAL OF FINITE-ELEMENTS IS STANDARD	75-0816
PLANAR	TRIPLET: A TWO-DIMENSIONAL, MULTIGROUP, TRIANGULAR MESH, PLANAR GEOMETRY, EXPLICIT TRANSPORT CODE	73-0955
PLANAR-FLOW	A FINITE-ELEMENT, PLANAR-FLOW MODEL OF CAMAS PRAIRIE, IDAHO	72-0980
PLANE	A " LOCAL " BASIS OF GENERALIZED SPLINES OVER RIGHT TRIANGLES DETERMINED FROM A NONUNIFORM PARTITION	72-0525
PLANE	A COMPUTER METHOD FOR THE SECOND ORDER ELASTIC ANALYSIS OF PLANE FRAMED STRUCTURES	73-0310
PLANE	A DATA-GENERATION PROCEDURE FOR THE FINITE ELEMENT ANALYSIS OF PLANE FRAMES AND GRILLAGES	68-0179
PLANE	A DISCRETE ELEMENT METHOD FOR ANALYSIS OF PLANE ELASTIC-PLASTIC STRESS PROBLEMS	65-0063
PLANE	A DISCRETE ELEMENT METHOD FOR THE ANALYSIS OF PLANE ELASTO-PLASTIC STRESS PROBLEMS	66-0062
PLANE	A FINITE ELEMENT ANALYSIS OF THE PLANE STRAIN SHEET DRAWING PROBLEM	71-0812
PLANE	A HYBRID-ELEMENT APPROACH TO CRACK PROBLEMS IN PLANE ELASTICITY	73-0569
PLANE	A LATTICE ANALOGY FOR THE SOLUTION OF PLANE STRESS	43-0002
PLANE	A METHOD OF ELASTIC-PLASTIC PLANE STRESS AND PLANE STRAIN ANALYSIS	66-0114
PLANE	A METHOD OF ELASTIC-PLASTIC PLANE STRESS AND PLANE STRAIN ANALYSIS	66-0114
PLANE	A QUADRILATERAL PLANE STRESS FINITE ELEMENT WITH QUADRATIC PROPERTIES	72-0969
PLANE	A STUDY OF THE PLANE STRESS OR STRAIN FINITE ELEMENT ANALYSIS FOR SOLUTION OF STRESS DISTRIBUTION IN	69-0487
PLANE	A STUDY OF THE PLANE STRESS OR STRAIN FINITE ELEMENT ANALYSIS FOR SOLUTION OF STRESS DISTRIBUTION IN	69-0487
PLANE	ALGORITHM FOR THE LARGE DEFLECTION GEOMETRICALLY NON-LINEAR BEHAVIOUR OF PLANE AND CURVED STRUCTURES	72-0083
PLANE	AN APPLICATION OF THE FINITE ELEMENT METHOD TO ELASTIC-PLASTIC PROBLEMS OF PLANE STRESS	70-0031
PLANE	AN AUTOMATIC MESH GENERATION SCHEME FOR PLANE AND CURVED SURFACES BY ISOPARAMETRIC COORDINATES	71-0271
PLANE	AN EVALUATION OF FINITE ELEMENTS WITH IMPROVED ASSURED CONVERGENCE FOR PLANE STRESS AND PLANE STRAIN	74-0429
PLANE	AN EVALUATION OF FINITE ELEMENTS WITH IMPROVED ASSURED CONVERGENCE FOR PLANE STRESS AND PLANE STRAIN	74-0429
PLANE	AN INVESTIGATION INTO THE ECONOMICS OF USING COMPLEX ELEMENTS IN PLANE STRESS ANALYSIS	74-0427
PLANE	ANALISI NUMERICA DEL COMPORTAMENTO ELASTICO DI UNA STRUTTURA BIDIMENSIONALE PIANA. (NUMERICAL ANALYS	74-1029
PLANE	ANALYSES OF AXISYMMETRIC UPSETTING AND PLANE STRAIN SIDE PRESSING OF SOLID CYLINDERS BY THE FINITE E	70-0083
PLANE	ANALYSIS OF ELASTIC-PLASTIC BEHAVIOR OF BEAMS AND COLUMNS BY PLANE FINITE ELEMENT MODEL	73-1073
PLANE	ANALYSIS OF PLANE STRAIN CONSOLIDATION IN SATURATED CLAYS BY FINITE ELEMENT METHOD	71-0815
PLANE	ANALYSIS OF PLANE STRAIN TEST USING THE FINITE ELEMENT METHOD	72-1002
PLANE	ANISOTROPIC WORK HARDENING IN METALS. FINITE ELEMENT APPLICATION TO PLANE STRESS AND PLANE STRAIN	73-0950
PLANE	ANISOTROPIC WORK HARDENING IN METALS. FINITE ELEMENT APPLICATION TO PLANE STRESS AND PLANE STRAIN	73-0950
PLANE	APPLICATION OF FINITE ELEMENT METHOD TO PLANE STRAIN EXTRUSION PROCESSES	74-1243
PLANE	APPLICATION OF THE DIRECT STIFFNESS METHOD TO PLANE PROBLEMS INVOLVING LARGE, TIME-DEPENDENT DEFORMA	66-0108
PLANE	APPLICATION TO PLANE STRESS AND PLANE STRAIN--LINEAR ELASTIC MATERIAL	68-0048
PLANE	APPLICATION TO PLANE STRESS AND PLANE STRAIN--LINEAR ELASTIC MATERIAL	68-0048
PLANE	APPLICATIONS OF FINITE-ELEMENT ANALYSIS TO NONLINEAR PLANE STRESS PROBLEMS	72-0976
PLANE	APPLICATIONS OF TRILINEAR COORDINATES TO SOME PROBLEMS IN PLANE ELASTICITY	72-0728
PLANE	AVOIDANCE OF PARASITIC SHEAR IN PLANE ELEMENT	75-0383
PLANE	BOUNDING PLANE STRESS SOLUTIONS BY FINITE ELEMENTS	66-0055
PLANE	BOUSSINESQ PROBLEM OF PLANE MICROPOLAR ELASTICITY	74-0197
PLANE	BRLESC FINITE ELEMENT PROGRAM FOR AXISYMMETRIC, PLANE STRAIN, AND PLANE STRESS, ORTHOTROPIC SOLIDS W	71-0749
PLANE	BRLESC FINITE ELEMENT PROGRAM FOR AXISYMMETRIC, PLANE STRAIN, AND PLANE STRESS, ORTHOTROPIC SOLIDS W	71-0749
PLANE	CAPE-2D, A COMPUTER PROGRAM FOR THE STRESS ANALYSIS OF PLANE AND AXISYMMETRIC COMPOSITE STRUCTURES.	69-0083
PLANE	CHISLENNOE RESHENIE PLOSKIKH ZADACH TEPLOPROVODNOSTI DLYA OBLASTEI SLOZHNOI FORMY (NUMERICAL SOLUTIO	74-0659
PLANE	COUPLE - A PSEUDO FINITE ELEMENT CONDUCTION CODE FOR THE TRANSIENT THERMAL RESPONE OF AXISYMMETRIC O	72-0778
PLANE	CRACK TIP FINITE ELEMENTS FOR PLANE ELASTICITY	71-0764
PLANE	DERIVATION OF STIFFNESS MATRICES FOR PROBLEMS IN PLANE ELASTICITY BY GALERKIN'S METHOD	69-0250
PLANE	DETERMINATION OF UPPER BOUNDS FOR PROBLEMS IN PLANE STRESS USING FINITE ELEMENT TECHNIQUES	67-0034
PLANE	DIGITAL COMPUTER PROGRAMS FOR STEADY-STATE OR TRANSIENT TEMPERATURE ANALYSIS OF PLANE AND AXISYMMETR	72-0915
PLANE	DISCRETE-ELEMENT PLASTIC ANALYSIS OF STRUCTURES IN A STATE OF MODIFIED PLANE STRAIN	69-0140
PLANE	DISCUSSION ON "PLANE STRESS LIMIT ANALYSIS BY FINITE ELEMENTS"	71-0792
PLANE	EIN NEUES VARIATIONSPRINZIP MIT ANWENDUNG AUF SCHWINGUNGEN FLACHER SCHALEN NACH DER METHODE DER FINI	73-0563
PLANE	ELASTIC AND PLASTIC INTERLAMINAR SHEAR DEFORMATION IN LAMINATED COMPOSITES UNDER GENERALIZED PLANE	71-0158
PLANE	ELASTIC PLASTIC CRACKED PLATES IN PLANE STRAIN	69-0321
PLANE	ELASTIC-PLASTIC STRESS ANALYSIS FOR A NOTCHED BAR IN PLANE STRAIN BENDING	71-0410
PLANE	ELASTO PLASTIC ANALYSIS OF PLANE STRUCTURES BY THE FINITE ELEMENT DISPLACEMENT METHOD	70-0299
PLANE	ELASTOPLASTIC ANALYSIS OF PLANE STRAIN AND AXISYMMETRIC FLAT PUNCH INDENTATION BY THE FINITE ELEMENT	70-0343
PLANE	ELEMENT STATIC FLEXIBILITY AND STIFFNESS MATRICES FOR THE FINITE ELEMENT ANALYSIS OF BEAM, PLATE, AN	67-0070
PLANE	ENGINEERING OPERATIONS REPORT: NERVA AXISYMMETRIC AND PLANE STRUCTURES FINITE ELEMENT STRESS ANALYSI	72-0917
PLANE	EXCITATION OF PLANE THERMOELASTIC WAVES IN A LAYER WITH A FINITE RATE OF HEAT PROPAGATION	74-0803
PLANE	EXPERIMENTAL AND THEORETICAL INVESTIGATION OF PLANE STRESS FRACTURE OF 2024 - T351 ALUMINUM ALLOY	70-0147
PLANE	FEATS, A COMPUTER PROGRAM FOR THE FINITE ELEMENT THERMAL STRESS ANALYSIS OF PLANE OR AXISYMMETRIC SO	68-0152
PLANE	FINITE ELEMENT ANALYSIS OF ELASTIC-PLASTIC PLANE STRESS PROBLEMS BASED UPON TRESCA YIELD CRITERION	70-0179
PLANE	FINITE ELEMENT ANALYSIS OF NOTCHED TENSILE SPECIMENS IN PLANE STRESS	71-0557
PLANE	FINITE ELEMENT ANALYSIS OF THE PLANE STRAIN DRAWING PROBLEM	74-1154
PLANE	FINITE ELEMENT ANALYSIS OF DENTAL STRUCTURES - AXISYMMETRIC AND PLANE STRESS IDEALIZATIONS	75-0525
PLANE	FINITE ELEMENT APPROACH TO PLANE MICROELASTICITY	72-0205
PLANE	FINITE ELEMENT APPROACH TO OPTIMAL DESIGN OF PLASTIC STRUCTURES IN PLANE STRESS.	72-0397
PLANE	FINITE ELEMENT SOLUTION TO SOME PLANE PROBLEMS IN SOIL DYNAMICS	74-0448
PLANE	FINITE ELEMENT SOLUTION OF THE INTERACTION OF A PLANE ACOUSTIC BLAST WAVE AND A CYLINDRIC STRUCTURE	74-0790
PLANE	FINITE ELEMENT STRESS ANALYSIS OF PLANE AND AXISYMMETRIC STRUCTURES	69-0062
PLANE	FINITE ELEMENT STUDY OF THE ELASTIC BEHAVIOR OF PLANE FRAMES WITH FILLED WALLS	71-0828
PLANE	FINITE ELEMENT TECHNIQUES FOR ORTHOTROPIC PLANE STRESS AND ORTHOTROPIC PLATE ANALYSIS	68-0274
PLANE	FINITE ELEMENT TECHNIQUES IN PLANE STRESS PROBLEMS	74-1206
PLANE	FINITE ELEMENTE ZUR BERECHNUNG DER WAERMEUEBERTRAGUNG FUER EBENE UND AXISYMMETRISCHE PROBLEME (FINIT	73-0449
PLANE	FINITE ELEMENTS FOR THE REPRESENTATION OF PLANE AND AXISYMMETRIC STRUCTURAL REINFORCEMENT	68-0037
PLANE	FINITE PLANE STRAIN OF INCOMPRESSIBLE ELASTIC SOLIDS BY THE FINITE ELEMENT METHOD	68-0128
PLANE	FORCE AND TORQUE ANALYSIS OF CONSTRAINED SPACE MECHANISMS AND PLANE MECHANISMS WITH OFFSET LINKS BY	74-0569
PLANE	GENERAL PURPOSE PROGRAM OF PLANE STRESS ANALYSIS BY FINITE ELEMENT METHOD, AND ITS APPLICATIONS	72-0186
PLANE	HIGH PRECISION FINITE ELEMENTS FOR PLANE ELASTIC PROBLEMS	72-0862
PLANE	HIGHER-ORDER FINITE ELEMENT FOR PLANE STRESS	67-0082
PLANE	HIGHER-ORDER FINITE ELEMENT FOR PLANE STRESS	68-0089
PLANE	HYBRID FINITE ELEMENTS IN THE PLANE STRESS PROBLEM	74-1071
PLANE	HYBRID-ELEMENT APPROACH TO CRACK PROBLEMS IN PLANE ELASTICITY	73-0508
PLANE	ISSLEDOVANIE PROCHNOSTI I ZHESTKOSTI PLASTIN, NAGRUZHENNYKH V SVOEI PLOSKOSTI, METHODOM KONECHNYKH E	73-0382
PLANE	LOWER BOUND LIMIT ANALYSIS OF PLANE STRESS PROBLEMS BY A FINITE ELEMENT METHOD	70-0428
PLANE	NONLINEAR IN-PLANE BENDING OF A PLANE STRESS-STRAIN FINITE ELEMENT MODEL	74-0211
PLANE	ON SOLUTION OF PLANE STRAIN CONSOLIDATION PROBLEMS BY FINITE ELEMENT METHOD	71-0532
PLANE	ON THE SOLUTION OF PLANE STRESS PROBLEMS BY FINITE ELEMENTS COMPUTER PROGRAMS	73-0978
PLANE	ON TRESCA YIELD CRITERION IN FINITE ELEMENT ANALYSIS OF PLANE STRESS PROBLEMS	73-0314
PLANE	ORTHOTROPIC FLAT PLATES SUPPORTED ON COLUMNS ELONGATED IN PLANE	72-0492
PLANE	OSCILLATION STABILITY OF INTERNALLY AND EXTERNALLY NOTCHED PLANE BARS OF EQUAL FORM FACTOR FROM 3943	73-0745

PLANE	PLACID: A GENERAL FINITE ELEMENT COMPUTER PROGRAM FOR STRESS ANALYSIS OF PLANE AND AXISYMMETRIC SOLI	74-0884
PLANE	PLANE ELASTICITY PROBLEMS AND THE DIRECT STIFFNESS METHOD	61-0009
PLANE	PLANE FRAMEWORK METHODS FOR PLATES IN EXTENSION	66-0126
PLANE	PLANE PLASTIC DEFORMATION OF SOIL.	71-0006
PLANE	PLANE POISEUILLE FLOW OF A RAREFIED GAS BASED ON FINITE ELEMENT METHOD	71-0575
PLANE	PLANE STRAIN CONSOLIDATION BY FINITE ELEMENTS	69-0471
PLANE	PLANE STRAIN CONSOLIDATION BY FINITE ELEMENTS	70-0043
PLANE	PLANE STRESS ANALYSIS USING THE FINITE ELEMENT METHOD. TRIANGULAR FINITE ELEMENT WITH SIX PARAMETERS	67-0106
PLANE	PLANE STRESS ANALYSIS OF RUBBER AT HIGH STRAINS USING FINITE ELEMENTS	71-0680
PLANE	PLANE STRESS AND PLANE STRAIN ANALYSIS BY FINITE ELEMENT METHOD	70-0388
PLANE	PLANE STRESS AND PLANE STRAIN ANALYSIS BY FINITE ELEMENT METHOD	70-0388
PLANE	PLANE STRESS CRACK TRAJECTORIES IN LOW MODULUS MATERIALS	75-0646
PLANE	PLANE STRESS ELEMENT OF VARYING THICKNESS APPLIED TO END BLOCK ANALYSIS	74-0396
PLANE	PLANE STRESS FINITE ELEMENTS FOR NONLINEAR ANALYSIS	69-0529
PLANE	PLANE STRESS LIMIT ANALYSIS BY FINITE ELEMENTS	70-0255
PLANE	PLANE STRESS REINFORCED CONCRETE FINITE ELEMENTS	74-0108
PLANE	PLANE STRESS SHAKEDOWN ANALYSIS BY FINITE ELEMENTS	72-0266
PLANE	PRECISION OF FINITE ELEMENT METHOD IN PLANE STRESS	69-0295
PLANE	QUADRILATERAL ELEMENTS IN PLANE ANALYSIS	66-0085
PLANE	QUADRILATERAL FINITE ELEMENT IN ELASTIC-PLASTIC PLANE STRESS ANALYSIS	67-0067
PLANE	QUADRILATERAL FINITE ELEMENT FOR PLANE STRESS AND PLANE STRAIN	70-0496
PLANE	QUADRILATERAL FINITE ELEMENT FOR PLANE STRESS AND PLANE STRAIN	70-0496
PLANE	RIGID PLATE ON NONLINEARLY DEFORMABLE COHESIVE BASE (PLANE PROBLEM)	75-0264
PLANE	SAAS 111: FINITE ELEMENT STRESS ANALYSIS OF AXISYMMETRIC AND PLANE SOLIDS WITH DIFFERENT ORTHOTROPIC	71-0741
PLANE	SAFE-CREEP: A COMPUTER PROGRAM FOR THE VISCOELASTIC ANALYSIS OF AXISYMMETRIC AND PLANE CONCRETE STRU	67-0148
PLANE	SASL - A FINITE ELEMENT CODE FOR THE STATIC ANALYSIS OF AXISYMMETRIC AND PLANE SOLIDS SUBJECTED TO A	72-0779
PLANE	SASL - A FINITE ELEMENT CODE FOR THE STATIC ANALYSIS OF AXISYMMETRIC AND PLANE SOLIDS SUBJECTED TO A	72-0779
PLANE	SECTOR FINITE ELEMENTS IN THE THEORY OF PLANE ELASTICITY	71-0766
PLANE	SIZE OF PLASTIC ZONE AT TIP OF A CRACK IN PLANE STRAIN SLATE BY FINITE ELEMENT METHOD	73-0006
PLANE	SMALL SCALE YIELDING NEAR A CRACK IN PLANE STRAIN-FINITE ELEMENT ANALYSIS	71-0355
PLANE	SMALL SCALE YIELDING ANALYSIS OF MIXED MODE PLANE STRAIN CRACK PROBLEMS	73-0801
PLANE	SOLUTION OF SOME PLANE STRAIN PROBLEMS IN SOIL MECHANICS USING THE METHOD OF FINITE ELEMENTS	69-0521
PLANE	SOME INVESTIGATIONS INTO THE FINITE ELEMENT METHOD WITH SPECIAL REFERENCE TO PLANE STRESS	66-0128
PLANE	SOME PLANE QUADRILATERAL HYBRID FINITE ELEMENTS	69-0082
PLANE	STATICAL CALCULATIONS OF PLANE AND SPACE BAR STRUCTURES AS WELL AS STUMPY FRAMEWORKS USING A MODIFIE	73-0747
PLANE	STRESS DISTRIBUTION UNDER UNIAXIAL ELASTIC PLANE STRAIN COMPRESSION WITH FRICTION	74-0526
PLANE	TAPERED PLANE STRESS ANNULAR SECTOR FINITE ELEMENT	75-0538
PLANE	THE APPLICATION OF THE MATRIX DISPLACEMENT METHOD IN PLANE ELASTO-PLASTIC STRESS PROBLEMS	64-0009
PLANE	THE EFFECT OF IN PLANE STATIC STRESSES ON THE VIBRATION CHARACTERISTICS OF CURVED SANDWICH PLATES	71-0588
PLANE	THE FINITE ELEMENT IN PLANE STRESS ANALYSIS	60-0001
PLANE	THE FINITE ELEMENT METHOD IN APPLICATION TO PLANE STRESS	68-0117
PLANE	THE FINITE ELEMENT METHOD IN PLANE STRESS ANALYSIS	69-0127
PLANE	THE METHOD OF INTERFACE STRESS ADJUSTMENT AND ITS USES IN SOME PLANE ELASTICITY PROBLEMS	61-0005
PLANE	THE PLANE ISOPARAMETRIC STRUCTURAL ELEMENT	70-0262
PLANE	THE RECTANGULAR PLANE STRESS ELEMENT BY TURNER, PIAN AND WILSON	74-1265
PLANE	THE ROLE OF FINITE DEFORMATION ANALYSIS IN PLANE STRESS AND STRAIN FRACTURES	74-0322
PLANE	THE SOLUTION OF STRUCTURAL EQUILIBRIUM EQUATIONS BY THE CONJUGATE GRADIENT METHOD WITH PARTICULAR RE	71-0268
PLANE	THE STRESS ANALYSIS OF AXISYMMETRIC SHELLS USING A PLANE FRAME ANALYSIS PROGRAM	69-0238
PLANE	THERMO-ELASTIC ANALYSIS OF RECTANGULAR PLATES IN PLANE STRESS BY THE FINITE-ELEMENT DISPLACEMENT MET	67-0087
PLANE	TRIANGULATION IN THE HYPERCIRCLE METHOD FOR PLANE PROBLEMS	52-0001
PLANE	USE OF SINGULARITY PROGRAMMING IN FINITE ELEMENT CALCULATIONS OF ELASTIC STRESS INTENSITY FACTORS. P	75-0252
PLANE-STRAIN	A COMPUTER PROGRAM FOR THE ANALYSIS OF LINEARLY ELASTIC PLANE-STRESS, PLANE-STRAIN PROBLEMS	68-0110
PLANE-STRAIN	ANALYSES OF AXISYMMETRIC UPSETTING AND PLANE-STRAIN SIDE-PRESSING OF SOLID CYLINDERS BY THE FINITE E	71-0794
PLANE-STRAIN	FINITE-ELEMENT PROGRAMME FOR THE PLANE-STRAIN ANALYSIS OF RUBBER	75-0362
PLANE-STRAIN	NON-UNIFORM DEFORMATION OF A BLOCK IN PLANE-STRAIN COMPRESSION CAUSED BY FRICTION. PART 4, ANALYSIS	71-0786
PLANE-STRAIN	ON THE NON-UNIFORM DEFORMATION OF A BLOCK IN PLANE-STRAIN COMPRESSION CAUSED BY FRICTION -3	71-0328
PLANE-STRESS	A COMPUTER PROGRAM FOR THE ANALYSIS OF LINEARLY ELASTIC PLANE-STRESS, PLANE-STRAIN PROBLEMS	68-0110
PLANE-STRESS	MIXED-DISPLACEMENT FINITE-ELEMENT ANALYSIS WITH PARTICULAR APPLICATION USING PLANE-STRESS TRIANGLES	72-0245
PLANES	APPLICATION OF THE FINITE ELEMENT METHOD TO CONVECTION HEAT TRANSFER BETWEEN PARALLEL PLANES	71-0244
PLANETARY	PLANETARY EXPLORER STRUCTURAL CONCEPT EVALUATION MODEL USING NASTRAN	71-0742
PLANFORM	PLANFORM INPUT GENERATOR (PING), A NASTRAN PRE-PROCESSOR FOR LIFTING SURFACES-THEORETICAL DEVELOPMEN	73-0776
PLANFORM	PLANFORM INPUT GENERATOR (PING), A NASTRAN PROCESSOR FOR LIFTING SURFACES-THEORETICAL DEVELOPMENT, U	73-0780
PLANS	PLANS - A FINITE ELEMENT PROGRAM FOR NONLINEAR ANALYSIS OF STRUCTURES	74-0966
PLANS	PLANS - A FINITE-ELEMENT PROGRAM FOR NONLINEAR ANALYSIS OF STRUCTURES	75-0011
PLANT	ANALYSIS FOR ATOMIC POWER PLANT BY FINITE ELEMENT METHOD	71-0198
PLANT	ANALYSIS OF BEHAVIOR OF EDWARD HYATT POWER PLANT	75-0055
PLANT	APPLICATION OF FINITE ELEMENT METHOD TO THE THREE DIMENSIONAL STRESS AND VIBRATION ANALYSIS OF BWR P	71-0180
PLANT	APPLICATION OF FINITE ELEMENT METHOD TO BOILING WATER REACTOR POWER PLANT (JAPANESE)	72-0816
PLANT	BERECHNUNGSMETHODEN ZUR ERMITTLUNG VON BELASTUNGSBEDINGUNGEN UND GRENZTRAGFAEHIGKEITSVERHALTEN VON K	73-0858
PLANT	EFFECTS OF FOUNDATION ROTATION ON SEISMIC INERTIA FORCES OF NUCLEAR POWER PLANT STRUCTURES	73-0177
PLANT	EVALUATION OF TORSIONAL EFFECTS IN THE SEISMIC ANALYSIS OF SYMMETRIC AND UNSYMMETRIC NUCLEAR POWER P	73-0175
PLANT	FINITE ELEMENT TREATMENT OF SOIL-STRUCTURE INTERACTION PROBLEM FOR NUCLEAR POWER PLANT UNDER SEISMIC	73-0834
PLANT	NON-LINEAR THERMAL STRESS ANALYSIS FOR NUCLEAR POWER PLANT BY FINITE ELEMENT METHOD	71-0181
PLANT	PWR DEVELOPMENT FOR NUCLEAR PROPULSION: VIBRATION ANALYSIS OF NUCLEAR POWER PLANT	71-0797
PLANT	STRUCTURAL DESIGN OF NUCLEAR PLANT FACILITIES	73-0812
PLANTS	FINITE ELEMENT ANALYSIS IN NUCLEAR POWER PLANTS (JAPANESE)	72-0685
PLANTS	PROBABILISTIC ASSESSMENT OF AIRCRAFT HAZARD FOR NUCLEAR POWER PLANTS	72-0235
PLAQUE	QUELQUES METHODES D'ELEMENTS FINIS POUR LE PROBLEME D'UNE PLAQUE ENCASTREE	74-0406
PLAQUES	APPLICATION DE LA METHODE DES ELEMENTS FINIS AU CALCUL DES PLAQUES FLECHIES (APPLICATION OF THE FINI	73-0019
PLAQUES	APPLICATIONS DE LA METHODE DES ELEMENTS FINIS A LA FLEXION DES PLAQUES	69-0511
PLAQUES	BORNES SUPERIEURES ET INFERIEURES DANS L'ANALYSE MATRICIELLE DES PLAQUES EN FLEXION TORSION	64-0015
PLAQUES	LES PLAQUES ET LES COQUES MINCES OU EPAISSES (THIN AND THICK PLATES AND SHELLS)	71-0432
PLASMA	ANALYSIS OF HYDROMAGNETIC PLASMA STABILITY BY FINITE-ELEMENT METHOD	72-0307
PLASMA	BIDIMENSIONAL COMPUTER-PROGRAM FOR CALCULATION OF GROWTH-RATES OF INCLUDED AXIALLY SYMMETRICAL PLASM	75-0420
PLASMA	DETERMINATION OF EIGENMODES OF HOMOGENEOUS PLASMA WITH ELLIPTICAL SECTION USING BI-DIMENSIONAL FINIT	74-0553
PLASMA	EIGENMODES OF CYLINDRICAL PLASMA USING FINITE ELEMENT METHOD	74-0174
PLASMA	FINITE-ELEMENT SOLUTIONS OF PLASMA EQUILIBRIUM	75-0455
PLASMA	NUMERICAL SOLUTIONS OF NONLINEAR PLASMA EQUATIONS BY THE FINITE-ELEMENT METHOD	75-0671
PLASMA	NUMERICAL STUDY OF PROPER MODES OF A HOMOGENEOUS CYLINDRICAL PLASMA BY METHOD OF FINITE ELEMENTS	74-0160
PLASMA	SEMIDISCRETE GALERKIN TECHNIQUES WITH TIME INTERPOLATION AND SPLITTING UP FOR PLASMA SIMULATION	75-0121
PLASMA-DIFFUSIO	FINITE-ELEMENT PLASMA-DIFFUSION CODE FOR ARBITRARY TOROIDAL GEOMETRY	75-0500
PLASTIC	A COMPARISON OF CURRENT WORK-HARDENING MODELS USED IN THE ANALYSIS OF PLASTIC DEFORMATIONS	73-0769
PLASTIC	A COMPUTER PROGRAMME FOR ELASTIC PLASTIC CALCULATIONS WITH DUAL PERTURBATION EXPANSION AND VARIATION	75-0882
PLASTIC	A FINITE ELEMENT METHOD FOR THE PLASTIC BENDING ANALYSIS OF STRUCTURES	68-0021
PLASTIC	A FINITE ELEMENT METHOD FOR THE PLASTIC BUCKLING ANALYSIS OF PLATES	69-0390
PLASTIC	A FINITE-ELEMENT METHOD FOR THE PLASTIC BUCKLING ANALYSIS OF PLATES	69-0214
PLASTIC	A METHOD OF FINITE DEFLECTION ELASTIC PLASTIC ANALYSIS OF FRAMED STRUCTURES SUBJECTED TO LARGE AXIAL	73-0239
PLASTIC	A NUMERICAL METHOD FOR THE ANALYSIS OF LONGITUDINAL ELASTIC PLASTIC STRESS WAVE PROPAGATION	73-0526
PLASTIC	AN ASSUMED STRESS HYBRID FINITE ELEMENT FOR THREE-DIMENSIONAL PLASTIC STRUCTURAL ANALYSIS	74-0717
PLASTIC	ANALYSIS OF SHELLS OF ANY FORM IN THE PLASTIC REGION BY THE FINITE ELEMENT METHOD	73-0940
PLASTIC	ANALYSIS OF THE ELASTIC PLASTIC PROBLEMS BY THE MATRIX DISPLACEMENT METHOD	68-0173
PLASTIC	ANALYSIS OF THERMAL ELASTIC PLASTIC STRESS AND STRAIN DURING WELDING BY FINITE ELEMENT METHOD	71-0382

320

PLASTIC	ANALYTICAL PERFORMANCE OF STEEL, ALUMINUM AND PLASTIC FOR A NEW DESIGN HIGH PRESSURE THIN-WALLED CAR	74-0756
PLASTIC	APPLICATION OF ELASTIC PLASTIC ANALYSIS TO THE DESIGN OF PRESSURE VESSELS	68-0157
PLASTIC	APPLICATION OF FINITE ELEMENT METHODS TO LARGE SCALE ELASTIC PLASTIC PROBLEMS OF FRACTURE MECHANICS	69-0372
PLASTIC	AXISYMMETRIC SOLUTION OF ELASTO PLASTIC PROBLEMS BY FINITE ELEMENT METHODS	69-0361
PLASTIC	BOPACE 3-D (THE BOEING PLASTIC ANALYSIS CAPABILITY FOR 3-DIMENSIONAL SOLIDS USING ISOPARAMETRIC FINI	75-0699
PLASTIC	BUCKLING PROBLEMS IN THE DESIGN OF FIBREGLASS-REINFORCED PLASTIC SHIPS	72-0164
PLASTIC	CALCULATION OF RESIDUAL STRESSES IN PLASTIC DEFORMATION PROCESSES	72-0241
PLASTIC	COMPARISON OF THE CAPABILITY OF OUR HARDENING RULES TO PREDICT A MATERIAL'S PLASTIC BEHAVIOR	75-0523
PLASTIC	COMPUTER PROGRAM JET 3 TO CALCULATE THE LARGE ELASTIC PLASTIC DYNAMICALLY INDUCED DEFORMATIONS OF FR	72-0609
PLASTIC	COMPUTER PROGRAMS FOR THE PLASTIC ANALYSIS OF STRUCTURES USING DISCRETE ELEMENT METHODS	67-0004
PLASTIC	CONTINUUM CALCULATION FROM ELASTIC PLASTIC REINFORCED MATERIAL USING A FINITE ELEMENT METHOD (DIE BE	72-0658
PLASTIC	CYCLIC PLASTIC ANALYSIS OF STRUCTURAL STEEL JOINTS	73-0799
PLASTIC	DEVELOPMENT OF FINITE ELEMENT COMPUTATIONAL METHODS FOR THREE DIMENSIONAL ELASTIC PLASTIC CRACK ANAL	70-0356
PLASTIC	DISCRETE ELEMENT METHODS FOR THE PLASTIC ANALYSIS OF STRUCTURES SUBJECTED TO CYCLIC LOADING	70-0023
PLASTIC	DISCRETE-ELEMENT METHODS FOR THE PLASTIC ANALYSIS OF STRUCTURES	67-0040
PLASTIC	DISCRETE-ELEMENT PLASTIC ANALYSIS OF STRUCTURES IN A STATE OF MODIFIED PLANE STRAIN	69-0140
PLASTIC	DYNAMIC PLASTIC LARGE DEFLECTIONS OF CLAMPED RECTANGULAR PLATE	74-0392
PLASTIC	EFFECTS OF LOCAL STRENGTH VARIATIONS ON PLASTIC ENCLAVE DEVELOPMENT AT CRACK TIPS	72-0822
PLASTIC	ELASTIC AND PLASTIC ANALYSIS BY FINITE ELEMENT METHOD	68-0119
PLASTIC	ELASTIC AND PLASTIC STRESS DISTRIBUTION ON BASE PLATES IN THE NEIGHBOURHOOD OF FILLET WELDS IN LAP J	70-0175
PLASTIC	ELASTIC AND PLASTIC INTERLAMINAR SHEAR DEFORMATION IN LAMINATED COMPOSITES UNDER GENERALIZED PLANE	71-0158
PLASTIC	ELASTIC PLASTIC ANALYSIS OF PRESSURE VESSEL COMPONENTS	69-0380
PLASTIC	ELASTIC PLASTIC ANALYSIS OF THREE DIMENSIONAL PROBLEMS BY THE FINITE ELEMENT METHOD	69-0387
PLASTIC	ELASTIC PLASTIC ANALYSIS OF NOTCHED SPECIMENS USING THE FINITE ELEMENT METHOD	70-0140
PLASTIC	ELASTIC PLASTIC ANALYSIS OF SAINT-VENANT TORSION PROBLEM BY A HYBRID STRESS MODEL	72-0385
PLASTIC	ELASTIC PLASTIC ANALYSIS BY ASSUMED STRESS HYBRID MODEL	74-0443
PLASTIC	ELASTIC PLASTIC ANALYSIS OF SHALLOW FOUNDATIONS	74-0446
PLASTIC	ELASTIC PLASTIC AND CREEP ANALYSIS OF PIPELINES BY FINITE ELEMENTS	73-0419
PLASTIC	ELASTIC PLASTIC BUCKLING OF STIFFENED RECTANGULAR PLATES	69-0122
PLASTIC	ELASTIC PLASTIC CRACKED PLATES IN PLANE STRAIN	69-0321
PLASTIC	ELASTIC PLASTIC DEFORMATION OF THICK WALLED CYLINDERS	69-0427
PLASTIC	ELASTIC PLASTIC DEFORMATION OF CYLINDRICAL PRESSURE VESSELS UNDER CYCLIC LOADING	74-0604
PLASTIC	ELASTIC PLASTIC DYNAMIC ANALYSIS OF AXISYMMETRIC SOLIDS	73-0657
PLASTIC	ELASTIC PLASTIC FINITE ELEMENT ANALYSIS OF PLATES	74-0093
PLASTIC	ELASTIC, PLASTIC BEHAVIOUR OF ROTO - 1	73-0458
PLASTIC	ELASTO PLASTIC ANALYSIS OF PLANE STRUCTURES BY THE FINITE ELEMENT DISPLACEMENT METHOD	70-0299
PLASTIC	ELASTO PLASTIC STRESS IN CRACKED PLATES	65-0042
PLASTIC	EVALUATION OF THE STABILITY OF NATURAL SLOPES IN PLASTIC CHAMPLAIN CLAYS	74-0264
PLASTIC	FINITE ELEMENT ANALYSIS OF STRUCTURES IN THE PLASTIC RANGE	71-0031
PLASTIC	FINITE ELEMENT ANALYSIS OF THE UNSTEADY FLOW OF A VISCOUS PLASTIC FLUID IN A CYLINDRICAL PIPE	74-0055
PLASTIC	FINITE ELEMENT ANALYSIS OF ELASTIC PLASTIC THICK-WALLED TUBES	74-0368
PLASTIC	FINITE ELEMENT APPROACH TO OPTIMAL DESIGN OF PLASTIC STRUCTURES IN PLANE STRESS.	72-0397
PLASTIC	FINITE ELEMENT ELASTIC PLASTIC CREEP ANALYSIS OF TWO DIMENSIONAL CONTINUUM WITH TEMPERATURE DEPENDEN	73-0226
PLASTIC	FINITE ELEMENT FORMULATION OF STRESS STRAIN MATRIX FOR AN ELASTICALLY PERFECTLY PLASTIC TRESCA MATER	70-0275
PLASTIC	FINITE ELEMENT METHOD FOR THE PLASTIC BENDING ANALYSIS OF STRUCTURES	68-0206
PLASTIC	FINITE ELEMENT METHOD FOR ELASTIC PLASTIC PLATES	72-0167
PLASTIC	FLOW OF PLASTIC AND VISCO-PLASTIC SOLIDS WITH SPECIAL REFERENCE TO EXTRUSION AND FORMING PROCESSES	74-0210
PLASTIC	INFLUENCE OF NON-SINGULAR STRESS TERMS AND SPECIMEN GEOMETRY ON SMALL SCALE YIELDING AT CRACK TIPS I	73-0203
PLASTIC	INVESTIGATION OF THE STATIONARY FLOW OF A RIGID PLASTIC MATERIAL BY THE NUMERICAL FINITE ELEMENT MET	73-0231
PLASTIC	J INTEGRAL ESTIMATES FOR STRAIN HARDENING MATERIALS USING FULLY PLASTIC SOLUTIONS	74-0690
PLASTIC	NONLINEAR THERMO ELASTIC PLASTIC AND CREEP ANALYSIS BY FINITE ELEMENT METHOD	74-0092
PLASTIC	ON NUMERICALLY ACCURATE FINITE ELEMENT SOLUTIONS IN THE FULLY PLASTIC RANGE	74-0529
PLASTIC	OPTIMUM PLASTIC DESIGN FOR MULTIPLE SETS OF LOADS	74-0980
PLASTIC	PENALTY FUNCTION APPROACH TO PROBLEMS OF PLASTIC FLOW OF METALS WITH LARGE SURFACE DEFORMATIONS	75-0282
PLASTIC	PLANE PLASTIC DEFORMATION OF SOIL.	71-0006
PLASTIC	PLASTIC ANALYSIS	74-0859
PLASTIC	PLASTIC ANALYSIS OF SHELLS BY FINITE ELEMENT METHODS	73-1017
PLASTIC	PLASTIC AND VISCOPLASTIC ANALYSIS OF AXISYMMETRIC SHELLS	75-0246
PLASTIC	PLASTIC COLLAPSE LOADS FOR PIPE ELBOWS USING INELASTIC ANALYSIS	73-0941
PLASTIC	PLASTIC COLLAPSE LOADS FOR PIPE ELBOWS USING INELASTIC ANALYSIS	74-0507
PLASTIC	PLASTIC FLOW AND FAILURE OF DISCONTINUOUS FIBRE COMPOSITE MATERIALS	70-0157
PLASTIC	PLASTIC INTENSITY FACTORS FOR CRACKED PLATES	71-0164
PLASTIC	PLASTIC SANDWICH BUCKLE-SHELL STRUCTURES	72-0119
PLASTIC	PLASTIC STRAINS ASSOCIATED WITH TORSION OF NONCIRCULAR BARS	72-0177
PLASTIC	PLASTIC STRESS STRAIN MATRIX OF ANISOTROPIC MATERIALS AND ITS APPLICATION	69-0383
PLASTIC	PLASTIC STRESS-STRAIN MATRIX AND ITS APPLICATION FOR THE SOLUTION OF ELASTIC-PLASTIC PROBLEMS BY THE	68-0171
PLASTIC	PRACTICAL SOLUTION OF PLASTIC DEFORMATION PROBLEMS IN THE ELASTIC PLASTIC RANGE	59-0004
PLASTIC	PRACTICAL SOLUTION OF PLASTIC DEFORMATION PROBLEMS IN THE ELASTIC PLASTIC RANGE	59-0004
PLASTIC	RADIAL INERTIA EFFECTS ON AN IDEALLY PLASTIC CIRCULAR PLATE UNDER IMPULSIVE AXIAL COMPRESSION	69-0324
PLASTIC	RESENI MAZNIHO STAVU ZELEZOBETONOVYCH DESEK (PLASTIC ANALYSIS OF REINFORCED CONCRETE PLATES)	74-0495
PLASTIC	SIMPLIFIED INELASTIC (PLASTIC AND CREEP) ANALYSIS OF PIPE ELBOWS SUBJECTED TO INPLANE AND OUT-OF-PLA	75-0654
PLASTIC	SIZE OF PLASTIC ZONE AT TIP OF A CRACK IN PLANE STRAIN SLATE BY FINITE ELEMENT METHOD	73-0006
PLASTIC	SOME RECENT FINITE ELEMENT APPROACHES IN ANALYSIS OF PLASTIC IMPACT	69-0299
PLASTIC	STABILITA STEN V PLASTICKEM OBORU (STABILITY IN WEBPLATES IN PLASTIC RANGE)	73-0387
PLASTIC	STIFFNESS ANALYSIS OF ELASTIC PLASTIC PLATES	65-0040
PLASTIC	STRESS ANALYSIS OF INDUSTRIAL COMPONENTS WITH PLASTIC AND FINITE ELEMENT MODELS	74-0512
PLASTIC	THE SIGNIFICANCE OF THE RATE OF PLASTIC WORK IN ELASTO-PLASTIC ANALYSIS	74-0437
PLASTIC	THE SIGNIFICANCE OF THE RATE OF PLASTIC WORK IN ELASTIC-PLASTIC ANALYSIS	74-0764
PLASTIC	THERMAL ELASTIC PLASTIC STRESS ANALYSIS BY THE FINITE ELEMENT METHOD	69-0391
PLASTIC	THREE DIMENSIONAL ELASTIC PLASTIC ANALYSIS FOR FRACTURE MECHANICS - 2.	71-0358
PLASTIC	VARIABLE TIME STEP METHOD FOR DETERMINING PLASTIC STRESS REFLECTIONS FROM BOUNDARIES	75-0641
PLASTIC	VISCOPLASTICITY, PLASTICITY AND PLASTIC FLOW	74-1278
PLASTIC	VISCOUS, INCOMPRESSIBLE FLOW WITH SPECIAL REFERENCE TO NON-NEWTONIAN (PLASTIC) FLUIDS	74-0053
PLASTIC-ELASTIC	STIFFNESS MATRIX IN PLASTIC-ELASTIC PROBLEMS OF CONTINUA	67-0116
PLASTICITY	A DESIGN ORIENTED APPROACH TO CREEP AND PLASTICITY IN FINITE ELEMENT PROGRAMS	70-0125
PLASTICITY	A FINITE ELEMENT MODEL FOR PLASTICITY PROBLEMS	74-1225
PLASTICITY	A GENERAL APPROACH TO PROBLEMS OF LARGE DEFORMATION, AND PLASTICITY USING ISOPARAMETRIC ELEMENT	71-0273
PLASTICITY	A MATHEMATICAL THEORY OF PLASTICITY FOR STRAIN HARDENING MATERIALS EXHIBITING INITIAL ANISOTROPY	72-0660
PLASTICITY	AN INVESTIGATION ON THE PLASTICITY MAGNIFICATION FACTOR IN DEEP SURFACE FLOWS	74-0766
PLASTICITY	ANALYSIS OF REAL STRUCTURES FOR CREEP, PLASTICITY AND OTHER COMPLEX CONSTITUTIVE LAWS.	69-0287
PLASTICITY	APPLICATION OF FINITE ELEMENT TECHNIQUES TO PLASTICITY PROBLEMS	69-0097
PLASTICITY	COMPUTER SOLUTIONS OF PLASTICITY PROBLEMS	72-0829
PLASTICITY	CONSTITUTIVE MATRIX AND FINITE ELEMENT FORMULATION FOR A DEFORMATION THEORY OF PLASTICITY	71-0700
PLASTICITY	DESIGN ORIENTED APPROACH TO CREEP AND PLASTICITY IN FINITE ELEMENT PROGRAMS	71-0281
PLASTICITY	EFFECT OF PLASTICITY AND CRACK BLUNTING ON THE STRESS DISTRIBUTION IN ORTHOTROPIC COMPOSITE MATERIAL	73-0145
PLASTICITY	FINITE ELEMENT ANALYSIS OF PLASTICITY PROBLEMS	68-0154
PLASTICITY	FINITE ELEMENT METHODS IN PLASTICITY	74-0801
PLASTICITY	GOVERNING EQUATIONS AND EXTREMUM PRINCIPLES OF ELASTICITY AND PLASTICITY GENERATED FROM A SINGLE FUN	73-0666
PLASTICITY	LIMITATIONS TO THE SMALL SCALE YIELDING APPROXIMATION FOR CRACK TIP PLASTICITY	74-0145
PLASTICITY	MATRIX ANALYSIS OF CREEP AND PLASTICITY PROBLEMS	65-0001
PLASTICITY	NUMERICAL ANALYSIS OF PLASTICITY IN SOILS	73-0611
PLASTICITY	PART 1. THEORY REPORT FOR CREEP-PLAST COMPUTER PROGRAM: ANALYSIS OF TWO-DIMENSIONAL PROBLEMS UNDER S	72-0765

PLASTICITY	PLASTICITY	70-0509
PLASTICITY	PLASTICITY AND LARGE DEFORMATION PROBLEMS BY FINITE ELEMENT METHOD	71-0529
PLASTICITY	PLASTICITY AND VISCOELASTICITY	72-0905
PLASTICITY	PLASTICITY OF POLYCRYSTALS AND GENERAL DISCUSSIONS ON PLASTICITY THEORY WITH IT	75-0312
PLASTICITY	PLASTICITY OF POLYCRYSTALS AND GENERAL DISCUSSIONS ON PLASTICITY THEORY WITH IT	75-0312
PLASTICITY	PLASTICITY THEORY AND FINITE ELEMENT APPLICATIONS	72-0318
PLASTICITY	PLASTICITY THEORY OF FIBROUS COMPOSITES - PART 11	73-0750
PLASTICITY	PLASTICITY THEORY OF FIBROUS COMPOSITES - PART 1	73-0751
PLASTICITY	RETSCP: A COMPUTER PROGRAM FOR ANALYSIS OF ROCKET ENGINE THERMAL STRAINS WITH CYCLIC PLASTICITY	74-0692
PLASTICITY	SOLUTION OF THE MIXED PROBLEM OF THE THEORY OF ELASTICITY AND THE THEORY OF PLASTICITY OF SOILS	74-1136
PLASTICITY	STRESS ANALYSIS OF TWO-DIMENSIONAL PROBLEMS UNDER SIMULTANEOUS CREEP AND PLASTICITY	71-0220
PLASTICITY	STRUCTURAL INELASTICITY. V1. A FINITE ELEMENT METHOD FOR PLASTICITY PROBLEMS	73-0887
PLASTICITY	VISCO-PLASTICITY, PLASTICITY AND CREEP IN ELASTIC SOLIDS. A UNIFIED NUMERICAL SOLUTION APPROACH	74-1279
PLASTICITY	VISCOPLASTICITY, PLASTICITY AND CREEP, A UNIFIED NUMERICAL SOLUTION APPROACH	74-1205
PLASTICITY	VISCOPLASTICITY, PLASTICITY AND PLASTIC FLOW	74-1278
PLASTICITY-AN	VISCO-PLASTICITY AND PLASTICITY-AN ALTERNATIVE FOR FINITE ELEMENT SOLUTION OF MATERIAL NONLINEARITIE	74-0410
PLASTICKEM	STABILITA STEN V PLASTICKEM OBORU (STABILITY IN WEBPLATES IN PLASTIC RANGE)	73-0387
PLASTIN	ISSLEDOVANIE PROCHNOSTI I ZHESTKOSTI PLASTIN, NAGRUZHENNYKH V SVOEI PLOSKOSTI, METHODOM KONECHNYKH E	73-0382
PLASTIN	MODIFICIROVANNYJ KONECHNYJ ELEMENT DIJA RASCHIOTA KOLEBANIJ TONKIKH PLASTIN	72-0891
PLASTIN	RASCHET KILEBANII PLASTIN SLOZHNOI FORMY S NEODNORODNYMI GRANICHNYMI USLOVIYAMI. (CALCULATION OF VIB	74-0988
PLATE	A COMPARATIVE STUDY OF VARIABLE AND CONSTANT THICKNESS HIGH PRECISION TRIANGULAR PLATE BENDING ELEME	74-0857
PLATE	A COMPATIBLE PLATE BENDING ANNULAR SECTOR FINITE ELEMENT OF NONUNIFORM THICKNESS	74-0430
PLATE	A COMPATIBLE QUADRILATERAL PLATE BENDING ELEMENT	72-0482
PLATE	A COMPATIBLE TRIANGULAR PLATE BENDING FINITE ELEMENT	68-0209
PLATE	A COMPATIBLE TRIANGULAR PLATE BENDING FINITE ELEMENT	70-0385
PLATE	A COMPUTER PROGRAM FOR THE FINITE ELEMENT ANALYSIS OF A CLASS OF PLATE BENDING PROBLEMS	69-0191
PLATE	A CONFORMING FINITE ELEMENT FOR PLATE BENDING	68-0071
PLATE	A CONFORMING QUARTIC TRIANGULAR ELEMENT FOR PLATE BENDING	69-0137
PLATE	A CONFORMING TRIANGULAR FINITE ELEMENT PLATE BENDING SOLUTION	70-0020
PLATE	A FAMILY OF EQUILIBRIUM PLATE BENDING ELEMENTS	74-0431
PLATE	A FINITE ELEMENT APPROACH TO PLATE VIBRATION PROBLEMS	65-0015
PLATE	A FINITE ELEMENT APPROACH TO PLATE BENDING PROBLEMS	68-0296
PLATE	A FINITE ELEMENT FORMULATION FOR LARGE DEFLECTION ELASTIC STRUCTURAL ANALYSIS APPLIED TO PLATE BENDI	70-0485
PLATE	A FINITE ELEMENT METHOD FOR RANDOM FOLDED PLATE STRUCTURES (EINE FINITE ELEMENT METHODE ZUR BERECHNU	73-0787
PLATE	A FINITE ELEMENT PLATE BENDING PROGRAMME	72-0579
PLATE	A FINITE ELEMENT SOLUTION FOR FOLDED PLATE STRUCTURES	67-0072
PLATE	A FINITE ELEMENT TECHNIQUE FOR ANALYZING FOLDED PLATE TYPE STRUCTURES WITH SKEWED SUPPORTS	69-0525
PLATE	A FINITE ELEMENT TENSOR APPROACH TO PLATE BUCKLING AND POST BUCKLING	73-0883
PLATE	A FULLY COMPATIBLE PLATE BENDING ELEMENT	67-0142
PLATE	A GENERAL QUADRILATERAL PLATE ELEMENT	73-0807
PLATE	A GENERAL SOLUTION OF THE VON KARMAN PLATE EQUATIONS BY THE FINITE ELEMENT METHOD	70-0487
PLATE	A HIGH PRECISION TRIANGULAR PLATE BENDING ELEMENT	68-0052
PLATE	A HIGH-ORDER FINITE ELEMENT FOR PRISMATIC AND NONPRISMATIC FOLDED PLATE STRUCTURES	75-0889
PLATE	A METHOD OF ELASTIC-PLASTIC ANALYSIS OF LARGELY DEFORMED PLATE PROBLEMS	72-0319
PLATE	A METHOD TO GENERATE BOTH UPPER AND LOWER BOUNDS TO PLATE EIGENVALUES BY CONFORMING DISPLACEMENT FIN	71-0166
PLATE	A MIXED FINITE ELEMENT FORMULATION FOR PLATE PROBLEMS	72-0556
PLATE	A NEW FAMILY OF CURVILINEAR PLATE BENDING ELEMENTS FOR VIBRATION AND STABILITY	72-0645
PLATE	A NEW FAMILY OF CURVILINEAR PLATE BENDING ELEMENTS FOR VIBRATION AND STABILITY	72-0999
PLATE	A NEW, FULLY COMPATIBLE FINITE ELEMENT FOR PLATE BENDING EIN NEUES VOLLVERTRAGLICHES ENDLICHES ELEME	68-0032
PLATE	A POLYGONAL FINITE FLEMENT FOR PLATE BENDING PROBLEMS USING THE ASSUMED STRESS APPROACH	69-0043
PLATE	A PROCEDURE FOR FINITE ELEMENT PLATE AND SHELL PRE AND POST BUCKLING ANALYSIS	71-0096
PLATE	A RAPIDLY CONVERGING TRIANGULAR PLATE ELEMENT	69-0245
PLATE	A RECTANGULAR PLATE BENDING ELEMENTS THE USE OF WHICH IS EQUIVALENT TO THE USE OF THE FINITE DIFFERE	71-0398
PLATE	A REFINED FINITE ELEMENT ANALYSIS OF FOLDED PLATE STRUCTURES	71-0671
PLATE	A REFINED MIXED-TYPE PLATE BENDING ELEMENT	69-0267
PLATE	A REFINED QUADRILATERAL ELEMENT FOR ANALYSIS OF PLATE BENDING	68-0047
PLATE	A REFINED TRIANGULAR PLATE BENDING FINITE ELEMENT	69-0056
PLATE	A SANDWICH PLATE FINITE ELEMENT	72-0077
PLATE	A SEQUEL TO TECHNICAL NOTE 14 ON THE TUBA FAMILY OF PLATE ELEMENTS	68-0015
PLATE	A SIMPLE ALGORITHM FOR THE STIFFNESS MATRIX OF TRIANGULAR PLATE BENDING ELEMENTS	71-0152
PLATE	A SIMPLE FINITE ELEMENT MODEL FOR ELASTIC-PLASTIC PLATE BENDING	72-0889
PLATE	A SIMPLE MIXED METHOD FOR PLATE AND SHELL PROBLEMS	73-1005
PLATE	A STUDY OF FINITE ELEMENTS APPLIED TO PLATE FLEXURE	66-0016
PLATE	A STUDY ON THE STRESS ANALYSIS OF BEAM PLATE STRUCTURE	69-0404
PLATE	A TECHNIQUE FOR CONNECTING BEAM ELEMENTS TO A PLATE MODEL OF A COMPLICATED BOX SECTION	74-0271
PLATE	A TECHNIQUE FOR MESH GRADING APPLIED TO CONFORMING PLATE BENDING FINITE ELEMENTS	73-1058
PLATE	A TRIANGULAR EQUILIBRIUM ELEMENT WITH LINEARLY VARYING BENDING MOMENTS FOR PLATE BENDING PROBLEMS	67-0054
PLATE	A TRIANGULAR FLAT PLATE BENDING ELEMENT	68-0252
PLATE	A TRIANGULAR PLATE BENDING ELEMENT FOR CONTACT PROBLEMS	73-0693
PLATE	ACCURACY OF THE FINITE ELEMENT ANALYSIS FOR THE ELASTIC PLATE WITH A CIRCULAR HOLE	73-0155
PLATE	AN ANNULAR SEGMENT FINITE ELEMENT FOR PLATE BENDING	71-0285
PLATE	AN EQUILIBRIUM MODEL FOR PLATE BENDING	68-0072
PLATE	ANALYSIS AND DESIGN OF AXISYMMETRIC REINFORCEMENT AROUND A CIRCULAR HOLE IN A THIN FLAT PLATE BY THE	74-1113
PLATE	ANALYSIS MODELS FOR PLATE ELEMENTS IN THE MEMBRANE STATE OF STRESS	67-0046
PLATE	ANALYSIS OF AN ORTHOTROPIC PLATE WITH TWO OPPOSITE EDGES SIMPLY SUPPORTED AND THE OTHER TWO FREE	73-0741
PLATE	ANALYSIS OF BOX GIRDERS BY FOLDED PLATE AND STRIP METHODS	73-0982
PLATE	ANALYSIS OF CURVED FOLDED PLATE STRUCTURES	70-0431
PLATE	ANALYSIS OF CURVED FOLDED PLATE STRUCTURES	71-0139
PLATE	ANALYSIS OF HYDRODYNAMIC AND PLATE STRUCTURES PROBLEMS BY THE FINITE ELEMENT METHOD	71-0091
PLATE	ANALYSIS OF PLATE AND SHELL STRUCTURES BY TRIANGULAR FINITE ELEMENTS	68-0303
PLATE	ANALYSIS OF PLATE AND SHELL STRUCTURES	69-0109
PLATE	ANALYSIS OF PLATE BENDING BY THE FINITE ELEMENT METHOD	60-0003
PLATE	ANALYSIS OF PLATE BENDING USING TRIANGULAR ELEMENTS	62-0010
PLATE	ANALYSIS OF PLATE BENDING BY TRIANGULAR ELEMENTS	68-0145
PLATE	ANALYSIS OF PLATE STRUCTURES BY A DUAL FINITE ELEMENT METHOD	69-0482
PLATE	ANALYSIS OF THREE-DIMENSIONAL ORTHOTROPIC SANDWICH PLATE STRUCTURES BY FINITE-ELEMENT METHOD	72-0945
PLATE	ANALYSIS ON THE LARGE DEFORMATION OF PLATE STRUCTURES	74-0435
PLATE	ANISOTROPIC PARAMETRIC PLATE DISCRETE ELEMENTS	73-1047
PLATE	ANNULAR AND CIRCULAR SECTOR FINITE ELEMENTS FOR PLATE BENDING	70-0100
PLATE	APPLICATION DE LA METHODE DES ELEMENTS FINIS AU CALCUL DES PLAQUES FLECHIES (APPLICATION OF THE FINI	73-0019
PLATE	APPLICATION OF A REFINED PLATE BENDING ELEMENT TO BUCKLING PROBLEMS	74-0937
PLATE	APPLICATION OF HERMITE POLYNOMIALS TO THE DETERMINATION OF THE STIFFNESS MATRIX OF PLATE BY FINITE E	71-0365
PLATE	APPLICATION OF HILL FUNCTIONS TO CIRCULAR PLATE PROBLEMS	75-0572
PLATE	APPLICATION OF SIMPLIFIED HYBRID DISPLACEMENT METHOD TO PLATE AND SHELL PROBLEMS	73-1037
PLATE	APPLICATION OF THE HIGH PRECISION TRIANGULAR PLATE BENDING ELEMENT TO PROBLEMS WITH CURVED BOUNDARIE	69-0071
PLATE	APPROXIMATE THREE-DIMENSIONAL STRESS SOLUTIONS FOR SOME LAMINATED PLATE PROBLEMS USING A HIGHER ORDE	70-0419
PLATE	AUTOMATED DATA GENERATION AND LARGE MATRIX OPERATION FOR 3 DIMENSIONAL PLATE STRUCTURE ANALYSIS	73-0339
PLATE	AXISYMMETRIC RING AND PLATE STRESS ANALYSIS PROGRAM - USERS GUIDE	70-0333
PLATE	BENDING ELEMENTS FOR PLATE AND SHELL NETWORKS	67-0043
PLATE	BERECHNUNG PRISMATISCHER FLAECHENTRAGWERKE MIT HILFE VON FINITEN STREIFENELEMENTEN. (CALCULATION OF	72-0133
PLATE	BOUNDING SOLUTIONS OF PLATE BENDING PROBLEMS	67-0084
PLATE	BUCKLING OF PLATE GIRDER WEBS UNDER PARTIAL EDGE LOADINGS	70-0283

322

PLATE	CALCULATION OF PLATE OF CONSTANT THICKNESS BY FINITE ELEMENT METHOD	69-0227
PLATE	CALCULATION OF STRESS CONCENTRATION DUE TO A HYPERBOLOID CAVITY IN A THIN PLATE USING NASTRAN	71-0260
PLATE	CERTAIN FINITE ELEMENTS FOR PLATE BENDING PROBLEMS	74-0395
PLATE	CLOSED FORM, FINITE ELEMENT SOLUTIONS FOR PLATE VIBRATIONS	69-0168
PLATE	CLOUGH-TOCHER TRIANGULAR PLATE BENDING ELEMENT IN VIBRATION	69-0093
PLATE	COMMENT ON A HIGHER ORDER CONFORMING RECTANGULAR PLATE ELEMENT	73-0999
PLATE	COMPARISON OF FINITE ELEMENTS FOR PLATE BENDING	72-0625
PLATE	COMPARISON OF PLATE BENDING FINITE ELEMENT FORMULATIONS	70-0288
PLATE	COMPARISON OF SOME PLATE BENDING FINITE ELEMENTS	72-0413
PLATE	COMPUTATION OF THE RESPONSE OF COUPLED PLATE ACOUSTIC SYSTEMS USING PLATE FINITE ELEMENTS AND ACOUST	71-0072
PLATE	COMPUTATION OF THE RESPONSE OF COUPLED PLATE ACOUSTIC SYSTEMS USING PLATE FINITE ELEMENTS AND ACOUST	71-0072
PLATE	COMPUTER PROGRAM FOR PRISMATIC FOLDED PLATES BY PLATE AND BEAM ELEMENTS	70-0371
PLATE	CONFORMING AND NONCONFORMING FINITE ELEMENT METHODS FOR SOLVING THE PLATE PROBLEM	73-0621
PLATE	CONFORMING FINITE ELEMENT METHODS FOR THE CLAMPED PLATE PROBLEM	75-0174
PLATE	CONFORMING RECTANGULAR AND TRIANGULAR PLATE BENDING ELEMENTS	71-0720
PLATE	CONTINUUM AND FINITE ELEMENT BRANCHING STUDIES OF THE CIRCULAR PLATE.	72-0427
PLATE	CONVERGENCE AND NUMERICAL ACCURACY WITH SPECIAL REFERENCE TO PLATE BENDING	69-0221
PLATE	CONVERGENCE OF A MIXED FINITE ELEMENT SCHEME FOR PLATE BENDING	73-0023
PLATE	CONVERGENCE OF A MIXED FINITE-ELEMENT METHOD FOR PLATE BENDING PROBLEMS	73-0107
PLATE	CONVERGENCE OF ANOTHER MIXED FINITE ELEMENT METHOD FOR PLATE BENDING PROBLEMS	72-0367
PLATE	CONVERGENCE OF EIGENVALUE SOLUTIONS IN CONFORMING PLATE BENDING FINITE ELEMENTS	72-0195
PLATE	CONVERGENCE OF SIMPLIFIED HYBRID DISPLACEMENT METHOD FOR PLATE BENDING	72-0412
PLATE	CONVERGENCE OF THE NASTRAN PLATE ELEMENTS FOR SHELL STRESS ANALYSIS	71-0261
PLATE	CONVERGENCE STUDIES OF EIGENVALUE SOLUTIONS USING TWO FINITE PLATE ELEMENTS	70-0085
PLATE	COUPLED RESPONSE OF SPATIAL VIBRATORY STRUCTURES MOUNTED TO ISOTROPIC PLATE ELEMENTS	71-0228
PLATE	CURVED FINITE ELEMENTS IN THE ANALYSIS OF SOLID SHELL AND PLATE STRUCTURES	69-0357
PLATE	DAS FLAECHENTRAGWERKSPROGRAM VON STRIP. (STRIP PROGRAM FOR PLATE STRUCTURES)	72-0150
PLATE	DEFORMATION AND FAILURE OF BORON-EPOXY PLATE WITH CIRCULAR HOLE	72-0536
PLATE	DERIVATION OF A STIFFNESS MATRIX FOR A RIGHT TRIANGULAR PLATE IN BENDING AND SUBJECTED TO INITIAL ST	62-0007
PLATE	DEVELOPMENT AND APPLICATION OF THE DIRECT STIFFNESS METHOD FOR OUT-OF-PLANE BENDING USING A TRIANGUL	62-0009
PLATE	DEVELOPMENT OF EXPLICIT STIFFNESS AND MASS MATRICES FOR A TRIANGULAR PLATE ELEMENT	69-0242
PLATE	DISCRETE METHODS IN THE ANALYSIS OF PLATE SYSTEMS	71-0835
PLATE	DISCRETE TRIANGULAR APPROXIMATION OF MOMENT AND DISPLACEMENT SURFACES FOR PLATE BENDING	71-0269
PLATE	DISTRIBUTED MASS MATRIX FOR PLATE ELEMENTS IN BENDING	65-0018
PLATE	DISTRIBUTED MASS MATRIX FOR PLATE ELEMENT BENDING (DISCUSSION)	66-0038
PLATE	DUGDALE MODEL SOLUTIONS FOR A SINGLE EDGE CRACKED PLATE	74-0221
PLATE	DYNAMIC ANALYSIS OF FRAME AND PLATE STRUCTURES WITH GEOMETRIC AND MATERIAL NONLINEARITIES	73-0649
PLATE	DYNAMIC PLASTIC LARGE DEFLECTIONS OF CLAMPED RECTANGULAR PLATE	74-0392
PLATE	EFFECTS OF A PEAK OVERLOAD ON FATIGUE CRACK PROPAGATION RATE-PLANE STRESS ANALYSIS OF CRACKED PLATE	74-1365
PLATE	EIGENVALUE PROBLEMS WITH A ' MIXED ' PLATE ELEMENT	69-0081
PLATE	ELASTIC PLATE SPALLATION	72-0682
PLATE	ELASTIC-PLASTIC PLATE BENDING WITH CONSTANT CURVATURE ELEMENTS	74-0308
PLATE	ELEMENT STATIC FLEXIBILITY AND STIFFNESS MATRICES FOR THE FINITE ELEMENT ANALYSIS OF BEAM, PLATE, AN	67-0070
PLATE	FATIGUE CRACK INITIATION IN A BORON EPOXY PLATE WITH A CIRCULAR HOLE	74-0820
PLATE	FINITE DEFLECTION STRUCTURAL ANALYSIS USING PLATE AND SHELL DISCRETE ELEMENTS	68-0143
PLATE	FINITE DEFLECTION STRUCTURAL ANALYSIS USING PLATE AND CYLINDRICAL SHELL DISCRETE ELEMENTS	68-0193
PLATE	FINITE DISPLACEMENT PLATE FLEXURE BY THE USE OF MATRIX INCREMENTAL APPROACH	72-0384
PLATE	FINITE ELEMENT ANALYSIS OF PLATE AND SHELL STRUCTURES	69-0360
PLATE	FINITE ELEMENT ANALYSIS OF PLATE BENDING WITH TRANSVERSE SHEAR DEFORMATION	70-0109
PLATE	FINITE ELEMENT ANALYSIS OF PLATE BUCKLING AND POST-BUCKLING	70-0332
PLATE	FINITE ELEMENT ANALYSIS OF PLATE BUCKLING USING A MIXED VARIATIONAL PRINCIPLE	71-0020
PLATE	FINITE ELEMENT ANALYSIS OF ALTERNATIVE AXIAL LOADING OF AN ELASTIC PLATE PRESSED BETWEEN TWO ELASTIC	71-0357
PLATE	FINITE ELEMENT ANALYSIS OF A PLATE ON A LAYERED CROSS-ANISOTROPIC FOUNDATION	74-0451
PLATE	FINITE ELEMENT ANALYSIS OF PRESTRESSED FOLDED PLATE STRUCTURES	74-1066
PLATE	FINITE ELEMENT ANALYSIS OF SPATIAL SANDWICH PLATE STRUCTURES	74-1132
PLATE	FINITE ELEMENT AND EXPERIMENTAL ANALYSIS OF AN INTERNAL FIXATION PLATE MODEL	75-0380
PLATE	FINITE ELEMENT APPLICATIONS OF A PRECISION TRIANGULAR PLATE BENDING ELEMENT TO LINEAR BOUNDARY VALUE	75-0788
PLATE	FINITE ELEMENT APPROACH TO THE CONTACT PROBLEM OF A PLATE ON THE ELASTIC HALF SPACE	72-0830
PLATE	FINITE ELEMENT DISPLACEMENT ANALYSIS OF PLATE BENDING BASED ON RECTANGULAR ELEMENTS	66-0032
PLATE	FINITE ELEMENT INCREMENTAL ANALYSIS OF ELASTOPLASTIC PLATE BENDING	73-0557
PLATE	FINITE ELEMENT METHOD FOR PLATE BENDING ANALYSIS BY DECOMPOSITION OF DIFFERENTIAL OPERATOR	71-0558
PLATE	FINITE ELEMENT METHOD IN PLATE AND SHELL INSTABILITY ANALYSIS	73-0983
PLATE	FINITE ELEMENT METHOD APPLIED TO CALCULATIONS OF NON-COPLANAR PLATE SYSTEMS	74-0241
PLATE	FINITE ELEMENT PLATE BENDING EQUILIBRIUM ANALYSIS	69-0045
PLATE	FINITE ELEMENT PROCEDURES IN THE SOLUTION OF PLATE AND SHELL PROBLEMS	65-0038
PLATE	FINITE ELEMENT SOLUTIONS FOR PLATE BENDING PROBLEMS BY SIMPLIFIED HYBRID DISPLACEMENT METHOD	72-0780
PLATE	FINITE ELEMENT STIFFNESS MATRICES FOR ANALYSIS OF PLATE BENDING	66-0021
PLATE	FINITE ELEMENT STRESS ANALYSIS OF A CRACK IN A BI-MATERIAL PLATE	72-0165
PLATE	FINITE ELEMENT TECHNIQUES APPLIED TO THE VIBRATIONS OF PLATE SYSTEMS	66-0033
PLATE	FINITE ELEMENT TECHNIQUES FOR ORTHOTROPIC PLANE STRESS AND ORTHOTROPIC PLATE ANALYSIS	68-0274
PLATE	FINITE ELEMENTS FOR PLATE BENDING	70-0325
PLATE	FINITE-ELEMENT METHOD FOR POTENTIAL CALCULATIONS IN A HALL PLATE	75-0454
PLATE	FLEXIBLE PLATE FINITE ELEMENT ON ELASTIC FOUNDATION	70-0128
PLATE	FOLDED PLATE STRUCTURES BY THE FINITE STRIP METHOD	69-0359
PLATE	FORMULATION OF A NEW TRIANGULAR PLATE BENDING ELEMENT	68-0051
PLATE	FURTHER APPLICATION OF QUADRATIC ISOPARAMETRIC FINITE-ELEMENTS TO LINEAR FRACTURE MECHANICS OF PLATE	75-0639
PLATE	GENERAL QUADRILATERAL PLATE ELEMENT	75-0339
PLATE	GENERALIZATION OF PLATE FINITE ELEMENTS TO SHELLS	72-0171
PLATE	HIGH-ORDER FINITE ELEMENT TECHNIQUE FOR PRISMATIC AND NONPRISMATIC FOLDED PLATE STRUCTURES	74-1130
PLATE	HIGH-PRECISION PLATE AND SHELL FINITE ELEMENTS	71-0437
PLATE	HYBRID PLATE BENDING ELEMENTS	72-0792
PLATE	IMPROVED GENERAL QUADRILATERAL PLATE ELEMENTS	75-0036
PLATE	IMPROVEMENT IN PLATE MODELING AND PLATE DESIGN BY THE FINITE ELEMENT METHOD	75-0791
PLATE	IMPROVEMENT IN PLATE MODELING AND PLATE DESIGN BY THE FINITE ELEMENT METHOD	75-0791
PLATE	INADEQUACY OF NODAL CONNECTIONS IN A STIFFNESS SOLUTION FOR PLATE BENDING	65-0021
PLATE	INCREMENTAL SOLUTION OF AXISYMMETRIC PLATE AND SHELL FINITE DEFORMATION	71-0776
PLATE	INFLUENCE OF CRACK CLOSURE AND ELASTO-PLASTIC FLOW ON THE BENDING OF A CRACKED PLATE	75-0735
PLATE	INTERACTIVE FINITE ELEMENT DATA GENERATION FOR THREE-DIMENSIONAL PLATE STRUCTURES	74-0909
PLATE	INTRODUCTION TO SHEARS DEFORMATION INTO A THIN PLATE DISPLACEMENTS FORMULATIONS	73-0617
PLATE	INVESTIGATION OF ELASTIC PLATE SPALLATION BY FINITE ELEMENTS	74-0382
PLATE	KINDS OF CONVERGENCE AND IMPROVED CONVERGENCE OF CONFORMING FINITE ELEMENT SOLUTIONS IN PLATE BENDIN	70-0323
PLATE	MACRO-APPROACH FOR RIBBED AND GRID PLATE SYSTEMS	75-0079
PLATE	MATRIX ANALYSIS OF A COMPOSITE STRUCTURE OF PLATE AND LINE ELEMENTS	71-0250
PLATE	MEMBRANE AND PLATE BENDING ELEMENTS	75-0405
PLATE	MODEL OF ELASTIC PLATE FROM FINITE ELEMENTS IN A SUPERSONIC FLOW	75-0038
PLATE	MODEL TEST AND ANALYSIS OF OBLIQUE FOLDED PLATE STRUCTURE	73-0230
PLATE	MULTIANGULAR ELEMENTS IN PLATE BENDING	73-0435
PLATE	NEW APPROACH FOR PLATE VIBRATIONS. COMBINATION OF TRANSFER MATRIX AND FINITE ELEMENT TECHNIQUE	71-0434
PLATE	NEW VARIATIONAL FUNCTIONAL FOR THE FINITE-ELEMENT METHOD AND ITS APPLICATION TO PLATE SHELL PROBLEMS	72-0108
PLATE	NONLINEAR ANALYSES OF THE STRESS FIELD NEAR A CRACK OR HOLE IN A PLATE	75-0694
PLATE	NONLINEAR BEAM AND PLATE ELEMENTS	72-0866
PLATE	NOTE ON STRESS AND STRAIN REDISTRIBUTION IN A NOTCHED PLATE SPECIMEN DURING CYCLIC LOADING	69-0194

PLATE	OBSERVATIONS REGARDING ASSUMED STRESS HYBRID PLATE ELEMENTS	74-0370
PLATE	ON ASSUMED DISPLACEMENTS FOR THE RECTANGULAR PLATE BENDING ELEMENT	67-0020
PLATE	ON THE APPLICATION OF THE FINITE ELEMENT METHOD TO THE VIBRATION ANALYSIS OF SPACE PLATE STRUCTURE	72-0630
PLATE	ON THE CONFORMING CUBIC TRIANGULAR ELEMENT FOR PLATE BENDING	72-0439
PLATE	ON THE DETERMINATION OF EFFECTIVE ELASTIC-PLASTIC PROPERTIES FOR THE EQUIVALENT SOLID PLATE ANALYSI	74-0494
PLATE	ON THE QUINTIC TRIANGULAR PLATE BENDING ELEMENT	72-0677
PLATE	ONE-SIDED APPROXIMATION AND PLATE BENDING	74-0336
PLATE	PAPA, STRUCTURAL ANALYSIS OF PLATES AND SHELLS USING TRAPEZOIDAL AND TRIANGULAR PLATE ELEMENTS	67-0059
PLATE	PARALLELOGRAM ELEMENT IN THE SOLUTION OF RHOMBIC CANTILEVER PLATE BENDING	66-0022
PLATE	PHOTOELASTIC INVESTIGATION ON THE CRACK-ARREST CAPABILITY OF A PRETENSIONED STIFFENED PLATE	75-0088
PLATE	PLATE AND SHELL PROBLEMS, FINITE ELEMENT DISPLACEMENT APPROACH	66-0005
PLATE	PLATE BENDING ANALYSIS USING ASSUMED STRESS METHODS	68-0135
PLATE	PLATE BENDING WITH NONCONFORMING ELEMENT: FINE-MESH CONVERGENCE TEST	65-0049
PLATE	PLATE BUCKLING ANALYSIS USING A FULLY COMPATIBLE FINITE ELEMENT	69-0065
PLATE	PLATE MEASUREMENTS USING FINITE ELEMENTS	70-0229
PLATE	PLATE VIBRATION PROBLEMS USING THE FINITE ELEMENT METHOD WITH ASSUMED STRESS DISTRIBUTION AND TRIANG	66-0068
PLATE	PRACTICAL ANALYSIS OF PLATE VIBRATIONS USING NASTRAN	71-0058
PLATE	PREDICTION OF PLATE VIBRATIONS USING A CONSISTENT MASS MATRIX	66-0043
PLATE	PREFABRICATED PRETENSIONED LIGHTWEIGHT CONCRETE FOLDED PLATE	73-0808
PLATE	PSEUDO ISOPARAMETRIC FINITE ELEMENTS FOR SHELL AND PLATE ANALYSIS	69-0040
PLATE	RADIAL INERTIA EFFECTS ON AN IDEALLY PLASTIC CIRCULAR PLATE UNDER IMPULSIVE AXIAL COMPRESSION	69-0324
PLATE	RECTANGULAR FINITE ELEMENTS FOR ANALYSIS OF PLATE VIBRATIONS	68-0113
PLATE	RECTANGULAR FINITE ELEMENT FOR PLATE BENDING ANALYSIS BASED ON HELLINGER-REISSNERS VARIATIONAL PRINC	71-0545
PLATE	REFINED FINITE ELEMENTS FOR FOLDED PLATE STRUCTURES	72-0574
PLATE	REFINED PLATE BENDING FINITE ELEMENT	74-1045
PLATE	RESIDUAL ENERGY BALANCING TECHNIQUE IN THE GENERATION OF PLATE BENDING FINITE ELEMENTS	74-0594
PLATE	RIGID PLATE ON NONLINEARLY DEFORMABLE COHESIVE BASE (PLANE PROBLEM)	75-0264
PLATE	SANDWICH PLATE STRUCTURE ANALYSIS BY FINITE ELEMENT	74-0163
PLATE	SECTOR ELEMENT FOR THIN PLATE FLEXURE	72-0174
PLATE	SHAPE FUNCTION SUBROUTINE FOR AN ISOPARAMETRIC THIN PLATE ELEMENT	73-0618
PLATE	SHEAR IN C AND C PLATE BENDING ELEMENTS	73-0692
PLATE	SHELL AND PLATE ANALYSIS BY FINITE ELEMENTS	73-0163
PLATE	SIMPLE "MIXED" METHOD FOR PLATE AND SHELL PROBLEMS	74-0952
PLATE	SIMPLE FINITE ELEMENT MODEL FOR ELASTIC-PLASTIC PLATE BENDING.	72-0447
PLATE	SIMPLIFIED HYBRID DISPLACEMENT METHOD APPLIED TO PLATE BUCKLING PROBLEMS	72-0781
PLATE	SINGLE ELEMENT PANEL FOR FLAT PLATE STRUCTURES	75-0279
PLATE	SOME CONFORMING, RECTANGULAR, PLATE ELEMENTS	75-0094
PLATE	SOME ELEMENTS FOR ANALYSIS OF PLATE BENDING	72-0626
PLATE	SOME FINITE ELEMENT SOLUTIONS FOR PLATE BENDING PROBLEMS BY SIMPLIFIED HYBRID DISPLACEMENT METHOD	72-0268
PLATE	SOME NONCONFORMING FINITE ELEMENTS FOR THE PLATE BENDING PROBLEM	75-0324
PLATE	STATIC AND DYNAMIC APPLICATIONS OF A HIGH-PRECISION TRIANGULAR PLATE BENDING ELEMENT	69-0086
PLATE	STATIC FLEXIBILITY MATRICES FOR A RECTANGULAR PLATE ELEMENT IN BENDING, TWISTING, AND SHEAR. UNIFORM	68-0217
PLATE	STATIC, VIBRATION, AND THERMAL STRESS ANALYSES OF LAMINATED PLATE AND SHELLS BY THE HYBRID-STRESS FI	72-0602
PLATE	STEEL PLATE ANALYSIS BY FINITE ELEMENTS	69-0100
PLATE	STIFFNESS MATRICES FOR PLATE BENDING ELEMENTS	69-0129
PLATE	STIFFNESS MATRIX FOR BENDING OF A RECTANGULAR PLATE ELEMENT WITH INITIAL MEMBRANE STRESSES	62-0006
PLATE	STRESS ANALYSIS OF THIN PLATE STRUCTURE BY THE FINITE ELEMENT METHOD	69-0403
PLATE	STRESS AND FAILURE ANALYSIS OF A GLASS-EPOXY COMPOSITE PLATE WITH A CIRCULAR HOLE	73-0276
PLATE	STRESS INTENSITY FACTORS IN PLATE STRUCTURES CALCULATED BY THE FINITE ELEMENT METHOD (IN JAPANESE)	71-0788
PLATE	STRESS INTENSITY FACTORS FOR CRACKS IN A PLATE CONTAINING A HOLE AND IN A SPINNING DISC	73-0467
PLATE	STRUCTURAL ANALYSIS OF A THREE-DIMENSIONAL PERFORATED PLATE BY THE FINITE ELEMENT METHOD WITH TEST C	74-0945
PLATE	STUDY ON FINITE ELEMENT METHOD FOR STRUCTURAL ANALYSIS - ISTRAN/PL (INH STRUCTURE ANALYSIS PLATE STR	73-0167
PLATE	TESTS OF GUSSET PLATE MODELS	71-0338
PLATE	THE ANALYSIS OF CLASSICAL PLATE BENDING PROBLEMS USING AN EQUILIBRIUM FINITE ELEMENT MODEL	68-0302
PLATE	THE APPLICATION OF FINITE ELEMENT TECHNIQUES TO PLATE AND SHELL PROBLEMS	65-0028
PLATE	THE CONSTANT BENDING MOMENT PLATE BENDING ELEMENT	71-0182
PLATE	THE ELASTIC STRESS ANALYSIS OF A BI-MATERIAL PLATE WITH A CRACK NORMAL TO THE INTERFACES	72-0819
PLATE	THE FINITE ELEMENT METHOD APPLIED TO PLATE FLEXURE	66-0095
PLATE	THE FINITE ELEMENT METHOD IN PLATE BUCKLING	69-0128
PLATE	THE FINITE ELEMENT METHOD APPLIED TO THE ANALYSIS OF SANDWICH PLATE AND SHELL STRUCTURES	72-0824
PLATE	THE FINITE ELEMENT METHOD IN PLATE AND SHELL STABILITY ANALYSIS	73-0644
PLATE	THE FUGA FAMILY FOR FOLDED PLATE STRUCTURES	69-0333
PLATE	THE FUGA FAMILY OF ELEMENTS FOR FOLDED PLATE STRUCTURES A FURTHER SEQUEL TO TECHNICAL NOTE 14	69-0446
PLATE	THE HYBRID DISPLACEMENT METHOD APPLIED TO PLATE AND SHELL PROBLEMS	75-0089
PLATE	THE TRIANGULAR EQUILIBRIUM ELEMENT IN THE SOLUTION OF PLATE BENDING PROBLEMS	68-0122
PLATE	THE TUBA FAMILY OF PLATE ELEMENTS FOR THE MATRIX DISPLACEMENT METHOD	68-0013
PLATE	THE VIBRATION CHARACTERISTICS OF A TENSIONED PLATE CONTAINING A FATIGUE CRACK	68-0134
PLATE	THEORETICAL AND EXPERIMENTAL INVESTIGATION ON A STEAM GENERATOR TUBE PLATE	69-0424
PLATE	THEORETICAL AND EXPERIMENTAL CONSIDERATIONS ON FLOW OF VISCOUS LIQUIDS PAST PLATE AT SMALL AND AVERA	70-0144
PLATE	THERMAL MECHANICAL ANALYSIS PROCEDURE FOR A SEMICIRCULAR PLATE WITH CLAMPED STRAIGHT EDGE	75-0651
PLATE	THERMAL STRESS DEVELOPED IN S PLATE, STRAIGHT PLATE, AND DEEP DISH WHEELS	73-0172
PLATE	THERMAL STRESS DEVELOPED IN S PLATE, STRAIGHT PLATE, AND DEEP DISH WHEELS	73-0172
PLATE	THREE DIMENSIONAL FINITE ELEMENT METHOD FOR PLATE BENDING	74-0243
PLATE	TRANSIENT RESPONSE OF A COUPLED PLATE-ACOUSTIC SYSTEM USING PLATE AND ACOUSTIC FINITE ELEMENTS	71-0463
PLATE	TRANSIENT RESPONSE OF A COUPLED PLATE-ACOUSTIC SYSTEM USING PLATE AND ACOUSTIC FINITE ELEMENTS	72-0275
PLATE	TRANSVERSE REINFORCEMENT OF CONTINUOUS, TROUGH-SHAPED CONCRETE RAILROAD STRUCTURES. AN IMPLEMENTATIO	71-0719
PLATE	TRIANGULAR ELEMENTS IN PLATE BENDING CONFORMING AND NON-CONFORMING SOLUTIONS	66-0011
PLATE	TRIANGULAR FINITE ELEMENTS FOR PLATE BENDING WITH CONSTANT AND LINEARLY VARYING BENDING MOMENTS	70-0019
PLATE	TRIANGULAR FINITE ELEMENT FOR PLATE BENDING ANALYSIS WITH TRANSVERSE SHEAR DEFORMATION	71-0555
PLATE	TRIANGULAR FINITE-ELEMENTS OF VARIABLE THICKNESS AND THEIR APPLICATION TO PLATE AND SHELL PROBLEMS	69-0095
PLATE	TRIANGULAR PLATE BENDING ELEMENTS	69-0057
PLATE	TRIANGULAR PLATE BENDING ELEMENT WITH ENFORCED COMPATIBILITY	71-0593
PLATE	TRIANGULAR PLATE ELEMENTS IN THE MATRIX FORCE METHOD OF STRUCTURAL ANALYSIS	63-0010
PLATE	TRIANGULAR THICK PLATE BENDING ELEMENTS	71-0167
PLATE	TRIGONOMETRIC FUNCTION REPRESENTATIONS FOR RECTANGULAR PLATE BENDING ELEMENTS	71-0765
PLATE	TWO DIMENSION, PLATE, SHELL AND FINITE PRISM ISOPARAMETRIC ELEMENTS AND THEIR APPLICATIONS	71-0772
PLATE	UNBONDED CONTACT PROBLEM OF A PLATE ON THE ELASTIC HALF SPACE	74-0208
PLATE	VIBRATION ANALYSIS OF A CANTILEVERED SQUARE PLATE BY THE STIFFNESS MATRIX METHOD	66-0012
PLATE	VIBRATION STUDIES OF A FLAT PLATE AND BUILT-UP WING	71-0238
PLATE	VIBRATIONS OF PLATE AND SHELL STRUCTURES USING TRIANGULAR FINITE ELEMENTS	67-0023
PLATE-ACOUSTIC	TRANSIENT RESPONSE OF A COUPLED PLATE-ACOUSTIC SYSTEM USING PLATE AND ACOUSTIC FINITE ELEMENTS	71-0463
PLATE-ACOUSTIC	TRANSIENT RESPONSE OF A COUPLED PLATE-ACOUSTIC SYSTEM USING PLATE AND ACOUSTIC FINITE ELEMENTS	72-0275
PLATE-BEAM	FREE VIBRATIONS OF PLATE-BEAM SYSTEMS	73-0322
PLATE-BENDING	DEPENDENCE OF PLATE-BENDING FINITE ELEMENT DEFLECTIONS AND EIGENVALUES ON POISSONS RATIO	74-0567
PLATE-BENDING	DEPENDENCE OF PLATE-BENDING FINITE ELEMENT DEFLECTIONS AND EIGENVALUES ON POISSONS RATIO	74-0583
PLATE-BENDING	FINITE PLATE-BENDING ELEMENTS IN POLAR COORDINATES	68-0234
PLATE-BENDING	NEW TRIANGULAR AND QUADRILATERAL PLATE-BENDING FINITE ELEMENTS	74-0833
PLATE-BENDING	NEW TRIANGULAR PLATE-BENDING FINITE ELEMENT WITH TRANSVERSE SHEAR FLEXIBILITY	74-0627
PLATE-BENDING	NEW TRIANGULAR PLATE-BENDING FINITE ELEMENT WITH TRANSVERSE SHEAR FLEXIBILITY	74-0964
PLATE-STIFFENER	LINEAR BUCKLING AND GEOMETRICALLY NONLINEAR ANALYSIS OF PLANAR PLATE-STIFFENER TYPE STRUCTURES BY TH	74-1097
PLATEAUS	NUMERICAL SOLUTION OF PLATEAUS PROBLEM BY A FINITE ELEMENT METHOD	74-0119
PLATED	DATA GENERATOR FOR THE IDEALIZATION FOR FINITE ELEMENT STRUCTURAL ANALYSIS OF NAVAL SHIP FLAT PLATED	72-0860

PLATED	MATHEMATICAL AND EXPERIMENTAL STUDIES ON THE MECHANICS OF PLATED TRANSVERSE FRACTURES	74-0656
PLATES	A BENDING ANALYSIS FOR PLATES	66-0035
PLATES	A COMPACT FINITE-ELEMENT METHOD FOR THE ANALYSIS OF ANISOTROPIC LAMINATED PLATES	73-1166
PLATES	A COMPARATIVE STUDY OF VARIABLE AND CONSTANT THICKNESS HIGH PRECISION TRIANGULAR PLATE BENDING ELEME	74-0857
PLATES	A COMPATIBLE ELEMENT FOR BUCKLING ANALYSIS OF SKEW PLATES WITH PARALLEL EDGES	73-0328
PLATES	A COMPUTER PROGRAM FOR ANALYSIS OF STIFFENED PLATES UNDER COMBINED INPLANE AND LATERAL LOADS	71-0703
PLATES	A CONFORMING QUADRILATERAL ELEMENT FOR BUCKLING ANALYSIS OF STIFFENED PLATES	73-0613
PLATES	A CONFORMING TRIANGULAR ELEMENT IN THE ANALYSIS OF THIN AND SEMI-THICK PLATES IN BENDING	70-0308
PLATES	A DISCRETE ELEMENT STRESS AND DISPLACEMENT ANALYSIS OF ELASTOPLASTIC PLATES	70-0108
PLATES	A FINITE ELEMENT ANALYSIS FOR CYLINDRICAL SHELLS AND PLATES	66-0034
PLATES	A FINITE ELEMENT ANALYSIS FOR ' MODERATELY THICK ' RECTANGULAR PLATES IN BENDING	68-0146
PLATES	A FINITE ELEMENT ANALYSIS FOR DYNAMIC BEHAVIOR OF STIFFENED PLATES	69-0392
PLATES	A FINITE ELEMENT ANALYSIS OF LAMINATED ORTHOTROPIC PLATES AND SHALLOW SHELLS	70-0123
PLATES	A FINITE ELEMENT ANALYSIS INCLUDING TRANSVERSE SHEAR EFFECTS FOR APPLICATIONS TO LAMINATED PLATES	71-0216
PLATES	A FINITE ELEMENT LINEAR PROGRAMMING METHOD FOR THE LIMIT ANALYSIS OF THIN PLATES	73-1064
PLATES	A FINITE ELEMENT METHOD FOR GEOMETRICALLY NONLINEAR LARGE DISPLACEMENT PROBLEMS IN THIN, ELASTIC PLA	69-0005
PLATES	A FINITE ELEMENT METHOD FOR THE PLASTIC BUCKLING ANALYSIS OF PLATES	69-0390
PLATES	A FINITE ELEMENT MODEL FOR LAMINATED PLATES INCLUDING TRANSVERSE SHEAR DEFORMATION	72-0818
PLATES	A FINITE ELEMENT PROCEDURE FOR PLATES WITH CURVED BOUNDARIES	75-0030
PLATES	A FINITE ELEMENT PROCEDURE FOR PLATES WITH CURVED BOUNDARIES	75-0750
PLATES	A FINITE-ELEMENT METHOD FOR TRANSVERSE VIBRATIONS OF BEAMS AND PLATES	67-0144
PLATES	A FINITE-ELEMENT METHOD FOR THE PLASTIC BUCKLING ANALYSIS OF PLATES	69-0214
PLATES	A FRAMEWORK METHOD FOR STRETCHED PLATES	71-0697
PLATES	A METHOD OF SOLUTION FOR THE COMPLICATED BUCKLING PROBLEMS OF ELASTIC PLATES WITH COMBINED USE OF RA	68-0097
PLATES	A MODIFIED FINITE ELEMENT METHOD FOR CALCULATION OF VIBRATION OF THIN PLATES	72-0598
PLATES	A STIFFNESS MATRIX FOR THE ANALYSIS OF THIN PLATES IN BENDING	61-0001
PLATES	A STUDY OF FINITE ELEMENTS FOR THE ANALYSIS OF SANDWICH PLATES	69-0503
PLATES	A STUDY OF STIFFNESS MATRICES FOR THE ANALYSIS OF FLAT PLATES	68-0105
PLATES	A TRIANGULAR ELEMENT FOR BENDING AND VIBRATION OF MULTILAYER SANDWICH PLATES	72-0489
PLATES	ACCURACY AND CONVERGENCE OF MIXED FINITE-ELEMENT APPROXIMATIONS OF THIN BARS, MEMBRANES, AND PLATES	73-1120
PLATES	ACCURATE CALCULATION OF STRESSES IN BEAMS, DISKS AND PLATES BY MEANS OF FINITE ELEMENT METHODS AND I	70-0133
PLATES	AN APPROXIMATE NON-LINEAR ANALYSIS OF THIN PLATES	68-0286
PLATES	AN ELASTIC FINITE-ELEMENT ANALYSIS OF THE BUCKLING OF ECCENTRICALLY STIFFENED PLATES AND SHELLS	73-1136
PLATES	AN ELASTIC-PLASTIC FINITE ELEMENT ANALYSIS FOR PLATES WITH EDGE BEAMS	69-0185
PLATES	AN INVESTIGATION OF THE EFFECTS OF SQUARE CUTOUTS ON THE NATURAL FREQUENCIES AND MODE SHAPES OF RECT	71-0638
PLATES	AN INVESTIGATION OF THE BENDING OF SYMMETRIC PLATES UNDER SYMMETRIC TRANSVERSE LOADING	72-0662
PLATES	ANALYSIS OF CIRCULAR AND ANGULAR LAMINATED THICK PLATES	74-0426
PLATES	ANALYSIS OF CLAMPED CIRCULAR PLATES OF VARIABLE THICKNESS SUBJECTED TO UNIFORMLY DISTRIBUTED LOAD BY	73-0583
PLATES	ANALYSIS OF ELASTIC PLATES IN FLEXURE BY A SIMPLIFIED FINITE ELEMENT METHOD	67-0035
PLATES	ANALYSIS OF ELASTIC PLATES BY A SIMPLIFIED FINITE ELEMENT METHOD	67-0156
PLATES	ANALYSIS OF ELASTIC-PLASTIC BUCKLING OF PLATES BY THE FINITE ELEMENT METHOD	69-0263
PLATES	ANALYSIS OF GEOMETRICALLY NONLINEAR PLATES AND SHELLS BY THE FINITE ELEMENT DISPLACEMENT METHOD	68-0050
PLATES	ANALYSIS OF LARGE DEFLECTION OF PLATES BY THE FINITE ELEMENT METHOD	69-0148
PLATES	ANALYSIS OF ORTHOTROPIC FOLDED PLATES	70-0272
PLATES	ANALYSIS OF ORTHOTROPIC FOLDED PLATES WITH ECCENTRIC STIFFENERS	70-0437
PLATES	ANALYSIS OF PLATES IN BENDING USING LARGE FINITE ELEMENTS	70-0185
PLATES	ANALYSIS OF PLATES ON ELASTIC FOUNDATIONS	73-0684
PLATES	ANALYSIS OF PLATES SUBJECTED TO IN-PLANE FORCES USING LARGE FINITE ELEMENTS	70-0271
PLATES	ANALYSIS OF RECTANGULAR PLATES IN TRANSIENT CREEP BENDING	69-0425
PLATES	ANALYSIS OF SIMPLY SUPPORTED THICK, LAYERED PLATES	71-0428
PLATES	ANALYSIS OF STIFFENED PLATES BY THE FINITE ELEMENT METHOD	68-0299
PLATES	ANALYSIS OF STIFFENED PLATES USING THE FINITE ELEMENT METHOD	70-0030
PLATES	ANALYSIS OF STIFFENED PLATES CONSIDERING NONLINEAR MATERIAL AND GEOMETRIC BEHAVIOUR	75-0327
PLATES	ANALYSIS OF STRESSES IN SKEWED PLATES BY THE METHOD OF FINITE ELEMENTS	73-0248
PLATES	ANALYSIS OF THIN PLATES IN BENDING USING TRIANGULAR FINITE ELEMENTS	68-0213
PLATES	APPLICATION OF CONFORMAL MAPPING AND VARIATIONAL METHOD TO THE STUDY OF HEAT CONDUCTION IN POLYGONAL	71-0444
PLATES	APPLICATION OF FINITE ELEMENT METHOD TO VIBRATIONS OF QUARTZ PLATES	74-0345
PLATES	APPLICATION OF THE DISCRETE ELEMENT METHOD TO THE BUCKLING ANALYSIS OF RECTANGULAR PLATES UNDER ARBI	68-0054
PLATES	APPLICATION OF THE FINITE ELEMENT METHOD TO THE EXTENSIONAL VIBRATIONS OF PIEZOELECTRIC PLATES	72-0073
PLATES	APPLICATION OF THE FINITE ELEMENT METHOD TO THE BUCKLING OF RECTANGULAR PLATES AND PLATES ON ELASTIC	73-0478
PLATES	APPLICATION OF THE FINITE ELEMENT METHOD TO THE BUCKLING OF RECTANGULAR PLATES AND PLATES ON ELASTIC	73-0478
PLATES	ARBITRARY SYMMETRIC LOADING PROBLEMS OF CENTRALLY CRACKED RECTANGULAR PLATES	75-0318
PLATES	ARCH DAMS ANALYSED BY A LINEAR FINITE ELEMENT SHELL THE FINITE STRIP METHOD IN THE ANALYSIS OF ELAST	68-0189
PLATES	ASYMMETRIC BENDING OF CIRCULAR PLATES USING THE FINITE ELEMENT METHOD	75-0175
PLATES	AXISYMMETRIC VIBRATIONS OF LINEARLY TAPERED ANNULAR PLATES	74-1266
PLATES	BENDING AND STRETCHING OF PLATES, SPECIAL MODELS FOR UPPER AND LOWER BOUNDS	66-0026
PLATES	BENDING AND VIBRATION OF MULTILAYER SANDWICH BEAMS AND PLATES	73-0272
PLATES	BENDING OF CIRCULAR PLATES OF HARDENING MATERIAL	67-0112
PLATES	BENDING OF MICROPOLAR PLATES	74-0263
PLATES	BENDING OF THIN PLATES - DISPLACEMENT MODELS	68-0049
PLATES	BIFURCATION OF ECCENTRICALLY LOADED MEMBERS AND OF STIFFENED PLATES	70-0057
PLATES	BUCKLING OF MULTILAYER PLATES BY FINITE ELEMENTS	67-0103
PLATES	CALCULATION OF PLATES BY MEANS OF METHOD OF FINITE ELEMENTS	70-0345
PLATES	CALCULATION OF RECTANGULAR PLATES BY THE FINITE ELEMENT METHOD	70-0506
PLATES	CALCULATION OF THE ELASTIC BUCKLING LOADS OF THIN FLAT REINFORCED PLATES USING TRIANGULAR FINITE ELE	73-1024
PLATES	CALCULATION OF THE ELASTIC BUCKLING LOADS OF THIN FLAT REINFORCED PLATES USING TRIANGULAR FINITE ELE	75-0177
PLATES	COLLAPSE LOAD OF REINFORCED CONCRETE PLATES.	72-0406
PLATES	COMBINED MATERIAL AND GEOMETRIC NONLINEARITY FOR THIN STEEL PLATES	75-0397
PLATES	COMPARISON OF PRACTICAL METHODS OF ANALYSING ELASTIC LATERALLY LOADED PLATES	71-0425
PLATES	COMPUTER PROGRAM FOR PRISMATIC FOLDED PLATES BY PLATE AND BEAM ELEMENTS	70-0371
PLATES	CONSTRUCTION OF NORMAL DISPLACEMENT FUNCTIONS OF TRIANGULAR FINITE ELEMENT OF PLATES AND SHELLS	72-0563
PLATES	CONVERGENCE OF SEMIANALYTICAL SOLUTION OF PLATES	73-0497
PLATES	CRACK PROPAGATION AND ARREST IN IMPACTED PLATES	72-0599
PLATES	CRACKING ANALYSIS OF REINFORCED CONCRETE PLATES	75-0085
PLATES	CREEP BENDING ANALYSIS OF PLATES BY THE FINITE ELEMENT METHOD	71-0415
PLATES	CREEP STRESS AND STRAIN ANALYSIS OF NOTCHED PLATES AND BARS BY MEANS OF THE FINITE ELEMENT METHOD	70-0141
PLATES	CURVED TRIANGULAR FINITE ELEMENTS FOR THE ANALYSIS OF THICK ELASTIC PLATES AND SHELLS	71-0814
PLATES	DEFLECTION FUNCTION FOR THE ASYMMETRICAL BENDING OF CIRCULAR PLATES	73-0468
PLATES	DER GERISSENE ZUSTAND ZWEISEITIG GELAGERTER PLATTEN UNTER EINZELLASTEN - NICHTLINEARE BERECHNUNG MIT	75-0858
PLATES	DERIVATION OF THE STIFFNESS MATRIX FOR PLATES BY GALERKIN METHOD	69-0402
PLATES	DETERMINATION OF STRESS INTENSITY FACTORS IN CRACKED PLATES BY THE FINITE ELEMENT METHOD	73-0148
PLATES	DETERMINATION OF STRESSES DUE TO DISCONTINUITIES IN FINITE PLATES OF ISOTROPIC AND ORTHOTROPIC MATER	73-1052
PLATES	DETERMINATION OF THE INITIAL BUCKLING STRESS OF FLAT PLATES USING THE DISPLACEMENT METHOD	69-0465
PLATES	DETERMINATION OF THE LOWEST CRITICAL BUCKLING STRESS OF VARIOUS SHAPED PLATES BY FINITE-ELEMENTS	73-1134
PLATES	DISCRETE MODEL ANALYSIS OF ELASTIC-PLASTIC PLATES	68-0009
PLATES	DISCUSSION OF "FINITE ELEMENT BENDING ANALYSIS OF REISSNER PLATES"	71-0826
PLATES	DISCUSSION OF "FINITE ELEMENT BENDING ANALYSIS OF REISSNER PLATES"	71-0832
PLATES	DOUBLE STRESS-WAVE DISCONTINUITIES IN FINITE, SHEAR-CORRECTED BEAMS AND PLATES	68-0210
PLATES	DYNAMIC ANALYSIS OF CIRCULAR AND SECTOR THICK, LAYERED PLATES	75-0198
PLATES	DYNAMIC BEHAVIOR OF ECCENTRICALLY STIFFENED PLATES	71-0796
PLATES	DYNAMIC BEHAVIOUR OF ECCENTRICALLY STIFFENED PLATES	71-0605
PLATES	DYNAMIC STABILITY OF PLATES BY FINITE ELEMENTS	71-0573

PLATES	DYNAMIC TRANSIENT NON-LINEAR BEHAVIOUR OF THICK AND THIN PLATES	75-0125
PLATES	DYNAMICAL FINITE ELEMENT ANALYSIS FOR ELASTIC WAVES IN BEAMS AND PLATES	68-0211
PLATES	EFFECT OF FIBRE DISTRIBUTION ON STRESS AND STRAIN CONCENTRATION AT HOLES IN COMPOSITE PLATES	72-0616
PLATES	EFFECTIVENESS OF EXCESSIVE NODAL CONTINUITIES IN THE FINITE ELEMENT ANALYSIS OF THIN RECTANGULAR AND	70-0346
PLATES	EFFICIENT FINITE ELEMENT ANALYSIS OF RECTANGULAR AND SKEW LAMINATED PLATES	73-0436
PLATES	ELASTIC AND PLASTIC STRESS DISTRIBUTION ON BASE PLATES IN THE NEIGHBOURHOOD OF FILLET WELDS IN LAP J	70-0175
PLATES	ELASTIC BENDING OF CIRCULAR PLATES OF VARIABLE THICKNESS. AN ANALYTICAL AND EXPERIMENTAL STUDY	70-0163
PLATES	ELASTIC PLASTIC BUCKLING OF STIFFENED RECTANGULAR PLATES	69-0122
PLATES	ELASTIC PLASTIC CRACKED PLATES IN PLANE STRAIN	69-0321
PLATES	ELASTIC PLASTIC FINITE ELEMENT ANALYSIS OF PLATES	74-0093
PLATES	ELASTIC SNAP THROUGH ANALYSIS OF CURVED PLATES USING DISCRETE ELEMENTS	72-0899
PLATES	ELASTIC SNAP-TRHOUGH ANALYSIS OF CURVED PLATES USING DISCRETE ELEMENTS	72-0927
PLATES	ELASTIC-PLASTIC ANALYSIS OF FLAT PLATES BY THE FINITE ELEMENT METHOD	68-0111
PLATES	ELASTIC-PLASTIC ANALYSIS OF AXISYMMETRIC PLATES AND SHELLS	71-0779
PLATES	ELASTIC-PLASTIC BUCKLING OF PLATES USING THE FINITE ELEMENT METHOD	67-0094
PLATES	ELASTIC-PLASTIC BUCKLING OF PLATES BY FINITE ELEMENT METHOD	69-0253
PLATES	ELASTIC-PLASTIC FINITE ELEMENT ANALYSIS OF PLATES - DISCUSSION	75-0139
PLATES	ELASTIC-PLASTIC FINITE ELEMENT ANALYSIS OF PLATES	75-0599
PLATES	ELASTO PLASTIC STRESS IN CRACKED PLATES	65-0042
PLATES	ELASTO-PLASTIC ANALYSIS OF ORTHOTROPIC PLATES AND SHELLS	68-0166
PLATES	ELASTO-PLASTIC ANALYSIS OF PLATES USING THE FINITE ELEMENT METHOD	71-0038
PLATES	ELASTO-PLASTIC ORTHOTROPIC PLATES AND SHELLS	69-0467
PLATES	ELASTODYNAMIC BEHAVIOR OF LAMINATED ORTHOTROPIC PLATES UNDER INITIAL STRESS	75-0249
PLATES	EQUIVALENT MASS MATRICES FOR RECTANGULAR PLATES IN BENDING	66-0063
PLATES	EVALUATION OF THE USE OF A SINGULARITY ELEMENT IN FINITE ELEMENT ANALYSIS OF CENTER-CRACKED PLATES	72-0870
PLATES	EXPERIMENTAL AND THEORETICAL BENDING DEFLECTIONS OF LAMINATED PLATES	72-0144
PLATES	EXPERIMENTAL AND THEORETICAL FATIGUE-CRACK PROPAGATION LIVES OF VARIOUSLY NOTCHED PLATES	74-0602
PLATES	EXPERIMENTAL VERIFICATION OF DISCRETE-ELEMENT SOLUTIONS FOR PLATES AND PAVEMENT SLABS	70-0411
PLATES	FINITE DEFLECTION DISCRETE ELEMENT ANALYSIS OF SANDWICH PLATES AND CYLINDRICAL SHELLS WITH LAMINATED	69-0236
PLATES	FINITE DEFORMATION OF ELASTIC PLATES, SHELLS, AND MEMBRANES BY THE FINITE ELEMENT METHOD	72-0805
PLATES	FINITE ELEMENT ANALYSIS OF PLYWOOD PLATES	67-0052
PLATES	FINITE ELEMENT ANALYSIS OF SKEW COMPOSITE PLATES	67-0160
PLATES	FINITE ELEMENT ANALYSIS OF SKEW PLATES IN BENDING	68-0120
PLATES	FINITE ELEMENT ANALYSIS OF SANDWICH PLATES AND CYLINDRICAL SHELLS WITH LAMINATED FACES	68-0121
PLATES	FINITE ELEMENT ANALYSIS OF STIFFENED PLATES	68-0240
PLATES	FINITE ELEMENT ANALYSIS OF LAMINATED ANISOTROPIC PLATES INCLUDING TRANSVERSE SHEAR DEFORMATIONS	70-0506
PLATES	FINITE ELEMENT ANALYSIS OF ELASTIC-PLASTIC PLATES AND ECCENTRICALLY STIFFENED PLATES	71-0585
PLATES	FINITE ELEMENT ANALYSIS OF ELASTIC-PLASTIC PLATES AND ECCENTRICALLY STIFFENED PLATES	71-0585
PLATES	FINITE ELEMENT ANALYSIS OF PLATES WITH CURVED EDGES.	72-0173
PLATES	FINITE ELEMENT ANALYSIS OF PLATES AND SHELLS	72-0409
PLATES	FINITE ELEMENT ANALYSIS OF PLATES ON A TWO PARAMETER FOUNDATION MODEL.	72-0425
PLATES	FINITE ELEMENT ANALYSIS OF VIBRATION OF THIN ELASTIC PLATES BY SIMPLIFIED HYBRID DISPLACEMENT METHOD	72-0565
PLATES	FINITE ELEMENT ANALYSIS OF FINITE DEFORMATIONS OF PLATES, SHELLS, AND MEMBRANES	72-0578
PLATES	FINITE ELEMENT ANALYSIS OF SKEW SANDWICH PLATES	72-0810
PLATES	FINITE ELEMENT ANALYSIS OF FINITE SIZED PLATES BOUNDED TO AN ELASTIC HALF SPACE	72-0880
PLATES	FINITE ELEMENT ANALYSIS OF PLATES AND ECCENTRICALLY STIFFENED PLATES	73-0691
PLATES	FINITE ELEMENT ANALYSIS OF PLATES AND ECCENTRICALLY STIFFENED PLATES	73-0691
PLATES	FINITE ELEMENT ANALYSIS OF PLATES ALLOWING FOR TRANSVERSE SHEAR DEFORMATION EFFECTS	73-0996
PLATES	FINITE ELEMENT ANALYSIS OF STATIC AND DYNAMIC STRESSES AROUND SUDDENLY PUNCHED HOLES IN PLATES AND S	73-1088
PLATES	FINITE ELEMENT ANALYSIS OF AXISYMMETRIC MULTILAYER SANDWICH PLATES AND SHELLS	74-0101
PLATES	FINITE ELEMENT ANALYSIS OF MULTILAYER SANDWICH PLATES AND SHELLS	74-0425
PLATES	FINITE ELEMENT ANALYSIS OF VIBRATION AND FLUTTER OF CANTILEVER ANISOTROPIC PLATES	74-0484
PLATES	FINITE ELEMENT ANALYSIS OF ANISOTROPIC PLATES USING Q-19 ELEMENT. PART 1. THEORETICAL DEVELOPMENT	74-0728
PLATES	FINITE ELEMENT ANALYSIS OF ANISOTROPIC PLATES USING Q-19 ELEMENT: PART 11. INSTRUCTIONS FOR USERS AN	74-0731
PLATES	FINITE ELEMENT ANALYSIS OF ANISOTROPIC PLATES USING Q-19 ELEMENT - 1, 2	74-0985
PLATES	FINITE ELEMENT ANALYSIS OF PERFORATED PLATES CONTAINING TRIANGULAR PENETRATION PATTERNS OF 5 AND 10	75-0216
PLATES	FINITE ELEMENT ANALYSIS OF PERFORATED PLATES CONTAINING TRIANGULAR PENETRATION PATTERNS OF 5 AND 10	75-0520
PLATES	FINITE ELEMENT APPROACH TO BENDING OF MICROPOLAR PLATES	74-0525
PLATES	FINITE ELEMENT BENDING ANALYSIS OF REISSNER PLATES	70-0214
PLATES	FINITE ELEMENT BENDING ANALYSIS OF MULTILAYER PLATES	74-0266
PLATES	FINITE ELEMENT BUCKLING ANALYSIS OF HOMOGENEOUS AND SANDWICH PLATES	75-0613
PLATES	FINITE ELEMENT DISPLACEMENT METHOD FOR LARGE AMPLITUDE FREE FLEXURAL VIBRATIONS OF BEAMS AND PLATES	73-0185
PLATES	FINITE ELEMENT FORMULATIONS FOR ELASTIC PLATES BY GENERAL VARIATIONAL STATEMENTS WITH DISCONTINUOUS	73-0559
PLATES	FINITE ELEMENT FORMULATIONS FOR ELASTIC PLATES BY GENERAL VARIATIONAL STATEMENTS WITH DISCONTINUOUS	73-112A
PLATES	FINITE ELEMENT LARGE DEFLECTION ANALYSIS OF PLATES	69-0208
PLATES	FINITE ELEMENT METHOD FOR CALCULATING STRESS INTENSITY FACTORS FOR CRACKED PLATES IN BENDING	71-0424
PLATES	FINITE ELEMENT METHOD FOR ELASTIC PLASTIC PLATES	72-0167
PLATES	FINITE ELEMENT NONLINEAR ANALYSIS FOR PLATES AND SHALLOW SHELLS	68-0216
PLATES	FINITE ELEMENT POST BUCKLING ANALYSIS OF THIN ELASTIC PLATES	68-0123
PLATES	FINITE ELEMENT PROCEDURE FOR LARGE DEFLECTION ANALYSIS OF PLATES WITH INITIAL DEFLECTIONS	71-0327
PLATES	FINITE ELEMENT PROCEDURE FOR POSTBUCKLING ANALYSIS OF INITIALLY CURVED PLATES	71-0421
PLATES	FINITE ELEMENT SOLUTIONS FOR LAMINATED THICK PLATES	72-0143
PLATES	FINITE ELEMENT STUDY OF VIBRATION OF TRAPEZOIDAL PLATES	73-0035
PLATES	FINITE ELEMENT THEORY OF PLATES AND SHELLS INCLUDING TRANSVERSE SHEAR STRAIN EFFECTS	70-0395
PLATES	FINITE ELEMENT VIBRATION ANALYSIS OF CRACKED PLATES IN TENSION	68-0133
PLATES	FINITE SEGMENT ANALYSIS OF FOLDED PLATES	69-0169
PLATES	FINITE-ELEMENT ANALYSIS OF ORTHOTROPIC PLATES WITH ECCENTRIC STIFFENERS	72-0958
PLATES	FINITE-ELEMENT ANALYSIS OF VIBRATION AND FLUTTER OF CANTILEVER ANISOTROPIC PLATES	74-0933
PLATES	FINITE-ELEMENT ANALYSIS OF SKEW SANDWICH PLATES	75-0185
PLATES	FINITE-ELEMENT BENDING ANALYSIS FOR PLATES	67-0036
PLATES	FINITE-ELEMENT LARGE DEFLECTION ANALYSIS OF PLATES	69-0196
PLATES	FINITE-ELEMENT TECHNIQUES FOR THE DYNAMIC ANALYSIS OF UNSYMMETRICALLY LAMINATED PLATES	73-1131
PLATES	FLEXURAL VIBRATIONS OF RECTANGULAR AND OTHER POLYGONAL PLATES	71-0417
PLATES	FORMULATION OF DISCRETE ELEMENT MODELS FOR STRESS AND VIBRATION ANALYSIS OF PLATES	70-0328
PLATES	FREE VIBRATION ANALYSIS OF STIFFENED PLATES	72-0303
PLATES	FREE VIBRATION AND TRANSIENT RESPONSE OF THICK AND THIN PLATES USING THE FINITE ELEMENT METHOD	74-0517
PLATES	FREE VIBRATION OF DOUBLY CURVED RECTANGULAR PLATES INCLUDING THE PRESENCE OF MEMBRANE STRESSES WITH	69-0468
PLATES	FREE VIBRATION OF ORTHOTROPIC VARIABLE THICKNESS, CONTINUOUS RECTANGULAR PLATES	70-0270
PLATES	FREE VIBRATION OF THICK, LAYERED RECTANGULAR PLATES BY FINITE LAYER METHOD	72-0440
PLATES	FREE VIBRATIONS OF FINITE ELEMENT PLATES SUBJECTED TO COMPLEX MIDDLE-PLANE FORCE SYSTEMS	72-0280
PLATES	FREE VIBRATIONS OF FINITE ELEMENT PLATES SUBJECTED TO COMPLEX MIDDLE-PLANE FORCE SYSTEMS	73-0097
PLATES	FREE VIBRATIONS OF FINITE ELEMENT PLATES SUBJECTED TO COMPLEX MIDDLE-PLANE FORCE SYSTEMS - REPLY	73-0122
PLATES	FREQUENCY ANALYSIS OF SKEW ORTHOTROPIC PLATES BY THE FINITE STRIP METHOD	71-0413
PLATES	FULL RANGE ANALYSIS OF ECCENTRICALLY STIFFENED PLATES	74-0149
PLATES	FULL-RANGE ANALYSIS OF STEEL PLATES AND STIFFENED PLATING UNDER UNIAXIAL COMPRESSION	75-0803
PLATES	FUNKCJA KSZTALTU NIESPELNIAPACA WARUNKOW CIAGLOSCI W ANALIZIE PLYT METODA ELEMENTOW SKONCZONYCH. (SH	75-0263
PLATES	FURTHER DEVELOPMENTS IN THE USE OF LINEAR DISPLACEMENT FUNCTIONS IN THE FINITE ELEMENT ANALYSIS OF O	71-0344
PLATES	GENERAL ANALYSIS OF FOLDED PLATES BY THE FINITE ELEMENT METHOD	69-0200
PLATES	GEOMETRIC NONLINEAR ANALYSIS OF THIN PLATES BY FINITE ELEMENTS	70-0492
PLATES	INELASTIC RESPONSE OF PLATES AND ECCENTRICALLY STIFFENED PLATES	74-0458
PLATES	INELASTIC RESPONSE OF PLATES AND ECCENTRICALLY STIFFENED PLATES	74-0458
PLATES	INVESTIGATION OF FLEXURAL STIFFNESS OF RECTANGULAR PLATES BY THE FINITE ELEMENT METHOD	73-0991

PLATES	ISSLEDOVANIE PROCHNOSTI I ZHESTKOSTI PLASTIN, NAGRUZHENNYKH V SVOEI PLOSKOSTI, METHODOM KONECHNYKH E	73-0382
PLATES	LAMINATED ORTHOTROPIC PLATES AND SHALLOW SHELLS	71-0400
PLATES	LARGE DEFLECTION ANALYSIS OF ELASTIC-PLASTIC PLATES AND SHELLS	69-0178
PLATES	LARGE DEFLECTION ANALYSIS OF PLATES AND SHALLOW SHELLS USING THE FINITE ELEMENT METHOD	73-0989
PLATES	LARGE DEFLECTION OF ELASTIC PLATES UNDER PATCH LOADING	72-0120
PLATES	LARGE DEFLECTIONS OF ELASTIC BEAMS AND PLATES USING THE FINITE ELEMENT METHOD	68-0101
PLATES	LARGE DISPLACEMENT ANALYSIS OF THIN PLATES	73-0261
PLATES	LARGE-DEFLECTION ELASTO-PLASTIC BUCKLING ANALYSIS OF PLATES USING FINITE ELEMENTS	73-0394
PLATES	LES PLAQUES ET LES COQUES MINCES OU EPAISSES (THIN AND THICK PLATES AND SHELLS)	71-0432
PLATES	LINEAR DYNAMIC ANALYSES OF LAMINATED PLATES AND SHELLS BY THE HYBRID-STRESS FINITE-ELEMENT METHOD	73-0786
PLATES	LUMPED FINITE ELEMENT BASES FOR BEAMS AND PLATES	72-0263
PLATES	MATRIX ANALYSIS OF LOCAL INSTABILITY IN PLATES, STIFFENED PANELS AND COLUMNS	72-0542
PLATES	MATRIX DISPLACEMENT ANALYSIS OF PLATES AND SHELLS	66-0010
PLATES	MATRIX DISPLACEMENT ANALYSIS OF PLATES AND SHELLS, PART 11. GENERAL FORMULATION OF THE LINEAR THEORY	71-0621
PLATES	MATRIX DISPLACEMENT ANALYSIS OF SHELLS AND PLATES INCLUDING TRANSVERSE SHEAR STRAIN EFFECTS	72-0643
PLATES	METHODS AND POSSIBILITIES FOR ELECTRONIC COMPUTATION OF PLATES IN FLEXURE	69-0329
PLATES	MIXED FINITE ELEMENT ANALYSIS OF PLATES IN BENDING, SMALL DEFLECTION THEORY OF ELASTIC AND ELASTO-PL	71-0783
PLATES	MIXED FINITE ELEMENT ANALYSIS OF PLATES IN BENDING, SMALL DEFLECTION THEORY OF ELASTIC AND ELASTO-PL	71-0783
PLATES	MIXED FINITE ELEMENT ANALYSIS OF ELASTO-PLASTIC PLATES IN BENDING	72-0135
PLATES	MIXED FINITE ELEMENT METHOD FOR BENDING OF PLATES	75-0229
PLATES	MIXED MODELS FOR PLATES	70-0384
PLATES	MULTIPLE STRESS WAVE DISCONTINUITIES IN FINITE BEAMS AND PLATES	67-0114
PLATES	NATURAL FREQUENCIES AND MODE SHAPES OF PLATES WITH INTERIOR CUT-OUTS	70-0402
PLATES	NON-LINEAR ANALYSIS OF PLATES CONSIDERING GEOMETRIC AND MATERIAL EFFECTS	71-0702
PLATES	NONCONSERVATIVE STABILITY OF PLATES BY FINITE ELEMENT METHOD	75-0789
PLATES	NONLINEAR ACOUSTIC RESPONSE ANALYSIS OF PLATES USING THE FINITE ELEMENT METHOD	72-0223
PLATES	NONLINEAR FINITE ELEMENT ANALYSIS OF REINFORCED CONCRETE PLATES	73-0530
PLATES	NONLINEAR FINITE ELEMENT ANALYSIS OF REINFORCED CONCRETE PLATES	73-0845
PLATES	NONLINEAR FINITE-ELEMENT METHOD FOR PLATES AND ITS APPLICATION TO DYNAMIC RESPONSE OF REACTOR FUEL S	74-0670
PLATES	NONLINEAR LAYERED ANALYSIS OF RC PLATES AND SHELLS	73-0208
PLATES	NONLINEAR LAYERED FINITE-ELEMENT ANALYSIS OF REINFORCED-CONCETE PLATES AND SHELLS	72-0946
PLATES	NONLINEAR RESPONSE OF PLATES TO IMPULSIVE LOADING USING HYBRID FINITE ELEMENTS	74-1018
PLATES	NUMERICAL METHODS FOR THE LIMIT ANALYSIS OF PLATES	68-0181
PLATES	NUMERICAL RESULTS FROM THE APPLICATION OF GRADIENT ITERATIVE TECHNIQUES TO THE FINITE ELEMENT VIBRAT	72-0138
PLATES	NUMERICAL SOLUTION OF THE VISCOUS FLOW IN THE ENTRANCE REGION OF PARALLEL PLATES	73-0022
PLATES	NUMERYCZNE ROZWIAZANIE PROBLEMU ORGAN WLASNYCH NA PRZYKLADZIE PLYT. (NUMERICAL SOLUTION OF THE PROBL	75-0296
PLATES	ON DERIVATION OF STIFFNESS MATRICES WITH CO ROTATION FIELDS FOR PLATES AND SHELLS	71-0254
PLATES	ON THE ENERGY FUNCTIONS FOR CURVED SANDWICH PLATES	71-0587
PLATES	ON THE GENERALIZATION OF STRESS FUNCTION PROCEDURE FOR DYNAMIC ANALYSIS OF PLATES	73-0561
PLATES	ON THE VIBRATION OF POINT-SUPPORTED PLATES	71-0602
PLATES	ON VIBRATION ANALYSIS OF RECTANGULAR CANTILEVER PLATES WITH A STEPPED CROSS-SECTION BY THE FINITE EL	71-0608
PLATES	ORTHOTROPIC FLAT PLATES SUPPORTED ON COLUMNS ELONGATED IN PLANE	72-0492
PLATES	PAPA, STRUCTURAL ANALYSIS OF PLATES AND SHELLS USING TRAPEZOIDAL AND TRIANGULAR PLATE ELEMENTS	67-0059
PLATES	PATTERN LOADING ON REINFORCED CONCRETE FLAT PLATES	71-0290
PLATES	PIECE-WISE CONTINUOUS FINITE ELEMENT FOR THE STRESS ANALYSIS OF PLATES AND SHELLS	75-0758
PLATES	PLANE FRAMEWORK METHODS FOR PLATES IN EXTENSION	66-0126
PLATES	PLASTIC INTENSITY FACTORS FOR CRACKED PLATES	71-0164
PLATES	PLATES AND BEAMS ON ELASTIC FOUNDATIONS: LINEAR AND NONLINEAR BEHAVIOUR	68-0042
PLATES	PLATES AND SHELLS, INELASTIC ANALYSIS, TIME DEPENDENT INELASTICITY, FIVE LECTURES	73-0633
PLATES	PLATES AND TANKS ON ELASTIC FOUNDATIONS, AN APPLICATION OF FINITE ELEMENT METHOD	65-0010
PLATES	POST-BUCKLING ANALYSIS OF SLIGHTLY CURVED PLATES BY THE FINITE ELEMENT METHOD	69-0365
PLATES	POST-ELASTIC FINITE-ELEMENT ANALYSIS OF REINFORCED PLATES	72-0979
PLATES	RANDOM VIBRATION RESPONSE OF CANTILEVER PLATES USING THE FINITE ELEMENT METHOD	70-0181
PLATES	RASCHET KILEBANII PLASTIN SLOZHNOI FORMY S NEODNORODNYMI GRANICHNYMI USLOVIYAMI. (CALCULATION OF VIB	74-0988
PLATES	REDUCED INTEGRATION TECHNIQUE IN GENERAL ANALYSIS OF PLATES AND SHELLS	71-0374
PLATES	RESENI MAZNIHO STAVU ZELEZOBETONOVYCH DESEK (PLASTIC ANALYSIS OF REINFORCED CONCRETE PLATES)	74-0495
PLATES	RESPONSE OF BEAMS AND PLATES TO MOVING LOADS	71-0277
PLATES	RESPONSE OF RECTANGULAR PLATES TO MOVING LOADS BY A FINITE ELEMENT PROCEDURE	69-0515
PLATES	RESPONSE OF RECTANGULAR ORTHOTROPIC PLATES TO RANDOM EXCITATIONS	73-0496
PLATES	RIESENIE VAZKOPRUZNYCH ANIZOTROPICKYCH DOASK METODOU KONECNYCH PRVKOV (SOLUTION OF VISCOELASTIC ANIS	74-0477
PLATES	SHEAR BUCKLING OF SQUARE PERFORATED PLATES	74-0713
PLATES	SHEAR DEFORMATION ANALYSIS OF BEAMS AND PLATES BY FINITE ELEMENT METHOD	75-0873
PLATES	SHELLS, PLATES AND MEMBRANES - A COMPUTER-AIDED DESIGN SYSTEM	74-1159
PLATES	SHORT TERM DEFLECTIONS OF CONCRETE FLAT PLATES	73-0277
PLATES	SIMPLE FINITE ELEMENT SOLUTION FOR PLATES OF HOMOGENEOUS, SANDWICH AND CELLULAR CONSTRUCTION	75-0140
PLATES	SOME APPROXIMATIONS IN THE NONLINEAR ANALYSIS OF RECTANGULAR PLATES USING FINITE ELEMENTS	74-1231
PLATES	SOME REMARKS CONCERNING HETEROGENEOUS ANISOTROPIC PLATES.	72-0430
PLATES	SOME RESULTS IN THE FINITE ELEMENT MODELING OF THIN PLATES	66-0018
PLATES	STABILITY CONTROL ANALYSIS OR ROTATING PLATES BY FINITE ELEMENT. EMPHASIS ON SLOTS AND HOLES	71-0284
PLATES	STABILITY INVESTIGATIONS OF PLATES USING FINITE ELEMENT METHOD BASED ON AN EXPANDED VARIATION PRINCI	71-0663
PLATES	STABILITY OF MODERATELY THICK RECTANGULAR PLATES USING A HIGH PRECISION TRIANGULAR FINITE ELEMENT	75-0334
PLATES	STABILITY OF PLATES USING THE FINITE ELEMENT METHOD	66-0081
PLATES	STABILITY OF PLATES USING THE FINITE ELEMENT METHOD	67-0030
PLATES	STABILITY OF PLATES USING A MIXED FINITE ELEMENT FORMULATION	74-0218
PLATES	STATIC AND DYNAMIC ANALYSIS OF MULTILAYERED SANDWICH PLATES	72-0493
PLATES	STATIC AND DYNAMIC BEHAVIOUR OF RECTANGULAR PLATES USING HIGHER ORDER FINITE STRIPS	72-0458
PLATES	STATIC, VIBRATION AND BUCKLING ANALYSIS OF AXISYMMETRIC CIRCULAR PLATES USING FINITE ELEMENTS	73-0157
PLATES	STIFFNESS ANALYSIS OF ELASTIC PLASTIC PLATES	65-0040
PLATES	STIFFNESS METHODS FOR PLATES BY GALERKIN'S METHOD	69-0249
PLATES	STRESS ANALYSIS OF FRAME STRUCTURE CONNECTED WITH PLATES	74-0265
PLATES	STRESS ANALYSIS OF LAMINATED CFRP PLATES CONTAINING A HOLE	74-1175
PLATES	STRESSES AND DEFORMATIONS IN MELTING PLATES	69-0476
PLATES	STUDY ON THE VIBRATION OF THIN PLATES IN A GAS STREAM BY THE FINITE ELEMENT METHOD	74-0454
PLATES	SYMMETRICAL BENDING OF CIRCULAR PLATES USING FINITE ELEMENTS	72-0428
PLATES	THE APPLICATION OF THE SMOOTH-SURFACE INTERPOLATION TECHNIQUE TO THE FINITE-ELEMENT ANALYSIS OF THIN	66-0120
PLATES	THE BEHAVIOR OF ADVANCED FILAMENTARY COMPOSITE PLATES WITH CUTOUTS	73-0891
PLATES	THE BEHAVIOUR OF ADVANCED FILAMENTARY COMPOSITE PLATES WITH CUTOUTS	70-0025
PLATES	THE DAMPED TRANSIENT BEHAVIOR OF FINITE BEAMS AND PLATES	69-0520
PLATES	THE EFFECT OF IN PLANE STATIC STRESSES ON THE VIBRATION CHARACTERISTICS OF CURVED SANDWICH PLATES	71-0588
PLATES	THE EFFECTIVENESS OF EXCESSIVE NODAL CONTINUITIES IN THE FINITE ELEMENT ANALYSIS OF THIN RECTANGULAR	70-0105
PLATES	THE ELASTO-PLASTIC AND LARGE-DISPLACEMENT RESPONSE OF PLATES TO BLAST LOADING	73-0731
PLATES	THE FINITE STRIP METHOD IN THE ANALYSIS OF ELASTIC PLATES WITH TWO SIMPLY SUPPORTED ENDS	68-0068
PLATES	THE VIBRATION OF STEPPED CANTILEVERED PLATES	67-0129
PLATES	THERMAL STRESS ANALYSIS OF PLATES AND SHALLOW SHELLS BY HYBRID FINITE-ELEMENT METHOD	74-0485
PLATES	THERMO-ELASTIC ANALYSIS OF RECTANGULAR PLATES IN PLANE STRESS BY THE FINITE-ELEMENT DISPLACEMENT MET	67-0087
PLATES	TRIANGULAR ELEMENT FOR MULTILAYER SANDWICH PLATES	72-0581
PLATES	TRIANGULAR FINITE ELEMENTS FOR THE BENDING ANALYSIS OF THICK ELASTIC PLATES	70-0466
PLATES	TWO HYBRID ELEMENTS FOR ANALYSIS OF THICK, THIN AND SANDWICH PLATES	72-0544
PLATES	UEBER DIE PRAKTISCHE ANWENDUNG VON FINITEN ELEMENTEN IN DER PLATTENSTATIK (APPLICATION OF THE FINITE	75-0805
PLATES	USE OF FINITE ELEMENT METHOD IN THE DYNAMICS OF PLATES AND SHELLS	70-0313
PLATES	USE OF NONCONFORMING MODES IN FINITE ELEMENT ANALYSIS OF PLATES AND SHELLS	73-0865
PLATES	USE OF THE FINITE ELEMENT METHOD IN EXAMINING ELASTIC BUCKLING OF THIN PLATES OF ANY SHAPE	75-0153

PLATES	VARIATION WITH TEMPERATURE OF RESONANT FREQUENCIES OF ANISOTROPIC PLATES	74-1046
PLATES	VARIATIONAL FORMULATIONS AND FINITE ELEMENT METHOD FOR THIN ELASTIC SHELLS AND PLATES	68-0064
PLATES	VELKE PRIEHYBY DOSIEK METODOU KONECNYCH PRVKOV (LARGE DEFLECTIONS OF PLATES ANALYZED BY FINITE ELEME	73-0373
PLATES	VIBRATION ANALYSIS OF CANTILEVERED CURVED PLATES USING A NEW CYLINDRICAL SHELL FINITE ELEMENT	68-0130
PLATES	VIBRATION ANALYSIS OF LAMINATED PLATES AND SHELLS BY A HYBRID STRESS ELEMENT	73-0120
PLATES	VIBRATION ANALYSIS OF PLATES, SHELLS AND BLADED RIMMED DISCS USING THE TRANSFER-MATRIX-FINITE-ELEMEN	73-1151
PLATES	VIBRATION AND BUCKLING ANALYSIS OF AXISYMMETRIC POLAR ORTHOTROPIC CIRCULAR PLATES	74-0574
PLATES	VIBRATION AND STABILITY OF PLATES USING FINITE ELEMENTS	68-0244
PLATES	VIBRATION OF CIRCULAR AND ANNULAR PLATES USING FINITE ELEMENTS	72-0194
PLATES	VIBRATION OF FOUR POINT-SUPPORTED PLATES BY A FINITE ELEMENT METHOD	71-0592
PLATES	VIBRATION OF RECTANGULAR PLATES WITH MIXED BOUNDARY CONDITIONS	73-1008
PLATES	VIBRATION OF SQUARE PLATES HAVING VARIOUS EDGE CONDITIONS AND LOADINGS USING A REFINED FINITE ELEMEN	75-0024
PLATES	VIBRATION OF STIFFENED PLATES USING FINITE-ELEMENTS	72-0951
PLATES	VIBRATIONS OF THICK PLATES USING A HIGH PRECISION TRIANGULAR ELEMENT	74-1002
PLATES	VIBRATORY BEHAVIOR OF TWISTED CANTILEVERED PLATES	75-0507
PLATES	VIBROACOUSTIC RESPONSE OF TURBULENCE EXCITED THIN RECTANGULAR FINITE PLATES IN HEAVY AND FLUID MEDIA	74-1309
PLATES	ZASTOSOWANIE ITERACJI SEIDLA W METODZIE ELEMENTOW SKONCZONYCH NA PRZYKLADZIE OBLICZEN STATYCZNYCH P	74-1031
PLATES	ZUR BERECHNUNG VON PLATTEN NACH DER THEORIE ZWEITER. ORDNUNG MIT HILFE EINES HYBRIDEN DEFORMATIONS M	73-0558
PLATES	ZUR BERECHNUNG, BEMESSUNG UND BEWEHRUNG RECHTECKIGER PLATTEN KONSTNATER DICKE MIT OEFFNUNGEN ODER AU	73-0664
PLATES	ZUR GENAUEN BERECHNUNG KONSTRUKTIV ORTHOTROPER PLATTEN (EXACT CALCULATION OF STRUCTURALLY ORTHOTROPI	72-0188
PLATES	ZUR WIRKLICHKEITSNAHEN BERECHNUNG VON STAHLBETONPLATTEN MIT DER FINITE-ELEMENT-METHODE. (APPROXIMATE	75-0832
PLATFORM	DESIGN, ANALYSIS AND CONSTRUCTION OF THE RESEARCH PLATFORM (NORDSEE)	75-0261
PLATFORM	DISCONTINUITY STRESS DECAY IN OFFSHORE PLATFORM JOINTS	74-0491
PLATFORM	OFFSHORE GRAVITY BASED OIL PRODUCTION PLATFORM INTERACTION WITH THE SEA BED	75-0221
PLATFORMS	DESIGN APPRAISAL OF OFFSHORE PLATFORMS OF WELDED TUBULAR CONSTRUCTION	74-1165
PLATFORMS	STOCHASTIC DYNAMIC RESPONSE ANALYSIS OF OFFSHORE PLATFORMS, WITH PARTICULAR REFERENCE TO GRAVITY-TYP	75-0220
PLATFORMS	STOCHASTIC DYNAMIC RESPONSE ANALYSIS OF OFFSHORE PLATFORMS, WITH PARTICULAR REFERENCE TO GRAVITY-TYP	75-0220
PLATING	FULL-RANGE ANALYSIS OF STEEL PLATES AND STIFFENED PLATING UNDER UNIAXIAL COMPRESSION	75-0803
PLATTE	THEORETICAL AND EXPERIMENTAL CONSIDERATIONS ON FLOW OF VISCOUS LIQUIDS PAST PLATE AT SMALL AND AVERA	70-0144
PLATTEN	A NEW, FULLY COMPATIBLE FINITE ELEMENT FOR PLATE BENDING EIN NEUES VOLLVERTRAGLICHES ENDLICHES ELEME	68-0032
PLATTEN	ACCURATE CALCULATION OF STRESSES IN BEAMS, DISKS AND PLATES BY MEANS OF FINITE ELEMENT METHODS AND I	70-0133
PLATTEN	BERECHNUNG VON PLATTEN NACH DEM RITZ' SCHEN VERFAHREN MIT HILFE DREIECKFOERMIGER MASCHENNETZE	66-0094
PLATTEN	DER GERISSENE ZUSTAND ZWEISEITIG GELAGERTER PLATTEN UNTER EINZELLASTEN - NICHTLINEARE BERECHNUNG MIT	75-0858
PLATTEN	STABILITY INVESTIGATIONS OF PLATES USING FINITE ELEMENT METHOD BASED ON AN EXPANDED VARIATION PRINCI	71-0663
PLATTEN	ZUR BERECHNUNG VON PLATTEN NACH DER THEORIE ZWEITER. ORDNUNG MIT HILFE EINES HYBRIDEN DEFORMATIONS M	73-0558
PLATTEN	ZUR BERECHNUNG, BEMESSUNG UND BEWEHRUNG RECHTECKIGER PLATTEN KONSTNATER DICKE MIT OEFFNUNGEN ODER AU	73-0664
PLATTEN	ZUR GENAUEN BERECHNUNG KONSTRUKTIV ORTHOTROPER PLATTEN (EXACT CALCULATION OF STRUCTURALLY ORTHOTROPI	72-0188
PLATTENSTATIK	UEBER DIE PRAKTISCHE ANWENDUNG VON FINITEN ELEMENTEN IN DER PLATTENSTATIK (APPLICATION OF THE FINITE	75-0805
PLOCHYCH	GEOMETRICKY NELINEARNE ULOHY PLOCHYCH SKRUPIN RIESENE PRIRASTKOVOU METODOU KONECNYCH PRVKOV. (GEOMET	74-1032
PLOSKIKH	CHISLENNOE RESHENIE PLOSKIKH ZADACH TEPLOPROVODNOSTI DLYA OBLASTEI SLOZHNOI FORMY (NUMERICAL SOLUTIO	74-0659
PLOSKOSTI	ISSLEDOVANIE PROCHNOSTI I ZHESTKOSTI PLASTIN, NAGRUZHENNYKH V SVOEI PLOSKOSTI, METHODOM KONECHNYKH E	73-0382
PLOT	MESH AND CONTOUR PLOT FOR TRIANGLE AND ISOPARAMETRIC ELEMENTS	70-0445
PLOTINY	RASCHET METODOM KONECHNYKH ELEMENTOV DINAMICHESKIKH KHARAKTEERISTIK ZEMLYANOI PLOTINY SOVMESTNO S OS	73-0437
PLOTS	PLOTS - THE PLOTTING SUBROUTINE INCORPORATED IN THE BALLISTIC RESEARCH LABORATORIES ELECTRONIC SCIEN	72-0803
PLOTTER	AN ALPHANUMERIC FINITE ELEMENT MESH PLOTTER	72-0543
PLOTTER	THE NASTRAN CONTOUR PLOTTER	72-0703
PLOTTING	A SPECIAL NASTRAN PROGRAM FOR INPUT CHECKING AND UNDEFORMED STRUCTURE PLOTTING	71-0234
PLOTTING	COMPUTER PROGRAMS FOR PLOTTING FINITE ELEMENT PATTERNS OF TWO-AND THREE-DIMENSIONAL STRUCTURES	70-0416
PLOTTING	CONPLOT: A GENERAL PURPOSE CONTOUR PLOTTING PROGRAM	74-0887
PLOTTING	MESH GENERATION CODE AND A CONTOUR PLOTTING ROUTINE	72-0802
PLOTTING	NASTRAN PLOTTING AT A REMOTE TERMINAL	74-0186
PLOTTING	NASTRAN PLOTTING CAPABILITIES	71-0172
PLOTTING	PLOTS - THE PLOTTING SUBROUTINE INCORPORATED IN THE BALLISTIC RESEARCH LABORATORIES ELECTRONIC SCIEN	72-0803
PLUG	COMPUTER MODELLING OF PLUG SEEDLING SURVIVAL	75-0176
PLUG	PHOTOELASTIC AND FINITE ELEMENT INVESTIGATION OF A NONSYMMETRICAL PLUG HATCH CONFIGURATION	73-0025
PLUG	STRESS DISTRIBUTIONS AROUND PLUG MATCHES WITH ANCHOR BLOCKS	69-0135
PLUMES	NUMERICAL ANALYSIS OF COOLING TOWER PLUMES AND FOG INCIDENCE	74-0040
PLUNGING	FINITE ELEMENT ANALYSIS AND COMPUTER GRAPHICS VISUALIZATION OF FLOW AROUND PITCHING AND PLUNGING AIR	73-0385
PLYT	FUNKCJA KSZTALTU NIESPELNIAPACA WARUNKOW CIAGLOSCI W ANALIZIE PLYT METODA ELEMENTOW SKONCZONYCH. (SH	75-0263
PLYT	NUMERYCZNE ROZWIAZANIE PROBLEMU DRGAN WLASNYCH NA PRZYKLADZIE PLYT. (NUMERICAL SOLUTION OF THE PROBL	75-0296
PLYT	ZASTOSOWANIE ITERACJI SEIDLA W METODZIE ELEMENTOW SKONCZONYCCH NA PRZYKLADZIE OBLICZEN STATYCZNYCH P	74-1031
PLYWOOD	FINITE ELEMENT ANALYSIS OF PLYWOOD PLATES	67-0052
PNEUMATIC	A STUDY OF INTERNAL STRESSES IN STATICALLY DEFORMED PNEUMATIC TIRES. A MATHEMATICAL MODEL FOR THE PN	70-0421
PNEUMATIC	A STUDY OF INTERNAL STRESSES IN STATICALLY DEFORMED PNEUMATIC TIRES. A MATHEMATICAL MODEL FOR THE PN	70-0421
PNEUMATIC	ANALYSIS OF PNEUMATIC SHELLS WITH OR WITHOUT CABLE FORMULATION	74-0591
PNEUMATIC	APPLICATION OF FINITE ELEMENT METHOD TO PNEUMATIC TYRE	75-0119
PNEUMATIC	NONLINEAR HYBRID STRESS FINITE ELEMENT ANALYSIS OF PNEUMATIC TIRES	75-0029
PNEUMATIC	NUMERICAL ANALYSIS OF NONLINEAR PNEUMATIC STRUCTURES	67-0060
POCKET	APPLICATION OF FINITE ELEMENT ANALYSIS TO HYDRODYNAMIC AND EXTERNALLY PRESSURIZED POCKET BEARINGS	72-0295
PODDED	THE APPLICATION OF THREE-DIMENSIONAL FINITE ELEMENTS TO THE ANALYSIS OF PODDED BOILER TYPE PRESTRESS	71-0047
PODKREPLENNYKH	CALCULATION OF THE REINFORCED THIN-WALLED STRUCTURES BY THE METHOD OF FINITE ELEMENTS (RASCHET PODKR	75-0685
PODZEMNOGO	RASCHET NAPRYAZHENIL L PEREMESHCHENII V SVODE PODZEMNOGO MASHINNOGO ZALA GES I V OKRUZHAYUSHCHEM SKA	74-0542
POETAPNOSTI	RASCHET NAPRYAZHENIL L PEREMESHCHENII V SVODE PODZEMNOGO MASHINNOGO ZALA GES I V OKRUZHAYUSHCHEM SKA	74-0542
POINT	A METHOD OF LIMIT POINT CALCULATION IN FINITE ELEMENT STRUCTURAL ANALYSIS	72-0786
POINT	APPLICATION OF FINITE ELEMENT TECHNIQUES TO POST-YIELD ANALYSIS OF PROPOSED STANDARD 3 POINT BEND FR	74-0118
POINT	DEVELOPMENT AND USE OF COMPUTER PROGRAMS FOR FINITE ELEMENT ANALYSIS. (LA MISE AU POINT ET L'UTILIZA	69-0289
POINT	ELASTIC DEFORMATION OF LIGHTWEIGHT MIRRORS METHOD OF LIMIT POINT CALCULATION IN FINITE ELEMENT STRUC	72-0232
POINT	EVALUATION OF NASTRAN FROM THE THE USER'S POINT OF VIEW	73-0596
POINT	FINITE ELEMENT METHOD - ACCURACY AT A POINT	74-0240
POINT	HERMITE INTERPOLATION TYPE RITZ METHODS FOR TWO POINT BOUNDARY VALUE PROBLEMS	66-0089
POINT	LOAD POINT COMPLIANCE OF THE CHARPY IMPACT SPECIMEN	73-0775
POINT	STRESS DISTRIBUTION NEAR A BROKEN POINT OF FILAMENT IN UNIDIRECTIONALLY FIBER REINFORCED COMPOSITES	73-0811
POINT-SUPPORTED	ON THE VIBRATION OF POINT-SUPPORTED PLATES	71-0602
POINT-SUPPORTED	VIBRATION OF FOUR POINT-SUPPORTED PLATES BY A FINITE ELEMENT METHOD	71-0592
POINTS	DETERMINATION OF STRESSES AT NODAL POINTS FROM FINITE ELEMENT SOLUTIONS TO ELASTIC PROBLEMS	71-0307
POINTS	LOCATION OF POINTS AT WHICH FINITE ELEMENT SOLUTIONS ARE VERY ACCURATE	74-0112
POINTS	ON FLOW NEAR SINGULAR POINTS OF A WALL BOUNDARY	74-0008
POINTWISE	POINTWISE CONVERGENCE OF FINITE-ELEMENTS	75-0821
POINTWISE	THE L2 AND LOO CONDITION NUMBERS OF THE FINITE ELEMENT MATRICES AND THE POINTWISE CONVERGENCE OF THE	75-0061
POISEUILLE	PLANE POISEUILLE FLOW OF A RAREFIED GAS BASED ON FINITE ELEMENT METHOD	71-0575
POISSON	LAGRANGIAN FINITE ELEMENT AND FINITE DIFFERENCE METHODS FOR POISSON PROBLEMS	74-1276
POISSON	NUMERICAL SOLUTION OF THE QUASI LINEAR POISSON EQUATION IN A NON-UNIFORM TRIANGLE MESH	67-0090
POISSON'S	DETERMINATION OF YOUNG'S MODULUS AND POISSON'S RATIO FOR WC-CO ALLOYS BY THE FINITE ELEMENT METHOD	72-0142
POISSON'S	FINITE ELEMENT TECHNIQUES FOR SOLUTION OF POISSON'S EQUATION	71-0494
POISSON'S	INFLUENCE OF POISSON'S RATIO ON THE CONDITION OF THE FINITE ELEMENT STIFFNESS MATRIX	71-0519
POISSON'S	RECTANGULAR AND TRIANGULAR FINITE ELEMENTS FOR POISSON'S EQUATION	75-0802
POISSON'S	SIMULATION OF ELASTIC MODULUS AND POISSON'S RATIO OF SPHEROIDAL GRAPHITE CAST IRON	74-0955
POISSON'S	TWO TYPES OF PIECEWISE QUADRATIC SPACES AND THEIR ORDER OF ACCURACY FOR POISSON'S EQUATION	72-0520
POISSONS	DEPENDENCE OF PLATE-BENDING FINITE ELEMENT DEFLECTIONS AND EIGENVALUES ON POISSONS RATIO	74-0567
POISSONS	DEPENDENCE OF PLATE-BENDING FINITE ELEMENT DEFLECTIONS AND EIGENVALUES ON POISSONS RATIO	74-0583
POISSONS	THE ECONOMICAL SOLUTION OF ELASTIC PROBLEMS FOR A RANGE OF POISSONS RATIO	74-0786
POLAR	FINITE ELEMENT ANALYSIS IN POLAR CO-ORDINATES OF THE SAINT VENANT TORSION PROBLEM	74-0500

328

POLAR	FINITE PLATE-BENDING ELEMENTS IN POLAR COORDINATES	68-0234
POLAR	POLAR GEOMETRY WAVEGUIDES BY FINITE-ELEMENT METHOD	74-0109
POLAR	VIBRATION AND BUCKLING ANALYSIS OF AXISYMMETRIC POLAR ORTHOTROPIC CIRCULAR PLATES	74-0574
POLARIZED	A FINITE-ELEMENT APPROACH TO THE ANALYSIS OF TANGENTIALLY POLARIZED PIEZOELECTRIC-CERAMIC FREE-FLOOD	74-0802
POLARIZED	3-DIMENSIONAL FINITE-ELEMENT ANALYSIS OF ELECTROELASTIC NORMAL MODES OF A TANGENTIALLY POLARIZED, SE	75-0591
POLEI	OPREDELENIE STATSIONARNYKH TEMPERATURNYKH POLEI V DETALYAKH DVIGATELEI VNUTRENNEGO SGORANIYA METODOM	73-0454
POLEI	RASCHET POLEI DEFORMATSII I NAPRYAZHENII V DETALYAKH DVIGATELEI VNUTRENNEGO SGORANIYA (CALCULATION O	73-0465
POLEI	RASHET TREKHMERNYKH MAGNITNYKE POLEI METODOM KONECHNYKH ELEMENTOV. (CALCULATION OF THREE-DIMENSIONAL	75-0766
POLEI	REALIZATSIYA METODA KONECHNYKH ELEMENTOV NA EVM DLYA RASCHETA DVUMERNYKH ELEKTRICHESKIKH I MAGNITNYK	74-0251
POLLUTANTS	A STUDY OF CONVECTIVE-DISPERSION EQUATION BY ISOPARAMETRIC FINITE ELEMENTS. TRANSPORT OF POLLUTANTS	75-0773
POLLUTION	GALERKIN FINITE ELEMENT SOLUTIONS FOR POLLUTION PROBLEMS IN PARTIALLY MIXED ESTUARIES	74-0047
POLYAXIAL	A STATISTICAL THEORY FOR THE FRACTURE OF BRITTLE STRUCTURES SUBJECTED TO NONUNIFORM POLYAXIAL STRESS	73-0928
POLYCRYSTALLINE	COMPUTER SIMULATION OF THE DEFORMATION OF COMPOSITE AND POLYCRYSTALLINE MATERIALS BY FINITE ELEMENT	72-0937
POLYCRYSTALLINE	ELASTIC DEFORMATION OF POLYCRYSTALLINE METAL - 1. INFLUENCE OF GRAIN SHAPE ON DISTRIBUTIONS OF STRES	72-0463
POLYCRYSTALLINE	ELASTIC DEFORMATION OF POLYCRYSTALLINE METAL - 2 EFFECT OF SLENDERNESS OF GRAINS ON DEFORMATION MODE	73-0275
POLYCRYSTALLINE	ELASTIC DEFORMATION OF POLYCRYSTALLINE METALS - 3	73-0361
POLYCRYSTALS	NON-UNIFORM FLOW OF POLYCRYSTALS BY GRAIN-BOUNDARY SLIDING ACCOMMODATED BY POWER-LAW CREEP	75-0562
POLYCRYSTALS	PLASTICITY OF POLYCRYSTALS AND GENERAL DISCUSSIONS ON PLASTICITY THEORY WITH IT	75-0312
POLYGONAL	A POLYGONAL FINITE ELEMENT FOR PLATE BENDING PROBLEMS USING THE ASSUMED STRESS APPROACH	69-0043
POLYGONAL	AN ADAPTIVE COMPUTER PROGRAM FOR DIV. (P(X,Y) GRAD U) EQUALS F(X,Y,U) IN A POLYGONAL REGION	75-0133
POLYGONAL	APPLICATION OF CONFORMAL MAPPING AND VARIATIONAL METHOD TO THE STUDY OF HEAT CONDUCTION IN POLYGONAL	71-0444
POLYGONAL	FLEXURAL VIBRATIONS OF RECTANGULAR AND OTHER POLYGONAL PLATES	71-0417
POLYGONAL	TWO NEARLY POLYGONAL HOLES. MATHEMATICAL CRACK PROBLEMS	69-0061
POLYGONS	PIECEWISE ANALYTIC INTERPOLATION AND APPROXIMATION IN TRIANGULATED POLYGONS	72-0508
POLYHEDRAL	BEHAVIOR OF TRIANGULAR SHELL ELEMENT STIFFNESS MATRICES ASSOCIATED WITH POLYHEDRAL DEFLECTION DISTRI	68-0159
POLYHEDRAL	FINITE-ELEMENT NONLINEAR BUCKLING ANALYSIS OF A PSEUDO-CYLINDRICAL CONCAVE POLYHEDRAL SHELL UNDER EX	73-1147
POLYMER	ANALYSIS OF BRIDGE DECKS USING POLYMER IMPREGNATED CONCRETE	73-0673
POLYMER	FINITE ELEMENT ANALYSIS OF DIAMETRAL TEST OF POLYMER MOLDINGS	73-0129
POLYMER	FINITE ELEMENT APPROACH TO POLYMER CONCRETE BRIDGE DECK DESIGN ANALYSIS	72-0918
POLYMER-IMPREGN	POLYMER-IMPREGNATED CONCRETE TUNNEL SUPPORT AND LINING	73-0736
POLYNOMIAL	AUTOMATIC GENERATION OF TRIANGULATIONS FOR PIECEWISE POLYNOMIAL APPROXIMATION	73-1118
POLYNOMIAL	COMPLETE POLYNOMIAL DISPLACEMENT FIELDS FOR FINITE ELEMENT METHODS	68-0062
POLYNOMIAL	COMPLETE POLYNOMIAL DISPLACEMENT FIELDS FOR FINITE ELEMENT METHODS	68-0092
POLYNOMIAL	CURVED BOUNDARY ELEMENTS, GENERAL FORMS OF POLYNOMIAL MAPPINGS	74-1256
POLYNOMIAL	HIGH ORDER POLYNOMIAL TRIANGULAR FINITE ELEMENTS FOR POTENTIAL PROBLEMS	69-0240
POLYNOMIAL	LEAST SQUARE POLYNOMIAL SPLINE APPROXIMATION	72-0526
POLYNOMIAL	PIECEWISE POLYNOMIAL INTERPOLATION AND APPROXIMATION	65-0058
POLYNOMIAL	PIECEWISE POLYNOMIAL INTERPOLATIONS AND THEIR APPLICATIONS TO PARTIAL DIFFERENTIAL EQUATIONS	70-0458
POLYNOMIAL	PIECEWISE POLYNOMIAL INTERPOLATIONS IN THE FINITE ELEMENT METHOD	73-1028
POLYNOMIAL	TETRAHEDRAL POLYNOMIAL FINITE ELEMENTS FOR THE HELMHOLTZ EQUATION.	72-0392
POLYNOMIAL	TRICUBIC POLYNOMIAL INTERPOLATION	71-0377
POLYNOMIALS	APPLICATION OF HERMITE POLYNOMIALS TO THE DETERMINATION OF THE STIFFNESS MATRIX OF PLATE BY FINITE E	71-0365
POLYNOMIALS	CONFORMING FINITE ELEMENTS BASED ON COMPLETE POLYNOMIALS	74-0302
POLYNOMIALS	DYNAMIC STIFFNESS MATRIX FORMULATION BY MEANS OF HERMITIAN POLYNOMIALS	66-0059
POLYNOMIALS	EXPERIENCES WITH ORTHOGONAL POLYNOMIALS AND "BEST" NUMERICAL INTEGRATION FORMULAS ON A TRIANGLE; WIT	74-0977
POLYNOMIALS	HELPFUL FORMULAS FOR INTEGRATING POLYNOMIALS IN THREE DIMENSIONS	64-0031
POLYNOMIALS	INTERPOLATION POLYNOMIALS ON THE TRIANGLE	70-0235
POLYNOMIALS	INTERPOLATION POLYNOMIALS ON THE TRIANGLE ELEMENT	70-0326
POLYNOMIALS	PIECEWISE POLYNOMIALS AND FINITE ELEMENT METHOD	73-0041
POLYNOMIALS	RAYLEIGH-RITZ APPROXIMATION OF PIECEWISE CUBIC POLYNOMIALS	66-0087
POLYNOMIALS	SURVEY OF UNDERLYING THEORY, APPROXIMATION BY PIECEWISE POLYNOMIALS, ILLEGAL ELEMENTS AND THE PATCH	73-0635
POLYNOMIALS	THE USE OF INTERPOLATORY POLYNOMIALS FOR A FINITE ELEMENT SOLUTION OF THE MULTIGROUP DIFFUSION EQUAT	72-0524
PONCTUELLE	ANALYSE NUMERIQUE DU CHAMP MAGNETIQUE D'UN ALTERNATEUR PAR ELEMENTS FINIS ET SURRELAXATION PONCTUELL	73-0625
POND	FINITE ELEMENT MODEL FOR TRANSIENT TWO-LAYER COOLING POND BEHAVIOR	75-0465
POND	MODELING AN OCEAN POND. A TWO-DIMENSIONAL, FINITE ELEMENT HYDRODYNAMIC MODEL OF NINIGRET POND, CHARL	75-0686
POND	MODELING AN OCEAN POND. A TWO-DIMENSIONAL, FINITE ELEMENT HYDRODYNAMIC MODEL OF NINIGRET POND, CHARL	75-0686
POND	SEEPAGE-ENVIRONMENTAL ANALYSIS OF THE SLIME ZONE OF A TAILINGS POND	74-0513
POND	ULTIMATE STRAIN CAPACITY AND TEMPERATURE RISE STUDIES, TRUMBULL POND DAM	72-0705
PONTS	CALCUL DES PONTS BIAIS A POUTRES MULTIPLES SANS ENTRETOISES PAR LA METHODE DES ELEMENTS (CALCULATION	72-0184
PONTS	CALCULATION OF THE BRIDGE DECKS BY USING NUMERICAL METHODS. (LE CALCUL DES TABLIERS DE PONTS PAR DES	71-0411
PONTS	LE CALCUL DES TABLIERS DE PONTS PAR DES METHODES NUMERIQUES	72-0197
PORE	STRESSES IN NEARLY INCOMPRESSIBLE MATERIALS BY FINITE ELEMENTS WITH APPLICATION TO THE CALCULATION O	74-0383
POROUS	A FINITE DIFFERENCE METHOD FOR UNSTEADY FLOW IN VARIABLY SATURATED POROUS MEDIA	71-0586
POROUS	A FINITE ELEMENT APPROACH TO TWO-PHASE FLOW IN POROUS MEDIA	75-0392
POROUS	AN APPLICATION OF THE FINITE ELEMENT METHOD TO TRANSIENT INTERFACE PROBLEMS IN POROUS MEDIA	73-0608
POROUS	AN INTRODUCTION TO THE FINITE ELEMENT ANALYSIS OF THE MOTION OF A GAS-LIQUID INTERFACE IN A POROUS M	72-0904
POROUS	APPLICATION OF FINITE ELEMENT METHODS TO LUBRICATION: SQUEEZE FILMS BETWEEN POROUS SURFACES	75-0764
POROUS	APPLICATION OF THE FINITE ELEMENT METHOD TO TRANSIENT FLOW IN POROUS MEDIA	68-0094
POROUS	CALCULATION OF FLOWS IN POROUS MEDIA BY THE FINITE ELEMENT METHOD	70-0473
POROUS	DYNAMIC STRESS ANALYSIS OF POROUS ELASTIC SOLIDS SATURATED WITH COMPRESSIBLE FLUIDS	71-0651
POROUS	FINITE ELEMENT ANALYSIS OF THE MOTION OF A GAS-LIQUID INTERFACE IN A POROUS MEDIUM	75-0672
POROUS	FINITE ELEMENT COMPUTER PROGRAMS FOR FLOW IN SATURATED UNSATURATED POROUS MEDIA	72-0364
POROUS	FINITE ELEMENT GALERKIN METHOD FOR ANALYSIS OF FLOW IN FRACTURED POROUS MEDIA	74-0066
POROUS	FINITE ELEMENT METHODS FOR FLOW IN POROUS MEDIA	74-0057
POROUS	FINITE ELEMENT MODELING OF FLOW THROUGH POROUS MEDIA	75-0706
POROUS	FINITE ELEMENT SOLUTION OF UNSTEADY AND UNSATURATED FLOW IN POROUS MEDIA	72-0064
POROUS	FINITE ELEMENT SOLUTION OF COUPLED ELECTROKINETIC AND HYDRODYNAMIC FLOW IN POROUS MEDIA	72-0398
POROUS	FINITE ELEMENT SOLUTION OF TWO DIMENSIONAL UNSTEADY AND UNSATURATED FLOW IN POROUS MEDIA	75-0113
POROUS	FINITE ELEMENT STRESS ANALYSIS OF POROUS MEDIA	69-0440
POROUS	FINITE ELEMENT STUDIES OF SOILS AND POROUS MEDIA	71-0504
POROUS	FINITE ELEMENT WEIGHTED RESIDUAL SOLUTION OF ONE-DIMENSIONAL, UNSTEADY AND UNSATURATED FLOWS IN PORO	73-0712
POROUS	FINITE ELEMENTS FOR TWO PHASE FLOW IN POROUS MEDIA	74-0060
POROUS	FLOW OF COMPRESSIBLE FLUID IN POROUS ELASTIC MEDIA	73-1004
POROUS	GALERKIN APPROACH TO SATURATED-UNSATURATED FLOW IN POROUS MEDIA	75-0393
POROUS	GENERALIZED VARIATIONAL PRINCIPLES FOR USE IN FLOW PROBLEMS THROUGH POROUS MEDIA	74-0067
POROUS	HIGHER ORDER FINITE ELEMENT METHODS FOR THE SOLUTION OF COMPRESSIBLE POROUS BEARING PROBLEMS	75-0341
POROUS	MEASUREMENT OF THE WATER DIFFUSIVITY OF POROUS MATERIALS	74-1138
POROUS	NON-LINEAR FLOW IN POROUS MEDIA BY FINITE ELEMENTS	69-0023
POROUS	ON THE SOLUTION OF TRANSIENT FREE-SURFACE FLOW PROBLEMS IN POROUS MEDIA BY THE FINITE ELEMENT METHOD	73-1019
POROUS	POROUS MEDIA - SOME PROBLEMS IN CIVIL AND PETROLEUM ENGINEERING	74-0349
POROUS	TRANSIENT FREE-SURFACE FLOW IN POROUS MEDIA BY FINITE ELEMENT METHOD	72-0297
POROUS	USE OF HYBRID COMPUTATION FOR SOLVING TRANSIENT FLOW PROBLEM IN POROUS MEDIA BY FINITE ELEMENT METHO	74-0068
POROUS	VARIATIONAL FORMULATION OF DYNAMICS OF FLUID-SATURATED POROUS ELASTIC SOLIDS	72-0017
POROUS	VARIATIONAL PRINCIPLES FOR FLUID FLOW IN POROUS MEDIA	71-0844
POROUS	WATER FLOW THROUGH SATURATED-UNSATURATED POROUS MEDIA USING FINITE-ELEMENT-GALERKIN METHODS	74-0879
POROUS	WATER MOVEMENT THROUGH SATURATED-UNSATURATED POROUS MEDIA: A FINITE-ELEMENT GALERKIN MODEL	75-0761
PORT	FINITE ELEMENT ANALYSES OF PORT ALLEN AND OLD RIVER LOCKS	69-0203
PORT	FINITE ELEMENT ANALYSIS OF PORT ALLEN LOCK	71-0293
PORTABILITY	PORTABILITY AND ADAPTABILITY OF CIVIL ENGINEERING SOFTWARE	73-0343
PORTAL	THE NATURAL FREQUENCIES OF PORTAL FRAMES II	65-0047
PORTANTE	ETUDE DE LA CAPACITE PORTANTE DES PIEUX PAR LA METHODE DES ELEMENTS FINIS. INFLUENCE DE LA DILATANCE	75-0624
PORTANTE	LA CAPACITE PORTANTE DES PIEUX. (BEARING CAPACITY OF PILES)	75-0463

PORTEUSE	STRIP PROGRAM FOR DESIGNING LOAD BEARING SURFACE STRUCTURES (PROGRAMME STRIP POUR LE CALCUL DES STRU	71-0408
PORV	THE SCANDINAVIAN PORV MODEL PROJECT: STRESS CALCULATION AND EXPERIMENTAL VERIFICATION	71-0021
POSITION	GALERKIN-FINITE ELEMENT TECHNIQUE FOR CALCULATING TRANSIENT POSITION OF SALTWATER FRONT	75-0135
POSITION	OPTIMUM POSITION OF THE CENTRAL CLAY CORE OF A ROCKFILL DAM IN RESPECT TO ARCHING AND HYDRAULIC FRAC	73-0447
POSITIONED	STRUCTURAL ANALYSIS OF TALL BUILDINGS HAVING IRREGULARLY POSITIONED SHEAR WALLS	73-0282
POSITIVE	A COMPARATIVE STUDY OF SEVERAL CORE STORAGE SCHEMES FOR LARGE SPARSE POSITIVE DEFINITE MATRICES WITH	72-0678
POSITIVE	A FINITE ELEMENT METHOD FOR FRIEDRICH'S SYMMETRIC POSITIVE SYSTEMS	71-0146
POSITIVE	RAYLEIGH RITZ GALERKIN METHODS FOR SYMMETRIC POSITIVE DIFFERENTIAL EQUATIONS APPLICATION TO FINITE E	73-1030
POST	A FINITE ELEMENT TENSOR APPROACH TO PLATE BUCKLING AND POST BUCKLING	73-0883
POST	A MATRIX DISPLACEMENT METHOD ON PRE AND POST BUCKLING ANALYSIS OF LINERS FOR REACTOR VESSELS	71-0267
POST	A PROCEDURE FOR FINITE ELEMENT PLATE AND SHELL PRE AND POST BUCKLING ANALYSIS	71-0096
POST	ANALYSIS OF CYLINDRICAL SHELL ROOFS WITH POST TENSIONED EDGEBEAMS	74-1042
POST	FINITE ELEMENT POST BUCKLING ANALYSIS OF THIN ELASTIC PLATES	68-0123
POST	REPORT ON POST CONF. MEETING ON COMPUTATIONAL ASPECTS OF ASPECTS OF FINITE ELEMENT METHOD ISD STUTTG	72-0475
POST	THE USE OF FINITE ELEMENT MID-INCREMENT STIFFNESS MATRICES IN THE POST BUCKLING ANALYSIS OF IMPERFEC	71-0383
POST	THEORETICAL POST YIELDING BEHAVIOUR OF COMPOSITE LAMINATES - 1. INELASTIC MICROMECHANICS	73-0168
POST	THEORETICAL POST YIELDING BEHAVIOUR OF COMPOSITE LAMINATES -2.	73-0408
POST	VEHICLE STRUCTURAL ANALYSIS, PRE AND POST PROCESSOR DATA HANDLING	71-0093
POST-BUCKLING	FINITE ELEMENT ANALYSIS OF PLATE BUCKLING AND POST-BUCKLING	70-0332
POST-BUCKLING	FINITE ELEMENT ANALYSIS OF POST-BUCKLING BEHAVIOR	71-0808
POST-BUCKLING	FINITE ELEMENT ANALYSIS OF BUCKLING AND POST-BUCKLING BEHAVIORS OF ARCHES WITH GEOMETRIC IMPERFECTIO	73-0186
POST-BUCKLING	FINITE ELEMENT ELASTIC THIN SHELL PRE-BUCKLING AND POST-BUCKLING ANALYSIS	71-0821
POST-BUCKLING	FINITE ELEMENT MATRIX FORMULATION OF POST-BUCKLING STABILITY AND IMPERFECTION SENSITIVITY	70-0246
POST-BUCKLING	POST-BUCKLING ANALYSIS OF SLIGHTLY CURVED PLATES BY THE FINITE ELFMENT METHOD	69-0365
POST-BUCKLING	POST-BUCKLING AND NON-LINEAR ANALYSIS BY FINITE ELEMENT METHOD AS A SUPPLEMENT TO A LINEAR ANALYSIS	75-0505
POST-BUCKLING	POST-BUCKLING RESPONSE OF STRUCTURES USING THE FINITE ELEMENT METHOD	69-0414
POST-ELASTIC	POST-ELASTIC FINITE-ELEMENT ANALYSIS OF REINFORCED PLATES	72-0979
POST-TENSIONED	ANCHORAGE ZONE STRESSES IN AXIALLY POST-TENSIONED I-SECTION MEMBERS WITH END BLOCKS	71-0305
POST-TENSIONED	POST-TENSIONED CONCRETE NUCLEAR STRUCTURES	70-0040
POST-TENSIONED	POST-TENSIONED SLAB BRIDGES ON ISOLATED SUPPORTS	75-0084
POST-YIELD	APPLICATION OF FINITE ELEMENT TECHNIQUES TO POST-YIELD ANALYSIS OF PROPOSED STANDARD 3 POINT BEND FR	74-0118
POST-YIELD	COMPUTATION OF POST-YIELD BEHAVIOUR IN NOTCH-BEND AND TENSION TESTPIECES.	72-0244
POST-YIELDED	BALANCED STRESSES IN POST-YIELDED MULTI-MATERIAL STRUCTURAL JOINTS	69-0271
POSTBUCKLING	FINITE DEFLECTIONS, BUCKLING AND POSTBUCKLING OF AN ARCH	69-0123
POSTBUCKLING	FINITE ELEMENT ANALYSIS OF POSTBUCKLING BEHAVIOR OF STRUCTURES	73-0099
POSTBUCKLING	FINITE ELEMENT PROCEDURE FOR POSTBUCKLING ANALYSIS OF INITIALLY CURVED PLATES	71-0421
POSTBUCKLING	FINITE ELEMENT PROCEDURE FOR NONLINEAR PREBUCKLING AND INITIAL POSTBUCKLING ANALYSIS	72-0170
POSTBUCKLING	POSTBUCKLING BEHAVIOUR OF A SECTION REPPRESENTATIVE OF THE B-1 AFT INTERMEDIATE FUSELAGE	73-0489
POSTBUCKLING	STABILITY AND POSTBUCKLING BEHAVIOUR OF HYPERELASTIC BODIES AT FINITE STRAIN BY THE FINITE ELEMENT M	73-0562
POSTBUCKLING	STABILITY AND POSTBUCKLING BEHAVIOUR OF HYPERELASTIC BODIES AT FINITE STRAIN BY THE FINITE ELEMENT ME	73-0838
POSTPROCESSOR	A NASTRAN POSTPROCESSOR FOR STRUCTURAL MODIFICATION REANALYSIS	71-0160
POSTPROCESSORS	A SURVEY OF PRE AND POSTPROCESSORS FOR NASTRAN	74-0114
POSTULATE	SIMPLIFIED MINIMUM INTERNAL ENERGY OF DEFORMATION POSTULATE AND ITS APPLICATION TO METALFORMING PROC	74-1155
POTASSIUM-APRES	FINITE-ELEMENT MODEL FOR THE MECHANICAL BEHAVIOUR OF THE LEFT VENTRICLE, PREDICTION OF DEFORMATION I	72-0304
POTEAUX	ETUDES DES POUTRES ET POTEAUX PAR LA METHODE DES ELEMENTS FINIS. (STABILITY ANALYSIS OF BEAMS AND C	74-1008
POTENTIAL	A MIXED FINITE ELEMENT MODEL FOR USE IN POTENTIAL FLOW PROBLEMS	73-0078
POTENTIAL	A VARIATIONAL METHOD IN POTENTIAL FLOWS WITH A FREE SURFACE	72-0723
POTENTIAL	AN APPLICATION OF THE FINITE ELEMENT TECHNIQUE TO BOUNDARY VALUE PROBLEMS OF POTENTIAL FLOW	70-0002
POTENTIAL	APPLICATION OF DUAL MINIMUM THEOREMS TO THE FINITE ELEMENT SOLUTION OF POTENTIAL PROBLEMS WITH SPECI	72-0036
POTENTIAL	CALCULATION OF POTENTIAL FLOW ABOUT ARBITRARY THREE-DIMENSIONAL LIFTING BODIES	72-0722
POTENTIAL	CURVED CUBIC TRIANGULAR FINITE ELEMENT FOR POTENTIAL FLOW PROBLEMS	73-0423
POTENTIAL	DIFFERENCE AND ELEMENT METHODS FOR POTENTIAL FLOW	74-0019
POTENTIAL	DUAL FORMULATION FOR POTENTIAL ENERGY AND COMPLEMENTARY ENERGY APPLICATION TO THE FINITE ELEMENT MET	72-0062
POTENTIAL	FINITE ELEMENT ANALYSIS OF PROBLEMS FORMULATED BY AN INTEGRAL EQUATION; APPLICATION TO POTENTIAL FLO	68-0191
POTENTIAL	FINITE ELEMENT ANALYSIS OF SHELLS OF REVOLUTION BY MINIMIZATION OF THE POTENTIAL ENERGY FUNCTIONAL	69-0114
POTENTIAL	FINITE ELEMENT ANALYSIS OF SHELLS OF REVOLUTION BY MINIMIZATION OF THE POTENTIAL ENERGY FUNCTIONAL	69-0354
POTENTIAL	FINITE ELEMENT ANALYSIS OF SHELLS OF REVOLUTION BY MINIMIZATION OF THE POTENTIAL ENERGY FUNCTIONAL	70-0481
POTENTIAL	FINITE ELEMENT ANALYSIS OF POTENTIAL FLOW PROBLEMS	75-0869
POTENTIAL	FINITE ELEMENT APPROACH TO THE INTEGRATED POTENTIAL FORMULATION OF GENERAL UNSTEADY SUPERSONIC AEROD	73-0910
POTENTIAL	FINITE ELEMENT METHOD VIA LOCAL POTENTIAL	74-0038
POTENTIAL	FINITE ELEMENT SOLUTION OF STEADY STATE POTENTIAL FLOW PROBLEMS	70-0480
POTENTIAL	FINITE-ELEMENT METHOD FOR POTENTIAL CALCULATIONS IN A HALL PLATE	75-0454
POTENTIAL	FLUTTER OF FLAT FINITE-ELEMENT PANELS IN A SUPERSONIC POTENTIAL FLOW	75-0814
POTENTIAL	FREE BOUNDARY POTENTIAL FLOW USING FINITE ELEMENTS	71-0829
POTENTIAL	GENERAL POTENTIAL ENERGY AND COMPLEMENTARY ENERGY MODELS BASED ON STRESS PARAMETERS	74-1240
POTENTIAL	HIGH ORDER POLYNOMIAL TRIANGULAR FINITE ELEMENTS FOR POTENTIAL PROBLEMS	69-0024
POTENTIAL	IMPORTANCE OF VARIATIONAL CALCULUS IN FLUID MECHANICS AND ITS APPLICATION TO COMPRESSIBLE POTENTIAL	73-1103
POTENTIAL	LOCAL POTENTIAL APPROACH TO FINITE ELEMENT METHOD IN UNSTEADY VISCOUS INCOMPRESSIBLE FLUID FLOW	73-0298
POTENTIAL	MODIFIED POTENTIAL ENERGY MASS REPRESENTATION FOR FREQUENCY PREDICTION	66-0053
POTENTIAL	ON THE APPLICATION OF FINITE-ELEMENT TECHNIQUES TO PROBLEMS CONCERNING POTENTIAL DISTRIBUTION AND ST	69-0268
POTENTIAL	OPTIMIZATION OF FINITE ELEMENT GRIDS BASED ON MINIMUM POTENTIAL ENERGY	71-0747
POTENTIAL	OPTIMIZATION OF FINITE ELEMENT GRIDS BASED ON MINIMUM POTENTIAL ENERGY	72-0109
POTENTIAL	POTENTIAL FLOW ANALYSIS BY FINITE ELEMENTS	72-0004
POTENTIAL	THE APPLICATION OF THE FINITE ELEMENT TECHNIQUE TO POTENTIAL FLOW PROBLEMS: PART I	69-0087
POTENTIAL	THE APPLICATION OF THE FINITE ELEMENT TECHNIQUE TO POTENTIAL FLOW PROBLEMS: PART II	69-0395
POTENTIAL	THE APPLICATION OF THE FINITE ELEMENT TECHNIQUE TO POTENTIAL FLOW PROBLEMS	71-0370
POTENTIAL	THREE DIMENSIONAL ANALYSIS OF COMPRESSIBLE POTENTIAL FLOWS WITH THE FINITE ELEMENT METHOD	73-0058
POTENTIAL	THREE DIMENSIONAL ANALYSIS OF COMPRESSIBLE POTENTIAL FLOWS	75-0625
POTENTIAL	TWO AND THREE DIMENSIONAL POTENTIAL FLOW BY THE METHOD OF SINGULARITIES	69-0049
POTENTIAL	TWO AND THREE DIMENSIONAL POTENTIAL FLOW ANALYSIS BY THE METHOD OF SINGULARITIES	70-0357
POTENTIAL	TWO AND THREE DIMENSIONAL ANALYSIS OF SUBSONIC COMPRESSIBLE POTENTIAL FLOWS IN DUCTS AND AROUND LIFT	74-0026
POTENTIALSTROEM	IMPORTANCE OF VARIATIONAL CALCULUS IN FLUID MECHANICS AND ITS APPLICATION TO COMPRESSIBLE POTENTIAL	73-1103
POTENTIELLES	FORMULATIONS DUALES DES ENERGIES POTENTIELLES ET COMPLEMENTAIRES. APPLICATION A LA METHODE DES ELEME	71-0752
POTOKE	RASCHIOT USTOICHIVOSTI IZGIBNO-KRUTILNYKH KOLEBANIJ KRULA V DOZVUKOVOM POTOKE METODOM KONECHNYKH ELE	72-0903
POUTDRES	ETUDES DES POUTDRES ET POTEAUX PAR LA METHODE DES ELEMENTS FINIS. (STABILITY ANALYSIS OF BEAMS AND C	74-1008
POUTRE	MATRICE DE RIGIDITE D'UN ELEMENT DE POUTRE SOLLICITE A LA TORSION-FLEXION (STIFFNESS MATRIX OF A BEA	73-0442
POUTRES	CALCUL DES PONTS BIAIS A POUTRES MULTIPLES SANS ENTRETOISES PAR LA METHODE DES ELEMENTS (CALCULATION	72-0184
POWDER	GRAVITY FLOW OF POWDER IN A LUNAR ENVIRONMENT. PT. 2-ANALYSIS OF FLOW INITIATION	71-0484
POWER	ANALYSIS FOR ATOMIC POWER PLANT BY FINITE ELEMENT METHOD	71-0198
POWER	ANALYSIS OF BEHAVIOR OF EDWARD HYATT POWER PLANT	75-0055
POWER	APPLICATION OF FINITE ELEMENT METHOD TO BOILING WATER REACTOR POWER PLANT (JAPANESE)	72-0816
POWER	BEPECHNUNGSMETHODEN ZUR ERMITTLUNG VON BELASTUNGSBEDINGUNGEN UND GRENZTRAGFAEHIGKEITSVERHALTEN VON K	73-0858
POWER	EFFECTS OF FOUNDATION ROTATION ON SEISMIC INERTIA FORCES OF NUCLEAR POWER PLANT STRUCTURES	73-0177
POWER	EVALUATION OF TORSIONAL EFFECTS IN THE SEISMIC ANALYSIS OF SYMMETRIC AND UNSYMMETRIC NUCLEAR POWER P	73-0175
POWER	FINITE ELEMENT ANALYSIS IN NUCLEAR POWER PLANTS (JAPANESE)	72-0685
POWER	FINITE ELEMENT TREATMENT OF SOIL-STRUCTURE INTERACTION PROBLEM FOR NUCLEAR POWER PLANT UNDER SEISMIC	73-0834
POWER	MODELLING HEAT FLOW FROM UNDERGROUND POWER CABLES	74-1158
POWER	NON-LINEAR THERMAL STRESS ANALYSIS FOR NUCLEAR POWER PLANT BY FINITE ELEMENT METHOD	71-0181
POWER	POWER SERIES EXPANSION OF THE GENERAL STIFFNESS MATRIX FOR BEAM ELEMENTS	75-0543
POWER	POWER SPECTRAL DENSITY ANALYSIS OF VEHICLE VIBRATION USING THE NASTRAN COMPUTER PROGRAM	74-0318
POWER	PROBABILISTIC ASSESSMENT OF AIRCRAFT HAZARD FOR NUCLEAR POWER PLANTS	72-0235
POWER	PWR DEVELOPMENT FOR NUCLEAR PROPULSION: VIBRATION ANALYSIS OF NUCLEAR POWER PLANT	71-0797

POWER	ROCK MECHANICS STUDIES FOR A HIGH CUT AT A NUCLEAR POWER STATION IN PENNSYLVANIA	70-0122
POWER	STRUCTURAL DESIGN OF CIRCULAR DIVERSION, POWER AND PRESSURE TUNNELS	71-0315
POWER	TRANSIENT HEAT CONDUCTION PROBLEMS IN POWER CABLES SOLVED BY FINITE ELEMENT METHOD	73-0007
POWER	V-BELT LIFE PREDICTION AND POWER RATING	75-0838
POWER-LAW	NON-UNIFORM FLOW OF POLYCRYSTALS BY GRAIN-BOUNDARY SLIDING ACCOMMODATED BY POWER-LAW CREEP	75-0562
POWERHOUSE	RASCHET NAPRYAZHENIL L PEREMESHCHENII V SVODE PODZEMNOGO MASHINNOGO ZALA GES I V OKRUZHAYUSHCHEM SKA	74-0542
PRACTICAL	A BOUND THEOREM IN EIGENVALUES AND ITS PRACTICAL APPLICATIONS	71-0128
PRACTICAL	A GENERAL APPROACH TO THE PRACTICAL SOLUTION OF CREEP PROBLEMS	59-0003
PRACTICAL	COMPARISON OF PRACTICAL METHODS OF ANALYSING ELASTIC LATERALLY LOADED PLATES	71-0425
PRACTICAL	IMPACT OF FINITE ELEMENT TECHNIQUES ON PRACTICAL DESIGN OF SHIP STRUCTURES	69-0344
PRACTICAL	PRACTICAL ANALYSIS OF PLATE VIBRATIONS USING NASTRAN	71-0058
PRACTICAL	PRACTICAL ANALYSIS OF STEEL STRUCTURES USING HIGHER-ORDER ELEMENTS	72-0326
PRACTICAL	PRACTICAL APPLICATION OF A COMPUTERIZED STRUCTURAL ANALYSIS SYSTEM WHICH ADOPTS A FINITE ELEMENT TEC	69-0225
PRACTICAL	PRACTICAL APPLICATION OF THE FINITE ELEMENT METHOD AND SOME NUMERICAL ASPECTS	69-0309
PRACTICAL	PRACTICAL APPLICATION OF FINITE ELEMENT METHODS IN THE DESIGN OF STRUCTURES	69-0420
PRACTICAL	PRACTICAL APPLICATIONS OF FINITE ELEMENT METHODS IN THE STRESS ANALYSIS OF THIN SHELL PRESSURE VESSE	71-0205
PRACTICAL	PRACTICAL APPLICATION OF METHOD OF FINITE ELEMENTS	74-0162
PRACTICAL	PRACTICAL CONSIDERATIONS IN THE APPLICATION OF FINITE ELEMENT TECHNIQUES TO SHIP STRUCTURES	69-0338
PRACTICAL	PRACTICAL FINITE ELEMENT METHOD OF FAILURE PREDICTION FOR COMPOSITE MATERIAL STRUCTURES	75-0644
PRACTICAL	PRACTICAL HANDLING OF FINITE-ELEMENT METHOD 2	75-0547
PRACTICAL	PRACTICAL HANDLING OF FINITE-ELEMENT METHOD 1	75-0551
PRACTICAL	PRACTICAL HINTS ON HOW TO UNDERSTAND, MODEL, USE AND INTERPRET FINITE ELEMENTS	73-0141
PRACTICAL	PRACTICAL METHOD FOR COMPUTER CALCULATION OF STRUCTURES	69-0098
PRACTICAL	PRACTICAL PROBLEMS ASSOCIATED WITH THE APPLICATION OF THE FINITE ELEMENT METHOD TO COMPOSITE MATERIA	74-0483
PRACTICAL	PRACTICAL SOLUTION OF PLASTIC DEFORMATION PROBLEMS IN THE ELASTIC PLASTIC RANGE	59-0004
PRACTICAL	PRAKTISCHE RESULTATEN, DOEL EN NUT VAN HET ONTWERPEN VAN MECHANISCHE ELEMENTEN MET BEHULP VAN COMPUT	73-0400
PRACTICAL	SOME PRACTICAL APPLICATIONS OF ACOUSTIC FINITE ELEMENTS	75-0414
PRACTICAL	SOME PRACTICAL SOLUTION TECHNIQUES FOR FINITE ELEMENT ANALYSIS	72-0082
PRACTICE	FINITE ELEMENT ANALYSIS WITH MATERIAL NONLINEARITIES: THEORY AND PRACTICE	69-0459
PRACTICE	FINITE ELEMENT ANALYSIS WITH MATERIAL NONLINEARITIES, THEORY AND PRACTICE	71-0168
PRACTICE	IMPACT OF COMPUTERS ON THE PRACTICE OF STRUCTURAL ENGINEERING IN CONCRETE	72-0414
PRACTICE	IMPACT OF COMPUTERS ON THE PRACTICE OF STRUCTURAL ENGINEERING IN CONCRETE	72-0415
PRACTICE	ROCK MECHANICS - THEORY AND PRACTICE	70-0107
PRACTICE	THEORY AND PRACTICE IN FINITE ELEMENT STRUCTURAL ANALYSIS	73-0700
PRACTICES	CURRENT FINITE ELEMENT ANALYSIS PRACTICES	73-0118
PRAIRIE	A FINITE-ELEMENT, PLANAR-FLOW MODEL OF CAMAS PRAIRIE, IDAHO	72-0980
PRAKTIKABLEN	UNTERSUCHUNGEN ZUM FESTIGKEITSVERHALTEN VON METALLKLEBUNGEN IM HIMBLICK AUF DIE ERSTELLUNG EINES PRA	74-0289
PRAKTISCHE	COMPUTER-UNTERSTUTZTES KONSTRUIEREN, PRAKTISCHE BEDEUTUNG DER FINIT-ELEMENT-VERFAHREN	71-0552
PRAKTISCHE	PRAKTISCHE BEISPIELE DER DIGITALEN BERECHNUNG VON GRUNDWASSERSTROMUNGS PROBLEMEN NACH DER METHODE EN	73-0081
PRAKTISCHE	PRAKTISCHE RESULTATEN, DOEL EN NUT VAN HET ONTWERPEN VAN MECHANISCHE ELEMENTEN MET BEHULP VAN COMPUT	73-0400
PRAKTISCHE	UEBER DIE PRAKTISCHE ANWENDUNG VON FINITEN ELEMENTEN IN DER PLATTENSTATIK (APPLICATION OF THE FINITE	75-0805
PRATICABLE	UNTERSUCHUNGEN ZUM FESTIGKEITSVERHALTEN VON METALLKLEBUNGEN IM HIMBLICK AUF DIE ERSTELLUNG EINES PRA	74-0289
PRE	A MATRIX DISPLACEMENT METHOD ON PRE AND POST BUCKLING ANALYSIS OF LINERS FOR REACTOR VESSELS	71-0267
PRE	A PROCEDURE FOR FINITE ELEMENT PLATE AND SHELL PRE AND POST BUCKLING ANALYSIS	71-0096
PRE	A SURVEY OF PRE AND POSTPROCESSORS FOR NASTRAN	74-0714
PRE	VEHICLE STRUCTURAL ANALYSIS, PRE AND POST PROCESSOR DATA HANDLING	71-0093
PRE-BUCKLING	FINITE ELEMENT ELASTIC THIN SHELL PRE-BUCKLING AND POST-BUCKLING ANALYSIS	71-0821
PRE-PROCESSOR	MISSILE BODY INPUT GENERATOR (BING), A NASTRAN PRE-PROCESSOR; THEORETICAL DEVELOPMENT, USER'S MANUAL	75-0717
PRE-PROCESSOR	PLANFORM INPUT GENERATOR (PING), A NASTRAN PRE-PROCESSOR FOR LIFTING SURFACES-THEORETICAL DEVELOPMEN	73-0776
PRE-STRESSED	RECENT DEVELOPMENTS IN THE FINITE ELEMENT ANALYSIS OF PRE-STRESSED CONCRETE REACTOR VESSELS	74-0582
PRE-TWISTED	SIMPLE FINITE ELEMENTS FOR PRE-TWISTED BLADING VIBRATION	74-0239
PREBUCKLING	FINITE ELEMENT PROCEDURE FOR NONLINEAR PREBUCKLING AND INITIAL POSTBUCKLING ANALYSIS	72-0170
PRECAST	CONSTRUCTION SELF-WEIGHT STRESSES OF PRECAST SHEAR WALLS	71-0203
PRECAST	DEVELOPMENT OF A DESIGN FOR A PRECAST CONCRETE BRIDGE BEAM OF U-SECTION	73-0392
PRECAST	EXPERIMENTAL AND FINITE ELEMENT ANALYSIS OF CABLE ROOF STRUCTURES INCLUDING PRECAST PANELS	74-1139
PRECAST	EXPERIMENTELL UNDERSOKNING AV ELEMENTFOGARS HALLFASTHETSEGENSKAPER (EXPERIMENTAL INVESTIGATION OF ST	72-0192
PRECAST	THEORETICAL STUDY OF THE SEQUENCE OF FAILURE IN PRECAST PANEL SHEAR WALLS	73-0490
PRECIS	MATRIX METHODS OF STRUCTURAL ANALYSIS, A PRECIS OF RECENT DEVELOPMENTS	63-0019
PRECISION	A COMPARATIVE STUDY OF VARIABLE AND CONSTANT THICKNESS HIGH PRECISION TRIANGULAR PLATE BENDING ELEME	74-0857
PRECISION	A HIGH PRECISION AXISYMMETRIC TRIANGULAR ELEMENT USED IN THE ANALYSIS OF HYDRAULIC TURBINE COMPONENT	70-0038
PRECISION	A HIGH PRECISION TRIANGULAR PLATE BENDING ELEMENT	68-0052
PRECISION	APPLICATION OF THE HIGH PRECISION TRIANGULAR PLATE BENDING ELEMENT TO PROBLEMS WITH CURVED BOUNDARIE	69-0071
PRECISION	FINITE ELEMENT APPLICATIONS OF A PRECISION TRIANGULAR PLATE BENDING ELEMENT TO LINEAR BOUNDARY VALUE	75-0788
PRECISION	HIGH PRECISION FINITE ELEMENT ANALYSIS OF SHELLS OF REVOLUTION	71-0524
PRECISION	HIGH PRECISION FINITE ELEMENTS FOR PLANE ELASTIC PROBLEMS	72-0862
PRECISION	PRECISION OF FINITE ELEMENT METHOD IN PLANE STRESS	69-0295
PRECISION	STABILITY OF MODERATELY THICK RECTANGULAR PLATES USING A HIGH PRECISION TRIANGULAR FINITE ELEMENT	75-0334
PRECISION	VIBRATIONS OF THICK PLATES USING A HIGH PRECISION TRIANGULAR ELEMENT	74-1002
PRECOMPRESSIONE	DISTRIBUZIONI OTTIMALI DI ARMATURE DI PRECOMPRESSIONE IN TRAVI-PARETE DI CALCESTRUZZO, (OPTIMAL DIST	75-0560
PRECONTRAINT	ANALYSE TRIDIMENSIONALE DU COMPORTEMENT NON-LINEAIRE D'UN CAISSON DE REACTEUR NUCLEAIRE EN BETON PRE	71-0144
PRECONTRAINT	NONLINEAR ANALYSIS OF PRESTRESSED CONCRETE FAST REACTOR PRESSURE VESSEL, (ANALYSE NON LINEAIRE DES C	70-0145
PREDICT	APPLICATION OF FINITE ELEMENT METHOD TO PREDICT STATIC RESPONSE OF AUTOMOTIVE BODY STRUCTURES	71-0340
PREDICT	APPLICATION OF THE FINITE ELEMENT METHOD TO PREDICT STATIC AND DYNAMIC RESPONSE OF AN UNSHROUDED CEN	71-0574
PREDICT	COMPARISON OF THE CAPABILITY OF OUR HARDENING RULES TO PREDICT A MATERIAL'S PLASTIC BEHAVIOR	75-0523
PREDICT	FULMOD - AN INELASTIC ANALYSIS PROGRAM TO PREDICT THE OPERATING BEHAVIOUR OF CANDU FUEL ELEMENTS	75-0574
PREDICT	THE USE OF EIGEN-DERIVATIVES TO PREDICT THE EFFECT OF DESIGN CHANGES UPON THE DYNAMIC CHARACTERISTIC	74-0737
PREDICT	TIME-DEPENDENT ANALYSIS TO PREDICT CLOSURE IN SALT CAVITIES	74-1149
PREDICT	VERIFICATION OF FRACTURE MECHANICS CONCEPTS TO PREDICT CRACKING OF FLEXIBLE PAVEMENTS	73-0772
PREDICTED	MEASURED AND PREDICTED LONGITUDINAL VIBRATIONS OF A LIQUID PROPELLANT TWO-STAGE LAUNCH VEHICLE	70-0366
PREDICTED	OBSERVED AND PREDICTED DEFORMATIONS IN A LARGE EMBANKMENT DAM DURING CONSTRUCTION	71-0301
PREDICTING	A COMPUTER BASED PROCEDURE FOR PREDICTING THE TRANSIENT RESPONSE AND FAILURE OF A TWO-DIMENSIONAL CO	71-0630
PREDICTING	A COMPUTER MODEL FOR PREDICTING THE LOAD-DEFLECTION RESPONSE OF EXPEDIENT SOIL SURFACING	73-0880
PREDICTING	A FINITE ELEMENT MODEL FOR PREDICTING LONG BONE FRACTURE FIXATION	71-0819
PREDICTING	A MULTI-LAYERED FINITE ELEMENT MODEL FOR PREDICTING MINE SUBSIDENCE	68-0293
PREDICTING	AN ANALYTIC METHOD FOR PREDICTING THE FLOW AND FRACTURE IN NOTCHED FIBER COMPOSITE MATERIALS	75-0689
PREDICTING	ANALYTICAL METHODS FOR PREDICTING SUBSIDENCE ABOVE SOLUTION-MINED CAVITIES	74-1145
PREDICTING	FINITE ELEMENT COMPUTER PROGRAM FOR PREDICTING THE NONLINEAR STATIC AND DYNAMIC BEHAVIOR OF VISCOELA	75-0191
PREDICTING	FINITE ELEMENT METHOD OF PREDICTING STRESS DISTRIBUTIONS AND SOIL DEFORMATION UNDER A TRACTIVE DEVIC	69-0389
PREDICTING	FINITE-ELEMENT METHOD APPLIED TO PREDICTING FATIGUE-CRACK GROWTH	73-0513
PREDICTING	METHODS FOR PREDICTING THE AERODYNAMIC AND STABILITY AND CONTROL CHARACTERISTICS OF STOL AIRCRAFT. V	73-0718
PREDICTING	METHODS FOR PREDICTING THE AERODYNAMIC AND STABILITY AND CONTROL CHARACTERISTICS OF STOL AIRCRAFT. V	73-0719
PREDICTING	METHODS OF PREDICTING THE AERODYNAMIC AND STABILITY AND CONTROL CHARACTERISTICS OF STOL AIRCRAFT. VO	73-0717
PREDICTING	NONLINEAR THEORY FOR PREDICTING THE PERFORMANCE OF FLEXIBLE HIGHWAY PAVEMENTS	70-0143
PREDICTING	PREDICTING MAGNETIC LEAKAGE FIELDS BY FINITE ELEMENT TECHNIQUES	75-0046
PREDICTING	PREDICTING THE BLOCK CAVING BEHAVIOUR OF OREBODIES	69-0044
PREDICTING	THE FINITE ELEMENT METHOD FOR PREDICTING STRESS DISTRIBUTION AND SOIL DEFORMATION UNDER A TRACTIVE D	69-0522
PREDICTING	THE FINITE ELEMENT METHOD FOR PREDICTING STRESS DISTRIBUTION AND SOIL DEFORMATION UNDER A TRACTIVE D	70-0801
PREDICTION	A FINITE ELEMENT COMPUTER PROGRAM FOR THE PREDICTION OF THE BEHAVIOR OF REINFORCED AND PRESTRESSED C	72-0799
PREDICTION	A METHOD FOR THE PREDICTION OF STRESSES IN AN ISOTROPIC INCLUSION OR OREBODY OF IRREGULAR SHAPE	72-0827
PREDICTION	ANALYTICAL FINITE ELEMENT SIMULATION MODEL FOR STRUCTURAL CRASHWORTHINESS PREDICTION	73-0778
PREDICTION	ANALYTICAL PREDICTION OF DEFECTS OCCURRENCE IN SIMPLE AND COMPLEX	69-0157
PREDICTION	COMPUTER PREDICTION OF FATIGUE CRACK PROPAGATION UNDER RANDOM LOADING	73-0851

PREDICTION	COUPLED ROTOR-AIRFRAME VIBRATION PREDICTION METHODS	74-0498
PREDICTION	FINITE ELEMENT METHOD FOR PREDICTION OF CRACK BEHAVIOUR	70-0342
PREDICTION	FINITE ELEMENT PREDICTION OF FIREPROOFING EFFECTIVENESS	72-0047
PREDICTION	FINITE-ELEMENT APPROACH FOR THE PREDICTION OF INELASTIC BEHAVIOR OF SHEAR WALL-FRAME SYSTEMS	72-0983
PREDICTION	FINITE-ELEMENT MODEL FOR THE MECHANICAL BEHAVIOUR OF THE LEFT VENTRICLE, PREDICTION OF DEFORMATION I	72-0304
PREDICTION	GAS TURBINE ENGINE DISK CYCLIC LIFE PREDICTION	75-0509
PREDICTION	GEOMETRIC METHOD FOR THE PREDICTION OF STRESSES IN INCLUSIONS, OREBODIES, AND MINING SYSTEMS	74-0219
PREDICTION	MATHEMATICAL PREDICTION OF DYNAMIC TIRE BEHAVIOR	73-0579
PREDICTION	MODELING AND ANALYSIS TECHNIQUES FOR PREDICTION OF STRUCTURAL AND BIODYNAMIC CRASH-IMPACT RESPONSE	75-0300
PREDICTION	MODIFIED POTENTIAL ENERGY MASS REPRESENTATION FOR FREQUENCY PREDICTION	66-0053
PREDICTION	MODULAR APPROACH TO STRUCTURAL SIMULATION FOR VEHICLE CRASHWORTHINESS PREDICTION	75-0737
PREDICTION	NUMERICAL MODEL FOR THE PREDICTION OF TRANSIENT WATER QUALITY IN ESTUARY NETWORKS	72-0635
PREDICTION	NUMERICAL PREDICTION OF TRACER CONCENTRATION IN AQUIFERS WITH TRANSIENT GROUNDWATER FLOW BY FINITE-E	75-0610
PREDICTION	PRACTICAL FINITE ELEMENT METHOD OF FAILURE PREDICTION FOR COMPOSITE MATERIAL STRUCTURES	75-0644
PREDICTION	PREDICTION OF ELASTIC CONSTANTS OF PARTICLEBOARD	74-0259
PREDICTION	PREDICTION OF LONG TERM DEFORMATION OF A COMPACTED COHESIVE SOIL EMBANKMENT OVER A SOFT FOUNDATION	74-0727
PREDICTION	PREDICTION OF LONG-TERM STRESS RANGES: USER'S MANUAL - BRIDGE LOAD GENERATOR	73-0737
PREDICTION	PREDICTION OF LONG-TERM STRESS RANGES: USER'S MANUAL - BRIDGE DYNAMIC STRESS ANALYSIS	74-0807
PREDICTION	PREDICTION OF MARINE PROPELLER DISTORTION AND STRESSES USING A SUPERPARAMETRIC THICK SHELL FINITE EL	73-0370
PREDICTION	PREDICTION OF PLATE VIBRATIONS USING A CONSISTENT MASS MATRIX	66-0043
PREDICTION	PREDICTION OF SLOW CRACK GROWTH BY A FINITE-ELEMENT METHOD	75-0416
PREDICTION	PREDICTION OF STATIC AND FATIGUE DAMAGE AND CRACK PROPAGATION IN COMPOSITE MATERIALS	75-0643
PREDICTION	PREDICTION OF TEMPERATURE DISTRIBUTIONS IN CARGOES WITH THE AID OF A COMPUTER PROGRAM USING THE METH	74-1215
PREDICTION	PREDICTION OF THE CREEP BEHAVIOUR OF A FLANGED JOINT	74-0757
PREDICTION	PREDICTION OF TRANSMISSION LOSS IN MUFFLERS BY THE FINITE ELEMENT METHOD	75-0160
PREDICTION	PREDICTION OF UNDRAINED MOVEMENTS CAUSED BY EMBANKMENTS ON MUSKEG	70-0070
PREDICTION	RECENT ADVANCES IN NUMERICAL WEATHER PREDICTION	74-0787
PREDICTION	SOLUTION OF WATER PREDICTION PROBLEMS BY FINITE ELEMENT METHODS	74-1345
PREDICTION	SOME APPLICATIONS OF FINITE ELEMENT ANALYSIS TO SHELL BUCKLING PREDICTION	68-0023
PREDICTION	STRESS - DEFORMATION PREDICTION IN CEMENT - TREATED SOIL PAVEMENTS	71-0286
PREDICTION	THE PREDICTION OF FULLY DEVELOPED TURBULENT FLOW IN DUCTS BY THE FINITE ELEMENT METHOD	75-0117
PREDICTION	THE PREDICTION OF REGULAR WAVE FORCES ON FLOATING BODIES WITH THE FINITE ELEMENT METHOD (IN DUTCH)	71-0743
PREDICTION	THEORETICAL PREDICTION OF DISC BRAKE TEMPERATURES AND A COMPARISON WITH EXPERIMENTAL DATA	72-0306
PREDICTION	THERMAL DEFORMATION PREDICTION IN ORBITING OBSERVATORY STRUCTURES	69-0292
PREDICTION	V-BELT LIFE PREDICTION AND POWER RATING	75-0838
PREDICTION	VEHICLE CRUSH PREDICTION USING FINITE-ELEMENT TECHNIQUES	73-0280
PREDICTIONS	AEROELASTIC LOADS PREDICTIONS USING FINITE-ELEMENT AERODYNAMICS	75-0815
PREDICTIONS	COMPARISON OF STRUCTURAL TEST RESULTS WITH PREDICTIONS OF FINITE ELEMENT ANALYSIS	73-0158
PREDICTIONS	COMPARISON OF TYPICAL INELASTIC ANALYSIS PREDICTIONS WITH BENCHMARK PROBLEM EXPERIMENTAL RESULTS	75-0696
PREDICTIONS	ELASTIC INSTABILITY PREDICTIONS FOR DOUBLY-CURVED SHELLS	68-0075
PREDICTIONS	ELASTIC WAVE PROPAGATION. A COMPARISON BETWEEN FINITE ELEMENT PREDICTIONS AND EXACT SOLUTION	67-0076
PREDICTIONS	EXPERIMENTAL ELASTIC STRESS ANALYSES OF CYLINDER-TO-CYLINDER SHELL MODELS AND COMPARISONS WITH THEOR	71-0065
PREDICTIONS	FINITE-ELEMENT PREDICTIONS OF TRANSIENT ELASTIC-PLASTIC LARGE DEFLECTIONS OF STIFFENED AND/OR UNSTIF	74-0825
PREDICTIONS	GROUNDWATER PREDICTIONS FOR A HIGHWAY CUT USING FINITE ELEMENTS	73-0346
PREDICTIONS	PREDICTIONS IN ENVIRONMENTAL HYDRODYNAMICS USING FINITE ELEMENT METHOD 1, THEORETICAL DEVELOPMENT	75-0003
PREDICTIONS	PREDICTIONS IN ENVIRONMENTAL HYDRODYNAMICS USING THE FINITE ELEMENT METHOD 2. APPLICATIONS	75-0064
PREFABRICATED	APPROXIMATE ANALYSIS OF PREFABRICATED TOWER STRUCTURES USING THIN SHELL FINITE ELEMENTS	73-0893
PREFABRICATED	ENTWURF UND SPANNUNGSBERECHNUNG EINES SPANNBETON DRUCKBEHAELTERS AUS FERTIGTEILEN. (DESIGN AND STRES	73-0856
PREFABRICATED	FIELD ANALYSIS FOR PREFABRICATED RIBBED CYLINDRICAL SHELLS	73-0899
PREFABRICATED	PREFABRICATED PRETENSIONED LIGHTWEIGHT CONCRETE FOLDED PLATE	73-0808
PREFABRICATED	SHAPE DETERMINATION OF PREFABRICATED PRESTRESSED NETWORKS	73-0889
PREFABRICATED	ZUSAMMENWIRKEN VON EINZELNEN FERTIGTEILEN ALS GROSSFLAECHIGE SCHEIBE (INTERACTION OF INDIVIDUAL PREF	73-0552
PRELOADED	SHOCK TUNNEL TESTS OF PRELOADED AND ARCHED WALL PANELS FINITE ELEMENT STRESS FORMULATION FOR WAVE PR	73-0892
PRENOTCHED	FATIGUE OF PRENOTCHED GRAPHITE FIBER COMPOSITES IN COMPRESSION	72-0215
PREPARATION	AUTOMATED INPUT DATA PREPARATION FOR NASTRAN	69-0080
PREPARATION	AUTOMATIC PREPARATION OF DATA CARDS FOR THE FINITE ELEMENT METHOD	66-0086
PREPARATION	USING INTERACTIVE GRAPHICS FOR THE PREPARATION AND MANAGEMENT OF FINITE ELEMENT DATA	74-0176
PREPARATIONS	STRESS ANALYSIS OF FIRST MOLARS WITH FULL CROWN PREPARATIONS BY THREE-DIMENSIONAL PHOTOELASTICITY AND	72-0939
PREPROCESSOR	FEMESH: A FINITE ELEMENT CODE PREPROCESSOR	73-0949
PREPROCESSORS	PREPROCESSORS FOR GENERAL PURPOSE FINITE ELEMENT PROGRAMS	74-1033
PRESCRIBED	ELEMENT STIFFNESS MATRICES FOR BOUNDARY COMPATIBILITY AND FOR PRESCRIBED BOUNDARY STRESSES	66-0061
PRESCRIBED	OPTIMAL DESIGN FOR PRESCRIBED BUCKLING LOADS	74-0568
PRESPONSE	VISCOELASTIC PRESPONSE OF THE CORNEO-SCLERAL SHELL UNDER TONOMETER LOADING	75-0855
PRESS	DER EINSATZ DER FINIT-ELEMENT-METHODE IN PRESSENBAU (APPLICATION OF THE FINITE ELEMENT METHOD FOR TH	74-0615
PRESSED	FINITE ELEMENT ANALYSIS OF ALTERNATIVE AXIAL LOADING OF AN ELASTIC PLATE PRESSED BETWEEN TWO ELASTIC	71-0357
PRESSENBAU	DER EINSATZ DER FINIT-ELEMENT-METHODE IN PRESSENBAU (APPLICATION OF THE FINITE ELEMENT METHOD FOR TH	74-0615
PRESSING	ANALYSES OF AXISYMMETRIC UPSETTING AND PLANE STRAIN SIDE PRESSING OF SOLID CYLINDERS BY THE FINITE E	70-0083
PRESSURE	ADVANCES IN THE ANALYSIS OF PRESTRESSED CONCRETE PRESSURE VESSELS	71-0162
PRESSURE	AN ASSUMED STRESS HYBRID FINITE ELEMENT MODEL FOR THE ANALYSIS OF AN AXISYMMETRIC THICK WALLED PRESS	71-0025
PRESSURE	ANALYSIS OF BAND PRESSURE IN A GUN TUBE USING FINITE ELEMENTS AND SHELL PUSHING DATA	72-0793
PRESSUPE	ANALYSIS OF MODERATELY THICK FLANGED REACTOR PRESSURE VESSELS	73-0821
PRESSURE	ANALYSIS OF THICK-WALLED PRESSURE VESSELS USING THE CONSTANT STRAIN AXISYMMETRIC FINITE ELEMENT	71-0653
PRESSURE	ANALYSIS OF VISCOPLASTIC EFFECTS IN PRESSURE VESSELS BY THE FINITE ELEMENT METHOD	72-0573
PRESSURE	ANALYTICAL PERFORMANCE OF STEEL, ALUMINUM AND PLASTIC FOR A NEW DESIGN HIGH PRESSURE THIN-WALLED CAR	74-0756
PRESSURE	APPLICATION OF ELASTIC PLASTIC ANALYSIS TO THE DESIGN OF PRESSURE VESSELS	68-0157
PRESSURE	APPLICATION OF FINITE ELEMENT METHOD TO AXISYMMETRIC BUCKLING OF SHALLOW SPHERICAL SHELLS UNDER EXTE	73-0111
PRESSURE	APPLICATION OF FRACTURE MECHANICS TO PRESTRESSED CONCRETE PRESSURE VESSELS	73-0228
PRESSURE	APPLICATION OF THE FINITE ELEMENT METHOD TO PRESSURE VESSEL DESIGN (JAPANESE)	70-0446
PRESSURE	APPLICATION OF THE FINITE ELEMENT METHOD TO THE CALCULATION OF REACTOR PRESSURE VESSELS	71-0645
PRESSURE	APPLICATION OF THE MATRIX DISPLACEMENT METHOD TO THE ANALYSIS OF PRESSURE VESSELS	69-0051
PRESSURE	AREAS OF CONTACT AND PRESSURE DISTRIBUTION IN BOLTED JOINTS	71-0329
PRESSURE	BERECHNUNGSMETHODEN ZUR ERMITTLUNG VON BELASTUNGSBEDINGUNGEN UND GRENZTRAGFAEHIGKEITSVERHALTEN VON K	73-0858
PRESSURE	BUCKLING OF CYLINDRICAL SHELLS BY WIND PRESSURE	74-0642
PRESSURE	CALCULATING STRESSES IN A PRESSURE VESSEL WITH FILL TUBE - COMPARISON OF THE FINITE ELEMENT AND PHOT	72-0782
PRESSURE	CALCULATIONS FOR PRESTRESSED CONCRETE PRESSURE VESSELS	71-0643
PRESSURE	CORRELATION BETWEEN PRESSURE AND THERMAL TRANSIENT STRESS LEVELS DERIVED BY THEORY AND EXPERIMENT ON	73-0469
PRESSURE	DEFORMATION OF THE ARTERIAL VASA VASORUM AT NORMAL AND HYPERTENSIVE ARTERIAL PRESSURE	71-0149
PRESSURE	DESIGN ASPECTS OF HIGH DUTY PRESSURE COMPONENTS	74-0199
PRESSURE	DISTRIBUTION OF THE CLAMPING PRESSURE IN FRICTION-GRIP BOLTED JOINTS	74-0250
PRESSURE	DYNAMIC RESPONSE OF SOIL/CONCRETE INTERFACES AT HIGH PRESSURE	74-0836
PRESSURE	EFFECT OF LATERAL PRESSURE ON NATURAL FREQUENCIES OF HYPERBOLOIDAL SHELLS OF REVOLUTION	71-0278
PRESSURE	EINFACHES NAEHERUNGSVERFAHREN FUR DIE GRENZTRAGFAEHIGKEITSANLYSE VON SPANNBETON-REAKTORDRUCKBEHAELTE	73-0854
PRESSURE	ELASTIC PLASTIC ANALYSIS OF PRESSURE VESSEL COMPONENTS	69-0380
PRESSURE	ELASTIC PLASTIC DEFORMATION OF CYLINDRICAL PRESSURE VESSELS UNDER CYCLIC LOADING	74-0604
PRESSURE	ELASTIC-PLASTIC ANALYSIS OF SOME PRESSURE VESSEL HEADS	69-0379
PRESSURE	ELASTIC-PLASTIC ANALYSIS OF THICK WALLED PRESSURE VESSELS WITH SHARP DISCONTINUITIES	71-0155
PRESSURE	ELASTIC-PLASTIC BEHAVIOR OF A LONGITUDINAL SEMI-ELLIPTIC CRACK IN A THICK PRESSURE VESSEL	73-0952
PRESSURE	ENTWURF UND SPANNUNGSBERECHNUNG EINES SPANNBETON DRUCKBEHAELTERS AUS FERTIGTEILEN. (DESIGN AND STRES	73-0856
PRESSURE	EURCYL: A PROGRAM TO GENERATE FINITE ELEMENT MESHES FOR PRESSURE VESSEL NOZZLES	74-1086
PRESSURE	FIELD INSTRUMENTATION FOR LONG TERM MEASUREMENT OF PILE LOAD AND RAFT CONTACT PRESSURE	73-0150
PRESSURE	FINITE ELEMENT ANALYSIS OF STRUCTURAL INTEGRITY OF A REACTOR PRESSURE VESSEL DURING EMERGENCY CORE C	70-0026
PRESSURE	FINITE ELEMENT ANALYSIS OF PRESSURE VESSELS	71-0348

332

PRESSURE	FINITE ELEMENT ANALYSIS OF PRESTRESSED CONCRETE PRESSURE VESSELS	74-1289
PRESSURE	FINITE ELEMENT APPROACH TO PRESSURE WAVE ATTENUATION BY REACTOR FUEL SUBASSEMBLIES	75-0522
PRESSURE	FINITE ELEMENT INCREMENTAL ELASTIC-PLASTIC ANALYSIS OF PRESSURE VESSELS	70-0253
PRESSURE	FINITE ELEMENT STUDY OF PRESSURE WAVE ATTENUATION BY REACTOR FUEL SUBASSEMBLIES	75-0215
PRESSURE	FINITE ELEMENTS FOR CALCULATION OF PRESTRESSED CONCRETE REACTOR PRESSURE VESSELS (FINITE ELEMENTE ZU	73-0797
PRESSURE	FINITE-ELEMENT CALCULATION OF PRESSURE HISTORY IN A GAS FIELD	75-0546
PRESSURE	FINITE-ELEMENT NONLINEAR BUCKLING ANALYSIS OF A PSEUDO-CYLINDRICAL CONCAVE POLYHEDRAL SHELL UNDER EX	73-1147
PRESSURE	FLOW BETWEEN LINES OF A GIVEN PRESSURE	74-0020
PRESSURE	FORECASTING THE ROCK PRESSURE AND ITS EFFECT ON THE LONGWALL FACES AND ROADWAYS IN THE BITUMINOUS CO	74-0374
PRESSURE	FORMING OF 7075-T6 ALUMINUM IN HIGH PRESSURE ENVIRONMENTS	71-0322
PRESSURE	FRACTURE AND SAFETY ANALYSIS OF NUCLEAR PRESSURE VESSELS	73-0257
PRESSURE	FREE VIBRATION AND RESPONSE TO RANDOM PRESSURE FIELD OF ANISOTROPIC THIN CYLINDRICAL SHELLS	74-0830
PRESSURE	FREE VIBRATIONS OF PRESSURE LOADED PARABOLOIDAL SHELLS	71-0601
PRESSURE	INELASTIC ANALYSIS OF A STRAIGHT TUBE UNDER COMBINED BENDING, PRESSURE AND THERMAL LOADS	75-0218
PRESSURE	INELASTIC ANALYSIS OF A STRAIGHT TUBE UNDER COMBINED BENDING, PRESSURE AND THERMAL LOADS	75-0513
PRESSURE	INELASTIC BUCKLING ANALYSIS OF PIPES SUBJECTED TO INTERNAL PRESSURE, FLEXURE AND AXIAL LOADING	74-0515
PRESSURE	INTERNAL PRESSURE EFFECTS ON THE VIBRATION OF PARTIALLY FILLED ELASTIC TANKS	75-0241
PRESSURE	INTERNAL SUPPORTING STRUCTURES OF PRESSURE TUBES TYPE REACTOR CORE, ANALYTICAL MODEL FOR STATIC ANAL	71-0049
PRESSURE	MATHEMATICAL MODEL OF THE CORNEO-SCLERAL SHELL AS APPLIED TO INTRAOCULAR PRESSURE VOLUME RELATIONS A	72-0471
PRESSURE	MODERN DEVELOPMENTS IN THE STRESS ANALYSIS OF PRESSURE VESSELS	69-0201
PRESSURE	NONLINEAR ANALYSIS OF PRESTRESSED CONCRETE FAST REACTOR PRESSURE VESSEL, (ANALYSE NON LINEAIRE DES C	70-0145
PRESSURE	NUMERICAL CALCULATIONS OF VELOCITY AND PRESSURE DISTRIBUTION AROUND OSCILLATING AIRFOILS	74-0847
PRESSURE	NUMERICAL SIMULATION OF FINITE-WIDTH THRUST BEARINGS, TAKING INTO ACCOUNT VISCOSITY VARIATION WITH T	75-0623
PRESSURE	NUMERICAL SOLUTIONS FOR DETERMINING WAVE-INDUCED PRESSURE DISTRIBUTIONS AROUND BURIED PIPELINES	74-1334
PRESSURE	ON THE USE OF FINITE ELEMENTS IN FRACTURE ANALYSIS OF PRESSURE VESSEL COMPONENTS	75-0514
PRESSURE	PENETRATION IN SHELLS UNDER EXTERNAL PRESSURE	70-0237
PRESSURE	PRACTICAL APPLICATIONS OF FINITE ELEMENT METHODS IN THE STRESS ANALYSIS OF THIN SHELL PRESSURE VESSE	71-0205
PRESSURE	PRESSURE DISTRIBUTION AND DEFORMATIONS OF MACHINED COMPONENTS IN CONTACT.	73-0372
PRESSURE	PRESSURE DISTRIBUTION AND BELT DEFORMATION IN V-BELT DRIVES	74-1028
PRESSURE	PRESSURE VESSEL CERTIFICATION BASED ON FRACTURE MECHANICS TECHNOLOGY	74-0948
PRESSURE	PRESSURE VESSEL WALL THICKNESS LIMITS BASED ON THERMAL STRESS RACHETTING AND CREEP FATIGUE INTERACTI	74-0968
PRESSURE	PRESSURE VESSELS ANALYSIS BY FINITE ELEMENT TECHNIQUES	67-0069
PRESSURE	PRESTRESSED CONCRETE PRESSURE VESSELS	69-0052
PRESSURE	PRESTRESSED CONCRETE REACTOR PRESSURE VESSEL FAILURE ANALYSIS, THE INCORPORATION OF STATISTICAL METH	71-0120
PRESSURE	RADIAL DISPLACEMENTS OF A HOLLOW SHAFT SUBJECTED TO A UNIFORM BAND OF EXTERNAL PRESSURE	72-0242
PRESSURE	SIMPLE FINITE ELEMENT METHOD FOR STRESS CONCENTRATION PROBLEMS IN PRESSURE VESSELS	71-0669
PRESSURE	SOLUTION OF THE PROBLEM OF PROCESSING METALS BY PRESSURE USING THE METHOD OF FINITE ELEMENTS	74-1363
PRESSURE	SOLVING THE PROBLEM OF PROCESSING METALS BY PRESSURE THROUGH THE METHOD OF FINITE ELEMENTS	74-1354
PRESSURE	SOME APPLICATIONS OF ENGINEERING MECHANICS IN PRESSURE VESSEL ANALYSIS	72-0892
PRESSURE	SOME CREEP EFFECTS IN STRESS ANALYSIS WITH PARTICULAR REFERENCE TO CONCRETE PRESSURE VESSELS	69-0284
PRESSURE	STRENGTH OF REINFORCED CONCRETE CHAMBERS UNDER EXTERNAL PRESSURE	75-0293
PRESSURE	STRENGTH OF REINFORCED CONCRETE CHAMBERS UNDER EXTERNAL PRESSURE	75-0512
PRESSURE	STRESS ANALYSIS OF A PRESTRESSED CONCRETE NUCLEAR PRESSURE VESSEL BY THE FINITE ELEMENT METHOD USING	71-0154
PRESSURE	STRESS ANALYSIS OF BWR PRESSURE VESSEL WITH BERSAFE AND FLHE	73-0406
PRESSURE	STRESS ANALYSIS OF PRESSURE VESSEL NOZZLES	73-0823
PRESSURE	STRESS DISTRIBUTIONS IN THIN-WALLED INTERSECTING CYLINDRICAL SHELLS SUBJECTED TO INTERNAL PRESSURE A	71-0023
PRESSURE	STRUCTURAL DESIGN OF CIRCULAR DIVERSION, POWER AND PRESSURE TUNNELS	71-0315
PRESSURE	THE APPLICATION OF THREE-DIMENSIONAL FINITE ELEMENTS TO THE ANALYSIS OF PODDED BOILER TYPE PRESTRESS	71-0047
PRESSURE	THE CHANGES IN THE (TWO-DIMENSIONAL) STRESS ROUNDED CORNERS, CAUSED BY A VARYING INTERNAL PRESSURE A	67-0141
PRESSURE	THE USE OF THE FINITE ELEMENT METHOD IN OBTAINING SOIL-WATER PRESSURE AND SOIL-WATER CONTENT DISTRIB	75-0112
PRESSURE	THERMAL SLEEVE DESIGN FOR PRESSURE VESSELS	74-1129
PRESSURE	THIN-WALLED PRESTRESSED CONCRETE PRESSURE VESSEL FOR HIGH TEMPERATURE REACTORS: EXPERIMENTAL INVESTI	73-0857
PRESSURE	THREE DIMENSIONAL ELASTIC AND ELASTO-PLASTIC ANALYSIS OF REACTOR PRESSURE VESSELS	69-0453
PRESSURE	THREE DIMENSIONAL RUPTURE ANALYSIS OF A PRESTRESSED CONCRETE PRESSURE VESSEL INCLUDING CREEP EFFECTS	73-0826
PRESSURE	THREE DIMENSIONAL RUPTURE ANALYSIS OF A PRESTRESSED CONCRETE PRESSURE VESSEL INCLUDING CREEP EFFECTS	74-0538
PRESSURE	THREE DIMENSIONAL STRESS ANALYSIS OF PENETRATIONS IN THICK WALLED PRESSURE VESSELS	74-0422
PRESSURE	THREE-DIMENSIONAL ANALYSIS OF REACTOR PRESSURE VESSEL NOZZLES	71-0221
PRESSURE	TRIANGULAR THICK SHELL FINITE ELEMENT FOR THE THREE-DIMENSIONAL STRESS ANALYSIS OF PRESSURE VESSELS	73-0860
PRESSURE	TWO-DIMENSIONAL TIME-DEPENDENT ANALYSIS OF PERFORATED END CAPS FOR NUCLEAR REACTOR PRESSURE VESSELS	71-0239
PRESSURE	VIBRATIONS OF PRESSURE LOADED HYPERBOLOIDAL SHELLS	72-0874
PRESSURE	VORAUSSAGE DES GEBIRGSDRUCKES UND SEINER AUSWIRKUNGEN AUF STREBEN UND STRECKEN IM STEINKOHLENBERGBAU	73-0679
PRESSURE-VESSEL	ANALYSIS OF A PRESSURE-VESSEL JUNCTION BY THE FINITE-ELEMENT METHOD	72-0949
PRESSURE-VESSEL	DISCONTINUITY STRESS ANALYSIS OF PRESSURE-VESSELS USING FINITE-ELEMENT METHOD	75-0346
PRESSUREMETER	FINITE ELEMENT PARAMETRIC STUDY OF VERTICAL STRAIN INFLUENCE FACTORS AND THE PRESSUREMETER TEST TO E	74-1127
PRESSURES	COMPARISON BETWEEN THEORETICAL AND MEASURED EARTH PRESSURES ACTING ON A LARGE MOTORWAY RETAINING WAL	74-0917
PRESSURES	COMPUTER BASED ANALYSES OF THE RESPONSE OF BOX TYPE STRUCTURES TO RANDOM PRESSURES	73-0227
PRESSURES	DYNAMIC ANALYSIS OF AXISYMMETRIC SHELLS UNDER ARBITRARY TRANSIENT PRESSURES	72-0467
PRESSURES	EARTH PRESSURES AGAINST THE ABUTMENTS AND WING WALLS OF STANDARD MOTORWAY BRIDGES	75-0629
PRESSURES	EARTH PRESSURES ON STRUCTURES DUE TO FAULT MOVEMENT	73-0403
PRESSURES	HYDRODYNAMIC PRESSURES ON HIGH DAMS DUE TO VERTICAL EARTHQUAKE MOTIONS	69-0438
PRESSURES	INFLUENCE OF CONSTRUCTION PROCEDURES ON GROUND PRESSURES AROUND A TUNNEL	74-0997
PRESSURES	ON THE ASYMPTOTICALLY SPHERICAL DEFORMATIONS OF ARBITRARY MEMBRANES OF REVOLUTION FIXED ALONG AN EDG	71-0640
PRESSURES	SEAL-SHELL-2, A COMPUTER PROGRAM FOR THE STRESS ANALYSIS OF A THICK SHELL OF REVOLUTION WITH AXISYMM	69-0107
PRESSURES	STRESSES IN NEARLY INCOMPRESSIBLE MATERIALS BY FINITE ELEMENTS WITH APPLICATION TO THE CALCULATION O	74-0383
PRESSURIZATION	EFFECT OF BACKPACKING AND INTERNAL PRESSURIZATION ON STRESSES TRANSMITTED TO BURIED CYLINDERS	73-0897
PRESSURIZED	ANALYSIS OF STRESSES IN PRESSURIZED WELDED PIPE IN THE CREEP RANGE	71-0404
PRESSURIZED	APPLICATION OF FINITE ELEMENT ANALYSIS TO HYDRODYNAMIC AND EXTERNALLY PRESSURIZED POCKET BEARINGS	72-0295
PRESSURIZED	FINITE ELEMENT STUDY OF PRESSURIZED CAVITIES IN GEOLOGIC MEDIA	71-0805
PRESSURIZED	STABILITY OF PRESSURIZED HYPERBOLOIDAL SHELLS	75-0585
PRESSURIZED	STRESS ANALYSIS OF PRESSURIZED CYLINDRICAL BORE IN RECTANGULAR BLOCK	69-0312
PRESSURIZED	STRESS CONCENTRATION FACTORS FOR CIRCULAR, REINFORCED PENETRATIONS IN PRESSURIZED CYLINDRICAL SHELLS	75-0726
PRESSURIZED	STRESS INTENSITY FACTOR OF A SURFACE CRACK IN A PRESSURIZED CYLINDER	75-0255
PRESSURIZED	TENSILE FIELD CALCULATION ABOUT A PRESSURIZED CAVITY: MONERO	72-0684
PRESSURIZED	VIBRATION CHARACTERISTICS OF THIN, PRESSURIZED SHELLS OF REVOLUTION PARTIALLY FILLED WITH LIQUID	72-0604
PRESTRESSED	A FINITE ELEMENT COMPUTER PROGRAM FOR THE PREDICTION OF THE BEHAVIOR OF REINFORCED AND PRESTRESSED C	72-0799
PRESTRESSED	A THREE-DIMENSIONAL FINITE ELEMENT ANALYSIS OF A PRESTRESSED CONCRETE REACTOR VESSEL MODEL	69-0084
PRESTRESSED	ADVANCES IN THE ANALYSIS OF PRESTRESSED CONCRETE PRESSURE VESSELS	71-0162
PRESTRESSED	AN ANALYSIS OF A REINFORCED AND A PRESTRESSED CONCRETE BEAM BY FINITE-ELEMENT METHOD	72-0940
PRESTRESSED	APPLICATION OF FRACTURE MECHANICS TO PRESTRESSED CONCRETE PRESSURE VESSELS	73-0228
PRESTRESSED	CALCULATIONS FOR PRESTRESSED CONCRETE PRESSURE VESSELS	71-0643
PRESTRESSED	CIRCUMFERENTIAL PRESTRESSING LOAD ANALYSIS DUE TO WIRE/STRAND WINDING SYSTEMS ON PRESTRESSED CONCRET	74-1202
PRESTRESSED	DESIGN OF A COMPOSITE STEEL AND CONCRETE CLOSURE FOR A PRESTRESSED CONCRETE REACTOR VESSEL.	72-0187
PRESTRESSED	DISCUSSION ON FINITE ELEMENT ANALYSIS OF PRESTRESSED CONCRETE REACTOR VESSELS	72-0648
PRESTRESSED	EINFACHES NAEHERUNGSVERFAHREN FUR DIE GRENZTRAGFAEHIGKEITSANLYSE VON SPANNBETON-REAKTORDRUCKBEHAELTE	73-0854
PRESTRESSED	ENTWURF UND SPANNUNGSBERECHNUNG EINES SPANNBETON DRUCKBEHAELTERS AUS FERTIGTEILEN. (DESIGN AND STRES	73-0856
PRESTRESSED	FINITE ELEMENT ANALYSIS OF HYPERBOLOIDAL PRESTRESSED CONCRETE CONTAINMENT VESSELS	69-0526
PRESTRESSED	FINITE ELEMENT ANALYSIS OF PRESTRESSED CONCRETE VOIDED BRIDGE DECKS	73-0106
PRESTRESSED	FINITE ELEMENT ANALYSIS OF PRESTRESSED FOLDED PLATE STRUCTURES	74-1066
PRESTRESSED	FINITE ELEMENT ANALYSIS OF PRESTRESSED CONCRETE PRESSURE VESSELS	74-1289
PRESTRESSED	FINITE ELEMENT METHOD IN ANALYSIS OF REINFORCED AND PRESTRESSED CONCRETE STRUCTURES	71-0270
PRESTRESSED	FINITE ELEMENTS FOR CALCULATION OF PRESTRESSED CONCRETE REACTOR PRESSURE VESSELS (FINITE ELEMENTE ZU	73-0797
PRESTRESSED	FINITE ELEMENTS METHOD IN DESIGN OF PRESTRESSED REACTOR CONCRETE CONSTRUCTION	73-0550

PRESTRESSED	LARGE DEFLECTION ANALYSIS OF PRESTRESSED NETWORKS	72-0169
PRESTRESSED	NONLINEAR ANALYSIS OF PRESTRESSED CONCRETE FAST REACTOR PRESSURE VESSEL. (ANALYSE NON LINEAIRE DES C	70-0145
PRESTRESSED	PRESTRESSED CONCRETE REACTOR VESSEL ANALYTICAL METHODS DEVELOPMENTS	68-0141
PRESTRESSED	PRESTRESSED CONCRETE PRESSURE VESSELS	69-0052
PRESTRESSED	PRESTRESSED CONCRETE REACTOR PRESSURE VESSEL FAILURE ANALYSIS. THE INCORPORATION OF STATISTICAL METH	71-0120
PRESTRESSED	PRESTRESSED CONCRETE DEEP SLABS WITH OPENINGS	73-0427
PRESTRESSED	PRINCIPAL STRESSES AT THE INTERMEDIATE SUPPORT OF PRESTRESSED CONCRETE CONTINUOUS BEAMS	70-0352
PRESTRESSED	RECENT DEVELOPMENTS IN FINITE ELEMENT ANALYSIS OF PRESTRESSED CONCRETE REACTOR VESSELS	74-0469
PRESTRESSED	RIDDLE OF BONDED AND UNBONDED TENDONS IN PRESTRESSED CONCRETE REACTOR VESSELS	74-0519
PRESTRESSED	SHAPE DETERMINATION OF PREFABRICATED PRESTRESSED NETWORKS	73-0889
PRESTRESSED	STEP AND LOADS EFFECT ON STRESSES IN PRESTRESSED CONCRETE SHORT BRACKETS	71-0334
PRESTRESSED	STRESS ANALYSIS AND SPECIAL PROBLEMS OF PRESTRESSED DAMS	61-0006
PRESTRESSED	STRESS ANALYSIS OF A PRESTRESSED CONCRETE NUCLEAR PRESSURE VESSEL BY THE FINITE ELEMENT METHOD USING	71-0154
PRESTRESSED	THE APPLICATION OF THREE-DIMENSIONAL FINITE ELEMENTS TO THE ANALYSIS OF PODDED BOILER TYPE PRESTRESS	71-0047
PRESTRESSED	THE INELASTIC ANALYSIS OF REINFORCED AND PRESTRESSED CONCRETE BEAMS	72-0667
PRESTRESSED	THIN-WALLED PRESTRESSED CONCRETE PRESSURE VESSEL FOR HIGH TEMPERATURE REACTORS: EXPERIMENTAL INVESTI	73-0857
PRESTRESSED	THREE DIMENSIONAL RUPTURE ANALYSIS OF A PRESTRESSED CONCRETE PRESSURE VESSEL INCLUDING CREEP EFFFCTS	73-0826
PRESTRESSED	THREE DIMENSIONAL RUPTURE ANALYSIS OF A PRESTRESSED CONCRETE PRESSURE VESSEL INCLUDING CREEP EFFECTS	74-0538
PRESTRESSED	VISCOELASTIC ANALYSIS OF PRESTRESSED CONCRETE SHELL OF REVOLUTION	73-0464
PRESTRESSED-CON	USE OF SMALL MODELS IN DESIGN AND ANALYSIS OF PRESTRESSED-CONCRETE REACTOR VESSELS	70-0433
PRESTRESSED-CON	USE OF SMALL MODELS IN DESIGN AND ANALYSIS OF PRESTRESSED-CONCRETE REACTOR VESSELS	71-0426
PRESTRESSING	CIRCUMFERENTIAL PRESTRESSING LOAD ANALYSIS DUE TO WIRE/STRAND WINDING SYSTEMS ON PRESTRESSED CONCRET	74-1202
PRESTRESSING	DISTRIBUZIONI OTTIMALI DI ARMATURE DI PRECOMPRESSIONE IN TRAVI-PARETE DI CALCESTRUZZO. (OPTIMAL DIST	75-0560
PRETENSIONED	PHOTOELASTIC INVESTIGATION ON THE CRACK-ARREST CAPABILITY OF A PRETENSIONED STIFFENED PLATE	75-0088
PRETENSIONED	PREFABRICATED PRETENSIONED LIGHTWEIGHT CONCRETE FOLDED PLATE	73-0808
PRIEHYBY	VELKE PRIEHYBY DOSIEK METODOU KONECNYCH PRVKOV (LARGE DEFLECTIONS OF PLATES ANALYZED BY FINITE ELEME	70-0373
PRIEREZU	BOX SHAPED GIRDERS. (NOSNIKY UZAVRETEHO PRIEREZU)	70-0169
PRIMAL	FORMULATION AND APPLICATION OF CERTAIN PRIMAL AND MIXED FINITE ELEMENT MODELS OF FINITE DEFORMATIONS	74-0344
PRIMARY	ANALYSIS OF THE DYNAMICS OF THIN PRIMARY MIRRORS FOR LARGE ASTRONOMICAL TELESCOPES	73-1084
PRIMARY	APPLICATION OF FINITE ELEMENT METHOD TO THE THREE DIMENSIONAL STRESS AND VIBRATION ANALYSIS OF BWR P	71-0180
PRIMARY	COMPUTERIZED STRUCTURAL ANALYSIS OF THE WORLD'S LARGEST LIGHT-GAGE STEEL PRIMARY STRUCTURAL SYSTEM	73-0503
PRIMARY	MODAL ANALYSIS OF A SATELLITE PRIMARY STRUCTURE USING A FINITE ELEMENT PROCEDURE	73-0624
PRIMARY	STRUCTURAL EVALUATION OF CANDIDATE DESIGNS FOR THE LARGE SPACE TELESCOPE PRIMARY MIRROR	75-0708
PRIMENENII	METOD KONECHNYKH ELEMENTOV V PRIMENENII K RASCHETU TSILINDRICHESKIKH OBOLOCHEK S PRYAMOUGOL'NYMI OTV	74-0536
PRIMITIVE	INTEGRATIONS OF PRIMITIVE EQUATIONS ON A SPHERE USING FINITE ELEMENT METHOD	74-0649
PRIMITIVE	INTEGRATIONS OF THE PRIMITIVE EQUATIONS ON A SPHERE USING THE FINITE ELEMENT METHOD	74-1347
PRIN	ASUPRA ANALIZEI PRIN METODA ELEMENTULUI FINIT A STARILOR DE EFORTURI SI DEFORMATII DIN MASIVELE DE P	70-0471
PRINCIPAL	CALCULATION OF THE NATURAL FREQUENCIES AND THE PRINCIPAL MODES OF HELICOPTER BLADES	72-0907
PRINCIPAL	PRINCIPAL STRESSES AT THE INTERMEDIATE SUPPORT OF PRESTRESSED CONCRETE CONTINUOUS BEAMS	70-0352
PRINCIPALS	FINITE ELEMENT METHODS BY DIFFERENT VARIATIONAL PRINCIPALS IN ELASTICITY	70-0407
PRINCIPLE	A SIMPLIFIED VERSION OF GURTIN'S VARIATIONAL PRINCIPLE	74-1244
PRINCIPLE	A SPECIAL VARIATIONAL PRINCIPLE FOR THE FINITE-ELEMENT METHOD	69-0181
PRINCIPLE	A USEFUL FORM OF THE VARIATIONAL PRINCIPLE OF IDEAL MHD FOR ONE-DIMENSIONAL NUMERICAL STABILITY CALC	73-0746
PRINCIPLE	A VARIATIONAL PRINCIPLE AND THE CONVERGENCE OF A FINITE-ELEMENT METHOD BASED ON ASSUMED STRESS DISTR	69-0260
PRINCIPLE	A VARIATIONAL PRINCIPLE WITH APPLICATIONS IN FINITE ELEMENTS	74-1285
PRINCIPLE	APPLICATION OF THE DUAL ANALYSIS PRINCIPLE	70-0306
PRINCIPLE	EIN NEUES VARIATIONSPRINZIP MIT ANWENDUNG AUF SCHWINGUNGEN FLACHER SCHALEN NACH DER METHODE DER FINI	73-0563
PRINCIPLE	FINITE ELEMENT ANALYSIS OF PLATE BUCKLING USING A MIXED VARIATIONAL PRINCIPLE	71-0020
PRINCIPLE	FINITE ELEMENT METHOD AS AN ASPECT OF THE PRINCIPLE OF MAXIMUM UNIFORMITY: NEW HYDRODYNAMICAL RAMIFA	74-0009
PRINCIPLE	FORMULATION OF A VARIATIONAL PRINCIPLE - FINITE ELEMENT METHOD FOR VISCOUS FLOW	72-0354
PRINCIPLE	HEAT TRANSFER WITH NONLINEAR BOUNDARY CONDITIONS VIA A VARIATIONAL PRINCIPLE	71-0443
PRINCIPLE	LEAST-SQUARES VARIATIONAL PRINCIPLE FOR FINITE-ELEMENT APPLICATIONS	75-0412
PRINCIPLE	MAXIMUM PRINCIPLE AND UNIFORM CONVERGENCE FOR THE FINITE ELEMENT METHOD	73-0852
PRINCIPLE	NEW VARIATIONAL PRINCIPLE FOR FINITE ELASTIC DISPLACEMENTS	72-0454
PRINCIPLE	NEW VARIATIONAL PRINCIPLE WITH APPLICATION TO VIBRATIONS OF SHALLOW SHELLS USING FINITE ELEMENTS MET	73-0354
PRINCIPLE	RECTANGULAR FINITE ELEMENT FOR PLATE BENDING ANALYSIS BASED ON HELLINGER-REISSNERS VARIATIONAL PRINC	71-0545
PRINCIPLE	SELF ADJOINT DIFFERENTIAL EQUATIONS AND VARIATIONAL PRINCIPLE - 1.	71-0388
PRINCIPLE	SELF ADJOINT DIFFERENTIAL EQUATIONS AND VARIATIONAL PRINCIPLE - 2.	72-0106
PRINCIPLE	SHELL FINITE ELEMENT METHOD VIA REISSNER'S PRINCIPLE	69-0216
PRINCIPLE	SOME APPLICATIONS OF THE PRINCIPLE OF VIRTUAL WORK	73-0930
PRINCIPLE	STABILITY INVESTIGATIONS OF PLATES USING FINITE ELEMENT METHOD BASED ON AN EXPANDED VARIATION PRINCI	71-0663
PRINCIPLE	THE DUAL PRINCIPLE OF ELASTODYNAMICS FINITE ELEMENT APPLICATIONS	71-0501
PRINCIPLE	TWO DUAL FINITE ELEMENT METHODS DERIVED FROM EXTENDED VARIATIONAL PRINCIPLE	73-0109
PRINCIPLE	USE OF VARIATIONAL PRINCIPLE IN STUDY ON RELATIVISTIC BOLTZMANN EQUATION	71-0442
PRINCIPLE	USEFUL FORM OF THE VARIATIONAL PRINCIPLE OF IDEAL MHD FOR ONE-DIMENSIONAL NUMERICAL STABILITY CALCUL	73-0946
PRINCIPLE	VARIATIONAL PRINCIPLE FOR INCOMPRESSIBLE AND NEARLY-INCOMPRESSIBLE ANISOTROPIC ELASTICITY	69-0319
PRINCIPLE	VARIATIONAL PRINCIPLE OF STEADY-STATE TRANSPORT PROCESSES	70-0190
PRINCIPLE	VARIATIONAL PRINCIPLE FOR MAGNETOHYDRODYNAMIC CHANNEL FLOW	70-0192
PRINCIPLE	VARIATIONAL PRINCIPLE FOR THE DYNAMIC ANALYSIS OF CONTINUA BY HYBRID FINITE ELEMENT METHOD	71-0492
PRINCIPLE	VARIATIONAL PRINCIPLE OF LINEAR DIFFERENTIAL EQUATIONS	72-0107
PRINCIPLE	VARIATIONAL PRINCIPLE FOR THE LAPLACE'S OPERATOR WITH APPLICATION IN THE TORSION OF COMPOSITE RODS	74-1044
PRINCIPLES	A MATHEMATICAL THEORY OF COMPLEMENTARY-DUAL VARIATIONAL PRINCIPLES AND MIXED FINITE ELEMENT APPROXIM	73-0804
PRINCIPLES	A UNIFIED THEORY OF VARIATIONAL PRINCIPLES IN MECHANICS	75-0155
PRINCIPLES	APPLICATION OF EXTENDED VARIATIONAL PRINCIPLES TO FINITE ELEMENT ANALYSIS	72-0529
PRINCIPLES	BASIC PRINCIPLES OF STRUCTURAL DYNAMICS	71-0505
PRINCIPLES	CONSTRAINED VARIATIONAL PRINCIPLES AND PENALTY FUNCTION METHODS IN FINITE ELEMENT ANALYSIS	72-0551
PRINCIPLES	DUALITY IN STRUCTURAL ANALYSIS BY FINITE ELEMENTS. STATIC GEOMETRIC ANALOGIES. THE DUAL PRINCIPLES O	71-0583
PRINCIPLES	FINITE ELEMENT APPLICATIONS OF A PRECISION TRIANGULAR PLATE BENDING ELEMENT TO LINEAR BOUNDARY VALUE	75-0788
PRINCIPLES	FINITE ELEMENT FORMULATION BY VARIATIONAL PRINCIPLES WITH RELAXED CONTINUITY REQUIREMENTS	72-0516
PRINCIPLES	FINITE ELEMENT METHODS AS APPLICATIONS OF VARIATIONAL PRINCIPLES	69-0121
PRINCIPLES	FINITE ELEMENT METHOD. BASIC PRINCIPLES AND FORMULATION	69-0328
PRINCIPLES	FINITE ELEMENT METHODS BY VARIATIONAL PRINCIPLES WITH RELAXED CONTINUITY REQUIREMENTS	72-0342
PRINCIPLES	FINITE ELEMENT STIFFNESS METHODS BY DIFFERENT VARIATIONAL PRINCIPLES IN ELASTICITY	70-0455
PRINCIPLES	FINITE-ELEMENT FORMULATION OF MIXED AND STRESS VARIATIONAL PRINCIPLES FOR STRUCTURAL DYNAMIC ANALYSI	72-0964
PRINCIPLES	GENERAL VARIATIONAL PRINCIPLES IN NONLINEAR AND LINEAR ELASTICITY WITH APPLICATIONS	72-0914
PRINCIPLES	GENERALIZED VARIATIONAL PRINCIPLES IN THE FINITE ELEMENT METHOD	69-0116
PRINCIPLES	GENERALIZED VARIATIONAL PRINCIPLES FOR USE IN FLOW PROBLEMS THROUGH POROUS MEDIA	74-0067
PRINCIPLES	GOVERNING EQUATIONS AND EXTREMUM PRINCIPLES OF ELASTICITY AND PLASTICITY GENERATED FROM A SINGLE FUN	73-0666
PRINCIPLES	HYDROLOGIC ENGINEERING METHODS FOR WATER RESOURCES DEVELOPMENT. VOLUME 10. PRINCIPLES OF GROUND-WATE	72-0696
PRINCIPLES	MIXED VARIATIONAL PRINCIPLES, FINITE ELEMENTS AND FINITE ELASTICITY	74-1254
PRINCIPLES	ON COMPLEMENTARY-DUAL VARIATIONAL PRINCIPLES IN MATHEMATICAL PHYSICS	74-0342
PRINCIPLES	ON THE APPLICATION OF GENERALIZED VARIATIONAL PRINCIPLES IN THE FINITE ELEMENT METHOD	68-0083
PRINCIPLES	ON THE ROLE OF VARIATIONAL PRINCIPLES AND METHOD OF WEIGHTED RESIDUALS IN THE FINITE ELEMENT METHODS	74-0551
PRINCIPLES	ON THE VARIATIONAL PRINCIPLES IN THE FINITE ELEMENT METHOD	74-1010
PRINCIPLES	ONE PARAMETER FAMILIES OF VARIATIONAL PRINCIPLES	72-0503
PRINCIPLES	PRINCIPLES OF THE FINITE ELEMENT METHOD	75-0303
PRINCIPLES	SOLUMN INSTABILITY UNDER NONCONSERVATIVE FORCES, WITH INTERNAL AND EXTERNAL DAMPING FINITE ELEMENT U	73-0566
PRINCIPLES	SOME TWO-FIELD VARIATIONAL PRINCIPLES FOR NONLINEAR DEFORMATION ANALYSIS OF SHELLS	75-0294
PRINCIPLES	TECHNICAL, PHYSICAL AND MATHEMATICAL PRINCIPLES OF THE FINITE ELEMENT METHOD	71-0676
PRINCIPLES	VARIATIONAL PRINCIPLES FOR LINEAR INITIAL VALUE PROBLEMS	64-0029
PRINCIPLES	VARIATIONAL PRINCIPLES AND THEIR APPLICATION TO FINITE ELEMENT METHODS	68-0137
PRINCIPLES	VARIATIONAL PRINCIPLES AND DUALISTIC SCHEME FOR INTERSECTION PROBLEMS IN ELASTICITY	69-0305

334

PRINCIPLES	VARIATIONAL PRINCIPLES FOR A RAREFIED GAS FLOW PAST AND ARBITRARY THREE-DIMENSIONAL BODY	70-0189
PRINCIPLES	VARIATIONAL PRINCIPLES AND FINITE ELEMENT METHOD IN PARTIAL DIFFERENTIAL EQUATIONS	71-0452
PRINCIPLES	VARIATIONAL PRINCIPLES FOR FLUID FLOW IN POROUS MEDIA	71-0844
PRINCIPLES	VARIATIONAL PRINCIPLES	72-0501
PRINCIPLES	VARIATIONAL PRINCIPLES, NONLINEAR OPERATORS, AND FINITE ELEMENT APPLICATIONS	72-0576
PRINCIPLES	VARIATIONAL PRINCIPLES OF ELASTOSTATICS WITH RELAXED CONTINUITY REQUIREMENTS AND THEIR APPLICABILITY	74-0758
PRINCIPLES	VARIATIONAL PRINCIPLES AND THE PATCH TEST	74-1282
PRINCIPLES	VARIATIONAL PRINCIPLES FOR ELASTOPLASTIC CONTINUA	75-0837
PRIRASTKOVOU	GEOMETRICKY NELINEARNE ULOHY PLOCHYCH SKRUPIN PRIRASTKOVOU METODOU KONECNYCH PRVKOV. (GEOMET	74-1032
PRISM	TWO DIMENSION, PLATE, SHELL AND FINITE PRISM ISOPARAMETRIC ELEMENTS AND THEIR APPLICATIONS	71-0772
PRISMATIC	A HIGH-ORDER FINITE ELEMENT FOR PRISMATIC AND NONPRISMATIC FOLDED PLATE STRUCTURES	75-0689
PRISMATIC	BERECHNUNG PRISMATISCHER FLAECHENTRAGWERKE MIT HILFE VON FINITEN STREIFENELEMENTEN. (CALCULATION OF	72-0133
PRISMATIC	COMPUTER PROGRAM FOR PRISMATIC FOLDED PLATES BY PLATE AND BEAM ELEMENTS	70-0371
PRISMATIC	DESIGN OF PRISMATIC SHELLS	72-0404
PRISMATIC	DYNAMICS OF PRISMATIC AND OPEN SECTION MEMBER GRIDS	69-0070
PRISMATIC	ELASTIC SHEAR ANALYSIS OF GENERAL PRISMATIC BEAMS	68-0114
PRISMATIC	HIGH-ORDER FINITE ELEMENT TECHNIQUE FOR PRISMATIC AND NONPRISMATIC FOLDED PLATE STRUCTURES	74-1130
PRISMATIC	MASS MATRIX FOR A PRISMATIC BEAM SEGMENT	74-0116
PRISMATIC	PRISMATIC SHELLS WITH HEXAGONAL CROSS-SECTION	68-0116
PRISMATIC	THERMAL AND STRESS ANALYSIS ON PRISMATIC NUCLEAR FUEL ELEMENTS	74-1043
PRISMATIC	THREE-DIMENSIONAL FINITE-ELEMENT ELASTIC ANALYSIS OF PRISMATIC SOLIDS	75-0120
PRISMATISCHER	BERECHNUNG PRISMATISCHER FLAECHENTRAGWERKE MIT HILFE VON FINITEN STREIFENELEMENTEN. (CALCULATION OF	73-1138
PRKLADNYKHH	THE DEVELOPMENT OF APPLIED METHODS IN PROBLEMS OF STATIC CALCULATION OF THIN-WALLED THREE-DIMENSIONA	72-0133
PROBABILISTIC	PROBABILISTIC ASSESSMENT OF AIRCRAFT HAZARD FOR NUCLEAR POWER PLANTS	73-0934
PROBABILISTIC	PROBABILISTIC BUCKLING BEHAVIOR OF STRUCTURES WITH RANDOM LACK-OF-FIT AND GEOMETRIC IMPERFECTIONS	72-0235
PROBABILISTIC	PROBABILISTIC EVALUATION OF THE RELATIVE BEHAVIOUR OF TUNGSTEN CARBIDE COMPONENTS IN STRUCTURAL APPL	73-0862
PROBABILISTIC	PROBABILISTIC FORMULATION FOR ANALYTICAL MODELING CC CONCRETE STRUCTURES	75-0579
PROBABILISTIC	PROBABILISTIC MODELING OF CONCRETE STRUCTURES	71-0709
PROBABILISTIC	PROBABILISTIC THEORY OF STRUCTURES USING FINITE ELEMENTS	72-0236
PROBABILISTIC	RESPONSE OF NUCLEAR STRUCTURAL SYSTEMS TO TRANSIENT AND RANDOM EXCITATIONS, USING BOTH DETERMINISTIC	71-0820
PROBABILISTIC	PROBALILISTIC MODEL FOR MATERIAL STRENGTH VARIATION AND SIZE EFFECT	74-1047
PROBED	BOX CAR BODY BOLSTER PROBED IN STRESS STUDY	72-0876
PROBLEME	ANALYSE NUMERIQUE D'UN PROBLEME DE STEFAN A DEUX PHASES PAR UNE METHODE D'ELEMENTS FINIS. (NUMERICAL	74-0654
PROBLEME	ANWENDUNG DER METHODEN DER FINITEN ELEMENTE MIT DEM PROGRAMMSYSTEM ASKA UNTER BESONDERER BERUCKSICHT	75-0387
PROBLEME	APPLICATION DE LAT METHODE DES ELEMENTS FINIS A LA RESOLUTION D'UN PROBLEME DE DOMAINE OPTIMAL	71-0624
PROBLEME	APPROXIMATIONS EXTERNES, PAR ELEMENTS FINIS DE LAGRANGE D ORDRE, UN ET DEUX, DU PROBLEME DE DIRICHLE	74-0412
PROBLEME	DIE LOSUNG NICHT-LINEARER PROBLEME NACH DER METHODE DER FINITEN ELEMENTE	72-0495
PROBLEME	EINIGE VERFAHREN ZUR BERECHNUNG VON EIGENWERTEN UND EIGENVEKTOREN VON NICHT HERMITESCHEN MATRIZEN U	74-0673
PROBLEME	FINITE ELEMENTE ZUR BERECHNUNG DER WAERMEUEBERTRAGUNG FUER EBENE UND AXISYMMETRISCHE PROBLEME (FINIT	72-0646
PROBLEME	KOMBINATION EINES INTEGRALGLEICHUNGS VERFAHRENS MIT METHODE DER FINITEN ELEMENTEN ZUR BERECHNUNG EBE	73-0449
PROBLEME	NICHTLINEARE DIRICHTLET PROBLEME UND EINFACHE FINITE ELEMENT VERFAHREN	74-0856
PROBLEME	QUELQUES METHODES D'ELEMENTS FINIS POUR DE PROBLEME D'UNE PLAQUE ENCASTREE	74-1263
PROBLEME	UNE METHODE VARIATIONNELLE ELEMENTS FINIS POUR LA RESOLUTION NUMERIQUE UN PROBLEME EXTERIEUR DANS R*	74-0406
PROBLEMEN	PRAKTISCHE BEISPIELE DER DIGITALEN BERECHNUNG VON GRUNDWASSERSTROMUNGS PROBLEMEN NACH DER METHODE EN	73-0576
PROBLEMES	APPROXIMATIONS EXTERNES, PAR ELEMENTS FINIS DE LAGRANGE D ORDRE, UN ET DEUX, DU PROBLEME DE DIRICHLE	73-0081
PROBLEMES	ELEMENTS FINIS ET PROBLEMES ELLIPTIQUES DEGENERES (FINITE ELEMENTS AND DEGENERATE ELLIPTIC PROBLEMS)	72-0495
PROBLEMES	LA METHODE DES ELEMENTS FINIS POUR LES PROBLEMES AUX LIMITES ELLIPTIQUES	73-0582
PROBLEMES	RESOLUTION NUMERIQUE PAR DES METHODES D ELEMENTS FINIS DE PROBLEMES ELLIPTIQUES DEGENERES	74-1228
PROBLEMES	SUR L APPROXIMATION, PAR ELEMENTS FINIS D ORDRE UN, ET LA RESOLUTION, PAR PENALISATION-DUALITE, D UN	72-0702
PROBLEMES	TOUR MAINE-MONTPARNASSE QUELQUES PROBLEMES DE STRUCTURE (SPECIFIC PROBLEMS ENCOUNTERED IN THE CONSTR	72-0845
PROBLEMI	IL METODO DEGLI ELEMENTI FINITI: POSSIBILITA DI APPLICAZIONE A PROBLEMI DI INTERESSE DEGLI ELETTROTE	74-0539
PROBLEMI	LA TECNICA DEGLI ELEMENTI FINITI PER PROBLEMI DI LUBRIFICAZIONE INCOMPRESSIBILE. (FINITE ELEMENT MET	75-0200
PROBLEMI	ON THE RESOLUTION OF NONLINEAR ELASTIC AND ELASTOPLASTIC PROBLEMS BY THE METHOD OF FINITE ELEMENTS.	75-0471
PROBLEMOW	OCENA DOKLADNOSCI ROZWIAZAN PROBLEMOW BRZEGOWYCH CIAL DYSKERTYZOWANYCH. (ASSESSING THE A	74-1326
PROBLEMU	APLIKACE MKP NA RESENI PROBLEMU DIFUZE V KOVECH. (APPLICATION OF THE FINITE ELEMENT METHOD TO THE SO	74-0921
PROBLEMU	NUMERYCZNE ROZWIAZANIE PROBLEMU ORGAN WLASNYCH NA PRZYKLADZIE PLYT. (NUMERICAL SOLUTION OF THE PROBL	74-0609
PROCEDE	APPLICATION DE LA METHODE DES ELEMENTS FINIS - UN PROCEDE DE SOUS-ASSEMBLAGE	75-0296
PROCEDIMENTO	PROCESS FOR THE AUTOMATIC GENERATION OF FLAT TRIANGULAR LATTICES FOR APPLICATION TO THE FINITE ELEME	74-0405
PROCESSING	APPLICATION OF THE DOUBLE LIGAMENT TENSILE TEST IN DEFORMATION PROCESSING	71-0366
PROCESSING	GRAPHICS APPLICATIONS FOR FINITE ELEMENT CODE PROCESSING	75-0827
PROCESSING	SOLUTION OF THE PROBLEM OF PROCESSING METALS BY PRESSURE USING THE METHOD OF FINITE ELEMENTS	74-0893
PROCESSING	SOLVING THE PROBLEM OF PROCESSING METALS BY PRESSURE THROUGH THE METHOD OF FINITE ELEMENTS	74-1363
PROCESSING	THERMAL ANALYSIS OF M552 EXPEIMENT FOR MATERIALS PROCESSING IN SPACE	74-1354
PROCESSING	THERMAL ANALYSIS OF M551 EXPERIMENT FOR MATERIALS PROCESSING IN SPACE	74-0704
PROCESSOR	PLANFORM INPUT GENERATOR (PING), A NASTRAN PROCESSOR FOR LIFTING SURFACES-THEORETICAL DEVELOPMENT, U	74-0725
PROCESSOR	VEHICLE STRUCTURAL ANALYSIS, PRE AND POST PROCESSOR DATA HANDLING	73-0780
PROCHNOSTI	ISSLEDOVANIE PROCHNOSTI I ZHESTKOSTI PLASTIN, NAGRUZHENNYKH V SVOEI PLOSKOSTI, METHODOM KONECHNYKH E	71-0093
PRODUCT	EXTRAPOLATED FINITE ELEMENT PROCEDURES OVER CARTESIAN PRODUCT DOMAINS	73-0382
PRODUCT	FINITE ELEMENT METHODS IN PRODUCT DESIGN	75-0044
PRODUCT	INTERFACING FINITE ELEMENT METHODS WITH PRODUCT DESIGN ENGINEERING	74-0117
PRODUCTIVITY	EFFECT OF PERFORATION DAMAGE ON WELL PRODUCTIVITY	73-0189
PRODUCTS	ANALYSIS OF VISCOELASTIC CONTACT STRESSES IN AGRICULTURAL PRODUCTS USING A FINITE ELEMENT METHOD	74-0936
PRODUCTS	ANALYSIS OF VISCOELASTIC CONTACT STRESSES IN AGRICULTURAL PRODUCTS USING A FINITE ELEMENT METHOD	74-0634
PROFILE	AN ALGORITHM FOR REDUCING THE BANDWITH AND PROFILE OF A SPARSE MATRIX	74-1199
PROFILE	STUDY OF UMBRELLA OSCILLATIONS OF TURBINE DISKS OF ARBITRARY PROFILE BY THE FINITE ELEMENT METHOD	74-0792
PROFILE	ZUR BERECHNUNG EINFACH-SYMMETRISCHER I-TRAEGER NACH DER THEORIE - 2. ORDNUNG UNTER BERUECKSICHTIGUNG	75-0332
PROFILES	FINITE ELEMENT OPTIMIZATION OF FINITE STEPPED SLIDER BEARING PROFILES	75-0510
PROGRAM	A COMPUTER PROGRAM FOR THE STATIC AND DYNAMIC FINITE ELEMENT ANALYSIS OF AXISYMMETRIC THIN SHELLS	74-0131
PROGRAM	A COMPUTER PROGRAM FOR THE ANALYSIS OF THIN SHELLS	66-0017
PROGRAM	A COMPUTER PROGRAM FOR THE ANALYSIS OF LINEARLY ELASTIC PLANE-STRESS, PLANE-STRAIN PROBLEMS	67-0041
PROGRAM	A COMPUTER PROGRAM FOR THE ANALYSIS OF TWO-DIMENSIONAL HEAT CONDUCTION USING THE FINITE ELEMENT TECH	68-0110
PROGRAM	A COMPUTER PROGRAM FOR THE FINITE ELEMENT ANALYSIS OF A CLASS OF PLATE BENDING PROBLEMS	69-0066
PROGRAM	A COMPUTER PROGRAM FOR THE CALCULATION OF TORSION FACTORS AND SHEAR STRESS FACTORS FOR BOX SECTIONS	69-0191
PROGRAM	A COMPUTER PROGRAM FOR THE ANALYSIS OF TUBULAR K JOINTS	69-0485
PROGRAM	A COMPUTER PROGRAM FOR EARTHQUAKE ANALYSIS OF DAMS	69-0488
PROGRAM	A COMPUTER PROGRAM FOR ANALYSIS OF STIFFENED PLATES UNDER COMBINED INPLANE AND LATERAL LOADS	70-0469
PROGRAM	A COMPUTER PROGRAM TO ANALYZE BEAM-COLUMN UNDER MOVEABLE LOADS	71-0703
PROGRAM	A COMPUTER PROGRAM TO DISPLAY THE DYNAMIC STRESS ANALYSIS OF STRUCTURES	68-0268
PROGRAM	A DIGITAL COMPUTER PROGRAM FOR THE FINITE ELEMENT ANALYSIS OF SOLIDS WITH NONLINEAR MATERIAL PROPERT	70-0077
PROGRAM	A FINITE ELEMENT COMPUTER PROGRAM FOR THE PREDICTION OF THE BEHAVIOR OF REINFORCED AND PRESTRESSED C	65-0032
PROGRAM	A FINITE ELEMENT PROGRAM FOR THE DYNAMIC LARGE DEFLECTION, ELASTIC-PLASTIC ANALYSIS OF STIFFENED SHE	72-0799
PROGRAM	A FINITE ELEMENT PROGRAM FOR DESIGN AND ANALYSIS OF THREE-DIMENSIONAL FLAT SPRINGS	72-0908
PROGRAM	A FINITE ELEMENT PROGRAM FOR FREE FORM SHELLS	74-0738
PROGRAM	A FINITE-ELEMENT PROGRAM FOR THE ANALYSIS OF EMBANKMENTS OVER SOFT, SATURATED SOILS INCLUDING CONSOL	75-0786
PROGRAM	A FORTRAN PROGRAM FOR CALCULATING THE FREQUENCIES AND MODE SHAPES OF AN ARBITRARY SHAPE CYLINDRICAL	73-1160
PROGRAM	A FORTRAN PROGRAM FOR CALCULATING THE FREQUENCIES AND MODE SHAPES OF A GUIDED CLAMPED, GUIDED PINNED	72-0594
PROGRAM	A FRONTAL SOLUTION PROGRAM FOR FINITE ELEMENT ANALYSIS	72-0890
PROGRAM	A GENERAL HIGH-ORDER FINITE-ELEMENT WAVEGUIDE ANALYSIS PROGRAM	70-0073
PROGRAM	A GENERAL PURPOSE THREE DIMENSIONAL STRESS ANALYSIS PROGRAM	69-0512
PROGRAM	A SELF-ORGANIZING MESH GENERATION PROGRAM	71-0686
PROGRAM	A SELF-ORGANIZING MESH GENERATION PROGRAM	74-1051
PROGRAM	A SPECIAL NASTRAN PROGRAM FOR INPUT CHECKING AND UNDEFORMED STRUCTURE PLOTTING	74-1249
		71-0234

335

PROGRAM	A TECHNICAL EVALUATION OF THE NASTRAN COMPUTER PROGRAM	71-0550
PROGRAM	A USER-ORIENTED PROGRAM FOR CRASH DYNAMICS	74-0321
PROGRAM	ADDITION OF AN ARBITRARY BODY ANALYSIS CAPABILITY TO THE BOEING TEA 236 FINITE ELEMENT COMPUTER PROG	73-0785
PROGRAM	AN ADAPTIVE COMPUTER PROGRAM FOR DIV. (P(X,Y) GRAD U) EQUALS F(X,Y,U) IN A POLYGONAL REGION	75-0133
PROGRAM	AN IMPROVED AUTOMATED STRUCTURAL OPTIMIZATION PROGRAM	74-0732
PROGRAM	AN INCORE FINITE ELEMENT PROGRAM FOR RESEARCH AND TEACHING PURPOSE	74-0666
PROGRAM	AN INTERACTIVE COMPUTER GRAPHICS PROGRAM FOR NASTRAN	71-0074
PROGRAM	AN INTERACTIVE GRAPHICS SHIP DESIGN ORIENTED PROGRAM PACKAGE FOR TIME SHARING SYSTEMS	73-0341
PROGRAM	ANALYSIS AND DESIGN CAPABILITY OF THE STRUDL PROGRAM	71-0054
PROGRAM	ANALYSIS OF SLOPE STABILITY BY FINITE ELEMENT METHOD AND ITS APPLICATION PROGRAM. SOIL MECHANICS	72-0919
PROGRAM	ANALYTIC MODELING OF ROCK-STRUCTURE INTERACTION. VOLUME 3. COMPUTER PROGRAM	73-0918
PROGRAM	ANALYTIC MODELING OF ROCK-STRUCTURE INTERACTION. VOLUME 2. USERS GUIDE FOR A COMPUTER PROGRAM	73-0919
PROGRAM	ANSYS (ENGINEERING ANALYSIS SYSTEM) PROGRAM	72-0156
PROGRAM	ANTRAS COMPUTER PROGRAM FOR LINEAR STRESS ANALYSIS BY FINITE ELEMENT METHOD	74-0601
PROGRAM	APACHE: A THREE-DIMENSIONAL FINITE ELEMENT PROGRAM FOR STEADY-STATE OR TRANSIENT HEAT CONDUCTION ANA	73-0971
PROGRAM	APPLICATION OF AN ' ELEMENT-METHOD ' PROGRAM TO SOLVING TEMPERATURE DISTRIBUTION PROBLEMS (ANWENDUNG	71-0464
PROGRAM	APPLICATION OF PROGRAM FATIQUE TEST TO MEMBER JOINTS OF HULLS	72-0814
PROGRAM	APPLICATION, TESTING, AND EVALUATION OF DOUBLY CURVED SHELL ELEMENTS FOR DYNAMIC ANALYSIS, PART III.	69-0256
PROGRAM	ARCH DAMS ANALYSED BY A LINEAR FINITE ELEMENT SHELL THE FINITE STRIP METHOD IN THE ANALYSIS OF ELAST	68-0189
PROGRAM	AXISYMMETRIC RING AND PLATE STRESS ANALYSIS PROGRAM - USERS GUIDE	70-0333
PROGRAM	AYER HEAT CONDUCTION COMPUTER PROGRAM	74-0885
PROGRAM	BASIS OF A WELL CONDITIONED FORCE PROGRAM FOR EQUILIBRIUM MODELS VIA THE SOUTHWELL SLAB ANALOGIES	67-0111
PROGRAM	BERGEN, AN AUTOMATIC FINITE ELEMENT MESH GENERATION PROGRAM FOR ARBITRARY STRUCTURES	72-0776
PROGRAM	BRLESC FINITE ELEMENT PROGRAM FOR AXISYMMETRIC, PLANE STRAIN, AND PLANE STRESS, ORTHOTROPIC SOLIDS W	71-0749
PROGRAM	CAPE-2D, A COMPUTER PROGRAM FOR THE STRESS ANALYSIS OF PLANE AND AXISYMMETRIC COMPOSITE STRUCTURES.	69-0083
PROGRAM	CHILES: A FINITE ELEMENT COMPUTER PROGRAM THAT CALCULATES THE INTENSITIES OF LINEAR ELASTIC SINGULAR	73-0959
PROGRAM	COMPANY TRAINING OF ENGINEERING IN STRUCTURAL ANALYSIS COMPUTER PROGRAM USAGE	72-0460
PROGRAM	COMPUTER PROGRAM CONSTRUCTION AND MAINTENANCE - FUTURE OF CENTRALIZED FINITE ELEMENT ACTIVITY	75-0014
PROGRAM	COMPUTER PROGRAM FOR ELASTIC-PLASTIC ANALYSIS OF AXISYMMETRICALLY LOADED SHELLS OF REVOLUTION	68-0239
PROGRAM	COMPUTER PROGRAM FOR PRISMATIC FOLDED PLATES BY PLATE AND BEAM ELEMENTS	70-0371
PROGRAM	COMPUTER PROGRAM FOR GENERATING FINITE ELEMENT MODELS OF MINE STRUCTURES	71-0493
PROGRAM	COMPUTER PROGRAM FOR PIT SLOPE STABILITY ANALYSIS BY THE FINITE ELEMENT STRESS ANALYSIS AND LIMITING	72-0225
PROGRAM	COMPUTER PROGRAM FOR STATIC AND DYNAMIC ANALYSIS OF LINEAR STRUCTURAL SYSTEMS	72-0683
PROGRAM	COMPUTER PROGRAM FOR SOLUTION OF LARGE, SPARSE, UNSYMMETRIC SYSTEMS OF LINEAR EQUATIONS	72-0719
PROGRAM	COMPUTER PROGRAM FOR AUTOMATIC MESH GENERATION FOR THE 2-D FINITE ELEMENT PROGRAM WILAX	73-0565
PROGRAM	COMPUTER PROGRAM FOR AUTOMATIC MESH GENERATION FOR THE 2-D FINITE ELEMENT PROGRAM WILAX	73-0565
PROGRAM	COMPUTER PROGRAM JET 3 TO CALCULATE THE LARGE ELASTIC PLASTIC DYNAMICALLY INDUCED DEFORMATIONS OF FR	72-0609
PROGRAM	CONIDA: A FINITE ELEMENT PROGRAM FOR THE STRESS ANALYSIS OF AXISYMMETRIC THIN SHELLS	74-1087
PROGRAM	CONPLOT: A GENERAL PURPOSE CONTOUR PLOTTING PROGRAM	74-0887
PROGRAM	DANUTA THREE DIMENSIONAL FINITE ELEMENT PROGRAM USED IN ANALYSIS OF TURBOMACHINERY	72-0296
PROGRAM	DAS FLAECHENTRAGWERKSPROGRAM VON STRIP. (STRIP PROGRAM FOR PLATE STRUCTURES)	72-0150
PROGRAM	DESCRIPTION AND INSTRUCTION FOR USE OF THE SAFE-SHELL. B FINITE ELEMENT PROGRAM IN THE IBM/360 VERSI	73-0945
PROGRAM	DESCRIPTION OF THE IBM-360 VERSION OF THE FINITE ELEMENTS COMPUTER PROGRAM SAFE-3D	72-0775
PROGRAM	DEVELOPMENT AND APPLICATIONS OF A GENERAL COMPUTER PROGRAM FOR EVALUATING STABILITY OF OPENINGS IN R	74-0994
PROGRAM	DIE-SWELL RECONSIDERED: SOME NUMERICAL SOLUTIONS USING A FINITE ELEMENT PROGRAM	73-0498
PROGRAM	DISCRETE-ELEMENT STATIC ANALYSIS OF BONDED, DOUBLE-LAYER, BRANCHED, THIN SHELLS OF REVOLUTION, PART	70-0081
PROGRAM	DYNAMIC STRUCTURAL ANALYSIS WITH THE NASTRAN COMPUTER PROGRAM	70-0213
PROGRAM	DYNAPLAS: A FINITE ELEMENT PROGRAM FOR THE DYNAMIC, LARGE DEFLECTION, ELASTIC-PLASTIC ANALYSIS OF ST	72-0773
PROGRAM	DYNASOR 11 - A FINITE ELEMENT PROGRAM FOR THE DYNAMIC NONLINEAR ANALYSIS OF SHELLS OF REVOLUTION (US	71-0666
PROGRAM	DYPLAS, A FINITE ELEMENT DYNAMIC ELASTIC-PLASTIC LARGE DEFORMATION ANALYSIS PROGRAM	74-0306
PROGRAM	ELAD PROGRAM MANUAL, A COMPUTER PROGRAM FOR THE ELASTIC ANALYSIS OF DOME STRUCTURES AND AXISYMMETRIC	69-0064
PROGRAM	ELAD PROGRAM MANUAL, A COMPUTER PROGRAM FOR THE ELASTIC ANALYSIS OF DOME STRUCTURES AND AXISYMMETRIC	69-0064
PROGRAM	ELAS - A GENERAL PURPOSE COMPUTER PROGRAM FOR THE EQUILIBRIUM PROBLEMS OF LINEAR STRUCTURES	68-0158
PROGRAM	ELAS: A GENERAL-PURPOSE COMPUTER PROGRAM FOR THE EQUILIBRIUM PROBLEMS OF LINEAR STRUCTURES. VOLUME 2	69-0462
PROGRAM	ELAS: A GENERAL-PURPOSE COMPUTER PROGRAM FOR THE EQUILIBRIUM PROBLEMS OF LINEAR STRUCTURES. VOLUME 2	69-0462
PROGRAM	ELASTIC-PLASTIC AND CREEP ANALYSIS VIA THE MARC 2 FINITE ELEMENT PROGRAM	71-0032
PROGRAM	ELASTOPLASTIC AND CREEP ANALYSIS WITH THE ASKA PROGRAM SYSTEM	74-0863
PROGRAM	ENGINEERING OPERATIONS REPORT: NERVA THIN SHELL FINITE ELEMENT STRESS ANALYSIS METHODOLOGY COMPUTER	72-0916
PROGRAM	EPAD, A COMPUTER PROGRAM FOR THE ELASTIC-PLASTIC ANALYSIS OF DOME STRUCTURES AND ARBITRARY SOLIDS OF	70-0036
PROGRAM	EURCYL: A COMPUTER PROGRAM TO GENERATE FINITE ELEMENT MESHES FOR CYLINDER-CYLINDER INTERSECTIONS	73-0962
PROGRAM	EURCYL: A PROGRAM TO GENERATE FINITE ELEMENT MESHES FOR PRESSURE VESSEL NOZZLES	74-1086
PROGRAM	EVALUATION OF A THREE-DIMENSIONAL STRESS ANALYSIS PROGRAM	72-0743
PROGRAM	EVALUATION OF THE COMPUTER PROGRAM CREEP-BLAST (CDC VERSION)	74-0684
PROGRAM	FAMSOR, A FINITE ELEMENT PROGRAM FOR THE FREQUENCIES AND MODE SHAPES OF SHELLS OF REVOLUTION	70-0090
PROGRAM	FAST MODAL EXTRACTION IN NASTRAN VIA THE FEER COMPUTER PROGRAM	73-0356
PROGRAM	FATIQUE STRENGTH OF LUGS CONTAINING LINERS. VOLUME II. COMPUTER PROGRAM USED FOR ANALYSIS	70-0418
PROGRAM	FEATS, A COMPUTER PROGRAM FOR THE FINITE ELEMENT THERMAL STRESS ANALYSIS OF PLANE OR AXISYMMETRIC SO	68-0152
PROGRAM	FEDGE - A GENERAL PURPOSE COMPUTER PROGRAM FOR FINITE ELEMENT DATA GENERATION. VOLUME 2 - PROGRAM MA	69-0492
PROGRAM	FEDGE - A GENERAL PURPOSE COMPUTER PROGRAM FOR FINITE ELEMENT DATA GENERATION. VOLUME 2 - PROGRAM MA	69-0492
PROGRAM	FEDGE - A GENERAL-PURPOSE COMPUTER PROGRAM FOR FINITE ELEMENT DATA GENERATION. VOLUME 1 - USER'S MAN	69-0491
PROGRAM	FEM2D: PROGRAM FOR SOLVING THE TWO-DIMENSIONAL DIFFUSION EQUATION BY THE METHOD OF FINITE ELEMENTS	74-1089
PROGRAM	FESTRAN: A FINITE ELEMENT COMPUTER PROGRAM	71-0811
PROGRAM	FETE PROGRAM PACKAGE	72-0161
PROGRAM	FINITE ELEMENT COMPUTER PROGRAM FOR THE ANALYSIS OF LAYERED ORTHOTROPIC STRUCTURES	74-0800
PROGRAM	FINITE ELEMENT COMPUTER PROGRAM FOR THE SOLUTION OF NONLINEAR AXISYMMETRIC CONTACT PROBLEMS WITH INT	74-1094
PROGRAM	FINITE ELEMENT COMPUTER PROGRAM FOR PREDICTING THE NONLINEAR STATIC AND DYNAMIC BEHAVIOR OF VISCOELA	75-0191
PROGRAM	FINITE ELEMENT MESH GENERATION PROGRAM FOR ARBITRARY TWO-AND THREE-DIMENSIONAL STRUCTURES	73-0966
PROGRAM	FINITE ELEMENT PROGRAM FEECON FOR UNDRAINED DEFORMATION ANALYSES OF GRANULAR EMBANKMENTS ON SOFT CLA	73-0875
PROGRAM	FINITE ELEMENT PROGRAM PACKAGE FOR AXISYMMETRIC SCALAR FIELD PROBLEMS	73-0113
PROGRAM	FINITE ELEMENT PROGRAM PACKAGE FOR AXISYMMETRIC VECTOR FIELD PROBLEMS	75-0138
PROGRAM	FINITE ELEMENT SOLUTION OF PARTIAL DIFFERENTIAL EQUATIONS FEMAC COMPUTER PROGRAM	73-0648
PROGRAM	FINITE ELEMENT SOLUTIONS OF SELECTED PARTIAL DIFFERENTIAL EQUATIONS "FEMAC COMPUTER PROGRAM"	74-0327
PROGRAM	FINPLT: A FINITE-ELEMENT FIELD-PLOTTING PROGRAM	72-0299
PROGRAM	FIRES-RC. A COMPUTER PROGRAM FOR THE FIRE RESPONSE OF STRUCTURES-REINFORCED CONCRETE FRAMES	74-0747
PROGRAM	FIRES-T. A COMPUTER PROGRAM FOR THE FIRE RESPONSE OF STRUCTURES - THERMAL	74-0850
PROGRAM	FUEL ROD DEFORMATIONS - A TIME DEPENDENT FINITE ELEMENT ANALYSIS ON AXISYMMETRIC FUEL RODS, INCLUDIN	70-0414
PROGRAM	FJLMOD - AN INELASTIC ANALYSIS PROGRAM TO PREDICT THE OPERATING BEHAVIOUR OF CANDU FUEL ELEMENTS	74-0574
PROGRAM	GENERAL DESCRIPTION OF SESAM-69, SUPER ELEMENT STRUCTURAL ANALYSIS (PROGRAM) MODULES	71-0612
PROGRAM	GENERAL PURPOSE COMPUTER PROGRAM FOR THE DYNAMIC SIMULATION OF VEHICLE GUIDEWAY INTERACTIONS	73-0260
PROGRAM	GENERAL PURPOSE PROGRAM OF PLANE STRESS ANALYSIS BY FINITE ELEMENT METHOD, AND ITS APPLICATIONS	72-0186
PROGRAM	GENERAL PURPOSE PROGRAM FOR FINITE ELEMENT ANALYSIS WITH REFERENCE TO INTERACTION BETWEEN SOFTWARE A	73-0334
PROGRAM	GENERAL PURPOSE PROGRAM FOR SHELL STRUCTURES: ISTRAN-S	75-0511
PROGRAM	GNATS: A FINITE ELEMENT COMPUTER PROGRAM FOR THE GENERAL NONLINEAR ANALYSIS OF TWO-DIMENSIONAL STRUC	74-0875
PROGRAM	HEAVY SECTION STEEL TECHNOLOGY PROGRAM. TECHNICAL REPORT NO. 4, OCTOBER 1969. A TWO-DIMENSIONAL ELAS	69-0474
PROGRAM	HEAVY SECTION STEEL TECHNOLOGY PROGRAM TECHNICAL OR PROGRAMMATIC MANUSCRIPT NO. 25. VARIABLE THICKNE	73-0948
PROGRAM	HETRAP: A HEAT TRANSFER ANALYSIS PROGRAM	74-0870
PROGRAM	HIGH-PERFORMANCE COMPOSITE MATERIALS FOR VEHICLE CONSTRUCTION: A FINITE ELEMENT COMPUTER PROGRAM FOR	73-0767
PROGRAM	IMPROVED ANALYTIC LONGITUDINAL RESPONSE ANALYSIS FOR AXISYMMETRIC LAUNCH VEHICLES, COMPUTER PROGRAM	65-0002
PROGRAM	INTEGRATED PROGRAM FOR THE DYNAMIC STRUCTURAL CALCULATION OF MOBIL OFFSHORE UNITS	74-0490
PROGRAM	INTERACTIVE PROGRAM IN DESIGN AND ANALYSIS OF COMPOSITE MATERIALS	74-0854
PROGRAM	LUSH: A COMPUTER PROGRAM FOR COMPLEX RESPONSE ANALYSIS OF SOIL-STRUCTURE SYSTEMS	74-0744
PROGRAM	MATHEMATICAL MODEL OF THE COLUMBIA RIVER FROM THE PACIFIC OCEAN TO BONNEVILLE DAM. PART 1. THEORY, P	69-0464

PROGRAM	MATUS: A THREE-DIMENSIONAL FINITE ELEMENT PROGRAM FOR SMALL-STRAIN ELASTIC ANALYSIS	73-0970
PROGRAM	MECHANICS OF CABLE MOORING SYSTEMS. VOLUME IV. A COMPUTER PROGRAM FOR ANALYZING THE STEADY STATE RES	72-0653
PROGRAM	MECHANICS OF CABLE MOORING SYSTEMS. VOLUME VI. A COMPUTER PROGRAM FOR ANALYZING THE STEADY STATE CON	73-0725
PROGRAM	MESH GENERATING COMPUTER PROGRAM FOR FINITE ELEMENT ANALYSES OF SIMPLE SLOPES IN HORIZONTALLY LAYERE	71-0748
PROGRAM	MESH GENERATION PROGRAM FOR HIGHWAY EXCAVATION CUTS	74-1250
PROGRAM	MESH 3D: A THREE-DIMENSIONAL FINITE ELEMENT MESH GENERATOR PROGRAM FOR EIGHT-NODE ISOPARAMETRIC ELEM	73-0972
PROGRAM	METHODS FOR PREDICTING THE AERODYNAMIC AND STABILITY AND CONTROL CHARACTERISTICS OF STOL AIRCRAFT. V	73-0718
PROGRAM	MISSILE BODY INPUT GENERATOR (BING), A NASTRAN PRE-PROCESSOR: THEORETICAL DEVELOPMENT, USER'S MANUAL	75-0717
PROGRAM	NASTRAN - A FINITE ELEMENT PROGRAM FOR STRUCTURAL ANALYSIS	72-0131
PROGRAM	NASTRAN, NASA'S GENERAL PURPOSE STRUCTURAL ANALYSIS PROGRAM	70-0370
PROGRAM	NONSAP - A GENERAL FINITE ELEMENT PROGRAM FOR NONLINEAR DYNAMIC ANALYSIS OF COMPLEX STRUCTURES	73-0740
PROGRAM	NONSAP - A NONLINEAR STRUCTURAL ANALYSIS PROGRAM	74-1006
PROGRAM	NONSAP: A STRUCTURAL ANALYSIS PROGRAM FOR STATIC AND DYNAMIC RESPONSE OF NONLINEAR SYSTEMS	74-0848
PROGRAM	ON IMPROVING EFFICIENCY OF AN ALGORITHM FOR STRUCTURAL OPTIMIZATION AND A USERS MANUAL FOR PROGRAM T	74-0778
PROGRAM	ON THE DEVELOPMENT OF THE "PASSAGE" PROGRAM	73-0342
PROGRAM	OPTIM 11: (MAGIC) COMPATIBLE LARGE SCALE AUTOMATED MINIMUM WEIGHT DESIGN PROGRAM - V 2. PROGRAMMER'S	74-1036
PROGRAM	OPTIM 11: A (MAGIC) COMPATIBLE LARGE SCALE AUTOMATED MINIMUM WEIGHT DESIGN PROGRAM - 1. ENGINEERS AN	74-1048
PROGRAM	OPTIM 11: A MAGIC COMPATIBLE LARGE-SCALE AUTOMATED MINIMUM WEIGHT DESIGN PROGRAM. VOLUME 1. ENGINEER	74-1096
PROGRAM	OPTIM 11: MAGIC COMPATIBLE LARGE SCALE AUTOMATED MINIMUM WEIGHT DESIGN PROGRAM. VOLUME 11. PROGRAMME	74-1093
PROGRAM	ORGANIZATION AND FUNCTIONAL PURPOSE OF A FINITE ELEMENT COMPUTER PROGRAM	74-0362
PROGRAM	ORGANIZATION OF GNATS: A GENERAL NONLINEAR ANALYSIS COMPUTER PROGRAM FOR TWO-DIMENSIONAL STRUCTURES	74-0871
PROGRAM	PART 1 - TASK 3.2.9. BENDING AND BUCKLING OF A ONE-DIMENSIONAL MODEL SIMULATING A CONICAL SHELL. PAR	71-0745
PROGRAM	PART 1. THEORY REPORT FOR CREEP-PLAST COMPUTER PROGRAM: ANALYSIS OF TWO-DIMENSIONAL PROBLEMS UNDER S	72-0765
PROGRAM	PLACID: A GENERAL FINITE ELEMENT COMPUTER PROGRAM FOR STRESS ANALYSIS OF PLANE AND AXISYMMETRIC SOLI	74-0884
PROGRAM	PLANFORM INPUT GENERATOR (PING), A NASTRAN PROCESSOR FOR LIFTING SURFACES-THEORETICAL DEVELOPMENT, U	73-0780
PROGRAM	PLANS - A FINITE ELEMENT PROGRAM FOR NONLINEAR ANALYSIS OF STRUCTURES	74-0966
PROGRAM	PLANS - A FINITE-ELEMENT PROGRAM FOR NONLINEAR ANALYSIS OF STRUCTURES	75-0011
PROGRAM	PLOTS - THE PLOTTING SUBROUTINE INCORPORATED IN THE BALLISTIC RESEARCH LABORATORIES ELECTRONIC SCIEN	72-0803
PROGRAM	POWER SPECTRAL DENSITY ANALYSIS OF VEHICLE VIBRATION USING THE NASTRAN COMPUTER PROGRAM	74-0318
PROGRAM	PREDICTION OF TEMPERATURE DISTRIBUTIONS IN CARGOES WITH THE AID OF A COMPUTER PROGRAM USING THE METH	74-1215
PROGRAM	PROGRAM DESIGN BY A MULTIDISCIPLINARY TEAM	75-0854
PROGRAM	PROGRAM FOR TRIANGULAR BENDING ELEMENTS WITH DERIVATIVES SMOOTHING	72-0473
PROGRAM	PROGRAM SPECIFICATION NV 300. GENERAL FINITE ELEMENT PROGRAM FOR 2- AND 3-DIMENSIONAL MEMBRANE STRUC	68-0184
PROGRAM	PROGRAM SPECIFICATION NV 300. GENERAL FINITE ELEMENT PROGRAM FOR 2- AND 3-DIMENSIONAL MEMBRANE STRUC	68-0184
PROGRAM	PROGRAM TO REDUCE THE SIZE OF STRUCTURAL MATRICES	72-0912
PROGRAM	QMESH: A SELF-ORGANIZING MESH GENERATION PROGRAM	74-0872
PROGRAM	QMESH: A SELF-ORGANIZING MESH GENERATION PROGRAM	74-0899
PROGRAM	QUAD-4: A COMPUTER PROGRAM FOR EVALUATING THE SEISMIC RESPONSE OF SOIL STRUCTURES BY VARIABLE DAMPIN	73-0771
PROGRAM	RANDOM VIBRATION ANALYSIS SYSTEM FOR COMPLEX STRUCTURES. PART 2. COMPUTER PROGRAM DESCRIPTION	69-0262
PROGRAM	RETSCP: A COMPUTER PROGRAM FOR ANALYSIS OF ROCKET ENGINE THERMAL STRAINS WITH CYCLIC PLASTICITY	74-0692
PROGRAM	REVIEW OF THE ASKA PROGRAM	71-0175
PROGRAM	SABOR 1. A FORTRAN PROGRAM FOR THE LINEAR AXISYMMETRIC LOADING BY USING THE MATRIX DISPLACEM	65-0027
PROGRAM	SAFE PCPS, A COMPUTER PROGRAM FOR THE STRESS ANALYSIS OF COMPOSITES BODIES OF REVOLUTION. INPUT INST	65-0013
PROGRAM	SAFE SHELL, A COMPUTER PROGRAM FOR THE STRESS ANALYSIS OF THIN SHELL BODIES OF REVOLUTION. A USER'S	67-0018
PROGRAM	SAFE-AXISYM, A COMPUTER PROGRAM FOR THE DESIGN OF COMPOSITE BODIES OF REVOLUTION. INPUT INSTRUCTIONS	65-0014
PROGRAM	SAFE-COLAPS: A FINITE ELEMENT PROGRAM FOR THE CREEP COLLAPSE ANALYSIS OF NUCLEAR FUEL RODS	75-0573
PROGRAM	SAFE-CREEP: A COMPUTER PROGRAM FOR THE VISCOELASTIC ANALYSIS OF AXISYMMETRIC AND PLANE CONCRETE STRU	67-0148
PROGRAM	SAFE-PLANE, A COMPUTER PROGRAM FOR THE STRESS ANALYSIS AND DESIGN OF TWO-DIMENSIONAL COMPOSITE BODIE	67-0017
PROGRAM	SAMMSOR 111: A FINITE ELEMENT PROGRAM TO DETERMINE STIFFNESS AND MASS MATRICES OF RING-STIFFENED SHE	72-0774
PROGRAM	SAP IV. A STRUCTURAL ANALYSIS PROGRAM FOR STATIC AND DYNAMIC RESPONSE OF LINEAR SYSTEMS	73-0650
PROGRAM	SAP-A GENERAL STRUCTURAL ANALYSIS PROGRAM FOR LINEAR SYSTEMS	72-0329
PROGRAM	SCALAR FINITE-ELEMENT PROGRAM PACKAGE FOR TWO-DIMENSIONAL FIELD PROBLEMS	71-0478
PROGRAM	SEAL-SHELL-2, A COMPUTER PROGRAM FOR THE STRESS ANALYSIS OF A THICK SHELL OF REVOLUTION WITH AXISYMM	69-0107
PROGRAM	SECOND GENERATION BOUNDARY INTEGRAL EQUATION PROGRAM FOR THREE-DIMENSIONAL ELASTIC ANALYSIS	75-0376
PROGRAM	SESAM 69 A GENERAL PURPOSE FINITE ELEMENT METHOD PROGRAM CDC/NASTRAN APPLICATIONS MANUAL	72-0552
PROGRAM	SESAM-69 - A GENERAL PURPOSE FINITE ELEMENT METHOD PROGRAM	74-0183
PROGRAM	SHIP (SIMPLIFIED-HELMHOLTZ-INTEGRAL PROGRAM): A FAST COMPUTER PROGRAM FOR CALCULATING THE ACOUSTIC R	72-0835
PROGRAM	SHIP (SIMPLIFIED-HELMHOLTZ-INTEGRAL PROGRAM): A FAST COMPUTER PROGRAM FOR CALCULATING THE ACOUSTIC R	72-0835
PROGRAM	SINGER: A COMPUTER CODE FOR GENERAL ANALYSIS OF TWO-DIMENSIONAL CONCRETE STRUCTURES. VOL. 11. PROGRA	75-0723
PROGRAM	SLADE D: A COMPUTER PROGRAM FOR THE DYNAMIC ANALYSIS OF THIN SHELLS	73-0964
PROGRAM	SLADE, A COMPUTER PROGRAM FOR THE STATIC ANALYSIS OF THIN SHELLS	70-0079
PROGRAM	SOLID SAP. A STATIC ANALYSIS PROGRAM FOR THREE DIMENSIONAL SOLID STRUCTURES	72-0855
PROGRAM	SOME PROGRAM DEVELOPMENTS OF MATRIX STRUCTURAL ANALYSIS IN JAPAN	72-0322
PROGRAM	SPARSE MATRICES IN FINITE ELEMENT PROGRAM DEVELOPMENT	73-0337
PROGRAM	STACUSS 1: A DISCRETE-ELEMENT PROGRAM FOR THE STATIC ANALYSIS OF SINGLE-LAYER CURVED STIFFENED SHELL	69-0160
PROGRAM	STRESS ANALYSIS PROGRAM BY FINITE ELEMENT METHOD FOR RADIAL FLOW IMPELLERS	74-1016
PROGRAM	STRIP PROGRAM FOR DESIGNING LOAD BEARING SURFACE STRUCTURES (PROGRAMME STRIP POUR LE CALCUL DES STRU	71-0408
PROGRAM	TACOS: A FINITE ELEMENT COMPUTER PROGRAM FOR THE TRANSIENT ANALYSIS OF CYLINDRICAL OBESE SHELLS	72-0588
PROGRAM	THALIA - ONE-DIMENSIONAL MAGNETOHYDRODYNAMIC STABILITY PROGRAM USING METHOD OF FINITE-ELEMENTS	75-0604
PROGRAM	THALIA: A ONE-DIMENSIONAL MAGNETOHYDRODYNAMIC STABILITY PROGRAM USING THE METHOD OF FINITE ELEMENTS	74-1085
PROGRAM	THE BANDIT COMPUTER PROGRAM FOR THE REDUCTION OF MATRIX BANDWIDTH FOR NASTRAN	72-0481
PROGRAM	THE MIDCOHV WEIGHT AND BALANCE COMPUTER PROGRAM (WTBAL)	74-0697
PROGRAM	THE NASTRAN COMPUTER PROGRAM FOR STRUCTURAL ANALYSIS	69-0183
PROGRAM	THE NASTRAN COMPUTER PROGRAM FOR STRUCTURAL ANALYSIS	71-0173
PROGRAM	THE NASTRAN COMPUTER PROGRAM FOR STRUCTURAL ANALYSIS	71-0682
PROGRAM	THE NASTRAN PROGRAM FOR STRUCTURAL ANALYSIS	72-0325
PROGRAM	THE PROGRAM " ASSEN " FOR STEADY FREE SURFACE FLOWS	74-0075
PROGRAM	THE STRESS ANALYSIS OF AXISYMMETRIC SHELLS USING A PLANE FRAME ANALYSIS PROGRAM	69-0238
PROGRAM	THE TRANSIENT LINEAR ELASTIC RESPONSE ANALYSIS COMPLEX THIN SHELLS OF REVOLUTION SUBJECTED TO ARBITR	70-0080
PROGRAM	THE USE OF A LASER FOR ARPA MILITARY GEOPHYSICS PROGRAM (ROCK MECHANICS AND RAPID EXCAVATION)	72-0800
PROGRAM	THEORETICAL BASIS FOR CBL/NONSAP, A NONLINEAR STRUCTURAL ANALYSIS PROGRAM	74-0740
PROGRAM	THERMAL AND STRESS ANALYSIS PROGRAM - HYBOS	72-0159
PROGRAM	THREE DIMENSIONAL SUPERTANKER COMPUTER PROGRAM	71-0661
PROGRAM	THREE-DIMENSIONAL FINITE-ELEMENT COMPUTER PROGRAM - USERS GUIDE	74-0755
PROGRAM	TRANSFORMER LEAKAGE FLUX PROGRAM BASED ON FINITE ELEMENT METHOD	73-0004
PROGRAM	TRANSPIRATION AND FILM COOLING BOUNDARY LAYER COMPUTER PROGRAM. VOLUME 2. COMPUTER PROGRAM AND USER'	71-0655
PROGRAM	TRANSPIRATION AND FILM COOLING BOUNDARY LAYER COMPUTER PROGRAM. VOLUME 2. COMPUTER PROGRAM AND USER'	71-0655
PROGRAM	TRIANGULAR MESH GENERATION PROGRAM	67-0037
PROGRAM	TRIANGULAR MESH GENERATION PROGRAM	67-0127
PROGRAM	USE OF FINITE ELEMENT METHOD IN A MARKER AND CELL PROGRAM FOR SOLVING TIME-DEPENDENT FLUID FLOW PROB	75-0043
PROGRAM	USER'S GUIDE FOR THE COMPUTER PROGRAM DESCRIBED IN TAM REPORT NO. 680 "DETERMINATION OF VOLUME CHANG	69-0473
PROGRAM	USER'S MANUAL FOR PROGRAM BEAM	73-1105
PROGRAM	USER'S MANUAL FOR UMR FINITE ELEMENT COMPUTER PROGRAM - STRATA	69-0429
PROGRAM	USERS MANUAL FOR CEL/NONSAP, A NONLINEAR STRUCTURAL ANALYSIS PROGRAM	74-0706
PROGRAM	USERS MANUAL FOR DIGITAL COMPUTER PROGRAM OK5 NUMERICAL ANALYSIS OF FINITE STRAINS AND DISPLACEMENTS	72-0718
PROGRAM	USERS MANUAL FOR PROGRAM BEAM	73-0714
PROGRAM	USERS MANUAL FOR QMESH, A SELF-ORGANIZING MESH GENERATION PROGRAM	74-0880
PROGRAM	USERS' MANUAL FOR DIGITAL COMPUTER PROGRAM OK1. NUMERICAL ANALYSIS OF FINITE AXISYMMETRIC DEFORMATIO	72-0861
PROGRAM	VERIFICATION OF A HYBRID GENERAL PURPOSE SHELL AND BODY OF REVOLUTION COMPUTER PROGRAM	71-0333
PROGRAM	VIEW FACTOR COMPUTER PROGRAM (PROGRAM VIEW) USER'S MANUAL	73-0789
PROGRAM	VIEW FACTOR COMPUTER PROGRAM (PROGRAM VIEW) USER'S MANUAL	73-0789
PROGRAM	VISCEL A GENERAL PURPOSE COMPUTER PROGRAM FOR ANALYSIS OF LINEAR VISCOELASTIC STRUCTURES, VOL. 2	72-0789

PROGRAMM	TURBAN VERSION 1.0. EIN SPEZIELLES PROGRAMM ZUR LINEAREN UND NICHTLINEAREN STATISCHEN UND DYNAMISCHE	71-0622
PROGRAMMATIC	HEAVY SECTION STEEL TECHNOLOGY PROGRAM TECHNICAL OR PROGRAMMATIC MANUSCRIPT NO. 25. VARIABLE THICKNE	73-0948
PROGRAMME	A COMPUTER PROGRAMME FOR ELASTIC-PLASTIC CALCULATIONS WITH USE OF TRIM-6 FINITE ELEMENTS	72-0993
PROGRAMME	A COMPUTER PROGRAMME FOR ELASTIC PLASTIC CALCULATIONS WITH DUAL PERTURBATION EXPANSION AND VARIATION	75-0882
PROGRAMME	A FINITE ELEMENT PLATE BENDING PROGRAMME	72-0579
PROGRAMME	FAST REACTOR PROGRAMME. FINITE ELEMENT TEMPERATURE ANALYSIS IN FUEL PINS	73-0939
PROGRAMME	FINITE-ELEMENT PROGRAMME FOR THE PLANE-STRAIN ANALYSIS OF RUBBER	75-0362
PROGRAMME	PROGRAMME ZUR METHODE DER FINITEN ELEMETE	69-0416
PROGRAMME	STRIP PROGRAM FOR DESIGNING LOAD BEARING SURFACE STRUCTURES (PROGRAMME STRIP POUR LE CALCUL DES STRU	71-0408
PROGRAMME	TEMPERATURE CALCULATIONS USING THE NASTRAN FINITE ELEMENT PROGRAMME	75-0640
PROGRAMMENTWICK	THEORIE, PROGRAMMENTWICKLUNG UND ERFAHRUNG ANVORGESPANNTEN NETZWERKKONSTRUKTIONEN	72-0640
PROGRAMMER'S	MAGIC 11: AN AUTOMATED GENERAL PURPOSE SYSTEM FOR STRUCTURAL ANALYSIS. VOLUME 111. PROGRAMMER'S MANU	71-0756
PROGRAMMER'S	MAGIC 111: AN AUTOMATED GENERAL PURPOSE SYSTEM FOR STRUCTURAL ANALYSIS. VOLUME 111. PROGRAMMER'S MAN	72-0726
PROGRAMMER'S	MAGIC, AN AUTOMATED GENERAL PURPOSE SYSTEM FOR STRUCTURAL ANALYSIS, VOLUME III. PROGRAMMER'S MANUAL	69-0090
PROGRAMMER'S	OPTIM 11: (MAGIC) COMPATIBLE LARGE SCALE AUTOMATED MINIMUM WEIGHT DESIGN PROGRAM - V 2. PROGRAMMER'S	74-1036
PROGRAMMER'S	OPTIM 11: MAGIC COMPATIBLE LARGE SCALE AUTOMATED MINIMUM WEIGHT DESIGN PROGRAM. VOLUME 11. PROGRAMME	74-1093
PROGRAMMER'S	PROGRAMMER'S MANUAL ADDITIONS AND DEMONSTRATION PROBLEMS FOR A THERMOSTRUCTURAL CAPABILITY FOR NASTR	75-0734
PROGRAMMER'S	THE NASTRAN PROGRAMMER'S MANUAL	72-0480
PROGRAMMES	APPLICATION OF AN ' ELEMENT-METHOD ' PROGRAM TO SOLVING TEMPERATURE DISTRIBUTION PROBLEMS (ANWENDUNG	71-0464
PROGRAMMES	DEVELOPMENT AND USE OF COMPUTER PROGRAMS FOR FINITE ELEMENT ANALYSIS. (LA MISE AU POINT ET L'UTILIZA	69-0289
PROGRAMMIERSPRA	ASKA - EINE PROGRAMMIERSPRACHE ZUR STATISCHEN UND DYNAMISCHEN ANALYSE	69-0457
PROGRAMMIERUNG	METHODE DER ELASTOSTATIK PROGRAMMIERUNG UND ANWENDUNG	70-0381
PROGRAMMIERUNG	ZUR LOESUNG DES KONTAKTPROBLEMS ELASTISCHER KOERPER MIT AUSGEDEHNTER KONTAKTFLAECHE DURCH QUADRATISC	74-1135
PROGRAMMING	A FINITE ELEMENT LINEAR PROGRAMMING METHOD FOR THE LIMIT ANALYSIS OF THIN PLATES	73-1064
PROGRAMMING	A MODIFIED GRADIENT METHOD FOR FINITE ELEMENT ELASTOPLASTIC ANALYSIS BY QUADRATIC PROGRAMMING	73-0998
PROGRAMMING	ASKA 105 A TECHNICAL MANUAL AND PROGRAMMING THEORY FOR THE ELEMENTS, FLA, TRIM, SPARTA AND TET	68-0254
PROGRAMMING	CALCULATION OF HARMONIC FIELDS IN WHICH SINGULARITIES ARE PRESENT USING FINITE ELEMENT AND FINITE DI	73-0979
PROGRAMMING	DYNAMIC PROGRAMMING APPROACH TO THE FORMULATION AND SOLUTION OF FINITE ELEMENT EQUATIONS	75-0059
PROGRAMMING	ELASTIC-PLASTIC ANALYSIS BY QUADRATIC PROGRAMMING	72-0238
PROGRAMMING	FINEL, UNIVERSELLES PROGRAMMSYSTEM ZUR BERECHNUNG DER ELASTISCHEN EIGENSCHAFTEN VON MASCHINENBAUTEIL	72-0132
PROGRAMMING	FINITE ELEMENT ELASTOPLASTIC ANALYSIS BY QUADRATIC PROGRAMMING: THE MULTISTAGE METHOD	73-0842
PROGRAMMING	FINITE ELEMENT ELASTOPLASTIC ANALYSIS OF UNDERGROUND OPENINGS BY QUADRATIC PROGRAMMING	74-1288
PROGRAMMING	FINITE ELEMENT LIMIT ANALYSIS USING LINEAR PROGRAMMING.	72-0461
PROGRAMMING	LINEAR PROGRAMMING APPROACH TO SHAKEDOWN ANALYSIS OF STRUCTURES	74-0181
PROGRAMMING	MATHEMATICAL PROGRAMMING	72-0100
PROGRAMMING	MESY - EIN PROGRAMMSYSTEM ZUR UNTERSUCHUNG VON TRAGWERKEN. (STRUCTURAL ANALYSIS BY MESY PROGRAMMING	75-0360
PROGRAMMING	NON-LINEAR EFFECTS OF HIGH TEMPERATURE ON THE VIBRATION OF A SIMPLY SUPPORTED BEAM WITH A CENTRAL MA	73-0838
PROGRAMMING	ORSAM - A PROGRAMMING SYSTEM FOR THE FINITE ELEMENT METHOD	73-0528
PROGRAMMING	QUASI-LINEAR PROGRAMMING ALGORITHM FOR OPTIMIZING FIBRE-REINFORCED STRUCTURES OF FIXED STIFFNESS	75-0225
PROGRAMMING	SESAM A PROGRAMMING SYSTEM FOR FINITE ELEMENT PROBLEMS	68-0276
PROGRAMMING	SHAKEDOWN THEORY IN PERFECT ELASTOPLASTICITY WITH ASSOCIATED AND NON ASSOCIATED FLOW LAWS - A FINITE	69-0172
PROGRAMMING	THE ELEMENT-CHAIN CONCEPT IN FINITE ELEMENT BASED PROGRAMMING	72-0934
PROGRAMMING	THE QUADRATIC PROGRAMMING APPROACH TO THE FINITE ELEMENT METHOD	73-0831
PROGRAMMING	USE OF SINGULARITY PROGRAMMING IN FINITE-DIFFERENCE AND FINITE ELEMENT COMPUTATIONS OF TEMPERATURE	73-0100
PROGRAMMING	USE OF SINGULARITY PROGRAMMING IN FINITE ELEMENT CALCULATIONS OF ELASTIC STRESS INTENSITY FACTORS. P	75-0252
PROGRAMMING	ZUR LOESUNG DES KONTAKTPROBLEMS ELASTISCHER KOERPER MIT AUSGEDEHNTER KONTAKTFLAECHE DURCH QUADRATISC	74-1135
PROGRAMMSYSTEM	ANWENDUNG DER METHODEN DER FINITEN ELEMENTE MIT DEM PROGRAMMSYSTEM ASKA UNTER BESONDERER BEPUCKSICHT	71-0624
PROGRAMMSYSTEM	DAS PROGRAMMSYSTEM ANTRAS EDV-SYSTEM ZUR BERECHNUNG BELIEBIGER STATISCHER SYSTEME NACH DER METHODE D	74-0918
PROGRAMMSYSTEM	FINEL, UNIVERSELLES PROGRAMMSYSTEM ZUR BERECHNUNG DER ELASTISCHEN EIGENSCHAFTEN VON MASCHINENBAUTEIL	72-0132
PROGRAMMSYSTEM	MESY - EIN PROGRAMMSYSTEM ZUR UNTERSUCHUNG VON TRAGWERKEN. (STRUCTURAL ANALYSIS BY MESY PROGRAMMING	75-0360
PROGRAMMSYSTEM	STRING EIN PROGRAMMSYSTEM FUR DIE MASCHINENUNABHANGIGE BEARBEITUNG VON ZEICHENKETTEN IN FORTRAN	68-0265
PROGRAMMSYSTEME	EINIGE ALLGEMEINE PROGRAMMSYSTEME FUR FINITE ELEMENTE	73-0581
PROGRAMS	A DESIGN ORIENTED APPROACH TO CREEP AND PLASTICITY IN FINITE ELEMENT PROGRAMS	70-0125
PROGRAMS	A SURVEY OF HEAT CONDUCTION COMPUTER PROGRAMS	71-0017
PROGRAMS	APPLICATION OF RESPONSE LIMITING TO FINITE-ELEMENT STRUCTURAL-ANALYSIS PROGRAMS	75-0424
PROGRAMS	ASKA SERIES OF PROGRAMS	72-0157
PROGRAMS	BASIS FOR ELEMENT INTERCHANGEABILITY IN FINITE ELEMENT PROGRAMS	71-0224
PROGRAMS	COMPUTER PROGRAMS FOR THE PLASTIC ANALYSIS OF STRUCTURES USING DISCRETE ELEMENT METHODS	67-0004
PROGRAMS	COMPUTER PROGRAMS FOR PLOTTING FINITE ELEMENT PATTERNS OF TWO-AND THREE-DIMENSIONAL STRUCTURES	70-0416
PROGRAMS	COMPUTER PROGRAMS FOR THE SOLUTION OF SYSTEMS OF LINEAR ALGEBRAIC EQUATIONS	73-0932
PROGRAMS	COMPUTER PROGRAMS FOR THE SOLUTION OF SYSTEMS OF LINEAR ALGEBRAIC EQUATIONS	73-1040
PROGRAMS	COMPUTER PROGRAMS FOR SIMULATING PHYSICAL PHENOMENA IN THREE SPACE DIMENSIONS AND TIME	74-0866
PROGRAMS	DESIGN ORIENTED APPROACH TO CREEP AND PLASTICITY IN FINITE ELEMENT PROGRAMS	71-0281
PROGRAMS	DETERMINATION OF STRESSES AND DEFORMATIONS ON PISTONS BY MEANS OF COMPUTER PROGRAMS AND PHOTOELASTIC	75-0553
PROGRAMS	DEVELOPMENT AND USE OF COMPUTER PROGRAMS FOR FINITE ELEMENT ANALYSIS. (LA MISE AU POINT ET L'UTILIZA	69-0289
PROGRAMS	DEVELOPMENT OF ADVANCED STRUCTURAL OPTIMIZATION PROGRAMS AND THEIR APPLICATION TO LARGE ORDER STRAIN	66-0024
PROGRAMS	DIGITAL COMPUTER PROGRAMS FOR STEADY-STATE OR TRANSIENT TEMPERATURE ANALYSIS OF PLANE AND AXISYMMETR	72-0915
PROGRAMS	DIGITAL COMPUTER PROGRAMS FOR STEADY-STATE OR TRANSIENT TEMPERATURE ANALYSIS OF PLANE AND AXISYMMETR	72-0915
PROGRAMS	DIRECT ITERATION METHOD FOR THE INCORPORATION OF PHASE CHANGE IN FINITE ELEMENT HEAT CONDUCTION PROG	74-0894
PROGRAMS	ELEMENTS FOR FINITE ELEMENT STRUCTURAL ANALYSIS PROGRAMS	71-0335
PROGRAMS	ENGINEERING OPERATIONS REPORT: NERVA AXISYMMETRIC AND PLANE STRUCTURES FINITE ELEMENT STRESS ANALYSI	72-0917
PROGRAMS	FESS AND FINESSE PROGRAMS	72-0158
PROGRAMS	FINITE ELEMENT COMPUTER PROGRAMS FOR FLOW IN SATURATED UNSATURATED POROUS MEDIA	72-0364
PROGRAMS	FINITE ELEMENT COMPUTER PROGRAMS FOR HEAT CONDUCTION PROBLEMS	73-0525
PROGRAMS	FINITE ELEMENT METHOD AND COMPUTER PROGRAMS	74-0450
PROGRAMS	FINITE ELEMENT PROGRAMS FOR SLAB BRIDGE DESIGN	71-0082
PROGRAMS	GENERAL PURPOSE PROGRAMS FOR NONLINEAR FINITE ELEMENT ANALYSIS	70-0247
PROGRAMS	GIFTS SYSTEM FOR THE PDP-15 AND OTHER CONSIDERATIONS IN DEVELOPING GRAPHICS AND CAPABILITIES FOR FIN	73-0671
PROGRAMS	HARDWARE AND SOFTWARE CONSIDERATIONS IN GENERAL PURPOSE PROGRAMS FOR FINITE ELEMENT ANALYSIS	70-0261
PROGRAMS	HYBRID FINITE ELEMENT OPTIMIZATION PROGRAMS FOR DISCS USING A DYNAMICS DESIGN TECHNOLOGY	74-0691
PROGRAMS	LARGE SCALE COMPUTER PROGRAMS FOR STRUCTURAL ANALYSIS	70-0211
PROGRAMS	LARGE USER-ORIENTATED SYSTEMS OF PROGRAMS FOR STRUCTURAL ANALYSIS AND DESIGN	74-0482
PROGRAMS	MARC SERIES OF PROGRAMS	72-0154
PROGRAMS	MARK5C/DAISY SERIES OF PROGRAMS	72-0154
PROGRAMS	MATHEMATICAL MODEL OF THE COLUMBIA RIVER FROM THE PACIFIC OCEAN TO BONNEVILLE DAM. PART 1. THEORY. P	69-0464
PROGRAMS	MODIFICATIONS TO ARL COMPUTER PROGRAMS USED FOR DESIGN OF AXIAL COMPRESSOR AIRFOILS	74-0508
PROGRAMS	ON GENERAL PURPOSE PROGRAMS FOR FINTIE ELEMENT ANALYSIS WITH SPECIAL REFERENCE TO GEOMETRIC AND MATE	70-0087
PROGRAMS	ON THE APPLICATION OF RESPONSE LIMITING TO FINITE ELEMENT STRUCTURAL ANALYSIS PROGRAMS	75-0192
PROGRAMS	ON THE APPLICATION OF RESPONSE LIMITING TO FINITE ELEMENT STRUCTURAL ANALYSIS PROGRAMS	75-0822
PROGRAMS	ON THE SOLUTION OF PLANE STRESS PROBLEMS BY FINITE ELEMENTS COMPUTER PROGRAMS	73-0978
PROGRAMS	PREPROCESSORS FOR GENERAL PURPOSE FINITE ELEMENT PROGRAMS	74-1033
PROGRAMS	ROHM AND HAAS THERMOSTRUCTURAL ANALYSIS SYSTEM. THERMAL STRUCTURAL ANALYSIS PROGRAMS - A SURVEY AND	72-0007
PROGRAMS	SHELL ANALYSIS WITH LARGE GENERAL PURPOSE PROGRAMS	71-0402
PROGRAMS	SOME ASPECTS OF NON-LINEAR FINITE ELEMENT COMPUTER PROGRAMS	75-0778
PROGRAMS	SPECIAL PURPOSE COMPUTER PROGRAMS	70-0154
PROGRAMS	STRESSES IN A PARTLY YIELDED NOTCHED BAR - AN ASSESSMENT OF THREE ALTERNATIVE PROGRAMS	73-0267
PROGRAMS	STRESSES IN A PARTLY YIELDED NOTCHED BAR EM DASH AN ASSESSMENT OF THREE ALTERNATIVE PROGRAMS	73-0268
PROGRAMS	STRUCTURAL DYNAMICS PROGRAMS	70-0412
PROGRAMS	STRUCTURAL MECHANICS COMPUTER PROGRAMS	74-0902
PROGRAMS	SURVEY OF GENERAL PURPOSE PROGRAMS FOR FINITE ELEMENT ANALYSIS	72-0323
PROGRAMS	TESTRAN PROGRAMS. A TOOL FOR TEMPERATURE AND STRESS ANALYSIS	73-0414
PROGRAMS	THE VERIFICATION AND ACCEPTANCE OF COMPUTER PROGRAMS FOR DESIGN ANALYSIS	70-0334

PROGRAMS	USE OF FINITE ELEMENT COMPUTER PROGRAMS IN FRACTURE MECHANICS	71-0022
PROGRAMS	USER'S GUIDE FOR ANALYSIS OF FINITE ELASTOPLASTIC DEFORMATION: THE FIPDEF AND FIPAX PROGRAMS FOR THE	74-0799
PROGRESS	ANNUAL PROGRESS REPORT, FY1973	73-0969
PROGRESS	PROGRESS IN COMPUTERIZED STRUCTURAL ANALYSIS OF LARGE VESSELS	74-0949
PROGRESS	PROGRESS IN NONLINEAR FINITE ELEMENT ANALYSIS USING ASYMPTOTIC SOLUTION TECHNIQUES	72-0316
PROGRESS	PROGRESS IN THREE-DIMENSIONAL ELASTIC-PLASTIC STRESS ANALYSIS FOR FRACTURE MECHANICS	71-0314
PROGRESS	PROGRESS REPORT ON DISCRETE-ELEMENT ELASTIC AND ELASTIC-PLASTIC ANALYSIS OF SHELLS OF REVOLUTION SUB	68-0170
PROGRESS	STRESS ANALYSIS OF B16.9 TEES BY THE FINITE ELEMENT METHOD, A PROGRESS REPORT	71-0215
PROGRESS	STRUCTURAL SYNTHESIS, 1959-1969, A DECADE OF PROGRESS	69-0237
PROGRESSIVE	APPLICATION OF GRIFFITH'S THEORY TO ANALYSIS OF PROGRESSIVE FRACTURE	75-0150
PROGRESSIVE	SIMULATION OF PROGRESSIVE FAILURE IN ROCK AROUND MINING EXCAVATIONS	70-0146
PROJECT	EMBANKMENT DAM FOR THE MARIMBONDO HYDRO PROJECT	75-0270
PROJECT	FINAL REPORT ON NASTRAN/GRAPHICS, 1971 FORD DOOR PROJECT	73-0595
PROJECT	M.I.T. TEST SECTION INSTRUMENTATION, MASSACHUSETTS BAY TRANSPORTATION AUTHORITY, HAYMARKET-NORTH EXT	72-0673
PROJECT	THE SCANDINAVIAN PORV MODEL PROJECT: STRESS CALCULATION AND EXPERIMENTAL VERIFICATION	71-0021
PROJECTILE	A FINITE ELEMENT STUDY OF THE PENETRATION OF A SOIL HALF-SPACE BY A RIGID PROJECTILE	69-0518
PROJECTILE	FINITE ELEMENT ANALYSIS OF A MULTI-COMPONENT KINETIC ENERGY PROJECTILE	74-1180
PROJECTILE	FINITE ELEMENT ANALYSIS OF A MULTI-COMPONENT KINETIC ENERGY PROJECTILE	75-0754
PROJECTILE	STRESS ANALYSIS OF 175 MM PROJECTILE, HE, M437	71-0736
PROJECTILE	STRUCTURAL RESPONSE TO MOVING PROJECTILE MASS BY THE FINITE ELEMENT METHOD	75-0693
PROJECTILES	FINITE ELEMENT STRESS ANALYSIS OF ARTILLERY PROJECTILES	73-0432
PROJECTIONAL	NON PROJECTIONAL METHODS	73-0632
PROJECTIONS	OVERVIEW, TRENDS AND PROJECTIONS: THEORY AND APPLICATIONS OF THE FINITE ELEMENT METHOD IN GEOTECHNIC	72-0988
PROJECTIONS	THEORY OF CONJUGATE PROJECTIONS IN FINITE ELEMENT ANALYSIS	71-0194
PROJECTIVE	PROJECTIVE METHOD APPLIED TO THREE-DIMENSIONAL ELASTICITY EQUATIONS	74-0617
PROJECTOR	APPLICATIONS OF THE FINITE-ELEMENT METHOD AND COMPUTER GRAPHICS. (A VIBRATIONAL ANALYSIS OF A SONAR	72-0729
PROJEKTOVANI	VYUZITI METODY KONECNYCH PRVKU PRO PROJEKTOVANI KOMBINOVANYCH KONSTRUCKCI (FINITE ELEMENT METHOD IDE	73-0667
PROPAGATE	FINITE ELEMENT APPROACH TO THE CRITICAL CYCLIC STRESS REQUIRED TO PROPAGATE A CRACK	73-0463
PROPAGATING	CYCLIC ANALYSIS OF A PROPAGATING CRACK AND ITS CORRELATION WITH FATIGUE CRACK GROWTH	75-0188
PROPAGATING	ELASTIC-PLASTIC ANALYSIS OF A PROPAGATING CRACK UNDER CYCLIC LOADING	75-0857
PROPAGATION	A FINITE ELEMENT ANALYSIS OF WAVE PROPAGATION IN HUMAN SPINE	73-0792
PROPAGATION	A FRACTURE MECHANICS APPROACH TO THE PROPAGATION OF FATIGUE CRACKS IN REACTOR COMPONENTS BY ACOUSTIC	71-0132
PROPAGATION	A NUMERICAL METHOD FOR THE ANALYSIS OF LONGITUDINAL ELASTIC PLASTIC STRESS WAVE PROPAGATION	73-0526
PROPAGATION	A STUDY OF A DYNAMICAL FINITE ELEMENT ANALYSIS FOR APPLICATION TO AXIAL WAVE PROPAGATION PROBLEMS IN	71-0589
PROPAGATION	ANALYSIS OF CRACK PROPAGATION IN WELDED STRUCTURES	74-1144
PROPAGATION	CALCULATION OF STRESS INTENSITY FACTORS AND FATIGUE CRACK PROPAGATION OF SEMI-ELLIPTICAL, PART-THROU	73-0360
PROPAGATION	COMPUTATION OF WAVE PROPAGATION BY MEANS OF THE FINITE ELEMENT METHOD	74-0032
PROPAGATION	COMPUTER PREDICTION OF FATIGUE CRACK PROPAGATION UNDER RANDOM LOADING	73-0851
PROPAGATION	COUPLING MATRICES FOR HIGH ORDER FINITE ELEMENT ANALYSIS OF ACOUSTIC - WAVE PROPAGATION	72-0040
PROPAGATION	CRACK PROPAGATION AND ARREST IN IMPACTED PLATES	72-0599
PROPAGATION	CRACK PROPAGATION IN SEA ICE - FINITE ELEMENT APPROACH	72-0509
PROPAGATION	CYCLIC LOADING AND CRACK PROPAGATION - AN ELASTOPLASTIC FINITE ELEMENT STUDY	74-1214
PROPAGATION	DISCRETIZATION-DISPERSION IN FINITE ELEMENT MODELLING OF WAVE PROPAGATION IN SOLIDS	74-1253
PROPAGATION	DYNAMIC FINITE ELEMENT ANALYSIS OF AXI-SYMMETRIC STRESS WAVE PROPAGATION IN SOIL-FILLED BIN	72-0622
PROPAGATION	DYNAMIC FINITE ELEMENT METHOD CALCULATION OF ELASTIC WAVE PROPAGATION IN TWO DIMENSIONAL STRUCTURES	74-0166
PROPAGATION	EFFECTS OF A PEAK OVERLOAD ON FATIGUE CRACK PROPAGATION RATE-PLANE STRESS ANALYSIS OF CRACKED PLATE	74-1365
PROPAGATION	ELASTIC WAVE PROPAGATION. A COMPARISON BETWEEN FINITE ELEMENT PREDICTIONS AND EXACT SOLUTION	67-0076
PROPAGATION	ELASTOPLASTIC CRACK PROPAGATION IN A TRANSVERSELY LOADED UNIDIRECTIONAL COMPOSITE	74-0217
PROPAGATION	EVALUATION OF NUMERICAL TIME INTEGRATION METHODS AS APPLIED TO ELASTIC-PLASTIC DYNAMIC PROBLEMS INVO	74-0858
PROPAGATION	EXCITATION OF PLANE THERMOELASTIC WAVES IN A LAYER WITH A FINITE RATE OF HEAT PROPAGATION	74-0803
PROPAGATION	EXPERIMENTAL AND THEORETICAL FATIGUE-CRACK PROPAGATION LIVES OF VARIOUSLY NOTCHED PLATES	74-0602
PROPAGATION	EXPLORATORY INVESTIGATION OF RAPID CRACK PROPAGATION AND CRACK ARREST	73-0757
PROPAGATION	FATIGUE CRACK FORMATION AND PROPAGATION IN PAVEMENTS CONTAINING SOIL-CEMENT BASES.	72-0435
PROPAGATION	FINITE ELEMENT ANALYSIS OF CRACK PROPAGATION IN THREE-DIMENSIONAL SOLIDS UNDER CYCLING LOADING	73-0529
PROPAGATION	FINITE ELEMENT ANALYSIS FOR WAVE PROPAGATION IN ELASTIC-PLASTIC SOLIDS	73-0990
PROPAGATION	FINITE ELEMENT ANALYSIS OF CRACK PROPAGATION IN 3-DIMENSIONAL SOLIDS UNDER CYCLIC LOADING	74-0479
PROPAGATION	FINITE ELEMENT ANALYSIS OF FATIGUE CRACK PROPAGATION - INCLUDING THE EFFECTS OF CRACK CLOSURE	74-1112
PROPAGATION	FINITE ELEMENT ANALYSIS OF WAVE PROPAGATION IN OPEN CHANNELS	75-0570
PROPAGATION	FINITE ELEMENT IN THE THEORY OF SH-WAVE PROPAGATION	73-1034
PROPAGATION	FINITE ELEMENT SOLUTION FOR WAVE PROPAGATION IN LAYERED MEDIA CAUSED BY A NUCLEAR DETONATION	67-0016
PROPAGATION	FINITE ELEMENT STRESS FORMULATION FOR WAVE PROPAGATION	71-0831
PROPAGATION	FINITE ELEMENT STRESS FORMULATION FOR WAVE PROPAGATION	73-1043
PROPAGATION	FINITE ELEMENT STUDY OF HARMONIC WAVE PROPAGATION IN PERIODIC STRUCTURES	74-0130
PROPAGATION	FINITE-ELEMENT ANALYSIS OF FRACTURE PROPAGATION IN TWO-DIMENSIONAL ELASTIC BRITTLE SOLIDS	72-0948
PROPAGATION	HIGH-PERFORMANCE COMPOSITE MATERIALS FOR VEHICLE CONSTRUCTION: A FINITE ELEMENT COMPUTER PROGRAM FOR	73-0767
PROPAGATION	LINEAR WAVE PROPAGATION PROBLEMS AND THE FINITE ELEMENT METHOD	74-1340
PROPAGATION	LINEAR WAVE PROPAGATION PROBLEMS AND THE FINITE ELEMENT METHOD	75-0395
PROPAGATION	MATHEMATICAL MODELS FOR GEOLOGICAL MATERIALS FOR WAVE PROPAGATION STUDIES	71-0740
PROPAGATION	MICROMECHANICAL ANALYSIS OF CRACK PROPAGATION IN AN ELASTOPLASTIC COMPOSITE MATERIAL	74-0565
PROPAGATION	MODELING WAVE PROPAGATION WITH FINITE ELEMENT METHODS	75-0015
PROPAGATION	NUMERICAL METHOD FOR INVESTIGATING THE PROPAGATION OF SURFACE WAVES ON DISSIPATIVE GUIDES	74-0613
PROPAGATION	ON SPARSE MATRICES IN FINITE ELEMENT WAVE PROPAGATION	73-0290
PROPAGATION	ON THE DEVELOPMENT OF A REFINED FINITE-ELEMENT FREE FROM DISCRETIZATION - DISPERSION FOR ELASTIC WAV	73-1148
PROPAGATION	PREDICTION OF STATIC AND FATIGUE DAMAGE AND CRACK PROPAGATION IN COMPOSITE MATERIALS	75-0643
PROPAGATION	PROPAGATION AND OPENING OF A THROUGH CRACK IN PIPE SUBJECT TO COMBINED CYCLIC THERMOMECHANICAL LOADI	75-0516
PROPAGATION	PROPAGATION IN ELLIPTICAL AND PARABOLIC WAVEGUIDES	74-1173
PROPAGATION	PROPAGATION OF ELLIPTICAL SURFACE CRACKS AND NONLINEAR FRACTURE MECHANICS BY THE FINITE ELEMENT METH	74-0906
PROPAGATION	PROPAGATION OF FAILURE IN A CIRCULAR CYLINDER OF ROCK SUBJECTED TO A COMPRESSIVE FORCE	73-0793
PROPAGATION	SHOCK TUNNEL TESTS OF PRELOADED AND ARCHED WALL PANELS FINITE ELEMENT STRESS FORMULATION FOR WAVE PR	73-0892
PROPAGATION	SOLUTION OF WAVE PROPAGATION PROBLEMS IN SATURATED MEDIUM	72-0784
PROPAGATION	STRESS-WAVE PROPAGATION AND DYNAMIC EFFECTS IN SHELLS (SPADES) MICROMECHANICAL STRESS-WAVE RESPONSE	72-0851
PROPAGATION	TENSILE CRACK PROPAGATION IN REINFORCED CONCRETE BEAMS BY THE FINITE ELEMENT TECHNIQUES	69-0264
PROPAGATION	THE APPLICATION OF THE FINITE ELEMENT METHOD TO A PROBLEM IN STRESS WAVE PROPAGATION	69-0505
PROPAGATION	THE PROPAGATION OF LOVE WAVES ACROSS NONHORIZONTALLY LAYERED STRUCTURES	71-0600
PROPAGATION	TIDAL AND LONG WAVE PROPAGATION - A FINITE ELEMENT APPROACH	73-1059
PROPAGATION	TIDAL PROPAGATION AND DISPERSION IN ESTUARIES	75-0390
PROPAGATION	WAVE PROPAGATION IN A VISCOELASTIC FIBER SUBJECTED TO TRANSVERSE IMPACT	71-0639
PROPAGATION	WAVE PROPAGATION IN A THIN HOLLOW CONE BY A FINITE ELEMENT METHOD	72-0045
PROPAGATION	WAVE PROPAGATION IN ANISOTROPIC ROCKS	72-0801
PROPAGATION	WAVE PROPAGATION IN ANISOTROPIC ROCKS	72-0873
PROPAGATION	WAVE PROPAGATION IN ANISOTROPIC ROCKS	73-0922
PROPAGATION	WAVE PROPAGATION IN FLUIDS	74-0034
PROPELLANT	ANALYSIS OF THE BIAXIAL STRIP AND SHEAR LAP TESTS FOR SOLID PROPELLANT CHARACTERIZATION	67-0019
PROPELLANT	APPLICATION OF THE FINITE ELEMENT METHOD TO STRESS ANALYSIS OF SOLID PROPELLANT ROCKET GRAINS	65-0009
PROPELLANT	CASE BOND STRESS CALCULATIONS FOR FLAPPED CYLINDRICAL ANALOGS OF SOLID PROPELLANT ROCKET MOTORS. BON	72-0842
PROPELLANT	DESIGN OF PROPELLANT FINS SUBJECTED TO AXIAL ACCELERATION	71-0734
PROPELLANT	DEVELOPMENT OF A SOLID ROCKET PROPELLANT NONLINEAR CONSTITUTTIVE THEORY	75-0722
PROPELLANT	HEAT CONDUCTION AND STRESS ANALYSIS OF SOLID PROPELLANT ROCKET MOTOR NOZZLES	69-0059
PROPELLANT	MEASURED AND PREDICTED LONGITUDINAL VIBRATIONS OF A LIQUID PROPELLANT TWO-STAGE LAUNCH VEHICLE	70-0366
PROPELLANT	STUDY OF PROPELLANT DYNAMICS IN A SHUTTLE TYPE LAUNCH VEHICLE	72-0699
PROPELLANT	THREE-DIMENSIONAL PHOTOELASTIC AND FINITE-ELEMENT ANALYSIS OF A PROPELLANT GRAIN	72-0031
PROPELLANTS	AN INVESTIGATION INTO THE METHODS OF DETECTING DEWETTED ZONES IN SOLID ROCKET PROPELLANTS USING HOLO	73-0890

339

PROPELLANTS	HANDBOOK FOR THE ENGINEERING STRUCTURAL ANALYSIS OF SOLID PROPELLANTS	71-0642
PROPELLANTS	MEASUREMENT OF THERMAL EXPANSION IN SOLID PROPELLANTS	73-0916
PROPELLER	INVESTIGATION ON FREE AND FORCED VIBRATIONS OF A LONG TANKER WITH OVERLAPPING PROPELLER ARRANGEMENT	73-0981
PROPELLER	LOADS AND RESPONSE OF LARGE DUCTED PROPELLER SYSTEMS	73-0810
PROPELLER	MARINE PROPELLER BLADES STRESSES AND DEFORMATIONS WITH GIVEN, FROZEN, HYDRODYNAMIC LOADING	73-0474
PROPELLER	PREDICTION OF MARINE PROPELLER DISTORTION AND STRESSES USING A SUPERPARAMETRIC THICK SHELL FINITE EL	73-0370
PROPELLER	PROPELLER BLADE STRESSES. APPLICATION OF FINITE ELEMENT METHODS	74-0207
PROPELLER-EXCIT	PROPELLER-EXCITED VIBRATION WITH PARTICULAR REFERENCE TO FULL-SCALE MEASUREMENTS	75-0018
PROPELLERS	ELASTIC STRENGTH OF PROPELLERS: AN ANALYSIS BY MATRIX METHODS	70-0468
PROPER	NUMERICAL STUDY OF PROPER MODES OF A HOMOGENEOUS CYLINDRICAL PLASMA BY METHOD OF FINITE ELEMENTS	74-0160
PROPER	ON THE DETERMINATION OF EFFECTIVE ELASTIC-PLASTIC PROPER TIES FOR THE EQUIVALENT SOLID PLATE ANALYSI	74-0494
PROPERTIES	A DIGITAL COMPUTER PROGRAM FOR THE FINITE ELEMENT ANALYSIS OF SOLIDS WITH NONLINEAR MATERIAL PROPERT	65-0032
PROPERTIES	A FINITE ELEMENT SIMULATION OF MECHANICAL PROPERTIES OF COMPOSITE MATERIALS	74-1294
PROPERTIES	A MODAL SYNTHESIS TECHNIQUE FOR DETERMINING DYNAMIC PROPERTIES FOR A STRUCTURE FOR MASS AND STIFFNES	74-0319
PROPERTIES	A QUADRILATERAL PLANE STRESS FINITE ELEMENT WITH QUADRATIC PROPERTIES	72-0969
PROPERTIES	APPLICATIONS OF FINITE ELEMENT STRESS ANALYSIS AND STRESS-STRAIN PROPERTIES IN DETERMINING NOTCH FAT	73-1114
PROPERTIES	ASAAS: ASYMMETRIC STRESS ANALYSIS OF AXISYMMETRIC SOLIDS WITH ORTHOTROPIC, TEMPERATURE-DEPENDENT MAT	71-0696
PROPERTIES	AVERAGE ELASTIC PROPERTIES AND STRESS FIELDS FOR COMPOSITE CONTINUA	69-0106
PROPERTIES	AVERAGE ELASTIC PROPERTIES AND STRESS FIELDS FOR COMPOSITE MATERIALS	69-0435
PROPERTIES	BRLESC FINITE ELEMENT PROGRAM FOR AXISYMMETRIC, PLANE STRAIN, AND PLANE STRESS, ORTHOTROPIC SOLIDS W	71-0749
PROPERTIES	CHARACTERISATION ELASTIQUE D'UN STRATIFIE RESINE-VERRE PAR LA METHODE DES DEPLACEMENTS (DETERMINATIO	75-0272
PROPERTIES	COMPARISON OF EXTRAPOLATION TECHNIQUES WITH HIGH ORDER FINITE ELEMENT METHODS FOR DIFFUSION EQUATION	73-0586
PROPERTIES	COMPLETENESS AND CONVERGENCE PROPERTIES OF FINITE ELEMENT DISPLACEMENT FUNCTIONS., A GENERAL TREATME	67-0053
PROPERTIES	EFFECT OF TIME-DEPENDENT MATERIAL PROPERTIES ON DYNAMIC RESPONSE	74-0900
PROPERTIES	ESTIMATION OF THE MECHANICAL PROPERTIES OF ROCK MASSES	73-0935
PROPERTIES	EXPERIMENTELL UNDERSOKNING AV ELEMENTFOGARS HALLFASTHETSEGENSKAPER (EXPERIMENTAL INVESTIGATION OF ST	72-0192
PROPERTIES	EXPLICIT INTEGRATION - STABILITY, SOLUTION PROPERTIES, COST	75-0298
PROPERTIES	FINEL, UNIVERSELLES PROGRAMMSYSTEM ZUR BERECHNUNG DER ELASTISCHEN EIGENSCHAFTEN VON MASCHINENBAUTEIL	72-0132
PROPERTIES	FINITE ELEMENT ANALYSIS FOR ARBITRARY AXISYMMETRIC STRUCTURES ORTHOTROPIC, TEMPERATURE-DEPENDENT MAT	68-0076
PROPERTIES	FINITE ELEMENT ANALYSIS OF COMPOSITE MATERIAL BEHAVIOURS FROM MECHANICAL PROPERTIES OF CONSTITUENT M	75-0181
PROPERTIES	FINITE ELEMENT ANALYSIS OF COMPOSITE MATERIAL BEHAVIORS FROM MECHANICAL PROPERTIES OF CONSTITUTENT M	75-0874
PROPERTIES	FINITE ELEMENT ELASTIC PLASTIC CREEP ANALYSIS OF TWO DIMENSIONAL CONTINUUM WITH TEMPERATURE DEPENDEN	73-0226
PROPERTIES	FINITE ELEMENT OF UNDERGROUND STRESSES UTILIZING STOCHASTICALLY SIMULATED MATERIAL PROPERTIES	70-0111
PROPERTIES	FINITE ELEMENT STRESS ANALYSIS OF AXISYMMETRIC SOLIDS WITH ORTHOTROPIC TEMPERATURE DEPENDENT MATERIA	67-0089
PROPERTIES	FREE VIBRATION OF A SIMPLY SUPPORTED BEAM WITH NONLINEAR MATERIAL PROPERTIES	71-0689
PROPERTIES	INFLUENCE OF ROCK PROPERTIES VARIABILITY ON MINE OPENING STABILITY ANALYSIS	74-1021
PROPERTIES	INFLUENCE OF TEMPERATURE DEPENDENT PROPERTIES ON THERMAL ROCK FRAGMENTATION	75-0222
PROPERTIES	INPUT SOIL PROPERTIES FOR THE FINITE ELEMENT CODE NTEDYNS	71-0694
PROPERTIES	LINEAR AND NON LINEAR ANALYSIS OF SHELLS OF REVOLUTION WITH ASYMMETRICAL STIFFNESS PROPERTIES	68-0151
PROPERTIES	MATRIX DISPLACEMENT ANALYSIS OF SHELLS WITH AN AXISYMMETRIC MIDSURFACE BUT HAVING CIRCUMFERENTIALLY	68-0031
PROPERTIES	MEASURED AND CALCULATED VIBRATION PROPERTIES OF RING-STIFFENED HONEY-COMB CYLINDERS	71-0834
PROPERTIES	MECHANICAL PROPERTIES OF TAPE COMPOSITES	69-0317
PROPERTIES	NONLINEAR DYNAMIC ANALYSIS OF SHELLS OF REVOLUTION WITH ASYMMETRIC PROPERTIES BY THE FINITE ELEMENT	75-0217
PROPERTIES	NONLINEAR DYNAMIC ANALYSIS OF SHELLS OF REVOLUTION WITH ASYMMETRIC PROPERTIES BY THE FINITE ELEMENT	75-0657
PROPERTIES	ON THE EXTREMAL PROPERTIES OF THE SOLUTION IN DYNAMICS OF RIGID VISCOPLASTIC BODIES ALLOWING FOR LAR	73-1080
PROPERTIES	ON THE FINITE ELEMENT ANALYSIS OF TWO-DIMENSIONAL STRESS PROBLEMS WITH TIME DEPENDENT PROPERTIES	65-0042
PROPERTIES	PROPERTIES OF SOLUTIONS OF ELLIPTIC BOUNDARY VALUE PROBLEMS	72-0499
PROPERTIES	PROPERTIES OF STABILITY AND MONOTONY WITH THE FINITE ELEMENT METHOD	69-0266
PROPERTIES	SAAS II: FINITE ELEMENT STRESS ANALYSIS OF AXISYMMETRIC SOLIDS WITH ORTHOTROPIC, TEMPERATURE DEPENDE	68-0096
PROPERTIES	SAAS 111: FINITE ELEMENT STRESS ANALYSIS OF AXISYMMETRIC AND PLANE SOLIDS WITH DIFFERENT ORTHOTROPIC	71-0741
PROPERTIES	SIMPLIFIED SOIL-STRUCTURE INTERACTION ANALYSIS WITH STRAIN DEPENDENT SOIL PROPERTIES	74-1052
PROPERTIES	SOME DISTRIBUTIONAL AND CONVERGENCE PROPERTIES OF THE FINITE ELEMENT METHOD, WITH APPLICATIONS IN NO	73-0598
PROPERTIES	SOME MATHEMATICAL PROPERTIES OF CERTAIN MIXED GALERKIN APPROXIMATIONS IN NONLINEAR ELASTICITY	74-1248
PROPERTIES	SOME OBSERVATIONS ON PROPERTIES OF CERTAIN MIXED FINITE ELEMENT APPROXIMATIONS	75-0617
PROPERTIES	SOME PROPERTIES OF LINEAR STRAIN TRIANGLES AND OPTIMAL FINITE ELEMENT MODELS	73-1044
PROPERTIES	STRAIN DISTRIBUTION AROUND UNDERGROUND OPENINGS. STATISTICAL RELATIONSHIPS FOR CERTAIN ROCK PROPERTI	71-0710
PROPERTIES	STRESS ANALYSIS AND DESIGN OF SILICON SOLAR CELL ARRAYS AND RELATED MATERIAL PROPERTIES	72-0840
PROPERTIES	STRESS ANALYSIS OF AXISYMMETRIC SOLIDS WITH ASYMMETRIC PROPERTIES	72-0140
PROPERTIES	STRESS-STRAIN PROPERTIES OF COMPACTED CLAYS	74-0650
PROPERTIES	STRUCTURAL DYNAMIC PROPERTIES OF TACTICAL MISSILE JOINTS - PHASE 1	70-0423
PROPERTIES	STRUCTURAL DYNAMIC PROPERTIES OF TACTICAL MISSILE JOINTS - PHASE 3	74-0811
PROPERTIES	THE GENERALIZED INVERSE - SOME PROPERTIES, A COMPUTATIONAL ALGORITHM AND A STRUCTURAL APPLICATION	68-0248
PROPERTIES	THE NONLINEAR DYNAMIC ANALYSIS OF SHELLS OF REVOLUTION WITH ASYMMETRIC PROPERTIES BY THE FINITE ELEM	71-0818
PROPERTIES	THEORY FOR DISTORTION STUDIES OF LUNG PARENCHYMA BASED ON ALVEOLAR MEMBRANE PROPERTIES	74-0222
PROPERTIES	TRANSIENT THERMAL STRESS ANALYSIS OF COMPOSITE STRUCTURES INCLUDING CONTINUOUSLY VARYING PROPERTIES	75-0681
PROPERTIES	TRANSVERSE PROPERTIES OF FIBROUS COMPOSITES	69-0069
PROPERTIES	TRANSVERSE PROPERTIES OF UNIDIRECTIONAL ALUMINUM MATRIX FIBROUS COMPOSITES	69-0167
PROPERTIES	TSAAS: FINITE ELEMENT THERMAL AND STRESS ANALYSIS OF AXISYMMETRIC SOLIDS WITH ORTHOTROPIC TEMPERATUR	74-0886
PROPERTIES	VARIATIONAL PROPERTIES OF STEADY FALL IN STOKES FLOW	72-0285
PROPERTIES	ZUSAMMENHANG ZWISCHEN OBERFLAECHENBESCHAFFENHEIT, VERBUND UND SPRENGWIRKUNG VON BEWEHRUNGSSTAEHLEN U	73-0450
PROPORTIONAL	ANWENDUNG DER MATRIZENVERSCHIEBUNGSMETHODE AUF ERZWUNGENE SCHWINGUNGEN (SCHWINGUNGSPROBLEME) (ARBEIT	71-0627
PROPRIE	SULLA DETERMINAZIONE DELLE FREGUENZE PROPRIE DI ALBERI SU PIU SUPPORTI ELASTICI. (DETERMINATION OF F	74-0091
PROPULSIF	COUPLAGE ENTRE L ECOULEMENT AUTOUR D UN ARRIERE-CORPS ET LE JET PROPULSIF EN THEORIE DE FLUIDE PARFA	74-1137
PROPULSION	FINITE ELEMENT ANALYSIS OF MERCURY SLOSH IN THE SOLAR ELECTRIC PROPULSION STAGE	75-0757
PROPULSION	OH - 58A PROPULSION SYSTEM VIBRATION INVESTIGATION	74-0733
PROPULSION	PWR DEVELOPMENT FOR NUCLEAR PROPULSION: VIBRATION ANALYSIS OF NUCLEAR POWER PLANT	71-0797
PROPULSION	TRANSVERSE VIBRATIONS OF SHIP'S PROPULSION SYSTEMS. PART 1: THEORETICAL ANALYSIS	74-1323
PROPULSION	TRANSVERSE VIBRATIONS OF SHIP'S PROPULSION SYSTEMS. PART 2: EXPERIMENTAL ANALYSIS	74-1324
PROPULSION	TRANSVERSE VIBRATIONS OF SHIP'S PROPULSION SYSTEMS - 1 THEORETICAL ANALYSIS	75-0291
PROPULSIVE	COUPLAGE ENTRE L ECOULEMENT AUTOUR D UN ARRIERE-CORPS ET LE JET PROPULSIF EN THEORIE DE FLUIDE PARFA	74-1137
PROSTHETIC	FINITE ELEMENT STRESS ANALYSIS OF PROSTHETIC TOOTH IMPLANTS	75-0073
PROSTHETIC	THE FINITE ELEMENT STRESS ANALYSIS OF THE HUMAN AORTIC VALVE TOWARD THE DEVELOPMENT OF A PROSTHETIC	74-1201
PROSTRANSTVENNY	THE DEVELOPMENT OF APPLIED METHODS IN PROBLEMS OF STATICAL CALCULATION OF THIN-WALLED THREE-DIMENSIONA	73-0934
PROTECTION	PRELIMINARY FINITE ELEMENT ANALYSIS, ATCHAFALAYA BASIN PROTECTION LEVEES, TEST SECTION 3	69-0463
PROTYPE	THE BOEING SST PROTYPE INTERNAL LOADS ANALYSIS SYSTEM AND PROCEDURES	71-0103
PRVKOV	GEOMETRICKY NELINEARNE ULOHY PLOCHYCH SKRUPIN RIESENE PRIRASTKOVOU METODOU KONECNYCH PRVKOV. (GEOMET	74-1032
PRVKOV	RIESENIE VAZKOPRUZNYCH ANIZOTROPICKYCH DOASK METODOU KONECNYCH PRVKOV (SOLUTION OF VISCOELASTIC ANIS	74-0477
PRVKOV	VELKE PRIEHYBY DOSIEK METODOU KONECNYCH PRVKOV (LARGE DEFLECTIONS OF PLATES ANALYZED BY FINITE ELEME	73-0373
PRVKU	FINITE ELEMENTS METHOD IN SOIL MECHANICS (METODA KONECNYCH PRVKU V MECHANICE SEMIN),	71-0468
PRVKU	VYUZITI METODY KONECNYCH PRVKU PRO PROJEKTOVANI KOMBINOVANYCH KONSTRUCKCI (FINITE ELEMENT METHOD IDE	73-0667
PRYAMOUGOL'NYMI	ISSLEDOVANIE NAPRYAZHENNOGO SOSTOYANIYA REBRISTYKH ISILINDRICHESKIKH OBOLOCHEK S PRYAMOUGOL'NYMI OTV	74-0944
PRYAMOUGOL'NYMI	METOD KONECHNYKH ELEMENTOV V PRIMENENII K RASCHETU TSILINDRICHESKIKH OBOLOCHEK S PRYAMOUGOL'NYMI OTV	73-0536
PRYING	HIGH STRENGTH BOLTS SUBJECT TO TENSION AND PRYING	74-0201
PRZEWODNICTWA	METODA ELEMENTOW SKONSZONYCH ROZWIAZYWANIA OSIOWO-SYMETRYCZNEGO PRZEWODNICTWA CIEPLA (FINITE-ELEMENT	73-0434
PRZYKLADZIE	NUMERYCZNE ROZWIAZANIE PROBLEMU DRGAN WLASNYCH NA PRZYKLADZIE PLYT. (NUMERICAL SOLUTION OF THE PROBL	75-0296
PRZYKLADZIE	ZASTOSOWANIE ITERACJI SEIDLA W METODZIE ELEMENTOW SKONCZONYCCH NA PRZYKLADZIE OBLICZEN STATYCZNYCH P	74-1031
PSEUDO	COUPLE - A PSEUDO FINITE ELEMENT CONDUCTION CODE FOR THE TRANSIENT THERMAL RESPONE OF AXISYMMETRIC O	72-0778
PSEUDO	PSEUDO ISOPARAMETRIC FINITE ELEMENTS FOR SHELL AND PLATE ANALYSIS	69-0040
PSEUDO	PSEUDO STATIC DEFORMATION AND FREQUENCIES OF ROTATING TURBOMACHINERY BLADES	72-0239
PSEUDO	PSEUDO 3-DIMENSIONAL FINITE-ELEMENT FORMULATION FOR ELASTOSTATIC PROBLEMS AND ITS GEOPHYSICAL APPLIC	75-0597
PSEUDO	THE PSEUDO FUNCTIONAL FINITE ELEMENT METHOD	74-0004
PSEUDO-CYLINDRI	FINITE-ELEMENT NONLINEAR BUCKLING ANALYSIS OF A PSEUDO-CYLINDRICAL CONCAVE POLYHEDRAL SHELL UNDER EX	73-1147

PSEUDO-FUNCTION	APPLICATION OF THE PSEUDO-FUNCTIONAL FINITE ELEMENT METHOD TO VISCO-PLASTIC TORSION	73-0002
PSEUDO-FUNCTION	APPLICATION OF THE PSEUDO-FUNCTIONAL FINITE ELEMENT METHOD TO NON-LINEAR PROBLEMS	74-0282
PSEUDO-FUNCTION	APPLICATION OF THE PSEUDO-FUNCTIONAL FINITE ELEMENT METHOD TO VISCOUS FLOW PROBLEMS	74-1183
PSEUDO-FUNCTION	APPLICATION OF THE PSEUDO-FUNCTIONAL FINITE ELEMENT METHOD TO COMPRESSIBLE FLOW PROBLEMS	74-1360
PSEUDO-FUNCTION	THE APPLICATION OF THE PSEUDO-FUNCTIONAL FINITE ELEMENT METHOD TO VISCOUS FLOW PROBLEMS	74-0287
PSEUDO-SOUND	RESPONSE OF STRUCTURE TO THE PSEUDO-SOUND FIELD OF A JET (USING A COMBINED CONTINUUM AND FINITE ELEM	67-0146
PSEUDO-SOUND	TEST RESULTS FROM THE BOUNDARY LAYER FACILITY - RESPONSE OF STRUCTURE TO THE PSEUDO-SOUND FIELD OF A	68-0262
PSEUDO-THREE-DI	PSEUDO-THREE-DIMENSIONAL FINITE ELEMENT FORMULATION FOR ELASTOSTATIC PROBLEMS	74-0167
PSEUDOCYLINDRIC	NUMERICAL AND EXPERIMENTAL ANALYSIS OF A PSEUDOCYLINDRICAL SHELL	74-1133
PUCHER'S	A FINITE SOLUTION OF PUCHER'S EQUATION	75-0782
PULSATILE	FINITE ELEMENT ANALYSIS METHODS IN DYNAMICS OF PULSATILE FLOW	69-0519
PULSATING	FINITE ELEMENT ANALYSIS OF STRUCTURAL INSTABILITY BY ASSOCIATION OF PULSATING EXCITATIONS	74-0453
PULSED	APPLICATIONS OF PULSED LASER INDUCED STRESS WAVES	73-0956
PUMP	CORRELATION BETWEEN PRESSURE AND THERMAL TRANSIENT STRESS LEVELS DERIVED BY THEORY AND EXPERIMENT ON	73-0469
PUMP	FESTIGKEITSUNTERSUCHUNGEN UND-BERECHNUNGEN AN PUMPENSPIRALGEHAEUSEN. (INVESTIGATIONS AND CALCULATION	75-0238
PUMP	REACTOR COOLANT PUMP FLYWHEELS OVERSPEED EVALUATION	74-0506
PUMP-TURBINE	ACTUAL STRESS OF HIGH HEAD PUMP-TURBINE RUNNER	75-0567
PUMPENSPIRALGEH	FESTIGKEITSUNTERSUCHUNGEN UND-BERECHNUNGEN AN PUMPENSPIRALGEHAEUSEN. (INVESTIGATIONS AND CALCULATION	75-0238
PUMPING	A FINITE ELEMENT APPROACH TO PHYSIOLOGICAL TRANSPORT PHENOMENA BY CILIARY AND PERISTALTIC PUMPING	72-0353
PUMPING	ANALYSIS OF PERISTALTIC PUMPING	72-0044
PUMPING	FINITE ELEMENT ANALYSIS OF TIME-DEPENDENT PUMPING TEST DATA IN AN UNCONFINED AQUIFER	75-0497
PUMPING	RESPONSE OF UNCONFINED AQUIFERS TO PUMPING	74-0073
PUNCH	ELASTOPLASTIC ANALYSIS OF PLANE STRAIN AND AXISYMMETRIC FLAT PUNCH INDENTATION BY THE FINITE ELEMENT	70-0343
PUNCH-INDENTATI	CONSTITUTIVE EQUATIONS AND PUNCH-INDENTATION OF CONCRETE	75-0286
PUNCH-INDENTATI	CONSTITUTIVE RELATIONS OF CONCRETE AND PUNCH-INDENTATION PROBLEMS	73-0877
PUNCHED	FINITE ELEMENT ANALYSIS OF STATIC AND DYNAMIC STRESSES AROUND SUDDENLY PUNCHED HOLES IN PLATES AND S	73-1088
PUNCHING	PUNCHING STRENGTH OF SLABS, A FLEXURAL APPROACH USING FINITE ELEMENTS	74-1307
PURE	PURE MOMENT LOADING OF AXISYMMETRIC FINITE ELEMENT MODELS	73-1007
PUSHING	ANALYSIS OF BAND PRESSURE IN A GUN TUBE USING FINITE ELEMENTS AND SHELL PUSHING DATA	72-0793
PWR	ANALYTICAL PROBLEMS ASSOCIATED WITH CORE SUPPORT STRUCTURES OF PWR	71-0039
PWR	PWR DEVELOPMENT FOR NUCLEAR PROPULSION: VIBRATION ANALYSIS OF NUCLEAR POWER PLANT	71-0797
PYLON	SPECTRUM TRUNCATION AND DAMAGE TOLERANCE STUDY ASSOCIATED WITH THE C-5A OUTBOARD PYLON AFT TRUSS LUG	74-0535
PYRAMID	MECHANICS OF CABLE MOORING SYSTEMS. VOLUME VII. THE STEADY-STATE BEHAVIOR OF A PYRAMID ARRAY SYSTEM	73-0724
PYRITIC	A FINITE ELEMENT ANALYSIS OF DISSOLVED OXYGEN DRAWDOWN AND SULFATE PRODUCTION IN STRIP MINE SPOIL DA	75-0795
PYROMETRIC	NONSTEADY PYROMETRIC CALCULATIONS AS THE BASIS FOR STRESS ANALYSIS OF STEAM GENERATORS AND HEAT EXCH	72-0451
Q-19	FINITE ELEMENT ANALYSIS OF ANISOTROPIC PLATES USING Q-19 ELEMENT. PART 1. THEORETICAL DEVELOPMENT	74-0728
Q-19	FINITE ELEMENT ANALYSIS OF ANISOTROPIC PLATES USING Q-19 ELEMENT - 1, 2	74-0985
QMESH	QMESH: A SELF-ORGANIZING MESH GENERATION PROGRAM	74-0872
QMESH	QMESH: A SELF-ORGANIZING MESH GENERATION PROGRAM	74-0899
QMESH	USERS MANUAL FOR QMESH, A SELF-ORGANIZING MESH GENERATION PROGRAM	74-0880
QUAD	MECHANICS OF CABLE MOORING SYSTEMS. VOLUME IV. A COMPUTER PROGRAM FOR ANALYZING THE STEADY STATE RES	72-0653
QUAD-4	QUAD-4: A COMPUTER PROGRAM FOR EVALUATING THE SEISMIC RESPONSE OF SOIL STRUCTURES BY VARIABLE DAMPIN	73-0771
QUADRATIC	A MODIFIED GRADIENT METHOD FOR FINITE ELEMENT ELASTOPLASTIC ANALYSIS BY QUADRATIC PROGRAMMING	73-0998
QUADRATIC	A PENALTY FUNCTION APPROACH TO THE MINIMIZATION OF QUADRATIC FUNCTIONALS IN FINITE ELEMENT ANALYSIS	74-0418
QUADRATIC	A QUADRILATERAL PLANE STRESS FINITE ELEMENT WITH QUADRATIC PROPERTIES	72-0969
QUADRATIC	APPLICATION OF QUADRATIC ISOPARAMETRIC FINITE ELEMENTS IN LINEAR FRACTURE MECHANICS	74-0521
QUADRATIC	ELASTIC-PLASTIC ANALYSIS BY QUADRATIC PROGRAMMING	72-0238
QUADRATIC	FINITE ELEMENT ELASTOPLASTIC ANALYSIS BY QUADRATIC PROGRAMMING: THE MULTISTAGE METHOD	73-0842
QUADRATIC	FINITE ELEMENT ELASTOPLASTIC ANALYSIS OF UNDERGROUND OPENINGS BY QUADRATIC PROGRAMMING	74-1288
QUADRATIC	FURTHER APPLICATION OF QUADRATIC ISOPARAMETRIC FINITE-ELEMENTS TO LINEAR FRACTURE MECHANICS OF PLATE	75-0639
QUADRATIC	QUADRATIC FINITE ELEMENT IN NEUTRON TRANSPORT	73-1039
QUADRATIC	QUADRATIC MATRIX EQUATIONS FOR DETERMINING VIBRATION MODES AND FREQUENCIES OF CONTINUOUS ELASTIC SYS	66-0064
QUADRATIC	THE QUADRATIC PROGRAMMING APPROACH TO THE FINITE ELEMENT METHOD	73-0831
QUADRATIC	TWO TYPES OF PIECEWISE QUADRATIC SPACES AND THEIR ORDER OF ACCURACY FOR POISSON'S EQUATION	72-0520
QUADRATIC	ZUR LOESUNG DES KONTAKTPROBLEMS ELASTISCHER KOERPER MIT AUSGEDEHNTER KONTAKTFLAECHE DURCH QUADRATISC	74-1135
QUADRATISCHE	ZUR LOESUNG DES KONTAKTPROBLEMS ELASTISCHER KOERPER MIT AUSGEDEHNTER KONTAKTFLAECHE DURCH QUADRATISC	74-1135
QUADRATURE	EFFECTS OF QUADRATURE ERRORS IN FINITE ELEMENT APPROXIMATION OF STEADY STATE, EIGENVALUE, AND PARABO	72-0343
QUADRATURE	GAUSSIAN QUADRATURE FORMULAS FOR TRIANGLES	73-0906
QUADRATURE	ON THE EFFECTS OF QUADRATURE ERRORS IN THE FINITE ELEMENT METHOD	72-0258
QUADRATURE	QUADRATURE RULES FOR BRICK BASED FINITE ELEMENTS	71-0531
QUADRILATERAL	A COMPATIBLE MEMBRANE QUADRILATERAL ELEMENT	72-0483
QUADRILATERAL	A COMPATIBLE QUADRILATERAL PLATE BENDING ELEMENT	72-0482
QUADRILATERAL	A CONFORMING QUADRILATERAL ELEMENT FOR BUCKLING ANALYSIS OF STIFFENED PLATES	73-0613
QUADRILATERAL	A GENERAL QUADRILATERAL PLATE ELEMENT	73-0807
QUADRILATERAL	A QUADRILATERAL PLANE STRESS FINITE ELEMENT WITH QUADRATIC PROPERTIES	72-0969
QUADRILATERAL	A REFINED QUADRILATERAL ELEMENT FOR ANALYSIS OF PLATE BENDING	68-0047
QUADRILATERAL	A SYNTHESIZED WARPED QUADRILATERAL ELEMENT GIVING SIMPLIFIED INPUT FOR EQUILIBRIUM MODELS	73-1013
QUADRILATERAL	A WARPED QUADRILATERAL STRAIN MEMBRANE ELEMENTS	75-0141
QUADRILATERAL	A WARPED QUADRILATERAL EQUILIBRIUM MEMBRANE ELEMENT	75-0885
QUADRILATERAL	AN EQUILIBRIUM MEMBRANE QUADRILATERAL ELEMENT	72-0484
QUADRILATERAL	ANALYSIS OF THIN SHELLS WITH A DOUBLY CURVED ARBITRARY QUADRILATERAL FINITE ELEMENT.	72-0443
QUADRILATERAL	ARBITRARY QUADRILATERAL SPAR WEBS FOR THE MATRIX DISPLACEMENT METHOD	66-0003
QUADRILATERAL	CURVED ISOPARAMETRIC QUADRILATERAL ELEMENTS FOR FINITE ANALYSIS	68-0008
QUADRILATERAL	CURVED ISOPARAMETRIC QUADRILATERAL ELEMENTS FOR FINITE ELEMENT ANALYSIS	68-0067
QUADRILATERAL	EFFECT OF OUT-OF-PLANENESS OF MEMBRANE QUADRILATERAL FINITE ELEMENTS	73-0031
QUADRILATERAL	ERROR ESTIMATE FOR QUADRILATERAL ISOPARAMETRIC FINITE-ELEMENTS OF TYPE Q1 WHICH CAN DEGENERATE INTO	75-0401
QUADRILATERAL	FLAT QUADRILATERAL EQUILIBRIUM MEMBRANE ELEMENTS	73-1116
QUADRILATERAL	GENERAL QUADRILATERAL PLATE ELEMENT	75-0339
QUADRILATERAL	IMPROVED GENERAL QUADRILATERAL PLATE ELEMENTS	75-0036
QUADRILATERAL	MIXED QUADRILATERAL ELEMENTS FOR BENDING	72-0623
QUADRILATERAL	NEW TRIANGULAR AND QUADRILATERAL PLATE-BENDING FINITE ELEMENTS	74-0833
QUADRILATERAL	QUADRILATERAL ELEMENTS IN PLANE ANALYSIS	66-0085
QUADRILATERAL	QUADRILATERAL FINITE ELEMENT IN ELASTIC-PLASTIC PLANE STRESS ANALYSIS	67-0067
QUADRILATERAL	QUADRILATERAL FINITE ELEMENT FOR PLANE STRESS AND PLANE STRAIN	70-0496
QUADRILATERAL	SOME PLANE QUADRILATERAL HYBRID FINITE ELEMENTS	69-0082
QUADRILATERAL	STATIC ANALYSIS OF SHELLS OF REVOLUTION USING DOUBLY-CURVED QUADRILATERAL ELEMENTS DERIVED FROM ALTE	69-0368
QUADRILATERAL	STRESS ANALYSIS OF AXISYMMETRIC SOLIDS UTILIZING HIGHER ORDER QUADRILATERAL FINITE ELEMENTS	69-0356
QUADRILATERAL	THE QUADRILATERAL SHEAR PANEL	51-0002
QUADRILATERAL	THERMAL DEFORMATION VECTOR FOR A BILINEAR TEMPERATURE DISTRIBUTION IN AN ANISOTROPIC QUADRILATERAL M	75-0670
QUADRILATERAL	VIBRATION OF CURVED STRUCTURES USING QUADRILATERAL FINITE ELEMENTS	72-0074
QUADRILATERAL	WARPED SEMI-MONOCOQUE QUADRILATERAL MEMBRANE ELEMENT FOR THE FINITE ELEMENT METHODS	70-0195
QUADRIPARTITE	PHOTOELASTIC AND FINITE-ELEMENT ANALYSIS OF A QUADRIPARTITE VAULT	73-0119
QUADRUPOLE	FINITE-ELEMENT TREATMENT OF QUADRUPOLE GROUND MOTIONS IN AXISYMMETRIC EARTH STRUCTURES	75-0434
QUALIFICATION	SOFTWARE USER: EDUCATION AND QUALIFICATION	72-0441
QUALIFICATION	THE QUALIFICATION OF SOLUTION METHODS	71-0549
QUALIFICATION	VERIFICATION AND QUALIFICATION OF A 3-D FINITE ELEMENT COMPUTER ANALYSIS	71-0487
QUALITY	A NUMERICAL MODEL OF TRANSIENT WATER QUALITY IN A ONE-DIMENSIONAL ESTUARY BASED ON THE FINITE ELEMEN	74-1369
QUALITY	NUMERICAL MODEL FOR THE PREDICTION OF TRANSIENT WATER QUALITY IN ESTUARY NETWORKS	72-0635
QUALITY	QUALITY IN STRUCTURAL ANALYSIS BY FINITE ELEMENTS	71-0500
QUALITY	TWO-DIMENSIONAL FINITE ELEMENT MODELING OF TRANSIENT FLOW IN REGIONAL AQUIFER SYSTEMS. ANALYSIS OF G	74-0743
QUANTITATIVE	QUANTITATIVE STRAIN-AND-STRESS STATE CRITERION FOR FAILURE IN THE VICINITY OF SHARP CRACKS	74-0639
QUANTITY	FINITE ELEMENT SOLUTION FOR QUANTITY OF STEADY SEEPAGE	71-0824

QUANTUM	FINITE ELEMENT METHOD FOR BOUND STATE CALCULATIONS IN QUANTUM MECHANICS	75-0005
QUARTER	FINITE ELEMENT MODELING OF A CIRCULAR RING USING HALF AND QUARTER SYMMETRY	66-0119
QUARTIC	A CONFORMING QUARTIC TRIANGULAR ELEMENT FOR PLATE BENDING	69-0137
QUARTIC	A QUARTIC, TWO DIMENSIONAL ISOPARAMETRIC ELEMENT	68-0187
QUARTZ	APPLICATION OF FINITE ELEMENT METHOD TO VIBRATIONS OF QUARTZ PLATES	74-0345
QUARTZ	FINITE ELEMENT CALCULATIONS RELEVANT TO AS-CUT QUARTZ RESONATORS	73-0672
QUARTZ	STRESS-WAVE PROPAGATION AND DYNAMIC EFFECTS IN SHELLS (SPADES) MICROMECHANICAL STRESS-WAVE RESPONSE	72-0851
QUASI	NUMERICAL SOLUTION OF THE QUASI LINEAR POISSON EQUATION IN A NON-UNIFORM TRIANGLE MESH	67-0090
QUASI-ANALYTICA	QUASI-ANALYTICAL FINITE ELEMENT PROCEDURES FOR AXISYMMETRIC ANISOTROPIC SHELLS AND SOLIDS	74-0297
QUASI-HARMONIC	FINITE ELEMENT ANALYSIS OF ENGINEERING SYSTEMS GOVERNED BY A NON-LINEAR QUASI-HARMONIC EQUATION	75-0664
QUASI-LINEAR	QUASI-LINEAR PROGRAMMING ALGORITHM FOR OPTIMIZING FIBRE-REINFORCED STRUCTURES OF FIXED STIFFNESS	75-0225
QUASILINEAR	A FINITE ELEMENT COLLOCATION METHOD FOR QUASILINEAR PARABOLIC EQUATIONS	73-1000
QUASILINEAR	FINITE ELEMENT COLLOCATION METHOD FOR QUASILINEAR PARABOLIC EQUATIONS	73-0047
QUASILINEARE	STABILITAT BEI DER METHODE DER FINITEN ELEMENTE FUR QUASILINEARE ELLIPTISCHE RANDWERTPROBLEME	74-1283
QUASILINEAREN	FINITE ELEMENT VERFAHREN BEI QUASILINEAREN ELLIPTISCHEN RANDWERTPROBLEM	74-1262
QUASIOPTIMAL	QUASIOPTIMAL FINITE-ELEMENT METHOD FOR ELLIPTIC INTERFACE PROBLEMS	75-0427
QUELQUES	QUELQUES METHODES D'ELEMENTS FINIS POUR DE PROBLEME D'UNE PLAQUE ENCASTREE	74-0406
QUELQUES	TOUR MAINE-MONTPARNASSE QUELQUES PROBLEMES DE STRUCTURE (SPECIFIC PROBLEMS ENCOUNTERED IN THE CONSTR	74-0539
QUERSCHNITT	TEMPERATURVERLAUF IM QUERSCHNITT GESCHWEISSTER ROHRWAENDE BEI UEBERWIEGEND KONVEKTIVEM WAERMEUEBERGA	75-0628
QUERSCHNITTE	EINE FINITE-ELEMENTE-LOESUNG FUR DIE TORSIONSSTEIFIGKEIT UND DEN SCHUBMITTELPUNKT BELIEBIGER QUERSCH	74-0256
QUERSCHNITTSVER	ZUR BERECHNUNG EINFACH-SYMMETRISCHER I-TRAEGER NACH DER THEORIE - 2. ORDNUNG UNTER BERUECKSICHTIGUNG	75-0510
QUICK	ANALYSIS OF QUICK SAND PHENOMENON BY FINITE ELEMENT METHOD	73-0191
QUICK	FINITE ELEMENT ANALYSES OF QUICK CAMP MODULE	75-0017
QUINTIC	ON THE QUINTIC TRIANGULAR PLATE BENDING ELEMENT	72-0677
RABBIT	EFFECT OF 2450-MHZ RADIATION ON THE RABBIT EYE	75-0365
RABBIT	FINITE ELEMENT MODEL SYSTEM OF A RABBIT CALVARIUM: AN INVESTIGATION OF OPTIMAL COMPRESSIVE LOADING	73-0206
RABBIT	MEASUREMENTS OF HEALING AT AN OSTEOTOMY IN A RABBIT CALVARIUM: THE INFLUENCE OF APPLIED COMPRESSIVE	74-0975
RACHETTING	PRESSURE VESSEL WALL THICKNESS LIMITS BASED ON THERMAL STRESS RACHETTING AND CREEP FATIGUE INTERACTI	74-0968
RADIAL	ANALYSIS OF RADIAL FLOW IMPELLERS BY THE MATRIX FINITE ELEMENT METHOD	71-0412
RADIAL	ANALYTIC INVESTIGATION OF COMPACT REINFORCEMENT FOR RADIAL NOZZLES IN SPHERICAL SHELLS	71-0107
RADIAL	RADIAL AND AXIAL TRANSMISSIBILITY CHARACTERISTICS OF TYPICAL ROCKET VEHICLE STRUCTURE TECHNICAL SUMM	69-0475
RADIAL	RADIAL DISPLACEMENTS OF A HOLLOW SHAFT SUBJECTED TO A UNIFORM BAND OF EXTERNAL PRESSURE	72-0242
RADIAL	RADIAL INERTIA EFFECTS ON AN IDEALLY PLASTIC CIRCULAR PLATE UNDER IMPULSIVE AXIAL COMPRESSION	69-0324
RADIAL	STRESS ANALYSIS PROGRAM BY FINITE ELEMENT METHOD FOR RADIAL FLOW IMPELLERS	74-1016
RADIAL	UNSTEADY RADIAL FLOW IN GAS RESERVOIRS BY FINITE ELEMENTS	74-0462
RADIALLY	A COMPARISON OF FINITE ELEMENT AND FOURIER SERIES SOLUTIONS AS APPLIED TO RADIALLY LOADED CIRCULAR R	68-0024
RADIALLY	EXACT SOLUTION TO THE RADIALLY END CONSTRAINED CIRCULAR CYLINDER UNDER TRIAXIAL LOADING	71-0433
RADIATING	FINITE ELEMENT ANALYSIS OF ACOUSTICALLY RADIATING STRUCTURES WITH APPLICATIONS TO SONAR TRANSDUCERS	73-0075
RADIATING	FINITE ELEMENT ANALYSIS OF ACOUSTICALLY RADIATING STRUCTURES WITH APPLICATIONS TO SONAR TRANSDUCERS	74-0524
RADIATION	APPLICATION OF THE FINITE-ELEMENT METHOD TO HEAT-TRANSFER PROBLEMS. PART 11. TRANSIENT TWO-DIMENSION	71-0759
RADIATION	EFFECT OF 2450-MHZ RADIATION ON THE RABBIT EYE	75-0365
RADIATION	FINITE ELEMENT APPROACH TO ACOUSTIC RADIATION FROM ELASTIC STRUCTURES	74-0121
RADIATION	FINITE ELEMENT METHOD APPLIED TO TRANSIENT TWO-DIMENSIONAL HEAT TRANSFER WITH CONVECTION AND RADIATI	70-0127
RADIATION	NASTRAN-GAP, AN INTEGRATED APPROACH TO THE ANALYSIS OF RADIATION PATTERNS OF DISTORTED REFLECTORS	71-0062
RADIATION	SHIP (SIMPLIFIED-HELMHOLTZ-INTEGRAL PROGRAM): A FAST COMPUTER PROGRAM FOR CALCULATING THE ACOUSTIC R	72-0835
RADIATION	SHIP (SIMPLIFIED-HELMHOLTZ-INTEGRAL PROGRAM): A FAST COMPUTER PROGRAM FOR CALCULATING THE ACOUSTIC R	72-0835
RADIATION	STRESS ANALYSIS OF A GRAPHITE FUEL ELEMENT INCLUDING RADIATION CREEP AND DIMENSIONAL CHANGES	71-0217
RADIATION-DAMAG	FINITE-ELEMENT ANALYSIS OF RADIATION-DAMAGE STRESSES IN THE GRAPHITE OF MATRIX FUEL ELEMENTS	70-0048
RADIATIVE	A RADIATIVE IGNITION MODEL OF A SOLID FUEL	74-1312
RADIO	NORMAL MODE ANALYSIS OF THE RADIO ASTRONOMY EXLPORER BOOMS AND SPACECRAFT	71-0130
RADIONUCLIDES	THE FATE OF RADIONUCLIDES IN A GROUNDWATER ENVIRONMENT	74-0061
RADIUS	EFFECTS OF SPECIMEN RADIUS ON THE STRESS STATE NEAR A ROOF BOLT ANCHOR: A FINITE ELEMENT DETERMINATI	73-0229
RADIUS	MEASUREMENT OF ACTUAL DEDENDUM STRAINS, AND THE INFLUENCE OF DEDENDUM RADIUS ON STRESSES AND DEDENDU	72-0118
RADOME	SHELL INSTABILITY ANALYSIS APPLIED TO A RADOME	74-0261
RADSAETZEN	EINSATZ MODERNER RECHENVERFAHREN ZUR ERMITTLUNG DER SPANNUNGEN UND VERFORMUNGEN BEI RADSAETZEN. (USE	75-0186
RADSAETZEN	UNTERSUCHUNGEN ZUR SPURMASSTABILITAET BEI RADSAETZEN MIT KLOTZGEBREMSTEN VOLLRAEDERN (INVESTIGATIONS	75-0810
RAFFINES	ELEMENTS FINIS RAFFINES EN ELASTICITE BIDIMENSIONNELLE	69-0411
RAFT	ANALYSIS OF A CIRCULAR RAFT IN ADHESIVE CONTACT WITH A THICK ELASTIC LAYER	74-1152
RAFT	ELASTIC SETTLEMENT OF A CIRCULAR RAFT IN ADHESIVE CONTACT WITH A TRANSVERSELY ISOTROPIC MEDIUM	75-0828
RAFT	FIELD INSTRUMENTATION FOR LONG TERM MEASUREMENT OF PILE LOAD AND RAFT CONTACT PRESSURE	73-0150
RAFTED	INTERACTIVE ANALYSIS OF A RAFTED MULTISTOREY SPACE FRAME RESSTING ON AN INHOMOGENEOUS CLAY STRATUM	74-0447
RAHMEN	VARIATIONAL PRINCIPLES OF ELASTOSTATICS WITH RELAXED CONTINUITY REQUIREMENTS AND THEIR APPLICABILITY	74-0758
RAHMENKORPERN	MODERNE VERFAHREN ZUR UNTERSUCHUNG DER STEIFIGKEITS-UND FESTIGKEITSVERHALTNISSE VON RAHMENKORPERN, I	69-0401
RAIL	ASSESSMENT OF DESIGN TOOLS AND CRITERIA FOR URBAN RAIL TRACK STRUCTURES, VOLUME 11. AT-GRADE SLAB TR	74-0813
RAILROAD	COMPUTER ANALYSIS OF A RAILROAD FREIGHT CAR BOLSTER UTILIZING THE FINITE ELEMENT METHOD	71-0283
RAILROAD	STRENGTH OF RAILROAD WHEELS - 3	73-0306
RAILROAD	TRANSVERSE REINFORCEMENT OF CONTINUOUS, TROUGH-SHAPED CONCRETE RAILROAD STRUCTURES. AN IMPLEMENTATIO	71-0719
RAILWAY	DESIGN OF INCLINED TIED ARCH RAILWAY BRIDGES OVER THE M56	71-0297
RAINFALL	CALCULATION OF AREAL RAINFALL USING FINITE ELEMENT TECHNIQUES WITH ALTITUDINAL CORRECTIONS	72-0344
RAINFALL	THE USE OF THE FINITE ELEMENT METHOD IN OBTAINING SOIL-WATER PRESSURE AND SOIL-WATER CONTENT DISTRIB	75-0112
RAM	AIRCRAFT FUEL TANK VULNERABILITY TO HYDRAULIC RAM: MODIFICATTION OF THE NORTHROP FINITE ELEMENT COMP	74-0112
RAMBERG	AN INVERSE FORM OF RAMBERG - OSGOOD FORMULA	69-0479
RAMIFACATIONS	FINITE ELEMENT METHOD AS AN ASPECT OF THE PRINCIPLE OF MAXIMUM UNIFORMITY: NEW HYDRODYNAMICAL RAMIFA	74-0009
RAMIFICATIONS	RAMIFICATIONS OF OPTIMUM IDEALIZATION GEOMETRY IN DISCRETE ELEMENT ANALYSIS	75-0182
RANDOM	A CONSISTENT FINITE ELEMENT METHOD FOR RANDOM RESPONSE PROBLEMS	71-0202
RANDOM	A CONSISTENT FINITE ELEMENT METHOD FOR RANDOM RESPONSE PROBLEMS	72-0896
RANDOM	A FINITE ELEMENT METHOD FOR RANDOM FOLDED PLATE STRUCTURES (EINE FINITE ELEMENT METHODE ZUR BERECHNU	73-0787
RANDOM	COMPUTER BASED ANALYSES OF THE RESPONSE OF BOX TYPE STRUCTURES TO RANDOM PRESSURES	73-0227
RANDOM	COMPUTER PREDICTION OF FATIQUE CRACK PROPAGATION UNDER RANDOM LOADING	73-0851
RANDOM	FINITE ELEMENT METHODS FOR THE RESPONSE OF STRUCTURES TO RANDOM EXCITATIONS	75-0230
RANDOM	FINITE-ELEMENT ANALYSIS OF COMPLEX PANEL RESPONSE TO RANDOM LOADS	68-0093
RANDOM	FREE VIBRATION AND RESPONSE TO RANDOM PRESSURE FIELD OF ANISOTROPIC THIN CYLINDRICAL SHELLS	74-0830
RANDOM	FREE VIBRATIONS AND RANDOM RESPONSE OF AN INTEGRALLY-STIFFENED PANEL	70-0138
RANDOM	METHOD FOR ESTIMATING RESPONSE OF PAYLOAD SECONDARY STRUCTURES TO RANDOM EXCITATION	68-0232
RANDOM	ON RESPONSE OF INITIALLY STRESSED STRUCTURES TO RANDOM EXCITATIONS	73-0278
RANDOM	PLANAR FINITE ELEMENT AT RANDOM TEMPERATURE	73-0140
RANDOM	PROBABILISTIC BUCKLING BEHAVIOR OF STRUCTURES WITH RANDOM LACK-OF-FIT AND GEOMETRIC IMPERFECTIONS	73-0862
RANDOM	RANDOM RESPONSE OF PERIODIC STRUCTURES BY A FINITE-ELEMENT TECHNIQUE	75-0426
RANDOM	RANDOM VIBRATION ANALYSIS SYSTEM FOR COMPLEX STRUCTURES, PART 2. COMPUTER PROGRAM DESCRIPTION	69-0262
RANDOM	RANDOM VIBRATION RESPONSE OF CANTILEVER PLATES USING THE FINITE ELEMENT METHOD	70-0181
RANDOM	RELABELING OF FINITE ELEMENT MESHES USING A RANDOM PROCESS	72-0754
RANDOM	RESPONSE OF NONLINEAR BEAM TO RANDOM EXCITATION	73-0285
RANDOM	RESPONSE OF NUCLEAR STRUCTURAL SYSTEMS TO TRANSIENT AND RANDOM EXCITATIONS, USING BOTH DETERMINISTIC	74-1047
RANDOM	RESPONSE OF RECTANGULAR ORTHOTROPIC PLATES TO RANDOM EXCITATIONS	73-0496
RANDOM	THE RESPONSE OF SHELLS TO DISTRIBUTED RANDOM LOADS USING NASTRAN	71-0134
RANDOM	VIBRATION MODES AND RANDOM RESPONSE OF A MULTI-BAY PANEL SYSTEM USING FINITE ELEMENTS	67-0128
RANDOM-VIBRATIO	RANDOM-VIBRATION ANALYSIS SYSTEM FOR COMPLEX STRUCTURES. PART 1. ENGINEERING USER'S GUIDE	68-0106
RANDOMLY	A FINITE ELEMENT APPROACH FOR THE ANALYSIS OF RANDOMLY EXCITED COMPLEX ELASTIC STRUCTURES	67-0057
RANDOMNESS	TREATMENT OF RANDOMNESS IN FINITE ELEMENT MODELING	70-0353
RANDWERTPROBLEM	FINITE ELEMENT VERFAHREN BEI QUASILINEAREN ELLIPTISCHEN RANDWERTPROBLEM	74-1262
RANDWERTPROBLEM	EINE EINFACHE KONSTRUKTION VON KOORDINATENFUNKTIONEN FUR DIE NUMERISCHE LOSUNG ZWEIDIMENSIONALER RAN	66-0134
RANDWERTPROBLEM	STABILITAT BEI DER METHODE DER FINITEN ELEMENTE FUR QUASILINEARE ELLIPTISCHE RANDWERTPROBLEME	74-1283

RANGE	ANALYSIS OF STRESSES IN PRESSURIZED WELDED PIPE IN THE CREEP RANGE	71-0404
RANGE	FINITE ELEMENT ANALYSIS OF STRUCTURES IN THE PLASTIC RANGE	71-0031
RANGE	FINITE-ELEMENT ANALYSIS OF A FRACTURE TOUGHNESS TEST SPECIMEN IN NON-LINEAR RANGE	72-0093
RANGE	FULL RANGE ANALYSIS OF ECCENTRICALLY STIFFENED PLATES	74-0149
RANGE	MATERIAL BEHAVIOUR CHARACTERISTICS FOR REINFORCED CONCRETE SHELLS STRESSED BEYOND THE ELASTIC RANGE	70-0166
RANGE	ON NUMERICALLY ACCURATE FINITE ELEMENT SOLUTIONS IN THE FULLY PLASTIC RANGE	74-0529
RANGE	PRACTICAL SOLUTION OF PLASTIC DEFORMATION PROBLEMS IN THE ELASTIC PLASTIC RANGE	59-0004
RANGE	STABILITA STEN V PLASTICKEM OBORU (STABILITY IN WEBPLATES IN PLASTIC RANGE)	73-0387
RANGE	THE ECONOMICAL SOLUTION OF ELASTIC PROBLEMS FOR A RANGE OF POISSONS RATIO	74-0786
RANGELY	EARTHQUAKE TRIGGERING BY FLUID INJECTION AT RANGELY, COLORADO	72-0634
RANGES	FINITE ELEMENT STUDY OF A REINFORCED CONCRETE CYLINDRICAL SHEL_ THROUGH ELASTIC, CRACKING, AND ULTIM	75-0317
RANGES	PREDICTION OF LONG-TERM STRESS RANGES: USER'S MANUAL - BRIDGE LOAD GENERATOR	73-0737
RANGES	PREDICTION OF LONG-TERM STRESS RANGES: USER'S MANUAL - BRIDGE DYNAMIC STRESS ANALYSIS	74-0807
RANK	DYNAMIC ANALYSIS OF STRUCTURES USING THE RANK FORCE METHOD	71-0553
RAPID	A STRUCTURAL EVALUATION OF RAPID METHODS OF BACKFILLING FOR BOMB DAMAGE REPAIR	74-0821
RAPID	A STRUCTURAL EVALUATION OF RAPID METHODS OF BACKFILLING FOR BOMB DAMAGE REPAIR - PHASE 11	75-0682
RAPID	EXPLORATORY INVESTIGATION OF RAPID CRACK PROPAGATION AND CRACK ARREST	73-0757
RAPID	LINEAR SHELL THEORY FOR RAPID VARIATION SOLUTIONS	71-0406
RAPID	RAPID CASE DESIGN USING FINITE ELEMENT STRESS ANALYSIS AND NASTRAN	73-0730
RAPID	THE USE OF A LASER FOR ARPA MILITARY GEOPHYSICS PROGRAM (ROCK MECHANICS AND RAPID EXCAVATION)	72-0800
RAPIDE	NONLINEAR ANALYSIS OF PRESTRESSED CONCRETE FAST REACTOR PRESSURE VESSEL, (ANALYSE NON LINEAIRE DES C	70-0145
RAPIDLY	A RAPIDLY CONVERGING TRIANGULAR PLATE ELEMENT	69-0245
RAREFIED	A FINITE ELEMENT ANALYSIS OF TRANSIENT RAREFIED GAS FLOW	73-0604
RAREFIED	ANALYSIS OF RAREFIED GAS FLOW THROUGH AN ARBITRARY CROSS SECTION BY THE FINITE ELEMENT METHOD	73-0365
RAREFIED	PLANE POISEUILLE FLOW OF A RAREFIED GAS BASED ON FINITE ELEMENT METHOD	71-0575
RAREFIED	THE FINITE ELEMENT ANALYSIS OF TRANSIENT RAREFIED GAS FLOW	74-0027
RAREFIED	VARIATIONAL PRINCIPLES FOR A RAREFIED GAS FLOW PAST AND ARBITRARY THREE-DIMENSIONAL BODY	70-0189
RASCHET	CALCULATION OF THE REINFORCED THIN-WALLED STRUCTURES BY THE METHOD OF FINITE ELEMENTS (RASCHET PODKR	75-0685
RASCHET	RASCHET KILEBANII PLASTIN SLOZHNDI FORMY S NEODNOPODNYMI GRANICHNYMI USLOVIYAMI. (CALCULATION OF VIB	74-0988
RASCHET	RASCHET METODOM KONECHNYKH ELEMENTOV DINAMICHESKIKH KHARAKTEERISTIK ZEMLYANOI PLOTINY SOVMESTNO S OS	73-0437
RASCHET	RASCHET NAPRYAZHENIL L PEREMESHCHENII V SVODE PODZEMNOGO MASHINNOGO ZALA GES I V OKRUZHAYUSHCHEM SKA	74-0542
RASCHET	RASCHET POLEI DEFORMATSII I NAPRYAZHENII V DETALYAKH DVIGATELEI VNUTRENNEGO SGORANIYA (CALCULATION O	73-0465
RASCHETA	APPLICATION OF THE METHOD OF FINITE ELEMENTS TO THE CALCULATION OF SHIP STRUCTURES (ISPOLZOVANIE MET	73-0933
RASCHETA	CHISLENNYI METOD RASCHETA SOBSTVENNYKH I VYNUZHDENNYKH KOLEBANII SOSTAVNYKH OBOLOCHECHNYKH KONSTRUKT	73-0461
RASCHETA	REALIZATSIYA METODA KONECHNYKH ELEMENTOV NA EVM DLYA RASCHETA DVUMERNYKH ELEKTRICHESKIKH I MAGNITNYK	74-0251
RASCHETA	THE DEVELOPMENT OF APPLIED METHODS IN PROBLEMS OF STATIC CALCULATION OF THIN-WALLED THREE-DIMENSIONA	73-0934
RASCHETAKH	METOD KONECHNYKH ELEMENTOV V RASCHETAKH KONSOLIDATSII VODONASYSHCHENNYKH GRUNTOV. (FINITE ELEMENT ME	75-0469
RASCHETNOGO	METODY RASCHETNOGO OPREDELENIYA TEMPERATURNYKH NAPRYAZHENII V KRYSHKAKH TSILINDROV DVIGATELEI VNUTRE	74-0179
RASCHETU	METOD KONECHNYKH ELEMENTOV V PRIMENENII K RASCHETU TSILINDRICHESKIKH OBOLOCHEK S PRYAMOUGOL'NYMI OTV	73-0536
RASCHIOT	RASCHIOT USTOICHIVOSTI IZGIBNO-KRUTILNYKH KOLEBANIJ KRULA V DOZVUKOVOM POTOKE METODOM KONECHNYKH ELE	72-0903
RASCHIOTA	MODIFICIROVANNYJ KONECHNYJ ELEMENT DIJA RASCHIOTA KOLEBANIJ TONKIKH PLASTIN	72-0891
RASHET	RASHET TREKHMERNYKH MAGNITNYKE POLEI METODOM KONECHNYKH ELEMENTOV. (CALCULATION OF THREE-DIMENSIONAL	75-0766
RAT	FINITE-ELEMENT MODEL FOR THE MECHANICAL BEHAVIOUR OF THE LEFT VENTRICLE, PREDICTION OF DEFORMATION I	72-0304
RATCHETTING	ELASTIC-PLASTIC-CREEP ANALYSIS OF THERMAL RATCHETTING IN STRAIGHT PIPE AND COMPARISONS WITH TEST RES	73-0393
RATE	EXCITATION OF PLANE THERMOELASTIC WAVES IN A LAYER WITH A FINITE RATE OF HEAT PROPAGATION	74-0803
RATE	RATE OF CHANGE OF DYNAMIC RESPONSE	72-1004
RATE	RATE OF CONVERGENCE OF THE FINITE ELEMENT METHOD	72-0502
RATE	RATE OF CONVERGENCE FOR THE ITERATIVE APPROACH IN ELASTIC-PLASTIC ANALYSIS OF CONTINUA	73-0518
RATE	THE RATE OF CONVERGENCE FOR THE FINITE ELEMENT METHOD	71-0034
RATE	THE SIGNIFICANCE OF THE RATE OF PLASTIC WORK IN ELASTO-PLASTIC ANALYSIS	74-0437
RATE	THE SIGNIFICANCE OF THE RATE OF PLASTIC WORK IN ELASTO-PLASTIC ANALYSIS	74-0764
RATE-PLANE	EFFECTS OF A PEAK OVERLOAD ON FATIGUE CRACK PROPAGATION RATE-PLANE STRESS ANALYSIS OF CRACKED PLATE	74-1365
RATES	DETERMINATION OF ENERGY RELEASE RATES AND STRESS - INTENSITY FACTORS BY THE FINITE-ELEMENT METHOD	72-0183
RATES	STRAIN ENERGY RELEASE RATES FOR TURNING CRACKS	72-0146
RATES	THEORETICAL AND NUMERICAL CONVERGENCE RATES FOR SOME CONFORMING FINITE ELEMENTS	74-0941
RATING	V-BELT LIFE PREDICTION AND POWER RATING	75-0878
RATIO	AEROELASTIC PROBLEMS OF LOW ASPECT RATIO WINGS (PART 1)	56-0006
RATIO	DEPENDENCE OF PLATE-BENDING FINITE ELEMENT DEFLECTIONS AND EIGENVALUES ON POISSONS RATIO	74-0567
RATIO	DEPENDENCE OF PLATE-BENDING FINITE ELEMENT DEFLECTIONS AND EIGENVALUES ON POISSONS RATIO	74-0583
RATIO	DETERMINATION OF YOUNG'S MODULUS AND POISSON'S RATIO FOR WC-CO ALLOYS BY THE FINITE ELEMENT METHOD	72-0142
RATIO	EFFECT OF ELASTIC MODULAR RATIO AND ANISOTROPY ON THE STRESS FIELD IN A TWO-DIMENSIONAL MATRIX INTO	71-0647
RATIO	INFLUENCE OF POISSON'S RATIO ON THE CONDITION OF THE FINITE ELEMENT STIFFNESS MATRIX	71-0519
RATIO	SIMULATION OF ELASTIC MODULUS AND POISSON'S RATIO OF SPHEROIDAL GRAPHITE CAST IRON	74-0955
RATIO	THE ECONOMICAL SOLUTION OF ELASTIC PROBLEMS FOR A RANGE OF POISSONS RATIO	74-0786
RATIO	THE EFFECT OF CAMBER ON THE NATURAL FREQUENCIES OF LOW ASPECT RATIO TURBOMACHINERY BLADES	73-1092
RATIONAL	A RATIONAL ANALYSIS AND DESIGN PROCEDURE FOR WOOD JOIST FLOOR SYSTEMS	74-0730
RATIONAL	RATIONAL DESIGN OF TUNNEL SUPPORTS: TUNNEL SUPPORT LOADING CAUSED BY ROCK FAILURE	75-0687
RATIONAL	THEORETICAL INVESTIGATION OF ROCK AND SUPPORT INTERACTION - DEVELOP MORE RATIONAL DESIGN METHODS AND	74-0782
RATIONALIZATION	RATIONALIZATION IN DERIVING ELEMENT STIFFNESS MATRIX BY ASSUMED STRESS APPROACH	68-0138
RATIOS	ON THE STRESS DISTRIBUTION IN NOZZLE TO CYLINDER CONNECTIONS FOR SMALL DIAMETER RATIOS	72-0426
RAUCHGASSEITE	TEMPERATURVERLAUF IM QUERSCHNITT GESCHWEISSTER ROHRWAENDE BEI UEBERWIEGEND KONVEKTIVEM WAERMEUEBERGA	75-0628
RAUM-UND	BERECHNUNG INSTATIONAERER TEMPERATUREFELDER MIT FINITEN ELEMENTEN IN RAUM-UND ZEITDIMENSION. (CALCUL	74-0987
RAUMFAHRT	EXPERIMENTS IN FORMING AND JOINTING OF TITANIUM AND TITANIUM ALLOYS FOR AIRCRAFT AND SPACE VEHICLES	70-0194
RAUMLICHER	BERECHNUNG RAUMLICHER SPANNUNGSVERTEILUNGEN IN STAUDAMMEN NACH DER MATRIZENVERSCHIEBUNGSMETHODE	67-0003
RAY-EFFECT	RAY-EFFECT MITIGATION IN DISCRETE ORDINATE-LIKE ANGULAR FINITE-ELEMENT APPROXIMATIONS IN NEUTRON-TRA	75-0496
RAYLEIGH	FINITE ELEMENT ANALOGUE OF THE MODIFIED RAYLEIGH RITZ METHOD FOR VIBRATION PROBLEMS	73-0181
RAYLEIGH	RAYLEIGH RITZ GALERKIN METHODS FOR SYMMETRIC POSITIVE DIFFERENTIAL EQUATIONS APPLICATION TO FINITE E	73-1030
RAYLEIGH	RAYLEIGH WAVES AT A CONTINENTAL BOUNDARY BY FINITE ELEMENT METHOD	72-0016
RAYLEIGH-RITZ	A FINITE ELEMENT ANALOGUE OF THE MODIFIED RAYLEIGH-RITZ METHOD FOR VIBRATION PROBLEMS	72-0478
RAYLEIGH-RITZ	A FINITE ELEMENT APPLICATION OF THE MODIFIED RAYLEIGH-RITZ METHOD	70-0091
RAYLEIGH-RITZ	AN ANALYSIS OF THE LARGE DEFLECTIONS OF BEAMS USING THE RAYLEIGH-RITZ FINITE ELEMENT METHOD	68-0126
RAYLEIGH-RITZ	DAS VERFAHREN VON RAYLEIGH-RITZ MIT HERMITESCHEN INTERPOLATIONSPOLYNOMEN	63-0018
RAYLEIGH-RITZ	EINE EINFACHE KONSTRUKTION VON KOORDINATENFUNKTIONEN FUR DIE NUMERISCHE LOSUNG ZWEIDIMENSIONALER RAN	66-0134
RAYLEIGH-RITZ	RAYLEIGH-RITZ AND GALERKIN FINITE ELEMENTS FOR DIFFUSION-CONVECTION PROBLEMS	73-0017
RAYLEIGH-RITZ	RAYLEIGH-RITZ APPROXIMATION OF PIECEWISE CUBIC POLYNOMIALS	66-0087
RAYLEIGH-RITZ'S	A METHOD OF SOLUTION FOR THE COMPLICATED BUCKLING PROBLEMS OF ELASTIC PLATES WITH COMBINED USE OF RA	68-0097
RAZRABOTKI	RASCHET NAPRYAZHENIL L PEREMESHCHENII V SVODE PODZEMNOGO MASHINNOGO ZALA GES I V OKRUZHAYUSHCHEM SKA	74-0542
RAZVITIE	THE DEVELOPMENT OF APPLIED METHODS IN PROBLEMS OF STATIC CALCULATION OF THIN-WALLED THREE-DIMENSIONA	73-0934
RE-ENTRY	MODAL SENSITIVITY STUDY OF A CONICAL RE-ENTRY VEHICLE SUBJECTED TO AERODYNAMICALLY INDUCED ACOUSTIC	69-0431
RE-ENTRY	RE-ENTRY THERMAL ANALYSIS OF VARIABLE THICKNESS SPHERICAL VEHICLES.	72-0468
RE-ENTRY	THE THERMO-ELASTO-PLASTIC ANALYSIS OF A RE-ENTRY BODY TYPE COMPOSITE SOLID STRUCTURE USING A NEW FIN	66-0093
RE-ENTRY	THERMAL STRESS ANALYSIS OF RE-ENTRY VEHICLE NOSETIPS AT ANGLE OF ATTACK	72-0852
REACTEUR	ANALYSE TRIDIMENSIONALE DU COMPORTEMENT NON-LINEAIRE D'UN CAISSON DE REACTEUR NUCLEAIRE EN BETON PRE	71-0144
REACTEUR	DETERMINATION DU CHAMP DES CONTRAINTES DANS LE CAISSON DU REACTEUR DE BUGEY L'	67-0121
REACTEUR	NONLINEAR ANALYSIS OF PRESTRESSED CONCRETE FAST REACTOR PRESSURE VESSEL, (ANALYSE NON LINEAIRE DES C	70-0145
REACTEUR	TENUE DES STRUCTURES DE GRAPHITE VIS-A-VIS DES SOLLICITATIONS RENCONTREES DANS UN COEUR DE REACTEUR	73-0817
REACTION	A FINITE ELEMENT ANALYSIS OF DISSOLVED OXYGEN DRAWDOWN AND SULFATE PRODUCTION IN STRIP MINE SPOIL DA	75-0795
REACTION	ANALYTICAL AND EXPERIMENTAL EVALUATION OF MODULUS OF SOIL REACTION	74-1168
REACTOR	A FRACTURE MECHANICS APPROACH TO THE PROPAGATION OF FATIGUE CRACKS IN REACTOR COMPONENTS BY ACOUSTIC	71-0132
REACTOR	A MATRIX DISPLACEMENT METHOD ON PRE AND POST BUCKLING ANALYSIS OF LINERS FOR REACTOR VESSELS	71-0267
REACTOR	A SEISMIC DESIGN OF NUCLEAR REACTOR BUILDING, STRESS ANALYSIS AND STIFFNESS EVALUATION OF THE ENTIRE	71-0251
REACTOR	A THREE-DIMENSIONAL FINITE ELEMENT ANALYSIS OF A PRESTRESSED CONCRETE REACTOR VESSEL MODEL	69-0084

REACTOR	ANALYSIS OF AN IN-PILE REACTOR TUBE	75-0193
REACTOR	ANALYSIS OF MODERATELY THICK FLANGED REACTOR PRESSURE VESSELS	73-0821
REACTOR	ANALYSIS OF MODERATELY THICK FLANGED REACTOR VESSELS	75-0320
REACTOR	APPLICATION OF FINITE ELEMENT METHOD TO BOILING WATER REACTOR POWER PLANT (JAPANESE)	72-0816
REACTOR	APPLICATION OF THE FINITE ELEMENT METHOD TO THE CALCULATION OF REACTOR PRESSURE VESSELS	71-0645
REACTOR	APPLICATION OF THE FINITE ELEMENT METHOD FOR THE SAFETY EVALUATION OF REACTOR COMPONENTS	73-0853
REACTOR	APPLICATION OF THE FINITE ELEMENT METHOD TO THE NONLINEAR ANALYSIS OF NUCLEAR REACTOR FUEL BEHAVIOR	74-1246
REACTOR	APPLICATION OF THREE-DIMENSIONAL FINITE ELEMENTS FOR ANALYSIS OF IRRADIATION INDUCED STRESSES IN NUC	72-0078
REACTOR	APPLICATIONS OF FINITE ELEMENT METHODS IN REACTOR MATHEMATICS - NUMERICAL SOLUTION OF THE NEUTRON DI	72-0535
REACTOR	APPLICATIONS OF FINITE ELEMENT METHODS IN REACTOR MATHEMATICS. NUMERICAL SOLUTION OF THE NEUTRON TRA	74-0853
REACTOR	CIRCUMFERENTIAL PRESTRESSING LOAD ANALYSIS DUE TO WIRE/STRAND WINDING SYSTEMS ON PRESTRESSED CONCRET	74-1202
REACTOR	COMPUTATIONAL ANALYSIS AND EVALUATION OF THE FINITE ELEMENT METHOD FOR A CLASS OF NUCLEAR REACTOR CO	74-0595
REACTOR	COMPUTATIONAL METHODS IN NUCLEAR REACTOR STRUCTURAL DESIGN FOR HIGH TEMPERATURE APPLICATIONS: AN INT	73-1001
REACTOR	CREEP DATA ACQUISITION AND APPLICATION TO REACTOR COMPONENT DESIGN	72-0113
REACTOR	DESIGN OF A COMPOSITE STEEL AND CONCRETE CLOSURE FOR A PRESTRESSED CONCRETE REACTOR VESSEL.	72-0187
REACTOR	DISCUSSION ON FINITE ELEMENT ANALYSIS OF PRESTRESSED CONCRETE REACTOR VESSELS	72-0648
REACTOR	EINFACHES NAEHERUNGSVERFAHREN FUR DIE GRENZTRAGFAEHIGKEITSANLYSE VON SPANNBETON-REAKTORDRUCKBEHAELTE	73-0854
REACTOR	FAST REACTOR PROGRAMME. FINITE ELEMENT TEMPERATURE ANALYSIS IN FUEL PINS	73-0939
REACTOR	FINITE ELEMENT ANALYSIS OF STRUCTURAL INTEGRITY OF A REACTOR PRESSURE VESSEL DURING EMERGENCY CORE C	70-0026
REACTOR	FINITE ELEMENT ANALYSIS OF STRESSES IN FUEL ELEMENTS OF A HIGH-TEMPERATURE GAS-COOLED REACTOR	71-0115
REACTOR	FINITE ELEMENT APPROACH TO PRESSURE WAVE ATTENUATION BY REACTOR FUEL SUBASSEMBLIES	75-0522
REACTOR	FINITE ELEMENT MATRIX DISPLACEMENT METHOD FOR TIME DEPENDENT STRESS ANALYSIS OF REACTOR MATERIALS IN	71-0674
REACTOR	FINITE ELEMENT METHOD APPLIED TO REACTOR PHYSICS PROBLEMS	71-0483
REACTOR	FINITE ELEMENT METHODS IN REACTOR DESIGN CALCULATIONS	71-0571
REACTOR	FINITE ELEMENT METHODS FOR SPACE-TIME REACTOR ANALYSIS	71-0673
REACTOR	FINITE ELEMENT METHODS IN ANALYSIS OF REACTOR VESSELS	72-0271
REACTOR	FINITE ELEMENT METHODS FOR REACTOR ANALYSIS	73-0108
REACTOR	FINITE ELEMENT METHODS IN REACTOR PHYSICS ANALYSIS	75-0020
REACTOR	FINITE ELEMENT SOLUTIONS OF A NONLINEAR REACTOR DYNAMICS PROBLEM	74-1268
REACTOR	FINITE ELEMENT STRESS ANALYSIS OF INTERACTING FUEL PELLET AND CANNING UNDER FAST REACTOR CONDITION	73-0814
REACTOR	FINITE ELEMENT STUDY OF PRESSURE WAVE ATTENUATION BY REACTOR FUEL SUBASSEMBLIES	75-0215
REACTOR	FINITE ELEMENTS FOR CALCULATION OF PRESTRESSED CONCRETE REACTOR PRESSURE VESSELS (FINITE ELEMENTE ZU	73-0797
REACTOR	FINITE ELEMENTS METHOD IN DESIGN OF PRESTRESSED REACTOR CONCRETE CONSTRUCTION	73-0550
REACTOR	FINITE-ELEMENT ELASTIC-PLASTIC ANALYSIS OF RESIDUAL STRESSES DUE TO CLAD WELDING IN REACTOR VESSELS	75-0779
REACTOR	FINITE-ELEMENT METHODS FOR REACTOR ANALYSIS - COMMENT	75-0009
REACTOR	FUEL ROD FINITE ELEMENT STRESS ANALYSIS: SEGMENTED FLIGHT REACTOR FUEL VERSUS SBDR FUEL	71-0798
REACTOR	INTERNAL SUPPORTING STRUCTURES OF PRESSURE TUBES TYPE REACTOR CORE. ANALYTICAL MODEL FOR STATIC ANAL	71-0049
REACTOR	INTERNAL SUPPORTING STRUCTURES OF THE (CIRENE) REACTOR CORE - RESULTS OF STATIC STRESS ANALYSIS FOR	73-0818
REACTOR	MATHEMATICAL AND COMPUTATIONAL PROBLEMS IN REACTOR CALCULATIONS	73-0806
REACTOR	MATHEMATICAL AND COMPUTATIONAL PROBLEMS IN REACTOR CALCULATIONS	73-0973
REACTOR	MATRIX DISPLACEMENT METHODS IN FRACTURE MECHANICS ANALYSIS OF REACTOR VESSELS	71-0226
REACTOR	NONLINEAR ANALYSIS OF PRESTRESSED CONCRETE FAST REACTOR PRESSURE VESSEL. (ANALYSE NON LINEAIRE DES C	70-0145
REACTOR	NONLINEAR FINITE-ELEMENT METHOD FOR PLATES AND ITS APPLICATION TO DYNAMIC RESPONSE OF REACTOR FUEL S	74-0670
REACTOR	ON THE APPLICATION OF THE FINITE ELEMENT METHOD IN REACTOR PHYSICS	75-0122
REACTOR	ON THE FINITE ELEMENT ANALYSIS OF THIN-WALLED STRUCTURES WITH APPLICATION TO STRUCTURAL PROBLEMS IN	73-0832
REACTOR	PRESTRESSED CONCRETE REACTOR VESSEL ANALYTICAL METHODS DEVELOPMENTS	68-0141
REACTOR	PRESTRESSED CONCRETE REACTOR PRESSURE VESSEL FAILURE ANALYSIS. THE INCORPORATION OF STATISTICAL METH	71-0120
REACTOR	REACTOR COOLANT PUMP FLYWHEELS OVERSPEED EVALUATION	74-0506
REACTOR	RECENT DEVELOPMENTS IN FINITE ELEMENT ANALYSIS OF PRESTRESSED CONCRETE REACTOR VESSELS	74-0469
REACTOR	RECENT DEVELOPMENTS IN THE FINITE ELEMENT ANALYSIS OF PRE-STRESSED CONCRETE REACTOR VESSELS	74-0582
REACTOR	RESPONSE OF REACTOR CORE SUBASSEMBLIES TO IMPULSIVE LOADINGS	73-0076
REACTOR	REVIEW OF CURRENT PROBLEMS FOR MULTIDIMENSIONAL REACTOR STATICS CALCULATIONS	73-0960
REACTOR	RIDDLE OF BONDED AND UNBONDED TENDONS IN PRESTRESSED CONCRETE REACTOR VESSELS	74-0519
REACTOR	SEISMIC EFFECTS ON A PROPOSED UNDERGROUND REACTOR FACILITY	73-0953
REACTOR	SPACE-TIME SOLUTIONS OF NONLINEAR REACTOR DYNAMICS BY FINITE-ELEMENT METHOD	75-0503
REACTOR	STARS - A COMPUTER CODE FOR STRUCTURAL DYNAMIC ANALYSIS OF REACTOR SYSTEM	71-0204
REACTOR	STRAP - A COMPUTER CODE FOR THE STATIC AND DYNAMIC ANALYSIS OF REACTOR STRUCTURAL SYSTEMS	71-0078
REACTOR	TENUE DES STRUCTURES DE GRAPHITE VIS-A-VIS DES SOLLICITATIONS RENCONTREES DANS UN COEUR DE REACTEUR	73-0817
REACTOR	THERMAL STRESSES IN FACT REACTOR BOILER TUBES AT THE SODIUM/GAS INTERFACE	73-0820
REACTOR	THREE DIMENSIONAL ELASTIC AND ELASTO-PLASTIC ANALYSIS OF REACTOR PRESSURE VESSELS	69-0453
REACTOR	THREE-DIMENSIONAL ANALYSIS OF REACTOR PRESSURE VESSEL NOZZLES	71-0221
REACTOR	TWO-DIMENSIONAL TIME-DEPENDENT ANALYSIS OF PERFORATED END CAPS FOR NUCLEAR REACTOR PRESSURE VESSELS	71-0239
REACTOR	USE OF SMALL MODELS IN DESIGN AND ANALYSIS OF PRESTRESSED-CONCRETE REACTOR VESSELS	70-0433
REACTOR	USE OF SMALL MODELS IN DESIGN AND ANALYSIS OF PRESTRESSED-CONCRETE REACTOR VESSELS	71-0426
REACTOR-PHYSICS	FINITE-ELEMENT METHODS IN REACTOR-PHYSICS ANALYSIS	75-0502
REACTOR-PHYSICS	2-DIMENSIONAL AND 3-DIMENSIONAL REACTOR-PHYSICS CALCULATIONS WITH FINITE-ELEMENT METHOD	75-0423
REACTORS	SOME MODELS FOR THE ANALYSIS OF STRESSES IN A TUBULAR FUEL PIN FOR HIGH TEMPERATURE REACTORS BY FINI	73-0816
REACTORS	STRESS DISTRIBUTION IN THE LIGAMENT ZONE OF PERFORATED CIRCULAR DISCS WITH SPECIAL REFERENCE TO THE	73-0827
REACTORS	THIN-WALLED PRESTRESSED CONCRETE PRESSURE VESSEL FOR HIGH TEMPERATURE REACTORS: EXPERIMENTAL INVESTI	73-0857
REACTORS	THREE-DIMENSIONAL MECHANICAL STRESSES IN TOROIDAL MAGNETS FOR CONTROLLED THERMONUCLEAR REACTORS	73-0669
REACTORS	TRANSIENT THERMAL STRESS TECHNIQUES IN NUCLEAR REACTORS. APPLIED MECHANICS ASPECTS OF NUCLEAR EFFECT	71-0474
REAKTORDRUCKBEH	ANWENDUNG DER FINITE-ELEMENT-METHODE BEI DER BERECHNUNG VON REAKTORDRUCKBEHALTERN	71-0036
REAL	ANALYSIS OF REAL STRUCTURES FOR CREEP, PLASTICITY AND OTHER COMPLEX CONSTITUTIVE LAWS.	69-0287
REALISTIC	BERECHNUNG VON STAHLBETONSCHEIBEN IM ZUSTAND II BEI ANNAHME EINES WIRKLICHKEITSNAHEN WERKSTOFFVERHAL	74-0978
REALIZATION	REALIZATSIYA METODA KONECHNYKH ELEMENTOV NA EVM DLYA RASCHETA DVUMERNYKH ELEKTRICHESKIKH I MAGNITNYK	74-0251
REALIZATSIYA	REALIZATSIYA METODA KONECHNYKH ELEMENTOV NA EVM DLYA RASCHETA DVUMERNYKH ELEKTRICHESKIKH I MAGNITNYK	74-0251
REALS	THE RING OF FINITE ELEMENTS IN A NON-STANDARD MODEL OF THE REALS	71-0450
REBO	REBO: COMPUTER CODE FOR THE GENERATION OF A THREE-DIMENSIONAL FINITE-ELEMENT GRID OF THE NOZZLE-CYLI	74-1310
REBOUND	FINITE-ELEMENT MODEL FOR POSSIBLE ISOSTATIC REBOUND IN GRAND CANYON	75-0037
REBRISTYKH	ISSLEDOVANIE NAPRYAZHENNOGO SOSTOYANIYA REBRISTYKH ISILINDRICHESKIKH OBOLOCHEK S PRYAMOUGOL'NYMI OTV	74-0944
RECESSES	ZUR BERECHNUNG, BEMESSUNG UND BEWEHRUNG RECHTECKIGER PLATTEN KONSTNATER DICKE MIT OEFFNUNGEN ODER AU	73-0664
RECHENVERFAHREN	EINSATZ MODERNER RECHENVERFAHREN ZUR ERMITTLUNG DER SPANNUNGEN UND VERFORMUNGEN BEI RADSAETZEN. (USE	75-0186
RECHENVERFAHREN	MESSUERFAHREN RECHENVERFAHREN ZUR ERMITTLUNG DREIACHSIGER EIGENSPANNUNGSZUSTAENDE	73-0507
RECHERCHES	TENDANCES ACTUELLES DES RECHERCHES SUR LES CONTRAINTES ET DEFORMATIONS DUES AU BRIDAGE EN RELATION A	74-0273
RECHNERISCHEN	NEUESTE ERFAHRUNGEN BEI DER RECHNERISCHEN UND KONSTRUKTIVEN AUSLEGUNG KOMPLIZIERTER SCHALENSYSTEME	71-0052
RECHTECKIGER	ZUR BERECHNUNG, BEMESSUNG UND BEWEHRUNG RECHTECKIGER PLATTEN KONSTNATER DICKE MIT OEFFNUNGEN ODER AU	73-0664
RECHTECKPLATTEN	DIE BEULLASTEN DIAGONAL VERSTEIFTER RECHTECKPLATTEN (BUCKLING LOADS IN DIAGONALLY STIFFENED RECTANGU	73-0405
RECONSTRUCTION	FINITE ELEMENT STUDIES OF FEMORAL ENDOPROSTHESES FOR HIP RECONSTRUCTION	75-0381
RECORD	RECORD CONCRETE SHELL ROOF FEATURED IN COLLISEUM	72-0218
RECORDING	AN AUTOMATIC RECORDING SCHEME FOR SIMULTANEOUS EQUATIONS DERIVED FROM NETWORK SYSTEMS	70-0285
RECTANGLE	HERMITEAN RECTANGLE FINITE ELEMENT FAMILY	73-1009
RECTANGULAR	A FINITE ELEMENT ANALYSIS FOR ' MODERATELY THICK ' RECTANGULAR PLATES IN BENDING	68-0146
RECTANGULAR	A FINITE ELEMENT SOLUTION OF GENERAL FLUID DYNAMICS PROBLEMS - NATURAL CONVECTION IN RECTANGULAR CAV	70-0008
RECTANGULAR	A HIGHER ORDER CONFORMING RECTANGULAR ELEMENT	73-0690
RECTANGULAR	A RECTANGULAR PLATE BENDING ELEMENTS THE USE OF WHICH IS EQUIVALENT TO THE USE OF THE FINITE DIFFERE	71-0398
RECTANGULAR	AN APPLICATION OF THE FINITE ELEMENT METHOD TO THE CALCULATION OF CYLINDRICAL SHELLS WITH RECTANGULA	74-1236
RECTANGULAR	AN INVESTIGATION OF THE EFFECTS OF SQUARE CUTOUTS ON THE NATURAL FREQUENCIES AND MODE SHAPES OF RECT	71-0638
RECTANGULAR	ANALYSIS OF ASYMMETRIC STIFFENED SHELL TYPE STRUCTURES BY THE FINITE ELEMENT METHOD: I FLAT RECTANGU	68-0283
RECTANGULAR	ANALYSIS OF CORNER SUPPORTED RECTANGULAR SLABS	70-0046
RECTANGULAR	ANALYSIS OF LONG RECTANGULAR TANKS	70-0276
RECTANGULAR	ANALYSIS OF RECTANGULAR TANKS BY FINITE ELEMENT TECHNIQUES	67-0044
RECTANGULAR	ANALYSIS OF RECTANGULAR PLATES IN TRANSIENT CREEP BENDING	69-0425

RECTANGULAR	APPLICATION OF THE DISCRETE ELEMENT METHOD TO THE BUCKLING ANALYSIS OF RECTANGULAR PLATES UNDER ARBI	68-0054
RECTANGULAR	APPLICATION OF THE FINITE ELEMENT METHOD TO THE BUCKLING OF RECTANGULAR PLATES AND PLATES ON ELASTIC	73-0478
RECTANGULAR	ARBITRARY SYMMETRIC LOADING PROBLEMS OF CENTRALLY CRACKED RECTANGULAR PLATES	75-0318
RECTANGULAR	CALCULATION OF RECTANGULAR PLATES BY THE FINITE ELEMENT METHOD	69-0506
RECTANGULAR	COMMENT ON A HIGHER ORDER CONFORMING RECTANGULAR PLATE ELEMENT	73-0999
RECTANGULAR	CONFORMING RECTANGULAR AND TRIANGULAR PLATE BENDING ELEMENTS	71-0720
RECTANGULAR	CREEPING FLOW IN RECTANGULAR DUCTS BY FINITE ELEMENT METHOD	75-0569
RECTANGULAR	DIE BEULLASTEN DIAGONAL VERSTEIFTER RECHTECKPLATTEN (BUCKLING LOADS IN DIAGONALLY STIFFENED RECTANGU	73-0405
RECTANGULAR	DISPLACEMENT METHOD IN STRUCTURAL ANALYSIS AS APPLIED TO RECTANGULAR SHELL BEAMS	72-0411
RECTANGULAR	DYNAMIC PLASTIC LARGE DEFLECTIONS OF CLAMPED RECTANGULAR PLATE	74-0392
RECTANGULAR	EFFECT OF ELASTIC MODULAR RATIO AND ANISOTROPY ON THE STRESS FIELD IN A TWO-DIMENSIONAL MATRIX INTO	71-0647
RECTANGULAR	EFFECTIVENESS OF EXCESSIVE NODAL CONTINUITIES IN THE FINITE ELEMENT ANALYSIS OF THIN RECTANGULAR AND	70-0346
RECTANGULAR	EFFICIENT FINITE ELEMENT ANALYSIS OF RECTANGULAR AND SKEW LAMINATED PLATES	73-0436
RECTANGULAR	ELASTIC PLASTIC BUCKLING OF STIFFENED RECTANGULAR PLATES	69-0122
RECTANGULAR	EQUIVALENT MASS MATRICES FOR RECTANGULAR PLATES IN BENDING	66-0063
RECTANGULAR	FINITE ELEMENT ANALYSIS OF CYLINDRICAL SHELLS. SOLUTION USING RECTANGULAR CIRCULAR CYLINDRICAL ELEME	70-0178
RECTANGULAR	FINITE ELEMENT ANALYSIS OF ALTERNATIVE AXIAL LOADING OF AN ELASTIC PLATE PRESSED BETWEEN TWO ELASTIC	71-0357
RECTANGULAR	FINITE ELEMENT DISPLACEMENT ANALYSIS OF PLATE BENDING BASED ON RECTANGULAR ELEMENTS	66-0032
RECTANGULAR	FLEXURAL VIBRATIONS OF RECTANGULAR AND OTHER POLYGONAL PLATES	71-0417
RECTANGULAR	FREE VIBRATION OF DOUBLY CURVED RECTANGULAR PLATES INCLUDING THE PRESENCE OF MEMBRANE STRESSES WITH	69-0468
RECTANGULAR	FREE VIBRATION OF ORTHOTROPIC VARIABLE THICKNESS, CONTINUOUS RECTANGULAR PLATES	70-0270
RECTANGULAR	FREE VIBRATION OF THICK, LAYERED RECTANGULAR PLATES BY FINITE LAYER METHOD	72-0440
RECTANGULAR	HIGH ORDER RECTANGULAR SHALLOW SHELL FINITE ELEMENT	73-0279
RECTANGULAR	HYPAR DISPLACEMENT FIELDS FOR RECTANGULAR FINITE ELEMENTS	73-0240
RECTANGULAR	INVESTIGATION OF FLEXURAL STIFFNESS OF RECTANGULAR PLATES BY THE FINITE ELEMENT METHOD	73-0991
RECTANGULAR	ISSLEDOVANIE NAPRYAZHENNOGO SOSTOYANIYA REBRISTYKH ISILINDRICHESKIKH OBOLOCHEK S PRYAMOUGOL'NYMI OTV	74-0944
RECTANGULAR	LOCAL MESH REFINEMENT USING RECTANGULAR BLENDED FINITE-ELEMENTS	75-0452
RECTANGULAR	MATHEMATICAL MODELS OF THE MASSACHUSETTS BAY. PART 1. FINITE ELEMENT MODELING OF TWO-DIMENSIONAL HYD	73-0800
RECTANGULAR	METOD KONECHNYKH ELEMENTOV V PRIMENENII K PASCHETU TSILINDRICHESKIKH OBOLOCHEK S PRYAMOUGOL'NYMI OTV	73-0536
RECTANGULAR	NEW RECTANGULAR FINITE ELEMENT FOR SHEAR WALL ANALYSIS	69-0182
RECTANGULAR	NEW RECTANGULAR FINITE ELEMENT FOR SHEAR WALL ANALYSIS	69-0362
RECTANGULAR	NOTE ON RECTANGULAR FINITE ELEMENTS WITH IN-PLANE FORCES	75-0080
RECTANGULAR	ON ASSUMED DISPLACEMENTS FOR THE RECTANGULAR PLATE BENDING ELEMENT	67-0020
RECTANGULAR	ON COMPUTATIONS INVOLVING STIFFNESS MATRICES STORED IN RECTANGULAR FORM	75-0576
RECTANGULAR	ON THE USE OF RECTANGULAR FINITE ELEMENTS	67-0162
RECTANGULAR	ON VIBRATION ANALYSIS OF RECTANGULAR CANTILEVER PLATES WITH A STEPPED CROSS-SECTION BY THE FINITE EL	71-0608
RECTANGULAR	RECTANGULAR AND TRIANGULAR FINITE ELEMENTS FOR POISSON'S EQUATION	75-0802
RECTANGULAR	RECTANGULAR FINITE ELEMENTS FOR ANALYSIS OF PLATE VIBRATIONS	68-0113
RECTANGULAR	RECTANGULAR FINITE ELEMENT FOR PLATE BENDING ANALYSIS BASED ON HELLINGER-REISSNERS VARIATIONAL PRINC	71-0545
RECTANGULAR	RESPONSE OF RECTANGULAR PLATES TO MOVING LOADS BY A FINITE ELEMENT PROCEDURE	69-0515
RECTANGULAR	RESPONSE OF RECTANGULAR ORTHOTROPIC PLATES TO RANDOM EXCITATIONS	73-0496
RECTANGULAR	SARD KERNEL THEOREMS ON TRIANGULAR AND RECTANGULAR DOMAINS WITH EXTENSIONS AND APPLICATIONS TO FINIT	72-0361
RECTANGULAR	SHEAR LAG ANALYSIS OF RECTANGULAR FULL-WIDTH TUBE JUNCTIONS	72-0087
RECTANGULAR	SOME APPROXIMATIONS IN THE NONLINEAR ANALYSIS OF RECTANGULAR PLATES USING FINITE ELEMENTS	74-1231
RECTANGULAR	SOME CONFORMING, RECTANGULAR, PLATE ELEMENTS	75-0094
RECTANGULAR	STABILITY OF MODERATELY THICK RECTANGULAR PLATES USING A HIGH PRECISION TRIANGULAR FINITE ELEMENT	75-0334
RECTANGULAR	STATIC AND DYNAMIC BEHAVIOUR OF RECTANGULAR PLATES USING HIGHER ORDER FINITE STRIPS	72-0458
RECTANGULAR	STATIC FLEXIBILITY MATRICES FOR A RECTANGULAR PLATE ELEMENT IN BENDING, TWISTING, AND SHEAR. UNIFORM	68-0217
RECTANGULAR	STIFFNESS MATRIX FOR BENDING OF A RECTANGULAR PLATE ELEMENT WITH INITIAL MEMBRANE STRESSES	62-0006
RECTANGULAR	STIFFNESS MATRIX FOR SHALLOW RECTANGULAR SHELL ELEMENT	67-0014
RECTANGULAR	STIFFNESS MATRIX FOR SHALLOW RECTANGULAR SHELL ELEMENT	68-0074
RECTANGULAR	STRESS ANALYSIS OF PRESSURIZED CYLINDRICAL BORE IN RECTANGULAR BLOCK	69-0312
RECTANGULAR	THE EFFECTIVENESS OF EXCESSIVE NODAL CONTINUITIES IN THE FINITE ELEMENT ANALYSIS OF THIN RECTANGULAR	70-0105
RECTANGULAR	THE RECTANGULAR PLANE STRESS ELEMENT BY TURNER, PIAN AND WILSON	74-1265
RECTANGULAR	THERMO-ELASTIC ANALYSIS OF RECTANGULAR PLATES IN PLANE STRESS BY THE FINITE-ELEMENT DISPLACEMENT MET	67-0087
RECTANGULAR	TRIGONOMETRIC FUNCTION REPRESENTATIONS FOR RECTANGULAR PLATE BENDING ELEMENTS	71-0765
RECTANGULAR	USE OF RECTANGULAR FINITE ELEMENTS	73-0104
RECTANGULAR	VARIATIONAL FINITE ELEMENT CALCULATION OF THE ACOUSTIC RESPONSE OF A RECTANGULAR PANEL	71-0100
RECTANGULAR	VIBRATION OF RECTANGULAR PLATES WITH MIXED BOUNDARY CONDITIONS	73-1086
RECTANGULAR	VIBROACOUSTIC RESPONSE OF TURBULENCE EXCITED THIN RECTANGULAR FINITE PLATES IN HEAVY AND FLUID MEDIA	74-1309
RECTANGULAR	ZUR BERECHNUNG, BEMESSUNG UND BEWEHRUNG RECHTECKIGER PLATTEN KONSTNATER DICKE MIT OEFFNUNGEN ODER AU	73-0664
RECYCLED	GUYAN REDUCTION SOLUTIONS RECYCLED FOR IMPROVED ACCURACY	71-0159
REDESIGN	ANALYSIS AND REDESIGN OF SONAR PIEZOELECTRIC RESONATORS USING THE FINITE ELEMENT METHOD.	72-0374
REDISTRIBUTION	NOTE ON STRESS AND STRAIN REDISTRIBUTION IN A NOTCHED PLATE SPECIMEN DURING CYCLIC LOADING	69-0194
REDUCTION	ALGORITHM FOR MATRIX BANDWIDTH REDUCTION	72-0895
REDUCTION	BANDWIDTH REDUCTION BY AUTOMATIC RENUMBERING	73-0993
REDUCTION	FINITE ELEMENT APPLICATION TO STRAIN GAGE DATA REDUCTION	73-0561
REDUCTION	GUYAN REDUCTION SOLUTIONS RECYCLED FOR IMPROVED ACCURACY	71-0159
REDUCTION	HIGH STRENGTH MATERIALS AND VEHICLE WEIGHT REDUCTION ANALYSIS	75-0157
REDUCTION	ON THE REDUCTION OF NUMERICAL ERROR IN THE MATRIX DISPLACEMENT METHOD	74-0417
REDUCTION	REDUCTION GEAR DAMAGES RELATED TO EXTERNAL INFLUENCES	75-0307
REDUCTION	REDUCTION OF STIFFNESS AND MASS MATRICES	65-0037
REDUCTION	REDUCTION OF STRESSES IN BURIED STRUCTURES	71-0698
REDUCTION	REDUCTION OF THE DEGREES OF FREEDOM IN SOLVING DYNAMIC PROBLEMS BY THE FINITE ELEMENT METHOD	71-0647
REDUCTION	STRUCTURAL BEHAVIOR OF A TWO SPAN REINFORCED CONCRETE BOX GIRDER BRIDGE MODEL. VOLUME 11. REDUCTION,	71-0659
REDUCTION	THE BANDIT COMPUTER PROGRAM FOR THE REDUCTION OF MATRIX BANDWIDTH FOR NASTRAN	72-0481
REDUCTION	USE OF THE FINITE ELEMENT DAMPED FORCED RESPONSE STRAIN ENERGY DISTRIBUTION FOR VIBRATION REDUCTION	72-0759
REDUCTION	VIBRATION REDUCTION BY USING BOTH THE FINITE ELEMENT STRAIN ENERGY DISTRIBUTION AND MOBILITY TECHNIQ	74-1303
REDUNDANCY	OPTIMIZATION OF REDUNDANCY SELECTION IN FINITE-ELEMENT FORCE METHOD	70-0228
REDUNDANT	A METHOD FOR REDUCING THE ANALYSIS OF COMPLEX REDUNDANT STRUCTURES TO A ROUTINE PROCEDURE	52-0002
REDUNDANT	REDUNDANT STRUCTURE CONCEPTS IN FINITE ELEMENT ANALYSES	75-0338
REDUNDANT	REDUNDANT STRUCTURE CONCEPTS IN FINITE ELEMENT ANALYSES	75-0626
REENTRY	THERMAL STRESS ANALYSIS OF REENTRY VEHICLE NOSETIPS AT ANGLE OF ATTACK	72-0456
REFERENCES	775 SELECTED REFERENCES ON THE FINITE ELEMENT METHOD AND MATRIX METHODS OF STRUCTURAL ANALYSIS	69-0241
REFINED	A REFINED AXISYMMETRIC FINITE ELEMENT FOR THE ANALYSIS OF NEARLY INCOMPRESSIBLE SOLIDS	70-0309
REFINED	A REFINED CURVED CYLINDRICAL SHELL FINITE ELEMENT	68-0038
REFINED	A REFINED CURVED ELEMENT FOR THIN SHELLS OF REVOLUTION	71-0214
REFINED	A REFINED FINITE ELEMENT ANALYSIS OF THIN SHELL STRUCTURES	67-0101
REFINED	A REFINED FINITE ELEMENT ANALYSIS OF THIN SHELL STRUCTURES INCLUDING DYNAMIC LOADINGS	67-0157
REFINED	A REFINED FINITE ELEMENT ANALYSIS OF FOLDED PLATE STRUCTURES	71-0671
REFINED	A REFINED MIXED-TYPE PLATE BENDING ELEMENT	69-0267
REFINED	A REFINED QUADRILATERAL ELEMENT FOR ANALYSIS OF PLATE BENDING	68-0047
REFINED	A REFINED TRIANGULAR PLATE BENDING FINITE ELEMENT	69-0056
REFINED	AN ASSUMED STRESS APPROACH TO REFINED ISOPARAMETRIC ELEMENTS IN THREE DIMENSIONS	74-0421
REFINED	APPLICATION OF A REFINED SHALLOW SHELL FINITE ELEMENT TO GEOMETRICALLY NONLINEAR STRUCTURES	73-0358
REFINED	APPLICATION OF A REFINED PLATE BENDING ELEMENT TO BUCKLING PROBLEMS	74-0937
REFINED	ON THE DEVELOPMENT OF A REFINED FINITE-ELEMENT FREE FROM DISCRETIZATION - DISPERSION FOR ELASTIC WAV	73-1148
REFINED	PANEL FLUTTER OPTIMIZATION - REFINED FINITE-ELEMENT APPROACH	75-0358
REFINED	REFINED CRACKED CONCRETE ANALYSIS OF CONCRETE CONTAINMENT STRUCTURES SUBJECT TO OPERATIONAL AND ENVI	74-0943
REFINED	REFINED FINITE ELEMENT ANALYSIS OF LINEAR AND NONLINEAR TWO-DIMENSIONAL STRUCTURES	66-0024
REFINED	REFINED FINITE ELEMENTS IN 2-DIMENSIONAL ELASTICITY	69-0320
REFINED	REFINED FINITE ELEMENT ANALYSIS OF ELASTIC-PLASTIC THIN SHELLS OF REVOLUTION	69-0497

345

REFINED	REFINED FINITE ELEMENTS FOR FOLDED PLATE STRUCTURES	72-0574
REFINED	REFINED FINITE ELEMENT STABILITY ANALYSIS OF THIN-WALLED MEMBERS	75-0784
REFINED	REFINED HEAT CONDUCTION FINITE ELEMENT FOR PERMAFROST ANALYSIS	73-0594
REFINED	REFINED MIXED METHOD FINITE ELEMENTS FOR SHELLS OF REVOLUTION	71-0101
REFINED	REFINED PLATE BENDING FINITE ELEMENT	74-1045
REFINED	REINVESTIGATION OF BUCKLING OF SHELLS OF REVOLUTION BY A REFINED FINITE ELEMENT	74-0097
REFINED	TWO-DIMENSIONAL AIRFOIL ANALYSIS USING A REFINED FINITE ELEMENT TECHNIQUE	73-0587
REFINED	VIBRATION OF SQUARE PLATES HAVING VARIOUS EDGE CONDITIONS AND LOADINGS USING A REFINED FINITE ELEMEN	75-0024
REFINEMENT	AUTOMATED MESH REFINEMENT STRATEGIES FOR FINITE ELEMENT COMPUTATIONS	75-0877
REFINEMENT	LOCAL MESH REFINEMENT WITH FINITE ELEMENTS FOR ELLIPTIC PROBLEMS	73-0073
REFINEMENT	LOCAL MESH REFINEMENT USING RECTANGULAR BLENDED FINITE-ELEMENTS	75-0452
REFINEMENTS	SELF-ADAPTIVE REFINEMENTS IN THE FINITE ELEMENT METHOD	75-0712
REFLECTION	STATIC ANALYSIS OF THE 64-M 210-PT ANTENNA REFLECTION STRUCTURE	71-0137
REFLECTIONS	A METHOD FOR THE THERMAL ANALYSIS OF SPACECRAFT INCLUDING ALL MULTIPLE REFLECTIONS AND SHADING AMONG	70-0424
REFLECTIONS	VARIABLE TIME STEP METHOD FOR DETERMINING PLASTIC STRESS REFLECTIONS FROM BOUNDARIES	75-0641
REFLECTORS	NASTRAN-GAP, AN INTEGRATED APPROACH TO THE ANALYSIS OF RADIATION PATTERNS OF DISTORTED REFLECTORS	71-0062
REFRACTORY	ELASTIC ANALYSIS OF A REFRACTORY SPRUNG ARCH	74-0386
REGION	AN ADAPTIVE COMPUTER PROGRAM FOR DIV. (P(X,Y) GRAD U) EQUALS F(X,Y,U) IN A POLYGONAL REGION	75-0133
REGION	ANALYSIS OF SHELLS OF ANY FORM IN THE PLASTIC REGION BY THE FINITE ELEMENT METHOD	73-0940
REGION	APPLICATION OF FINITE ELEMENT METHOD TO A MICROZONATION STUDY OF SALONICA REGION	73-0610
REGION	COMMENT ON "SOLUTION OF LAPLACE'S EQUATION IN A REGION WITH DIELECTRIC INTERFACES OF ARBITRARY SHAPE	70-0238
REGION	CONTRIBUTION TO NUMERICAL-SOLUTION OF DEVELOPING LAMINAR-FLOW IN ENTRANCE REGION OF CONCENTRIC ANNUL	75-0448
REGION	FINITE ELEMENT ANALYSIS OF CRACKS IN REGION OF A BI-MATERIAL INTERFACE	74-0164
REGION	FORTRAN SUBROUTINES FOR THE SOLUTION OF LAPLACE'S EQUATION OVER A GENERAL REGION IN TWO DIMENSIONS	70-0425
REGION	INTERACTIVE PROCEDURE FOR CURVILINEAR COORDINATION OF A PLANAR REGION WITH APPLICATIONS TO FINITE-EL	73-0040
REGION	NUMERICAL SOLUTION OF THE VISCOUS FLOW IN THE ENTRANCE REGION OF PARALLEL PLATES	73-0022
REGION	RESPONSE OF BURIED CAPSULES IN THE HIGH-OVERPRESSURE REGION	70-0093
REGION	STRESS DISTRIBUTION IN THE LIGAMENT ZONE OF PERFORATED CIRCULAR DISCS WITH SPECIAL REFERENCE TO THE	73-0827
REGIONAL	SENSITIVITY ANALYSIS OF INPUT PARAMETERS IN NUMERICAL MODELING OF STEADY STATE REGIONAL GROUNDWATER	74-0926
REGIONAL	TECHNIQUES FOR HANDLING INPUT DATA FOR FINITE ELEMENT ANALYSIS OF REGIONAL GROUNDWATER FLOW	75-0806
REGIONAL	TWO-DIMENSIONAL FINITE ELEMENT MODELING OF TRANSIENT FLOW IN REGIONAL AQUIFER SYSTEMS. ANALYSIS OF G	74-0743
REGIONS	ACOUSITC FINITE ELEMENT ANALYSIS OF AXISYMMETRIC FLUID REGIONS	73-0763
REGIONS	APPLICATION OF INTEGRAL EQUATIONS TO FLUID FLOWS IN UNBOUNDED REGIONS	74-0029
REGIONS	AUTOMATIC TRIANGULATION OF TWO DIMENSIONAL REGIONS	72-0697
REGIONS	CHISLENNOE RESHENIE PLOSKIKH ZADACH TEPLOPROVODNOSTI DLYA OBLASTEI SLOZHNOI FORMY (NUMERICAL SOLUTIO	74-0659
REGIONS	INCOMPRESSIBLE FLOW IN MULTIPLY CONNECTED REGIONS	74-0904
REGIONS	ORDER OF CONVERGENCE OF A FINITE-ELEMENT METHOD IN REGIONS WITH CURVED BOUNDARIES	75-0817
REGIONS	PHOTOELASTIC INVESTIGATION OF STRESS CONCENTRATIONS IN SPHERE-CYLINDER TRANSITION REGIONS: INCLUDING	68-0258
REGISTER	SHIP STRUCTURAL ANALYSIS IN LLOYD'S REGISTER OF SHIPPING	75-0285
REGRESSION	INTERPRETATION OF COMPLEX STRESS ANALYSIS BY MULTIPLE REGRESSION	74-1203
REGULAR	COMPLEXITY BOUNDS FOR REGULAR FINITE DIFFERENCE AND FINITE ELEMENT GRIDS	72-0872
REGULAR	COMPLEXITY BOUNDS FOR REGULAR FINITE DIFFERENCE AND FINITE ELEMENT GRIDS	73-0043
REGULAR	NESTED DISSECTION OF A REGULAR FINITE ELEMENT MESH	73-0042
REGULAR	THE PREDICTION OF REGULAR WAVE FORCES ON FLOATING BODIES WITH THE FINITE ELEMENT METHOD (IN DUTCH)	71-0743
REGULARITY	FINITE ELEMENT WITH A HIGH DEGREE OF REGULARITY	70-0052
REGULARIZATION	APPROXIMATION BY FINITE ELEMENT FUNCTIONS USING REGULARIZATION	74-1229
REGULARIZATION	APPROXIMATION BY FINITE-ELEMENT FUNCTIONS USING LOCAL REGULARIZATION	75-0436
REIBUNGSLOSEN	BERECHNUNG DES REIBUNGSLOSEN STROEMUNGSFELDES VON DREIDIMENSIONALEN AUFTRIEBSBEHAFTETENTRAGFLUEGELN	73-0533
REINFORCED	A FINITE ELEMENT COMPUTER PROGRAM FOR THE PREDICTION OF THE BEHAVIOR OF REINFORCED AND PRESTRESSED C	72-0799
REINFORCED	A FINITE-ELEMENT MODEL TO SIMULATE THE NONLINEAR RESPONSE OF REINFORCED CONCRETE FRAMES WITH MASONRY	73-1168
REINFORCED	A FINITE-ELEMENT STUDY ON THE NONLINEAR BEHAVIOR OF REINFORCED CONCRETE FLEXURAL MEMBERS DUE TO SHOR	73-1154
REINFORCED	A STUDY OF BEHAVIOUR OF REINFORCED CONCRETE ELEMENTS USING FINITE ELEMENTS	70-0335
REINFORCED	AN ANALYSIS OF A REINFORCED AND A PRESTRESSED CONCRETE BEAM BY FINITE-ELEMENT METHOD	72-0940
REINFORCED	AN INVESTIGATION OF INITIAL YIELD SURFACES FOR UNIDIRECTIONAL REINFORCED COMPOSITES	72-0837
REINFORCED	ANALYSIS OF COMPOSITE REINFORCED CUTOUTS AND CRACKS	74-0824
REINFORCED	ANALYSIS OF NONLINEAR REINFORCED CONCRETE FRAME BY FINITE ELEMENT METHOD	74-1104
REINFORCED	ANALYSIS OF STRENGTH AND RESIDUAL STRESSES IN FILAMENT REINFORCED ALUMINUM CYLINDERS	75-0517
REINFORCED	BERECHNUNG VON STAHLBETONSCHEIBEN IM ZUSTAND II BEI ANNAHME EINES WIRKLICHKEITSNAHEN WERKSTOFFVERHAL	74-0978
REINFORCED	CALCULATION OF THE ELASTIC BUCKLING LOADS OF THIN FLAT REINFORCED PLATES USING TRIANGULAR FINITE ELE	73-1024
REINFORCED	CALCULATION OF THE ELASTIC BUCKLING LOADS OF THIN FLAT REINFORCED PLATES USING TRIANGULAR FINITE ELE	75-0177
REINFORCED	CALCULATION OF THE REINFORCED THIN-WALLED STRUCTURES BY THE METHOD OF FINITE ELEMENTS (RASCHET PODKR	75-0685
REINFORCED	CALCULATION OF THE REINFORCED THIN-WALLED CONSTRUCTION BY THE FINITE ELEMENT	75-0756
REINFORCED	COLLAPSE LOAD OF REINFORCED CONCRETE PLATES.	74-0406
REINFORCED	COMPARISON OF FINITE ELEMENT AND UNIQUE LIMIT ANALYSIS SOLUTIONS FOR CERTAIN REINFORCED CONCRETE SLA	69-0184
REINFORCED	COMPUTER SIMULATION OF LAMINATED ROOF REINFORCED WITH GROUTED BOLTS	73-1045
REINFORCED	CONTINUUM CALCULATION FROM ELASTIC PLASTIC REINFORCED MATERIAL USING A FINITE ELEMENT METHOD (DIE BE	72-0658
REINFORCED	CRACKING ANALYSIS OF REINFORCED CONCRETE PLATES	75-0085
REINFORCED	DETERMINATION OF VOLUME CHANGE STRESSES IN PLAIN AND REINFORCED CONCRETE USING FINITE ELEMENT ANALYS	69-0477
REINFORCED	DIRECT MEASUREMENT OF BOND SLIP IN REINFORCED CONCRETE	69-0428
REINFORCED	DYNAMIC RESPONSE OF REINFORCED OPENINGS IN ROCK	71-0311
REINFORCED	ELASTIC-PLASTIC ANALYSIS OF REINFORCED CONCRETE MEMBERS BY FEM	73-0292
REINFORCED	FINITE ELEMENT ANALYSIS OF REINFORCED CONCRETE BEAMS	67-0058
REINFORCED	FINITE ELEMENT ANALYSIS OF REINFORCED CONCRETE	67-0161
REINFORCED	FINITE ELEMENT ANALYSIS OF REINFORCED CONCRETE SLABS	71-0133
REINFORCED	FINITE ELEMENT ANALYSIS OF REINFORCED CONCRETE STRUCTURES	72-0051
REINFORCED	FINITE ELEMENT ANALYSIS OF REINFORCED CONCRETE MEMBERS	72-0583
REINFORCED	FINITE ELEMENT ANALYSIS OF A DIAGONAL CRACKING BEHAVIOUR IN REINFORCED CONCRETE	73-0462
REINFORCED	FINITE ELEMENT ANALYSIS OF REINFORCED CONCRETE	73-0494
REINFORCED	FINITE ELEMENT ANALYSIS OF REINFORCED CONCRETE	74-0356
REINFORCED	FINITE ELEMENT ANALYSIS OF NONLINEAR CABLE REINFORCED MAMBRANES	74-1073
REINFORCED	FINITE ELEMENT ANALYSIS OF SHEAR STRENGTH OF REINFORCED CONCRETE BEAMS	74-1306
REINFORCED	FINITE ELEMENT FOR REINFORCED CONCRETE FRAME STUDY	73-0204
REINFORCED	FINITE ELEMENT FOR REINFORCED CONCRETE FRAME STUDY	74-0110
REINFORCED	FINITE ELEMENT FOR REINFORCED CONCRETE FRAME STUDY - SPACE DISCUSSION	74-0246
REINFORCED	FINITE ELEMENT FOR REINFORCED CONCRETE FRAME STUDY - CLOSURE	75-0087
REINFORCED	FINITE ELEMENT METHOD IN ANALYSIS OF REINFORCED AND PRESTRESSED CONCRETE STRUCTURES	71-0270
REINFORCED	FINITE ELEMENT STUDY OF REINFORCED CONCRETE BEAMS WITH DIAGONAL TENSION CRACKS	74-1308
REINFORCED	FINITE ELEMENT STUDY OF A REINFORCED CONCRETE CYLINDRICAL SHELL THROUGH ELASTIC, CRACKING, AND ULTIM	75-0317
REINFORCED	FRACTURE OF GRAPHITE FIBER REINFORCED COMPOSITES	74-0695
REINFORCED	INELASTIC ANALYSIS OF REINFORCED CONCRETE FRAMES	74-0587
REINFORCED	INELASTIC FINITE ELEMENT ANALYSIS OF REINFORCED CONCRETE PANELS UNDER IN-PLANE LOADS	70-0482
REINFORCED	INELASTIC MODEL FOR CYCLIC BIAXIAL LOADING OF REINFORCED CONCRETE	74-0503
REINFORCED	INITIAL YIELD SURFACE OF A UNIDIRECTIONALLY REINFORCED COMPOSITE	71-0280
REINFORCED	LIMIT ANALYSIS OF REINFORCED CONCRETE SLABS BY FINITE ELEMENT METHOD	71-0802
REINFORCED	LIMIT ANALYSIS OF REINFORCED CONCRETE SLABS BY A FINITE ELEMENT METHOD	72-0985
REINFORCED	MATERIAL BEHAVIOUR CHARACTERISTICS FOR REINFORCED CONCRETE SHELLS STRESSED BEYOND THE ELASTIC RANGE	70-0166
REINFORCED	MODIFIED APPROACH FOR ESTIMATING THE CRACKING MOMENT OF REINFORCED CONCRETE BEAMS	75-0485
REINFORCED	NON LINEAR ANALYSIS OF REINFORCED CONCRETE MEMBERS BY FINITE ELEMENT METHOD, PART 1: ANALYSIS, PART	71-0313
REINFORCED	NON-LINEAR ANALYSIS OF REINFORCED CONCRETE SLABS	72-0191
REINFORCED	NON-LINEAR FINITE ELEMENT ANALYSIS OF REINFORCED CONCRETE BY THE FINITE ELEMENT METHOD	68-0241
REINFORCED	NONLINEAR ANALYSIS OF REINFORCED CONCRETE BY THE FINITE ELEMENT METHOD	68-0126
REINFORCED	NONLINEAR ANALYSIS OF REINFORCED CONCRETE FRAMES AND PANELS	70-0436

346

REINFORCED	NONLINEAR ANALYSIS OF REINFORCED CONCRETE BUILDINGS	73-0324
REINFORCED	NONLINEAR ANALYSIS OF PLANAR REINFORCED CONCRETE STRUCTURES	74-0533
REINFORCED	NONLINEAR BEHAVIOR OF REINFORCED CONCRETE BEAMS BY THE FINITE-ELEMENT METHOD	72-0954
REINFORCED	NONLINEAR FINITE ELEMENT ANALYSIS OF REINFORCED CONCRETE PLATES	73-0530
REINFORCED	NONLINEAR FINITE ELEMENT ANALYSIS OF REINFORCED CONCRETE PLATES	73-0845
REINFORCED	NONLINEAR FINITE ELEMENT ANALYSIS OF REINFORCED CONCRETE STRUCTURES	74-0442
REINFORCED	NONLINEAR FINITE ELEMENT ANALYSIS OF REINFORCED CONCRETE FRAMES	74-1124
REINFORCED	PATTERN LOADING ON REINFORCED CONCRETE FLAT PLATES	71-0290
REINFORCED	PAVEMENT DESIGN AND CONTINUOUSLY REINFORCED CONCRETE PAVEMENT PERFORMANCE	73-0749
REINFORCED	PHOTOELASTIC TESTS ON COMPLEX INTERSECTIONS OF REINFORCED THIN-SHELL CYLINDERS	75-0363
REINFORCED	PLANE STRESS REINFORCED CONCRETE FINITE ELEMENTS	74-0108
REINFORCED	POST-ELASTIC FINITE-ELEMENT ANALYSIS OF REINFORCED PLATES	72-0979
REINFORCED	PROBLEM OF USING THE METHOD OF FINITE ELEMENTS TO CONSTRUCT AN ALGORITHM FOR CALCULATING A THIN-WALL	74-0851
REINFORCED	REINFORCED CONCRETE - A FINITE-ELEMENT FORMULATION	72-0929
REINFORCED	REINFORCED CONCRETE ON CONSTITUTIVE RELATIONS	75-0753
REINFORCED	REINFORCED FIELDS OF TRIANGULAR ELEMENTS WITH LINEARLY VARYING STRAIN: EFFECT OF INITIAL STRAINS	65-0004
REINFORCED	RESENI MAZNIHO STAVU ZELEZOBETONOVYCH DESEK (PLASTIC ANALYSIS OF REINFORCED CONCRETE PLATES)	74-0495
REINFORCED	SINGER: A COMPUTER CODE FOR GENERAL ANALYSIS OF TWO-DIMENSIONAL REINFORCED CONCRETE STRUCTURES. VOLU	75-0724
REINFORCED	SINGER: A COMPUTER CODE FOR GENERAL ANALYSIS OF TWO-DIMENSIONAL REINFORCED CONCRETE STRUCTURES. VOLU	75-0731
REINFORCED	STATICKE RESENI FYZIKALNE NELINEARNIHO CHOVANI ZELEZOBETO- NOVYCH DESEK (ANALYSIS OF PHYSICALLY NONL	74-0514
REINFORCED	STRENGTH OF REINFORCED CONCRETE CHAMBERS UNDER EXTERNAL PRESSURE	75-0293
REINFORCED	STRENGTH OF REINFORCED CONCRETE CHAMBERS UNDER EXTERNAL PRESSURE	75-0512
REINFORCED	STRESS ANALYSIS OF RIBBON REINFORCED COMPOSITES	70-0221
REINFORCED	STRESS ANALYSIS OF TAPE REINFORCED COMPOSITES	69-0068
REINFORCED	STRESS CONCENTRATION FACTORS FOR CIRCULAR, REINFORCED PENETRATIONS IN PRESSURIZED CYLINDRICAL SHELLS	75-0726
REINFORCED	STRESS DISTRIBUTION NEAR A BROKEN POINT OF FILAMENT IN UNIDIRECTIONALLY FIBER REINFORCED COMPOSITES	73-0811
REINFORCED	STRUCTURAL BEHAVIOR OF A TWO SPAN REINFORCED CONCRETE BOX GIRDER BRIDGE MODEL. VOLUME 11. REDUCTION,	71-0659
REINFORCED	STRUCTURAL CAPACITY OF REINFORCED CONCRETE COLUMNS SUBJECTED TO FIRE INDUCED THERMAL GRADIENTS	73-0913
REINFORCED	STUDY OF ELASTIC-PLASTIC MATRIX ANALYSIS FOR FOUNDATION AND REINFORCED CONCRETE IN JAPAN	69-0278
REINFORCED	STUDY OF FORCE DISPLACEMENT RELATIONSHIPS FOR THE FINITE ELEMENT ANALYSIS OF REINFORCED CONCRETE	73-0645
REINFORCED	TENSILE CRACK PROPAGATION IN REINFORCED CONCRETE BEAMS BY THE FINITE ELEMENT TECHNIQUES	69-0264
REINFORCED	THE DETERMINATION OF STRESSES AND DEFORMATIONS OF REINFORCED CONCRETE AFTER CRACKING	64-0004
REINFORCED	THE INELASTIC ANALYSIS OF REINFORCED AND PRESTRESSED CONCRETE BEAMS	72-0567
REINFORCED	THE INELASTIC ANALYSIS OF REINFORCED CONCRETE SLABS	74-1333
REINFORCED	THICK WALLED MULTIPLE OPENING REINFORCED CONCRETE CONDUITS	72-0666
REINFORCED	TIME DEPENDENT REINFORCED CONCRETE SLAB DEFLECTIONS	74-0586
REINFORCED	TWO DIMENSIONAL NON LINEAR ANALYSIS OF REINFORCED CONCRETE MEMBERS BY THE FINITE ELEMENT METHOD. PAR	71-0210
REINFORCED	USER'S GUIDE FOR THE COMPUTER PROGRAM DESCRIBED IN TAM REPORT NO. 680 "DETERMINATION OF VOLUME CHANG	69-0473
REINFORCED	YIELD CRITERION FOR REINFORCED CONCRETE SLABS	73-0985
REINFORCED	ZUR WIRKLICHKEITSNAHEN BERECHNUNG VON STAHLBETONPLATTEN MIT DER FINITE-ELEMENT-METHODE. (APPROXIMATE	75-0832
REINFORCED-CONC	NONLINEAR LAYERED FINITE-ELEMENT ANALYSIS OF REINFORCED-CONCRETE PLATES AND SHELLS	72-0946
REINFORCEMENT	ANALYSIS AND DESIGN OF AXISYMMETRIC REINFORCEMENT AROUND A CIRCULAR HOLE IN A THIN FLAT PLATE BY THE	74-1113
REINFORCEMENT	ANALYSIS AND MODEL STUDIES OF UNDERGROUND OPENINGS IN JOINTED ROCK. RESEARCH ON ROCK BOLT REINFORCEM	71-0438
REINFORCEMENT	ANALYSIS AND MODEL STUDIES OF UNDERGROUND OPENINGS IN JOINTED ROCK. RESEARCH ON ROCK BOLT REINFORCEM	71-0750
REINFORCEMENT	ANALYTIC INVESTIGATION OF COMPACT REINFORCEMENT FOR RADIAL NOZZLES IN SPHERICAL SHELLS	71-0107
REINFORCEMENT	COMPOSITE-OVERLAY REINFORCEMENT OF CUTOUTS AND CRACKS IN METAL SHEET	73-0878
REINFORCEMENT	FINITE ELEMENTS FOR THE REPRESENTATION OF PLANE AND AXISYMMETRIC STRUCTURAL REINFORCEMENT	68-0037
REINFORCEMENT	NUMERICAL AND PHYSICAL MODELLING OF REINFORCEMENT SYSTEMS FOR TUNNELS IN JOINTED ROCK	73-0872
REINFORCEMENT	TRANSVERSE REINFORCEMENT OF CONTINUOUS, TROUGH-SHAPED CONCRETE RAILROAD STRUCTURES. AN IMPLEMENTATIO	71-0719
REINFORCEMENT	ZUR BERECHNUNG, BEMESSUNG UND BEWEHRUNG RECHTECKIGER PLATTEN KONSTNATER DICKE MIT OEFFNUNGEN ODER AU	73-0664
REINFORCEMENTS	A METHOD FOR SIMULATING CONCENTRATED FORCES AND LOCAL REINFORCEMENTS IN STRESS COMPUTATIONS	72-0081
REINFORCING	ANALYSIS OF STRESSES IN CONCRETE NEAR A REINFORCING BAR DUE TO BOND AND TRANSVERSE CRACKING	70-0351
REINFORCING	REINFORCING REQUIREMENTS FOR CONCRETE BEAMS WITH LARGE WEB OPENINGS	73-0198
REINFORCING	ZUSAMMENHANG ZWISCHEN OBERFLAECHENBESCHAFFENHEIT, VERBUND UND SPRENGWIRKUNG VON BEWEHRUNGSSTAEHLEN U	73-0450
REINVESTIGATION	REINVESTIGATION OF BUCKLING OF SHELLS OF REVOLUTION BY A REFINED FINITE ELEMENT	74-0097
REISSNER	DISCUSSION OF "FINITE ELEMENT BENDING ANALYSIS OF REISSNER PLATES"	71-0826
REISSNER	DISCUSSION OF "FINITE ELEMENT BENDING ANALYSIS OF REISSNER PLATES"	71-0832
REISSNER	FINITE ELEMENT BENDING ANALYSIS OF REISSNER PLATES	70-0214
REISSNER'S	SHELL FINITE ELEMENT METHOD VIA REISSNER'S PRINCIPLE	69-0216
RELABELING	AN AUTOMATIC RELABELING ALGORITHM FOR BANDWIDTH MINIMIZATION	75-0763
RELABELING	RELABELING OF FINITE ELEMENT MESHES USING A RANDOM PROCESS	72-0754
RELABELLING	AN AUTOMATIC NODE RELABELLING SCHEME FOR BANDWIDTH MINIMIZATION OF STIFFNESS MATRICES	68-0256
RELATION	RECENT TRENDS OF RESEARCH INTO RESTRAINT STRESSES AND STRAINS IN RELATION TO WELD CRACKING	73-0483
RELATION	TENDANCES ACTUELLES DES RECHERCHES SUR LES CONTRAINTES ET DEFORMATIONS DUES AU BRIDAGE EN RELATION A	74-0273
RELATION	TENDANCES ACTUELLES DES RECHERCHES SUR LES CONTRAINTES ET DEFORMATIONS DUES AU BRIDAGE EN RELATION A	74-0273
RELATION	THE RELATION BETWEEN THE GALERKIN AND COLLOCATION METHODS USING SMOOTH SPLINES	74-0912
RELATION	TWO-DIMENSIONAL FINITE ELEMENT MODELING OF TRANSIENT FLOW IN REGIONAL AQUIFER SYSTEMS. ANALYSIS OF G	74-0743
RELATION	VOLLSTAENDIGE SPANNUNGS-DEHNUNGS-TEMPERATURAENDERUNGSBEZIEHUNG FUER DIE SCHWEISSEIGENSPANNUNGSBERECH	75-0807
RELATION	WEIGHTED RESIDUAL METHODS AND THEIR RELATION TO FINITE ELEMENT METHODS IN FLOW PROBLEMS	74-0003
RELATIONSHIPS	RELATIONSHIPS BETWEEN MATRIX METHODS IN FINITE ELEMENTS	67-0086
RELATIONSHIPS	SOIL RESISTANCE MOVEMENT RELATIONSHIPS FOR LATERALLY LOADED PILES IN CLAYS AT GREAT DEPTHS	70-0162
RELATIONSHIPS	STRAIN DISTRIBUTION AROUND UNDERGROUND OPENINGS. STATISTICAL RELATIONSHIPS FOR CERTAIN ROCK PROPERTI	71-0710
RELATIONSHIPS	STUDY OF FORCE DISPLACEMENT RELATIONSHIPS FOR THE FINITE ELEMENT ANALYSIS OF REINFORCED CONCRETE	73-0645
RELATIVE	DYNAMIC TORSIONAL RESPONSE OF THE HUMAN LEG RELATIVE TO SKIING INJURIES	73-0471
RELATIVE	PROBABILISTIC EVALUATION OF THE RELATIVE BEHAVIOUR OF TUNGSTEN CARBIDE COMPONENTS IN STRUCTURAL APPL	75-0579
RELATIVE	RELATIVE EFFICIENCY OF FINITE DIFFERENCE AND FINITE ELEMENT METHODS - 1. HYPERBOLIC PROBLEMS AND SPL	74-0986
RELATIVE	REVIEW OF NASTRAN DEVELOPMENT RELATIVE TO EFFICIENCY OF EXECUTION	73-0938
RELATIVE	THE RELATIVE EFFICIENCY OF FINITE DIFFERENCE AND FINITE ELEMENT METHODS. 1. HYPERBOLIC PROBLEMS AND	74-0911
RELATIVISTIC	USE OF VARIATIONAL PRINCIPLE IN STUDY ON RELATIVISTIC BOLTZMANN EQUATION	71-0442
RELAXATION	A RELAXATION PROCEDURE FOR COMPUTING STRESSES IN BUILT-UP STRUCTURES	69-0063
RELAXATION	THE APPLICATION OF DYNAMIC RELAXATION TO THE FINITE ELEMENT METHOD OF STRUCTURAL ANALYSIS	68-0108
RELAXED	FINITE ELEMENT FORMULATION BY VARIATIONAL PRINCIPLES WITH RELAXED CONTINUITY REQUIREMENTS	72-0516
RELAXED	FINITE ELEMENT METHODS BY VARIATIONAL PRINCIPLES WITH RELAXED CONTINUITY REQUIREMENTS	72-0342
RELAXED	FINITE ELEMENTS WITH RELAXED CONTINUITY - AN INVESTIGATION OF THE VARIATIONAL BASIS AND NUMERICAL PE	74-1065
RELAXED	STRUCTURAL RESPONSE OF RELAXED AND CONSTRICTED ARTERIOLES	68-0200
RELAXED	VARIATIONAL PRINCIPLES OF ELASTOSTATICS WITH RELAXED CONTINUITY REQUIREMENTS AND THEIR APPLICABILITY	74-0758
RELAYS	PARAMETRIC ANALYSIS OF CONTACT CHATTER IN RELAYS	74-0487
RELEASE	DETERMINATION OF ENERGY RELEASE RATES AND STRESS - INTENSITY FACTORS BY THE FINITE-ELEMENT METHOD	72-0183
RELEASE	STRAIN ENERGY RELEASE RATES FOR TURNING CRACKS	72-0146
RELEVANT	FINITE ELEMENT CALCULATIONS RELEVANT TO AS-CUT QUARTZ RESONATORS	73-0672
RELIABILITY	A NEW MEASURING METHOD OF RESIDUAL STRESSES WITH THE AID OF FINITE ELEMENT METHOD AND RELIABILITY OF	75-0875
RELIABILITY	RELIABILITY STUDY OF LINEAR VARIATION TRIANGULAR FINITE ELEMENT	72-0281
RELIEF	TWO-DIMENSIONAL PHOTOELASTIC STUDY OF STEAM-TURBINE CASING FLANGES WITH INTERFACE RELIEF	74-0603
RELIEVING	ANALYSIS OF WELDING STRESS RELIEVING BY ANNEALING BASED ON FINITE ELEMENT METHOD	75-0777
RELUCTANCE	FINITE-ELEMENT FIELD ANALYSIS OF AN INHOMOGENEOUS ANISOTROPIC RELUCTANCE MACHINE ROTOR	73-0020
RELUCTIVITY	SIMPLE EQUATIONS FOR THE MAGNETIZATION AND RELUCTIVITY CURVES OF STEEL	75-0156
REMOTE	ERROR ANALYSIS OF A REMOTE TRANSIENT HEAT FLUX SENSOR	74-0753
REMOTE	NASTRAN PLOTTING AT A REMOTE TERMINAL	74-0186
REMOVAL	REMOVAL OF TRUNCATION ERROR IN FINITE ELEMENT ANALYSIS	75-0131
REMPLISSAGE	CONTRIBUTION A L'ETUDE DU COMPLEXE BARRAGE-FONDATION ET EXPLICATION DES SEISMES DUS AU REMPLISSAGE D	72-0554
RENCONTREES	TENUE DES STRUCTURES DE GRAPHITE VIS-A-VIS DES SOLLICITATIONS RENCONTREES DANS UN COEUR DE REACTEUR	73-0817
RENUMBERING	BANDWIDTH REDUCTION BY AUTOMATIC RENUMBERING	73-0993

REORDERING	EMPIRICAL INVESTIGATION OF REORDERING AND DATA MANAGEMENT FOR FINITE ELEMENT SYSTEMS OF EQUATIONS	73-0944
REPAIR	A STRUCTURAL EVALUATION OF RAPID METHODS OF BACKFILLING FOR BOMB DAMAGE REPAIR	74-0821
REPAIR	A STRUCTURAL EVALUATION OF RAPID METHODS OF BACKFILLING FOR BOMB DAMAGE REPAIR - PHASE 11	75-0682
REPRESENTATION	FINITE ELEMENT REPRESENTATION OF STABLE CRACK-GROWTH	73-0086
REPRESENTATION	FINITE ELEMENTS FOR THE REPRESENTATION OF PLANE AND AXISYMMETRIC STRUCTURAL REINFORCEMENT	68-0037
REPRESENTATION	FINITE-ELEMENT CURRENT-DENSITY REPRESENTATION IN NUMERICAL-SOLUTION OF MICROSTRIP PROBLEMS	75-0400
REPRESENTATION	INTEGRAL REPRESENTATION OF FIELD VARIABLES FOR THE FINITE ELEMENT SOLUTION OF VISCOUS FLOW PROBLEMS	74-0468
REPRESENTATION	MODIFIED POTENTIAL ENERGY MASS REPRESENTATION FOR FREQUENCY PREDICTION	66-0053
REPRESENTATION	ON THE EXPLICIT REPRESENTATION OF TRANSFER-MATRICES OF CURVED BEAMS AND THEIR APPLICATIONS	75-0565
REPRESENTATION	REPRESENTATION OF SINGULARITIES WITH ISOPARAMETRIC FINITE ELEMENTS	74-1075
REPRESENTATION	SHELL DEFORMATION UNDER TOOL LOADS REPRESENTATION OF SINGULARITIES WITH ISOPARAMETRIC FINITE ELEMENT	74-0365
REPRESENTATION	SOME PROBLEMS IN THE DISCRETE ELEMENT REPRESENTATION OF AIRCRAFT STRUCTURES	64-0021
REPRESENTATIONS	EIGENVALUES OF COLLINEAR BEAM STRUCTURES USING FINITE ELEMENT TECHNIQUES AND VARIOUS DYNAMIC REPRESE	67-0124
REPRESENTATIONS	INVESTIGATION OF FINITE ELEMENT REPRESENTATIONS OF THE GEOPOTENTIAL	75-0290
REPRESENTATIONS	ON THE ACCURACY OF CERTAIN C DEGREE CONTINUOUS FINITE ELEMENT REPRESENTATIONS	74-0367
REPRESENTATIONS	SOME REMARKS ON INFINITE VENEZIANO REPRESENTATIONS	69-0472
REPRESENTATIONS	TRIGONOMETRIC FUNCTION REPRESENTATIONS FOR RECTANGULAR PLATE BENDING ELEMENTS	71-0765
RESENI	APLIKACE MKP NA RESENI PROBLEMU DIFUZE V KOVECH. (APPLICATION OF THE FINITE ELEMENT METHOD TO THE SO	74-0609
RESENI	RESENI MAZNIHO STAVU ZELEZOBETONOVYCH DESEK (PLASTIC ANALYSIS OF REINFORCED CONCRETE PLATES)	74-0495
RESENI	STATICKE RESENI FYZIKALNE NELINEARNIHO CHOVANI ZELEZOBETO- NOVYCH DESEK (ANALYSIS OF PHYSICALLY NONL	74-0514
RESEQUENCING	ELEMENT RESEQUENCING FOR FRONTAL SOLUTIONS	75-0132
RESERVOIR	EARTHQUAKE RESPONSE OF GRAVITY DAMS INCLUDING RESERVOIR INTERACTION EFFECTS	72-0675
RESERVOIR	EFFECT OF RESERVOIR FILLING ON STRESSES AND MOVEMENT IN EARTH AND ROCKFILL DAMS. A REPORT OF AN INVE	72-0836
RESERVOIR	ERROR BOUNDS FOR SEMIDISCRETE GALERKIN APPROXIMATIONS OF PARABOLIC PROBLEMS WITH APPLICATIONS TO PET	70-0291
RESERVOIR	FINITE ELEMENT METHOD IN PETROLEUM RESERVOIR ANALYSIS	72-0049
RESERVOIR	FINITE ELEMENT MODEL FOR LOWER GRANITE RESERVOIR	73-0065
RESERVOIR	FRONT DISPLACEMENT MODEL IN A FRACTURED RESERVOIR	74-0071
RESERVOIR	GALERKIN FINITE-ELEMENT SIMULATION OF A GEOTHERMAL RESERVOIR	73-0900
RESERVOIR	INFLUENCE OF RESERVOIR WATER ON EARTHQUAKE RESPONSE OF EARTH DAMS	70-0375
RESERVOIR	OBLICZANIE KONSTRUKCJI TOROIDALNEGO ZBIORNIKA NA WODE. (STATICAL ANALYSIS OF A TOROIDAL WATER RESERV	75-0211
RESERVOIR	RESERVOIR SIMULATION BY FINITE ELEMENT METHOD	73-0062
RESERVOIR	RESERVOIR SIMULATION BY GALERKIN'S METHOD	73-0662
RESERVOIR	SEISMIC ANALYSIS OF EARTH DAM RESERVOIR SYSTEMS	73-0412
RESERVOIR-DAM	EARTHQUAKE BEHAVIOUR OF RESERVOIR-DAM SYSTEMS	68-0043
RESERVOIR-DAM	EARTHQUAKE BEHAVIOUR OF RESERVOIR-DAM SYSTEMS (DISCUSSION)	69-0283
RESERVOIRS	CONTRIBUTION A L'ETUDE DU COMPLEXE BARRAGE-FONDATION ET EXPLICATION DES SEISMES DUS AU REMPLISSAGE D	72-0554
RESERVOIRS	CONTRIBUTION A L'ETUDE DU COMPLEXE BARRAGE-FONDATION ET EXPLICATION DES SEISMES DUS AU REMPLISSAGE D	72-0554
RESERVOIRS	UNE METHODE DE CALCUL PAR ELEMENTS FINIS DES MOUVEMENTS DE LIQUIDES DANS DES RESERVOIRS RIGIDES OU D	74-1171
RESERVOIRS	UNSTEADY RADIAL FLOW IN GAS RESERVOIRS BY FINITE ELEMENTS	74-0462
RESHENIE	CHISLENNOE RESHENIE PLOSKIKH ZADACH TEPLOPROVODNOSTI DLYA OBLASTEI SLOZHNOI FORMY (NUMERICAL SOLUTIO	74-0659
RESIDUAL	A NEW MEASURING METHOD OF RESIDUAL STRESSES WITH THE AID OF FINITE ELEMENT METHOD AND RELIABILITY OF	75-0875
RESIDUAL	ANALYSIS OF STRENGTH AND RESIDUAL STRESSES IN FILAMENT REINFORCED ALUMINUM CYLINDERS	75-0517
RESIDUAL	APPLICATION OF THE FINITE ELEMENT METHOD ON RESIDUAL STRESSES AND DEFORMATIONS IN WELDS	68-0178
RESIDUAL	CALCULATION OF RESIDUAL STRESSES IN PLASTIC DEFORMATION PROCESSES	72-0241
RESIDUAL	CREEP OF B/AL COMPOSITES AS INFLUENCED BY RESIDUAL STRESSES, BOND STRENGTH, AND FIBER PACKING GEOMET	74-1208
RESIDUAL	EFFECT OF THE VARIATION OF CONSISTENCY IN RESIDUAL SOILS ON LOAD TRANSFER BY BORED PILES	75-0812
RESIDUAL	FINITE ELEMENT ANALYSIS OF TRANSIENT HEAT CONDUCTION. APPLICATION OF THE WEIGHTED RESIDUAL PROCESS	74-0277
RESIDUAL	FINITE ELEMENT ANALYSIS OF INCOMPRESSIBLE MATERIAL BY RESIDUAL ENERGY BALANCING	74-1041
RESIDUAL	FINITE ELEMENT ANALYSIS OF THIN ELASTIC SHELLS WITH RESIDUAL ENERGY BALANCING AND THE ROLE OF THE RI	75-0034
RESIDUAL	FINITE ELEMENT WEIGHTED RESIDUAL SOLUTION TO ONE DIMENSIONAL FIELD PROBLEMS	73-0388
RESIDUAL	FINITE ELEMENT WEIGHTED RESIDUAL SOLUTION OF ONE-DIMENSIONAL, UNSTEADY AND UNSATURATED FLOWS IN PORO	73-0712
RESIDUAL	FINITE-ELEMENT ANALYSIS OF THIN ELASTIC SHELLS WITH RESIDUAL ENERGY BALANCING AND ROLE OF RIGID BODY	75-0048
RESIDUAL	FINITE-ELEMENT ELASTIC-PLASTIC ANALYSIS OF RESIDUAL STRESSES DUE TO CLAD WELDING IN REACTOR VESSELS	75-0779
RESIDUAL	RESIDUAL ENERGY BALANCING TECHNIQUE IN THE GENERATION OF PLATE BENDING FINITE ELEMENTS	74-0594
RESIDUAL	RESIDUAL STRESS PRODUCED BY METAL CUTTING	71-0414
RESIDUAL	STABILITY AND OSCILLATION CHARACTERISTICS OF FINITE ELEMENT, FINITE DIFFERENCE AND WEIGHTED RESIDUAL	73-0139
RESIDUAL	VARIATIONAL AND WEIGHTED RESIDUAL METHODS	73-0630
RESIDUAL	VOLLSTAENDIGE SPANNUNGS-DEHNUNGS-TEMPERATURAENDERUNGSBEZIEHUNG FUER DIE SCHWEISSEIGENSPANNUNGSBERECH	75-0807
RESIDUAL	WEIGHTED RESIDUAL METHODS AND THEIR RELATION TO FINITE ELEMENT METHODS IN FLOW PROBLEMS	74-0003
RESIDUAL	WEIGHTED RESIDUAL PROCESSES IN FINITE ELEMENT WITH PARTICULAR REFERENCE TO SOME TRANSIENT AND COUPLE	71-0503
RESIDUALS	METHOD FOR FINITE ELEMENTS AS SPECIAL CASE OF METHOD OF WEIGHTED RESIDUALS	72-0141
RESIDUALS	ON THE ROLE OF VARIATIONAL PRINCIPLES AND METHOD OF WEIGHTED RESIDUALS IN THE FINITE ELEMENT METHODS	74-0551
RESIDUEN	FEHLERBETRACHTUNGEN AUFGRUND DER METHODE DER GEWICHTETEN RESIDUEN FUR DAS VERFAHREN DER FINITEN ELEM	72-0368
RESILIENT	THE EFFECT OF STRESS HISTORY ON THE RESILIENT RESPONSE OF SOILS	73-0914
RESIN	ANALYSIS OF MULTI-AXIAL ANISOTROPIC CREEP, ROOF RESIN BOLTS, AND BED SEPARATION IN ROCK STRUCTURES U	74-1122
RESIN-GLASS	CHARACTERISATION ELASTIQUE D'UN STRATIFIE RESINE-VERRE PAR LA METHODE DES DEPLACEMENTS (DETERMINATIO	75-0272
RESINE-VERRE	CHARACTERISATION ELASTIQUE D'UN STRATIFIE RESINE-VERRE PAR LA METHODE DES DEPLACEMENTS (DETERMINATIO	75-0272
RESISTANCE	FAST FRACTURE RESISTANCE AND CRACK ARREST IN STRUCTURAL STEELS	73-0766
RESISTANCE	LATERAL RESISTANCE AND DEFLECTION OF VERTICAL PILES	73-0912
RESISTANCE	METHODS OF CALCULATION OF THREE-DIMENSIONAL PROBLEMS OF PERCOLATION OF FISSURED ROCK BY FINITE ELEME	70-0472
RESISTANCE	SOIL RESISTANCE MOVEMENT RELATIONSHIPS FOR LATERALLY LOADED PILES IN CLAYS AT GREAT DEPTHS	70-0162
RESISTANCE	USE OF THE FINITE ELEMENT TECHNIQUE TO DETERMINE THE FIRE RESISTANCE CAPABILITIES OF INSULATED STRUC	71-0147
RESIZING	ON THE ACCURACY OF THE TAYLOR APPROXIMATION FOR STRUCTURE RESIZING	74-0147
RESONANCE	EFFECTS OF SHEAR DEFORMATION, ROTARY INERTIA, AND ELASTICITY OF THE SUPPORT ON THE RESONANCE FREQUEN	75-0369
RESONANT	VARIATION WITH TEMPERATURE OF RESONANT FREQUENCIES OF ANISOTROPIC PLATES	74-1046
RESONATOR	PARAMETRIC STUDY OF A BEAM WITH A COMPOUND SIDE-BRANCH RESONATOR AS A DEVICE TO EVALUATE PRELIMINARY	70-0465
RESONATORS	ANALYSIS AND REDESIGN OF SONAR PIEZOELECTRIC RESONATORS USING THE FINITE ELEMENT METHOD.	72-0374
RESONATORS	FINITE ELEMENT CALCULATIONS RELEVANT TO AS-CUT QUARTZ RESONATORS	73-0672
RESONATORS	FINITE ELEMENT SIMULATION OF TWO-DIMENSIONAL ELECTROMECHANICAL RESONATORS	74-1014
RESONATORS	FINITE ELEMENT SIMULATION OF ENERGY-TRAPPED ELECTROMECHANICAL RESONATORS	75-0142
RESONATORS	FINITE ELEMENT SIMULATION OF ELECTROMECHANICAL RESONATORS WITH ARBITRARY ELECTRODES	75-0492
RESONATORS	VIBRATION ANALYSIS OF AXISYMMETRIC RESONATORS	75-0199
RESOURCES	HYDROLOGIC ENGINEERING METHODS FOR WATER RESOURCES DEVELOPMENT. VOLUME 10. PRINCIPLES OF GROUND-WATE	72-0696
RESOURCES	THE EFFICIENT USE OF COMPUTER RESOURCES	72-0099
RESPONCE	INELASTIC RESPONCE OF FRAMES TO DYNAMIC LOADS	71-0611
RESPONE	COUPLE - A PSEUDO FINITE ELEMENT CONDUCTION CODE FOR THE TRANSIENT THERMAL RESPONE OF AXISYMMETRIC O	72-0778
RESPONSE	A COMPUTER BASED PROCEDURE FOR PREDICTING THE TRANSIENT RESPONSE AND FAILURE OF A TWO-DIMENSIONAL CO	71-0630
RESPONSE	A COMPUTER MODEL FOR PREDICTING THE LOAD-DEFLECTION RESPONSE OF EXPEDIENT SOIL SURFACING	73-0880
RESPONSE	A CONSISTENT FINITE ELEMENT METHOD FOR RANDOM RESPONSE PROBLEMS	71-0202
RESPONSE	A CONSISTENT FINITE ELEMENT METHOD FOR RANDOM RESPONSE PROBLEMS	72-0896
RESPONSE	A FINITE ELEMENT APPROACH TO THE DETERMINATION OF THE DYNAMIC RESPONSE OF BEAMS SUPPORTING MOVING MA	74-1114
RESPONSE	A FINITE-ELEMENT MODEL TO SIMULATE THE NONLINEAR RESPONSE OF REINFORCED CONCRETE FRAMES WITH MASONRY	73-1168
RESPONSE	A NONLINEAR FINITE ELEMENT CODE FOR ANALYZING THE BLAST RESPONSE OF UNDERGROUND STRUCTURES	70-0054
RESPONSE	ACOUSTIC RESPONSE OF A SPACECRAFT SOLAR PANEL USING NORMAL MODE METHOD	69-0422
RESPONSE	AN INVESTIGATION OF THE DYNAMIC RESPONSE CHARACTERISTICS OF THE BON TEMPE DAM	66-0047
RESPONSE	ANALYSIS OF DYNAMIC RESPONSE OF SAFEGUARD STRUCTURES	70-0462
RESPONSE	ANALYSIS OF STRUCTURAL VIBRATIONS AND DYNAMIC RESPONSE	71-0059
RESPONSE	APPLICATION OF FINITE ELEMENT METHOD TO PREDICT STATIC RESPONSE OF AUTOMOTIVE BODY STRUCTURES	71-0340
RESPONSE	APPLICATION OF RESPONSE LIMITING TO FINITE-ELEMENT STRUCTURAL-ANALYSIS PROGRAMS	75-0424
RESPONSE	APPLICATION OF THE FINITE ELEMENT METHOD TO PREDICT STATIC AND DYNAMIC RESPONSE OF AN UNSHROUDED CEN	71-0574
RESPONSE	APPROXIMATE METHOD FOR CALCULATING THE RESPONSE OF EMPLACEMENT STRUCTURES SUBJECTED TO GROUND SHOCK	73-0654
RESPONSE	BASELINE SOLUTIONS FOR THE SMOOTH WALL THERMOCHEMICAL ABLATION RESPONSE OF GRAPHITE AND CARBON PHENO	71-0728

348

RESPONSE	COMPARISON OF THE RESULTS OF THE DYNAMIC RESPONSE OF CYLINDRICAL SHELLS BY CHARACTERISTIC AND BY FIN	74-0781
RESPONSE	COMPUTATION OF THE RESPONSE OF COUPLED PLATE ACOUSTIC SYSTEMS USING PLATE FINITE ELEMENTS AND ACOUST	71-0072
RESPONSE	COMPUTER BASED ANALYSES OF THE RESPONSE OF BOX TYPE STRUCTURES TO RANDOM PRESSURES	73-0227
RESPONSE	CONSIDERATION ON THE SEISMIC RESPONSE ANALYSIS OF UNDERGROUND STRUCTURES	75-0470
RESPONSE	COUPLED RESPONSE OF SPATIAL VIBRATORY STRUCTURES MOUNTED TO ISOTROPIC PLATE ELEMENTS	71-0228
RESPONSE	CRASH: A COMPUTER SIMULATION OF NONLINEAR TRANSIENT RESPONSE OF STRUCTURES	72-0569
RESPONSE	DECONVOLUTION OF SEISMIC RESPONSE FOR LINEAR SYSTEMS	73-0774
RESPONSE	DYNAMIC ELASTOPLASTIC RESPONSE OF GEOMETRICALLY NONLINEAR ARBITRARY SHELLS OF REVOLUTION UNDER IMPUL	70-0098
RESPONSE	DYNAMIC RESPONSE ANALYSIS OF GEOMETRICALLY NON-LINEAR STRUCTURES SUBJECTED TO HIGH IMPACT	72-0193
RESPONSE	DYNAMIC RESPONSE ANALYSIS OF GEOMETRICALLY NON-LINEAR STRUCTURES SUBJECTED TO HIGH IMPACT	72-0555
RESPONSE	DYNAMIC RESPONSE ANALYSIS OF THE IRREGULAR GROUND BY MEANS OF THE FINITE ELEMENT METHOD	75-0860
RESPONSE	DYNAMIC RESPONSE BY A STEP BY STEP ANALYSIS	62-0014
RESPONSE	DYNAMIC RESPONSE BY LARGE STEP INTEGRATION	73-0567
RESPONSE	DYNAMIC RESPONSE OF EARTH DAMS	67-0024
RESPONSE	DYNAMIC RESPONSE OF COHESIVE EARTH DAMS TO EARTHQUAKES	67-0131
RESPONSE	DYNAMIC RESPONSE OF ARBITRARY SHELLS OF REVOLUTION UNDER COMBINED IMPULSIVE NONISOTHERMAL AND IRREGU	70-0365
RESPONSE	DYNAMIC RESPONSE OF A CONSTRAINED FIBROUS SYSTEM SUBJECTED TO TRANSVERSE IMPACT. II. A MECHANICAL MO	70-0369
RESPONSE	DYNAMIC RESPONSE OF REINFORCED OPENINGS IN ROCK	71-0311
RESPONSE	DYNAMIC RESPONSE OF NONLINEAR SPACE FRAMES BY FINITE ELEMENT METHODS	71-0584
RESPONSE	DYNAMIC RESPONSE OF STRUCTURES SUBJECTED TO TIME-DEPENDENT BOUNDARY CONDITIONS USING THE FINITE ELEM	72-0129
RESPONSE	DYNAMIC RESPONSE OF MODEL PAVEMENT STRUCTURE	72-0226
RESPONSE	DYNAMIC RESPONSE OF A CYLINDER BURIED IN AN EARTH BEAM - RESULTS OF A FINITE ELEMENT ANALYSIS	72-0592
RESPONSE	DYNAMIC RESPONSE OF THE HUMAN ARM AS A COMPOSITE STRUCTURE	72-0611
RESPONSE	DYNAMIC RESPONSE OF VISCOELASTIC MATERIALS TO SINUSOIDALLY VARYING LOADINGS	73-1072
RESPONSE	DYNAMIC RESPONSE OF VISCOELASTIC MATERIALS	73-1075
RESPONSE	DYNAMIC RESPONSE OF FAST-REACTOR CORE SUBASSEMBLIES	74-0540
RESPONSE	DYNAMIC RESPONSE OF STRUCTURES IN LAYERED SOILS	74-0761
RESPONSE	DYNAMIC RESPONSE OF SOIL/CONCRETE INTERFACES AT HIGH PRESSURE	74-0836
RESPONSE	DYNAMIC RESPONSE OF ELASTIC HEAD MODELS	75-0378
RESPONSE	DYNAMIC RESPONSE ON NONLINEAR MEDIA AT LARGE STRAINS	72-0237
RESPONSE	DYNAMIC SHELL THEORY COUPLING THICKNESS STRESS WAVE EFFECTS WITH GROSS STRUCTURAL RESPONSE	73-0379
RESPONSE	DYNAMIC TORSIONAL RESPONSE OF THE HUMAN LEG RELATIVE TO SKIING INJURIES	73-0471
RESPONSE	EARTHQUAKE RESPONSE OF EARTH DAMS	67-0010
RESPONSE	EARTHQUAKE RESPONSE OF CONCRETE GRAVITY DAMS	70-0042
RESPONSE	EARTHQUAKE RESPONSE OF ARCH DAMS	70-0113
RESPONSE	EARTHQUAKE RESPONSE OF GRAVITY DAMS INCLUDING RESERVOIR INTERACTION EFFECTS	72-0675
RESPONSE	EFFECT OF GEOLOGIC IRREGULARITIES OF SEISMIC RESPONSE	74-0898
RESPONSE	EFFECT OF NONLINEAR MATERIAL RESPONSE ON THE BEHAVIOUR OF PAVEMENTS UNDER TRAFFIC	70-0171
RESPONSE	EFFECT OF TIME-DEPENDENT MATERIAL PROPERTIES ON DYNAMIC RESPONSE	74-0900
RESPONSE	EFFECTS OF FLEXIBLE CONNECTIONS ON BODY STRUCTURAL RESPONSE	74-0140
RESPONSE	EFFECTS OF RISE TIME AND DAMPING ON FINITE ELEMENT ANALYSIS OF RESPONSE OF STRUCTURES	71-0087
RESPONSE	EFFECTS OF SHEAR WALLS ON DYNAMIC RESPONSE OF FRAMES	69-0230
RESPONSE	EFFECTS OF SOURCE STRUCTURE AND SITE RESPONSE ON GROUND MOTION IN THE NEAR FIELD OF EARTHQUAKES	73-0732
RESPONSE	ELASTIC DYNAMIC RESPONSE OF AXISYMMETRIC STRUCTURES	69-0275
RESPONSE	ELASTIC-PLASTIC EARTHQUAKE RESPONSE OF SOIL-BUILDING SYSTEMS	72-0717
RESPONSE	ELASTIC-PLASTIC LARGE DEFORMATION RESPONSE OF CLAY TO FOOTING LOADS	74-0699
RESPONSE	ELASTO-PLASTIC SEISMIC RESPONSE ANALYSIS OF FRAMED STRUCTURES BY MODE SUPERPOSITION	73-0325
RESPONSE	EVALUATION OF DISCRETE METHODS FOR THE LINEAR DYNAMIC RESPONSE OF ELASTIC AND VISCOELASTIC SOLIDS	70-0439
RESPONSE	FEATHER: FINITE ELEMENT ANALYSIS FOR THREE-DIMENSIONAL ELASTIC RESPONSE	72-0607
RESPONSE	FINITE ELEMENT ANALYSIS FOR SEISMIC RESPONSE OF EARTH DAMS	68-0091
RESPONSE	FINITE ELEMENT ANALYSES OF SHELL RESPONSE FOR ARBITRARY EXCITATIONS	70-0348
RESPONSE	FINITE ELEMENT ANALYSIS OF DYNAMIC RESPONSE	72-0681
RESPONSE	FINITE ELEMENT ANALYSIS OF LOAD CELL RESPONSE IN HIGH SPEED TENSILE TESTING	73-1071
RESPONSE	FINITE ELEMENT BASED PROCEDURE FOR SIMULATING THE TRANSIENT RESPONSE AND FAILURE OF A TWO-DIMENSIONA	72-0123
RESPONSE	FINITE ELEMENT METHODS FOR THE RESPONSE OF STRUCTURES TO RANDOM EXCITATIONS	75-0230
RESPONSE	FINITE ELEMENT PERTURBATION ANALYSIS OF NONLINEAR DYNAMIC RESPONSE OF ELASTIC CONTINUA	74-0452
RESPONSE	FINITE ELEMENT PROCEDURE FOR THE LARGE DEFORMATION DYNAMIC RESPONSE OF AXISYMMETRIC SOLIDS	74-0527
RESPONSE	FINITE ELEMENT PROCEDURE FOR THE LARGE DEFORMATION DYNAMIC RESPONSE OF AXISYMMETRIC SOLIDS	74-0967
RESPONSE	FINITE ELEMENT PROCEDURE FOR LARGE DEFORMATION DYNAMIC RESPONSE OF AXISYMMETRIC SOLIDS	75-0013
RESPONSE	FINITE-ELEMENT ANALYSIS OF COMPLEX PANEL RESPONSE TO RANDOM LOADS	68-0093
RESPONSE	FINITE-ELEMENT ANALYSIS OF STRUCTURAL RESPONSE IN HCDA	75-0407
RESPONSE	FIRES-RC. A COMPUTER PROGRAM FOR THE FIRE RESPONSE OF STRUCTURES-REINFORCED CONCRETE FRAMES	74-0747
RESPONSE	FIRES-T. A COMPUTER PROGRAM FOR THE FIRE RESPONSE OF STRUCTURES - THERMAL	74-0850
RESPONSE	FREE VIBRATION AND RESPONSE TO RANDOM PRESSURE FIELD OF ANISOTROPIC THIN CYLINDRICAL SHELLS	74-0830
RESPONSE	FREE VIBRATION AND TRANSIENT RESPONSE OF THICK AND THIN PLATES USING THE FINITE ELEMENT METHOD	74-0517
RESPONSE	FREE VIBRATIONS AND RANDOM RESPONSE OF AN INTEGRALLY-STIFFENED PANEL	70-0138
RESPONSE	GRAPHICAL DISPLAY OF COMPUTER SIMULATED UNBALANCED ROTOR RESPONSE	71-0486
RESPONSE	IMPROVED ANALYTIC LONGITUDINAL RESPONSE ANALYSIS FOR AXISYMMETRIC LAUNCH VEHICLES, COMPUTER PROGRAM	65-0002
RESPONSE	IMPROVED LINEAR AXISYMMETRIC SHELL FLUID MODEL FOR LAUNCH VEHICLE LONGITUDINAL RESPONSE ANALYSIS	65-0044
RESPONSE	IN-VACUO MODAL DYNAMIC RESPONSE OF THE HUMAN SKULL	73-0460
RESPONSE	INELASTIC RESPONSE OF PLATES AND ECCENTRICALLY STIFFENED PLATES	74-0458
RESPONSE	INELASTIC RESPONSE OF NUCLEAR PIPING SUBJECTED TO RUPTURE FORCES	75-0515
RESPONSE	INFLUENCE OF RESERVOIR WATER ON EARTHQUAKE RESPONSE OF EARTH DAMS	70-0375
RESPONSE	INFLUENCE OF SOIL/STRUCTURE INTERACTION PARAMETERS ON FLOOR RESPONSE SPECTRA	73-0833
RESPONSE	LARGE DEFLECTION ELASTIC-PLASTIC DYNAMIC RESPONSE OF STIFFENED SHELLS OF REVOLUTION	72-0608
RESPONSE	LARGE DEFLECTION ELASTIC-PLASTIC DYNAMIC RESPONSE OF STIFFENED SHELLS OF REVOLUTION	74-0496
RESPONSE	LOADS AND RESPONSE OF LARGE DUCTED PROPELLER SYSTEMS	73-0810
RESPONSE	LUSH: A COMPUTER PROGRAM FOR COMPLEX RESPONSE ANALYSIS OF SOIL-STRUCTURE SYSTEMS	74-0744
RESPONSE	MAGAZINE HEADWALL RESPONSE TO EXPLOSIVE BLAST	74-0754
RESPONSE	MAGNETOTELLURIC RESPONSE OF A 2-DIMENSIONAL SLOPING CONTACT BY FINITE ELEMENT METHOD	73-0130
RESPONSE	MAGNETOTELLURIC RESPONSE OF LATERALLY INHOMOGENEOUS AND ANISOTROPIC MEDIA	75-0850
RESPONSE	MECHANICS OF CABLE MOORING SYSTEMS. VOLUME IV. A COMPUTER PROGRAM FOR ANALYZING THE STEADY STATE RES	72-0653
RESPONSE	MECHANICS OF CABLE MOORING SYSTEMS. VOLUME 1. THREE DIMENSIONAL RESPONSE OF DEEP WATER MOORING LINES	72-0654
RESPONSE	METHOD FOR ESTIMATING RESPONSE OF PAYLOAD SECONDARY STRUCTURES TO RANDOM EXCITATION	68-0232
RESPONSE	MODAL RESPONSE ANALYSIS OF CONSTRAINED FINITE-ELEMENTS ELASTIC CONTINUUM	75-0767
RESPONSE	MODAL RESPONSE OF FREE ROCKETS UNDER THRUST LOADING	74-0855
RESPONSE	MODELING AND ANALYSIS TECHNIQUES FOR PREDICTION OF STRUCTURAL AND BIODYNAMIC CRASH-IMPACT RESPONSE	75-0300
RESPONSE	NON-LINEAR RESPONSE OF A BEAM TO PERIODIC LOADING.	72-0407
RESPONSE	NONLINEAR ACOUSTIC RESPONSE ANALYSIS OF PLATES USING THE FINITE ELEMENT METHOD	72-0223
RESPONSE	NONLINEAR FINITE-ELEMENT METHOD FOR PLATES AND ITS APPLICATION TO DYNAMIC RESPONSE OF REACTOR FUEL S	74-0670
RESPONSE	NONLINEAR RESPONSE OF PLATES TO IMPULSIVE LOADING USING HYBRID FINITE ELEMENTS	74-1018
RESPONSE	NONLINEAR STATIC AND DYNAMIC RESPONSE OF UNDERWATER CABLE STRUCTURES USING THE FINITE ELEMENT METHOD	75-0460
RESPONSE	NONSAP: A STRUCTURAL ANALYSIS PROGRAM FOR STATIC AND DYNAMIC RESPONSE OF NONLINEAR SYSTEMS	74-0848
RESPONSE	ON RESPONSE OF INITIALLY STRESSED STRUCTURES TO RANDOM EXCITATIONS	73-0278
RESPONSE	ON THE APPLICATION OF RESPONSE LIMITING TO FINITE ELEMENT STRUCTURAL ANALYSIS PROGRAMS	75-0192
RESPONSE	ON THE APPLICATION OF RESPONSE LIMITING TO FINITE ELEMENT STRUCTURAL ANALYSIS PROGRAMS	75-0822
RESPONSE	PART 1 - TASK 3.2.9. BENDING AND BUCKLING OF A ONE-DIMENSIONAL MODEL SIMULATING A CONICAL SHELL. PAR	71-0745
RESPONSE	POST-BUCKLING RESPONSE OF STRUCTURES USING THE FINITE ELEMENT METHOD	69-0414
RESPONSE	QUAD-4: A COMPUTER PROGRAM FOR EVALUATING THE SEISMIC RESPONSE OF SOIL STRUCTURES BY VARIABLE DAMPIN	73-0771
RESPONSE	RANDOM RESPONSE OF PERIODIC STRUCTURES BY A FINITE-ELEMENT TECHNIQUE	75-0426
RESPONSE	RANDOM VIBRATION RESPONSE OF CANTILEVER PLATES USING THE FINITE ELEMENT METHOD	70-0181
RESPONSE	RATE OF CHANGE OF DYNAMIC RESPONSE	72-1004

RESPONSE	RC COLUMN EARTHQUAKE RESPONSE IN TWO DIMENSIONS	74-0651
RESPONSE	RESPONSE OF BEAMS AND PLATES TO MOVING LOADS	71-0277
RESPONSE	RESPONSE OF BOX-LIKE STRUCTURES TO WEAK EXPLOSIONS	75-0197
RESPONSE	RESPONSE OF BURIED CAPSULES IN THE HIGH-OVERPRESSURE REGION	70-0093
RESPONSE	RESPONSE OF EARTH BANKS DURING EARTHQUAKES	67-0039
RESPONSE	RESPONSE OF EARTH DAMS TO TRAVELLING SEISMIC WAVES	69-0092
RESPONSE	RESPONSE OF NONLINEAR BEAM TO RANDOM EXCITATION	73-0285
RESPONSE	RESPONSE OF NUCLEAR STRUCTURAL SYSTEMS TO TRANSIENT AND RANDOM EXCITATIONS, USING BOTH DETERMINISTIC	74-1047
RESPONSE	RESPONSE OF REACTOR CORE SUBASSEMBLIES TO IMPULSIVE LOADINGS	73-0076
RESPONSE	RESPONSE OF RECTANGULAR PLATES TO MOVING LOADS BY A FINITE ELEMENT PROCEDURE	69-0515
RESPONSE	RESPONSE OF RECTANGULAR ORTHOTROPIC PLATES TO RANDOM EXCITATIONS	73-0496
RESPONSE	RESPONSE OF SIMPLY SUPPORTED TIMOSHENKO BEAMS	69-0489
RESPONSE	RESPONSE OF STRUCTURE TO THE PSEUDO-SOUND FIELD OF A JET (USING A COMBINED CONTINUUM AND FINITE ELEM	67-0146
RESPONSE	RESPONSE OF STRUCTURES TO COMBINED BLAST EFFECTS	73-0424
RESPONSE	RESPONSE OF UNCONFINED AQUIFERS TO PUMPING	74-0073
RESPONSE	SAP IV. A STRUCTURAL ANALYSIS PROGRAM FOR STATIC AND DYNAMIC RESPONSE OF LINEAR SYSTEMS	73-0650
RESPONSE	SEISMIC RESPONSE BY VARIABLE DAMPING FINITE ELEMENTS	74-0143
RESPONSE	SEISMIC RESPONSE EFFECTS ON EMBEDDED STRUCTURES	72-0902
RESPONSE	SEISMIC RESPONSE OF SOIL DEPOSITS UNDERLAIN BY SLOPING ROCK BOUNDARIES	69-0484
RESPONSE	SEISMIC RESPONSE OF HIGHWAY BRIDGES	75-0879
RESPONSE	SKIN-STRINGER PANEL NORMAL MODE RESPONSE EXPERIMENTAL DATA	71-0691
RESPONSE	SOIL-STRUCTURE INTERACTION ANALYSES FOR EVALUATING SEISMIC RESPONSE	74-0745
RESPONSE	STOCHASTIC DYNAMIC RESPONSE ANALYSIS OF OFFSHORE PLATFORMS, WITH PARTICULAR REFERENCE TO GRAVITY-TYP	75-0220
RESPONSE	STOCHASTIC FINITE ELEMENT SIMULATION OF THE NONLINEAR STRUCTURAL RESPONSE OF FIBROUS COMPOSITE MATER	75-0762
RESPONSE	STRESS STRAIN RESPONSE AND FRACTURE OF A CONCRETE MODEL IN BIAXIAL LOADING	71-0295
RESPONSE	STRESS-WAVE PROPAGATION AND DYNAMIC EFFECTS IN SHELLS (SPADES) MICROMECHANICAL STRESS-WAVE RESPONSE	72-0851
RESPONSE	STRUCTURAL RESPONSE OF RELAXED AND CONSTRICTED ARTERIOLES	68-0200
RESPONSE	STRUCTURAL RESPONSE OF UNSTIFFENED TOROIDAL SHELLS	69-0198
RESPONSE	STRUCTURAL RESPONSE AND LOADING OF WALL PANELS	71-0735
RESPONSE	STRUCTURAL RESPONSE TO IMPACT DAMAGE	75-0325
RESPONSE	STRUCTURAL RESPONSE TO MOVING PROJECTILE MASS BY THE FINITE ELEMENT METHOD	75-0693
RESPONSE	STUDIES OF SEISMIC RESPONSE OF CLAY BANKS	71-0833
RESPONSE	STUDY OF EFFECTS OF DESIGN DETAILS ON STRUCTURAL RESPONSE TO ACOUSTIC EXCITATION	72-0864
RESPONSE	TEST RESULTS FROM THE BOUNDARY LAYER FACILITY - RESPONSE OF STRUCTURE TO THE PSEUDO-SOUND FIELD OF A	68-0262
RESPONSE	THE COMPARISON OF THE RESPONSE OF A HIGHWAY BRIDGE TO UNIFORM GROUND SHOCK AND MOVING GROUND EXCITAT	72-0737
RESPONSE	THE EFFECT OF STRESS HISTORY ON THE RESILIENT RESPONSE OF SOILS	73-0914
RESPONSE	THE EFFECTS OF THIN CORES AND NONHOMOGENEITY ON DYNAMIC RESPONSE OF EARTH DAMS	67-0021
RESPONSE	THE ELASTO-PLASTIC AND LARGE-DISPLACEMENT RESPONSE OF PLATES TO BLAST LOADING	73-0731
RESPONSE	THE RESPONSE OF SHELLS TO DISTRIBUTED RANDOM LOADS USING NASTRAN	71-0134
RESPONSE	THE STOCHASTIC FINITE ELEMENT SIMULATION OF THE NONLINEAR STRUCTURAL RESPONSE OF FIBROUS COMPOSITE M	75-0798
RESPONSE	THE TRANSIENT LINEAR ELASTIC RESPONSE ANALYSIS COMPLEX THIN SHELLS OF REVOLUTION SUBJECTED TO ARBITR	70-0080
RESPONSE	THERMOMECHANICAL RESPONSE OF INELASTIC FIBRE COMPOSITES	75-0235
RESPONSE	THREE-DIMENSIONAL FINITE ELEMENT ANALYSIS INCLUDING ELASTIC-PLASTIC MATERIAL RESPONSE	74-1126
RESPONSE	TRANSIENT DYNAMIC RESPONSE OF VISCOELASTIC STRUCTURES	72-0234
RESPONSE	TRANSIENT RESPONSE OF LINEAR ELASTIC STRUCTURES DETERMINED BY THE MATRIX EXPONENTAL METHOD	71-0240
RESPONSE	TRANSIENT RESPONSE OF A COUPLED PLATE-ACOUSTIC SYSTEM USING PLATE AND ACOUSTIC FINITE ELEMENTS	71-0463
RESPONSE	TRANSIENT RESPONSE OF A COUPLED PLATE-ACOUSTIC SYSTEM USING PLATE AND ACOUSTIC FINITE ELEMENTS	72-0275
RESPONSE	TRANSIENT RESPONSE OF NON-UNIFORM, NON-HOMOGENEOUS BEAMS	73-0162
RESPONSE	TRANSIENT RESPONSE OF INELASTICALLY CONSTRAINED RIGID BODY SYSTEMS	73-0291
RESPONSE	TRANSIENT RESPONSE OF ELASTIC, VISCO-PLASTIC BEAMS	73-0380
RESPONSE	TRANSIENT SHELL RESPONSE BY NUMERICAL TIME INTEGRATION	72-0305
RESPONSE	TRANSIENT TEMPERATURE RESPONSE OF COMPOSITE SLABS	74-1038
RESPONSE	USE OF THE FINITE ELEMENT DAMPED FORCED RESPONSE STRAIN ENERGY DISTRIBUTION FOR VIBRATION REDUCTION	72-0759
RESPONSE	USER'S MANUAL FOR NTEDYNS - A NONLINEAR FINITE ELEMENT CODE FOR ANALYZING THE BLAST RESPONSE OF UNDE	71-0695
RESPONSE	VARIATIONAL FINITE ELEMENT CALCULATION OF THE ACOUSTIC RESPONSE OF A RECTANGULAR PANEL	71-0100
RESPONSE	VERTICAL RESPONSE OF A RIGID BASE TO DYNAMIC LOADING	69-0316
RESPONSE	VIBRATION MODES AND RANDOM RESPONSE OF A MULTI-BAY PANEL SYSTEM USING FINITE ELEMENTS	67-0128
RESPONSE	VIBRATIONAL RESPONSE OF SONAR TRANSDUCERS USING PIEZO-ELECTRIC FINITE ELEMENTS	74-0522
RESPONSE	VIBROACOUSTIC RESPONSE OF TURBULENCE EXCITED THIN RECTANGULAR FINITE PLATES IN HEAVY AND FLUID MEDIA	74-1309
RESPONSE	VISCOELASTIC RESPONSE OF GRAPHITE MATERIALS IN IRRADIATION ENVIRONMENTS	70-0314
RESPONSE-BOUND	APPLICATION OF RESPONSE-BOUND METHOD IN SHOCK AND VIBRATION ANALYSIS OF TELEPHONE STRUCTURES	74-0655
RESPONSES	APPLICATION OF THE COLLISION-IMPARTED VELOCITY METHOD FOR ANALYZING THE RESPONSES OF CONTAINMENT AND	73-0770
RESPONSES	APPLICATIONS OF NASTRAN TO COUPLED STRUCTURAL AND HYDRODYNAMIC RESPONSES IN AIRCRAFT HYDRAULIC SYSTE	71-0744
RESPONSES	DYNAMIC RESPONSES OF CYLINDRICAL SHELLS INCLUDING GEOMETRIC AND MATERIAL NONLINEARITIES	74-0150
RESPONSES	NONLINEAR TRANSIENT RESPONSES OF STRUCTURES BY SPATIAL FINITE-ELEMENT METHOD	73-0138
RESPONSES	SUBTERRANEAN GUIDEWAY RESPONSES TO LEVITATION VEHICLE LOADING	74-0927
RESSTING	INTERACTIVE ANALYSIS OF A RAFTED MULTISTOREY SPACE FRAME RESSTING ON AN INHOMOGENEOUS CLAY STRATUM	74-0447
RESTART	STRUCTURAL DYNAMIC ANALYSIS OF ELECTRONIC ASSEMBLIES USING NASTRAN RESTART/FORMAT CHANGE CAPABILITY	71-0230
RESTING	FINITE ELEMENT ANALYSIS OF THE SURFACE DEFORMATION DUE TO A UNIFORM LOADING ON A LAYER OF GIBSON SOI	75-0240
RESTORED	FINITE ELEMENT STRESS ANALYSIS OF A RESTORED AXISYMMETRIC FIRST MOLAR	73-0053
RESTORED	PHOTOELASTIC AND FINITE ELEMENT STRESS ANALYSIS OF A RESTORED AXISYMMETRIC FIRST MOLAR	73-0142
RESTRAINED	COMPUTER PROGRAM JET 3 TO CALCULATE THE LARGE ELASTIC PLASTIC DYNAMICALLY INDUCED DEFORMATIONS OF FR	72-0609
RESTRAINED	STRESSES DUE TO CONCENTRATED LOADS IN RESTRAINED BENDING TESTS	67-0097
RESTRAINT	DESIGN ANALYSIS FOR LMFBR CORE RESTRAINT	71-0208
RESTRAINT	RECENT TRENDS OF RESEARCH INTO RESTRAINT STRESSES AND STRAINS IN RELATION TO WELD CRACKING	73-0483
RESTRAINT	TENDANCES ACTUELLES DES RECHERCHES SUR LES CONTRAINTES ET DEFORMATIONS DUES AU BRIDAGE EN RELATION A	74-0273
RESTRAINT	THE EFFECT OF END RESTRAINT CONDITION ON THE FRACTURE MECHANISM AND STRENGTH OF ROCK-LIKE MATERIALS	72-0134
RESTRAINTS	STRESS ANALYSIS OF CURVED PIPES WITH END RESTRAINTS	75-0667
RESTRICTIONS	RESTRICTIONS IN THE APPLICATION OF AUTOMATIC GENERATION SCHEMES BY ISOPARAMETRIC CO-ORDINATES	74-1377
RESULTANTS	COMPUTATION OF STRESS RESULTANTS IN FINITE ELEMENT ANALYSIS	66-0056
RESULTANTS	COMPUTATION OF STRESS RESULTANTS FROM ELEMENT STIFFNESS MATRICES	66-0125
RESULTANTS	STRAIN RESULTANTS IN CERTAIN FINITE ELEMENTS	69-0079
RESULTANTS	STRESS RESULTANTS IN SHELLS BY AN AUGMENTED FINITE ELEMENT METHOD	73-0585
RESULTATEN	PRAKTISCHE RESULTATEN, DOEL EN NUT VAN HET ONTWERPEN VAN MECHANISCHE ELEMENTEN MET BEHULP VAN COMPUT	73-0400
RESOLUTION	UNE METHODE VARIATIONNELLE ELEMENTS FINIS POUR LA RESOLUTION NUMERIQUE UN PROBLEME EXTERIEUR DANS R*	73-0576
RETAINING	COMPARISON BETWEEN THEORETICAL AND MEASURED EARTH PRESSURES ACTING ON A LARGE MOTORWAY RETAINING WAL	74-0917
RETAINING	FINITE ELEMENT ANALYSES OF RETAINING WALL BEHAVIOUR	71-0320
RETAINING	TIED-BACK RETAINING WALLS	75-0308
RETARDATION	MANAGEMENT OF RETARDATION OF SALT WATER INTRUSION IN COASTAL AQUIFERS	75-0692
RETICOLI	OPTIMIZZAZIONE DI RETICOLI TRIANGOLARI PER IL METODO DEGLI ELEMENTI FINITI IN IDROLOGIA	72-0987
RETICOLI	PROCESS FOR THE AUTOMATIC GENERATION OF FLAT TRIANGULAR LATTICES FOR APPLICATION TO THE FINITE ELEME	71-0366
RETICULATED	NONLINEAR ANALYSIS OF RETICULATED SPACE TRUSSES	75-0287
RETICULATED	SNAP-THROUGH BUCKLING OF RETICULATED SHELLS	75-0640
RETORT	COMPUTER SIMULATION OF AN IN SITU SHALE-OIL RETORT	74-1059
RETSCP	RETSCP: A COMPUTER PROGRAM FOR ANALYSIS OF ROCKET ENGINE THERMAL STRAINS WITH CYCLIC PLASTICITY	74-0692
REVERSE	COMPARATIVE ANALYSIS OF THE CUTHILL-MCKEE AND THE REVERSE CUTHILL-MCKEE ORDERING ALGORITHMS FOR SPAR	73-1099
REVERSERS	ANALYSIS OF CURVED TARGET-TYPE THRUST REVERSERS	75-0152
REVIEW	A CRITICAL REVIEW OF NASTRAN	71-0246
REVIEW	A REVIEW OF SOME ASPECTS OF STRUCTURAL OPTIMIZATION	75-0467
REVIEW	CONTINUUM MECHANICS - A BRIEF REVIEW	71-0407
REVIEW	ISOPARAMETRIC AND ALLIED NUMERICALLY INTEGRATED ELEMENTS - A REVIEW	71-0763
REVIEW	REVIEW OF BASIC THEORY AND METHODS OF ANALYSIS OF CONTINUOUS STRUCTURES	73-0697

REVIEW	REVIEW OF CONTINUUM, FINITE ELEMENT AND HYBRID TECHNIQUES IN THE ANALYSIS OF STRESS CONCENTRATIONS I	73-1051
REVIEW	REVIEW OF CONTINUUM, FINITE ELEMENT AND HYBRID TECHNIQUES IN ANALYSIS OF STRESS CONCENTRATIONS IN ST	74-0914
REVIEW	REVIEW OF CURRENT PROBLEMS FOR MULTIDIMENSIONAL REACTOR STATICS CALCULATIONS	73-0960
REVIEW	REVIEW OF NASTRAN DEVELOPMENT RELATIVE TO EFFICIENCY OF EXECUTION	73-0938
REVIEW	REVIEW OF OPTIMALITY CRITERIA APPROACHES TO STRUCTURAL OPTIMIZATION	74-0652
REVIEW	REVIEW OF THE ASKA PROGRAM	71-0175
REVIVAL	FINITE ELEMENT METHOD - MATHEMATICAL REVIVAL	74-0169
REVOLUTION	A DYNAMIC ELASTOPLASTIC ANALYSIS OF THIN SHELLS OF REVOLUTION UNDER ASYMMETRIC MECHANICAL AND THERMA	72-0068
REVOLUTION	A FINITE ELEMENT ANALYSIS OF SHELLS OF REVOLUTION	73-0759
REVOLUTION	A FINITE ELEMENT ANALYSIS OF BODIES OF REVOLUTION. A COMPARISON BETWEEN CURRENTLY USED TYPES OF ELEM	74-1319
REVOLUTION	A FINITE ELEMENT FOR THERMAL STRESS ANALYSIS OF SHELLS OF REVOLUTION	73-0798
REVOLUTION	A FINITE ELEMENT PROGRAM FOR THE DYNAMIC LARGE DEFLECTION, ELASTIC-PLASTIC ANALYSIS OF STIFFENED SHE	72-0908
REVOLUTION	A METHOD FOR COMPUTATION OF VIBRATION MODES AND FREQUENCIES OF ORTHOTROPIC THIN SHELLS OF REVOLUTION	69-0037
REVOLUTION	A NOTE ON THE FINITE DEFORMATIONS OF A THICK ELASTIC SHELL OF REVOLUTION	71-0193
REVOLUTION	A REFINED CURVED ELEMENT FOR THIN SHELLS OF REVOLUTION	71-0214
REVOLUTION	A STUDY OF A DYNAMICAL FINITE ELEMENT ANALYSIS FOR APPLICATION TO AXIAL WAVE PROPAGATION PROBLEMS IN	71-0589
REVOLUTION	A STUDY OF ELASTIC-PLASTIC ANALYSIS BY THE ASSUMED-STRESS HYBRID FINITE-ELEMENT MODEL, WITH APPLICAT	74-0718
REVOLUTION	A STUDY OF THE MATRIX DISPLACEMENT METHOD AS APPLIED TO SHELLS OF REVOLUTION	66-0111
REVOLUTION	ACCURACY OF FINITE-ELEMENT SOLUTIONS FOR MODAL CHARACTERISTICS OF SHELLS OF REVOLUTION	70-0231
REVOLUTION	AN IMPROVED SHELL OF REVOLUTION ELEMENT UTILISING CUBIC DISPLACEMENT FUNCTIONS	70-0451
REVOLUTION	ANALYSES OF SHELLS OF REVOLUTION BY FINITE ELEMENTS	67-0149
REVOLUTION	ANALYSIS OF BUCKLING AND VIBRATION OF RING-STIFFENED, SEGMENTED SHELLS OF REVOLUTION	69-0496
REVOLUTION	ANALYSIS OF ELASTIC STABILITY OF SHELLS OF REVOLUTION BY THE FINITE ELEMENT METHOD	67-0056
REVOLUTION	ANALYSIS OF ELASTIC SHELLS OF REVOLUTION WITH MEMBRANE AND FLEXURE STRESSES UNDER ARBITRARY LOADING	68-0289
REVOLUTION	ANALYSIS OF ELASTIC-PLASTIC SHELLS OF REVOLUTION UNDER AXI-SYMMETRIC LOADING BY THE FINITE ELEMENT M	67-0102
REVOLUTION	ANALYSIS OF ELASTIC-PLASTIC SHELLS OF REVOLUTION	67-0220
REVOLUTION	ANALYSIS OF LAMINATED SHELL FOR REVOLUTION	66-0023
REVOLUTION	ANALYSIS OF NONLINEAR THERMOELASTIC AND THERMOPLASTIC BEHAVIOUR OF SOLIDS OF REVOLUTION BY THE FINIT	71-0192
REVOLUTION	ANALYSIS OF RING-STIFFENED SHELLS OF REVOLUTION UNDER COMBINED THERMAL AND MECHANICAL LOADING	71-0342
REVOLUTION	ANALYSIS OF SHELLS OF REVOLUTION BY A CURVED FINITE ELEMENT	75-0680
REVOLUTION	APPLICATION OF MATRIX DISPLACEMENT METHOD TO LINEAR ELASTIC ANALYSIS OF SHELLS OF REVOLUTION	65-0025
REVOLUTION	APPLICATION OF THE METHOD OF FINITE ELEMENTS TO CALCULATIONS OF THE STRESSED AND STRAINED STATE OF I	75-0775
REVOLUTION	AXISYMMETRIC VIBRATION OF SHELLS OF REVOLUTION BY THE FINITE ELEMENT METHOD	67-0100
REVOLUTION	CALCULATION OF NON-LINEAR AXISYMMETRIC VIBRATIONS OF THIN SHELLS OF REVOLUTION BY A FINITE ELEMENT M	75-0067
REVOLUTION	CALCULATION OF TEMPERATURE DISTRIBUTIONS IN THIN SHELLS OF REVOLUTION BY THE FINITE ELEMENT METHOD	71-0018
REVOLUTION	COMPUTER PROGRAM FOR ELASTIC-PLASTIC ANALYSIS OF AXISYMMETRICALLY LOADED SHELLS OF REVOLUTION	68-0239
REVOLUTION	CONICAL SEGMENT METHOD FOR ANALYZING OPEN CROWN SHELLS OF REVOLUTION FOR EDGE LOADING	63-0007
REVOLUTION	CRITERIA FOR FINITE ELEMENT DISCRETIZATION OF SHELLS OF REVOLUTION	73-0152
REVOLUTION	DIRECT STIFFNESS METHOD OF ANALYSIS OF SHELLS OF REVOLUTION UTILIZING CURVED ELEMENTS	66-0042
REVOLUTION	DISCRETE-ELEMENT ANALYSIS OF SHELLS OF REVOLUTION	68-0168
REVOLUTION	DISCRETE-ELEMENT MODELLING AND SENSITIVITY STUDIES OF STATICALLY LOADED SINGLE-LAYER AND BOUNDED DOU	70-0129
REVOLUTION	DISCRETE-ELEMENT STATIC ANALYSIS OF BONDED, DOUBLE-LAYER, BRANCHED, THIN SHELLS OF REVOLUTION, PART	68-0102
REVOLUTION	DISCRETE-ELEMENT STATIC ANALYSIS OF CORE-STIFFENED SHELLS OF REVOLUTION FOR ASYMMETRIC MECHANICAL LO	68-0148
REVOLUTION	DISCRETE-ELEMENT STATIC ANALYSIS OF BONDED, DOUBLE-LAYER, BRANCHED, THIN SHELLS OF REVOLUTION, PART	70-0081
REVOLUTION	DYNAMIC ELASTOPLASTIC RESPONSE OF GEOMETRICALLY NONLINEAR ARBITRARY SHELLS OF REVOLUTION UNDER IMPUL	70-0098
REVOLUTION	DYNAMIC RESPONSE OF ARBITRARY SHELLS OF REVOLUTION UNDER COMBINED IMPULSIVE NONISOTHERMAL AND IRREGU	70-0365
REVOLUTION	DYNAPLAS: A FINITE ELEMENT PROGRAM FOR THE DYNAMIC, LARGE DEFLECTION, ELASTIC-PLASTIC ANALYSIS OF ST	72-0773
REVOLUTION	DYNASOR 11 - A FINITE ELEMENT PROGRAM FOR THE DYNAMIC NONLINEAR ANALYSIS OF SHELLS OF REVOLUTION (US	71-0666
REVOLUTION	EFFECT OF LATERAL PRESSURE ON NATURAL FREQUENCIES OF HYPERBOLOIDAL SHELLS OF REVOLUTION	71-0278
REVOLUTION	ELAD PROGRAM MANUAL, A COMPUTER PROGRAM FOR THE ELASTIC ANALYSIS OF DOME STRUCTURES AND AXISYMMETRIC	69-0064
REVOLUTION	ELASTIC STABILITY OF SHELLS OF REVOLUTION BY THE VARIATIONAL APPROACH USING DISCRETE ELEMENTS	66-0106
REVOLUTION	ELASTIC-PLASTIC DISCRETE ELEMENT ANALYSIS OF AXISYMMETRICALLY LOADED SHELLS OF REVOLUTION	67-0099
REVOLUTION	EPAD, A COMPUTER PROGRAM FOR THE ELASTIC-PLASTIC ANALYSIS OF DOME STRUCTURES AND ARBITRARY SOLIDS OF	70-0036
REVOLUTION	FAMSOR, A FINITE ELEMENT PROGRAM FOR THE FREQUENCIES AND MODE SHAPES OF SHELLS OF REVOLUTION	70-0090
REVOLUTION	FINITE ELEMENT ANALYSIS FOR SOLIDS OF REVOLUTION	68-0153
REVOLUTION	FINITE ELEMENT ANALYSIS OF OPEN SHELLS OF REVOLUTION UNDER NONSYMMETRICAL LOADING	68-0295
REVOLUTION	FINITE ELEMENT ANALYSIS OF SHELLS OF REVOLUTION BY MINIMIZATION OF THE POTENTIAL ENERGY FUNCTIONAL	69-0114
REVOLUTION	FINITE ELEMENT ANALYSIS OF SHELLS OF REVOLUTION BY MINIMIZATION OF THE POTENTIAL ENERGY FUNCTIONAL	69-0354
REVOLUTION	FINITE ELEMENT ANALYSIS OF SHELLS OF REVOLUTION BY MINIMIZATION OF THE POTENTIAL ENERGY FUNCTIONAL	70-0481
REVOLUTION	FINITE ELEMENT ANALYSIS OF AN AXISYMMETRICALLY LOADED ORTHOTROPIC SHELL OF REVOLUTION.	72-0438
REVOLUTION	FINITE ELEMENT ANALYSIS OF VIBRATIONS OF INITIALLY STRESSED THIN SHELLS OF REVOLUTION	74-1056
REVOLUTION	FINITE ELEMENT LARGE DEFLECTION ANALYSIS OF ELASTIC-PLASTIC SHELLS OF REVOLUTION SUBJECTED TO AXISYM	72-0137
REVOLUTION	FINITE ELEMENT METHOD FOR THE ANALYSIS OF NOZZLE OPENINGS IN SHELLS OF REVOLUTION	69-0254
REVOLUTION	FINITE ELEMENT METHOD FOR STRESS ANALYSIS OF BODIES OF REVOLUTION SUBJECTED TO AXISYMMETRIC LOADING	71-0409
REVOLUTION	FINITE ELEMENT METHOD FOR DETERMINATION OF THERMAL STRESSES INTERACTIVE GRAPHICS FOR FINITE ELEMENT	74-0155
REVOLUTION	FINITE ELEMENT MODELS FOR SHELLS OF REVOLUTION	68-0027
REVOLUTION	FINITE ELEMENT SOLUTION FOR THIN SHELLS OF REVOLUTION	63-0024
REVOLUTION	FINITE ELEMENT SOLUTION FOR THIN SHELLS OF REVOLUTION	64-0005
REVOLUTION	FINITE ELEMENT SOLUTION OF AXISYMMETRICAL DYNAMIC PROBLEMS OF SHELLS OF REVOLUTION	66-0020
REVOLUTION	FINITE-ELEMENT METHOD FOR THE DETERMINATION OF THERMAL STRESSES IN ANISOTROPIC SOLIDS OF REVOLUTION	74-0262
REVOLUTION	FREE VIBRATION OF SHELLS OF REVOLUTION USING RING ELEMENTS	67-0088
REVOLUTION	FREE VIBRATION OF SHELLS OF REVOLUTION USING FINITE ELEMENT METHOD	74-0232
REVOLUTION	HIGH PRECISION FINITE ELEMENT ANALYSIS OF SHELLS OF REVOLUTION	71-0524
REVOLUTION	HIGH-PRECISION FINITE-ELEMENT ANALYSIS OF SHELLS OF REVOLUTION	73-1164
REVOLUTION	IMPROVEMENTS ON THE ANALYSIS OF SHELLS OF REVOLUTION BY THE MATRIX DISPLACEMENT METHOD	66-0070
REVOLUTION	INCREMENTAL ANALYSIS OF LARGE DEFLECTIONS OF SHELLS OF REVOLUTION	71-0296
REVOLUTION	INEXTENSIONAL VIBRATIONS OF PARABOLOIDAL SHELLS OF REVOLUTION. DOMES AND SHELLS	71-0427
REVOLUTION	INTERFACE CORE DISCRETE-ELEMENT FOR STIFFENED SHELLS OF REVOLUTION	67-0092
REVOLUTION	LARGE DEFLECTION ANALYSIS OF ELASTIC-PLASTIC SHELLS OF REVOLUTION	69-0174
REVOLUTION	LARGE DEFLECTION ANALYSIS OF ELASTIC-PLASTIC SHELLS OF REVOLUTION	70-0507
REVOLUTION	LARGE DEFLECTION ELASTIC-PLASTIC ANALYSIS OF SHELLS OF REVOLUTION	71-0263
REVOLUTION	LARGE DEFLECTION ELASTIC-PLASTIC DYNAMIC RESPONSE OF STIFFENED SHELLS OF REVOLUTION	72-0608
REVOLUTION	LARGE DEFLECTION ELASTIC-PLASTIC DYNAMIC RESPONSE OF STIFFENED SHELLS OF REVOLUTION	74-0496
REVOLUTION	LINEAR AND NON LINEAR ANALYSIS OF SHELLS OF REVOLUTION WITH ASYMMETRICAL STIFFNESS PROPERTIES	68-0151
REVOLUTION	LINEAR AND NONLINEAR STATIC ANALYSIS OF AXISYMMETRICALLY LOADED THIN SHELLS OF REVOLUTION	69-0215
REVOLUTION	METHOD OF FINITE ELEMENTS FOR CALCULATION OF HETEROGENEOUS BODIES OF REVOLUTION IN ELASTO-PLASTIC ME	75-0144
REVOLUTION	NATURAL VIBRATIONS OF LAMINATED ORTHOTROPIC SHELLS OF REVOLUTION	70-0367
REVOLUTION	NEW FINITE ELEMENT METHOD FOR ANALYZING SYMMETRICALLY LOADED THIN SHELLS OF REVOLUTION	73-0389
REVOLUTION	NONLINEAR ANALYSIS OF SHELLS OF REVOLUTION BY THE MATRIX DISPLACEMENT METHOD	67-0079
REVOLUTION	NONLINEAR ANALYSIS OF ELASTIC SHELLS OF REVOLUTION UNDER AXISYMMETRIC LOADING	70-0161
REVOLUTION	NONLINEAR BEHAVIOR OF SHELLS OF REVOLUTION UNDER CYCLIC LOADING	73-0242
REVOLUTION	NONLINEAR DYNAMIC ANALYSIS OF SHELLS OF REVOLUTION BY MATRIX DISPLACEMENT METHOD	70-0110
REVOLUTION	NONLINEAR DYNAMIC ANALYSIS OF SHELLS OF REVOLUTION WITH ASYMMETRIC PROPERTIES BY THE FINITE ELEMENT	75-0217
REVOLUTION	NONLINEAR DYNAMIC ANALYSIS OF SHELLS OF REVOLUTION WITH ASYMMETRIC PROPERTIES BY THE FINITE ELEMENT	75-0657
REVOLUTION	NONLINEAR FINITE ELEMENT ANALYSIS OF SANDWICH SHELLS OF REVOLUTION	72-0180
REVOLUTION	NONLINEAR TRANSIENT ANALYSIS OF SHELLS AND SOLIDS OF REVOLUTION BY CONVECTED ELEMENTS	73-0519
REVOLUTION	NUMERICAL ANALYSIS FOR FINITE AXISYMMETRIC DEFORMATIONS OF INCOMPRESSIBLE ELASTIC SOLIDS OF REVOLUTI	70-0094
REVOLUTION	ON THE ASYMPTOTICALLY SPHERICAL DEFORMATIONS OF ARBITRARY MEMBRANES OF REVOLUTION FIXED ALONG AN EDG	71-0640
REVOLUTION	ON THE DYNAMIC BUCKLING OF SHELLS OF REVOLUTION	73-1102
REVOLUTION	PART 1 - TASK 3.2.9. BENDING AND BUCKLING OF A ONE-DIMENSIONAL MODEL SIMULATING A CONICAL SHELL. PAR	71-0745
REVOLUTION	PROGRESS REPORT ON DISCRETE-ELEMENT ELASTIC AND ELASTIC-PLASTIC ANALYSIS OF SHELLS OF REVOLUTION SUB	68-0170
REVOLUTION	REFINED FINITE ELEMENT ANALYSIS OF ELASTIC-PLASTIC THIN SHELLS OF REVOLUTION	69-0497

351

REVOLUTION	REFINED MIXED METHOD FINITE ELEMENTS FOR SHELLS OF REVOLUTION	71-0101
REVOLUTION	REINVESTIGATION OF BUCKLING OF SHELLS OF REVOLUTION BY A REFINED FINITE ELEMENT	74-0097
REVOLUTION	SAFE PCPS, A COMPUTER PROGRAM FOR THE STRESS ANALYSIS OF COMPOSITES BODIES OF REVOLUTION. INPUT INST	65-0013
REVOLUTION	SAFE SHELL, A COMPUTER PROGRAM FOR THE STRESS ANALYSIS OF THIN SHELL BODIES OF REVOLUTION. A USER'S	67-0018
REVOLUTION	SAFE-AXISYM, A COMPUTER PROGRAM FOR THE DESIGN OF COMPOSITE BODIES OF REVOLUTION. INPUT INSTRUCTIONS	65-0014
REVOLUTION	SAMMSOR 111: A FINITE ELEMENT PROGRAM TO DETERMINE STIFFNESS AND MASS MATRICES OF RING-STIFFENED SHE	72-0774
REVOLUTION	SEAL-SHELL-2, A COMPUTER PROGRAM FOR THE STRESS ANALYSIS OF A THICK SHELL OF REVOLUTION WITH AXISYMM	69-0107
REVOLUTION	SNAP-THROUGH BUCKLING ANALYSIS OF SHELLS OF REVOLUTION USING INCREMENTAL STIFFNESS MATRICES	74-1349
REVOLUTION	STABILITY ANALYSIS OF SHELLS OF REVOLUTION BY THE FINITE ELEMENT METHOD	68-0124
REVOLUTION	STABILITY ANALYSIS OF SHELLS OF REVOLUTION	68-0136
REVOLUTION	STATIC ANALYSIS OF SHELLS OF REVOLUTION USING DOUBLY-CURVED QUADRILATERAL ELEMENTS DERIVED FROM ALTE	69-0368
REVOLUTION	STATIC, STABILITY, AND DYNAMIC ANALYSIS OF SHELLS OF REVOLUTION BY NUMERICAL INTEGRATION - A COMPARI	74-0389
REVOLUTION	STRESS, STABILITY AND VIBRATION OF COMPLEX SHELLS OF REVOLUTION: ANALYSIS AND USERS MANUAL FOR BOSOR	69-0441
REVOLUTION	THE LINEAR ELASTIC DYNAMIC ANALYSIS OF SHELLS OF REVOLUTION BY THE MATRIX DISPLACEMENT METHOD	66-0046
REVOLUTION	THE NONLINEAR DYNAMIC ANALYSIS OF SHELLS OF REVOLUTION WITH ASYMMETRIC PROPERTIES BY THE FINITE ELEM	71-0818
REVOLUTION	THE TRANSIENT LINEAR ELASTIC RESPONSE ANALYSIS COMPLEX THIN SHELLS OF REVOLUTION SUBJECTED TO ARBITR	70-0080
REVOLUTION	THEORETICAL ANALYSIS OF THICK DOME STRUCTURES AND AXISYMMETRIC SOLIDS OF REVOLUTION SUBJECTED TO UNSYM	70-0037
REVOLUTION	UNSYMMETRIC BUCKLING OF SHELLS OF REVOLUTION BY FINITE ELEMENT METHOD	70-0500
REVOLUTION	USE OF HIGHER ORDER DISPLACEMENT FUNCTIONS IN THE FREE VIBRATION ANALYSIS OF SHELLS OF REVOLUTION HA	74-1055
REVOLUTION	USERS' MANUAL FOR DIGITAL COMPUTER PROGRAM OK1. NUMERICAL ANALYSIS OF FINITE AXISYMMETRIC DEFORMATIO	72-0861
REVOLUTION	VERIFICATION OF A HYBRID GENERAL PURPOSE SHELL AND BODY OF REVOLUTION COMPUTER PROGRAM	71-0333
REVOLUTION	VIBRATION CHARACTERISTICS OF THIN, PRESSURIZED SHELLS OF REVOLUTION PARTIALLY FILLED WITH LIQUID	72-0604
REVOLUTION	VISCOELASTIC ANALYSIS OF PRESTRESSED CONCRETE SHELL OF REVOLUTION	73-0464
REVOLUTTION	NONLINEAR TRANSIENT ANALYSIS OF SHELLS AND SOLIDS OF REVOLUTTION BY CONVECTED ELEMENTS	74-0476
REYNOLD'S	FINITE ELEMENT SOLUTION OF THE THREE DIMENSIONAL FLOW PROBLEMS AND OF REYNOLD'S EQUATION FOR INCOMPR	72-0011
REYNOLDS	A FINITE ELEMENT METHOD FOR LOW REYNOLDS NUMBER FLOW AND ITS APPLICATION TO BIOMECHANICS	69-0304
REYNOLDS	LOW REYNOLDS NUMBER DEVELOPING FLOW	69-0020
REYNOLDS	NUMERICAL SOLUTION FOR THE FLOW AROUND A CYLINDER AT REYNOLDS NUMBERS OF 40, 200 AND 500	69-0378
REYNOLDSZAHLEN	THEORETICAL AND EXPERIMENTAL CONSIDERATIONS ON FLOW OF VISCOUS LIQUIDS PAST PLATE AT SMALL AND AVERA	70-0144
RHEOLOGICAL	RHEOLOGICAL THEORY OF STRENGTH OF MATERIALS	69-0313
RHODE	MODELING AN OCEAN POND. A TWO-DIMENSIONAL, FINITE ELEMENT HYDRODYNAMIC MODEL OF NINIGRET POND, CHARL	75-0686
RHOMBIC	PARALLELOGRAM ELEMENT IN THE SOLUTION OF RHOMBIC CANTILEVER PLATE BENDING	66-0022
RIBBED	FIELD ANALYSIS FOR PREFABRICATED RIBBED CYLINDRICAL SHELLS	73-0899
RIBBED	ISSLEDOVANIE NAPRYAZHENNOGO SOSTOYANIYA REBRISTYKH ISILINDRICHESKIKH OBOLOCHEK S PRYAMOUGOL'NYMI OTV	74-0944
RIBBED	MACRO-APPROACH FOR RIBBED AND GRID PLATE SYSTEMS	75-0079
RIBBON	STRESS ANALYSIS OF RIBBON REINFORCED COMPOSITES	70-0221
RICHARDSON	COMBINED APPLICATION OF FINITE ELEMENT METHODS AND RICHARDSON EXTRAPOLATION TO THE TORSION PROBLEM	72-0066
RICHARDSON	COMPILED APPLICATION OF FINITE ELEMENT METHODS AND RICHARDSON EXTRAPOLATION TO THE TORSION PROBLEM	73-1012
RICHARDSON	RICHARDSON EXTRAPOLATION FOR PARABOLIC GALERKIN METHODS	72-0522
RIDDLE	RIDDLE OF BONDED AND UNBONDED TENDONS IN PRESTRESSED CONCRETE REACTOR VESSELS	74-0519
RIDGE	THEORETICAL COMPUTATIONS ON RIDGE ACOUSTIC SURFACE WAVES USING THE FINITE-ELEMENT METHOD	71-0011
RIESENE	GEOMETRICKY NELINEARNE ULOHY PLOCHYCH SKRUPIN RIESENE PRIRASTKOVOU METODOU KONECNYCH PRVKOV. (GEOMET	74-1032
RIESENIE	RIESENIE VAZKOPRUZNYCH ANIZOTROPICKYCH DOASK METODOU KONECNYCH PRVKOV (SOLUTION OF VISCOELASTIC ANIS	74-0477
RIGHT	A " LOCAL " BASIS OF GENERALIZED SPLINES OVER RIGHT TRIANGLES DETERMINED FROM A NONUNIFORM PARTITION	72-0525
RIGHT	DERIVATION OF A STIFFNESS MATRIX FOR A RIGHT TRIANGULAR PLATE IN BENDING AND SUBJECTED TO INITIAL ST	62-0007
RIGHT	ORTHOTROPIC RIGHT BRIDGES BY THE FINITE STRIP METHOD	69-0028
RIGHT	TEMPERATURE RISE IN THE HIGH SPEED COMPRESSION OF RIGHT CYLINDRICAL BILLETS	74-1153
RIGID	A FINITE ELEMENT STUDY OF THE PENETRATION OF A SOIL HALF-SPACE BY A RIGID PROJECTILE	69-0518
RIGID	ALGORYT OBLICZEN ORGAN WYMUSZONYCH METODA SZTYWNYCH ELEMENTOW SKONCZONYCH (ALGORITHM FOR CALCULATIN	74-0581
RIGID	AN ADVANCED ANALYSIS OF RIGID PAVEMENTS	72-0417
RIGID	AUTOMATIC EXTRACTION OF RIGID BODY MODES FROM STRESS AND STRAIN ELEMENTS	72-0913
RIGID	CALCULATION METHOD FOR THE STUDY OF CRACKED RIGID STRUCTURES (METODO DI CALCOLO PER LO STUDIO DI STR	71-0637
RIGID	COMPUTER ANALYSIS OF RIGID FRAMES	71-0401
RIGID	DYNAMIC ANALYSIS OF A SYSTEM OF HINGE-CONNECTED RIGID BODIES WITH NONRIGID APPENDAGES	74-0852
RIGID	EFFECT OF OPENINGS ON STRESSES IN RIGID PAVEMENTS	73-0344
RIGID	ELASTIC-PLASTIC ANALYSIS OF A WHEEL ROLLING ON A RIGID TRACK	74-1040
RIGID	ELASTOPLASTIC INDENTATION OF HALF-SPACE BY AN INFINITELY LONG RIGID CIRCULAR CYLINDER	71-0395
RIGID	EVALUATION OF RIGID PAVEMENTS BY NONDESTRUCTIVE TESTS.	72-0459
RIGID	EVALUATION OF STRESSES IN SOIL UNDER RIGID WHEELS BY FINITE ELEMENT METHOD	73-1109
RIGID	FINITE ELEMENT ANALYSIS OF CYLINDRICAL SHELLS UTILIZING CURVED ELEMENTS CONTAINING RIGID BODY DEFORM	71-0823
RIGID	FINITE ELEMENT ANALYSIS OF THIN ELASTIC SHELLS WITH RESIDUAL ENERGY BALANCING AND THE ROLE OF THE RI	75-0034
RIGID	FINITE ELEMENT ANALYSIS OF THE SURFACE DEFORMATION DUE TO A UNIFORM LOADING ON A LAYER OF GIBSON SOI	75-0240
RIGID	FINITE-ELEMENT ANALYSIS OF CONCRETE SLABS AND ITS IMPLICATIONS FOR RIGID PAVEMENT DESIGN	73-0538
RIGID	FINITE-ELEMENT ANALYSIS OF THIN ELASTIC SHELLS WITH RESIDUAL ENERGY BALANCING AND ROLE OF RIGID BODY	75-0048
RIGID	FORCES ON RIGID CULVERTS UNDER HIGH FILLS	67-0008
RIGID	IMPLICIT RIGID BODY MOTION IN CURVED FINITE ELEMENTS	71-0174
RIGID	INVESTIGATION OF THE STATIONARY FLOW OF A RIGID PLASTIC MATERIAL BY THE NUMERICAL FINITE ELEMENT MET	73-0231
RIGID	MECHANICS OF CABLE MOORING SYSTEMS. VOLUME V1. A COMPUTER PROGRAM FOR ANALYZING THE STEADY STATE CON	73-0725
RIGID	ON THE EXTREMAL PROPERTIES OF THE SOLUTION IN DYNAMICS OF RIGID VISCOPLASTIC BODIES ALLOWING FOR LAR	73-1080
RIGID	ON THE RIGID DISPLACEMENT CONDITION.	72-0405
RIGID	RIGID BODY DISPLACEMENTS OF CURVED ELEMENTS IN THE ANALYSIS OF SHELLS BY THE MATRIX-DISPLACEMENT MET	67-0031
RIGID	RIGID BODY MOTIONS AND EQUILIBRIUM IN FINITE ELEMENTS	74-0102
RIGID	RIGID BODY MOTIONS IN CURVED FINITE ELEMENTS	70-0035
RIGID	RIGID PAVEMENT STRESSES UNDER AIRCRAFT LOADING	71-0423
RIGID	RIGID PLATE ON NONLINEARLY DEFORMABLE COHESIVE BASE (PLANE PROBLEM)	75-0264
RIGID	STEADY STATE FLOW IN RIGID NETWORKS OF FRACTURES	74-0278
RIGID	STRESS DISTRIBUTIONS AROUND SHALLOW BURIED RIGID PIPES	74-0142
RIGID	STRUCTURAL INELASTICITY. V11. ELASTIC-PLASTIC ANALYSIS OF A WHEEL ROLLING ON A RIGID TRACK	73-0886
RIGID	THE ELASTO-PLASTIC INDENTATION OF A HALF-SPACE BY A RIGID SPHERE	71-0113
RIGID	TRANSIENT RESPONSE OF INELASTICALLY CONSTRAINED RIGID BODY SYSTEMS	73-0291
RIGID	VERTICAL RESPONSE OF A RIGID BASE TO DYNAMIC LOADING	69-0316
RIGID-BODY	EXPLICIT ADDITION OF RIGID-BODY MOTIONS IN CURVED FINITE ELEMENTS	73-0037
RIGID-BODY	RIGID-BODY MOTIONS AND STRAIN-DISPLACEMENT EQUATIONS OF CURVED SHELL FINITE ELEMENTS	72-0267
RIGID-FINITE-EL	OBLICZANIE ORGAN METODA SZTYWNYCH ELEMENTOW SKONCZONYCH. (CALCULATING VIBRATIONS BY THE RIGID-FINITE	75-0649
RIGIDES	UNE METHODE DE CALCUL PAR ELEMENTS FINIS DES MOUVEMENTS DE LIQUIDES DANS DES RESERVOIRS RIGIDES OU D	74-1171
RIGIDITE	DEPLACEMENTS DANS LES STRUCTURES MINCES A FAIBLE RIGIDITE GEOMETRIQUE-DETERMINATION NUMERIQUE (NUMER	72-0213
RIGIDITE	MATRICE DE RIGIDITE D'UN ELEMENT DE POUTRE SOLLICITE A LA TORSION-FLEXION (STIFFNESS MATRIX OF A BEA	73-0442
RIGIDITIES	RELATIONSHIP BETWEEN FINITE ELEMENT AND MEASURED JOINT-FLOOR CONNECTION RIGIDITIES	75-0738
RIGIDITY	EINE FINITE-ELEMENTE-LOESUNG FUR DIE TORSIONSSTEIFIGKEIT UND DEN SCHUBMITTELPUNKT BELIEBIGER QUERSCH	74-0256
RIGIDITY	FINITE ELEMENTS SOLUTION FOR TORSION RIGIDITY AND SHEAR CENTER OF ARBITRARY CROSS-SECTIONS	74-0168
RIGIDITY	FLEXURAL RIGIDITY OF A THIN WALLED BUILD-UP ROTOR FOR JET ENGINE. MEASUREMENT BY STATIC LOAD TEST AND	73-0756
RIGIDITY	ISSLEDOVANIE PROCHNOSTI I ZHESTKOSTI PLASTIN. NAGRUZHENNYKH V SVOEI PLOSKOSTI, METHODOM KONECNYKH E	73-0382
RIMMED	VIBRATION ANALYSIS OF PLATES. SHELLS AND BLADED RIMMED DISCS USING THE TRANSFER-MATRIX-FINITE-ELEMEN	73-1151
RING	AXISYMMETRIC RING AND PLATE STRESS ANALYSIS PROGRAM - USERS GUIDE	70-0333
RING	FINITE ELEMENT MODELING OF A CIRCULAR RING USING HALF AND QUARTER SYMMETRY	66-0119
RING	FINITE-ELEMENT METHOD WITH ELEMENTS WITH CIRCULAR RING SECTOR FORM	73-0105
RING	FORMULATION AND EVALUATION OF A TOROIDAL RING DISCRETE ELEMENT	66-0082
RING	FORMULATION AND EVALUATION OF A TRIANGULAR CROSS-SECTION RING DISCRETE ELEMENT	66-0083
RING	FORMULATION AND EVALUATION OF A TRAPEZOIDAL CROSS-SECTION RING DISCRETE ELEMENT	66-0084
RING	FREE VIBRATION OF RING STIFFENED CYCLINDRICAL SHELLS	70-0172
RING	FREE VIBRATION OF SHELLS OF REVOLUTION USING RING ELEMENTS	67-0088
RING	MASS LOADING EFFECTS ON VIBRATION OF RING AND SHELL STRUCTURES	68-0245

RING	ON MAXIMUM FILLET STRESSES IN BREECH RING	72-0736
RING	RING FINITE ELEMENTS FOR AXISYMMETRIC AND NON-AXISYMMETRIC THIN SHELL ANALYSIS	72-0922
RING	THE ANALYSIS OF COOLING TOWER RING BEAMS	68-0063
RING	THE RING OF FINITE ELEMENTS IN A NON-STANDARD MODEL OF THE REALS	71-0450
RING-STIFFENED	ANALYSIS OF BUCKLING AND VIBRATION OF RING-STIFFENED, SEGMENTED SHELLS OF REVOLUTION	69-0496
RING-STIFFENED	ANALYSIS OF RING-STIFFENED SHELLS OF REVOLUTION UNDER COMBINED THERMAL AND MECHANICAL LOADING	71-0342
RING-STIFFENED	MEASURED AND CALCULATED VIBRATION PROPERTIES OF RING-STIFFENED HONEY-COMB CYLINDERS	71-0834
RING-STIFFENED	SAMMSOR 111: A FINITE ELEMENT PROGRAM TO DETERMINE STIFFNESS AND MASS MATRICES OF RING-STIFFENED SHE	72-0774
RINGS	A COMPARISON OF FINITE ELEMENT AND FOURIER SERIES SOLUTIONS AS APPLIED TO RADIALLY LOADED CIRCULAR R	68-0024
RINGS	FINITE-ELEMENT PREDICTIONS OF TRANSIENT ELASTIC-PLASTIC LARGE DEFLECTIONS OF STIFFENED AND/OR UNSTIF	74-0825
RINGS	THEORETICAL AND EXPERIMENTAL STUDIES OF TRANSIENT ELASTIC-PLASTIC LARGE DEFLECTIONS OF GEOMETRICALLY	75-0830
RINGS	THEORETICAL AND EXPERIMENTAL STUDIES OF TRANSIENT ELASTIC-PLASTIC LARGE DEFLECTIONS OF GEOMETRICALLY	75-0831
RISE	EFFECTS OF RISE TIME AND DAMPING ON FINITE ELEMENT ANALYSIS OF RESPONSE OF STRUCTURES	71-0087
RISE	MEMBRANE ANALYSIS OF UNFRAMED HIGH RISE BUILDINGS	73-0451
RISE	TEMPERATURE RISE IN THE HIGH SPEED COMPRESSION OF RIGHT CYLINDRICAL BILLETS	74-1153
RISE	ULTIMATE STRAIN CAPACITY AND TEMPERATURE RISE STUDIES, TRUMBULL POND DAM	72-0705
RISK	ROCK MECHANICS AND RISK IN OPEN PIT MINING	73-0544
RISOLUZIONE	ON THE RESOLUTION OF NONLINEAR ELASTIC AND ELASTOPLASTIC PROBLEMS BY THE METHOD OF FINITE ELEMENTS.	74-1326
RITZ	FINITE ELEMENT ANALOGUE OF THE MODIFIED RAYLEIGH RITZ METHOD FOR VIBRATION PROBLEMS	73-0181
RITZ	HERMITE INTERPOLATION TYPE RITZ METHODS FOR TWO POINT BOUNDARY VALUE PROBLEMS	66-0089
RITZ	RAYLEIGH RITZ GALERKIN METHODS FOR SYMMETRIC POSITIVE DIFFERENTIAL EQUATIONS APPLICATION TO FINITE E	73-1030
RITZ	STRESS WAVE ANALYSIS IN LAYERED THERMOVISCOELASTIC MATERIALS BY THE EXTENDED RITZ METHOD, VOLUME II	68-0125
RITZ	THE FINITE DIFFERENCE AND LOCALIZED RITZ METHODS	71-0447
RITZ	UNIFICATION OF THE FINITE ELEMENT AND CLASSICAL RITZ METHODS	73-1139
RITZ	UNIFORM CONVERGENCE OF VARIATIONAL FINITE ELEMENT AND RITZ METHODS	74-0420
RITZ-GALERKIN	FOURIER ANALYSIS OF THE FINITE ELEMENT METHOD IN RITZ-GALERKIN THEORY	69-0297
RITZ'	BERECHNUNG VON PLATTEN NACH DEM RITZ' SCHEN VERFAHREN MIT HILFE DREIECKFOERMIGER MASCHENNETZE	66-0094
RITZ'S	APPLICATION OF PITZ'S METHOD TO THIN ELASTIC SHELL ANALYSIS	71-0083
RIVER	FINITE ELEMENT ANALYSES OF PORT ALLEN AND OLD RIVER LOCKS	69-0203
RIVER	MATHEMATICAL MODEL OF THE COLUMBIA RIVER FROM THE PACIFIC OCEAN TO BONNEVILLE DAM. PART 1. THEORY, P	69-0464
RIVER	MATHEMATICAL MODEL OF THE COLUMBIA RIVER FROM THE PACIFIC OCEAN TO BONNEVILLE DAM. PART II: INPUT-OU	70-0417
RIVER	PROCEEDINGS OF INTERNATIONAL ASSOCIATION FOR HYDRAULIC RESEARCH AND PROCEEDINGS OF INTERNATIONAL SYM	73-0534
RIVER	SEEPAGE IN MISSISSIPPI RIVER BANKS. REPORT 1. ANALYSIS OF TRANSIENT SEEPAGE USING VISCOUS FLOW MODEL	70-0405
RIVER	SEEPAGE IN MISSISSIPPI RIVER BANKS. REPORT 1. ANALYSIS OF TRANSIENT SEEPAGE USING A VISCOUS-FLOW MOD	73-0909
ROAD	APPLICATION DES CALCULUS DE FONDATIONS A L'EVALUATION DU COMPORTEMENT DES CHAUSSEES EM DASH 1, 2. (A	71-0429
ROADBUILDING	SIGNIFICANCE OF VARIABLY CONFINED TRIAXIAL TESTING ROADBUILDING MATERIALS	74-0554
ROADWAYS	FORECASTING THE ROCK PRESSURE AND ITS EFFECT ON THE LONGWALL FACES AND ROADWAYS IN THE BITUMINOUS CO	74-0374
ROADWAYS	VORAUSSAGE DES GEBIRGSDRUCKES UND SEINER AUSQIRKUNGEN AUF STREBEN UND STRECKEN IM STEINKOHLENBERGBAU	73-0679
ROCHE	ANALYSE TRIDIMENSIONELLE DE LA STABILITE D'UN TUNNEL AU VOISINAGE DU FRONT DE TAILLE DANS UNE ROCHE	74-0995
ROCHES	CALCUL DE MASSIFS EN ELASTOPLASTICITE APPLICATIONS A LA MECANIQUE DES ROCHES ET DES SOLS. (APPLICATI	74-0223
ROCHEUX	ANALYSE DE LA STABILITY DES TALUS ROCHEUX PAR UNE METHODE GENERALISEE DE L EQUILIBRE LIMITE (ROCK SL	74-1019
ROCK	A CASE STUDY OF IN-SITU ROCK DEFORMATION BEHAVIOR FOR THE DESIGN OF GROUND SUPPORT SYSTEM	72-0668
ROCK	A DESIGN PROCEDURE FOR HIGH CUTS IN JOINTED HARD ROCK. THREE DIMENSIONAL SOLUTIONS. VOLUME 1	71-0656
ROCK	A FINITE ELEMENT MODEL FOR ANISOTROPIC YIELDING IN GRAVITY LOADED ROCK	69-0012
ROCK	A MODEL FOR THE MECHANICS OF JOINTED ROCK	68-0079
ROCK	A THEORETICAL METHOD FOR EVALUATING STABILITY OF OPENINGS IN ROCK	72-0877
ROCK	AN APPLICATION OF THE FINITE ELEMENT METHOD FOR ROCK SLOPE STABILITY ANALYSIS	71-0813
ROCK	ANALYSE DE LA STABILITY DES TALUS ROCHEUX PAR UNE METHODE GENERALISEE DE L EQUILIBRE LIMITE (ROCK SL	74-1019
ROCK	ANALYSE TRIDIMENSIONELLE DE LA STABILITE D'UN TUNNEL AU VOISINAGE DU FRONT DE TAILLE DANS UNE ROCHE	74-0995
ROCK	ANALYSIS AND MODEL STUDIES OF UNDERGROUND OPENINGS IN JOINTED ROCK. RESEARCH ON ROCK BOLT REINFORCEM	71-0438
ROCK	ANALYSIS AND MODEL STUDIES OF UNDERGROUND OPENINGS IN JOINTED ROCK. RESEARCH ON ROCK BOLT REINFORCEM	71-0438
ROCK	ANALYSIS AND MODEL STUDIES OF UNDERGROUND OPENINGS IN JOINTED ROCK. RESEARCH ON ROCK BOLT REINFORCEM	71-0750
ROCK	ANALYSIS AND MODEL STUDIES OF UNDERGROUND OPENINGS IN JOINTED ROCK. RESEARCH ON ROCK BOLT REINFORCEM	71-0750
ROCK	ANALYSIS OF MULTI-AXIAL ANISOTROPIC CREEP, ROOF RESIN BOLTS, AND BED SEPARATION IN ROCK STRUCTURES U	74-1122
ROCK	ANALYSIS OF NONELASTIC AND NONLINEAR VIBRATION OF ROCK DAMS USING FINITE ELEMENT METHOD	73-0680
ROCK	ANALYSIS OF NONLINEAR PROBLEMS IN ROCK MECHANICS WITH PARTICULAR REFERENCE TO JOINTED ROCK SYSTEMS	70-0282
ROCK	ANALYSIS OF NONLINEAR PROBLEMS IN ROCK MECHANICS WITH PARTICULAR REFERENCE TO JOINTED ROCK SYSTEMS	70-0282
ROCK	ANALYSIS OF ROCK MASS DEFORMATIONS DUE TO EXCAVATIONS	73-0395
ROCK	ANALYSIS OF ROCK SLOPES, USING THE FINITE ELEMENT METHOD	70-0184
ROCK	ANALYSIS OF UNDERGROUND OPENINGS IN ROCK BY FINITE ELEMENT METHODS	73-0895
ROCK	APPLICATION OF ELASTO-PLASTIC ANALYSIS IN ROCK MECHANICS BY FINITE ELEMENT METHOD	71-0349
ROCK	APPLICATION OF THE FINITE ELEMENT METHOD OF ANALYSIS IN SOLVING BOUNDARY VALUE PROBLEMS IN ROCK MECH	66-0013
ROCK	APPLICATION OF THE FINITE ELEMENT METHOD TO PROBLEMS OF ROCK MECHANICS	66-0088
ROCK	APPLICATION OF THE FINITE ELEMENT METHOD TO PROBLEMS IN SOIL AND ROCK MECHANICS	67-0118
ROCK	APPLICATION OF THE FINITE ELEMENT METHOD TO THE ANALYSIS OF ROCK SAMPLES UNDER UNIAXIAL COMPRESSION	70-0182
ROCK	ASPECTS OF MECHANICAL BEHAVIOR OF ROCK UNDER STATIC AND CYCLIC LOADING. PART A. MECHANICAL BEHAVIOR	72-0761
ROCK	ASPECTS OF MECHANICAL BEHAVIOR OF ROCK UNDER STATIC AND CYCLIC LOADING. PART A. MECHANICAL BEHAVIOR	72-0761
ROCK	ASPECTS OF MECHANICAL BEHAVIOR OF ROCK UNDER STATIC AND CYCLIC LOADING. PART A: MECHANICAL BEHAVIOR	72-0762
ROCK	ASPECTS OF MECHANICAL BEHAVIOR OF ROCK UNDER STATIC AND CYCLIC LOADING. PART A: MECHANICAL BEHAVIOR	72-0762
ROCK	BEHAVIOUR OF THE ROCK MASS AT THE SITE OF A ROCKFILL DAM	74-0992
ROCK	BIT PENETRATION INTO ROCK - A FINITE ELEMENT STUDY	75-0800
ROCK	BUTTRESS DAMS ON COMPLEX ROCK FOUNDATIONS	64-0013
ROCK	CALCUL DE MASSIFS EN ELASTOPLASTICITE APPLICATIONS A LA MECANIQUE DES ROCHES ET DES SOLS. (APPLICATI	74-0223
ROCK	COMPUTER APPLICATIONS IN ROCK MECHANICS	72-0190
ROCK	CONTINUUM CHARACTERIZATION OF JOINTED ROCK MASSES - 1, 2	73-0196
ROCK	CONTINUUM MECHANICS AS AN APPROACH TO ROCK MASS PROBLEMS	68-0172
ROCK	CONTROLLED FAILURE OF HOLLOW ROCK CYLINDERS IN UNIAXIAL COMPRESSION	72-0196
ROCK	DESIGN OF ROOF BOLT PATTERNS FOR JOINTED ROCK	74-1318
ROCK	DEVELOPMENT AND APPLICATIONS OF THEORETICAL METHODS FOR EVALUATING STABILITY OF OPENINGS IN ROCK	74-0791
ROCK	DEVELOPMENT AND APPLICATIONS OF A GENERAL COMPUTER PROGRAM FOR EVALUATING STABILITY OF OPENINGS IN R	74-0994
ROCK	DIE BERECHNUNG DER BEWEGUNGEN UND SPANNUNGEN DES GEBIRGES VOM ABBAU BIS ZUR TAGESOBERFLAECHE NACH DE	73-0456
ROCK	DIRECT AND INVERSE HEAT CONDUCTION IN ROCK MATERIALS USING THE FINITE ELEMENT METHOD	74-1098
ROCK	DYNAMIC RESPONSE OF REINFORCED OPENINGS IN ROCK	71-0311
ROCK	EFFECT OF DISCONTINUITIES IN STRESSES AND DEFORMATIONS OF ROCK AROUND A TUNNEL, (EINFLUSS VON DISKON	70-0170
ROCK	EFFECT OF SHAPE AND HETEROGENEITY OF ROCK SAMPLES ON THEIR COMPRESSIVE STRENGTH	69-0315
ROCK	ESTIMATION OF THE MECHANICAL PROPERTIES OF ROCK MASSES	73-0935
ROCK	EXCAVATION IN ROCK, AN APPRECIATION OF THE FINITE ELEMENT METHOD OF ANALYSIS	73-0313
ROCK	FINITE ELEMENT ANALYSES OF SLOPES IN JOINTED ROCK	68-0269
ROCK	FINITE ELEMENT FOR ROCK JOINTS AND INTERFACES	73-0411
ROCK	FINITE ELEMENT METHOD AND ROCK MECHANICS	73-0551
ROCK	FINITE ELEMENT MODELING CRITERIA FOR UNDERGROUND OPENINGS IN ROCK	74-0630
ROCK	FINITE ELEMENT TECHNIQUES IN ROCK MECHANICS	68-0028
ROCK	FORECASTING THE ROCK PRESSURE AND ITS EFFECT ON THE LONGWALL FACES AND ROADWAYS IN THE BITUMINOUS CO	74-0374
ROCK	INFLUENCE OF ROCK PROPERTIES VARIABILITY ON MINE OPENING STABILITY ANALYSIS	74-1021
ROCK	INFLUENCE OF TEMPERATURE DEPENDENT PROPERTIES ON THERMAL ROCK FRAGMENTATION	75-0222
ROCK	INTERACTION OF ROCK AND TUNNEL LINING	73-1057
ROCK	LOAD TRANSFER MECHANISM FOR PIERS SOCKETTED IN HARD SOILS OR ROCK	74-0969
ROCK	METHODS OF CALCULATION OF THREE-DIMENSIONAL PROBLEMS OF PERCOLATION OF FISSURED ROCK BY FINITE ELEME	70-0472
ROCK	METODA ELEMENTOW SKONSZONYCH W MECHANICE GRUNTOW I MECHANICE GOROTWORU (FINITE ELEMENT METHOD IN SOI	73-0500
ROCK	NEW HORIZONS IN ROCK MECHANICS	73-0428
ROCK	NEW HORIZONS IN ROCK MECHANICS	73-0446
ROCK	NON-LINEAR STRESS ANALYSIS OF TWO DIMENSIONAL PROBLEMS WITH SPECIAL REFERENCE TO ROCK AND SOIL MECHA	68-0183

ROCK	NONLINEAR FINITE ELEMENT ANALYSIS OF MULTI JOINTED ROCK MASSES	72-0956
ROCK	NUMERICAL ANALYSES OF DEFORMABILITY TESTS IN JOINTED ROCK. ' JOINT PERTURBATION ' AND ' NO TENSION '	71-0439
ROCK	NUMERICAL AND PHYSICAL MODELLING OF REINFORCEMENT SYSTEMS FOR TUNNELS IN JOINTED ROCK	73-0872
ROCK	NUMERICAL DETERMINATION OF STRESSES AND DEFORMATIONS IN ROCK TAKING INTO ACCOUNT DISCONTINUITIES	70-0066
ROCK	NUMERICAL DETERMINATION OF STRESSES AND DEFORMATIONS IN ROCK TAKING INTO ACCOUNT DISCONTINUITIES	70-0167
ROCK	NUMERICAL STUDY OF UNIAXIAL AND TRIAXIAL ROCK COMPRESSION TESTS	73-0896
ROCK	PROBLEMS IN ROCK MECHANICS	65-0033
ROCK	PROPAGATION OF FAILURE IN A CIRCULAR CYLINDER OF ROCK SUBJECTED TO A COMPRESSIVE FORCE	73-0793
ROCK	RASCHET NAPRYAZHENII L PEREMESHCHENII V SVODE PODZEMNOGO MASHINNOGO ZALA GES I V OKRUZHAYUSHCHEM SKA	74-0542
ROCK	RATIONAL DESIGN OF TUNNEL SUPPORTS: TUNNEL SUPPORT LOADING CAUSED BY ROCK FAILURE	75-0687
ROCK	ROCK FAILURE AROUND A CIRCULAR OPENING IN A GRAVITY FIELD WITH TECTONIC FORCES	70-0071
ROCK	ROCK FRAGMENTATION BY SUBSURFACE THERMAL INCLUSIONS - A FINITE ELEMENT STUDY	73-1159
ROCK	ROCK MECHANICS - THEORY AND PRACTICE	70-0107
ROCK	ROCK MECHANICS AND RISK IN OPEN PIT MINING	73-0544
ROCK	ROCK MECHANICS STUDIES FOR A HIGH CUT AT A NUCLEAR POWER STATION IN PENNSYLVANIA	70-0122
ROCK	SEEPAGE IN FRACTURED ROCK: FINITE ELEMENT ANALYSIS OF ROCK MASS BEHAVIOR UNDER COUPLED ACTION OF BOD	71-0825
ROCK	SEEPAGE IN FRACTURED ROCK: FINITE ELEMENT ANALYSIS OF ROCK MASS BEHAVIOR UNDER COUPLED ACTION OF BOD	71-0825
ROCK	SEISMIC RESPONSE OF SOIL DEPOSITS UNDERLAIN BY SLOPING ROCK BOUNDARIES	69-0484
ROCK	SIMULATION OF PROGRESSIVE FAILURE IN ROCK AROUND MINING EXCAVATIONS	70-0146
ROCK	SOLUTION OF STRESS-DEFORMATION PROBLEMS IN SOIL AND ROCK MECHANICS USING FINITE ELEMENT METHODS	69-0272
ROCK	SOME NON LINEAR PROBLEMS IN SOIL AND ROCK MECHANICS - FINITE ELEMENT SOLUTION	69-0363
ROCK	STRAIN DISTRIBUTION AROUND UNDERGROUND OPENINGS. STATISTICAL RELATIONSHIPS FOR CERTAIN ROCK PROPERTI	71-0710
ROCK	STRESS ANALYSIS OF ROCK AS A NO-TENSION MATERIAL	68-0174
ROCK	STRESS AND DISPLACEMENTS AROUND OPENINGS IN ROCK CONTAINING AN INELASTIC DISCONTINUITY	75-0170
ROCK	STRESS DISTRIBUTION DUE TO GRAVITY IN A VERTICAL ROCK BANK	68-0090
ROCK	STRESS DISTRIBUTION AND ROCK FRACTURE ZONES IN THE ROOF OF LONGWALL FACE IN COAL MINE	73-0247
ROCK	STRESSES AND DISPLACEMENTS AROUND OPENINGS IN HOMOGENEOUS ROCK	75-0168
ROCK	STRESSES AND DISPLACEMENTS AROUND OPENINGS IN ROCK CONTAINING AN ELASTIC DISCONTINUITY	75-0169
ROCK	STRESSES IN ANISOTROPIC MEDIA WITH PARTICULAR REFERENCE TO PROBLEMS OF ROCK MECHANICS	66-0079
ROCK	STRESSES IN ANISOTROPIC MEDIA OF ROCK MECHANICS	66-0127
ROCK	THE USE OF A LASER FOR ARPA MILITARY GEOPHYSICS PROGRAM (ROCK MECHANICS AND RAPID EXCAVATION)	72-0800
ROCK	THE USE OF ROCK BOLTS OR WIRE ROPE TO INCREASE THE STRENGTH OF FRACTURED MODEL PILLARS	71-0722
ROCK	THEORETICAL INVESTIGATION OF ROCK AND SUPPORT INTERACTION - DEVELOP MORE RATIONAL DESIGN METHODS AND	74-0782
ROCK	THERMAL-MECHANICAL FRAGMENTATION OF HARD ROCK - FIELD TESTS	74-0998
ROCK	THREE DIMENSIONAL FINITE ELEMENT ANALYSIS OF JOINTED ROCK SLOPES	70-0502
ROCK	THREE DIMENSIONAL FINITE ELEMENT STRESS ANALYSIS APPLIED TO TWO PROBLEMS IN ROCK MECHANICS	72-0189
ROCK	VORAUSSAGE DES GEBIRGSDRUCKES UND SEINER AUSWIRKUNGEN AUF STREBEN UND STRECKEN IM STEINKOHLENBERGBAU	73-0679
ROCK	WECHSELWIRKUNG ZWISCHEN DEFORMATION UND DURCHSTROEMUNG IM KLUEFTIGEN, ANISOTROPEN GEBIRGE. (INTERACT	72-0612
ROCK-LIKE	THE EFFECT OF END RESTRAINT CONDITION ON THE FRACTURE MECHANISM AND STRENGTH OF ROCK-LIKE MATERIALS	72-0134
ROCK-SALT	FINITE ELEMENT ANALYSIS OF CREEP IN ROCK-SALT PILLAR MODELS	72-0935
ROCK-STRUCTURE	ANALYTIC MODELING OF ROCK-STRUCTURE INTERACTION	72-0808
ROCK-STRUCTURE	ANALYTIC MODELING OF ROCK-STRUCTURE INTERACTION. VOLUME 3. COMPUTER PROGRAM	73-0918
ROCK-STRUCTURE	ANALYTIC MODELING OF ROCK-STRUCTURE INTERACTION. VOLUME 2. USERS GUIDE FOR A COMPUTER PROGRAM	73-0919
ROCK-STRUCTURE	ANALYTIC MODELING OF ROCK-STRUCTURE INTERACTION. VOLUME 1	73-0920
ROCKET	ACOUSTIC ANALYSIS OF SOLID ROCKET MOTOR CAVITIES BY A FINITE ELEMENT METHOD	71-0119
ROCKET	AN INVESTIGATION INTO THE METHODS OF DETECTING DEWETTED ZONES IN SOLID ROCKET PROPELLANTS USING HOLO	73-0890
ROCKET	APPLICATION OF THE FINITE ELEMENT METHOD TO STRESS ANALYSIS OF SOLID PROPELLANT ROCKET GRAINS	65-0009
ROCKET	AUTOMATED EVALUATION OF INTERCEPTOR ROCKET MOTOR DESIGNS UNDER COMBINED OPERATIONAL AND NUCLEAR EFFE	74-0698
ROCKET	CASE BOND STRESS CALCULATIONS FOR FLAPPED CYLINDRICAL ANALOGS OF SOLID PROPELLANT ROCKET MOTORS. BON	72-0842
ROCKET	CYCLIC FATIGUE ANALYSIS OF ROCKET THRUST CHAMBERS. VOLUME 2: ATTITUDE CONTROL THRUSTER HIGH CYCLE FA	74-0693
ROCKET	CYCLIC FATIGUE ANALYSIS OF ROCKET THRUST CHAMBERS. VOLUME 1: OFHC COPPER CHAMBER LOW CYCLE FATIGUE	74-0694
ROCKET	DEVELOPMENT OF A SOLID ROCKET PROPELLANT NONLINEAR CONSTITUTTIVE THEORY	75-0722
ROCKET	DEVELOPMENT OF TECHNOLOGY FOR MODELING OF A 1/8-SCALE DYNAMIC MODEL OF THE SHUTTLE SOLID ROCKET BOOS	74-0703
ROCKET	FRACTURE OF ROCKET GRAIN MODELS SUBJECTED TO COOLDOWN: A STUDY OF TEMPERATURE-INITIATED FRACTURES OF	74-0536
ROCKET	FRACTURE OF ROCKET GRAIN MODELS SUBJECTED TO COOLDOWN: A STUDY OF TEMPERATURE-INITIATED FRACTURES OF	74-0536
ROCKET	HEAT CONDUCTION AND STRESS ANALYSIS OF SOLID PROPELLANT ROCKET MOTOR NOZZLES	69-0059
ROCKET	METHODS FOR DETERMINING CHARACTERISTICS OF ACOUSTIC WAVES IN ROCKET MOTORS	75-0075
ROCKET	RADIAL AND AXIAL TRANSMISSIBILITY CHARACTERISTICS OF TYPICAL ROCKET VEHICLE STRUCTURE TECHNICAL SUMM	69-0454
ROCKET	RETSCP: A COMPUTER PROGRAM FOR ANALYSIS OF ROCKET ENGINE THERMAL STRAINS WITH CYCLIC PLASTICITY	74-0692
ROCKET	WATER IMPACT ANALYSIS OF SPACE SHUTTLE SOLID ROCKET BY THE FINITE ELEMENT METHOD	74-0760
ROCKETS	MODAL RESPONSE OF FREE ROCKETS UNDER THRUST LOADING	74-0855
ROCKFILL	BEHAVIOUR OF THE ROCK MASS AT THE SITE OF A ROCKFILL DAM	74-0992
ROCKFILL	CONSTRUCTIONAL DEFORMATIONS IN ROCKFILL DAM	73-0281
ROCKFILL	CRACKING OF EARTH AND ROCKFILL DAMS. A THEORETICAL INVESTIGATION BY MEANS OF THE FINITE ELEMENT METH	69-0085
ROCKFILL	CRACKING OF EARTH AND ROCKFILL DAMS	69-0294
ROCKFILL	CRACKING OF EARTH AND ROCKFILL DAMS: COMPARISON OF OBSERVED AND THEORETICAL TENSILE STRAINS IN THE C	71-0699
ROCKFILL	CRACKING OF EARTH AND ROCKFILL DAMS: COMPARISON OF OBSERVED AND THEORETICAL TENSILE STRAINS IN THE C	71-0699
ROCKFILL	EFFECT OF RESERVOIR FILLING ON STRESSES AND MOVEMENT IN EARTH AND ROCKFILL DAMS. A REPORT OF AN INVE	70-0836
ROCKFILL	ELASTIC ANALYSIS FOR BEHAVIOR OF ROCKFILL	70-0032
ROCKFILL	HYDRAULIC FRACTURING IN ZONED EARTH AND ROCKFILL DAMS: A REPORT OF AN INVESTIGATION	73-0873
ROCKFILL	OPTIMUM POSITION OF THE CENTRAL CLAY CORE OF A ROCKFILL DAM IN RESPECT TO ARCHING AND HYDRAULIC FRAC	73-0447
ROCKFILL	TENSION ZONES IN DECKED ROCKFILL DAMS	73-0540
ROCKS	CALCULATION OF DISPLACEMENTS AND STRESSES IN ANISOTROPIC ROCKS BY USING METHOD OF FINITE ELEMENTS	73-0265
ROCKS	DAMS ON ROCKS OF VARYING ELASTICITY	74-0671
ROCKS	DIRECT TENSILE TESTING OF ANISOTROPIC ROCKS	74-0989
ROCKS	DUPLICATION OF DILATANT BEHAVIOR IN THE ANALYSIS OF JOINTED ROCKS	72-0757
ROCKS	FLOW IN ROCKS WITH DEFORMABLE FRACTURES	74-0065
ROCKS	INDIRECT TENSILE TESTING OF ANISOTROPIC ROCKS	73-0376
ROCKS	ON THE DISTRIBUTION OF STRESSES AROUND CIRCULAR TUNNELS IN NON-HOMOGENEOUS ROCKS	66-0029
ROCKS	STRESSES AND DEFORMATIONS OF VERTICAL SLOPES IN ELASTO-PLASTIC ROCKS	70-0114
ROCKS	SURFACE FOOTINGS ON SOFT ROCKS	74-0990
ROCKS	WAVE PROPAGATION IN ANISOTROPIC ROCKS	72-0801
ROCKS	WAVE PROPAGATION IN ANISOTROPIC ROCKS	72-0873
ROCKS	WAVE PROPAGATION IN ANISOTROPIC ROCKS	73-0922
ROD	ANALYSIS OF NUCLEAR FUEL ROD LOCALIZED STRESSES	73-0708
ROD	BLOWDOWN EFFECTS ON A FUEL ROD	75-0575
ROD	FUEL ROD DEFORMATIONS - A TIME DEPENDENT FINITE ELEMENT ANALYSIS ON AXISYMMETRIC FUEL RODS, INCLUDIN	70-0414
ROD	FUEL ROD FINITE ELEMENT STRESS ANALYSIS: SEGMENTED FLIGHT REACTOR FUEL VERSUS S8DR FUEL	71-0798
ROD	HIGH STRENGTH END FITTINGS FOR FRP ROD AND ROPE	74-0605
ROD	STRESS ANALYSIS OF A BOLT-NUT JOINT BY THE FINITE ELEMENT METHOD AND THE COPPER-ELECTROPLATING METHO	73-0220
ROD	THERMAL AND STRESS ANALYSIS OF A FUEL ROD	75-0713
ROD	THREE-DIMENSIONAL STUDY OF NUCLEAR FUEL ROD BEHAVIOUR DURING STARTUP	73-0815
ROD	THREE-DIMENSIONAL STUDY OF NUCLEAR FUEL ROD BEHAVIOR DURING STARTUP	74-1003
ROD	USE OF INTEGRAL PARAMETERS IN ROD PROBLEMS	74-1194
RODS	CALCULATIONAL PROCEDURE FOR DETERMINING CREEP COLLAPSE OF LWR FUEL RODS	74-1001
RODS	FUEL ROD DEFORMATIONS - A TIME DEPENDENT FINITE ELEMENT ANALYSIS ON AXISYMMETRIC FUEL RODS, INCLUDIN	70-0414
RODS	MATHEMATICAL MODELING AND ANALYSIS OF FUEL RODS	74-0930
RODS	SAFE-COLAPS: A FINITE ELEMENT PROGRAM FOR THE CREEP COLLAPSE ANALYSIS OF NUCLEAR FUEL RODS	75-0573
RODS	VARIATIONAL PRINCIPLE FOR THE LAPLACE'S OPERATOR WITH APPLICATION IN THE TORSION OF COMPOSITE RODS	74-1044
RODS	VIBRATORY WEAR OF FUEL RODS	75-0771
RODS	ZUR BERECHNUNG DER BIEGESTEIFIGKEIT ABGESETZTER STAEBE UND WELLEN UNTER ANWENDUNG VON FINITEN ELEMEN	75-0481
ROHRWAENDE	TEMPERATURVERLAUF IM QUERSCHNITT GESCHWEISSTER ROHRWAENDE BEI UEBERWIEGEND KONVEKTIVEM WAERMEUEBERGA	75-0628

ROLLED	THEORETICAL AND EXEPIMENTAL ANALYSIS OF ELASTOPLASTIC DEFORMATION IN A UNIAXIALLY LOADED STRIP OF 2	73-0720
ROLLER	HOLLOW-ENDED ROLLER. A SOLUTION FOR IMPROVING FATIGUE LIFE IN ASYMMETRICALLY LOADED CYLINDRICAL ROLL	71-0393
ROLLER	HOLLOW-ENDED ROLLER. A SOLUTION FOR IMPROVING FATIGUE LIFE IN ASYMMETRICALLY LOADED CYLINDRICAL ROLL	71-0393
ROLLING	A FINITE ELEMENT METHOD OF VISCOELASTIC STRESS ANALYSES WITH APPLICATION TO ROLLING CONTACT PROBLEMS	69-0330
ROLLING	ELASTIC-PLASTIC ANALYSIS OF A WHEEL ROLLING ON A RIGID TRACK	74-1040
ROLLING	FINITE ELEMENT METHOD IN ROLLING BEARING ENGINEERING	73-0054
ROLLING	NONLINEAR ANALYSIS OF A ROLLING CONTACT PROBLEM BY FINITE ELEMENTS	74-1216
ROLLING	SHAKEDOWN OF ROLLING WHEEL UNDER HUB LOADING	74-0663
ROLLING	STRUCTURAL INELASTICITY. V11. ELASTIC-PLASTIC ANALYSIS OF A WHEEL ROLLING ON A RIGID TRACK	73-0886
ROOF	ANALYSIS OF HIPPED ROOF HYPERBOLIC PARABOLOID STRUCTURES	72-0216
ROOF	ANALYSIS OF MULTI-AXIAL ANISOTROPIC CREEP, ROOF RESIN BCLTS, AND BED SEPARATION IN ROCK STRUCTURES U	74-1122
ROOF	COMPUTER SIMULATION OF LAMINATED ROOF REINFORCED WITH GROUTED BOLTS	73-1045
ROOF	DESIGN OF ROOF BOLT PATTERNS FOR JOINTED ROCK	74-1318
ROOF	EFFECTS OF SPECIMEN RADIUS ON THE STRESS STATE NEAR A ROOF BOLT ANCHOR: A FINITE ELEMENT DETERMINATI	73-0229
ROOF	EXPERIMENTAL AND FINITE ELEMENT ANALYSIS OF CABLE ROOF STRUCTURES INCLUDING PRECAST PANELS	74-1139
ROOF	RECORD CONCRETE SHELL ROOF FEATURED IN COLLISEUM	72-0218
ROOF	STRESS DISTRIBUTION AND ROCK FRACTURE ZONES IN THE ROOF OF LONGWALL FACE IN COAL MINE	73-0247
ROOF	STRESSES IN GABLE ROOF H. P. SHELLS.	72-0394
ROOFS	ANALYSIS OF CYLINDRICAL SHELL ROOFS WITH POST TENSIONED EDGEBEAMS	74-1042
ROOFS	DYNAMIC BEHAVIOUR OF SHELL ROOFS	69-0228
ROOM	IN SITU CREEP ANALYSIS OF ROOM IN FROZEN SOIL	72-0128
ROOMS	ANALYSIS OF ACOUSTIC FIELD IN IRREGULARLY SHAPED ROOMS BY FINITE ELEMENT METHOD	73-0133
ROOT	NEW APPROACH TO GEAR TOOTH ROOT STRESSES	72-0117
ROOT	STRESS ANALYSIS OF A BOLT-NUT JOINT BY THE FINITE ELEMENT METHOD AND THE COPPER-ELECTROPLATING METHO	73-0220
ROOT	STRESS ANALYSIS OF A BOLT-NUT JOINT BY FINITE FLEMENT METHOD AND COPPER-ELECTROPLATING METHOD - 2. S	74-0153
ROOT	STRESS AND STRAIN DISTRIBUTIONS AT THE ROOT OF CRACKS	69-0290
ROOTS	FINITE ELEMENT ANALYSIS OF 2-DIMENSIONAL FLOW IN SOILS CONSIDERING WATER UPTAKE BY ROOTS .1. THEORY	75-0137
ROOTS	FINITE ELEMENT ANALYSIS OF 2-DIMENSIONAL FLOW IN SOILS CONSIDERING WATER UPTAKE BY RONTS .2. FIELD A	75-0147
ROPE	HIGH STRENGTH END FITTINGS FOR FRP ROD AND ROPE	74-0605
ROPE	THE USE OF ROCK BOLTS OR WIRE ROPE TO INCREASE THE STRENGTH OF FRACTURED MODEL PILLARS	71-0722
ROSENBERG	SIMPLIFIED CALCULATION OF THE INTERACTION FORCES BETWEEN HULL AND TANK SYSTEM FOR A MOSS ROSENBERG T	75-0561
ROTARY	EFFECT OF SHEAR DEFORMATIONS AND ROTARY INERTIA ON OPTIMUM BEAM FREQUENCIES	75-0231
ROTARY	EFFECTS OF SHEAR DEFORMATION, ROTARY INERTIA, AND ELASTICITY OF THE SUPPORT ON THE RESONANCE FREQUEN	75-0369
ROTARY	GALERKIN FINITE ELEMENT ANALYSIS OF A UNIFORM BEAM CARRYING A CONCENTRATED MASS AND ROTARY INERTIA W	74-0552
ROTARY	ROTARY INERTIA CORRECTION FOR MASS MATRICES	63-0011
ROTATING	BERECHNUNG VON LAUFRAEDERN MIT DER METHODE DER FINITEN ELEMENTE (DESIGN OF ROTATING WHEELS BY THE FI	74-0547
ROTATING	CERAMIC ROTATING BLADES: SOME CRITICAL DESIGN PARAMETERS FOR GAS TURBINE APPLICATIONS	74-0146
ROTATING	CONTRIBUTION TO NUMERICAL-SOLUTION OF DEVELOPING LAMINAR-FLOW IN ENTRANCE REGION OF CONCENTRIC ANNUL	75-0448
ROTATING	ELASTIC DEFLECTION IN THIN WALLED CONES ROTATING ABOUT THEIR AXES OF SYMMETRY	73-0545
ROTATING	ELASTO-PLASTIC ANALYSIS OF A ROTATING MODEL CONICAL TURBINE DISC	75-0281
ROTATING	FINITE ELEMENT ANALYSIS OF ROTATING SHELLS	74-0096
ROTATING	FINITE ELEMENT ANALYSIS OF THE MAGNETIC FIELD DISTRIBUTION INSIDE A ROTATING FERROMAGNETIC BAR	74-1037
ROTATING	GENERAL METHOD FOR STABILITY ANALYSIS OF ROTATING SHAFTS	75-0528
ROTATING	GEOMETRIC STIFFNESS CHARACTERISTICS OF A ROTATING ELASTIC APPENDAGE	74-1296
ROTATING	PSEUDO STATIC DEFORMATION AND FREQUENCIES OF ROTATING TURBOMACHINERY BLADES	72-0239
ROTATING	STABILITY CONTROL ANALYSIS OR ROTATING PLATES BY FINITE ELEMENT. EMPHASIS ON SLOTS AND HOLES	71-0284
ROTATING	STRESSES IN A ROTATING DISK WITH CENTRAL AND NONCENTRAL HOLES (STRESS ANALYSIS BY THE FINITE ELEMENT	74-1337
ROTATING	TRAPEZOIDAL FINITE ELEMENTS: THEIR DERIVATION AND USE FOR AXISYMMETRIC ROTATING BODIES	73-0192
ROTATING	UNSYMMETRICAL TRANSIENT HEAT CONDUCTION: ROTATING DISK APPLICATIONS	70-0005
ROTATING	VIBRATION ANALYSIS OF ROTATING COMPRESSOR BLADES	73-0417
ROTATING	VIBRATION ANALYSIS OF ROTATING TURBINE BLADES	74-0959
ROTATION	EFFECTS OF FOUNDATION ROTATION ON SEISMIC INERTIA FORCES OF NUCLEAR POWER PLANT STRUCTURES	73-0177
ROTATION	ON DERIVATION OF STIFFNESS MATRICES WITH CO ROTATION FIELDS FOR PLATES AND SHELLS	71-0254
ROTATIONAL	CURVED ROTATIONAL SHELL ELEMENTS BY THE CONSTRAINT METHOD	72-0924
ROTATIONAL	NON-LINEAR ANALYSIS OF THE ROTATIONAL STIFFNESS OF A NUCLEAR STEAM GENERATOR FOUNDATION	69-0286
ROTATIONAL	ROTATIONAL STIFFNESS OF CONCRETE SLABS	74-0660
ROTATIONS	FINITE ELEMENTS, FINITE ROTATIONS AND SMALL STRAINS OF FLEXIBLE SHELLS	69-0516
ROTATIONSSYMMET	ROTATIONSSYMMETRISCHE BERECHNUNG DES 1:5 THTR SPANNBETON-BEHAL-BEHAL TERMODELLS	71-0629
ROTATOPY	TRANSVERSE VIBRATION OF BEAMS WITH INTERNAL DAMPING, ROTATORY INERTIA AND SHEAR	63-0020
ROTIERENDEN	TURBAN VERSION 1.0. EIN SPEZIELLES PROGRAMM ZUR LINEAREN UND NICHTLINEAREN STATISCHEN UND DYNAMISCHE	71-0622
ROTO	ELASTIC, PLASTIC BEHAVIOUR OF ROTO - 1	73-0458
ROTOR	APPLICATION OF THE COLLISION-IMPARTED VELOCITY METHOD FOR ANALYZING THE RESPONSES OF CONTAINMENT AND	73-0770
ROTOR	DYNAMICS OF DISTRIBUTED PARAMETER ROTOR SYSTEMS: TRANSFER MATRIX AND FINITE ELEMENT TECHNIQUES	70-0461
ROTOR	FINITE ELEMENT ANALYSIS OF LARGE TRANSIENT ELASTIC-PLASTIC DEFORMATIONS OF SIMPLE STRUCTURES, WITH A	72-0849
ROTOR	FINITE-ELEMENT FIELD ANALYSIS OF AN INHOMOGENEOUS ANISOTROPIC RELUCTANCE MACHINE ROTOR	73-0020
ROTOR	FLEXUAL RIGIDITY OF A THIN WALLED BUILD-UP ROTOR FOR JET ENGINE. MEASUREMENT BY STATIC LOAD TEST AND	73-0756
ROTOR	FREQUENCY TUNING OF COMPRESSOR ROTOR BLADES	66-0112
ROTOR	GRAPHICAL DISPLAY OF COMPUTER SIMULATED UNBALANCED ROTOR RESPONSE	71-0485
ROTOR	ROTOR BLADE VIBRATIONS BY GALERKIN FINITE-ELEMENT METHOD	75-0389
ROTOR	SENSITIVITY ANALYSIS OF TORSIONAL VIBRATION CHARACTERISTICS OF HELICOPTER ROTOR BLADES. PART 1: STRU	74-0846
ROTOR	SENSITIVITY OF ROTOR BLADE VIBRATION CHARACTERISTICS TO TORSIONAL OSCILLATIONS	74-0611
ROTOR-AIRFRAME	COUPLED ROTOR-AIRFRAME VIBRATION PREDICTION METHODS	74-0498
ROTOR-AND-BEARI	ON THE STABILITY OF ROTOR-AND-BEARING SYSTEMS AND ON THE CALCULATION OF SLIDING BEARINGS	72-0037
ROTORBLATTERN	SPEKTRALANALYSE VON ROTORBLATTERN NACH DERMATRIZENVERS CHIEBUNGSMETHODE UND MATRIZENKRAFTEMETHODE	66-0123
ROTORS	FINITE-ELEMENT ANALYSIS OF AXISYMMETRIC ROTORS	69-0244
ROTORS	UNBALANCE GENERATED VIBRATIONS AND STABILITY OF TURBINE ROTORS	73-0224
ROUGH	EFFECTS OF INSERTS ON THE ELASTIC BEHAVIOUR OF CYLINDRICAL MATERIALS LOADED BETWEEN ROUGH END-PLATES	71-0318
ROUGHNESS	INFLUENCE OF COOLING, FREE-STREAM TURBULENCE AND SURFACE ROUGHNESS ON THE AERODYNAMIC BEHAVIOR OF CA	74-0970
ROUND	STRAIN CONCENTRATION BEHAVIOUR IN A NOTCHED ROUND BAR SUBJECTED TO CREEP	73-0440
ROUND	THEORETICAL AND EXPERIMENTAL ANALYSIS OF ELASTOPLASTIC DEFORMATION IN A UNIAXIALLY LOADED STRIP OF 2	73-0720
ROUND-OFF	DISCPETIZATION AND ROUND-OFF ERROR IN THE FINITE ELEMENT ANALYSIS OF ELLIPTIC BOUNDARY VALUE PROBLEM	71-0514
ROUND-OFF	SOME CONSIDERATIONS ON ROUND-OFF ERRORS OF THE FINITE ELEMENT METHOD	72-0260
ROUNDED	THE CHANGES IN THE (TWO-DIMENSIONAL) STRESS ROUNDED CORNERS, CAUSED BY A VARYING INTERNAL PRESSURE A	67-0141
ROUNDING-OFF	NUMERICAL ACCURACY AND ROUNDING-OFF ERRORS	69-0348
ROUNDOFF	ROUNDOFF CRITERIA IN DIRECT STIFFNESS SOLUTIONS	68-0204
ROUTH	SIMPLIFICATION OF HIGH ORDER MECHANICAL SYSTEMS USING THE ROUTH APPROXIMATION	75-0853
ROUTINE	A METHOD FOR REDUCING THE ANALYSIS OF COMPLEX REDUNDANT STRUCTURES TO A ROUTINE PROCEDURE	52-0002
ROUTINE	MESH GENERATION CODE AND A CONTOUR PLOTTING ROUTINE	72-0802
ROUTINES	GENERAL SOLUTION ROUTINES FOR SYMMETRIC EQUATION SYSTEMS	75-0129
ROUTINES	THREE DIMENSIONAL FINITE ELEMENT STUDIES. PART ONE: SERVICE ROUTINES	72-0721
ROZWIAZAN	OCENA DOKLADNOSCI ROZWIAZAN PROBLEMOW BRZEGOWYCH SPRZYSTYCH CIAL DYSKERTYZOWANYCH. (ASSESSING THE A	74-0921
ROZWIAZANIE	NUMERYCZNE ROZWIAZANIE PROBLEMU ORGAN WLASNYCH NA PRZYKLADZIE PLYT. (NUMERICAL SOLUTION OF THE PROBL	75-0296
ROZWIAZYWANIA	METODA ELEMENTOW SKONSZONYCH ROZWIAZYWANIA OSIOWO-SYMETRYCZNEGO PRZEWODNICTWA CIEPLA (FINITE-FLEMENT	73-0434
RUBBER	ENERGY FOR CRACK GROWTH IN MODEL RUBBER COMPONENTS	72-0568
RUBBER	FINITE-ELEMENT PROGRAMME FOR THE PLANE-STRAIN ANALYSIS OF RUBBER	75-0362
RUBBER	PLANE STRESS ANALYSIS OF RUBBER AT HIGH STRAINS USING FINITE ELEMENTS	71-0680
RUBBER	RUBBER IN ENGINEERING	73-0367
RUBBER	SOME APPLICATIONS OF FINITE ELEMENT ANALYSIS IN RUBBER SPRING DESIGN	72-0539
RUBBER	THE COMPUTATION OF TEARING ENERGY OF NICKED RUBBER STRIPS IN EXTENSION	74-0315
RUBBERS	APPLICATION OF FINITE ELASTICITY THEORY TO THE DEFORMATION OF RUBBERS	62-0005
RUEMPFEN	BERECHNUNG DES REIBUNGSLOSEN STROEMUNGSFELDES VON DREIDIMENSIONALEN AUFTRIEBSBEHAFTETENTRAGFLUEGELN	73-0533
RULE	STRESS ANALYSIS IN VISCOELASTIC BODIES USING FINITE ELEMENTS AND A CORRESPONDENCE RULE WITH ELASTICI	69-0270

RULES	COMPARISON OF THE CAPABILITY OF OUR HARDENING RULES TO PREDICT A MATERIAL'S PLASTIC BEHAVIOR	75-0523
RULES	QUADRATURE RULES FOR BRICK BASED FINITE ELEMENTS	71-0531
RULES	RULES GOVERNING THE NUMBER OF NODES AND ELEMENTS IN A FINITE ELEMENT MESH	70-0053
RULES	THE APPLICATION OF NODE-ELEMENT RULES FOR FORECASTING PROBLEMS IN THE GENERATION OF FINITE ELEMENT M	75-0614
RUN	FINITE ELEMENT SOLUTIONS OF THE SHALLOW WATER EQUATIONS SURFACE RUN OFF	74-0046
RUN-OFF	AN ANALYSIS OF TWO-DIMENSIONAL SURFACE RUN-OFF BY FINITE ELEMENTS	74-1336
RUN-UP	NUMERICAL MODEL FOR TSUNAMI RUN-UP	70-0200
RUNNER	ACTUAL STRESS OF HIGH HEAD PUMP-TURBINE RUNNER	75-0567
RUNOFF	FINITE ELEMENT APPROACH TO WATERSHED RUNOFF	74-0234
RUNOFF	SIMULATION OF RUNOFF HYDROGRAPHS FROM NATURAL WATERSHEDS BY FINITE-ELEMENT METHOD	73-1140
RUNOFF-	FINITE-ELEMENT APPROACH TO WATERSHED RUNOFF- DISCUSSION	75-0402
RUPTURE	ESTIMATES OF THE CREEP RUPTURE LIFETIME OF STRUCTURES USING THE FINITE ELEMENT METHOD	75-0833
RUPTURE	FINITE ELEMENT FRACTURE MECHANICS ANALYSIS OF CREEP RUPTURE OF FUEL ELEMENT CLADDING	74-0929
RUPTURE	INELASTIC RESPONSE OF NUCLEAR PIPING SUBJECTED TO RUPTURE FORCES	75-0515
RUPTURE	THREE DIMENSIONAL RUPTURE ANALYSIS OF A PRESTRESSED CONCRETE PRESSURE VESSEL INCLUDING CREEP EFFECTS	73-0826
RUPTURE	THREE DIMENSIONAL RUPTURE ANALYSIS OF A PRESTRESSED CONCRETE PRESSURE VESSEL INCLUDING CREEP EFFECTS	74-0538
SAAS	SAAS II: FINITE ELEMENT STRESS ANALYSIS OF AXISYMMETRIC SOLIDS WITH ORTHOTROPIC, TEMPERATURE DEPENDE	68-0096
SAAS	SAAS 111: FINITE ELEMENT STRESS ANALYSIS OF AXISYMMETRIC AND PLANE SOLIDS WITH DIFFERENT ORTHOTROPIC	71-0741
SABOR	DISCRETE-ELEMENT STATIC ANALYSIS OF BONDED, DOUBLE-LAYER, BRANCHED, THIN SHELLS OF REVOLUTION, PART	70-0081
SABOR	SABOR CS-2 USER'S MANUAL	70-0408
SABOR	SABOR 1, A FORTRAN PROGRAM FOR THE LINEAR ELASTIC AXISYMMETRIC LOADING BY USING THE MATRIX DISPLACEM	65-0027
SABOR	THE TRANSIENT LINEAR ELASTIC RESPONSE ANALYSIS COMPLEX THIN SHELLS OF REVOLUTION SUBJECTED TO ARBITR	70-0080
SACS	A MODEL STUDY OF GAS DIFFUSION IN ALVEOLAR SACS	73-0029
SAFE	SAFE PCPS, A COMPUTER PROGRAM FOR THE STRESS ANALYSIS OF COMPOSITES BODIES OF REVOLUTION, INPUT INST	65-0013
SAFE	SAFE SHELL, A COMPUTER PROGRAM FOR THE STRESS ANALYSIS OF THIN SHELL BODIES OF REVOLUTION, A USER'S	67-0018
SAFE-AXISYM	SAFE-AXISYM, A COMPUTER PROGRAM FOR THE DESIGN OF COMPOSITE BODIES OF REVOLUTION, INPUT INSTRUCTIONS	65-0014
SAFE-COLAPS	SAFE-COLAPS: A FINITE ELEMENT PROGRAM FOR THE CREEP COLLAPSE ANALYSIS OF NUCLEAR FUEL RODS	75-0573
SAFE-CREEP	SAFE-CREEP: A COMPUTER PROGRAM FOR THE VISCOELASTIC ANALYSIS OF AXISYMMETRIC AND PLANE CONCRETE STRU	67-0148
SAFE-PLANE	SAFE-PLANE, A COMPUTER PROGRAM FOR THE STRESS ANALYSIS AND DESIGN OF TWO-DIMENSIONAL COMPOSITE BODIE	67-0017
SAFE-SHELL	DESCRIPTION AND INSTRUCTION FOR USE OF THE SAFE-SHELL, 8 FINITE ELEMENT PROGRAM IN THE IBM/360 VERSI	73-0945
SAFE-SHELL	VERIFICATION OF THE PEC CONTAINMENT VESSEL FOR SOME OPERATING CONDITIONS BY USE OF THE FINITE ELEMEN	74-0883
SAFE-2D	VERIFICATION OF THE FINITE ELEMENT COMPUTER CODE SAFE-2D	74-0873
SAFE-3D	DESCRIPTION OF THE IBM-360 VERSION OF THE FINITE ELEMENTS COMPUTER PROGRAM SAFE-3D	72-0775
SAFEGUARD	ANALYSIS OF DYNAMIC RESPONSE OF SAFEGUARD STRUCTURES	70-0462
SAFETY	APPLICATION OF THE FINITE ELEMENT METHOD FOR THE SAFETY EVALUATION OF REACTOR COMPONENTS	73-0853
SAFETY	FACTOR OF SAFETY APPROACH FOR EVALUATING SEISMIC STABILITY OF SLOPES	75-0645
SAFETY	FRACTURE AND SAFETY ANALYSIS OF NUCLEAR PRESSURE VESSELS	73-0257
SAFETY	USE OF FRACTURE MECHANICS AND FINITE ELEMENT TECHNIQUES FOR DETERMINING THE SAFETY OF OFFSHORE STRUC	74-1211
SAINT	FINITE ELEMENT ANALYSIS IN POLAR CO-ORDINATES OF THE SAINT VENANT TORSION PROBLEM	74-0500
SAINT	FINITE ELEMENT STRESS FIELD SOLUTION OF THE PROBLEM OF SAINT VENANT TORSION	73-1042
SAINT-VENANT	A FINITE ELEMENT SOLUTION FOR SAINT-VENANT TORSION	69-0161
SAINT-VENANT	ANALYSIS OF SAINT-VENANT TORSION PROBLEM BY A HYBRID STRESS MODEL	69-0277
SAINT-VENANT	ELASTIC PLASTIC ANALYSIS OF SAINT-VENANT TORSION PROBLEM BY A HYBRID STRESS MODEL	72-0385
SAINT-VENANT	FINITE ELEMENT SOLUTION FOR SAINT-VENANT BENDING	71-0346
SALONICA	APPLICATION OF FINITE ELEMENT METHOD TO A MICROZONATION STUDY OF SALONICA REGION	73-0610
SALT	FINITE ELEMENT ANALYSIS OF SALT PILLAR MODELS	68-0291
SALT	FINITE ELEMENT ANALYSES OF SALT DOMES WITH STORED HOT WASTES	74-1148
SALT	MANAGEMENT OF RETARDATION OF SALT WATER INTRUSION IN COASTAL AQUIFERS	75-0692
SALT	TIME-DEPENDENT ANALYSIS TO PREDICT CLOSURE IN SALT CAVITIES	74-1149
SALTWATER	GALERKIN-FINITE ELEMENT TECHNIQUE FOR CALCULATING TRANSIENT POSITION OF SALTWATER FRONT	75-0135
SAMECS	SAMECS STRUCTURAL ANALYSIS SYSTEM - USER'S DOCUMENT	69-0495
SAMIS	APPLICATION OF SAMIS TO THE ANALYSIS OF A FRONT SUSPENSION UPPER ARM	70-0260
SAMIS	STRUCTURAL ANALYSIS AND MATRIX INTERPRETIVE SYSTEM (SAMIS)	66-0098
SAMIS	STRUCTURAL ANALYSIS AND MATRIX INTERPRETIVE SYSTEM (SAMIS) USER REPORT	67-0125
SAMMSOR	SAMMSOR 111: A FINITE ELEMENT PROGRAM TO DETERMINE STIFFNESS AND MASS MATRICES OF RING-STIFFENED SHE	72-0774
SAN	DYNAMIC ANALYSIS OF THE SLIDE IN THE LOWER SAN FERNANDO DAM DURING THE EARTHQUAKE OF FEBRUARY 9, 197	75-0277
SAND	AN EXPERIMENTAL AND THEORETICAL STUDY OF HOLLOW CYLINDER TESTS ON SAND	71-0008
SAND	ANALYSIS OF QUICK SAND PHENOMENON BY FINITE ELEMENT METHOD	73-0191
SAND	FINITE ELEMENT PARAMETRIC STUDY OF VERTICAL STRAIN INFLUENCE FACTORS AND THE PRESSUREMETER TEST TO E	74-1127
SAND	SETTLEMENTS AND STRENGTHENING OF SOFT CLAY ACCELERATED BY SAND DRAINS	73-0438
SANDER	SUR L ELEMENT DE FRAEIJS DE VEUBEKE ET SANDER (FRAEIJS DE VEUBEKE-SANDER ELEMENT)	74-0924
SANDS	NUMERICAL DESIGN-ANALYSIS FOR PILES IN SANDS	74-0333
SANDS	THE ANALYSIS OF FLEXIBLE BULKHEADS IN SANDS	74-0369
SANDWICH	A SANDWICH PLATE FINITE ELEMENT	72-0077
SANDWICH	A STUDY OF FINITE ELEMENTS FOR THE ANALYSIS OF SANDWICH PLATES	69-0503
SANDWICH	A TRIANGULAR ELEMENT FOR BENDING AND VIBRATION OF MULTILAYER SANDWICH PLATES	72-0489
SANDWICH	ANALYSIS OF THREE-DIMENSIONAL ORTHOTROPIC SANDWICH PLATE STRUCTURES BY FINITE-ELEMENT METHOD	72-0945
SANDWICH	APPLICATIONS OF CURVED FINITE ELEMENT TO SANDWICH BEAMS USING SHEAR STRAIN FORMULATION	74-0523
SANDWICH	APPLICATIONS OF CURVED FINITE ELEMENTS TO SANDWICH SHELLS USING SHEAR STRAIN FORMULATION	75-0433
SANDWICH	BENDING AND VIBRATION OF MULTILAYER SANDWICH BEAMS AND PLATES	73-0272
SANDWICH	DISCRETE ELEMENT FINITE DISPLACEMENT ANALYSIS OF ANISOTROPIC SANDWICH SHELLS	70-0504
SANDWICH	DYNAMIC ANALYSIS OF SANDWICH BEAMS	72-0586
SANDWICH	FINITE DEFLECTION DISCRETE ELEMENT ANALYSIS OF SANDWICH PLATES AND CYLINDRICAL SHELLS WITH LAMINATED	69-0236
SANDWICH	FINITE ELEMENT ANALYSIS OF SANDWICH PLATES AND CYLINDRICAL SHELLS WITH LAMINATED FACES	68-0121
SANDWICH	FINITE ELEMENT ANALYSIS OF SANDWICH PANELS	70-0449
SANDWICH	FINITE ELEMENT ANALYSIS OF SANDWICH PANELS	71-0662
SANDWICH	FINITE ELEMENT ANALYSIS OF SKEW SANDWICH PLATES	72-0810
SANDWICH	FINITE ELEMENT ANALYSIS OF AXISYMMETRIC MULTILAYER SANDWICH PLATES AND SHELLS	74-0101
SANDWICH	FINITE ELEMENT ANALYSIS OF MULTILAYER SANDWICH PLATES AND SHELLS	74-0425
SANDWICH	FINITE ELEMENT ANALYSIS OF SPATIAL SANDWICH PLATE STRUCTURES	74-1132
SANDWICH	FINITE ELEMENT BUCKLING ANALYSIS OF HOMOGENEOUS AND SANDWICH PLATES	75-0613
SANDWICH	FINITE-ELEMENT ANALYSIS OF MULTILAYER SANDWICH STRUCTURES	72-0953
SANDWICH	FINITE-ELEMENT ANALYSIS OF SANDWICH SHELLS	73-1130
SANDWICH	FINITE-ELEMENT ANALYSIS OF SKEW SANDWICH PLATES	75-0185
SANDWICH	FREE VIBRATION OF CURVED SANDWICH BEAMS BY METHOD OF FINITE ELEMENTS	71-0351
SANDWICH	FREE VIBRATIONS OF CURVED SANDWICH BEAMS	71-0729
SANDWICH	MATRIX ANALYSIS OF MULTILAYERED AND SANDWICH SHELLS BY FINITE ELEMENT METHOD	73-0101
SANDWICH	NONLINEAR BUCKLING ANALYSIS OF SANDWICH ARCHES	71-0292
SANDWICH	NONLINEAR FINITE ELEMENT ANALYSIS OF SANDWICH SHELLS OF REVOLUTION	72-0180
SANDWICH	NUMERICAL STUDY OF DAMPING IN VISCOELASTIC SANDWICH BEAMS	73-0443
SANDWICH	ON THE ENERGY FUNCTIONS FOR CURVED SANDWICH PLATES	71-0587
SANDWICH	PLASTIC SANDWICH BUCKLE-SHELL STRUCTURES	72-0119
SANDWICH	SANDWICH AND COMPOSITE ELEMENTS	75-0136
SANDWICH	SANDWICH PLATE STRUCTURE ANALYSIS BY FINITE ELEMENT	74-0163
SANDWICH	SIMPLE FINITE ELEMENT SOLUTION FOR PLATES OF HOMOGENEOUS, SANDWICH AND CELLULAR CONSTRUCTION	75-0140
SANDWICH	SONIC FATIGUE OF DIFFUSION BOUNDED TITANIUM SANDWICH STRUCTURES	72-0290
SANDWICH	STABILITY ANALYSIS OF MULTILAYER SANDWICH STRUCTRES	73-0345
SANDWICH	STATIC ANALYSIS OF SANDWICH STRUCTURES BY THE FINITE ELEMENT METHOD	71-0730
SANDWICH	STATIC AND DYNAMIC ANALYSIS OF SANDWICH STRUCTURES BY METHOD OF FINITE ELEMENTS	71-0352
SANDWICH	STATIC AND DYNAMIC ANALYSIS OF MULTILAYERED SANDWICH PLATES	72-0493
SANDWICH	STATIC AND DYNAMIC FINITE ELEMENT ANALYSIS OF SANDWICH STRUCTURES	68-0003
SANDWICH	STIFFNESS MATRIX FOR A TRIANGULAR SANDWICH ELEMENT IN BENDING	67-0050

SANDWICH	THE EFFECT OF IN PLANE STATIC STRESSES ON THE VIBRATION CHARACTERISTICS OF CURVED SANDWICH PLATES	71-0588
SANDWICH	THE FINITE ELEMENT ANALYSIS OF SANDWICH BUCKLE-SHELL STRUCTURES	72-0968
SANDWICH	THE FINITE ELEMENT METHOD APPLIED TO THE ANALYSIS OF SANDWICH PLATE AND SHELL STRUCTURES	72-0824
SANDWICH	TRIANGULAR ELEMENT FOR MULTILAYER SANDWICH PLATES	72-0581
SANDWICH	TWO HYBRID ELEMENTS FOR ANALYSIS OF THICK, THIN AND SANDWICH PLATES	72-0544
SANDY	GROUNDWATER FLOW IN A SANDY TIDAL BEACH: II. 2-DIMENSIONAL FINITE ELEMENT ANALYSIS	72-0300
SANDY	GROUNDWATER FLOWN IN A SANDY TIDAL BEACH .1. ONE-DIMENSIONAL FINITE ELEMENT ANALYSIS	71-0567
SAP	SAP IV. A STRUCTURAL ANALYSIS PROGRAM FOR STATIC AND DYNAMIC RESPONSE OF LINEAR SYSTEMS	73-0650
SAP	SOLID SAP. A STATIC ANALYSIS PROGRAM FOR THREE DIMENSIONAL SOLID STRUCTURES	72-0855
SAP-A	SAP-A GENERAL STRUCTURAL ANALYSIS PROGRAM FOR LINEAR SYSTEMS	72-0329
SARD	SARD KERNEL THEOREMS ON TRIANGULAR AND RECTANGULAR DOMAINS WITH EXTENSIONS AND APPLICATIONS TO FINIT	72-0361
SARD	THE EXTENSION AND APPLICATION OF SARD KERNEL THEOREMS TO COMPUTE FINITE ELEMENT ERROR BOUNDS	72-0519
SASL	SASL - A FINITE ELEMENT CODE FOR THE STATIC ANALYSIS OF AXISYMMETRIC AND PLANE SOLIDS SUBJECTED TO A	72-0779
SATELLITE	MODAL ANALYSIS OF A SATELLITE PRIMARY STRUCTURE USING A FINITE ELEMENT PROCEDURE	73-0624
SATURABLE	COMPARAISON DES DIVERSES MODELISATIONS NUMERIQUES D'UN SYSTEME MAGNETIQUE SATURABLE. (COMPARISON OF	75-0295
SATURABLE	COMPARAISON DES DIVERSES MODELISATIONS NUMERIQUES D'UN SYSTEME MAGNETIQUE SATURABLE. (COMPARISON OF	75-0295
SATURABLE	CURVILINEAR FINITE ELEMENTS FOR 2-DIMENSIONAL SATURABLE MAGNETIC FIELDS	74-0562
SATURABLE	FINITE ELEMENT SOLUTION OF SATURABLE MAGNETIC FIELD PROBLEMS	70-0205
SATURATED	A FINITE DIFFERENCE METHOD FOR UNSTEADY FLOW IN VARIABLY SATURATED POROUS MEDIA	71-0586
SATURATED	A FINITE-ELEMENT PROGRAM FOR THE ANALYSIS OF EMBANKMENTS OVER SOFT, SATURATED SOILS INCLUDING CONSOL	73-1160
SATURATED	ANALYSIS OF PLANE STRAIN CONSOLIDATION IN SATURATED CLAYS BY FINITE ELEMENT METHOD	71-0815
SATURATED	ANWENDUNG DER FINITE-ELEMENT-METHODE ZUR BERECHNUNG VON SPANNUNGEN IN WASSERGESAETTIGTEN BOEDEN (FIN	73-0502
SATURATED	DYNAMIC STRESS ANALYSIS OF POROUS ELASTIC SOLIDS SATURATED WITH COMPRESSIBLE FLUIDS	71-0651
SATURATED	FINITE ELEMENT ANALYSIS OF MAGNETICALLY SATURATED D-C MACHINES	71-0566
SATURATED	FINITE ELEMENT COMPUTER PROGRAMS FOR FLOW IN SATURATED UNSATURATED POROUS MEDIA	72-0364
SATURATED	FINITE ELEMENT SOLUTION OF SATURATED TRAVELLING MAGNETIC FIELD PROBLEMS	75-0600
SATURATED	GROUNDWATER FLOW IN PARTIALLY SATURATED SOILS	72-0745
SATURATED	METOD KONECHNYKH ELEMENTOV V RASCHETAKH KONSOLIDATSII VODONASYSHCHENNYKH GRUNTOV. (FINITE ELEMENT ME	75-0469
SATURATED	SATURATED MAGNETIC ENERGY FUNCTIONAL FOR FINITE-ELEMENT ANALYSIS OF ELECTRIC MACHINES	75-0636
SATURATED	SOLUTION OF WAVE PROPAGATION PROBLEMS IN SATURATED MEDIUM	72-0784
SATURATED	THEORETICAL TRANSIENT BEHAVIOUR OF SATURATED AND UNSATURATED SOILS UNDER LOAD AND CHANGING MOISTURE	73-0901
SATURATED-UNSAT	GALERKIN APPROACH TO SATURATED-UNSATURATED FLOW IN POROUS MEDIA	75-0393
SATURATED-UNSAT	SATURATED-UNSATURATED SEEPAGE BY FINITE ELEMENTS	73-0459
SATURATED-UNSAT	WATER FLOW THROUGH SATURATED-UNSATURATED POROUS MEDIA USING FINITE-ELEMENT-GALERKIN METHODS	74-0879
SATURATED-UNSAT	WATER MOVEMENT THROUGH SATURATED-UNSATURATED POROUS MEDIA: A FINITE-ELEMENT GALERKIN MODEL	75-0761
SCALAR	AXISYMETRIC TRIANGULAR ELEMENT FOR THE SCALAR HELMHOLTZ EQUATION	73-0688
SCALAR	FINITE ELEMENT PROGRAM PACKAGE FOR AXISYMMETRIC SCALAR FIELD PROBLEMS	73-0113
SCALAR	SCALAR ALGORITHM OF DETERMINATION OF STIFFNESS MATRIX AND OF ANOTHER FINITE ELEMENT CHARACTERISTIC	74-1013
SCALAR	SCALAR FINITE-ELEMENT PROGRAM PACKAGE FOR TWO-DIMENSIONAL FIELD PROBLEMS	71-0478
SCALE	A TORSIONAL STRENGTH ANALYSIS ON THE CONTAINER SHIP BY MEANS OF THE FINITE ELEMENT PROCEDURE AND FUL	71-0243
SCALE	ANALYTICAL AND EXPERIMENTAL INVESTIGATION OF A 1/8 SCALE DYNAMIC MODEL OF THE SHUTTLE ORBITER. VOLUM	74-1186
SCALE	APPLICATION OF FINITE ELEMENT METHODS TO LARGE SCALE ELASTIC PLASTIC PROBLEMS OF FRACTURE MECHANICS	69-0372
SCALE	ASKA, A LARGE SCALE SOFTWARE SYSTEM FOR FINITE ELEMENT ANALYSIS	69-0331
SCALE	BLOCKING TECHNIQUE FOR LARGE SCALE STRUCTURAL ANALYSIS	73-0243
SCALE	COMPUTER SYSTEM FOR THE ANALYSIS OF LARGE SCALE PROBLEMS IN SOLID, FLUID AND THERMODYNAMICS	73-0554
SCALE	EXAMPLE OF LARGE-SCALE STRUCTURAL ANALYSIS. COMPARISON BETWEEN FINITE ELEMENT CALCULATION AND FULL S	74-0177
SCALE	EXPERIMENTAL AND THEORETICAL STUDY ON A 1/10TH SCALE MODEL PCPV	73-0829
SCALE	INFLUENCE OF NON-SINGULAR STRESS TERMS AND SPECIMEN GEOMETRY ON SMALL SCALE YIELDING AT CRACK TIPS I	73-0203
SCALE	LARGE SCALE ANALYSIS OF CURRENT AIRCRAFT	70-0383
SCALE	LARGE SCALE COMPUTER PROGRAMS FOR STRUCTURAL ANALYSIS	70-0211
SCALE	LIMITATIONS TO THE SMALL SCALE YIELDING APPROXIMATION FOR CRACK TIP PLASTICITY	74-0145
SCALE	MODAL COUPLING PROCEDURES ADAPTED TO NASTRAN ANALYSIS OF THE 1/8 - SCALE SHUTTLE STRUCTURAL DYNAMICS	75-0677
SCALE	NONLINEAR FINITE ELEMENT AND ITS APPLICATION TO LARGE SCALE STRUCTURAL PROBLEMS	75-0386
SCALE	OPTIM 11: (MAGIC) COMPATIBLE LARGE SCALE AUTOMATED MINIMUM WEIGHT DESIGN PROGRAM - V 2. PROGRAMMER'S	74-1036
SCALE	OPTIM 11: A (MAGIC) COMPATIBLE LARGE SCALE AUTOMATED MINIMUM WEIGHT DESIGN PROGRAM - 1. ENGINEERS AN	74-1044
SCALE	OPTIM 11: MAGIC COMPATIBLE LARGE SCALE AUTOMATED MINIMUM WEIGHT DESIGN PROGRAM. VOLUME 11. PROGRAMME	74-1093
SCALE	SMALL SCALE YIELDING NEAR A CRACK IN PLANE STRAIN-FINITE ELEMENT ANALYSIS	71-0355
SCALE	SMALL SCALE YIELDING ANALYSIS OF MIXED MODE PLANE STRAIN CRACK PROBLEMS	73-0801
SCALE	THE DEVELOPMENT OF COMPUTER GRAPHICS FOR LARGE SCALE FINITE ELEMENT CODES	71-0042
SCALE	THE DEVELOPMENT OF LARGE SCALE DIGITAL COMPUTER CODES FOR PRODUCTION STRUCTURAL ANALYSIS	71-0721
SCALE	THIN-WALLED PRESTRESSED CONCRETE PRESSURE VESSEL FOR HIGH TEMPERATURE REACTORS: EXPERIMENTAL INVESTI	73-0857
SCANDINAVIAN	THE SCANDINAVIAN PCRV MODEL PROJECT: STRESS CALCULATION AND EXPERIMENTAL VERIFICATION	71-0021
SCATTERED-LIGHT	THREE DIMENSIONAL SCATTERED-LIGHT STRESS ANALYSIS OF DISCONTINUOUS FIBER-REINFORCED COMPOSITES	73-0218
SCATTERING	A FINITE ELEMENT APPROACH TO SCATTERING FROM ELASTIC SPHERICAL SHELLS	74-0702
SCATTERING	FINITE ELEMENT ANALYSIS OF ELASTIC WAVE SCATTERING FROM DISCONTINUITIES	73-0547
SCATTERING	FINITE-ELEMENT APPROACH TO ACOUSTIC SCATTERING FROM ELASTIC STRUCTURES	75-0069
SCATTERING	OSCILLATIONS AND WAVE FORCES IN AN OFFSHORE HARBOR - APPLICATIONS OF HYBRID FINITE ELEMENT METHOD TO	74-1024
SCATTERING	SCATTERING BY BURIED OBSTACLES	74-0739
SCATTERING	SCATTERING BY DIELECTRIC CYLINDERS	74-1025
SCATTERING	UNIMOMENT METHOD OF SOLVING ANTENNA AND SCATTERING PROBLEMS	74-1020
SCHALEN	COMPUTATION OF SHELLS WITH FINITE DISPLACEMENTS: MIXED FINITE ELEMENTS (BERECHNUNG VON SCHALEN MIT E	73-0788
SCHALEN	EIN NEUES VARIATIONSPRINZIP MIT ANWENDUNG AUF SCHWINGUNGEN FLACHER SCHALEN NACH DER METHODE DER FINI	73-0563
SCHALENELEMENT	DESH, EIN EINFACHES FLACHES SCHALENELEMENT UEBER DREIECKIGEM GRUNDRISS. (DESH, A SIMPLE TRIANGULAR S	75-0304
SCHALENELEMENT	HYBRIDE SCHALENELEMENTE MIT EINER ANWENDUNG AUF WENDELSCHALEN. (HYBRID SHELL ELEMENTS IN APPLICATION	75-0609
SCHALENSYSTEME	NEUESTE ERFAHRUNGEN BEI DER RECHNERISCHEN UND KONSTRUKTIVEN AUSLEGUNG KOMPLIZIERTER SCHALENSYSTEME	71-0052
SCHALENTRAGWERK	TURBAN VERSION 1.0. EIN SPEZIELLES PROGRAMM ZUR LINEAREN UND NICHTLINEAREN STATISCHEN UND DYNAMISCHE	71-0622
SCHALENVERFORMU	UEBER PAUSCHALE FEHLERSCHRANKEN FUER APPROXIMATIONEN VON SCHALENVERFORMUNGEN. (GLOBAL BOUNDARY CONDI	75-0292
SCHEIBE	ZUSAMMENWIRKEN VON EINZELNEN FERTIGTEILEN ALS GROSSFLAECHIGE SCHEIBE (INTERACTION OF INDIVIDUAL PREF	73-0552
SCHEIBEN	ACCURATE CALCULATION OF STRESSES IN BEAMS, DISKS AND PLATES BY MEANS OF FINITE ELEMENT METHODS AND I	70-0133
SCHEN	BERECHNUNG VON PLATTEN NACH DEM RITZ' SCHEN VERFAHREN MIT HILFE DREIECKFOERMIGER MASCHENNETZE	66-0094
SCHRAUBENFEDERN	SCHWINGUNGEN ZYLINDRISCHER SCHRAUBENFEDERN (VIBRATIONS OF CYLINDRICAL COIL SPRINGS)	74-0331
SCHRODINGER	A METHOD TO GALERKIN TYPE ACHIEVING OPTIMUM L(2) ACCURACY FOR FIRST ORDER HYPERBOLICS AND EQUATIONS	72-0521
SCHUBMITTELPUNK	EINE FINITE-ELEMENTE-LOESUNG FUR DIE TORSIONSSTEIFIGKEIT UND DEN SCHUBMITTELPUNKT BELIEBIGER QUERSCH	74-0256
SCHWEISSEIGENSP	BERECHNUNG DER SCHWEISSEIGENSPANNUNGEN UND SCHWEISSFORMAENDERUNGEN MIT ELASTISCH-PLASTISCHEN FINITEN	75-0271
SCHWEISSEIGENSP	VOLLSTAENDIGE SPANNUNGS-DEHNUNGS-TEMPERATURAENDERUNGSBEZIEHUNG FUER DIE SCHWEISSEIGENSPANNUNGSBERECH	75-0807
SCHWEISSFORMAEN	BERECHNUNG DER SCHWEISSEIGENSPANNUNGEN UND SCHWEISSFORMAENDERUNGEN MIT ELASTISCH-PLASTISCHEN FINITEN	75-0271
SCHWEISSKONSTRU	ANWENDUNG DES FINITE ELEMENT VERFAHRENS SUR BESTIMMUNG DER FESTIGKEIT VON SCHWEISSKONSTRUKTIONEN. (	72-0185
SCHWINGFESTIGKE	OSCILLATION STABILITY OF INTERNALLY AND EXTERNALLY NOTCHED PLANE BARS OF EQUAL FORM FACTOR FROM 3943	73-0745
SCHWINGFESTIGKE	FORM FACTORS OF FLAT BARS FOR FATIGUE STRENGTH TESTS. (FORMZAHLEN VON FLACHSTAEBEN FUER SCHWINGFESTI	71-0641
SCHWINGUNGEN	ANWENDUNG DER MATRIZENVERSCHIEBUNGSMETHODE AUF ERZWUNGENE SCHWINGUNGEN (SCHWINGSPROBLEME) (ARBEIT	71-0627
SCHWINGUNGEN	EIN NEUES VARIATIONSPRINZIP MIT ANWENDUNG AUF SCHWINGUNGEN FLACHER SCHALEN NACH DER METHODE DER FINI	73-0563
SCHWINGUNGEN	SCHWINGUNGEN ZYLINDRISCHER SCHRAUBENFEDERN (VIBRATIONS OF CYLINDRICAL COIL SPRINGS)	74-0331
SCHWINGUNGSMAES	SCHWINGUNGSMAESSIGE AUSLEGUNG DER BESCHAUFELUNG AXIALER TURBOMASCHINEN (DESIGN OF DISPLACEMENT-FREE	74-0253
SCHWINGUNGSPROB	ANWENDUNG DER MATRIZENVERSCHIEBUNGSMETHODE AUF ERZWUNGENE SCHWINGUNGEN (SCHWINGSPROBLEME) (ARBEIT	71-0627
SCIENTIFIC	FINITE ELEMENTS AND SCIENTIFIC COMPUTING	75-0042
SCIENTIFIC	PLOTS - THE PLOTTING SUBROUTINE INCORPORATED IN THE BALLISTIC RESEARCH LABORATORIES ELECTRONIC SCIEN	72-0803
SCIENTIFIC	USING THE NASTRAN THERMAL ANALYZER TO SIMULATE A FLIGHT SCIENTIFIC INSTRUMENT PACKAGE	74-1011
SCREENING	EDDY CURRENTS SCREENING PROBLEMS SOLVED BY MEANS OF FINITE ELEMENT DIFFERENCE AND FINITE ELEMENT MET	75-0552
SCREW	NUMERICAL STUDY OF EIGEN OSCILLATIONS OF A SCREW PINCH WITH FINITE ELEMENTS	74-0159
SEA	CRACK PROPAGATION IN SEA ICE - FINITE ELEMENT APPROACH	72-0029
SEA	DEEP SEA ENGINEERING	71-0739
SEA	NUMERICAL DETERMINATION OF SEA WATER INTRUSION IN AQUIFERS	74-1181

SEA	OFFSHORE GRAVITY BASED OIL PRODUCTION PLATFORM INTERACTION WITH THE SEA BED	75-0221
SEA	OSCILLATIONS AND WAVE FORCES IN A MAN-MADE HARBOUR IN THE OPEN SEA	74-0151
SEAL	COMPRESSIBLE SEAL FLOW ANALYSIS USING THE FINITE ELEMENT METHOD WITH GALERKIN SOLUTION TECHNIQUE	74-0688
SEAL-SHELL-2	SEAL-SHELL-2, A COMPUTER PROGRAM FOR THE STRESS ANALYSIS OF A THICK SHELL OF REVOLUTION WITH AXISYMM	69-0107
SEAWATER	ON SEAWATER ENCROACHMENT IN COASTAL AQUIFERS	74-0577
SEAWAY	A GENERAL CALCULATION OF THE STRUCTURAL LOADS ON A SHIP IN A SEAWAY	71-0345
SECOND	A COMPARISON OF FIRST AND SECOND ORDER AXIALLY SYMMETRIC FINITE ELEMENTS	73-0997
SECOND	A COMPUTER METHOD FOR THE SECOND ORDER ELASTIC ANALYSIS OF PLANE FRAMED STRUCTURES	73-0310
SECOND	A FINITE ELEMENT PROCEDURE OF THE SECOND ORDER ACCURACY	70-0277
SECOND	A SECOND ORDER FINITE ELEMENT METHOD FOR THE ONE-DIMENSIONAL STEFAN PROBLEM	74-1222
SECOND	APPROXIMATION OF STEKLOV EIGENVALUES OF NON-SELFADJOINT SECOND ORDER ELLIPTIC OPERATORS	72-0509
SECOND	COMPUTER SIMULATION OF SECOND ORDER FAULTS	71-0010
SECOND	ELASTIC-PLASTIC DEFORMATIONS OF FLEXURALLY STIFF FRAMEWORKS FROM SECOND ORDER STRESS THEORY	73-1061
SECOND	FEASIBILITY OF USING FINITE ELEMENTS IN ANALYSIS OF SECOND BREAKDOWN IN SEMICONDUCTOR DEVICES	73-0874
SECOND	FINITE ELEMENTS IN CURVILINEAR COORDINATES - SECOND ORDER SURFACES	73-0753
SECOND	ON SOME FINITE ELEMENT PROCEDURES FOR SOLVING SECOND ORDER BOUNDARY VALUE PROBLEMS	69-0288
SECOND	SECOND GENERATION BOUNDARY INTEGRAL EQUATION PROGRAM FOR THREE-DIMENSIONAL ELASTIC ANALYSIS	75-0376
SECOND	SECOND ORDER THEORY OF OSCILLATING CYLINDERS	70-0430
SECOND	THE SECOND VARIAION TEST WITH ALGEBRAIC AND DIFFERENTIAL CONTRASTS	69-0342
SECOND	ZUR BERECHNUNG VON PLATTEN NACH DER THEORIE ZWEITER. ORDNUNG MIT HILFE EINES HYBRIDEN DEFORMATIONS M	73-0558
SECOND-ORDER	ZUR BERECHNUNG EINFACH-SYMMETRISCHER I-TRAEGER NACH DER THEORIE - 2. ORDNUNG UNTER BERUECKSICHTIGUNG	75-0510
SECONDARY	FINITE ELEMENT INVESTIGATION OF SECONDARY CRACKING DURING BRITTLE FRACTURE	71-0804
SECONDARY	METHOD FOR ESTIMATING RESPONSE OF PAYLOAD SECONDARY STRUCTURES TO RANDOM EXCITATION	68-0232
SECTION	A TECHNIQUE FOR CONNECTING BEAM ELEMENTS TO A PLATE MODEL OF A COMPLICATED BOX SECTION	74-0271
SECTION	ANALYSIS OF RAREFIED GAS FLOW THROUGH AN ARBITRARY CROSS SECTION BY THE FINITE ELEMENT METHOD	73-0365
SECTION	CLASSIFICATION OF FINITE ELEMENT STRESSES ACCORDING TO ASME SECTION 111 STRESS CATEGORIES	74-0610
SECTION	COUPLED VIBRATIONS OF THIN-WALLED BEAMS OF OPEN SECTION USING FINITE ELEMENT METHOD	70-0230
SECTION	COUPLED VIBRATIONS OF THIN-WALLED BEAMS OF OPEN SECTION USING FINITE ELEMENT METHOD	71-0568
SECTION	DETERMINATION OF EIGENMODES OF HOMOGENEOUS PLASMA WITH ELLIPTICAL SECTION USING BI-DIMENSIONAL FINIT	74-0553
SECTION	DYNAMICS OF PRISMATIC AND OPEN SECTION MEMBER GRIDS	69-0070
SECTION	EINE FINITE-ELEMENTE-LOESUNG FUR DIE TORSIONSSTEIFIGKEIT UND DEN SCHUBMITTELPUNKT BELIEBIGER QUERSCH	74-0256
SECTION	FINITE ELEMENT ANALYSIS OF AN EXPANDED I - SECTION BEAM AND AN AXISYMMETRIC FLANGED CYLINDER	68-0002
SECTION	FINITE ELEMENT ANALYSIS ON THE TORSION OF A BAR WITH UNIFORM CROSS SECTION	68-0186
SECTION	FINITE ELEMENT ANALYSIS OF THIN-WALLED MEMBERS OF OPEN SECTION	72-0966
SECTION	FINITE ELEMENT SIMULATION OF ACOUSTIC FILTERS WITH VARIABLE CIRCULAR CROSS SECTION	75-0871
SECTION	HEAVY SECTION STEEL TECHNOLOGY PROGRAM. TECHNICAL REPORT NO. 4, OCTOBER 1969. A TWO-DIMENSIONAL ELAS	69-0474
SECTION	HEAVY SECTION STEEL TECHNOLOGY PROGRAM TECHNICAL OR PROGRAMMATIC MANUSCRIPT NO. 25. VARIABLE THICKNE	73-0948
SECTION	INTERPRETATION OF FINITE ELEMENT STRESSES ACCORDING TO ASME SECTION 111	75-0655
SECTION	M.I.T. TEST SECTION INSTRUMENTATION, MASSACHUSETTS BAY TRANSPORTATION AUTHORITY, HAYMARKET-NORTH EXT	72-0673
SECTION	MATRICE DE RIGIDITE D'UN ELEMENT DE POUTRE SOLLICITE A LA TORSION-FLEXION (STIFFNESS MATRIX OF A BEA	73-0442
SECTION	POSTBUCKLING BEHAVIOUR OF A SECTION REPRESENTATIVE OF THE B-1 AFT INTERMEDIATE FUSELAGE	73-0489
SECTION	PRELIMINARY FINITE ELEMENT ANALYSIS, ATCHAFALAYA BASIN PROTECTION LEVEES, TEST SECTION 3	69-0463
SECTION	THREE DIMENSIONAL INELASTIC ANALYSIS OF A SHELL APPURTENANCE PER SECTION 111 -'APPENDIX F.	74-0716
SECTION	TRANSVERSE STRENGTH AT FORE AND AFT SECTION OF TANKER CARGO HOLD	74-0618
SECTIONAL	APPLICATION OF FINITE ELEMENT METHOD WITH SECTIONAL LINEARIZATION TO FLOW PROBLEMS	75-0611
SECTIONS	A COMPUTER PROGRAM FOR THE CALCULATION OF TORSION FACTORS AND SHEAR STRESS FACTORS FOR BOX SECTIONS	69-0485
SECTIONS	APPROXIMATION METHOD FOR COMPUTATION OF NATURAL FREQUENCIES OF SPIRAL SECTIONS	71-0775
SECTIONS	FRACTURE TOUGHNESS TESTING WITH SECTIONS OF CYLINDERS	74-0999
SECTIONS	MODELING TECHNIQUES OF THIN-WALLED BEAMS WITH OPEN CROSS SECTIONS	72-0573
SECTIONS	TRANSVERSE VIBRATIONS OF CANTILEVER BARS OF VARIABLE CROSS SECTIONS	66-0001
SECTOR	A COMPATIBLE PLATE BENDING ANNULAR SECTOR FINITE ELEMENT OF NONUNIFORM THICKNESS	74-0430
SECTOR	ANNULAR AND CIRCULAR SECTOR FINITE ELEMENTS FOR PLATE BENDING	70-0100
SECTOR	COMMENT ON STIFFNESS MATRICES FOR SECTOR ELEMENTS	70-0233
SECTOR	DYNAMIC ANALYSIS OF CIRCULAR AND SECTOR THICK, LAYERED PLATES	75-0198
SECTOR	FINITE-ELEMENT METHOD WITH ELEMENTS WITH CIRCULAR RING SECTOR FORM	73-0105
SECTOR	SECTOR ELEMENT FOR THIN PLATE FLEXURE	72-0174
SECTOR	SECTOR ELEMENTS FOR MATRIX DISPLACEMENT ANALYSIS	73-1049
SECTOR	SECTOR FINITE ELEMENTS IN THE THEORY OF PLANE ELASTICITY	71-0766
SECTOR	STIFFNESS MATRICES FOR SECTOR ELEMENTS	69-0219
SECTOR	TAPERED PLANE STRESS ANNULAR SECTOR STRESS ELEMENT	75-0538
SEDIMENT	A FINITE ELEMENT MODEL FOR SEDIMENT TRANSPORT IN ESTUARIES	74-1102
SEDIMENTS	OPTIMAL DESIGN OF WATER WELLS IN UNCONSOLIDATED SEDIMENTS	74-1182
SEEDLING	COMPUTER MODELLING OF PLUG SEEDLING SURVIVAL	75-1076
SEEPAGE	A FINITE ELEMENT MODEL FOR STABILITY ANALYSIS OF MINE WASTE EMBANKMENTS SUBJECTED TO SEEPAGE	74-1166
SEEPAGE	ANALYSIS OF SEEPAGE THROUGH DAMS AND IN FOUNDATION BY FINITE ELEMENT METHOD	73-1110
SEEPAGE	ANALYSIS OF THE SEEPAGE THROUGH AN EARTH DAM BY FINITE ELEMENT METHOD	72-0585
SEEPAGE	APPLICATION OF DUAL MINIMUM THEOREMS TO THE FINITE ELEMENT SOLUTION OF POTENTIAL PROBLEMS WITH SPECI	72-0036
SEEPAGE	APPROXIMATE SOLUTION FOR UNCONFINED SEEPAGE	73-0348
SEEPAGE	DETERMINING SEEPAGE CHARACTERISTICS OF MILL-TAILING DAMS BY THE FINITE ELEMENT METHOD	71-0477
SEEPAGE	DISPERSION IN STEADY UNSTEADY SEEPAGE FLOWFIELD	74-0072
SEEPAGE	EFFECT OF SEEPAGE ON EMBANKMENT DEFORMATIONS DUE TO WATER LOADING	74-1329
SEEPAGE	FIELD SOLUTIONS FOR TURBULENT SEEPAGE FLOW	71-0206
SEEPAGE	FINITE ELEMENT ANALYSIS OF SEEPAGE	67-0130
SEEPAGE	FINITE ELEMENT ANALYSIS OF SEEPAGE THROUGH DAMS	67-0152
SEEPAGE	FINITE ELEMENT ANALYSIS OF SEEPAGE IN ELASTIC MEDIA	69-0024
SEEPAGE	FINITE ELEMENT ANALYSIS OF UNCONFINED SEEPAGE PROBLEMS	71-0767
SEEPAGE	FINITE ELEMENT ANALYSIS OF A NONSTEADY GROUNDWATER SEEPAGE PROBLEM DAM, DRAINAGE	74-1375
SEEPAGE	FINITE ELEMENT METHOD OF ANALYZING STEADY SEEPAGE WITH A FREE SURFACE	70-0092
SEEPAGE	FINITE ELEMENT METHODS: TWO-DIMENSIONAL SEEPAGE WITH A FREE SURFACE	71-0475
SEEPAGE	FINITE ELEMENT PROCEDURES FOR SEEPAGE ANALYSIS USING AN ISOPARAMETRIC ELEMENT	72-0366
SEEPAGE	FINITE ELEMENT SOLUTION FOR QUANTITY OF STEADY SEEPAGE	71-0824
SEEPAGE	FINITE-ELEMENT ANALYSIS OF SEEPAGE THROUGH DAMS	67-0025
SEEPAGE	NUMERICAL ANALYSIS OF FREE SURFACE SEEPAGE PROBLEMS	71-0469
SEEPAGE	ON SOME FREE SURFACE ON TRANSIENT FLOW PROBLEMS OF SEEPAGE AND IRROTATIONAL FLOW	72-0070
SEEPAGE	SATURATED-UNSATURATED SEEPAGE BY FINITE ELEMENTS	73-0459
SEEPAGE	SEEPAGE ANALYSIS OF EARTH BANKS UNDER DRAWDOWN	72-0014
SEEPAGE	SEEPAGE AND GROUNDWATER EFFECTS ASSOCIATED WITH EXPLOSIVE CRATERING	72-0656
SEEPAGE	SEEPAGE IN FRACTURED ROCK: FINITE ELEMENT ANALYSIS OF ROCK MASS BEHAVIOR UNDER COUPLED ACTION OF BOD	71-0825
SEEPAGE	SEEPAGE IN MISSISSIPPI RIVER BANKS. REPORT 1. ANALYSIS OF TRANSIENT SEEPAGE USING VISCOUS FLOW MODEL	70-0405
SEEPAGE	SEEPAGE IN MISSISSIPPI RIVER BANKS. REPORT 1. ANALYSIS OF TRANSIENT SEEPAGE USING VISCOUS FLOW MODEL	70-0405
SEEPAGE	SEEPAGE IN MISSISSIPPI RIVER BANKS. REPORT 1. ANALYSIS OF TRANSIENT SEEPAGE USING A VISCOUS-FLOW MOD	73-0909
SEEPAGE	SEEPAGE IN MISSISSIPPI RIVER BANKS. REPORT 1. ANALYSIS OF TRANSIENT SEEPAGE USING A VISCOUS-FLOW MOD	73-0909
SEEPAGE	SEEPAGE INTO AN EXCAVATION IN A MEDIUM POSSESSING STRESS-DEPENDENT PERMEABILITY	72-0629
SEEPAGE	SOLUTION OF ANISOTROPIC SEEPAGE PROBLEMS BY FINITE ELEMENTS	66-0080
SEEPAGE	SOLUTION OF ANISOTROPIC SEEPAGE BY FINITE ELEMENTS	67-0071
SEEPAGE	THE APPLICATION OF THE FINITE ELEMENT METHOD TO SEEPAGE AND STRESS-DEFORMATION PROBLEMS IN SOIL MECH	70-0267
SEEPAGE	TWO AND THREE DIMENSIONAL GROUNDWATER SEEPAGE BY FINITE ELEMENTS	74-0463
SEEPAGE	UNCONFINED TRANSIENT SEEPAGE IN SLOPING BANKS	71-0378
SEEPAGE	UTILISATION DE MODELES MATHEMATIQUES MULTIDIMENSIONNELS POUR L'ETUDE DES ECOULEMENTS DANS LES DIGUES	73-0381
SEEPAGE	VARIABLE TIME STEP ANALYSIS OF UNCONFINED SEEPAGE	74-0064
SEEPAGE-ENVIRON	SEEPAGE-ENVIRONMENTAL ANALYSIS OF THE SLIME ZONE OF A TAILINGS POND	74-0513
SEEPAGE-LINE	STUDIES ON THE DETERMINATION OF SEEPAGE-LINE OF FILL DAM BY THE METHOD OF FINITE ELEMENT. WATER FLOW	74-1348

358

SEGMENT	A FULLY COMPATIBLE ANNULAR SEGMENT FINITE ELEMENT	72-0582
SEGMENT	A FULLY COMPATIBLE ANNULAR SEGMENT FINITE ELEMENT	72-0663
SEGMENT	AN ANNULAR SEGMENT FINITE ELEMENT FOR PLATE BENDING	71-0285
SEGMENT	CONICAL SEGMENT METHOD FOR ANALYZING OPEN CROWN SHELLS OF REVOLUTION FOR EDGE LOADING	63-0007
SEGMENT	FINITE SEGMENT ANALYSIS OF FOLDED PLATES	69-0169
SEGMENT	HORIZONTALLY CURVED HIGHWAY BRIDGES. A FULLY COMPATIBLE ANNULAR SEGMENT FINITE ELEMENT	72-0834
SEGMENT	MASS MATRIX FOR A PRISMATIC BEAM SEGMENT	68-0116
SEGMENTALLY	COMPUTER ANALYSIS OF SEGMENTALLY ERECTED BRIDGES	75-0172
SEGMENTED	ANALYSIS OF BUCKLING AND VIBRATION OF RING-STIFFENED, SEGMENTED SHELLS OF REVOLUTION	69-0496
SEGMENTED	FUEL ROD FINITE ELEMENT STRESS ANALYSIS: SEGMENTED FLIGHT REACTOR FUEL VERSUS SRDR FUEL	71-0798
SEGMENTED	VIBRATIONS OF SEGMENTED SHELLS	73-0161
SEGMENTED	3-DIMENSIONAL FINITE-ELEMENT ANALYSIS OF ELECTROELASTIC NORMAL MODES OF A TANGENTIALLY POLARIZED, SE	75-0591
SEGMENTS	BUCKLING OF SEGMENTS OF TOROIDAL SHELLS	73-0515
SEIDEL'S	ZASTOSOWANIE ITERACJI SEIDLA W METODZIE ELEMENTOW SKONCZONYCH NA PRZYKLADZIE OBLICZEN STATYCZNYCH P	74-1031
SEIDLA	ZASTOSOWANIE ITERACJI SEIDLA W METODZIE ELEMENTOW SKONCZONYCH NA PRZYKLADZIE OBLICZEN STATYCZNYCH P	74-1031
SEINER	VORAUSSAGE DES GEBIRGSDRUCKES UND SEINER AUSQIRKUNGEN AUF STREBEN UND STRECKEN IM STEINKOHLENBERGBAU	73-0679
SEISMES	CONTRIBUTION A L'ETUDE DU COMPLEXE BARRAGE-FONDATION ET EXPLICATION DES SEISMES DUS AU REMPLISSAGE D	72-0554
SEISMIC	A SEISMIC DESIGN OF NUCLEAR REACTOR BUILDING. STRESS ANALYSIS AND STIFFNESS EVALUATION OF THE ENTIRE	71-0251
SEISMIC	A SEISMIC STRENGTH OF KOLKEWADI DAM	73-0197
SEISMIC	ANALYSIS OF BUILDINGS ON ELASTIC MEDIUM UNDER DYNAMIC OR SEISMIC LOADS	74-0456
SEISMIC	ANALYSIS OF SOIL-STRUCTURE INTERACTION EFFECTS UNDER SEISMIC EXCITATION	71-0064
SEISMIC	COMPORTAREA ELASTICA A DIGURILOR DE PAMINT SOLICITATE SEISMIC (ELASTIC BEHAVIOR OF THE EARTH DAMS SEI	74-1185
SEISMIC	COMPUTATIONAL METHOD FOR DYNAMIC ANALYSIS OF STRUCTURES WITH LARGE SIZE STIFFNESS AND MASS MATRICES.	73-0830
SEISMIC	CONSIDERATION ON THE SEISMIC RESPONSE ANALYSIS OF UNDERGROUND STRUCTURES	75-0470
SEISMIC	DECONVOLUTION OF SEISMIC RESPONSE FOR LINEAR SYSTEMS	73-0774
SEISMIC	EFFECT OF GEOLOGIC IRREGULARITIES OF SEISMIC RESPONSE	74-0898
SEISMIC	EFFECTS OF FOUNDATION ROTATION ON SEISMIC INERTIA FORCES OF NUCLEAR POWER PLANT STRUCTURES	73-0177
SEISMIC	ELASTO-PLASTIC SEISMIC RESPONSE ANALYSIS OF FRAMED STRUCTURES BY MODE SUPERPOSITION	73-0325
SEISMIC	EVALUATION OF TORSIONAL EFFECTS IN THE SEISMIC ANALYSIS OF SYMMETRIC AND UNSYMMETRIC NUCLEAR POWER P	73-0175
SEISMIC	FACTOR OF SAFETY APPROACH FOR EVALUATING SEISMIC STABILITY OF SLOPES	75-0645
SEISMIC	FINITE ELEMENT ANALYSIS FOR SEISMIC RESPONSE OF EARTH DAMS	68-0091
SEISMIC	FINITE ELEMENT ANALYSIS FOR THE SEISMIC STABILITY OF EARTH STRUCTURES	74-1106
SEISMIC	FINITE ELEMENT TREATMENT OF SOIL-STRUCTURE INTERACTION PROBLEM FOR NUCLEAR POWER PLANT UNDER SEISMIC	73-0834
SEISMIC	NONLINEAR SEISMIC ANALYSIS OF CURRENT DIVIDER SUPPORT SYSTEM FOR PACIFIC DC INTERTIE	73-0200
SEISMIC	QUAD-4: A COMPUTER PROGRAM FOR EVALUATING THE SEISMIC RESPONSE OF SOIL STRUCTURES BY VARIABLE DAMPIN	73-0771
SEISMIC	RESPONSE OF EARTH DAMS TO TRAVELLING SEISMIC WAVES	69-0092
SEISMIC	SEISMIC ANALYSIS OF EARTH DAM RESERVOIR SYSTEMS	73-0412
SEISMIC	SEISMIC ANALYSIS OF A THREE-PHASE BREAKER INSTALLATION	74-0522
SEISMIC	SEISMIC BEHAVIOUR OF FRAMED TUBES	75-0306
SEISMIC	SEISMIC EFFECTS ON A PROPOSED UNDERGROUND REACTOR FACILITY	73-0953
SEISMIC	SEISMIC RESPONSE BY VARIABLE DAMPING FINITE ELEMENTS	74-0143
SEISMIC	SEISMIC RESPONSE EFFECTS ON EMBEDDED STRUCTURES	72-0902
SEISMIC	SEISMIC RESPONSE OF SOIL DEPOSITS UNDERLAIN BY SLOPING ROCK BOUNDARIES	69-0484
SEISMIC	SEISMIC RESPONSE OF HIGHWAY BRIDGES	75-0879
SEISMIC	SOIL-STRUCTURE INTERACTION ANALYSES FOR EVALUATING SEISMIC RESPONSE	74-0745
SEISMIC	STATIC AND SEISMIC BEHAVIOUR OF AN EARTH DAM	67-0026
SEISMIC	STUDIES OF SEISMIC RESPONSE OF CLAY BANKS	71-0833
SEISMIC	STUDY OF NUMERICAL METHODS IN SEISMIC PROBLEMS	72-0738
SEISMIC	THE SEISMIC DESIGN STUDY OF A DOUBLE CURVATURE ARCH DAM	69-0053
SEISMIC	THREE DIMENSIONAL SEISMIC ANALYSIS OF STRUCTURE-GROUND SYSTEMS BY THE FINITE ELEMENT METHOD	74-0268
SEISMIC	3-D SEISMIC CODE FOR ILLIAC-1V	73-0868
SEISMICALLY	COMPORTAREA ELASTICA A DIGURILOR DE PAMINT SOLICITATE SEISMIC (ELASTIC BEHAVIOR OF THE EARTH DAMS SEI	74-1185
SEISMICALLY	ENVIRONMENTAL CONSIDERATIONS OF DAM CONSTRUCTION AND OPERATION IN SEISMICALLY ACTIVE URBAN AREAS	73-0470
SEISMOLOGICAL	DYNAMIC FINITE ELEMENT ANALYSIS WITH APPLICATIONS TO SEISMOLOGICAL PROBLEMS	75-0785
SEISMOLOGY	SEISMOLOGY AND ACOUSTIC-GRAVITY WAVES	73-0744
SELECTING	CRITERION FOR SELECTING STIFFNESS MATRICES	65-0061
SELECTING	GUIDELINES FOR SELECTING FINITE ELEMENT GRIDS BASED ON AN OPTIMIZATION STUDY	74-0303
SELECTING	PROCEDURES FOR SELECTING NEAR OPTIMUM FINITE ELEMENT GRIDS	73-0537
SELECTION	APPLICATION OF GRID SELECTION PROCEDURES FOR IMPROVED FINITE ELEMENT STRESS ANALYSIS	74-0326
SELECTION	FURTHER DEVELOPMENTS IN GRID SELECTION PROCEDURES IN THE FINITE ELEMENT METHOD	75-0028
SELECTION	LUBRICATION PROBLEMS - THE SELECTION OF MATHEMATICAL MODELS	75-0881
SELECTION	ON THE AUTOMATIC SELECTION OF INDEPENDENT DISPLACEMENT VARIABLES IN GENERAL PURPOSE FINITE ELEMENT C	73-0676
SELECTION	OPTIMIZATION OF REDUNDANCY SELECTION IN FINITE-ELEMENT FORCE METHOD	70-0228
SELECTION	SELECTION OF LOCATIONS FOR LONG TERM STORAGE OF WHEAT	75-0323
SELF	CLOSURE ANALYSIS USING SELF ENERGIZING, DOUBLE CONICAL GASKETS	74-0675
SELF	PROBLEMS OF SELF ADAPTIVENESS IN THE FINITE ELEMENT METHOD	75-0097
SELF	SELF ADJOINT DIFFERENTIAL EQUATIONS AND VARIATIONAL PRINCIPLE - 1.	71-0388
SELF	SELF ADJOINT DIFFERENTIAL EQUATIONS AND VARIATIONAL PRINCIPLE - 2.	72-0106
SELF-ADAPTIVE	SELF-ADAPTIVE REFINEMENTS IN THE FINITE ELEMENT METHOD	75-0712
SELF-ORGANIZING	A SELF-ORGANIZING MESH GENERATION PROGRAM	74-1051
SELF-ORGANIZING	A SELF-ORGANIZING MESH GENERATION PROGRAM	74-1249
SELF-ORGANIZING	QMESH: A SELF-ORGANIZING MESH GENERATION PROGRAM	74-0872
SELF-ORGANIZING	QMESH: A SELF-ORGANIZING MESH GENERATION PROGRAM	74-0899
SELF-ORGANIZING	USERS MANUAL FOR QMESH. A SELF-ORGANIZING MESH GENERATION PROGRAM	74-0880
SELF-SEALING	STRESS ANALYSIS OF SELF-SEALING TYPE PISTON HEAD BY FINITE-ELEMENT METHOD	72-0259
SELF-SEALING	STRESS ANALYSIS OF THE "SELF-SEALING" TYPE PISTON HEAD BY THE FINITE ELEMENT METHOD	72-0617
SELF-STRESS	SELF-STRESS CONCENTRATIONS	71-0317
SELF-WEIGHT	CONSTRUCTION SELF-WEIGHT STRESSES OF PRECAST SHEAR WALLS	71-0203
SEMI-ANALYTIC	SEMI-ANALYTIC SOLUTION OF SKEW BOX GIRDER BRIDGES	75-0578
SEMI-ANALYTICAL	SEMI-ANALYTICAL FINITE ELEMENT PROCEDURE FOR CONDUCTION IN ANISOTROPIC AXISYMMETRIC SOLIDS	74-0270
SEMI-AUTOMATIC	MASS CONDENSATION, A SEMI-AUTOMATIC METHOD FOR REDUCING THE SIZE OF VIBRATION PROBLEMS	69-0220
SEMI-CONDUCTOR	TWO-DIMENSIONAL FINITE ELEMENT ANALYSIS OF SEMI-CONDUCTOR STEADY STATE TRANSPORT EQUATIONS	74-1224
SEMI-DISCRETE	ON THE SEMI-DISCRETE GALERKIN METHOD FOR HYPERBOLIC PROBLEMS AND ITS APPLICATION TO PROBLEMS IN ELAS	72-0457
SEMI-ELLIPTIC	ELASTIC-PLASTIC BEHAVIOR OF A LONGITUDINAL SEMI-ELLIPTIC CRACK IN A THICK PRESSURE VESSEL	73-0952
SEMI-ELLIPTICAL	CALCULATION OF STRESS INTENSITY FACTORS AND FATIGUE CRACK PROPAGATION OF SEMI-ELLIPTICAL, PART-THROU	73-0360
SEMI-ELLIPTICAL	THREE-DIMENSIONAL ELASTIC ANALYSIS OF SEMI-ELLIPTICAL SURFACE CRACKS SUBJECT TO THERMAL SHOCK	75-0256
SEMI-INFINITE	A STUDY OF A DYNAMICAL FINITE ELEMENT ANALYSIS FOR APPLICATION TO AXIAL WAVE PROPAGATION PROBLEMS IN	71-0589
SEMI-INFINITE	ANALYSIS OF FOUNDATIONS ON A SEMI-INFINITE ELASTIC MASS UNDER DYNAMIC LOADS	72-0763
SEMI-INFINITE	END DISPLACEMENTS OF SEMI-INFINITE CYLINDERS DUE TO ANNULAR LOADINGS	69-0399
SEMI-LOOF	THE SEMI-LOOF SHELL ELEMENT	74-0629
SEMI-MONOCOQUE	ELASTIC AND EQUILIBRIUM MATRICES OF GENERAL SEMI-MONOCOQUE MEMBRANE ELEMENTS	70-0159
SEMI-MONOCOQUE	AN EVALUATION OF CURRENT NASTRAN DISCRETE ELEMENT MODELS FOR MONOCOQUE AND SEMI-MONOCOQUE STRUCTURES	71-0236
SEMI-MONOCOQUE	EVALUATION OF CALAC SEMI-MONOCOQUE MEMBRANE ELEMENT	70-0278
SEMI-MONOCOQUE	WARPED SEMI-MONOCOQUE QUADRILATERAL MEMBRANE ELEMENT FOR THE FINITE ELEMENT METHODS	70-0195
SEMI-RIGID	FINITE ELEMENT ANALYSIS FOR BUCKLING OF SEMI-RIGID SPACE FRAMES	68-0270
SEMI-THICK	A CONFORMING TRIANGULAR ELEMENT IN THE ANALYSIS OF THIN AND SEMI-THICK PLATES IN BENDING	70-0308
SEMIANALYTICAL	CONVERGENCE OF SEMIANALYTICAL SOLUTION OF PLATES	73-0497
SEMICIRCULAR	THERMAL MECHANICAL ANALYSIS PROCEDURE FOR A SEMICIRCULAR PLATE WITH CLAMPED STRAIGHT EDGE	75-0651
SEMICONDUCTOR	FEASIBILITY OF USING FINITE ELEMENTS IN ANALYSIS OF SECOND BREAKDOWN IN SEMICONDUCTOR DEVICES	73-0874
SEMICONDUCTOR	FINITE DIFFERENCE AND FINITE ELEMENT METHODS FOR THE SEMICONDUCTOR FLOW EQUATIONS	74-0028
SEMICONDUCTOR	TWO-DIMENSIONAL FINITE ELEMENT SIMULATION OF SEMICONDUCTOR DEVICES	74-0260
SEMICONDUCTOR-D	2-DIMENSIONAL SIMULATION OF SEMICONDUCTOR-DEVICES USING FINITE ELEMENT METHODS	75-0548

359

SEMICONDUCTORS	ON A VARIATIONAL INEQUALITY AND ITS APPROXIMATION, IN THE THEORY OF SEMICONDUCTORS	75-0829
SEMIDISCRETE	ERROR BOUNDS FOR SEMIDISCRETE GALERKIN APPROXIMATIONS OF PARABOLIC PROBLEMS WITH APPLICATIONS TO PET	70-0291
SEMIDISCRETE	SEMIDISCRETE GALERKIN TECHNIQUES WITH TIME INTERPOLATION AND SPLITTING UP FOR PLASMA SIMULATION	75-0121
SEMILOOP	THE SEMILOOP SHELL ELEMENT	75-0536
SEMILOOP	A BEAM ELEMENT FOR USE WITH THE SEMILOOP SHELL ELEMENT	73-0620
SEMILOOP	AN ANALYSIS OF AN INTERSECTION PROBLEM USING THE SEMILOOP SHELL ELEMENT	74-0401
SEMIN	FINITE ELEMENTS METHOD IN SOIL MECHANICS (METODA KONECNYCH PRVKU V MECHANICE SEMIN).	71-0468
SENSING	CRACK GROWTH MONITORING BY STRAIN SENSING	73-0486
SENSITIVITY	AN ANALYSIS OF SOME FACTORS THAT AFFECT HYDROPHONE SENSITIVITY	72-0739
SENSITIVITY	DISCRETE-ELEMENT MODELLING AND SENSITIVITY STUDIES OF STATICALLY LOADED SINGLE-LAYER AND BOUNDED DOU	70-0129
SENSITIVITY	FINITE ELEMENT MATRIX FORMULATION OF POST-BUCKLING STABILITY AND IMPERFECTION SENSITIVITY	70-0246
SENSITIVITY	MODAL SENSITIVITY STUDY OF A CONICAL RE-ENTRY VEHICLE SUBJECTED TO AERODYNAMICALLY INDUCED ACOUSTIC	69-0431
SENSITIVITY	SENSITIVITY ANALYSIS OF TORSIONAL VIBRATION CHARACTERISTICS OF HELICOPTER ROTOR BLADES. PART 1: STRU	74-0846
SENSITIVITY	SENSITIVITY ANALYSIS OF INPUT PARAMETERS IN NUMERICAL MODELING OF STEADY STATE REGIONAL GROUNDWATER	74-0926
SENSITIVITY	SENSITIVITY OF ROTOR BLADE VIBRATION CHARACTERISTICS TO TORSIONAL OSCILLATIONS	74-0611
SENSOR	ERROR ANALYSIS OF A REMOTE TRANSIENT HEAT FLUX SENSOR	74-0753
SEPARATION	ANALYSIS OF MULTI-AXIAL ANISOTROPIC CREEP, ROOF RESIN BOLTS, AND BED SEPARATION IN ROCK STRUCTURES U	74-1122
SEPARATION	INFLUENCE OF ASCENT HEATING ON THE SEPARATION DYNAMICS OF A SPACECRAFT FAIRING	72-0906
SEPARATION	LEAST SQUARES FINITE ELEMENT ANALYSIS OF TWO-DIMENSIONAL LAMINAR BOUNDARY-LAYER EQUATION WITH SEPARA	75-0796
SEPARATION	STRESS SEPARATION IN PHOTOELASTICIMETRIC MEASUREMENT BY THE FINITE ELEMENT METHOD	71-0807
SEPARATION	THE NUMERICAL CALCULATION OF LAMINAR BOUNDARY-LAYER SEPARATION	74-0785
SEQUENCE	ON A COMBINED STURM SEQUENCE AND INVERSE ITERATION TECHNIQUE FOR EIGENPROBLEM SOLUTION OF SPINNING S	73-0413
SEQUENCE	SOLUTION OF EIGENVALUE PROBLEMS BY STURM SEQUENCE METHOD.	72-0400
SEQUENCE	THEORETICAL STUDY OF THE SEQUENCE OF FAILURE IN PRECAST PANEL SHEAR WALLS	73-0490
SEQUENCING	SAVINGS IN NASTRAN DECOMPOSITION TIME BY SEQUENCING TO REDUCE ACTIVE COLUMNS	71-0161
SEQUENTIAL	INTEGRATED SEQUENTIAL SOLVER FOR LARGE MATRIX EQUATIONS	74-0276
SERENDIPEAN	GENERATION OF SERENDIPEAN SHAPE FUNCTIONS IN FINITE ELEMENT METHOD	74-1012
SERENDIPITY	A REMARK ON THE SERENDIPITY FAMILY	73-1081
SERIES	A COMPARISON OF FINITE ELEMENT AND FOURIER SERIES SOLUTIONS AS APPLIED TO RADIALLY LOADED CIRCULAR R	68-0024
SERIES	ASKA SERIES OF PROGRAMS	72-0157
SERIES	MARC SERIES OF PROGRAMS	72-0154
SERIES	MARK5C/DAISY SERIES OF PROGRAMS	72-0160
SERIES	POWER SERIES EXPANSION OF THE GENERAL STIFFNESS MATRIX FOR BEAM ELEMENTS	75-0543
SERIES	THE DEVELOPMENT OF A SERIES OF HYBRID-STRESS FINITE ELEMENTS	75-0335
SERIES	USE OF FOURIER SERIES IN THE FINITE ELEMENT METHOD	68-0103
SERVICE	DESIGN AND ANALYSIS EXPERIENCE WITH A LIQUID METAL HEAT EXCHANGER FOR PFTF SERVICE	72-0772
SERVICE	THREE DIMENSIONAL FINITE ELEMENT STUDIES. PART ONE: SERVICE ROUTINES	72-0721
SESAM	SESAM A PROGRAMMING SYSTEM FOR FINITE ELEMENT PROBLEMS	68-0276
SESAM	SESAM 69 A GENERAL PURPOSE FINITE ELEMENT METHOD PROGRAM CDC/NASTRAN APPLICATIONS MANUAL	72-0552
SESAM-69	GENERAL DESCRIPTION OF SESAM-69, SUPER ELEMENT STRUCTURAL ANALYSIS (PROGRAM) MODULES	71-0612
SESAM-69	SESAM-69 - A GENERAL PURPOSE FINITE ELEMENT METHOD PROGRAM	74-0183
SESAM-69	SHELL ANALYSIS BY SESAM-69	70-0249
SETS	EINSATZ MODERNER RECHENVERFAHREN ZUR ERMITTLUNG DER SPANNUNGEN UND VERFORMUNGEN BEI RADSAETZEN. (USE	75-0186
SETS	ON THE SOLUTION OF SETS OF NONLINEAR EQUATIONS ARISING IN THE APPLICATION OF FINITE ELEMENT METHODS	75-0127
SETS	OPTIMUM PLASTIC DESIGN FOR MULTIPLE SETS OF LOADS	74-0980
SETS	UNTERSUCHUNGEN ZUR SPURMASSTABILITAET BEI RADSAETZEN MIT KLOTZGEBREMSTEN VOLLRAEDERN (INVESTIGATIONS	75-0810
SETTING	K VOPROSU OB AVTOMATIZATSII ZADANIYA INFORMATSII V METODE KONECHNYKH ELEMENTOV. (PROBLEM OF AUTOMATI	75-0382
SETTLEMENT	A TECHNIQUE FOR THE SETTLEMENT AND STRESS ANALYSIS OF MAT FOUNDATIONS USING THE FINITE ELEMENT METHO	72-0950
SETTLEMENT	ELASTIC SETTLEMENT OF A CIRCULAR RAFT IN ADHESIVE CONTACT WITH A TRANSVERSELY ISOTROPIC MEDIUM	75-0828
SETTLEMENT	FINITE ELEMENT PARAMETRIC STUDY OF VERTICAL STRAIN INFLUENCE FACTORS AND THE PRESSUREMETER TEST TO E	74-1127
SETTLEMENT	FINITE ELEMENT PERMAFROST THAW SETTLEMENT MODEL	74-0543
SETTLEMENT	INITIAL SETTLEMENT OF STRUCTURES ON CLAY	71-0304
SETTLEMENT	NONLINEAR SETTLEMENT ANALYSIS BY FINITE ELEMENT	75-0486
SETTLEMENT	SETTLEMENT ANALYSIS OF A STRIP FOOTING USING A CRITICAL STATE SOIL MODEL IN CONJUNCTION WITH FINITE	71-0533
SETTLEMENT	SETTLEMENT ANALYSIS OF DRIVEN PILES	75-0319
SETTLEMENTS	IMMEDIATE SETTLEMENTS OF THREE TEST EMBANKMENTS ON CHAMPLAIN CLAY	74-0275
SETTLEMENTS	SETTLEMENTS AND STRENGTHENING OF SOFT CLAY ACCELERATED BY SAND DRAINS	73-0438
SEVENTIES	ASKA. SOFTWARE ENGINEERING FOR THE SEVENTIES IN STRUCTURAL ANALYSIS	73-0704
SGORANIYA	METODY RASCHETNOGO OPREDELENIYA TEMPERATURNYKH NAPRYAZHENII V KRYSHKAKH TSILINDROV DVIGATELEI VNUTRE	74-0179
SGORANIYA	OPREDELENIE STATSIONARNYKH TEMPERATURNYKH POLEI V DETALYAKH DVIGATELEI VNUTRENNEGO SGORANIYA METODOM	73-0454
SGORANIYA	RASCHET POLEI DEFORMATSII I NAPRYAZHENII V DETALYAKH DVIGATELEI VNUTRENNEGO SGORANIYA (CALCULATION O	73-0465
SH-WAVE	FINITE ELEMENT IN THE THEORY OF SH-WAVE PROPAGATION	73-1034
SHADING	A METHOD FOR THE THERMAL ANALYSIS OF SPACECRAFT INCLUDING ALL MULTIPLE REFLECTIONS AND SHADING AMONG	70-0424
SHAFT	FAILURE OF THE CONCRETE LINING IN AN INCLINED SHAFT AT GREAT DEPTH.	74-0996
SHAFT	FINITE ELEMENT APPROACH TO THE VIBRATION ANALYSIS OF ELASTIC DISKS ON A FLEXIBLE SHAFT	75-0368
SHAFT	ON THE STATES OF STRESS AND DEFORMATION AROUND THE BOTTOM OF A CIRCULAR VERTICAL SHAFT	69-0323
SHAFT	RADIAL DISPLACEMENTS OF A HOLLOW SHAFT SUBJECTED TO A UNIFORM BAND OF EXTERNAL PRESSURE	72-0242
SHAFT	RECENT DEVELOPMENTS IN HULL AND SHAFT VIBRATION ANALYSIS	72-0176
SHAFTS	COMPUTER METHOD FOR COMPUTING CRITICAL SPEEDS OF NON-UNIFORM SHAFTS GEARED SYSTEMS AND CURVED ASSEMB	73-0675
SHAFTS	FINITE ELEMENT ANALYSIS OF STRESSES IN SHAFTS DUE TO INTERFERENCE-FIT HUBS	69-0353
SHAFTS	GENERAL METHOD FOR STABILITY ANALYSIS OF ROTATING SHAFTS	75-0528
SHAFTS	STRESSES IN SHAFTS	67-0104
SHAFTS	SULLA DETERMINAZIONE DELLE FREQUENZE PROPRIE DI ALBERI SU PIU SUPPORTI ELASTICI. (DETERMINATION OF F	74-0916
SHAFTS	ZASTOSOWANIE METODY SZTYWNYCH ELEMENTOW SKONCZONYCH DO OBLICZEN ORGAN WALOW OKRETOYCH Z UWZGLEDNIENI	74-0962
SHAFTS	ZUR BERECHNUNG DER BIEGESTEIFIGKEIT ABGESETZTER STAEBE UND WELLEN UNTER ANWENDUNG VON FINITEN ELEMEN	75-0481
SHAKEDOWN	LINEAR PROGRAMMING APPROACH TO SHAKEDOWN ANALYSIS OF STRUCTURES	74-0181
SHAKEDOWN	PLANE STRESS SHAKEDOWN ANALYSIS BY FINITE ELEMENTS	72-0266
SHAKEDOWN	SHAKEDOWN OF ROLLING WHEEL UNDER HUB LOADING	74-0663
SHAKEDOWN	SHAKEDOWN THEORY IN PERFECT ELASTOPLASTICITY WITH ASSOCIATED AND NON ASSOCIATED FLOW LAWS - A FINITE	69-0172
SHAKING	COMPUTER ANALYSIS OF FRUIT DETACHMENT DURING TREE SHAKING	74-1162
SHALE-OIL	COMPUTER SIMULATION OF AN IN SITU SHALE-OIL RETORT	74-1059
SHALES	INITIATION OF FAILURE IN SLOPES IN OVERCONSOLIDATED CLAYS AND CLAY SHALES.	70-0186
SHALES	SLOPES IN STIFF FISSURED CLAYS AND SHALES	69-0094
SHALLOW	A COMPARISON OF RECENT SHALLOW SHELL FINITE-ELEMENT ANALYSES	70-0232
SHALLOW	A FINITE ELEMENT ANALYSIS OF LAMINATED ORTHOTROPIC PLATES AND SHALLOW SHELLS	70-0123
SHALLOW	A FINITE-DEFLECTION ANALYSIS OF SHALLOW ARCHES BY THE DISCRETE ELEMENT METHOD	71-0077
SHALLOW	A MIXED FINITE-ELEMENT SHALLOW SHELL FORMULATION	69-0078
SHALLOW	A NONLINEAR FINITE ELEMENT ANALYSIS OF SHALLOW CIRCULAR ARCHES	69-0269
SHALLOW	A NUMERICAL MODEL FOR THE SIMULATION OF TWO DIMENSIONAL CONVECTIVE DISPERSION IN SHALLOW ESTUARIES	74-0052
SHALLOW	A SHALLOW SHELL FINITE ELEMENT OF TRIANGULAR SHAPE	70-0044
SHALLOW	A STIFFNESS MATRIX FOR SHALLOW SHELL FINITE ELEMENTS	69-0229
SHALLOW	AN EXTENSION OF THE SHALLOW TO THE NON-SHALLOW STIFFNESS MATRIX FOR A CYLINDRICAL SHELL FINITE ELEME	70-0104
SHALLOW	ANALYSIS OF ARBITRARY SHELLS USING SHALLOW SHELL FINITE ELEMENTS	71-0717
SHALLOW	ANALYSIS OF ARBITRARY SHELLS USING SHALLOW SHELL FINITE ELEMENTS	74-1374
SHALLOW	ANALYSIS OF SHALLOW SHELL STRUCTURES BY A DISCRETE ELEMENT SYSTEM	66-0113
SHALLOW	APPLICATION OF A REFINED SHALLOW SHELL FINITE ELEMENT TO GEOMETRICALLY NONLINEAR STRUCTURES	73-0358
SHALLOW	APPLICATION OF FINITE ELEMENT METHOD TO AXISYMMETRIC BUCKLING OF SHALLOW SPHERICAL SHELLS UNDER EXTE	73-0111
SHALLOW	COMPARISON OF RECENT SHALLOW SHELL FINITE ELEMENT ANALYSES	71-0362
SHALLOW	CURVED FINITE ELEMENTS FOR THE ANALYSIS OF SHALLOW AND DEEP ARCHES	74-0332
SHALLOW	DESH. EIN EINFACHES FLACHES SCHALENELEMENT UEBER DREIECKIGEM GRUNDRISS. (DESH, A SIMPLE TRIANGULAR S	75-0304
SHALLOW	DYNAMIC ANALYSIS OF SHALLOW SHELLS WITH A DOUBLY-CURVED TRIANGULAR FINITE ELEMENT	71-0353
SHALLOW	ELASTIC PLASTIC ANALYSIS OF SHALLOW FOUNDATIONS	74-0446

360

SHALLOW	FINITE ELEMENT ANALYSIS OF SKEWED SHALLOW SHELLS	68-0250
SHALLOW	FINITE ELEMENT ANALYSIS OF SKEWED SHALLOW SHELLS	69-0211
SHALLOW	FINITE ELEMENT CIRCULATION ANALYSIS OF VARIABLE-DEPTH SHALLOW LAKES	73-0727
SHALLOW	FINITE ELEMENT CREEP BUCKLING ANALYSIS OF VISCO-ELASTIC SHALLOW SPHERICAL SHELL	75-0159
SHALLOW	FINITE ELEMENT DYNAMIC ANALYSIS OF SHALLOW SHELL STRUCTURES	70-0164
SHALLOW	FINITE ELEMENT LARGE DEFLECTION ANALYSIS OF SHALLOW SHELLS	75-0356
SHALLOW	FINITE ELEMENT NONLINEAR ANALYSIS FOR PLATES AND SHALLOW SHELLS	68-0216
SHALLOW	FINITE ELEMENT SOLUTIONS OF THE SHALLOW WATER EQUATIONS SURFACE RUN OFF	74-0046
SHALLOW	FINITE-ELEMENT ANALYSIS OF SHALLOW TEMPERATURE ANOMALIES	71-0471
SHALLOW	GEOMETRICKY NELINEARNE ULOHY PLOCHYCH SKRUPIN RIESENE PRIRASTKOVOU METODOU KONECNYCH PRVKOV. (GEOMET	74-1032
SHALLOW	HIGH ORDER RECTANGULAR SHALLOW SHELL FINITE ELEMENT	73-0279
SHALLOW	LAMINATED ORTHOTROPIC PLATES AND SHALLOW SHELLS	71-0400
SHALLOW	LARGE DEFLECTION ANALYSIS OF ARBITRARY SHALLOW SHELLS BY THE FINITE ELEMENT METHOD	73-0937
SHALLOW	LARGE DEFLECTION ANALYSIS OF PLATES AND SHALLOW SHELLS USING THE FINITE ELEMENT METHOD	73-0989
SHALLOW	LINEAR AND NON-LINEAR ANALYSIS OF THIN SHALLOW SHELLS BY MIXED FINITE ELEMENTS	75-0665
SHALLOW	MATHEMATICAL MODELS OF THE MASSACHUSETTS BAY. PART 1. FINITE ELEMENT MODELING OF TWO-DIMENSIONAL HYD	73-0800
SHALLOW	MIXED FINITE ELEMENT METHOD FOR VIBRATION ANALYSIS OF SHALLOW SHELLS	73-0593
SHALLOW	NEW VARIATIONAL PRINCIPLE WITH APPLICATION TO VIBRATIONS OF SHALLOW SHELLS USING FINITE ELEMENTS MET	73-0354
SHALLOW	STIFFNESS MATRIX FOR SHALLOW RECTANGULAR SHELL ELEMENT	67-0014
SHALLOW	STIFFNESS MATRIX FOR SHALLOW RECTANGULAR SHELL ELEMENT	68-0074
SHALLOW	STRESS DISTRIBUTIONS AROUND SHALLOW BURIED RIGID PIPES	74-0142
SHALLOW	THE FINITE-ELEMENT METHOD AS APPLIED TO SHALLOW FOUNDATIONS	71-0074
SHALLOW	THERMAL STRESS ANALYSIS OF PLATES AND SHALLOW SHELLS BY HYBRID FINITE-ELEMENT METHOD	74-0485
SHAPE	A FORTRAN PROGRAM FOR CALCULATING THE FREQUENCIES AND MODE SHAPES OF AN ARBITRARY SHAPE CYLINDRICAL	72-0594
SHAPE	A GENERAL METHOD FOR THE SHAPE FINDING OF LIGHTWEIGHT TENSION STRUCTURES	74-1217
SHAPE	A METHOD FOR THE PREDICTION OF STRESSES IN AN ISOTROPIC INCLUSION OR OREBODY OF IRREGULAR SHAPE	72-0827
SHAPE	A SHALLOW SHELL FINITE ELEMENT OF TRIANGULAR SHAPE	70-0044
SHAPE	CALCULATION OF STIFFNESS MATRICES FOR FINITE ELEMENTS OF THIN SHELLS OF ARBITRARY SHAPE	68-0127
SHAPE	CALCULATION OF THE FREE OSCILLATIONS OF A LIQUID IN AXISYMMETRIC MOTIONLESS CONTAINERS OF ARBITRARY	74-0387
SHAPE	COMMENT ON "SOLUTION OF LAPLACE'S EQUATION IN A REGION WITH DIELECTRIC INTERFACES OF ARBITRARY SHAPE	70-0238
SHAPE	CURSHL: A HIGH-PRECISION FINITE ELEMENT FOR SHELLS OF ARBITRARY SHAPE	71-0430
SHAPE	DOUBLY CURVED TRIANGULAR FINITE ELEMENTS FOR SHELLS OF ARBITRARY SHAPE	73-0881
SHAPE	EFFECT OF SHAPE AND HETEROGENEITY OF ROCK SAMPLES ON THEIR COMPRESSIVE STRENGTH	69-0315
SHAPE	ELASTIC DEFORMATION OF POLYCRYSTALLINE METAL - 1. INFLUENCE OF GRAIN SHAPE ON DISTRIBUTIONS OF STRES	72-0463
SHAPE	FINESSE - A FINITE ELEMENT SYSTEM. PART V FINITE ELEMENT DESIGN - DERIVATIVES FOR STRUCTURAL SHAPE D	72-0963
SHAPE	FINITE ELEMENT ANALYSIS OF UNSTEADY INCOMPRESSIBLE FLOW AROUND AN OSCILLATING OBSTACLE OF ARBITRARY	73-0067
SHAPE	FREQUENCY-DEPENDENT SHAPE FUNCTIONS FOR FREE AND FORCED STRUCTURAL VIBRATION CALCULATION	73-1091
SHAPE	FUNKCJA KSZTALTU NIESPELNIAJACA WARUNKOW CIAGLOSCI W ANALIZIE PLYT METODA ELEMENTOW SKONCZONYCH. (SH	75-0263
SHAPE	GENERATION OF SERENDIPEAN SHAPE FUNCTIONS IN FINITE ELEMENT METHOD	74-1012
SHAPE	INVESTIGATION OF THE EFFECT OF FOUNDATION YIELD ON THE FREQUENCY AND SHAPE OF INHERENT VIBRATIONS OF	72-0419
SHAPE	ON COMPLETENESS OF SHAPE FUNCTIONS FOR FINITE ELEMENT ANALYSIS	72-0105
SHAPE	ON THE COMPLETENESS OF SHAPE FUNCTIONS FOR FINITE ELEMENT ANALYSIS	72-0885
SHAPE	SHAPE DETERMINATION OF PREFABRICATED PRESTRESSED NETWORKS	73-0889
SHAPE	SHAPE FUNCTION FORMULATIONS FOR ELEMENTS OTHER THAN DISPLACEMENT MODELS	72-0472
SHAPE	SHAPE FUNCTION SUBROUTINE FOR AN ISOPARAMETRIC THIN PLATE ELEMENT	73-0618
SHAPE	SHAPE FUNCTIONS AND ACCURACY OF ARCH FINITE ELEMENTS	73-0038
SHAPE	SLAB BRIDGES WITH ARBITRARY SHAPE AND SUPPORT CONDITIONS - A GENERAL METHOD OF ANALYSIS BASED ON FIN	68-0233
SHAPE	STEADY STATE EQUATIONS OF MOTION, EQUILIBRIUM SHAPE AND STABILITY DERIVATIVES OF ELASTIC AIRPLANES E	72-0541
SHAPE	THE ENERGY METHOD OF NETWORKS OF ARBITRARY SHAPE IN PROBLEMS OF THE THEORY OF ELASTICITY	59-0001
SHAPE	USE OF THE FINITE ELEMENT METHOD IN EXAMINING ELASTIC BUCKLING OF THIN PLATES OF ANY SHAPE	75-0153
SHAPED	ANALYSIS OF ACOUSTIC FIELD IN IRREGULARLY SHAPED ROOMS BY FINITE ELEMENT METHOD	73-0133
SHAPED	ARBITRARILY SHAPED INHOMOGENEOUS OPTICAL FIBER OR INTEGRATED OPTICAL WAVEGUIDES	75-0372
SHAPED	BOX SHAPED GIRDERS. (NOSNIKY UZAVRETEHO PRIEREZU)	70-0169
SHAPED	DETERMINATION OF THE LOWEST CRITICAL BUCKLING STRESS OF VARIOUS SHAPED PLATES BY FINITE-ELEMENTS	73-1134
SHAPED	FINITE ELEMENT ANALYSIS OF IRREGULARLY SHAPED ACOUSTIC FIELD	71-0711
SHAPES	USE OF SIMPLE THREE-DIMENSIONAL ACOUSTIC FINITE ELEMENTS FOR DETERMINING NATURAL MODES AND FREQUENCI	72-0010
SHAPES	A FORTRAN PROGRAM FOR CALCULATING THE FREQUENCIES AND MODE SHAPES OF AN ARBITRARY SHAPE CYLINDRICAL	72-0594
SHAPES	A FORTRAN PROGRAM FOR CALCULATING THE FREQUENCIES AND MODE SHAPES OF A GUIDED CLAMPED, GUIDED PINNED	72-0890
SHAPES	AN INVESTIGATION OF THE EFFECTS OF SQUARE CUTOUTS ON THE NATURAL FREQUENCIES AND MODE SHAPES OF RECT	71-0638
SHAPES	COMPUTATION OF VIBRATION MODAL SHAPES AND NATURAL FREQUENCIES OF A GENERAL ORTHOTROPIC LAMINATED, TH	71-0309
SHAPES	ELASTIC TORSIONAL ANALYSIS OF IRREGULAR SHAPES	65-0020
SHAPES	ELASTIC TORSIONAL ANALYSIS OR IRREGULAR SHAPES	66-0078
SHAPES	FAMSOR, A FINITE ELEMENT PROGRAM FOR THE FREQUENCIES AND MODE SHAPES OF SHELLS OF REVOLUTION	70-0090
SHAPES	FORBIDDEN SHAPES IN THE FINITE ELEMENT METHOD	71-0705
SHAPES	MODE SHAPES AND FREQUENCIES BY FINITE ELEMENT METHOD USING CONSISTENT AND LUMPED MASSES	71-0610
SHAPES	NATURAL FREQUENCIES AND MODE SHAPES OF TRANSFORMER CORES	55-0048
SHAPES	NATURAL FREQUENCIES AND MODE SHAPES OF PLATES WITH INTERIOR CUT-OUTS	70-0402
SHAPES	RASCHET KILEBANII PLASTIN SLOZHNOI FORMY S NEODNORODNYMI GRANICHNYMI USLOVIYAMI. (CALCULATION OF VIB	74-0988
SHAPES	STRESS ANALYSIS IN HEATED, COMPLEX SHAPES	62-0002
SHAPES	STRESS CONCENTRATIONS IN HULL SHAPES FROM SURFACE DISCONTINUITIES	69-0130
SHARP	ELASTIC-PLASTIC ANALYSIS OF THICK WALLED PRESSURE VESSELS WITH SHARP DISCONTINUITIES	71-0155
SHARP	QUANTITATIVE STRAIN-AND-STRESS STATE CRITERION FOR FAILURE IN THE VICINITY OF SHARP CRACKS	74-0639
SHEAR	A COMPUTER PROGRAM FOR THE CALCULATION OF TORSION FACTORS AND SHEAR STRESS FACTORS FOR BOX SECTIONS	69-0485
SHEAR	A DISCRETE-ELEMENT METHOD OF ANALYSIS FOR COMBINED BENDING AND SHEAR DEFORMATION OF A BEAM	69-0494
SHEAR	A FINITE ELEMENT ANALYSIS INCLUDING TRANSVERSE SHEAR EFFECTS FOR APPLICATIONS TO LAMINATED PLATES	71-0216
SHEAR	A FINITE ELEMENT MODEL FOR LAMINATED PLATES INCLUDING TRANSVERSE SHEAR DEFORMATION	72-0818
SHEAR	A SIX NODE TRIANGULAR SHEAR PANEL FOR EQUILIBRIUM MODELS	73-1011
SHEAR	ANALYSIS OF FRAMES WITH SHEAR WALLS BY FINITE ELEMENTS	69-0199
SHEAR	ANALYSIS OF FRAMES WITH SHEAR WALLS WITH FINITE ELEMENTS	69-0500
SHEAR	ANALYSIS OF IN-PLANE VIBRATION OF SHEAR WALLS BY A FINITE ELEMENT METHOD	72-0090
SHEAR	ANALYSIS OF LIGHT GATE STEEL SHEAR DIAPHRAGMS	73-0733
SHEAR	ANALYSIS OF SHEAR WALLS WITH OPENINGS	69-0111
SHEAR	ANALYSIS OF SHEAR WALLS BY FINITE ELEMENT METHOD	69-0112
SHEAR	ANALYSIS OF SHEAR WALLS BY FINITE ELEMENTS	71-0569
SHEAR	ANALYSIS OF STRESS DISTRIBUTION IN SHEAR WALLS BY THE FINITE ELEMENT DISPLACEMENT METHOD	67-0033
SHEAR	ANALYSIS OF THE BIAXIAL STRIP AND SHEAR LAP TESTS FOR SOLID PROPELLANT CHARACTERIZATION	67-0019
SHEAR	APPLICATIONS OF CURVED FINITE ELEMENT TO SANDWICH BEAMS USING SHEAR STRAIN FORMULATION	74-0523
SHEAR	APPLICATIONS OF CURVED FINITE ELEMENTS TO SANDWICH SHELLS USING SHEAR STRAIN FORMULATION	75-0433
SHEAR	AVOIDANCE OF PARASITIC SHEAR IN PLANE ELEMENT	75-0383
SHEAR	BEHAVIOR OF ISOLATED SHEAR WALLS SUBJECTED TO LATERAL LOADS	75-0704
SHEAR	CALCULATION OF STRESS INTENSITY FACTORS FOR COMBINED TENSILE AND SHEAR LOADING	75-0205
SHEAR	COMPARISON OF THE THREE-DIMENSIONAL ANALYSIS OF A SHEAR WALL BUILDING BY TWO DIFFERENT METHODS	75-0705
SHEAR	COMPARISON OF THEORETICAL AND EXPERIMENTAL SHEAR STRESS IN THE ADHESIVE LAYER OF A LAP JOINT MODEL	74-1176
SHEAR	CONSTRAINED AND SHEAR MODULI FOR FINITE ELEMENTS	74-0774
SHEAR	CONSTRAINED AND SHEAR MODULI FOR FINITE ELEMENTS	75-0228
SHEAR	CONSTRUCTION SELF-WEIGHT STRESSES OF PRECAST SHEAR WALLS	71-0203
SHEAR	COUPLED SHEAR WALL ANALYSIS BY LAGRANGE MULTIPLIERS	75-0202
SHEAR	COUPLED SHEAR WALLS. IMPROVED METHODS FOR ELASTIC ANALYSIS	74-0818
SHEAR	DESIGN, MANUFACTURE, DEVELOPMENT, TEST, AND EVALUATION OF BORON/ALUMINUM STRUCTURAL COMPONENTS FOR S	74-0798
SHEAR	DIRECT BEAM STIFFNESS MATRIX CALCULATIONS INCLUDING SHEAR EFFECTS	62-0003
SHEAR	DYNAMIC BEHAVIOR OF IN-LINE SHEAR WALLS CONNECTED BY FLOOR SLABS	72-0689
SHEAR	EFFECT OF SHEAR DEFORMATIONS AND ROTARY INERTIA ON OPTIMUM BEAM FREQUENCIES	75-0231

SHEAR	EFFECT OF TRANSVERSE SHEAR DEFORMATION IN VIBRATIONS OF PLANAR STRUCTURES COMPOSED OF BEAM TYPE ELEM	73-0402
SHEAR	EFFECTS OF SHEAR DEFORMATION, ROTARY INERTIA, AND ELASTICITY OF THE SUPPORT ON THE RESONANCE FREQUEN	75-0369
SHEAR	EFFECTS OF SHEAR WALLS ON DYNAMIC RESPONSE OF FRAMES	69-0230
SHEAR	EINE FINITE-ELEMENTE-LOESUNG FUR DIE TORSIONSSTEIFIGKEIT UND DEN SCHUBMITTELPUNKT BELIEBIGER QUERSCH	74-0256
SHEAR	ELASTIC ANALYSIS OF SPATIAL SYSTEMS OF INTERCONNECTED SHEAR WALLS AND FRAMES	74-1195
SHEAR	ELASTIC AND PLASTIC INTERLAMINAR SHEAR DEFORMATION IN LAMINATED COMPOSITIES UNDER GENERALIZED PLANE	71-0158
SHEAR	ELASTIC SHEAR ANALYSIS OF GENERAL PRISMATIC BEAMS	68-0114
SHEAR	EVALUATION OF COMPOSITE SHEAR PANELS WITH CUTOUTS	72-0615
SHEAR	FINITE ELEMENT ANALYSIS OF PLATE BENDING WITH TRANSVERSE SHEAR DEFORMATION	70-0109
SHEAR	FINITE ELEMENT ANALYSIS OF LAMINATED ANISOTROPIC PLATES INCLUDING TRANSVERSE SHEAR DEFORMATIONS	70-0506
SHEAR	FINITE ELEMENT ANALYSIS OF INTERLAMINAR SHEAR FIBROUS COMPOSITES	71-0129
SHEAR	FINITE ELEMENT ANALYSIS OF PLATES ALLOWING FOR TRANSVERSE SHEAR DEFORMATION EFFECTS	73-0996
SHEAR	FINITE ELEMENT ANALYSIS OF METAL DECK SHEAR DIAPHRAGMS	74-0128
SHEAR	FINITE ELEMENT ANALYSIS OF SHEAR STRENGTH OF REINFORCED CONCRETE BEAMS	74-1306
SHEAR	FINITE ELEMENT FOR THE INTERACTION OF FRAMES AND SHEAR WALLS	73-0681
SHEAR	FINITE ELEMENT THEORY OF PLATES AND SHELLS INCLUDING TRANSVERSE SHEAR STRAIN EFFECTS	70-0395
SHEAR	FINITE ELEMENTS FOR VIBRATION OF FRAMED SHEAR WALLS	73-0036
SHEAR	FINITE ELEMENTS SOLUTION FOR TORSION RIGIDITY AND SHEAR CENTER OF ARBITRARY CROSS-SECTIONS	74-0168
SHEAR	FINITE-ELEMENT APPROACH FOR THE PREDICTION OF INELASTIC BEHAVIOR OF SHEAR WALL-FRAME SYSTEMS	72-0983
SHEAR	FINITE-ELEMENT FOR BUCKLING OF CURVED BEAMS AND SHELLS WITH SHEAR	75-0607
SHEAR	HIGHER ORDER FINITE ELEMENT METHOD FOR STATIC AND DYNAMIC ANALYSIS OF SHEAR WALLS	71-0368
SHEAR	INCLUSION OF SHEAR DEFLECTION IN THE STIFFNESS MATRIX OF A BEAM ELEMENT	70-0307
SHEAR	INCLUSION OF TRANSVERSE SHEAR DEFORMATION IN FINITE ELEMENT DISPLACEMENT FORMULATIONS	74-0662
SHEAR	INTERACTION OF FRAME AND SHEAR WALL WITH OPENINGS	74-1190
SHEAR	INTERACTION OF SOLID SHEAR WALLS AND THEIR SUPPORTING STRUCTURES.	72-0450
SHEAR	MATRIX DISPLACEMENT ANALYSIS OF SHELLS AND PLATES INCLUDING TRANSVERSE SHEAR STRAIN EFFECTS	72-0643
SHEAR	NEW RECTANGULAR FINITE ELEMENT FOR SHEAR WALL ANALYSIS	69-0182
SHEAR	NEW RECTANGULAR FINITE ELEMENT FOR SHEAR WALL ANALYSIS	69-0362
SHEAR	NEW TRIANGULAR PLATE-BENDING FINITE ELEMENT WITH TRANSVERSE SHEAR FLEXIBILITY	74-0627
SHEAR	NEW TRIANGULAR PLATE-BENDING FINITE ELEMENT WITH TRANSVERSE SHEAR FLEXIBILITY	74-0964
SHEAR	NUMERICAL PROCEDURES FOR THE DETERMINATION OF THE BEHAVIOUR OF A SHEAR WALL FRAME SYSTEM	73-0216
SHEAR	SHEAR BUCKLING OF SQUARE PERFORATED PLATES	74-0713
SHEAR	SHEAR DEFORMATION ANALYSIS OF BEAMS AND PLATES BY FINITE ELEMENT METHOD	75-0873
SHEAR	SHEAR IN C AND C PLATE BENDING ELEMENTS	73-0692
SHEAR	SHEAR LAG ANALYSIS OF RECTANGULAR FULL-WIDTH TUBE JUNCTIONS	72-0087
SHEAR	SHEAR LAG IN STEEL BOX GIRDER BRIDGES	75-0310
SHEAR	SHEAR LAG IN STIFFENED BOX GIRDERS	70-0243
SHEAR	SHEAR STRESSES IN BEAMS WITH DISCONTINUITY IN DEPTH	75-0487
SHEAR	STATIC FLEXIBILITY MATRICES FOR A RECTANGULAR PLATE ELEMENT IN BENDING, TWISTING, AND SHEAR. UNIFORM	68-0217
SHEAR	STRUCTURAL ANALYSIS OF TALL BUILDINGS HAVING IRREGULARLY POSITIONED SHEAR WALLS	73-0282
SHEAR	THE ANALYSIS OF THIN SHELLS WITH TRANSVERSE SHEAR STRAINS BY THE FINITE ELEMENT METHOD	68-0098
SHEAR	THE BEHAVIOUR OF SQUARE SHEAR WEBS HAVING A CIRCULAR HOLE	67-0108
SHEAR	THE QUADRILATERAL SHEAR PANEL	51-0002
SHEAR	THEORETICAL STUDY OF THE SEQUENCE OF FAILURE IN PRECAST PANEL SHEAR WALLS	73-0490
SHEAR	THREE-DIMENSIONAL ANALYSIS OF SHEAR WALL BUILDINGS TO LATERAL LOAD	74-0220
SHEAR	TRANSVERSE VIBRATION OF BEAMS WITH INTERNAL DAMPING, ROTATORY INERTIA AND SHEAR	63-0020
SHEAR	TRIANGULAR FINITE ELEMENT FOR PLATE BENDING ANALYSIS WITH TRANSVERSE SHEAR DEFORMATION	71-0555
SHEAR	WOOD SHEAR PANELS BONDED WITH FLEXIBLE ADHESIVES	75-0086
SHEAR	ZUM TRAGVERHALTEN VON WAENDEN AUS FERTIGTEILTAFELN (BEHAVIOR OF LARGE PANEL SHEAR WALLS)	74-0299
SHEAR	ZUSAMMENWIRKEN VON EINZELNEN FERTIGTEILEN ALS GROSSFLAECHIGE SCHEIBE (INTERACTION OF INDIVIDUAL PREF	—
SHEAR-CORRECTED	DOUBLE STRESS-WAVE DISCONTINUITIES IN FINITE, SHEAR-CORRECTED BEAMS AND PLATES	68-0210
SHEARS	INTRODUCTION TO SHEARS DEFORMATION INTO A THIN PLATE DISPLACEMENTS FORMULATIONS	73-0617
SHEBA	A SEQUEL TO TECHNICAL NOTE 15: THE SHEBA FAMILY OF SHELL ELEMENTS FOR THE MATRIX DISPLACEMENT METHOD	69-0444
SHEBA	ON THE APPLICATION OF THE SHEBA SHELL ELEMENT	72-0651
SHEBA	THE SHEBA FAMILY OF SHELL ELEMENTS FOR THE MATRIX DISPLACEMENT METHOD	68-0014
SHEBA	THE SHEBA FAMILY OF SHELL ELEMENTS FOR THE MATRIX DISPLACEMENT METHOD. 111, LARGE DISPLACEMENTS	69-0499
SHEELS	ANALYSIS OF TOTALLY SUBMERGED THIN SHEELS BY FINITE ELEMENT TECHNIQUE	71-0616
SHEET	A FINITE ELEMENT ANALYSIS OF THE PLANE STRAIN SHEET DRAWING PROBLEM	71-0812
SHEET	AN ELASTOPLASTIC ANALYSIS OF A UNIAXIALLY LOADED SHEET WITH AN INTERFERENCE - FIT BOLT	74-0687
SHEET	ANALYTICAL AND EXPERIMENTAL INVESTIGATION OF FATIGUE IN A SHEET SPECIMEN WITH AN INTERFERENCE-FIT BO	75-0701
SHEET	COMPOSITE-OVERLAY REINFORCEMENT OF CUTOUTS AND CRACKS IN METAL SHEET	73-0878
SHEET	EXPERIMENTAL AND ANALYTICAL STRAINS IN A EDGE-CRACKED SHEET	69-0483
SHEET	FINITE ELEMENT ANALYSIS AND EXPERIMENTAL INVESTIGATION OF SHEET METAL STRETCHING	72-0229
SHEET	FINITE ELEMENT ANALYSIS OF GENERAL DEFORMATION OF SHEET METALS	74-0501
SHEET	FINITE ELEMENT METHOD OF ANALYSIS OF THE HYDROSTATIC BULGING OF A SHEET METAL - 1.	74-0958
SHEET	STIFFNESS ANALYSIS OF SHEET METAL SHELLS UNDER CONCENTRATED LOADS	74-0325
SHEET	THEORETICAL AND EXPERIMENTAL ANALYSIS OF ELASTOPLASTIC DEFORMATION IN A UNIAXIALLY LOADED STRIP OF 2	73-0720
SHEET-METAL	APPLICATION OF FINITE-ELEMENT METHOD AND MOIRE TECHNIQUE TO SHEET-METAL FORMING PROBLEMS	71-0784
SHEETS	CRACK BRANCHING IN HOMALITE-100 SHEETS	72-0843
SHEETS	CRACK BRANCHING IN HOMALITE-100 SHEETS	74-0291
SHEETS	ELASTO-PLASTIC ANALYSIS OF METAL SHEETS WITH A CENTRAL CRACK	67-0137
SHEETS	FINITE ELEMENT ANALYSIS OF CUT-GROWTH IN SHEETS OF HIGHLY ELASTIC MATERIALS	73-0514
SHEETS	FINITE ELEMENT METHOD FOR OPTIMAL DESIGN OF VARIABLE THICKNESS SHEETS	73-0131
SHEETS	FRACTURE DYNAMICS OF HOMALITE-100 SHEETS	73-0809
SHEETS	ON THE DETERMINATION OF EFFECTIVE ELASTIC-PLASTIC PROPERTIES FOR THE EQUIVALENT SOLID PLATE ANALYSI	74-0494
SHELL	A BEAM ELEMENT FOR USE WITH THE SEMILOOP SHELL ELEMENT	73-0620
SHELL	A COMPARISON OF RECENT SHALLOW SHELL FINITE-ELEMENT ANALYSES	70-0232
SHELL	A COMPARISON OF SEVERAL COMPUTER SOLUTIONS TO THREE STRUCTURAL SHELL ANALYSIS PROBLEMS	73-0884
SHELL	A COMPARISON OF SEVERAL COMPUTER SOLUTIONS TO THREE STRUCTURAL SHELL ANALYSIS PROBLEMS	73-0907
SHELL	A CURVED CIRCULAR CYLINDRICAL SHELL FINITE ELEMENT FOR HIGHER ORDER SHELL THEORY	69-0528
SHELL	A CURVED CIRCULAR CYLINDRICAL SHELL FINITE ELEMENT FOR HIGHER ORDER SHELL THEORY	69-0528
SHELL	A CURVED CYLINDRICAL SHELL DISCRETE ELEMENT	67-0009
SHELL	A CURVED, CYLINDRICAL SHELL, FINITE ELEMENT	68-0039
SHELL	A CURVILINEAR SHELL FINITE ELEMENT	69-0153
SHELL	A CYLINDRICAL SHELL DISCRETE ELEMENT	67-0007
SHELL	A DOUBLY CURVED TRIANGULAR SHELL ELEMENT	68-0149
SHELL	A FEASIBILITY STUDY OF THE ANALYSIS OF ASYMMETRIC SHELL BY THE DISCRETE-ELEMENT METHOD	68-0115
SHELL	A FINITE CYLINDRICAL SHELL ELEMENT AND THE VIBRATIONS OF A CURVED FAN BLADE	68-0131
SHELL	A FINITE ELEMENT FORMULATION FOR STABILITY ANALYSIS OF DOUBLY CURVED THIN SHELL STRUCTURES	69-0513
SHELL	A FINITE ELEMENT TECHNIQUE FOR THE IDEALIZATION OF CONTINUOUS SHELL STRUCTURES	68-0290
SHELL	A FLAT TRIANGULAR SHELL ELEMENT STIFFNESS MATRIX	66-0052
SHELL	A FORTRAN PROGRAM FOR CALCULATING THE FREQUENCIES AND MODE SHAPES OF AN ARBITRARY SHAPE CYLINDRICAL	72-0594
SHELL	A FORTRAN PROGRAM FOR CALCULATING THE FREQUENCIES AND MODE SHAPES OF A GUIDED CLAMPED, GUIDED PINNED	72-0890
SHELL	A FURTHER MODIFICATION TO AHMAD'S SHELL ELEMENT	73-0619
SHELL	A HIGH-PRECISION TRIANGULAR CYLINDRICAL SHELL FINITE ELEMENT	71-0163
SHELL	A HYBRID STRESS ELEMENT FOR THIN SHELL ANALYSIS	74-0433
SHELL	A MATHEMATICAL MODEL OF THE CORNEO-SCLERAL SHELL	71-0150
SHELL	A MIXED FINITE ELEMENT SHELL FORMULATION	71-0063
SHELL	A MIXED FINITE-ELEMENT SHALLOW SHELL FORMULATION	69-0078
SHELL	A NEW HYBRID CYLINDRICAL SHELL FINITE ELEMENT	71-0118
SHELL	A NOTE ON THE FINITE DEFORMATIONS OF A THICK ELASTIC SHELL OF REVOLUTION	71-0193
SHELL	A PROCEDURE FOR FINITE ELEMENT PLATE AND SHELL PRE AND POST BUCKLING ANALYSIS	71-0096

SHELL	A REFINED CURVED CYLINDRICAL SHELL FINITE ELEMENT	68-0038
SHELL	A REFINED FINITE ELEMENT ANALYSIS OF THIN SHELL STRUCTURES	67-0101
SHELL	A REFINED FINITE ELEMENT ANALYSIS OF THIN SHELL STRUCTURES INCLUDING DYNAMIC LOADINGS	67-0157
SHELL	A SEQUEL TO TECHNICAL NOTE 15: THE SHEBA FAMILY OF SHELL ELEMENTS FOR THE MATRIX DISPLACEMENT METHOD	69-0444
SHELL	A SHALLOW SHELL FINITE ELEMENT OF TRIANGULAR SHAPE	70-0044
SHELL	A SIMPLE MATRIX-VECTOR HANDLING SCHEME FOR THREE-DIMENSIONAL AND SHELL ANALYSIS	70-0317
SHELL	A SIMPLE MIXED METHOD FOR PLATE AND SHELL PROBLEMS	73-1005
SHELL	A STATIC AND DYNAMIC FINITE ELEMENT SHELL ANALYSIS WITH EXPERIMENTAL VERIFICATION	69-0156
SHELL	A STATIC AND DYNAMIC FINITE ELEMENT SHELL ANALYSIS WITH EXPERIMENTAL VERIFICATION	71-0769
SHELL	A STIFFNESS MATRIX FOR A CURVED MEMBRANE SHELL	69-0055
SHELL	A STIFFNESS MATRIX FOR SHALLOW SHELL FINITE ELEMENTS	69-0229
SHELL	A TRIANGULAR THIN SHELL FINITE ELEMENT: NONLINEAR ANALYSIS	75-0697
SHELL	A TRIANGULAR THIN SHELL FINITE ELEMENT: LINEAR ANALYSIS	75-0698
SHELL	A TWO-DIMENSIONAL AXISYMMETRIC SHELL FLUID MODEL FOR LAUNCH VEHICLE LONGITUDINAL DYNAMIC ANALYSIS	69-0046
SHELL	AN ANALYSIS OF AN INTERSECTION PROBLEM USING THE SEMILOOP SHELL ELEMENT	74-0401
SHELL	AN ASSESSMENT OF CURRENT CAPABILITY FOR COMPUTER ANALYSIS OF SHELL STRUCTURES	71-0706
SHELL	AN EFFICIENT TRIANGULAR SHELL ELEMENT	70-0382
SHELL	AN EFFICIENT TRIANGULAR SHELL ELEMENT	70-0441
SHELL	AN EXTENSION OF THE SHALLOW TO THE NON-SHALLOW STIFFNESS MATRIX FOR A CYLINDRICAL SHELL FINITE ELEME	70-0104
SHELL	AN IMPROVED SHELL OF REVOLUTION ELEMENT UTILISING CUBIC DISPLACEMENT FUNCTIONS	70-0451
SHELL	ANALYSIS OF ARBITRARY SHELLS USING SHALLOW SHELL FINITE ELEMENTS	71-0717
SHELL	ANALYSIS OF ARBITRARY SHELLS USING SHALLOW SHELL FINITE ELEMENTS	74-1374
SHELL	ANALYSIS OF ASYMMETRIC STIFFENED SHELL TYPE STRUCTURES BY THE FINITE ELEMENT METHOD: I FLAT RECTANGU	68-0283
SHELL	ANALYSIS OF ASYMMETRIC STIFFENED SHELL TYPE STRUCTURES BY FINITE ELEMENT METHOD II. SIX DEGREE-OF-FR	69-0373
SHELL	ANALYSIS OF BAND PRESSURE IN A GUN TUBE USING FINITE ELEMENTS AND SHELL PUSHING DATA	72-0793
SHELL	ANALYSIS OF CYLINDRICAL SHELL ROOFS WITH POST TENSIONED EDGEBEAMS	74-1042
SHELL	ANALYSIS OF DOUBLE CORRUGATED SHELL STRUCTURES BY THE FINITE ELEMENT METHOD	74-1116
SHELL	ANALYSIS OF LAMINATED SHELL FOR REVOLUTION	66-0023
SHELL	ANALYSIS OF PLATE AND SHELL STRUCTURES BY TRIANGULAR FINITE ELEMENTS	68-0303
SHELL	ANALYSIS OF PLATE AND SHELL STRUCTURES	69-0109
SHELL	ANALYSIS OF SHALLOW SHELL STRUCTURES BY A DISCRETE ELEMENT SYSTEM	66-0113
SHELL	ANALYSIS OF SHELL STRUCTURES BY THE FINITE ELEMENT METHOD	61-0008
SHELL	ANALYSIS OF THE CORNEO-SCLERAL SHELL BY THE METHOD OF DIRECT STIFFNESS	71-0151
SHELL	ANALYSIS OF THICK AND THIN SHELL STRUCTURE BY CURVED FINITE ELEMENTS	70-0016
SHELL	ANALYTICAL AND EXPERIMENTAL INVESTIGATION OF A NOZZLE TO CYLINDRICAL SHELL JUNCTION	74-0638
SHELL	APPLICATION OF A REFINED SHALLOW SHELL FINITE ELEMENT TO GEOMETRICALLY NONLINEAR STRUCTURES	73-0358
SHELL	APPLICATION OF RITZ'S METHOD TO THIN ELASTIC SHELL ANALYSIS	71-0083
SHELL	APPLICATION OF SIMPLIFIED HYBRID DISPLACEMENT METHOD TO PLATE AND SHELL PROBLEMS	73-1037
SHELL	APPLICATION OF THE STIFFNESS METHOD TO THE ANALYSIS OF SHELL STRUCTURES	61-0004
SHELL	APPLICATION, TESTING AND EVALUATION OF DOUBLY CURVED SHELL ELEMENTS FOR DYNAMIC ANALYSIS. PART II. F	69-0117
SHELL	APPLICATION, TESTING, AND EVALUATION OF DOUBLY CURVED SHELL ELEMENTS FOR DYNAMIC ANALYSIS. PART 1. T	69-0246
SHELL	APPLICATION, TESTING, AND EVALUATION OF DOUBLY CURVED SHELL ELEMENTS FOR DYNAMIC ANALYSIS, PART III.	69-0256
SHELL	APPLICATION, TESTING, AND EVALUATION OF DOUBLY CURVED SHELL ELEMENTS FOR DYNAMIC ANALYSIS, PART III.	69-0256
SHELL	APPLIED FINITE ELEMENT DISPLACEMENT ANALYSIS OF A SIMPLE INTEGRAL CONSTRUCTION VEHICLE BODY SHELL	72-0264
SHELL	APPROXIMATE ANALYSIS OF PREFABRICATED TOWER STRUCTURES USING THIN SHELL FINITE ELEMENTS	73-0893
SHELL	ARCH DAMS ANALYSED BY A LINEAR FINITE ELEMENT SHELL THE FINITE STRIP METHOD IN THE ANALYSIS OF ELAST	68-0189
SHELL	ASYMMETRICAL BENDING AND VIBRATION OF A CONICAL SHELL FINITE ELEMENT	74-0085
SHELL	AUTOMATIC DESIGN OF SHELL STRUCTURES	71-0336
SHELL	BEHAVIOR OF TRIANGULAR SHELL ELEMENT STIFFNESS MATRICES ASSOCIATED WITH POLYHEDRAL DEFLECTION DISTRI	68-0159
SHELL	BENDING ELEMENTS FOR PLATE AND SHELL NETWORKS	67-0043
SHELL	CHISLENNYI METOD RASCHETA SOBSTVENNYKH I VYNUZHDENNYKH KOLEBANII SOSTAVNYKH OBOLOCHECHNYKH KONSTRUKT	73-0461
SHELL	COMPARISON OF RECENT SHALLOW SHELL FINITE ELEMENT ANALYSES	71-0362
SHELL	COMPUTATION OF VIBRATION MODAL SHAPES AND NATURAL FREQUENCIES OF A GENERAL ORTHOTROPIC LAMINATED, TH	71-0309
SHELL	COMPUTER ANALYSIS OF SHELL STRUCTURES	69-0060
SHELL	COMPUTER ORIENTED ANALYSIS OF SHELL STRUCTURES	71-0559
SHELL	COMPUTER ORIENTED ANALYSIS OF SHELL STRUCTURES	71-0762
SHELL	COMPUTER-AIDED DESIGN OF SHELL STRUCTURES USING THE FINITE ELEMENT METHOD	72-0998
SHELL	CONFORMING FINITE ELEMENT METHOD FOR THE SHELL PROBLEM	75-0096
SHELL	CONICAL SHELL FINITE ELEMENT	74-0593
SHELL	CONVERGENCE CHARACTERISTICS OF CYLINDRICAL SHELL FINITE ELEMENTS	72-0943
SHELL	CONVERGENCE OF CONFORMING FINITE-ELEMENT METHODS FOR SHELL PROBLEMS	75-0494
SHELL	CONVERGENCE OF THE NASTRAN PLATE ELEMENTS FOR SHELL STRESS ANALYSIS	71-0261
SHELL	CORRECTED ASSESSMENT OF CYLINDRICAL SHELL FINITE ELEMENT OF BOGNER, FOX AND SCHMIT WHEN APPLIED TO A	73-0024
SHELL	CURVED FINITE ELEMENTS IN THE ANALYSIS OF SOLID SHELL AND PLATE STRUCTURES	69-0357
SHELL	CURVED FINITE ELEMENTS IN SHELL CALCULATIONS	70-0227
SHELL	CURVED FINITE ELEMENTS IN THE ANALYSIS OF SHELL STRUCTURES	71-0076
SHELL	CURVED ROTATIONAL SHELL ELEMENTS BY THE CONSTRAINT METHOD	72-0924
SHELL	CURVED THICK SHELL AND MEMBRANE ELEMENTS WITH PARTICULAR REFERENCE TO AXISYMMETRIC PROBLEMS	68-0005
SHELL	CURVED, CYLINDRICAL SHELL, FINITE ELEMENT	72-0096
SHELL	DEFICIENCY IN CURRENT FINITE ELEMENTS FOR THIN SHELL APPLICATIONS	73-0184
SHELL	DELINQUENT FINITE ELEMENTS FOR SHELL IDEALIZATION	75-0478
SHELL	DESH, EIN EINFACHES FLACHES SCHALENELEMENT UEBER DREIECKIGEM GRUNDRISS. (DESH, A SIMPLE TRIANGULAR S	75-0304
SHELL	DEVELOPMENT AND EVALUATION OF A TRIANGULAR THIN SHELL ELEMENT BASED UPON THE HYBRID ASSUMED STRESS M	69-0252
SHELL	DEVELOPMENT OF A FINITE ELEMENT MODEL FOR THE CLASS V FLEXTENSIONAL UNDERWATER TRANSDUCER SHELL	71-0754
SHELL	DEVELOPMENT OF A FINITE ELEMENT MODEL FOR THE CLASS 5 FLEXTENSIONAL UNDERWATER TRANSDUCER SHELL	71-0840
SHELL	DEVELOPMENT OF A FINITE ELEMENT MODEL FOR THE CLASS IV FLEXTENSIONAL UNDERWATER TRANSDUCER SHELL	72-0593
SHELL	DEVELOPMENT OF TWO SIMPLE SHELL ELEMENTS	72-0676
SHELL	DISCRETE ELEMENT PROCEDURE FOR THIN SHELL INSTABILITY ANALYSIS	67-0028
SHELL	DISPLACEMENT METHOD IN STRUCTURAL ANALYSIS AS APPLIED TO RECTANGULAR SHELL BEAMS	72-0411
SHELL	DOUBLE CURVATURE SHELL FINITE-ELEMENT AND ITS USE IN THE DYNAMIC ANALYSIS OF COOLING TOWERS AND OTHE	70-0312
SHELL	DOUBLE CURVATURE SHELL FINITE-ELEMENT AND ITS USE IN THE DYNAMIC ANALYSIS OF COOLING TOWERS AND OTHE	70-0312
SHELL	DYNAMIC ANALYSIS OF A COMPLEX SHELL STRUCTURE UNDER IMPULSIVE ASYMMETRIC LOADING	68-0226
SHELL	DYNAMIC BEHAVIOUR OF SHELL ROOFS	69-0228
SHELL	DYNAMIC LOADING OF A FLUID-FILLED SPHERICAL SHELL.	72-0444
SHELL	DYNAMIC SHELL THEORY COUPLING THICKNESS STRESS WAVE EFFECTS WITH GROSS STRUCTURAL RESPONSE	73-0379
SHELL	ELEMENT STATIC FLEXIBILITY AND STIFFNESS MATRICES FOR THE FINITE ELEMENT ANALYSIS OF BEAM, PLATE, AN	67-0070
SHELL	ENGINEERING OPERATIONS REPORT: NERVA THIN SHELL FINITE ELEMENT STRESS ANALYSIS METHODOLOGY COMPUTER	72-0916
SHELL	ERRORS IN SHELL FINITE-ELEMENT MODELS FOR VIBRATION OF CIRCULAR-CYLINDERS	75-0203
SHELL	EXPERIMENTAL ELASTIC STRESS ANALYSES OF CYLINDER-TO-CYLINDER SHELL MODELS AND COMPARISONS WITH THEOR	71-0065
SHELL	EXTENSION OF THREE DIMENSIONAL ANALYSIS TO SHELL STRUCTURES USING THE FINITE ELEMENT IDEALIZATION	74-0849
SHELL	FINITE DEFLECTION STRUCTURAL ANALYSIS USING PLATE AND SHELL DISCRETE ELEMENTS	68-0143
SHELL	FINITE DEFLECTION STRUCTURAL ANALYSIS USING PLATE AND CYLINDRICAL SHELL DISCRETE ELEMENTS	68-0193
SHELL	FINITE ELEMENT ANALYSIS OF DYNAMIC SHELL BEHAVIOUR	67-0013
SHELL	FINITE ELEMENT ANALYSIS OF SHELL STRUCTURES IN JAPAN	69-0151
SHELL	FINITE ELEMENT ANALYSIS OF PLATE AND SHELL STRUCTURES	69-0360
SHELL	FINITE ELEMENT ANALYSES OF SHELL RESPONSE FOR ARBITRARY EXCITATIONS	70-0348
SHELL	FINITE ELEMENT ANALYSIS OF AN AXISYMMETRICALLY LOADED ORTHOTROPIC SHELL OF REVOLUTION.	72-0438
SHELL	FINITE ELEMENT CREEP BUCKLING ANALYSIS OF VISCO-ELASTIC SHALLOW SPHERICAL SHELL	75-0159
SHELL	FINITE ELEMENT DYNAMIC ANALYSIS OF SHALLOW SHELL STRUCTURES	70-0164
SHELL	FINITE ELEMENT ELASTIC THIN SHELL PRE-BUCKLING AND POST-BUCKLING ANALYSIS	71-0821
SHELL	FINITE ELEMENT METHOD IN SHELL STABILITY ANALYSIS	73-0250
SHELL	FINITE ELEMENT METHOD IN PLATE AND SHELL INSTABILITY ANALYSIS	73-0983

363

SHELL	FINITE ELEMENT METHOD OF MIXED TYPE AND ITS CONVERGENCE IN LINEAR SHELL PROBLEMS	73-1041
SHELL	FINITE ELEMENT PROCEDURES IN THE SOLUTION OF PLATE AND SHELL PROBLEMS	65-0038
SHELL	FINITE ELEMENT STUDY OF A REINFORCED CONCRETE CYLINDRICAL SHELL THROUGH ELASTIC, CRACKING, AND ULTIM	75-0317
SHELL	FINITE ELEMENTE FUR GEKRUEMMTE FLAECHENTRAGWERKE. (FINITE ELEMENTS SOLUTIONS FOR CURVED SHELL STRUCT	71-0391
SHELL	FINITE-DIFFERENCE ENERGY METHOD FOR NONLINEAR SHELL ANALYSIS	71-0399
SHELL	FINITE-ELEMENT NONLINEAR BUCKLING ANALYSIS OF A PSEUDO-CYLINDRICAL CONCAVE POLYHEDRAL SHELL UNDER EX	73-1147
SHELL	FLAT FINITE ELEMENT FOR THIN SHELL ANALYSIS DERIVED BY ASSUMED STRESS APPROACH	73-0147
SHELL	FREQUENCY COEFFICIENTS OF A SPHERICAL SHELL FOR VARYING THICKNESS AND OPENING ANGLE WITH GUIDED PINN	72-0595
SHELL	GENERAL PURPOSE PROGRAM FOR SHELL STRUCTURES: ISTRAN-S	75-0511
SHELL	HIGH ORDER RECTANGULAR SHALLOW SHELL FINITE ELEMENT	73-0279
SHELL	HIGH-PRECISION CURVED SHELL FINITE ELEMENT	71-0422
SHELL	HIGH-PRECISION PLATE AND SHELL FINITE ELEMENTS	71-0437
SHELL	HYBRIDE SCHALENELEMENTE MIT EINER ANWENDUNG AUF WENDELSCHALEN. (HYBRID SHELL ELEMENTS IN APPLICATION	75-0609
SHELL	IMPROVED FINITE-ELEMENT ANALYSIS OF NUCLEAR-CONTAINMENT SHELL STRUCTURES	73-0209
SHELL	IMPROVED LINEAR AXISYMMETRIC SHELL FLUID MODEL FOR LAUNCH VEHICLE LONGITUDINAL RESPONSE ANALYSIS	65-0044
SHELL	IMPROVED NUMERICAL INTEGRATION OF THICK SHELL FINITE ELEMENTS	71-0396
SHELL	INCORPORATION OF FOUNDATION DEFORMATION IN A FINITE ELEMENT SHELL ANALYSIS	68-0227
SHELL	INCREMENTAL SOLUTION OF AXISYMMETRIC PLATE AND SHELL FINITE DEFORMATION	71-0776
SHELL	INELASTIC ANALYSIS OF A BRANCH SHELL JUNCTION	74-0682
SHELL	INFLUENCE OF THE UNPRECISION OF EXECUTION ON THE BEHAVIOUR OF A SPHERICAL SHELL WITH CONCENTRATED LO	73-0802
SHELL	LINEAR SHELL THEORY FOR RAPID VARIATION SOLUTIONS	71-0406
SHELL	MASS LOADING EFFECTS ON VIBRATION OF RING AND SHELL STRUCTURES	68-0245
SHELL	MATHEMATICAL MODEL OF THE CORNEO-SCLERAL SHELL AS APPLIED TO INTRAOCULAR PRESSURE VOLUME RELATIONS A	72-0471
SHELL	MATRIX ANALYSIS OF SHELL STRUCTURES	64-0003
SHELL	MEASUREMENT OF SLOPE CHANGE ON CYLINDRICAL SHELL SURFACES	73-0439
SHELL	MIXED FORMULATIONS FOR FINITE ELEMENT SHELL ANALYSIS	71-0222
SHELL	MIXED ISOPARAMETRIC LAMINATED COMPOSITE SHELL ELEMENTS	75-0450
SHELL	MODELING EARDRUM AS A DOUBLY CURVED SHELL USING FINITE-ELEMENT METHOD	75-0047
SHELL	NASTRAN BUCKLING ANALYSIS OF A LARGE STIFFENED CYLINDRICAL SHELL WITH A CUTOUT	71-0045
SHELL	NEW CYLINDRICAL SHELL FINITE ELEMENT BASED ON SIMPLE INDEPENDENT STRAIN FUNCTIONS	72-0094
SHELL	NEW HYBRID CYLINDRICAL SHELL FINITE ELEMENT	72-0092
SHELL	NEW VARIATIONAL FUNCTIONAL FOR THE FINITE-ELEMENT METHOD AND ITS APPLICATION TO PLATE SHELL PROBLEMS	72-0108
SHELL	NONLINEAR MATERIAL AND GEOMETRIC BEHAVIOUR OF SHELL STRUCTURES	71-0084
SHELL	NUMERICAL ANALYSIS OF SHELL STRUCTURES	67-0077
SHELL	NUMERICAL AND EXPERIMENTAL ANALYSIS OF A PSEUDOCYLINDRICAL SHELL	74-1133
SHELL	ON SHELL EFFECTS ON FERRO-CEMENT VESSELS	73-0455
SHELL	ON THE APPLICATION OF THE SHEBA SHELL ELEMENT	72-0651
SHELL	ON THE PARABOLIC VELAROIDAL SHELL	69-0314
SHELL	ON THE VALIDITY OF AN APPROXIMATION AVAILABLE IN THE FINITE ELEMENT SHELL ANALYSIS	75-0663
SHELL	PART 1 - TASK 3.2.9. BENDING AND BUCKLING OF A ONE-DIMENSIONAL MODEL SIMULATING A CONICAL SHELL. PAR	71-0745
SHELL	PLANAR AND CURVED SHELL ELEMENTS	69-0186
SHELL	PLATE AND SHELL PROBLEMS. FINITE ELEMENT DISPLACEMENT APPROACH	66-0005
SHELL	PRACTICAL APPLICATIONS OF FINITE ELEMENT METHODS IN THE STRESS ANALYSIS OF THIN SHELL PRESSURE VESSE	71-0205
SHELL	PREDICTION OF MARINE PROPELLER DISTORTION AND STRESSES USING A SUPERPARAMETRIC THICK SHELL FINITE EL	73-0370
SHELL	PROBLEM OF USING THE METHOD OF FINITE ELEMENTS TO CONSTRUCT AN ALGORITHM FOR CALCULATING A THIN-WALL	74-0851
SHELL	PSEUDO ISOPARAMETRIC FINITE ELEMENTS FOR SHELL AND PLATE ANALYSIS	69-0040
SHELL	RECORD CONCRETE SHELL ROOF FEATURED IN COLLISEUM	72-0218
SHELL	RIGID-BODY MOTIONS AND STRAIN-DISPLACEMENT EQUATIONS OF CURVED SHELL FINITE ELEMENTS	72-0267
SHELL	RING FINITE ELEMENTS FOR AXISYMMETRIC AND NON-AXISYMMETRIC THIN SHELL ANALYSIS	72-0922
SHELL	SAFE SHELL. A COMPUTER PROGRAM FOR THE STRESS ANALYSIS OF THIN SHELL BODIES OF REVOLUTION. A USER'S	67-0018
SHELL	SAFE SHELL. A COMPUTER PROGRAM FOR THE STRESS ANALYSIS OF THIN SHELL BODIES OF REVOLUTION. A USER'S	67-0018
SHELL	SEAL-SHELL-2. A COMPUTER PROGRAM FOR THE STRESS ANALYSIS OF A THICK SHELL OF REVOLUTION WITH AXISYMM	69-0107
SHELL	SHELL ANALYSIS BY SESAM-69	70-0249
SHELL	SHELL ANALYSIS BY THE FINITE ELEMENT METHOD	69-0376
SHELL	SHELL ANALYSIS USING PLANAR TRIANGULAR ELEMENTS	69-0041
SHELL	SHELL ANALYSIS USING A SIMPLE FACET ELEMENT	72-0986
SHELL	SHELL ANALYSIS WITH LARGE GENERAL PURPOSE PROGRAMS	71-0402
SHELL	SHELL AND PLATE ANALYSIS BY FINITE ELEMENTS	73-0163
SHELL	SHELL DEFORMATION UNDER TOOL LOADS REPRESENTATION OF SINGULARITIES WITH ISOPARAMETRIC FINITE ELEMENT	74-0365
SHELL	SHELL ELEMENTS	75-0545
SHELL	SHELL FINITE ELEMENT METHOD VIA REISSNER'S PRINCIPLE	69-0216
SHELL	SHELL INSTABILITY ANALYSIS APPLIED TO A RADOME	74-0261
SHELL	SHELL STRUCTURES AND CLIMATIC INFLUENCES	72-0389
SHELL	SHELL STRUCTURES AND CLIMATIC INFLUENCES	72-0399
SHELL	SIMPLE "MIXED" METHOD FOR PLATE AND SHELL PROBLEMS	74-0952
SHELL	SOLUTION OF A SINGLE HYPERBOLOID SHELL BY FINITE ELEMENT METHOD	73-0151
SHELL	SOME APPLICATIONS OF FINITE ELEMENT ANALYSIS TO SHELL BUCKLING PREDICTION	68-0023
SHELL	SOME MODIFICATIONS OF AN ISOPARAMETRIC SHELL ELEMENT	73-1010
SHELL	STABILITY INVESTIGATION OF A CYLINDRICAL SHELL WITH FINITE ELEMENTS	72-0127
SHELL	STATIC ANALYSIS OF A COMPLEX SHELL STRUCTURE UNDER ASYMMETRIC LOADING	67-0120
SHELL	STIFFNESS MATRIX FOR SHALLOW RECTANGULAR SHELL ELEMENT	67-0014
SHELL	STIFFNESS MATRIX FOR SHALLOW RECTANGULAR SHELL ELEMENT	68-0074
SHELL	STIFFNESS MATRIX FOR CURVED TAPERED AND STRAIGHT TAPERED SHELL STIFFENERS	69-0074
SHELL	STRONGLY CURVED FINITE ELEMENT FOR SHELL ANALYSIS	73-0244
SHELL	STRUCTURAL ANALYSIS AND EXPERIMENTAL CORRELATION OF A COMPLEX SHELL UNDER ASYMMETRIC LOADS	69-0155
SHELL	STRUCTURAL ANALYSIS OF SHELL INTERSECTIONS	69-0217
SHELL	STUDIES IN DOUBLY CURVED SHELL ELEMENTS	74-0133
SHELL	THE APPLICATION OF A CURVED MIXED TYPE SHELL ELEMENT	70-0118
SHELL	THE APPLICATION OF A CURVED MIXED-TYPE SHELL ELEMENT	70-0440
SHELL	THE APPLICATION OF FINITE ELEMENT TECHNIQUES TO PLATE AND SHELL PROBLEMS	65-0028
SHELL	THE APPLICATION OF GRADIENT MINIMIZATION METHODS AND HIGHER ORDER DISCRETE ELEMENTS TO SHELL BUCKLIN	71-0237
SHELL	THE EVOLUTION OF SHELL STRUCTURES IN OCEAN TECHNOLOGY	69-0042
SHELL	THE FINITE ELEMENT METHOD IN SHELL ANALYSIS	69-0058
SHELL	THE FINITE ELEMENT METHOD APPLIED TO THE ANALYSIS OF SANDWICH PLATE AND SHELL STRUCTURES	72-0824
SHELL	THE FINITE ELEMENT METHOD IN PLATE AND SHELL STABILITY ANALYSIS	73-0644
SHELL	THE HYBRID DISPLACEMENT METHOD APPLIED TO PLATE AND SHELL PROBLEMS	75-0089
SHELL	THE SEMI-LOOF SHELL ELEMENT	74-0629
SHELL	THE SEMILOOF SHELL ELEMENT	75-0536
SHELL	THE SHEBA FAMILY OF SHELL ELEMENTS FOR THE MATRIX DISPLACEMENT METHOD	68-0014
SHELL	THE SHEBA FAMILY OF SHELL ELEMENTS FOR THE MATRIX DISPLACEMENT METHOD. 111. LARGE DISPLACEMENTS	69-0499
SHELL	THIN SHELL THEORY IN CARTESIAN COORDINATES - METHODS FOR FINITE ELEMENTS	70-0359
SHELL	THREE DIMENSIONAL INELASTIC ANALYSIS OF A SHELL APPURTENANCE PER SECTION 111 - APPENDIX F.	74-0716
SHELL	TRANSIENT SHELL RESPONSE BY NUMERICAL TIME INTEGRATION	72-0305
SHELL	TRIANGULAR FINITE-ELEMENTS OF VARIABLE THICKNESS AND THEIR APPLICATION TO PLATE AND SHELL PROBLEMS	69-0095
SHELL	TRIANGULAR THICK SHELL FINITE ELEMENT FOR THE THREE-DIMENSIONAL STRESS ANALYSIS OF PRESSURE VESSELS	73-0860
SHELL	TRIANGULAR THIN SHELL FINITE ELEMENT: LINEAR ANALYSIS	75-0602
SHELL	TRIANGULAR THIN SHELL FINITE ELEMENT: NONLINEAR ANALYSIS	75-0668
SHELL	TWO DIMENSION, PLATE, SHELL AND FINITE PRISM ISOPARAMETRIC ELEMENTS AND THEIR APPLICATIONS	71-0772
SHELL	UEBER PAUSCHALE FEHLERSCHRANKEN FUER APPROXIMATIONEN VON SCHALENVERFORMUNGEN. (GLOBAL BOUNDARY CONDI	75-0292
SHELL	USE OF FINITE ELEMENT METHOD FOR SHELL PROBLEMS	75-0598
SHELL	USE OF GROUP-THEORETIC METHODS IN THE DEVELOPMENT OF NONLINEAR SHELL FINITE ELEMENTS	71-0704
SHELL	VERIFICATION OF A HYBRID GENERAL PURPOSE SHELL AND BODY OF REVOLUTION COMPUTER PROGRAM	71-0333

SHELL	VIBRATION ANALYSIS OF CANTILEVERED CURVED PLATES USING A NEW CYLINDRICAL SHELL FINITE ELEMENT	68-0130
SHELL	VIBRATION PROBLEMS USING ISOPARAMETRIC SHELL ELEMENTS	72-0549
SHELL	VIBRATIONS OF PLATE AND SHELL STRUCTURES USING TRIANGULAR FINITE ELEMENTS	67-0023
SHELL	VISCOELASTIC ANALYSIS OF PRESTRESSED CONCRETE SHELL OF REVOLUTION	73-0464
SHELL	VISCOELASTIC PRESPONSE OF THE CORNEO-SCLERAL SHELL UNDER TONOMETER LOADING	75-0855
SHELL	3-DIMENSIONAL FINITE-ELEMENT ANALYSIS OF ELECTROELASTIC NORMAL MODES OF A TANGENTIALLY POLARIZED, SE	75-0591
SHELL-TYPE	FINITE ELEMENT ANALYSIS OF SHELL-TYPE STRUCTURES	71-0789
SHELLED	EXPERIMENTAL AND FINITE ELEMENT STRESS ANALYSIS OF A THIN SHELLED CYLINDER TO CYLINDER MODEL	71-0066
SHELLS	A COMPUTER PROGRAM FOR THE STATIC AND DYNAMIC FINITE ELEMENT ANALYSIS OF AXISYMMETRIC THIN SHELLS	66-0017
SHELLS	A COMPUTER PROGRAM FOR THE ANALYSIS OF THIN SHELLS	67-0041
SHELLS	A CONICAL ELEMENT FOR DISPLACEMENT ANALYSIS OF AXI-SYMMETRIC SHELLS	69-0120
SHELLS	A CURVED ELEMENT APPROXIMATION IN THE ANALYSIS OF AXISYMMETRIC THIN SHELLS	70-0063
SHELLS	A CURVED FINITE ELEMENT FOR CYLINDRICAL SHELLS	68-0294
SHELLS	A CURVED FINITE ELEMENT FOR THIN ELASTIC SHELLS	69-0486
SHELLS	A DYNAMIC ELASTOPLASTIC ANALYSIS OF THIN SHELLS OF REVOLUTION UNDER ASYMMETRIC MECHANICAL AND THERMA	72-0068
SHELLS	A FINITE ELEMENT ANALYSIS FOR CYLINDRICAL SHELLS AND PLATES	66-0034
SHELLS	A FINITE ELEMENT ANALYSIS OF ECCENTRICALLY STIFFENED-CIRCULAR CYLINDRICAL SHELLS	69-0159
SHELLS	A FINITE ELEMENT ANALYSIS OF LAMINATED ORTHOTROPIC PLATES AND SHALLOW SHELLS	70-0123
SHELLS	A FINITE ELEMENT ANALYSIS OF SHELLS OF REVOLUTION	73-0759
SHELLS	A FINITE ELEMENT APPROXIMATION FOR THE ANALYSIS OF THIN SHELLS	68-0046
SHELLS	A FINITE ELEMENT APPROACH TO SCATTERING FROM ELASTIC SPHERICAL SHELLS	74-0702
SHELLS	A FINITE ELEMENT DISPLACEMENT APPROACH FOR THE ELASTO-PLASTIC ANALYSIS OF THIN CYLINDRICAL SHELLS	74-1077
SHELLS	A FINITE ELEMENT FOR THERMAL STRESS ANALYSIS OF SHELLS OF REVOLUTION	73-0798
SHELLS	A FINITE ELEMENT METHOD FOR GEOMETRICALLY NONLINEAR LARGE DISPLACEMENT PROBLEMS IN THIN, ELASTIC PLA	69-0005
SHELLS	A FINITE ELEMENT PROGRAM FOR THE DYNAMIC LARGE DEFLECTION, ELASTIC-PLASTIC ANALYSIS OF STIFFENED SHE	72-0908
SHELLS	A FINITE ELEMENT PROGRAM FOR FREE FORM SHELLS	75-0786
SHELLS	A FINITE-ELEMENT ANALYSIS FOR THIN SHELLS	68-0087
SHELLS	A METHOD FOR COMPUTATION OF VIBRATION MODES AND FREQUENCIES OF ORTHOTROPIC THIN SHELLS OF REVOLUTION	69-0037
SHELLS	A MIXED FINITE ELEMENT METHOD FOR THIN SHELLS ANALYSIS	68-0182
SHELLS	A REFINED CURVED ELEMENT FOR THIN SHELLS OF REVOLUTION	71-0214
SHELLS	A STRUCTURAL ANALYSIS OF CYLINDER-CONE-CYLINDER SHELLS BY THE FINITE ELEMENT METHOD	72-0796
SHELLS	A STUDY OF A DYNAMICAL FINITE ELEMENT ANALYSIS FOR APPLICATION TO AXIAL WAVE PROPAGATION PROBLEMS IN	71-0589
SHELLS	A STUDY OF ELASTIC-PLASTIC ANALYSIS BY THE ASSUMED-STRESS HYBRID FINITE-ELEMENT MODEL, WITH APPLICAT	74-0718
SHELLS	A STUDY OF FREE-VIBRATIONS OF THIN ELASTIC SHELLS SUBJECTED TO INITIAL LOADS USING FINITE-ELEMENTS	73-1143
SHELLS	A STUDY OF THE MATRIX DISPLACEMENT METHOD AS APPLIED TO SHELLS OF REVOLUTION	66-0111
SHELLS	A SURVEY OF THE ANALYSIS OF SHELLS BY THE DISPLACEMENT METHOD	66-0041
SHELLS	ACCURACY OF FINITE-ELEMENT SOLUTIONS FOR MODAL CHARACTERISTICS OF SHELLS OF REVOLUTION	70-0231
SHELLS	AN APPLICATION OF THE FINITE ELEMENT METHOD TO THE CALCULATION OF CYLINDRICAL SHELLS WITH RECTANGULA	74-1236
SHELLS	AN ELASTIC FINITE-ELEMENT ANALYSIS OF THE BUCKLING OF ECCENTRICALLY STIFFENED PLATES AND SHELLS	73-1136
SHELLS	AN EVALUATION OF FINITE DIFFERENCE AND FINITE ELEMENT TECHNIQUES FOR ANALYSIS OF GENERAL SHELLS	70-0254
SHELLS	ANALYSES OF SHELLS OF REVOLUTION BY FINITE ELEMENTS	67-0149
SHELLS	ANALYSIS OF ARBITRARY SHELLS USING SHALLOW SHELL FINITE ELEMENTS	71-0717
SHELLS	ANALYSIS OF ARBITRARY SHELLS USING SHALLOW SHELL FINITE ELEMENTS	74-1374
SHELLS	ANALYSIS OF AXISYMMETRIC SHELLS BY THE DIRECT STIFFNESS METHOD	63-0006
SHELLS	ANALYSIS OF BUCKLING AND VIBRATION OF RING-STIFFENED, SEGMENTED SHELLS OF REVOLUTION	69-0496
SHELLS	ANALYSIS OF ECCENTRICALLY STIFFENED CYLINDRICAL SHELLS	72-0217
SHELLS	ANALYSIS OF ELASTIC STABILITY OF SHELLS OF REVOLUTION BY THE FINITE ELEMENT METHOD	67-0056
SHELLS	ANALYSIS OF ELASTIC SHELLS OF REVOLUTION WITH MEMBRANE AND FLEXURE STRESSES UNDER ARBITRARY LOADING	68-0289
SHELLS	ANALYSIS OF ELASTIC-PLASTIC SHELLS OF REVOLUTION UNDER AXI-SYMMETRIC LOADING BY THE FINITE ELEMENT M	67-0102
SHELLS	ANALYSIS OF ELASTIC-PLASTIC SHELLS OF REVOLUTION	70-0220
SHELLS	ANALYSIS OF ELASTIC-PLASTIC CREEP BUCKLING OF AXISYMMETRIC SHELLS BY FINITE-ELEMENT METHOD	75-0355
SHELLS	ANALYSIS OF ELASTICALLY COUPLED SHELLS ON ELASTIC FOUNDATION BY HYBRID METHOD	71-0276
SHELLS	ANALYSIS OF GEOMETRICALLY NONLINEAR PLATES AND SHELLS BY THE FINITE ELEMENT DISPLACEMENT METHOD	68-0050
SHELLS	ANALYSIS OF HYPERBOLIC PARABOLOID SHELLS	70-0324
SHELLS	ANALYSIS OF INFLATABLE SHELLS BY THE FINITE ELEMENT METHOD	71-0822
SHELLS	ANALYSIS OF LARGE ELASTIC-PLASTIC CREEP DEFORMATIONS OF AXISYMMETRIC SHELLS BY THE FINITE ELEMENT ME	74-0563
SHELLS	ANALYSIS OF PNEUMATIC SHELLS WITH OR WITHOUT CABLE FORMULATION	74-0591
SHELLS	ANALYSIS OF RING-STIFFENED SHELLS OF REVOLUTION UNDER COMBINED THERMAL AND MECHANICAL LOADING	71-0342
SHELLS	ANALYSIS OF SHELLS BY THE FINITE ELEMENT DISPLACEMENT METHOD	68-0185
SHELLS	ANALYSIS OF SHELLS OF ANY FORM IN THE PLASTIC REGION BY THE FINITE ELEMENT METHOD	73-0940
SHELLS	ANALYSIS OF SHELLS OF REVOLUTION BY A CURVED FINITE ELEMENT	75-0680
SHELLS	ANALYSIS OF THIN SHELLS USING CURVED FINITE ELEMENTS	69-0523
SHELLS	ANALYSIS OF THIN SHELLS WITH A DOUBLY CURVED ARBITRARY QUADRILATERAL FINITE ELEMENT.	72-0443
SHELLS	ANALYSIS OF THIN-SHELL HYPERBOLIC PARABOLOID SHELLS	72-0121
SHELLS	ANALYSIS OF VISCOELASTOPLASTIC STRUCTURAL BEHAVIOUR OF ANISOTROPIC SHELLS BY THE FINITE ELEMENT METH	71-0056
SHELLS	ANALYTIC INVESTIGATION OF COMPACT REINFORCEMENT FOR RADIAL NOZZLES IN SPHERICAL SHELLS	71-0107
SHELLS	APPLICATION OF FINITE ELEMENT METHOD TO AXISYMMETRIC BUCKLING OF SHALLOW SPHERICAL SHELLS UNDER EXTE	73-0111
SHELLS	APPLICATION OF MATRIX DISPLACEMENT METHOD TO LINEAR ELASTIC ANALYSIS OF SHELLS OF REVOLUTION	65-0025
SHELLS	APPLICATION OF THE FINITE ELEMENT METHOD TO THE ANALYSIS OF CYLINDRICAL SHELLS	73-0828
SHELLS	APPLICATIONS OF CURVED FINITE ELEMENTS TO SANDWICH SHELLS USING SHEAR STRAIN FORMULATION	75-0433
SHELLS	AXISYMMETRIC SHELLS	69-0481
SHELLS	AXISYMMETRIC VIBRATION OF SHELLS OF REVOLUTION BY THE FINITE ELEMENT METHOD	67-0100
SHELLS	AXISYMMETRIC VIBRATION OF CONICAL SHELLS	70-0350
SHELLS	BASIC COMPUTATIONAL PROBLEMS IN THE FINITE ELEMENT ANALYSIS OF SHELLS	71-0678
SHELLS	BENDING OF CYLINDRICAL SHELLS - SOLUTION USING CYLINDRICAL STRIP ELEMENTS - CONTRIBUTION TO FINITE	70-0363
SHELLS	BUCKLING OF COOLING-TOWER SHELLS: BIFURCATION RESULTS	75-0384
SHELLS	BUCKLING OF COOLING-TOWER SHELLS: STATE-OF-THE-ART	75-0385
SHELLS	BUCKLING OF CYLINDRICAL SHELLS BY WIND PRESSURE	74-0642
SHELLS	BUCKLING OF DEEP SHELLS	73-0849
SHELLS	BUCKLING OF OPEN CYLINDRICAL SHELLS	67-0115
SHELLS	BUCKLING OF SEGMENTS OF TOROIDAL SHELLS	73-0515
SHELLS	BUCKLING OF SHELLS BY FINITE ELEMENT METHOD	74-0091
SHELLS	CALCULATION OF NON-LINEAR AXISYMMETRIC VIBRATIONS OF THIN SHELLS OF REVOLUTION BY A FINITE ELEMENT M	75-0067
SHELLS	CALCULATION OF STIFFNESS MATRICES FOR FINITE ELEMENTS OF THIN SHELLS OF ARBITRARY SHAPE	68-0127
SHELLS	CALCULATION OF TEMPERATURE DISTRIBUTIONS IN THIN SHELLS OF REVOLUTION BY THE FINITE ELEMENT METHOD	71-0018
SHELLS	CALCULATION OF THE CHARACTERISTIC FREQUENCIES OF FLUID-FILLED SPHERICAL SHELLS	73-0511
SHELLS	COMPARISON OF SOME FINITE-ELEMENT MODELS FOR CIRCULAR CYLINDRICAL SHELLS	73-1158
SHELLS	COMPARISON OF THE RESULTS OF THE DYNAMIC RESPONSE OF CYLINDRICAL SHELLS BY CHARACTERISTIC AND BY FIN	74-0781
SHELLS	COMPARISON OF TWO HIGH-PRECISION TRIANGULAR FINITE ELEMENTS FOR ARBITRARY DEEP SHELLS	71-0071
SHELLS	COMPUTATION OF SHELLS WITH FINITE DISPLACEMENTS: MIXED FINITE ELEMENTS (BERECHNUNG VON SCHALEN MIT E	73-0788
SHELLS	COMPUTATIONAL EFFICIENCY ASPECTS OF THE FINITE ELEMENT NONLINEAR ANALYSIS OF SHELLS	75-0031
SHELLS	COMPUTER PROGRAM FOR ELASTIC-PLASTIC ANALYSIS OF AXISYMMETRICALLY LOADED SHELLS OF REVOLUTION	68-0239
SHELLS	CONICAL SEGMENT METHOD FOR ANALYZING OPEN CROWN SHELLS OF REVOLUTION FOR EDGE LOADING	63-0007
SHELLS	CONIDA: A FINITE ELEMENT PROGRAM FOR THE STRESS ANALYSIS OF AXISYMMETRIC THIN SHELLS	74-1087
SHELLS	CONSISTENT STIFFNESS MATRICES IN THE ANALYSIS OF SHELLS	67-0078
SHELLS	CONSTRUCTION OF NORMAL DISPLACEMENT FUNCTIONS OF TRIANGULAR FINITE ELEMENT OF PLATES AND SHELLS	72-0563
SHELLS	CRITERIA FOR FINITE ELEMENT DISCRETIZATION OF SHELLS OF REVOLUTION	73-0152
SHELLS	CURSHL: A HIGH-PRECISION FINITE ELEMENT FOR SHELLS OF ARBITRARY SHAPE	71-0430
SHELLS	CURVED FINITE ELEMENT FOR THIN ELASTIC SHELLS	70-0173
SHELLS	CURVED TRIANGULAR ELEMENTS FOR THE ANALYSIS OF SHELLS	68-0030
SHELLS	CURVED TRIANGULAR FINITE ELEMENTS FOR THE ANALYSIS OF THICK ELASTIC PLATES AND SHELLS	71-0814
SHELLS	DESIGN OF PRISMATIC SHELLS	72-0404

SHELLS	DIRECT STIFFNESS METHOD OF ANALYSIS OF SHELLS OF REVOLUTION UTILIZING CURVED ELEMENTS	66-0042
SHELLS	DISCRETE ELEMENT FINITE DISPLACEMENT ANALYSIS OF ANISOTROPIC SANDWICH SHELLS	70-0504
SHELLS	DISCRETE-ELEMENT ANALYSIS OF SHELLS OF REVOLUTION	68-0168
SHELLS	DISCRETE-ELEMENT MODELLING AND SENSITIVITY STUDIES OF STATICALLY LOADED SINGLE-LAYER AND BOUNDED DOU	70-0129
SHELLS	DISCRETE-ELEMENT STATIC ANALYSIS OF BONDED, DOUBLE-LAYER, BRANCHED, THIN SHELLS OF REVOLUTION, PART	68-0102
SHELLS	DISCRETE-ELEMENT STATIC ANALYSIS OF CORE-STIFFENED SHELLS OF REVOLUTION FOR ASYMMETRIC MECHANICAL LO	68-0148
SHELLS	DISCRETE-ELEMENT STATIC ANALYSIS OF BONDED, DOUBLE-LAYER, BRANCHED, THIN SHELLS OF REVOLUTION, PART	70-0081
SHELLS	DISCUSSION ON FINITE ELEMENT SOLUTION FOR AXISYMMETRICAL SHELLS	65-0062
SHELLS	DOUBLY CURVED TRIANGULAR FINITE ELEMENTS FOR SHELLS OF ARBITRARY SHAPE	73-0881
SHELLS	DUALITY IN THE ANALYSIS OF SHELLS BY THE FINITE ELEMENT METHOD	71-0124
SHELLS	DYNAMIC ANALYSIS OF SHELLS USING DOUBLY-CURVED FINITE ELEMENTS	68-0082
SHELLS	DYNAMIC ANALYSIS OF SHELLS USING CURVED FINITE ELEMENTS	69-0296
SHELLS	DYNAMIC ANALYSIS OF SHALLOW SHELLS WITH A DOUBLY-CURVED TRIANGULAR FINITE ELEMENT	71-0353
SHELLS	DYNAMIC ANALYSIS OF AXISYMMETRIC SHELLS UNDER ARBITRARY TRANSIENT PRESSURES.	72-0467
SHELLS	DYNAMIC ANALYSIS OF VISCOELASTOPLASTIC ANISOTROPIC SHELLS	73-0253
SHELLS	DYNAMIC ANALYSIS OF COLUMN-SUPPORTED HYPERBOLOIDAL SHELLS	74-0188
SHELLS	DYNAMIC ELASTOPLASTIC RESPONSE OF GEOMETRICALLY NONLINEAR ARBITRARY SHELLS OF REVOLUTION UNDER IMPUL	70-0098
SHELLS	DYNAMIC FINITE ELEMENT ANALYSIS OF ARBITRARY THIN SHELLS	71-0061
SHELLS	DYNAMIC RESPONSE OF ARBITRARY SHELLS OF REVOLUTION UNDER COMBINED IMPULSIVE NONISOTHERMAL AND IRREGU	70-0365
SHELLS	DYNAMIC RESPONSES OF CYLINDRICAL SHELLS INCLUDING GEOMETRIC AND MATERIAL NONLINEARITIES	74-0150
SHELLS	DYNAMIC STABILITY OF CYLINDRICAL SHELLS BY FINITE ELEMENTS	68-0292
SHELLS	DYNAMIC STABILITY OF CONICAL SHELLS BY FINITE ELEMENTS	70-0467
SHELLS	DYNAMICAL ANALYSIS OF AXIALLY NON-UNIFORM, THIN CYLINDRICAL SHELLS. PART 1 - MATRIX FORMULATION	70-0413
SHELLS	DYNAMICAL ANALYSIS OF AXIALLY NONUNIFORM, THIN CYLINDRICAL SHELLS. PART 2 FREE VIBRATION	71-0716
SHELLS	DYNAPLAS: A FINITE ELEMENT PROGRAM FOR THE DYNAMIC, LARGE DEFLECTION, ELASTIC-PLASTIC ANALYSIS OF ST	72-0773
SHELLS	DYNASOR 11 - A FINITE ELEMENT PROGRAM FOR THE DYNAMIC NONLINEAR ANALYSIS OF SHELLS OF REVOLUTION (US	71-0666
SHELLS	EFFECT OF FILLETS ON STRESS CONCENTRATIONS IN CYLINDRICAL SHELLS WITH STEP CHANGES IN OUTSIDE DIAMET	71-0108
SHELLS	EFFECT OF LATERAL PRESSURE ON NATURAL FREQUENCIES OF HYPERBOLOIDAL SHELLS OF REVOLUTION	71-0278
SHELLS	EIN NEUES VARIATIONSPRINZIP MIT ANWENDUNG AUF SCHWINGUNGEN FLACHER SCHALEN NACH DER METHODE DER FINI	73-0563
SHELLS	ELASTIC ANALYSIS OF CRACKED THIN SHELLS BY THE FINITE ELEMENT METHOD	71-0048
SHELLS	ELASTIC INSTABILITY PREDICTIONS FOR DOUBLY-CURVED SHELLS	68-0075
SHELLS	ELASTIC STABILITY OF SHELLS OF REVOLUTION BY THE VARIATIONAL APPROACH USING DISCRETE ELEMENTS	66-0106
SHELLS	ELASTIC-PLASTIC ANALYSIS OF AXISYMMETRIC PLATES AND SHELLS	71-0779
SHELLS	ELASTIC-PLASTIC DISCRETE ELEMENT ANALYSIS OF AXISYMMETRICALLY LOADED SHELLS OF REVOLUTION	67-0099
SHELLS	ELASTO-PLASTIC ANALYSIS OF ORTHOTROPIC PLATES AND SHELLS	68-0166
SHELLS	ELASTO-PLASTIC ORTHOTROPIC PLATES AND SHELLS	69-0467
SHELLS	ELEMENT STIFFNESS MATRICES OF THICK-WALLED ORTHOTROPIC SHELLS WITH APPLICATIONS	69-0261
SHELLS	FAMSOR, A FINITE ELEMENT PROGRAM FOR THE FREQUENCIES AND MODE SHAPES OF SHELLS OF REVOLUTION	70-0090
SHELLS	FIELD ANALYSIS FOR PREFABRICATED RIBBED CYLINDRICAL SHELLS	73-0899
SHELLS	FINITE DEFLECTION DISCRETE ELEMENT ANALYSIS OF SANDWICH PLATES AND CYLINDRICAL SHELLS WITH LAMINATED	69-0236
SHELLS	FINITE DEFLECTION, DISCRETE ELEMENT ANALYSIS OF SHELLS	68-0029
SHELLS	FINITE DEFORMATION OF ELASTIC PLATES, SHELLS, AND MEMBRANES BY THE FINITE ELEMENT METHOD	72-0805
SHELLS	FINITE DIFFERENCE SOLUTIONS FOR LARGE DEFORMATIONS OF CYLINDRICAL SHELLS - A COMPARISON WITH FINITE	72-0270
SHELLS	FINITE ELEMENT ANALYSIS OF SANDWICH PLATES AND CYLINDRICAL SHELLS WITH LAMINATED FACES	68-0121
SHELLS	FINITE ELEMENT ANALYSIS OF SKEWED SHALLOW SHELLS	68-0250
SHELLS	FINITE ELEMENT ANALYSIS OF OPEN SHELLS OF REVOLUTION UNDER NONSYMMETRICAL LOADING	68-0295
SHELLS	FINITE ELEMENT ANALYSIS OF DISCONTINUOUS SHELLS	68-0301
SHELLS	FINITE ELEMENT ANALYSIS OF SHELLS UNDER NONUNIFORM SNOW LOADS	69-0113
SHELLS	FINITE ELEMENT ANALYSIS OF SHELLS OF REVOLUTION BY MINIMIZATION OF THE POTENTIAL ENERGY FUNCTIONAL	69-0114
SHELLS	FINITE ELEMENT ANALYSIS OF THIN SHELLS (DISCUSSION)	69-0187
SHELLS	FINITE ELEMENT ANALYSIS OF SKEWED SHALLOW SHELLS	69-0211
SHELLS	FINITE ELEMENT ANALYSIS OF SHELLS OF REVOLUTION BY MINIMIZATION OF THE POTENTIAL ENERGY FUNCTIONAL	69-0354
SHELLS	FINITE ELEMENT ANALYSIS OF CYLINDRICAL SHELLS. SOLUTION USING RECTANGULAR CIRCULAR CYLINDRICAL ELEME	70-0178
SHELLS	FINITE ELEMENT ANALYSIS OF ARBITRARY THIN SHELLS	70-0329
SHELLS	FINITE ELEMENT ANALYSIS OF THIN SHELLS	70-0457
SHELLS	FINITE ELEMENT ANALYSIS OF SHELLS OF REVOLUTION BY MINIMIZATION OF THE POTENTIAL ENERGY FUNCTIONAL	70-0481
SHELLS	FINITE ELEMENT ANALYSIS OF AXISYMMETRICAL SHELLS OF COMPOSITE MATERIAL	70-0483
SHELLS	FINITE ELEMENT ANALYSIS OF HYPERBOLIC PARABOLOID SHELLS	71-0245
SHELLS	FINITE ELEMENT ANALYSIS OF CYLINDRICAL SHELLS UTILIZING CURVED ELEMENTS CONTAINING RIGID BODY DEFORM	71-0823
SHELLS	FINITE ELEMENT ANALYSIS OF PERFORATED SHELLS	72-0228
SHELLS	FINITE ELEMENT ANALYSIS OF TRANSLATIONAL SHELLS	72-0348
SHELLS	FINITE ELEMENT ANALYSIS OF PLATES AND SHELLS	72-0409
SHELLS	FINITE ELEMENT ANALYSIS OF FINITE DEFORMATIONS OF PLATES, SHELLS, AND MEMBRANES	72-0578
SHELLS	FINITE ELEMENT ANALYSIS OF HYPERBOLIC PARABOLOID SHELLS	72-0887
SHELLS	FINITE ELEMENT ANALYSIS OF INFLATABLE SHELLS	73-0241
SHELLS	FINITE ELEMENT ANALYSIS OF STATIC AND DYNAMIC STRESSES AROUND SUDDENLY PUNCHED HOLES IN PLATES AND S	73-1088
SHELLS	FINITE ELEMENT ANALYSIS OF ROTATING SHELLS	74-0096
SHELLS	FINITE ELEMENT ANALYSIS OF AXISYMMETRIC MULTILAYER SANDWICH PLATES AND SHELLS	74-0101
SHELLS	FINITE ELEMENT ANALYSIS OF ELASTO-PLASTIC SHELLS	74-0249
SHELLS	FINITE ELEMENT ANALYSIS OF MULTILAYER SANDWICH PLATES AND SHELLS	74-0425
SHELLS	FINITE ELEMENT ANALYSIS OF INFLATABLE SHELLS	74-0559
SHELLS	FINITE ELEMENT ANALYSIS OF VIBRATIONS OF INITIALLY STRESSED THIN SHELLS OF REVOLUTION	74-1056
SHELLS	FINITE ELEMENT ANALYSIS OF THIN ELASTIC SHELLS WITH RESIDUAL ENERGY BALANCING AND THE ROLE OF THE RI	75-0034
SHELLS	FINITE ELEMENT LARGE DEFLECTION ANALYSIS OF ELASTIC-PLASTIC SHELLS OF REVOLUTION SUBJECTED TO AXISYM	72-0137
SHELLS	FINITE ELEMENT LARGE DEFLECTION ANALYSIS OF SHALLOW SHELLS	75-0356
SHELLS	FINITE ELEMENT METHOD OF ANALYSIS FOR ARCH DAMS SHELLS AND COMPARISON WITH FINITE DIFFERENCE PROCEDU	65-0036
SHELLS	FINITE ELEMENT METHOD FOR THE ANALYSIS OF NOZZLE OPENINGS IN SHELLS OF REVOLUTION	69-0254
SHELLS	FINITE ELEMENT MODELS FOR SHELLS OF REVOLUTION	68-0027
SHELLS	FINITE ELEMENT MODEL FOR SHELLS BASED ON THE DISCRETE KIRCHHOFF HYPOTHESIS.	72-0388
SHELLS	FINITE ELEMENT NONLINEAR ANALYSIS FOR PLATES AND SHALLOW SHELLS	68-0216
SHELLS	FINITE ELEMENT SOLUTION FOR THIN SHELLS OF REVOLUTION	63-0024
SHELLS	FINITE ELEMENT SOLUTION FOR THIN SHELLS OF REVOLUTION	64-0005
SHELLS	FINITE ELEMENT SOLUTION FOR AXISYMMETRIC SHELLS	64-0010
SHELLS	FINITE ELEMENT SOLUTION FOR AXISYMMETRIC SHELLS	65-0023
SHELLS	FINITE ELEMENT SOLUTION OF AXISYMMETRICAL DYNAMIC PROBLEMS OF SHELLS OF REVOLUTION	66-0020
SHELLS	FINITE ELEMENT STRESS ANALYSIS OF THIN SHELLS	65-0024
SHELLS	FINITE ELEMENT THEORY OF PLATES AND SHELLS INCLUDING TRANSVERSE SHEAR STRAIN EFFECTS	70-0395
SHELLS	FINITE ELEMENTS, FINITE ROTATIONS AND SMALL STRAINS OF FLEXIBLE SHELLS	69-0516
SHELLS	FINITE-ELEMENT ANALYSIS OF THIN SHELLS	68-0164
SHELLS	FINITE-ELEMENT ANALYSIS FOR FREE VIBRATION OF GENERAL ANISOTROPIC LAMINATED THIN SHELLS	72-0618
SHELLS	FINITE-ELEMENT ANALYSIS OF SANDWICH SHELLS	73-1130
SHELLS	FINITE-ELEMENT ANALYSIS OF THIN ELASTIC SHELLS WITH RESIDUAL ENERGY BALANCING AND ROLE OF RIGID BODY	75-0048
SHELLS	FINITE-ELEMENT ANALYSIS OF MODERATELY THICK SHELLS	75-0588
SHELLS	FINITE-ELEMENT FOR BUCKLING OF CURVED BEAMS AND SHELLS WITH SHEAR	75-0607
SHELLS	FINITE-ELEMENT PREDICTIONS OF TRANSIENT ELASTIC-PLASTIC LARGE DEFLECTIONS OF STIFFENED AND/OR UNSTIF	74-0825
SHELLS	FREE VIBRATION AND RESPONSE TO RANDOM PRESSURE FIELD OF ANISOTROPIC THIN CYLINDRICAL SHELLS	74-0830
SHELLS	FREE VIBRATION OF CYLINDRICAL SHELLS PARTIALLY FILLED WITH LIQUID	71-0839
SHELLS	FREE VIBRATION OF RING STIFFENED CYLINDRICAL SHELLS	70-0172
SHELLS	FREE VIBRATION OF SHELLS OF REVOLUTION USING RING ELEMENTS	67-0088
SHELLS	FREE VIBRATION OF SHELLS OF REVOLUTION USING FINITE ELEMENT METHOD	74-0232
SHELLS	FREE VIBRATION OF THIN CYLINDRICAL SHELLS WITH A DISCONTINUITY IN THE THICKNESS	69-0439
SHELLS	FREE VIBRATIONS OF CYLINDRICAL SHELLS WITH VARIOUS END CLOSURES	74-0571

SHELLS	FREE VIBRATIONS OF PRESSURE LOADED PARABOLOIDAL SHELLS	71-0601
SHELLS	FURTHER APPLICATION OF QUADRATIC ISOPARAMETRIC FINITE-ELEMENTS TO LINEAR FRACTURE MECHANICS OF PLATE	75-0639
SHELLS	FURTHER DEVELOPMENTS IN THE USE OF LINEAR DISPLACEMENT FUNCTIONS IN THE FINITE ELEMENT ANALYSIS OF O	71-0344
SHELLS	GENERALIZATION OF PLATE FINITE ELEMENTS TO SHELLS	72-0171
SHELLS	GEOMETRICKY NELINEARNE ULOHY PLOCHYCH SKRUPIN RIESENE PRIRASTKOVOU METODOU KONECNYCH PRVKOV. (GEOMET	74-1032
SHELLS	HIGH PRECISION FINITE ELEMENT ANALYSIS OF SHELLS OF REVOLUTION	71-0524
SHELLS	HIGH-PRECISION FINITE-ELEMENT ANALYSIS OF SHELLS OF REVOLUTION	73-1164
SHELLS	HYBRIDE SCHALENELEMENTE MIT EINER ANWENDUNG AUF WENDELSCHALEN. (HYBRID SHELL ELEMENTS IN APPLICATION	75-0609
SHELLS	IMPROVEMENTS ON THE ANALYSIS OF SHELLS OF REVOLUTION BY THE MATRIX DISPLACEMENT METHOD	66-0070
SHELLS	INCREMENTAL ANALYSIS OF LARGE DEFLECTIONS OF SHELLS OF REVOLUTION	71-0296
SHELLS	INCREMENTAL VARIATIONAL METHOD FOR THE LARGE DISPLACEMENT ANALYSIS OF SHELLS WITH GEOMETRIC IMPERFEC	72-0894
SHELLS	INEXTENSIONAL VIBRATIONS OF PARABOLOIDAL SHELLS OF REVOLUTION. DOMES AND SHELLS	71-0427
SHELLS	INEXTENSIONAL VIBRATIONS OF PARABOLOIDAL SHELLS OF REVOLUTION. DOMES AND SHELLS	71-0427
SHELLS	INSTABILITY OF THIN SHELLS BY THE FINITE ELEMENT METHOD	70-0187
SHELLS	INSTABILITY OF THIN SHELLS BY THE FINITE ELEMENT METHOD	70-0477
SHELLS	INTERFACE CORE DISCRETE-ELEMENT FOR STIFFENED SHELLS OF REVOLUTION	67-0092
SHELLS	INVESTIGATION OF THE STRESS-STRAINED STATE OF SPHERICAL SHELLS WITH AN ECCENTRIC CUTOUT ON THE BASIS	73-0222
SHELLS	ISSLEDOVANIE NAPRYAZHENNOGO SOSTOYANIYA REBRISTYKH ISILINDRICHESKIKH OBOLOCHEK S PRYAMOUGOL'NYMI OTV	74-0944
SHELLS	LAMINATED ORTHOTROPIC PLATES AND SHALLOW SHELLS	71-0400
SHELLS	LARGE DEFLECTION ANALYSIS OF ELASTIC-PLASTIC SHELLS OF REVOLUTION	69-0174
SHELLS	LARGE DEFLECTION ANALYSIS OF ELASTIC-PLASTIC PLATES AND SHELLS	69-0178
SHELLS	LARGE DEFLECTION ANALYSIS OF ELASTIC-PLASTIC SHELLS OF REVOLUTION	70-0507
SHELLS	LARGE DEFLECTION ANALYSIS OF ARBITRARY SHALLOW SHELLS BY THE FINITE ELEMENT METHOD	73-0937
SHELLS	LARGE DEFLECTION ANALYSIS OF PLATES AND SHALLOW SHELLS USING THE FINITE ELEMENT METHOD	73-0989
SHELLS	LARGE DEFLECTION DYNAMIC ANALYSIS OF THIN SHELLS USING FINITE ELEMENT METHOD	70-0379
SHELLS	LARGE DEFLECTION ELASTIC-PLASTIC ANALYSIS OF SHELLS OF REVOLUTION	71-0263
SHELLS	LARGE DEFLECTION ELASTIC-PLASTIC DYNAMIC RESPONSE OF STIFFENED SHELLS OF REVOLUTION	72-0608
SHELLS	LARGE DEFLECTION ELASTIC-PLASTIC DYNAMIC RESPONSE OF STIFFENED SHELLS OF REVOLUTION	74-0496
SHELLS	LARGE DISPLACEMENT ANALYSIS OF ASIXYMMETRIC SHELLS	69-0276
SHELLS	LES PLAQUES ET LES COQUES MINCES OU EPAISSES (THIN AND THICK PLATES AND SHELLS)	71-0432
SHELLS	LINEAR AND NON LINEAR ANALYSIS OF SHELLS OF REVOLUTION WITH ASYMMETRICAL STIFFNESS PROPERTIES	68-0151
SHELLS	LINEAR AND NON-LINEAR ANALYSIS OF THIN SHALLOW SHELLS BY MIXED FINITE ELEMENTS	75-0665
SHELLS	LINEAR AND NONLINEAR STATIC ANALYSIS OF AXISYMMETRICALLY LOADED THIN SHELLS OF REVOLUTION	69-0215
SHELLS	LINEAR AND NONLINEAR ANALYSIS OF SHELLS	72-0832
SHELLS	LINEAR DYNAMIC ANALYSES OF LAMINATED PLATES AND SHELLS BY THE HYBRID-STRESS FINITE-ELEMENT METHOD	73-0786
SHELLS	LUMPED PARAMETER ELEMENT FOR THE ANALYSIS OF HYPERBOLIC PARABOLOID SHELLS	72-0148
SHELLS	MATERIAL BEHAVIOUR CHARACTERISTICS FOR REINFORCED CONCRETE SHELLS STRESSED BEYOND THE ELASTIC RANGE	70-0166
SHELLS	MATRIX ANALYSIS OF AXISYMMETRIC SHELLS UNDER GENERAL LOADING	64-0075
SHELLS	MATRIX ANALYSIS OF MULTILAYERED AND SANDWICH SHELLS BY FINITE ELEMENT METHOD	73-0101
SHELLS	MATRIX ANALYSIS OF SHELLS - SMALL AND LARGE DISPLACEMENTS	68-0016
SHELLS	MATRIX ANALYSIS OF THIN SHELLS USING FINITE ELEMENTS	69-0527
SHELLS	MATRIX DISPLACEMENT ANALYSIS OF ANISOTROPIC SHELLS BY TRIANGULAR ELEMENTS	65-0005
SHELLS	MATRIX DISPLACEMENT ANALYSIS OF PLATES AND SHELLS	66-0010
SHELLS	MATRIX DISPLACEMENT ANALYSIS OF SHELLS WITH AN AXISYMMETRIC MIDSURFACE BUT HAVING CIRCUMFERENTIALLY	68-0031
SHELLS	MATRIX DISPLACEMENT ANALYSIS OF PLATES AND SHELLS. PART 11. GENERAL FORMULATION OF THE LINEAR THEORY	71-0621
SHELLS	MATRIX DISPLACEMENT ANALYSIS OF SHELLS AND PLATES INCLUDING TRANSVERSE SHEAR STRAIN EFFECTS	72-0643
SHELLS	METOD KONECHNYKH ELEMENTOV V PRIMENENII K RASCHETU TSILINDRICHESKIKH OBOLOCHEK S PRYAMOUGOL'NYMI OTV	73-0536
SHELLS	MIXED FINITE ELEMENT METHOD FOR AXISYMMETRIC SHELLS	72-0149
SHELLS	MIXED FINITE ELEMENT METHOD FOR VIBRATION ANALYSIS OF SHALLOW SHELLS	73-0593
SHELLS	NATURAL GEOMETRY OF SURFACE WITH SPECIFIC REFERENCE TO THE MATRIX DISPLACEMENT ANALYSIS OF SHELLS. 1	73-0709
SHELLS	NATURAL VIBRATIONS OF LAMINATED ORTHOTROPIC SHELLS OF REVOLUTION	70-0367
SHELLS	NEW CONCEPTS FOR FINITE ELEMENTS OF SHELLS	68-0195
SHELLS	NEW DEVELOPMENTS IN THE FINITE ELEMENT ANALYSIS OF SHELLS	69-0334
SHELLS	NEW FINITE ELEMENT METHOD FOR ANALYZING SYMMETRICALLY LOADED THIN SHELLS OF REVOLUTION	73-0389
SHELLS	NEW FINITE-ELEMENT SCHEME FOR INSTABILITY ANALYSIS OF THIN SHELLS	75-0357
SHELLS	NEW VARIATIONAL PRINCIPLE WITH APPLICATION TO VIBRATIONS OF SHALLOW SHELLS USING FINITE ELEMENTS MET	73-0354
SHELLS	NON-LINEAR DYNAMIC ANALYSIS OF AXISYMMETRIC SHELLS	75-0474
SHELLS	NONCONFORMING FINITE ELEMENT ANALYSIS OF SHELLS	75-0539
SHELLS	NONLINEAR ANALYSIS OF SHELLS OF REVOLUTION BY THE MATRIX DISPLACEMENT METHOD	67-0079
SHELLS	NONLINEAR ANALYSIS OF ELASTIC SHELLS OF REVOLUTION UNDER AXISYMMETRIC LOADING	70-0161
SHELLS	NONLINEAR ANALYSIS OF BEAMS AND SHELLS	73-1074
SHELLS	NONLINEAR ANALYSIS OF ARBITRARY, HYPERELASTIC MEMBRANE SHELLS BY THE FINITE-ELEMENT METHOD	73-1132
SHELLS	NONLINEAR ANALYSIS OF RC SHELLS OF GENERAL FORM	75-0053
SHELLS	NONLINEAR BEHAVIOR OF SHELLS OF REVOLUTION UNDER CYCLIC LOADING	73-0242
SHELLS	NONLINEAR BUCKLING ANALYSIS OF COMPOSITE SHELLS	74-0973
SHELLS	NONLINEAR BUCKLING ANALYSIS OF COMPOSITE SHELLS	75-0214
SHELLS	NONLINEAR DYNAMIC ANALYSIS OF SHELLS OF REVOLUTION BY MATRIX DISPLACEMENT METHOD	70-0110
SHELLS	NONLINEAR DYNAMIC ANALYSIS OF SHELLS OF REVOLUTION WITH ASYMMETRIC PROPERTIES BY THE FINITE ELEMENT	75-0217
SHELLS	NONLINEAR DYNAMIC ANALYSIS OF SHELLS OF REVOLUTION WITH ASYMMETRIC PROPERTIES BY THE FINITE ELEMENT	75-0657
SHELLS	NONLINEAR DYNAMIC ANALYSIS OF TORSIONAL SHELLS BY COMBINED USE OF FINITE ELEMENT AND MODE SUPERPOSIT	75-0859
SHELLS	NONLINEAR FINITE ELEMENT ANALYSIS OF SANDWICH SHELLS OF REVOLUTION	72-0180
SHELLS	NONLINEAR FINITE ELEMENT ANALYSIS OF LAMINATED COMPOSITE SHELLS	74-0751
SHELLS	NONLINEAR FINITE-ELEMENT ANALYSIS OF THIN SHELLS	73-1169
SHELLS	NONLINEAR HYBRID STRESS FINITE ELEMENT ANALYSIS OF LAMINATED SHELLS	74-1350
SHELLS	NONLINEAR LAYERED ANALYSIS OF RC PLATES AND SHELLS	73-0208
SHELLS	NONLINEAR LAYERED FINITE-ELEMENT ANALYSIS OF REINFORCED-CONCRETE PLATES AND SHELLS	72-0946
SHELLS	NONLINEAR TRANSIENT ANALYSIS OF SHELLS AND SOLIDS OF REVOLUTION BY CONVECTED ELEMENTS	73-0519
SHELLS	NONLINEAR TRANSIENT ANALYSIS OF SHELLS AND SOLIDS OF REVOLUTTION BY CONVECTED ELEMENTS	74-0476
SHELLS	NUMERICAL ANALYSIS OF THIN SHELLS BY CURVED TRIANGULAR ELEMENTS BASED ON DISCRETE - KIRCHHOFF HYPOTH	69-0091
SHELLS	ON DERIVATION OF STIFFNESS MATRICES WITH CO ROTATION FIELDS FOR PLATES AND SHELLS	71-0254
SHELLS	ON SHELLS OF CONSTANT STRENGTH	72-0431
SHELLS	ON THE DYNAMIC BUCKLING OF SHELLS OF REVOLUTION	73-1102
SHELLS	OPTIMUM DESIGN OF COMPOSITE SHELLS SUBJECT TO NATURAL FREQUENCY CONSTRAINTS	73-0154
SHELLS	PAPA, STRUCTURAL ANALYSIS OF PLATES AND SHELLS USING TRAPEZOIDAL AND TRIANGULAR PLATE ELEMENTS	67-0059
SHELLS	PART 1 - TASK 3.2.9. BENDING AND BUCKLING OF A ONE-DIMENSIONAL MODEL SIMULATING A CONICAL SHELL. PAR	71-0745
SHELLS	PENETRATION IN SHELLS UNDER EXTERNAL PRESSURE	70-0237
SHELLS	PIECE-WISE CONTINUOUS FINITE ELEMENT FOR THE STRESS ANALYSIS OF PLATES AND SHELLS	75-0758
SHELLS	PLASTIC ANALYSIS OF SHELLS BY FINITE ELEMENT METHODS	73-1017
SHELLS	PLASTIC AND VISCOPLASTIC ANALYSIS OF AXISYMMETRIC SHELLS	75-0244
SHELLS	PLATES AND SHELLS, INELASTIC ANALYSIS. TIME DEPENDENT INELASTICITY, FIVE LECTURES	73-0633
SHELLS	PRISMATIC SHELLS WITH HEXAGONAL CROSS-SECTION	74-1043
SHELLS	PROGRESS REPORT ON DISCRETE-ELEMENT ELASTIC AND ELASTIC-PLASTIC ANALYSIS OF SHELLS OF REVOLUTION SUB	68-0170
SHELLS	QUASI-ANALYTICAL FINITE ELEMENT PROCEDURES FOR AXISYMMETRIC ANISOTROPIC SHELLS AND SOLIDS	74-0297
SHELLS	REDUCED INTEGRATION TECHNIQUE IN GENERAL ANALYSIS OF PLATES AND SHELLS	71-0374
SHELLS	REFINED FINITE ELEMENT ANALYSIS OF ELASTIC-PLASTIC THIN SHELLS OF REVOLUTION	69-0497
SHELLS	REFINED MIXED METHOD FINITE ELEMENTS FOR SHELLS OF REVOLUTION	71-0101
SHELLS	REINVESTIGATION OF BUCKLING OF SHELLS OF REVOLUTION BY A REFINED FINITE ELEMENT	74-0097
SHELLS	RIGID BODY DISPLACEMENTS OF CURVED ELEMENTS IN THE ANALYSIS OF SHELLS BY THE MATRIX-DISPLACEMENT MET	67-0031
SHELLS	SAMMSOR 111: A FINITE ELEMENT PROGRAM TO DETERMINE STIFFNESS AND MASS MATRICES OF RING-STIFFENED SHE	72-0774
SHELLS	SHELLS, PLATES AND MEMBRANES - A COMPUTER-AIDED DESIGN SYSTEM	74-1159
SHELLS	SIMPLIFIED HYBRID DISPLACEMENT METHOD FOR LINEAR FINITE ELEMENT ANALYSIS OF GENERAL SHELLS	74-0614
SHELLS	SLADE D: A COMPUTER PROGRAM FOR THE DYNAMIC ANALYSIS OF THIN SHELLS	73-0964

SHELLS	SLADE, A COMPUTER PROGRAM FOR THE STATIC ANALYSIS OF THIN SHELLS	70-0079
SHELLS	SNAP-THROUGH BUCKLING ANALYSIS OF SHELLS OF REVOLUTION USING INCREMENTAL STIFFNESS MATRICES	74-1349
SHELLS	SNAP-THROUGH BUCKLING OF RETICULATED SHELLS	75-0640
SHELLS	SOME TWO-FIELD VARIATIONAL PRINCIPLES FOR NONLINEAR DEFORMATION ANALYSIS OF SHELLS	75-0294
SHELLS	STABILITY ANALYSIS OF SHELLS OF REVOLUTION BY THE FINITE ELEMENT METHOD	68-0124
SHELLS	STABILITY ANALYSIS OF SHELLS OF REVOLUTION	68-0136
SHELLS	STABILITY ANALYSIS OF AXISYMMETRIC SHELLS	69-0131
SHELLS	STABILITY OF HYPERBOLOIDAL SHELLS	75-0488
SHELLS	STABILITY OF PRESSURIZED HYPERBOLOIDAL SHELLS	75-0585
SHELLS	STACUSS 1: A DISCRETE-ELEMENT PROGRAM FOR THE STATIC ANALYSIS OF SINGLE-LAYER CURVED STIFFENED SHELL	69-0160
SHELLS	STATIC ANALYSIS OF SHELLS OF REVOLUTION USING DOUBLY-CURVED QUADRILATERAL ELEMENTS DERIVED FROM ALTE	69-0368
SHELLS	STATIC ANALYSIS OF STIFFENED SHELLS BY THE FINITE ELEMENT METHOD	71-0803
SHELLS	STATIC AND DYNAMIC ANALYSIS OF VISCOELASTOPLASTIC FIBER-REINFORCED COMPOSITE SHELLS IN MISSILE STRUC	73-0651
SHELLS	STATIC, STABILITY, AND DYNAMIC ANALYSIS OF SHELLS OF REVOLUTION BY NUMERICAL INTEGRATION - A COMPARI	74-0389
SHELLS	STATIC, VIBRATION, AND THERMAL STRESS ANALYSES OF LAMINATED PLATE AND SHELLS BY THE HYBRID-STRESS FI	72-0602
SHELLS	STATICS AND DYNAMICS OF HYDROSTATICALLY LOADED SHELLS BY FINITE ELEMENT METHOD	71-0044
SHELLS	STIFFNESS ANALYSIS OF SHEET METAL SHELLS UNDER CONCENTRATED LOADS	74-0325
SHELLS	STRESS CONCENTRATION FACTORS FOR CIRCULAR, REINFORCED PENETRATIONS IN PRESSURIZED CYLINDRICAL SHELLS	75-0726
SHELLS	STRESS DISTRIBUTIONS IN THIN-WALLED INTERSECTING CYLINDRICAL SHELLS SUBJECTED TO INTERNAL PRESSURE A	71-0023
SHELLS	STRESS RESULTANTS IN SHELLS BY AN AUGMENTED FINITE ELEMENT METHOD	73-0585
SHELLS	STRESS-WAVE PROPAGATION AND DYNAMIC EFFECTS IN SHELLS (SPADES) MICROMECHANICAL STRESS-WAVE RESPONSE	72-0851
SHELLS	STRESS, STABILITY AND VIBRATION OF COMPLEX SHELLS OF REVOLUTION: ANALYSIS AND USERS MANUAL FOR BOSOR	69-0441
SHELLS	STRESSES AND DEFLECTIONS IN COOLING TOWER SHELLS DUE TO WIND LOADING	68-0088
SHELLS	STRESSES IN GABLE ROOF H. P. SHELLS.	72-0394
SHELLS	STRUCTURAL RESPONSE OF UNSTIFFENED TOROIDAL SHELLS	69-0198
SHELLS	TACOS: A FINITE ELEMENT COMPUTER PROGRAM FOR THE TRANSIENT ANALYSIS OF CYLINDRICAL OBESE SHELLS	72-0588
SHELLS	THE ANALYSIS OF MODERATELY THICK TO THIN SHELLS BY THE FINITE ELEMENT METHOD	70-0315
SHELLS	THE ANALYSIS OF THIN SHELLS BY A FINITE ELEMENT PROCEDURE	67-0123
SHELLS	THE ANALYSIS OF THIN SHELLS WITH TRANSVERSE SHEAR STRAINS BY THE FINITE ELEMENT METHOD	68-0098
SHELLS	THE ANALYSIS OF THIN SHELLS BY THE FINITE ELEMENT METHOD	70-0078
SHELLS	THE APPLICATION OF FINITE ELEMENTS TO THE LARGE DEFLECTION GEOMETRICALLY NONLINEAR BEHAVIOR OF CYLIN	72-0815
SHELLS	THE APPLICATION OF THE FINITE ELEMENT METHOD TO THE ANALYSIS OF FRACTURE OF CYLINDRICAL SHELLS	73-0728
SHELLS	THE APPROXIMATE ANALYSIS OF THIN SHELLS BY THE FINITE ELEMENT METHOD	66-0131
SHELLS	THE DEVELOPMENT OF APPLIED METHODS IN PROBLEMS OF STATIC CALCULATION OF THIN-WALLED THREE-DIMENSIONA	73-0934
SHELLS	THE EFFECT OF AN INITIAL AXISYMMETRIC IMPERFECTION ON THE NATURAL FREQUENCIES OF CONICAL SHELLS	70-0368
SHELLS	THE FINITE ELEMENT ANALYSIS OF THIN ELASTIC SHELLS	73-1076
SHELLS	THE FINITE ELEMENT DYNAMIC STABILITY ANALYSIS OF THIN SHELLS SUBMERGED IN FLUIDS	71-0057
SHELLS	THE FINITE ELEMENT METHOD APPLIED TO THIN SHELLS AND BOX STRUCTURES	70-0444
SHELLS	THE LINEAR ELASTIC DYNAMIC ANALYSIS OF SHELLS OF REVOLUTION BY THE MATRIX DISPLACEMENT METHOD	66-0046
SHELLS	THE NONLINEAR DYNAMIC ANALYSIS OF SHELLS OF REVOLUTION WITH ASYMMETRIC PROPERTIES BY THE FINITE ELEM	71-0818
SHELLS	THE RESPONSE OF SHELLS TO DISTRIBUTED RANDOM LOADS USING NASTRAN	71-0134
SHELLS	THE STRESS ANALYSIS OF AXISYMMETRIC SHELLS USING A PLANE FRAME ANALYSIS PROGRAM	69-0238
SHELLS	THE TRANSIENT DYNAMIC ANALYSIS OF THIN SHELLS BY THE FINITE ELEMENT METHODS	71-0142
SHELLS	THE TRANSIENT LINEAR ELASTIC RESPONSE ANALYSIS COMPLEX THIN SHELLS OF REVOLUTION SUBJECTED TO ARBITR	70-0080
SHELLS	THE USE OF FINITE ELEMENT METHODS FOR STATIC AND DYNAMIC ANALYSES OF COMPLEX SHELLS	68-0247
SHELLS	THEORETICAL FORMULATION OF FINITE ELEMENT METHODS IN LINEAR ELASTIC ANALYSIS OF GENERAL SHELLS	72-0496
SHELLS	THERMAL STRESS ANALYSIS OF PLATES AND SHALLOW SHELLS BY HYBRID FINITE-ELEMENT METHOD	74-0485
SHELLS	THERMAL STRESSES IN OPEN CYLINDRICAL SHELLS OF ARBITRARY BOUNDARY CONDITIONS BY A FINITE STRIP METHO	72-0371
SHELLS	THERMAL STRESSES IN THIN CYLINDRICAL SHELLS	73-0822
SHELLS	TRANSIENT ANALYSIS OF THICK CYLINDRICAL SHELLS	71-0590
SHELLS	TREUGOL'NY I ELEMENT DLYA KRUGOVOI TSILINDRICHESKOI OBOLOCHKI (TRIANGULAR FINITE ELEMENT FOR CIRCULA	73-0493
SHELLS	TRIANGULAR FINITE ELEMENT FOR CYLINDRICAL SHELLS	70-0488
SHELLS	UNSYMMETRIC BUCKLING OF SHELLS OF REVOLUTION BY FINITE ELEMENT METHOD	70-0500
SHELLS	USE OF FINITE ELEMENT METHOD IN THE DYNAMICS OF PLATES AND SHELLS	70-0313
SHELLS	USE OF HIGHER ORDER DISPLACEMENT FUNCTIONS IN THE FREE VIBRATION ANALYSIS OF SHELLS OF REVOLUTION HA	74-1055
SHELLS	USE OF NONCONFORMING MODES IN FINITE ELEMENT ANALYSIS OF PLATES AND SHELLS	73-0865
SHELLS	VARIATIONAL FORMULATIONS AND FINITE ELEMENT METHOD FOR THIN ELASTIC SHELLS AND PLATES	68-0064
SHELLS	VIBRATION ANALYSIS OF LAMINATED PLATES AND SHELLS BY A HYBRID STRESS ELEMENT	73-0120
SHELLS	VIBRATION ANALYSIS OF PLATES, SHELLS AND BLADED RIMMED DISCS USING THE TRANSFER-MATRIX-FINITE-ELEMEN	73-1151
SHELLS	VIBRATION CHARACTERISTICS OF THIN, PRESSURIZED SHELLS OF REVOLUTION PARTIALLY FILLED WITH LIQUID	72-0604
SHELLS	VIBRATION CHARACTERISTICS OF FREE THIN CYLINDRICAL SHELLS	73-0396
SHELLS	VIBRATION OF CYLINDRICAL SHELLS BY HYBRID FINITE ELEMENT METHOD	72-0253
SHELLS	VIBRATION OF CYLINDRICAL SHELLS BY A HYBRID FINITE ELEMENT METHOD	72-0795
SHELLS	VIBRATION OF THICK, CURVED, SHELLS WITH PARTICULAR REFERENCE TO TURBINE BLADES	70-0017
SHELLS	VIBRATIONAL ANALYSIS OF GENERAL SHELLS UTILIZING FINITE-ELEMENT TECHNIQUES	73-1133
SHELLS	VIBRATIONS OF PRESSURE LOADED HYPERBOLOIDAL SHELLS	72-0874
SHELLS	VIBRATIONS OF SEGMENTED SHELLS	73-0161
SHELLS	2-DIMENSIONAL FINITE-ELEMENT ANALYSIS OF ELECTROELASTIC NORMAL MODES OF FERROELECTRIC DISKS AND SHEL	75-0594
SHIFTING	SHIFTING ROLES OF COMPUTERS, EXPERIMENTS AND ANALYSIS IN APPLIED MECHANICS	71-0604
SHIP	A GENERAL CALCULATION OF THE STRUCTURAL LOADS ON A SHIP IN A SEAWAY	71-0345
SHIP	A NUMERICAL APPROACH TO THE STUDY OF IRREGULAR SHIP MOTIONS	74-0030
SHIP	A TORSIONAL STRENGTH ANALYSIS ON THE CONTAINER SHIP BY MEANS OF THE FINITE ELEMENT PROCEDURE AND FUL	71-0243
SHIP	ACCURACY AND OPTIMIZATION IN THE ANALYSIS OF SHIP STRUCTURES, USING FINITE ELEMENT METHODS	71-0418
SHIP	AN INTERACTIVE GRAPHICS SHIP DESIGN ORIENTED PROGRAM PACKAGE FOR TIME SHARING SYSTEMS	73-0341
SHIP	ANALYSIS FOR CRACKS IN SHIP STRUCTURE USING THE FINITE ELEMENT METHOD WITH THE CONCEPT OF CRACK SUB-	74-0248
SHIP	ANALYSIS FOR CRACKS IN SHIP STRUCTURES USING THE FINITE-ELEMENT METHOD WITH THE CONCEPT OF CRACK SUB	74-1367
SHIP	ANALYSIS OF LARGE CLOSED SHIP STRUCTURES: APPLICATIONS TO TANKER STRUCTURAL DESIGN	70-0289
SHIP	ANALYSIS OF SHIP STRUCTURES BY THE FINITE ELEMENT METHOD	69-0336
SHIP	APPLICATION OF FINITE ELEMENT COMPUTATIONS TO SOME SHIP STRUCTURAL PROBLEMS	69-0337
SHIP	APPLICATION OF FINITE ELEMENT METHOD TO ADDED VIRTUAL MASS OF SHIP HULL VIBRATION	70-0463
SHIP	APPLICATION OF FINITE ELEMENT METHODS TO SHIP STRUCTURES	73-0429
SHIP	APPLICATION OF THE FINITE ELEMENT METHOD TO SHIP STRUCTURES	71-0403
SHIP	APPLICATION OF THE FINITE ELEMENT METHOD TO THE CALCULATION OF SHIP HULLS	71-0431
SHIP	APPLICATION OF THE METHOD OF FINITE ELEMENTS TO THE CALCULATION OF SHIP STRUCTURES (ISPOLZOVANIE MET	73-0933
SHIP	APPLICATIONS OF FINITE ELEMENT METHOD TO SHIP STRUCTURAL ANALYSIS	69-0210
SHIP	DATA GENERATION FOR FINITE-ELEMENT STRUCTURAL ANALYSIS OF THREE-DIMENSIONAL NAVAL SHIP STRUCTURES	75-0707
SHIP	DATA GENERATOR FOR THE IDEALIZATION FOR FINITE ELEMENT STRUCTURAL ANALYSIS OF NAVAL SHIP FLAT PLATED	72-0860
SHIP	DESIGN OF NEW SHIP TYPES BY MEANS OF FINITE ELEMENT METHOD	72-0562
SHIP	FINITE ELEMENT ANALYSIS OF SHIP STRUCTURES	69-0226
SHIP	FINITE ELEMENT ANALYSIS OF THE FORCED OSCILLATION OF SHIP HULL FORMS	74-0814
SHIP	FINITE ELEMENT TECHNIQUE IN SHIP VIBRATION ANALYSIS	68-0180
SHIP	FINITE ELEMENT TECHNIQUES IN SHIP STRUCTURE DESIGN IN FINITE ELEMENT TECHNIQUES	69-0335
SHIP	FUNDAMENTAL STUDIES ON STEADY SHIP WAVE PROBLEMS BY THE FINITE ELEMENT METHOD	74-0564
SHIP	HYBRYDOWA METODA ELEMENTOW SKONCZONYCH W ZASTOSOWANIU DO OBLICZEN ORGAN URZADZEN OKRETOWYCH. (HYBRID	74-0963
SHIP	IMPACT OF FINITE ELEMENT TECHNIQUES ON PRACTICAL DESIGN OF SHIP STRUCTURES	59-0344
SHIP	MATRIX METHOD OF ANALYSIS OF SHIP STRUCTURES 3RD REPORT	68-0225
SHIP	METHOD SZTYWNYCH ELEMENTOW SKONCZONYCH W OBLICZENIACH KONSTRUCJI OKRETOWYCH. (STIFF FINITE ELEMENT M	74-0923
SHIP	MODEL EXPERIMENTS AND FINITE ELEMENT ANALYSIS OF STRESS IN AN OPEN SHIP	66-0054
SHIP	NUMERICAL ANALYSIS OF SHIP STRUCTURAL ELEMENTS BY MEANS OF FINITE ELEMENT METHOD AND PHOTO-ELASTIC E	69-0125
SHIP	NUMERICAL HULL VIBRATION ANALYSIS OF A FAR EAST CONTAINER SHIP	74-1325
SHIP	PRACTICAL CONSIDERATIONS IN THE APPLICATION OF FINITE ELEMENT TECHNIQUES TO SHIP STRUCTURES	69-0338
SHIP	PROCEEDINGS OF THE NAVY-NASTRAN COLLOQUIUM (4TH) HELD AT NAVY SHIP RESEARCH AND DEVELOPMENT CENTER,	73-0888

369

SIMPLE	SIMPLE APPROACH TO THE STRENGTH ANALYSIS OF TANKERS	71-0319
SIMPLE	SIMPLE EQUATIONS FOR THE MAGNETIZATION AND RELUCTIVITY CURVES OF STEEL	75-0156
SIMPLE	SIMPLE FINITE ELEMENT METHOD FOR STRESS CONCENTRATION PROBLEMS IN PRESSURE VESSELS	71-0669
SIMPLE	SIMPLE FINITE ELEMENT MODEL FOR ELASTIC-PLASTIC PLATE BENDING.	72-0447
SIMPLE	SIMPLE FINITE ELEMENTS FOR PRE-TWISTED BLADING VIBRATION	74-0239
SIMPLE	SIMPLE FINITE ELEMENT SOLUTION FOR PLATES OF HOMOGENEOUS, SANDWICH AND CELLULAR CONSTRUCTION	75-0140
SIMPLE	SIMPLE TEACHING EXAMPLES FOR FINITE ELEMENT STRESS ANALYSIS	72-0898
SIMPLE	SIMPLE TRIANGULAR MEMBRANE FINITE ELEMENT	74-0881
SIMPLE	STUDY AND DEVELOPMENT OF SIMPLE MATRIX METHODS OF INELASTIC STRUCTURES	66-0115
SIMPLE	THE SIMPLE METHOD OF EVALUATION OF STRESS INTENSITY FACTOR USING THE FINITE ELEMENT METHOD	74-0829
SIMPLE	THREE-DIMENSIONAL ELASTIC-PLASTIC STRESS AND STRAIN ANALYSIS FOR FRACTURE MECHANICS. PHASE 1. SIMPLE	70-0188
SIMPLE	USE OF SIMPLE THREE-DIMENSIONAL ACOUSTIC FINITE ELEMENTS FOR DETERMINING NATURAL MODES AND FREQUENCI	72-0010
SIMPLIFICATION	SIMPLIFICATION OF HIGH ORDER MECHANICAL SYSTEMS USING THE ROUTH APPROXIMATION	75-0853
SIMPLIFIED	A COMPARATIVE STUDY OF THE CONSISTENT AND SIMPLIFIED FINITE ELEMENTS ANALYSIS OF EIGENVALUE PROBLEMS	70-0034
SIMPLIFIED	A COMPARATIVE STUDY OF THE CONSISTENT AND SIMPLIFIED FINITE ELEMENT ANALYSES OF EIGENVALUE PROBLEMS	70-0373
SIMPLIFIED	A SIMPLIFIED VERSION OF GURTIN'S VARIATIONAL PRINCIPLE	74-1244
SIMPLIFIED	A SYNTHESIZED WARPED QUADRILATERAL ELEMENT GIVING SIMPLIFIED INPUT FOR EQUILIBRIUM MODELS	73-1013
SIMPLIFIED	ANALYSIS OF ELASTIC PLATES IN FLEXURE BY A SIMPLIFIED FINITE ELEMENT METHOD	67-0035
SIMPLIFIED	ANALYSIS OF ELASTIC PLATES BY A SIMPLIFIED FINITE ELEMENT METHOD	67-0156
SIMPLIFIED	APPLICATION OF SIMPLIFIED HYBRID DISPLACEMENT METHOD TO PLATE AND SHELL PROBLEMS	73-1037
SIMPLIFIED	CONVERGENCE OF SIMPLIFIED HYBRID DISPLACEMENT METHOD FOR PLATE BENDING	72-0412
SIMPLIFIED	FINITE ELEMENT ANALYSIS OF VIBRATION OF THIN ELASTIC PLATES BY SIMPLIFIED HYBRID DISPLACEMENT METHOD	72-0565
SIMPLIFIED	FINITE ELEMENT SOLUTIONS FOR PLATE BENDING PROBLEMS BY SIMPLIFIED HYBRID DISPLACEMENT METHOD	72-0780
SIMPLIFIED	ON THE CONVERGENCE OF FINITE ELEMENT SCHEMES BASED ON NON-CONFORMING AND SIMPLIFIED HYBRID DISPLACEM	73-0841
SIMPLIFIED	SIMPLIFIED CALCULATION OF THE INTERACTION FORCES BETWEEN HULL AND TANK SYSTEM FOR A MOSS ROSENBERG T	75-0561
SIMPLIFIED	SIMPLIFIED HYBRID DISPLACEMENT METHOD APPLIED TO PLATE BUCKLING PROBLEMS	72-0781
SIMPLIFIED	SIMPLIFIED HYBRID DISPLACEMENT METHOD FOR LINEAR FINITE ELEMENT ANALYSIS OF GENERAL SHELLS	74-0614
SIMPLIFIED	SIMPLIFIED INELASTIC HIGH TEMPERATURE STRUCTURAL ANALYSIS OF MODERATELY COMPLEX SPATIALLY THREE DIME	74-0445
SIMPLIFIED	SIMPLIFIED INELASTIC (PLASTIC AND CREEP) ANALYSIS OF PIPE ELBOWS SUBJECTED TO INPLANE AND OUT-OF-PLA	75-0654
SIMPLIFIED	SIMPLIFIED MINIMUM INTERNAL ENERGY OF DEFORMATION POSTULATE AND ITS APPLICATION TO METALFORMING PROC	74-1155
SIMPLIFIED	SIMPLIFIED SOIL-STRUCTURE INTERACTION ANALYSIS WITH STRAIN DEPENDENT SOIL PROPERTIES	74-1052
SIMPLIFIED	SOME FINITE ELEMENT SOLUTIONS FOR PLATE BENDING PROBLEMS BY SIMPLIFIED HYBRID DISPLACEMENT METHOD	72-0268
SIMPLIFIED-HELM	SHIP (SIMPLIFIED-HELMHOLTZ-INTEGRAL PROGRAM): A FAST COMPUTER PROGRAM FOR CALCULATING THE ACOUSTIC R	72-0835
SIMPLY	ANALYSIS OF AN ORTHOTROPIC PLATE WITH TWO OPPOSITE EDGES SIMPLY SUPPORTED AND THE OTHER TWO FREE	73-0741
SIMPLY	ANALYSIS OF SIMPLY SUPPORTED THICK, LAYERED PLATES	71-0428
SIMPLY	ARCH DAMS ANALYSED BY A LINEAR FINITE ELEMENT SHELL THE FINITE STRIP METHOD IN THE ANALYSIS OF ELAST	68-0189
SIMPLY	FREE VIBRATION OF A SIMPLY SUPPORTED BEAM WITH NONLINEAR MATERIAL PROPERTIES	71-0689
SIMPLY	NON-LINEAR EFFECTS OF HIGH TEMPERATURE ON THE VIBRATION OF A SIMPLY SUPPORTED BEAM WITH A CENTRAL MA	73-0838
SIMPLY	RESPONSE OF SIMPLY SUPPORTED TIMOSHENKO BEAMS	69-0489
SIMPLY	THE FINITE STRIP METHOD IN THE ANALYSIS OF ELASTIC PLATES WITH TWO SIMPLY SUPPORTED ENDS	68-0068
SIMULATE	A FINITE-ELEMENT MODEL TO SIMULATE THE NONLINEAR RESPONSE OF REINFORCED CONCRETE FRAMES WITH MASONRY	73-1168
SIMULATE	FINITE-ELEMENT CODE TO SIMULATE ANISOTROPIC-PLASMA DIFFUSION IN LINEAR MULTIPOLES	75-0409
SIMULATE	USING THE NASTRAN THERMAL ANALYZER TO SIMULATE A FLIGHT SCIENTIFIC INSTRUMENT PACKAGE	74-1011
SIMULATED	FINITE ELEMENT OF UNDERGROUND STRESSES UTILIZING STOCHASTICALLY SIMULATED MATERIAL PROPERTIES	70-0111
SIMULATED	GRAPHICAL DISPLAY OF COMPUTER SIMULATED UNBALANCED ROTOR RESPONSE	71-0486
SIMULATING	A METHOD FOR SIMULATING CONCENTRATED FORCES AND LOCAL REINFORCEMENTS IN STRESS COMPUTATIONS	72-0081
SIMULATING	COMPUTER PROGRAMS FOR SIMULATING PHYSICAL PHENOMENA IN THREE SPACE DIMENSIONS AND TIME	74-0866
SIMULATING	FINITE ELEMENT BASED PROCEDURE FOR SIMULATING THE TRANSIENT RESPONSE AND FAILURE OF A TWO-DIMENSIONA	72-0123
SIMULATING	PART 1 - TASK 3.2.9. BENDING AND BUCKLING OF A ONE-DIMENSIONAL MODEL SIMULATING A CONICAL SHELL. PAR	71-0745
SIMULATING	SIMULATING BEHAVIOR OF EARDRUM BY FINITE-ELEMENT METHOD	74-0511
SIMULATION	A FINITE ELEMENT SIMULATION OF MECHANICAL PROPERTIES OF COMPOSITE MATERIALS	74-1294
SIMULATION	A NUMERICAL MODEL FOR THE SIMULATION OF TWO DIMENSIONAL CONVECTIVE DISPERSION IN SHALLOW ESTUARIES	74-0052
SIMULATION	A SURVEY OF ANALYTICAL METHODS FOR SIMULATION OF CAB	73-0936
SIMULATION	ANALYTICAL DYNAMICS AND NONRIGID SPACECRAFT SIMULATION	74-0771
SIMULATION	ANALYTICAL FINITE ELEMENT SIMULATION MODEL FOR STRUCTURAL CRASHWORTHINESS PREDICTION	73-0778
SIMULATION	APPLICATION OF THE FINITE-ELEMENT METHOD FOR SIMULATION OF SURFACE WATER TRANSPORT PROBLEMS	72-0826
SIMULATION	BEITRAG ZUR SIMULATION NICHTSTATIONARER HORIZONTAL-EBENER GRUNDWASSERSTROMUNGSFELDER AUF DER GRUNDLA	74-0088
SIMULATION	COMPUTER SIMULATION OF SECOND ORDER FAULTS	71-0010
SIMULATION	COMPUTER SIMULATION OF DYNAMIC STRESS, DEFORMATION, AND FRACTURE OF GEAR TEETH	72-0114
SIMULATION	COMPUTER SIMULATION OF THE DEFORMATION OF COMPOSITE AND POLYCRYSTALLINE MATERIALS BY FINITE ELEMENT	72-0937
SIMULATION	COMPUTER SIMULATION OF LAMINATED ROOF REINFORCED WITH GROUTED BOLTS	73-1045
SIMULATION	COMPUTER SIMULATION OF AN IN SITU SHALE-OIL RETORT	74-1059
SIMULATION	COMPUTER SIMULATION OF VIBRATION OF WHOLE TREE SYSTEMS USING FINITE ELEMENT METHODS	74-1161
SIMULATION	CRASH: A COMPUTER SIMULATION OF NONLINEAR TRANSIENT RESPONSE OF STRUCTURES	72-0569
SIMULATION	DEVELOPMENT OF METHODS FOR COMPUTER SIMULATION OF BEAM-COLUMNS AND GRID-BEAM AND SLAB SYSTEMS	73-1101
SIMULATION	DYNAMIC SIMULATION OF COMPLEX SYSTEMS UTILIZING EXPERIMENTAL AND ANALYTICAL TECHNIQUES	75-0864
SIMULATION	FINITE ELEMENT SIMULATION OF TWO-DIMENSIONAL ELECTROMECHANICAL RESONATORS	74-1014
SIMULATION	FINITE ELEMENT SIMULATION OF THERMALLY INDUCED FLOW FIELDS	74-1235
SIMULATION	FINITE ELEMENT SIMULATION OF ENERGY-TRAPPED ELECTROMECHANICAL RESONATORS	75-0142
SIMULATION	FINITE ELEMENT SIMULATION OF ELECTROMECHANICAL RESONATORS WITH ARBITRARY ELECTRODES	75-0492
SIMULATION	FINITE ELEMENT SIMULATION OF ACOUSTIC FILTERS WITH VARIABLE CIRCULAR CROSS SECTION	75-0871
SIMULATION	FINITE-ELEMENT SIMULATION OF FLOOD HYDROGRAPHS	75-0632
SIMULATION	GALERKIN FINITE-ELEMENT SIMULATION OF A GEOTHERMAL RESERVOIR	73-0900
SIMULATION	GALERKIN-FINITE ELEMENT SIMULATION OF GROUNDWATER CONTAMINATION ON LONG-ISLAND, NEW-YORK	73-0093
SIMULATION	GENERAL PURPOSE COMPUTER PROGRAM FOR THE DYNAMIC SIMULATION OF VEHICLE GUIDEWAY INTERACTIONS	73-0260
SIMULATION	LEAST SQUARE OPTIMAL APPROXIMATION SIMULATION TECHNIQUES	73-0871
SIMULATION	MATHEMATICAL SIMULATION OF THE SUBSIDENCE OF VENICE 1 THEORY	73-0992
SIMULATION	MODULAR APPROACH TO STRUCTURAL SIMULATION FOR VEHICLE CRASHWORTHINESS PREDICTION	75-0737
SIMULATION	NUMERICAL SIMULATION OF FINITE-WIDTH THRUST BEARINGS, TAKING INTO ACCOUNT VISCOSITY VARIATION WITH T	75-0623
SIMULATION	RESERVOIR SIMULATION BY FINITE ELEMENT METHOD	73-0062
SIMULATION	RESERVOIR SIMULATION BY GALERKIN'S METHOD	73-0662
SIMULATION	SEMIDISCRETE GALERKIN TECHNIQUES WITH TIME INTERPOLATION AND SPLITTING UP FOR PLASMA SIMULATION	75-0121
SIMULATION	SIMULATION OF ELASTIC MODULUS AND POISSON'S RATIO OF SPHEROIDAL GRAPHITE CAST IRON	74-0955
SIMULATION	SIMULATION OF FINITE ELEMENT METHOD FOR LOADING PROCESS OF GRAIN CARGO AND INTERACTION WITH SHIP'S S	75-0161
SIMULATION	SIMULATION OF GROUND WATER CONTAMINATION USING A GALERKIN FINITE ELEMENT TECHNIQUE	74-1362
SIMULATION	SIMULATION OF PROGRESSIVE FAILURE IN ROCK AROUND MINING EXCAVATIONS	70-0146
SIMULATION	SIMULATION OF RUNOFF HYDROGRAPHS FROM NATURAL WATERSHEDS BY FINITE-ELEMENT METHOD	73-1140
SIMULATION	SIMULATION OF VIBRATION OF WHOLE TREE SYSTEMS USING FINITE ELEMENTS	75-0765
SIMULATION	SIMULATION ON THE INITIATION AND GROWTH OF DUCTILE FRACTURE VOIDS	75-0836
SIMULATION	SIMULATION ON THE TENSILE BEHAVIOUR OF TWO-PHASE ALLOY CONTAINING A FIBER UNDER TENSILE LOAD	73-0542
SIMULATION	SINGLE-TRACK SUBWAY ENVIRONMENTAL SIMULATION MODEL. PHASE 1	71-0693
SIMULATION	STOCHASTIC FINITE ELEMENT SIMULATION OF THE NONLINEAR STRUCTURAL RESPONSE OF FIBROUS COMPOSITE MATER	75-0762
SIMULATION	STUDIES ON TWO PHASE COMPOSITE MATERIALS THROUGH COMPUTER SIMULATION	73-0315
SIMULATION	THE STOCHASTIC FINITE ELEMENT SIMULATION OF THE NONLINEAR STRUCTURAL RESPONSE OF FIBROUS COMPOSITE M	75-0798
SIMULATION	THERMAL, THERMOELASTIC, AND WEAR SIMULATION OF A HIGH-ENERGY SLIDING CONTACT PROBLEM	73-0430
SIMULATION	TIDAL SIMULATION GY TIME SPACE ELEMENT	75-0867
SIMULATION	TWO-DIMENSIONAL FINITE ELEMENT SIMULATION OF SEMICONDUCTOR DEVICES	74-0260
SIMULATION	USE OF THE OVER-RELAXATION TECHNIQUE IN THE SIMULATION OF LARGE GROUNDWATER BASINS BY THE FINITE ELE	75-0606
SIMULATION	2-DIMENSIONAL SIMULATION OF SEMICONDUCTOR-DEVICES USING FINITE ELEMENT METHODS	75-0548
SIMULTANEOUS	A COMPACT STORAGE SCHEME FOR THE SOLUTION OF SYMMETRICAL LINEAR SIMULTANEOUS EQUATIONS	66-0116
SIMULTANEOUS	A TEST ON CONVERGENCE OF EIGENSOLUTIONS BY SIMULTANEOUS ITERATION METHOD	73-0317
SIMULTANEOUS	AN AUTOMATIC RECORDING SCHEME FOR SIMULTANEOUS EQUATIONS DERIVED FROM NETWORK SYSTEMS	70-0285

SIMULTANEOUS	APPLICATION OF SIMULTANEOUS ITERATION METHOD TO TORSIONAL VIBRATION PROBLEMS	73-0660
SIMULTANEOUS	BANDED MATRIX TECHNIQUES AND SOLUTIONS OF LINEAR SIMULTANEOUS EQUATIONS OF FINITE ELEMENT ANALYSES	70-0302
SIMULTANEOUS	PART 1. THEORY REPORT FOR CREEP-PLAST COMPUTER PROGRAM: ANALYSIS OF TWO-DIMENSIONAL PROBLEMS UNDER S	72-0765
SIMULTANEOUS	SIMULTANEOUS EIGENVECTOR ITERATION APPLIED TO LARGE SYSTEMS	74-0680
SIMULTANEOUS	STRESS ANALYSIS OF TWO-DIMENSIONAL PROBLEMS UNDER SIMULTANEOUS CREEP AND PLASTICITY	71-0220
SINGER	SINGER: A COMPUTER CODE FOR GENERAL ANALYSIS OF TWO-DIMENSIONAL CONCRETE STRUCTURES. VOL. 11. PROGRA	75-0723
SINGER	SINGER: A COMPUTER CODE FOR GENERAL ANALYSIS OF TWO-DIMENSIONAL REINFORCED CONCRETE STRUCTURES.	75-0724
SINGER	SINGER: A COMPUTER CODE FOR GENERAL ANALYSIS TO TWO-DIMENSIONAL CONCRETE STRUCTURES. VOLUME 4. DEMON	75-0730
SINGER	SINGER: A COMPUTER CODE FOR GENERAL ANALYSIS OF TWO-DIMENSIONAL REINFORCED CONCRETE STRUCTURES. VOLU	75-0731
SINGLE	DUGDALE MODEL SOLUTIONS FOR A SINGLE EDGE CRACKED PLATE	74-0221
SINGLE	FREQUENCY ANALYSIS OF CERTAIN SINGLE AND CONTINUOUS SPAN BRIDGES	71-0337
SINGLE	GOVERNING EQUATIONS AND EXTREMUM PRINCIPLES OF ELASTICITY AND PLASTICITY GENERATED FROM A SINGLE FUN	73-0666
SINGLE	SINGLE ELEMENT PANEL FOR FLAT PLATE STRUCTURES	75-0279
SINGLE	SOLUTION OF A SINGLE HYPERBOLOID SHELL BY FINITE ELEMENT METHOD	73-0151
SINGLE	STABILITY ANALYSIS OF AN EXPERIMENTALLY DEFORMED SINGLE LAYER OF INDIANA LIMESTONE USING FINITE ELEM	74-0212
SINGLE	STRENGTH DISTRIBUTIONS OF SINGLE FILAMENTS	75-0305
SINGLE-	DIAPHRAGMS IN SINGLE- AND DOUBLE-CELL BOX GIRDER BRIDGES WITH VARYING ANGLE OF SKEW	72-0130
SINGLE-EDGE	APPLICATION OF THE SINGLE-EDGE NOTCHED BEAM TO FRACTURE TOUGHNESS TESTING OF CERAMICS	74-0623
SINGLE-LAYER	DISCRETE-ELEMENT MODELLING AND SENSITIVITY STUDIES OF STATICALLY LOADED SINGLE-LAYER AND BOUNDED DOU	70-0129
SINGLE-LAYER	STACUSS 1: A DISCRETE-ELEMENT PROGRAM FOR THE STATIC ANALYSIS OF SINGLE-LAYER CURVED STIFFENED SHELL	69-0160
SINGLE-SYMMETRI	ZUR BERECHNUNG EINFACH-SYMMETRISCHER I-TRAEGER NACH DER THEORIE - 2. ORDNUNG UNTER BERUECKSICHTIGUNG	75-0510
SINGLE-TRACK	SINGLE-TRACK SUBWAY ENVIRONMENTAL SIMULATION MODEL. PHASE 1	71-0693
SINGULAR	ACCURATE FINITE ELEMENT APPROXIMATION FOR PROBLEMS WITH SINGULAR DATA	75-0098
SINGULAR	ADJOINING APPROPRIATE SINGULAR ELEMENTS TO TRANSPORT THEORY COMPUTATIONS	74-0946
SINGULAR	FINITE ELEMENT CONVERGENCE FOR SINGULAR DATA	73-0079
SINGULAR	ON FLOW NEAR SINGULAR POINTS OF A WALL BOUNDARY	74-0008
SINGULAR	SINGULAR ELEMENTS IN VARIATIONAL AND FINITE ELEMENT TRANSPORT COMPUTATIONS	73-0902
SINGULAR	USE OF SINGULAR FUNCTIONS WITH FINITE ELEMENT APPROXIMATIONS	73-0102
SINGULARITIES	CALCULATION OF HARMONIC FIELDS IN WHICH SINGULARITIES ARE PRESENT USING FINITE ELEMENT AND FINITE DI	73-0979
SINGULARITIES	CHILES: A FINITE ELEMENT COMPUTER PROGRAM THAT CALCULATES THE INTENSITIES OF LINEAR ELASTIC SINGULAR	73-0959
SINGULARITIES	CORNER SINGULARITIES IN ELLIPTIC PROBLEMS BY FINITE ELEMENT METHODS	71-0456
SINGULARITIES	ELASTIC ANALYSIS OF CYLINDRICAL CONFIGURATIONS WITH STRESS SINGULARITIES	71-0325
SINGULARITIES	FINITE ELEMENT ANALYSIS OF STRESS CONCENTRATIONS AND SINGULARITIES	72-0841
SINGULARITIES	FINITE ELEMENT METHODS FOR ELLIPTIC BOUNDARY VALUE PROBLEMS CONTAINING SINGULARITIES	72-0360
SINGULARITIES	FINITE ELEMENT SOLUTION OF BOUNDARY-VALUE PROBLEMS WITH NON-REMOVABLE SINGULARITIES	73-0124
SINGULARITIES	FINITE ELEMENT TECHNIQUES APPLIED TO CRACKS INTERACTING WITH SELECTED SINGULARITIES	75-0196
SINGULARITIES	FINITE ELEMENT TREATMENT OF SINGULARITIES OF BOUNDARY VALUE PROBLEMS AND ITS APPLICATION TO ANALYSIS	73-0306
SINGULARITIES	HIGHER ORDER SINGULARITIES FOR INTERFACE PROBLEMS	72-0513
SINGULARITIES	NUMERICAL SOLUTION OF STEADY STATE DIFFUSION PROBLEMS CONTAINING SINGULARITIES	74-0036
SINGULARITIES	REPRESENTATION OF SINGULARITIES WITH ISOPARAMETRIC FINITE ELEMENTS	74-1075
SINGULARITIES	SHELL DEFORMATION UNDER TOOL LOADS REPRESENTATION OF SINGULARITIES WITH ISOPARAMETRIC FINITE ELEMENT	74-0365
SINGULARITIES	SINGULARITIES IN ELECTROMAGNETIC TRANSMISSION LINES	72-0072
SINGULARITIES	SINGULARITIES IN THE FINITE ELEMENT APPROXIMATION OF TWO DIMENSIONAL DIFFUSION PROBLEMS	73-1015
SINGULARITIES	SOLUTION OF PROBLEMS WITH INTERFACES AND SINGULARITIES	74-0822
SINGULARITIES	SOLUTION OF PROBLEMS WITH INTERFACES AND SINGULARITIES	74-0891
SINGULARITIES	SOME EFFECTS OF SYSTEM IDEALIZATIONS, SINGULARITIES AND MESH PATTERNS ON FINITE ELEMENT SOLUTIONS	74-0939
SINGULARITIES	SPECIAL FINITE ELEMENTS FOR THE ANALYSIS OF STRESS CONCENTRATIONS AND SINGULARITIES	71-0219
SINGULARITIES	STRESS CONCENTRATIONS AND SINGULARITIES AT INTERFACE CORNERS	71-0419
SINGULARITIES	STRESS SINGULARITIES AT DIAPHRAGM-WEB-FLANGE JUNCTIONS IN BOX GIRDERS	75-0242
SINGULARITIES	TWO AND THREE DIMENSIONAL POTENTIAL FLOW BY THE METHOD OF SINGULARITIES	69-0449
SINGULARITIES	TWO AND THREE DIMENSIONAL POTENTIAL FLOW ANALYSIS BY THE METHOD OF SINGULARITIES	70-0357
SINGULARITIES	USE OF HIGHER ORDER DISPLACEMENT FUNCTIONS IN THE FREE VIBRATION ANALYSIS OF SHELLS OF REVOLUTION HA	74-1055
SINGULARITY	BEST FINITE ELEMENTS DISTRIBUTION AROUND A SINGULARITY	72-0252
SINGULARITY	CALCULATION OF HARMONIC FIELDS IN WHICH SINGULARITIES ARE PRESENT USING FINITE ELEMENT AND FINITE DI	73-0979
SINGULARITY	DEVELOPMENT OF AN ANISOTROPIC SINGULARITY FINITE ELEMENT UTILIZING THE HYBRID DISPLACEMENT METHOD	74-1099
SINGULARITY	EVALUATION OF THE USE OF A SINGULARITY ELEMENT IN FINITE ELEMENT ANALYSIS OF CENTER-CRACKED PLATES	72-0870
SINGULARITY	ON THE CONVERGENCE OF THE FINITE ELEMENT METHOD FOR PROBLEMS WITH SINGULARITY	73-0173
SINGULARITY	THREE-DIMENSIONAL HARMONIC FUNCTIONS NEAR TERMINATION OR INTERSECTION OF GRADIENT SINGULARITY LINES:	74-0178
SINGULARITY	USE OF SINGULARITY PROGRAMMING IN FINITE-DIFFERENCE AND FINITE ELEMENT COMPUTATIONS OF TEMPERATURE	73-0100
SINGULARITY	USE OF SINGULARITY PROGRAMMING IN FINITE ELEMENT CALCULATIONS OF ELASTIC STRESS INTENSITY FACTORS. P	75-0252
SINUSOIDAL	GENERAL SINUSOIDAL STIFFNESS MATRICES FOR BUCKLING AND VIBRATION ANALYSES OF THIN FLAT-WALLED STRUCT	68-0278
SINUSOIDALLY	DYNAMIC RESPONSE OF VISCOELASTIC MATERIALS TO SINUSOIDALLY VARYING LOADINGS	73-1072
SITE	BEHAVIOUR OF THE ROCK MASS AT THE SITE OF A ROCKFILL DAM	74-0992
SITE	EFFECTS OF SOURCE STRUCTURE AND SITE RESPONSE ON GROUND MOTION IN THE NEAR FIELD OF EARTHQUAKES	73-0732
SITU	COMPUTER SIMULATION OF AN IN SITU SHALE-OIL RETORT	74-1059
SITU	IN SITU CREEP ANALYSIS OF ROOM IN FROZEN SOIL	72-0128
SIX	A SIX NODE TRIANGULAR EQUILIBRIUM (CONSTANT STRESS) MEMBRANE ELEMENT	73-0783
SIX	A SIX NODE TRIANGULAR SHEAR PANEL FOR EQUILIBRIUM MODELS	73-1011
SIX	ANALYSIS OF ASYMMETRIC STIFFENED SHELL TYPE STRUCTURES BY FINITE ELEMENT METHOD IT. SIX DEGREE-OF-FR	69-0373
SIX	ANALYSIS OF OUT-OF-PLANE STRESS BY THE SIX NODES, FOURTH ORDER ACCURACY TRIANGULAR ELEMENTS	69-0375
SIX	PLANE STRESS ANALYSIS USING THE FINITE ELEMENT METHOD. TRIANGULAR FINITE ELEMENT WITH SIX PARAMETERS	67-0106
SIZED	FINITE ELEMENT ANALYSIS OF FINITE SIZED PLATES BOUNDED TO AN ELASTIC HALF SPACE	72-0880
SIZED	NOUVELLE CONCEPTION DES DRAINS DANS LES BARRAGES EN TERRE HOMOGENE DE PETITE ET MOYENNE DIMENSIONS (	73-0404
SKALNOM	RASCHET NAPRYAZHENIL L PEREMESHCHENII V SVODE PODZEMNOGO MASHINNOGO ZALA GES I V OKRUZHAYUSHCHEM SKA	74-0542
SKEW	A COMPATIBLE ELEMENT FOR BUCKLING ANALYSIS OF SKEW PLATES WITH PARALLEL EDGES	73-0328
SKEW	ANALYSIS AND DESIGN OF SKEW BOX GIRDER BRIDGES	72-0659
SKEW	ANALYSIS OF A SKEW SLAB-BEAM SYSTEM BY FINITE ELEMENT METHOD	73-0233
SKEW	ANALYSIS OF CORNER SUPPORTED SKEW SLABS	68-0203
SKEW	CALCUL DES PONTS BIAIS A POUTRES MULTIPLES SANS ENTRETOISES PAR LA METHODE DES ELEMENTS (CALCULATION	72-0184
SKEW	DIAPHRAGMS IN SINGLE- AND DOUBLE-CELL BOX GIRDER BRIDGES WITH VARYING ANGLE OF SKEW	72-0130
SKEW	EFFECTIVENESS OF EXCESSIVE NODAL CONTINUITIES IN THE FINITE ELEMENT ANALYSIS OF THIN RECTANGULAR AND	70-0346
SKEW	EFFICIENT FINITE ELEMENT ANALYSIS OF RECTANGULAR AND SKEW LAMINATED PLATES	73-0436
SKEW	FINITE ELEMENT ANALYSIS OF SKEW COMPOSITE GIRDER BRIDGES	67-0150
SKEW	FINITE ELEMENT ANALYSIS OF SKEW COMPOSITE PLATES	67-0160
SKEW	FINITE ELEMENT ANALYSIS OF SKEW PLATES IN BENDING	68-0120
SKEW	FINITE ELEMENT ANALYSIS OF SKEW, CURVED BOX-GIRDER BRIDGE	70-0241
SKEW	FINITE ELEMENT ANALYSIS OF SKEW BOX GIRDER BRIDGES	72-0209
SKEW	FINITE ELEMENT ANALYSIS OF SKEW SANDWICH PLATES	72-0810
SKEW	FINITE ELEMENT OF SKEW VAULT BRIDGES	72-0434
SKEW	FINITE STRIP ANALYSIS OF SKEW SLABS	72-0369
SKEW	FINITE-ELEMENT ANALYSIS OF SKEW SANDWICH PLATES	75-0185
SKEW	FLUTTER OF SKEW PANELS BY THE MATRIX DISPLACEMENT APPROACH	70-0201
SKEW	FREQUENCY ANALYSIS OF SKEW ORTHOTROPIC PLATES BY THE FINITE STRIP METHOD	71-0413
SKEW	MATRIX-DISPLACEMENT DIRECT-ELEMENT METHOD FOR FORCE AND TORQUE ANALYSIS OF DETERMINATE AND INDETERMI	74-0531
SKEW	NUMERICAL RESULTS FROM THE APPLICATION OF GRADIENT ITERATIVE TECHNIQUES TO THE FINITE ELEMENT VIBRAT	72-0138
SKEW	SEMI-ANALYTIC SOLUTION OF SKEW BOX GIRDER BRIDGES	75-0578
SKEW	THE ANALYSIS OF SKEW BRIDGE DECKS, A NEW FINITE ELEMENT APPROACH	69-0234
SKEW	THE EFFECTIVENESS OF EXCESSIVE NODAL CONTINUITIES IN THE FINITE ELEMENT ANALYSIS OF THIN RECTANGULAR	70-0105
SKEW	THEORY OF EQUIVALENCE AND ITS APPLICATION TO VARIOUS PROBLEMS IN ELASTICITY AND DESIGN OF SKEW BRIDG	72-0470
SKEWED	A FINITE ELEMENT TECHNIQUE FOR ANALYZING FOLDED PLATE TYPE STRUCTURES WITH SKEWED SUPPORTS	69-0525
SKEWED	A HIGHER ORDER IN-PLANE PARALLELOGRAM ELEMENT AND ITS APPLICATION TO SKEWED CURVED BOX GIRDER BRIDGE	71-0002
SKEWED	ANALYSIS OF STRESSES IN SKEWED PLATES BY THE METHOD OF FINITE ELEMENTS	73-0248

371

SKEWED	FINITE ELEMENT ANALYSIS OF SKEWED SHALLOW SHELLS	68-0250
SKEWED	FINITE ELEMENT ANALYSIS OF SKEWED SHALLOW SHELLS	69-0211
SKIING	DYNAMIC TORSIONAL RESPONSE OF THE HUMAN LEG RELATIVE TO SKIING INJURIES	73-0471
SKIN	FINITE ELEMENTS FOR IMPROVED SKIN STRINGER STRUCTURAL ANALYSIS	70-0397
SKIN	OPTIMUM DESIGN OF STRESSED SKIN STRUCTURES	73-0472
SKIN-RIB	A FINITE ELEMENT STUDY OF THE VIBRATION OF SKIN-RIB STRUCTURES	74-0537
SKIN-STRINGER	NODAL ANALYSIS OF A NINE BAY SKIN-STRINGER PANEL	71-0102
SKIN-STRINGER	SKIN-STRINGER PANEL NORMAL MODE RESPONSE EXPERIMENTAL DATA	71-0691
SKLADKI	THE DEVELOPMENT OF APPLIED METHODS IN PROBLEMS OF STATIC CALCULATION OF THIN-WALLED THREE-DIMENSIONA	73-0934
SKONCZONYCCH	ZASTOSOWANIE ITERACJI SEIDLA W METODZIE ELEMENTOW SKONCZONYCCH NA PRZYKLADZIE OBLICZEN STATYCZNYCH P	74-1031
SKONCZONYCH	ALGORYTM OBLICZEN ORGAN WYMUSZONYCH METODA SZTYWNYCH ELEMENTOW SKONCZONYCH (ALGORITHM FOR CALCULATIN	74-0581
SKONCZONYCH	ANALIZA ORGAN WYMUSZONYCH ZLOZONYCH UKLADOW LINIOWYCH METODA SZTYWNYCH ELEMENTOW SKONCZONYCH (ANALYS	72-0246
SKONCZONYCH	APPLICATION OF HERMITE POLYNOMIALS TO THE DETERMINATION OF THE STIFFNESS MATRIX OF PLATE BY FINITE E	71-0365
SKONCZONYCH	FUNKCJA KSZTALTU NIESPELNIAPACA WARUNKOW,CIAGLOSCI W ANALIZIE PLYT METODA ELEMENTOW SKONCZONYCH. (SH	75-0263
SKONCZONYCH	HYBRYDOWA METODA ELEMENTOW SKONCZONYCH W ZASTOSOWANIU DO OBLICZEN ORGAN URZADZEN OKRETOWYCH. (HYBRID	74-0963
SKONCZONYCH	METHOD SZTYWNYCH ELEMENTOW SKONCZONYCH W OBLICZENIACH KONSTRUCJI OKRETOWYCH. (STIFF FINITE ELEMENT M	74-0923
SKONCZONYCH	OBLICZANIE ORGAN METODA SZTYWNYCH ELEMENTOW SKONCZONYCH. (CALCULATING VIBRATIONS BY THE RIGID-FINITE	75-0649
SKONCZONYCH	ZASTOSOWANIE METODY SZTYWNYCH ELEMENTOW SKONCZONYCH DO OBLICZEN ORGAN WALOW OKRETOYCH Z UWZGLEDNIENI	74-0962
SKONSZONYCH	METODA ELEMENTOW SKONSZONYCH ROZWIAZYWANIA OSIOWO-SYMETRYCZNEGO PRZEWODNICTWA CIEPLA (FINITE-ELEMENT	73-0434
SKONSZONYCH	METODA ELEMENTOW SKONSZONYCH W MECHANICE GRUNTOW I MECHANICE GOROTWORU (FINITE ELEMENT METHOD IN SOI	73-0500
SKRUPIN	GEOMETRICKY NELINEARNE ULOHY PLOCHYCH SKRUPIN RIESENE PRIRASTKOVOU METODOU KONECNYCH PRVKOV. (GEOMET	74-1032
SKULL	ELASTIC ANALYSIS OF A SKULL	73-0415
SKULL	IN-VACUO MODAL DYNAMIC RESPONSE OF THE HUMAN SKULL	73-0460
SKYLAB	COMPLEX EIGENVALUE SOLUTION TO A SPINNING SKYLAB PROBLEM	71-0207
SKYLINE-STORED	SOLUTION OF LINEAR EQUATIONS WITH SKYLINE-STORED SYMMETRIC MATRIX	75-0662
SL-7	STRUCTURAL ANALYSIS OF SL-7 CONTAINERSHIP UNDER COMBINED LOADING OF VERTICAL, LATERAL AND TORSIONAL	74-0734
SLAB	A DISCRETE-ELEMENT METHOD OF ANALYSIS FOR ORTHOGONAL SLAB AND GRID BRIDGE FLOOR SYSTEMS	72-0690
SLAB	A THEORETICAL AND EXPERIMENTAL STUDY ON FREE VIBRATION OF CURVED SLAB BRIDGES	71-0496
SLAB	ANALYSIS OF SLAB AND GIRDER BRIDGES BY THE FINITE STRIP METHOD	70-0210
SLAB	ANALYSIS OF SLAB BEAM BRIDGES	69-0077
SLAB	ANALYSIS OF SLAB BRIDGES BY FINITE STRIP METHOD	73-0687
SLAB	ANALYSIS OF SLAB BRIDGES BY FINITE ELEMENTS	73-0695
SLAB	ASSESSMENT OF DESIGN TOOLS AND CRITERIA FOR URBAN RAIL TRACK STRUCTURES, VOLUME 11. AT-GRADE SLAB TR	74-0813
SLAB	BASIS OF A WELL CONDITIONED FORCE PROGRAM FOR EQUILIBRIUM MODELS VIA THE SOUTHWELL SLAB ANALOGIES	67-0111
SLAB	COMPUTATIONAL EXPERIMENTATION ON FINITE ELEMENT METHOD IN BARE SLAB CRITICALITY CALCULATIONS	73-0015
SLAB	DESIGN OF CONCRETE TRACK SLAB ON ELASTIC FOUNDATION USING THE FINITE ELEMENT METHOD	73-0479
SLAB	DEVELOPMENT OF METHODS FOR COMPUTER SIMULATION OF BEAM-COLUMNS AND GRID-BEAM AND SLAB SYSTEMS	73-1101
SLAB	ETUDE THEORIQUE DU COMPORTEMENT D'UNE DALLE DE BETON. (THEORETICAL STUDY OF THE BEHAVIOR OF A CONCRE	75-0273
SLAB	FINITE ELEMENT ANALYSIS OF A BROAD ARCH BRIDGE WITH A SLAB DECK	71-0724
SLAB	FINITE ELEMENT PROGRAMS FOR SLAB BRIDGE DESIGN	71-0082
SLAB	FREE VIBRATION OF CURVED AND STRAIGHT BEAM SLAB OR BOX GIRDER BRIDGES	72-0724
SLAB	INFLUENCE CHARACTERISTICS FOR SLAB BRIDGES	71-0188
SLAB	POST-TENSIONED SLAB BRIDGES ON ISOLATED SUPPORTS	75-0084
SLAB	SLAB BRIDGE DESIGN AND DRAWING, AN AUTOMATED PROCESS	70-0347
SLAB	SLAB BRIDGES WITH ARBITRARY SHAPE AND SUPPORT CONDITIONS - A GENERAL METHOD OF ANALYSIS BASED ON FIN	68-0233
SLAB	STRAIN-ENERGY BOUNDS IN FINITE-ELEMENT ANALYSIS BY SLAB ANALOGY	67-0027
SLAB	STRESS CONCENTRATION FACTORS AROUND A PENETRATION IN A STRIP OR SLAB OF FINITE DIMENSIONS	69-0409
SLAB	SUBSTRUCTURE ANALYSIS OF MULTISTOREY FLAT SLAB BUILDINGS	74-0394
SLAB	THREE-DIMENSIONAL STRESS ANALYSIS OF A FINITE SLAB CONTAINING A TRANSVERSE CENTRAL CRACK	75-0732
SLAB	TIME DEPENDENT REINFORCED CONCRETE SLAB DEFLECTIONS	74-0586
SLAB-BEAM	ANALYSIS OF A SKEW SLAB-BEAM SYSTEM BY FINITE ELEMENT METHOD	73-0233
SLABS	A DISCRETE-ELEMENT METHOD OF MULTIPLE-LOADING ANALYSIS FOR TWO-WAY BRIDGE FLOOR SLABS	70-0435
SLABS	A SUMMARY OF DISCRETE-ELEMENT METHODS OF ANALYSIS FOR PAVEMENT SLABS	72-0674
SLABS	ANALYSIS OF CORNER SUPPORTED SKEW SLABS	68-0203
SLABS	ANALYSIS OF CORNER SUPPORTED RECTANGULAR SLABS	70-0046
SLABS	ANALYSIS OF SLABS WITH EDGE BEAMS	71-0075
SLABS	COMPARISON OF FINITE ELEMENT AND UNIQUE LIMIT ANALYSIS SOLUTIONS FOR CERTAIN REINFORCED CONCRETE SLA	69-0184
SLABS	DUAL ANALYSIS OF COLUMN SUPPORTED SLABS	74-0399
SLABS	DYNAMIC BEHAVIOR OF IN-LINE SHEAR WALLS CONNECTED BY FLOOR SLABS	72-0689
SLABS	EXPERIMENTAL VERIFICATION OF DISCRETE-ELEMENT SOLUTIONS FOR PLATES AND PAVEMENT SLABS	70-0411
SLABS	FINITE ELEMENT ANALYSIS OF REINFORCED CONCRETE SLABS	71-0133
SLABS	FINITE ELEMENT ANALYSIS OF SLABS ON ELASTIC SOLIDS	74-0120
SLABS	FINITE STRIP ANALYSIS OF SKEW SLABS	72-0369
SLABS	FINITE STRIP METHOD OF ANALYSIS OF ELASTIC SLABS	68-0190
SLABS	FINITE-ELEMENT ANALYSIS OF CONCRETE SLABS AND ITS IMPLICATIONS FOR RIGID PAVEMENT DESIGN	73-0538
SLABS	LIMIT ANALYSIS OF REINFORCED CONCRETE SLABS BY FINITE ELEMENT METHOD	71-0802
SLABS	LIMIT ANALYSIS OF REINFORCED CONCRETE SLABS BY A FINITE ELEMENT METHOD	72-0985
SLABS	NON-LINEAR ANALYSIS OF REINFORCED CONCRETE SLABS	72-0191
SLABS	ON THE BENDING STIFFNESS OF SLABS IN CROSS-WALLED STRUCTURES	74-0398
SLABS	PRESTRESSED CONCRETE DEEP SLABS WITH OPENINGS	73-0602
SLABS	PUNCHING STRENGTH OF SLABS, A FLEXURAL APPROACH USING FINITE ELEMENTS	74-1307
SLABS	ROTATIONAL STIFFNESS OF CONCRETE SLABS	74-0660
SLABS	STATICKE RESENI FYZIKALNE NELINEARNIHO CHOVANI ZELEZOBETO- NOVYCH DESEK (ANALYSIS OF PHYSICALLY NONL	74-0514
SLABS	STRESS ANALYSIS OF CONCRETE TRACK SLABS ON AN ELASTIC FOUNDATION BY THE FINITE ELEMENT METHOD	74-1049
SLABS	THE FINITE ELEMENT METHOD FOR ANALYSIS OF ELASTIC ISOTROPIC AND ORTHOTROPIC SLABS	64-0014
SLABS	THE FINITE ELEMENT METHOD FOR FLEXURE OF SLABS WHEN STRESS DISTRIBUTIONS ARE ASSUMED	66-0069
SLABS	THE INELASTIC ANALYSIS OF REINFORCED CONCRETE SLABS	74-1333
SLABS	TRANSIENT TEMPERATURE RESPONSE OF COMPOSITE SLABS	74-1038
SLABS	YIELD CRITERION FOR REINFORCED CONCRETE SLABS	73-0985
SLADE	SLADE D: A COMPUTER PROGRAM FOR THE DYNAMIC ANALYSIS OF THIN SHELLS	73-0964
SLADE	SLADE, A COMPUTER PROGRAM FOR THE STATIC ANALYSIS OF THIN SHELLS	70-0079
SLAMCODE	SLAMCODE: FINITE ELEMENT STRESS WAVE ANALYSIS	72-0603
SLATE	SIZE OF PLASTIC ZONE AT TIP OF A CRACK IN PLANE STRAIN SLATE BY FINITE ELEMENT METHOD	73-0006
SLEEVE	THERMAL SLEEVE DESIGN FOR PRESSURE VESSELS	74-1129
SLENDERNESS	ELASTIC DEFORMATION OF POLYCRYSTALLINE METAL - 2 EFFECT OF SLENDERNESS OF GRAINS ON DEFORMATION MODE	73-0275
SLIDE	DYNAMIC ANALYSIS OF THE SLIDE IN THE LOWER SAN FERNANDO DAM DURING THE EARTHQUAKE OF FEBRUARY 9, 197	75-0277
SLIDER	FINITE ELEMENT OPTIMIZATION OF FINITE STEPPED SLIDER BEARING PROFILES	74-0131
SLIDER	OIL-FILM TEMPERATURE DISTRIBUTION IN AN INFINITELY WIDE SLIDER BEARING: AN APPLICATION OF THE FINITE	73-1128
SLIDING	NON-UNIFORM FLOW OF POLYCRYSTALS BY GRAIN-BOUNDARY SLIDING ACCOMMODATED BY POWER-LAW CREEP	75-0562
SLIDING	ON THE STABILITY OF ROTOR-AND-BEARING SYSTEMS AND ON THE CALCULATION OF SLIDING BEARINGS	72-0037
SLIDING	THERMAL, THERMOELASTIC, AND WEAR SIMULATION OF A HIGH-ENERGY SLIDING CONTACT PROBLEM	73-0430
SLIGHTLY	POST-BUCKLING ANALYSIS OF SLIGHTLY CURVED PLATES BY THE FINITE ELEMENT METHOD	69-0365
SLIME	SEEPAGE-ENVIRONMENTAL ANALYSIS OF THE SLIME ZONE OF A TAILINGS POND	74-0513
SLIP	DIRECT MEASUREMENT OF BOND SLIP IN REINFORCED CONCRETE	69-0428
SLOPE	ACCURACY OF EQUILIBRIUM SLOPE STABILITY ANALYSIS	73-0517
SLOPE	AN APPLICATION OF THE FINITE ELEMENT METHOD FOR ROCK SLOPE STABILITY ANALYSIS	71-0813
SLOPE	ANALYSE DE LA STABILITY DES TALUS ROCHEUX PAR UNE METHODE GENERALISEE DE L EQUILIBRE LIMITE (ROCK SL	72-1010
SLOPE	ANALYSIS OF SLOPE STABILITY BY FINITE ELEMENT METHOD AND ITS APPLICATION PROGRAM. SOIL MECHANICS	72-0919
SLOPE	COMPUTER PROGRAM FOR PIT SLOPE STABILITY ANALYSIS BY THE FINITE ELEMENT STRESS ANALYSIS AND LIMITING	72-0225
SLOPE	CONTROLLING OPEN-PIT SLOPE FAILURES AT SHIRLEY BASIN	73-0149
SLOPE	CORRELATION BETWEEN MODEL STUDY AND FINITE ELEMENT ANALYSIS IN AN INVESTIGATION OF SLOPE STABILITY	72-0227
SLOPE	DER SPANNUNGS UND BEANSPRUCHUNGSZUSTAND DES BODENMATERIALS IN DER NACHBARSCHAFT EINER BODENBOESCHUNG	75-0371

SLOPE	FINITE ELEMENT METHOD WITH CONSIDERATION TO THE CONTINUITY OF NORMAL SLOPE	72-0752
SLOPE	FINITE ELEMENT METHOD WITH CONSIDERATION TO THE CONTINUITY OF NORMAL SLOPE	72-0813
SLOPE	FINITE ELEMENT METHOD AND DENDIXEN-OSTENFELD SLOPE DEFLECTION METHOD	74-0557
SLOPE	MEASUREMENT OF SLOPE CHANGE ON CYLINDRICAL SHELL SURFACES	73-0439
SLOPE	SLOPE STABILITY ANALYSIS BY THE FINITE ELEMENT STRESS ANALYSIS AND LIMITING EQUILIBRIUM METHOD	70-0222
SLOPES	ANALYSES OF SLOPES BY THE FINITE ELEMENT METHOD	68-0297
SLOPES	ANALYSIS OF ROCK SLOPES, USING THE FINITE ELEMENT METHOD	70-0184
SLOPES	APPLICATION OF FINITE ELEMENT METHOD IN DETERMINING STABILITY OF CRATER SLOPES, PRELIMINARY REPORT	68-0026
SLOPES	APPLICATION OF FINITE ELEMENT METHOD IN DETERMINING STABILITY OF CRATER SLOPES	72-0888
SLOPES	DEVELOPMENT OF FAILURE AROUND EXCAVATED SLOPES	70-0051
SLOPES	EVALUATION OF THE STABILITY OF NATURAL SLOPES IN PLASTIC CHAMPLAIN CLAYS	74-0264
SLOPES	FACTOR OF SAFETY APPROACH FOR EVALUATING SEISMIC STABILITY OF SLOPES	75-0645
SLOPES	FINITE ELEMENT ANALYSIS OF SLOPES IN SOIL	68-0061
SLOPES	FINITE ELEMENT ANALYSES OF SLOPES IN JOINTED ROCK	68-0269
SLOPES	FINITE ELEMENT ANALYSIS OF CENTRIFUGED AND BUILT-UP SLOPES	74-0415
SLOPES	FINITE ELEMENT ANALYSIS OF CENTRIFUGED AND BUILT UP SLOPES	74-1376
SLOPES	INELASTIC ANALYSIS OF SLOPES BY THE FINITE ELEMENT METHOD	68-0298
SLOPES	INITIATION OF FAILURE IN SLOPES IN OVERCONSOLIDATED CLAYS AND CLAY SHALES.	70-0186
SLOPES	MESH GENERATING COMPUTER PROGRAM FOR FINITE ELEMENT ANALYSES OF SIMPLE SLOPES IN HORIZONTALLY LAYERE	71-0748
SLOPES	SLOPES IN STIFF FISSURED CLAYS AND SHALES	69-0094
SLOPES	STATIC AND DYNAMIC STRESSES IN SLOPES	66-0025
SLOPES	STEADY STATE AND TRANSIENT GROUNDWATER FLOW IN SLOPES	70-0273
SLOPES	STRESSES AND DEFORMATIONS OF VERTICAL SLOPES IN ELASTO-PLASTIC ROCKS	70-0114
SLOPES	THE BEHAVIOR OF STATISTICALLY HETEROGENEOUS EXCAVATED EARTH SLOPES	74-1330
SLOPES	THREE DIMENSIONAL FINITE ELEMENT ANALYSIS OF JOINTED ROCK SLOPES	70-0502
SLOPES	THREE-DIMENSIONAL CONSIDERATION OF THE STRESSES SURROUNDING OPEN-PIT MINE SLOPES	73-0492
SLOPING	MAGNETOTELLURIC RESPONSE OF A 2-DIMENSIONAL SLOPING CONTACT BY FINITE ELEMENT METHOD	73-0130
SLOPING	SEISMIC RESPONSE OF SOIL DEPOSITS UNDERLAIN BY SLOPING ROCK BOUNDARIES	69-0484
SLOPING	UNCONFINED TRANSIENT SEEPAGE IN SLOPING BANKS	71-0378
SLOSH	FINITE ELEMENT ANALYSIS OF MERCURY SLOSH IN THE SOLAR ELECTRIC PROPULSION STAGE	75-0757
SLOSHING	A COMPARISON OF SOME FINITE ELEMENT AND FINITE-DIFFERENCE METHODS FOR A SIMPLE SLOSHING PROBLEM	71-0055
SLOSHING	FINITE ELEMENT ANALYSIS FOR LIQUID SLOSHING PROBLEMS	69-0029
SLOSHING	FINITE ELEMENT SOLUTION OF LOW BOND NUMBER SLOSHING	75-0728
SLOSHING	LIQUID SLOSHING IN AN ELASTIC CONTAINER	66-0096
SLOSHING	SLOSHING OF LIQUID IN CIRCULAR CHANNELS AND SPHERICAL TANKS	60-0005
SLOTS	STABILITY CONTROL ANALYSIS OR ROTATING PLATES BY FINITE ELEMENT, EMPHASIS ON SLOTS AND HOLES	71-0284
SLOW	A NOTE ON SOME VARIATIONAL STATEMENTS FOR THE SLOW FLOW OF A NAVIER-POISSON FLUID	74-1054
SLOW	FINITE ELEMENT ANALYSIS OF TWO-DIMENSIONAL SLOW NON-NEWTONIAN FLOWS	72-0035
SLOW	FINITE ELEMENT ANALYSIS OF SLOW NON-NEWTONIAN CHANNEL FLOW	72-0308
SLOW	FINITE ELEMENT METHOD FOR INCOMPRESSIBLE SLOW VISCOUS FLOW WITH A FREE SURFACE	69-0303
SLOW	FINITE-ELEMENT ANALYSIS OF SLOW INCOMPRESSIBLE VISCOUS FLUID MOTION	74-0154
SLOW	PREDICTION OF SLOW CRACK GROWTH BY A FINITE-ELEMENT METHOD	75-0416
SLOW	SLOW FINITE DEFORMATIONS OF VISCOUS SOLIDS	69-0423
SLOW	SLOW PARTICULATE VISCOUS FLOW IN CHANNELS AND TUBES, APPLICATION TO BIOMECHANICS	71-0248
SLOW	SLOW VISCOUS FLOW AND ITS APPLICATION TO BIOMECHANICS	71-0619
SLOZHNOI	RASCHET KILEBANII PLASTIN SLOZHNOI FORMY S NEODNORODNYMI GRANICHNYMI USLOVIYAMI. (CALCULATION OF VIB	74-0988
SLOZHNOI	CHISLENNOE RESHENIE PLOSKIKH ZADACH TEPLOPROVODNOSTI DLYA OBLASTEI SLOZHNOI FORMY (NUMERICAL SOLUTIO	74-0659
SLUICE	FINITE ELEMENT ANALYSIS OF SLUICE GATE FLOW	71-0540
SLURRY	ANALYTICAL PROBLEMS IN MODELING SLURRY WALL CONSTRUCTION	73-0764
SMALL	ANALYSIS OF NONLINEAR CONCRETE FRAMES USING SMALL GYPSUM MODELS AND THE FINITE ELEMENT METHOD	70-0498
SMALL	APPLICATION OF NATURAL KINEMATIC MODES TO MATRIX ANALYSIS: SMALL AND LARGE DISPLACEMENTS AND STRAINS	67-0138
SMALL	APPLICATION OF SMALL AND LARGE-DISPLACEMENT THEORY TO A LARGE FLEXIBLE SOLAR ARRAY AND COMPARISON WI	71-0231
SMALL	DEPLACEMENTS DANS LES STRUCTURES MINCES A FAIBLE RIGIDITE GEOMETRIQUE-DETERMINATION NUMERIQUE (NUMER	72-0213
SMALL	FINITE ELEMENT APPROACH TO HYDROTHERMAL ANALYSIS OF SMALL LAKES	72-0363
SMALL	FINITE ELEMENTS, FINITE ROTATIONS AND SMALL STRAINS OF FLEXIBLE SHELLS	69-0516
SMALL	HOW FINITE ELEMENT METHODS ARE INTRODUCED IN LARGE AND SMALL ORGANIZATIONS	74-0366
SMALL	INFLUENCE OF NON-SINGULAR STRESS TERMS AND SPECIMEN GEOMETRY ON SMALL SCALE YIELDING AT CRACK TIPS I	73-0203
SMALL	LIMITATIONS TO THE SMALL SCALE YIELDING APPROXIMATION FOR CRACK TIP PLASTICITY	74-0145
SMALL	MATRIX ANALYSIS OF SHELLS - SMALL AND LARGE DISPLACEMENTS	68-0016
SMALL	MATRIX ANALYSIS OF THREE-DIMENSIONAL ELASTIC MEDIA: SMALL AND LARGE DISPLACEMENTS	65-0003
SMALL	MIXED FINITE ELEMENT ANALYSIS OF PLATES IN BENDING, SMALL DEFLECTION THEORY OF ELASTIC AND ELASTO-PL	71-0783
SMALL	NONLINEAR CONSIDERATION OF GRAVITY IN A STIFFNESS TEST OF A WEAK STRUCTURE AT SMALL STRAINS	71-0035
SMALL	NOUVELLE CONCEPTION DES DRAINS DANS LES BARRAGES EN TERRE HOMOGENE DE PETITE ET MOYENNE DIMENSIONS (	73-0404
SMALL	ON SMALL VIBRATIONS AND PERTURBATIONS OF FLEXIBLE BODIES UNDERGOING ARBITRARY NOMINAL MOTION	70-0376
SMALL	ON THE STRESS DISTRIBUTION IN NOZZLE TO CYLINDER CONNECTIONS FOR SMALL DIAMETER RATIOS	72-0426
SMALL	SMALL SCALE YIELDING NEAR A CRACK IN PLANE STRAIN-FINITE ELEMENT ANALYSIS	71-0355
SMALL	SMALL SCALE YIELDING ANALYSIS OF MIXED MODE PLANE STRAIN CRACK PROBLEMS	73-0801
SMALL	STRESS ANALYSIS OF CONFIGURATIONS INVOLVING SMALL FILLETS	75-0361
SMALL	THEORETICAL AND EXPERIMENTAL CONSIDERATIONS ON FLOW OF VISCOUS LIQUIDS PAST PLATE AT SMALL AND AVERA	70-0144
SMALL	THREE DIMENSIONAL ANISOTROPIC AND INHOMOGENEOUS ELASTIC MEDIA MATRIX ANALYSIS FOR SMALL AND LARGE DI	65-0008
SMALL	TRANSIENT HYDROTHERMAL ANALYSIS OF SMALL LAKES	73-0445
SMALL	USE OF SMALL MODELS IN DESIGN AND ANALYSIS OF PRESTRESSED-CONCRETE REACTOR VESSELS	70-0433
SMALL	USE OF SMALL MODELS IN DESIGN AND ANALYSIS OF PRESTRESSED-CONCRETE REACTOR VESSELS	71-0426
SMALL-SCALE	AN ANALYSIS OF THE SMALL-SCALE STRENGTH TESTING OF ICE	72-0881
SMALL-STRAIN	MATUS: A THREE-DIMENSIONAL FINITE ELEMENT PROGRAM FOR SMALL-STRAIN ELASTIC ANALYSIS	73-0970
SMALLEST	AN ECONOMICAL METHOD FOR DETERMINING THE SMALLEST EIGENVALUES OF LARGE LINEAR SYSTEMS	71-0551
SMOOTH	BASELINE SOLUTIONS FOR THE SMOOTH WALL THERMOCHEMICAL ABLATION RESPONSE OF GRAPHITE AND CARBON PHENO	71-0728
SMOOTH	FINITE ELEMENT ANALYSIS OF THE SURFACE DEFORMATION DUE TO A UNIFORM LOADING ON A LAYER OF GIBSON SOI	75-0240
SMOOTH	FINITE ELEMENT METHOD FOR NON SMOOTH DOMAINS AND COEFFICIENTS	72-0504
SMOOTH	THE RELATION BETWEEN THE GALERKIN AND COLLOCATION METHODS USING SMOOTH SPLINES	74-0912
SMOOTH-SURFACE	APPLICATION OF THE SMOOTH-SURFACE INTERPOLATION TO THE FINITE ELEMENT ANALYSIS	69-0088
SMOOTH-SURFACE	THE APPLICATION OF THE SMOOTH-SURFACE INTERPOLATION TECHNIQUE TO THE FINITE-ELEMENT ANALYSIS OF THIN	66-0120
SMOOTHING	LEAST SQUARES SMOOTHING OF EXPERIMENTAL DATA USING FINITE ELEMENTS	68-0224
SMOOTHING	LEAST SQUARES SURFACE FITTING BY FINITE ELEMENTS AND AN APPLICATION TO STRESS SMOOTHING	67-0154
SMOOTHING	LOCAL AND GLOBAL SMOOTHING OF DISCONTINUOUS FINITE ELEMENT FUNCTIONS USING A LEAST SQUARES METHOD	74-0378
SMOOTHING	LOCAL LEAST SQUARES STRESS SMOOTHING FOR PARABOLIC ISOPARAMETRIC ELEMENTS	75-0593
SMOOTHING	PROGRAM FOR TRIANGULAR BENDING ELEMENTS WITH DERIVATIVES SMOOTHING	72-0473
SNAP	ELASTIC SNAP THROUGH ANALYSIS OF CURVED PLATES USING DISCRETE ELEMENTS	72-0899
SNAP	SNAP DYNAMICS	72-0911
SNAP-THROUGH	ADAPTION OF KOITER'S METHOD TO FINITE ELEMENT ANALYSIS OF SNAP-THROUGH BUCKLING BEHAVIOUR	71-0109
SNAP-THROUGH	APPLICATION OF THE STIFFNESS METHOD TO A SNAP-THROUGH PROBLEM	60-0006
SNAP-THROUGH	SNAP-THROUGH BUCKLING ANALYSIS OF SHELLS OF REVOLUTION USING INCREMENTAL STIFFNESS MATRICES	74-1349
SNAP-THROUGH	SNAP-THROUGH BUCKLING OF RETICULATED SHELLS	75-0640
SNAP-TRHOUGH	ELASTIC SNAP-TRHOUGH ANALYSIS OF CURVED PLATES USING DISCRETE ELEMENTS	72-0927
SNOW	ELASTIC STRESSES IN LAYERED SNOW PACKS	72-0925
SNOW	FINITE ELEMENT ANALYSIS OF SHELLS UNDER NONUNIFORM SNOW LOADS	69-0113
SOBOLEV	METHODE DES ELEMENTS FINIS; UNE FAMILE D'APPROXIMATION DES ESPACES DE SOBOLEV PAR LES TRANSLATES DE	70-0301
SOBRE	SOBRE TORRES ESPIADAS - METODO DE CALCULO E UM EXEMPLO DE APLICACAO. (METHOD AND AN EXAMPLE OF CALCU	74-0915
SOBSTVENNYKH	CHISLENNYI METOD RASCHETA SOBSTVENNYKH I VYNUZHDENNYKH KOLEBANII SOSTAVNYKH OBOLOCHECHNYKH KONSTRUKT	73-0461
SOCKETTED	LOAD TRANSFER MECHANISM FOR PIERS SOCKETTED IN HARD SOILS OR ROCK	74-0969
SODIUM	THERMAL STRESSES IN FACT REACTOR BOILER TUBES AT THE SODIUM/GAS INTERFACE	73-0820
SOFT	A FINITE-ELEMENT PROGRAM FOR THE ANALYSIS OF EMBANKMENTS OVER SOFT, SATURATED SOILS INCLUDING CONSOL	73-1160

SOFT	ANALYSIS OF LOAD-BEARING FILLS OVER SOFT SUBSOILS	71-0330
SOFT	FINITE ELEMENT PROGRAM FEECON FOR UNDRAINED DEFORMATION ANALYSES OF GRANULAR EMBANKMENTS ON SOFT CLA	72-0875
SOFT	NUMERICAL ANALYSIS OF EMBANKMENTS OVER SOFT SOILS	75-0719
SOFT	PHOTOELASTIC AND FINITE ELEMENT ANALYSIS OF EMBANKMENTS CONSTRUCTED OVER SOFT SOILS	70-0151
SOFT	PREDICTION OF LONG TERM DEFORMATION OF A COMPACTED COHESIVE SOIL EMBANKMENT OVER A SOFT FOUNDATION	74-0727
SOFT	SETTLEMENTS AND STRENGTHENING OF SOFT CLAY ACCELERATED BY SAND DRAINS	73-0438
SOFT	SURFACE FOOTINGS ON SOFT ROCKS	74-0990
SOFT-BONDED	VIBRATION AND STRESS ANALYSIS OF SOFT-BONDED SHUTTLE INSULATION TILES, MODAL ANALYSIS WITH COMPACT W	74-0708
SOFTENING	ELASTO-PLASTIC STRESS ANALYSIS. A GENERALIZATION FOR VARIOUS CONSTITUTIVE RELATIONS INCLUDING STRAIN	72-0547
SOFTWARE	ASKA, A LARGE SCALE SOFTWARE SYSTEM FOR FINITE ELEMENT ANALYSIS	69-0331
SOFTWARE	ASKA, SOFTWARE ENGINEERING FOR THE SEVENTIES IN STRUCTURAL ANALYSIS	73-0704
SOFTWARE	DEVELOPMENT AND MAINTENANCE OF LARGE FINITE ELEMENT SOFTWARE SYSTEMS	74-0865
SOFTWARE	FINITE ELEMENT SOFTWARE IN THE NEXT DECADE	75-0259
SOFTWARE	GENERAL PURPOSE PROGRAM FOR FINITE ELEMENT ANALYSIS WITH REFERENCE TO INTERACTION BETWEEN SOFTWARE A	73-0334
SOFTWARE	HARDWARE AND SOFTWARE CONSIDERATIONS IN GENERAL PURPOSE PROGRAMS FOR FINITE ELEMENT ANALYSIS	70-0261
SOFTWARE	PORTABILITY AND ADAPTABILITY OF CIVIL ENGINEERING SOFTWARE	73-0343
SOFTWARE	SOFTWARE USER: EDUCATION AND QUALIFICATION	72-0441
SOFTWARE	STRUCTURAL MECHANICS SOFTWARE. A BIBLIOGRAPHY WITH ABSTRACTS	74-0842
SOIL	A COMPUTER MODEL FOR PREDICTING THE LOAD-DEFLECTION RESPONSE OF EXPEDIENT SOIL SURFACING	73-0880
SOIL	A FINITE ELEMENT STUDY OF THE PENETRATION OF A SOIL HALF-SPACE BY A RIGID PROJECTILE	69-0518
SOIL	ADAPTATION OF CRITICAL STATE SOIL MECHANICS FOR USE IN FINITE ELEMENTS	71-0534
SOIL	ANALYSIS OF SLOPE STABILITY BY FINITE ELEMENT METHOD AND ITS APPLICATION PROGRAM. SOIL MECHANICS	72-0919
SOIL	ANALYSIS OF SOIL MOVEMENT AROUND A DEEP EXCAVATION	70-0039
SOIL	ANALYTICAL AND EXPERIMENTAL EVALUATION OF MODULUS OF SOIL REACTION	74-1168
SOIL	APPLICATION OF THE FINITE ELEMENT METHOD TO STRESS DISTRIBUTION IN SOIL MEDIA	67-0075
SOIL	APPLICATION OF THE FINITE ELEMENT METHOD TO PROBLEMS IN SOIL AND ROCK MECHANICS	67-0118
SOIL	APPLICATION OF THE THREE-DIMENSIONAL FINITE ELEMENT METHOD TO SOME PROBLEMS IN SOIL MECHANICS	70-0510
SOIL	ASUPRA ANALIZEI PRIN METODA ELEMENTULUI FINIT A STARILOR DE EFORTURI SI DEFORMATII DIN MASIVELE DE P	70-0471
SOIL	CALCUL DE MASSIFS EN ELASTOPLASTICITE APPLICATIONS A LA MECANIQUE DES ROCHES ET DES SOLS. (APPLICATI	74-0223
SOIL	DER SPANNUNGS UND BEANSPRUCHUNGSZUSTAND DES BODENMATERIALS IN DER NACHBARSCHAFT EINER BODENBOESCHUNG	75-0371
SOIL	DYNAMIC ANALYSIS OF SOIL - STRUCTURE SYSTEM USING FINITE ELEMENT METHOD	67-0001
SOIL	DYNAMIC RESPONSE OF SOIL/CONCRETE INTERFACES AT HIGH PRESSURE	74-0836
SOIL	DYNAMICS OF BUILDING - SOIL INTERACTION	72-0856
SOIL	ELASTO-PLASTIC FINITE ELEMENT ANALYSIS OF SOIL DEFORMATION	73-1157
SOIL	EVALUATION OF STRESSES IN SOIL UNDER RIGID WHEELS BY FINITE ELEMENT METHOD	73-1109
SOIL	FINITE ELEMENT ANALYSIS OF SLOPES IN SOIL	68-0061
SOIL	FINITE ELEMENT ANALYSIS OF NONLINEAR SOIL MEDIA	69-0132
SOIL	FINITE ELEMENT ANALYSIS OF SOIL DEFORMATIONS	69-0134
SOIL	FINITE ELEMENT ANALYSES OF SOIL MOVEMENTS CAUSED BY DEEP EXCAVATION AND DEWATERING	69-0530
SOIL	FINITE ELEMENT ANALYSIS OF CREEP PROBLEMS IN SOIL MECHANICS	71-0809
SOIL	FINITE ELEMENT ANALYSIS OF CONSOLIDATION OF UNSATURATED SOIL	74-0082
SOIL	FINITE ELEMENT ANALYSIS OF SOIL COMPACTION	74-0931
SOIL	FINITE ELEMENT ANALYSIS OF THE SURFACE DEFORMATION DUE TO A UNIFORM LOADING ON A LAYER OF GIBSON SOI	75-0240
SOIL	FINITE ELEMENT METHOD APPLIED TO PROBLEMS IN STRESSES AND DEFORMATION OF SOIL	67-0085
SOIL	FINITE ELEMENT METHOD FOR PROBLEMS IN SOIL MECHANICS	68-0078
SOIL	FINITE ELEMENT METHOD OF PREDICTING STRESS DISTRIBUTIONS AND SOIL DEFORMATION UNDER A TRACTIVE DEVIC	69-0389
SOIL	FINITE ELEMENT METHOD FOR SOIL DEFORMATIONS	70-0495
SOIL	FINITE ELEMENT SOLUTION TO SOME PLANE PROBLEMS IN SOIL DYNAMICS	74-0448
SOIL	FINITE ELEMENTS METHOD IN SOIL MECHANICS (METODA KONECNYCH PRVKU V MECHANICE SEMIN),	71-0468
SOIL	FINITE-ELEMENT METHOD APPLIED TO SOME PROBLEMS IN SOIL MECHANICS	68-0077
SOIL	FRACTURE IN SOIL CEMENT USING THE FINITE ELEMENT METHOD	74-0434
SOIL	GRAVITY FLOW BIN DESIGN FOR LUNAR SOIL	72-0240
SOIL	HYSTERETIC DAMPING SYSTEMS APPLIED TO SOIL STRUCTURE INTERACTION BY FINITE ELEMENT METHOD	73-0663
SOIL	IN SITU CREEP ANALYSIS OF ROOM IN FROZEN SOIL	72-0128
SOIL	INCREMENTAL NUMERICAL SOLUTION OF A SIMPLE DEFORMATION PROBLEM IN SOIL MECHANICS	70-0380
SOIL	INFLUENCE OF SOIL/STRUCTURE INTERACTION PARAMETERS ON FLOOR RESPONSE SPECTRA	73-0833
SOIL	INPUT SOIL PROPERTIES FOR THE FINITE ELEMENT CODE NTEDYNS	71-0694
SOIL	LUMPED PARAMETER APPROACH TO AXISYMMETRIC DYNAMIC SOIL DEFORMATIONS	69-0170
SOIL	METODA ELEMENTOW SKONSZONYCH W MECHANICE GRUNTOW I MECHANICE GOROTWORU (FINITE ELEMENT METHOD IN SOI	73-0500
SOIL	NON-LINEAR STRESS ANALYSIS OF TWO DIMENSIONAL PROBLEMS WITH SPECIAL REFERENCE TO ROCK AND SOIL MECHA	68-0183
SOIL	NUMERICAL COMPUTATION OF BOUNDARY VALUE PROBLEMS IN SOIL MECHANICS	73-0269
SOIL	PLANE PLASTIC DEFORMATION OF SOIL,	71-0006
SOIL	PREDICTION OF LONG TERM DEFORMATION OF A COMPACTED COHESIVE SOIL EMBANKMENT OVER A SOFT FOUNDATION	74-0727
SOIL	QUAD-4: A COMPUTER PROGRAM FOR EVALUATING THE SEISMIC RESPONSE OF SOIL STRUCTURES BY VARIABLE DAMPIN	73-0771
SOIL	SEISMIC RESPONSE OF SOIL DEPOSITS UNDERLAIN BY SLOPING ROCK BOUNDARIES	69-0484
SOIL	SETTLEMENT ANALYSIS OF A STRIP FOOTING USING A CRITICAL STATE SOIL MODEL IN CONJUNCTION WITH FINITE	71-0533
SOIL	SIMPLIFIED SOIL-STRUCTURE INTERACTION ANALYSIS WITH STRAIN DEPENDENT SOIL PROPERTIES	74-1052
SOIL	SOIL RESISTANCE MOVEMENT RELATIONSHIPS FOR LATERALLY LOADED PILES IN CLAYS AT GREAT DEPTHS	70-0162
SOIL	SOLUTION OF SOME PLANE STRAIN PROBLEMS IN SOIL MECHANICS USING THE METHOD OF FINITE ELEMENTS	69-0521
SOIL	SOLUTION OF STRESS-DEFORMATION PROBLEMS IN SOIL AND ROCK MECHANICS USING FINITE ELEMENT METHODS	69-0272
SOIL	SOME NON LINEAR PROBLEMS IN SOIL AND ROCK MECHANICS - FINITE ELEMENT SOLUTION	69-0363
SOIL	STEADY STATE IN SOIL SUBJECTED TO SEVERAL COUPLED FORCES: A FINITE ELEMENT FORMULATION	70-0207
SOIL	STRESS - DEFORMATION PREDICTION IN CEMENT - TREATED SOIL PAVEMENTS	71-0286
SOIL	STRESS ANALYSIS OF CONTRACTIVE OR DILATIVE SOIL	71-0579
SOIL	TEST OF VALIDITY OF DARCY'S LAW FOR A CLAY SOIL	74-0070
SOIL	THE APPLICATION OF THE FINITE ELEMENT METHOD TO SEEPAGE AND STRESS-DEFORMATION PROBLEMS IN SOIL MECH	70-0267
SOIL	THE EFFECTS OF SOIL PARAMETERS AND BOUNDARY CONDITIONS ON THE CONSOLIDATION OF AN ELASTIC LAYER	70-0422
SOIL	THE FINITE ELEMENT METHOD AND ITS APPLICATION TO SOIL DYNAMIC PROBLEMS	68-0288
SOIL	THE FINITE ELEMENT METHOD FOR PREDICTING STRESS DISTRIBUTION AND SOIL DEFORMATION UNDER A TRACTIVE D	69-0522
SOIL	THE FINITE ELEMENT METHOD FOR PREDICTING STRESS DISTRIBUTION AND SOIL DEFORMATION UNDER A TRACTIVE D	71-0801
SOIL	THE USES OF FINITE ELEMENT TECHNIQUES IN SOIL MECHANICS WITH PARTICULAR REFERENCE TO DEFORMATION PLA	72-0002
SOIL	UNDRAINED VISCO-ELASTIC ANALYSIS OF SOIL DEFORMATION	72-0424
SOIL-BUILDING	ELASTIC-PLASTIC EARTHQUAKE RESPONSE OF SOIL-BUILDING SYSTEMS	72-0717
SOIL-CEMENT	FATIGUE CRACK FORMATION AND PROPAGATION IN PAVEMENTS CONTAINING SOIL-CEMENT BASES.	72-0435
SOIL-CULVERT	FINITE-ELEMENT SOLUTION OF STRESSES AND DISPLACEMENTS IN A SOIL-CULVERT SYSTEM	73-1156
SOIL-FILLED	DYNAMIC FINITE ELEMENT ANALYSIS OF AXI-SYMMETRIC STRESS WAVE PROPAGATION IN SOIL-FILLED BIN	72-0622
SOIL-FOUNDATION	APPLICATION OF FINITE ELEMENT METHOD TO SOIL-FOUNDATION INTERACTION ANALYSES	74-1080
SOIL-STRUCTURE	A FINITE ELEMENT APPROACH TO ELASTIC SOIL-STRUCTURE INTERACTION	70-0013
SOIL-STRUCTURE	ANALYSIS OF SOIL-STRUCTURE INTERACTION EFFECTS UNDER SEISMIC EXCITATION	71-0064
SOIL-STRUCTURE	FINITE ELEMENT ANALYSES OF SOIL-STRUCTURE INTERACTION IN U-FRAME LOCKS	69-0531
SOIL-STRUCTURE	FINITE ELEMENT TREATMENT OF SOIL-STRUCTURE INTERACTION PROBLEM FOR NUCLEAR POWER PLANT UNDER SEISMIC	73-0834
SOIL-STRUCTURE	INVESTIGATION OF SOIL-STRUCTURE INTERACTION: FLEXIBLE CULVERT PIPES	72-0661
SOIL-STRUCTURE	LUSH: A COMPUTER PROGRAM FOR COMPLEX RESPONSE ANALYSIS OF SOIL-STRUCTURE SYSTEMS	74-0744
SOIL-STRUCTURE	SIMPLIFIED SOIL-STRUCTURE INTERACTION ANALYSIS WITH STRAIN DEPENDENT SOIL PROPERTIES	74-1052
SOIL-STRUCTURE	SOIL-STRUCTURE INTERACTION ANALYSES FOR EVALUATING SEISMIC RESPONSE	74-0745
SOIL-STRUCTURE	SOIL-STRUCTURE INTERACTION - CONTINUUM OR FINITE ELEMENT	74-0954
SOIL-STRUCTURE	SUMMARY OF SOIL-STRUCTURE INTERACTION	72-0817
SOIL-WATER	THE USE OF THE FINITE ELEMENT METHOD IN OBTAINING SOIL-WATER PRESSURE AND SOIL-WATER CONTENT DISTRIB	75-0112
SOIL-WATER	THE USE OF THE FINITE ELEMENT METHOD IN OBTAINING SOIL-WATER PRESSURE AND SOIL-WATER CONTENT DISTRIB	75-0112
SOILS	A FINITE-ELEMENT PROGRAM FOR THE ANALYSIS OF EMBANKMENTS OVER SOFT, SATURATED SOILS INCLUDING CONSOL	73-1160
SOILS	ANALYSIS OF CIRCULAR FOOTINGS ON LAYERED SOILS	70-0281
SOILS	ANWENDUNG DER FINITE-ELEMENT-METHODE ZUR BERECHNUNG VON SPANNUNGEN IN WASSERGESAETTIGTEN BOEDEN (FIN	73-0502

SOILS	BEHAVIOUR OF STRIP FOOTINGS ON LAYERED COHESIVE SOILS	69-0223
SOILS	CALCULATION OF STEADY-STATE WATER-TABLE HEIGHTS IN DRAINED SOILS BY MEANS OF FINITE-ELEMENT METHOD	75-0457
SOILS	COUPLED HEAT AND MOISTURE TRANSPORT MODEL FOR ARCTIC SOILS	74-0578
SOILS	DYNAMIC RESPONSE OF STRUCTURES IN LAYERED SOILS	74-0761
SOILS	EFFECT OF DENSIFICATION ON THE ENGINEERING CHARACTERISTICS OF ORGANIC SOILS - VOL. 1 AND 2.	73-0805
SOILS	EFFECT OF THE VARIATION OF CONSISTENCY IN RESIDUAL SOILS ON LOAD TRANSFER BY BORED PILES	75-0812
SOILS	EFFECTIVE STRESS-STRAIN-STRENGTH MODEL FOR SOILS	75-0056
SOILS	EVALUATION OF A THREE-DIMENSIONAL STRESS CELL FOR GRANULAR SOILS	72-0882
SOILS	FINITE ELEMENT ANALYSIS OF ELASTO-PLASTIC SOILS. REPORT NO. 4, FINITE ELEMENT ANALYSIS OF ELASTO-PLA	72-0788
SOILS	FINITE ELEMENT ANALYSIS OF 2-DIMENSIONAL FLOW IN SOILS CONSIDERING WATER UPTAKE BY ROOTS .1. THEORY	75-0137
SOILS	FINITE ELEMENT ANALYSIS OF 2-DIMENSIONAL FLOW IN SOILS CONSIDERING WATER UPTAKE BY ROOTS .2. FIELD A	75-0147
SOILS	FINITE ELEMENT STUDIES OF SOILS AND POROUS MEDIA	71-0504
SOILS	GALERKIN APPROACH TO UNSATURATED FLOW IN SOILS	74-0058
SOILS	GROUNDWATER FLOW IN PARTIALLY SATURATED SOILS	72-0745
SOILS	LOAD TRANSFER MECHANISM FOR PIERS SOCKETTED IN HARD SOILS OR ROCK	74-0969
SOILS	MECHANICAL CHARACTERIZATION AND FINITE-ELEMENT ANALYSIS OF ELASTIC-PLASTIC, WORK-HARDENING SOILS	72-0974
SOILS	METOD KONECHNYKH ELEMENTOV V RASCHETAKH KONSOLIDATSII VODONASYSHCHENNYKH GRUNTOV. (FINITE ELEMENT ME	75-0469
SOILS	NONLINEAR ANALYSIS OF STRESS AND STRAIN IN SOILS	70-0050
SOILS	NUMERICAL ANALYSIS OF ELECTRO-OSMOTIC FLOW IN SOILS	73-0444
SOILS	NUMERICAL ANALYSIS OF PLASTICITY IN SOILS	73-0611
SOILS	NUMERICAL ANALYSIS OF EMBANKMENTS OVER SOFT SOILS	75-0719
SOILS	PHOTOELASTIC AND FINITE ELEMENT ANALYSIS OF EMBANKMENTS CONSTRUCTED OVER SOFT SOILS	70-0151
SOILS	SOLUTION OF THE MIXED PROBLEM OF THE THEORY OF ELASTICITY AND THE THEORY OF PLASTICITY OF SOILS	74-1136
SOILS	THE EFFECT OF STRESS HISTORY ON THE RESILIENT RESPONSE OF SOILS	73-0914
SOILS	THE USE OF THE FINITE ELEMENT METHOD IN THE SOLUTION OF THE FLOW EQUATIONS IN SOILS	74-0449
SOILS	THEORETICAL TRANSIENT BEHAVIOUR OF SATURATED AND UNSATURATED SOILS UNDER LOAD AND CHANGING MOISTURE	73-0901
SOILS	ULTIMATE CAPACITY OF CIRCULAR FOOTINGS ON LAYERED SOILS	70-0280
SOLAR	ACOUSTIC RESPONSE OF A SPACECRAFT SOLAR PANEL USING NORMAL MODE METHOD	69-0422
SOLAR	ADVANCED INTERCONNECT SYSTEMS FOR LIGHTWEIGHT SOLAR ARRAYS	73-0696
SOLAR	APPLICATION OF SMALL AND LARGE-DISPLACEMENT THEORY TO A LARGE FLEXIBLE SOLAR ARRAY AND COMPARISON WI	71-0231
SOLAR	DEVELOPMENT OF ADVANCED INTERCONNECTORS FOR SOLAR CELLS	74-0371
SOLAR	FINITE ELEMENT ANALYSIS OF MERCURY SLOSH IN THE SOLAR ELECTRIC PROPULSION STAGE	75-0757
SOLAR	STRESS ANALYSIS AND DESIGN OF SILICON SOLAR CELL ARRAYS AND RELATED MATERIAL PROPERTIES	72-0840
SOLDER	APPLICABILITY OF APPROXIMATE AND EXACT TRANSIENT HEAT TRANSFER ANALYSES TO HEATING PROCESSES USED TO	75-0244
SOLICIATE	COMPORTAREA ELASTICA A DIGURILOR DE PAMINT SOLICIATE SEISMIC (ELASTIC BEHAVIOR OF THE EARTH DAMS SEI	74-1185
SOLID	A DERIVATION OF PERTURBED EQUATIONS OF MOTION AND HEAT CONDUCTION OF A THERMOELASTIC SOLID	72-0735
SOLID	A NEW ASSUMED STRESS HYBRID FINITE ELEMENT MODEL FOR SOLID CONTINUA	71-0371
SOLID	A NEW FINITE ELEMENT MODEL FOR SOLID CONTINUA	68-0156
SOLID	A RADIATIVE IGNITION MODEL OF A SOLID FUEL	74-1312
SOLID	A THREE DIMENSIONAL "VARIABLE NODE" ISOPARAMETRIC SOLID ELEMENT	74-0779
SOLID	ACOUSTIC ANALYSIS OF SOLID ROCKET MOTOR CAVITIES BY A FINITE ELEMENT METHOD	71-0119
SOLID	AN INVESTIGATION INTO THE METHODS OF DETECTING DEWETTED ZONES IN SOLID ROCKET PROPELLANTS USING HOLO	73-0890
SOLID	ANALYSES OF AXISYMMETRIC UPSETTING AND PLANE STRAIN SIDE PRESSING OF SOLID CYLINDERS BY THE FINITE E	70-0083
SOLID	ANALYSES OF AXISYMMETRIC UPSETTING AND PLANE-STRAIN SIDE-PRESSING OF SOLID CYLINDERS BY THE FINITE E	71-0794
SOLID	ANALYSIS OF THE BIAXIAL STRIP AND SHEAR LAP TESTS FOR SOLID PROPELLANT CHARACTERIZATION	67-0019
SOLID	APPLICATION OF NONLINEAR ANALYSIS TO SOLID STRUCTURES	72-0212
SOLID	APPLICATION OF THE FINITE ELEMENT METHOD TO STRESS ANALYSIS OF SOLID PROPELLANT ROCKET GRAINS	65-0009
SOLID	BASIS OF FINITE ELEMENT METHODS FOR SOLID CONTINUA	69-0213
SOLID	CASE BOND STRESS CALCULATIONS FOR FLAPPED CYLINDRICAL ANALOGS OF SOLID PROPELLANT ROCKET MOTORS. BON	72-0842
SOLID	COMPUTATIONAL METHODS IN SOLID MECHANICS AND STRESS ANALYSIS	69-0006
SOLID	COMPUTER AIDED DESIGN AND CONSTRUCTION BY MEANS OF CURVED SURFACE AND SOLID ELEMENTS	73-0165
SOLID	COMPUTER SYSTEM FOR THE ANALYSIS OF LARGE SCALE PROBLEMS IN SOLID, FLUID AND THERMODYNAMICS	73-0554
SOLID	CURVED FINITE ELEMENTS IN THE ANALYSIS OF SOLID SHELL AND PLATE STRUCTURES	69-0357
SOLID	DEVELOPMENT OF A SOLID ROCKET PROPELLANT NONLINEAR CONSTITUTTIVE THEORY	75-0722
SOLID	DEVELOPMENT OF TECHNOLOGY FOR MODELING OF A 1/8-SCALE DYNAMIC MODEL OF THE SHUTTLE SOLID ROCKET BOOS	74-0703
SOLID	DIFFUSION ON SOLID SPHERICAL PARTICLE IN VISCOUS FLUID FLOW AT FINITE VALUES OF RE NUMBERS	69-0419
SOLID	DISCUSSION ON THE ADAPTION OF CRITICAL STATE SOLID MECHANICS THEORY FOR USE IN FINITE ELEMENTS	71-0007
SOLID	ECONOMICAL COMPUTER TECHNIQUES FOR SOLID FINITE ELEMENTS	68-0086
SOLID	FINITE ELEMENT METHOD FOR CONTACT PROBLEMS OF SOLID BODIES - 2.	71-0435
SOLID	FINITE ELEMENT METHOD FOR A SOLID ENCLOSING AN INVISCID INCOMPRESSIBLE FLUID	73-0090
SOLID	FINITE ELEMENT VISCOELASTIC STRESS ANALYSIS STUDY USING THE GENERALIZED KELVIN SOLID MODEL	73-0957
SOLID	FORMULATIONS OF FINITE ELEMENT METHODS FOR SOLID CONTINUA	71-0679
SOLID	FOUNDATIONS OF FINITE ELEMENT METHODS FOR SOLID CONTINUA	71-0213
SOLID	FUNCTIONAL MINIMIZATION IN NONLINEAR SOLID MECHANICS	74-1251
SOLID	HANDBOOK FOR THE ENGINEERING STRUCTURAL ANALYSIS OF SOLID PROPELLANTS	71-0642
SOLID	HEAT CONDUCTION AND STRESS ANALYSIS OF SOLID PROPELLANT ROCKET MOTOR NOZZLES	69-0059
SOLID	INTERACTION OF SOLID SHEAR WALLS AND THEIR SUPPORTING STRUCTURES.	72-0450
SOLID	MEASUREMENT OF THERMAL EXPANSION IN SOLID PROPELLANTS	73-0916
SOLID	NEW DISPLACEMENT HYBRID FINITE ELEMENT MODELS FOR SOLID CONTINUA	70-0115
SOLID	NOTES ON THE FINITE ELEMENT ANALYSIS OF THE AXISYMMETRIC ELASTIC SOLID	74-0185
SOLID	ON THE DETERMINATION OF EFFECTIVE ELASTIC-PLASTIC PROPER TIES FOR THE EQUIVALENT SOLID PLATE ANALYSI	74-0494
SOLID	PERIODIC HEAT CONDUCTION IN A TWO PHASE, TWO DIMENSIONAL SOLID DOMAIN	74-1164
SOLID	SOLID ELEMENT REQUIREMENT FOR NASTRAN	71-0256
SOLID	SOLID SAP. A STATIC ANALYSIS PROGRAM FOR THREE DIMENSIONAL SOLID STRUCTURES	72-0855
SOLID	SOLID SAP. A STATIC ANALYSIS PROGRAM FOR THREE DIMENSIONAL SOLID STRUCTURES	72-0855
SOLID	SOLID WEDGE FINITE ELEMENT FOR STRESS ANALYSIS	69-0124
SOLID	SOLVING A NONLINEAR PROBLEM OF NONSTEADY THERMAL CONDUCTIVITY IN SOLID BODIES OF ARBITRARY FORM	68-0019
SOLID	SOME CONTRIBUTIONS TO NONLINEAR SOLID MECHANICS	73-0705
SOLID	SOME CONTRIBUTIONS TO NONLINEAR SOLID MECHANICS	74-0674
SOLID	STRAIN HARDENING TORSION OF SOLID BARS AS A MINIMIZATION PROBLEM	70-0174
SOLID	STRUCTURALLY COMPATIBLE FLUID FINITE-ELEMENT FOR SOLID FLUID INTERACTION STUDIES	75-0344
SOLID	THE EFFECT OF GRAVITY UPON THE MELT-THROUGH TIME OF A SOLID SUBJECTED TO A HIGH INTENSITY LASER	72-0693
SOLID	THE FINITE ELEMENT METHOD IN SOLID MECHANICS	70-0295
SOLID	THE THERMO-ELASTO-PLASTIC ANALYSIS OF A RE-ENTRY BODY TYPE COMPOSITE SOLID STRUCTURE USING A NEW FIN	66-0093
SOLID	UNTERSUCHUNGEN ZUR SPURMASSTABILITAET BEI RADSAETZEN MIT KLOTZGEBREMSTEN VOLLRAEDERN (INVESTIGATIONS	75-0810
SOLID	VARIATIONAL FORMULATIONS OF NUMERICAL METHODS IN SOLID CONTINUA	71-0670
SOLID	WATER IMPACT ANALYSIS OF SPACE SHUTTLE SOLID ROCKET BY THE FINITE ELEMENT METHOD	74-0760
SOLID	WELDED JOINT STIFFNESS OBTAINED USING SOLID FINITE ELEMENTS	75-0190
SOLIDIFICATION	CALCULATION OF SOLIDIFICATION OF MOLTEN METAL BY FINITE-ELEMENT METHOD	74-1357
SOLIDIFICATION	FINITE-ELEMENT MODELING OF DIRECTIONAL SOLIDIFICATION OF SUPER ALLOY EUTECTICS	75-0350
SOLIDS	A DIGITAL COMPUTER PROGRAM FOR THE FINITE ELEMENT ANALYSIS OF SOLIDS WITH NONLINEAR MATERIAL PROPERT	65-0032
SOLIDS	A NEW APPROACH TO THE FINITE-ELEMENT FORMULATION AND SOLUTION OF A CLASS OF PROBLEMS IN COUPLED THER	72-0732
SOLIDS	A NEW APPROACH TO THE FINITE-ELEMENT FORMULATION AND SOLUTION OF A CLASS OF PROBLEMS IN COUPLED THER	73-0600
SOLIDS	A REFINED AXISYMMETRIC FINITE ELEMENT FOR THE ANALYSIS OF NEARLY INCOMPRESSIBLE SOLIDS	70-0309
SOLIDS	ANALYSIS OF AXISYMMETRIC SOLIDS BY THE DIRECT STIFFNESS METHOD	66-0049
SOLIDS	ANALYSIS OF FINITE DEFORMATIONS OF ELASTIC SOLIDS BY THE FINITE ELEMENT METHOD	70-0096
SOLIDS	ANALYSIS OF NONLINEAR THERMOELASTIC AND THERMOPLASTIC BEHAVIOUR OF SOLIDS OF REVOLUTION BY THE FINIT	71-0192
SOLIDS	APPLICATION OF THE FINITE ELEMENT METHOD TO HEAT CONDUCTION IN SOLIDS	67-0005
SOLIDS	APPLICATION OF THE FINITE-ELEMENT METHOD TO HEAT-TRANSFER PROBLEMS. PART 1. FINITE-ELEMENT METHOD AP	71-0760
SOLIDS	APPLICATION OF THE METHOD OF FINITE ELEMENTS TO CALCULATIONS OF, THE STRESSED AND STRAINED STATE OF I	75-0775
SOLIDS	APPROXIMATIONS AND NUMERICAL ANALYSIS OF FINITE DEFORMATIONS OF ELASTIC SOLIDS	74-0343
SOLIDS	ASAAS: ASYMMETRIC STRESS ANALYSIS OF AXISYMMETRIC SOLIDS WITH ORTHOTROPIC, TEMPERATURE-DEPENDENT MAT	71-0696

SOLIDS	BASIC RESEARCH IN METHODS OF APPROXIMATION IN THE NONLINEAR MECHANICS OF SOLIDS AND STRUCTURES	72-0857
SOLIDS	BASIC RESEARCH IN METHODS OF APPROXIMATION IN THE NONLINEAR MECHANICS OF SOLIDS AND STRUCTURES, VOLU	74-1091
SOLIDS	BOPACE 3-D (THE BOEING PLASTIC ANALYSIS CAPABILITY FOR 3-DIMENSIONAL SOLIDS USING ISOPARAMETRIC FINI	75-0699
SOLIDS	BRLESC FINITE ELEMENT PROGRAM FOR AXISYMMETRIC, PLANE STRAIN, AND PLANE STRESS, ORTHOTROPIC SOLIDS W	71-0749
SOLIDS	CHARACTERIZATION OF ELASTIC SOLIDS USING FINITE ELEMENT METHODS	72-0712
SOLIDS	DISCRETE APPROXIMATIONS RELATED TO NONLINEAR THEORIES OF SOLIDS	71-0259
SOLIDS	DISCRETIZATION-DISPERSION IN FINITE ELEMENT MODELLING OF WAVE PROPAGATION IN SOLIDS	74-1253
SOLIDS	DYNAMIC STRESS ANALYSIS OF POROUS ELASTIC SOLIDS SATURATED WITH COMPRESSIBLE FLUIDS	71-0651
SOLIDS	ELAD PROGRAM MANUAL. A COMPUTER PROGRAM FOR THE ELASTIC ANALYSIS OF DOME STRUCTURES AND AXISYMMETRIC	69-0064
SOLIDS	ELASTIC PLASTIC DYNAMIC ANALYSIS OF AXISYMMETRIC SOLIDS	73-0657
SOLIDS	EPAD, A COMPUTER PROGRAM FOR THE ELASTIC-PLASTIC ANALYSIS OF DOME STRUCTURES AND ARBITRARY SOLIDS OF	70-0036
SOLIDS	EVALUATION OF DISCRETE METHODS FOR THE LINEAR DYNAMIC RESPONSE OF ELASTIC AND VISCOELASTIC SOLIDS	70-0439
SOLIDS	FEATS, A COMPUTER PROGRAM FOR THE FINITE ELEMENT THERMAL STRESS ANALYSIS OF PLANE OR AXISYMMETRIC SO	68-0152
SOLIDS	FINITE ELEMENT ANALYSIS OF AXISYMMETRIC SOLIDS	65-0011
SOLIDS	FINITE ELEMENT ANALYSIS OF AXISYMMETRIC SOLIDS WITH ARBITRARY LOADINGS	67-0145
SOLIDS	FINITE ELEMENT ANALYSIS FOR SOLIDS OF REVOLUTION	68-0153
SOLIDS	FINITE ELEMENT ANALYSIS OF HEAT CONDUCTION IN ANISOTROPIC SOLIDS	68-0161
SOLIDS	FINITE ELEMENT ANALYSIS OF CRACK PROPAGATION IN THREE-DIMENSIONAL SOLIDS UNDER CYCLING LOADING	73-0529
SOLIDS	FINITE ELEMENT ANALYSIS FOR WAVE PROPAGATION IN ELASTIC-PLASTIC SOLIDS	73-0990
SOLIDS	FINITE ELEMENT ANALYSIS OF SLABS ON ELASTIC SOLIDS	74-0120
SOLIDS	FINITE ELEMENT ANALYSIS OF CRACK PROPAGATION IN 3-DIMENSIONAL SOLIDS UNDER CYCLIC LOADING	74-0479
SOLIDS	FINITE ELEMENT APPLICATIONS IN THE CHARACTERIZATION OF ELASTIC SOLIDS	71-0138
SOLIDS	FINITE ELEMENT METHOD FOR DETERMINATION OF THERMAL STRESSES INTERACTIVE GRAPHICS FOR FINITE ELEMENT	74-0155
SOLIDS	FINITE ELEMENT PROCEDURE FOR THE LARGE DEFORMATION DYNAMIC RESPONSE OF AXISYMMETRIC SOLIDS	74-0527
SOLIDS	FINITE ELEMENT PROCEDURE FOR THE LARGE DEFORMATION DYNAMIC RESPONSE OF AXISYMMETRIC SOLIDS	74-0967
SOLIDS	FINITE ELEMENT PROCEDURE FOR LARGE DEFORMATION DYNAMIC RESPONSE OF AXISYMMETRIC SOLIDS	75-0013
SOLIDS	FINITE ELEMENT STRESS ANALYSIS OF AXISYMMETRIC SOLIDS WITH ORTHOTROPIC TEMPERATURE DEPENDENT MATERIA	67-0089
SOLIDS	FINITE ELEMENT STRESS ANALYSIS FOR ANISOTROPIC SOLIDS	70-0478
SOLIDS	FINITE ELEMENT TECHNIQUE APPLIED TO HEAT CONDUCTION IN SOLIDS WITH TEMPERATURE DEPENDENT THERMAL CON	69-0004
SOLIDS	FINITE ELEMENTS FOR AXISYMMETRIC SOLIDS UNDER ARBITRARY LOADINGS WITH NODES ON ORIGIN.	72-0251
SOLIDS	FINITE ELEMENTS FOR AXISYMMETRIC SOLIDS UNDER ARBITRARY LOADINGS WITH NODES ON ORIGIN	73-0094
SOLIDS	FINITE PLANE STRAIN OF INCOMPRESSIBLE ELASTIC SOLIDS BY THE FINITE ELEMENT METHOD	68-0128
SOLIDS	FINITE-ELEMENT ANALYSIS OF FRACTURE PROPAGATION IN TWO-DIMENSIONAL ELASTIC BRITTLE SOLIDS	72-0948
SOLIDS	FINITE-ELEMENT METHOD APPLIED TO HEAT CONDUCTION IN SOLIDS WITH NONLINEAR BOUNDARY CONDITIONS	73-0005
SOLIDS	FINITE-ELEMENT METHOD FOR THE DETERMINATION OF THERMAL STRESSES IN ANISOTROPIC SOLIDS OF REVOLUTION	74-0262
SOLIDS	FLOW OF PLASTIC AND VISCO-PLASTIC SOLIDS WITH SPECIAL REFERENCE TO EXTRUSION AND FORMING PROCESSES	74-0210
SOLIDS	IDENTIFICATION OF NONLINEAR ELASTIC SOLIDS BY A FINITE ELEMENT METHOD	74-0530
SOLIDS	INCREMENTAL THERMOMECHANICAL THEORY OF VISCOELASTOPLASTIC SOLIDS AND SOLUTION BY FINITE ELEMENTS	73-0861
SOLIDS	MECHANICS OF DEFORMABLE SOLIDS - EURASIAN COMMUNIST COUNTRIES	73-0794
SOLIDS	NEW APPROACH TO FINITE ELEMENT FORMULATION AND SOLUTION OF A CLASS OF PROBLEMS IN COUPLED THERMOELAS	73-0013
SOLIDS	NONLINEAR DYNAMIC ANALYSIS OF AXISYMMETRIC SOLIDS BY THE FINITE ELEMENT METHOD	73-0942
SOLIDS	NONLINEAR DYNAMIC ANALYSIS OF AXISYMMETRIC SOLIDS BY THE FINITE ELEMENT METHOD	74-0510
SOLIDS	NONLINEAR DYNAMICS OF SOLIDS BY FINITE ELEMENT METHODS	69-0490
SOLIDS	NONLINEAR DYNAMICS OF SOLIDS BY THE FINITE ELEMENT METHOD	71-0116
SOLIDS	NONLINEAR FINITE ELEMENT DYNAMIC ANALYSIS OF AXISYMMETRIC SOLIDS	74-0775
SOLIDS	NONLINEAR TRANSIENT ANALYSIS OF SHELLS AND SOLIDS OF REVOLUTION BY CONVECTED ELEMENTS	73-0519
SOLIDS	NONLINEAR TRANSIENT ANALYSIS OF SHELLS AND SOLIDS OF REVOLUTION BY CONVECTED ELEMENTS	74-0476
SOLIDS	NUMERICAL ANALYSIS FOR FINITE AXISYMMETRIC DEFORMATIONS OF INCOMPRESSIBLE ELASTIC SOLIDS OF REVOLUTI	70-0094
SOLIDS	PLACID: A GENERAL FINITE ELEMENT COMPUTER PROGRAM FOR STRESS ANALYSIS OF PLANE AND AXISYMMETRIC SOLI	74-0884
SOLIDS	QUASI-ANALYTICAL FINITE ELEMENT PROCEDURES FOR AXISYMMETRIC ANISOTROPIC SHELLS AND SOLIDS	74-0297
SOLIDS	SAAS II: FINITE ELEMENT STRESS ANALYSIS OF AXISYMMETRIC SOLIDS WITH ORTHOTROPIC, TEMPERATURE DEPENDE	68-0096
SOLIDS	SAAS 111: FINITE ELEMENT STRESS ANALYSIS OF AXISYMMETRIC AND PLANE SOLIDS WITH DIFFERENT ORTHOTROPIC	71-0741
SOLIDS	SASL - A FINITE ELEMENT CODE FOR THE STATIC ANALYSIS OF AXISYMMETRIC AND PLANE SOLIDS SUBJECTED TO A	72-0779
SOLIDS	SEMI-ANALYTICAL FINITE ELEMENT PROCEDURE FOR CONDUCTION IN ANISOTROPIC AXISYMMETRIC SOLIDS	74-0270
SOLIDS	SHORT COMMUNICATIONS: TORSION OF AXI-SYMMETRIC SOLIDS OF VARIABLE DIAMETER - INCLUDING ACCELERATION	74-0203
SOLIDS	SLOW FINITE DEFORMATIONS OF VISCOUS SOLIDS	69-0423
SOLIDS	STABILITY AND OSCILLATION CHARACTERISTICS OF FINITE ELEMENT, FINITE-DIFFERENCE, AND WEIGHTED-RESIDUA	75-0615
SOLIDS	STATIC STRESS ANALYSIS OF AXISYMMETRIC SOLIDS WITH MATERIAL AND GEOMETRIC NONLINEARITIES BY THE FINI	73-0958
SOLIDS	STRESS ANALYSIS OF AXISYMMETRIC SOLIDS UTILIZING HIGHER ORDER QUADRILATERAL FINITE ELEMENTS	69-0356
SOLIDS	STRESS ANALYSIS OF AXISYMMETRIC SOLIDS WITH ASYMMETRIC PROPERTIES	72-0140
SOLIDS	STRUCTURAL ANALYSIS OF SOLIDS	63-0009
SOLIDS	STRUCTURAL ANALYSIS OF AXISYMMETRIC SOLIDS	65-0031
SOLIDS	STRUCTURAL ANALYSIS OF AXISYMMETRIC SOLIDS	65-0056
SOLIDS	THE FINITE ELEMENT ANALYSIS OF PHYSICALLY AND KINEMATICALLY NONLINEAR ELASTIC SOLIDS	69-0146
SOLIDS	THEMIS FINAL REPORT. VOL. II. NONLINEAR MECHANICS IN SOLIDS AND STRUCTURES	74-0722
SOLIDS	THEMIS FINAL REPORT. VOL. 1. BASIC RESEARCH IN METHODS OF APPROXIMATION IN THE NONLINEAR MECHANICS O	74-0723
SOLIDS	THEORETICAL ANALYSIS OF THICK DOME STRUCTURES AND ARBITRARY SOLIDS OF REVOLUTION SUBJECTED TO UNSYMM	70-0037
SOLIDS	THERMOMECHANICAL ANALYSIS OF VISCOELASTIC SOLIDS	70-0240
SOLIDS	THERMOMECHANICAL DISSIPATION ANALYSIS OF THERMOVISCOELASTIC SOLIDS BY FINITE ELEMENTS	72-0403
SOLIDS	THREE-DIMENSIONAL ANALYSIS OF ELASTIC SOLIDS I. ANALYSIS PROCEDURE	69-0222
SOLIDS	THREE-DIMENSIONAL FINITE-ELEMENT ELASTIC ANALYSIS OF PRISMATIC SOLIDS	73-1138
SOLIDS	TRANSIENT HEAT FLOW IN SOLIDS	74-0461
SOLIDS	TSAAS: FINITE ELEMENT THERMAL AND STRESS ANALYSIS OF AXISYMMETRIC SOLIDS WITH ORTHOTROPIC TEMPERATUR	72-0886
SOLIDS	USERS' MANUAL FOR DIGITAL COMPUTER PROGRAM DK1. NUMERICAL ANALYSIS OF FINITE AXISYMMETRIC DEFORMATIO	72-0861
SOLIDS	VARIATIONAL FORMULATION OF DYNAMICS OF FLUID-SATURATED POROUS ELASTIC SOLIDS	72-0017
SOLIDS	VIBRATIONAL CHARACTERISTICS OF THREE-DIMENSIONAL SOLIDS, WITH APPLICATIONS TO EARTH DAMS	69-0432
SOLIDS	VISCO-PLASTICITY, PLASTICITY AND CREEP IN ELASTIC SOLIDS. A UNIFIED NUMERICAL SOLUTION APPROACH	74-1279
SOLLICITATIONS	TENUE DES STRUCTURES DE GRAPHITE VIS-A-VIS DES SOLLICITATIONS RENCONTREES DANS UN COEUR DE REACTEUR	73-0817
SOLLICITE	MATRICE DE RIGIDITE D'UN ELEMENT DE POUTRE SOLLICITE A LA TORSION-FLEXION (STIFFNESS MATRIX OF A BEA	73-0442
SOLS	CALCUL DE MASSIFS EN ELASTOPLASTICITE APPLICATIONS A LA MECANIQUE DES ROCHES ET DES SOLS. (APPLICATI	74-0223
SOLUMN	SOLUMN INSTABILITY UNDER NONCONSERVATIVE FORCES, WITH INTERNAL AND EXTERNAL DAMPING FINITE ELEMENT U	74-0566
SOLUTION-MINED	ANALYTICAL METHODS FOR PREDICTING SUBSIDENCE ABOVE SOLUTION-MINED CAVITIES	74-1145
SOLVE	SIGNIFICANT APPLICATIONS OF FINITE ELEMENTS AND FINITE DIFFERENCES METHODS TO SOLVE PROBLEMS OF THER	73-0864
SOLVENCY	OCENA DOKLADNOSCI ROZWIAZAN PROBLEMOW BRZEGOWYCH SPREZYSTYCH CIAL DYSKRETYZOWANYCH. (ASSESSING THE A	74-0921
SOLVER	AN EQUATION SOLVER OF VERY LARGE CAPACITY	71-0677
SOLVER	INTEGRATED SEQUENTIAL SOLVER FOR LARGE MATRIX EQUATIONS	74-0276
SOLVER	LARGE CAPACITY EQUATION SOLVER FOR STRUCTURAL ANALYSIS	74-0983
SOLVER	PARAMETRIC STUDIES OF LAYERED PAVEMENT SYSTEMS USING AFPAV CODE WITH FAST EQUATION SOLVER	75-0747
SONAR	ANALYSIS AND REDESIGN OF SONAR PIEZOELECTRIC RESONATORS USING THE FINITE ELEMENT METHOD.	72-0374
SONAR	APPLICATIONS OF THE FINITE-ELEMENT METHOD AND COMPUTER GRAPHICS. (A VIBRATIONAL ANALYSIS OF A SONAR	72-0729
SONAR	FINITE ELEMENT ANALYSIS OF ACOUSTICALLY RADIATING STRUCTURES WITH APPLICATIONS TO SONAR TRANSDUCERS	73-0075
SONAR	FINITE ELEMENT ANALYSIS OF ACOUSTICALLY RADIATING STRUCTURES WITH APPLICATIONS TO SONAR TRANSDUCERS	74-0524
SONAR	VIBRATIONAL RESPONSE OF SONAR TRANSDUCERS USING PIEZO-ELECTRIC FINITE ELEMENTS	74-0522
SONIC	SONIC FATIGUE OF DIFFUSION BOUNDED TITANIUM SANDWICH STRUCTURES	72-0290
SOR	ON THE CONVERGENCE OF SOR ITERATIONS FOR FINITE ELEMENT APPROXIMATIONS TO ELLIPTIC BOUNDARY VALUE PR	71-0449
SOSTAVNYKH	CHISLENNYI METOD RASCHETA SOBSTVENNYKH I VYNUZHDENNYKH KOLEBANII SOSTAVNYKH OBOLOCHECHNYKH KONSTRUKT	73-0461
SOSTOYANIYA	ISSLEDOVANIE NAPRYAZHENNOGO SOSTOYANIYA REBRISTYKH ISILINDRICHESKIKH OBOLOCHEK S PRYAMOUGOL'NYMI OTV	74-0944
SOSTOYANIYA	ISSLEDOVANIE NAPRYAZHENNO-DEFORMIROVANNOGO SOSTOYANIYA SIMMETRICHNYKH DISKOV SO STUPITSEI. (INVESTIG	74-0965
SOUDURES	TENDANCES ACTUELLES DES RECHERCHES SUR LES CONTRAINTES ET DEFORMATIONS DUES AU BRIDAGE EN RELATION A	74-0273
SOUND	ACOUSTIC FINITE ELEMENT APPROACH FOR STUDYING BOUNDARY FLEXIBILITY AND SOUND TRANSMISSION BETWEEN IR	73-0096
SOUND	AN ACOUSTIC FINITE ELEMENT APPROACH FOR STUDYING BOUNDARY FLEXIBILITY AND SOUND TRANSMISSION BETWEEN	73-1087
SOURCE	EFFECTS OF SOURCE STRUCTURE AND SITE RESPONSE ON GROUND MOTION IN THE NEAR FIELD OF EARTHQUAKES	73-0732

SOURCE	MIMIC SOURCE APPROACH TO VISCOPLASTIC ANALYSIS	59-0164
SOURCES	SHIP (SIMPLIFIED-HELMHOLTZ-INTEGRAL PROGRAM): A FAST COMPUTER PROGRAM FOR CALCULATING THE ACOUSTIC P	72-0835
SOUS-ASSEMBLAGE	APPLICATION DE LA METHODE DES ELEMENTS FINIS - UN PROCEDE DE SOUS-ASSEMBLAGE	74-0405
SOUTHWELL	BASIS OF A WELL CONDITIONED FORCE PROGRAM FOR EQUILIBRIUM MODELS VIA THE SOUTHWELL SLAB ANALOGIES	67-0111
SOVMESTNO	RASCHET METODOM KONECHNYKH ELEMENTOV DINAMICHESKIKH KHARAKTEERISTIK ZEMLYANOI PLOTINY SOVMESTNO S OS	73-0437
SP-222	SP-222 NASTRAN THEORETICAL MANUAL	70-0387
SP-222	SP-222 NASTRAN USERS MANUAL	70-0386
SPACE	ADAPTATION OF NASTRAN TO THE ANALYSIS OF THE VIKING SPACE VEHICLE	71-0135
SPACE	ANALYSIS OF SPACE CLOSING SPRINGS IN ORTHODONTICS	74-0236
SPACE	APPLICATION OF NASTRAN TO THE ANALYSIS OF A SPACE SHUTTLE ORBITER STRUCTURE	71-0099
SPACE	APPLICATION OF NASTRAN TO A SPACE SHUTTLE DYNAMICS MODEL	71-0247
SPACE	APPLICATION OF PHASE SPACE FINITE ELEMENTS TO THE TWO DIMENSIONAL NEUTRON TRANSPORT EQUATION IN X-Y	73-0453
SPACE	APPLICATIONS OF FINITE ELEMENTS IN SPACE AND TIME	72-0147
SPACE	APPROXIMATION IN ELASTICITY BASED ON THE CONCEPT OF FUNCTION SPACE	47-0001
SPACE	BERECHNUNG INSTATIONAERER TEMPERATUREFELDER MIT FINITEN ELEMENTEN IN RAUM-UND ZEITDIMENSION. (CALCUL	74-0987
SPACE	BIFURCATION OF SPACE FRAMES	70-0055
SPACE	CALCULATION OF TRANSIENT TEMPERATURE FIELDS WITH FINITE ELEMENTS IN SPACE AND TIME DIMENSIONS	74-0379
SPACE	CODE TAFEST: NUMERICAL SOLUTION TO TRANSIENT HEAT-CONDUCTION PROBLEMS USING FINITE ELEMENTS IN SPACE	74-0901
SPACE	COMPUTER PROGRAMS FOR SIMULATING PHYSICAL PHENOMENA IN THREE SPACE DIMENSIONS AND TIME	74-0866
SPACE	CONTINUOUS CONCEPT OF A SPACE GRID	73-0755
SPACE	DESIGN, MANUFACTURE, DEVELOPMENT, TEST, AND EVALUATION OF BORON/ALUMINUM STRUCTURAL COMPONENTS FOR S	74-0798
SPACE	DYNAMIC RESPONSE OF NONLINEAR SPACE FRAMES BY FINITE ELEMENT METHODS	71-0584
SPACE	EXPERIMENTS IN FORMING AND JOINTING OF TITANIUM AND TITANIUM ALLOYS FOR AIRCRAFT AND SPACE VEHICLES	70-0134
SPACE	FINITE ELEMENT ANALYSIS FOR BUCKLING OF SEMI-RIGID SPACE FRAMES	68-0270
SPACE	FINITE ELEMENT ANALYSIS OF FINITE SIZED PLATES BOUNDED TO AN ELASTIC HALF SPACE	72-0880
SPACE	FINITE ELEMENT APPROACH TO THE CONTACT PROBLEM OF A PLATE ON THE ELASTIC HALF SPACE	72-0830
SPACE	FINITE ELEMENT FOR REINFORCED CONCRETE FRAME STUDY - SPACE DISCUSSION	74-0246
SPACE	FINITE ELEMENTS IN TIME AND SPACE	69-0442
SPACE	FINITE ELEMENTS IN TIME AND SPACE	69-0450
SPACE	FORCE AND TORQUE ANALYSIS OF CONSTRAINED SPACE MECHANISMS AND PLANE MECHANISMS WITH OFFSET LINKS BY	74-0569
SPACE	INTERACTIVE ANALYSIS OF A RAFTED MULTISTOREY SPACE FRAME RESSTING ON AN INHOMOGENEOUS CLAY STRATUM	74-0447
SPACE	INVESTIGATION OF THE STRESS-STRAINED STATE OF SPHERICAL SHELLS WITH AN ECCENTRIC CUTOUT ON THE BASIS	73-0222
SPACE	MACHINERY SPACE DEFLECTION AND VIBRATION ANALYSES USING FINITE ELEMENT METHODS	72-0474
SPACE	MATRIX-DISPLACEMENT DIRECT-ELEMENT METHOD FOR FORCE AND TORQUE ANALYSIS OF DETERMINATE AND INDETERMI	74-0531
SPACE	NODAL ANALYSIS OF THE MATED SPACE SHUTTLE CONFIGURATION	71-0097
SPACE	NONLINEAR ANALYSIS OF RETICULATED SPACE TRUSSES	75-0287
SPACE	ON THE APPLICATION OF THE FINITE ELEMENT METHOD TO THE VIBRATION ANALYSIS OF SPACE PLATE STRUCTURE	72-0630
SPACE	SPACE FINITE-ELEMENTS 2-DIMENSIONAL MULTIGROUP TRANSPORT CALCULATIONS USING PHASE-SPACE FINITE-ELEME	75-0504
SPACE	SPACE SHUTTLE, THE NEED FOR SUBSTRUCTURING	71-0104
SPACE	SPACE TIME FINITE ELEMENTS IN DYNAMIC PROBLEMS	72-0402
SPACE	STATIC AND DYNAMIC ELASTO-PLASTIC ANALYSIS BY THE METHOD OF FINITE ELEMENTS IN SPACE AND TIME	72-0642
SPACE	STATICAL CALCULATIONS OF PLANE AND SPACE BAR STRUCTURES AS WELL AS STUMPY FRAMEWORKS USING A MODIFIE	73-0747
SPACE	STRUCTURAL DESIGN AND ANALYSIS OF THE SPACE SHUTTLE	70-0336
SPACE	STRUCTURAL EVALUATION OF CANDIDATE DESIGNS FOR THE LARGE SPACE TELESCOPE PRIMARY MIRROR	75-0708
SPACE	STUDY OF HYPERVELOCITY METEOROID IMPACT ON ORBITAL SPACE STATIONS	73-0905
SPACE	TANGENT STIFFNESS MATRIX FOR SPACE FRAME MEMBERS	69-0397
SPACE	THEORETICAL AND EXPERIMENTAL VIBRATION ANALYSIS OF AN OBLIQUE SPACE FRAME	70-0460
SPACE	THERMAL ANALYSIS OF M552 EXPERIMENT FOR MATERIALS PROCESSING IN SPACE	74-0704
SPACE	THERMAL ANALYSIS OF M551 EXPERIMENT FOR MATERIALS PROCESSING IN SPACE	74-0725
SPACE	THERMOSTRUCTURAL DESIGN CONSIDERATIONS TO ACHIEVE THE LARGE SPACE TELESCOPE LINE-OF-SIGHT REQUIREMEN	75-0839
SPACE	TIDAL SIMULATION GY TIME SPACE ELEMENT	75-0867
SPACE	TRANSPORT SOLUTIONS USING FINITE ELEMENTS IN SPACE-ANGLE PHASE SPACE	71-0480
SPACE	UNBONDED CONTACT PROBLEM OF A PLATE ON THE ELASTIC HALF SPACE	74-0208
SPACE	VIBRATION ANALYSIS OF A 1/15-SCALE DYNAMIC MODAL OF A SPACE SHUTTLE CONFIGURATION	71-0609
SPACE	VIBRATIONAL CHARACTERISTICS OF SPACE FRAMES	69-0434
SPACE	WATER IMPACT ANALYSIS OF SPACE SHUTTLE SOLID ROCKET BY THE FINITE ELEMENT METHOD	74-0760
SPACE-ANGLE	TRANSPORT SOLUTIONS USING FINITE ELEMENTS IN SPACE-ANGLE PHASE SPACE	71-0480
SPACE-FRAME	ANALYSIS OF ARCH DAMS USING A SPACE-FRAME MODEL	73-0176
SPACE-TIME	ERROR ESTIMATES FOR FINITE ELEMENT METHOD FOR HEAT TRANSFER PROBLEMS IN THE SPACE-TIME DOMAIN	75-0104
SPACE-TIME	FINITE ELEMENT METHODS FOR SPACE-TIME REACTOR ANALYSIS	71-0673
SPACE-TIME	METHOD OF FINITE SPACE-TIME ELEMENTS IN DYNAMICS OF STRUCTURES	75-0823
SPACE-TIME	SPACE-TIME SOLUTIONS OF NONLINEAR REACTOR DYNAMICS BY FINITE-ELEMENT METHOD	75-0503
SPACEBOUND	ELASTIC ANALYSIS OF LARGE SPACEBOUND MIRRORS	72-0619
SPACECRAFT	A METHOD FOR THE THERMAL ANALYSIS OF SPACECRAFT INCLUDING ALL MULTIPLE REFLECTIONS AND SHADING AMONG	70-0424
SPACECRAFT	ACOUSTIC RESPONSE OF A SPACECRAFT SOLAR PANEL USING NORMAL MODE METHOD	69-0422
SPACECRAFT	ANALYTICAL DYNAMICS AND NONRIGID SPACECRAFT SIMULATION	74-0771
SPACECRAFT	INFLUENCE OF ASCENT HEATING ON THE SEPARATION DYNAMICS OF A SPACECRAFT FAIRING	72-0906
SPACECRAFT	LARGE STRUCTURES FOR MANNED SPACECRAFT: MATHEMATICAL ANALYSIS, DESIGN, CONSTRUCTION AND TESTS	74-1327
SPACECRAFT	NORMAL MODE ANALYSIS OF THE RADIO ASTRONOMY EXLPORER BOOMS AND SPACECRAFT	71-0130
SPACECRAFT	STRUCTURAL ANALYSIS AND DESIGN OPTIMIZATION OF SPACECRAFT USING NASTRAN	71-0068
SPACED	VIBRATION AND STRESS ANALYSIS OF SOFT-BONDED SHUTTLE INSULATION TILES, MODAL ANALYSIS WITH COMPACT W	74-0708
SPACES	STABILITY IN LQ OF L2-PROJECTION INTO FINITE ELEMENT FUNCTION SPACES	75-0071
SPACES	TWO TYPES OF PIECEWISE QUADRATIC SPACES AND THEIR ORDER OF ACCURACY FOR POISSON'S EQUATION	72-0520
SPADES	STRESS-WAVE PROPAGATION AND DYNAMIC EFFECTS IN SHELLS (SPADES) MICROMECHANICAL STRESS-WAVE RESPONSE	72-0851
SPAENDINGSHYBRI	SPAENDINGSHYBRIDE FINITE ELEMENTER TIL SVINGNINGSPROBLEMER. (STRESS HYBRID FINITE ELEMENT METHOD FOR	75-0208
SPALLATION	ELASTIC PLATE SPALLATION	72-0682
SPALLATION	INVESTIGATION OF ELASTIC PLATE SPALLATION BY FINITE ELEMENTS	74-0382
SPAN	FREQUENCY ANALYSIS OF CERTAIN SINGLE AND CONTINUOUS SPAN BRIDGES	71-0337
SPAN	STRUCTURAL BEHAVIOR OF A TWO SPAN REINFORCED CONCRETE BOX GIRDER BRIDGE MODEL. VOLUME 11. REDUCTION,	71-0659
SPANNBETON	ENTWURF UND SPANNUNGSBEPECHNUNG EINES SPANNBETON DRUCKBEHAELTERS AUS FERTIGTEILEN. (DESIGN AND STRES	73-0856
SPANNBETON-BEHA	POTATIONSSYMMETRISCHE BERECHNUNG DES 1:5 THTR SPANNBETON-BEHAL-BEHAL TERMODELLS	71-0629
SPANNBETON-REAK	EINFACHES NAEHERUNGSVERFAHREN FUR DIE GRENZTRAGFAEHIGKEITSANLYSE VON SPANNBETON-REAKTORDRUCKBEHAELTE	73-0854
SPANNBETON-REAK	FINITE ELEMENTS FOR CALCULATION OF PRESTRESSED CONCRETE REACTOR PRESSURE VESSELS (FINITE ELEMENTE ZU	73-0797
SPANNBETONREAKT	FINITE ELEMENTE ZUR BERECHNUNG VON SPANNBETONREAKTORDRUCKBEHALTERN	73-0702
SPANNEBETONDRUC	UEBER DAS VERHALTEN VON SPANNEBETONDRUCKBEHAELTRERN IM GRENZTRAGFAEHIGKEITSZUSTAND UND DESSEN EINFUL	73-0855
SPANNUNGEN	ANWENDUNG DER FINITE-ELEMENT-METHODE ZUR BERECHNUNG VON SPANNUNGEN IN WASSERGESAETTIGTEN BOEDEN (FIN	73-0502
SPANNUNGEN	BERECHNUNG VON SPANNUNGEN UND TEMPERATURFELDERN IN FLOSSENWAENDEN UND AUFGESCHWEISSTEN BAUTEILEN. (C	74-0544
SPANNUNGEN	DIE BERECHNUNG DER BEWEGUNGEN UND SPANNUNGEN DES GEBIRGES VOM ABBAU BIS ZUR TAGESOBERFLAECHE NACH DE	73-0456
SPANNUNGEN	EFFECT OF DISCONTINUITIES ON STRESSES AND DEFORMATIONS OF ROCK AROUND A TUNNEL. (EINFLUSS VON DISKON	70-0170
SPANNUNGEN	EINSATZ MODERNER RECHENVERFAHREN ZUR ERMITTLUNG DER SPANNUNGEN UND VERFORMUNGEN BEI RADSAETZEN. (USE	75-0186
SPANNUNGEN	SPANNUNGEN IN MEMBRANWAENDEN DURCH WAERMEBEANSPRUCHUNG UND MECHANISCHE BELASTUNG. (STRESSES IN MEMBR	75-0066
SPANNUNGS	DER SPANNUNGS UND BEANSPRUCHUNGSZUSTAND DES BODENMATERIALS IN DER NACHBARSCHAFT EINER BODENBOESCHUNG	75-0371
SPANNUNGS-DEHNU	VOLLSTAENDIGE SPANNUNGS-DEHNUNGS-TEMPERATURAENDERUNGSBEZIEHUNG.FUER DIE SCHWEISSEIGENSPANNUNGSBERECH	75-0807
SPANNUNGSBERECH	ACCURATE CALCULATION OF STRESSES IN BEAMS, DISKS AND PLATES BY MEANS OF FINITE ELEMENT METHODS AND I	70-0133
SPANNUNGSBERECH	ENTWURF UND SPANNUNGSBERECHNUNG EINES SPANNBETON DRUCKBEHAELTERS AUS FERTIGTEILEN. (DESIGN AND STRES	73-0856
SPANNUNGSBERECH	MODERN TECHNIQUES FOR CALCULATING STRESSES IN TANKS, AND THEIR LIMITATIONS. (MODERNE VERFAHREN ZUR S	70-0153
SPANNUNGSKOMPON	DIE VERWENDUNG DER MATRIZENVERSCHIEBUNGSMETHODE ZUR BERECHNUNG VON SPANNUNGSKOMPONENTEN	67-0140
SPANNUNGSKONZEN	BERECHNUNG VON SPANNUNGSKONZENTRATIONEN AN TRAGWERKEN UNTER THERMO-MECHANISCHER BELASTUNG	73-0686
SPANNUNGSKONZEN	DIE ANWENDUNG DER MATRIZENVERSCHIEBUNGSMETHODE ZUM BERECHNEN VON SPANNUNGSKONZENTRATIONEN	68-0208
SPANNUNGSKONZEN	DIE VERWENDUNG DER MATRIZENVERSCHIEBUNGSMETHODE ZUR BERECHNUNG VON SPANNUNGSKONZENTRATIONEN	70-0400
SPANNUNGSKONZZE	CALCULATION OF STRESS CONCENTRATIONS ON SUPPORT STRUCTURES UNDER THERMOMECHANICAL LOADING (BERECHNUN	73-0796
SPANNUNGSUMLAGE	BERECHNUNG VON SPANNUNGSUMLAGERUNGEN IN FELS UND BODEN MIT HILFE DER ELEMENTENMETHODE	69-0364

SPANNUNGSVERTEI	BERECHNUNG RAUMLICHER SPANNUNGSVERTEILUNGEN IN STAUDAMMEN NACH DER MATRIZENVERSCHIEBUNGSMETHODE	67-0003
SPANWISE	ANALYSIS OF THE EFFECTS ON LIFE OF LEADING-EDGE HOLES IN AN AIRFOIL SUBJECTED TO ARBITRARY SPANWISE	75-0700
SPAR	ARBITRARY QUADRILATERAL SPAR WEBS FOR THE MATRIX DISPLACEMENT METHOD	66-0003
SPARSE	A COMPARATIVE STUDY OF SEVERAL CORE STORAGE SCHEMES FOR LARGE SPARSE POSITIVE DEFINITE MATRICES WITH	72-0678
SPARSE	A SURVEY OF SPARSE MATRIX TECHNOLOGY	72-0989
SPARSE	AN ALGORITHM FOR REDUCING THE BANDWITH AND PROFILE OF A SPARSE MATRIX	74-0792
SPARSE	AN IMPROVED METHOD FOR REDUCING THE BANDWIDTH OF SPARSE SYMMETRIC MATRICES	72-0990
SPARSE	COMPARATIVE ANALYSIS OF THE CUTHILL-MCKEE AND THE REVERSE CUTHILL-MCKEE ORDERING ALGORITHMS FOR SPAR	73-1099
SPARSE	COMPARISON BETWEEN SPARSE STIFFNESS MATRIX AND SUB-STRUCTURE METHODS	73-1068
SPARSE	COMPUTER PROGRAM FOR SOLUTION OF LARGE, SPARSE, UNSYMMETRIC SYSTEMS OF LINEAR EQUATIONS	72-0719
SPARSE	COMPUTING WITH SPARSE MATRICES	73-0926
SPARSE	NOTE ON MINIMIZING THE BANDWIDTH OF SPARSE, SYMMETRIC MATRICES	73-1107
SPARSE	ON SPARSE MATRICES IN FINITE ELEMENT WAVE PROPAGATION	73-0290
SPARSE	REDUCING THE BANDWIDTH OF SPARSE SYMMETRIC MATRICES	69-0469
SPARSE	REDUCING THE BANDWIDTH OF SPARSE SYMMETRIC MATRICES	69-0502
SPARSE	SOLVING A CLASS OF LARGE SPARSE LINEAR SYSTEMS OF EQUATIONS BY PARTITIONING	73-0580
SPARSE	SPARSE MATRICES IN FINITE ELEMENT PROGRAM DEVELOPMENT	73-0337
SPARSE	SPARSE MATRIX OPERATIONS IN NASTRAN	73-0336
SPARSE	SPARSE MATRIX TECHNIQUES IN DIRECT SOLUTION OF FINITE ELEMENT EQUATIONS	74-0115
SPARSE	SURVEY OF SPARSE MATRIX METHODS IN THE DIRECT SOLUTION OF FINITE ELEMENT EQUATIONS	73-0572
SPARSE	THE ANALYSIS AND APPLICATION OF SPARSE MATRIX ALGORITHMS IN THE FINITE ELEMENT METHOD	72-0079
SPARSELY	PROCEDURE FOR SOLVING LINEAR EQUATION SYSTEMS WITH SPARSELY OCCUPIED COEFFICIENT MATRICES ASSOCIATED	74-0352
SPARSITY	USE OF SPARSITY IN THE SOLUTION OF FINITE ELEMENT SYSTEMS OF EQUATIONS BY GAUSSIAN ELIMINATION-TYPE	74-0868
SPARTA	ASKA 105 A TECHNICAL MANUAL AND PROGRAMMING THEORY FOR THE ELEMENTS, ELA, TRIM, SPARTA AND TET	68-0254
SPATIAL	COUPLED RESPONSE OF SPATIAL VIBRATORY STRUCTURES MOUNTED TO ISOTROPIC PLATE ELEMENTS	71-0228
SPATIAL	ELASTIC ANALYSIS OF SPATIAL SYSTEMS OF INTERCONNECTED SHEAR WALLS AND FRAMES	74-1195
SPATIAL	EXACT THEORY OF SPATIAL STRUCTURES	73-0781
SPATIAL	FINITE ELEMENT ANALYSIS OF SPATIAL SANDWICH PLATE STRUCTURES	74-1132
SPATIAL	LARGE DEFLECTION SPATIAL BUCKLING OF THIN WALLED BEAMS AND FRAMES	73-1085
SPATIAL	MULTI-ELEMENT SPATIAL STRUCTURES	73-0768
SPATIAL	NONLINEAR TRANSIENT RESPONSES OF STRUCTURES BY SPATIAL FINITE-ELEMENT METHOD	73-0138
SPATIALLY	SIMPLIFIED INELASTIC HIGH TEMPERATURE STRUCTURAL ANALYSIS OF MODERATELY COMPLEX SPATIALLY THREE DIME	74-0445
SPECIALISTS	TECHNICAL EVALUATION REPORT ON AGARD SPECIALISTS	71-0737
SPECIFIC	NATURAL GEOMETRY OF SURFACE WITH SPECIFIC REFERENCE TO THE MATRIX DISPLACEMENT ANALYSIS OF SHELLS, 1	73-0709
SPECIFIC	TOUR MAINE-MONTPARNASSE QUELQUES PROBLEMES DE STRUCTURE (SPECIFIC PROBLEMS ENCOUNTERED IN THE CONSTR	74-0539
SPECIFICATION	PROGRAM SPECIFICATION NV 300. GENERAL FINITE ELEMENT PROGRAM FOR 2- AND 3-DIMENSIONAL MEMBRANE STRUC	68-0184
SPECIFICATION	SPECIFICATION OF GEOMETRICAL PARAMETERS FOR ELEMENTS WITH CUBIC MAPPING FUNCTIONS	73-1082
SPECIFIED	A FINITE ELEMENT FORMULATION FOR MATERIALS WITH SPECIFIED CHANGE OF VOLUME	68-0044
SPECIMEN	A DYNAMIC PHOTOELASTIC ANALYSIS OF DYNAMIC TEAR TEST SPECIMEN	74-0736
SPECIMEN	ANALYTICAL AND EXPERIMENTAL INVESTIGATION OF FATIGUE IN A SHEET SPECIMEN WITH AN INTERFERENCE-FIT 30	75-0701
SPECIMEN	APPLICATION OF FINITE ELEMENT ELASTIC-PLASTIC STRESS ANALYSIS TO NOTCHED FATIGUE SPECIMEN BEHAVIOUR	71-0185
SPECIMEN	APPLICATIONS OF FINITE ELEMENT STRESS ANALYSIS AND STRESS-STRAIN PROPERTIES IN DETERMINING NOTCH FAT	73-1114
SPECIMEN	COMPLIANCE OF A TAPERED DCB SPECIMEN - CONFIGURATIONS BY A FINITE ELEMENT METHOD	74-0545
SPECIMEN	EFFECTS OF SPECIMEN RADIUS ON THE STRESS STATE NEAR A ROOF BOLT ANCHOR: A FINITE ELEMENT DETERMINATI	73-0229
SPECIMEN	ESTABLISHMENT OF A VALID FRACTURE TOUGHNESS SPECIMEN GEOMETRY USING A FINITE ELEMENT ANALYSIS	74-0135
SPECIMEN	FINITE ELEMENT ANALYSIS ON UNIAXIAL COMPRESSIVE STRENGTH OF CYLINDRICAL BRITTLE SPECIMEN	70-0155
SPECIMEN	FINITE ELEMENT AND CUMULATIVE DAMAGE ANALYSIS OF A KEYHOLE TEST SPECIMEN	75-0163
SPECIMEN	FINITE ELEMENT AND CUMULATIVE DAMAGE ANALYSIS OF A KEYHOLE TEST SPECIMEN	75-0774
SPECIMEN	FINITE ELEMENT CALCULATIONS OF THE COMPLIANCE OF A TAPERED DOUBLE CANTILEVER BEAM SPECIMEN FOR DIFFE	72-0665
SPECIMEN	FINITE-ELEMENT ANALYSIS OF A FRACTURE TOUGHNESS TEST SPECIMEN IN NON-LINEAR RANGE	72-0091
SPECIMEN	HEAVY SECTION STEEL TECHNOLOGY PROGRAM TECHNICAL OR PROGRAMMATIC MANUSCRIPT NO. 25. VARIABLE THICKNE	73-0948
SPECIMEN	INFLUENCE OF NON-SINGULAR STRESS TERMS AND SPECIMEN GEOMETRY ON SMALL SCALE YIELDING AT CRACK TIPS I	73-0203
SPECIMEN	LOAD POINT COMPLIANCE OF THE CHARPY IMPACT SPECIMEN	73-0775
SPECIMEN	NOTE ON STRESS AND STRAIN REDISTRIBUTION IN A NOTCHED PLATE SPECIMEN DURING CYCLIC LOADING	69-0194
SPECIMEN	NUMERICAL ANALYSIS ON STRESS DISTRIBUTION IN PARTIAL COMPRESSED SPECIMEN (HARDBOARD) BY FINITE ELEME	72-0920
SPECIMEN	STRESS CONCENTRATION FACTORS FOR A FATIGUE TEST SPECIMEN WITH A U-TYPE NOTCH IN ONE EDGE BY FINITE E	74-1344
SPECIMENS	A FINITE ELEMENT STUDY OF ELASTIC-PLASTIC STRESS DISTRIBUTIONS IN NOTCHED SPECIMENS UNDER TENSION	68-0177
SPECIMENS	COMPARISON OF FINITE-ELEMENT AND EXPERIMENTAL STUDIES ON DEFORMATIONS OF ZIRCONIUM NOTCHED BEND SPEC	71-0359
SPECIMENS	CREEP BENDING TESTS OF LARGE WELD OVERLAY AND BASE METAL SPECIMENS AT 1000 DEGREE F. (537.7 DEG. C)	75-0521
SPECIMENS	CRUCIFORM SPECIMENS FOR BIAXIAL FATIGUE TESTS - AN INVESTIGATION USING FINITE-ELEMENT ANALYSIS AND P	71-0795
SPECIMENS	ELASTIC PLASTIC ANALYSIS OF NOTCHED SPECIMENS USING THE FINITE ELEMENT METHOD	70-0140
SPECIMENS	FINITE ELEMENT ANALYSIS OF NOTCHED TENSILE SPECIMENS IN PLANE STRESS	71-0557
SPECIMENS	HEAVY SECTION STEEL TECHNOLOGY PROGRAM. TECHNICAL REPORT NO. 4, OCTOBER 1969. A TWO-DIMENSIONAL ELAS	69-0474
SPECIMENS	ON THE STATES OF STRESS AND DEFORMATION OF CYLINDRICAL SPECIMENS OF BRITTLE MATERIAL UNDER UNIAXIAL	69-0150
SPECIMENS	STRESS INTENSITY FACTORS FOR HEXAGONAL FRACTURE TOUGHNESS SPECIMENS	71-0799
SPECIMENS	STRESS INTENSITY FACTORS FOR ANISOTROPIC FRACTURE TEST SPECIMENS OF SEVERAL GEOMETRIES	74-0293
SPECIMENS	STUDY ON STATE OF STRESS AND DEFORMATION OF CYLINDRICAL SPECIMENS OF BRITTLE MATERIAL UNDER UNIAXIAL	69-0152
SPECIMENS	THREE-DIMENSIONAL ELASTIC-PLASTIC STRESS AND STRAIN ANALYSIS FOR FRACTURE MECHANICS. PHASE 1. SIMPLE	70-0188
SPECKLE	INTERFERENCE-FIT FASTENER DISPLACEMENT MEASUREMENT BY SPECKLE PHOTOGRAPHY	74-1178
SPECTRA	FINITE ELEMENT ANALYSIS OF FREQUENCY SPECTRA FOR ELASTIC WAVEGUIDES	75-0247
SPECTRA	INFLUENCE OF SOIL/STRUCTURE INTERACTION PARAMETERS ON FLOOR RESPONSE SPECTRA	73-0823
SPECTRAL	BOUNDS ON EXTERNAL EIGENVALUES OF FINITE ELEMENT STIFFNESS AND MASS MATRICES AND THEIR SPECTRAL COND	72-0256
SPECTRAL	BOUNDS ON THE SPECTRAL AND MAXIMUM NORMS OF THE FINITE ELEMENT STIFFNESS, FLEXIBILITY AND MASS MATRI	73-0409
SPECTRAL	ON A SPECTRAL PROBLEM IN VIBRATION MECHANICS: COMPUTATION OF ELASTIC TANKS PARTIALLY FILLED WITH LIQ	75-0267
SPECTRAL	POWER SPECTRAL DENSITY ANALYSIS OF VEHICLE VIBRATION USING THE NASTRAN COMPUTER PROGRAM	74-0318
SPECTRUM	FINITE AND DISCONNECTED SUBGROUPS OF SU(3) AND THEIR APPLICATION TO THE ELEMENTARY PARTICLE SPECTRUM	66-0129
SPECTRUM	SPECTRUM TRUNCATION AND DAMAGE TOLERANCE STUDY ASSOCIATED WITH THE C-5A OUTBOARD PYLON AFT TRUSS LUG	74-0535
SPEED	AN AUTOMATIC SYSTEM FOR KINEMATIC ANALYSIS ASKA HIGH SPEED COMPUTING OF ELASTIC STRUCTURES	70-0358
SPEED	FINITE ELEMENT ANALYSIS OF LOAD CELL RESPONSE IN HIGH SPEED TENSILE TESTING	73-1071
SPEED	TEMPERATURE RISE IN THE HIGH SPEED COMPRESSION OF RIGHT CYLINDRICAL BILLETS	74-1153
SPEED	THE FINITE ELEMENT METHOD IN LOW SPEED AERODYNAMICS	75-0702
SPEEDS	COMPUTER METHOD FOR COMPUTING CRITICAL SPEEDS OF NON-UNIFORM SHAFTS GEARED SYSTEMS AND CURVED ASSEMB	73-0675
SPEKTRALANALYSE	SPEKTRALANALYSE VON ROTORBLATTERN NACH DER MATRIZENVERSCHIEBUNGSMETHODE UND MATRIZENKRAFTEMETHODE	66-0123
SPENNUNGS	KOMBINATION EINES INTEGRALGLEICHUNGS VERFAHRENS MIT METHODE DER FINITEN ELEMENTEN ZUR BERECHNUNG EBE	74-0856
SPEZIELLES	TURBAN VERSION 1.0. EIN SPEZIELLES PROGRAMM ZUR LINEAREN UND NICHTLINEAREN STATISCHEN UND DYNAMISCHE	71-0622
SPHERE	INTEGRATIONS OF PRIMITIVE EQUATIONS ON A SPHERE USING FINITE ELEMENT METHOD	74-0649
SPHERE	INTEGRATIONS OF THE PRIMITIVE EQUATIONS ON A SPHERE USING THE FINITE ELEMENT METHOD	74-1347
SPHERE	THE ELASTO-PLASTIC INDENTATION OF A HALF-SPACE BY A RIGID SPHERE	71-0113
SPHERE-CONE	STATIC AND HEAT TRANSFER ANALYSIS OF A SPHERE-CONE INTERSECTION IN A NUCLEAR CONTAINMENT VESSEL.	75-0166
SPHERE-CYLINDER	PHOTOELASTIC INVESTIGATION OF STRESS CONCENTRATIONS IN SPHERE-CYLINDER TRANSITION REGIONS: INCLUDING	68-0258
SPHERICAL	A FINITE ELEMENT APPROACH TO SCATTERING FROM ELASTIC SPHERICAL SHELLS	74-0702
SPHERICAL	A FORTRAN PROGRAM FOR CALCULATING THE FREQUENCIES AND MODE SHAPES OF A GUIDED CLAMPED, GUIDED PINNED	72-0890
SPHERICAL	ANALYTIC INVESTIGATION OF COMPACT REINFORCEMENT FOR RADIAL NOZZLES IN SPHERICAL SHELLS	71-0107
SPHERICAL	APPLICATION OF FINITE ELEMENT METHOD TO AXISYMMETRIC BUCKLING OF SHALLOW SPHERICAL SHELLS UNDER EXTE	73-0111
SPHERICAL	CALCULATION OF THE CHARACTERISTIC FREQUENCIES OF FLUID-FILLED SPHERICAL SHELLS	73-0511
SPHERICAL	DIFFUSION ON SOLID SPHERICAL PARTICLE IN VISCOUS FLUID FLOW AT FINITE VALUES OF RE NUMBERS	69-0419
SPHERICAL	DYNAMIC LOADING OF A FLUID-FILLED SPHERICAL SHELL.	72-0444
SPHERICAL	FINITE ELEMENT CREEP BUCKLING ANALYSIS OF VISCO-ELASTIC SHALLOW SPHERICAL SHELL	75-0159
SPHERICAL	FREQUENCY COEFFICIENTS OF A SPHERICAL SHELL FOR VARYING THICKNESS AND OPENING ANGLE WITH GUIDED PINN	72-0595
SPHERICAL	INFLUENCE OF THE UNPRECISION OF EXECUTION ON THE BEHAVIOUR OF A SPHERICAL SHELL WITH CONCENTRATED LO	73-0802
SPHERICAL	INVESTIGATION OF THE STRESS-STRAINED STATE OF SPHERICAL SHELLS WITH AN ECCENTRIC CUTOUT ON THE BASIS	73-0222
SPHERICAL	ON THE ASYMPTOTICALLY SPHERICAL DEFORMATIONS OF ARBITRARY MEMBRANES OF REVOLUTION FIXED ALONG AN EDG	71-0640

SPHERICAL	RE-ENTRY THERMAL ANALYSIS OF VARIABLE THICKNESS SPHERICAL VEHICLES.	72-0468
SPHERICAL	SLOSHING OF LIQUID IN CIRCULAR CHANNELS AND SPHERICAL TANKS	60-0005
SPHERICAL	SPHERICAL WAVES IN INELASTIC MATERIAL	70-0075
SPHERICAL	THREE-DIMENSIONAL FINITE ELEMENT ANALYSIS OF SPHERICAL PARTICLE COMPOSITES	74-0137
SPHEROIDAL	SIMULATION OF ELASTIC MODULUS AND POISSON'S RATIO OF SPHEROIDAL GRAPHITE CAST IRON	74-0955
SPILLWAY	FINITE ELEMENT METHOD APPLIED TO ANALYSIS OF FLOW OVER A SPILLWAY CREST	73-0300
SPILLWAY	SPECIAL STUDY ON THE STABILITY OF A DAM SPILLWAY	72-0469
SPINE	A FINITE ELEMENT ANALYSIS OF WAVE PROPAGATION IN HUMAN SPINE	73-0792
SPINE	ANALYSIS OF SPINE BEAM BRIDGES USING FINITE ELEMENTS	70-0236
SPINNING	COMPLEX EIGENVALUE SOLUTION TO A SPINNING SKYLAB PROBLEM	71-0207
SPINNING	FREE VIBRATION ANALYSIS OF SPINNING STRUCTURAL SYSTEMS	73-1077
SPINNING	MATHEMATICAL MODELING OF SPINNING ELASTIC BODIES FOR MODAL ANALYSIS	73-0441
SPINNING	ON A COMBINED STURM SEQUENCE AND INVERSE ITERATION TECHNIQUE FOR EIGENPROBLEM SOLUTION OF SPINNING S	73-0413
SPINNING	STRESS INTENSITY FACTORS FOR CRACKS IN A PLATE CONTAINING A HOLE AND IN A SPINNING DISC	73-0467
SPIRAL	APPROXIMATION METHOD FOR COMPUTATION OF NATURAL FREQUENCIES OF SPIRAL SECTIONS	71-0775
SPIRAL	FESTIGKEITSUNTERSUCHUNGEN UND-BERECHNUNGEN AN PUMPENSPIRALGEHAEUSEN. (INVESTIGATIONS AND CALCULATION	75-0238
SPLINE	CUBIC SPLINE GALERKIN APPROXIMATIONS TO PARABOLIC SYSTEMS WITH COUPLED NON-LINEAR BOUNDARY CONDITION	75-0232
SPLINE	FONCTIONS SPLINES ET METHODE D ELEMENTS FINIS (SPLINE FUNCTIONS AND FINITE ELEMENT METHOD)	75-0846
SPLINE	LEAST SQUARE POLYNOMIAL SPLINE APPROXIMATION	72-0526
SPLINE	NONLINEAR ANALYSES USING SPLINE FUNCTIONS	71-0466
SPLINE	SOME INTERIOR SUPER CONVERGENCE ESTIMATES FOR SPLINE APPROXIMATION	73-0641
SPLINE	SPLINE APPROXIMATION AND DIFFERENCE SCHEMES FOR THE HEAT EQUATION	72-0518
SPLINE	SPLINE FUNCTIONS AND FINITE-ELEMENT METHOD	75-0430
SPLINE	SPLINE INTERPOLATION TECHNIQUES FOR VARIATIONAL METHODS	73-1079
SPLINES	A " LOCAL " BASIS OF GENERALIZED SPLINES OVER RIGHT TRIANGLES DETERMINED FROM A NONUNIFORM PARTITION	72-0525
SPLINES	FONCTIONS SPLINES ET METHODE D ELEMENTS FINIS (SPLINE FUNCTIONS AND FINITE ELEMENT METHOD)	75-0846
SPLINES	RELATIVE EFFICIENCY OF FINITE DIFFERENCE AND FINITE ELEMENT METHODS - 1. HYPERBOLIC PROBLEMS AND SPL	74-0986
SPLINES	SPLINES AND FINITE ELEMENTS	73-0629
SPLINES	THE RELATION BETWEEN THE GALERKIN AND COLLOCATION METHODS USING SMOOTH SPLINES	74-0912
SPLINES	THE RELATIVE EFFICIENCY OF FINITE DIFFERENCE AND FINITE ELEMENT METHODS, 1. HYPERBOLIC PROBLEMS AND	74-0911
SPLITTING	SEMIDISCRETE GALERKIN TECHNIQUES WITH TIME INTERPOLATION AND SPLITTING UP FOR PLASMA SIMULATION	75-0121
SPOIL	A FINITE ELEMENT ANALYSIS OF DISSOLVED OXYGEN DRAWDOWN AND SULFATE PRODUCTION IN STRIP MINE SPOIL DA	75-0795
SPONTANEOUS	SPONTANEOUS IGNITION - FINITE ELEMENT SOLUTIONS FOR STEADY AND TRANSIENT CONDITIONS	74-0353
SPRENGWIRKUNG	ZUSAMMENHANG ZWISCHEN OBERFLAECHENBESCHAFFENHEIT, VERBUND UND SPRENGWIRKUNG VON BEWEHRUNGSSTAEHLEN U	74-0450
SPREZYSTYCH	OCENA DOKLADNOSCI ROZWIAZAN PROBLEMOW. BRZEGOWYCH SPREZYSTYCH CIAL DYSKERTYZOWANYCH. (ASSESSING THE A	74-0921
SPRING	GALERKIN FINITE ELEMENT ANALYSIS OF A UNIFORM BEAM CARRYING A CONCENTRATED MASS AND ROTARY INERTIA W	74-0552
SPRING	SOME APPLICATIONS OF FINITE ELEMENT ANALYSIS IN RUBBER SPRING DESIGN	72-0539
SPRINGS	A FINITE ELEMENT PROGRAM FOR DESIGN AND ANALYSIS OF THREE-DIMENSIONAL FLAT SPRINGS	74-0738
SPRINGS	ANALYSIS OF SPACE CLOSING SPRINGS IN ORTHODONTICS	74-0236
SPRINGS	SCHWINGUNGEN ZYLINDRISCHER SCHRAUBENFEDERN (VIBRATIONS OF CYLINDRICAL COIL SPRINGS)	74-0331
SPRUNG	ELASTIC ANALYSIS OF A REFRACTORY SPRUNG ARCH	74-0386
SPUR	EVALUATION OF STRESSES AND DEFLECTION OF SPUR GEAR TEETH UNDER STRAIN	72-0115
SPURMASSTABILIT	UNTERSUCHUNGEN ZUR SPURMASSTABILITAET BEI RADSAETZEN MIT KLOTZGEBREMSTEN VOLLRAEDERN (INVESTIGATIONS	75-0810
SQUARE	AN INVESTIGATION OF THE EFFECTS OF SQUARE CUTOUTS ON THE NATURAL FREQUENCIES AND MODE SHAPES OF RECT	71-0638
SQUARE	ELASTIC DEFORMATION OF SQUARE EXTRUSION DIES DUE TO LOAD AND TEMPERATURE	75-0581
SQUARE	LEAST SQUARE OPTIMAL APPROXIMATION SIMULATION TECHNIQUES	73-0871
SQUARE	LEAST SQUARE POLYNOMIAL SPLINE APPROXIMATION	72-0526
SQUARE	SHEAR BUCKLING OF SQUARE PERFORATED PLATES	74-0713
SQUARE	THE BEHAVIOUR OF SQUARE SHEAR WEBS HAVING A CIRCULAR HOLE	67-0108
SQUARE	VIBRATION ANALYSIS OF A CANTILEVERED SQUARE PLATE BY THE STIFFNESS MATRIX METHOD	66-0012
SQUARE	VIBRATION OF SQUARE PLATES HAVING VARIOUS EDGE CONDITIONS AND LOADINGS USING A REFINED FINITE ELEMEN	75-0024
SQUARE-FINITE	LEAST SQUARE-FINITE ELEMENT FOR ELASTO-STATIC PROBLEMS	74-0279
SQUARES	A DISCRETE LEAST SQUARES AND FINITE ELEMENT METHOD	69-0322
SQUARES	A LEAST SQUARES FINITE ELEMENT SOLUTION OF NONLINEAR OPERATORS	72-0060
SQUARES	A LEAST SQUARES FINITE ELEMENT ANALYSIS OF NONLINEAR BIOPHYSICAL DIFFUSION-KINETICS EQUATIONS	74-1257
SQUARES	A TWO-DIMENSIONAL FINITE ELEMENT - LEAST SQUARES SOLUTION	75-0025
SQUARES	AN APPLICATION OF LEAST SQUARES TO ONE-DIMENSIONAL TRANSIENT PROBLEMS	74-0347
SQUARES	AN APPLICATION OF LEAST SQUARES FINITE ELEMENT METHODS TO TWO DIMENSIONAL STEADY INVISCID COMPRESSIB	75-0114
SQUARES	FINITE ELEMENT APPLICATIONS OF THE LEAST SQUARES METHOD	72-0052
SQUARES	FINITE ELEMENTS FORMULATED BY THE WEIGHTED DISCRETE LEAST SQUARES METHOD	74-0200
SQUARES	LEAST SQUARES ANALYSIS USING FINITE ELEMENTS	68-0287
SQUARES	LEAST SQUARES FINITE ELEMENT ANALYSIS OF LAMINAR BOUNDARY LAYER FLOWS	74-0616
SQUARES	LEAST SQUARES FINITE ELEMENT ANALYSIS OF TWO-DIMENSIONAL LAMINAR BOUNDARY-LAYER EQUATION WITH SEPARA	75-0796
SQUARES	LEAST SQUARES SMOOTHING OF EXPERIMENTAL DATA USING FINITE ELEMENTS	68-0224
SQUARES	LEAST SQUARES SURFACE FITTING BY FINITE ELEMENTS AND AN APPLICATION TO STRESS SMOOTHING	67-0154
SQUARES	LOCAL AND GLOBAL SMOOTHING OF DISCONTINUOUS FINITE ELEMENT FUNCTIONS USING A LEAST SQUARES METHOD	74-0378
SQUARES	LOCAL LEAST SQUARES STRESS SMOOTHING FOR PARABOLIC ISOPARAMETRIC ELEMENTS	75-0593
SQUARES	TRANSIENT TWO-DIMENSIONAL PROBLEMS UTILIZING THE LEAST SQUARES ALGORITHM	75-0219
SQUARES	USE OF THE LEAST SQUARES CRITERION IN THE FINITE ELEMENT FORMULATION	73-0270
SQUEEZE	APPLICATION OF FINITE ELEMENT METHODS TO LUBRICATION: SQUEEZE FILMS BETWEEN POROUS SURFACES	75-0764
SST	THE BOEING SST PROTYPE INTERNAL LOADS ANALYSIS SYSTEM AND PROCEDURES	71-0103
STABILITA	STABILITA STEN V PLASTICKEM OBORU (STABILITY IN WEBPLATES IN PLASTIC RANGE)	73-0387
STABILITAETSUNT	STABILITY INVESTIGATIONS OF PLATES USING FINITE ELEMENT METHOD BASED ON AN EXPANDED VARIATION PRINCI	71-0663
STABILITAT	STABILITAT BEI DER METHODE DER FINITEN ELEMENTE FUR QUASILINEARE ELLIPTISCHE RANDWERTPROBLEME	74-1283
STABILITATSPROB	UBER DIE ANWENDUNG DER ENERGETISCHEN METHODE AUF STABILITATSPROBLEME	38-0001
STABILITE	ANALYSE TRIDIMENSIONELLE DE LA STABILITE D'UN TUNNEL AU VOISINAGE DU FRONT DE TAILLE DANS UNE ROCHE	74-0995
STABILITY	A FINITE ELEMENT APPROACH TO STABILITY ANALYSIS IN FRAMES	74-0455
STABILITY	A FINITE ELEMENT FORMULATION FOR STABILITY ANALYSIS OF DOUBLY CURVED THIN SHELL STRUCTURES	69-0513
STABILITY	A FINITE ELEMENT FORMULATION FOR THE GENERAL STABILITY ANALYSIS OF THIN WALLED MEMBERS	70-0337
STABILITY	A FINITE ELEMENT MODEL FOR STABILITY ANALYSIS OF MINE WASTE EMBANKMENTS SUBJECTED TO SEEPAGE	74-1166
STABILITY	A FINITE ELEMENT SOLUTION OF THE STABILITY OF SUPERPOSED FLUIDS	74-1226
STABILITY	A MATRIX METHOD FOR VIBRATION AND STABILITY PROBLEMS	66-0091
STABILITY	A NEW FAMILY OF CURVILINEAR PLATE BENDING ELEMENTS FOR VIBRATION AND STABILITY	72-0645
STABILITY	A NEW FAMILY OF CURVILINEAR PLATE BENDING ELEMENTS FOR VIBRATION AND STABILITY	72-0999
STABILITY	A STABILITY ANALYSIS OF CYLINDRICAL PANELS USING A FINITE ELEMENT FORMULATION	71-0636
STABILITY	A SURVEY OF FINITE ELEMENT FRAMEWORK STABILITY ANALYSIS	69-0418
STABILITY	A THEORETICAL METHOD FOR EVALUATING STABILITY OF OPENINGS IN ROCK	72-0877
STABILITY	A USEFUL FORM OF THE VARIATIONAL PRINCIPLE OF IDEAL MHD FOR ONE-DIMENSIONAL NUMERICAL STABILITY CALC	73-0746
STABILITY	ACCURACY OF EQUILIBRIUM SLOPE STABILITY ANALYSIS	73-0517
STABILITY	ACCURACY, STABILITY AND OSCILLATION CHARACTERISTICS OF FINITE ELEMENT METHOD FOR SOLVING HEAT CONDUC	69-0007
STABILITY	ACCURACY, STABILITY, AND OSCILLATION CHARACTERISTICS OF TRANSIENT TWO-DIMENSIONAL HEAT CONDUCTION	75-0849
STABILITY	AN APPLICATION OF THE FINITE ELEMENT METHOD FOR ROCK SLOPE STABILITY ANALYSIS	71-0813
STABILITY	AN INVESTIGATION OF THE STABILITY OF NUMERICAL SOLUTIONS OF BIOT'S EQUATIONS OF CONSOLIDATION	74-0746
STABILITY	ANALYSE DE LA STABILITY DES TALUS ROCHEUX PAR UNE METHODE GENERALISEE DE L EQUILIBRE LIMITE (ROCK SL	74-1019
STABILITY	ANALYSE TRIDIMENSIONELLE DE LA STABILITE D'UN TUNNEL AU VOISINAGE DU FRONT DE TAILLE DANS UNE ROCHE	74-0995
STABILITY	ANALYSIS OF ELASTIC STABILITY OF SHELLS OF REVOLUTION BY THE FINITE ELEMENT METHOD	67-0056
STABILITY	ANALYSIS OF HYDROMAGNETIC PLASMA STABILITY BY FINITE-ELEMENT METHOD	72-0307
STABILITY	ANALYSIS OF SLOPE STABILITY BY FINITE ELEMENT METHOD AND ITS APPLICATION PROGRAM. SOIL MECHANICS	72-0919
STABILITY	APPLICATION OF FINITE ELEMENT METHOD IN DETERMINING STABILITY OF CRATER SLOPES. PRELIMINARY REPORT	68-0026
STABILITY	APPLICATION OF FINITE ELEMENT METHOD IN DETERMINING STABILITY OF CRATER SLOPES	72-0888
STABILITY	APPLICATION OF THE FINITE ELEMENT STIFFNESS METHOD TO NONCONSERVATIVE STABILITY PROBLEMS	67-0158
STABILITY	APPLICATION OF THE FINITE ELEMENT METHOD TO NONCONSERVATIVE STABILITY PROBLEMS WITH DAMPING	72-0610

STABILITY	AUTOMATED LARGE DEFLECTION AND STABILITY ANALYSIS OF THREE DIMENSIONAL BAR STRUCTURES	69-0393
STABILITY	AUTOMATED OPTIMUM DESIGN OF AIRCRAFT WINGS TO SATISFY STRENGTH, STABILITY, FREQUENCY AND FLUTTER REQ	72-0605
STABILITY	BERECHNUNG UND BEURTEILUNG DER STANDSICHERHEIT VON ABBAUHOHLRAEUMEN IN STEILEN GANGERZLAGERSTAETTEN	74-0597
STABILITY	BOTTOM STABILITY UNDER WATER INDUCED LOADING	72-0373
STABILITY	BOUNDS FOR EIGENVALUES IN SOME VIBRATION AND STABILITY PROBLEMS.	72-0390
STABILITY	CAVITY STABILITY: FINITE ELEMENT ANALYSIS FOR STEADY AND TRANSIENT CREEP	74-1082
STABILITY	COMPARISON OF CALCULATION BY THE FINITE ELEMENT METHOD WITH EXPERIMENT AND AN EXAMPLE OF A DYNAMIC S	72-0597
STABILITY	COMPUTATIONAL STABILITY IN NONLINEAR DISCRETE STRUCTURAL SYSTEMS	69-0145
STABILITY	COMPUTER PROGRAM FOR PIT SLOPE STABILITY ANALYSIS BY THE FINITE ELEMENT STRESS ANALYSIS AND LIMITING	72-0225
STABILITY	CONVERGENCE AND STABILITY OF NONLINEAR FINITE-ELEMENT EQUATIONS	75-0441
STABILITY	CONVERGENCE, ACCURACY AND STABILITY OF FINITE ELEMENT APPROXIMATIONS OF A CLASS OF NON-LINEAR HYPERB	73-0159
STABILITY	CORRELATION BETWEEN MODEL STUDY AND FINITE ELEMENT ANALYSIS IN AN INVESTIGATION OF SLOPE STABILITY	72-0227
STABILITY	DERIVATION OF STIFFNESS MATRICES FOR THE ANALYSIS OF LARGE DEFLECTION AND STABILITY PROBLEMS	65-0066
STABILITY	DEVELOPMENT AND APPLICATIONS OF THEORETICAL METHODS FOR EVALUATING STABILITY OF OPENINGS IN ROCK	73-0791
STABILITY	DEVELOPMENT AND APPLICATIONS OF A GENERAL COMPUTER PROGRAM FOR EVALUATING STABILITY OF OPENINGS IN R	74-0994
STABILITY	DISCRETE-ELEMENT METHODS FOR STABILITY ANALYSIS OF COMPLEX STRUCTURES	68-0140
STABILITY	DYNAMIC STABILITY ANALYSIS BY THE CONJUGATE GRADIENT METHOD	72-0749
STABILITY	DYNAMIC STABILITY OF CYLINDRICAL SHELLS BY FINITE ELEMENTS	68-0292
STABILITY	DYNAMIC STABILITY OF CONICAL SHELLS BY FINITE ELEMENTS	70-0467
STABILITY	DYNAMIC STABILITY OF PLATES BY FINITE ELEMENTS	71-0573
STABILITY	DYNAMIC STABILITY OF A NONLINEAR BEAM SUBJECTED TO BOTH LONGITUDINAL AND TRANSVERSE EXCITATION	73-0553
STABILITY	DYNAMIC STABILITY OF TIMOSHENKO BEAMS BY FINITE ELEMENT METHOD	75-0194
STABILITY	DYNAMIC STABILITY OF TIMOSHENKO BEAMS BY FINITE ELEMENT METHOD	75-0422
STABILITY	EFFECT OF INITIAL DISPLACEMENT ON PROBLEM OF LARGE DEFLECTION AND STABILITY	67-0113
STABILITY	ELASTIC EQUILIBRIUM MODELS, STRESS FUNCTIONS, AND DUALITY NON LINEARITIES AND STABILITY, FIVE LECTUR	73-0631
STABILITY	ELASTIC STABILITY OF SHELLS OF REVOLUTION BY THE VARIATIONAL APPROACH USING DISCRETE ELEMENTS	66-0106
STABILITY	ELASTIC-PLASTIC STABILITY ANALYSIS OF MINE-WASTE EMBANKMENTS	75-0835
STABILITY	ETUDES DES POUTRES ET POTEAUX PAR LA METHODE DES ELEMENTS FINIS. (STABILITY ANALYSIS OF BEAMS AND C	74-1008
STABILITY	EVALUATION OF THE STABILITY OF NATURAL SLOPES IN PLASTIC CHAMPLAIN CLAYS	74-0264
STABILITY	EXPLICIT INTEGRATION - STABILITY, SOLUTION PROPERTIES, COST	75-0298
STABILITY	FACTOR OF SAFETY APPROACH FOR EVALUATING SEISMIC STABILITY OF SLOPES	75-0645
STABILITY	FINITE ELEMENT ANALYSIS OF FREE VIBRATION AND STABILITY OF STRUCTURES SUBJECTED TO INITIAL STRESS	68-0300
STABILITY	FINITE ELEMENT ANALYSIS OF TORSIONAL AND TORSIONAL-FLEXURAL STABILITY PROBLEMS	70-0028
STABILITY	FINITE ELEMENT ANALYSIS OF VIBRATION AND STABILITY PROBLEMS	72-0477
STABILITY	FINITE ELEMENT ANALYSIS OF PROBLEMS IN NONCONSERVATIVE ELASTIC STABILITY	72-0957
STABILITY	FINITE ELEMENT ANALYSIS FOR THE SEISMIC STABILITY OF EARTH STRUCTURES	74-1106
STABILITY	FINITE ELEMENT MATRIX FORMULATION OF POST-BUCKLING STABILITY AND IMPERFECTION SENSITIVITY	70-0246
STABILITY	FINITE ELEMENT METHOD APPLIED TO THE PROBLEM OF STABILITY OF A NON-CONSERVATIVE SYSTEM	71-0291
STABILITY	FINITE ELEMENT METHOD IN SHELL STABILITY ANALYSIS	73-0250
STABILITY	FINITE ELEMENT SCHEMES: STABILITY AND CONVERGENCE	72-0302
STABILITY	FINITE ELEMENT SOLUTION TO DYNAMIC STABILITY OF BARS	68-0519
STABILITY	GALERKIN FINITE ELEMENT SOLUTION FOR THE STABILITY OF CANTILEVER COLUMNS SUBJECTED TO TANGENTIAL LOA	75-0534
STABILITY	GALERKIN FINITE-ELEMENT SOLUTION FOR STABILITY OF CANTILIVER COLUMNS SUBJECTED TO TANGENTIAL LOADS	75-0618
STABILITY	GENERAL METHOD FOR STABILITY ANALYSIS OF ROTATING SHAFTS	75-0528
STABILITY	HYDROMAGNETIC STABILITY STUDIES USING THE FINITE ELEMENT METHOD	75-0878
STABILITY	INFLUENCE OF ROCK PROPERTIES VARIABILITY ON MINE OPENING STABILITY ANALYSIS	74-1021
STABILITY	LINEAR STABILITY ANALYSIS OF BEAM-COLUMNS	73-0499
STABILITY	MATRIX FORMULATION OF STRUCTURAL STABILITY PROBLEMS	65-0039
STABILITY	METHODS FOR PREDICTING THE AERODYNAMIC AND STABILITY AND CONTROL CHARACTERISTICS OF STOL AIRCRAFT. V	73-0718
STABILITY	METHODS FOR PREDICTING THE AERODYNAMIC AND STABILITY AND CONTROL CHARACTERISTICS OF STOL AIRCRAFT. V	73-0719
STABILITY	METHODS OF PREDICTING THE AERODYNAMIC AND STABILITY AND CONTROL CHARACTERISTICS OF STOL AIRCRAFT. VO	73-0717
STABILITY	MINIMUM WEIGHT DESIGN WITH STABILITY CONSTRAINT	70-0274
STABILITY	NONCONSERVATIVE STABILITY BY FINITE ELEMENT	71-0572
STABILITY	NONCONSERVATIVE STABILITY OF PLATES BY FINITE ELEMENT METHOD	75-0789
STABILITY	NUMERICAL RESULTS FROM THE APPLICATION OF GRADIENT ITERATIVE TECHNIQUES TO THE FINITE ELEMENT VIBRAT	72-0138
STABILITY	NUMERICAL STUDY OF HYDROGMAGNETIC STABILITY USING FINITE ELEMENT METHOD	73-0092
STABILITY	ON THE ACCURACY AND NUMERICAL STABILITY OF THE FINITE ELEMENT APPROXIMATION FOR PARABOLIC AND HYPERB	75-0102
STABILITY	ON THE DERIVATION OF STIFFNESS MATRICES FOR THE ANALYSIS OF LARGE DEFLECTION AND STABILITY PROBLEMS	66-0050
STABILITY	ON THE INTERRELATION BETWEEN STABILITY AND COMPUTATIONS	73-0674
STABILITY	ON THE STABILITY OF ROTOR-AND-BEARING SYSTEMS AND ON THE CALCULATION OF SLIDING BEARINGS	72-0037
STABILITY	ON THE STABILITY OF EXPLICIT METHODS FOR THE NUMERICAL INTEGRATION OF THE EQUATIONS OF MOTION IN FIN	72-0172
STABILITY	ON THE UNCONDITIONAL STABILITY OF AN IMPLICIT ALGORITHM FOR NONLINEAR STRUCTURAL DYNAMICS	75-0825
STABILITY	OPTIMIZATION OF STRUCTURES FOR STRENGTH AND STABILITY REQUIREMENTS	73-0575
STABILITY	OSCILLATION STABILITY OF INTERNALLY AND EXTERNALLY NOTCHED PLANE BARS OF EQUAL FORM FACTOR FROM 3943	73-0745
STABILITY	PERTURBATION METHOD IN THE ANALYSIS OF GEOMETRICALLY NON LINEAR AND STABILITY PROBLEMS	72-0321
STABILITY	PROPERTIES OF STABILITY AND MONOTONY WITH THE FINITE ELEMENT METHOD	69-0266
STABILITY	REFINED FINITE ELEMENT STABILITY ANALYSIS OF THIN-WALLED MEMBERS	75-0784
STABILITY	SLOPE STABILITY ANALYSIS BY THE FINITE ELEMENT STRESS ANALYSIS AND LIMITING EQUILIBRIUM METHOD	70-0222
STABILITY	SPECIAL STUDY ON THE STABILITY OF A DAM SPILLWAY	72-0469
STABILITY	STABILITA STEN V PLASTICKEM OBORU (STABILITY IN WEBPLATES IN PLASTIC RANGE)	73-0387
STABILITY	STABILITY ANALYSIS BY FINITE ELEMENTS	70-0056
STABILITY	STABILITY ANALYSIS OF COMPLEX STRUCTURES USING DISCRETE ELEMENT TECHNIQUE	67-0068
STABILITY	STABILITY ANALYSIS OF SHELLS OF REVOLUTION BY THE FINITE ELEMENT METHOD	68-0124
STABILITY	STABILITY ANALYSIS OF SHELLS OF REVOLUTION	68-0136
STABILITY	STABILITY ANALYSIS OF AXISYMMETRIC SHELLS	69-0131
STABILITY	STABILITY ANALYSIS OF COLUMNS USING FINITE ELEMENTS	69-0478
STABILITY	STABILITY ANALYSIS OF STRUCTURES BY A REDUCED SYSTEM OF GENERALIZED COORDINATES	70-0168
STABILITY	STABILITY ANALYSIS OF MULTILAYER SANDWICH STRUCTRES	70-0345
STABILITY	STABILITY ANALYSIS OF AN EXPERIMENTALLY DEFORMED SINGLE LAYER OF INDIANA LIMESTONE USING FINITE ELEM	74-0212
STABILITY	STABILITY ANALYSIS OF THE TORSIONAL BENDING VIBRATIONS OF WING IN A SUBSONIC FLOW BY METHOD OF THE F	75-0755
STABILITY	STABILITY AND ACCURACY OF NUMERICAL APPROXIMATIONS TO TIME DEPENDENT FLOWS	75-0183
STABILITY	STABILITY AND OSCILLATION CHARACTERISTICS OF FINITE ELEMENT, FINITE DIFFERENCE AND WEIGHTED RESIDUAL	73-0139
STABILITY	STABILITY AND OSCILLATION CHARACTERISTICS OF FINITE ELEMENT, FINITE-DIFFERENCE, AND WEIGHTED-RESIDUA	75-0615
STABILITY	STABILITY AND POSTBUCKLING BEHAVIOUR OF HYPERELASTIC BODIES AT FINITE STRAIN BY THE FINITE ELEMENT M	73-0562
STABILITY	STABILITY AND POSTBUCKLING BEHAVIOR OF HYPERELASTIC BODIES AT FINITE STRAIN BY THE FINITE ELEMENT ME	74-0838
STABILITY	STABILITY CONTROL ANALYSIS OR ROTATING PLATES BY FINITE ELEMENT, EMPHASIS ON SLOTS AND HOLES	71-0284
STABILITY	STABILITY CRITERIA FOR TWO FINITE ELEMENT SCHEMES FOR PARABOLIC EQUATION	75-0568
STABILITY	STABILITY IN LQ OF L2-PROJECTION INTO FINITE ELEMENT FUNCTION SPACES	75-0071
STABILITY	STABILITY INVESTIGATIONS OF PLATES USING FINITE ELEMENT METHOD BASED ON AN EXPANDED VARIATION PRINCI	71-0663
STABILITY	STABILITY INVESTIGATION OF A CYLINDRICAL SHELL WITH FINITE ELEMENTS	72-0127
STABILITY	STABILITY OF FINITE ELEMENT SCHEMES FOR VIBRATION PROBLEMS OF THE EQUATION OF ELASTICITY THEORY	71-0521
STABILITY	STABILITY OF HYPERBOLOIDAL SHELLS	75-0488
STABILITY	STABILITY OF MASONRY PIERS	75-0248
STABILITY	STABILITY OF MODERATELY THICK RECTANGULAR PLATES USING A HIGH PRECISION TRIANGULAR FINITE ELEMENT	75-0334
STABILITY	STABILITY OF PLATES USING THE FINITE ELEMENT METHOD	66-0081
STABILITY	STABILITY OF PLATES USING THE FINITE ELEMENT METHOD	67-0030
STABILITY	STABILITY OF PLATES USING A MIXED FINITE ELEMENT FORMULATION	74-0218
STABILITY	STABILITY OF PRESSURIZED HYPERBOLOIDAL SHELLS	75-0585
STABILITY	STABILITY OF TILTING-PAD JOURNAL BEARINGS	71-0687
STABILITY	STATIC, STABILITY, AND DYNAMIC ANALYSIS OF SHELLS OF REVOLUTION BY NUMERICAL INTEGRATION - A COMPARI	72-0389
STABILITY	STATISTICAL ASPECTS OF THE STABILITY OF COMPOUND SYSTEMS WITH BILINEAR ELEMENTS	73-0986
STABILITY	STEADY STATE EQUATIONS OF MOTION, EQUILIBRIUM SHAPE AND STABILITY DERIVATIVES OF ELASTIC AIRPLANES E	72-0541

STABILITY   STRESS, STABILITY AND VIBRATION OF COMPLEX SHELLS OF REVOLUTION: ANALYSIS AND USERS MANUAL FOR BOSOR   69-0441
STABILITY   STUDIES OF HYDROMAGNETIC STABILITY   74-0035
STABILITY   THALIA - ONE-DIMENSIONAL MAGNETOHYDRODYNAMIC STABILITY PROGRAM USING METHOD OF FINITE-ELEMENTS   75-0458
STABILITY   THALIA: A ONE-DIMENSIONAL MAGNETOHYDRODYNAMIC STABILITY PROGRAM USING THE METHOD OF FINITE ELEMENTS   74-1085
STABILITY   THE EFFECT OF INITIAL DISPLACEMENTS OF PROBLEMS OF LARGE DEFLECTION AND STABILITY   67-0048
STABILITY   THE FINITE ELEMENT DYNAMIC STABILITY ANALYSIS OF THIN SHELLS SUBMERGED IN FLUIDS   71-0057
STABILITY   THE FINITE ELEMENT METHOD OF THIN-SHELL STABILITY ANALYSIS   72-0725
STABILITY   THE FINITE ELEMENT METHOD IN PLATE AND SHELL STABILITY ANALYSIS   73-0644
STABILITY   UNBALANCE GENERATED VIBRATIONS AND STABILITY OF TURBINE ROTORS   73-0224
STABILITY   UNTERSUCHUNGEN ZUR SPURMASSTABILITAET BEI RADSAETZEN MIT KLOTZGEBREMSTEN VOLLRAEDERN (INVESTIGATIONS   75-0810
STABILITY   USEFUL FORM OF THE VARIATIONAL PRINCIPLE OF IDEAL MHD FOR ONE-DIMENSIONAL NUMERICAL STABILITY CALCUL   73-0946
STABILITY   VIBRATION AND STABILITY OF PLATES USING FINITE ELEMENTS   68-0244
STABILITY-FINIT   COUPLED, NONCONSERVATIVE STABILITY-FINITE ELEMENT   72-0220
STABILIZATION   FINITE ELEMENT APPROACH TO HYDRODYNAMICS AND MESH STABILIZATION   74-1298
STABILIZED   ANALYSIS OF FEEDBACK CONTROL STABILIZED VIBRATION BY THE FINITE ELEMENT METHOD   74-1188
STABILIZER   DEVELOPMENT OF A GRAPHITE HORIZONTAL STABILIZER   73-0931
STABILIZER   STATIC AND DYNAMIC ANALYSIS OF F-14A BORON HORIZONTAL STABILIZER   71-0123
STABLE   A HIGHLY STABLE EXPLICIT INTEGRATION TECHNIQUE FOR COMPUTATIONAL CONTINUUM MECHANICS   74-1277
STABLE   FINITE ELEMENT REPRESENTATION OF STABLE CRACK-GROWTH   73-0086
STABLE   FINITE ELEMENT SOLUTION FOR ENERGY CONSERVATION USING A HIGHLY STABLE EXPLICIT INTEGRATION ALGORITHM   72-0633
STABLE   UNCONDITIONALLY STABLE FINITE ELEMENT SCHEMES FOR PARABOLIC EQUATIONS   74-1286
STABLY   A NUMERICAL INVESTIGATION OF STABLY STRATIFIED, WIND DRIVEN CAVITY FLOW BY THE FINITE ELEMENT METHOD   74-1372
STACKS   NON-LINEAR ELASTIC ANALYSIS OF UNGUYED TOWERS AND STACKS   74-1157
STACUSS   STACUSS 1: A DISCRETE-ELEMENT PROGRAM FOR THE STATIC ANALYSIS OF SINGLE-LAYER CURVED STIFFENED SHELL   69-0160
STAEBE   ZUR BERECHNUNG DER BIEGESTEIFIGKEIT ABGESETZTER STAEBE UND WELLEN UNTER ANWENDUNG VON FINITEN ELEMEN   75-0481
STAGE   FINITE ELEMENT ANALYSIS OF MERCURY SLOSH IN THE SOLAR ELECTRIC PROPULSION STAGE   75-0757
STAGS   EXTENSIONS TO THE STAGS COMPUTER CODE   73-0364
STAGS   EXTENSIONS TO THE STAGS COMPUTER CODE   73-0917
STAHLBETONPLATT   ZUR WIRKLICHKEITSNAHEN BERECHNUNG VON STAHLBETONPLATTEN MIT DER FINITE-ELEMENT-METHODE. (APPROXIMATE   75-0332
STAHLBETONSCHEI   BERECHNUNG VON STAHLBETONSCHEIBEN IM ZUSTAND II BEI ANNAHME EINES WIRKLICHKEITSNAHEN WERKSTOFFVERHAL   74-0978
STAHLDRAHT-KUNS   BERECHNUNG EINES EINACHSIGEN STAHLDRAHT-KUNSTSTOFFVERBUNDES UNTER EINSATZ DER FINIT-ELEMENT METHODE   72-0644
STANDARD   APPLICATION OF FINITE ELEMENT TECHNIQUES TO POST-YIELD ANALYSIS OF PROPOSED STANDARD 3 POINT BEND FR   74-0118
STANDARD   BOOLEAN MATRICES AS A STANDARD FACILITY IN A MATRIX CODE   65-0057
STANDARD   COMPUTER - AIDED DESIGN - STANDARD OIL OF INDIANA BUILDING   73-0182
STANDARD   EARTH PRESSURES AGAINST THE ABUTMENTS AND WING WALLS OF STANDARD MOTORWAY BRIDGES   75-0529
STANDARD   SOME STRUCTURAL PROBLEMS - STANDARD OIL OF INDIANA BUILDING   73-0180
STANDARD   STRONGLY PLANAR AC-LATTICES IN WHICH IDEAL OF FINITE-ELEMENTS IS STANDARD   75-0816
STANDARDS   JOURNAL OF RESEARCH OF THE NATIONAL BUREAU OF STANDARDS: C. ENGINEERING AND INSTRUMENTATION   71-0564
STANDPIPE   STRESS DISTRIBUTION IN THE LIGAMENT ZONE OF PERFORATED CIRCULAR DISCS WITH SPECIAL REFERENCE TO THE   73-0827
STANDSICHERHEIT   BERECHNUNG UND BEURTEILUNG DER STANDSICHERHEIT VON ABBAUHOHLRAEUMEN IN STEILEN GANGERZLAGERSTAETTEN   74-0597
STANZIONARI   STUDIO DI CAMPI ELETTRICI E MAGNETICI STANZIONARI CON IL METODO DEGLI ELEMENTI FINITI. APPLICAZIONE   75-0207
STAR-100   HYPERMATRIX SCHEME FOR FINITE-ELEMENT SYSTEMS ON CDC STAR-100 COMPUTER   75-0818
STAR-100   IMPACT OF CDC STAR-100 COMPUTER ON FINITE ELEMENT SYSTEMS   75-0039
STARDYNE   MRI/STARDYNE USER INFORMATION MANUAL   73-0541
STARDYNE   STARDYNE STRUCTURAL ANALYSIS SYSTEM, ANALYTICAL CAPABILITY MANUAL   68-0235
STARFIGHTER   F 104 G FINAL MOTIVE: GROUND VIBRATION CALCULATIONS FOR THE STARFIGHTER F 104 G   69-0456
STARILOR   ASUPRA ANALIZEI PRIN METODA ELEMENTULUI FINIT A STARILOR DE EFORTURI SI DEFORMATII DIN MASIVELE DE P   70-0471
STARR-PLASTISCH   STARR-PLASTISCHE TRAGLASTBERECHNUNGEN MITTELS DER METHODE DER FINITEN ELEMENTE   71-0635
STARS   STARS - A COMPUTER CODE FOR STRUCTURAL DYNAMIC ANALYSIS OF REACTOR SYSTEM   71-0204
STARTUP   THREE-DIMENSIONAL STUDY OF NUCLEAR FUEL ROD BEHAVIOUR DURING STARTUP   73-0815
STARTUP   THREE-DIMENSIONAL STUDY OF NUCLEAR FUEL ROD BEHAVIOR DURING STARTUP   74-1003
STATE   ADAPTATION OF CRITICAL STATE SOIL MECHANICS FOR USE IN FINITE ELEMENTS   71-0534
STATE   ANALYSIS MODELS FOR PLATE ELEMENTS IN THE MEMBRANE STATE OF STRESS   67-0044
STATE   APPLICATION OF THE METHOD OF FINITE ELEMENTS TO CALCULATIONS OF THE STRESSED AND STRAINED STATE OF I   75-0775
STATE   DER SPANNUNGS UND BEANSPRUCHUNGSZUSTAND DES BODENMATERIALS IN DER NACHBARSCHAFT EINER BODENBOESCHUNG   75-0371
STATE   DETERMINATION OF THE STRESS-STRAINED STATE WHEN CUTTING METAL WITH A WEDGE   73-0237
STATE   DISCRETE-ELEMENT PLASTIC ANALYSIS OF STRUCTURES IN A STATE OF MODIFIED PLANE STRAIN   69-0140
STATE   DISCUSSION ON THE ADAPTION OF CRITICAL STATE SOLID MECHANICS THEORY FOR USE IN FINITE ELEMENTS   71-0007
STATE   EFFECT OF INITIAL STATE FORCES ON THE HYDROELASTIC VIBRATION OF TUBES   73-0589
STATE   EFFECTS OF QUADRATURE ERRORS IN FINITE ELEMENT APPROXIMATION OF STEADY STATE, EIGENVALUE, AND PARABO   72-0343
STATE   EFFECTS OF SPECIMEN RADIUS ON THE STRESS STATE NEAR A ROOF BOLT ANCHOR: A FINITE ELEMENT DETERMINATI   73-0229
STATE   FINITE ELEMENT METHOD FOR BOUND STATE CALCULATIONS IN QUANTUM MECHANICS   75-0005
STATE   FINITE ELEMENT SOLUTION OF THE STEADY STATE COMPRESSIBLE LUBRICATION PROBLEM   70-0007
STATE   FINITE ELEMENT SOLUTION OF STEADY STATE POTENTIAL FLOW PROBLEMS   70-0480
STATE   FINITE ELEMENT STUDY OF THE TRIAXIAL STRESS STATE AROUND AN INSPECTION TUNNEL IN AN ARCH DAM   74-0831
STATE   GENERALIZED STEADY STATE FIELD PROBLEM AND ITS SOLUTION   74-0059
STATE   INFLUENCE OF FLUID INJECTION ON THE STATE OF STRESS IN THE EARTH'S CRUST   72-0624
STATE   INVESTIGATION OF THE STRESS-STRAINED STATE OF SPHERICAL SHELLS WITH AN ECCENTRIC CUTOUT ON THE BASIS   73-0222
STATE   ISSLEDOVANIE NAPRYAZHENNOGO SOSTOYANIYA REBRISTYKH ISILINDRICHESKIKH OBOLOCHEK S PRYAMOUGOL'NYMI OTV   74-0944
STATE   ISSLEDOVANIE NAPRYAZHENNO-DEFORMIROVANNOGO SOSTOYANIYA SIMMETRICHNYKH DISKOV SO STUPITSEI. (INVESTIG   74-0965
STATE   MECHANICS OF CABLE MOORING SYSTEMS. VOLUME IV. A COMPUTER PROGRAM FOR ANALYZING THE STEADY STATE RES   72-0653
STATE   MECHANICS OF CABLE MOORING SYSTEMS. VOLUME 1. THREE DIMENSIONAL RESPONSE OF DEEP WATER MOORING LINES   72-0654
STATE   MECHANICS OF CABLE MOORING SYSTEMS. VOLUME V1. A COMPUTER PROGRAM FOR ANALYZING THE STEADY STATE CON   73-0725
STATE   NUMERICAL SOLUTION OF STEADY STATE DIFFUSION PROBLEMS CONTAINING SINGULARITIES   74-0036
STATE   QUANTITATIVE STRAIN-AND-STRESS STATE CRITERION FOR FAILURE IN THE VICINITY OF SHARP CRACKS   74-0639
STATE   SENSITIVITY ANALYSIS OF INPUT PARAMETERS IN NUMERICAL MODELING OF STEADY STATE REGIONAL GROUNDWATER   74-0926
STATE   SETTLEMENT ANALYSIS OF A STRIP FOOTING USING A CRITICAL STATE SOIL MODEL IN CONJUNCTION WITH FINITE   71-0533
STATE   STATISTICAL STATE OF STRESS STUDIED BY GRID ANALYSIS: "NUMERICAL METHODS OF ANALYSIS"   49-0001
STATE   STEADY STATE AND TRANSIENT GROUNDWATER FLOW IN SLOPES   70-0273
STATE   STEADY STATE AND TRANSIENT HEAT CONDUCTION AND STRESS ANALYSES OF TWO AND THREE DIMENSIONAL STRUCTUR   71-0675
STATE   STEADY STATE EQUATIONS OF MOTION, EQUILIBRIUM SHAPE AND STABILITY DERIVATIVES OF ELASTIC AIRPLANES E   72-0541
STATE   STEADY STATE FLOW IN RIGID NETWORKS OF FRACTURES   74-0278
STATE   STEADY STATE IN SOIL SUBJECTED TO SEVERAL COUPLED FORCES: A FINITE ELEMENT FORMULATION   70-0207
STATE   STUDY ON STATE OF STRESS AND DEFORMATION OF CYLINDRICAL SPECIMENS OF BRITTLE MATERIAL UNDER UNIAXIAL   69-0152
STATE   TWO-DIMENSIONAL FINITE ELEMENT ANALYSIS OF SEMI-CONDUCTOR STEADY STATE TRANSPORT EQUATIONS   74-1224
STATE-OF-THE-AR   BUCKLING OF COOLING-TOWER SHELLS: STATE-OF-THE-ART   75-0385
STATES   FINITE ELEMENT CALCULATION OF LEFT VENTRICULAR WALL STRESS IN NORMAL AND DISEASE STATES   73-0588
STATES   NONLINEAR STRESS ANALYSIS OF INDETERMINATE STRUCTURAL ELEMENTS SUBJECTED TO BIAXIAL STRESS STATES US   73-1152
STATES   ON THE STATES OF STRESS AND DEFORMATION OF CYLINDRICAL SPECIMENS OF BRITTLE MATERIAL UNDER UNIAXIAL   69-0150
STATES   ON THE STATES OF STRESS AND DEFORMATION AROUND THE BOTTOM OF A CIRCULAR VERTICAL SHAFT   69-0323
STATIC   A COMPUTER PROGRAM FOR THE STATIC AND DYNAMIC FINITE ELEMENT ANALYSIS OF AXISYMMETRIC THIN SHELLS   66-0017
STATIC   A STATIC AND DYNAMIC FINITE ELEMENT SHELL ANALYSIS WITH EXPERIMENTAL VERIFICATION   69-0156
STATIC   A STATIC AND DYNAMIC FINITE ELEMENT SHELL ANALYSIS WITH EXPERIMENTAL VERIFICATION   71-0769
STATIC   APPLICATION OF FINITE ELEMENT METHOD TO PREDICT STATIC RESPONSE OF AUTOMOTIVE BODY STRUCTURES   71-0340
STATIC   APPLICATION OF SMALL AND LARGE-DISPLACEMENT THEORY TO A LARGE FLEXIBLE SOLAR ARRAY AND COMPARISON WI   71-0231
STATIC   APPLICATION OF THE FINITE ELEMENT METHOD TO PREDICT STATIC AND DYNAMIC RESPONSE OF AN UNSHROUDED CEN   71-0574
STATIC   ASPECTS OF MECHANICAL BEHAVIOR OF ROCK UNDER STATIC AND CYCLIC LOADING. PART A. MECHANICAL BEHAVIOR   72-0761
STATIC   ASPECTS OF MECHANICAL BEHAVIOR OF ROCK UNDER STATIC AND CYCLIC LOADING. PART A. MECHANICAL BEHAVIOR   72-0761
STATIC   ASPECTS OF MECHANICAL BEHAVIOR OF ROCK UNDER STATIC AND CYCLIC LOADING. PART A: MECHANICAL BEHAVIOR   72-0762
STATIC   ASPECTS OF MECHANICAL BEHAVIOR OF ROCK UNDER STATIC AND CYCLIC LOADING. PART A: MECHANICAL BEHAVIOR   72-0762
STATIC   COMPARISON OF STRUCTURAL LOADS: STATIC VERSUS DYNAMIC   70-0464
STATIC   COMPUTER PROGRAM FOR STATIC AND DYNAMIC ANALYSIS OF LINEAR STRUCTURAL SYSTEMS   72-0683

STATIC	DISCRETE-ELEMENT STATIC ANALYSIS OF BONDED, DOUBLE-LAYER, BRANCHED, THIN SHELLS OF REVOLUTION, PART	68-0102
STATIC	DISCRETE-ELEMENT STATIC ANALYSIS OF CORE-STIFFENED SHELLS OF REVOLUTION FOR ASYMMETRIC MECHANICAL LO	68-0148
STATIC	DISCRETE-ELEMENT STATIC ANALYSIS OF BONDED, DOUBLE-LAYER, BRANCHED, THIN SHELLS OF REVOLUTION, PART	70-0081
STATIC	DUALITY BETWEEN DISPLACEMENT AND EQUILIBRIUM METHOD WITH A VIEW TO OBTAINING UPPER AND LOWER BOUNDS	62-0013
STATIC	DUALITY IN STRUCTURAL ANALYSIS BY FINITE ELEMENTS. STATIC GEOMETRIC ANALOGIES, THE DUAL PRINCIPLES O	71-0583
STATIC	ELEMENT STATIC FLEXIBILITY AND STIFFNESS MATRICES FOR THE FINITE ELEMENT ANALYSIS OF BEAM, PLATE, AN	67-0070
STATIC	FINITE ELEMENT ANALYSIS OF STATIC AND DYNAMIC STRESSES AROUND SUDDENLY PUNCHED HOLES IN PLATES AND S	73-1088
STATIC	FINITE ELEMENT COMPUTER PROGRAM FOR PREDICTING THE NONLINEAR STATIC AND DYNAMIC BEHAVIOR OF VISCOELA	75-0191
STATIC	FLEXUAL RIGIDITY OF A THIN WALLED BUILD-UP ROTOR FOR JET ENGINE. MEASUREMENT BY STATIC LOAD TEST AND	73-0756
STATIC	HIGHER ORDER FINITE ELEMENT METHOD FOR STATIC AND DYNAMIC ANALYSIS OF SHEAR WALLS	71-0368
STATIC	INTERNAL SUPPORTING STRUCTURES OF PRESSURE TUBES TYPE REACTOR CORE. ANALYTICAL MODEL FOR STATIC ANAL	71-0049
STATIC	INTERNAL SUPPORTING STRUCTURES OF THE (CIRENE) REACTOR CORE - RESULTS OF STATIC STRESS ANALYSIS FOR	73-0818
STATIC	INVERSION TECHNIQUE FOR STATIC FINITE ELEMENT METHOD	74-0170
STATIC	LINEAR AND NONLINEAR STATIC ANALYSIS OF AXISYMMETRICALLY LOADED THIN SHELLS OF REVOLUTION	69-0215
STATIC	NONLINEAR STATIC AND DYNAMIC RESPONSE OF UNDERWATER CABLE STRUCTURES USING THE FINITE ELEMENT METHOD	75-0460
STATIC	NONSAP: A STRUCTURAL ANALYSIS PROGRAM FOR STATIC AND DYNAMIC RESPONSE OF NONLINEAR SYSTEMS	74-0848
STATIC	NUMERICAL SOLUTION SCHEMES FOR HIGHLY NONLINEAR STATIC STRUCTURAL BEHAVIOR	75-0679
STATIC	OPTIMAL SYNTHESIS OF FLEXIBLE LINK MECHANISMS WITH LARGE STATIC DEFLECTIONS	74-0525
STATIC	OPTIMIZATION OF COMPLEX STRUCTURES TO SATISFY STATIC, DYNAMIC AND AEROELASTIC REQUIREMENTS	74-0272
STATIC	OPTIMUM DESIGN OF ADVANCED COMPOSITE STRUCTURES FOR STATIC LOADS	72-0613
STATIC	PART 1 - TASK 3.2.9. BENDING AND BUCKLING OF A ONE-DIMENSIONAL MODEL SIMULATING A CONICAL SHELL. PAR	71-0745
STATIC	PREDICTION OF STATIC AND FATIGUE DAMAGE AND CRACK PROPAGATION IN COMPOSITE MATERIALS	75-0643
STATIC	PSEUDO STATIC DEFORMATION AND FREQUENCIES OF ROTATING TURBOMACHINERY BLADES	72-0239
STATIC	SAP IV. A STRUCTURAL ANALYSIS PROGRAM FOR STATIC AND DYNAMIC RESPONSE OF LINEAR SYSTEMS	73-0650
STATIC	SASL - A FINITE ELEMENT CODE FOR THE STATIC ANALYSIS OF AXISYMMETRIC AND PLANE SOLIDS SUBJECTED TO A	72-0779
STATIC	SLADE, A COMPUTER PROGRAM FOR THE STATIC ANALYSIS OF THIN SHELLS	70-0079
STATIC	SOLID SAP. A STATIC ANALYSIS PROGRAM FOR THREE DIMENSIONAL SOLID STRUCTURES	72-0855
STATIC	STACUSS 1: A DISCRETE-ELEMENT PROGRAM FOR THE STATIC ANALYSIS OF SINGLE-LAYER CURVED STIFFENED SHELL	69-0160
STATIC	STATIC ANALYSIS OF A COMPLEX SHELL STRUCTURE UNDER ASYMMETRIC LOADING	67-0120
STATIC	STATIC ANALYSIS OF SHELLS OF REVOLUTION USING DOUBLY-CURVED QUADRILATERAL ELEMENTS DERIVED FROM ALTE	69-0368
STATIC	STATIC ANALYSIS OF SANDWICH STRUCTURES BY THE FINITE ELEMENT METHOD	71-0730
STATIC	STATIC ANALYSIS OF STIFFENED SHELLS BY THE FINITE ELEMENT METHOD	71-0803
STATIC	STATIC ANALYSIS OF THE 64-M 210-FT ANTENNA REFLECTION STRUCTURE	71-0137
STATIC	STATIC ANALYSIS VIA SUBSTRUCTURING OF AN EXPERIMENTAL VEHICLE FRONT-END BODY STRUCTURE	74-0313
STATIC	STATIC AND DYNAMIC APPLICATIONS OF A HIGH-PRECISION TRIANGULAR PLATE BENDING ELEMENT	69-0086
STATIC	STATIC AND DYNAMIC ANALYSIS OF F-14A BORON HORIZONTAL STABILIZER	71-0123
STATIC	STATIC AND DYNAMIC ANALYSIS OF SANDWICH STRUCTURES BY METHOD OF FINITE ELEMENTS	71-0352
STATIC	STATIC AND DYNAMIC ANALYSIS OF MULTILAYERED SANDWICH PLATES	72-0493
STATIC	STATIC AND DYNAMIC ANALYSIS OF VISCOELASTOPLASTIC FIBER-REINFORCED COMPOSITE SHELLS IN MISSILE STRUC	73-0651
STATIC	STATIC AND DYNAMIC ANALYSIS OF WOOD-JOIST FLOORS BY THE FINITE-ELEMENT METHOD	73-1161
STATIC	STATIC AND DYNAMIC ANALYSIS OF AN ARCH DAM	74-1101
STATIC	STATIC AND DYNAMIC ANALYSIS OF NONLINEAR STRUCTURES	75-0718
STATIC	STATIC AND DYNAMIC BEHAVIOUR OF GUYED MASTS	68-0271
STATIC	STATIC AND DYNAMIC BUCKLING OF THIN-WALLED COLUMNS USING FINITE ELEMENTS	70-0505
STATIC	STATIC AND DYNAMIC BEHAVIOUR OF RECTANGULAR PLATES USING HIGHER ORDER FINITE STRIPS	72-0458
STATIC	STATIC AND DYNAMIC ELASTO-PLASTIC ANALYSIS BY THE METHOD OF FINITE ELEMENTS IN SPACE AND TIME	72-0642
STATIC	STATIC AND DYNAMIC FINITE ELEMENT ANALYSIS OF SANDWICH STRUCTURES	68-0003
STATIC	STATIC AND DYNAMIC GEOMETRIC AND MATERIAL NONLINEAR ANALYSIS	74-1302
STATIC	STATIC AND DYNAMIC NONLINEAR STRUCTURAL PROBLEMS	72-0848
STATIC	STATIC AND DYNAMIC STRESSES IN SLOPES	66-0025
STATIC	STATIC AND HEAT TRANSFER ANALYSIS OF A SPHERE-CONE INTERSECTION IN A NUCLEAR CONTAINMENT VESSEL	75-0166
STATIC	STATIC AND SEISMIC BEHAVIOUR OF AN EARTH DAM	67-0026
STATIC	STATIC FLEXIBILITY MATRICES FOR A RECTANGULAR PLATE ELEMENT IN BENDING, TWISTING, AND SHEAR. UNIFORM	68-0217
STATIC	STATIC GEOMETRIC AND MATERIAL NONLINEAR ANALYSIS	72-0314
STATIC	STATIC STRESS ANALYSIS OF AXISYMMETRIC SOLIDS WITH MATERIAL AND GEOMETRIC NONLINEARITIES BY THE FINI	73-0958
STATIC	STATIC STRESSES BY LINEAR AND NONLINEAR METHODS	75-0278
STATIC	STATIC STRUCTURAL ANALYSIS OF SUSPENSION BRIDGES	66-0110
STATIC	STATIC THREE DIMENSIONAL ANALYSIS OF ELEVATED STEEL TRANSPORTATION STRUCTURES	74-1187
STATIC	STATIC, STABILITY, AND DYNAMIC ANALYSIS OF SHELLS OF REVOLUTION BY NUMERICAL INTEGRATION - A COMPARI	74-0389
STATIC	STATIC, VIBRATION AND BUCKLING ANALYSIS OF AXISYMMETRIC CIRCULAR PLATES USING FINITE ELEMENTS	73-0157
STATIC	STATIC, VIBRATION, AND THERMAL STRESS ANALYSES OF LAMINATED PLATE AND SHELLS BY THE HYBRID-STRESS FI	72-0602
STATIC	STRAP - A COMPUTER CODE FOR THE STATIC AND DYNAMIC ANALYSIS OF REACTOR STRUCTURAL SYSTEMS	71-0078
STATIC	STRAP: COMPUTER CODE FOR STATIC AND DYNAMIC STRUCTURAL ANALYSIS AND STUDIES MADE USING THE CODE	70-0409
STATIC	SURVEY OF SOLUTION PROCEDURES FOR NONLINEAR STATIC AND DYNAMIC ANALYSES	74-0307
STATIC	SURVEY OF STATIC GEOMETRIC AND MATERIAL NONLINEAR ANALYSIS BY THE FINITE ELEMENT METHOD	72-0783
STATIC	THE ANALYSIS OF ARCH DAMS UNDER STATIC LOADING BY THE FINITE ELEMENT METHOD	68-0059
STATIC	THE CALCULATION OF STATIC AND DYNAMIC CHARACTERISTICS OF FLIGHT VEHICLE STRUCTURES BY THE FINITE ELE	73-1018
STATIC	THE DEVELOPMENT OF APPLIED METHODS IN PROBLEMS OF STATIC CALCULATION OF THIN-WALLED THREE-DIMENSIONA	73-0934
STATIC	THE EFFECT OF IN PLANE STATIC STRESSES ON THE VIBRATION CHARACTERISTICS OF CURVED SANDWICH PLATES	71-0588
STATIC	THE FINITE ELEMENT METHOD APPLIED TO THE STATIC AND DYNAMIC BUCKLING OF CIRCULAR ARCHES	70-0497
STATIC	THE STATIC CONDENSATION ALGORITHM	74-0328
STATIC	THE USE OF FINITE ELEMENT METHODS FOR STATIC AND DYNAMIC ANALYSES OF COMPLEX SHELLS	68-0247
STATIC	USE OF PAS-1 FINITE ELEMENT COMPUTER SYSTEM FOR STATIC AND DYNAMIC ANALYSIS OF 2-DIMENSIONAL AND 3-D	75-0070
STATIC-GEOMETRI	STATIC-GEOMETRIC ANALOGIES AND FINITE ELEMENT MODELS	71-0499
STATICAL	COMPUTER-DEVELOPED STATICAL METHODS	69-0374
STATICAL	OBLICZANIE KONSTRUKCJI TOROIDALNEGO ZBIORNIKA NA WODE. (STATICAL ANALYSIS OF A TOROIDAL WATER RESERV	75-0211
STATICAL	STATICAL CALCULATIONS OF PLANE AND SPACE BAR STRUCTURES AS WELL AS STUMPY FRAMEWORKS USING A MODIFIE	73-0747
STATICAL	THREE-DIMENSIONAL ISOPARAMETRIC ELEMENTS AND THEIR USE FOR STATICAL ANALYSIS OF MASSIVE WATER-POWER	71-0364
STATICAL	ZASTOSOWANIE ITERACJI SEIDLA W METODZIE ELEMENTOW SKONCZONYCH NA PRZYKLADZIE OBLICZEN STATYCZNYCH P	74-1031
STATICALLY	A GENERAL DIGITAL COMPUTER ANALYSIS OF STATICALLY INDETERMINATE STRUCTURES	59-0007
STATICALLY	A STUDY OF INTERNAL STRESSES IN STATICALLY DEFORMED PNEUMATIC TIRES. A MATHEMATICAL MODEL FOR THE PN	70-0421
STATICALLY	DISCRETE-ELEMENT MODELLING AND SENSITIVITY STUDIES OF STATICALLY LOADED SINGLE-LAYER AND BOUNDED DOU	70-0129
STATICHESK	THE DEVELOPMENT OF APPLIED METHODS IN PROBLEMS OF STATIC CALCULATION OF THIN-WALLED THREE-DIMENSIONA	73-0934
STATICKE	STATICKE RESENI FYZIKALNE NELINEARNIHO CHOVANI ZELEZOBETO- NOVYCH DESEK (ANALYSIS OF PHYSICALLY NONL	74-0514
STATICS	ILLUSTRATIONS OF AUTOMOTIVE FINITE ELEMENT MODELS - STATICS	74-0373
STATICS	REVIEW OF CURRENT PROBLEMS FOR MULTIDIMENSIONAL REACTOR STATICS CALCULATIONS	73-0960
STATICS	STATICS AND DYNAMICS OF HYDROSTATICALLY LOADED SHELLS BY FINITE ELEMENT METHOD	71-0044
STATICS	THE MATRIX THEORY OF STATICS	57-0002
STATICS	ZUR SYSTEMATIK NUMERISCHER NAEHERUNGSVERFAHREN IN DER STATIK UND DYNAMIK DER KONSTRUKTIONEN (METHODS	72-0557
STATIK	DIE MATRIZENTHEORIE DER STATIK	57-0003
STATIK	MATRIZENTHEORIE DER STATIK	60-0009
STATIK	ZUR SYSTEMATIK NUMERISCHER NAEHERUNGSVERFAHREN IN DER STATIK UND DYNAMIK DER KONSTRUKTIONEN (METHODS	72-0557
STATION	PHOTOELASTIC DETERMINATION OF THE STRESSES IN THE FOUNDATION OF THE NUCLEAR-POWER STATION BUGEY 2	75-0834
STATION	ROCK MECHANICS STUDIES FOR A HIGH CUT AT A NUCLEAR POWER STATION IN PENNSYLVANIA	70-0122
STATIONARY	CONFORMING AND NONCONFORMING FINITE ELEMENT METHODS FOR SOLVING THE STATIONARY STOKES EQUATIONS - 1	73-0390
STATIONARY	CONVERGENCE AND ERROR ESTIMATES FOR FINITE ELEMENT APPROXIMATIONS OF STATIONARY NAVIER-STOKES EQUATI	74-0122
STATIONARY	FINITE-ELEMENT METHOD TO COMPUTE STATIONARY SUBCRITICAL FLOWS	75-0439
STATIONARY	INVESTIGATION OF THE STATIONARY FLOW OF A RIGID PLASTIC MATERIAL BY THE NUMERICAL FINITE ELEMENT MET	74-0231
STATIONARY	NUMERICAL SOLUTION OF THE STATIONARY NAVIER-STOKES EQUATIONS BY FINITE ELEMENT METHODS	74-0408
STATIONARY	OPREDELENIE STATSIONARNYKH TEMPERATURNYKH POLEI V DETALYAKH DVIGATELEI VNUTRENNEGO SGORANIYA METODOM	73-0454
STATIONARY	USE OF METHOD OF FINITE ELEMENTS FOR STUDY OF STATIONARY METAL FORMING TECHNIQUES	72-0024
STATIONS	STUDY OF HYPERVELOCITY METEOROID IMPACT ON ORBITAL SPACE STATIONS	73-0905

STATIQUE	ANALYSE STATIQUE ET DYNAMIQUE DES COQUES PAR LA METHODE DES ELEMENTS FINIS	74-0509
STATISCHEN	ASKA - EINE PROGRAMMIERSPRACHE ZUR STATISCHEN UND DYNAMISCHEN ANALYSE	69-0457
STATISCHEN	TURBAN VERSION 1.0. EIN SPEZIELLES PROGRAMM ZUR LINEAREN UND NICHTLINEAREN STATISCHEN UND DYNAMISCHE	71-0622
STATISCHER	DAS PROGRAMMSYSTEM ANTRAS EDV-SYSTEM ZUR BERECHNUNG BELIEBIGER STATISCHER SYSTEME NACH DER METHODE D	74-0918
STATISTICAL	A STATISTICAL THEORY FOR THE FRACTURE OF BRITTLE STRUCTURES SUBJECTED TO NONUNIFORM POLYAXIAL STRESS	73-0928
STATISTICAL	LOAD LIMIT ANALYSIS. COMBINATION OF STATISTICAL METHODS (MONTE CARLO METHOD) AND FINITE ELEMENT METH	71-0768
STATISTICAL	PRESTRESSED CONCRETE REACTOR PRESSURE VESSEL FAILURE ANALYSIS, THE INCORPORATION OF STATISTICAL METH	71-0120
STATISTICAL	STATISTICAL ASPECTS OF THE STABILITY OF COMPOUND SYSTEMS WITH BILINEAR ELEMENTS	73-0986
STATISTICAL	STATISTICAL IDENTIFICATION OF STRUCTURES	74-0141
STATISTICAL	STATISTICAL STATE OF STRESS STUDIED BY GRID ANALYSIS: "NUMERICAL METHODS OF ANALYSIS"	49-0001
STATISTICAL	STRAIN DISTRIBUTION AROUND UNDERGROUND OPENINGS. STATISTICAL RELATIONSHIPS FOR CERTAIN ROCK PROPERTI	71-0710
STATISTICALLY	THE BEHAVIOR OF STATISTICALLY HETEROGENEOUS EXCAVATED EARTH SLOPES	74-1330
STATOR	DESIGN AND ANALYSIS OF A CERAMIC STATOR VANE	75-0366
STATSIONARNYKH	OPREDELENIE STATSIONARNYKH TEMPERATURNYKH POLEI V DETALYAKH DVIGATELEI VNUTRENNEGO SGORANIYA METODOM	73-0454
STATYCZNYCH	ZASTOSOWANIE ITERACJI SEIDLA W METODZIE ELEMENTOW SKONCZONYCH NA PRZYKLADZIE OBLICZEN STATYCZNYCH P	74-1031
STAUDAMMEN	BERECHNUNG RAUMLICHER SPANNUNGSVERTEILUNGEN IN STAUDAMMEN NACH DER MATRIZENVERSCHIEBUNGSMETHODE	67-0003
STAVU	RESENI MAZNIHO STAVU ZELEZOBETONOVYCH DESEK (PLASTIC ANALYSIS OF REINFORCED CONCRETE PLATES)	74-0495
STAYED	FINITE ELEMENT ANALYSIS OF CABLE STAYED BRIDGES	72-0494
STEADILY	STEADILY GROWING ELASTIC-PLASTIC CRACK TIP IN A FINITE ELEMENT TREATMENT	73-0087
STEADY	A FINITE ELEMENT ANALYSIS FOR STEADY AND OSCILLATORY SUBSONIC FLOW AROUND COMPLEX CONFIGURATIONS	74-1317
STEADY	A FINITE ELEMENT MODEL FOR TWO-DIMENSIONAL STEADY FLOW THROUGH CONTRACTIONS IN NATURAL CHANNELS	71-0810
STEADY	A FINITE-ELEMENT ANALYSIS OF STEADY, TWO-DIMENSIONAL, INCOMPRESSIBLE, LAMINAR FLOW	72-0981
STEADY	A FINITE-ELEMENT METHOD FOR LIFTING SURFACES IN STEADY INCOMPRESSIBLE SUBSONIC FLOW	74-1321
STEADY	A VARIATIONAL FINITE ELEMENT METHOD FOR TWO DIMENSIONAL STEADY VISCOUS FLOWS	72-0355
STEADY	AN APPLICATION OF LEAST SQUARES FINITE ELEMENT METHODS TO TWO DIMENSIONAL STEADY INVISCID COMPRESSIB	75-0114
STEADY	ANALYSIS OF STEADY ELECTRICAL AND MAGNETIC-FIELDS BY FINITE-ELEMENT METHOD - APPLICATION TO TRANSFOR	75-0550
STEADY	ANALYSIS OF STEADY INCOMPRESSIBLE VISCOUS FLOW	74-0013
STEADY	CAVITY STABILITY: FINITE ELEMENT ANALYSIS FOR STEADY AND TRANSIENT CREEP	74-1082
STEADY	COMPARISON OF THE FINITE ELEMENT AND FINITE DIFFERENCE METHODS FOR THE ANALYSIS OF STEADY TWO DIMENS	75-0223
STEADY	DISPERSION IN STEADY UNSTEADY SEEPAGE FLOWFIELD	74-0072
STEADY	EFFECTS OF QUADRATURE ERRORS IN FINITE ELEMENT APPROXIMATION OF STEADY STATE, EIGENVALUE, AND PARABO	72-0343
STEADY	FINITE ELEMENT ANALYSIS OF STEADY FLUID AND METAL FLOW	74-0054
STEADY	FINITE ELEMENT ANALYSIS OF THE STEADY FLOW OF NON NEWTONIAN FLUIDS THROUGH PARALLEL SIDED CONDUITS	74-0056
STEADY	FINITE ELEMENT ANALYSIS OF STEADY FLOW OF NON-NEWTONIAN FLUIDS	74-0465
STEADY	FINITE ELEMENT ANALYSIS OF STEADY FLUID AND METAL FLOW	75-0178
STEADY	FINITE ELEMENT FORMULATION FOR SOLUTION OF STEADY FLOW OF INCOMPRESSIBLE VISCOUS FLUID	73-0299
STEADY	FINITE ELEMENT METHOD OF ANALYZING STEADY SEEPAGE WITH A FREE SURFACE	70-0092
STEADY	FINITE ELEMENT SOLUTION OF THE STEADY STATE COMPRESSIBLE LUBRICATION PROBLEM	70-0007
STEADY	FINITE ELEMENT SOLUTION OF STEADY STATE POTENTIAL FLOW PROBLEMS	70-0480
STEADY	FINITE ELEMENT SOLUTION FOR QUANTITY OF STEADY SEEPAGE	71-0824
STEADY	FUNDAMENTAL STUDIES ON STEADY SHIP WAVE PROBLEMS BY THE FINITE ELEMENT METHOD	74-0564
STEADY	GALERKIN METHOD FOR A CLASS OF STEADY, 2-DIMENSIONAL, INCOMPRESSIBLE, LAMINAR BOUNDARY-LAYER FLOWS	75-0437
STEADY	GENERALIZED STEADY STATE FIELD PROBLEM AND ITS SOLUTION	74-0059
STEADY	MECHANICS OF CABLE MOORING SYSTEMS. VOLUME IV. A COMPUTER PROGRAM FOR ANALYZING THE STEADY STATE RES	72-0653
STEADY	MECHANICS OF CABLE MOORING SYSTEMS. VOLUME 1. THREE DIMENSIONAL RESPONSE OF DEEP WATER MOORING LINES	72-0654
STEADY	MECHANICS OF CABLE MOORING SYSTEMS. VOLUME V1. A COMPUTER PROGRAM FOR ANALYZING THE STEADY STATE CON	73-0725
STEADY	NUMERICAL SOLUTION OF STEADY STATE DIFFUSION PROBLEMS CONTAINING SINGULARITIES	74-0036
STEADY	NUMERICAL SOLUTION OF THE STEADY INTERFACE IN A CONFINED COASTAL AQUIFER	74-1007
STEADY	SENSITIVITY ANALYSIS OF INPUT PARAMETERS IN NUMERICAL MODELING OF STEADY STATE REGIONAL GROUNDWATER	74-0926
STEADY	SPONTANEOUS IGNITION - FINITE ELEMENT SOLUTIONS FOR STEADY AND TRANSIENT CONDITIONS	74-0353
STEADY	STEADY CONDUCTION OF HEAT IN LINEAR AND NONLINEAR FULLY ANISOTROPIC MEDIA BY FINITE ELEMENTS	74-0359
STEADY	STEADY FLOW ANALYSIS OF INCOMPRESSIBLE VISCOUS FLOW BY THE FINITE ELEMENT METHOD	73-0332
STEADY	STEADY STATE AND TRANSIENT GROUNDWATER FLOW IN SLOPES	70-0273
STEADY	STEADY STATE AND TRANSIENT HEAT CONDUCTION AND STRESS ANALYSES OF TWO AND THREE DIMENSIONAL STRUCTUR	71-0675
STEADY	STEADY STATE EQUATIONS OF MOTION. EQUILIBRIUM SHAPE AND STABILITY DERIVATIVES OF ELASTIC AIRPLANES E	72-0541
STEADY	STEADY STATE FLOW IN RIGID NETWORKS OF FRACTURES	74-0278
STEADY	STEADY STATE IN SOIL SUBJECTED TO SEVERAL COUPLED FORCES: A FINITE ELEMENT FORMULATION	70-0207
STEADY	STUDIO DI CAMPI ELETTRICI E MAGNETICI STANZIONARI CON IL METODO DEGLI ELEMENTI FINITI. APPLICAZIONE	75-0207
STEADY	THE NUMERICAL SOLUTION OF THE STEADY INTERFACE IN A COASTAL CONFINED AQUIFER	73-0711
STEADY	THE PROGRAM " ASSEN " FOR STEADY FREE SURFACE FLOWS	74-0075
STEADY	TWO-DIMENSIONAL FINITE ELEMENT ANALYSIS OF SEMI-CONDUCTOR STEADY STATE TRANSPORT EQUATIONS	74-1224
STEADY	VARIATIONAL PROPERTIES OF STEADY FALL IN STOKES FLOW	72-0285
STEADY-AND	A FINITE-ELEMENT ANALYSIS FOR STEADY-AND OSCILLATORY SUPERSONIC FLOWS AROUND COMPLEX CONFIGURATIONS	74-1316
STEADY-STATE	APACHE: A THREE-DIMENSIONAL FINITE ELEMENT PROGRAM FOR STEADY-STATE OR TRANSIENT HEAT CONDUCTION ANA	73-0971
STEADY-STATE	CALCULATION BY FINITE ELEMENT TECHNIQUE OF STEADY-STATE OR TRANSIENT HEAT-FLOW CONDUCTION	74-0637
STEADY-STATE	CALCULATION OF STEADY-STATE WATER-TABLE HEIGHTS IN DRAINED SOILS BY MEANS OF FINITE-ELEMENT METHOD	75-0457
STEADY-STATE	DIGITAL COMPUTER PROGRAMS FOR STEADY-STATE OR TRANSIENT TEMPERATURE ANALYSIS OF PLANE AND AXISYMMETR	72-0915
STEADY-STATE	EVALUATION OF A CLASS OF METHODS FOR BOUNDING STEADY-STATE CREEP DEFORMATION	74-0892
STEADY-STATE	FINITE ELEMENT ANALYSIS OF STEADY-STATE NONLINEAR HEAT TRANSFER PROBLEMS	74-0572
STEADY-STATE	MECHANICS OF CABLE MOORING SYSTEMS. VOLUME V11. THE STEADY-STATE BEHAVIOR OF A PYRAMID ARRAY SYSTEM	73-0724
STEADY-STATE	VARIATIONAL PRINCIPLE OF STEADY-STATE TRANSPORT PROCESSES	70-0190
STEAM	NON-LINEAR ANALYSIS OF THE ROTATIONAL STIFFNESS OF A NUCLEAR STEAM GENERATOR FOUNDATION	69-0286
STEAM	NONSTEADY PYROMETRIC CALCULATIONS AS THE BASIS FOR STRESS ANALYSIS OF STEAM GENERATORS AND HEAT EXCH	72-0451
STEAM	THEORETICAL AND EXPERIMENTAL INVESTIGATION ON A STEAM GENERATOR TUBE PLATE	69-0424
STEAM	THREE-DIMENSIONAL FINITE ELEMENT ANALYSIS OF A STEAM GENERATOR CHANNEL-HEAD-COMPLEX - COMPARISON OF	75-0653
STEAM-TURBINE	TWO-DIMENSIONAL PHOTOELASTIC STUDY OF STEAM-TURBINE CASING FLANGES WITH INTERFACE RELIEF	74-0603
STEEL	ANALYSIS OF LIGHT GATE STEEL SHEAR DIAPHRAGMS	73-0733
STEEL	ANALYTICAL PERFORMANCE OF STEEL, ALUMINUM AND PLASTIC FOR A NEW DESIGN HIGH PRESSURE THIN-WALLED CAR	74-0756
STEEL	COMBINED MATERIAL AND GEOMETRIC NONLINEARITY FOR THIN STEEL PLATES	75-0397
STEEL	COMPUTER ANALYSIS OF STEEL FRAME IN FIRE	75-0173
STEEL	COMPUTERIZED STRUCTURAL ANALYSIS OF THE WORLD'S LARGEST LIGHT-GAGE STEEL PRIMARY STRUCTURAL SYSTEM	73-0503
STEEL	CYCLIC MECHANICAL TESTS AND AN APPROPRIATE ANALYTICAL STRESS-STRAIN MODEL FOR A36 STEEL	74-0815
STEEL	CYCLIC PLASTIC ANALYSIS OF STRUCTURAL STEEL JOINTS	73-0799
STEEL	DEFLECTION ANALYSIS OF EXPANDED OPEN-WEB STEEL BEAMS	74-0224
STEEL	DESIGN OF A COMPOSITE STEEL AND CONCRETE CLOSURE FOR A PRESTRESSED CONCRETE REACTOR VESSEL.	72-0187
STEEL	DYNAMIC BEHAVIOR OF STEEL FOUNDATIONS FOR TURBO-ALTERNATORS	71-0837
STEEL	ELASTIC LATERAL BUCKLING OF STEEL BEAMS	70-0364
STEEL	FINITE ELEMENT SOLUTION OF HEAT TRANSMISSION IN STEEL INGOTS	72-0311
STEEL	FULL-RANGE ANALYSIS OF STEEL PLATES AND STIFFENED PLATING UNDER UNIAXIAL COMPRESSION	75-0803
STEEL	HEAVY SECTION STEEL TECHNOLOGY PROGRAM. TECHNICAL REPORT NO. 4, OCTOBER 1969. A TWO-DIMENSIONAL ELAS	69-0474
STEEL	HEAVY SECTION STEEL TECHNOLOGY PROGRAM TECHNICAL OR PROGRAMMATIC MANUSCRIPT NO. 25. VARIABLE THICKNE	73-0948
STEEL	INELASTIC BUCKLING OF STEEL BEAMS UNDER NON UNIFORM MOMENT	75-0149
STEEL	INFLATION FORMING OF STEEL FIBER-REINFORCED CONCRETE DOMES	74-0705
STEEL	LIGHT GAGE STEEL INFILL PANELS IN MULTISTORY STEEL FRAMES	74-0229
STEEL	LIGHT GAGE STEEL INFILL PANELS IN MULTISTORY STEEL FRAMES	74-0229
STEEL	LOAD DISTRIBUTION IN A COMPOSITE STEEL BOX GIRDER BRIDGE	73-0765
STEEL	NOTCH EFFECT IN LOW CYCLE FATIGUE - REPORT 1: LOW-CYCLE FATIGUE STRENGTH OF 0.48 PERCENT CARBON STEE	73-1090
STEEL	PRACTICAL ANALYSIS OF STEEL STRUCTURES USING HIGHER-ORDER ELEMENTS	72-0326
STEEL	SHEAR LAG IN STEEL BOX GIRDER BRIDGES	75-0310
STEEL	SIMPLE EQUATIONS FOR THE MAGNETIZATION AND RELUCTIVITY CURVES OF STEEL	75-0156
STEEL	STATIC THREE DIMENSIONAL ANALYSIS OF ELEVATED STEEL TRANSPORTATION STRUCTURES	74-1187
STEEL	STEEL PLATE ANALYSIS BY FINITE ELEMENTS	69-0100

STEEL	THEORETICAL STUDY OF THE INFLUENCE OF INCLUSIONS UPON THE INITIATION AND GROWTH OF FATIGUE CRACKS IN	71-0394
STEEL-COMPOSITE	FIELD TESTS OF A STEEL-COMPOSITE BOX-GIRDER BRIDGE	75-0633
STEEL-CONCRETE	THERMAL STRESSES IN STEEL-CONCRETE COMPOSITE BRIDGES	75-0154
STEEL-WORK	DEFORMABILITY OF THE HULL STEEL-WORK AND DEFORMATIONS OF THE ENGINE-ROOM OF LARGE TANKERS	74-0532
STEELS	FAST FRACTURE RESISTANCE AND CRACK ARREST IN STRUCTURAL STEELS	73-0766
STEELS	PRELIMINARY INVESTIGATION OF THE DIE-LESS DRAWING OF TITANIUM AND SOME STEELS	75-0770
STEELS	ZUSAMMENHANG ZWISCHEN OBERFLAECHENBESCHAFFENHEIT, VERBUND UND SPRENGWIRKUNG VON BEWEHRUNGSSTAEHLEN U	73-0450
STEEP-PITCH	BERECHNUNG UND BEURTEILUNG DER STANDSICHERHEIT VON ABBAUHOHLRAEUMEN IN STEILEN GANGERZLAGERSTAETTEN	74-0597
STEEPLY	UNDERGROUND MEASUREMENTS IN A STEEPLY DIPPING OREBODY	68-0085
STEFAN	A SECOND ORDER FINITE ELEMENT METHOD FOR THE ONE-DIMENSIONAL STEFAN PROBLEM	74-1222
STEFAN	ANALYSE NUMERIQUE D'UN PROBLEME DE STEFAN A DEUX PHASES PAR UNE METHODE D'ELEMENTS FINIS. (NUMERICAL	75-0387
STEFAN	ANALYSE NUMERIQUE D'UN PROBLEME DE STEFAN A DEUX PHASES PAR UNE METHODE D'ELEMENTS FINIS. (NUMERICAL	75-0387
STEFAN	NUMERICAL ANALYSIS OF 2-PHASE STEFAN PROBLEM BY METHOD OF FINITE ELEMENTS	75-0616
STEFAN	THE MULTIDIMENSIONAL STEFAN PROBLEM: A FINITE ELEMENT APPROACH	74-0464
STEIFIGKEITS-UN	MODERNE VERFAHREN ZUR UNTERSUCHUNG DER STEIFIGKEITS-UND FESTIGKEITSVERHALTNISSE VON RAHMENKORPERN, I	69-0401
STEILEN	BERECHNUNG UND BEURTEILUNG DER STANDSICHERHEIT VON ABBAUHOHLRAEUMEN IN STEILEN GANGERZLAGERSTAETTEN	74-0597
STEINKOHLENBERG	VORAUSSAGE DES GEBIRGSDRUCKES UND SEINER AUSQUIRKUNG AUF STREBEN UND STRECKEN IM STEINKOHLENBERGBAU	73-0679
STEKLOV	APPROXIMATION OF STEKLOV EIGENVALUES OF NON-SELFADJOINT SECOND ORDER ELLIPTIC OPERATORS	72-0509
STEN	STABILITA STEN V PLASTICKEM OBORU (STABILITY IN WEBPLATES IN PLASTIC RANGE)	73-0387
STEP	DYNAMIC RESPONSE BY A STEP BY STEP ANALYSIS	62-0014
STEP	DYNAMIC RESPONSE BY A STEP BY STEP ANALYSIS	62-0014
STEP	DYNAMIC RESPONSE BY LARGE STEP INTEGRATION	73-0567
STEP	EFFECT OF FILLETS ON STRESS CONCENTRATIONS IN CYLINDRICAL SHELLS WITH STEP CHANGES IN OUTSIDE DIAMET	71-0108
STEP	NUMERICAL SOLUTION TO THE DYNAMICS OF VISCOUS FLUID FLOW BY A FINITE ELEMENT ALGORITHM: A FIRST STEP	71-0004
STEP	NUMERICAL SOLUTION TO THE DYNAMICS OF VISCOUS FLUID FLOW BY A FINITE ELEMENT ALGORITHM: A FIRST STEP	71-0791
STEP	RASCHET NAPRYAZHENIL L PEREMESHCHENII V SVODE PODZEMNOGO MASHINNOGO ZALA GES I OKRUZHAYUSHCHEM SKA	74-0542
STEP	STEP AND LOADS EFFECT ON STRESSES IN PRESTRESSED CONCRETE SHORT BRACKETS	71-0334
STEP	VARIABLE TIME STEP ANALYSIS OF UNCONFINED SEEPAGE	74-0064
STEP	VARIABLE TIME STEP METHOD FOR DETERMINING PLASTIC STRESS REFLECTIONS FROM BOUNDARIES	75-0641
STEP-BY	RASCHET NAPRYAZHENIL L PEREMESHCHENII V SVODE PODZEMNOGO MASHINNOGO ZALA GES I OKRUZHAYUSHCHEM SKA	74-0542
STEPPED	FINITE ELEMENT OPTIMIZATION OF FINITE ELEMENT SLIDER BEARING PROFILES	74-0131
STEPPED	ON VIBRATION ANALYSIS OF RECTANGULAR CANTILEVER PLATES WITH A STEPPED CROSS-SECTION BY THE FINITE EL	71-0608
STEPPED	STIFFNESS OF STEPPED BARS	74-0711
STEPPED	THE VIBRATION OF STEPPED CANTILEVERED PLATES	67-0129
STEPPING	AN ANALYSIS OF VARIOUS TIME STEPPING SCHEMES FOR INITIAL VALUE PROBLEMS	73-0607
STEPWISE	VARIATIONAL SOLUTIONS BY USE OF STEPWISE CONSTANT FUNCTIONS: ONE LINEAR CASE	72-0286
STETIGKEITSFORD	VARIATIONAL PRINCIPLES OF ELASTOSTATICS WITH RELAXED CONTINUITY REQUIREMENTS AND THEIR APPLICABILITY	74-0758
STIFF	ANALIZA ORGAN WYMUSZONYCH ZLOZONYCH UKLADOW LINIOWYCH METODA SZTYWNYCH ELEMENTOW SKONCZONYCH (ANALYS	72-0246
STIFF	ELASTIC-PLASTIC STIFF FRAMEWORKS FROM SECOND ORDER STRESS THEORY	73-1061
STIFF	METHOD SZTYWNYCH ELEMENTOW SKONCZONYCH W OBLICZENIACH KONSTRUCJI OKRETOWYCH. (STIFF FINITE ELEMENT M	74-0923
STIFF	SLOPES IN STIFF FISSURED CLAYS AND SHALES	69-0094
STIFF	ZASTOSOWANIE METODY SZTYWNYCH ELEMENTOW SKONCZONYCH DO OBLICZEN ORGAN WALOW OKRETOYCH Z UWZGLEDNIENI	74-0962
STIFFENED	A COMPUTER PROGRAM FOR ANALYSIS OF STIFFENED PLATES UNDER COMBINED INPLANE AND LATERAL LOADS	71-0703
STIFFENED	A CONFORMING QUADRILATERAL ELEMENT FOR BUCKLING ANALYSIS OF STIFFENED PLATES	73-0613
STIFFENED	A FINITE ELEMENT ANALYSIS FOR DYNAMIC BEHAVIOR OF STIFFENED PLATES	69-0392
STIFFENED	A FINITE ELEMENT PROGRAM FOR THE DYNAMIC LARGE DEFLECTION, ELASTIC-PLASTIC ANALYSIS OF STIFFENED SHE	72-0908
STIFFENED	ACCURATE FINITE ELEMENT MODELLING OF FLAT AND CURVED STIFFENED PANELS	72-0538
STIFFENED	AN ELASTIC FINITE-ELEMENT ANALYSIS OF THE BUCKLING OF ECCENTRICALLY STIFFENED PLATES AND SHELLS	73-1136
STIFFENED	ANALYSIS OF ASYMMETRIC STIFFENED SHELL TYPE STRUCTURES BY THE FINITE ELEMENT METHOD: I FLAT RECTANGU	68-0283
STIFFENED	ANALYSIS OF ASYMMETRIC STIFFENED SHELL TYPE STRUCTURES BY FINITE ELEMENT METHOD II. SIX DEGREE-OF-FR	69-0373
STIFFENED	ANALYSIS OF ECCENTRICALLY STIFFENED CYLINDRICAL SHELLS	72-0217
STIFFENED	ANALYSIS OF STIFFENED PLATES BY THE FINITE ELEMENT METHOD	68-0299
STIFFENED	ANALYSIS OF STIFFENED PLATES USING THE FINITE ELEMENT METHOD	70-0030
STIFFENED	ANALYSIS OF STIFFENED PLATES CONSIDERING NONLINEAR MATERIAL AND GEOMETRIC BEHAVIOUR	75-0327
STIFFENED	BIFURCATION OF ECCENTRICALLY LOADED MEMBERS AND OF STIFFENED PLATES	70-0613
STIFFENED	DIE BEULLASTEN DIAGONAL VERSTEIFTER RECHTECKPLATTEN (BUCKLING LOADS IN DIAGONALLY STIFFENED RECTANGU	73-0405
STIFFENED	DYNAMIC ANALYSIS OF STIFFENED PANEL STRUCTURES	71-0522
STIFFENED	DYNAMIC BEHAVIOR OF ECCENTRICALLY STIFFENED PLATES	71-0796
STIFFENED	DYNAMIC BEHAVIOUR OF ECCENTRICALLY STIFFENED PLATES	71-0605
STIFFENED	DYNAPLAS: A FINITE ELEMENT PROGRAM FOR THE DYNAMIC, LARGE DEFLECTION, ELASTIC-PLASTIC ANALYSIS OF ST	72-0773
STIFFENED	ELASTIC PLASTIC BUCKLING OF STIFFENED RECTANGULAR PLATES	69-0122
STIFFENED	FINITE ELEMENT ANALYSIS OF STIFFENED PLATES	68-0240
STIFFENED	FINITE ELEMENT ANALYSIS OF ELASTIC-PLASTIC PLATES AND ECCENTRICALLY STIFFENED PLATES	71-0585
STIFFENED	FINITE ELEMENT ANALYSIS OF PLATES AND ECCENTRICALLY STIFFENED PLATES	73-0691
STIFFENED	FINITE-ELEMENT PREDICTIONS OF TRANSIENT ELASTIC-PLASTIC LARGE DEFLECTIONS OF STIFFENED AND/OR UNSTIF	74-0825
STIFFENED	FREE VIBRATION ANALYSIS OF STIFFENED PLATES	72-0303
STIFFENED	FREE VIBRATION OF RING STIFFENED CYLINDRICAL SHELLS	70-0172
STIFFENED	FULL RANGE ANALYSIS OF ECCENTRICALLY STIFFENED PLATES	74-0196
STIFFENED	FULL-RANGE ANALYSIS OF STEEL PLATES AND STIFFENED PLATING UNDER UNIAXIAL COMPRESSION	75-0803
STIFFENED	INELASTIC RESPONSE OF PLATES AND ECCENTRICALLY STIFFENED PLATES	74-0458
STIFFENED	INTERFACE CORE DISCRETE-ELEMENT FOR STIFFENED SHELLS OF REVOLUTION	67-0092
STIFFENED	JET NOISE EXCITATION OF AN INTEGRALLY STIFFENED PANEL	71-0546
STIFFENED	LARGE DEFLECTION ELASTIC-PLASTIC DYNAMIC RESPONSE OF STIFFENED SHELLS OF REVOLUTION	72-0608
STIFFENED	LARGE DEFLECTION ELASTIC-PLASTIC DYNAMIC RESPONSE OF STIFFENED SHELLS OF REVOLUTION	74-0496
STIFFENED	LIMIT ANALYSIS OF FRAME SYSTEMS STIFFENED BY PANELS	74-0584
STIFFENED	MATRIX ANALYSIS OF LOCAL INSTABILITY IN PLATES, STIFFENED PANELS AND COLUMNS	72-0542
STIFFENED	NASTRAN BUCKLING ANALYSIS OF A LARGE STIFFENED CYLINDRICAL SHELL WITH A CUTOUT	71-0045
STIFFENED	PHOTOELASTIC INVESTIGATION ON THE CRACK-ARREST CAPABILITY OF A PRETENSIONED STIFFENED PLATE	75-0088
STIFFENED	SHEAR LAG IN STIFFENED BOX GIRDERS	70-0243
STIFFENED	STACUSS 1: A DISCRETE-ELEMENT PROGRAM FOR THE STATIC ANALYSIS OF SINGLE-LAYER CURVED STIFFENED SHELL	69-0160
STIFFENED	STATIC ANALYSIS OF STIFFENED SHELLS BY THE FINITE ELEMENT METHOD	71-0803
STIFFENED	STRESS DIFFUSION IN THICK STIFFENED PANELS	74-0620
STIFFENED	STRESSES AROUND HOLES IN STIFFENED COMPOSITE PANELS USING LAURENT-SERIES AND FINITE ELEMENT METHODS	74-1252
STIFFENED	STRESSES AROUND HOLES IN STIFFENED COMPOSITE PANELS USING LAURENT-SERIES AND FINITE ELEMENT METHODS	75-0733
STIFFENED	THEORETICAL AND EXPERIMENTAL ANALYSIS OF STIFFENED PANEL UNDER DYNAMIC CONDITIONS	70-0130
STIFFENED	THEORETICAL AND EXPERIMENTAL STUDIES OF TRANSIENT ELASTIC-PLASTIC LARGE DEFLECTIONS OF GEOMETRICALLY	75-0830
STIFFENED	THEORETICAL AND EXPERIMENTAL STUDIES OF TRANSIENT ELASTIC-PLASTIC LARGE DEFLECTIONS OF GEOMETRICALLY	75-0831
STIFFENED	VIBRATION OF STIFFENED PLATES USING FINITE-ELEMENTS	72-0951
STIFFENED-CIRCU	A FINITE ELEMENT ANALYSIS OF ECCENTRICALLY STIFFENED-CIRCULAR CYLINDRICAL SHELLS	69-0159
STIFFENER	FORCES AND DEFORMATIONS AT THE LATERAL STIFFENER OF THIN WALLED U-BARS	73-0236
STIFFENERS	ANALYSIS OF ORTHOTROPIC FOLDED PLATES WITH ECCENTRIC STIFFENERS	70-0437
STIFFENERS	FINITE-ELEMENT ANALYSIS OF ORTHOTROPIC PLATES WITH ECCENTRIC STIFFENERS	72-0958
STIFFENERS	ON BUCKLING OF UNIDIRECTIONAL BORON-ALUMINUM STIFFENERS - A CAUTION TO DESIGNERS	75-0634
STIFFENERS	STIFFNESS MATRIX FOR CURVED TAPERED AND STRAIGHT TAPERED SHELL STIFFENERS	69-0074
STIFFENING	STRESS ANALYSIS OF A TUBULAR CROSS-JOINT WITHOUT ANY INTERNAL STIFFENING FOR OFFSHORE STRUCTURES	74-0549
STIFFNESS	A CONVERGENCE INVESTIGATION OF THE DIRECT STIFFNESS METHOD	66-0102
STIFFNESS	A FLAT TRIANGULAR SHELL ELEMENT STIFFNESS MATRIX	66-0102
STIFFNESS	A FORMULA FOR CERTAIN TYPES OF STIFFNESS MATRICES OF STRUCTURAL ELEMENTS	63-0003
STIFFNESS	A GENERAL FORMULA FOR STIFFNESS MATRICES OF STRUCTURAL ELEMENTS	63-0002
STIFFNESS	A METHOD OF COMPUTING NUMERICALLY INTEGRATED STIFFNESS MATRICES	72-0546
STIFFNESS	A MODAL SYNTHESIS TECHNIQUE FOR DETERMINING DYNAMIC PROPERTIES FOR A STRUCTURE FOR MASS AND STIFFNES	74-0319
STIFFNESS	A NEW APPROACH FOR DERIVING GOOD ELEMENT STIFFNESS MATRICES	75-0128

384

STIFFNESS	A SEISMIC DESIGN OF NUCLEAR REACTOR BUILDING, STRESS ANALYSIS AND STIFFNESS EVALUATION OF THE ENTIRE	71-0251
STIFFNESS	A SIMPLE ALGORITHM FOR THE STIFFNESS MATRIX OF TRIANGULAR PLATE BENDING ELEMENTS	71-0152
STIFFNESS	A STIFFNESS MATRIX FOR THE ANALYSIS OF THIN PLATES IN BENDING	61-0001
STIFFNESS	A STIFFNESS MATRIX FOR A CURVED MEMBRANE SHELL	69-0055
STIFFNESS	A STIFFNESS MATRIX FOR SHALLOW SHELL FINITE ELEMENTS	69-0229
STIFFNESS	A STIFFNESS MATRIX METHOD OF NATURAL MODE ANALYSIS	58-0002
STIFFNESS	A STRAIN ENERGY BASIS FOR STUDIES OF ELEMENT STIFFNESS MATRICES	72-0621
STIFFNESS	A STUDY OF STIFFNESS MATRICES FOR THE ANALYSIS OF FLAT PLATES	68-0105
STIFFNESS	AN AUTOMATIC NODE RELABELLING SCHEME FOR BANDWIDTH MINIMIZATION OF STIFFNESS MATRICES	68-0256
STIFFNESS	AN EXTENSION OF THE SHALLOW TO THE NON-SHALLOW STIFFNESS MATRIX FOR A CYLINDRICAL SHELL FINITE ELEME	70-0104
STIFFNESS	ANALYSIS OF AXISYMMETRIC SHELLS BY THE DIRECT STIFFNESS METHOD	63-0006
STIFFNESS	ANALYSIS OF AXISYMMETRIC SOLIDS BY THE DIRECT STIFFNESS METHOD	66-0049
STIFFNESS	ANALYSIS OF ELEMENT STIFFNESS	67-0117
STIFFNESS	ANALYSIS OF THE CORNEO-SCLERAL SHELL BY THE METHOD OF DIRECT STIFFNESS	71-0151
STIFFNESS	APPLICATION OF GENERALIZED CONSTRAINTS IN THE STIFFNESS METHOD OF STRUCTURAL ANALYSIS	65-0030
STIFFNESS	APPLICATION OF HERMITE POLYNOMIALS TO THE DETERMINATION OF THE STIFFNESS MATRIX OF PLATE BY FINITE E	71-0365
STIFFNESS	APPLICATION OF THE DIRECT STIFFNESS METHOD TO PLANE PROBLEMS INVOLVING LARGE, TIME-DEPENDENT DEFORMA	66-0108
STIFFNESS	APPLICATION OF THE FINITE ELEMENT STIFFNESS METHOD TO NONCONSERVATIVE STABILITY PROBLEMS	67-0158
STIFFNESS	APPLICATION OF THE STIFFNESS METHOD TO A SNAP-THROUGH PROBLEM	60-0006
STIFFNESS	APPLICATION OF THE STIFFNESS METHOD TO THE ANALYSIS OF SHELL STRUCTURES	61-0004
STIFFNESS	AUTOMATIC GENERATION AND CONDENSATION OF LARGE MASS AND STIFFNESS MATRICES	72-0466
STIFFNESS	AUTOMOBILE PANEL SWEEP STIFFNESS ANALYSIS	74-0175
STIFFNESS	BANDWIDTH MINIMIZATION OF STIFFNESS MATRICES	70-0045
STIFFNESS	BASIS OF DERIVATION OF MATRICES FOR THE DIRECT STIFFNESS METHOD	63-0008
STIFFNESS	BEHAVIOR OF TRIANGULAR SHELL ELEMENT STIFFNESS MATRICES ASSOCIATED WITH POLYHEDRAL DEFLECTION DISTRI	68-0159
STIFFNESS	BOUNDS ON EXTERNAL EIGENVALUES OF FINITE ELEMENT STIFFNESS AND MASS MATRICES AND THEIR SPECTRAL COND	72-0256
STIFFNESS	BOUNDS ON THE SPECTRAL AND MAXIMUM NORMS OF THE FINITE ELEMENT STIFFNESS, FLEXIBILITY AND MASS MATRI	73-0409
STIFFNESS	CALCULATION OF GEOMETRIC STIFFNESS MATRICES FOR COMPLEX STRUCTURES	66-0058
STIFFNESS	CALCULATION OF STIFFNESS MATRICES FOR FINITE ELEMENTS OF THIN SHELLS OF ARBITRARY SHAPE	68-0127
STIFFNESS	COMMENT ON MATRICES FOR THE DIRECT STIFFNESS METHOD BY R. J. MELOSH	64-0022
STIFFNESS	COMMENT ON STIFFNESS MATRICES FOR SECTOR ELEMENTS	70-0233
STIFFNESS	COMMENTS ON DERIVATION OF ELEMENT STIFFNESS MATRICES BY T. H. H. PIAN	65-0060
STIFFNESS	COMPARISON AND EVALUATION OF STIFFNESS MATRICES	66-0045
STIFFNESS	COMPARISON BETWEEN SPARSE STIFFNESS MATRIX AND SUB-STRUCTURE METHODS	73-1068
STIFFNESS	COMPUTATION OF STRESS RESULTANTS FROM ELEMENT STIFFNESS MATRICES	66-0125
STIFFNESS	COMPUTATIONAL METHOD FOR DYNAMIC ANALYSIS OF STRUCTURES WITH LARGE SIZE STIFFNESS AND MASS MATRICES.	73-0830
STIFFNESS	CONSISTENT STIFFNESS MATRICES IN THE ANALYSIS OF SHELLS	67-0078
STIFFNESS	CONSISTENT STRESSES FOR THE FINITE ELEMENT STIFFNESS METHOD	72-0698
STIFFNESS	CRITERION FOR SELECTING STIFFNESS MATRICES	65-0061
STIFFNESS	DEPLACEMENTS DANS LES STRUCTURES MINCES A FAIBLE RIGIDITE GEOMETRIQUE-DETERMINATION NUMERIQUE (NUMER	72-0213
STIFFNESS	DERIVATION OF A STIFFNESS MATRIX FOR A RIGHT TRIANGULAR PLATE IN BENDING AND SUBJECTED TO INITIAL ST	62-0007
STIFFNESS	DERIVATION OF COMPATIBLE STIFFNESS MATRICES FOR LARGE DISCRETE ELEMENTS	67-0080
STIFFNESS	DERIVATION OF ELEMENT STIFFNESS MATRICES BY ASSUMED STRESS DISTRIBUTIONS	64-0007
STIFFNESS	DERIVATION OF ELEMENT STIFFNESS MATRICES	64-0008
STIFFNESS	DERIVATION OF GEOMETRIC STIFFNESS AND MASS MATRICES FOR FINITE ELEMENT HYBRID MODELS	74-0844
STIFFNESS	DERIVATION OF STIFFNESS MATRICES FOR THE ANALYSIS OF LARGE DEFLECTION AND STABILITY PROBLEMS	65-0066
STIFFNESS	DERIVATION OF STIFFNESS MATRICES FOR PROBLEMS IN PLANE ELASTICITY BY GALERKIN'S METHOD	69-0250
STIFFNESS	DERIVATION OF THE STIFFNESS MATRIX FOR PLATES BY GALERKIN METHOD	69-0402
STIFFNESS	DETERMINATION OF THE APPLICABILITY OF SOME THEORIES ABOUT STRENGTH AND STIFFNESS OF THIN-WALLED BEAM	69-0318
STIFFNESS	DEVELOPMENT AND APPLICATION OF THE DIRECT STIFFNESS METHOD FOR OUT-OF-PLANE BENDING USING A TRIANGUL	62-0009
STIFFNESS	DEVELOPMENT OF EXPLICIT STIFFNESS AND MASS MATRICES FOR A TRIANGULAR PLATE ELEMENT	69-0242
STIFFNESS	DEVELOPMENT OF THE STIFFNESS METHOD TO DEFINE BOUNDS ON ELASTIC BEHAVIOUR OF STRUCTURES	62-0011
STIFFNESS	DIRECT BEAM STIFFNESS MATRIX CALCULATIONS INCLUDING SHEAR EFFECTS	62-0003
STIFFNESS	DIRECT STIFFNESS METHOD OF ANALYSIS OF SHELLS OF REVOLUTION UTILIZING CURVED ELEMENTS	66-0042
STIFFNESS	DYNAMIC STIFFNESS MATRIX FORMULATION BY MEANS OF HERMITIAN POLYNOMIALS	66-0059
STIFFNESS	EFFECT OF OPENINGS ON THE LATERAL STIFFNESS OF INFILLED FRAMES	71-0300
STIFFNESS	ELEMENT STATIC FLEXIBILITY AND STIFFNESS MATRICES FOR THE FINITE ELEMENT ANALYSIS OF BEAM, PLATE, AN	67-0070
STIFFNESS	ELEMENT STIFFNESS MATRICES FOR BOUNDARY COMPATIBILITY AND FOR PRESCRIBED BOUNDARY STRESSES	66-0061
STIFFNESS	ELEMENT STIFFNESS MATRICES FOR CONSTANT AND LINEAR STRESS - TRIANGULAR, CONSTANT AND LINEAR STRESS T	67-0006
STIFFNESS	ELEMENT STIFFNESS MATRICES OF THICK-WALLED ORTHOTROPIC SHELLS WITH APPLICATIONS	69-0261
STIFFNESS	ELEMENT MATRIX GENERATOR	71-0106
STIFFNESS	ENGINEERING APPLICATIONS OF NUMERICAL INTEGRATION IN STIFFNESS METHOD	66-0040
STIFFNESS	EXPLICIT EXPRESSIONS FOR TRIANGULAR TOWA ELEMENT STIFFNESS MATRIX	68-0160
STIFFNESS	FINITE ELEMENT STIFFNESS MATRICES FOR ANALYSIS OF PLATE BENDING	66-0021
STIFFNESS	FINITE ELEMENT STIFFNESS METHODS BY DIFFERENT VARIATIONAL PRINCIPLES IN ELASTICITY	70-0455
STIFFNESS	FLEXIBILITY AND STIFFNESS MATRICES FOR AN OPEN-TUBE WARPING CONSTRAINT FINITE ELEMENT	72-0076
STIFFNESS	FURTHER DEVELOPMENT AND APPLICATIONS OF THE STIFFNESS METHOD	64-0020
STIFFNESS	GENERAL SINUSOIDAL STIFFNESS MATRICES FOR BUCKLING AND VIBRATION ANALYSES OF THIN FLAT-WALLED STRUCT	68-0278
STIFFNESS	GENERALIZED STIFFNESS MATRIX OF A CURVED-BEAM ELEMENT	69-0165
STIFFNESS	GEOMETRIC STIFFNESS CHARACTERISTICS OF A ROTATING ELASTIC APPENDAGE	74-1296
STIFFNESS	GEOMETRICALLY NONLINEAR ANALYSIS BY THE DIRECT STIFFNESS METHOD	70-0331
STIFFNESS	GEOMETRICALLY NONLINEAR STRUCTURAL ANALYSIS BY THE DIRECT STIFFNESS METHOD	71-0560
STIFFNESS	INADEQUACY OF NODAL CONNECTIONS IN A STIFFNESS SOLUTION FOR PLATE BENDING	65-0021
STIFFNESS	INCLUSION OF SHEAR DEFLECTION IN THE STIFFNESS MATRIX OF A BEAM ELEMENT	70-0307
STIFFNESS	INCREMENTAL STIFFNESS METHOD FOR FINITE ELEMENT ANALYSIS OF THE NONLINEAR DYNAMIC PROBLEM	71-0525
STIFFNESS	INFLUENCE OF END-CLOSURE STIFFNESS ON BEHAVIOR OF CONCRETE CYLINDRICAL HULLS SUBJECTED TO HYDROSTATI	71-0725
STIFFNESS	INFLUENCE OF POISSON'S RATIO ON THE CONDITION OF THE FINITE ELEMENT STIFFNESS MATRIX	71-0519
STIFFNESS	INVESTIGATION OF FLEXURAL STIFFNESS OF RECTANGULAR PLATES BY THE FINITE ELEMENT METHOD	73-0991
STIFFNESS	LINEAR AND NON LINEAR ANALYSIS OF SHELLS OF REVOLUTION WITH ASYMMETRICAL STIFFNESS PROPERTIES	68-0151
STIFFNESS	MATRICE DE RIGIDITE D'UN ELEMENT DE POUTRE SOLLICITE A LA TORSION-FLEXION (STIFFNESS MATRIX OF A BEA	73-0442
STIFFNESS	NASTRAN DIFFERENTIAL STIFFNESS ANALYSIS OF AN AIRCRAFT CANOPY	71-0037
STIFFNESS	NATURAL FACTOR FORMULATION OF THE STIFFNESS FOR THE MATRIX DISPLACEMENT METHOD	75-0057
STIFFNESS	NON-LINEAR ANALYSIS OF THE ROTATIONAL STIFFNESS OF A NUCLEAR STEAM GENERATOR FOUNDATION	69-0286
STIFFNESS	NONLINEAR CONSIDERATION OF GRAVITY IN A STIFFNESS TEST OF A WEAK STRUCTURE AT SMALL STRAINS	71-0035
STIFFNESS	NUMERICAL CONDITIONING OF STIFFNESS MATRIX FORMULATIONS FOR FRAME STRUCTURES	68-0022
STIFFNESS	ON COMPUTATIONS INVOLVING STIFFNESS MATRICES STORED IN RECTANGULAR FORM	75-0576
STIFFNESS	ON DERIVATION OF STIFFNESS MATRICES WITH CO ROTATION FIELDS FOR PLATES AND SHELLS	71-0254
STIFFNESS	ON SOME GENERALIZATIONS OF THE INCREMENTAL STIFFNESS RELATIONS FOR FINITE DEFORMATIONS OF COMPRESSIB	71-0189
STIFFNESS	ON THE BENDING STIFFNESS OF SLABS IN CROSS-WALLED STRUCTURES	74-0398
STIFFNESS	ON THE DERIVATION OF STIFFNESS MATRICES FOR THE ANALYSIS OF LARGE DEFLECTION AND STABILITY PROBLEMS	66-0050
STIFFNESS	ON THE SOLUTION OF TWO-DIMENSIONAL PROBLEMS OF ELASTOPLASTICITY BY THE FINITE ELEMENT AND DIRECT STI	67-0002
STIFFNESS	PLANE ELASTICITY PROBLEMS AND THE DIRECT STIFFNESS METHOD	61-0009
STIFFNESS	POSSIBILITIES OF THE METHOD OF FINITE ELEMENTS IN THE CALCULATION OF THE STRENGTH AND STIFFNESS OF T	70-0344
STIFFNESS	POWER SERIES EXPANSION OF THE GENERAL STIFFNESS MATRIX FOR BEAM ELEMENTS	75-0543
STIFFNESS	PRACTICAL APPLICATION OF A COMPUTERIZED STRUCTURAL ANALYSIS SYSTEM WHICH ADOPTS A FINITE ELEMENT TEC	69-0225
STIFFNESS	PROBLEMS IN BUILDING AND MANIPULATING LARGE STIFFNESS MATRICES	71-0596
STIFFNESS	QUASI-LINEAR PROGRAMMING ALGORITHM FOR OPTIMIZING FIBRE-REINFORCED STRUCTURES OF FIXED STIFFNESS	75-0225
STIFFNESS	RATIONALIZATION IN DERIVING ELEMENT STIFFNESS MATRIX BY ASSUMED STRESS APPROACH	68-0138
STIFFNESS	REDUCTION OF STIFFNESS AND MASS MATRICES	65-0037
STIFFNESS	ROTATIONAL STIFFNESS OF CONCRETE SLABS	74-0660
STIFFNESS	ROUNDOFF CRITERIA IN DIRECT STIFFNESS SOLUTIONS	68-0204
STIFFNESS	SAMMSOR 111: A FINITE ELEMENT PROGRAM TO DETERMINE STIFFNESS AND MASS MATRICES OF RING-STIFFENED SHE	72-0774

385

STIFFNESS	SCALAR ALGORITHM OF DETERMINATION OF STIFFNESS MATRIX AND OF ANOTHER FINITE ELEMENT CHARACTERISTIC	74-1013
STIFFNESS	SNAP-THROUGH BUCKLING ANALYSIS OF SHELLS OF REVOLUTION USING INCREMENTAL STIFFNESS MATRICES	74-1349
STIFFNESS	STIFFNESS ANALYSIS OF ELASTIC PLASTIC PLATES	65-0040
STIFFNESS	STIFFNESS ANALYSIS OF SHEET METAL SHELLS UNDER CONCENTRATED LOADS	74-0325
STIFFNESS	STIFFNESS AND DEFLECTION ANALYSIS OF COMPLEX STRUCTURES	56-0001
STIFFNESS	STIFFNESS COEFFICIENTS FOR EMBEDDED FOOTINGS	75-0540
STIFFNESS	STIFFNESS DERIVATIVE FINITE ELEMENT TECHNIQUE FOR DETERMINATION OF CRACK TIP STRESS INTENSITY FACTOR	74-0956
STIFFNESS	STIFFNESS DERIVATIVE FINITE ELEMENT TECHNIQUE FOR DETERMINATION OF CRACK TIP STRESS INTENSITY FACTOR	74-1358
STIFFNESS	STIFFNESS MATRICES FOR THIN TRIANGULAR ELEMENTS OF NONZERO GAUSSIAN CURVATURE	66-0073
STIFFNESS	STIFFNESS MATRICES FOR PLATE BENDING ELEMENTS	69-0129
STIFFNESS	STIFFNESS MATRICES FOR SECTOR ELEMENTS	69-0219
STIFFNESS	STIFFNESS MATRICES FOR THE GENERAL DEFORMATION (OUT-OF-PLANE AND INPLANE) OF CURVED BEAM MEMBERS BAS	75-0124
STIFFNESS	STIFFNESS MATRIX FOR BENDING OF A RECTANGULAR PLATE ELEMENT WITH INITIAL MEMBRANE STRESSES	62-0006
STIFFNESS	STIFFNESS MATRIX FOR SHALLOW RECTANGULAR SHELL ELEMENT	67-0014
STIFFNESS	STIFFNESS MATRIX FOR A TRIANGULAR SANDWICH ELEMENT IN BENDING	67-0050
STIFFNESS	STIFFNESS MATRIX FOR SHALLOW RECTANGULAR SHELL ELEMENT	68-0074
STIFFNESS	STIFFNESS MATRIX FOR CURVED TAPERED AND STRAIGHT TAPERED SHELL STIFFENERS	69-0074
STIFFNESS	STIFFNESS MATRIX IN PLASTIC-ELASTIC PROBLEMS OF CONTINUA	67-0116
STIFFNESS	STIFFNESS METHODS FOR PLATES BY GALERKIN'S METHOD	69-0249
STIFFNESS	STIFFNESS OF STEPPED BARS	74-0711
STIFFNESS	STRUCTURAL ANALYSIS OF AN ARTERIOLE BY THE DIRECT STIFFNESS METHOD	66-0109
STIFFNESS	TANGENT STIFFNESS MATRIX FOR SPACE FRAME MEMBERS	69-0397
STIFFNESS	THE ECONOMICAL COMPUTATION OF STIFFNESS OF LARGE STRUCTURAL ELEMENTS	66-0090
STIFFNESS	THE GENERATION OF INTERELEMENT, COMPATIBLE STIFFNESS AND MASS MATRICES BY THE USE OF INTERPOLATION F	66-0014
STIFFNESS	THE USE OF FINITE ELEMENT MID-INCREMENT STIFFNESS MATRICES IN THE POST BUCKLING ANALYSIS OF IMPERFEC	71-0383
STIFFNESS	TRUSS ANALYSIS BY STIFFNESS CONSIDERATIONS	58-0001
STIFFNESS	VIBRATION ANALYSIS OF A CANTILEVERED SQUARE PLATE BY THE STIFFNESS MATRIX METHOD	66-0012
STIFFNESS	VIBRATION OF FRAMES AND OTHER STRUCTURES WITH BANDED STIFFNESS MATRIX	70-0322
STIFFNESS	WELDED JOINT STIFFNESS OBTAINED USING SOLID FINITE ELEMENTS	75-0190
STIFFNESS	WIELOSTOPNIOWA SYNTEZA MACIERZY SZTYWNOSCI (MULTI-STAGE SYNTHESIS OF THE STIFFNESS MATRICES)	73-0668
STINGER	OPTIMIZATION OF TENSION LEVEL AND STINGER LENGTH FOR OFFSHORE PIPELINE INSTALLATION	74-0504
STOCHASTIC	DYNAMIC AND STOCHASTIC ASPECTS IN THE STRUCTURAL ANALYSIS OF VLCCS AND ULCCS	74-0298
STOCHASTIC	MULTIDEGREE DYNAMIC ANALYSIS OF TALL BUILDINGS SUBJECTED TO WIND AS A STOCHASTIC PROCESS	74-1196
STOCHASTIC	OPTIMIZATION OF STRUCTURES UNDERGOING HARMONIC OR STOCHASTIC EXCITATION	75-0711
STOCHASTIC	STOCHASTIC DYNAMIC RESPONSE ANALYSIS OF OFFSHORE PLATFORMS, WITH PARTICULAR REFERENCE TO GRAVITY-TYP	75-0220
STOCHASTIC	STOCHASTIC FINITE ELEMENT SIMULATION OF THE NONLINEAR STRUCTURAL RESPONSE OF FIBROUS COMPOSITE MATER	75-0762
STOCHASTIC	THE STOCHASTIC FINITE ELEMENT SIMULATION OF THE NONLINEAR STRUCTURAL RESPONSE OF FIBROUS COMPOSITE M	75-0798
STOCHASTIC	TWO DIMENSIONAL STOCHASTIC MODEL OF A HETEROGENEOUS GEOLOGIC SYSTEM	73-0715
STOCHASTICALLY	FINITE ELEMENT OF UNDERGROUND STRESSES UTILIZING STOCHASTICALLY SIMULATED MATERIAL PROPERTIES	70-0111
STOFFANSATZES	PARAMETER EINES ELASTO-PLASTISCHEN STOFFANSATZES FUER KOERNIGE ERDSTOFFE (DETERMINATION OF PARAMETER	74-0657
STOKES	A HYBRID FINITE ELEMENT METHOD FOR TWO-DIMENSIONAL NAVIER STOKES EQUATIONS	75-0027
STOKES	A NON VARIATIONAL FINITE ELEMENT ANALYSIS FOR THE NAVIER STOKES EQUATIONS	74-0010
STOKES	AN APPROXIMATE NUMERICAL SOLUTION OF THE NAVIER STOKES EQUATIONS BY GALERKIN METHOD	74-0011
STOKES	CONFORMING AND NONCONFORMING FINITE ELEMENT METHODS FOR SOLVING THE STATIONARY STOKES EQUATIONS - 1	73-0390
STOKES	NAVIER STOKES EQUATIONS USING MIXED INTERPOLATION	74-0014
STOKES	THE SOLUTION OF THE NAVIER STOKES EQUATIONS	74-1220
STOKES	VARIATIONAL FINITE ELEMENT METHODS FOR TWO DIMENSIONAL AND AXISYMMETRIC NAVIER STOKES EQUATIONS	74-0012
STOKES	VARIATIONAL PROPERTIES OF STEADY FALL IN STOKES FLOW	72-0285
STOL	METHODS FOR PREDICTING THE AERODYNAMIC AND STABILITY AND CONTROL CHARACTERISTICS OF STOL AIRCRAFT. V	73-0718
STOL	METHODS FOR PREDICTING THE AERODYNAMIC AND STABILITY AND CONTROL CHARACTERISTICS OF STOL AIRCRAFT. V	73-0718
STOL	METHODS FOR PREDICTING THE AERODYNAMIC AND STABILITY AND CONTROL CHARACTERISTICS OF STOL AIRCRAFT. V	73-0719
STOL	METHODS OF PREDICTING THE AERODYNAMIC AND STABILITY AND CONTROL CHARACTERISTICS OF STOL AIRCRAFT. VO	73-0717
STORAGE	A COMPACT STORAGE SCHEME FOR THE SOLUTION OF SYMMETRICAL LINEAR SIMULTANEOUS EQUATIONS	66-0116
STORAGE	A COMPARATIVE STUDY OF SEVERAL CORE STORAGE SCHEMES FOR LARGE SPARSE POSITIVE DEFINITE MATRICES WITH	72-0678
STORAGE	ANALYSIS OF BOTTOM ENTRY NOZZLES FOR ATMOSPHERIC STORAGE TANKS	75-0331
STORAGE	APPLICATION OF FINITE ELEMENT ANALYSIS TO UNDERGROUND STORAGE OF NATURAL GAS	70-0198
STORAGE	ORGANIZATION TECHNIQUES OF THE FINITE ELEMENT METHOD FOR REDUCING CORE STORAGE REQUIREMENTS	74-0074
STORAGE	ORGANIZATION TECHNIQUES OF THE FINITE ELEMENT METHOD FOR REDUCING CORE STORAGE REQUIREMENTS	74-1120
STORAGE	SELECTION OF LOCATIONS FOR LONG TERM STORAGE OF WHEAT	75-0323
STORED	FINITE ELEMENT ANALYSES OF SALT DOMES WITH STORED HOT WASTES	74-1148
STORED	ON COMPUTATIONS INVOLVING STIFFNESS MATRICES STORED IN RECTANGULAR FORM	75-0576
STORM	EVALUATION OF TECHNIQUES FOR NUMERICAL CALCULATION OF STORM SURGES	75-0554
STORM	NUMERICAL CALCULATION OF STORM SURGES: AN EVALUATION OF TECHNIQUES	75-0461
STRAIGHT	BENDING AND TORSION OF THIN-WALLED STRAIGHT AND CURVED BEAMS WITH VARIABLE OPEN CROSS-SECTION	69-0493
STRAIGHT	ELASTIC-PLASTIC PROBLEMS OF COMPRESSIBLE STRAIGHT BAR UNDER COMBINED LOADING CONDITION	75-0313
STRAIGHT	ELASTIC-PLASTIC-CREEP ANALYSIS OF THERMAL RATCHETTING IN STRAIGHT PIPE AND COMPARISONS WITH TEST RES	73-0393
STRAIGHT	FREE VIBRATION OF CURVED AND STRAIGHT BEAM SLAB OR BOX GIRDER BRIDGES	72-0487
STRAIGHT	INELASTIC ANALYSIS OF A STRAIGHT TUBE UNDER COMBINED BENDING, PRESSURE AND THERMAL LOADS	75-0218
STRAIGHT	INELASTIC ANALYSIS OF A STRAIGHT TUBE UNDER COMBINED BENDING, PRESSURE AND THERMAL LOADS	75-0513
STRAIGHT	STIFFNESS MATRIX FOR CURVED TAPERED AND STRAIGHT TAPERED SHELL STIFFENERS	69-0074
STRAIGHT	STRESS INTENSITY FACTOR FOR A STRAIGHT CRACK APPROACHING A HOLE	69-0158
STRAIGHT	THERMAL MECHANICAL ANALYSIS PROCEDURE FOR A SEMICIRCULAR PLATE WITH CLAMPED STRAIGHT EDGE	75-0651
STRAIGHT	THERMAL STRESS DEVELOPED IN S PLATE, STRAIGHT PLATE, AND DEEP DISH WHEELS	73-0172
STRAIGHT	USE OF STRAIGHT BEAM FINITE ELEMENTS FOR ANALYSIS OF VIBRATIONS OF CURVED BEAMS	73-0616
STRAIN	A FINITE ELEMENT ANALYSIS OF THE PLANE STRAIN SHEET DRAWING PROBLEM	71-0812
STRAIN	A FINITE ELEMENT ANALYSIS OF SHOCK AND FINITE-AMPLITUDE WAVES IN ONE-DIMENSIONAL HYPERELASTIC BODIES	72-0731
STRAIN	A FINITE ELEMENT FORMULATION FOR PROBLEMS OF LARGE STRAIN AND LARGE DISPLACEMENT	70-0069
STRAIN	A MATHEMATICAL THEORY OF PLASTICITY FOR STRAIN HARDENING MATERIALS EXHIBITING INITIAL ANISOTROPY	72-0660
STRAIN	A METHOD FOR DETERMINING LOCAL ELASTOPLASTIC STRESS AND STRAIN IN METALLURGICALLY BONDED NOTCHED LAM	74-0686
STRAIN	A METHOD OF ELASTIC-PLASTIC PLANE STRESS AND PLANE STRAIN ANALYSIS	66-0114
STRAIN	A STRAIN ENERGY BASIS FOR STUDIES OF ELEMENT STIFFNESS MATRICES	72-0621
STRAIN	A STUDY OF THE PLANE STRESS OR STRAIN FINITE ELEMENT ANALYSIS FOR SOLUTION OF STRESS DISTRIBUTION IN	69-0487
STRAIN	A WARPED QUADRILATERAL STRAIN MEMBRANE ELEMENTS	75-0141
STRAIN	AN EVALUATION OF FINITE ELEMENTS WITH IMPROVED ASSURED CONVERGENCE FOR PLANE STRESS AND PLANE STRAIN	74-0429
STRAIN	AN INCREMENTAL FINITE ELEMENT ANALYSIS OF TWO DIMENSIONAL LARGE STRAIN AND LARGE DISPLACEMENT PROBLE	72-0338
STRAIN	ANALYSES OF AXISYMMETRIC UPSETTING AND PLANE STRAIN SIDE PRESSING OF SOLID CYLINDERS BY THE FINITE E	70-0083
STRAIN	ANALYSIS OF PLANE STRAIN CONSOLIDATION IN SATURATED CLAYS BY FINITE ELEMENT METHOD	71-0815
STRAIN	ANALYSIS OF PLANE STRAIN TEST USING THE FINITE ELEMENT METHOD	72-1002
STRAIN	ANALYSIS OF STRESS AND STRAIN DISTRIBUTIONS AT CRACK TIP BY FINITE ELEMENT METHOD	69-0193
STRAIN	ANALYSIS OF STRESS AND STRAIN DISTRIBUTIONS IN TRIAXIAL TESTS USING THE METHOD OF FINITE ELEMENTS	73-0903
STRAIN	ANALYSIS OF THERMAL ELASTIC PLASTIC STRESS AND STRAIN DURING WELDING BY FINITE ELEMENT METHOD	71-0382
STRAIN	ANALYSIS OF THICK-WALLED PRESSURE VESSELS USING THE CONSTANT STRAIN AXISYMMETRIC FINITE ELEMENT	71-0653
STRAIN	ANISOTROPIC WORK HARDENING IN METALS. FINITE ELEMENT APPLICATION TO PLANE STRESS AND PLANE STRAIN	73-0950
STRAIN	APPLICATION OF FINITE ELEMENT METHOD TO PLANE STRAIN EXTRUSION PROCESSES	74-1243
STRAIN	APPLICATIONS OF CURVED FINITE ELEMENT TO SANDWICH BEAMS USING SHEAR STRAIN FORMULATION	74-0523
STRAIN	APPLICATIONS OF CURVED FINITE ELEMENTS TO SANDWICH SHELLS USING SHEAR STRAIN FORMULATION	75-0433
STRAIN	AUTOMATIC EXTRACTION OF RIGID BODY MODES FROM STRESS AND STRAIN ELEMENTS	72-0913
STRAIN	BRAIN TISSUE FRAGILITY - A FINITE STRAIN ANALYSIS BY A HYBRID FINITE-ELEMENT METHOD	75-0162
STRAIN	BRAIN TISSUE FRAGILITY - A FINITE STRAIN ANALYSIS BY A HYBRID FINITE-ELEMENT METHOD	75-0243
STRAIN	BRAIN-TISSUE FRAGILITY - FINITE STRAIN ANALYSIS BY A HYBRID FINITE-ELEMENT METHOD	75-0631
STRAIN	BRLESC FINITE ELEMENT PROGRAM FOR AXISYMMETRIC, PLANE STRAIN, AND PLANE STRESS, ORTHOTROPIC SOLIDS W	71-0749
STRAIN	CRACK GROWTH MONITORING BY STRAIN SENSING	73-0486

STRAIN	CRACK TIP STRAIN - A COMPARISON OF FEM CALCULATIONS AND MEASUREMENTS	75-0721
STRAIN	CREEP STRESS AND STRAIN ANALYSIS OF NOTCHED PLATES AND BARS BY MEANS OF THE FINITE ELEMENT METHOD	70-0141
STRAIN	DER SPANNUNGS UND BEANSPRUCHUNGSZUSTAND DES BODENMATERIALS IN DER NACHBARSCHAFT EINER BODENBOESCHUNG	75-0371
STRAIN	DISCRETE-ELEMENT PLASTIC ANALYSIS OF STRUCTURES IN A STATE OF MODIFIED PLANE STRAIN	69-0140
STRAIN	DISPLACEMENT AND STRAIN FIELDS FROM UNDERGROUND EXPLOSIONS USING ANALYTICAL AND FINITE ELEMENT METHO	72-0587
STRAIN	EFFECT OF FIBRE DISTRIBUTION ON STRESS AND STRAIN CONCENTRATION AT HOLES IN COMPOSITE PLATES	72-0616
STRAIN	ELASTIC DEFORMATION OF POLYCRYSTALLINE METAL - 1. INFLUENCE OF GRAIN SHAPE ON DISTRIBUTIONS OF STRES	72-0463
STRAIN	ELASTIC PLASTIC CRACKED PLATES IN PLANE STRAIN	69-0321
STRAIN	ELASTIC-PLASTIC STRESS ANALYSIS FOR A NOTCHED BAR IN PLANE STRAIN BENDING	71-0410
STRAIN	ELASTO-PLASTIC STRESS ANALYSIS. A GENERALIZATION FOR VARIOUS CONSTITUTIVE RELATIONS INCLUDING STRAIN	72-0547
STRAIN	ELASTOPLASTIC ANALYSIS OF PLANE STRAIN AND AXISYMMETRIC FLAT PUNCH INDENTATION BY THE FINITE ELEMENT	70-0343
STRAIN	EVALUATION OF STRESSES AND DEFLECTION OF SPUR GEAR TEETH UNDER STRAIN	72-0115
STRAIN	FINITE ELEMENT ANALYSIS OF THE PLANE STRAIN DRAWING PROBLEM	74-1154
STRAIN	FINITE ELEMENT APPLICATION TO STRAIN GAGE DATA REDUCTION	73-0561
STRAIN	FINITE ELEMENT FORMULATION OF STRESS STRAIN MATRIX FOR AN ELASTICALLY PERFECTLY PLASTIC TRESCA MATER	70-0275
STRAIN	FINITE ELEMENT FORMULATIONS FOR LARGE DISPLACEMENT AND LARGE STRAIN ANALYSIS	73-0457
STRAIN	FINITE ELEMENT PARAMETRIC STUDY OF VERTICAL STRAIN INFLUENCE FACTORS AND THE PRESSUREMETER TEST TO E	74-1127
STRAIN	FINITE ELEMENT THEORY OF PLATES AND SHELLS INCLUDING TRANSVERSE SHEAR STRAIN EFFECTS	70-0395
STRAIN	FINITE PLANE STRAIN OF INCOMPRESSIBLE ELASTIC SOLIDS BY THE FINITE ELEMENT METHOD	68-0128
STRAIN	FORMULATION OF STRESS STRAIN INDUCED DIFFUSION OF HYDROGEN AND ITS SOLUTION BY COMPUTER-AIDED FINITE	75-0326
STRAIN	FRACTURE ANALYSIS IN AREAS OF HIGH NOMINAL STRAIN	73-0495
STRAIN	IMPROVED APPROXIMATION OF CONSTITUTIVE ELASTO-PLASTIC STRESS STRAIN RELATIONSHIP FOR FINITE ELEMENT	74-0566
STRAIN	INCREMENTAL THEORY OF LARGE STRAIN AND LARGE DISPLACEMENT PROBLEMS AND ITS FINITE ELEMENT FORMULATIO	72-0274
STRAIN	J INTEGRAL ESTIMATES FOR STRAIN HARDENING MATERIALS USING FULLY PLASTIC SOLUTIONS	74-0690
STRAIN	LAGRANGIAN AND EULERIAN FINITE ELEMENT FORMULATION FOR LARGE STRAIN ELASTO-PLASTICITY	75-0209
STRAIN	LARGE STRAIN ELASTO-PLASTIC FINITE ELEMENT ANALYSIS	71-0121
STRAIN	LARGE STRAIN, ELASTO-PLASTIC NUMERICAL ANALYSIS BY MEANS OF FINITE ELEMENT METHOD	71-0790
STRAIN	LARGE STRAIN, LARGE DISPLACEMENT ANALYSIS	71-0508
STRAIN	LARGE STRAIN, VISCOELASTIC NUMERICAL ANALYSIS BY MEANS OF FINITE ELEMENT METHOD	72-0768
STRAIN	LARGE STRAIN, VISCOELASTIC AND ELASTO-VISCOPLASTIC NUMERICAL-ANALYSIS BY MEANS OF FINITE-ELEMENT MET	75-0435
STRAIN	LARGE STRAIN, VISCOELASTIC AND ELASTO VISCOPLASTIC NUMERICAL ANALYSIS BY MEANS OF THE FINITE ELEMENT	75-0563
STRAIN	MATRIX DISPLACEMENT ANALYSIS OF SHELLS AND PLATES INCLUDING TRANSVERSE SHEAR STRAIN EFFECTS	72-0643
STRAIN	MEMBRANE PARALLELOGRAM ELEMENT WITH LINEARLY VARYING EDGE STRAIN FOR THE MATRIX DISPLACEMENT METHOD	66-0006
STRAIN	MIXED DISPLACEMENT 3D ISOPARAMETRIC FINITE ELEMENTS AND A STRAIN ENERGY BASIS FOR STUDIES OF ELEMENT	70-0286
STRAIN	NEW CYLINDRICAL SHELL FINITE ELEMENT BASED ON SIMPLE INDEPENDENT STRAIN FUNCTIONS	72-0094
STRAIN	NONLINEAR ANALYSIS OF STRESS AND STRAIN IN SOILS	70-0050
STRAIN	NONLINEAR FINITE ELEMENT ANALYSIS INCLUDING HIGHER ORDER STRAIN ENERGY TERMS	70-0065
STRAIN	NOTE ON STRESS AND STRAIN REDISTRIBUTION IN A NOTCHED PLATE SPECIMEN DURING CYCLIC LOADING	69-0194
STRAIN	ON SOLUTION OF PLANE STRAIN CONSOLIDATION PROBLEMS BY FINITE ELEMENT METHOD	71-0532
STRAIN	ON THE EFFECT OF THE FORM OF THE STRAIN ENERGY FUNCTION ON THE SOLUTION OF A BOUNDARY-VALUE PROBLEM	72-0429
STRAIN	PLANE STRAIN CONSOLIDATION BY FINITE ELEMENTS	69-0471
STRAIN	PLANE STRAIN CONSOLIDATION BY FINITE ELEMENTS	70-0043
STRAIN	PLANE STRESS AND PLANE STRAIN ANALYSIS BY FINITE ELEMENT METHOD	70-0388
STRAIN	PLASTIC STRESS STRAIN MATRIX OF ANISOTROPIC MATERIALS AND ITS APPLICATION	69-0383
STRAIN	QUADRILATERAL FINITE ELEMENT FOR PLANE STRESS AND PLANE STRAIN	70-0496
STRAIN	REINFORCED FIELDS OF TRIANGULAR ELEMENTS WITH LINEARLY VARYING STRAIN: EFFECT OF INITIAL STRAINS	65-0004
STRAIN	SIMPLIFIED SOIL-STRUCTURE INTERACTION ANALYSIS WITH STRAIN DEPENDENT SOIL PROPERTIES	74-1052
STRAIN	SIZE OF PLASTIC ZONE AT TIP OF A CRACK IN PLANE STRAIN SLATE BY FINITE ELEMENT METHOD	73-0006
STRAIN	SMALL SCALE YIELDING ANALYSIS OF MIXED MODE PLANE STRAIN CRACK PROBLEMS	73-0801
STRAIN	SOLUTION OF SOME PLANE STRAIN PROBLEMS IN SOIL MECHANICS USING THE METHOD OF FINITE ELEMENTS	69-0521
STRAIN	SOME PROPERTIES OF LINEAR STRAIN TRIANGLES AND OPTIMAL FINITE ELEMENT MODELS	73-1044
STRAIN	STABILITY AND POSTBUCKLING BEHAVIOUR OF HYPERELASTIC BODIES AT FINITE STRAIN BY THE FINITE ELEMENT M	73-0562
STRAIN	STABILITY AND POSTBUCKLING BEHAVIOR OF HYPERELASTIC BODIES AT FINITE STRAIN BY THE FINITE ELEMENT ME	74-0838
STRAIN	STIFFNESS MATRICES FOR THE GENERAL DEFORMATION (OUT-OF-PLANE AND INPLANE) OF CURVED BEAM MEMBERS BAS	75-0124
STRAIN	STRAIN CONCENTRATION ANALYSIS USING THE FINITE ELEMENT METHOD	71-0090
STRAIN	STRAIN CONCENTRATION BEHAVIOUR IN A NOTCHED ROUND BAR SUBJECTED TO CREEP	73-0440
STRAIN	STRAIN DISTRIBUTION AROUND UNDERGROUND OPENINGS	69-0212
STRAIN	STRAIN DISTRIBUTION AROUND UNDERGROUND OPENINGS. STATISTICAL RELATIONSHIPS FOR CERTAIN ROCK PROPERTI	71-0710
STRAIN	STRAIN DISTRIBUTION AROUND UNDERGROUND OPENINGS. FINITE ELEMENT ANALYSIS OF JOINTED SYSTEMS	72-0838
STRAIN	STRAIN DISTRIBUTION IN AN ALUMINA BICRYSTAL STUDIES BY THE FINITE ELEMENT METHOD AND X-RAY DIFFRACTI	74-1070
STRAIN	STRAIN ENERGY FUNCTIONS AND MATHEMATICAL MODELS OF LUNG ELASTICITY	73-0591
STRAIN	STRAIN ENERGY RELEASE RATES FOR TURNING CRACKS	72-0146
STRAIN	STRAIN HARDENING TORSION OF SOLID BARS AS A MINIMIZATION PROBLEM	70-0174
STRAIN	STRAIN RESULTANTS IN CERTAIN FINITE ELEMENTS	69-0079
STRAIN	STRESS ANALYSIS OF A PRESTRESSED CONCRETE NUCLEAR PRESSURE VESSEL BY THE FINITE ELEMENT METHOD USING	71-0154
STRAIN	STRESS AND STRAIN DISTRIBUTIONS AT THE ROOT OF CRACKS	69-0290
STRAIN	STRESS AND STRAIN DISTRIBUTION IN THE VICINITY OF INTERFERENCE FIT FASTENERS	73-0363
STRAIN	STRESS DISTRIBUTION UNDER UNIAXIAL ELASTIC PLANE STRAIN COMPRESSION WITH FRICTION	74-0526
STRAIN	STRESS STRAIN MATRIX AND ITS APPLICATIONS FOR THE SOLUTION OF ELASTIC-PLASTIC PROBLEMS BY THE FINITE	68-0282
STRAIN	STRESS STRAIN RELATIONSHIP FROM COMPRESSION TESTS ON NONLINEAR MATERIALS	69-0189
STRAIN	STRESS STRAIN RESPONSE AND FRACTURE OF A CONCRETE MODEL IN BIAXIAL LOADING	71-0295
STRAIN	TETRAHEDRON ELEMENTS WITH LINEARLY VARYING STRAIN FOR THE MATRIX DISPLACEMENT METHOD	65-0006
STRAIN	THE ROLE OF FINITE DEFORMATION ANALYSIS IN PLANE STRESS AND STRAIN FRACTURES	74-0322
STRAIN	THE USES OF FINITE ELEMENT TECHNIQUES IN SOIL MECHANICS WITH PARTICULAR REFERENCE TO DEFORMATION PLA	72-0002
STRAIN	THE VALIDITY OF THE INITIAL STRAIN CONCEPT	61-0010
STRAIN	THREE-DIMENSIONAL ELASTIC-PLASTIC STRESS AND STRAIN ANALYSIS FOR FRACTURE MECHANICS. PHASE 1. SIMPLE	70-0188
STRAIN	TRIANGULAR ELEMENTS WITH LINEARLY VARYING STRAIN FOR THE MATRIX DISPLACEMENT METHOD	65-0053
STRAIN	ULTIMATE STRAIN CAPACITY AND TEMPERATURE RISE STUDIES. TRUMBULL POND DAM	72-0705
STRAIN	USE OF THE FINITE ELEMENT DAMPED FORCED RESPONSE STRAIN ENERGY DISTRIBUTION FOR VIBRATION REDUCTION	72-0759
STRAIN	VIBRATION REDUCTION BY USING BOTH THE FINITE ELEMENT STRAIN ENERGY DISTRIBUTION AND MOBILITY TECHNIQ	74-1303
STRAIN--LINEAR	APPLICATION TO PLANE STRESS AND STRAIN--LINEAR ELASTIC MATERIAL	68-0048
STRAIN-AND-STRE	QUANTITATIVE STRAIN-AND-STRESS STATE CRITERION FOR FAILURE IN THE VICINITY OF SHARP CRACKS	74-0639
STRAIN-BASED	BEHAVIOR WITH DIMINISHING CURVATURE OF STRAIN-BASED ARCH FINITE ELEMENTS	73-0088
STRAIN-CONSTANT	ON THE UNITY OF THE CONSTANT STRAIN-CONSTANT MOMENT FINITE ELEMENT METHODS	73-0164
STRAIN-DISPLACE	RIGID-BODY MOTIONS AND STRAIN-DISPLACEMENT EQUATIONS OF CURVED SHELL FINITE ELEMENTS	72-0267
STRAIN-ENERGY	STRAIN-ENERGY BOUNDS IN FINITE-ELEMENT ANALYSIS BY SLAB ANALOGY	67-0027
STRAIN-FINITE	SMALL SCALE YIELDING NEAR A CRACK IN PLANE STRAIN-FINITE ELEMENT ANALYSIS	71-0355
STRAIN-GRADIENT	ON THE NUMERICAL SOLUTION OF A CLASS OF PROBLEMS IN A LINEAR FIRST STRAIN-GRADIENT THEORY OF ELASTIC	70-0318
STRAIN-HARDENIN	NEAR-CRACK DEFORMATION FIELDS IN STRAIN-HARDENING MATERIAL	74-0938
STRAIN-SOFTENIN	FINITE ELEMENT ANALYSIS OF STRAIN-SOFTENING CLAY	72-0028
STRAINED	APPLICATION OF THE METHOD OF FINITE ELEMENTS TO CALCULATIONS OF THE STRESSED AND STRAINED STATE OF I	75-0775
STRAINED	ISSLEDOVANIE NAPRYAZHENNO-DEFORMIROVANNOGO SOSTOYANIYA SIMMETRICHNYKH DISKOV SO STUPITSEI. (INVESTIG	74-0965
STRAINS	A NEW METHOD OF ANALYSIS TO OBTAIN EXACT SOLUTIONS FOR STRESSES AND STRAINS IN CIRCULAR INCLUSIONS	74-1063
STRAINS	APPLICATION OF NATURAL KINEMATIC MODES TO MATRIX ANALYSIS: SMALL AND LARGE DISPLACEMENTS AND STRAINS	67-0138
STRAINS	CALCULATION OF THERMALLY INDUCED STRAINS IN MACHINE TOOL ELEMENTS	72-0464
STRAINS	CONTINUUM BASES FOR CONSISTENT NUMERICAL FORMULATIONS OF FINITE STRAINS IN ELASTIC AND INELASTIC STR	75-0299
STRAINS	CRACKING OF EARTH AND ROCKFILL DAMS: COMPARISON OF OBSERVED AND THEORETICAL TENSILE STRAINS IN THE C	71-0699
STRAINS	CURVED FINITE ELEMENTS BY METHODS OF INITIAL STRAINS	73-0030
STRAINS	DEVELOPMENT OF ADVANCED STRUCTURAL OPTIMIZATION PROGRAMS AND THEIR APPLICATION TO LARGE ORDER STRAIN	66-0028
STRAINS	DYNAMIC RESPONSE ON NONLINEAR MEDIA AT LARGE STRAINS	72-0237
STRAINS	DYNAMIC STRAINS IN A FOUR BAR MECHANISM	73-0677
STRAINS	EXPERIMENTAL AND ANALYTICAL STRAINS IN A EDGE-CRACKED SHEET	69-0483

STRAINS	FINITE ELEMENTS, FINITE ROTATIONS AND SMALL STRAINS OF FLEXIBLE SHELLS	69-0516
STRAINS	FINITE STRAINS AND DISPLACEMENTS OF ELASTIC MEMBRANES BY THE FINITE ELEMENT METHOD	67-0062
STRAINS	INITIAL STRAINS IN THE MATRIX FORCE METHOD OF STRUCTURAL ANALYSIS	60-0008
STRAINS	LARGE NATURAL STRAINS AND SOME SPECIAL DIFFICULTIES DUE TO NONLINEARITY AND INCOMPRESSIBILITY IN FIN	74-0528
STRAINS	LARGE NATURAL STRAINS AND SOME SPECIAL DIFFICULTIES DUE TO NONLINEARITY AND INCOMPRESSIBILITY IN FIN	74-1027
STRAINS	MEASUREMENT OF ACTUAL DEDENDUM STRAINS, AND THE INFLUENCE OF DEDENDUM RADIUS ON STRESSES AND DEDENDU	72-0118
STRAINS	NATRUAL STRAINS AND STRESSES FOR TRAPEZOIDAL STRUCTURES ANALYSIS	69-0101
STRAINS	NONLINEAR CONSIDERATION OF GRAVITY IN A STIFFNESS TEST OF A WEAK STRUCTURE AT SMALL STRAINS	71-0035
STRAINS	PLANE STRESS ANALYSIS OF RUBBER AT HIGH STRAINS USING FINITE ELEMENTS	71-0680
STRAINS	PLASTIC STRAINS ASSOCIATED WITH TORSION OF NONCIRCULAR BARS	72-0177
STRAINS	RECENT TRENDS OF RESEARCH INTO RESTRAINT STRESSES AND STRAINS IN RELATION TO WELD CRACKING	73-0483
STRAINS	REINFORCED FIELDS OF TRIANGULAR ELEMENTS WITH LINEARLY VARYING STRAIN: EFFECT OF INITIAL STRAINS	65-0004
STRAINS	RETSCP: A COMPUTER PROGRAM FOR ANALYSIS OF ROCKET ENGINE THERMAL STRAINS WITH CYCLIC PLASTICITY	74-0692
STRAINS	TENDANCES ACTUELLES DES RECHERCHES SUR LES CONTRAINTES ET DEFORMATIONS DUES AU BRIDAGE EN RELATION A	74-0273
STRAINS	THE ANALYSIS OF THIN SHELLS WITH TRANSVERSE SHEAR STRAINS BY THE FINITE ELEMENT METHOD	68-0098
STRAINS	USERS MANUAL FOR DIGITAL COMPUTER PROGRAM OK5 NUMERICAL ANALYSIS OF FINITE STRAINS AND DISPLACEMENTS	72-0718
STRAND	CIRCUMFERENTIAL PRESTRESSING LOAD ANALYSIS DUE TO WIRE/STRAND WINDING SYSTEMS ON PRESTRESSED CONCEPT	74-1202
STRAP	STRAP - A COMPUTER CODE FOR THE STATIC AND DYNAMIC ANALYSIS OF REACTOR STRUCTURAL SYSTEMS	71-0078
STRAP	STRAP: COMPUTER CODE FOR STATIC AND DYNAMIC STRUCTURAL ANALYSIS AND STUDIES MADE USING THE CODE	70-0409
STRATA	DIE BERECHNUNG DER BEWEGUNGEN UND SPANNUNGEN DES GEBIRGES VOM ABBAU BIS ZUR TAGESOBERFLAECHE NACH DE	73-0456
STPATA	USER'S MANUAL FOR UMR FINITE ELEMENT COMPUTER PROGRAM - STRATA	69-0429
STRATEGIES	AUTOMATED MESH REFINEMENT STRATEGIES FOR FINITE ELEMENT COMPUTATIONS	75-0877
STRATEGIES	SEVERAL STRATEGIES FOR REDUCING THE BANDWIDTH OF MATRICES	72-0996
STRATIFIE	CHARACTERISATION ELASTIQUE D'UN STRATIFIE RESINE-VERRE PAR LA METHODE DES DEPLACEMENTS (DETERMINATIO	75-0272
STRATIFIED	A FINITE ELEMENT SOLUTION FOR TWO-DIMENSIONAL DENSITY STRATIFIED FLOW	73-0911
STRATIFIED	A FINITE ELEMENT SOLUTION FOR TWO-DIMENSIONAL STRATIFIED FLOW PROBLEMS	75-0391
STRATIFIED	A NUMERICAL INVESTIGATION OF STABLY STRATIFIED, WIND DRIVEN CAVITY FLOW BY THE FINITE ELEMENT METHOD	74-1372
STRATIFIED	FINITE ELEMENT MODEL FOR TWO-DIMENSIONAL DENSITY STRATIFIED FLOW	73-0064
STRATIFIES	DETERMINATION OF THE TECHNICAL CONSTANTS OF LAMINATES IN OBLIQUE DIRECTIONS. THEORY AND EXPERIMENTAL	73-0748
STRATUM	INTERACTIVE ANALYSIS OF A RAFTED MULTISTOREY SPACE FRAME RESSTING ON AN INHOMOGENEOUS CLAY STRATUM	74-0447
STREAM	FINITE ELEMENT ANALYSIS OF VISCOUS FLOW BY STREAM FUNCTION	75-0526
STREAM	FINITE ELEMENT SOLUTION OF THE INCOMPRESSIBLE FLOW OVER AN AIRFOIL IN A NON UNIFORM STREAM	74-0432
STREAM	STUDY ON THE VIBRATION OF THIN PLATES IN A GAS STREAM BY THE FINITE ELEMENT METHOD	74-0544
STREBEN	VORAUSSAGE DES GEBIRGSDRUCKES UND SEINER AUSWIRKUNGEN AUF STREBEN UND STRECKEN IM STEINKOHLENBERGBAU	73-0679
STRECKEN	VORAUSSAGE DES GEBIRGSDRUCKES UND SEINER AUSWIRKUNGEN AUF STREBEN UND STRECKEN IM STEINKOHLENBERGBAU	73-0679
STREIFENELEMENT	BERECHNUNG PRISMATISCHER FLAECHENTRAGWERKE MIT HILFE VON FINITEN STREIFENELEMENTEN. (CALCULATION OF	72-0133
STRENGTH	A FINITE ELEMENT ANALYSIS FOR THE STRENGTH OF BEAM COLUMNS	66-0037
STRENGTH	A SEISMIC STRENGTH OF KOLKEWADI DAM	73-0197
STRENGTH	A TORSIONAL STRENGTH ANALYSIS ON THE CONTAINER SHIP BY MEANS OF THE FINITE ELEMENT PROCEDURE AND FUL	71-0243
STRENGTH	AN ANALYSIS OF THE SMALL-SCALE STRENGTH TESTING OF ICE	72-0881
STRENGTH	ANALYSIS OF STRENGTH AND RESIDUAL STRESSES IN FILAMENT REINFORCED ALUMINUM CYLINDERS	75-0517
STRENGTH	ANWENDUNG DES FINITE ELEMENT VERFAHRENS ZUR BESTIMMUNG DER FESTIGKEIT VON SCHWEISSKONSTRUKTIONEN. (	72-0185
STRENGTH	APPLICATION OF FINITE ELEMENT METHOD TO THE STUDY OF STRENGTH OF CAR BODY STRUCTURE AND WHEEL	74-0281
STRENGTH	AUTOMATED OPTIMUM DESIGN OF AIRCRAFT WINGS TO SATISFY STRENGTH, STABILITY, FREQUENCY AND FLUTTER REQ	72-0605
STRENGTH	AUTOMATED PROCEDURE FOR DESIGN OF WING STRUCTURES TO SATISFY STRENGTH AND FLUTTER REQUIREMENTS	73-0867
STRENGTH	CREEP OF B/AL COMPOSITES AS INFLUENCED BY RESIDUAL STRESSES, BOND STRENGTH, AND FIBER PACKING GEOMET	74-1208
STRENGTH	DETERMINATION OF THE APPLICABILITY OF SOME THEORIES ABOUT STRENGTH AND STIFFNESS OF THIN-WALLED BEAM	69-0318
STRENGTH	DIAGONAL TENSILE STRENGTH OF BRICKWORK	70-0158
STRENGTH	EFFECT OF SHAPE AND HETEROGENEITY OF ROCK SAMPLES ON THEIR COMPRESSIVE STRENGTH	69-0315
STRENGTH	EFFECTS OF LOCAL STRENGTH VARIATIONS ON PLASTIC ENCLAVE DEVELOPMENT AT CRACK TIPS	72-0822
STRENGTH	ELASTIC STRENGTH OF PROPELLERS: AN ANALYSIS BY MATRIX METHODS	70-0468
STRENGTH	EXPERIMENTELL UNDERSOKNING AV ELEMENTFOGARS HALLFASTHETSEGENSKAPER (EXPERIMENTAL INVESTIGATION OF ST	72-0192
STRENGTH	FATIQUE STRENGTH OF LUGS CONTAINING LINERS. VOLUME II. COMPUTER PROGRAM USED FOR ANALYSIS	70-0418
STRENGTH	FESTIGKEITSUNTERSUCHUNG MIT DER METHODE DER FINITEN ELEMENTE. (STRENGTH OF THE MATERIALS INVESTIGATE	74-1009
STRENGTH	FESTIGKEITSUNTERSUCHUNGEN UND-BERECHNUNGEN AN PUMPENSPIRALGEHAEUSEN. (INVESTIGATIONS AND CALCULATION	75-0238
STRENGTH	FINITE ELEMENT ANALYSIS ON UNIAXIAL COMPRESSIVE STRENGTH OF CYLINDRICAL BRITTLE SPECIMEN	70-0155
STRENGTH	FINITE ELEMENT ANALYSIS OF SHEAR STRENGTH OF REINFORCED CONCRETE BEAMS	74-1306
STRENGTH	FINITE-ELEMENT METHOD FOR DETERMINING STRENGTH OF WELDED STRUCTURES	72-0272
STRENGTH	FORM FACTORS OF FLAT BARS FOR FATIGUE STRENGTH TESTS. (FORMZAHLEN VON FLACHSTAEBEN FUER SCHWINGFESTI	71-0641
STRENGTH	FRACTURE (ULTIMATE STRENGTH) ANALYSES OF ASPHALT PAVEMENT LAYERS RESULTING FROM TRAFFIC LOADING	73-0548
STRENGTH	HIGH STRENGTH BOLTS SUBJECT TO TENSION AND PRYING	74-0201
STRENGTH	HIGH STRENGTH CERAMICS	72-0878
STRENGTH	HIGH STRENGTH END FITTINGS FOR FRP ROD AND ROPE	74-0605
STRENGTH	HIGH STRENGTH MATERIALS AND VEHICLE WEIGHT REDUCTION ANALYSIS	75-0157
STRENGTH	ISSLEDOVANIE PROCHNOSTI I ZHESTKOSTI PLASTIN, NAGRUZHENNYKH V SVOEI PLOSKOSTI, METHODOM KONECHNYKH E	73-0382
STRENGTH	LOW CYCLE FATIGUE STRENGTH OF PIPING COMPONENTS	73-0366
STRENGTH	MEASUREMENT OF ACTUAL DEDENDUM STRAINS, AND THE INFLUENCE OF DEDENDUM RADIUS ON STRESSES AND DEDENDU	72-0118
STRENGTH	NOTCH EFFECT IN LOW CYCLE FATIGUE - REPORT 1: LOW-CYCLE FATIGUE STRENGTH OF 0.48 PERCENT CARBON STEE	73-1090
STRENGTH	ON SHELLS OF CONSTANT STRENGTH	72-0431
STRENGTH	ON STRENGTH OF TRANSVERSE WEBS AND THEIR JOINTS OF LARGE TANKER	69-0327
STRENGTH	ON THE THEORY OF CALCULATION OF THE OVERALL STRENGTH OF A FUSELAGE BY THE METHOD OF FINITE ELEMENTS	72-0692
STRENGTH	OPTIMIZATION OF STRUCTURES FOR STRENGTH AND STABILITY REQUIREMENTS	73-0575
STRENGTH	POSSIBILITIES OF THE METHOD OF FINITE ELEMENTS IN THE CALCULATION OF THE STRENGTH AND STIFFNESS OF T	70-0344
STRENGTH	PROBALILISTIC MODEL FOR MATERIAL STRENGTH VARIATION AND SIZE EFFECT	72-0876
STRENGTH	PROBLEMS OF LONGITUDINAL STRENGTH IN THE CONSTRUCTION OF VERY LARGE TANK TRUCKS	73-0251
STRENGTH	PUNCHING STRENGTH OF SLABS. A FLEXURAL APPROACH USING FINITE ELEMENTS	74-1307
STRENGTH	RHEOLOGICAL THEORY OF STRENGTH OF MATERIALS	69-0313
STRENGTH	SIMPLE APPROACH TO THE STRENGTH ANALYSIS OF TANKERS	71-0319
STRENGTH	STRENGTH DISTRIBUTIONS OF SINGLE FILAMENTS	75-0305
STRENGTH	STRENGTH OF RAILROAD WHEELS - 2	71-0306
STRENGTH	STRENGTH OF REINFORCED CONCRETE CHAMBERS UNDER EXTERNAL PRESSURE	75-0293
STRENGTH	STRENGTH OF REINFORCED CONCRETE CHAMBERS UNDER EXTERNAL PRESSURE	75-0512
STRENGTH	STRENGTH OF SHIP STRUCTURAL ELEMENTS	75-0051
STRENGTH	STRENGTH OF TRANSVERSE FILLET WELDED JOINTS	75-0144
STRENGTH	STRESS DETERMINATION IN 2 CONDENSER HEAT EXCHANGER BAFFLE CAPS BY METHOD OF FINITE ELEMENTS AND A ST	74-0661
STRENGTH	STUDY ON MOTION AND STRENGTH OF FLOATING MARINE STRUCTURES IN WAVES	74-0489
STRENGTH	TANKER TRANSVERSE STRENGTH ANALYSIS: USER'S MANUAL	72-0750
STRENGTH	THE EFFECT OF END RESTRAINT CONDITION ON THE FRACTURE MECHANISM AND STRENGTH OF ROCK-LIKE MATERIALS	72-0134
STRENGTH	THE FATIGUE STRENGTH OF FILLET WELDED CONNECTIONS	71-0684
STRENGTH	THE USE OF ROCK BOLTS OR WIRE ROPE TO INCREASE THE STRENGTH OF FRACTURED MODEL PILLARS	71-0722
STRENGTH	TRAGLAST VON T-TRAEGERN. EXPERIMENT UND F.E. BERECHNUNGEN (ULTIMATE STRENGTH OF T-BEAMS. EXPERIMENTS	74-0935
STRENGTH	TRANSVERSE STRENGTH OF TANKERS, FINITE ELEMENT APPLICATIONS	68-0142
STRENGTH	TRANSVERSE STRENGTH AT FORE AND AFT SECTION OF TANKER CARGO HOLD	74-0618
STRENGTH	ULTIMATE STRENGTH OF BOX GIRDERS BY FINITE ELEMENT METHOD	75-0794
STRENGTH	UNTERSUCHUNGEN ZUM FESTIGKEITSVERHALTEN VON METALLKLEBUNGEN IM HIMBLICK AUF DIE ERSTELLUNG EINES PRA	74-0289
STRENGTH	ZUR BERECHNUNG DER BIEGESTEIFIGKEIT ABGESETZTER STAEBE UND WELLEN UNTER ANWENDUNG VON FINITEN ELEMEN	75-0481
STRENGTHENING	SETTLEMENTS AND STRENGTHENING OF SOFT CLAY ACCELERATED BY SAND DRAINS	75-0438
STRENGTHS	ELASTIC-PLASTIC STRESS-STRAIN RELATIONS FOR MATERIALS WITH DIFFERENT YIELD STRENGTHS IN TENSION AND	72-0764
STRESS	A COMPUTER PROGRAM FOR THE CALCULATION OF TORSION FACTORS AND SHEAR STRESS FACTORS FOR BOX SECTIONS	69-0485
STRESS	A COMPUTER PROGRAM TO DISPLAY THE DYNAMIC STRESS ANALYSIS OF STRUCTURES	70-0077
STRESS	A CONSIDERATION ON THE FINITE ELEMENT METHOD FOR CALCULATING STRESS INTENSITY FACTOR KI USING THE J	75-0863
STRESS	A DEEP HOLE STRESS MEASUREMENT DEVICE	74-0804

STRESS	A DISCRETE ELEMENT METHOD FOR ANALYSIS OF PLANE ELASTIC-PLASTIC STRESS PROBLEMS	65-0063
STRESS	A DISCRETE ELEMENT METHOD FOR THE ANALYSIS OF PLANE ELASTO-PLASTIC STRESS PROBLEMS	66-0062
STRESS	A DISCRETE ELEMENT STRESS AND DISPLACEMENT ANALYSIS OF ELASTOPLASTIC PLATES	70-0108
STRESS	A FINITE ELEMENT FOR THERMAL STRESS ANALYSIS OF SHELLS OF REVOLUTION	73-0798
STRESS	A FINITE ELEMENT METHOD OF VISCOELASTIC STRESS ANALYSES WITH APPLICATION TO ROLLING CONTACT PROBLEMS	69-0330
STRESS	A FINITE ELEMENT STRESS ANALYSIS OF THE VERTICAL BUTTRESSES OF A NUCLEAR CONTAINMENT VESSEL	69-0280
STRESS	A FINITE ELEMENT STUDY OF ELASTIC-PLASTIC STRESS DISTRIBUTIONS IN NOTCHED SPECIMENS UNDER TENSION	68-0177
STRESS	A GENERAL PURPOSE THREE DIMENSIONAL STRESS ANALYSIS PROGRAM	71-0686
STRESS	A HYBRID FINITE ELEMENT FOR DETERMINATION OF STRESS INTENSITY FACTORS AT A CRACK TIP	73-0643
STRESS	A HYBRID STRESS ELEMENT FOR THIN SHELL ANALYSIS	74-0433
STRESS	A LATTICE ANALOGY FOR THE SOLUTION OF PLANE STRESS	43-0002
STRESS	A METHOD FOR DETERMINING LOCAL ELASTOPLASTIC STRESS AND STRAIN IN METALLURGICALLY BONDED NOTCHED LAM	74-0686
STRESS	A METHOD FOR SIMULATING CONCENTRATED FORCES AND LOCAL REINFORCEMENTS IN STRESS COMPUTATIONS	72-0081
STRESS	A METHOD OF ELASTIC-PLASTIC PLANE STRESS AND PLANE STRAIN ANALYSIS	66-0114
STRESS	A NEW ASSUMED STRESS HYBRID FINITE ELEMENT MODEL FOR SOLID CONTINUA	71-0371
STRESS	A NUMERICAL METHOD FOR THE ANALYSIS OF LONGITUDINAL ELASTIC PLASTIC STRESS WAVE PROPAGATION	73-0526
STRESS	A NUMERICAL METHOD OF VISCO-ELASTIC STRESS ANALYSIS	68-0175
STRESS	A POLYGONAL FINITE ELEMENT FOR PLATE BENDING PROBLEMS USING THE ASSUMED STRESS APPROACH	69-0043
STRESS	A QUADRILATERAL PLANE STRESS FINITE ELEMENT WITH QUADRATIC PROPERTIES	72-0969
STRESS	A SEISMIC DESIGN OF NUCLEAR REACTOR BUILDING, STRESS ANALYSIS AND STIFFNESS EVALUATION OF THE ENTIRE	71-0251
STRESS	A SIX NODE TRIANGULAR EQUILIBRIUM (CONSTANT STRESS) MEMBRANE ELEMENT	73-0783
STRESS	A STUDY OF THE PLANE STRESS OR STRAIN FINITE ELEMENT ANALYSIS FOR SOLUTION OF STRESS DISTRIBUTION IN	69-0487
STRESS	A STUDY OF THE PLANE STRESS OR STRAIN FINITE ELEMENT ANALYSIS FOR SOLUTION OF STRESS DISTRIBUTION IN	69-0487
STRESS	A STUDY ON THE STRESS ANALYSIS OF BEAM PLATE STRUCTURE	69-0404
STRESS	A SURVEY OF THE APPLICATION OF FINITE ELEMENT METHODS TO STRESS ANALYSIS PROBLEMS WITH PARTICULAR EM	71-0632
STRESS	A TECHNIQUE FOR THE SETTLEMENT AND STRESS ANALYSIS OF MAT FOUNDATIONS USING THE FINITE ELEMENT METHO	72-0950
STRESS	A TETRA-CORE STRESS ANALYSIS MODEL	72-0845
STRESS	A TRIANGULAR EQUILIBRIUM (LINEAR STRESS) MEMBRANE ELEMENT	75-0883
STRESS	A VARIATIONAL PRINCIPLE AND THE CONVERGENCE OF A FINITE-ELEMENT METHOD BASED ON ASSUMED STRESS DISTR	69-0260
STRESS	ACCURACY OF MODAL STRESS CALCULATIONS BY FINITE ELEMENT METHOD	70-0223
STRESS	ACTUAL STRESS OF HIGH HEAD PUMP-TURBINE RUNNER	75-0567
STRESS	ALTERNATE HYBRID STRESS FINITE ELEMENT MODELS	74-0476
STRESS	AN APPLICATION OF THE FINITE ELEMENT METHOD TO ELASTIC-PLASTIC PROBLEMS OF PLANE STRESS	70-0031
STRESS	AN APPROXIMATE METHOD FOR THERMO-VISCOELASTIC STRESS ANALYSIS	66-0100
STRESS	AN ASSUMED STRESS APPROACH TO REFINED ISOPARAMETRIC ELEMENTS IN THREE DIMENSIONS	74-0421
STRESS	AN ASSUMED STRESS HYBRID FINITE ELEMENT METHOD FOR AN INCOMPRESSIBLE AND NEAR INCOMPRESSIBLE MATERIA	69-0259
STRESS	AN ASSUMED STRESS HYBRID FINITE ELEMENT MODEL FOR THE ANALYSIS OF AN AXISYMMETRIC THICK WALLED PRESS	71-0025
STRESS	AN ASSUMED STRESS HYBRID FINITE ELEMENT FOR THREE-DIMENSIONAL PLASTIC STRUCTURAL ANALYSIS	74-0717
STRESS	AN ASSUMED STRESS VERSION OF THE WILSON 8-NODE ISOPARAMETRIC BRICK	72-0532
STRESS	AN EQUILIBRIUM STRESS FIELD MODEL FOR FINITE ELEMENT SOLUTIONS OF TWO-DIMENSIONAL ELASTIC PROBLEMS	68-0163
STRESS	AN EVALUATION OF FINITE ELEMENTS WITH IMPROVED ASSURED CONVERGENCE FOR PLANE STRESS AND PLANE STRAIN	74-0429
STRESS	AN INCREMENTAL COMPLEMENTARY ENERGY METHOD OF NONLINEAR STRESS ANALYSIS	69-0281
STRESS	AN INVESTIGATION INTO THE USE OF FINITE ELEMENT METHODS FOR ANALYSING STRESS DISTRIBUTIONS IN BLOCK	71-0530
STRESS	AN INVESTIGATION INTO THE ECONOMICS OF USING COMPLEX ELEMENTS IN PLANE STRESS ANALYSIS	74-0427
STRESS	ANALYSIS MODELS FOR PLATE ELEMENTS IN THE MEMBRANE STATE OF STRESS	67-0046
STRESS	ANALYSIS OF DISLOCATION STRESS FIELD BY FINITE ELEMENT METHOD	75-0862
STRESS	ANALYSIS OF ONE-DIMENSIONAL STRESS WAVE BY THE FINITE ELEMENT METHOD	71-0523
STRESS	ANALYSIS OF OUT-OF-PLANE STRESS BY THE SIX NODES, FOURTH ORDER ACCURACY TRIANGULAR ELEMENTS	69-0375
STRESS	ANALYSIS OF SAINT-VENANT TORSION PROBLEM BY A HYBRID STRESS MODEL	69-0277
STRESS	ANALYSIS OF STRESS AND STRAIN DISTRIBUTIONS AT CRACK TIP BY FINITE ELEMENT METHOD	69-0193
STRESS	ANALYSIS OF STRESS AND STRAIN DISTRIBUTIONS IN TRIAXIAL TESTS USING THE METHOD OF FINITE ELEMENTS	73-0903
STRESS	ANALYSIS OF STRESS DISTRIBUTION IN SHEAR WALLS BY THE FINITE ELEMENT DISPLACEMENT METHOD	67-0033
STRESS	ANALYSIS OF THERMAL ELASTIC PLASTIC STRESS AND STRAIN DURING WELDING BY FINITE ELEMENT METHOD	71-0382
STRESS	ANALYSIS OF WELDING STRESS RELIEVING BY ANNEALING BASED ON FINITE ELEMENT METHOD	75-0777
STRESS	ANCHOR ZONE STRESS ANALYSIS BY FINITE ELEMENT METHOD	68-0147
STRESS	ANISOTROPIC WORK HARDENING IN METALS. FINITE ELEMENT APPLICATION TO PLANE STRESS AND PLANE STRAIN	73-0950
STRESS	ANTRAS COMPUTER PROGRAM FOR LINEAR STRESS ANALYSIS BY FINITE ELEMENT METHOD	74-0601
STRESS	APPLICATION OF FINITE ELEMENT TECHNIQUES TO STRESS AND STRUCTURAL ANALYSIS	70-0362
STRESS	APPLICATION OF FINITE ELEMENT METHOD TO THE THREE DIMENSIONAL STRESS AND VIBRATION ANALYSIS OF BWR P	71-0180
STRESS	APPLICATION OF FINITE ELEMENT ELASTIC-PLASTIC STRESS ANALYSIS TO NOTCHED FATIGUE SPECIMEN BEHAVIOUR	71-0185
STRESS	APPLICATION OF FINITE ELEMENT METHOD – STRESS CALCULATIONS ON 90 DEGREES NOZZLES OF THIN-WALLED CYLI	71-0253
STRESS	APPLICATION OF FINITE ELEMENT METHODS TO STRESS ANALYSIS	73-0524
STRESS	APPLICATION OF FRACTURE MECHANICS TO GRAPHITE UNDER COMPLEX STRESS CONDITIONS	72-0771
STRESS	APPLICATION OF GRID SELECTION PROCEDURES FOR IMPROVED FINITE ELEMENT STRESS ANALYSIS	74-0326
STRESS	APPLICATION OF THE BOUNDARY-INTEGRAL EQUATION METHOD TO THREE DIMENSIONAL STRESS ANALYSIS	73-0252
STRESS	APPLICATION OF THE FINITE ELEMENT METHOD TO STRESS ANALYSIS OF SOLID PROPELLANT ROCKET GRAINS	65-0009
STRESS	APPLICATION OF THE FINITE ELEMENT METHOD TO STRESS DISTRIBUTION IN SOIL MEDIA	67-0075
STRESS	APPLICATION OF THE FINITE ELEMENT TECHNIQUE TO THE STUDY OF AREAS OF HIGH STRESS GRADIENTS WITHIN A	73-0259
STRESS	APPLICATION TO PLANE STRESS AND PLANE STRAIN--LINEAR ELASTIC MATERIAL	68-0048
STRESS	APPLICATIONS OF FINITE-ELEMENT ANALYSIS TO NONLINEAR PLANE STRESS PROBLEMS	72-0976
STRESS	APPLICATIONS OF FINITE ELEMENT STRESS ANALYSIS AND STRESS-STRAIN PROPERTIES IN DETERMINING NOTCH FAT	73-1114
STRESS	APPLICATIONS OF PULSED LASER INDUCED STRESS WAVES	73-0956
STRESS	APPLICATIONS OF THE FINITE ELEMENT METHOD TO THE STRESS ANALYSIS OF MATERIALS SUBJECT TO CREEP	71-0548
STRESS	APPLICATIONS OF THE FINITE ELEMENT METHOD TO THE CALCULATIONS OF STRESS INTENSITY FACTORS	72-0900
STRESS	APPROXIMATE THREE-DIMENSIONAL STRESS SOLUTIONS FOR SOME LAMINATED PLATE PROBLEMS USING A HIGHER ORDE	70-0410
STRESS	ASAAS: ASYMMETRIC STRESS ANALYSIS OF AXISYMMETRIC SOLIDS WITH ORTHOTROPIC, TEMPERATURE-DEPENDENT MAT	71-0696
STRESS	ASSUMED STRESS HYBRID FINITE ELEMENT METHOD FOR FRACTURE MECHANICS AND ELASTIC-PLASTIC ANALYSIS	72-0584
STRESS	ASSUMED STRESS HYBRID FINITE ELEMENT MODEL FOR LINEAR ELASTODYNAMIC ANALYSIS	73-0089
STRESS	AUTOMATED STRESS ANALYSIS USING SUBSTRUCTURES	66-0071
STRESS	AUTOMATIC EXTRACTION OF RIGID BODY MODES FROM STRESS AND STRAIN ELEMENTS	72-0913
STRESS	AVERAGE ELASTIC PROPERTIES AND STRESS FIELDS FOR COMPOSITE CONTINUA	69-0106
STRESS	AVERAGE ELASTIC PROPERTIES AND STRESS FIELDS FOR COMPOSITE MATERIALS	69-0435
STRESS	AXISYMMETRIC RING AND PLATE STRESS ANALYSIS PROGRAM - USERS GUIDE	70-0333
STRESS	BASIS FOR ISOPARAMETRIC STRESS ELEMENTS	73-0532
STRESS	BERSAFE - A COMPUTER SYSTEM FOR STRESS ANALYSIS BY FINITE ELEMENTS	70-0212
STRESS	BOUNDING PLANE STRESS SOLUTIONS BY FINITE ELEMENTS	66-0055
STRESS	BOUNDS TO THE INFLUENCE COEFFICIENTS BY THE ASSUMED STRESS METHOD	70-0165
STRESS	BOX CAR BODY BOLSTER PROBED IN STRESS STUDY	74-0654
STRESS	BRLESC FINITE ELEMENT PROGRAM FOR AXISYMMETRIC, PLANE STRAIN, AND PLANE STRESS, ORTHOTROPIC SOLIDS W	71-0749
STRESS	CALCULATION OF STRESS INTENSITY FACTORS USING FINITE ELEMENT METHOD WITH CRACKED ELEMENTS	70-0224
STRESS	CALCULATION OF STRESS INTENSITY FACTORS USING THE FINITE ELEMENT METHOD WITH CRACKED ELEMENTS	70-0475
STRESS	CALCULATION OF STRESS CONCENTRATION DUE TO A HYPERBOLOID CAVITY IN A THIN PLATE USING NASTRAN	71-0260
STRESS	CALCULATION OF STRESS INTENSITY FACTORS AT CRACK TIPS USING SPECIAL FINITE ELEMENTS	72-0071
STRESS	CALCULATION OF STRESS INTENSITY FACTORS AND FATIGUE CRACK PROPAGATION OF SEMI-ELLIPTICAL, PART-THROU	73-0360
STRESS	CALCULATION OF STRESS CONCENTRATIONS ON SUPPORT STRUCTURES UNDER THERMOMECHANICAL LOADING (BERECHNUN	73-0796
STRESS	CALCULATION OF STRESS INTENSITY FACTORS FOR COMBINED TENSILE AND SHEAR LOADING	75-0205
STRESS	CALCULATION OF STRESS INTENSITY FACTORS IN TWO AND THREE DIMENSIONS USING FINITE ELEMENTS	75-0254
STRESS	CALCULATION OF STRESS INTENSITY FACTORS IN ANISOTROPIC MATERIALS BY A CONTOUR INTEGRAL METHOD	75-0257
STRESS	CAPE-2D, A COMPUTER PROGRAM FOR THE STRESS ANALYSIS OF PLANE AND AXISYMMETRIC COMPOSITE STRUCTURES.	69-0083
STRESS	CASE BOND STRESS CALCULATIONS FOR FLAPPED CYLINDRICAL ANALOGS OF SOLID PROPELLANT ROCKET MOTORS. BON	72-0842
STRESS	CENTRIFUGAL MODEL STUDY AND FINITE ELEMENT ANALYSIS OF STRESS NEAR A FAULT	67-0159
STRESS	CLASSIFICATION OF FINITE ELEMENT STRESSES ACCORDING TO ASME SECTION 111 STRESS CATEGORIES	74-0610

STRESS	CODE FOR STRESS ANALYSIS BY THE THREE DIMENSIONAL FINITE ELEMENT METHOD (JAPANESE)	69-0509
STRESS	COMPARATIVE CONSIDERATION OF PARAMETRIC STRESS CONCENTRATION INVESTIGATIONS WITH FINITE ELEMENTS AND	72-0423
STRESS	COMPARATIVE CONSIDERATION OF PARAMETRIC STRESS CONCENTRATION INVESTIGATIONS WITH FINITE ELEMENTS AND	72-0423
STRESS	COMPARATIVE OBSERVATION ON PARAMETRIC STRESS STUDIES WITH FINITE ELEMENTS WITH COMPLEX STRESS FUNCTI	72-0250
STRESS	COMPARATIVE OBSERVATION ON PARAMETRIC STRESS STUDIES WITH FINITE ELEMENTS WITH COMPLEX STRESS FUNCTI	72-0250
STRESS	COMPARISON OF NUMERICAL METHODS FOR DETERMINING STRESS INTENSITY FACTORS	72-0176
STRESS	COMPARISON OF THEORETICAL AND EXPERIMENTAL SHEAR STRESS IN THE ADHESIVE LAYER OF A LAP JOINT MODEL	74-1176
STRESS	COMPUTATION OF STRESS RESULTANTS IN FINITE ELEMENT ANALYSIS	66-0056
STRESS	COMPUTATION OF STRESS RESULTANTS FROM ELEMENT STIFFNESS MATRICES	66-0125
STRESS	COMPUTATION OF STRESS INTENSITY FACTORS BY A SPECIAL FINITE ELEMENT TECHNIQUE	71-0287
STRESS	COMPUTATIONAL METHODS IN SOLID MECHANICS AND STRESS ANALYSIS	69-0006
STRESS	COMPUTATIONAL PROCEDURES FOR INELASTIC STRESS ANALYSIS	73-0859
STRESS	COMPUTER PROGRAM FOR PIT SLOPE STABILITY ANALYSIS BY THE FINITE ELEMENT STRESS ANALYSIS AND LIMITING	72-0225
STRESS	COMPUTER SIMULATION OF DYNAMIC STRESS, DEFORMATION, AND FRACTURE OF GEAR TEETH	72-0114
STRESS	COMPUTER STRESS ANALYSIS	68-0099
STRESS	CONFERENCE ON APPLICATION OF FINITE ELEMENT METHODS TO STRESS ANALYSIS PROBLEMS IN NUCLEAR ENGINEERI	71-0774
STRESS	CONICAL FINITE ELEMENT STRESS ANALYSIS OF AN AXISYMMETRIC TANK ON AN ELASTIC FOUNDATION	70-0494
STRESS	CONIDA: A FINITE ELEMENT PROGRAM FOR THE STRESS ANALYSIS OF AXISYMMETRIC THIN SHELLS	74-1087
STRESS	CONVERGENCE OF THE NASTRAN PLATE ELEMENTS FOR SHELL STRESS ANALYSIS	71-0261
STRESS	CORRELATION BETWEEN PRESSURE AND THERMAL TRANSIENT STRESS LEVELS DERIVED BY THEORY AND EXPERIMENT ON	73-0469
STRESS	CREEP STRESS AND STRAIN ANALYSIS OF NOTCHED PLATES AND BARS BY MEANS OF THE FINITE ELEMENT METHOD	70-0141
STRESS	DAS PROGRAMMSYSTEM ANTRAS EDV-SYSTEM ZUR BERECHNUNG BELIEBIGER STATISCHER SYSTEME NACH DER METHODE D	74-0918
STRESS	DER SPANNUNGS UND BEANSPRUCHUNGSZUSTAND DES BODENMATERIALS IN DER NACHBARSCHAFT EINER BODENBOESCHUNG	75-0371
STRESS	DERIVATION OF ELEMENT STIFFNESS MATRICES BY ASSUMED STRESS DISTRIBUTIONS	64-0007
STRESS	DETERMINATION OF ENERGY RELEASE RATES AND STRESS - INTENSITY FACTORS BY THE FINITE-ELEMENT METHOD	72-0183
STRESS	DETERMINATION OF STRESS INTENSITY FACTORS IN CRACKED PLATES BY THE FINITE ELEMENT METHOD	73-0148
STRESS	DETERMINATION OF STRESS CONCENTRATIONS AT NOZZLE TO CYLINDER INTERSECTIONS AND COMPARISON WITH EXPER	73-0491
STRESS	DETERMINATION OF THE INITIAL BUCKLING STRESS OF FLAT PLATES USING THE DISPLACEMENT METHOD	69-0465
STRESS	DETERMINATION OF THE LOWEST CRITICAL BUCKLING STRESS OF VARIOUS SHAPED PLATES BY FINITE-ELEMENTS	73-1134
STRESS	DETERMINATION OF UPPER BOUNDS FOR PROBLEMS IN PLANE STRESS USING FINITE ELEMENT TECHNIQUES	67-0034
STRESS	DEVELOPMENT AND APPLICATION OF HYBRID STRESS FINITE ELEMENTS TO AN AUTOMOTIVE STRUCTURE	74-1221
STRESS	DEVELOPMENT AND EVALUATION OF A TRIANGULAR THIN SHELL ELEMENT BASED UPON THE HYBRID ASSUMED STRESS M	69-0252
STRESS	DIE BERECHNUNG DER BEWEGUNGEN UND SPANNUNGEN DES GEBIRGES VOM ABBAU BIS ZUR TAGESOBERFLAECHE NACH DE	73-0456
STRESS	DISCONTINUITY STRESS DECAY IN OFFSHORE PLATFORM JOINTS	74-0491
STRESS	DISCONTINUITY STRESS ANALYSIS OF PRESSURE-VESSELS USING FINITE-ELEMENT METHOD	75-0346
STRESS	DISCUSSION ON "PLANE STRESS LIMIT ANALYSIS BY FINITE ELEMENTS"	71-0792
STRESS	DISTRIBUTION OF STRESS AROUND A TUNNEL CAVITY, USING THE FINITE ELEMENT METHOD	71-0843
STRESS	DYNAMIC FINITE ELEMENT ANALYSIS OF AXI-SYMMETRIC STRESS WAVE PROPAGATION IN SOIL-FILLED BIN	72-0622
STRESS	DYNAMIC SHELL THEORY COUPLING THICKNESS STRESS WAVE EFFECTS WITH GROSS STRUCTURAL RESPONSE	73-0379
STRESS	DYNAMIC STRESS ANALYSIS OF AXISYMMETRIC STRUCTURES UNDER ARBITRARY LOADING	69-0433
STRESS	DYNAMIC STRESS ANALYSIS OF POROUS ELASTIC SOLIDS SATURATED WITH COMPRESSIBLE FLUIDS	71-0651
STRESS	EARTHQUAKE STRESS ANALYSIS IN EARTH DAMS	65-0043
STRESS	EFFECT OF ELASTIC MODULAR RATIO AND ANISOTROPY ON THE STRESS FIELD IN A TWO-DIMENSIONAL MATRIX INTO	71-0647
STRESS	EFFECT OF FIBRE DISTRIBUTION ON STRESS AND STRAIN CONCENTRATION AT HOLES IN COMPOSITE PLATES	72-0616
STRESS	EFFECT OF FILLETS ON STRESS CONCENTRATIONS IN CYLINDRICAL SHELLS WITH STEP CHANGES IN OUTSIDE DIAMET	71-0108
STRESS	EFFECT OF PLASTICITY AND CRACK BLUNTING ON THE STRESS DISTRIBUTION IN ORTHOTROPIC COMPOSITE MATERIAL	73-0145
STRESS	EFFECTS OF A PEAK OVERLOAD ON FATIGUE CRACK PROPAGATION RATE-PLANE STRESS ANALYSIS OF CRACKED PLATE	74-1365
STRESS	EFFECTS OF SPECIMEN RADIUS ON THE STRESS STATE NEAR A ROOF BOLT ANCHOR: A FINITE ELEMENT DETERMINATI	73-0229
STRESS	EFFICIENT FINITE ELEMENT METHODS FOR STRESS INTENSITY FACTOR	75-0148
STRESS	EFFICIENT FINITE-ELEMENT METHODS FOR STRESS INTENSITY FACTORS USING WEIGHT FUNCTIONS	75-0411
STRESS	EFFICIENT FINITE-ELEMENT METHODS FOR STRESS INTENSITY FACTORS USING WEIGHT FUNCTIONS - REPLY	75-0421
STRESS	ELASTIC ANALYSIS OF CYLINDRICAL CONFIGURATIONS WITH STRESS SINGULARITIES	71-0325
STRESS	ELASTIC AND PLASTIC STRESS DISTRIBUTION ON BASE PLATES IN THE NEIGHBOURHOOD OF FILLET WELDS IN LAP J	70-0175
STRESS	ELASTIC AND PLASTIC INTERLAMINAR SHEAR DEFORMATION IN LAMINATED COMPOSITES UNDER GENERALIZED PLANE	71-0158
STRESS	ELASTIC DEFORMATION OF POLYCRYSTALLINE METAL - 1. INFLUENCE OF GRAIN SHAPE ON DISTRIBUTIONS OF STRES	72-0463
STRESS	ELASTIC EQUILIBRIUM MODELS, STRESS FUNCTIONS, AND DUALITY NON _INEARITIES AND STABILITY, FIVE LECTUR	73-0631
STRESS	ELASTIC PLASTIC ANALYSIS OF SAINT-VENANT TORSION PROBLEM BY A HYBRID STRESS MODEL	72-0385
STRESS	ELASTIC PLASTIC ANALYSIS BY ASSUMED STRESS HYBRID MODEL	74-0443
STRESS	ELASTIC-PLASTIC ANALYSIS OF TWO-DIMENSIONAL STRESS SYSTEMS BY THE FINITE ELEMENT METHOD	67-0049
STRESS	ELASTIC-PLASTIC AND THERMAL STRESS ANALYSIS	68-0169
STRESS	ELASTIC-PLASTIC DEFORMATIONS OF FLEXURALLY STIFF FRAMEWORKS FROM SECOND ORDER STRESS THEORY	73-1061
STRESS	ELASTIC-PLASTIC STRESS ANALYSIS FOR A NOTCHED BAR IN PLANE STRAIN BENDING	71-0410
STRESS	ELASTO PLASTIC STRESS IN CRACKED PLATES	65-0042
STRESS	ELASTO-PLASTIC ANALYSIS OF STRESS IN A GAS-TURBINE WHEEL	74-0323
STRESS	ELASTO-PLASTIC SOLUTIONS OF ENGINEERING PROBLEMS ' INITIAL STRESS ' FINITE ELEMENT APPROACH	69-0282
STRESS	ELASTO-PLASTIC STRESS ANALYSIS. A GENERALIZATION FOR VARIOUS CONSTITUTIVE RELATIONS INCLUDING STRAIN	72-0547
STRESS	ELASTODYNAMIC BEHAVIOR OF LAMINATED ORTHOTROPIC PLATES UNDER INITIAL STRESS	75-0249
STRESS	ELEMENT STIFFNESS MATRICES FOR CONSTANT AND LINEAR STRESS - TRIANGULAR, CONSTANT AND LINEAR STRESS T	67-0006
STRESS	ELEMENT STIFFNESS MATRICES FOR CONSTANT AND LINEAR STRESS - TRIANGULAR, CONSTANT AND LINEAR STRESS T	67-0006
STRESS	ELEMENTMETODEN OG DEFORMATIONSMETODEN (FINITE ELEMENT AND DEFORMATION METHODS IN STRESS ANALYSIS)	74-0580
STRESS	ENGINEERING OPERATIONS REPORT: NERVA THIN SHELL FINITE ELEMENT STRESS ANALYSIS METHODOLOGY COMPUTER	72-0916
STRESS	ENGINEERING OPERATIONS REPORT: NERVA AXISYMMETRIC AND PLANE STRUCTURES FINITE ELEMENT STRESS ANALYSI	72-0917
STRESS	ENTWURF UND SPANNUNGSBERECHNUNG EINES SPANNBETON DRUCKBEHAELTERS AUS FERTIGTEILEN. (DESIGN AND STRES	73-0856
STRESS	EQUILIBRIUM STRESS FIELD MODELS IN FINITE ELEMENTS	67-0011
STRESS	EVALUATION OF A THREE-DIMENSIONAL STRESS ANALYSIS PROGRAM	72-0743
STRESS	EVALUATION OF A THREE-DIMENSIONAL STRESS CELL FOR GRANULAR SOILS	72-0882
STRESS	EVALUATION OF FINITE ELEMENT METHODS FOR THE COMPUTATION OF ELASTIC STRESS INTENSITY FACTORS	72-0249
STRESS	EVALUATION OF STRESS CELL PERFORMANCE	74-0664
STRESS	EXAMPLES OF STRESS ANALYSIS OF STRUCTURES BY THE FINITE ELEMENT METHOD. (JAPANESE)	70-0450
STRESS	EXPERIENCE IN THE USE OF THE CONSTANT STRESS TRIANGLE FOR FINITE ELEMENT STRESS ANALYSIS	69-0407
STRESS	EXPERIENCE IN THE USE OF THE CONSTANT STRESS TRIANGLE FOR FINITE ELEMENT STRESS ANALYSIS	69-0407
STRESS	EXPERIENCES FROM STRESS ANALYSIS OF AXI-SYMMETRIC PROBLEMS IN MACHINE DESIGN	69-0247
STRESS	EXPERIMENTAL AND FINITE ELEMENT STRESS ANALYSIS OF A THIN SHELLED CYLINDER TO CYLINDER MODEL	71-0066
STRESS	EXPERIMENTAL AND THEORETICAL INVESTIGATION OF PLANE STRESS FRACTURE OF 2024 - T351 ALUMINUM ALLOY	70-0147
STRESS	EXPERIMENTAL ELASTIC STRESS ANALYSES OF CYLINDER-TO-CYLINDER SHELL MODELS AND COMPARISONS WITH THEOR	71-0065
STRESS	FACTORS OF STRESS CONCENTRATION AT A HOLE IN FINITE STRIPS UNDER BENDING	72-0744
STRESS	FEATS, A COMPUTER PROGRAM FOR THE FINITE ELEMENT THERMAL STRESS ANALYSIS OF PLANE OR AXISYMMETRIC SO	68-0152
STRESS	FIESTA - FINITE ELEMENT STRESS AND TEMPERATURE ANALYSIS. A DESCRIPTION OF THE THEORY USED FOR THE CO	71-0718
STRESS	FINITE ELEMENT ANALYSIS OF TWO-DIMENSIONAL TIME-DEPENDENT STRESS PROBLEMS	65-0022
STRESS	FINITE ELEMENT ANALYSIS OF FREE VIBRATION AND STABILITY OF STRUCTURES SUBJECTED TO INITIAL STRESS	68-0300
STRESS	FINITE ELEMENT ANALYSIS OF ELASTIC-PLASTIC PLANE STRESS PROBLEMS BASED UPON TRESCA YIELD CRITERION	70-0179
STRESS	FINITE ELEMENT ANALYSIS OF NOTCHED TENSILE SPECIMENS IN PLANE STRESS	71-0557
STRESS	FINITE ELEMENT ANALYSIS OF STRESS CONCENTRATIONS AND SINGULARITIES	72-0841
STRESS	FINITE ELEMENT ANALYSIS OF DENTAL STRUCTURES - AXISYMMETRIC AND PLANE STRESS IDEALIZATIONS	75-0525
STRESS	FINITE ELEMENT APPROACH TO STRESS WAVE PROBLEMS	67-0015
STRESS	FINITE ELEMENT APPROACH TO OPTIMAL DESIGN OF PLASTIC STRUCTURES IN PLANE STRESS.	72-0397
STRESS	FINITE ELEMENT APPROACH TO THE CRITICAL CYCLIC STRESS REQUIRED TO PROPAGATE A CRACK	73-0463
STRESS	FINITE ELEMENT CALCULATION OF LEFT VENTRICULAR WALL STRESS IN NORMAL AND DISEASE STATES	73-0588
STRESS	FINITE ELEMENT CALCULATION OF STRESS INTENSITY FACTORS USING ENERGY TECHNIQUES	73-0825
STRESS	FINITE ELEMENT CALCULATION OF STRESS DISTRIBUTIONS IN ARCHES	74-0094
STRESS	FINITE ELEMENT CALCULATION OF STRESS INTENSITY FACTORS USING SUPERPOSITION	75-0251
STRESS	FINITE ELEMENT FORMULATION OF STRESS STRAIN MATRIX FOR AN ELASTICALLY PERFECTLY PLASTIC TRESCA MATER	70-0275

STRESS	FINITE ELEMENT MATRIX DISPLACEMENT METHOD FOR TIME DEPENDENT STRESS ANALYSIS OF REACTOR MATERIALS IN	71-0674
STRESS	FINITE ELEMENT METHODS IN STRESS ANALYSIS	69-0126
STRESS	FINITE ELEMENT METHOD OF PREDICTING STRESS DISTRIBUTIONS AND SOIL DEFORMATION UNDER A TRACTIVE DEVIC	69-0389
STRESS	FINITE ELEMENT METHODS IN STRESS ANALYSIS	69-0410
STRESS	FINITE ELEMENT METHOD IN STRESS ANALYSIS	69-0413
STRESS	FINITE ELEMENT METHOD FOR STRESS ANALYSIS OF BODIES OF REVOLUTION SUBJECTED TO AXISYMMETRIC LOADING	71-0409
STRESS	FINITE ELEMENT METHOD FOR CALCULATING STRESS INTENSITY FACTORS FOR CRACKED PLATES IN BENDING	71-0424
STRESS	FINITE ELEMENT METHODS IN STRESS ANALYSIS	71-0576
STRESS	FINITE ELEMENT METHOD - STRESS ANALYSIS AND MUCH MORE	74-0095
STRESS	FINITE ELEMENT METHOD FOR CALCULATING STRESS INTENSITY FACTORS AND ITS APPLICATION TO COMPOSITES	74-1000
STRESS	FINITE ELEMENT STRESS ANALYSIS OF THIN SHELLS	65-0024
STRESS	FINITE ELEMENT STRESS ANALYSIS OF AXISYMMETRIC SOLIDS WITH ORTHOTROPIC TEMPERATURE DEPENDENT MATERIA	67-0089
STRESS	FINITE ELEMENT STRESS ANALYSIS OF ANISOTROPIC BODIES	68-0033
STRESS	FINITE ELEMENT STRESS ANALYSIS OF PLANE AND AXISYMMETRIC STRUCTURES	69-0062
STRESS	FINITE ELEMENT STRESS ANALYSIS OF POROUS MEDIA	69-0440
STRESS	FINITE ELEMENT STRESS ANALYSIS FOR ANISOTROPIC SOLIDS	70-0478
STRESS	FINITE ELEMENT STRESS ANALYSIS OF TURBINE BLADE HOOKS	70-0479
STRESS	FINITE ELEMENT STRESS ANALYSIS, PART 2	71-0209
STRESS	FINITE ELEMENT STRESS ANALYSIS PART 1	71-0440
STRESS	FINITE ELEMENT STRESS FORMULATION FOR WAVE PROPAGATION	71-0831
STRESS	FINITE ELEMENT STRESS ANALYSIS OF A CRACK IN A BI-MATERIAL PLATE	72-0165
STRESS	FINITE ELEMENT STRESS ANALYSIS OF A RESTORED AXISYMMETRIC FIRST MOLAR	73-0053
STRESS	FINITE ELEMENT STRESS ANALYSIS OF A MULTICONDUCTOR ARMOR CABLE	73-0095
STRESS	FINITE ELEMENT STRESS ANALYSIS OF ARTILLERY PROJECTILES	73-0432
STRESS	FINITE ELEMENT STRESS ANALYSIS OF INTERACTING FUEL PELLET AND CANNING UNDER FAST REACTOR CONDITION	73-0814
STRESS	FINITE ELEMENT STRESS FIELD SOLUTION OF THE PROBLEM OF SAINT VENANT TORSION	73-1042
STRESS	FINITE ELEMENT STRESS FORMULATION FOR WAVE PROPAGATION	73-1043
STRESS	FINITE ELEMENT STRESS ANALYSIS OF AN INTERVERTEBRAL DISC	74-0244
STRESS	FINITE ELEMENT STRESS ANALYSIS OF PROSTHETIC TOOTH IMPLANTS	75-0073
STRESS	FINITE ELEMENT STRESS FORMULATION FOR DYNAMIC ELASTIC-PLASTIC ANALYSIS	75-0484
STRESS	FINITE ELEMENT STUDY OF THE TRIAXIAL STRESS STATE AROUND AN INSPECTION TUNNEL IN AN ARCH DAM	74-0831
STRESS	FINITE ELEMENT TECHNIQUES FOR ORTHOTROPIC PLANE STRESS AND ORTHOTROPIC PLATE ANALYSIS	68-0274
STRESS	FINITE ELEMENT TREATMENT OF SINGULARITIES OF BOUNDARY VALUE PROBLEMS AND ITS APPLICATION TO ANALYSIS	73-0306
STRESS	FINITE ELEMENT VISCOELASTIC STRESS ANALYSIS STUDY USING THE GENERALIZED KELVIN SOLID MODEL	73-0957
STRESS	FINITE ELEMENTS FOR DETERMINATION OF CRACK TIP ELASTIC STRESS INTENSITY FACTORS	71-0436
STRESS	FINITE ELEMENTS FOR DETERMINATION OF CRACK-TIP ELASTIC STRESS INTENSITY FACTORS	71-0785
STRESS	FINITE-ELEMENT FORMULATION OF MIXED AND STRESS VARIATIONAL PRINCIPLES FOR STRUCTURAL DYNAMIC ANALYSI	72-0964
STRESS	FINITE-ELEMENT STRESS ANALYSIS OF A MULTIPIECE PISTON	69-0274
STRESS	FINITE-ELEMENT STRESS ANALYSIS OF A NON-LINEAR PROBLEM. A CONNECTING-ROD EYE LOADED BY MEANS OF A PI	70-0239
STRESS	FLAT FINITE ELEMENT FOR THIN SHELL ANALYSIS DERIVED BY ASSUMED STRESS APPROACH	73-0147
STRESS	FORMULATION OF DISCRETE ELEMENT MODELS FOR STRESS AND VIBRATION ANALYSIS OF PLATES	70-0328
STRESS	FORMULATION OF STRESS STRAIN INDUCED DIFFUSION OF HYDROGEN AND ITS SOLUTION BY COMPUTER-AIDED FINITE	75-0326
STRESS	FRACTURE SURFACE ENERGIES BY FINITE ELEMENT STRESS ANALYSIS	73-0915
STRESS	FUEL ROD FINITE ELEMENT STRESS ANALYSIS: SEGMENTED FLIGHT REACTOR FUEL VERSUS S8DR FUEL	71-0798
STRESS	GENERAL POTENTIAL ENERGY AND COMPLEMENTARY ENERGY MODELS BASED ON STRESS PARAMETERS	74-1240
STRESS	GENERAL PURPOSE PROGRAM OF PLANE STRESS ANALYSIS BY FINITE ELEMENT METHOD, AND ITS APPLICATIONS	72-0186
STRESS	GENERALIZED HYBRID STRESS FINITE-ELEMENT MODELS	73-0027
STRESS	GENERALIZED STRESS MODELS FOR FINITE ELEMENT ANALYSIS	74-1067
STRESS	HEAT CONDUCTION AND STRESS ANALYSIS OF SOLID PROPELLANT ROCKET MOTOR NOZZLES	69-0059
STRESS	HIGHER-ORDER FINITE ELEMENT FOR PLANE STRESS	67-0082
STRESS	HIGHER-ORDER FINITE ELEMENT FOR PLANE STRESS	68-0089
STRESS	HYBRID FINITE ELEMENTS IN THE PLANE STRESS PROBLEM	74-1071
STRESS	IMPROVED APPROXIMATION OF CONSTITUTIVE ELASTO-PLASTIC STRESS STRAIN RELATIONSHIP FOR FINITE ELEMENT	74-0566
STRESS	IMPROVED FINITE ELEMENT METHOD FOR STRESS CONCENTRATIONS	73-0127
STRESS	IMPROVED THERMAL STRESS DETERMINATION BY FINITE ELEMENT METHODS	74-0247
STRESS	INEXPENSIVE, INTERACTIVE COMPUTER-AIDED DESIGN APPLIED TO FINITE ELEMENT STRESS ANALYSIS	74-0592
STRESS	INFLUENCE OF FLUID INJECTION ON THE STATE OF STRESS IN THE EARTH'S CRUST	72-0624
STRESS	INFLUENCE OF NON-SINGULAR STRESS TERMS AND SPECIMEN GEOMETRY ON SMALL SCALE YIELDING AT CRACK TIPS I	73-0203
STRESS	INTERNAL SUPPORTING STRUCTURES OF THE (CIRENE) REACTOR CORE - RESULTS OF STATIC STRESS ANALYSIS FOR	73-0818
STRESS	INTERPRETATION OF COMPLEX STRESS ANALYSIS BY MULTIPLE REGRESSION	74-1203
STRESS	INTERPRETATION OF FINITE ELEMENT PROCEDURE AS STRESS ERROR MINIMIZATION PROCEDURE	72-0984
STRESS	INVESTIGATION OF TECHNIQUES OF THREE-DIMENSIONAL FINITE ELEMENT STRESS ANALYSIS	70-0029
STRESS	ISSLEDOVANIE NAPRYAZHENNO-DEFORMIROVANNOGO SOSTOYANIYA SIMMETRICHNYKH DISKOV SO STUPITSEI. (INVESTIG	74-0965
STRESS	ITERATIVE TECHNIQUE FOR THE STRESS ANALYSIS OF PCR STRUCTURES USING THE FINITE ELEMENT METHOD	64-0026
STRESS	LEAST SQUARES SURFACE FITTING BY FINITE ELEMENTS AND AN APPLICATION TO STRESS SMOOTHING	67-0154
STRESS	LOCAL LEAST SQUARES STRESS SMOOTHING FOR PARABOLIC ISOPARAMETRIC ELEMENTS	75-0593
STRESS	LOWER BOUND LIMIT ANALYSIS OF PLANE STRESS PROBLEMS BY A FINITE ELEMENT METHOD	70-0428
STRESS	MATHEMATICAL ANALYSIS OF STRESS CRACKS. VOLUME 2.	75-0742
STRESS	MATHEMATICAL ANALYSIS OF STRESS CRACKS. VOLUME 1	75-0743
STRESS	MEASUREMENTS OF HEALING AT AN OSTEOTOMY IN A RABBIT CALVARIUM: THE INFLUENCE OF APPLIED COMPRESSIVE	74-0975
STRESS	METHOD OF EVALUATION OF THREE-DIMENSIONAL STRESS INTENSITY FACTOR USING THE FINITE ELEMENT METHOD	75-0316
STRESS	MICA DAM EMBANKMENT STRESS ANALYSIS	75-0054
STRESS	MINUTEMAN STV STRESS ANALYSIS AND TESTING	74-0809
STRESS	MODEL EXPERIMENTS AND FINITE ELEMENT ANALYSIS OF STRESS IN AN OPEN SHIP	66-0054
STRESS	MODERN DEVELOPMENTS IN THE STRESS ANALYSIS OF PRESSURE VESSELS	69-0201
STRESS	MULTIPLE STRESS WAVE DISCONTINUITIES IN FINITE BEAMS AND PLATES	67-0114
STRESS	NON-LINEAR STRESS ANALYSIS OF TWO DIMENSIONAL PROBLEMS WITH SPECIAL REFERENCE TO ROCK AND SOIL MECHA	68-0183
STRESS	NON-LINEAR THERMAL STRESS ANALYSIS FOR NUCLEAR POWER PLANT BY FINITE ELEMENT METHOD	71-0181
STRESS	NONLINEAR ANALYSES OF THE STRESS FIELD NEAR A CRACK OR HOLE IN A PLATE	75-0694
STRESS	NONLINEAR ANALYSIS BY ASSUMED STRESS HYBRID MODELS	74-1004
STRESS	NONLINEAR ANALYSIS OF STRESS AND STRAIN IN SOILS	70-0050
STRESS	NONLINEAR CREEP ANALYSIS BY ASSUMED STRESS FINITE ELEMENT METHODS	74-0631
STRESS	NONLINEAR CREEP ANALYSIS BY ASSUMED STRESS FINITE ELEMENT METHODS	74-0922
STRESS	NONLINEAR HYBRID STRESS FINITE ELEMENT ANALYSIS OF LAMINATED SHELLS	74-1350
STRESS	NONLINEAR HYBRID STRESS FINITE ELEMENT ANALYSIS OF PNEUMATIC TIRES	75-0029
STRESS	NONLINEAR STRESS ANALYSIS OF INDETERMINATE STRUCTURAL ELEMENTS SUBJECTED TO BIAXIAL STRESS STATES US	73-1152
STRESS	NONLINEAR STRESS ANALYSIS OF INDETERMINATE STRUCTURAL ELEMENTS SUBJECTED TO BIAXIAL STRESS STATES US	73-1152
STRESS	NONLINEAR STRESS DEFORMATION ANALYSES OF LAKE AGASSIZ CLAYS USING FINITE ELEMENT METHOD	74-1072
STRESS	NONSTEADY PYROMETRIC CALCULATIONS AS THE BASIS FOR STRESS ANALYSIS OF STEAM GENERATORS AND HEAT EXCH	72-0451
STRESS	NOTCH EFFECT IN LOW-CYCLE FATIGUE - 2. EFFECT OF STRESS CHANGE AND MEAN STRESS	75-0537
STRESS	NOTCH EFFECT IN LOW-CYCLE FATIGUE - 2. EFFECT OF STRESS CHANGE AND MEAN STRESS	75-0537
STRESS	NOTE ON STRESS AND STRAIN REDISTRIBUTION IN A NOTCHED PLATE SPECIMEN DURING CYCLIC LOADING	69-0194
STRESS	NOTE ON THE STRESS CONCENTRATIONS AT THE END OF A CYLINDRICAL HOLE	70-0139
STRESS	NUMERICAL ANALYSIS ON STRESS DISTRIBUTION IN PARTIAL COMPRESSED SPECIMEN (HARDBOARD) BY FINITE ELEME	72-0920
STRESS	NUMERICAL EXPERIMENTS ON THE DETERMINATION OF STRESS CONCENTRATION FACTORS	74-0292
STRESS	NUMERICAL STRESS ANALYSIS OF INTERVERTEBRAL DISK	72-0125
STRESS	OBSERVATIONS REGARDING ASSUMED STRESS HYBRID PLATE ELEMENTS	74-0370
STRESS	ON THE APPLICATION OF FINITE-ELEMENT TECHNIQUES TO PROBLEMS CONCERNING POTENTIAL DISTRIBUTION AND ST	69-0268
STRESS	ON THE CALCULATION OF CONSISTENT STRESS DISTRIBUTIONS IN FINITE ELEMENT APPROXIMATIONS	71-0191
STRESS	ON THE FINITE ELEMENT ANALYSIS OF TWO-DIMENSIONAL STRESS PROBLEMS WITH TIME DEPENDENT PROPERTIES	65-0052
STRESS	ON THE FINITE ELEMENT STRESS ANALYSIS OF OIL TANKER STRUCTURES	74-0606
STRESS	ON THE GENERALIZATION OF STRESS FUNCTION PROCEDURE FOR DYNAMIC ANALYSIS OF PLATES	73-0661

STRESS	ON THE HYBRID STRESS FINITE ELEMENT MODEL FOR INCREMENTAL ANALYSIS OF LARGE DEFLECTION PROBLEMS	73-0371
STRESS	ON THE MATHEMATICAL ANALYSIS OF STRESS IN THE HUMAN FEMUR	72-0207
STRESS	ON THE SOLUTION OF PLANE STRESS PROBLEMS BY FINITE ELEMENTS COMPUTER PROGRAMS	73-0978
STRESS	ON THE STATES OF STRESS AND DEFORMATION OF CYLINDRICAL SPECIMENS OF BRITTLE MATERIAL UNDER UNIAXIAL	69-0150
STRESS	ON THE STATES OF STRESS AND DEFORMATION AROUND THE BOTTOM OF A CIRCULAR VERTICAL SHAFT	69-0323
STRESS	ON THE STRESS ANALYSIS AT JOINTS OF DEEP BEAMS	67-0055
STRESS	ON THE STRESS COMPUTATION IN FINITE ELEMENT MODELS BASED UPON DISPLACEMENT APPROXIMATIONS	74-0598
STRESS	ON THE STRESS DISTRIBUTION IN NOZZLE TO CYLINDER CONNECTIONS FOR SMALL DIAMETER RATIOS	72-0426
STRESS	ON THE USE OF FINITE ELEMENTS AND COMPLIANCE CALCULATIONS FOR DETERMINING CRACK TIP STRESS INTENSITY	69-0448
STRESS	ON TRESCA YIELD CRITERION IN FINITE ELEMENT ANALYSIS OF PLANE STRESS PROBLEMS	73-0314
STRESS	PARAMETRISCHE KERBSPANNUNGSUNTERSUCHUNGEN AN DER LOCHSCHEIBE MIT DER METHODE DER FINITEN ELEMENTE. (	75-0603
STRESS	PHOTOELASTIC AND FINITE ELEMENT STRESS ANALYSIS OF A RESTORED AXISYMMETRIC FIRST MOLAR	73-0142
STRESS	PHOTOELASTIC INVESTIGATION OF STRESS CONCENTRATIONS IN SPHERE-CYLINDER TRANSITION REGIONS: INCLUDING	68-0258
STRESS	PIECE-WISE CONTINUOUS FINITE ELEMENT FOR THE STRESS ANALYSIS OF PLATES AND SHELLS	75-0758
STRESS	PIPELINE STRESS ANALYSIS	75-0078
STRESS	PLACID: A GENERAL FINITE ELEMENT COMPUTER PROGRAM FOR STRESS ANALYSIS OF PLANE AND AXISYMMETRIC SOLI	74-0884
STRESS	PLANE STRESS ANALYSIS USING THE FINITE ELEMENT METHOD. TRIANGULAR FINITE ELEMENT WITH SIX PARAMETERS	67-0106
STRESS	PLANE STRESS ANALYSIS OF RUBBER AT HIGH STRAINS USING FINITE ELEMENTS	71-0680
STRESS	PLANE STRESS AND PLANE STRAIN ANALYSIS BY FINITE ELEMENT METHOD	70-0388
STRESS	PLANE STRESS CRACK TRAJECTORIES IN LOW MODULUS MATERIALS	75-0646
STRESS	PLANE STRESS ELEMENT OF VARYING THICKNESS APPLIED TO END BLOCK ANALYSIS	74-0396
STRESS	PLANE STRESS FINITE ELEMENTS FOR NONLINEAR ANALYSIS	69-0529
STRESS	PLANE STRESS LIMIT ANALYSIS BY FINITE ELEMENTS	70-0255
STRESS	PLANE STRESS REINFORCED CONCRETE FINITE ELEMENTS	74-0108
STRESS	PLANE STRESS SHAKEDOWN ANALYSIS BY FINITE ELEMENTS	72-0266
STRESS	PLASTIC STRESS STRAIN MATRIX OF ANISOTROPIC MATERIALS AND ITS APPLICATION	69-0383
STRESS	PLATE BENDING ANALYSIS USING ASSUMED STRESS METHODS	68-0135
STRESS	PLATE VIBRATION PROBLEMS USING THE FINITE ELEMENT METHOD WITH ASSUMED STRESS DISTRIBUTION AND TRIANG	66-0068
STRESS	PRACTICAL APPLICATIONS OF FINITE ELEMENT METHODS IN THE STRESS ANALYSIS OF THIN SHELL PRESSURE VESSE	71-0205
STRESS	PRECISION OF FINITE ELEMENT METHOD IN PLANE STRESS	69-0295
STRESS	PREDICTION OF LONG-TERM STRESS RANGES: USER'S MANUAL - BRIDGE LOAD GENERATOR	73-0737
STRESS	PREDICTION OF LONG-TERM STRESS RANGES: USER'S MANUAL - BRIDGE DYNAMIC STRESS ANALYSIS	74-0807
STRESS	PREDICTION OF LONG-TERM STRESS RANGES: USER'S MANUAL - BRIDGE DYNAMIC STRESS ANALYSIS	74-0807
STRESS	PRESSURE VESSEL WALL THICKNESS LIMITS BASED ON THERMAL STRESS RACHETTING AND CREEP FATIGUE INTERACTI	74-0968
STRESS	PROCEEDINGS OF THE NOSETIP STRESS ANALYSIS TECHNICAL INTERCHANGE MEETING	67-0042
STRESS	PROGRESS IN THREE-DIMENSIONAL ELASTIC-PLASTIC STRESS ANALYSIS FOR FRACTURE MECHANICS	71-0314
STRESS	QUADRILATERAL FINITE ELEMENT IN ELASTIC-PLASTIC PLANE STRESS ANALYSIS	67-0067
STRESS	QUADRILATERAL FINITE ELEMENT FOR PLANE STRESS AND PLANE STRAIN	70-0496
STRESS	RAPID CASE DESIGN USING FINITE ELEMENT STRESS ANALYSIS AND NASTRAN	73-0730
STRESS	RASCHET POLEI DEFORMATSII I NAPRYAZHENII V DETALYAKH DVIGATELEI VNUTRENNEGO SGORANIYA (CALCULATION O	73-0465
STRESS	RATIONALIZATION IN DERIVING ELEMENT STIFFNESS MATRIX BY ASSUMED STRESS APPROACH	68-0138
STRESS	RESIDUAL STRESS PRODUCED BY METAL CUTTING	71-0414
STRESS	REVIEW OF CONTINUUM, FINITE ELEMENT AND HYBRID TECHNIQUES IN THE ANALYSIS OF STRESS CONCENTRATIONS I	73-1051
STRESS	REVIEW OF CONTINUUM, FINITE ELEMENT AND HYBRID TECHNIQUES IN ANALYSIS OF STRESS CONCENTRATIONS IN ST	74-0914
STRESS	SAAS II: FINITE ELEMENT STRESS ANALYSIS OF AXISYMMETRIC SOLIDS WITH ORTHOTROPIC, TEMPERATURE DEPENDE	68-0096
STRESS	SAAS III: FINITE ELEMENT STRESS ANALYSIS OF AXISYMMETRIC AND PLANE SOLIDS WITH DIFFERENT ORTHOTROPIC	71-0741
STRESS	SAFE PCPS, A COMPUTER PROGRAM FOR THE STRESS ANALYSIS OF COMPOSITES BODIES OF REVOLUTION. INPUT INST	65-0013
STRESS	SAFE SHELL, A COMPUTER PROGRAM FOR THE STRESS ANALYSIS OF THIN SHELL BODIES OF REVOLUTION. A USER'S	67-0018
STRESS	SAFE-PLANE, A COMPUTER PROGRAM FOR THE STRESS ANALYSIS AND DESIGN OF TWO-DIMENSIONAL COMPOSITE BODIE	67-0017
STRESS	SEAL-SHELL-2, A COMPUTER PROGRAM FOR THE STRESS ANALYSIS OF A THICK SHELL OF REVOLUTION WITH AXISYMM	69-0107
STRESS	SHOCK TUNNEL TESTS OF PRELOADED AND ARCHED WALL PANELS FINITE ELEMENT STRESS FORMULATION FOR WAVE PR	73-0892
STRESS	SIMPLE FINITE ELEMENT METHOD FOR STRESS CONCENTRATION PROBLEMS IN PRESSURE VESSELS	71-0669
STRESS	SIMPLE TEACHING EXAMPLES FOR FINITE ELEMENT STRESS ANALYSIS	72-0898
STRESS	SLAMCODE: FINITE ELEMENT STRESS WAVE ANALYSIS	72-0603
STRESS	SLOPE STABILITY ANALYSIS BY THE FINITE ELEMENT STRESS ANALYSIS AND LIMITING EQUILIBRIUM METHOD	70-0222
STRESS	SOLID WEDGE FINITE ELEMENT FOR STRESS ANALYSIS	69-0124
STRESS	SOLUTION BOUNDS IN SOME STRESS PROBLEMS BY THE HYPERCIRCLE AND FINITE ELEMENT METHODS	72-0706
STRESS	SOME CONSIDERATIONS ON THE DYNAMIC STRESS ANALYSIS OF A STRIP BY FINITE ELEMENT METHOD	71-0591
STRESS	SOME CONSIDERATIONS OF THE FINITE ELEMENT METHOD IN STRESS ANALYSIS	72-0997
STRESS	SOME CONSIDERATIONS ON THE CONVERGENCE OF HYBRID STRESS METHOD	73-0303
STRESS	SOME CREEP EFFECTS IN STRESS ANALYSIS WITH PARTICULAR REFERENCE TO CONCRETE PRESSURE VESSELS	69-0284
STRESS	SOME INVESTIGATIONS INTO THE FINITE ELEMENT METHOD WITH SPECIAL REFERENCE TO PLANE STRESS	66-0128
STRESS	SOME RECENT STUDIES IN ASSUMED STRESS HYBRID MODELS	72-0262
STRESS	SOME STRESS INTENSITY CALCULATIONS USING FINITE ELEMENTS	74-0423
STRESS	SPAENDINGSHYBRIDE FINITE ELEMENTER TIL SVINGNINGSPROBLEMER. (STRESS HYBRID FINITE ELEMENT METHOD FOR	75-0003
STRESS	SPANNUNGEN IN MEMBRANWAENDEN DURCH WAERMEBEANSPRUCHUNG UND MECHANISCHE BELASTUNG. (STRESSES IN MEMBR	75-0066
STRESS	SPECIAL FINITE ELEMENTS FOR THE ANALYSIS OF STRESS CONCENTRATIONS AND SINGULARITIES	71-0219
STRESS	STATIC STRESS ANALYSIS OF AXISYMMETRIC SOLIDS WITH MATERIAL AND GEOMETRIC NONLINEARITIES BY THE FINI	73-0958
STRESS	STATIC, VIBRATION, AND THERMAL STRESS ANALYSES OF LAMINATED PLATE AND SHELLS BY THE HYBRID-STRESS FI	72-0602
STRESS	STATISTICAL STATE OF STRESS STUDIED BY GRID ANALYSIS: "NUMERICAL METHODS OF ANALYSIS"	49-0001
STRESS	STEADY STATE AND TRANSIENT HEAT CONDUCTION AND STRESS ANALYSES OF TWO AND THREE DIMENSIONAL STRUCTUR	71-0675
STRESS	STIFFNESS DERIVATIVE FINITE ELEMENT TECHNIQUE FOR DETERMINATION OF CRACK TIP STRESS INTENSITY FACTOR	74-0956
STRESS	STIFFNESS DERIVATIVE FINITE ELEMENT TECHNIQUE FOR DETERMINATION OF CRACK TIP STRESS INTENSITY FACTOR	74-1358
STRESS	STRESS - DEFORMATION PREDICTION IN CEMENT - TREATED SOIL PAVEMENTS	71-0286
STRESS	STRESS ANALYSIS	65-0055
STRESS	STRESS ANALYSIS AND SPECIAL PROBLEMS OF PRESTRESSED DAMS	61-0006
STRESS	STRESS ANALYSIS AND DESIGN OF SILICON SOLAR CELL ARRAYS AND RELATED MATERIAL PROPERTIES	72-0840
STRESS	STRESS ANALYSIS BY FINITE ELEMENTS	68-0165
STRESS	STRESS ANALYSIS BY SUPERPOSITION OF THE ANALYTICAL SOLUTION AND FINITE ELEMENT SOLUTION	72-1000
STRESS	STRESS ANALYSIS BY THE FINITE ELEMENT METHOD - THERMAL EFFECTS	57-0022
STRESS	STRESS ANALYSIS BY THE FINITE ELEMENT DISPLACEMENT METHOD	68-0036
STRESS	STRESS ANALYSIS FOR STRUCTURES UNDER CONSTRUCTION	69-0209
STRESS	STRESS ANALYSIS FOR CLEAN LEAN DESIGN	71-0386
STRESS	STRESS ANALYSIS IN HEATED, COMPLEX SHAPES	62-0002
STRESS	STRESS ANALYSIS IN HULL STRUCTURAL MEMBERS BY PHOTOELASTIC EXPERIMENTS AND FINITE ELEMENT METHODS -	70-0341
STRESS	STRESS ANALYSIS IN SOME COLD FORGING TOOL COMPONENTS AND THEIR DESIGN GUIDES	74-0589
STRESS	STRESS ANALYSIS IN VISCOELASTIC BODIES USING FINITE ELEMENTS AND A CORRESPONDENCE RULE WITH ELASTICI	69-0270
STRESS	STRESS ANALYSIS OF A GRAVITY DAM BY THE FINITE ELEMENT METHOD	63-0004
STRESS	STRESS ANALYSIS OF AXISYMMETRIC SOLIDS UTILIZING HIGHER ORDER QUADRILATERAL FINITE ELEMENTS	69-0356
STRESS	STRESS ANALYSIS OF A CONICAL ACRYLIC VIEWPORT	70-0106
STRESS	STRESS ANALYSIS OF AN AUTOMOBILE CHASSIS FRAME	70-0137
STRESS	STRESS ANALYSIS OF A PRESTRESSED CONCRETE NUCLEAR PRESSURE VESSEL BY THE FINITE ELEMENT METHOD USING	71-0154
STRESS	STRESS ANALYSIS OF A GRAPHITE FUEL ELEMENT INCLUDING RADIATION CREEP AND DIMENSIONAL CHANGES	71-0217
STRESS	STRESS ANALYSIS OF AXISYMMETRIC SOLIDS WITH ASYMMETRIC PROPERTIES	72-0140
STRESS	STRESS ANALYSIS OF A BOLT-NUT JOINT BY THE FINITE ELEMENT METHOD AND THE COPPER-ELECTROPLATING METHO	73-0220
STRESS	STRESS ANALYSIS OF A BOLT-NUT JOINT BY THE FINITE ELEMENT METHOD AND THE COPPER-ELECTROPLATING METHO	73-0220
STRESS	STRESS ANALYSIS OF A CERAMIC BLADE DESIGN SUBJECTED TO CENTRIFUGAL AND THERMAL LOADS	73-0790
STRESS	STRESS ANALYSIS OF A BOLT-NUT JOINT BY FINITE ELEMENT METHOD AND COPPER-ELECTROPLATING METHOD - 2. S	74-0153
STRESS	STRESS ANALYSIS OF A BOLT-NUT JOINT BY FINITE ELEMENT METHOD AND COPPER-ELECTROPLATING METHOD - 2. S	74-0153
STRESS	STRESS ANALYSIS OF ADHESIVE-BONDED LAP JOINTS	74-0548
STRESS	STRESS ANALYSIS OF A TUBULAR CROSS-JOINT WITHOUT ANY INTERNAL STIFFENING FOR OFFSHORE STRUCTURES	74-0549
STRESS	STRESS ANALYSIS OF B16.9 TEES BY THE FINITE ELEMENT METHOD	71-0060

STRESS	STRESS ANALYSIS OF B16.9 TEES BY THE FINITE ELEMENT METHOD, A PROGRESS REPORT	71-0215
STRESS	STRESS ANALYSIS OF BWR PRESSURE VESSEL WITH BERSAFE AND FLHE	73-0406
STRESS	STRESS ANALYSIS OF CONTRACTIVE OR DILATIVE SOIL	71-0579
STRESS	STRESS ANALYSIS OF COMPLEX SHIP COMPONENTS BY A NUMERICAL PROCEDURE USING CURVED FINITE ELEMENTS	73-0879
STRESS	STRESS ANALYSIS OF COLDWORKED FASTENER HOLES	74-0783
STRESS	STRESS ANALYSIS OF CONCRETE TRACK SLABS ON AN ELASTIC FOUNDATION BY THE FINITE ELEMENT METHOD	74-1049
STRESS	STRESS ANALYSIS OF CONFIGURATIONS INVOLVING SMALL FILLETS	75-0361
STRESS	STRESS ANALYSIS OF CURVED PIPES WITH END RESTRAINTS	75-0667
STRESS	STRESS ANALYSIS OF COMPOSITE MATERIALS BY THE HYBRID FINITE ELEMENT METHOD	75-0799
STRESS	STRESS ANALYSIS OF FIRST MOLARS WITH FULL CROWN PREPARATIONS BY THREE-DIMENSIONAL PHOTOELASTICITY AND	72-0939
STRESS	STRESS ANALYSIS OF FRAME STRUCTURE CONNECTED WITH PLATES	74-0265
STRESS	STRESS ANALYSIS OF HUMAN FEMUR	73-0327
STRESS	STRESS ANALYSIS OF HUMAN TEETH	73-0505
STRESS	STRESS ANALYSIS OF INDUSTRIAL COMPONENTS WITH PLASTIC AND FINITE ELEMENT MODELS	74-0512
STRESS	STRESS ANALYSIS OF LAMINATED CFRP PLATES CONTAINING A HOLE	74-1175
STRESS	STRESS ANALYSIS OF MULTI-CELLULAR CAISSONS	74-0338
STRESS	STRESS ANALYSIS OF PRESSURIZED CYLINDRICAL BORE IN RECTANGULAR BLOCK	69-0312
STRESS	STRESS ANALYSIS OF PARACHUTES USING FINITE ELEMENTS	71-0481
STRESS	STRESS ANALYSIS OF PRESSURE VESSEL NOZZLES	73-0823
STRESS	STRESS ANALYSIS OF ROCK AS A NO-TENSION MATERIAL	68-0174
STRESS	STRESS ANALYSIS OF RIBBON REINFORCED COMPOSITES	70-0221
STRESS	STRESS ANALYSIS OF SELF-SEALING TYPE PISTON HEAD BY FINITE-ELEMENT METHOD	72-0259
STRESS	STRESS ANALYSIS OF TAPE REINFORCED COMPOSITES	69-0068
STRESS	STRESS ANALYSIS OF THIN PLATE STRUCTURE BY THE FINITE ELEMENT METHOD	69-0403
STRESS	STRESS ANALYSIS OF TWO-DIMENSIONAL PROBLEMS UNDER SIMULTANEOUS CREEP AND PLASTICITY	71-0220
STRESS	STRESS ANALYSIS OF THE "SELF-SEALING" TYPE PISTON HEAD BY THE FINITE ELEMENT METHOD	72-0617
STRESS	STRESS ANALYSIS OF TWO AND THREE DIMENSIONAL STRUCTURES BY THE FINITE ELEMENT METHOD	72-0794
STRESS	STRESS ANALYSIS OF TWO JOINTS BY THE FINITE-ELEMENT METHOD	72-0942
STRESS	STRESS ANALYSIS OF THE HUMAN AORTIC VALVE	73-0166
STRESS	STRESS ANALYSIS OF TUBULAR JOINTS FOR OFFSHORE STRUCTURES	73-0207
STRESS	STRESS ANALYSIS OF 175 MM PROJECTILE, HE, M437	71-0736
STRESS	STRESS ANALYSIS PROGRAM BY FINITE ELEMENT METHOD FOR RADIAL FLOW IMPELLERS	74-1016
STRESS	STRESS AND DEFLECTION ANALYSIS OF MECHANICALLY FASTENED JOINTS	70-0067
STRESS	STRESS AND DEFORMATION ANALYSIS OF IRRADIATION INDUCED SWELLING	74-0305
STRESS	STRESS AND DEFORMATION ANALYSIS OF IRRADIATION INDUCED SWELLING	74-0882
STRESS	STRESS AND DISPLACEMENTS AROUND OPENINGS IN ROCK CONTAINING AN INELASTIC DISCONTINUITY	75-0170
STRESS	STRESS AND FAILURE ANALYSIS OF A GLASS-EPOXY COMPOSITE PLATE WITH A CIRCULAR HOLE	73-0276
STRESS	STRESS AND STRAIN DISTRIBUTIONS AT THE ROOT OF CRACKS	69-0290
STRESS	STRESS AND STRAIN DISTRIBUTION IN THE VICINITY OF INTERFERENCE FIT FASTENERS	73-0363
STRESS	STRESS AND VIBRATION ANALYSIS ON THE SYSTEM OF FRAME AND FOUNDATION	69-0218
STRESS	STRESS AND VIBRATION ANALYSIS OF TURBOPUMP INDUCER BLADES BY FINITE ELEMENTS	71-0363
STRESS	STRESS CONCENTRATIONS IN TIMBER BEAMS	67-0038
STRESS	STRESS CONCENTRATIONS IN HULL SHAPES FROM SURFACE DISCONTINUITIES	69-0130
STRESS	STRESS CONCENTRATION FACTORS AROUND A PENETRATION IN A STRIP OR SLAB OF FINITE DIMENSIONS	69-0409
STRESS	STRESS CONCENTRATIONS NEAR A DISCONTINUITY IN FIBROUS COMPOSITES	71-0316
STRESS	STRESS CONCENTRATIONS AND SINGULARITIES AT INTERFACE CORNERS	71-0419
STRESS	STRESS CONCENTRATION FACTORS FOR BONDED LAP JOINTS	71-0842
STRESS	STRESS CONCENTRATIONS NEAR HOLES IN LAMINATES	74-0329
STRESS	STRESS CONCENTRATION FACTORS FOR CIRCULAR NOTCHES	74-0791
STRESS	STRESS CONCENTRATION FACTORS FOR A FATIGUE TEST SPECIMEN WITH A U-TYPE NOTCH IN ONE EDGE BY FINITE E	74-1344
STRESS	STRESS CONCENTRATION FACTORS FOR CIRCULAR, REINFORCED PENETRATIONS IN PRESSURIZED CYLINDRICAL SHELLS	75-0726
STRESS	STRESS CRACK PHENOMENA	75-0744
STRESS	STRESS DETERMINATION IN 2 CONDENSER HEAT EXCHANGER BAFFLE CAPS BY METHOD OF FINITE ELEMENTS AND A ST	74-0661
STRESS	STRESS DIFFUSION IN THICK STIFFENED PANELS	74-0620
STRESS	STRESS DISTRIBUTION DUE TO GRAVITY IN A VERTICAL ROCK BANK	68-0090
STRESS	STRESS DISTRIBUTION AND ROCK FRACTURE ZONES IN THE ROOF OF LONGWALL FACE IN COAL MINE	73-0247
STRESS	STRESS DISTRIBUTION IN BONE ARISING FROM LOADING ON ENDOSTEAL DENTAL IMPLANTS	73-0501
STRESS	STRESS DISTRIBUTION NEAR A BROKEN POINT OF FILAMENT IN UNIDIRECTIONALLY FIBER REINFORCED COMPOSITES	73-0811
STRESS	STRESS DISTRIBUTION IN THE LIGAMENT ZONE OF PERFORATED CIRCULAR DISCS WITH SPECIAL REFERENCE TO THE	73-0827
STRESS	STRESS DISTRIBUTION UNDER UNIAXIAL ELASTIC PLANE STRAIN COMPRESSION WITH FRICTION	74-0526
STRESS	STRESS DISTRIBUTIONS AROUND PLUG MATCHES WITH ANCHOR BLOCKS	69-0135
STRESS	STRESS DISTRIBUTIONS IN THIN-WALLED INTERSECTING CYLINDRICAL SHELLS SUBJECTED TO INTERNAL PRESSURE A	71-0023
STRESS	STRESS DISTRIBUTIONS AROUND SHALLOW BURIED RIGID PIPES	74-0142
STRESS	STRESS FIELDS AROUND MULTIPLE INCLUSIONS	69-0067
STRESS	STRESS FUNCTION APPROACH	75-0016
STRESS	STRESS INTENSITY FACTOR FOR A STRAIGHT CRACK APPROACHING A HOLE	69-0158
STRESS	STRESS INTENSITY FACTORS CALCULATED GENERALLY BY FINITE ELEMENT TECHNIQUE	69-0369
STRESS	STRESS INTENSITY FACTORS IN PLATE STRUCTURES CALCULATED BY THE FINITE ELEMENT METHOD (IN JAPANESE)	71-0788
STRESS	STRESS INTENSITY FACTORS FOR HEXAGONAL FRACTURE TOUGHNESS SPECIMENS	71-0799
STRESS	STRESS INTENSITY FACTORS FOR CRACKS IN A PLATE CONTAINING A HOLE AND IN A SPINNING DISC	73-0467
STRESS	STRESS INTENSITY FACTORS FOR ANISOTROPIC FRACTURE TEST SPECIMENS OF SEVERAL GEOMETRIES	74-0293
STRESS	STRESS INTENSITY FACTORS FOR THICK-WALLED CYLINDERS	74-0300
STRESS	STRESS INTENSITY FACTOR OF A SURFACE CRACK IN A PRESSURIZED CYLINDER	75-0255
STRESS	STRESS INTENSITY FACTORS FOR A NOTCHED BEAM ON AN ELASTIC FOUNDATION	75-0566
STRESS	STRESS RESULTANTS IN SHELLS BY AN AUGMENTED FINITE ELEMENT METHOD	73-0585
STRESS	STRESS SEPARATION IN PHOTOELASTICIMETRIC MEASUREMENT BY THE FINITE ELEMENT METHOD	71-0807
STRESS	STRESS SINGULARITIES AT DIAPHRAGM-WEB-FLANGE JUNCTIONS IN BOX GIRDERS	75-0242
STRESS	STRESS STRAIN MATRIX AND ITS APPLICATIONS FOR THE SOLUTION OF ELASTIC-PLASTIC PROBLEMS BY THE FINITE	68-0282
STRESS	STRESS STRAIN RELATIONSHIP FROM COMPRESSION TESTS ON NONLINEAR MATERIALS	69-0189
STRESS	STRESS STRAIN RESPONSE AND FRACTURE OF A CONCRETE MODEL IN BIAXIAL LOADING	71-0295
STRESS	STRESS WAVE ANALYSIS IN LAYERED THERMOVISCOELASTIC MATERIALS BY THE EXTENDED RITZ METHOD, VOLUME II	68-0125
STRESS	STRESS, DISPLACEMENT, LINE INTEGRAL AND CLOSURE ENERGY DETERMINATIONS OF CRACK TIP STRESS INTENSITY	74-0621
STRESS	STRESS, DISPLACEMENT, LINE INTEGRAL AND CLOSURE ENERGY DETERMINATIONS OF CRACK TIP STRESS INTENSITY	74-0621
STRESS	STRESS, STABILITY AND VIBRATION OF COMPLEX SHELLS OF REVOLUTION: ANALYSIS AND USERS MANUAL FOR BOSOR	69-0441
STRESS	STRESSES IN A ROTATING DISK WITH CENTRAL AND NONCENTRAL HOLES (STRESS ANALYSIS BY THE FINITE ELEMENT	74-1337
STRESS	STUDY ON STATE OF STRESS AND DEFORMATION OF CYLINDRICAL SPECIMENS OF BRITTLE MATERIAL UNDER UNIAXIAL	69-0152
STRESS	SYSTEMATIC ENFORCEMENT OF STRESS BOUNDARY CONDITIONS IN THE ASSUMED STRESS HYBRID MODEL BASED ON THE	71-0771
STRESS	SYSTEMATIC ENFORCEMENT OF STRESS BOUNDARY CONDITIONS IN THE ASSUMED STRESS HYBRID MODEL BASED ON THE	71-0771
STRESS	TAPERED PLANE STRESS ANNULAR SECTOR FINITE ELEMENT	75-0538
STRESS	TESTRAN PROGRAMS, A TOOL FOR TEMPERATURE AND STRESS ANALYSIS	73-0414
STRESS	THE ANALYSIS OF STRESS DISTRIBUTION AROUND THE TIP OF A CRACK BY FINITE ELEMENT METHOD WITH THE AID	71-0517
STRESS	THE APPLICATION OF FINITE ELEMENT METHODS TO THE SOLUTION OF STRESS AND DIFFUSION PROBLEMS IN CONTIN	68-0034
STRESS	THE APPLICATION OF THE MATRIX DISPLACEMENT METHOD IN PLANE ELASTO-PLASTIC STRESS PROBLEMS	64-0009
STRESS	THE APPLICATION OF THE FINITE ELEMENT METHOD TO A PROBLEM IN STRESS WAVE PROPAGATION	69-0505
STRESS	THE CHANGES IN THE (TWO-DIMENSIONAL) STRESS ROUNDED CORNERS, CAUSED BY A VARYING INTERNAL PRESSURE A	67-0141
STRESS	THE COMPARISON OF ITERATIVE AND DIRECT SOLUTION TECHNIQUES IN THE ANALYSIS OF TIME-DEPENDENT STRESS	72-0080
STRESS	THE CONSTRAINT METHOD FOR FINITE ELEMENT STRESS ANALYSIS	75-0432
STRESS	THE EFFECT OF STRESS HISTORY ON THE RESILIENT RESPONSE OF SOILS	73-0914
STRESS	THE ELASTIC STRESS ANALYSIS OF A BI-MATERIAL PLATE WITH A CRACK NORMAL TO THE INTERFACES	72-0819
STRESS	THE FINITE ELEMENT CALCULATION OF STRESS INTENSITY FACTORS USING SUPERPOSITION	75-0886
STRESS	THE FINITE ELEMENT IN PLANE STRESS ANALYSIS	60-0001
STRESS	THE FINITE ELEMENT METHOD FOR FLEXURE OF SLABS WHEN STRESS DISTRIBUTIONS ARE ASSUMED	66-0069

STRESS	THE FINITE ELEMENT METHOD IN APPLICATION TO PLANE STRESS	68-0117
STRESS	THE FINITE ELEMENT METHOD IN PLANE STRESS ANALYSIS	69-0127
STRESS	THE FINITE ELEMENT METHOD FOR PREDICTING STRESS DISTRIBUTION AND SOIL DEFORMATION UNDER A TRACTIVE D	69-0522
STRESS	THE FINITE ELEMENT METHOD FOR PREDICTING STRESS DISTRIBUTION AND SOIL DEFORMATION UNDER A TRACTIVE D	71-0801
STRESS	THE FINITE ELEMENT STRESS ANALYSIS OF THE HUMAN AORTIC VALVE TOWARD THE DEVELOPMENT OF A PROSTHETIC	74-1201
STRESS	THE HEAT CONDUCTION AND THERMAL STRESS ANALYSIS BY THE FINITE ELEMENT METHOD	68-0073
STRESS	THE METHOD OF INTERFACE STRESS ADJUSTMENT AND ITS USES IN SOME PLANE ELASTICITY PROBLEMS	61-0005
STRESS	THE RECTANGULAR PLANE STRESS ELEMENT BY TURNER, PIAN AND WILSON	74-1265
STRESS	THE ROLE OF FINITE DEFORMATION ANALYSIS IN PLANE STRESS AND STRAIN FRACTURES	74-0322
STRESS	THE SCANDINAVIAN PORV MODEL PROJECT: STRESS CALCULATION AND EXPERIMENTAL VERIFICATION	71-0021
STRESS	THE SIMPLE METHOD OF EVALUATION OF STRESS INTENSITY FACTOR USING THE FINITE ELEMENT METHOD	74-0829
STRESS	THE SOLUTION OF STRUCTURAL EQUILIBRIUM EQUATIONS BY THE CONJUGATE GRADIENT METHOD WITH PARTICULAR RE	71-0268
STRESS	THE STRESS ANALYSIS OF AXISYMMETRIC SHELLS USING A PLANE FRAME ANALYSIS PROGRAM	69-0238
STRESS	THE STRESS DISTRIBUTION OF NORFOLK DAM	62-0001
STRESS	THEORETICAL AND EXPERIMENTAL STRESS ANALYSES OF ORNL THIN-SHELL CYLINDER-TO-CYLINDER MODEL 3	75-0690
STRESS	THEORETICAL AND EXPERIMENTAL STRESS ANALYSES OF ORNL THIN-SHELL CYLINDER-TO-CYLINDER MODEL 4	75-0695
STRESS	THEORETICAL BASIS OF THE BOREHOLE DEEPENING METHOD OF ABSOLUTE STRESS MEASUREMENT	70-0047
STRESS	THERMAL AND STRESS ANALYSIS BY NASTRAN	72-0153
STRESS	THERMAL AND STRESS ANALYSIS PROGRAM - HYBOS	72-0159
STRESS	THERMAL AND STRESS ANALYSIS ON PRISMATIC NUCLEAR FUEL ELEMENTS	75-0120
STRESS	THERMAL AND STRESS ANALYSIS OF A FUEL ROD	75-0713
STRESS	THERMAL ELASTIC PLASTIC STRESS ANALYSIS BY THE FINITE ELEMENT METHOD	69-0391
STRESS	THERMAL STRESS ANALYSIS BY THE FINITE ELEMENT METHOD	70-0124
STRESS	THERMAL STRESS ANALYSIS OF REENTRY VEHICLE NOSETIPS AT ANGLE OF ATTACK	72-0456
STRESS	THERMAL STRESS ANALYSIS OF RE-ENTRY VEHICLE NOSETIPS AT ANGLE OF ATTACK	72-0852
STRESS	THERMAL STRESS ANALYSIS OF PLATES AND SHALLOW SHELLS BY HYBRID FINITE-ELEMENT METHOD	74-0485
STRESS	THERMAL STRESS AND TEMPERATURE FIELDS - A FINITE-ELEMENT APPROACH	71-0782
STRESS	THERMAL STRESS DEVELOPED IN S PLATE, STRAIGHT PLATE, AND DEEP DISH WHEELS	73-0172
STRESS	THERMO-ELASTIC ANALYSIS OF RECTANGULAR PLATES IN PLANE STRESS BY THE FINITE-ELEMENT DISPLACEMENT MET	67-0087
STRESS	THERMOVISCOELASTIC AXISYMMETRIC STRESS ANALYSIS BY FINITE ELEMENT AND FINITE DIFFERENCE COMPUTER TEC	71-0105
STRESS	THREE DIMENSIONAL ANALYSIS OF STRESS CONCENTRATIONS NEAR HOLES IN LAMINATED COMPOSITES	72-0713
STRESS	THREE DIMENSIONAL FINITE ELEMENT STRESS ANALYSIS APPLIED TO TWO PROBLEMS IN ROCK MECHANICS	72-0189
STRESS	THREE DIMENSIONAL SCATTERED-LIGHT STRESS ANALYSIS OF DISCONTINUOUS FIBER-REINFORCED COMPOSITES	73-0218
STRESS	THREE DIMENSIONAL STRESS ANALYSIS	70-0086
STRESS	THREE DIMENSIONAL STRESS ANALYSIS OF PENETRATIONS IN THICK WALLED PRESSURE VESSELS	74-0422
STRESS	THREE NODE ISOPARAMETRIC STRESS AXIAL ELEMENT	72-0992
STRESS	THREE-DIMENSIONAL ANALYSIS FOR THE STRESS DISTRIBUTION NEAR CIRCULAR HOLES IN LAMINATED COMPOSITES	74-0793
STRESS	THREE-DIMENSIONAL BAR CELL FOR ELASTIC STRESS ANALYSIS	70-0219
STRESS	THREE-DIMENSIONAL ELASTIC-PLASTIC STRESS AND STRAIN ANALYSIS FOR FRACTURE MECHANICS. PHASE 1. SIMPLE	70-0188
STRESS	THREE-DIMENSIONAL GRAIN STRESS ANALYSIS USING THE FINITE ELEMENT METHOD	71-0732
STRESS	THREE-DIMENSIONAL STRESS ANALYSIS OF ARCH DAMS BY THE FINITE ELEMENT METHOD: PART I, ARCH DAM NO. 1,	66-0099
STRESS	THREE-DIMENSIONAL STRESS ANALYSIS AND FIELD PROBLEMS BY THE FINITE ELEMENT METHOD	70-0183
STRESS	THREE-DIMENSIONAL STRESS ANALYSES OF MITERED ELBOWS	75-0650
STRESS	THREE-DIMENSIONAL STRESS ANALYSIS OF A FINITE SLAB CONTAINING A TRANSVERSE CENTRAL CRACK	75-0732
STRESS	THREE-DIMENSIONAL THERMAL AND STRESS ANALYSIS OF A PIPING TEE	75-0715
STRESS	TRANSIENT THERMAL STRESS TECHNIQUES IN NUCLEAR REACTORS. APPLIED MECHANICS ASPECTS OF NUCLEAR EFFECT	71-0474
STRESS	TRANSIENT THERMAL STRESS ANALYSIS OF COMPOSITE STRUCTURES INCLUDING CONTINUOUSLY VARYING PROPERTIES	75-0681
STRESS	TRESS - A FINITE ELEMENT CODE FOR STRESS ANALYSIS PROBLEMS	71-0715
STRESS	TRIANGULAR THICK SHELL FINITE ELEMENT FOR THE THREE-DIMENSIONAL STRESS ANALYSIS OF PRESSURE VESSELS	73-0860
STRESS	TSAAS: FINITE ELEMENT THERMAL AND STRESS ANALYSIS OF AXISYMMETRIC SOLIDS WITH ORTHOTROPIC TEMPERATUR	74-0886
STRESS	TWO-DIMENSIONAL FINITE ELEMENT COMPUTER CODES FOR STRESS ANALYSIS AND FIELD PROBLEMS	73-0055
STRESS	UNDRAINED STRESS DISTRIBUTION BY NUMERICAL METHODS	68-0045
STRESS	USE OF SINGULARITY PROGRAMMING IN FINITE ELEMENT CALCULATIONS OF ELASTIC STRESS INTENSITY FACTORS. P	75-0252
STRESS	USE OF SINGULARITY PROGRAMMING IN FINITE ELEMENT CALCULATIONS OF ELASTIC STRESS INTENSITY FACTORS. P	75-0252
STRESS	VARIABLE TIME STEP METHOD FOR DETERMINING PLASTIC STRESS REFLECTIONS FROM BOUNDARIES	75-0641
STRESS	VIBRATION ANALYSIS OF LAMINATED PLATES AND SHELLS BY A HYBRID STRESS ELEMENT	73-0120
STRESS	VIBRATION AND STRESS ANALYSIS OF SOFT-BONDED SHUTTLE INSULATION TILES, MODAL ANALYSIS WITH COMPACT W	74-0708
STRESS	VISCOELASTIC STRESS ANALYSIS	68-0080
STRESS	3-D STRESS WAVE CODE FOR THE ILLTAC IV	74-0790
STRESS-ANALYSIS	USE OF THE ' TITUS ' FINITE ELEMENT CODE FOR THE NUCLEAR STRUCTURES AND INTERNAL STRESS-ANALYSIS	73-0840
STRESS-DEFORMAT	SOLUTION OF STRESS-DEFORMATION PROBLEMS IN SOIL AND ROCK MECHANICS USING FINITE ELEMENT METHODS	69-0272
STRESS-DEFORMAT	THE APPLICATION OF THE FINITE ELEMENT METHOD TO SEEPAGE AND STRESS-DEFORMATION PROBLEMS IN SOIL MECH	70-0267
STRESS-DEPENDEN	SEEPAGE INTO AN EXCAVATION IN A MEDIUM POSSESSING STRESS-DEPENDENT PERMEABILITY	72-0629
STRESS-HYBRID	STRESS-HYBRID FINITE-ELEMENT	75-0619
STRESS-PATH	STRESS-PATH DEPENDENT DEFORMATION MODULI OF CLAY	75-0050
STRESS-STRAIN	ANALYTICAL EVALUATION OF STRESS-STRAIN TEST DATA.	72-0442
STRESS-STRAIN	APPLICATIONS OF FINITE ELEMENT STRESS ANALYSIS AND STRESS-STRAIN PROPERTIES IN DETERMINING NOTCH FAT	73-1114
STRESS-STRAIN	BIAXIAL STRESS-STRAIN RELATIONS FOR CONCRETE	72-0203
STRESS-STRAIN	CYCLIC MECHANICAL TESTS AND AN APPROPRIATE ANALYTICAL STRESS-STRAIN MODEL FOR A36 STEEL	74-0815
STRESS-STRAIN	ELASTIC-PLASTIC STRESS-STRAIN RELATIONS FOR MATERIALS WITH DIFFERENT YIELD STRENGTHS IN TENSION AND	72-0764
STRESS-STRAIN	ELASTIC-PLASTIC STRESS-STRAIN RELATIONS USED IN CODE EPSOLA	73-0784
STRESS-STRAIN	NONLINEAR IN-PLANE BENDING OF A PLANE STRESS-STRAIN FINITE ELEMENT MODEL	74-0211
STRESS-STRAIN	PLASTIC STRESS-STRAIN MATRIX AND ITS APPLICATION FOR THE SOLUTION OF ELASTIC-PLASTIC PROBLEMS BY THE	68-0171
STRESS-STRAIN	STRESS-STRAIN PROPERTIES OF COMPACTED CLAYS	74-0650
STRESS-STRAIN-S	EFFECTIVE STRESS-STRAIN-STRENGTH MODEL FOR SOILS	75-0056
STRESS-STRAIN-T	VOLLSTAENDIGE SPANNUNGS-DEHNUNGS-TEMPERATURAENDERUNGSBEZIEHUNG FUER DIE SCHWEISSEIGENSPANNUNGSBERECH	75-0807
STRESS-STRAINED	DETERMINATION OF THE STRESS-STRAINED STATE WHEN CUTTING METAL WITH A WEDGE	73-0237
STRESS-STRAINED	INVESTIGATION OF THE STRESS-STRAINED STATE OF SPHERICAL SHELLS WITH AN ECCENTRIC CUTOUT ON THE BASIS	73-0222
STRESS-WAVE	DOUBLE STRESS-WAVE DISCONTINUITIES IN FINITE, SHEAR-CORRECTED BEAMS AND PLATES	68-0210
STRESS-WAVE	STRESS-WAVE PROPAGATION AND DYNAMIC EFFECTS IN SHELLS (SPADES) MICROMECHANICAL STRESS-WAVE RESPONSE	72-0851
STRESS-WAVE	STRESS-WAVE PROPAGATION AND DYNAMIC EFFECTS IN SHELLS (SPADES) MICROMECHANICAL, STRESS-WAVE RESPONSE	72-0851
STRESSED	APPLICATION OF FRACTURE MECHANICS TO THERMALLY STRESSED CYLINDERS USING A FINITE ELEMENT METHOD	69-0089
STRESSED	APPLICATION OF THE METHOD OF FINITE ELEMENTS TO CALCULATIONS OF THE STRESSED AND STRAINED STATE OF I	75-0775
STRESSED	COMPORTAREA ELASTICA A DIGUILOR DE PAMINT SOLICIATE SEISMIC (ELASTIC BEHAVIOR OF THE EARTH DAMS SEI	74-1185
STRESSED	FINITE ELEMENT ANALYSIS OF VIBRATIONS OF INITIALLY STRESSED THIN SHELLS OF REVOLUTION	74-1056
STRESSED	ISSLEDOVANIE NAPRYAZHENNOGO SOSTOYANIYA REBRISTYKH ISILINDRICHESKIKH OBOLOCHEK S PRYAMOUGOL'NYMI OTV	74-0944
STRESSED	MATERIAL BEHAVIOUR CHARACTERISTICS FOR REINFORCED CONCRETE SHELLS STRESSED BEYOND THE ELASTIC RANGE	70-0166
STRESSED	ON RESPONSE OF INITIALLY STRESSED STRUCTURES TO RANDOM EXCITATIONS	73-0278
STRESSED	OPTIMUM DESIGN OF STRESSED SKIN STRUCTURES	73-0472
STRESSES	A FRACTURE MECHANICS APPROACH TO THE PROPAGATION OF FATIGUE CRACKS IN REACTOR COMPONENTS BY ACOUSTIC	71-0132
STRESSES	A METHOD FOR DETERMINING THE SURFACE CONTACT STRESSES RESULTING FROM INTERFERENCE FITS	70-0101
STRESSES	A METHOD FOR THE PREDICTION OF STRESSES IN AN ISOTROPIC INCLUSION OR OREBODY OF IRREGULAR SHAPE	72-0827
STRESSES	A NEW MEASURING METHOD OF RESIDUAL STRESSES WITH THE AID OF FINITE ELEMENT METHOD AND RELIABILITY OF	75-0875
STRESSES	A NEW METHOD OF ANALYSIS TO OBTAIN EXACT SOLUTIONS FOR STRESSES AND STRAINS IN CIRCULAR INCLUSIONS	74-1063
STRESSES	A RELAXATION PROCEDURE FOR COMPUTING STRESSES IN BUILT-UP STRUCTURES	69-0063
STRESSES	A STATISTICAL THEORY FOR THE FRACTURE OF BRITTLE STRUCTURES SUBJECTED TO NONUNIFORM POLYAXIAL STRESS	73-0928
STRESSES	A STUDY OF INTERNAL STRESSES IN STATICALLY DEFORMED PNEUMATIC TIRES. A MATHEMATICAL MODEL FOR THE PN	70-0421
STRESSES	ACCURATE CALCULATION OF STRESSES IN BEAMS, DISKS AND PLATES BY MEANS OF FINITE ELEMENT METHODS. AND I	70-0133
STRESSES	AN AXISYMMETRIC FINITE ELEMENT ANALYSIS OF THE MECHANICAL AND THERMAL STRESSES IN BRAKE DRUMS	74-0311
STRESSES	ANALYSIS OF ELASTIC SHELLS OF REVOLUTION WITH MEMBRANE AND FLEXURE STRESSES UNDER ARBITRARY LOADING	68-0289
STRESSES	ANALYSIS OF ELASTO-PLASTIC THERMAL STRESSES BY THE FINITE ELEMENT METHOD	70-0142
STRESSES	ANALYSIS OF EMBANKMENT STRESSES AND DEFORMATIONS	67-0012

STRESSES	ANALYSIS OF NUCLEAR FUEL ROD LOCALIZED STRESSES	73-0708
STRESSES	ANALYSIS OF STRENGTH AND RESIDUAL STRESSES IN FILAMENT REINFORCED ALUMINUM CYLINDERS	75-0517
STRESSES	ANALYSIS OF STRESSES IN CONCRETE NEAR A REINFORCING BAR DUE TO BOND AND TRANSVERSE CRACKING	70-0351
STRESSES	ANALYSIS OF STRESSES IN PRESSURIZED WELDED PIPE IN THE CREEP RANGE	71-0404
STRESSES	ANALYSIS OF STRESSES IN SKEWED PLATES BY THE METHOD OF FINITE ELEMENTS	73-0248
STRESSES	ANALYSIS OF STRESSES AND DEFLECTIONS IN FRAME SUPPORTED TENTS	74-0689
STRESSES	ANALYSIS OF THERMAL STRESSES IN CORED CERAMIC BRICKS USED IN HYPERSONIC WIND TUNNEL HEATERS	72-0655
STRESSES	ANALYSIS OF THERMAL STRESSES AND METAL MOVEMENT DURING WELDING 1. ANALYTICAL STUDY	74-1026
STRESSES	ANALYSIS OF VISCOELASTIC CONTACT STRESSES IN AGRICULTURAL PRODUCTS USING A FINITE ELEMENT METHOD	74-0634
STRESSES	ANALYSIS OF VISCOELASTIC CONTACT STRESSES IN AGRICULTURAL PRODUCTS USING A FINITE ELEMENT METHOD	74-1199
STRESSES	ANCHORAGE ZONE STRESSES IN AXIALLY POST-TENSIONED I-SECTION MEMBERS WITH END BLOCKS	71-0305
STRESSES	ANWENDUNG DER FINITE-ELEMENT-METHODE ZUR BERECHNUNG VON SPANNUNGEN IN WASSERGESAETTIGTEN BOEDEN (FIN	73-0502
STRESSES	APPLICATION OF FINITE ELEMENTS TO THE ANALYSIS OF GEAR TOOTH STRESSES	72-0116
STRESSES	APPLICATION OF FINITE ELEMENT METHOD FOR THE DETAILED ANALYSIS OF HATCH CORNER STRESSES	73-0735
STRESSES	APPLICATION OF THE FINITE ELEMENT METHOD ON RESIDUAL STRESSES AND DEFORMATIONS IN WELDS	68-0178
STRESSES	APPLICATION OF THREE-DIMENSIONAL FINITE ELEMENTS FOR ANALYSIS OF IRRADIATION INDUCED STRESSES IN NUC	72-0078
STRESSES	ARCH DAM CONSTRUCTION STRESSES DUE TO HYDROSTATIC AND GRAVITY LOADS	75-0658
STRESSES	ASUPRA ANALIZEI PRIN METODA ELEMENTULUI FINIT A STARILOR DE EFORTURI SI DEFORMATII DIN MASIVELE DE P	70-0471
STRESSES	BALANCED STRESSES IN POST-YIELDED MULTI-MATERIAL STRUCTURAL JOINTS	69-0271
STRESSES	BERECHNUNG DER SCHWEISSEIGENSPANNUNGEN UND SCHWEISSFORMAENDERUNGEN MIT ELASTISCH-PLASTISCHEN FINITEN	75-0271
STRESSES	BERECHNUNG VON SPANNUNGEN UND TEMPERATURFELDERN IN FLOSSENWAENDEN UND AUFGESCHWEISSTEN BAUTEILEN. (C	74-0544
STRESSES	CALCULATING STRESSES IN A PRESSURE VESSEL WITH FILL TUBE - COMPARISON OF THE FINITE ELEMENT AND PHOT	72-0782
STRESSES	CALCULATION OF DISPLACEMENTS AND STRESSES IN ANISOTROPIC ROCKS BY USING METHOD OF FINITE ELEMENTS	73-0265
STRESSES	CALCULATION OF RESIDUAL STRESSES IN PLASTIC DEFORMATION PROCESSES	72-0241
STRESSES	CALCULATION OF THERMAL STRESSES IN A CYLINDER-CYLINDER INTERSECTION BY MEANS OF FINITE ELEMENTS	73-0571
STRESSES	CALCULATION OF THERMAL STRESSES IN TUBULAR CERAMIC COMPONENTS	73-0670
STRESSES	CLASSIFICATION OF FINITE ELEMENT STRESSES ACCORDING TO ASME SECTION 111 STRESS CATEGORIES	74-0610
STRESSES	COMPUTATION OF STRESSES IN TRIANGULAR FINITE ELEMENTS	66-0072
STRESSES	COMPUTER CALCULATIONS OF STRESSES IN AXISYMMETRIC THERMALLY LOADED COMPONENTS	68-0069
STRESSES	CONSISTENT STRESSES FOR THE FINITE ELEMENT STIFFNESS METHOD	72-0698
STRESSES	CONSTRUCTION SELF-WEIGHT STRESSES OF PRECAST SHEAR WALLS	71-0203
STRESSES	CONSTRUCTION STRESSES IN DWORSHAK DAM	65-0029
STRESSES	CREEP OF B/AL COMPOSITES AS INFLUENCED BY RESIDUAL STRESSES, BOND STRENGTH, AND FIBER PACKING GEOMET	74-1208
STRESSES	DERIVATION OF A STIFFNESS MATRIX FOR A RIGHT TRIANGULAR PLATE IN BENDING AND SUBJECTED TO INITIAL ST	62-0007
STRESSES	DETERMINATION OF STRESSES AT NODAL POINTS FROM FINITE ELEMENT SOLUTIONS TO ELASTIC PROBLEMS	71-0307
STRESSES	DETERMINATION OF STRESSES DUE TO DISCONTINUITIES IN FINITE PLATES OF ISOTROPIC AND ORTHOTROPIC MATER	73-1052
STRESSES	DETERMINATION OF STRESSES AND DEFORMATIONS ON PISTONS BY MEANS OF COMPUTER PROGRAMS AND PHOTOELASTIC	75-0553
STRESSES	DETERMINATION OF VOLUME CHANGE STRESSES IN PLAIN AND REINFORCED CONCRETE USING FINITE ELEMENT ANALYS	69-0477
STRESSES	DIGITAL COMPUTATION OF STRESSES AND DEFLECTIONS IN A BOX BEAM	67-0073
STRESSES	DIGITAL COMPUTATION OF STRESSES AND DEFLECTIONS IN A BOX BEAM - A PERFORMANCE COMPARISON BETWEEN FIN	69-0355
STRESSES	EFFECT OF BACKPACKING AND INTERNAL PRESSURIZATION ON STRESSES TRANSMITTED TO BURIED CYLINDERS	73-0897
STRESSES	EFFECT OF DISCONTINUITIES ON STRESSES AND DEFORMATIONS OF ROCK AROUND A TUNNEL. (EINFLUSS VON DISKON	70-0170
STRESSES	EFFECT OF OPENINGS ON STRESSES IN RIGID PAVEMENTS	73-0344
STRESSES	EFFECT OF RESERVOIR FILLING ON STRESSES AND MOVEMENT IN EARTH AND ROCKFILL DAMS. A REPORT OF AN INVE	72-0836
STRESSES	EINSATZ MODERNER RECHENVERFAHREN ZUR ERMITTLUNG DER SPANNUNGEN UND VERFORMUNGEN BEI RADSAETZEN. (USE	75-0186
STRESSES	ELASTIC STRESSES IN LAYERED SNOW PACKS	72-0925
STRESSES	ELEMENT STIFFNESS MATRICES FOR BOUNDARY COMPATIBILITY AND FOR PRESCRIBED BOUNDARY STRESSES	66-0061
STRESSES	EVALUATION OF STRESSES AND DEFLECTION OF SPUR GEAR TEETH UNDER STRAIN	72-0115
STRESSES	EVALUATION OF STRESSES IN SOIL UNDER RIGID WHEELS BY FINITE ELEMENT METHOD	73-1109
STRESSES	EXPERIMENTAL STUDY OF FLEXIBILITY FACTORS AND STRESSES OF WELDING ELBOWS WITH END EFFECTS	73-0422
STRESSES	FINITE ELEMENT ANALYSIS OF STRESSES IN SHAFTS DUE TO INTERFERENCE-FIT HUBS	69-0353
STRESSES	FINITE ELEMENT ANALYSIS OF STRESSES IN FUEL ELEMENTS OF A HIGH-TEMPERATURE GAS-COOLED REACTOR	71-0115
STRESSES	FINITE ELEMENT ANALYSIS SYSTEM FOR THE MECHANICAL BEHAVIOR OF ORIENTED FIBER COMPOSITE MATERIALS UND	73-0869
STRESSES	FINITE ELEMENT ANALYSIS OF STATIC AND DYNAMIC STRESSES AROUND SUDDENLY PUNCHED HOLES IN PLATES AND S	73-1088
STRESSES	FINITE ELEMENT ANALYSIS OF LEFT VENTRICULAR MYOCARDIAL STRESSES	74-0669
STRESSES	FINITE ELEMENT METHOD APPLIED TO PROBLEMS IN STRESSES AND DEFORMATION OF SOIL	67-0085
STRESSES	FINITE ELEMENT METHOD FOR DETERMINATION OF THERMAL STRESSES INTERACTIVE GRAPHICS FOR FINITE ELEMENT	74-0155
STRESSES	FINITE ELEMENT OF UNDERGROUND STRESSES UTILIZING STOCHASTICALLY SIMULATED MATERIAL PROPERTIES	70-0111
STRESSES	FINITE ELEMENT SOLUTION FOR AXISYMMETRIC TRANSIENT THERMAL STRESSES	72-0823
STRESSES	FINITE-ELEMENT ANALYSIS OF RADIATION-DAMAGE STRESSES IN THE GRAPHITE OF MATRIX FUEL ELEMENTS	70-0048
STRESSES	FINITE-ELEMENT ELASTIC-PLASTIC ANALYSIS OF RESIDUAL STRESSES DUE TO CLAD WELDING IN REACTOR VESSELS	75-0779
STRESSES	FINITE-ELEMENT METHOD FOR THE DETERMINATION OF THERMAL STRESSES IN ANISOTROPIC SOLIDS OF REVOLUTION	74-0262
STRESSES	FINITE-ELEMENT SOLUTION OF STRESSES AND DISPLACEMENTS IN A SOIL-CULVERT SYSTEM	73-1156
STRESSES	FREE VIBRATION OF DOUBLY CURVED RECTANGULAR PLATES INCLUDING THE PRESENCE OF MEMBRANE STRESSES WITH	69-0468
STRESSES	FUEL ELEMENT STRESSES AT DISCONTINUITIES AND INTERACTIONS BY FINITE ELEMENTS IN TWO DIMENSIONS	73-0083
STRESSES	GEOMETRIC METHOD FOR THE PREDICTION OF STRESSES IN INCLUSIONS, OREBODIES, AND MINING SYSTEMS	74-0219
STRESSES	HEAT TRANSFER PROBLEMS AND TEMPERATURE STRESSES	69-0326
STRESSES	IN VIVO STRESSES IN THE HUMAN LEFT VENTRICULAR WALL: ANALYSIS ACCOUNTING FOR THE IRREGULAR THREE DIM	72-0136
STRESSES	IN-SITU HORIZONTAL STRESSES IN LATERITE BY FINITE ELEMENT ANALYSIS	73-1048
STRESSES	INELASTIC MICROMECHANICS OF CURING STRESSES IN COMPOSITES	75-0284
STRESSES	INFLUENCE OF DEFECTS ON FUEL ELEMENT CLADDING STRESSES	73-0863
STRESSES	INFLUENCE OF END-EFFECTS ON STRESSES AND FLEXIBILITY OF A PIPING ELBOW WITH IN-PLANE MOMENT	74-0534
STRESSES	INTERACTION STRESSES IN COMPOSITE SYSTEMS	70-0215
STRESSES	INTERPRETATION OF FINITE ELEMENT STRESSES ACCORDING TO ASME SECTION 111	75-0655
STRESSES	MARINE PROPELLER BLADES STRESSES AND DEFORMATIONS WITH GIVEN, FROZEN, HYDRODYNAMIC LOADING	73-0474
STRESSES	MATRICE DE RIGIDITE D'UN ELEMENT DE POUTRE SOLLICITE A LA TORSION-FLEXION (STIFFNESS MATRIX OF A BEA	73-0442
STRESSES	MATRIX AND INTERFACE STRESSES IN A DISCONTINUOUS FIBER COMPOSITE MODEL	68-0040
STRESSES	MEASUREMENT OF ACTUAL DEDENDUM STRAINS, AND THE INFLUENCE OF DEDENDUM RADIUS ON STRESSES AND DEDENDU	72-0118
STRESSES	METHOD FOR THE COMPUTATION OF CYCLIC STRESSES IN A GAS-TURBINE WHEEL	74-0972
STRESSES	METHOD OF MINIMUM WEIGHT DESIGN WITH REQUIREMENTS IMPOSED ON STRESSES AND NATURAL FREQUENCIES	70-0149
STRESSES	METODY RASCHETNOGO OPREDELENIYA TEMPERATURNYKH NAPRYAZHENII V KRYSHKAKH TSILINDROV DVIGATELEI VNUTRE	74-0179
STRESSES	MODERN TECHNIQUES FOR CALCULATING STRESSES IN TANKS, AND THEIR LIMITATIONS. (MODERNE VERFAHREN ZUR S	70-0153
STRESSES	NATRUAL STRAINS AND STRESSES FOR TRAPEZOIDAL STRUCTURES ANALYSIS	69-0101
STRESSES	NEW APPROACH TO GEAR TOOTH ROOT STRESSES	72-0117
STRESSES	NOTCH STRESSES AT THE LOAD CARRYING ELASTIC CORE	73-0266
STRESSES	NOTE ON AN APPROXIMATE METHOD FOR COMPUTING CONSISTENT CONJUGATE STRESSES IN ELASTIC FINITE ELEMENTS	73-0050
STRESSES	NUMERICAL DETERMINATION OF STRESSES AND DEFORMATIONS IN ROCK TAKING INTO ACCOUNT DISCONTINUITIES	70-0066
STRESSES	NUMERICAL DETERMINATION OF STRESSES AND DEFORMATIONS IN ROCK TAKING INTO ACCOUNT DISCONTINUITIES	70-0167
STRESSES	ON MAXIMUM FILLET STRESSES IN BREECH RING	72-0736
STRESSES	ON THE DISTRIBUTION OF STRESSES AROUND CIRCULAR TUNNELS IN NON-HOMOGENEOUS ROCKS	66-0029
STRESSES	PARAMETRISCHE KERBSPANNUNGSUNTERSUCHUNG AM ELASTISCHEN KERN.(PARAMETRICAL INVESTIGATION OF NOTCH STR	72-0208
STRESSES	PHOTOELASTIC DETERMINATION OF THE STRESSES IN THE FOUNDATION OF THE NUCLEAR-POWER STATION BUGEY 2	75-0834
STRESSES	PREDICTION OF MARINE PROPELLER DISTORTION AND STRESSES USING A SUPERPARAMETRIC THICK SHELL FINITE EL	73-0370
STRESSES	PRINCIPAL STRESSES AT THE INTERMEDIATE SUPPORT OF PRESTRESSED CONCRETE CONTINUOUS BEAMS	70-0352
STRESSES	PROPELLER BLADE STRESSES, APPLICATION OF FINITE ELEMENT METHODS	74-0207
STRESSES	RASCHET NAPRYAZHENII L PEREMESHCHENII V SVODE PODZEMNOGO MASHINNOGO ZALA GES I V OKRUZHAYUSHCHEM SKA	74-0542
STRESSES	RECENT TRENDS OF RESEARCH INTO RESTRAINT STRESSES AND STRAINS IN RELATION TO WELD CRACKING	73-0483
STRESSES	REDUCTION OF STRESSES IN BURIED STRUCTURES	71-0698
STRESSES	RIGID PAVEMENT STRESSES UNDER AIRCRAFT LOADING	71-0423
STRESSES	SHEAR STRESSES IN BEAMS WITH DISCONTINUITY IN DEPTH	75-0487
STRESSES	SIGNIFICANT APPLICATIONS OF FINITE ELEMENTS AND FINITE DIFFERENCES METHODS TO SOLVE PROBLEMS OF THER	73-0864
STRESSES	SOLUTION OF STRESSES AND TRANSFORMATIONS OF EARTH DAMS BY THE METHOD OF FINITE ELEMENTS	69-0426

STRESSES	SOME MODELS FOR THE ANALYSIS OF STRESSES IN A TUBULAR FUEL PIN FOR HIGH TEMPERATURE REACTORS BY FINI	73-0816
STRESSES	SPANNUNGEN IN MEMBRANWAENDEN DURCH WAERMEBEANSPRUCHUNG UND MECHANISCHE BELASTUNG. (STRESSES IN MEMBR	75-0066
STRESSES	SPECIAL FINITE ELEMENTS FOR FLAW AND DISCONTINUITY STRESSES IN COMPOSITES	73-0846
STRESSES	STATIC AND DYNAMIC STRESSES IN SLOPES	66-0025
STRESSES	STATIC STRESSES BY LINEAR AND NONLINEAR METHODS	75-0278
STRESSES	STEP AND LOADS EFFECT ON STRESSES IN PRESTRESSED CONCRETE SHORT BRACKETS	71-0334
STRESSES	STIFFNESS MATRIX FOR BENDING OF A RECTANGULAR PLATE ELEMENT WITH INITIAL MEMBRANE STRESSES	62-0006
STRESSES	STRESSES ANALYSIS USING FINITE ELEMENT METHODS	70-0349
STRESSES	STRESSES AND DEFLECTIONS IN COOLING TOWER SHELLS DUE TO WIND LOADING	68-0088
STRESSES	STRESSES AND DEFORMATIONS IN MELTING PLATES	69-0476
STRESSES	STRESSES AND DEFORMATIONS OF VERTICAL SLOPES IN ELASTO-PLASTIC ROCKS	70-0114
STRESSES	STRESSES AND DEFORMATIONS IN JAIL GULCH EMBANKMENT	73-0391
STRESSES	STRESSES AND DISPLACEMENTS AROUND A CIRCULAR TUNNEL IN A THREE LAYER MEDIUM	69-0038
STRESSES	STRESSES AND DISPLACEMENTS AROUND TWO ADJACENT CIRCULAR OPENINGS NEAR TO THE GROUND SURFACE	74-0993
STRESSES	STRESSES AND DISPLACEMENTS AROUND OPENINGS IN HOMOGENEOUS ROCK	75-0168
STRESSES	STRESSES AND DISPLACEMENTS AROUND OPENINGS IN ROCK CONTAINING AN ELASTIC DISCONTINUITY	75-0169
STRESSES	STRESSES AND MOVEMENTS IN OROVILLE DAM	72-0200
STRESSES	STRESSES AND VIBRATION ANALYSIS OF INDUCER BLADES USING FINITE ELEMENT TECHNIQUE	70-0305
STRESSES	STRESSES AROUND FLEXIBLE ELLIPTIC PIPES	73-0215
STRESSES	STRESSES AROUND HOLES IN STIFFENED COMPOSITE PANELS USING LAURENT-SERIES AND FINITE ELEMENT METHODS	74-1252
STRESSES	STRESSES AROUND HOLES IN STIFFENED COMPOSITE PANELS USING LAURENT-SERIES AND FINITE ELEMENT METHODS	75-0733
STRESSES	STRESSES DUE TO CONCENTRATED LOADS IN RESTRAINED BENDING TESTS	67-0097
STRESSES	STRESSES IN A PARTLY YIELDED NOTCHED BAR - AN ASSESSMENT OF THREE ALTERNATIVE PROGRAMS	73-0267
STRESSES	STRESSES IN A PARTLY YIELDED NOTCHED BAR EM DASH AN ASSESSMENT OF THREE ALTERNATIVE PROGRAMS	73-0268
STRESSES	STRESSES IN A ROTATING DISK WITH CENTRAL AND NONCENTRAL HOLES (STRESS ANALYSIS BY THE FINITE ELEMENT	74-1337
STRESSES	STRESSES IN ANISOTROPIC MEDIA WITH PARTICULAR REFERENCE TO PROBLEMS OF ROCK MECHANICS	66-0079
STRESSES	STRESSES IN ANISOTROPIC MEDIA OF ROCK MECHANICS	66-0127
STRESSES	STRESSES IN BUTTRESS DAMS	65-0035
STRESSES	STRESSES IN CONTINUOUS CONCRETE BOX GIRDER BRIDGES	71-0233
STRESSES	STRESSES IN GABLE ROOF H. P. SHELLS.	72-0394
STRESSES	STRESSES IN NEARLY INCOMPRESSIBLE MATERIALS BY FINITE ELEMENTS WITH APPLICATION TO THE CALCULATION O	74-0383
STRESSES	STRESSES IN SHAFTS	67-0104
STRESSES	STRESSES IN SHORT THICK COMPOUND CYLINDERS	69-0385
STRESSES	TEMPERATURES AND STRESSES IN A BOILER MEMBRANE WALL TUBE	74-1053
STRESSES	TENDANCES ACTUELLES DES RECHERCHES SUR LES CONTRAINTES ET DEFORMATIONS DUES AU BRIDAGE EN RELATION A	74-0273
STRESSES	THE CALCULATION OF TIME-DEPENDENT STRESSES IN PCPVS BY AN APPROXIMATE METHOD	71-0086
STRESSES	THE CALCULATION OF THREE-DIMENSIONAL TEMPERATURE DISTRIBUTIONS AND THERMAL STRESSES USING FINITE ELE	71-0187
STRESSES	THE COMPUTATION OF TEMPERATURE DISTRIBUTIONS AND THERMAL STRESSES USING FINITE ELEMENT TECHNIQUES	71-0092
STRESSES	THE DETERMINATION OF STRESSES AND DEFORMATIONS OF REINFORCED CONCRETE AFTER CRACKING	64-0004
STRESSES	THE EARTHQUAKE EXPERIENCE AT KOYNA DAM AND STRESSES IN CONCRETE GRAVITY DAMS	72-0589
STRESSES	THE EFFECT OF IN PLANE STATIC STRESSES ON THE VIBRATION CHARACTERISTICS OF CURVED SANDWICH PLATES	71-0588
STRESSES	THE STRESSES IN AN ADHESIVE LAYER	72-0910
STRESSES	THERMAL STRESSES IN OPEN CYLINDRICAL SHELLS OF ARBITRARY BOUNDARY CONDITIONS BY A FINITE STRIP METHO	72-0371
STRESSES	THERMAL STRESSES IN FAST REACTOR BOILER TUBES AT THE SODIUM/GAS INTERFACE	73-0820
STRESSES	THERMAL STRESSES IN THIN CYLINDRICAL SHELLS	73-0822
STRESSES	THERMAL STRESSES IN A NOZZLE-DRUM INTERSECTION	73-0824
STRESSES	THERMAL STRESSES IN STEEL-CONCRETE COMPOSITE BRIDGES	75-0154
STRESSES	THREE-DIMENSIONAL CONSIDERATION OF THE STRESSES SURROUNDING OPEN-PIT MINE SLOPES	73-0492
STRESSES	THREE-DIMENSIONAL MECHANICAL STRESSES IN TOROIDAL MAGNETS FOR CONTROLLED THERMONUCLEAR REACTORS	73-0669
STRESSES	USER'S GUIDE FOR THE COMPUTER PROGRAM DESCRIBED IN TAM REPORT NO. 680 "DETERMINATION OF VOLUME CHANG	69-0473
STRESSES	VOLLSTAENDIGE SPANNUNGS-DEHNUNGS-TEMPERATURAENDERUNGSBEZIEHUNG FUER DIE SCHWEISSEIGENSPANNUNGSBERECH	75-0807
STRESSING	THE INELASTIC BEHAVIOUR OF STRUCTURES SUBJECTED TO CYCLIC, THERMAL AND MECHANICAL STRESSING CONDITIO	60-0004
STRETCHED	A FRAMEWORK METHOD FOR STRETCHED PLATES	71-0697
STRETCHING	BENDING AND STRETCHING OF PLATES, SPECIAL MODELS FOR UPPER AND LOWER BOUNDS	66-0026
STRETCHING	FINITE ELEMENT ANALYSIS AND EXPERIMENTAL INVESTIGATION OF SHEET METAL STRETCHING	72-0229
STRING	STRING EIN PROGRAMMSYSTEM FUR DIE MASCHINENUNABHANGIGE BEARBEITUNG VON ZEICHENKETTEN IN FORTRAN	68-0265
STRINGER	FINITE ELEMENTS FOR IMPROVED SKIN STRINGER STRUCTURAL ANALYSIS	70-0297
STRINGERS	VIBRATION AND STRESS ANALYSIS OF SOFT-BONDED SHUTTLE INSULATION TILES. MODAL ANALYSIS WITH COMPACT W	74-0708
STRIP	A FINITE ELEMENT ANALYSIS OF DISSOLVED OXYGEN DRAWDOWN AND SULFATE PRODUCTION IN STRIP MINE SPOIL DA	75-0795
STRIP	ANALYSIS OF BOX GIRDER BRIDGES BY FINITE STRIP METHOD	69-0050
STRIP	ANALYSIS OF BOX GIRDERS BY FOLDED PLATE AND STRIP METHODS	73-0982
STRIP	ANALYSIS OF CURVED BOX GIRDER BRIDGES BY FINITE STRIP METHOD	71-0485
STRIP	ANALYSIS OF SLAB AND GIRDER BRIDGES BY THE FINITE STRIP METHOD	70-0210
STRIP	ANALYSIS OF SLAB BRIDGES BY FINITE STRIP METHOD	73-0687
STRIP	ANALYSIS OF THE BIAXIAL STRIP AND SHEAR LAP TESTS FOR SOLID PROPELLANT CHARACTERIZATION	67-0019
STRIP	ARCH DAMS ANALYSED BY A LINEAR FINITE ELEMENT SHELL THE FINITE STRIP METHOD IN THE ANALYSIS OF ELAST	68-0189
STRIP	BEHAVIOUR OF STRIP FOOTINGS ON LAYERED COHESIVE SOILS	69-0223
STRIP	BENDING OF CYLINDRICAL SHELLS - SOLUTION USING CYLINDRICAL STRIP ELEMENTS - CONTRIBUTION TO FINITE	70-0363
STRIP	BUCKLING UNDER COMBINED LOADING OF THIN FLAT-WALLED STRUCTURES BY A COMPLEX FINITE STRIP METHOD	74-0788
STRIP	CONSOLIDATION OF A LAYER UNDER A STRIP LOAD	72-0222
STRIP	DAS FLAECHENTRAGWERKSPROGRAM VON STRIP. (STRIP PROGRAM FOR PLATE STRUCTURES)	72-0150
STRIP	DAS FLAECHENTRAGWERKSPROGRAM VON STRIP. (STRIP PROGRAM FOR PLATE STRUCTURES)	72-0150
STRIP	FINITE ELEMENT AND FINITE STRIP ANALYSES OF BRIDGES	72-0490
STRIP	FINITE ELEMENT METHOD AND FINITE STRIP METHOD	73-0347
STRIP	FINITE STRIP ANALYSIS OF SKEW SLABS	72-0369
STRIP	FINITE STRIP METHOD OF ANALYSIS OF ELASTIC SLABS	68-0190
STRIP	FOLDED PLATE STRUCTURES BY THE FINITE STRIP METHOD	69-0359
STRIP	FREQUENCY ANALYSIS OF SKEW ORTHOTROPIC PLATES BY THE FINITE STRIP METHOD	71-0413
STRIP	FREQUENCY ANALYSIS OF ORTHOTROPIC CIRCULAR CYLINDRICAL PANELS BY THE FINITE STRIP METHOD	73-0274
STRIP	ORTHOTROPIC RIGHT BRIDGES BY THE FINITE STRIP METHOD	69-0028
STRIP	SETTLEMENT ANALYSIS OF A STRIP FOOTING USING A CRITICAL STATE SOIL MODEL IN CONJUNCTION WITH FINITE	71-0533
STRIP	SOME CONSIDERATIONS ON THE DYNAMIC STRESS ANALYSIS OF A STRIP BY FINITE ELEMENT METHOD	71-0591
STRIP	STRESS CONCENTRATION FACTORS AROUND A PENETRATION IN A STRIP OR SLAB OF FINITE DIMENSIONS	69-0409
STRIP	STRIP FOOTING ANALYSIS	70-0755
STRIP	STRIP PROGRAM FOR DESIGNING LOAD BEARING SURFACE STRUCTURES (PROGRAMME STRIP POUR LE CALCUL DES STRU	71-0408
STRIP	STRIP PROGRAM FOR DESIGNING LOAD BEARING SURFACE STRUCTURES (PROGRAMME STRIP POUR LE CALCUL DES STRU	71-0408
STRIP	THE FINITE STRIP METHOD IN THE ANALYSIS OF ELASTIC PLATES WITH TWO SIMPLY SUPPORTED ENDS	68-0068
STRIP	THE FINITE STRIP METHOD AND ITS APPLICATIONS	73-0642
STRIP	THEORETICAL AND EXPERIMENTAL ANALYSIS OF ELASTOPLASTIC DEFORMATION IN A UNIAXIALLY LOADED STRIP OF 2	73-0720
STRIP	THERMAL STRESSES IN OPEN CYLINDRICAL SHELLS OF ARBITRARY BOUNDARY CONDITIONS BY A FINITE STRIP METHO	72-0371
STRIPS	FACTORS OF STRESS CONCENTRATION AT A HOLE IN FINITE STRIPS UNDER BENDING	72-0744
STRIPS	PARAMETRISCHE KERBSPANNUNGSUNTERSUCHUNGEN AN DER LOCHSCHEIBE MIT DER METHODE DER FINITEN ELEMENTE. (	75-0603
STRIPS	STATIC AND DYNAMIC BEHAVIOUR OF RECTANGULAR PLATES USING HIGHER ORDER FINITE STRIPS	72-0458
STRIPS	THE COMPUTATION OF TEARING ENERGY OF NICKED RUBBER STRIPS IN EXTENSION	74-0315
STRIPS	VYPOCET KRUHOVYCH LOMENICOVYCH KONSTRUKCII METODOU KONECNYCH PASOV. (CALCULATION OF CIRCULAR FOLDED-	74-0612
STROEMUNG	THEORETICAL AND EXPERIMENTAL CONSIDERATIONS ON FLOW OF VISCOUS LIQUIDS PAST PLATE AT SMALL AND AVERA	70-0144
STROEMUNGSFELDE	BERECHNUNG DES REIBUNGSLOSEN STROEMUNGSFELDES VON DREIDIMENSIONALEN AUFTRIEBSBEHAFTETENTRAGFLUEGELN	73-0533
STROEMUNGSMECHA	IMPORTANCE OF VARIATIONAL CALCULUS IN FLUID MECHANICS AND ITS APPLICATION TO COMPRESSIBLE POTENTIAL	73-1103
STRONGEST	STRONGEST COLUMN BY FINITE ELEMENT DISPLACEMENT METHOD	73-0136
STRONGLY	STRONGLY CURVED FINITE ELEMENT FOR SHELL ANALYSIS	73-0244
STRONGLY	STRONGLY PLANAR AC-LATTICES IN WHICH IDEAL OF FINITE-ELEMENTS IS STANDARD	75-0816
STRUCTRES	STABILITY ANALYSIS OF MULTILAYER SANDWICH STRUCTRES	73-0345

STRUCTURAL	A COMPARATIVE STUDY OF DIFFERENT NUMERICAL SOLUTION TECHNIQUES AS APPLIED TO A NONLINEAR STRUCTURAL	73-1117
STRUCTURAL	A COMPARISON OF SEVERAL COMPUTER SOLUTIONS TO THREE STRUCTURAL SHELL ANALYSIS PROBLEMS	73-0884
STRUCTURAL	A COMPARISON OF SEVERAL COMPUTER SOLUTIONS TO THREE STRUCTURAL SHELL ANALYSIS PROBLEMS	73-0907
STRUCTURAL	A CORRELATION STUDY OF METHODS OF MATRIX STRUCTURAL ANALYSIS	69-0110
STRUCTURAL	A FINITE ELEMENT APPROACH TO THE STRUCTURAL INSTABILITY OF BEAM COLUMNS, FRAMES, AND ARCHES	70-0434
STRUCTURAL	A FINITE ELEMENT FORMULATION FOR LARGE DEFLECTION ELASTIC STRUCTURAL ANALYSIS APPLIED TO PLATE BENDI	70-0485
STRUCTURAL	A FINITE ELEMENT METHOD FOR BENDING ANALYSIS OF LAYERED STRUCTURAL SYSTEMS	67-0134
STRUCTURAL	A FINITE-ELEMENT APPROACH TO OPTIMAL STRUCTURAL DESIGN	73-1162
STRUCTURAL	A FINITE-ELEMENT METHOD OF SOLUTION FOR STRUCTURAL FRAMES	67-0133
STRUCTURAL	A FORMULA FOR CERTAIN TYPES OF STIFFNESS MATRICES OF STRUCTURAL ELEMENTS	63-0003
STRUCTURAL	A GENERAL CALCULATION OF THE STRUCTURAL LOADS ON A SHIP IN A SEAWAY	71-0345
STRUCTURAL	A GENERAL FORMULA FOR STIFFNESS MATRICES OF STRUCTURAL ELEMENTS	63-0002
STRUCTURAL	A GENERAL SYSTEM FOR STRUCTURAL FINITE ELEMENT ANALYSIS	68-0107
STRUCTURAL	A GENERALIZATION OF THE DIRECT-STIFFNESS METHOD OF STRUCTURAL ANALYSIS	64-0002
STRUCTURAL	A KOITER TYPE METHOD FOR FINITE ELEMENT ANALYSIS OF NONLINEAR STRUCTURAL BEHAVIOUR	70-0064
STRUCTURAL	A KOITER-TYPE METHOD FOR FINITE ELEMENT ANALYSIS OF NONLINEAR STRUCTURAL BEHAVIOR	70-0493
STRUCTURAL	A METHOD OF COMPUTATION FOR STRUCTURAL DYNAMICS	59-0002
STRUCTURAL	A METHOD OF LIMIT POINT CALCULATION IN FINITE ELEMENT STRUCTURAL ANALYSIS	72-0786
STRUCTURAL	A NASTRAN POSTPROCESSOR FOR STRUCTURAL MODIFICATION REANALYSIS	71-0160
STRUCTURAL	A NEW LOOK AT STRUCTURAL ENERGY DISSIPATION	74-0816
STRUCTURAL	A REVIEW OF SOME ASPECTS OF STRUCTURAL OPTIMIZATION	75-0467
STRUCTURAL	A STRUCTURAL ANALYSIS OF CYLINDER-CONE-CYLINDER SHELLS BY THE FINITE ELEMENT METHOD	72-0796
STRUCTURAL	A STRUCTURAL EVALUATION OF RAPID METHODS OF BACKFILLING FOR BOMB DAMAGE REPAIR	74-0821
STRUCTURAL	A STRUCTURAL EVALUATION OF RAPID METHODS OF BACKFILLING FOR BOMB DAMAGE REPAIR - PHASE 11	75-0682
STRUCTURAL	A STUDY OF NUMERICAL ERROR IN STRUCTURAL SYSTEMS	70-0394
STRUCTURAL	A TECHNIQUE FOR THE FINITE ELEMENT ANALYSIS OF LARGE COMPLEX STRUCTURAL SYSTEMS	69-0514
STRUCTURAL	ACCURACY OF FINITE ELEMENT APPROXIMATIONS TO STRUCTURAL PROBLEMS	70-0120
STRUCTURAL	ACCURACY OF FINITE ELEMENT APPROXIMATIONS TO STRUCTURAL PROBLEMS	70-0438
STRUCTURAL	AFPAV COMPUTER CODE FOR STRUCTURAL ANALYSIS OF AIRFIELD PAVEMENTS	75-0684
STRUCTURAL	AN ASSUMED STRESS HYBRID FINITE ELEMENT FOR THREE-DIMENSIONAL PLASTIC STRUCTURAL ANALYSIS	74-0717
STRUCTURAL	AN IMPROVED AUTOMATED STRUCTURAL OPTIMIZATION PROGRAM	74-0732
STRUCTURAL	AN INTERACTIVE HYBRID TECHNIQUE FOR CRASHWORTHY DESIGN OF COMPLEX VEHICULAR STRUCTURAL SYSTEMS	74-0317
STRUCTURAL	AN ITERATIVE METHOD FOR LARGE SYSTEMS OF LINEAR STRUCTURAL EQUATIONS	73-1020
STRUCTURAL	ANALYSIS OF LARGE CLOSED SHIP STRUCTURES: APPLICATIONS TO TANKER STRUCTURAL DESIGN	70-0289
STRUCTURAL	ANALYSIS OF STRUCTURAL VIBRATIONS AND DYNAMIC RESPONSE	71-0059
STRUCTURAL	ANALYSIS OF STRUCTURAL ACOUSTIC INTERACTIONS IN METAL CERAMIC TRANSDUCERS	73-0401
STRUCTURAL	ANALYSIS OF THERMALLY INDUCED STRUCTURAL VIBRATIONS OF FINITE ELEMENT TECHNIQUES	68-0272
STRUCTURAL	ANALYSIS OF VISCOELASTOPLASTIC STRUCTURAL BEHAVIOUR OF ANISOTROPIC SHELLS BY THE FINITE ELEMENT METH	71-0056
STRUCTURAL	ANALYTICAL FINITE ELEMENT SIMULATION MODEL FOR STRUCTURAL CRASHWORTHINESS PREDICTION	73-0778
STRUCTURAL	APPLICATION OF FINITE ELEMENT COMPUTATIONS TO SOME SHIP STRUCTURAL PROBLEMS	69-0337
STRUCTURAL	APPLICATION OF FINITE ELEMENT TECHNIQUES TO STRESS AND STRUCTURAL ANALYSIS	70-0362
STRUCTURAL	APPLICATION OF FINITE ELEMENT METHODS TO COMPLETE AUTOMOBILE STRUCTURAL DESIGN EVALUATION	74-0312
STRUCTURAL	APPLICATION OF GENERALIZED CONSTRAINTS IN THE STIFFNESS METHOD OF STRUCTURAL ANALYSIS	66-0030
STRUCTURAL	APPLICATION OF INTERACTIVE GRAPHICS TO THE NUMERICAL METHODS USED IN STRUCTURAL ANALYSIS	73-0384
STRUCTURAL	APPLICATION OF MATRIX METHODS IN STRUCTURAL THEORY	69-0454
STRUCTURAL	APPLICATIONS OF CONTINUOUS TONE COMPUTER-GENERATED IMAGES IN STRUCTURAL MECHANICS	74-1035
STRUCTURAL	APPLICATIONS OF FINITE ELEMENT METHOD TO SHIP STRUCTURAL ANALYSIS	69-0210
STRUCTURAL	APPLICATIONS OF NASTRAN TO COUPLED STRUCTURAL AND HYDRODYNAMIC RESPONSES IN AIRCRAFT HYDRAULIC SYSTE	71-0744
STRUCTURAL	ASKA A COMPUTER SYSTEM FOR STRUCTURAL ENGINEERS	69-0345
STRUCTURAL	ASKA 105 - A PROBLEM ORIENTED LANGUAGE FOR STRUCTURAL ENGINEERS, VOLUME 11, APPLICATION	69-0461
STRUCTURAL	ASKA 105, A PROBLEM ORIENTED LANGUAGE FOR STRUCTURAL ENGINEERS, VOLUME 1, DESIGN	69-0460
STRUCTURAL	ASKA, SOFTWARE ENGINEERING FOR THE SEVENTIES IN STRUCTURAL ANALYSIS	73-0704
STRUCTURAL	ASTRA-BOEING'S ADVANCED STRUCTURAL ANALYZER	69-0118
STRUCTURAL	AUTOMATED STRUCTURAL DESIGN AND ANALYSIS OF ADVANCED COMPOSITE WING MODELS	73-0221
STRUCTURAL	AUTOMATED STRUCTURAL SYNTHESIS USING A RE NUMBER OF DESIGN COORDINATES	73-0584
STRUCTURAL	BALANCED STRESSES IN POST-YIELDED MULTI-MATERIAL STRUCTURAL JOINTS	69-0271
STRUCTURAL	BASIC PRINCIPLES OF STRUCTURAL DYNAMICS	71-0505
STRUCTURAL	BERECHNUNG VON STAHLBETONSCHEIBEN IM ZUSTAND II BEI ANNAHME EINES WIRKLICHKEITSNAHEN WERKSTOFFVERHAL	74-0978
STRUCTURAL	BERECHNUNGSMETHODEN ZUR ERMITTLUNG VON BELASTUNGSBEDINGUNGEN UND GRENZTRAGFAEHIGKEITSVERHALTEN VON K	73-0858
STRUCTURAL	BESTRAN: A TECHNIQUE FOR PERFORMING STRUCTURAL ANALYSES	75-0759
STRUCTURAL	BLOCKING TECHNIQUE FOR LARGE SCALE STRUCTURAL ANALYSIS	73-0243
STRUCTURAL	COMPANY TRAINING OF ENGINEERING IN STRUCTURAL ANALYSIS COMPUTER PROGRAM USAGE	72-0460
STRUCTURAL	COMPARISON OF STRUCTURAL LOADS: STATIC VERSUS DYNAMIC	70-0464
STRUCTURAL	COMPARISON OF STRUCTURAL TEST RESULTS WITH PREDICTIONS OF FINITE ELEMENT ANALYSIS	73-0158
STRUCTURAL	COMPUTATIONAL METHODS IN NUCLEAR REACTOR STRUCTURAL DESIGN FOR HIGH TEMPERATURE APPLICATIONS: AN INT	73-1001
STRUCTURAL	COMPUTATIONAL STABILITY IN NONLINEAR DISCRETE STRUCTURAL SYSTEMS	69-0145
STRUCTURAL	COMPUTATIONAL TECHNIQUES IN FINITE ELEMENT ANALYSIS OF STRUCTURAL PROBLEMS	71-0556
STRUCTURAL	COMPUTER AIDED STRUCTURAL ANALYSIS: THE MACHINE INDEPENDENT SYSTEM ASKA	70-0397
STRUCTURAL	COMPUTER GRAPHICS FOR STRUCTURAL ANALYSIS	74-0832
STRUCTURAL	COMPUTER METHODS FOR THE STRUCTURAL ANALYSIS OF MACHINE TOOLS	69-0325
STRUCTURAL	COMPUTER PROGRAM FOR STATIC AND DYNAMIC ANALYSIS OF LINEAR STRUCTURAL SYSTEMS	72-0683
STRUCTURAL	COMPUTER-AIDED INPUT/OUTPUT FOR USE WITH FINITE ELEMENT METHOD OF STRUCTURAL ANALYSIS	70-0330
STRUCTURAL	COMPUTER-AIDED INPUT/OUTPUT FOR THE USE WITH FINITE ELEMENT STRUCTURAL ANALYSES	73-0923
STRUCTURAL	COMPUTERIZED STRUCTURAL ANALYSIS OF THE WORLD'S LARGEST LIGHT-GAGE STEEL PRIMARY STRUCTURAL SYSTEM	73-0503
STRUCTURAL	COMPUTERIZED STRUCTURAL ANALYSIS OF THE WORLD'S LARGEST LIGHT-GAGE STEEL PRIMARY STRUCTURAL SYSTEM	73-0503
STRUCTURAL	COMPUTING SYSTEM FOR STRUCTURAL ANALYSIS OF CAR BODIES	72-0324
STRUCTURAL	CONF. ON MATRIX METHODS IN STRUCTURAL MECHANICS	68-0284
STRUCTURAL	CONFERENCE ON MATRIX METHODS IN STRUCTURAL MECHANICS	65-0065
STRUCTURAL	CONFERENCE ON MATRIX METHODS IN STRUCTURAL MECHANICS	69-0470
STRUCTURAL	CONFERENCE ON MATRIX METHODS IN STRUCTURAL MECHANICS	71-0773
STRUCTURAL	CONSISTENT MATRIX FORMULATIONS FOR STRUCTURAL ANALYSIS USING INFLUENCE COEFFICIENT TECHNIQUES	64-0017
STRUCTURAL	CORRELATION OF FINITE ELEMENT STRUCTURAL DYNAMIC ANALYSIS WITH MEASURED FREE VIBRATION CHARACTERISTI	74-0486
STRUCTURAL	CYCLIC PLASTIC ANALYSIS OF STRUCTURAL STEEL JOINTS	73-0799
STRUCTURAL	DATA GENERATION FOR FINITE-ELEMENT STRUCTURAL ANALYSIS OF THREE-DIMENSIONAL NAVAL SHIP STRUCTURES	75-0707
STRUCTURAL	DATA GENERATOR FOR THE IDEALIZATION FOR FINITE ELEMENT STRUCTURAL ANALYSIS OF NAVAL SHIP FLAT PLATED	72-0860
STRUCTURAL	DESIGN, MANUFACTURE, DEVELOPMENT, TEST, AND EVALUATION OF BORON/ALUMINUM STRUCTURAL COMPONENTS FOR S	74-0798
STRUCTURAL	DETAILED STRUCTURAL ANALYSIS OF THE HUMAN TIBIA	75-0379
STRUCTURAL	DEVELOPMENT OF A GUN-TURRET MODEL AND OF THE EQUATIONS FOR COUPLING IT WITH A HELICOPTER FINITE-ELEM	73-0752
STRUCTURAL	DEVELOPMENT OF ADVANCED STRUCTURAL OPTIMIZATION PROGRAMS AND THEIR APPLICATION TO LARGE ORDER STRAIN	66-0028
STRUCTURAL	DEVELOPMENTS IN DISCRETE ELEMENT FINITE DEFLECTION STRUCTURAL ANALYSIS BY FUNCTION MINIMIZATION	68-0144
STRUCTURAL	DEVELOPMENTS IN DISCRETE ELEMENT FINITE DEFLECTION STRUCTURAL ANALYSIS BY FUNCTION MINIMIZATION	68-0243
STRUCTURAL	DEVELOPMENTS IN STRUCTURAL ANALYSIS BY DIRECT ENERGY MINIMIZATION	68-0070
STRUCTURAL	DEVELOPMENTS IN STRUCTURAL ANALYSIS AND SYNTHESIS BY FUNCTION MINIMIZATION	71-0384
STRUCTURAL	DISCRETE ELEMENT APPROACH TO STRUCTURAL INSTABILITY ANALYSIS	63-0005
STRUCTURAL	DISCRETE ELEMENT STRUCTURAL THEORY OF FLUIDS	70-0009
STRUCTURAL	DISCUSSION OF FINITE ELEMENT TECHNIQUES IN STRUCTURAL MECHANICS	70-0263
STRUCTURAL	DISPLACEMENT METHOD IN STRUCTURAL ANALYSIS AS APPLIED TO RECTANGULAR SHELL BEAMS	72-0411
STRUCTURAL	DUALITY IN STRUCTURAL ANALYSIS BY FINITE ELEMENTS. STATIC GEOMETRIC ANALOGIES, THE DUAL PRINCIPLES O	71-0583
STRUCTURAL	DYNAMIC ANALYSIS OF STRUCTURAL SYSTEMS USING COMPONENT MODES	64-0024
STRUCTURAL	DYNAMIC ANALYSIS OF AUTOMOTIVE STRUCTURAL SYSTEMS	70-0340
STRUCTURAL	DYNAMIC ANALYSIS OF STRUCTURAL DEFORMATION AND METAL FORMING	75-0058
STRUCTURAL	DYNAMIC AND STOCHASTIC ASPECTS IN THE STRUCTURAL ANALYSIS OF VLCCS AND ULCCS	74-0298

STRUCTURAL	DYNAMIC SHELL THEORY COUPLING THICKNESS STRESS WAVE EFFECTS WITH GROSS STRUCTURAL RESPONSE	73-0379
STRUCTURAL	DYNAMIC STRUCTURAL ANALYSIS WITH THE NASTRAN COMPUTER PROGRAM	70-0213
STRUCTURAL	EFFECTS OF FLEXIBLE CONNECTIONS ON BODY STRUCTURAL RESPONSE	74-0140
STRUCTURAL	EIGENVALUES OF COLLINEAR BEAM STRUCTURES USING FINITE ELEMENT TECHNIQUES AND VARIOUS DYNAMIC REPRESE	67-0124
STRUCTURAL	ELASTIC DEFORMATION OF LIGHTWEIGHT MIRRORS METHOD OF LIMIT POINT CALCULATION IN FINITE ELEMENT STRUC	72-0232
STRUCTURAL	ELASTO-PLASTIC CYCLIC ANALYSIS OF STRUCTURAL MEMBERS	72-0202
STRUCTURAL	ELASTODYNAMICS OF COMPLEX STRUCTURAL SYSTEMS	71-0563
STRUCTURAL	ELEMENTS FOR FINITE ELEMENT STRUCTURAL ANALYSIS PROGRAMS	71-0335
STRUCTURAL	ENERGY THEOREMS AND STRUCTURAL ANALYSIS	54-0001
STRUCTURAL	ENERGY THEOREMS AND STRUCTURAL ANALYSIS	67-0136
STRUCTURAL	ESTIMATIONS OF THE EFFECT OF DESIGN CHANGES ON THE MODES AND FREQUENCIES OF VIBRATING STRUCTURAL ELE	74-0819
STRUCTURAL	EVALUATION OF ALGORITHMS FOR STRUCTURAL MODIFICATION	69-0243
STRUCTURAL	EVALUATION OF SOLUTION PROCEDURES FOR MATERIAL AND/OR GEOMETRICALLY NONLINEAR STRUCTURAL ANALYSIS	73-0362
STRUCTURAL	EXAMPLE OF LARGE-SCALE STRUCTURAL ANALYSIS. COMPARISON BETWEEN FINITE ELEMENT CALCULATION AND FULL S	74-0177
STRUCTURAL	EXPERIENCE WITH FINITE ELEMENT COMPUTATIONS FOR NONLINEAR STRUCTURAL PROBLEMS	74-1239
STRUCTURAL	FAST FRACTURE RESISTANCE AND CRACK ARREST IN STRUCTURAL STEELS	73-0766
STRUCTURAL	FINEL, UNIVERSELLES PROGRAMMSYSTEM ZUR BERECHNUNG DER ELASTISCHEN EIGENSCHAFTEN VON MASCHINENBAUTEIL	72-0132
STRUCTURAL	FINESSE - A FINITE ELEMENT SYSTEM. PART V FINITE ELEMENT DESIGN - DERIVATIVES FOR STRUCTURAL SHAPE D	72-0963
STRUCTURAL	FINITE DEFLECTION STRUCTURAL ANALYSIS USING PLATE AND SHELL DISCRETE ELEMENTS	68-0143
STRUCTURAL	FINITE DEFLECTION STRUCTURAL ANALYSIS USING PLATE AND CYLINDRICAL SHELL DISCRETE ELEMENTS	68-0193
STRUCTURAL	FINITE ELEMENT ANALYSIS OF STRUCTURAL FRAMES	67-0032
STRUCTURAL	FINITE ELEMENT ANALYSIS OF STRUCTURAL INTEGRITY OF A REACTOR PRESSURE VESSEL DURING EMERGENCY CORE C	70-0026
STRUCTURAL	FINITE ELEMENT ANALYSIS OF WOOD STRUCTURAL MEMBERS	70-0180
STRUCTURAL	FINITE ELEMENT ANALYSIS OF STRUCTURAL INSTABILITY BY ASSOCIATION OF PULSATING EXCITATIONS	74-0453
STRUCTURAL	FINITE ELEMENT ANALYSIS OF TRANSIENT NONLINEAR STRUCTURAL BEHAVIOR	75-0297
STRUCTURAL	FINITE ELEMENT APPLICATIONS IN NONLINEAR STRUCTURAL ANALYSIS	69-0204
STRUCTURAL	FINITE ELEMENT METHOD FOR MACHINE TOOL STRUCTURAL ANALYSIS	71-0136
STRUCTURAL	FINITE ELEMENT STRUCTURAL ANALYSIS USING COMPUTERS	70-0216
STRUCTURAL	FINITE ELEMENT STRUCTURAL ANALYSIS OF LOCAL INSTABILITY	72-0179
STRUCTURAL	FINITE ELEMENT STRUCTURAL ANALYSIS OF LOCAL INSTABILITY	74-0098
STRUCTURAL	FINITE ELEMENT STRUCTURAL ANALYSIS OF LOCAL INSTABILITY - REPLY	74-0099
STRUCTURAL	FINITE ELEMENT STRUCTURAL ANALYSIS AS APPLIED TO AN AUTOMOTIVE DOOR STRUCTURE	74-0310
STRUCTURAL	FINITE ELEMENT TECHNIQUES IN STRUCTURAL MECHANICS	70-0116
STRUCTURAL	FINITE ELEMENTS FOR THE REPRESENTATION OF PLANE AND AXISYMMETRIC STRUCTURAL REINFORCEMENT	68-0037
STRUCTURAL	FINITE ELEMENTS FOR IMPROVED SKIN STRINGER STRUCTURAL ANALYSIS	70-0297
STRUCTURAL	FINITE ELEMENTS IN STRUCTURAL ANALYSIS	75-0739
STRUCTURAL	FINITE-ELEMENT ANALYSIS OF STRUCTURAL PROBLEMS	74-0636
STRUCTURAL	FINITE-ELEMENT ANALYSIS OF STRUCTURAL RESPONSE IN HCDA	75-0407
STRUCTURAL	FINITE-ELEMENT FORMULATION OF MIXED AND STRESS VARIATIONAL PRINCIPLES FOR STRUCTURAL DYNAMIC ANALYSI	72-0964
STRUCTURAL	FORMULATION AND EXPERIMENTAL VERIFICATION OF AN AXISYMMETRIC FINITE-ELEMENT STRUCTURAL ANALYSIS	71-0354
STRUCTURAL	FORMULATION, COMPUTATION, AND SOLUTION PROCEDURES FOR MATERIAL AND/OR GEOMETRIC NONLINEAR STRUCTURAL	72-0346
STRUCTURAL	FREE VIBRATION ANALYSIS OF SPINNING STRUCTURAL SYSTEMS	73-1077
STRUCTURAL	FREQUENCY-DEPENDENT SHAPE FUNCTIONS FOR FREE AND FORCED STRUCTURAL VIBRATION CALCULATION	73-1091
STRUCTURAL	FUNDAMENTALS OF FINITE ELEMENT TECHNIQUES FOR STRUCTURAL ENGINEERS	73-1050
STRUCTURAL	FUTURE DEVELOPMENTS IN STRUCTURAL ANALYSIS	74-0390
STRUCTURAL	GENERAL DESCRIPTION OF SESAM-69, SUPER ELEMENT STRUCTURAL ANALYSIS (PROGRAM) MODULES	71-0612
STRUCTURAL	GENERAL TREATMENT OF STRUCTURAL MODIFICATIONS	72-0166
STRUCTURAL	GENERALISED MATRIX FORCE AND DISPLACEMENT METHODS FOR LINEAR STRUCTURAL ANALYSIS	65-0064
STRUCTURAL	GEOMETRICALLY NONLINEAR STRUCTURAL ANALYSIS BY THE DIRECT STIFFNESS METHOD	71-0560
STRUCTURAL	GLOBAL-LOCAL FINITE ELEMENT STRUCTURAL ANALYSIS, ANALYTICAL CONSIDERATIONS/ELEMENT FORMULATION	74-1090
STRUCTURAL	HABEAS- A STRUCTURAL DYNAMICS ANALYSIS SYSTEM	70-0121
STRUCTURAL	HANDBOOK FOR THE ENGINEERING STRUCTURAL ANALYSIS OF SOLID PROPELLANTS	71-0642
STRUCTURAL	HORIZONS IN THE STRUCTURAL USE OF ADVANCED MATERIALS - THRAUSTICS	70-0415
STRUCTURAL	IMPACT OF COMPUTERS ON THE PRACTICE OF STRUCTURAL ENGINEERING IN CONCRETE	72-0414
STRUCTURAL	IMPACT OF COMPUTERS ON THE PRACTICE OF STRUCTURAL ENGINEERING IN CONCRETE	72-0415
STRUCTURAL	IMPACT OF FINITE ELEMENT METHOD ON STRUCTURAL ENGINEERING	75-0007
STRUCTURAL	IMPROVEMENTS OF FINITE ELEMENT SOLUTIONS FOR STRUCTURAL AND NON-STRUCTURAL APPLICATIONS	71-0229
STRUCTURAL	IMPROVEMENTS OF FINITE ELEMENT SOLUTIONS FOR STRUCTURAL AND NONSTRUCTURAL APPLICATIONS	72-0671
STRUCTURAL	INCREMENTAL SOLUTION PROCEDURE FOR NONLINEAR STRUCTURAL DYNAMICS PROBLEMS	75-0656
STRUCTURAL	INFLUENCE OF JOINTS IN PANELIZED STRUCTURAL SYSTEMS	72-0204
STRUCTURAL	INITIAL STRAINS IN THE MATRIX FORCE METHOD OF STRUCTURAL ANALYSIS	60-0008
STRUCTURAL	INTEGRATED PROGRAM FOR THE DYNAMIC STRUCTURAL CALCULATION OF MOBIL OFFSHORE UNITS	74-0490
STRUCTURAL	INTERACTION OF AREA COORDINATES IN MATRIX STRUCTURAL ANALYSIS	68-0150
STRUCTURAL	INTERACTIVE COMPUTER GRAPHICS IN STRUCTURAL ANALYSIS: ONLINE 72	72-0639
STRUCTURAL	INTERACTIVE GRAPHICS AIDED STRUCTURAL ANALYSIS	74-0867
STRUCTURAL	INTRODUCTION TO FINITE ELEMENT METHODS OF STRUCTURAL ANALYSIS	68-0231
STRUCTURAL	ISOTROPIC-KINEMATIC HARDENING MODEL FOR ELASTIC-PLASTIC CYCLIC STRUCTURAL ANALYSIS	75-0210
STRUCTURAL	ITERATIVE METHOD FOR THE ANALYSIS OF LARGE STRUCTURAL SYSTEMS	68-0236
STRUCTURAL	KRON'S METHOD: AN ALGORITHM FOR THE EIGENVALUE ANALYSIS OF LARGE-SCALE STRUCTURAL SYSTEMS	73-0758
STRUCTURAL	LAGRANGE MULTIPLIER TECHNIQUES IN STRUCTURAL ANALYSIS	65-0045
STRUCTURAL	LARGE CAPACITY EQUATION SOLVER FOR STRUCTURAL ANALYSIS	74-0983
STRUCTURAL	LARGE SCALE COMPUTER PROGRAMS FOR STRUCTURAL ANALYSIS	70-0211
STRUCTURAL	LARGE USER-ORIENTATED SYSTEMS OF PROGRAMS FOR STRUCTURAL ANALYSIS AND DESIGN	74-0482
STRUCTURAL	LIMITATIONS OF THE FINITE ELEMENT METHOD IN STRUCTURAL ANALYSIS	73-0420
STRUCTURAL	LINEAR AND NONLINEAR FINITE ELEMENT ANALYSIS OF WOOD STRUCTURAL MEMBERS	73-0117
STRUCTURAL	LINEAR METHODS OF STRUCTURAL ANALYSIS	71-0028
STRUCTURAL	MAGIC II. AN AUTOMATED GENERAL PURPOSE SYSTEM FOR STRUCTURAL ANALYSIS	71-0495
STRUCTURAL	MAGIC II: AN AUTOMATED GENERAL PURPOSE SYSTEM FOR STRUCTURAL ANALYSIS. VOLUME III. PROGRAMMER'S MANU	71-0756
STRUCTURAL	MAGIC II: AN AUTOMATED GENERAL PURPOSE SYSTEM FOR STRUCTURAL ANALYSIS. VOLUME 1. ENGINEER'S MANUAL (	71-0757
STRUCTURAL	MAGIC III: AN AUTOMATED GENERAL PURPOSE SYSTEM FOR STRUCTURAL ANALYSIS. VOLUME III. PROGRAMMER'S MAN	72-0726
STRUCTURAL	MAGIC III: AN AUTOMATED GENERAL PURPOSE SYSTEM FOR STRUCTURAL ANALYSIS VOLUME 1. ENGINEER'S MANUAL	72-0727
STRUCTURAL	MAGIC, AN AUTOMATED GENERAL PURPOSE SYSTEM FOR STRUCTURAL ANALYSIS, VOLUME III. PROGRAMMER'S MANUAL	69-0090
STRUCTURAL	MAGIC, AN AUTOMATED GENERAL PURPOSE SYSTEM FOR STRUCTURAL ANALYSIS, VOLUME II, USER'S MANUAL	69-0144
STRUCTURAL	MAGIC, AN AUTOMATED GENERAL PURPOSE SYSTEM FOR STRUCTURAL ANALYSIS	69-0173
STRUCTURAL	MAN MACHINE INTERACTIVE STRUCTURAL ANALYSIS AS A PRELIMINARY DESIGN AID	69-0293
STRUCTURAL	MANIPULATION ERRORS IN COMPUTER SOLUTION OF CRITICAL SIZE STRUCTURAL EQUATIONS	71-0367
STRUCTURAL	MATHEMATICAL OBSERVATIONS IN STRUCTURAL DYNAMICS	73-0156
STRUCTURAL	MATRIX AND DIGITAL COMPUTER METHODS IN STRUCTURAL ANALYSIS	69-0143
STRUCTURAL	MATRIX DISPLACEMENT ANALYSIS OF SHELLS WITH AN AXISYMMETRIC MIDSURFACE BUT HAVING CIRCUMFERENTIALLY	68-0031
STRUCTURAL	MATRIX FORMULATION OF STRUCTURAL STABILITY PROBLEMS	65-0039
STRUCTURAL	MATRIX ITERATION AND ACCELERATION PROCESSES IN FINITE ELEMENT PROBLEMS OF STRUCTURAL MECHANICS	70-0132
STRUCTURAL	MATRIX METHODS AND USE OF COMPUTERS IN STRUCTURAL ENGINEERING	71-0615
STRUCTURAL	MATRIX METHODS IN STRUCTURAL MECHANICS	66-0065
STRUCTURAL	MATRIX METHODS OF AEROSPACE STRUCTURAL ANALYSIS	68-0198
STRUCTURAL	MATRIX METHODS OF STRUCTURAL ANALYSIS	62-0004
STRUCTURAL	MATRIX METHODS OF STRUCTURAL ANALYSIS	63-0014
STRUCTURAL	MATRIX METHODS OF STRUCTURAL ANALYSIS. A PRECIS OF RECENT DEVELOPMENTS	63-0019
STRUCTURAL	MATRIX METHODS OF STRUCTURAL ANALYSIS	71-0620
STRUCTURAL	MATRIX STRUCTURAL ANALYSIS	71-0614
STRUCTURAL	MESY - EIN PROGRAMMSYSTEM ZUR UNTERSUCHUNG VON TRAGWERKEN. (STRUCTURAL ANALYSIS BY MESY PROGRAMMING	75-0360
STRUCTURAL	MODAL COUPLING PROCEDURES ADAPTED TO NASTRAN ANALYSIS OF THE 1/8 - SCALE SHUTTLE STRUCTURAL DYNAMICS	75-0677
STRUCTURAL	MODAL COUPLING PROCEDURES ADAPTED TO NASTRAN ANALYSIS OF THE 1/8-SCALE SHUTTLE STRUCTURAL DYNAMICS M	75-0678

STRUCTURAL	MODELING AND ANALYSIS TECHNIQUES FOR PREDICTION OF STRUCTURAL AND BIODYNAMIC CRASH-IMPACT RESPONSE	75-0300
STRUCTURAL	MODULAR APPROACH TO STRUCTURAL SIMULATION FOR VEHICLE CRASHWORTHINESS PREDICTION	75-0737
STRUCTURAL	NASTRAN - A FINITE ELEMENT PROGRAM FOR STRUCTURAL ANALYSIS	72-0131
STRUCTURAL	NASTRAN AS ANALYSIS TOOL IN A STRUCTURAL DESIGN OPTIMIZATION PROCESS	71-0235
STRUCTURAL	NASTRAN STRUCTURAL/THERMAL ANALYZER	72-0152
STRUCTURAL	NASTRAN, NASA'S GENERAL PURPOSE STRUCTURAL ANALYSIS PROGRAM	70-0370
STRUCTURAL	NODE NUMBERING OPTIMIZATION IN STRUCTURAL ANALYSIS	75-0077
STRUCTURAL	NONDETERMINISTIC PROBLEMS IN STRUCTURAL MECHANICS	71-0746
STRUCTURAL	NONLINEAR FINITE ELEMENT AND ITS APPLICATION TO LARGE SCALE STRUCTURAL PROBLEMS	75-0386
STRUCTURAL	NONLINEAR METHODS OF STRUCTURAL ANALYSIS	71-0029
STRUCTURAL	NONLINEAR STRESS ANALYSIS OF INDETERMINATE STRUCTURAL ELEMENTS SUBJECTED TO BIAXIAL STRESS STATES US	73-1152
STRUCTURAL	NONLINEAR STRUCTURAL ANALYSIS; TENSOR FORMULATION	65-0046
STRUCTURAL	NONSAP - A NONLINEAR STRUCTURAL ANALYSIS PROGRAM	74-1006
STRUCTURAL	NONSAP: A STRUCTURAL ANALYSIS PROGRAM FOR STATIC AND DYNAMIC RESPONSE OF NONLINEAR SYSTEMS	74-0848
STRUCTURAL	NOTE ON THE THEORY OF AIRCRAFT STRUCTURAL ANALYSIS	59-0005
STRUCTURAL	NUMERICAL ANALYSIS OF SHIP STRUCTURAL ELEMENTS BY MEANS OF FINITE ELEMENT METHOD AND PHOTO-ELASTIC F	69-0125
STRUCTURAL	NUMERICAL METHODS FOR THE SOLUTION OF NONLINEAR PROBLEMS IN STRUCTURAL ANALYSIS	73-0506
STRUCTURAL	NUMERICAL SOLUTION SCHEMES FOR HIGHLY NONLINEAR STATIC STRUCTURAL BEHAVIOR	75-0679
STRUCTURAL	NUMERICAL-PERTURBATION METHOD FOR THE NONLINEAR ANALYSIS OF STRUCTURAL VIBRATIONS	74-0499
STRUCTURAL	ON IMPROVING EFFICIENCY OF AN ALGORITHM FOR STRUCTURAL OPTIMIZATION AND A USERS MANUAL FOR PROGRAM T	74-0778
STRUCTURAL	ON THE APPLICATION OF RESPONSE LIMITING TO FINITE ELEMENT STRUCTURAL ANALYSIS PROGRAMS	75-0192
STRUCTURAL	ON THE APPLICATION OF RESPONSE LIMITING TO FINITE ELEMENT STRUCTURAL ANALYSIS PROGRAMS	75-0822
STRUCTURAL	ON THE EFFECT OF APPROXIMATING HIGHER ORDER TERMS IN CERTAIN FINITE ELEMENT MODELS OF NONLINEAR STRU	73-0603
STRUCTURAL	ON THE FINITE ELEMENT ANALYSIS OF THIN-WALLED STRUCTURES WITH APPLICATION TO STRUCTURAL PROBLEMS IN	73-0832
STRUCTURAL	ON THE STRUCTURAL DESIGN OF TUBULAR JOINTS	74-0492
STRUCTURAL	ON THE UNCONDITIONAL STABILITY OF AN IMPLICIT ALGORITHM FOR NONLINEAR STRUCTURAL DYNAMICS	75-0925
STRUCTURAL	OPTIMALITY CRITERIA IN STRUCTURAL DESIGN	71-0758
STRUCTURAL	OPTIMIZATION OF STRUCTURAL DESIGN 1	73-0904
STRUCTURAL	PAPA, STRUCTURAL ANALYSIS OF PLATES AND SHELLS USING TRAPEZOIDAL AND TRIANGULAR PLATE ELEMENTS	67-0059
STRUCTURAL	PERFORMANCE OF STRUCTURAL WOOD MEMBERS EXPOSED TO FIRE	75-0060
STRUCTURAL	PLANETARY EXPLORER STRUCTURAL CONCEPT EVALUATION MODEL USING NASTRAN	71-0742
STRUCTURAL	PRACTICAL APPLICATION OF A COMPUTERIZED STRUCTURAL ANALYSIS SYSTEM WHICH ADOPTS A FINITE ELEMENT TEC	69-0225
STRUCTURAL	PROBABILISTIC EVALUATION OF THE RELATIVE BEHAVIOUR OF TUNGSTEN CARBIDE COMPONENTS IN STRUCTURAL APPL	75-0579
STRUCTURAL	PROGRAM TO REDUCE THE SIZE OF STRUCTURAL MATRICES	72-0912
STRUCTURAL	PROGRESS IN COMPUTERIZED STRUCTURAL ANALYSIS OF LARGE VESSELS	74-0949
STRUCTURAL	QUALITY IN STRUCTURAL ANALYSIS BY FINITE ELEMENTS	71-0500
STRUCTURAL	REANALYSIS FOR LIMITED STRUCTURAL DESIGN MODIFICATIONS	72-0650
STRUCTURAL	RECENT ADVANCES IN COMPUTATIONAL METHODS IN STRUCTURAL MECHANICS AND DESIGN,	72-0575
STRUCTURAL	RECENT ADVANCES IN MATRIX METHODS OF STRUCTURAL ANALYSIS	64-0016
STRUCTURAL	RECENT ADVANCES IN MATRIX METHODS IN STRUCTURAL ANALYSIS AND DESIGN	70-0290
STRUCTURAL	RECENT ADVANCES IN NUMERICAL ANALYSIS OF STRUCTURAL EIGENVALUE PROBLEMS	73-0316
STRUCTURAL	RECENT TRENDS IN STUDIES AT INTERNATIONAL CONFERENCE ON STRUCTURAL ENGINEERING	71-0347
STRUCTURAL	RESPONSE OF NUCLEAR STRUCTURAL SYSTEMS TO TRANSIENT AND RANDOM EXCITATIONS, USING BOTH DETERMINISTIC	74-1047
STRUCTURAL	RESULTS ON THE CONVERGENCE OF THE FINITE ELEMENT METHOD IN STRUCTURAL AND NONSTRUCTURAL CASES	74-0416
STRUCTURAL	REVIEW OF OPTIMALITY CRITERIA APPROACHES TO STRUCTURAL OPTIMIZATION	74-0652
STRUCTURAL	ROHM AND HAAS THERMOSTRUCTURAL ANALYSIS SYSTEM, THERMAL STRUCTURAL ANALYSIS PROGRAMS - A SURVEY AND	72-0007
STRUCTURAL	SAMECS STRUCTURAL ANALYSIS SYSTEM - USER'S DOCUMENT	69-0495
STRUCTURAL	SAP IV, A STRUCTURAL ANALYSIS PROGRAM FOR STATIC AND DYNAMIC RESPONSE OF LINEAR SYSTEMS	73-0650
STRUCTURAL	SAP-A GENERAL STRUCTURAL ANALYSIS PROGRAM FOR LINEAR SYSTEMS	72-0329
STRUCTURAL	SENSITIVITY ANALYSIS OF TORSIONAL VIBRATION CHARACTERISTICS OF HELICOPTER ROTOR BLADES. PART 1: STRU	74-0846
STRUCTURAL	SHIP STRUCTURAL ANALYSIS IN LLOYD'S REGISTER OF SHIPPING	75-0285
STRUCTURAL	SHOCK ANALYSIS OF STRUCTURAL SYSTEMS	70-0426
STRUCTURAL	SIMPLIFIED INELASTIC HIGH TEMPERATURE STRUCTURAL ANALYSIS OF MODERATELY COMPLEX SPATIALLY THREE DIME	74-0445
STRUCTURAL	SOLUTION METHODS FOR EIGENVALUE PROBLEMS IN STRUCTURAL MECHANICS	73-1095
STRUCTURAL	SOLUTION OF AUTOMOTIVE STRUCTURAL PROBLEMS USING THE FINITE ELEMENT METHOD AND COMPUTER GRAPHICS	71-0341
STRUCTURAL	SOLUTION PROCEDURE FOR NONLINEAR STRUCTURAL PROBLEMS	73-0425
STRUCTURAL	SOLUTION SCHEMES FOR PROBLEMS OF NONLINEAR STRUCTURAL DYNAMICS	74-0505
STRUCTURAL	SOME CONSIDERATIONS REGARDING STRUCTURAL OPTIMIZATION AND FINITE ELEMENT ANALYSIS	74-0424
STRUCTURAL	SOME PROGRAM DEVELOPMENTS OF MATRIX STRUCTURAL ANALYSIS IN JAPAN	72-0322
STRUCTURAL	SOME STRUCTURAL PROBLEMS - STANDARD OIL OF INDIANA BUILDING	73-0180
STRUCTURAL	SPECIAL STRUCTURAL ELEMENTS FOR PIPING ANALYSIS	74-0502
STRUCTURAL	STARDYNE STRUCTURAL ANALYSIS SYSTEM, ANALYTICAL CAPABILITY MANUAL	68-0235
STRUCTURAL	STARS - A COMPUTER CODE FOR STRUCTURAL DYNAMIC ANALYSIS OF REACTOR SYSTEM	71-0204
STRUCTURAL	STATIC AND DYNAMIC NONLINEAR STRUCTURAL PROBLEMS	72-0848
STRUCTURAL	STATIC STRUCTURAL ANALYSIS OF SUSPENSION BRIDGES	66-0110
STRUCTURAL	STOCHASTIC FINITE ELEMENT SIMULATION OF THE NONLINEAR STRUCTURAL RESPONSE OF FIBROUS COMPOSITE MATER	75-0762
STRUCTURAL	STRAP - A COMPUTER CODE FOR THE STATIC AND DYNAMIC ANALYSIS OF REACTOR STRUCTURAL SYSTEMS	71-0078
STRUCTURAL	STRAP: COMPUTER CODE FOR STATIC AND DYNAMIC STRUCTURAL ANALYSIS AND STUDIES MADE USING THE CODE	70-0409
STRUCTURAL	STRENGTH OF SHIP STRUCTURAL ELEMENTS	75-0051
STRUCTURAL	STRESS ANALYSIS IN HULL STRUCTURAL MEMBERS BY PHOTOELASTIC EXPERIMENTS AND FINITE ELEMENT METHODS -	70-0341
STRUCTURAL	STRUCTURAL ANALYSIS OF SWEPT-BACK WINGS BY MATRIX-TRANSFORMATION	51-0001
STRUCTURAL	STRUCTURAL ANALYSIS BY THE MATRIX FORCE METHOD WITH APPLICATION TO AIRCRAFT WINGS	56-0004
STRUCTURAL	STRUCTURAL ANALYSIS BY THE MATRIX DISPLACEMENT METHOD	61-0003
STRUCTURAL	STRUCTURAL ANALYSIS OF SOLIDS	63-0009
STRUCTURAL	STRUCTURAL ANALYSIS OF AXISYMMETRIC SOLIDS	65-0031
STRUCTURAL	STRUCTURAL ANALYSIS OF AXISYMMETRIC SOLIDS	65-0056
STRUCTURAL	STRUCTURAL ANALYSIS AND MATRIX INTERPRETIVE SYSTEM (SAMIS)	66-0098
STRUCTURAL	STRUCTURAL ANALYSIS OF AN ARTERIOLE BY THE DIRECT STIFFNESS METHOD	66-0109
STRUCTURAL	STRUCTURAL ANALYSIS OF AERODYNAMIC DECELERATION SYSTEMS BY THE FINITE ELEMENT METHOD	67-0064
STRUCTURAL	STRUCTURAL ANALYSIS AND MATRIX INTERPRETIVE SYSTEM (SAMIS) USER REPORT	67-0125
STRUCTURAL	STRUCTURAL ANALYSIS AND EXPERIMENTAL CORRELATION OF A COMPLEX SHELL UNDER ASYMMETRIC LOADS	69-0155
STRUCTURAL	STRUCTURAL ANALYSIS OF SHELL INTERSECTIONS	69-0217
STRUCTURAL	STRUCTURAL ANALYSIS OF AIRCRAFT IMPACT ON A NUCLEAR CONTAINMENT VESSEL AND ASSOCIATED STRUCTURES	69-0279
STRUCTURAL	STRUCTURAL ANALYSIS AND DESIGN OPTIMIZATION OF SPACECRAFT USING NASTRAN	71-0068
STRUCTURAL	STRUCTURAL ANALYSIS OF SYMMETRICAL PENSTOCK BIFURCATIONS BY THE FINITE ELEMENT METHOD	71-0310
STRUCTURAL	STRUCTURAL ANALYSIS OF ARAVA FUSELAGE USING FINITE ELEMENT METHOD	71-0578
STRUCTURAL	STRUCTURAL ANALYSIS BY PROBLEM ORIENTED LANGUAGES	71-0623
STRUCTURAL	STRUCTURAL ANALYSIS OF LONGITUDINALLY FRAMED SHIPS	72-0436
STRUCTURAL	STRUCTURAL ANALYSIS OF A PARAWING DURING DEPLOYMENT	72-0560
STRUCTURAL	STRUCTURAL ANALYSIS AND DESIGN OF A CATAMARAN CROSS-STRUCTURE BY FINITE ELEMENT METHOD	73-0052
STRUCTURAL	STRUCTURAL ANALYSIS OF TALL BUILDINGS HAVING IRREGULARLY POSITIONED SHEAR WALLS	73-0282
STRUCTURAL	STRUCTURAL ANALYSIS OF PETROCHEMICAL EQUIPMENT BY THE FINITE ELEMENT METHOD	73-0377
STRUCTURAL	STRUCTURAL ANALYSIS OF A COAL MINE OPENING IN ELASTIC, MULTILAYERED MATERIAL	74-0235
STRUCTURAL	STRUCTURAL ANALYSIS OF MAST BY FINITE ELEMENT TECHNIQUE	74-0385
STRUCTURAL	STRUCTURAL ANALYSIS OF WELDED WALL FOR BOILER FURNACE	74-0653
STRUCTURAL	STRUCTURAL ANALYSIS OF SL-7 CONTAINERSHIP UNDER COMBINED LOADING OF VERTICAL, LATERAL AND TORSIONAL	74-0734
STRUCTURAL	STRUCTURAL ANALYSIS OF A COAL MINE OPENING IN ELASTIC, MULTILAYERED MATERIAL	74-0840
STRUCTURAL	STRUCTURAL ANALYSIS OF A THREE-DIMENSIONAL PERFORATED PLATE BY THE FINITE ELEMENT METHOD WITH TEST C	74-0945
STRUCTURAL	STRUCTURAL ANALYSIS: A UNIFIED CLASSICAL AND MATRIX APPROACH	72-0055
STRUCTURAL	STRUCTURAL AND SYSTEM MODELS	75-0669
STRUCTURAL	STRUCTURAL BEHAVIOR OF A TWO SPAN REINFORCED CONCRETE BOX GIRDER BRIDGE MODEL. VOLUME 11. REDUCTION,	71-0659
STRUCTURAL	STRUCTURAL BEHAVIOR OF A FLEXIBLE METAL CULVERT UNDER A DEEP EARTH EMBANKMENT USING METHOD A BACKFIL	73-0739

STRUCTURAL	STRUCTURAL CAPACITY OF REINFORCED CONCRETE COLUMNS SUBJECTED TO FIRE INDUCED THERMAL GRADIENTS	73-0913
STRUCTURAL	STRUCTURAL DESIGN AND ANALYSIS OF THE SPACE SHUTTLE	70-0336
STRUCTURAL	STRUCTURAL DESIGN DATA FOR CONCRETE DRIFT LININGS IN BLOCK CAVING	73-0773
STRUCTURAL	STRUCTURAL DESIGN OF SUPERTANKERS	70-0152
STRUCTURAL	STRUCTURAL DESIGN OF CIRCULAR DIVERSION, POWER AND PRESSURE TUNNELS	71-0315
STRUCTURAL	STRUCTURAL DESIGN OF NUCLEAR PLANT FACILITIES	73-0812
STRUCTURAL	STRUCTURAL DYNAMIC ANALYSIS OF ELECTRONIC ASSEMBLIES USING NASTRAN RESTART/FORMAT CHANGE CAPABILITY	71-0230
STRUCTURAL	STRUCTURAL DYNAMIC PROPERTIES OF TACTICAL MISSILE JOINTS - PHASE 1	70-0423
STRUCTURAL	STRUCTURAL DYNAMIC PROPERTIES OF TACTICAL MISSILE JOINTS - PHASE 3	74-0811
STRUCTURAL	STRUCTURAL DYNAMICS PROGRAMS	70-0412
STRUCTURAL	STRUCTURAL DYNAMICS BY FINITE ELEMENTS: MODAL AND FOURIER ANALYSIS	73-0836
STRUCTURAL	STRUCTURAL EFFECTS OF CONFINED DYNAMIC LOADS	72-0151
STRUCTURAL	STRUCTURAL EIGENVALUE PROBLEMS: ELIMINATION OF UNWANTED VARIABLES	65-0050
STRUCTURAL	STRUCTURAL EVALUATION OF CANDIDATE DESIGNS FOR THE LARGE SPACE TELESCOPE PRIMARY MIRROR	75-0708
STRUCTURAL	STRUCTURAL FINITE ELEMENT ANALYSIS AIDED BY COMPUTER GRAPHICS	74-0304
STRUCTURAL	STRUCTURAL INELASTICITY. V11. ELASTIC-PLASTIC ANALYSIS OF A WHEEL ROLLING ON A RIGID TRACK	73-0886
STRUCTURAL	STRUCTURAL INELASTICITY. V1. A FINITE ELEMENT METHOD FOR PLASTICITY PROBLEMS	73-0887
STRUCTURAL	STRUCTURAL MECHANICS SOFTWARE. A BIBLIOGRAPHY WITH ABSTRACTS	74-0842
STRUCTURAL	STRUCTURAL MECHANICS COMPUTER PROGRAMS	74-0902
STRUCTURAL	STRUCTURAL MODEL FOR THE MECHANICAL BEHAVIOUR OF TRABECULAR BONE	73-0473
STRUCTURAL	STRUCTURAL MODEL STUDIES OF LARGE PIPE BIFURCATION	73-0283
STRUCTURAL	STRUCTURAL MODELING OF A WELDED JOINT TYPICAL FOR VEHICLE FRAMES	73-0539
STRUCTURAL	STRUCTURAL OPTIMIZATION USING THE FINITE ELEMENT METHOD APPLIED TO A BEAM	75-0233
STRUCTURAL	STRUCTURAL OPTIMIZATION USING THE FORCE METHOD	75-0440
STRUCTURAL	STRUCTURAL RESPONSE OF RELAXED AND CONSTRICTED ARTERIOLES	68-0200
STRUCTURAL	STRUCTURAL RESPONSE OF UNSTIFFENED TOROIDAL SHELLS	69-0198
STRUCTURAL	STRUCTURAL RESPONSE AND LOADING OF WALL PANELS	71-0735
STRUCTURAL	STRUCTURAL RESPONSE TO IMPACT DAMAGE	75-0325
STRUCTURAL	STRUCTURAL RESPONSE TO MOVING PROJECTILE MASS BY THE FINITE ELEMENT METHOD	75-0693
STRUCTURAL	STRUCTURAL SYNTHESIS, 1959-1969, A DECADE OF PROGRESS	69-0237
STRUCTURAL	STRUCTURAL VIBRATION ANALYSIS USING TRIANGULAR FINITE ELEMENTS	66-0060
STRUCTURAL	STRUCTURAL VIBRATION - A MATRIX APPROACH	71-0385
STRUCTURAL	STRUDL-A COMPUTER SYSTEM FOR STRUCTURAL DESIGN	66-0118
STRUCTURAL	STUDY OF EFFECTS OF DESIGN DETAILS ON STRUCTURAL RESPONSE TO ACOUSTIC EXCITATION	72-0864
STRUCTURAL	STUDY ON FINITE ELEMENT METHOD FOR STRUCTURAL ANALYSIS - ISTRAN/PL (INH STRUCTURE ANALYSIS PLATE STR	73-0167
STRUCTURAL	STUDY ON WEIGHT MATRIX IN STRUCTURAL ANALYSIS	75-0062
STRUCTURAL	SURVEY OF ADVANCED STRUCTURAL DESIGN ANALYSIS TECHNIQUES	69-0306
STRUCTURAL	SURVEY OF STRUCTURAL ANALYSIS SYSTEMS	74-1290
STRUCTURAL	TETRAHEDRON ELEMENTS IN THE MATRIX FORCE METHOD OF STRUCTURAL ANALYSIS	64-0011
STRUCTURAL	THE APPLICATION OF DYNAMIC RELAXATION TO THE FINITE ELEMENT METHOD OF STRUCTURAL ANALYSIS	68-0108
STRUCTURAL	THE APPLICATION OF THE FINITE ELEMENT METHOD TO MULTIHULLED SHIP STRUCTURAL ANALYSIS	72-0886
STRUCTURAL	THE AUTOMATIC COMPUTATION OF THE NATURAL FREQUENCIES OF STRUCTURAL FRAMES USING AN EXACT MATRIX TECH	73-0318
STRUCTURAL	THE DEVELOPMENT OF LARGE SCALE DIGITAL COMPUTER CODES FOR PRODUCTION STRUCTURAL ANALYSIS	71-0721
STRUCTURAL	THE ECONOMICAL COMPUTATION OF STIFFNESS OF LARGE STRUCTURAL ELEMENTS	66-0090
STRUCTURAL	THE FINITE ELEMENT METHOD IN STRUCTURAL MECHANICS	65-0012
STRUCTURAL	THE FINITE ELEMENT METHOD IN STRUCTURAL MECHANICS	66-0051
STRUCTURAL	THE FINITE ELEMENT METHOD IN STRUCTURAL AND CONTINUUM MECHANICS	67-0093
STRUCTURAL	THE FINITE ELEMENT TECHNIQUE - A NEW TOOL IN STRUCTURAL ANALYSIS	69-0195
STRUCTURAL	THE GENERALIZED INVERSE - SOME PROPERTIES, A COMPUTATIONAL ALGORITHM AND A STRUCTURAL APPLICATION	68-0248
STRUCTURAL	THE IMPACT OF THE DIGITAL COMPUTER ON STRUCTURAL ANALYSIS	69-0447
STRUCTURAL	THE MATRIC SOLUTION OF CERTAIN NONLINEAR PROBLEMS IN STRUCTURAL ANALYSIS	56-0002
STRUCTURAL	THE MATRIX DEFORMATION METHOD OF STRUCTURAL ANALYSIS	71-0079
STRUCTURAL	THE MATRIX EIGENVALUE PROBLEM FOR DAMPED STRUCTURAL VIBRATION	66-0121
STRUCTURAL	THE MATRIX FORCE METHOD OF STRUCTURAL ANALYSIS AND SOME NEW APPLICATIONS	56-0005
STRUCTURAL	THE NASTRAN COMPUTER PROGRAM FOR STRUCTURAL ANALYSIS	69-0183
STRUCTURAL	THE NASTRAN COMPUTER PROGRAM FOR STRUCTURAL ANALYSIS	71-0173
STRUCTURAL	THE NASTRAN COMPUTER PROGRAM FOR STRUCTURAL ANALYSIS	71-0682
STRUCTURAL	THE NASTRAN PROGRAM FOR STRUCTURAL ANALYSIS	72-0325
STRUCTURAL	THE PLANE ISOPARAMETRIC STRUCTURAL ELEMENT	70-0262
STRUCTURAL	THE ROLE OF COMPUTER GRAPHICS IN THE STRUCTURAL DESIGN PROCESS	68-0215
STRUCTURAL	THE ROLE OF MECHANICS IN DESIGN - STRUCTURAL JOINTS	74-0770
STRUCTURAL	THE SOLUTION OF STRUCTURAL EQUILIBRIUM EQUATIONS BY THE CONJUGATE GRADIENT METHOD WITH PARTICULAR RE	71-0268
STRUCTURAL	THE STOCHASTIC FINITE ELEMENT SIMULATION OF THE NONLINEAR STRUCTURAL RESPONSE OF FIBROUS COMPOSITE M	75-0798
STRUCTURAL	THE USE OF EIGEN-DERIVATIVES TO PREDICT THE EFFECT OF DESIGN CHANGES UPON THE DYNAMIC CHARACTERISTIC	74-0737
STRUCTURAL	THEORETICAL BASIS FOR CBL/NONSAP, A NONLINEAR STRUCTURAL ANALYSIS PROGRAM	74-0740
STRUCTURAL	THEORY AND PRACTICE IN FINITE ELEMENT STRUCTURAL ANALYSIS	73-0700
STRUCTURAL	THEORY OF MATRIX STRUCTURAL ANALYSIS	68-0139
STRUCTURAL	THERMAL-STRESS CONCENTRATION CAUSED BY STRUCTURAL DISCONTINUITIES	69-0308
STRUCTURAL	TRIANGULAR PLATE ELEMENTS IN THE MATRIX FORCE METHOD OF STRUCTURAL ANALYSIS	63-0010
STRUCTURAL	TWO-DIMENSIONAL AUTOMATIC MESH GENERATION FOR STRUCTURAL ANALYSIS	70-0058
STRUCTURAL	UEBER DIE PRAKTISCHE ANWENDUNG VON FINITEN ELEMENTEN IN DER PLATTENSTATIK (APPLICATION OF THE FINITE	75-0805
STRUCTURAL	UPPER AND LOWER BOUNDS IN MATRIX STRUCTURAL ANALYSIS	64-0018
STRUCTURAL	UPPER AND LOWER BOUNDS TO STRUCTURAL DEFORMATIONS BY DUAL ANALYSIS IN FINITE ELEMENTS	64-0017
STRUCTURAL	USE OF THE FINITE ELEMENT TECHNIQUE TO DETERMINE THE FIRE RESISTANCE CAPABILITIES OF INSULATED STRUC	71-0147
STRUCTURAL	USERS MANUAL FOR CEL/NONSAP, A NONLINEAR STRUCTURAL ANALYSIS PROGRAM	74-0706
STRUCTURAL	VEHICLE STRUCTURAL ANALYSIS, PRE AND POST PROCESSOR DATA HANDLING	71-0093
STRUCTURAL	ZUR SYSTEMATIK NUMERISCHER NAEHERUNGSVERFAHREN IN DER STATIK UND DYNAMIK DER KONSTRUKTIONEN (METHODS	72-0557
STRUCTURAL	775 SELECTED REFERENCES ON THE FINITE ELEMENT METHOD AND MATRIX METHODS OF STRUCTURAL ANALYSIS	69-0241
STRUCTURAL-ACOU	ANALYSIS OF STRUCTURAL-ACOUSTIC INTERACTIONS IN METAL-CERAMIC TRANSDUCERS	72-0850
STRUCTURAL-ANAL	APPLICATION OF RESPONSE LIMITING TO FINITE-ELEMENT STRUCTURAL-ANALYSIS PROGRAMS	75-0424
STRUCTURALLY	STRUCTURALLY COMPATIBLE FLUID FINITE-ELEMENT FOR SOLID FLUID INTERACTION STUDIES	75-0344
STRUCTURALLY	ZUR GENAUEN BERECHNUNG KONSTRUKTIV ORTHOTROPER PLATTEN (EXACT CALCULATION OF STRUCTURALLY ORTHOTROPI	72-0188
STRUCTURE	A MODAL SYNTHESIS TECHNIQUE FOR DETERMINING DYNAMIC PROPERTIES FOR A STRUCTURE FOR MASS AND STIFFNES	74-0319
STRUCTURE	A NUMERICAL THERMO-MECHANICAL MODEL FOR THE WELDING AND SUBSEQUENT LOADING OF A FABRICATED STRUCTURE	72-0847
STRUCTURE	A SPECIAL NASTRAN PROGRAM FOR INPUT CHECKING AND UNDEFORMED STRUCTURE PLOTTING	71-0234
STRUCTURE	A STUDY ON THE STRESS ANALYSIS OF BEAM PLATE STRUCTURE	69-0404
STRUCTURE	ANALISI NUMERICA DEL COMPORTAMENTO ELASTICO DI UNA STRUTTURA BIDIMENSIONALE PIANA. (NUMERICAL ANALYS	74-1029
STRUCTURE	ANALYSIS FOR CRACKS IN SHIP STRUCTURE USING THE FINITE ELEMENT METHOD WITH THE CONCEPT OF CRACK SUB-	74-0248
STRUCTURE	ANALYSIS OF THICK AND THIN SHELL STRUCTURE BY CURVED FINITE ELEMENTS	70-0016
STRUCTURE	ANALYSIS OF VEHICLE BODY STRUCTURE BY THE FINITE ELEMENT METHOD	71-0324
STRUCTURE	APPLICATION OF FINITE ELEMENT TECHNIQUES TO THE ANALYSIS OF AN AUTOMOBILE STRUCTURE	70-0136
STRUCTURE	APPLICATION OF FINITE ELEMENT METHOD TO THE STUDY OF STRENGTH OF CAR BODY STRUCTURE AND WHEEL	74-0281
STRUCTURE	APPLICATION OF NASTRAN TO THE ANALYSIS OF A SPACE SHUTTLE ORBITER STRUCTURE	71-0099
STRUCTURE	AUTOMATED DATA GENERATION AND LARGE MATRIX OPERATION FOR 3 DIMENSIONAL PLATE STRUCTURE ANALYSIS	73-0339
STRUCTURE	COMPUTATIONAL ASPECTS OF FINITE ELEMENT METHOD FOR A FRAME STRUCTURE	72-0175
STRUCTURE	COMPUTER AIDED DESIGN APPLIED TO A MODEL OF A CHASSIS TYPE STRUCTURE USING FINITE ELEMENT TECHNIQUES	69-0421
STRUCTURE	COMPUTER AIDED DESIGN OF MACHINE TOOL STRUCTURE	74-1189
STRUCTURE	CONTRIBUTION A L'ANALYSE DES VIBRATIONS LIBRES D'UNE STRUCTURE. APPLICATION AUX AUBAGES DE TURBOMACH	75-0804
STRUCTURE	CONTRIBUTION A L'ANALYSE DES VIBRATIONS LIBRES D'UNE STRUCTURE. APPLICATION AUX AUBAGES DE TURBOMACH	75-0804
STRUCTURE	COUPLED VIBRATIONS OF A STRUCTURE SUBMERGED IN A COMPRESSIBLE FLUID	69-0001
STRUCTURE	COUPLED VIBRATIONS OF A STRUCTURE SUBMERGED IN A COMPRESSIBLE FLUID	69-0507
STRUCTURE	CREEP DEFORMATION OF A MULTI-STOREY STRUCTURE IN A FIRE	74-1197

STRUCTURE	CRUSTAL STRUCTURE FROM SURFACE LOAD TILTS USING A FINITE ELEMENT MODEL	72-0005
STRUCTURE	DESIGN OF LOW-THERMAL-DISTORTION LST METERING STRUCTURE	75-0840
STRUCTURE	DEVELOPMENT AND APPLICATION OF HYBRID STRESS FINITE ELEMENTS TO AN AUTOMOTIVE STRUCTURE	74-1221
STRUCTURE	DYNAMIC ANALYSIS OF SOIL - STRUCTURE SYSTEM USING FINITE ELEMENT METHOD	67-0001
STRUCTURE	DYNAMIC ANALYSIS OF A COMPLEX SHELL STRUCTURE UNDER IMPULSIVE ASYMMETRIC LOADING	68-0226
STRUCTURE	DYNAMIC RESPONSE OF MODEL PAVEMENT STRUCTURE	72-0226
STRUCTURE	DYNAMIC RESPONSE OF THE HUMAN ARM AS A COMPOSITE STRUCTURE	72-0611
STRUCTURE	EFFECTS OF SOURCE STRUCTURE AND SITE RESPONSE ON GROUND MOTION IN THE NEAR FIELD OF EARTHQUAKES	73-0732
STRUCTURE	ELASTO-PLASTIC ANALYSIS OF AUTOMOBILE BODY STRUCTURE BY THE FINITE ELEMENT METHOD	74-0196
STRUCTURE	FINITE ELEMENT ANALYSIS OF A WING STRUCTURE	72-0333
STRUCTURE	FINITE ELEMENT SOLUTION OF THE INTERACTION OF A PLANE ACOUSTIC BLAST WAVE AND A CYLINDRIC STRUCTURE	74-0790
STRUCTURE	FINITE ELEMENT STRUCTURAL ANALYSIS AS APPLIED TO AN AUTOMOTIVE DOOR STRUCTURE	74-0310
STRUCTURE	FINITE ELEMENT TECHNIQUES IN SHIP STRUCTURE DESIGN IN FINITE ELEMENT TECHNIQUES	69-0335
STRUCTURE	HYSTERETIC DAMPING SYSTEMS APPLIED TO SOIL STRUCTURE INTERACTION BY FINITE ELEMENT METHOD	73-0563
STRUCTURE	INFLUENCE OF SOIL/STRUCTURE INTERACTION PARAMETERS ON FLOOR RESPONSE SPECTRA	73-0833
STRUCTURE	MATRIX ANALYSIS OF A COMPOSITE STRUCTURE OF PLATE AND LINE ELEMENTS	71-0250
STRUCTURE	MODAL ANALYSIS OF A SATELLITE PRIMARY STRUCTURE USING A FINITE ELEMENT PROCEDURE	73-0624
STRUCTURE	MODEL TEST AND ANALYSIS OF OBLIQUE FOLDED PLATE STRUCTURE	73-0230
STRUCTURE	NAVIGATION LOCK STRUCTURE ANALYSIS BY FINITE ELEMENT METHOD	71-0816
STRUCTURE	NONLINEAR CONSIDERATION OF GRAVITY IN A STIFFNESS TEST OF A WEAK STRUCTURE AT SMALL STRAINS	71-0035
STRUCTURE	NORMAL MODE ANALYSIS OF A FUZE SUPPORT STRUCTURE USING NASTRAN. PART 1	72-0853
STRUCTURE	NUMERICAL, THERMO-MECHANICAL MODEL FOR THE WELDING AND SUBSEQUENT LOADING OF A FABRICATED STRUCTURE	73-0418
STRUCTURE	ON THE ACCURACY OF THE TAYLOR APPROXIMATION FOR STRUCTURE RESIZING	74-0147
STRUCTURE	ON THE APPLICATION OF THE FINITE ELEMENT METHOD TO THE VIBRATION ANALYSIS OF SPACE PLATE STRUCTURE	72-0630
STRUCTURE	PRODUCTION OF MOTION PICTURES OF A CONTINUOUSLY DEFORMING STRUCTURE USING NASTRAN	71-0223
STRUCTURE	RADIAL AND AXIAL TRANSMISSIBILITY CHARACTERISTICS OF TYPICAL ROCKET VEHICLE STRUCTURE TECHNICAL SUMM	69-0475
STRUCTURE	REDUNDANT STRUCTURE CONCEPTS IN FINITE ELEMENT ANALYSES	75-0338
STRUCTURE	REDUNDANT STRUCTURE CONCEPTS IN FINITE ELEMENT ANALYSES	75-0526
STRUCTURE	RESPONSE OF STRUCTURE TO THE PSEUDO-SOUND FIELD OF A JET (USING A COMBINED CONTINUUM AND FINITE ELEM	67-0146
STRUCTURE	SANDWICH PLATE STRUCTURE ANALYSIS BY FINITE ELEMENT	74-0163
STRUCTURE	SIMULATION OF FINITE ELEMENT METHOD FOR LOADING PROCESS OF GRAIN CARGO AND INTERACTION WITH SHIP'S S	75-0161
STRUCTURE	STATIC ANALYSIS OF A COMPLEX SHELL STRUCTURE UNDER ASYMMETRIC LOADING	67-0120
STRUCTURE	STATIC ANALYSIS OF THE 64-M 210-FT ANTENNA REFLECTION STRUCTURE	71-0137
STRUCTURE	STATIC ANALYSIS VIA SUBSTRUCTURING OF AN EXPERIMENTAL VEHICLE FRONT-END BODY STRUCTURE	74-0313
STRUCTURE	STRESS ANALYSIS OF FRAME STRUCTURE CONNECTED WITH PLATES	74-0265
STRUCTURE	STRESS ANALYSIS OF THIN PLATE STRUCTURE BY THE FINITE ELEMENT METHOD	69-0403
STRUCTURE	STUDY ON FINITE ELEMENT METHOD FOR STRUCTURAL ANALYSIS - ISTRAN/PL (INH STRUCTURE ANALYSIS PLATE STR	73-0167
STRUCTURE	STUDY ON FINITE ELEMENT METHOD FOR STRUCTURAL ANALYSIS - ISTRAN/PL (INH STRUCTURE ANALYSIS PLATE STR	73-0167
STRUCTURE	TEST RESULTS FROM THE BOUNDARY LAYER FACILITY - RESPONSE OF STRUCTURE TO THE PSEUDO-SOUND FIELD OF A	68-0262
STRUCTURE	THE THERMO-ELASTO-PLASTIC ANALYSIS OF A RE-ENTRY BODY TYPE COMPOSITE SOLID STRUCTURE USING A NEW FIN	66-0093
STRUCTURE	TOUR MAINE-MONTPARNASSE QUELQUES PROBLEMES DE STRUCTURE (SPECIFIC PROBLEMS ENCOUNTERED IN THE CONSTR	74-0539
STRUCTURE	VIBRATION OF A BOX-TYPE STRUCTURE - 1	71-0405
STRUCTURE	VIBRATION OF A BOX-TYPE STRUCTURE - 2.	71-0420
STRUCTURE	3-DIMENSIONAL ANALYSIS OF HULL STRUCTURE BY FINITE ELEMENT METHOD	73-0160
STRUCTURE-	STRUCTURE- MEDIUM INTERACTION PHENOMENOLOGY	72-0863
STRUCTURE-FOUND	EARTHQUAKE ANALYSIS OF STRUCTURE-FOUNDATION SYSTEMS	73-0876
STRUCTURE-FOUND	EARTHQUAKE FINITE ELEMENT ANALYSIS OF STRUCTURE-FOUNDATION SYSTEMS	74-0558
STRUCTURE-FOUND	EARTHQUAKE FINITE-ELEMENT ANALYSIS OF STRUCTURE-FOUNDATION SYSTEMS	75-0410
STRUCTURE-GROUN	THREE DIMENSIONAL SEISMIC ANALYSIS OF STRUCTURE-GROUND SYSTEMS BY THE FINITE ELEMENT METHOD	74-0268
STRUCTURE-MEDIU	STRUCTURE-MEDIUM INTERACTION AND DESIGN PROCEDURES STUDY. VOLUME I. ANALYSIS METHOD, THEORY, VERIFIC	69-0166
STRUCTURE-MEDIU	STRUCTURE-MEDIUM INTERACTION AND DESIGN PROCEDURES STUDY. VOLUME IV. FEAT CODE USER'S GUIDE	69-0265
STRUCTURES	A COMPUTER METHOD FOR THE SECOND ORDER ELASTIC ANALYSIS OF PLANE FRAMED STRUCTURES	73-0310
STRUCTURES	A COMPUTER PROGRAM TO DISPLAY THE DYNAMIC STRESS ANALYSIS OF STRUCTURES	70-0077
STRUCTURES	A CRUCIFORM ELEMENT FOR SURFACES OF INFLATABLE STRUCTURES	75-0118
STRUCTURES	A FINITE ELEMENT APPROACH FOR THE ANALYSIS OF RANDOMLY EXCITED COMPLEX ELASTIC STRUCTURES	67-0057
STRUCTURES	A FINITE ELEMENT COMPUTER PROGRAM FOR THE PREDICTION OF THE BEHAVIOR OF REINFORCED AND PRESTRESSED C	72-0799
STRUCTURES	A FINITE ELEMENT FORMULATION FOR STABILITY ANALYSIS OF DOUBLY CURVED THIN SHELL STRUCTURES	69-0513
STRUCTURES	A FINITE ELEMENT METHOD FOR THE PLASTIC BENDING ANALYSIS OF STRUCTURES	68-0021
STRUCTURES	A FINITE ELEMENT METHOD FOR RANDOM FOLDED PLATE STRUCTURES (EINE FINITE ELEMENT METHODE ZUR BERECHNU	73-0787
STRUCTURES	A FINITE ELEMENT SOLUTION FOR FOLDED PLATE STRUCTURES	67-0072
STRUCTURES	A FINITE ELEMENT STUDY OF THE VIBRATION OF SKIN-RIB STRUCTURES	74-0537
STRUCTURES	A FINITE ELEMENT TECHNIQUE FOR THE IDEALIZATION OF CONTINUOUS SHELL STRUCTURES	68-0290
STRUCTURES	A FINITE ELEMENT TECHNIQUE FOR ANALYZING FOLDED PLATE TYPE STRUCTURES WITH SKEWED SUPPORTS	69-0525
STRUCTURES	A GENERAL DIGITAL COMPUTER ANALYSIS OF STATICALLY INDETERMINATE STRUCTURES	59-0007
STRUCTURES	A GENERAL METHOD FOR THE SHAPE FINDING OF LIGHTWEIGHT TENSION STRUCTURES	74-1217
STRUCTURES	A HIGH-ORDER FINITE ELEMENT FOR PRISMATIC AND NONPRISMATIC FOLDED PLATE STRUCTURES	75-0889
STRUCTURES	A METHOD FOR REDUCING THE ANALYSIS OF COMPLEX REDUNDANT STRUCTURES TO A ROUTINE PROCEDURE	52-0002
STRUCTURES	A METHOD OF ANALYSIS OF LINE STRUCTURES BY TRANSFER MATRICES DERIVED FROM FINITE ELEMENTS	74-0808
STRUCTURES	A METHOD OF FINITE DEFLECTION ELASTIC PLASTIC ANALYSIS OF FRAMED STRUCTURES SUBJECTED TO LARGE AXIAL	69-0239
STRUCTURES	A NONLINEAR FINITE ELEMENT CODE FOR ANALYZING THE BLAST RESPONSE OF UNDERGROUND STRUCTURES	70-0054
STRUCTURES	A REFINED FINITE ELEMENT ANALYSIS OF THIN SHELL STRUCTURES	67-0101
STRUCTURES	A REFINED FINITE ELEMENT ANALYSIS OF THIN SHELL STRUCTURES INCLUDING DYNAMIC LOADINGS	67-0157
STRUCTURES	A REFINED FINITE ELEMENT ANALYSIS OF FOLDED PLATE STRUCTURES	71-0671
STRUCTURES	A RELAXATION PROCEDURE FOR COMPUTING STRESSES IN BUILT-UP STRUCTURES	69-0063
STRUCTURES	A STATISTICAL THEORY FOR THE FRACTURE OF BRITTLE STRUCTURES SUBJECTED TO NONUNIFORM POLYAXIAL STRESS	73-0928
STRUCTURES	A STUDY OF MATRIX ANALYSIS METHODS FOR INELASTIC STRUCTURES	63-0022
STRUCTURES	A STUDY OF VARIOUS IDEALIZATIONS FOR WING STRUCTURES AND NUMERICAL PROCEDURES INVOLVED USING MATRIX	69-0257
STRUCTURES	A STUDY ON COMPOSITE STRUCTURES EMPLOYING DIFFERENT TYPES OF FINITE ELEMENTS	69-0231
STRUCTURES	A THREE-DIMENSIONAL FINITE ELEMENT COMPUTER CODE FOR THE ANALYSIS OF COMPLEX STRUCTURES	71-0275
STRUCTURES	ACCURACY AND OPTIMIZATION IN THE ANALYSIS OF SHIP STRUCTURES. USING FINITE ELEMENT METHODS	74-0418
STRUCTURES	ALGORITHM FOR THE LARGE DEFLECTION GEOMETRICALLY NON-LINEAR BEHAVIOUR OF PLANE AND CURVED STRUCTURES	72-0083
STRUCTURES	AN ASSESSMENT OF CURRENT CAPABILITY FOR COMPUTER ANALYSIS OF SHELL STRUCTURES	71-0706
STRUCTURES	AN AUTOMATIC SYSTEM FOR KINEMATIC ANALYSIS ASKA HIGH SPEED COMPUTING OF ELASTIC STRUCTURES	70-0358
STRUCTURES	AN EVALUATION OF CURRENT NASTRAN DISCRETE ELEMENT MODELS FOR MONOCOQUE AND SEMI-MONOCOQUE STRUCTURES	71-0236
STRUCTURES	AN INTERACTIVE COMPUTER-AIDED MODELING SYSTEM FOR THE DISCRETIZATION OF BRIDGE STRUCTURES FOR FINITE	73-1165
STRUCTURES	AN ITERATION METHOD FOR THE NONLINEAR BUCKLING OF FRAME STRUCTURES	66-0107
STRUCTURES	ANALYSE DES CONTRAINTES DAN DES STRUCTURES COMPLEXES EN PRESENCE DE FLUAGE, DE VARIATIONS DE DIMENSI	71-0081
STRUCTURES	ANALYSE DYNAMIQUE DUALE DES STRUCTURES PAR LA METHODE DES ELEMENTS FINIS	73-0623
STRUCTURES	ANALYSIS FOR CRACKS IN SHIP STRUCTURES USING THE FINITE-ELEMENT METHOD WITH THE CONCEPT OF CRACK SUB	74-1367
STRUCTURES	ANALYSIS OF ASYMMETRIC STIFFENED SHELL TYPE STRUCTURES BY THE FINITE ELEMENT METHOD: I FLAT RECTANGU	68-0283
STRUCTURES	ANALYSIS OF ASYMMETRIC STIFFENED SHELL TYPE STRUCTURES BY FINITE ELEMENT METHOD II. SIX DEGREE-OF-FR	69-0373
STRUCTURES	ANALYSIS OF AXISYMMETRIC COMPOSITE STRUCTURES BY THE FINITE ELEMENT METHOD	66-0066
STRUCTURES	ANALYSIS OF BIOLOGICAL STRUCTURES	71-0148
STRUCTURES	ANALYSIS OF BODY STRUCTURES - 2.	71-0339
STRUCTURES	ANALYSIS OF BONDED JOINTS IN VEHICULAR STRUCTURES	73-0375
STRUCTURES	ANALYSIS OF CRACK PROPAGATION IN WELDED STRUCTURES	74-1144
STRUCTURES	ANALYSIS OF CURVED FOLDED PLATE STRUCTURES	70-0431
STRUCTURES	ANALYSIS OF CURVED FOLDED PLATE STRUCTURES	71-0139
STRUCTURES	ANALYSIS OF DOUBLE CORRUGATED SHELL STRUCTURES BY THE FINITE ELEMENT METHOD	74-1116
STRUCTURES	ANALYSIS OF DYNAMIC RESPONSE OF SAFEGUARD STRUCTURES	70-0462
STRUCTURES	ANALYSIS OF HIPPED ROOF HYPERBOLIC PARABOLOID STRUCTURES	72-0216
STRUCTURES	ANALYSIS OF HYDRODYNAMIC AND PLATE STRUCTURES PROBLEMS BY THE FINITE ELEMENT METHOD	71-0091

STRUCTURES	ANALYSIS OF INPLANE VIBRATION OF BOX-TYPE STRUCTURES BY A FINITE ELEMENT METHOD	72-0091
STRUCTURES	ANALYSIS OF LARGE CLOSED SHIP STRUCTURES: APPLICATIONS TO TANKER STRUCTURAL DESIGN	70-0289
STRUCTURES	ANALYSIS OF MULTI-AXIAL ANISOTROPIC CREEP, ROOF RESIN BOLTS, AND BED SEPARATION IN ROCK STRUCTURES U	74-1122
STRUCTURES	ANALYSIS OF OPEN STRUCTURES BY FINITE ELEMENT RESOLUTION OF EQUIVALENT CLOSED-BOUNDARY PROBLEM	74-0111
STRUCTURES	ANALYSIS OF PLATE AND SHELL STRUCTURES BY TRIANGULAR FINITE ELEMENTS	68-0303
STRUCTURES	ANALYSIS OF PLATE AND SHELL STRUCTURES	69-0109
STRUCTURES	ANALYSIS OF PLATE STRUCTURES BY A DUAL FINITE ELEMENT METHOD	69-0482
STRUCTURES	ANALYSIS OF REAL STRUCTURES FOR CREEP, PLASTICITY AND OTHER COMPLEX CONSTITUTIVE LAWS.	69-0287
STRUCTURES	ANALYSIS OF SHALLOW SHELL STRUCTURES BY A DISCRETE ELEMENT SYSTEM	66-0113
STRUCTURES	ANALYSIS OF SHELL STRUCTURES BY THE FINITE ELEMENT METHOD	61-0008
STRUCTURES	ANALYSIS OF SHIP STRUCTURES BY THE FINITE ELEMENT METHOD	69-0336
STRUCTURES	ANALYSIS OF STRUCTURES USING FINITE ELEMENT TECHNIQUES	71-0356
STRUCTURES	ANALYSIS OF TENSION STRUCTURES BY MATRIX METHODS	67-0119
STRUCTURES	ANALYSIS OF THE DYNAMICS OF ELASTIC STRUCTURES BY THE FINITE ELEMENT METHOD	72-0893
STRUCTURES	ANALYSIS OF THREE-DIMENSIONAL THIN WALLED STRUCTURES	69-0188
STRUCTURES	ANALYSIS OF THREE-DIMENSIONAL ORTHOTROPIC SANDWICH PLATE STRUCTURES BY FINITE-ELEMENT METHOD	72-0945
STRUCTURES	ANALYSIS ON THE LARGE DEFORMATION OF PLATE STRUCTURES	74-0435
STRUCTURES	ANALYTICAL MODEL FOR DETERMINING ENERGY DISSIPATION IN DYNAMICALLY LOADED STRUCTURES	75-0328
STRUCTURES	ANALYTICAL PROBLEMS ASSOCIATED WITH CORE SUPPORT STRUCTURES OF PWR	71-0039
STRUCTURES	ANWENDUNG DES FINITE ELEMENT VERFAHRENS SUR BESTIMMUNG DER FESTIGKERIT VON SCHWEISSKONSTRUKTIONEN. (	72-0185
STRUCTURES	APPLICATION OF A CONJUGATED-GRADIENT-PROCESS IN FINITE-ELEMENT-CALCULATION OF NON-LINEAR STRUCTURES	74-0214
STRUCTURES	APPLICATION OF A REFINED SHALLOW SHELL FINITE ELEMENT TO GEOMETRICALLY NONLINEAR STRUCTURES	73-0358
STRUCTURES	APPLICATION OF FINITE ELEMENT METHOD TO ANALYSIS OF AIRCRAFT WING STRUCTURES	69-0232
STRUCTURES	APPLICATION OF FINITE ELEMENT METHOD TO THE DYNAMIC ANALYSIS OF TALL STRUCTURES	71-0111
STRUCTURES	APPLICATION OF FINITE ELEMENT METHOD TO PREDICT STATIC RESPONSE OF AUTOMOTIVE BODY STRUCTURES	71-0340
STRUCTURES	APPLICATION OF FINITE ELEMENT METHOD TO CYCLIC LOADING OF ELASTIC-PLASTIC STRUCTURES CONTAINING EFFE	73-0103
STRUCTURES	APPLICATION OF FINITE ELEMENT METHODS TO SHIP STRUCTURES	73-0429
STRUCTURES	APPLICATION OF FLOWGRAPHS TO FREE VIBRATION OF STRUCTURES	68-0196
STRUCTURES	APPLICATION OF NONLINEAR ANALYSIS TO SOLID STRUCTURES	72-0212
STRUCTURES	APPLICATION OF RESPONSE-BOUND METHOD IN SHOCK AND VIBRATION ANALYSIS OF TELEPHONE STRUCTURES	74-0655
STRUCTURES	APPLICATION OF THE COLLISION-IMPARTED VELOCITY METHOD FOR ANALYZING THE RESPONSES OF CONTAINMENT AND	73-0770
STRUCTURES	APPLICATION OF THE FINITE ELEMENT METHOD FOR ANALYSIS AND OPTIMIZATION OF SUBMARINE STRUCTURES	71-0350
STRUCTURES	APPLICATION OF THE FINITE ELEMENT METHOD TO SHIP STRUCTURES	71-0403
STRUCTURES	APPLICATION OF THE FINITE ELEMENT METHOD TO OFF-SHORE STRUCTURES	74-0206
STRUCTURES	APPLICATION OF THE METHOD OF FINITE ELEMENTS TO THE CALCULATION OF SHIP STRUCTURES (ISPOLZOVANIE MET	73-0933
STRUCTURES	APPLICATION OF THE STIFFNESS METHOD TO THE ANALYSIS OF SHELL STRUCTURES	61-0004
STRUCTURES	APPROXIMATE ALGORITHM FOR THE REANALYSIS OF STRUCTURES BY THE FINITE ELEMENT METHOD.	72-0445
STRUCTURES	APPROXIMATE ANALYSIS OF PREFABRICATED TOWER STRUCTURES USING THIN SHELL FINITE ELEMENTS	73-0893
STRUCTURES	APPROXIMATE METHOD FOR CALCULATING THE RESPONSE OF EMPLACEMENT STRUCTURES SUBJECTED TO GROUND SHOCK	73-0654
STRUCTURES	ASPECTS OF FINITE ELEMENT METHOD APPLIED TO AEROSPACE STRUCTURES	73-0703
STRUCTURES	ASPECTS OF THE FINITE ELEMENT METHOD AS APPLIED TO AERO-SPACE STRUCTURES	72-0649
STRUCTURES	ASSESSMENT OF COMMERCIALLY AVAILABLE FINITE ELEMENT CODES FOR THE DYNAMIC ANALYSIS OF STRUCTURES	73-0954
STRUCTURES	ASSESSMENT OF DESIGN TOOLS AND CRITERIA FOR URBAN RAIL TRACK STRUCTURES. VOLUME 11. AT-GRADE SLAB TR	74-0813
STRUCTURES	AUTOMATED DESIGN OPTIMIZATION OF SUPERSONIC AIRPLANE WING STRUCTURES UNDER DYNAMIC CONSTRAINTS	72-0178
STRUCTURES	AUTOMATED LARGE DEFLECTION AND STABILITY ANALYSIS OF THREE DIMENSIONAL BAR STRUCTURES	69-0393
STRUCTURES	AUTOMATED PROCEDURE FOR DESIGN OF WING STRUCTURES TO SATISFY STRENGTH AND FLUTTER REQUIREMENTS	73-0867
STRUCTURES	AUTOMATIC CALCULATION OF THIN STRUCTURES	73-0201
STRUCTURES	AUTOMATIC DESIGN OF SHELL STRUCTURES	71-0336
STRUCTURES	AUTOMATIC MESH GENERATION FOR THREE DIMENSIONAL STRUCTURES BASED ON THEIR THREE VIEWS	73-0340
STRUCTURES	BALLISTIC IMPACT OF TEXTILE STRUCTURES.	72-0433
STRUCTURES	BASIC RESEARCH IN METHODS OF APPROXIMATION IN THE NONLINEAR MECHANICS OF SOLIDS AND STRUCTURES	72-0857
STRUCTURES	BASIC RESEARCH IN METHODS OF APPROXIMATION IN THE NONLINEAR MECHANICS OF SOLIDS AND STRUCTURES. VOLU	74-1091
STRUCTURES	BERECHNUNG PRISMATISCHER FLAECHENTRAGWERKE MIT HILFE VON FINITEN STREIFENELEMENTEN. (CALCULATION OF	72-0133
STRUCTURES	BERECHNUNGSMETHODEN ZUR ERMITTLUNG VON BELASTUNGSBEDINGUNGEN UND GRENZTRAGFAEHIGKEITSVERHALTEN VON K	73-0858
STRUCTURES	BERGEN, AN AUTOMATIC FINITE ELEMENT MESH GENERATION PROGRAM FOR ARBITRARY STRUCTURES	72-0776
STRUCTURES	BOUNDARIES AND FRACTURES IN FINITE ELEMENT MODELS OF GEOLOGICAL STRUCTURES	73-0352
STRUCTURES	BUCKLING UNDER COMBINED LOADING OF THIN FLAT-WALLED STRUCTURES BY A COMPLEX FINITE STRIP METHOD	74-0788
STRUCTURES	CALCULATION METHOD FOR THE STUDY OF CRACKED RIGID STRUCTURES (METODO DI CALCOLO PER LO STUDIO DI STR	71-0637
STRUCTURES	CALCULATION OF GEOMETRIC STIFFNESS MATRICES FOR COMPLEX STRUCTURES	66-0058
STRUCTURES	CALCULATION OF STRESS CONCENTRATIONS ON SUPPORT STRUCTURES UNDER THERMOMECHANICAL LOADING (BERECHNUN	73-0796
STRUCTURES	CALCULATION OF THE REINFORCED THIN-WALLED STRUCTURES BY THE METHOD OF FINITE ELEMENTS (RASCHET PODKR	75-0685
STRUCTURES	CALCULATION OF VIBRATIONS OF STRUCTURES CAUSED BY EARTHQUAKES USING THE FINITE ELEMENT METHOD	71-0690
STRUCTURES	CAPE-2D, A COMPUTER PROGRAM FOR THE STRESS ANALYSIS OF PLANE AND AXISYMMETRIC COMPOSITE STRUCTURES.	69-0083
STRUCTURES	CELLULAR STRUCTURES OF ARBITRARY PLAN GEOMETRY	72-0201
STRUCTURES	CHISLENNYI METOD RASCHETA SOBSTVENNYKH I VYNUZHDENNYKH KOLEBANII SOSTAVNYKH OBOLOCHECHNYKH KONSTRUKT	73-0461
STRUCTURES	COMPARISON OF SOME FORMULATIONS IN THERMAL ANALYSIS OF STRUCTURES	75-0888
STRUCTURES	COMPARISON OF THE NUMERICAL INTEGRATION TECHNIQUE AND THE FINITE ELEMENT METHOD IN THE ANALYSIS OF T	73-0848
STRUCTURES	COMPUTATION OF INFLUENCE COEFFICIENTS FOR AIRCRAFT STRUCTURES WITH DISCONTINUITIES AND SWEEPBACK	47-0002
STRUCTURES	COMPUTATIONAL METHOD FOR DYNAMIC ANALYSIS OF STRUCTURES WITH LARGE SIZE STIFFNESS AND MASS MATRICES.	73-0830
STRUCTURES	COMPUTER ANALYSIS OF SHELL STRUCTURES	69-0060
STRUCTURES	COMPUTER BASED ANALYSES OF THE RESPONSE OF BOX TYPE STRUCTURES TO RANDOM PRESSURES	73-0227
STRUCTURES	COMPUTER GENERATED DISPLAYS OF STRUCTURES IN VIBRATION	74-1305
STRUCTURES	COMPUTER ORIENTED ANALYSIS OF SHELL STRUCTURES	71-0559
STRUCTURES	COMPUTER ORIENTED ANALYSIS OF SHELL STRUCTURES	71-0762
STRUCTURES	COMPUTER PROGRAM FOR GENERATING FINITE ELEMENT MODELS OF MINE STRUCTURES	71-0493
STRUCTURES	COMPUTER PROGRAMS FOR THE PLASTIC ANALYSIS OF STRUCTURES USING DISCRETE ELEMENT METHODS	67-0004
STRUCTURES	COMPUTER PROGRAMS FOR PLOTTING FINITE ELEMENT PATTERNS OF TWO-AND THREE-DIMENSIONAL STRUCTURES	70-0416
STRUCTURES	COMPUTER-AIDED DESIGN OF SHELL STRUCTURES USING THE FINITE ELEMENT METHOD	72-0998
STRUCTURES	COMPUTERIZED ANALYSIS OF NON-ISOTROPIC STRUCTURES	69-0192
STRUCTURES	CONSIDERATION ON THE SEISMIC RESPONSE ANALYSIS OF UNDERGROUND STRUCTURES	75-0470
STRUCTURES	CONTINUUM BASES FOR CONSISTENT NUMERICAL FORMULATIONS OF FINITE STRAINS IN ELASTIC AND INELASTIC STR	75-0299
STRUCTURES	COUPLED RESPONSE OF SPATIAL VIBRATORY STRUCTURES MOUNTED TO ISOTROPIC PLATE ELEMENTS	71-0228
STRUCTURES	CRASH: A COMPUTER SIMULATION OF NONLINEAR TRANSIENT RESPONSE OF STRUCTURES	72-0569
STRUCTURES	CREEP OF FABRICATED STRUCTURES	68-0001
STRUCTURES	CURRENT COMPUTATION ON COMPLEX STRUCTURES BY FINITE-ELEMENT METHOD	75-0456
STRUCTURES	CURVED FINITE ELEMENTS IN THE ANALYSIS OF SOLID SHELL AND PLATE STRUCTURES	69-0357
STRUCTURES	CURVED FINITE ELEMENTS IN THE ANALYSIS OF SHELL STRUCTURES	71-0076
STRUCTURES	DAS FLAECHENTRAGWERKSPROGRAM VON STRIP. (STRIP PROGRAM FOR PLATE STRUCTURES)	72-0150
STRUCTURES	DATA GENERATION FOR FINITE-ELEMENT STRUCTURAL ANALYSIS OF THREE-DIMENSIONAL NAVAL SHIP STRUCTURES	75-0707
STRUCTURES	DEPLACEMENTS DANS LES STRUCTURES MINCES A FAIBLE RIGIDITE GEOMETRIQUE-DETERMINATION NUMERIQUE (NUMER	72-0213
STRUCTURES	DEPLACEMENTS DANS LES STRUCTURES MINCES A FAIBLE RIGIDITE GEOMETRIQUE-DETERMINATION NUMERIQUE (NUMER	72-0213
STRUCTURES	DESIGN OF OPTICAL MIRROR STRUCTURES	71-0658
STRUCTURES	DESIGN OF OPTIMUM STRUCTURES FOR DYNAMIC LOADS	71-0255
STRUCTURES	DESIGN OF TWO AND THREE DIMENSIONAL STRUCTURES BY THE FINITE ELEMENT METHOD (IN GERMAN)	74-1364
STRUCTURES	DEVELOPMENT OF THE STIFFNESS METHOD TO DEFINE BOUNDS ON ELASTIC BEHAVIOUR OF STRUCTURES	62-0011
STRUCTURES	DIE ANWENDUNG VON INTERAKTIVER COMPUTER-GRAPHIK BEI DER BERECHNUNG VON TRAGWERKEN. (APPLICATION OF I	72-0102
STRUCTURES	DIGITAL ANALYSIS OF NON-LINEAR STRUCTURES BY THE FORCE METHOD	64-0023
STRUCTURES	DISCRETE ELEMENT ANALYSIS OF COMPOSITE STRUCTURES	75-0201
STRUCTURES	DISCRETE ELEMENT METHODS FOR THE PLASTIC ANALYSIS OF STRUCTURES SUBJECTED TO CYCLIC LOADING	70-0023
STRUCTURES	DISCRETE-ELEMENT METHODS FOR THE PLASTIC ANALYSIS OF STRUCTURES	67-0040
STRUCTURES	DISCRETE-ELEMENT METHODS FOR STABILITY ANALYSIS OF COMPLEX STRUCTURES	68-0140
STRUCTURES	DISCRETE-ELEMENT PLASTIC ANALYSIS OF STRUCTURES IN A STATE OF MODIFIED PLANE STRAIN	69-0140

STRUCTURES	DOUBLE CURVATURE SHELL FINITE-ELEMENT AND ITS USE IN THE DYNAMIC ANALYSIS OF COOLING TOWERS AND OTHE	70-0312
STRUCTURES	DYNAMIC ANALYSIS OF CIVIL ENGINEERING STRUCTURES	71-0264
STRUCTURES	DYNAMIC ANALYSIS OF MECHANICAL STRUCTURES	71-0381
STRUCTURES	DYNAMIC ANALYSIS OF MACHINE TOOL STRUCTURES BY THE FINITE ELEMENT METHOD	71-0416
STRUCTURES	DYNAMIC ANALYSIS OF STIFFENED PANEL STRUCTURES	71-0522
STRUCTURES	DYNAMIC ANALYSIS OF STRUCTURES USING THE RANK FORCE METHOD	71-0553
STRUCTURES	DYNAMIC ANALYSIS OF STRUCTURES	72-0298
STRUCTURES	DYNAMIC ANALYSIS OF FRAME STRUCTURES BY THE FORCE METHOD	72-0313
STRUCTURES	DYNAMIC ANALYSIS OF FRAME AND PLATE STRUCTURES WITH GEOMETRIC AND MATERIAL NONLINEARITIES	73-0649
STRUCTURES	DYNAMIC ANALYSIS OF THE DEFORMATION OF STRUCTURES AND OF METAL FORMING	74-0823
STRUCTURES	DYNAMIC ELASTIC-PLASTIC ANALYSIS OF AXISYMMETRIC STRUCTURES BY FINITE ELEMENT METHOD COMPUTER CODE	73-0844
STRUCTURES	DYNAMIC FINITE ELEMENT METHOD CALCULATION OF ELASTIC WAVE PROPAGATION IN TWO DIMENSIONAL STRUCTURES	74-0166
STRUCTURES	DYNAMIC MATRIX ANALYSIS OF FRAMED STRUCTURES	62-0012
STRUCTURES	DYNAMIC RESPONSE ANALYSIS OF GEOMETRICALLY NON-LINEAR STRUCTURES SUBJECTED TO HIGH IMPACT	72-0193
STRUCTURES	DYNAMIC RESPONSE ANALYSIS OF GEOMETRICALLY NON-LINEAR STRUCTURES SUBJECTED TO HIGH IMPACT	72-0555
STRUCTURES	DYNAMIC RESPONSE OF STRUCTURES SUBJECTED TO TIME-DEPENDENT BOUNDARY CONDITIONS USING THE FINITE ELEM	72-0129
STRUCTURES	DYNAMIC RESPONSE OF STRUCTURES IN LAYERED SOILS	74-0761
STRUCTURES	DYNAMIC STRESS ANALYSIS OF AXISYMMETRIC STRUCTURES UNDER ARBITRARY LOADING	69-0433
STRUCTURES	DYNAMICS OF CRACKED STRUCTURES USING FINITE ELEMENTS	74-1115
STRUCTURES	DYNAMICS OF SUBMARINE STRUCTURES, PART 2	68-0025
STRUCTURES	DYNAMICS OF SUBMARINE STRUCTURES, VOLUME 1	68-0251
STRUCTURES	DYNAMICS OF SUBMARINE STRUCTURES, PART 1.	69-0054
STRUCTURES	EARTH PRESSURES ON STRUCTURES DUE TO FAULT MOVEMENT	73-0403
STRUCTURES	EARTHQUAKE-RESISTANT DESIGN OF ENGINEERING STRUCTURES	73-0284
STRUCTURES	EFFECT OF TRANSVERSE SHEAR DEFORMATION IN VIBRATIONS OF PLANAR STRUCTURES COMPOSED OF BEAM TYPE ELEM	73-0402
STRUCTURES	EFFECTS OF FOUNDATION ROTATION ON SEISMIC INERTIA FORCES OF NUCLEAR POWER PLANT STRUCTURES	73-0177
STRUCTURES	EFFECTS OF RISE TIME AND DAMPING ON FINITE ELEMENT ANALYSIS OF RESPONSE OF STRUCTURES	71-0087
STRUCTURES	EIGENVALUES OF COLLINEAR BEAM STRUCTURES USING FINITE ELEMENT TECHNIQUES AND VARIOUS DYNAMIC REPRESE	67-0124
STRUCTURES	ELAD PROGRAM MANUAL, A COMPUTER PROGRAM FOR THE ELASTIC ANALYSIS OF DOME STRUCTURES AND AXISYMMETRIC	69-0064
STRUCTURES	ELAS - A GENERAL PURPOSE COMPUTER PROGRAM FOR THE EQUILIBRIUM PROBLEMS OF LINEAR STRUCTURES	68-0158
STRUCTURES	ELAS: A GENERAL-PURPOSE COMPUTER PROGRAM FOR THE EQUILIBRIUM PROBLEMS OF LINEAR STRUCTURES, VOLUME 2	69-0462
STRUCTURES	ELASTIC DYNAMIC RESPONSE OF AXISYMMETRIC STRUCTURES	69-0275
STRUCTURES	ELASTIC NONLINEAR FINITE ELEMENT ANALYSIS OF PLANAR STRUCTURES	71-0830
STRUCTURES	ELASTIC-PLASTIC ANALYSIS OF FRAMED STRUCTURES	69-0197
STRUCTURES	ELASTO PLASTIC ANALYSIS OF PLANE STRUCTURES BY THE FINITE ELEMENT DISPLACEMENT METHOD	70-0299
STRUCTURES	ELASTO-PLASTIC ANALYSIS OF THREE-DIMENSIONAL STRUCTURES USING THE ISOPARAMETRIC ELEMENT	72-0248
STRUCTURES	ELASTO-PLASTIC SEISMIC RESPONSE ANALYSIS OF FRAMED STRUCTURES BY MODE SUPERPOSITION	73-0325
STRUCTURES	ELASTODYNAMICS OF CRACKED STRUCTURES USING FINITE ELEMENTS	74-0625
STRUCTURES	ELEMENT STATIC FLEXIBILITY AND STIFFNESS MATRICES FOR THE FINITE ELEMENT ANALYSIS OF BEAM, PLATE, AN	67-0070
STRUCTURES	ENGINEERING OPERATIONS REPORT: NERVA AXISYMMETRIC AND PLANE STRUCTURES FINITE ELEMENT STRESS ANALYSI	72-0917
STRUCTURES	EPAD, A COMPUTER PROGRAM FOR THE ELASTIC-PLASTIC ANALYSIS OF DOME STRUCTURES AND ARBITRARY SOLIDS OF	70-0036
STRUCTURES	ESTIMATES OF THE CREEP RUPTURE LIFETIME OF STRUCTURES USING THE FINITE ELEMENT METHOD	75-0833
STRUCTURES	EVALUATION OF ATTENUATION FROM LOSSLESS TRIANGULAR-FINITE-ELEMENT SOLUTIONS FOR INHOMOGENEOUSLY FILL	73-0059
STRUCTURES	EVALUATION OF COMPUTER AIDED METHODS FOR THE DESIGN OF MACHINE TOOL STRUCTURES	73-0577
STRUCTURES	EVALUATION OF DAMAGE TOLERANCE IN AIRCRAFT STRUCTURES	75-0559
STRUCTURES	EVALUATION OF TORSIONAL EFFECTS IN THE SEISMIC ANALYSIS OF SYMMETRIC AND UNSYMMETRIC NUCLEAR POWER P	73-0175
STRUCTURES	EXACT THEORY OF SPATIAL STRUCTURES	73-0781
STRUCTURES	EXAMPLES OF STRESS ANALYSIS OF STRUCTURES BY THE FINITE ELEMENT METHOD, (JAPANESE)	70-0450
STRUCTURES	EXPERIMENTAL AND FINITE ELEMENT ANALYSIS OF CABLE ROOF STRUCTURES INCLUDING PRECAST PANELS	74-1139
STRUCTURES	EXTENSION OF THREE DIMENSIONAL ANALYSIS TO SHELL STRUCTURES USING THE FINITE ELEMENT IDEALIZATION	74-0849
STRUCTURES	FAMILLE D ELEMENTS FINIS DE COQUE PLAN CINEMATIQUEMENT ADMISSIBLES POUR L'ANALYSE DES STRUCTURES	74-0488
STRUCTURES	FINITE ELEMENT ANALYSIS OF TWO DIMENSIONAL STRUCTURES	63-0015
STRUCTURES	FINITE ELEMENT ANALYSIS OF AXISYMMETRIC COMPOSITE STRUCTURES	65-0030
STRUCTURES	FINITE ELEMENT ANALYSIS FOR ARBITRARY AXISYMMETRIC STRUCTURES ORTHOTROPIC, TEMPERATURE-DEPENDENT MAT	68-0076
STRUCTURES	FINITE ELEMENT ANALYSIS OF NONLINEAR STRUCTURES	68-0109
STRUCTURES	FINITE ELEMENT ANALYSIS OF FREE VIBRATION AND STABILITY OF STRUCTURES SUBJECTED TO INITIAL STRESS	68-0300
STRUCTURES	FINITE ELEMENT ANALYSIS OF SHELL STRUCTURES IN JAPAN	69-0151
STRUCTURES	FINITE ELEMENT ANALYSIS OF NONLINEAR STRUCTURES	69-0205
STRUCTURES	FINITE ELEMENT ANALYSIS OF INELASTIC STRUCTURES	69-0224
STRUCTURES	FINITE ELEMENT ANALYSIS OF SHIP STRUCTURES	69-0226
STRUCTURES	FINITE ELEMENT ANALYSIS OF CELLULAR STRUCTURES	69-0350
STRUCTURES	FINITE ELEMENT ANALYSIS OF PLATE AND SHELL STRUCTURES	69-0360
STRUCTURES	FINITE ELEMENT ANALYSIS OF STRUCTURES WITH SURFACES OF ARBITRARY CURVATURE	70-0102
STRUCTURES	FINITE ELEMENT ANALYSIS OF STRUCTURES IN THE PLASTIC RANGE	71-0031
STRUCTURES	FINITE ELEMENT ANALYSIS OF THIN-WALLED STRUCTURES BASED ON THE MODERN ENGINEERING THEORY OF BEAMS	71-0140
STRUCTURES	FINITE ELEMENT ANALYSIS OF SHELL-TYPE STRUCTURES	71-0789
STRUCTURES	FINITE ELEMENT ANALYSIS OF REINFORCED CONCRETE STRUCTURES	72-0051
STRUCTURES	FINITE ELEMENT ANALYSIS OF MINE STRUCTURES	72-0687
STRUCTURES	FINITE ELEMENT ANALYSIS OF LARGE TRANSIENT ELASTIC-PLASTIC DEFORMATIONS OF SIMPLE STRUCTURES, WITH A	72-0849
STRUCTURES	FINITE ELEMENT ANALYSIS OF NONLINEAR MEMBRANE STRUCTURES	72-0909
STRUCTURES	FINITE ELEMENT ANALYSIS OF NONLINEAR MEMBRANE STRUCTURES	72-0944
STRUCTURES	FINITE ELEMENT ANALYSIS OF ACOUSTICALLY RADIATING STRUCTURES WITH APPLICATIONS TO SONAR TRANSDUCERS	73-0075
STRUCTURES	FINITE ELEMENT ANALYSIS OF POSTBUCKLING BEHAVIOR OF STRUCTURES	73-0099
STRUCTURES	FINITE ELEMENT ANALYSIS OF CLAD STRUCTURES	73-0353
STRUCTURES	FINITE ELEMENT ANALYSIS OF PAVEMENT STRUCTURES USING AFPAV CODE LINEAR ELASTIC ANALYSIS	73-0658
STRUCTURES	FINITE ELEMENT ANALYSIS OF NONLINEAR STRUCTURES	73-0975
STRUCTURES	FINITE ELEMENT ANALYSIS OF PILE-SUPPORTED PAVEMENT STRUCTURES	74-0157
STRUCTURES	FINITE ELEMENT ANALYSIS OF AUTOMOBILE STRUCTURES	74-0309
STRUCTURES	FINITE ELEMENT ANALYSIS OF HUMAN CARDIAC STRUCTURES	74-0438
STRUCTURES	FINITE ELEMENT ANALYSIS OF ACOUSTICALLY RADIATING STRUCTURES WITH APPLICATIONS TO SONAR TRANSDUCERS	74-0524
STRUCTURES	FINITE ELEMENT ANALYSIS OF PRESTRESSED FOLDED PLATE STRUCTURES	74-1066
STRUCTURES	FINITE ELEMENT ANALYSIS FOR THE SEISMIC STABILITY OF EARTH STRUCTURES	74-1106
STRUCTURES	FINITE ELEMENT ANALYSIS OF SPATIAL SANDWICH PLATE STRUCTURES	74-1132
STRUCTURES	FINITE ELEMENT ANALYSIS OF THREE-DIMENSIONAL CABLE STRUCTURES	74-1275
STRUCTURES	FINITE ELEMENT ANALYSIS OF PILE-SUPPORTED PAVEMENT STRUCTURES	75-0145
STRUCTURES	FINITE ELEMENT ANALYSIS OF CRACKED STRUCTURES SUBJECTED TO SHOCK LOADS	75-0258
STRUCTURES	FINITE ELEMENT ANALYSIS OF DENTAL STRUCTURES - AXISYMMETRIC AND PLANE STRESS IDEALIZATIONS	75-0525
STRUCTURES	FINITE ELEMENT APPROACH TO OPTIMAL DESIGN OF PLASTIC STRUCTURES IN PLANE STRESS.	72-0397
STRUCTURES	FINITE ELEMENT APPROACH TO ACOUSTIC RADIATION FROM ELASTIC STRUCTURES	74-0121
STRUCTURES	FINITE ELEMENT COMPUTER PROGRAM FOR THE ANALYSIS OF LAYERED ORTHOTROPIC STRUCTURES	74-0800
STRUCTURES	FINITE ELEMENT DYNAMIC ANALYSIS OF SHALLOW SHELL STRUCTURES	70-0164
STRUCTURES	FINITE ELEMENT FLUTTER ANALYSIS OF MULTI-WEB WING STRUCTURES	75-0004
STRUCTURES	FINITE ELEMENT FRACTURE MECHANICS ANALYSIS OF TWO-DIMENSIONAL AND AXISYMMETRIC ELASTIC AND ELASTIC-P	74-1092
STRUCTURES	FINITE ELEMENT MESH GENERATION PROGRAM FOR ARBITRARY TWO-AND THREE-DIMENSIONAL STRUCTURES	73-0966
STRUCTURES	FINITE ELEMENT METHOD FOR THE PLASTIC BENDING ANALYSIS OF STRUCTURES	68-0206
STRUCTURES	FINITE ELEMENT METHOD IN ANALYSIS OF REINFORCED AND PRESTRESSED CONCRETE STRUCTURES	71-0270
STRUCTURES	FINITE ELEMENT METHOD FOR DETERMINATION OF MACHINE TOOL STRUCTURES	72-0095
STRUCTURES	FINITE ELEMENT METHOD FOR THE ANALYSIS OF MACHINE TOOL STRUCTURES.	72-0448
STRUCTURES	FINITE ELEMENT METHODS FOR ANALYSIS OF MULTICELLULAR STRUCTURES	72-0561
STRUCTURES	FINITE ELEMENT METHODS FOR THE RESPONSE OF STRUCTURES TO RANDOM EXCITATIONS	75-0230
STRUCTURES	FINITE ELEMENT MODELLING OF THREE-DIMENSIONAL STRUCTURES	68-0267
STRUCTURES	FINITE ELEMENT MODELING AND OPTIMIZATION OF AEROSPACE STRUCTURES	72-0416

403

STRUCTURES	FINITE ELEMENT MODELING AND OPTIMIZATION OF AEROSPACE STRUCTURES	72-0734
STRUCTURES	FINITE ELEMENT SOLUTION OF NONLINEAR STRUCTURES BY PERTURBATION TECHNIQUE	74-1273
STRUCTURES	FINITE ELEMENT STRESS ANALYSIS OF PLANE AND AXISYMMETRIC STRUCTURES	69-0062
STRUCTURES	FINITE ELEMENT STUDY OF HARMONIC WAVE PROPAGATION IN PERIODIC STRUCTURES	74-0130
STRUCTURES	FINITE ELEMENT ULTIMATE LOAD ANALYSIS OF THREE DIMENSIONAL CONCRETE STRUCTURES	74-0672
STRUCTURES	FINITE ELEMENTE FUR GEKRUEMMTE FLAECHENTRAGWERKE. (FINITE ELEMENTS SOLUTIONS FOR CURVED SHELL STRUCT	71-0391
STRUCTURES	FINITE-ELEMENT ANALYSIS OF LARGE ELASTIC-PLASTIC TRANSIENT DEFORMATIONS OF SIMPLE STRUCTURES	71-0326
STRUCTURES	FINITE-ELEMENT ANALYSIS OF MULTILAYER SANDWICH STRUCTURES	72-0953
STRUCTURES	FINITE-ELEMENT ANALYSIS OF PAVEMENT STRUCTURES USING AFPAV CODE (NONLINEAR ELASTIC ANALYSIS)	75-0748
STRUCTURES	FINITE-ELEMENT APPROACH TO ACOUSTIC SCATTERING FROM ELASTIC STRUCTURES	75-0069
STRUCTURES	FINITE-ELEMENT METHOD FOR DETERMINING STRENGTH OF WELDED STRUCTURES	72-0272
STRUCTURES	FINITE-ELEMENT SOLUTION OF EDDY-CURRENT PROBLEM IN MAGNETIC STRUCTURES	74-0106
STRUCTURES	FINITE-ELEMENT TREATMENT OF QUADRUPOLE GROUND MOTIONS IN AXISYMMETRIC EARTH STRUCTURES	75-0434
STRUCTURES	FIRES-T. A COMPUTER PROGRAM FOR THE FIRE RESPONSE OF STRUCTURES - THERMAL	74-0850
STRUCTURES	FOLDED PLATE STRUCTURES BY THE FINITE STRIP METHOD	69-0359
STRUCTURES	FREE VIBRATIONS OF FLUID-BORNE STRUCTURES: INVESTIGATIONS ON A SIMPLE MODEL	74-0976
STRUCTURES	FREE-FREE STRUCTURES IN FINITE ELEMENT ANALYSIS	70-0442
STRUCTURES	GENERAL PURPOSE PROGRAM FOR SHELL STRUCTURES: ISTRAN-S	75-0511
STRUCTURES	GENERAL SINUSOIDAL STIFFNESS MATRICES FOR BUCKLING AND VIBRATION ANALYSES OF THIN FLAT-WALLED STRUCT	68-0278
STRUCTURES	GNATS: A FINITE ELEMENT COMPUTER PROGRAM FOR THE GENERAL NONLINEAR ANALYSIS OF TWO-DIMENSIONAL STRUC	74-0875
STRUCTURES	HIGH-ORDER FINITE ELEMENTS FOR INHOMOGENEOUS ACOUSTIC GUIDING STRUCTURES	73-0137
STRUCTURES	HIGH-ORDER FINITE ELEMENT TECHNIQUE FOR PRISMATIC AND NONPRISMATIC FOLDED PLATE STRUCTURES	74-1130
STRUCTURES	HYBRID FINITE ELEMENT ANALYSIS WITH PARTICULAR REFERENCE TO AXISYMMETRIC STRUCTURES	70-0248
STRUCTURES	HYBRYDOWA METODA ELEMENTOW SKONCZONYCH W ZASTOSOWANIU DO OBLICZEN ORGAN URZADZEN OKRETOWYCH. (HYBRID	74-0963
STRUCTURES	IMPACT OF FINITE ELEMENT TECHNIQUES ON PRACTICAL DESIGN OF SHIP STRUCTURES	69-0344
STRUCTURES	IMPACT ON COMPLEX MECHANICAL STRUCTURES	71-0598
STRUCTURES	IMPROVED FINITE-ELEMENT ANALYSIS OF NUCLEAR-CONTAINMENT SHELL STRUCTURES	73-0239
STRUCTURES	INDETS - INTEGRATED DESIGN OF TANKER STRUCTURES	75-0524
STRUCTURES	INERTIA LOADING IN FINITE-ELEMENT ANALYSIS OF STRUCTURES SUBJECT TO COMPOUND MOTION	75-0351
STRUCTURES	INHERITED ERROR IN FINITE ELEMENT ANALYSES OF STRUCTURES	73-0452
STRUCTURES	INITIAL SETTLEMENT OF STRUCTURES ON CLAY	71-0304
STRUCTURES	INTERACTION OF SOLID SHEAR WALLS AND THEIR SUPPORTING STRUCTURES.	72-0450
STRUCTURES	INTERACTIVE FINITE ELEMENT DATA GENERATION FOR THREE-DIMENSIONAL PLATE STRUCTURES	74-0909
STRUCTURES	INTERNAL SUPPORTING STRUCTURES OF PRESSURE TUBES TYPE REACTOR CORE. ANALYTICAL MODEL FOR STATIC ANAL	71-0049
STRUCTURES	INTERNAL SUPPORTING STRUCTURES OF THE (CIRENE) REACTOR CORE - RESULTS OF STATIC STRESS ANALYSIS FOR	73-0818
STRUCTURES	INTRODUCTION TO FINITE ELEMENT METHOD AND ITS APPLICATION TO MACHINE TOOL STRUCTURES	72-0112
STRUCTURES	ITERATIVE TECHNIQUE FOR THE STRESS ANALYSIS OF PCR STRUCTURES USING THE FINITE ELEMENT METHOD	64-0026
STRUCTURES	LARGE DEFLECTIONS OF STRUCTURES SUBJECT TO HEATING AND EXTERNAL LOADS	60-0002
STRUCTURES	LARGE DEFORMATION AND CRITICAL LOADS ANALYSIS OF FRAMED STRUCTURES	74-1293
STRUCTURES	LARGE STRUCTURES FOR MANNED SPACECRAFT: MATHEMATICAL ANALYSIS, DESIGN, CONSTRUCTION AND TESTS	74-1327
STRUCTURES	LINEAR BUCKLING AND GEOMETRICALLY NONLINEAR ANALYSIS OF PLANAR PLATE-STIFFENER TYPE STRUCTURES BY TH	74-1097
STRUCTURES	LINEAR PROGRAMMING APPROACH TO SHAKEDOWN ANALYSIS OF STRUCTURES	74-0181
STRUCTURES	LOVE WAVE DISPERSION THROUGH IRREGULAR STRUCTURES USING FINITE-ELEMENT ALGORITHMS	75-0431
STRUCTURES	MANIPULATION ERRORS IN FINITE ELEMENT ANALYSIS OF STRUCTURES	69-0349
STRUCTURES	MASS LOADING EFFECTS ON VIBRATION OF RING AND SHELL STRUCTURES	68-0245
STRUCTURES	MATHEMATICAL THEORY OF LINEAR STRUCTURES	71-0498
STRUCTURES	MATRIX ANALYSIS METHODS FOR INELASTIC STRUCTURES	65-0051
STRUCTURES	MATRIX ANALYSIS METHODS FOR ANISOTROPIC INELASTIC STRUCTURES	66-0104
STRUCTURES	MATRIX ANALYSIS OF SHELL STRUCTURES	64-0003
STRUCTURES	MATRIX DIFFERENCE EQUATION ANALYSIS OF VIBRATING PERIODIC STRUCTURES	75-0151
STRUCTURES	MATRIX METHOD OF ANALYSIS OF SHIP STRUCTURES 3RD REPORT	68-0225
STRUCTURES	METHOD FOR ESTIMATING RESPONSE OF PAYLOAD SECONDARY STRUCTURES TO RANDOM EXCITATION	68-0232
STRUCTURES	METHOD OF FINITE SPACE-TIME ELEMENTS IN DYNAMICS OF STRUCTURES	75-0823
STRUCTURES	METHOD SZTYWNYCH ELEMENTOW SKONCZONYCH W OBLICZENIACH KONSTRUCJI OKRETOWYCH. (STIFF FINITE ELEMENT M	74-0923
STRUCTURES	METHODS OF MODELING AND ANALYZING VISCOELASTICALLY DAMPED STRUCTURES	71-0473
STRUCTURES	METHODS OF MODELING AND ANALYZING VISCOELASTICALLY DAMPED STRUCTURES	71-0836
STRUCTURES	MINIMUM WEIGHT DESIGN OF COMPLEX STRUCTURES SUBJECT TO A FREQUENCY CONSTRAINT	70-0374
STRUCTURES	MINIMUM WEIGHT DESIGN OF FINITE ELEMENT STRUCTURES	74-0148
STRUCTURES	MULTI-ELEMENT SPATIAL STRUCTURES	73-0768
STRUCTURES	NATRUAL STRAINS AND STRESSES FOR TRAPEZOIDAL STRUCTURES ANALYSIS	69-0101
STRUCTURES	NATURAL FREQUENCIES OF COMPLEX FREE OR SUBMERGED STRUCTURES BY THE FINITE ELEMENT METHOD	65-0026
STRUCTURES	NATURAL FREQUENCIES OF BOX-TYPE STRUCTURES BY A FINITE-ELEMENT METHOD	74-0193
STRUCTURES	NATURAL VIBRATIONS OF THIN, FLAT WALLED STRUCTURES WITH DIFFERENT BOUNDARY CONDITIONS	71-0509
STRUCTURES	NEW METHOD OF BULKHEAD ANALYSIS IN AIRCRAFT STRUCTURES	74-0334
STRUCTURES	NONLINEAR ANALYSIS OF ELASTIC FRAMED STRUCTURES	68-0222
STRUCTURES	NONLINEAR ANALYSIS OF STRUCTURES	74-0363
STRUCTURES	NONLINEAR ANALYSIS OF PLANAR REINFORCED CONCRETE STRUCTURES	74-0533
STRUCTURES	NONLINEAR DYNAMIC ANALYSIS OF COMPLEX STRUCTURES	73-0193
STRUCTURES	NONLINEAR FINITE ELEMENT ANALYSIS OF REINFORCED CONCRETE STRUCTURES	74-0442
STRUCTURES	NONLINEAR FINITE-ELEMENT FORMULATION FOR TRANSIENT ANALYSIS OF THREE DIMENSIONAL THIN STRUCTURES	74-0874
STRUCTURES	NONLINEAR MATERIAL AND GEOMETRIC BEHAVIOUR OF SHELL STRUCTURES	71-0084
STRUCTURES	NONLINEAR STATIC AND DYNAMIC RESPONSE OF UNDERWATER CABLE STRUCTURES USING THE FINITE ELEMENT METHOD	75-0460
STRUCTURES	NONLINEAR STRUCTURES FOR SYSTEM IDENTIFICATION	74-1030
STRUCTURES	NONLINEAR TRANSIENT RESPONSES OF STRUCTURES BY SPATIAL FINITE-ELEMENT METHOD	73-0138
STRUCTURES	NONLINEAR TRANSIENT ANALYSIS OF AUTOMOTIVE STRUCTURES BY FINITE ELEMENTS	75-0301
STRUCTURES	NONPROPORTIONAL LOADING LIMIT FOR STRUCTURES	74-0228
STRUCTURES	NONSAP - A GENERAL FINITE ELEMENT PROGRAM FOR NONLINEAR DYNAMIC ANALYSIS OF COMPLEX STRUCTURES	73-0740
STRUCTURES	NUMERICAL ANALYSIS OF NONLINEAR PNEUMATIC STRUCTURES	67-0060
STRUCTURES	NUMERICAL ANALYSIS OF SHELL STRUCTURES	67-0077
STRUCTURES	NUMERICAL ANALYSIS OF VISCO-ELASTIC STRUCTURES BY THE FINITE ELEMENT METHOD	70-0177
STRUCTURES	NUMERICAL ANALYSIS OF AUTOMOBILE STRUCTURES	72-0334
STRUCTURES	NUMERICAL CONDITIONING OF STIFFNESS MATRIX FORMULATIONS FOR FRAME STRUCTURES	68-0022
STRUCTURES	NUMERICAL PROCEDURE FOR THE ANALYSIS OF STRUCTURES SUPPORTED BY VISCOELASTIC FOUNDATIONS	73-0564
STRUCTURES	ON A COMBINED STURM SEQUENCE AND INVERSE ITERATION TECHNIQUE FOR EIGENPROBLEM SOLUTION OF SPINNING S	73-0413
STRUCTURES	ON AN APPROACH TO TECHNIQUES FOR THE ANALYSIS OF THE STRUCTURES OF LARGE SYSTEMS OF EQUATIONS	62-0015
STRUCTURES	ON CRACKS IN WELDS AND WELDED STRUCTURES	71-0688
STRUCTURES	ON RESPONSE OF INITIALLY STRESSED STRUCTURES TO RANDOM EXCITATIONS	73-0278
STRUCTURES	ON THE BENDING STIFFNESS OF SLABS IN CROSS-WALLED STRUCTURES	74-0398
STRUCTURES	ON THE DIMENSIONING OF UNDERGROUND STRUCTURES	69-0147
STRUCTURES	ON THE FINITE ELEMENT ANALYSIS OF THIN-WALLED STRUCTURES WITH APPLICATION TO STRUCTURAL PROBLEMS IN	73-0832
STRUCTURES	ON THE FINITE ELEMENT STRESS ANALYSIS OF OIL TANKER STRUCTURES	74-0606
STRUCTURES	ON THE MATRIX THEORY OF STRUCTURES	60-0007
STRUCTURES	ON THE ROLE OF DEFECTS IN CRACK INITIATION IN WELDED STRUCTURES	74-1142
STRUCTURES	OPTIMAL DESIGN OF STRUCTURES WITH BUCKLING CONSTRAINTS	73-0246
STRUCTURES	OPTIMIZATION OF COMPLEX STRUCTURES TO SATISFY STATIC, DYNAMIC AND AEROELASTIC REQUIREMENTS	74-0272
STRUCTURES	OPTIMIZATION OF STRUCTURES TO SATISFY FLUTTER REQUIREMENTS	69-0311
STRUCTURES	OPTIMIZATION OF STRUCTURES FOR STRENGTH AND STABILITY REQUIREMENTS	73-0575
STRUCTURES	OPTIMIZATION OF STRUCTURES UNDERGOING HARMONIC OR STOCHASTIC EXCITATION	75-0711
STRUCTURES	OPTIMUM CONFIGURATIONAL AND DIMENSIONAL DESIGN OF TRUSS STRUCTURES	74-0588
STRUCTURES	OPTIMUM CONFIGURATION OF TRUSS STRUCTURES	74-1315
STRUCTURES	OPTIMUM DESIGN OF ADVANCED COMPOSITE STRUCTURES FOR STATIC LOADS	72-0613
STRUCTURES	OPTIMUM DESIGN OF CONCRETE STRUCTURES	74-0979

STRUCTURES	OPTIMUM DESIGN OF FRAMED STRUCTURES	69-0251
STRUCTURES	OPTIMUM DESIGN OF STRESSED SKIN STRUCTURES	73-0472
STRUCTURES	ORGANIZATION OF GNATS: A GENERAL NONLINEAR ANALYSIS COMPUTER PROGRAM FOR TWO-DIMENSIONAL STRUCTURES	74-0871
STRUCTURES	PERFORMANCE OF EARTH AND EARTH-SUPPORTED STRUCTURES	72-0449
STRUCTURES	PERIODIC SHOCK EXCITATION OF ELASTIC STRUCTURES	69-0163
STRUCTURES	PLANS - A FINITE ELEMENT PROGRAM FOR NONLINEAR ANALYSIS OF STRUCTURES	74-0966
STRUCTURES	PLANS - A FINITE-ELEMENT PROGRAM FOR NONLINEAR ANALYSIS OF STRUCTURES	75-0011
STRUCTURES	PLASTIC SANDWICH BUCKLE-SHELL STRUCTURES	72-0119
STRUCTURES	POSSIBILITY OF USING FINITE ELEMENT METHOD IN THE ANALYSIS OF JOINT PERFORMANCE OF THE HYDRAULIC STR	72-0418
STRUCTURES	POST-BUCKLING RESPONSE OF STRUCTURES USING THE FINITE ELEMENT METHOD	69-0414
STRUCTURES	POST-TENSIONED CONCRETE NUCLEAR STRUCTURES	70-0040
STRUCTURES	PRACTICAL ANALYSIS OF STEEL STRUCTURES USING HIGHER-ORDER ELEMENTS	72-0326
STRUCTURES	PRACTICAL APPLICATION OF FINITE ELEMENT METHODS IN THE DESIGN OF STRUCTURES	69-0420
STRUCTURES	PRACTICAL CONSIDERATIONS IN THE APPLICATION OF FINITE ELEMENT TECHNIQUES TO SHIP STRUCTURES	69-0338
STRUCTURES	PRACTICAL FINITE ELEMENT METHOD OF FAILURE PREDICTION FOR COMPOSITE MATERIAL STRUCTURES	75-0644
STRUCTURES	PRACTICAL METHOD FOR COMPUTER CALCULATION OF STRUCTURES	69-0098
STRUCTURES	PROBABILISTIC BUCKLING BEHAVIOR OF STRUCTURES WITH RANDOM LACK-OF-FIT AND GEOMETRIC IMPERFECTIONS	73-0862
STRUCTURES	PROBABILISTIC FORMULATION FOR ANALYTICAL MODELING OC CONCRETE STRUCTURES	71-0709
STRUCTURES	PROBABILISTIC MODELING OF CONCRETE STRUCTURES	72-0236
STRUCTURES	PROBABILISTIC THEORY OF STRUCTURES USING FINITE ELEMENTS	71-0820
STRUCTURES	PROCEDURE FOR EFFICIENT GENERATING, CHECKING, AND DISPLAYING NASTRAN INPUT AND OUTPUT DATA FOR ANALY	71-0098
STRUCTURES	PROGRAM SPECIFICATION NV 300. GENERAL FINITE ELEMENT PROGRAM FOR 2- AND 3-DIMENSIONAL MEMBRANE STRUC	68-0184
STRUCTURES	QUAD-4: A COMPUTER PROGRAM FOR EVALUATING THE SEISMIC RESPONSE OF SOIL STRUCTURES BY VARIABLE DAMPIN	73-0771
STRUCTURES	QUASI-LINEAR PROGRAMMING ALGORITHM FOR OPTIMIZING FIBRE-REINFORCED STRUCTURES OF FIXED STIFFNESS	75-0225
STRUCTURES	RANDOM RESPONSE OF PERIODIC STRUCTURES BY A FINITE-ELEMENT TECHNIQUE	75-0426
STRUCTURES	RANDOM VIBRATION ANALYSIS SYSTEM FOR COMPLEX STRUCTURES. PART 2. COMPUTER PROGRAM DESCRIPTION	69-0262
STRUCTURES	RANDOM-VIBRATION ANALYSIS SYSTEM FOR COMPLEX STRUCTURES. PART 1. ENGINEERING USER'S GUIDE	68-0106
STRUCTURES	RECENT DEVELOPMENTS OF MATRIX THEORY OF STRUCTURES	59-0006
STRUCTURES	REDUCTION OF STRESSES IN BURIED STRUCTURES	71-0698
STRUCTURES	REFINED CRACKED CONCRETE ANALYSIS OF CONCRETE CONTAINMENT STRUCTURES SUBJECT TO OPERATIONAL AND ENVI	74-0943
STRUCTURES	REFINED FINITE ELEMENT ANALYSIS OF LINEAR AND NONLINEAR TWO-DIMENSIONAL STRUCTURES	66-0024
STRUCTURES	REFINED FINITE ELEMENTS FOR FOLDED PLATE STRUCTURES	72-0574
STRUCTURES	RESPONSE OF BOX-LIKE STRUCTURES TO WEAK EXPLOSIONS	75-0197
STRUCTURES	RESPONSE OF STRUCTURES TO COMBINED BLAST EFFECTS	73-0424
STRUCTURES	REVIEW OF BASIC THEORY AND METHODS OF ANALYSIS OF CONTINUOUS STRUCTURES	73-0697
STRUCTURES	REVIEW OF CONTINUUM, FINITE ELEMENT AND HYBRID TECHNIQUES IN THE ANALYSIS OF STRESS CONCENTRATIONS I	73-1051
STRUCTURES	REVIEW OF CONTINUUM, FINITE ELEMENT AND HYBRID TECHNIQUES IN ANALYSIS OF STRESS CONCENTRATIONS IN ST	74-0914
STRUCTURES	SAFE-CREEP: A COMPUTER PROGRAM FOR THE VISCOELASTIC ANALYSIS OF AXISYMMETRIC AND PLANE CONCRETE STRU	67-0148
STRUCTURES	SEISMIC RESPONSE EFFECTS ON EMBEDDED STRUCTURES	72-0902
STRUCTURES	SHELL STRUCTURES AND CLIMATIC INFLUENCES	72-0389
STRUCTURES	SHELL STRUCTURES AND CLIMATIC INFLUENCES	72-0399
STRUCTURES	SINGER: A COMPUTER CODE FOR GENERAL ANALYSIS OF TWO-DIMENSIONAL CONCRETE STRUCTURES. VOL. 11. PROGRA	75-0723
STRUCTURES	SINGER: A COMPUTER CODE FOR GENERAL ANALYSIS OF TWO-DIMENSIONAL REINFORCED CONCRETE STRUCTURES. VOLU	75-0724
STRUCTURES	SINGER: A COMPUTER CODE FOR GENERAL ANALYSIS TO TWO-DIMENSIONAL CONCRETE STRUCTURES. VOLUME 4. DEMON	75-0730
STRUCTURES	SINGER: A COMPUTER CODE FOR GENERAL ANALYSIS OF TWO-DIMENSIONAL REINFORCED CONCRETE STRUCTURES. VOLU	75-0731
STRUCTURES	SINGLE ELEMENT PANEL FOR FLAT PLATE STRUCTURES	75-0279
STRUCTURES	SOLID SAP. A STATIC ANALYSIS PROGRAM FOR THREE DIMENSIONAL SOLID STRUCTURES	72-0855
STRUCTURES	SOME APPLICATIONS OF FINITE ELEMENTS TO MODEL GEOLOGICAL STRUCTURES	73-0143
STRUCTURES	SOME DEVELOPMENTS IN THE ANALYSIS OF COMPLEX SHIP STRUCTURES	72-0332
STRUCTURES	SOME EXAMPLES OF THE APPLICATION OF THE FINITE ELEMENT METHOD SHIP STRUCTURES	74-0189
STRUCTURES	SOME PROBLEMS IN THE DISCRETE ELEMENT REPRESENTATION OF AIRCRAFT STRUCTURES	64-0021
STRUCTURES	SOME RESULTS ON THE FREE-FREE OSCILLATIONS OF AIRCRAFT TYPE STRUCTURES	65-0007
STRUCTURES	SONIC FATIGUE OF DIFFUSION BOUNDED TITANIUM SANDWICH STRUCTURES	72-0290
STRUCTURES	STABILITY ANALYSIS OF COMPLEX STRUCTURES USING DISCRETE ELEMENT TECHNIQUE	67-0068
STRUCTURES	STABILITY ANALYSIS OF STRUCTURES BY A REDUCED SYSTEM OF GENERALIZED COORDINATES	70-0168
STRUCTURES	STATIC ANALYSIS OF SANDWICH STRUCTURES BY THE FINITE ELEMENT METHOD	71-0730
STRUCTURES	STATIC AND DYNAMIC ANALYSIS OF SANDWICH STRUCTURES BY METHOD OF FINITE ELEMENTS	71-0352
STRUCTURES	STATIC AND DYNAMIC ANALYSIS OF VISCOELASTOPLASTIC FIBER-REINFORCED COMPOSITE SHELLS IN MISSILE STRUC	73-0651
STRUCTURES	STATIC AND DYNAMIC ANALYSIS OF NONLINEAR STRUCTURES	75-0718
STRUCTURES	STATIC AND DYNAMIC FINITE ELEMENT ANALYSIS OF SANDWICH STRUCTURES	68-0003
STRUCTURES	STATIC THREE DIMENSIONAL ANALYSIS OF ELEVATED STEEL TRANSPORTATION STRUCTURES	74-1187
STRUCTURES	STATICAL CALCULATIONS OF PLANE AND SPACE BAR STRUCTURES AS WELL AS STUMPY FRAMEWORKS USING A MODIFIE	73-0747
STRUCTURES	STATISTICAL IDENTIFICATION OF STRUCTURES	74-0141
STRUCTURES	STEADY STATE AND TRANSIENT HEAT CONDUCTION AND STRESS ANALYSES OF TWO AND THREE DIMENSIONAL STRUCTUR	71-0675
STRUCTURES	STIFFNESS AND DEFLECTION ANALYSIS OF COMPLEX STRUCTURES	56-0001
STRUCTURES	STRESS ANALYSIS FOR STRUCTURES UNDER CONSTRUCTION	69-0209
STRUCTURES	STRESS ANALYSIS OF A TUBULAR CROSS-JOINT WITHOUT ANY INTERNAL STIFFENING FOR OFFSHORE STRUCTURES	74-0549
STRUCTURES	STRESS ANALYSIS OF TWO AND THREE DIMENSIONAL STRUCTURES BY THE FINITE ELEMENT METHOD	72-0794
STRUCTURES	STRESS ANALYSIS OF TUBULAR JOINTS FOR OFFSHORE STRUCTURES	73-0207
STRUCTURES	STRESS INTENSITY FACTORS IN PLATE STRUCTURES CALCULATED BY THE FINITE ELEMENT METHOD (IN JAPANESE)	71-0788
STRUCTURES	STRIP PROGRAM FOR DESIGNING LOAD BEARING SURFACE STRUCTURES (PROGRAMME STRIP POUR LE CALCUL DES STRU	71-0408
STRUCTURES	STRIP PROGRAM FOR DESIGNING LOAD BEARING SURFACE STRUCTURES (PROGRAMME STRIP POUR LE CALCUL DES STRU	71-0408
STRUCTURES	STRUCTURAL ANALYSIS OF AIRCRAFT IMPACT ON A NUCLEAR CONTAINMENT VESSEL AND ASSOCIATED STRUCTURES	69-0279
STRUCTURES	STUDY AND DEVELOPMENT OF SIMPLE MATRIX METHODS OF INELASTIC STRUCTURES	66-0115
STRUCTURES	STUDY OF MOTION AND STRENGTH OF FLOATING MARINE STRUCTURES IN WAVES	74-0489
STRUCTURES	STUDY OF THE ACOUSTIC EXCITATION OF STRUCTURES	74-1332
STRUCTURES	TENSILE CREEP ANALYSIS OF CONCRETE STRUCTURES	74-0439
STRUCTURES	TENUE DES STRUCTURES DE GRAPHITE VIS-A-VIS DES SOLLICITATIONS RENCONTREES DANS UN COEUR DE REACTEUR	73-0817
STRUCTURES	TENUE DES STRUCTURES DE GRAPHITE VIS-A-VIS DES SOLLICITATIONS RENCONTREES DANS UN COEUR DE REACTEUR	73-0817
STRUCTURES	THE ANALYSIS OF COMPLEX SHIP STRUCTURES BY THE FINITE ELEMENT TECHNIQUE	64-0006
STRUCTURES	THE APPLICATION OF THE SUPER ELEMENT METHOD IN ANALYSIS AND DESIGN OF SHIP STRUCTURES AND MACHINERY	72-0657
STRUCTURES	THE CALCULATION OF STATIC AND DYNAMIC CHARACTERISTICS OF FLIGHT VEHICLE STRUCTURES BY THE FINITE ELE	73-1018
STRUCTURES	THE EFFECT OF COMPOSITE ACTION ON THE ELASTOPLASTIC ANALYSIS OF COMPLETE BUILDING STRUCTURES	74-0441
STRUCTURES	THE ELASTO-PLASTIC CALCULATION OF GENERAL STRUCTURES AND CONTINUA	68-0207
STRUCTURES	THE EVOLUTION OF SHELL STRUCTURES IN OCEAN TECHNOLOGY	69-0042
STRUCTURES	THE FINITE ELEMENT ANALYSIS OF SANDWICH BUCKLE-SHELL STRUCTURES	72-0968
STRUCTURES	THE FINITE ELEMENT METHOD APPLIED TO THIN SHELLS AND BOX STRUCTURES	70-0444
STRUCTURES	THE FINITE ELEMENT METHOD APPLIED TO THE ANALYSIS OF SANDWICH PLATE AND SHELL STRUCTURES	72-0824
STRUCTURES	THE FINITE ELEMENT METHOD AND LOCAL BOUNDS FOR BOUNDARY VALUE PROBLEMS OF ELASTIC STRUCTURES	73-1056
STRUCTURES	THE FUGA FAMILY FOR FOLDED PLATE STRUCTURES	69-0333
STRUCTURES	THE FUGA FAMILY OF ELEMENTS FOR FOLDED PLATE STRUCTURES A FURTHER SEQUEL TO TECHNICAL NOTE 14	69-0446
STRUCTURES	THE INELASTIC BEHAVIOUR OF STRUCTURES SUBJECTED TO CYCLIC, THERMAL AND MECHANICAL STRESSING CONDITIO	60-0004
STRUCTURES	THE KANSAS TEST TRACK. NON-CONVENTIONAL TRACK STRUCTURES. DESIGN REPORT	72-0797
STRUCTURES	THE MATRIX ANALYSIS OF STRUCTURES WITH CUT-OUTS AND MODIFICATIONS	56-0003
STRUCTURES	THE PIECEWISE LINEAR ANALYSIS OF TWO CONNECTED STRUCTURES INCLUDING THE EFFECT OF CLEARANCE AT THE C	71-0379
STRUCTURES	THE PROPAGATION OF LOVE WAVES ACROSS NONHORIZONTALLY LAYERED STRUCTURES	71-0600
STRUCTURES	THE USE OF FINITE ELEMENT MID-INCREMENT STIFFNESS MATRICES IN THE POST BUCKLING ANALYSIS OF IMPERFEC	71-0383
STRUCTURES	THE USE OF NASTRAN IN THE ANALYSIS OF LARGE COMPLEX AIRFRAME STRUCTURES. MODELING TECHNIQUES AND ORG	71-0110
STRUCTURES	THEMIS FINAL REPORT, VOL. II. NONLINEAR MECHANICS IN SOLIDS AND STRUCTURES	74-0722
STRUCTURES	THEMIS FINAL REPORT, VOL. 1. BASIC RESEARCH IN METHODS OF APPROXIMATION IN THE NONLINEAR MECHANICS O	74-0723
STRUCTURES	THEORETICAL ANALYSIS OF THICK DOME STRUCTURES AND ARBITRARY SOLIDS OF REVOLUTION SUBJECTED TO UNSYMM	70-0037
STRUCTURES	THERMAL AND MECHANICAL ANALYSIS OF WELDED STRUCTURES	73-0819

405

STRUCTURES	THERMAL DEFORMATION PREDICTION IN ORBITING OBSERVATORY STRUCTURES	69-0292
STRUCTURES	THERMAL NONLINEAR BEHAVIOUR OF STRUCTURES	72-0317
STRUCTURES	THERMOMECHANICAL ANALYSIS OF STRUCTURES	71-0618
STRUCTURES	THERMOMECHANICAL ANALYSIS OF STRUCTURES	73-0368
STRUCTURES	THIN-SHELL STRUCTURES: THEORY, EXPERIMENT, AND DESIGN	74-0828
STRUCTURES	THREE DIMENSIONAL STRUCTURES/MEDIA INTERACTION. DYNAX DEVELOPMENT AND APPLICATIONS.	70-0378
STRUCTURES	THREE-DIMENSIONAL ISOPARAMETRIC ELEMENTS AND THEIR USE FOR STATICAL ANALYSIS OF MASSIVE WATER-POWER	71-0364
STRUCTURES	TRANSIENT DYNAMIC RESPONSE OF VISCOELASTIC STRUCTURES	72-0234
STRUCTURES	TRANSIENT RESPONSE OF LINEAR ELASTIC STRUCTURES DETERMINED BY THE MATRIX EXPONENTAL METHOD	71-0240
STRUCTURES	TRANSIENT THERMAL STRESS ANALYSIS OF COMPOSITE STRUCTURES INCLUDING CONTINUOUSLY VARYING PROPERTIES	75-0681
STRUCTURES	TRANSVERSE REINFORCEMENT OF CONTINUOUS, TROUGH-SHAPED CONCRETE RAILROAD STRUCTURES. AN IMPLEMENTATIO	71-0719
STRUCTURES	UNBOUNDED ERRORS IN NUMERICAL ANALYSIS OF STRUCTURES	74-1210
STRUCTURES	USE OF FRACTURE MECHANICS AND FINITE ELEMENT TECHNIQUES FOR DETERMINING THE SAFETY OF OFFSHORE STRUC	74-1211
STRUCTURES	USE OF PAS-1 FINITE ELEMENT COMPUTER SYSTEM FOR STATIC AND DYNAMIC ANALYSIS OF 2-DIMENSIONAL AND 3-D	75-0070
STRUCTURES	USE OF THE ' TITUS ' FINITE ELEMENT CODE FOR THE NUCLEAR STRUCTURES AND INTERNAL STRESS-ANALYSIS	73-0640
STRUCTURES	USER'S MANUAL FOR NTEDYNS - A NONLINEAR FINITE ELEMENT CODE FOR ANALYZING THE BLAST RESPONSE OF UNDE	71-0695
STRUCTURES	VIBRATION ANALYSIS OF STRUCTURES BY COMPONENT MODE SUBSTITUTION	71-0303
STRUCTURES	VIBRATION OF CURVED STRUCTURES USING QUADRILATERAL FINITE ELEMENTS	72-0074
STRUCTURES	VIBRATION OF FRAMES AND OTHER STRUCTURES WITH BANDED STIFFNESS MATRIX	70-0322
STRUCTURES	VIBRATION OF SUBMERGED STRUCTURES AS COMPUTED BY THE FINITE ELEMENT METHOD	75-0227
STRUCTURES	VIBRATIONS OF PLATE AND SHELL STRUCTURES USING TRIANGULAR FINITE ELEMENTS	67-0023
STRUCTURES	VISCEL A GENERAL PURPOSE COMPUTER PROGRAM FOR ANALYSIS OF LINEAR VISCOELASTIC STRUCTURES. VOL. 2	72-0789
STRUCTURES	VYPOCET KRUHOVYCH LOMENICOVYCH KONSTRUKCII METODOU KONECNYCH PASOV. (CALCULATION OF CIRCULAR FOLDED-	74-0612
STRUCTURES-REIN	FIRES-RC. A COMPUTER PROGRAM FOR THE FIRE RESPONSE OF STRUCTURES-REINFORCED CONCRETE FRAMES	74-0747
STRUDL	ANALYSIS AND DESIGN CAPABILITY OF THE STRUDL PROGRAM	71-0054
STRUDL	THE DEVELOPMENT OF ICES STRUDL	72-0327
STRUDL-A	STRUDL-A COMPUTER SYSTEM FOR STRUCTURAL DESIGN	66-0118
STRUDL-II	ICES STRUDL-II ENGINEERING USERS MANUAL. VOLS. 1, 11, 111	69-0398
STRUKTURDYNAMIS	EINIGE VERFAHREN ZUR BERECHNUNG VON EIGENWERTEN UND EIGENVEKTOREN VON NICHT HERMITESCHEN MATRIZEN U	72-0646
STRUKTUREN	BERECHNUNG DER ERDBEBENSCHWINGUNGEN VON STRUKTUREN MIT DER FINITE ELEMENT METHODE, MECHANISCHE MODELL	71-0170
STRUTTURA	ANALISI NUMERICA DEL COMPORTAMENTO ELASTICO DI UNA STRUTTURA BIDIMENSIONALE PIANA. (NUMERICAL ANALYS	74-1029
STRUTURE	CALCULATION METHOD FOR THE STUDY OF CRACKED RIGID STRUCTURES (METODO DI CALCOLO PER LO STUDIO DI STR	71-0637
STUB	ANALYSIS OF A STUB END BY FINITE ELEMENT METHOD	72-0210
STUCTURAL	MATHEMATICAL MODELLING AND STUCTURAL ANALYSIS OF THE MANDIBLE	72-0124
STUDIED	STATISTICAL STATE OF STRESS STUDIED BY GRID ANALYSIS: "NUMERICAL METHODS OF ANALYSIS"	49-0001
STUDIO	CALCULATION METHOD FOR THE STUDY OF CRACKED RIGID STRUCTURES (METODO DI CALCOLO PER LO STUDIO DI STR	71-0637
STUDIO	STUDIO DI CAMPI ELETTRICI E MAGNETICI STANZIONARI CON IL METODO DEGLI ELEMENTI FINITI. APPLICAZIONE	75-0207
STUMPY	STATICAL CALCULATIONS OF PLANE AND SPACE BAR STRUCTURES AS WELL AS STUMPY FRAMEWORKS USING A MODIFIE	73-0747
STUPITSEI	ISSLEDOVANIE NAPRYAZHENNO-DEFORMIROVANNOGO SOSTOYANIYA SIMMETRICHNYKH DISKOV SO STUPITSEI. (INVESTIG	74-0965
STURM	ON A COMBINED STURM SEQUENCE AND INVERSE ITERATION TECHNIQUE FOR EIGENPROBLEM SOLUTION OF SPINNING S	73-0413
STURM	SOLUTION OF EIGENVALUE PROBLEMS BY STURM SEQUENCE METHOD.	72-0400
STUTTGART	REPORT ON POST CONF. MEETING ON COMPUTATIONAL ASPECTS OF ASPECTS OF FINITE ELEMENT METHOD ISD STUTTG	72-0475
STV	MINUTEMAN STV STRESS ANALYSIS AND TESTING	74-0809
SUB-STRUCTURE	ANALYSIS FOR CRACKS IN SHIP STRUCTURE USING THE FINITE ELEMENT METHOD WITH THE CONCEPT OF CRACK SUB-	74-0248
SUB-STRUCTURE	ANALYSIS FOR CRACKS IN SHIP STRUCTURES USING THE FINITE-ELEMENT METHOD WITH THE CONCEPT OF CRACK SUB	74-1367
SUB-STRUCTURE	CALCULATION OF TWO AND THREE DIMENSIONAL FLOW PROBLEMS ACCORDING TO SUB-STRUCTURE THEORY WITH ISOPAR	74-0215
SUB-STRUCTURE	COMPARISON BETWEEN SPARSE STIFFNESS MATRIX AND SUB-STRUCTURE METHODS	73-1068
SUB-STRUCTURING	INTERACTIVE FINITE ELEMENT DATA GENERATION USING THE SUB-STRUCTURING TECHNIQUE	74-1291
SUBASSEMBLIES	DYNAMIC RESPONSE OF FAST-REACTOR CORE SUBASSEMBLIES	74-0540
SUBASSEMBLIES	FINITE ELEMENT APPROACH TO PRESSURE WAVE ATTENUATION BY REACTOR FUEL SUBASSEMBLIES	75-0522
SUBASSEMBLIES	FINITE ELEMENT STUDY OF PRESSURE WAVE ATTENUATION BY REACTOR FUEL SUBASSEMBLIES	75-0215
SUBASSEMBLIES	NONLINEAR FINITE-ELEMENT METHOD FOR PLATES AND ITS APPLICATION TO DYNAMIC RESPONSE OF REACTOR FUEL S	74-0670
SUBASSEMBLIES	RESPONSE OF REACTOR CORE SUBASSEMBLIES TO IMPULSIVE LOADINGS	73-0076
SUBCRITICAL	FINITE ELEMENT CALCULATION OF AXISYMMETRIC SUBCRITICAL COMPRESSIBLE FLOW	74-1064
SUBCRITICAL	FINITE-ELEMENT METHOD TO COMPUTE STATIONARY SUBCRITICAL FLOWS	75-0439
SUBGRADE	AIRCRAFT-PAVEMENT INTERACTION STUDIES, PHASE 1: A FINITE ELEMENT MODEL OF A JOINTED CONCRETE PAVEMEN	73-0894
SUBGROUPS	FINITE AND DISCONNECTED SUBGROUPS OF SU(3) AND THEIR APPLICATION TO THE ELEMENTARY PARTICLE SPECTRUM	66-0129
SUBMARINE	APPLICATION OF THE FINITE ELEMENT METHOD FOR ANALYSIS AND OPTIMIZATION OF SUBMARINE STRUCTURES	71-0350
SUBMARINE	DYNAMICS OF SUBMARINE STRUCTURES, PART 2	68-0025
SUBMARINE	DYNAMICS OF SUBMARINE STRUCTURES, VOLUME 1	68-0251
SUBMARINE	DYNAMICS OF SUBMARINE STRUCTURES, PART 1.	69-0053
SUBMERGED	ANALYSIS OF TOTALLY SUBMERGED THIN SHEELS BY FINITE ELEMENT TECHNIQUE	71-0616
SUBMERGED	COUPLED VIBRATIONS OF A STRUCTURE SUBMERGED IN A COMPRESSIBLLE FLUID	69-0006
SUBMERGED	COUPLED VIBRATIONS OF A STRUCTURE SUBMERGED IN A COMPRESSIBLE FLUID	69-0507
SUBMERGED	EARTHQUAKE ANALYSIS OF AXISYMMETRIC TOWERS PARTIALLY SUBMERGED IN WATER	75-0076
SUBMERGED	NATURAL FREQUENCIES OF COMPLEX FREE OR SUBMERGED STRUCTURES BY THE FINITE ELEMENT METHOD	65-0026
SUBMERGED	THE FINITE ELEMENT DYNAMIC STABILITY ANALYSIS OF THIN SHELLS SUBMERGED IN FLUIDS	71-0057
SUBMERGED	VIBRATION OF SUBMERGED STRUCTURES AS COMPUTED BY THE FINITE ELEMENT METHOD	75-0227
SUBROUTINE	PLOTS - THE PLOTTING SUBROUTINE INCORPORATED IN THE BALLISTIC RESEARCH LABORATORIES ELECTRONIC SCIEN	72-0803
SUBROUTINES	ANALOG SUBROUTINES FOR THE FINITE ELEMENT METHOD	72-0652
SUBROUTINES	FORTRAN SUBROUTINES FOR THE SOLUTION OF LAPLACE'S EQUATION OVER A GENERAL REGION IN TWO DIMENSIONS	70-0425
SUBROUTINES	FORTRAN SUBROUTINES FOR FINITE ELEMENT ANALYSIS	72-0694
SUBROUTING	SHAPE FUNCTION SUBROUTING FOR AN ISOPARAMETRIC THIN PLATE ELEMENT	73-0618
SUBSEQUENT	A NUMERICAL THERMO-MECHANICAL MODEL FOR THE WELDING AND SUBSEQUENT LOADING OF A FABRICATED STRUCTURE	72-0484
SUBSEQUENT	NUMERICAL, THERMO-MECHANICAL MODEL FOR THE WELDING AND SUBSEQUENT LOADING OF A FABRICATED STRUCTURE	73-0418
SUBSIDENCE	A MULTI-LAYERED FINITE ELEMENT MODEL FOR PREDICTING MINE SUBSIDENCE	68-0293
SUBSIDENCE	ANALYTICAL METHODS FOR PREDICTING SUBSIDENCE ABOVE SOLUTION-MINED CAVITIES	74-1145
SUBSIDENCE	FINITE ELEMENT ANALYSIS OF LAND SUBSIDENCE	69-0233
SUBSIDENCE	MATHEMATICAL SIMULATION OF THE SUBSIDENCE OF VENICE 1 THEORY	73-0992
SUBSOILS	ANALYSIS OF LOAD-BEARING FILLS OVER SOFT SUBSOILS	71-0330
SUBSONIC	A FINITE ELEMENT ANALYSIS FOR STEADY AND OSCILLATORY SUBSONIC FLOW AROUND COMPLEX CONFIGURATIONS	74-1317
SUBSONIC	A FINITE ELEMENT METHOD FOR CALCULATING AERODYNAMIC COEFFICIENTS OF A SUBSONIC AIRPLANE	73-0655
SUBSONIC	A FINITE-ELEMENT FORMULATION FOR SUBSONIC FLOWS AROUND COMPLEX CONFIGURATIONS	73-0742
SUBSONIC	A FINITE-ELEMENT METHOD FOR LIFTING SURFACES IN STEADY INCOMPRESSIBLE SUBSONIC FLOW	74-1321
SUBSONIC	INTEGRAL EQUATION METHOD FOR SUBSONIC FLOW PAST AIRFOILS IN VENTILATED WIND TUNNELS	75-0065
SUBSONIC	STABILITY ANALYSIS OF THE TORSIONAL BENDING VIBRATIONS OF WING IN A SUBSONIC FLOW BY METHOD OF THE F	75-0755
SUBSONIC	SUBSONIC AND SUPERSONIC INDICIAL AERODYNAMICS AND AERODYNAMIC TRANSFER FUNCTION FOR COMPLEX CONFIGUR	74-1320
SUBSONIC	THE MBB-UFE SUBSONIC PANEL METHOD. PART 2 - THE LIFT-CONNECTED DISPLACEMENT PROBLEM IN COMPRESSIBLE	72-0865
SUBSONIC	TWO AND THREE DIMENSIONAL ANALYSIS OF SUBSONIC COMPRESSIBLE POTENTIAL FLOWS IN DUCTS AND AROUND LIFT	74-0026
SUBSONIC	UNSTEADY SUBSONIC COLLOCATION METHOD FOR WINGS WITH AND WITHOUT CONTROL SURFACES	72-0012
SUBSPACES	ON DIRICHLET PROBLEMS USING SUBSPACES WITH NEARLY ZERO BOUNDARY CONDITIONS	72-0514
SUBSPACES	SUBSPACES WITH ACCURATELY INTERPOLATED BOUNDARY CONDITIONS	72-0527
SUBSTITUTION	VIBRATION ANALYSIS OF STRUCTURES BY COMPONENT MODE SUBSTITUTION	71-0303
SUBSTRUCTURE	COMPUTERIZED SUBSTRUCTURE ANALYSIS	75-0419
SUBSTRUCTURE	SUBSTRUCTURE ANALYSIS TECHNIQUES AND AUTOMATION	73-0597
SUBSTRUCTURE	SUBSTRUCTURE ANALYSIS OF MULTISTOREY FLAT SLAB BUILDINGS	74-0394
SUBSTRUCTURES	AUTOMATED STRESS ANALYSIS USING SUBSTRUCTURES	66-0071
SUBSTRUCTURES	DYNAMIC SUBSTRUCTURES METHOD FOR SHOCK ANALYSIS	68-0285
SUBSTRUCTURES	SUBSTRUCTURES ANALYSIS OF IMPELLER VIBRATION MODES	75-0283
SUBSTRUCTURING	ADVANCED SUBSTRUCTURING TECHNIQUES	71-0838
SUBSTRUCTURING	APPLICATION OF A SUBSTRUCTURING TECHNIQUE TO THE PROBLEM OF CRACK EXTENSION AND CLOSURE	74-0772
SUBSTRUCTURING	RELATIONSHIP OF SUBSTRUCTURING TO IMPROVEMENT OF DYNAMIC MODELING	71-0323

SUBSTRUCTURING	SPACE SHUTTLE, THE NEED FOR SUBSTRUCTURING	71-0104
SUBSTRUCTURING	STATIC ANALYSIS VIA SUBSTRUCTURING OF AN EXPERIMENTAL VEHICLE FRONT-END BODY STRUCTURE	74-0313
SUBSTRUCTURING	SUBSTRUCTURING TECHNIQUES	71-0692
SUBSTRUCTURING	SYSTEMATIC SUBSTRUCTURING	74-0895
SUBSURFACE	NUMERICAL METHODS IN SUBSURFACE HYDROLOGY: WITH AN INTRODUCTION TO THE FINITE ELEMENT METHOD	71-0800
SUBSURFACE	ROCK FRAGMENTATION BY SUBSURFACE THERMAL INCLUSIONS - A FINITE ELEMENT STUDY	73-1159
SUBTERRANEAN	SUBTERRANEAN GUIDEWAY RESPONSES TO LEVITATION VEHICLE LOADING	74-0927
SUBVOLUMES	FINITE ELEMENT CYCLIC THERMOPLASTICITY ANALYSIS BY THE METHOD OF SUBVOLUMES	75-0787
SUBWAY	SINGLE-TRACK SUBWAY ENVIRONMENTAL SIMULATION MODEL. PHASE 1	71-0693
SUCCESSIVE	EXTENSIONS OF THE SUCCESSIVE OVER-RELAXATION THEORY WITH APPLICATIONS TO FINITE ELEMENT APPROXIMATIO	73-0980
SUCCESSIVE	FINITE ELEMENT SOLUTION OF A NON-LINEAR DIFFERENTIAL EQUATION USING KACHANOV'S AND SUCCESSIVE APPROX	75-0108
SUCHETOM	RASCHET NAPRYAZHENIL L PEREMESHCHENII V SVODE PODZEMNOGO MASHINNOGO ZALA GES I V OKRUZHAYUSHCHEM SKA	74-0542
SUCTION	ON THE FICTION OF LEADING EDGE SUCTION	71-0712
SUDDENLY	FINITE ELEMENT ANALYSIS OF STATIC AND DYNAMIC STRESSES AROUND SUDDENLY PUNCHED HOLES IN PLATES AND S	73-1088
SUDOVYKH	APPLICATION OF THE METHOD OF FINITE ELEMENTS TO THE CALCULATION OF SHIP STRUCTURES (ISPOLZOVANIE MET	73-0933
SUFFICIENCY	NUMERICAL SUFFICIENCY TEST FOR MONOTONIC CONVERGENCE OF FINITE ELEMENT MODELS	75-0620
SUFFICIENCY	ON A NUMERICAL SUFFICIENCY TEST FOR MONOTONIC CONVERGENCE OF FINITE ELEMENT MODELS	75-0533
SUFFICIENT	SUFFICIENT CONDITIONS FOR CONVERGENCE IN THE FINITE ELEMENT METHOD FOR ANY SOLUTION OF FINITE ENERGY	72-0065
SUITABILITY	SUITABILITY OF THE FINITE ELEMENT METHOD FOR ANALYSIS OF UNSTEADY FLOW AROUND OSCILLATING AIRFOILS	73-0069
SUL	CHARACTERISTICS OF A HYBRID CALCULATION SYSTEM BASED ON THE FINITE ELEMENT METHOD. (CARATTERISTICHE	72-0715
SULFATE	A FINITE ELEMENT ANALYSIS OF DISSOLVED OXYGEN DRAWDOWN AND SULFATE PRODUCTION IN STRIP MINE SPOIL DA	75-0795
SULLA	ON THE RESOLUTION OF NONLINEAR ELASTIC AND ELASTOPLASTIC PROBLEMS BY THE METHOD OF FINITE ELEMENTS.	74-1326
SULLA	SULLA DETERMINAZIONE DELLE FREQUENZE PROPRIE DI ALBERI SU PIU SUPPORTI ELASTICI. (DETERMINATION OF F	74-0916
SUMMARY	A SUMMARY OF DISCRETE-ELEMENT METHODS OF ANALYSIS FOR PAVEMENT SLABS	72-0674
SUMMARY	ANALYTICAL AND EXPERIMENTAL INVESTIGATION OF A 1/8 SCALE DYNAMIC MODEL OF THE SHUTTLE ORBITER. VOLUM	74-1186
SUMMARY	RADIAL AND AXIAL TRANSMISSIBILITY CHARACTERISTICS OF TYPICAL ROCKET VEHICLE STRUCTURE TECHNICAL SUMM	69-0475
SUMMARY	SUMMARY OF SOIL-STRUCTURE INTERACTION	72-0317
SUPER	FINITE-ELEMENT MODELING OF DIRECTIONAL SOLIDIFICATION OF SUPER ALLOY EUTECTICS.	75-0350
SUPER	GENERAL DESCRIPTION OF SESAM-69, SUPER ELEMENT STRUCTURAL ANALYSIS (PROGRAM) MODULES	71-0512
SUPER	SOME INTERIOR SUPER CONVERGENCE ESTIMATES FOR SPLINE APPROXIMATION	73-0541
SUPER	THE APPLICATION OF THE SUPER ELEMENT METHOD IN ANALYSIS AND DESIGN OF SHIP STRUCTURES AND MACHINERY	72-0657
SUPER-ELEMENT	A SUPER-ELEMENT FOR CRACK ANALYSIS	73-0754
SUPERCONVERGENC	A SUPERCONVERGENCE RESULT FOR THE APPROXIMATE SOLUTION OF THE HEAT EQUATION BY A COLLOCATION METHOD	72-0510
SUPERCONVERGENC	SOME SUPERCONVERGENCE RESULTS FOR AN H-GALERKIN PROCEDURE FOR THE HEAT EQUATION	74-0411
SUPERIEURES	BORNES SUPERIEURES ET INFERIEURES DANS L'ANALYSE MATRICIELLE DES PLAQUES EN FLEXION TORSION	64-0015
SUPERPARAMETRIC	PREDICTION OF MARINE PROPELLER DISTORTION AND STRESSES USING A SUPERPARAMETRIC THICK SHELL FINITE EL	73-0370
SUPERPATCH	THE SUPERPATCH THEOREM AND OTHER PROPOSITIONS RELATING TO THE PATCH TEST	75-0026
SUPERPOSED	A FINITE ELEMENT SOLUTION OF THE STABILITY OF SUPERPOSED FLUIDS	74-1226
SUPERPOSITION	ANALYSIS OF WATER WAVE PROBLEMS BY THE METHOD OF SUPERPOSITION OF ANALYTICAL AND FINITE ELEMENT SOLU	75-0870
SUPERPOSITION	ELASTIC-PLASTIC FINITE ELEMENT ANALYSIS USING SUPERPOSITION	75-0314
SUPERPOSITION	ELASTIC-PLASTIC FINITE-ELEMENT ANALYSIS USING SUPERPOSITION	75-0429
SUPERPOSITION	ELASTO-PLASTIC SEISMIC RESPONSE ANALYSIS OF FRAMED STRUCTURES BY MODE SUPERPOSITION	73-0325
SUPERPOSITION	FINITE ELEMENT CALCULATION OF STRESS INTENSITY FACTORS USING SUPERPOSITION	75-0251
SUPERPOSITION	NONLINEAR DYNAMIC ANALYSIS OF TORSIONAL SHELLS BY COMBINED USE OF FINITE ELEMENT AND MODE SUPERPOSIT	75-0869
SUPERPOSITION	STRESS ANALYSIS BY SUPERPOSITION OF THE ANALYTICAL SOLUTION AND FINITE ELEMENT SOLUTION	72-1000
SUPERPOSITION	THE FINITE ELEMENT CALCULATION OF STRESS INTENSITY FACTORS USING SUPERPOSITION	75-0886
SUPERSONIC	A FINITE-ELEMENT ANALYSIS FOR STEADY-AND OSCILLATORY SUPERSONIC FLOWS AROUND COMPLEX CONFIGURATIONS	74-1316
SUPERSONIC	AUTOMATED DESIGN OPTIMIZATION OF SUPERSONIC AIRPLANE WING STRUCTURES UNDER DYNAMIC CONSTRAINTS	72-0178
SUPERSONIC	CYCLIC-STRESS ANALYSIS OF NOTCHES FOR SUPERSONIC TRANSPORT CONDITIONS	74-0834
SUPERSONIC	DEVELOPMENT AND APPLICATIONS OF SUPERSONIC UNSTEADY CONSISTENT AERODYNAMICS FOR INTERFERING PARALLEL	72-0716
SUPERSONIC	FINITE ELEMENT ANALYSIS OF SUPERSONIC PANEL FLUTTER	73-0481
SUPERSONIC	FINITE ELEMENT APPROACH TO THE INTEGRATED POTENTIAL FORMULATION OF GENERAL UNSTEADY SUPERSONIC AEROD	73-0910
SUPERSONIC	FLUTTER OF FLAT FINITE-ELEMENT PANELS IN A SUPERSONIC POTENTIAL FLOW	75-0814
SUPERSONIC	FURTHER DEVELOPMENTS IN CONSISTENT UNSTEADY SUPERSONIC AERODYNAMIC COEFFICIENTS	72-0559
SUPERSONIC	KERNEL FUNCTION FOR NONPLANAR OSCILLATING SURFACES IN SUPERSONIC FLOW	71-0472
SUPERSONIC	KINEMATICALLY CONSISTENT UNSTEADY AERODYNAMIC COEFFICIENTS IN SUPERSONIC FLOW	70-0014
SUPERSONIC	KINEMATICALLY CONSISTENT UNSTEADY AERODYNAMIC COEFFICIENTS IN SUPERSONIC FLOW	70-0316
SUPERSONIC	MODEL OF ELASTIC PLATE FROM FINITE ELEMENTS IN A SUPERSONIC FLOW	75-0038
SUPERSONIC	OPTIMIZATION OF A SUPERSONIC PANEL SUBJECT TO A FLUTTER CONSTRAINT. A FINITE ELEMENT SOLUTION	71-0073
SUPERSONIC	SUBSONIC AND SUPERSONIC INDICIAL AERODYNAMICS AND AERODYNAMIC TRANSFER FUNCTION FOR COMPLEX CONFIGUR	74-1320
SUPERSTRUCTURE	HULL AND SUPERSTRUCTURE VIBRATIONS DESIGN CALCULATIONS BY FINITE ELEMENTS	75-0556
SUPERTANKER	THREE DIMENSIONAL SUPERTANKER COMPUTER PROGRAM	71-0661
SUPERTANKERS	STRUCTURAL DESIGN OF SUPERTANKERS	70-0152
SUPPLEMENT	POST-BUCKLING AND NON-LINEAR ANALYSIS BY FINITE ELEMENT METHOD AS A SUPPLEMENT TO A LINEAR ANALYSIS	75-0505
SUPPORT	A CASE STUDY OF IN-SITU ROCK DEFORMATION BEHAVIOR FOR THE DESIGN OF GROUND SUPPORT SYSTEM	72-0668
SUPPORT	ANALYTICAL PROBLEMS ASSOCIATED WITH CORE SUPPORT STRUCTURES OF PWR	71-0039
SUPPORT	CALCULATION OF STRESS CONCENTRATIONS ON SUPPORT STRUCTURES UNDER THERMOMECHANICAL LOADING (BERECHNUN	73-0796
SUPPORT	DESIGN, DEVELOPMENT, DOCUMENTATION AND SUPPORT OF A MAJOR FINITE ELEMENT SYSTEM	73-0205
SUPPORT	DYNAMIC ANALYSIS OF A FLEXIBLE VEHICLE MOVING ALONG A FLEXIBLE SUPPORT	71-0597
SUPPORT	EFFECTS OF SHEAR DEFORMATION, ROTARY INERTIA, AND ELASTICITY OF THE SUPPORT ON THE RESONANCE FREQUEN	75-0369
SUPPORT	FINITE ELEMENT STUDIES OF PILE DRIVER TUNNELS INCLUDING CONSIDERATIONS OF SUPPORT REQUIREMENTS	72-0679
SUPPORT	NONLINEAR SEISMIC ANALYSIS OF CURRENT DIVIDER SUPPORT SYSTEM FOR PACIFIC DC INTERTIE	73-0200
SUPPORT	NORMAL MODE ANALYSIS OF A FUZE SUPPORT STRUCTURE USING NASTRAN. PART 1	72-0853
SUPPORT	POLYMER-IMPREGNATED CONCRETE TUNNEL SUPPORT AND LINING	73-0736
SUPPORT	PRINCIPAL STRESSES AT THE INTERMEDIATE SUPPORT OF PRESTRESSED CONCRETE CONTINUOUS BEAMS	70-0352
SUPPORT	RATIONAL DESIGN OF TUNNEL SUPPORTS: TUNNEL SUPPORT LOADING CAUSED BY ROCK FAILURE	75-0687
SUPPORT	RESEARCH TO IMPROVE TUNNEL SUPPORT SYSTEMS	74-0777
SUPPORT	SLAB BRIDGES WITH ARBITRARY SHAPE AND SUPPORT CONDITIONS - A GENERAL METHOD OF ANALYSIS BASED ON FIN	68-0233
SUPPORT	STUDIES TO IMPROVE TUNNEL SUPPORT SYSTEMS	74-0957
SUPPORT	SUPPORT PERFORMANCE OF HYDRAULIC BACKFILL	74-0257
SUPPORT	THEORETICAL INVESTIGATION OF ROCK AND SUPPORT INTERACTION - DEVELOP MORE RATIONAL DESIGN METHODS AND	74-0782
SUPPORT	THEORETICAL INVESTIGATION OF ROCK AND SUPPORT INTERACTION - DEVELOP MORE RATIONAL DESIGN METHODS AND	74-0782
SUPPORTED	ANALYSIS OF AN ORTHOTROPIC PLATE WITH TWO OPPOSITE EDGES SIMPLY SUPPORTED AND THE OTHER TWO FREE	73-0741
SUPPORTED	ANALYSIS OF CORNER SUPPORTED SKEW SLABS	68-0203
SUPPORTED	ANALYSIS OF CORNER SUPPORTED RECTANGULAR SLABS	70-0046
SUPPORTED	ANALYSIS OF SIMPLY SUPPORTED THICK, LAYERED PLATES	71-0428
SUPPORTED	ANALYSIS OF STRESSES AND DEFLECTIONS IN FRAME SUPPORTED TENTS	74-0689
SUPPORTED	ARCH DAMS ANALYSED BY A LINEAR FINITE ELEMENT SHELL THE FINITE STRIP METHOD IN THE ANALYSIS OF ELAST	68-0189
SUPPORTED	DER GERISSENE ZUSTAND ZWEISEITIG GELAGERTER PLATTEN UNTER EINZELLASTEN - NICHTLINEARE BERECHNUNG MIT	75-0858
SUPPORTED	DUAL ANALYSIS OF COLUMN SUPPORTED SLABS	74-0399
SUPPORTED	FLEXURE OF A BEAM SUPPORTED ON AN ELASTIC FOUNDATION	72-0871
SUPPORTED	FREE VIBRATION OF A SIMPLY SUPPORTED BEAM WITH NONLINEAR MATERIAL PROPERTIES	71-0689
SUPPORTED	NON-LINEAR EFFECTS OF HIGH TEMPERATURE ON THE VIBRATION OF A SIMPLY SUPPORTED BEAM WITH A CENTRAL MA	73-0838
SUPPORTED	NUMERICAL PROCEDURE FOR THE ANALYSIS OF STRUCTURES SUPPORTED BY VISCOELASTIC FOUNDATIONS	73-0564
SUPPORTED	ORTHOTROPIC FLAT PLATES SUPPORTED ON COLUMNS ELONGATED IN PLANE	72-0492
SUPPORTED	RESPONSE OF SIMPLY SUPPORTED TIMOSHENKO BEAMS	69-0489
SUPPORTED	THE FINITE STRIP METHOD IN THE ANALYSIS OF ELASTIC PLATES WITH TWO SIMPLY SUPPORTED ENDS	68-0068
SUPPORTI	SULLA DETERMINAZIONE DELLE FREQUENZE PROPRIE DI ALBERI SU PIU SUPPORTI ELASTICI. (DETERMINATION OF F	74-0916
SUPPORTING	A FINITE ELEMENT APPROACH TO THE DETERMINATION OF THE DYNAMIC RESPONSE OF BEAMS SUPPORTING MOVING MA	74-1114
SUPPORTING	HIGHER-ORDER FINITE-ELEMENT ANALYSIS OF TOPOGRAPHIC GUIDES SUPPORTING ELASTIC SURFACE WAVES	73-0114
SUPPORTING	INTERACTION OF SOLID SHEAR WALLS AND THEIR SUPPORTING STRUCTURES.	72-0450
SUPPORTING	INTERNAL SUPPORTING STRUCTURES OF PRESSURE TUBES TYPE REACTOR CORE. ANALYTICAL MODEL FOR STATIC ANAL	71-0049

SUPPORTING	INTERNAL SUPPORTING STRUCTURES OF THE (CIRENE) REACTOR CORE - RESULTS OF STATIC STRESS ANALYSIS FOR	73-0818
SUPPORTING	MODAL COUPLING PROCEDURES ADAPTED TO NASTRAN ANALYSIS OF THE 1/8-SCALE SHUTTLE STRUCTURAL DYNAMICS M	75-0678
SUPPORTS	A FINITE ELEMENT TECHNIQUE FOR ANALYZING FOLDED PLATE TYPE STRUCTURES WITH SKEWED SUPPORTS	69-0525
SUPPORTS	POST-TENSIONED SLAB BRIDGES ON ISOLATED SUPPORTS	75-0084
SUPPORTS	RATIONAL DESIGN OF TUNNEL SUPPORTS: TUNNEL SUPPORT LOADING CAUSED BY ROCK FAILURE	75-0687
SURFACE	A FINITE ELEMENT APPROACH TO SURFACE DEFINITION	69-0030
SURFACE	A METHOD FOR DETERMINING THE SURFACE CONTACT STRESSES RESULTING FROM INTERFERENCE FITS	70-0101
SURFACE	A THERMO-ELASTIC ANALYSIS OF THE PARALLEL SURFACE THRUST BEARING	74-1069
SURFACE	A VARIATIONAL METHOD IN POTENTIAL FLOWS WITH A FREE SURFACE	72-0723
SURFACE	AN ANALYSIS OF TWO-DIMENSIONAL SURFACE RUN-OFF BY FINITE ELEMENTS	74-1336
SURFACE	AN INVESTIGATION ON THE PLASTICITY MAGNIFICATION FACTOR IN DEEP SURFACE FLOWS	74-0766
SURFACE	ANALYSIS OF NONSTEADY FLOW WITH A FREE SURFACE USING FINITE ELEMENT METHOD	71-0482
SURFACE	APPLICATION OF THE FINITE-ELEMENT METHOD FOR SIMULATION OF SURFACE WATER TRANSPORT PROBLEMS	72-0826
SURFACE	APPLICATION OF THE FINITE ELEMENT METHOD TO SURFACE WAVE ANALYSIS	74-0913
SURFACE	CALCULATION OF STRESS INTENSITY FACTORS AND FATIGUE CRACK PROPAGATION OF SEMI-ELLIPTICAL, PART-THROU	73-0360
SURFACE	COMPUTER AIDED DESIGN AND CONSTRUCTION BY MEANS OF CURVED SURFACE AND SOLID ELEMENTS	73-0165
SURFACE	CONTACT OF AN INFLATED TOROIDAL MEMBRANE WITH A FLAT SURFACE AS AN APPROACH TO THE TIRE DEFLECTION P	75-0164
SURFACE	CRUSTAL STRUCTURE FROM SURFACE LOAD TILTS USING A FINITE ELEMENT MODEL	72-0005
SURFACE	DARCY FLOW SOLUTIONS WITH A FREE SURFACE	67-0081
SURFACE	DEFORMATION MECHANICS OF DEEP SURFACE FLAW CRACKS	72-0785
SURFACE	DIE BERECHNUNG DER BEWEGUNGEN UND SPANNUNGEN DES GEBIRGES VOM ABBAU BIS ZUR TAGESOBERFLAECHE NACH DE	73-0456
SURFACE	ERROR ESTIMATES FOR A FINITE-ELEMENT APPROXIMATION OF A MINIMAL SURFACE	75-0491
SURFACE	FINITE ELEMENT ANALYSIS OF DISPERSION-FREE GUIDES FOR ELASTIC SURFACE WAVES	73-0009
SURFACE	FINITE ELEMENT ANALYSIS OF SURFACE WAVE PROBLEMS USING INTEGRAL EQUATIONS	74-0624
SURFACE	FINITE ELEMENT ANALYSIS OF THE SURFACE DEFORMATION DUE TO A UNIFORM LOADING ON A LAYER OF GIBSON SOI	75-0240
SURFACE	FINITE ELEMENT METHOD FOR INCOMPRESSIBLE SLOW VISCOUS FLOW WITH A FREE SURFACE	69-0303
SURFACE	FINITE ELEMENT METHOD OF ANALYZING STEADY SEEPAGE WITH A FREE SURFACE	70-0092
SURFACE	FINITE ELEMENT METHODS: TWO-DIMENSIONAL SEEPAGE WITH A FREE SURFACE	71-0475
SURFACE	FINITE ELEMENT SOLUTIONS OF THE SHALLOW WATER EQUATIONS SURFACE RUN OFF	74-0046
SURFACE	FINITE ELEMENT TECHNIQUE IN LIFTING SURFACE PROBLEMS	74-0022
SURFACE	FINITE-ELEMENT MODELING OF SURFACE DEFORMATION ASSOCIATED WITH VOLCANISM	75-0596
SURFACE	FORCE AND MOMENTS ON ASYMMETRIC AND YAWED BODIES IN A FREE SURFACE	73-0908
SURFACE	FRACTURE SURFACE ENERGIES BY FINITE ELEMENT STRESS ANALYSIS	73-0915
SURFACE	HIGHER-ORDER FINITE-ELEMENT ANALYSIS OF TOPOGRAPHIC GUIDES SUPPORTING ELASTIC SURFACE WAVES	73-0114
SURFACE	INFLUENCE OF COOLING, FREE-STREAM TURBULENCE AND SURFACE ROUGHNESS ON THE AERODYNAMIC BEHAVIOR OF CA	74-0970
SURFACE	INITIAL YIELD SURFACE OF A UNIDIRECTIONALLY REINFORCED COMPOSITE	71-0280
SURFACE	JET-WING LIFTING - SURFACE THEORY USING ELEMENTARY VORTEX DISTRIBUTIONS	75-0245
SURFACE	LEAST SQUARES SURFACE FITTING BY FINITE ELEMENTS AND AN APPLICATION TO STRESS SMOOTHING	67-0154
SURFACE	LIFTING SURFACE PROBLEMS ANALYSIS	73-0333
SURFACE	MECHANISM OF PARALLEL SURFACE THRUST BEARING	75-0322
SURFACE	NATURAL GEOMETRY OF SURFACE WITH SPECIFIC REFERENCE TO THE MATRIX DISPLACEMENT ANALYSIS OF SHELLS, 1	73-0709
SURFACE	NATURAL VIBRATION OF A FREE SURFACE	62-0008
SURFACE	NUMERICAL ANALYSIS OF FREE SURFACE SEEPAGE PROBLEMS	71-0469
SURFACE	NUMERICAL METHOD FOR INVESTIGATING THE PROPAGATION OF SURFACE WAVES ON DISSIPATIVE GUIDES	74-0613
SURFACE	ON SOME FREE SURFACE ON TRANSIENT FLOW PROBLEMS OF SEEPAGE AND IRROTATIONAL FLOW	72-0070
SURFACE	PENALTY FUNCTION APPROACH TO PROBLEMS OF PLASTIC FLOW OF METALS WITH LARGE SURFACE DEFORMATIONS	75-0282
SURFACE	PROPAGATION OF ELLIPTICAL SURFACE CRACKS AND NONLINEAR FRACTURE MECHANICS BY THE FINITE ELEMENT METH	74-0906
SURFACE	STRESS CONCENTRATIONS IN HULL SHAPES FROM SURFACE DISCONTINUITIES	69-0130
SURFACE	STRESS INTENSITY FACTOR OF A SURFACE CRACK IN A PRESSURIZED CYLINDER	75-0255
SURFACE	STRESSES AND DISPLACEMENTS AROUND TWO ADJACENT CIRCULAR OPENINGS NEAR TO THE GROUND SURFACE	74-0993
SURFACE	STRIP PROGRAM FOR DESIGNING LOAD BEARING SURFACE STRUCTURES (PROGRAMME STRIP POUR LE CALCUL DES STRU	71-0408
SURFACE	STRIP PROGRAM FOR DESIGNING LOAD BEARING SURFACE STRUCTURES (PROGRAMME STRIP POUR LE CALCUL DES STRU	71-0408
SURFACE	SURFACE FOOTINGS ON SOFT ROCKS	74-0990
SURFACE	THE APPLICATION OF FINITE ELEMENT METHOD TO THE LIFTING SURFACE PROBLEM	75-0868
SURFACE	THE PROGRAM " ASSEN " FOR STEADY FREE SURFACE FLOWS	74-0075
SURFACE	THE USE OF THE FINITE ELEMENT METHOD IN OBTAINING SOIL-WATER PRESSURE AND SOIL-WATER CONTENT DISTRIB	75-0112
SURFACE	THEORETICAL COMPUTATIONS ON RIDGE ACOUSTIC SURFACE WAVES USING THE FINITE-ELEMENT METHOD	71-0011
SURFACE	THERMO-ELASTIC ANALYSIS OF THE PARALLEL SURFACE THRUST BEARING	74-1172
SURFACE	THREE-DIMENSIONAL ELASTIC ANALYSIS OF SEMI-ELLIPTICAL SURFACE CRACKS SUBJECT TO THERMAL SHOCK	75-0256
SURFACE	VARIATIONAL APPROACH TO THE LIFTING SURFACE PROBLEM	73-0779
SURFACE	ZUSAMMENHANG ZWISCHEN OBERFLAECHENBESCHAFFENHEIT, VERBUND UND SPRENGWIRKUNG VON BEWEHRUNGSSTAEHLEN U	73-0450
SURFACES	A CRUCIFORM ELEMENT FOR SURFACES OF INFLATABLE STRUCTURES	75-0118
SURFACES	A FINITE-ELEMENT METHOD FOR LIFTING SURFACES IN STEADY INCOMPRESSIBLE SUBSONIC FLOW	74-1321
SURFACES	A METHOD FOR THE THERMAL ANALYSIS OF SPACECRAFT INCLUDING ALL MULTIPLE REFLECTIONS AND SHADING AMONG	70-0424
SURFACES	AN AUTOMATIC MESH GENERATION SCHEME FOR PLANE AND CURVED SURFACES BY ISOPARAMETRIC COORDINATES	71-0271
SURFACES	AN INVESTIGATION OF INITIAL YIELD SURFACES FOR UNIDIRECTIONAL REINFORCED COMPOSITES	72-0837
SURFACES	APPLICATION OF FINITE ELEMENT METHODS TO LUBRICATION: SQUEEZE FILMS BETWEEN POROUS SURFACES	75-0764
SURFACES	DISCRETE TRIANGULAR APPROXIMATION OF MOMENT AND DISPLACEMENT SURFACES FOR PLATE BENDING	71-0269
SURFACES	FINITE ELEMENT ANALYSIS OF STRUCTURES WITH SURFACES OF ARBITRARY CURVATURE	70-0102
SURFACES	FINITE ELEMENT METHODS FOR THE SOLUTION OF SOME INCOMPRESSIBLE NON-NEWTONIAN FLUID MECHANICS PROBLEM	75-0224
SURFACES	FINITE ELEMENTS IN CURVILINEAR COORDINATES - SECOND ORDER SURFACES	73-0753
SURFACES	KERNEL FUNCTION FOR NONPLANAR OSCILLATING SURFACES IN SUPERSONIC FLOW	71-0472
SURFACES	MEASUREMENT OF SLOPE CHANGE ON CYLINDRICAL SHELL SURFACES	73-0439
SURFACES	TECHNIQUES IN TRIANGULAR MESH GENERATION FLAT AND CURVED SURFACES	69-0394
SURFACES	UNSTEADY SUBSONIC COLLOCATION METHOD FOR WINGS WITH AND WITHOUT CONTROL SURFACES	72-0012
SURFACES	USE OF FINITE ELEMENT METHOD IN A MARKER AND CELL PROGRAM FOR SOLVING TIME-DEPENDENT FLUID FLOW PROB	75-0043
SURFACES-THEORE	PLANFORM INPUT GENERATOR (PING), A NASTRAN PRE-PROCESSOR FOR LIFTING SURFACES-THEORETICAL DEVELOPMEN	73-0776
SURFACES-THEORE	PLANFORM INPUT GENERATOR (PING), A NASTRAN PROCESSOR FOR LIFTING SURFACES-THEORETICAL DEVELOPMENT, U	73-0780
SURFACING	A COMPUTER MODEL FOR PREDICTING THE LOAD-DEFLECTION RESPONSE OF EXPEDIENT SOIL SURFACING	73-0880
SURGES	EVALUATION OF TECHNIQUES FOR NUMERICAL CALCULATION OF STORM SURGES	75-0554
SURGES	NUMERICAL CALCULATION OF STORM SURGES: AN EVALUATION OF TECHNIQUES	75-0461
SURRELAXATION	ANALYSE NUMERIQUE DU CHAMP MAGNETIQUE D'UN ALTERNATEUR PAR ELEMENTS FINIS ET SURRELAXATION PONCTUELL	73-0625
SURROUNDING	RASCHET NAPRYAZHENIL L PEREMESHCHENII V SVODE PODZEMNOGO MASHINNOGO ZALA GES I V OKRUZHAYUSHCHEM SKA	74-0542
SURROUNDING	THREE-DIMENSIONAL CONSIDERATION OF THE STRESSES SURROUNDING OPEN-PIT MINE SLOPES	73-0492
SURVEY	A SURVEY OF ANALYTICAL METHODS FOR DYNAMIC SIMULATION OF CAB	73-0936
SURVEY	A SURVEY OF FINITE ELEMENT FRAMEWORK STABILITY ANALYSIS	69-0418
SURVEY	A SURVEY OF FINITE ELEMENT FORMULATIONS OF GEOMETRICALLY NONLINEAR PROBLEMS	71-0171
SURVEY	A SURVEY OF FINITE-ELEMENT METHODS IN CONTINUUM MECHANICS	72-0751
SURVEY	A SURVEY OF HEAT CONDUCTION COMPUTER PROGRAMS	71-0017
SURVEY	A SURVEY OF PRE AND POSTPROCESSORS FOR NASTRAN	74-0714
SURVEY	A SURVEY OF SPARSE MATRIX TECHNOLOGY	72-0989
SURVEY	A SURVEY OF THE ANALYSIS OF SHELLS BY THE DISPLACEMENT METHOD	66-0041
SURVEY	A SURVEY OF THE APPLICATION OF FINITE ELEMENT METHODS TO STRESS ANALYSIS PROBLEMS WITH PARTICULAR EM	71-0632
SURVEY	FINITE ELEMENT ANALYSIS OF FLUID FILM LUBRICATION - A SURVEY	74-0041
SURVEY	FINITE ELEMENT ANALYSIS OF FLUID FILM LUBRICATION - A SURVEY	74-1100
SURVEY	FINITE ELEMENT ANALYSIS OF FLUID FILM LUBRICATION SURVEY	75-0880
SURVEY	FINITE ELEMENT FORMULATION OF FINITE DEFORMATIONS AND IRREVERSIBLE THERMODYNAMICS OF NONLINEAR CONTI	69-0207
SURVEY	FINITE ELEMENT FORMULATION OF PROBLEMS OF FINITE DEFORMATION AND IRREVERSIBLE THERMODYNAMICS OF NONL	71-0190
SURVEY	MESH GENERATION - SURVEY	73-0048
SURVEY	ROHM AND HAAS THERMOSTRUCTURAL ANALYSIS SYSTEM, THERMAL STRUCTURAL ANALYSIS PROGRAMS - A SURVEY AND	72-0007
SURVEY	SURVEY AND EVALUATION OF THE FINITE ELEMENT METHOD IN LINEAR FRACTURE MECHANICS ANALYSIS	71-0094
SURVEY	SURVEY LECTURES ON THE MATHEMATICAL FOUNDATIONS OF THE FINITE ELEMENT METHOD	72-0670

SURVEY	SURVEY OF ADVANCED STRUCTURAL DESIGN ANALYSIS TECHNIQUES	69-0306
SURVEY	SURVEY OF GENERAL PURPOSE PROGRAMS FOR FINITE ELEMENT ANALYSIS	72-0323
SURVEY	SURVEY OF SOLUTION PROCEDURES FOR NONLINEAR STATIC AND DYNAMIC ANALYSES	74-0307
SURVEY	SURVEY OF SOME FINITE ELEMENT METHOD PROPOSED FOR TREATING DIRICHLET PROBLEM	75-0612
SURVEY	SURVEY OF SPARSE MATRIX METHODS IN THE DIRECT SOLUTION OF FINITE ELEMENT EQUATIONS	73-0572
SURVEY	SURVEY OF STATIC GEOMETRIC AND MATERIAL NONLINEAR ANALYSIS BY THE FINITE ELEMENT METHOD	72-0783
SURVEY	SURVEY OF STRUCTURAL ANALYSIS SYSTEMS	74-1290
SURVEY	SURVEY OF UNDERLYING THEORY, APPROXIMATION BY PIECEWISE POLYNOMIALS, ILLEGAL ELEMENTS AND THE PATCH	73-0635
SURVIVAL	COMPUTER MODELLING OF PLUG SEEDLING SURVIVAL	75-0176
SUSPENSION	APPLICATION OF SAMIS TO THE ANALYSIS OF A FRONT SUSPENSION UPPER ARM	70-0260
SUSPENSION	STATIC STRUCTURAL ANALYSIS OF SUSPENSION BRIDGES	66-0110
SVINGNINGSPROBL	SPAENDINGSHYBRIDE FINITE ELEMENTER TIL SVINGNINGSPROBLEMER. (STRESS HYBRID FINITE ELEMENT METHOD FOR	75-0208
SVODE	RASCHET NAPRYAZHENIL L PEREMESHCHENII V SVODE PODZEMNOGO MASHINNOGO ZALA GES I V OKRUZHAYUSHCHEM SKA	74-0542
SVOEI	ISSLEDOVANIE PROCHNOSTI I ZHESTKOSTI PLASTIN, NAGRUZHENNYKH V SVOEI PLOSKOSTI, METODOM KONECHNYKH E	73-0382
SWAGE	FINITE ELEMENT ANALYSIS OF SEVERAL SWAGE MANDREL DESIGNS	74-0794
SWEEP	AUTOMOBILE PANEL SWEEP STIFFNESS ANALYSIS	74-0175
SWEEPBACK	COMPUTATION OF INFLUENCE COEFFICIENTS FOR AIRCRAFT STRUCTURES WITH DISCONTINUITIES AND SWEEPBACK	47-0002
SWELLING	APPLICATION OF FINITE ELEMENT METHODS FOR THE ANALYSIS OF THERMAL CREEP, IRRADIATION INDUCED CREEP,	71-0242
SWELLING	STRESS AND DEFORMATION ANALYSIS OF IRRADIATION INDUCED SWELLING	74-0305
SWELLING	STRESS AND DEFORMATION ANALYSIS OF IRRADIATION INDUCED SWELLING	74-0882
SWEPT-BACK	STRUCTURAL ANALYSIS OF SWEPT-BACK WINGS BY MATRIX-TRANSFORMATION	51-0001
SYMBOLIC	USE OF COMPUTERIZED SYMBOLIC INTEGRATION IN FINITE ELEMENT DEVELOPMENT	74-0520
SYMBOLIC	USE OF SYMBOLIC MANIPULATION IN DEVELOPMENT OF TWO-DIMENSIONAL FINITE ELEMENTS	74-0103
SYMMETRIC	A COMPARISON OF FIRST AND SECOND ORDER AXIALLY SYMMETRIC FINITE ELEMENTS	73-0997
SYMMETRIC	A FINITE ELEMENT METHOD FOR FRIEDRICH'S SYMMETRIC POSITIVE SYSTEMS	71-0146
SYMMETRIC	AN IMPROVED METHOD FOR REDUCING THE BANDWIDTH OF SPARSE SYMMETRIC MATRICES	72-0990
SYMMETRIC	AN INVESTIGATION OF THE BENDING OF SYMMETRIC PLATES UNDER SYMMETRIC TRANSVERSE LOADING	72-0662
SYMMETRIC	AN INVESTIGATION OF THE BENDING OF SYMMETRIC PLATES UNDER SYMMETRIC TRANSVERSE LOADING	72-0662
SYMMETRIC	APPROXIMATE THREE-DIMENSIONAL SOLUTIONS FOR SYMMETRIC LAMINATES UNDER INPLANE LOADING	71-0392
SYMMETRIC	ARBITRARY SYMMETRIC LOADING PROBLEMS OF CENTRALLY CRACKED RECTANGULAR PLATES	75-0318
SYMMETRIC	COMPARIOSN OF RESULTS OBTAINED FOR THE PROBLEM OF SYMMETRIC FLUID OSCILLATIONS IN TANKS BY THE FINIT	71-0761
SYMMETRIC	EIGENVALUE PROBLEMS (A - LAMBDA B) X EQUALS 0 FOR SYMMETRIC MATRICES OF HIGH ORDER	74-0205
SYMMETRIC	EVALUATION OF TORSIONAL EFFECTS IN THE SEISMIC ANALYSIS OF SYMMETRIC AND UNSYMMETRIC NUCLEAR POWER P	73-0175
SYMMETRIC	FINITE ELEMENT METHODS FOR SYMMETRIC HYPERBOLIC EQUATIONS	73-0116
SYMMETRIC	GENERAL SOLUTION ROUTINES FOR SYMMETRIC EQUATION SYSTEMS	75-0129
SYMMETRIC	NON-UNIFORM DEFORMATION OF MATERIAL IN AXIALLY SYMMETRIC COMPRESSION CAUSED BY FRICTION, PART 3, THE	72-0923
SYMMETRIC	NOTE ON MINIMIZING THE BANDWIDTH OF SPARSE, SYMMETRIC MATRICES	73-1107
SYMMETRIC	ON THE NONUNIFORM DEFORMATION OF MATERIAL IN AXIALLY SYMMETRIC COMPRESSION CAUSED BY FRICTION - 2	71-0321
SYMMETRIC	ON THE SOLUTION OF THE NAVIER-STOKES EQUATION IN TWO DIMENSIONAL AND AXIAL SYMMETRIC PROBLEMS	74-0007
SYMMETRIC	RAYLEIGH RITZ GALERKIN METHODS FOR SYMMETRIC POSITIVE DIFFERENTIAL EQUATIONS APPLICATION TO FINITE E	73-1030
SYMMETRIC	REDUCING THE BANDWIDTH OF SPARSE SYMMETRIC MATRICES	69-0469
SYMMETRIC	REDUCING THE BANDWIDTH OF SPARSE SYMMETRIC MATRICES	69-0502
SYMMETRIC	SOLUTION OF LINEAR EQUATIONS WITH SKYLINE-STORED SYMMETRIC MATRIX	75-0662
SYMMETRIC	THE THERMO-ELASTO-PLASTIC ANALYSIS OF A RE-ENTRY BODY TYPE COMPOSITE SOLID STRUCTURE USING A NEW FIN	66-0093
SYMMETRIC	VERGLEICH NUMERISCHER LOESUNGEN NACH DEM DIFFERENZENVERFAHREN UND NACH DER FINITE-ELEMENT-METHODE AM	71-0016
SYMMETRICAL	A COMPACT STORAGE SCHEME FOR THE SOLUTION OF SYMMETRICAL LINEAR SIMULTANEOUS EQUATIONS	66-0116
SYMMETRICAL	AN ALGORITHM FOR REDUCING THE BANDWIDTH OF A MATRIX OF SYMMETRICAL CONFIGURATION	65-0067
SYMMETRICAL	BIDIMENSIONAL COMPUTER-PROGRAM FOR CALCULATION OF GROWTH-RATES OF INCLUDED AXIALLY SYMMETRICAL PLASM	75-0420
SYMMETRICAL	ISSLEDOVANIE NAPRYAZHENNO-DEFORMIROVANNOGO SOSTOYANIYA SIMMETRICHNYKH DISKOV SO STUPITSEI. (INVESTIG	74-0965
SYMMETRICAL	STRUCTURAL ANALYSIS OF SYMMETRICAL PENSTOCK BIFURCATIONS BY THE FINITE ELEMENT METHOD	71-0310
SYMMETRICAL	SYMMETRICAL BENDING OF CIRCULAR PLATES USING FINITE ELEMENTS	72-0428
SYMMETRICALLY	NEW FINITE ELEMENT METHOD FOR ANALYZING SYMMETRICALLY LOADED THIN SHELLS OF REVOLUTION	73-0389
SYMMETRISCHER	VERGLEICH NUMERISCHER LOESUNGEN NACH DEM DIFFERENZENVERFAHREN UND NACH DER FINITE-ELEMENT-METHODE AM	71-0016
SYMMETRY	ELASTIC DEFLECTION IN THIN WALLED CONES ROTATING ABOUT THEIR AXES OF SYMMETRY	73-0545
SYMMETRY	FINITE ELEMENT MODELING OF A CIRCULAR RING USING HALF AND QUARTER SYMMETRY	66-0119
SYMMETRY	SIGNIFICANCE OF SYMMETRY IN MULTIPLE FRACTURE ANALYSIS	74-0869
SYNTEZA	WIELOSTOPNIOWA SYNTEZA MACIERZY SZTYWNOSCI (MULTI-STAGE SYNTHESIS OF THE STIFFNESS MATRICES)	73-0668
SYNTHESIS	A HYBRID METHOD OF COMPONENT MODE SYNTHESIS	71-0681
SYNTHESIS	A MODAL SYNTHESIS TECHNIQUE FOR DETERMINING DYNAMIC PROPERTIES FOR A STRUCTURE FOR MASS AND STIFFNES	74-0319
SYNTHESIS	ANALYTICAL METHODS IN COMPONENT MODAL SYNTHESIS	75-0824
SYNTHESIS	AUTOMATED STRUCTURAL SYNTHESIS USING A RE NUMBER OF DESIGN COORDINATES	73-0584
SYNTHESIS	DEVELOPMENTS IN STRUCTURAL ANALYSIS AND SYNTHESIS BY FUNCTION MINIMIZATION	71-0384
SYNTHESIS	FINITE ELEMENT MODELING STUDIES IN THE NORMAL-MODE METHOD AND SYNTHESIS	71-0067
SYNTHESIS	MEASUREMENTS OF HEALING AT AN OSTEOTOMY IN A RABBIT CALVARIUM: THE INFLUENCE OF APPLIED COMPRESSIVE	74-0975
SYNTHESIS	NASTRAN MODELING STUDIES IN THE NORMAL-MODE AND NORMAL-MODE SYNTHESIS	71-0069
SYNTHESIS	OPTIMAL SYNTHESIS OF FLEXIBLE LINK MECHANISMS WITH LARGE STATIC DEFLECTIONS	74-0525
SYNTHESIS	STRUCTURAL SYNTHESIS, 1959-1969, A DECADE OF PROGRESS	69-0237
SYNTHESIS	WIELOSTOPNIOWA SYNTEZA MACIERZY SZTYWNOSCI (MULTI-STAGE SYNTHESIS OF THE STIFFNESS MATRICES)	73-0668
SYNTHESIZED	A SYNTHESIZED WARPED QUADRILATERAL ELEMENT GIVING SIMPLIFIED INPUT FOR EQUILIBRIUM MODELS	73-1013
SYSTEM	A CASE STUDY OF IN-SITU ROCK DEFORMATION BEHAVIOR FOR THE DESIGN OF GROUND SUPPORT SYSTEM	72-0668
SYSTEM	A FINITE ELEMENT EIGENVALUE SYSTEM	68-0192
SYSTEM	A GENERAL SYSTEM FOR STRUCTURAL FINITE ELEMENT ANALYSIS	68-0107
SYSTEM	AN AUTOMATIC SYSTEM FOR KINEMATIC ANALYSIS ASKA HIGH SPEED COMPUTING OF ELASTIC STRUCTURES	70-0358
SYSTEM	AN AUTOMATIC SYSTEM FOR KINEMATIC ANALYSIS, ASKA PART 1	70-0396
SYSTEM	AN AUTOMATIC SYSTEM FOR KINEMATIC ANALYSIS ASKA PART 1	71-0626
SYSTEM	AN INTERACTIVE COMPUTER-AIDED MODELING SYSTEM FOR THE DISCRETIZATION OF BRIDGE STRUCTURES FOR FINITE	73-1165
SYSTEM	ANALYSIS OF A SKEW SLAB-BEAM SYSTEM BY FINITE ELEMENT METHOD	73-0233
SYSTEM	ANALYSIS OF SHALLOW SHELL STRUCTURES BY A DISCRETE ELEMENT SYSTEM	66-0113
SYSTEM	ANSYS (ENGINEERING ANALYSIS SYSTEM) PROGRAM	72-0156
SYSTEM	ASAS - A COMMERCIAL FINITE ELEMENT SYSTEM WITH DISPLACEMENT AND FORCE CONTINUITY OPTIONS	73-0947
SYSTEM	ASKA - AUTOMATIC SYSTEM FOR KINEMATIC ANALYSIS	65-0059
SYSTEM	ASKA A COMPUTER SYSTEM FOR STRUCTURAL ENGINEERS	69-0345
SYSTEM	ASKA AUTOMATIC SYSTEM FOR KINEMATIC ANALYSIS	69-0307
SYSTEM	ASKA, A LARGE SCALE SOFTWARE SYSTEM FOR FINITE ELEMENT ANALYSIS	69-0331
SYSTEM	AUTOMATIC SYSTEM FOR KINEMATIC ANALYSIS, NASKA	65-0054
SYSTEM	BERSAFE - A COMPUTER SYSTEM FOR STRESS ANALYSIS BY FINITE ELEMENTS	70-0212
SYSTEM	BERSAFE FINITE ELEMENT SYSTEM	74-0190
SYSTEM	CHARACTERISTICS OF A HYBRID CALCULATION SYSTEM BASED ON THE FINITE ELEMENT METHOD. (CARATTERISTICHE	72-0715
SYSTEM	COMBINING EXPERIMENTAL AND ANALYTICAL TECHNIQUES FOR DYNAMIC SYSTEM ANALYSIS	73-0321
SYSTEM	COMPARAISON DES DIVERSES MODELISATIONS NUMERIQUES D'UN SYSTEME MAGNETIQUE SATURABLE. (COMPARISON OF	75-0295
SYSTEM	COMPUTER AIDED STRUCTURAL ANALYSIS: THE MACHINE INDEPENDENT SYSTEM ASKA	70-0397
SYSTEM	COMPUTER SYSTEM FOR ANALYSIS OF LARGE FRAMEWORKS	67-0122
SYSTEM	COMPUTER SYSTEM FOR THE ANALYSIS OF LARGE SCALE PROBLEMS IN SOLID, FLUID AND THERMODYNAMICS	73-0554
SYSTEM	COMPUTERIZED STRUCTURAL ANALYSIS OF THE WORLD'S LARGEST LIGHT-GAGE STEEL PRIMARY STRUCTURAL SYSTEM	73-0503
SYSTEM	COMPUTERUNTERSTUETZTES SYSTEM FUER DIE CONSTRUKTION DES WERKZEUGMASCHINENAUFBAUS (COMPUTER-AIDED SYS	73-0482
SYSTEM	COMPUTERUNTERSTUETZTES SYSTEM FUER DIE CONSTRUKTION DES WERKZEUGMASCHINENAUFBAUS (COMPUTER-AIDED SYS	73-0482
SYSTEM	COMPUTING SYSTEM FOR STRUCTURAL ANALYSIS OF CAR BODIES	72-0324
SYSTEM	CONTRIBUTION A L'ETUDE DU COMPLEXE BARRAGE-FONDATION ET EXPLICATION DES SEISMES DUS AU REMPLISSAGE D	72-0554
SYSTEM	DATA GENERATION FOR A 3-D FINITE ELEMENT SYSTEM	74-0361
SYSTEM	DATA GENERATION FOR A 3-D FINITE ELEMENT SYSTEM	74-0475
SYSTEM	DESIGN, DEVELOPMENT, DOCUMENTATION AND SUPPORT OF A MAJOR FINITE ELEMENT SYSTEM	73-0205
SYSTEM	DYNAMIC ANALYSIS OF SOIL - STRUCTURE SYSTEM USING FINITE ELEMENT METHOD	67-0001

SYSTEM	DYNAMIC ANALYSIS OF A SYSTEM OF HINGE-CONNECTED RIGID BODIES WITH NONRIGID APPENDAGES	74-0852
SYSTEM	DYNAMIC RESPONSE OF A CONSTRAINED FIBROUS SYSTEM SUBJECTED TO TRANSVERSE IMPACT. II. A MECHANICAL MO	70-0369
SYSTEM	EFFECT OF SYSTEM PARAMETER VARIATIONS ON NATURAL FREQUENCIES	75-0661
SYSTEM	ELASTOPLASTIC AND CREEP ANALYSIS WITH THE ASKA PROGRAM SYSTEM	74-0863
SYSTEM	FINEL, UNIVERSELLES PROGRAMMSYSTEM ZUR BERECHNUNG DER ELASTISCHEN EIGENSCHAFTEN VON MASCHINENBAUTEIL	72-0132
SYSTEM	FINESSE - A FINITE ELEMENT SYSTEM, PART V FINITE ELEMENT DESIGN - DERIVATIVES FOR STRUCTURAL SHAPE D	72-0963
SYSTEM	FINITE ELEMENT - ANALOGUE METHOD FOR DETERMINING THE DYNAMIC CHARACTERISTICS OF AN ARCH DAM-RESERVOI	73-0678
SYSTEM	FINITE ELEMENT ANALYSIS SYSTEM FOR THE MECHANICAL BEHAVIOR OF ORIENTED FIBER COMPOSITE MATERIALS UND	73-0869
SYSTEM	FINITE ELEMENT ANALYSIS OF A BONE-PLATE-SCREW SYSTEM	74-0920
SYSTEM	FINITE ELEMENT METHOD APPLIED TO THE PROBLEM OF STABILITY OF A NON-CONSERVATIVE SYSTEM	71-0291
SYSTEM	FINITE ELEMENT MODEL SYSTEM OF A RABBIT CALVARIUM: AN INVESTIGATION OF OPTIMAL COMPRESSIVE LOADING	73-0206
SYSTEM	FINITE ELEMENT SOLUTION SYSTEM	69-0358
SYSTEM	FINITE-ELEMENT SOLUTION OF STRESSES AND DISPLACEMENTS IN A SOIL-CULVERT SYSTEM	73-1156
SYSTEM	FORMAT, A GENERAL PURPOSE MATRIX ANALYSIS SYSTEM IN FORTRAN	68-0266
SYSTEM	GALERKIN-FINITE ELEMENT ANALYSIS OF HYDROTHERMAL SYSTEM AT WAIRAKEI, NEW-ZEALAND	75-0601
SYSTEM	GIFTS SYSTEM	74-1060
SYSTEM	GIFTS SYSTEM FOR THE PDP-15 AND OTHER CONSIDERATIONS IN DEVELOPING GRAPHICS AND CAPABILITIES FOR FIN	73-0671
SYSTEM	GRIP: AN INFORMATION SYSTEM FOR DESIGN AUTOMATION	73-0210
SYSTEM	HABEAS- A STRUCTURAL DYNAMICS ANALYSIS SYSTEM	70-0121
SYSTEM	ICON: THE INTERACTIVE CREATION OF NASTRAN DATA. A SYSTEM DESCRIPTION	75-0274
SYSTEM	ISOPARAMETRIC FINITE ELEMENT SYSTEM: A NEW CONCEPT IN FINITE ELEMENT ANALYSIS	68-0223
SYSTEM	LAYERED PAVEMENT SYSTEMS. PART 1. LAYERED SYSTEM DESIGN. PART II. FATIGUE OF PLAIN CONCRETE	72-0867
SYSTEM	LONG TERM CREEP CLOSURE OF SOLUTION CAVITY SYSTEM	74-1150
SYSTEM	MAGIC II. AN AUTOMATED GENERAL PURPOSE SYSTEM FOR STRUCTURAL ANALYSIS	71-0495
SYSTEM	MAGIC II: AN AUTOMATED GENERAL PURPOSE SYSTEM FOR STRUCTURAL ANALYSIS. VOLUME III. PROGRAMMER'S MANU	71-0756
SYSTEM	MAGIC II: AN AUTOMATED GENERAL PURPOSE SYSTEM FOR STRUCTURAL ANALYSIS. VOLUME I. ENGINEER'S MANUAL (	71-0757
SYSTEM	MAGIC III: AN AUTOMATED GENERAL PURPOSE SYSTEM FOR STRUCTURAL ANALYSIS. VOLUME III. PROGRAMMER'S MAN	72-0726
SYSTEM	MAGIC III: AN AUTOMATED GENERAL PURPOSE SYSTEM FOR STRUCTURAL ANALYSIS VOLUME I. ENGINEER'S MANUAL	72-0727
SYSTEM	MAGIC, AN AUTOMATED GENERAL PURPOSE SYSTEM FOR STRUCTURAL ANALYSIS, VOLUME III. PROGRAMMER'S MANUAL	69-0090
SYSTEM	MAGIC, AN AUTOMATED GENERAL PURPOSE SYSTEM FOR STRUCTURAL ANALYSIS. VOLUME II. USER'S MANUAL	69-0144
SYSTEM	MAGIC, AN AUTOMATED GENERAL PURPOSE SYSTEM FOR STRUCTURAL ANALYSIS	69-0173
SYSTEM	MECHANICS OF CABLE MOORING SYSTEMS. VOLUME VII. THE STEADY-STATE BEHAVIOR OF A PYRAMID ARRAY SYSTEM	73-0724
SYSTEM	MESY - EIN PROGRAMMSYSTEM ZUR UNTERSUCHUNG VON TRAGWERKEN. (STRUCTURAL ANALYSIS BY MESY PROGRAMMING	75-0360
SYSTEM	MODELING AN INPUT-OUTPUT GEOKINETIC SYSTEM UTILIZING A FINITE ELEMENT APPROACH	75-0704
SYSTEM	NONLINEAR SEISMIC ANALYSIS OF CURRENT DIVIDER SUPPORT SYSTEM FOR PACIFIC DC INTERTIE	73-0200
SYSTEM	NONLINEAR STRUCTURES FOR SYSTEM IDENTIFICATION	74-1030
SYSTEM	NUMERICAL PROCEDURES FOR THE DETERMINATION OF THE BEHAVIOUR OF A SHEAR WALL FRAME SYSTEM	73-0216
SYSTEM	OH - 58A PROPULSION SYSTEM VIBRATION INVESTIGATION	74-0733
SYSTEM	ORSAM - A PROGRAMMING SYSTEM FOR THE FINITE ELEMENT METHOD	73-0528
SYSTEM	PRACTICAL APPLICATION OF A COMPUTERIZED STRUCTURAL ANALYSIS SYSTEM WHICH ADOPTS A FINITE ELEMENT TEC	69-0225
SYSTEM	RANDOM VIBRATION ANALYSIS SYSTEM FOR COMPLEX STRUCTURES, PART 2, COMPUTER PROGRAM DESCRIPTION	69-0262
SYSTEM	RANDOM-VIBRATION ANALYSIS SYSTEM FOR COMPLEX STRUCTURES. PART 1, ENGINEERING USER'S GUIDE	68-0106
SYSTEM	ROHM AND HAAS THERMOSTRUCTURAL ANALYSIS SYSTEM. THERMAL STRUCTURAL ANALYSIS PROGRAMS - A SURVEY AND	72-0007
SYSTEM	SAMECS STRUCTURAL ANALYSIS SYSTEM - USER'S DOCUMENT	69-0495
SYSTEM	SESAM A PROGRAMMING SYSTEM FOR FINITE ELEMENT PROBLEMS	68-0276
SYSTEM	SHELLS, PLATES AND MEMBRANES - A COMPUTER-AIDED DESIGN SYSTEM	74-1159
SYSTEM	SIMPLIFIED CALCULATION OF THE INTERACTION FORCES BETWEEN HULL AND TANK SYSTEM FOR A MOSS ROSENBERG T	75-0463
SYSTEM	SIMPLIFIED INELASTIC HIGH TEMPERATURE STRUCTURAL ANALYSIS OF MODERATELY COMPLEX SPATIALLY THREE DIME	74-0445
SYSTEM	SOME ASPECTS OF THE NATURAL COORDINATE SYSTEM IN THE FINITE-ELEMENT METHOD	69-0105
SYSTEM	SOME EFFECTS OF SYSTEM IDEALIZATIONS, SINGULARITIES AND MESH PATTERNS ON FINITE ELEMENT SOLUTIONS	74-0939
SYSTEM	STABILITY ANALYSIS OF STRUCTURES BY A REDUCED SYSTEM OF GENERALIZED COORDINATES	70-0168
SYSTEM	STARDYNE STRUCTURAL ANALYSIS SYSTEM. ANALYTICAL CAPABILITY MANUAL	68-0235
SYSTEM	STARS - A COMPUTER CODE FOR STRUCTURAL DYNAMIC ANALYSIS OF REACTOR SYSTEM	71-0204
SYSTEM	STRESS AND VIBRATION ANALYSIS ON THE SYSTEM OF FRAME AND FOUNDATION	69-0218
SYSTEM	STRUCTURAL ANALYSIS AND MATRIX INTERPRETIVE SYSTEM (SAMIS)	66-0098
SYSTEM	STRUCTURAL ANALYSIS AND MATRIX INTERPRETIVE SYSTEM (SAMIS) USER REPORT	67-0125
SYSTEM	STRUCTURAL AND SYSTEM MODELS	75-0669
SYSTEM	STRUDL-A COMPUTER SYSTEM FOR STRUCTURAL DESIGN	66-0118
SYSTEM	THE APPLICATION OF THE BERSAFE FINITE ELEMENT SYSTEM TO NUCLEAR DESIGN PROBLEMS	71-0117
SYSTEM	THE BOEING SST PROTYPE INTERNAL LOADS ANALYSIS SYSTEM AND PROCEDURES	71-0103
SYSTEM	THE ISOPARAMETRIC FINITE ELEMENT SYSTEM - A NEW CONCEPT IN FINITE ELEMENT ANALYSIS	69-0139
SYSTEM	THEORY OF FLOW IN A CONFINED TWO AQUIFER SYSTEM	69-0382
SYSTEM	TRANSIENT RESPONSE OF A COUPLED PLATE-ACOUSTIC SYSTEM USING PLATE AND ACOUSTIC FINITE ELEMENTS	71-0463
SYSTEM	TRANSIENT RESPONSE OF A COUPLED PLATE-ACOUSTIC SYSTEM USING PLATE AND ACOUSTIC FINITE ELEMENTS	72-0275
SYSTEM	TWO DIMENSIONAL STOCHASTIC MODEL OF A HETEROGENEOUS GEOLOGIC SYSTEM	73-0715
SYSTEM	USE OF PAS-1 FINITE ELEMENT COMPUTER SYSTEM FOR STATIC AND DYNAMIC ANALYSIS OF 2-DIMENSIONAL AND 3-D	75-0070
SYSTEM	VIBRATION MODES AND RANDOM RESPONSE OF A MULTI-BAY PANEL SYSTEM USING FINITE ELEMENTS	67-0128
SYSTEMATIC	SYSTEMATIC ENFORCEMENT OF STRESS BOUNDARY CONDITIONS IN THE ASSUMED STRESS HYBRID MODEL BASED ON THE	71-0771
SYSTEMATIC	SYSTEMATIC SUBSTRUCTURING	74-0895
SYSTEMATIK	ZUR SYSTEMATIK NUMERISCHER NAEHERUNGSVERFAHREN IN DER STATIK UND DYNAMIK DER KONSTRUKTIONEN (METHODS	72-0557
SYSTEME	ANWENDUNG DER MATRIZENVERSCHIEBUNGSMETHODE AUF ERZWUNGENE SCHWINGUNGEN (SCHWINGUNGSPROBLEME) (ARBEIT	71-0627
SYSTEME	COMPARAISON DES DIVERSES MODELISATIONS NUMERIQUES D'UN SYSTEME MAGNETIQUE SATURABLE. (COMPARISON OF	75-0295
SYSTEME	DAS PROGRAMMSYSTEM ANTRAS EDV-SYSTEM ZUR BERECHNUNG BELIEBIGER STATISCHER SYSTEME NACH DER METHODE D	74-0918
SYSTEMES	LA METHODE FRONTALE POUR LA RESOLUTION DES SYSTEMES LINEAIRESS. (FRONTAL METHOD FOR SOLVING LINEAR SY	73-1097
SYSTEMS	A DISCRETE-ELEMENT METHOD OF ANALYSIS FOR ORTHOGONAL SLAB AND GRID BRIDGE FLOOR SYSTEMS	72-0690
SYSTEMS	A FINITE ELEMENT METHOD FOR BENDING ANALYSIS OF LAYERED STRUCTURAL SYSTEMS	67-0134
SYSTEMS	A FINITE ELEMENT METHOD FOR FRIEDRICH'S SYMMETRIC POSITIVE SYSTEMS	71-0146
SYSTEMS	A RATIONAL ANALYSIS AND DESIGN PROCEDURE FOR WOOD JOIST FLOOR SYSTEMS	74-0730
SYSTEMS	A STUDY OF NUMERICAL ERROR IN STRUCTURAL SYSTEMS	70-0394
SYSTEMS	A TECHNIQUE FOR THE FINITE ELEMENT ANALYSIS OF LARGE COMPLEX STRUCTURAL SYSTEMS	69-0514
SYSTEMS	ADI METHODS FOR FINITE ELEMENT SYSTEMS	71-0652
SYSTEMS	ADVANCED INTERCONNECT SYSTEMS FOR LIGHTWEIGHT SOLAR ARRAYS	73-0696
SYSTEMS	AIRCRAFT-PAVEMENT INTERACTION STUDIES. PHASE 1: A FINITE ELEMENT MODEL OF A JOINTED CONCRETE PAVEMEN	73-0894
SYSTEMS	AN ANALYSIS OF GENERAL CHAIN SYSTEMS	72-0787
SYSTEMS	AN AUTOMATIC RECORDING SCHEME FOR SIMULTANEOUS EQUATIONS DERIVED FROM NETWORK SYSTEMS	70-0285
SYSTEMS	AN ECONOMICAL METHOD FOR DETERMINING THE SMALLEST EIGENVALUES OF LARGE LINEAR SYSTEMS	71-0551
SYSTEMS	AN INTERACTIVE GRAPHICS SHIP DESIGN ORIENTED PROGRAM PACKAGE FOR TIME SHARING SYSTEMS	73-0341
SYSTEMS	AN INTERACTIVE HYBRID TECHNIQUE FOR CRASHWORTHY DESIGN OF COMPLEX VEHICULAR STRUCTURAL SYSTEMS	74-0317
SYSTEMS	AN ITERATIVE METHOD FOR LARGE SYSTEMS OF LINEAR STRUCTURAL EQUATIONS	73-1020
SYSTEMS	ANALIZA DRGAN WYMUSZONYCH ZLOZONYCH UKLADOW LINIOWYCH METODA SZTYWNYCH ELEMENTOW SKONCZONYCH (ANALYS	72-0246
SYSTEMS	ANALYSIS OF LAYERED SYSTEMS SUBJECTED TO WHEEL LOADS	72-0868
SYSTEMS	ANALYSIS OF NONLINEAR PROBLEMS IN ROCK MECHANICS WITH PARTICULAR REFERENCE TO JOINTED ROCK SYSTEMS	70-0282
SYSTEMS	APPLICATION OF FINITE ELEMENT METHOD TO THE THREE DIMENSIONAL STRESS AND VIBRATION ANALYSIS OF BWR P	71-0180
SYSTEMS	APPLICATION OF FINITE ELEMENT METHODS FOR CALUCLATION OF TRANSVERSE OSCILLATIONS IN ONE DIMENSIONAL	71-0599
SYSTEMS	APPLICATION OF METHOD OF FINITE ELEMENTS TO ANALYSIS OF COMPOSITE FLOOR SYSTEMS	75-0418
SYSTEMS	APPLICATION OF THE FAST FOURIER TRANSFORM TO LINEAR SYSTEMS IN CIVIL ENGINEERING	73-1119
SYSTEMS	APPLICATIONS OF NASTRAN TO COUPLED STRUCTURAL AND HYDRODYNAMIC RESPONSES IN AIRCRAFT HYDRAULIC SYSTE	71-0744
SYSTEMS	BLOCK ELIMINATION ON FINITE ELEMENT SYSTEMS OF EQUATIONS	72-0530
SYSTEMS	CALCULATION OF AXISYMMETRIC LONGITUDINAL MODES FOR FLUID-ELASTIC TANK-ULLAGE GAS SYSTEMS AND COMPARI	64-0019
SYSTEMS	CIRCUMFERENTIAL PRESTRESSING LOAD ANALYSIS DUE TO WIRE/STRAND WINDING SYSTEMS ON PRESTRESSED CONCRET	74-1202
SYSTEMS	COMPLIANCES OF LAYERED ELASTIC SYSTEMS	74-1204
SYSTEMS	COMPUTATION OF THE RESPONSE OF COUPLED PLATE ACOUSTIC SYSTEMS USING PLATE FINITE ELEMENTS AND ACOUST	71-0072

SYSTEMS	COMPUTATIONAL STABILITY IN NONLINEAR DISCRETE STRUCTURAL SYSTEMS	69-0145
SYSTEMS	COMPUTER METHOD FOR COMPUTING CRITICAL SPEEDS OF NON-UNIFORM SHAFTS GEARED SYSTEMS AND CURVED ASSEMB	73-0675
SYSTEMS	COMPUTER PROGRAM FOR STATIC AND DYNAMIC ANALYSIS OF LINEAR STRUCTURAL SYSTEMS	72-0683
SYSTEMS	COMPUTER PROGRAM FOR SOLUTION OF LARGE, SPARSE, UNSYMMETRIC SYSTEMS OF LINEAR EQUATIONS	72-0719
SYSTEMS	COMPUTER PROGRAMS FOR THE SOLUTION OF SYSTEMS OF LINEAR ALGEBRAIC EQUATIONS	73-0932
SYSTEMS	COMPUTER PROGRAMS FOR THE SOLUTION OF SYSTEMS OF LINEAR ALGEBRAIC EQUATIONS	73-1040
SYSTEMS	COMPUTER SIMULATION OF VIBRATION OF WHOLE TREE SYSTEMS USING FINITE ELEMENT METHODS	74-1161
SYSTEMS	CONSISTENT MASS MATRIX FOR DISTRIBUTED SYSTEMS	63-0001
SYSTEMS	CONSTRUCTION OF CURVILINEAR COORDINATE SYSTEMS AND APPLICATIONS TO MESH GENERATION	74-1295
SYSTEMS	CUBIC SPLINE GALERKIN APPROXIMATIONS TO PARABOLIC SYSTEMS WITH COUPLED NON-LINEAR BOUNDARY CONDITION	75-0232
SYSTEMS	DECONVOLUTION OF SEISMIC RESPONSE FOR LINEAR SYSTEMS	73-0774
SYSTEMS	DESIGN OPTIMIZATION IN UNDERGROUND COAL SYSTEMS	74-1095
SYSTEMS	DEVELOPMENT AND MAINTENANCE OF LARGE FINITE ELEMENT SOFTWARE SYSTEMS	74-0865
SYSTEMS	DEVELOPMENT OF METHODS FOR COMPUTER SIMULATION OF BEAM-COLUMNS AND GRID-BEAM AND SLAB SYSTEMS	73-1101
SYSTEMS	DISCRETE METHODS IN THE ANALYSIS OF PLATE SYSTEMS	71-0835
SYSTEMS	DISPLAYS OF KINEMATIC AND ELASTIC SYSTEMS	71-0053
SYSTEMS	DYNAMIC ANALYSIS OF STRUCTURAL SYSTEMS USING COMPONENT MODES	64-0024
SYSTEMS	DYNAMIC ANALYSIS OF AUTOMOTIVE STRUCTURAL SYSTEMS	70-0340
SYSTEMS	DYNAMIC ANALYSIS OF HYDROELASTIC SYSTEMS USING THE FINITE-ELEMENT METHOD	73-1137
SYSTEMS	DYNAMIC LOADS ON GROUTED EMPLACEMENT SYSTEMS	70-0406
SYSTEMS	DYNAMIC SIMULATION OF COMPLEX SYSTEMS UTILIZING EXPERIMENTAL AND ANALYTICAL TECHNIQUES	75-0864
SYSTEMS	DYNAMICS OF DISTRIBUTED PARAMETER ROTOR SYSTEMS: TRANSFER MATRIX AND FINITE ELEMENT TECHNIQUES	70-0461
SYSTEMS	DYNAMICS OF GRAVITY DAM-RESERVOIR SYSTEMS	73-0225
SYSTEMS	EARTHQUAKE ANALYSIS OF STRUCTURE-FOUNDATION SYSTEMS	73-0876
SYSTEMS	EARTHQUAKE BEHAVIOUR OF RESERVOIR-DAM SYSTEMS	68-0043
SYSTEMS	EARTHQUAKE BEHAVIOUR OF RESERVOIR-DAM SYSTEMS (DISCUSSION)	69-0283
SYSTEMS	EARTHQUAKE FINITE ELEMENT ANALYSIS OF STRUCTURE-FOUNDATION SYSTEMS	74-0558
SYSTEMS	EARTHQUAKE FINITE-ELEMENT ANALYSIS OF STRUCTURE-FOUNDATION SYSTEMS	75-0410
SYSTEMS	EFFICIENT COMPUTER ANALYSIS OF LARGE THREE-DIMENSIONAL PIPING SYSTEMS BY THE FINITE ELEMENT METHOD	72-0378
SYSTEMS	ELASTIC ANALYSIS OF SPATIAL SYSTEMS OF INTERCONNECTED SHEAR WALLS AND FRAMES	74-1195
SYSTEMS	ELASTIC-PLASTIC ANALYSIS OF TWO-DIMENSIONAL STRESS SYSTEMS BY THE FINITE ELEMENT METHOD	67-0049
SYSTEMS	ELASTIC-PLASTIC EARTHQUAKE RESPONSE OF SOIL-BUILDING SYSTEMS	72-0717
SYSTEMS	ELASTODYNAMICS OF COMPLEX STRUCTURAL SYSTEMS	71-0563
SYSTEMS	EMPIRICAL INVESTIGATION OF REORDERING AND DATA MANAGEMENT FOR FINITE ELEMENT SYSTEMS OF EQUATIONS	73-0944
SYSTEMS	FAILURE MECHANISMS IN COMPOSITE SYSTEMS	75-0688
SYSTEMS	FINITE ELEMENT ANALYSIS OF LONG WAVES IN OPEN CHANNEL SYSTEMS	74-0033
SYSTEMS	FINITE ELEMENT ANALYSIS OF HYDROTHERMAL SYSTEMS	74-0048
SYSTEMS	FINITE ELEMENT ANALYSIS OF ELASTO-PLASTIC MATERIALS BY USE OF DISLOCATION DIPOLE SYSTEMS	74-0269
SYSTEMS	FINITE ELEMENT ANALYSIS OF NON-LINEAR COMPOSITE MATERIALS BY USE OF OVERLAY SYSTEMS	74-0928
SYSTEMS	FINITE ELEMENT ANALYSIS OF LAYERED WOOD SYSTEMS	75-0288
SYSTEMS	FINITE ELEMENT ANALYSIS OF ENGINEERING SYSTEMS GOVERNED BY A NON-LINEAR QUASI-HARMONIC EQUATION	75-0664
SYSTEMS	FINITE ELEMENT APPROACH TO THE MODELING OF HYDROTHERMAL SYSTEMS	73-0011
SYSTEMS	FINITE ELEMENT METHOD APPLIED TO CALCULATIONS OF NON-COPLANAR PLATE SYSTEMS	74-0241
SYSTEMS	FINITE ELEMENT MODEL FOR DISTRIBUTED PARAMETER TURBOMOTOR SYSTEMS	71-0332
SYSTEMS	FINITE ELEMENT SOLUTION SYSTEMS (FESS).	67-0098
SYSTEMS	FINITE ELEMENT TECHNIQUES APPLIED TO THE VIBRATIONS OF PLATE SYSTEMS	66-0033
SYSTEMS	FINITE-ELEMENT APPROACH FOR THE PREDICTION OF INELASTIC BEHAVIOR OF SHEAR WALL-FRAME SYSTEMS	72-0983
SYSTEMS	FREE VIBRATION ANALYSIS OF SPINNING STRUCTURAL SYSTEMS	73-1077
SYSTEMS	FREE VIBRATIONS OF FINITE ELEMENT PLATES SUBJECTED TO COMPLEX MIDDLE-PLANE FORCE SYSTEMS	72-0280
SYSTEMS	FREE VIBRATIONS OF FINITE ELEMENT PLATES SUBJECTED TO COMPLEX MIDDLE-PLANE FORCE SYSTEMS	73-0097
SYSTEMS	FREE VIBRATIONS OF FINITE ELEMENT PLATES SUBJECTED TO COMPLEX MIDDLE-PLANE FORCE SYSTEMS - REPLY	73-0122
SYSTEMS	FREE VIBRATIONS OF PLATE-BEAM SYSTEMS	73-0322
SYSTEMS	GENERAL SOLUTION ROUTINES FOR SYMMETRIC EQUATION SYSTEMS	75-0129
SYSTEMS	GEOMETRIC ASPECTS OF THE FINITE ELEMENT METHOD: CONSTRUCTION OF CURVILINEAR COORDINATE SYSTEMS AND T	72-0534
SYSTEMS	GEOMETRIC METHOD FOR THE PREDICTION OF STRESSES IN INCLUSIONS, OREBODIES, AND MINING SYSTEMS	74-0219
SYSTEMS	HYDROELASTIC ANALYSIS OF AXISYMMETRIC SYSTEMS BY A FINITE ELEMENT METHOD	68-0084
SYSTEMS	HYPERMATRIX SCHEME FOR FINITE-ELEMENT SYSTEMS ON CDC STAR-100 COMPUTER	75-0183
SYSTEMS	HYSTERETIC DAMPING SYSTEMS APPLIED TO SOIL STRUCTURE INTERACTION BY FINITE ELEMENT METHOD	73-0663
SYSTEMS	IMPACT OF CDC STAR-100 COMPUTER ON FINITE ELEMENT SYSTEMS	75-0039
SYSTEMS	IMPROVED CALCULATION OF ELASTO-PLASTIC SYSTEMS BY THE FINITE ELEMENT METHOD	70-0076
SYSTEMS	IMPROVED DEFORMATION FUNCTIONS FOR THE FINITE ELEMENT ANALYSIS OF BEAM SYSTEMS	69-0075
SYSTEMS	INFLUENCE OF JOINTS IN PANELIZED STRUCTURAL SYSTEMS	72-0204
SYSTEMS	INTERACTION STRESSES IN COMPOSITE SYSTEMS	70-0215
SYSTEMS	ITERATIVE METHOD FOR THE ANALYSIS OF LARGE STRUCTURAL SYSTEMS	68-0236
SYSTEMS	ITERATIVE METHODS FOR FINITE ELEMENT SYSTEMS	70-0196
SYSTEMS	ITERATIVE SOLUTION OF LINEAR SYSTEMS ARISING FROM FINITE ELEMENT TECHNIQUES	75-0126
SYSTEMS	KRON'S METHOD: AN ALGORITHM FOR THE EIGENVALUE ANALYSIS OF LARGE-SCALE STRUCTURAL SYSTEMS	73-0758
SYSTEMS	LA METHODE FRONTALE POUR LA RESOLUTION DES SYSTEMES LINEARESS. (FRONTAL METHOD FOR SOLVING LINEAR SY	73-1097
SYSTEMS	LARGE USER-ORIENTATED SYSTEMS OF PROGRAMS FOR STRUCTURAL ANALYSIS AND DESIGN	74-0482
SYSTEMS	LAYERED PAVEMENT SYSTEMS. PART 1. LAYERED SYSTEM DESIGN. PART 11. FATIQUE OF PLAIN CONCRETE	72-0867
SYSTEMS	LIMIT ANALYSIS OF FRAME SYSTEMS STIFFENED BY PANELS	74-0584
SYSTEMS	LOADS AND RESPONSE OF LARGE DUCTED PROPELLER SYSTEMS	73-0810
SYSTEMS	LUSH: A COMPUTER PROGRAM FOR COMPLEX RESPONSE ANALYSIS OF SOIL-STRUCTURE SYSTEMS	74-0744
SYSTEMS	MACRO-APPROACH FOR RIBBED AND GRID PLATE SYSTEMS	75-0079
SYSTEMS	MATHEMATICAL MODELS OF THE MASSACHUSETTS BAY. PART 1. FINITE ELEMENT MODELING OF TWO-DIMENSIONAL HYD	73-0800
SYSTEMS	MECHANICS OF CABLE MOORING SYSTEMS. VOLUME IV. A COMPUTER PROGRAM FOR ANALYZING THE STEADY STATE RES	72-0653
SYSTEMS	MECHANICS OF CABLE MOORING SYSTEMS. VOLUME 1. THREE DIMENSIONAL RESPONSE OF DEEP WATER MOORING LINES	72-0654
SYSTEMS	MECHANICS OF CABLE MOORING SYSTEMS. VOLUME V11. THE STEADY-STATE BEHAVIOR OF A PYRAMID ARRAY SYSTEM	73-0724
SYSTEMS	MECHANICS OF CABLE MOORING SYSTEMS. VOLUME V1. A COMPUTER PROGRAM FOR ANALYZING THE STEADY STATE CON	73-0725
SYSTEMS	MULTIDEGREE-OF-FREEDOM ELASTIC SYSTEMS HAVING MULTIPLE CLEARANCES	73-0653
SYSTEMS	NATURAL COORDINATE SYSTEMS. AN AUTOMATIC INPUT DATA GENERATION SCHEME FOR A FINITE ELEMENT METHOD	70-0018
SYSTEMS	NONSAP: A STRUCTURAL ANALYSIS PROGRAM FOR STATIC AND DYNAMIC RESPONSE OF NONLINEAR SYSTEMS	74-0848
SYSTEMS	NOTE ON THE CONVERGENCE OF FINITE ELEMENT APPROXIMATIONS FOR PROBLEMS FORMULATED IN CURVILINEAR COOR	74-0267
SYSTEMS	NUMERICAL AND PHYSICAL MODELLING OF REINFORCEMENT SYSTEMS FOR TUNNELS IN JOINTED ROCK	73-0872
SYSTEMS	ON AN APPROACH TO TECHNIQUES FOR THE ANALYSIS OF THE STRUCTURES OF LARGE SYSTEMS OF EQUATIONS	62-0015
SYSTEMS	ON THE STABILITY OF ROTOR-AND-BEARING SYSTEMS AND ON THE CALCULATION OF SLIDING BEARINGS	72-0037
SYSTEMS	PARAMETRIC STUDIES OF LAYERED PAVEMENT SYSTEMS USING AFPAV CODE WITH FAST EQUATION SOLVER	75-0747
SYSTEMS	PROCEDURE FOR SOLVING LINEAR EQUATION SYSTEMS WITH SPARSELY OCCUPIED COEFFICIENT MATRICES ASSOCIATED	74-0352
SYSTEMS	QUADRATIC MATRIX EQUATIONS FOR DETERMINING VIBRATION MODES AND FREQUENCIES OF CONTINUOUS ELASTIC SYS	66-0064
SYSTEMS	RESEARCH TO IMPROVE TUNNEL SUPPORT SYSTEMS	74-0777
SYSTEMS	RESPONSE OF NUCLEAR STRUCTURAL SYSTEMS TO TRANSIENT AND RANDOM EXCITATIONS. USING BOTH DETERMINISTIC	74-1047
SYSTEMS	SAP IV. A STRUCTURAL ANALYSIS PROGRAM FOR STATIC AND DYNAMIC RESPONSE OF LINEAR SYSTEMS	73-0650
SYSTEMS	SAP-A GENERAL STRUCTURAL ANALYSIS PROGRAM FOR LINEAR SYSTEMS	72-0329
SYSTEMS	SEISMIC ANALYSIS OF EARTH DAM RESERVOIR SYSTEMS	73-0412
SYSTEMS	SHOCK ANALYSIS OF STRUCTURAL SYSTEMS	70-0426
SYSTEMS	SHOCK LOADS ON PIPING SYSTEMS	72-0760
SYSTEMS	SIMPLIFICATION OF HIGH ORDER MECHANICAL SYSTEMS USING THE ROUTH APPROXIMATION	75-0853
SYSTEMS	SIMULATION OF VIBRATION OF WHOLE TREE SYSTEMS USING FINITE ELEMENTS	75-0765
SYSTEMS	SIMULTANEOUS EIGENVECTOR ITERATION APPLIED TO LARGE SYSTEMS	74-0680
SYSTEMS	SOLUTION OF LARGE SYSTEMS OF EQUATIONS	75-0608
SYSTEMS	SOLVING A CLASS OF LARGE SPARSE LINEAR SYSTEMS OF EQUATIONS BY PARTITIONING	73-0580
SYSTEMS	STATISTICAL ASPECTS OF THE STABILITY OF COMPOUND SYSTEMS WITH BILINEAR ELEMENTS	73-0986

411

SYSTEMS	STRAIN DISTRIBUTION AROUND UNDERGROUND OPENINGS. FINITE ELEMENT ANALYSIS OF JOINTED SYSTEMS	72-0838
SYSTEMS	STRAP - A COMPUTER CODE FOR THE STATIC AND DYNAMIC ANALYSIS OF REACTOR STRUCTURAL SYSTEMS	71-0078
SYSTEMS	STRUCTURAL ANALYSIS OF AERODYNAMIC DECELERATION SYSTEMS BY THE FINITE ELEMENT METHOD	67-0064
SYSTEMS	STUDIES TO IMPROVE TUNNEL SUPPORT SYSTEMS	74-0957
SYSTEMS	SURVEY OF STRUCTURAL ANALYSIS SYSTEMS	74-1290
SYSTEMS	THE DEVELOPMENT OF APPLIED METHODS IN PROBLEMS OF STATIC CALCULATION OF THIN-WALLED THREE-DIMENSIONA	73-0934
SYSTEMS	THE THERMO-ELASTO-PLASTIC ANALYSIS OF A RE-ENTRY BODY TYPE COMPOSITE SOLID STRUCTURE USING A NEW FIN	66-0093
SYSTEMS	THEORETICAL INVESTIGATION OF ROCK AND SUPPORT INTERACTION - DEVELOP MORE RATIONAL DESIGN METHODS AND	74-0782
SYSTEMS	THREE DIMENSIONAL SEISMIC ANALYSIS OF STRUCTURE-GROUND SYSTEMS BY THE FINITE ELEMENT METHOD	74-0268
SYSTEMS	TRANSIENT FLOW OF GROUNDWATER TO WELLS IN MULTIPLE-AQUIFER SYSTEMS	69-0501
SYSTEMS	TRANSIENT RESPONSE OF INELASTICALLY CONSTRAINED RIGID BODY SYSTEMS	73-0291
SYSTEMS	TRANSIENT VIBRATION ANALYSIS OF LINEAR SYSTEMS USING TRANSITION MATRICES	68-0053
SYSTEMS	TRANSMISSION OF VIBRATION IN BEAM SYSTEMS	69-0405
SYSTEMS	TRANSVERSE VIBRATIONS OF SHIP'S PROPULSION SYSTEMS. PART 1: THEORETICAL ANALYSIS	74-1323
SYSTEMS	TRANSVERSE VIBRATIONS OF SHIP'S PROPULSION SYSTEMS. PART 2: EXPERIMENTAL ANALYSIS	74-1324
SYSTEMS	TRANSVERSE VIBRATIONS OF SHIP'S PROPULSION SYSTEMS - 1 THEORETICAL ANALYSIS	75-0291
SYSTEMS	TWO-DIMENSIONAL FINITE ELEMENT MODELING OF TRANSIENT FLOW IN REGIONAL AQUIFER SYSTEMS. ANALYSIS OF G	74-0743
SYSTEMS	USE OF SPARSITY IN THE SOLUTION OF FINITE ELEMENT SYSTEMS OF EQUATIONS BY GAUSSIAN ELIMINATION-TYPE	74-0868
SYSTEMS	USE OF THE FINITE ELEMENT METHOD IN SOLVING TRANSIENT FLOW PROBLEMS IN AQUIFER SYSTEMS	68-0167
SYSTEMS	USING NASTRAN FOR DYNAMIC ANALYSIS OF VEHICLE SYSTEMS	74-0316
SYSTEMS	VIBRATION ANALYSIS OF FINITE ELEMENT SYSTEMS	71-0506
SYSTEMS	VIBRATION STUDY OF COMPLETE TREE SYSTEMS USING FINITE ELEMENT METHOD	75-0790
SZTYWNOSCI	APPLICATION OF HERMITE POLYNOMIALS TO THE DETERMINATION OF THE STIFFNESS MATRIX OF PLATE BY FINITE E	71-0365
SZTYWNOSCI	WIELOSTOPNIOWA SYNTEZA MACIERZY SZTYWNOSCI (MULTI-STAGE SYNTHESIS OF THE STIFFNESS MATRICES)	73-0668
SZTYWNYCH	ALGORYTM OBLICZEN DRGAN WYMUSZONYCH METODA SZTYWNYCH ELEMENTOW SKONCZONYCH (ALGORITHM FOR CALCULATIN	74-0581
SZTYWNYCH	ANALIZA DRGAN WYMUSZONYCH ZLOZONYCH UKLADOW LINIOWYCH METODA SZTYWNYCH ELEMENTOW SKONCZONYCH (ANALYS	72-0246
SZTYWNYCH	METHOD SZTYWNYCH ELEMENTOW SKONCZONYCH W OBLICZENIACH KONSTRUCJI OKRETOWYCH. (STIFF FINITE ELEMENT M	74-0923
SZTYWNYCH	OBLICZANIE DRGAN METODA SZTYWNYCH ELEMENTOW SKONCZONYCH. (CALCULATING VIBRATIONS BY THE RIGID-FINITE	75-0649
SZTYWNYCH	ZASTOSOWANIE METODY SZTYWNYCH ELEMENTOW SKONCZONYCH DO OBLICZEN DRGAN WALOW OKRETOYCH Z UWZGLEDNIENI	74-0962
S8DR	FUEL ROD FINITE ELEMENT STRESS ANALYSIS: SEGMENTED FLIGHT REACTOR FUEL VERSUS S8DR FUEL	71-0798
T-BEAMS	TRAGLAST VON T-TRAEGERN. EXPERIMENT UND F.E. BERECHNUNGEN (ULTIMATE STRENGTH OF T-BEAMS. EXPEDIMENTS	74-0935
T-CONNECTIONS	ANALYSIS OF TUBULAR T-CONNECTIONS BY THE FINITE ELEMENT METHOD: COMPARISON WITH EXPERIMENTS	72-0380
T-TRAEGERN	TRAGLAST VON T-TRAEGERN. EXPERIMENT UND F.E. BERECHNUNGEN (ULTIMATE STRENGTH OF T-BEAMS. EXPERIMENTS	74-0935
TABLE	INVESTIGATION OF THE WATER TABLE IN A TIDAL BEACH	71-0648
TABLIERS	CALCULATION OF THE BRIDGE DECKS BY USING NUMERICAL METHODS. (LE CALCUL DES TABLIERS DE PONTS PAR DES	71-0411
TABLIERS	LE CALCUL DES TABLIERS DE PONTS PAR DES METHODES NUMERIQUES	72-0197
TACOS	TACOS: A FINITE ELEMENT COMPUTER PROGRAM FOR THE TRANSIENT ANALYSIS OF CYLINDRICAL OBESE SHELLS	72-0588
TACTICAL	STRUCTURAL DYNAMIC PROPERTIES OF TACTICAL MISSILE JOINTS - PHASE 1	70-0423
TACTICAL	STRUCTURAL DYNAMIC PROPERTIES OF TACTICAL MISSILE JOINTS - PHASE 3	74-0811
TAFEST	CODE TAFEST: NUMERICAL SOLUTION TO TRANSIENT HEAT-CONDUCTION PROBLEMS USING FINITE ELEMENTS IN SPACE	74-0901
TAGESOBERFLAECH	DIE BERECHNUNG DER BEWEGUNGEN UND SPANNUNGEN DES GEBIRGES VOM ABBAU BIS ZUR TAGESOBERFLAECHE NACH DE	73-0456
TAILINGS	SEEPAGE-ENVIRONMENTAL ANALYSIS OF THE SLIME ZONE OF A TAILINGS POND	74-0513
TAILLE	ANALYSE TRIDIMENSIONELLE DE LA STABILITE D'UN TUNNEL AU VOISINAGE DU FRONT DE TAILLE DANS UNE ROCHE	74-0995
TALL	APPLICATION OF FINITE ELEMENT METHOD TO THE DYNAMIC ANALYSIS OF TALL STRUCTURES	71-0111
TALL	FINITE-ELEMENT ANALYSIS OF TALL BUILDINGS	72-0959
TALL	MULTIDEGREE DYNAMIC ANALYSIS OF TALL BUILDINGS SUBJECTED TO WIND AS A STOCHASTIC PROCESS	74-1196
TALL	STRUCTURAL ANALYSIS OF TALL BUILDINGS HAVING IRREGULARLY POSITIONED SHEAR WALLS	73-0282
TALUS	ANALYSE DE LA STABILITY DES TALUS ROCHEUX PAR UNE METHODE GENERALISEE DE L EQUILIBRE LIMITE (ROCK SL	74-1019
TANGENT	TANGENT STIFFNESS MATRIX FOR SPACE FRAME MEMBERS	69-0397
TANGENTIAL	GALERKIN FINITE ELEMENT SOLUTION FOR THE STABILITY OF CANTILEVER COLUMNS SUBJECTED TO TANGENTIAL LOA	75-0534
TANGENTIAL	GALERKIN FINITE-ELEMENT SOLUTION FOR STABILITY OF CANTILIVER COLUMNS SUBJECTED TO TANGENTIAL LOADS	75-0618
TANGENTIALLY	A FINITE-ELEMENT APPROACH TO THE ANALYSIS OF TANGENTIALLY POLARIZED PIEZOELECTRIC-CERAMIC FREE-FLOOD	74-0802
TANGENTIALLY	3-DIMENSIONAL FINITE-ELEMENT ANALYSIS OF ELECTROELASTIC NORMAL MODES OF A TANGENTIALLY POLARIZED, SE	75-0591
TANK	AIRCRAFT FUEL TANK VULNERABILITY TO HYDRAULIC RAM: MODIFICATION OF THE NORTHROP FINITE ELEMENT COMP	74-0724
TANK	CONICAL FINITE ELEMENT STRESS ANALYSIS OF AN AXISYMMETRIC TANK ON AN ELASTIC FOUNDATION	70-0494
TANK	EINSATZ DES COMPUTERS BEI DER ENTWICKLUNG EINER PANZERWANNE (THE USE OF A COMPUTER FOR DEVELOPING A	72-0211
TANK	PROBLEMS OF LONGITUDINAL STRENGTH IN THE CONSTRUCTION OF VERY LARGE TANK TRUCKS	73-0251
TANK	SIMPLIFIED CALCULATION OF THE INTERACTION FORCES BETWEEN HULL AND TANK SYSTEM FOR A MOSS ROSENBERG T	75-0561
TANK-ULLAGE	CALCULATION OF AXISYMMETRIC LONGITUDINAL MODES FOR FLUID-ELASTIC TANK-ULLAGE GAS SYSTEMS AND COMPARI	64-0019
TANKAGE	LEAKAGE THROUGH CRACKS IN LNG TANKAGE	74-0749
TANKER	ANALYSIS OF LARGE CLOSED SHIP STRUCTURES: APPLICATIONS TO TANKER STRUCTURAL DESIGN	70-0289
TANKER	EXAMPLE OF LARGE-SCALE STRUCTURAL ANALYSIS. COMPARISON BETWEEN FINITE ELEMENT CALCULATION AND FULL S	74-0177
TANKER	INDETS - INTEGRATED DESIGN OF TANKER STRUCTURES	75-0524
TANKER	INVESTIGATION ON FREE AND FORCED VIBRATIONS OF A LONG TANKER WITH OVERLAPPING PROPELLER ARRANGEMENT	73-0981
TANKER	ON STRENGTH OF TRANSVERSE WEBS AND THEIR JOINTS OF LARGE TANKER	69-0327
TANKER	ON THE FINITE ELEMENT STRESS ANALYSIS OF OIL TANKER STRUCTURES	74-0606
TANKER	TANKER TRANSVERSE STRENGTH ANALYSIS: USER'S MANUAL	72-0750
TANKER	TRANSVERSE STRENGTH AT FORE AND AFT SECTION OF TANKER CARGO HOLD	74-0618
TANKERS	DEFORMABILITY OF THE HULL STEEL-WORK AND DEFORMATIONS OF THE ENGINE-ROOM OF LARGE TANKERS	74-0532
TANKERS	SIMPLE APPROACH TO THE STRENGTH ANALYSIS OF TANKERS	71-0319
TANKERS	TRANSVERSE STRENGTH OF TANKERS. FINITE ELEMENT APPLICATIONS	68-0142
TANKS	ANALYSIS OF BOTTOM ENTRY NOZZLES FOR ATMOSPHERIC STORAGE TANKS	75-0331
TANKS	ANALYSIS OF LONG RECTANGULAR TANKS	70-0276
TANKS	ANALYSIS OF RECTANGULAR TANKS BY FINITE ELEMENT TECHNIQUES	67-0044
TANKS	BENDING MOMENTS IN LONG WALL TANKS	67-0110
TANKS	COMPARIOSN OF RESULTS OBTAINED FOR THE PROBLEM OF SYMMETRIC FLUID OSCILLATIONS IN TANKS BY THE FINIT	71-0761
TANKS	DYNAMIC BEHAVIOR OF LIQUIDS IN ELASTIC TANKS	71-0141
TANKS	INTERNAL PRESSURE EFFECTS ON THE VIBRATION OF PARTIALLY FILLED ELASTIC TANKS	75-0241
TANKS	MODERN TECHNIQUES FOR CALCULATING STRESSES IN TANKS. AND THEIR LIMITATIONS. (MODERNE VERFAHREN ZUR S	70-0153
TANKS	ON A SPECTRAL PROBLEM IN VIBRATION MECHANICS: COMPUTATION OF ELASTIC TANKS PARTIALLY FILLED WITH LIQ	75-0267
TANKS	PLATES AND TANKS ON ELASTIC FOUNDATIONS, AN APPLICATION OF FINITE ELEMENT METHOD	65-0010
TANKS	SLOSHING OF LIQUID IN CIRCULAR CHANNELS AND SPHERICAL TANKS	60-0005
TANKS	VERGLEICH NUMERISCHER LOESUNGEN NACH DEM DIFFERENZENVERFAHREN UND NACH DER FINITE-ELEMENT-METHODE AM	71-0016
TANKS	VIBRATIONS OF TANKS PARTIALLY FILLED WITH LIQUIDS	74-0089
TAPE	MECHANICAL PROPERTIES OF TAPE COMPOSITES	69-0317
TAPE	STRESS ANALYSIS OF TAPE REINFORCED COMPOSITES	69-0068
TAPE	STUDY ON FOIL BEARINGS - AN APPLICATION TO TAPE MEMORY DEVICES	75-0311
TAPERED	A TAPERED THICKNESS TRIM 6 ELEMENT FOR THE MATRIX DISPLACEMENT METHOD	66-0007
TAPERED	AN INHOMOGENEOUS, TAPERED FINITE ELEMENT	73-1038
TAPERED	ANISOTROPIC TAPERED ELEMENTS USING DISPLACEMENT MODELS	72-0391
TAPERED	AXISYMMETRIC VIBRATIONS OF LINEARLY TAPERED ANNULAR PLATES	74-1266
TAPERED	COMPLIANCE OF A TAPERED DCB SPECIMEN - CONFIGURATIONS BY A FINITE ELEMENT METHOD	74-0545
TAPERED	DYNAMIC COUPLING IN AN UNSYMMETRICALLY TAPERED BEAM	74-1301
TAPERED	FINITE ELEMENT ANALYSIS OF COUPLED VIBRATION OF TAPERED TWISTED BLADES.	72-0766
TAPERED	FINITE ELEMENT CALCULATIONS OF THE COMPLIANCE OF A TAPERED DOUBLE CANTILEVER BEAM SPECIMEN FOR DIFFE	72-0665
TAPERED	FINITE ELEMENT LARGE DEFORMATION ANALYSIS OF A TAPERED AORTA	75-0856
TAPERED	FINITE-ELEMENT ANALYSIS OF TAPERED VISCOUS FLOW	72-0961
TAPERED	IMPROVED FINITE ELEMENTS FOR VIBRATION ANALYSIS OF TAPERED BEAMS	73-0033
TAPERED	STIFFNESS MATRIX FOR CURVED TAPERED AND STRAIGHT TAPERED SHELL STIFFENERS	69-0074
TAPERED	STIFFNESS MATRIX FOR CURVED TAPERED AND STRAIGHT TAPERED SHELL STIFFENERS	69-0074
TAPERED	TAPERED PLANE STRESS ANNULAR SECTOR FINITE ELEMENT	75-0538
TAPERED	VIBRATIONS OF TAPERED AND CURVED MEMBERS BY FINITE ELEMENT METHODS	69-0517

TARCZY   APPLICATION OF HERMITE POLYNOMIALS TO THE DETERMINATION OF THE STIFFNESS MATRIX OF PLATE BY FINITE E   71-0365
TARGET-TYPE   ANALYSIS OF CURVED TARGET-TYPE THRUST REVERSERS   75-0152
TARGETS   THIN WINDOWS FOR GASEOUS AND LIQUID TARGETS: AN OPTIMIZATION PROCEDURE   75-0741
TASK   PART 1 - TASK 3.2.9. BENDING AND BUCKLING OF A ONE-DIMENSIONAL MODEL SIMULATING A CONICAL SHELL. PAR   71-0745
TASK   PART 1 - TASK 3.2.9. BENDING AND BUCKLING OF A ONE-DIMENSIONAL MODEL SIMULATING A CONICAL SHELL. PAR   71-0745
TAYLOR   MULTIPOINT TAYLOR FORMULAS AND APPLICATIONS TO THE FINITE ELEMENT METHOD   71-0446
TAYLOR   ON THE ACCURACY OF THE TAYLOR APPROXIMATION FOR STRUCTURE RESIZING   74-0147
TEA   ADDITION OF AN ARBITRARY BODY ANALYSIS CAPABILITY TO THE BOEING TEA 236 FINITE ELEMENT COMPUTER PROG   73-0785
TEA   THE TET 20 AND TEA 8 ELEMENTS FOR THE MATRIX DISPLACEMENT METHOD   68-0012
TEACHING   AN INCORE FINITE ELEMENT PROGRAM FOR RESEARCH AND TEACHING PURPOSE   74-0666
TEACHING   COMPUTER AIDED TEACHING OF THE FINITE ELEMENT DISPLACEMENT METHOD   69-0430
TEACHING   SIMPLE TEACHING EXAMPLES FOR FINITE ELEMENT STRESS ANALYSIS   72-0898
TEAM   PROGRAM DESIGN BY A MULTIDISCIPLINARY TEAM   75-0854
TEAR   A DYNAMIC PHOTOELASTIC ANALYSIS OF DYNAMIC TEAR TEST SPECIMEN   74-0736
TEARING   LAMELLAR TEARING IN WELDED GIRDERS   70-0218
TEARING   THE COMPUTATION OF TEARING ENERGY OF NICKED RUBBER STRIPS IN EXTENSION   74-0315
TEC   A SEQUEL TO TECHNICAL NOTE 13 - THE CURVED TETRAHEDRONAL AND TRIANGULAR ELEMENTS TEC AND TRIC FOR TH   69-0048
TEC   THE CURVED TETRAHEDRONAL AND TRIANGULAR ELEMENTS TEC AND TRIC FOR THE MATRIX DISPLACEMENT METHOD   69-0047
TECHNICAL   A SEQUEL TO TECHNICAL NOTE 14 ON THE TUBA FAMILY OF PLATE ELEMENTS   68-0015
TECHNICAL   A SEQUEL TO TECHNICAL NOTE 13 - THE CURVED TETRAHEDRONAL AND TRIANGULAR ELEMENTS TEC AND TRIC FOR TH   69-0048
TECHNICAL   A SEQUEL TO TECHNICAL NOTE 15: THE SHEBA FAMILY OF SHELL ELEMENTS FOR THE MATRIX DISPLACEMENT METHOD   69-0444
TECHNICAL   A TECHNICAL EVALUATION OF THE NASTRAN COMPUTER PROGRAM   71-0550
TECHNICAL   A TECHNICAL MANUAL FOR THE BEAM ELEMENTS   69-0455
TECHNICAL   ASKA 105 - A TECHNICAL MANUAL FOR THE DIRECT SOLUTION AND INVERSION PACKAGE   68-0255
TECHNICAL   ASKA 105 A TECHNICAL MANUAL AND PROGRAMMING THEORY FOR THE ELEMENTS, FLA, TRIM, SPARTA AND TET   68-0254
TECHNICAL   BULLETIN DE L ASSOCIATION TECHNIQUE MARITIME ET AERONAUTIQUE (BULLETIN OF THE TECHNICAL MARITIME AND   74-1108
TECHNICAL   DETERMINATION OF THE TECHNICAL CONSTANTS OF LAMINATES IN OBLIQUE DIRECTIONS. THEORY AND EXPERIMENTAL   73-0748
TECHNICAL   HEAVY SECTION STEEL TECHNOLOGY PROGRAM. TECHNICAL REPORT NO. 4, OCTOBER 1969. A TWO-DIMENSIONAL ELAS   69-0474
TECHNICAL   HEAVY SECTION STEEL TECHNOLOGY PROGRAM TECHNICAL OR PROGRAMMATIC MANUSCRIPT NO. 25. VARIABLE THICKNE   73-0948
TECHNICAL   MODAL COUPLING PROCEDURES ADAPTED TO NASTRAN ANALYSIS OF THE 1/8 - SCALE SHUTTLE STRUCTURAL DYNAMICS   75-0677
TECHNICAL   PROCEEDINGS OF THE NOSETIP STRESS ANALYSIS TECHNICAL INTERCHANGE MEETING   67-0042
TECHNICAL   RADIAL AND AXIAL TRANSMISSIBILITY CHARACTERISTICS OF TYPICAL ROCKET VEHICLE STRUCTURE TECHNICAL SUMM   69-0475
TECHNICAL   TECHNICAL EVALUATION REPORT ON AGARD SPECIALISTS   71-0737
TECHNICAL   TECHNICAL, PHYSICAL AND MATHEMATICAL PRINCIPLES OF THE FINITE ELEMENT METHOD   71-0676
TECHNICAL   THE FUGA FAMILY OF ELEMENTS FOR FOLDED PLATE STRUCTURES A FURTHER SEQUEL TO TECHNICAL NOTE 14   69-0446
TECHNOLOGY   A SURVEY OF SPARSE MATRIX TECHNOLOGY   72-0989
TECHNOLOGY   DEVELOPMENT OF TECHNOLOGY FOR MODELING OF A 1/8-SCALE DYNAMIC MODEL OF THE SHUTTLE SOLID ROCKET BOOS   74-0703
TECHNOLOGY   HEAVY SECTION STEEL TECHNOLOGY PROGRAM. TECHNICAL REPORT NO. 4, OCTOBER 1969. A TWO-DIMENSIONAL ELAS   69-0474
TECHNOLOGY   HEAVY SECTION STEEL TECHNOLOGY PROGRAM TECHNICAL OR PROGRAMMATIC MANUSCRIPT NO. 25. VARIABLE THICKNE   73-0948
TECHNOLOGY   HYBRID FINITE ELEMENT OPTIMIZATION PROGRAMS FOR DISCS USING A DYNAMICS DESIGN TECHNOLOGY   74-0691
TECHNOLOGY   ON THE FINITE ELEMENT ANALYSIS OF THIN-WALLED STRUCTURES WITH APPLICATION TO STRUCTURAL PROBLEMS IN   73-0832
TECHNOLOGY   PRESSURE VESSEL CERTIFICATION BASED ON FRACTURE MECHANICS TECHNOLOGY   74-0948
TECHNOLOGY   THE EVOLUTION OF SHELL STRUCTURES IN OCEAN TECHNOLOGY   69-0042
TECHNOLOGY   USE OF COMPUTERS IN WELDING TECHNOLOGY   74-1160
TECNICA   LA TECNICA DEGLI ELEMENTI FINITI PER PROBLEMI DI LUBRIFICAZIONE INCOMPRESSIBILE. (FINITE ELEMENT MET   75-0441
TECTONIC   ROCK FAILURE AROUND A CIRCULAR OPENING IN A GRAVITY FIELD WITH TECTONIC FORCES   70-0071
TEE   FINITE ELEMENT ANALYSIS OF ELASTO-PLASTIC TEE JOINTS   74-1313
TEE   THREE-DIMENSIONAL THERMAL AND STRESS ANALYSIS OF A PIPING TEE   75-0703
TEES   STRESS ANALYSIS OF B16.9 TEES BY THE FINITE ELEMENT METHOD   71-0060
TEES   STRESS ANALYSIS OF B16.9 TEES BY THE FINITE ELEMENT METHOD, A PROGRESS REPORT   71-0215
TEETH   COMPUTER SIMULATION OF DYNAMIC STRESS, DEFORMATION, AND FRACTURE OF GEAR TEETH   72-0114
TEETH   EVALUATION OF STRESSES AND DEFLECTION OF SPUR GEAR TEETH UNDER STRAIN   72-0115
TEETH   STRESS ANALYSIS OF HUMAN TEETH   73-0505
TELEPHONE   APPLICATION OF RESPONSE-BOUND METHOD IN SHOCK AND VIBRATION ANALYSIS OF TELEPHONE STRUCTURES   74-0655
TELESCOPE   STRUCTURAL EVALUATION OF CANDIDATE DESIGNS FOR THE LARGE SPACE TELESCOPE PRIMARY MIRROR   75-0708
TELESCOPE   THERMOSTRUCTURAL DESIGN CONSIDERATIONS TO ACHIEVE THE LARGE SPACE TELESCOPE LINE-OF-SIGHT REQUIREMEN   75-0839
TELESCOPES   ANALYSIS OF THE DYNAMICS OF THIN PRIMARY MIRRORS FOR LARGE ASTRONOMICAL TELESCOPES   73-1084
TEMPE   AN INVESTIGATION OF THE DYNAMIC RESPONSE CHARACTERISTICS OF THE BON TEMPE DAM   66-0047
TEMPERATURE   A FINITE ELEMENT METHOD FOR THE DETERMINATION OF NON-STATIONARY TEMPERATURE DISTRIBUTIONS AND THERMA   66-0074
TEMPERATURE   ANALYSE DES CONTRAINTES DAN DES STRUCTURES COMPLEXES EN PRESENCE DE FLUAGE, DE VARIATIONS DE DIMENSI   71-0081
TEMPERATURE   ANALYSIS OF THE EFFECTS ON LIFE OF LEADING-EDGE HOLES IN AN AIRFOIL SUBJECTED TO ARBITRARY SPANWISE   75-0700
TEMPERATURE   APPLICATION OF AN ' ELEMENT-METHOD ' PROGRAM TO SOLVING TEMPERATURE DISTRIBUTION PROBLEMS (ANWENDUNG   71-0464
TEMPERATURE   APPLICATION OF CONFORMAL MAPPING AND VARIATIONAL METHOD TO THE STUDY OF HEAT CONDUCTION IN POLYGONAL   71-0444
TEMPERATURE   APPLICATION OF THE FINITE ELEMENT METHOD TO PROBLEMS OF TEMPERATURE DISTRIBUTION ANALYSIS FOR BODIES   74-1117
TEMPERATURE   BERECHNUNG INSTATIONAERER TEMPERATUREFELDER MIT FINITEN ELEMENTEN IN RAUM-UND ZEITDIMENSION. (CALCUL   74-0987
TEMPERATURE   BERECHNUNG VON SPANNUNGEN UND TEMPERATURFELDERN IN FLOSSENWAENDEN UND AUFGESCHWEISSIEN BAUTEILEN. (C   74-0544
TEMPERATURE   CALCULATION OF TEMPERATURE DISTRIBUTIONS IN THIN SHELLS OF REVOLUTION BY THE FINITE ELEMENT METHOD   71-0018
TEMPERATURE   CALCULATION OF TEMPERATURE DISTRIBUTION WITHIN A MELTING-FREEZING MATERIAL USING FINITE ELEMENT TECH   72-0043
TEMPERATURE   CALCULATION OF TRANSIENT TEMPERATURE FIELDS WITH FINITE ELEMENTS IN SPACE AND TIME DIMENSIONS   74-0379
TEMPERATURE   COMPUTATIONAL METHODS IN NUCLEAR REACTOR STRUCTURAL DESIGN FOR HIGH TEMPERATURE APPLICATIONS: AN INT   73-1001
TEMPERATURE   DIGITAL COMPUTER PROGRAMS FOR STEADY-STATE OR TRANSIENT TEMPERATURE ANALYSIS OF PLANE AND AXISYMMETR   72-0915
TEMPERATURE   ELASTIC DEFORMATION OF SQUARE EXTRUSION DIES DUE TO LOAD AND TEMPERATURE   75-0581
TEMPERATURE   EVALUATION OF THE USE OF FINITE ELEMENT METHOD IN COMPUTING TEMPERATURE   69-0017
TEMPERATURE   EVALUATION OF USE OF THE FINITE-ELEMENT METHOD IN COMPUTATION OF TEMPERATURE   71-0467
TEMPERATURE   FAST REACTOR PROGRAMME. FINITE ELEMENT TEMPERATURE ANALYSIS IN FUEL PINS   73-0939
TEMPERATURE   FIESTA - FINITE ELEMENT STRESS AND TEMPERATURE ANALYSIS. A DESCRIPTION OF THE THEORY USED FOR THE CO   71-0718
TEMPERATURE   FINITE ELEMENT ELASTIC PLASTIC CREEP ANALYSIS OF TWO DIMENSIONAL CONTINUUM WITH TEMPERATURE DEPENDEN   73-0226
TEMPERATURE   FINITE ELEMENT STRESS ANALYSIS OF AXISYMMETRIC SOLIDS WITH ORTHOTROPIC TEMPERATURE DEPENDENT MATERIA   67-0089
TEMPERATURE   FINITE ELEMENT TECHNIQUE APPLIED TO HEAT CONDUCTION IN SOLIDS WITH TEMPERATURE DEPENDENT THERMAL CON   69-0004
TEMPERATURE   FINITE-ELEMENT ANALYSIS OF SHALLOW TEMPERATURE ANOMALIES   71-0471
TEMPERATURE   HEAT TRANSFER PROBLEMS AND TEMPERATURE STRESSES   69-0326
TEMPERATURE   HIGH-ORDER FINITE ELEMENT ANALYSIS FOR TRANSIENT TEMPERATURE ANALYSIS OF INHOMOGENEOUS MATERIALS   69-0301
TEMPERATURE   INELASTIC ANALYSIS AND SATISFACTION OF DESIGN CRITERIA OF A HIGH TEMPERATURE COMPONENT   75-0265
TEMPERATURE   INFLUENCE OF INTERLAMINATE FLAWS ON TRANSIENT TEMPERATURE DISTRIBUTIONS   73-0223
TEMPERATURE   INFLUENCE OF TEMPERATURE DEPENDENT PROPERTIES ON THERMAL ROCK FRAGMENTATION   75-0222
TEMPERATURE   NON-LINEAR EFFECTS OF HIGH TEMPERATURE ON THE VIBRATION OF A SIMPLY SUPPORTED BEAM WITH A CENTRAL MA   73-0838
TEMPERATURE   NONLINEAR CREEP ANALYSIS AT ELEVATED TEMPERATURE   71-0050
TEMPERATURE   NUMERICAL SIMULATION OF FINITE-WIDTH THRUST BEARINGS. TAKING INTO ACCOUNT VISCOSITY VARIATION WITH T   75-0623
TEMPERATURE   NUMERISCHE BERECHNUNG INSTATIONAERER TEMPERATURFELDER UND THERMISCH BEDINGTER VERFORMUNGEN VON WERKZ   75-0847
TEMPERATURE   OIL-FILM TEMPERATURE DISTRIBUTION IN AN INFINITELY WIDE SLIDER BEARING: AN APPLICATION OF THE FINITE   73-1128
TEMPERATURE   OPREDELENIE STATSIONARNYKH TEMPERATURNYKH POLEI V DETALYAKH DVIGATELEI VNUTRENNEGO SGORANIYA METODOM   73-0454
TEMPERATURE   PLANAR FINITE ELEMENT AT RANDOM TEMPERATURE   73-0140
TEMPERATURE   PREDICTION OF TEMPERATURE DISTRIBUTIONS IN CARGOES WITH THE AID OF A COMPUTER PROGRAM USING THE METH   74-1215
TEMPERATURE   RESEARCH NOTE: A THREE-DIMENSIONAL OIL FILM TEMPERATURE DISTRIBUTION IN TILTING THRUST BEARINGS   74-0608
TEMPERATURE   SAAS II: FINITE ELEMENT STRESS ANALYSIS OF AXISYMMETRIC SOLIDS WITH ORTHOTROPIC, TEMPERATURE DEPENDE   68-0096
TEMPERATURE   SIMPLIFIED INELASTIC HIGH TEMPERATURE STRUCTURAL ANALYSIS OF MODERATELY COMPLEX SPATIALLY THREE DIME   74-0445
TEMPERATURE   SOME MODELS FOR THE ANALYSIS OF STRESSES IN A TUBULAR FUEL PIN FOR HIGH TEMPERATURE REACTORS BY FINI   73-0816
TEMPERATURE   TEMPERATURE CALCULATIONS USING THE NASTRAN FINITE ELEMENT PROGRAMME   75-0642
TEMPERATURE   TEMPERATURE RISE IN THE HIGH SPEED COMPRESSION OF RIGHT CYLINDRICAL BILLETS   74-1153
TEMPERATURE   TEMPERATURVERLAUF IM QUERSCHNITT GESCHWEISSTER ROHRWAENDE BEI JEBERWIEGEND KONVEKTIVEM WAERMEUEBERGA   75-0628
TEMPERATURE   TENUE DES STRUCTURES DE GRAPHITE VIS-A-VIS DES SOLLICITATIONS RENCONTREES DANS UN COEUR DE REACTEUR   73-0817
TEMPERATURE   TENUE DES STRUCTURES DE GRAPHITE VIS-A-VIS DES SOLLICITATIONS RENCONTREES DANS UN COEUR DE REACTEUR   73-0817

413

TEMPERATURE	TESTRAN PROGRAMS, A TOOL FOR TEMPERATURE AND STRESS ANALYSIS	73-0414
TEMPERATURE	THE CALCULATION OF THREE-DIMENSIONAL TEMPERATURE DISTRIBUTIONS AND THERMAL STRESSES USING FINITE ELE	71-0187
TEMPERATURE	THE CHANGES IN THE (TWO-DIMENSIONAL) STRESS ROUNDED CORNERS, CAUSED BY A VARYING INTERNAL PRESSURE A	67-0141
TEMPERATURE	THE COMPUTATION OF TEMPERATURE DISTRIBUTIONS AND THERMAL STRESSES USING FINITE ELEMENT TECHNIQUES	71-0092
TEMPERATURE	THERMAL DEFORMATION VECTOR FOR A BILINEAR TEMPERATURE DISTRIBUTION IN AN ANISOTROPIC QUADRILATERAL M	75-0670
TEMPERATURE	THERMAL STRESS AND TEMPERATURE FIELDS - A FINITE-ELEMENT APPROACH	71-0782
TEMPERATURE	THIN-WALLED PRESTRESSED CONCRETE PRESSURE VESSEL FOR HIGH TEMPERATURE REACTORS: EXPERIMENTAL INVESTI	73-0857
TEMPERATURE	TRANSIENT TEMPERATURE RESPONSE OF COMPOSITE SLABS	74-1038
TEMPERATURE	ULTIMATE STRAIN CAPACITY AND TEMPERATURE RISE STUDIES, TRUMBULL POND DAM	72-0705
TEMPERATURE	USE OF SINGULARITY PROGRAMMING IN FINITE-DIFFERENCE AND FINITE ELEMENT COMPUTATIONS OF TEMPERATURE	73-0100
TEMPERATURE	USING THE FINITE ELEMENT METHOD TO DETERMINE TEMPERATURE DISTRIBUTIONS IN ORTHOGONAL MACHINING	74-1212
TEMPERATURE	VARIATION WITH TEMPERATURE OF RESONANT FREQUENCIES OF ANISOTROPIC PLATES	74-1046
TEMPERATURE	YIELDING IN UNIDIRECTIONAL COMPOSITES UNDER EXTERNAL LOADS AND TEMPERATURE CHANGES.	73-0169
TEMPERATURE-DEP	ASAAS: ASYMMETRIC STRESS ANALYSIS OF AXISYMMETRIC SOLIDS WITH ORTHOTROPIC, TEMPERATURE-DEPENDENT MAT	71-0696
TEMPERATURE-DEP	BRLESC FINITE ELEMENT PROGRAM FOR AXISYMMETRIC, PLANE STRAIN, AND PLANE STRESS, ORTHOTROPIC SOLIDS W	71-0749
TEMPERATURE-DEP	FINITE ELEMENT ANALYSIS FOR ARBITRARY AXISYMMETRIC STRUCTURES ORTHOTROPIC, TEMPERATURE-DEPENDENT MAT	68-0076
TEMPERATURE-DEP	SAAS 111: FINITE ELEMENT STRESS ANALYSIS OF AXISYMMETRIC AND PLANE SOLIDS WITH DIFFERENT ORTHOTROPIC	71-0741
TEMPERATURE-DEP	TSAAS: FINITE ELEMENT THERMAL AND STRESS ANALYSIS OF AXISYMMETRIC SOLIDS WITH ORTHOTROPIC TEMPERATUR	74-0886
TEMPERATURE-INI	FRACTURE OF ROCKET GRAIN MODELS SUBJECTED TO COOLDOWN: A STUDY OF TEMPERATURE-INITIATED FRACTURES OF	74-0536
TEMPERATUREFELD	BERECHNUNG INSTATIONAERER TEMPERATUREFELDER MIT FINITEN ELEMENTEN IN RAUM-UND ZEITDIMENSION. (CALCUL	74-0987
TEMPERATURES	TEMPERATURES AND STRESSES IN A BOILER MEMBRANE WALL TUBE	74-1053
TEMPERATURES	THEORETICAL PREDICTION OF DISC BRAKE TEMPERATURES AND A COMPARISON WITH EXPERIMENTAL DATA	72-0306
TEMPERATURFELDE	NUMERISCHE BERECHNUNG INSTATIONAERER TEMPERATURFELDER UND THERMISCH BEDINGTER VERFORMUNGEN VON WERKZ	75-0847
TEMPERATURFELDE	BERECHNUNG VON SPANNUNGEN UND TEMPERATURFELDERN IN FLOSSENWAENDEN UND AUFGESCHWEISSTEN BAUTEILEN. (C	74-0544
TEMPERATURNYKH	METODY RASCHETNOGO OPREDELENIYA TEMPERATURNYKH NAPRYAZHENII V KRYSHKAKH TSILINDROV DVIGATELEI VNUTRE	74-0179
TEMPERATURNYKH	OPREDELENIE STATSIONARNYKH TEMPERATURNYKH POLEI V DETALYAKH DVIGATELEI VNUTRENNEGO SGORANIYA METODOM	73-0454
TEMPERATURVERLA	TEMPERATURVERLAUF IM QUERSCHNITT GESCHWEISSTER ROHRWAENDE BEI UEBERWIEGEND KONVEKTIVEM WAERMEUEBERGA	75-0628
TEMPERATURVERTE	APPLICATION OF AN ' ELEMENT-METHOD ' PROGRAM TO SOLVING TEMPERATURE DISTRIBUTION PROBLEMS (ANWENDUNG	74-0464
TEMPERATURVETEI	DIE BERECHNUNG ACHSENSY METRISCHER TRAGWERKE UNTER ACHSENSYMETRISCHER UND NICHT-SYMMETRISCHER BELAST	69-0452
TEMPORARY	TEMPORARY EXCAVATION IN VARVED CLAY	75-0049
TENDONS	RIDDLE OF BONDED AND UNBONDED TENDONS IN PRESTRESSED CONCRETE REACTOR VESSELS	74-0519
TENSILE	APPLICATION OF THE DOUBLE LIGAMENT TENSILE TEST IN DEFORMATION PROCESSING	75-0827
TENSILE	CALCULATION OF STRESS INTENSITY FACTORS FOR COMBINED TENSILE AND SHEAR LOADING	75-0205
TENSILE	CRACKING OF EARTH AND ROCKFILL DAMS: COMPARISON OF OBSERVED AND THEORETICAL TENSILE STRAINS IN THE C	71-0699
TENSILE	DIAGONAL TENSILE STRENGTH OF BRICKWORK	70-0158
TENSILE	DIRECT TENSILE TESTING OF ANISOTROPIC ROCKS	74-0989
TENSILE	FINITE ELEMENT ANALYSIS OF NOTCHED TENSILE SPECIMENS IN PLANE STRESS	71-0557
TENSILE	FINITE ELEMENT ANALYSIS OF LOAD CELL RESPONSE IN HIGH SPEED TENSILE TESTING	73-1071
TENSILE	INDIRECT TENSILE TESTING OF ANISOTROPIC ROCKS	73-0376
TENSILE	SIMULATION ON THE TENSILE BEHAVIOUR OF TWO-PHASE ALLOY CONTAINING A FIBER UNDER TENSILE LOAD	73-0542
TENSILE	SIMULATION ON THE TENSILE BEHAVIOUR OF TWO-PHASE ALLOY CONTAINING A FIBER UNDER TENSILE LOAD	73-0542
TENSILE	STRESS ANALYSIS OF A BOLT-NUT JOINT BY THE FINITE ELEMENT METHOD AND THE COPPER-ELECTROPLATING METHO	73-0220
TENSILE	STRESS ANALYSIS OF A BOLT-NUT JOINT BY FINITE ELEMENT METHOD AND COPPER-ELECTROPLATING METHOD - 2. S	74-0153
TENSILE	TENSILE CRACK PROPAGATION IN REINFORCED CONCRETE BEAMS BY THE FINITE ELEMENT TECHNIQUES	69-0264
TENSILE	TENSILE CREEP ANALYSIS OF CONCRETE STRUCTURES	74-0439
TENSILE	TENSILE FIELD CALCULATION ABOUT A PRESSURIZED CAVITY: MONERO	72-0684
TENSION	A FINITE ELEMENT STUDY OF ELASTIC-PLASTIC STRESS DISTRIBUTIONS IN NOTCHED SPECIMENS UNDER TENSION	68-0177
TENSION	A GENERAL METHOD FOR THE SHAPE FINDING OF LIGHTWEIGHT TENSION STRUCTURES	74-1217
TENSION	ANALYSIS OF TENSION STRUCTURES BY MATRIX METHODS	67-0119
TENSION	APPLICATION OF THE FINITE ELEMENT METHOD TO CALCULATE WOOD TENSION	75-0768
TENSION	COMPUTATION OF POST-YIELD BEHAVIOUR IN NOTCH-BEND AND TENSION TESTPIECES.	72-0244
TENSION	ELASTIC-PLASTIC STRESS-STRAIN RELATIONS FOR MATERIALS WITH DIFFERENT YIELD STRENGTHS IN TENSION AND	72-0764
TENSION	FINITE ELEMENT STUDY OF REINFORCED CONCRETE BEAMS WITH DIAGONAL TENSION CRACKS	74-1308
TENSION	FINITE ELEMENT VIBRATION ANALYSIS OF CRACKED PLATES IN TENSION	68-0133
TENSION	FREE VIBRATIONS OF CIRCULAR MEMBRANES UNDER ARBITRARY TENSION BY THE FINITE ELEMENT METHOD	69-0190
TENSION	HIGH STRENGTH BOLTS SUBJECT TO TENSION AND PRYING	74-0201
TENSION	LES FONCTIONS DE TENSION DANS LA METHODE DES ELEMENTS FINIS	73-1113
TENSION	NONLINEAR DYNAMICS OF CABLES WITH LOW INITIAL TENSION	72-0740
TENSION	NUMERICAL ANALYSES OF DEFORMABILITY TESTS IN JOINTED ROCK. ' JOINT PERTURBATION ' AND ' NO TENSION '	71-0439
TENSION	OPTIMIZATION OF TENSION LEVEL AND STINGER LENGTH FOR OFFSHORE PIPELINE INSTALLATION	74-0504
TENSION	SAAS 111: FINITE ELEMENT STRESS ANALYSIS OF AXISYMMETRIC AND PLANE SOLIDS WITH DIFFERENT ORTHOTROPIC	71-0741
TENSION	TENSION ZONES IN DECKED ROCKFILL DAMS	73-0540
TENSIONED	ANALYSIS OF CYLINDRICAL SHELL ROOFS WITH POST TENSIONED EDGEBEAMS	74-1042
TENSIONED	THE VIBRATION CHARACTERISTICS OF A TENSIONED PLATE CONTAINING A FATIGUE CRACK	68-0134
TENSOR	A FINITE ELEMENT TENSOR APPROACH TO PLATE BUCKLING AND POST BUCKLING	73-0883
TENSOR	NONLINEAR STRUCTURAL ANALYSIS: TENSOR FORMULATION	65-0004
TENTS	ANALYSIS OF STRESSES AND DEFLECTIONS IN FRAME SUPPORTED TENTS	74-0689
TENUE	TENUE DES STRUCTURES DE GRAPHITE VIS-A-VIS DES SOLLICITATIONS RENCONTREES DANS UN COEUR DE REACTEUR	73-0817
TEPLOPROVODNOST	CHISLENNOE RESHENIE PLOSKIKH ZADACH TEPLOPROVODNOSTI DLYA OBLASTEI SLOZHNOI FORMY (NUMERICAL SOLUTIO	74-0659
TERM	FIELD INSTRUMENTATION FOR LONG TERM MEASUREMENT OF PILE LOAD AND RAFT CONTACT PRESSURE	73-0150
TERM	LONG TERM CREEP CLOSURE OF SOLUTION CAVITY SYSTEM	74-1150
TERM	PREDICTION OF LONG TERM DEFORMATION OF A COMPACTED COHESIVE SOIL EMBANKMENT OVER A SOFT FOUNDATION	74-0727
TERM	SELECTION OF LOCATIONS FOR LONG TERM STORAGE OF WHEAT	75-0323
TERM	SHORT TERM DEFLECTIONS OF CONCRETE FLAT PLATES	73-0277
TERMINAL	NASTRAN PLOTTING AT A REMOTE TERMINAL	74-0186
TERMINATION	THREE-DIMENSIONAL HARMONIC FUNCTIONS NEAR TERMINATION OR INTERSECTION OF GRADIENT SINGULARITY LINES:	74-0178
TERMODELLS	ROTATIONSSYMMETRISCHE BERECHNUNG DES 1:5 THTR SPANNBETON-BEHAL-BEHAL TERMODELLS	71-0629
TERMS	EFFECT OF HIGHER ORDER TERMS IN CERTAIN NONLINEAR FINITE ELEMENT MODELS	73-0125
TERMS	EFFECT OF HIGHER ORDER TERMS IN CERTAIN NONLINEAR FINITE ELEMENT MODELS - COMMENT	74-0473
TERMS	INFLUENCE OF HIGHER ORDER TERMS IN THE LARGE DEFLECTION ANALYSIS OF FRAMEWORKS	68-0259
TERMS	INFLUENCE OF NON-SINGULAR STRESS TERMS AND SPECIMEN GEOMETRY ON SMALL SCALE YIELDING AT CRACK TIPS I	73-0200
TERMS	NONLINEAR FINITE ELEMENT ANALYSIS INCLUDING HIGHER ORDER STRAIN ENERGY TERMS	70-0065
TERMS	ON THE EFFECT OF APPROXIMATING HIGHER ORDER TERMS IN CERTAIN FINITE ELEMENT MODELS OF NONLINEAR STRU	73-0603
TERMS	USE OF TRIGONOMETRIC TERMS IN FINITE ELEMENT METHOD WITH APPLICATION TO VIBRATING MEMBRANES	74-0127
TERMS	USE OF TRIGONOMETRIC TERMS IN FINITE ELEMENT METHOD WITH APPLICATION TO VIBRATING MEMBRANES	74-0555
TERRE	NOUVELLE CONCEPTION DES DRAINS DANS LES BARRAGES EN TERRE HOMOGENE DE PETITE ET MOYENNE DIMENSIONS (	73-0404
TERRE	UTILISATION DE MODELES MATHEMATIQUES MULTIDIMENSIONNELS POUR L'ETUDE DES ECOULEMENTS DANS LES DIGUES	73-0381
TEST	A DYNAMIC PHOTOELASTIC ANALYSIS OF DYNAMIC TEAR TEST SPECIMEN	74-0736
TEST	A TEST ON CONVERGENCE OF EIGENSOLUTIONS BY SIMULTANEOUS ITERATION METHOD	73-0317
TEST	ANALYSIS OF PLANE STRAIN TEST USING THE FINITE ELEMENT METHOD	72-1002
TEST	ANALYTICAL EVALUATION OF STRESS-STRAIN TEST DATA.	72-0442
TEST	APPLICATION OF FINITE ELEMENT TECHNIQUES TO POST-YIELD ANALYSIS OF PROPOSED STANDARD 3 POINT BEND FR	74-0118
TEST	APPLICATION OF PROGRAM FATIQUE TEST TO MEMBER JOINTS OF HULLS	72-0814
TEST	APPLICATION OF SMALL AND LARGE-DISPLACEMENT THEORY TO A LARGE FLEXIBLE SOLAR ARRAY AND COMPARISON WI	71-0231
TEST	APPLICATION OF THE DOUBLE LIGAMENT TENSILE TEST IN DEFORMATION PROCESSING	75-0827
TEST	BERECHNUNG DES KEGELDRUCKVERSUCHES MIT EINER FINITE-ELEMENT-METHODE. (CALCULATION OF THE CONE INDENT	75-0187
TEST	CALCULATION OF CONE INDENTATION TEST USING A FINITE-ELEMENT METHOD	75-0438
TEST	CH-53A FLEXIBLE FRAME VIBRATION ANALYSIS/TEST CORRELATION	73-1093
TEST	COMPARISON OF STRUCTURAL TEST RESULTS WITH PREDICTIONS OF FINITE ELEMENT ANALYSIS	73-0158
TEST	DESIGN, MANUFACTURE, DEVELOPMENT, TEST, AND EVALUATION OF BORON/ALUMINUM STRUCTURAL COMPONENTS FOR S	74-0798
TEST	DESIGN, MANUFACTURE, DEVELOPMENT, TEST, AND EVALUATION OF BORON/ALUMINUM STRUCTURAL COMPONENTS FOR S	74-0798
TEST	DESTRESSING TEST AT THE GALENA MINE, WALLACE, IDAHO	72-0126

414

TEST	ELASTIC-PLASTIC-CREEP ANALYSIS OF THERMAL PATCHETTING IN STRAIGHT PIPE AND COMPARISONS WITH TEST RES	73-0393
TEST	EXPERIENCE WITH THE PATCH TEST FOR CONVERGENCE OF FINITE ELEMENTS	72-0365
TEST	FINITE ELEMENT ANALYSIS OF DIAMETRAL TEST OF POLYMER MOLDINGS	73-0129
TEST	FINITE ELEMENT ANALYSIS OF TIME-DEPENDENT PUMPING TEST DATA IN AN UNCONFINED AQUIFER	75-0497
TEST	FINITE ELEMENT AND CUMULATIVE DAMAGE ANALYSIS OF A KEYHOLE TEST SPECIMEN	75-0163
TEST	FINITE ELEMENT AND CUMULATIVE DAMAGE ANALYSIS OF A KEYHOLE TEST SPECIMEN	75-0774
TEST	FINITE ELEMENT PARAMETRIC STUDY OF VERTICAL STRAIN INFLUENCE FACTORS AND THE PRESSUREMETER TEST TO E	74-1127
TEST	FINITE-ELEMENT ANALYSIS OF A FRACTURE TOUGHNESS TEST SPECIMEN IN NON-LINEAR RANGE	72-0093
TEST	FINITE-ELEMENT METHOD FOR CALCULATING THE CONE-INDENTATION TEST	75-0813
TEST	FLEXUAL RIGIDITY OF A THIN WALLED BUILD-UP ROTOR FOR JET ENGINE. MEASUREMENT BY STATIC LOAD TEST AND	73-0756
TEST	FLEXUAL RIGIDITY OF A THIN WALLED BUILD-UP ROTOR FOR JET ENGINE. MEASUREMENT BY STATIC LOAD TEST AND	73-0756
TEST	HEAVY SECTION STEEL TECHNOLOGY PROGRAM. TECHNICAL REPORT NO. 4, OCTOBER 1969. A TWO-DIMENSIONAL ELAS	69-0474
TEST	IMMEDIATE SETTLEMENTS OF THREE TEST EMBANKMENTS ON CHAMPLAIN CLAY	74-0275
TEST	M.I.T. TEST SECTION INSTRUMENTATION. MASSACHUSETTS BAY TRANSPORTATION AUTHORITY, HAYMARKET-NORTH EXT	72-0673
TEST	MODEL TEST AND ANALYSIS OF OBLIQUE FOLDED PLATE STRUCTURE	73-0230
TEST	NASTRAN ANALYSIS OF A TURBINE BLADE AND COMPARISON WITH TEST AND FIELD DATA	75-0367
TEST	NON-CONFORMING FINITE ELEMENT METHODS AND THE PATCH TEST	73-0640
TEST	NONLINEAR CONSIDERATION OF GRAVITY IN A STIFFNESS TEST OF A WEAK STRUCTURE AT SMALL STRAINS	71-0035
TEST	NUMERICAL SUFFICIENCY TEST FOR MONOTONIC CONVERGENCE OF FINITE ELEMENT MODELS	75-0620
TEST	ON A NUMERICAL SUFFICIENCY TEST FOR MONOTONIC CONVERGENCE OF FINITE ELEMENT MODELS	75-0533
TEST	PLATE BENDING WITH NONCONFORMING ELEMENT: FINE-MESH CONVERGENCE TEST	65-0049
TEST	PRELIMINARY FINITE ELEMENT ANALYSIS, ATCHAFALAYA BASIN PROTECTION LEVEES, TEST SECTION 3	69-0424
TEST	STRESS CONCENTRATION FACTORS FOR A FATIGUE TEST SPECIMEN WITH A U-TYPE NOTCH IN ONE EDGE BY FINITE E	74-1344
TEST	STRESS INTENSITY FACTORS FOR ANISOTROPIC FRACTURE TEST SPECIMENS OF SEVERAL GEOMETRIES	74-0293
TEST	STRUCTURAL ANALYSIS OF A THREE-DIMENSIONAL PERFORATED PLATE BY THE FINITE ELEMENT METHOD WITH TEST C	74-0945
TEST	SURVEY OF UNDERLYING THEORY, APPROXIMATION BY PIECEWISE POLYNOMIALS, ILLEGAL ELEMENTS AND THE PATCH	73-0635
TEST	TEST OF VALIDITY OF DARCY'S LAW FOR A CLAY SOIL	74-0070
TEST	TEST RESULTS FROM THE BOUNDARY LAYER FACILITY - RESPONSE OF STRUCTURE TO THE PSEUDO-SOUND FIELD OF A	68-0262
TEST	THE KANSAS TEST TRACK. NON-CONVENTIONAL TRACK STRUCTURES. DESIGN REPORT	72-0797
TEST	THE PATCH TEST	74-0339
TEST	THE SECOND VARIATION TEST WITH ALGEBRAIC AND DIFFERENTIAL CONTRASTS	69-0342
TEST	THE SUPERPATCH THEOREM AND OTHER PROPOSITIONS RELATING TO THE PATCH TEST	75-0026
TEST	VARIATIONAL PRINCIPLES AND THE PATCH TEST	74-1282
TESTING	A TORSIONAL STRENGTH ANALYSIS ON THE CONTAINER SHIP BY MEANS OF THE FINITE ELEMENT PROCEDURE AND FUL	71-0243
TESTING	AN ANALYSIS OF THE SMALL-SCALE STRENGTH TESTING OF ICE	72-0881
TESTING	ANALYSIS FOR THE DESIGN OF BEAM TESTING APPARATUS	75-0527
TESTING	APPLICATION OF THE SINGLE-EDGE NOTCHED BEAM TO FRACTURE TOUGHNESS TESTING OF CERAMICS	74-0623
TESTING	APPLICATION, TESTING AND EVALUATION OF DOUBLY CURVED SHELL ELEMENTS FOR DYNAMIC ANALYSIS. PART II. F	69-0117
TESTING	APPLICATION, TESTING, AND EVALUATION OF DOUBLY CURVED SHELL ELEMENTS FOR DYNAMIC ANALYSIS. PART 1, T	69-0246
TESTING	APPLICATION, TESTING, AND EVALUATION OF DOUBLY CURVED SHELL ELEMENTS FOR DYNAMIC ANALYSIS, PART III.	69-0256
TESTING	DIRECT TENSILE TESTING OF ANISOTROPIC ROCKS	74-0989
TESTING	DOSTIGNUCA I TENDENCIJE U ISPITIVANJIMA ALATNIH MASINA (ACHIEVEMENTS AND TENDENCIES IN MACHINE TOOLS	73-0426
TESTING	DYNAMIC TESTING AND COMPUTER ANALYSIS OF AUTOMOTIVE FRAMES	72-0089
TESTING	FINITE ELEMENT ANALYSIS OF LOAD CELL RESPONSE IN HIGH SPEED TENSILE TESTING	73-1071
TESTING	FOUNDATION TESTING FOR AUBURN DAM	70-0119
TESTING	FRACTURE TOUGHNESS TESTING WITH SECTIONS OF CYLINDERS	74-0999
TESTING	INDIRECT TENSILE TESTING OF ANISOTROPIC ROCKS	73-0376
TESTING	MINUTEMAN STV STRESS ANALYSIS AND TESTING	74-0809
TESTING	SIGNIFICANCE OF VARIABLY CONFINED TRIAXIAL TESTING ROADBUILDING MATERIALS	74-0554
TESTING	TESTING AND ASSESSING FINITE ELEMENTS BY AN EIGENVALUE TECHNIQUE	68-0188
TESTPIECES	COMPUTATION OF POST-YIELD BEHAVIOUR IN NOTCH-BEND AND TENSION TESTPIECES.	72-0244
TESTPAN	TESTPAN PROGRAMS, A TOOL FOR TEMPERATURE AND STRESS ANALYSIS	73-0414
TESTS	AN EXPERIMENTAL AND THEORETICAL STUDY OF HOLLOW CYLINDER TESTS ON SAND	71-0008
TESTS	ANALYSIS OF STRESS AND STRAIN DISTRIBUTIONS IN TRIAXIAL TESTS USING THE METHOD OF FINITE ELEMENTS	73-0903
TESTS	ANALYSIS OF THE BIAXIAL STRIP AND SHEAR LAP TESTS FOR SOLID PROPELLANT CHARACTERIZATION	67-0019
TESTS	CREEP BENDING TESTS OF LARGE WELD OVERLAY AND BASE METAL SPECIMENS AT 1000 DEGREE F. (537.7 DEG. C)	75-0521
TESTS	CRUCIFORM SPECIMENS FOR BIAXIAL FATIGUE TESTS - AN INVESTIGATION USING FINITE-ELEMENT ANALYSIS AND P	71-0795
TESTS	CYCLIC MECHANICAL TESTS AND AN APPROPRIATE ANALYTICAL STRESS-STRAIN MODEL FOR A36 STEEL	74-0815
TESTS	EVALUATION OF RIGID PAVEMENTS BY NONDESTRUCTIVE TESTS.	72-0459
TESTS	FIELD TESTS OF A STEEL-COMPOSITE BOX-GIRDER BRIDGE	75-0633
TESTS	FORM FACTORS OF FLAT BARS FOR FATIGUE STRENGTH TESTS. (FORMZAHLEN VON FLACHSTAEBEN FUER SCHWINGFESTI	71-0641
TESTS	IDENTIFICATION OF AQUIFER CONSTANTS: TESTS OF TWO INVERSE METHODS	74-0051
TESTS	LARGE STRUCTURES FOR MANNED SPACECRAFT: MATHEMATICAL ANALYSIS, DESIGN, CONSTRUCTION AND TESTS	74-1327
TESTS	MULTI-CELL BRIDGE DECKS - MODEL TESTS AND THEORETICAL ANALYSIS	72-0381
TESTS	NUMERICAL ANALYSES OF DEFORMABILITY TESTS IN JOINTED ROCK. ' JOINT PERTURBATION ' AND ' NO TENSION '	71-0439
TESTS	NUMERICAL STUDY OF UNIAXIAL AND TRIAXIAL ROCK COMPRESSION TESTS	73-0896
TESTS	PARAMETRISCHE KERBSPANNUNGSUNTERSUCHUNGEN AN DER LOCHSCHEIBE MIT DER METHODE DER FINITEN ELEMENTE. (	75-0603
TESTS	PHOTOELASTIC TESTS ON COMPLEX INTERSECTIONS OF REINFORCED THIN-SHELL CYLINDERS	75-0363
TESTS	SHOCK TUNNEL TESTS OF PRELOADED AND ARCHED WALL PANELS FINITE ELEMENT STRESS FORMULATION FOR WAVE PR	73-0892
TESTS	STRESS STRAIN RELATIONSHIP FROM COMPRESSION TESTS ON NONLINEAR MATERIALS	69-0189
TESTS	STRESSES DUE TO CONCENTRATED LOADS IN RESTRAINED BENDING TESTS	67-0097
TESTS	TESTS OF GUSSET PLATE MODELS	71-0338
TESTS	THERMAL-MECHANICAL FRAGMENTATION OF HARD ROCK - FIELD TESTS	74-0998
TET	ASKA 105 A TECHNICAL MANUAL AND PROGRAMMING THEORY FOR THE ELEMENTS, FLA, TRIM, SPARTA AND TET	68-0254
TET	THE TET 20 AND TEA 8 ELEMENTS FOR THE MATRIX DISPLACEMENT METHOD	68-0012
TETRA-CORE	A TETRA-CORE STRESS ANALYSIS MODEL	72-0845
TETRAHEDRAL	TETRAHEDRAL POLYNOMIAL FINITE ELEMENTS FOR THE HELMHOLTZ EQUATION.	72-0392
TETRAHEDRON	ELEMENT STIFFNESS MATRICES FOR CONSTANT AND LINEAR STRESS - TRIANGULAR, CONSTANT AND LINEAR STRESS T	67-0006
TETRAHEDRON	TETRAHEDRON ELEMENTS IN THE MATRIX FORCE METHOD OF STRUCTURAL ANALYSIS	64-0011
TETRAHEDRON	TETRAHEDRON ELEMENTS WITH LINEARLY VARYING STRAIN FOR THE MATRIX DISPLACEMENT METHOD	65-0006
TETRAHEDRONAL	A SEQUEL TO TECHNICAL NOTE 13 - THE CURVED TETRAHEDRONAL AND TRIANGULAR ELEMENTS TEC AND TRIC FOR TH	69-0048
TETRAHEDRONAL	THE CURVED TETRAHEDRONAL AND TRIANGULAR ELEMENTS TEC AND TRIC FOR THE MATRIX DISPLACEMENT METHOD	69-0047
TETRAPOLAIRE	ETUDE NUMERIQUE DU CHAMP MAGNETIQUE DANS UN ALTERNATEUR TETRAPOLAIRE PAR LA METHODE DES ELEMENTS FIN	74-0413
TEXAS	PAVEMENT CRACKING IN WEST TEXAS DUE TO FREEZE-THAW CYCLING	75-0276
TEXTILE	BALLISTIC IMPACT OF TEXTILE STRUCTURES.	72-0433
THALIA	THALIA - ONE-DIMENSIONAL MAGNETOHYDRODYNAMIC STABILITY PROGRAM USING METHOD OF FINITE-ELEMENTS	75-0458
THALIA	THALIA: A ONE-DIMENSIONAL MAGNETOHYDRODYNAMIC STABILITY PROGRAM USING THE METHOD OF FINITE ELEMENTS	74-1085
THAW	FINITE ELEMENT PERMAFROST THAW SETTLEMENT MODEL	74-0543
THEMIS	THEMIS FINAL REPORT, VOL. IV, RECENT DEVELOPMENTS IN THE NUMERICAL SOLUTION OF CERTAIN PROBLEMS IN N	74-0721
THEMIS	THEMIS FINAL REPORT, VOL. II, NONLINEAR MECHANICS IN SOLIDS AND STRUCTURES	74-0722
THEMIS	THEMIS FINAL REPORT, VOL. 1, BASIC RESEARCH IN METHODS OF APPROXIMATION IN THE NONLINEAR MECHANICS O	74-0723
THEOREM	A BOUND THEOREM IN EIGENVALUES AND ITS PRACTICAL APPLICATIONS	71-0128
THEOREM	A FINITE ELEMENT APPLICATION OF THE HELLINGER-REISSNER VARIATIONAL THEOREM	68-0060
THEOREM	A THEOREM FOR OPTIMUM IDEALIZATIONS IN FINITE-ELEMENT ANALYSIS	72-0791
THEOREM	ELASTICITY EQUATIONS FOR INCOMPRESSIBLE AND NEARLY INCOMPRESSIBLE MATERIALS BY A VARATIONAL THEOREM	65-0019
THEOREM	FINITE ELEMENT FORMULATION FOR PIEZOELECTRIC CONTINUA BY A VARIATIONAL THEOREM	68-0230
THEOREM	ON A VARIATIONAL THEOREM FOR INCOMPRESSIBLE AND NEARLY INCOMPRESSIBLE ORTHOTROPIC ELASTICITY	68-0249
THEOREM	THE SUPERPATCH THEOREM AND OTHER PROPOSITIONS RELATING TO THE PATCH TEST	75-0026
THEOREM	THEOREM FOR OPTIMUM FINITE-ELEMENT IDEALIZATIONS	73-0255
THEOREMS	APPLICATION OF DUAL MINIMUM THEORFMS TO THE FINITE ELEMENT SOLUTION OF POTENTIAL PROBLEMS WITH SPECI	72-0036
THEOREMS	ENERGY THEOREMS AND STRUCTURAL ANALYSIS	54-0001
THEOREMS	ENERGY THEOREMS AND STRUCTURAL ANALYSIS	67-0136

415

THEOREMS	SARD KERNEL THEOREMS ON TRIANGULAR AND RECTANGULAR DOMAINS WITH EXTENSIONS AND APPLICATIONS TO FINIT	72-0361
THEOREMS	THE EXTENSION AND APPLICATION OF SARD KERNEL THEOREMS TO COMPUTE FINITE ELEMENT ERROR BOUNDS	72-0519
THEORETICAL	A THEORETICAL AND EXPERIMENTAL STUDY ON THE FREE VIBRATION OF CURVED SLAB BRIDGES	71-0496
THEORETICAL	A THEORETICAL AND NUMERICAL COMPARISON OF ELASTIC NONLINEAR FINITE ELEMENT METHODS	72-0709
THEORETICAL	A THEORETICAL METHOD FOR EVALUATING STABILITY OF OPENINGS IN ROCK	72-0877
THEORETICAL	AN EXPERIMENTAL AND THEORETICAL STUDY OF HOLLOW CYLINDER TESTS ON SAND	71-0008
THEORETICAL	APPLICATION, TESTING, AND EVALUATION OF DOUBLY CURVED SHELL ELEMENTS FOR DYNAMIC ANALYSIS. PART 1, T	69-0246
THEORETICAL	COMPARISON BETWEEN THEORETICAL AND MEASURED EARTH PRESSURES ACTING ON A LARGE MOTORWAY RETAINING WAL	74-0917
THEORETICAL	COMPARISON OF THEORETICAL AND EXPERIMENTAL SHEAR STRESS IN THE ADHESIVE LAYER OF A LAP JOINT MODEL	74-1176
THEORETICAL	CRACKING OF EARTH AND ROCKFILL DAMS. A THEORETICAL INVESTIGATION BY MEANS OF THE FINITE ELEMENT METH	69-0085
THEORETICAL	CRACKING OF EARTH AND ROCKFILL DAMS: COMPARISON OF OBSERVED AND THEORETICAL TENSILE STRAINS IN THE C	71-0699
THEORETICAL	DEVELOPMENT AND APPLICATIONS OF THEORETICAL METHODS FOR EVALUATING STABILITY OF OPENINGS IN ROCK	73-0791
THEORETICAL	ETUDE THEORIQUE DU COMPORTEMENT D'UNE DALLE DE BETON. (THEORETICAL STUDY OF THE BEHAVIOR OF A CONCRE	75-0273
THEORETICAL	ETUDE THEORIQUE DU COMPORTEMENT DES PIEUX SOUS CHARGE VERTICALE. INTRODUCTION DE LA DILATANCE. (THEO	75-0309
THEORETICAL	EXPERIMENTAL AND THEORETICAL INVESTIGATION OF PAVEMENT DEFLECTIONS	69-0154
THEORETICAL	EXPERIMENTAL AND THEORETICAL INVESTIGATION OF PLANE STRESS FRACTURE OF 2024 - T351 ALUMINUM ALLOY	70-0147
THEORETICAL	EXPERIMENTAL AND THEORETICAL STUDIES OF HULL-DECKHOUSE INTERACTION - 2-LEVEL DECKHOUSE	71-0672
THEORETICAL	EXPERIMENTAL AND THEORETICAL BENDING DEFLECTIONS OF LAMINATED PLATES	72-0144
THEORETICAL	EXPERIMENTAL AND THEORETICAL STUDY ON A 1/10TH SCALE MODEL PCPV	73-0829
THEORETICAL	EXPERIMENTAL AND THEORETICAL FATIGUE-CRACK PROPAGATION LIVES OF VARIOUSLY NOTCHED PLATES	74-0602
THEORETICAL	EXPERIMENTAL AND THEORETICAL INVESTIGATION OF THE BEHAVIOR OF BOX GIRDERS CURVED IN PLAN	75-0659
THEORETICAL	EXPERIMENTAL ELASTIC STRESS ANALYSES OF CYLINDER-TO-CYLINDER SHELL MODELS AND COMPARISONS WITH THEOR	71-0065
THEORETICAL	FINITE ELEMENT ANALYSIS OF ANISOTROPIC PLATES USING Q-19 ELEMENT. PART 1. THEORETICAL DEVELOPMENT	74-0728
THEORETICAL	MATHEMATICAL ASPECTS OF FINITE ELEMENT APPROXIMATIONS IN THEORETICAL MECHANICS	75-0852
THEORETICAL	METHODS FOR PREDICTING THE AERODYNAMIC AND STABILITY AND CONTROL CHARACTERISTICS OF STOL AIRCRAFT. V	73-0719
THEORETICAL	MISSILE BODY INPUT GENERATOR (BING), A NASTRAN PRE-PROCESSOR: THEORETICAL DEVELOPMENT, USER'S MANUAL	75-0717
THEORETICAL	MULTI-CELL BRIDGE DECKS - MODEL TESTS AND THEORETICAL ANALYSIS	72-0381
THEORETICAL	NON-UNIFORM DEFORMATION OF MATERIAL IN AXIALLY SYMMETRIC COMPRESSION CAUSED BY FRICTION. PART 3, THE	72-0923
THEORETICAL	PREDICTIONS IN ENVIRONMENTAL HYDRODYNAMICS USING FINITE ELEMENT METHOD 1, THEORETICAL DEVELOPMENT	75-0003
THEORETICAL	SP-222 NASTRAN THEORETICAL MANUAL	70-0387
THEORETICAL	THE NASTRAN THEORETICAL MANUAL	70-0298
THEORETICAL	THEORETICAL ANALYSIS OF THICK DOME STRUCTURES AND ARBITRARY SOLIDS OF REVOLUTION SUBJECTED TO UNSYMM	70-0037
THEORETICAL	THEORETICAL AND EXPERIMENTAL INVESTIGATION ON A STEAM GENERATOR TUBE PLATE	69-0424
THEORETICAL	THEORETICAL AND EXPERIMENTAL ANALYSIS OF STIFFENED PANEL UNDER DYNAMIC CONDITIONS	70-0130
THEORETICAL	THEORETICAL AND EXPERIMENTAL CONSIDERATIONS ON FLOW OF VISCOUS PAST PLATE AT SMALL AND AVERA	70-0144
THEORETICAL	THEORETICAL AND EXPERIMENTAL VIBRATION ANALYSIS OF AN OBLIQUE SPACE FRAME	70-0460
THEORETICAL	THEORETICAL AND EXPERIMENTAL ANALYSIS OF ELASTOPLASTIC DEFORMATION IN A UNIAXIALLY LOADED STRIP OF 2	73-0720
THEORETICAL	THEORETICAL AND EXPERIMENTAL STRESS ANALYSES OF ORNL THIN-SHELL CYLINDER-TO-CYLINDER MODEL 3	75-0690
THEORETICAL	THEORETICAL AND EXPERIMENTAL STRESS ANALYSES OF ORNL THIN-SHELL CYLINDER-TO-CYLINDER MODEL 4	75-0695
THEORETICAL	THEORETICAL AND EXPERIMENTAL STUDIES OF TRANSIENT ELASTIC-PLASTIC LARGE DEFLECTIONS OF GEOMETRICALLY	75-0830
THEORETICAL	THEORETICAL AND EXPERIMENTAL STUDIES OF TRANSIENT ELASTIC-PLASTIC LARGE DEFLECTIONS OF GEOMETRICALLY	75-0831
THEORETICAL	THEORETICAL AND NUMERICAL COMPARISON OF ELASTIC NONLINEAR FINITE ELEMENT METHODS.	72-0452
THEORETICAL	THEORETICAL AND NUMERICAL CONVERGENCE RATES FOR SOME CONFORMING FINITE ELEMENTS	74-0941
THEORETICAL	THEORETICAL BASIS FOR CBL/NONSAP, A NONLINEAR STRUCTURAL ANALYSIS PROGRAM	74-0740
THEORETICAL	THEORETICAL BASIS FOR HYBRID FINITE ELEMENTS IN DYNAMIC PROBLEMS	75-0489
THEORETICAL	THEORETICAL BASIS OF THE BOREHOLE DEEPENING METHOD OF ABSOLUTE STRESS MEASUREMENT	70-0047
THEORETICAL	THEORETICAL COMPUTATIONS ON RIDGE ACOUSTIC SURFACE WAVES USING THE FINITE-ELEMENT METHOD	71-0011
THEORETICAL	THEORETICAL FORMULATION OF FINITE ELEMENT METHODS IN LINEAR ELASTIC ANALYSIS OF GENERAL SHELLS	72-0049
THEORETICAL	THEORETICAL FOUNDATIONS ON THE FINITE ELEMENT METHOD	68-0055
THEORETICAL	THEORETICAL FOUNDATIONS OF THE FINITE ELEMENT METHOD (DISCUSSION)	70-0074
THEORETICAL	THEORETICAL FOUNDATIONS OF THE FINITE ELEMENT METHOD	70-0453
THEORETICAL	THEORETICAL INVESTIGATION OF ROCK AND SUPPORT INTERACTION - DEVELOP MORE RATIONAL DESIGN METHODS AND	74-0782
THEORETICAL	THEORETICAL POST YIELDING BEHAVIOUR OF COMPOSITE LAMINATES - 1. INELASTIC MICROMECHANICS	73-0168
THEORETICAL	THEORETICAL POST YIELDING BEHAVIOUR OF COMPOSITE LAMINATES -2.	73-0408
THEORETICAL	THEORETICAL PREDICTION OF DISC BRAKE TEMPERATURES AND A COMPARISON WITH EXPERIMENTAL DATA	72-0306
THEORETICAL	THEORETICAL STUDY OF THE INFLUENCE OF INCLUSIONS UPON THE INITIATION AND GROWTH OF FATIGUE CRACKS IN	71-0394
THEORETICAL	THEORETICAL STUDY OF THE EFFECT OF THE INTERFACE ON COMPOSITE TOUGHNESS	73-0386
THEORETICAL	THEORETICAL STUDY OF THE SEQUENCE OF FAILURE IN PRECAST PANEL SHEAR WALLS	73-0490
THEORETICAL	THEORETICAL TRANSIENT BEHAVIOUR OF SATURATED AND UNSATURATED SOILS UNDER LOAD AND CHANGING MOISTURE	73-0901
THEORETICAL	TRANSVERSE VIBRATIONS OF SHIP'S PROPULSION SYSTEMS. PART 1: THEORETICAL ANALYSIS	74-1323
THEORETICAL	TRANSVERSE VIBRATIONS OF SHIP'S PROPULSION SYSTEMS - 1 THEORETICAL ANALYSIS	75-0291
THEORETICAL	ULTIMATE LOAD BEHAVIOUR OF PCRV TOP CLOSURES. THEORETICAL AND EXPERIMENTAL INVESTIGATIONS	70-0835
THEORETISCHE	THEORETICAL AND EXPERIMENTAL CONSIDERATIONS ON FLOW OF VISCOUS LIQUIDS PAST PLATE AT SMALL AND AVERA	70-0144
THEORIE	COUPLAGE ENTRE L ECOULEMENT AUTOUR D UN ARRIERE-CORPS ET LE JET PROPULSIF EN THEORIE DE FLUIDE PARFA	74-1137
THEORIE	THEORIE UND EXPERIMENT ZU EINEM NEUEN INJEKTIONSVERFAHREN. (THEORY AND EXPERIMENT OF A NEW GROUTING	74-0991
THEORIE	THEORIE, PROGRAMMENTWICKLUNG UND ERFAHRUNG ANVORGESPANNTEN NETZWERKKONSTRUKTIONEN	72-0640
THEORIE	ZUR BERECHNUNG EINFACH-SYMMETRISCHER I-TRAEGER NACH DER THEORIE - 2. ORDNUNG UNTER BERUECKSICHTIGUNG	75-0510
THEORIE	ZUR BERECHNUNG VON PLATTEN NACH DER THEORIE ZWEITER. ORDNUNG MIT HILFE EINES HYBRIDEN DEFORMATIONS M	73-0558
THEORIES	COMBINED ANALYTICAL - EXPERIMENTAL FRACTURE STUDY OF THE TWO LEADING THEORIES OF ELASTIC-PLASTIC FRA	73-0951
THEORIES	DETERMINATION OF THE APPLICABILITY OF SOME THEORIES ABOUT STRENGTH AND STIFFNESS OF THIN-WALLED BEAM	69-0318
THEORIES	DISCRETE APPROXIMATIONS RELATED TO NONLINEAR THEORIES OF SOLIDS	71-0259
THEORIES	UNIFIED APPROACH TO THREE FINITE ELEMENT THEORIES FOR GEOMETRIC NONLINEARITY	74-0599
THEORIQUE	ETUDE THEORIQUE DU COMPORTEMENT D'UNE DALLE DE BETON. (THEORETICAL STUDY OF THE BEHAVIOR OF A CONCRE	75-0273
THEORIQUE	ETUDE THEORIQUE DU COMPORTEMENT DES PIEUX SOUS CHARGE VERTICALE. INTRODUCTION DE LA DILATANCE. (THEO	75-0309
THERMAL	A DYNAMIC ELASTOPLASTIC ANALYSIS OF THIN SHELLS OF REVOLUTION UNDER ASYMMETRIC MECHANICAL AND THERMA	72-0068
THERMAL	A FINITE ELEMENT FOR THERMAL STRESS ANALYSIS OF SHELLS OF REVOLUTION	73-0798
THERMAL	A FINITE ELEMENT METHOD FOR THE DETERMINATION OF NON-STATIONARY TEMPERATURE DISTRIBUTIONS AND THERMA	66-0014
THERMAL	A METHOD FOR THE THERMAL ANALYSIS OF SPACECRAFT INCLUDING ALL MULTIPLE REFLECTIONS AND SHADING AMONG	70-0424
THERMAL	A THERMAL, THERMOELASTIC, AND WEAR ANALYSIS OF HIGH-ENERGY DISK BRAKES	74-0839
THERMAL	AN AXISYMMETRIC FINITE ELEMENT ANALYSIS OF THE MECHANICAL AND THERMAL STRESSES IN BRAKE DRUMS	74-0311
THERMAL	ANALYSIS OF ELASTO-PLASTIC THERMAL STRESSES BY THE FINITE ELEMENT METHOD	70-0142
THERMAL	ANALYSIS OF LOADING AND THERMAL EFFECTS ON FUELLED GRAPHITE BRICKS FOR A HIGH-TEMPERATURE GAS COOLED	71-0085
THERMAL	ANALYSIS OF RING-STIFFENED SHELLS OF REVOLUTION UNDER COMBINED THERMAL AND MECHANICAL LOADING	71-0342
THERMAL	ANALYSIS OF THERMAL ELASTIC PLASTIC STRESS AND STRAIN DURING WELDING BY FINITE ELEMENT METHOD	71-0382
THERMAL	ANALYSIS OF THERMAL DEFORMATION OF MACHINE TOOL BY THE FINITE ELEMENT METHOD.	72-0432
THERMAL	ANALYSIS OF THERMAL STRESSES IN CORED CERAMIC BRICKS USED IN HYPERSONIC WIND TUNNEL HEATERS	72-0655
THERMAL	ANALYSIS OF THERMAL STRESSES AND METAL MOVEMENT DURING WELDING 1. ANALYTICAL STUDY	74-1026
THERMAL	APPLICATION OF FINITE ELEMENT METHODS FOR THE ANALYSIS OF THERMAL CREEP, IRRADIATION INDUCED CREEP,	71-0242
THERMAL	APPLICATION OF LAPLACE TRANSFORM AND FINITE ELEMENT METHOD TO SOLUTION OF NONSTATIONARY THERMAL COND	74-0171
THERMAL	CALCULATION OF THERMAL STRESSES IN A CYLINDER-CYLINDER INTERSECTION BY MEANS OF FINITE ELEMENTS	73-0571
THERMAL	CALCULATION OF THERMAL STRESSES IN TUBULAR CERAMIC COMPONENTS	73-0670
THERMAL	CHISLENNOE RESHENIE PLOSKIKH ZADACH TEPLOPROVODNOSTI DLYA OBLASTEI SLOZHNOI FORMY (NUMERICAL SOLUTIO	74-0659
THERMAL	COMOC THERMAL ANALYSIS VARIANT, USER'S MANUAL	72-0701
THERMAL	COMPARISON OF SOME FORMULATIONS IN THERMAL ANALYSIS OF STRUCTURES	72-0888
THERMAL	CORRELATION BETWEEN PRESSURE AND THERMAL TRANSIENT STRESS LEVELS DERIVED BY THEORY AND EXPERIMENT ON	73-0469
THERMAL	COUPLE - A PSEUDO FINITE ELEMENT CONDUCTION CODE FOR THE TRANSIENT THERMAL RESPONSE OF AXISYMMETRIC O	72-0778
THERMAL	DISCONTINUOUS FINITE ELEMENTS IN THERMAL ANALYSIS	70-0256
THERMAL	DYNAMIC ELASTOPLASTIC RESPONSE OF GEOMETRICALLY NONLINEAR ARBITRARY SHELLS OF REVOLUTION UNDER IMPUL	70-0098
THERMAL	ELASTIC-PLASTIC AND THERMAL STRESS ANALYSIS	68-0169
THERMAL	ELASTIC-PLASTIC-CREEP ANALYSIS OF THERMAL RATCHETTING IN STRAIGHT PIPE AND COMPARISONS WITH TEST RES	73-0393
THERMAL	FEATS, A COMPUTER PROGRAM FOR THE FINITE ELEMENT THERMAL STRESS ANALYSIS OF PLANE OR AXISYMMETRIC SO	68-0152
THERMAL	FINITE ELEMENT LAKE CIRCULATION AND THERMAL ANALYSIS	74-0043

THERMAL	FINITE ELEMENT METHOD FOR DETERMINATION OF THERMAL STRESSES INTERACTIVE GRAPHICS FOR FINITE ELEMENT	74-0155
THERMAL	FINITE ELEMENT SOLUTION FOR AXISYMMETRIC TRANSIENT THERMAL STRESSES	72-0823
THERMAL	FINITE ELEMENT TECHNIQUE APPLIED TO HEAT CONDUCTION IN SOLIDS WITH TEMPERATURE DEPENDENT THERMAL CON	69-0004
THERMAL	FINITE ELEMENTE ZUR BERECHNUNG DER WAERMEUEBERTRAGUNG FUER EBENE UND AXISYMMETRISCHE PROBLEME (FINIT	73-0449
THERMAL	FINITE-ELEMENT METHOD FOR THE DETERMINATION OF THERMAL STRESSES IN ANISOTROPIC SOLIDS OF REVOLUTION	74-0262
THERMAL	FIRES-T. A COMPUTER PROGRAM FOR THE FIRE RESPONSE OF STRUCTURES - THERMAL	74-0850
THERMAL	GENERAL VARIATIONAL ANALYSIS OF HYDRODYNAMIC, THERMAL AND DIFFUSIONAL BOUNDARY LAYERS	70-0191
THERMAL	IMPROVED THERMAL STRESS DETERMINATION BY FINITE ELEMENT METHODS	74-0247
THERMAL	INELASTIC ANALYSIS OF A STRAIGHT TUBE UNDER COMBINED BENDING, PRESSURE AND THERMAL LOADS	75-0218
THERMAL	INELASTIC ANALYSIS OF A STRAIGHT TUBE UNDER COMBINED BENDING, PRESSURE AND THERMAL LOADS	75-0513
THERMAL	INFLUENCE OF TEMPERATURE DEPENDENT PROPERTIES ON THERMAL ROCK FRAGMENTATION	75-0222
THERMAL	MEASUREMENT OF THERMAL EXPANSION IN SOLID PROPELLANTS	73-0916
THERMAL	METODY RASCHETNOGO OPREDELENIYA TEMPERATURNYKH NAPRYAZHENII V KRYSHKAKH TSILINDROV DVIGATELEI VNUTRE	74-0179
THERMAL	NASTRAN STRUCTURAL/THERMAL ANALYZER	72-0152
THERMAL	NON-LINEAR THERMAL STRESS ANALYSIS FOR NUCLEAR POWER PLANT BY FINITE ELEMENT METHOD	71-0181
THERMAL	PRESSURE VESSEL WALL THICKNESS LIMITS BASED ON THERMAL STRESS RACHETTING AND CREEP FATIGUE INTERACTI	74-0968
THERMAL	RE-ENTRY THERMAL ANALYSIS OF VARIABLE THICKNESS SPHERICAL VEHICLES.	72-0468
THERMAL	RETSCP: A COMPUTER PROGRAM FOR ANALYSIS OF ROCKET ENGINE THERMAL STRAINS WITH CYCLIC PLASTICITY	74-0692
THERMAL	ROCK FRAGMENTATION BY SUBSURFACE THERMAL INCLUSIONS - A FINITE ELEMENT STUDY	73-1159
THERMAL	ROHM AND HAAS THERMOSTRUCTURAL ANALYSIS SYSTEM, THERMAL STRUCTURAL ANALYSIS PROGRAMS - A SURVEY AND	72-0007
THERMAL	SIGNIFICANT APPLICATIONS OF FINITE ELEMENTS AND FINITE DIFFERENCES METHODS TO SOLVE PROBLEMS OF THER	73-0864
THERMAL	SOLVING A NONLINEAR PROBLEM OF NONSTEADY THERMAL CONDUCTIVITY IN SOLID BODIES OF ARBITRARY FORM	68-0019
THERMAL	SPANNUNGEN IN MEMBRANWAENDEN DURCH WAERMEBEANSPRUCHUNG UND MECHANISCHE BELASTUNG. (STRESSES IN MEMBR	75-0066
THERMAL	STACUSS 1: A DISCRETE-ELEMENT PROGRAM FOR THE STATIC ANALYSIS OF SINGLE-LAYER CURVED STIFFENED SHELL	69-0160
THERMAL	STATIC, VIBRATION, AND THERMAL STRESS ANALYSES OF LAMINATED PLATE AND SHELLS BY THE HYBRID-STRESS FI	72-0602
THERMAL	STRESS ANALYSIS BY THE FINITE ELEMENT METHOD - THERMAL EFFECTS	67-0022
THERMAL	STRESS ANALYSIS OF A CERAMIC BLADE DESIGN SUBJECTED TO CENTRIFUGAL AND THERMAL LOADS	73-0790
THERMAL	STRUCTURAL CAPACITY OF REINFORCED CONCRETE COLUMNS SUBJECTED TO FIRE INDUCED THERMAL GRADIENTS	73-0913
THERMAL	THE CALCULATION OF THREE-DIMENSIONAL TEMPERATURE DISTRIBUTIONS AND THERMAL STRESSES USING FINITE ELE	71-0187
THERMAL	THE COMPUTATION OF TEMPERATURE DISTRIBUTIONS AND THERMAL STRESSES USING FINITE ELEMENT TECHNIQUES	71-0092
THERMAL	THE HEAT CONDUCTION AND THERMAL STRESS ANALYSIS BY THE FINITE ELEMENT METHOD	68-0073
THERMAL	THE INELASTIC BEHAVIOUR OF STRUCTURES SUBJECTED TO CYCLIC, THERMAL AND MECHANICAL STRESSING CONDITIO	60-0004
THERMAL	THERMAL ANALYSIS OF M552 EXPERIMENT FOR MATERIALS PROCESSING IN SPACE	74-0704
THERMAL	THERMAL ANALYSIS OF M551 EXPERIMENT FOR MATERIALS PROCESSING IN SPACE	74-0725
THERMAL	THERMAL AND MECHANICAL ANALYSIS OF WELDED STRUCTURES	73-0819
THERMAL	THERMAL AND STRESS ANALYSIS BY NASTRAN	72-0153
THERMAL	THERMAL AND STRESS ANALYSIS PROGRAM - HYBOS	72-0159
THERMAL	THERMAL AND STRESS ANALYSIS ON PRISMATIC NUCLEAR FUEL ELEMENTS	75-0120
THERMAL	THERMAL AND STRESS ANALYSIS OF A FUEL ROD	75-0713
THERMAL	THERMAL DEFORMATION PREDICTION IN ORBITING OBSERVATORY STRUCTURES	69-0292
THERMAL	THERMAL DEFORMATION VECTOR FOR A BILINEAR TEMPERATURE DISTRIBUTION IN AN ANISOTROPIC QUADRILATERAL M	75-0670
THERMAL	THERMAL ELASTIC PLASTIC STRESS ANALYSIS BY THE FINITE ELEMENT METHOD	69-0391
THERMAL	THERMAL LOADING OF THIN-SHELL CONCRETE COOLING TOWERS	74-0667
THERMAL	THERMAL MECHANICAL ANALYSIS PROCEDURE FOR A SEMICIRCULAR PLATE WITH CLAMPED STRAIGHT EDGE	75-0651
THERMAL	THERMAL NONLINEAR BEHAVIOUR OF STRUCTURES	72-0317
THERMAL	THERMAL SLEEVE DESIGN FOR PRESSURE VESSELS	74-1129
THERMAL	THERMAL STRESS ANALYSIS BY THE FINITE ELEMENT METHOD	70-0124
THERMAL	THERMAL STRESS ANALYSIS OF REENTRY VEHICLE NOSETIPS AT ANGLE OF ATTACK	72-0456
THERMAL	THERMAL STRESS ANALYSIS OF RE-ENTRY VEHICLE NOSETIPS AT ANGLE OF ATTACK	72-0852
THERMAL	THERMAL STRESS ANALYSIS OF PLATES AND SHALLOW SHELLS BY HYBRID FINITE-ELEMENT METHOD	74-0485
THERMAL	THERMAL STRESS AND TEMPERATURE FIELDS - A FINITE-ELEMENT APPROACH	71-0782
THERMAL	THERMAL STRESS DEVELOPED IN S PLATE, STRAIGHT PLATE, AND DEEP DISH WHEELS	73-0172
THERMAL	THERMAL STRESSES IN OPEN CYLINDRICAL SHELLS OF ARBITRARY BOUNDARY CONDITIONS BY A FINITE STRIP METHO	72-0371
THERMAL	THERMAL STRESSES IN FAST REACTOR BOILER TUBES AT THE SODIUM/GAS INTERFACE	73-0820
THERMAL	THERMAL STRESSES IN THIN CYLINDRICAL SHELLS	73-0822
THERMAL	THERMAL STRESSES IN A NOZZLE-DRUM INTERSECTION	73-0824
THERMAL	THERMAL STRESSES IN STEEL-CONCRETE COMPOSITE BRIDGES	75-0154
THERMAL	THERMAL, THERMOELASTIC, AND WEAR SIMULATION OF A HIGH-ENERGY SLIDING CONTACT PROBLEM	73-0430
THERMAL	THREE-DIMENSIONAL ELASTIC ANALYSIS OF SEMI-ELLIPTICAL SURFACE CRACKS SUBJECT TO THERMAL SHOCK	75-0256
THERMAL	THREE-DIMENSIONAL THERMAL AND STRESS ANALYSIS OF A PIPING TEE	75-0703
THERMAL	TRANSIENT THERMAL STRESS TECHNIQUES IN NUCLEAR REACTORS. APPLIED MECHANICS ASPECTS OF NUCLEAR EFFECT	71-0474
THERMAL	TRANSIENT THERMAL STRESS ANALYSIS OF COMPOSITE STRUCTURES INCLUDING CONTINUOUSLY VARYING PROPERTIES	75-0681
THERMAL	TSAAS: FINITE ELEMENT THERMAL AND STRESS ANALYSIS OF AXISYMMETRIC SOLIDS WITH ORTHOTROPIC TEMPERATUR	74-0886
THERMAL	USE OF SINGULARITY PROGRAMMING IN FINITE ELEMENT CALCULATIONS OF ELASTIC STRESS INTENSITY FACTORS. P	75-0252
THERMAL	USING THE NASTRAN THERMAL ANALYZER TO SIMULATE A FLIGHT SCIENTIFIC INSTRUMENT PACKAGE	74-1011
THERMAL-MECHANI	THERMAL-MECHANICAL FRAGMENTATION OF HARD ROCK - FIELD TESTS	74-0998
THERMAL-STRESS	THERMAL-STRESS CONCENTRATION CAUSED BY STRUCTURAL DISCONTINUITIES	69-0308
THERMALLY	ANALYSIS OF THERMALLY INDUCED STRUCTURAL VIBRATIONS OF FINITE ELEMENT TECHNIQUES	68-0272
THERMALLY	APPLICATION OF FRACTURE MECHANICS TO THERMALLY STRESSED CYLINDERS USING A FINITE ELEMENT METHOD	69-0089
THERMALLY	CALCULATION OF THERMALLY INDUCED STRAINS IN MACHINE TOOL ELEMENTS	72-0464
THERMALLY	COMPUTER AIDED DESIGN OF THERMALLY LOADED AXISYMMETRIC DIESEL ENGINE COMPONENTS	74-0596
THERMALLY	COMPUTER CALCULATIONS OF STRESSES IN AXISYMMETRIC THERMALLY LOADED COMPONENTS	68-0069
THERMALLY	FINITE ELEMENT SIMULATION OF THERMALLY INDUCED FLOW FIELDS	74-1235
THERMISCH	NUMERISCHE BERECHNUNG INSTATIONAERER TEMPERATURFELDER UND THERMISCH BEDINGTER VERFORMUNGEN VON WERKZ	75-0847
THERMO	NONLINEAR THERMO ELASTIC PLASTIC AND CREEP ANALYSIS BY FINITE ELEMENT METHOD	74-0092
THERMO-ELASTIC	A THERMO-ELASTIC ANALYSIS OF THE PARALLEL SURFACE THRUST BEARING	74-1069
THERMO-ELASTIC	THERMO-ELASTIC ANALYSIS OF RECTANGULAR PLATES IN PLANE STRESS BY THE FINITE-ELEMENT DISPLACEMENT MET	67-0087
THERMO-ELASTIC	THERMO-ELASTIC ANALYSIS OF THE PARALLEL SURFACE THRUST BEARING	74-1172
THERMO-ELASTIC-	THERMO-ELASTIC-PLASTIC CYCLIC ANALYSIS BY FINITE ELEMENT METHOD	73-1003
THERMO-ELASTICI	FINITE ELEMENT FORMULATION OF GENERAL THERMO-ELASTICITY PROBLEMS	71-0001
THERMO-ELASTO-H	ON THERMO-ELASTO-HYDRODYNAMIC FLUID FILM BEARINGS	75-0532
THERMO-ELASTO-P	THE THERMO-ELASTO-PLASTIC ANALYSIS OF A RE-ENTRY BODY TYPE COMPOSITE SOLID STRUCTURE USING A NEW FIN	66-0004
THERMO-ELASTOPL	FINITE-ELEMENT METHOD IN CYCLIC THERMO-ELASTOPLASTICITY	73-1111
THERMO-MECHANIC	A NUMERICAL THERMO-MECHANICAL MODEL FOR THE WELDING AND SUBSEQUENT LOADING OF A FABRICATED STRUCTURE	72-0847
THERMO-MECHANIC	NUMERICAL, THERMO-MECHANICAL MODEL FOR THE WELDING AND SUBSEQUENT LOADING OF A FABRICATED STRUCTURE	73-0418
THERMO-MECHANIS	BERECHNUNG VON SPANNUNGSKONZENTRATIONEN AN TRAGWERKEN UNTER THERMO-MECHANISCHER BELASTUNG	73-0686
THERMO-PLASTIC	THERMO-PLASTIC CONSTITUTIVE RELATIONS AND A VARIATIONAL SOLUTION TECHNIQUE USING FINITE ELEMENTS	70-0486
THERMO-STRUCTUR	THERMO-STRUCTURAL CAPABILITY OF ANSYS	72-0155
THERMO-VISCOELA	AN APPROXIMATE METHOD FOR THERMO-VISCOELASTIC STRESS ANALYSIS	66-0100
THERMO-VISCOELA	FINITE ELEMENT ANALYSIS OF THERMO-VISCOELASTIC PROBLEMS	71-0526
THERMO-VISCOELA	FINITE ELEMENT ANALYSIS OF THERMO-VISCOELASTIC PROBLEMS	71-0781
THERMOCHEMICAL	BASELINE SOLUTIONS FOR THE SMOOTH WALL THERMOCHEMICAL ABLATION RESPONSE OF GRAPHITE AND CARBON PHENO	71-0728
THERMODYNAMICS	A FINITE ELEMENT COMPUTATIONAL THEORY FOR THE MECHANICS AND THERMODYNAMICS OF A VISCOUS, COMPRESSIBL	71-0787
THERMODYNAMICS	COMPUTER SYSTEM FOR THE ANALYSIS OF LARGE SCALE PROBLEMS IN SOLID, FLUID AND THERMODYNAMICS	73-0554
THERMODYNAMICS	FINITE ELEMENT FORMULATION OF FINITE DEFORMATIONS AND IRREVERSIBLE THERMODYNAMICS OF NONLINEAR CONTI	69-0207
THERMODYNAMICS	FINITE ELEMENT FORMULATION OF PROBLEMS OF FINITE DEFORMATION AND IRREVERSIBLE THERMODYNAMICS OF NONL	71-0190
THERMODYNAMICS	FINITE ELEMENT THEORY FOR THE MECHANICS AND THERMODYNAMICS OF A VISCOUS, COMPRESSIBLE MULTI-SPECIE F	71-0005
THERMODYNAMICS	NONLINEAR THERMOELASTICITY, IRREVERSIBLE THERMODYNAMICS AND ELASTIC INSTABILITY	73-0762
THERMOELASTIC	A DERIVATION OF PERTURBED EQUATIONS OF MOTION AND HEAT CONDUCTION OF A THERMOELASTIC SOLID	72-0735
THERMOELASTIC	A METHOD FOR THE TRANSIENT NONLINEAR COUPLED THERMOELASTIC ANALYSIS OF AXISYMMETRIC BODIES USING FIN	74-1140
THERMOELASTIC	A THERMAL, THERMOELASTIC, AND WEAR ANALYSIS OF HIGH-ENERGY DISK BRAKES	74-0839
THERMOELASTIC	ANALYSIS OF NONLINEAR THERMOELASTIC AND THERMOPLASTIC BEHAVIOUR OF SOLIDS OF REVOLUTION BY THE FINIT	71-0192

417

THERMOELASTIC	EXCITATION OF PLANE THERMOELASTIC WAVES IN A LAYER WITH A FINITE RATE OF HEAT PROPAGATION	74-0803
THERMOELASTIC	THE TRIDIMENSIONAL THERMOELASTIC COMPUTER CODE TITUS	71-0156
THERMOELASTIC	THERMAL, THERMOELASTIC, AND WEAR SIMULATION OF A HIGH-ENERGY SLIDING CONTACT PROBLEM	73-0430
THERMOELASTICIT	ANALYSIS OF GENERAL COUPLED THERMOELASTICITY PROBLEMS BY THE FINITE ELEMENT METHOD	68-0129
THERMOELASTICIT	APPROXIMATE SOLUTIONS IN LINEAR COUPLED THERMOELASTICITY	68-0220
THERMOELASTICIT	FINITE ELEMENT ANALYSIS OF NONLINEAR PROBLEMS IN THE DYNAMICAL THEORY OF COUPLED THERMOELASTICITY	69-0202
THERMOELASTICIT	NONLINEAR THERMOELASTICITY, IRREVERSIBLE THERMODYNAMICS AND ELASTIC INSTABILITY	73-0762
THERMOELASTICIT	ON THE NUMERICAL SOLUTION OF A CLASS OF NONLINEAR PROBLEMS IN DYNAMIC COUPLED THERMOELASTICITY	71-0441
THERMOELASTO-DY	MEAN-SQUARE ERRORS OF FINITE ELEMENT APPROXIMATIONS ON LINEAR ELASTOSTATICS, ELASTODYNAMICS, AND THE	74-0107
THERMOELASTOPLA	FINITE ELEMENT THERMOELASTOPLASTIC ANALYSIS	72-0139
THERMOELASTOPLA	INCREMENTAL THEORY OF THREE DIMENSIONAL TRANSIENT THERMOELASTOPLASTICITY - FORMULATION AND SOLUTION	73-0555
THERMOELASTOPLA	THREE-DIMENSIONAL FINITE ELEMENT METHOD OF THERMOELASTOPLASTICITY WITH CREEP EFFECT	73-0837
THERMOELASTOVIS	A NEW APPROACH TO THE FINITE-ELEMENT FORMULATION AND SOLUTION OF A CLASS OF PROBLEMS IN COUPLED THER	72-0732
THERMOELASTOVIS	A NEW APPROACH TO THE FINITE-ELEMENT FORMULATION AND SOLUTION OF A CLASS OF PROBLEMS IN COUPLED THER	73-0600
THERMOELASTOVIS	NEW APPROACH TO FINITE ELEMENT FORMULATION AND SOLUTION OF A CLASS OF PROBLEMS IN COUPLED THERMOELAS	73-0013
THERMOHYDRODYNA	APPLICATION OF FINITE ELEMENT METHODS TO THERMOHYDRODYNAMIC LUBRICATION	73-0074
THERMOMECHANICA	CALCULATION OF STRESS CONCENTRATIONS ON SUPPORT STRUCTURES UNDER THERMOMECHANICAL LOADING (BERECHNUN	73-0796
THERMOMECHANICA	FINITE ELEMENT ANALYSIS OF THERMOMECHANICAL PROBLEMS	71-0030
THERMOMECHANICA	FORMULATION OF GENERAL DISCRETE MODELS OF THERMOMECHANICAL BEHAVIOR OF MATERIALS WITH MEMORY	69-0381
THERMOMECHANICA	INCREMENTAL THERMOMECHANICAL THEORY OF VISCOELASTOPLASTIC SOLIDS AND SOLUTION BY FINITE ELEMENTS	73-0861
THERMOMECHANICA	PROPAGATION AND OPENING OF A THROUGH CRACK IN PIPE SUBJECT TO COMBINED CYCLIC THERMOMECHANICAL LOADI	75-0516
THERMOMECHANICA	THERMOMECHANICAL ANALYSIS OF VISCOELASTIC SOLIDS	70-0240
THERMOMECHANICA	THERMOMECHANICAL ANALYSIS OF STRUCTURES	71-0618
THERMOMECHANICA	THERMOMECHANICAL ANALYSIS OF STRUCTURES	73-0368
THERMOMECHANICA	THERMOMECHANICAL ANALYSIS OF THE WELDING PROCESS USING THE FINITE ELEMENT METHOD	75-0213
THERMOMECHANICA	THERMOMECHANICAL ANALYSIS OF THE WELDING PROCESS USING THE FINITE ELEMENT METHOD	75-0518
THERMOMECHANICA	THERMOMECHANICAL ANALYSIS OF THE WELDING PROCESS USING THE FINITE ELEMENT METHOD	75-0776
THERMOMECHANICA	THERMOMECHANICAL CREEP OF AGEING CONCRETE - A UNIFIED APPROACH	74-0607
THERMOMECHANICA	THERMOMECHANICAL DISSIPATION ANALYSIS OF THERMOVISCOELASTIC SOLIDS BY FINITE ELEMENTS	72-0403
THERMOMECHANICA	THERMOMECHANICAL RESPONSE OF INELASTIC FIBRE COMPOSITES	75-0235
THERMOMECHANICS	MECHANICS AND THERMOMECHANICS	73-0929
THERMOMECHANISC	CALCULATION OF STRESS CONCENTRATIONS ON SUPPORT STRUCTURES UNDER THERMOMECHANICAL LOADING (BERECHNUN	73-0796
THERMONUCLEAR	THREE-DIMENSIONAL MECHANICAL STRESSES IN TOROIDAL MAGNETS FOR CONTROLLED THERMONUCLEAR REACTORS	73-0669
THERMOPLASTIC	ANALYSIS OF NONLINEAR THERMOELASTIC AND THERMOPLASTIC BEHAVIOUR OF SOLIDS OF REVOLUTION BY THE FINIT	71-0192
THERMOPLASTICIT	FINITE ELEMENT CYCLIC THERMOPLASTICITY ANALYSIS BY THE METHOD OF SUBVOLUMES	75-0787
THERMOPLASTICIT	SOLUTION OF A THREE DIMENSIONAL AXISYMMETRICAL PROBLEM OF THERMOPLASTICITY AS APPLIED TO THICK TURBI	74-0384
THERMOSTRUCTURA	PROGRAMMER'S MANUAL ADDITIONS AND DEMONSTRATION PROBLEMS FOR A THERMOSTRUCTURAL CAPABILITY FOR NASTR	75-0734
THERMOSTRUCTURA	ROHM AND HAAS THERMOSTRUCTURAL ANALYSIS SYSTEM, THERMAL STRUCTURAL ANALYSIS PROGRAMS - A SURVEY AND	72-0007
THERMOSTRUCTURA	THERMOSTRUCTURAL DESIGN CONSIDERATIONS TO ACHIEVE THE LARGE SPACE TELESCOPE LINE-OF-SIGHT REQUIREMEN	75-0839
THERMOVISCO	FINITE ELEMENT FORMULATION FOR LINEAR THERMOVISCO ELASTIC MATERIALS	69-0351
THERMOVISCOELAS	STRESS WAVE ANALYSIS IN LAYERED THERMOVISCOELASTIC MATERIALS BY THE EXTENDED RITZ METHOD, VOLUME II	68-0125
THERMOVISCOELAS	THERMOMECHANICAL DISSIPATION ANALYSIS OF THERMOVISCOELASTIC SOLIDS BY FINITE ELEMENTS	72-0403
THERMOVISCOELAS	THERMOVISCOELASTIC AXISYMMETRIC STRESS ANALYSIS BY FINITE ELEMENT AND FINITE DIFFERENCE COMPUTER TEC	71-0105
THERMOVISCOELAS	ANALYSIS OF NONLINEAR DYNAMIC, COUPLED THERMOVISCOELASTICITY PROBLEMS BY THE FINITE ELEMENT METHOD	71-0527
THERMOVISCOELAS	ANALYSIS OF NONLINEAR THERMOVISCOELASTICITY PROBLEMS BY THE FINITE ELEMENT METHOD	71-0528
THERMOVISCOELAS	FINITE ELEMENT APPLICATIONS IN LINEAR AND NONLINEAR THERMOVISCOELASTICITY	67-0065
THERMOVISCOELAS	FINITE ELEMENT APPROXIMATIONS IN NONLINEAR THERMOVISCOELASTICITY	71-0459
THERMOVISCOELAS	SOME METHODS OF SOLVING PROBLEMS OF NON-LINEAR THERMOVISCOELASTICITY	74-1207
THICK	A COMPUTER PROGRAM FOR THE CALCULATION OF TORSION FACTORS AND SHEAR STRESS FACTORS FOR BOX SECTIONS	69-0485
THICK	A FINITE ELEMENT ANALYSIS OF ' MODERATELY THICK ' RECTANGULAR PLATES IN BENDING	68-0146
THICK	A NOTE ON THE FINITE DEFORMATIONS OF A THICK ELASTIC SHELL OF REVOLUTION	71-0193
THICK	A STUDY OF ELASTIC-PLASTIC ANALYSIS BY THE ASSUMED-STRESS HYBRID FINITE-ELEMENT MODEL, WITH APPLICAT	74-0718
THICK	A UNIFIED TREATMENT OF THICK AND THIN FILM ELASTOHYDRODYNAMIC PROBLEMS USING HIGHER ORDER ELEMENT PR	75-0890
THICK	AN ASSUMED STRESS HYBRID FINITE ELEMENT MODEL FOR THE ANALYSIS OF AN AXISYMMETRIC THICK WALLED PRESS	71-0025
THICK	ANALYSIS OF A CIRCULAR RAFT IN ADHESIVE CONTACT WITH A THICK ELASTIC LAYER	74-1152
THICK	ANALYSIS OF CIRCULAR AND ANGULAR LAMINATED THICK PLATES	74-0426
THICK	ANALYSIS OF MODERATELY THICK FLANGED REACTOR PRESSURE VESSELS	73-0821
THICK	ANALYSIS OF MODERATELY THICK FLANGED REACTOR VESSELS	75-0320
THICK	ANALYSIS OF SIMPLY SUPPORTED THICK, LAYERED PLATES	71-0428
THICK	ANALYSIS OF THICK AND THIN SHELL STRUCTURE BY CURVED FINITE ELEMENTS	70-0016
THICK	APPLICATION OF FRACTURE MECHANICS TO FAILURE OF THICK WALLED CYLINDERS	71-0511
THICK	CURVED THICK SHELL AND MEMBRANE ELEMENTS WITH PARTICULAR REFERENCE TO AXISYMMETRIC PROBLEMS	68-0005
THICK	CURVED TRIANGULAR FINITE ELEMENTS FOR THE ANALYSIS OF THICK ELASTIC PLATES AND SHELLS	71-0814
THICK	DYNAMIC ANALYSIS OF CIRCULAR AND SECTOR THICK, LAYERED PLATES	75-0198
THICK	DYNAMIC TRANSIENT NON-LINEAR BEHAVIOUR OF THICK AND THIN PLATES	75-0125
THICK	ELASTIC PLASTIC DEFORMATION OF THICK WALLED CYLINDERS	69-0427
THICK	ELASTIC-PLASTIC ANALYSIS OF THICK WALLED PRESSURE VESSELS WITH SHARP DISCONTINUITIES	71-0155
THICK	ELASTIC-PLASTIC BEHAVIOR OF A LONGITUDINAL SEMI-ELLIPTIC CRACK IN A THICK PRESSURE VESSEL	73-0952
THICK	FINITE ELEMENT ANALYSIS OF A THICK WALLED MICROPOLAR CYLINDER LOADED AXISYMMETRICALLY	73-0847
THICK	FINITE ELEMENT SOLUTIONS FOR LAMINATED THICK PLATES	72-0143
THICK	FINITE-ELEMENT ANALYSIS OF MODERATELY THICK SHELLS	75-0588
THICK	FREE VIBRATION AND TRANSIENT RESPONSE OF THICK AND THIN PLATES USING THE FINITE ELEMENT METHOD	74-0517
THICK	FREE VIBRATION OF THICK, LAYERED RECTANGULAR PLATES BY FINITE LAYER METHOD	72-0440
THICK	FREE VIBRATION OF THICK, LAYERED CYLINDERS HAVING FINITE LENGTH WITH VARIOUS BOUNDARY CONDITIONS	72-0491
THICK	IMPROVED NUMERICAL INTEGRATION OF THICK SHELL FINITE ELEMENTS	71-0396
THICK	LES PLAQUES ET LES COQUES MINCES OU EPAISSES (THIN AND THICK PLATES AND SHELLS)	71-0412
THICK	PREDICTION OF MARINE PROPELLER DISTORTION AND STRESSES USING A SUPERPARAMETRIC THICK SHELL FINITE EL	73-0370
THICK	SEAL-SHELL-2, A COMPUTER PROGRAM FOR THE STRESS ANALYSIS OF A THICK SHELL OF REVOLUTION WITH AXISYMM	69-0100
THICK	SOLUTION OF A THREE DIMENSIONAL AXISYMMETRICAL PROBLEM OF THERMOPLASTICITY AS APPLIED TO THICK TURBI	74-0384
THICK	STABILITY OF MODERATELY THICK RECTANGULAR PLATES USING A HIGH PRECISION TRIANGULAR FINITE ELEMENT	75-0334
THICK	STRESS DIFFUSION IN THICK STIFFENED PANELS	74-0620
THICK	STRESSES IN SHORT THICK COMPOUND CYLINDERS	69-0385
THICK	THE ANALYSIS OF MODERATELY THICK TO THIN SHELLS BY THE FINITE ELEMENT METHOD	70-0315
THICK	THEORETICAL ANALYSIS OF THICK DOME STRUCTURES AND ARBITRARY SOLIDS OF REVOLUTION SUBJECTED TO UNSYMM	70-0037
THICK	THICK WALLED MULTIPLE OPENING REINFORCED CONCRETE CONDUITS	72-0666
THICK	THREE DIMENSIONAL FINITE ELEMENT ANALYSIS OF THICK LAMINATED COMPOSITES - INCLUDING INTERLAMINAR AND	73-0882
THICK	THREE DIMENSIONAL STRESS ANALYSIS OF PENETRATIONS IN THICK WALLED PRESSURE VESSELS	74-0422
THICK	THREE-DIMENSIONAL FINITE ELEMENT ANALYSIS OF THICK LAMINATED COMPOSITES - INCLUDING INTERLAMINAR AND	74-1151
THICK	TRANSIENT ANALYSIS OF THICK CYLINDRICAL SHELLS	71-0590
THICK	TRIANGULAR FINITE ELEMENTS FOR THE BENDING ANALYSIS OF THICK ELASTIC PLATES	70-0466
THICK	TRIANGULAR THICK PLATE BENDING ELEMENTS	71-0167
THICK	TRIANGULAR THICK SHELL FINITE ELEMENT FOR THE THREE-DIMENSIONAL STRESS ANALYSIS OF PRESSURE VESSELS	73-0860
THICK	TWO HYBRID ELEMENTS FOR ANALYSIS OF THICK, THIN AND SANDWICH PLATES	72-0544
THICK	VIBRATION OF THICK, CURVED, SHELLS WITH PARTICULAR REFERENCE TO TURBINE BLADES	70-0017
THICK	VIBRATIONS OF THICK PLATES USING A HIGH PRECISION TRIANGULAR ELEMENT	74-1002
THICK-WALLED	ANALYSIS OF THICK-WALLED PRESSURE VESSELS USING THE CONSTANT STRAIN AXISYMMETRIC FINITE ELEMENT	71-0653
THICK-WALLED	ELEMENT STIFFNESS MATRICES OF THICK-WALLED ORTHOTROPIC SHELLS WITH APPLICATIONS	69-0261
THICK-WALLED	FINITE ELEMENT ANALYSIS OF ELASTIC PLASTIC THICK-WALLED TUBES	74-0368
THICK-WALLED	STRESS INTENSITY FACTORS FOR THICK-WALLED CYLINDERS	74-0300
THICK-WALLED	THE FINITE ELEMENT ANALYSIS OF ELASTIC-PLASTIC THICK-WALLED TUBES	74-0768
THICK-WALLED	THREE-DIMENSIONAL FINITE ELEMENT ANALYSIS OF THICK-WALLED PIPE-NOZZLE JUNCTIONS WITH CURVED TRANSITI	71-0153
THICKNESS	A COMPARATIVE STUDY OF VARIABLE AND CONSTANT THICKNESS HIGH PRECISION TRIANGULAR PLATE BENDING ELEME	74-0857

THICKNESS	A COMPARATIVE STUDY OF VARIABLE AND CONSTANT THICKNESS HIGH PRECISION TRIANGULAR PLATE BENDING ELEME	74-0857
THICKNESS	A COMPATIBLE PLATE BENDING ANNULAR SECTOR FINITE ELEMENT OF NONUNIFORM THICKNESS	74-0430
THICKNESS	A TAPERED THICKNESS TRIM 6 ELEMENT FOR THE MATRIX DISPLACEMENT METHOD	66-0007
THICKNESS	ANALYSIS OF CLAMPED CIRCULAR PLATES OF VARIABLE THICKNESS SUBJECTED TO UNIFORMLY DISTRIBUTED LOAD BY	73-0583
THICKNESS	CALCULATION OF PLATE OF CONSTANT THICKNESS BY FINITE ELEMENT METHOD	69-0227
THICKNESS	DYNAMIC SHELL THEORY COUPLING THICKNESS STRESS WAVE EFFECTS WITH GROSS STRUCTURAL RESPONSE	73-0379
THICKNESS	ELASTIC BENDING OF CIRCULAR PLATES OF VARIABLE THICKNESS. AN ANALYTICAL AND EXPERIMENTAL STUDY	70-0163
THICKNESS	FINITE ELEMENT METHOD FOR OPTIMAL DESIGN OF VARIABLE THICKNESS SHEETS	73-0131
THICKNESS	FREE VIBRATION OF ORTHOTROPIC VARIABLE THICKNESS, CONTINUOUS RECTANGULAR PLATES	70-0270
THICKNESS	FREE VIBRATION OF THIN CYLINDRICAL SHELLS WITH A DISCONTINUITY IN THE THICKNESS	69-0439
THICKNESS	FREQUENCY COEFFICIENTS OF A SPHERICAL SHELL FOR VARYING THICKNESS AND OPENING ANGLE WITH GUIDED PINN	72-0595
THICKNESS	HEAVY SECTION STEEL TECHNOLOGY PROGRAM TECHNICAL OR PROGRAMMATIC MANUSCRIPT NO. 25. VARIABLE THICKNE	73-0948
THICKNESS	PLANE STRESS ELEMENT OF VARYING THICKNESS APPLIED TO END BLOCK ANALYSIS	74-0396
THICKNESS	PRESSURE VESSEL WALL THICKNESS LIMITS BASED ON THERMAL STRESS RACHETTING AND CREEP FATIGUE INTERACTI	74-0968
THICKNESS	RE-ENTRY THERMAL ANALYSIS OF VARIABLE THICKNESS SPHERICAL VEHICLES.	72-0468
THICKNESS	TRIANGULAR FINITE-ELEMENTS OF VARIABLE THICKNESS AND THEIR APPLICATION TO PLATE AND SHELL PROBLEMS	69-0005
THICKNESS	ZUR BERECHNUNG, BEMESSUNG UND BEWEHRUNG RECHTECKIGER PLATTEN KONSTNATER DICKE MIT OEFFNUNGEN ODER AU	73-0664
THIN	A COMPUTER PROGRAM FOR THE STATIC AND DYNAMIC FINITE ELEMENT ANALYSIS OF AXISYMMETRIC THIN SHELLS	66-0017
THIN	A COMPUTER PROGRAM FOR THE ANALYSIS OF THIN SHELLS	67-0041
THIN	A CONFORMING TRIANGULAR ELEMENT IN THE ANALYSIS OF THIN AND SEMI-THICK PLATES IN BENDING	70-0308
THIN	A CURVED ELEMENT APPROXIMATION IN THE ANALYSIS OF AXISYMMETRIC THIN SHELLS	70-0063
THIN	A CURVED FINITE ELEMENT FOR THIN ELASTIC SHELLS	69-0486
THIN	A DYNAMIC ELASTOPLASTIC ANALYSIS OF THIN SHELLS OF REVOLUTION UNDER ASYMMETRIC MECHANICAL AND THERMA	72-0068
THIN	A FINITE ELEMENT APPROXIMATION FOR THE ANALYSIS OF THIN SHELLS	68-0046
THIN	A FINITE ELEMENT DISPLACEMENT APPROACH FOR THE ELASTO-PLASTIC ANALYSIS OF THIN CYLINDRICAL SHELLS	74-1077
THIN	A FINITE ELEMENT FORMULATION FOR STABILITY ANALYSIS OF DOUBLY CURVED THIN SHELL STRUCTURES	69-0513
THIN	A FINITE ELEMENT FORMULATION FOR THE GENERAL STABILITY ANALYSIS OF THIN WALLED MEMBERS	70-0337
THIN	A FINITE ELEMENT LINEAR PROGRAMMING METHOD FOR THE LIMIT ANALYSIS OF THIN PLATES	73-1064
THIN	A FINITE ELEMENT METHOD FOR GEOMETRICALLY NONLINEAR LARGE DISPLACEMENT PROBLEMS IN THIN, ELASTIC PLA	69-0005
THIN	A FINITE-ELEMENT ANALYSIS FOR THIN SHELLS	68-0087
THIN	A HYBRID STRESS ELEMENT FOR THIN SHELL ANALYSIS	74-0433
THIN	A METHOD FOR COMPUTATION OF VIBRATION MODES AND FREQUENCIES OF ORTHOTROPIC THIN SHELLS OF REVOLUTION	69-0037
THIN	A MIXED FINITE ELEMENT METHOD FOR THIN SHELLS ANALYSIS	68-0182
THIN	A MODIFIED FINITE ELEMENT METHOD FOR CALCULATION OF VIBRATION OF THIN PLATES	72-0598
THIN	A REFINED CURVED ELEMENT FOR THIN SHELLS OF REVOLUTION	71-0214
THIN	A REFINED FINITE ELEMENT ANALYSIS OF THIN SHELL STRUCTURES	67-0101
THIN	A REFINED FINITE ELEMENT ANALYSIS OF THIN SHELL STRUCTURES INCLUDING DYNAMIC LOADINGS	67-0157
THIN	A STIFFNESS MATRIX FOR THE ANALYSIS OF THIN PLATES IN BENDING	61-0001
THIN	A STUDY OF FREE-VIBRATIONS OF THIN ELASTIC SHELLS SUBJECTED TO INITIAL LOADS USING FINITE-ELEMENTS	73-1143
THIN	A TRIANGULAR THIN SHELL FINITE ELEMENT: NONLINEAR ANALYSIS	75-0697
THIN	A TRIANGULAR THIN SHELL FINITE ELEMENT: LINEAR ANALYSIS	75-0698
THIN	A UNIFIED TREATMENT OF THICK AND THIN FILM ELASTOHYDRODYNAMIC PROBLEMS USING HIGHER ORDER ELEMENT PR	75-0890
THIN	ACCURACY AND CONVERGENCE OF MIXED FINITE-ELEMENT APPROXIMATIONS OF THIN BARS, MEMBRANES, AND PLATES	73-1120
THIN	AN APPROXIMATE NON-LINEAR ANALYSIS OF THIN PLATES	68-0286
THIN	ANALYSIS AND DESIGN OF AXISYMMETRIC REINFORCEMENT AROUND A CIRCULAR HOLE IN A THIN FLAT PLATE BY THE	74-1113
THIN	ANALYSIS OF THE DYNAMICS OF THIN PRIMARY MIRRORS FOR LARGE ASTRONOMICAL TELESCOPES	73-1084
THIN	ANALYSIS OF THICK AND THIN SHELL STRUCTURE BY CURVED FINITE ELEMENTS	70-0016
THIN	ANALYSIS OF THIN ARCH DAMS BY THE FINITE ELEMENT METHOD	64-0001
THIN	ANALYSIS OF THIN PLATES IN BENDING USING TRIANGULAR FINITE ELEMENTS	68-0213
THIN	ANALYSIS OF THIN SHELLS USING CURVED FINITE ELEMENTS	69-0523
THIN	ANALYSIS OF THIN SHELLS WITH A DOUBLY CURVED ARBITRARY QUADRILATERAL FINITE ELEMENT.	72-0443
THIN	ANALYSIS OF THREE-DIMENSIONAL THIN WALLED STRUCTURES	69-0188
THIN	ANALYSIS OF TOTALLY SUBMERGED THIN SHELLS BY FINITE ELEMENT TECHNIQUE	71-0616
THIN	APPLICATION OF RITZ'S METHOD TO THIN ELASTIC SHELL ANALYSIS	71-0083
THIN	APPROXIMATE ANALYSIS OF PREFABRICATED TOWER STRUCTURES USING THIN SHELL FINITE ELEMENTS	73-0893
THIN	AUTOMATIC CALCULATION OF THIN STRUCTURES	73-0201
THIN	BENDING OF THIN PLATES - DISPLACEMENT MODELS	68-0049
THIN	BUCKLING UNDER COMBINED LOADING OF THIN FLAT-WALLED STRUCTURES BY A COMPLEX FINITE STRIP METHOD	74-0788
THIN	CALCULATION OF NON-LINEAR AXISYMMETRIC VIBRATIONS OF THIN SHELLS OF REVOLUTION BY A FINITE ELEMENT M	75-0067
THIN	CALCULATION OF STIFFNESS MATRICES FOR FINITE ELEMENTS OF THIN SHELLS OF ARBITRARY SHAPE	68-0127
THIN	CALCULATION OF STRESS CONCENTRATION DUE TO A HYPERBOLOID CAVITY IN A THIN PLATE USING NASTRAN	71-0260
THIN	CALCULATION OF TEMPERATURE DISTRIBUTIONS IN THIN SHELLS OF REVOLUTION BY THE FINITE ELEMENT METHOD	71-0018
THIN	CALCULATION OF THE CRITICAL LOAD OF ELASTIC THIN WALLED BARS BY THE METHOD OF FINITE ELEMENTS	72-0446
THIN	CALCULATION OF THE ELASTIC BUCKLING LOADS OF THIN FLAT REINFORCED PLATES USING TRIANGULAR FINITE ELE	73-1024
THIN	CALCULATION OF THE ELASTIC BUCKLING LOADS OF THIN FLAT REINFORCED PLATES USING TRIANGULAR FINITE ELE	75-0177
THIN	COMBINED MATERIAL AND GEOMETRIC NONLINEARITY FOR THIN STEEL PLATES	75-0397
THIN	COMPUTATION OF VIBRATION MODAL SHAPES AND NATURAL FREQUENCIES OF A GENERAL ORTHOTROPIC LAMINATED, TH	71-0309
THIN	CONIDA: A FINITE ELEMENT PROGRAM FOR THE STRESS ANALYSIS OF AXISYMMETRIC THIN SHELLS	74-1087
THIN	CURVED FINITE ELEMENT FOR THIN ELASTIC SHELLS	70-0173
THIN	DEFICIENCY IN CURRENT FINITE ELEMENTS FOR THIN SHELL APPLICATIONS	73-0184
THIN	DEPLACEMENTS DANS LES STRUCTURES MINCES A FAIBLE RIGIDITE GEOMETRIQUE-DETERMINATION NUMERIQUE (NUMER	72-0213
THIN	DEVELOPMENT AND EVALUATION OF A TRIANGULAR THIN SHELL ELEMENT BASED UPON THE HYBRID ASSUMED STRESS M	69-0252
THIN	DISCRETE ELEMENT PROCEDURE FOR THIN SHELL INSTABILITY ANALYSIS	67-0028
THIN	DISCRETE-ELEMENT MODELLING AND SENSITIVITY STUDIES OF STATICALLY LOADED SINGLE-LAYER AND BOUNDED DOU	70-0129
THIN	DISCRETE-ELEMENT STATIC ANALYSIS OF BONDED, DOUBLE-LAYER, BRANCHED, THIN SHELLS OF REVOLUTION. PART	68-0102
THIN	DISCRETE-ELEMENT STATIC ANALYSIS OF BONDED, DOUBLE-LAYER, BRANCHED, THIN SHELLS OF REVOLUTION, PART	70-0081
THIN	DYNAMIC FINITE ELEMENT ANALYSIS OF ARBITRARY THIN SHELLS	71-0061
THIN	DYNAMIC TRANSIENT NON-LINEAR BEHAVIOUR OF THICK AND THIN PLATES	75-0125
THIN	DYNAMICAL ANALYSIS OF AXIALLY NON-UNIFORM, THIN CYLINDRICAL SHELLS. PART 1 - MATRIX FORMULATION	70-0413
THIN	DYNAMICAL ANALYSIS OF AXIALLY NONUNIFORM, THIN CYLINDRICAL SHELLS. PART 2 FREE VIBRATION	71-0716
THIN	EFFECTIVENESS OF EXCESSIVE NODAL CONTINUITIES IN THE FINITE ELEMENT ANALYSIS OF THIN RECTANGULAR AND	70-0346
THIN	ELASTIC ANALYSIS OF CRACKED THIN SHELLS BY THE FINITE ELEMENT METHOD	71-0048
THIN	ELASTIC DEFLECTION IN THIN WALLED CONES ROTATING ABOUT THEIR AXES OF SYMMETRY	73-0545
THIN	ENGINEERING OPERATIONS REPORT: NERVA THIN SHELL FINITE ELEMENT STRESS ANALYSIS METHODOLOGY COMPUTER	72-0916
THIN	EXPERIMENTAL AND FINITE ELEMENT STRESS ANALYSIS OF A THIN SHELLED CYLINDER TO CYLINDER MODEL	71-0066
THIN	FINITE ELEMENT ANALYSIS OF THIN SHELLS (DISCUSSION)	69-0187
THIN	FINITE ELEMENT ANALYSIS OF ARBITRARY THIN SHELLS	70-0329
THIN	FINITE ELEMENT ANALYSIS OF THIN SHELLS	70-0457
THIN	FINITE ELEMENT ANALYSIS OF VIBRATION OF THIN ELASTIC PLATES BY SIMPLIFIED HYBRID DISPLACEMENT METHOD	72-0565
THIN	FINITE ELEMENT ANALYSIS OF VIBRATIONS OF INITIALLY STRESSED THIN SHELLS OF REVOLUTION	74-1056
THIN	FINITE ELEMENT ANALYSIS OF THIN ELASTIC SHELLS WITH RESIDUAL ENERGY BALANCING AND THE ROLE OF THE RI	75-0034
THIN	FINITE ELEMENT ELASTIC THIN SHELL PRE-BUCKLING AND POST-BUCKLING ANALYSIS	71-0821
THIN	FINITE ELEMENT FOR TORSION OF THIN WALLED OPEN TUBES	71-0308
THIN	FINITE ELEMENT POST BUCKLING ANALYSIS OF THIN ELASTIC PLATES	68-0123
THIN	FINITE ELEMENT SOLUTION FOR THIN SHELLS OF REVOLUTION	63-0024
THIN	FINITE ELEMENT SOLUTION FOR THIN SHELLS OF REVOLUTION	64-0005
THIN	FINITE ELEMENT STRESS ANALYSIS OF THIN SHELLS	65-0024
THIN	FINITE-ELEMENT ANALYSIS OF THIN SHELLS	68-0164
THIN	FINITE-ELEMENT ANALYSIS FOR FREE VIBRATION OF GENERAL ANISOTROPIC LAMINATED THIN SHELLS	72-0618
THIN	FINITE-ELEMENT ANALYSIS OF THIN ELASTIC SHELLS WITH RESIDUAL ENERGY BALANCING AND ROLE OF RIGID BODY	75-0048
THIN	FLAT FINITE ELEMENT FOR THIN SHELL ANALYSIS DERIVED BY ASSUMED STRESS APPROACH	73-0147
THIN	FLEXURAL RIGIDITY OF A THIN WALLED BUILD-UP ROTOR FOR JET ENGINE. MEASUREMENT BY STATIC LOAD TEST AND	73-0756

THIN	FORCES AND DEFORMATIONS AT THE LATERAL STIFFENER OF THIN WALLED U-BARS	73-0236
THIN	FREE VIBRATION AND RESPONSE TO RANDOM PRESSURE FIELD OF ANISOTROPIC THIN CYLINDRICAL SHELLS	74-0830
THIN	FREE VIBRATION AND TRANSIENT RESPONSE OF THICK AND THIN PLATES USING THE FINITE ELEMENT METHOD	74-0517
THIN	FREE VIBRATION OF THIN CYLINDRICAL SHELLS WITH A DISCONTINUITY IN THE THICKNESS	69-0439
THIN	GENERAL SINUSOIDAL STIFFNESS MATRICES FOR BUCKLING AND VIBRATION ANALYSES OF THIN FLAT-WALLED STRUCT	68-0278
THIN	GEOMETRIC NONLINEAR ANALYSIS OF THIN PLATES BY FINITE ELEMENTS	70-0492
THIN	INSTABILITY OF THIN SHELLS BY THE FINITE ELEMENT METHOD	70-0187
THIN	INSTABILITY OF THIN SHELLS BY THE FINITE ELEMENT METHOD	70-0447
THIN	INTRODUCTION TO SHEARS DEFORMATION INTO A THIN PLATE DISPLACEMENTS FORMULATIONS	73-0617
THIN	LARGE DEFLECTION DYNAMIC ANALYSIS OF THIN SHELLS USING FINITE ELEMENT METHOD	70-0379
THIN	LARGE DEFLECTION SPATIAL BUCKLING OF THIN WALLED BEAMS AND FRAMES	73-1085
THIN	LARGE DISPLACEMENT ANALYSIS OF THIN PLATES	73-0261
THIN	LES PLAQUES ET LES COQUES MINCES OU EPAISSES (THIN AND THICK PLATES AND SHELLS)	71-0432
THIN	LINEAR AND NON-LINEAR ANALYSIS OF THIN SHALLOW SHELLS BY MIXED FINITE ELEMENTS	75-0665
THIN	LINEAR AND NONLINEAR STATIC ANALYSIS OF AXISYMMETRICALLY LOADED THIN SHELLS OF REVOLUTION	69-0215
THIN	MATRIX ANALYSIS OF THIN SHELLS USING FINITE ELEMENTS	69-0527
THIN	NATURAL VIBRATIONS OF THIN, FLAT WALLED STRUCTURES WITH DIFFERENT BOUNDARY CONDITIONS	71-0509
THIN	NEW FINITE ELEMENT METHOD FOR ANALYZING SYMMETRICALLY LOADED THIN SHELLS OF REVOLUTION	73-0389
THIN	NEW FINITE-ELEMENT SCHEME FOR INSTABILITY ANALYSIS OF THIN SHELLS	75-0357
THIN	NONLINEAR FINITE-ELEMENT ANALYSIS OF THIN SHELLS	73-1169
THIN	NONLINEAR FINITE-ELEMENT FORMULATION FOR TRANSIENT ANALYSIS OF THREE DIMENSIONAL THIN STRUCTURES	74-0874
THIN	NUMERICAL ANALYSIS OF THIN SHELLS BY CURVED TRIANGULAR ELEMENTS BASED ON DISCRETE - KIRCHHOFF HYPOTH	69-0091
THIN	PRACTICAL APPLICATIONS OF FINITE ELEMENT METHODS IN THE STRESS ANALYSIS OF THIN SHELL PRESSURE VESSE	71-0205
THIN	REFINED FINITE ELEMENT ANALYSIS OF ELASTIC-PLASTIC THIN SHELLS OF REVOLUTION	69-0497
THIN	RING FINITE ELEMENTS FOR AXISYMMETRIC AND NON-AXISYMMETRIC THIN SHELL ANALYSIS	72-0922
THIN	SAFE SHELL, A COMPUTER PROGRAM FOR THE STRESS ANALYSIS OF THIN SHELL BODIES OF REVOLUTION. A USER'S	67-0018
THIN	SECTOR ELEMENT FOR THIN PLATE FLEXURE	72-0174
THIN	SHAPE FUNCTION SUBROUTING FOR AN ISOPARAMETRIC THIN PLATE ELEMENT	73-0618
THIN	SLADE D: A COMPUTER PROGRAM FOR THE DYNAMIC ANALYSIS OF THIN SHELLS	73-0964
THIN	SLADE, A COMPUTER PROGRAM FOR THE STATIC ANALYSIS OF THIN SHELLS	70-0079
THIN	SOME RESULTS IN THE FINITE ELEMENT MODELING OF THIN PLATES	66-0018
THIN	STIFFNESS MATRICES FOR THIN TRIANGULAR ELEMENTS OF NONZERO GAUSSIAN CURVATURE	66-0073
THIN	STRESS ANALYSIS OF THIN PLATE STRUCTURE BY THE FINITE ELEMENT METHOD	69-0403
THIN	STUDY ON THE VIBRATION OF THIN PLATES IN A GAS STREAM BY THE FINITE ELEMENT METHOD	74-0454
THIN	THE ANALYSIS OF MODERATELY THICK TO THIN SHELLS BY THE FINITE ELEMENT METHOD	70-0315
THIN	THE ANALYSIS OF THIN SHELLS BY A FINITE ELEMENT PROCEDURE	67-0123
THIN	THE ANALYSIS OF THIN SHELLS WITH TRANSVERSE SHEAR STRAINS BY THE FINITE ELEMENT METHOD	68-0098
THIN	THE ANALYSIS OF THIN SHELLS BY THE FINITE ELEMENT METHOD	70-0078
THIN	THE APPLICATION OF THE SMOOTH-SURFACE INTERPOLATION TECHNIQUE TO THE FINITE-ELEMENT ANALYSIS OF THIN	66-0120
THIN	THE APPROXIMATE ANALYSIS OF THIN SHELLS BY THE FINITE ELEMENT METHOD	66-0131
THIN	THE EFFECTIVENESS OF EXCESSIVE NODAL CONTINUITIES IN THE FINITE ELEMENT ANALYSIS OF THIN RECTANGULAR	70-0105
THIN	THE EFFECTS OF THIN CORES AND NONHOMOGENEITY ON DYNAMIC RESPONSE OF EARTH DAMS	67-0021
THIN	THE FINITE ELEMENT ANALYSIS OF THIN ELASTIC SHELLS	73-1076
THIN	THE FINITE ELEMENT DYNAMIC STABILITY ANALYSIS OF THIN SHELLS SUBMERGED IN FLUIDS	71-0057
THIN	THE FINITE ELEMENT METHOD APPLIED TO THIN SHELLS AND BOX STRUCTURES	70-0444
THIN	THE TRANSIENT DYNAMIC ANALYSIS OF THIN SHELLS BY THE FINITE ELEMENT METHODS	71-0142
THIN	THE TRANSIENT LINEAR ELASTIC RESPONSE ANALYSIS COMPLEX THIN SHELLS OF REVOLUTION SUBJECTED TO ARBITR	70-0080
THIN	THERMAL STRESSES IN THIN CYLINDRICAL SHELLS	73-0822
THIN	THIN SHELL THEORY IN CARTESIAN COORDINATES - METHODS FOR FINITE ELEMENTS	70-0359
THIN	THIN WINDOWS FOR GASEOUS AND LIQUID TARGETS: AN OPTIMIZATION PROCEDURE	75-0741
THIN	TRIANGULAR THIN SHELL FINITE ELEMENT: LINEAR ANALYSIS	75-0602
THIN	TRIANGULAR THIN SHELL FINITE ELEMENT: NONLINEAR ANALYSIS	75-0668
THIN	TWO HYBRID ELEMENTS FOR ANALYSIS OF THICK, THIN AND SANDWICH PLATES	72-0544
THIN	USE OF THE FINITE ELEMENT METHOD IN EXAMINING ELASTIC BUCKLING OF THIN PLATES OF ANY SHAPE	75-0153
THIN	VARIATIONAL FORMULATIONS AND FINITE ELEMENT METHOD FOR THIN ELASTIC SHELLS AND PLATES	68-0064
THIN	VIBRATION CHARACTERISTICS OF THIN, PRESSURIZED SHELLS OF REVOLUTION PARTIALLY FILLED WITH LIQUID	72-0504
THIN	VIBRATION CHARACTERISTICS OF FREE THIN CYLINDRICAL SHELLS	73-0396
THIN	VIBROACOUSTIC RESPONSE OF TURBULENCE EXCITED THIN RECTANGULAR FINITE PLATES IN HEAVY AND FLUID MEDIA	74-1309
THIN	WAVE PROPAGATION IN A THIN HOLLOW CONE BY A FINITE ELEMENT METHOD	72-0045
THIN-SHELL	A DISCRETE ELEMENT PROCEDURE FOR THIN-SHELL INSTABILITY ANALYSIS	67-0029
THIN-SHELL	ANALYSIS OF THIN-SHELL HYPERBOLIC PARABOLOID SHELLS	72-0121
THIN-SHELL	COMPARISON OF THE NUMERICAL INTEGRATION TECHNIQUE AND THE FINITE ELEMENT METHOD IN THE ANALYSIS OF T	73-0848
THIN-SHELL	PHOTOELASTIC TESTS ON COMPLEX INTERSECTIONS OF REINFORCED THIN-SHELL CYLINDERS	75-0363
THIN-SHELL	THE FINITE ELEMENT METHOD OF THIN-SHELL STABILITY ANALYSIS	72-0725
THIN-SHELL	THEORETICAL AND EXPERIMENTAL STRESS ANALYSES OF ORNL THIN-SHELL CYLINDER-TO-CYLINDER MODEL 3	75-0690
THIN-SHELL	THEORETICAL AND EXPERIMENTAL STRESS ANALYSES OF ORNL THIN-SHELL CYLINDER-TO-CYLINDER MODEL 4	75-0695
THIN-SHELL	THERMAL LOADING OF THIN-SHELL CONCRETE COOLING TOWERS	74-0667
THIN-SHELL	THIN-SHELL STRUCTURES: THEORY, EXPERIMENT, AND DESIGN	74-0828
THIN-WALLED	A CONSISTENT DISCRETE ELEMENT TECHNIQUE FOR THIN-WALLED ASSEMBLAGES	68-0104
THIN-WALLED	ANALYSIS OF THIN-WALLED CURVED BEAMS BY THE TRANSFER MATRIX METHOD	72-0335
THIN-WALLED	ANALYTICAL PERFORMANCE OF STEEL, ALUMINUM AND PLASTIC FOR A NEW DESIGN HIGH PRESSURE THIN-WALLED CAR	74-0756
THIN-WALLED	APPLICATION OF FINITE ELEMENT METHOD - STRESS CALCULATIONS ON 90 DEGREES NOZZLES OF THIN-WALLED CYLI	71-0253
THIN-WALLED	BENDING AND TORSION OF THIN-WALLED STRAIGHT AND CURVED BEAMS WITH VARIABLE OPEN CROSS-SECTION	69-0493
THIN-WALLED	CALCULATION OF THE REINFORCED THIN-WALLED STRUCTURES BY THE METHOD OF FINITE ELEMENTS (RASCHET PODKR	75-0685
THIN-WALLED	CALCULATION OF THE REINFORCED THIN-WALLED CONSTRUCTION BY THE FINITE ELEMENT	75-0756
THIN-WALLED	COUPLED VIBRATIONS OF THIN-WALLED BEAMS OF OPEN SECTION USING FINITE ELEMENT METHOD	70-0230
THIN-WALLED	COUPLED VIBRATIONS OF THIN-WALLED BEAMS OF OPEN SECTION USING FINITE ELEMENT METHOD	71-0568
THIN-WALLED	DETERMINATION OF THE APPLICABILITY OF SOME THEORIES ABOUT STRENGTH AND STIFFNESS OF THIN-WALLED BEAM	69-0318
THIN-WALLED	FAILURE OF THIN-WALLED MEMBERS UNDER PATCH LOADINGS	72-0230
THIN-WALLED	FINITE ELEMENT ANALYSIS OF THIN-WALLED STRUCTURES BASED ON THE MODERN ENGINEERING THEORY OF BEAMS	71-0140
THIN-WALLED	FINITE ELEMENT ANALYSIS OF THIN-WALLED MEMBERS OF OPEN SECTION	72-0966
THIN-WALLED	FINITE ELEMENT LARGE DEFLECTION ANALYSIS OF THIN-WALLED BEAMS	74-0444
THIN-WALLED	GEOMETRICALLY NONLINEAR BEHAVIOUR OF THIN-WALLED MEMBERS USING FINITE-ELEMENTS	73-1142
THIN-WALLED	MODELING TECHNIQUES OF THIN-WALLED BEAMS WITH OPEN CROSS SECTIONS	72-0573
THIN-WALLED	ON THE FINITE ELEMENT ANALYSIS OF THIN-WALLED STRUCTURES WITH APPLICATION TO STRUCTURAL PROBLEMS IN	73-0832
THIN-WALLED	POSSIBILITIES OF THE METHOD OF FINITE ELEMENTS IN THE CALCULATION OF THE STRENGTH AND STIFFNESS OF T	70-0344
THIN-WALLED	PROBLEM OF USING THE METHOD OF FINITE ELEMENTS TO CONSTRUCT AN ALGORITHM FOR CALCULATING A THIN-WALL	74-0851
THIN-WALLED	REFINED FINITE ELEMENT STABILITY ANALYSIS OF THIN-WALLED MEMBERS	75-0784
THIN-WALLED	STATIC AND DYNAMIC BUCKLING OF THIN-WALLED COLUMNS USING FINITE ELEMENTS	75-0505
THIN-WALLED	STRESS DISTRIBUTIONS IN THIN-WALLED INTERSECTING CYLINDRICAL SHELLS SUBJECTED TO INTERNAL PRESSURE A	71-0023
THIN-WALLED	THE DEVELOPMENT OF APPLIED METHODS IN PROBLEMS OF STATIC CALCULATION OF THIN-WALLED THREE-DIMENSIONA	73-0934
THIN-WALLED	THIN-WALLED PRESTRESSED CONCRETE PRESSURE VESSEL FOR HIGH TEMPERATURE REACTORS: EXPERIMENTAL INVESTI	73-0857
THRAUSTICS	HORIZONS IN THE STRUCTURAL USE OF ADVANCED MATERIALS - THRAUSTICS	70-0415
THREAD	STRESS ANALYSIS OF A BOLT-NUT JOINT BY FINITE ELEMENT METHOD AND COPPER-ELECTROPLATING METHOD - 2. S	74-0153
THREADED	FINITE ELEMENT ANALYSIS OF THREADED CONNECTIONS	74-1179
THREADS	FINITE ELEMENT ANALYSIS OF CASING THREADS	75-0330
THREADS	FINITE-ELEMENT ANALYSIS OF CASING THREADS	75-0417
THREE-	INCIPIENT CHIP FORMATION IN METAL CUTTING - A THREE- DIMENSION FINITE-ELEMENT ANALYSIS	73-1146
THREE-DIMENSION	A FINITE ELEMENT PROGRAM FOR DESIGN AND ANALYSIS OF THREE-DIMENSIONAL FLAT SPRINGS	74-0738
THREE-DIMENSION	A SIMPLE MATRIX-VECTOR HANDLING SCHEME FOR THREE-DIMENSIONAL AND SHELL ANALYSIS	70-0317
THREE-DIMENSION	A THREE-DIMENSIONAL FINITE ELEMENT ANALYSIS OF A PRESTRESSED CONCRETE REACTOR VESSEL MODEL	69-0008A
THREE-DIMENSION	A THREE-DIMENSIONAL FINITE ELEMENT COMPUTER CODE FOR THE ANALYSIS OF COMPLEX STRUCTURES	71-0275

THREE-DIMENSION	AN ASSUMED STRESS HYBRID FINITE ELEMENT FOR THREE-DIMENSIONAL PLASTIC STRUCTURAL ANALYSIS	74-0717
THREE-DIMENSION	ANALYSIS OF THREE-DIMENSIONAL THIN WALLED STRUCTURES	69-0188
THREE-DIMENSION	ANALYSIS OF THREE-DIMENSIONAL ORTHOTROPIC SANDWICH PLATE STRUCTURES BY FINITE-ELEMENT METHOD	72-0945
THREE-DIMENSION	ANALYSIS OF THREE-DIMENSIONAL UNSTEADY VISCOUS FLOW AROUND OSCILLATING WINGS	74-0982
THREE-DIMENSION	ANALYSIS OF UNSTEADY FLOW TOWARD ARTESIAN WELLS BY THREE-DIMENSIONAL FINITE ELEMENTS	74-0742
THREE-DIMENSION	ANALYSIS OF UNSTEADY FLOW TOWARD AN ARTESIAN WELL BY THREE-DIMENSIONAL FINITE ELEMENTS	74-1351
THREE-DIMENSION	APACHE: A THREE-DIMENSIONAL FINITE ELEMENT PROGRAM FOR STEADY-STATE OR TRANSIENT HEAT CONDUCTION ANA	73-0971
THREE-DIMENSION	APPLICATION OF THE THREE-DIMENSIONAL FINITE ELEMENT METHOD TO SOME PROBLEMS IN SOIL MECHANICS	70-0510
THREE-DIMENSION	APPLICATION OF THREE-DIMENSIONAL FINITE ELEMENTS FOR ANALYSIS OF IRRADIATION INDUCED STRESSES IN NUC	72-0078
THREE-DIMENSION	APPLICATION OF THREE-DIMENSIONAL FINITE ELEMENT ANALYSES TO THE MICROMECHANICS OF FIBROUS COMPOSITE	74-1079
THREE-DIMENSION	APPROXIMATE THREE-DIMENSIONAL STRESS SOLUTIONS FOR SOME LAMINATED PLATE PROBLEMS USING A HIGHER ORDE	70-0419
THREE-DIMENSION	APPROXIMATE THREE-DIMENSIONAL SOLUTIONS FOR SYMMETRIC LAMINATES UNDER INPLANE LOADING	71-0392
THREE-DIMENSION	AUTOMATED METHOD FOR THE LARGE DEFLECTION AND INSTABILITY ANALYSIS OF THREE-DIMENSIONAL TRUSS AND FR	66-0105
THREE-DIMENSION	BODY ORIENTED (NATURAL) CO-ORDINATES FOR GENERATING THREE-DIMENSIONAL MESHES	74-0908
THREE-DIMENSION	CALCULATION OF POTENTIAL FLOW ABOUT ARBITRARY THREE-DIMENSIONAL LIFTING BODIES	72-0722
THREE-DIMENSION	COMPARISON OF THE THREE-DIMENSIONAL ANALYSIS OF A SHEAR WALL BUILDING BY TWO DIFFERENT METHODS	75-0705
THREE-DIMENSION	COMPUTER PROGRAMS FOR PLOTTING FINITE ELEMENT PATTERNS OF TWO-AND THREE-DIMENSIONAL STRUCTURES	70-0416
THREE-DIMENSION	DATA GENERATION FOR FINITE-ELEMENT STRUCTURAL ANALYSIS OF THREE-DIMENSIONAL NAVAL SHIP STRUCTURES	75-0707
THREE-DIMENSION	EFFICIENT COMPUTER ANALYSIS OF LARGE THREE-DIMENSIONAL PIPING SYSTEMS BY THE FINITE ELEMENT METHOD	72-0378
THREE-DIMENSION	ELASTIC-PLASTIC ANALYSIS OF THREE-DIMENSIONAL MEDIA	66-0002
THREE-DIMENSION	ELASTO-PLASTIC ANALYSIS OF THREE-DIMENSIONAL STRUCTURES USING THE ISOPARAMETRIC ELEMENT	72-0248
THREE-DIMENSION	ELASTO-PLASTIC MATRIX DISPLACEMENT ANALYSIS OF THREE-DIMENSIONAL CONTINUA	66-0004
THREE-DIMENSION	EVALUATION OF A THREE-DIMENSIONAL STRESS ANALYSIS PROGRAM	72-0743
THREE-DIMENSION	EVALUATION OF A THREE-DIMENSIONAL STRESS CELL FOR GRANULAR SOILS	72-0882
THREE-DIMENSION	FEATHER: FINITE ELEMENT ANALYSIS FOR THREE-DIMENSIONAL ELASTIC RESPONSE	72-0607
THREE-DIMENSION	FINITE ELEMENT ANALYSIS OF CRACK PROPAGATION IN THREE-DIMENSIONAL SOLIDS UNDER CYCLING LOADING	73-0529
THREE-DIMENSION	FINITE ELEMENT ANALYSIS OF THREE-DIMENSIONAL GROUNDWATER FLOW PROBLEMS	74-0254
THREE-DIMENSION	FINITE ELEMENT ANALYSIS OF THREE-DIMENSIONAL CABLE STRUCTURES	74-1275
THREE-DIMENSION	FINITE ELEMENT COMPUTATIONAL THEORY FOR THREE-DIMENSIONAL BOUNDARY LAYER FLOW	72-0632
THREE-DIMENSION	FINITE ELEMENT MESH GENERATION PROGRAM FOR ARBITRARY TWO-AND THREE-DIMENSIONAL STRUCTURES	73-0966
THREE-DIMENSION	FINITE ELEMENT MODELLING OF THREE-DIMENSIONAL STRUCTURES	68-0267
THREE-DIMENSION	FINITE ELEMENT SOLUTION THEORY FOR THREE-DIMENSIONAL BOUNDARY FLOWS	74-0579
THREE-DIMENSION	FINITE ELEMENTS FOR THREE-DIMENSIONAL ELASTIC CRACK ANALYSIS	74-0136
THREE-DIMENSION	GENERATION OF TWO- AND THREE-DIMENSIONAL FINITE ELEMENTS	71-0040
THREE-DIMENSION	INTERACTIVE FINITE ELEMENT DATA GENERATION FOR THREE-DIMENSIONAL PLATE STRUCTURES	74-0909
THREE-DIMENSION	INVESTIGATION OF TECHNIQUES OF THREE-DIMENSIONAL FINITE ELEMENT STRESS ANALYSIS	70-0029
THREE-DIMENSION	ISOPARAMETRIC FINITE ELEMENTS IN TWO AND THREE-DIMENSIONAL ANALYSIS	68-0237
THREE-DIMENSION	MATRIX ANALYSIS OF THREE-DIMENSIONAL ELASTIC MEDIA: SMALL AND LARGE DISPLACEMENTS	65-0003
THREE-DIMENSION	MATUS: A THREE-DIMENSIONAL FINITE ELEMENT PROGRAM FOR SMALL-STRAIN ELASTIC ANALYSIS	73-0970
THREE-DIMENSION	MESH 3D: A THREE-DIMENSIONAL FINITE ELEMENT MESH GENERATOR PROGRAM FOR EIGHT-NODE ISOPARAMETRIC ELEM	73-0972
THREE-DIMENSION	METHOD OF EVALUATION OF THREE-DIMENSIONAL STRESS INTENSITY FACTOR USING THE FINITE ELEMENT METHOD	75-0316
THREE-DIMENSION	METHODS OF CALCULATION OF THREE-DIMENSIONAL PROBLEMS OF PERCOLATION OF FISSURED ROCK BY FINITE ELEME	74-0472
THREE-DIMENSION	ON FURTHER APPLICATION OF THE FINITE ELEMENT METHOD TO THREE-DIMENSIONAL ELASTIC ANALYSIS	70-0252
THREE-DIMENSION	ON THE DETERMINATION OF EFFECTIVE MODULI OF COMPOSITE MATERIALS BY A THREE-DIMENSIONAL FINITE ELEMEN	74-1123
THREE-DIMENSION	PROGRESS IN THREE-DIMENSIONAL ELASTIC-PLASTIC STRESS ANALYSIS FOR FRACTURE MECHANICS	71-0314
THREE-DIMENSION	PROJECTIVE METHOD APPLIED TO THREE-DIMENSIONAL PLASTICITY EQUATIONS	74-0617
THREE-DIMENSION	RASHET TREKHMERNYKH MAGNITNYKE POLEI METODOM KONECHNYKH ELEMENTOV. (CALCULATION OF THREE-DIMENSIONAL	75-0766
THREE-DIMENSION	REBO: COMPUTER CODE FOR THE GENERATION OF A THREE-DIMENSIONAL FINITE-ELEMENT GRID OF THE NOZZLE-CYLI	74-1310
THREE-DIMENSION	RESEARCH NOTE: A THREE-DIMENSIONAL OIL FILM TEMPERATURE DISTRIBUTION IN TILTING THRUST BEARINGS	74-0508
THREE-DIMENSION	SECOND GENERATION BOUNDARY INTEGRAL EQUATION PROGRAM FOR THREE-DIMENSIONAL ELASTIC ANALYSIS	75-0376
THREE-DIMENSION	SOLUTION OF THREE-DIMENSIONAL FIELD PROBLEMS BY THE FINITE ELEMENT METHOD	67-0091
THREE-DIMENSION	STRESS ANALYSIS OF FIRST MOLARS WTH FULL CROWN PREPARATIONS BY THREE-DIMENSIONAL PHOTOELASTICITY AND	72-0939
THREE-DIMENSION	STRUCTURAL ANALYSIS OF A THREE-DIMENSIONAL PERFORATED PLATE BY THE FINITE ELEMENT METHOD WITH TEST C	74-0945
THREE-DIMENSION	THE APPLICATION OF THREE-DIMENSIONAL FINITE ELEMENTS TO A CYLINDER-CYLINDER INTERSECTION	70-0068
THREE-DIMENSION	THE APPLICATION OF THREE-DIMENSIONAL FINITE ELEMENTS TO THE ANALYSIS OF PODDED BOILER TYPE PRESTRESS	71-0047
THREE-DIMENSION	THE CALCULATION OF THREE-DIMENSIONAL TEMPERATURE DISTRIBUTIONS AND THERMAL STRESSES USING FINITE ELE	71-0187
THREE-DIMENSION	THE DEVELOPMENT OF APPLIED METHODS IN PROBLEMS OF STATIC CALCULATION OF THIN-WALLED THREE-DIMENSIONA	73-0934
THREE-DIMENSION	THREE-DIMENSIONAL ANALYSIS OF ELASTIC SOLIDS I. ANALYSIS PROCEDURE	69-0222
THREE-DIMENSION	THREE-DIMENSIONAL ANALYSIS OF REACTOR PRESSURE VESSEL NOZZLES	71-0221
THREE-DIMENSION	THREE-DIMENSIONAL ANALYSIS OF SHEAR WALL BUILDINGS TO LATERAL LOAD	74-0220
THREE-DIMENSION	THREE-DIMENSIONAL ANALYSIS FOR THE STRESS DISTRIBUTION NEAR CIRCULAR HOLES IN LAMINATED COMPOSITES	74-0793
THREE-DIMENSION	THREE-DIMENSIONAL APPLICATION OF THE FINITE ELEMENT METHOD	70-0250
THREE-DIMENSION	THREE-DIMENSIONAL BAR CELL FOR ELASTIC STRESS ANALYSIS	70-0219
THREE-DIMENSION	THREE-DIMENSIONAL BEHAVIOR OF A CENTRAL CORE DAM	72-0714
THREE-DIMENSION	THREE-DIMENSIONAL CONSIDERATION OF THE STRESSES SURROUNDING OPEN-PIT MINE SLOPES	73-0492
THREE-DIMENSION	THREE-DIMENSIONAL CRACK ANALYSIS APPLIED TO AN LWR NOZZLE-CYLINDER INTERSECTION	75-0165
THREE-DIMENSION	THREE-DIMENSIONAL ELASTIC-PLASTIC STRESS AND STRAIN ANALYSIS FOR FRACTURE MECHANICS. PHASE 1. SIMPLE	70-0188
THREE-DIMENSION	THREE-DIMENSIONAL ELASTO-STATIC PROBLEMS USING ISOPARAMETRIC FINITE ELEMENTS	71-0751
THREE-DIMENSION	THREE-DIMENSIONAL ELASTIC FINITE ELEMENT ANALYSIS USING GRADIENT DEGREES OF FREEDOM	73-0843
THREE-DIMENSION	THREE-DIMENSIONAL ELASTO-PLASTIC FINITE ELEMENT ANALYSIS	75-0236
THREE-DIMENSION	THREE-DIMENSIONAL ELASTIC ANALYSIS OF SEMI-ELLIPTICAL SURFACE CRACKS SUBJECT TO THERMAL SHOCK	75-0256
THREE-DIMENSION	THREE-DIMENSIONAL FIELD PROBLEMS BY HIGHER ORDER FINITE ELEMENTS	70-0257
THREE-DIMENSION	THREE-DIMENSIONAL FINITE ELEMENT ANALYSIS OF THICK-WALLED PIPE-NOZZLE JUNCTIONS WITH CURVED TRANSITI	71-0153
THREE-DIMENSION	THREE-DIMENSIONAL FINITE ELEMENT ANALYSIS OF DAMS	71-0733
THREE-DIMENSION	THREE-DIMENSIONAL FINITE ELEMENT ANALYSIS FOR FRACTURE MECHANICS	72-0162
THREE-DIMENSION	THREE-DIMENSIONAL FINITE ELEMENT ANALYSIS OF LAMINATED COMPOSITES	72-0345
THREE-DIMENSION	THREE-DIMENSIONAL FINITE ELEMENT ANALYSIS FOR FRACTURE MECHANICS	72-0921
THREE-DIMENSION	THREE-DIMENSIONAL FINITE ELEMENT ANALYSES OF DAMS	73-0202
THREE-DIMENSION	THREE-DIMENSIONAL FINITE ELEMENT METHOD OF THERMOELASTOPLASTICITY WITH CREEP EFFECT	73-0837
THREE-DIMENSION	THREE-DIMENSIONAL FINITE-ELEMENT ELASTIC ANALYSIS OF PRISMATIC SOLIDS	73-1138
THREE-DIMENSION	THREE-DIMENSIONAL FINITE ELEMENT ANALYSIS OF SPHERICAL PARTICLE COMPOSITES	74-0137
THREE-DIMENSION	THREE-DIMENSIONAL FINITE-ELEMENT COMPUTER PROGRAM - USERS GUIDE	74-0755
THREE-DIMENSION	THREE-DIMENSIONAL FINITE ELEMENT ANALYSIS INCLUDING ELASTIC-PLASTIC MATERIAL RESPONSE	74-1126
THREE-DIMENSION	THREE-DIMENSIONAL FINITE ELEMENT ANALYSIS OF THICK LAMINATED COMPOSITES - INCLUDING INTERLAMINAR AND	74-1151
THREE-DIMENSION	THREE-DIMENSIONAL FINITE ELEMENT ANALYSIS OF VERTICALLY LOADED PILE GROUPS	75-0239
THREE-DIMENSION	THREE-DIMENSIONAL FINITE ELEMENT ANALYSIS OF A STEAM GENERATOR CHANNEL-HEAD-COMPLEX - COMPARISON OF	75-0653
THREE-DIMENSION	THREE-DIMENSIONAL GRAIN STRESS ANALYSIS USING THE FINITE ELEMENT METHOD	71-0732
THREE-DIMENSION	THREE-DIMENSIONAL HARMONIC FUNCTIONS NEAR TERMINATION OR INTERSECTION OF GRADIENT SINGULARITY LINES:	74-0178
THREE-DIMENSION	THREE-DIMENSIONAL ISOPARAMETRIC ELEMENTS AND THEIR USE FOR STATICAL ANALYSIS OF MASSIVE WATER-POWER	71-0364
THREE-DIMENSION	THREE-DIMENSIONAL LINEAR FRACTURE MECHANICS ANALYSIS BY A DISPLACEMENT-HYBRID FINITE-ELEMENT MODEL	75-0884
THREE-DIMENSION	THREE-DIMENSIONAL MECHANICAL STRESSES IN TOROIDAL MAGNETS FOR CONTROLLED THERMONUCLEAR REACTORS	73-0669
THREE-DIMENSION	THREE-DIMENSIONAL MESH GENERATOR FOR FINITE ELEMENT COMPUTER CODES	74-0864
THREE-DIMENSION	THREE-DIMENSIONAL PHOTOELASTIC AND FINITE-ELEMENT ANALYSIS OF A PROPELLANT GRAIN	72-0031
THREE-DIMENSION	THREE-DIMENSIONAL STRESS ANALYSIS OF ARCH DAMS BY THE FINITE ELEMENT METHOD: PART I, ARCH DAM NO. 1,	66-0099
THREE-DIMENSION	THREE-DIMENSIONAL STRESS ANALYSIS AND FIELD PROBLEMS BY THE FINITE ELEMENT METHOD	70-0183
THREE-DIMENSION	THREE-DIMENSIONAL STUDY OF NUCLEAR FUEL ROD BEHAVIOUR DURING STARTUP	73-0815
THREE-DIMENSION	THREE-DIMENSIONAL STUDY OF NUCLEAR FUEL ROD BEHAVIOR DURING STARTUP	74-1003
THREE-DIMENSION	THREE-DIMENSIONAL STRESS ANALYSES OF MITERED ELBOWS	75-0650
THREE-DIMENSION	THREE-DIMENSIONAL STRESS ANALYSIS OF A FINITE SLAB CONTAINING A TRANSVERSE CENTRAL CRACK	75-0732
THREE-DIMENSION	THREE-DIMENSIONAL THERMAL AND STRESS ANALYSIS OF A PIPING TEE	75-0703
THREE-DIMENSION	TRANSIENT FIELD PROBLEMS: TWO-DIMENSIONAL AND THREE-DIMENSIONAL ANALYSIS BY ISO-PARAMETRIC FINITE EL	70-0001
THREE-DIMENSION	TRIANGULAR THICK SHELL FINITE ELEMENT FOR THE THREE-DIMENSIONAL STRESS ANALYSIS OF PRESSURE VESSELS	73-0860

421

THREE-DIMENSION	TWO AND THREE-DIMENSIONAL FLOW USING FINITE ELEMENTS	69-0008
THREE-DIMENSION	TWO-AND THREE-DIMENSIONAL FLOW ANALYSIS USING FINITE ELEMENTS	70-0021
THREE-DIMENSION	USE OF SIMPLE THREE-DIMENSIONAL ACOUSTIC FINITE ELEMENTS FOR DETERMINING NATURAL MODES AND FREQUENCI	72-0010
THREE-DIMENSION	VARIATIONAL PRINCIPLES FOR A RAREFIED GAS FLOW PAST AND ARBITRARY THREE-DIMENSIONAL BODY	70-0189
THREE-DIMENSION	VIBRATIONAL CHARACTERISTICS OF THREE-DIMENSIONAL SOLIDS, WITH APPLICATIONS TO EARTH DAMS	69-0432
THREE-PHASE	SEISMIC ANALYSIS OF A THREE-PHASE BREAKER INSTALLATION	74-0622
THROUGH-FLOW	THE MERIDIONAL THROUGH-FLOW CALCULATION IN AN AXIAL FLOW MACHINE BY THE FINITE ELEMENT METHOD	75-0115
THROUGHFLOW	A VARIATIONAL FORMULATION OF THE COMPRESSIBLE THROUGHFLOW PROBLEM	74-1343
THROUGHFLOW	A VARIATIONAL FORMULATION OF THE COMPRESSIBLE THROUGHFLOW PROBLEM	75-0021
THRUST	A THERMO-ELASTIC ANALYSIS OF THE PARALLEL SURFACE THRUST BEARING	74-1069
THRUST	ANALYSIS OF CURVED TARGET-TYPE THRUST REVERSERS	75-0152
THRUST	CYCLIC FATIGUE ANALYSIS OF ROCKET THRUST CHAMBERS. VOLUME 2: ATTITUDE CONTROL THRUSTER HIGH CYCLE FA	74-0693
THRUST	CYCLIC FATIGUE ANALYSIS OF ROCKET THRUST CHAMBERS. VOLUME 1: OFHC COPPER CHAMBER LOW CYCLE FATIGUE	74-0694
THRUST	DYNAMIC BEHAVIOR OF HYDROSTATIC THRUST BEARING	75-0189
THRUST	MECHANISM OF PARALLEL SURFACE THRUST BEARING	75-0322
THRUST	MODAL RESPONSE OF FREE ROCKETS UNDER THRUST LOADING	74-0855
THRUST	NUMERICAL SIMULATION OF FINITE-WIDTH THRUST BEARINGS, TAKING INTO ACCOUNT VISCOSITY VARIATION WITH T	75-0623
THRUST	RESEARCH NOTE: A THREE-DIMENSIONAL OIL FILM TEMPERATURE DISTRIBUTION IN TILTING THRUST BEARINGS	74-0608
THRUST	THERMO-ELASTIC ANALYSIS OF THE PARALLEL SURFACE THRUST BEARING	74-1172
THRUSTER	CYCLIC FATIGUE ANALYSIS OF ROCKET THRUST CHAMBERS. VOLUME 2: ATTITUDE CONTROL THRUSTER HIGH CYCLE FA	74-0693
THTR	ROTATIONSSYMMETRISCHE BERECHNUNG DES 1:5 THTR SPANNBETON-BEHAL-BEHAL TERMODELLS	71-0629
THTR	UEBER DAS VERHALTEN VON SPANNBETONDRUCKBEHAELTRERN IM GRENZTRAGFAEHIGKEITSZUSTAND UND DESSEN EINFUL	73-0855
THTR-VESSEL	UEBER DAS VERHALTEN VON SPANNBETONDRUCKBEHAELTRERN IM GRENZTRAGFAEHIGKEITSZUSTAND UND DESSEN EINFUL	73-0855
TIBIA	DETAILED STRUCTURAL ANALYSIS OF THE HUMAN TIBIA	75-0379
TIDAL	GROUNDWATER FLOW IN A SANDY TIDAL BEACH: II. 2-DIMENSIONAL FINITE ELEMENT ANALYSIS	72-0300
TIDAL	GROUNDWATER FLOWN IN A SANDY TIDAL BEACH .1. ONE-DIMENSIONAL FINITE ELEMENT ANALYSIS	71-0567
TIDAL	INVESTIGATION OF THE WATER TABLE IN A TIDAL BEACH	71-0648
TIDAL	TIDAL AND LONG WAVE PROPAGATION - A FINITE ELEMENT APPROACH	73-1059
TIDAL	TIDAL PROPAGATION AND DISPERSION IN ESTUARIES	75-0390
TIDAL	TIDAL SIMULATION GY TIME SPACE ELEMENT	75-0867
TIDES	A FINITE ELEMENT MODEL OF TIDES IN ESTUARIES	74-0044
TIED	DESIGN OF INCLINED TIED ARCH RAILWAY BRIDGES OVER THE M56	71-0297
TIED-BACK	PERFORMANCE OF TIED-BACK WALLS IN CLAY	74-0665
TIED-BACK	TIED-BACK RETAINING WALLS	75-0308
TIES	ON THE DETERMINATION OF EFFECTIVE ELASTIC-PLASTIC PROPER TIES FOR THE EQUIVALENT SOLID PLATE ANALYSI	74-0494
TIL	SPAENDINGSHYBRIDE FINITE ELEMENTER TIL SVINGNINGSPROBLEMER. (STRESS HYBRID FINITE ELEMENT METHOD FOR	75-0208
TILES	VIBRATION AND STRESS ANALYSIS OF SOFT-BONDED SHUTTLE INSULATION TILES, MODAL ANALYSIS WITH COMPACT W	74-0708
TILTING	RESEARCH NOTE: A THREE-DIMENSIONAL OIL FILM TEMPERATURE DISTRIBUTION IN TILTING THRUST BEARINGS	74-0608
TILTING-PAD	STABILITY OF TILTING-PAD JOURNAL BEARINGS	71-0687
TILTS	CRUSTAL STRUCTURE FROM SURFACE LOAD TILTS USING A FINITE ELEMENT MODEL	72-0005
TIMBER	STRESS CONCENTRATIONS IN TIMBER BEAMS	67-0038
TIME	A FINITE ELEMENT SOLUTION OF TIME DEPENDENT FIELD PROBLEMS	67-0155
TIME	A FINITE ELEMENT SOLUTION OF TIME DEPENDENT FIELD PROBLEMS USING A REDUCED VARIATIONAL FORMULATION	69-0504
TIME	ACCURACY AND CONVERGENCE OF FINITE ELEMENT GALERKIN APPROXIMATIONS OF TIME DEPENDENT PROBLEMS WITH E	73-0987
TIME	ACCURACY AND CONVERGENCE OF FINITE ELEMENT/GALERKIN APPROXIMATIONS OF TIME DEPENDENT PROBLEMS WITH E	75-0006
TIME	AN ANALYSIS OF VARIOUS TIME STEPPING SCHEMES FOR INITIAL VALUE PROBLEMS	73-0607
TIME	AN INTERACTIVE GRAPHICS SHIP DESIGN ORIENTED PROGRAM PACKAGE FOR TIME SHARING SYSTEMS	73-0341
TIME	APPLICATIONS OF FINITE ELEMENTS IN SPACE AND TIME	72-0147
TIME	AUTOMATED SOLUTION OF TIME DEPENDENT PROBLEMS	72-0084
TIME	BERECHNUNG INSTATIONAERER TEMPERATUREFELDER MIT FINITEN ELEMENTEN IN RAUM-UND ZEITDIMENSION. (CALCUL	74-0887
TIME	CALCULATION OF TRANSIENT TEMPERATURE FIELDS WITH FINITE ELEMENTS IN SPACE AND TIME DIMENSIONS	74-0379
TIME	CODE TAFEST: NUMERICAL SOLUTION TO TRANSIENT HEAT-CONDUCTION PROBLEMS USING FINITE ELEMENTS IN SPACE	74-0901
TIME	COMPARISON OF TIME MARCHING SCHEMES FOR THE TRANSIENT HEAT CONDUCTION EQUATION	75-0477
TIME	COMPUTER PROGRAMS FOR SIMULATING PHYSICAL PHENOMENA IN THREE SPACE DIMENSIONS AND TIME	74-0866
TIME	CREEP ANALYSIS OF TRANSVERSELY ISOTROPIC BODIES SUBJECTED TO TIME DEPENDENT LOADING	73-0170
TIME	CURVED ELEMENTS, ERROR BOUNDS, TIME DEPENDENT PROBLEMS FIVE LECTURES	73-0636
TIME	EFFECTS OF RISE TIME AND DAMPING ON FINITE ELEMENT ANALYSIS OF RESPONSE OF STRUCTURES	71-0087
TIME	EVALUATION OF NUMERICAL TIME INTEGRATION METHODS AS APPLIED TO ELASTIC-PLASTIC DYNAMIC PROBLEMS INVO	74-0858
TIME	FINITE ELEMENT MATRIX DISPLACEMENT METHOD FOR TIME DEPENDENT STRESS ANALYSIS OF REACTOR MATERIALS IN	71-0674
TIME	FINITE ELEMENT METHOD FOR TIME DEPENDENT PROBLEMS	72-0821
TIME	FINITE ELEMENTS IN TIME AND SPACE	69-0442
TIME	FINITE ELEMENTS IN TIME AND SPACE	69-0450
TIME	FUEL ROD DEFORMATIONS - A TIME DEPENDENT FINITE ELEMENT ANALYSIS ON AXISYMMETRIC FUEL RODS, INCLUDIN	70-0414
TIME	GALERKIN APPROXIMATION OF THE TIME DERIVATIVE IN THE FINITE ELEMENT ANALYSIS	73-0573
TIME	GALERKIN APPROXIMATION OF THE TIME DERIVATIVE IN THE FINITE ELEMENT ANALYSIS OF GROUNDWATER FLOW	74-0658
TIME	NUMERICAL METHODS FOR TIME DEPENDENT NONLINEAR BOUNDARY VALUE PROBLEMS	70-0296
TIME	ON FINITE ELEMENT APPROXIMATIONS ON TIME DEPENDENT PROBLEMS	72-0528
TIME	ON THE DISCRETIZATION OF THE TIME VARIABLE FOR PARABOLIC PARTIAL DIFFERENTIAL EQUATIONS	75-0103
TIME	ON THE FINITE ELEMENT ANALYSIS OF TWO-DIMENSIONAL STRESS PROBLEMS WITH TIME DEPENDENT PROPERTIES	65-0052
TIME	PLATES AND SHELLS, INELASTIC ANALYSIS, TIME DEPENDENT INELASTICITY, FIVE LECTURES	73-0633
TIME	SAVINGS IN NASTRAN DECOMPOSITION TIME BY SEQUENCING TO REDUCE ACTIVE COLUMNS	71-0161
TIME	SEMIDISCRETE GALERKIN TECHNIQUES WITH TIME INTERPOLATION AND SPLITTING UP FOR PLASMA SIMULATION	75-0121
TIME	SOLUTION OF NONLINEAR EQUATIONS BY EXPLICIT TIME INTEGRATION - A SHORT COMMUNICATION	73-0606
TIME	SOME REMARKS ON FINITE ELEMENT ANALYSIS OF TIME DEPENDENT FIELD PROBLEMS	73-0307
TIME	SPACE TIME FINITE ELEMENTS IN DYNAMIC PROBLEMS	72-0402
TIME	STABILITY AND ACCURACY OF NUMERICAL APPROXIMATIONS TO TIME DEPENDENT FLOWS	75-0183
TIME	STATIC AND DYNAMIC ELASTO-PLASTIC ANALYSIS BY THE METHOD OF FINITE ELEMENTS IN SPACE AND TIME	72-0642
TIME	THE EFFECT OF GRAVITY UPON THE MELT-THROUGH TIME OF A SOLID SUBJECTED TO A HIGH INTENSITY LASER	72-0693
TIME	THE FINITE ELEMENT METHOD FOR TIME DEPENDENT PROBLEMS	72-0507
TIME	TIDAL SIMULATION GY TIME SPACE ELEMENT	75-0867
TIME	TIME DEPENDENT REINFORCED CONCRETE SLAB DEFLECTIONS	74-0586
TIME	TRANSIENT SHELL RESPONSE BY NUMERICAL TIME INTEGRATION	72-0305
TIME	VARIABLE TIME STEP ANALYSIS OF UNCONFINED SEEPAGE	74-0064
TIME	VARIABLE TIME STEP METHOD FOR DETERMINING PLASTIC STRESS REFLECTIONS FROM BOUNDARIES	75-0641
TIME-DEPENDENT	APPLICATION OF THE DIRECT STIFFNESS METHOD TO PLANE PROBLEMS INVOLVING LARGE, TIME-DEPENDENT DEFORMA	66-0108
TIME-DEPENDENT	DYNAMIC RESPONSE OF STRUCTURES SUBJECTED TO TIME-DEPENDENT BOUNDARY CONDITIONS USING THE FINITE ELEM	72-0129
TIME-DEPENDENT	EFFECT OF TIME-DEPENDENT MATERIAL PROPERTIES ON DYNAMIC RESPONSE	74-0900
TIME-DEPENDENT	EVALUATION AND ANALYSES OF SOME FINITE ELEMENT AND FINITE DIFFERENCE PROCEDURES FOR TIME-DEPENDENT P	75-0745
TIME-DEPENDENT	FINITE ELEMENT ANALYSIS OF TWO-DIMENSIONAL TIME-DEPENDENT STRESS PROBLEMS	65-0022
TIME-DEPENDENT	FINITE ELEMENT ANALYSIS FOR TIME-DEPENDENT PROBLEMS	74-0192
TIME-DEPENDENT	FINITE ELEMENT ANALYSIS OF TIME-DEPENDENT PUMPING TEST DATA IN AN UNCONFINED AQUIFER	75-0497
TIME-DEPENDENT	FINITE ELEMENT ANALYSIS FOR THE TIME-DEPENDENT INELASTIC MATERIAL BEHAVIOR	75-0749
TIME-DEPENDENT	FINITE ELEMENT APPROXIMATIONS TO TIME-DEPENDENT PROBLEMS	72-0269
TIME-DEPENDENT	FINITE-ELEMENT ANALYSIS OF TIME-DEPENDENT PHENOMENA	69-0104
TIME-DEPENDENT	LINEAR DISCONTINUOUS FINITE-ELEMENT SCHEME IN TIME-DEPENDENT TRANSPORT CALCULATIONS	75-0099
TIME-DEPENDENT	THE CALCULATION OF TIME-DEPENDENT STRESSES IN PCPVS BY AN APPROXIMATE METHOD	71-0086
TIME-DEPENDENT	THE COMPARISON OF ITERATIVE AND DIRECT SOLUTION TECHNIQUES IN THE ANALYSIS OF TIME-DEPENDENT STRESS	72-0080
TIME-DEPENDENT	TIME-DEPENDENT ANALYSIS TO PREDICT CLOSURE IN SALT CAVITIES	74-1149
TIME-DEPENDENT	TWO-DIMENSIONAL TIME-DEPENDENT ANALYSIS OF PERFORATED END CAPS FOR NUCLEAR REACTOR PRESSURE VESSELS	71-0239
TIME-DEPENDENT	USE OF FINITE ELEMENT METHOD IN A MARKER AND CELL PROGRAM FOR SOLVING TIME-DEPENDENT FLUID FLOW PROB	75-0043
TIME-SPACE	FINITE TIME-SPACE ELEMENTS APPLIED TO NONLINEAR PROBLEMS	70-0398
TIME-STEP	FINITE ELEMENT ANALYSIS OF UNCONFINED FLOW WITH A VARIABLE TIME-STEP PROCEDURE	74-0776

TIMING	TIMING COMPARISON STUDY FOR SOME HIGH ORDER FINITE ELEMENT PROCEDURES AND A LOW ORDER FINITE DIFFERE	72-0767
TIMOSHENKO	A TIMOSHENKO BEAM ELEMENT	72-0590
TIMOSHENKO	CONVERGENCE OF CONSISTENTLY DERIVED TIMOSHENKO BEAM FINITE ELEMENTS	72-0396
TIMOSHENKO	DYNAMIC STABILITY OF TIMOSHENKO BEAMS BY FINITE ELEMENT METHOD	75-0194
TIMOSHENKO	DYNAMIC STABILITY OF TIMOSHENKO BEAMS BY FINITE ELEMENT METHOD	75-0422
TIMOSHENKO	FINITE ELEMENT FOR TIMOSHENKO BEAM BASED ON MECHANICAL IMPEDANCE	71-0507
TIMOSHENKO	FINITE ELEMENT MODEL FOR DYNAMIC ANALYSIS OF TIMOSHENKO BEAM	75-0275
TIMOSHENKO	RESPONSE OF SIMPLY SUPPORTED TIMOSHENKO BEAMS	69-0489
TIMOSHENKO	TIMOSHENKO BEAM FINITE ELEMENTS	73-0032
TIMOSHENKO	VIBRATIONS OF A TIMOSHENKO BEAM, USING FINITE-ELEMENT APPROACH	66-0044
TINY	FRACTURE MECHANICS OF TINY CRACKS NEAR FASTENERS	74-0735
TIP	A HYBRID FINITE ELEMENT FOR DETERMINATION OF STRESS INTENSITY FACTORS AT A CRACK TIP	73-0643
TIP	AN ELASTIC-PLASTIC FINITE ELEMENT ANALYSIS OF CRACK TIP FIELDS UNDER BIAXIAL LOADING CONDITIONS	73-0760
TIP	ANALYSIS OF STRESS AND STRAIN DISTRIBUTIONS AT CRACK TIP BY FINITE ELEMENT METHOD	69-0193
TIP	CRACK TIP FINITE ELEMENTS FOR PLANE ELASTICITY	71-0764
TIP	CRACK TIP FINITE ELEMENTS ARE UNNECESSARY	73-1016
TIP	CRACK TIP FINITE ELEMENTS ARE UNNECESSARY	75-0472
TIP	CRACK TIP STRAIN - A COMPARISON OF FEM CALCULATIONS AND MEASUREMENTS	75-0721
TIP	ELASTIC-PLASTIC FINITE ELEMENT ANALYSIS OF CRACK TIP FIELDS UNDER BIAXIAL LOADING CONDITIONS	74-0296
TIP	FINITE ELEMENT CRACK TIP MODELING WHEN EVALUATING J-INTEGRAL	75-0637
TIP	FINITE ELEMENT TREATMENT OF A UNIFORMLY MOVING ELASTIC-PLASTIC CRACK TIP	74-0470
TIP	FINITE ELEMENTS FOR DETERMINATION OF CRACK TIP ELASTIC STRESS INTENSITY FACTORS	71-0436
TIP	LIMITATIONS TO THE SMALL SCALE YIELDING APPROXIMATION FOR CRACK TIP PLASTICITY	74-0145
TIP	ON THE USE OF FINITE ELEMENTS AND COMPLIANCE CALCULATIONS FOR DETERMINING CRACK TIP STRESS INTENSITY	69-0448
TIP	SIZE OF PLASTIC ZONE AT TIP OF A CRACK IN PLANE STRAIN SLATE BY FINITE ELEMENT METHOD	73-0006
TIP	STEADILY GROWING ELASTIC-PLASTIC CRACK TIP IN A FINITE ELEMENT TREATMENT	73-0087
TIP	STIFFNESS DERIVATIVE FINITE ELEMENT TECHNIQUE FOR DETERMINATION OF CRACK TIP STRESS INTENSITY FACTOR	74-0956
TIP	STIFFNESS DERIVATIVE FINITE ELEMENT TECHNIQUE FOR DETERMINATION OF CRACK TIP STRESS INTENSITY FACTOR	74-1358
TIP	STRESS, DISPLACEMENT, LINE INTEGRAL AND CLOSURE ENERGY DETERMINATIONS OF CRACK TIP STRESS INTENSITY	74-0621
TIP	THE ANALYSIS OF STRESS DISTRIBUTION AROUND THE TIP OF A CRACK BY FINITE ELEMENT METHOD WITH THE AID	71-0517
TIPS	CALCULATION OF STRESS INTENSITY FACTORS AT CRACK TIPS USING SPECIAL FINITE ELEMENTS	72-0071
TIPS	EFFECTS OF LOCAL STRENGTH VARIATIONS ON PLASTIC ENCLAVE DEVELOPMENT AT CRACK TIPS	72-0822
TIPS	INFLUENCE OF NON-SINGULAR STRESS TERMS AND SPECIMEN GEOMETRY ON SMALL SCALE YIELDING AT CRACK TIPS I	73-0203
TIRE	A STUDY OF INTERNAL STRESSES IN STATICALLY DEFORMED PNEUMATIC TIRES. A MATHEMATICAL MODEL FOR THE PN	70-0421
TIRE	CONTACT OF AN INFLATED TOROIDAL MEMBRANE WITH A FLAT SURFACE AS AN APPROACH TO THE TIRE DEFLECTION P	75-0164
TIRE	MATHEMATICAL PREDICTION OF DYNAMIC TIRE BEHAVIOR	73-0579
TIRES	A STUDY OF INTERNAL STRESSES IN STATICALLY DEFORMED PNEUMATIC TIRES. A MATHEMATICAL MODEL FOR THE PN	70-0421
TIRES	NONLINEAR HYBRID STRESS FINITE ELEMENT ANALYSIS OF PNEUMATIC TIRES	75-0029
TISSUE	BRAIN TISSUE FRAGILITY - A FINITE STRAIN ANALYSIS BY A HYBRID FINITE-ELEMENT METHOD	75-0162
TISSUE	BRAIN TISSUE FRAGILITY - A FINITE STRAIN ANALYSIS BY A HYBRID FINITE-ELEMENT METHOD	75-0243
TISSUE	MECHANICS OF BRAIN TISSUE FRAGILITY	73-0214
TITAN	EXPERIMENTS IN FORMING AND JOINTING OF TITANIUM AND TITANIUM ALLOYS FOR AIRCRAFT AND SPACE VEHICLES	70-0134
TITANIUM	EXPERIMENTS IN FORMING AND JOINTING OF TITANIUM AND TITANIUM ALLOYS FOR AIRCRAFT AND SPACE VEHICLES	70-0134
TITANIUM	EXPERIMENTS IN FORMING AND JOINTING OF TITANIUM AND TITANIUM ALLOYS FOR AIRCRAFT AND SPACE VEHICLES	70-0134
TITANIUM	PRELIMINARY INVESTIGATION OF THE DIE-LESS DRAWING OF TITANIUM AND SOME STEELS	75-0770
TITANIUM	SONIC FATIGUE OF DIFFUSION BOUNDED TITANIUM SANDWICH STRUCTURES	72-0290
TITANLEGIERUNGE	EXPERIMENTS IN FORMING AND JOINTING OF TITANIUM AND TITANIUM ALLOYS FOR AIRCRAFT AND SPACE VEHICLES	70-0134
TITUS	THE TRIDIMENSIONAL THERMOELASTIC COMPUTER CODE TITUS	71-0156
TITUS	USE OF THE ' TITUS ' FINITE ELEMENT CODE FOR THE NUCLEAR STRUCTURES AND INTERNAL STRESS-ANALYSIS	73-0840
TOCHER	SUR L ELEMENT DE CLOUGH ET TOCHER (CLOUGH-TOCHER ELEMENT)	74-1022
TOLERANCE	EVALUATION OF DAMAGE TOLERANCE IN AIRCRAFT STRUCTURES	75-0559
TOLERANCE	SPECTRUM TRUNCATION AND DAMAGE TOLERANCE STUDY ASSOCIATED WITH THE C-5A OUTBOARD PYLON AFT TRUSS LUG	74-0535
TONE	APPLICATIONS OF CONTINUOUS TONE COMPUTER-GENERATED IMAGES IN STRUCTURAL MECHANICS	74-1035
TONKIKH	MODIFICIROVANNYJ KONECHNYJ ELEMENT DIJA PASCHIOTA KOLEBANIJ TONKIKH PLASTIN	72-0891
TONKOSTENNYKH	CALCULATION OF THE REINFORCED THIN-WALLED STRUCTURES BY THE METHOD OF FINITE ELEMENTS (RASCHET PODKR	75-0685
TONKOSTENNYKH	THE DEVELOPMENT OF APPLIED METHODS IN PROBLEMS OF STATIC CALCULATION OF THIN-WALLED THREE-DIMENSIONA	73-0934
TONOMETER	MECHANICS OF MACKAY-MARG TONOMETER	73-0433
TONOMETER	VISCOELASTIC PRESPONSE OF THE CORNEO-SCLERAL SHELL UNDER TONOMETER LOADING	75-0855
TONOMETRY	MATHEMATICAL MODEL OF THE CORNEO-SCLERAL SHELL AS APPLIED TO INTRAOCULAR PRESSURE VOLUME RELATIONS A	72-0471
TOOL	ANALYSIS OF THERMAL DEFORMATION OF MACHINE TOOL BY THE FINITE ELEMENT METHOD.	72-0432
TOOL	CALCULATION OF THERMALLY INDUCED STRAINS IN MACHINE TOOL ELEMENTS	72-0464
TOOL	COMPUTER AIDED DESIGN OF MACHINE TOOL STRUCTURE	74-1189
TOOL	COMPUTERUNTERSTUETZTES SYSTEM FUER DIE CONSTRUKTION DES WERKZEUGMASCHINENAUFBAUS (COMPUTER-AIDED SYS	73-0482
TOOL	DYNAMIC ANALYSIS OF MACHINE TOOL STRUCTURES BY THE FINITE ELEMENT METHOD	71-0416
TOOL	EVALUATION OF COMPUTER AIDED METHODS FOR THE DESIGN OF MACHINE TOOL STRUCTURES	73-0577
TOOL	FINITE ELEMENT ANALYSIS, AN AUTOMOBILE ENGINEER'S TOOL	74-0238
TOOL	FINITE ELEMENT METHOD FOR MACHINE TOOL STRUCTURAL ANALYSIS	71-0136
TOOL	FINITE ELEMENT METHOD FOR DETERMINATION OF MACHINE TOOL STRUCTURES	72-0095
TOOL	FINITE ELEMENT METHOD FOR THE ANALYSIS OF MACHINE TOOL STRUCTURES.	72-0448
TOOL	FINITE ELEMENT METHOD - A VERSATILE TOOL FOR CIVIL ENGINEERING	72-0627
TOOL	INTERNATIONAL MACHINE TOOL DESIGN AND RESEARCH CONFERENCE, 12TH PROCEEDINGS, 1971.	72-0224
TOOL	INTERNATIONAL MACHINE TOOL DESIGN AND RESEARCH CONFERENCE, 13TH PROCEEDINGS, 1972	73-0504
TOOL	INTRODUCTION TO FINITE ELEMENT METHOD AND ITS APPLICATION TO MACHINE TOOL STRUCTURES	72-0112
TOOL	NASTRAN AS ANALYSIS TOOL IN A STRUCTURAL DESIGN OPTIMIZATION PROCESS	71-0235
TOOL	NUMERISCHE BERECHNUNG INSTATIONAERER TEMPERATURFELDER UND THERMISCH BEDINGTER VERFORMUNGEN VON WERKZ	75-0847
TOOL	SHELL DEFORMATION UNDER TOOL LOADS REPRESENTATION OF SINGULARITIES WITH ISOPARAMETRIC FINITE ELEMENT	74-0365
TOOL	STRESS ANALYSIS IN SOME COLD FORGING TOOL COMPONENTS AND THEIR DESIGN GUIDES	74-0589
TOOL	TESTRAN PROGRAMS, A TOOL FOR TEMPERATURE AND STRESS ANALYSIS	73-0414
TOOL	THE FINITE ELEMENT TECHNIQUE - A NEW TOOL IN STRUCTURAL ANALYSIS	69-0195
TOOL	THE GROWTH OF INTERACTIVE GRAPHICS AS AN ENGINEERING TOOL.	69-0347
TOOLS	ANWENDUNG DER METHODE FINITER ELEMENTE BEI DER ANALYSE DES DYNAMISCHEN VERHALTENS GEDAEMPFTER WERKZE	75-0580
TOOLS	ASSESSMENT OF DESIGN TOOLS AND CRITERIA FOR URBAN RAIL TRACK STRUCTURES, VOLUME 11. AT-GRADE SLAB TR	74-0813
TOOLS	COMPUTER METHODS FOR THE STRUCTURAL ANALYSIS OF MACHINE TOOLS	69-0325
TOOLS	DOSTIGNUCA I TENDENCIJE U ISPITIVANJIMA ALATNIH MASINA (ACHIEVEMENTS AND TENDENCIES IN MACHINE TOOLS	73-0426
TOOTH	APPLICATION OF FINITE ELEMENTS TO THE ANALYSIS OF GEAR TOOTH STRESSES	72-0116
TOOTH	FINITE ELEMENT STRESS ANALYSIS OF PROSTHETIC TOOTH IMPLANTS	75-0073
TOOTH	NEW APPROACH TO GEAR TOOTH ROOT STRESSES	72-0117
TOP	ULTIMATE LOAD BEHAVIOUR OF PCRV TOP CLOSURES. THEORETICAL AND EXPERIMENTAL INVESTIGATIONS	73-0835
TOPOGRAPHIC	FINITE ELEMENT ANALYSIS OF THE PIEZOELECTRIC TOPOGRAPHIC WAVEGUIDE	74-1111
TOPOGRAPHIC	HIGHER-ORDER FINITE-ELEMENT ANALYSIS OF TOPOGRAPHIC GUIDES SUPPORTING ELASTIC SURFACE WAVES	73-0114
TOPOLOGICAL	SOME APPLICATIONS OF TOPOLOGICAL CONSIDERATION AND WEIGHT MATRIX METHOD TO FINITE ELEMENT ANALYSIS	73-0305
TOROIDAL	BUCKLING OF SEGMENTS OF TOROIDAL SHELLS	73-0515
TOROIDAL	CONTACT OF AN INFLATED TOROIDAL MEMBRANE WITH A FLAT SURFACE AS AN APPROACH TO THE TIRE DEFLECTION P	75-0164
TOROIDAL	FINITE-ELEMENT PLASMA-DIFFUSION CODE FOR ARBITRARY TOROIDAL GEOMETRY	75-0500
TOROIDAL	FORMULATION AND EVALUATION OF A TOROIDAL RING DISCRETE ELEMENT	66-0082
TOROIDAL	OBLICZANIE KONSTRUKCJI TOROIDALNEGO ZBIORNIKA NA WODE. (STATICAL ANALYSIS OF A TOROIDAL WATER RESERV	75-0211
TOROIDAL	STRUCTURAL RESPONSE OF UNSTIFFENED TOROIDAL SHELLS	69-0198
TOROIDAL	THREE-DIMENSIONAL MECHANICAL STRESSES IN TOROIDAL MAGNETS FOR CONTROLLED THERMONUCLEAR REACTORS	73-0669
TOROIDALNEGO	OBLICZANIE KONSTRUKCJI TOROIDALNEGO ZBIORNIKA NA WODE. (STATICAL ANALYSIS OF A TOROIDAL WATER RESERV	75-0211
TORQUE	A GENERAL EQUILIBRIUM TORQUE ELEMENT	74-0187
TORQUE	FORCE AND TORQUE ANALYSIS OF CONSTRAINED SPACE MECHANISMS AND PLANE MECHANISMS WITH OFFSET LINKS BY	74-0569
TORQUE	MATRIX-DISPLACEMENT DIRECT-ELEMENT METHOD FOR FORCE AND TORQUE ANALYSIS OF DETERMINATE AND INDETERMI	74-0531

TORRES	SOBRE TORRES ESPIADAS - METODO DE CALCULO E UM EXEMPLO DE APLICACAO. (METHOD AND AN EXAMPLE OF CALCU	74-0915
TORSION	A COMPUTER PROGRAM FOR THE CALCULATION OF TORSION FACTORS AND SHEAR STRESS FACTORS FOR BOX SECTIONS	69-0485
TORSION	A FINITE ELEMENT SOLUTION FOR SAINT-VENANT TORSION	69-0161
TORSION	ANALYSIS OF SAINT-VENANT TORSION PROBLEM BY A HYBRID STRESS MODEL	69-0277
TORSION	APPLICATION OF THE PSEUDO-FUNCTIONAL FINITE ELEMENT METHOD TO VISCO-PLASTIC TORSION	73-0002
TORSION	BENDING AND TORSION OF THIN-WALLED STRAIGHT AND CURVED BEAMS WITH VARIABLE OPEN CROSS-SECTION	69-0493
TORSION	BORNES SUPERIEURES ET INFERIEURES DANS L'ANALYSE MATRICIELLE DES PLAQUES EN FLEXION TORSION	64-0015
TORSION	COMBINED APPLICATION OF FINITE ELEMENT METHODS AND RICHARDSON EXTRAPOLATION TO THE TORSION PROBLEM	72-0066
TORSION	COMPILED APPLICATION OF FINITE ELEMENT METHODS AND RICHARDSON EXTRAPOLATION TO THE TORSION PROBLEM	73-1012
TORSION	ELASTIC PLASTIC ANALYSIS OF SAINT-VENANT TORSION PROBLEM BY A HYBRID STRESS MODEL	72-0385
TORSION	ELASTIC-PLASTIC ANALYSIS OF A BAR WITH UNIFORM CROSS-SECTION UNDER COMBINED LOADINGS 1 - COMBINATION	71-0778
TORSION	FINITE ELEMENT ANALYSIS ON THE TORSION OF A BAR WITH UNIFORM CROSS SECTION	68-0186
TORSION	FINITE ELEMENT ANALYSIS IN POLAR CO-ORDINATES OF THE SAINT VENANT TORSION PROBLEM	74-0500
TORSION	FINITE ELEMENT FOR TORSION OF THIN WALLED OPEN TUBES	71-0308
TORSION	FINITE ELEMENT STRESS FIELD SOLUTION OF THE PROBLEM OF SAINT VENANT TORSION	73-1042
TORSION	FINITE ELEMENTS SOLUTION FOR TORSION RIGIDITY AND SHEAR CENTER OF ARBITRARY CROSS-SECTIONS	74-0168
TORSION	PLASTIC STRAINS ASSOCIATED WITH TORSION OF NONCIRCULAR BARS	72-0177
TORSION	SHORT COMMUNICATIONS: TORSION OF AXI-SYMMETRIC SOLIDS OF VARIABLE DIAMETER - INCLUDING ACCELERATION	74-0203
TORSION	STRAIN HARDENING TORSION OF SOLID BARS AS A MINIMIZATION PROBLEM	70-0174
TORSION	VARIATIONAL PRINCIPLE FOR THE LAPLACE'S OPERATOR WITH APPLICATION IN THE TORSION OF COMPOSITE RODS	74-1044
TORSION-BENDING	FLEXIBILITY MATRIX FOR AN EXACT TORSION-BENDING FINITE ELEMENT	72-0214
TORSION-FLEXION	MATRICE DE RIGIDITE D'UN ELEMENT DE POUTRE SOLLICITE A LA TORSION-FLEXION (STIFFNESS MATRIX OF A BEA	73-0442
TORSIONAL	A TORSIONAL STRENGTH ANALYSIS ON THE CONTAINER SHIP BY MEANS OF THE FINITE ELEMENT PROCEDURE AND FUL	71-0243
TORSIONAL	APPLICATION OF SIMULTANEOUS ITERATION METHOD TO TORSIONAL VIBRATION PROBLEMS	73-0660
TORSIONAL	CURVED BEAM FINITE ELEMENTS FOR COUPLED BENDING AND TORSIONAL VIBRATION	72-0720
TORSIONAL	DYNAMIC TORSIONAL RESPONSE OF THE HUMAN LEG RELATIVE TO SKIING INJURIES	73-0471
TORSIONAL	EINE FINITE-ELEMENTE-LOESUNG FUR DIE TORSIONSSTEIFIGKEIT UND DEN SCHUBMITTELPUNKT BELIEBIGER QUERSCH	74-0256
TORSIONAL	ELASTIC TORSIONAL ANALYSIS OF IRREGULAR SHAPES	65-0020
TORSIONAL	ELASTIC TORSIONAL ANALYSIS OR IRREGULAR SHAPES	66-0078
TORSIONAL	EVALUATION OF TORSIONAL EFFECTS IN THE SEISMIC ANALYSIS OF SYMMETRIC AND UNSYMMETRIC NUCLEAR POWER P	73-0175
TORSIONAL	FINITE ELEMENT ANALYSIS OF TORSIONAL AND TORSIONAL-FLEXURAL STABILITY PROBLEMS	70-0028
TORSIONAL	MATRICE DE RIGIDITE D'UN ELEMENT DE POUTRE SOLLICITE A LA TORSION-FLEXION (STIFFNESS MATRIX OF A BEA	73-0442
TORSIONAL	NONLINEAR DYNAMIC ANALYSIS OF TORSIONAL SHELLS BY COMBINED USE OF FINITE ELEMENT AND MODE SUPERPOSIT	75-0859
TORSIONAL	SENSITIVITY ANALYSIS OF TORSIONAL VIBRATION CHARACTERISTICS OF HELICOPTER ROTOR BLADES. PART 1: STRU	74-0846
TORSIONAL	SENSITIVITY OF ROTOR BLADE VIBRATION CHARACTERISTICS TO TORSIONAL OSCILLATIONS	74-0611
TORSIONAL	STABILITY ANALYSIS OF THE TORSIONAL BENDING VIBRATIONS OF WING IN A SUBSONIC FLOW BY METHOD OF THE F	75-0755
TORSIONAL	STRUCTURAL ANALYSIS OF SL-7 CONTAINERSHIP UNDER COMBINED LOADING OF VERTICAL, LATERAL AND TORSIONAL	74-0734
TORSIONAL-FLEXU	FINITE ELEMENT ANALYSIS OF TORSIONAL AND TORSIONAL-FLEXURAL STABILITY PROBLEMS	70-0028
TORSIONSSTEIFIG	EINE FINITE-ELEMENTE-LOESUNG FUR DIE TORSIONSSTEIFIGKEIT UND DEN SCHUBMITTELPUNKT BELIEBIGER QUERSCH	74-0256
TORT	AN ANALYSIS OF FINITE ELEMENT CODE TORT 11	75-0752
TOUGHNESS	APPLICATION OF THE SINGLE-EDGE NOTCHED BEAM TO FRACTURE TOUGHNESS TESTING OF CERAMICS	74-0623
TOUGHNESS	ESTABLISHMENT OF A VALID FRACTURE TOUGHNESS SPECIMEN GEOMETRY USING A FINITE ELEMENT ANALYSIS	74-0135
TOUGHNESS	FINITE-ELEMENT ANALYSIS OF A FRACTURE TOUGHNESS TEST SPECIMEN IN NON-LINEAR RANGE	72-0093
TOUGHNESS	FRACTURE TOUGHNESS PARAMETERS AND ELASTIC-PLASTIC ANALYSIS OF NON-MODERATE FRACTURE CONDITIONS USING	73-0258
TOUGHNESS	FRACTURE TOUGHNESS PARAMETERS AND ELASTIC-PLASTIC ANALYSIS OF NON-MODERATE FRACTURE CONDITIONS USING	73-1126
TOUGHNESS	FRACTURE TOUGHNESS TESTING WITH SECTIONS OF CYLINDERS	74-0999
TOUGHNESS	STRESS INTENSITY FACTORS FOR HEXAGONAL FRACTURE TOUGHNESS SPECIMENS	71-0799
TOUGHNESS	STUDY ON FRACTURE TOUGHNESS EVALUATION BY ELASTO-PLASTIC ANALYSIS AROUND NOTCHES	74-0619
TOUGHNESS	THEORETICAL STUDY OF THE EFFECT OF THE INTERFACE ON COMPOSITE TOUGHNESS	73-0386
TOUR	TOUR MAINE-MONTPARNASSE QUELQUES PROBLEMES DE STRUCTURE (SPECIFIC PROBLEMS ENCOUNTERED IN THE CONSTR	74-0539
TOWA	EXPLICIT EXPRESSIONS FOR TRIANGULAR TOWA ELEMENT STIFFNESS MATRIX	68-0160
TOWED	APPLICATION OF THE FINITE-ELEMENT METHOD TO TOWED CABLE DYNAMICS	73-1141
TOWED	APPLICATION OF THE FINITE-ELEMENT METHOD TO TOWED CABLE DYNAMICS	75-0809
TOWER	APPROXIMATE ANALYSIS OF PREFABRICATED TOWER STRUCTURES USING THIN SHELL FINITE ELEMENTS	73-0893
TOWER	FOUNDATION INSTRUMENTATION FOR THE NATIONAL WESTMINISTER BANK TOWER	74-1147
TOWER	NONLINEAR DYNAMIC ANALYSIS OF COOLING TOWER.	72-0219
TOWER	NUMERICAL ANALYSIS OF COOLING TOWER PLUMES AND FOG INCIDENCE	74-0040
TOWER	STRESSES AND DEFLECTIONS IN COOLING TOWER SHELLS DUE TO WIND LOADING	68-0088
TOWER	THE ANALYSIS OF COOLING TOWER RING BEAMS	68-0063
TOWERS	ANALYSIS OF COOLING TOWERS BY THE MATRIX FINITE ELEMENT METHOD- 2	70-0148
TOWERS	DOUBLE CURVATURE SHELL FINITE-ELEMENT AND ITS USE IN THE DYNAMIC ANALYSIS OF COOLING TOWERS AND OTHE	70-0312
TOWERS	EARTHQUAKE ANALYSIS OF AXISYMMETRIC TOWERS PARTIALLY SUBMERGED IN WATER	75-0076
TOWERS	NON-LINEAR ELASTIC ANALYSIS OF UNGUYED TOWERS AND STACKS	74-1157
TOWERS	SOBRE TORRES ESPIADAS - METODO DE CALCULO E UM EXEMPLO DE APLICACAO. (METHOD AND AN EXAMPLE OF CALCU	74-0915
TOWERS	THERMAL LOADING OF THIN-SHELL CONCRETE COOLING TOWERS	74-0667
TRABECULAR	STRUCTURAL MODEL FOR THE MECHANICAL BEHAVIOUR OF TRABECULAR BONE	73-0473
TRACER	NUMERICAL PREDICTION OF TRACER CONCENTRATION IN AQUIFERS WITH TRANSIENT GROUNDWATER FLOW BY FINITE-E	75-0610
TRACER	SOLUTION OF A TWO DIMENSIONAL DIFFUSION CONVECTION EQUATION DESCRIBING THE MASS TRANSFER OF A NEUTRA	74-0037
TRACK	ASSESSMENT OF DESIGN TOOLS AND CRITERIA FOR URBAN RAIL TRACK STRUCTURES. VOLUME 11. AT-GRADE SLAB TR	74-0813
TRACK	ASSESSMENT OF DESIGN TOOLS AND CRITERIA FOR URBAN RAIL TRACK STRUCTURES. VOLUME 11. AT-GRADE SLAB TR	74-0813
TRACK	DESIGN OF CONCRETE TRACK SLAB ON ELASTIC FOUNDATION USING THE FINITE ELEMENT METHOD	73-0479
TRACK	ELASTIC-PLASTIC ANALYSIS OF A WHEEL ROLLING ON A RIGID TRACK	74-1040
TRACK	STRESS ANALYSIS OF CONCRETE TRACK SLABS ON AN ELASTIC FOUNDATION BY THE FINITE ELEMENT METHOD	74-1049
TRACK	STRUCTURAL INELASTICITY. V11. ELASTIC-PLASTIC ANALYSIS OF A WHEEL ROLLING ON A RIGID TRACK	73-0886
TRACK	THE KANSAS TEST TRACK. NON-CONVENTIONAL TRACK STRUCTURES. DESIGN REPORT	72-0797
TRACK	THE KANSAS TEST TRACK. NON-CONVENTIONAL TRACK STRUCTURES. DESIGN REPORT	72-0797
TRACK	UNTERSUCHUNGEN ZUR SPURMASSTABILITAET BEI RADSAETZEN MIT KLOTZGEBREMSTEN VOLLRAEDERN (INVESTIGATIONS	75-0810
TRACKED	CONCEPT FEASIBILITY COMBAT TRACKED VEHICLE SIGNATURE DUPLICATOR	64-0028
TRACTIVE	FINITE ELEMENT METHOD OF PREDICTING STRESS DISTRIBUTIONS AND SOIL DEFORMATION UNDER A TRACTIVE DEVIC	69-0389
TRACTIVE	THE FINITE ELEMENT METHOD FOR PREDICTING STRESS DISTRIBUTION AND SOIL DEFORMATION UNDER A TRACTIVE D	69-0522
TRACTIVE	THE FINITE ELEMENT METHOD FOR PREDICTING STRESS DISTRIBUTION AND SOIL DEFORMATION UNDER A TRACTIVE D	71-0801
TRAFFIC	EFFECT OF NONLINEAR MATERIAL RESPONSE ON THE BEHAVIOUR OF PAVEMENTS UNDER TRAFFIC	70-0171
TRAFFIC	FRACTURE (ULTIMATE STRENGTH) ANALYSES OF ASPHALT PAVEMENT LAYERS RESULTING FROM TRAFFIC LOADING	73-0634
TRAGLAST	TRAGLAST VON T-TRAEGERN. EXPERIMENT UND F.E. BERECHNUNGEN (ULTIMATE STRENGTH OF T-BEAMS. EXPERIMENTS	74-0935
TRAGLASTBERECHN	STARR-PLASTISCHE TRAGLASTBERECHNUNGEN MITTELS DER METHODE DER FINITEN ELEMENTE	71-0635
TRAGVERHALTEN	ZUM TRAGVERHALTEN VON WAENDEN AUS FERTIGTEILTAFELN (BEHAVIOR OF LARGE PANEL SHEAR WALLS)	74-0299
TRAGWERKE	DIE BERECHNUNG ACHSENSYMETRISCHER TRAGWERKE UNTER ACHSENSYMETRISCHER UND NICHT-SYMMETRISCHER BELAST	69-0452
TRAGWERKEN	BERECHNUNG VON SPANNUNGSKONZENTRATIONEN AN TRAGWERKEN UNTER THERMO-MECHANISCHER BELASTUNG	73-0686
TRAGWERKEN	CALCULATION OF STRESS CONCENTRATIONS ON SUPPORT STRUCTURES UNDER THERMOMECHANICAL LOADING (BERECHNUN	73-0796
TRAGWERKEN	DIE ANWENDUNG VON INTERAKTIVER COMPUTER-GRAPHIK BEI DER BERECHNUNG VON TRAGWERKEN. (APPLICATION OF I	72-0102
TRAGWERKEN	DIE ELASTO-PLASTISCHE BERECHNUNG VON ALLGEMEINEN TRAGWERKEN UND KONTINUA	69-0443
TRAGWERKEN	MALEN DIMENSIONIERUNG VON TRAGWERKEN	74-0683
TRAGWERKEN	MESY - EIN PROGRAMMSYSTEM ZUR UNTERSUCHUNG VON TRAGWERKEN. (STRUCTURAL ANALYSIS BY MESY PROGRAMMING	75-0360
TRAINING	COMPANY TRAINING OF ENGINEERING IN STRUCTURAL ANALYSIS COMPUTER PROGRAM USAGE	72-0460
TRAJECTORIES	PLANE STRESS CRACK TRAJECTORIES IN LOW MODULUS MATERIALS	75-0646
TRANAL	TRANAL: A 3-D FINITE ELEMENT CODE FOR TRANSIENT NONLINEAR ANALYSIS	74-0707
TRANSACTIONS	TRANSACTIONS OF ARMY MATHEMATICIANS RESEARCH CONFERENCE	73-0925
TRANSDUCER	APPLICATIONS OF THE FINITE-ELEMENT METHOD AND COMPUTER GRAPHICS. (A VIBRATIONAL ANALYSIS OF A SONAR	72-0729
TRANSDUCER	DEVELOPMENT OF A FINITE ELEMENT MODEL FOR THE CLASS V FLEXTENSIONAL UNDERWATER TRANSDUCER SHELL	71-0754
TRANSDUCER	DEVELOPMENT OF A FINITE ELEMENT MODEL FOR THE CLASS 5 FLEXTENSIONAL UNDERWATER TRANSDUCER SHELL	71-0840
TRANSDUCER	DEVELOPMENT OF A FINITE ELEMENT MODEL FOR THE CLASS IV FLEXTENSIONAL UNDERWATER TRANSDUCER SHELL	72-0593
TRANSDUCER	IMPEDANCE OF INTERSTITIAL GAPS IN TRANSDUCER ARRAYS	70-0410

424

TRANSDUCER	VIBRATIONAL ANALYSIS OF THE FLEXING HEAD OF THE BQS-6 TRANSDUCER	71-0708
TRANSDUCERS	A FINITE-ELEMENT APPPOACH TO THE ANALYSIS OF TANGENTIALLY POLARIZED PIEZOELECTRIC-CERAMIC FREE-FLOOD	74-0802
TRANSDUCERS	ANALYSIS OF STRUCTURAL-ACOUSTIC INTERACTIONS IN METAL-CERAMIC TRANSDUCERS	72-0850
TRANSDUCERS	ANALYSIS OF STRUCTURAL ACOUSTIC INTERACTIONS IN METAL CERAMIC TRANSDUCERS	73-0401
TRANSDUCERS	FINITE ELEMENT ANALYSIS OF FLEXURE - TYPE VIBRATORS WITH ELECTROSTRICTIVE TRANSDUCERS	70-0338
TRANSDUCERS	FINITE ELEMENT ANALYSIS OF ACOUSTICALLY RADIATING STRUCTURES WITH APPLICATIONS TO SONAR TRANSDUCERS	73-0075
TRANSDUCERS	FINITE ELEMENT ANALYSIS OF ACOUSTICALLY RADIATING STRUCTURES WITH APPLICATIONS TO SONAR TRANSDUCERS	74-0524
TRANSDUCERS	VIBRATIONAL RESPONSE OF SONAR TRANSDUCERS USING PIEZO-ELECTRIC FINITE ELEMENTS	74-0522
TRANSFER	A FINITE ELEMENT FORMULATION OF HEAT CONDUCTION AND HEAT AND MASS TRANSFER PROBLEMS	74-0350
TRANSFER	A METHOD OF ANALYSIS OF LINE STRUCTURES BY TRANSFER MATRICES DERIVED FROM FINITE ELEMENTS	74-0808
TRANSFER	ANALYSIS OF THIN-WALLED CURVED BEAMS BY THE TRANSFER MATRIX METHOD	72-0335
TRANSFER	ANALYTIC METHODS IN CONDUCTION HEAT TRANSFER	71-0186
TRANSFER	APPLICABILITY OF APPROXIMATE AND EXACT TRANSIENT HEAT TRANSFER ANALYSES TO HEATING PROCESSES USED TO	75-0244
TRANSFER	APPLICATION OF THE FINITE ELEMENT METHOD TO CONVECTION HEAT TRANSFER BETWEEN PARALLEL PLANES	71-0244
TRANSFER	APPLICATION OF THE FINITE-ELEMENT METHOD TO HEAT-TRANSFER PROBLEMS. PART 11. TRANSIENT TWO-DIMENSION	71-0759
TRANSFER	COUPLED CONVECTIVE AND CONDUCTIVE HEAT TRANSFER BY FINITE ELEMENT METHODS	74-0050
TRANSFER	COUPLED CONVECTIVE/CONDUCTIVE HEAT TRANSFER INCLUDING VELOCITY FIELD EVALUATION	75-0116
TRANSFER	DYNAMICS OF DISTRIBUTED PARAMETER ROTOR SYSTEMS: TRANSFER MATRIX AND FINITE ELEMENT TECHNIQUES	70-0461
TRANSFER	EFFECT OF THE VARIATION OF CONSISTENCY IN RESIDUAL SOILS ON LOAD TRANSFER BY BORED PILES	75-0812
TRANSFER	ERROR ESTIMATES FOR FINITE ELEMENT METHOD FOR HEAT TRANSFER PROBLEMS IN THE SPACE-TIME DOMAIN	75-0104
TRANSFER	FINITE CONDUCTANCE ELEMENT METHOD OF CONDUCTION HEAT TRANSFER	73-0963
TRANSFER	FINITE ELEMENT ANALYSIS OF STEADY-STATE NONLINEAR HEAT TRANSFER PROBLEMS	74-0572
TRANSFER	FINITE ELEMENT ANALYSIS OF LINEAR AND NONLINEAR HEAT TRANSFER	74-0626
TRANSFER	FINITE ELEMENT ANALYSIS OF LAMINAR CONVECTIVE HEAT TRANSFER IN VERTICAL DUCTS WITH ARBITRARY CROSS-S	75-0555
TRANSFER	FINITE ELEMENT METHOD IN FLUID DYNAMICS AND HEAT TRANSFER	67-0151
TRANSFER	FINITE ELEMENT METHOD APPLIED TO TRANSIENT TWO-DIMENSIONAL HEAT TRANSFER WITH CONVECTION AND RADIATI	70-0127
TRANSFER	HEAT TRANSFER PROBLEMS AND TEMPERATURE STRESSES	69-0326
TRANSFER	HEAT TRANSFER WITH NONLINEAR BOUNDARY CONDITIONS VIA A VARIATIONAL PRINCIPLE	71-0443
TRANSFER	HEATRAN, A FINITE ELEMENT CODE FOR HEAT TRANSFER PROBLEMS	69-0076
TRANSFER	HETRAP: A HEAT TRANSFER ANALYSIS PROGRAM	74-0870
TRANSFER	LOAD TRANSFER MECHANISM FOR PIERS SOCKETTED IN HARD SOILS OR ROCK	74-0969
TRANSFER	MECHANICS OF LOAD TRANSFER AT THE FIBER/MATRIX INTERFACE	72-0879
TRANSFER	NEW APPROACH FOR PLATE VIBRATIONS. COMBINATION OF TRANSFER MATRIX AND FINITE ELEMENT TECHNIQUE	71-0434
TRANSFER	SOLUTION OF A TWO DIMENSIONAL DIFFUSION CONVECTION EQUATION DESCRIBING THE MASS TRANSFER OF A NEUTRA	74-0037
TRANSFER	STATIC AND HEAT TRANSFER ANALYSIS OF A SPHERE-CONE INTERSECTION IN A NUCLEAR CONTAINMENT VESSEL	75-0166
TRANSFER	SUBSONIC AND SUPERSONIC INDICIAL AERODYNAMICS AND AERODYNAMIC TRANSFER FUNCTION FOR COMPLEX CONFIGUR	74-1320
TRANSFER	TEMPERATURVERLAUF IM QUERSCHNITT GESCHWEISSTER ROHRWAENDE BEI UEBERWIEGEND KONVEKTIVEM WAERMEUEBERGA	75-0628
TRANSFER	THE FINITE ELEMENT SOLUTION OF FLOW AND HEAT TRANSFER IN ELLIPTICAL DUCTS	74-0039
TRANSFER-MATRIC	ON THE EXPLICIT REPRESENTATION OF TRANSFER-MATRICES OF CURVED BEAMS AND THEIR APPLICATIONS	75-0565
TRANSFER-MATRIX	VIBRATION ANALYSIS OF PLATES, SHELLS AND BLADED RIMMED DISCS USING THE TRANSFER-MATRIX-FINITE-ELEMEN	73-1151
TRANSFINITE	DISCRETIZATION ERROR BOUNDS FOR TRANSFINITE ELEMENTS	72-0533
TRANSFINITE	TRANSFINITE ELEMENT METHODS: BLENDING FUNCTION INTERPOLATION OVER ARBITRARY CURVED ELEMENT DOMAINS	73-1006
TRANSFORM	APPLICATION OF LAPLACE TRANSFORM AND FINITE ELEMENT METHOD TO SOLUTION OF NONSTATIONARY THERMAL COND	74-0171
TRANSFORM	APPLICATION OF THE FAST FOURIER TRANSFORM TO LINEAR SYSTEMS IN CIVIL ENGINEERING	73-1119
TRANSFORM	FINITE ELEMENT METHOD AND LAPLACE TRANSFORM COMPARATIVE SOLUTIONS OF TRANSIENT HEAT CONDUCTION PROBL	74-0575
TRANSFORMATIONS	SOLUTION OF STRESSES AND TRANSFORMATIONS OF EARTH DAMS BY THE METHOD OF FINITE ELEMENTS	69-0426
TRANSFORMATIONS	TRANSFORMATIONS FOR ELLIPTICAL HOLES APPLIED TO FINITE ELEMENT ANALYSIS	74-0132
TRANSFORMED	COMPUTER INTEGRATION OF CONGRUENTLY TRANSFORMED MATRICES	71-0580
TRANSFORMER	ANALYSIS OF TRANSFORMER LEAKAGE PHENOMENA BY HIGH-ORDER FINITE ELEMENTS	73-0135
TRANSFORMER	NATURAL FREQUENCIES AND MODE SHAPES OF TRANSFORMER CORES	65-0048
TRANSFORMER	TRANSFORMER LEAKAGE FLUX PROGRAM BASED ON FINITE ELEMENT METHOD	73-0004
TRANSFORMERS	ANALYSIS OF STEADY ELECTRICAL AND MAGNETIC-FIELDS BY FINITE-ELEMENT METHOD - APPLICATION TO TRANSFOR	75-0550
TRANSFORMERS	STUDIO DI CAMPI ELETTRICI E MAGNETICI STANZIONARI CON IL METODO DEGLI FLEMENTI FINITI. APPLICAZIONE	75-0207
TRANSFORMS	THE NUMERICAL SOLUTION OF VISCO-ELASTIC PROBLEMS USING LAPLACE TRANSFORMS	74-0436
TRANSIENT	A COMPUTER BASED PROCEDURE FOR PREDICTING THE TRANSIENT RESPONSE AND FAILURE OF A TWO-DIMENSIONAL CO	71-0630
TRANSIENT	A DISCRETE LEAST-SQUARES FINITE ELEMENT METHOD FOR TRANSIENT HEAT CONDUCTION PROBLEMS	74-1191
TRANSIENT	A FINITE ELEMENT ANALYSIS OF TRANSIENT RAREFIED GAS FLOW	73-0604
TRANSIENT	A FINITE ELEMENT SOLUTION OF THE TRANSIENT MAGNETOHYDRODYNAMIC COUETTE FLOW	74-1103
TRANSIENT	A METHOD FOR THE TRANSIENT NONLINEAR COUPLED THERMOELASTIC ANALYSIS OF AXISYMMETRIC BODIES USING FIN	74-1140
TRANSIENT	A METHOD OF ANALYZING TRANSIENT FLUID FLOW IN MULTILAYERED AQUIFERS	69-0142
TRANSIENT	A NUMERICAL MODEL OF TRANSIENT WATER QUALITY IN A ONE-DIMENSIONAL ESTUARY BASED ON THE FINITE ELEMEN	74-1369
TRANSIENT	A THREE DIMENSIONAL SOLUTION OF THE TRANSIENT FIELD PROBLEM USING ISOPARAMETRIC FINITE ELEMENTS	72-0601
TRANSIENT	ACCURACY, STABILITY, AND OSCILLATION CHARACTERISTICS OF TRANSIENT TWO-DIMENSIONAL HEAT CONDUCTION	75-0849
TRANSIENT	AN APPLICATION OF LEAST SQUARES TO ONE-DIMENSIONAL TRANSIENT PROBLEMS	74-0347
TRANSIENT	AN APPLICATION OF THE FINITE ELEMENT METHOD TO TRANSIENT INTERFACE PROBLEMS IN POROUS MEDIA	73-0608
TRANSIENT	ANALYSIS OF RECTANGULAR PLATES IN TRANSIENT CREEP BENDING	69-0425
TRANSIENT	APACHE: A THREE-DIMENSIONAL FINITE ELEMENT PROGRAM FOR STEADY-STATE OR TRANSIENT HEAT CONDUCTION ANA	73-0971
TRANSIENT	APPLICABILITY OF APPROXIMATE AND EXACT TRANSIENT HEAT TRANSFER ANALYSES TO HEATING PROCESSES USED TO	75-0244
TRANSIENT	APPLICATION OF THE FINITE ELEMENT METHOD TO TRANSIENT FLOW IN POROUS MEDIA	68-0094
TRANSIENT	APPLICATION OF THE FINITE-ELEMENT METHOD TO HEAT-TRANSFER PROBLEMS. PART 11. TRANSIENT TWO-DIMENSION	71-0759
TRANSIENT	BASIC RESEARCH IN METHODS OF APPROXIMATION IN THE NONLINEAR MECHANICS OF SOLIDS AND STRUCTURES. VOLU	74-1091
TRANSIENT	BERECHNUNG INSTATIONAERER TEMPERATUREFELDER MIT FINITEN ELEMENTEN IN RAUM-UND ZEITDIMENSION. (CALCUL	74-0987
TRANSIENT	CALCULATION BY FINITE ELEMENT TECHNIQUE OF STEADY-STATE OR TRANSIENT HEAT-FLOW CONDUCTION	74-0637
TRANSIENT	CALCULATION OF TRANSIENT TEMPERATURE FIELDS WITH FINITE ELEMENTS IN SPACE AND TIME DIMENSIONS	74-0379
TRANSIENT	CAVITY STABILITY: FINITE ELEMENT ANALYSIS FOR STEADY AND TRANSIENT CREEP	74-1082
TRANSIENT	CODE TAFEST: NUMERICAL SOLUTION TO TRANSIENT HEAT-CONDUCTION PROBLEMS USING FINITE ELEMENTS IN SPACE	74-0901
TRANSIENT	COMPARISON OF FINITE-DIFFERENCE AND FINITE-ELEMENT SOLUTION TECHNIQUE IN TRANSIENT HEAT-CONDUCTION	75-0348
TRANSIENT	COMPARISON OF TIME MARCHING SCHEMES FOR THE TRANSIENT HEAT CONDUCTION EQUATION	75-0477
TRANSIENT	CONVECTED COORDINATES FOR TRANSIENT NONLINEAR FINITE-ELEMENT ANALYSIS	73-0850
TRANSIENT	CORRELATION BETWEEN PRESSURE AND THERMAL TRANSIENT STRESS LEVELS DERIVED BY THEORY AND EXPERIMENT ON	73-0469
TRANSIENT	COUPLE - A PSEUDO FINITE ELEMENT CONDUCTION CODE FOR THE TRANSIENT THERMAL RESPONE OF AXISYMMETRIC 0	72-0778
TRANSIENT	CRASH: A COMPUTER SIMULATION OF NONLINEAR TRANSIENT RESPONSE OF STRUCTURES	72-0569
TRANSIENT	DIGITAL COMPUTER PROGRAMS FOR STEADY-STATE OR TRANSIENT TEMPERATURE ANALYSIS OF PLANE AND AXISYMMETR	72-0915
TRANSIENT	DISCRETE VARIABLE METHOD APPLIED TO TRANSIENT HEAT CONDUCTION PROBLEMS	74-1163
TRANSIENT	DYNAMIC ANALYSIS OF AXISYMMETRIC SHELLS UNDER ARBITRARY TRANSIENT PRESSURES.	72-0467
TRANSIENT	DYNAMIC TRANSIENT NON-LINEAR BEHAVIOUR OF THICK AND THIN PLATES	75-0125
TRANSIENT	EFFICIENT SOLUTION PROCESSES FOR FINITE ELEMENT ANALYSIS OF TRANSIENT HEAT CONDUCTION	69-0014
TRANSIENT	ERROR ANALYSIS OF A REMOTE TRANSIENT HEAT FLUX SENSOR	74-0753
TRANSIENT	FINITE ELEMENT ANALYSIS OF LARGE TRANSIENT ELASTIC-PLASTIC DEFORMATIONS OF SIMPLE STRUCTURES, WITH A	72-0849
TRANSIENT	FINITE ELEMENT ANALYSIS OF TRANSIENT HEAT CONDUCTION. APPLICATION OF THE WEIGHTED RESIDUAL PROCESS	74-0277
TRANSIENT	FINITE ELEMENT ANALYSIS OF TRANSIENT NONLINEAR STRUCTURAL BEHAVIOR	75-0297
TRANSIENT	FINITE ELEMENT BASED PROCEDURE FOR SIMULATING THE TRANSIENT RESPONSE AND FAILURE OF A TWO-DIMENSIONA	72-0123
TRANSIENT	FINITE ELEMENT METHOD APPLIED TO TRANSIENT TWO-DIMENSIONAL HEAT TRANSFER WITH CONVECTION AND RADIATI	70-0127
TRANSIENT	FINITE ELEMENT METHOD FOR TRANSIENT LINEAR VISCOUS FLOW PROBLEMS	74-0474
TRANSIENT	FINITE ELEMENT METHOD AND LAPLACE TRANSFORM COMPARATIVE SOLUTIONS OF TRANSIENT HEAT CONDUCTION PROBL	74-0575
TRANSIENT	FINITE ELEMENT MODEL FOR TRANSIENT TWO-LAYER COOLING POND BEHAVIOR	75-0465
TRANSIENT	FINITE ELEMENT SOLUTION FOR AXISYMMETRIC TRANSIENT THERMAL STRESSES	72-0823
TRANSIENT	FINITE ELEMENT SOLUTIONS OF TRANSIENT, TWO-REGIME FLOW TOWARDS WELLS	74-0466
TRANSIENT	FINITE-ELEMENT ANALYSIS OF LARGE ELASTIC-PLASTIC TRANSIENT DEFORMATIONS OF SIMPLE STRUCTURES	71-0326
TRANSIENT	FINITE-ELEMENT PREDICTIONS OF TRANSIENT ELASTIC-PLASTIC LARGE DEFLECTIONS OF STIFFENED AND/OR UNSTIF	74-0825
TRANSIENT	FINITE-ELEMENT SOLUTION OF NAVIER-STOKES EQUATIONS FOR TRANSIENT 2-DIMENSIONAL INCOMPRESSIBLE FLOW	75-0134
TRANSIENT	FOURIER ANALYSIS OF TRANSIENT PROBLEMS BY THE FINITE ELEMENT METHOD	74-1331

425

TRANSIENT	FREE VIBRATION AND TRANSIENT RESPONSE OF THICK AND THIN PLATES USING THE FINITE ELEMENT METHOD	74-0517
TRANSIENT	GALERKIN-FINITE ELEMENT TECHNIQUE FOR CALCULATING TRANSIENT POSITION OF SALTWATER FRONT	75-0135
TRANSIENT	HIGH-ORDER FINITE ELEMENT ANALYSIS FOR TRANSIENT TEMPERATURE ANALYSIS OF INHOMOGENEOUS MATERIALS	69-0301
TRANSIENT	INCREMENTAL THEORY OF THREE DIMENSIONAL TRANSIENT THERMOELASTOPLASTICITY - FORMULATION AND SOLUTION	73-0555
TRANSIENT	INFLUENCE OF INTERLAMINATE FLAWS ON TRANSIENT TEMPERATURE DISTRIBUTIONS	73-0223
TRANSIENT	LARGE DISPLACEMENT, NONLINEAR TRANSIENT ANALYSIS BY FINITE ELEMENTS	74-0324
TRANSIENT	NON-LINEAR TRANSIENT FINITE ELEMENT ANALYSIS WITH CONVECTED CO-ORDINATES	73-0378
TRANSIENT	NONLINEAR FINITE-ELEMENT FORMULATION FOR TRANSIENT ANALYSIS OF THREE DIMENSIONAL THIN STRUCTURES	74-0874
TRANSIENT	NONLINEAR TRANSIENT RESPONSES OF STRUCTURES BY SPATIAL FINITE-ELEMENT METHOD	73-0138
TRANSIENT	NONLINEAR TRANSIENT ANALYSIS OF SHELLS AND SOLIDS OF REVOLUTION BY CONVECTED ELEMENTS	73-0519
TRANSIENT	NONLINEAR TRANSIENT ANALYSIS OF SHELLS AND SOLIDS OF REVOLUTTION BY CONVECTED ELEMENTS	74-0476
TRANSIENT	NONLINEAR TRANSIENT ANALYSIS OF AUTOMOTIVE STRUCTURES BY FINITE ELEMENTS	75-0301
TRANSIENT	NUMERICAL MODEL FOR THE PREDICTION OF TRANSIENT WATER QUALITY IN ESTUARY NETWORKS	72-0635
TRANSIENT	NUMERICAL PREDICTION OF TRACER CONCENTRATION IN AQUIFERS WITH TRANSIENT GROUNDWATER FLOW BY FINITE-E	75-0610
TRANSIENT	ON SOME FREE SURFACE ON TRANSIENT FLOW PROBLEMS OF SEEPAGE AND IRROTATIONAL FLOW	72-0070
TRANSIENT	ON THE ACCURACY OF FINITE ELEMENT SOLUTIONS TO THE TRANSIENT HEAT CONDUCTION EQUATION	74-0182
TRANSIENT	ON THE SOLUTION OF TRANSIENT FREE-SURFACE FLOW PROBLEMS IN POROUS MEDIA BY THE FINITE ELEMENT METHOD	73-1019
TRANSIENT	RESPONSE OF NUCLEAR STRUCTURAL SYSTEMS TO TRANSIENT AND RANDOM EXCITATIONS, USING BOTH DETERMINISTIC	74-1047
TRANSIENT	SEEPAGE IN MISSISSIPPI RIVER BANKS. REPORT 1. ANALYSIS OF TRANSIENT SEEPAGE USING VISCOUS FLOW MODEL	70-0405
TRANSIENT	SEEPAGE IN MISSISSIPPI RIVER BANKS. REPORT 1. ANALYSIS OF TRANSIENT SEEPAGE USING A VISCOUS-FLOW MOD	73-0909
TRANSIENT	SOLUTION OF TRANSIENT GROUNDWATER FLOW PROBLEMS BY FINITE ELEMENT METHOD	72-0046
TRANSIENT	SPONTANEOUS IGNITION - FINITE ELEMENT SOLUTIONS FOR STEADY AND TRANSIENT CONDITIONS	74-0353
TRANSIENT	STABILITY AND OSCILLATION CHARACTERISTICS OF FINITE ELEMENT, FINITE-DIFFERENCE, AND WEIGHTED-RESIDUA	75-0615
TRANSIENT	STEADY STATE AND TRANSIENT GROUNDWATER FLOW IN SLOPES	70-0273
TRANSIENT	STEADY STATE AND TRANSIENT HEAT CONDUCTION AND STRESS ANALYSES OF TWO AND THREE DIMENSIONAL STRUCTUR	71-0675
TRANSIENT	TACOS: A FINITE ELEMENT COMPUTER PROGRAM FOR THE TRANSIENT ANALYSIS OF CYLINDRICAL OBESE SHELLS	72-0588
TRANSIENT	THE DAMPED TRANSIENT BEHAVIOR OF FINITE BEAMS AND PLATES	69-0520
TRANSIENT	THE FINITE ELEMENT ANALYSIS OF TRANSIENT RAREFIED GAS FLOW	74-0027
TRANSIENT	THE TRANSIENT DYNAMIC ANALYSIS OF THIN SHELLS BY THE FINITE ELEMENT METHODS	71-0142
TRANSIENT	THE TRANSIENT LINEAR ELASTIC RESPONSE ANALYSIS COMPLEX THIN SHELLS OF REVOLUTION SUBJECTED TO ARBITR	70-0080
TRANSIENT	THEORETICAL AND EXPERIMENTAL STUDIES OF TRANSIENT ELASTIC-PLASTIC LARGE DEFLECTIONS OF GEOMETRICALLY	75-0830
TRANSIENT	THEORETICAL AND EXPERIMENTAL STUDIES OF TRANSIENT ELASTIC-PLASTIC LARGE DEFLECTIONS OF GEOMETRICALLY	75-0831
TRANSIENT	THEORETICAL TRANSIENT BEHAVIOUR OF SATURATED AND UNSATURATED SOILS UNDER LOAD AND CHANGING MOISTURE	73-0901
TRANSIENT	TRANAL: A 3-D FINITE ELEMENT CODE FOR TRANSIENT NONLINEAR ANALYSIS	74-0707
TRANSIENT	TRANSIENT ANALYSIS OF THICK CYLINDRICAL SHELLS	71-0590
TRANSIENT	TRANSIENT AND NONLINEAR-ANALYSIS OF FIELD PROBLEMS BY FINITE ELEMENTS	72-0975
TRANSIENT	TRANSIENT DYNAMIC RESPONSE OF VISCOELASTIC STRUCTURES	72-0234
TRANSIENT	TRANSIENT ELASTOHYDRODYNAMIC LUBRICATION BY FINITE ELEMENT METHODS	72-0087
TRANSIENT	TRANSIENT FIELD PROBLEMS: TWO-DIMENSIONAL AND THREE-DIMENSIONAL ANALYSIS BY ISO-PARAMETRIC FINITE EL	70-0001
TRANSIENT	TRANSIENT FLOW OF GROUNDWATER TO WELLS IN MULTIPLE-AQUIFER SYSTEMS	69-0501
TRANSIENT	TRANSIENT FREE-SURFACE FLOW IN POROUS MEDIA BY FINITE ELEMENT METHOD	72-0297
TRANSIENT	TRANSIENT HEAT CONDUCTION PROBLEMS IN POWER CABLES SOLVED BY FINITE ELEMENT METHOD	73-0007
TRANSIENT	TRANSIENT HEAT FLOW IN SOLIDS	74-0461
TRANSIENT	TRANSIENT HYDROTHERMAL ANALYSIS OF SMALL LAKES	73-0445
TRANSIENT	TRANSIENT LAMINAR FLOW IN DUCTS OF ARBITRARY CROSS-SECTION BY FINITE ELEMENT METHODS	72-0013
TRANSIENT	TRANSIENT NATURAL CONVECTION COOLING OF A VERTICAL CIRCULAR CYLINDER	74-1169
TRANSIENT	TRANSIENT RESPONSE OF LINEAR ELASTIC STRUCTURES DETERMINED BY THE MATRIX EXPONENTAL METHOD	71-0240
TRANSIENT	TRANSIENT RESPONSE OF A COUPLED PLATE-ACOUSTIC SYSTEM USING PLATE AND ACOUSTIC FINITE ELEMENTS	71-0463
TRANSIENT	TRANSIENT RESPONSE OF A COUPLED PLATE-ACOUSTIC SYSTEM USING PLATE AND ACOUSTIC FINITE ELEMENTS	72-0275
TRANSIENT	TRANSIENT RESPONSE OF NON-UNIFORM, NON-HOMOGENEOUS BEAMS	73-0162
TRANSIENT	TRANSIENT RESPONSE OF INELASTICALLY CONSTRAINED RIGID BODY SYSTEMS	73-0291
TRANSIENT	TRANSIENT RESPONSE OF ELASTIC, VISCO-PLASTIC BEAMS	73-0380
TRANSIENT	TRANSIENT SHELL RESPONSE BY NUMERICAL TIME INTEGRATION	72-0305
TRANSIENT	TRANSIENT TEMPERATURE RESPONSE OF COMPOSITE SLABS	74-1038
TRANSIENT	TRANSIENT THERMAL STRESS TECHNIQUES IN NUCLEAR REACTORS. APPLIED MECHANICS ASPECTS OF NUCLEAR EFFECT	71-0474
TRANSIENT	TRANSIENT THERMAL STRESS ANALYSIS OF COMPOSITE STRUCTURES INCLUDING CONTINUOUSLY VARYING PROPERTIES	75-0681
TRANSIENT	TRANSIENT TWO DIMENSIONAL UNCONFINED AQUIFER PROBLEMS	74-0069
TRANSIENT	TRANSIENT TWO DIMENSIONAL HEAT CONDUCTION PROBLEMS SOLVED BY THE FINITE ELEMENT METHOD	74-0364
TRANSIENT	TRANSIENT TWO-DIMENSIONAL PROBLEMS UTILIZING THE LEAST SQUARES ALGORITHM	75-0219
TRANSIENT	TRANSIENT VIBRATION ANALYSIS OF LINEAR SYSTEMS USING TRANSITION MATRICES	68-0053
TRANSIENT	TWO-DIMENSIONAL FINITE ELEMENT MODELING OF TRANSIENT FLOW IN REGIONAL AQUIFER SYSTEMS. ANALYSIS OF G	74-0743
TRANSIENT	UNCONFINED TRANSIENT SEEPAGE IN SLOPING BANKS	71-0378
TRANSIENT	UNSYMMETRICAL TRANSIENT HEAT CONDUCTION: ROTATING DISK APPLICATIONS	70-0005
TRANSIENT	USE OF FINITE ELEMENT METHODS IN SOLUTION OF TRANSIENT HEAT CONDUCTION PROBLEMS	69-0016
TRANSIENT	USE OF HYBRID COMPUTATION FOR SOLVING TRANSIENT FLOW PROBLEM IN POROUS MEDIA BY FINITE ELEMENT METHO	74-0068
TRANSIENT	USE OF THE FINITE ELEMENT METHOD IN SOLVING TRANSIENT FLOW PROBLEMS IN AQUIFER SYSTEMS	68-0167
TRANSIENT	WEIGHTED RESIDUAL PROCESSES IN FINITE ELEMENT WITH PARTICULAR REFERENCE TO SOME TRANSIENT AND COUPLE	71-0503
TRANSITION	PHOTOELASTIC INVESTIGATION OF STRESS CONCENTRATIONS IN SPHERE-CYLINDER TRANSITION REGIONS: INCLUDING	68-0258
TRANSITION	TRANSIENT VIBRATION ANALYSIS OF LINEAR SYSTEMS USING TRANSITION MATRICES	68-0053
TRANSITIONS	THREE-DIMENSIONAL FINITE ELEMENT ANALYSIS OF THICK-WALLED PIPE-NOZZLE JUNCTIONS WITH CURVED TRANSITI	71-0153
TRANSITORY	NUMERICAL STUDY OF TRANSITORY OR NONLINEAR WAVES	74-0031
TRANSLATES	METHODE DES ELEMENTS FINIS; UNE FAMILE D'APPROXIMATION DES ESPACES DE SOBOLEV PAR LES TRANSLATES DE	70-0301
TRANSLATIONAL	FINITE ELEMENT ANALYSIS OF TRANSLATIONAL SHELLS	72-0348
TRANSMISSIBILIT	RADIAL AND AXIAL TRANSMISSIBILITY CHARACTERISTICS OF TYPICAL ROCKET VEHICLE STRUCTURE TECHNICAL SUMM	69-0475
TRANSMISSION	ACOUSTIC FINITE ELEMENT APPROACH FOR STUDYING BOUNDARY FLEXIBILITY AND SOUND TRANSMISSION BETWEEN IR	73-0096
TRANSMISSION	AN ACOUSTIC FINITE ELEMENT APPROACH FOR STUDYING BOUNDARY FLEXIBILITY AND SOUND TRANSMISSION BETWEEN	73-1087
TRANSMISSION	FINITE ELEMENT SOLUTION OF HEAT TRANSMISSION IN STEEL INGOTS	72-0311
TRANSMISSION	FINITE ELEMENT VIBRATION ANALYSIS OF MULTICONDUCTOR ELECTRICAL TRANSMISSION LINES	73-0590
TRANSMISSION	PREDICTION OF TRANSMISSION LOSS IN MUFFLERS BY THE FINITE ELEMENT METHOD	75-0160
TRANSMISSION	SINGULARITIES IN ELECTROMAGNETIC TRANSMISSION LINES	72-0072
TRANSMISSION	TRANSMISSION OF VIBRATION IN BEAM SYSTEMS	69-0405
TRANSMISSION	ZASTOSOWANIE METODY SZTYWNYCH ELEMENTOW SKONCZONYCH DO OBLICZEN ORGAN WALOW OKRETOYCH Z UWZGLEDNIENI	74-0962
TRANSMITTED	EFFECT OF BACKPACKING AND INTERNAL PRESSURIZATION ON STRESSES TRANSMITTED TO BURIED CYLINDERS	73-0897
TRANSONIC	COMPOSITE WING FOR TRANSONIC IMPROVEMENT. VOLUME 11. ADVANCED ANALYSIS EVALUATION	71-0657
TRANSONIC	FINITE ELEMENT ANALYSIS OF TRANSONIC FLOW	74-1223
TRANSPIRATION	TRANSPIRATION AND FILM COOLING BOUNDARY LAYER COMPUTER PROGRAM. VOLUME 2, COMPUTER PROGRAM AND USER'	71-0655
TRANSPORT	A FINITE ELEMENT APPROACH TO PHYSIOLOGICAL TRANSPORT PHENOMENA BY CILIARY AND PERISTALTIC PUMPING	72-0353
TRANSPORT	A FINITE ELEMENT MODEL FOR SEDIMENT TRANSPORT IN ESTUARIES	74-1102
TRANSPORT	A STUDY OF CONVECTIVE-DISPERSION EQUATION BY ISOPARAMETRIC FINITE ELEMENTS. TRANSPORT OF POLLUTANTS	75-0773
TRANSPORT	ADJOINING APPROPRIATE SINGULAR ELEMENTS TO TRANSPORT THEORY COMPUTATIONS	74-0946
TRANSPORT	APPLICATION OF FINITE ELEMENT TECHNIQUES FOR THE NUMERICAL SOLUTION OF THE NEUTRON TRANSPORT AND DIF	71-0013
TRANSPORT	APPLICATION OF FINITE ELEMENT SOLUTION TECHNIQUE TO NEUTRON DIFFUSION AND TRANSPORT EQUATIONS	71-0015
TRANSPORT	APPLICATION OF FINITE ELEMENT METHOD TO MASS TRANSPORT LIMITED EPITAXIAL GROWTH PROCESSES	71-0488
TRANSPORT	APPLICATION OF PHASE-SPACE FINITE ELEMENTS TO ONE DIMENSIONAL NEUTRON TRANSPORT EQUATION	73-0123
TRANSPORT	APPLICATION OF PHASE SPACE FINITE ELEMENTS TO THE TWO DIMENSIONAL NEUTRON TRANSPORT EQUATION IN X-Y	73-0453
TRANSPORT	APPLICATION OF THE FINITE-ELEMENT METHOD FOR SIMULATION OF SURFACE WATER TRANSPORT PROBLEMS	72-0826
TRANSPORT	APPLICATIONS OF FINITE ELEMENT METHODS IN REACTOR MATHEMATICS. NUMERICAL SOLUTION OF THE NEUTRON TRA	74-0853
TRANSPORT	CONTINUOUS AND DISCONTINUOUS FINITE ELEMENT METHODS FOR SOLVING THE TRANSPORT EQUATION	75-0099
TRANSPORT	CONVECTIVE TRANSPORT FINITE ELEMENT ANALOG	75-0289
TRANSPORT	COUPLED HEAT AND MOISTURE TRANSPORT MODEL FOR ARCTIC SOILS	74-0578
TRANSPORT	CURVED LOW ORDER ANGULAR ELEMENTS IN TWO DIMENSIONAL NEUTRON TRANSPORT PROBLEMS	74-1322
TRANSPORT	CYCLIC-STRESS ANALYSIS OF NOTCHES FOR SUPERSONIC TRANSPORT CONDITIONS	74-0834

TRANSPORT	FINITE ELEMENT INTEGRAL NEUTRON TRANSPORT	73-1032
TRANSPORT	FINITE ELEMENT METHODS FOR THE TRANSPORT EQUATION	74-0925
TRANSPORT	FINITE ELEMENT METHOD FOR SOLVING NEUTRON TRANSPORT PROBLEMS IN TWO-DIMENSIONAL CYLINDRICAL GEOMETRY	74-1081
TRANSPORT	FINITE ELEMENT SOLUTION TECHNIQUE FOR NEUTRON TRANSPORT EQUATION	72-0730
TRANSPORT	FINITE ELEMENT SOLUTION OF MULTIGROUP TRANSPORT EQUATION IN 2 DIMENSIONAL GEOMETRIES	74-0648
TRANSPORT	FINITE ELEMENT SOLUTION OF THE WIND DRIVEN CURRENTS AND ITS MASS TRANSPORT IN LAKES	74-1368
TRANSPORT	FINITE ELEMENTS IN NEUTRON TRANSPORT THEORY	73-0961
TRANSPORT	ISOPARAMETRIC FINITE-ELEMENT METHODS FOR 2-DIMENSIONAL TRANSPORT CALCULATIONS	75-0354
TRANSPORT	LINEAR DISCONTINUOUS FINITE-ELEMENT SCHEME IN TIME-DEPENDENT TRANSPORT CALCULATIONS	75-0498
TRANSPORT	NEW TRANSPORT METHODS AND CODES	74-0878
TRANSPORT	NON-SELF-ADJOINT VARIATIONAL PROCEDURE FOR FINITE-ELEMENT APPROXIMATION OF TRANSPORT EQUATION	75-0072
TRANSPORT	ON A FINITE ELEMENT METHOD FOR SOLVING THE NEUTRON TRANSPORT EQUATION	74-1255
TRANSPORT	ON A FINITE ELEMENT METHOD FOR SOLVING THE NEUTRON TRANSPORT EQUATION	74-1355
TRANSPORT	QUADRATIC FINITE ELEMENT IN NEUTRON TRANSPORT	73-1039
TRANSPORT	SINGULAR ELEMENTS IN VARIATIONAL AND FINITE ELEMENT TRANSPORT COMPUTATIONS	73-0902
TRANSPORT	SOLUTION OF MULTI-DIMENSIONAL NEUTRON TRANSPORT EQUATION BY FINITE ELEMENT METHOD	72-0564
TRANSPORT	SOLUTION OF TRANSPORT PROBLEMS BY USE OF TRIAL FUNCTION TECHNIQUES	72-0991
TRANSPORT	SPACE FINITE-ELEMENTS 2-DIMENSIONAL MULTIGROUP TRANSPORT CALCULATIONS USING PHASE-SPACE FINITE-ELEME	75-0504
TRANSPORT	TRANSPORT SOLUTIONS USING FINITE ELEMENTS IN SPACE-ANGLE PHASE SPACE	71-0480
TRANSPORT	TRANSPORT THEORY NUMERICAL METHODS	72-0769
TRANSPORT	TRIANGULAR MESH METHODS FOR THE NEUTRON TRANSPORT EQUATION	73-1062
TRANSPORT	TRIPLET: A TWO-DIMENSIONAL, MULTIGOUP, TRIANGULAR MESH, PLANAR GEOMETRY, EXPLICIT TRANSPORT CODE	73-0955
TRANSPORT	TWO-DIMENSIONAL FINITE ELEMENT ANALYSIS OF SEMI-CONDUCTOR STEADY STATE TRANSPORT EQUATIONS	74-1224
TRANSPORT	TWO-DIMENSIONAL TRANSPORT CALCULATIONS USING PHASE-SPACE FINITE ELEMENTS	72-0027
TRANSPORT	VARIATIONAL PRINCIPLE OF STEADY-STATE TRANSPORT PROCESSES	70-0190
TRANSPORTATION	M.I.T. TEST SECTION INSTRUMENTATION, MASSACHUSETTS BAY TRANSPORTATION AUTHORITY, HAYMARKET-NORTH EXT	72-0673
TRANSPORTATION	STATIC THREE DIMENSIONAL ANALYSIS OF ELEVATED STEEL TRANSPORTATION STRUCTURES	74-1187
TRANSVERSE	A FINITE ELEMENT ANALYSIS INCLUDING TRANSVERSE SHEAR EFFECTS FOR APPLICATIONS TO LAMINATED PLATES	71-0215
TRANSVERSE	A FINITE ELEMENT MODEL FOR LAMINATED PLATES INCLUDING TRANSVERSE SHEAR DEFORMATION	72-0818
TRANSVERSE	A FINITE-ELEMENT METHOD FOR TRANSVERSE VIBRATIONS OF BEAMS AND PLATES	67-0144
TRANSVERSE	AN INVESTIGATION OF THE BENDING OF SYMMETRIC PLATES UNDER SYMMETRIC TRANSVERSE LOADING	72-0662
TRANSVERSE	ANALYSIS OF MULTI-CELL BRIDGES WITHOUT TRANSVERSE DIAPHRAGMS, A FINITE ELEMENT APPROACH	69-0235
TRANSVERSE	ANALYSIS OF STRESSES IN CONCRETE NEAR A REINFORCING BAR DUE TO BOND AND TRANSVERSE CRACKING	70-0351
TRANSVERSE	APPLICATION OF FINITE ELEMENT METHODS FOR CALUCLATION OF TRANSVERSE OSCILLATIONS IN ONE DIMENSIONAL	71-0599
TRANSVERSE	DYNAMIC RESPONSE OF A CONSTRAINED FIBROUS SYSTEM SUBJECTED TO TRANSVERSE IMPACT. II. A MECHANICAL MO	70-0369
TRANSVERSE	DYNAMIC STABILITY OF A NONLINEAR BEAM SUBJECTED TO BOTH LONGITUDINAL AND TRANSVERSE EXCITATION	73-0553
TRANSVERSE	EFFECT OF TRANSVERSE SHEAR DEFORMATION IN VIBRATIONS OF PLANAR STRUCTURES COMPOSED OF BEAM TYPE ELEM	73-0402
TRANSVERSE	FINITE ELEMENT ANALYSIS OF PLATE BENDING WITH TRANSVERSE SHEAR DEFORMATION	70-0109
TRANSVERSE	FINITE ELEMENT ANALYSIS OF LAMINATED ANISOTROPIC PLATES INCLUDING TRANSVERSE SHEAR DEFORMATIONS	70-0506
TRANSVERSE	FINITE ELEMENT ANALYSIS OF PLATES ALLOWING FOR TRANSVERSE SHEAR DEFORMATION EFFECTS	73-0996
TRANSVERSE	FINITE ELEMENT ANALYSIS OF TRANSVERSE CRACKING IN LOW-EMBANKMENT DAMS	74-0696
TRANSVERSE	FINITE ELEMENT THEORY OF PLATES AND SHELLS INCLUDING TRANSVERSE SHEAR STRAIN EFFECTS	70-0395
TRANSVERSE	FINITE ELEMENTS IN TRANSVERSE FRAME ANALYSIS	72-0182
TRANSVERSE	INCLUSION OF TRANSVEPSE SHEAR DEFORMATION IN FINITE ELEMENT DISPLACEMENT FORMULATIONS	74-0662
TRANSVERSE	INELASTIC ANALYSIS OF A UNIDIRECTIONAL COMPOSITE SUBJECTED TO TRANSVERSE NORMAL LOADING	70-0015
TRANSVERSE	MATHEMATICAL AND EXPERIMENTAL STUDIES ON THE MECHANICS OF PLATED TRANSVERSE FRACTURES	74-0656
TRANSVERSE	MATRIX DISPLACEMENT ANALYSIS OF SHELLS AND PLATES INCLUDING TRANSVERSE SHEAR STRAIN EFFECTS	72-0643
TRANSVERSE	NEW TRIANGULAR PLATE-BENDING FINITE ELEMENT WITH TRANSVERSE SHEAR FLEXIBILITY	74-0627
TRANSVERSE	NEW TRIANGULAR PLATE-BENDING FINITE ELEMENT WITH TRANSVERSE SHEAR FLEXIBILITY	74-0964
TRANSVERSE	ON STRENGTH OF TRANSVERSE WEBS AND THEIR JOINTS OF LARGE TANKER	69-0327
TRANSVERSE	STRENGTH OF TRANSVERSE FILLET WELDED JOINTS	74-0144
TRANSVERSE	TANKER TRANSVERSE STRENGTH ANALYSIS: USER'S MANUAL	72-0750
TRANSVERSE	THE ANALYSIS OF THIN SHELLS WITH TRANSVERSE SHEAR STRAINS BY THE FINITE ELEMENT METHOD	68-0098
TRANSVERSE	THREE-DIMENSIONAL STRESS ANALYSIS OF A FINITE SLAB CONTAINING A TRANSVERSE CENTRAL CRACK	75-0732
TRANSVERSE	TRANSVERSE IMPACT PROBLEMS BY HIGHER-ORDER BEAM FINITE-ELEMENTS	75-0819
TRANSVERSE	TRANSVERSE LOADING OF UNIDIRECTIONAL FIBER COMPOSITES	69-0119
TRANSVERSE	TRANSVERSE PROPERTIES OF FIBROUS COMPOSITES	69-0069
TRANSVERSE	TRANSVERSE PROPERTIES OF UNIDIRECTIONAL ALUMINUM MATRIX FIBROUS COMPOSITES	69-0167
TRANSVERSE	TRANSVERSE REINFORCEMENT OF CONTINUOUS, TROUGH-SHAPED CONCRETE RAILROAD STRUCTURES. AN IMPLEMENTATIO	71-0719
TRANSVERSE	TRANSVERSE STRENGTH OF TANKERS, FINITE ELEMENT APPLICATIONS	68-0142
TRANSVERSE	TRANSVERSE STRENGTH AT FORE AND AFT SECTION OF TANKER CARGO HOLD	74-0618
TRANSVERSE	TRANSVERSE VIBRATION OF BEAMS WITH INTERNAL DAMPING, ROTATORY INERTIA AND SHEAR	63-0020
TRANSVERSE	TRANSVERSE VIBRATIONS OF CANTILEVER BARS OF VARIABLE CROSS SECTIONS	66-0001
TRANSVERSE	TRANSVERSE VIBRATIONS OF NON-UNIFORM BEAMS	74-1058
TRANSVERSE	TRANSVERSE VIBRATIONS OF SHIP'S PROPULSION SYSTEMS. PART 1: THEORETICAL ANALYSIS	74-1323
TRANSVERSE	TRANSVERSE VIBRATIONS OF SHIP'S PROPULSION SYSTEMS. PART 2: EXPERIMENTAL ANALYSIS	74-1324
TRANSVERSE	TRANSVERSE VIBRATIONS OF SHIP'S PROPULSION SYSTEMS - 1 THEORETICAL ANALYSIS	75-0291
TRANSVERSE	TRIANGULAR FINITE ELEMENT FOR PLATE BENDING ANALYSIS WITH TRANSVERSE SHEAR DEFORMATION	71-0555
TRANSVERSE	WAVE PROPAGATION IN A VISCOELASTIC FIBER SUBJECTED TO TRANSVERSE IMPACT	71-0639
TRANSVERSE	ZUSAMMENHANG ZWISCHEN OBERFLAECHENBESCHAFFENHEIT, VERBUND UND SPRENGWIRKUNG VON BEWEHRUNGSSTAEHLEN U	73-0450
TRANSVERSE-SHEA	INCLUSION OF TRANSVERSE-SHEAR DEFORMATION IN FINITE-ELEMENT DISPLACEMENT FORMULATIONS - COMMENT	75-0406
TRANSVERSELY	CREEP ANALYSIS OF TRANSVERSELY ISOTROPIC BODIES SUBJECTED TO TIME DEPENDENT LOADING	73-0170
TRANSVERSELY	ELASTIC SETTLEMENT OF A CIRCULAR RAFT IN ADHESIVE CONTACT WITH A TRANSVERSELY ISOTROPIC MEDIUM	75-0828
TRANSVERSELY	ELASTOPLASTIC CRACK PROPAGATION IN A TRANSVERSELY LOADED UNIDIRECTIONAL COMPOSITE	74-0217
TRANSVERSELY	IMPACT ON A TRANSVERSELY ANISOTROPIC HALF-SPACE	74-0950
TRAPEZOIDAL	ANALYSIS OF ELASTIC SHELLS OF REVOLUTION WITH MEMBRANE AND FLEXURE STRESSES UNDER ARBITRARY LOADING	68-0289
TRAPEZOIDAL	FINITE ELEMENT STUDY OF VIBRATION OF TRAPEZOIDAL PLATES	73-0035
TRAPEZOIDAL	FORMULATION AND EVALUATION OF A TRAPEZOIDAL CROSS-SECTION RING DISCRETE ELEMENT	66-0084
TRAPEZOIDAL	NATRUAL STRAINS AND STRESSES FOR TRAPEZOIDAL STRUCTURES ANALYSIS	69-0101
TRAPEZOIDAL	PAPA, STRUCTURAL ANALYSIS OF PLATES AND SHELLS USING TRAPEZOIDAL AND TRIANGULAR PLATE ELEMENTS	67-0059
TRAPEZOIDAL	TRAPEZOIDAL FINITE ELEMENTS: THEIR DERIVATION AND USE FOR AXISYMMETRIC ROTATING BODIES	73-0192
TRASFORMATORI	STUDIO DI CAMPI ELETTRICI E MAGNETICI STANZIONARI CON IL METODO DEGLI ELEMENTI FINITI. APPLICAZIONE	75-0207
TRAVELLING	FINITE ELEMENT SOLUTION OF SATURATED TRAVELLING MAGNETIC FIELD PROBLEMS	75-0600
TRAVELLING	RESPONSE OF EARTH DAMS TO TRAVELLING SEISMIC WAVES	69-0092
TRAVI-PARETE	DISTRIBUZIONI OTTIMALI DI ARMATURE DI PRECOMPRESSIONE IN TRAVI-PARETE DI CALCESTRUZZO. (OPTIMAL DIST	75-0560
TREATED	GEOMETRICKY NELINEARNE ULOHY PLOCHYCH SKRUPIN RIESENE PRIRASTKOVOU METODOU KONECNYCH PRVKOV. (GEOMET	74-1032
TREATED	STRESS - DEFORMATION PREDICTION IN CEMENT - TREATED SOIL PAVEMENTS	71-0286
TREATING	SURVEY OF SOME FINITE ELEMENT METHOD PROPOSED FOR TREATING DIRICHLET PROBLEM	75-0612
TREATISE	A TREATISE OF NONLINEAR FINITE ELEMENT ANALYSIS	73-1063
TREATMENT	A FINITE ELEMENT TREATMENT OF NEUTRON DIFFUSION	71-0373
TREATMENT	A UNIFIED TREATMENT OF THICK AND THIN FILM ELASTOHYDRODYNAMIC PROBLEMS USING HIGHER ORDER ELEMENT PR	75-0890
TREATMENT	COMPLETENESS AND CONVERGENCE PROPERTIES OF FINITE ELEMENT DISPLACEMENT FUNCTIONS, A GENERAL TREATME	67-0053
TREATMENT	FINITE ELEMENT TREATMENT OF SINGULARITIES OF BOUNDARY VALUE PROBLEMS AND ITS APPLICATION TO ANALYSIS	73-0306
TREATMENT	FINITE ELEMENT TREATMENT OF SOIL-STRUCTURE INTERACTION PROBLEM FOR NUCLEAR POWER PLANT UNDER SEISMIC	73-0834
TREATMENT	FINITE ELEMENT TREATMENT OF A UNIFORMLY MOVING ELASTIC-PLASTIC CRACK TIP	74-0470
TREATMENT	FINITE-ELEMENT TREATMENT OF QUADRUPOLE GROUND MOTIONS IN AXISYMMETRIC EARTH STRUCTURES	75-0434
TREATMENT	GENERAL TREATMENT OF STRUCTURAL MODIFICATIONS	72-0166
TREATMENT	HARBOUR OSCILLATION - A NUMERICAL TREATMENT FOR UNDAMPED NATURAL MODES	70-0208
TREATMENT	NUMERICAL TREATMENT OF ALGEBRAIC INTEGRAL EQUATIONS BY VARIATIONAL METHODS	75-0374
TREATMENT	ON THE NUMERICAL TREATMENT OF THE NAVIER-STOKES EQUATIONS FOR AN INCOMPRESSIBLE FLUID	73-0809
TREATMENT	STEADILY GROWING ELASTIC-PLASTIC CRACK TIP IN A FINITE ELEMENT TREATMENT	73-0087
TREATMENT	THE TREATMENT OF NATURAL BOUNDARY CONDITIONS IN THE FINITE ELEMENT AND FINITE DIFFERENCE METHODS	73-1098

TREATMENT	TREATMENT OF RANDOMNESS IN FINITE ELEMENT MODELING	70-0353
TREE	COMPUTER ANALYSIS OF FRUIT DETACHMENT DURING TREE SHAKING	74-1162
TREE	COMPUTER SIMULATION OF VIBRATION OF WHOLE TREE SYSTEMS USING FINITE ELEMENT METHODS	74-1161
TREE	SIMULATION OF VIBRATION OF WHOLE TREE SYSTEMS USING FINITE ELEMENTS	75-0765
TREE	VIBRATION STUDY OF COMPLETE TREE SYSTEMS USING FINITE ELEMENT METHOD	75-0790
TREKHMERNYKH	RASHET TREKHMERNYKH MAGNITNYKE POLEI METODOM KONECHNYKH ELEMENTOV. (CALCULATION OF THREE-DIMENSIONAL	75-0766
TRENCHES	GRAVITY FLOW TO EXCAVATIONS AND DRAINAGE TRENCHES IN LAYERED AQUIFERS	71-0014
TRENDS	OVERVIEW, TRENDS AND PROJECTIONS: THEORY AND APPLICATIONS OF THE FINITE ELEMENT METHOD IN GEOTECHNIC	72-0988
TRENDS	RECENT TRENDS IN STUDIES AT INTERNATIONAL CONFERENCE ON STRUCTURAL ENGINEERING	71-0347
TRENDS	RECENT TRENDS OF RESEARCH INTO RESTRAINT STRESSES AND STRAINS IN RELATION TO WELD CRACKING	73-0483
TRENDS	TENDANCES ACTUELLES DES RECHERCHES SUR LES CONTRAINTES ET DEFORMATIONS DUES AU BRIDAGE EN RELATION A	74-0273
TRENDS	TRENDS AND DIRECTIONS IN THE APPLICATIONS OF NUMERICAL ANALYSIS.	71-0667
TRESCA	FINITE ELEMENT ANALYSIS OF ELASTIC-PLASTIC PLANE STRESS PROBLEMS BASED UPON TRESCA YIELD CRITERION	70-0179
TRESCA	FINITE ELEMENT FORMULATION OF STRESS STRAIN MATRIX FOR AN ELASTICALLY PERFECTLY PLASTIC TRESCA MATER	70-0275
TRESCA	ON TRESCA YIELD CRITERION IN FINITE ELEMENT ANALYSIS OF PLANE STRESS PROBLEMS	73-0314
TRESS	TRESS - A FINITE ELEMENT CODE FOR STRESS ANALYSIS PROBLEMS	71-0715
TREUGOL'NY	TREUGOL'NY I ELEMENT DLYA KRUGOVOI TSILINDRICHESKOI OBOLOCHKI (TRIANGULAR FINITE ELEMENT FOR CIRCULA	73-0493
TRI-MOORED	MECHANICS OF CABLE MOORING SYSTEMS. VOLUME V1. A COMPUTER PROGRAM FOR ANALYZING THE STEADY STATE CON	73-0725
TRIAL	SOLUTION OF TRANSPORT PROBLEMS BY USE OF TRIAL FUNCTION TECHNIQUES	72-0991
TRIAL	SOME ELEMENT TYPES AND TRIAL FUNCTIONS IN THE FINITE ELEMENT METHOD	70-0242
TRIANGLE	AN IRREGULAR TRIANGLE MESH GENERATOR	64-0032
TRIANGLE	EXPERIENCE IN THE USE OF THE CONSTANT STRESS TRIANGLE FOR FINITE ELEMENT STRESS ANALYSIS	69-0407
TRIANGLE	EXPERIENCES WITH ORTHOGONAL POLYNOMIALS AND "BEST" NUMERICAL INTEGRATION FORMULAS ON A TRIANGLE; WIT	74-0977
TRIANGLE	FINITE ELEMENT SOLUTIONS FOR AN EQUILATERAL TRIANGLE	74-1281
TRIANGLE	INTERPOLATION POLYNOMIALS ON THE TRIANGLE	70-0235
TRIANGLE	INTERPOLATION POLYNOMIALS ON THE TRIANGLE ELEMENT	70-0326
TRIANGLE	MESH AND CONTOUR PLOT FOR TRIANGLE AND ISOPARAMETRIC ELEMENTS	70-0445
TRIANGLE	NUMERICAL SOLUTION OF THE QUASI LINEAR POISSON EQUATION IN A NON-UNIFORM TRIANGLE MESH	67-0090
TRIANGLES	A " LOCAL " BASIS OF GENERALIZED SPLINES OVER RIGHT TRIANGLES DETERMINED FROM A NONUNIFORM PARTITION	72-0525
TRIANGLES	BICUBIC INTERPOLATION OVER TRIANGLES	69-0367
TRIANGLES	ERROR ANALYSIS OF FINITE ELEMENT METHODS WITH TRIANGLES FOR ELLIPTIC BOUNDARY VALUE PROBLEMS	73-0072
TRIANGLES	ERROR ESTIMATE FOR QUADRILATERAL ISOPARAMETRIC FINITE-ELEMENTS OF TYPE Q1 WHICH CAN DEGENERATE INTO	75-0401
TRIANGLES	GAUSSIAN QUADRATURE FORMULAS FOR TRIANGLES	73-0906
TRIANGLES	MIXED-DISPLACEMENT FINITE-ELEMENT ANALYSIS WITH PARTICULAR APPLICATION USING PLANE-STRESS TRIANGLES	72-0245
TRIANGLES	SOME PROPERTIES OF LINEAR STRAIN TRIANGLES AND OPTIMAL FINITE ELEMENT MODELS	73-1044
TRIANGOLARI	OPTIMIZZAZIONE DI RETICOLI TRIANGOLARI PER IL METODO DEGLI ELEMENTI FINITI IN IDROLOGIA	72-0987
TRIANGOLARI	PROCESS FOR THE AUTOMATIC GENERATION OF FLAT TRIANGULAR LATTICES FOR APPLICATION TO THE FINITE ELEME	71-0366
TRIANGULAR	A COMPARATIVE STUDY OF VARIABLE AND CONSTANT THICKNESS HIGH PRECISION TRIANGULAR PLATE BENDING ELEME	74-0857
TRIANGULAR	A COMPATIBLE TRIANGULAR PLATE BENDING FINITE ELEMENT	68-0209
TRIANGULAR	A COMPATIBLE TRIANGULAR PLATE BENDING FINITE ELEMENT	70-0385
TRIANGULAR	A CONFORMING QUARTIC TRIANGULAR ELEMENT FOR PLATE BENDING	69-0137
TRIANGULAR	A CONFORMING TRIANGULAR FINITE ELEMENT PLATE BENDING SOLUTION	70-0020
TRIANGULAR	A CONFORMING TRIANGULAR ELEMENT IN THE ANALYSIS OF THIN AND SEMI-THICK PLATES IN BENDING	70-0308
TRIANGULAR	A DERIVATION PROCEDURE FOR THE DYNAMIC FLEXIBILITY MATRIX OF A TRIANGULAR BENDING ELEMENT	70-0103
TRIANGULAR	A DOUBLY CURVED TRIANGULAR SHELL ELEMENT	68-0149
TRIANGULAR	A FLAT TRIANGULAR SHELL ELEMENT STIFFNESS MATRIX	66-0052
TRIANGULAR	A HIGH ORDER TRIANGULAR FIELD ELEMENTS	71-0535
TRIANGULAR	A HIGH PRECISION AXISYMMETRIC TRIANGULAR ELEMENT USED IN THE ANALYSIS OF HYDRAULIC TURBINE COMPONENT	70-0038
TRIANGULAR	A HIGH PRECISION TRIANGULAR PLATE BENDING ELEMENT	68-0052
TRIANGULAR	A HIGH-PRECISION TRIANGULAR CYLINDRICAL SHELL FINITE ELEMENT	71-0163
TRIANGULAR	A HIGHER ORDER TRIANGULAR FINITE ELEMENT FOR THE SOLUTION OF FIELD PROBLEMS IN ORTHOTROPIC MEDIA	72-0370
TRIANGULAR	A RAPIDLY CONVERGING TRIANGULAR PLATE ELEMENT	69-0245
TRIANGULAR	A REFINED TRIANGULAR PLATE BENDING FINITE ELEMENT	69-0056
TRIANGULAR	A SEQUEL TO TECHNICAL NOTE 13 - THE CURVED TETRAHEDRONAL AND TRIANGULAR ELEMENTS TEC AND TRIC FOR TH	69-0048
TRIANGULAR	A SHALLOW SHELL FINITE ELEMENT OF TRIANGULAR SHAPE	70-0044
TRIANGULAR	A SIMPLE ALGORITHM FOR THE STIFFNESS MATRIX OF TRIANGULAR PLATE BENDING ELEMENTS	71-0152
TRIANGULAR	A SIX NODE TRIANGULAR EQUILIBRIUM (CONSTANT STRESS) MEMBRANE ELEMENT	73-0783
TRIANGULAR	A SIX NODE TRIANGULAR SHEAR PANEL FOR EQUILIBRIUM MODELS	73-1011
TRIANGULAR	A TRIANGULAR ELEMENT FOR BENDING AND VIBRATION OF MULTILAYER SANDWICH PLATES	72-0489
TRIANGULAR	A TRIANGULAR EQUILIBRIUM ELEMENT WITH LINEARLY VARYING BENDING MOMENTS FOR PLATE BENDING PROBLEMS	67-0054
TRIANGULAR	A TRIANGULAR EQUILIBRIUM (LINEAR STRESS) MEMBRANE ELEMENT	75-0883
TRIANGULAR	A TRIANGULAR FLAT PLATE BENDING ELEMENT	68-0252
TRIANGULAR	A TRIANGULAR PLATE BENDING ELEMENT FOR CONTACT PROBLEMS	73-0693
TRIANGULAR	A TRIANGULAR THIN SHELL FINITE ELEMENT: NONLINEAR ANALYSIS	75-0697
TRIANGULAR	A TRIANGULAR THIN SHELL FINITE ELEMENT: LINEAR ANALYSIS	75-0698
TRIANGULAR	ALGORITHMS FOR THE PRODUCTION OF CONTOUR MAPS OVER AN IRREGULAR TRIANGULAR MESH	72-0825
TRIANGULAR	AN EFFICIENT TRIANGULAR SHELL ELEMENT	70-0382
TRIANGULAR	AN EFFICIENT TRIANGULAR SHELL ELEMENT	70-0441
TRIANGULAR	ANALYSIS OF OUT-OF-PLANE STRESS BY THE SIX NODES, FOURTH ORDER ACCURACY TRIANGULAR ELEMENTS	69-0375
TRIANGULAR	ANALYSIS OF PLATE AND SHELL STRUCTURES BY TRIANGULAR FINITE ELEMENTS	68-0303
TRIANGULAR	ANALYSIS OF PLATE BENDING USING TRIANGULAR ELEMENTS	62-0010
TRIANGULAR	ANALYSIS OF PLATE BENDING BY TRIANGULAR ELEMENTS	68-0145
TRIANGULAR	ANALYSIS OF THIN PLATES IN BENDING USING TRIANGULAR FINITE ELEMENTS	68-0213
TRIANGULAR	APPLICATION OF THE HIGH PRECISION TRIANGULAR PLATE BENDING ELEMENT TO PROBLEMS WITH CURVED BOUNDARIE	69-0071
TRIANGULAR	AXISYMETRIC TRIANGULAR ELEMENT FOR THE SCALAR HELMHOLTZ EQUATION	73-0688
TRIANGULAR	BEHAVIOR OF TRIANGULAR SHELL ELEMENT STIFFNESS MATRICES ASSOCIATED WITH POLYHEDRAL DEFLECTION DISTRI	68-0159
TRIANGULAR	CALCULATION OF THE ELASTIC BUCKLING LOADS OF THIN FLAT REINFORCED PLATES USING TRIANGULAR FINITE ELE	73-1024
TRIANGULAR	CALCULATION OF THE ELASTIC BUCKLING LOADS OF THIN FLAT REINFORCED PLATES USING TRIANGULAR FINITE ELE	75-0177
TRIANGULAR	CLOUGH-TOCHER TRIANGULAR PLATE BENDING ELEMENT IN VIBRATION	69-0093
TRIANGULAR	COMPARISON OF TWO HIGH-PRECISION TRIANGULAR FINITE ELEMENTS FOR ARBITRARY DEEP SHELLS	71-0071
TRIANGULAR	COMPATIBLE TRIANGULAR FINITE ELEMENTS	74-0330
TRIANGULAR	COMPUTATION OF STRESSES IN TRIANGULAR FINITE ELEMENTS	66-0072
TRIANGULAR	CONFORMING RECTANGULAR AND TRIANGULAR PLATE BENDING ELEMENTS	71-0720
TRIANGULAR	CONSTRUCTION OF NORMAL DISPLACEMENT FUNCTIONS OF TRIANGULAR FINITE ELEMENT OF PLATES AND SHELLS	72-0563
TRIANGULAR	CURVED CUBIC TRIANGULAR FINITE ELEMENT FOR POTENTIAL FLOW PROBLEMS	73-0423
TRIANGULAR	CURVED TRIANGULAR ELEMENTS FOR THE ANALYSIS OF SHELLS	73-0030
TRIANGULAR	CURVED TRIANGULAR FINITE ELEMENTS FOR THE ANALYSIS OF THICK ELASTIC PLATES AND SHELLS	71-0814
TRIANGULAR	DERIVATION OF A STIFFNESS MATRIX FOR A RIGHT TRIANGULAR PLATE IN BENDING AND SUBJECTED TO INITIAL ST	62-0007
TRIANGULAR	DESH, EIN EINFACHES FLACHES SCHALENELEMENT UEBER DREIECKIGEM GRUNDRISS. (DESH, A SIMPLE TRIANGULAR S	75-0304
TRIANGULAR	DEVELOPMENT AND APPLICATION OF THE DIRECT STIFFNESS METHOD FOR OUT-OF-PLANE BENDING USING A TRIANGUL	62-0009
TRIANGULAR	DEVELOPMENT AND EVALUATION OF A TRIANGULAR THIN SHELL ELEMENT BASED UPON THE HYBRID ASSUMED STRESS M	69-0252
TRIANGULAR	DEVELOPMENT OF EXPLICIT STIFFNESS AND MASS MATRICES FOR A TRIANGULAR PLATE ELEMENT	69-0242
TRIANGULAR	DISCRETE TRIANGULAR APPROXIMATION OF MOMENT AND DISPLACEMENT SURFACES FOR PLATE BENDING	71-0269
TRIANGULAR	DOUBLY CURVED TRIANGULAR FINITE ELEMENTS FOR SHELLS OF ARBITRARY SHAPE	73-0881
TRIANGULAR	DYNAMIC ANALYSIS OF SHALLOW SHELLS WITH A DOUBLY-CURVED TRIANGULAR FINITE ELEMENT	71-0353
TRIANGULAR	ELEMENT STIFFNESS MATRICES FOR CONSTANT AND LINEAR STRESS - TRIANGULAR, CONSTANT AND LINEAR STRESS T	67-0006
TRIANGULAR	EXPLICIT EXPRESSIONS FOR TRIANGULAR TOWA ELEMENT STIFFNESS MATRIX	68-0160
TRIANGULAR	EXPLICIT TRIANGULAR BENDING ELEMENT MATRIX	74-0288
TRIANGULAR	FINITE ELEMENT ANALYSIS OF PERFORATED PLATES CONTAINING TRIANGULAR PENETRATION PATTERNS OF 5 AND 10	75-0216
TRIANGULAR	FINITE ELEMENT ANALYSIS OF PERFORATED PLATES CONTAINING TRIANGULAR PENETRATION PATTERNS OF 5 AND 10	75-0520
TRIANGULAR	FINITE ELEMENT APPLICATIONS OF A PRECISION TRIANGULAR PLATE BENDING ELEMENT TO LINEAR BOUNDARY VALUE	75-0788
TRIANGULAR	FORMULATION AND EVALUATION OF A TRIANGULAR CROSS-SECTION RING DISCRETE ELEMENT	66-0083

428

TRIANGULAR	FORMULATION OF A NEW TRIANGULAR PLATE BENDING ELEMENT	68-0051
TRIANGULAR	HIGH ORDER POLYNOMIAL TRIANGULAR FINITE ELEMENTS FOR POTENTIAL PROBLEMS	69-0240
TRIANGULAR	HIGHER ORDER COMPATIBLE TRIANGULAR FINITE ELEMENTS	74-0152
TRIANGULAR	MATRIX DISPLACEMENT ANALYSIS OF ANISOTROPIC SHELLS BY TRIANGULAR ELEMENTS	65-0005
TRIANGULAR	NEW TRIANGULAR AND QUADRILATERAL PLATE-BENDING FINITE ELEMENTS	74-0833
TRIANGULAR	NEW TRIANGULAR ELEMENT FOR FINITE DIFFERENCE SOLUTION OF AXISYMMETRIC CONDUCTION PROBLEMS IN CYLINDR	74-0381
TRIANGULAR	NEW TRIANGULAR PLATE-BENDING FINITE ELEMENT WITH TRANSVERSE SHEAR FLEXIBILITY	74-0627
TRIANGULAR	NEW TRIANGULAR PLATE-BENDING FINITE ELEMENT WITH TRANSVERSE SHEAR FLEXIBILITY	74-0964
TRIANGULAR	NUMERICAL ANALYSIS OF THIN SHELLS BY CURVED TRIANGULAR ELEMENTS BASED ON DISCRETE - KIRCHHOFF HYPOTH	69-0091
TRIANGULAR	ON THE CONFORMING CUBIC TRIANGULAR ELEMENT FOR PLATE BENDING	72-0439
TRIANGULAR	ON THE QUINTIC TRIANGULAR PLATE BENDING ELEMENT	72-0677
TRIANGULAR	PAPA, STRUCTURAL ANALYSIS OF PLATES AND SHELLS USING TRAPEZOIDAL AND TRIANGULAR PLATE ELEMENTS	67-0059
TRIANGULAR	PLANE STRESS ANALYSIS USING THE FINITE ELEMENT METHOD. TRIANGULAR FINITE ELEMENT WITH SIX PARAMETERS	67-0106
TRIANGULAR	PLATE VIBRATION PROBLEMS USING THE FINITE ELEMENT METHOD WITH ASSUMED STRESS DISTRIBUTION AND TRIANG	66-0068
TRIANGULAR	PROCESS FOR THE AUTOMATIC GENERATION OF FLAT TRIANGULAR LATTICES FOR APPLICATION TO THE FINITE ELEME	71-0366
TRIANGULAR	PROGRAM FOR TRIANGULAR BENDING ELEMENTS WITH DERIVATIVES SMOOTHING	72-0473
TRIANGULAR	RECTANGULAR AND TRIANGULAR FINITE ELEMENTS FOR POISSON'S EQUATION	75-0802
TRIANGULAR	REINFORCED FIELDS OF TRIANGULAR ELEMENTS WITH LINEARLY VARYING STRAIN: EFFECT OF INITIAL STRAINS	65-0004
TRIANGULAR	RELIABILITY STUDY OF LINEAR VARIATION TRIANGULAR FINITE ELEMENT	72-0281
TRIANGULAR	SARD KERNEL THEOREMS ON TRIANGULAR AND RECTANGULAR DOMAINS WITH EXTENSIONS AND APPLICATIONS TO FINIT	72-0361
TRIANGULAR	SHELL ANALYSIS USING PLANAR TRIANGULAR ELEMENTS	69-0041
TRIANGULAR	SIMPLE TRIANGULAR MEMBRANE FINITE ELEMENT	74-0881
TRIANGULAR	STABILITY OF MODERATELY THICK RECTANGULAR PLATES USING A HIGH PRECISION TRIANGULAR FINITE ELEMENT	75-0334
TRIANGULAR	STATIC AND DYNAMIC APPLICATIONS OF A HIGH-PRECISION TRIANGULAR PLATE BENDING ELEMENT	69-0086
TRIANGULAR	STIFFNESS MATRICES FOR THIN TRIANGULAR ELEMENTS OF NONZERO GAUSSIAN CURVATURE	66-0073
TRIANGULAR	STIFFNESS MATRIX FOR A TRIANGULAR SANDWICH ELEMENT IN BENDING	67-0050
TRIANGULAR	STRUCTURAL VIBRATION ANALYSIS USING TRIANGULAR FINITE ELEMENTS	66-0060
TRIANGULAR	TECHNIQUES IN TRIANGULAR MESH GENERATION FLAT AND CURVED SURFACES	69-0394
TRIANGULAR	THE CURVED TETRAHEDRONAL AND TRIANGULAR ELEMENTS TEC AND TRIC FOR THE MATRIX DISPLACEMENT METHOD	69-0047
TRIANGULAR	THE TRIANGULAR EQUILIBRIUM ELEMENT IN THE SOLUTION OF PLATE BENDING PROBLEMS	68-0122
TRIANGULAR	TREUGOL'NY I ELEMENT DLYA KRUGOVOI TSILINDRICHESKOI OBOLOCHKI (TRIANGULAR FINITE ELEMENT FOR CIRCULA	73-0493
TRIANGULAR	TRIANGULAR ELEMENT FOR MULTILAYER SANDWICH PLATES	72-0581
TRIANGULAR	TRIANGULAR ELEMENTS WITH LINEARLY VARYING STRAIN FOR THE MATRIX DISPLACEMENT METHOD	65-0053
TRIANGULAR	TRIANGULAR ELEMENTS IN PLATE BENDING CONFORMING AND NON-CONFORMING SOLUTIONS	66-0011
TRIANGULAR	TRIANGULAR ELEMENTS IN THE FINITE ELEMENT METHOD	70-0194
TRIANGULAR	TRIANGULAR FINITE ELEMENTS FOR PLATE BENDING WITH CONSTANT AND LINEARLY VARYING BENDING MOMENTS	70-0019
TRIANGULAR	TRIANGULAR FINITE ELEMENTS FOR THE BENDING ANALYSIS OF THICK ELASTIC PLATES	70-0466
TRIANGULAR	TRIANGULAR FINITE ELEMENT FOR CYLINDRICAL SHELLS	70-0488
TRIANGULAR	TRIANGULAR FINITE ELEMENT FOR PLATE BENDING ANALYSIS WITH TRANSVERSE SHEAR DEFORMATION	71-0555
TRIANGULAR	TRIANGULAR FINITE ELEMENTS FOR THE GENERALIZED BESSEL EQUATION OF ORDER M	73-1094
TRIANGULAR	TRIANGULAR FINITE-ELEMENTS OF VARIABLE THICKNESS AND THEIR APPLICATION TO PLATE AND SHELL PROBLEMS	69-0095
TRIANGULAR	TRIANGULAR MESH GENERATION PROGRAM	67-0037
TRIANGULAR	TRIANGULAR MESH GENERATION PROGRAM	67-0127
TRIANGULAR	TRIANGULAR MESH METHODS FOR THE NEUTRON TRANSPORT EQUATION	73-1062
TRIANGULAR	TRIANGULAR PLATE BENDING ELEMENTS	69-0057
TRIANGULAR	TRIANGULAR PLATE BENDING ELEMENT WITH ENFORCED COMPATIBILITY	71-0593
TRIANGULAR	TRIANGULAR PLATE ELEMENTS IN THE MATRIX FORCE METHOD OF STRUCTURAL ANALYSIS	63-0010
TRIANGULAR	TRIANGULAR THICK PLATE BENDING ELEMENTS	71-0167
TRIANGULAR	TRIANGULAR THICK SHELL FINITE ELEMENT FOR THE THREE-DIMENSIONAL STRESS ANALYSIS OF PRESSURE VESSELS	73-0860
TRIANGULAR	TRIANGULAR THIN SHELL FINITE ELEMENT: LINEAR ANALYSIS	75-0602
TRIANGULAR	TRIANGULAR THIN SHELL FINITE ELEMENT: NONLINEAR ANALYSIS	75-0668
TRIANGULAR	TRIPLET: A TWO-DIMENSIONAL, MULTIGROUP, TRIANGULAR MESH, PLANAR GEOMETRY, EXPLICIT TRANSPORT CODE	73-0955
TRIANGULAR	VIBRATIONS OF PLATE AND SHELL STRUCTURES USING TRIANGULAR FINITE ELEMENTS	67-0063
TRIANGULAR	VIBRATIONS OF THICK PLATES USING A HIGH PRECISION TRIANGULAR ELEMENT	74-1002
TRIANGULAR-FINI	EVALUATION OF ATTENUATION FROM LOSSLESS TRIANGULAR-FINITE-ELEMENT SOLUTIONS FOR INHOMOGENEOUSLY FILL	73-0059
TRIANGULATED	PIECEWISE ANALYTIC INTERPOLATION AND APPROXIMATION IN TRIANGULATED POLYGONS	72-0508
TRIANGULATION	AUTOMATIC NODAL TRIANGULATION FOR FINITE ELEMENTS	73-0178
TRIANGULATION	AUTOMATIC TRIANGULATION OF TWO DIMENSIONAL REGIONS	72-0697
TRIANGULATION	AUTOMATIC TRIANGULATION OF ARBITRARY PLANAR DOMAINS FOR THE FINITE ELEMENT METHOD	74-0480
TRIANGULATION	TRIANGULATION IN THE HYPERCIRCLE METHOD FOR PLANE PROBLEMS	52-0001
TRIANGULATIONS	AUTOMATIC GENERATION OF TRIANGULATIONS FOR PIECEWISE POLYNOMIAL APPROXIMATION	73-1118
TRIAX	RESULT OF CALCULATIONS WITH TRIM 6 AND TRIAX 6 ELASTIC-PLASTIC ELEMENTS	73-1122
TRIAX	THE TRIAX 6 ELEMENT FOR AXISYMMETRIC ANALYSIS BY THE MATRIX DISPLACEMENT METHOD: PART 1, FOUNDATIONS	66-0124
TRIAX	TRIAX 6 ELEMENT FOR AXISYMMETRIC ANALYSIS BY THE MATRIX DISPLACEMENT METHOD	66-0009
TRIAXIAL	ANALYSIS OF STRESS AND STRAIN DISTRIBUTIONS IN TRIAXIAL TESTS USING THE METHOD OF FINITE ELEMENTS	73-0903
TRIAXIAL	EXACT SOLUTION TO THE RADIALLY END CONSTRAINED CIRCULAR CYLINDER UNDER TRIAXIAL LOADING	71-0433
TRIAXIAL	FINITE ELEMENT STUDY OF THE TRIAXIAL STRESS STATE AROUND AN INSPECTION TUNNEL IN AN ARCH DAM	74-0831
TRIAXIAL	NUMERICAL STUDY OF UNIAXIAL AND TRIAXIAL ROCK COMPRESSION TESTS	73-0896
TRIAXIAL	SIGNIFICANCE OF VARIABLY CONFINED TRIAXIAL TESTING ROADBUILDING MATERIALS	74-0554
TRIC	A SEQUEL TO TECHNICAL NOTE 13 - THE CURVED TETRAHEDRONAL AND TRIANGULAR ELEMENTS TEC AND TRIC FOR TH	69-0048
TRIC	THE CURVED TETRAHEDRONAL AND TRIANGULAR ELEMENTS TEC AND TRIC FOR THE MATRIX DISPLACEMENT METHOD	69-0047
TRICUBIC	TRICUBIC POLYNOMIAL INTERPOLATION	71-0377
TRIDIMENSIONAL	THE TRIDIMENSIONAL THERMOELASTIC COMPUTER CODE TITUS	71-0156
TRIDIMENSIONALE	ANALYSE TRIDIMENSIONALE DU COMPORTEMENT NON-LINEAIRE D'UN CAISSON DE REACTEUR NUCLEAIRE EN BETON PRE	71-0144
TRIDIMENSIONELL	ANALYSE TRIDIMENSIONELLE DE LA STABILITE D'UN TUNNEL AU VOISINAGE DU FRONT DE TAILLE DANS UNE ROCHE	74-0995
TRIGGERING	EARTHQUAKE TRIGGERING BY FLUID INJECTION AT RANGELY, COLORADO	72-0634
TRIGONOMETRIC	TRIGONOMETRIC FUNCTION REPRESENTATIONS FOR RECTANGULAR PLATE BENDING ELEMENTS	71-0765
TRIGONOMETRIC	USE OF TRIGONOMETRIC TERMS IN FINITE ELEMENT METHOD WITH APPLICATION TO VIBRATING MEMBRANES	74-0127
TRIGONOMETRIC	USE OF TRIGONOMETRIC TERMS IN FINITE ELEMENT METHOD WITH APPLICATION TO VIBRATING MEMBRANES	74-0555
TRILEAFLET	THE FINITE ELEMENT STRESS ANALYSIS OF THE HUMAN AORTIC VALVE TOWARD THE DEVELOPMENT OF A PROSTHETIC	74-1201
TRILINEAR	APPLICATIONS OF TRILINEAR COORDINATES TO SOME PROBLEMS IN PLANE ELASTICITY	72-0728
TRILINEAR	SOME NONCONFORMING VARIANTS ON THE TRILINEAR ISOPARAMETRIC BRICK FOR LINEAR ELASTIC PROBLEMS	72-0531
TRIM	A TAPERED THICKNESS TRIM 6 ELEMENT FOR THE MATRIX DISPLACEMENT METHOD	66-0007
TRIM	ASKA 105 A TECHNICAL MANUAL AND PROGRAMMING THEORY FOR THE ELEMENTS, FLA, TRIM, SPARTA AND TET	68-0254
TRIM	RESULT OF CALCULATIONS WITH TRIM 6 AND TRIAX 6 ELASTIC-PLASTIC ELEMENTS	73-1122
TRIM-6	A COMPUTER PROGRAMME FOR ELASTIC-PLASTIC CALCULATIONS WITH USE OF TRIM-6 FINITE ELEMENTS	72-0993
TRIPLET	TRIPLET: A TWO-DIMENSIONAL, MULTIGROUP, TRIANGULAR MESH, PLANAR GEOMETRY, EXPLICIT TRANSPORT CODE	73-0955
TRIPOD	WEAPON-MOUNT INTERFACE STUDY - TRIPOD MOUNT M3	72-0846
TROUGH-SHAPED	TRANSVERSE REINFORCEMENT OF CONTINUOUS, TROUGH-SHAPED CONCRETE RAILROAD STRUCTURES. AN IMPLEMENTATIO	71-0719
TRUCKS	PROBLEMS OF LONGITUDINAL STRENGTH IN THE CONSTRUCTION OF VERY LARGE TANK TRUCKS	73-0251
TRUMBULL	ULTIMATE STRAIN CAPACITY AND TEMPERATURE RISE STUDIES, TRUMBULL POND DAM	72-0705
TRUNCATION	REMOVAL OF TRUNCATION ERROR IN FINITE ELEMENT ANALYSIS	75-0131
TRUNCATION	SPECTRUM TRUNCATION AND DAMAGE TOLERANCE STUDY ASSOCIATED WITH THE C-5A OUTBOARD PYLON AFT TRUSS LUG	74-0535
TRUSS	AUTOMATED METHOD FOR THE LARGE DEFLECTION AND INSTABILITY ANALYSIS OF THREE-DIMENSIONAL TRUSS AND FR	66-0105
TRUSS	OPTIMUM CONFIGURATIONAL AND DIMENSIONAL DESIGN OF TRUSS STRUCTURES	74-0588
TRUSS	OPTIMUM CONFIGURATION OF TRUSS STRUCTURES	74-1315
TRUSS	SPECTRUM TRUNCATION AND DAMAGE TOLERANCE STUDY ASSOCIATED WITH THE C-5A OUTBOARD PYLON AFT TRUSS LUG	74-0535
TRUSS	TRUSS ANALYSIS BY STIFFNESS CONSIDERATIONS	58-0001
TRUSSES	A NEW MODEL FOR ELASTIC-PLASTIC TRUSSES	74-0805
TRUSSES	NONLINEAR ANALYSIS OF RETICULATED SPACE TRUSSES	75-0287
TRUSSOPT2	ON IMPROVING EFFICIENCY OF AN ALGORITHM FOR STRUCTURAL OPTIMIZATION AND A USERS MANUAL FOR PROGRAM T	74-0778
TSAAS	TSAAS: FINITE ELEMENT THERMAL AND STRESS ANALYSIS OF AXISYMMETRIC SOLIDS WITH ORTHOTROPIC TEMPERATUR	74-0886

TSILINDRICHESKI	METOD KONECHNYKH ELEMENTOV V PRIMENENII K RASCHETU TSILINDRICHESKIKH OBOLOCHEK S PRYAMOUGOL'NYMI OTV	73-0536
TSILINDRICHESKO	TREUGOL'NY I ELEMENT DLYA KRUGOVOI TSILINDRICHESKOI OBOLOCHKI (TRIANGULAR FINITE ELEMENT FOR CIRCULA	73-0493
TSILINDROV	METODY RASCHETNOGO OPREDELENIYA TEMPERATURNYKH NAPRYAZHENII V KRYSHKAKH TSILINDROV DVIGATELEI VNUTRE	74-0179
TSUNAMI	NUMERICAL MODEL FOR TSUNAMI RUN-UP	70-0200
TUBA	A SEQUEL TO TECHNICAL NOTE 14 ON THE TUBA FAMILY OF PLATE ELEMENTS	68-0015
TUBA	THE TUBA FAMILY OF PLATE ELEMENTS FOR THE MATRIX DISPLACEMENT METHOD	68-0013
TUBE	ANALYSIS OF AN IN-PILE REACTOR TUBE	75-0193
TUBE	ANALYSIS OF BAND PRESSURE IN A GUN TUBE USING FINITE ELEMENTS AND SHELL PUSHING DATA	72-0793
TUBE	CALCULATING STRESSES IN A PRESSURE VESSEL WITH FILL TUBE - COMPARISON OF THE FINITE ELEMENT AND PHOT	72-0782
TUBE	ELASTIC-PLASTIC SOLUTION OF A TWO-DIMENSIONAL TUBE PROBLEM BY THE FINITE ELEMENT METHOD	74-0769
TUBE	INELASTIC ANALYSIS OF A STRAIGHT TUBE UNDER COMBINED BENDING, PRESSURE AND THERMAL LOADS	75-0218
TUBE	INELASTIC ANALYSIS OF A STRAIGHT TUBE UNDER COMBINED BENDING, PRESSURE AND THERMAL LOADS	75-0513
TUBE	ON THE DETERMINATION OF EFFECTIVE ELASTIC-PLASTIC PROPER TIES FOR THE EQUIVALENT SOLID PLATE ANALYSI	72-0494
TUBE	SHEAR LAG ANALYSIS OF RECTANGULAR FULL-WIDTH TUBE JUNCTIONS	72-0087
TUBE	TEMPERATURES AND STRESSES IN A BOILER MEMBRANE WALL TUBE	74-1053
TUBE	TEMPERATURVERLAUF IM QUERSCHNITT GESCHWEISSTER ROHRWAENDE BEI UEBERWIEGEND KONVEKTIVEM WAERMEUEBERGA	75-0628
TUBE	THEORETICAL AND EXPERIMENTAL INVESTIGATION ON A STEAM GENERATOR TUBE PLATE	69-0424
TUBES	A FINITE ELEMENT ANALYSIS OF LAMINATED ANISOTROPIC TUBES	70-0319
TUBES	CALCULATION OF NATURAL FREQUENCIES OF HEAT EXCHANGER TUBES	73-0665
TUBES	EFFECT OF INITIAL STATE FORCES ON THE HYDROELASTIC VIBRATION OF TUBES	73-0589
TUBES	FINITE ELEMENT PLASTIC ANALYSIS OF ELASTIC PLASTIC THICK-WALLED TUBES	74-0368
TUBES	FINITE ELEMENT FOR TORSION OF THIN WALLED OPEN TUBES	71-0308
TUBES	FULLY DEVELOPED VISCOUS FLOW IN INTERNALLY FINNED TUBES	75-0648
TUBES	INTERNAL SUPPORTING STRUCTURES OF PRESSURE TUBES TYPE REACTOR CORE, ANALYTICAL MODEL FOR STATIC ANAL	71-0049
TUBES	SEISMIC BEHAVIOUR OF FRAMED TUBES	75-0306
TUBES	SLOW PARTICULATE VISCOUS FLOW IN CHANNELS AND TUBES, APPLICATION TO BIOMECHANICS	71-0248
TUBES	THE FINITE ELEMENT ANALYSIS OF ELASTIC-PLASTIC THICK-WALLED TUBES	74-0768
TUBES	THERMAL STRESSES IN FACT REACTOR BOILER TUBES AT THE SODIUM/GAS INTERFACE	73-0820
TUBESHEETS	INELASTIC ANALYSIS OF TUBESHEETS BY THE FINITE ELEMENT METHOD	75-0652
TUBULAR	A COMPUTER PROGRAM FOR THE ANALYSIS OF TUBULAR K JOINTS	69-0488
TUBULAR	ANALYSIS OF TUBULAR T-CONNECTIONS BY THE FINITE ELEMENT METHOD: COMPARISON WITH EXPERIMENTS	72-0380
TUBULAR	CALCULATION OF THERMAL STRESSES IN TUBULAR CERAMIC COMPONENTS	73-0670
TUBULAR	DESIGN APPRAISAL OF OFFSHORE PLATFORMS OF WELDED TUBULAR CONSTRUCTION	74-1165
TUBULAR	FINITE ELEMENT ANALYSIS OF TUBULAR JOINTS: A REPORT ON A FEASIBILITY STUDY	67-0147
TUBULAR	FINITE ELEMENT ANALYSIS OF TUBULAR K JOINTS	70-0427
TUBULAR	ON THE STRUCTURAL DESIGN OF TUBULAR JOINTS	74-0492
TUBULAR	SOME MODELS FOR THE ANALYSIS OF STRESSES IN A TUBULAR FUEL PIN FOR HIGH TEMPERATURE REACTORS BY FINI	73-0816
TUBULAR	STRESS ANALYSIS OF A TUBULAR CROSS-JOINT WITHOUT ANY INTERNAL STIFFENING FOR OFFSHORE STRUCTURES	74-0549
TUBULAR	STRESS ANALYSIS OF TUBULAR JOINTS FOR OFFSHORE STRUCTURES	73-0207
TUBULAR	THE FLEXIBILITY OF A TUBULAR WELDED JOINT IN A VEHICLE FRAME	74-0209
TUNGSTEN	ADVANCED WIRE-WOUND TUNGSTEN NOZZLES	66-0031
TUNGSTEN	PROBABILISTIC EVALUATION OF THE RELATIVE BEHAVIOUR OF TUNGSTEN CARBIDE COMPONENTS IN STRUCTURAL APPL	75-0579
TUNING	FREQUENCY TUNING OF COMPRESSOR ROTOR BLADES	66-0112
TUNNEL	ANALYSE TRIDIMENSIONELLE DE LA STABILITE D'UN TUNNEL AU VOISINAGE DU FRONT DE TAILLE DANS UNE ROCHE	74-0995
TUNNEL	ANALYSE TRIDIMENSIONELLE DE LA STABILITE D'UN TUNNEL AU VOISINAGE DU FRONT DE TAILLE DANS UNE ROCHE	74-0995
TUNNEL	ANALYSIS OF THERMAL STRESSES IN CORED CERAMIC BRICKS USED IN HYPERSONIC WIND TUNNEL HEATERS	72-0655
TUNNEL	DISTRIBUTION OF STRESS AROUND A TUNNEL CAVITY, USING THE FINITE ELEMENT METHOD	71-0843
TUNNEL	EFFECT OF DISCONTINUITIES ON STRESSES AND DEFORMATIONS OF ROCK AROUND A TUNNEL. (EINFLUSS VON DISKON	70-0170
TUNNEL	FINITE ELEMENT STUDY OF THE TRIAXIAL STRESS STATE AROUND AN INSPECTION TUNNEL IN AN ARCH DAM	74-0831
TUNNEL	INFLUENCE OF CONSTRUCTION PROCEDURES ON GROUND PRESSURES AROUND A TUNNEL	74-0997
TUNNEL	INTERACTION OF ROCK AND TUNNEL LINING	73-1057
TUNNEL	POLYMER-IMPREGNATED CONCRETE TUNNEL SUPPORT AND LINING	73-0736
TUNNEL	RATIONAL DESIGN OF TUNNEL SUPPORTS: TUNNEL SUPPORT LOADING CAUSED BY ROCK FAILURE	75-0687
TUNNEL	RATIONAL DESIGN OF TUNNEL SUPPORTS: TUNNEL SUPPORT LOADING CAUSED BY ROCK FAILURE	75-0687
TUNNEL	RESEARCH TO IMPROVE TUNNEL SUPPORT SYSTEMS	74-0777
TUNNEL	SHOCK TUNNEL TESTS OF PRELOADED AND ARCHED WALL PANELS FINITE ELEMENT STRESS FORMULATION FOR WAVE PR	73-0892
TUNNEL	STRESSES AND DISPLACEMENTS AROUND A CIRCULAR TUNNEL IN A THREE LAYER MEDIUM	69-0038
TUNNEL	STUDIES TO IMPROVE TUNNEL SUPPORT SYSTEMS	74-0957
TUNNEL	TUNNEL CONSTRUCTION - A BIBLIOGRAPHY WITH ABSTRACTS	74-0748
TUNNEL	TUNNEL DESIGN CONSIDERATIONS: ANALYSIS OF MEDIUM-SUPPORT INTERACTION	74-1062
TUNNEL	UNUSUAL FOUNDATIONS OVER THE MERSEY TUNNEL	70-0197
TUNNELLING	APPLICATION OF NEW AUSTRIAN TUNNELLING METHOD IN DIFFICULT BUILTOVER AREAS IN FRANKFURT/MAIN METRO	73-1025
TUNNELROEHRE	EFFECT OF DISCONTINUITIES ON STRESSES AND DEFORMATIONS OF ROCK AROUND A TUNNEL. (EINFLUSS VON DISKON	70-0170
TUNNELS	FINITE ELEMENT STUDIES OF PILE DRIVER TUNNELS INCLUDING CONSIDERATIONS OF SUPPORT REQUIREMENTS	72-0679
TUNNELS	INTEGRAL EQUATION METHOD FOR SUBSONIC FLOW PAST AIRFOILS IN VENTILATED WIND TUNNELS	75-0065
TUNNELS	LINER-MEDIUM INTERACTION IN TUNNELS	75-0052
TUNNELS	NUMERICAL AND PHYSICAL MODELLING OF REINFORCEMENT SYSTEMS FOR TUNNELS IN JOINTED ROCK	73-0872
TUNNELS	ON THE DISTRIBUTION OF STRESSES AROUND CIRCULAR TUNNELS IN NON-HOMOGENEOUS ROCKS	66-0029
TUNNELS	STRUCTURAL DESIGN OF CIRCULAR DIVERSION, POWER AND PRESSURE TUNNELS	71-0315
TURBAN	TURBAN VERSION 1.0. EIN SPEZIELLES PROGRAMM ZUR LINEAREN UND NICHTLINEAREN STATISCHEN UND DYNAMISCHE	71-0622
TURBINE	A HIGH PRECISION AXISYMMETRIC TRIANGULAR ELEMENT USED IN THE ANALYSIS OF HYDRAULIC TURBINE COMPONENT	70-0038
TURBINE	CALCULATION OF AXISYMMETRICAL OSCILLATION OF FLEXIBLE TURBINE DISCS. (CALCOLO DELLE OSCILLAZIONI ASS	74-1328
TURBINE	CALCULATION OF AXISYMMETRICAL OSCILLATION OF FLEXIBLE TURBINE DISCS. (CALCOLO DELLE OSCILLAZIONI ASS	74-1328
TURBINE	CERAMIC ROTATING BLADES: SOME CRITICAL DESIGN PARAMETERS FOR GAS TURBINE APPLICATIONS	74-0146
TURBINE	ELASTO-PLASTIC ANALYSIS OF A ROTATING MODEL CONICAL TURBINE DISC	75-0281
TURBINE	FINITE ELEMENT ANALYSIS OF A GAS TURBINE BLADE	74-0971
TURBINE	FINITE ELEMENT STRESS ANALYSIS OF TURBINE BLADE HOOKS	70-0479
TURBINE	GAS TURBINE ENGINE DISK CYCLIC LIFE PREDICTION	75-0509
TURBINE	HOLOGRAPHIC INSTRUMENTATION OF TURBINE BLADES	73-0466
TURBINE	MODELING OF GAS TURBINE ENGINE COMPRESSOR BLADES FOR VIBRATION ANALYSIS	75-0508
TURBINE	NASTRAN ANALYSIS OF A TURBINE BLADE AND COMPARISON WITH TEST AND FIELD DATA	75-0367
TURBINE	SOLUTION OF A THREE DIMENSIONAL AXISYMMETRICAL PROBLEM OF THERMOPLASTICITY AS APPLIED TO THICK TURBI	75-0384
TURBINE	STUDY OF UMBRELLA OSCILLATIONS OF TURBINE DISKS OF ARBITRARY PROFILE BY THE FINITE ELEMENT METHOD	75-0332
TURBINE	UNBALANCE GENERATED VIBRATIONS AND STABILITY OF TURBINE ROTORS	73-0224
TURBINE	VIBRATION ANALYSIS OF ROTATING TURBINE BLADES	74-0959
TURBINE	VIBRATION ANALYSIS OF AXIAL FLOW TURBINE DISKS USING FINITE ELEMENTS	75-0195
TURBINE	VIBRATION ANALYSIS OF AXIAL-FLOW TURBINE DISKS USING FINITE-ELEMENTS	75-0415
TURBINE	VIBRATION OF THICK, CURVED, SHELLS WITH PARTICULAR REFERENCE TO TURBINE BLADES	70-0017
TURBINES	APPLICATION OF THE FINITE ELEMENT METHOD TO THE VIBRATION ANALYSIS OF AXIAL FLOW TURBINES	73-1125
TURBINES	BUILDING THREE NEW MILLION-HORSEPOWER HYDRAULIC TURBINES FOR GRAND COULEE	74-0981
TURBO-ALTERNATO	DYNAMIC BEHAVIOR OF STEEL FOUNDATIONS FOR TURBO-ALTERNATORS	71-0837
TURBOALTERNATOR	ANALYSIS OF TURBOALTERNATOR MAGNETIC FIELDS BY FINITE ELEMENTS	71-0460
TURBOMACHINE	CONTRIBUTION A L'ANALYSE DES VIBRATIONS LIBRES D'UNE STRUCTURE. APPLICATION AUX AUBAGES DE TURBOMACH	75-0804
TURBOMACHINE	FINITE-ELEMENT METHOD FOR AXISYMMETRIC FLOW COMPUTATION IN A TURBOMACHINE	75-0359
TURBOMACHINERY	DANUTA THREE DIMENSIONAL FINITE ELEMENT PROGRAM USED IN ANALYSIS OF TURBOMACHINERY	72-0296
TURBOMACHINERY	PSEUDO STATIC DEFORMATION AND FREQUENCIES OF ROTATING TURBOMACHINERY BLADES	72-0239
TURBOMACHINERY	SCHWINGUNGSMAESSIGE AUSLEGUNG DER BESCHAUFELUNG AXIALER TURBOMASCHINEN (DESIGN OF DISPLACEMENT-FREE	74-0253
TURBOMACHINERY	THE EFFECT OF CAMBER ON THE NATURAL FREQUENCIES OF LOW ASPECT RATIO TURBOMACHINERY BLADES	73-1092
TURBOMACHINES	A FINITE ELEMENT METHOD FOR FLOW CALCULATIONS IN TURBOMACHINES	74-0726
TURBOMACHINES	COMPUTATION OF FLOWS IN TURBOMACHINES	74-0076
TURBOMACHINES	CONTRIBUTION A L'ANALYSE DES VIBRATIONS LIBRES D'UNE STRUCTURE. APPLICATION AUX AUBAGES DE TURBOMACH	75-0804
TURBOMACHINES	NUMERICAL CALCULATION OF MERIDIONAL FLOW FIELD IN TURBOMACHINES USING FINITE ELEMENTS METHOD	74-0481

TURBOMASCHINEN	SCHWINGUNGSMAESSIGE AUSLEGUNG DER BESCHAUFELUNG AXIALER TURBOMASCHINEN (DESIGN OF DISPLACEMENT-FREE	74-0253
TURBOMOTOR	FINITE ELEMENT MODEL FOR DISTRIBUTED PARAMETER TURBOMOTOR SYSTEMS	71-0332
TURBOPUMP	STRESS AND VIBRATION ANALYSIS OF TURBOPUMP INDUCER BLADES BY FINITE ELEMENTS	71-0363
TURBULENCE	EFFECTS OF TURBULENCE ON FLOOD FLOW EXPLORED BY FINITE ELEMENT ANALYSIS.	72-0410
TURBULENCE	INFLUENCE OF COOLING, FREE-STREAM TURBULENCE AND SURFACE ROUGHNESS ON THE AERODYNAMIC BEHAVIOR OF CA	74-0970
TURBULENCE	VIBROACOUSTIC RESPONSE OF TURBULENCE EXCITED THIN RECTANGULAR FINITE PLATES IN HEAVY AND FLUID MEDIA	74-1309
TURBULENT	FIELD SOLUTIONS FOR TURBULENT SEEPAGE FLOW	71-0206
TURBULENT	THE PREDICTION OF FULLY DEVELOPED TURBULENT FLOW IN DUCTS BY THE FINITE ELEMENT METHOD	75-0117
TURNER	THE RECTANGULAR PLANE STRESS ELEMENT BY TURNER, PIAN AND WILSON	74-1265
TURNING	STRAIN ENERGY RELEASE RATES FOR TURNING CRACKS	72-0146
TWISTED	ANALYSIS OF BENT AND TWISTED BARS USING THE FINITE ELEMENT METHOD	67-0045
TWISTED	FINITE ELEMENT ANALYSIS OF COUPLED VIBRATION OF TAPERED TWISTED BLADES	72-0766
TWISTED	TWISTED BEAM ELEMENT MATRICES FOR BENDING	70-0082
TWISTED	VIBRATORY BEHAVIOR OF TWISTED CANTILEVERED PLATES	75-0507
TWISTING	STATIC FLEXIBILITY MATRICES FOR A RECTANGULAR PLATE ELEMENT IN BENDING, TWISTING, AND SHEAR. UNIFORM	68-0217
TWO-	GENERATION OF TWO- AND THREE-DIMENSIONAL FINITE ELEMENTS	71-0040
TWO-AND	COMPUTER PROGRAMS FOR PLOTTING FINITE ELEMENT PATTERNS OF TWO-AND THREE-DIMENSIONAL STRUCTURES	70-0416
TWO-AND	FINITE ELEMENT MESH GENERATION PROGRAM FOR ARBITRARY TWO-AND THREE-DIMENSIONAL STRUCTURES	73-0966
TWO-AND	TWO-AND THREE-DIMENSIONAL FLOW ANALYSIS USING FINITE ELEMENTS	70-0021
TWO-AND-THREE	COMPUTATIONAL METHODS FOR INVISCID AND VISCOUS TWO-AND-THREE DIMENSIONAL FLOW FIELDS	75-0716
TWO-COMPONENT	DISCUSSION OF PROCEEDINGS, PAPER 6612. ELASTIC BUCKLING OF TWO-COMPONENT COLUMNS	70-0321
TWO-DIMENSIONAL	A COMPUTER BASED PROCEDURE FOR PREDICTING THE TRANSIENT RESPONSE AND FAILURE OF A TWO-DIMENSIONAL CO	71-0630
TWO-DIMENSIONAL	A COMPUTER PROGRAM FOR THE ANALYSIS OF TWO-DIMENSIONAL HEAT CONDUCTION USING THE FINITE ELEMENT TECH	69-0066
TWO-DIMENSIONAL	A FINITE ELEMENT MODEL FOR TWO-DIMENSIONAL STEADY FLOW THROUGH CONTRACTIONS IN NATURAL CHANNELS	71-0810
TWO-DIMENSIONAL	A FINITE ELEMENT SOLUTION FOR TWO-DIMENSIONAL DENSITY STRATIFIED FLOW	73-0911
TWO-DIMENSIONAL	A FINITE ELEMENT SOLUTION FOR TWO-DIMENSIONAL STRATIFIED FLOW PROBLEMS	75-0391
TWO-DIMENSIONAL	A FINITE-ELEMENT ANALYSIS OF STEADY, TWO-DIMENSIONAL, INCOMPRESSIBLE, LAMINAR FLOW	72-0981
TWO-DIMENSIONAL	A GENERAL METHOD OF ANALYSIS FOR TWO-DIMENSIONAL PILE GROUPS	71-0298
TWO-DIMENSIONAL	A GENERAL NUMERICAL SOLUTION OF THE TWO-DIMENSIONAL DIFFUSION-CONVECTION EQUATION BY THE FINITE ELEM	70-0004
TWO-DIMENSIONAL	A HIGH-ORDER FINITE ELEMENT FOR TWO-DIMENSIONAL CRACK PROBLEMS	74-1128
TWO-DIMENSIONAL	A HYBRID FINITE ELEMENT METHOD FOR TWO-DIMENSIONAL NAVIER STOKES EQUATIONS	75-0027
TWO-DIMENSIONAL	A NUMERICAL SOLUTION TECHNIQUE FOR A CLASS OF TWO-DIMENSIONAL PROBLEMS IN FLUID DYNAMICS FORMULATED	69-0019
TWO-DIMENSIONAL	A TWO-DIMENSIONAL AXISYMMETRIC SHELL FLUID MODEL FOR LAUNCH VEHICLE LONGITUDINAL DYNAMIC ANALYSIS	69-0046
TWO-DIMENSIONAL	A TWO-DIMENSIONAL FINITE ELEMENT - LEAST SQUARES SOLUTION	75-0025
TWO-DIMENSIONAL	ACCURACY, STABILITY, AND OSCILLATION CHARACTERISTICS OF TRANSIENT TWO-DIMENSIONAL HEAT CONDUCTION	75-0849
TWO-DIMENSIONAL	AN ANALYSIS OF TWO-DIMENSIONAL SURFACE RUN-OFF BY FINITE ELEMENTS	74-1336
TWO-DIMENSIONAL	AN EQUILIBRIUM STRESS FIELD MODEL FOR FINITE ELEMENT SOLUTIONS OF TWO-DIMENSIONAL ELASTIC PROBLEMS	68-0163
TWO-DIMENSIONAL	ANALISI NUMERICA DEL COMPORTAMENTO ELASTICO DI UNA STRUTTURA BIDIMENSIONALE PIANA. (NUMERICAL ANALYS	74-1029
TWO-DIMENSIONAL	APPLICATION OF AN ASSUMED DISPLACEMENT HYBRID FINITE ELEMENT PROCEDURE TO TWO-DIMENSIONAL PROBLEMS I	74-0729
TWO-DIMENSIONAL	APPLICATION OF FINITE ELEMENT ANALYSIS METHOD TO TWO-DIMENSIONAL PROBLEMS IN FRACTURE MECHANICS	69-0033
TWO-DIMENSIONAL	APPLICATION OF FINITE-ELEMENTS TO ONE-DIMENSIONAL AND TWO-DIMENSIONAL NEUTRON-TRANSPORT	73-1150
TWO-DIMENSIONAL	APPLICATION OF FINITE DIFFERENCES FOR SOLVING THE TWO-DIMENSIONAL ELASTICITY PROBLEM BY MEANS OF THE	75-0482
TWO-DIMENSIONAL	APPLICATION OF THE FINITE-ELEMENTS METHOD TO HEAT-TRANSFER PROBLEMS. PART 11. TRANSIENT TWO-DIMENSION	71-0759
TWO-DIMENSIONAL	APPLICATION OF THE FINITE ELEMENTS METHOD IN TWO-DIMENSIONAL HYDRODYNAMICS USING THE LAGRANGE VARIAB	73-0976
TWO-DIMENSIONAL	APPLICATION OF THE METHOD OF FINITE ELEMENT ANALYSIS TO TWO-DIMENSIONAL PROBLEMS IN FRACTURE MECHANI	68-0100
TWO-DIMENSIONAL	APPLICATION OF THE TWO-DIMENSIONAL INTEGRAL EQUATION METHOD TO ENGINEERING PROBLEMS	75-0377
TWO-DIMENSIONAL	COMBINED ELASTIC-PLASTIC CREEP ANALYSIS OF TWO-DIMENSIONAL BODIES	71-0232
TWO-DIMENSIONAL	EFFECT OF ELASTIC MODULAR RATIO AND ANISOTROPY ON THE STRESS FIELD IN A TWO-DIMENSIONAL MATRIX INTO	71-0647
TWO-DIMENSIONAL	EFFECT OF ELASTIC MODULAR RATIO AND ANISOTROPY ON THE STRESS FIELD IN A TWO-DIMENSIONAL MATRIX INTO	71-0647
TWO-DIMENSIONAL	EFFICIENCY EVALUATION OF A TWO-DIMENSIONAL INCOMPATIBLE FINITE ELEMENT	73-0988
TWO-DIMENSIONAL	ELASTIC-PLASTIC ANALYSIS OF TWO-DIMENSIONAL STRESS SYSTEMS BY THE FINITE ELEMENT METHOD	67-0049
TWO-DIMENSIONAL	ELASTIC-PLASTIC SOLUTION OF A TWO-DIMENSIONAL TUBE PROBLEM BY THE FINITE ELEMENT METHOD	74-0769
TWO-DIMENSIONAL	EQUIPOTENTIAL ZONING OF TWO-DIMENSIONAL MESHES	63-0025
TWO-DIMENSIONAL	EQUIPOTENTIAL ZONING OF TWO-DIMENSIONAL MESHES	64-0033
TWO-DIMENSIONAL	ESTIMATING MATRIX DISPLACEMENT SOLUTIONS OF TWO-DIMENSIONAL PROBLEMS BY LARGE ELEMENT TECHNIQUE	70-0160
TWO-DIMENSIONAL	FEM2D: PROGRAM FOR SOLVING THE TWO-DIMENSIONAL DIFFUSION EQUATION BY THE METHOD OF FINITE ELEMENTS	74-1089
TWO-DIMENSIONAL	FINITE ELEMENT ANALYSIS OF TWO-DIMENSIONAL TIME-DEPENDENT STRESS PROBLEMS	65-0022
TWO-DIMENSIONAL	FINITE ELEMENT ANALYSIS OF TWO-DIMENSIONAL SLOW NON-NEWTONIAN FLOWS	72-0035
TWO-DIMENSIONAL	FINITE ELEMENT ANALYSIS OF TWO-DIMENSIONAL ADDED MASS AND DAMPING	75-0394
TWO-DIMENSIONAL	FINITE ELEMENT BASED PROCEDURE FOR SIMULATING THE TRANSIENT RESPONSE AND FAILURE OF A TWO-DIMENSIONA	72-0123
TWO-DIMENSIONAL	FINITE ELEMENT FRACTURE MECHANICS ANALYSIS OF TWO-DIMENSIONAL AND AXISYMMETRIC ELASTIC AND ELASTIC-P	74-1092
TWO-DIMENSIONAL	FINITE ELEMENT METHOD APPLIED TO TRANSIENT TWO-DIMENSIONAL HEAT TRANSFER WITH CONVECTION AND RADIATI	70-0127
TWO-DIMENSIONAL	FINITE ELEMENT METHODS: TWO-DIMENSIONAL SEEPAGE WITH A FREE SURFACE	71-0475
TWO-DIMENSIONAL	FINITE ELEMENT METHOD FOR SOLVING NEUTRON TRANSPORT PROBLEMS IN TWO-DIMENSIONAL CYLINDRICAL GEOMETRY	74-1081
TWO-DIMENSIONAL	FINITE ELEMENT MODEL FOR TWO-DIMENSIONAL DENSITY STRATIFIED FLOW	73-0064
TWO-DIMENSIONAL	FINITE ELEMENT SIMULATION OF TWO-DIMENSIONAL ELECTROMECHANICAL RESONATORS	74-1014
TWO-DIMENSIONAL	FINITE ELEMENT SOLUTION OF THE FLOW AROUND A TWO-DIMENSIONAL DESCOID	69-0298
TWO-DIMENSIONAL	FINITE-ELEMENT ANALYSIS OF FRACTURE PROPAGATION IN TWO-DIMENSIONAL ELASTIC BRITTLE SOLIDS	72-0948
TWO-DIMENSIONAL	FRACTURE OF ROCKET GRAIN MODELS SUBJECTED TO COOLDOWN: A STUDY OF TEMPERATURE-INITIATED FRACTURES OF	74-0536
TWO-DIMENSIONAL	GNATS: A FINITE ELEMENT COMPUTER PROGRAM FOR THE GENERAL NONLINEAR ANALYSIS OF TWO-DIMENSIONAL STRUC	74-0875
TWO-DIMENSIONAL	HEAVY SECTION STEEL TECHNOLOGY PROGRAM. TECHNICAL REPORT NO. 4, OCTOBER 1969. A TWO-DIMENSIONAL ELAS	69-0474
TWO-DIMENSIONAL	IMPROVED TWO-DIMENSIONAL FINITE ELEMENT	74-0355
TWO-DIMENSIONAL	ISOFINEL: ISOPARAMETRIC FINITE ELEMENT CODE FOR ELASTIC ANALYSIS OF TWO-DIMENSIONAL BODIES	75-0676
TWO-DIMENSIONAL	LEAST SQUARES FINITE ELEMENT ANALYSIS OF TWO-DIMENSIONAL LAMINAR BOUNDARY-LAYER EQUATION WITH SEPARA	75-0796
TWO-DIMENSIONAL	MATHEMATICAL MODELS OF THE MASSACHUSETTS BAY. PART 1. FINITE ELEMENT MODELING OF TWO-DIMENSIONAL HYD	73-0800
TWO-DIMENSIONAL	MODELING AN OCEAN POND. A TWO-DIMENSIONAL, FINITE ELEMENT HYDRODYNAMIC MODEL OF NINIGRET POND. CHARL	75-0686
TWO-DIMENSIONAL	NONLINEAR ANALYSIS OF TWO-DIMENSIONAL PROBLEMS IN CONCRETE CREEP	71-0299
TWO-DIMENSIONAL	ON THE FINITE ELEMENT ANALYSIS OF TWO-DIMENSIONAL STRESS PROBLEMS WITH TIME DEPENDENT PROPERTIES	65-0052
TWO-DIMENSIONAL	ON THE SOLUTION OF TWO-DIMENSIONAL PROBLEMS OF ELASTOPLASTICITY BY THE FINITE ELEMENT AND DIRECT STI	67-0002
TWO-DIMENSIONAL	ORGANIZATION OF GNATS: A GENERAL NONLINEAR ANALYSIS COMPUTER PROGRAM FOR TWO-DIMENSIONAL STRUCTURES	74-0871
TWO-DIMENSIONAL	PART 1. THEORY REPORT FOR CREEP-PLAST COMPUTER PROGRAM: ANALYSIS OF TWO-DIMENSIONAL PROBLEMS UNDER S	72-0765
TWO-DIMENSIONAL	REALIZATSIYA METODA KONECHNYKH ELEMENTOV NA EVM DLYA RASCHETA DVUMERNYKH ELEKTRICHESKIKH I MAGNITNYK	74-0251
TWO-DIMENSIONAL	REFINED FINITE ELEMENT ANALYSIS OF LINEAR AND NONLINEAR TWO-DIMENSIONAL STRUCTURES	66-0024
TWO-DIMENSIONAL	SAFE-PLANE, A COMPUTER PROGRAM FOR THE STRESS ANALYSIS AND DESIGN OF TWO-DIMENSIONAL COMPOSITE BODIE	67-0017
TWO-DIMENSIONAL	SCALAR FINITE-ELEMENT PROGRAM PACKAGE FOR TWO-DIMENSIONAL FIELD PROBLEMS	71-0478
TWO-DIMENSIONAL	SINGER: A COMPUTER CODE FOR GENERAL ANALYSIS OF TWO-DIMENSIONAL CONCRETE STRUCTURES. VOL. 11. PROGRA	75-0723
TWO-DIMENSIONAL	SINGER: A COMPUTER CODE FOR GENERAL ANALYSIS OF TWO-DIMENSIONAL REINFORCED CONCRETE STRUCTURES. VOLU	75-0724
TWO-DIMENSIONAL	SINGER: A COMPUTER CODE FOR GENERAL ANALYSIS TO TWO-DIMENSIONAL CONCRETE STRUCTURES. VOLUME 4. DEMON	75-0730
TWO-DIMENSIONAL	SINGER: A COMPUTER CODE FOR GENERAL ANALYSIS OF TWO-DIMENSIONAL REINFORCED CONCRETE STRUCTURES. VOLU	75-0731
TWO-DIMENSIONAL	SOME ASPECTS OF THE FINITE ELEMENT METHOD IN TWO-DIMENSIONAL PROBLEMS	69-0141
TWO-DIMENSIONAL	STABILITY AND OSCILLATION CHARACTERISTICS OF FINITE ELEMENT, FINITE-DIFFERENCE, AND WEIGHTED-RESIDUA	75-0615
TWO-DIMENSIONAL	STRESS ANALYSIS OF TWO-DIMENSIONAL PROBLEMS UNDER SIMULTANEOUS CREEP AND PLASTICITY	71-0220
TWO-DIMENSIONAL	THE APPLICATION OF FINITE ELEMENT ANALYSIS TO THE STUDY OF TWO-DIMENSIONAL CRACKED BODIES	72-0695
TWO-DIMENSIONAL	THE CHANGES IN THE (TWO-DIMENSIONAL) STRESS ROUNDED CORNERS, CAUSED BY A VARYING INTERNAL PRESSURE A	67-0141
TWO-DIMENSIONAL	TRANSIENT FIELD PROBLEMS: TWO-DIMENSIONAL AND THREE-DIMENSIONAL ANALYSIS BY ISO-PARAMETRIC FINITE EL	70-0001
TWO-DIMENSIONAL	TRANSIENT TWO-DIMENSIONAL PROBLEMS UTILIZING THE LEAST SQUARES ALGORITHM	75-0219
TWO-DIMENSIONAL	TRIPLET: A TWO-DIMENSIONAL, MULTIGROUP, TRIANGULAR MESH, PLANAR GEOMETRY, EXPLICIT TRANSPORT CODE	73-0955
TWO-DIMENSIONAL	TWO-DIMENSIONAL AIRFOIL ANALYSIS USING A REFINED FINITE ELEMENT TECHNIQUE	73-0587
TWO-DIMENSIONAL	TWO-DIMENSIONAL ANALYSIS OF BACKWATER AT BRIDGES	74-0184
TWO-DIMENSIONAL	TWO-DIMENSIONAL AUTOMATIC MESH GENERATION FOR STRUCTURAL ANALYSIS	70-0058
TWO-DIMENSIONAL	TWO-DIMENSIONAL FINITE ELEMENT MODEL OF A CERAMIC BODY	73-0010
TWO-DIMENSIONAL	TWO-DIMENSIONAL FINITE ELEMENT COMPUTER CODES FOR STRESS ANALYSIS AND FIELD PROBLEMS	73-0055

TWO-DIMENSIONAL	TWO-DIMENSIONAL FINITE ELEMENT SIMULATION OF SEMICONDUCTOR DEVICES	74-0260
TWO-DIMENSIONAL	TWO-DIMENSIONAL FINITE ELEMENT MODELING OF TRANSIENT FLOW IN REGIONAL AQUIFER SYSTEMS. ANALYSIS OF G	74-0743
TWO-DIMENSIONAL	TWO-DIMENSIONAL FINITE ELEMENT ANALYSIS OF SEMI-CONDUCTOR STEADY STATE TRANSPORT EQUATIONS	74-1224
TWO-DIMENSIONAL	TWO-DIMENSIONAL HEATED PIPELINE IN PERMAFROST PIPELINES	74-0541
TWO-DIMENSIONAL	TWO-DIMENSIONAL PHOTOELASTIC STUDY OF STEAM-TURBINE CASING FLANGES WITH INTERFACE RELIEF	74-0603
TWO-DIMENSIONAL	TWO-DIMENSIONAL TIME-DEPENDENT ANALYSIS OF PERFORATED END CAPS FOR NUCLEAR REACTOR PRESSURE VESSELS	71-0239
TWO-DIMENSIONAL	TWO-DIMENSIONAL TRANSPORT CALCULATIONS USING PHASE-SPACE FINITE ELEMENTS	72-0027
TWO-DIMENSIONAL	USE OF SYMBOLIC MANIPULATION IN DEVELOPMENT OF TWO-DIMENSIONAL FINITE ELEMENTS	74-0103
TWO-FIELD	SOME TWO-FIELD VARIATIONAL PRINCIPLES FOR NONLINEAR DEFORMATION ANALYSIS OF SHELLS	75-0294
TWO-GROUP	AN OPTIMUM IDEALIZATION IN FINITE ELEMENT METHOD FOR A TWO-GROUP NEUTRON DIFFUSION PROBLEM	74-1143
TWO-HIGH-PRECIS	A COMPARISON BETWEEN TWO-HIGH-PRECISION FINITE ELEMENTS	75-0033
TWO-LAYER	FINITE ELEMENT MODEL FOR TRANSIENT TWO-LAYER COOLING POND BEHAVIOR	75-0465
TWO-LAYERED	FINITE-ELEMENT SOLUTION FOR A CRACKED TWO-LAYERED ELASTIC CYLINDER	71-0397
TWO-PHASE	A FINITE ELEMENT APPROACH TO TWO-PHASE FLOW IN POROUS MEDIA	75-0392
TWO-PHASE	ANALYSE NUMERIQUE D'UN PROBLEME DE STEFAN A DEUX PHASES PAR UNE METHODE D'ELEMENTS FINIS. (NUMERICAL	75-0387
TWO-PHASE	SIMULATION ON THE TENSILE BEHAVIOUR OF TWO-PHASE ALLOY CONTAINING A FIBER UNDER TENSILE LOAD	73-0542
TWO-POINT	AN ERROR EXPANSION FOR A FINITE-ELEMENT SOLUTION OF A TWO-POINT BOUNDARY-VALUE PROBLEM	75-0801
TWO-REGIME	FINITE ELEMENT SOLUTIONS OF TRANSIENT, TWO-REGIME FLOW TOWARDS WELLS	74-0466
TWO-REGIME	FINITE ELEMENT SOLUTION OF TWO-REGIME FLOW TOWARDS WELLS	74-1352
TWO-STAGE	MEASURED AND PREDICTED LONGITUDINAL VIBRATIONS OF A LIQUID PROPELLANT TWO-STAGE LAUNCH VEHICLE	70-0366
TWO-WAY	A DISCRETE-ELEMENT METHOD OF MULTIPLE-LOADING ANALYSIS FOR TWO-WAY BRIDGE FLOOR SLABS	70-0435
TYPICAL	COMPARISON OF TYPICAL INELASTIC ANALYSIS PREDICTIONS WITH BENCHMARK PROBLEM EXPERIMENTAL RESULTS	75-0696
TYPICAL	RADIAL AND AXIAL TRANSMISSIBILITY CHARACTERISTICS OF TYPICAL ROCKET VEHICLE STRUCTURE TECHNICAL SUMM	69-0475
TYPICAL	STRUCTURAL MODELING OF A WELDED JOINT TYPICAL FOR VEHICLE FRAMES	73-0539
TYRE	APPLICATION OF FINITE ELEMENT METHOD TO PNEUMATIC TYRE	75-0119
U-BARS	FORCES AND DEFORMATIONS AT THE LATERAL STIFFENER OF THIN WALLED U-BARS	73-0236
U-FRAME	FINITE ELEMENT ANALYSES OF SOIL-STRUCTURE INTERACTION IN U-FRAME LOCKS	69-0531
U-SECTION	DEVELOPMENT OF A DESIGN FOR A PRECAST CONCRETE BRIDGE BEAM OF U-SECTION	73-0392
U-TYPE	STRESS CONCENTRATION FACTORS FOR A FATIGUE TEST SPECIMEN WITH A U-TYPE NOTCH IN ONE EDGE BY FINITE E	74-1344
UEBERWIEGEND	TEMPERATURVERLAUF IM QUERSCHNITT GESCHWEISSTER ROHRWAENDE BEI UEBERWIEGEND KONVEKTIVEM WAERMEUEBERGA	75-0628
UKLADOW	ANALIZA ORGAN WYMUSZONYCH ZLOZONYCH UKLADOW LINIOWYCH METODA SZTYWNYCH ELEMENTOW SKONCZONYCH (ANALYS	72-0246
ULCCS	DYNAMIC AND STOCHASTIC ASPECTS IN THE STRUCTURAL ANALYSIS OF VLCCS AND ULCCS	72-0298
ULOHY	GEOMETRICKY NELINEARNE ULOHY PLOCHYCH SKRUPIN PIESENE PRIRASTKOVOU METODOU KONECNYCH PRVKOV. (GEOMET	74-1032
ULTIMATE	EINFACHES NAEHERUNGSVERFAHREN FUR DIE GRENZTRAGFAEHIGKEITSANLYSE VON SPANNBETON-REAKTORDRUCKBEHAELTE	73-0854
ULTIMATE	FINITE ELEMENT STUDY OF A REINFORCED CONCRETE CYLINDRICAL SHELL THROUGH ELASTIC, CRACKING, AND ULTIM	74-0317
ULTIMATE	FINITE ELEMENT ULTIMATE LOAD ANALYSIS OF THREE DIMENSIONAL CONCRETE STRUCTURES	74-0672
ULTIMATE	FRACTURE (ULTIMATE STRENGTH) ANALYSES OF ASPHALT PAVEMENT LAYERS RESULTING FROM TRAFFIC LOADING	73-0548
ULTIMATE	TRAGLAST VON T-TRAEGERN. EXPERIMENT UND F.E. BERECHNUNGEN (ULTIMATE STRENGTH OF T-BEAMS. EXPERIMENTS	74-0935
ULTIMATE	UEBER DAS VERHALTEN VON SPANNBETONDRUCKBEHAELTERN IM GRENZTRAGFAEHIGKEITSZUSTAND UND DESSEN EINFLU	73-0855
ULTIMATE	ULTIMATE CAPACITY OF CIRCULAR FOOTINGS ON LAYERED SOILS	70-0280
ULTIMATE	ULTIMATE LOAD BEHAVIOUR OF PCRV TOP CLOSURES. THEORETICAL AND EXPERIMENTAL INVESTIGATIONS	73-0835
ULTIMATE	ULTIMATE STRAIN CAPACITY AND TEMPERATURE RISE STUDIES, TRUMBULL POND DAM	72-0705
ULTIMATE	ULTIMATE STRENGTH OF BOX GIRDERS BY FINITE ELEMENT METHOD	75-0794
UMBRELLA	STUDY OF UMBRELLA OSCILLATIONS OF TURBINE DISKS OF ARBITRARY PROFILE BY THE FINITE ELEMENT METHOD	75-0332
UMGEBUNG	EFFECT OF DISCONTINUITIES ON STRESSES AND DEFORMATIONS OF ROCK AROUND A TUNNEL. (EINFLUSS VON DISKON	70-0170
UMR	USER'S MANUAL FOR UMR FINITE ELEMENT COMPUTER PROGRAM - STRATA	69-0429
UNBALANCE	UNBALANCE GENERATED VIBRATIONS AND STABILITY OF TURBINE ROTORS	73-0224
UNBALANCED	GRAPHICAL DISPLAY OF COMPUTER SIMULATED UNBALANCED ROTOR RESPONSE	71-0486
UNBONDED	RIDDLE OF BONDED AND UNBONDED TENDONS IN PRESTRESSED CONCRETE REACTOR VESSELS	74-0519
UNBONDED	UNBONDED CONTACT PROBLEM OF A PLATE ON THE ELASTIC HALF SPACE	74-0208
UNBOUNDED	APPLICATION OF INTEGRAL EQUATIONS TO FLUID FLOWS IN UNBOUNDED REGIONS	74-0029
UNBOUNDED	FINITE-ELEMENT SOLUTION OF UNBOUNDED FIELD PROBLEMS	72-0025
UNBOUNDED	UNBOUNDED ERRORS IN NUMERICAL ANALYSIS OF STRUCTURES	74-1210
UNCONDITIONAL	ON THE UNCONDITIONAL STABILITY OF AN IMPLICIT ALGORITHM FOR NONLINEAR STRUCTURAL DYNAMICS	75-0825
UNCONDITIONALLY	UNCONDITIONALLY STABLE FINITE ELEMENT SCHEMES FOR PARABOLIC EQUATIONS	74-1286
UNCONFINED	APPROXIMATE SOLUTION FOR UNCONFINED SEEPAGE	73-0348
UNCONFINED	FINITE ELEMENT ANALYSIS OF UNCONFINED SEEPAGE PROBLEMS	71-0767
UNCONFINED	FINITE ELEMENT ANALYSIS OF UNCONFINED FLOW WITH A VARIABLE TIME-STEP PROCEDURE	74-0776
UNCONFINED	FINITE ELEMENT ANALYSIS OF TIME-DEPENDENT PUMPING TEST DATA IN AN UNCONFINED AQUIFER	75-0497
UNCONFINED	RESPONSE OF UNCONFINED AQUIFERS TO PUMPING	74-0073
UNCONFINED	TRANSIENT TWO DIMENSIONAL UNCONFINED AQUIFER PROBLEMS	74-0069
UNCONFINED	UNCONFINED TRANSIENT SEEPAGE IN SLOPING BANKS	71-0378
UNCONFINED	VARIABLE TIME STEP ANALYSIS OF UNCONFINED SEEPAGE	74-0064
UNCONSOLIDATED	OPTIMAL DESIGN OF WATER WELLS IN UNCONSOLIDATED SEDIMENTS	74-1182
UND-BERECHNUNGE	FESTIGKEITSUNTERSUCHUNGEN UND-BERECHNUNGEN AN PUMPENSPIRALGEHAEUSEN. (INVESTIGATIONS AND CALCULATION	75-0238
UNDAMPED	HARBOUR OSCILLATION - A NUMERICAL TREATMENT FOR UNDAMPED NATURAL MODES	70-0208
UNDEFORMED	A SPECIAL NASTRAN PROGRAM FOR INPUT CHECKING AND UNDEFORMED STRUCTURE PLOTTING	71-0234
UNDERGOING	ON SMALL VIBRATIONS AND PERTURBATIONS OF FLEXIBLE BODIES UNDERGOING ARBITRARY NOMINAL MOTION	70-0376
UNDERGOING	OPTIMIZATION OF STRUCTURES UNDERGOING HARMONIC OR STOCHASTIC EXCITATION	75-0711
UNDERGROUND	A NONLINEAR FINITE ELEMENT CODE FOR ANALYZING THE BLAST RESPONSE OF UNDERGROUND STRUCTURES	70-0054
UNDERGROUND	ANALYSIS AND MODEL STUDIES OF UNDERGROUND OPENINGS IN JOINTED ROCK. RESEARCH ON ROCK BOLT REINFORCEM	71-0438
UNDERGROUND	ANALYSIS AND MODEL STUDIES OF UNDERGROUND OPENINGS IN JOINTED ROCK. RESEARCH ON ROCK BOLT REINFORCEM	71-0750
UNDERGROUND	ANALYSIS OF UNDERGROUND OPENINGS IN ROCK BY FINITE ELEMENT METHODS	73-0895
UNDERGROUND	APPLICATION OF FINITE ELEMENT ANALYSIS TO UNDERGROUND STORAGE OF NATURAL GAS	70-0198
UNDERGROUND	APPLICATION OF THE FINITE ELEMENT METHOD TO THE DESIGN OF UNDERGROUND BULKHEADS	72-0387
UNDERGROUND	APPROXIMATE METHOD FOR CALCULATING THE RESPONSE OF EMPLACEMENT STRUCTURES SUBJECTED TO GROUND SHOCK	73-0654
UNDERGROUND	CONSIDERATION ON THE SEISMIC RESPONSE ANALYSIS OF UNDERGROUND STRUCTURES	75-0470
UNDERGROUND	DESIGN OPTIMIZATION IN UNDERGROUND COAL SYSTEMS	74-1095
UNDERGROUND	DISPLACEMENT AND STRAIN FIELDS FROM UNDERGROUND EXPLOSIONS USING ANALYTICAL AND FINITE ELEMENT METHO	72-0587
UNDERGROUND	ELASTO-PLASTIC ANALYSIS OF UNDERGROUND OPENINGS BY THE FINITE ELEMENT METHOD	66-0067
UNDERGROUND	ELECTRO-OSMOSIS AND THE CONTAMINATION OF UNDERGROUND FLUIDS	73-0609
UNDERGROUND	FINITE ELEMENT ELASTOPLASTIC ANALYSIS OF UNDERGROUND OPENINGS BY QUADRATIC PROGRAMMING	74-1288
UNDERGROUND	FINITE ELEMENT MODELING CRITERIA FOR UNDERGROUND OPENINGS IN ROCK	74-0630
UNDERGROUND	FINITE ELEMENT OF UNDERGROUND STRESSES UTILIZING STOCHASTICALLY SIMULATED MATERIAL PROPERTIES	70-0111
UNDERGROUND	MODELLING HEAT FLOW FROM UNDERGROUND POWER CABLES	74-1158
UNDERGROUND	ON THE DIMENSIONING OF UNDERGROUND STRUCTURES	69-0147
UNDERGROUND	RASCHET NAPRYAZHENII L PEREMESHCHENII V SVODE PODZEMNOGO MASHINNOGO ZALA GES I V OKRUZHAYUSHCHEM SKA	74-0542
UNDERGROUND	SEISMIC EFFECTS ON A PROPOSED UNDERGROUND REACTOR FACILITY	73-0953
UNDERGROUND	STRAIN DISTRIBUTION AROUND UNDERGROUND OPENINGS	69-0212
UNDERGROUND	STRAIN DISTRIBUTION AROUND UNDERGROUND OPENINGS. STATISTICAL RELATIONSHIPS FOR CERTAIN ROCK PROPERTI	71-0710
UNDERGROUND	STRAIN DISTRIBUTION AROUND UNDERGROUND OPENINGS. FINITE ELEMENT ANALYSIS OF JOINTED SYSTEMS	72-0838
UNDERGROUND	UNDERGROUND MEASUREMENTS IN A STEEPLY DIPPING OREBODY	68-0085
UNDERGROUND	USER'S MANUAL FOR NTEDYNS - A NONLINEAR FINITE ELEMENT CODE FOR ANALYZING THE BLAST RESPONSE OF UNDE	71-0695
UNDERLAIN	SEISMIC RESPONSE OF SOIL DEPOSITS UNDERLAIN BY SLOPING ROCK BOUNDARIES	69-0484
UNDERLYING	SURVEY OF UNDERLYING THEORY. APPROXIMATION BY PIECEWISE POLYNOMIALS. ILLEGAL ELEMENTS AND THE PATCH	73-0635
UNDERSOKNING	EXPERIMENTELL UNDERSOKNING AV ELEMENTFOGARS HALLFASTHETSEGENSKAPER (EXPERIMENTAL INVESTIGATION OF ST	
UNDERWATER	DEVELOPMENT OF A FINITE ELEMENT MODEL FOR THE CLASS V FLEXTENSIONAL UNDERWATER TRANSDUCER SHELL	71-0754
UNDERWATER	DEVELOPMENT OF A FINITE ELEMENT MODEL FOR THE CLASS 5 FLEXTENSIONAL UNDERWATER TRANSDUCER SHELL	71-0840
UNDERWATER	DEVELOPMENT OF A FINITE ELEMENT MODEL FOR THE CLASS IV FLEXTENSIONAL UNDERWATER TRANSDUCER SHELL	72-0593
UNDERWATER	FUNDAMENTAL STUDY OF UNDERWATER WELDING	72-0382
UNDERWATER	NONLINEAR STATIC AND DYNAMIC RESPONSE OF UNDERWATER CABLE STRUCTURES USING THE FINITE ELEMENT METHOD	75-0460
UNDRAINED	FINITE ELEMENT PROGRAM FEECON FOR UNDRAINED DEFORMATION ANALYSES OF GRANULAR EMBANKMENTS ON SOFT CLA	72-0875

UNDRAINED	PREDICTION OF UNDRAINED MOVEMENTS CAUSED BY EMBANKMENTS ON MUSKEG	70-0070
UNDRAINED	UNDRAINED STRESS DISTRIBUTION BY NUMERICAL METHODS	68-0045
UNDRAINED	UNDRAINED VISCO-ELASTIC ANALYSIS OF SOIL DEFORMATION	72-0424
UNFRAMED	MEMBRANE ANALYSIS OF UNFRAMED HIGH RISE BUILDINGS	73-0451
UNGUYED	NON-LINEAR ELASTIC ANALYSIS OF UNGUYED TOWERS AND STACKS	74-1157
UNIAXIAL	APPLICATION OF THE FINITE ELEMENT METHOD TO THE ANALYSIS OF ROCK SAMPLES UNDER UNIAXIAL COMPRESSION	70-0182
UNIAXIAL	CONTROLLED FAILURE OF HOLLOW ROCK CYLINDERS IN UNIAXIAL COMPRESSION	72-0196
UNIAXIAL	CREEP OF A UNIAXIAL METAL MATRIX COMPOSITE SUBJECTED TO AXIAL AND NORMAL LATERAL LOADS	71-0041
UNIAXIAL	CREEP OF A UNIAXIAL METAL MATRIX COMPOSITE SUBJECTED TO AXIAL AND NORMAL LATERAL LOADS	71-0727
UNIAXIAL	FINITE ELEMENT ANALYSIS ON UNIAXIAL COMPRESSIVE STRENGTH OF CYLINDRICAL BRITTLE SPECIMEN	70-0155
UNIAXIAL	FULL-RANGE ANALYSIS OF STEEL PLATES AND STIFFENED PLATING UNDER UNIAXIAL COMPRESSION	75-0803
UNIAXIAL	NUMERICAL STUDY OF UNIAXIAL COMPRESSION IN CIRCULAR ELASTIC-PLASTIC COLUMNS	73-0238
UNIAXIAL	NUMERICAL STUDY OF UNIAXIAL AND TRIAXIAL ROCK COMPRESSION TESTS	73-0896
UNIAXIAL	ON THE STATES OF STRESS AND DEFORMATION OF CYLINDRICAL SPECIMENS OF BRITTLE MATERIAL UNDER UNIAXIAL	69-0150
UNIAXIAL	STRESS DISTRIBUTION UNDER UNIAXIAL ELASTIC PLANE STRAIN COMPRESSION WITH FRICTION	74-0526
UNIAXIAL	STUDY ON STATE OF STRESS AND DEFORMATION OF CYLINDRICAL SPECIMENS OF BRITTLE MATERIAL UNDER UNIAXIAL	69-0152
UNIAXIALLY	AN ELASTOPLASTIC ANALYSIS OF A UNIAXIALLY LOADED SHEET WITH AN INTERFERENCE - FIT BOLT	74-0687
UNIAXIALLY	THEORETICAL AND EXPERIMENTAL ANALYSIS OF ELASTOPLASTIC DEFORMATION IN A UNIAXIALLY LOADED STRIP OF 2	73-0720
UNIDIRECTIONAL	AN INVESTIGATION OF INITIAL YIELD SURFACES FOR UNIDIRECTIONAL REINFORCED COMPOSITES	72-0837
UNIDIRECTIONAL	ELASTOPLASTIC CRACK PROPAGATION IN A TRANSVERSELY LOADED UNIDIRECTIONAL COMPOSITE	74-0217
UNIDIRECTIONAL	HIGH-PERFORMANCE COMPOSITE MATERIALS FOR VEHICLE CONSTRUCTION: A FINITE ELEMENT COMPUTER PROGRAM FOR	73-0767
UNIDIRECTIONAL	INELASTIC ANALYSIS OF A UNIDIRECTIONAL COMPOSITE SUBJECTED TO TRANSVERSE NORMAL LOADING	70-0015
UNIDIRECTIONAL	ON BUCKLING OF UNIDIRECTIONAL BORON-ALUMINUM STIFFENERS - A CAUTION TO DESIGNERS	75-0634
UNIDIRECTIONAL	TRANSVERSE LOADING OF UNIDIRECTIONAL FIBER COMPOSITES	69-0154
UNIDIRECTIONAL	TRANSVERSE PROPERTIES OF UNIDIRECTIONAL ALUMINUM MATRIX FIBROUS COMPOSITES	69-0167
UNIDIRECTIONAL	YIELDING IN UNIDIRECTIONAL COMPOSITES UNDER EXTERNAL LOADS AND TEMPERATURE CHANGES.	73-0169
UNIDIRECTIONALL	INITIAL YIELD SURFACE OF A UNIDIRECTIONALLY REINFORCED COMPOSITE	71-0280
UNIDIRECTIONALL	STRESS DISTRIBUTION NEAR A BROKEN POINT OF FILAMENT IN UNIDIRECTIONALLY FIBER REINFORCED COMPOSITES	73-0811
UNIFICATION	UNIFICATION OF THE FINITE ELEMENT AND CLASSICAL RITZ METHODS	73-1139
UNIFIED	A UNIFIED FINITE ELEMENT APPROACH TO COLUMNS SUBJECTED TO NONCONSERVATIVE LOADS, WITH DAMPING EFFECT	74-0795
UNIFIED	A UNIFIED THEORY OF VARIATIONAL PRINCIPLES IN MECHANICS	75-0155
UNIFIED	A UNIFIED TREATMENT OF THICK AND THIN FILM ELASTOHYDRODYNAMIC PROBLEMS USING HIGHER ORDER ELEMENT PR	75-0890
UNIFIED	STRUCTURAL ANALYSIS: A UNIFIED CLASSICAL AND MATRIX APPROACH	72-0055
UNIFIED	THERMOMECHANICAL CREEP OF AGEING CONCRETE - A UNIFIED APPROACH	74-0607
UNIFIED	UNIFIED APPROACH TO THREE FINITE ELEMENT THEORIES FOR GEOMETRIC NONLINEARITY	74-0599
UNIFIED	VISCO-PLASTICITY, PLASTICITY AND CREEP IN ELASTIC SOLIDS. A UNIFIED NUMERICAL SOLUTION APPROACH	74-1279
UNIFIED	VISCOPLASTICITY, PLASTICITY AND CREEP. A UNIFIED NUMERICAL SOLUTION APPROACH	74-1205
UNIFORM	APPLICATIONS OF FINITE ELEMENT METHODS TO UNSTEADY VISCOUS FLOW AROUND A BOX-GIRDER BRIDGE OSCILLATI	74-0827
UNIFORM	ELASTIC-PLASTIC ANALYSIS OF BEAMS WITH UNIFORM CROSS-SECTION UNDER COMBINED LOADINGS	71-0777
UNIFORM	ELASTIC-PLASTIC ANALYSIS OF A BAR WITH UNIFORM CROSS-SECTION UNDER COMBINED LOADINGS 1 - COMBINATION	71-0778
UNIFORM	ELASTIC-PLASTIC ANALYSIS OF BEAMS WITH UNIFORM CROSS-SECTION UNDER COMBINED LOADINGS (2)	73-1070
UNIFORM	FINITE ELEMENT ANALYSIS ON THE TORSION OF A BAR WITH UNIFORM CROSS SECTION	68-0186
UNIFORM	FINITE ELEMENT ANALYSIS OF THE SURFACE DEFORMATION DUE TO A UNIFORM LOADING ON A LAYER OF GIBSON SOI	75-0240
UNIFORM	FINITE ELEMENT SOLUTION OF THE INCOMPRESSIBLE FLOW OVER AN AIRFOIL IN A NON UNIFORM STREAM	74-0432
UNIFORM	GALERKIN FINITE ELEMENT ANALYSIS OF A UNIFORM BEAM CARRYING A CONCENTRATED MASS AND ROTARY INERTIA W	74-0552
UNIFORM	INELASTIC BUCKLING OF STEEL BEAMS UNDER NON UNIFORM MOMENT	75-0149
UNIFORM	MAXIMUM PRINCIPLE AND UNIFORM CONVERGENCE FOR THE FINITE ELEMENT METHOD	73-0852
UNIFORM	RADIAL DISPLACEMENTS OF A HOLLOW SHAFT SUBJECTED TO A UNIFORM BAND OF EXTERNAL PRESSURE	72-0242
UNIFORM	STATIC FLEXIBILITY MATRICES FOR A RECTANGULAR PLATE ELEMENT IN BENDING, TWISTING, AND SHEAR. UNIFORM	68-0217
UNIFORM	THE COMPARISON OF THE RESPONSE OF A HIGHWAY BRIDGE TO UNIFORM GROUND SHOCK AND MOVING GROUND EXCITAT	72-0737
UNIFORM	THE USE OF THE FINITE ELEMENT METHOD IN OBTAINING SOIL-WATER PRESSURE AND SOIL-WATER CONTENT DISTRIB	75-0112
UNIFORM	UNIFORM CONVERGENCE OF VARIATIONAL FINITE ELEMENT AND RITZ METHODS	74-0420
UNIFORMITY	FINITE ELEMENT METHOD AS AN ASPECT OF THE PRINCIPLE OF MAXIMUM UNIFORMITY: NEW HYDRODYNAMICAL RAMIFA	74-0009
UNIFORMLY	ANALYSIS OF CLAMPED CIRCULAR PLATES OF VARIABLE THICKNESS SUBJECTED TO UNIFORMLY DISTRIBUTED LOAD BY	73-0583
UNIFORMLY	FINITE ELEMENT TREATMENT OF A UNIFORMLY MOVING ELASTIC-PLASTIC CRACK TIP	74-0470
UNIMOD	UNIMOD - APPLICATIONS ORIENTED FINITE ELEMENT SCHEME FOR THE ANALYSIS OF FRACTURE MECHANICS PROBLEMS	74-1356
UNIMOMENT	APPLICATION OF A UNIMOMENT TECHNIQUE TO A BICONICAL ANTENNA WITH INHOMOGENEOUS DIELECTRIC LOADING	75-0375
UNIMOMENT	UNIMOMENT METHOD OF SOLVING ANTENNA AND SCATTERING PROBLEMS	74-1020
UNIQUE	COMPARISON OF FINITE ELEMENT AND UNIQUE LIMIT ANALYSIS SOLUTIONS FOR CERTAIN REINFORCED CONCRETE SLA	69-0184
UNIT	VIBRATION ANALYSIS OF AN INTERNAL REFERENCE UNIT	69-0176
UNITS	INTEGRATED PROGRAM FOR THE DYNAMIC STRUCTURAL CALCULATION OF MOBIL OFFSHORE UNITS	74-0490
UNITY	ON THE UNITY OF THE CONSTANT STRAIN-CONSTANT MOMENT FINITE ELEMENT METHODS	73-0164
UNIVERSAL	FINEL, UNIVERSELLES PROGRAMMSYSTEM ZUR BERECHNUNG DER ELASTISCHEN EIGENSCHAFTEN VON MASCHINENBAUTEIL	72-0132
UNIVERSELLES	FINEL, UNIVERSELLES PROGRAMMSYSTEM ZUR BERECHNUNG DER ELASTISCHEN EIGENSCHAFTEN VON MASCHINENBAUTEIL	72-0132
UNKNOWN	FINITE ELEMENT SOLUTION OF PROBLEMS WITH UNKNOWN AND MOVING BOUNDARIES	74-0460
UNNECESSARY	CRACK TIP FINITE ELEMENTS ARE UNNECESSARY	73-1016
UNNECESSARY	CRACK TIP FINITE ELEMENTS ARE UNNECESSARY	75-0472
UNPRECISION	INFLUENCE OF THE UNPRECISION OF EXECUTION ON THE BEHAVIOUR OF A SPHERICAL SHELL WITH CONCENTRATED LO	73-0802
UNREGELMADIGER	BEITRAG ZUR SIMULATION NICHTSTATIONARER HORIZONTAL-EVENER GRUNDWASSERSTROMUNGSFELDER AUF DER GRUNDLA	74-0088
UNSATUPATED	FINITE ELEMENT ANALYSIS OF CONSOLIDATION OF UNSATURATED SOIL	74-0082
UNSATUPATED	FINITE ELEMENT COMPUTER PROGRAMS FOR FLOW IN SATURATED UNSATURATED POROUS MEDIA	72-0364
UNSATUPATED	FINITE ELEMENT SOLUTION OF UNSTEADY AND UNSATURATED FLOW IN POROUS MEDIA	72-0064
UNSATURATED	FINITE ELEMENT SOLUTION OF TWO DIMENSIONAL UNSTEADY AND UNSATURATED FLOW IN POROUS MEDIA	75-0113
UNSATURATED	FINITE ELEMENT WEIGHTED RESIDUAL SOLUTION OF ONE-DIMENSIONAL, UNSTEADY AND UNSATURATED FLOWS IN PORO	73-0712
UNSATURATED	GALERKIN APPROACH TO UNSATURATED FLOW IN SOILS	74-0058
UNSATURATED	THEORETICAL TRANSIENT BEHAVIOUR OF SATURATED AND UNSATURATED SOILS UNDER LOAD AND CHANGING MOISTURE	73-0901
UNSHROUDED	APPLICATION OF THE FINITE ELEMENT METHOD TO PREDICT STATIC AND DYNAMIC RESPONSE OF AN UNSHROUDED CEN	71-0574
UNSTEADY	A FINITE DIFFERENCE METHOD FOR UNSTEADY FLOW IN VARIABLY SATURATED POROUS MEDIA	71-0586
UNSTEADY	A LAGRANGIAN FINITE ELEMENT SOLUTION TO UNSTEADY FLOW	74-1264
UNSTEADY	ANALYSIS OF THREE-DIMENSIONAL UNSTEADY VISCOUS FLOW AROUND OSCILLATING WINGS	74-0982
UNSTEADY	ANALYSIS OF UNSTEADY FLOW TOWARD AN ARTESIAN WELL BY 3-DIMENSIONAL FINITE ELEMENTS	74-0351
UNSTEADY	ANALYSIS OF UNSTEADY FLOW TOWARD ARTESIAN WELLS BY THREE-DIMENSIONAL FINITE ELEMENTS	74-0742
UNSTEADY	ANALYSIS OF UNSTEADY FLOW TOWARD AN ARTESIAN WELL BY THREE-DIMENSIONAL FINITE ELEMENTS	74-1351
UNSTEADY	APPLICATION OF THE FINITE ELEMENT METHOD TO UNSTEADY FLOW PROBLEMS	70-0310
UNSTEADY	APPLICATIONS OF FINITE ELEMENT METHOD IN UNSTEADY AERODYNAMICS	74-0105
UNSTEADY	APPLICATIONS OF FINITE ELEMENT METHODS TO UNSTEADY VISCOUS FLOW AROUND A BOX-GIRDER BRIDGE OSCILLATI	74-0827
UNSTEADY	COMPUTATIONAL CONSIDERATIONS IN APPLICATION OF THE FINITE ELEMENT METHOD FOR ANALYSIS OF UNSTEADY FL	73-0068
UNSTEADY	DEVELOPMENT AND APPLICATIONS OF SUPERSONIC UNSTEADY CONSISTENT AERODYNAMICS FOR INTERFERING PARALLEL	72-0716
UNSTEADY	DEVELOPMENT OF A FINITE ELEMENT APPROACH FOR APPROXIMATE ANALYSIS OF UNSTEADY COMPRESSIBLE FLUID FLO	72-0828
UNSTEADY	DISPERSION IN STEADY UNSTEADY SEEPAGE FLOWFIELD	74-0072
UNSTEADY	FINITE ELEMENT ANALYSIS OF THE UNSTEADY, LINEARIZED BOLTZMANN EQUATION	70-0033
UNSTEADY	FINITE ELEMENT ANALYSIS OF UNSTEADY INCOMPRESSIBLE FLOW AROUND AN OSCILLATING OBSTACLE OF ARBITRARY	73-0067
UNSTEADY	FINITE ELEMENT ANALYSIS OF THE UNSTEADY FLOW OF A VISCOUS PLASTIC FLUID IN A CYLINDRICAL PIPE	74-0055
UNSTEADY	FINITE ELEMENT APPROACH TO THE INTEGRATED POTENTIAL FORMULATION OF GENERAL UNSTEADY SUPERSONIC AEROD	73-0910
UNSTEADY	FINITE ELEMENT METHOD APPLIED TO UNSTEADY NON-DARCY FLOW	70-0088
UNSTEADY	FINITE ELEMENT METHODS FOR UNSTEADY COMPRESSIBLE GAS DYNAMICS	75-0045
UNSTEADY	FINITE ELEMENT SOLUTION OF UNSTEADY AND UNSATURATED FLOW IN POROUS MEDIA	72-0064
UNSTEADY	FINITE ELEMENT SOLUTION OF BOUSSINESQ'S EQUATION FOR UNSTEADY GROUNDWATER FLOW	74-0459
UNSTEADY	FINITE ELEMENT SOLUTION OF TWO DIMENSIONAL UNSTEADY AND UNSATURATED FLOW IN POROUS MEDIA	75-0113
UNSTEADY	FINITE ELEMENT WEIGHTED RESIDUAL SOLUTION OF ONE-DIMENSIONAL, UNSTEADY AND UNSATURATED FLOWS IN PORO	73-0712
UNSTEADY	FURTHER DEVELOPMENTS IN CONSISTENT UNSTEADY SUPERSONIC AERODYNAMIC COEFFICIENTS	72-0559
UNSTEADY	KINEMATICALLY CONSISTENT UNSTEADY AERODYNAMIC COEFFICIENTS IN SUPERSONIC FLOW	70-0014

433

UNSTEADY	KINEMATICALLY CONSISTENT UNSTEADY AERODYNAMIC COEFFICIENTS IN SUPERSONIC FLOW	70-0316
UNSTEADY	LOCAL POTENTIAL APPROACH TO FINITE ELEMENT METHOD IN UNSTEADY VISCOUS INCOMPRESSIBLE FLUID FLOW	73-0298
UNSTEADY	NONLINEAR EQUATION OF UNSTEADY GROUND-WATER FLOW	73-0262
UNSTEADY	NUMERICAL METHODS FOR UNSTEADY NON-DARCY FLOW	74-0062
UNSTEADY	NUMERISCHE BERECHNUNG INSTATIONAERER TEMPERATURFELDER UND THERMISCH BEDINGTER VERFORMUNGEN VON WERKZ	75-0847
UNSTEADY	SUITABILITY OF THE FINITE ELEMENT METHOD FOR ANALYSIS OF UNSTEADY FLOW AROUND OSCILLATING AIRFOILS	73-0069
UNSTEADY	UNSTEADY MHD DUCT FLOW BY THE FINITE ELEMENT METHOD	73-0263
UNSTEADY	UNSTEADY RADIAL FLOW IN GAS RESERVOIRS BY FINITE ELEMENTS	74-0462
UNSTEADY	UNSTEADY SUBSONIC COLLOCATION METHOD FOR WINGS WITH AND WITHOUT CONTROL SURFACES	72-0012
UNSTIFFENED	FINITE-ELEMENT PREDICTIONS OF TRANSIENT ELASTIC-PLASTIC LARGE DEFLECTIONS OF STIFFENED AND/OR UNSTIF	70-0825
UNSTIFFENED	STRUCTURAL RESPONSE OF UNSTIFFENED TOROIDAL SHELLS	69-0198
UNSYMMETRIC	ANALYTICAL DISPLACEMENTS AND VIBRATIONS OF CANTILEVERED UNSYMMETRIC FIBER COMPOSITE LAMINATES	75-0736
UNSYMMETRIC	COMPUTER PROGRAM FOR SOLUTION OF LARGE, SPARSE, UNSYMMETRIC SYSTEMS OF LINEAR EQUATIONS	72-0719
UNSYMMETRIC	EVALUATION OF TORSIONAL EFFECTS IN THE SEISMIC ANALYSIS OF SYMMETRIC AND UNSYMMETRIC NUCLEAR POWER P	73-0175
UNSYMMETRIC	UNSYMMETRIC BUCKLING OF SHELLS OF REVOLUTION BY FINITE ELEMENT METHOD	70-0500
UNSYMMETRICAL	ELAD PROGRAM MANUAL, A COMPUTER PROGRAM FOR THE ELASTIC ANALYSIS OF DOME STRUCTURES AND AXISYMMETRIC	69-0064
UNSYMMETRICAL	EPAD, A COMPUTER PROGRAM FOR THE ELASTIC-PLASTIC ANALYSIS OF DOME STRUCTURES AND ARBITRARY SOLIDS OF	70-0036
UNSYMMETRICAL	THEORETICAL ANALYSIS OF THICK DOME STRUCTURES AND ARBITRARY SOLIDS OF REVOLUTION SUBJECTED TO UNSYMM	70-0037
UNSYMMETRICAL	UNSYMMETRICAL TRANSIENT HEAT CONDUCTION: ROTATING DISK APPLICATIONS	70-0005
UNSYMMETRICALLY	DYNAMIC COUPLING IN AN UNSYMMETRICALLY TAPERED BEAM	74-1301
UNSYMMETRICALLY	FINITE-ELEMENT TECHNIQUES FOR THE DYNAMIC ANALYSIS OF UNSYMMETRICALLY LAMINATED PLATES	73-1131
UNTERSPEZIELLER	UEBER DAS VERHALTEN VON SPANNBETONDRUCKBEHAELTRERN IM GRENZTRAGFAEHIGKEITSZUSTAND UND DESSEN EINFLU	73-0855
UNTERSUCHUNG	MESY - EIN PROGRAMMSYSTEM ZUR UNTERSUCHUNG VON TRAGWERKEN. (STRUCTURAL ANALYSIS BY MESY PROGRAMMING	75-0360
UNTERSUCHUNG	MODERNE VERFAHREN ZUR UNTERSUCHUNG DER STEIFIGKEITS-UND FESTIGKEITSVERHALTNISSE VON RAHMENKORPERN, I	69-0401
UNTERSUCHUNGEN	PARAMETRISCHE UNTERSUCHUNGEN VON KERBEN MIT DER MATRIZENVERSCHIEBUNGSMETHODE UND BILDSCHIRMUNTERSTUT	72-0638
UNTERSUCHUNGEN	UNTERSUCHUNGEN ZUM FESTIGKEITSVERHALTEN VON METALLKLEBUNGEN IN HINBLICK AUF DIE ERSTELLUNG EINES PRA	74-0289
UNTERSUCHUNGEN	UNTERSUCHUNGEN ZUR SPURMASSTABILITAET BEI RADSAETZEN MIT KLOTZGEBREMSTEN VOLLRAEDERN (INVESTIGATIONS	75-0810
UNTERSUCHUNGSME	NEUARTIGE ANWENDUNG GEOMECHANISCHER UNTERSUCHUNGSMETHODEN ZUR BEURTEILUNG DER ABBAUWIRKUNGEN IM GANG	75-0530
UPPER	A METHOD TO GENERATE BOTH UPPER AND LOWER BOUNDS TO PLATE EIGENVALUES BY CONFORMING DISPLACEMENT FIN	70-0166
UPPER	APPLICATION OF SAMIS TO THE ANALYSIS OF A FRONT SUSPENSION UPPER ARM	70-0260
UPPER	BENDING AND STRETCHING OF PLATES, SPECIAL MODELS FOR UPPER AND LOWER BOUNDS	66-0026
UPPER	DETERMINATION OF UPPER BOUNDS FOR PROBLEMS IN PLANE STRESS USING FINITE ELEMENT TECHNIQUES	67-0034
UPPER	DUALITY BETWEEN DISPLACEMENT AND EQUILIBRIUM METHOD WITH A VIEW TO OBTAINING UPPER AND LOWER BOUNDS	62-0013
UPPER	UPPER AND LOWER BOUNDS IN MATRIX STRUCTURAL ANALYSIS	64-0018
UPPER	UPPER AND LOWER BOUNDS TO STRUCTURAL DEFORMATIONS BY DUAL ANALYSIS IN FINITE ELEMENTS	67-0074
UPSETTING	ANALYSES OF AXISYMMETRIC UPSETTING AND PLANE STRAIN SIDE PRESSING OF SOLID CYLINDERS BY THE FINITE E	70-0083
UPSETTING	ANALYSES OF AXISYMMETRIC UPSETTING AND PLANE-STRAIN UPSETTING OF SOLID CYLINDERS BY THE FINITE E	71-0794
UPTAKE	FINITE ELEMENT ANALYSIS OF 2-DIMENSIONAL FLOW IN SOILS CONSIDERING WATER UPTAKE BY ROOTS .1. THEORY	75-0137
UPTAKE	FINITE ELEMENT ANALYSIS OF 2-DIMENSIONAL FLOW IN SOILS CONSIDERING WATER UPTAKE BY ROOTS .2. FIELD A	75-0147
URBAN	ASSESSMENT OF DESIGN TOOLS AND CRITERIA FOR URBAN RAIL TRACK STRUCTURES, VOLUME 11. AT-GRADE SLAB TR	74-0813
URBAN	ENVIRONMENTAL CONSIDERATIONS OF DAM CONSTRUCTION AND OPERATION IN SEISMICALLY ACTIVE URBAN AREAS	73-0470
URZADZEN	HYBRYDOWA METODA ELEMENTOW SKONCZONYCH W ZASTOSOWANIU DO OBLICZEN ORGAN URZADZEN OKRETOWYCH. (HYBRID	74-0963
USER	EAC/EASE USER INFORMATION MANUAL	70-0320
USER	MARC-CDC USER INFORMATION MANUALS	71-0562
USER	MRI/STARDYNE USER INFORMATION MANUAL	73-0541
USER	NASTRAN USER INTERFACES, AUTOMATED INPUT INNOVATIONS	71-0126
USER	NASTRAN: USER EXPERIENCE WITH FOUR EXAMPLE PROBLEMS	72-0804
USER	SOFTWARE USER: EDUCATION AND QUALIFICATION	72-0441
USER	STRUCTURAL ANALYSIS AND MATRIX INTERPRETIVE SYSTEM (SAMIS) USER REPORT	67-0125
USER	USER EDUCATION: THE FINITE ELEMENT SHORT COURSE	72-0437
USER	USER INFORMATION MANUAL	73-0543
USER-ORIENTATED	LARGE USER-ORIENTATED SYSTEMS OF PROGRAMS FOR STRUCTURAL ANALYSIS AND DESIGN	74-0482
USER-ORIENTED	A USER-ORIENTED PROGRAM FOR CRASH DYNAMICS	74-0321
USER'S	ANSYS USER'S MANUAL	71-0561
USER'S	CAPE-2D, A COMPUTER PROGRAM FOR THE STRESS ANALYSIS OF PLANE AND AXISYMMETRIC COMPOSITE STRUCTURES.	69-0083
USER'S	COMOC THERMAL ANALYSIS VARIANT, USER'S MANUAL	72-0701
USER'S	DEFINIT - A NEW ELEMENT DEFINITION CAPABILITY FOR NASTRAN: USER'S MANUAL	73-0738
USER'S	DYNASOR 11 A FINITE ELEMENT PROGRAM FOR THE DYNAMIC NONLINEAR ANALYSIS OF SHELLS OF REVOLUTION (US	71-0666
USER'S	EVALUATION OF NASTRAN FROM THE THE USER'S POINT OF VIEW	73-0596
USER'S	FEABL (FINITE ELEMENT ANALYSIS BASIC LIBRARY) USER'S GUIDE	72-0748
USER'S	FEDGE - A GENERAL-PURPOSE COMPUTER PROGRAM FOR FINITE ELEMENT DATA GENERATION. VOLUME 1 - USER'S MAN	69-0491
USER'S	MAGIC, AN AUTOMATED GENERAL PURPOSE SYSTEM FOR STRUCTURAL ANALYSIS, VOLUME II, USER'S MANUAL	69-0144
USER'S	MISSILE BODY INPUT GENERATOR (BING), A NASTRAN PRE-PROCESSOR: THEORETICAL DEVELOPMENT, USER'S MANUAL	75-0717
USER'S	NASTRAN: USER'S EXPERIENCES	75-0675
USER'S	PLANFORM INPUT GENERATOR (PING), A NASTRAN PROCESSOR FOR LIFTING SURFACES-THEORETICAL DEVELOPMENT, U	73-0780
USER'S	PREDICTION OF LONG-TERM STRESS RANGES: USER'S MANUAL - BRIDGE LOAD GENERATOR	73-0737
USER'S	PREDICTION OF LONG-TERM STRESS RANGES: USER'S MANUAL - BRIDGE DYNAMIC STRESS ANALYSIS	74-0807
USER'S	RANDOM-VIBRATION ANALYSIS SYSTEM FOR COMPLEX STRUCTURES. PART 1, ENGINEERING USER'S GUIDE	68-0106
USER'S	SABOR CS-2 USER'S MANUAL	70-0408
USER'S	SAFE SHELL, A COMPUTER PROGRAM FOR THE STRESS ANALYSIS OF THIN SHELL BODIES OF REVOLUTION. A USER'S	67-0018
USER'S	SAMECS STRUCTURAL ANALYSIS SYSTEM - USER'S DOCUMENT	69-0455
USER'S	SINGER: A COMPUTER CODE FOR GENERAL ANALYSIS OF TWO-DIMENSIONAL REINFORCED CONCRETE STRUCTURES. VOLU	75-0731
USER'S	STRUCTURE-MEDIUM INTERACTION AND DESIGN PROCEDURES STUDY. VOLUME IV. FEAT CODE USER'S GUIDE	69-0265
USER'S	TANKER TRANSVERSE STRENGTH ANALYSIS: USER'S MANUAL	72-0750
USER'S	THE NASTRAN USER'S MANUAL	72-0479
USER'S	TRANSPIRATION AND FILM COOLING BOUNDARY LAYER COMPUTER PROGRAM. VOLUME 2, COMPUTER PROGRAM AND USER'	71-0655
USER'S	USER'S GUIDE FOR ANALYSIS OF FINITE ELASTOPLASTIC DEFORMATION: THE FIPDEF AND FIPAX PROGRAMS FOR THE	74-0799
USER'S	USER'S GUIDE FOR THE COMPUTER PROGRAM DESCRIBED IN TAM REPORT NO. 680 "DETERMINATION OF VOLUME CHANG	69-0473
USER'S	USER'S MANUAL FOR NTEDYNS - A NONLINEAR FINITE ELEMENT CODE FOR ANALYZING THE BLAST RESPONSE OF UNDE	71-0695
USER'S	USER'S MANUAL FOR PROGRAM BEAM	73-1105
USER'S	USER'S MANUAL FOR UMP FINITE ELEMENT COMPUTER PROGRAM - STRATA	69-0429
USER'S	VIEW FACTOR COMPUTER PROGRAM (PROGRAM VIEW) USER'S MANUAL	73-0789
USERS	AIRCRAFT FUEL TANK VULNERABILITY TO HYDRAULIC RAM: MODIFICATTION OF THE NORTHROP FINITE ELEMENT COMP	74-0724
USERS	ANALYTIC MODELING OF ROCK-STRUCTURE INTERACTION. VOLUME 2. USERS GUIDE FOR A COMPUTER PROGRAM	73-0919
USERS	AXISYMMETRIC RING AND PLATE STRESS ANALYSIS PROGRAM - USERS GUIDE	70-0333
USERS	DEVELOPMENT AND APPLICATIONS OF SUPERSONIC UNSTEADY CONSISTENT AERODYNAMICS FOR INTERFERING PARALLEL	74-0716
USERS	DYNAN USERS MANUEL	71-0625
USERS	FINITE ELEMENT ANALYSIS OF ANISOTROPIC PLATES USING 0-19 ELEMENT: PART 11. INSTRUCTIONS FOR USERS AN	74-0731
USERS	ICES STRUDL-II ENGINEERING USERS MANUAL. VOLS. 1, 11, 111	69-0398
USERS	NASTRAN USERS EXPERIENCES	71-0751
USERS	NASTRAN USERS EXPERIENCES	72-0753
USERS	ON IMPROVING EFFICIENCY OF AN ALGORITHM FOR STRUCTURAL OPTIMIZATION AND A USERS MANUAL FOR PROGRAM T	74-0778
USERS	OPTIM 11: A (MAGIC) COMPATIBLE LARGE AUTOMATED MINIMUM WEIGHT DESIGN PROGRAM - 1. ENGINEERS AN	74-1048
USERS	OPTIM 11: A MAGIC COMPATIBLE LARGE-SCALE AUTOMATED MINIMUM WEIGHT DESIGN PROGRAM. VOLUME 1. ENGINEER	74-1096
USERS	SAFE-CREEP: A COMPUTER PROGRAM FOR THE VISCOELASTIC ANALYSIS OF AXISYMMETRIC AND PLANE CONCRETE STRU	67-0068
USERS	SP-222 NASTRAN USERS MANUAL	70-0386
USERS	STRESS, STABILITY AND VIBRATION OF COMPLEX SHELLS OF REVOLUTION: ANALYSIS AND USERS MANUAL FOR BOSOR	69-0441
USERS	THREE-DIMENSIONAL FINITE-ELEMENT COMPUTER PROGRAM - USERS GUIDE	74-0755
USERS	USERS MANUAL FOR CEL/NONSAP, A NONLINEAR STRUCTURAL ANALYSIS PROGRAM	74-0706
USERS	USERS MANUAL FOR DIGITAL COMPUTER PROGRAM OK5 NUMERICAL ANALYSIS OF FINITE STRAINS AND DISPLACEMENTS	72-0718
USERS	USERS MANUAL FOR PROGRAM BEAM	73-0714
USERS	USERS MANUAL FOR QMESH, A SELF-ORGANIZING MESH GENERATION PROGRAM	74-0880

434

USERS'	USERS' MANUAL FOR DIGITAL COMPUTER PROGRAM OK1. NUMERICAL ANALYSIS OF FINITE AXISYMMETRIC DEFORMATIO	72-0861
USLOVIYAMI	RASCHET KILEBANII PLASTIN SLOZHNOI FORMY S NEODNORODNYMI GRANICHNYMI USLOVIYAMI. (CALCULATION OF VIB	74-0988
USTOICHIVOSTI	RASCHIOT USTOICHIVOSTI IZGIBNO-KRUTILNYKH KOLEBANIJ KRULA V DOZVUKOVOM POTOKE METODOM KONECHNYKH ELE	72-0903
UTILISING	AN IMPROVED SHELL OF REVOLUTION ELEMENT UTILISING CUBIC DISPLACEMENT FUNCTIONS	70-0451
UWZGLEDNIENIEM	ZASTOSOWANIE METODY SZTYWNYCH ELEMENTOW SKONCZONYCH DO OBLICZEN ORGAN WALOW OKRETOYCH Z UWZGLEDNIENI	74-0962
V-BELT	PRESSURE DISTRIBUTION AND BELT DEFORMATION IN V-BELT DRIVES	74-1028
V-BELT	V-BELT LIFE PREDICTION AND POWER RATING	75-0838
V-GROOVED	STRESS ANALYSIS OF A BOLT-NUT JOINT BY THE FINITE ELEMENT METHOD AND THE COPPER-ELECTROPLATING METHO	73-0220
VALID	ESTABLISHMENT OF A VALID FRACTURE TOUGHNESS SPECIMEN GEOMETRY USING A FINITE ELEMENT ANALYSIS	74-0135
VALIDITY	ON THE VALIDITY OF AN APPROXIMATION AVAILABLE IN THE FINITE ELEMENT SHELL ANALYSIS	75-0663
VALIDITY	ON THE VALIDITY OF THE FINITE ELEMENT ANALYSIS OF CIRCULAR ARCHES REPRESENTED BY AN ASSEMBLAGE OF BE	75-0479
VALIDITY	TEST OF VALIDITY OF DARCY'S LAW FOR A CLAY SOIL	74-0070
VALIDITY	THE VALIDITY OF THE INITIAL STRAIN CONCEPT	61-0010
VALLEYS	APPROACH TO ANALYSIS OF ARCH DAMS IN WIDE VALLEYS	73-0153
VALUE	A FINITE ELEMENT METHOD FOR INITIAL VALUE PROBLEMS	71-0145
VALUE	A MATHEMATICAL THEORY OF COMPLEMENTARY-DUAL VARIATIONAL PRINCIPLES AND MIXED FINITE ELEMENT APPROXIM	73-0804
VALUE	A THEORY OF MIXED FINITE ELEMENT APPROXIMATIONS OF NON-SELF-ADJOINT BOUNDARY VALUE PROBLEMS	74-0715
VALUE	AN ANALYSIS OF VARIOUS TIME STEPPING SCHEMES FOR INITIAL VALUE PROBLEMS	73-0607
VALUE	AN APPLICATION OF THE FINITE ELEMENT TECHNIQUE TO BOUNDARY VALUE PROBLEMS OF POTENTIAL FLOW	70-0002
VALUE	APPLICATION OF THE FINITE ELEMENT METHOD OF ANALYSIS IN SOLVING BOUNDARY VALUE PROBLEMS IN ROCK MECH	66-0013
VALUE	APPLICATIONS OF VARIATIONAL EQUATIONS TO ORDINARY AND PARTIAL DIFFERENTIAL EQUATIONS - MULTIPLE SOLU	75-0010
VALUE	COMPUTABLE ERROR BOUNDS FOR THE FINITE ELEMENT METHOD FOR ELLIPTIC BOUNDARY VALUE PROBLEMS	74-1219
VALUE	CONNECTION BETWEEN THE FINITE DIFFERENCE LIKE METHODS AND THE METHODS BASED ON INITIAL VALUE PROBLEM	74-0877
VALUE	CONVERGENCE OF A FINITE ELEMENT PROCEDURE FOR SOLVING BOUNDARY VALUE PROBLEMS OF THE FOURTH ORDER	70-0131
VALUE	CONVERGENCE OF LUMPED FINITE ELEMENT SCHEMES FOR SELECTED INITIAL VALUE PROBLEMS	73-1026
VALUE	CONVERGENCE OF MIXED FINITE ELEMENT APPROXIMATIONS OF A CLASS OF LINEAR-BOUNDARY VALUE PROBLEMS	73-0599
VALUE	DISCRETIZATION AND ROUND-OFF ERROR IN THE FINITE ELEMENT ANALYSIS OF ELLIPTIC BOUNDARY VALUE PROBLEM	71-0514
VALUE	ELLIPTIC BOUNDARY VALUE PROBLEMS WITH NONVARIATIONAL PERTURBATION AND THE FINITE ELEMENT METHOD	73-1022
VALUE	ERROR ANALYSIS OF FINITE ELEMENT METHODS WITH TRIANGLES FOR ELLIPTIC BOUNDARY VALUE PROBLEMS	73-0072
VALUE	FINITE ELEMENT APPLICATIONS OF A PRECISION TRIANGULAR PLATE BENDING ELEMENT TO LINEAR BOUNDARY VALUE	75-0788
VALUE	FINITE ELEMENT FORMULATIONS OF NONLINEAR BOUNDARY VALUE PROBLEMS	71-0497
VALUE	FINITE ELEMENT GALERKIN METHOD FOR MIXED INITIAL-BOUNDARY VALUE PROBLEMS IN ELASTICITY THEORY	71-0520
VALUE	FINITE ELEMENT METHODS FOR INITIAL VALUE PROBLEMS	72-0059
VALUE	FINITE ELEMENT METHODS FOR ELLIPTIC BOUNDARY VALUE PROBLEMS CONTAINING SINGULARITIES	72-0360
VALUE	FINITE ELEMENT METHODS FOR SOLVING INITIAL BOUNDARY VALUE PROBLEMS	74-1125
VALUE	FINITE ELEMENT TREATMENT OF SINGULARITIES OF BOUNDARY VALUE PROBLEMS AND ITS APPLICATION TO ANALYSIS	73-0306
VALUE	FUNCTIONAL ANALYSIS AND THE FINITE ELEMENT METHOD APPLIED TO ELLIPTIC BOUNDARY VALUE PROBLEMS	72-0056
VALUE	FUNCTIONAL APPROXIMATIONS FOR SOLVING BOUNDARY VALUE PROBLEMS BY COMPUTER	69-0013
VALUE	GENERALIZED CONJUGATE FUNCTIONS FOR MIXED FINITE ELEMENT APPROXIMATIONS OF BOUNDARY VALUE PROBLEMS	72-0515
VALUE	HERMITE INTERPOLATION TYPE RITZ METHODS FOR TWO POINT BOUNDARY VALUE PROBLEMS	66-0089
VALUE	LOCAL ERROR BOUNDS IN BOUNDARY VALUE PROBLEMS USING FINITE ELEMENT DISCRETIZATION	72-0982
VALUE	NONLINEAR INITIAL-BOUNDARY VALUE SOLUTIONS BY THE FINITE ELEMENT METHOD	74-0812
VALUE	NUMERICAL COMPUTATION OF BOUNDARY VALUE PROBLEMS IN SOIL MECHANICS	73-0269
VALUE	NUMERICAL METHODS FOR TIME DEPENDENT NONLINEAR BOUNDARY VALUE PROBLEMS	70-0296
VALUE	NUMERICAL METHODS OF HIGH-ORDER ACCURACY FOR NONLINEAR BOUNDARY VALUE PROBLEMS	68-0194
VALUE	NUMERICAL SOLUTION OF 2-POINT BOUNDARY VALUE PROBLEMS	74-1141
VALUE	ON SOME FINITE ELEMENT PROCEDURES FOR SOLVING SECOND ORDER BOUNDARY VALUE PROBLEMS	69-0288
VALUE	ON THE CONVERGENCE OF SOR ITERATIONS FOR FINITE ELEMENT APPROXIMATIONS TO ELLIPTIC BOUNDARY VALUE PR	71-0449
VALUE	PROPERTIES OF SOLUTIONS OF ELLIPTIC BOUNDARY VALUE PROBLEMS	72-0499
VALUE	SOLUTION OF A FREE-SURFACE BOUNDARY VALUE PROBLEM USING AN INVERSE FORMULATION AND THE FINITE ELEMEN	75-0466
VALUE	SOLUTION OF ELASTO-STATIC BOUNDARY VALUE PROBLEMS BY THE FINITE ELEMENT METHOD	64-0025
VALUE	SOLUTION OF ELASTO-STATIC BOUNDARY VALUE PROBLEMS BY THE FINITE ELEMENT METHOD	64-0030
VALUE	SOLUTION OF MIXED BOUNDARY VALUE PROBLEMS IN FINITE DOMAINS BY FOURIERS METHOD	72-0790
VALUE	SOLUTION OF MIXED BOUNDARY VALUE PROBLEMS WITH LOCAL ERROR BOUND BY THE FINITE ELEMENT METHOD	73-0795
VALUE	THE FINITE ELEMENT METHOD AND LOCAL BOUNDS FOR BOUNDARY VALUE PROBLEMS OF ELASTIC STRUCTURES	73-1056
VALUE	THREE DIMENSIONAL NONLINEAR MAGNETIC FIELD BOUNDARY VALUE PROBLEM AND ITS NUMERICAL SOLUTION	70-0060
VALUE	VARIATIONAL PRINCIPLES FOR LINEAR INITIAL VALUE PROBLEMS	64-0029
VALUES	A NEW MEASURING METHOD OF RESIDUAL STRESSES WITH THE AID OF FINITE ELEMENT METHOD AND RELIABILITY OF	75-0875
VALUES	DIFFUSION ON SOLID SPHERICAL PARTICLE IN VISCOUS FLUID FLOW AT FINITE VALUES OF RE NUMBERS	69-0419
VALVE	FINITE-ELEMENT ANALYSIS OF HIGH-PRESSURE, LARGE VALVE	75-0329
VALVE	FINITE-ELEMENT ANALYSIS OF HIGH-PRESSURE LARGE VALVE GATES	75-0413
VALVE	STRESS ANALYSIS OF THE HUMAN AORTIC VALVE	73-0166
VALVE	THE FINITE ELEMENT STRESS ANALYSIS OF THE HUMAN AORTIC VALVE TOWARD THE DEVELOPMENT OF A PROSTHETIC	74-1201
VALVE	THE FINITE ELEMENT STRESS ANALYSIS OF THE HUMAN AORTIC VALVE TOWARD THE DEVELOPMENT OF A PROSTHETIC	74-1201
VANE	DESIGN AND ANALYSIS OF A CERAMIC STATOR VANE	75-0656
VARATIONAL	ELASTICITY EQUATIONS FOR INCOMPRESSIBLE AND NEARLY INCOMPRESSIBLE MATERIALS BY A VARATIONAL THEOREM	65-0019
VARIABILITY	INFLUENCE OF ROCK PROPERTIES VARIABILITY ON MINE OPENING STABILITY ANALYSIS	74-1021
VARIABLE	A COMPARATIVE STUDY OF VARIABLE AND CONSTANT THICKNESS HIGH PRECISION TRIANGULAR PLATE BENDING ELEME	74-0857
VARIABLE	A COMPARATIVE STUDY OF VARIABLE AND CONSTANT THICKNESS HIGH PRECISION TRIANGULAR PLATE BENDING ELEME	74-0857
VARIABLE	A THREE DIMENSIONAL "VARIABLE NODE" ISOPARAMETRIC SOLID ELEMENT	74-0779
VARIABLE	ANALYSIS OF CLAMPED CIRCULAR PLATES OF VARIABLE THICKNESS SUBJECTED TO UNIFORMLY DISTRIBUTED LOAD BY	73-0583
VARIABLE	BENDING AND TORSION OF THIN-WALLED STRAIGHT AND CURVED BEAMS WITH VARIABLE OPEN CROSS-SECTION	69-0493
VARIABLE	CREEP AND SHRINKAGE LAW FOR CONCRETE AT VARIABLE HUMIDITY	74-0668
VARIABLE	DERIVATION OF CRACK ELEMENT-STIFFNESS MATRIX BY THE COMPLEX VARIABLE APPROACH	73-0870
VARIABLE	DISCRETE VARIABLE METHOD APPLIED TO TRANSIENT HEAT CONDUCTION PROBLEMS	74-1163
VARIABLE	ELASTIC BENDING OF CIRCULAR PLATES OF VARIABLE THICKNESS. AN ANALYTICAL AND EXPERIMENTAL STUDY	70-0163
VARIABLE	FINITE ELEMENT ANALYSIS OF UNCONFINED FLOW WITH A VARIABLE TIME-STEP PROCEDURE	74-0776
VARIABLE	FINITE ELEMENT METHOD FOR OPTIMAL DESIGN OF VARIABLE THICKNESS SHEETS	73-0131
VARIABLE	FINITE ELEMENT SIMULATION OF ACOUSTIC FILTERS WITH VARIABLE CIRCULAR CROSS SECTION	75-0871
VARIABLE	FREE VIBRATION OF ORTHOTROPIC VARIABLE THICKNESS, CONTINUOUS RECTANGULAR PLATES	70-0270
VARIABLE	HEAVY SECTION STEEL TECHNOLOGY PROGRAM TECHNICAL OR PROGRAMMATIC MANUSCRIPT NO. 25. VARIABLE THICKNE	73-0948
VARIABLE	ON THE DISCRETIZATION OF THE TIME VARIABLE FOR PARABOLIC PARTIAL DIFFERENTIAL EQUATIONS	75-0103
VARIABLE	QUAD-4: A COMPUTER PROGRAM FOR EVALUATING THE SEISMIC RESPONSE OF SOIL STRUCTURES BY VARIABLE DAMPIN	73-0771
VARIABLE	RE-ENTRY THERMAL ANALYSIS OF VARIABLE THICKNESS SPHERICAL VEHICLES.	72-0468
VARIABLE	SEISMIC RESPONSE BY VARIABLE DAMPING FINITE ELEMENTS	74-0143
VARIABLE	SHORT COMMUNICATIONS: TORSION OF AXI-SYMMETRIC SOLIDS OF VARIABLE DIAMETER - INCLUDING ACCELERATION	74-0203
VARIABLE	STRESS ANALYSIS OF A PRESTRESSED CONCRETE NUCLEAR PRESSURE VESSEL BY THE FINITE ELEMENT METHOD USING	71-0154
VARIABLE	TRANSVERSE VIBRATIONS OF CANTILEVER BARS OF VARIABLE CROSS SECTIONS	66-0001
VARIABLE	TRIANGULAR FINITE-ELEMENTS OF VARIABLE THICKNESS AND THEIR APPLICATION TO PLATE AND SHELL PROBLEMS	69-0095
VARIABLE	VARIABLE GRID FINITE DIFFERENCE-ELEMENT SOLUTION OF ELLIPTIC PARTIAL DIFFERENTIAL EQUATIONS	74-0862
VARIABLE	VARIABLE TIME STEP ANALYSIS OF UNCONFINED SEEPAGE	74-0064
VARIABLE	VARIABLE TIME STEP METHOD FOR DETERMINING PLASTIC STRESS REFLECTIONS FROM BOUNDARIES	75-0641
VARIABLE-DEPTH	FINITE ELEMENT CIRCULATION ANALYSIS OF VARIABLE-DEPTH SHALLOW LAKES	73-0727
VARIABLES	APPLICATION OF THE FINITE ELEMENTS METHOD IN TWO-DIMENSIONAL HYDRODYNAMICS USING THE LAGRANGE VARIAB	73-0976
VARIABLES	ERROR BOUNDS FOR EIGENVALUE ANALYSIS BY ELIMINATION OF VARIABLES	71-0594
VARIABLES	INTEGRAL REPRESENTATION OF FIELD VARIABLES FOR THE FINITE ELEMENT SOLUTION OF VISCOUS FLOW PROBLEMS	74-0468
VARIABLES	NODAL VARIABLES FOR ARBITRARY ORDER FINITE ELEMENTS	75-0002
VARIABLES	ON LAGRANGIAN INTERPOLATION IN N VARIABLES	70-0452
VARIABLES	ON THE AUTOMATIC SELECTION OF INDEPENDENT DISPLACEMENT VARIABLES IN GENERAL PURPOSE FINITE ELEMENT C	73-0676
VARIABLES	PIECEWISE HERMITE INTERPOLATION IN ONE AND TWO VARIABLES WITH APPLICATIONS TO PARTIAL DIFFERENTIAL E	68-0201
VARIABLES	ROLE OF PART-INVERSION IN FLUID-STRUCTURE PROBLEMS WITH MIXED VARIABLES	70-0209
VARIABLES	STRUCTURAL EIGENVALUE PROBLEMS: ELIMINATION OF UNWANTED VARIABLES	65-0050

VARIABLY	A FINITE DIFFERENCE METHOD FOR UNSTEADY FLOW IN VARIABLY SATURATED POROUS MEDIA	71-0586
VARIABLY	SIGNIFICANCE OF VARIABLY CONFINED TRIAXIAL TESTING ROADBUILDING MATERIALS	74-0554
VARIAION	THE SECOND VARIAION TEST WITH ALGEBRAIC AND DIFFERENTIAL CONTRASTS	69-0342
VARIANT	COMOC THERMAL ANALYSIS VARIANT, USER'S MANUAL	72-0701
VARIANTS	SOME NONCONFORMING VARIANTS ON THE TRILINEAR ISOPARAMETRIC BRICK FOR LINEAR ELASTIC PROBLEMS	72-0531
VARIATION	EFFECT OF THE VARIATION OF CONSISTENCY IN RESIDUAL SOILS ON LOAD TRANSFER BY BORED PILES	75-0812
VARIATION	LINEAR SHELL THEORY FOR RAPID VARIATION SOLUTIONS	71-0406
VARIATION	NUMERICAL SIMULATION OF FINITE-WIDTH THRUST BEARINGS, TAKING INTO ACCOUNT VISCOSITY VARIATION WITH T	75-0623
VARIATION	PROBALILISTIC MODEL FOR MATERIAL STRENGTH VARIATION AND SIZE EFFECT	72-0876
VARIATION	RELIABILITY STUDY OF LINEAR VARIATION TRIANGULAR FINITE ELEMENT	72-0281
VARIATION	STABILITY INVESTIGATIONS OF PLATES USING FINITE ELEMENT METHOD BASED ON AN EXPANDED VARIATION PRINCI	71-0663
VARIATION	TEMPERATURVERLAUF IM QUERSCHNITT GESCHWEISSTER ROHRWAENDE BEI UEBERWIEGEND KONVEKTIVEM WAERMEUEBERGA	75-0628
VARIATION	VARIATION METHOD ON FINITE ELEMENTS FOR NUMERICAL RESOLUTION TO PROBLEMS WITH EXTERNAL R3	73-0359
VARIATION	VARIATION WITH TEMPERATURE OF RESONANT FREQUENCIES OF ANISOTROPIC PLATES	74-1046
VARIATIONAL	A COMPUTER PROGRAMME FOR ELASTIC PLASTIC CALCULATIONS WITH DUAL PERTURBATION EXPANSION AND VARIATION	75-0882
VARIATIONAL	A FINITE ELEMENT APPLICATION OF THE HELLINGER-REISSNER VARIATIONAL THEOREM	68-0060
VARIATIONAL	A FINITE ELEMENT SOLUTION OF TIME DEPENDENT FIELD PROBLEMS USING A REDUCED VARIATIONAL FORMULATION	69-0504
VARIATIONAL	A GENERAL THEORY OF VARIATIONAL METHODS WITH APPLICATION TO FINITE ELEMENTS	72-0897
VARIATIONAL	A MATHEMATICAL THEORY OF COMPLEMENTARY-DUAL VARIATIONAL PRINCIPLES AND MIXED FINITE ELEMENT APPROXIM	73-0804
VARIATIONAL	A NON VARIATIONAL FINITE ELEMENT ANALYSIS FOR THE NAVIER STOKES EQUATIONS	74-0010
VARIATIONAL	A NOTE ON SOME VARIATIONAL STATEMENTS FOR THE SLOW FLOW OF A NAVIER-POISSON FLUID	74-1054
VARIATIONAL	A SIMPLIFIED VERSION OF GURTIN'S VARIATIONAL PRINCIPLE	74-1244
VARIATIONAL	A SPECIAL VARIATIONAL PRINCIPLE FOR THE FINITE-ELEMENT METHOD	69-0181
VARIATIONAL	A THEORY OF VARIATIONAL METHODS AND ITS APPLICATION TO FINITE ELEMENTS	73-0639
VARIATIONAL	A UNIFIED THEORY OF VARIATIONAL PRINCIPLES IN MECHANICS	75-0155
VARIATIONAL	A USEFUL FORM OF THE VARIATIONAL PRINCIPLE OF IDEAL MHD FOR ONE-DIMENSIONAL NUMERICAL STABILITY CALC	73-0746
VARIATIONAL	A VARIATIONAL APPROACH TO GRID OPTIMIZATION IN THE FINITE ELEMENT METHOD	72-0572
VARIATIONAL	A VARIATIONAL FINITE ELEMENT METHOD FOR TWO DIMENSIONAL STEADY VISCOUS FLOWS	72-0355
VARIATIONAL	A VARIATIONAL FORMULATION OF DAMPED ACOUSTO-STRUCTURAL PROBLEMS	66-0133
VARIATIONAL	A VARIATIONAL FORMULATION FOR DISCONTINUOUS FIELDS IN LINEARLY ELASTIC FINITE ELEMENT APPLICATIONS	72-0962
VARIATIONAL	A VARIATIONAL FORMULATION OF THE COMPRESSIBLE THROUGHFLOW PROBLEM	74-1343
VARIATIONAL	A VARIATIONAL FORMULATION OF THE COMPRESSIBLE THROUGHFLOW PROBLEM	75-0021
VARIATIONAL	A VARIATIONAL METHOD FOR LARGE DEFLECTION ELASTO-PLASTIC THEORY	69-0115
VARIATIONAL	A VARIATIONAL METHOD IN POTENTIAL FLOWS WITH A FREE SURFACE	72-0723
VARIATIONAL	A VARIATIONAL METHOD FOR FREE BOUNDARY PROBLEMS	75-0105
VARIATIONAL	A VARIATIONAL PRINCIPLE AND THE CONVERGENCE OF A FINITE-ELEMENT METHOD BASED ON ASSUMED STRESS DISTR	69-0260
VARIATIONAL	A VARIATIONAL PRINCIPLE WITH APPLICATIONS IN FINITE ELEMENTS	74-1285
VARIATIONAL	AN INTRODUCTION TO VARIATIONAL AND FINITE ELEMENT METHODS	72-0994
VARIATIONAL	APPLICATION OF CONFORMAL MAPPING AND VARIATIONAL METHOD TO THE STUDY OF HEAT CONDUCTION IN POLYGONAL	71-0444
VARIATIONAL	APPLICATION OF EXTENDED VARIATIONAL PRINCIPLES TO FINITE ELEMENT ANALYSIS	72-0529
VARIATIONAL	APPLICATIONS OF VARIATIONAL EQUATIONS TO ORDINARY AND PARTIAL DIFFERENTIAL EQUATIONS - MULTIPLE SOLU	75-0010
VARIATIONAL	CLASSICAL AND NATURAL CONDUCTION AND DEFORMATION MODES CONCEPTS FOR AXISYMMETRIC VARIATIONAL METHODS	72-1003
VARIATIONAL	CONSTRAINED VARIATIONAL PRINCIPLES AND PENALTY FUNCTION METHODS IN FINITE ELEMENT ANALYSIS	72-0551
VARIATIONAL	DIFFERENTIAL OF VARIATIONAL FORMS	74-0445
VARIATIONAL	DISCRETE APPROXIMATIONS OF ELASTIC-PLASTIC BODIES BY VARIATIONAL METHODS	72-0710
VARIATIONAL	EIN NEUES VARIATIONSPRINZIP MIT ANWENDUNG AUF SCHWINGUNGEN FLACHER SCHALEN NACH DER METHODE DER FINI	73-0563
VARIATIONAL	ELASTIC STABILITY OF SHELLS OF REVOLUTION BY THE VARIATIONAL APPROACH USING DISCRETE ELEMENTS	66-0106
VARIATIONAL	FINITE ELEMENT ANALYSIS OF PLATE BUCKLING USING A MIXED VARIATIONAL PRINCIPLE	71-0020
VARIATIONAL	FINITE ELEMENT FORMULATION FOR PIEZOELECTRIC CONTINUA BY A VARIATIONAL THEOREM	68-0230
VARIATIONAL	FINITE ELEMENT FORMULATION BY VARIATIONAL PRINCIPLES WITH RELAXED CONTINUITY REQUIREMENTS	72-0516
VARIATIONAL	FINITE ELEMENT FORMULATIONS FOR ELASTIC PLATES BY GENERAL VARIATIONAL STATEMENTS WITH DISCONTINUOUS	73-0559
VARIATIONAL	FINITE ELEMENT FORMULATIONS FOR ELASTIC PLATES BY GENERAL VARIATIONAL STATEMENTS WITH DISCONTINUOUS	73-1124
VARIATIONAL	FINITE ELEMENT METHODS AS APPLICATIONS OF VARIATIONAL PRINCIPLES	69-0121
VARIATIONAL	FINITE ELEMENT METHODS BY DIFFERENT VARIATIONAL PRINCIPALS IN ELASTICITY	70-0407
VARIATIONAL	FINITE ELEMENT METHODS BY VARIATIONAL PRINCIPLES WITH RELAXED CONTINUITY REQUIREMENTS	72-0342
VARIATIONAL	FINITE ELEMENT STIFFNESS METHODS BY DIFFERENT VARIATIONAL PRINCIPLES IN ELASTICITY	70-0455
VARIATIONAL	FINITE ELEMENTS FOR FLUID DYNAMICS. VARIATIONAL FORMULATIONS FOR NONLINEAR, INITIAL-VALUE PROBLEMS	74-0710
VARIATIONAL	FINITE ELEMENTS WITH RELAXED CONTINUITY - AN INVESTIGATION OF THE VARIATIONAL BASIS AND NUMERICAL PE	74-1065
VARIATIONAL	FINITE-ELEMENT FORMULATION OF MIXED AND STRESS VARIATIONAL PRINCIPLES FOR STRUCTURAL DYNAMIC ANALYSI	72-0964
VARIATIONAL	FORMULATION OF A VARIATIONAL PRINCIPLE - FINITE ELEMENT METHOD FOR VISCOUS FLOW	72-0354
VARIATIONAL	FUNCTIONAL ANALYSIS DERIVATION AND GENERALIZATION OF HYBRID VARIATIONAL METHODS	73-1104
VARIATIONAL	GENERAL VARIATIONAL ANALYSIS OF HYDRODYNAMIC, THERMAL AND DIFFUSIONAL BOUNDARY LAYERS	70-0191
VARIATIONAL	GENERAL VARIATIONAL PRINCIPLES IN NONLINEAR AND LINEAR ELASTICITY WITH APPLICATIONS	72-0914
VARIATIONAL	GENERALIZED VARIATIONAL PRINCIPLES IN THE FINITE ELEMENT METHOD	69-0116
VARIATIONAL	GENERALIZED VARIATIONAL PRINCIPLES FOR USE IN FLOW PROBLEMS THROUGH POROUS MEDIA	74-0067
VARIATIONAL	GENERATION OF FINITE ELEMENTS BY EULER'S VARIATIONAL METHOD	72-0596
VARIATIONAL	HEAT TRANSFER WITH NONLINEAR BOUNDARY CONDITIONS VIA A VARIATIONAL PRINCIPLE	71-0443
VARIATIONAL	IMPORTANCE OF VARIATIONAL CALCULUS IN FLUID MECHANICS AND ITS APPLICATION TO COMPRESSIBLE POTENTIAL	73-1103
VARIATIONAL	INCREMENTAL VARIATIONAL METHOD FOR THE LARGE DISPLACEMENT ANALYSIS OF SHELLS WITH GEOMETRIC IMPERFEC	72-0894
VARIATIONAL	LEAST-SQUARES VARIATIONAL PRINCIPLE FOR FINITE-ELEMENT APPLICATIONS	75-0412
VARIATIONAL	MIXED VARIATIONAL PRINCIPLES, FINITE ELEMENTS AND FINITE ELASTICITY	74-1254
VARIATIONAL	NEW VARIATIONAL FUNCTIONAL FOR THE FINITE-ELEMENT METHOD AND ITS APPLICATION TO PLATE SHELL PROBLEMS	72-0108
VARIATIONAL	NEW VARIATIONAL PRINCIPLE FOR FINITE ELASTIC DISPLACEMENTS	72-0454
VARIATIONAL	NEW VARIATIONAL PRINCIPLE WITH APPLICATION TO VIBRATIONS OF SHALLOW SHELLS USING FINITE ELEMENTS MET	73-0354
VARIATIONAL	NON-SELF-ADJOINT VARIATIONAL PROCEDURE FOR FINITE-ELEMENT APPROXIMATION OF TRANSPORT EQUATION	75-0072
VARIATIONAL	NUMERICAL TREATMENT OF ALGEBRAIC INTEGRAL EQUATIONS BY VARIATIONAL METHODS	75-0374
VARIATIONAL	ON A VARIATIONAL INEQUALITY AND ITS APPROXIMATION, IN THE THEORY OF SEMICONDUCTORS	75-0829
VARIATIONAL	ON A VARIATIONAL THEOREM FOR INCOMPRESSIBLE AND NEARLY INCOMPRESSIBLE ORTHOTROPIC ELASTICITY	68-0249
VARIATIONAL	ON COMPLEMENTARY-DUAL VARIATIONAL PRINCIPLES IN MATHEMATICAL PHYSICS	74-0342
VARIATIONAL	ON THE APPLICATION OF GENERALIZED VARIATIONAL PRINCIPLES IN THE FINITE ELEMENT METHOD	68-0083
VARIATIONAL	ON THE ROLE OF VARIATIONAL PRINCIPLES AND METHOD OF WEIGHTED RESIDUALS IN THE FINITE ELEMENT METHODS	74-0551
VARIATIONAL	ON THE VARIATIONAL PRINCIPLES IN THE FINITE ELEMENT METHOD	74-1010
VARIATIONAL	ONE PARAMETER FAMILIES OF VARIATIONAL PRINCIPLES	72-0503
VARIATIONAL	ONE-SIDED APPROXIMATION AND VARIATIONAL INEQUALITIES	74-0335
VARIATIONAL	PERTURBATION VARIATIONAL SOLUTION OF COMPRESSIBLE FLOWS	74-0025
VARIATIONAL	RECTANGULAR FINITE ELEMENT FOR PLATE BENDING ANALYSIS BASED ON HELLINGER-REISSNERS VARIATIONAL PRINC	71-0545
VARIATIONAL	SELF ADJOINT DIFFERENTIAL EQUATIONS AND VARIATIONAL PRINCIPLE - 1.	71-0388
VARIATIONAL	SELF ADJOINT DIFFERENTIAL EQUATIONS AND VARIATIONAL PRINCIPLE - 2.	72-0106
VARIATIONAL	SINGULAR ELEMENTS IN VARIATIONAL AND FINITE ELEMENT TRANSPORT COMPUTATIONS	73-0902
VARIATIONAL	SOLUMN INSTABILITY UNDER NONCONSERVATIVE FORCES, WITH INTERNAL AND EXTERNAL DAMPING FINITE ELEMENT U	73-0565
VARIATIONAL	SOME TWO-FIELD VARIATIONAL PRINCIPLES FOR NONLINEAR DEFORMATION ANALYSIS OF SHELLS	75-0294
VARIATIONAL	SPLINE INTERPOLATION TECHNIQUES FOR VARIATIONAL METHODS	73-1079
VARIATIONAL	STATIC ANALYSIS OF SHELLS OF REVOLUTION USING DOUBLY-CURVED QUADRILATERAL ELEMENTS DERIVED FROM ALTE	69-0368
VARIATIONAL	THERMO-PLASTIC CONSTITUTIVE RELATIONS AND A VARIATIONAL SOLUTION TECHNIQUE USING FINITE ELEMENTS	70-0486
VARIATIONAL	TWO DUAL FINITE ELEMENT METHODS DERIVED FROM EXTENDED VARIATIONAL PRINCIPLE	73-0109
VARIATIONAL	UNE METHODE VARIATIONNELLE ELEMENTS FINIS POUR LA RESULUTION NUMERIQUE UN PROBLEME EXTERIEUR DANS R*	73-0576
VARIATIONAL	UNIFORM CONVERGENCE OF VARIATIONAL FINITE ELEMENT AND RITZ METHODS	74-0420
VARIATIONAL	USE OF VARIATIONAL PRINCIPLE IN STUDY ON RELATIVISTIC BOLTZMANN EQUATION	71-0442
VARIATIONAL	USEFUL FORM OF THE VARIATIONAL PRINCIPLE OF IDEAL MHD FOR ONE-DIMENSIONAL NUMERICAL STABILITY CALCUL	73-0946
VARIATIONAL	VARIATIONAL AND WEIGHTED RESIDUAL METHODS	73-0630
VARIATIONAL	VARIATIONAL APPROACH TO NON-DARCY FLOW	70-0089
VARIATIONAL	VARIATIONAL APPROACH TO THE LIFTING SURFACE PROBLEM	73-0779

VARIATIONAL	VARIATIONAL CRIMES IN THE FINITE ELEMENT METHOD	72-0517
VARIATIONAL	VARIATIONAL FINITE ELEMENT CALCULATION OF THE ACOUSTIC RESPONSE OF A RECTANGULAR PANEL	71-0100
VARIATIONAL	VARIATIONAL FINITE ELEMENT METHODS FOR TWO DIMENSIONAL AND AXISYMMETRIC NAVIER STOKES EQUATIONS	74-0012
VARIATIONAL	VARIATIONAL FORMULATIONS AND FINITE ELEMENT METHOD FOR THIN ELASTIC SHELLS AND PLATES	68-0064
VARIATIONAL	VARIATIONAL FORMULATION OF FINITE DISPLACEMENT ANALYSIS	70-0284
VARIATIONAL	VARIATIONAL FORMULATIONS OF NUMERICAL METHODS	70-0392
VARIATIONAL	VARIATIONAL FORMULATIONS OF NUMERICAL METHODS IN SOLID CONTINUA	71-0670
VARIATIONAL	VARIATIONAL FORMULATION OF DYNAMICS OF FLUID-SATURATED POROUS ELASTIC SOLIDS	72-0017
VARIATIONAL	VARIATIONAL FORMULATION FOR A CLASS OF FREE BOUNDARY PROBLEMS ARISING IN HYDRODYNAMIC LUBRICATION	75-0531
VARIATIONAL	VARIATIONAL METHODS FOR THE SOLUTION OF PROBLEMS OF EQUILIBRIUM AND VIBRATION	43-0001
VARIATIONAL	VARIATIONAL PRINCIPLES FOR LINEAR INITIAL VALUE PROBLEMS	64-0029
VARIATIONAL	VARIATIONAL PRINCIPLES AND THEIR APPLICATION TO FINITE ELEMENT METHODS	68-0137
VARIATIONAL	VARIATIONAL PRINCIPLES AND DUALISTIC SCHEME FOR INTERSECTION PROBLEMS IN ELASTICITY	69-0305
VARIATIONAL	VARIATIONAL PRINCIPLE FOR INCOMPRESSIBLE AND NEARLY-INCOMPRESSIBLE ANISOTROPIC ELASTICITY	69-0319
VARIATIONAL	VARIATIONAL PRINCIPLE FOR A RAREFIED GAS FLOW PAST AND ARBITRARY THREE-DIMENSIONAL BODY	70-0189
VARIATIONAL	VARIATIONAL PRINCIPLE OF STEADY-STATE TRANSPORT PROCESSES	70-0190
VARIATIONAL	VARIATIONAL PRINCIPLE FOR MAGNETOHYDRODYNAMIC CHANNEL FLOW	70-0192
VARIATIONAL	VARIATIONAL PRINCIPLES AND FINITE ELEMENT METHOD IN PARTIAL DIFFERENTIAL EQUATIONS	71-0452
VARIATIONAL	VARIATIONAL PRINCIPLE FOR THE DYNAMIC ANALYSIS OF CONTINUA BY HYBRID FINITE ELEMENT METHOD	71-0492
VARIATIONAL	VARIATIONAL PRINCIPLES FOR FLUID FLOW IN POROUS MEDIA	71-0844
VARIATIONAL	VARIATIONAL PRINCIPLE OF LINEAR DIFFERENTIAL EQUATIONS	72-0107
VARIATIONAL	VARIATIONAL PRINCIPLES	72-0501
VARIATIONAL	VARIATIONAL PRINCIPLES, NONLINEAR OPERATORS, AND FINITE ELEMENT APPLICATIONS	72-0576
VARIATIONAL	VARIATIONAL PRINCIPLES OF ELASTOSTATICS WITH RELAXED CONTINUITY REQUIREMENTS AND THEIR APPLICABILITY	74-0758
VARIATIONAL	VARIATIONAL PRINCIPLE FOR THE LAPLACE'S OPERATOR WITH APPLICATION IN THE TORSION OF COMPOSITE RODS	74-1044
VARIATIONAL	VARIATIONAL PRINCIPLES AND THE PATCH TEST	74-1282
VARIATIONAL	VARIATIONAL PRINCIPLES FOR ELASTOPLASTIC CONTINUA	75-0837
VARIATIONAL	VARIATIONAL PROCEDURES AND CONVERGENCE OF FINITE ELEMENT METHODS	71-0070
VARIATIONAL	VARIATIONAL PROPERTIES OF STEADY FALL IN STOKES FLOW	72-0285
VARIATIONAL	VARIATIONAL SOLUTIONS BY USE OF STEPWISE CONSTANT FUNCTIONS: ONE LINEAR CASE	72-0286
VARIATIONAL	VARIATIONAL SOLUTION OF TWO DIMENSIONAL FLUID FLOW PROBLEMS	73-1014
VARIATIONAL-DIF	THE COMPARISON OF THE FINITE ELEMENT METHOD WITH THE VARIATIONAL-DIFFERENCE METHOD IN THE THEORY OF	67-0096
VARIATIONNELLE	UNE METHODE VARIATIONNELLE ELEMENTS FINIS POUR LA RESOLUTION NUMERIQUE UN PROBLEME EXTERIEUR DANS R*	73-0576
VARIATIONS	ANALYSE DES CONTRAINTES DAN DES STRUCTURES COMPLEXES EN PRESENCE DE FLUAGE, DE VARIATIONS DE DIMENSI	71-0081
VARIATIONS	EFFECT OF SYSTEM PARAMETER VARIATIONS ON NATURAL FREQUENCIES	75-0661
VARIATIONS	EFFECTS OF LOCAL STRENGTH VARIATIONS ON PLASTIC ENCLAVE DEVELOPMENT AT CRACK TIPS	72-0822
VARIATIONSPRINZ	EIN NEUES VARIATIONSPRINZIP MIT ANWENDUNG AUF SCHWINGUNGEN FLACHER SCHALEN NACH DER METHODE DER FINI	73-0563
VARIATIONSPRINZ	VARIATIONAL PRINCIPLES OF ELASTOSTATICS WITH RELAXED CONTINUITY REQUIREMENTS AND THEIR APPLICABILITY	74-0758
VARIATIONSPRINZ	STABILITY INVESTIGATIONS OF PLATES USING FINITE ELEMENT METHOD BASED ON AN EXPANDED VARIATION PRINCI	71-0663
VARIATIONSRECHN	IMPORTANCE OF VARIATIONAL CALCULUS IN FLUID MECHANICS AND ITS APPLICATION TO COMPRESSIBLE POTENTIAL	73-1103
VARVED	BEHAVIOR OF EMBANKMENTS ON NEW LISKHEARD VARVED CLAY	73-0721
VARVED	PERFORMANCE OF AN EMBANKMENT CONSTRUCTED ON VARVED CLAY	73-0743
VARVED	TEMPORARY EXCAVATION IN VARVED CLAY	75-0049
VARY	ASAAS: ASYMMETRIC STRESS ANALYSIS OF AXISYMMETRIC SOLIDS WITH ORTHOTROPIC, TEMPERATURE-DEPENDENT MAT	71-0696
VARYING	A TRIANGULAR EQUILIBRIUM ELEMENT WITH LINEARLY VARYING BENDING MOMENTS FOR PLATE BENDING PROBLEMS	67-0054
VARYING	DAMS ON ROCKS OF VARYING ELASTICITY	74-0671
VARYING	DIAPHRAGMS IN SINGLE- AND DOUBLE-CELL BOX GIRDER BRIDGES WITH VARYING ANGLE OF SKEW	72-0130
VARYING	DYNAMIC RESPONSE OF VISCOELASTIC MATERIALS TO SINUSOIDALLY VARYING LOADINGS	73-1072
VARYING	FREQUENCY COEFFICIENTS OF A SPHERICAL SHELL FOR VARYING THICKNESS AND OPENING ANGLE WITH GUIDED PINN	72-0595
VARYING	MATRIX DISPLACEMENT ANALYSIS OF SHELLS WITH AN AXISYMMETRIC MIDSURFACE BUT HAVING CIRCUMFERENTIALLY	68-0031
VARYING	MEMBRANE PARALLELOGRAM ELEMENT WITH LINEARLY VARYING EDGE STRAIN FOR THE MATRIX DISPLACEMENT METHOD	66-0006
VARYING	PLANE STRESS ELEMENT OF VARYING THICKNESS APPLIED TO END BLOCK ANALYSIS	74-0396
VARYING	REINFORCED FIELDS OF TRIANGULAR ELEMENTS WITH LINEARLY VARYING STRAIN: EFFECT OF INITIAL STRAINS	65-0004
VARYING	TETRAHEDRON ELEMENTS WITH LINEARLY VARYING STRAIN FOR THE MATRIX DISPLACEMENT METHOD	65-0006
VARYING	THE CHANGES IN THE (TWO-DIMENSIONAL) STRESS ROUNDED CORNERS, CAUSED BY A VARYING INTERNAL PRESSURE A	67-0141
VARYING	TRANSIENT THERMAL STRESS ANALYSIS OF COMPOSITE STRUCTURES INCLUDING CONTINUOUSLY VARYING PROPERTIES	75-0681
VARYING	TRIANGULAR ELEMENTS WITH LINEARLY VARYING STRAIN FOR THE MATRIX DISPLACEMENT METHOD	65-0053
VARYING	TRIANGULAR FINITE ELEMENTS FOR PLATE BENDING WITH CONSTANT AND LINEARLY VARYING BENDING MOMENTS	70-0019
VASA	DEFORMATION OF THE ARTERIAL VASA VASORUM AT NORMAL AND HYPERTENSIVE ARTERIAL PRESSURE	71-0149
VASORUM	DEFORMATION OF THE ARTERIAL VASA VASORUM AT NORMAL AND HYPERTENSIVE ARTERIAL PRESSURE	71-0149
VAULT	A FINITE ELEMENT ANALYSIS OF A DOUBLY CURVED GOTHIC VAULT	71-0827
VAULT	FINITE ELEMENT OF SKEW VAULT BRIDGES	72-0434
VAULT	PHOTOELASTIC AND FINITE-ELEMENT ANALYSIS OF A QUADRIPARTITE VAULT	73-0119
VAULT	RASCHET NAPRYAZHENIL L PEREMESHCHENII V SVODE PODZEMNOGO MASHINNOGO ZALA GES I V OKRUZHAYUSHCHEM SKA	74-0542
VAZKOPRUZNYCH	RIESENIE VAZKOPRUZNYCH ANIZOTROPICKYCH DOASK METODOU KONECNYCH PRVKOV (SOLUTION OF VISCOELASTIC ANIS	74-0477
VECTOR	FINITE ELEMENT ANALYSIS OF COUPLED IRREVERSIBLE VECTOR PROCESSES	71-0701
VECTOR	FINITE ELEMENT PROGRAM PACKAGE FOR AXISYMMETRIC VECTOR FIELD PROBLEMS	75-0138
VECTOR	THERMAL DEFORMATION VECTOR FOR A BILINEAR TEMPERATURE DISTRIBUTION IN AN ANISOTROPIC QUADRILATERAL M	75-0670
VEHICLE	A TWO-DIMENSIONAL AXISYMMETRIC SHELL FLUID MODEL FOR LAUNCH VEHICLE LONGITUDINAL DYNAMIC ANALYSIS	69-0046
VEHICLE	ADAPTATION OF NASTRAN TO THE ANALYSIS OF THE VIKING SPACE VEHICLE	71-0135
VEHICLE	ANALYSIS OF VEHICLE BODY STRUCTURE BY THE FINITE ELEMENT METHOD	71-0324
VEHICLE	APPLIED FINITE ELEMENT DISPLACEMENT ANALYSIS OF A SIMPLE INTEGRAL CONSTRUCTION VEHICLE BODY SHELL	72-0264
VEHICLE	CONCEPT FEASIBILITY COMBAT TRACKED VEHICLE SIGNATURE DUPLICATOR	64-0028
VEHICLE	DYNAMIC ANALYSIS OF A FLEXIBLE VEHICLE MOVING ALONG A FLEXIBLE SUPPORT	71-0597
VEHICLE	GENERAL PURPOSE COMPUTER PROGRAM FOR THE DYNAMIC SIMULATION OF VEHICLE GUIDEWAY INTERACTIONS	73-0260
VEHICLE	HIGH STRENGTH MATERIALS AND VEHICLE WEIGHT REDUCTION ANALYSIS	75-0157
VEHICLE	HIGH-PERFORMANCE COMPOSITE MATERIALS FOR VEHICLE CONSTRUCTION: A FINITE ELEMENT COMPUTER PROGRAM FOR	73-0767
VEHICLE	IMPROVED LINEAR AXISYMMETRIC SHELL FLUID MODEL FOR LAUNCH VEHICLE LONGITUDINAL RESPONSE ANALYSIS	65-0044
VEHICLE	MEASURED AND PREDICTED LONGITUDINAL VIBRATIONS OF A LIQUID PROPELLANT TWO-STAGE LAUNCH VEHICLE	70-0366
VEHICLE	MODAL SENSITIVITY STUDY OF A CONICAL RE-ENTRY VEHICLE SUBJECTED TO AERODYNAMICALLY INDUCED ACOUSTIC	69-0431
VEHICLE	MODULAR APPROACH TO STRUCTURAL SIMULATION FOR VEHICLE CRASHWORTHINESS PREDICTION	75-0737
VEHICLE	POWER SPECTRAL DENSITY ANALYSIS OF VEHICLE VIBRATION USING THE NASTRAN COMPUTER PROGRAM	74-0318
VEHICLE	PROCEDURE FOR EFFICIENT GENERATING, CHECKING, AND DISPLAYING NASTRAN INPUT AND OUTPUT DATA FOR ANALY	71-0098
VEHICLE	RADIAL AND AXIAL TRANSMISSIBILITY CHARACTERISTICS OF TYPICAL ROCKET VEHICLE STRUCTURE TECHNICAL SUMM	69-0475
VEHICLE	STATIC ANALYSIS VIA SUBSTRUCTURING OF AN EXPERIMENTAL VEHICLE FRONT-END BODY STRUCTURE	74-0313
VEHICLE	STRUCTURAL MODELING OF A WELDED JOINT TYPICAL FOR VEHICLE FRAMES	73-0539
VEHICLE	STUDY OF PROPELLANT DYNAMICS IN A SHUTTLE TYPE LAUNCH VEHICLE	72-0699
VEHICLE	SUBTERRANEAN GUIDEWAY RESPONSES TO LEVITATION VEHICLE LOADING	74-0927
VEHICLE	THE CALCULATION OF STATIC AND DYNAMIC CHARACTERISTICS OF FLIGHT VEHICLE STRUCTURES BY THE FINITE ELE	73-1018
VEHICLE	THE FLEXIBILITY OF A TUBULAR WELDED JOINT IN A VEHICLE FRAME	74-0209
VEHICLE	THERMAL STRESS ANALYSIS OF REENTRY VEHICLE NOSETIPS AT ANGLE OF ATTACK	72-0456
VEHICLE	THERMAL STRESS ANALYSIS OF RE-ENTRY VEHICLE NOSETIPS AT ANGLE OF ATTACK	72-0852
VEHICLE	USING NASTRAN FOR DYNAMIC ANALYSIS OF VEHICLE SYSTEMS	74-0316
VEHICLE	VEHICLE CRUSH PREDICTION USING FINITE-ELEMENT TECHNIQUES	73-0280
VEHICLE	VEHICLE STRUCTURAL ANALYSIS, PRE AND POST PROCESSOR DATA HANDLING	71-0093
VEHICLES	EVALUATION OF A FINITE ELEMENT ANALYSIS FOR THE LONGITUDINAL VIBRATIONS OF LIQUID-PROPELLANT LAUNCH	70-0204
VEHICLES	EXPERIMENTS IN FORMING AND JOINTING OF TITANIUM AND TITANIUM ALLOYS FOR AIRCRAFT AND SPACE VEHICLES	70-0134
VEHICLES	IMPROVED ANALYTIC LONGITUDINAL RESPONSE ANALYSIS FOR AXISYMMETRIC LAUNCH VEHICLES, COMPUTER PROGRAM	65-0002
VEHICLES	RE-ENTRY THERMAL ANALYSIS OF VARIABLE THICKNESS SPHERICAL VEHICLES.	72-0468
VEHICULAR	AN INTERACTIVE HYBRID TECHNIQUE FOR CRASHWORTHY DESIGN OF COMPLEX VEHICULAR STRUCTURAL SYSTEMS	74-0317
VEHICULAR	ANALYSIS OF BONDED JOINTS IN VEHICULAR STRUCTURES	73-0375
VEIN	NEUARTIGE ANWENDUNG GEOMECHANISCHER UNTERSUCHUNGSMETHODEN ZUR BEURTEILUNG DER ABBAUWIRKUNGEN IM GANG	75-0530

VELAROIDAL	ON THE PARABOLIC VELAROIDAL SHELL	69-0314
VELKE	VELKE PRIEHYBY DOSIEK METODOU KONECNYCH PRVKOV (LARGE DEFLECTIONS OF PLATES ANALYZED BY FINITE ELEME	73-0373
VELOCIMETER	ANALYSIS OF THE MUTUAL INDUCTANCE PARTICLE VELOCIMETER	74-0712
VELOCITY	APPLICATION OF THE COLLISION-IMPARTED VELOCITY METHOD FOR ANALYZING THE RESPONSES OF CONTAINMENT AND	73-0770
VELOCITY	COUPLED CONVECTIVE/CONDUCTIVE HEAT TRANSFER INCLUDING VELOCITY FIELD EVALUATION	75-0116
VELOCITY	NUMERICAL CALCULATIONS OF VELOCITY AND PRESSURE DISTRIBUTION AROUND OSCILLATING AIRFOILS	74-0847
VENANT	FINITE ELEMENT ANALYSIS IN POLAR CO-ORDINATES OF THE SAINT VENANT TORSION PROBLEM	74-0500
VENANT	FINITE ELEMENT STRESS FIELD SOLUTION OF THE PROBLEM OF SAINT VENANT TORSION	73-1042
VENEER	INVESTIGATION OF VENEER CUTTING BY ELASTIC FINITE ELEMENT MODELS	74-1110
VENEZIANO	SOME REMARKS ON INFINITE VENEZIANO REPRESENTATIONS	69-0472
VENICE	MATHEMATICAL SIMULATION OF THE SUBSIDENCE OF VENICE 1 THEORY	73-0992
VENTILATED	INTEGRAL EQUATION METHOD FOR SUBSONIC FLOW PAST AIRFOILS IN VENTILATED WIND TUNNELS	75-0065
VENTRICLE	DEFORMATION OF THE DIASTOLIC LEFT VENTRICLE - 2. NONLINEAR GEOMETRIC EFFECTS	74-0974
VENTRICLE	DEFORMATION OF THE END DIASTOLIC LEFT VENTRICLE, COMPARISON OF THEORY WITH EXPERIMENT	71-0131
VENTRICLE	FINITE-ELEMENT MODEL FOR THE MECHANICAL BEHAVIOUR OF THE LEFT VENTRICLE. PREDICTION OF DEFORMATION I	72-0304
VENTRICULAR	FINITE ELEMENT ANALYSIS OF LEFT VENTRICULAR MYOCARDIAL STRESSES	74-0669
VENTRICULAR	FINITE ELEMENT CALCULATION OF LEFT VENTRICULAR WALL STRESS IN NORMAL AND DISEASE STATES	73-0588
VENTRICULAR	IN VIVO STRESSES IN THE HUMAN LEFT VENTRICLE: ANALYSIS ACCOUNTING FOR THE IRREGULAR THREE DIM	72-0136
VEON	EIN VERFAHREN DUR DIE FORMFINDUNG VEON BELIEBIGEN VORGESPANNTEN NETZWERKKONSTRUKTIONEN	72-0641
VERALLGEMEINERU	ACCURATE CALCULATION OF STRESSES IN BEAMS, DISKS AND PLATES BY MEANS OF FINITE ELEMENT METHODS AND I	70-0133
VERBUND	ZUSAMMENHANG ZWISCHEN OBERFLAECHENBESCHAFFENHEIT, VERBUND UND SPRENGWIRKUNG VON BEWEHRUNGSSTAEHLEN U	73-0450
VEREFAHREN	EINIGE VERFAHREN ZUR BERECHNUNG VON EIGENWERTEN UND EIGENVEKTOREN VON NICHT HERMITESCHEN MATRIZEN U	72-0646
VERFAHREN	BERECHNUNG VON PLATTEN NACH DEM RITZ' SCHEN VERFAHREN MIT HILFE DREIECKFOERMIGER MASCHENNETZE	66-0094
VERFAHREN	DAS VERFAHREN VON RAYLEIGH-RITZ MIT HERMITESCHEN INTERPOLATIONSPOLYNOMEN	63-0018
VERFAHREN	EIN VERFAHREN DUR DIE FORMFINDUNG VEON BELIEBIGEN VORGESPANNTEN NETZWERKKONSTRUKTIONEN	72-0641
VERFAHREN	FEHLERBETRACHTUNGEN AUFGRUND DER METHODE DER GEWICHTETEN RESIDUEN FUR DAS VERFAHREN DER FINITEN ELEM	72-0368
VERFAHREN	FINITE ELEMENT VERFAHREN BEI QUASILINEAREN ELLIPTISCHEN RANDWERTPROBLEM	74-1262
VERFAHREN	HYBRIDE FINITE VERFAHREN FUER LINEARE UND NICHTLINEARE INSTATIONAERE WAERMELEITPROBLEME (HYBRID FINI	73-0484
VERFAHREN	MODERN TECHNIQUES FOR CALCULATING STRESSES IN TANKS, AND THEIR LIMITATIONS. (MODERNE VERFAHREN ZUR S	70-0153
VERFAHREN	MODERNE VERFAHREN ZUR UNTERSUCHUNG DER STEIFIGKEITS-UND FESTIGKEITSVERHALTNISSE VON RAHMENKOERPERN. I	69-0401
VERFAHREN	NICHTLINEARE DIRICHTLET PROBLEME UND EINFACHE FINITE ELEMENT VERFAHREN	74-1263
VERFAHRENS	ANWENDUNG DES FINITE ELEMENT VERFAHRENS SUR BESTIMMUNG DER FESTIGKERIT VON SCHWEISSKONSTRUKTIONEN. (	72-0185
VERFAHRENS	KOMBINATION EINES INTEGRALGLEICHUNGS VERFAHRENS MIT METHODE DER FINITEN ELEMENTEN ZUR BERECHNUNG EBE	74-0856
VERFESTINGUNG	CONTINUUM CALCULATION FROM ELASTIC PLASTIC REINFORCED MATERIAL USING A FINITE ELEMENT METHOD (DIE BE	72-0658
VERFORMUNGEN	EINSATZ MODERNER RECHENVERFAHREN ZUR ERMITTLUNG DER SPANNUNGEN UND VERFORMUNGEN BEI RADSAETZEN	75-0186
VERFORMUNGEN	NUMERISCHE BERECHNUNG INSTATIONAERER TEMPERATURFELDER UND THERMISCH BEDINGTER VERFORMUNGEN VON WERKZ	75-0847
VERGLEICH	VERGLEICH NUMERISCHER LOESUNGEN NACH DEM DIFFERENZENVERFAHREN UND NACH DER FINITE-ELEMENT-METHODE AM	71-0016
VERHALTEN	UEBER DAS VERHALTEN VON SPANNBETONDRUCKBEHAELTRERN IM GRENZTRAGFAEHIGKEITSZUSTAND UND DESSEN EINFLU	73-0855
VERHALTENS	ANWENDUNG DER METHODE FINITER ELEMENTE BEI DER ANALYSE DES DYNAMISCHEN VERHALTENS GEDAEMPFTER WERKZE	75-0580
VERIFICATION	A STATIC AND DYNAMIC FINITE ELEMENT ANALYSIS WITH EXPERIMENTAL VERIFICATION	69-0156
VERIFICATION	A STATIC AND DYNAMIC FINITE ELEMENT SHELL ANALYSIS WITH EXPERIMENTAL VERIFICATION	71-0769
VERIFICATION	EXPERIMENTAL VERIFICATION OF DISCRETE-ELEMENT SOLUTIONS FOR PLATES AND PAVEMENT SLABS	70-0354
VERIFICATION	FORMULATION AND EXPERIMENTAL VERIFICATION OF AN AXISYMMETRIC FINITE-ELEMENT STRUCTURAL ANALYSIS	71-0354
VERIFICATION	MATHEMATICAL MODEL OF THE COLUMBIA RIVER FROM THE PACIFIC OCEAN TO BONNEVILLE DAM. PART II: INPUT-OU	70-0417
VERIFICATION	STRUCTURE-MEDIUM INTERACTION AND DESIGN PROCEDURES STUDY. VOLUME I. ANALYSIS METHOD, THEORY, VERIFIC	69-0166
VERIFICATION	THE SCANDINAVIAN PORV MODEL PROJECT: STRESS CALCULATION AND EXPERIMENTAL VERIFICATION	71-0021
VERIFICATION	THE VERIFICATION AND ACCEPTANCE OF COMPUTER PROGRAMS FOR DESIGN ANALYSIS	70-0334
VERIFICATION	VERIFICATION AND QUALIFICATION OF A 3-D FINITE ELEMENT COMPUTER ANALYSIS	74-0467
VERIFICATION	VERIFICATION OF A HYBRID GENERAL PURPOSE SHELL AND BODY OF REVOLUTION COMPUTER PROGRAM	71-0333
VERIFICATION	VERIFICATION OF FRACTURE MECHANICS CONCEPTS TO PREDICT CRACKING OF FLEXIBLE PAVEMENTS	73-0772
VERIFICATION	VERIFICATION OF THE FINITE ELEMENT COMPUTER CODE SAFE-2D	74-0873
VERIFICATION	VERIFICATION OF THE PEC CONTAINMENT VESSEL FOR SOME OPERATING CONDITIONS BY USE OF THE FINITE ELEMEN	74-0883
VERSCHIEBUNGEN	COMPUTATION OF SHELLS WITH FINITE DISPLACEMENTS: MIXED FINITE ELEMENTS (BERECHNUNG VON SCHALEN MIT E	73-0788
VERSION	A SIMPLIFIED VERSION OF GURTIN'S VARIATIONAL PRINCIPLE	74-1244
VERSION	AN ASSUMED STRESS VERSION OF THE WILSON 8-NODE ISOPARAMETRIC BRICK	72-0532
VERSION	DESCRIPTION AND INSTRUCTION FOR USE OF THE SAFE-SHELL, 8 FINITE ELEMENT PROGRAM IN THE IBM/360 VERSI	73-0945
VERSION	DESCRIPTION OF THE IBM-360 VERSION OF THE FINITE ELEMENTS COMPUTER PROGRAM SAFE-3D	72-0775
VERSION	EVALUATION OF THE COMPUTER PROGRAM CREEP-BLAST (CDC VERSION)	74-0684
VERSION	TURBAN VERSION 1.0. EIN SPEZIELLES PROGRAMM ZUR LINEAREN UND NICHTLINEAREN STATISCHEN UND DYNAMISCHE	71-0522
VERSTEIFTER	DIE BEULLASTEN DIAGONAL VERSTEIFTER RECHTECKPLATTEN (BUCKLING LOADS IN DIAGONALLY STIFFENED RECTANGU	73-0405
VERTICAL	A FINITE ELEMENT STRESS ANALYSIS OF THE VERTICAL BUTTRESSES OF A NUCLEAR CONTAINMENT VESSEL	69-0280
VERTICAL	ETUDE THEORIQUE DU COMPORTEMENT DU PIEUX SOUS CHARGE VERTICALE. INTRODUCTION DE LA DILATANCE. (THEO	75-0309
VERTICAL	FINITE ELEMENT ANALYSIS OF LAMINAR CONVECTIVE HEAT TRANSFER IN VERTICAL DUCTS WITH ARBITRARY CROSS-S	75-0555
VERTICAL	FINITE ELEMENT PARAMETRIC STUDY OF VERTICAL STRAIN INFLUENCE FACTORS AND THE PRESSUREMETER TEST TO E	74-1127
VERTICAL	HYDRODYNAMIC PRESSURES ON HIGH DAMS DUE TO VERTICAL EARTHQUAKE MOTIONS	69-0438
VERTICAL	LATERAL RESISTANCE AND DEFLECTION OF VERTICAL PILES	73-0912
VERTICAL	ON THE STATES OF STRESS AND DEFORMATION AROUND THE BOTTOM OF A CIRCULAR VERTICAL SHAFT	69-0323
VERTICAL	STRESS DISTRIBUTION DUE TO GRAVITY IN A VERTICAL ROCK BANK	68-0090
VERTICAL	STRESSES AND DEFORMATIONS OF VERTICAL SLOPES IN ELASTO-PLASTIC ROCKS	70-0114
VERTICAL	STRUCTURAL ANALYSIS OF SL-7 CONTAINERSHIP UNDER COMBINED LOADING OF VERTICAL, LATERAL AND TORSIONAL	74-0734
VERTICAL	THE IMPORTANCE OF THE VERTICAL COMPONENT OF EARTHQUAKE MOTIONS	66-0019
VERTICAL	TRANSIENT NATURAL CONVECTION COOLING OF A VERTICAL CIRCULAR CYLINDER	74-1169
VERTICAL	VERTICAL RESPONSE OF A RIGID BASE TO DYNAMIC LOADING	69-0316
VERTICAL	VERTICAL VIBRATION OF EMBEDDED FOOTINGS	72-0231
VERTICAL	VERTICAL VIBRATIONS OF FOOTINGS EMBEDDED IN LAYERED MEDIA	69-0436
VERTICALE	ETUDE THEORIQUE DU COMPORTEMENT DES PIEUX SOUS CHARGE VERTICALE. INTRODUCTION DE LA DILATANCE. (THEO	75-0309
VERTICALLY	FINITE-ELEMENT ANALYSES OF INITIAL ELASTIC DEFORMATION OF VERTICALLY DEFORMED CRYSTAL BLOCKS	72-0970
VERTICALLY	THREE-DIMENSIONAL FINITE ELEMENT ANALYSIS OF VERTICALLY LOADED PILE GROUPS	75-0239
VERWENDBARKEIET	VARIATIONAL PRINCIPLES OF ELASTOSTATICS WITH RELAXED CONTINUITY REQUIREMENTS AND THEIR APPLICABILITY	74-0758
VERWENDUNG	DIE VERWENDUNG DER MATRIZENVERSCHIEBUNGSMETHODE ZUR BERECHNUNG VON SPANNUNGSKOMPONENTEN	67-0140
VERWENDUNG	DIE VERWENDUNG DER MATRIZENVERSCHIEBUNGSMETHODE ZUR BERECHNUNG VON SPANNUNGSKONZENTRATIONEN	70-0400
VESSEL	A FINITE ELEMENT STRESS ANALYSIS OF THE VERTICAL BUTTRESSES OF A NUCLEAR CONTAINMENT VESSEL	69-0280
VESSEL	A THREE-DIMENSIONAL FINITE ELEMENT ANALYSIS OF A PRESTRESSED CONCRETE REACTOR VESSEL MODEL	69-0084
VESSEL	AN ASSUMED STRESS HYBRID FINITE ELEMENT MODEL FOR THE ANALYSIS OF AN AXISYMMETRIC THICK WALLED PRESS	71-0025
VESSEL	APPLICATION OF THE FINITE ELEMENT METHOD TO PRESSURE VESSEL DESIGN (JAPANESE)	70-0446
VESSEL	CALCULATING STRESSES IN A PRESSURE VESSEL WITH FILL TUBE - COMPARISON OF THE FINITE ELEMENT AND PHOT	72-0782
VESSEL	DESIGN OF A COMPOSITE STEEL AND CONCRETE CLOSURE FOR A PRESTRESSED CONCRETE REACTOR VESSEL.	72-0187
VESSEL	ELASTIC PLASTIC ANALYSIS OF PRESSURE VESSEL COMPONENTS	69-0380
VESSEL	ELASTIC-PLASTIC ANALYSIS OF SOME PRESSURE VESSEL HEADS	69-0379
VESSEL	ELASTIC-PLASTIC BEHAVIOR OF A LONGITUDINAL SEMI-ELLIPTIC CRACK IN A THICK PRESSURE VESSEL	73-0952
VESSEL	ENTWURF UND SPANNUNGSBERECHNUNG EINES SPANNBETON DRUCKBEHAELTERS AUS FERTIGTEILEN. (DESIGN AND STRES	73-0856
VESSEL	EURCYL: A PROGRAM TO GENERATE FINITE ELEMENT MESHES FOR PRESSURE VESSEL NOZZLES	74-1086
VESSEL	FINITE ELEMENT ANALYSIS OF STRUCTURAL INTEGRITY OF A REACTOR PRESSURE VESSEL DURING EMERGENCY CORE C	70-0026
VESSEL	FINITE-ELEMENT ANALYSIS OF FLOW IN BLOOD VESSEL WITH ARBITRARY CROSS-SECTION	72-0936
VESSEL	NONLINEAR ANALYSIS OF PRESTRESSED CONCRETE FAST REACTOR PRESSURE VESSEL. (ANALYSE NON LINEAIRE DES C	70-0145
VESSEL	ON THE USE OF FINITE ELEMENTS IN FRACTURE ANALYSIS OF PRESSURE VESSEL COMPONENTS	75-0514
VESSEL	PRESSURE VESSEL CERTIFICATION BASED ON FRACTURE MECHANICS TECHNOLOGY	74-0948
VESSEL	PRESSURE VESSEL WALL THICKNESS LIMITS BASED ON THERMAL STRESS RACHETTING AND CREEP FATIGUE INTERACTI	74-0968
VESSEL	PRESTRESSED CONCRETE REACTOR VESSEL ANALYTICAL METHODS DEVELOPMENTS	68-0141
VESSEL	PRESTRESSED CONCRETE REACTOR PRESSURE VESSEL FAILURE ANALYSIS. THE INCORPORATION OF STATISTICAL METH	71-0120
VESSEL	REBO: COMPUTER CODE FOR THE GENERATION OF A THREE-DIMENSIONAL FINITE-ELEMENT GRID OF THE NOZZLE-CYLI	74-1310
VESSEL	SOME APPLICATIONS OF ENGINEERING MECHANICS IN PRESSURE VESSEL ANALYSIS	72-0892

VESSEL	STATIC AND HEAT TRANSFER ANALYSIS OF A SPHERE-CONE INTERSECTION IN A NUCLEAR CONTAINMENT VESSEL	75-0166
VESSEL	STRESS ANALYSIS OF A PRESTRESSED CONCRETE NUCLEAR PRESSURE VESSEL BY THE FINITE ELEMENT METHOD USING	71-0154
VESSEL	STRESS ANALYSIS OF BWR PRESSURE VESSEL WITH BERSAFE AND FLHE	73-0406
VESSEL	STRESS ANALYSIS OF PRESSURE VESSEL NOZZLES	73-0823
VESSEL	STRUCTURAL ANALYSIS OF AIRCRAFT IMPACT ON A NUCLEAR CONTAINMENT VESSEL AND ASSOCIATED STRUCTURES	69-0279
VESSEL	THIN-WALLED PRESTRESSED CONCRETE PRESSURE VESSEL FOR HIGH TEMPERATURE REACTORS: EXPERIMENTAL INVESTI	73-0857
VESSEL	THREE DIMENSIONAL RUPTURE ANALYSIS OF A PRESTRESSED CONCRETE PRESSURE VESSEL INCLUDING CREEP EFFECTS	73-0826
VESSEL	THREE DIMENSIONAL RUPTURE ANALYSIS OF A PRESTRESSED CONCRETE PRESSURE VESSEL INCLUDING CREEP EFFECTS	74-0538
VESSEL	THREE-DIMENSIONAL ANALYSIS OF REACTOR PRESSURE VESSEL NOZZLES	71-0221
VESSEL	VERIFICATION OF THE PEC CONTAINMENT VESSEL FOR SOME OPERATING CONDITIONS BY USE OF THE FINITE ELEMEN	74-0883
VESSELS	A MATRIX DISPLACEMENT METHOD ON PRE AND POST BUCKLING ANALYSIS OF LINERS FOR REACTOR VESSELS	71-0267
VESSELS	ADVANCES IN THE ANALYSIS OF PRESTRESSED CONCRETE PRESSURE VESSELS	71-0162
VESSELS	ANALYSIS OF MODERATELY THICK FLANGED REACTOR PRESSURE VESSELS	73-0821
VESSELS	ANALYSIS OF MODERATELY THICK FLANGED REACTOR VESSELS	75-0320
VESSELS	ANALYSIS OF THICK-WALLED PRESSURE VESSELS USING THE CONSTANT STRAIN AXISYMMETRIC FINITE ELEMENT	71-0653
VESSELS	ANALYSIS OF VISCOPLASTIC EFFECTS IN PRESSURE VESSELS BY THE FINITE ELEMENT METHOD	74-0573
VESSELS	APPLICATION OF ELASTIC PLASTIC ANALYSIS TO THE DESIGN OF PRESSURE VESSELS	68-0157
VESSELS	APPLICATION OF FINITE ELEMENT METHOD - STRESS CALCULATIONS ON 90 DEGREES NOZZLES OF THIN-WALLED CYLI	71-0253
VESSELS	APPLICATION OF FRACTURE MECHANICS TO PRESTRESSED CONCRETE PRESSURE VESSELS	73-0228
VESSELS	APPLICATION OF THE FINITE ELEMENT METHOD TO THE CALCULATION OF REACTOR PRESSURE VESSELS	71-0645
VESSELS	APPLICATION OF THE MATRIX DISPLACEMENT METHOD TO THE ANALYSIS OF PRESSURE VESSELS	69-0051
VESSELS	CALCULATIONS FOR PRESTRESSED CONCRETE PRESSURE VESSELS	71-0643
VESSELS	CIRCUMFERENTIAL PRESTRESSING LOAD ANALYSIS DUE TO WIRE/STRAND WINDING SYSTEMS ON PRESTRESSED CONCRET	74-1202
VESSELS	DISCUSSION ON FINITE ELEMENT ANALYSIS OF PRESTRESSED CONCRETE REACTOR VESSELS	72-0648
VESSELS	EINFACHES NAEHERUNGSVERFAHREN FUR DIE GRENZTRAGFAEHIGKEITSANLYSE VON SPANNBETON-REAKTORDRUCKBEHAELTE	73-0854
VESSELS	ELASTIC PLASTIC DEFORMATION OF CYLINDRICAL PRESSURE VESSELS UNDER CYCLIC LOADING	74-0604
VESSELS	ELASTIC-PLASTIC ANALYSIS OF THICK WALLED PRESSURE VESSELS WITH SHARP DISCONTINUITIES	71-0155
VESSELS	FINITE ELEMENT ANALYSIS OF HYPERBOLOIDAL PRESTRESSED CONCRETE CONTAINMENT VESSELS	69-0526
VESSELS	FINITE ELEMENT ANALYSIS OF CONTAINMENT VESSELS	71-0080
VESSELS	FINITE ELEMENT ANALYSIS OF PRESSURE VESSELS	71-0348
VESSELS	FINITE ELEMENT ANALYSIS OF PRESTRESSED CONCRETE PRESSURE VESSELS	74-1289
VESSELS	FINITE ELEMENT INCREMENTAL ELASTIC-PLASTIC ANALYSIS OF PRESSURE VESSELS	70-0253
VESSELS	FINITE ELEMENT METHODS IN ANALYSIS OF REACTOR VESSELS	72-0271
VESSELS	FINITE ELEMENTS FOR CALCULATION OF PRESTRESSED CONCRETE REACTOR PRESSURE VESSELS (FINITE ELEMENTE ZU	73-0797
VESSELS	FINITE-ELEMENT ELASTIC-PLASTIC ANALYSIS OF RESIDUAL STRESSES DUE TO CLAD WELDING IN REACTOR VESSELS	75-0779
VESSELS	FRACTURE AND SAFETY ANALYSIS OF NUCLEAR PRESSURE VESSELS	73-0257
VESSELS	MATRIX DISPLACEMENT METHODS IN FRACTURE MECHANICS ANALYSIS OF REACTOR VESSELS	71-0226
VESSELS	MODERN DEVELOPMENTS IN THE STRESS ANALYSIS OF PRESSURE VESSELS	69-0201
VESSELS	ON SHELL EFFECTS ON FERRO-CEMENT VESSELS	73-0455
VESSELS	PRACTICAL APPLICATIONS OF FINITE ELEMENT METHODS IN THE STRESS ANALYSIS OF THIN SHELL PRESSURE VESSE	71-0205
VESSELS	PRESSURE VESSELS ANALYSIS BY FINITE ELEMENT TECHNIQUES	67-0069
VESSELS	PRESTRESSED CONCRETE PRESSURE VESSELS	69-0052
VESSELS	PROGRESS IN COMPUTERIZED STRUCTURAL ANALYSIS OF LARGE VESSELS	74-0949
VESSELS	RECENT DEVELOPMENTS IN FINITE ELEMENT ANALYSIS OF PRESTRESSED CONCRETE REACTOR VESSELS	74-0469
VESSELS	RECENT DEVELOPMENTS IN THE FINITE ELEMENT ANALYSIS OF PRE-STRESSED CONCRETE REACTOR VESSELS	74-0582
VESSELS	RIDDLE OF BONDED AND UNBONDED TENDONS IN PRESTRESSED CONCRETE REACTOR VESSELS	74-0519
VESSELS	SIMPLE FINITE ELEMENT METHOD FOR STRESS CONCENTRATION PROBLEMS IN PRESSURE VESSELS	71-0669
VESSELS	SOME CREEP EFFECTS IN STRESS ANALYSIS WITH PARTICULAR REFERENCE TO CONCRETE PRESSURE VESSELS	69-0284
VESSELS	THE APPLICATION OF THREE-DIMENSIONAL FINITE ELEMENTS TO THE ANALYSIS OF PODDED BOILER TYPE PRESTRESS	71-0047
VESSELS	THERMAL SLEEVE DESIGN FOR PRESSURE VESSELS	74-1129
VESSELS	THREE DIMENSIONAL ELASTIC AND ELASTO-PLASTIC ANALYSIS OF REACTOR PRESSURE VESSELS	69-0453
VESSELS	THREE DIMENSIONAL STRESS ANALYSIS OF PENETRATIONS IN THICK WALLED PRESSURE VESSELS	74-0422
VESSELS	TRIANGULAR THICK SHELL FINITE ELEMENT FOR THE THREE-DIMENSIONAL STRESS ANALYSIS OF PRESSURE VESSELS	73-0860
VESSELS	TWO-DIMENSIONAL TIME-DEPENDENT ANALYSIS OF PERFORATED END CAPS FOR NUCLEAR REACTOR PRESSURE VESSELS	71-0239
VESSELS	USE OF SMALL MODELS IN DESIGN AND ANALYSIS OF PRESTRESSED-CONCRETE REACTOR VESSELS	70-0433
VESSELS	USE OF SMALL MODELS IN DESIGN AND ANALYSIS OF PRESTRESSED-CONCRETE REACTOR VESSELS	71-0426
VEUBEKE	SUR L ELEMENT DE FRAEIJS DE VEUBEKE ET SANDER (FRAEIJS DE VEUBEKE-SANDER ELEMENT)	74-0924
VEUBEKE-SANDER	SUR L ELEMENT DE FRAEIJS DE VEUBEKE ET SANDER (FRAEIJS DE VEUBEKE-SANDER ELEMENT)	74-0924
VIBRATING	ESTIMATIONS OF THE EFFECT OF DESIGN CHANGES ON THE MODES AND FREQUENCIES OF VIBRATING STRUCTURAL ELE	74-0819
VIBRATING	MATRIX DIFFERENCE EQUATION ANALYSIS OF VIBRATING PERIODIC STRUCTURES	75-0151
VIBRATING	USE OF TRIGONOMETRIC TERMS IN FINITE ELEMENT METHOD WITH APPLICATION TO VIBRATING MEMBRANES	74-0127
VIBRATING	USE OF TRIGONOMETRIC TERMS IN FINITE ELEMENT METHOD WITH APPLICATION TO VIBRATING MEMBRANES	74-0555
VIBRATION	A COMPARISON OF CURVED BEAM FINITE ELEMENTS WHEN USED IN VIBRATION PROBLEMS	71-0227
VIBRATION	A FINITE ELEMENT ANALOGUE OF THE MODIFIED RAYLEIGH-RITZ METHOD FOR VIBRATION PROBLEMS	72-0478
VIBRATION	A FINITE ELEMENT APPROACH TO PLATE VIBRATION PROBLEMS	65-0015
VIBRATION	A FINITE ELEMENT STUDY OF THE VIBRATION OF SKIN-RIB STRUCTURES	74-0537
VIBRATION	A MATRIX METHOD FOR VIBRATION AND STABILITY PROBLEMS	66-0091
VIBRATION	A METHOD FOR COMPUTATION OF VIBRATION MODES AND FREQUENCIES OF ORTHOTROPIC THIN SHELLS OF REVOLUTION	69-0037
VIBRATION	A MODIFIED FINITE ELEMENT METHOD FOR CALCULATION OF VIBRATION OF THIN PLATES	72-0598
VIBRATION	A NASTRAN VIBRATION MODEL OF THE AH-IG HELICOPTER AIRFRAME. VOLUME 1	74-1338
VIBRATION	A NASTRAN VIBRATION MODEL OF THE AH-IG HELICOPTER AIRFRAME. VOLUME 11	74-1339
VIBRATION	A NEW FAMILY OF CURVILINEAR PLATE BENDING ELEMENTS FOR VIBRATION AND STABILITY	72-0645
VIBRATION	A NEW FAMILY OF CURVILINEAR PLATE BENDING ELEMENTS FOR VIBRATION AND STABILITY	72-0999
VIBRATION	A THEORETICAL AND EXPERIMENTAL STUDY ON THE FREE VIBRATION OF CURVED SLAB BRIDGES	71-0496
VIBRATION	A TRIANGULAR ELEMENT FOR BENDING AND VIBRATION OF MULTILAYER SANDWICH PLATES	72-0489
VIBRATION	AN IMPROVED MATRIX DISPLACEMENT METHOD FOR VIBRATION OF NONUNIFORM BEAMS	71-0841
VIBRATION	ANALYSIS OF BUCKLING AND VIBRATION OF RING-STIFFENED, SEGMENTED SHELLS OF REVOLUTION	69-0496
VIBRATION	ANALYSIS OF FEEDBACK CONTROL STABILIZED VIBRATION BY THE FINITE ELEMENT METHOD	74-1188
VIBRATION	ANALYSIS OF IN-PLANE VIBRATION OF SHEAR WALLS BY A FINITE ELEMENT METHOD	72-0090
VIBRATION	ANALYSIS OF INPLANE VIBRATION OF BOX-TYPE STRUCTURES BY A FINITE ELEMENT METHOD	72-0091
VIBRATION	ANALYSIS OF NONELASTIC AND NONLINEAR VIBRATION OF ROCK DAMS USING FINITE ELEMENT METHOD	73-0680
VIBRATION	APPLICATION OF FINITE ELEMENT METHOD TO ADDED VIRTUAL MASS OF SHIP HULL VIBRATION	70-0463
VIBRATION	APPLICATION OF FINITE ELEMENT METHOD TO THE THREE DIMENSIONAL STRESS AND VIBRATION ANALYSIS OF BWR P	71-0180
VIBRATION	APPLICATION OF FLOWGRAPHS TO FREE VIBRATION OF STRUCTURES	68-0196
VIBRATION	APPLICATION OF RESPONSE-BOUND METHOD IN SHOCK AND VIBRATION ANALYSIS OF TELEPHONE STRUCTURES	74-0655
VIBRATION	APPLICATION OF SIMULTANEOUS ITERATION METHOD TO TORSIONAL VIBRATION PROBLEMS	73-0660
VIBRATION	APPLICATION OF THE FINITE ELEMENT METHOD TO THE VIBRATION ANALYSIS OF AXIAL FLOW TURBINES	73-1125
VIBRATION	ASYMMETRICAL BENDING AND VIBRATION OF A CONICAL SHELL FINITE ELEMENT	74-0085
VIBRATION	AXIAL-FLEXURAL VIBRATION OF FRAMEWORKS USING FINITE-ELEMENT APPROACH	73-0051
VIBRATION	AXISYMMETRIC VIBRATION OF SHELLS OF REVOLUTION BY THE FINITE ELEMENT METHOD	67-0100
VIBRATION	AXISYMMETRIC VIBRATION OF CONICAL SHELLS	70-0350
VIBRATION	BENDING AND VIBRATION OF MULTILAYER SANDWICH BEAMS AND PLATES	73-0272
VIBRATION	BOUNDS FOR EIGENVALUES IN SOME VIBRATION AND STABILITY PROBLEMS.	72-0390
VIBRATION	CH-53A FLEXIBLE FRAME VIBRATION ANALYSIS/TEST CORRELATION	73-1093
VIBRATION	CLOUGH-TOCHER TRIANGULAR PLATE BENDING ELEMENT IN VIBRATION	69-0093
VIBRATION	COMMENT ON DISCRETE ELEMENT IDEALIZATION OF AN INCOMPRESSIBLE LIQUID FOR VIBRATION ANALYSIS	73-0694
VIBRATION	COMPUTATION OF VIBRATION MODAL SHAPES AND NATURAL FREQUENCIES OF A GENERAL ORTHOTROPIC LAMINATED, TH	71-0309
VIBRATION	COMPUTER GENERATED DISPLAYS OF STRUCTURES IN VIBRATION	74-1305
VIBRATION	COMPUTER SIMULATION OF VIBRATION OF WHOLE TREE SYSTEMS USING FINITE ELEMENT METHODS	74-1161
VIBRATION	CONSTANT CURVATURE BEAM FINITE ELEMENTS FOR IN-PLANE VIBRATION	72-0254
VIBRATION	CORRELATION OF FINITE ELEMENT STRUCTURAL DYNAMIC ANALYSIS WITH MEASURED FREE VIBRATION CHARACTERISTI	74-0486
VIBRATION	COUPLED ROTOR-AIRFRAME VIBRATION PREDICTION METHODS	74-0498

439

VIBRATION	CURVED BEAM FINITE ELEMENTS FOR COUPLED BENDING AND TORSIONAL VIBRATION	72-0720
VIBRATION	DEFLECTION AND VIBRATION ANALYSIS OF A FLEXIBLE FAN BLADE	73-0556
VIBRATION	DISCRETE ELEMENT IDEALIZATION OF AN INCOMPRESSIBLE LIQUID FOR VIBRATION ANALYSIS	70-0391
VIBRATION	DYNAMICAL ANALYSIS OF AXIALLY NONUNIFORM, THIN CYLINDRICAL SHELLS. PART 2 FREE VIBRATION	71-0716
VIBRATION	EARTH VIBRATION EFFECTS AND ABATEMENT FOR MILITARY FACILITIES. REPORT 3. ANALYSIS METHOD FOR FOOTING	72-0806
VIBRATION	EFFECT OF INITIAL STATE FORCES ON THE HYDROELASTIC VIBRATION OF TUBES	73-0589
VIBRATION	EFFECTS OF MATERIAL NONLINEARITIES ON VIBRATION	73-0323
VIBRATION	EIGENVALUE ECONOMISERS IN VIBRATION PROBLEMS	63-0013
VIBRATION	ERRORS IN SHELL FINITE-ELEMENT MODELS FOR VIBRATION OF CIRCULAR-CYLINDERS	75-0203
VIBRATION	F 104 G FINAL MOTIVE: GROUND VIBRATION CALCULATIONS FOR THE STARFIGHTER F 104 G	69-0456
VIBRATION	FINITE ELEMENT ANALOGUE OF THE MODIFIED RAYLEIGH RITZ METHOD FOR VIBRATION PROBLEMS	73-0181
VIBRATION	FINITE ELEMENT ANALYSIS OF FREE VIBRATION AND STABILITY OF STRUCTURES SUBJECTED TO INITIAL STRESS	68-0300
VIBRATION	FINITE ELEMENT ANALYSIS OF VIBRATION AND STABILITY PROBLEMS	72-0477
VIBRATION	FINITE ELEMENT ANALYSIS OF VIBRATION OF THIN ELASTIC PLATES BY SIMPLIFIED HYBRID DISPLACEMENT METHOD	72-0565
VIBRATION	FINITE ELEMENT ANALYSIS OF COUPLED VIBRATION OF TAPERED TWISTED BLADES	72-0766
VIBRATION	FINITE ELEMENT ANALYSIS OF NONLINEAR VIBRATION OF BEAM COLUMNS	73-0028
VIBRATION	FINITE ELEMENT ANALYSIS OF VIBRATION AND FLUTTER OF CANTILEVER ANISOTROPIC PLATES	74-0484
VIBRATION	FINITE ELEMENT APPROACH TO THE VIBRATION ANALYSIS OF ELASTIC DISKS ON A FLEXIBLE SHAFT	75-0368
VIBRATION	FINITE ELEMENT METHOD FOR PIEZOELECTRIC VIBRATION	70-0012
VIBRATION	FINITE ELEMENT MODELLING OF NATURAL VIBRATION PROBLEMS	73-0320
VIBRATION	FINITE ELEMENT STUDY OF VIBRATION OF TRAPEZOIDAL PLATES	73-0035
VIBRATION	FINITE ELEMENT TECHNIQUE IN SHIP VIBRATION ANALYSIS	68-0180
VIBRATION	FINITE ELEMENT VIBRATION ANALYSIS OF CRACKED PLATES IN TENSION	68-0133
VIBRATION	FINITE ELEMENT VIBRATION ANALYSIS OF MULTICONDUCTOR ELECTRICAL TRANSMISSION LINES	73-0590
VIBRATION	FINITE ELEMENTS FOR VIBRATION OF FRAMED SHEAR WALLS	73-0036
VIBRATION	FINITE ELEMENTS FOR THE VIBRATION OF CONES AND CYLINDERS	75-0340
VIBRATION	FINITE-ELEMENT ANALYSIS FOR FREE VIBRATION OF GENERAL ANISOTROPIC LAMINATED THIN SHELLS	72-0618
VIBRATION	FINITE-ELEMENT ANALYSIS OF VIBRATION AND FLUTTER OF CANTILEVER ANISOTROPIC PLATES	74-0933
VIBRATION	FLEXUAL RIGIDITY OF A THIN WALLED BUILD-UP ROTOR FOR JET ENGINE. MEASUREMENT BY STATIC LOAD TEST AND	73-0756
VIBRATION	FORMULATION OF DISCRETE ELEMENT MODELS FOR STRESS AND VIBRATION ANALYSIS OF PLATES	70-0328
VIBRATION	FREE VIBRATION ANALYSIS OF STIFFENED PLATES	72-0303
VIBRATION	FREE VIBRATION ANALYSIS OF SPINNING STRUCTURAL SYSTEMS	73-1077
VIBRATION	FREE VIBRATION AND RESPONSE TO RANDOM PRESSURE FIELD OF ANISOTROPIC THIN CYLINDRICAL SHELLS	74-0830
VIBRATION	FREE VIBRATION AND TRANSIENT RESPONSE OF THICK AND THIN PLATES USING THE FINITE ELEMENT METHOD	74-0517
VIBRATION	FREE VIBRATION OF A CURVED BEAM	71-0606
VIBRATION	FREE VIBRATION OF A SIMPLY SUPPORTED BEAM WITH NONLINEAR MATERIAL PROPERTIES	71-0689
VIBRATION	FREE VIBRATION OF CURVED SANDWICH BEAMS BY METHOD OF FINITE ELEMENTS	71-0351
VIBRATION	FREE VIBRATION OF CURVED AND STRAIGHT BEAM SLAB OR BOX GIRDER BRIDGES	72-0487
VIBRATION	FREE VIBRATION OF CYLINDRICAL SHELLS PARTIALLY FILLED WITH LIQUID	71-0839
VIBRATION	FREE VIBRATION OF DOUBLY CURVED RECTANGULAR PLATES INCLUDING THE PRESENCE OF MEMBRANE STRESSES WITH	69-0468
VIBRATION	FREE VIBRATION OF MODEL OF CAR BODY	71-0312
VIBRATION	FREE VIBRATION OF ORTHOTROPIC VARIABLE THICKNESS, CONTINUOUS RECTANGULAR PLATES	70-0270
VIBRATION	FREE VIBRATION OF RING STIFFENED CYLINDRICAL SHELLS	70-0172
VIBRATION	FREE VIBRATION OF SHELLS OF REVOLUTION USING RING ELEMENTS	67-0088
VIBRATION	FREE VIBRATION OF SHELLS OF REVOLUTION USING FINITE ELEMENT METHOD	74-0232
VIBRATION	FREE VIBRATION OF THIN CYLINDRICAL SHELLS WITH A DISCONTINUITY IN THE THICKNESS	69-0439
VIBRATION	FREE VIBRATION OF THICK, LAYERED RECTANGULAR PLATES BY FINITE LAYER METHOD	72-0440
VIBRATION	FREE VIBRATION OF THICK, LAYERED CYLINDERS HAVING FINITE LENGTH WITH VARIOUS BOUNDARY CONDITIONS	72-0491
VIBRATION	FREQUENCY-DEPENDENT SHAPE FUNCTIONS FOR FREE AND FORCED STRUCTURAL VIBRATION CALCULATION	73-1091
VIBRATION	GALERKIN FINITE ELEMENT METHOD FOR VIBRATION PROBLEMS	73-0026
VIBRATION	GALERKIN FINITE ELEMENT METHOD FOR VIBRATION PROBLEMS	73-0659
VIBRATION	GENERAL SINUSOIDAL STIFFNESS MATRICES FOR BUCKLING AND VIBRATION ANALYSES OF THIN FLAT-WALLED STRUCT	68-0278
VIBRATION	IMPROVED FINITE ELEMENTS FOR VIBRATION ANALYSIS OF TAPERED BEAMS	73-0033
VIBRATION	INTERNAL PRESSURE EFFECTS ON THE VIBRATION OF PARTIALLY FILLED ELASTIC TANKS	75-0241
VIBRATION	LARGE AMPLITUDE, FREE VIBRATION GEOMETRICALLY NONLINEAR CIRCULAR ARCHES	72-0606
VIBRATION	MACHINERY SPACE DEFLECTION AND VIBRATION ANALYSES USING FINITE ELEMENT METHODS	72-0474
VIBRATION	MASS CONDENSATION, A SEMI-AUTOMATIC METHOD FOR REDUCING THE SIZE OF VIBRATION PROBLEMS	69-0220
VIBRATION	MASS LOADING EFFECTS ON VIBRATION OF RING AND SHELL STRUCTURES	68-0245
VIBRATION	MATHEMATICAL MODELING VIA DIRECT USE OF VIBRATION DATA	69-0310
VIBRATION	MEASURED AND CALCULATED VIBRATION PROPERTIES OF RING-STIFFENED HONEY-COMB CYLINDERS	71-0834
VIBRATION	MIXED FINITE ELEMENT METHOD FOR VIBRATION ANALYSIS OF SHALLOW SHELLS	73-0593
VIBRATION	MODELING OF GAS TURBINE ENGINE COMPRESSOR BLADES FOR VIBRATION ANALYSIS	75-0508
VIBRATION	NATURAL VIBRATION OF A FREE SURFACE	62-0008
VIBRATION	NON-LINEAR EFFECTS OF HIGH TEMPERATURE ON THE VIBRATION OF A SIMPLY SUPPORTED BEAM WITH A CENTRAL MA	73-0838
VIBRATION	NUMERICAL HULL VIBRATION ANALYSIS OF A FAR EAST CONTAINER SHIP	74-1325
VIBRATION	NUMERICAL RESULTS FROM THE APPLICATION OF GRADIENT ITERATIVE TECHNIQUES TO THE FINITE ELEMENT VIBRAT	72-0138
VIBRATION	OH - 58A PROPULSION SYSTEM VIBRATION INVESTIGATION	74-0733
VIBRATION	ON A SPECTRAL PROBLEM IN VIBRATION MECHANICS: COMPUTATION OF ELASTIC TANKS PARTIALLY FILLED WITH LIQ	75-0267
VIBRATION	ON THE APPLICATION OF THE FINITE ELEMENT METHOD TO THE VIBRATION ANALYSIS OF SPACE PLATE STRUCTURE	72-0630
VIBRATION	ON THE VIBRATION OF POINT-SUPPORTED PLATES	71-0602
VIBRATION	ON VIBRATION ANALYSIS OF RECTANGULAR CANTILEVER PLATES WITH A STEPPED CROSS-SECTION BY THE FINITE EL	71-0608
VIBRATION	PLATE VIBRATION PROBLEMS USING THE FINITE ELEMENT METHOD WITH ASSUMED STRESS DISTRIBUTION AND TRIANG	66-0004
VIBRATION	POWER SPECTRAL DENSITY ANALYSIS OF VEHICLE VIBRATION USING THE NASTRAN COMPUTER PROGRAM	74-0318
VIBRATION	PROPELLER-EXCITED VIBRATION WITH PARTICULAR REFERENCE TO FULL-SCALE MEASUREMENTS	75-0018
VIBRATION	PWR DEVELOPMENT FOR NUCLEAR PROPULSION: VIBRATION ANALYSIS OF NUCLEAR POWER PLANT	71-0797
VIBRATION	QUADRATIC MATRIX EQUATIONS FOR DETERMINING VIBRATION MODES AND FREQUENCIES OF CONTINUOUS ELASTIC SYS	66-0064
VIBRATION	RANDOM VIBRATION ANALYSIS SYSTEM FOR COMPLEX STRUCTURES, PART 2, COMPUTER PROGRAM DESCRIPTION	69-0262
VIBRATION	RANDOM VIBRATION RESPONSE OF CANTILEVER PLATES USING THE FINITE ELEMENT METHOD	70-0181
VIBRATION	RECENT DEVELOPMENTS IN HULL AND SHAFT VIBRATION ANALYSIS	70-0176
VIBRATION	SENSITIVITY ANALYSIS OF TORSIONAL VIBRATION CHARACTERISTICS OF HELICOPTER ROTOR BLADES. PART 1: STRU	74-0845
VIBRATION	SENSITIVITY OF ROTOR BLADE VIBRATION CHARACTERISTICS TO TORSIONAL OSCILLATIONS	74-0611
VIBRATION	SIMPLE FINITE ELEMENTS FOR PRE-TWISTED BLADING VIBRATION	74-0239
VIBRATION	SIMULATION OF VIBRATION OF WHOLE TREE SYSTEMS USING FINITE ELEMENTS	75-0765
VIBRATION	SPAENDINGSHYBRIDE FINITE ELEMENTER TIL SVINGNINGSPROBLEMER. (STRESS HYBRID FINITE ELEMENT METHOD FOR	75-0208
VIBRATION	STABILITY OF FINITE ELEMENT SCHEMES FOR VIBRATION PROBLEMS OF THE EQUATION OF ELASTICITY THEORY	71-0521
VIBRATION	STATIC, VIBRATION AND BUCKLING ANALYSIS OF AXISYMMETRIC CIRCULAR PLATES USING FINITE ELEMENTS	73-0157
VIBRATION	STATIC, VIBRATION, AND THERMAL STRESS ANALYSES OF LAMINATED PLATE AND SHELLS BY THE HYBRID-STRESS FI	72-0602
VIBRATION	STRESS AND VIBRATION ANALYSIS ON THE SYSTEM OF FRAME AND FOUNDATION	69-0218
VIBRATION	STRESS AND VIBRATION ANALYSIS OF TURBOPUMP INDUCER BLADES BY FINITE ELEMENTS	71-0363
VIBRATION	STRESS, STABILITY AND VIBRATION OF COMPLEX SHELLS OF REVOLUTION: ANALYSIS AND USERS MANUAL FOR BOSOR	69-0441
VIBRATION	STRESSES AND VIBRATION ANALYSIS OF INDUCER BLADES USING FINITE ELEMENT TECHNIQUE	70-0305
VIBRATION	STRUCTURAL VIBRATION ANALYSIS USING TRIANGULAR FINITE ELEMENTS	66-0060
VIBRATION	STRUCTURAL VIBRATION - A MATRIX APPROACH	71-0385
VIBRATION	STUDY ON THE VIBRATION OF THIN PLATES IN A GAS STREAM BY THE FINITE ELEMENT METHOD	74-0454
VIBRATION	SUBSTRUCTURES ANALYSIS OF IMPELLER VIBRATION MODES	75-0283
VIBRATION	THE APPLICATION OF GRADIENT MINIMIZATION METHODS AND HIGHER ORDER DISCRETE ELEMENTS TO SHELL BUCKLIN	71-0237
VIBRATION	THE EFFECT OF IN PLANE STATIC STRESSES ON THE VIBRATION CHARACTERISTICS OF CURVED SANDWICH PLATES	71-0588
VIBRATION	THE FINITE ELEMENT METHOD AND ITS APPLICATION IN VIBRATION ANALYSIS	66-0077
VIBRATION	THE FINITE ELEMENT METHOD IN VIBRATION ANALYSIS	67-0095
VIBRATION	THE MATRIX EIGENVALUE PROBLEM FOR DAMPED STRUCTURAL VIBRATION	66-0121
VIBRATION	THE VIBRATION CHARACTERISTICS OF A TENSIONED PLATE CONTAINING A FATIGUE CRACK	68-0134
VIBRATION	THE VIBRATION OF STEPPED CANTILEVERED PLATES	67-0129

VIBRATION	THEORETICAL AND EXPERIMENTAL VIBRATION ANALYSIS OF AN OBLIQUE SPACE FRAME	70-0460
VIBRATION	TRANSIENT VIBRATION ANALYSIS OF LINEAR SYSTEMS USING TRANSITION MATRICES	68-0053
VIBRATION	TRANSMISSION OF VIBRATION IN BEAM SYSTEMS	69-0405
VIBRATION	TRANSVERSE VIBRATION OF BEAMS WITH INTERNAL DAMPING, ROTATORY INERTIA AND SHEAR	63-0020
VIBRATION	USE OF HIGHER ORDER DISPLACEMENT FUNCTIONS IN THE FREE VIBRATION ANALYSIS OF SHELLS OF REVOLUTION HA	74-1055
VIBRATION	USE OF THE FINITE ELEMENT DAMPED FORCED RESPONSE STRAIN ENERGY DISTRIBUTION FOR VIBRATION REDUCTION	72-0759
VIBRATION	VARIATIONAL METHODS FOR THE SOLUTION OF PROBLEMS OF EQUILIBRIUM AND VIBRATION	43-0001
VIBRATION	VERTICAL VIBRATION OF EMBEDDED FOOTINGS	72-0231
VIBRATION	VIBRATION ANALYSIS OF A CANTILEVERED SQUARE PLATE BY THE STIFFNESS MATRIX METHOD	66-0012
VIBRATION	VIBRATION ANALYSIS OF CANTILEVERED CURVED PLATES USING A NEW CYLINDRICAL SHELL FINITE ELEMENT	68-0130
VIBRATION	VIBRATION ANALYSIS OF AN INTERNAL REFERENCE UNIT	69-0176
VIBRATION	VIBRATION ANALYSIS OF STRUCTURES BY COMPONENT MODE SUBSTITUTION	71-0303
VIBRATION	VIBRATION ANALYSIS OF FINITE ELEMENT SYSTEMS	71-0506
VIBRATION	VIBRATION ANALYSIS OF A 1/15-SCALE DYNAMIC MODEL OF A SPACE SHUTTLE CONFIGURATION	71-0609
VIBRATION	VIBRATION ANALYSIS OF CYLINDRICAL PANELS	72-0485
VIBRATION	VIBRATION ANALYSIS OF LAMINATED PLATES AND SHELLS BY A HYBRID STRESS ELEMENT	73-0120
VIBRATION	VIBRATION ANALYSIS OF ROTATING COMPRESSOR BLADES	73-0417
VIBRATION	VIBRATION ANALYSIS OF PLATES, SHELLS AND BLADED RIMMED DISCS USING THE TRANSFER-MATRIX-FINITE-ELEMEN	73-1151
VIBRATION	VIBRATION ANALYSIS OF ROTATING TURBINE BLADES	74-0959
VIBRATION	VIBRATION ANALYSIS OF AXIAL FLOW TURBINE DISKS USING FINITE ELEMENTS	75-0195
VIBRATION	VIBRATION ANALYSIS OF AXISYMMETRIC RESONATORS	75-0199
VIBRATION	VIBRATION ANALYSIS OF AXIAL-FLOW TURBINE DISKS USING FINITE-ELEMENTS	75-0415
VIBRATION	VIBRATION AND BUCKLING ANALYSIS OF AXISYMMETRIC POLAR ORTHOTROPIC CIRCULAR PLATES	74-0574
VIBRATION	VIBRATION AND STABILITY OF PLATES USING FINITE ELEMENTS	68-0244
VIBRATION	VIBRATION AND STRESS ANALYSIS OF SOFT-BONDED SHUTTLE INSULATION TILES, MODAL ANALYSIS WITH COMPACT W	74-0708
VIBRATION	VIBRATION CHARACTERISTICS OF THIN, PRESSURIZED SHELLS OF REVOLUTION PARTIALLY FILLED WITH LIQUID	72-0604
VIBRATION	VIBRATION CHARACTERISTICS OF FREE THIN CYLINDRICAL SHELLS	73-0396
VIBRATION	VIBRATION CHARACTERISTICS OF THE NORTH FORK DAM MODEL	74-0845
VIBRATION	VIBRATION MODES AND RANDOM RESPONSE OF A MULTI-BAY PANEL SYSTEM USING FINITE ELEMENTS	67-0128
VIBRATION	VIBRATION OF A BOX-TYPE STRUCTURE - 1	71-0405
VIBRATION	VIBRATION OF A BOX-TYPE STRUCTURE - 2.	71-0420
VIBRATION	VIBRATION OF AN ELASTIC BODY IMMERSED IN A FLUID	72-0312
VIBRATION	VIBRATION OF BEAMS WITH OVERHANGS	73-0488
VIBRATION	VIBRATION OF CIRCULAR AND ANNULAR PLATES USING FINITE ELEMENTS	72-0194
VIBRATION	VIBRATION OF CURVED STRUCTURES USING QUADRILATERAL FINITE ELEMENTS	72-0074
VIBRATION	VIBRATION OF CYLINDRICAL SHELLS BY HYBRID FINITE ELEMENT METHOD	72-0253
VIBRATION	VIBRATION OF CYLINDRICAL SHELLS BY A HYBRID FINITE ELEMENT METHOD	72-0795
VIBRATION	VIBRATION OF FOUR POINT-SUPPORTED PLATES BY A FINITE ELEMENT METHOD	71-0592
VIBRATION	VIBRATION OF FRAMES AND OTHER STRUCTURES WITH BANDED STIFFNESS MATRIX	70-0322
VIBRATION	VIBRATION OF MISSILES	61-0007
VIBRATION	VIBRATION OF MULTISUPPORTED CURVED BEAMS	74-1297
VIBRATION	VIBRATION OF NON-UNIFORM BEAMS	63-0021
VIBRATION	VIBRATION OF RECTANGULAR PLATES WITH MIXED BOUNDARY CONDITIONS	73-1086
VIBRATION	VIBRATION OF SQUARE PLATES HAVING VARIOUS EDGE CONDITIONS AND LOADINGS USING A REFINED FINITE ELEMEN	75-0024
VIBRATION	VIBRATION OF STIFFENED PLATES USING FINITE-ELEMENTS	72-0951
VIBRATION	VIBRATION OF SUBMERGED STRUCTURES AS COMPUTED BY THE FINITE ELEMENT METHOD	75-0227
VIBRATION	VIBRATION OF THICK, CURVED, SHELLS WITH PARTICULAR REFERENCE TO TURBINE BLADES	70-0017
VIBRATION	VIBRATION PROBLEMS USING ISOPARAMETRIC SHELL ELEMENTS	72-0549
VIBRATION	VIBRATION REDUCTION BY USING BOTH THE FINITE ELEMENT STRAIN ENERGY DISTRIBUTION AND MOBILITY TECHNIQ	74-1303
VIBRATION	VIBRATION STUDIES OF A FLAT PLATE AND BUILT-UP WING	71-0238
VIBRATION	VIBRATION STUDY OF COMPLETE TREE SYSTEMS USING FINITE ELEMENT METHOD	75-0790
VIBRATIONAL	APPLICATIONS OF THE FINITE-ELEMENT METHOD AND COMPUTER GRAPHICS. (A VIBRATIONAL ANALYSIS OF A SONAR	72-0729
VIBRATIONAL	FUNDAMENTAL VIBRATIONAL FREQUENCY BY A 3D - FEM PROCEDURE	75-0710
VIBRATIONAL	GENERAL FINITE ELEMENT VIBRATIONAL ANALYSIS OF PLANAR MECHANISMS	74-1107
VIBRATIONAL	VIBRATIONAL ANALYSIS OF THE FLEXING HEAD OF THE BQS-6 TRANSDUCER	71-0708
VIBRATIONAL	VIBRATIONAL ANALYSIS OF GENERAL SHELLS UTILIZING FINITE-ELEMENT TECHNIQUES	73-1133
VIBRATIONAL	VIBRATIONAL CHARACTERISTICS OF THREE-DIMENSIONAL SOLIDS, WITH APPLICATIONS TO EARTH DAMS	69-0432
VIBRATIONAL	VIBRATIONAL CHARACTERISTICS OF SPACE FRAMES	69-0434
VIBRATIONAL	VIBRATIONAL RESPONSE OF SONAR TRANSDUCERS USING PIEZO-ELECTRIC FINITE ELEMENTS	74-0522
VIBRATIONS	A FINITE CYLINDRICAL SHELL ELEMENT AND THE VIBRATIONS OF A CURVED FAN BLADE	68-0131
VIBRATIONS	A FINITE-ELEMENT METHOD FOR TRANSVERSE VIBRATIONS OF BEAMS AND PLATES	67-0144
VIBRATIONS	A NASTRAN CORRELATION STUDY FOR VIBRATIONS OF A CROSS-STIFFENED SHIP'S DECK	72-0756
VIBRATIONS	ALGORYTM OBLICZEN ORGAN WYMUSZONYCH METODA SZTYWNYCH ELEMENTOW SKONCZONYCH (ALGORITHM FOR CALCULATIN	74-0581
VIBRATIONS	ANALIZA ORGAN WYMUSZONYCH ZLOZONYCH UKLADOW LINIOWYCH METODA SZTYWNYCH ELEMENTOW SKONCZONYCH (ANALYS	72-0246
VIBRATIONS	ANALYSIS OF STRUCTURAL VIBRATIONS AND DYNAMIC RESPONSE	71-0059
VIBRATIONS	ANALYSIS OF THERMALLY INDUCED STRUCTURAL VIBRATIONS OF FINITE ELEMENT TECHNIQUES	68-0272
VIBRATIONS	ANALYTICAL DISPLACEMENTS AND VIBRATIONS OF CANTILEVERED UNSYMMETRIC FIBER COMPOSITE LAMINATES	75-0736
VIBRATIONS	APPLICATION OF FINITE ELEMENT METHOD TO VIBRATIONS OF QUARTZ PLATES	74-0345
VIBRATIONS	APPLICATION OF THE FINITE ELEMENT METHOD TO THE EXTENSIONAL VIBRATIONS OF PIEZOELECTRIC PLATES	72-0073
VIBRATIONS	AXISYMMETRIC VIBRATIONS OF LINEARLY TAPERED ANNULAR PLATES	74-1266
VIBRATIONS	BUILDING VIBRATIONS AND ACOUSTICS	73-0656
VIBRATIONS	CALCULATION OF NON-LINEAR AXISYMMETRIC VIBRATIONS OF THIN SHELLS OF REVOLUTION BY A FINITE ELEMENT M	75-0067
VIBRATIONS	CALCULATION OF VIBRATIONS OF STRUCTURES CAUSED BY EARTHQUAKES USING THE FINITE ELEMENT METHOD	71-0690
VIBRATIONS	CHISLENNYI METOD RASCHETA SOBSTVENNYKH I VYNUZHDENNYKH KOLEBANII SOSTAVNYKH OBOLOCHECHNYKH KONSTRUKT	73-0461
VIBRATIONS	CLOSED FORM, FINITE ELEMENT SOLUTIONS FOR PLATE VIBRATIONS	69-0168
VIBRATIONS	CONTRIBUTION A L'ANALYSE DES VIBRATIONS LIBRES D'UNE STRUCTURE. APPLICATION AUX AUBAGES DE TURBOMACH	75-0804
VIBRATIONS	CONTRIBUTION A L'ANALYSE DES VIBRATIONS LIBRES D'UNE STRUCTURE. APPLICATION AUX AUBAGES DE TURBOMACH	75-0804
VIBRATIONS	COUPLED VIBRATIONS OF A STRUCTURE SUBMERGED IN A COMPRESSIBLE FLUID	69-0001
VIBRATIONS	COUPLED VIBRATIONS OF A STRUCTURE SUBMERGED IN A COMPRESSIBLE FLUID	69-0507
VIBRATIONS	COUPLED VIBRATIONS OF THIN-WALLED BEAMS OF OPEN SECTION USING FINITE ELEMENT METHOD	70-0230
VIBRATIONS	COUPLED VIBRATIONS OF THIN-WALLED BEAMS OF OPEN SECTION USING FINITE ELEMENT METHOD	71-0568
VIBRATIONS	EARTH VIBRATION EFFECTS AND ABATEMENT FOR MILITARY FACILITIES. REPORT 3. ANALYSIS METHOD FOR FOOTING	72-0806
VIBRATIONS	EFFECT OF TRANSVERSE SHEAR DEFORMATION IN VIBRATIONS OF PLANAR STRUCTURES COMPOSED OF BEAM TYPE ELEM	73-0402
VIBRATIONS	EIN NEUES VARIATIONSPRINZIP MIT ANWENDUNG AUF SCHWINGUNGEN FLACHER SCHALEN NACH DER METHODE DER FINI	73-0563
VIBRATIONS	EVALUATION OF A FINITE ELEMENT ANALYSIS FOR THE LONGITUDINAL VIBRATIONS OF LIQUID-PROPELLANT LAUNCH	70-0204
VIBRATIONS	FINITE ELEMENT ANALYSIS OF AXISYMMETRIC VIBRATIONS OF CYLINDERS	72-0255
VIBRATIONS	FINITE ELEMENT ANALYSIS OF VIBRATIONS OF INITIALLY STRESSED THIN SHELLS OF REVOLUTION	74-1056
VIBRATIONS	FINITE ELEMENT DISPLACEMENT METHOD FOR LARGE AMPLITUDE FREE FLEXURAL VIBRATIONS OF BEAMS AND PLATES	73-0185
VIBRATIONS	FINITE ELEMENT MODELLING TECHNIQUES FOR THE ANALYSIS OF BUILDING VIBRATIONS	72-0839
VIBRATIONS	FINITE ELEMENT TECHNIQUES APPLIED TO THE VIBRATIONS OF PLATE SYSTEMS	66-0033
VIBRATIONS	FLEXURAL VIBRATIONS OF RECTANGULAR AND OTHER POLYGONAL PLATES	71-0417
VIBRATIONS	FORCED VIBRATIONS OF CIRCULAR FOUNDATIONS ON LAYERED MEDIA	74-0762
VIBRATIONS	FREE VIBRATIONS AND RANDOM RESPONSE OF AN INTEGRALLY-STIFFENED PANEL	70-0138
VIBRATIONS	FREE VIBRATIONS OF CIRCULAR MEMBRANES UNDER ARBITRARY TENSION BY THE FINITE ELEMENT METHOD	69-0190
VIBRATIONS	FREE VIBRATIONS OF CURVED SANDWICH BEAMS	71-0729
VIBRATIONS	FREE VIBRATIONS OF CYLINDRICAL SHELLS WITH VARIOUS END CLOSURES	74-0571
VIBRATIONS	FREE VIBRATIONS OF FINITE ELEMENT PLATES SUBJECTED TO COMPLEX MIDDLE-PLANE FORCE SYSTEMS	72-0280
VIBRATIONS	FREE VIBRATIONS OF FINITE ELEMENT PLATES SUBJECTED TO COMPLEX MIDDLE-PLANE FORCE SYSTEMS	73-0097
VIBRATIONS	FREE VIBRATIONS OF FINITE ELEMENT PLATES SUBJECTED TO COMPLEX MIDDLE-PLANE FORCE SYSTEMS - REPLY	73-0122
VIBRATIONS	FREE VIBRATIONS OF FLUID-BORNE STRUCTURES: INVESTIGATIONS ON A SIMPLE MODEL	74-0976
VIBRATIONS	FREE VIBRATIONS OF PRESSURE LOADED PARABOLOIDAL SHELLS	71-0601
VIBRATIONS	FREE VIBRATIONS OF PLATE-BEAM SYSTEMS	73-0322

VIBRATIONS	HULL AND SUPERSTRUCTURE VIBRATIONS DESIGN CALCULATIONS BY FINITE ELEMENTS	75-0556
VIBRATIONS	HYBRYDOWA METODA ELEMENTOW SKONCZONYCH W ZASTOSOWANIU DO OBLICZEN ORGAN URZADZEN OKRETOWYCH. (HYBRID	74-0963
VIBRATIONS	INEXTENSIONAL VIBRATIONS OF PARABOLOIDAL SHELLS OF REVOLUTION. DOMES AND SHELLS	71-0427
VIBRATIONS	INVESTIGATION OF THE EFFECT OF FOUNDATION YIELD ON THE FREQUENCY AND SHAPE OF INHERENT VIBRATIONS OF	72-0419
VIBRATIONS	INVESTIGATION ON FREE AND FORCED VIBRATIONS OF A LONG TANKER WITH OVERLAPPING PROPELLER ARRANGEMENT	73-0981
VIBRATIONS	LUMPED PARAMETER METHODS APPLIED TO ELASTIC VIBRATIONS	63-0016
VIBRATIONS	MEASURED AND PREDICTED LONGITUDINAL VIBRATIONS OF A LIQUID PROPELLANT TWO-STAGE LAUNCH VEHICLE	70-0366
VIBRATIONS	NATURAL VIBRATIONS OF LAMINATED ORTHOTROPIC SHELLS OF REVOLUTION	70-0367
VIBRATIONS	NATURAL VIBRATIONS OF THIN, FLAT WALLED STRUCTURES WITH DIFFERENT BOUNDARY CONDITIONS	71-0509
VIBRATIONS	NEW APPROACH FOR PLATE VIBRATIONS. COMBINATION OF TRANSFER MATRIX AND FINITE ELEMENT TECHNIQUE	71-0434
VIBRATIONS	NEW VARIATIONAL PRINCIPLE WITH APPLICATION TO VIBRATIONS OF SHALLOW SHELLS USING FINITE ELEMENTS MET	73-0354
VIBRATIONS	NONLINEAR VIBRATIONS OF NONUNIFORM BEAMS WITH CONCENTRATED MASSES	74-1300
VIBRATIONS	NUMERICAL-PERTURBATION METHOD FOR THE NONLINEAR ANALYSIS OF STRUCTURAL VIBRATIONS	74-0499
VIBRATIONS	NUMERYCZNE ROZWIAZANIE PROBLEMU ORGAN WLASNYCH NA PRZYKLADZIE PLYT. (NUMERICAL SOLUTION OF THE PROBL	75-0296
VIBRATIONS	OBLICZANIE ORGAN METODA SZTYWNYCH ELEMENTOW SKONCZONYCH. (CALCULATING VIBRATIONS BY THE RIGID-FINITE	75-0649
VIBRATIONS	ON SMALL VIBRATIONS AND PERTURBATIONS OF FLEXIBLE BODIES UNDERGOING ARBITRARY NOMINAL MOTION	70-0376
VIBRATIONS	PRACTICAL ANALYSIS OF PLATE VIBRATIONS USING NASTRAN	71-0058
VIBRATIONS	PREDICTION OF PLATE VIBRATIONS USING A CONSISTENT MASS MATRIX	66-0043
VIBRATIONS	RASCHET KILEBANII PLASTIN SLOZHNOI FORMY S NEODNORODNYMI GRANICHNYMI USLOVIYAMI. (CALCULATION OF VIB	74-0988
VIBRATIONS	RECTANGULAR FINITE ELEMENTS FOR ANALYSIS OF PLATE VIBRATIONS	68-0113
VIBRATIONS	ROTOR BLADE VIBRATIONS BY GALERKIN FINITE-ELEMENT METHOD	75-0389
VIBRATIONS	SCHWINGUNGEN ZYLINDRISCHER SCHRAUBENFEDERN (VIBRATIONS OF CYLINDRICAL COIL SPRINGS)	74-0331
VIBRATIONS	SIGNIFICANT APPLICATIONS OF FINITE ELEMENTS AND FINITE DIFFERENCES METHODS TO SOLVE PROBLEMS OF THER	73-0864
VIBRATIONS	STABILITY ANALYSIS OF THE TORSIONAL BENDING VIBRATIONS OF WING IN A SUBSONIC FLOW BY METHOD OF THE F	75-0755
VIBRATIONS	TRANSVERSE VIBRATIONS OF CANTILEVER BARS OF VARIABLE CROSS SECTIONS	66-0001
VIBRATIONS	TRANSVERSE VIBRATIONS OF NON-UNIFORM BEAMS	74-1058
VIBRATIONS	TRANSVERSE VIBRATIONS OF SHIP'S PROPULSION SYSTEMS. PART 1: THEORETICAL ANALYSIS	74-1323
VIBRATIONS	TRANSVERSE VIBRATIONS OF SHIP'S PROPULSION SYSTEMS. PART 2: EXPERIMENTAL ANALYSIS	74-1324
VIBRATIONS	TRANSVERSE VIBRATIONS OF SHIP'S PROPULSION SYSTEMS - 1 THEORETICAL ANALYSIS	75-0291
VIBRATIONS	UNBALANCE GENERATED VIBRATIONS AND STABILITY OF TURBINE ROTORS	73-0224
VIBRATIONS	USE OF STRAIGHT BEAM FINITE ELEMENTS FOR ANALYSIS OF VIBRATIONS OF CURVED BEAMS	73-0616
VIBRATIONS	VERTICAL VIBRATIONS OF FOOTINGS EMBEDDED IN LAYERED MEDIA	69-0436
VIBRATIONS	VIBRATIONS OF A NON-UNIFORM BEAM	68-0246
VIBRATIONS	VIBRATIONS OF A TIMOSHENKO BEAM, USING FINITE-ELEMENT APPROACH	66-0044
VIBRATIONS	VIBRATIONS OF GRIDS BY THE FINITE ELEMENT METHOD	73-1089
VIBRATIONS	VIBRATIONS OF PLATE AND SHELL STRUCTURES USING TRIANGULAR FINITE ELEMENTS	67-0023
VIBRATIONS	VIBRATIONS OF PRESSURE LOADED HYPERBOLOIDAL SHELLS	72-0874
VIBRATIONS	VIBRATIONS OF SEGMENTED SHELLS	73-0161
VIBRATIONS	VIBRATIONS OF TANKS PARTIALLY FILLED WITH LIQUIDS	74-0089
VIBRATIONS	VIBRATIONS OF TAPERED AND CURVED MEMBERS BY FINITE ELEMENT METHODS	69-0517
VIBRATIONS	VIBRATIONS OF THICK PLATES USING A HIGH PRECISION TRIANGULAR ELEMENT	74-1002
VIBRATORS	FINITE ELEMENT ANALYSIS OF FLEXURE - TYPE VIBRATORS WITH ELECTROSTRICTIVE TRANSDUCERS	70-0338
VIBRATORY	COUPLED RESPONSE OF SPATIAL VIBRATORY STRUCTURES MOUNTED TO ISOTROPIC PLATE ELEMENTS	71-0228
VIBRATORY	VIBRATORY BEHAVIOR OF TWISTED CANTILEVERED PLATES	75-0507
VIBRATORY	VIBRATORY WEAR OF FUEL RODS	75-0771
VIBROACOUSTIC	VIBROACOUSTIC RESPONSE OF TURBULENCE EXCITED THIN RECTANGULAR FINITE PLATES IN HEAVY AND FLUID MEDIA	74-1309
VICINITY	DER SPANNUNGS UND BEANSPRUCHUNGSZUSTAND DES BODENMATERIALS IN DER NACHBARSCHAFT EINER BODENBOESCHUNG	75-0371
VICINITY	QUANTITATIVE STRAIN-AND-STRESS STATE CRITERION FOR FAILURE IN THE VICINITY OF SHARP CRACKS	74-0639
VICINITY	STRESS AND STRAIN DISTRIBUTION IN THE VICINITY OF INTERFERENCE FIT FASTENERS	73-0363
VIERENDEEL	VIERENDEEL BENT CAPS	72-0758
VIEWPORT	STRESS ANALYSIS OF A CONICAL ACRYLIC VIEWPORT	70-0106
VIKING	ADAPTATION OF NASTRAN TO THE ANALYSIS OF THE VIKING SPACE VEHICLE	71-0135
VIKING	VIKING DYNAMICS - AN OVERVIEW	74-1304
VIOLATING	FUNKCJA KSZTALTU NIESPELNIAPACA WARUNKOW CIAGLOSCI W ANALIZIE PLYT METODA ELEMENTOW SKONCZONYCH. (SH	75-0263
VIRTUAL	APPLICATION OF FINITE ELEMENT METHOD TO ADDED VIRTUAL MASS OF SHIP HULL VIBRATION	70-0463
VIRTUAL	CALCULATION OF THE VIRTUAL MASS, MOVEMENT AND INERTIA BY THE FINITE ELEMENT METHOD	68-0221
VIRTUAL	ON THE METHOD OF VIRTUAL CRACK EXTENSIONS	75-0237
VIRTUAL	SOME APPLICATIONS OF THE PRINCIPLE OF VIRTUAL WORK	73-0930
VISCEL	VISCEL A GENERAL PURPOSE COMPUTER PROGRAM FOR ANALYSIS OF LINEAR VISCOELASTIC STRUCTURES. VOL. 2	72-0789
VISCO-ELASTIC	A NUMERICAL METHOD OF VISCO-ELASTIC STRESS ANALYSIS	68-0175
VISCO-ELASTIC	ANALYSIS OF LONGITUDINAL WAVE IN A BAR EXHIBITING VISCO-ELASTIC BEHAVIOR OR WITH MATERIAL DAMPING	71-0780
VISCO-ELASTIC	FINITE ELEMENT CREEP BUCKLING ANALYSIS OF VISCO-ELASTIC SHALLOW SPHERICAL SHELL	75-0159
VISCO-ELASTIC	NUMERICAL ANALYSIS OF VISCO-ELASTIC STRUCTURES BY THE FINITE ELEMENT METHOD	70-0177
VISCO-ELASTIC	THE NUMERICAL SOLUTION OF VISCO-ELASTIC PROBLEMS USING LAPLACE TRANSFORMS	74-0436
VISCO-ELASTIC	UNDRAINED VISCO-ELASTIC ANALYSIS OF SOIL DEFORMATION	72-0424
VISCO-PLASTIC	APPLICATION OF THE PSEUDO-FUNCTIONAL FINITE ELEMENT METHOD TO VISCO-PLASTIC TORSION	73-0002
VISCO-PLASTIC	FLOW OF PLASTIC AND VISCO-PLASTIC SOLIDS WITH SPECIAL REFERENCE TO EXTRUSION AND FORMING PROCESSES	74-0210
VISCO-PLASTIC	TRANSIENT RESPONSE OF ELASTIC, VISCO-PLASTIC BEAMS	73-0380
VISCO-PLASTICIT	VISCO-PLASTICITY AND PLASTICITY-AN ALTERNATIVE FOR FINITE ELEMENT SOLUTION OF MATERIAL NONLINEARITIE	74-0410
VISCO-PLASTICIT	VISCO-PLASTICITY SOLUTION BY FINITE ELEMENT PROCESS	72-0048
VISCO-PLASTICIT	VISCO-PLASTICITY, PLASTICITY AND CREEP IN ELASTIC SOLIDS. A UNIFIED NUMERICAL SOLUTION APPROACH	74-1279
VISCOELASTIC	A FINITE ELEMENT METHOD OF VISCOELASTIC STRESS ANALYSES WITH APPLICATION TO ROLLING CONTACT PROBLEMS	69-0330
VISCOELASTIC	A GENERALIZATION OF FINITE ELEMENT METHOD TO DYNAMIC VISCOELASTIC ANALYSIS	75-0106
VISCOELASTIC	ANALYSIS OF VISCOELASTIC CONTACT STRESSES IN AGRICULTURAL PRODUCTS USING A FINITE ELEMENT METHOD	74-0634
VISCOELASTIC	ANALYSIS OF VISCOELASTIC CONTACT STRESSES IN AGRICULTURAL PRODUCTS USING A FINITE ELEMENT METHOD	74-1199
VISCOELASTIC	ANALYSIS OF VISCOELASTIC FLUID BY FINITE ELEMENT METHOD	75-0865
VISCOELASTIC	DYNAMIC RESPONSE OF VISCOELASTIC MATERIALS TO SINUSOIDALLY VARYING LOADINGS	73-1072
VISCOELASTIC	DYNAMIC RESPONSE OF VISCOELASTIC MATERIALS	73-1075
VISCOELASTIC	EFFICIENT METHOD FOR SOLUTION OF VISCOELASTIC PROBLEMS	73-0232
VISCOELASTIC	EVALUATION OF DISCRETE METHODS FOR THE LINEAR DYNAMIC RESPONSE OF ELASTIC AND VISCOELASTIC SOLIDS	70-0439
VISCOELASTIC	FINITE ELEMENT COMPUTER PROGRAM FOR PREDICTING THE NONLINEAR STATIC AND DYNAMIC BEHAVIOR OF VISCOELA	75-0191
VISCOELASTIC	FINITE ELEMENT VISCOELASTIC STRESS ANALYSIS STUDY USING THE GENERALIZED KELVIN SOLID MODEL	73-0957
VISCOELASTIC	ICE ENGINEERING: VISCOELASTIC FINITE ELEMENT FORMULATION	74-0861
VISCOELASTIC	LARGE DISPLACEMENT ANALYSIS OF VISCOELASTIC BEAMS AND FRAMES BY FINITE-ELEMENT METHOD	74-0644
VISCOELASTIC	LARGE STRAIN, VISCOELASTIC NUMERICAL ANALYSIS BY MEANS OF FINITE ELEMENT METHOD	72-0768
VISCOELASTIC	LARGE STRAIN, VISCOELASTIC AND ELASTO-VISCOPLASTIC NUMERICAL-ANALYSIS BY MEANS OF FINITE-ELEMENT MET	75-0435
VISCOELASTIC	LARGE STRAIN, VISCOELASTIC AND ELASTO VISCOPLASTIC NUMERICAL ANALYSIS BY MEANS OF THE FINITE ELEMENT	75-0563
VISCOELASTIC	MIMIC SOURCE APPROACH TO VISCOELASTIC ANALYSIS	69-0164
VISCOELASTIC	NUMERICAL PROCEDURE FOR THE ANALYSIS OF STRUCTURES SUPPORTED BY VISCOELASTIC FOUNDATIONS	73-0564
VISCOELASTIC	NUMERICAL STUDY OF DAMPING IN VISCOELASTIC SANDWICH BEAMS	73-0443
VISCOELASTIC	RIESENIE VAZKOPRUZNYCH ANIZOTROPICKYCH DOASK METODOU KONECNYCH PRVKOV (SOLUTION OF VISCOELASTIC ANIS	74-0477
VISCOELASTIC	SAFE-CREEP: A COMPUTER PROGRAM FOR THE VISCOELASTIC ANALYSIS OF AXISYMMETRIC AND PLANE CONCRETE STRU	67-0148
VISCOELASTIC	STRESS ANALYSIS IN VISCOELASTIC BODIES USING FINITE ELEMENTS AND A CORRESPONDENCE RULE WITH ELASTICI	69-0270
VISCOELASTIC	THERMOMECHANICAL ANALYSIS OF VISCOELASTIC SOLIDS	70-0240
VISCOELASTIC	TRANSIENT DYNAMIC RESPONSE OF VISCOELASTIC STRUCTURES	72-0234
VISCOELASTIC	VISCEL A GENERAL PURPOSE COMPUTER PROGRAM FOR ANALYSIS OF LINEAR VISCOELASTIC STRUCTURES. VOL. 2	72-0789
VISCOELASTIC	VISCOELASTIC ANALYSIS OF P C CONTAINMENT USING FINITE ELEMENT METHOD	73-0034
VISCOELASTIC	VISCOELASTIC ANALYSIS OF PRESTRESSED CONCRETE SHELL OF REVOLUTION	73-0464
VISCOELASTIC	VISCOELASTIC FINITE ELEMENTS	74-0440
VISCOELASTIC	VISCOELASTIC RESPONSE OF THE CORNEO-SCLERAL SHELL UNDER TONOMETER LOADING	75-0855
VISCOELASTIC	VISCOELASTIC RESPONSE OF GRAPHITE MATERIALS IN IRRADIATION ENVIRONMENTS	70-0314
VISCOELASTIC	VISCOELASTIC STRESS ANALYSIS	68-0080

442

VISCOELASTIC	WAVE PROPAGATION IN A VISCOELASTIC FIBER SUBJECTED TO TRANSVERSE IMPACT	71-0639
VISCOELASTICALL	METHODS OF MODELING AND ANALYZING VISCOELASTICALLY DAMPED STRUCTURES	71-0473
VISCOELASTICALL	METHODS OF MODELING AND ANALYZING VISCOELASTICALLY DAMPED STRUCTURES	71-0836
VISCOELASTICITY	A GENERALIZATION OF THE FINITE ELEMENT CONCEPT AND ITS APPLICATION TO A CLASS OF PROBLEMS IN NONLINE	68-0212
VISCOELASTICITY	APPROXIMATE SOLUTION IN LINEAR VISCOELASTICITY	66-0101
VISCOELASTICITY	CONVERGENCE OF FINITE ELEMENT SOLUTIONS IN VISCOELASTICITY	74-0628
VISCOELASTICITY	FINITE ELEMENT METHOD IN LINEAR VISCOELASTICITY	74-0213
VISCOELASTICITY	FINITE ELEMENTS AND DYNAMIC VISCOELASTICITY	71-0479
VISCOELASTICITY	FINITE ELEMENTS IN LINEAR VISCOELASTICITY	68-0281
VISCOELASTICITY	GALERKIN METHOD AS APPLIED TO PROBLEMS IN VISCOELASTICITY	74-1039
VISCOELASTICITY	MATRIX DIFFERENTIAL EQUATION AND HIGHER-ORDER NUMERICAL METHODS FOR PROBLEMS OF NON-LINEAR CREEP, VI	72-0198
VISCOELASTICITY	NUMERICAL FORMULATION OF A CLASS OF PROBLEMS IN NONLINEAR VISCOELASTICITY	67-0061
VISCOELASTICITY	ON A GENERALIZATION OF THE FINITE ELEMENT CONCEPT AND ITS APPLICATION TO A CLASS OF PROBLEMS IN NONL	69-0206
VISCOELASTICITY	PLASTICITY AND VISCOELASTICITY	72-0905
VISCOELASTOPLAS	ANALYSIS OF VISCOELASTOPLASTIC STRUCTURAL BEHAVIOUR OF ANISOTROPIC SHELLS BY THE FINITE ELEMENT METH	71-0056
VISCOELASTOPLAS	DYNAMIC ANALYSIS OF VISCOELASTOPLASTIC ANISOTROPIC SHELLS	73-0253
VISCOELASTOPLAS	INCREMENTAL THERMOMECHANICAL THEORY OF VISCOELASTOPLASTIC SOLIDS AND SOLUTION BY FINITE ELEMENTS	73-0861
VISCOELASTOPLAS	STATIC AND DYNAMIC ANALYSIS OF VISCOELASTOPLASTIC FIBER-REINFORCED COMPOSITE SHELLS IN MISSILE STRUC	73-0651
VISCOPLASTIC	ANALYSIS OF VISCOPLASTIC EFFECTS IN PRESSURE VESSELS BY THE FINITE ELEMENT METHOD	74-0573
VISCOPLASTIC	LARGE STRAIN, VISCOELASTIC AND ELASTO VISCOPLASTIC NUMERICAL ANALYSIS BY MEANS OF THE FINITE ELEMENT	75-0563
VISCOPLASTIC	NUMERICAL INVESTIGATION OF AXI-SYMMETRIC VISCOPLASTIC FLOWS BY FINITE-ELEMENT PROCEDURE	73-0132
VISCOPLASTIC	ON THE EXTREMAL PROPERTIES OF THE SOLUTION IN DYNAMICS OF RIGID VISCOPLASTIC BODIES ALLOWING FOR LAR	73-1080
VISCOPLASTIC	PLASTIC AND VISCOPLASTIC ANALYSIS OF AXISYMMETRIC SHELLS	75-0246
VISCOPLASTICITY	VISCOPLASTICITY, PLASTICITY AND CREEP. A UNIFIED NUMERICAL SOLUTION APPROACH	74-1205
VISCOPLASTICITY	VISCOPLASTICITY, PLASTICITY AND VISCOUS FLOW	74-1278
VISCOSITY	NUMERICAL SIMULATION OF FINITE-WIDTH THRUST BEARINGS, TAKING INTO ACCOUNT VISCOSITY VARIATION WITH T	75-0623
VISCOUS	A FINITE ELEMENT COMPUTATIONAL THEORY FOR THE MECHANICS AND THERMODYNAMICS OF A VISCOUS, COMPRESSIBL	71-0787
VISCOUS	A FINITE ELEMENT SOLUTION FOR A CLASS OF TWO DIMENSIONAL VISCOUS FLOW DYNAMICS PROBLEMS	71-0544
VISCOUS	A VARIATIONAL FINITE ELEMENT METHOD FOR TWO DIMENSIONAL STEADY VISCOUS FLOWS	72-0355
VISCOUS	AIRCRAFT-PAVEMENT INTERACTION STUDIES, PHASE 1: A FINITE ELEMENT MODEL OF A JOINTED CONCRETE PAVEMEN	73-0894
VISCOUS	AN ITERATIVE FINITE ELEMENT METHOD OF VISCOUS FLOW	74-1353
VISCOUS	ANALYSIS OF FLOW OF VISCOUS FLUIDS BY FINITE-ELEMENT METHOD	72-0032
VISCOUS	ANALYSIS OF STEADY INCOMPRESSIBLE VISCOUS FLOW	74-0013
VISCOUS	ANALYSIS OF THREE-DIMENSIONAL UNSTEADY VISCOUS FLOW AROUND OSCILLATING WINGS	74-0982
VISCOUS	APPLICATION OF THE PSEUDO-FUNCTIONAL FINITE ELEMENT METHOD TO VISCOUS FLOW PROBLEMS	74-1183
VISCOUS	APPLICATIONS OF FINITE ELEMENT METHODS TO UNSTEADY VISCOUS FLOW AROUND A BOX-GIRDER BRIDGE OSCILLATI	74-0827
VISCOUS	COMPUTATIONAL METHODS FOR INVISCID AND VISCOUS TWO-AND-THREE DIMENSIONAL FLOW FIELDS	75-0716
VISCOUS	COMPUTATIONALLY EFFICIENT FINITE ELEMENT ANALYSIS OF VISCOUS FLOW PROBLEMS	74-1050
VISCOUS	COMPUTATIONALLY EFFICIENT FINITE ELEMENT ANALYSIS OF VISCOUS FLOW PROBLEMS	74-1258
VISCOUS	CONVERGENCE OF FINITE ELEMENT SOLUTIONS IN VISCOUS FLOW PROBLEMS	75-0337
VISCOUS	DIFFUSION ON SOLID SPHERICAL PARTICLE IN VISCOUS FLUID FLOW AT FINITE VALUES OF RE NUMBERS	69-0419
VISCOUS	FINITE ELEMENT ANALYSIS OF THE UNSTEADY FLOW OF A VISCOUS PLASTIC FLUID IN A CYLINDRICAL PIPE	74-0055
VISCOUS	FINITE ELEMENT ANALYSIS OF SEVERAL COMPRESSIBLE AND INCOMPRESSIBLE VISCOUS FLOW PROBLEMS	74-1119
VISCOUS	FINITE ELEMENT ANALYSIS OF VISCOUS FLOW BY STREAM FUNCTION	75-0526
VISCOUS	FINITE ELEMENT ANALYSIS OF VISCOUS, INCOMPRESSIBLE FLUID FLOW	75-0674
VISCOUS	FINITE ELEMENT ANALYSIS OF VISCOUS, INCOMPRESSIBLE FLUID FLOW	75-0760
VISCOUS	FINITE ELEMENT ANALYSIS OF INCOMPRESSIBLE MICROPOLAR VISCOUS FLUID FLOW	75-0866
VISCOUS	FINITE ELEMENT FORMULATION FOR SOLUTION OF STEADY FLOW OF INCOMPRESSIBLE VISCOUS FLUID	73-0299
VISCOUS	FINITE ELEMENT METHOD FOR INCOMPRESSIBLE SLOW VISCOUS FLOW WITH A FREE SURFACE	69-0303
VISCOUS	FINITE ELEMENT METHOD FOR TRANSIENT LINEAR VISCOUS FLOW PROBLEMS	74-0474
VISCOUS	FINITE ELEMENT SOLUTION OF VISCOUS FLOW IN AN ARBITRARY INTERNAL PASSAGE	71-0461
VISCOUS	FINITE ELEMENT SOLUTION ALGORITHM FOR VISCOUS INCOMPRESSIBLE FLUID DYNAMICS	73-0063
VISCOUS	FINITE ELEMENT THEORY FOR THE MECHANICS AND THERMODYNAMICS OF A VISCOUS, COMPRESSIBLE MULTI-SPECIE F	71-0005
VISCOUS	FINITE-ELEMENT ANALYSIS OF TAPERED VISCOUS FLOW	72-0961
VISCOUS	FINITE-ELEMENT ANALYSIS OF SLOW INCOMPRESSIBLE VISCOUS FLUID MOTION	74-0154
VISCOUS	FORMULATION OF A VARIATIONAL PRINCIPLE – FINITE ELEMENT METHOD FOR VISCOUS FLOW	72-0354
VISCOUS	FULLY DEVELOPED VISCOUS FLOW IN INTERNALLY FINNED TUBES	75-0648
VISCOUS	INTEGRAL REPRESENTATION OF FIELD VARIABLES FOR THE FINITE ELEMENT SOLUTION OF VISCOUS FLOW PROBLEMS	74-0468
VISCOUS	LOCAL POTENTIAL APPROACH TO FINITE ELEMENT METHOD IN UNSTEADY VISCOUS INCOMPRESSIBLE FLUID FLOW	73-0298
VISCOUS	NEWTONIAN AND NON-NEWTONIAN VISCOUS FLOW	75-0109
VISCOUS	NUMERICAL SOLUTION OF THE VISCOUS FLOW IN THE ENTRANCE REGION OF PARALLEL PLATES	73-0022
VISCOUS	NUMERICAL SOLUTION TO THE DYNAMICS OF VISCOUS FLUID FLOW BY A FINITE ELEMENT ALGORITHM: A FIRST STEP	71-0004
VISCOUS	NUMERICAL SOLUTION TO THE DYNAMICS OF VISCOUS FLUID FLOW BY A FINITE ELEMENT ALGORITHM: A FIRST STEP	71-0791
VISCOUS	SEEPAGE IN MISSISSIPPI RIVER BANKS. REPORT 1. ANALYSIS OF TRANSIENT SEEPAGE USING VISCOUS FLOW MODEL	70-0405
VISCOUS	SLOW FINITE DEFORMATIONS OF VISCOUS SOLIDS	69-0423
VISCOUS	SLOW PARTICULATE VISCOUS FLOW IN CHANNELS AND TUBES, APPLICATION TO BIOMECHANICS	71-0248
VISCOUS	SLOW VISCOUS FLOW AND ITS APPLICATION TO BIOMECHANICS	71-0619
VISCOUS	SOLUTION OF VISCOUS INCOMPRESSIBLE JET AND FREE-SURFACE FLOWS USING FINITE ELEMENT METHODS	74-0357
VISCOUS	STEADY FLOW ANALYSIS OF INCOMPRESSIBLE VISCOUS FLOW BY THE FINITE ELEMENT METHOD	73-0332
VISCOUS	THE APPLICATION OF THE PSEUDO-FUNCTIONAL FINITE ELEMENT METHOD TO VISCOUS FLOW PROBLEMS	74-0287
VISCOUS	THEORETICAL AND EXPERIMENTAL CONSIDERATIONS ON FLOW OF VISCOUS LIQUIDS PAST PLATE AT SMALL AND AVERA	70-0144
VISCOUS	VISCOUS, INCOMPRESSIBLE FLOW WITH SPECIAL REFERENCE TO NON-NEWTONIAN (PLASTIC) FLUIDS	74-0053
VISCOUS	WHIRLING CABLE SUBJECTED TO VISCOUS DRAG	74-0457
VISCOUS-FLOW	FINITE-ELEMENT ANALYSIS OF COMPRESSIBLE AND INCOMPRESSIBLE VISCOUS-FLOW AND HEAT-TRANSFER PROBLEMS	75-0347
VISCOUS-FLOW	SEEPAGE IN MISSISSIPPI RIVER BANKS. REPORT 1. ANALYSIS OF TRANSIENT SEEPAGE USING A VISCOUS-FLOW MOD	73-0909
VISUALIZATION	FINITE ELEMENT ANALYSIS AND COMPUTER GRAPHICS VISUALIZATION OF FLOW AROUND PITCHING AND PLUNGING AIR	73-0385
VIVO	IN VIVO STRESSES IN THE HUMAN LEFT VENTRICULAR WALL: ANALYSIS ACCOUNTING FOR THE IRREGULAR THREE DIM	72-0136
VLCCS	DYNAMIC AND STOCHASTIC ASPECTS IN THE STRUCTURAL ANALYSIS OF VLCCS AND ULCCS	74-0298
VNUTRENNEGO	METODY RASCHETNOGO OPREDELENIYA TEMPERATURNYKH NAPRYAZHENII V KRYSHKAKH TSILINDROV DVIGATELEI VNUTRE	74-0179
VNUTRENNEGO	OPREDELENIE STATSIONARNYKH TEMPERATURNYKH POLEI V DETALYAKH DVIGATELEI VNUTRENNEGO SGORANIYA METODOM	73-0454
VNUTRENNEGO	RASCHET POLEI DEFORMATSII I NAPRYAZHENII V DETALYAKH DVIGATELEI VNUTRENNEGO SGORANIYA (CALCULATION O	73-0465
VODONASYSHCHENN	METOD KONECHNYKH ELEMENTOV V RASCHETAKH KONSOLIDATSII VODONASYSHCHENNYKH GRUNTOV. (FINITE ELEMENT ME	75-0469
VOGT	VOGT BOUNDARY FOR FINITE ELEMENT ARCH DAM ANALYSIS	74-0156
VOID	YIELDING OF A TWO DIMENSIONAL VOID ASSEMBLY IN AN ORGANIC GLASS	73-0416
VOID-GROWTH	DEFORMATION MODE OF VOID-GROWTH AND COALESCENCE IN THE PROCESS OF DUCTILE FRACTURE	75-0683
VOIDED	FINITE ELEMENT ANALYSIS OF PRESTRESSED CONCRETE VOIDED BRIDGE DECKS	73-0106
VOIDS	SIMULATION ON THE INITIATION AND GROWTH OF DUCTILE FRACTURE VOIDS	75-0836
VOILES	ANALYSE DES VOILES MINCES DANS LES DOMAINES LINEAIRES ET GEOMETRIQUEMENT NONLINEAIRES PAR LA METHODE	71-0122
VOISINAGE	ANALYSE TRIDIMENSIONELLE DE LA STABILITE D'UN TUNNEL AU VOISINAGE DU FRONT DE TAILLE DANS UNE ROCHE	74-0995
VOLCANISM	FINITE-ELEMENT MODELING OF SURFACE DEFORMATION ASSOCIATED WITH VOLCANISM	75-0596
VOLLRAEDERN	UNTERSUCHUNGEN ZUR SPURMASSTABILITAET BEI RADSAETZEN MIT KLOTZGEBREMSTEN VOLLRAEDERN (INVESTIGATIONS	75-0810
VOLLSTAENDIGE	VOLLSTAENDIGE SPANNUNGS-DEHNUNGS-TEMPERATURAENDERUNGSBEZIEHUNG FUER DIE SCHWEISSEIGENSPANNUNGSBERECH	75-0807
VOLLVERTRAGLICH	A NEW, FULLY COMPATIBLE FINITE ELEMENT FOR PLATE BENDING EIN NEUES VOLLVERTRAGLICHES ENDLICHES ELEME	68-0032
VOLS	ICES STRUDL-II ENGINEERING USERS MANUAL. VOLS. 1, 11, 111	69-0398
VOLUME	A DESIGN PROCEDURE FOR HIGH CUTS IN JOINTED HARD ROCK. THREE DIMENSIONAL SOLUTIONS. VOLUME 1	71-0656
VOLUME	A FINITE ELEMENT FORMULATION FOR MATERIALS WITH SPECIFIED CHANGE OF VOLUME	68-0044
VOLUME	A NASTRAN VIBRATION MODEL OF THE AH-IG HELICOPTER AIRFRAME. VOLUME 1	74-1338
VOLUME	A NASTRAN VIBRATION MODEL OF THE AH-IG HELICOPTER AIRFRAME. VOLUME 11	74-1339
VOLUME	ADVANCES IN APPLIED MECHANICS, VOLUME 12.	72-0375
VOLUME	ANALYTIC MODELING OF ROCK-STRUCTURE INTERACTION. VOLUME 3. COMPUTER PROGRAM	73-0918
VOLUME	ANALYTIC MODELING OF ROCK-STRUCTURE INTERACTION. VOLUME 2. USERS GUIDE FOR A COMPUTER PROGRAM	73-0919
VOLUME	ANALYTIC MODELING OF ROCK-STRUCTURE INTERACTION. VOLUME 1	73-0920

VOLUME	ANALYTICAL AND EXPERIMENTAL INVESTIGATION OF A 1/8 SCALE DYNAMIC MODEL OF THE SHUTTLE ORBITER. VOLUM	74-1186
VOLUME	ASKA 105 - A PROBLEM ORIENTED LANGUAGE FOR STRUCTURAL ENGINEERS, VOLUME 11, APPLICATION	69-0461
VOLUME	ASKA 105, A PROBLEM ORIENTED LANGUAGE FOR STRUCTURAL ENGINEERS, VOLUME 1, DESIGN	69-0460
VOLUME	ASSESSMENT OF DESIGN TOOLS AND CRITERIA FOR URBAN RAIL TRACK STRUCTURES, VOLUME 11. AT-GRADE SLAB TR	74-0813
VOLUME	BASIC RESEARCH IN METHODS OF APPROXIMATION IN THE NONLINEAR MECHANICS OF SOLIDS AND STRUCTURES, VOLU	74-1091
VOLUME	COMPOSITE WING FOR TRANSONIC IMPROVEMENT. VOLUME 11. ADVANCED ANALYSIS EVALUATION	71-0657
VOLUME	CYCLIC FATIGUE ANALYSIS OF ROCKET THRUST CHAMBERS. VOLUME 2: ATTITUDE CONTROL THRUSTER HIGH CYCLE FA	74-0693
VOLUME	CYCLIC FATIGUE ANALYSIS OF ROCKET THRUST CHAMBERS. VOLUME 1: OFHC COPPER CHAMBER LOW CYCLE FATIGUE	74-0694
VOLUME	DESIGN, MANUFACTURE, DEVELOPMENT, TEST, AND EVALUATION OF BORON/ALUMINUM STRUCTURAL COMPONENTS FOR S	74-0798
VOLUME	DETERMINATION OF VOLUME CHANGE STRESSES IN PLAIN AND REINFORCED CONCRETE USING FINITE ELEMENT ANALYS	69-0477
VOLUME	DEVELOPMENT OF AN UNDERSTANDING OF THE FATIQUE PHENOMENA OF BONDED AND BOLTED JOINTS IN ADVANCED FIL	72-0798
VOLUME	DYNAMICS OF SUBMARINE STRUCTURES. VOLUME 1	68-0251
VOLUME	ELAS: A GENERAL-PURPOSE COMPUTER PROGRAM FOR THE EQUILIBRIUM PROBLEMS OF LINEAR STRUCTURES. VOLUME 2	69-0462
VOLUME	FATIQUE STRENGTH OF LUGS CONTAINING LINERS. VOLUME II. COMPUTER PROGRAM USED FOR ANALYSIS	70-0418
VOLUME	FEDGE - A GENERAL PURPOSE COMPUTER PROGRAM FOR FINITE ELEMENT DATA GENERATION. VOLUME 2 - PROGRAM MA	69-0492
VOLUME	FEDGE - A GENERAL-PURPOSE COMPUTER PROGRAM FOR FINITE ELEMENT DATA GENERATION. VOLUME 1 - USER'S MAN	69-0491
VOLUME	HYDROLOGIC ENGINEERING METHODS FOR WATER RESOURCES DEVELOPMENT. VOLUME 10. PRINCIPLES OF GROUND-WATE	72-0696
VOLUME	INTEGRATION OF AREA AND VOLUME COORDINATES IN FINITE-ELEMENT METHOD	69-0370
VOLUME	MAGIC 11: AN AUTOMATED GENERAL PURPOSE SYSTEM FOR STRUCTURAL ANALYSIS. VOLUME 111. PROGRAMMER'S MANU	71-0756
VOLUME	MAGIC 11: AN AUTOMATED GENERAL PURPOSE SYSTEM FOR STRUCTURAL ANALYSIS. VOLUME 1. ENGINEER'S MANUAL (	71-0757
VOLUME	MAGIC 111: AN AUTOMATED GENERAL PURPOSE SYSTEM FOR STRUCTURAL ANALYSIS. VOLUME 111. PROGRAMMER'S MAN	72-0726
VOLUME	MAGIC 111: AN AUTOMATED GENERAL PURPOSE SYSTEM FOR STRUCTURAL ANALYSIS. VOLUME 1. ENGINEER'S MANUAL	72-0727
VOLUME	MAGIC, AN AUTOMATED GENERAL PURPOSE SYSTEM FOR STRUCTURAL ANALYSIS. VOLUME III. PROGRAMMER'S MANUAL	69-0090
VOLUME	MAGIC, AN AUTOMATED GENERAL PURPOSE SYSTEM FOR STRUCTURAL ANALYSIS. VOLUME II. USER'S MANUAL	69-0144
VOLUME	MATHEMATICAL ANALYSIS OF STRESS CRACKS. VOLUME 2.	75-0742
VOLUME	MATHEMATICAL ANALYSIS OF STRESS CRACKS. VOLUME 1	75-0743
VOLUME	MATHEMATICAL MODEL OF THE CORNEO-SCLERAL SHELL AS APPLIED TO INTRAOCULAR PRESSURE VOLUME RELATIONS A	72-0471
VOLUME	MECHANICS OF CABLE MOORING SYSTEMS. VOLUME IV. A COMPUTER PROGRAM FOR ANALYZING THE STEADY STATE RES	72-0653
VOLUME	MECHANICS OF CABLE MOORING SYSTEMS. VOLUME 1. THREE DIMENSIONAL RESPONSE OF DEEP WATER MOORING LINES	72-0654
VOLUME	MECHANICS OF CABLE MOORING SYSTEMS. VOLUME V11. THE STEADY-STATE BEHAVIOR OF A PYRAMID ARRAY SYSTEM	73-0724
VOLUME	MECHANICS OF CABLE MOORING SYSTEMS. VOLUME V1. A COMPUTER PROGRAM FOR ANALYZING THE STEADY STATE CON	73-0725
VOLUME	METHODS FOR PREDICTING THE AERODYNAMIC AND STABILITY AND CONTROL CHARACTERISTICS OF STOL AIRCRAFT. V	73-0718
VOLUME	METHODS FOR PREDICTING THE AERODYNAMIC AND STABILITY AND CONTROL CHARACTERISTICS OF STOL AIRCRAFT. V	73-0719
VOLUME	METHODS OF PREDICTING THE AERODYNAMIC AND STABILITY AND CONTROL CHARACTERISTICS OF STOL AIRCRAFT. VO	73-0717
VOLUME	MODAL COUPLING PROCEDURES ADAPTED TO NASTRAN ANALYSIS OF THE 1/8 - SCALE SHUTTLE STRUCTURAL DYNAMICS	75-0677
VOLUME	MODAL COUPLING PROCEDURES ADAPTED TO NASTRAN ANALYSIS OF THE 1/8-SCALE SHUTTLE STRUCTURAL DYNAMICS M	75-0678
VOLUME	OPTIM 11: A MAGIC COMPATIBLE LARGE-SCALE AUTOMATED MINIMUM WEIGHT DESIGN PROGRAM. VOLUME 1. ENGINEER	74-1096
VOLUME	OPTIM 11: MAGIC COMPATIBLE LARGE SCALE AUTOMATED MINIMUM WEIGHT DESIGN PROGRAM. VOLUME 11. PROGRAMME	74-1093
VOLUME	SINGER: A COMPUTER CODE FOR GENERAL ANALYSIS OF TWO-DIMENSIONAL REINFORCED CONCRETE STRUCTURES. VOLU	75-0724
VOLUME	SINGER: A COMPUTER CODE FOR GENERAL ANALYSIS TO TWO-DIMENSIONAL CONCRETE STRUCTURES. VOLUME 4. DEMON	75-0730
VOLUME	SINGER: A COMPUTER CODE FOR GENERAL ANALYSIS OF TWO-DIMENSIONAL REINFORCED CONCRETE STRUCTURES. VOLU	75-0731
VOLUME	STRESS WAVE ANALYSIS IN LAYERED THERMOVISCOELASTIC MATERIALS BY THE EXTENDED RITZ METHOD. VOLUME II	68-0125
VOLUME	STRUCTURAL BEHAVIOR OF A TWO SPAN REINFORCED CONCRETE BOX GIRDER BRIDGE MODEL. VOLUME 11. REDUCTION,	71-0659
VOLUME	STRUCTURE-MEDIUM INTERACTION AND DESIGN PROCEDURES STUDY. VOLUME 1. ANALYSIS METHOD, THEORY, VERIFIC	69-0166
VOLUME	STRUCTURE-MEDIUM INTERACTION AND DESIGN PROCEDURES STUDY. VOLUME IV. FEAT CODE USER'S GUIDE	69-0265
VOLUME	TRANSPIRATION AND FILM COOLING BOUNDARY LAYER COMPUTER PROGRAM. VOLUME 2. COMPUTER PROGRAM AND USER'	71-0655
VOLUME	USER'S GUIDE FOR THE COMPUTER PROGRAM DESCRIBED IN TAM REPORT NO. 680 "DETERMINATION OF VOLUME CHANG	69-0473
VOLUME-DISPLACE	COMPUTATION OF THE RESPONSE OF COUPLED PLATE ACOUSTIC SYSTEMS USING PLATE FINITE ELEMENTS AND ACOUST	71-0072
VOPROSU	K VOPROSU OB AVTOMATIZATSII ZADANIYA INFORMATSII V METODE KONECHNYKH ELEMENTOV. (PROBLEM OF AUTOMATI	75-0382
VORAUSSAGE	VORAUSSAGE DES GEBIRGSDRUCKES UND SEINER AUSWIRKUNGEN AUF STREBEN UND STRECKEN IM STEINKOHLENBERGBAU	73-0679
VORGESPANNTEN	EIN VERFAHREN ZUR DIE FORMFINDUNG VEON BELIEBIGEN VORGESPANNTEN NETZWERKKONSTRUKTIONEN	72-0641
VORGESPANNTER	BERECHNUNG VORGESPANNTER NETZWERKE	70-0399
VORTEX	JET-WING LIFTING - SURFACE THEORY USING ELEMENTARY VORTEX DISTRIBUTIONS	75-0245
VORTICITY	AN APPLICATION OF LEAST SQUARES FINITE ELEMENT METHODS TO TWO DIMENSIONAL STEADY INVISCID COMPRESSIB	75-0114
VULNERABILITY	AIRCRAFT FUEL TANK VULNERABILITY TO HYDRAULIC RAM: MODIFICATION OF THE NORTHROP FINITE ELEMENT COMP	74-0724
VYNUZHDENNYKH	CHISLENNYI METOD RASCHETA SOBSTVENNYKH I VYNUZHDENNYKH KOLEBANII SOSTAVNYKH OBOLOCHECHNYKH KONSTRUKT	73-0461
VYPOCET	VYPOCET KRUHOVYCH LOMENICOVYCH KONSTRUKCII METODOU KONECNYCH PASOV. (CALCULATION OF CIRCULAR FOLDED-	74-0612
VYUZITI	VYUZITI METODY KONECNYCH PRVKU PRO PROJEKTOVANI KOMBINOVANYCH KONSTRUCKCI (FINITE ELEMENT METHOD IDE	73-0667
V11	MECHANICS OF CABLE MOORING SYSTEMS. VOLUME V11. THE STEADY-STATE BEHAVIOR OF A PYRAMID ARRAY SYSTEM	73-0724
V11	STRUCTURAL INELASTICITY. V11. ELASTIC-PLASTIC ANALYSIS OF A WHEEL ROLLING ON A RIGID TRACK	73-0886
WAENDEN	ZUM TRAGVERHALTEN VON WAENDEN AUS FERTIGTEILTAFELN (BEHAVIOR OF LARGE PANEL SHEAR WALLS)	74-0299
WAERMEBEANSPRUC	SPANNUNGEN IN MEMBRANWAENDEN DURCH WAERMEBEANSPRUCHUNG UND MECHANISCHE BELASTUNG. (STRESSES IN MEMBR	75-0066
WAERMELEITPROBL	HYBRIDE FINITE VERFAHREN FUER LINEARE UND NICHTLINEARE INSTATIONAERE WAERMELEITPROBLEME (HYBRID FINI	73-0484
WAERMEUEBERGANG	TEMPERATURVERLAUF IM QUERSCHNITT GESCHWEISSTER ROHRWAENDE BEI UEBERWIEGEND KONVEKTIVEM WAERMEUEBERGA	75-0628
WAERMEUEBERTRAG	FINITE ELEMENTE ZUR BERECHNUNG DER WAERMEUEBERTRAGUNG FUER EBENE UND AXISYMMETRISCHE PROBLEME (FINIT	73-0449
WAIRAKEI	GALERKIN-FINITE ELEMENT ANALYSIS OF HYDROTHERMAL SYSTEM AT WAIRAKEI, NEW-ZEALAND	75-0601
WALL	ANALYTICAL PROBLEMS IN MODELING SLURRY WALL CONSTRUCTION	73-0764
WALL	BASELINE SOLUTIONS FOR THE SMOOTH WALL THERMOCHEMICAL ABLATION RESPONSE OF GRAPHITE AND CARBON PHENO	71-0728
WALL	BENDING MOMENTS IN LONG WALL TANKS	67-0110
WALL	COMPARISON BETWEEN THEORETICAL AND MEASURED EARTH PRESSURES ACTING ON A LARGE MOTORWAY RETAINING WAL	74-0917
WALL	COMPARISON OF THE THREE-DIMENSIONAL ANALYSIS OF A SHEAR WALL BUILDING BY TWO DIFFERENT METHODS	75-0705
WALL	COUPLED SHEAR WALL ANALYSIS BY LAGRANGE MULTIPLIERS	75-0202
WALL	DISTRIBUZIONI OTTIMALI DI ARMATURE DI PRECOMPRESSIONE IN TRAVI-PARETE DI CALCESTRUZZO. (OPTIMAL DIST	75-0560
WALL	FINITE ELEMENT ANALYSES OF RETAINING WALL BEHAVIOUR	71-0320
WALL	FINITE ELEMENT CALCULATION OF LEFT VENTRICULAR WALL STRESS IN NORMAL AND DISEASE STATES	73-0588
WALL	IN VIVO STRESSES IN THE HUMAN LEFT VENTRICULAR WALL: ANALYSIS ACCOUNTING FOR THE IRREGULAR THREE DIM	72-0136
WALL	INTERACTION OF FRAME AND SHEAR WALL WITH OPENINGS	74-1190
WALL	NEW RECTANGULAR FINITE ELEMENT FOR SHEAR WALL ANALYSIS	69-0182
WALL	NEW RECTANGULAR FINITE ELEMENT FOR SHEAR WALL ANALYSIS	69-0362
WALL	NUMERICAL PROCEDURES FOR THE DETERMINATION OF THE BEHAVIOUR OF A SHEAR WALL FRAME SYSTEM	73-0216
WALL	ON FLOW NEAR SINGULAR POINTS OF A WALL BOUNDARY	74-0008
WALL	PRESSURE VESSEL WALL THICKNESS LIMITS BASED ON THERMAL STRESS RACHETTING AND CREEP FATIGUE INTERACTI	74-0968
WALL	SHOCK TUNNEL TESTS OF PRELOADED AND ARCHED WALL PANELS FINITE ELEMENT STRESS FORMULATION FOR WAVE PR	73-0892
WALL	STRUCTURAL ANALYSIS OF WELDED WALL FOR BOILER FURNACE	74-0653
WALL	STRUCTURAL RESPONSE AND LOADING OF WALL PANELS	71-0735
WALL	TEMPERATURES AND STRESSES IN A BOILER MEMBRANE WALL TUBE	74-1053
WALL	THREE DIMENSIONAL ANALYSIS OF BUILDINGS COMPOSED OF FLOOR AND WALL PANELS	71-0387
WALL	THREE-DIMENSIONAL ANALYSIS OF SHEAR WALL BUILDINGS TO LATERAL LOAD	74-0220
WALL-FRAME	FINITE-ELEMENT APPROACH FOR THE PREDICTION OF INELASTIC BEHAVIOR OF SHEAR WALL-FRAME SYSTEMS	72-0983
WALLACE	DESTRESSING TEST AT THE GALENA MINE, WALLACE, IDAHO	72-0126
WALLED	A FINITE ELEMENT FORMULATION FOR THE GENERAL STABILITY ANALYSIS OF THIN WALLED MEMBERS	70-0337
WALLED	AN ASSUMED STRESS HYBRID FINITE ELEMENT MODEL FOR THE ANALYSIS OF AN AXISYMMETRIC THICK WALLED PRESS	71-0025
WALLED	ANALYSIS OF THREE-DIMENSIONAL THIN WALLED STRUCTURES	69-0188
WALLED	APPLICATION OF FRACTURE MECHANICS TO FAILURE OF THICK WALLED CYLINDERS	71-0511
WALLED	CALCULATION OF THE CRITICAL LOAD OF ELASTIC THIN WALLED BARS BY THE METHOD OF FINITE ELEMENTS	72-0446
WALLED	ELASTIC DEFLECTION IN THIN WALLED CONES ROTATING ABOUT THEIR AXES OF SYMMETRY	73-0545
WALLED	ELASTIC PLASTIC DEFORMATION OF THICK WALLED CYLINDERS	69-0427
WALLED	ELASTIC-PLASTIC ANALYSIS OF THICK WALLED PRESSURE VESSELS WITH SHARP DISCONTINUITIES	71-0155
WALLED	FINITE ELEMENT ANALYSIS OF A THICK WALLED MICROPOLAR CYLINDER LOADED AXISYMMETRICALLY	73-0847
WALLED	FINITE ELEMENT FOR TORSION OF THIN WALLED OPEN TUBES	71-0308
WALLED	FLEXUAL RIGIDITY OF A THIN WALLED BUILD-UP ROTOR FOR JET ENGINE. MEASUREMENT BY STATIC LOAD TEST AND	73-0756
WALLED	FORCES AND DEFORMATIONS AT THE LATERAL STIFFENER OF THIN WALLED U-BARS	73-0236

444

WALLED	LARGE DEFLECTION SPATIAL BUCKLING OF THIN WALLED BEAMS AND FRAMES	73-1085
WALLED	NATURAL VIBRATIONS OF THIN, FLAT WALLED STRUCTURES WITH DIFFERENT BOUNDARY CONDITIONS	71-0509
WALLED	THICK WALLED MULTIPLE OPENING REINFORCED CONCRETE CONDUITS	72-0666
WALLED	THREE DIMENSIONAL STRESS ANALYSIS OF PENETRATIONS IN THICK WALLED PRESSURE VESSELS	74-0422
WALLS	A COMPUTER PROGRAM FOR THE CALCULATION OF TORSION FACTORS AND SHEAR STRESS FACTORS FOR BOX SECTIONS	69-0485
WALLS	A FINITE-ELEMENT METHOD FOR ANALYSIS OF ANCHORED BULKHEADS AND ANCHOR WALLS	66-0130
WALLS	A FINITE-ELEMENT MODEL TO SIMULATE THE NONLINEAR RESPONSE OF REINFORCED CONCRETE FRAMES WITH MASONRY	73-1168
WALLS	ANALYSIS OF FRAMES WITH SHEAR WALLS BY FINITE ELEMENTS	69-0199
WALLS	ANALYSIS OF FRAMES WITH SHEAR WALLS WITH FINITE ELEMENTS	69-0500
WALLS	ANALYSIS OF IN-PLANE VIBRATION OF SHEAR WALLS BY A FINITE ELEMENT METHOD	72-0090
WALLS	ANALYSIS OF SHEAR WALLS WITH OPENINGS	69-0111
WALLS	ANALYSIS OF SHEAR WALLS BY FINITE ELEMENT METHOD	69-0112
WALLS	ANALYSIS OF SHEAR WALLS BY FINITE ELEMENTS	71-0569
WALLS	ANALYSIS OF STRESS DISTRIBUTION IN SHEAR WALLS BY THE FINITE ELEMENT DISPLACEMENT METHOD	67-0033
WALLS	BEHAVIOR OF ISOLATED SHEAR WALLS SUBJECTED TO LATERAL LOADS	75-0704
WALLS	BERECHNUNG VON SPANNUNGEN UND TEMPERATURFELDERN IN FLOSSENWAENDEN UND AUFGESCHWEISSTEN BAUTEILEN. (C	74-0544
WALLS	CONSTRUCTION SELF-WEIGHT STRESSES OF PRECAST SHEAR WALLS	71-0203
WALLS	CONTRIBUTION TO NUMERICAL-SOLUTION OF DEVELOPING LAMINAR-FLOW IN ENTRANCE REGION OF CONCENTRIC ANNUL	75-0448
WALLS	COUPLED SHEAR WALLS. IMPROVED METHODS FOR ELASTIC ANALYSIS	74-0818
WALLS	DYNAMIC BEHAVIOR OF IN-LINE SHEAR WALLS CONNECTED BY FLOOR SLABS	72-0689
WALLS	EARTH PRESSURES AGAINST THE ABUTMENTS AND WING WALLS OF STANDARD MOTORWAY BRIDGES	75-0529
WALLS	EFFECTS OF SHEAR WALLS ON DYNAMIC RESPONSE OF FRAMES	69-0230
WALLS	ELASTIC ANALYSIS OF SPATIAL SYSTEMS OF INTERCONNECTED SHEAR WALLS AND FRAMES	74-1195
WALLS	FINITE ELEMENT FOR THE INTERACTION OF FRAMES AND SHEAR WALLS	73-0681
WALLS	FINITE ELEMENT STUDY OF COMPOSITE ACTION IN WALLS	69-0291
WALLS	FINITE ELEMENT STUDY OF THE ELASTIC BEHAVIOR OF PLANE FRAMES WITH FILLER WALLS	71-0828
WALLS	FINITE ELEMENTS FOR VIBRATION OF FRAMED SHEAR WALLS	73-0036
WALLS	HIGHER ORDER FINITE ELEMENT METHOD FOR STATIC AND DYNAMIC ANALYSIS OF SHEAR WALLS	71-0368
WALLS	INTERACTION OF SOLID SHEAR WALLS AND THEIR SUPPORTING STRUCTURES.	72-0450
WALLS	LINEAR ELASTIC FINITE ELEMENT ANALYSIS OF MASONRY WALLS ON BUILDINGS	74-1015
WALLS	PERFORMANCE OF TIED-BACK WALLS IN CLAY	74-0665
WALLS	SPANNUNGEN IN MEMBRANWAENDEN DURCH WAERMEBEANSPRUCHUNG UND MECHANISCHE BELASTUNG. (STRESSES IN MEMBR	75-0066
WALLS	STRUCTURAL ANALYSIS OF TALL BUILDINGS HAVING IRREGULARLY POSITIONED SHEAR WALLS	73-0282
WALLS	TEMPERATURVERLAUF IM QUERSCHNITT GESCHWEISSTER ROHRWAENDE BEI UEBERWIEGEND KONVEKTIVEM WAERMEUEBERGA	75-0628
WALLS	THEORETICAL STUDY OF THE SEQUENCE OF FAILURE IN PRECAST PANEL SHEAR WALLS	73-0490
WALLS	TIED-BACK RETAINING WALLS	75-0308
WALLS	ZUM TRAGVERHALTEN VON WAENDEN AUS FERTIGTEILTAFELN (BEHAVIOR OF LARGE PANEL SHEAR WALLS)	74-0299
WALOW	ZASTOSOWANIE METODY SZTYWNYCH ELEMENTOW SKONCZONYCH DO OBLICZEN ORGAN WALOW OKRETOYCH Z UWZGLEDNIENI	74-0962
WARPED	A SYNTHESIZED WARPED QUADRILATERAL ELEMENT GIVING SIMPLIFIED INPUT FOR EQUILIBRIUM MODELS	73-1013
WARPED	A WARPED QUADRILATERAL STRAIN MEMBRANE ELEMENTS	75-0141
WARPED	A WARPED QUADRILATERAL EQUILIBRIUM MEMBRANE ELEMENT	75-0885
WARPED	WARPED SEMI-MONOCOQUE QUADRILATERAL MEMBRANE ELEMENT FOR THE FINITE ELEMENT METHODS	70-0195
WARPING	FLEXIBILITY AND STIFFNESS MATRICES FOR AN OPEN-TUBE WARPING CONSTRAINT FINITE ELEMENT	72-0076
WARPING	WARPING AND DISTORTION AT I-SECTION JOINTS	73-0722
WARUNKOW	FUNKCJA KSZTALTU NIESPELNIAPACA WARUNKOW CIAGLOSCI W ANALIZIE PLYT METODA ELEMENTOW SKONCZONYCH. (SH	75-0263
WASHINGTON	NAVY-NASTRAN COLLOQUIUM, WASHINGTON, D.C.	70-0403
WASHINGTON	NAVY-NASTRAN COLLOQUIM AT WASHINGTON, D.C.	72-0669
WASHINGTON	NSRDC: NASTRAN COLLOQUIUM, WASHINGTON, D.C.	70-0404
WASHINGTON	PROCEEDINGS OF THE NAVY-NASTRAN COLLOQUIUM (4TH) HELD AT NAVY SHIP RESEARCH AND DEVELOPMENT CENTER,	73-0888
WASSERGESAETTIG	ANWENDUNG DER FINITE-ELEMENT-METHODE ZUR BERECHNUNG VON SPANNUNGEN IN WASSERGESAETTIGTEN BOEDEN (FIN	73-0502
WASTE	A FINITE ELEMENT MODEL FOR STABILITY ANALYSIS OF MINE WASTE EMBANKMENTS SUBJECTED TO SEEPAGE	74-1166
WASTES	FINITE ELEMENT ANALYSES OF SALT DOMES WITH STORED HOT WASTES	74-1148
WATER	A NUMERICAL MODEL OF TRANSIENT WATER QUALITY IN A ONE-DIMENSIONAL ESTUARY BASED ON THE FINITE ELEMEN	74-1369
WATER	ANALYSIS OF WATER WAVE PROBLEMS BY THE METHOD OF SUPERPOSITION OF ANALYTICAL AND FINITE ELEMENT SOLU	75-0870
WATER	APPLICATION OF FINITE ELEMENT METHOD TO BOILING WATER REACTOR POWER PLANT (JAPANESE)	72-0816
WATER	APPLICATION OF THE FINITE-ELEMENT METHOD FOR SIMULATION OF SURFACE WATER TRANSPORT PROBLEMS	72-0826
WATER	APPLICATION OF THE FINITE ELEMENT METHOD TO PROBLEMS OF TEMPERATURE DISTRIBUTION ANALYSIS FOR BODIES	74-1117
WATER	BOTTOM STABILITY UNDER WATER INDUCED LOADING	72-0373
WATER	COMPARATIVE STUDY OF FRESH-SALT WATER INTERFACES USING FINITE ELEMENT AND SIMPLE APPROACHES	75-0468
WATER	EARTHQUAKE ANALYSIS OF AXISYMMETRIC TOWERS PARTIALLY SUBMERGED IN WATER	75-0076
WATER	EFFECT OF SEEPAGE ON EMBANKMENT DEFORMATIONS DUE TO WATER LOADING	74-1329
WATER	EMBANKMENT DEFORMATIONS DUE TO WATER LOADS	73-0782
WATER	FINITE ELEMENT ANALYSIS OF LONG-PERIOD WATER WAVES	73-1008
WATER	FINITE ELEMENT ANALYSIS OF 2-DIMENSIONAL FLOW IN SOILS CONSIDERING WATER UPTAKE BY ROOTS .1. THEORY	75-0137
WATER	FINITE ELEMENT ANALYSIS OF 2-DIMENSIONAL FLOW IN SOILS CONSIDERING WATER UPTAKE BY ROOTS .2. FIELD A	75-0147
WATER	FINITE ELEMENT SOLUTIONS OF THE SHALLOW WATER EQUATIONS SURFACE RUN OFF	74-0046
WATER	HYBRID-ELEMENT METHOD FOR WATER WAVES	75-0808
WATER	HYDROLOGIC ENGINEERING METHODS FOR WATER RESOURCES DEVELOPMENT. VOLUME 10. PRINCIPLES OF GROUND-WATE	72-0696
WATER	INFLUENCE OF RESERVOIR WATER ON EARTHQUAKE RESPONSE OF EARTH DAMS	70-0375
WATER	INVESTIGATION OF THE WATER TABLE IN A TIDAL BEACH	71-0648
WATER	MANAGEMENT OF RETARDATION OF SALT WATER INTRUSION IN COASTAL AQUIFERS	75-0692
WATER	MATHEMATICAL MODEL OF HEATED EFFLUENTS IN COASTAL WATER	73-0921
WATER	MATHEMATICAL MODELING OF MOVEMENT OF DISSOLVED CONSTITUENTS IN GROUND WATER AQUIFERS BY THE FINITE E	70-0490
WATER	MATHEMATICAL MODELS OF THE MASSACHUSETTS BAY. PART 1. FINITE ELEMENT MODELING OF TWO-DIMENSIONAL HYD	73-0800
WATER	MEASUREMENT OF THE WATER DIFFUSIVITY OF POROUS MATERIALS	74-1138
WATER	MECHANICS OF CABLE MOORING SYSTEMS. VOLUME 1. THREE DIMENSIONAL RESPONSE OF DEEP WATER MOORING LINES	72-0654
WATER	METOD KONECHNYKH ELEMENTOV V RASCHETAKH KONSOLIDATSII VODONASYSHCHENNYKH GRUNTOV. (FINITE ELEMENT ME	75-0469
WATER	NUMERICAL DETERMINATION OF SEA WATER INTRUSION IN AQUIFERS	74-1181
WATER	NUMERICAL MODEL FOR THE PREDICTION OF TRANSIENT WATER QUALITY IN ESTUARY NETWORKS	72-0635
WATER	OBLICZANIE KONSTRUKCJI TOROIDALNEGO ZBIORNIKA NA WODE. (STATICAL ANALYSIS OF A TOROIDAL WATER RESERV	75-0211
WATER	OPTIMAL DESIGN OF WATER WELLS IN UNCONSOLIDATED SEDIMENTS	74-1182
WATER	SIMULATION OF GROUND WATER CONTAMINATION USING A GALERKIN FINITE ELEMENT TECHNIQUE	74-1362
WATER	SOLUTION OF WATER PREDICTION PROBLEMS BY FINITE ELEMENT METHODS	74-1345
WATER	STUDIES ON THE DETERMINATION OF SEEPAGE-LINE OF FILL DAM BY THE METHOD OF FINITE ELEMENT. WATER FLOW	74-1348
WATER	THE DETERMINATION OF A WATER/GAS INTERFACE USING FINITE ELEMENTS	74-0346
WATER	TWO-DIMENSIONAL FINITE ELEMENT MODELING OF TRANSIENT FLOW IN REGIONAL AQUIFER SYSTEMS. ANALYSIS OF G	74-0743
WATER	WATER FLOW THROUGH SATURATED-UNSATURATED POROUS MEDIA USING FINITE-ELEMENT-GALERKIN METHODS	74-0879
WATER	WATER IMPACT ANALYSIS OF SPACE SHUTTLE SOLID ROCKET BY THE FINITE ELEMENT METHOD	74-0760
WATER	WATER MOVEMENT THROUGH SATURATED-UNSATURATED POROUS MEDIA: A FINITE-ELEMENT GALERKIN MODEL	75-0761
WATER-DISTRIBUT	FINITE-ELEMENT METHOD FOR WATER-DISTRIBUTION NETWORKS	75-0269
WATER-POWER	THREE-DIMENSIONAL ISOPARAMETRIC ELEMENTS AND THEIR USE FOR STATICAL ANALYSIS OF MASSIVE WATER-POWER	71-0364
WATER-TABLE	CALCULATION OF STEADY-STATE WATER-TABLE HEIGHTS IN DRAINED SOILS BY MEANS OF FINITE-ELEMENT METHOD	75-0457
WATER-WAVE	OSCILLATIONS AND WAVE FORCES IN AN OFFSHORE HARBOR - APPLICATIONS OF HYBRID FINITE ELEMENT METHOD TO	74-1024
WATERS	HYDRODYNAMICAL-NUMERICAL MODELS FOR COASTAL WATERS AND OPEN OCEAN AREAS	75-0571
WATERSHED	FINITE ELEMENT APPROACH TO WATERSHED RUNOFF	74-0234
WATERSHED	FINITE-ELEMENT APPROACH TO WATERSHED RUNOFF- DISCUSSION	75-0402
WATERSHEDS	SIMULATION OF RUNOFF HYDROGRAPHS FROM NATURAL WATERSHEDS BY FINITE-ELEMENT METHOD	73-1140
WAVE	A FINITE ELEMENT ANALYSIS OF WAVE PROPAGATION IN HUMAN SPINE	73-0792
WAVE	A NUMERICAL METHOD FOR THE ANALYSIS OF LONGITUDINAL ELASTIC PLASTIC STRESS WAVE PROPAGATION	73-0526
WAVE	A STUDY OF A DYNAMICAL FINITE ELEMENT ANALYSIS FOR APPLICATION TO AXIAL WAVE PROPAGATION PROBLEMS IN	71-0589
WAVE	ANALYSIS OF LONGITUDINAL WAVE IN A BAR EXHIBITING VISCO-ELASTIC BEHAVIOR OR WITH MATERIAL DAMPING	71-0780
WAVE	ANALYSIS OF ONE-DIMENSIONAL STRESS WAVE BY THE FINITE ELEMENT METHOD	71-0523

445

WAVE	ANALYSIS OF WATER WAVE PROBLEMS BY THE METHOD OF SUPERPOSITION OF ANALYTICAL AND FINITE ELEMENT SOLU	75-0870
WAVE	APPLICATION OF THE FINITE ELEMENT METHOD TO SURFACE WAVE ANALYSIS	74-0913
WAVE	COMPUTATION OF WAVE PROPAGATION BY MEANS OF THE FINITE ELEMENT METHOD	74-0032
WAVE	COUPLING MATRICES FOR HIGH ORDER FINITE ELEMENT ANALYSIS OF ACOUSTIC - WAVE PROPAGATION	72-0040
WAVE	DISCRETIZATION-DISPERSION IN FINITE ELEMENT MODELLING OF WAVE PROPAGATION IN SOLIDS	74-1253
WAVE	DYNAMIC FINITE ELEMENT ANALYSIS OF AXI-SYMMETRIC STRESS WAVE PROPAGATION IN SOIL-FILLED BIN	72-0622
WAVE	DYNAMIC FINITE ELEMENT METHOD CALCULATION OF ELASTIC WAVE PROPAGATION IN TWO DIMENSIONAL STRUCTURES	74-0166
WAVE	DYNAMIC SHELL THEORY COUPLING THICKNESS STRESS WAVE EFFECTS WITH GROSS STRUCTURAL RESPONSE	73-0379
WAVE	ELASTIC WAVE PROPAGATION, A COMPARISON BETWEEN FINITE ELEMENT PREDICTIONS AND EXACT SOLUTION	67-0076
WAVE	EVALUATION OF NUMERICAL TIME INTEGRATION METHODS AS APPLIED TO ELASTIC-PLASTIC DYNAMIC PROBLEMS INVO	74-0858
WAVE	FINITE ELEMENT ANALYSIS OF ELASTIC WAVE SCATTERING FROM DISCONTINUITIES	73-0547
WAVE	FINITE ELEMENT ANALYSIS FOR WAVE PROPAGATION IN ELASTIC-PLASTIC SOLIDS	73-0990
WAVE	FINITE ELEMENT ANALYSIS OF SURFACE WAVE PROBLEMS USING INTEGRAL EQUATIONS	74-0624
WAVE	FINITE ELEMENT ANALYSIS OF WAVE PROPAGATION IN OPEN CHANNELS	75-0570
WAVE	FINITE ELEMENT APPROACH TO STRESS WAVE PROBLEMS	67-0015
WAVE	FINITE ELEMENT APPROACH TO WAVE MOTIONS IN ELASTIC-PLASTIC CONTINUA	68-0041
WAVE	FINITE ELEMENT APPROACH TO PRESSURE WAVE ATTENUATION BY REACTOR FUEL SUBASSEMBLIES	75-0522
WAVE	FINITE ELEMENT METHOD FOR WAVE GUIDE PROBLEMS	75-0872
WAVE	FINITE ELEMENT SOLUTION FOR WAVE PROPAGATION IN LAYERED MEDIA CAUSED BY A NUCLEAR DETONATION	67-0016
WAVE	FINITE ELEMENT SOLUTIONS OF THE WAVE EQUATION	74-0216
WAVE	FINITE ELEMENT SOLUTION OF THE INTERACTION OF A PLANE ACOUSTIC BLAST WAVE AND A CYLINDRIC STRUCTURE	74-0790
WAVE	FINITE ELEMENT STRESS FORMULATION FOR WAVE PROPAGATION	71-0831
WAVE	FINITE ELEMENT STRESS FORMULATION FOR WAVE PROPAGATION	73-1043
WAVE	FINITE ELEMENT STUDY OF HARMONIC WAVE PROPAGATION IN PERIODIC STRUCTURES	74-0130
WAVE	FINITE ELEMENT STUDY OF PRESSURE WAVE ATTENUATION BY REACTOR FUEL SUBASSEMBLIES	75-0215
WAVE	FUNDAMENTAL STUDIES ON STEADY SHIP WAVE PROBLEMS BY THE FINITE ELEMENT METHOD	74-0564
WAVE	LINEAR WAVE PROPAGATION PROBLEMS AND THE FINITE ELEMENT METHOD	74-1340
WAVE	LINEAR WAVE PROPAGATION PROBLEMS AND THE FINITE ELEMENT METHOD	75-0395
WAVE	LOVE WAVE DISPERSION THROUGH IRREGULAR STRUCTURES USING FINITE-ELEMENT ALGORITHMS	75-0431
WAVE	MATHEMATICAL MODELS FOR GEOLOGICAL MATERIALS FOR WAVE PROPAGATION STUDIES	71-0740
WAVE	MODELING WAVE PROPAGATION WITH FINITE ELEMENT METHODS	75-0015
WAVE	MULTIPLE STRESS WAVE DISCONTINUITIES IN FINITE BEAMS AND PLATES	67-0114
WAVE	ON SPARSE MATRICES IN FINITE ELEMENT WAVE PROPAGATION	73-0290
WAVE	ON THE DEVELOPMENT OF A REFINED FINITE-ELEMENT FREE FROM DISCRETIZATION - DISPERSION FOR ELASTIC WAV	73-1148
WAVE	OSCILLATIONS AND WAVE FORCES IN A MAN-MADE HARBOUR IN THE OPEN SEA	74-0151
WAVE	OSCILLATIONS AND WAVE FORCES IN AN OFFSHORE HARBOR - APPLICATIONS OF HYBRID FINITE ELEMENT METHOD TO	74-1024
WAVE	SHOCK TUNNEL TESTS OF PRELOADED AND ARCHED WALL PANELS FINITE ELEMENT STRESS FORMULATION FOR WAVE PR	72-0892
WAVE	SLAMCODE: FINITE ELEMENT STRESS WAVE ANALYSIS	72-0603
WAVE	SOLUTION OF WAVE PROPAGATION PROBLEMS IN SATURATED MEDIUM	72-0784
WAVE	STRESS WAVE ANALYSIS IN LAYERED THERMOVISCOELASTIC MATERIALS BY THE EXTENDED RITZ METHOD. VOLUME II	68-0125
WAVE	THE APPLICATION OF THE FINITE ELEMENT METHOD TO A PROBLEM IN STRESS WAVE PROPAGATION	69-0505
WAVE	THE PREDICTION OF REGULAR WAVE FORCES ON FLOATING BODIES WITH THE FINITE ELEMENT METHOD (IN DUTCH)	71-0743
WAVE	TIDAL AND LONG WAVE PROPAGATION - A FINITE ELEMENT APPROACH	73-1059
WAVE	WAVE PROPAGATION IN A VISCOELASTIC FIBER SUBJECTED TO TRANSVERSE IMPACT	71-0639
WAVE	WAVE PROPAGATION IN A THIN HOLLOW CONE BY A FINITE ELEMENT METHOD	72-0045
WAVE	WAVE PROPAGATION IN ANISOTROPIC ROCKS	72-0801
WAVE	WAVE PROPAGATION IN ANISOTROPIC ROCKS	72-0873
WAVE	WAVE PROPAGATION IN ANISOTROPIC ROCKS	73-0922
WAVE	WAVE PROPAGATION IN FLUIDS	74-0034
WAVE	3-D STRESS WAVE CODE FOR THE ILLTAC 1V	74-0780
WAVE-INDUCED	NUMERICAL SOLUTIONS FOR DETERMINING WAVE-INDUCED PRESSURE DISTRIBUTIONS AROUND BURIED PIPELINES	74-1334
WAVE-PROPAGATIO	APPLICATION OF FINITE-ELEMENT ANALYSIS TO BODY WAVE-PROPAGATION PROBLEMS	75-0446
WAVEGUIDE	A GENERAL HIGH-ORDER FINITE-ELEMENT WAVEGUIDE ANALYSIS PROGRAM	69-0512
WAVEGUIDE	APPLICATION OF FINITE ELEMENT ANALYSIS TO COPLANAR WAVEGUIDE	73-0924
WAVEGUIDE	FINITE BOUNDARY CORRECTIONS TO THE COPLANAR WAVEGUIDE ANALYSIS	73-0398
WAVEGUIDE	FINITE ELEMENT ANALYSIS OF THE PIEZOELECTRIC TOPOGRAPHIC WAVEGUIDE	74-1111
WAVEGUIDE	FINITE ELEMENT SOLUTION OF THE HOMOGENEOUS WAVEGUIDE PROBLEMS	69-0010
WAVEGUIDE	FINITE-ELEMENT METHOD FOR WAVEGUIDE PROBLEMS	68-0006
WAVEGUIDE	WAVEGUIDE SOLUTIONS BY THE FINITE ELEMENT METHOD	69-0011
WAVEGUIDES	ARBITRARILY SHAPED INHOMOGENEOUS OPTICAL FIBER OR INTEGRATED OPTICAL WAVEGUIDES	75-0372
WAVEGUIDES	FINITE ELEMENT ANALYSIS OF PIEZOELECTRIC ELASTIC WAVEGUIDES	73-0115
WAVEGUIDES	FINITE ELEMENT ANALYSIS OF FREQUENCY SPECTRA FOR ELASTIC WAVEGUIDES	75-0247
WAVEGUIDES	FINITE ELEMENT METHODS FOR INHOMOGENEOUS WAVEGUIDES	69-0498
WAVEGUIDES	FINITE ELEMENT SOLUTION OF DIELECTRIC LOADED WAVEGUIDES	70-0206
WAVEGUIDES	FINITE-ELEMENT METHODS FOR INHOMOGENEOUS WAVEGUIDES	69-0039
WAVEGUIDES	NUMERICAL SOLUTION OF DIELECTRIC LOADED WAVEGUIDES: I. FINITE-ELEMENT ANALYSIS	70-0199
WAVEGUIDES	POLAR GEOMETRY WAVEGUIDES BY FINITE-ELEMENT METHOD	74-0109
WAVEGUIDES	PROPAGATION IN ELLIPTICAL AND PARABOLIC WAVEGUIDES	74-1173
WAVES	A FINITE ELEMENT ANALYSIS OF SHOCK AND FINITE-AMPLITUDE WAVES IN ONE-DIMENSIONAL HYPERELASTIC BODIES	72-0731
WAVES	APPLICATIONS OF PULSED LASER INDUCED STRESS WAVES	73-0956
WAVES	BERECHNUNGSMETHODEN ZUR ERMITTLUNG VON BELASTUNGSBEDINGUNGEN UND GRENZTRAGFAEHIGKEITSVERHALTEN VON K	73-0858
WAVES	CATCHING WAVES WITH COMPUTERS	74-0280
WAVES	DIFFRACTION OF OBLIQUE WAVES BY AN INFINITE CYLINDER	75-0605
WAVES	DYNAMICAL FINITE ELEMENT ANALYSIS FOR ELASTIC WAVES IN BEAMS AND PLATES	68-0211
WAVES	EXCITATION OF PLANE THERMOELASTIC WAVES IN A LAYER WITH A FINITE RATE OF HEAT PROPAGATION	74-0803
WAVES	FINITE ELEMENT ANALYSIS OF DISPERSION-FREE GUIDES FOR ELASTIC SURFACE WAVES	73-0009
WAVES	FINITE ELEMENT ANALYSIS OF LONG-PERIOD WATER WAVES	73-1008
WAVES	FINITE ELEMENT ANALYSIS OF LONG WAVES IN OPEN CHANNEL SYSTEMS	74-0033
WAVES	FINITE ELEMENT ANALYSIS OF WAVES IN NONLINEAR MATERIALS	75-0110
WAVES	HIGHER-ORDER FINITE-ELEMENT ANALYSIS OF TOPOGRAPHIC GUIDES SUPPORTING ELASTIC SURFACE WAVES	73-0114
WAVES	HYBRID-ELEMENT METHOD FOR WATER WAVES	75-0808
WAVES	METHODS FOR DETERMINING CHARACTERISTICS OF ACOUSTIC WAVES IN ROCKET MOTORS	75-0075
WAVES	NUMERICAL METHOD FOR INVESTIGATING THE PROPAGATION OF SURFACE WAVES ON DISSIPATIVE GUIDES	74-0613
WAVES	NUMERICAL STUDY OF TRANSITORY OR NONLINEAR WAVES	74-0031
WAVES	RAYLEIGH WAVES AT A CONTINENTAL BOUNDARY BY FINITE ELEMENT METHOD	72-0016
WAVES	RESPONSE OF EARTH DAMS TO TRAVELLING SEISMIC WAVES	69-0092
WAVES	SEISMOLOGY AND ACOUSTIC-GRAVITY WAVES	73-0744
WAVES	SOME RECENT DEVELOPMENTS IN THE THEORY AND APPLICATION OF FINITE ELEMENTS TO MIXED PROBLEMS AND SHOC	73-0637
WAVES	SPHERICAL WAVES IN INELASTIC MATERIAL	70-0075
WAVES	STUDY OF MOTION AND STRENGTH OF FLOATING MARINE STRUCTURES IN WAVES	74-0489
WAVES	THE PROPAGATION OF LOVE WAVES ACROSS NONHORIZONTALLY LAYERED STRUCTURES	71-0600
WAVES	THEORETICAL COMPUTATIONS ON RIDGE ACOUSTIC SURFACE WAVES USING THE FINITE-ELEMENT METHOD	71-0011
WC-CO	DETERMINATION OF YOUNG'S MODULUS AND POISSON'S RATIO FOR WC-CO ALLOYS BY THE FINITE ELEMENT METHOD	72-0142
WEAK	METHOD OF WEAK ELEMENTS	74-1084
WEAK	NONLINEAR CONSIDERATION OF GRAVITY IN A STIFFNESS TEST OF A WEAK STRUCTURE AT SMALL STRAINS	71-0035
WEAK	RESPONSE OF BOX-LIKE STRUCTURES TO WEAK EXPLOSIONS	75-0197
WEAK-WEAK	AN IMBEDDING TECHNIQUE FOR THE GENERATION OF WEAK-WEAK FINITE ELEMENT APPROXIMATIONS OF LINEAR AND N	72-0577
WEAKENED	THEORETICAL AND EXPERIMENTAL ANALYSIS OF ELASTOPLASTIC DEFORMATION IN A UNIAXIALLY LOADED STRIP OF 2	73-0720
WEAPON-MOUNT	WEAPON-MOUNT INTERFACE STUDY - TRIPOD MOUNT M3	72-0846
WEAR	A THERMAL, THERMOELASTIC, AND WEAR ANALYSIS OF HIGH-ENERGY DISK BRAKES	74-0839
WEAR	THERMAL, THERMOELASTIC, AND WEAR SIMULATION OF A HIGH-ENERGY SLIDING CONTACT PROBLEM	73-0430
WEAR	VIBRATORY WEAR OF FUEL RODS	75-0771

446

WEATHER	RECENT ADVANCES IN NUMERICAL WEATHER PREDICTION	74-0787
WEB	REINFORCING REQUIREMENTS FOR CONCRETE BEAMS WITH LARGE WEB OPENINGS	73-0198
WEBPLATES	STABILITA STEN V PLASTICKEM OBORU (STABILITY IN WEBPLATES IN PLASTIC RANGE)	73-0387
WEBS	ARBITRARY QUADRILATERAL SPAR WEBS FOR THE MATRIX DISPLACEMENT METHOD	66-0003
WEBS	BUCKLING OF PLATE GIRDER WEBS UNDER PARTIAL EDGE LOADINGS	70-0283
WEBS	ON STRENGTH OF TRANSVERSE WEBS AND THEIR JOINTS OF LARGE TANKER	69-0327
WEBS	THE BEHAVIOUR OF SQUARE SHEAR WEBS HAVING A CIRCULAR HOLE	67-0108
WECHSELWIRKUNG	WECHSELWIRKUNG ZWISCHEN DEFORMATION UND DURCHSTROIMUNG IM KLUEFTIGEN, ANISOTROPEN GEBIRGE. (INTERACT	72-0512
WEDGE	DETERMINATION OF THE STRESS-STRAINED STATE WHEN CUTTING METAL WITH A WEDGE	73-0237
WEDGE	SOLID WEDGE FINITE ELEMENT FOR STRESS ANALYSIS	69-0124
WEIGHT	EFFICIENT FINITE-ELEMENT METHODS FOR STRESS INTENSITY FACTORS USING WEIGHT FUNCTIONS	75-0411
WEIGHT	EFFICIENT FINITE-ELEMENT METHODS FOR STRESS INTENSITY FACTORS USING WEIGHT FUNCTIONS - REPLY	75-0421
WEIGHT	HIGH STRENGTH MATERIALS AND VEHICLE WEIGHT REDUCTION ANALYSIS	75-0157
WEIGHT	METHOD OF MINIMUM WEIGHT DESIGN WITH REQUIREMENTS IMPOSED ON STRESSES AND NATURAL FREQUENCIES	70-0149
WEIGHT	MINIMUM WEIGHT DESIGN WITH STABILITY CONSTRAINT	70-0274
WEIGHT	MINIMUM WEIGHT DESIGN OF COMPLEX STRUCTURES SUBJECT TO A FREQUENCY CONSTRAINT	70-0374
WEIGHT	MINIMUM WEIGHT DESIGN OF FINITE ELEMENT STRUCTURES	74-0148
WEIGHT	OPTIM 11: (MAGIC) COMPATIBLE LARGE SCALE AUTOMATED MINIMUM WEIGHT DESIGN PROGRAM - V 2. PROGRAMMER'S	74-1036
WEIGHT	OPTIM 11: A (MAGIC) COMPATIBLE LARGE SCALE AUTOMATED MINIMUM WEIGHT DESIGN PROGRAM - 1. ENGINEERS AN	74-1048
WEIGHT	OPTIM 11: A MAGIC COMPATIBLE LARGE-SCALE AUTOMATED MINIMUM WEIGHT DESIGN PROGRAM. VOLUME 1. ENGINEER	74-1096
WEIGHT	OPTIM 11: MAGIC COMPATIBLE LARGE SCALE AUTOMATED MINIMUM WEIGHT DESIGN PROGRAM. VOLUME 11. PROGRAMME	74-1093
WEIGHT	SOME APPLICATIONS OF TOPOLOGICAL CONSIDERATION AND WEIGHT MATRIX METHOD TO FINITE ELEMENT ANALYSIS	73-0305
WEIGHT	STUDY ON WEIGHT MATRIX IN STRUCTURAL ANALYSIS	75-0062
WEIGHT	THE MIDCOHV WEIGHT AND BALANCE COMPUTER PROGRAM (WTBAL)	74-0697
WEIGHTED	FINITE ELEMENT ANALYSIS OF TRANSIENT HEAT CONDUCTION. APPLICATION OF THE WEIGHTED RESIDUAL PROCESS	74-0277
WEIGHTED	FINITE ELEMENT WEIGHTED RESIDUAL SOLUTION TO ONE DIMENSIONAL FIELD PROBLEMS	73-0388
WEIGHTED	FINITE ELEMENT WEIGHTED RESIDUAL SOLUTION OF ONE-DIMENSIONAL, UNSTEADY AND UNSATURATED FLOWS IN PORO	73-0712
WEIGHTED	FINITE ELEMENTS FORMULATED BY THE WEIGHTED DISCRETE LEAST SQUARES METHOD	74-0200
WEIGHTED	METHOD FOR FINITE ELEMENTS AS SPECIAL CASE OF METHOD OF WEIGHTED RESIDUALS	72-0141
WEIGHTED	ON THE ROLE OF VARIATIONAL PRINCIPLES AND METHOD OF WEIGHTED RESIDUALS IN THE FINITE ELEMENT METHODS	74-0551
WEIGHTED	STABILITY AND OSCILLATION CHARACTERISTICS OF FINITE ELEMENT, FINITE DIFFERENCE AND WEIGHTED RESIDUAL	73-0139
WEIGHTED	VARIATIONAL AND WEIGHTED RESIDUAL METHODS	73-0630
WEIGHTED	WEIGHTED RESIDUAL METHODS AND THEIR RELATICN TO FINITE ELEMENT METHODS IN FLOW PROBLEMS	74-0003
WEIGHTED	WEIGHTED RESIDUAL PROCESSES IN FINITE ELEMENT WITH PARTICULAR REFERENCE TO SOME TRANSIENT AND COUPLE	71-0503
WEIGHTED-RESIDU	STABILITY AND OSCILLATION CHARACTERISTICS OF FINITE ELEMENT, FINITE-DIFFFRENCE, AND WEIGHTED-RESIDUA	75-0615
WELD	CREEP BENDING TESTS OF LARGE WELD OVERLAY AND BASE METAL SPECIMENS AT 1000 DEGREE F. (537,7 DEG. C)	75-0521
WELD	RECENT TRENDS OF RESEARCH INTO RESTRAINT STRESSES AND STRAINS IN RELATION TO WELD CRACKING	73-0483
WELD	TENDANCES ACTUELLES DES RECHERCHES SUR LES CONTRAINTES ET DEFORMATIONS DUES AU BRIDAGE EN RELATION A	74-0273
WELDED	ANALYSIS OF CRACK PROPAGATION IN WELDED STRUCTURES	74-1144
WELDED	ANALYSIS OF STRESSES IN PRESSURIZED WELDED PIPE IN THE CREEP RANGE	71-0404
WELDED	ANWENDUNG DES FINITE ELEMENT VERFAHRENS SUR BESTIMMUNG DER FESTIGKEIT VON SCHWEISSKONSTRUKTIONEN. (	72-0185
WELDED	DESIGN APPRAISAL OF OFFSHORE PLATFORMS OF WELDED TUBULAR CONSTRUCTION	74-1165
WELDED	FINITE-ELEMENT METHOD FOR DETERMINING STRENGTH OF WELDED STRUCTURES	72-0272
WELDED	LAMELLAR TEARING IN WELDED GIRDERS	70-0218
WELDED	ON CRACKS IN WELDS AND WELDED STRUCTURES	71-0688
WELDED	ON THE ROLE OF DEFECTS IN CRACK INITIATION IN WELDED STRUCTURES	74-1142
WELDED	STRENGTH OF TRANSVERSE FILLET WELDED JOINTS	74-0144
WELDED	STRUCTURAL ANALYSIS OF WELDED WALL FOR BOILER FURNACE	74-0653
WELDED	STRUCTURAL MODELING OF A WELDED JOINT TYPICAL FOR VEHICLE FRAMES	73-0539
WELDED	TEMPERATURVERLAUF IM QUERSCHNITT GESCHWEISSTER ROHRWAENDE BEI UEBERWIEGEND KONVEKTIVEM WAERMEUEBERGA	75-0628
WELDED	THE FATIGUE STRENGTH OF FILLET WELDED CONNECTIONS	71-0684
WELDED	THE FLEXIBILITY OF A TUBULAR WELDED JOINT IN A VEHICLE FRAME	74-0209
WELDED	THERMAL AND MECHANICAL ANALYSIS OF WELDED STRUCTURES	73-0819
WELDED	WELDED JOINT STIFFNESS OBTAINED USING SOLID FINITE ELEMENTS	75-0190
WELDED-ON	BERECHNUNG VON SPANNUNGEN UND TEMPERATURFELDERN IN FLOSSENWAENDEN UND AUFGESCHWEISSTEN BAUTEILEN. (C	74-0544
WELDING	A NUMERICAL THERMO-MECHANICAL MODEL FOR THE WELDING AND SUBSEQUENT LOADING OF A FABRICATED STRUCTURE	72-0847
WELDING	ANALYSIS OF THERMAL ELASTIC PLASTIC STRESS AND STRAIN DURING WELDING BY FINITE ELEMENT METHOD	71-0382
WELDING	ANALYSIS OF THERMAL STRESSES AND METAL MOVEMENT DURING WELDING 1. ANALYTICAL STUDY	74-1026
WELDING	ANALYSIS OF WELDING STRESS RELIEVING BY ANNEALING BASED ON FINITE ELEMENT METHOD	75-0777
WELDING	BERECHNUNG DER SCHWEISSEIGENSPANNUNGEN UND SCHWEISSFORMAENDERUNGEN MIT ELASTISCH-PLASTISCHEN FINITEN	75-0271
WELDING	EXPERIMENTAL STUDY OF FLEXIBILITY FACTORS AND STRESSES OF WELDING ELBOWS WITH END EFFECTS	73-0422
WELDING	FINITE-ELEMENT ELASTIC-PLASTIC ANALYSIS OF RESIDUAL STRESSES DUE TO CLAD WELDING IN REACTOR VESSELS	75-0779
WELDING	FUNDAMENTAL STUDY OF UNDERWATER WELDING	72-0382
WELDING	NUMERICAL, THERMO-MECHANICAL MODEL FOR THE WELDING AND SUBSEQUENT LOADING OF A FABRICATED STRUCTURE	73-0418
WELDING	THERMOMECHANICAL ANALYSIS OF THE WELDING PROCESS USING THE FINITE ELEMENT METHOD	75-0213
WELDING	THERMOMECHANICAL ANALYSIS OF THE WELDING PROCESS USING THE FINITE ELEMENT METHOD	75-0518
WELDING	THERMOMECHANICAL ANALYSIS OF THE WELDING PROCESS USING THE FINITE ELEMENT METHOD	75-0776
WELDING	USE OF COMPUTERS IN WELDING TECHNOLOGY	74-1160
WELDING	VOLLSTAENDIGE SPANNUNGS-DEHNUNGS-TEMPERATURAENDERUNGSBEZIEHUNG FUER DIE SCHWEISSEIGENSPANNUNGSBERECH	75-0807
WELDS	APPLICATION OF THE FINITE ELEMENT METHOD ON RESIDUAL STRESSES AND DEFORMATIONS IN WELDS	68-0178
WELDS	DIFFUSION OF HYDROGEN IN FILLET WELDS	75-0373
WELDS	ELASTIC AND PLASTIC STRESS DISTRIBUTION ON BASE PLATES IN THE NEIGHBOURHOOD OF FILLET WELDS IN LAP J	70-0175
WELDS	KERBSPANNUNGSANALYSE NACH DER FINITE-ELEMENT-METHODE AN KEHLNAEHTEN. (NOTCH EFFECT ANALYSIS USING TH	75-0673
WELDS	ON CRACKS IN WELDS AND WELDED STRUCTURES	71-0688
WELL	ANALYSIS OF UNSTEADY FLOW TOWARD AN ARTESIAN WELL BY 3-DIMENSIONAL FINITE ELEMENTS	74-0351
WELL	ANALYSIS OF UNSTEADY FLOW TOWARD AN ARTESIAN WELL BY THREE-DIMENSIONAL FINITE ELEMENTS	74-1351
WELL	BASIS OF A WELL CONDITIONED FORCE PROGRAM FOR EQUILIBRIUM MODELS VIA THE SOUTHWELL SLAB ANALOGIES	67-0111
WELL	EFFECT OF PERFORATION DAMAGE ON WELL PRODUCTIVITY	74-0936
WELL	STATICAL CALCULATIONS OF PLANE AND SPACE BAR STRUCTURES AS WELL AS STUMPY FRAMEWORKS USING A MODIFIE	73-0747
WELLEN	ZUR BERECHNUNG DER BIEGESTEIFIGKEIT ABGESETZTER STAEBE UND WELLEN UNTER ANWENDUNG VON FINITEN ELEMEN	75-0481
WELLS	ANALYSIS OF LIQUID-WASTE INJECTION WELLS IN ILLINOIS BY MATHEMATICAL MODELS	74-0518
WELLS	ANALYSIS OF UNSTEADY FLOW TOWARD ARTESIAN WELLS BY THREE-DIMENSIONAL FINITE ELEMENTS	74-0742
WELLS	FINITE ELEMENT SOLUTIONS OF TRANSIENT, TWO-REGIME FLOW TOWARDS WELLS	74-0466
WELLS	FINITE ELEMENT SOLUTION OF TWO-REGIME FLOW TOWARDS WELLS	74-1352
WELLS	OPTIMAL DESIGN OF WATER WELLS IN UNCONSOLIDATED SEDIMENTS	74-1182
WELLS	TRANSIENT FLOW OF GROUNDWATER TO WELLS IN MULTIPLE-AQUIFER SYSTEMS	69-0501
WENDELSCHALEN	HYBRIDE SCHALENELEMENTE MIT EINER ANWENDUNG AUF WENDELSCHALEN. (HYBRID SHELL ELEMENTS IN APPLICATION	75-0609
WERKSTOFFVERHAL	BERECHNUNG VON STAHLBETONSCHEIBEN IM ZUSTAND II BEI ANNAHME EINES WIRKLICHKEITSNAHEN WERKSTOFFVERHAL	74-0978
WERKZEUGMASCHIN	COMPUTERUNTERSTUETZTES SYSTEM FUER DIE CONSTRUKTION DES WERKZEUGMASCHINENAUFBAUS (COMPUTER-AIDED SYS	73-0482
WERKZEUGMASCHIN	NUMERISCHE BERECHNUNG INSTATIONAERER TEMPERATURFELDER UND THERMISCH BEDINGTER VERFORMUNGEN VON WERKZ	75-0847
WERKZEUGMASCHIN	ANWENDUNG DER METHODE FINITER ELEMENTE BEI DER ANALYSE DES DYNAMISCHEN VERHALTENS GEDAEMPFTER WERKZE	75-0580
WEST	PAVEMENT CRACKING IN WEST TEXAS DUE TO FREEZE-THAW CYCLING	75-0276
WESTMINISTER	FOUNDATION INSTRUMENTATION FOR THE NATIONAL WESTMINISTER BANK TOWER	74-1147
WHEAT	SELECTION OF LOCATIONS FOR LONG TERM STORAGE OF WHEAT	75-0323
WHEEL	ANALYSIS OF LAYERED SYSTEMS SUBJECTED TO WHEEL LOADS	72-0868
WHEEL	APPLICATION OF FINITE ELEMENT METHOD TO THE STUDY OF STRENGTH OF CAR BODY STRUCTURE AND WHEEL	74-0281
WHEEL	EINSATZ MODERNER RECHENVERFAHREN ZUR ERMITTLUNG DER SPANNUNGEN UND VERFORMUNGEN BEI RADSAETZEN. (USE	75-0186
WHEEL	ELASTIC-PLASTIC ANALYSIS OF A WHEEL ROLLING ON A RIGID TRACK	74-1040
WHEEL	ELASTO-PLASTIC ANALYSIS OF STRESS IN A GAS-TURBINE WHEEL	74-0323
WHEEL	METHOD FOR THE COMPUTATION OF CYCLIC STRESSES IN A GAS-TURBINE WHEEL	74-0972
WHEEL	SHAKEDOWN OF ROLLING WHEEL UNDER HUB LOADING	74-0663
WHEEL	STRUCTURAL INELASTICITY. V11. ELASTIC-PLASTIC ANALYSIS OF A WHEEL ROLLING ON A RIGID TRACK	73-0886

WHEEL	UNTERSUCHUNGEN ZUR SPURMASSTABILITAET BEI RADSAETZEN MIT KLOTZGEBREMSTEN VOLLRAEDERN (INVESTIGATIONS	75-0810
WHEELS	BERECHNUNG VON LAUFRAEDERN MIT DER METHODE DER FINITEN ELEMENTE (DESIGN OF ROTATING WHEELS BY THE FI	74-0547
WHEELS	EVALUATION OF STRESSES IN SOIL UNDER RIGID WHEELS EY FINITE ELEMENT METHOD	73-1109
WHEELS	STRENGTH OF RAILROAD WHEELS - 3	71-0306
WHEELS	THERMAL STRESS DEVELOPED IN S PLATE, STRAIGHT PLATE, AND DEEP DISH WHEELS	73-0172
WHEELS	UNTERSUCHUNGEN ZUR SPURMASSTABILITAET BEI RADSAETZEN MIT KLOTZGEBREMSTEN VOLLRAEDERN (INVESTIGATIONS	75-0810
WHIRL	APPLICATION OF A FINITE ELEMENT METHOD TO EVALUATION OF OIL WHIRL CHARACTERISTICS	74-0590
WHIRLING	WHIRLING CABLE SUBJECTED TO VISCOUS DRAG	74-1161
WHOLE	COMPUTER SIMULATION OF VIBRATION OF WHOLE TREE SYSTEMS USING FINITE ELEMENT METHODS	74-1161
WHOLE	SIMULATION OF VIBRATION OF WHOLE TREE SYSTEMS USING FINITE ELEMENTS	75-0765
WIDE	APPROACH TO ANALYSIS OF ARCH DAMS IN WIDE VALLEYS	73-0153
WIDE	NATURAL FREQUENCIES OF WIDE CHORD COMPRESSOR BLADES	74-0231
WIDE	OIL-FILM TEMPERATURE DISTRIBUTION IN AN INFINITELY WIDE SLIDER BEARING: AN APPLICATION OF THE FINITE	73-1128
WIDE-FLANGE	COUPLED LOCAL BUCKLING IN WIDE-FLANGE BEAM-COLUMNS	73-0235
WIDELY	VIBRATION AND STRESS ANALYSIS OF SOFT-BONDED SHUTTLE INSULATION TILES, MODAL ANALYSIS WITH COMPACT W	74-0708
WIDTH	APPLICATION OF FINITE ELEMENT METHOD TO HYDRODYNAMIC LUBRICATION PROBLEMS (PART 2, " FINITE WIDTH BE	71-0538
WIE	BERECHNUNGSMETHODEN ZUR ERMITTLUNG VON BELASTUNGSBEDINGUNGEN UND GRENZTRAGFAEHIGKEITSVERHALTEN VON K	73-0858
WIELOMIANOW	APPLICATION OF HERMITE POLYNOMIALS TO THE DETERMINATION OF THE STIFFNESS MATRIX OF PLATE BY FINITE E	71-0365
WIELOSTOPNIOWA	WIELOSTOPNIOWA SYNTEZA MACIERZY SZTYWNOSCI (MULTI-STAGE SYNTHESIS OF THE STIFFNESS MATRICES)	73-0668
WILAX	COMPUTER PROGRAM FOR AUTOMATIC MESH GENERATION FOR THE 2-D FINITE ELEMENT PROGRAM WILAX	73-0565
WILLIAMS	DISCUSSION OF A PAPER BY F. W. WILLIAMS	73-1078
WILSON	AN ASSUMED STRESS VERSION OF THE WILSON 8-NODE ISOPARAMETRIC BRICK	72-0532
WILSON	THE RECTANGULAR PLANE STRESS ELEMENT BY TURNER, PIAN AND WILSON	74-1265
WIND	A NUMERICAL INVESTIGATION OF STABLY STRATIFIED, WIND DRIVEN CAVITY FLOW BY THE FINITE ELEMENT METHOD	74-1372
WIND	ANALYSIS OF THERMAL STRESSES IN CORED CERAMIC BRICKS USED IN HYPERSONIC WIND TUNNEL HEATERS	72-0655
WIND	BUCKLING OF CYLINDRICAL SHELLS BY WIND PRESSURE	74-0642
WIND	FINITE ELEMENT SOLUTION OF THE WIND DRIVEN CURRENTS AND ITS MASS TRANSPORT IN LAKES	74-1368
WIND	INTEGRAL EQUATION METHOD FOR SUBSONIC FLOW PAST AIRFOILS IN VENTILATED WIND TUNNELS	75-0065
WIND	MULTIDEGREE DYNAMIC ANALYSIS OF TALL BUILDINGS SUBJECTED TO WIND AS A STOCHASTIC PROCESS	74-1196
WIND	STRESSES AND DEFLECTIONS IN COOLING TOWER SHELLS DUE TO WIND LOADING	68-0088
WIND	WIND DRIVEN LAKE CIRCULATION BY THE FINITE ELEMENT METHOD	70-0041
WIND-DRIVEN	FINITE-DIFFERENCE AND FINITE-ELEMENT METHODS IN GLOBAL WIND-DRIVEN OCEAN CIRCULATION	75-0345
WINDING	CIRCUMFERENTIAL PRESTRESSING LOAD ANALYSIS DUE TO WIRE/STRAND WINDING SYSTEMS ON PRESTRESSED CONCRET	74-1202
WINDOWS	THIN WINDOWS FOR GASEOUS AND LIQUID TARGETS: AN OPTIMIZATION PROCEDURE	75-0741
WING	A STUDY OF VARIOUS IDEALIZATIONS FOR WING STRUCTURES AND NUMERICAL PROCEDURES INVOLVED USING MATRIX	69-0257
WING	ANALYSIS OF THE 747 AIRCRAFT WING BODY INTERSECTION	68-0275
WING	APPLICATION OF FINITE ELEMENT METHOD TO ANALYSIS OF AIRCRAFT WING STRUCTURES	69-0232
WING	AUTOMATED DESIGN OPTIMIZATION OF SUPERSONIC AIRPLANE WING STRUCTURES UNDER DYNAMIC CONSTRAINTS	72-0178
WING	AUTOMATED PROCEDURE FOR DESIGN OF WING STRUCTURES TO SATISFY STRENGTH AND FLUTTER REQUIREMENTS	73-0867
WING	AUTOMATED STRUCTURAL DESIGN AND ANALYSIS OF ADVANCED COMPOSITE WING MODELS	73-0221
WING	COMPOSITE WING FOR TRANSONIC IMPROVEMENT. VOLUME 11. ADVANCED ANALYSIS EVALUATION	71-0657
WING	EARTH PRESSURES AGAINST THE ABUTMENTS AND WING WALLS OF STANDARD MOTORWAY BRIDGES	75-0629
WING	FINITE ELEMENT ANALYSIS OF A WING STRUCTURE	72-0333
WING	FINITE ELEMENT FLUTTER ANALYSIS OF MULTI-WEB WING STRUCTURES	75-0004
WING	STABILITY ANALYSIS OF THE TORSIONAL BENDING VIBRATIONS OF WING IN A SUBSONIC FLOW BY METHOD OF THE F	75-0755
WING	VIBRATION STUDIES OF A FLAT PLATE AND BUILT-UP WING	71-0238
WING-FUSELAGE	BERECHNUNG DES REIBUNGSLOSEN STROEMUNGSFELDES VON DREIDIMENSIONALEN AUFTRIEBSBEHAFTETENTRAGFLUEGELN	73-0533
WINGS	AEROELASTIC PROBLEMS OF LOW ASPECT RATIO WINGS (PART 1)	56-0006
WINGS	ANALYSIS OF THREE-DIMENSIONAL UNSTEADY VISCOUS FLOW AROUND OSCILLATING WINGS	74-0982
WINGS	AUTOMATED OPTIMUM DESIGN OF AIRCRAFT WINGS TO SATISFY STRENGTH, STABILITY, FREQUENCY AND FLUTTER REQ	72-0605
WINGS	BERECHNUNG DES REIBUNGSLOSEN STROEMUNGSFELDES VON DREIDIMENSIONALEN AUFTRIEBSBEHAFTETENTRAGFLUEGELN	73-0533
WINGS	DEVELOPMENT AND APPLICATIONS OF SUPERSONIC UNSTEADY CONSISTENT AERODYNAMICS FOR INTERFERING PARALLEL	72-0716
WINGS	STRUCTURAL ANALYSIS OF SWEPT-BACK WINGS BY MATRIX-TRANSFORMATION	51-0001
WINGS	STRUCTURAL ANALYSIS BY THE MATRIX FORCE METHOD WITH APPLICATION TO AIRCRAFT WINGS	56-0004
WINGS	UNSTEADY SUBSONIC COLLOCATION METHOD FOR WINGS WITH AND WITHOUT CONTROL SURFACES	72-0012
WINSCAR	PERFORMANCE OF CULVERT UNDER WINSCAR DAM	75-0638
WIRE	CIRCUMFERENTIAL PRESTRESSING LOAD ANALYSIS DUE TO WIRE/STRAND WINDING SYSTEMS ON PRESTRESSED CONCRET	74-1202
WIRE	NUMERICAL ANALYSIS OF THE LINE CAPACITANCE AND CROSSTALK FACTOR FOR INSULATED WIRE PAIRS	74-0685
WIRE	THE USE OF ROCK BOLTS OR WIRE ROPE TO INCREASE THE STRENGTH OF FRACTURED MODEL PILLARS	71-0722
WIRE-WOUND	ADVANCED WIRE-WOUND TUNGSTEN NOZZLES	66-0031
WIRKLICHKEITSNA	BERECHNUNG VON STAHLBETONSCHEIBEN IM ZUSTAND II BEI ANNAHME EINES WIRKLICHKEITSNAHEN WERKSTOFFVERHAL	74-0978
WIRKLICHKEITSNA	ZUR WIRKLICHKEITSNAHEN BERECHNUNG VON STAHLBETONPLATTEN MIT DER FINITE-ELEMENT-METHODE. (APPROXIMATE	75-0832
WLASNYCH	NUMERYCZNE ROZWIAZANIE PROBLEMU DRGAN WLASNYCH NA PRZYKLADZIE PLYT. (NUMERICAL SOLUTION OF THE PROBL	75-0296
WODE	OBLICZANIE KONSTRUKCJI TOROIDALNEGO ZBIORNIKA NA WODE. (STATICAL ANALYSIS OF A TOROIDAL WATER RESERV	75-0211
WOOD	A DISCUSSION ON A MODIFIED CONJUGATE GRADIENT METHOD BY W.L. WOOD	74-0402
WOOD	A RATIONAL ANALYSIS AND DESIGN PROCEDURE FOR WOOD JOIST FLOOR SYSTEMS	74-0730
WOOD	APPLICATION OF THE FINITE ELEMENT METHOD TO CALCULATE WOOD TENSION	75-0768
WOOD	FINITE ELEMENT ANALYSIS OF WOOD STRUCTURAL MEMBERS	70-0180
WOOD	FINITE ELEMENT ANALYSIS OF WOOD BEAMS	70-0503
WOOD	FINITE ELEMENT ANALYSIS OF LAYERED WOOD SYSTEMS	75-0288
WOOD	FINITE ELEMENT METHOD FOR WOOD MECHANICS	71-0294
WOOD	FINITE ELEMENT METHOD FOR WOOD MECHANICS	71-0613
WOOD	LINEAR AND NONLINEAR FINITE ELEMENT ANALYSIS OF WOOD STRUCTURAL MEMBERS	73-0117
WOOD	PERFORMANCE OF STRUCTURAL WOOD MEMBERS EXPOSED TO FIRE	75-0060
WOOD	WOOD SHEAR PANELS BONDED WITH FLEXIBLE ADHESIVES	75-0086
WOOD-JOIST	STATIC AND DYNAMIC ANALYSIS OF WOOD-JOIST FLOORS BY THE FINITE-ELEMENT METHOD	73-1161
WORD	COMPUTER WORD LENGTH AND SOLUTION ACCURACY. EXPERIENCE WITH ELAS AND THE IBM 360	73-0335
WORK	ANISOTROPIC WORK HARDENING IN METALS. FINITE ELEMENT APPLICATION TO PLANE STRESS AND PLANE STRAIN	73-0950
WORK	FINITE ELEMENT SOLUTION OF EXTRUSION OF ELASTO-PLASTIC WORK HARDENING MATERIALS	73-1127
WORK	SOME APPLICATIONS OF THE PRINCIPLE OF VIRTUAL WORK	73-0930
WORK	THE SIGNIFICANCE OF THE RATE OF PLASTIC WORK IN ELASTO-PLASTIC ANALYSIS	74-0437
WORK	THE SIGNIFICANCE OF THE RATE OF PLASTIC WORK IN ELASTO-PLASTIC ANALYSIS	74-0764
WORK-HARDENING	A COMPARISON OF CURRENT WORK-HARDENING MODELS USED IN THE ANALYSIS OF PLASTIC DEFORMATIONS	73-0769
WORK-HARDENING	MECHANICAL CHARACTERIZATION AND FINITE-ELEMENT ANALYSIS OF ELASTIC-PLASTIC, WORK-HARDENING SOILS	72-0974
WORK-SOFTENING	ON APPLICATION OF THE FINITE ELEMENT METHOD TO LIMIT ANALYSIS CONSISTENT FINITE ELEMENT TECHNIQUE FO	74-1234
WORKSOFTENING	A CONSISTENT FINITE ELEMENT TECHNIQUE FOR WORKSOFTENING BEHAVIOUR	74-1233
WORTH	NASTRAN EXPERIENCES OF FORT WORTH OPERATION, CONVAIR AEROSPACE DIVISION OF GENERAL DYNAMICS	71-0019
WORTHINESS	STUDY ON AUTOMOBILE CRASH WORTHINESS	70-0217
WTBAL	THE MIDCOHV WEIGHT AND BALANCE COMPUTER PROGRAM (WTBAL)	74-0697
WTH	STRESS ANALYSIS OF FIRST MOLARS WTH FULL CROWN PREPARATIONS BY THREE-DIMENSIONAL PHOTOELASTICITY AND	72-0939
WYMUSZONYCH	ALGORYTM OBLICZEN DRGAN WYMUSZONYCH METODA SZTYWNYCH ELEMENTOW SKONCZONYCH (ALGORITHM FOR CALCULATIN	74-0581
WYMUSZONYCH	ANALIZA DRGAN WYMUSZONYCH ZLOZONYCH UKLADOW LINIOWYCH METODA SZTYWNYCH ELEMENTOW SKONCZONYCH (ANALYS	72-0246
WYZNACZANIA	APPLICATION OF HERMITE POLYNOMIALS TO THE DETERMINATION OF THE STIFFNESS MATRIX OF PLATE BY FINITE E	71-0365
X-RAY	STRAIN DISTRIBUTION IN AN ALUMINA BICRYSTAL STUDIES BY THE FINITE ELEMENT METHOD AND X-RAY DIFFRACTI	74-1070
YAWED	FORCE AND MOMENTS ON ASYMMETRIC AND YAWED BODIES IN A FREE SURFACE	73-0908
YIELD	AN INVESTIGATION OF INITIAL YIELD SURFACES FOR UNIDIRECTIONAL REINFORCED COMPOSITES	72-0837
YIELD	ELASTIC-PLASTIC STRESS-STRAIN RELATIONS FOR MATERIALS WITH DIFFERENT YIELD STRENGTHS IN TENSION AND	72-0764
YIELD	FINITE ELEMENT ANALYSIS OF ELASTIC-PLASTIC PLANE STRESS PROBLEMS BASED UPON TRESCA YIELD CRITERION	70-0179
YIELD	INITIAL YIELD SURFACE OF A UNIDIRECTIONALLY REINFORCED COMPOSITE	71-0280
YIELD	INVESTIGATION OF THE EFFECT OF FOUNDATION YIELD ON THE FREQUENCY AND SHAPE OF INHERENT VIBRATIONS OF	72-0419
YIELD	ON TRESCA YIELD CRITERION IN FINITE ELEMENT ANALYSIS OF PLANE STRESS PROBLEMS	73-0314
YIELD	YIELD CRITERION FOR REINFORCED CONCRETE SLABS	73-0985

YIELDED   STRESSES IN A PARTLY YIELDED NOTCHED BAR - AN ASSESSMENT OF THREE ALTERNATIVE PROGRAMS   73-0267
YIELDED   STRESSES IN A PARTLY YIELDED NOTCHED BAR EM DASH AN ASSESSMENT OF THREE ALTERNATIVE PROGRAMS   73-0268
YIELDING   A FINITE ELEMENT MODEL FOR ANISOTROPIC YIELDING IN GRAVITY LOADED ROCK   69-0012
YIELDING   INFLUENCE OF NON-SINGULAR STRESS TERMS AND SPECIMEN GEOMETRY ON SMALL SCALE YIELDING AT CRACK TIPS I   73-0203
YIELDING   LIMITATIONS TO THE SMALL SCALE YIELDING APPROXIMATION FOR CRACK TIP PLASTICITY   74-0145
YIELDING   SMALL SCALE YIELDING NEAR A CRACK IN PLANE STRAIN-FINITE ELEMENT ANALYSIS   71-0355
YIELDING   SMALL SCALE YIELDING ANALYSIS OF MIXED MODE PLANE STRAIN CRACK PROBLEMS   73-0801
YIELDING   THEORETICAL POST YIELDING BEHAVIOUR OF COMPOSITE LAMINATES - 1. INELASTIC MICROMECHANICS   73-0168
YIELDING   THEORETICAL POST YIELDING BEHAVIOUR OF COMPOSITE LAMINATES -2.   73-0408
YIELDING   YIELDING IN UNIDIRECTIONAL COMPOSITES UNDER EXTERNAL LOADS AND TEMPERATURE CHANGES.   73-0169
YIELDING   YIELDING OF A TWO DIMENSIONAL VOID ASSEMBLY IN AN ORGANIC GLASS   73-0416
YOUNG'S   DETERMINATION OF YOUNG'S MODULUS AND POISSON'S RATIO FOR WC-CO ALLOYS BY THE FINITE ELEMENT METHOD   72-0142
ZADACH   CHISLENNOE RESHENIE PLOSKIKH ZADACH TEPLOPROVODNOSTI DLYA OBLASTEI SLOZHNOI FORMY (NUMERICAL SOLUTIO   74-0659
ZADACHAKH   THE DEVELOPMENT OF APPLIED METHODS IN PROBLEMS OF STATIC CALCULATION OF THIN-WALLED THREE-DIMENSIONA   73-0934
ZADANIYA   K VOPROSU OB AVTOMATIZATSII ZADANIYA INFORMATSII V METODE KONECHNYKH ELEMENTOV. (PROBLEM OF AUTOMATI   75-0382
ZAEHER   THEORETICAL AND EXPERIMENTAL CONSIDERATIONS ON FLOW OF VISCOUS LIQUIDS PAST PLATE AT SMALL AND AVERA   70-0144
ZALA   RASCHET NAPRYAZHENII L PEREMESHCHENII V SVODE PODZEMNOGO MASHINNOGO ZALA GES I V OKRUZHAYUSHCHEM SKA   74-0542
ZASTOSOWANIE   APPLICATION OF HERMITE POLYNOMIALS TO THE DETERMINATION OF THE STIFFNESS MATRIX OF PLATE BY FINITE E   71-0365
ZASTOSOWANIE   ZASTOSOWANIE ITERACJI SEIDLA W METODZIE ELEMENTOW SKONCZONYCH NA PRZYKLADZIE OBLICZEN STATYCZNYCH P   74-1031
ZASTOSOWANIE   ZASTOSOWANIE METODY SZTYWNYCH ELEMENTOW SKONCZONYCH DO OBLICZEN ORGAN WALOW OKRETOYCH Z UWZGLEDNIENI   74-0962
ZASTOSOWANIU   HYBRYDOWA METODA ELEMENTOW SKONCZONYCH W ZASTOSOWANIU DO OBLICZEN ORGAN URZADZEN OKRETOWYCH. (HYBRID   74-0963
ZBIORNIKA   OBLICZANIE KONSTRUKCJI TOROIDALNEGO ZBIORNIKA NA WODE. (STATICAL ANALYSIS OF A TOROIDAL WATER RESERV   75-0211
ZEICHENKETTEN STRING EIN PROGRAMMSYSTEM FUR DIE MASCHINENUNABHANGIGE BEARBEITUNG VON ZEICHENKETTEN IN FORTRAN   68-0265
ZEITDIMENSION BERECHNUNG INSTATIONAERER TEMPERATUREFELDER MIT FINITEN ELEMENTEN IN RAUM-UND ZEITDIMENSION. (CALCUL   74-0987
ZELEZOBETO-   STATICKE RESENI FYZIKALNE NELINEARNIHO CHOVANI ZELEZOBETO- NOVYCH DESEK (ANALYSIS OF PHYSICALLY NONL   74-0514
ZELEZOBETONOVYC RESENI MAZNIHO STAVU ZELEZOBETONOVYCH DESEK (PLASTIC ANALYSIS OF REINFORCED CONCRETE PLATES)   74-0495
ZEMLYANOI   RASCHET METODOM KONECHNYKH ELEMENTOV DINAMICHESKIKH KHARAKTEERISTIK ZEMLYANOI PLOTINY SOVMESTNO S OS   73-0437
ZERO   ON DIRICHLET PROBLEMS USING SUBSPACES WITH NEARLY ZERO BOUNDARY CONDITIONS   72-0514
ZHESTKOSTI   ISSLEDOVANIE PROCHNOSTI I ZHESTKOSTI PLASTIN. NAGRUZHENNYKH V SVOEI PLOSKOSTI. METHODOM KONECHNYKH E   73-0382
ZIRCONIUM   COMPARISON OF FINITE-ELEMENT AND EXPERIMENTAL STUDIES ON DEFORMATIONS OF ZIRCONIUM NOTCHED BEND SPEC   71-0359
ZLOZONYCH   ANALIZA ORGAN WYMUSZONYCH ZLOZONYCH UKLADOW LINIOWYCH METODA SZTYWNYCH ELEMENTOW SKONCZONYCH (ANALYS   72-0246
ZONE   ANCHOR ZONE STRESS ANALYSIS BY FINITE ELEMENT METHOD   68-0147
ZONE   ANCHORAGE ZONE STRESSES IN AXIALLY POST-TENSIONED I-SECTION MEMBERS WITH END BLOCKS   71-0305
ZONE   REBO: COMPUTER CODE FOR THE GENERATION OF A THREE-DIMENSIONAL FINITE-ELEMENT GRID OF THE NOZZLE-CYLI   74-1310
ZONE   SEEPAGE-ENVIRONMENTAL ANALYSIS OF THE SLIME ZONE OF A TAILINGS POND   74-0513
ZONE   SIZE OF PLASTIC ZONE AT TIP OF A CRACK IN PLANE STRAIN SLATE BY FINITE ELEMENT METHOD   73-0006
ZONE   STRESS DISTRIBUTION IN THE LIGAMENT ZONE OF PERFORATED CIRCULAR DISCS WITH SPECIAL REFERENCE TO THE   73-0827
ZONED   HYDRAULIC FRACTURING IN ZONED EARTH AND ROCKFILL DAMS: A REPORT OF AN INVESTIGATION   73-0873
ZONES   AN INVESTIGATION INTO THE METHODS OF DETECTING DEWETTED ZONES IN SOLID ROCKET PROPELLANTS USING HOLO   73-0890
ZONES   STRESS DISTRIBUTION AND ROCK FRACTURE ZONES IN THE ROOF OF LONGWALL FACE IN COAL MINE   73-0247
ZONES   TENSION ZONES IN DECKED ROCKFILL DAMS   73-0540
ZONING   EQUIPOTENTIAL ZONING OF TWO-DIMENSIONAL MESHES   63-0025
ZONING   EQUIPOTENTIAL ZONING OF TWO-DIMENSIONAL MESHES   64-0033
ZUSAMMENHANG   ZUSAMMENHANG ZWISCHEN OBERFLAECHENBESCHAFFENHEIT, VERBUND UND SPRENGWIRKUNG VON BEWEHRUNGSSTAEHLEN U   73-0450
ZUSAMMENWIRKEN ZUSAMMENWIRKEN VON EINZELNEN FERTIGTEILEN ALS GROSSFLAECHIGE SCHEIBE (INTERACTION OF INDIVIDUAL PREF   73-0552
ZUSTAND   BERECHNUNG VON STAHLBETONSCHEIBEN IM ZUSTAND II BEI ANNAHME EINES WIRKLICHKEITSNAHEN WERKSTOFFVERHAL   74-0978
ZUSTAND   DER GERISSENE ZUSTAND ZWEISEITIG GELAGERTER PLATTEN UNTER EINZELLASTEN - NICHTLINEARE BERECHNUNG MIT   75-0858
ZWEIDIMENSIONAL EINE EINFACHE KONSTRUKTION VON KOORDINATENFUNKTIONEN FUR DIE NUMERISCHE LOSUNG ZWEIDIMENSIONALER RAN   66-0134
ZWEISEITIG   DER GERISSENE ZUSTAND ZWEISEITIG GELAGERTER PLATTEN UNTER EINZELLASTEN - NICHTLINEARE BERECHNUNG MIT   75-0858
ZWEITER   ZUR BERECHNUNG VON PLATTEN NACH DER THEORIE ZWEITER. ORDNUNG MIT HILFE EINES HYBRIDEN DEFORMATIONS M   73-0558
ZYLINDRISCHER SCHWINGUNGEN ZYLINDRISCHER SCHRAUBENFEDERN (VIBRATIONS OF CYLINDRICAL COIL SPRINGS)   74-0331

Citation Listing

38-0001 MARGUERRE, K.
UBER DIE ANWENDUNG DER ENERGETISCHEN METHODE AUF
STABILITATSPROBLEME
HJHRB., D.V.L., 1938, PP. 252-262.

41-0001 HRENNIKOFF, A.
SOLUTION OF PROBLEMS IN ELASTICITY BY THE
FRAMEWORK METHOD
J. APPL. MECH., DECEMBER 1941, PP. A-169-A-175.

43-0001 COURANT, R.
VARIATIONAL METHODS FOR THE SOLUTION OF PROBLEMS
OF EQUILIBRIUM AND VIBRATION
BULL. AM. MATH. SOC., 49, 1-23, 1943.

43-0002 MCHENRY, D.
A LATTICE ANALOGY FOR THE SOLUTION OF PLANE STRESS
J. INST. CIVIL ENG. VOL. 21, 1943, PP: 59-82

47-0001 SYNGE, J. L.
PRAGER, W.
APPROXIMATION IN ELASTICITY BASED ON THE CONCEPT
OF FUNCTION SPACE
QUART. APPL. MATH., 5,241-69,1947.

47-0002 LEVY, S.
COMPUTATION OF INFLUENCE COEFFICIENTS FOR AIRCRAFT
STRUCTURES WITH DISCONTINUITIES AND SWEEPBACK
J. AERO. SOC., VOL. 14, NO. 10, OCTOBER 1947, PP.
547-560.

49-0001 GINTER, L. E.
STATISTICAL STATE OF STRESS STUDIED BY GRID
ANALYSIS: "NUMERICAL METHODS OF ANALYSIS"
NUMERICAL METH. OF ANALYSIS IN L.E. GINTER,
ENGRG., MACMILLAN, NEW YORK, 1949.

51-0001 LANGEFORS, B.
STRUCTURAL ANALYSIS OF SWEPT-BACK WINGS BY
MATRIX-TRANSFORMATION
SAAB TN 3, LINKOPING, 1951.

51-0002 GARVEY, S. J.
THE QUADRILATERAL SHEAR PANEL
AIRCRAFT ENG., VOL. 23, NO. 264, MAY 1951, PP.
134, 135, 144.

52-0001 SYNGE, J. L.
TRIANGULATION IN THE HYPERCIRCLE METHOD FOR PLANE
PROBLEMS
PROC. ROY. IRISH ACAD., VOL. 54A, 1952, PP.
341-367.

52-0002 WEHLE, L. B.
LANSING, W.
A METHOD FOR REDUCING THE ANALYSIS OF COMPLEX
REDUNDANT STRUCTURES TO A ROUTINE PROCEDURE
J. AERO. SC., VOL. 19, NO. 10, OCTOBER 1952, PP.
677-684.

54-0001 ARGYRIS, J. H.
KELSEY, S.
ENERGY THEOREMS AND STRUCTURAL ANALYSIS
AIRCRAFT ENGINEERING, 1954-1955

56-0001 TURNER, M. J.
CLOUGH, R. W.
MARTIN, H. C.
TOPP, L. J.
STIFFNESS AND DEFLECTION ANALYSIS OF COMPLEX
STRUCTURES
J. AEROSPACE SCIENCE, VOL. 23, 1956, PP. 805-823.

56-0002 DENKE, P. H.
THE MATRIC SOLUTION OF CERTAIN NONLINEAR PROBLEMS
IN STRUCTURAL ANALYSIS
J. AERONAUT. SCI. VOL. 23, 1956, PP. 3, 231.

56-0003 ARGYRIS, J. H.
THE MATRIX ANALYSIS OF STRUCTURES WITH CUT-OUTS
AND MODIFICATIONS
COMMUNICATION TO THE CONGRESS OF APPLIED
MECHANICS, BRUSSELS, SEPTEMBER 1956, PP: 131-142.

56-0004 ARGYRIS, J. H.
KELSEY, S.
STRUCTURAL ANALYSIS BY THE MATRIX FORCE METHOD
WITH APPLICATION TO AIRCRAFT WINGS
WISSENSCHAFTLICHE GESELLSCHAFT FUR LUFTFAHRT,
JAHRBUCH, 1956, PP: 78-98.

56-0005 ARGYRIS, J. H.
KELSEY, S.
THE MATRIX FORCE METHOD OF STRUCTURAL ANALYSIS AND
SOME NEW APPLICATIONS
AERONAUTICAL RESEARCH COUNCIL, R. AND M. 3034,
FEBRUARY, 1956 AND MAY, 1957.

56-0006 THOMANN, G. E. A.
AEROELASTIC PROBLEMS OF LOW ASPECT RATIO WINGS
(PART 1)
AIRCRAFT ENG., VOL. 28, NO. 324, FEBRUARY 1956,
PP. 36-42.

57-0001 SYNGE, J. L.
THE HYPERCIRCLE IN MATHEMATICAL PHYSICS
CAMBRIDGE, UNIVERSITY PRESS, 1957.

57-0002 ARGYRIS, J. H.
THE MATRIX THEORY OF STATICS
INGENIEUR-ARCHIV. VOL. 25, 1957, P: 174

57-0003 ARGYRIS, J. H.
DIE MATRIZENTHEORIE DER STATIK
INGENIEUR ARCHIV., VOL. XXV, MAY 1957, PP: 174-192

58-0001 MARTIN, H. C.
TRUSS ANALYSIS BY STIFFNESS CONSIDERATIONS
TRANS. AM. SOC. CIVIL. ENGR., VOL. 123, 1958, PP.
1182-1194.

58-0002 ARCHER, J. S.
A STIFFNESS MATRIX METHOD OF NATURAL MODE ANALYSIS
PROC. I.A.S. NATL. MEETING ON DYNAMICS AND
AEROELASTICITY, NOVEMBER 6-7, 1958, PP. 88-97.

59-0001 SZMELTER, J.
THE ENERGY METHOD OF NETWORKS OF ARBITRARY SHAPE
IN PROBLEMS OF THE THEORY OF ELASTICITY
PROC. I.U.T.A.M., SYMPOSIUM ON NON-HOMOGENEITY IN
ELASTICITY AND PLASTICITY, ED. BY W. OLSZAK,
PERGAMON PRESS, 1959.

59-0002 NEWMARK, N. M.
A METHOD OF COMPUTATION FOR STRUCTURAL DYNAMICS
PROCEEDINGS OF ASCE, VOL. EM 3, JULY 1959, PP.
67-94.

59-0003 MENDELSON, A.
HISCHBERG, M. H.
MANSON, S. S.
A GENERAL APPROACH TO THE PRACTICAL SOLUTION OF
CREEP PROBLEMS
J. OF BASIC ENG., TRANS., ASME SERIES D, VOL. 81,
1959, PP. 585-598.

59-0004 MANSON, S. S.
PRACTICAL SOLUTION OF PLASTIC DEFORMATION PROBLEMS
IN THE ELASTIC PLASTIC RANGE
NATL. AERON. SPACE ADMIN. TR R28, 1959

59-0005 ARGYRIS, J. H.
NOTE ON THE THEORY OF AIRCRAFT STRUCTURAL ANALYSIS
ZEITSCHRIFT FUR FLUGWISSENSCHAFTEN, VOL. 7, MARCH
1959, PP: 73-77.

59-0006 ARGYRIS, J. H.
RECENT DEVELOPMENTS OF MATRIX THEORY OF STRUCTURES
ADVISORY GROUP FOR AERONAUTICAL RESEARCH AND
DEVELOPMENTS, 10TH MEETING OF THE STRUCT. AND
MATER. PANEL, AACHEN, SEPTEMBER 1959.

59-0007 DENKE, P. H.
A GENERAL DIGITAL COMPUTER ANALYSIS OF STATICALLY
INDETERMINATE STRUCTURES
STRUCT. AND MATER. PANEL OF AGARD, AACHEN,
GERMANY, SEPTEMBER 1959. (ALSO NASA TN-1666,
DECEMBER 1963)

60-0001 CLOUGH, R. W.
THE FINITE ELEMENT IN PLANE STRESS ANALYSIS
PROC. 2ND A.S.C.E. CONF. ON ELECTRONIC
COMPUTATION, PITTSBURGH, PA., SEPTEMBER 1960.

60-0002 TURNER, M. J.
DILL, E. H.
MARTIN, H. C.
MELOSH, R. J.
LARGE DEFLECTIONS OF STRUCTURES SUBJECT TO HEATING
AND EXTERNAL LOADS
J. AEROSPACE SCI., VOL. 27, FEBRUARY 1960, PP.
97-106

60-0003 ADINI, A.
CLOUGH, R. W.
ANALYSIS OF PLATE BENDING BY THE FINITE ELEMENT
METHOD
NAT. SCI. FOUNDATION GRANT G7337, 1960.

60-0004 PADLOG, J.
HUFF, R. D.
HOLLOWAY, G. F.
THE INELASTIC BEHAVIOUR OF STRUCTURES SUBJECTED TO
CYCLIC, THERMAL AND MECHANICAL STRESSING
CONDITIONS
BELL AEROSYSTEMS CO. REPT., WADD TR 60-271, 1960.

60-0005 BUDIANSKY, B.
SLOSHING OF LIQUID IN CIRCULAR CHANNELS AND
SPHERICAL TANKS
J. AEROSPACE SCIENCE, VOL. 27, 1960, PP. 161-173.

60-0006 ORTEGA, M. A.
APPLICATION OF THE STIFFNESS METHOD TO A
SNAP-THROUGH PROBLEM
STRUCTURAL ANALYSIS RESEARCH MEMO., NO. 16, THE
BOEING COMPANY, AEROSPACE DIVISION, SEATTLE,
WASHINGTON, MAY 1960.

60-0007 ARGYRIS, J. H.
KELSEY, S.
ON THE MATRIX THEORY OF STRUCTURES
ZEITSCHRIFT FUR FLUGWISSENSCHAFTEN, VOL. 8, JUNE
1960, PP: 169-172.

60-0008 ARGYRIS, J. H.
INITIAL STRAINS IN THE MATRIX FORCE METHOD OF
STRUCTURAL ANALYSIS
J. OF ROY. AERO. SOC., VOL. 64, 1960, PP: 493-495.

60-0009 ARGYRIS, J. H.
MATRIZENTHEORIE DER STATIK
TECHNIKA CHRONIKA, VOL. 37, 1960, NO. 1, PP. 1-12,
NO. 2, PP. 75-89.

61-0001 MELOSH, R. J.
A STIFFNESS MATRIX FOR THE ANALYSIS OF THIN PLATES
IN BENDING
J. AEROSPACE SCI., VOL. 28, 1961, PP. 34-42.

61-0002 BLAND, D. R.
SOLUTIONS OF LAPLACE'S EQUATIONS
ROUTLEDGE AND KEGAN PAUL, LONDON, 1961.

61-0003 TAIG, I. C.
STRUCTURAL ANALYSIS BY THE MATRIX DISPLACEMENT
METHOD
ENG. ELECTRIC AVIATION REPORT NO. S017, 1961.

61-0004 GREENE, B. E.
STROME, D. R.
WEIKEL, R. C.
APPLICATION OF THE STIFFNESS METHOD TO THE
ANALYSIS OF SHELL STRUCTURES
PROC. AVIATION CONF., AMER. SOC. MECH. ENG., LOS
ANGELES, MARCH, 1961.

61-0005 ZIENKIEWICZ, O. C.
GERSTNER, R. W.
THE METHOD OF INTERFACE STRESS ADJUSTMENT AND ITS
USES IN SOME PLANE ELASTICITY PROBLEMS
INT. J. MECH. SCI., VOL. 2, 1961, PP. 267-276.

61-0006 ZIENKIEWICZ, O. C.
GERSTNER, R. W.
STRESS ANALYSIS AND SPECIAL PROBLEMS OF
PRESTRESSED DAMS
PROC. AM. SOC. CIV. ENG., VOL. 87, POI, 1961, PP.
7-43.

61-0007 COX, H. L.
VIBRATION OF MISSILES
AIRCRAFT ENG., VOL. 32, 1961, PP. 2-7, AND 48-55.

61-0008 ADINI, A.
ANALYSIS OF SHELL STRUCTURES BY THE FINITE ELEMENT
METHOD
PH. D. DISSERTATION, DEPT. OF CIVIL ENG., UNIV. OF
CALIF., BERKELEY, CALIF., 1961.

61-0009 MARTIN, H. C.
PLANE ELASTICITY PROBLEMS AND THE DIRECT STIFFNESS
METHOD
THE TREND IN ENGINEERING, VOL. 13, JANUARY 1961.

61-0010 ARGYRIS, J. H.
KELSEY, S.
THE VALIDITY OF THE INITIAL STRAIN CONCEPT
J. ROY. AERO. SOC., VOL. 65, 1961, PP: 129-138.

62-0001 CLOUGH, R. W.
THE STRESS DISTRIBUTION OF NORFOLK DAM
AD-701 898, AUGUST 1962, 15 PP.

62-0002 GALLAGHER, R. H.
STRESS ANALYSIS IN HEATED, COMPLEX SHAPES
J. AERO-SPACE SCIENCE, VOL. 29, 1962, PP. 700-707.

62-0003 MEISSNER, C. J.
DIRECT BEAM STIFFNESS MATRIX CALCULATIONS
INCLUDING SHEAR EFFECTS
J. AEROSPACE SCI., VOL. 29, 1962, PP. 247-248.

62-0004 MELOSH, R. J.
MATRIX METHODS OF STRUCTURAL ANALYSIS
J. AEROSPACE SCI., VOL. 29, 1962, PP. 365-366.

62-0005 BLATZ, P
KO, W. L.
APPLICATION OF FINITE ELASTICITY THEORY TO THE
DEFORMATION OF RUBBERS
TRANSACTIONS OF THE SOCIETY OF RHEOLOGY, VOL. VI,
PP 223-2511962.

62-0006 GREENE, B. E.
STIFFNESS MATRIX FOR BENDING OF A RECTANGULAR
PLATE ELEMENT WITH INITIAL MEMBRANE STRESSES
STRUCTURAL ANALYSIS RESEARCH MEMORANDUM, NO. 45,
THE BOEING COMPANY, SEATTLE, AUGUST, 1962.

62-0007 LUNDER, C. A.
DERIVATION OF A STIFFNESS MATRIX FOR A RIGHT
TRIANGULAR PLATE IN BENDING AND SUBJECTED TO
INITIAL STRESSES
M.SC., DEPT. OF AERON. AND ASTRON., UNIV. OF
WASHINGTON, SEATTLE, 1962.

62-0008 JENNING, A.
NATURAL VIBRATION OF A FREE SURFACE
AIRCRAFT ENG., VOL. 34, 1962, PP. 81-83.

62-0009 MCMAHAN, L. L.
DEVELOPMENT AND APPLICATION OF THE DIRECT
STIFFNESS METHOD FOR OUT-OF-PLANE BENDING USING A
TRIANGULAR PLATE ELEMENT
M.SC. THESIS, DEPT. OF AERON. AND ASTRON. UNIV. OF
WASHINGTON, SEATTLE, 1962.

62-0010 TOCHER, J. L.
ANALYSIS OF PLATE BENDING USING TRIANGULAR
ELEMENTS
PH. D., CIVIL ENG. DEPT., UNIV. OF CALIF.,
BERKELEY, 1962.

62-0011 MELOSH, R. J.
DEVELOPMENT OF THE STIFFNESS METHOD TO DEFINE
BOUNDS ON ELASTIC BEHAVIOUR OF STRUCTURES
PH.D. DISSERTATION, CIVIL ENG. DEPT., UNIVERSITY
OF WASHINGTON, SEATTLE, 1962.

62-0012 LAURSEN, J. I.
SHUBINSKI, R. P.
CLOUGH, R. W.
DYNAMIC MATRIX ANALYSIS OF FRAMED STRUCTURES
PROC. 4TH U.S. NAT. CONGR. APPL. MECH., VOL. 1,
1962, PP. 99-105.

62-0013 FRAEIJS DE VEUBEKE, B.
DUALITY BETWEEN DISPLACEMENT AND EQUILIBRIUM
METHOD WITH A VIEW TO OBTAINING UPPER AND LOWER
BOUNDS TO STATIC INFLUENCE COEFFICIENTS
PROC. OF FOURTEENTH MEETING OF THE AGARD
STRUCTURES AND MATERIALS PANEL, PARIS, JULY 1962.

62-0014 WILSON, E. L.
CLOUGH, R. W.
DYNAMIC RESPONSE BY A STEP BY STEP ANALYSIS
PROC. SYMP. USE OF COMPUTERS IN CIVIL ENG.,
LISBON, PAPER NO. 29, 1962.

62-0015 STEWARD, D. V.
ON AN APPROACH TO TECHNIQUES FOR THE ANALYSIS OF
THE STRUCTURES OF LARGE SYSTEMS OF EQUATIONS
S.I.A.M. REVIEW, VOL. 4, 1962.

63-0001 ARCHER, J. S.
CONSISTENT MASS MATRIX FOR DISTRIBUTED SYSTEMS
PROC. AMER. SOC. CIV. ENG., VOL. 89, ST4, 1963, P.
161.

63-0002 BEST, G. C.
A GENERAL FORMULA FOR STIFFNESS MATRICES OF
STRUCTURAL ELEMENTS
A.I.A.A. J., VOL. 1, 1963, PP. 1920-1921.

63-0003 BEST, G. C.
A FORMULA FOR CERTAIN TYPES OF STIFFNESS MATRICES
OF STRUCTURAL ELEMENTS
A.I.A.A. J., VOL. 1, 1963, PP. 212-213.

63-0004 CLOUGH, R. W.
WILSON, E. L.
STRESS ANALYSIS OF A GRAVITY DAM BY THE FINITE
ELEMENT METHOD
RILEM BULLETIN, NO. 19, 1963, PP. 45-54.

63-0005 GALLAGHER, R. H.
PADLOG, J.
DISCRETE ELEMENT APPROACH TO STRUCTURAL
INSTABILITY ANALYSIS
A.I.A.A. J., VOL. 1, NO. 6, 1963, PP. 1437-1439.

63-0006 GRAFTON, P. E.
STROME, D. R.
ANALYSIS OF AXISYMMETRIC SHELLS BY THE DIRECT
STIFFNESS METHOD
A.I.A.A. J., VOL. 1, 1963, PP. 2342-2347.

63-0007 MEYER, R. R.
HARMON, M. B.
CONICAL SEGMENT METHOD FOR ANALYZING OPEN CROWN
SHELLS OF REVOLUTION FOR EDGE LOADING
A.I.A.A.J., VOL. 1, NO. 4, APRIL 1963, PP.
886-891.

63-0008 MELOSH, R. J.
BASIS OF DERIVATION OF MATRICES FOR THE DIRECT
STIFFNESS METHOD
A.I.A.A. J., VOL. 1, 1963, PP. 1631-1637.

63-0009 MELOSH, R. J.
STRUCTURAL ANALYSIS OF SOLIDS
J. STRUCT. DIV., A.S.C.E., VOL. 89, NO. ST4, 1963,
PP. 205-223.

63-0010 PRZEMIENIECKI, J. S.
TRIANGULAR PLATE ELEMENTS IN THE MATRIX FORCE
METHOD OF STRUCTURAL ANALYSIS
A.I.A.A. J., VOL. 1, AUGUST 1963, PP. 1895-1897.

63-0011 MCCALLEY, R. B.
ROTARY INERTIA CORRECTION FOR MASS MATRICES
REPORT NO. DIG/SA 63-73, GENERAL ELECTRIC COMPANY,
KNOLLS ATOMIC POWER LAB., SCHENECTADY, NEW YORK,
1963.

63-0012 LECKIE, F. A.
LINDBERG, G. M.
THE EFFECT OF LUMPED PARAMETERS ON BEAM
FREQUENCIES
THE AERO. QUART., VOL. 14, 1963, P. 234.

63-0013 IRONS, B. M.
EIGENVALUE ECONOMISERS IN VIBRATION PROBLEMS
J. ROY. AERO. SOC., VOL. 67, 1963, PP. 526.

63-0014 ARGYRIS, J. H.
KELSEY, S.
KAMEL, H. A.
MATRIX METHODS OF STRUCTURAL ANALYSIS
AGARD-AGRAPH 72, PERGAMON PRESS, 1963.

63-0015 WILSON, E. L.
FINITE ELEMENT ANALYSIS OF TWO DIMENSIONAL
STRUCTURES
DIV. STRUCT. ENGR. STRUCT. MECH. REPT. 63-2, UNIV.
OF CALIF., BERKELEY, 1963, ALSO PH.D. THESIS,
UNIVERSITY OF CALIF., BERKLEY 1963.

63-0016 LINDBERG, G. M.
LUMPED PARAMETER METHODS APPLIED TO ELASTIC
VIBRATIONS
PH.D. THESIS, DEPARTMENT OF ENGINEERING,
UNIVERSITY OF CAMBRIDGE, NOVEMBER 1963.

63-0017 WISSMANN, J. W.
NUMERISCHE BERECHNUNG NICHTLINEARER ELASTISCHER
KOERPER
DISSERTATION, HANNOVER, 1963.

63-0018 FALK, S.
DAS VERFAHREN VON RAYLEIGH-RITZ MIT HERMITESCHEN
INTERPOLATIONSPOLYNOMEN
ZAMM, VOL. 43, 1963, PP. 149-166.

63-0019 ARGYRIS, J. H.
KELSEY, S.
KAMEL, W. H.
MATRIX METHODS OF STRUCTURAL ANALYSIS. A PRECIS OF
RECENT DEVELOPMENTS
PROC. 14TH MEETING OF STRUCTURES AND MATERIALS
PANEL, AGARD, 1963.

63-0020 SNOWDON, J. C.
TRANSVERSE VIBRATION OF BEAMS WITH INTERNAL
DAMPING, ROTATORY INERTIA AND SHEAR
J ACOUST. SOC. AM., VOL. 35, 1963, PP. 1997-2006.

63-0021 LINDBERG, G. M.
VIBRATION OF NON-UNIFORM BEAMS
THE AERONAUTICAL QUARTERLY, VOL. 14, NOVEMBER
1963.

63-0022 PERCY, J. H.
LODEN, W. A.
NAVARATNA, D. R.
A STUDY OF MATRIX ANALYSIS METHODS FOR INELASTIC
STRUCTURES
AIR FORCE SYSTEMS COMMAND, TECH. DOC. REP. NO.
RTD-TDR-63-4032, 1963.

63-0023 ARGYRIS, J. H.
KELSEY, S.
MODERN FUSELAGE ANALYSIS AND THE ELASTIC AIRCRAFT
BUTTERWORTHS SCIENTIFIC PUB., LONDON, 1963.

63-0024 LU, Z. A.
PENZIEN, J.
POPOV, E. P.
FINITE ELEMENT SOLUTION FOR THIN SHELLS OF
REVOLUTION
I.E.P. TECHNICAL REPORT SESM 63-3, U. OF CALIF.,
BERKELEY, CALIF., SEPTEMBER , 1963.

63-0025 WINSLOW, A. M.
EQUIPOTENTIAL ZONING OF TWO-DIMENSIONAL MESHES
LAWRENCE LIVERMORE LAB., REPORT UCRL-7312, JUNE
1963.

64-0001 CLOUGH, R. W.
TOCHER, J. L.
ANALYSIS OF THIN ARCH DAMS BY THE FINITE ELEMENT
METHOD
PROC. INTERN. SYMP. THEORY ARCH DAMS, SOUTHAMPTON,
UNIVERSITY, ENGLAND, APRIL 1964, PERGAMON PRESS,
NEW YORK, 1964.

64-0002 JONES, R. E.
A GENERALIZATION OF THE DIRECT-STIFFNESS METHOD OF
STRUCTURAL ANALYSIS
A.I.A.A. J., VOL. 2, 1964, PP. 821-826.

64-0003 KLEIN, S.
MATRIX ANALYSIS OF SHELL STRUCTURES
AD-616 907, JUNE 1964, 174 P.

64-0004 LOOV, R. E.
THE DETERMINATION OF STRESSES AND DEFORMATIONS OF
REINFORCED CONCRETE AFTER CRACKING
INT. CONF. ON STRUCT. SOLID MECH. IN ENGR., DESIGN
OF C.E. MAT., SOUTHAMPTON UNIV., APRIL 1964, PAPER
103, P. 4.

64-0005 PENZIEN, J.
LU, Z. A.
POPOV, E. P.
FINITE ELEMENT SOLUTION FOR THIN SHELLS OF
REVOLUTION
NASA CP-37, JULY 1964.

64-0006 PAULLING, J. R.
THE ANALYSIS OF COMPLEX SHIP STRUCTURES BY THE
FINITE ELEMENT TECHNIQUE
J. SHIP RESEARCH, VOL. 8, NO. 3, 1964, PP. 1-19.

64-0007 PIAN, T. H. H.
DERIVATION OF ELEMENT STIFFNESS MATRICES BY
ASSUMED STRESS DISTRIBUTIONS
A.I.A.A. J., VOL. 2, JULY 1964, PP. 1333-1336.

64-0008 PIAN, T. H. H.
DERIVATION OF ELEMENT STIFFNESS MATRICES
A.I.A.A. J., VOL. 2, 1964, PP. 576-577.

64-0009 POPE, G. G.
THE APPLICATION OF THE MATRIX DISPLACEMENT METHOD
IN PLANE ELASTO-PLASTIC STRESS PROBLEMS
PROC. CONF. MATRIX METH. STRUCT. MECH., AD-646
300, NOVEMBER 1964, PP. 635-654.

64-0010 POPOV, E. P.
PENZIEN, J.
LU, Z. A.
FINITE ELEMENT SOLUTION FOR AXISYMMETRIC SHELLS
PROC. A.S.C.E., 1964, PP. 119-145.

64-0011 PRZEMIENIECKI, J. S.
TETRAHEDRON ELEMENTS IN THE MATRIX FORCE METHOD OF
STRUCTURAL ANALYSIS
A.I.A.A. J., VOL. 2, 1964, PP. 1152-1154.

64-0012 SMITH, R. G.
WEBSTER, J. A.
MATRIX ANALYSIS OF BEAM COLUMNS BY THE METHOD OF
FINITE ELEMENTS
PROC. ROY. SOC., EDINBURGH, SEC. A. VOL. 67,
1964-65, PP. 156-173.

64-0013 ZIENKIEWICZ, O. C.
CHEUNG, Y. K.
BUTTRESS DAMS ON COMPLEX ROCK FOUNDATIONS
WATER POWER, VOL. 16, 1964, P. 193.

64-0014 ZIENKIEWICZ, O. C.
CHEUNG, Y. K.
THE FINITE ELEMENT METHOD FOR ANALYSIS OF ELASTIC
ISOTROPIC AND ORTHOTROPIC SLABS
PROC. INST. CIV. ENGRS., VOL. 28, 1964, PP.
471-488.

64-0015 SANDER, C.
BORNES SUPERIEURES ET INFERIEURES DANS L'ANALYSE
MATRICIELLE DES PLAQUES EN FLEXION TORSION
BULL. SOC. ROYALE DES SC. DE LIEGE, VOL. 33, 1964,
PP. 456-494.

64-0016 ARGYRIS, J. H.
RECENT ADVANCES IN MATRIX METHODS OF STRUCTURAL
ANALYSIS
PROGR. AERON. SCI., VOL. 4, PERGAMON PRESS, NEW
YORK, 1964.

64-0017 ARCHER, J. S.
CONSISTENT MATRIX FORMULATIONS FOR STRUCTURAL
ANALYSIS USING INFLUENCE COEFFICIENT TECHNIQUES
FIRST AM. INST. AERON. ASTRON. ANNUAL MEETING,
PAPER NO. 64-488, JUNE 29-JULY 2, 1964., ALSO
IN AIAA J., VOL. 3, NO. 10, OCTOBER 1965.

64-0018 FRAEIJS DE VEUBEKE, B.
UPPER AND LOWER BOUNDS IN MATRIX STRUCTURAL
ANALYSIS
MATRIX METHODS OF STRUCTURAL ANALYSES, EDITED BY
B. M. FRAEIJS DE VEUBEKE, PERGAMON PRESS, LONDON,
1964, PP. 165-201.

64-0019 PALMER, J. H.
ASHER, G. W.
CALCULATION OF AXISYMMETRIC LONGITUDINAL MODES FOR
FLUID-ELASTIC TANK-ULLAGE GAS SYSTEMS AND
COMPARISON WITH MODAL RESULTS
PROC. AIAA SYMP. ON STRUCT. DYN., AND
AERO-ELASTICITY, BOSTON, 1964.

64-0020 TURNER, M. J.
MARTIN, H. C.
WEIKEL, R. C.
FURTHER DEVELOPMENT AND APPLICATIONS OF THE
STIFFNESS METHOD
PRESENTED AT AGARD STRUCTURES AND MATERIALS PANEL,
PARIS, FRANCE, JULY 1962, AND PUBLISHED IN
AGARD-OGRAPH, 72, PERGAMON PRESS, 1964, PP.
203-266.

64-0021 TAIG, I. C.
KERR, R. I.
SOME PROBLEMS IN THE DISCRETE ELEMENT
REPRESENTATION OF AIRCRAFT STRUCTURES
MATRIX METHODS OF STRUCTURAL ANALYSES, EDITED BY
B. M. FRAEIJS DE VEUBEKE, PERGAMON PRESS, LONDON,
1964.

64-0022 IRONS, B. M.
BARLOW, J.
COMMENT ON MATRICES FOR THE DIRECT STIFFNESS
METHOD BY R. J. MELOSH
J. A.I.A.A., 1964, P. 403.

64-0023 DENKE, P. H.
DIGITAL ANALYSIS OF NON-LINEAR STRUCTURES BY THE
FORCE METHOD
AGARDOGRAPH 72, PERGAMON PRESS, OXFORD, 1964, PP.
317-342.

64-0024 HURTY, C. H.
DYNAMIC ANALYSIS OF STRUCTURAL SYSTEMS USING
COMPONENT MODES
A.I.A.A. J., VOL. 3, NO. 3, 1964.

64-0025 RASHID, Y.
SOLUTION OF ELASTO-STATIC BOUNDARY VALUE PROBLEMS
BY THE FINITE ELEMENT METHOD
PH.D. THESIS, UNIV. OF CALIF., BERKELEY, 1964.

64-0026 CORNELL, D. C.
ITERATIVE TECHNIQUE FOR THE STRESS ANALYSIS OF PCR
STRUCTURES USING THE FINITE ELEMENT METHOD
JOHN JAY HOPKINS LAB. FOR PURE AND APPLIED SCI.,
GENERAL ATOMIC, SAN DIEGO, CALIF., NOVEMBER 30,
1964, 21 P. (N.T.I.S. - GAMD-5926)

64-0027 KAMEL, H. A.
AUTOMATIC ANALYSIS OF FUSELAGES AND PROBLEMS OF
CONDITIONING
RESEARCH REPORT NO. 3, INSTITUT FUR STATIK UND
DYNAMIK DER LUFTUND RAUMFAHRTKONSTRUKTIONEN, DEC.
1964.

64-0028 JONES, B. D.
COHRON, G. T.
CONCEPT FEASIBILITY COMBAT TRACKED VEHICLE
SIGNATURE DUPLICATOR
DEFENCE ENG., CHRYLSTER CORP., DETROIT, MICH.,
FINAL TECH. REPORT, MAY 23, 1973 - MARCH 15, 1974,
133 P. (N.T.I.S.-AD-780 127/1)

64-0029 GURTIN, M. E.
VARIATIONAL PRINCIPLES FOR LINEAR INITIAL VALUE
PROBLEMS
QUART. APPL. MATH., VOL. 22, 1964, PP. 252-256.

64-0030 RASHID, Y. R.
SOLUTION OF ELASTO-STATIC BOUNDARY VALUE PROBLEMS
BY THE FINITE ELEMENT METHOD
PH.D., UNIV. OF CALIF., BERKELEY, 1964.

64-0031 BEST, G. C.
HELPFUL FORMULAS FOR INTEGRATING POLYNOMIALS IN
THREE DIMENSIONS
MATH. COMP., VOL. 18, 1964, PP. 310-312.

64-0032 WINSLOW, A. M.
AN IRREGULAR TRIANGLE MESH GENERATOR
LAWRENCE RADIATION LAB., UNIV. OF CALIF.,
LIVERMORE, REPORT NO. UCRL-7880, 1964.

64-0033 WINSLOW, A. M.
EQUIPOTENTIAL ZONING OF TWO-DIMENSIONAL MESHES
LAWRENCE RADIATION LAB., UNIV. OF CALIF.,
LIVERMORE, REPORT NO. UCRL-7312, 1964.

65-0001 BESSELING, J. F.
MATRIX ANALYSIS OF CREEP AND PLASTICITY PROBLEMS
CONF. ON MATRIX METH. IN STRUCT. MECH.,
WRIGHT-PATTERSON AIR FORCE BASE, OHIO, OCTOBER
1965, PP. 26-28.

65-0002 ARCHER, J. S.
RUBIN, C. P.
IMPROVED ANALYTIC LONGITUDINAL RESPONSE ANALYSIS
FOR AXISYMMETRIC LAUNCH VEHICLES, COMPUTER PROGRAM
DESCRIPTION
VOL. 2, NASA CR-346, WASHINGTON, D.C., DECEMBER
1965.

65-0003 ARGYRIS, J. H.
MATRIX ANALYSIS OF THREE-DIMENSIONAL ELASTIC
MEDIA: SMALL AND LARGE DISPLACEMENTS
A.I.A.A. J., VOL. 3, NO. 1, JANUARY 1965, PP.
45-51, ALSO IN J. ROY. AERO. SOC., VOL. 69, MARCH
1966, P. 452.

65-0004 ARGYRIS, J. H.
REINFORCED FIELDS OF TRIANGULAR ELEMENTS WITH
LINEARLY VARYING STRAIN: EFFECT OF INITIAL
STRAINS
J. ROY. AERO. SOC., VOL. 69, NOVEMBER 1965, PP.
799-801.

65-0005 ARGYRIS, J. H.
MATRIX DISPLACEMENT ANALYSIS OF ANISOTROPIC SHELLS
BY TRIANGULAR ELEMENTS
J. ROY. AERO. SOC., VOL. 69, NOVEMBER 1965, PP.
801-805.

65-0006 ARGYRIS, J. H.
TETRAHEDRON ELEMENTS WITH LINEARLY VARYING STRAIN
FOR THE MATRIX DISPLACEMENT METHOD
J. ROY. AERO. SOC., VOL. 69, DECEMBER 1965, PP.
877-880.

65-0007 ARGYRIS, J. H.
SOME RESULTS ON THE FREE-FREE OSCILLATIONS OF
AIRCRAFT TYPE STRUCTURES
REVUE DE LA SOCIETE FRANCAISE DE MECANIQUE, NO. 3,
1965.

65-0008 ARGYRIS, J. H.
THREE DIMENSIONAL ANISOTROPIC AND INHOMOGENEOUS
ELASTIC MEDIA MATRIX ANALYSIS FOR SMALL AND LARGE
DISPLACEMENTS
INGENIEUR ARCHIV, VOL. 34, NO. 1, 1965, PP. 33-35.

65-0009 BECKER, E. B.
BRISBANE, J. J.
APPLICATION OF THE FINITE ELEMENT METHOD TO STRESS
ANALYSIS OF SOLID PROPELLANT ROCKET GRAINS
SPECIAL REPORT NO. S-76, ROHM AND HAAS CO.,
HUNTSVILLE, ALA., NOVEMBER 1965, VOL. 1, 92 PP.,
AD-476 515, VOL. 2, PART 1, 96 PP. AD-474 031,
VOL. 2, PART 2, 104 PP. AD-476 735.

65-0010 CHEUNG, Y. K.
ZIENKIEWICZ, O. C.
PLATES AND TANKS ON ELASTIC FOUNDATIONS, AN
APPLICATION OF FINITE ELEMENT METHOD
INT. J. SOLIDS STRUCT., VOL. 1, 1965, PP. 451-461.

65-0011 CLOUGH, R. W.
RASHID, Y. R.
FINITE ELEMENT ANALYSIS OF AXISYMMETRIC SOLIDS
J. ENG. MECH. DIV., A.S.C.E., VOL. 91, NO. EM 1,
FEBRUARY 1965, PP. 71-85.

65-0012 CLOUGH, R. W.
THE FINITE ELEMENT METHOD IN STRUCTURAL MECHANICS
STRESS ANALYSIS, O.C. ZIENKIEWICZ AND G.S.
HOLISTER, (EDS) WILEY, CHAPTER 7, 1965.

65-0013 CORNELL, D. C.
SAFE PCPS, A COMPUTER PROGRAM FOR THE STRESS
ANALYSIS OF COMPOSITES BODIES OF REVOLUTION.
INPUT INSTRUCTIONS
GENERAL ATOMIC, SAN DIEGO, CALIF. JOHN JAY HOPKINS
LAB. FOR PURE AND APPLIED SCIENCE, AUGUST 1, 1965,
P. 43, GA-6588.

65-0014 CORNELL, D. C.
SAFE-AXISYM, A COMPUTER PROGRAM FOR THE DESIGN OF
COMPOSITE BODIES OF REVOLUTION. INPUT
INSTRUCTIONS
GENERAL ATOMIC, SAN DIEGO, CALIF., JOHN JAY
HOPKINS LAB FOR PURE AND APPLIED SCIENCE, OCTOBER
15, 1965, P. 37, GA-6697.

65-0015 DAWE, D. J.
A FINITE ELEMENT APPROACH TO PLATE VIBRATION
PROBLEMS
J. MECH. ENG. SCI., VOL. 7, 1965, P. 28.

65-0016 FRAEIJS DE VEUBEKE, B.
DISPLACEMENT AND EQUILIBRIUM MODELS IN THE FINITE
ELEMENT METHOD
STRESS ANALYSIS, O.C. ZIENKIEWICZ AND G.S.
HOLISTER, (EDS.) WILEY, CHAPTER 9, 1965.

65-0017 GLADWELL, G. M. L.
A FINITE ELEMENT METHOD FOR ACOUSTICS
CONGRESS INTERN. D'ACOUSTIQUE, 5TH LIEGE, BELGIUM,
SEPTEMBER, 1965.

65-0018 GUYAN, R. J.
DISTRIBUTED MASS MATRIX FOR PLATE ELEMENTS IN
BENDING
A.I.A.A. J., VOL. 3, 1965, PP. 567.

65-0019 HERRMANN, L. R.
ELASTICITY EQUATIONS FOR INCOMPRESSIBLE AND NEARLY
INCOMPRESSIBLE MATERIALS BY A VARATIONAL THEOREM
A.I.A.A. J., VOL. 3, NO. 10, OCTOBER 1965, PP.
1896-1900.

65-0020 HERRMANN, L. R.
ELASTIC TORSIONAL ANALYSIS OF IRREGULAR SHAPES
J. ENGR. MECH. DIV., A.S.C.E., VOL. 91, NO. EM 6,
DECEMBER 1965, PP. 11-20.

65-0021 IRONS, B. M.
DRAPER, J. K.
INADEQUACY OF NODAL CONNECTIONS IN A STIFFNESS
SOLUTION FOR PLATE BENDING
A.I.A.A. J., VOL. 3, NO. 5, 1965, P. 961.

65-0022 KING, I. P.
FINITE ELEMENT ANALYSIS OF TWO-DIMENSIONAL
TIME-DEPENDENT STRESS PROBLEMS
AD-701 894, JANUARY 1965, 109 P.

65-0023 LIPP, R.
FINITE ELEMENT SOLUTION FOR AXISYMMETRIC SHELLS
J. ENGR. MECH. DIV., A.S.C.E., VOL. 91, NO. EM3,
PP. 262-264., DISCUSSION BY S. KLEIN, PP. 264-268,
JUNE 1965, CLOSURE TO DISCUSSION, E. P. POPOV, ET
AL., VOL. 91, NO. EM 6, DECEMBER 1965, PP.
178-181.

65-0024 PARFITT, V. R.
FINITE ELEMENT STRESS ANALYSIS OF THIN SHELLS
AD-475 772, NOVEMBER 1965, P. 65.

65-0025 PERCY, J. H.
PIAN, T. H. H.
KLEIN, S.
NAVARATNA, D. R.
APPLICATION OF MATRIX DISPLACEMENT METHOD TO
LINEAR ELASTIC ANALYSIS OF SHELLS OF REVOLUTION
A.I.A.A. J., VOL. 3, NO. 11, NOVEMBER 1965, PP.
2138-2145.

65-0026 ZIENKIEWICZ, O. C.
IRONS, B. M.
NATH, P.
NATURAL FREQUENCIES OF COMPLEX FREE OR SUBMERGED
STRUCTURES BY THE FINITE ELEMENT METHOD
SYMP. ON VIBRATION IN CIV. ENG., INST. CIV. ENG.,
(BUTTERWORTH), LONDON, 1965.

65-0027 PERCY, J. H.
NAVARATNA, D. R.
KLEIN, S.
SABOR 1, A FORTRAN PROGRAM FOR THE LINEAR ELASTIC
AXISYMMETRIC LOADING BY USING THE MATRIX
DISPLACEMENT
AD-616 725, APRIL 1965, P. 104.

65-0028 PETYT, M.
THE APPLICATION OF FINITE ELEMENT TECHNIQUES TO
PLATE AND SHELL PROBLEMS
SOUTHHAMPTON UNIVERSITY, ENGLAND, AD-640 839,
FEBRUARY 1965, P. 58.

65-0029 RAPHAEL, J. M.
CLOUGH, R. W.
CONSTRUCTION STRESSES IN DWORSHAK DAM
AD-701 895, APRIL 1965, P. 90.

65-0030 RASHID, Y. R.
FINITE ELEMENT ANALYSIS OF AXISYMMETRIC COMPOSITE
STRUCTURES
GENERAL ATOMIC, SAN DIEGO, CALIF., JOHN JAY
HOPKINS LAB FOR PURE AND APPLIED SCIENCE, JUNE 4,
1965, CONTRACT AT(04-3)-167, P. 81, GA-6303.

65-0031 WILSON, E. L.
STRUCTURAL ANALYSIS OF AXISYMMETRIC SOLIDS
A.I.A.A. PAPER NO. 65-143, JANUARY 1965.

65-0032 WILSON, E. L.
A DIGITAL COMPUTER PROGRAM FOR THE FINITE ELEMENT
ANALYSIS OF SOLIDS WITH NONLINEAR MATERIAL
PROPERTIES
AEROJET-GENERAL CORP., TECHNICAL MEMO. NO. 23,
SACRAMENTO, CALIF., JULY 1965.

65-0033 ZIENKIEWICZ, O. C.
PROBLEMS IN ROCK MECHANICS
THE ENGINEER, FEBRUARY 12, 1965.

65-0034 ZIENKIEWICZ, O. C.
CHEUNG, Y. K.
FINITE ELEMENTS IN THE SOLUTION OF FIELD PROBLEMS
THE ENGINEER, SEPTEMBER 23, 1965, PP. 507-510.

65-0035 ZIENKIEWICZ, O. C.
CHEUNG, Y. K.
STRESSES IN BUTTRESS DAMS
WATER POWER, VOL. 17, 1965, P. 69.

65-0036 ZIENKIEWICZ, O. C.
CHEUNG, Y. K.
FINITE ELEMENT METHOD OF ANALYSIS FOR ARCH DAMS
SHELLS AND COMPARISON WITH FINITE DIFFERENCE
PROCEDURES
THEORY OF ARCH DAMS, PERGAMON PRESS, EDITOR
RYDZEWSKI, 1965. ALSO IN PROC OF INT. SYMP. ON THE
THEORY OF ARCH DAMS, SOUTHAMPTON, UNIVERSITY,
ENGLAND, APRIL 1964.

65-0037 GUYAN, R. J.
REDUCTION OF STIFFNESS AND MASS MATRICES
AIAA J., VOL. 3, 1965, P. 380.

65-0038 ZIENKIEWICZ, O. C.
FINITE ELEMENT PROCEDURES IN THE SOLUTION OF PLATE
AND SHELL PROBLEMS
STRESS ANALYSIS, O.C. ZIENKIEWICZ AND G.S.
HOLISTER, (EDS) WILEY, CHAPTER 8, 1965.

65-0039 HARTZ, B. J.
MATRIX FORMULATION OF STRUCTURAL STABILITY
PROBLEMS
PROC. A.S.C.E., J. STRUCT. DIV., VOL. 91, NO. ST6,
DECEMBER 1965.

65-0040 SWEDLOW, J. L.
YANG, W. H.
STIFFNESS ANALYSIS OF ELASTIC PLASTIC PLATES
GRADUATE AERONAUTICAL LAB., CALIFORNIA INST. OF
TECH., SM 65-10, 1965.

65-0041 MASSONNET, C. E.
NUMERICAL USE OF INTEGRAL PROCEDURES
CHAPTER 10 OF STRESS ANALYSIS, ED. O. C.
ZIENKIEWICZ AND G. S. HOLISTER, J. WILEY AND SON,
1965.

65-0042 SWEDLOW, J. L.
WILLIAMS, M. L.
YANG, W. M.
ELASTO PLASTIC STRESS IN CRACKED PLATES
CALCIT., REPORT SM, 65-19, CALIF. INST. OF TECH.,
1965.

65-0043 CLOUGH, R. W.
CHOPRA, A. K.
EARTHQUAKE STRESS ANALYSIS IN EARTH DAMS
STRUCTURES AND MATERIALS RESEARCH REPORT NO. 65-8,
UNIV. OF CALIFORNIA, BERKELEY, CALIF., 1965, ALSO
IN J. OF ENG. MECH. DIV., A.S.C.E., VOL. 92, NO.
EM2, PROC. PAPER 4793, APRIL 1967, PP. 197-212.

65-0044 ARCHER, J. S.
RUBIN, C. P.
IMPROVED LINEAR AXISYMMETRIC SHELL FLUID MODEL FOR
LAUNCH VEHICLE LONGITUDINAL RESPONSE ANALYSIS
PROC. CONF. ON MATRIX METHODS IN STRUCT. MECH.,
AIR FORCE INST. OF TECH., WRIGHT PATTERSON AIR
FORCE BASE, OHIO, OCTOBER, 1965.

65-0045 IRONS, B. M.
DRAPER, K.
LAGRANGE MULTIPLIER TECHNIQUES IN STRUCTURAL
ANALYSIS
J. AIAA., VOL. 3, 1965, PP. 1172-1176.

65-0046 WISSMANN, J. W.
NONLINEAR STRUCTURAL ANALYSIS; TENSOR FORMULATION
PROC. CONF. MATRIX METH. STRUCT. MECH., AF FLIGHT
DYNAMICS LAB., TECH. REPORT 66-80, DAYTON, OHIO,
1965, PP. 670-696.

65-0047 RIEGER, N.F.
MCCALLION, H.
THE NATURAL FREQUENCIES OF PORTAL FRAMES II
INT. J. MECH. SCI., VOL. 7, 1965, PP. 263-276.

65-0048 HENSHELL, R. D.
BENNETT, P.
MCCALLION, H.
MILNER, M.
NATURAL FREQUENCIES AND MODE SHAPES OF TRANSFORMER
CORES
PROC. INSTN. ELEC. ENGRS., VOL. 112, 1965, PP.
2133-2139.

65-0049 IRONS, B. M.
PLATE BENDING WITH NONCONFORMING ELEMENT:
FINE-MESH CONVERGENCE TEST
UNPUBLISHED, 1965

65-0050 DRAPER, K.
IRONS, B. M.
STRUCTURAL EIGENVALUE PROBLEMS: ELIMINATION OF
UNWANTED VARIABLES
AIAA J., VOL. 3, 1965, PP. 961-962.

65-0051 JENSEN, W. R.
LANSING, W.
FALBY, W. E.
MATRIX ANALYSIS METHODS FOR INELASTIC STRUCTURES
CONF. ON MATRIX METHODS IN STRUCT. MECH.,
WRIGHT-PATTERSON AFB, OHIO, OCTOBER 1965, PP.
26-28.

65-0052 KING, I. P.
ON THE FINITE ELEMENT ANALYSIS OF TWO-DIMENSIONAL
STRESS PROBLEMS WITH TIME DEPENDENT PROPERTIES
PH.D., UNIV. OF CALIF., BERKELEY, 1965.

65-0053 ARGYRIS, J. H.
TRIANGULAR ELEMENTS WITH LINEARLY VARYING STRAIN
FOR THE MATRIX DISPLACEMENT METHOD
J. ROY. AERO. SOC., TECH. NOTE, VOL. 69, OCTOBER
1965, PP. 711-713.

65-0054 ARGYRIS, J. H.
KAMEL, H.
AUTOMATIC SYSTEM FOR KINEMATIC ANALYSIS, NASKA
RESEARCH REPORT NO. 8, ISSUED BY THE INSTITUT FUR
STATIK UND DYNAMIK DER LUFT UND
RAUMFAHRTKONSTRUKTIONEN, OCTOBER 1965.

65-0055   ZIENKIEWICZ, O. C.
HOLISTER, G. S.
STRESS ANALYSIS
(EDS.) FRAEIJS DE VEUBEKE, B., ZIENKIEWICZ, O.
CLOUGH, C., HOLISTER, G. S., CHEUNG, Y.K., WILEY,
NEW YORK, 1965.

65-0056   WILSON, E. L.
STRUCTURAL ANALYSIS OF AXISYMMETRIC SOLIDS
A.I.A.A. J. VOL. 3, NO. 12, DECEMBER 1965, PP.
2269-2274.

65-0057   ARGYRIS, J. H.
KAMEL, H. A.
BOOLEAN MATRICES AS A STANDARD FACILITY IN A
MATRIX CODE
RESEARCH REPORT NO. 7, INSTITUT FUR STATIK UND
DYNAMIK DER LUFT UND RAUMFAHRTKONSTRUKTIONEN,
FEBRUARY 1965.

65-0058   BIRKHOFF, G.
DE BOOR, C.
PIECEWISE POLYNOMIAL INTERPOLATION AND
APPROXIMATION
IN APPROXIMATION OF FUNCTIONS, H. L. GARABEDIAN
(ED.), ELSEVIER, NEW YORK, 1965.

65-0059   ANON
ASKA - AUTOMATIC SYSTEM FOR KINEMATIC ANALYSIS
INST. STATIK. DYN. LUFT. RAUMFAHRTKONSTRUKTIONEN,
UNIV. STUTTGART, 1965.

65-0060   GALLAGHER, R. H.
COMMENTS ON DERIVATION OF ELEMENT STIFFNESS
MATRICES BY T. H. H. PIAN
A.I.A.A. J., VOL. 3, 1965, PP. 186-187.

65-0061   KHANNA, J.
CRITERION FOR SELECTING STIFFNESS MATRICES
A.I.A.A. J., VOL. 3, 1965, P. 1976.

65-0062   KLEIN, S.
DISCUSSION ON FINITE ELEMENT SOLUTION FOR
AXISYMMETRICAL SHELLS
J. ENG. MECH. DIV., A.S.C.E., VOL. 91, EMX, 1965,
PP. 262-268.

65-0063   POPE, G.
A DISCRETE ELEMENT METHOD FOR ANALYSIS OF PLANE
ELASTIC-PLASTIC STRESS PROBLEMS
ROY. AERON. ESTABL., TR65028, 1965.

65-0064   YETTRAM, A. L.
HUSAIN, H. M.
GENERALISED MATRIX FORCE AND DISPLACEMENT METHODS
FOR LINEAR STRUCTURAL ANALYSIS
A.I.A.A. J., VOL. 3, 1965, PP. 1154-1156.

65-0065   ANON
CONFERENCE ON MATRIX METHODS IN STRUCTURAL
MECHANICS
PROC. 1ST CONF., WRIGHT-PATTERSON AFB, OHIO,
REPORT AFFDL-TR-66-80, 1965.

65-0066   MARTIN, H. C.
DERIVATION OF STIFFNESS MATRICES FOR THE ANALYSIS
OF LARGE DEFLECTION AND STABILITY PROBLEMS
PROC. 1ST CONF. MATRIX METH. IN STRUCT. MECH.,
WRIGHT-PATTERSON AFB, OHIO, REPORT AFFDL TR66-80,
1965.

65-0067   ALWAY, G. G.
MARTIN, S. W.
AN ALGORITHM FOR REDUCING THE BANDWIDTH OF A
MATRIX OF SYMMETRICAL CONFIGURATION
COMP. J., VOL. 8, 1965, PP. 264-272.

66-0001   GAINS, J. H.
VOLTERRA, E.
TRANSVERSE VIBRATIONS OF CANTILEVER BARS OF
VARIABLE CROSS SECTIONS
J. ACOUST. SOC. AM., VOL. 39, 1966, PP.4.

66-0002   ARGYRIS, J. H.
ELASTIC-PLASTIC ANALYSIS OF THREE-DIMENSIONAL
MEDIA
ACTA TECHNICA ACAD., SCIENT., HUNGARICAE, VOL. 54,
1966.

66-0003   ARGYRIS, J. H.
ARBITRARY QUADRILATERAL SPAR WEBS FOR THE MATRIX
DISPLACEMENT METHOD
J. ROY. AERO. SOC., VOL. 70, FEBRUARY 1966, PP.
359-362.

66-0004   ARGYRIS, J. H.
ELASTO-PLASTIC MATRIX DISPLACEMENT ANALYSIS OF
THREE-DIMENSIONAL CONTINUA
J. ROY. AERO. SOC., VOL. 69, MARCH 1966, P. 452.

66-0005   ZIENKIEWICZ, O. C.
CHEUNG, Y. K.
PLATE AND SHELL PROBLEMS, FINITE ELEMENT
DISPLACEMENT APPROACH
PROCEEDINGS OF THE SYMPOSIUM ON THE USE OF
ELECTRONIC DIGITAL COMPUTERS IN STRUCTURAL
ENGINEERING, JULY 1966, UNIVERSITY OF
NEWCASTLE-UPON-TYNE.

66-0006   ARGYRIS, J. H.
MEMBRANE PARALLELOGRAM ELEMENT WITH LINEARLY
VARYING EDGE STRAIN FOR THE MATRIX DISPLACEMENT
METHOD
J. ROY. AERO. SOC., VOL. 70, MAY 1966, PP.
599-604.

66-0007   ARGYRIS, J. H.
A TAPERED THICKNESS TRIM 6 ELEMENT FOR THE MATRIX
DISPLACEMENT METHOD
J. OF ROY. AERO. SOC., VOL. 70, NOVEMBER 1966, PP.
1040-1043.

66-0008   ARGYRIS, J. H.
CONTINUA AND DISCONTINUA
PROC. CONF. MATRIX METH. STRUCT. MECH., AD-646
300, NOVEMBER 1966, PP. 11-190.

66-0009   ARGYRIS, J. H.
TRIAX 6 ELEMENT FOR AXISYMMETRIC ANALYSIS BY THE
MATRIX DISPLACEMENT METHOD
J. ROY. AERO. SOC., VOL. 3, DECEMBER 1966, PP.
1098-1102.

66-0010   ARGYRIS, J. H.
MATRIX DISPLACEMENT ANALYSIS OF PLATES AND SHELLS
INGENIEUR ARCHIV, VOL. 35, NO. 2, 1966, PP.
102-142. ALSO IN AD-626 836.

66-0011   BAZELEY, G. P.
CHEUNG, Y. K.
IRONS, B. M.
ZIENKIEWICZ, O. C.
TRIANGULAR ELEMENTS IN PLATE BENDING CONFORMING
AND NON-CONFORMING SOLUTIONS
PROC. CONF. MATRIX METH. STRUCT. MECH., AD-646
300, NOVEMBER, 1966, PP. 547-576.

66-0012   BEST, G. C.
VIBRATION ANALYSIS OF A CANTILEVERED SQUARE PLATE
BY THE STIFFNESS MATRIX METHOD
PROC. CONF. MATRIX METH. STRUCT. MECH., AD-646
300, NOVEMBER, 1966, PP. 849-861.

66-0013   BLAKE, W.
APPLICATION OF THE FINITE ELEMENT METHOD OF
ANALYSIS IN SOLVING BOUNDARY VALUE PROBLEMS IN
ROCK MECHANICS
INT. J. ROCK. MECH. AND MIN. SCI., VOL. 3, NO. 3,
1966, PP. 169-180.

66-0014   BOGNER, F. K.
FOX, R. L.
SCHMIT, L. A.
THE GENERATION OF INTERELEMENT, COMPATIBLE
STIFFNESS AND MASS MATRICES BY THE USE OF
INTERPOLATION FORMULAS
PROC. CONF. MATRIX METH. STRUCT. MECH., AD-646
300, NOVEMBER 1966, PP. 397-444. ALSO IN PROC. OF
THE FIRST AIR FORCE CONFERENCE ON MATRIX METHODS
IN STRUCTURAL MECHANICS, AFFDL-TR-66-80, NOVEMBER
1965.

66-0015   BROWN, C. B.
KING, I. P.
AUTOMATIC EMBANKMENT ANALYSIS, EQUILIBRIUM AND
INSTABILITY CONDITIONS
GEOTECHNIQUE, VOL. 16, NO. 3, 1966, PP. 209-219.

66-0016   BUTLIN, G. A.
LECKIE, F. A.
A STUDY OF FINITE ELEMENTS APPLIED TO PLATE
FLEXURE
NUMERICAL METHODS FOR VIBRATION PROBLEMS,
I.S.V.R., VOL. 2, JULY 1966, PP. 26-37.

66-0017   CHEN, H. Y.
A COMPUTER PROGRAM FOR THE STATIC AND DYNAMIC
FINITE ELEMENT ANALYSIS OF AXISYMMETRIC THIN
SHELLS
UNIV. OF CALIFORNIA, NO. 67-18095, OCTOBER 1966,
PP. 146.

66-0018   CHEN, Y.
TORIDIS, T. G.
SOME RESULTS IN THE FINITE ELEMENT MODELING OF
THIN PLATES
NO. 67-22753, GEORGE WASHINGTON UNIVERSITY, 1966.

66-0019   CHOPRA, A. K.
THE IMPORTANCE OF THE VERTICAL COMPONENT OF
EARTHQUAKE MOTIONS
BULL. SEIS. SOC. AMER., VOL. 56, NO. 5, OCTOBER
1966.

66-0020 CHOW, H. Y.
POPOV, E. P.
FINITE ELEMENT SOLUTION OF AXISYMMETRICAL DYNAMIC
PROBLEMS OF SHELLS OF REVOLUTION
UNIV. OF CALIF., NASA-CR-76100, SESM-66-3, GRANT
NSG-274, NO. 66-30175, APRIL 1966, 52 PP.

66-0021 CLOUGH, R. W.
TOCHER, J. L.
FINITE ELEMENT STIFFNESS MATRICES FOR ANALYSIS OF
PLATE BENDING
PROC. CONF. MATRIX METH. STRUCT. MECH. AD-646 300,
NOVEMBER 1966, PP. 515-546.

66-0022 DAWE, D. J.
PARALLELOGRAM ELEMENT IN THE SOLUTION OF RHOMBIC
CANTILEVER PLATE BENDING
J. STRAIN ANALYSIS, VOL. 3, 1966.

66-0023 DONG, S. B.
ANALYSIS OF LAMINATED SHELL FOR REVOLUTION
J. ENG. MECH. DIV., A.S.C.E., VOL. 92, NO. EM 6,
PROC. PAPER 5044, DECEMBER 1966, PP. 135-155.

66-0024 FELIPPA, C. A.
REFINED FINITE ELEMENT ANALYSIS OF LINEAR AND
NONLINEAR TWO-DIMENSIONAL STRUCTURES
STRUCTURAL ENGR. LAB., UNIV. OF CALIF., BERKLEY,
REPORT NO. SESM-66-22, OCTOBER 1966.

66-0025 FINN, W. D.
STATIC AND DYNAMIC STRESSES IN SLOPES
PROC. FIRST CONF. INT. SOC. OF ROCK. MECH.,
LISBON, PORTUGAL, 1966.

66-0026 FRAEIJS DE VEUBEKE, B.
BENDING AND STRETCHING OF PLATES, SPECIAL MODELS
FOR UPPER AND LOWER BOUNDS
PROC. CONF. MATRIX METH. STRUCT. MECH., AD-646300,
NOVEMBER 1966, PP. 863-886.

66-0027 FULTON, R. E.
THE ACCURACY OF THE FINITE ELEMENT METHODS IN
CONTINUUM PROBLEMS
FIFTH U.S. NATIONAL CONGRESS OF APPLIED MECHANICS,
JUNE 1966

66-0028 GELLATLY, R. A.
GALLAGHER, R. H.
DEVELOPMENT OF ADVANCED STRUCTURAL OPTIMIZATION
PROGRAMS AND THEIR APPLICATION TO LARGE ORDER
STRAINS
PROC. CONF. MATRIX METH. STRUCT. MECH., AD-646
300, NOVEMBER 1966, PP. 231-253.

66-0029 GOODMAN, R. E.
ON THE DISTRIBUTION OF STRESSES AROUND CIRCULAR
TUNNELS IN NON-HOMOGENEOUS ROCKS
PROC. FIRST CONG. INT. SOC. OF ROCK MECHANICS,
LISBON, PORTUGAL, 1966.

66-0030 GREENE, B. E.
APPLICATION OF GENERALIZED CONSTRAINTS IN THE
STIFFNESS METHOD OF STRUCTURAL ANALYSIS
AD-807 049, FEBRUARY 1966, 49 PP.

66-0031 GREENING, T. A.
ADVANCED WIRE-WOUND TUNGSTEN NOZZLES
AD-489 658, SEPTEMBER 1966, 67 PP.

66-0032 HANSTEEN, H.
FINITE ELEMENT DISPLACEMENT ANALYSIS OF PLATE
BENDING BASED ON RECTANGULAR ELEMENTS
INT. SYMP. ON USE OF ELECT. DIG. COMP. IN STRUCT.
ENGR., NEWCASTLE, 1966.

66-0033 HENSHELL, R. D.
WARBURTON, G. B.
FINITE ELEMENT TECHNIQUES APPLIED TO THE
VIBRATIONS OF PLATE SYSTEMS
NUM. METHODS FOR VIB. PROBLEMS, I.S.V.R., VOL. 2,
JULY 1966, PP. 18-25.

66-0034 HERRMANN, L. R.
A FINITE ELEMENT ANALYSIS FOR CYLINDRICAL SHELLS
AND PLATES
AD-703 705, AUGUST 1966, 71 PP.

66-0035 HERRMANN, L. R.
A BENDING ANALYSIS FOR PLATES
PROC. CONF. MATRIX METH. STRUCT. MECH., AD-646
300, NOVEMBER 1966, PP. 577-603.

66-0036 HICKS, G. W. JR.
A GENERAL TECHNIQUE FOR BUCKLING ANALYSIS WITH A
FINITE ELEMENT SOLUTION APPLICATION
AD-675 798, FEBRUARY 1966, 219 P.

66-0037 HICKS, G. W. JR.
A FINITE ELEMENT ANALYSIS FOR THE STRENGTH OF BEAM
COLUMNS
AD-675 797, SEPTEMBER 1966, 116 P.

66-0038 IRONS, B. M.
DISTRIBUTED MASS MATRIX FOR PLATE ELEMENT BENDING
(DISCUSSION)
A.I.A.A. J., VOL. 4, 1966, P. 189.

66-0039 IRONS, B. M.
NUMERICAL INTEGRATION APPLIED TO FINITE ELEMENT
METHODS
CONF. ON USE OF DIGITAL COMPUTERS IN STRUCTURAL
ENGINEERING, UNIV. OF NEWCASTLE, JULY 1966.

66-0040 IRONS, B. M.
ENGINEERING APPLICATIONS OF NUMERICAL INTEGRATION
IN STIFFNESS METHOD
A.I.A.A. J., VOL. 4, 1966, PP. 2035-2037.

66-0041 JONES, R. E.
STROME, D. R.
A SURVEY OF THE ANALYSIS OF SHELLS BY THE
DISPLACEMENT METHOD
PROC. CONF. MATRIX METH. STRUCT. MECH., AD-646
300, NOVEMBER 1966, PP. 205-230.

66-0042 JONES, R. E.
STROME, D. R.
DIRECT STIFFNESS METHOD OF ANALYSIS OF SHELLS OF
REVOLUTION UTILIZING CURVED ELEMENTS
A.I.A.A. J., 1966, PP. 1519-1525.

66-0043 KAPUR, K. K.
PREDICTION OF PLATE VIBRATIONS USING A CONSISTENT
MASS MATRIX
A.I.A.A. J., VOL. 4, 1966, PP. 565-566.

66-0044 KAPUR, K. K.
VIBRATIONS OF A TIMOSHENKO BEAM, USING
FINITE-ELEMENT APPROACH
J. ACOUST. SOC. OF AMER., VOL. 40, NO. 5, FEBRUARY
1966, PP. 1058-1063.

66-0045 KHANNA, J.
HOOLEY, R. F.
COMPARISON AND EVALUATION OF STIFFNESS MATRICES
A.I.A.A. J., VOL. 4, 1966, PP. 2105-2111.

66-0046 KLEIN, S.
SYLVESTER, R. J.
THE LINEAR ELASTIC DYNAMIC ANALYSIS OF SHELLS OF
REVOLUTION BY THE MATRIX DISPLACEMENT METHOD
PROC. CONF. MATRIX METH. STRUCT. MECH., AD-646
300, NOVEMBER 1966, PP. 299-328.

66-0047 MARTIN, G. P.
SEED, H. B.
AN INVESTIGATION OF THE DYNAMIC RESPONSE
CHARACTERISTICS OF THE BON TEMPE DAM
SOIL MECH. AND BITUMINOUS MAT. RES. LAB., UNIV. OF
CALIF., BERKELY, REPORT NO. TE-66-2, FEBRUARY
1966.

66-0048 MASON, V.
THE FINITE ELEMENT TECHNIQUE APPLIED TO AN
ACOUSTIC DAMPING PROBLEM
NUM. METHODS FOR VIB. PROBLEMS, VOL. 2, I.S.V.R.,
JULY 1966, PP. 51-66.

66-0049 MCLAY, R. W.
SUTHERLAND, W. H.
GREENE, B. E.
JONES, R. E.
ANALYSIS OF AXISYMMETRIC SOLIDS BY THE DIRECT
STIFFNESS METHOD
AD-864 333, FEBRUARY 1966, P. 61.

66-0050 MARTIN, H. C.
ON THE DERIVATION OF STIFFNESS MATRICES FOR THE
ANALYSIS OF LARGE DEFLECTION AND STABILITY
PROBLEMS
PROC. 1ST CONF. ON MATRIX METHODS IN STRUCT.
MECH., NOVEMBER 1966, PP. 697-716, ALSO IN
AFFDL-TR-66-80, NOVEMBER 1965.

66-0051 MEEK, J. L.
THE FINITE ELEMENT METHOD IN STRUCTURAL MECHANICS
CIV. ENG. TRANS. I.E.AUST., VOL. E8, NO. 2,
OCTOBER 1966, PP. 166-178.

66-0052 MELOSH, R. J.
A FLAT TRIANGULAR SHELL ELEMENT STIFFNESS MATRIX
PROC. CONF. MATRIX METH. STRUCT. MECH., AD-646
300, NOVEMBER 1966, PP. 503-514.

66-0053 MELOSH, R. J.
LANG, T. E.
MODIFIED POTENTIAL ENERGY MASS REPRESENTATION FOR
FREQUENCY PREDICTION
PROC. CONF. MATRIX METH. STRUCT. MECH., AD-646
300, NOVEMBER 1966, PP. 445-456.

66-0054 MOE, J.
TONNESSEN, A.
MODEL EXPERIMENTS AND FINITE ELEMENT ANALYSIS OF
STRESS IN AN OPEN SHIP
EUROPEAN SHIPBUILDING, VOL. 15, NO. 5, 1966, PP.
76-81.

66-0055 MOOLEY, R. F.
HIBBERT, P. D.
BOUNDING PLANE STRESS SOLUTIONS BY FINITE ELEMENTS
J. STRUCT. DIV. A.S.C.E., VOL. 92, NO. ST 1, PAPER
9663, FEBRUARY 1966, PP. 39-48.

66-0056 NAVARATNA, D. R.
COMPUTATION OF STRESS RESULTANTS IN FINITE ELEMENT
ANALYSIS
A.I.A.A. J., VOL. 4, NO. 11, NOVEMBER 1966, PP.
2058-2060.

66-0057 ODEN, J. T.
ANALYSIS OF LARGE DEFORMATIONS OF ELASTIC
MEMBRANES BY THE FINITE ELEMENT METHOD
IASS SYMP., LENINGRAD, USSR, SEPTEMBER 1966.

66-0058 ODEN, J. T.
CALCULATION OF GEOMETRIC STIFFNESS MATRICES FOR
COMPLEX STRUCTURES
A.I.A.A. J., VOL. 4, 1966, PP. 1480-1482.

66-0059 PESTEL, E.
DYNAMIC STIFFNESS MATRIX FORMULATION BY MEANS OF
HERMITIAN POLYNOMIALS
PROC. CONF. MATRIX METH. STRUCT. MECH., AD-646
300, NOVEMBER 1966, PP. 479-502.

66-0060 PETYT, M.
STRUCTURAL VIBRATION ANALYSIS USING TRIANGULAR
FINITE ELEMENTS
NUMERICAL METHODS FOR VIBRATION PROBLEMS,
I.S.V.R., VOL. 2, JULY 1966, PP. 55-56.

66-0061 PIAN, T. H. H.
ELEMENT STIFFNESS MATRICES FOR BOUNDARY
COMPATIBILITY AND FOR PRESCRIBED BOUNDARY STRESSES
PROC. CONF. MATRIX METH. STRUCT. MECH., AD-646
300, NOVEMBER 1966, PP. 457-478.

66-0062 POPE, G. G.
A DISCRETE ELEMENT METHOD FOR THE ANALYSIS OF
PLANE ELASTO-PLASTIC STRESS PROBLEMS
AERON. QUART., VOL. 17, PART 1, FEBRUARY 1966, PP.
83-104. ALSO IN R.A.E. FARNBOROUGH, T. R. 65028,
1965.

66-0063 PRZEMIENIECKI, J. S.
EQUIVALENT MASS MATRICES FOR RECTANGULAR PLATES IN
BENDING
A.I.A.A. J., VOL. 1, NO. 5, MAY 1966, PP. 949-950.

66-0064 PRZEMIENIECKI, J. S.
QUADRATIC MATRIX EQUATIONS FOR DETERMINING
VIBRATION MODES AND FREQUENCIES OF CONTINUOUS
ELASTIC SYSTEMS
PROC. CONF. MATRIX METH. STRUCT. MECH., AD-646
300, NOVEMBER 1966, PP. 779-902.

66-0065 PRZEMIENIECKI, J. S.
BADER, R. M.
BOZICH, W. F.
JOHNSON, J. R.
MYKTOW, W. J.
MATRIX METHODS IN STRUCTURAL MECHANICS
PROC. OF CONF. MATRIX METH. STRUCT. MECH.,
AFFDL-TR-66-80, AD-646 300, NOVEMBER 1966.

66-0066 RASHID, Y. R.
ANALYSIS OF AXISYMMETRIC COMPOSITE STRUCTURES BY
THE FINITE ELEMENT METHOD
NUCLEAR ENGR. AND DESIGN, VOL. 3, NO. 1, JANUARY
1966, PP. 163-182.

66-0067 REYES, S. F.
DEERE, D. U.
ELASTO-PLASTIC ANALYSIS OF UNDERGROUND OPENINGS BY
THE FINITE ELEMENT METHOD
FIRST INT. CONGRESS ON ROCK MECHANICS, VOL. 2,
LISBON, 1966, PP. 477-486., ALSO IN PH.D. THESIS
DEPARTMENT OF CIVIL ENGINEERING, UNIVERSITY OF
ILLINOIS, 1966, PP. 16-36.

66-0068 SEVERN, R. T.
TAYLOR, P.
PLATE VIBRATION PROBLEMS USING THE FINITE ELEMENT
METHOD WITH ASSUMED STRESS DISTRIBUTION AND
TRIANGULAR ELEMENTS
NUM. METH. FOR VIB. PROBLEMS, ISVR., VOL. 2, JULY
1966, PP. 38-54.

66-0069 SEVERN, R. T.
TAYLOR, P.
THE FINITE ELEMENT METHOD FOR FLEXURE OF SLABS
WHEN STRESS DISTRIBUTIONS ARE ASSUMED
PROC. INSTN. OF CIVIL ENGINEERS, 1966, VOL. 34, P.
153.

66-0070 STRICKLIN, J. A.
NAVARATNA, D. R.
PIAN, T. H. H.
IMPROVEMENTS ON THE ANALYSIS OF SHELLS OF
REVOLUTION BY THE MATRIX DISPLACEMENT METHOD
A.I.A.A. J. VOL. 4, NO. 11, NOVEMBER 1966, PP.
2069-2071.

66-0071 TAIG, I. C.
AUTOMATED STRESS ANALYSIS USING SUBSTRUCTURES
PROC. CONF. MATRIX METH. STRUCT. MECH., AD-646
300, NOVEMBER 1966, PP. 275-298.

66-0072 UTKU, S.
COMPUTATION OF STRESSES IN TRIANGULAR FINITE
ELEMENTS
NASA-CR-76292, JPL-TR-32-948, CONTRACT NAS7-100,
JUNE 1966, P. 37, N66-30787.

66-0073 UTKU, S.
STIFFNESS MATRICES FOR THIN TRIANGULAR ELEMENTS OF
NONZERO GAUSSIAN CURVATURE
4TH AEROSPACE SCIENCES MEET., A.I.A.A. J., LOS
ANGELES, CALIF., PAPER 66-530, JUNE 1966.

66-0074 VISSER, W.
A FINITE ELEMENT METHOD FOR THE DETERMINATION OF
NON-STATIONARY TEMPERATURE DISTRIBUTIONS AND
THERMAL
PROC. CONF. MATRIX METH. STRUCT. MECH., AD-646
300, NOVEMBER 1966, PP. 925-944.

66-0075 WALLACE, C. D. JR.
MATRIX ANALYSIS OF AXISYMMETRIC SHELLS UNDER
GENERAL LOADING
AD-639 448, MAY 1966, P. 111.

66-0076 WILSON, E. L.
NICKELL, R. E.
APPLICATION OF THE FINITE ELEMENT METHOD TO HEAT
CONDUCTION ANALYSIS
NUCLEAR ENGR. AND DESIGN, VOL. 4, NO. 3, OCTOBER
1966, PP. 276-286.

66-0077 ZIENKIEWICZ, O. C.
THE FINITE ELEMENT METHOD AND ITS APPLICATION IN
VIBRATION ANALYSIS
NUM. METHODS FOR VIBRATION PROBLEMS, I.S.V.R.,
VOL. 2, JULY 1966, PP. 1-17.

66-0078 ZIENKIEWICZ, O. C.
ELASTIC TORSIONAL ANALYSIS OR IRREGULAR SHAPES
J. ENG. MECH. DIV., A.S.C.E., VOL. 92, NO. EM 4,
AUGUST 1966, PP. 78-79.

66-0079 ZIENKIEWICZ, O. C.
CHEUNG, Y. K.
STAGG, K. G.
STRESSES IN ANISOTROPIC MEDIA WITH PARTICULAR
REFERENCE TO PROBLEMS OF ROCK MECHANICS
J. STRAIN ANALYSIS, VOL. 1, 1966, PP. 172-182.

66-0080 ZIENKIEWICZ, O. C.
MAYER, P.
CHEUNG, Y. K.
SOLUTION OF ANISOTROPIC SEEPAGE PROBLEMS BY FINITE
ELEMENTS
PROC. AMER. SOC., CIV. ENG., VOL. 98, NO. EM 1,
1966, PP. 111-120.

66-0081 KAPUR, K. K.
HARTZ, B. J.
STABILITY OF PLATES USING THE FINITE ELEMENT
METHOD
PROC. A.S.C.E., J., ENG. MECH. DIV., VOL. 92, NO.
EM2, APRIL 1966.

66-0082 MALLETT, R. H.
HELLE, E.
FORMULATION AND EVALUATION OF A TOROIDAL RING
DISCRETE ELEMENT
BELL AEROSYSTEMS CORPORATION REPORT NO.
9500-941-001, MAY 1966.

66-0083 HELLE, E.
FORMULATION AND EVALUATION OF A TRIANGULAR
CROSS-SECTION RING DISCRETE ELEMENT
BELL AEROSYSTEMS CORPORATION REPORT NO.
9500-941-003, JUNE 1966.

66-0084 MALLETT, R. H.
JORDAN, S.
FORMULATION AND EVALUATION OF A TRAPEZOIDAL
CROSS-SECTION RING DISCRETE ELEMENT
BELL AEROSYSTEMS CORP. REPORT NO. 9500-941-004,
NOVEMBER 1966.

66-0085 ERGATOUDIS, J.
QUADRILATERAL ELEMENTS IN PLANE ANALYSIS
M. SC. THESIS, UNIV. OF WALES, SWANSEA, 1966.

66-0086 CHEUNG, Y. K.
PEDRO, J. O.
AUTOMATIC PREPARATION OF DATA CARDS FOR THE FINITE
ELEMENT METHOD
RES. REP., UNIVERSITY COLLEGE OF SWANSEA, MARCH
1966.

66-0087 BIRKHOFF, G.
DEBOOR, C.
SWARTZ, B.
WENDROFF, B.
RAYLEIGH-RITZ APPROXIMATION OF PIECEWISE CUBIC
POLYNOMIALS
SIAM JOURNAL OF NUMERICAL ANALYSIS, VOL. 3, NO. 2,
1966, PP. 188-203.

66-0088 CHEUNG, Y. K.
APPLICATION OF THE FINITE ELEMENT METHOD TO
PROBLEMS OF ROCK MECHANICS
PROC. OF THE INT. CONGRESS ON ROCK MECHANICS,
SEPTEMBER 1966. NATIONAL LABORATORY OF CIVIL
ENGINEERING, LISBON, PORTUGAL

66-0089 VARGA, R. S.
HERMITE INTERPOLATION TYPE RITZ METHODS FOR TWO
POINT BOUNDARY VALUE PROBLEMS
IN J. H. BRAMBLE (ED.), NUMERICAL SOLUTION OF
PARTIAL DIFFERENTIAL EQUATIONS, ACADEMIC PRESS,
NEW YORK, 1966.

66-0090 LOOF, H. W.
THE ECONOMICAL COMPUTATION OF STIFFNESS OF LARGE
STRUCTURAL ELEMENTS
INT. SYMP. ON THE USE OF ELECT. DIGITAL COMPUTERS
IN STRUCTURAL ENGINEERING, UNIVERSITY OF
NEWCASTLE-UPON-TYNE, DEPT. OF CIVIL ENGINEERING,
1966.

66-0091 RAJU, S. P. G.
RAO, A. K.
A MATRIX METHOD FOR VIBRATION AND STABILITY
PROBLEMS
J. AERONAUT. SOC. INDIA, VOL. 8, 1966, PP. 90-99.

66-0092 RAJU, S. P. G.
SOME APPLICATIONS OF MATRIX FORCE METHOD
M.SC. THESIS, FACULTY OF ENGINEERING, INDIAN
INSTITUTE OF SCIENCE, BANGALOR, INDIA, 1966.

66-0093 ARGYRIS, J. H.
SPOONER, J. H.
THE THERMO-ELASTO-PLASTIC ANALYSIS OF A RE-ENTRY
BODY TYPE COMPOSITE SOLID STRUCTURE USING A NEW
FINITE ELEMENT DEVELOPED FOR AXIALLY SYMMETRIC
SYSTEMS
PROC. 17TH INT. ASTRON. CONGR., MADRID, 1966, PP.
307-326.

66-0094 WITHUM, D.
BERECHNUNG VON PLATTEN NACH DEM RITZ' SCHEN
VERFAHREN MIT HILFE DREIECKFOERMIGER MASCHENNETZE
MITTEILUNGEN DES INSTITUTS FUR STATIK DER
TECHNISCHEN HOCHSCHULE HANOVER, VOL. 9, 1966.

66-0095 BUTLIN, G. A.
THE FINITE ELEMENT METHOD APPLIED TO PLATE FLEXURE
PH.D. THESIS, CAMBRIDGE UNIVERSITY, 1966.

66-0096 TONG, P.
LIQUID SLOSHING IN AN ELASTIC CONTAINER
A.F.O.S.R., 66-0943, JUNE 1966.

66-0097 BECKER, E. B.
A NUMERICAL SOLUTION OF A CLASS OF PROBLEMS OF
FINITE ELASTIC DEFORMATION
PH.D. THESIS, UNIVERSITY OF CALIFORNIA, 1966.

66-0098 MELOSH, R. J.
CHRISTIANSEN, H. N.
STRUCTURAL ANALYSIS AND MATRIX INTERPRETIVE SYSTEM
(SAMIS)
TECH. REPORT TM 33-307, JET PROPULSION LAB, DEC.,
1966.

66-0099 IRONS, B. M.
ZIENKIEWICZ, O. C.
ERGATOUDIS, J.
THREE-DIMENSIONAL STRESS ANALYSIS OF ARCH DAMS BY
THE FINITE ELEMENT METHOD: PART I. ARCH DAM NO. 1,
OCT. 1966; PART II. ARCH DAM NO. 5, DEC. 1966;
AD/1735 AND AD/1745
REPORTS TO ARCH DAMS COMMITTEE OF INST. OF CIV.
ENGRS. LONDON, 1966, RESEARCH REPORT NO. C.R.58/66
UNIV. OF WALES, SWANSEA.

66-0100 TAYLOR, R. L.
CHANG, T. Y.
AN APPROXIMATE METHOD FOR THERMO-VISCOELASTIC
STRESS ANALYSIS
NUCL. ENG. DES., VOL. 4, 1966, PP. 21-28.

66-0101 CHANG, T. Y.
APPROXIMATE SOLUTION IN LINEAR VISCOELASTICITY
STRUCT. ENGNG. LAB. REP. NO. 66-8, UNIVERSITY OF
CALIFORNIA, 1966.

66-0102 KEY, S. W.
A CONVERGENCE INVESTIGATION OF THE DIRECT
STIFFNESS METHOD
PH.D THESIS, UNIVERSITY OF WASHINGTON, SEATTLE,
1966.

66-0103 YAMAMOTO, Y.
A FORMULATION OF MATRIX DISPLACEMENT METHOD
DEPT. OF AERON. AND ASTRON., M.I.T., NOVEMBER
1966.

66-0104 JENSEN, W. R.
FALBY, W. E.
PRINCE, N.
MATRIX ANALYSIS METHODS FOR ANISOTROPIC INELASTIC
STRUCTURES
AFFDL-TR-65-220, 1966.

66-0105 MALLETT, R. H.
BERKE, L.
AUTOMATED METHOD FOR THE LARGE DEFLECTION AND
INSTABILITY ANALYSIS OF THREE-DIMENSIONAL TRUSS
AND FRAME ASSEMBLIES
AFFDL-TR-66-102, DECEMBER 1966.

66-0106 NAVARATNA, D. R.
ELASTIC STABILITY OF SHELLS OF REVOLUTION BY THE
VARIATIONAL APPROACH USING DISCRETE ELEMENTS
ASRL-TR-139-1, MIT, JUNE 1966.

66-0107 TEZCAN, S. S.
OVUNC, B.
AN ITERATION METHOD FOR THE NONLINEAR BUCKLING OF
FRAME STRUCTURES
PROC. INT. CONF. ON SPACE STRUCTURES, LONDON,
SEPTEMBER 1966

66-0108 GERSTENKORN, G. F.
KOBAYASHI, A. S.
APPLICATION OF THE DIRECT STIFFNESS METHOD TO
PLANE PROBLEMS INVOLVING LARGE, TIME-DEPENDENT
DEFORMATION
J. BASIC ENGINEERING, TRANS. OF ASME, VOL. 88,
SERIES D, NO.4, DECEMBER 1966, PP. 771-776.

66-0109 GERSTENKORN, G. F.
KOBAYASHI, A. S.
WIEDERHIELM, C. A.
RUSHMER, R. F.
STRUCTURAL ANALYSIS OF AN ARTERIOLE BY THE DIRECT
STIFFNESS METHOD
J. ENG. FOR IND., TRANS., AM. SOC. MECH. ENGRS.,
VOL. 88, SERIES B, NO. 4, 1966, PP. 363-368.

66-0110 FUJINO, T.
OHSAKA, K.
STATIC STRUCTURAL ANALYSIS OF SUSPENSION BRIDGES
MITSUBISHI NIPPON HEAVY-IND., TECH. REV., VOL. 3,
NO. 6, 1966, PP. 17-23.

66-0111 KLEIN, S.
A STUDY OF THE MATRIX DISPLACEMENT METHOD AS
APPLIED TO SHELLS OF REVOLUTION
MATRIX METH. STRUCT. MECH., 1966, PP. 275-298.

66-0112 FULLARD, K.
FREQUENCY TUNING OF COMPRESSOR ROTOR BLADES
SYMP. PAP. IN NUMERICAL METHODS FOR VIBRATION
PROBLEMS, JULY 12-15, 1966, PP. 57-65.

66-0113 MOHRAZ, B.
SCHNOBRICH, W. C.
ANALYSIS OF SHALLOW SHELL STRUCTURES BY A DISCRETE
ELEMENT SYSTEM
CIV. ENG. STUDIES, STRUCT., RESEARCH SERIES NO.
304, UNIV. OF ILL. 1966.

66-0114 TUBA, I. S.
A METHOD OF ELASTIC-PLASTIC PLANE STRESS AND PLANE
STRAIN ANALYSIS
J. OF STRAIN ANALYSIS, VOL. 1, 1966.

66-0115 MENTEL, T.J.
STUDY AND DEVELOPMENT OF SIMPLE MATRIX METHODS OF
INELASTIC STRUCTURES
J. SPACECRAFT ROCKETS VOL. 3, 1966, PP. 4, 449.

66-0116 JENNINGS, A.
A COMPACT STORAGE SCHEME FOR THE SOLUTION OF
SYMMETRICAL LINEAR SIMULTANEOUS EQUATIONS
COMPUTER J., VOL. 9, 1966, PP. 281-285.

66-0117 MATLOCK, H.
HALIBURTON, T.
A FINITE ELEMENT METHOD OF SOLUTION FOR LINEARLY
ELASTIC BEAM-COLUMNS
BUREAU OF PUBLIC RDS., TEXAS HWY DEPT., WASH.,
D.C. REPORT NO: RR-56-1, 1966, 177 PP. (N.T.I.S. -
PB-173 727)

66-0118 LOGCHER, R. D.
STURMAN, G. M.
STRUDL-A COMPUTER SYSTEM FOR STRUCTURAL DESIGN
J. OF STUCTURES, A.S.C.E., VOL. 92, NO. ST6,
DECEMBER 1966.

66-0119 COOK, W. L.
FINITE ELEMENT MODELING OF A CIRCULAR RING USING
HALF AND QUARTER SYMMETRY
GODDARD SPACE FLIGHT CENTER, NATL. AERON. AND
SPACE ADMIN., GREENBELT, MD., REPORT NO:
NASA-TM-X-55746, X-321-66-587, DECEMBER 1966, 22
PP. (N.T.I.S. - N67-23317)

66-0120 DEAK, A. L.
PIAN, T. H. H.
THE APPLICATION OF THE SMOOTH-SURFACE
INTERPOLATION TECHNIQUE TO THE FINITE-ELEMENT
ANALYSIS OF THIN PLATES
AEROELASTIC AND STRUCT. RESEARCH LAB., MASS. INST.
OF TECH., CAMBRIDGE, REPORT NO: MIT-ASRL-136-2,
AUGUST 1966, 56 PP. (N.T.I.S. - AD-643 870)

66-0121 PATTON, P. C.
KARIAPPA, S.
THE MATRIX EIGENVALUE PROBLEM FOR DAMPED
STRUCTURAL VIBRATION
RESEARCH REPORT NO. 27, INSTITUT FUR STATIK UND
DYNAMIK DER LUFTUND RAUMFAHRTKONSTRUKTIONEN, JAN.
1966.

66-0122 HICKS, G. W.
A FINITE ELEMENT ANALYSIS OF THE ELASTIC BEHAVIOR
OF BEAM COLUMNS
AIRPLANE DIV., BOEING CO., RENTON, WASH., REPORT
NO. D6-15505, SEPTEMBER 19, 1966, 39 PP. (N.T.I.S.
- AD-675 795)

66-0123 ARGYRIS, J. H.
LOCHNER, N. M.
SCHARPF, D. W.
SPEKTRALANALYSE VON ROTORBLATTERN NACH
DERMATRIZENVERS CHIEBUNGSMETHODE UND
MATRIZENKRAFTEMETHODE
WGLR TAGUNG UBER SCHWINGUNG VON ROTORBLATTERN,
MUNICH, 1965 (PUBL. BY WGLR, STUTTGART, NOVEMBER,
1966).

66-0124 ARGYRIS, J. H.
THE TRIAX 6 ELEMENT FOR AXISYMMETRIC ANALYSIS BY
THE MATRIX DISPLACEMENT METHOD: PART 1,
FOUNDATIONS
J. ROY. AERO. SOC., VOL. 70, NO. 12, DECEMBER
1966, PP: 1102-1105.

66-0125 STRICKLIN, J. A.
COMPUTATION OF STRESS RESULTANTS FROM ELEMENT
STIFFNESS MATRICES
A.I.A.A. J., VOL. 4, 1966, PP. 1095-1096.

66-0126 YETTRAM, A. L.
HUSAIN, H. M.
PLANE FRAMEWORK METHODS FOR PLATES IN EXTENSION
J. ENG. MECH. DIV., A.S.C.E., VOL. 92, 1966, PP.
157-168.

66-0127 ZIENKIEWICZ, O. C.
CHEUNG, Y. K.
STAGG, K. G.
STRESSES IN ANISOTROPIC MEDIA OF ROCK MECHANICS
J. STRAIN ANALYSIS, VOL. 1, 1966, PP. 172-182.

66-0128 KHANNA, J.
SOME INVESTIGATIONS INTO THE FINITE ELEMENT METHOD
WITH SPECIAL REFFRENCE TO PLANE STRESS
PH.D., THE UNIV. OF BRITISH COLUMBIA, CANADA,
1966.

66-0129 KLINK, W. H.
FINITE AND DISCONNECTED SUBGROUPS OF SU(3) AND
THEIR APPLICATION TO THE ELEMENTARY PARTICLE
SPECTRUM
PH.D., THE JOHNS HOPKINS UNIV., 1966.

66-0130 RAUHUT, J. B.
A FINITE-ELEMENT METHOD FOR ANALYSIS OF ANCHORED
BULKHEADS AND ANCHOR WALLS
PH.D., THE UNIV. OF TEXAS, AUSTIN, 1966.

66-0131 VISSER, C.
THE APPROXIMATE ANALYSIS OF THIN SHELLS BY THE
FINITE ELEMENT METHOD
PH.D., THE OHIO STATE UNIV., 1966.

66-0132 WATWOOD, V. B. JR.
A STUDY OF THE EQUILIBRIUM MODEL FOR USE IN FINITE
ELEMENT ANALYSIS
PH.D., UNIV. OF WASHINGTON, 1966.

66-0133 GLADWELL, G. M. L.
A VARIATIONAL FORMULATION OF DAMPED
ACOUSTO-STRUCTURAL PROBLEMS
J. SOUND AND VIB., VOL. 3, 1966, P. 233.

66-0134 SCHAEFER, H.
EINE EINFACHE KONSTRUKTION VON
KOORDINATENFUNKTIONEN FUR DIE NUMERISCHE LOSUNG
ZWEIDIMENSIONALER RANDWERTPROBLEME NACH
RAYLEIGH-RITZ
ING. ARCH., VOL. 35, 1966, PP. 73-81.

67-0001 DUNS, C. S.
BUTTERFIELD, R.
DYNAMIC ANALYSIS OF SOIL - STRUCTURE SYSTEM USING
FINITE ELEMENT METHOD
PROC., INT. SYMP. ON WAVE PROPAGATION AND DYNAMIC
PROPERTIES OF EARTH MATERIALS, AUGUST 23-25, 1967,
HELD UNIV. OF NEW MEXICO WITH NEW MEXICO SECT.
ASCE. ET AL., ALBURQUERQUE. PP. 615-629.

67-0002 AKYUZ, F. A.
ON THE SOLUTION OF TWO-DIMENSIONAL PROBLEMS OF
ELASTOPLASTICITY BY THE FINITE ELEMENT AND DIRECT
STIFFNESS METHOD
PROC. 5TH A.I.A.A. AEROSPACE SCIENCES MEETING,
A.I.A.A., PAPER 67-144, JANUARY 23-26, 1967.

67-0003 ARGYRIS, J. H.
BERECHNUNG RAUMLICHER SPANNUNGSVERTEILUNGEN IN
STAUDAMMEN NACH DER MATRIZENVERSCHIEBUNGSMETHODE
INGENIEUR ARCHIV, VOL. 36, 1967, PP. 320-334.

67-0004 ARMEN, H.
PIFKO, A. B.
COMPUTER PROGRAMS FOR THE PLASTIC ANALYSIS OF
STRUCTURES USING DISCRETE ELEMENT METHODS
NASA CP-66364, GRUMMAN RESEARCH DEPT., REPORT
PE-292, GRUMMAN AIRCRAFT ENGINEERING CORPORATION,
BETHPAGE, NEW YORK, JUNE1967.

67-0005 BECKER, E. B.
PARR, C. H.
APPLICATION OF THE FINITE ELEMENT METHOD TO HEAT
CONDUCTION IN SOLIDS
AD-823 105, NOVEMBER 1967, 125 PP.

67-0006 BLACKMAN, K.
ELEMENT STIFFNESS MATRICES FOR CONSTANT AND LINEAR
STRESS - TRIANGULAR, CONSTANT AND LINEAR STRESS
TETRAHEDRON
AD-675 794, JANUARY 1967, P. 54.

67-0007 BOGNER, F. K.
FOX, R. L.
SCHMIT, L. A.
A CYLINDRICAL SHELL DISCRETE ELEMENT
A.I.A.A. J., VOL. 5, NO. 4, APRIL, 1967, PP.
745-750.

67-0008 BROWN, C. B.
FORCES ON RIGID CULVERTS UNDER HIGH FILLS
J. STRUCT. DIV., A.S.C.E., VOL. 93, NO. 5, PROC.
PAPER 5501, OCTOBER 1967, PP. 195-218.

67-0009 CANTIN, G.
CLOUGH, R. W.
A CURVED CYLINDRICAL SHELL DISCRETE ELEMENT
A.I.A.A. J., VOL. 5, NO. 4, APRIL 1967.

67-0010 CHOPRA, A. K.
EARTHQUAKE RESPONSE OF EARTH DAMS
J. SOIL MECH. FOUNDATIONS DIV., A.S.C.E., VOL. 93,
NO. SM2, PROC. PAPER 5134, MARCH 1967, PP. 65-81.

67-0011 HARTZ, B. J.
WATWOOD, V. B.
EQUILIBRIUM STRESS FIELD MODELS IN FINITE ELEMENTS
PROC. 4TH INT. CONGR. ON THE APPLICATION OF
MATHEMATICS IN ENG., WEIMAR, EAST GERMANY, JUNE
1967.

67-0012 CLOUGH, R. W.
WOODWARD, R. J.
ANALYSIS OF EMBANKMENT STRESSES AND DEFORMATIONS
J. SOIL MECH. AND FOUNDATION DIV., A.S.C.E., VOL.
93, NO. SM 4, PROC. PAPER 5329, JUNE 1967, PP.
529-549.

67-0013 CLOUGH, R. W.
CARR, A. J.
FINITE ELEMENT ANALYSIS OF DYNAMIC SHELL BEHAVIOUR
DISCRETE AND CONTINUUM CONCEPTS IN MICRO AND MACRO
MECHANICS, A.S.C.E., NOVEMBER 1967, PP. 99-102.

67-0014 CONNOR, J. J.
BREBBIA, C.
STIFFNESS MATRIX FOR SHALLOW RECTANGULAR SHELL
ELEMENT
J. ENG. MECH. DIV., A.S.C.E., VOL. 93, NO. EM 5,
OCTOBER 1967, PP. 46-66.

67-0015 CONSTANTINO, C. J.
FINITE ELEMENT APPROACH TO STRESS WAVE PROBLEMS
J. ENGR. MECH. DIV., A.S.C.E., VOL. 93, NO. EM 2,
PROC PAPER 5206, APRIL 1967, PP. 153-176.

67-0016 CONSTANTINO, C. J.
WACHOWSKI, A.
BARNWELL, U. L.
FINITE ELEMENT SOLUTION FOR WAVE PROPAGATION IN
LAYERED MEDIA CAUSED BY A NUCLEAR DETONATION
PROC. INT. SYMP. WAVE PROPAGATION AND DYNAMIC
PROPERTIES OF EARTH MATERIALS, NEW MEXICO, AUGUST
1967.

67-0017 CORNELL, D. C.
SAFE-PLANE, A COMPUTER PROGRAM FOR THE STRESS
ANALYSIS AND DESIGN OF TWO-DIMENSIONAL COMPOSITE
BODIES
GENERAL DYNAMICS CORP, SAN DIEGO, CALIF., A USER'S
MANUAL, JUNE 1967, P. 51, GA-7851.

67-0018 CORNELL, D. C.
SAFE SHELL, A COMPUTER PROGRAM FOR THE STRESS
ANALYSIS OF THIN SHELL BODIES OF REVOLUTION. A
USER'S MANUAL
GULF GENERAL ATOMIC, INC., SAN DIEGO, CALIF., JUNE
30, 1967, P. 43, GA-7852.

67-0019 COST, T. L.
PARR, C. H.
ANALYSIS OF THE BIAXIAL STRIP AND SHEAR LAP TESTS
FOR SOLID PROPELLANT CHARACTERIZATION
AD-813 719, MAY 1967, 60 PP.

67-0020 DAWE, D. J.
ON ASSUMED DISPLACEMENTS FOR THE RECTANGULAR PLATE
BENDING ELEMENT
J. ROY. AERO. SOC., VOL. 71, OCTOBER 1967, PP.
722-724.

67-0021 DIBAJ, M.
PENZIEN, J.
THE EFFECTS OF THIN CORES AND NONHOMOGENEITY ON
DYNAMIC RESPONSE OF EARTH DAMS
SOIL MECH. AND BITUMINOUS MAT. RES. LAB., UNIV. OF
CALIF., BERKELY, REPORT NO. TE-67-2, SEPTEMBER
1967.

67-0022 ZIENKIEWICZ, O. C.
CHEUNG, Y. K.
WATSON, M.
STRESS ANALYSIS BY THE FINITE ELEMENT METHOD -
THERMAL EFFECTS
PROC. OF CONFERENCE ON PRESTRESSED CONCRETE
PRESSURE VESSELS, MARCH 1967., INST. OF CIV.
ENGRS., LONDON

67-0023 DUNGAR, R.
VIBRATIONS OF PLATE AND SHELL STRUCTURES USING
TRIANGULAR FINITE ELEMENTS
J. STRAIN ANALYSIS, VOL. 2, NO. 1, JANUARY 1967,
PP. 73-83.

67-0024 FINN, W. D.
KHANNA, J.
DYNAMIC RESPONSE OF EARTH DAMS
TRANS. NINTH INTERN. CONG. ON LARGE DAMS, VOL. 4,
ISTANBUL, 1967.

67-0025 FINN, W. D.
FINITE-ELEMENT ANALYSIS OF SEEPAGE THROUGH DAMS
J. SOIL MECH. AND FOUNDATION DIV., A.S.C.E., VOL.
93, NO. SM 6, PROC. PAPER 5552, NOVEMBER 1967, PP.
41-48.

67-0026 FINN, W. D.
STATIC AND SEISMIC BEHAVIOUR OF AN EARTH DAM
CANADIAN GEOTECHNICAL J., VOL. 4, NO. 1, 1967.

67-0027 FRAEIJS DE VEUBEKE, B.
STRAIN-ENERGY BOUNDS IN FINITE-ELEMENT ANALYSIS BY
SLAB ANALOGY
J. OF STRAIN ANALYSIS, VOL. 2, NO. 4, JULY 1967,
PP. 265-271.

67-0028 GALLAGHER, R. H.
GELLATLY, R. A.
PADLOG, J.
MALLETT, R. H.
DISCRETE ELEMENT PROCEDURE FOR THIN SHELL
INSTABILITY ANALYSIS
A.I.A.A. J., VOL. 5, NO. 1, JANUARY 1967, PP.
138-145.

67-0029 GALLAGHER, R. H.
A DISCRETE ELEMENT PROCEDURE FOR THIN-SHELL
INSTABILITY ANALYSIS
A.I.A.A. J., VOL. 5, NO. 2, FEBRUARY 1967, PP.
367.

67-0030 GALLAGHER, R. H.
STABILITY OF PLATES USING THE FINITE ELEMENT
METHOD
J. ENGR. MECH. DIV., A.S.C.E., VOL. 93, NO. EM 1,
FEBRUARY 1967, PP. 81-83. CLOSURE TO DISCUSSION,
K. K. KAPUR AND B. J. HARTZ, VOL. 93, NO. EM 4,
AUGUST 1967, PP. 175-176.

67-0031 HAISLER, W. E.
STRICKLIN, J. A.
RIGID BODY DISPLACEMENTS OF CURVED ELEMENTS IN THE
ANALYSIS OF SHELLS BY THE MATRIX-DISPLACEMENT
METHOD
A.I.A.A. J., VOL. 5, NO. 8, AUGUST 1967, PP.
1525-1527.

67-0032 HALIBURTON, T. A.
MATLOCK, H.
FINITE ELEMENT ANALYSIS OF STRUCTURAL FRAMES
ASCE STRUCTURAL ENGINEERING CONF., SEATTLE,
WASHINGTON, PREPRINT 441, MAY 1967.

67-0033 HANSTEEN, O. E.
ANALYSIS OF STRESS DISTRIBUTION IN SHEAR WALLS BY
THE FINITE ELEMENT DISPLACEMENT METHOD
4TH INTER. CONG. ON APPLICATION OF MATH IN
PRACTICE OF ENGR., WEIMAR, GERMANY, 1967.

67-0034 HAYES, D. J.
MARCAL, P. V.
DETERMINATION OF UPPER BOUNDS FOR PROBLEMS IN
PLANE STRESS USING FINITE ELEMENT TECHNIQUES
INT. J. MECH. SCIENCES, VOL. 9, NO. 5, MAY 1967,
PP. 245-252.

67-0035 HELLEN, T. K.
ANALYSIS OF ELASTIC PLATES IN FLEXURE BY A
SIMPLIFIED FINITE ELEMENT METHOD
ACTA POLTECHNICA SCANDINAVIAN, NO. 46, 1967.

67-0036 HERRMANN, L. R.
FINITE-ELEMENT BENDING ANALYSIS FOR PLATES
J. ENGR. MECH. DIV., A.S.C.E., VOL. 93, NO. EM 5,
OCTOBER 1967, PP. 13-26.

67-0037 ZIENKIEWICZ, O. C.
TRIANGULAR MESH GENERATION PROGRAM
REPORT NO. 2, UNIV. OF WALES, SWANSEA, 1967.

67-0038 MOOLEY, R. F.
HIBBERT, P. D.
STRESS CONCENTRATIONS IN TIMBER BEAMS
J. STRUCT. DIV., A.S.C.E., VOL. 93, NO. ST 2,
PAPER 5180, APRIL 1967, PP. 127-139.

67-0039 IDRISS, I. M.
SEED, H. B.
RESPONSE OF EARTH BANKS DURING EARTHQUAKES
J. SOIL MECH. AND FOUNDATION DIV., A.S.C.E., VOL.
93, NO. SM 3, PROC. PAPER 5232, MAY 1967, PP.
61-82.

67-0040 ISAKSON, G.
DISCRETE-ELEMENT METHODS FOR THE PLASTIC ANALYSIS
OF STRUCTURES
NASA-CR-803, OCTOBER 1967, P. 215.

67-0041 JOHNSON, C. P.
SMITH, O.
A COMPUTER PROGRAM FOR THE ANALYSIS OF THIN SHELLS
AD-701 896, AUGUST 1967, P. 92.

67-0042 JONES, R. M.
PROCEEDINGS OF THE NOSETIP STRESS ANALYSIS
TECHNICAL INTERCHANGE MEETING
AD-824 285, NOVEMBER 1967, P. 364.

67-0043 KAUFMAN, S.
HALL, D. B.
BENDING ELEMENTS FOR PLATE AND SHELL NETWORKS
A.I.A.A. J., VOL. 5, NO. 3, MARCH 1967, PP.
402-405.

67-0044 CHEUNG, Y. K.
DAVIES, J. D.
ANALYSIS OF RECTANGULAR TANKS BY FINITE ELEMENT
TECHNIQUES
"CONCRETE", MAY 1967

67-0045 KRAHULA, J. L.
ANALYSIS OF BENT AND TWISTED BARS USING THE FINITE
ELEMENT METHOD
A.I.A.A. J., VOL. 5, JUNE 1967, PP. 1194-1197.

67-0046 LLOYD, J. R.
ANALYSIS MODELS FOR PLATE ELEMENTS IN THE MEMBRANE
STATE OF STRESS
AD-856 604, DECEMBER 1967, 32 P.

67-0047 MARCAL, P. V.
A COMPARATIVE STUDY OF NUMERICAL METHODS OF
ELASTIC-PLASTIC ANALYSIS
AIAA/ASME 8TH STRUCTURES, STRUCTURAL DYNAMICS AND
MATERIALS CONFERENCE, MARCH 1967, PP. 212-216.

67-0048 MARCAL, P. V.
THE EFFECT OF INITIAL DISPLACEMENTS OF PROBLEMS OF
LARGE DEFLECTION AND STABILITY
AD-664 979, NOVEMBER 1967, 20 P.

67-0049 MARCAL, P. V.
KING, I. P.
ELASTIC-PLASTIC ANALYSIS OF TWO-DIMENSIONAL STRESS
SYSTEMS BY THE FINITE ELEMENT METHOD
INT. J. MECH. SCI., VOL. 9, 1967, PP. 143-155.

67-0050 MARTIN, H. C.
STIFFNESS MATRIX FOR A TRIANGULAR SANDWICH ELEMENT
IN BENDING
INST. OF TECH., JET PROPULSION LAB., CALIF., TECH.
REPORT 32-1158, 1967.

67-0051 MCCALLEY, R. B. JR.
SHOCK ANALYSIS BY MATRIX METHODS
GENERAL ELECTRIC CO., SCHENECTADY, N. Y., MAY
1967. CONTRACT W-31-109-ENG-52, P. 101.
KAPL-P-3343.

67-0052 MCGLEEN, J.
HARTZ, B. J.
FINITE ELEMENT ANALYSIS OF PLYWOOD PLATES
ASCE STRUCTURAL ENGINEERING CONF., SEATTLE,
WASHINGTON, PREPRINT 518, MAY 1967.

67-0053 MCLAY, R. W.
COMPLETENESS AND CONVERGENCE PROPERTIES OF FINITE
ELEMENT DISPLACEMENT FUNCTIONS., A GENERAL
TREATMENT
A.I.A.A. PAPER 67-143, JANUARY 1967.

67-0054 MORLEY, L. S. D.
A TRIANGULAR EQUILIBRIUM ELEMENT WITH LINEARLY
VARYING BENDING MOMENTS FOR PLATE BENDING PROBLEMS
J. ROY. AERO. SOC., VOL. 71, OCTOBER 1967, PP.
715-719.

67-0055 NAGAMOTO, R.
ON THE STRESS ANALYSIS AT JOINTS OF DEEP BEAMS
SOC. NAV. ARCH. WEST JAPAN, PT1, J. NO. 33, PP.
1-14, 1967. PT2, J. NO. 34, 1967, PP. 87-96, PT.
3, J. NO. 65, 1968.

67-0056 NAVARATNA, D. R.
ANALYSIS OF ELASTIC STABILITY OF SHELLS OF
REVOLUTION BY THE FINITE ELEMENT METHOD
AIAA/ASME 8TH STRUCTURES, STRUCTURAL DYNAMICS AND
MATERIALS CONFERENCE, MARCH 1967, PP. 176-185.

463

67-0057 NEWSOM, C. D.
A FINITE ELEMENT APPROACH FOR THE ANALYSIS OF
RANDOMLY EXCITED COMPLEX ELASTIC STRUCTURES
AIAA/ASME 8TH STRUCTURES, STRUCTURAL DYNAMICS AND
MATERIALS CONFERENCE, MARCH 1967, PP. 125-134.

67-0058 NGO, D.
SCORDELIS, A. C.
FINITE ELEMENT ANALYSIS OF REINFORCED CONCRETE
BEAMS
A.C.I., J., VOL. 64, NO. 3, MARCH 1967, PP.
152-163.

67-0059 NORTH, E. L.
PAPA, STRUCTURAL ANALYSIS OF PLATES AND SHELLS
USING TRAPEZOIDAL AND TRIANGULAR PLATE ELEMENTS
GENERAL ELECTRIC CO., SAN JOSE, CALIF., ATOMIC
POWER EQUIPMENT DEPT., MARCH 1967, CONTRACT
AT(04-3)-189, P. 46, GEAP-5471.

67-0060 ODEN, J. T.
KUBITZA, W. K.
NUMERICAL ANALYSIS OF NONLINEAR PNEUMATIC
STRUCTURES
PROC. 1ST INTER. COLL. ON PNEUMATIC STRUCTURES,
IASS, MAY 1967, PP. 87-107.

67-0061 ODEN, J. T.
NUMERICAL FORMULATION OF A CLASS OF PROBLEMS IN
NONLINEAR VISCOELASTICITY
ADV. IN ASTRON. SCIENCES, JUNE 1967.

67-0062 ODEN, J. T.
SATO, T.
FINITE STRAINS AND DISPLACEMENTS OF ELASTIC
MEMBRANES BY THE FINITE ELEMENT METHOD
INT. J. OF SOLIDS AND STRUCT., VOL. 3, NO. 2, MAY
1967, PP. 471-488.

67-0063 ODEN, J. T.
NUMERICAL FORMULATION OF NONLINEAR ELASTICITY
PROBLEMS
J. STRUCTURED DIV., A.S.C.E., VOL. 93, NO. ST 3,
JUNE 1967, PP. 235-255.

67-0064 ODEN, J. T.
SATO, T.
STRUCTURAL ANALYSIS OF AERODYNAMIC DECELERATION
SYSTEMS BY THE FINITE ELEMENT METHOD
ADV. IN ASTRONOMICAL SCIENCES, VOL. 24, PART III,
JUNE 1967.

67-0065 ODEN, J. T.
FINITE ELEMENT APPLICATIONS IN LINEAR AND
NONLINEAR THERMOVISCOELASTICITY
DISCRETE AND CONTINUUM CONCEPTS IN MICRO AND MACRO
MECHANICS, A.S.C.E., NOVEMBER 1967, PP. 32-36.

67-0066 OLSON, M. D.
FINITE ELEMENTS APPLIED TO PANEL FLUTTER
A.I.A.A. J., VOL. 5, NO. 12, DECEMBER 1967, PP.
2267-2269.

67-0067 PERCY, J. H.
QUADRILATERAL FINITE ELEMENT IN ELASTIC-PLASTIC
PLANE STRESS ANALYSIS
A.I.A.A. J., VOL. 5, NO. 2, FEBRUARY 1967, P. 367.

67-0068 PRZEMIENIECKI, J. S.
STABILITY ANALYSIS OF COMPLEX STRUCTURES USING
DISCRETE ELEMENT TECHNIQUE
SYMP. ON STRUCTURAL STABILITY AND OPTIMIZATION,
ROY. AERON. SOC. AND LOUGHBOROUGH UNIVERSITY OF
TECHNOLOGY, ENGLAND, MARCH 23-24, 1967.

67-0069 RASHID, Y. R.
ROCKENHAUSER, W.
PRESSURE VESSELS ANALYSIS BY FINITE ELEMENT
TECHNIQUES
GENERAL DYNAMICS CORP., SAN DIEGO, CALIF., GENERAL
ATOMIC DIV., FEBRUARY 27, 1967, CONTRACT
AT(04-3)-167, GA-7810, P. 32, ALSO IN PROC. CONF.
ON PRESTRESSED CONCRETE PRESSURE VESSELS, INST.
CIV. ENG., 1968.

67-0070 ROBINSON, J.
ELEMENT STATIC FLEXIBILITY AND STIFFNESS MATRICES
FOR THE FINITE ELEMENT ANALYSIS OF BEAM, PLATE,
AND SHELL STRUCTURES, PART 1: PLANE ELEMENTS
ISVR-196, SEPTEMBER 1967, P. 85.

67-0071 RUSHTON, K. R.
SOLUTION OF ANISOTROPIC SEEPAGE BY FINITE ELEMENTS
J. ENGR. MECH. DIV., A.S.C.E., VOL. 92, NO. EM 5,
OCTOBER 1966, P. 141, CLOSURE TO DISCUSSION, O. C.
ZIENKIEWICZ, ET AL. VOL. 93, NO. EM 2, APRIL 1967,
PP. 235-236.

67-0072 ROCKEY, K. C.
EVANS, H. R.
A FINITE ELEMENT SOLUTION FOR FOLDED PLATE
STRUCTURES
SPACE STRUCTURES, BLACKWELL SCIENTIFIC PUB., 1967,
PP. 155-189.

67-0073 SANDER, G.
BECKERS, P.
NGUYEN, H. D.
DIGITAL COMPUTATION OF STRESSES AND DEFLECTIONS IN
A BOX BEAM
AD-664 930, 1967, P. 54.

67-0074 FRAEIJS DE VEUBEKE, B.
SANDER, G.
UPPER AND LOWER BOUNDS TO STRUCTURAL DEFORMATIONS
BY DUAL ANALYSIS IN FINITE ELEMENTS
AFFDL-TR-66-199, 1967.

67-0075 SHERIF, M. A.
CHEN, J. T.
APPLICATION OF THE FINITE ELEMENT METHOD TO STRESS
DISTRIBUTION IN SOIL MEDIA
THE TREND IN ENGINEERING, VOL. 19, NO. 4, OCTOBER
1967.

67-0076 SHIPLEY, S. A.
LEISTNER, H. G.
JONES, R. E.
ELASTIC WAVE PROPAGATION, A COMPARISON BETWEEN
FINITE ELEMENT PREDICTIONS AND EXACT SOLUTION
PROC. INTERN. SYMP. WAVE PROPAGATION AND DYNAMIC
PROPERTIES OF EARTH MATERIALS, NEW MEXICO, AUGUST
1967.

67-0077 STERN, P.
NUMERICAL ANALYSIS OF SHELL STRUCTURES
DISCRETE AND CONTINUUM CONCEPTS IN MICRO AND MACRO
MECHANICS, A.S.C.E., NOVEMBER 1967, PP. 107-110.

67-0078 STRICKLIN, J. A.
CONSISTENT STIFFNESS MATRICES IN THE ANALYSIS OF
SHELLS
AIAA/ASME 8TH STRUCTURES, STRUCTURAL DYNAMICS AND
MATERIALS, CONFERENCE, MARCH 1967, PP. 162-174.

67-0079 STRICKLIN, J. A.
HAISLER, W. E.
NONLINEAR ANALYSIS OF SHELLS OF REVOLUTION BY THE
MATRIX DISPLACEMENT METHOD
SANDIA, AUGUST 1967, P. 11, SC-DC-67-2054, ALSO IN
J. AIAA, VOL. 6, NO. 12, DECEMBER 1968, PP.
2306-2311.

67-0080 SZILARD, R.
WEST, A. S.
DERIVATION OF COMPATIBLE STIFFNESS MATRICES FOR
LARGE DISCRETE ELEMENTS
DISCRETE AND CONTINUUM CONCEPTS IN MICRO AND MACRO
MECHANICS, A.S.C.E., NOVEMBER 1967, PP. 111-114.

67-0081 TAYLOR, R. L.
BROWN, C. B.
DARCY FLOW SOLUTIONS WITH A FREE SURFACE
J. HYDRAULIC DIV., A.S.C.E., VOL. 93, NO. HY 2,
PROC. PAPER 5126, MARCH 1967, PP. 25-33.

67-0082 TOCHER, J. L.
HARTZ, B. J.
HIGHER-ORDER FINITE ELEMENT FOR PLANE STRESS
J. ENGR. MECH. DIV., A.S.C.E., VOL. 93, NO. EM 4,
PROC. PAPER 5402, AUGUST, 1967, PP. 149-172.

67-0083 TONG, P.
PIAN, T. H. H.
THE CONVERGENCE OF FINITE ELEMENT METHOD IN
SOLVING LINEAR ELASTIC PROBLEMS
INT. J. SOLIDS AND STRUCTURES, VOL. 3, NO. 5,
SEPTEMBER 1967, PP. 865-880.

67-0084 UTKU, S.
BOUNDING SOLUTIONS OF PLATE BENDING PROBLEMS
DISCRETE AND CONTINUUM CONCEPTS IN MICRO AND MACRO
MECHANICS, A.S.C.E., NOVEMBER 1967, PP. 63-106.

67-0085 VALLABHAN, C. V. G.
REESE, L. C.
FINITE ELEMENT METHOD APPLIED TO PROBLEMS IN
STRESSES AND DEFORMATION OF SOIL
JANUARY 1967, P. 178, NASA-CR-6638, N67-2-1441.

67-0086 WATWOOD, V. B.
RELATIONSHIPS BETWEEN MATRIX METHODS IN FINITE
ELEMENTS
DISCRETE AND CONTINUUM CONCEPTS IN MICRO AND MACRO
MECHANICS, A.S.C.E., NOVEMBER 1967, PP. 115-118.

67-0087 WEBBER, J. P. H.
THERMO-ELASTIC ANALYSIS OF RECTANGULAR PLATES IN
PLANE STRESS BY THE FINITE-ELEMENT DISPLACEMENT
METHOD
J. STRAIN ANALYSIS, VOL. 2, NO. 1, JANUARY 1967,
PP. 43-51.

67-0088 WEBSTER, J. J.
FREE VIBRATION OF SHELLS OF REVOLUTION USING RING
ELEMENTS
INT. J. MECH. SCI., VOL. 9, 1967, P. 559.

67-0089 WILSON, E. L.
JONES, R. M.
FINITE ELEMENT STRESS ANALYSIS OF AXISYMMETRIC
SOLIDS WITH ORTHOTROPIC TEMPERATURE DEPENDENT
MATERIAL PROPERTIES
AD-820 991, SEPTEMBER 1967, P. 89.

67-0090 WINSLOW, A. M.
NUMERICAL SOLUTION OF THE QUASI LINEAR POISSON
EQUATION IN A NON-UNIFORM TRIANGLE MESH
J. COMP. PHYSICS, VOL. 1, 1967, PP. 149-172.

67-0091 ZIENKIEWICZ, O. C.
BAHRANI, A. K.
ARLETT, P. L.
SOLUTION OF THREE-DIMENSIONAL FIELD PROBLEMS BY
THE FINITE ELEMENT METHOD
ENGINEER, OCTOBER 27, 1967, PP. 547-550.

67-0092 YIN, F. C. P.
INTERFACE CORE DISCRETE-ELEMENT FOR STIFFENED
SHELLS OF REVOLUTION
A.I.A.A. J., VOL. 5, NO. 12, DECEMBER 1967, PP.
2270-2273.

67-0093 ZIENKIEWICZ, O. C.
CHEUNG, Y. K.
THE FINITE ELEMENT METHOD IN STRUCTURAL AND
CONTINUUM MECHANICS
MCGRAW HILL PUB. CO. LTD., LONDON, 1967.

67-0094 TERAZAWA, K.
UEDA, Y.
MATSUISHI, M.
ELASTIC-PLASTIC BUCKLING OF PLATES USING THE
FINITE ELEMENT METHOD
J. SOC. NAV. ARCH. OF JAPAN, VOL. 122, 1967, PP.
129-136.

67-0095 ZIENKIEWICZ, O. C.
THE FINITE ELEMENT METHOD IN VIBRATION ANALYSIS
SYMP. ON NUM. METHODS IN VIB. PROBLEMS, UNIVERSITY
OF SOUTHHAMPTON, 1967.

67-0096 KORNEEV, V. G.
THE COMPARISON OF THE FINITE ELEMENT METHOD WITH
THE VARIATIONAL-DIFFERENCE METHOD IN THE THEORY OF
ELASTICITY
IZVESTIA REPORTS OF THE WHOLE-UNION SCIENTIFIC
INVESTIGATION OF THE HYDRO-TECHNICAL INSTITUTE
VOL. 83, 1967, PP. 286-307.

67-0097 CHEUNG, Y. K.
DAVIES, J. D.
NATH, P.
STRESSES DUE TO CONCENTRATED LOADS IN RESTRAINED
BENDING TESTS
"STRAIN", APRIL 1967

67-0098 KING, I. P.
FINITE ELEMENT SOLUTION SYSTEMS (FESS),
INTERNAL REPORT CENTRE FOR NUMERICAL METHODS IN
ENGINEERING, UNIVERSITY OF WALES SWANSEA, 1967.

67-0099 FOWLER, J. N.
ELASTIC-PLASTIC DISCRETE ELEMENT ANALYSIS OF
AXISYMMETRICALLY LOADED SHELLS OF REVOLUTION
MIT ASRL TR 146-3, JULY 1967.

67-0100 DELPAK, R.
AXISYMMETRIC VIBRATION OF SHELLS OF REVOLUTION BY
THE FINITE ELEMENT METHOD
M.SC. THESIS, UNIV. OF WALES, SWANSEA, 1967.

67-0101 CARR, A. J.
A REFINED FINITE ELEMENT ANALYSIS OF THIN SHELL
STRUCTURES
INCLUDING DYNAMIC LOADING, SEL REPORT NO. 67-9,
UNIV. OF CALIFORNIA, BERKELEY, 1967.

67-0102 KHOJASTEH-BAKHT, M.
ANALYSIS OF ELASTIC-PLASTIC SHELLS OF REVOLUTION
UNDER AXI-SYMMETRIC LOADING BY THE FINITE
ELEMENT METHOD
DEPT. CIV. ENG. UNIV. OF CALIFORNIA, SE SA 67-8,
1967.

67-0103 LUNDGREN, H. R.
BUCKLING OF MULTILAYER PLATES BY FINITE ELEMENTS
PH.D. DISS., OKLAHOMA STATE UNIVERSITY, 1967.

67-0104 ZIENKIEWICZ, O. C.
CHEUNG, Y. K.
STRESSES IN SHAFTS
THE ENGINEER, NOVEMBER 1967.

67-0105 ODEN, J. T.
SATO, T.
FINITE DEFORMATION OF ELASTIC MEMBRANES BY THE
FINITE ELEMENT METHOD
INTL. J. SOLIDS AND STRUCT., VOL. 3, 1967, PP.
471-488.

67-0106 BERGAN, P. G.
PLANE STRESS ANALYSIS USING THE FINITE ELEMENT
METHOD. TRIANGULAR FINITE ELEMENT WITH SIX
PARAMETERS IN EACH NODE
DIVISION OF STRUCTURAL MECHANICS, THE TECHNICAL
UNIVERSITY OF NORWAY, 1967.

67-0107 HICKS, G. W. JR.
FINITE ELEMENT ELASTIC BUCKLING ANALYSIS
PROC. A.S.C.E., J. STRUCT. DIV., PAPER 5662, VOL.
93, 1967, PP. 71-86.

67-0108 ROCKEY, K. C.
ANDERSON, R. G.
CHEUNG, Y. K.
THE BEHAVIOUR OF SQUARE SHEAR WEBS HAVING A
CIRCULAR HOLE
PROC. OF THE SYMPOSIUM ON THIN WALLED STEEL
STRUCTURES, SEPTEMBER 1967, UNIVERSITY COLLEGE OF
SWANSEA.

67-0109 ZIENKIEWICZ, O. C.
CHEUNG, Y. K.
RIBEIRO, A. A.
AZEVEDO, M. C.
FERREIRA, M. J. F.
PEDRO, J. O.
COMPARISON OF ANALYTICAL AND EXPERIMENTAL RESULTS
IN THE DESIGN OF A HOLLOW GRAVITY DAM
PROC. 9TH INT. CONGRESS ON LARGE DAMS, ISTANBUL,
1967.

67-0110 DAVIES, J. D.
CHEUNG, Y. K.
BENDING MOMENTS IN LONG WALL TANKS
ACI JOURNAL, PROCEEDINGS V. 64, NO. 10, OCTOBER
1967, PP. 685-690

67-0111 FRAEIJS DE VEUBEKE, B.
BASIS OF A WELL CONDITIONED FORCE PROGRAM FOR
EQUILIBRIUM MODELS VIA THE SOUTHWELL SLAB
ANALOGIES
TECH. REPORT: AFFDL-TR-67-10, MARCH 1967.

67-0112 YAGHMAI, S. K. M.
POPOV, E. P.
KHOJASTEH-BAKHT, M.
BENDING OF CIRCULAR PLATES OF HARDENING MATERIAL
INT. J. SOL. STRUCT., VOL. 3, 1967, PP. 975-988.

67-0113 MARCAL, P. V.
EFFECT OF INITIAL DISPLACEMENT ON PROBLEM OF LARGE
DEFLECTION AND STABILITY
TECH. REP. ARPA E54, BROWN UNIV., 1967.

67-0114 KOENIG, H. A.
MULTIPLE STRESS WAVE DISCONTINUITIES IN FINITE
BEAMS AND PLATES
A THESIS IN ENGINEERING MECHANICS, PENN. STATE
UNIV., 1967.

67-0115 CHUM K. H.
KRISHNAMURTHY, G.
BUCKLING OF OPEN CYLINDRICAL SHELLS
PROCEEDINGS OF ASCE, J. ENGNG. MECH. DIV., VOL.
93, NO. EM2, APRIL, 1967.

67-0116 YAMADA, Y.
STIFFNESS MATRIX IN PLASTIC-ELASTIC PROBLEMS OF
CONTINUA
SEISAN KENKYU, VOL. 19, NO. 3, 1967, PP. 75-76 (IN
JAPANESE)

67-0117 FELIPPA, C. A.
ANALYSIS OF ELEMENT STIFFNESS
ENGINEERING EXTENSION COURSE ON FINITE ELEMENT
METHOD IN STRUCTURAL MECHANICS, UNIVERSITY OF
CALIFORNIA, 1967.

67-0118 GIRIJAVALLABHAN, C. V.
APPLICATION OF THE FINITE ELEMENT METHOD TO
PROBLEMS IN SOIL AND ROCK MECHANICS
PH.D. THESIS, UNIV. OF TEXAS, AUSTIN, 1967.

67-0119 THUBOI, Y.
OHYAMA, H.
ANALYSIS OF TENSION STRUCTURES BY MATRIX METHODS
REPRINT OF THE 17TH JAPAN NAT. CONGR. APPLI.
MECH., 1967. PP. 95-96

67-0120 KLEIN, S.
STATIC ANALYSIS OF A COMPLEX SHELL STRUCTURE UNDER
ASYMMETRIC LOADING
STRUCT. ANALYSIS OF A REENTRY VEHICLE TYPE SHELL,
AEROSPACE CORP., SAN BERNARDINO, CALIF., VOL. 1,
1967.

67-0121 BOISSERIE, J. M.
BENJAMIN, C.
LIDA, M.
DETERMINATION DU CHAMP DES CONTRAINTES DANS LE
CAISSON DU REACTEUR DE BUGEY L'
BULLETIN DU CENTRE DE RECHERCHES ET D'ESSAIS DE
CHATOU, NO. 3, 1967.

67-0122 BROOKS, D.F.
BROTTON, D.M.
COMPUTER SYSTEM FOR ANALYSIS OF LARGE FRAMEWORKS
J. STRUCT. DIV. AM. SOC. CIV. ENGRS., VOL. 93,
1967, PP. 1-23.

67-0123 JOHNSON, C. P.
THE ANALYSIS OF THIN SHELLS BY A FINITE ELEMENT
PROCEDURE
STRUCT. AND MATER. RESEARCH, DEPT. OF CIV. ENG.,
UNIV. OF CALIF., BERKELEY, REPORT NO. 67-22, 1967.

67-0124 ROBINSON, J.
EIGENVALUES OF COLLINEAR BEAM STRUCTURES USING
FINITE ELEMENT TECHNIQUES AND VARIOUS DYNAMIC
REPRESENTATIONS FOR THE STRUCTURAL ELEMENTS
WISS. Z. HOCHSCH. ARCHIT. BAUW. WEIMAR., VOL. 3,
1967, PP. 279.

67-0125 LANG, T. E.
STRUCTURAL ANALYSIS AND MATRIX INTERPRETIVE SYSTEM
(SAMIS) USER REPORT
TM 33-305, JET PROPULSION LAB, MARCH, 1967.

67-0126 DE ARANTES E OLIVEIRA, E. R.
MATHEMATICAL FOUNDATION OF THE FINITE ELEMENT
METHOD
DIV. OF APPL. MATH., LABORATORIO NACIONAL DE
ENGENHARIA CIVIL, LISBON, 1967.

67-0127 KING, I. P.
TRIANGULAR MESH GENERATION PROGRAM
CENTRE FOR NUMERICAL METHODS IN ENGINEERING,
SWANSEA, C.P.R.2, 1967.

67-0128 LINDBERG, G. M.
OLSON, M. D.
VIBRATION MODES AND RANDOM RESPONSE OF A MULTI-BAY
PANEL SYSTEM USING FINITE ELEMENTS
NAT. RES. COUNCIL OF CANADA, AERONAUT. REP.
LR-492, DECEMBER 1967.

67-0129 LINDBERG, G. M.
OLSON, M. D.
THE VIBRATION OF STEPPED CANTILEVERED PLATES
NAT. RES. COUNCIL OF CANADA, AERONAUT. REP.
LR-494, DECEMBER 1967.

67-0130 FINN, W. D. L.
FINITE ELEMENT ANALYSIS OF SEEPAGE
PROC. ASCE., VOL. 93, NO. SM6, 1967.

67-0131 MARTIN, G. R.
DYNAMIC RESPONSE OF COHESIVE EARTH DAMS TO
EARTHQUAKES
PROC. 5TH AUSTRALIA-NEW ZEALAND CONF. ON SOIL
MECH. AND FOUND., ENG., FEBRUARY 13-17, 1967, PP.
121-131.

67-0132 KUZANEK, J. F.
FINITE ELEMENT MODELING OF A CANTILVER BEAM
CLEARINGHOUSE NO. 67-22881, MARCH 1967.

67-0133 MATLOCK, H.
GRUBBS, B. R.
A FINITE-ELEMENT METHOD OF SOLUTION FOR STRUCTURAL
FRAMES
BUREAU OF PUBLIC ROADS, TEXAS HIGHWAY DEPT., WASH.
D.C., REPORT NO: RR-56-3, MAY 1967, 105 PP.
(N.T.I.S. - PB-175 748) SEE ALSO PB-173 727

67-0134 MATLOCK, H.
INGRAM, W. B.
A FINITE ELEMENT METHOD FOR BENDING ANALYSIS OF
LAYERED STRUCTURAL SYSTEMS
BUREAU OF PUBLIC ROADS, TEXAS HIGHWAY DEPT.,
WASH., D.C., REPORT NO.: RR-56-5, JUNE 1967, 239
PP. (N.T.I.S. - PB-175 749) SEE ALSO PB-173 727.

67-0135 OLSON, M. D.
ON APPLYING FINITE ELEMENTS TO PANEL FLUTTER
NATL. RESEARCH COUNCIL OF CANADA, OTTAWA, ONT.,
REPORT NO: LR-476, NRC-9642, MARCH 1967, 26 PP.
(N.T.I.S. - 67-34429)

67-0136 ARGYRIS, J. H.
ENERGY THEOREMS AND STRUCTURAL ANALYSIS
BUTTERWORTHS SCIENTIFIC PUBL., LONDON, THIRD
EDITION, 1967.

67-0137 SCHARPF, D. W.
WEBER, J.
ELASTO-PLASTIC ANALYSIS OF METAL SHEETS WITH A
CENTRAL CRACK
INSTITUT REPORT NO. 37, JANUARY 1967.

67-0138 ARGYRIS, J. H.
APPLICATION OF NATURAL KINEMATIC MODES TO MATRIX
ANALYSIS: SMALL AND LARGE DISPLACEMENTS AND
STRAINS
ACADEMY OF SCI., BELGRADE, DECEMBER 22 1965, PUBL.
IN COMPTES RENDUES DE L'ACADEMIE DE SCI. DE
SERVIE, 1967.

67-0139 ARGYRIS, J. H.
PATTON, P. C.
A LOOK INTO THE FUTURE: HOW COMPUTERS WILL
INFLUENCE ENGINEERING
J. ROY. AERO. SOC., VOL. 71, NO. 676, APRIL 1967,
PP: 244-252.

67-0140 ARGYRIS, J. H.
SPOONER, J. B.
DIE VERWENDUNG DER MATRIZENVERSCHIEBUNGSMETHODE
ZUR BERECHNUNG VON SPANNUNGSKOMPONENTEN
RESEARCH REPORT NO. 43, INSTITUT FUR STATIK UND
DYNAMIK DER LUFTUND RAUMFAHRTKONSTRUKTIONEN,
DECEMBER 1967.

67-0141 ARGYRIS, J. H.
SPOONER, J. B.
BRONLUND, O. E.
FRIED, I.
THE CHANGES IN THE (TWO-DIMENSIONAL) STRESS
ROUNDED CORNERS, CAUSED BY A VARYING INTERNAL
PRESSURE AND TEMPERATURE
PROC. 9TH INT. ASTRON. CONF., BELGRAD, OCTOBER
1967.

67-0142 ARGYRIS, J. H.
BOSSHARD, W.
FRIED, I.
HILBER, H.
A FULLY COMPATIBLE PLATE BENDING ELEMENT
RESEARCH REPORT NO. 42, INSTITUT FUR STATIK UND
DYNAMIK DER LUFT UND RAUMFAHRTKONSTRUKTIONEN,
DECEMBER 1967.

67-0143 ELIAS, Z. M.
DUALITY IN FINITE ELEMENT METHODS
DEPT. OF CIVIL ENG., MASS. INST. OF TECH.,
CAMBRIDGE, REPORT NO: NASA-CR-90500, 367-16, MAY
1967, 77 PP. (N.T.I.S. - N68-11612)

67-0144 SALANI, H.
A FINITE-ELEMENT METHOD FOR TRANSVERSE VIBRATIONS
OF BEAMS AND PLATES
BUREAU OF PUBLIC RDS., WASH., D.C., REPORT NO:
RR-56-8, JUNE 1967, 112 PP. (N.T.I.S. - PB-176
388)

67-0145 DUNHAM, R. S.
NICKELL, R. E.
FINITE ELEMENT ANALYSIS OF AXISYMMETRIC SOLIDS
WITH ARBITRARY LOADINGS
STANFORD RESEARCH INST., MENLO PARK, CALIF., AND
ROHM AND HAAS CO., HUNTSVILLE, ALA., REPORT NO:
SEL-67-6, JUNE 1967, 125 P. (N.T.I.S. - AD-655
253)

67-0146 MAESTRELLO, K.
GEDGE, M. R.
REDDAWAY, A. R. F.
RESPONSE OF STRUCTURE TO THE PSEUDO-SOUND FIELD OF
A JET (USING A COMBINED CONTINUUM AND FINITE
ELEMENT METHOD) PART 1
FLIGHT SCI. LAB., BOEING SCIENTIFIC RESEARCH
LABS., SEATTLE, WASH. REPORT NO: D1-82-3652, 118,
SEPTEMBER 1967, 46 PP. (N.T.I.S. - AD-663 662)

67-0147 GRESTE, O.
CLOUGH, R. W.
FINITE ELEMENT ANALYSIS OF TUBULAR JOINTS: A
REPORT ON A FEASIBILITY STUDY
STRUCT. ENG. LAB., UNIV. OF CALIF., BERKELEY,
REPORT NO: SESM-67-7, APRIL 1967, 53 PP. (N.T.I.S.
- PB-189 497)

67-0148 DEARRIAGA, F. J.
RASHID, Y. R.
SAFE-CREEP: A COMPUTER PROGRAM FOR THE
VISCOELASTIC ANALYSIS OF AXISYMMETRIC AND PLANE
CONCRETE STRUCTURES: USERS MANUAL
GULF GENERAL ATOMIC INC., SAN DIEGO, CALIF., JULY
31, 1967, 43 PP. (N.T.I.S. - GA-8111)

67-0149 POPOV, E. P.
ANALYSES OF SHELLS OF REVOLUTION BY FINITE
ELEMENTS
CONF. AT INTERN. CONGR. OF SHELL STRUCT. IN
ARCHITECTURE, MEXICO CITY, SEPTEMBER 3-7, 1967,
REPORT NO: NASA-CR-37648, 1967, 11 PP. (N.T.I.S. -
N68-25746)

67-0150 MEHRAIN, M.
FINITE ELEMENT ANALYSIS OF SKEW COMPOSITE GIRDER
BRIDGES
STRUCT. ENG. LAB., UNIV. OF CALIF., BERKELEY,
REPORT NO: SEL-67-28, NOVEMBER 1967, 151 PP.
(N.T.I.S. - PE-177 815).

67-0151 FRIED, I.
FINITE ELEMENT METHOD IN FLUID DYNAMICS AND HEAT
TRANSFER
INSTITUT FUR STATIK UND DYNAMIK DER LUFT UND
RAUMFAHRTKONSTRUKTIONEN, UNIVERSITAT STUTTGART,
REPORT 38, 1967.

67-0152 LIAM FINN, W. D.
FINITE ELEMENT ANALYSIS OF SEEPAGE THROUGH DAMS
PROC. J. SOIL MECH. AND FOUNDATIONS DIV.,
A.S.C.E., REPORT SM6, 1967, PP. 41-48.

67-0153 MARCAL, P. V.
COMPARATIVE STUDY OF NUMERICAL METHODS OF
ELASTIC-PLASTIC ANALYSIS
A.I.A.A. J., VOL. 6, NO. 1, 1967, PP. 157-158.

67-0154 IRONS, B. M.
LEAST SQUARES SURFACE FITTING BY FINITE ELEMENTS
AND AN APPLICATION TO STRESS SMOOTHING
AERO. STRESS MEMO, ROLLS-ROYCE, NO. ASM 1524,
1967.

67-0155 PAREKH, C. J.
A FINITE ELEMENT SOLUTION OF TIME DEPENDENT FIELD
PROBLEMS
M.SC. THESIS, UNIV. OF WALES, SWANSEA, 1967.

67-0156 HELLAN, K.
ANALYSIS OF ELASTIC PLATES BY A SIMPLIFIED FINITE
ELEMENT METHOD
ACTA POLYTECHNICA SCANDINAVICA, CIVIL ENG. AND
BUILDING CONST., SERIES NO. 46, 1967.

67-0157 CARR, A. J.
A REFINED FINITE ELEMENT ANALYSIS OF THIN SHELL
STRUCTURES INCLUDING DYNAMIC LOADINGS
PH.D., UNIV. OF CALIF., BERKELEY, 1967.

67-0158 GEBHARDT, L.
APPLICATION OF THE FINITE ELEMENT STIFFNESS METHOD
TO NONCONSERVATIVE STABILITY PROBLEMS
PH.D., UNIV. OF WASHINGTON, 1967.

67-0159 HAYCOCKS, C.
CENTRIFUGAL MODEL STUDY AND FINITE ELEMENT
ANALYSIS OF STRESS NEAR A FAULT
PH.D., UNIV. OF MISSOURI, ROLLA, 1967.

67-0160 MEHRAIN, M.
FINITE ELEMENT ANALYSIS OF SKEW COMPOSITE PLATES
PH.D., UNIV. OF CALIF., BERKELEY, 1967.

67-0161 NILSON, A. H.
FINITE ELEMENT ANALYSIS OF REINFORCED CONCRETE
PH.D., UNIV. OF CALIF., BERKELEY, 1967.

67-0162 MASON, V.
ON THE USE OF RECTANGULAR FINITE ELEMENTS
INST. OF SOUND AND VIB., UNIV. OF SOUTHAMPTON,
REPORT NO: 161, 1967.

68-0001 WARREN, J. W. L.
CREEP OF FABRICATED STRUCTURES
ROY. AERON. SOC. CONF. ON BEHALF OF JT. BRIT.
COMM. FOR STRESS ANAL. MARCH 26-29, 1968, LONDON,
ENGLAND, SESS. 2, PP. 15-26.

68-0002 HUMPHREY, A. T.
SUNLEY, V. K.
FINITE ELEMENT ANALYSIS OF AN EXPANDED I - SECTION
BEAM AND AN AXISYMMETRIC FLANGED CYLINDER
ROY. AERON. SOC., CONF. ON BEHALF OF JT. BRIT.
COMM. FOR STRESS ANAL., MARCH 26-29, 1968, LONDON,
ENGLAND, SESS. 3, PP. 14-23.

68-0003 ABEL, J. F.
POPOV, E. P.
STATIC AND DYNAMIC FINITE ELEMENT ANALYSIS OF
SANDWICH STRUCTURES
PROC. 2ND CONF. ON MATRIX METHODS IN STRUCTURAL
MECHANICS, AFFDL-TR-68-150, 1968, PP. 213-546.

68-0004 ADLER, A.
GALLAGHER, R. H.
THE USE OF FINITE ELEMENT METHODS IN HEAT FLOW
ANALYSIS
BELL AEROSYSTEMS CO., REPORT NO. 9500-920134,
AUGUST 1968.

68-0005 AHMAD, S.
IRONS, B. M.
ZIENKIEWICZ, O. C.
CURVED THICK SHELL AND MEMBRANE ELEMENTS WITH
PARTICULAR REFERENCE TO AXISYMMETRIC PROBLEMS
PROC. 2ND CONF. ON MATRIX METHODS IN STRUCTURAL
MECHANICS, AFFDL-TR-68-150, 1968, PP. 539-572.

68-0006 AHMED, S.
FINITE-ELEMENT METHOD FOR WAVEGUIDE PROBLEMS
ELECTRONIC LETTERS, VOL. 4, SEPTEMBER 1968, PP.
387-389.

68-0007 AKYUZ, F. A.
MERWIN, J. E.
SOLUTION OF NONLINEAR PROBLEMS OF ELASTOPLASTICITY
BY THE FINITE ELEMENT METHOD
A.I.A.A. J., VOL. 6, NO. 10, OCTOBER 1968, PP.
1825-1831.

68-0008 ERGATOUDIS, I.
IRONS, B. M.
ZIENKIEWICZ, O. C.
CURVED ISOPARAMETRIC QUADRILATERAL ELEMENTS FOR
FINITE ANALYSIS
INT. J. SOLIDS AND STRUCTS., VOL. 4, 1968.

68-0009 ANG, A. H.
LOPEZ, L. A.
DISCRETE MODEL ANALYSIS OF ELASTIC-PLASTIC PLATES
J. ENG. MECH. DIV., A.S.C.E., VOL. 94, NO. EM1,
PROC. PAPER 5809, 1968, PP. 271-293.

68-0010 ARGYRIS, J. H.
FRIED, I.
THE LUMINA ELEMENT FOR THE MATRIX DISPLACEMENT
METHOD
J. ROY. AERO. SOC., VOL. 72, JUNE 1968, PP.
514-517.

68-0011 ARGYRIS, J. H.
THE HERMES 8 ELEMENT FOR THE MATRIX DISPLACEMENT
METHOD
J. ROY. AERO. SOC., VOL. 72, JULY 1968, PP.
613-617.

68-0012 ARGYRIS, J. H.
THE TET 20 AND TEA 8 ELEMENTS FOR THE MATRIX
DISPLACEMENT METHOD
J. ROY. AERO. SOC., VOL. 72, JULY 1968, PP.
618-623.

68-0013 ARGYRIS, J. H.
THE TUBA FAMILY OF PLATE ELEMENTS FOR THE MATRIX
DISPLACEMENT METHOD
J. ROY. AERO. SOC., VOL. 72, AUGUST 1968, PP.
701-709.

68-0014 ARGYRIS, J. H.
SCHARPF, D. W.
THE SHEBA FAMILY OF SHELL ELEMENTS FOR THE MATRIX
DISPLACEMENT METHOD
J ROY. AERO. SOC. VOL. 72, OCTOBER 1968, PP.
873-883.

68-0015 ARGYRIS, J. H.
VUCKY, K. E.
A SEQUEL TO TECHNICAL NOTE 14 ON THE TUBA FAMILY
OF PLATE ELEMENTS
J. ROY. AERO. SOC., VOL. 72, NOVEMBER 1968, PP.
977-983.

68-0016 ARGYRIS, J. H.
MATRIX ANALYSIS OF SHELLS - SMALL AND LARGE
DISPLACEMENTS
AD-681 153, NOVEMBER 1968, 220 PP.

68-0017 ARGYRIS, J. H.
BUCK, K. E.
FRIED, I.
HILBER, H. M.
MARECZEK, G.
SCHARPF, D. W.
SOME NEW ELEMENTS FOR THE MATRIX DISPLACEMENT
METHOD
PROC. 2ND CONF. ON MATRIX METHODS IN STRUCTURAL
MECHANICS, AFFDL-TR-68-150, 1968, PP. 333-398.

68-0018 ARGYRIS, J. H.
REDSHAW, S. C.
THREE DIMENSIONAL ANALYSIS OF TWO ARCH DAMS BY THE
FINITE ELEMENT METHOD
SYMP. ON ARCH DAMS, I.C.E., 1968.

68-0019 ARKALEV, B. A.
VARVASINSKAYA, L. I.
SHATROVSKAYA, G. N.
SOLVING A NONLINEAR PROBLEM OF NONSTEADY THERMAL
CONDUCTIVITY IN SOLID BODIES OF ARBITRARY FORM
THERMAL ENGINEERING, VOL. 15, NO. 12, DECEMBER
1968, PP. 82-84.

68-0020 ARLETT, P. L.
BAHRANI, A. K.
ZIENKIEWICZ, O. C.
APPLICATION OF FINITE ELEMENTS TO THE SOLUTION OF
HELMHOLT'S EQUATION
PROC. I.E.E., VOL. 15, NO. 12, DECEMBER 1968, PP.
1762-1766.

68-0021 ARMEN, H.
PIFKO, A. B.
LEVINE, H. S.
A FINITE ELEMENT METHOD FOR THE PLASTIC BENDING
ANALYSIS OF STRUCTURES
PROC. 2ND CONF. ON MATRIX METHODS IN STRUCT.
MECH., REPORT NO. AFFDL-TR-68-150, OCTOBER 1968,
PP. 1301-1340.

68-0022 GLOUDEMAN, J. F.
ROSANOFF, R.
LEVY, S.
NUMERICAL CONDITIONING OF STIFFNESS MATRIX
FORMULATIONS FOR FRAME STRUCTURES
PROC. OF THE 2ND CONF. ON MATRIX METHODS IN
STRUCTURAL MECHANICS, AFFDL-TR-68-150, OCTOBER
1968.

68-0023 MELLO, R. M.
BACKUS, W. E.
SOME APPLICATIONS OF FINITE ELEMENT ANALYSIS TO
SHELL BUCKLING PREDICTION
PROC. 2ND CONF. ON MATRIX METHODS IN STRUCTURAL
MECHANICS, AFFDL-TR-68-150, 1968, PP. 869-894.

68-0024 BAKER, O. C.
A COMPARISON OF FINITE ELEMENT AND FOURIER SERIES
SOLUTIONS AS APPLIED TO RADIALLY LOADED CIRCULAR
RINGS
AD-844 167, JUNE 1968, 76 PP.

68-0025 BALTRUKONIS, J. H.
DYNAMICS OF SUBMARINE STRUCTURES, PART 2
AD-689 201, 1968, 27 PP.

68-0026 BANKS, D. C.
         PALMERTON, J. B.
         APPLICATION OF FINITE ELEMENT METHOD IN
         DETERMINING STABILITY OF CRATER SLOPES,
         PRELIMINARY REPORT
         AD-676 836, MAY 1968, 44 PP.

68-0027 BIRCHLER, W. D.
         FINITE ELEMENT MODELS FOR SHELLS OF REVOLUTION
         E.E.S. SERIES REPORT NO. 21, UNIV. OF ARIZONA,
         OCTOBER 1968.

68-0028 BLAKEY, L. H.
         FINITE ELEMENT TECHNIQUES IN ROCK MECHANICS
         AD-837 103, 1968, 13 PP.

68-0029 BOGNER, F. K.
         FINITE DEFLECTION, DISCRETE ELEMENT ANALYSIS OF
         SHELLS
         AFFDL-TR-67-185, JUNE 1968, P. 170.

68-0030 DHATT, G. S.
         BONNES, G.
         GIROUX, Y. M.
         ROBICHAUD, L. P. A.
         CURVED TRIANGULAR ELEMENTS FOR THE ANALYSIS OF
         SHELLS
         PROC. 2ND CONF. ON MATRIX METHODS IN STRUCTURAL
         MECHANICS, AFFDL-TR-68-150, 1968, PP. 617-640.

68-0031 BORING, L. M.
         MATRIX DISPLACEMENT ANALYSIS OF SHELLS WITH AN
         AXISYMMETRIC MIDSURFACE BUT HAVING
         CIRCUMFERENTIALLY VARYING STRUCTURAL PROPERTIES
         AD-841 165, JULY 1968, 93 PP.

68-0032 BOSSHARD, W.
         A NEW, FULLY COMPATIBLE FINITE ELEMENT FOR PLATE
         BENDING EIN NEUES VOLLVERTRAGLICHES ENDLICHES
         ELEMENT FUR PLATTEN BIEGUNG
         PUBLICATIONS I, ASSOC. FOR BRIDGE AND STRUCTURAL
         ENGR., VOL. 28, PART I, 1968, PP. 27-40.

68-0033 BRISBANE, J. J.
         FINITE ELEMENT STRESS ANALYSIS OF ANISOTROPIC
         BODIES
         AD-839 957, AUGUST 1968, 138 PP.

68-0034 BRISBANE, J. J.
         BECKER, E. B.
         PARR, C. H.
         THE APPLICATION OF FINITE ELEMENT METHODS TO THE
         SOLUTION OF STRESS AND DIFFUSION PROBLEMS IN
         CONTINUA
         AD-686 921, 1968, 19 PP.

68-0035 BROWN, J. E.
         FINITE ELEMENT SOLUTION TO DYNAMIC STABILITY OF
         BARS
         A.I.A.A. J. VOL. 6, NO. 7, JULY 1968, PP.
         1423-1424.

68-0036 CAMPBELL, J. S.
         STRESS ANALYSIS BY THE FINITE ELEMENT DISPLACEMENT
         METHOD
         BERKELEY NUCLEAR LABS. CENTRAL ELECTRICITY
         GENERATING BOARD, BERKELEY, ENGLAND, DECEMBER
         1968, P. 15, RD/B/N-8349

68-0037 CAMPBELL, J. S.
         FINITE ELEMENTS FOR THE REPRESENTATION OF PLANE
         AND AXISYMMETRIC STRUCTURAL REINFORCEMENT
         BERKELEY NUCLEAR LABS, CENTRAL ELECTRICITY
         GENERATING BOARD, BERKELEY, ENGLAND, DECEMBER
         1968, P. 16, RD-B-N-1195.

68-0038 CANTIN, G.
         CLOUGH, R. W.
         A REFINED CURVED CYLINDRICAL SHELL FINITE ELEMENT
         A.I.A.A. 6TH AEROSPACE SCIENCES MEETING, JANUARY
         1968, 10 PP. PAPER 68-176.

68-0039 CANTIN, G.
         CLOUGH, R. W.
         A CURVED, CYLINDRICAL SHELL, FINITE ELEMENT
         A.I.A.A. J. VOL. 6, NO. 6, JUNE 1968, PP.
         1057-1062.

68-0040 CARRARA, S.
         MCGARRY, F.
         MATRIX AND INTERFACE STRESSES IN A DISCONTINUOUS
         FIBER COMPOSITE MODEL
         J. OF COMPOS. MAT., VOL. 2, NO. 2, APRIL 1968, PP.
         222-243.

68-0041 CHANG, G. C.
         FLAIR, G. S.
         FINITE ELEMENT APPROACH TO WAVE MOTIONS IN
         ELASTIC-PLASTIC CONTINUA
         A.I.A.A. 6TH AEROSPACE SCIENCES MEETING, PAPER NO.
         68-145, NEW YORK, JANUARY 1968.

68-0042 CHEUNG, Y. K.
         NAG, D. K.
         PLATES AND BEAMS ON ELASTIC FOUNDATIONS: LINEAR
         AND NONLINEAR BEHAVIOUR
         GEOTECHNIQUE, VOL. 18, NO. 2, JUNE 1968, PP.
         250-260.

68-0043 CHOPRA, A. K.
         EARTHQUAKE BEHAVIOUR OF RESERVOIR-DAM SYSTEMS
         J. ENG. MECH. DIV., A.S.C.E., VOL. 94, NO. EM6,
         DECEMBER 1968, PP. 1475-1500.

68-0044 CHRISTIAN, J. T.
         A FINITE ELEMENT FORMULATION FOR MATERIALS WITH
         SPECIFIED CHANGE OF VOLUME
         M.I.T. SCHOOL OF ENGR., JUNE 1968, P. 14.

68-0045 CHRISTIAN, J. T.
         UNDRAINED STRESS DISTRIBUTION BY NUMERICAL METHODS
         J. SOIL MECH. AND FOUNDATIONS DIV., A.S.C.E., VOL.
         94, NO. SM 6, NOVEMBER 1968, PP. 1333-1345.

68-0046 JOHNSON, C. P.
         CLOUGH, R. W.
         A FINITE ELEMENT APPROXIMATION FOR THE ANALYSIS OF
         THIN SHELLS
         INT. J. SOLIDS AND STRUCT., VOL. 4, NO. 1, JANUARY
         1968, PP. 43-60.

68-0047 CLOUGH, R. W.
         FELIPPA, C. A.
         A REFINED QUADRILATERAL ELEMENT FOR ANALYSIS OF
         PLATE BENDING
         PROC. 2ND CONF. ON MATRIX METHODS IN STRUCTURAL
         MECHANICS, AFFDL-TR-68-150, 1968, PP. 399-440.

68-0048 CONNOR, J. J.
         APPLICATION TO PLANE STRESS AND PLANE
         STRAIN--LINEAR ELASTIC MATERIAL
         M.I.T. SCHOOL OF ENGR., JUNE 1968, P. 67.

68-0049 CONNOR, J. J.
         BENDING OF THIN PLATES - DISPLACEMENT MODELS
         M.I.T. SCHOOL OF ENGR., JUNE 1968, P. 67.

68-0050 CONNOR, J. J.
         ANALYSIS OF GEOMETRICALLY NONLINEAR PLATES AND
         SHELLS BY THE FINITE ELEMENT DISPLACEMENT METHOD
         M.I.T. SCHOOL OF ENGR., JUNE 1968, P. 36.

68-0051 COWPER, G. R.
         KOSKO, E.
         FORMULATION OF A NEW TRIANGULAR PLATE BENDING
         ELEMENT
         CASI TRANS., VOL. 1, SEPTEMBER 1968, PP. 86-90.

68-0052 COWPER, G. R.
         KOSKO, E.
         LINDBERG, G. M.
         OLSON, M. D.
         A HIGH PRECISION TRIANGULAR PLATE BENDING ELEMENT
         NAT. RES. COUNCIL OF CANADA, AERON. REP. LR-514,
         DECEMBER 1968, ALSO AD-685 576

68-0053 CRAGGS, A.
         TRANSIENT VIBRATION ANALYSIS OF LINEAR SYSTEMS
         USING TRANSITION MATRICES
         NASA-CR-1237, NOVEMBER 1968, P. 26.

68-0054 DAWE, D. J.
         APPLICATION OF THE DISCRETE ELEMENT METHOD TO THE
         BUCKLING ANALYSIS OF RECTANGULAR PLATES UNDER
         ARBITRARY MEMBRANE LOADING
         AD-840 106, FEBRUARY 1968, 25 PP.

68-0055 IRONS, B. M.
         ZIENKIEWICZ, O. C.
         DE ARANTES E OLIVEIRA, E. R.
         THEORETICAL FOUNDATIONS ON THE FINITE ELEMENT
         METHOD
         INT. J. SOLIDS AND STRUCT., VOL. 4, NO. 10, 1968,
         PP. 929-952.

68-0056 DE ARANTES E OLIVEIRA, E. R.
         COMPLETENESS AND CONVERGENCE IN THE FINITE ELEMENT
         METHOD
         PROC. 2ND CONF. ON MATRIX METHODS IN STRUCTURAL
         MECHANICS, AFFDL-TR-68-150, 1968, PP. 1061-4090,
         ALSO IN TECNIA (LISBON) VOL. 33, NO. 43, 1970, PP.
         108-124.

68-0057 DONG, S. B.
         HASSON, C.
         WESTMANN, R. A.
         EMBANKMENT ANALYSIS AND FIELD CORRELATION
         HIGHWAY RESEARCH BOARD, HIGHWAY RESEARCH REPORT
         NO. 228, 1968.

68-0058 DUNCAN, J. M.
         MONISMITH, C. L.
         WILSON, E. L.
         FINITE ELEMENT ANALYSIS OF PAVEMENTS
         HIGHWAY RESEARCH RECORD NUMBER 228, HIGHWAY
         RESEARCH BOARD, 1968.

68-0059 DUNGAR, R.
         SEVERN, R. T.
         THE ANALYSIS OF ARCH DAMS UNDER STATIC LOADING BY
         THE FINITE ELEMENT METHOD
         SYMP. ON ARCH DAMS, I.C.E., 1968.

68-0060 DUNHAM, R. S.
PISTER, K. S.
A FINITE ELEMENT APPLICATION OF THE
HELLINGER-REISSNER VARIATIONAL THEOREM
PROC. 2ND CONF. ON MATRIX METHODS IN STRUCTURAL
MECHANICS, AFFDL-TR-68-150, 1968, PP. 471-488.

68-0061 DUNLOP, P.
DUNCAN, J. M.
SEED, H. B.
FINITE ELEMENT ANALYSIS OF SLOPES IN SOIL
AD-687 176, MAY 1968, 229 PP.

68-0062 DUNNE, P. C.
COMPLETE POLYNOMIAL DISPLACEMENT FIELDS FOR FINITE
ELEMENT METHODS
THE AERON. J., VOL. 72, AUGUST 1968, PP. 245-247.

68-0063 DAVIES, J. D.
CHEUNG, Y. K.
THE ANALYSIS OF COOLING TOWER RING BEAMS
PROC. INST. CIV. ENGRS., V. 39, PP. 567-579, APRIL
1968

68-0064 ELIAS, Z. M.
VARIATIONAL FORMULATIONS AND FINITE ELEMENT METHOD
FOR THIN ELASTIC SHELLS AND PLATES
M.I.T. SCHOOL OF ENGR., JUNE 1968, P. 43.

68-0065 ELIAS, Z. M.
DUALITY IN FINITE ELEMENTS METHODS
J. ENG. MECH. DIV., A.S.C.E., VOL. 94, NO. EM4,
AUGUST 1968, PP. 931-946.

68-0066 ERGATOUDIS, J. J.
THREE DIMENSIONAL ANALYSIS OF ARCH DAMS AND THEIR
FOUNDATIONS
SYMP. ON ARCH DAMS, I.C.E., 1968.

68-0067 ERGATOUDIS, J. J.
CURVED ISOPARAMETRIC QUADRILATERAL ELEMENTS FOR
FINITE ELEMENT ANALYSIS
INT. J. SOLIDS STRUCT., VOL. 4, 1968, PP. 31-42.

68-0068 CHEUNG, Y. K.
THE FINITE STRIP METHOD IN THE ANALYSIS OF ELASTIC
PLATES WITH TWO SIMPLY SUPPORTED ENDS
PROC. INST. CIV. ENGRS., V. 40, MAY 1968

68-0069 FISHKAA, G.
IVERSEN, P. A.
SARSTEN, A.
COMPUTER CALCULATIONS OF STRESSES IN AXISYMMETRIC
THERMALLY LOADED COMPONENTS
SYMP. ON COMPUTERS IN I.C. ENGINE DESIGN, INST.
MECH. ENGR., APRIL 1968.

68-0070 FOX, R. L.
STANTON, E. L.
DEVELOPMENTS IN STRUCTURAL ANALYSIS BY DIRECT
ENERGY MINIMIZATION
A.I.A.A. J., VOL. 6, NO. L, JUNE 1968, PP.
1036-1042.

68-0071 FRAEIJS DE VEUBEKE, B.
A CONFORMING FINITE ELEMENT FOR PLATE BENDING
INT. J. SOLIDS AND STRUCT., VOL. 4, NO. 1, JANUARY
1968, PP. 65-108.

68-0072 FRAEIJS DE VEUBEKE, F.
AN EQUILIBRIUM MODEL FOR PLATE BENDING
INTL. J. SOLIDS AND STRUCT., VOL. 4, NO. 4, APRIL
1968, PP. 447-468.

68-0073 FUJINO, T.
OHSAKA, K.
THE HEAT CONDUCTION AND THERMAL STRESS ANALYSIS BY
THE FINITE ELEMENT METHOD
PROC. 2ND CONF. ON MATRIX METHODS IN STRUCTURAL
MECHANICS, AFFDL-TR-68-150, 1968, PP. 1121-1164.

68-0074 GALLAGHER, R. H.
STIFFNESS MATRIX FOR SHALLOW RECTANGULAR SHELL
ELEMENT
J. ENGR. MECH. DIV., A.S.C.E., VOL. 94, NO. EM 2,
APRIL 1968, PP. 708-709.

68-0075 GALLAGHER, R. H.
ELASTIC INSTABILITY PREDICTIONS FOR DOUBLY-CURVED
SHELLS
PROC. 2ND CONF. ON MATRIX METH. STRUCT. MECH.,
AFFDL-TR-68-150, 1968, PP. 711-742.

68-0076 GIFFORD, L. N. JR.
FINITE ELEMENT ANALYSIS FOR ARBITRARY AXISYMMETRIC
STRUCTURES ORTHOTROPIC, TEMPERATURE-DEPENDENT
MATERIAL PROPERTIES
AD-835 243, MARCH 1968, 156 PP.

68-0077 GIRIJAVALLABHAN, C. V.
REESE, L. C.
FINITE-ELEMENT METHOD APPLIED TO SOME PROBLEMS IN
SOIL MECHANICS
J. SOIL MECH. AND FOUNDATIONS DIV., A.S.C.E., VOL.
92, NO. SM 2, PROC. PAPER 5864, MARCH 1968.

68-0078 GIRIJAVALLABHAN, C. V.
REESE, L. C.
FINITE ELEMENT METHOD FOR PROBLEMS IN SOIL
MECHANICS
J. OF SOIL MECH. AND FOUNDATIONS DIV., A.S.C.E.,
VOL. 94, NO. SM 2, MARCH 1968, PP. 473-496.

68-0079 GOODMAN, R. E.
TAYLOR, P. L.
BREKKE, T.
A MODEL FOR THE MECHANICS OF JOINTED ROCK
J. OF SOIL MECH. AND FOUND. DIV., A.S.C.E., VOL.
94, 1968, PP. 637-659.

68-0080 GRADOWCZYK, M. H.
VISCOELASTIC STRESS ANALYSIS
M.I.T. SCHOOL OF ENGR., JUNE 1968, P. 19.

68-0081 GREENBAUM, G. A.
RUBINSTEIN, M. F.
CREEP ANALYSIS OF AXISYMMETRIC BODIES
NUCLEAR ENGR. AND DESIGN, VOL. 7, NO. 4, APRIL
1968, PP. 379-397.

68-0082 GREENE, B. E.
JONES, R. E.
MCLAY, R. W.
STROME, D. R.
DYNAMIC ANALYSIS OF SHELLS USING DOUBLY-CURVED
FINITE ELEMENTS
PROC. 2ND CONF. ON MATRIX METHODS IN STRUCTURAL
MECHANICS, AFFDL-TR-68-150, 1968, PP. 185-212.

68-0083 GREENE, B. E.
JONES, R. E.
MCLAY, R. W.
STROME, D. R.
ON THE APPLICATION OF GENERALIZED VARIATIONAL
PRINCIPLES IN THE FINITE ELEMENT METHOD
A.I.A.A./A.S.M.E. 9TH STRUCTURES, STRUCTURAL
DYNAMICS AND MATERIALS CONF., A.I.A.A. PAPER NO.
68-290, PALMS SPRINGS, CALIF., 1968.

68-0084 GUYAN, R. J.
UJIHARA, B. H.
WELCH, P. W.
HYDROELASTIC ANALYSIS OF AXISYMMETRIC SYSTEMS BY A
FINITE ELEMENT METHOD
PROC. 2ND CONF. ON MATRIX METHODS IN STRUCTURAL
MECHANICS, AFFDL-TR-68-150, 1968, PP. 1165-1206.

68-0085 HEDLEY, D. G. F.
ZAHARY, G.
SODERLAND, H. W.
COATES, D. F.
UNDERGROUND MEASUREMENTS IN A STEEPLY DIPPING
OREBODY
PROC. 5TH CANADIAN ROCK MECH. SYMP., DECEMBER
1968, PP. 105-125.

68-0086 HELLEN, T. K.
ECONOMICAL COMPUTER TECHNIQUES FOR SOLID FINITE
ELEMENTS
CENTRAL ELECTRICITY GENERATING BOARD, BERKELEY,
ENGLAND, BERKELEY NUCLEAR LABS., DECEMBER 1968, P.
13, RD/B/N-1305.

68-0087 HERRMANN, L. R.
CAMPBELL, D. M.
A FINITE-ELEMENT ANALYSIS FOR THIN SHELLS
A.I.A.A. J., VOL. 6, NO. 10, OCTOBER 1968, PP.
1842-1846.

68-0088 HILL, D. W.
COFFIN, G. K.
STRESSES AND DEFLECTIONS IN COOLING TOWER SHELLS
DUE TO WIND LOADING
BULLETIN I.A.S.S., NO. 35, SEPTEMBER 1968, PP.
43-51.

68-0089 HOLAND, I.
BERGAN, P. G.
HIGHER-ORDER FINITE ELEMENT FOR PLANE STRESS
J. ENGR. MECH. DIV., A.S.C.E., NO. EM 2, APRIL
1968, PP. 698-702. (DISCUSSION)

68-0090 HOYAUX, B.
LADANYI, B.
STRESS DISTRIBUTION DUE TO GRAVITY IN A VERTICAL
ROCK BANK
TENTH SYMP. ON ROCK MECHANICS, AUSTIN TEXAS, 1968.

68-0091 IDRISS, I. M.
FINITE ELEMENT ANALYSIS FOR SEISMIC RESPONSE OF
EARTH DAMS
J. SOIL MECH. AND FOUNDATION DIV., A.S.C.E., VOL.
93, NO. SM4, 1968.

68-0092 IRONS, B. M.
ERGATOUDIS, J. G.
ZIENKIEWICZ, O. C.
COMPLETE POLYNOMIAL DISPLACEMENT FIELDS FOR FINITE
ELEMENT METHODS
TRANS. ROYAL AERO. SOC., VOL. 72, 1968, PP.
709-711.

68-0093 JACOBS, L. D.
LAGERQUIST, D. R.
FINITE-ELEMENT ANALYSIS OF COMPLEX PANEL RESPONSE
TO RANDOM LOADS
AD-845 309, OCTOBER 1968, 135 P.

68-0094 JAVANDEL, I.
WITHERSPOON, P. A.
APPLICATION OF THE FINITE ELEMENT METHOD TO
TRANSIENT FLOW IN POROUS MEDIA
SOC. PET. ENGR. J., VOL. 8, NO. 3, SEPTEMBER 1968,
P. 241.

68-0095 JOHNSON, M. W.
MCLAY, R. W.
CONVERGENCE OF THE FINITE ELEMENT METHOD IN THE
THEORY OF ELASTICITY
J. APPLIED MECH., A.S.M.E., VOL. 35, JUNE 1968,
PP. 274-278.

68-0096 JONES, R. M.
CROSS, J. G.
SAAS II: FINITE ELEMENT STRESS ANALYSIS OF
AXISYMMETRIC SOLIDS WITH ORTHOTROPIC, TEMPERATURE
DEPENDENT MATERIAL PROPERTIES
AD-679 983, SEPTEMBER 1968, 204 P.

68-0097 KAWAI, T.
OHTSUBO, H.
A METHOD OF SOLUTION FOR THE COMPLICATED BUCKLING
PROBLEMS OF ELASTIC PLATES WITH COMBINED USE OF
RAYLEIGH-RITZ'S PROCEDURE IN THE FINITE ELEMENT
METHOD
PROC. 2ND CONF. ON MATRIX METHODS IN STRUCTURAL
MECHANICS, AFFDL-TR-68-150, 1968, PP. 967-994.

68-0098 BEISINGER, Z. E.
KEY, S. W.
THE ANALYSIS OF THIN SHELLS WITH TRANSVERSE SHEAR
STRAINS BY THE FINITE ELEMENT METHOD
PROC. 2ND CONF. ON MATRIX METHODS IN STRUCTURAL
MECHANICS, OHIO, 1968.

68-0099 KHAL, R.
COMPUTER STRESS ANALYSIS
MACHINE DESIGN, VOL. 40, NOVEMBER 21, 1968, PP.
136-145.

68-0100 KOBAYASHI, A. S.
MAIDEN, D. E.
SIMON, B. J.
IIDA, S.
APPLICATION OF THE METHOD OF FINITE ELEMENT
ANALYSIS TO TWO-DIMENSIONAL PROBLEMS IN FRACTURE
MECHANICS
TECH. REPORT NO. 5, DEPT. OF MECH. ENGR., UNIV. OF
WASHINGTON, SEATTLE, 1968.

68-0101 VON RIESEMANN, W.A .
LARGE DEFLECTIONS OF ELASTIC BEAMS AND PLATES
USING THE FINITE ELEMENT METHOD
PH.D. THESIS, STANFORD UNIV., 1968.

68-0102 KOTANCHIK, J. J.
DISCRETE-ELEMENT STATIC ANALYSIS OF BONDED,
DOUBLE-LAYER, BRANCHED, THIN SHELLS OF REVOLUTION,
PART I. ANALYSIS AND EVALUATION
AD-856 452, MAY 1968, 218 P.

68-0103 KRAHULA, J. L.
POLHEMUS, J. F.
USE OF FOURIER SERIES IN THE FINITE ELEMENT METHOD
A.I.A.A. J., VOL. 6, NO. 4, APRIL 1968, PP.
726-727.

68-0104 KRAJCINOVIC, D.
A CONSISTENT DISCRETE ELEMENT TECHNIQUE FOR
THIN-WALLED ASSEMBLAGES
AD-694 593, NOVEMBER 1968, P. 27.

68-0105 KROSS, D. A.
A STUDY OF STIFFNESS MATRICES FOR THE ANALYSIS OF
FLAT PLATES
NASA-TN-D-4927, DECEMBER 1968, P. 47.

68-0106 LAGERQUIST, D. R.
JACOBS, L. D.
RANDOM-VIBRATION ANALYSIS SYSTEM FOR COMPLEX
STRUCTURES. PART 1, ENGINEERING USER'S GUIDE
AD-845 604, NOVEMBER 1968, 209 P.

68-0107 LOGCHER, R. D.
A GENERAL SYSTEM FOR STRUCTURAL FINITE ELEMENT
ANALYSIS
M.I.T. SCHOOL OF ENGR., JUNE 1968, P. 47.

68-0108 LYNCH, R. D.
KELSEY, S.
SAXE, H. C.
THE APPLICATION OF DYNAMIC RELAXATION TO THE
FINITE ELEMENT METHOD OF STRUCTURAL ANALYSIS
AD-676 566, SEPTEMBER 1968, 125 P.

68-0109 MALLETT, R. H.
MARCAL, P. V.
FINITE ELEMENT ANALYSIS OF NONLINEAR STRUCTURES
J. STRUCT. DIV., A.S.C.E., VOL., 94, NO. ST 9,
SEPTEMBER 1968, PP. 2081-2106.

68-0110 MALONE, J. P.
A COMPUTER PROGRAM FOR THE ANALYSIS OF LINEARLY
ELASTIC PLANE-STRESS, PLANE-STRAIN PROBLEMS
AD-844 095, SEPTEMBER 1968 132 P.

68-0111 MARCAL, P. V.
MALLETT, R. H.
ELASTIC-PLASTIC ANALYSIS OF FLAT PLATES BY THE
FINITE ELEMENT METHOD
PAPER NO. 68-WA/PVP-10, DECEMBER 1968.

68-0112 MARTIN, H. C.
FINITE ELEMENT ANALYSIS OF FLUID FLOWS
PROC. 2ND CONF. ON MATRIX METHODS IN STRUCTURAL
MECHANICS, AFFDL-TR-68-150, 1968, PP. 517-538.

68-0113 MASON, V.
RECTANGULAR FINITE ELEMENTS FOR ANALYSIS OF PLATE
VIBRATIONS
J. SOUND VIB., VOL. 7, NO. 3, 1968 PP. 437-448.

68-0114 MASON, WM. E. JR.
ELASTIC SHEAR ANALYSIS OF GENERAL PRISMATIC BEAMS
J. ENGR. MECH. DIV., A.S.C.E., VOL. 94, NO. EM4,
AUGUST 1968, PP. 965-983.

68-0115 MCCALLUM, R. B.
WITMER, E. A.
A FEASIBILITY STUDY OF THE ANALYSIS OF ASYMMETRIC
SHELL BY THE DISCRETE-ELEMENT METHOD
AD-844 813, SEPTEMBER 1968, 115 P.

68-0116 MCCALLEY, R. B. JR.
MASS MATRIX FOR A PRISMATIC BEAM SEGMENT
KNOLLS ATOMIC POWER LAB., SCHENECTADY, N. Y.,
MARCH 1968. CONTRACT W-31-109-ENG-52, P. 73,
KAPG-M-6913.

68-0117 HRENNIKOFF, A.
THE FINITE ELEMENT METHOD IN APPLICATION TO PLANE
STRESS
PUBLICATION, I.A.B.S.E., VOL. 28, PART II, 1968,
PP. 49-68.

68-0118 MEEK, J. L.
FIELD PROBLEMS SOLUTIONS BY FINITE METHODS
CIVIL ENGINEERING TRANS., I. E. AUST., OCTOBER
1968, PP. 173-180.

68-0119 MIYAMOTO, H.
MIYOSHI, T.
ELASTIC AND PLASTIC ANALYSIS BY FINITE ELEMENT
METHOD
SOC. OF STEEL CONSTRUCTION OF JAPAN, 1968.

68-0120 MONFORTON, G. R.
SCHMIT, L. A.
FINITE ELEMENT ANALYSIS OF SKEW PLATES IN BENDING
A.I.A.A. J., VOL. 6, NO. 6, JUNE 1968, PP.
1150-1151.

68-0121 MONFORTON, G. R.
SCHMIT, L. A.
FINITE ELEMENT ANALYSIS OF SANDWICH PLATES AND
CYLINDRICAL SHELLS WITH LAMINATED FACES
PROC. 2ND CONF. ON MATRIX METHODS IN STRUCTURAL
MECHANICS, AFFDL-TR-68-150, 1968, PP. 573-616.

68-0122 MORLEY, L. S. D.
THE TRIANGULAR EQUILIBRIUM ELEMENT IN THE SOLUTION
OF PLATE BENDING PROBLEMS
AERON. QUART., VOL. 19, MAY 1968, PP. 149-169.

68-0123 MURRAY, D. W.
WILSON, E. L.
FINITE ELEMENT POST BUCKLING ANALYSIS OF THIN
ELASTIC PLATES
PROC. SECOND CONF. ON MATRIX METHODS IN STRUCTURAL
MECHANICS, AFFDL-TR-69, WRIGHT PATTERSON AIR FORCE
BASE, OHIO, OCTOBER 1968. ALSO IN A.I.A.A. J.,
VOL. 7, NO. 10, 1969, PP. 1915-1920.

68-0124 NAVARATNA, D. R.
STABILITY ANALYSIS OF SHELLS OF REVOLUTION BY THE
FINITE ELEMENT METHOD
A.I.A.A. J., VOL. 6, NO. 2, FEBRUARY 1968, PP.
355-360.

68-0125 NICKELL, R. E.
STRESS WAVE ANALYSIS IN LAYERED THERMOVISCOELASTIC
MATERIALS BY THE EXTENDED RITZ METHOD, VOLUME II
AD-841 224, OCTOBER 1968, P. 108.

68-0126 NILSON, A. H.
NONLINEAR ANALYSIS OF REINFORCED CONCRETE BY THE
FINITE ELEMENT METHOD
A.C.I.J., VOL. 65, NO. 9, SEPTEMBER 1968, PP.
757-766.

68-0127 ODEN, J. T.
CALCULATION OF STIFFNESS MATRICES FOR FINITE
ELEMENTS OF THIN SHELLS OF ARBITRARY SHAPE
A.I.A.A. J., VOL. 6, NO. 5, MAY 1968, PP. 969-971.

470

68-0128 ODEN, J. T.
FINITE PLANE STRAIN OF INCOMPRESSIBLE ELASTIC
SOLIDS BY THE FINITE ELEMENT METHOD
AERON. QUART., VOL. 19, AUGUST 1968, PP. 259-264.

68-0129 ODEN, J. T.
KROSS, D. A.
ANALYSIS OF GENERAL COUPLED THERMOELASTICITY
PROBLEMS BY THE FINITE ELEMENT METHOD
PROC. 2ND CONF. ON MATRIX METHODS IN STRUCTURAL
MECHANICS, AFFDL-TR-68-150, 1968, PP. 1091-1120.

68-0130 OLSON, M. D.
LINDBERG, G. M.
VIBRATION ANALYSIS OF CANTILEVERED CURVED PLATES
USING A NEW CYLINDRICAL SHELL FINITE ELEMENT
PROC. 2ND CONF. ON MATRIX METHODS IN STRUCTURAL
MECHANICS, AFFDL-TR-68-150, 1968, PP. 247-270.

68-0131 OLSON, M. D.
LINDBERG, G. M.
A FINITE CYLINDRICAL SHELL ELEMENT AND THE
VIBRATIONS OF A CURVED FAN BLADE
NRC-10198, CANADA, (N68-33842), FEBRUARY 1968, P.
45.

68-0132 PAULLING, J. R.
PAYER, H. G.
HULL-DECKHOUSE INTERACTION BY FINITE ELEMENT
CALCULATIONS
SOC. NAV. ARCH. MAR. ENGRS., ANNUAL MEETING, NEW
YORK, 1968.

68-0133 PETYT, M.
FINITE ELEMENT VIBRATION ANALYSIS OF CRACKED
PLATES IN TENSION
AD-666 448, JANUARY 1968, P. 165.

68-0134 PETYT, M.
THE VIBRATION CHARACTERISTICS OF A TENSIONED PLATE
CONTAINING A FATIGUE CRACK
ISRAEL J. TECH., VOL. 6, NO. 1/2, JANUARY 1968,
PP. 58-66.

68-0135 PIAN, T. H. H.
PLATE BENDING ANALYSIS USING ASSUMED STRESS
METHODS
M.I.T. SCHOOL OF ENGR., JUNE 1968 P. 26.

68-0136 PIAN, T. H. H.
STABILITY ANALYSIS OF SHELLS OF REVOLUTION
M.I.T. SCHOOL OF ENGR., JUNE 1968, P. 26.

68-0137 PIAN, T. H. H.
VARIATIONAL PRINCIPLES AND THEIR APPLICATION TO
FINITE ELEMENT METHODS
M.I.T. SCHOOL OF ENGR., JUNE 1968.

68-0138 PIAN, T. H. H.
TONG, P.
RATIONALIZATION IN DERIVING ELEMENT STIFFNESS
MATRIX BY ASSUMED STRESS APPROACH
PROC. 2ND CONF. ON MATRIX METHODS IN STRUCTURAL
MECHANICS, AFFDL-TR-68-150, 1968 PP. 441-470.

68-0139 PRZEMIENIECKI, J. S.
THEORY OF MATRIX STRUCTURAL ANALYSIS
MCGRAW HILL PUB. CO., 1968.

68-0140 PRZEMIENIECKI, J. S.
DISCRETE-ELEMENT METHODS FOR STABILITY ANALYSIS OF
COMPLEX STRUCTURES
J. ROY. AERON. SOC., VOL. 72, DECEMBER 1968, PP.
1077-1086.

68-0141 RASHID, Y. R.
PRESTRESSED CONCRETE REACTOR VESSEL ANALYTICAL
METHODS DEVELOPMENTS
GULF GENERAL ATOMIC INC., SAN DIEGO, CALIF., FINAL
PROGRESS REPORT, JULY 1, 1968-JUNE 30, 1969, JULY
1969, P. 25.

68-0142 ROREN, E. M. Q.
TRANSVERSE STRENGTH OF TANKERS, FINITE ELEMENT
APPLICATIONS
EUROPEAN SHIPBUILDING, PT. 1, VOL. 17, NO. 3,
1968, PP. 42-49, PT. 2, VOL. 17, NO. 4, 1968, PP.
58-65.

68-0143 SCHMIT, L. A.
FINITE DEFLECTION STRUCTURAL ANALYSIS USING PLATE
AND SHELL DISCRETE ELEMENTS
A.I.A.A. J., VOL. 6, NO. 5, MAY 1968 PP. 781-791.

68-0144 SCHMIT, L. A.
GOBLE, G. G.
STANTON, E. L.
GIBSON, W.
SWARTS, J. D.
CAMPBELL, F. S.
FOX, R. L.
DEVELOPMENTS IN DISCRETE ELEMENT FINITE DEFLECTION
STRUCTURAL ANALYSIS BY FUNCTION MINIMIZATION
TECHNICAL REPORT AFFDL-TR-68-126, WRIGHT-PATTERSON
AFB, OHIO, SEPTEMBER 1968.

68-0145 SHIEH, W. Y. J.
ANALYSIS OF PLATE BENDING BY TRIANGULAR ELEMENTS
J. ENGR. MECH. DIV., A.S.C.E., VOL. 94, NO. EM 5,
OCTOBER 1968, PP. 1089-1108.

68-0146 SMITH, I. M.
A FINITE ELEMENT ANALYSIS FOR ' MODERATELY THICK '
RECTANGULAR PLATES IN BENDING
INT. J. OF MECH. SCI., VOL. 10, JULY 1968, PP.
563-570.

68-0147 SOM, P. K.
GHOSH, K.
MUKHERJEE, A. K.
ANCHOR ZONE STRESS ANALYSIS BY FINITE ELEMENT
METHOD
J. INST. ENGR., INDIA, C.E. DIV., VOL. 48, NO.
CI6, JULY 1968, PP. 1536-1549.

68-0148 SPEARE, J. C.
DISCRETE-ELEMENT STATIC ANALYSIS OF CORE-STIFFENED
SHELLS OF REVOLUTION FOR ASYMMETRIC MECHANICAL
LOADING
AD-847 384, SEPTEMBER 1968, P. 251.

68-0149 STRICKLAND, G. E.
LODEN, W. A.
A DOUBLY CURVED TRIANGULAR SHELL ELEMENT
PROC. 2ND CONF. ON MATRIX METHODS IN STRUCTURAL
MECHANICS, AFFDL-TR-68-150, 1968, PP. 667-710.

68-0150 STRICKLIN, J. A.
INTERACTION OF AREA COORDINATES IN MATRIX
STRUCTURAL ANALYSIS
A.I.A.A. J., VOL. 6, NO. 10, OCTOBER 1968, PP.
2023.

68-0151 STRICKLIN, J. A.
DE ANDRADE, J. C.
LINEAR AND NON LINEAR ANALYSIS OF SHELLS OF
REVOLUTION WITH ASYMMETRICAL STIFFNESS PROPERTIES
PROC. SECOND CONF. MATRIX METHODS STRUCT. MECH.,
AIR FORCE INST. OF TECHN., WRIGHT PATTERSON AIR
FORCE BASE, OHIO, 1968.

68-0152 SWANSON, J. A.
FEATS, A COMPUTER PROGRAM FOR THE FINITE ELEMENT
THERMAL STRESS ANALYSIS OF PLANE OR
AXISYMMETRIC SOLIDS
WESTINGHOUSE ELECTRIC CORP., PITTSBURGH, PA.,
ASTRONUCLEAR LAB., DECEMBER 1968, P. 142,
WANL-TME-1888.

68-0153 TAKAHASHI, S. K.
DONG, S. B.
FINITE ELEMENT ANALYSIS FOR SOLIDS OF REVOLUTION
AD-667 235, MARCH 1968, P. 164.

68-0154 TOCHER, J. L.
VARANASI, S. R.
FINITE ELEMENT ANALYSIS OF PLASTICITY PROBLEMS
AMER. SOC. CIVIL ENGRS., JOINT SPECIALITY
CONFERENCE, CHICAGO 1968.

68-0155 TONG, P.
ON THE CONVERGENCE OF FINITE ELEMENT METHODS
M.I.T. SCHOOL OF ENGR., JUNE 1968, P. 35.

68-0156 TONG, P.
A NEW FINITE ELEMENT MODEL FOR SOLID CONTINUA
MIT., ASRL-TR-144-2, JULY 1968, P. 31.

68-0157 TOWNLEY, C. H. A.
APPLICATION OF ELASTIC PLASTIC ANALYSIS TO THE
DESIGN OF PRESSURE VESSELS
CENTRAL ELECTRICITY GENERATING BOARD, BERKELEY,
ENGLAND, BERKELEY NUCLEAR LABS, JULY 1968, P. 33,
RD/B/N-1105.

68-0158 UTKU, S.
AKYUZ, F. A.
ELAS - A GENERAL PURPOSE COMPUTER PROGRAM FOR THE
EQUILIBRIUM PROBLEMS OF LINEAR STRUCTURES
TECH. REP. 32-1240, JET PROPULSION LAB., CALIF.,
INST. OF TECHNOLOGY, FEBRUARY 1968.

68-0159 UTKU, S.
MELOSH, R. J.
BEHAVIOR OF TRIANGULAR SHELL ELEMENT STIFFNESS
MATRICES ASSOCIATED WITH POLYHEDRAL DEFLECTION
DISTRIBUTIONS
A.I.A.A. J., VOL. 6, NO. 2, FEBRUARY 1968, PP.
374-375.

68-0160 UTKU, S.
EXPLICIT EXPRESSIONS FOR TRIANGULAR TORUS ELEMENT
STIFFNESS MATRIX
A.I.A.A. J., VOL. 6, NO. 6, JUNE 1968, PP.
1174-1176.

68-0161 VISSER, C.
FINITE ELEMENT ANALYSIS OF HEAT CONDUCTION IN
ANISOTROPIC SOLIDS
LWBR DEVELOPMENT PROGRAM, WESTINGHOUSE RESEARCH
LABS, PITTSBURGH, PA., JULY 1968, P. 64,
WERL-1114-4.

68-0162 WALKER, A. C.
HALL, D. G.
AN ANALYSIS OF THE LARGE DEFLECTIONS OF BEAMS
USING THE RAYLEIGH-RITZ FINITE ELEMENT METHOD
AERON. QUARTERLY, VOL. 19, NOVEMBER 1968, PP.
357-367.

68-0163 WATWOOD, V. B.
HARTZ, B. J.
AN EQUILIBRIUM STRESS FIELD MODEL FOR FINITE
ELEMENT SOLUTIONS OF TWO-DIMENSIONAL ELASTIC
PROBLEMS
INT. J. SOLIDS AND STRUCTURES, VOL. 4, NO. 9,
SEPTEMBER 1968, PP. 854-874.

68-0164 WEMPNER, G. A.
ODEN, J. T.
KROSS, D. A.
FINITE-ELEMENT ANALYSIS OF THIN SHELLS
J. ENG. MECH. DIV., A.S.C.E., VOL. 94, NO. EM6,
DECEMBER 1968, PP. 1273-1294.

68-0165 WESTMANN, R. A.
STRESS ANALYSIS BY FINITE ELEMENTS
HIGHWAY RESEARCH BOARD, HIGHWAY RESEARCH RECORD
NO. 228, 1968.

68-0166 WHANG, B.
ELASTO-PLASTIC ANALYSIS OF ORTHOTROPIC PLATES AND
SHELLS
AD-688 471, OCTOBER 1968, P. 319.

68-0167 WITHERSPOON, P. A.
JAVANDEL, I.
NEUMAN, S. P.
USE OF THE FINITE ELEMENT METHOD IN SOLVING
TRANSIENT FLOW PROBLEMS IN AQUIFER SYSTEMS
THE USE OF ANALOG AND DIGITAL COMPUTERS IN
HYDROLOGY, A.I.H.S., PUBLICATION 81, 1968.

68-0168 WITMER, E. A.
DISCRETE-ELEMENT ANALYSIS OF SHELLS OF REVOLUTION
M.I.T. SCHOOL OF ENGR., JUNE 1968 P. 100.

68-0169 WITMER, E. A.
ELASTIC-PLASTIC AND THERMAL STRESS ANALYSIS
M.I.T. SCHOOL OF ENGR., JUNE 1968 P. 62.

68-0170 WITMER, E. A.
KOTANCHIK, J. J.
PROGRESS REPORT ON DISCRETE-ELEMENT ELASTIC AND
ELASTIC-PLASTIC ANALYSIS OF SHELLS OF REVOLUTION
SUBJECTED TO AXISYMMETRIC AND ASSYMMETRIC LOADING
PROC. 2ND CONF. ON MATRIX METHODS IN STRUCTURAL
MECHANICS, AFFDL-TR-68-150, 1968, P. 1341.

68-0171 YAMADA, Y.
PLASTIC STRESS-STRAIN MATRIX AND ITS APPLICATION
FOR THE SOLUTION OF ELASTIC-PLASTIC PROBLEMS BY
THE FINITE ELEMENT METHOD
INT. J. MECH. SCI., VOL. 10, MAY 1968, PP.
343-354.

68-0172 ZIENKIEWICZ, O. C.
CONTINUUM MECHANICS AS AN APPROACH TO ROCK MASS
PROBLEMS
CHAPTER 8, ROCK MECHANICS IN ENGINEERING PRACTICE,
STAGG, K. G., AND ZIENKIEWICZ, O. C., EDS., JOHN
WILEY AND SONS, 1968.

68-0173 YAMADA, Y.
KAWAI, T.
YOSHIMURA, N.
SAKURAI, T.
ANALYSIS OF THE ELASTIC PLASTIC PROBLEMS BY THE
MATRIX DISPLACEMENT METHOD
PROC. 2ND CONF. MATRIX METHODS STRUCT. MECH.,
WRIGHT-PATTERSON AIR FORCE BASE, OHIO,
AFFDL-TR-68-150, 1968, PP. 1271-1299.

68-0174 ZIENKIEWICZ, O. C.
VALLIAPPAN, S.
KING, I. P.
STRESS ANALYSIS OF ROCK AS A NO-TENSION MATERIAL
GEOTECHNIQUE, VOL. 18, 1968, PP. 56-66.

68-0175 ZIENKIEWICZ, O. C.
WATSON, M.
KING, I. P.
A NUMERICAL METHOD OF VISCO-ELASTIC STRESS
ANALYSIS
INT. J. MECH. SCI., VOL. 10, 1968, PP. 807-827.

68-0176 ZLAMAL, M.
ON THE FINITE ELEMENT METHOD
NUMERISCHE MATHEMATIK, VOL. 12, 1968, PP. 394-409.

68-0177 KAPKOWSKI, J.
A FINITE ELEMENT STUDY OF ELASTIC-PLASTIC STRESS
DISTRIBUTIONS IN NOTCHED SPECIMENS UNDER TENSION
MEDDELELSE SKB II/M 13, DEPARTMENT OF SHIP
STRUCTURES, THE TECHNICAL UNIVERSITY OF NORWAY,
TRONDHEIM, 1968.

68-0178 WESTBY, O.
APPLICATION OF THE FINITE ELEMENT METHOD ON
RESIDUAL STRESSES AND DEFORMATIONS IN WELDS
DEPARTMENT OF METALLURGY AND METALS WORKING, THE
TECHNICAL UNIVERSITY OF NORWAY, TRONDHEIM, 1968

68-0179 KAYE, G.
A DATA-GENERATION PROCEDURE FOR THE FINITE ELEMENT
ANALYSIS OF PLANE FRAMES AND GRILLAGES
N.C.R.E. TECHNICAL NOTE (UNPUBLISHED), 1968.

68-0180 HYLARIDES, S.
FINITE ELEMENT TECHNIQUE IN SHIP VIBRATION
ANALYSIS
INT. SHIPBUILDING PROGRESS, VOL. 15, SEPTEMBER
1968.

68-0181 HODGE, P. G.
BELYTSCHKO, T.
NUMERICAL METHODS FOR THE LIMIT ANALYSIS OF PLATES
J. APPL. MECH., VOL. 35, 1968, PP. 796-802.

68-0182 PRATO, C. A.
A MIXED FINITE ELEMENT METHOD FOR THIN SHELLS
ANALYSIS
PH. D. DISSERTATION, MASS. INST. OF TECH., 1968.

68-0183 VALLIAPPAN, S.
NON-LINEAR STRESS ANALYSIS OF TWO DIMENSIONAL
PROBLEMS WITH SPECIAL REFERENCE TO ROCK AND SOIL
MECHANICS
PH.D. THESIS, UNIVERSITY OF WALES, 1968.

68-0184 EGELAND, O.
PROGRAM SPECIFICATION NV 300. GENERAL FINITE
ELEMENT PROGRAM FOR 2- AND 3-DIMENSIONAL MEMBRANE
STRUCTURES
DET NORSKE VERITAS RES. DEPT. REPORT NO. 68-22-S,
OSLO 1968.

68-0185 YAO, F. C-C.
ANALYSIS OF SHELLS BY THE FINITE ELEMENT
DISPLACEMENT METHOD
PH.D. THESIS, MASS. INST. OF TECH., 1968.

68-0186 KAWAI, T.
YOSHIMURA, N.
FINITE ELEMENT ANALYSIS ON THE TORSION OF A BAR
WITH UNIFORM CROSS SECTION
SEISAN KENKYU, VOL. 20, NO. 5, MAY 1968.

68-0187 SCOTT, F. C.
A QUARTIC, TWO DIMENSIONAL ISOPARAMETRIC ELEMENT
UNDERGRADUATE PROJECT, UNIV. OF WALES, SWANSEA,
1968.

68-0188 IRONS, B. M.
TESTING AND ASSESSING FINITE ELEMENTS BY AN
EIGENVALUE TECHNIQUE
PROC. CONF. ON RECENT DEVELOPMENTS IN STRESS
ANALYSIS, J. BR. SOC. ST. AN., ROYAL AERO
SOC., 1968.

68-0189 ZIENKIEWICZ, O. C.
PAREKH, C. J.
KING, I. P.
CHEUNG, Y. K.
ARCH DAMS ANALYSED BY A LINEAR FINITE ELEMENT
SHELL THE FINITE STRIP METHOD IN THE ANALYSIS OF
ELASTIC PLATES SOLUTION PROGRAM WITH TWO OPPOSITE
SIMPLY SUPPORTED ENDS
PROC. SYMP. ARCH DAMS, INST. CIV. ENG., LONDON,
VOL. 40, 1968, PP. 1-7.

68-0190 CHEUNG, Y. K.
FINITE STRIP METHOD OF ANALYSIS OF ELASTIC SLABS
PROC. AM. SOC. CIV. ENG., 94, EM6, 1365-78, 1968.

68-0191 FRIED, I.
FINITE ELEMENT ANALYSIS OF PROBLEMS FORMULATED BY
AN INTEGRAL EQUATION; APPLICATION TO POTENTIAL
FLOW
INST. FUR STATIK UND DYNAMIK, LUFT UND
RAUMFAHRTSANSTALT, STUTTGART, 1968.

68-0192 ANDERSON, R. G.
A FINITE ELEMENT EIGENVALUE SYSTEM
PH.D. THESIS, UNIVERSITY OF WALES, SWANSEA, 1968.

68-0193 SCHMIT, L. A.
BOGNER, F. K.
FOX, R. L.
FINITE DEFLECTION STRUCTURAL ANALYSIS USING PLATE
AND CYLINDRICAL SHELL DISCRETE ELEMENTS
PROC. AIAA/ASME 8TH STRUCT. AND STRESS DYNAMIC
CONFERENCE, PALM SPRINGS, CALIFORNIA, 197-211,
MARCH 1967. ALSO J.A.I.A.A., 5, 1525-7,
1968.

68-0194 CIARLET, P. G.
SCHULTZ, M. H.
VARGA, R. S.
NUMERICAL METHODS OF HIGH-ORDER ACCURACY FOR
NONLINEAR BOUNDARY VALUE PROBLEMS
NUMER. MATH., VOL. 9, 1967, VOLS. 11, 12, 1968.

68-0195 WEMPNER, G. A.
NEW CONCEPTS FOR FINITE ELEMENTS OF SHELLS
ZEITSCHRIFT FUR ANGEWANDTE MATHEMATIK UND MECHANIK
VOL. N8, NO. 48, 1968, PP. T174-T180.

68-0196 WRIGHT, P. M.
FENG, C. C.
APPLICATION OF FLOWGRAPHS TO FREE VIBRATION OF
STRUCTURES
U.S. NAVAL RESEARCH LAB.-SHOCK & VIBRATION BUL.
38, PT. 1, AUG. 1968, PP. 99-108.

68-0197 CARD, C. C. H.
FINITE ELEMENT APPLICATIONS TO NEWTONIAN FLUIDS
PH.D. THESIS, UNIVERSITY OF WALES, SWANSEA, 1968.

68-0198 DENKE, P. H.
MATRIX METHODS OF AEROSPACE STRUCTURAL ANALYSIS
PROCEEDINGS OF THE 2ND CONF. ON MATRIX METHODS IN
STRUCTURAL MECHANICS, AFFDL-TR-68-150, OCTOBER
1968.

68-0199 TOCHER, J. L.
VARANASI, S. R.
NON-LINEAR MATERIAL ANALYSIS WITH FINITE ELEMENTS
AND INCREMENTAL METHODS
THE BOEING COMPANY, DOC. ADS., CAD RENTON 70-36,
DOC. DE-29460, JANUARY 1968.

68-0200 WIEDERHIELM, C. A.
KOBAYASHI, A. S.
STROMBERG, D. D.
WOO, S. L. Y.
STRUCTURAL RESPONSE OF RELAXED AND CONSTRICTED
ARTERIOLES
JOURNAL OF BIOMECHANICS, VOL. 1, 1968, PP.
259-270.

68-0201 BIRKHOFF, G.
SCHULTZ, M. H.
VARGA, R. S.
PIECEWISE HERMITE INTERPOLATION IN ONE AND TWO
VARIABLES WITH APPLICATIONS TO PARTIAL
DIFFERENTIAL EQUATIONS
NUMER. MATH. VOL. 11, 1968, PP. 232-256.

68-0202 ROBINSON, J.
HAGGENMACHER, G. W.
A CURVED BEAM ELEMENT
LOCKHEED - CALIFORNIA COMPANY, REPORT NO. 22096,
1968.

68-0203 DAVIES, J. D.
MA, S. P. K.
CHEUNG, Y. K.
ANALYSIS OF CORNER SUPPORTED SKEW SLABS
"BUILDING SCIENCES", NOVEMBER 1968.

68-0204 IRONS, B. M.
ROUNDOFF CRITERIA IN DIRECT STIFFNESS SOLUTIONS
AIAA J., VOL. 6, NO. 7, 1968, PP. 1308-1312.

68-0205 AKESSON, B. A.
FINITE ELEMENT METHOD
INTERNAL REPORT NO. 60, CHALMERS TECHNICAL
UNIVERSITY, 1968.

68-0206 ARMEN, H.
PIFKO, A. B.
LEVINE, H. S.
FINITE ELEMENT METHOD FOR THE PLASTIC BENDING
ANALYSIS OF STRUCTURES
PROC. OF THE 2ND CONF. ON MATRIX METHODS IN
STRUCTURAL MECHANICS, AFFDL-TR-68-140, 1968, PP.
1301-1339.

68-0207 ARGYRIS, J. H.
SCHARPF, D. W.
SPOONER, J. H.
THE ELASTO-PLASTIC CALCULATION OF GENERAL
STRUCTURES AND CONTINUA
PROC. 3RD CONF. DIMENSIONING, BUDAPEST, 1968, PP.
345-384.

68-0208 ARGYRIS, J. H.
SPOONER, J. H.
DIE ANWENDUNG DER MATRIZENVERSCHIEBUNGSMETHODE ZUM
BERECHNEN VON SPANNUNGSKONZENTRATIONEN
SONDERDRUCK, KOLLOQUIUM TRAGFAHIGKEITSERMITTLUNG
BEI SCHWEISSVERBINDUNGEN, BRAUNSCHWEIG, 1968, PP.
201-219.

68-0209 BUTLIN, G. A.
FORD, R.
A COMPATIBLE TRIANGULAR PLATE BENDING FINITE
ELEMENT
REPORT 68-15, OCTOBER 1968, DEPARTMENT OF
ENGINEERING, UNIVERSITY OF LEICESTER.

68-0210 DAVIDS, N.
KOENIG, H. A.
DOUBLE STRESS-WAVE DISCONTINUITIES IN FINITE,
SHEAR-CORRECTED BEAMS AND PLATES
13TH MIDWESTERN MECHANICS CONF., AUGUST 1967.,
ALSO IN DEV. MECH., VOL. 4, AUGUST 1968.

68-0211 DAVIDS, N.
KOENIG, H. A.
DYNAMICAL FINITE ELEMENT ANALYSIS FOR ELASTIC
WAVES IN BEAMS AND PLATES
INT. J. SOLIDS AND STRUCT., VOL. 4, 1968, PP.
643-660.

68-0212 ODEN, J. T.
A GENERALIZATION OF THE FINITE ELEMENT CONCEPT AND
ITS APPLICATION TO A CLASS OF PROBLEMS IN
NONLINEAR VISCOELASTICITY
DEV. THEOR. AND APPL. MECH., VOL. 4, PERGAMON
PRESS, OXFORD, 1968.

68-0213 BELL, K.
ANALYSIS OF THIN PLATES IN BENDING USING
TRIANGULAR FINITE ELEMENTS
DIVISION OF STRUCTURAL MECHANICS, THE TECHNICAL
UNIVERSITY OF NORWAY, TRONDHEIM, FEBRUARY 1968.

68-0214 PICKARD, J.
FORMAT-FORTRAN MATRIX ABSTRACTION TECHNIQUE
ENGINEERING USER AND TECHNICAL REPORT, VOL. 5,
AFFDL-TR-66-207, DECEMBER 1968.

68-0215 BATDORF, N.
KAPUR, S.
SAYER, R.
THE ROLE OF COMPUTER GRAPHICS IN THE STRUCTURAL
DESIGN PROCESS
PROC. 2ND CONF. ON MATRIX METHODS IN STRUCTURAL
MECHANICS, AFFDL-TR-68-150, OCTOBER 1968.

68-0216 RODRIGUEZ, L.
FINITE ELEMENT NONLINEAR ANALYSIS FOR PLATES AND
SHALLOW SHELLS
PH.D. THESIS, DEPT. OF CIVIL ENG., M.I.T.,
SEPTEMBER 1968.

68-0217 ROBINSON, J.
STATIC FLEXIBILITY MATRICES FOR A RECTANGULAR
PLATE ELEMENT IN BENDING, TWISTING, AND SHEAR.
UNIFORM AND NON-UNIFORM DISTRIBUTED LOADING
PAPER PRESENTED AT INT. SYMP. ON EXPERIENCES WITH
COMPUTATION TECHNIQUES FOR BRIDGE ERECTION,
RESEARCH INST. OF CIV. ENGNG., BRATISLAVA,
CZECHOSLOVAKIA, OCTOBER 1968.

68-0218 ARGYRIS, J. H.
INTRODUCTION TO FINITE ELEMENT ANALYSIS
UNIVERSITY OF STUTTGART, MAY 1968.

68-0219 STROH, G. B.
LAWRENCE, H. H.
DEIBEL, W. T.
EFFECTS OF SHOE FORCE GEOMETRY ON HEAVY DUTY
INTERNAL SHOE BRAKE PERFORMANCE
SAE PAPER 68-0432.

68-0220 NICKELL, R. E.
SACKMAN, J. J.
APPROXIMATE SOLUTIONS IN LINEAR COUPLED
THERMOELASTICITY
J. APPL. MECH., VOL. 35, SERIES E., NO. 2, JUNE
1968, PP. 255-266.

68-0221 MATSUURA, Y.
KAWASAKI, H.
CALCULATION OF THE VIRTUAL MASS, MOVEMENT AND
INERTIA BY THE FINITE ELEMENT METHOD
J. SOC. NAV. ARCH. OF JAPAN, TECH. PAPER NO. 124,
1968 36 PP.

68-0222 CONNOR, J. J.
LOGCHER J.
CHAN, S. C.
NONLINEAR ANALYSIS OF ELASTIC FRAMED STRUCTURES
ASCE., VOL. 94, NO. ST6, PROC. PAPER 6011, JUNE
1968.

68-0223 IRONS, B. M.
ZIENKIEWICZ, O. C.
ISOPARAMETRIC FINITE ELEMENT SYSTEM: A NEW
CONCEPT IN FINITE ELEMENT ANALYSIS
ROYAL AERO. SOC., CONF. ON BEHALF OF JT. BRIT.
COMM. FOR STRESS ANAL. RECENT ADVANCES IN STRESS
ANALYSIS, MARCH 26-29, 1968, LONDON, ENGLAND,
SESS. 3, PP. 35-40.

68-0224 IRONS, B. M.
HINTON, E.
LEAST SQUARES SMOOTHING OF EXPERIMENTAL DATA USING
FINITE ELEMENTS
STRAIN, VOL. 4, 1968, PP. 1-4.

68-0225 YOSHIKII, M.
KAWAI, T.
YOSHIMURA, N.
MATRIX METHOD OF ANALYSIS OF SHIP STRUCTURES 3RD
REPORT
J. SOC. NAV. ARCHIT., JAPAN, VOL. 123, 1968, PP.
188-195.

68-0226 KLEIN, S.
DYNAMIC ANALYSIS OF A COMPLEX SHELL STRUCTURE
UNDER IMPULSIVE ASYMMETRIC LOADING
STRUCTURAL ANALYSIS OF A REENTRY VEHICLE TYPE
SHELL VOL. 2, THE AEROSPACE CORPORATION, SAN
BERNARDINO, CALIF. 1968.

68-0227
HOLAND, I.
INCORPORATION OF FOUNDATION DEFORMATION IN A
FINITE ELEMENT SHELL ANALYSIS
PROC. SYMP. ARCH DAMS, INST. CIV. ENGRS., LONDON,
1968.

68-0228
FULTON, R. E.
CYRUS, N. J.
WALZ, J. E.
ACCURACY AND CONVERGENCE OF FINITE ELEMENT
APPROXIMATION
PROC. 2ND CONF. MATRIX METH. STRUCT. MECH.
WRIGHT-PATTERSON AIR FORCE BASE, OHIO, 1968.

68-0229
FRIED, I.
SCHARPF, D. W.
A COMPUTATIONAL PROCEDURE FOR GRADIENT ITERATIVE
TECHNIQUES IN THE FINITE ELEMENT METHOD
LECTURE 11TH CONF. EUROMECH, STUTTGART, APRIL
1968.

68-0230
HUGHES, T. J. R.
ALLIK, H.
FINITE ELEMENT FORMULATION FOR PIEZOELECTRIC
CONTINUA BY A VARIATIONAL THEOREM
GD/ELECTRIC BOAT DIVISION REPORT P411-68-056,
1968.

68-0231
TOLEFSON, D. C.
BRAND, L.
INTRODUCTION TO FINITE ELEMENT METHODS OF
STRUCTURAL ANALYSIS
MAR. TECH-NOL., VOL. 5, 1968, PP. 4, 331-345.

68-0232
KAPUR, K. K.
SLUPEK, B. A.
METHOD FOR ESTIMATING RESPONSE OF PAYLOAD
SECONDARY STRUCTURES TO RANDOM EXCITATION
INST. ENVIRONMENTAL SCIENCES - ANNUAL TECH.
MEETING, 14TH ST. LOUIS, APRIL 29-MAY 1, 1968, PP.
185-190.

68-0233
CHEUNG, Y. K.
KING, I. P.
ZIENKIEWICZ, O. C.
SLAB BRIDGES WITH ARBITRARY SHAPE AND SUPPORT
CONDITIONS - A GENERAL METHOD OF ANALYSIS BASED ON
FINITE ELEMENTS
PROC. INST. CIV. ENGRS., VOL. 40, 1968, PP. 9-36.

68-0234
OLSON, M. D.
LINDBERG, G. M.
TULLOCH, H. A.
FINITE PLATE-BENDING ELEMENTS IN POLAR COORDINATES
NATIONAL RESEARCH COUNCIL OF CANADA, AERONAUTICS
REPORT LR-512, 1968.

68-0235
ROSEN, R.
RAGLE, R.
MISEL, J.
CURTIS, R.
FALCONER, R.
VAN COUVERING, D.
STARDYNE STRUCTURAL ANALYSIS SYSTEM, ANALYTICAL
CAPABILITY MANUAL
MECHANICS RESEARCH, INC., DEC., 1968.

68-0236
VENKAYYA, V. B.
ITERATIVE METHOD FOR THE ANALYSIS OF LARGE
STRUCTURAL SYSTEMS
AIR FORCE DYNAMICS LAB., AIR FORCE SYSTEMS
COMMAND, WRIGHT-PATTERSON AIR FORCE BASE, OHIO,
TECH. REP. AFFDL-TR-67-194, 1968, PP. 42.

68-0237
ERGATOUDIS, J.
ISOPARAMETRIC FINITE ELEMENTS IN TWO AND
THREE-DIMENSIONAL ANALYSIS
PH.D. THESIS, UNIV. OF WALES, SWANSEA, 1968.

68-0238
VISSER, M.
THE FINITE ELEMENT METHOD IN DEFORMATION AND HEAT
CONDUCTION PROBLEMS
DELFT, 1968.

68-0239
KHOJASTEH-BAKHT, M.
COMPUTER PROGRAM FOR ELASTIC-PLASTIC ANALYSIS OF
AXISYMMETRICALLY LOADED SHELLS OF REVOLUTION
REPORT STRUCTURAL ENGINEERING AND STRUCTURAL
MECHANICS 68-3, STRUCT. ENGINEERING LABORATORY,
UNIV. OF CALIF. BERKELEY, 1968.

68-0240
MCBEAN, R. P.
FINITE ELEMENT ANALYSIS OF STIFFENED PLATES
ANN. CONF. OF ENG. INST. OF CANADA, OTTAWA, 1970.
(ALSO PH.D. THESIS, STANFORD UNIV., 1968)

68-0241
NILSON, A. H.
NON-LINEAR FINITE ELEMENT ANALYSIS OF REINFORCED
CONCRETE BY THE FINITE ELEMENT METHOD
J. AMERICAN CONCRETE INSTITUTE, VOL. 65, NO. 9,
1968, PP. 757-766.

68-0242
PARSONS, B.
ANALYSIS OF AXISYMMETRIC COMPOSITE BODIES BY THE
FINITE ELEMENT METHOD
RECENT ADVANCES IN ENG. SCI., VOL. 5, PT. 2, 1970,
PROC. 6TH ANNUAL MEETING SOC. ENG. SCI., NOVEMBER
11-13, 1968, PP. 295-310.

68-0243
ANON
DEVELOPMENTS IN DISCRETE ELEMENT FINITE DEFLECTION
STRUCTURAL ANALYSIS BY FUNCTION MINIMIZATION
AIR FORCE FLIGHT DYNAMICS LAB., AIR FORCE SYSTEMS
COMMAND, WRIGHT-PATTERSON AIR FORCE BASE, OHIO,
TECH. REP. AFFDL-TR- 68-126, SEPTEMBER 1968, 294
PP.

68-0244
ANDERSON, R. G.
ZIENKIEWICZ, O. C.
IRONS, B. M.
VIBRATION AND STABILITY OF PLATES USING FINITE
ELEMENTS
INT. J. SOLIDS AND STRUCTS., VOL. 4, 1968, PP.
1031-1055.

68-0245
LEE, S. Y.
TANG, S. S.
LIYEOS, J. G.
MASS LOADING EFFECTS ON VIBRATION OF RING AND
SHELL STRUCTURES
REPORT NASA-CR-98508, FEBRUARY 1968.

68-0246
MALTBAEK, J. C.
VIBRATIONS OF A NON-UNIFORM BEAM
ENGINEER, VOL. 220, 1965, PP. 714-16. ALSO IN
VOL. 226, 1968, PP. 906-909.

68-0247
SINGHAL, A. C.
THE USE OF FINITE ELEMENT METHODS FOR STATIC AND
DYNAMIC ANALYSES OF COMPLEX SHELLS
REPORT: S-13, DECEMBER 1968, LAVAL UNIV., QUEBEC,
CANADA

68-0248
KLEEMANN, P. H.
THE GENERALIZED INVERSE - SOME PROPERTIES, A
COMPUTATIONAL ALGORITHM AND A STRUCTURAL
APPLICATION
INSTITUT FUR STATIK UND DYNAMIK DER LUFT UND
RAUMFAHRTKONSTRUKTIONEN, RESEARCH REPORT NO. 47,
FEBRUARY 1968.

68-0249
TAYLOR, R. L.
PISTER, K.
HERRMANN, L. R.
ON A VARIATIONAL THEOREM FOR INCOMPRESSIBLE AND
NEARLY INCOMPRESSIBLE ORTHOTROPIC ELASTICITY
INT. J. SOLIDS STRUCT., VOL. 4, 1968, PP. 875-883.

68-0250
PECKNOLD, D. A. W.
SCHNOBRICH, W. C.
FINITE ELEMENT ANALYSIS OF SKEWED SHALLOW SHELLS
DOCTORAL THESIS, DEPT. OF CIVIL ENGNG., UNIV. OF
ILL., URBANA, REPORT NO: STRUCT. RESEARCH SER-332,
JANUARY 1968, 171 PP. (N.T.I.S. - AD-664 822)

68-0251
AUTHOR
DYNAMICS OF SUBMARINE STRUCTURES. VOLUME 1
CATHOLIC UNIV. OF AMERICA, WASH. D.C. 1968, 27 PP.
(N.T.I.S. - AD-689 204) SEE ALSO VOL. 2, PART 1,
MARCH 1969, AD-689 203

68-0252
CONNOR, J.
WILL, G.
A TRIANGULAR FLAT PLATE BENDING ELEMENT
TR 68-3, DEPT. OF CIVIL ENG., M.I.T., CAMBRIDGE,
MASS., 1968

68-0253
MELOSH, R. J.
MANIPULATION ERRORS IN FINITE ELEMENT ANALYSIS
REPORT NO. 1385, NASA CONTRACTORS, PHILCO-FORD,
1968.

68-0254
SORENSON, M.
BALMER, H.
ASKA 105 A TECHNICAL MANUAL AND PROGRAMMING THEORY
FOR THE ELEMENTS, FLA, TRIM, SPARTA AND TET
INSTITUT FUR STATIK UND DYNAMIK DER LUFT UND
RAUMFAHRTKONSTRUKTIONEN, RESEARCH REPORT NO. 49,
APRIL 1968.

68-0255
SORENSON, M.
BALMER, H.
ASKA 105 - A TECHNICAL MANUAL FOR THE DIRECT
SOLUTION AND INVERSION PACKAGE
INSTITUT FUR STATIK UND DYNAMIK DER LUFT UND
RAUMFAHRTKONSTRUKTIONEN, RESEARCH REPORT NO. 59,
NOVEMBER 1968.

68-0256
AKYUZ, F. A.
UTKU, S.
AN AUTOMATIC NODE RELABELLING SCHEME FOR BANDWIDTH
MINIMIZATION OF STIFFNESS MATRICES
A.I.A.A. JOURNAL, VOL. 6, NO. 4, APRIL 1968, PP:
728-730

68-0257 LOGCHER, R. D.
FLACHSBART, B. B.
HALL, F. J.
POWER, C. M.
WELLS, A.
FERRANTE, A. J.
ICES-STRUDL-11
MASSACHUSETTS INSTITUTE OF TECH., CAMBRIDGE,
MASS., 1968

68-0258 TAKAHASHI, S. K.
MARK, R.
PHOTOELASTIC INVESTIGATION OF STRESS
CONCENTRATIONS IN SPHERE-CYLINDER TRANSITION
REGIONS: INCLUDING A COMPARISON OF RESULTS FROM
PHOTOELASTIC AND FINITE ELEMENT ANALYSES
NAVAL CIVIL ENG. LAB., PORT HUENEME, CALIF.,
REPORT NO: NCEL-TR-572, APRIL 1968, 108 PP.
(N.T.I.S. - AD-667 834)

68-0259 PURDY, M. D.
PRZEMIENIECKI, J. S.
INFLUENCE OF HIGHER ORDER TERMS IN THE LARGE
DEFLECTION ANALYSIS OF FRAMEWORKS
PROC. AMER. SOC. CIVIL ENG., JOINT SPECIALTY
CONF., OPTIMIZATION AND NONLINEAR PROBLEMS, 1968,
PP. 142-152.

68-0260 VERRUIJT, A.
GROUNDWATER FLOW - THE FINITE ELEMENT METHOD
REPORT NO. 75-3, TECHNOLOGICAL U. OF DELFT,
DECEMBER 1968.

68-0261 ARGYRIS, J. H.
PART 11: LUMINA AND HEXE ELEMENTS
PROC. OF THE SYMP. ON ARCH DAMS, INST. CIVIL ENG.,
MARCH 20-21, 1968.

68-0262 MAESTRELLO, L.
TEST RESULTS FROM THE BOUNDARY LAYER FACILITY -
RESPONSE OF STRUCTURE TO THE PSEUDO-SOUND FIELD OF
A JET (USING COMBINED CONTINUUM AND FINITE ELEMENT
METHOD)
COMMERCIAL AIRPLANE DIV., BOEING CO., RENTON,
WASH., REPORT NO: D6-9944-VOL-4, JANUARY 2, 1968,
49 PP. (N.T.I.S. - AD-669 215) SEE ALSO VOL. 3,
AD-669 217

68-0263 TAYLOR, T. P.
MATLOCK, H.
A FINITE-ELEMENT METHOD OF ANALYSIS FOR COMPOSITE
BEAMS
TEXAS HIGHWAY DEPT., BUREAU OF PUBLIC ROADS, WASH.
D.C., REPORT NO: RR-56-10, JANUARY 1968, 159 PP.
(N.T.T.S. - PB-178 397)

68-0264 ARGYRIS, J. H.
GROSRECHENANLAGEN IN DER FORSCHUNG
BILD DER WISSENSCHAFT, STUTTGART, AUGUST 1968.

68-0265 ARGYRIS, J. H.
VON FUCHS, G.
SCHREM, E.
STRING EIN PROGRAMMSYSTEM FUR DIE
MASCHINENUNABHANGIGE BEARBEITUNG VON ZEICHENKETTEN
IN FORTRAN
INSTITUT FUR STATIK UND DYNAMIK DER LUFT UND
RAUMFAHRTKONSTRUKTIONEN, RESEARCH REPORT NO. 45,
JANUARY 1968.

68-0266 ALMOND, J.
BOTTERUD, O. H.
LEDERER, E.
FORMAT, A GENERAL PURPOSE MATRIX ANALYSIS SYSTEM
IN FORTRAN
INSTITUT FUR STATIK UND DYNAMIK DER LUFT UND
RAUMFAHRTKONSTRUKTIONEN, RESEARCH REPORT NO. 44,
FEBRUARY 1968.

68-0267 CASE, W. R. JR.
FINITE ELEMENT MODELLING OF THREE-DIMENSIONAL
STRUCTURES
GODDARD SPACE FLIGHT CENTER, NATL. AERON. SPACE
ADMIN., GREENBELT, NASA-TM-X-63238, X-321-68-158,
MAY 1968, 36 PP. (N.T.I.S. - N68-27454)

68-0268 MATLOCK, H.
TAYLOR, T. P.
A COMPUTER PROGRAM TO ANALYZE BEAM-COLUMN UNDER
MOVEABLE LOADS
BUREAU OF PUBLIC ROADS, WASH., D.C., REPORT NO:
RR-56-4, JUNE 1968, 191 PP. (N.T.I.S. - PN-179
901)

68-0269 DUNCAN, J. M.
GOODMAN, R. E.
FINITE ELEMENT ANALYSES OF SLOPES IN JOINTED ROCK
OFFICE OF RESEARCH SERVICES, UNIV. OF CALIF.,
BERKELEY, REPORT NO: TE-68-1, FEBRUARY 1968, 273
PP. (N.T.I.S. - AD-678 632)

68-0270 RAMSEY, J. W. JR.
FINITE ELEMENT ANALYSIS FOR BUCKLING OF SEMI-RIGID
SPACE FRAMES
LANGLEY RESEARCH CENTER, NATL. AERON. SPACE
ADMIN., LANGLEY STATION, VA., REPORT NO:
NASA-TM-X-61465, JUNE 1968, 74 PP. (N.T.I.S. -
N69-19885)

68-0271 SHEARS, M.
STATIC AND DYNAMIC BEHAVIOUR OF GUYED MASTS
STRUCT. ENGN. LAB., UNIV. OF CALIF., BERKELEY,
REPORT NO: SESM-68-6, JUNE 1968, 173 P. (N.T.I.S.
- PB-189 500)

68-0272 MASON, J. B.
ANALYSIS OF THERMALLY INDUCED STRUCTURAL
VIBRATIONS OF FINITE ELEMENT TECHNIQUES
GODDARD SPACE FLIGHT CENTER, NATL. AERON. SPACE
ADMIN., GREENBELT, MD., REPORT NO:
NASA-TM-X-63488, X-321-68-333, AUGUST 1968, 31 PP.
(N.T.I.S. - N69-21010)

68-0273 SANBONGI, S.
APPLICATIONS OF THE FINITE ELEMENT METHOD TO BOX
BEAMS
NATL. AEROSPACE LAB., TOKYO, JAPAN, REPORT NO:
NAL-TR-165, 1968, 11 P. (N.T.I.S. - N69-37376)

68-0274 MAKI, A. C.
FINITE ELEMENT TECHNIQUES FOR ORTHOTROPIC PLANE
STRESS AND ORTHOTROPIC PLATE ANALYSIS
FOREST PRODUCTS LAB., MADISON WIS., REPORT NO:
FSRP-FPL-37, JUNE 1968, 43 PP. (N.T.I.S. - AD-672
016)

68-0275 ANDERTON, G. L.
CONNACHER, N. E.
DOUGHERTY, C. S.
HANSEN, S. D.
ANALYSIS OF THE 747 AIRCRAFT WING BODY
INTERSECTION
PROC. 2ND CONF. MATRIX METH. IN STRUCT. MECH.,
WRIGHT-PATTERSON AFB., OHIO, AFFDL-TR-68-150,
1968.

68-0276 BERGAN, P. G.
ALDSTEDT, E.
SESAM A PROGRAMMING SYSTEM FOR FINITE ELEMENT
PROBLEMS
TECH. UNIV. OF NORWAY, TRONDHEIM, 1968.

68-0277 CHEUNG, Y. K.
KING, I. P.
ZIENKIEWICZ, O. C.
GENERAL METHOD OF ANALYSIS BASED ON FINITE
ELEMENTS
PROC. INST. CIVIL ENG., VOL. 40, 1968, PP. 9-36.

68-0278 WITTRICK, W. H.
GENERAL SINUSOIDAL STIFFNESS MATRICES FOR BUCKLING
AND VIBRATION ANALYSES OF THIN FLAT-WALLED
STRUCTURES
INTL. J. MECH. SCI., VOL. 10, 1968, PP. 949-966.

68-0279 DE ARANTES E OLIVEIRA, E. R.
COMPLETENESS AND CONVERGENCE IN THE FINITE ELEMENT
METHOD
PROC. 2ND CONF. MATRIX METH. STRUCT. MECH.,
WRIGHT-PATTERSON AFB., OHIO, AFFDL-TR-68-150,
1968.

68-0280 STURGIS, J. D.
A FINITE ELEMENT ANALYSIS OF AN AIRCRAFT LANDING
MAT ON AN ELASTIC FOUNDATION
PH.D. THESIS, UNIV. OF ALABAMA, 1968.

68-0281 WHITE, J. L.
FINITE ELEMENTS IN LINEAR VISCOELASTICITY
PROC. 2ND CONF. MATRIX METH. IN STRUCT. MECH.,
WRIGHT-PATTERSON AFB. OHIO, REPORT
AFFDL-TR-68-150, 1968.

68-0282 YAMADA, Y.
YASHIMURA, N.
SAKURAI, T.
STRESS STRAIN MATRIX AND ITS APPLICATIONS FOR THE
SOLUTION OF ELASTIC-PLASTIC PROBLEMS BY THE FINITE
ELEMENT METHOD
INT. J. MECH. SCI., VOL. 10, 1968, PP. 343-354.

68-0283 ZUDANS, Z.
ANALYSIS OF ASYMMETRIC STIFFENED SHELL TYPE
STRUCTURES BY THE FINITE ELEMENT METHOD: I FLAT
RECTANGULAR ELEMENTS
NUCL. ENG. DESIGN, VOL. 8, 1968, PP. 367-379.

68-0284 ANON
CONF. ON MATRIX METHODS IN STRUCTURAL MECHANICS
PROC. 2ND CONF., WRIGHT-PATTERSON AFB, OHIO,
REPORT AFFDL-TR-68-150, 1968.

68-0285 PAKSTYS, M. JR.
DYNAMIC SUBSTRUCTURES METHOD FOR SHOCK ANALYSIS
SHOCK AND VIB. BULL., VOL. 38, NO. 2, 1968, PP.
11-22.

68-0286 MURRAY, D. W.
WILSON, E. L.
AN APPROXIMATE NON-LINEAR ANALYSIS OF THIN PLATES
PROC. 2ND AIR FORCE CONF. ON MATRIX METH. IN
STRUCT. MECH., WRIGHT-PATTERSON AIR FORCE BASE,
OHIO, OCTOBER 1968.

68-0287 HINTON, E.
LEAST SQUARES ANALYSIS USING FINITE ELEMENTS
M.SC. THESIS, CIVIL ENG. DEPT., UNIV. COLLEGE OF
SWANSEA, 1968.

68-0288 LYSMER, J.
THE FINITE ELEMENT METHOD AND ITS APPLICATION TO
SOIL DYNAMIC PROBLEMS
SUMMER SHORT COURSE ON VIBRATIONS OF SOILS AND
FOUNDATIONS, UNIV. OF MICH., 1968, PP. 1-30.

68-0289 AGRAWAL, K. M.
ANALYSIS OF ELASTIC SHELLS OF REVOLUTION WITH
MEMBRANE AND FLEXURE STRESSES UNDER ARBITRARY
LOADING USING TRAPEZOIDAL FINITE ELEMENTS
PH.D. THESIS, UNIV. OF BRITISH COLUMBIA, CANADA,
1968.

68-0290 BAYER, D. M.
A FINITE ELEMENT TECHNIQUE FOR THE IDEALIZATION OF
CONTINUOUS SHELL STRUCTURES
PH.D. THESIS, VANDERBILT UNIV., 1968.

68-0291 BERGERON, W. J.
FINITE ELEMENT ANALYSIS OF SALT PILLAR MODELS
PH.D. THESIS, LOUISIANA STATE UNIV. AND AGRI. AND
MECH. COLL., 1968.

68-0292 BLACK, D. E.
DYNAMIC STABILITY OF CYLINDRICAL SHELLS BY FINITE
ELEMENTS
PH.D. THESIS, OKLAHOMA STATE UNIV., 1968.

68-0293 BROWN, R. E.
A MULTI-LAYERED FINITE ELEMENT MODEL FOR
PREDICTING MINE SUBSIDENCE
PH.D. THESIS, CARNEGIE-MELLON UNIV., 1968.

68-0294 CANTIN, G.
A CURVED FINITE ELEMENT FOR CYLINDRICAL SHELLS
PH.D. THESIS, UNIV. OF CALIF., BERKELEY, 1968.

68-0295 COFFIN, G. K.
FINITE ELEMENT ANALYSIS OF OPEN SHELLS OF
REVOLUTION UNDER NONSYMMETRICAL LOADING
PH.D. THESIS, DUKE UNIV., 1968.

68-0296 CRUZ., L. V.
A FINITE ELEMENT APPROACH TO PLATE BENDING
PROBLEMS
PH.D. THESIS, DUKE UNIV., 1968.

68-0297 DUNLOP, P.
ANALYSES OF SLOPES BY THE FINITE ELEMENT METHOD
PH.D. THESIS, UNIV. OF CALIF., BERKELEY, 1968.

68-0298 GATES, R. H.
INELASTIC ANALYSIS OF SLOPES BY THE FINITE ELEMENT
METHOD
PH.D. THESIS, UNIV. OF ILLINOIS, URBANA,
CHAMPAIGN, 1968.

68-0299 MCBEAN, R. P.
ANALYSIS OF STIFFENED PLATES BY THE FINITE ELEMENT
METHOD
PH.D. THESIS, STANFORD UNIV., 1968

68-0300 MEI, C.
FINITE ELEMENT ANALYSIS OF FREE VIBRATION AND
STABILITY OF STRUCTURES SUBJECTED TO INITIAL
STRESS
PH.D. THESIS, CORNELL UNIV., 1968.

68-0301 RICHARDS, A. M.
FINITE ELEMENT ANALYSIS OF DISCONTINUOUS SHELLS
PH.D. THESIS, UNIV. OF CINCINNATI, 1968.

68-0302 SCHWAEGLER, R. T.
THE ANALYSIS OF CLASSICAL PLATE BENDING PROBLEMS
USING AN EQUILIBRIUM FINITE ELEMENT MODEL
PH.D. THESIS, UNIV. OF WASHINGTON, 1968.

68-0303 SHIEH, W. Y-J.
ANALYSIS OF PLATE AND SHELL STRUCTURES BY
TRIANGULAR FINITE ELEMENTS
PH.D. THESIS, NORTHWESTERN UNIV., 1968.

69-0001 NEWTON, R. E.
COUPLED VIBRATIONS OF A STRUCTURE SUBMERGED IN A
COMPRESSIBLLE FLUID
PROC. SYMP. FINITE ELEMENT TECHN., INSTITUT FUR
STATIK UND DYNAMIK DER LUFT UND
RAUMFAHRKONSTRUKTIONEN, UNIV. OF STUTTGART,
GERMANY, JUNE 10-12, 1969.

69-0002 ODEN, J. T.
FINITE ELEMENT ANALOGUE OF NAVIER-STOKES EQUATIONS
J. ENG. MECH. DIV., PROC. ASCE 96, NO. EM-3, 1969,
PP. 529-534.

69-0003 ODEN, J. T.
SONNOGYI, D.
FINITE ELEMENT APPLICATIONS IN FLUID DYNAMICS
J. ENG. MECH. DIV., PROC. ASCE 95, NO. EM-3, 1969,
PP. 821-826.

69-0004 AGUIRRE-RAMIREZ, G.
ODEN, J. T.
FINITE ELEMENT TECHNIQUE APPLIED TO HEAT
CONDUCTION IN SOLIDS WITH TEMPERATURE DEPENDENT
THERMAL CONDUCTIVITY
PAPER 69-WA/HT-34, ASME WINTER ANN. MEETING, LOS
ANGELES, NOVEMBER 16-20, 1969.

69-0005 MELLIERE, R. A.
A FINITE ELEMENT METHOD FOR GEOMETRICALLY
NONLINEAR LARGE DISPLACEMENT PROBLEMS IN THIN,
ELASTIC PLATES AND SHELLS
PH.D. THESIS, UNIV. OF MISSOURI, ROLLA, 1969.

69-0006 RASHID, Y. R.
COMPUTATIONAL METHODS IN SOLID MECHANICS AND
STRESS ANALYSIS
GULF GENERAL ATOMIC INC., SAN DIEGO, CALIFORNIA,
GA-8298, 1969, P. 27.

69-0007 LEMMON, E. C.
HEATON, H. S.
ACCURACY, STABILITY AND OSCILLATION
CHARACTERISTICS OF FINITE ELEMENT METHOD FOR
SOLVING HEAT CONDUCTION EQUATION
PREPRINT 69-WA/HT-35, ASME WINTER ANN. MEETING,
LOS ANGELES, NOVEMBER 16-20, 1969.

69-0008 ARGYRIS, J. H.
MARECZEK, G.
SCHARPF, D. W.
TWO AND THREE-DIMENSIONAL FLOW USING FINITE
ELEMENTS
AERO. J. ROY. SOC. 73, 1969, PP. 961-964.

69-0009 CHEUNG, Y. K.
MEDWELL, O.
FINITE ELEMENT METHOD APPLIED TO THE SOLUTION OF
HEAT CONDUCTION PROBLEMS
REV. ROUM. SCI. TECH., MECH. APPL. 14, NO. 2,
1969, PP. 361-372.

69-0010 SILVESTER, P.
FINITE ELEMENT SOLUTION OF THE HOMOGENEOUS
WAVEGUIDE PROBLEMS
ALTA FREQUENZA 38 (SPECIAL ISSUE), 1969, PP.
313-317.

69-0011 AHMED, S.
DALY, P.
ARCHER, J. S.
WAVEGUIDE SOLUTIONS BY THE FINITE ELEMENT METHOD
RADIO ELECTRON, ENG., VOL. 38, NO. 4, 1969, PP.
217-223.

69-0012 DAHL, H. D.
A FINITE ELEMENT MODEL FOR ANISOTROPIC YIELDING IN
GRAVITY LOADED ROCK
PH.D. THESIS, PENN. STATE UNIV., 1969.

69-0013 THOMAS, D. T.
FUNCTIONAL APPROXIMATIONS FOR SOLVING BOUNDARY
VALUE PROBLEMS BY COMPUTER
IEEE TRANS. MICROWAVE THEORY TECH. MIT-17, NO. 8,
1969, PP. 447-454.

69-0014 GALLAGHER, R. H.
MALLETT, R. H.
EFFICIENT SOLUTION PROCESSES FOR FINITE ELEMENT
ANALYSIS OF TRANSIENT HEAT CONDUCTION
PAPER 69-WA/HT-32, ASME WINTER ANN. MEETING, LOS
ANGELES, NOVEMBER 16-20, 1969.

69-0015 BROCCI, R. A.
ANALYSIS OF AXISYMMETRIC LINEAR HEAT CONDUCTION
PROBLEMS BY FINITE ELEMENT METHOD
PAPER 69-WA/HT-37, ASME WINTER ANN. MEETING, LOS
ANGELES, NOVEMBER 16-20, 1969.

69-0016 RICHARDSON, P. D.
SHUM, Y. M.
USE OF FINITE ELEMENT METHODS IN SOLUTION OF
TRANSIENT HEAT CONDUCTION PROBLEMS
PAPER 69-WA/HT-36, ASME WINTER ANN. MEETING, LOS
ANGELES, NOVEMBER 16-20, 1969.

69-0017 CARSON, W. W.
EMERY, A. F.
EVALUATION OF THE USE OF FINITE ELEMENT METHOD IN
COMPUTING TEMPERATURE
PAPER 69-WA/HT-38, ASME WINTER ANN. MEETING, LOS
ANGELES, NOVEMBER 16-20, 1969.

69-0018 TONG, P.
THE FINITE ELEMENT METHOD FOR FLUID FLOW
PAPER U.S. 5-4, JAPAN - U.S. SEMINAR ON MATRIX
METHODS OF STRUCTURAL ANALYSIS AND DESIGN,
TOKYO, JAPAN, AUGUST 25-30, 1969, ALSO IN MATRIX
METHODS OF STRUCTURAL ANALYSIS AND DESIGN, R.
GALLAGHER, Y. YAMADA, AND J. T. ODEN (EDITORS), U.
OF ALABAMA PRESS, 1971, PP. 737-808.

69-0019 BAKER, A. J.
A NUMERICAL SOLUTION TECHNIQUE FOR A CLASS OF
TWO-DIMENSIONAL PROBLEMS IN FLUID DYNAMICS
FORMULATED WITH THE USE OF DISCRETE ELEMENTS
TECH. NOTE TCTN-1005, BELL AEROSYSTEMS CO.,
BUFFALO, NEW YORK, JANUARY 1969.

69-0020   ATKINSON, B.
           BROCKLEBANK, M. P.
           CARD, C. C. H.
           SMITH, J. M.
           LOW REYNOLDS NUMBER DEVELOPING FLOW
           AICHE J. 15, NO. 4, 1969, PP. 548-553.

69-0021   REDDI, M. M.
           FINITE ELEMENT SOLUTION OF THE INCOMPRESSIBLE
           LUBRICATION PROBLEM
           J. LUB. TECH. TRANS. A.S.M.E., VOL. 91, SER. F,
           NO. 3, 1969, PP. 524-533. ALSO IN J. MECH. ENG.
           VOL. 91, NO. 10, 1969, P. 86.

69-0022   ARGYRIS, J. H.
           SCHARPF, D. W.
           THE INCOMPRESSIBLE LUBRICATION PROBLEM
           AERO. J. ROY. AERO. SOC. 73, 1969, PP. 1044-1046.

69-0023   VOLKER, R. E.
           NON-LINEAR FLOW IN POROUS MEDIA BY FINITE ELEMENTS
           J. HYD. DIV., ASCE 95, PAPER 6927P, 1969, PP.
           2093-2114.

69-0024   SANDHU, R. S.
           WILSON, E. L.
           FINITE ELEMENT ANALYSIS OF SEEPAGE IN ELASTIC
           MEDIA
           J. ENG. MECH. DIV., ASCE 95, 1969, PP. 641-651.

69-0025   LEONARD, J. W.
           LINEARIZED COMPRESSIBLE FLOW BY THE FINITE ELEMENT
           METHOD
           BELL AEROSYSTEMS CO., TECH. NOTE TCTN-9500-920156,
           DECEMBER 1969.

69-0026   BRAMLETTE, T. T.
           MALLETT, R. H.
           A FINITE ELEMENT SOLUTION TECHNIQUE FOR THE
           BOLTZMANN EQUATION
           REPORT NO. 9500-920163, BELL AEROSPACE CO.,
           BUFFALO, N.Y., JUNE 1969. ALSO IN J. FLUID MECH.,
           VOL. 42, PART 2, 1970, PP. 177-191.

69-0027   FUJINO, T.
           ANALYSIS OF HYDRODYNAMIC PROBLEMS BY THE FINITE
           ELEMENT METHOD
           JAPAN-U.S. SEMINAR ON MATRIX METHODS OF STRUCTURAL
           ANALYSIS AND DESIGN, TOKYO, JAPAN, AUGUST 25-30,
           1969.

69-0028   CHEUNG, Y. K.
           ORTHOTROPIC RIGHT BRIDGES BY THE FINITE STRIP
           METHOD
           PROC. 2ND INT. SYMP. ON CONCRETE BRIDGE DESIGN,
           CHICAGO, MARCH 1969. ALSO IN ACI PUB. SP-26, PP.
           182-205.

69-0029   LUK, C. H.
           FINITE ELEMENT ANALYSIS FOR LIQUID SLOSHING
           PROBLEMS
           S. M. THESIS, MASSACHUSSETS INST. OF TECH., 1969.
           ALSO IN AFOSR 69-1504 TR; ASRL TR 144-3,
           MAY 1969.

69-0030   THROSBY, P. W.
           A FINITE ELEMENT APPROACH TO SURFACE DEFINITION
           COMPUTING J. 12, 1969, PP. 385-387.

69-0031   ROWAN, W. H.
           HACKETT, R. M.
           APPLICATION OF FINITE ELEMENT METHODS IN CIVIL
           ENGINEERING
           PROC. A.S.C.E. SYMP., VANDERBILT UNIV., NASHVILLE,
           TENNESSEE, NOVEMBER 13-14, 1969.

69-0032   CHAN, S. K.
           TUBA, I. S.
           WILSON, W. K.
           ON THE FINITE ELEMENT METHOD IN LINEAR FRACTURE
           MECHANICS
           PROC. NAT. SYMP. ON FRACTURE MECH., 2ND LEHIGH
           UNIV., JUNE 17-19, 1969.

69-0033   KOBAYASHI, A. S.
           MAIDEN, D. E.
           SIMON, B. J.
           LIDA, S.
           APPLICATION OF FINITE ELEMENT ANALYSIS METHOD TO
           TWO-DIMENSIONAL PROBLEMS IN FRACTURE MECHANICS
           PAPER 69-WA/PVP-12, ASME WINTER ANN. MEETING, LOS
           ANGELES, NOVEMBER 16-20, 1969.

69-0034   ODEN, J. T.
           A GENERAL THEORY OF FINITE ELEMENTS
           INT. J. NUM. METHODS ENG. (PART 1) 1, PP. 205-221;
           (PART 2) 1, PP. 247-259, 1969.

69-0035   CHARGIN, A. K.
           MAGNET FIELD AND FORCE BY FINITE ELEMENT
           TECHNIQUES
           PAPER G4, PROC. THIRD SYMP. ENG. PROBLEMS FUSION
           RES., LOS ALAMOS, NEW MEXICO, APRIL 8-11, 1969.

69-0036   ABSI, E.
           FINITE ELEMENT METHOD
           ANNALES DE L'INSTITUT TECHNIQUE DU BATIMENT ET DES
           TRAVAUX PUBLICS N 262, PP. 1593-162, OCT. 1969 (IN
           FRENCH)

69-0037   ADELMAN, H. M.
           A METHOD FOR COMPUTATION OF VIBRATION MODES AND
           FREQUENCIES OF ORTHOTROPIC THIN SHELLS OF
           REVOLUTION HAVING GENERAL MERIDIONAL CURVATURE
           NASA - TN - D - 4972, JANUARY 1969, P. 88.

69-0038   AGARWAL, R. K.
           BOSHKOV, S. H.
           STRESSES AND DISPLACEMENTS AROUND A CIRCULAR
           TUNNEL IN A THREE LAYER MEDIUM
           PROC. I.E.E., VOL. 116, NO. 10, OCTOBER 1969, PP.
           1661-1664.

69-0039   AHMAD, S.
           DALY, P.
           FINITE-ELEMENT METHODS FOR INHOMOGENEOUS
           WAVEGUIDES
           PROC. I.E.E., PART 1, VOL. 6, PP. 519-528,
           NOVEMBER 1969, PART 2, VOL. 6, NO. 6, NOVEMBER
           1969, PP. 529-540.

69-0040   AHMAD, S.
           PSEUDO ISOPARAMETRIC FINITE ELEMENTS FOR SHELL AND
           PLATE ANALYSIS
           RECENT ADVANCES IN STRESS ANALYSIS, ROY. AERON.
           SOC., PP. 6-21, 1969.

69-0041   ALDSTEDT, E.
           SHELL ANALYSIS USING PLANAR TRIANGULAR ELEMENTS
           FINITE ELEMENT METHODS, TAPIR, 1969, PP. 255-286.

69-0042   ALLMENDINGER, E. E.
           SZILARD, R.
           HARRENSTINE, H.
           THE EVOLUTION OF SHELL STRUCTURES IN OCEAN
           TECHNOLOGY
           INTER. COLLOQUIUM ON PROGRESS OF SHELL STRUCTURES,
           I.A.S.S., VOL. 3, 1969.

69-0043   ALLWOOD, R. J.
           CORNES, G. M. M.
           A POLYGONAL FINITE ELEMENT FOR PLATE BENDING
           PROBLEMS USING THE ASSUMED STRESS APPROACH
           INTERN. J. NUM. METH. ENG. V. 1, N. 1, PP.
           135-150, JANUARY 1969.

69-0044   MCMAHON, B. K.
           KENDRICK, R. F.
           PREDICTING THE BLOCK CAVING BEHAVIOUR OF OREBODIES
           AIME, PREPRINT NO. 69-AU-51, 1969, 15 PP.

69-0045   ANDERHEGGEN, E.
           FINITE ELEMENT PLATE BENDING EQUILIBRIUM ANALYSIS
           J. ENG. MECH. DIV., VOL. 95, NO. 4, AUGUST 1969.

69-0046   ARCHER, J. S.
           PINSON, L. D.
           RUBIN, C. P.
           A TWO-DIMENSIONAL AXISYMMETRIC SHELL FLUID MODEL
           FOR LAUNCH VEHICLE LONGITUDINAL DYNAMIC ANALYSIS
           JAPAN-U.S. SEMINAR ON MATRIX METHODS OF STRUCTURAL
           ANALYSIS AND DESIGN, TOKYO, JAPAN, AUGUST 1969,
           PAPER US 3-3.

69-0047   ARGYRIS, J. H.
           SCHARPF, D. W.
           THE CURVED TETRAHEDRONAL AND TRIANGULAR ELEMENTS
           TEC AND TRIC FOR THE MATRIX DISPLACEMENT METHOD
           THE AERONAUTICAL JOURNAL, VOL. 73, 1969.

69-0048   ARGYRIS, J. H.
           SCHARPF, D. W.
           A SEQUEL TO TECHNICAL NOTE 13 - THE CURVED
           TETRAHEDRONAL AND TRIANGULAR ELEMENTS TEC AND TRIC
           FOR THE MATRIX DISPLACEMENT METHOD
           J. ROY. AERO. SOC., VOL. 73, PP. 55-65, JANUARY
           1969.

69-0049   ARGYRIS, J. H.
           SCHARPF, D. W.
           SOME GENERAL CONSIDERATIONS ON THE NATURAL MODE
           TECHNIQUE
           THE AERONAUTICAL JOURNAL, VOL. 73, NO. 699, PP.
           218-226, MARCH 1969.

69-0050   CHEUNG, Y. K.
           ANALYSIS OF BOX GIRDER BRIDGES BY FINITE STRIP
           METHOD
           PROC. 2ND INT. SYMP. ON CONCRETE BRIDGE DESIGN,
           CHICAGO, MARCH 1969. ALSO IN ACI PUB. SP-26, PP.
           357-378.

69-0051   ARGYRIS, J. H.
           APPLICATION OF THE MATRIX DISPLACEMENT METHOD TO
           THE ANALYSIS OF PRESSURE VESSELS
           A.S.M.E. MEETING, PAPER NO. 69-WA/PVP-3, 1969,
           ALSO IN J. ENG. IND., VOL. 92, NO. 2, 1970, PP.
           317-329.

69-0052 ARTHUR, I. P.
PRESTRESSED CONCRETE PRESSURE VESSELS
CHEM. ENG PROGRESS, VOL. 65, NO. 5, PP. 84-88, MAY
1969.

69-0053 BACK, P. A. A.
CASSELL, A. C.
DUNGAR, R.
GAUKROGER, D. R.
SEVERN, R. T.
THE SEISMIC DESIGN STUDY OF A DOUBLE CURVATURE
ARCH DAM
PROC. INST. CIV. ENGRS., VOL. 43, PP. 217-248,
1969.

69-0054 BALTRUKONIS, J. H.
DYNAMICS OF SUBMARINE STRUCTURES, PART 1.
AD-689 203, MARCH 1969, 97 PP.

69-0055 BARLOW, J.
A STIFFNESS MATRIX FOR A CURVED MEMBRANE SHELL
RECENT ADVANCES IN STRESS ANALYSIS, ROY. AERO.
SOC., PP. 1-20, 1-23, 1969.

69-0056 BELL, K.
A REFINED TRIANGULAR PLATE BENDING FINITE ELEMENT
INT. J. NUM. METH. ENGNG., VOL. 1, NO. 1, JANUARY
1969, PP. 101-122.

69-0057 BELL, K.
TRIANGULAR PLATE BENDING ELEMENTS
FINITE ELEMENT METHODS, TAPIR, 1969, PP. 213-254.

69-0058 BREBBIA, C.
TOTTENHAM, H.
TAHBILDAR, A.
THE FINITE ELEMENT METHOD IN SHELL ANALYSIS
INT. COLLOQUIUM ON PROGRESS OF SHELL STRUCTURES,
I.A.S.S., VOL. III, 1969.

69-0059 BRISBANE, J. J.
HEAT CONDUCTION AND STRESS ANALYSIS OF SOLID
PROPELLANT ROCKET MOTOR NOZZLES
AD-848 594, FEBRUARY 1969, 194 PP.

69-0060 BUSHNELL, D.
COMPUTER ANALYSIS OF SHELL STRUCTURES
A.S.M.E., 69-WA/PVP-13, 1969.

69-0061 BYSKOV, E.
TWO NEARLY POLYGONAL HOLES, MATHEMATICAL CRACK
PROBLEMS
AD-700 848, 1969, 7 PP.

69-0062 CAMPBELL, J. S.
FINITE ELEMENT STRESS ANALYSIS OF PLANE AND
AXISYMMETRIC STRUCTURES
CENTRAL ELECTRICITY GENERATING BOARD, BERKELEY,
ENGLAND, BERKELEY NUCLEAR LABS., JANUARY 1969, P.
24, RD/B/N-1209.

69-0063 CAPEY, E. C.
A RELAXATION PROCEDURE FOR COMPUTING STRESSES IN
BUILT-UP STRUCTURES
RECENT ADVANCES IN STRESS ANALYSIS, ROY. AERO.
SOC., PP. 3-24, 3-34, 1969.

69-0064 CAPPELLI, A. P.
AGRAWAL, G. L.
PAULEY, K. E.
ELAD PROGRAM MANUAL, A COMPUTER PROGRAM FOR THE
ELASTIC ANALYSIS OF DOME STRUCTURES AND
AXISYMMETRIC SOLIDS OF REVOLUTION SUBJECTED TO
UNSYMMETRICAL LOADINGS
AD-862 676, OCTOBER 1969, 143 PP.

69-0065 CARSON, W. G.
NEWTON, R. E.
PLATE BUCKLING ANALYSIS USING A FULLY COMPATIBLE
FINITE ELEMENT
A.I.A.A. J. VOL. 7, NO. 3, PP. 527-529, MARCH
1969.

69-0066 CHALOUPKA, A. B.
A COMPUTER PROGRAM FOR THE ANALYSIS OF
TWO-DIMENSIONAL HEAT CONDUCTION USING THE FINITE
ELEMENT TECHNIQUE
AD-690 450, JUNE 1969, 85 PP.

69-0067 CHEN, P. E.
LAVENGOOD, R. E.
STRESS FIELDS AROUND MULTIPLE INCLUSIONS
AD-846 907, JANUARY 1969, 42 PP.

69-0068 CHEN, P. E.
LEWIS, T. B.
STRESS ANALYSIS OF TAPE REINFORCED COMPOSITES
AD-846 680, JANUARY 1969, 40 PP.

69-0069 CHEN, P. E.
LIN, J. M.
TRANSVERSE PROPERTIES OF FIBROUS COMPOSITES
MATER. RES. STAND., VOL. 9, PP. 29-33, AUGUST
1969.

69-0070 CHENG, P. Y.
DYNAMICS OF PRISMATIC AND OPEN SECTION MEMBER
GRIDS
APPLICATION OF FINITE ELEMENT METHODS IN CIVIL
ENGINEERING, A.S.C.E., NOVEMBER 1969, PP. 339-374.

69-0071 CHERNUKA, M. W.
COWPER, G. R.
LINDBERG, G. M.
OLSON, M. D.
APPLICATION OF THE HIGH PRECISION TRIANGULAR PLATE
BENDING ELEMENT TO PROBLEMS WITH CURVED BOUNDARIES
AD-708 668, OCTOBER 1969, 57 PP.

69-0072 CHOPRA, A. K.
DIBAJ, M.
CLOUGH, R. W.
PENZIEN, J.
SEED, H. B.
EARTHQUAKE ANALYSIS OF EARTH DAMS
PROC. FOURTH WORLD CONF. EARTHQUAKE ENGINEERING,
SANTIAGO, CHILE, JANUARY 1969.

69-0073 CLOUGH, R. W.
COMPARISON OF THREE DIMENSIONAL FINITE ELEMENTS
APPLICATION OF FINITE ELEMENT METHODS IN CIVIL
ENGINEERING, A.S.C.E., NOVEMBER 1969, PP. 1-26.

69-0074 COCO, R. H.
STIFFNESS MATRIX FOR CURVED TAPERED AND STRAIGHT
TAPERED SHELL STIFFENERS
AD-689 435, MARCH 1969, P. 66.

69-0075 COHEN, E.
MCCALLION, H.
IMPROVED DEFORMATION FUNCTIONS FOR THE FINITE
ELEMENT ANALYSIS OF BEAM SYSTEMS
INT. J. NUM. METHODS ENG., VOL. 1, NO. 2, PP.
163-168, APRIL, 1969.

69-0076 COLLIER, W. D.
HEATRAN, A FINITE ELEMENT CODE FOR HEAT TRANSFER
PROBLEMS
UNITED KINGDOM ATOMIC ENERGY AUTHORITY, RISLEY,
ENGLAND, 1969, P. 49, TRG-REPORT-1807.

69-0077 CHEUNG, Y. K.
CHEUNG, M. S.
ANALYSIS OF SLAB BEAM BRIDGES
PROCEEDINGS OF THE 2ND AUSTRALASIAN CONFERENCE ON
THE MECHANICS OF STRUCTURES AND MATERIALS,
UNIVERSITY OF ADELAIDE, AUGUST 1969.

69-0078 CONNOR, J. J.
WILL, G. T.
A MIXED FINITE-ELEMENT SHALLOW SHELL FORMULATION
PROC. OF THE U.S.-JAPAN SEMINAR ON MATRIX METHODS
OF STRUCTURAL ANALYSIS AND DESIGN, TOKYO,
SEPTEMBER 1969.

69-0079 COOK, R. D.
STRAIN RESULTANTS IN CERTAIN FINITE ELEMENTS
A.I.A.A. J., VOL. 7, NO. 3, PP. 535, MARCH 1969.

69-0080 COOK, W. L.
AUTOMATED INPUT DATA PREPARATION FOR NASTRAN
(NASA-TM-X-63607) NASA, APRIL 1969, PP. 40,
(N69-32112)

69-0081 COOK, R. D.
EIGENVALUE PROBLEMS WITH A * MIXED * PLATE ELEMENT
A.I.A.A. J., VOL. 7, NO. 5, MAY 1969, P. 982.

69-0082 COOK, R. D.
AL-ABDULLA, J. K.
SOME PLANE QUADRILATERAL HYBRID FINITE ELEMENTS
A.I.A.A. J., VOL. 7, P. 11, PP. 2184-2185,
NOVEMBER 1969.

69-0083 CORNELL, D. C.
CAPE-2D, A COMPUTER PROGRAM FOR THE STRESS
ANALYSIS OF PLANE AND AXISYMMETRIC COMPOSITE
STRUCTURES. A USER'S MANUAL.
GULF GENERAL ATOMIC, INC., SAN DIEGO, CALIF.,
FEBRUARY 12, 1969, P. 43, GA-9076.

69-0084 CORUM, J. M.
KRISHNAMURTHY, N.
A THREE-DIMENSIONAL FINITE ELEMENT ANALYSIS OF A
PRESTRESSED CONCRETE REACTOR VESSEL MODEL
APPLICATION OF FINITE ELEMENT METHODS IN CIVIL
ENGINEERING, A.S.C.E., NOVEMBER 1969, PP. 63-94.

69-0085 COVARRUBIAS, S. W.
CRACKING OF EARTH AND ROCKFILL DAMS. A
THEORETICAL INVESTIGATION BY MEANS OF THE FINITE
ELEMENT METHOD
AD-708 935, APRIL 1969, 166 PP.

69-0086 COWPER, G. R.
KOSKO, E.
LINDBERG, G. M.
OLSON, M. D.
STATIC AND DYNAMIC APPLICATIONS OF A
HIGH-PRECISION TRIANGULAR PLATE BENDING ELEMENT
A.I.A.A. J., VOL. 7, PP. 1957-1965, OCTOBER 1969.

69-0087 DE VRIES, G.
NORRIE, D. H.
THE APPLICATION OF THE FINITE ELEMENT TECHNIQUE TO
POTENTIAL FLOW PROBLEMS: PART I
REPORT NO. 7, MECHANICAL ENGINEERING DEPARTMENT,
UNIVERSITY OF CALGARY, CANADA, AUGUST 1969

69-0088 DEAK, A.
PIAN, T. H. H.
APPLICATION OF THE SMOOTH-SURFACE INTERPOLATION TO
THE FINITE ELEMENT ANALYSIS
A.I.A.A. J., VOL. 5, NO. 1, JANUARY 1969.
ALSO IN J. AMERICAN INST. AERON. ASTRON.,
JANUARY 1967.

69-0089 DESALVO, G. J.
APPLICATION OF FRACTURE MECHANICS TO THERMALLY
STRESSED CYLINDERS USING A FINITE ELEMENT METHOD
ASTRONUCLEAR LAB., WESTINGHOUSE ELECTRIC CORP.,
PITTSBURGH, PA., MARCH 1969, P. 36, WANL-TME-1897.

69-0090 DESANTIS, D.
MAGIC, AN AUTOMATED GENERAL PURPOSE SYSTEM FOR
STRUCTURAL ANALYSIS, VOLUME III, PROGRAMMER'S
MANUAL
AD-685 191, JANUARY 1969, 376 PP.

69-0091 DHATT, G. S.
NUMERICAL ANALYSIS OF THIN SHELLS BY CURVED
TRIANGULAR ELEMENTS BASED ON DISCRETE - KIRCHHOFF
HYPOTHESIS
APPLICATION OF FINITE ELEMENT METHODS IN CIVIL
ENGINEERING, A.S.C.E., NOVEMBER 1969, PP. 255-278.

69-0092 DIBAJ, M.
PENZIEN, J.
RESPONSE OF EARTH DAMS TO TRAVELLING SEISMIC WAVES
J. SOIL MECH. + FOUNDATIONS DIV., A.S.C.E., VOL.
95, NO. SM2, PP. 541-561, 1969.

69-0093 DICKINSON, S. M.
HENSHELL, R. D.
CLOUGH-TOCHER TRIANGULAR PLATE BENDING ELEMENT IN
VIBRATION
A.I.A.A. J., VOL. 7, NO. 3, PP. 560-561, MARCH
1969.

69-0094 DUNCAN, J. M.
DUNLOP, P.
SLOPES IN STIFF FISSURED CLAYS AND SHALES
J. SOIL MECH. + FOUNDATION DIV., A.S.C.E., VOL.
95, NO. SM2, 1969.

69-0095 DUNGAR, R.
SEVERN, R. T.
TRIANGULAR FINITE-ELEMENTS OF VARIABLE THICKNESS
AND THEIR APPLICATION TO PLATE AND SHELL PROBLEMS
J. STRAIN ANAL., VOL. 4, NO. 1, PP. 10-21, JANUARY
1969.

69-0096 EGEBERG, J. L.
MESHGEN, A COMPUTER CODE FOR AUTOMATIC FINITE
ELEMENT MESH GENERATION
SANDIA LABORATORIES, REPORT SCL-DR-69-49, JULY
1969.

69-0097 EGELAND, O.
APPLICATION OF FINITE ELEMENT TECHNIQUES TO
PLASTICITY PROBLEMS
FINITE ELEMENT METHODS, TAPIR, 1969, PP. 435-450.

69-0098 FARGETTE, F.
MAURY, J. F.
PRACTICAL METHOD FOR COMPUTER CALCULATION OF
STRUCTURES
ANNALES DE L'INSTITUT TECHNIQUE DU BATIMENT ET DES
TRAVAUX PUBLICS, NO. 259-260, PP. 1195-1228, JULY
1969.

69-0099 FJELD, S. A.
THREE DIMENSIONAL THEORY OF ELASTICITY
FINITE ELEMENT METHODS, TAPIR, 1969, PP. 333-364.

69-0100 FIRKINS, N. L.
STEEL PLATE ANALYSIS BY FINITE ELEMENTS
A.I.S.C. ENG. J., OCTOBER 1969, PP. 130-137.

69-0101 HUNG, N. D.
FRAEIJS DE VEUBEKE, B.
NATRUAL STRAINS AND STRESSES FOR TRAPEZOIDAL
STRUCTURES ANALYSIS
AD-687 186, FEBRUARY 1969, 46 PP.

69-0102 FRIED, I.
MORE ON GRADIENT ITERATIVE METHODS IN
FINITE-ELEMENT ANALYSIS
A.I.A.A. J., VOL. 7, NO. 3, PP. 565-566, MARCH
1969.

69-0103 FRIED, I.
GRADIENT METHODS FOR FINITE ELEMENT EIGENPROBLEMS
A.I.A.A. J., VOL. 7, NO. 4, PP. 739-740, APRIL
1969.

69-0104 FRIED, I.
FINITE-ELEMENT ANALYSIS OF TIME-DEPENDENT
PHENOMENA
A.I.A.A. J., VOL. 7, PP. 1170-1173, JUNE 1969.

69-0105 FRIED, I.
SOME ASPECTS OF THE NATURAL COORDINATE SYSTEM IN
THE FINITE-ELEMENT METHOD
A.I.A.A. J., VOL. 7, PP. 1366-1368, JULY 1969.

69-0106 GAONKAR, G. H.
AVERAGE ELASTIC PROPERTIES AND STRESS FIELDS FOR
COMPOSITE CONTINUA
AIAA/ASME 10TH STRUCTURES, STRUCTURAL DYNAMICS AND
MATERIALS CONFERENCE, PP. 172-182, APRIL 1969.

69-0107 FRIEDRICH, C. M.
SEAL-SHELL-2, A COMPUTER PROGRAM FOR THE STRESS
ANALYSIS OF A THICK SHELL OF REVOLUTION WITH
AXISYMMETRIC PRESSURES
BETTIS ATOMIC POWER LAB., JUNE 1969, P. 25,
WAPD-TM-398.

69-0108 GALLAGHER, R. H.
FINITE ELEMENT METHOD IN ELASTIC INSTABILITY
ANALYSIS
PROC. OF ISD/ISSC SYMP. ON FINITE ELEMENT
TECHNIQUES, STUTTGART, WEST GERMANY, JUNE 1969.

69-0109 GALLAGHER, R. H.
ANALYSIS OF PLATE AND SHELL STRUCTURES
APPLICATION OF FINITE ELEMENT METHODS IN CIVIL
ENGINEERING, A.S.C.E., NOVEMBER 1969, PP. 155-206.

69-0110 GALLAGHER, R. H.
A CORRELATION STUDY OF METHODS OF MATRIX
STRUCTURAL ANALYSIS
AGARD-OGRAPH 69, PERGAMON PRESS, 1969.

69-0111 GIRIJAVALLABHAN, C. V.
ANALYSIS OF SHEAR WALLS WITH OPENINGS
J. STRUC. DIV., A.S.C.E., VOL. 95, NO. ST10,
OCTOBER 1969.

69-0112 GIRIJAVALLABHAN, C. V.
ANALYSIS OF SHEAR WALLS BY FINITE ELEMENT METHOD
APPLICATION OF FINITE ELEMENT METHODS IN CIVIL
ENGINEERING, A.S.C.E., NOVEMBER 1969, PP. 631-641.

69-0113 GIROUX, Y. M.
BONNES, G.
FINITE ELEMENT ANALYSIS OF SHELLS UNDER NONUNIFORM
SNOW LOADS
INT. COLLOQUIUM ON PROGRESS OF SHELL STRUCTURES,
I.A.S.S., VOL. III., 1969.

69-0114 GOULD, L.
FINITE ELEMENT ANALYSIS OF SHELLS OF REVOLUTION BY
MINIMIZATION OF THE POTENTIAL ENERGY FUNCTIONAL
APPLICATION OF FINITE ELEMENT METHODS IN CIVIL
ENGINEERING, A.S.C.E., NOVEMBER 1969, PP. 279-308.

69-0115 GRAVES SMITH, T. R.
A VARIATIONAL METHOD FOR LARGE DEFLECTION
ELASTO-PLASTIC THEORY
INT. CONF. ON STRUCT. SOLID MECH. IN ENG., DESIGN
OF CIVIL ENGG., SOUTHAMPTON UNIV., APRIL 1969,
PP. 7, PAPER 102.

69-0116 GREENE, B. E.
JONES, R. E.
MCLAY, R. W.
STROME, D. R.
GENERALIZED VARIATIONAL PRINCIPLES IN THE FINITE
ELEMENT METHOD
A.I.A.A. J., VOL. 7, NO. 7, PP. 1254-1260, JULY
1969.

69-0117 GREENE, B. E.
JONES, R. E.
MCLAY, R. W.
APPLICATION, TESTING AND EVALUATION OF DOUBLY
CURVED SHELL ELEMENTS FOR DYNAMIC ANALYSIS, PART
II. FINITE ELEMENT DERIVATIONS
AD-857 297, JULY 1969, 133 PP.

69-0118 GREENE, B. E.
HERNES, M. W.
SALUS, W. L.
STROME, D. R.
ASTRA-BOEING'S ADVANCED STRUCTURAL ANALYZER
STRUCTURES TECH. FOR LARGE RADIO AND RADAR
TELESCOPE SYSTEMS, MIT PRESS, PP. 465-498, 1969.

69-0119 HAENER, J.
PUPPO, A.
FENG, M.
TRANSVERSE LOADING OF UNIDIRECTIONAL FIBER
COMPOSITES
AD-862 147, JULY 1969, 131 PP.

69-0120 HANSTEEN, O. E.
A CONICAL ELEMENT FOR DISPLACEMENT ANALYSIS OF
AXI-SYMMETRIC SHELLS
FINITE ELEMENT METHODS, TAPIR, 1969, PP. 319-332.

69-0121 HANSTEEN, O. E.
FINITE ELEMENT METHODS AS APPLICATIONS OF
VARIATIONAL PRINCIPLES
FINITE ELEMENT METHODS, TAPIR, 1969, PP. 451-474.

69-0122 HARRIS, H. G.
PIFKO, A. B.
ELASTIC PLASTIC BUCKLING OF STIFFENED RECTANGULAR
PLATES
APPLICATION OF FINITE ELEMENT METHODS IN CIVIL
ENGINEERING, A.S.C.E., NOVEMBER 1969, PP. 207-254.

69-0123 WEMPNER, G. A.
PATRICK, G. E.
FINITE DEFLECTIONS, BUCKLING AND POSTBUCKLING OF
AN ARCH
DEVELOPMENTS IN MECHANICS, VOL. 5, PROC. OF 11TH
MIDWESTERN MECH. CONF., AUGUST 18-20, 1969, AT
IOWA STATE UNIV., AMES, PP. 439-450.

69-0124 HELLEN, T. K.
SOLID WEDGE FINITE ELEMENT FOR STRESS ANALYSIS
CENTRAL ELECTRICITY GENERATING BOARD, BERKELY,
ENGLAND, BERKELEY NUCLEAR LABS., FEBRUARY
1969, P. 15, RD/B/N-1305.

69-0125 HIGUCHI, M.
NUMERICAL ANALYSIS OF SHIP STRUCTURAL ELEMENTS BY
MEANS OF FINITE ELEMENT METHOD AND PHOTO-ELASTIC
EXPERIMENT
JAPAN-U.S. SEMINAR ON MATRIX METHODS OF STRUCTURAL
ANALYSIS AND DESIGN, TOKYO, AUGUST 1969, PAPER J
4-6.

69-0126 HOLAND, I.
BELL, K.
FINITE ELEMENT METHODS IN STRESS ANALYSIS
TAPIR, TECHNICAL UNIV. OF NORWAY, N-7034
TRONDHEIM-NTH, NORWAY, 1969.

69-0127 HOLAND, I.
THE FINITE ELEMENT METHOD IN PLANE STRESS ANALYSIS
FINITE ELEMENT METHODS, TAPIR, 1969, PP. 43-92.

69-0128 HOLAND, I.
MOAN, T.
THE FINITE ELEMENT METHOD IN PLATE BUCKLING
FINITE ELEMENT METHODS, TAPIR, 1969, PP 475-500.

69-0129 HOLAND, I.
STIFFNESS MATRICES FOR PLATE BENDING ELEMENTS
FINITE ELEMENT METHODS, TAPIR, 196, PP. 159-178.

69-0130 HOYT, P.
CHANG, B.
STRESS CONCENTRATIONS IN HULL SHAPES FROM SURFACE
DISCONTINUITIES
AD-698 303, APRIL 1969, 81 PP.

69-0131 HSUEH, T.
WILSON, E. L.
STABILITY ANALYSIS OF AXISYMMETRIC SHELLS
AD-866 839, SEPTEMBER 1969, 107 PP.

69-0132 HUANG, Y. H.
FINITE ELEMENT ANALYSIS OF NONLINEAR SOIL MEDIA
APPLICATION OF FINITE ELEMENT METHODS IN CIVIL
ENGINEERING, A.S.C.E., NOVEMBER 1969, PP. 662-689.

69-0133 HUGHES, T. J. R.
ALLIK, H.
FINITE ELEMENTS FOR COMPRESSIBLE AND
INCOMPRESSIBLE CONTINUA
APPLICATION OF FINITE ELEMENT METHODS IN CIVIL
ENGINEERING, NOVEMBER 1969, PP. 27-62, A.S.C.E.

69-0134 HWANG, C. T.
HO, M. K.
WILSON, N. E.
FINITE ELEMENT ANALYSIS OF SOIL DEFORMATIONS
APPLICATION OF FINITE ELEMENT METHODS IN CIVIL
ENGINEERING, A.S.C.E., NOVEMBER 1969, PP. 729-746.

69-0135 HUNTER, A. R.
SKOGH, J.
STRESS DISTRIBUTIONS AROUND PLUG MATCHES WITH
ANCHOR BLOCKS
AD-861 451, AUGUST 1969, 83 PP.

69-0136 ICHIKAWA, T.
SOME CONSIDERATIONS ON METHODS FOR FLUTTER
ANALYSIS
JAPAN-U.S. SEMINAR ON MATRIX METHODS OF STRUCTURAL
ANALYSIS AND DESIGN, TOKYO, JAPAN, AUGUST 1969,
PAPER J 3/4.

69-0137 IRONS, B. M.
A CONFORMING QUARTIC TRIANGULAR ELEMENT FOR PLATE
BENDING
INT. J. NUM. METHODS ENG., VOL. 1, NO. 1, JANUARY
1969.

69-0138 IRONS, B. M.
ECONOMICAL COMPUTER TECHNIQUES FOR NUMERICALLY
INTEGRATED FINITE ELEMENTS
INT. J. NUM. METHODS ENG., VOL. 1, NO. 2, PP.
201-204, APRIL 1969.

69-0139 IRONS, B. M.
ZIENKIEWICZ, O. C.
THE ISOPARAMETRIC FINITE ELEMENT SYSTEM - A NEW
CONCEPT IN FINITE ELEMENT ANALYSIS
RECENT ADV. IN STRESS ANALYSIS, ROY. AERO. SOC.,
PP. 3-35, 3-40, 1969

69-0140 ISAKSON, G.
DISCRETE-ELEMENT PLASTIC ANALYSIS OF STRUCTURES IN
A STATE OF MODIFIED PLANE STRAIN
A.I.A.A. J., VOL. 7, NO. 1, PP. 545-546, MARCH
1969.

69-0141 IVERSEN, P. A.
SOME ASPECTS OF THE FINITE ELEMENT METHOD IN
TWO-DIMENSIONAL PROBLEMS
FINITE ELEMENT METHODS, TAPIR, 1969, PP. 93-114.

69-0142 JAVANDEL, I.
WITHERSPOON, P. A.
A METHOD OF ANALYZING TRANSIENT FLUID FLOW IN
MULTILAYERED AQUIFERS
WATER RESOUR. RES., VOL. 5, NO. 4, PP. 856-869,
AUGUST 1969.

69-0143 JENKINS, W. M.
MATRIX AND DIGITAL COMPUTER METHODS IN STRUCTURAL
ANALYSIS
MCGRAW-HILL BOOK CO., 1969.

69-0144 JORDAN, S.
MADDUX, G. E.
MALLETT, R. H.
MAGIC, AN AUTOMATED GENERAL PURPOSE SYSTEM FOR
STRUCTURAL ANALYSIS, VOLUME II, USER'S MANUAL
AD-691 893, JULY 1969, 37 PP.

69-0145 KAMEL, H. A.
SACK, R. L.
COMPUTATIONAL STABILITY IN NONLINEAR DISCRETE
STRUCTURAL SYSTEMS
AIAA/ASME 10TH STRUCTURES, STRUCTURAL DYNAMICS AND
MATERIALS CONFERENCE, PP. 114-124, APRIL 1969.

69-0146 KAVANAGH, K. T.
THE FINITE ELEMENT ANALYSIS OF PHYSICALLY AND
KINEMATICALLY NONLINEAR ELASTIC SOLIDS
AD-701 897, APRIL 1969, 138 PP.

69-0147 KAVARI, K.
ON THE DIMENSIONING OF UNDERGROUND STRUCTURES
SCHWEIZERISCHE BAUZEITUNG, VOL. 87, NO. 37, PP.
687-697, SEPTEMBER 1969 (IN GERMAN).

69-0148 KAWAI, T.
YOSHIMURA, N.
ANALYSIS OF LARGE DEFLECTION OF PLATES BY THE
FINITE ELEMENT METHOD
INT. J. NUM. METHODS ENG., VOL. 1, NO. 1, PP.
123-134, JANUARY 1969.

69-0149 KAWAI, T.
FINITE ELEMENT ANALYSIS OF THE GEOMETRICALLY
NONLINEAR PROBLEMS
PROC. JAPAN-U.S. SEMINAR ON MATRIX METHODS IN
STRUCTURAL ANALYSIS AND DESIGN, TOKYO, 1969.

69-0150 KAWAI, T.
OHTSUBO, H.
ON THE STATES OF STRESS AND DEFORMATION OF
CYLINDRICAL SPECIMENS OF BRITTLE MATERIAL UNDER
UNIAXIAL COMPRESSION
PROC. JAPAN-U.S. SEMINAR ON MATRIX METHODS IN
STRUCTURAL ANALYSIS AND DESIGN, TOKYO, 1969.

69-0151 KAWAI, T.
FINITE ELEMENT ANALYSIS OF SHELL STRUCTURES IN
JAPAN
JAPAN-U.S. SEMINAR ON MATRIX METHODS OF STRUCTURAL
ANALYSIS AND DESIGN, TOKYO, JAPAN, AUGUST 1969,
PAPER J 1-3.

69-0152 KAWAMOTO, T.
KAJITA, T.
STUDY ON STATE OF STRESS AND DEFORMATION OF
CYLINDRICAL SPECIMENS OF BRITTLE MATERIAL UNDER
UNIAXIAL COMPRESSION
JAPAN-U.S. SEMINAR ON MATRIX METHODS OF STRUCTURAL
ANALYSIS AND DESIGN, TOKYO, JAPAN, AUGUST 1969,
PAPER J 2-6.

69-0153 KIESS, D. W.
A CURVILINEAR SHELL FINITE ELEMENT
AD-704 514, DECEMBER 1969, 85 PP.

69-0154 KIRWAN, R. W.
GLYNN, T. E.
EXPERIMENTAL AND THEORETICAL INVESTIGATION OF
PAVEMENT DEFLECTIONS
AD-703 902, NOVEMBER 1969, 147 PP.

69-0155 KLEIN, S.
STRUCTURAL ANALYSIS AND EXPERIMENTAL CORRELATION
OF A COMPLEX SHELL UNDER ASYMMETRIC LOADS
AIAA/ASME 13TH STRUCTURES, STRUCTURAL DYNAMICS AND
MATERIALS CONFERENCE, APRIL 1969, PP. 31-39.

69-0156 KLEIN, S.
A STATIC AND DYNAMIC FINITE ELEMENT SHELL ANALYSIS
WITH EXPERIMENTAL VERIFICATION
JAPAN-U.S. SEMINAR ON MATRIX METHODS OF STRUCTURAL
ANALYSIS AND DESIGN, TOKYO, JAPAN, AUGUST 1969,
PAPER US 3-2. ALSO IN INT. J. NUMER. METH.
ENGNG., VOL. 3, NO. 3, JULY 1971, PP. 299-316.

69-0157 KOBAYASHI, A. S.
SHABAIK, A. H.
LEE, C. H.
ANALYTICAL PREDICTION OF DEFECTS OCCURRENCE IN
SIMPLE AND COMPLEX
AD-700 225, JULY 1969, 91 PP.

69-0158 KOBAYASHI, A. S.
MAIDEN, D. E.
STRESS INTENSITY FACTOR FOR A STRAIGHT CRACK
APPROACHING A HOLE
AD-700 277, DECEMBER 1969, 20 PP.

69-0159 KOHNKE, P. C.
SCHNOBRICH, W. C.
A FINITE ELEMENT ANALYSIS OF ECCENTRICALLY
STIFFENED-CIRCULAR CYLINDRICAL SHELLS
AD-695 743, OCTOBER 1969, 94 PP.

69-0160 KOTANCHIK, J. J.
BERG, B. A.
STACUSS 1: A DISCRETE-ELEMENT PROGRAM FOR THE
STATIC ANALYSIS OF SINGLE-LAYER CURVED STIFFENED
SHELLS SUBJECTED TO MECHANICAL AND THERMAL LOADS
AD-707 838, DECEMBER 1969, 702 PP.

69-0161 KRAHULA, J. L.
A FINITE ELEMENT SOLUTION FOR SAINT-VENANT TORSION
A.I.A.A. J., VOL. 7, PP. 220-2203, DECEMBER 1969.

69-0162 CHEUNG, Y. K.
THE ANALYSIS OF CURVILINEAR ORTHOTROPIC CURVED
BRIDGE DECKS
PUBLICATIONS, INTERNATIONAL ASSOCIATION OF BRIDGES
AND STRUCTURAL ENGINEERING, VOL. 29, NO. 11,
DECEMBER 1969.

69-0163 KRAJCINOVIC, D.
HERRMANN, G.
PERIODIC SHOCK EXCITATION OF ELASTIC STRUCTURES
SHOCK VIB. BULL., VOL. 40, PART 2, 1969, PP.
57-66. ALSO IN AD-708 261, 1969, 13 PP.

69-0164 KRIZEK, R. J.
MIMIC SOURCE APPROACH TO VISCOELASTIC ANALYSIS
APPLICATION OF FINITE ELEMENT METHODS IN CIVIL
ENGINEERING, A.S.C.E., NOVEMBER 1969, PP. 517-528.

69-0165 LEE, K. P.
GENERALIZED STIFFNESS MATRIX OF A CURVED-BEAM
ELEMENT
A.I.A.A. J., VOL. 7, PP. 2043-2045, OCTOBER 1969.

69-0166 LEISTNER, H. G.
JONES, R. E.
WALKER, W. J.
STRUCTURE-MEDIUM INTERACTION AND DESIGN PROCEDURES
STUDY. VOLUME I. ANALYSIS METHOD, THEORY,
VERIFICATION AND APPLICABILITY
AD-863 248, OCTOBER 1969, 245 PP.

69-0167 LIN, J. M.
CHEN, P. E.
DIBENEDETTO, A. T.
TRANSVERSE PROPERTIES OF UNIDIRECTIONAL ALUMINUM
MATRIX FIBROUS COMPOSITES
AD-861 188, AUGUST 1969, 43 PP.
ALSO IN POLYM. ENG. SCI., VOL. 11, NO. 4,
JULY 1971, PP. 344-352.

69-0168 LINDBERG, G. M.
OLSON, M. D.
TULLOCH, H. A.
CLOSED FORM, FINITE ELEMENT SOLUTIONS FOR PLATE
VIBRATIONS
AD-690 836, FEBRUARY 1969, 53 PP.

69-0169 LO, K. S.
SCORDELIS, A. C.
FINITE SEGMENT ANALYSIS OF FOLDED PLATES
J. STRUCT. DIV., A.S.C.E., VOL. 95, NO. ST 5, PP.
831-852, MAY 1969.

69-0170 LUMING, H.
LUMPED PARAMETER APPROACH TO AXISYMMETRIC DYNAMIC
SOIL DEFORMATIONS
APPLICATION OF FINITE ELEMENT METHODS IN CIVIL
ENGINEERING, A.S.C.E., NOVEMBER 1969, PP. 642-661.

69-0171 LYSMER, J.
KUHLEMEYER, R. L.
FINITE DYNAMIC MODEL FOR INFINITE MEDIA
J. ENG. MECH. DIV., A.S.C.E., VOL. 95, NO. EM4,
PP. 859-877, AUGUST 1969.

69-0172 MAIER, G.
SHAKEDOWN THEORY IN PERFECT ELASTOPLASTICITY WITH
ASSOCIATED AND NON ASSOCIATED FLOW LAWS - A FINITE
ELEMENT LINEAR PROGRAMMING APPROACH
MECCANICA, VOL. 4, SEPTEMBER 1969, PP. 250-260.

69-0173 MALLETT, R. H.
JORDAN, S.
MAGIC, AN AUTOMATED GENERAL PURPOSE SYSTEM FOR
STRUCTURAL ANALYSIS
BELL AEROSYSTEMS CO., BUFFALO, N. Y., JANUARY
1969, PROJ. AF-1467, AD 685 190.

69-0174 MARCAL, P. V.
LARGE DEFLECTION ANALYSIS OF ELASTIC-PLASTIC
SHELLS OF REVOLUTION
PROC. 10TH ASME/AIAA STRUCTURES, STRUCT. DYNAMICS,
AND MAT. CONF., APRIL 1969.

69-0175 MARCAL, P. V.
FINITE ELEMENT ANALYSIS OF COMBINED PROBLEMS ON
NONLINEAR MATERIAL AND GEOMETRIC BEHAVIOUR
CONF. ON COMPUTATIONAL APPROACHES IN APPLIED
MECH., A.S.M.E., CHICAGO, JUNE 1969 (ALSO AD-689
877)

69-0176 ISHINO, B.
VIBRATION ANALYSIS OF AN INTERNAL REFERENCE UNIT
U.S. NAVAL RES. LAB., SHOCK VIB., BULL., VOL. 40,
PT 3, DECEMBER, 1969, PP. 257-266.

69-0177 MARCAL, P. V.
FINITE-ELEMENT ANALYSIS WITH MATERIAL
NONLINEARITIES
PROC. JAPAN-U.S. SEMINAR ON MATRIX METHODS IN
STRUCTURAL ANALYSIS AND DESIGN, TOKYO, 1969.

69-0178 MARCAL, P. V.
LARGE DEFLECTION ANALYSIS OF ELASTIC-PLASTIC
PLATES AND SHELLS
INT. CONF. ON PRESSURE VESSEL TECH., A.S.M.E., PP.
75-87, 1969.

69-0179 MARTIN, H. C.
FINITE ELEMENTS AND THE ANALYSIS OF GEOMETRICALLY
NONLINEAR PROBLEMS
JAPAN-U.S. SEMINAR ON MATRIX METHODS OF STRUCTURAL
ANALYSIS AND DESIGN, TOKYO, JAPAN, AUGUST 1969.
PAPER US 2-2. ALSO IN RECENT ADVANCES IN
MATRIX METHODS OF STRUCTURAL ANALYSIS AND DESIGN,
U. OF ALABAMA PRESS, 1971, PP. 343-381.

69-0180 MARTIN, H. C.
FINITE ELEMENT FORMULATIONS OF GEOMETRICALLY
NONLINEAR PROBLEMS
PROC. OF U.S.-JAPAN SEMINAR ON MATRIX METHODS OF
STRUCTURAL ANALYSIS AND DESIGN, TOKYO, JAPAN,
SEPTEMBER 1969.

69-0181 MCLAY, R. W.
A SPECIAL VARIATIONAL PRINCIPLE FOR THE
FINITE-ELEMENT METHOD
A.I.A.A. J., VOL. 7, NO. 3, PP. 533-539, MARCH
1969.

69-0182 MCLEON, I. A.
NEW RECTANGULAR FINITE ELEMENT FOR SHEAR WALL
ANALYSIS
J. STRUCT. DIV., A.S.C.E., VOL. 95, NO. ST 3, PP.
399-410, MARCH 1969.

69-0183 MCNEAL, R. H.
MCCORMICK, C. W.
THE NASTRAN COMPUTER PROGRAM FOR STRUCTURAL
ANALYSIS
SAE, PAPER 690612, P. 15, OCTOBER 1969,
ALSO IN COMPUT. STRUCT., VOL. 1, NO. 3,
OCTOBER 1971, PP. 389-412.

69-0184 MCNEICE, G. M.
KEMP, K. O.
COMPARISON OF FINITE ELEMENT AND UNIQUE LIMIT
ANALYSIS SOLUTIONS FOR CERTAIN REINFORCED CONCRETE
SLABS
INST. CIVIL ENG., PROC., VOL. 43, PP. 629-640,
AUGUST 1969.

69-0185 MCNEICE, G. M.
AN ELASTIC-PLASTIC FINITE ELEMENT ANALYSIS FOR
PLATES WITH EDGE BEAMS
APPLICATION OF FINITE ELEMENT METHODS IN CIVIL
ENGINEERING, A.S.C.E., NOVEMBER 196., PP. 529-536.

69-0186 MEGARD, G.
PLANAR AND CURVED SHELL ELEMENTS
FINITE ELEMENT METHODS, TAPIR, 1969, PP. 287-318.

69-0187 MEHROTRA, B. L.
MUFTI, A. A.
FINITE ELEMENT ANALYSIS OF THIN SHELLS
(DISCUSSION)
J. ENG. MECH. DIV., A.S.C.E., VOL. 95, NO. EM4,
PP. 1021-1024, AUGUST 1969.

69-0188 MEHROTRA, B. L.
MUFTI, A. A.
REDWOOD, R. G.
ANALYSIS OF THREE-DIMENSIONAL THIN WALLED
STRUCTURES
J. STRUCT. DIV., A.S.C.E., VOL. 95, NO. ST12, PP.
2863-2872, DECEMBER 1969.

69-0189 MEHTA, K. C.
STRESS STRAIN RELATIONSHIP FROM COMPRESSION TESTS
ON NONLINEAR MATERIALS
APPLICATION OF FINITE ELEMENT METHODS IN CIVIL
ENGINEERING, A.S.C.E. NOVEMBER 1969, PP. 457-480.

69-0190 MEI, C.
FREE VIBRATIONS OF CIRCULAR MEMBRANES UNDER
ARBITRARY TENSION BY THE FINITE ELEMENT METHOD
J. ACOUST. SOC. AMER., VOL. 46, NO. 3, PART 2, PP.
693-700, SEPTEMBER 1969.

69-0191 MERRIFIELD, B.
A COMPUTER PROGRAM FOR THE FINITE ELEMENT ANALYSIS
OF A CLASS OF PLATE BENDING PROBLEMS
AD-870 895, DECEMBER 1969, 72 PP.

69-0192 MIKKELSON, P. T.
COMPUTERIZED ANALYSIS OF NON-ISOTROPIC STRUCTURES
APPLICATION OF FINITE ELEMENT METHODS IN CIVIL
ENGINEERING, A.S.C.E., NOVEMBER 1969, PP. 372-418.

69-0193 MIYAMOTO, H.
ANALYSIS OF STRESS AND STRAIN DISTRIBUTIONS AT
CRACK TIP BY FINITE ELEMENT METHOD
PROC. JAPAN-U.S. SEMINAR ON MATRIX METHODS IN
STRUCTURAL ANALYSIS AND DESIGN, TOKYO, 1969
ALSO IN RECENT ADVANCES IN
MATRIX METH. OF STRUCT. ANALYSIS AND DESIGN, 1971,
PP. 317-342.

69-0194 MOWBRAY, D. F.
SLOT, T.
NOTE ON STRESS AND STRAIN REDISTRIBUTION IN A
NOTCHED PLATE SPECIMEN DURING CYCLIC LOADING
J. BASIC ENG., A.S.M.E., VOL. 91, NO. 3, PP.
379-382, SEPTEMBER 1969.

69-0195 MOE, J.
THE FINITE ELEMENT TECHNIQUE - A NEW TOOL IN
STRUCTURAL ANALYSIS
FINITE ELEMENT METHODS, TAPIR, 1969, PP. 7-42.

69-0196 MURRAY, D. W.
WILSON, E. L.
FINITE-ELEMENT LARGE DEFLECTION ANALYSIS OF PLATES
J. ENG. MECH. DIV., A.S.C.E., VOL. 95, AM 1, PP.
143-166, FEBRUARY 1969.

69-0197 NAKAMURA, T.
ELASTIC-PLASTIC ANALYSIS OF FRAMED STRUCTURES
JAPAN-U.S. SEMINAR ON MATRIX METHODS OF STRUCTURAL
ANALYSIS AND DESIGN, TOKYO, JAPAN, AUGUST 1969,
PAPER J 4-2.

69-0198 NORDELL, W. J.
CRAWFORD, J. E.
BEARD, R. M.
STRUCTURAL RESPONSE OF UNSTIFFENED TOROIDAL SHELLS
AD-697 274, NOVEMBER 1969, 24 PP.

69-0199 OAKBERG, R. G.
ANALYSIS OF FRAMES WITH SHEAR WALLS BY FINITE
ELEMENTS
APPLICATION OF FINITE ELEMENT METHODS IN CIVIL
ENGINEERING, A.S.C.E., NOVEMBER 1969, PP. 567-608.

69-0200 DAUGHTRIDGE, A. S.
GENERAL ANALYSIS OF FOLDED PLATES BY THE FINITE
ELEMENT METHOD
PH.D. THESIS, NORTH CAROLINA STATE UNIV., RALEIGH,
1969.

69-0201 ARGYRIS, J. H.
BUCK, K. E.
MODERN DEVELOPMENTS IN THE STRESS ANALYSIS OF
PRESSURE VESSELS
ASME, FIRST INT. CONF. ON PRESSURE VESSEL TECH.,
DELFT, NETHERLANDS, SEPTEMBER 29-OCTOBER 2, 1969,
PT. 3, PP. 33-49.

69-0202 ODEN, J. T.
FINITE ELEMENT ANALYSIS OF NONLINEAR PROBLEMS IN
THE DYNAMICAL THEORY OF COUPLED THERMOELASTICITY
NUCL. ENG. DESIGN, AMSTERDAM, JULY 1969.

69-0203 CLOUGH, G. W.
DUNCAN, J. M.
FINITE ELEMENT ANALYSES OF PORT ALLEN AND OLD
RIVER LOCKS
U.S. ARMY ENG. WATERWAYS EXP. STA., CORPS ENG.,
CONTRACT REP. TE 69-3(S-69-6) SEPTEMBER 1969, 264
PP.

69-0204 ODEN, J. T.
FINITE ELEMENT APPLICATIONS IN NONLINEAR
STRUCTURAL ANALYSIS
APPLICATION OF FINITE ELEMENT METHODS IN CIVIL
ENGINEERING, A.S.C.E., NOVEMBER 1969, PP. 419-456.

69-0205 ODEN, J. T.
FINITE ELEMENT ANALYSIS OF NONLINEAR STRUCTURES
J. STRUCT. DIV., (DISCUSSION BY R. H. MALLETT AND
P. V. MARCAL), A.S.C.E., VOL. 95, NO. ST6, PP.
1379-1381, 1969.

69-0206 ODEN, J. T.
ON A GENERALIZATION OF THE FINITE ELEMENT CONCEPT
AND ITS APPLICATION TO A CLASS OF PROBLEMS IN
NONLINEAR VISCOELASTICITY
DEVELOPMENTS IN THEORETICAL AND APPLIED MECHANICS,
VOL. IV, PERGAMON PRESS, LONDON, PP. 581-593,
1969.

69-0207 ODEN, J. T.
FINITE ELEMENT FORMULATION OF FINITE DEFORMATIONS
AND IRREVERSIBLE THERMODYNAMICS OF NONLINEAR
CONTINUA - A SURVEY AND EXTENSIONS OF RECENT
DEVELOPMENTS
PROC. JAPAN-U.S. SEMINAR ON MATRIX METHODS IN
STRUCTURAL ANALYSIS AND DESIGN, TOKYO, 1969.

69-0208 ODEN, J. T.
FINITE ELEMENT LARGE DEFLECTION ANALYSIS OF PLATES
J. ENG. MECH. DIV., PROC., A.S.C.E., VOL. 95, P.
143, 1969.

69-0209 OHCHI, Y.
STRESS ANALYSIS FOR STRUCTURES UNDER CONSTRUCTION
JAPAN-U.S. SEMINAR ON MATRIX METH. STRUCT.
ANALYSIS AND DESIGN, TOKYO, JAPAN, PAPER J 4-3,
AUGUST 1969.

69-0210 PAULLING, J. R.
APPLICATIONS OF FINITE ELEMENT METHOD TO SHIP
STRUCTURAL ANALYSIS
JAPAN-U.S. SEMINAR ON MATRIX METHODS OF STRUCTURAL
ANALYSIS AND DESIGN, TOKYO, JAPAN, AUGUST 1969,
PAPER US 4-4.

69-0211 PECKNOLD, D. A.
SCHNOBRICH, W. C.
FINITE ELEMENT ANALYSIS OF SKEWED SHALLOW SHELLS
J. STRUCT. DIV., A.S.C.E., VOL. 95, ST 4, PP.
715-745, APRIL 1969.

69-0212 PERLOFF, W. H.
STRAIN DISTRIBUTION AROUND UNDERGROUND OPENINGS
AD-701 764, JUNE 1969, 241 PP.

69-0213 PIAN, T. H. H.
TONG, P.
BASIS OF FINITE ELEMENT METHODS FOR SOLID CONTINUA
INT. J. NUM. METHODS ENG., VOL. 1, NO. 1, JANUARY
1969.

69-0214 PIFKO, A. B.
ISAKSON, G.
A FINITE-ELEMENT METHOD FOR THE PLASTIC BUCKLING
ANALYSIS OF PLATES
A.I.A.A. J., VOL. 7, PP. 1950-1957, OCTOBER 1969.

69-0215 POPOV, E. P.
YAGHMAI, S.
LINEAR AND NONLINEAR STATIC ANALYSIS OF
AXISYMMETRICALLY LOADED THIN SHELLS OF REVOLUTION
INT. CONF. ON PRESSURE VESSEL TECH., A.S.M.E., PP
237-244, 1969.

69-0216 PRATO, C. A.
SHELL FINITE ELEMENT METHOD VIA REISSNER'S
PRINCIPLE
INT. J. SOLIDS STRUCT., VOL. 5, PP. 1119-1133,
OCTOBER 1969.

69-0217 PRINCE, N.
RASHID, Y. R.
STRUCTURAL ANALYSIS OF SHELL INTERSECTIONS
INT. CONF. ON PRESSURE VESSEL TECHNOL., 1969, PP.
245-254. ALSO IN GULF GENERAL ATOMIC INC., SAN
DIEGO, CALIF., MARCH 1969, P. 28, GA-9184.

69-0218 HATTORI, T.
TOMINO, H.
STRESS AND VIBRATION ANALYSIS ON THE SYSTEM OF
FRAME AND FOUNDATION
PROC., NAT. SYMP. ON MATRIX METHODS OF STRUCTURAL
ANALYSIS AND DESIGN, J.S.S.C., MAY 1969, PP.
459-466 (IN JAPANESE)

69-0219 RAJU, I. S.
RAO, A. K.
STIFFNESS MATRICES FOR SECTOR ELEMENTS
A.I.A.A. J., VOL. 7, NO. 1, P. 156, JANUARY 1969.

69-0220 RAMSDEN, J. N.
STOKER, J. R.
MASS CONDENSATION, A SEMI-AUTOMATIC METHOD FOR
REDUCING THE SIZE OF VIBRATION PROBLEMS
INT. J. NUM. METHODS ENG., VOL. 1, PP. 333-350,
OCTOBER 1969.

69-0221 RAMSTAD, H.
CONVERGENCE AND NUMERICAL ACCURACY WITH SPECIAL
REFERENCE TO PLATE BENDING
FINITE ELEMENT METHODS, TAPIR, 1969, PP. 179-212.

69-0222 RASHID, Y. R.
THREE-DIMENSIONAL ANALYSIS OF ELASTIC SOLIDS I.
ANALYSIS PROCEDURE
INT. J. SOLIDS STRUCT., VOL. 5, NO. 12, PP.
1311-1331, DECEMBER 1969.

69-0223 REESE, L. C.
BEHAVIOUR OF STRIP FOOTINGS ON LAYERED COHESIVE
SOILS
APPLICATION OF FINITE ELEMENT METHODS IN CIVIL
ENGINEERING, A.S.C.E., NOVEMBER 1969, PP. 690-728.

69-0224 RICHARD, R. M.
BLACKLOCK, J. R.
FINITE ELEMENT ANALYSIS OF INELASTIC STRUCTURES
A.I.A.A. J., VOL. 7, NO. 3, MARCH 1969, PP.
432-438.

69-0225 ROBINSON, J.
PRACTICAL APPLICATION OF A COMPUTERIZED STRUCTURAL
ANALYSIS SYSTEM WHICH ADOPTS A FINITE ELEMENT
TECHNIQUE DIRECT STIFFNESS METHOD
RECENT ADV. IN STRESS ANALYSIS, ROY. AERO. SOC.,
PP. 1-33, 1-43, 1969.

69-0226 ROREN, E. M. Q.
FINITE ELEMENT ANALYSIS OF SHIP STRUCTURES
FINITE ELEMENT METHODS, TAPIR, 1969, PP. 115-158.

69-0227 HOLUSA, L.
KRATOCHVIL, J.
CALCULATION OF PLATE OF CONSTANT THICKNESS BY
FINITE ELEMENT METHOD
STAVEBNICKY CASPOIS, VOL. 17, NO. 10, 1969, PP.
779-783.

69-0228 CARR, A. J.
CLOUGH, R. W.
DYNAMIC BEHAVIOUR OF SHELL ROOFS
PROC. FOURTH WORLD CONF. ON EARTHQUAKE
ENGINEERING, SANTIAGO, CHILE, 1969.

69-0229 SABIR, A. B.
ASHWELL, D. G.
A STIFFNESS MATRIX FOR SHALLOW SHELL FINITE
ELEMENTS
INT. J. MECH. SCI., VOL. II, PP. 269-279, MARCH
1969.

69-0230 SAGHERA, S. S.
EFFECTS OF SHEAR WALLS ON DYNAMIC RESPONSE OF
FRAMES
APPLICATION OF FINITE ELEMENT METHODS IN CIVIL
ENGINEERING, A.S.C.E., NOVEMBER 1969, PP. 609-630.

69-0231 SAKURAI, T.
A STUDY ON COMPOSITE STRUCTURES EMPLOYING
DIFFERENT TYPES OF FINITE ELEMENTS
JAPAN-U.S. SEMINAR ON MATRIX METHODS OF STRUCTURAL
ANALYSIS AND DESIGN, TOKYO, JAPAN, AUGUST 1969,
PAPER J 1-7.

69-0232 SANBONGI, S.
APPLICATION OF FINITE ELEMENT METHOD TO ANALYSIS
OF AIRCRAFT WING STRUCTURES
JAPAN-U.S. SEMINAR ON MATRIX METHODS OF STRUCTURAL
ANALYSIS AND DESIGN, TOKYO, JAPAN, AUGUST 1969,
PAPER J 5-2.

69-0233 SANDHU, R. S.
WILSON, E. L.
FINITE ELEMENT ANALYSIS OF LAND SUBSIDENCE
INT. SYMP. LAND SUBSIDENCE, TOKYO, SEPTEMBER 1969.

69-0234 SAWKO, F.
COPE, R. J.
THE ANALYSIS OF SKEW BRIDGE DECKS, A NEW FINITE
ELEMENT APPROACH
STRUCTURAL ENGINEER, VOL. 47, NO. 6, PP. 215-224,
JUNE 1969.

69-0235 SAWKO, F.
COPE, R. J.
ANALYSIS OF MULTI-CELL BRIDGES WITHOUT TRANSVERSE
DIAPHRAGMS, A FINITE ELEMENT APPROACH
STRUCTURAL ENGINEER, VOL. 47, NO. 11, PP. 455-460,
NOVEMBER 1969.

69-0236 SCHMIT, L. A.
MONFORTON, G. R.
FINITE DEFLECTION DISCRETE ELEMENT ANALYSIS OF
SANDWICH PLATES AND CYLINDRICAL SHELLS WITH
LAMINATED FACES
AIAA/ASME 10TH STRUCTURES, STRUCTURAL DYNAMICS AND
MATERIALS

69-0237 SCHMIT, L. A.
STRUCTURAL SYNTHESIS, 1959-1969, A DECADE OF
PROGRESS
JAPAN-U.S. SEMINAR ON MATRIX METHODS OF STRUCTURAL
ANALYSIS AND DESIGN, TOKYO, JAPAN, AUGUST 1969,
PAPER US. 4-1.

69-0238 SCHOLES, A.
STROVER, E. M.
THE STRESS ANALYSIS OF AXISYMMETRIC SHELLS USING A
PLANE FRAME ANALYSIS PROGRAM
RECENT ADV. IN STRESS ANALYSIS, ROY. AERO. SOC.,
PP. 3-7, 3-13, 1969.

69-0239 FUJITA, Y.
KAWAI, T.
OTSUBO, H.
YUHARA, T.
A METHOD OF FINITE DEFLECTION ELASTIC PLASTIC
ANALYSIS OF FRAMED STRUCTURES SUBJECTED TO LARGE
AXIAL FORCES
PROC. 3RD ANN. NAT. SYMP. ON MATRIX METHODS OF
STRUCTURAL ANALYSIS AND DESIGN, J.S.S.C., TOKYO,
MAY 1969, PP. 311-318. (IN JAPANESE)

69-0240 SILVESTER, P.
HIGH ORDER POLYNOMIAL TRIANGULAR FINITE ELEMENTS
FOR POTENTIAL PROBLEMS
INT. J. ENG. SCI., VOL. 7, NO. 8, PP. 849-861,
AUGUST 1969.

69-0241 SINGHAL, A. C.
775 SELECTED REFERENCES ON THE FINITE ELEMENT
METHOD AND MATRIX METHODS OF STRUCTURAL ANALYSIS
CIVIL ENG. DEPT., LAVAL UNIV., QUEBEC, REPORT
S-12, JANUARY 1969. (OUT OF PRINT)

69-0242 SLYPER, H. A.
DEVELOPMENT OF EXPLICIT STIFFNESS AND MASS
MATRICES FOR A TRIANGULAR PLATE ELEMENT
INT. J. SOLIDS STRUCT., VOL. 5, PP. 241-249, MARCH
1969.

69-0243 SOBIESZCZANSKI, J.
EVALUATION OF ALGORITHMS FOR STRUCTURAL
MODIFICATION
APPLICATION OF FINITE ELEMENT METHODS IN CIVIL
ENGINEERING, A.S.C.E., NOVEMBER 1969, PP. 129-154.

69-0244 STORDAHL, H.
CHRISTENSEN, H.
FINITE-ELEMENT ANALYSIS OF AXISYMMETRIC ROTORS
J. STRAIN ANAL., VOL. 4, PP. 163-168, JULY 1969

69-0245 STRICKLIN, J. A.
A RAPIDLY CONVERGING TRIANGULAR PLATE ELEMENT
A.I.A.A. J., VOL. 7, NO. 1, P. 180, JANUARY 1969.

69-0246 STROME, D. R.
GREENE, B. E.
JONES, R. E.
APPLICATION, TESTING, AND EVALUATION OF DOUBLY
CURVED SHELL ELEMENTS FOR DYNAMIC ANALYSIS. PART
1. THEORY, TECHNIQUES, THEORETICAL AND
EXPERIMENTAL CORRELATION
AD-857 296, JULY 1969, P. 152.

69-0247 STORDAHL, H.
CHRISTENSEN, H.
EXPERIENCES FROM STRESS ANALYSIS OF AXI-SYMMETRIC
PROBLEMS IN MACHINE DESIGN
FINITE ELEMENT METHODS, TAPIR, 1969, PP. 365-382.

69-0248 SUTHERLAND, W. H.
A FINITE ELEMENT COMPUTER CODE, AXICAP FOR CREEP
ANALYSIS
BATTELLE-NORTHWEST, RICHLAND, WASHINGTON, PACIFIC
NORTHWEST LAB., OCTOBER 1969, P. 59, BNWL-1142.
ALSO IN NUCL. ENG. AND DESIGN,
VOL. 11, 1970, PP. 269-285.

69-0249 SZABO, B. A.
LEE, G. C.
STIFFNESS METHODS FOR PLATES BY GALERKIN'S METHOD
J. ENG. MECH. DIV., A.S.C.E., VOL. 95, EM 3, PP.
571-586, JUNE 1969

69-0250 SZABO, B. A.
LEE, G. C.
DERIVATION OF STIFFNESS MATRICES FOR PROBLEMS IN
PLANE ELASTICITY BY GALERKIN'S METHOD
INT. J. NUM. METHODS ENG., VOL. 1, PP. 301-309,
JULY 1969.

69-0251 TANAKA, H.
OPTIMUM DESIGN OF FRAMED STRUCTURES
JAPAN-U.S. SEMINAR ON MATRIX METHODS OF STRUCTURAL
ANALYSIS AND DESIGN, TOKYO, JAPAN, AUGUST 1969,
PAPER J 4-1.

69-0252 TANAKA, M.
DEVELOPMENT AND EVALUATION OF A TRIANGULAR THIN
SHELL ELEMENT BASED UPON THE HYBRID ASSUMED STRESS
MODEL
AD-702 444, DECEMBER 1969, 122 PP.

69-0253 TERAZAWA, K.
UEDA, Y.
MATSUISHI, M.
ELASTIC-PLASTIC BUCKLING OF PLATES BY FINITE
ELEMENT METHOD
A.S.C.E., NATIONAL MEETING, NEW ORLEANS,
LOUISIANA, FEBRUARY 3-7, 1969, MEETING PREPRINT
NO. 845.

69-0254 MOHRAZ, B.
SCHNOBRICH, W. C.
FINITE ELEMENT METHOD FOR THE ANALYSIS OF NOZZLE
OPENINGS IN SHELLS OF REVOLUTION
UNIV-DEPT. CIV. ENG.-STRUCTURAL RESEARCH SER. 355,
ILLINOIS, DECEMBER 1969, 68 PP.

69-0255 THOMPSON, J. M. T.
HUNT, G. W.
COMPARATIVE PERTURBATION STUDIES OF THE ELASTICA
INT. J. MECH. SCI., VOL. 11, PP. 999-1014,
DECEMBER 1969.

69-0256 THORNE, R. G.
DURBIN, C. F.
TRIPP, L.
KIERSKY, L. B.
APPLICATION, TESTING, AND EVALUATION OF DOUBLY
CURVED SHELL ELEMENTS FOR DYNAMIC ANALYSIS, PART
III. THE SHELL COMPUTER PROGRAM - ITS USAGE, FLOW
AND ORGANIZATION
AD-857 493, JULY 1969, 82 PP.

69-0257   TINAWI, R. A.
A STUDY OF VARIOUS IDEALIZATIONS FOR WING
STRUCTURES AND NUMERICAL PROCEDURES INVOLVED USING
MATRIX METHODS
AIAA/ASME 10TH STRUCTURES, STRUCTURAL DYNAMICS AND
MATERIALS CONFERENCE, PP. 9-17, APRIL 1969.

69-0258   TONG, P.
EXACT SOLUTION OF CERTAIN PROBLEMS BY THE
FINITE-ELEMENT METHOD
A.I.A.A. J., VOL. 7, NO. 1, P. 178, JANUARY 1969.
(ALSO AD-703 988).

69-0259   TONG, P.
AN ASSUMED STRESS HYBRID FINITE ELEMENT METHOD FOR
AN INCOMPRESSIBLE AND NEAR INCOMPRESSIBLE MATERIAL
INT. J. SOLIDS AND STRUCTURES, VOL. 5, NO. 5, MAY
1969.   PP. 455-462, ALSO AD-692 808.

69-0260   TONG, P.
PIAN, T. H. H.
A VARIATIONAL PRINCIPLE AND THE CONVERGENCE OF A
FINITE-ELEMENT METHOD BASED ON ASSUMED STRESS
DISTRIBUTION
INT. J. SOLIDS STRUCT., VOL. 5, NO. 5, PP.
463-472, MAY 1969. (ALSO AD-667 813)

69-0261   TSUI, E. Y. W.
MASSARD, J. M.
LODEN, W. A.
ELEMENT STIFFNESS MATRICES OF THICK-WALLED
ORTHOTROPIC SHELLS WITH APPLICATIONS
APPLICATION OF FINITE ELEMENT METHODS IN CIVIL
ENGINEERING, A.S.C.E., NOVEMBER 1969, PP. 95-128.

69-0262   TSURUSAKI, K.
WALLACE, F. S.
RANDOM VIBRATION ANALYSIS SYSTEM FOR COMPLEX
STRUCTURES, PART 2, COMPUTER PROGRAM DESCRIPTION
AD-849 017, JANUARY 1969, P. 49.

69-0263   UEDA, Y.
MATSUISKI, M.
ANALYSIS OF ELASTIC-PLASTIC BUCKLING OF PLATES BY
THE FINITE ELEMENT METHOD
PROC., JAPAN-U.S. SEMINAR ON MATRIX METHODS IN
STRUCTURAL ANALYSIS AND DESIGN, TOKYO, 1969.

69-0264   VALLIAPPAN, S.
NATH, P.
TENSILE CRACK PROPAGATION IN REINFORCED CONCRETE
BEAMS BY THE FINITE ELEMENT TECHNIQUES
INT. CONF. ON SHEAR TORSION, AND BOND IN
REINFORCED CONCRETE, COIMBATORE, INDIA, JANUARY
1969

69-0265   VANDERLINDEN, J.
FAIR, G. S.
STRUCTURE-MEDIUM INTERACTION AND DESIGN PROCEDURES
STUDY. VOLUME IV. FEAT CODE USER'S GUIDE
AD-863 249, OCTOBER 1969, 153 PP.

69-0266   VILLAGGIO, P.
PROPERTIES OF STABILITY AND MONOTONY WITH THE
FINITE ELEMENT METHOD
AEROTECHNICA, VOL. 49, NO. 316, JUNE - DECEMBER,
1969, PP. 94-102. (IN ITALIAN)

69-0267   VISSER, W.
A REFINED MIXED-TYPE PLATE BENDING ELEMENT
A.I.A.A. J., VOL. 7, PP. 1801-1803, SEPTEMBER
1969.

69-0268   VOIGHT, B.
SAMUELSON, A. C.
ON THE APPLICATION OF FINITE-ELEMENT TECHNIQUES TO
PROBLEMS CONCERNING POTENTIAL DISTRIBUTION AND
STRESS ANALYSIS IN THE EARTH SCIENCES
PURE AND APPLIED GEOPHYSICS, VOL. 76, NO. 5, 1969,
PP. 40-55.

69-0269   WALKER, A. C.
A NONLINEAR FINITE ELEMENT ANALYSIS OF SHALLOW
CIRCULAR ARCHES
INT. J. SOLIDS STRUCT., VOL. 5, NO. 2, PP. 97-108,
FEBRUARY 1969.

69-0270   WEBBER, J. P. H.
STRESS ANALYSIS IN VISCOELASTIC BODIES USING
FINITE ELEMENTS AND A CORRESPONDENCE RULE WITH
ELASTICITY
J. STRAIN ANAL., VOL. 4, PP. 236-243, JULY 1969.

69-0271   WEI, B. C. F.
BALANCED STRESSES IN POST-YIELDED MULTI-MATERIAL
STRUCTURAL JOINTS
APPLICATION OF FINITE ELEMENT METHODS IN CIVIL
ENGINEERING, A.S.C.E., NOVEMBER 1969, PP. 309-328.

69-0272   DESAI, C. S.
SOLUTION OF STRESS-DEFORMATION PROBLEMS IN SOIL
AND ROCK MECHANICS USING FINITE ELEMENT METHODS
PH.D. THESIS, UNIV. OF TEXAS, AUSTIN, 1969.

69-0273   DARIO, N. P.
A COMPARISON OF FINITE ELEMENT AND FINITE
DIFFERENCE METHODS IN ELASTOSTATIC PROBLEMS
PH.D. THESIS, MICHIGAN STATE UNIV., 1969.

69-0274   WHITE, D. J.
ENDERBY, L. R.
FINITE-ELEMENT STRESS ANALYSIS OF A MULTIPIECE
PISTON
J. STRAIN ANAL., VOL. 4, NO. 1, PP. 33-39, JANUARY
1969.

69-0275   WILSON, E. L.
ELASTIC DYNAMIC RESPONSE OF AXISYMMETRIC
STRUCTURES
AD-702 989, JANUARY 1969, 62 PP.

69-0276   WILSON, E. A.
JONES, L.
HSUEH, T.
LARGE DISPLACEMENT ANALYSIS OF ASIXYMMETRIC SHELLS
AD-866 841, MARCH 1969, 79 PP.

69-0277   YAMADA, Y.
NAKAGIRI, S.
TAKATSUKA, K.
ANALYSIS OF SAINT-VENANT TORSION PROBLEM BY A
HYBRID STRESS MODEL
1ST U.S.-JAPAN SEMINAR PAPER ON MATRIX METH. OF
STRUCT. ANALYSIS AND DESIGN, TOKYO, JAPAN, PAPER J
1-5, AUGUST 1969.

69-0278   YAMAGUCHI, H.
STUDY OF ELASTIC-PLASTIC MATRIX ANALYSIS FOR
FOUNDATION AND REINFORCED CONCRETE IN JAPAN
JAPAN-U.S. SEMINAR ON MATRIX METHODS OF STRUCTURAL
ANALYSIS AND DESIGN, TOKYO, JAPAN, AUGUST 1969,
PAPER J 2-2.

69-0279   YANG, H. T. Y.
GODFREY, D. A.
STRUCTURAL ANALYSIS OF AIRCRAFT IMPACT ON A
NUCLEAR CONTAINMENT VESSEL AND ASSOCIATED
STRUCTURES
NUCL. ENG. DESIGN, VOL. 11, NO. 2, PP. 295-307,
1969.

69-0280   YANG, H. T. Y.
A FINITE ELEMENT STRESS ANALYSIS OF THE VERTICAL
BUTTRESSES OF A NUCLEAR CONTAINMENT VESSEL
NUCL. ENG. DESIGN, VOL. 11, NO. 2, PP. 255-268,
1969.

69-0281   RYBICKI, E. F.
SCHMIT, L. A.
AN INCREMENTAL COMPLEMENTARY ENERGY METHOD OF
NONLINEAR STRESS ANALYSIS
AIAA PAPER NO. 69-119, PRESENTED AT THE 7TH
AEROSPACE SCIENCES MEETING, NEW YORK, JANUARY
1969.

69-0282   ZIENKIEWICZ, O. C.
VALLIAPPAN, S.
KING, I. P.
ELASTO-PLASTIC SOLUTIONS OF ENGINEERING PROBLEMS '
INITIAL STRESS ' FINITE ELEMENT APPROACH
INT. J. NUM. METHODS ENG., VOL. 1, NO. 1, PP.
75-100, JANUARY 1969. ALSO IN UNICIV REPORT NO.
R-70, U. OF NEW SOUTH WALES, AUSTRALIA, AUGUST
1971.

69-0283   ZIENKIEWICZ, O. C.
EARTHQUAKE BEHAVIOUR OF RESERVOIR-DAM SYSTEMS
(DISCUSSION)
J. ENG. MECH. DIV., A.S.C.E., VOL. 95, NO. EM3,
PP. 801-803, JUNE 1969.

69-0284   ZIENKIEWICZ, O. C.
WATSON, M.
SOME CREEP EFFECTS IN STRESS ANALYSIS WITH
PARTICULAR REFERENCE TO CONCRETE PRESSURE VESSELS
NUCL. ENG. DESIGN, VOL. 4, NO. 4, PP. 406-412,
NOVEMBER 1969.

69-0285   ZIENKIEWICZ, O. C.
IRONS, B. M.
ERGATOUDIS, J.
AHMAD, S.
SCOTT, F. C.
ISO-PARAMETRIC AND ASSOCIATED ELEMENT FAMILIES FOR
TWO AND THREE DIMENSIONAL ANALYSIS
FINITE ELEMENT METHODS, TAPIR, 1969, PP. 383-434.

69-0286   YANG, H. T. Y.
NON-LINEAR ANALYSIS OF THE ROTATIONAL STIFFNESS OF
A NUCLEAR STEAM GENERATOR FOUNDATION
NUCL. ENG. AND DES., VOL. 10, NO. 3, JULY 1969,
PP. 339-348.

69-0287   ZIENKIEWICZ, O. C.
VALLIAPPAN, S.
ANALYSIS OF REAL STRUCTURES FOR CREEP, PLASTICITY
AND OTHER COMPLEX CONSTITUTIVE LAWS.
CONF. ON MATERIALS IN CIV. ENG., UNIV. OF
SOUTHAMPTON, 1969, WILEY (1970).

69-0288   ZLAMAL, M.
ON SOME FINITE ELEMENT PROCEDURES FOR SOLVING
SECOND ORDER BOUNDARY VALUE PROBLEMS
NUMERISCHE MATHEMATIK, VOL. 14, NO. 1, PP. 42-48,
1969.

69-0289 YU, Y. S.
COATES, D. F.
DEVELOPMENT AND USE OF COMPUTER PROGRAMS FOR
FINITE ELEMENT ANALYSIS. (LA MISE AU POINT ET
L'UTILIZATION DE PROGRAMMES SUR ORDINATEUR EN VUE
DE L'ANALYSE DE LA METHODS DES ELEMENTS FINIS)
CAN. DEP. ENERGY, MINES RESOURCES, MINES CR., RES.
REP. 198, JULY 1969, P. 97.

69-0290 MIYAMOTO, H.
MIYOSHI, T.
STRESS AND STRAIN DISTRIBUTIONS AT THE ROOT OF
CRACKS
J. FAC. ENG., UNIV. TOKYO, SER. B., VOL. 30, NO.
2, SEPTEMBER 1969, PP. 139-153.

69-0291 MALE, D. J.
ARBON, P. F.
FINITE ELEMENT STUDY OF COMPOSITE ACTION IN WALLS
PROC. 2ND AUSTRALASIAN CONF. ON MECH. STRUCTURES
MATER., ADELAIDE, SOUTH AUSTRALIA, AUGUST 25-27,
1969, PAPER 14, 23 PP.

69-0292 MASON, J. B.
THERMAL DEFORMATION PREDICTION IN ORBITING
OBSERVATORY STRUCTURES
NASA SPEC. PUBL. 233 (OPTICAL TELESCOPE TECHNOL.),
APRIL 29-MAY 1 1969, HUNTSVILLE, ALA., PP.
401-415.

69-0293 BUTLIN, G. A.
MAN MACHINE INTERACTIVE STRUCTURAL ANALYSIS AS A
PRELIMINARY DESIGN AID
AGARD CONF. PROC. , NO. 36, SYMP. ON STRUCT.
OPTIMIZATION BY STRUCT. AND MATER. PANEL, OCTOBER
8-10, 1969, ISTANBUL, TURKEY, PAPER 15, P. 25.

69-0294 COVARRUBIAS, S. W.
CRACKING OF EARTH AND ROCKFILL DAMS
HARVARD UNIV., HARVARD SOIL MECH. SER., NO. 82,
APRIL 1969, 156 PP.

69-0295 HRENNIKOFF, A.
PRECISION OF FINITE ELEMENT METHOD IN PLANE STRESS
INT. ASS. BRIDGE STRUCT. ENG., PUBL. VOL. 29, PART
2, 1969, PP. 125-137.

69-0296 TOTTENHAM, H.
TAHBILDAR, U. C.
DYNAMIC ANALYSIS OF SHELLS USING CURVED FINITE
ELEMENTS
BULL. INT. ASS. SHELL STRUCT., NO. 40, DECEMBER
1969, PP. 15-34.

69-0297 FIX, G. J.
STRANG, G.
FOURIER ANALYSIS OF THE FINITE ELEMENT METHOD IN
RITZ-GALERKIN THEORY
STUD. APPL. MATH. VOL. 48, NO. 3, 1969, PP.
265-273.

69-0298 BUG, G.
BLAIR, P.
FINITE ELEMENT SOLUTION OF THE FLOW AROUND A
TWO-DIMENSIONAL DESCOID
A.S.C.E. NAT. MEETING, NEW ORLEANS, LOUISIANA,
FEBRUARY 1969.

69-0299 PUTTAIAH, G.
GUPTA, B. P.
SOME RECENT FINITE ELEMENT APPROACHES IN ANALYSIS
OF PLASTIC IMPACT
ISPAEL J. TECH., VOL. 7, NO . 6, 1969, PP.
479-483.

69-0300 DALY, P.
FINITE ELEMENT COUPLING MATRICES
ELECTRON. LETT., VOL. 5, 1969, PP. 613-615.

69-0301 RYBICKI, E. F.
HOPPER, A. T.
HIGH-ORDER FINITE ELEMENT ANALYSIS FOR TRANSIENT
TEMPERATURE ANALYSIS OF INHOMOGENEOUS MATERIALS
PREPRINT 69-WA/HT-33, ASME. WINTER ANN. MEETING,
LOS ANGELES, NOVEMBER 16-20, 1969.

69-0302 SOOSAAR, K.
APPLICATION OF FINITE ELEMENT TECHNIQUE TO DESIGN
AND MANUFACTURE OF ASTRONOMICAL MIRRORS
OPTICAL TELESCOPE TECHNOL., HUNTSVILLE, ALA.,
APRIL 29-MAY 1, NASA SPEC. PUBL. 233, 1969, PP.
249-255.

69-0303 THOMPSON, E. G.
MACK, L. R.
LIN, E. S.
FINITE ELEMENT METHOD FOR INCOMPRESSIBLE SLOW
VISCOUS FLOW WITH A FREE SURFACE
PROC. MIDWESTERN CONF., DEVELOP. MECH., 11TH, VOL.
5, PP. 93-111, IOWA STATE UNIV. PRESS, AMES, IOWA,
1969.

69-0304 TONG, P.
A FINITE ELEMENT METHOD FOR LOW REYNOLDS NUMBER
FLOW AND ITS APPLICATION TO BIOMECHANICS
CAN. CONGR. OF APPL. MECH., WATERLOO, ONTARIO,
CANADA, 2ND, MAY 1969, PP. 20-23.

69-0305 YAMAMOTO, Y.
ISSHIKI, M.
VARIATIONAL PRINCIPLES AND DUALISTIC SCHEME FOR
INTERSECTION PROBLEMS IN ELASTICITY
J. FAC. ENG. UNIV. TOKYO, SER. B., VOL. 30, NO. 1,
1969, PP. 17-30.

69-0306 ZUDANS, Z.
SURVEY OF ADVANCED STRUCTURAL DESIGN ANALYSIS
TECHNIQUES
NUCL. ENG. DESIGN, VOL. 10, 1969, PP. 400-440.
ALSO IN PAPER 69-DE-13 FOR ASME MEETING, MAY 5-8,
1969, 27 PP.

69-0307 ARGYRIS, J. H.
ASKA AUTOMATIC SYSTEM FOR KINEMATIC ANALYSIS
NUCL. ENG. DESIGN, VOL. 10, 1969, PP. 441-455.

69-0308 EMERY, A. F.
WILLIAMS, J. A.
THERMAL-STRESS CONCENTRATION CAUSED BY STRUCTURAL
DISCONTINUITIES
EXP. MECH., VOL. 9, NO. 12, DECEMBER 1969, PP.
558-564.

69-0309 VAN BOMMEL, P.
PRACTICAL APPLICATION OF THE FINITE ELEMENT METHOD
AND SOME NUMERICAL ASPECTS
INGENIEUR (HAGUE), VOL. 81, NO. 27, JULY 4, 1969,
PP. 83-91.

69-0310 YOUNG, J. P.
ON, F. J.
MATHEMATICAL MODELING VIA DIRECT USE OF VIBRATION
DATA
SAE-PAPER 690615, FOR MEETING OCTOBER 6-10, 1969,
9 PP.

69-0311 TURNER, M. J.
OPTIMIZATION OF STRUCTURES TO SATISFY FLUTTER
REQUIREMENTS
AIAA J., VOL. 7, NO. 5, MAY 1969, PP. 945-951.

69-0312 GANGAL, M. B.
PAUL, B.
STRESS ANALYSIS OF PRESSURIZED CYLINDRICAL BORE IN
RECTANGULAR BLOCK
ASME PAPER 69-WA/PVP11 FOR MEETING NOVEMBER 16-20,
1969, 4 PP.

69-0313 HURTADO, J.
RHEOLOGICAL THEORY OF STRENGTH OF MATERIALS
GENIE CIVIL VOL. 146, NO. 5, MAY 1969, PP.
293-296.

69-0314 BREBBIA, C.
ON THE PARABOLIC VELAROIDAL SHELL
A.J. FERRANTE, INT. ASSN. FOR SHELL STRUCTURES-
BUL. NO. 38, JUNE 1969, PP. 31-43.

69-0315 TEMPLIER, A.
FINE, J.
EFFECT OF SHAPE AND HETEROGENEITY OF ROCK SAMPLES
ON THEIR COMPRESSIVE STRENGTH
REV. DE L'INDUSTRIE MINERALE, VOL. 52, NO. 4,
APRIL 1969, PP. 365-375.

69-0316 DUNS, C. S.
VERTICAL RESPONSE OF A RIGID BASE TO DYNAMIC
LOADING
CIV. ENG. (LONDON), VOL. 64, NO. 760, NOV. 1969,
PP. 1091-1095.

69-0317 CHEN, P. E.
NIELSEN, L. E.
MECHANICAL PROPERTIES OF TAPE COMPOSITES
KOLLOID-Z AND Z, POLYM., VOL. 235, NO. 1, NOV.
1969, PP. 1174-1181.

69-0318 JANSSEN, J. D.
DETERMINATION OF THE APPLICABILITY OF SOME
THEORIES ABOUT STRENGTH AND STIFFNESS OF
THIN-WALLED BEAMS
INGENIEUR (HAGUE), VOL. 81, NO. 42, OCTOBER 17,
1969, PP. 119-127.

69-0319 KEY, S. W.
VARIATIONAL PRINCIPLE FOR INCOMPRESSIBLE AND
NEARLY-INCOMPRESSIBLE ANISOTROPIC ELASTICITY
INT. J. SOLIDS AND STRUCTURES, VOL. 5, NO. 9,
SEPTEMBER 1969, PP. 951-964.

69-0320 DUPUIS, G.
GOEL, J. J.
REFINED FINITE ELEMENTS IN 2-DIMENSIONAL
ELASTICITY
ZEITSCHRIFT FUR ANGEWANDTE MATHEMATIK UND PHYSIK,
VOL. 20, NO. 6, 1969, PP. 858-881.

69-0321 SWEDLOW, J. L.
ELASTIC PLASTIC CRACKED PLATES IN PLANE STRAIN
INT. J. FRACTURE MECH., VOL. 5, 1969, PP. 33-44.

69-0322 BRANDES, K.
A DISCRETE LEAST SQUARES AND FINITE ELEMENT METHOD
NUCL. ENG. AND DES., VOL. 10, NO. 4, 1969, PP.
503-504.

69-0323 AKIMOTO, M.
KAWAMOTO, T.
ON THE STATES OF STRESS AND DEFORMATION AROUND THE
BOTTOM OF A CIRCULAR VERTICAL SHAFT
J. MINING MET. INST. JAP., VOL. 85, NO. 979,
NOVEMBER 1969, PP. 917-922.

69-0324 DAVIDS, N.
PUTTAIAH, G.
RADIAL INERTIA EFFECTS ON AN IDEALLY PLASTIC
CIRCULAR PLATE UNDER IMPULSIVE AXIAL COMPRESSION
INT. J. SOLIDS AND STRUCTURES, VOL. 5, NO. 11,
NOVEMBER 1969, PP. 1221-1230.

69-0325 TAYLOR, S.
TOBIAS, S. A.
COMPUTER METHODS FOR THE STRUCTURAL ANALYSIS OF
MACHINE TOOLS
C.I.R.P.-ANNALS, VOL. 17, NO. 4, AUGUST 1969, PP.
519-531.

69-0326 VISSER, W.
HEAT TRANSFER PROBLEMS AND TEMPERATURE STRESSES
INGENIEUR, VOL. 81, NO. 25, JUNE 20, 1969, PP.
73-81.

69-0327 NAGAMOTO, R.
FUNAOKA, K.
ON STRENGTH OF TRANSVERSE WEBS AND THEIR JOINTS OF
LARGE TANKER
MITSUBISHI HEAVY IND., LTD-TECH. REV., VOL. 6, NO.
2, MAY 1969, PP. 154-168.

69-0328 BLAAUWENDRAAD, J.
FINITE ELEMENT METHOD. BASIC PRINCIPLES AND
FORMULATION
INGENIEUR, VOL. 81, NO. 22, MAY 30, 1969, PP.
49-58.

69-0329 BLAAUWENDRAAD, J.
METHODS AND POSSIBILITIES FOR ELECTRONIC
COMPUTATION OF PLATES IN FLEXURE
INGENIEUR, VOL. 81, NO. 6, FEBRUARY 7, 1969, PP.
21-31.

69-0330 LYNCH, F.
A FINITE ELEMENT METHOD OF VISCOELASTIC STRESS
ANALYSES WITH APPLICATION TO ROLLING CONTACT
PROBLEMS
INT. J. NUM. METH. ENGNG., VOL. 1, NO. 4,
OCTOBER-DECEMBER 1969, PP. 379-394.

69-0331 BERGMAN, H. W.
ASKA, A LARGE SCALE SOFTWARE SYSTEM FOR FINITE
ELEMENT ANALYSIS
JAPAN-U.S. SEMINAR ON MATRIX METHODS OF STRUCTURAL
ANALYSIS AND DESIGN, TOKYO, JAPAN, AUGUST 1969.

69-0332 BREBBIA, C.
CONNOR, J. J.
GEOMETRICALLY NONLINEAR FINITE-ELEMENT ANALYSIS
J. ENG. MECH. DIV., ASCE., VOL. 98, EM2, APRIL
1969, PP. 463-483.

69-0333 ARGYRIS, J. H.
SCHARPF, D. W.
THE FUGA FAMILY FOR FOLDED PLATE STRUCTURES
AERON. J. ROY. AERON. SOC., VOL. 73, JUNE 1969.

69-0334 LINDBERG, G. M.
OLSON, M. D.
COWPER, G. R.
NEW DEVELOPMENTS IN THE FINITE ELEMENT ANALYSIS OF
SHELLS
DME/NAE QUARTERLY BULLETIN, VOL. 4, 1969, PP.
1-38.

69-0335 MOE, J.
FINITE ELEMENT TECHNIQUES IN SHIP STRUCTURE DESIGN
IN FINITE ELEMENT TECHNIQUES
PROC. SYMP. ON FINITE ELEMENT TECHNIQUES, INSTITUT
FUR STATIK UND DYNAMIK DER LUFT UND,
RAUMFAHRTKONSTRUKTIONEN, UNIV. OF STUTTGART,
GERMANY, EDITED BY M. SORENSEN, FROM AIR FORCE
FLIGHT DYNAMICS LAB., TECH. REPORT, JUNE 10-12,
1969, PP. 1-59.

69-0336 DIXON, J. R.
DUKES, T. P.
ANALYSIS OF SHIP STRUCTURES BY THE FINITE ELEMENT
METHOD
IN FINITE ELEMENT TECHNIQUES, EDITED BY M.
SORENSEN, FROM PROCEEDINGS OF THE SYMPOSIUM ON
FINITE ELEMENT TECHNIQUES, HELD AT THE INSTITUT
FUR STATIK UND DYNAMIK DER LUFT-UND
RAUMFAHRTKONSTRUKTIONEN, UNIVERSITY OF STUTTGART,
GERMANY, JUNE 10-12, 1969, PP. 61-98.

69-0337 PAULLING, J. R.
APPLICATION OF FINITE ELEMENT COMPUTATIONS TO SOME
SHIP STRUCTURAL PROBLEMS
IN FINITE ELEMENT TECHNIQUES, EDITED BY M.
SORENSEN, FROM PROCEEDINGS OF THE SYMPOSIUM ON
FINITE ELEMENT TECHNIQUES, HELD AT THE INSTITUT
FUR STATIK UND DYNAMIK DER LUFT-UND
RAUMFAHRTKONSTRUKTIONEN, UNIVERSITY OF STUTTGART,
GERMANY, JUNE 10-12, 1969, PP. 101-141.

69-0338 SMITH, C. S.
MITCHELL, G. C.
PRACTICAL CONSIDERATIONS IN THE APPLICATION OF
FINITE ELEMENT TECHNIQUES TO SHIP STRUCTURES
IN FINITE ELEMENT TECHNIQUES, EDITED BY M.
SORENSEN, FROM PROCEEDINGS OF THE SYMPOSIUM ON
FINITE ELEMENT TECHNIQUES, HELD AT THE INSTITUT
FUR STATIK UND DYNAMIK DER LUFT-UND
RAUMFAHRTKONSTRUKTIONEN, UNIVERSITY OF STUTTGART,
GERMANY, JUNE 10-12, 1969, PP. 143-185.

69-0339 GALLAGHER, R. H.
THE FINITE ELEMENT METHOD IN ELASTIC INSTABILITY
ANALYSIS
IN FINITE ELEMENT TECHNIQUES, EDITED BY M.
SORENSEN, FROM PROCEEDINGS OF THE SYMPOSIUM ON
FINITE ELEMENT TECHNIQUES, HELD AT THE INSTITUT
FUR STATIK UND DYNAMIK DER LUFT-UND
RAUMFAHRTKONSTRUKTIONEN, UNIVERSITY OF STUTTGART,
GERMANY, JUNE 10-12, 1969, PP. 187-239.

69-0340 ARGYRIS, J. H.
BUCK, K. E.
GLOUDEMAN, J. F.
SCHARPF, D. W.
SOME ASPECTS OF FINITE ELEMENT TECHNIQUES
IN FINITE ELEMENT TECHNIQUES, EDITED BY M.
SORENSEN, FROM PROCEEDINGS OF THE SYMPOSIUM ON
FINITE ELEMENT TECHNIQUES, HELD AT THE INSTITUT
FUR STATIK UND DYNAMIK DER LUFT-UND
RAUMFAHRTKONSTRUKTIONEN, UNIVERSITY OF STUTTGART,
GERMANY, JUNE 10-12, 1969, PP. 241-301.

69-0341 BRONLUND, O. E.
EIGENVALUES OF LARGE MATRICES
IN FINITE ELEMENT TECHNIQUES, EDITED BY M.
SORENSEN, FROM PROCEEDINGS OF THE SYMPOSIUM ON
FINITE ELEMENT TECHNIQUES, HELD AT THE INSTITUT
FUR STATIK UND DYNAMIK DER LUFT-UND
RAUMFAHRTKONSTRUKTIONEN, UNIVERSITY OF STUTTGART,
GERMANY, JUNE 10-12, 1969, PP. 303-357.

69-0342 FRAEIJS DE VEUBEKE, B.
THE SECOND VARIAION TEST WITH ALGEBRAIC AND
DIFFERENTIAL CONTRASTS
ADVANCED PROBLEMS AND METH. FOR SPACE FLIGHT
OPTIMIZATION, PERGAMON PRESS, 1969.

69-0343 ARGYRIS, J. H.
SCHARPF, D. W.
METHODS OF ELASTOPLASTIC ANALYSIS
IN FINITE ELEMENT TECHNIQUES, EDITED BY M.
SORENSEN, FROM PROCEEDINGS OF THE SYMPOSIUM ON
FINITE ELEMENT TECHNIQUES, HELD AT THE INSTITUT
FUR STATIK UND DYNAMIK DER LUFT-UND
RAUMFAHRTKONSTRUKTIONEN, UNIVERSITY OF STUTTGART,
GERMANY, JUNE 10-12, 1969, PP. 381-418.

69-0344 ROREN, E. M. Q.
IMPACT OF FINITE ELEMENT TECHNIQUES ON PRACTICAL
DESIGN OF SHIP STRUCTURES
IN FINITE ELEMENT TECHNIQUES, EDITED BY M.
SORENSEN, FROM PROCEEDINGS OF THE SYMPOSIUM ON
FINITE ELEMENT TECHNIQUES, HELD AT THE INSTITUT
FUR STATIK UND DYNAMIK DER LUFT-UND
RAUMFAHRTKONSTRUKTIONEN, UNIVERSITY OF STUTTGART,
GERMANY, JUNE 10-12, 1969, PP. 419-466.

69-0345 VON FUCHS, G.
SCHREM, E.
ASKA A COMPUTER SYSTEM FOR STRUCTURAL ENGINEERS
PROC. SYMP. ON FINITE ELEMENT TECHNIQUES, INSTITUT
FUR STATIK UND DYNAMIK DER LUFT UND
RAUMFAHRTKONSTRUKTIONEN, UNIV. OF STUTTGART,
GERMANY, M. SORENSEN (ED.), JUNE 10-12, 1969, PP.
467-507.

69-0346 HOLAND, I.
FINITE ELEMENTS FOR THE COMPUTATIONS OF
HYDRODYNAMIC MASS
IN FINITE ELEMENT TECHNIQUES, EDITED BY M.
SORENSEN, FROM PROCEEDINGS OF THE SYMPOSIUM ON
FINITE ELEMENT TECHNIQUES, HELD AT THE INSTITUT
FUR STATIK UND DYNAMIK DER LUFT-UND
RAUMFAHRTKONSTRUKTIONEN, UNIVERSITY OF STUTTGART,
GERMANY, JUNE 10-12, 1969, PP. 509-530.

69-0347 GLOUDEMAN, J. F.
THE GROWTH OF INTERACTIVE GRAPHICS AS AN
ENGINEERING TOOL.
IN FINITE ELEMENT TECHNIQUES, EDITED BY M.
SORENSEN, FROM PROCEEDINGS OF THE SYMPOSIUM ON
FINITE ELEMENT TECHNIQUES, HELD AT THE INSTITUT
FUR STATIK UND DYNAMIK DER LUFT-UND
RAUMFAHRTKONSTRUKTIONEN, UNIVERSITY OF STUTTGART,
GERMANY, JUNE 10-12, 1969, PP. 532-540.

69-0348 HOLAND, I.
NUMERICAL ACCURACY AND ROUNDING-OFF ERRORS
IN FINITE ELEMENT TECHNIQUES, EDITED BY M.
SORENSEN. FROM PROCEEDINGS OF THE SYMPOSIUM ON
FINITE ELEMENT TECHNIQUES, HELD AT THE INSTITUT
FUR STATIK UND DYNAMIK DER LUFT-UND
RAUMFAHRTKONSTRUKTIONEN, UNIVERSITY OF STUTTGART,
GERMANY, JUNE 10-12, 1969, PP. 550-551.

69-0349 MELOSH, R. J.
PALACOL, E. L.
MANIPULATION ERRORS IN FINITE ELEMENT ANALYSIS OF
STRUCTURES
NASA CR-1385, PHILCO-FORD, 1969.

69-0350 WILLIAM, K.
FINITE ELEMENT ANALYSIS OF CELLULAR STRUCTURES
PH.D. DISSERTATION, DEPT. OF CIVIL ENG., UNIV. OF
CALIF., BERKELEY, CALIF., 1969.

69-0351 HEER, E.
CHEN, J. C.
FINITE ELEMENT FORMULATION FOR LINEAR THERMOVISCO
ELASTIC MATERIALS
NASA TECHNICAL REPORT, 32-1381, JPL, 1969.

69-0352 FUJINO, T.
ANALYSIS OF CONTINUUM BY THE FINITE ELEMENT METHOD
MITSUBIS HEAVY INDUSTRIES, TECHNICAL REVIEW, VOL.
6, NO. 2, 1969, (IN JAPANESE)

69-0353 WHITE, D. J.
HUMPHERSON, J.
FINITE ELEMENT ANALYSIS OF STRESSES IN SHAFTS DUE
TO INTERFERENCE-FIT HUBS
J. STRAIN ANALYSIS 4(2), APRIL 1969, PP. 105-14.

69-0354 BROMBOLICH, L. J.
GOULD, P. L.
FINITE ELEMENT ANALYSIS OF SHELLS OF REVOLUTION BY
MINIMIZATION OF THE POTENTIAL ENERGY
FUNCTIONAL
PROC. OF SYMPOSIUM OF APPLICATION OF FINITE
ELEMENT METHODS IN CIVIL ENGINEERING,
VANDERBILT UNIVERSITY, NOVEMBER 1969, PP.
279-308.

69-0355 SANDER, G.
BECKERS, P.
NGUYEN, H. D.
DIGITAL COMPUTATION OF STRESSES AND DEFLECTIONS IN
A BOX BEAM - A PERFORMANCE COMPARISON BETWEEN
FINITE ELEMENT MODELS AND IDEALIZATION
PATTERNS
AFFDL-TR-69-4, 1969.

69-0356 DOHERTY, W. P.
WILSON, E. L.
TAYLOR, R. L.
STRESS ANALYSIS OF AXISYMMETRIC SOLIDS UTILIZING
HIGHER ORDER QUADRILATERAL FINITE ELEMENTS
STRUCT. ENG. LAB., UNIV. OF CALIF. BERKELEY, 1969.

69-0357 AHMAD, S.
CURVED FINITE ELEMENTS IN THE ANALYSIS OF SOLID
SHELL AND PLATE STRUCTURES
PH.D. THESIS, UNIV. OF WALES, SWANSEA, 1969

69-0358 PAREKH, C. J.
FINITE ELEMENT SOLUTION SYSTEM
PH.D. THESIS, UNIV. OF WALES, SWANSEA, 1969.

69-0359 CHEUNG, Y. K.
FOLDED PLATE STRUCTURES BY THE FINITE STRIP METHOD
PROC. AM. SOC. CIV. ENG., 95 ST, 2963-79, 1969.

69-0360 GALLAGHER, R. H.
FINITE ELEMENT ANALYSIS OF PLATE AND SHELL
STRUCTURES
SYMP. ON APPLIED FINITE ELEMENT METHODS,
VANDERBILT UNIV., NASHVILLE, TENN., NOVEMBER
13-14, 1969.

69-0361 MEEK, J. L.
CAREY, G. F.
AXISYMMETRIC SOLUTION OF ELASTO PLASTIC PROBLEMS
BY FINITE ELEMENT METHODS
CIVIL ENGR. DEPT. UNIV. OF QUEENSLAND, AUSTRALIA,
BULL. NO. 11, MAY 1969

69-0362 MACLEAD, I. A.
NEW RECTANGULAR FINITE ELEMENT FOR SHEAR WALL
ANALYSIS
J. STRUCT. DIV., ASCE, VOL. 95, NO. ST3, 1969, PP.
399-409.

69-0363 ZIENKIEWICZ, O. C.
BEST, B.
SOME NON LINEAR PROBLEMS IN SOIL AND ROCK
MECHANICS - FINITE ELEMENT SOLUTION
CONF. ON ROCK MECHANICS, UNIV. OF QUEENSLAND,
TOWNSVILLE. JUNE 1969.

69-0364 MALINA, H.
BERECHNUNG VON SPANNUNGSUMLAGERUNGEN IN FELS UND
BODEN MIT HILFE DER ELEMENTENMETHODE
VEROFFENTLICHUNGEN UNIV. KARLSRUHE, 40, 1-90,
1969.

69-0365 ROBERTS, T. M.
ASHWELL, D. G.
POST-BUCKLING ANALYSIS OF SLIGHTLY CURVED PLATES
BY THE FINITE ELEMENT METHOD
DEPT. OF CIVIL AND STRUCT. ENG., UNIV. OF WALES,
CARDIFF, REPORT 2, 1969.

69-0366 MCCORQUODALE, J. A.
NON-DARCY FLOW SOLVED BY FINITE ELEMENT ANALYSIS
PROC. OF THE 13TH CONG. IAHR, VOL. 4, KYOTO,
JAPAN, SEPTEMBER 1969, PP. 347-355.

69-0367 HALL, C. A.
BICUBIC INTERPOLATION OVER TRIANGLES
J. MATH. MECH., VOL. 19, 1969, PP. 1-11.

69-0368 ATLURI, S.
STATIC ANALYSIS OF SHELLS OF REVOLUTION USING
DOUBLY-CURVED QUADRILATERAL ELEMENTS DERIVED FROM
ALTERNATE VARIATIONAL MODELS
DOCTORAL DISSERTATION, DEPT. OF AERON. AND ASTRON.
M.I.T., JUNE 1969.

69-0369 DIXON, J. R.
POOK, L. P.
STRESS INTENSITY FACTORS CALCULATED GENERALLY BY
FINITE ELEMENT TECHNIQUE
NATURE VOL. 224, ISSUE N5215, 1969, PP. 166-167.

69-0370 BREBBIA, C.
INTEGRATION OF AREA AND VOLUME COORDINATES IN
FINITE-ELEMENT METHOD
A.I.A.A. J., VOL. 7, NO. 6, 1969, P. 1212.

69-0371 PRATO, C. A.
COMMENT ON EXACT SOLUTION OF CERTAIN PROBLEMS BY
FINITE ELEMENT METHOD
AMERICAN INST. AERONAUTICS AND ASTRONAUTICS. JOUR.
VOL. 7, ISSUE N6, 1969, PP 1215-1216.

69-0372 LEVY, N.
APPLICATION OF FINITE ELEMENT METHODS TO LARGE
SCALE ELASTIC PLASTIC PROBLEMS OF FRACTURE
MECHANICS
PH.D. DISSERTATION, BROWN UNIVERSITY, JULY 1969.

69-0373 ZUDANS, Z.
REDDI, M. M.
ANALYSIS OF ASYMMETRIC STIFFENED SHELL TYPE
STRUCTURES BY FINITE ELEMENT METHOD II. SIX
DEGREE-OF-FREEDOM MODEL
NUCLEAR ENGINEERING AND DESIGN VOL. 9, ISSUE N3,
1969, PP. 302-310, REF 5.

69-0374 WOLF, J. P.
COMPUTER-DEVELOPED STATICAL METHODS
SCHWEIZ BAUZTG. V. 87, N.6, FEB 6, 1969, PP. 93-6.

69-0375 FUJINO, T.
ARAI, K.
ANALYSIS OF OUT-OF-PLANE STRESS BY THE SIX NODES,
FOURTH ORDER ACCURACY TRIANGULAR ELEMENTS
PREPRINT, 46TH ANN. MEETING OF JAPAN SOC. OF MECH.
ENGRS., 1969, PP. 139-142. (IN JAPANESE)

69-0376 FUJINO, T.
ARAI, K.
SHELL ANALYSIS BY THE FINITE ELEMENT METHOD
PROC. SYMP. ON MATRIX METHODS OF STRUCT. ANALYSIS
AND DESIGN, SOC. OF STEEL CONSTRUCTION OF JAPAN,
TOKYO, 1969, PP. 100-107 (IN JAPANESE)

69-0377 KURIBAYASHI, E.
IWASAKI, T.
EIGENVALUE ANALYSIS BY FINITE ELEMENT METHODS
PROC. NAT. SYMP. ON MATRIX METHODS OF STRUCTURAL
ANALYSIS AND DESIGN, SOCIETY OF STEEL
CONSTRUCTION OF JAPAN, MAY 1969, PP. 489-496 (IN
JAPANESE)

69-0378 SON, J. S.
HANRATTY, T. J.
NUMERICAL SOLUTION FOR THE FLOW AROUND A CYLINDER
AT REYNOLDS NUMBERS OF 40, 200 AND 500
J. FLUID MECHANICS, V. 35, PT. 2, FEB 3, 1969, PP.
369-86.

69-0379 POPOV, E. P.
KHOJASTEH-BAKHT, M.
ELASTIC-PLASTIC ANALYSIS OF SOME PRESSURE VESSEL
HEADS
ASME PAP. 69-WA/PVP-7 FOR MEETING NO. 16-20 1969,
8P. J. ENGOND., TRANS. ASME., V. 92, SER. B, N. 2,
MAY 1970, PP. 309-16.

69-0380 MARCAL, P. V.
ELASTIC PLASTIC ANALYSIS OF PRESSURE VESSEL
COMPONENTS
FIRST PRESSURE VESSEL AND PIPING CONFERENCE,
AMERICAN SOCIETY OF MECHANICAL ENGINEERS, COMPUTER
SEMINAR, DALLAS, TEXAS, 1968, ARPA E 62, DIV. OF
ENGNG., BROWN UNIV., JANUARY 1969, 23 PP.

69-0381 ODEN, J. T.
FORMULATION OF GENERAL DISCRETE MODELS OF
THERMOMECHANICAL BEHAVIOR OF MATERIALS WITH
MEMORY
G. AGUIRRE RAMIREZ, UNIV. OF ALABAMA, HUNTSVILLE.
INT. J SOLIDS & STRUCTURES, V. 5, N.10, OCT
1969, PP. 1077-93.

69-0382 NEUMAN, S. P.
THEORY OF FLOW IN A CONFINED TWO AQUIFER SYSTEM
WITHERSPOON PA, WATER RESOUR RES., V. 5, N.4, AUG
1969, PP. 803-16.

69-0383 YAMADA, Y.
NAKAGIRI, S.
YOKOUCHI, Y.
PLASTIC STRESS STRAIN MATRIX OF ANISOTROPIC
MATERIALS AND ITS APPLICATION
PROC. NATL. SYMP. ON MATRIX METH. IN STRUCT.
ANALYSIS AND DESIGN, J.S.S.C. PAPER, TOKYO, 1969,
PP. 371-376. (IN JAPANESE)

69-0384 SUHARA, J.
FUKUDA, J.
ON THE APPLICATIONS OF COLLOCATION OF ANALYTIC
SOLUTION TO FINITE ELEMENT METHOD
PROC. OF NATIONAL SYMP. ON MATRIX METHODS OF
STRUCTURAL ANALYSIS AND DESIGN, TOKYO, SOCIETY OF
STEEL CONSTRUCTION OF JAPAN, 1969, PP. 28-39 (IN
JAPANESE)

69-0385 DIMITRIOU, C.
STRESSES IN SHORT THICK COMPOUND CYLINDERS
S. AFR. MECH. ENG., V. 19, N. 5, DEC 1969, PP.
127-30.

69-0386 DEIST, F. H.
DIMITRIOU, C.
FINITE ELEMENT METHOD
S. AFR. MECH. ENG., V. 19, N. 5, DEC 1969, PP.
124-6.

69-0387 MIYAMOTO, H.
MIYOSHI, T.
ELASTIC PLASTIC ANALYSIS OF THREE DIMENSIONAL
PROBLEMS BY THE FINITE ELEMENT METHOD
PROCEEDINGS NATIONAL SYMPOSIUM ON MATRIX METHODS
OF STRUCTURAL ANALYSIS AND DESIGN, SOCIETY OF
STEEL CONSTRUCTION IN JAPAN, TOKYO, 1969, PP.
404-411 (IN JAPANESE)

69-0388 YAMADA, Y.
YOKOUCHI, Y.
SASAOKA, G.
ANALYSIS OF CONTACT PROBLEMS BY THE MATRIX METHOD
PROCEEDINGS, NATIONAL SYMPOSIUM ON MATRIX METHODS
OF STRUCTURAL ANALYSIS AND DESIGN, SOCIETY OF
STEEL CONSTRUCTION OF JAPAN, TOKYO, 1969, PP.
412-421 (IN JAPANESE)

69-0389 PERUMPRAL, J. V.
LILJEDAHL, J. B.
PERLOFF, W. H.
FINITE ELEMENT METHOD OF PREDICTING STRESS
DISTRIBUTIONS AND SOIL DEFORMATION UNDER A
TRACTIVE DEVICE
AGRIC. ENG., VOL. 50, NO. 10, 1969, PP. 612-638,
ALSO IN INT. J. NUM. METHODS ENG., VOL. 3, NO. 4,
OCTOBER 1971, PP. 575-586, AND IN TRANS. OF AM.
SOC. OF AGRICULTURAL ENGINEERS, VOL. 14, NO. 6,
1971, PP. 1184-1210.

69-0390 ANON
A FINITE ELEMENT METHOD FOR THE PLASTIC BUCKLING
ANALYSIS OF PLATES
GRUMMAN AIRCRAFT ENG., CO., AD-870 706L, OCTOBER
1969, 41 PP

69-0391 UEDA, H.
INOUE, T.
TAIRA, S.
THERMAL ELASTIC PLASTIC STRESS ANALYSIS BY THE
FINITE ELEMENT METHOD
PROCEEDINGS, NATIONAL SYMPOSIUM ON MATRIX METHODS
OF STRUCTURAL ANALYSIS AND DESIGN, SOCIETY OF
STEEL CONSTRUCTION OF JAPAN, TOKYO, 1969, PP.
396-403. (IN JAPANESE)

69-0392 DAMLE, S. K.
A FINITE ELEMENT ANALYSIS FOR DYNAMIC BEHAVIOR OF
STIFFENED PLATES
PH.D. THESIS, UNIVERSITY OF COLORADO, 1969.

69-0393 BERKE, L.
MALLETT, R. H.
AUTOMATED LARGE DEFLECTION AND STABILITY ANALYSIS
OF THREE DIMENSIONAL BAR STRUCTURES
PROC. STRUCTURES TECHNOLOGY FOR LARGE RADIO AND
RADAR TELESCOPE SYSTEMS, MIT PRESS, 1969.

69-0394 PHILLIPS, D. V.
TECHNIQUES IN TRIANGULAR MESH GENERATION FLAT AND
CURVED SURFACES
UNIV. OF WALES, SWANSEA, 1969.

69-0395 DE VRIES, G.
NORRIE, D. H.
THE APPLICATION OF THE FINITE ELEMENT TECHNIQUE TO
POTENTIAL FLOW PROBLEMS: PART II
REPORT NO. 8, MECHANICAL ENGINEERING DEPARTMENT,
UNIVERSITY OF CALGARY, CANADA, JULY 1969

69-0396 MELOSH, R. J.
BAMFORD, R. M.
EFFICIENT SOLUTION OF LOAD-DEFLECTING EQUATIONS
J. AM. SOC. CIV. ENGRS. STRUCT. DIV., PAPER NO.
6510, 1969, PP. 661-676.

69-0397 TEZCAN, S. S.
MAHAPATRA, B. C.
TANGENT STIFFNESS MATRIX FOR SPACE FRAME MEMBERS
JOURNAL OF THE STRUCTURAL DIVISION, ASCE, VOL. 95,
NO. ST6, JUNE 1969, PP. 1257-1270.

69-0398 LOGCHER, R. D.
ICES STRUDL-II ENGINEERING USERS MANUAL. VOLS. 1,
11, 111
DEPT. OF CIVIL ENG., M.I.T., CAMBRIDGE, MASS.,
1969.

69-0399 WILSON, E. A.
END DISPLACEMENTS OF SEMI-INFINITE CYLINDERS DUE
TO ANNULAR LOADINGS
INT. J. NUMER. METHOD ENG., VOL. 1, NO. 4,
OCTOBER-DECEMBER, 1969, P. 395.

69-0400 ODEN, J. T.
SOMOGYI, D.
ON THE APPLICATION OF THE FINITE ELEMENT METHOD TO
A CLASS OF PROBLEMS IN FLUID DYNAMICS
J. ENGNG. MECH. DIV., ASCE., VOL. 95, NO. EM3,
1969.

69-0401 VAN WINSEN, F.H.
ZIMMER, A.
MODERNE VERFAHREN ZUR UNTERSUCHUNG DER
STEIFIGKEITS-UND FESTIGKEITSVERHALTNISSE VON
RAHMENKORPERN, INSBESONDERE VON KAROSSERIEN
AUTOMOBIL-INDUSTRIE, VOL. 69, NO. 1, 1969, PP. 45.

69-0402 SZABO, B. A.
LEE, G. C.
DERIVATION OF THE STIFFNESS MATRIX FOR PLATES BY
GALERKIN METHOD
INTL. J. NUMER. METH. ENG., VOL. 1, 1969.

69-0403 YOSHIMURA, N.
MORITA, J.
STRESS ANALYSIS OF THIN PLATE STRUCTURE BY THE
FINITE ELEMENT METHOD
PROC. OF NAT. SYMP. ON MATRIX METHODS OF
STRUCTURAL ANALYSIS AND DESIGN, SOC. OF STEEL
CONSTRUCTION OF JAPAN, TOKYO, 1969, PP. 188-294
(IN JAPANESE)

69-0404 NAKAO, K.
YAMASHITA, M.
KAWASHIMA, M.
A STUDY ON THE STRESS ANALYSIS OF BEAM PLATE
STRUCTURE
PROC. OF NAT. SYMP. ON MATRIX METHODS OF
STRUCTURAL ANALYSIS AND DESIGN, SOC. OF STEEL
CONSTRUCTION OF JAPAN, TOKYO, 1969, PP. 224-231
(IN JAPANESE)

69-0405 HENSHELL, R. D.
WARBURTON, G. B.
TRANSMISSION OF VIBRATION IN BEAM SYSTEMS
INT. J. NUMER. METHODS ENG., VOL. 1, NO. 1,
JANUARY-MARCH 1969, PP 47-66

69-0406 KARIAPPA, V.
SOMASHEKAR, B. R.
APPLICATION OF MATRIX DISPLACEMENT METHODS IN THE
STUDY OF PANEL FLUTTER
A.I.A.A.J., VOL. 7, 1969, PP. 50-53.

69-0407 FULLARD, K.
EXPERIENCE IN THE USE OF THE CONSTANT STRESS
TRIANGLE FOR FINITE ELEMENT STRESS ANALYSIS
CENTRAL ELECTRICITY GENERATING BOARD, REPORT
RD/B/N1435, 1969.

69-0408 WHETSTONE, W. D.
COMPUTER ANALYSIS OF LARGE LINEAR FRAMES
J. STRUCT. DIV. AM. SOC. CIV. ENGRS. VOL. 95,
1969, PP. 2401-2417.

69-0409 O'CONNELL, J. M.
STRESS CONCENTRATION FACTORS AROUND A PENETRATION
IN A STRIP OR SLAB OF FINITE DIMENSIONS
NUCL. ENGNG. DESIGN, VOL. 10, 1969, PP. 356-360.

69-0410 RAMSTAD, H.
FINITE ELEMENT METHODS IN STRESS ANALYSIS
TAPIR, TRONDHEIM, NORWAY, 1969, PP. 179-211.

69-0411 GOEL, J. J.
DUPUIS, G.
ELEMENTS FINIS RAFFINES EN ELASTICITE BIDIMENSIONNELLE
ZAMP, VOL. 20, 1969, PP. 858-881.

69-0412 HELLEN, T. K.
A FRONT SOLUTION FOR FINITE ELEMENT TECHNIQUES
CENTRAL ELECTRICITY GENERATING BOARD, REPORT RD/B/N1459, 1969.

69-0413 ZIENKIEWICZ, O. C.
IRONS, B. M.
SCOTT, F. C.
ERGATOUDIS, J.
AHMAD, S.
FINITE ELEMENT METHOD IN STRESS ANALYSIS
(EDS. I. HOLAND AND K. BELL), TAPIR, TRONDHEIM, 1969 CHAP 13

69-0414 LANG, T. E.
POST-BUCKLING RESPONSE OF STRUCTURES USING THE FINITE ELEMENT METHOD
PH.D. THESIS, UNIV. OF WASHINGTON, 1969.

69-0415 ZIMMER, A
BAUSINGER, R.
ELASTOSTATIK-ELEMENTMETHODE (ESEM)
IBM-NACHRICHTEN, VOL. 19, NO. 195, 1969, PP. 701-709., NO. 196, PP. 779-785, AND NO. 197, PP. 860-866.

69-0416 ANDERHEGGEN, E.
PROGRAMME ZUR METHODE DER FINITEN ELEMETE
INSTITUT FUR BAUSTATIK, EIDGENOSSISCHE TECHNISCHE HOCHSCHULE, ZURICH, 1969.

69-0417 LEONARD, J. W.
COMPRESSIBLE FLOW BY THE FINITE ELEMENT METHOD
TECH. NOTE TCTN-9500-920156, BELL AEROSYSTEMS COMPANY, NIAGARA FALLS, NEW YORK, 1969.

69-0418 GALLAGHER, R. H.
A SURVEY OF FINITE ELEMENT FRAMEWORK STABILITY ANALYSIS
AM. SOC. CIV. ENGRS, NATIONAL MEETING, NEW ORLEANS, 1969.

69-0419 GUPALP, Y. U. P.
RYAZANTSEV, Y. U. S.
DIFFUSION ON SOLID SPHERICAL PARTICLE IN VISCOUS FLUID FLOW AT FINITE VALUES OF RE NUMBERS
IZV. AKAD. NAUK. MEKH. ZHIDK. GASA., NO. 6, NOVEMBER-DECEMBER 1969, PP. 127-130.

69-0420 FLEMMING, M.
PRACTICAL APPLICATION OF FINITE ELEMENT METHODS IN THE DESIGN OF STRUCTURES
AGARD CONF. PROC. NO. 36, SYMP. ON STRUCT. OPTIMIZATION BY STRUCT. AND MATER. PANEL OCTOBER 8-10, 1969, ISTANBUL, TURKEY, PAPER 8, 28 PP.

69-0421 ALI, R.
HEDGES, J. L.
COMPUTER AIDED DESIGN APPLIED TO A MODEL OF A CHASSIS TYPE STRUCTURE USING FINITE ELEMENT TECHNIQUES
INST. MECH. ENG., PROC. (PART 2A), AUTO. DIV. PROC., VOL. 184, NO. 1, 1969-70, PP. 15-24.

69-0422 STROEVE, A.
ACOUSTIC RESPONSE OF A SPACECRAFT SOLAR PANEL USING NORMAL MODE METHOD
U.S. NAVAL RES. LAB., SHOCK VIB. BULL. 40, PT. 3, DECEMBER 1969, PP. 49-55.

69-0423 DIETERICH, J. H.
ONAT, E. T.
SLOW FINITE DEFORMATIONS OF VISCOUS SOLIDS
J. GEOPHYS. RES., VOL. 74, NO. 8, APRIL 15, 1969, PP. 2081-2088.

69-0424 ARAV, F.
FRANCKEN, A. J.
THEORETICAL AND EXPERIMENTAL INVESTIGATION ON A STEAM GENERATOR TUBE PLATE
ASME, FIRST INT. CONF. ON PRESSURE VESSEL TECHNOL, DELFT, NETHERLANDS, SEPTEMBER 29-OCTOBER 2, 1969, PT. 2, PP. 1123-1134.

69-0425 HELLAN, K.
ANALYSIS OF RECTANGULAR PLATES IN TRANSIENT CREEP BENDING
ACTA POLTECH SCAND., MECH. ENG. SER. ME. 46, 1969, 20 PP.

69-0426 KRATOCHVIL, J.
HANSLIAN, J.
SOLUTION OF STRESSES AND TRANSFORMATIONS OF EARTH DAMS BY THE METHOD OF FINITE ELEMENTS
INZENYRSKE STAVBY., VOL. P7, NO. 1, JANUARY 1969, PP. 7-10.

69-0427 MEIJERS, P.
ELASTIC PLASTIC DEFORMATION OF THICK WALLED CYLINDERS
ASME, FIRST INT. CONF. ON PRESSURE VESSEL TECHNOL, DELFT NETHERLANDS, SEPTEMBER 29-OCTOBER 2, 1969, PT. 1, PP. 19-34.

69-0428 WAHLA, M. I.
SCOTT, N. R.
DIRECT MEASUREMENT OF BOND SLIP IN REINFORCED CONCRETE
AMER. SOC. AGR. ENG. WINTER MEETING CHICAGO, ILL., DECEMBER 9-12, 1969, PAPER 69-921, 24 PP.

69-0429 DAVIS, R. L.
KEITH, H. D.
USER'S MANUAL FOR UMR FINITE ELEMENT COMPUTER PROGRAM - STRATA
UNIVERSITY OF MISSOURI-ROLLA, U.S.A., 1969.

69-0430 CONNOR, J.
WILL, G.
COMPUTER AIDED TEACHING OF THE FINITE ELEMENT DISPLACEMENT METHOD
M.I.T., CAMBRIDGE, MASS., U.S.A., 1969.

69-0431 CHAUMP, L. E.
MERCHANT, J. K.
MODAL SENSITIVITY STUDY OF A CONICAL RE-ENTRY VEHICLE SUBJECTED TO AERODYNAMICALLY INDUCED ACOUSTIC LOADINGS
U.S. NAVAL RES. LAB., SHOCK VIB. BULL., 40, PART 3, DECEMBER 1969, PP. 31-48.

69-0432 FRAZIER, G. A.
VIBRATIONAL CHARACTERISTICS OF THREE-DIMENSIONAL SOLIDS, WITH APPLICATIONS TO EARTH DAMS
MONTANA STATE UNIV., 1969, 200 PP., UNIV. MICROFILMS 70-10, 023.

69-0433 GHOSH, S.
DYNAMIC STRESS ANALYSIS OF AXISYMMETRIC STRUCTURES UNDER ARBITRARY LOADING
UNIV. CALIF., 1969, UNIV. MICROFILMS 70-13, 054, 1969.

69-0434 WHETSTONE, W. D.
JONES, C. E.
VIBRATIONAL CHARACTERISTICS OF SPACE FRAMES
J. STRUCT. DIV., PROC. ASCE., VOL. 95, NO. ST10, OCTOBER 1969, PP. 2077-2091.

69-0435 GOANKAR, G. H.
AVERAGE ELASTIC PROPERTIES AND STRESS FIELDS FOR COMPOSITE MATERIALS
AIAA.ASME 10TH STRUCTURES, STRUCTURAL DYNAMICS AND MATERIALS CONF., TECH. PAPERS FOR MEETING, NEW ORLEANS, LA., APRIL 14 1969, PP. 172-178.

69-0436 KUHLEMEYER, R. L.
VERTICAL VIBRATIONS OF FOOTINGS EMBEDDED IN LAYERED MEDIA
UNIV. CALIF., 1969, 253 PP.

69-0437 MAINS, R. M.
COMPARISONS OF CONSISTENT MASS MATRIX SCHEMES
SHOCK AND VIBRATION BULL., VOL. 39, NO. 3, 1969, PP. 129-142.

69-0438 NATH, B.
HYDRODYNAMIC PRESSURES ON HIGH DAMS DUE TO VERTICAL EARTHQUAKE MOTIONS
INST. CIV. ENGRS., PROC. VOL. 42, MARCH 1969, PP. 413-421.

69-0439 WARBURTON, G. B.
AL-NAJAFI, A. M.
FREE VIBRATION OF THIN CYLINDRICAL SHELLS WITH A DISCONTINUITY IN THE THICKNESS
J. SOUND AND VIB., VOL. 9, NO. 3, MAY 1969, PP. 373-382.

69-0440 CROSE, J. G.
FINITE ELEMENT STRESS ANALYSIS OF POROUS MEDIA
SAN BERNARDINO OPERATIONS, AEROSPACE CORP., SAN BERNARDINO, CALIF., REPORT NO: TR-0200(S4816-76)-1, MAY 23, 1969, 78 P, (N.T.I.S. - AD-688 250)

69-0441 BUSHNELL, D.
STRESS, STABILITY AND VIBRATION OF COMPLEX SHELLS OF REVOLUTION: ANALYSIS AND USERS MANUAL FOR BOSOR 3.
LOCKHEED MISSILES AND SPACE COMPANY, SUNNYVALE, CALIF., 1969

69-0442 ARGYRIS, J. H.
SCHARPF, D. W.
FINITE ELEMENTS IN TIME AND SPACE
NUCLEAR ENG. AND DESIGN, VOL. 10, NO. 4, 1969, PP: 456-464.

69-0443 ARGYRIS, J. H.
SCHARPF, D. W.
SPOONER, J. B.
DIE ELASTO-PLASTISCHE BERECHNUNG VON ALLGEMEINEN TRAGWERKEN UND KONTINUA
INGENIEUR ARCHIV. BD. 37, 5. HEFT, 1969, PP: 326-352

69-0444 ARGYRIS, J. H.
SCHARPF, D. W.
A SEQUEL TO TECHNICAL NOTE 15: THE SHEBA FAMILY OF
SHELL ELEMENTS FOR THE MATRIX DISPLACEMENT METHOD,
PART III: LARGE DISPLACEMENTS
J. ROY. AERO. SOC., VOL. 73, NO. 701, MAY 1969,
PP. 423-426.

69-0445 ARGYRIS, J. H.
SCHARPF, D. W.
SOME GENERAL CONSIDERATIONS ON THE NATURAL MODE
TECHNIQUE. PART II: LARGE DISPLACEMENTS
J. ROY. AERO. SOC., VOL. 73, NO. 700, APRIL 1969,
PP. 361-368.

69-0446 ARGYRIS, J. H.
SCHARPF, D. W.
THE FUGA FAMILY OF ELEMENTS FOR FOLDED PLATE
STRUCTURES A FURTHER SEQUEL TO TECHNICAL NOTE 14
J. ROY. AERO. SOC., VOL. 73, NO. 703, JULY 1969,
PP. 594-602.

69-0447 ARGYRIS, J. H.
THE IMPACT OF THE DIGITAL COMPUTER ON STRUCTURAL
ANALYSIS
NJCL. ENG. DES., VOL. 10, NO. 4, AUGUST 1969.

69-0448 ANDREWS, W. R.
MOWBRAY, D. F.
ON THE USE OF FINITE ELEMENTS AND COMPLIANCE
CALCULATIONS FOR DETERMINING CRACK TIP STRESS
INTENSITY FACTORS
KNOLLS ATOMIC POWER LAB., SCHENECTADY, N.Y. REPORT
NO: KAPL-P-3815, MARCH 13, 1969, 13 P. (N.T.I.S. -
PB-184 350)

69-0449 ARGYRIS, J. H.
SCHARPF, D. W.
TWO AND THREE DIMENSIONAL POTENTIAL FLOW BY THE
METHOD OF SINGULARITIES
J. ROY. AERO. SOC., VOL. 73, NO. 707, NOVEMBER
1969, PP. 959-961.

69-0450 ARGYRIS, J. H.
SCHARPF, D. W.
FINITE ELEMENTS IN TIME AND SPACE
J. ROY. AERO. SOC., VOL. 73, NO. 708, DECEMBER
1969, PP. 1041-1044.

69-0451 ARGYRIS, J. H.
BUCK, K. E.
EINIGE ENTWICKLUNGEN IN DER
MATRIZEN-VERSCHIEBUNGSMETHODE
PLENUMVORTRAG DGLR-JAHRESTAGUNG 1969, BREMEN,
SEPTEMBER JAHRBUCH 1969, DER DGLR PP: 140-153.

69-0452 ARGYRIS, J. H.
BUCK, K. E.
DIE BERECHNUNG ACHSENSYMETRISCHER TRAGWERKE UNTER
ACHSENSYMETRISCHER UND NIGHT-SYMMETRISHCER
BELASTUNG UND TEMPERATURVETEILUNG
VORTRAG NR. 62, DGLR-JAHRESTAGUNG 1969, BREMEN,
SEPTEMBER JAHRBUCH 1969 DER DGLR PP: 154-162.

69-0453 ARGYRIS, J. H.
GRIEGER, I.
THREE DIMENSIONAL ELASTIC AND ELASTO-PLASTIC
ANALYSIS OF REACTOR PRESSURE VESSELS
SWEITE INFORMATIONSTAGUNG UBER
REAKTORDRUCKBEHALTER AUS SPANNEBETON UND IHRE
WARMEISOLIERUNG, KOMMISSON DEREUROPAISCHEN
BRUSSELS, NOVEMBER 1969.

69-0454 SCHILLING, C. H.
SORENSEN, M.
APPLICATION OF MATRIX METHODS IN STRUCTURAL THEORY
INSTITUT FUR STATIK UND DYNAMIK DER LUFT UND
RAUMFAHRTKONSTRUKTIONEN, RESEARCH REPORT NO. 66,
JUNE 1969.

69-0455 SORENSEN, M.
A TECHNICAL MANUAL FOR THE BEAM ELEMENTS
INSTITUT FUR STATIK UND DYNAMIK DER LUFT UND
RAUMFAHRTKONSTRUKTIONEN, RESEARCH REPORT NO: 67,
JUNE 1969.

69-0456 MLEJNEK, H. P.
MAI, M.
BUHLMEIER, J.
F 104 G FINAL MOTIVE: GROUND VIBRATION
CALCULATIONS FOR THE STARFIGHTER F 104 G
INSTITUT FUR STATIK UND DYNAMIK DER LUFT UND
RAUMFAHRTKONSTRUKTIONEN, RESEARCH REPORT NO. 68,
JUNE 1969.

69-0457 GRIEGER, I.
ASKA - EINE PROGRAMMIEPSPRACHE ZUR STATISCHEN UND
DYNAMISCHEN ANALYSE
INSTITUT FUR STATIK UND DYNAMIK DER LUFT UND
RAUMFAHRTKONSTRUKTIONEN, SEPTEMBER 1969.

69-0458 BALMER, H.
GENERAL BEAM ELEMENT FOR ASKA
INSTITUT FUR STATIK UND DYNAMIK DER LUFT UND
RAUMFAHRTKONSTRUKTIONEN, RESEARCH REPORT NO: 75,
DECEMBER 1969.

69-0459 MARCAL, P. V.
FINITE ELEMENT ANALYSIS WITH MATERIAL
NONLINEARTIES: THEORY AND PRACTICE
SEMINAR ON MATRIX METH. OF STPUCT. ANALYSIS AND
DESIGN AND JSSC SYMP. ON MATRIX METH. OF STRUCT.
ANALYSIS, TOKYO, JAPAN, REPORT NO: TR-3, AUGUST
1969, 39 P. (N.T.I.S. - AD-693 992) SEE ALSO TECH.
REPT. NO. 2, AD-691 782.

69-0460 SORENSEN, M.
ASKA 105, A PROBLEM ORIENTED LANGUAGE FOR
STRUCTURAL ENGINEERS. VOLUME 1, DESIGN
INSTITUT FUR STATIK UND DYNAMIK DER LUFT UND
RAUMFAHRTKONSTRUKTIONEN, RESEARCH REPORT NO. 81,
JANUARY 1969.

69-0461 SORENSEN, M.
ASKA 105 - A PROBLEM ORIENTED LANGUAGE FOR
STRUCTURAL ENGINEERS, VOLUME 11, APPLICATION
INSTITUT FUR STATIK UND DYNAMIK DER LUFT UND
RAUMFAHRTKONSTRUKTIONEN, RESEARCH REPORT NO. 82,
JANUARY 1969.

69-0462 UTKU, S.
ELAS: A GENERAL-PURPOSE COMPUTER PROGRAM FOR THE
EQUILIBRIUM PROBLEMS OF LINEAR STRUCTURES. VOLUME
2: DOCUMENTATION OF THE PROGRAM
CALIF. INST. OF TECH., JET PROPULSION LAB.,
PASADENA, REPORT NO: NASA-CR-137236,
JPL-TR-1240-VOL-2-ADD, OCTOBER 15, 1969, 175 P.
(N.T.I.S. - N74-19549/6)

69-0463 PALMERTON, J. B.
PRELIMINARY FINITE ELEMENT ANALYSIS, ATCHAFALAYA
BASIN PROTECTION LEVEES, TEST SECTION 3
ARMY ENGR. WATERWAYS EXPERIMENT STATION,
VICKSBURG, MISS., REPORT NO:
AEWES-MISC-PAPER-S-69-53, DECEMBER 1969, 37 P.
(N.T.I.S. - AD-735 782)

69-0464 CALLAWAY, R. J.
BYRAN, K. V.
DITSWORTH, G. R.
MATHEMATICAL MODEL OF THE COLUMBIA RIVER FROM THE
PACIFIC OCEAN TO BONNEVILLE DAM. PART 1. THEORY,
PROGRAM NOTES AND PROGRAMS
PACIFIC NORTHWEST WATER LAB., CORVALLIS, OREG.,
NOVEMBER 1969, 157 P. (N.T.I.S. - PB-202 422) SEF
ALSO PART 2, PB-202 464.

69-0465 MOLHOEK, F. A.
VAN BATEN, T. J.
DETERMINATION OF THE INITIAL BUCKLING STRESS OF
FLAT PLATES USING THF DISPLACEMENT METHOD
VLIEGTUIGBOUWKUNDE, TECHNISCHE HOGESCHOOL, DELFT,
NETHERLANDS, REPORT NO: VTH-161, DECEMBER 1969, 54
P. (N.T.I.S. - N71-30041)

69-0466 JOHNSON, R. L. JR.
BEHAVIOR OF A BI-MATERIAL CYLINDER INCLUDING THE
EFFECTS OF FRACTURE
DEPT. OF CIVIL ENGNG. UNIV. OF NEW MEXICO,
ALBUQUERQUE, REPORT NO: CE-16(69)S-093-2. REPORT
NO: SC-CR-69-3246, AUGUST 1969, 65 PP. (N.T.I.S. -
PB-186 062)

69-0467 WHANG, B.
ELASTO-PLASTIC ORTHOTROPIC PLATES AND SHELLS
PROC. A.S.C.E. SYMP. ON APPLICATION OF FINITE
ELEMENT METHODS IN CIVIL ENG., SCHOOL OF ENG.,
VANDERBILT UNIV., 1969, PP. 481-516.

69-0468 NATH, J. M.
PETYT, M.
FREE VIBRATION OF DOUBLY CURVED RECTANGULAR PLATES
INCLUDING THE PRESENCE OF MEMBRANE STRESSES WITH
SPECIAL REFERENCE TO THE FINITE ELEMENT TECHNIQUE
INST. OF SOUND AND VIB. RESEARCH, SOUTHAMPTON
UNIV., ENGLAND, REPORT NO: ISVR-TR-22, JULY 1969,
114 PP. (N.T.I.S. - N70-17460)

69-0469 CUTHILL, E.
MCKEE, J.
REDUCING THE BANDWIDTH OF SPARSE SYMMETRIC
MATRICES
NAVAL SHIP RESEARCH AND DEV. CENTER, WASH., D.C.,
AML-40-69, JUNE 1969.

69-0470 AUTHORS
CONFERENCE ON MATRIX METHODS IN STRUCTURAL
MECHANICS
PROC. 2ND. CONF. ON MATRIX METH. IN STRUCT. MECH.,
WRIGHT-PATTERSON AFB, OHIO, OCTOBER 15-17, 1968,
REPORT NO: AFFDL-TR-68-150. DECEMBER 1969, 1466
PP. (N.T.I.S. - AD-703 685) SEE ALSO PROC. OF 1ST.
CONF., AD-646 300

69-0471 BOEHMER, J. W.
CHRISTIAN, J. T.
PLANE STRAIN CONSOLIDATION BY FINITE ELEMENTS
DEPT. OF CIVIL ENGNG., MASS. INST. OF TECH.,
CAMBRIDGE, REPORT NO: R69-60, SOIL MECH. PUB/243,
AUGUST 1969, 183 PP. (N.T.I.S. - PB-188 023)

69-0472 FERRARA, S.
GRILLO, A. F.
SOME REMARKS ON INFINITE VENEZIANO REPRESENTATIONS
COMITATO NAZIONALE PER L ENERGIA NUCLEARE,
FRASCATI, ITALY, REPORT NO: LNF-69/79, DECEMBER
23, 1969, 6 P. (N.T.I.S. - N70-33452)

69-0473 REJALI, H. M.
USER'S GUIDE FOR THE COMPUTER PROGRAM DESCRIBED IN
TAM REPORT NO. 680 "DETERMINATION OF VOLUME CHANGE
STRESSES IN PLAIN AND REINFORCED CONCRETE USING
FINITE ELEMENT ANALYSIS"
DEPT. OF THEORETICAL AND APPLIED MECH., UNIV. OF
ILLINOIS, URBANA, AUGUST 1969, 51 PP. (N.T.I.S. /
PB-288 188)

69-0474 VISSER, C.
GABRIELSE, S. E.
VANBUREN, W.
HEAVY SECTION STEEL TECHNOLOGY PROGRAM. TECHNICAL
REPORT NO. 4, OCTOBER 1969. A TWO-DIMENSIONAL
ELASTIC-PLASTIC ANALYSIS OF FRACTURE TEST
SPECIMENS
OAK RIDGE NAT. LAB., TENN., AUGUST 1969, 178 P.
(N.T.I.S. - WCAP-7368)

69-0475 LEE, S. Y.
RADIAL AND AXIAL TRANSMISSIBILITY CHARACTERISTICS
OF TYPICAL ROCKET VEHICLE STRUCTURE TECHNICAL
SUMMARY REPORT
SPACE DIV., NORTH AMER. ROCKWELL CORP., DOWNEY,
CALIF., REPORT NO: NASA-CR-102725, SD-69-766,
DECEMBER 1969, 431 P. (N.T.I.S. - N70-31061)

69-0476 BOLEY, B. A.
FRIEDMAN, E.
STRESSES AND DEFORMATIONS IN MELTING PLATES
GENERAL ELECTRIC CO., PHILADELPHIA, PA. AND
COLUMBIA UNIV., NEW YORK, CONTRACT NONR-4259(07),
REVISION OF REPORT JULY 28, 1969. OCTOBER 31,
1969, 10 P. (N.T.I.S. - AD-710 116) SEE ALSO J. OF
SPACECRAFT AND ROCKETS, VOL. 7, NO. 3, MARCH 1970,
PP. 324-333.

69-0477 REJALI, H. M.
LOTT, L. J.
KESLER, C. E.
DETERMINATION OF VOLUME CHANGE STRESSES IN PLAIN
AND REINFORCED CONCRETE USING FINITE ELEMENT
ANALYSIS
DEPT. OF THEORETICAL AND APPLIED MECH., UNIV. OF
ILL., URBANA, REPORT NO: T/AM-680, AUGUST 1969, 51
P. (N.T.I.S. - PB-188 281)

69-0478 MURTY, A. V.
RAO, G. V.
STABILITY ANALYSIS OF COLUMNS USING FINITE
ELEMENTS
DEPT. OF AERON. ENGNG., INDIAN INST. OF SCI.,
BANGALORE, REPPORT NO: AE-238-S, MARCH 1969, 13 P.
(N.T.I.S. - N70-29708)

69-0479 MURTY, A. V.
RAO, G. V.
AN INVERSE FORM OF RAMBERG - OSGOOD FORMULA
DEPT. OF AERON. ENGNG., INDIAN INST. OF SCI.,
BANGALORE, REPORT NO: AE-239-S, MARCH 1969, 9 P.
(N.T.I.S. N70-29706)

69-0480 CULVER, C.
NASIR, G.
INSTABILITY OF HORIZONTALLY CURVED MEMBERS.
FLANGE BUCKLING STUDIES
DEPT. OF CIVIL ENGNG., CARNEGIE-MELLON UNIV.,
PITTSBURGH, PA., NOVEMBER 1969, 107 P. (N.T.I.S. -
PB-192 901)

69-0481 KAUFMAN, S.
AXISYMMETRIC SHELLS
BELLCOMM. INC., WASH., D.C., REPORT NO:
NASA-CR-106583, TM-69-2031-3, JULY 24, 1969, 36 P.
(N.T.I.S. - N70-10515)

69-0482 HO, P. K.
ANALYSIS OF PLATE STRUCTURES BY A DUAL FINITE
ELEMENT METHOD
DEPT. OF CIVIL ENGNG., MASS. INST. OF TECH.,
CAMBRIDGE, REPORT NO: NASA-CR-106969, R-69-58,
SEPTEMBER 1969, 197 P. (N.T.I.S. - N70-11950)

69-0483 KENDALL, D. P.
SWEDLOW, J. L.
UNDERWOOD, J. H.
EXPERIMENTAL AND ANALYTICAL STRAINS IN A
EDGE-CRACKED SHEET
BENET R AND E LABS., WATERVLIET ARSENAL, N.Y.,
REPORT NO: NASA-CR-106904, WVT-6933, SEPTEMBER
1969, 38 PP. (N.T.I.S. - N70-11952)

69-0484 DEZFULIAN, H.
SEED, B. H.
SEISMIC RESPONSE OF SOIL DEPOSITS UNDERLAIN BY
SLOPING ROCK BOUNDARIES
EARTHQUAKE ENGNG. RESEARCH CENTER, UNIV. OF
CALIF., BERKELEY, REPORT NO: EERC-69-9, AUGUST
1969, 44 PP. (N.T.I.S. /PB-189 114)

69-0485 STIEDA, C. K. A.
A COMPUTER PROGRAM FOR THE CALCULATION OF TORSION
FACTORS AND SHEAR STRESS FACTORS FOR BOX SECTIONS
WITH THICK WALLS
FOREST PRODUCTS LAB., VANCOUVER, B.C., REPORT NO:
VP-X-55, NOVEMBER 1969, 58 PP. (N.T.I.S. - PB-189
379)

69-0486 DUPUIS, G.
GOEL, J. J.
A CURVED FINITE ELEMENT FOR THIN ELASTIC SHELLS
DIV. OF ENGNG., BROWN UNIV., PROVIDENCE, R.I.,
REPORT NO: TR-4, DECEMBER 1969, 32 PP. (N.T.I.S. -
AD-701 325)

69-0487 HELBING, T. C.
A STUDY OF THE PLANE STRESS OR STRAIN FINITE
ELEMENT ANALYSIS FOR SOLUTION OF STRESS
DISTRIBUTION IN PLANE ELASTIC CONTINUA
COLL. OF ARCH. AND DESIGN, LANGLEY RESEARCH
CENTER, NAT. AERON. AND SPACE ADMIN., LANGLEY
STATION, VA., REPORT NO: NASA-TM-X-62848, 1969,
161 P. (N.T.I.S. - N70-25364)

69-0488 GRESTE, O.
A COMPUTER PROGRAM FOR THE ANALYSIS OF TUBULAR K
JOINTS
STRUCT. ENG. LAB., UNIV. OF CALIF., BERKELEY,
REPORT NO: SES-69-19, NOVEMBER 1969, 88 PP.
(N.T.I.S. - PB189 501)

69-0489 RAVERA, R. J.
RESPONSE OF SIMPLY SUPPORTED TIMOSHENKO BEAMS
BELLCOMM. INC., WASH., D.C., REPORT NO:
NASA-CR-107588, TM-69-1022-11, NOVEMBER 14, 1969,
28 PP. (N.T.I.S. - N70-15643)

69-0490 HARTZMAN, M.
NONLINEAR DYNAMICS OF SOLIDS BY FINITE ELEMENT
METHODS
LAWRENCE RADIATION LAB., UNIV. OF CALIF.,
LIVERMORE, DECEMBER 19, 1969, 101 P. (N.T.I.S. -
UCRL-50779)

69-0491 AKYUZ, F. A.
FEDGE - A GENERAL-PURPOSE COMPUTER PROGRAM FOR
FINITE ELEMENT DATA GENERATION. VOLUME 1 - USER'S
MANUAL
INST. OF TECH., JET PROPULSION LAB., PASADENA,
CALIF., REPORT NO: NASA-CR-109530,
JPL-TM-33-431-VOL-1, SEPTEMBER 15, 1969, 41 P.
(N.T.I.S. - N70-24901)

69-0492 AKYUZ, F. A.
FEDGE - A GENERAL PURPOSE COMPUTER PROGRAM FOR
FINITE ELEMENT DATA GENERATION. VOLUME 2 - PROGRAM
MANUAL
INST. OF TECH., JET PROPULSION LAB., PASADENA,
CALIF., REPORT NO: NASA-CR-109467,
JPL-TM-33-431-VOL-2, SEPTEMBER 15, 1969, 81 PP.
(N.T.I.S. - N70-24900)

69-0493 LONKAR, S.
BENDING AND TORSION OF THIN-WALLED STRAIGHT AND
CURVED BEAMS WITH VARIABLE OPEN CROSS-SECTION
EIDGENOSSISCHE TECHNISCHE HOCHSCHULE, ZURICH,
SWITZERLAND, 1969, 82 P. (N.T.I.S. - N70-24699)

69-0494 TANKERSLEY, D. F.
DAWKINS, W. P.
A DISCRETE-ELEMENT METHOD OF ANALYSIS FOR COMBINED
BENDING AND SHEAR DEFORMATION OF A BEAM
CENTER FOR HIGHWAY RESEARCH, TEXAS UNIV., AUSTIN,
REPORT NO: RR-56-12, DECEMBER 1969, 239 PP.
(N.T.I.S. - PB-190 840)

69-0495 CONNACHER, N.
MCELROY, M.
HANSEN, S. D.
SAMECS STRUCTURAL ANALYSIS SYSTEM - USER'S
DOCUMENT
COMMERCIAL AIRPLANE DIV., BOEING CO., RENTON,
WASH., REPORT NO: D6-23727-1TN, FEBRUARY 21, 1969,
241 PP. (N.T.I.S. - AD-702 472)

69-0496 BUSHNELL, D.
ANALYSIS OF BUCKLING AND VIBRATION OF
RING-STIFFENED, SEGMENTED SHELLS OF REVOLUTION
INT. J. SOLIDS STRUCT., VOL. 6, 1970, PP. 157-181,
MAY 8, 1968, 25 PP. (N.T.I.S. - AD-702 652)
REVISION NOV. 8, 1968.

69-0497 SHARIFI, P.
REFINED FINITE ELEMENT ANALYSIS OF ELASTIC-PLASTIC
THIN SHELLS OF REVOLUTION
STRUCT. ENGNG. LAB., UNIV. OF CALIF., BERKELEY,
REPORT NO: SESM-69-28, DECEMBER 1969, 122 P.
(N.T.I.S. - AD703 908)

69-0498 AHMED, S.
DALY, P.
FINITE ELEMENT METHODS FOR INHOMOGENEOUS
WAVEGUIDES
PROC. I.E.E., VOL. 116, 1969, PP. 1661-1664.

69-0499 ARGYRIS, J. H.
SCHARPF, D. W.
THE SHEBA FAMILY OF SHELL ELEMENTS FOR THE MATRIX
DISPLACEMENT METHOD. 111, LARGE DISPLACEMENTS
J. ROY. AERON. SOC., VOL. 73, 1969, PP. 423-426.

69-0500 BRANTLY, E.
ANALYSIS OF FRAMES WITH SHEAR WALLS WITH FINITE
ELEMENTS
PROC. SYMP. APPL. OF FINITE ELEMENT METH. IN CIVIL
ENG., VANDERBILT UNIV., 1969, PP. 567-608.

69-0501 NEUMAN, S. P.
WITHERSPOON, P. A.
TRANSIENT FLOW OF GROUNDWATER TO WELLS IN
MULTIPLE-AQUIFER SYSTEMS
DEPT. CIVIL ENG., UNIV. OF CALIF., BERKELEY, 1969.
(PUBLICATION 69-1)

69-0502 CUTHILL, E. H.
MCKEE, J. M.
REDUCING THE BANDWIDTH OF SPARSE SYMMETRIC
MATRICES
PROC. 24TH NATL. CONF. ASSOC. COMPUT. MACH., NEW
YORK, 1969. (ACM PUBLICATION P-69)

69-0503 MAPLE, J. A.
A STUDY OF FINITE ELEMENTS FOR THE ANALYSIS OF
SANDWICH PLATES
PH.D. THESIS, PURDUE UNIV., 1969.

69-0504 KIDBIRIGE, S. L.
A FINITE ELEMENT SOLUTION OF TIME DEPENDENT FIELD
PROBLEMS USING A REDUCED VARIATIONAL FORMULATION
B.SC. THESIS, UNIV. OF WALES, SWANSEA, 1969.

69-0505 MORGAN, B. J.
THE APPLICATION OF THE FINITE ELEMENT METHOD TO A
PROBLEM IN STRESS WAVE PROPAGATION
PH.D. THESIS, UNIV. OF NOTRE DAME, 1969.

69-0506 KRATOCHVIL, J.
ZENISEK, A.
CALCULATION OF RECTANGULAR PLATES BY THE FINITE
ELEMENT METHOD
STAVEBNICKY CASOPIS, VOL. 17, 1969, PP. 641-653.

69-0507 ZIENKIEWICZ, O. C.
NEWTON, R. E.
COUPLED VIBRATIONS OF A STRUCTURE SUBMERGED IN A
COMPRESSIBLE FLUID
INT. SYMP. ON FINITE ELEMENT TECHNIQUES IN
SHIPBUILDING, STUTTGART, 1969.

69-0508 MARCAL, P. V.
FINITE ELEMENT ANALYSIS OF COMBINED PROBLEMS OF
MATERIAL AND GEOMETRIC BEHAVIOUR
BROWN UNIV., TECH. REPORT, 1 ONY, 1969.

69-0509 MORI, K.
CODE FOR STRESS ANALYSIS BY THE THREE DIMENSIONAL
FINITE ELEMENT METHOD (JAPANESE)
FAPIG (TOKYO), VOL. 55, 1969, PP. 186-192.

69-0510 NIELSON, H. B.
A FINITE ELEMENT METHOD FOR CALCULATING OPEN
CHANNEL FLOW
TECH. UNIV. OF DENMARK, REPORT 19, 1969.

69-0511 SANDER, G.
APPLICATIONS DE LA METHODE DES ELEMENTS FINIS A LA
FLEXION DES PLAQUES
THESIS, UNIV. OF LIEGE, FACULTY OF APPL. SCI., NO.
15, 1969.

69-0512 SILVESTER, P.
A GENERAL HIGH-ORDER FINITE-ELEMENT WAVEGUIDE
ANALYSIS PROGRAM
I.E.E.E., TRANS. MTT-A, 1969, PP. 204-210.

69-0513 YANG, H. T-Y.
A FINITE ELEMENT FORMULATION FOR STABILITY
ANALYSIS OF DOUBLY CURVED THIN SHELL STRUCTURES
PH.D., CORNELL UNIV., 1969.

69-0514 TARPY, T. S. JR.
A TECHNIQUE FOR THE FINITE ELEMENT ANALYSIS OF
LARGE COMPLEX STRUCTURAL SYSTEMS
PH.D., VANDERBILT UNIV., 1969.

69-0515 DAMLE, S. K.
FEESER, L. J.
RESPONSE OF RECTANGULAR PLATES TO MOVING LOADS BY
A FINITE ELEMENT PROCEDURE
SHOCK AND VIB. BULL., VOL. 40, NO. 3, DECEMBER
1969, PP. 99-109.

69-0516 WEMPNER, G. A.
FINITE ELEMENTS, FINITE ROTATIONS AND SMALL
STRAINS OF FLEXIBLE SHELLS
INT. J. SOLIDS STRUCT., VOL. 5, 1969, PP. 117-153.

69-0517 SAMLER, R. G.
VIBRATIONS OF TAPERED AND CURVED MEMBERS BY FINITE
ELEMENT METHODS

69-0518 REEVES, G. N.
A FINITE ELEMENT STUDY OF THE PENETRATION OF A
SOIL HALF-SPACE BY A RIGID PROJECTILE
PH.D., TEXAS A AND M UNIV., 1969.

69-0519 RAY, G.
FINITE ELEMENT ANALYSIS METHODS IN DYNAMICS OF
PULSATILE FLOW
PH.D., PENNSYLVANIA STATE UNIV., 1969.

69-0520 KOENIG, H. A.
DAVIDS, N.
THE DAMPED TRANSIENT BEHAVIOR OF FINITE BEAMS AND
PLATES
INT. J. NUMER. METH. ENGR., VOL. 1, NO. 2, APRIL -
JUNE, 1969, PP. 151-162.

69-0521 RADHAKRISHNAN, N.
SOLUTION OF SOME PLANE STRAIN PROBLEMS IN SOIL
MECHANICS USING THE METHOD OF FINITE ELEMENTS
PH.D. THESIS, UNIV. OF TEXAS, AUSTIN, 1969.

69-0522 PERUMPPAL, J. V.
THE FINITE ELEMENT METHOD FOR PREDICTING STRESS
DISTRIBUTION AND SOIL DEFORMATION UNDER A TRACTIVE
DEVICE
THESIS, PURDUE UNIV., ANN ARBOR, 1973, 117-120.

69-0523 MEGARD, G.
ANALYSIS OF THIN SHELLS USING CURVED FINITE
ELEMENTS
DIV. OF STRUCT. MECH., THE NORWEGIAN INST. OF
TECH., TRONDHEIM, REPORT NO. 69-1, 1969.

69-0524 BONNES, G.
ANALYSE DES BOILES MINCES PAR ELEMENTS FINIS
COURBES
D.SC. DISSERTATION, DEPT. OF CIVIL ENG., UNIV.
LAVAL, QUEBEC, 1969.

69-0525 DEASON, D. H.
A FINITE ELEMENT TECHNIQUE FOR ANALYZING FOLDED
PLATE TYPE STRUCTURES WITH SKEWED SUPPORTS
PH.D. DISSERTATION, VANDERBILT UNIV., NASHVILLE,
TENNESSEE.

69-0526 PATTERSON, G. J.
FINITE ELEMENT ANALYSIS OF HYPERBOLOIDAL
PRESTRESSED CONCRETE CONTAINMENT VESSELS
PH.D., UNIV. OF ILLINOIS, URBANA, CHAMPAIGN, 1969.

69-0527 MUFTI, A. A.
MATRIX ANALYSIS OF THIN SHELLS USING FINITE
ELEMENTS
PH.D., MCGILL UNIV., CANADA, 1969.

69-0528 AL-ADEEB, A. M.
A CURVED CIRCULAR CYLINDRICAL SHELL FINITE ELEMENT
FOR HIGHER ORDER SHELL THEORY
PH.D., UNIV. OF COLORADO, 1969.

69-0529 BLACKLOCK, J. R.
PLANE STRESS FINITE ELEMENTS FOR NONLINEAR
ANALYSIS
PH.D., THE UNIV. OF ARIZONA, 1969.

69-0530 CHANG, C-Y.
FINITE ELEMENT ANALYSES OF SOIL MOVEMENTS CAUSED
BY DEEP EXCAVATION AND DEWATERING
PH.D., UNIV. OF CALIF., BERKELEY, 1969.

69-0531 CLOUGH, G. W.
FINITE ELEMENT ANALYSES OF SOIL-STRUCTURE
INTERACTION IN U-FRAME LOCKS
PH.D., UNIV. OF CALIF., BERKELEY, 1969.

70-0001 ZIENKIEWICZ, O. C.
PAREKH, C. J.
TRANSIENT FIELD PROBLEMS: TWO-DIMENSIONAL AND
THREE-DIMENSIONAL ANALYSIS BY ISO-PARAMETRIC
FINITE ELEMENTS
INT. J. NUM. METHODS ENG. 2, 1970, PP. 61-71

70-0002 DOCTORS, L. J.
AN APPLICATION OF THE FINITE ELEMENT TECHNIQUE TO
BOUNDARY VALUE PROBLEMS OF POTENTIAL FLOW
INT. J. NUM. METHODS ENG. 2, 1970, PP. 243-252.

70-0003 ARGYRIS, J. H.
THE IMPACT OF THE DIGITAL COMPUTER ON THE
ENGINEERING SCIENCES - PART I
AERO. J. ROY. SOC. 74, 1970, PP. 13-41.

70-0004 GUYMON, G. L.
SCOTT, V. H.
HERRMAN, L. R.
A GENERAL NUMERICAL SOLUTION OF THE
TWO-DIMENSIONAL DIFFUSION-CONVECTION EQUATION BY
THE FINITE ELEMENT METHOD
WATER RESOURCES RES. 6, NO. 6, 1970, PP 1611-1617.

70-0005 MOTE, C. D.
UNSYMMETRICAL TRANSIENT HEAT CONDUCTION: ROTATING
DISK APPLICATIONS
J. ENG. IND., TRANS. ASME 92, SER. B, NO. 1, 1970,
PP. 181-190.

70-0006 LEONARD, J. W.
GALERKIN FINITE ELEMENT FORMULATION FOR
INCOMPRESSIBLE FLOW
REP. NO. 9500-920181, BELL AEROSYSTEMS CO.,
BUFFALO, NEW YORK, APRIL 1970.

70-0007 REDDI, M. M.
CHU, T. Y.
FINITE ELEMENT SOLUTION OF THE STEADY STATE
COMPRESSIBLE LUBRICATION PROBLEM
J. LUB. TECH. 92, SER. F., NO. 3, 1970, PP.
495-503.

70-0008 SKIBA, E.
A FINITE ELEMENT SOLUTION OF GENERAL FLUID
DYNAMICS PROBLEMS - NATURAL CONVECTION IN
RECTANGULAR CAVITIES
M. APPL. SCI. CIVIL. ENG. THESIS, UNIV. OF
WATERLOO, ONTARIO, CANADA, APRIL 1970.

70-0009 HUNT, D. A.
DISCRETE ELEMENT STRUCTURAL THEORY OF FLUIDS
PAPER 70-23, AIAA 8TH AEROSPACE SCIENCES MEETING,
JANUARY 19-21, 1970.

70-0010 ZIENKIEWICZ, O. C.
THE FINITE ELEMENT METHOD: FROM INTUITION TO
GENERALITY
APPL. MECH. REV. 23, 1970, PP. 249-256.

70-0011 OLSON, M. D.
SOME FLUTTER SOLUTIONS USING FINITE ELEMENTS
AIAA J. 8, NO. 4, 1970, PP. 747-752.

70-0012 ALLIK, H.
HUGHES, T. J. R.
FINITE ELEMENT METHOD FOR PIEZOELECTRIC VIBRATION
INT. J. NUM. METHODS ENG. 2, 1970, PP. 151-157.

70-0013 SMITH, I. M.
A FINITE ELEMENT APPROACH TO ELASTIC
SOIL-STRUCTURE INTERACTION
CAN. GEOTECH. J. 7, 1970, PP. 95-105.

70-0014 KARIAPPA, V.
KINEMATICALLY CONSISTENT UNSTEADY AERODYNAMIC
COEFFICIENTS IN SUPERSONIC FLOW
INT. J. NUM. METHODS ENG. 2, 1970, PP. 495-507.

70-0015 ADAMS, D. F.
INELASTIC ANALYSIS OF A UNIDIRECTIONAL COMPOSITE
SUBJECTED TO TRANSVERSE NORMAL LOADING
AD-707 505, MAY 70, 53 P.

70-0016 AHMAD, S.
IRONS, B. M.
ZIENKIEWICZ, O. C.
ANALYSIS OF THICK AND THIN SHELL STRUCTURE BY
CURVED FINITE ELEMENTS
INT. J. NUM. METHODS ENG. VOL. 2, NO. 3, PP.
429-452, JULY-SEPTEMBER 1970.

70-0017 AHMAD, S.
ANDERSON, R. G.
ZIENKIEWICZ, O. C.
VIBRATION OF THICK, CURVED, SHELLS WITH PARTICULAR
REFERENCE TO TURBINE BLADES
J. STRAIN ANALYSIS, V. 5, PP. 200-206, 1970.

70-0018 AKYUZ, A.
NATURAL COORDINATE SYSTEMS, AN AUTOMATIC INPUT
DATA GENERATION SCHEME FOR A FINITE
ELEMENT METHOD
NUCLEAR ENGINEERING AND DESIGN, V. 11, 1970, PP.
195-207.

70-0019 ALLMAN, D. J.
TRIANGULAR FINITE ELEMENTS FOR PLATE BENDING WITH
CONSTANT AND LINEARLY VARYING BENDING MOMENTS
IUTAM SYMP. HIGH SPEED COMPUTING OF ELASTIC
STRUCTURES, LIEGE, 1970.

70-0020 ANDERHEGGEN, E.
A CONFORMING TRIANGULAR FINITE ELEMENT PLATE
BENDING SOLUTION
INT. J. NUM. METHODS ENG., VOL. 2, NO. 2, PP.
259-264, APRIL-JUNE, 1970.

70-0021 ARGYRIS, J. H.
MARECZEK, G.
SCHARPF, D. W.
TWO-AND THREE-DIMENSIONAL FLOW ANALYSIS USING
FINITE ELEMENTS
NUCL. ENG. AND DESIGN, 1970, PP. 230-236.
ALSO IN AERO. J. OF THE ROY. AERON. SOC.,
VOL. 73, 1969, PP. 961-964.

70-0022 ARGYRIS, J. H.
SCHARPF, D. W.
FINITE ELEMENT FORMULATION OF THE INCOMPRESSIBLE
LUBRICATION PROBLEM
NUCL. ENG. AND DESIGN, 1970, PP. 225-229.

70-0023 ARMEN, H.
ISAKSON, G.
PIFKO, A. B.
DISCRETE ELEMENT METHODS FOR THE PLASTIC ANALYSIS
OF STRUCTURES SUBJECTED TO CYCLIC LOADING
INT. J. NUM. METHODS ENG., VOL. 2, NO. 2,
APRIL-JUNE 1970, PP. 189-206.

70-0024 ATKINSON, B.
CARD, C. C. M.
IRONS, B. M.
APPLICATION OF THE FINITE ELEMENT METHOD TO
CREEPING FLOW PROBLEMS
TRANS. INSTRN. CHEM. ENGRS., VOL. 48, T276-T284,
1970.

70-0025 AUGUST, A.
HADCOCK, R.
THE BEHAVIOUR OF ADVANCED FILAMENTARY COMPOSITE
PLATES WITH CUTOUTS
AD-871 939, MAY 1970, 48 PP.

70-0026 AYRES, D. J.
SIDDALL, W. F.
FINITE ELEMENT ANALYSIS OF STRUCTURAL INTEGRITY OF
A REACTOR PRESSURE VESSEL DURING EMERGENCY CORE
COOLING
COMBUSTION ENGINEERING, INC., WINDSOR, CONN.,
JANUARY 1970, PP. 54, NP-18198.

70-0027 BABUSKA, I.
THE FINITE ELEMENT METHOD FOR ELLIPTIC
DIFFERENTIAL EQUATIONS
NUMERICAL SOLUTION OF PARTIAL DIFFERENTIAL
EQUATIONS, PP. 69-106, 1970. ALSO IN TECH. NOTE
BN-653, INST. FOR FLUID DYNAMICS AND APPLIED
MATHEMATICS, UNIV. OF MARYLAND, 1970.

70-0028 BARSOUM, R. S.
GALLAGHER, R. H.
FINITE ELEMENT ANALYSIS OF TORSIONAL AND
TORSIONAL-FLEXURAL STABILITY PROBLEMS
INT. J. NUM. METHODS ENG. VOL. 2, NO. 3, PP.
335-352, JULY-SEPTEMBER 1970.

70-0029 BRISBANE, J. J.
BECKER, E. B.
SCHKADE, A. F. (JR.)
INVESTIGATION OF TECHNIQUES OF THREE-DIMENSIONAL
FINITE ELEMENT STRESS ANALYSIS
AD-866 971, MARCH 1970, 125 PP.

70-0030 BERGAN, P. G.
CLOUGH, R. W.
ANALYSIS OF STIFFENED PLATES USING THE FINITE
ELEMENT METHOD
AD-873 395, JANUARY 1970, 79 PP.

70-0031 SANDHU, R. S.
SALMON, M.
AN APPLICATION OF THE FINITE ELEMENT METHOD TO
ELASTIC-PLASTIC PROBLEMS OF PLANE STRESS
AD-708 516, MAY 1970, 55 PP.

70-0032 BOUGHTON, N. O.
ELASTIC ANALYSIS FOR BEHAVIOR OF ROCKFILL
J. SOIL MECH. AND FOUND., A.S.C.E., VOL. 96, NO.
SM5, SEPTEMBER 1970, PP. 1715-1733.

70-0033 BRAMLETTE, T. T.
LEONARD, J. W.
FINITE ELEMENT ANALYSIS OF THE UNSTEADY,
LINEARIZED BOLTZMANN EQUATION
BELL AEROSYSTEMS CO., REPORT NO. 9500-920180,
APRIL 1970.

70-0034 VENKATESWARA RAO, G.
KRISHNA MURTY, A. V.
A COMPARATIVE STUDY OF THE CONSISTENT AND
SIMPLIFIED FINITE ELEMENTS ANALYSIS OF EIGENVALUE
PROBLEMS
J. AERONAUT., SOC. INDIA., VOL. 22, 1970, PP.
183-188.

70-0035 CANTIN, G.
RIGID BODY MOTIONS IN CURVED FINITE ELEMENTS
A.I.A.A. J. VOL. 8, NO. 7, PP. 1252-1255, JULY
1970.

70-0036 CAPPELLI, A. P.
AGRAWAL, G. L.
ROMERO, V. E.
EPAD, A COMPUTER PROGRAM FOR THE ELASTIC-PLASTIC
ANALYSIS OF DOME STRUCTURES AND ARBITRARY SOLIDS
OF REVOLUTION SUBJECTED TO UNSYMMETRICAL LOADINGS
AD-866 588, FEBRUARY 1970, 125 PP.

70-0037 CAPPELLI, A. P.
THEORETICAL ANALYSIS OF THICK DOME STRUCTURES AND
ARBITRARY SOLIDS OF REVOLUTION SUBJECTED TO
UNSYMMETRICAL LOADINGS
AD-868 329, APRIL 1970, 31 PP.

70-0038   CHACOUR, S.
A HIGH PRECISION AXISYMMETRIC TRIANGULAR ELEMENT
USED IN THE ANALYSIS OF HYDRAULIC TURBINE
COMPONENTS
A.S.M.E. PAPER 70-FE-19, ALSO J. BASIC
ENGINEERING, 1970.

70-0039   CHANG, C.
DUNCAN, J. M.
ANALYSIS OF SOIL MOVEMENT AROUND A DEEP EXCAVATION
J. SOIL MECH. AND FOUNDATIONS DIV., A.S.C.E., VOL.
96, NO. SM5, PP. 1655-1681, SEPTEMBER 1970.

70-0040   CHAUVIN, G. A.
SMALL, R. J.
WALSER, A.
POST-TENSIONED CONCRETE NUCLEAR STRUCTURES
J. POWER DIV., A.S.C.E., VOL. 93, NO. PO3, PP.
401-414, JUNE, 1970.

70-0041   CHENG, R. T.
TUNG, C.
WIND DRIVEN LAKE CIRCULATION BY THE FINITE ELEMENT
METHOD
PROC. 13TH CONF. GREAT LAKES RES., BUFFALO, NEW
YORK, 1970.

70-0042   CHOPRA, A. K.
EARTHQUAKE RESPONSE OF CONCRETE GRAVITY DAMS
AD-709 640, JANUARY 1970, 40 PP.

70-0043   CHRISTIAN, J. T.
BOEHMER, J. W.
PLANE STRAIN CONSOLIDATION BY FINITE ELEMENTS
J. SOIL MECH. AND FOUNDATION DIV., A.S.C.E., VOL.
96, NO. SM4, PP. 1435-1457, JULY 1970.

70-0044   COWPER, G. R.
LINDBERG, G. M.
OLSON, M. D.
A SHALLOW SHELL FINITE ELEMENT OF TRIANGULAR SHAPE
INT. J. SOLIDS STRUCT., VOL. 6, NO. 8, PP.
1133-1156, AUGUST 1970.

70-0045   CROSE, J. G.
BANDWIDTH MINIMIZATION OF STIFFNESS MATRICES
AD-705 579, MAY 1970, 14 PP.

70-0046   DAVIES, J. D.
ANALYSIS OF CORNER SUPPORTED RECTANGULAR SLABS
STRUCTURAL ENGINEER, VOL. 48, NO. 2, PP. 75-82,
FEBRUARY 1970.

70-0047   DE LA CRUZ, R. V.
GOODMAN, R. E.
THEORETICAL BASIS OF THE BOREHOLE DEEPENING METHOD
OF ABSOLUTE STRESS MEASUREMENT
ROCK MECHANICS--THEORY AND PRACTICE: CHAPTER 20,
W. H. SOMERTON (EDITOR), PORT CITY PRESS, 1970.

70-0048   DONEA, J.
FINITE-ELEMENT ANALYSIS OF RADIATION-DAMAGE
STRESSES IN THE GRAPHITE OF MATRIX FUEL ELEMENTS
EUROPEAN ATOMIC ENERGY COMMUNITY, ISPRA, ITALY,
JOINT NUCLEAR RESEARCH CENTER, 1970. P. 18,
EUR-4472E.

70-0049   DOUGLAS, A.
FINITE ELEMENTS FOR GEOLOGICAL MODELLING
NATURE, VOL. 226, MAY 16, 1970.

70-0050   DUNCAN, J. M.
CHANG, C.
NONLINEAR ANALYSIS OF STRESS AND STRAIN IN SOILS
J. SOIL MECH. FOUNDATION DIV., A.S.C.E., VOL. 96,
NO. SM5, SEPTEMBER 1970, PP. 1629-1653.

70-0051   DUNLOP, P.
DUNCAN, J. M.
DEVELOPMENT OF FAILURE AROUND EXCAVATED SLOPES
J. SOIL MECH. + FOUNDATION DIV., A.S.C.E., VOL.
96, NO. SM2, MARCH 1970.

70-0052   DUPUIS, G.
GOEL, J. J.
FINITE ELEMENT WITH A HIGH DEGREE OF REGULARITY
INT. J. NUM. METHODS ENG., VOL. 2, PP. 563-576,
OCTOBER 1970.

70-0053   EWING, D. J. F.
FAWKES, A. J.
GRIFFITHS, J. R.
RULES GOVERNING THE NUMBER OF NODES AND ELEMENTS
IN A FINITE ELEMENT MESH
INT. J. NUM. METHODS ENG., VOL. 2, PP. 595-602,
OCTOBER 1970.

70-0054   WILSON, E. A.
FARHOOMAND, I.
A NONLINEAR FINITE ELEMENT CODE FOR ANALYZING THE
BLAST RESPONSE OF UNDERGROUND STRUCTURES
AD-703 920, JANUARY 1970, 73 PP.

70-0055   FRAEIJS DE VEUBEKE, F.
KABAILA, A. P.
BIFURCATION OF SPACE FRAMES
AFFDL-TR-70-36, MARCH 1970, 33 PP.

70-0056   KABAILA, A. P.
FRAEIJS DE VEUBEKE, B.
STABILITY ANALYSIS BY FINITE ELEMENTS
AFFDL-TR-70-35, MARCH 1970, 31 PP.

70-0057   KABAILA, A. P.
BIFURCATION OF ECCENTRICALLY LOADED MEMBERS AND OF
STIFFENED PLATES
AFFDL-TR-70-37, MARCH 1970, 35 PP.

70-0058   FREDERICK, C. O.
WONG, Y. C.
EDGE, F. W.
TWO-DIMENSIONAL AUTOMATIC MESH GENERATION FOR
STRUCTURAL ANALYSIS
INT. J. NUM. METHODS ENG., VOL. 2, NO. 1, PP.
133-144, JANUARY 1970.

70-0059   FRIED, I.
A GRADIENT COMPUTATIONAL PROCEDURE FOR THE
SOLUTION OF LARGE PROBLEMS ARISING FROM THE FINITE
ELEMENT DISCRETIZATION METHOD
INT. J. NUM. METHODS, VOL. 2, OCTOBER 1970, , PP.
477-494.

70-0060   KOZAKOFF, D. J.
SIMONS, JR. F. O.
THREE DIMENSIONAL NONLINEAR MAGNETIC FIELD
BOUNDARY VALUE PROBLEM AND ITS NUMERICAL SOLUTION
IEEE. TRANS. MAGN. VOL. MAG-6, NO. 4, DECEMBER
1970, PP. 828-833.

70-0061   FU, C. C.
A METHOD FOR THE NUMERICAL INTEGRATION OF THE
EQUATIONS OF MOTION ARISING FROM A FINITE ELEMENT
ANALYSIS
J. APPL. MECH., VOL. 37, SEPTEMBER 1970.

70-0062   GALLAGHER, R. H.
LEE, C. H.
MATRIX DYNAMIC AND INSTABILITY ANALYSIS WITH
NON-UNIFORM ELEMENTS
INT. J. NUM. METHODS ENG., VOL. 2, NO. 2, PP.
265-276, APRIL- JUNE 1970.

70-0063   GIANNINI, M.
MILES, G. A.
A CURVED ELEMENT APPROXIMATION IN THE ANALYSIS OF
AXISYMMETRIC THIN SHELLS
INT. J. NUM. METHODS ENG., VOL. 2, PP. 459-476,
OCTOBER 1970.

70-0064   HAFTKA, R. T.
MALLETT, R. H.
NACHBAR, W.
A KOITER TYPE METHOD FOR FINITE ELEMENT ANALYSIS
OF NONLINEAR STRUCTURAL BEHAVIOUR
AFFDL TR 70-130, NOVEMBER 1970.

70-0065   HAISLER, W. E.
STRICKLIN, J. A.
NONLINEAR FINITE ELEMENT ANALYSIS INCLUDING HIGHER
ORDER STRAIN ENERGY TERMS
A.I.A.A. J., VOL. 8, NO. 6, P. 1158, JUNE 1970.

70-0066   HALINA, H.
NUMERICAL DETERMINATION OF STRESSES AND
DEFORMATIONS IN ROCK TAKING INTO ACCOUNT
DISCONTINUITIES
ROCK MECH., VOL. 2, NO. 1, PP. 1-16, MAY 1970.

70-0067   HARRIS, J. G.
OJALVO, I. U.
HODSON, R. E.
STRESS AND DEFLECTION ANALYSIS OF MECHANICALLY
FASTENED JOINTS
AD-709 221, MAY 1970, 288 PP.

70-0068   HELLEN, T. K.
MONEY, H. A.
THE APPLICATION OF THREE-DIMENSIONAL FINITE
ELEMENTS TO A CYLINDER-CYLINDER INTERSECTION
INT. J. NUM. METHODS ENG., VOL. 2, NO. 3, PP.
415-418, JULY-SEPTEMBER 1970.

70-0069   HIBBITT, H. D.
MARCAL, P. V.
RICE, J. R.
A FINITE ELEMENT FORMULATION FOR PROBLEMS OF LARGE
STRAIN AND LARGE DISPLACEMENT
INT. J. SOLIDS STRUCT., VOL. 6, NO. 8, PP.
1069-1086, AUGUST 1970. (ALSO AD-691 782).

70-0070   HOLLINGSHEAD, G.
RAYMOND, G.
PREDICTION OF UNDRAINED MOVEMENTS CAUSED BY
EMBANKMENTS ON MUSKEG
AD-870 043, MAY 1970, 23 PP.

70-0071   HOYAUX, B.
LADANYI, B.
ROCK FAILURE AROUND A CIRCULAR OPENING IN A
GRAVITY FIELD WITH TECTONIC FORCES
ROCK MECH., THEORY AND PRACTICE; CHAPTER 17, W. H.
SOMERTON (EDITOR), PORT CITY PRESS, 1970.

70-0072   HUBBARD, B.
NUMERICAL SOLUTION OF PARTIAL DIFFERENTIAL
EQUATIONS, 11
ACADEMIC PRESS, 1970.

70-0073   IRONS, B. M.
A FRONTAL SOLUTION PROGRAM FOR FINITE ELEMENT
ANALYSIS
INT. J. NUM. METHODS ENG., VOL. 2, NO. 1, PP.
5-32, JANUARY 1970.

70-0074   IRONS, B. M.
ZIENKIEWICZ, O. C.
DE APANTES E OLIVEIRA, E. R.
THEORETICAL FOUNDATIONS OF THE FINITE ELEMENT
METHOD (DISCUSSION)
INTL. J. SOLIDS STRUCT., VOL. 6, NO. 5, MAY 1970,
P. 695. ALSO IN INTL. J. SOLIDS STRUCT., VOL. 4,
1968, PP. 929-952.

70-0075   ISENBERG, J.
BHAUNIK, A. K.
WONG, P. S.
SPHERICAL WAVES IN INELASTIC MATERIAL
AD-703 295, MARCH 1970, 62 PP.

70-0076   JORDAN, W. B.
IMPROVED CALCULATION OF ELASTO-PLASTIC SYSTEMS BY
THE FINITE ELEMENT METHOD
KNOLLS ATOMIC POWER LAB., SCHENECTADY, N. Y.,
JANUARY 1970, P. 25, KAPL-M-7108.

70-0077   KAITZ, M. J.
SHALLON, S. M.
A COMPUTER PROGRAM TO DISPLAY THE DYNAMIC STRESS
ANALYSIS OF STRUCTURES
AD-871 627L, APRIL 1970, 99 PP.

70-0078   KEY, S. W.
BEISINGER, Z. E.
THE ANALYSIS OF THIN SHELLS BY THE FINITE ELEMENT
METHOD
I.U.T.A.M. SYMP. HIGH SPEED COMPUTING OF ELASTIC
STRUCTURES LIEGE, BELGIUM, AUGUST 1970.

70-0079   KEY, S. W.
BEISINGER, Z. E.
SLADE, A COMPUTER PROGRAM FOR THE STATIC ANALYSIS
OF THIN SHELLS
SANDIA LABS., REPORT SC-RR-69-369, NOVEMBER 1970.

70-0080   KOTANCHIK, J. J.
YEGHIAYAN, R. P.
WITMER, E. A.
BERG, B. A.
THE TRANSIENT LINEAR ELASTIC RESPONSE ANALYSIS
COMPLEX THIN SHELLS OF REVOLUTION SUBJECTED TO
ARBITRARY EXTERNAL LOADING BY THE FINITE ELEMENT
PROGRAM SABOR 5-DRASTIC
AD-709 189, APRIL 1970, 500 PP.

70-0081   KOTANCHIK, J. J.
DISCRETE-ELEMENT STATIC ANALYSIS OF BONDED,
DOUBLE-LAYER, BRANCHED, THIN SHELLS OF REVOLUTION,
PART 2: THE SABOR 5 PROGRAM
AD-861 128L, MAY 1970, 470 PP.

70-0082   LAWRENCE, K. L.
TWISTED BEAM ELEMENT MATRICES FOR BENDING
A.I.A.A. J., VOL. 8, NO. 6, PP. 1160-1161, JUNE
1970.

70-0083   KOBAYASHI, A. S.
LEE, C. H.
ANALYSES OF AXISYMMETRIC UPSETTING AND PLANE
STRAIN SIDE PRESSING OF SOLID CYLINDERS BY THE
FINITE ELEMENT METHOD
A.S.M.E. PAPER NO. 70-WA/PROD-4, 1970.

70-0084   BRAMLETTE, T. T.
LEONARD, J. W.
FINITE ELEMENT SOLUTIONS TO DIFFERENTIAL EQUATIONS
J. ENG. MECH. DIV., A.S.C.E., VOL. 96, NO. EM6,
DECEMBER 1970, PP. 1277-1283.

70-0085   LINDBERG, G. M.
OLSON, M. D.
CONVERGENCE STUDIES OF EIGENVALUE SOLUTIONS USING
TWO FINITE PLATE ELEMENTS
INT. J. NUM. METHODS ENG., VOL. 2, NO. 1, PP.
99-116, JANUARY, 1970.

70-0086   ZIENKIEWICZ, O. C.
IRONS, B. M.
SCOTT, F. C.
CAMPBELL, J. S.
THREE DIMENSIONAL STRESS ANALYSIS
PROC. IUTAM SYMP. HIGH SPEED COMPUT., LIEGE,
BELGIUM, 1970.

70-0087   MARCAL, P. V.
ON GENERAL PURPOSE PROGRAMS FOR FINTIE ELEMENT
ANALYSIS WITH SPECIAL REFERENCE TO GEOMETRIC AND
MATERIAL NONLINEARITIES
NUMERICAL SOLUTION OF PARTIAL DIFFERENTIAL
EQUATIONS, PP. 433-468, 1970. ALSO AD-707, 792,
MAY 1970, 34 PP.

70-0088   MCCORQUODALE, J.A.
FINITE ELEMENT METHOD APPLIED TO UNSTEADY
NON-DARCY FLOW
A.S.C.E. HYDRAULICS DIV., SPECIALTY CONF.,
MINNEAPOLIS, MINN., AUGUST 1970.

70-0089   MCCORQUODALE, J.A.
VARIATIONAL APPROACH TO NON-DARCY FLOW
J. HYDRAULICS DIV., A.S.C.E., VOL. 96, NO. HY11,
NOV. 1970, PP. 2265-2278.

70-0090   MCWHORTER, L. B.
HAISLER, W. E.
FAMSOR, A FINITE ELEMENT PROGRAM FOR THE
FREQUENCIES AND MODE SHAPES OF SHELLS OF
REVOLUTION
SANDIA LABS., REPORT SC-CR-715125, OCTOBER 1970.

70-0091   MORLEY, L. S. D.
A FINITE ELEMENT APPLICATION OF THE MODIFIED
RAYLEIGH-RITZ METHOD
INT. J. NUM. METHODS ENG., VOL. 2, NO. 1, PP.
85-98, JANUARY, 1970.

70-0092   NEUMAN, S. P.
WITHERSPOON, P. A.
FINITE ELEMENT METHOD OF ANALYZING STEADY SEEPAGE
WITH A FREE SURFACE
WATER RESOUR. RES. VOL. 6, NO. 3, P. 829, 1970.

70-0093   ODELLO, R. J.
ALLGOOD, J. R.
RESPONSE OF BURIED CAPSULES IN THE
HIGH-OVERPRESSURE REGION
AD-705 990, APRIL 1970, 71 PP.

70-0094   ODEN, J. T.
KEY, J. E.
NUMERICAL ANALYSIS FOR FINITE AXISYMMETRIC
DEFORMATIONS OF INCOMPRESSIBLE ELASTIC SOLIDS OF
REVOLUTION
INT. J. SOLIDS STRUCT., VOL. 6, NO. 5, MAY 1970,
PP. 497-518.

70-0095   ODEN, J. T.
AGUIRRE-RAMIREZ, G.
WU, T. S.
A NUMERICAL SOLUTION OF THE BOLTZMANN EQUATION BY
THE FINITE ELEMENT TECHNIQUE
PROC. 7TH INT. SYMP. ON RAREFIED GAS DYNAMICS,
PISA, ITALY, JUNE 1970.

70-0096   ODEN, J. T.
KEY, J. E.
ANALYSIS OF FINITE DEFORMATIONS OF ELASTIC SOLIDS
BY THE FINITE ELEMENT METHOD
I.U.T.A.M. SYMP. HIGH SPEED COMPUTING OF ELASTIC
STRUCTURES, LIEGE, BELGIUM, AUGUST 1970.

70-0097   ODEN, J. T.
NOTE ON AN APPROXIMATE METHOD FOR COMPUTING
NONCONSERVATIVE GENERALIZED FORCES ON FINITELY
DEFORMED FINITE ELEMENTS
A.I.A.A. J., VOL. 8, NO. 11, PP. 2088-2090,
NOVEMBER 1970.

70-0098   ODEN, J. T.
CHUNG, T. J.
EIDSON, R. L.
DYNAMIC ELASTOPLASTIC RESPONSE OF GEOMETRICALLY
NONLINEAR ARBITRARY SHELLS OF REVOLUTION UNDER
IMPULSIVE MECHANICAL AND THERMAL LOADINGS
THE SHOCK AND VIBRATION BULLETIN, NO. 41, PART 7,
DECEMBER 1970.

70-0099   ODEN, J. T.
A FINITE ELEMENT ANALOGUE OF THE NAVIER-STOKES
EQUATION
J. ENG. MECH. DIV., A.S.C.E., VOL. 96, NO. EM4,
1970.

70-0100   OLSON, M. D.
LINDBERG, G. M.
ANNULAR AND CIRCULAR SECTOR FINITE ELEMENTS FOR
PLATE BENDING
INT. J. MECH. SCI., VOL. 12, NO. 1, PP. 17-34,
JANUARY 1970.

70-0101   PARSONS, B.
WILSON, E. A.
A METHOD FOR DETERMINING THE SURFACE CONTACT
STRESSES RESULTING FROM INTERFERENCE FITS
J. ENG. IND., FEBRUARY 1970, PP. 208-218.

70-0102   SEGUI, W. T.
FINITE ELEMENT ANALYSIS OF STRUCTURES WITH
SURFACES OF ARBITRARY CURVATURE
PH.D. THESIS, UNIV. OF SOUTH CAROLINA, 1970.

70-0103 ROBINSON, J.
A DERIVATION PROCEDURE FOR THE DYNAMIC FLEXIBILITY
MATRIX OF A TRIANGULAR BENDING ELEMENT
AERON. J., VOL. 74, P. 327, 1970.

70-0104 SABIR, A. B.
AN EXTENSION OF THE SHALLOW TO THE NON-SHALLOW
STIFFNESS MATRIX FOR A CYLINDRICAL SHELL FINITE
ELEMENT
INT. J. MECH. SCI., VOL. 11, NO. 3, PP. 287-292,
MARCH 1970.

70-0105 SMITH, I. M.
DUNCAN, W.
THE EFFECTIVENESS OF EXCESSIVE NODAL CONTINUITIES
IN THE FINITE ELEMENT ANALYSIS OF THIN RECTANGULAR
AND SKEW PLATES IN BENDING
INT. J. NUM. METHODS ENG., VOL. 2, NO. 2, PP.
253-258, APRIL-JUNE 1970.

70-0106 SNOEY, M. R.
CRAWFORD, J. E.
STRESS ANALYSIS OF A CONICAL ACRYLIC VIEWPORT
AD-708 009, APRIL 1970, 69 PP.

70-0107 SOMERTON, W. H. (ED.)
ROCK MECHANICS - THEORY AND PRACTICE
11TH SYMP. ON ROCK MECHANICS, AMER. INST. MINING,
MET. AND PETR. ENGRS., NEW YORK, 1970.

70-0108 STANTON, E. L.
SCHMIT, L. A.
A DISCRETE ELEMENT STRESS AND DISPLACEMENT
ANALYSIS OF ELASTOPLASTIC PLATES
A.I.A.A. J., VOL. 8, NO. 7, PP. 1245-1251, JULY
1970.

70-0109 GREIMANN, L. F.
LYNN, P. P.
FINITE ELEMENT ANALYSIS OF PLATE BENDING WITH
TRANSVERSE SHEAR DEFORMATION
NUCL. ENGNG. DES., VOL. 14, 1970, PP. 223-230.

70-0110 STRICKLIN, J. A.
MARTINEZ, J. E.
TILLERSON, J. R.
HONG, J. H.
NONLINEAR DYNAMIC ANALYSIS OF SHELLS OF REVOLUTION
BY MATRIX DISPLACEMENT METHOD
TEXAS ENGINEERING EXPERIMENT STATION, FEBRUARY
1970, P. 102, 38-4 UNLTD DIST, SC-CR-70-6070. ALSO
A.I.A.A. J., VOL. 9, NO. 4, APRIL 1971, PP.
629-636.

70-0111 SU, Y. L.
WANG, Y. J.
STEFANKO, R.
FINITE ELEMENT OF UNDERGROUND STRESSES UTILIZING
STOCHASTICALLY SIMULATED MATERIAL PROPERTIES
ROCK MECHANICS--THEORY AND PRACTICE, CHAPTER 15,
W. H. SOMERTON (EDITOR), PORT CITY PRESS, 1970.

70-0112 TADA, Y.
LEE, G. C.
FINITE ELEMENT SOLUTION TO AN ELASTICA PROBLEM OF
BEAMS
INT. J. NUM. METHODS ENG., VOL. 2, NO. 2, PP.
229-242, APRIL-JUNE 1970.

70-0113 TAHBILDAR, U. C.
TOTTENHAM, H.
EARTHQUAKE RESPONSE OF ARCH DAMS
J. STRUCT. DIV., A.S.C.E., VOL. 96, NO. ST11,
NOVEMBER 1970, PP. 2321-2336.

70-0114 TAMULY-PHUKAN, A. L.
LO, K. V.
LA-ROCHELLE, P.
STRESSES AND DEFORMATIONS OF VERTICAL SLOPES IN
ELASTO-PLASTIC ROCKS
ROCK MECHANICS--THEORY AND PRACTICE, CHAPTER 11,
W. H. SOMERTON (EDITOR), PORT CITY PRESS, 1970.

70-0115 TONG, P.
NEW DISPLACEMENT HYBRID FINITE ELEMENT MODELS FOR
SOLID CONTINUA
AD-703 989, MARCH 1969, 13 PP. ALSO INT. J. NUM.
METHODS ENG., VOL. 2, PP. 73-83, 1970.

70-0116 BREBBIA, C.
TOTTENHAM, H.
FINITE ELEMENT TECHNIQUES IN STRUCTURAL MECHANICS
UNIVERSITY OF SOUTHAMPTON PRESS, 1970.

70-0117 VERRUIJT, A.
THEORY OF GROUNDWATER FLOW
GORDON AND BREACH SCIENCE PUBLISHERS, 1970.

70-0118 VISSER, W.
THE APPLICATION OF A CURVED MIXED TYPE SHELL
ELEMENT
I.U.T.A.M. SYMP. HIGH SPEED COMPUTING OF ELASTIC
STRUCTURES, LIEGE, BELGIUM, AUGUST 1970.

70-0119 WALLACE, G. B.
SLEBIR, E. J.
ANDERSON, F. A.
FOUNDATION TESTING FOR AUBURN DAM
ROCK MECHANICS--THEORY AND PRACTICE, CHAPTER 25,
W. H. SOMERTON (EDITOR), PORT CITY PRESS, 1970.

70-0120 WALZ, J. E.
ACCURACY OF FINITE ELEMENT APPROXIMATIONS TO
STRUCTURAL PROBLEMS
N.A.S.A., TN D-5728, MARCH 1970.
ALSO IN PROC. OF 2ND CONF. OF MATRIX
METHODS IN STRUCTURAL MECHANICS, WRIGHT-PATTERSON
AIR FORCE BASE, OHIO, 1969, PP. 995-1028.

70-0121 WANG, S. N.
POTTS, J. S.
NEWMAN, M.
HABEAS- A STRUCTURAL DYNAMICS ANALYSIS SYSTEM
AD-704-099, MARCH 1970, 66 PP.

70-0122 WEST, L. J.
PERRY, P. M.
ROCK MECHANICS STUDIES FOR A HIGH CUT AT A NUCLEAR
POWER STATION IN PENNSYLVANIA
ROCK MECHANICS--THEORY AND PRACTICE, CHAPTER 24,
W. H. SOMERTON (EDITOR), PORT CITY PRESS, 1970.

70-0123 WHANG, B.
A FINITE ELEMENT ANALYSIS OF LAMINATED ORTHOTROPIC
PLATES AND SHALLOW SHELLS
AD-704 495, MARCH 1970, 92 PP.

70-0124 WHANG, B.
THERMAL STRESS ANALYSIS BY THE FINITE ELEMENT
METHOD
AD-872 958, JULY 1970, 99 PP.

70-0125 WILSON, E. A.
A DESIGN ORIENTED APPROACH TO CREEP AND PLASTICITY
IN FINITE ELEMENT PROGRAMS
A.S.M.E. PAPER NO. 70-WA/DE-4, 1970.
ALSO IN J. ENG. IND., VOL. 93, NO. 3,
1971, PP. 793-798.

70-0126 WILSON, E. A.
PARSONS, B.
FINITE ELEMENT ANALYSIS OF ELASTIC CONTACT
PROBLEMS USING DIFFERENTIAL DISPLACEMENTS
INT. J. NUMER. METH. IN ENG., VOL. 2, NO. 3,
JULY-SEPTEMBER 1970, PP. 387-396.

70-0127 CHU, S. C.
YALAMANCHILI, R. S.
FINITE ELEMENT METHOD APPLIED TO TRANSIENT
TWO-DIMENSIONAL HEAT TRANSFER WITH CONVECTION AND
RADIATION BOUNDARY CONDITIONS
AD-709 604, JUNE 1970, 44 PP.

70-0128 YANG, H. T. Y.
FLEXIBLE PLATE FINITE ELEMENT ON ELASTIC
FOUNDATION
J. STRUCT. DIV., A.S.C.E., VOL. 96, NO. ST10, PP.
2083-2101, OCTOBER 1970.

70-0129 YEGHIAYAN, R. P.
WITMER, E. A.
FRENCH, S. B.
DISCRETE-ELEMENT MODELLING AND SENSITIVITY STUDIES
OF STATICALLY LOADED SINGLE-LAYER AND BONDED
DOUBLE-LAYER THIN SHELLS OF REVOLUTION
AD-702 445, JANUARY 1970, P. 253.

70-0130 ZAK, A. R.
YURKOVICH, R. N.
SCHMIDT, J. H.
THEORETICAL AND EXPERIMENTAL ANALYSIS OF STIFFENED
PANEL UNDER DYNAMIC CONDITIONS
AD-866 658, JANUARY 1970, 50 PP.

70-0131 ZENISEK, A.
ZLAMAL, M.
CONVERGENCE OF A FINITE ELEMENT PROCEDURE FOR
SOLVING BOUNDARY VALUE PROBLEMS OF THE FOURTH
ORDER
INT. J. NUMER. METH. IN ENG., VOL. 2, NO. 3,
JULY-SEPTEMBER 1970, PP. 307-310.

70-0132 ZIENKIEWICZ, O. C.
IRONS, B. M.
MATRIX ITERATION AND ACCELERATION PROCESSES IN
FINITE ELEMENT PROBLEMS OF STRUCTURAL MECHANICS
NUMERICAL METHODS FOR NON-LINEAR ALGEBRAIC
EQUATIONS, CHAPTER 9, P. RABINOWITZ (ED.), GORDON
AND BREACH, 1970.

70-0133 BUSSE, L.
ACCURATE CALCULATION OF STRESSES IN BEAMS, DISKS
AND PLATES BY MEANS OF FINITE ELEMENT METHODS AND
ITS GENERALIZATION TO ELASTIC BODIES. (DIE GENAUE
SPANNUNGSBERECHNUNG VON BALKEN, SCHEIBEN UND
PLATTEN MIT FINITEN ELEMENTEN UND IHRE
VERALLGEMEINERUNG AUF ELASTISCHE KOERPER)
VDI FORTSCHR BER PT 4, NO. 19, DECEMBER 1970, 98
PP.

70-0134 DORN, L.
EXPERIMENTS IN FORMING AND JOINTING OF TITANIUM
AND TITANIUM ALLOYS FOR AIRCRAFT AND SPACE
VEHICLES - 2. (ERFAHRUNGEN ZUR FORMUNGS UND
EUEGETECHNIK VON TITAN UND TITANLEGIERUNGEN AUS
DER LUFT UND RAUMFAHRT)
BAENDER BLECHE ROHRE, VOL. 11, NO. 10, OCTOBER
1970, PP. 509-515.

70-0135 ALI, R.
HEDGES, J. L.
DYNAMIC ANALYSIS OF AN AUTOMOBILE CHASSIS FRAME
INST. MECH. ENG., PROC. (PART 1), GEN. PROC., VOL.
185, NO. 44, 1970-1971, PP. 683-690.

70-0136 ALI, R.
HEDGES, J. L.
APPLICATION OF FINITE ELEMENT TECHNIQUES TO THE
ANALYSIS OF AN AUTOMOBILE STRUCTURE
INST. MECH. ENG., PROC. (PART 1), GEN. PROC., VOL.
185, NO. 44, 1970-1971, PP. 665-674.

70-0137 HEDGES, J. L.
NORVILLE, C. C.
STRESS ANALYSIS OF AN AUTOMOBILE CHASSIS FRAME
INST. MECH. ENG., PROC. (PART 1), GEN. PROC., VOL.
185, NO. 44, 1970-1971, PP. 675-682.

70-0138 OLSON, M. D.
LINDBERG, G. M.
FREE VIBRATIONS AND RANDOM RESPONSE OF AN
INTEGRALLY-STIFFENED PANEL
NAT. RES. COUNC. CAN., AERONAUT. REP. LR-544,
OCTOBER 1970, 59 PP.

70-0139 COATES, D. F.
YU, Y. S.
NOTE ON THE STRESS CONCENTRATIONS AT THE END OF A
CYLINDRICAL HOLE
INT. J. ROCK. MECH. MINING SCI. . VOL. 7, NO. 6,
NOVEMBER 1970, PP. 583-588.

70-0140 OHJI, K.
OGURA, K.
ELASTIC PLASTIC ANALYSIS OF NOTCHED SPECIMENS
USING THE FINITE ELEMENT METHOD
PROC. 14TH JAP. CONGR. ON MATER. RES., KYOTO,
JAPAN, SEPTEMBER 1970, PP. 170-173.

70-0141 TAIRA, S.
OHTANI, R.
CREEP STRESS AND STRAIN ANALYSIS OF NOTCHED PLATES
AND BARS BY MEANS OF THE FINITE ELEMENT METHOD
PROC. 14TH JAP. CONGR. ON MATER. RES., KYOTO,
JAPAN, SEPTEMBER 1970 (1971), PP. 158-164.

70-0142 TAIRA, S.
INOUE, T.
ANALYSIS OF ELASTO-PLASTIC THERMAL STRESSES BY THE
FINITE ELEMENT METHOD
PROC. 14TH JAP. CONGR. ON MATER. RES., KYOTO,
JAPAN, SEPTEMBER 1970 (1971), PP. 165-169.

70-0143 BARKSDALE, R. D.
NONLINEAR THEORY FOR PREDICTING THE PERFORMANCE OF
FLEXIBLE HIGHWAY PAVEMENTS
HIGHW. RES. REC. NO. 337, 1970, PP. 22-39.

70-0144 DUMITRESCU, D.
CAZACU, M. D.
THEORETICAL AND EXPERIMENTAL CONSIDERATIONS ON
FLOW OF VISCOUS LIQUIDS PAST PLATE AT SMALL AND
AVERAGE RE NUMBERS. (THEORETISCHE UND
EXPERIMENTELLE BETRACHTUNGEN UEBER DIE STROEMUNG
ZAEHER FLUESSIGKEITEN UM EINE PLATTE BEI KLEINEN
UND MITTLEREN REYNOLDSZAHLEN)
Z. ANGEW MATH. MECH., VOL. 50, NO. 5, MAY 1970,
PP. 257-280.

70-0145 KHAN, M. H.
SAUGY, B.
NONLINEAR ANALYSIS OF PRESTRESSED CONCRETE FAST
REACTOR PRESSURE VESSEL. (ANALYSE NON LINEAIRE DES
CONTRAINTES DANS LE FUT D'UN CAISSON DE BETON
PRECONTRAINT POUR REACTEUR RAPIDE)
BULL. TECH. SUISSE ROMANDE, VOL. 96, NO. 10, MAY
16, 1970, PP. 141-145.

70-0146 SUZUKI, K.
ISHIJIMA, Y.
SIMULATION OF PROGRESSIVE FAILURE IN ROCK AROUND
MINING EXCAVATIONS
J. MINING MET. INST. JAP., VOL. 86, NO. 982,
FEBRUARY 1970, PP. 69-74.

70-0147 FEDDERSEN, C. E.
SIMONEN, F. A.
EXPERIMENTAL AND THEORETICAL INVESTIGATION OF
PLANE STRESS FRACTURE OF 2024 - T351 ALUMINUM
ALLOY
NASA CONTRACT REP. CR-1679, SEPTEMBER 1970, P. 78.

70-0148 CHAN, A. S. L.
FIRMIN, A.
ANALYSIS OF COOLING TOWERS BY THE MATRIX FINITE
ELEMENT METHOD- 2
AERON. J., VOL. 74, NO. 718, OCTOBER 1970, PP.
826-835.

70-0149 YASAKA, T.
METHOD OF MINIMUM WEIGHT DESIGN WITH REQUIREMENTS
IMPOSED ON STRESSES AND NATURAL FREQUENCIES
INST. SPACE AERONAUT. SCI., UNIV. TOKYO, REP. 452,
JULY 1970, PP. 259-316.

70-0150 JOUNG, K. S.
HO, B. P. C.
APPLYING FINITE ELEMENT ANALYSIS
ALLIS-CHALMERS ENG. REV., VOL. 35, NO. 1, 1970,
PP. 4-8.

70-0151 THOMS, R. L.
ARMAN, A.
PHOTOELASTIC AND FINITE ELEMENT ANALYSIS OF
EMBANKMENTS CONSTRUCTED OVER SOFT SOILS
HIGHW. RES. REC. NO. 323, 1970, PP. 71-86.

70-0152 KENDRICK, S.
STRUCTURAL DESIGN OF SUPERTANKERS
ROY. INST. NAV. ARCHITECTS, QUART. TRANS., VOL.
112, NO. 4, OCTOBER 1970, PP. 391-420.

70-0153 SCHNELL, W.
MODERN TECHNIQUES FOR CALCULATING STRESSES IN
TANKS, AND THEIR LIMITATIONS. (MODERNE VERFAHREN
ZUR SPANNUNGSBERECHNUNG IM BEHAELTERBAU UND IHRE
GRENZEN)
CHEM-ING-TECH., VOL. 42, NO. 11, JUNE 1970, PP.
764-771.

70-0154 RODABAUGH, E. C.
SPECIAL PURPOSE COMPUTER PROGRAMS
ASME, 2ND PRESSURE VESSELS AND PIPING CONF.,
DENVER, COLORADO, SEPTEMBER 16, 1970, P. 13.

70-0155 KAJITA, T.
KAWAMOTO, T.
FINITE ELEMENT ANALYSIS ON UNIAXIAL COMPRESSIVE
STRENGTH OF CYLINDRICAL BRITTLE SPECIMEN
PROC. JAP. SOC. CIV. ENG., NO. 177, 1970, PP.
71-76. (IN JAPANESE).

70-0156 PUPPO, A.
EVENSEN, H.
CALCULATION AND DESIGN OF JOINTS MADE FROM
COMPOSITE MATERIALS
AD-872 159, MAY 1970, 108 PP.

70-0157 IREMONGER, M. J.
WOOD, W. G.
PLASTIC FLOW AND FAILURE OF DISCONTINUOUS FIBRE
COMPOSITE MATERIALS
J. STRAIN ANAL. VOL. 5, NO. 3, JULY 1970, PP.
212-222.

70-0158 SMITH, B. S.
CARTER, C.
DIAGONAL TENSILE STRENGTH OF BRICKWORK
STRUCT. ENG., VOL. 48, NO. 6, JUNE 1970, PP.
219-225.

70-0159 ROBINSON, J.
WILLIAMS, R.
ELASTIC AND EQUILIBRIUM MATRICES OF GENERAL
SEMI-MONOCOGUE MEMBRANE ELEMENTS
AERON. J., VOL. 74, NO. 714, JUNE 1970, PP.
502-506.

70-0160 SZILARD, R.
ESTIMATING MATRIX DISPLACEMENT SOLUTIONS OF
TWO-DIMENSIONAL PROBLEMS BY LARGE ELEMENT
TECHNIQUE
ACTA TECH., VOL. 68, NO. 3-4, 1970, PP. 293-310.

70-0161 OVUNC, B.
NONLINEAR ANALYSIS OF ELASTIC SHELLS OF REVOLUTION
UNDER AXISYMMETRIC LOADING
BULL. INT. ASS. SHELL STRUCT., NO. 43, SEPTEMBER
1970, PP. 9-16.

70-0162 RADHAKRISHNAN, N.
REESE, L. C.
SOIL RESISTANCE MOVEMENT RELATIONSHIPS FOR
LATERALLY LOADED PILES IN CLAYS AT GREAT DEPTHS
J. INDIAN NATL. SOC. SOIL MECH. FOUND. ENG., VOL.
9, NO. 4, OCTOBER 1970, PP. 387-402.

70-0163 LORD, H. W.
YOUSEF, S. S.
ELASTIC BENDING OF CIRCULAR PLATES OF VARIABLE
THICKNESS. AN ANALYTICAL AND EXPERIMENTAL STUDY
INT. J. MECH. SCI., VOL. 12, NO. 5, MAY 1970, PP.
417-434.

70-0164 LINDBERG, G. M.
OLSON, M. D.
SARAZIN, A. C.
FINITE ELEMENT DYNAMIC ANALYSIS OF SHALLOW SHELL
STRUCTURES
NAT. RES. COUNCIL CANADA, AERONAUT. REP. LR-540,
JULY 1970, 39 PP.

70-0165 TONG, P.
PIAN, T. H. H.
BOUNDS TO THE INFLUENCE COEFFICIENTS BY THE
ASSUMED STRESS METHOD
INT. J. SOLIDS STRUCT., VOL. 6, NO. 11, NOVEMBER
1970, PP. 1429-1432.

70-0166 SCHNOBRICH, W. C.
MIKKOLA, M.
MATERIAL BEHAVIOUR CHARACTERISTICS FOR REINFORCED
CONCRETE SHELLS STRESSED BEYOND THE ELASTIC RANGE
ILL. UNIV. CIV. ENG. STUD., STRUCT. RES. SER. 367,
AUGUST 1970, 41 PP.

70-0167 MALINA, H.
NUMERICAL DETERMINATION OF STRESSES AND
DEFORMATIONS IN ROCK TAKING INTO ACCOUNT
DISCONTINUITIES
ROCK MECH., FELSMECH, MEC ROCHES, VOL. 2, NO. 1,
MAY 1970, PP. 1-16.

70-0168 WOLF, J. A.
DONG, S. B.
STABILITY ANALYSIS OF STRUCTURES BY A REDUCED
SYSTEM OF GENERALIZED COORDINATES
INT. J. SOLIDS STRUCT., VOL. 6, NO. 10, OCTOBER
1970, PP. 1377-1388.

70-0169 DJUBEK, J.
BOX SHAPED GIRDERS, (NOSNIKY UZAVRETEHO PRIEREZU)
STAVEBNICKY CAS., VOL. 18, NO. 7, 1970, PP.
542-557.

70-0170 BAUDENDISTEL, M.
MALINA, H.
EFFECT OF DISCONTINUITIES ON STRESSES AND
DEFORMATIONS OF ROCK AROUND A TUNNEL, (EINFLUSS
VON DISKONTINUITAETEN AUF DIE SPANNUNGEN UND
DEFORMATIONEN IN DER UMGEBUNG EINER TUNNELROEHRE)
ROCK MECH., FELSMECH., MEC. ROCHES, VOL. 2, NO. 1,
MAY 1970, PP. 17-40.

70-0171 DEHLEN, G. L.
MONISMITH, C. L.
EFFECT OF NONLINEAR MATERIAL RESPONSE ON THE
BEHAVIOUR OF PAVEMENTS UNDER TRAFFIC
HIGHW. RES. REC., NO. 310, 1970, PP. 1-16.

70-0172 AL-NAJAFI, A. M. J.
WARBURTON, G. B.
FREE VIBRATION OF RING STIFFENED CYLINDRICAL
SHELLS
J. SOUND VIB., VOL. 13, NO. 1, SEPTEMBER 1970, PP.
9-25.

70-0173 DUPUIS, G.
GOEL, J. J.
CURVED FINITE ELEMENT FOR THIN ELASTIC SHELLS
INT. J. SOLIDS STRUCT., VOL. 6, NO. 11, NOVEMBER
1970, PP. 1413-1428.

70-0174 HERAKOVICH, C. T.
STRAIN HARDENING TORSION OF SOLID BARS AS A
MINIMIZATION PROBLEM
INT. J. MECH. SCI., VOL. 12, NO. 11, NOVEMBER
1970, PP. 985-995.

70-0175 HORIKAWA, J.
OKUMURA, T.
ELASTIC AND PLASTIC STRESS DISTRIBUTION ON BASE
PLATES IN THE NEIGHBOURHOOD OF FILLET WELDS IN LAP
JOINTS
TRANS. JAP. WELD. SOC., VOL. 1, NO. 1, APRIL 1970,
PP. 130-133.

70-0176 HYLARIDES, S.
RECENT DEVELOPMENTS IN HULL AND SHAFT VIBRATION
ANALYSIS
INT. SHIPBLDG. PROGR., VOL. 17, NO. 190, JUNE
1970, PP. 185-190.

70-0177 HORII, K.
KAWAHARA, M.
NUMERICAL ANALYSIS OF VISCO-ELASTIC STRUCTURES BY
THE FINITE ELEMENT METHOD
PROC. JAP. SOC. CIV. ENG., NO. 179, 1970, PP.
23-25.

70-0178 JIROUSEK, J.
FINITE ELEMENT ANALYSIS OF CYLINDRICAL SHELLS.
SOLUTION USING RECTANGULAR CIRCULAR CYLINDRICAL
ELEMENTS
ACTA POLYTECH SCAND., MATH. COMPUT. MACH. SER. MA
20, 1970, PP. 37-57.

70-0179 ANAND, S. C.
LEE, S. L.
FINITE ELEMENT ANALYSIS OF ELASTIC-PLASTIC PLANE
STRESS PROBLEMS BASED UPON TRESCA YIELD CRITERION
ING-ARCH VOL. 39, NO. 2, 1970, PP. 73-86.

70-0180 MAGHSOOD, J.
SCOTT, N. R.
FINITE ELEMENT ANALYSIS OF WOOD STRUCTURAL MEMBERS
A.S.A.E., CHICAGO, ILL., PAPER 70-922, DECMEBER
1970, 41 PP. (ALSO IN AGRICULTURAL ENG., VOL. 51,
NO. 10, 1970, P. 590.

70-0181 JONES, A. T.
BEADLE, C. W.
RANDOM VIBRATION RESPONSE OF CANTILEVER PLATES
USING THE FINITE ELEMENT METHOD
AIAA J., VOL. 8, NO. 10, OCTOBER 1970, PP.
1905-1907.

70-0182 DINIS DA GAMA, C.
APPLICATION OF THE FINITE ELEMENT METHOD TO THE
ANALYSIS OF ROCK SAMPLES UNDER UNIAXIAL
COMPRESSION
TECHNICAN (LISBON), VOL. 45, (NO) 403, DECEMBER
1970, PP. 143-155.

70-0183 SKJOLINGSTAD, L.
THREE-DIMENSIONAL STRESS ANALYSIS AND FIELD
PROBLEMS BY THE FINITE ELEMENT METHOD
M.SC. THESIS, THE UNIVERSITY OF CALGARY, 1970.

70-0184 YU, Y. S.
COATES, D. F.
ANALYSIS OF ROCK SLOPES, USING THE FINITE ELEMENT
METHOD
CAN. DEP. ENERGY, MINES RESOUR., MINES BR., RES.
REP. 229, OCTOBER 1970, 74 PP.

70-0185 GHALI, A.
BATHE, K. J.
ANALYSIS OF PLATES IN BENDING USING LARGE FINITE
ELEMENTS
PUBLICATIONS, INT. ASSOC. FOR BRIDGE AND STRUCT.
ENGR., VOL.30, NO. 2, ZURICH 1970, PP. 29-40.

70-0186 CONSTANTOPOULOS, U. V.
CHRISTIAN, J. T.
INITIATION OF FAILURE IN SLOPES IN
OVERCONSOLIDATED CLAYS AND CLAY SHALES.
U.S. ARMY ENG. NUCL. CRATERING GROUP, NCG TECH.
REP. NO. 29,NOVEMBER 1970, 151 PP.

70-0187 DHATT, G.
INSTABILITY OF THIN SHELLS BY THE FINITE ELEMENT
METHOD
INT. SYMP. OF THE ASSOC. OF SHELL STRUCT., VIENNA,
1970.

70-0188 LEVY, N.
MARCAL, P. V.
THREE-DIMENSIONAL ELASTIC-PLASTIC STRESS AND
STRAIN ANALYSIS FOR FRACTURE MECHANICS. PHASE 1.
SIMPLE FLAWED SPECIMENS
BROWN UNIV., DIV. ENG., PROVIDENCE, RI, HSST.
TECH., REP. NO. 12, DECEMBER 1970, 48 PP.

70-0189 SU, C. L.
VARIATIONAL PRINCIPLES FOR A RAREFIED GAS FLOW
PAST AND ARBITRARY THREE-DIMENSIONAL BODY
PHYS. FLUID, VOL. 13, NO. 1, 1970, PP. 71-78.

70-0190 WANG, C. F.
VARIATIONAL PRINCIPLE OF STEADY-STATE TRANSPORT
PROCESSES
AIAA J. VOL. 8, NO. 2, 1970, PP. 335-338.

70-0191 WEIHS, D.
GAL-OR, B.
GENERAL VARIATIONAL ANALYSIS OF HYDRODYNAMIC,
THERMAL AND DIFFUSIONAL BOUNDARY LAYERS
INT. J. ENG. SCI., VOL. 8, NO. 3, 1970, PP.
231-249.

70-0192 WENGER, N. C.
VARIATIONAL PRINCIPLE FOR MAGNETOHYDRODYNAMIC
CHANNEL FLOW
J. FLUID MECH., VOL. 43, PT. 1, 1970, PP. 211-224.

70-0193 BABUSKA, I.
FINITE ELEMENT METHOD FOR ELLIPTIC EQUATIONS WITH
DISCONTINUOUS COEFFICIENTS
COMPUTING, VOL 5, NO. 3, 1970, PP. 207-213.

70-0194 BRAMBLE, J. H.
ZLAMAL, M.
TRIANGULAR ELEMENTS IN THE FINITE ELEMENT METHOD
MATH. COMPUT., VOL. 24, NO. 112, 1970, PP.
809-820.

70-0195 ROBINSON, J.
HAGGENMACHER, G. W.
WARPED SEMI-MONOCOQUE QUADRILATERAL MEMBRANE
ELEMENT FOR THE FINITE ELEMENT METHODS
LOCKHEED - CALIFORNIA COMPANY REPORT NO. 23676,
1970.

70-0196 FIX, G. J.
ITERATIVE METHODS FOR FINITE ELEMENT SYSTEMS
SIAM MEETING, BOSTON, MASS., OCTOBER 12-14, 1970.

70-0197 BINGHAM, C. M.
SOANE, A. J. M.
UNUSUAL FOUNDATIONS OVER THE MERSEY TUNNEL
STRUCT. ENG., VOL. 48, NO. 4, APRIL 1970, PP.
153-163.

70-0198 CHUGH, Y. P.
HARDY, H. R.
APPLICATION OF FINITE ELEMENT ANALYSIS TO
UNDERGROUND STORAGE OF NATURAL GAS
TRANS. AMER. GEOPHYS. UN. VOL. 51, NO. 4, 1970, P.
432.

70-0199 CZENDES, Z. J.
SILVESTER, P.
NUMERICAL SOLUTION OF DIELECTRIC LOADED
WAVEGUIDES: I. FINITE-ELEMENT ANALYSIS
IEEE TRANS. MICROWAVE THEORY TECH. MTT-18, NO. 12,
1970, PP. 1124-1131.

70-0200 HEITNER, K. L.
HOUSNER, G. W.
NUMERICAL MODEL FOR TSUNAMI RUN-UP
J. ASCE. WATERWAYS, HARBOURS DIV., VOL. 96. NO.
WW3, PAPER 7491, 1970, PP. 701-719.

70-0201 KARIAPPA, V.
SOMASHEKAR, B. R.
FLUTTER OF SKEW PANELS BY THE MATRIX DISPLACEMENT
APPROACH
AERO. J. VOL. 74, NO. 716, 1970, PP. 672-675.

70-0202 MILES, G. A.
WHITE, D. J.
FINITE-ELEMENT METHOD AND ITS APPLICATIONS IN AN
ENGINEERING LABORATORY
J. SCI. TECH., VOL. 37, NO. 3, 1970, PP. 127-136.

70-0203 PARISEAU, W. G.
VOIGHT, B
FINITE ELEMENT ANALYSES OF ELASTIC-PLASTIC
PROBLEMS IN THE MECHANICS OF GEOLOGIC MEDIA: AN
OVERVIEW
PROC. 2ND CONGR. OF INT. SOC. ROCK. MECH.,
BELGRADE, YUGOSLAVIA, 1970, PAPER 45, VOL. 2,
THEME 3, PP. 311-323.

70-0204 PINSON, L. D.
EVALUATION OF A FINITE ELEMENT ANALYSIS FOR THE
LONGITUDINAL VIBRATIONS OF LIQUID-PROPELLANT
LAUNCH VEHICLES
NASA REP. NO. TN D-5803, JUNE 1970.

70-0205 SILVESTER, P.
MADABUSHI, V. J.
FINITE ELEMENT SOLUTION OF SATURABLE MAGNETIC
FIELD PROBLEMS
IEEE TRANS. POWER APPARATUS SYSTEM, VOL. PA-89,
1970, PP. 1642-1651.

70-0206 SILVESTER, P.
CSENDES, Z.
FINITE ELEMENT SOLUTION OF DIELECTRIC LOADED
WAVEGUIDES
IEEE G-MTT INT. MICROWAVE SYMP., NEWPORT BEACH,
CALIFORNIA, DIGEST TECH. PAPER, MAY 11-14, 1970,
PP. 150-155.

70-0207 SMIRNOFF, T. P.
HRIBAR, J. A.
STEADY STATE IN SOIL SUBJECTED TO SEVERAL COUPLED
FORCES: A FINITE ELEMENT FORMULATION
SOIL SCI., VOL 110, NO. 3, 1970, PP. 183-190.

70-0208 WILSON, B. W.
FISHER, S. M.
TAYLOR, C.
PATIL, B. S.
ZIENKIEWICZ, O. C.
HARBOUR OSCILLATION - A NUMERICAL TREATMENT FOR
UNDAMPED NATURAL MODES
PROC. INST. CIVIL ENG., VOL. 46, 1970, PP.
203-211.

70-0209 IRONS, B. M.
ROLE OF PART-INVERSION IN FLUID-STRUCTURE PROBLEMS
WITH MIXED VARIABLES
AIAA J. VOL. 8, NO. 3, 1970, P. 568.

70-0210 CHEUNG, M. S.
CHEUNG, Y. K.
GHALI, A.
ANALYSIS OF SLAB AND GIRDER BRIDGES BY THE FINITE
STRIP METHOD
BUILD. SCI., VOL. 5, 1970, PERGAMON PRESS, G.
BRITAIN, PP. 95-104

70-0211 GALLAGHER, R. H.
LARGE SCALE COMPUTER PROGRAMS FOR STRUCTURAL
ANALYSIS
ASME CONF., GENERAL PURPOSE FINITE ELEMENT COMPUT.
PROGRAMS, NEW YORK, NOVEMBER 30, 1970, PP. 1-14.

70-0212 HELLEN, T. K.
BERSAFE - A COMPUTER SYSTEM FOR STRESS ANALYSIS BY
FINITE ELEMENTS
PROC. CONF. 'STRESS ANAL. TODAY' UNIVERSITY OF
SURREY, SEPTEMBER 23-25, 1970.

70-0213 MACNEAL, R. H.
DYNAMIC STRUCTURAL ANALYSIS WITH THE NASTRAN
COMPUTER PROGRAM
ASME CONF. GENERAL PURPOSE FINITE ELEMENT COMPUT.
PROGRAMS, NEW YORK, NOVEMBER 30, 1970, PP. 78-97.

70-0214 PRYOR, C. W. JR.
BARKER, R. M.
FINITE ELEMENT BENDING ANALYSIS OF REISSNER PLATES
ASCE J. ENG. MECH. DIV., VOL. 96, NO. EM6,
DECEMBER 1970, PAPER 7760, PP. 967-983.

70-0215 IQBAL, M. A.
KROKOSKY, E. M.
INTERACTION STRESSES IN COMPOSITE SYSTEMS
ASCE J. ENG. MECH. DIV., VOL. 96, NO. EM6,
DECEMBER 1970, PAPER 7738, PP. 825-845.

70-0216 COOK, E. L.
GOLDBERG, D. R.
FINITE ELEMENT STRUCTURAL ANALYSIS USING COMPUTERS
SAE-PAPER 700218, FOR MEETING MARCH 18-20, 1970,
10 PP.

70-0217 TANI, M.
EMORI, R. I.
STUDY ON AUTOMOBILE CRASH WORTHINESS
SAE-PAPER 700175 FOR MEETING JANUARY 12-16, 1970,
5 PP.

70-0218 SWEETSUR, G.
LAMELLAR TEARING IN WELDED GIRDERS
METAL CONSTR. BRIT. WELD. J. VOL. 2, NO. 3, MARCH
1970, PP. 103-107.

70-0219 HRENNIKOFF, A.
GANTAYAT, A. N.
THREE-DIMENSIONAL BAR CELL FOR ELASTIC STRESS
ANALYSIS
ASCE J. ENG. MECH. DIV., VOL. 96, NO. EM3, JUNE
1970, PAPER 7352, PP. 313-326.

70-0220 KHOJASTEH-BAKHT, M.
POPOV, E. P.
ANALYSIS OF ELASTIC-PLASTIC SHELLS OF REVOLUTION
ASCE J. ENG. MECH. DIV., VOL. 96, NO. EM3, JUNE
1970, PAPER 7354, PP. 327-340.

70-0221 CHEN, P. E.
LEWIS, T. B.
STRESS ANALYSIS OF RIBBON REINFORCED COMPOSITES
POLYM. ENG. SCI. VOL. 10, NO. 1, JANUARY 1970, PP.
43-47.

70-0222 WANG, F. D.
SUN, M. C.
SLOPE STABILITY ANALYSIS BY THE FINITE ELEMENT
STRESS ANALYSIS AND LIMITING EQUILIBRIUM METHOD
U.S. BUR. MINES-REPORT INVESTIAGTIONS 7341,
JANUARY 1970, 16PP.

70-0223 ADELMAN, H. M.
CATHERINES, D. S.
WALTON, W. C.
ACCURACY OF MODAL STRESS CALCULATIONS BY FINITE
ELEMENT METHOD
J. AIAA, VOL. 8, NO. 3, 1970, PP. 462-468, ALSO IN
J. AIAA., VOL. 7, 1969, PP. 156-157.

70-0224 BYSKOV, E.
CALCULATION OF STRESS INTENSITY FACTORS USING
FINITE ELEMENT METHOD WITH CRACKED ELEMENTS
INT. J. FRACTURE MECH., VOL. 6, NO. 2, 1970, PP.
159-167.

70-0225 DOUGLAS, M. R.
LUNNISS, R. C.
AN APPLICATION OF FINITE ELEMENT TECHNIQUE IN
BRIDGE DESIGN
CONCRETE, VOL. 4, NO. 5, 1970, PP. 197-200.

70-0226 KOLAR, V.
INFLUENCE FUNCTIONS IN FINITE ELEMENT METHOD
ZEITSCHRIFT FUR ANGEWANDTE MATHEMATIK UND
MECHANIK, VOL. 50, NO. 1/4, 1970, PP. T129-T131.

70-0227 FETTE, H.
CURVED FINITE ELEMENTS IN SHELL CALCULATIONS
ZEITSCHRIFT FUR ANGEWANDTE MATHEMATIK UND
MECHANIK, VOL. 50, NO. 1/4, 1970, PP. T115-T117.

70-0228 ROBINSON, J.
HAGGENMACHER, G. W.
OPTIMIZATION OF REDUNDANCY SELECTION IN
FINITE-ELEMENT FORCE METHOD
A.I.A.A.J., VOL. 8, NO. 8, 1970, PP. 1429-1433.

70-0229 BUFLER, H.
STEIN, E.
PLATE MEASUREMENTS USING FINITE ELEMENTS
INGENIEUR-ARCHIV, VOL. 39, NO. 4, 1970, PP.
248-260.

70-0230 MEI, C.
COUPLED VIBRATIONS OF THIN-WALLED BEAMS OF OPEN
SECTION USING FINITE ELEMENT METHOD
INT. J. MECH. SCI., VOL. 12, NO. 10, 1970, PP.
883-891.

70-0231 WEBSTER, J. J.
ACCURACY OF FINITE-ELEMENT SOLUTIONS FOR MODAL
CHARACTERISTICS OF SHELLS OF REVOLUTION
INT. J. MECH. SCI., VOL. 12, NO. 2, 1970, PP.
157-168.

70-0232 BREBBIA, C.
DEBNATH, J. M.
A COMPARISON OF RECENT SHALLOW SHELL
FINITE-ELEMENT ANALYSES
INT. J. MECH. SCI., VOL. 12, NO. 10, 1970, PP.
849-857.

70-0233 IRONS, B. M.
COMMENT ON STIFFNESS MATRICES FOR SECTOR ELEMENTS
AIAA J., VOL. 8, NO. 1, 1970, P. 191, ALSO IN AIAA
J., VOL. 7, 1969, P. 156-157.

70-0234 MURRAY, K. H.
COMMENTS ON CONVERGENCE OF FINITE ELEMENT
SOLUTIONS
A.I.A.A.J., VOL. 8, NO. 4, 1970, PP. 815-816.

70-0235 ZENISEK, A.
INTERPOLATION POLYNOMIALS ON THE TRIANGLE
NUMER. MATH., VOL. 15, 1970, PP. 283-296.

70-0236 SAWKO, F.
ANALYSIS OF SPINE BEAM BRIDGES USING FINITE
ELEMENTS
CIV. ENG. PUB. WORKS, REV., VOL. 65, NO. 763,
1970, PP. 146-147.

70-0237 DEHART, R. C.
GREIMANN, L. F.
PENETRATION IN SHELLS UNDER EXTERNAL PRESSURE
ASME PAPER 69-WA/UNT-4, 6 PP. ALSO IN J. ENG.
IND., TRANS. ASME, VOL. 92, SER. B., NO. 2, MAY
1970, PP. 269-274.

70-0238 BAHRANI, A. K.
COMMENT ON "SOLUTION OF LAPLACE'S EQUATION IN A
REGION WITH DIELECTRIC INTERFACES OF ARBITRARY
SHAPE"
PROC. IEEE, VOL. 58, NO. 5, MAY 1970, PP. 794-795.

70-0239 WHITE, D. J.
ENDERBY, L. R.
FINITE-ELEMENT STRESS ANALYSIS OF A NON-LINEAR
PROBLEM. A CONNECTING-ROD EYE LOADED BY MEANS OF
A PIN
J. STRAIN ANAL., VOL. 5, NO. 1, JANUARY 1970, PP.
41-49.

70-0240 TAYLOR, R. L.
PISTER, K. S.
THERMOMECHANICAL ANALYSIS OF VISCOELASTIC SOLIDS
INT. J. NUMER. METHODS ENG., VOL. 2, NO. 1,
JANUARY-MARCH 1970, PP. 45-59.

70-0241 SISODIYA, R. G.
CHEUNG, Y. K.
GHALI, A.
FINITE ELEMENT ANALYSIS OF SKEW, CURVED BOX-GIRDER
BRIDGE
"PUBLICATIONS", INTERNATIONAL ASSOCIATION FOR
BRIDGE AND STRUCTURAL ENGINEERING, VOL. 30 NO. 2,
ZURICH 1970, PP. 191-199

70-0242 ZENISEK, A.
SOME ELEMENT TYPES AND TRIAL FUNCTIONS IN THE
FINITE ELEMENT METHOD
STAVEBNICKY CAS., VOL. 18, NO. 1, 1970, PP. 48-62.

70-0243 MALCOM, D. J.
REDWOOD, R. G.
SHEAR LAG IN STIFFENED BOX GIRDERS
ASCE J. STRUCT. DIV., VOL. 96, NO. ST7, JULY 1970,
PAPER 7409, PP. 1403-1419.

70-0244 BABUSKA, I.
FINITE ELEMENT METHOD FOR DOMAINS WITH CORNERS
COMPUTING, VOL. 6, 1970, PP. 264-273.

70-0245 MELKES, F.
THE FINITE ELEMENT METHOD FOR NONLINEAR PROBLEMS
APLIKACE MATEMATIKY, VOL. 15, 1970, PP. 177-189.

70-0246 LANG, T. E.
HARTZ, B. J.
FINITE ELEMENT MATRIX FORMULATION OF POST-BUCKLING
STABILITY AND IMPERFECTION SENSITIVITY
PRESENTED AT THE IUTAM COLLOQUIUM ON HIGH SPEED
COMPUTING OF ELASTIC STRUCTURES, LIEGE, BELGIUM,
AUGUST 1970.

70-0247 HIBBITT, H. D.
LEVY, N.
MARCAL, P. V.
GENERAL PURPOSE PROGRAMS FOR NONLINEAR FINITE
ELEMENT ANALYSIS
PROC. 1ST A.S.M.E. ANN. MEETING, NEW YORK, 1970,
PP. 98-122.

70-0248 HIBBITT, H. D.
MARCAL, P. V.
HYBRID FINITE ELEMENT ANALYSIS WITH PARTICULAR
REFERENCE TO AXISYMMETRIC STRUCTURES
PROC. AIAA EIGHTH AEROSPACE SCIENCES MEETING,
PAPER NO. 70-137, 1970.

70-0249 AAMODT, B.
SHELL ANALYSIS BY SESAM-69
SEMINAR ON THE PRACTICAL APPLICATION OF THE FINITE
ELEMENT METHOD, NORWAY, JANUARY 1970.

70-0250 RASHID, Y. R.
THREE-DIMENSIONAL APPLICATION OF THE FINITE
ELEMENT METHOD
IUTAM COLLOQUIUM ON HIGH SPEED COMPUTING OF
ELASTIC STRUCTURES, UNIVERSITY OF LIEGE, BELGIUM,
AUGUST 23-28, 1970.

70-0251 KAMEL, H. A.
AUTOMATIC MESH GENERATION IN TWO AND THREE
DIMENSIONAL INTERCONNECTED DOMAINS
SYMP. ON HIGH SPEED COMPUT. OF ELASTIC STRUCT.,
LIEGE. (IUTAM), PAPER PRESENTED BY H. A. KAMEL ET
AL., 1970, PP. 1-40.

70-0252 RASHID, Y. R.
PRINCE, N.
ON FURTHER APPLICATION OF THE FINITE ELEMENT
METHOD TO THREE-DIMENSIONAL ELASTIC ANALYSIS
PROC. OF SYMPOSIUM ON HIGH SPEED COMPUTING OF
ELASTIC STRUCTURES, LEIGE, BELGIUM, 1970.

70-0253 SMITH, P. D.
ZUDANS, Z.
FINITE ELEMENT INCREMENTAL ELASTIC-PLASTIC
ANALYSIS OF PRESSURE VESSELS
TRANS. ASME, 92(2), J. ENGR. IND., MAY 1970, PP.
293-302.

70-0254 FORSBERG, K.
AN EVALUATION OF FINITE DIFFERENCE AND FINITE
ELEMENT TECHNIQUES FOR ANALYSIS OF GENERAL
SHELLS
PROC. OF SYMPOSIUM ON HIGH SPEED COMPUTING OF
ELASTIC STRUCTURES, LIEGE, BELGIUM, 1970.

70-0255 BELYTSCHKO, T.
HODGE, P. G.
PLANE STRESS LIMIT ANALYSIS BY FINITE ELEMENTS
PROC. ASCE, JOURNAL OF STRUCTURAL DIVISION, VOL.
96, EM6, DECEMBER 1970, PP. 931-44.

70-0256 THOMPSON, J. J.
CHEN, P. Y. P.
DISCONTINUOUS FINITE ELEMENTS IN THERMAL ANALYSIS
NUCL. ENGINEERING AND DESIGN, VOL. 14, 1970, PP.
211-22.

70-0257 SKJOLINGSTAD, L.
CHEUNG, Y. K.
THREE-DIMENSIONAL FIELD PROBLEMS BY HIGHER ORDER
FINITE ELEMENTS
REPORT, DEPT. OF CIVIL ENGR., UNIVERSITY OF
CALGARY, FEBRUARY 1970.

70-0258 GUYMON, G. L.
A FINITE ELEMENT SOLUTION OF THE ONE DIMENSIONAL
DIFFUSION CONVECTION EQUATION
WATER RESOURCES RESEARCH, VOL. 6, NO. 1, FEBRUARY
1970, PP. 204-210.

70-0259 YAMADA, Y.
THE STATUS AND OUTLOOK FOR THE SOLUTION OF
NONLINEAR PROBLEMS BY THE FINITE ELEMENT
METHOD
SEISAN-KENKYU, VOL. 22, NO. 1, 1970, PP. 6-14. (IN
JAPANESE)

70-0260 CONNOLLY, P. L.
APPLICATION OF SAMIS TO THE ANALYSIS OF A FRONT
SUSPENSION UPPER ARM
FORD REPORT 69-2001, JULY 1970.

70-0261 MCCORMICK, C. W.
HARDWARE AND SOFTWARE CONSIDERATIONS IN GENERAL
PURPOSE PROGRAMS FOR FINITE ELEMENT ANALYSIS
PROC. 1ST A.S.M.E. ANN. MEETING, NEW YORK, 1970,
PP. 68-77.

70-0262 JORDAN, W. B.
THE PLANE ISOPARAMETRIC STRUCTURAL ELEMENT
GENERAL ELECTRIC COMPANY, KAPL-M-7112,
SCHENECTADY, N. Y., FEB, 1970.

70-0263 IRONS, B. M.
DISCUSSION OF FINITE ELEMENT TECHNIQUES IN
STRUCTURAL MECHANICS
SOUTHAMPTON UNIV. PRESS, H. TOTTENHAM AND C.
BREBBIA (EDS.), 1970, PP. 328-331.

70-0264 TONG, P.
ON THE NUMERICAL PROBLEMS BY THE FINITE ELEMENT
METHODS
COMPUTER AIDED ENGINEERING, EDITED BY GLADWALL, U.
OF WATERLOO, WATERLOO, CANADA, 1970, PP.
539-560.

70-0265 HUSSEY, M. J. L., ET AL
ON THE CONSTRUCTION AND USE OF FINITE ELEMENTS
JOURNAL OF THE INSTITUTE OF MATHEMATICS AND ITS
APPLICATIONS, VOL. 6, NO. 3, 1970, PP
263-282.

70-0266 SKJOLINGSTAD, L.
CHEUNG, Y. K.
VENART, J. E. S.
THE APPLICATION OF HIGHER ORDER FINITE ELEMENTS TO
THREE DIMENSIONAL HEAT CONDUCTION PROBLEMS
PROC. 3RD WESTERN CAN. CONF. ON HEAT TRANSFER,
EDMONTON, MAY, 1970.

70-0267 CHOWDHURY, R. N.
THE APPLICATION OF THE FINITE ELEMENT METHOD TO
SEEPAGE AND STRESS-DEFORMATION PROBLEMS IN SOIL
MECHANICS
PH.D. THESIS, UNIVERSITY OF LIVERPOOL, ENGLAND,
1970.

70-0268 MCCORQUODALE, J. A.
FINITE ELEMENT ANALYSIS OF NON-DARCY FLOW
PH. D. THESIS, UNIVERSITY OF WINDSOR, WINDSOR,
CANADA, 1970.

70-0269 GUYMON, G. L.
SCOTT, V. H.
APPLICATION OF FINITE ELEMENT METHOD TO GENERAL
NUMERICAL SOLUTION OF 2-DIMENSIONAL
DIFFUSION-CONVECTION EQUATION
TRANSACTIONS, AMERICAN GEOPHYSICAL UNION VOL. 51,
ISSUE N4, 1970, PAGE 282.

70-0270 CHEUNG, Y. K.
CHEUNG, M. S.
FREE VIBRATION OF ORTHOTROPIC VARIABLE THICKNESS,
CONTINUOUS RECTANGULAR PLATES
REVUE ROMAINE DES SCIENCES TECHNIQUE MECANIQUE
APPLIQUEE, NO. 3, 1970.

70-0271 GHALI, A.
BATHE, K. J.
ANALYSIS OF PLATES SUBJECTED TO IN-PLANE FORCES
USING LARGE FINITE ELEMENTS
PUBLICATIONS, INT. ASSOC. FOR BRIDGE AND STRUCT.
ENGR., VOL. 30, NO. 1, ZURICH 1970, PP. 61-72.

70-0272 CHEUNG, Y. K.
ANALYSIS OF ORTHOTROPIC FOLDED PLATES
PAPER PRESENTED AT IASS SYMPOSIUM ON FOLDED PLATE
CONSTRUCTION, VIENNA, SEPTEMBER 1970.

70-0273 CHEUNG, Y. K.
WATSON, M.
SKJOLINGSTAD, L.
STEADY STATE AND TRANSIENT GROUNDWATER FLOW IN
SLOPES
PAPER PRESENTED AT THE 23RD ANNUAL CANADIAN
GEOTECHNICAL CONFERENCE, BANFF, NOVEMBER 1970.

70-0274 ZARGHAMEE, M. S.
MINIMUM WEIGHT DESIGN WITH STABILITY CONSTRAINT
ASCE, J. STRUCT. DIV., V. 96, N. ST8, AUG 1970,
PAPER 7486, PP. 1697-1710.

70-0275 LEVY, N.
FINITE ELEMENT FORMULATION OF STRESS STRAIN MATRIX
FOR AN ELASTICALLY PERFECTLY PLASTIC TRESCA
MATERIAL
BROWN UNIVERSITY, TECH. REP., NGL40-002-080/2,
1970.

70-0276 DAVIES, J. D.
CHEUNG, Y. K.
GORECKI, A.
ANALYSIS OF LONG RECTANGULAR TANKS
WATER AND WATER ENGINEERING, DECEMBER 1970

70-0277 ZLAMAL, M.
A FINITE ELEMENT PROCEDURE OF THE SECOND ORDER
ACCURACY
NUMER. MATH., VOL. 14, 1970, PP. 394-402.

70-0278 ROBINSON, J.
HAGGENMACHER, G. W.
EVALUATION OF CALAC SEMI-MONOCOQUE MEMBRANE
ELEMENT
LOCKHEED - CALIFORNIA COMPANY REPORT NO. 24055,
1970.

70-0279 CREAGER, M.
DEVELOPMENT OF A CRACKED FINITE ELEMENT
LOCKHEED CALIFORNIA COMPANY REPORT, NO. 23996,
1970.

70-0280 DESAI, C. S.
ULTIMATE CAPACITY OF CIRCULAR FOOTINGS ON LAYERED
SOILS
J. INDIAN NAT. SOC. SOIL MECH. FOUND., ENG., V. 9,
N. 1, JAN 1970, PP. 41-50.

70-0281 DESAI, C. S.
REESE, L. C.
ANALYSIS OF CIRCULAR FOOTINGS ON LAYERED SOILS
ASCE, J. SOIL MECH. FOUND. DIV., V. 96, N. SM4,
JULY 1970, PAPER 7408, PP. 1289-1310.

70-0282 ZIENKIEWICZ, O. C.
BEST, B.
DULLAGE, C.
STAGG, K. G.
ANALYSIS OF NONLINEAR PROBLEMS IN ROCK MECHANICS
WITH PARTICULAR REFERENCE TO JOINTED ROCK SYSTEMS
PROC. 2ND INT. CONG. ON ROCK MECHANICS, BELGRADE,
1970

70-0283 ROCKEY, K. C.
BAGCHI, D. K.
BUCKLING OF PLATE GIRDER WEBS UNDER PARTIAL EDGE
LOADINGS
INT. J. MECH. SCI., VOL. 12, 1970, PP. 61-76.

70-0284 PIAN, T. H. H.
TONG, P.
VARIATIONAL FORMULATION OF FINITE DISPLACEMENT
ANALYSIS
SYMP. INT. UN. TH. APPL. MECH. ON HIGH SPEED
COMPUTING OF ELASTIC STRUCTURES, LIEGE, 1970.

70-0285 KING, I. P.
AN AUTOMATIC RECORDING SCHEME FOR SIMULTANEOUS
EQUATIONS DERIVED FROM NETWORK SYSTEMS
INT. J. NUM. METH. ENG., 1970.

70-0286 RIGBY, G. L.
MIXED DISPLACEMENT 3D ISOPARAMETRIC FINITE
ELEMENTS AND A STRAIN ENERGY BASIS FOR STUDIES OF
ELEMENT GEOMETRY
M.A.SC. THESIS, UNIVERSITY OF WATERLOO, 1970.

70-0287 WHITESTONE, W. D.
YEN, C. L.
COMPARISON OF MEMBRANE FINITE ELEMENT FORMULATIONS
LMSC-HREC D162553, LOCKHEED MISSILES AND SPACE
CO., HUNTSVILLE, ALA., SEPTEMBER 1970.

70-0288 YEN, C. L.
COMPARISON OF PLATE BENDING FINITE ELEMENT
FORMULATIONS
LOCKHEED MISSILES AND SPACE CO., HUNTSVILLE, ALA.,
LMSC-HREC D162811, DECEMBER 1970.

70-0289 KAMEL, H. A.
ANALYSIS OF LARGE CLOSED SHIP STRUCTURES:
APPLICATIONS TO TANKER STRUCTURAL DESIGN
SEM. ON FINITE ELEMENT METHODS OF SHIP STRUCTURE
ANALYSIS, WEBB INST. OF NAVAL ARCHITECTURE, NEW
YORK, JUNE 1970.

70-0290 GALLAGHER, R. H.
YAMADA, Y.
ODEN, J. T.
RECENT ADVANCES IN MATRIX METHODS IN STRUCTURAL
ANALYSIS AND DESIGN
PROC. JAPAN-U.S. SEMINAR MATRIX METHODS STRUCT.
ANALYSIS AND DESIGN, TOKYO, AUGUST 1969, UNIV. OF
ALABAMA PRESS, ALABAMA 1970.

70-0291 PRICE, H. S.
VARGA, R. S.
ERROR BOUNDS FOR SEMIDISCRETE GALERKIN
APPROXIMATIONS OF PARABOLIC PROBLEMS WITH
APPLICATIONS TO PETROLEUM RESERVOIR MECHANICS
IN G. BIRKHOFF AND A. S. VARGA (EDS.), NUMERICAL
SOLUTION OF FIELD PROBLEMS IN CONTINUUM PHYSICS,
PROC. SOC. IND. APPL. MECH.-AM. MATH. SOC., VOL.
2, 1970.

70-0292 BRYOR, C. W.
BARKER, R. M.
FINITE ELEMENT ANALYSIS OF BENDING-EXTENSIONAL
COUPLING IN LAMINATED COMPOSITES
J. OF COMP. MATER., VOL. 4, 1970, PP. 549-552.

70-0293 ANON
FINITE ELEMENT COMPUTER ANALYSIS OF KOCHMUS HULL
520.
CHEVRON SHIPPING COMPANY, SAN FRANCISCO, CALIF.,
MARCH 1970.

70-0294 ANON
FINITE ELEMENT COMPUTER ANALYSIS OF MITSUBISHI
HULL 1670
CHEVRON SHIPPING COMPANY, SAN FRANCISCO, CALIF.,
JULY 1970.

70-0295 FELIPPA, C. A.
CLOUGH, R. W.
THE FINITE ELEMENT METHOD IN SOLID MECHANICS
NUMERICAL SOLUTION OF FIELD PROBLEMS IN CONTINUUM
PHYSICS SIAM-AMS PROCEEDINGS, VOL. II, AMER.
MATH. SOC., PROVIDENCE,R.I., 1970.

70-0296 CULHAM, W. E.
VARGA, R. S.
NUMERICAL METHODS FOR TIME DEPENDENT NONLINEAR
BOUNDARY VALUE PROBLEMS
PROC. 2ND SYMP. ON NUMERICAL SIMULATION RESERVOIR
PERFORMANCE, SOC. PETROL. ENG., A.I.M.E., 1970.

70-0297 BARLOW, J.
FINITE ELEMENTS FOR IMPROVED SKIN STRINGER
STRUCTURAL ANALYSIS
AERO STRESS MEMO. 2637, ROLLS-ROYCE LTD., DERBY,
DECEMBER 1970.

70-0298 MACNEAL, R. H. (ED)
THE NASTRAN THEORETICAL MANUAL
NASA SP-221, SEPTEMBER, 1970.

70-0299 VARANASI, S. R.
CAREY, G. F.
ELASTO PLASTIC ANALYSIS OF PLANE STRUCTURES BY THE
FINITE ELEMENT DISPLACEMENT METHOD
THE BOEING CO., DOC. ADS., CAD RENTON 70-36, DOC.
D6-24524, 1970.

70-0300 BABUSKA, I.
A REMARK ON THE FINITE ELEMENT METHOD
COMMENT. MATH. UNIV. CAROLINA, VOL 12, NO. 1,
1971, PP. 367-375. TECH. NOTE BN-675, NOVEMBER
1970, INST. FOR FLUID DYNAMICS AND APPLIED
MATHEMATICS, UNIV. OF MARYLAND.

70-0301 GUGLIELMO, F. D.
METHODE DES ELEMENTS FINIS; UNE FAMILE
D'APPROXIMATION DES ESPACES DE SOBOLEV PAR LES
TRANSLATES DE P FUNCTIONS
CALCOLO, VOL. 7, 1970, PP. 185-234.

70-0302 KAWAZURA, A.
WAKESUGI, T.
BANDED MATRIX TECHNIQUES AND SOLUTIONS OF LINEAR
SIMULTANEOUS EQUATIONS OF FINITE ELEMENT
ANALYSES
PROCEEDINGS OF 20TH JAPAN NATIONAL CONGRESS FOR
THEORETICAL AND APPLIED MECHANICS, 1970, PP.
275-280.

70-0303 BABUSKA, I.
SEGETHOVA, J.
SEGETH, K.
NUMERICAL EXPERIMENT WITH FINITE ELEMENT I
TECH. NOTE BN-669, AUGUST 1970, INSTITUTE FOR
FLUID DYNAMICS AND APPLIED MATHEMATICS, UNIVERSITY
OF MARYLAND.

70-0304 BABUSKA, I.
COMPUTATION OF DERIVATIVES IN THE FINITE ELEMENT
METHOD
COMMENT MATH. UNIV. CAROLINAE, VOL. 11, NO. 3,
1970, PP. 545-558. TECH. NOTE BN-650, APRIL 1970,
INSTITUTE FOR FLUID DYNAMICS AND APPLIED
MATHEMATICS, UNIVERSITY OF MARYLAND

70-0305 BARTEN, H. J.
SCHEURENBRAND, J. A.
SCHEER, D. D.
STRESSES AND VIBRATION ANALYSIS OF INDUCER BLADES
USING FINITE ELEMENT TECHNIQUE
AIAA, 6TH PROP. JOINT SPEC. CONF., AIAA PAPER NO.
70-630, 1970.

70-0306 SANDER, G.
APPLICATION OF THE DUAL ANALYSIS PRINCIPLE
PROC. IUTAM COLLOQIUM ON HIGH SPEED COMPUTING OF
ELASTIC STRUCTURES, LIEGE, BELGIUM, AUGUST
1970.

70-0307 SEVERN, R. T.
INCLUSION OF SHEAR DEFLECTION IN THE STIFFNESS
MATRIX OF A BEAM ELEMENT
J. STRAIN ANALYSIS, VOL. 5, 1970.

70-0308 RAZZAQUE, A.
A CONFORMING TRIANGULAR ELEMENT IN THE ANALYSIS OF
THIN AND SEMI-THICK PLATES IN BENDING
M. SC. THESIS, UNIVERSITY OF WALES, 1970.

70-0309 SCHKADE, A. F.
A REFINED AXISYMMETRIC FINITE ELEMENT FOR THE
ANALYSIS OF NEARLY INCOMPRESSIBLE SOLIDS
PH.D. THESIS, UNIV. OF TEXAS, AUSTIN, 1970.

70-0310 NORRIE, D. H.
DE VRIES, G.
APPLICATION OF THE FINITE ELEMENT METHOD TO
UNSTEADY FLOW PROBLEMS
N.R.C. BLADE SEMINAR, OTTAWA, 1970

70-0311 DE VRIES, G.
BERARD, G. P.
NORRIE, D. H.
APPLICATION OF THE FINITE ELEMENT TECHNIQUE TO
COMPRESSIBLE FLOW PROBLEMS
REPORT NO. 18, MECHANICAL ENGINEERING DEPARTMENT,
UNIVERSITY OF CALGARY, CANADA, AUGUST, 1970

70-0312 WOODMAN, N. J.
SEVERN, R. T.
DOUBLE CURVATURE SHELL FINITE-ELEMENT AND ITS USE
IN THE DYNAMIC ANALYSIS OF COOLING TOWERS AND
OTHER SHELL STRUCTURES
SYMP. ON STRUCTURAL DYNAMICS, PROC. MARCH 23-25,
1970, LOUGHBOROUGH UNIV. OF TECHNOL., DEPT. OF
TRANSPORT TECHNOL., ENGL., BY BRIT. ACOUSTICAL
SOC., ET. AL., VOL. 1, PAPER A.5, 18 PP.

70-0313 REDDY, D. V.
USE OF FINITE ELEMENT METHOD IN THE DYNAMICS OF
PLATES AND SHELLS
PROC. 15TH CONGRESS OF THEORETICAL AND APPLIED
MECHANICS, SINDRI, INDIA, 1970, PP. 413-424.

70-0314 CHANG, T. Y.
RASHID, Y. R.
VISCOELASTIC RESPONSE OF GRAPHITE MATERIALS IN
IRRADIATION ENVIRONMENTS
NUCLEAR ENGINEERING AND DESIGN, VOL. 14, 1970, PP.
181-190.

70-0315 PAWSEY, S. F.
THE ANALYSIS OF MODERATELY THICK TO THIN SHELLS BY
THE FINITE ELEMENT METHOD
SESM REPORT 70-12, STRUCTURAL ENGINEERING
LABORATORY, UNIV. OF CALIFORNIA, BERKELEY,
1970 P. 102

70-0316 APPA, K.
KINEMATICALLY CONSISTENT UNSTEADY AERODYNAMIC
COEFFICIENTS IN SUPERSONIC FLOW
INT. J. FOR NUM. METH. IN ENG., VOL. 2, NO. 4,
1970, PP. 495-509.

70-0317 IRONS, B. M.
ZIENKIEWCIZ, O. C.
AHMAD, S.
A SIMPLE MATRIX-VECTOR HANDLING SCHEME FOR
THREE-DIMENSIONAL AND SHELL ANALYSIS
INT. J. FOR NUM. METH. IN ENG., VOL. 2, NO. 4,
1970, PP. 509-523.

70-0318 ODEN, J. T.
RIGSBY, D. M.
CORNETT, D.
ON THE NUMERICAL SOLUTION OF A CLASS OF PROBLEMS
IN A LINEAR FIRST STRAIN-GRADIENT THEORY OF
ELASTICITY
INT. J. FOR NUM. METH. IN ENG., VOL. 2, NO. 2,
1970, PP. 159-175.

70-0319 RIZZO, R. R.
A FINITE ELEMENT ANALYSIS OF LAMINATED ANISOTROPIC
TUBES
JOURNAL OF COMPOSITE MATERIALS VOL. 4, 1970, P.
344.

70-0320 ANON
EAC/EASE USER INFORMATION MANUAL
MINNEAPOLIS, MINNESOTA, CONTROL DATA CORP. DATA
SERVICES DIVISION, PUBLICATION NO.
D0001200602,1970.

70-0321 BARSOUM, R. S.
DISCUSSION OF PROCEEDINGS, PAPER 6612, ELASTIC
BUCKLING OF TWO-COMPONENT COLUMNS
J. ENG. MECH. DIV. PROC. AM. SOC. CIV. ENGRS.,
VOL. 96, 1970

70-0322 GUPTA, K. K.
VIBRATION OF FRAMES AND OTHER STRUCTURES WITH
BANDED STIFFNESS MATRIX
INT. J. NUM. METH. ENGNG., VOL. 2, 1970, PP.
221-228.

70-0323 LYNN, P. O.
BORESI, A. P.
KINDS OF CONVERGENCE AND IMPROVED CONVERGENCE OF
CONFORMING FINITE ELEMENT SOLUTIONS IN PLATE
BENDING
NUCL. ENGNG. DES., VOL. 11, 1970, PP. 159-176.

70-0324 SCHNOBRICH, W. C.
ANALYSIS OF HYPERBOLIC PARABOLOID SHELLS
PAPER PRESENTED AT THE AMERICAN CONCRETE INST.
MEETING, NEW YORK, 1970.

70-0325 OLSEN, M. D.
LINDBERG, G. M.
FINITE ELEMENTS FOR PLATE BENDING
INT. J. MECH. SCI., VOL. 2, 1970, PP. 17-33.

70-0326 ZENISEK, A.
INTERPOLATION POLYNOMIALS ON THE TRIANGLE ELEMENT
NUM. MATH., VOL. 45, 1970, PP. 283-296.

70-0327 HOLUSA, L.
KRATOCHVIL, J.
ZLAMAL, M.
ZENISEK, A.
THE FINITE ELEMENT METHOD
REPORT, IN CZECH, COMPUTING CENTRE OF THE
TECHNICAL UNIV., BRNO, 1970.

70-0328 MOTE, C. D.
FORMULATION OF DISCRETE ELEMENT MODELS FOR STRESS
AND VIBRATION ANALYSIS OF PLATES
FOREST PRODUCTS LAB., REPORT 35.01.77, UNIVERSITY
OF CALIF., BERKELEY, CALIF., 1970.

70-0329 CLOUGH, R. W.
JOHNSON, C. P.
FINITE ELEMENT ANALYSIS OF ARBITRARY THIN SHELLS
PROCEEDINGS ACI SYMPOSIUM ON CONCRETE THIN SHELLS,
NEW YORK, 1970.

70-0330 ROCKWELL, R. D.
COMPUTER-AIDED INPUT/OUTPUT FOR USE WITH FINITE
ELEMENT METHOD OF STRUCTURAL ANALYSIS
IEEE COMPUTER GROUP NEWS, VOL. 3, NO. 3, 1970, P.
66.

70-0331 STRICKLIN, J. A.
HAISLER, W. E.
VON RIESEMANN, W. A.
GEOMETRICALLY NONLINEAR ANALYSIS BY THE DIRECT
STIFFNESS METHOD
REPORT 70-16, AEROSPACE ENGINEERING DEPT., TEXAS A
& M UNIV. COLLEGE STATION, TEXAS, AUGUST 1970.

70-0332 VOS, R. F.
FINITE ELEMENT ANALYSIS OF PLATE BUCKLING AND
POST-BUCKLING
PH.D. THESIS, RICE UNIV., HOUSTON, TEXAS, 1970.

70-0333 ANON
AXISYMMETRIC RING AND PLATE STRESS ANALYSIS
PROGRAM - USERS GUIDE
STRUCTURAL SCIENCES RESEARCH INC. LOS ANGELES,
CALIF. 1970.

70-0334 GRIFFIN, D. S.
THE VERIFICATION AND ACCEPTANCE OF COMPUTER
PROGRAMS FOR DESIGN ANALYSIS
IN: SYMPOSIUM ON GENERAL PURPOSE FINITE ELEMENT
COMPUTER PROGRAMS, NEW YORK: AM. SOC. MECH. ENG.,
1970, PP. 1443-1450.

70-0335 MUFTI, A. A.
MIRZA, M. S.
MCCUTCHEON, J. O.
HOUDE, J.
A STUDY OF BEHAVIOUR OF REINFORCED CONCRETE
ELEMENTS USING FINITE ELEMENTS
STRUCTURAL CONCRETE SERIES NO. 70-5, MCGILL
UNIVERSITY, MONTREAL, P. 23, 1970.

70-0336 COWPER, G. R.
KOSKO, E.
LINDBERG, G. M.
OLSON, M. D.
STRUCTURAL DESIGN AND ANALYSIS OF THE SPACE
SHUTTLE
STRUCT. DESIGN TECH. SESSION, SPACE SHUTTLE TECH.
CONF., NASA LEWIS RESEARCH CENTER, CLEVELAND,
OHIO, JULY 15-17, 1970.

70-0337 BARSOUM, R. S.
A FINITE ELEMENT FORMULATION FOR THE GENERAL
STABILITY ANALYSIS OF THIN WALLED MEMBERS
PH.D. THESIS, CORNELL UNIV., 1970.

70-0338 KAGAWA, Y.
GLADWELL, G. M. L.
FINITE ELEMENT ANALYSIS OF FLEXURE - TYPE
VIBRATORS WITH ELECTROSTRICTIVE TRANSDUCERS
IEEE TRANS. SONICS ULTRASONICS VOL. SU-17, NO. 1,
JANUARY 1970, PP. 41-49.

70-0339 CARMICHAEL, R. L.
RECENT EXPERIENCE IN USING FINITE ELEMENT METHODS
FOR THE SOLUTION OF PROBLEMS IN AERODYNAMIC
INTERFERENCE
AGARD CONF. PROC. NO. 71, FLUID DYNAMICS PANEL
SPECIALISTS' MEETING, SILVER SPINGS, MD.,
SEPTEMBER 28-30, 1970, PAPER 4, 5 PP.

70-0340 SELNA, L.
DYNAMIC ANALYSIS OF AUTOMOTIVE STRUCTURAL SYSTEMS
SAE PAP. 700844 FOR MEETING OCTOBER 5-9, 1970, 8
PP.

70-0341 HIGUCHI, M.
HORIOKA, M.
STRESS ANALYSIS IN HULL STRUCTURAL MEMBERS BY
PHOTOELASTIC EXPERIMENTS AND FINITE ELEMENT
METHODS - 1
NIPPON KOKAN TECH. REP., OVERSEAS NO. 11, DECEMBER
1970, PP. 97-105.

70-0342 HAMID, M. S.
WATWOOD, V. B.
FINITE ELEMENT METHOD FOR PREDICTION OF CRACK
BEHAVIOUR
NUCL. ENG. DES., VOL. 11, NO. 2, 1969, PP.
323-332.

70-0343 LEE, C. H.
KOBAYASHI, A. S.
ELASTOPLASTIC ANALYSIS OF PLANE STRAIN AND
AXISYMMETRIC FLAT PUNCH INDENTATION BY THE FINITE
ELEMENT METHOD
INT. J. MECH. SCI., VOL. 12, NO. 4, APRIL 1970,
PP. 349-370.

70-0344 JANSSEN, J. D.
POSSIBILITIES OF THE METHOD OF FINITE ELEMENTS IN
THE CALCULATION OF THE STRENGTH AND STIFFNESS OF
THIN-WALLED BEAMS
INGENIEUR (HAGUE) VOL. 82, NO. 43, OCTOBER 23,
1970, PP. 121-129.

70-0345 BUFLER, H.
STEIN, E.
CALCULATION OF PLATES BY MEANS OF METHOD OF FINITE
ELEMENTS
ING-ARCH. VOL. 39, NO. 4, 1970, PP. 248-260.

70-0346 SMITH, I. M.
DUNCAN, W.
EFFECTIVENESS OF EXCESSIVE NODAL CONTINUITIES IN
THE FINITE ELEMENT ANALYSIS OF THIN RECTANGULAR
AND SKEW PLATES IN BENDING
J. NUMER. METHODS ENG., VOL. 2, NO. 2, APRIL-JUNE
1970, PP. 253-257.

70-0347 APERGHIS, G. G.
SLAB BRIDGE DESIGN AND DRAWING. AN AUTOMATED
PROCESS
PROC. INST. CIV. ENG., VOL. 46, MAY 1970, PP.
55-75.

70-0348 TOTTENHAM, H.
TAHBILDAR, U. C.
FINITE ELEMENT ANALYSES OF SHELL RESPONSE FOR
ARBITRARY EXCITATIONS
SYMP. ON STRUCTURAL DYNAMICS, PROC. MARCH 23-25,
1973, LOUGHBOROUGH, ENGLAND, VOL. 1, PAPER A.4,
25 PP.

70-0349 MALLETT, R. H.
JORDAN, S.
STRESSES ANALYSIS USING FINITE ELEMENT METHODS
ASME, ON GENERAL PURPOSE FINITE ELEMENT COMPUTER
PROGRAMS, NEW YORK, NEW YORK, NOVEMBER 30, 1970,
PP. 30-67.

70-0350 HARTUNG, R. F.
LODEN, W. A.
AXISYMMETRIC VIBRATION OF CONICAL SHELLS
J. SPACECRAFT ROCKETS, VOL. 7, NO. 10, OCTOBER
1970, PP. 1153-1159.

70-0351 LUTZ, L. A.
ANALYSIS OF STRESSES IN CONCRETE NEAR A
REINFORCING BAR DUE TO BOND AND TRANSVERSE
CRACKING
J. AMER. CONCRETE INST. VOL. 67, NO. 10, OCTOBER
1970, PP. 778-787.

70-0352 SARGIOUS, M.
PRINCIPAL STRESSES AT THE INTERMEDIATE SUPPORT OF
PRESTRESSED CONCRETE CONTINUOUS BEAMS
J. AMER. CONCRETE INST., VOL. 67, NO. 10, OCTOBER
1970, PP. 828-836.

70-0353 HART, G. C.
TREATMENT OF RANDOMNESS IN FINITE ELEMENT MODELING
SAE PAP 700842 FOR MEETING OCTOBER 5-9, 1970, 12
PP.

70-0354 BORONKAY, T. G.
MEI, C.
ANALYSIS AND DESIGN OF MULTIPLE INPUT FLEXIBLE
LINK MECHANISMS
J. MECHANISMS, VOL. 5, NO. 1, 1970, PP. 29-40.

70-0355 MELOSH, R. J.
CHARACTERISTICS OF MANIPULATION ERRORS IN SOLVING
LOAD DEFLECTION EQUATIONS
ASME, ON GENERAL PURPOSE FINITE ELEMENT COMPUTER
PROGRAMS, NEW YORK, NOVEMBER 1970, PP. 123-142.

70-0356 LEVY, M. J.
MARCAL, P. V.
RICE, J. R.
DEVELOPMENT OF FINITE ELEMENT COMPUTATIONAL
METHODS FOR THREE DIMENSIONAL ELASTIC PLASTIC
CRACK ANALYSIS
NUCLEAR ENGINEERING AND DESIGN, VOL. 14, NO. 2,
1970, P. 356

70-0357 ARGYRIS, J. H.
SCHARPF, D. W.
TWO AND THREE DIMENSIONAL POTENTIAL FLOW ANALYSIS
BY THE METHOD OF SINGULARITIES
NUCL. ENG. DES., VOL. 11, NO. 2, 1969, PP.
237-240.

70-0358 SCHREM, R.
AN AUTOMATIC SYSTEM FOR KINEMATIC ANALYSIS ASKA
HIGH SPEED COMPUTING OF ELASTIC STRUCTURES
PROC. OF SYMP. IUTAM-AUGUST 1970.

70-0359 DEPRES, J.
THIN SHELL THEORY IN CARTESIAN COORDINATES -
METHODS FOR FINITE ELEMENTS
BULLETIN DE LA CLASSE DES SCI. ACAD. ROYALE DE
BELGIQUE, VOL. 56, NO. 10, 1970, P. 1155.

70-0360 GALLAGHER, R. H.
FINITE ELEMENT ANALYSIS IN BRITTLE MATERIAL DESIGN
J. FRANKLIN INSTITUTE, VOL. 290, NO. 6, 1970, P.
523.

70-0361 HODGE, P. G.
CONSISTENT FINITE ELEMENT MODEL FOR TWO
DIMENSIONAL CONTINUUM
INGENIEUR ARCHIV, VOL. 39, NO. 6, 1970, P. 375.

70-0362 HUMPHREY, A. T.
APPLICATION OF FINITE ELEMENT TECHNIQUES TO STRESS
AND STRUCTURAL ANALYSIS
MARCONI REVIEW, VOL. 33, NO. 179, 1970, P. 315.

70-0363 JIROUSEK, J.
BENDING OF CYLINDRICAL SHELLS - SOLUTION USING
CYLINDRICAL STRIP ELEMENTS - CONTRIBUTION TO
FINITE ELEMENT METHODS
ACTA POLYTECHNICA SCANDINAVICA. MATHEMATICS AND
COMPUT., VOL. 1970, NO. 20, 1970, P. 3.

70-0364 POWELL, G.
KLINGNER, R.
ELASTIC LATERAL BUCKLING OF STEEL BEAMS
ASCE J. STRUCT. DIV., VOL. 96, NO. SR9, SEPTEMBER
1970, PAPER 7555, PP. 1919-1932.

70-0365 CHUNG, T. J.
MASTERS, A. E.
DYNAMIC RESPONSE OF ARBITRARY SHELLS OF REVOLUTION
UNDER COMBINED IMPULSIVE NONISOTHERMAL AND
IRREGULAR MECHANICAL LOADINGS
SHOCK AND VIBRATION BULLETIN, VOL. 41, NO. 7,
DECEMBER 1970.

70-0366 CLARY, R. R.
TURNER, L. J.
MEASURED AND PREDICTED LONGITUDINAL VIBRATIONS OF
A LIQUID PROPELLANT TWO-STAGE LAUNCH VEHICLE
NASA-TN-D-5943 NASA-LANGLEY, AUGUST 1970, 61 PP.

70-0367 DONG, S. B.
SELNA, L. G.
NATURAL VIBRATIONS OF LAMINATED ORTHOTROPIC SHELLS
OF REVOLUTION
J. COMPOS. MATER., VOL. 4, 1970, PP. 2-19.

70-0368 FURUIKE, T.
THE EFFECT OF AN INITIAL AXISYMMETRIC IMPERFECTION
ON THE NATURAL FREQUENCIES OF CONICAL SHELLS
UNIV. MICROFILMS 71-7710, UNIVERSITY OF SOUTH
CALIF., 257 PP., 1970.

70-0369 LYNCH, F. D.
DYNAMIC RESPONSE OF A CONSTRAINED FIBROUS SYSTEM
SUBJECTED TO TRANSVERSE IMPACT. II. A MECHANICAL
MODEL
REPORT AMMRC-TR-70-16, JULY 1970.

70-0370 MCCOMAS, R. L.
NASTRAN, NASA'S GENERAL PURPOSE STRUCTURAL
ANALYSIS PROGRAM
N71-29099 NASA-MARSHALL SPACE FLIGHT CTR.,
NOVEMBER 1, 1970, 4 PP.

70-0371 MEYER, C.
SCORDELIS, A. C.
COMPUTER PROGRAM FOR PRISMATIC FOLDED PLATES BY
PLATE AND BEAM ELEMENTS
REPORT: SESM-70-3, FEBRUARY 1970.

70-0372 NEUBERT, V. H.
LEE, H.
FINITE BEAM ELEMENTS FOR DYNAMIC ANALYSIS
SHOCK, AND VIBRATION BULL., VOL. 41, NO. 6,
DECEMBER 1970.

70-0373 RAO, G. V.
MURTY, A. V. K.
A COMPARATIVE STUDY OF THE CONSISTENT AND
SIMPLIFIED FINITE ELEMENT ANALYSES OF EIGENVALUE
PROBLEMS
J. AERONAUT. SOC. INDIA, VOL. 22, NO. 3, AUGUST
1970, PP. 183-186.

70-0374 RUBIN, C. P.
MINIMUM WEIGHT DESIGN OF COMPLEX STRUCTURES
SUBJECT TO A FREQUENCY CONSTRAINT
AIAA J., VOL. 8, NO. 5, MAY 1970, PP. 923-927.

70-0375 SAINI, S. S.
CHANDRASEKARAN, A. R.
INFLUENCE OF RESERVOIR WATER ON EARTHQUAKE
RESPONSE OF EARTH DAMS
J. INST. ENG. INDIA, CIV. ENG. DIV., VOL. 50, NO.
11, JULY 6, 1970, PP. 354-358.

70-0376 UTKU, S.
ON SMALL VIBRATIONS AND PERTURBATIONS OF FLEXIBLE
BODIES UNDERGOING ARBITRARY NOMINAL MOTION
CALIF. INST. TECHNOL., PASADENA, U.S.A., REPORT
NASA-CR-110435, JUNE 1970.

70-0377 TABANDEH, N.
CONVERGENCE OF THE FINITE ELEMENT METHOD
PH.D. THESIS, LOUISIANA STATE UNIV. AND AGRI. AND
MECH. COLL., 1970.

70-0378 WELCH, R. E.
THREE DIMENSIONAL STRUCTURES/MEDIA INTERACTION,
DYNAX DEVELOPMENT AND APPLICATIONS,
IIT. RES. INST., MAY 1973, VO. 1, 250 PP.

70-0379 YEH, C. H.
LARGE DEFLECTION DYNAMIC ANALYSIS OF THIN SHELLS
USING FINITE ELEMENT METHOD
UNIV. CALIF., UNIV. MICROFILMS 71-20, 1970, PP.
118.

70-0380 SMITH, I. M.
DUNCAN, W.
INCREMENTAL NUMERICAL SOLUTION OF A SIMPLE
DEFORMATION PROBLEM IN SOIL MECHANICS
GEOTECHNIQUE, VOL. 20, NO. 4, 1970, PP. 357-372.

70-0381 GROTH, P.
METHODE DER ELASTOSTATIK PROGRAMMIERUNG UND
ANWENDUNG
MUNCHEN, WIEN: R. OLDENBOURG, 1970.

70-0382 DHATT, G.
AN EFFICIENT TRIANGULAR SHELL ELEMENT
A.I.A.A. J., VOL. 8, 1970, PP. 2100-2102.

70-0383 MILLER, R. E.
HANSEN, S. D.
LARGE SCALE ANALYSIS OF CURRENT AIRCRAFT
A.S.M.E. SPECIAL PUBLICATION, NEW YORK, N.Y.,
1970.

70-0384 CONNOR, J.
MIXED MODELS FOR PLATES
PROC. OF SEMINAR, FINITE ELEMENT TECHNIQUES IN
STRUCTURAL MECHANICS, UNIV. OF SOUTHAMPTON, 1970.

70-0385 BUTLIN, G.
FORD, R.
A COMPATIBLE TRIANGULAR PLATE BENDING FINITE
ELEMENT
INT. J. SOLIDS STRUCT., VOL. 6, 1970, PP: 323-332.

70-0386 ANON
SP-222 NASTRAN USERS MANUAL
COSMIC, U. OF GEORGIA, ATHENS, GA., 1970

70-0387 ANON
SP-222 NASTRAN THEORETICAL MANUAL
COSMIC, U. OF GEORGIA, ATHENS, GA., 1970

70-0388 CHANG, D. C.
PLANE STRESS AND PLANE STRAIN ANALYSIS BY FINITE
ELEMENT METHOD
PROGRAM NO. 0308 BSR 2907, BENDIX CORP., ANN
ARBOR, MICH., AUGUST 1970.

70-0389 SCHMID, G.
THE HARMONIC FINITE ELEMENT MODEL FOR USE IN FIELD
PROBLEMS
PH.D. THESIS, UNIV. OF WASHINGTON, 1970.

70-0390 IRONS, B. M.
NUMERICAL INTEGRATION APPLIED TO FINITE ELEMENT
METHODS
ROLLS ROYCE LTD. AND CIVIL ENG. DEPT., UNIV. OF
WALES, SWANSEA, 1970.

70-0391 HUNT, D. A.
DISCRETE ELEMENT IDEALIZATION OF AN INCOMPRESSIBLE
LIQUID FOR VIBRATION ANALYSIS
A.I.A.A. JOL., VOL. 8, NO. 6, JUNE, 1970, PP:
1001-1004

70-0392 PIAN, T. H. H.
VARIATIONAL FORMULATIONS OF NUMERICAL METHODS
IN "COMPUTER AIDED ENG.", EDITED BY G. M. L.
GLADWELL, UNIV. OF WATERLOO, 1970, P. 421.

70-0393 KELSEY, S.
LEE, K. N.
MAK, C. K. K.
THE CONDITION OF SOME FINITE ELEMENT COEFFICIENT
MATRICES
IN "COMPUTER AIDED ENG.", EDITED BY G. M. L.
GLADWELL, UNIV. OF WATERLOO, 1970, P. 267.

70-0394 ROY, J. R.
A STUDY OF NUMERICAL ERROR IN STRUCTURAL SYSTEMS
RESEARCH REPORT NO. 77, INSTITUT FUR STATIK UND
DYNAMIK DER LUFT UND RAUMFAHRTKONSTRUKTIONEN,
FEBRUARY 1970.

70-0395 ARGYRIS, J. H.
SCHARPF, D. W.
FINITE ELEMENT THEORY OF PLATES AND SHELLS
INCLUDING TRANSVERSE SHEAR STRAIN EFFECTS
COLL. OF INT. UNION OF THEORETICAL AND APPL.
MECH., UNIV. OF LIEGE, BELGIUM, AUGUST 23-28,
1970.

70-0396 SCHREM, E.
ROY, J. R.
AN AUTOMATIC SYSTEM FOR KINEMATIC ANALYSIS, ASKA
PART 1
PAPER PRESENTED AT THE COLLOQUIUM OF THE
INTERNATIONAL UNION THEORETICAL AND APPLIED
MECHANICS, U. OF LIEGE BELGIUM, AUGUST
23-28, 1970.

70-0397 ARGYRIS, J. H.
BRONLUND, O. E.
SORENSEN, M.
COMPUTER AIDED STRUCTURAL ANALYSIS: THE MACHINE
INDEPENDENT SYSTEM ASKA
NORD DATA-70 CONF., COPENHAGEN, AUGUST 1970.

70-0398 ARGYRIS, J. H.
GLOUDEMAN, J. F.
FINITE TIME-SPACE ELEMENTS APPLIED TO NONLINEAR
PROBLEMS
REPORT SD 70-577, SPACE DIV., NORTH AMERICAN
ROCKWELL, OCTOBER, 1970.

70-0399 ARGYRIS, J. H.
SCHARPF, D. W.
BERECHNUNG VORGESPANNTER NETZWERKE
BAYERISCHE   AKADEMIE   DER   WISSENSCHAFTEN,
SONDERDRUCK   4   AUS   DEN   SITZUNGSBERICHTEN S.
25-58, 1970.

70-0400 ARGYRIS, J. H.
SPOONER, J. B.
DIE   VERWENDUNG   DER   MATRIZENVERSCHIEBUNGSMETHODE
ZUR BERECHNUNG VON SPANNUNGSKONZENTRATIONEN
MISCELLANY, DEDICATED TO THE MEMORY OF THE LATE
ACADEMICIAN JAKOB M. KLITSCHIEFF, THE SERBIAN
ACADEMY OF SCIENCES AND ARTS, SECTION OF THE
TECH. SCI. BEOGRAD, 1970, PP: 1-20

70-0401 MUNCH, K. V.
A COLLECTION OF PROBLEMS SOLVED WITH ASKA
RESEARCH REPORT NO. 91, INSTITUT FUR STATIK UND
DYNAMIK   DER   LUFT   UND   RAUMFAHRTKONSTRUKTIONEN,
AUGUST 1970.

70-0402 MONAHAN, J.
NATURAL FREQUENCIES AND MODE SHAPES OF PLATES WITH
INTERIOR CUT-OUTS
MASTER'S THESIS, SCHOOL OF ENGNG., AIR FORCE INST.
OF TECH., WRIGHT/PATTERSON AFB, OHIO, REPORT NO:
GAM/MC/71-1,   SEPTEMBER,   1970,   94   P.
(N.T.I.S.-AD-770 090/9)

70-0403 AUTHORS
NAVY-NASTRAN COLLOQUIUM, WASHINGTON, D.C.
PROC. 2ND. NAVY-NASTRAN COLLOQ., NAVAL SHIP
RESEARCH   AND   DEVELOPMENT   CENTER,   BETHESDA,   MD.,
DECEMBER 9, 1970, 18 P. (N.T.I.S. - AD-764 507)
SEE ALSO AD-764-299)

70-0404 AUTHORS
NSRDC: NASTRAN COLLOQUIUM, WASHINGTON, D.C.
PROC. 1ST. NSRDC: NASTRAN COLLOQ., NAVAL SHIP
RESEARCH AND DEVOLPMENT CENTER, BETHESDA, MD.,
JANUARY 1970, 194 P. (N.T.I.S. - AD-764 298)

70-0405 DESAI, C. S.
SEEPAGE IN MISSISSIPPI RIVER BANKS. REPORT 1.
ANALYSIS OF TRANSIENT SEEPAGE USING VISCOUS FLOW
MODEL AND NUMERICAL METHODS
ARMY   ENGR.   WATERWAYS   EXPERIMENT   STATION,
VICKSBURG,   MISS.,   REPORT   NO:
AEWES-MISC-PAPER-S-70-3-1, FEBRUARY 1970, 71 P.
(N.T.I.S. - AD-757 396)

70-0406 NORRIS, D. M, JR.,
STAATS, J. R.
DYNAMIC LOADS ON GROUTED EMPLACEMENT SYSTEMS
LAWRENCE LIVERMORE LAB., UNIV. OF CALIF.,
LIVERMORE,   DECEMBER,   1973,   28   P.
(N.T.I.S.-UCID-15994) SEE ALSO NSA 3102, NO. 03296

70-0407 PIAN, T. H. H.
FINITE ELEMENT METHODS BY DIFFERENT VARIATIONAL
PRINCIPALS IN ELASTICITY
PROC. OF SIAM-AMS, AMER. MATH. SOC., PROVIDENCE,
R.I., VOL. 2, 1970.

70-0408 CHANE, H. L.
SABOR CS-2 USER'S MANUAL
MCDONNELL DOUGLAS ASTRONAUTICS CO-WEST, HUNTINGTON
BEACH, CALIF., REPORT NO: MDC-G0519, JANUARY 1970,
193 P. (N.T.I.S. - AD-747 028)

70-0409 DEARIEN, J. A.
NOBLE, C.
ULDRICH, E. D.
STRAP: COMPUTER CODE FOR STATIC AND DYNAMIC
STRUCTURAL ANALYSIS AND STUDIES MADE USING THE
CODE
IDAHO NUCL. CORP., IDAHO FALLS, JUNE 1970, 43 PP.
(N.T.I.S. IN-1362)

70-0410 SHERMAN, C. H.
IMPEDANCE OF INTERSTITIAL GAPS IN TRANSDUCER
ARRAYS
PARKE MATH. LABS. INC., CARLISLE, MASS, REPORT NO:
0424-TM-4, NOVEMBER 1970, 9 P. (N.T.I.S. - AD-739
722)

70-0411 AGARWAL, S. L.
HUDSON, W. R.
EXPERIMENTAL VERIFICATION OF DISCRETE-ELEMENT
SOLUTIONS FOR PLATES AND PAVEMENT SLABS
CENTER FOR HIGHWAY RESEARCH, UNIV. OF TEXAS,
AUSTIN, REPORT NO: RR-56-15, APRIL 1970, 268 PP.
(N.T.I.S. - PB-193 825)

70-0412 MURRAY, M. T.
STRUCTURAL DYNAMICS PROGRAMS
ADMIRALTY RESEARCH LAB., TEDDINGTON, ENGLAND,
REPORT NO: ARL/M/P28, MAY 1970, 33 PP. (N.T.I.S. -
AD-711 519)

70-0413 LAKIS, A. A.
PAIDOUSSIS, M. P.
DYNAMICAL ANALYSIS OF AXIALLY NON-UNIFORM, THIN
CYLINDRICAL SHELLS. PART 1 - MATRIX FORMULATION
DEPT. OF MECH. ENGNG., MCGILL UNIV., MONTREAL,
QUE., REPORT NO: MERL-70-9-PT-1, DECEMBER 1970, 49
P. (N.T.I.S. - N71-33819)

70-0414 VISSER, C.
VANBUREN, W.
WILSON, W. K.
GABRIELSE, S. E.
FUEL ROD DEFORMATIONS - A TIME DEPENDENT FINITE
ELEMENT ANALYSIS ON AXISYMMETRIC FUEL RODS,
INCLUDING PELLET END EFFECTS (LWBR DEVELOPMENT
PROGRAM)
WESTINGHOUSE RESEARCH LABS., PITTSBURGH, PA.,
AUGUST 1970, 273 P. (N.T.I.S. - WERL-ELPLA-1)

70-0415 POMERANTZ, M. A.
HORIZONS IN THE STRUCTURAL USE OF ADVANCED
MATERIALS - THRAUSTICS
FRANKLIN INST. RESEARCH LABS., PHILADELPHIA, PA.,
1970, 133 P. (N.T.I.S. - AD-730 586) SEE ALSO J.
OF FRANKLIN INSTITUTE, VOL. 290, NO. 6, DECEMBER
1970, PP. 483-604.

70-0416 DAILEY, G.
COMPUTER PROGRAMS FOR PLOTTING FINITE ELEMENT
PATTERNS OF TWO-AND THREE-DIMENSIONAL STRUCTURES
APPLIED PHYSICS LAB., JOHNS HOPKINS UNIV., SILVER
SPRING, MD., REPORT NO: APL-TG-1140, OCTOBER 1970,
32 P. (N.T.I.S. - AD-730 490)

70-0417 CALLAWAY, R. J.
BYRAM, K. V.
MATHEMATICAL MODEL OF THE COLUMBIA RIVER FROM THE
PACIFIC   OCEAN   TO   BONNEVILLE   DAM.   PART   II:
INPUT-OUTPUT AND INITIAL VERIFICATION PROCEDURES
PACIFIC NORTHWEST WATER LAB., CORVALLIS, OREG.,
DECEMBER 1970, 126 PP., (N.T.I.S. - PB-202 423)
SEE ALSO PART I, PB-202 423.

70-0418 MAYERJAK, R. J.
FATIGUE STRENGTH OF LUGS CONTAINING LINERS. VOLUME
II. COMPUTER PROGRAM USED FOR ANALYSIS
KAMAN AEROSPACE CORP., BLOOMFIELD, CONN., REPORT
NO. R-850-VOL.-2, NOVEMBER 1970, 93 PP., (N.T.I.S.
- AD-880 290) SEE ALSO VOL. 1, AD-880 289.

70-0419 RYBICKI, E. F.
APPROXIMATE THREE-DIMENSIONAL STRESS SOLUTIONS FOR
SOME LAMINATED PLATE PROBLEMS USING A HIGHER ORDER
EQUILIBRIUM FINITE ELEMENT MODEL. PART 1.
PRELIMINARY INVESTIGATIONS
BATTELLE MEMORIAL INST., COLUMBUS, OHIO, REPORT
NO. BAT-G731-1, AUGUST 1970, 51 PP. (N.T.I.S. -
AD-711 399)

70-0420 SMITH, C. V.
FINITE ELEMENT MODEL WITH APPLICATIONS TO
BUILDINGS K-33 AND K-31
COMPUTING TECH. CENTER, UNION CARBIDE CORP., OAK
RIDGE, TENN., JUNE 1, 1973, 109 PP. (N.T.I.S. -
CTC-29)

70-0421 ZOROWSKI, C. F.
DUNN, S. E. JR.
A STUDY OF INTERNAL STRESSES IN STATICALLY
DEFORMED PNEUMATIC TIRES. A MATHEMATICAL MODEL
FOR THE PNEUMATIC TIRE
DEPT. OF MECH. AND AEROSPACE ENGNG., NORTH
CAROLINA STATE UNIV., RALEIGH, FINAL REPORT,
NOVEMBER 1970, 111 P. (N.T.I.S. - PB-201 551) SEE
ALSO PB-201 546.

70-0422 CHRISTIAN, J. T.
BOEHMER, J. W.
THE EFFECTS OF SOIL PARAMETERS AND BOUNDARY
CONDITIONS ON THE CONSOLIDATION OF AN ELASTIC
LAYER
DEPT. OF CIVIL ENGNG., MASS. INST. OF TECH.,
CAMBRIDGE, REPORT NO: R70-50, AUGUST 1970, 124 P.
(N.T.I.S. - PB-201 550)

70-0423 MALONEY, J. G.
SHELTON, M. T.
UNDERHILL, D. A.
STRUCTURAL DYNAMIC PROPERTIES OF TACTICAL MISSILE
JOINTS - PHASE 1
ELECTRO DYNAMIC DIV., GENERAL DYNAMICS, POMONA,
CALIF., REPORT NO: CR-6-348-945-001, JUNE 1970,
133 P. (N.T.I.S. - AD-727 617)

70-0424 SWOFFORD, D. P.
A METHOD FOR THE THERMAL ANALYSIS OF SPACECRAFT
INCLUDING ALL MULTIPLE REFLECTIONS AND SHADING
AMONG DIFFUSE, GRAY SURFACES
LANGLEY RESEARCH CENTER, NAT. AERON. AND SPACE
ADMIN., LANGLEY STATION, VA., REPORT NO:
NASA-TN-D-5910, L-6063, JULY 1970, 64 PP.
(N.T.I.S. - N70-33123)

70-0425 REID, J. K.
TURNER, A. B.
FORTRAN SUBROUTINES FOR THE SOLUTION OF LAPLACE'S
EQUATION OVER A GENERAL REGION IN TWO DIMENSIONS
ATOMIC ENERGY RESEARCH ESTABLISHMENT, HARWELL,
ENGLAND, SEPTEMBER 1970, 30 P. (N.T.I.S. -TP-422)
SEE ALSO NASA 2802, NO. 04990.

70-0426 NEUBERT, V. H.
RANGAIAH, V. P.
VOGEL, W. H.
SHOCK ANALYSIS OF STRUCTURAL SYSTEMS
DEPT. OF ENGNG. MECH., PENN. STATE UNIV.,
UNIVERSITY PARK, FEBRUARY 1970, 161 PP. (N.T.I.S.
- AD-707 055) SEE ALSO INTERIM REPORT NO. 8,
AD-687 429.

70-0427 GRESTE, O.
CLOUGH, R. W.
FINITE ELEMENT ANALYSIS OF TUBULAR K JOINTS
STRUCT. ENGNG. LAB., UNIV. OF CALIF., BERKELEY,
REPORT NO: SESM-70-11B, JUNE 1970, 169 PP.
(N.T.I.S. - PB-193 560)

70-0428 BELYTSCHKO, T.
HODGE, P. G. JR.
LOWER BOUND LIMIT ANALYSIS OF PLANE STRESS
PROBLEMS BY A FINITE ELEMENT METHOD
DEPT. OF MECH., ILLINOIS INST. OF TECH., CHICAGO,
REPORT NO: DOMIIT-1-42, MARCH 1970, 26 P.
(N.T.I.S. - AD-702 403)

70-0429 FRAEIJS DE VEUBEKE, B.
BAUDOUIN, M.
LARGE DISPLACEMENT FORMULATION FOR ELASTIC BODIES
LOUVAIN UNIV., BELGIUM, REPORT NO: SA-15, MARCH
1970, 41 PP. (N.T.I.S. - AD-705 628)

70-0430 POTASH, P. L.
SECOND ORDER THEORY OF OSCILLATING CYLINDERS
COLL. OF ENGNG., UNIV. OF CALIF., BERKELEY, REPORT
NO: NA-70-3, JUNE 1970, 167 PP. (N.T.I.S. - AD-710
767)

70-0431 MEYER, C.
SCORDELIS, A. C.
ANALYSIS OF CURVED FOLDED PLATE STRUCTURES
COLL. OF ENGNG., UNIV. OF CALIF., BERKELEY, REPORT
NO: SESM--70-8, JUNE 1970, 84 P. (N.T.I.S. -
PB-193 535)

70-0432 RICE, J. R.
NUMERICAL ANALYSIS
DEPT. OF COMPUTER SCI., PURDUE UNIV., LAFAYETTE,
IND., REPORT NO: CSD-TR-49, SEPTEMBER 1, 1970, 12
PP. (N.T.I.S. - AD-710 431)

70-0433 CORUM, J. M.
SMITH, J. E.
USE OF SMALL MODELS IN DESIGN AND ANALYSIS OF
PRESTRESSED-CONCRETE REACTOR VESSELS
OAK RIDGE NAT. LAB., TENN. MAY 1970, 279 PP.
(N.T.I.S. - ORNL-4346)

70-0434 THOMAS, J. M.
A FINITE ELEMENT APPROACH TO THE STRUCTURAL
INSTABILITY OF BEAM COLUMNS, FRAMES, AND ARCHES
MARSHALL SPACE FLIGHT CENTER, NAT. AERON. AND
SPACE ADMIN., HUNTSVILLE, ALA., REPORT NO:
NASA-TN-D-5782, M-351, MAY 1970, 55 PP. (N.T.I.S.
- N70-26568)

70-0435 PANAK, J. J.
MATLOCK, H.
A DISCRETE-ELEMENT METHOD OF MULTIPLE-LOADING
ANALYSIS FOR TWO-WAY BRIDGE FLOOR SLABS
CENTER FOR HIGHWAY RESEARCH, UNIV. OF TEXAS,
AUSTIN, REPORT NO: RR-56-13, JANUARY 1970, 203 PP.
(N.T.I.S. - PB-191 978)

70-0436 FRANKLIN, H. A.
NONLINEAR ANALYSIS OF REINFORCED CONCRETE FRAMES
AND PANELS
COLL. OF ENGNG., UNIV. OF CALIF., BERKELEY, REPORT
NO: SESM-70-5, MARCH 1970, 275 P. (N.T.I.S. -
PB-191 937)

70-0437 WILLAM, K. J.
SCORDELIS, A. C.
ANALYSIS OF ORTHOTROPIC FOLDED PLATES WITH
ECCENTRIC STIFFENERS
COLL. OF ENGNG., UNIV. OF CALIF., BERKELEY, REPORT
NO: SESM-70-28 FEBRUARY 1970, 142 P. (N.T.I.S. -
PB-191 051)

70-0438 CYRUS, N. J.
EPPINK, R. T.
FULTON, R. E.
WALZ, J. E.
ACCURACY OF FINITE ELEMENT APPROXIMATIONS TO
STRUCTURAL PROBLEMS
LANGLEY RESEARCH CENTER, NAT. AERON. AND SPACE
ADMIN., LANGLEY STATION, VA., REPORT NO:
NASA-TN-D-5728,L-5656, MARCH 1970, 54 P. (N.T.I.S.
- N70-21708)

70-0439 GOUDREAU, G. L.
EVALUATION OF DISCRETE METHODS FOR THE LINEAR
DYNAMIC RESPONSE OF ELASTIC AND VISCOELASTIC
SOLIDS
DEPT. CIVIL ENG., UNIV. OF CALIF., BERKELEY,
CALIF., REPORT NO. 69-15, 1970.

70-0440 VISSER, K.
THE APPLICATION OF A CURVED MIXED-TYPE SHELL
ELEMENT
SYMP. CN HIGH SPEED COMPUT. OF ELASTIC STRUCT.,
INT. UNION OF THEOR. APPL. MECH., LIEGE, AUGUST
1970.

70-0441 DHATT, G. S.
AN EFFICIENT TRIANGULAR SHELL ELEMENT
A.I.A.A. J., VOL. 8, 1970, PP. 2100-2102.

70-0442 BERKOVIC, M.
FREE-FREE STRUCTURES IN FINITE ELEMENT ANALYSIS
10TH YUGOSLAV. CONG. ON MECH., 1970.

70-0443 BARFIELD, W. D.
NUMERICAL METHOD FOR GENERATING ORTHOGONAL
CURVILINEAR MESHES
J. COMP. PHYS., VOL. 5, 1970, PP. 23-33.

70-0444 TODD, M.
THE FINITE ELEMENT METHOD APPLIED TO THIN SHELLS
AND BOX STRUCTURES
PH.D. THESIS, LOUGHBOROUGH UNIV. OF TECH., 1970.

70-0445 KAN, D.
MESH AND CONTOUR PLOT FOR TRIANGLE AND
ISOPARAMETRIC ELEMENTS
DEPT. OF CIVIL ENG., UNIV. OF WALES, SWANSEA,
COMPUT. REPORT CNME/CR/39, 1970.

70-0446 KAWAI, T.
APPLICATION OF THE FINITE ELEMENT METHOD TO
PRESSURE VESSEL DESIGN (JAPANESE)
KOATSURYOKU, VOL. 8, 1970, PP. 1877-1885.

70-0447 KOLAR, V.
THE INFLUENCE OF DIVISION ON THE RESULTS IN THE
FINITE ELEMENT METHOD
Z.A.M.M., GAMM-TAGUNG, DELFT, 1973.

70-0448 DOUGLAS, J.
DUPONT, T.
GALERKIN METHODS FOR PARABOLIC EQUATIONS
S.I.A.M., J. ON NUMER. ANALYSIS, VOL. 7, NO. 4,
1970, PP. 575-626.

70-0449 BARTELDS, B.
OTTENS, H. H.
FINITE ELEMENT ANALYSIS OF SANDWICH PANELS
IUTAM SYMP. ON THE HIGH SPEED COMPUTING OF ELASTIC
STRUCT., LIEGE, 1970.

70-0450 TAKENAKA, Y.
MORI, K.
NAKAI, Y.
NAGATO, K.
SASAKI, T.
EXAMPLES OF STRESS ANALYSIS OF STRUCTURES BY THE
FINITE ELEMENT METHOD, (JAPANESE)
FAPIG (TOKYO), VOL. 59, 1970, PP. 182-187.

70-0451 MEBANE, P. M.
AN IMPROVED SHELL OF REVOLUTION ELEMENT UTILISING
CUBIC DISPLACEMENT FUNCTIONS
M.SC. THESIS, TEXAS A AND M UNIV., 1970.

70-0452 NICOLAIDES, R. A.
ON LAGRANGIAN INTERPOLATION IN N VARIABLES
INST. COMP. SCI., UNIV. OF LONDON, TECH. NOTE 274,
1970.

70-0453 DE ARANTES E OLIVEIRA, E. R.
THEORETICAL FOUNDATIONS OF THE FINITE ELEMENT
METHOD
TECNICA., VOL. 32, 1970, PP. 399-400.

70-0454 PARIKH, S. K.
ANALYSIS OF EARTHEN DAMS BY THE METHOD OF FINITE
ELEMENTS
J. SOIL MECH. AND FOUNDATION ENG., A.S.C.E., VOL.
96, NO. SM2, 1970, PP. 155-169.

70-0455 PIAN, T. H. H.
FINITE ELEMENT STIFFNESS METHODS BY DIFFERENT
VARIATIONAL PRINCIPLES IN ELASTICITY
PROC. SIAM-AMS, "IN NUMERICAL SOLUTION OF FIELD
PROBLEMS IN CONTINUUM MECHANICS", BIRKHOFF AND
VARGA (EDS.), PROVIDENCE, R.I., VOL. 11, 1970.

70-0456 WAIT, R.
THE FINITE ELEMENT METHOD IN PARTIAL DIFFERENTIAL
EQUATIONS
PH.D. THESIS, UNIV. OF DUNDEE, 1970.

70-0457 WEMPNER, G.
ODEN, J. T.
KROSS, D.
FINITE ELEMENT ANALYSIS OF THIN SHELLS
PROC. J. ENG. MECH. DIV., A.S.C.E., VOL. 96, NO.
EM6, 1970, PP. 967-983.

506

70-0458
KOUKAL, S.
PIECEWISE POLYNOMIAL INTERPOLATIONS AND THEIR
APPLICATIONS TO PARTIAL DIFFERENTIAL EQUATIONS
CZECH. SBORNIK VAAZ, BRUNO, 1970, PP. 29-38.

70-0459
YAMADA, YOSHIAKI
MATRIX METHODS IN MECHANICS OF MATERIALS
BAIFUKAN, 1970 (JAPANESE).

70-0460
SIDDALL, J. N.
DOKAINISH, M. A.
RAGHAVA, R. S.
TIWARI, S. K.
THEORETICAL AND EXPERIMENTAL VIBRATION ANALYSIS OF
AN OBLIQUE SPACE FRAME
EXPERIMENTAL MECH., VOL. 10, NO. 9, SEPTEMBER
1970, PP. 353-361.

70-0461
RUHL, R. L.
DYNAMICS OF DISTRIBUTED PARAMETER ROTOR SYSTEMS:
TRANSFER MATRIX AND FINITE ELEMENT TECHNIQUES
CORNELL UNIV., 1970, 335 PP.

70-0462
WERNER, S. D.
ANALYSIS OF DYNAMIC RESPONSE OF SAFEGUARD
STRUCTURES
SHOCK AND VIB. BULL., VOL. 41, NO. 2, DECEMBER
1970.

70-0463
MATSUMOTO, K.
APPLICATION OF FINITE ELEMENT METHOD TO ADDED
VIRTUAL MASS OF SHIP HULL VIBRATION
J. SOC. OF NAVAL ARCHITECTS OF JAPAN, VOL. 127,
1970, PP. 83-90.

70-0464
JONES, P. J.
KACENA, W. J.
COMPARISON OF STRUCTURAL LOADS: STATIC VERSUS
DYNAMIC
SHOCK AND VIB. BULL., VOL. 41, NO. 6, DECEMBER
1970.

70-0465
RAVENSCRAFT, R.
PARAMETRIC STUDY OF A BEAM WITH A COMPOUND
SIDE-BRANCH RESONATOR AS A DEVICE TO EVALUATE
PRELIMINARY DESIGN LOADS
SHOCK AND VIB. BULL., VOL. 41, NO. 6, DECEMBER
1970.

70-0466
DHILLON, B. S.
TRIANGULAR FINITE ELEMENTS FOR THE BENDING
ANALYSIS OF THICK ELASTIC PLATES
UNIV. OF COLO., ORDER NO. 70-23,702, 1970, 190 PP.

70-0467
PROMSIT, S.
DYNAMIC STABILITY OF CONICAL SHELLS BY FINITE
ELEMENTS
OKLA. STATE UNIV., 1970, 133 PP.

70-0468
GENALIS, P.
ELASTIC STRENGTH OF PROPELLERS: AN ANALYSIS BY
MATRIX METHODS
NAVAL SHIP R AND D CTR., JULY 1970, 270 PP.

70-0469
CHOPRA, A. K.
CHAKRAEARTI, P.
A COMPUTER PROGRAM FOR EARTHQUAKE ANALYSIS OF DAMS
CALIF. UNIV., SEPTEMBER 1970, 78 PP.

70-0470
SAWADA, T.
HASEGAWA, T.
AOYAMA, S.
FINITE ELEMENT ANALYSIS OF SOME MECHANICAL
PROBLEMS OF HOLLOW GRAVITY DAM
TRANS. AGRIC. ENG. SOC., JAPAN, VOL. 33, AUGUST
1970, PP. 62-75.

70-0471
COMSA, R.
POPOVICI, L.
ASUPRA ANALIZEI PRIN METODA ELEMENTULUI FINIT A
STARILOR DE EFORTURI SI DEFORMATII DIN MASIVELE DE
PAMINT. (ON THE ANALYSIS OF STRESSES INSIDE SOIL
MASSES BY THE FINITE ELEMENT METHOD)
STUD GEOTEH FUNDATII CONSTR HIDROTEH, VOL. 15,
1970, PP. 299-336.

70-0472
WITTKE, W.
METHODS OF CALCULATION OF THREE-DIMENSIONAL
PROBLEMS OF PERCOLATION OF FISSURED ROCK BY FINITE
ELEMENTS AND RESISTANCE NETWORKS
PROC. 2ND CONGRESS INT. SOC. OF ROCK MECH.,
BELGRADE, VOL. 3, 1970.

70-0473
GUELLEC, P.
CALCULATION OF FLOWS IN POROUS MEDIA BY THE FINITE
ELEMENT METHOD
LABORATORIES DES PONTS ET CHAUSSEES, RAPPORT DE
RECHERCHE NO. 11, NOVEMBER 1970.

70-0474
ANON
NOTE ON THE FINITE ELEMENT METHOD IN LINEAR
FRACTURE MECHANICS
ENG. FRACT. MECH., VOL. 2, NO. 2, NOVEMBER 1970,
PP. 173-176.

70-0475
ANON
CALCULATION OF STRESS INTENSITY FACTORS USING THE
FINITE ELEMENT METHOD WITH CRACKED ELEMENTS
INT. J. FRACT. MECH., VOL. 6, NO. 2, JUNE 1970,
PP. 159-167.

70-0476
CHAN, S. K.
TUBA, I. S.
FINITE ELEMENT METHOD IN LINEAR FRACTURE MECHANICS
ENG. FRACT. MECH., VOL. 2, NO. 1, JULY 1970, PP.
1-17.

70-0477
DHATT, G. S.
INSTABILITY OF THIN SHELLS BY THE FINITE ELEMENT
METHOD
PROC. I.A.S.S. SYMP. FOLDED PLATES AND PRISMATIC
STRUCT., VIENNA, 1970.

70-0478
AL-DABBAGH, A. M. A.
FINITE ELEMENT STRESS ANALYSIS FOR ANISOTROPIC
SOLIDS
PH.D., COLORADO STATE UNIV., 1970.

70-0479
BALASUBRAMANIAN, T. V.
FINITE ELEMENT STRESS ANALYSIS OF TURBINE BLADE
HOOKS
PH.D., CORNELL UNIV., 1970.

70-0480
ANON
FINITE ELEMENT SOLUTION OF STEADY STATE POTENTIAL
FLOW PROBLEMS
HYDROLOGIC ENG. CENTER, U.S.A. ENG. DISTRICT,
SACRAMENTO, CALIF., NO. 723-G2L2440, NOVEMBER
1970.

70-0481
BROMBOLICH, L. J.
FINITE ELEMENT ANALYSIS OF SHELLS OF REVOLUTION BY
MINIMIZATION OF THE POTENTIAL ENERGY FUNCTIONAL
SC.D., WASHINGTON UNIV., 1970.

70-0482
CERVENKA, V.
INELASTIC FINITE ELEMENT ANALYSIS OF REINFORCED
CONCRETE PANELS UNDER IN-PLANE LOADS
PH.D., UNIV. OF COLORADO, 1970.

70-0483
CHAUDHARY, V. P.
FINITE ELEMENT ANALYSIS OF AXISYMMETRICAL SHELLS
OF COMPOSITE MATERIAL
PH.D., THE UNIV. OF NEBRASKA, LINCOLN, 1970.

70-0484
CHARI, M. V.
FINITE-ELEMENT ANALYSIS OF NONLINEAR MAGNETIC
FIELDS IN ELECTRIC MACHINES
PH.D., MCGILL UNIV., CANADA, 1970.

70-0485
CHEN, L. K-C.
A FINITE ELEMENT FORMULATION FOR LARGE DEFLECTION
ELASTIC STRUCTURAL ANALYSIS APPLIED TO PLATE
BENDING
PH.D., UNIV. OF CINCINNATI, 1970.

70-0486
CHARLWOOD, R. G.
THERMO-PLASTIC CONSTITUTIVE RELATIONS AND A
VARIATIONAL SOLUTION TECHNIQUE USING FINITE
ELEMENTS
PH.D., THE UNIV. OF BRITISH COLUMBIA, CANADA,
1970.

70-0487
COLVILLE, J.
A GENERAL SOLUTION OF THE VON KARMAN PLATE
EQUATIONS BY THE FINITE ELEMENT METHOD
PH.D., THE UNIV. OF TEXAS, AUSTIN, 1970.

70-0488
CYR, N. A.
TRIANGULAR FINITE ELEMENT FOR CYLINDRICAL SHELLS
PH.D., STANFORD UNIV., 1970.

70-0489
MOWBRAY, D. F.
A NOTE ON THE FINITE ELEMENT METHOD IN LINEAR
FRACTURE MECHANICS
ENG. FRACT. MECH., VOL. 2, 1970, PP. 173-176.

70-0490
GUYMON, G. L.
MATHEMATICAL MODELING OF MOVEMENT OF DISSOLVED
CONSTITUENTS IN GROUND WATER AQUIFERS BY THE
FINITE ELEMENT METHOD
PH.D., UNIV. OF CALIF., DAVIS, 1970.

70-0491
HARVEY, J. W.
CONSTRAINT TECHNIQUE IN FINITE ELEMENT ANALYSIS
PH.D., UNIV. OF NOTRE DAME, 1970.

70-0492
HASKELL, W. E. JR.
GEOMETRIC NONLINEAR ANALYSIS OF THIN PLATES BY
FINITE ELEMENTS
PH.D., UNIV. OF MASSACHUSETTS, 1970.

70-0493
HAFTKA, R. T.
A KOITER-TYPE METHOD FOR FINITE ELEMENT ANALYSIS
OF NONLINEAR STRUCTURAL BEHAVIOR
PH.D., UNIV. OF CALIF., SAN DIEGO, 1970.

70-0494
HORN, H. R. JR.
CONICAL FINITE ELEMENT STRESS ANALYSIS OF AN
AXISYMMETRIC TANK ON AN ELASTIC FOUNDATION
PH.D., TEXAS TECH. UNIV., 1970.

70-0495
HWANG, C. T.
FINITE ELEMENT METHOD FOR SOIL DEFORMATIONS
PH.D., MCMASTER UNIV., CANADA, 1970.

507

70-0496 KIDDER, R. L.
QUADRILATERAL FINITE ELEMENT FOR PLANE STRESS AND
PLANE STRAIN
PH.D., STANFORD UNIV., 1970.

70-0497 KNOWLES, N. C.
THE FINITE ELEMENT METHOD APPLIED TO THE STATIC
AND DYNAMIC BUCKLING OF CIRCULAR ARCHES
PH.D., UNIV. OF NOTRE DAME, 1970.

70-0498 KROENKE, W. C.
ANALYSIS OF NONLINEAR CONCRETE FRAMES USING SMALL
GYPSUM MODELS AND THE FINITE ELEMENT METHOD
PH.D. THESIS, PURDUE UNIV., 1970.

70-0499 KULHAWY, F. H.
FINITE ELEMENT ANALYSIS OF THE BEHAVIOR OF
EMBANKMENTS
PH.D., UNIV. OF CALIF., BERKELEY, 1970.

70-0500 KUO, C. T.
UNSYMMETRIC BUCKLING OF SHELLS OF REVOLUTION BY
FINITE ELEMENT METHOD
PH.D. THESIS, UNIV. OF CALIF., BERKELEY, 1970.

70-0501 LEVY, N.
APPLICATION OF THE FINITE ELEMENT METHOD TO 2 AND
3 DIMENSIONAL ELASTIC-PLASTIC PROBLEMS OF FRACTURE
MECHANICS
PH.D., BROWN UNIV., 1970.

70-0502 MAHTAB, M. A.
THREE DIMENSIONAL FINITE ELEMENT ANALYSIS OF
JOINTED ROCK SLOPES
PH.D., UNIV. OF CALIF., BERKELEY, 1970.

70-0503 MAGHSOOD, J.
FINITE ELEMENT ANALYSIS OF WOOD BEAMS
PH.D., CORNELL UNIV., 1970.

70-0504 MONFORTON, G. R.
DISCRETE ELEMENT FINITE DISPLACEMENT ANALYSIS OF
ANISOTROPIC SANDWICH SHELLS
PH.D., CASE WESTERN RESERVE UNIV., 1970.

70-0505 PARDOEN, G. C.
STATIC AND DYNAMIC BUCKLING OF THIN-WALLED COLUMNS
USING FINITE ELEMENTS
PH.D., STANFORD UNIV., 1970.

70-0506 PRYOR, C. W. JR.
FINITE ELEMENT ANALYSIS OF LAMINATED ANISOTROPIC
PLATES INCLUDING TRANSVERSE SHEAR DEFORMATIONS
PH.D., VIRGINIA POLYTECHNIC INST. AND STATE UNIV.,
1970.

70-0507 GERDEEN, J. C.
SIMONEN, F. A.
HUNTER, D. T.
LARGE DEFLECTION ANALYSIS OF ELASTIC-PLASTIC
SHELLS OF REVOLUTION
A.I.A.A./A.S.M.E., 11TH CONF. STRUCT. DYNAMICS AND
MATER., DENVER, COLORADO, APRIL 1970, PP. 239-249.

70-0508 RANGASWAMY, T.
FINITE ELEMENT ANALYSIS OF ORTHOGONAL FRAME-PLATE
INTERACTION
PH.D., UNIV. OF KENTUCKY, 1970.

70-0509 ARMEN, H.
PIFKO, A. B.
LEVINE, H. S.
ISAKSON, G.
PLASTICITY
IN "FINITE ELEMENT TECHNIQUES IN STRUCTURAL
MECHANICS", H. TOTTENHAM, ET AL (EDS.),
SOUTHAMPTON UNIV. PRESS, 1970.

70-0510 RUSER, J. R.
APPLICATION OF THE THREE-DIMENSIONAL FINITE
ELEMENT METHOD TO SOME PROBLEMS IN SOIL MECHANICS
PH.D. THESIS, UNIV. OF TEXAS, AUSTIN, 1970.

71-0001 ODEN, J. T.
KELLEY, B. E.
FINITE ELEMENT FORMULATION OF GENERAL
THERMO-ELASTICITY PROBLEMS
INT. J. NUM. METHODS ENG. 3, 1971, PP. 161-179.

71-0002 SISODIYA, R. G.
CHEUNG, Y. K.
A HIGHER ORDER IN-PLANE PARALLELOGRAM ELEMENT AND
ITS APPLICATION TO SKEWED CURVED BOX GIRDER
BRIDGES
PAPER PRESENTED AT CONF. ON DEVELOPMENTS IN BRIDGE
DESIGN AND CONSTRUCTION, UNIVERSITY COLLEGE,
CARDIFF, MARCH 1971.

71-0003 GELDER, D.
SOLUTION OF THE COMPRESSIBLE FLOW EQUATIONS
INT. J. NUM. METHODS ENG., VOL. 3, 1971, PP.
35-43.

71-0004 BAKER, A. J.
NUMERICAL SOLUTION TO THE DYNAMICS OF VISCOUS
FLUID FLOW BY A FINITE ELEMENT ALGORITHM: A FIRST
STEP TOWARDS COMPUTATIONAL CONTINUUM MECHANICS
REPORT BELL AEROSPACE CO., BUFFALO, NEW YORK,
1971.

71-0005 BAKER, A. J.
FINITE ELEMENT THEORY FOR THE MECHANICS AND
THERMODYNAMICS OF A VISCOUS, COMPRESSIBLE
MULTI-SPECIE FLUID
REP. NO. 9500-920200, BELL AEROSPACE CO., BUFFALO,
NEW YORK, MARCH 1971.

71-0006 SMITH, I. M.
PLANE PLASTIC DEFORMATION OF SOIL,
ROSCOE MEMORIAL SYMP., ENG. DEPT., CAMBRIDGE
UNIV., MARCH 29-31, 1971.

71-0007 ZIENKIEWICZ, O. C.
NAYLOR, D. J.
DISCUSSION ON THE ADAPTION OF CRITICAL STATE SOLID
MECHANICS THEORY FOR USE IN FINITE ELEMENTS
ROSCOE MEMORIAL SYMP., ENG. DEPT., CAMBRIDGE
UNIV., MARCH 29-31, 1971.

71-0008 KAY, S.
AN EXPERIMENTAL AND THEORETICAL STUDY OF HOLLOW
CYLINDER TESTS ON SAND
M.SC. THESIS, UNIV. OF MANCHESTER, 1971.

71-0009 ALLAN, T.
THE EFFECT OF DEFORMATION ON THE BEHAVIOUR OF
HYDRODYNAMIC JOURNAL BEARINGS
TRIBOLOGY 4, NO. 3, AUGUST 1971, PP. 164-169.

71-0010 BOCK, H.
COMPUTER SIMULATION OF SECOND ORDER FAULTS
ROCK MECH., FELSMECH., MEC. ROCHES 3, NO. 4,
DECEMBER 1971, PP. 225-238.

71-0011 BURRIDGE, R.
SABINA, F. J.
THEORETICAL COMPUTATIONS ON RIDGE ACOUSTIC SURFACE
WAVES USING THE FINITE-ELEMENT METHOD
ELECTRON LETT. 7, NO. 24, DECEMBER 2, 1971, PP.
720-722.

71-0012 HOLZLOEHNER, U.
DIE BERECHNUNG VON KONSOLIDATIONSVORGAENGEN MIT
DER FINITE-ELEMENT-METHODE, (CALCULATION OF
CONSOLIDATION PROCESS BY MEANS OF FINITE ELEMENT
METHOD)
BAUINGENIEUR 46, NO. 7, JULY 1971, PP. 238-242.

71-0013 KAPER, H. G.
LEAF, G. K.
LINDEMAN, A. J.
APPLICATION OF FINITE ELEMENT TECHNIQUES FOR THE
NUMERICAL SOLUTION OF THE NEUTRON TRANSPORT AND
DIFFUSION EQUATIONS
SECOND CONFERENCE ON TRANSPORT THEORY, LOS ALAMOS
SCIENTIFIC LAB., NEW MEXICO, JANUARY 26-29, 1971.

71-0014 MCLEAN, F. G.
KRIZEK, R. J.
GRAVITY FLOW TO EXCAVATIONS AND DRAINAGE TRENCHES
IN LAYERED AQUIFERS
HIGHW. RES. REC. 360, 1971, PP. 65-76.

71-0015 OHNISHI, T.
APPLICATION OF FINITE ELEMENT SOLUTION TECHNIQUE
TO NEUTRON DIFFUSION AND TRANSPORT EQUATIONS
PROC. AMER. NUCL. SOC. CONF-710302, IDAHO FALLS,
MARCH 29-31, 1971.

71-0016 SCHIFFNER, K.
VERGLEICH NUMERISCHER LOESUNGEN NACH DEM
DIFFERENZENVERFAHREN UND NACH DER
FINITE-ELEMENT-METHODE AM BEISPIEL SYMMETRISCHER
FLUESSIGKEITSSCHWINGUNGEN IN BEHAELTERN,
(COMPARISON OF RESULTS OBTAINED FOR THE PROBLEM OF
SYMMETRIC FLUID OSCILLATIONS IN TANKS BY THE
FINITE DIFFERENCES TECHNIQUE AND THE FINITE
ELEMENT METHOD)
DEUT. LUFT RAUMFAHRT, FORSCHUNGSBER. DLR FB 71-69,
AUGUST 1971.

71-0017 SHUKER, W. A.
A SURVEY OF HEAT CONDUCTION COMPUTER PROGRAMS
NUCL. SAFETY 12, NO. 6, NOVEMBER-DECEMBER 1971,
PP. 569-582.

71-0018 ADELMAN, H. M.
CATHERINES, D. S.
CALCULATION OF TEMPERATURE DISTRIBUTIONS IN THIN
SHELLS OF REVOLUTION BY THE FINITE ELEMENT METHOD
NASA TN D-6100, FEBRUARY 1971.

71-0019 ALLEN, M.
NASTRAN EXPERIENCES OF FORT WORTH OPERATION,
CONVAIR AEROSPACE DIVISION OF GENERAL DYNAMICS
NASTRAN USERS EXPERIENCES, NASA TM X-2378,
SEPTEMBER, 1971.

71-0020 ALLMAN, D. J.
FINITE ELEMENT ANALYSIS OF PLATE BUCKLING USING A
MIXED VARIATIONAL PRINCIPLE
THIRD CONF. ON MATRIX METHODS IN STRUCTURAL
MECHANICS, DAYTON, OHIO, OCTOBER, 1971.

71-0021 ANDERSEN, S. I.
THE SCANDINAVIAN PORV MODEL PROJECT: STRESS
CALCULATION AND EXPERIMENTAL VERIFICATION
FIRST INT. CONF. STRUCT. MECH. IN REACTOR TECH.,
SEPTEMBER 1971, PAPER NO. H 5/3.

71-0022 ANDERSON, G. P.
RUGGLES, V. L.
STIBOR, G. S.
USE OF FINITE ELEMENT COMPUTER PROGRAMS IN
FRACTURE MECHANICS
INT. J. FRACTURE MECHANICS, VOL. 7, NO. 1, PP.
63-76, MARCH, 1971.

71-0023 ANDO, Y.
YAGAWA, G.
KIKUCHI, F.
STRESS DISTRIBUTIONS IN THIN-WALLED INTERSECTING
CYLINDRICAL SHELLS SUBJECTED TO INTERNAL PRESSURE
AND IN-PLANE FORCE
FIRST INT. CONF. STRUCT. MECH. IN REACTOR TECH.,
SEPTEMBER 1971, PAPER NO. G 2/2.

71-0024 ANEJA, I. K.
ROLL, F.
EXPERIMENTAL AND ANALYTICAL INVESTIGATION OF A
HORIZONTALLY CURVED BOX-BEAM HIGHWAY BRIDGE MODEL
SEC. INT. SYMP. CONCRETE BRIDGE DESIGN, ACI,
SEPTEMBER 16-26, 1971.

71-0025 APPA RAO, T. V. S. R.
AN ASSUMED STRESS HYBRID FINITE ELEMENT MODEL FOR
THE ANALYSIS OF AN AXISYMMETRIC THICK WALLED
PRESSURE VESSEL
FIRST INT. CONF. STRUCT. MECH. REACTOR TECH.,
SEPTEMBER 1971, PAPER NO. M 6/3.

71-0026 ARAL, M. M.
APPLICATION OF FINITE ELEMENT ANALYSIS IN FLUID
MECHANICS
GEORGIA INSTITUTE OF TECHNOLOGY, PH. D. THESIS,
ATLANTA, GEORGIA, SEPTEMBER 1971.

71-0027 ARAL, M. M.
MAYER, P. G.
SMITH, C. V. (JR.)
FINITE ELEMENT GALERKIN METHOD SOLUTIONS TO
SELECTED ELLIPTIC AND PARABOLIC DIFFERENTIAL
EQUATIONS
THIRD CONF. MATRIC METHODS IN STRUCTURAL
MECHANICS, DAYTON, OHIO, OCTOBER 1971.

71-0028 ARGYRIS, J. H.
BUCK, K. E.
SCHARPF, D. W.
WILLAM, K.
LINEAR METHODS OF STRUCTURAL ANALYSIS
FIRST INT. CONF. STRUCT. MECH. REACTOR TECH.,
SEPTEMBER 1971, PAPER NO. M 2/1. ALSO NUCL. ENG.
DES., VOL. 19, NO. 1, 1972, PP. 139-167.

71-0029 ARGYRIS, J. H.
BUCK, K. E.
SCHARPF, D. W.
WILLAM, K.
NONLINEAR METHODS OF STRUCTURAL ANALYSIS
FIRST INT. CONF. STRUCT. MECH. REACTOR TECH.,
SEPTEMBER 1971, PAPER NO. M 2/2. ALSO NUCL. ENG.
DES., VOL. 19, NO. 1, 1972, PP. 169-191.

71-0030 ARGYRIS, J. H.
BALMER, H.
DOLTSINIS, J.
WILLAM, K.
FINITE ELEMENT ANALYSIS OF THERMOMECHANICAL
PROBLEMS
THIRD CONF. ON MATRIX METHODS IN STRUCTURAL
MECHANICS, DAYTON, OHIO, OCTOBER 1971

71-0031 ARMEN, H.
LEVINE, H. S.
PIFKO, A. B.
FINITE ELEMENT ANALYSIS OF STRUCTURES IN THE
PLASTIC RANGE
NASA CP-1649, 295 PP., FEBRUARY 1971.

71-0032 AYRES, D. J.
ELASTIC-PLASTIC AND CREEP ANALYSIS VIA THE MARC 2
FINITE ELEMENT PROGRAM
INT. SYMP. NUMERICAL AND COMPUTER METH. IN
STRUCTURAL MECHANICS, U.S. OFFICE OF NAVAL
RESEARCH, URBANA, ILLINOIS, SEPTEMBER 1971.

71-0033 BABUSKA, I.
ERROR BOUNDS FOR FINITE ELEMENT METHOD
1971.
NUMERISCHE MATHEMATIK, VOL. 16, NO.4, PP. 322-333,
JANUARY,

71-0034 BABUSKA, I.
THE RATE OF CONVERGENCE FOR THE FINITE ELEMENT
METHOD
J. NUMERICAL ANALYSIS, S.I.A.M., VOL. 8, NO. 2,
PP. 304-315, JUNE 1971.

71-0035 BARCLAY, R. G.
HANLEIN, S. L.
SWEEN, J. (JR.)
KING, P. H.
NONLINEAR CONSIDERATION OF GRAVITY IN A STIFFNESS
TEST OF A WEAK STRUCTURE AT SMALL STRAINS
NASTRAN USERS EXPERIENCES, NASA TM X-2378,
SEPTEMBER 1971.

71-0036 BAZANT, E.
ANWENDUNG DER FINITE-ELEMENT-METHODE BEI DER
BERECHNUNG VON REAKTORDRUCKBEHALTERN
FIRST INT. CONF. STRUCT. MECH IN REACTOR TECH.,
SEPTEMBER 1971, PAPER NO. G 1/4.

71-0037 BENNETT, R. D.
NASTRAN DIFFERENTIAL STIFFNESS ANALYSIS OF AN
AIRCRAFT CANOPY
NASTRAN USERS EXPERIENCES, NASA TM X-2378,
SEPTEMBER 1971.

71-0038 BERGAN, P. G.
CLOUGH, R. W.
ELASTO-PLASTIC ANALYSIS OF PLATES USING THE FINITE
ELEMENT METHOD
THIRD CONF. ON MATRIX METHODS IN STRUCTURAL
MECHANICS, DAYTON, OHIO, OCTOBER 1971.

71-0039 BOHM, G. J.
ANALYTICAL PROBLEMS ASSOCIATED WITH CORE SUPPORT
STRUCTURES OF PWR
FIRST INT. CONF. STRUCT. MECH. IN REACTOR TECH.,
SEPTEMBER 1971, PAPER NO F 1/1.

71-0040 BOISSERIE, J. M.
GENERATION OF TWO- AND THREE-DIMENSIONAL FINITE
ELEMENTS
INT. J. NUM. METHODS ENGR., VOL. 3, NO. 3, JULY
1971, PP. 327-348.

71-0041 BRANCA, T. R.
BORESI, A. P.
CREEP OF A UNIAXIAL METAL MATRIX COMPOSITE
SUBJECTED TO AXIAL AND NORMAL LATERAL LOADS
FIRST INT. CONF. STRUCT. MECH. REACTOR TECH.,
SEPTEMBER 1971, PAPER NO. L 3/1.

71-0042 BOUSQUET, R. D.
YATES, D. N.
SABLE, W. W.
VINSON, T. J.
THE DEVELOPMENT OF COMPUTER GRAPHICS FOR LARGE
SCALE FINITE ELEMENT CODES
THIRD CONF. ON MATRIX METHODS IN STRUCTURAL
MECHANICS, DAYTON, OHIO, OCTOBER 1971. ALSO
MISSILES SYSTEMS DIV., LOCKHEED MISSILES AND SPACE
CO., AFLC-WPAFB OCTOBER 1971, 33 PP.

71-0043 BRAUCHLI, H. J.
ODEN, J. T.
CONJUGATE APPROXIMATION FUNCTIONS IN FINITE
ELEMENT ANALYSIS
J. APPL. MATH., 1971.

71-0044 BREBBIA, C.
SABANATHAN, S.
TAHBILDAR, U. C.
TOTTENHAM, H.
STATICS AND DYNAMICS OF HYDROSTATICALLY LOADED
SHELLS BY FINITE ELEMENT METHOD
I.A.S.S. SYMP. HYDROMECHANICALLY LOADED SHELLS,
HONOLULU, HAWAII, OCTOBER 1971.

71-0045 BROLLIAR, R. H.
NASTRAN BUCKLING ANALYSIS OF A LARGE STIFFENED
CYLINDRICAL SHELL WITH A CUTOUT
NASTRAN USERS EXPERIENCES, NASA TM X-2378,
SEPTEMBER 1971.

71-0046 BUSHNELL, D.
ENERGY APPROACHES TO FINITE-DIFFERENCE AND FINITE
ELEMENT METHODS
INT. SYMP. NUMERICAL AND COMPUTER METH. IN
STRUCTURAL MECHANICS, U. S. OFFICE OF NAVAL
RESEARCH, URBANA, ILLINOIS, SEPTEMBER 1971.

71-0047 CARMICHAEL, G. D. T.
THE APPLICATION OF THREE-DIMENSIONAL FINITE
ELEMENTS TO THE ANALYSIS OF PODDED BOILER TYPE
PRESTRESSED CONCRETE PRESSURE VESSELS
NUCL. ENG. AND DESIGN, VOL. 16, NO. 1, PP. 35-44,
MAY, 1971.

71-0048 CARMIGNANI, C.
CELLA, A.
ELASTIC ANALYSIS OF CRACKED THIN SHELLS BY THE
FINITE ELEMENT METHOD
FIRST INT. CONF. STRUCT. MECH. REACTOR TECH.,
SEPTEMBER 1971, PAPER NO. J 1/6.

509

71-0049 CARUSO, S.
INTERNAL SUPPORTING STRUCTURES OF PRESSURE TUBES
TYPE REACTOR CORE, ANALYTICAL MODEL FOR STATIC
ANALYSIS
FIRST INT. CONF. STRUCT. MECH. IN REACTOR TECH.,
SEPTEMBER 1971, PAPER NO. F 2/1.

71-0050 CHANG, T. Y.
RASHID, Y. R.
NONLINEAR CREEP ANALYSIS AT ELEVATED TEMPERATURE
FIRST INT. CONF. STRUCT. MECH. REACTOR TECH.,
SEPTEMBER 1971, PAPER NO. L 4/2.

71-0051 CHATTOPADHYAY, A.
SETLUR, A. V.
APPLICATION OF FINITE ELEMENT METHOD TO CONTINUUM
MECHANICS PROBLEMS
THIRD CONF. ON MATRIX METHODS IN STRUCTURAL
MECHANICS, DAYTON, OHIO, OCTOBER 1971.

71-0052 CHRIST, G.
FISCHER, D.
ZWINGENBERGER, L.
NEUESTE ERFAHRUNGEN BEI DER RECHNERISCHEN UND
KONSTRUKTIVEN AUSLEGUNG KOMPLIZIERTER
SCHALENSYSTEME
FIRST INT. CONF. STRUCT. MECH. REACTOR TECH.,
SEPTEMBER 1971, PAPER NO. J 1/2.

71-0053 CHRISTIANSEN, H. N.
DISPLAYS OF KINEMATIC AND ELASTIC SYSTEMS
THIRD CONF. ON MATRIX METHODS IN STRUCTURAL
MECHANICS, DAYTON, OHIO, OCTOBER 1971.

71-0054 CHU, S. L.
ANALYSIS AND DESIGN CAPABILITY OF THE STRUDL
PROGRAM
INT. SYMP. NUMERICAL AND COMPUTER METH. IN
STRUCTURAL MECHANICS, U.S. OFFICE OF NAVAL
RESEARCH, URBANA, ILLINOIS, SEPTEMBER 1971.

71-0055 CHU, W. H.
A COMPARISON OF SOME FINITE ELEMENT AND
FINITE-DIFFERENCE METHODS FOR A SIMPLE SLOSHING
PROBLEM
A.I.A.A. J., VOL. 9, NO. 10, PP. 2094-2095,
OCTOBER 1971.

71-0056 CHUNG, T. J.
EIDSON, R. L.
ANALYSIS OF VISCOELASTOPLASTIC STRUCTURAL
BEHAVIOUR OF ANISOTROPIC SHELLS BY THE FINITE
ELEMENT METHOD
FIRST INT. CONF. STRUCT. MECH. REACTOR TECH.,
SEPTEMBER 1971, PAPER NO. J 1/7.

71-0057 CHUNG, T. J.
JENKINS, J. F.
THE FINITE ELEMENT DYNAMIC STABILITY ANALYSIS OF
THIN SHELLS SUBMERGED IN FLUIDS
I.A.S.S. SYMP. HYDROMECHANICALLY LOADED SHELLS,
HONOLULU, HAWAII, OCTOBER 1971.

71-0058 CLARY, R. R.
PRACTICAL ANALYSIS OF PLATE VIBRATIONS USING
NASTRAN
NASTRAN USERS EXPERIENCES, NASA TM X-2378,
SEPTEMBER 1971.

71-0059 CLOUGH, R. W.
ANALYSIS OF STRUCTURAL VIBRATIONS AND DYNAMIC
RESPONSE
RECENT ADV. IN MATRIX METH. OF STRUCT. ANALYSIS
AND DESIGN, PP. 441-486, 1971.

71-0060 CLOUGH, R. W.
POWELL, G. H.
GANTAYAT, A. N.
STRESS ANALYSIS OF B16.9 TEES BY THE FINITE
ELEMENT METHOD
FIRST INT. CONF. STRUCT. MECH. IN REACTOR TECH.,
SEPTEMBER 1971, PAPER NO. F 4/7.

71-0061 CLOUGH, R. W.
WILSON, E. L.
DYNAMIC FINITE ELEMENT ANALYSIS OF ARBITRARY THIN
SHELLS
COMPUTERS AND STRUCTURES J., VOL. 1, 1971.

71-0062 COOK, W. L.
NASTRAN-GAP, AN INTEGRATED APPROACH TO THE
ANALYSIS OF RADIATION PATTERNS OF DISTORTED
REFLECTORS
NASTRAN USERS EXPERIENCES, NASA TM X-2378,
SEPTEMBER 1971.

71-0063 CONNOR, J. J.
WILL, G. T.
A MIXED FINITE ELEMENT SHELL FORMULATION
RECENT ADV. IN MATRIX METH. OF STRUCT. ANALYSIS
AND DESIGN, PP. 105-138, 1971.

71-0064 CONSTANTINO, C. J.
ANALYSIS OF SOIL-STRUCTURE INTERACTION EFFECTS
UNDER SEISMIC EXCITATION
FIRST INT. CONF. STRUCT. MECH. REACTOR TECH.,
SEPTEMBER 1971, PAPER NO. K 3/4.

71-0065 CORUM, J. M.
GREENSTREET, W. L.
EXPERIMENTAL ELASTIC STRESS ANALYSES OF
CYLINDER-TO-CYLINDER SHELL MODELS AND COMPARISONS
WITH THEORETICAL PREDICTIONS
FIRST INT. CONF. STRUCT. MECH. IN REACTOR TECH.,
SEPTEMBER 1971, PAPER NO. G 2/5.

71-0066 CORUM, J. M.
BOLT, S. E.
GREENSTREET, W. L.
GWALTNEY, W. L.
EXPERIMENTAL AND FINITE ELEMENT STRESS ANALYSIS OF
A THIN SHELLED CYLINDER TO CYLINDER MODEL
A.S.M.E. PAPER NO. 71-PVP-36, 1971.

71-0067 COURTNEY, R. L.
FINITE ELEMENT MODELING STUDIES IN THE NORMAL-MODE
METHOD AND NORMAL-MODE SYNTHESIS
N.A.S.A. TN D-6326, MAY 1971.

71-0068 COURTNEY, R. L.
STRUCTURAL ANALYSIS AND DESIGN OPTIMIZATION OF
SPACECRAFT USING NASTRAN
NASTRAN USERS EXPERIENCES, NASA TM X-2378,
SEPTEMBER 1971.

71-0069 COURTNEY, R. L.
NASTRAN MODELING STUDIES IN THE NORMAL-MODE AND
NORMAL-MODE SYNTHESIS
NASTRAN USERS EXPERIENCES, NASA TM X-2378,
SEPTEMBER 1971.

71-0070 COWPER, G. R.
VARIATIONAL PROCEDURES AND CONVERGENCE OF FINITE
ELEMENT METHODS
INT. SYMP. NUMER. AND COMPUTER METH. IN STRUCT.
MECH., U.S. OFFICE OF NAVAL RESEARCH, URBANA,
ILLINOIS, SEPTEMBER 1971.

71-0071 COWPER, G. R.
LINDBERG, G. M.
OLSON, M. D.
COMPARISON OF TWO HIGH-PRECISION TRIANGULAR FINITE
ELEMENTS FOR ARBITRARY DEEP SHELLS
THIRD CONF. ON MATRIX METHODS IN STRUCTURAL
MECHANICS, DAYTON, OHIO, OCTOBER 1971.

71-0072 CRAGGS, A.
COMPUTATION OF THE RESPONSE OF COUPLED PLATE
ACOUSTIC SYSTEMS USING PLATE FINITE ELEMENTS AND
ACOUSTIC VOLUME-DISPLACEMENT THEORY
J. SOUND VIB., VOL. 18, NO. 2, PP. 235-246,
SEPTEMBER 1971.

71-0073 CRAIG, R. R.
OPTIMIZATION OF A SUPERSONIC PANEL SUBJECT TO A
FLUTTER CONSTRAINT, A FINITE ELEMENT SOLUTION
AIAA PAPER NO. 71-330, APRIL 1971.

71-0074 CRONK, M.
AN INTERACTIVE COMPUTER GRAPHICS PROGRAM FOR
NASTRAN
NASTRAN USERS EXPERIENCES, NASA TM X-2378,
SEPTEMBER 1971.

71-0075 DAVIES, J. D.
PAREKH, C. J.
ZIENKIEWICZ, O. C.
ANALYSIS OF SLABS WITH EDGE BEAMS
SEC. INT. SYMP. CONCRETE BRIDGE DESIGN, ACI, SP
26-6, 1971.

71-0076 DAWE, D. J.
CURVED FINITE ELEMENTS IN THE ANALYSIS OF SHELL
STRUCTURES
FIRST INT. CONF. STRUCT. MECH. REACTOR TECH.,
SEPTEMBER 1971, PAPER NO J 1/4.

71-0077 DAWE, D. J.
A FINITE-DEFLECTION ANALYSIS OF SHALLOW ARCHES BY
THE DISCRETE ELEMENT METHOD
INT. J. NUM. METHODS ENG., VOL. 3, NO. 4, PP.
529-552, OCTOBER 1971.

71-0078 DEARIEN, J. A.
STRAP - A COMPUTER CODE FOR THE STATIC AND DYNAMIC
ANALYSIS OF REACTOR STRUCTURAL SYSTEMS
FIRST INT. CONF. STRUCT. MECH. IN REACTOR TECH.
SEPTEMBER 1971, PAPER NO. F 3/2.

71-0079 DENKE, P. H.
EIDE, G. R.
THE MATRIX DEFORMATION METHOD OF STRUCTURAL
ANALYSIS
THIRD CONF. ON MATRIX METHODS IN STRUCTURAL
MECHANICS, DAYTON, OHIO, OCTOBER 1971.

71-0080 DHATT, G. S.
VENKATASUBBU, S.
FINITE ELEMENT ANALYSIS OF CONTAINMENT VESSELS
FIRST INT. CONF. STRUCT. MECH. REACTOR TECH.,
SEPTEMBER 1971, PAPER NO. J 3/6.

71-0081 DONEA, J.
GIULIANI, S.
ANALYSE DES CONTRAINTES DAN DES STRUCTURES
COMPLEXES EN PRESENCE DE FLUAGE, DE VARIATIONS DE
DIMENSIONS ET D'UN CHAMP DE TEMPERATURE
FIRST INT. CONF. STRUCT. MECH. REACTOR TECH.,
SEPTEMBER 1971, PAPER NO. M 5/7.

71-0082 DOUGLAS, M. R.
PAREKH, C. J.
ZIENKIEWICZ, O. C.
FINITE ELEMENT PROGRAMS FOR SLAB BRIDGE DESIGN
SEC. INT. SYMP. CONCRETE BRIDGE DESIGN, ACI, SP
26-5, 1971.

71-0083 DUPUIS, G.
APPLICATION OF RITZ'S METHOD TO THIN ELASTIC SHELL
ANALYSIS
A.S.M.E. PAPER 71-APM-32, 1971.

71-0084 DUPUIS, G. A.
HIBBITT, H. D.
MCNAMARA, S. F.
MARCAL, P. V.
NONLINEAR MATERIAL AND GEOMETRIC BEHAVIOUR OF
SHELL STRUCTURES
COMPUTERS AND STRUCTURES, J., VOL. 1, 1971.

71-0085 EAGLE, J.
BECKETT, V.
ANALYSIS OF LOADING AND THERMAL EFFECTS ON FUELLED
GRAPHITE BRICKS FOR A HIGH-TEMPERATURE GAS COOLED
REACTOR
FIRST INT. CONF. STRUCT. MECH. IN REACTOR TECH.,
SEPTEMBER, 1971, PAPER NO. D 2/4.

71-0086 ENGLAND, G. L.
THE CALCULATION OF TIME-DEPENDENT STRESSES IN
PCPVS BY AN APPROXIMATE METHOD
FIRST INT. CONF. STRUCT. MECH. IN REACTOR TECH.,
SEPTEMBER 1971, PAPER NO. H 3/8.

71-0087 FARRELL, J. J.
DAI, P. K.
EFFECTS OF RISE TIME AND DAMPING ON FINITE ELEMENT
ANALYSIS OF RESPONSE OF STRUCTURES
THIRD CONF. ON MATRIX METHODS IN STRUCTURAL
MECHANICS, DAYTON, OHIO, OCTOBER 1971.

71-0088 FRIED, I.
ACCURACY OF FINITE ELEMENT EIGENPROBLEMS
J. SOUND VIB., VOL. 18, NO. 2, PP. 289, SEPTEMBER
1971.

71-0089 FRIED, I.
DISCRETIZATION AND COMPUTATIONAL ERRORS IN
HIGHER-ORDER FINITE ELEMENTS
A.I.A.A. J., VOL. 9, NO. 10, PP. 2071-2072,
OCTOBER 1971.

71-0090 FRIEDMAN, F.
STRAIN CONCENTRATION ANALYSIS USING THE FINITE
ELEMENT METHOD
A.S.M.E. PAPER NO. 71-PVP-39, 1971.

71-0091 FUJINO, T.
ANALYSIS OF HYDRODYNAMIC AND PLATE STRUCTURES
PROBLEMS BY THE FINITE ELEMENT METHOD
RECENT ADV. IN MATRIX METH. OF STRUCT. ANALYSIS
AND DESIGN, 1971, PP. 725-786.

71-0092 FULLARD, K.
THE COMPUTATION OF TEMPERATURE DISTRIBUTIONS AND
THERMAL STRESSES USING FINITE ELEMENT TECHNIQUES
FIRST INT. CONF. STRUCT. MECH. REACTOR TECH.,
SEPTEMBER 1971, PAPER NO. M 5/3.

71-0093 FURUIKE, T.
YAHATA, S.
VEHICLE STRUCTURAL ANALYSIS, PRE AND POST
PROCESSOR DATA HANDLING
NASTRAN USERS EXPERIENCES, NASA TM X-2378,
SEPTEMBER 1971.

71-0094 GALLAGHER, R. H.
SURVEY AND EVALUATION OF THE FINITE ELEMENT METHOD
IN LINEAR FRACTURE MECHANICS ANALYSIS
FIRST INT. CONF. STRUCT. MECH. REACTOR TECH.,
SEPTEMBER 1971, PAPER NO. L 6/9.

71-0095 GALLAGHER, R. H.
DHALLA, A. K.
DIRECT FLEXIBILITY FINITE ELEMENT ELASTOPLASTIC
ANALYSIS
FIRST INT. CONF. STRUCT. MECH. REACTOR TECH.,
SEPTEMBER 1971, PAPER NO. M 6/9.

71-0096 GALLAGHER, R. H.
LIEN, S.
MAU, S. T.
A PROCEDURE FOR FINITE ELEMENT PLATE AND SHELL PRE
AND POST BUCKLING ANALYSIS
THIRD CONF. ON MATRIX METHODS IN STRUCTURAL
MECHANICS, DAYTON, OHIO, OCTOBER 1971.

71-0097 GIESKE, R. K.
NODAL ANALYSIS OF THE MATED SPACE SHUTTLE
CONFIGURATION
NASTRAN USERS EXPERIENCES, NASA TM X-2378,
SEPTEMBER 1971.

71-0098 GILES, G. L.
BLACKBURN, C. L.
PROCEDURE FOR EFFICIENT GENERATING, CHECKING, AND
DISPLAYING NASTRAN INPUT AND OUTPUT DATA FOR
ANALYSIS OF AEROSPACE VEHICLE STRUCTURES
NASTRAN USERS EXPERIENCES, NASA TM X-2378,
SEPTEMBER 1971.

71-0099 GILES, G. L.
DUTTON, J. H.
APPLICATION OF NASTRAN TO THE ANALYSIS OF A SPACE
SHUTTLE ORBITER STRUCTURE
NASTRAN USERS EXPERIENCES, NASA TM X-2378,
SEPTEMBER 1971.

71-0100 GLADWELL, G. M. L.
MASON, V.
VARIATIONAL FINITE ELEMENT CALCULATION OF THE
ACOUSTIC RESPONSE OF A RECTANGULAR PANEL
J. SOUND VIB., VOL. 14, NO. 1, JANUARY 1971, PP.
115-136.

71-0101 GOULD, P. L.
SEN, S. K.
REFINED MIXED METHOD FINITE ELEMENTS FOR SHELLS OF
REVOLUTION
THIRD CONF. ON MATRIX METHODS IN STRUCTURAL
MECHANICS, DAYTON, OHIO, OCTOBER 1971.

71-0102 GRANDLE, R. E.
RUCKER, C. E.
NODAL ANALYSIS OF A NINE BAY SKIN-STRINGER PANEL
NASTRAN USERS EXPERIENCES, NASA TM X-2378,
SEPTEMBER 1971.

71-0103 GRISHAM, A. F.
THE BOEING SST PROTYPE INTERNAL LOADS ANALYSIS
SYSTEM AND PROCEDURES
THIRD CONF. ON MATRIX METHODS IN STRUCTURAL
MECHANICS, DAYTON, OHIO, OCTOBER 1971.

71-0104 GROOMS, H. R.
YAHATA, S.
SPACE SHUTTLE, THE NEED FOR SUBSTRUCTURING
NASTRAN USERS EXPERIENCES, NASA TM X-2378,
SEPTEMBER 1971.

71-0105 GROSS, H.
THERMOVISCOELASTIC AXISYMMETRIC STRESS ANALYSIS BY
FINITE ELEMENT AND FINITE DIFFERENCE COMPUTER
TECHNIQUES
FIRST INT. CONF. STRUCT. MECH. REACTOR TECH.,
SEPTEMBER 1971, PAPER NO. M 5/5.

71-0106 GUNDERSON, R. H.
CETINER, A.
ELEMENT STIFFNESS MATRIX GENERATOR
J. STRUCT. DIV., A.S.C.E., VOL. 97, NO. ST1, PP.
363-375, JANUARY 1971.

71-0107 GWALTNEY, R. C.
CORUM, J. M.
ANALYTIC INVESTIGATION OF COMPACT REINFORCEMENT
FOR RADIAL NOZZLES IN SPHERICAL SHELLS
A.S.M.E. PAPER NO. 71-PVP-26, 1971.

71-0108 GWALTNEY, R. C.
CORUM, J. M.
GREENSTREET, W. L.
EFFECT OF FILLETS ON STRESS CONCENTRATIONS IN
CYLINDRICAL SHELLS WITH STEP CHANGES IN OUTSIDE
DIAMETER
A.S.M.E. PAPER NO. 71-PVP-27, 1971.

71-0109 HAFTKA, R. T.
MALLETT, R. H.
NACHBAR, W.
ADAPTION OF KOITER'S METHOD TO FINITE ELEMENT
ANALYSIS OF SNAP-THROUGH BUCKLING BEHAVIOUR
INT. J. SOLIDS STRUCT., VOL. 7, NO. 10, OCTOBER
1971, PP. 1427-1446.

71-0110 HAGGENMACHER, G. W.
THE USE OF NASTRAN IN THE ANALYSIS OF LARGE
COMPLEX AIRFRAME STRUCTURES, MODELING TECHNIQUES
AND ORGANIZATION
NASTRAN USERS EXPERIENCES, NASA TM X-2378,
SEPTEMBER 1971.

71-0111 HANDA, K. N.
CLARKSON, B. L.
APPLICATION OF FINITE ELEMENT METHOD TO THE
DYNAMIC ANALYSIS OF TALL STRUCTURES
J. SOUND VIB., VOL. 18, NO. 3, OCTOBER 1971.

71-0112 HARDER, R. L.
MCNEAL, R. H.
DOGGERT, R. V. (JR.)
A DESIGN STUDY FOR THE INCORPORATION OF
AEROELASTIC CAPABILITY INTO NASTRAN
NASTRAN USERS EXPERIENCES, NASA TM X-2378,
SEPTEMBER 1971.

71-0113 HARDY, C.
BARONET, C. N.
TORDION, G. V.
THE ELASTO-PLASTIC INDENTATION OF A HALF-SPACE BY
A RIGID SPHERE
INT. J. NUM. METHODS ENG., VOL. 3, NO. 4, PP.
451-463, OCTOBER 1971.

71-0114 HARTZ, B. J.
NATHAN, N. D.
FINITE ELEMENT FORMULATION OF GEOMETRICALLY
NONLINEAR PROBLEMS OF ELASTICITY
RECENT ADV. IN MATRIX METH. OF STRUCT. ANALYSIS
AND DESIGN, PP. 415-437, 1971.

71-0115 HEAD, J. L.
JEZERNIK, A.
FINITE ELEMENT ANALYSIS OF STRESSES IN FUEL
ELEMENTS OF A HIGH-TEMPERATURE GAS-COOLED REACTOR
FIRST INT. CONF. STRUCT. MECH. IN REACTOR TECH.,
SEPTEMBER 1971, PAPER NO. D 2/2.

71-0116 HARTZMAN, M.
HUTCHINSON, J. R.
NONLINEAR DYNAMICS OF SOLIDS BY THE FINITE ELEMENT
METHOD
COMPUTERS AND STRUCTURES, VOL. 1, 1971.

71-0117 HELLEN, T. K.
THE APPLICATION OF THE BERSAFE FINITE ELEMENT
SYSTEM TO NUCLEAR DESIGN PROBLEMS
FIRST INT. CONF. STRUCT. MECH REACTOR TECH.,
SEPTEMBER, 1971, PAPER NO. M 6/1.

71-0118 HENSHELL, R. D.
NEALE, B. K.
WARBURTON, G. B.
A NEW HYBRID CYLINDRICAL SHELL FINITE ELEMENT
J. SOUND VIB., VOL. 16, NO. 4, PP. 519-532, JUNE
1971.

71-0119 HERTING, D. N.
JOSEPH, J. A.
KUUSINEN, L. R.
MCNEAL, R. H.
ACOUSTIC ANALYSIS OF SOLID ROCKET MOTOR CAVITIES
BY A FINITE ELEMENT METHOD
NASTRAN USERS EXPERIENCES, NASA TM X-2378,
SEPTEMBER 1971.

71-0120 HOFMANN, H. H.
PRESTRESSED CONCRETE REACTOR PRESSURE VESSEL
FAILURE ANALYSIS, THE INCORPORATION OF STATISTICAL
METHODS MONTE CARLO METHOD IN FINITE ELEMENT
CALCULATIONS
FIRST INT. CONF. STRUCT. MECH. IN REACTOR TECH.,
SEPTEMBER 1971, PAPER NO. H 3/6.

71-0121 HOFMEISTER, L. D.
GREENBAUM, G. A.
EVENSEN, D. A.
LARGE STRAIN ELASTO-PLASTIC FINITE ELEMENT
ANALYSIS
A.I.A.A. J., VOL. 9, NO. 7, PP. 1248-1254, JULY
1971.

71-0122 TAHIANI, C.
ANALYSE DES VOILES MINCES DANS LES DOMAINES
LINEAIRES ET GEOMETRIQUEMENT NONLINEAIRES PAR LA
METHODE DES ELEMENTS FINIS MIXTES
PH.D. THESIS, DEPT. CIVIL ENG., LAVAL UNIV.,
QUEBEC, 1971.

71-0123 HUANG, S. L.
RUBIN, H.
STATIC AND DYNAMIC ANALYSIS OF F-14A BORON
HORIZONTAL STABILIZER
NASTRAN USERS EXPERIENCES, NASA TM X-2378,
SEPTEMBER 1971.

71-0124 HUNG, N. D.
DUALITY IN THE ANALYSIS OF SHELLS BY THE FINITE
ELEMENT METHOD
INT. J. SOLIDS STRUCT., VOL. 7, NO. 3, PP.
281-300, MARCH 1971.

71-0125 HURWITZ, M. M.
GOLDEN, M. E.
NEW ELEMENT DEFINITION CAPABILITY FOR NASTRAN
NASTRAN USERS EXPERIENCES, NASA TM X-2378,
SEPTEMBER 1971.

71-0126 ICE, M. W.
NASTRAN USER INTERFACES, AUTOMATED INPUT
INNOVATIONS
NASTRAN USERS EXPERIENCES, NASA TM X-2378,
SEPTEMBER 1971.

71-0127 IRONS, B. M.
KAN, D. K-Y.
EQUATION SOLVING ALGORITHMS FOR THE FINITE ELEMENT
METHOD
INT. SYMP. NUMERICAL AND COMPUTER METH. IN
STRUCTURAL MECHANICS, U.S. OFFICE OF NAVAL
RESEARCH, URBANA, ILLINOIS, SEPTEMBER 1971.

71-0128 IRONS, B. M.
TREHARNE, G.
A BOUND THEOREM IN EIGENVALUES AND ITS PRACTICAL
APPLICATIONS
THIRD CONF. ON MATRIX METHODS IN STRUCTURAL
MECHANICS, DAYTON, OHIO, OCTOBER 1971.

71-0129 ISAKSON, G.
LEVY, A.
FINITE ELEMENT ANALYSIS OF INTERLAMINAR SHEAR
FIBROUS COMPOSITES
J. COMPOS. MATER., VOL. 5, PP. 273-276, APRIL
1971.

71-0130 JABBOUR, K. N.
NORMAL MODE ANALYSIS OF THE RADIO ASTRONOMY
EXLPORER BOOMS AND SPACECRAFT
NASTRAN USERS EXPERIENCES, NASA TM X-2378,
SEPTEMBER 1971.

71-0131 JANZ, R. F.
GRIMM, A. F.
DEFORMATION OF THE END DIASTOLIC LEFT VENTRICLE,
COMPARISON OF THEORY WITH EXPERIMENT
24TH ANNUAL CONF. ENGINEERING IN MEDICINE AND
BIOLOGY, LAS VEGAS, NEVADA, PAPER NO. 21.9,
OCTOBER 1971.

71-0132 JERRAM, K.
A FRACTURE MECHANICS APPROACH TO THE PROPAGATION
OF FATIGUE CRACKS IN REACTOR COMPONENTS BY
ACOUSTICALLY INDUCED STRESSES
FIRST INT. CONF. STRUCT. MECH. REACTOR TECH.,
SEPTEMBER 1971, PAPER NO. L 6/3.

71-0133 JOFRIET, J. C.
MCNEICE, G. M.
FINITE ELEMENT ANALYSIS OF REINFORCED CONCRETE
SLABS
J. STRUCT. DIV., A.S.C.E., VOL. 97, NO. ST3, MARCH
1971, PP. 785-806.

71-0134 JONES, G. K.
THE RESPONSE OF SHELLS TO DISTRIBUTED RANDOM LOADS
USING NASTRAN
NASTRAN USERS EXPERIENCES, NASA TM X-2378,
SEPTEMBER 1971.

71-0135 JONES, T. C.
PINSON, L. D.
ADAPTATION OF NASTRAN TO THE ANALYSIS OF THE
VIKING SPACE VEHICLE
NASTRAN USERS EXPERIENCES, NASA TM X-2378,
SEPTEMBER 1971.

71-0136 COWLEY, A.
HINDUJA, S.
FINITE ELEMENT METHOD FOR MACHINE TOOL STRUCTURAL
ANALYSIS
ANN CIRP. VOL. 19, NO. 1, APRIL 1971, PP. 171-181.

71-0137 KATOW, M. S.
STATIC ANALYSIS OF THE 64-M 210-FT ANTENNA
REFLECTION STRUCTURE
NASTRAN USERS EXPERIENCES, NASA TM X-2378,
SEPTEMBER 1971.

71-0138 KAVANAGH, K. T.
CLOUGH, R. W.
FINITE ELEMENT APPLICATIONS IN THE
CHARACTERIZATION OF ELASTIC SOLIDS
INT. J. SOLIDS STRUCT., VOL. 7, NO. 1, PP. 11-24,
JANUARY 1971.

71-0139 MEYER, C.
SCORDELIS, A. C.
ANALYSIS OF CURVED FOLDED PLATE STRUCTURES
ASCE J. STRUCT. DIV., VOL. 97, NO. ST10, OCTOBER
1971, PAPER 8434, PP. 2459-2480.

71-0140 KAWAI, T.
MURAKI, T.
TANAKA, N.
IWAKI, T.
FINITE ELEMENT ANALYSIS OF THIN-WALLED STRUCTURES
BASED ON THE MODERN ENGINEERING THEORY OF BEAMS
THIRD CONF. ON MATRIX METHODS IN STRUCTURAL
MECHANICS, DAYTON, OHIO, OCTOBER 1971.

71-0141 KHABBAZ, G. R.
DYNAMIC BEHAVIOR OF LIQUIDS IN ELASTIC TANKS
A.I.A.A. J., VOL. 9, NO. 10, OCTOBER 1971, PP.
1985-1990.

71-0142 KEY, S. W.
BEISINGER, Z. E.
THE TRANSIENT DYNAMIC ANALYSIS OF THIN SHELLS BY
THE FINITE ELEMENT METHODS
THIRD CONF. ON MATRIX METHODS IN STRUCTURAL
MECHANICS, DAYTON, OHIO, OCTOBER 1971.

71-0143 KEY, S. W.
KRIEG, R. D.
COMPARISON OF FINITE ELEMENT AND FINITE DIFFERENCE
METHODS
O.N.R. SYMP. NUMERICAL AND COMPUTER METH. IN
STRUCTURAL MECHANICS, URBANA, ILLINOIS, SEPTEMBER
1971.

71-0144 KHAN, M. H.
SAUGY, B.
ZIMMERMANN, T.
ANALYSE TRIDIMENSIONALE DU COMPORTEMENT
NON-LINEAIRE D'UN CAISSON DE REACTEUR NUCLEAIRE EN
BETON PRECONTRAINT
FIRST INT. CONF. STRUCT. MECH. IN REACTOR TECH.,
SEPTEMBER 1971, PAPER NO. H 3/5.

71-0145 KIKUCHI, F.
ANDO, Y.
A FINITE ELEMENT METHOD FOR INITIAL VALUE PROBLEMS
THIRD CONF. ON MATRIX METHODS IN STRUCTURAL
MECHANICS, DAYTON, OHIO, OCTOBER 1971.

71-0146 KIKUCHI, F.
ANDO, Y.
A FINITE ELEMENT METHOD FOR FRIEDRICH'S SYMMETRIC
POSITIVE SYSTEMS
21ST JAPAN NAT. CONG. APPL. MECH. TOKYO, 1971.

71-0147 WADE, L. V.
USE OF THE FINITE ELEMENT TECHNIQUE TO DETERMINE
THE FIRE RESISTANCE CAPABILITIES OF INSULATED
STRUCTURAL MEMBERS
DISS ABSTR. INT. SEC. B., UNIV. MICROFILMS, ANN
ARBOR, MICHIGAN, (ORDER NO. 71-24, 873,
83PP.) VOL 32(4), OCTOBER 1971, PP.
2154--8.

71-0148 KOBAYASHI, A. S.
WOO, S. L. Y.
ANALYSIS OF BIOLOGICAL STRUCTURES
RECENT ADV. IN MATRIX METH. OF STRUCT. ANALYSIS
AND DESIGN, PP. 837-853, 1971.

71-0149 KOBAYASHI, A. S.
SIMON, B. R.
WIEDERHIELM, C. A.
STRANDNESS, D. E.
DEFORMATION OF THE ARTERIAL VASA VASORUM AT NORMAL
AND HYPERTENSIVE ARTERIAL PRESSURE
24TH ANNUAL CONF. ENGINEERING IN MEDICINE AND
BIOLOGY, LAS VEGAS, NEVADA, PAPER NO. 36.1,
OCTOBER 1971, ALSO IN J. BIOMECH., VOL. 6, NO. 4,
JULY 1973, PP. 349-359.

71-0150 KOBAYASHI, A. S.
WOO, S. L. Y.
LAWRENCE, C.
SCHLEGEL, W. A.
A MATHEMATICAL MODEL OF THE CORNEO-SCLERAL SHELL
24TH ANNUAL CONF. ENGINEERING IN MEDICINE AND
BIOLOGY, LAS VEGAS, NEVADA, PAPER NO. 39.7,
OCTOBER 1971.

71-0151 KOBAYASHI, A. S.
WOO, S. L. Y.
LAWRENCE, C.
SCHLEGEL, W. A.
ANALYSIS OF THE CORNEO-SCLERAL SHELL BY THE METHOD
OF DIRECT STIFFNESS
J. BIOMECH., VOL. 4, NO. 5, PP. 323-330, OCTOBER
1971.

71-0152 KRATOCHVIL, J.
ZENISEK, A.
ZLAMAL, M.
A SIMPLE ALGORITHM FOR THE STIFFNESS MATRIX OF
TRIANGULAR PLATE BENDING ELEMENTS
INT. J. NUM. METHODS ENG., VOL. 3, NO. 4, PP.
553-564, OCTOBER 1971.

71-0153 KRISHNAMURTHY, N.
THREE-DIMENSIONAL FINITE ELEMENT ANALYSIS OF
THICK-WALLED PIPE-NOZZLE JUNCTIONS WITH CURVED
TRANSITIONS
FIRST INT. CONF. STRUCT. MECH. IN REACTOR TECH.,
SEPTEMBER 1971, PAPER NO. G 2/7.

71-0154 KURODA, T.
MURAKAMI, H.
YAMAMOTO, S.
STRESS ANALYSIS OF A PRESTRESSED CONCRETE NUCLEAR
PRESSURE VESSEL BY THE FINITE ELEMENT METHOD USING
VARIABLE STRAIN ELEMENTS
FIRST INT. CONF. STRUCT. MECH. IN REACTOR TECH.,
SEPTEMBER 1971, PAPER NO. H 3/2.

71-0155 LARSEN, P. K.
POPOV, E. P.
ELASTIC-PLASTIC ANALYSIS OF THICK WALLED PRESSURE
VESSELS WITH SHARP DISCONTINUITIES
A.S.M.E. PAPER NO. 71-PVP-23, 1971.

71-0156 LAUNAY, P.
CHARPENET, G.
VOUILLON, C. C.
THE TRIDIMENSIONAL THERMOELASTIC COMPUTER CODE
TITUS
FIRST INT. CONF. STRUCT. MECH. REACTOR TECH.,
SEPTEMBER 1971, PAPER NO. M 5/4.

71-0157 LEONARD, J. W.
MELFI, D.
3-D FINITE ELEMENT MODEL FOR LAKE CIRCULATION
THIRD CONF. ON MATRIX METHODS IN STRUCTURAL
MECHANICS, DAYTON, OHIO, OCTOBER 1971.

71-0158 LEVY, A.
ARMEN, H.
WHITESIDE, J.
ELASTIC AND PLASTIC INTERLAMINAR SHEAR DEFORMATION
IN LAMINATED COMPOSITIES UNDER GENERALIZED PLANE
STRESS
THIRD CONF. ON MATRIX METHODS IN STRUCTURAL
MECHANICS, DAYTON, OHIO, OCTOBER 1971.

71-0159 LEVY, R.
GUYAN REDUCTION SOLUTIONS RECYCLED FOR IMPROVED
ACCURACY
NASTRAN USERS EXPERIENCES, NASA TM X-2378,
SEPTEMBER 1971.

71-0160 LEVY, R.
A NASTRAN POSTPROCESSOR FOR STRUCTURAL
MODIFICATION REANALYSIS
NASTRAN USERS EXPERIENCES, NASA TM X-2378,
SEPTEMBER 1971.

71-0161 LEVY, R.
WALL, S.
SAVINGS IN NASTRAN DECOMPOSITION TIME BY
SEQUENCING TO REDUCE ACTIVE COLUMNS
NASTRAN USERS EXPERIENCES, NASA TM X-2378,
SEPTEMBER 1971.

71-0162 LEWIS, D. J.
IRVING, J.
CARMICHAEL, G. D. T.
ADVANCES IN THE ANALYSIS OF PRESTRESSED CONCRETE
PRESSURE VESSELS
FIRST INT. CONF. STRUCT. MECH. IN REACTOR TECH.,
SEPTEMBER 1971, PAPER NO. H 3/1. ALSO NUCL. ENG.
DES. VOL. 20, NO. 2, 1972, PP. 543-573.

71-0163 LINDBERG, G. M.
OLSON, M. D.
A HIGH-PRECISION TRIANGULAR CYLINDRICAL SHELL
FINITE ELEMENT
A.I.A.A. J., VOL. 9, NO. 3, PP 530-532, MARCH
1971.

71-0164 HILTON, P. D.
HUTCHINSON, J. W.
PLASTIC INTENSITY FACTORS FOR CRACKED PLATES
ENG. FRACT. MECH., VOL. 3, 1971, PP. 435-451.

71-0165 LUFT, R. W.
ROESSET, J. M.
CONNOR, J. J.
AUTOMATIC GENERATION OF FINITE ELEMENT MATRICES
J. STRUCT. DIV., A.S.C.E., VOL. 97, NO. ST 1, PP.
349-962, JANUARY 1971.

71-0166 LYNN, P. P.
RAMEY, G. E.
A METHOD TO GENERATE BOTH UPPER AND LOWER BOUNDS
TO PLATE EIGENVALUES BY CONFORMING DISPLACEMENT
FINITE ELEMENTS
FIRST INT. CONF. STRUCT. MECH. REACTOR TECH.,
SEPTEMBER 1971, PAPER NO. M 6/4.

71-0167 LYNN, P. P.
DHILLON, B. S.
TRIANGULAR THICK PLATE BENDING ELEMENTS
FIRST INT. CONF. STRUCT. MECH. REACTOR TECH.,
SEPTEMBER 1971, PAPER NO. M 6/5.

71-0168 MARCAL, P. V.
FINITE ELEMENT ANALYSIS WITH MATERIAL
NONLINEARITIES, THEORY AND PRACTICE
RECENT ADV. IN MATRIX METH. OF STRUCT. ANALYSIS
AND DESIGN, PP. 257-282, 1971.

71-0169 MARCAL, P. V.
NONLINEAR ANALYSIS
INT. SYMP. NUMERICAL + COMPUTER METH. IN
STRUCTURAL MECHANICS, U.S. OFFICE OF NAVAL
RESEARCH, URBANA, ILLINOIS, SEPTEMBER 1971.

71-0170 MARGUERRE, K.
SCHALK, M.
WOLFEL, H.
BERECHNUNG DER ERDBEBENSCHWINGUNGEN VON STRUKTUREN
MIT DER FINITE ELEMENT METHODE, MECHANISCHE
MODELLE VON KERNKRAFTWERKEN MIT EINBAUTEN
FIRST INT. CONF. STRUCT. MECH. REACTOR TECH.,
SEPTEMBER 1971, PAPER NO. K 2/7.

71-0171 MARTIN, H. C.
A SURVEY OF FINITE ELEMENT FORMULATIONS OF
GEOMETRICALLY NONLINEAR PROBLEMS
RECENT ADV. IN MATRIX METH. OF STRUCT. ANALYSIS
AND DESIGN, PP. 343-382, 1971.

71-0172 MCDONOUGH, J. R.
NASTRAN PLOTTING CAPABILITIES
NASTRAN USERS EXPERIENCES, NASA TM X-2378,
SEPTEMBER 1971.

71-0173 MCNEAL, R. H.
THE NASTRAN COMPUTER PROGRAM FOR STRUCTURAL
ANALYSIS
COMPUT. AND STRUCT. J., VOL. 1, 1971.

71-0174 MEBANE, P. M.
STRICKLIN, J. A.
IMPLICIT RIGID BODY MOTION IN CURVED FINITE
ELEMENTS
A.I.A.A. J., VOL. 9, NO. 2, PP. 344, FEBRUARY
1971.

71-0175 MEIJERS, P.
REVIEW OF THE ASKA PROGRAM
INT. SYMP. NUMERICAL + COMPUTER METH. IN
STRUCTURAL MECHANICS, U.S. OFFICE OF NAVAL
RESEARCH, URBANA, ILLINOIS, SEPTEMBER 1971.

71-0176 MELOSH, R. J.
MANIPULATION ERRORS IN THE FINITE ELEMENT ANALYSES
RECENT ADV. IN MATRIX METH. OF STRUCT. ANALYSIS
AND DESIGN, PP. 857-877, 1971.

71-0177 MELOSH, R. J.
THE OPTIMUM APPROACH TO ANALYSIS OF ELASTIC
CONTINUA
COMPUTERS AND STRUCTURES J., VOL. 1, 1971.

71-0178 MIYAMOTO, H.
APPLICATION OF FINITE ELEMENT METHOD TO FRACTURE
MECHANICS
PROC. 1ST INT. CONF. ON STRUCT. MECH. REACT.
TECHNOL., TOKYO, SEPTEMBER 1971, PAPER NO. 6/4,
ALSO IN J. FRACT. ENG., VOL. 31, NO. 19, 1971, PP.
217-270.

71-0179 PUTTER, S.
BEAM BENDING FINITE ELEMENT CONTAINING INTEGRATION
PARAMETERS
ISRAEL J. TECH., VOL. 9, NO. 5, 1971, PP. 483-488.

71-0180 MIZUMACHI, W.
APPLICATION OF FINITE ELEMENT METHOD TO THE THREE
DIMENSIONAL STRESS AND VIBRATION ANALYSIS OF BWR
PRIMARY PLANT SYSTEMS
FIRST INT. CONF. STRUCT. MECH. REACTOR TECH.,
SEPTEMBER 1971, PAPER NO. M 6/7.

71-0181 MIZUMACHI, W.
NON-LINEAR THERMAL STRESS ANALYSIS FOR NUCLEAR
POWER PLANT BY FINITE ELEMENT METHOD
THIRD CONF. ON MATRIX METHODS IN STRUCTURAL
MECHANICS, DAYTON, OHIO, OCTOBER 1971.

71-0182 MORLEY, L. S. D.
THE CONSTANT BENDING MOMENT PLATE BENDING ELEMENT
J. STRAIN ANAL., VOL. 6, NO. 1, PP. 20-24,
JANUARY, 1971.

71-0183 MORLEY, L. S. D.
EXTENDED INTERPOLATION IN FINITE ELEMENT ANALYSIS
THIRD CONF. ON MATRIX METHODS IN STRUCTURAL
MECHANICS, DAYTON, OHIO, OCTOBER 1971.

71-0184 MOTE, C. D.
GLOBAL-LOCAL FINITE ELEMENTS
INT. J. NUM. METHODS ENG., VOL. 3, NO. 4, PP.
565-574, OCTOBER 1971.

71-0185 MOWBRAY, D. F.
MCCONNELLEE, J. E.
APPLICATION OF FINITE ELEMENT ELASTIC-PLASTIC
STRESS ANALYSIS TO NOTCHED FATIGUE SPECIMEN
BEHAVIOUR
FIRST INT. CONF. STRUCT. MECH. REACTOR TECH.,
SEPTEMBER 1971, PAPER NO. M 6/8.

71-0186 MYERS, G. E.
ANALYTIC METHODS IN CONDUCTION HEAT TRANSFER
MC GRAW-HILL BOOK CO., 1971.

71-0187 NAEHRIG, TH. K.
GASCHEN, J. P.
THE CALCULATION OF THREE-DIMENSIONAL TEMPERATURE
DISTRIBUTIONS AND THERMAL STRESSES USING FINITE
ELEMENT METHODS
FIRST INT. CONF. STRUCT. MECH. REACTOR TECH.,
SEPTEMBER 1971, PAPER NO. M 5/2.

71-0188 NAYAK, G. C.
DAVIES, J. D.
INFLUENCE CHARACTERISTICS FOR SLAB BRIDGES
SEC. INT. SYMP. CONCRETE BRIDGE DESIGN, ACI, SP
26-1, 1971.

71-0189 ODEN, J. T.
KEY, J. N.
ON SOME GENERALIZATIONS OF THE INCREMENTAL
STIFFNESS RELATIONS FOR FINITE DEFORMATIONS OF
COMPRESSIBLE AND INCOMPRESSIBLE FINITE ELEMENTS
NUCL. ENG. DESIGN, VOL. 15, NO. 2, PP. 121-134,
APRIL 1971.

71-0190 ODEN, J. T.
FINITE ELEMENT FORMULATION OF PROBLEMS OF FINITE
DEFORMATION AND IRREVERSIBLE THERMODYNAMICS OF
NONLINEAR CONTINUA, A SURVEY AND EXTENSION OF
RECENT DEVELOPMENTS
RECENT ADV. IN MATRIX METH. OF STRUCT. ANALYSIS
AND DESIGN, PP. 693-724, 1971.

71-0191 ODEN, J. T.
BRAUCHLI, H. J.
ON THE CALCULATION OF CONSISTENT STRESS
DISTRIBUTIONS IN FINITE ELEMENT APPROXIMATIONS
INT. J. NUM. METHODS ENG., VOL. 3, NO. 3, PP.
317-326, JULY 1971.

71-0192 ODEN, J. T.
CHUNG, T. J.
KEY, J. E.
ANALYSIS OF NONLINEAR THERMOELASTIC AND
THERMOPLASTIC BEHAVIOUR OF SOLIDS OF REVOLUTION BY
THE FINITE ELEMENT METHOD
FIRST INT. CONF. STRUCT. MECH. REACTOR TECH.,
SEPTEMBER 1971, PAPER NO. M 5/6.

71-0193 ODEN, J. T.
KEY, J. E.
A NOTE ON THE FINITE DEFORMATIONS OF A THICK
ELASTIC SHELL OF REVOLUTION
FIRST INT. CONF. STRUCT. MECH. REACTOR TECH.,
SEPTEMBER 1971, PAPER NO. J 2/5.

71-0194 ODEN, J. T.
THEORY OF CONJUGATE PROJECTIONS IN FINITE ELEMENT
ANALYSIS
N.A.T.O. ADVANCED STUDY INST., PORTUGAL, SEPTEMBER
1971. ALSO IN THE NATO LECTURES ON FINITE
ELEMENT METHODS IN CONTINUUM MECHANICS, UAH PRESS,
HUNTSVILLE, 1972.

71-0195 NAGANO, H.
SOME TECHNIQUES FOR FINITE ELEMENT ANALYSIS
PREPRINTS, ANNUAL MEETING OF JAPAN SOCIETY OF
CIVIL ENGINEERS, 1971, PP. 317-320, (IN JAPANESE)

71-0196 FUJINO, T.
YAMADA, YOSHIAKI
ANALYSIS OF COMPRESSIBLE HYDRODYNAMICS BY THE
FINITE ELEMENT METHOD
2ND J.S.S.C. CONF. ON MATRIX METHODS AND STRUCT.
ANALYSIS, 1971. (IN JAPANESE)

71-0197 ODEN, J. T.
THE FINITE ELEMENT METHOD IN FLUID MECHANICS
N.A.T.O. ADVANCED STUDY INST., PORTUGAL, SEPTEMBER
1971.

71-0198 MIZUMACHI, W.
ANALYSIS FOR ATOMIC POWER PLANT BY FINITE ELEMENT
METHOD
PROC. NATIONAL SYMPOSIUM ON MATRIX METHODS OF
STRUCTURAL ANALYSIS AND DESIGN, SOC. OF STEEL
CONSTRUCTION OF JAPAN, TOKYO, 1971, PP. 506-513.
(IN JAPANESE)

71-0199 ODEN, J. T.
FINITE ELEMENT MODELS OF NONLINEAR OPERATOR
EQUATIONS
THIRD CONF. ON MATRIX METHODS IN STRUCTURAL
MECHANICS, DAYTON, OHIO, OCTOBER 1971.

71-0200 ODEN, J. T.
FINITE ELEMENTS IN NONLINEAR CONTINUA
MC GRAW-HILL BOOK PUBLISHERS, NEW YORK, 1971.

71-0201 SCHREM, E.
IMPLEMENTATION OF THE FINITE ELEMENT PROCEDURE
INT. SYMP. NUMERICAL + COMPUTER METH. IN
STRUCTURAL MECHANICS, U.S. OFFICE OF NAVAL
RESEARCH, URBANA, ILLINOIS, SEPTEMBER, 1971.

71-0202 OLSON, M. D.
A CONSISTENT FINITE ELEMENT METHOD FOR RANDOM
RESPONSE PROBLEMS
COMPUTERS AND STRUCTURES J., VOL. 1, 1971.

71-0203 ORME, D. H.
SMITH, R. G.
CONSTRUCTION SELF-WEIGHT STRESSES OF PRECAST SHEAR
WALLS
J. STRUCT. DIV., A.S.C.E., VOL. 97, NO. ST1, PP.
393-406, JANUARY 1971.

71-0204 PAREKH, R. J.
STARS - A COMPUTER CODE FOR STRUCTURAL DYNAMIC
ANALYSIS OF REACTOR SYSTEM
FIRST INT. CONF. STRUCT. MECH. IN REACTOR TECH.,
SEPTEMBER 1971, PAPER NO. E 4/4.

71-0205 PARKER, J. V.
PRACTICAL APPLICATIONS OF FINITE ELEMENT METHODS
IN THE STRESS ANALYSIS OF THIN SHELL PRESSURE
VESSELS
FIRST INT. CONF. STRUCT. MECH. REACTOR TECH.,
SEPTEMBER 1971, PAPER NO. J 1/3.

71-0206 PARKING, A. K.
FIELD SOLUTIONS FOR TURBULENT SEEPAGE FLOW
J. SOIL MECH., + FOUNDATIONS DIV., A.S.C.E., VOL.
97, SM 1 P. 208-218, JANUARY 1971.

71-0207 PATEL, J. S.
SELTZER, S. M.
COMPLEX EIGENVALUE SOLUTION TO A SPINNING SKYLAB
PROBLEM
NASTRAN USERS EXPERIENCES, NASA TM X-2378,
SEPTEMBER 1971.

71-0208 PATTERSON, J. F.
YANT, H. W.
DESIGN ANALYSIS FOR LMFBR CORE RESTRAINT
FIRST INT. CONF. STRUCT. MECH. IN REACTOR TECH.,
SEPTEMBER 1971, PAPER NO. F 1/6.

71-0209 PAULSEN, W. C.
FINITE ELEMENT STRESS ANALYSIS, PART 2
MACHINE DESIGN, PART 1, SEPTEMBER 30, 1971, PART
2, OCTOBER 14, 1971, PART 3, OCTOBER 28, 1971.

71-0210 KOKUSHO, S.   ET. AL.
TWO DIMENSIONAL NON LINEAR ANALYSIS OF REINFORCED
CONCRETE MEMBERS BY THE FINITE ELEMENT METHOD,
PART 1: ASSUMPTION AND ANALYTICAL PROCEDURES
TRANS. OF THE ARCH. INST. OF JAPAN, NO. 189,
NOVEMBER 1971, PP. 51-57.

71-0211 PIAN, T. H. H.
HYBRID MODELS
INT. SYMP. NUMERICAL + COMPUTER METH. IN
STRUCTURAL MECHANICS, U.S. OFFICE OF NAVAL
RESEARCH, URBANA, ILLINOIS, SEPTEMBER, 1971.

71-0212 PIAN, T. H. H.
TONG, P.
LUK, C. H.
ELASTIC CRACK ANALYSIS BY A FINITE ELEMENT HYBRID
METHOD
THIRD CONF. ON MATRIX METHODS IN STRUCTURAL
MECHANICS, DAYTON, OHIO, OCTOBER 1971.

71-0213 PIAN, T. H. H.
FOUNDATIONS OF FINITE ELEMENT METHODS FOR SOLID
CONTINUA
RECENT ADV. IN MATRIX METH. OF STRUCT. ANALYSIS
AND DESIGN, PP. 49-84, 1971.

71-0214 POPOV, E. P.
SHARIFI, P.
A REFINED CURVED ELEMENT FOR THIN SHELLS OF
REVOLUTION
INT. J. NUM. METHODS ENG., VOL. 3, NO. 4, PP.
495-508, OCTOBER 1971.

71-0215 POWELL, G. H.
CLOUGH, R. W.
GANTAYAT, A. N.
STRESS ANALYSIS OF B16.9 TEES BY THE FINITE
ELEMENT METHOD, A PROGRESS REPORT
A.S.M.E. PAPER NO. 71-PVP-40, 1971.

71-0216 PRYOR, C. W.
BARKER, R. M.
A FINITE ELEMENT ANALYSIS INCLUDING TRANSVERSE
SHEAR EFFECTS FOR APPLICATIONS TO LAMINATED PLATES
A.I.A.A. J., VOL. 9, NO. 5, PP. 912-916, MAY 1971.

71-0217 PUTZEYS, J.
REYNEN, J.
STRESS ANALYSIS OF A GRAPHITE FUEL ELEMENT
INCLUDING RADIATION CREEP AND DIMENSIONAL CHANGES
FIRST INT. CONF. STRUCT. MECH. IN REACTOR TECH.,
SEPTEMBER 1971, PAPER NO. D 2/3.

71-0218 RAO, A. K.
RAJU, I. S.
KRISHNA MURTY, A. V.
A POWERFUL HYBRID METHOD IN FINITE ELEMENT
ANALYSIS
INTL. J. NUMER. METH. IN ENG., VOL. 3, NO. 3, JULY
1971, PP. 289-403.

71-0219 RAO, A. K.
RAJU, I. S.
KRISHNA MURTY, A. V.
SPECIAL FINITE ELEMENTS FOR THE ANALYSIS OF STRESS
CONCENTRATIONS AND SINGULARITIES
FIRST INT. CONF. STRUCT. MECH. REACTOR TECH.,
SEPTEMBER 1971, PAPER NO. M 6/6.

71-0220 RASHID, Y. R.
CHANG, T. Y.
STRESS ANALYSIS OF TWO-DIMENSIONAL PROBLEMS UNDER
SIMULTANEOUS CREEP AND PLASTICITY
FIRST INT. CONF. STRUCT. MECH. REACTOR TECH.,
SEPTEMBER 1971, PAPER NO. L 4/5.

71-0221 RASHID, Y. R.
GILMAN, J. D.
THREE-DIMENSIONAL ANALYSIS OF REACTOR PRESSURE
VESSEL NOZZLES
FIRST INT. CONF. STRUCT. MECH. IN REACTOR TECH.,
SEPTEMBER 1971, PAPER NO. G 2/6.

71-0222 HERMANN, L. R.
MASON, W. E.
MIXED FORMULATIONS FOR FINITE ELEMENT SHELL
ANALYSIS
CONF. ON COMPUTER-ORIENTED ANALYSIS OF SHELL
STRUCT., REPORT NO. AFFDL-TR-71-79, JUNE 1971.

71-0223 REYNOLDS, R. J.
PRODUCTION OF MOTION PICTURES OF A CONTINUOUSLY
DEFORMING STRUCTURE USING NASTRAN
NASTRAN USERS EXPERIENCES, NASA TM X-2378,
SEPTEMBER 1971.

71-0224 ROBINSON, J.
HAGGENMACHER, G. W.
BASIS FOR ELEMENT INTERCHANGEABILITY IN FINITE
ELEMENT PROGRAMS
THIRD CONF. ON MATRIX METHODS IN STRUCTURAL
MECHANICS, DAYTON, OHIO, OCTOBER 1971.

71-0225 ROBINSON, J.
HAGGENMACHER, G. W.
SOME NEW DEVELOPMENTS IN MATRIX FORCE ANALYSIS
RECENT ADV. IN MATRIX METH. OF STRUCT. ANALYSIS
AND DESIGN, PP. 183-228, 1971.

71-0226 ROWE, G. H.
MATRIX DISPLACEMENT METHODS IN FRACTURE MECHANICS
ANALYSIS OF REACTOR VESSELS
PROC. 1ST INT. CONF. STRUCT. MECH. IN REACTOR
TECH., SEPTEMBER 1971, PAPER NO. G 6/1, ALSO IN
NUCL. ENG. DES., VOL. 20, NO. 1, 1972, P. 251.

71-0227 SABIR, A. B.
ASHWELL, D. G.
A COMPARISON OF CURVED BEAM FINITE ELEMENTS WHEN
USED IN VIBRATION PROBLEMS
J. SOUND VIB., VOL. 18, NO. 4, PP. 555-564,
OCTOBER 1971.

71-0228 SACZALSKI, K. J.
HUANG, T. C.
COUPLED RESPONSE OF SPATIAL VIBRATORY STRUCTURES
MOUNTED TO ISOTROPIC PLATE ELEMENTS
A.S.M.E. PAPER 71-VIBR-3, 1971.

71-0229 SANDER, G.
BECKERS, P.
IMPROVEMENTS OF FINITE ELEMENT SOLUTIONS FOR
STRUCTURAL AND NON-STRUCTURAL APPLICATIONS
THIRD CONF. ON MATRIX METHODS IN STRUCTURAL
MECHANICS,   DAYTON, OHIO, OCTOBER 1971.

71-0230 SCHMITZ, R. P.
STRUCTURAL DYNAMIC ANALYSIS OF ELECTRONIC
ASSEMBLIES USING NASTRAN RESTART/FORMAT CHANGE
CAPABILITY
NASTRAN USERS EXPERIENCES, NASA TM X-2378,
SEPTEMBER 1971.

71-0231 SCHMITZ, R. P.
APPLICATION OF SMALL AND LARGE-DISPLACEMENT THEORY
TO A LARGE FLEXIBLE SOLAR ARRAY AND COMPARISON
WITH STATIC TEST RESULTS
NASTRAN USERS EXPERIENCES, NASA TM X-2378,
SEPTEMBER 1971.

71-0232 SCHULTZ, C. C.
VAN POSSEN, D. B.
COMBINED ELASTIC-PLASTIC CREEP ANALYSIS OF
TWO-DIMENSIONAL BODIES
A.S.M.E. PAPER NO. 71-PVP-30, 1971.

71-0233 SCORDELIS, A. C.
DAVIS, D. B.
STRESSES IN CONTINUOUS CONCRETE BOX GIRDER BRIDGES
SEC. INT. SYMP. CONCRETE BRIDGE DESIGN, ACI, SP
26-13, 1971.

71-0234 SMITH, W. W.
A SPECIAL NASTRAN PROGRAM FOR INPUT CHECKING AND
UNDEFORMED STRUCTURE PLOTTING
NASTRAN USERS EXPERIENCES, NASA TM X-2378,
SEPTEMBER 1971.

71-0235 SOBIESZCZANSKI, J. E.
LOENDORF, D. D.
NASTRAN AS ANALYSIS TOOL IN A STRUCTURAL DESIGN
OPTIMIZATION PROCESS
NASTRAN USERS EXPERIENCES, NASA TM X-2378,
SEPTEMBER 1971.

71-0236 STANTON, E. L.
BOZICH, W. F.
AN EVALUATION OF CURRENT NASTRAN DISCRETE ELEMENT
MODELS FOR MONOCOQUE AND SEMI-MONOCOQUE STRUCTURES
NASTRAN USERS EXPERIENCES, NASA TM X-2378,
SEPTEMBER 1971.

71-0237 MCGOVERN, D. J.
STANTON, E. L.
THE APPLICATION OF GRADIENT MINIMIZATION METHODS
AND HIGHER ORDER DISCRETE ELEMENTS TO SHELL
BUCKLING AND VIBRATION EIGENPROBLEMS
COMPUT. AND STRUCT. J., VOL. 1, 1971.

71-0238 STARNES, J. H. JR.
VIBRATION STUDIES OF A FLAT PLATE AND BUILT-UP
WING
NASTRAN USERS EXPERIENCES, NASA TM X-2378,
SEPTEMBER 1971.

71-0239 STEFANOU, G. D.
YU, C. W.
ENGLAND, G. L.
TWO-DIMENSIONAL TIME-DEPENDENT ANALYSIS OF
PERFORATED END CAPS FOR NUCLEAR REACTOR PRESSURE
VESSELS BY THE FINITE ELEMENT METHOD
FIRST INT. CONF. STRUCT. MECH. IN REACTOR TECH.,
SEPTEMBER 1971, PAPER NO. H 3/3.

71-0240   STODDART, W. C. T.
TRANSIENT RESPONSE OF LINEAR ELASTIC STRUCTURES
DETERMINED BY THE MATRIX EXPONENTAL METHOD
FIRST INT. CONF. STRUCT. MECH. IN REACTOR TECH.,
SEPTEMBER 1971, PAPER NO. E 1/6.

71-0241   SUHARA, J.
FUKUDA, J.
NEW APPROACHES TO THE FINITE ELEMENT ANALYSIS
RECENT ADV. IN MATRIX METH. OF STRUCT. ANALYSIS
AND DESIGN, PP. 229-254, 1971.

71-0242   SWANSON, J. A.
PATTERSON, J. F.
APPLICATION OF FINITE ELEMENT METHODS FOR THE
ANALYSIS OF THERMAL CREEP, IRRADIATION INDUCED
CREEP, AND SWELLING FOR LMFBR DESIGN
FIRST INT. CONF. STRUCT. MECH. REACTOR TECH.,
SEPTEMBER 1971, PAPER NO. L 4/3.

71-0243   TANAKA, N.
SANBONGI, S.
KAWAI, T.
A TORSIONAL STRENGTH ANALYSIS ON THE CONTAINER
SHIP BY MEANS OF THE FINITE ELEMENT PROCEDURE AND
FULL SCALE TESTING
INT. SHIPBUILDING PROG., VOL. 18, NO. 201, MAY
1971.

71-0244   TAY, A. O.
DE VAHL DAVIS, G.
APPLICATION OF THE FINITE ELEMENT METHOD TO
CONVECTION HEAT TRANSFER BETWEEN PARALLEL PLANES
INT. J. HEAT AND MASS TRANSFER, VOL. 14, NO. 8,
PP. 1057-1070, AUGUST 1971.

71-0245   TEZCAN, S. S.
AGRAWAL, K. M.
KOSTRO, C.
FINITE ELEMENT ANALYSIS OF HYPERBOLIC PARABOLOID
SHELLS
J. STRUCT. DIV., A.S.C.E., VOL. 94, NO. ST1, PP.
407-424, JANUARY 1971.

71-0246   TOCHER, J. L.
HERNESS, E. D.
A CRITICAL REVIEW OF NASTRAN
INT. SYMP. NUMERICAL + COMPUTER METH. IN
STRUCTURAL MECHANICS, U.S. OFFICE OF NAVAL
RESEARCH, URBANA, ILLINOIS, SEPTEMBER 1971.

71-0247   THORNTON, E. A.
APPLICATION OF NASTRAN TO A SPACE SHUTTLE DYNAMICS
MODEL
NASTRAN USERS EXPERIENCES, NASA   TM   X-2378,
SEPTEMBER 1971.

71-0248   TONG, P.
FUNG, Y. C.
SLOW PARTICULATE VISCOUS FLOW IN CHANNELS AND
TUBES, APPLICATION TO BIOMECHANICS
A.S.M.E. PAPER NO. 71-APM-R, 1971.

71-0249   TONG, P.
THE FINITE ELEMENT METHOD IN FLUID FLOW ANALYSIS
RECENT ADV. IN MATRIX METH. OF STRUCT. ANALYSIS
AND DESIGN, PP. 787-808, 1971.

71-0250   TSUBOI, Y.
KAWAMATA, S.
SHIOYA, S.
MATRIX ANALYSIS OF A COMPOSITE STRUCTURE OF PLATE
AND LINE ELEMENTS
RECENT ADV. IN MATRIX METH. OF STRUCT. ANALYSIS
AND DESIGN, PP. 655-690, 1971.

71-0251   TSUSHIMA, Y.
HAYAMIZU, Y.
NISHIYAMA, K.
A SEISMIC DESIGN OF NUCLEAR REACTOR BUILDING,
STRESS ANALYSIS AND STIFFNESS EVALUATION OF THE
ENTIRE BUILDING BY THE FINITE ELEMENT METHOD
FIRST INT. CONF. STRUCT. MECH. REACTOR TECH.,
SEPTEMBER 1971, PAPER NO. K 2/5.

71-0252   YAMAMOTO, Y.
TOKUDA, N.
A NOTE ON CONVERGENCE OF FINITE ELEMENT SOLUTIONS
INT. J. NUM. METHODS ENG., VOL. 8, NO. 4, PP.
485-494, OCTOBER 1971.

71-0253   SCHULTE, D.
APPLICATION OF FINITE ELEMENT METHOD - STRESS
CALCULATIONS ON 90 DEGREES NOZZLES OF THIN-WALLED
CYLINDRICAL VESSELS
TECH. MITT. KRUPP, FORSCHUNGSBERICHTE, VOL. 29,
NO. 2, 1971, PP. 89-99.

71-0254   UTKU, S.
ON DERIVATION OF STIFFNESS MATRICES WITH C0
ROTATION FIELDS FOR PLATES AND SHELLS
THIRD CONF. ON MATRIX METHODS IN STRUCTURAL
MECHANICS, DAYTON, OHIO, OCTOBER 1971.

71-0255   VENKAYYA, V. B.
KHOT, N. S.
TISCHLER, V. A.
TAYLOR, R. F.
DESIGN OF OPTIMUM STRUCTURES FOR DYNAMIC LOADS
THIRD CONF. ON MATRIX METHODS IN STRUCTURAL
MECHANICS, DAYTON, OHIO, OCTOBER 1971.

71-0256   WAHLSTROM, S.
SOLID ELEMENT REQUIREMENT FOR NASTRAN
NASTRAN USERS EXPERIENCES, NASA   TM   X-2378,
SEPTEMBER 1971.

71-0257   WASHIZU, K.
SOME REMARKS ON BASIC THEORY FOR FINITE ELEMENT
METHOD
RECENT ADV. IN MATRIX METH. OF STRUCT. ANALYSIS
AND DESIGN, PP. 25-48, 1971.

71-0258   WEISSHAAR, T. A.
AN APPLICATION OF FINITE ELEMENT METHODS TO PANEL
FLUTTER OPTIMIZATION
THIRD CONF. ON MATRIX METHODS IN STRUCTURAL
MECHANICS, DAYTON, OHIO, OCTOBER 1971.

71-0259   WEMPNER, G. A.
DISCRETE APPROXIMATIONS RELATED TO NONLINEAR
THEORIES OF SOLIDS
INT. J. SOLIDS STRUCT., VOL. 7, NO. 11, PP.
1581-1599, NOVEMBER 1971.

71-0260   WILCOX, P. R.
CALCULATION OF STRESS CONCENTRATION DUE TO A
HYPERBOLOID CAVITY IN A THIN PLATE USING NASTRAN
NASTRAN USERS EXPERIENCES, NASA   TM   X-2378,
SEPTEMBER 1971.

71-0261   WILKINSON, M. T.
FULTON, R. E.
CONVERGENCE OF THE NASTRAN PLATE ELEMENTS FOR
SHELL STRESS ANALYSIS
NASTRAN USERS EXPERIENCES, NASA   TM   X-2378,
SEPTEMBER 1971.

71-0262   WILSON, E. L.
TAYLOR, R. L.
INCOMPATIBLE DISPLACEMENT MODELS
INT. SYMP. NUMERICAL + COMPUTER METH. IN
STRUCTURAL MECHANICS, U.S. OFFICE OF NAVAL
RESEARCH, URBANA, ILLINOIS, SEPTEMBER, 1971.

71-0263   YAGHMAI, S.
POPOV, E. P.
LARGE DEFLECTION ELASTIC-PLASTIC ANALYSIS OF
SHELLS OF REVOLUTION
FIRST INT. CONF. STRUCT. MECH. REACTOR TECH.,
SEPTEMBER 1971, PAPER NO. J. 2/2.

71-0264   YAMADA, YOSHIKAZU
DYNAMIC ANALYSIS OF CIVIL ENGINEERING STRUCTURES
RECENT ADV. IN MATRIX METH. OF STRUCT. ANALYSIS
AND DESIGN, 1971, PP. 487-514.

71-0265   YAMADA, Y.
RECENT DEVELOPMENTS IN MATRIX DISPLACEMENT METHOD
FOR ELASTIC-PLASTIC PROBLEMS IN JAPAN
RECENT ADV. IN MATRIX METH. OF STRUCT. ANALYSIS
AND DESIGN, 1971, PP. 283-316.

71-0266   YAMAMOTO, Y.
FINITE ELEMENT APPROACHES WITH THE AID OF ANALYTIC
SOLUTIONS
RECENT ADV. IN MATRIX METH. OF STRUCT. ANALYSIS
AND DESIGN, PP. 85-104, 1971.

71-0267   YANG, T. Y.
A MATRIX DISPLACEMENT METHOD ON PRE AND POST
BUCKLING ANALYSIS OF LINERS FOR REACTOR VESSELS
FIRST INT. CONF. STRUCT. MECH. IN REACTOR TECH.,
SEPTEMBER 1971, PAPER NO. H 6/2.

71-0268   YETTRAM, A. L.
HIRST, M. J. S.
THE SOLUTION OF STRUCTURAL EQUILIBRIUM EQUATIONS
BY THE CONJUGATE GRADIENT METHOD WITH PARTICULAR
REFERENCE TO PLANE STRESS ANALYSIS
INT. J. NUM. METHODS ENG., VOL. 3, NO. 3, PP.
349-360, JULY 1971.

71-0269   YOSHIDA, Y.
DISCRETE TRIANGULAR APPROXIMATION OF MOMENT AND
DISPLACEMENT SURFACES FOR PLATE BENDING
RECENT ADV. IN MATRIX METH. OF STRUCT. ANALYSIS
AND DESIGN, PP. 139-182, 1971.

71-0270   ZIENKIEWICZ, O. C.
OWEN, D. R. J.
PHILLIPS, D. V.
FINITE ELEMENT METHOD IN ANALYSIS OF REINFORCED
AND PRESTRESSED CONCRETE STRUCTURES
FIRST INT. CONF. STRUCT. MECH. REACTOR TECH.,
SEPTEMBER 1971, PAPER NO. M 5/1.

71-0271   ZIENKIEWICZ, O. C.
PHILLIPS, D. V.
AN AUTOMATIC MESH GENERATION SCHEME FOR PLANE AND
CURVED SURFACES BY ISOPARAMETRIC COORDINATES
INT. J. NUM. METHODS ENG., VOL. 3, NO. 4, PP.
519-528, OCTOBER 1971.

71-0272   ZIENKIEWICZ, O. C.
IRONS, B. M.
ISOPARAMETRIC AND OTHER NUMERICALLY INTEGRATED
ELEMENTS
INT. SYMP. NUMERICAL + COMPUTER METH. IN
STRUCTURAL MECHANICS, U.S. OFFICE OF NAVAL
RESEARCH, URBANA, ILLINOIS, SEPTEMBER 1971.

71-0273   ZIENKIEWICZ, O. C.
NAYAK, G. C.
A GENERAL APPROACH TO PROBLEMS OF LARGE
DEFORMATION, AND PLASTICITY USING ISOPARAMETRIC
ELEMENT
THIRD CONF. ON MATRIX METHODS IN STRUCTURAL
MECHANICS, DAYTON, OHIO, OCTOBER 1971.

71-0274   ZIENKIEWICZ, O. C.
THE FINITE ELEMENT METHOD IN ENGINEERING SCIENCE
MCGRAW HILL PUB. CO., 1971.

71-0275   ZUDANS, Z.
REDDI, M. M.
FISHMAN, H. M.
GRAY, D.
A THREE-DIMENSIONAL FINITE ELEMENT COMPUTER CODE
FOR THE ANALYSIS OF COMPLEX STRUCTURES
PROC. 1ST INT. CONF. STRUCT. MECH. IN REACTOR
TECH., SEPTEMBER, 1971, PAPER NO. F 5/1, ALSO IN
NUCL. ENG. AND DES., VOL. 20, NO. 1, 1972, PP.
149-167.

71-0276   ZUDANS, Z.
ANALYSIS OF ELASTICALLY COUPLED SHELLS ON ELASTIC
FOUNDATION BY HYBRID METHOD
FIRST INT. CONF. STRUCT. MECH. IN REACTOR TECH.,
SEPTEMBER 1971, PAPER NO. J 1/1.

71-0277   YOSHIDA, D. M.
WEAVER, J. R. W.
RESPONSE OF BEAMS AND PLATES TO MOVING LOADS
ASCE, NAT. STRUCT. ENG. MEETING, BALTIMORE,
MARYLAND, APRIL 19-23 1971, PAPER 1366, 30 PP.

71-0278   LASHKARI, M.
WEINGARTEN, V. I.
EFFECT OF LATERAL PRESSURE ON NATURAL FREQUENCIES
OF HYPERBOLOIDAL SHELLS OF REVOLUTION
ASCE, NAT. STRUCT. ENG. MEETING, BALTIMORE,
MARYLAND, APRIL 19-23 1971, PAPER 1390, 30 PP.

71-0279   ASHWELL, D. G.
SABIR, A. B.
LIMITATIONS OF CERTAIN CURVED FINITE ELEMENTS WHEN
APPLIED TO ARCHES
INT. J. MECH. SCI., VOL. 13, NO. 2, FEBRUARY 1971,
PP. 133-140.

71-0280   LIN, T. H.
SALINAS, D.
INITIAL YIELD SURFACE OF A UNIDIRECTIONALLY
REINFORCED COMPOSITE
ASME PAPER 71-APMW-19 FOR MEETING AUGUST 23-25,
1971, 6 PP.

71-0281   WILSON, E. A.
DESIGN ORIENTED APPROACH TO CREEP AND PLASTICITY
IN FINITE ELEMENT PROGRAMS
J. ENG. IND., TRANS. ASME, VOL. 93, SER. B, NO. 3,
AUGUST 1971, PP. 793-798. PAPER 70-WA/DE-4.

71-0282   ANEJA, I. K.
ROLL, F.
MODEL ANALYSIS OF CURVED BOX-BEAM HIGHWAY BRIDGE
ASCE, J. STRUCT. DIV., VOL. 97, NO. ST12, DECEMBER
1971, PAPER 8603, PP. 2861-2878.

71-0283   MOEHLING, C.
MATE, D. P.
COMPUTER ANALYSIS OF A RAILROAD FREIGHT CAR
BOLSTER UTILIZING THE FINITE ELEMENT METHOD
ASME, 1971, RAIL TRANSPORTATION PROC., JOINT
IEEE/ASME RAILROAD CONF., NEW YORK, NEW YORK,
APRIL 20-21, 1971, 7 PP.

71-0284   MOTE, C. D.
STABILITY CONTROL ANALYSIS OR ROTATING PLATES BY
FINITE ELEMENT. EMPHASIS ON SLOTS AND HOLES
ASME PAPER 71-WA/AUT-2, FOR MEETING NOVEMBER
28-DECEMBER 2 1971, 7 PP.

71-0285   SAWKO, F.
MERIMAN, P. A.
AN ANNULAR SEGMENT FINITE ELEMENT FOR PLATE
BENDING
INT. J. NUMER. METHODS ENG., VOL. 3, NO. 1,
JANUARY-MARCH 1971, PP. 119-129.

71-0286   WANG, M. C.
MITCHELL, J. K.
STRESS - DEFORMATION PREDICTION IN CEMENT -
TREATED SOIL PAVEMENTS
HIGHW. RES. REC. NO. 351, 1971, PP. 93-111.

71-0287   WALSH, P. F.
COMPUTATION OF STRESS INTENSITY FACTORS BY A
SPECIAL FINITE ELEMENT TECHNIQUE
INT. J. SOLIDS STRUCT., VOL. 7, NO. 10, OCTOBER
1971, PP. 1333-1342.

71-0288   KHAN, M. H.
FINITE ELEMENT METHOD IN NONLINEAR DOMAIN,
(METHODE DES ELEMENTS FINIS DANS LE DOMAINE NON
LINEAIRE).
BULL. TECH. SUISSE ROMANDE, VOL. 97, NO. 3,
FEBRUARY 6, 1971, PP. 23-33.

71-0289   HADDADIN, M. J.
MATS AND COMBINED FOOTINGS. ANALYSIS BY THE
FINITE ELEMENT METHOD
J. AMER. CONCRETE INST., VOL. 68, NO. 12, DECEMBER
1971, P. 9459.

71-0290   JOFRIET, J. C.
MCNEICE, G. M.
PATTERN LOADING ON REINFORCED CONCRETE FLAT PLATES
J. AMER. CONCRETE INST., VOL. 68, NO. 12, DECEMBER
1971, PP. 968-972.

71-0291   BARSOUM, R. S.
FINITE ELEMENT METHOD APPLIED TO THE PROBLEM OF
STABILITY OF A NON-CONSERVATIVE SYSTEM
INT. J. NUMER. METHODS ENG., VOL. 3, NO. 1,
JANUARY-MARCH 1971, PP. 63-87.

71-0292   SHARIFI, P.
POPOV, E. P.
NONLINEAR BUCKLING ANALYSIS OF SANDWICH ARCHES
ASCE J. ENG. MECH. DIV., VOL. 97, NO. EM5, OCTOBER
1971, PAPER 8417, PP. 1397-1412.

71-0293   DUNCAN, J. M.
CLOUGH, G. W.
FINITE ELEMENT ANALYSIS OF PORT ALLEN LOCK
ASCE J. SOIL MECH. FOUND. DIV., VOL. 97, NO. SM8,
AUGUST 1971, PAPER 8317, PP. 1053-1068.

71-0294   AL-DABBAGH, A. M. A.
GOODMAN, J. R.
FINITE ELEMENT METHOD FOR WOOD MECHANICS
A.S.C.E. NAT. STRUCT. ENG. MEETING, BALTIMORE,
MARYLAND, APRIL 19-23, 1973, PAPER NO. 1389, 28
PP. ALSO IN A.S.C.E. J., STRUCT. DIV., PAPER 8784,
VOL. 98, NO. ST3, MARCH 1972, PP. 569-586.

71-0295   BUYUKOZTURK, O.
NILSON, A. H.
STRESS STRAIN RESPONSE AND FRACTURE OF A CONCRETE
MODEL IN BIAXIAL LOADING
J. AMER. CONCRETE INST., VOL. 68, NO. 8, AUGUST
1971, PP. 590-599.

71-0296   YAGHMAI, S.
POPOV, E. P.
INCREMENTAL ANALYSIS OF LARGE DEFLECTIONS OF
SHELLS OF REVOLUTION
INT. J. SOLIDS STRUCT., VOL. 7, NO. 10, OCTOBER
1971, PP. 1375-1393.

71-0297   MAJID, K. I.
SPINDEL, J. E.
DESIGN OF INCLINED TIED ARCH RAILWAY BRIDGES OVER
THE M56
PROC. INST. CIV. ENG., VOL. 50, OCTOBER 1971, PP.
139-160.

71-0298   ANDERSON, H.
A GENERAL METHOD OF ANALYSIS FOR TWO-DIMENSIONAL
PILE GROUPS
PROC., INST. CIV. ENG., VOL. 49, MAY 1971, PP.
37-51.

71-0299   RASHID, Y. R.
NONLINEAR ANALYSIS OF TWO-DIMENSIONAL PROBLEMS IN
CONCRETE CREEP
ASME, PAPER 71-APMW-25, FOR MEETING AUGUST 23-25,
1971, 8 PP.

71-0300   MALLICK, D. V.
GARG, R. P.
EFFECT OF OPENINGS ON THE LATERAL STIFFNESS OF
INFILLED FRAMES
PROC. INST. CIV. ENG., VOL. 49, JUNE 1971, PP.
193-209.

71-0301   PENMAN, A. D. M.
BURLAND, J. B.
OBSERVED AND PREDICTED DEFORMATIONS IN A LARGE
EMBANKMENT DAM DURING CONSTRUCTION
PROC. INST. CIV. ENG., VOL. 49, MAY 1971, PP.
1-21.

71-0302 WOLF, J. A.
NATURAL FREQUENCIES OF CIRCULAR ARCHES
ASCE J. STRUCT. DIV., VOL. 97, NO. ST9, SEPTEMBER
1971, PAPER 8393, PP. 2337-2350.

71-0303 BENFIELD, W. A.
HRUDA, R. F.
VIBRATION ANALYSIS OF STRUCTURES BY COMPONENT MODE
SUBSTITUTION
AIAA J., VOL. 9 NO. 7, JULY 1971, PP. 1255-1261.

71-0304 D'APPOLONIA, D. J.
POULOS, H. G.
INITIAL SETTLEMENT OF STRUCTURES ON CLAY
ASCE J. SOIL MECH. FOUND. DIV., VOL. 97, NO. SM10,
OCTOBER 1971, PAPER 8438, PP. 1359-1377.

71-0305 YETTRAM, A. L.
ROBBINS, K.
ANCHORAGE ZONE STRESSES IN AXIALLY POST-TENSIONED
I-SECTION MEMBERS WITH END BLOCKS
MAG. CONGR. RES., VOL. 23, NO. 74, MARCH 1971, PP.
37-42.

71-0306 NISHIOKA, K.
MORITA, Y.
STRENGTH OF RAILROAD WHEELS - 3
BULL. ASME., VOL. 14, NO. 67, JANUARY 1971, PP.
11-19.

71-0307 JOBSON, D. A.
DETERMINATION OF STRESSES AT NODAL POINTS FROM
FINITE ELEMENT SOLUTIONS TO ELASTIC PROBLEMS
AERON. J., VOL. 75, NO. 723, MARCH 1971, PP.
194-196.

71-0308 BOLLAND, G. B.
FINITE ELEMENT FOR TORSION OF THIN WALLED OPEN
TUBES
AERON. J., VOL. 75, NO. 724, APRIL 1971, PP.
282-284.

71-0309 THOMPSON, G. L.
COMPUTATION OF VIBRATION MODAL SHAPES AND NATURAL
FREQUENCIES OF A GENERAL ORTHOTROPIC LAMINATED,
THIN SHELL USING FINITE ELEMENTS
DISS. ABSTR. INT. SEC. B., UNIV. MICROFILMS, ANN
ARBOR, MICHIGAN, (DISSERTATION ORDER NO.
71-268561, 170 PP), VOL. 32, NO. 4, OCTOBER 1971,
PP. 2120-2128.

71-0310 SHUGAR, T. A.
STRUCTURAL ANALYSIS OF SYMMETRICAL PENSTOCK
BIFURCATIONS BY THE FINITE ELEMENT METHOD
DISS. ABSTR. INT. SEC. B, UNIV. MICROFILMS, ANN
ARBOR, MICHIGAN, (DISSERTATION ORDER NO.
71-22,151, 181 PP.), VOL. 32, NO. 3, SEPTEMBER
1971, P. 1555-B.

71-0311 BLAKEY, L. H.
DYNAMIC RESPONSE OF REINFORCED OPENINGS IN ROCK
DISS. ABSTR. INT. SEC. B, UNIV. MICROFILMS, ANN
ARBOR, MICHIGAN, (DISSERTATION ORDER NO.
71-22,758, 283 PP.), VOL. 32, NO. 3, SEPTEMBER
1971, P. 1545-B.

71-0312 NOSSEIR, T. A.
DICKINSON, S. M.
FREE VIBRATION OF MODEL OF CAR BODY
J. SOUND VIB., VOL. 15, NO. 2, MARCH 22, 1971, PP.
257-268.

71-0313 MUTO, K.
NON LINEAR ANALYSIS OF REINFORCED CONCRETE MEMBERS
BY FINITE ELEMENT METHOD, PART 1: ANALYSIS, PART
2: SIMPLE BEAMS
PROC. OF ANN. MEET. OF THE ARCH. INST. OF JAPAN,
TOKYO, K. MUTO ET AL (EDS.), 1971, PP. 735-738.

71-0314 LEVY, N.
MARCAL, P. V.
PROGRESS IN THREE-DIMENSIONAL ELASTIC-PLASTIC
STRESS ANALYSIS FOR FRACTURE MECHANICS
NUCL. ENG. DES., VOL. 17, NO. 1, 1971, PP. 64-75.

71-0315 BANGASH, Y.
STRUCTURAL DESIGN OF CIRCULAR DIVERSION, POWER AND
PRESSURE TUNNELS
TUNNELS TUNNELLING, VOL. 3, NO. 3, 4, 5, MAY-JUNE
1971, PP. 161-171, JULY-AUGUST, PP. 271-273, 275,
SEPTEMBER-OCTOBER, PP. 356-359.

71-0316 BARKER, R. M.
MACLAUGHLIN, T. F.
STRESS CONCENTRATIONS NEAR A DISCONTINUITY IN
FIBROUS COMPOSITES
J. COMPOS. MATER., VOL. 5, OCTOBER 1971, PP.
492-503.

71-0317 TODD, P. H.
FUCHS, H. O.
SELF-STRESS CONCENTRATIONS
EXP. MECH., VOL. 11, NO. 12, DECEMBER 1971, PP.
548-553.

71-0318 BRADY, B. T.
EFFECTS OF INSERTS ON THE ELASTIC BEHAVIOUR OF
CYLINDRICAL MATERIALS LOADED BETWEEN ROUGH
END-PLATES
INT. J. ROCK. MECH. MINING SCI., VOL. 8, NO. 4,
JULY 1971, PP. 357-369.

71-0319 NIELSEN, R.
CHANG, P. Y.
SIMPLE APPROACH TO THE STRENGTH ANALYSIS OF
TANKERS
SOC. NAV. ARCHITECTS MAR. ENG., NEW YORK, PAPER
NO. 6, FOR MEETING NOVEMBER 11-12, 1971, 14 PP.

71-0320 CLOUGH, G. W.
DUNCAN, J. M.
FINITE ELEMENT ANALYSES OF RETAINING WALL
BEHAVIOUR
ASCE J. SOIL MECH. FOUND. DIV., VOL. 97, NO. SM12,
DECEMBER 1971, PAPER 8583, PP. 1657-1673.

71-0321 NAGAMATSU, A.
MUROTA, T.
ON THE NONUNIFORM DEFORMATION OF MATERIAL IN
AXIALLY SYMMETRIC COMPRESSION CAUSED BY FRICTION -
2
BULL JSME, VOL. 14, NO. 70, APRIL 1971, PP.
339-347.

71-0322 PENDLETON, R. L.
DAVIS, R. L.
FORMING OF 7075-T6 ALUMINUM IN HIGH PRESSURE
ENVIRONMENTS
PAPER 71-WA/PT-11, A.S.M.E. MEETING, NOVEMBER
28-DECEMBER 2, 1971, P. 4.

71-0323 NEUBERT, V. H.
RELATIONSHIP OF SUBSTRUCTURING TO IMPROVEMENT OF
DYNAMIC MODELING
ASME, SYNTHESIS OF VIBRATING SYST. WINTER ANNUAL
MEETING, WASHINGTON, D.C., NOVEMBER 30, 1971, PP.
49-56.

71-0324 OKUBO, Y.
KURATA, T.
ANALYSIS OF VEHICLE BODY STRUCTURE BY THE FINITE
ELEMENT METHOD
BULL JSAE, NO. 3, 1971, PP. 102-108.

71-0325 ZAK, A. R.
ELASTIC ANALYSIS OF CYLINDRICAL CONFIGURATIONS
WITH STRESS SINGULARITIES
ASME PAPER 71-WA/APM-18, FOR MEETING NOVEMBER
28-DECEMBER 2,1971, 14 PP.

71-0326 WU, R. W. H.
WITMER, E. A.
FINITE-ELEMENT ANALYSIS OF LARGE ELASTIC-PLASTIC
TRANSIENT DEFORMATIONS OF SIMPLE STRUCTURES
AIAA J., VOL. 9, NO. 9, SEPTEMBER 1971, PP.
1719-1724.

71-0327 YANG, T. Y.
FINITE ELEMENT PROCEDURE FOR LARGE DEFLECTION
ANALYSIS OF PLATES WITH INITIAL DEFLECTIONS
AIAA J. VOL. 9, NO. 8, AUGUST 1971, PP. 1468-1473.

71-0328 NAGAMATSU, A.
MUROTA, T.
ON THE NON-UNIFORM DEFORMATION OF A BLOCK IN
PLANE-STRAIN COMPRESSION CAUSED BY FRICTION -3
BULL JSME, VOL. 14, NO. 70, APRIL 1971, PP.
314-321.

71-0329 GOULD, H. H.
MIKIC, B. B.
AREAS OF CONTACT AND PRESSURE DISTRIBUTION IN
BOLTED JOINTS
PAPER 71-WA/DE-3, A.S.M.E. MEETING, NOVEMBER
28-DECEMBER 2, 1971, 7 PP.

71-0330 MITCHELL, J. K.
GARDNER, W. S.
ANALYSIS OF LOAD-BEARING FILLS OVER SOFT SUBSOILS
ASCE J. SOIL MECH. FOUND. DIV., VOL. 97, NO. SM11,
NOVEMBER 1971, PAPER 8522, PP. 1549-1571.

71-0331 WADDELL, G. G.
CROCKER, T. J.
TECHNIQUE OF MEASURING INITIAL DEFORMATION AROUND
AN OPENING
U.S. BUR. MINES, REP. INVEST., 7505, APRIL 1971,
60 PP.

71-0332 RUHL, R. L.
BOOKER, J. F.
FINITE ELEMENT MODEL FOR DISTRIBUTED PARAMETER
TURBOMOTOR SYSTEMS
PAPER 71-VIBR-56, A.S.M.E. MEETING, SEPTEMBER
8-10, 1971. ALSO IN J. ENG. IND., VOL. 94, NO. 1,
1972, PP. 126-132.

71-0333 ZUDANS, Z.
CHOW, T. Y.
VERIFICATION OF A HYBRID GENERAL PURPOSE SHELL AND
BODY OF REVOLUTION COMPUTER PROGRAM
ASME, SYMP. ON ENG. COMPUTER SOFTWARE, SAN
FRANCISCO, CALIF., MAY 1971, PP. 26-35.

71-0334 SARGIOUS, M.
TADROS, G.
STEP AND LOADS EFFECT ON STRESSES IN PRESTRESSED
CONCRETE SHORT BRACKETS
J. AMER. CONCRETE INST., VOL. 68, NO. 11, NOVEMBER
1971, PP. 861-866.

71-0335 ROBINSON, J.
HAGGENMACHER, G. W.
ELEMENTS FOR FINITE ELEMENT STRUCTURAL ANALYSIS
PROGRAMS
LOCKHEED CALIFORNIA COMPANY REPORT NO. 24910,
1971.

71-0336 SMITH, P. G.
WILSON, E. L.
AUTOMATIC DESIGN OF SHELL STRUCTURES
ASCE J. STRUCT. DIV., VOL. 97, NO. ST6, JANUARY
1971, PAPER 7825, PP. 191-201.

71-0337 CHEUNG, Y. K.
REDDY, D. V.
FREQUENCY ANALYSIS OF CERTAIN SINGLE AND
CONTINUOUS SPAN BRIDGES
CONF. ON DEVELOPMENTS IN BRIDGE DESIGN AND
CONSTRUCTION, UNIV. COLL., CARDIFF, MARCH 1971.

71-0338 VASARHELYI, D. D.
TESTS OF GUSSET PLATE MODELS
ASCE J. STRUCT. DIV., VOL. 97, NO. ST2, FEBRUARY
1971, PAPER 7901, PP. 665-678.

71-0339 KIRIOKA, K.
OHKUBO, Y.
ANALYSIS OF BODY STRUCTURES - 2.
PAPER 710157, A.S.M.E. MEETING, JANUARY 11-15,
1971, 11 PP.

71-0340 PETERSEN, W.
APPLICATION OF FINITE ELEMENT METHOD TO PREDICT
STATIC RESPONSE OF AUTOMOTIVE BODY STRUCTURES
PAPER 710263, A.S.M.E. MEETING, JANUARY 11-15,
1971, 17 PP.

71-0341 HESSEL, J. J.
LAMMERS, S. J.
SOLUTION OF AUTOMOTIVE STRUCTURAL PROBLEMS USING
THE FINITE ELEMENT METHOD AND COMPUTER GRAPHICS
PAPER 710243, A.S.M.E. MEETING, JANUARY 11-15,
1971, 10 PP.

71-0342 BUSHNELL, D.
ANALYSIS OF RING-STIFFENED SHELLS OF REVOLUTION
UNDER COMBINED THERMAL AND MECHANICAL LOADING
AIAA J. VOL. 9, NO. 3, MARCH 1971, PP. 401-410.

71-0343 CHOPRA, A. K.
PERUMALSWAMI, P. R.
DYNAMICS OF EARTH DAMS WITH FOUNDATION INTERACTION
ASCE J. ENG. MECH. DIV., VOL. 9M, NO. EM2, APRIL
1971, PAPER 8049, PP 181-191.

71-0344 MUFTI, A. A.
HARRIS, P. J.
FURTHER DEVELOPMENTS IN THE USE OF LINEAR
DISPLACEMENT FUNCTIONS IN THE FINITE ELEMENT
ANALYSIS OF ORTHOTROPIC PLATES AND SHELLS
ENG. J. (MONTREAL), VOL. 54, NO. 1-2,
JANUARY-FEBRUARY, 1971, PP. I-X.

71-0345 YPEREZ, L. P.
A GENERAL CALCULATION OF THE STRUCTURAL LOADS ON A
SHIP IN A SEAWAY
MAR. TECHNOL., VOL. 8, NO. 2, APRIL 1971, PP.
186-200.

71-0346 LAUTERBACH, G. F.
PEPPIN, P. J.
FINITE ELEMENT SOLUTION FOR SAINT-VENANT BENDING
AIAA, VOL. 9, NO. 3, MARCH 1971, PP. 525-527.

71-0347 MAEDA, Y.
RECENT TRENDS IN STUDIES AT INTERNATIONAL
CONFERENCE ON STRUCTURAL ENGINEERING
J. JAP. SOC. CIV. ENG., VOL. 56, NO. 8, AUGUST
1971, P. 637.

71-0348 DAVIS, R. L.
KEITH, H. D.
FINITE ELEMENT ANALYSIS OF PRESSURE VESSELS
A.S.M.E. MEETING, NOVEMBER 28-DECEMBER 2, 1971,
PAPER NO. 71-WA/PT-10, 5 PP. ALSO IN J. BASIC
ENG., VOL. 94, NO. 2, 1972, PP. 401-406.

71-0349 BAKER, L. E.
SANDHU, R. S.
APPLICATION OF ELASTO-PLASTIC ANALYSIS IN ROCK
MECHANICS BY FINITE ELEMENT METHOD
SOC. MINING ENG., AIME, PROC. 11TH SYMP. ON ROCK
MECHANICS, JUNE 16-19, 1969, BERKELEY, CALIF.,
1970, PP. 237-251.

71-0350 VINSON, T. J.
YATES, D. N.
APPLICATION OF THE FINITE ELEMENT METHOD FOR
ANALYSIS AND OPTIMIZATION OF SUBMARINE STRUCTURES
A.S.M.E. MEETING, NOVEMBER 28-DECEMBER 2, 1971,
PAPER NO. 71-WA/UNT-7, 19 PP. ALSO IN J. MECH.
ENG., VOL. 94, NO. 4, 1972, 72 PP.

71-0351 AHMED, K. M.
FREE VIBRATION OF CURVED SANDWICH BEAMS BY METHOD
OF FINITE ELEMENTS
J. SOUND AND VIBRATION, VOL. 18, NO. 1, 1971, PP.
61-74.

71-0352 AHMED, K. M.
STATIC AND DYNAMIC ANALYSIS OF SANDWICH STRUCTURES
BY METHOD OF FINITE ELEMENTS
J. SOUND AND VIBRATION, VOL. 18, NO. 1, 1971, PP.
75-91.

71-0353 OLSON, M. D.
LINDBERG, G. M.
DYNAMIC ANALYSIS OF SHALLOW SHELLS WITH A
DOUBLY-CURVED TRIANGULAR FINITE ELEMENT
J. SOUND AND VIBRATION, VOL. 19, NO. 3, 1971, PP.
299-318.

71-0354 MITCHELL, R. A.
WOOLLEY, R. M.
FISHER, C. R.
FORMULATION AND EXPERIMENTAL VERIFICATION OF AN
AXISYMMETRIC FINITE-ELEMENT STRUCTURAL ANALYSIS
J. RESEARCH NBS SECTION C, ENGINEERING, VOL. C
75, NO. 3-4, 1971, PP. 155-163.

71-0355 LEVY, N.
OSTERGREN, W. J.
RICE, J. R.
SMALL SCALE YIELDING NEAR A CRACK IN PLANE
STRAIN-FINITE ELEMENT ANALYSIS
INT. J. FRACTURE MECHANICS, VOL. 7, NO. 2, 1971,
PP. 143-156.

71-0356 ROSS, C. T.
ANALYSIS OF STRUCTURES USING FINITE ELEMENT
TECHNIQUES
ENG. MATERIALS AND DESIGN, VOL. 14, NO. 4, 1971,
PP. 369-372.

71-0357 WRIGHT, G. P.
O'CONNOR, J. J.
FINITE ELEMENT ANALYSIS OF ALTERNATIVE AXIAL
LOADING OF AN ELASTIC PLATE PRESSED BETWEEN TWO
ELASTIC RECTANGULAR BLOCKS WITH FINITE FRICTION
INT. J. ENG. SCIENCE, VOL. 9, NO. 6, 1971, PP.
555-570.

71-0358 LEVY, N.
MARCAL, P. V.
THREE DIMENSIONAL ELASTIC PLASTIC ANALYSIS FOR
FRACTURE MECHANICS - 2.
HEAVY SECT. STEEL TECHNOL. PROGRAM, TECH. REP. NO.
17, NOVEMBER 1971, P. 29.

71-0359 LUXMOORE, A. R.
GARDNER, N. A.
WYATT, P. J.
COMPARISON OF FINITE-ELEMENT AND EXPERIMENTAL
STUDIES ON DEFORMATIONS OF ZIRCONIUM NOTCHED BEND
SPECIMENS
J. OF THE MECHANICS AND PHYSICS OF SOLIDS, VOL.
19, NO. 6, 1971, PP. 395-406.

71-0360 ASHWELL, D. G.
SABIR, A. B.
ROBERTS, T. M.
APPLICATION OF CURVED FINITE ELEMENTS TO CIRCULAR
ARCHES
INT. J. MECH. SCIENCES, VOL. 13, NO. 6, 1971, PP.
507-517.

71-0361 NETHERCOT, D. A.
ROCKEY, K. C.
FINITE ELEMENT SOLUTIONS FOR BUCKLING OF COLUMNS
AND BEAMS
INT. J. MECH. SCIENCES, VOL. 13, NO. 11, 1971, PP.
945-949.

71-0362 HENSHELL, R. D.
NEALE, B. K.
WARBURTON, G. B.
COMPARISON OF RECENT SHALLOW SHELL FINITE ELEMENT
ANALYSES
INT. J. MECH. SCIENCES, VOL. 13, NO. 3, 1971, PP.
283-284.

71-0363 BARTEN, H. J.
SCHEURENBRAND, J. A.
SCHEER, D. D.
STRESS AND VIBRATION ANALYSIS OF TURBOPUMP INDUCER
BLADES BY FINITE ELEMENTS
J. SPACECRAFT AND ROCKETS, VOL. 8, NO. 6, 1971,
PP. 657-664.

71-0364 KRATOCHVIL, J.
LEITNER, F.
THREE-DIMENSIONAL ISOPARAMETRIC ELEMENTS AND THEIR
USE FOR STATICAL ANALYSIS OF MASSIVE WATER-POWER
STRUCTURES
STAVEBNICKLY CAS. VOL. 19, NO. 7, 1971, PP.
481-503.

71-0365 DEMS, K.
APPLICATION OF HERMITE POLYNOMIALS TO THE
DETERMINATION OF THE STIFFNESS MATRIX OF PLATE BY
FINITE ELEMENT METHOD. (ZASTOSOWANIE WIELOMIANOW
HERMITE'A DO WYZNACZANIA MACIERZY SZTYWNOSCI
ELEMENTU TARCZY W METODZIE ELEMENTOW SKONCZONYCH)
MECH. TEOR, I STOSOW, VOL. 9, NO. 3, 1971, PP.
355-373.

71-0366 DI MONACO, A.
PROCESS FOR THE AUTOMATIC GENERATION OF FLAT
TRIANGULAR LATTICES FOR APPLICATION TO THE FINITE
ELEMENT METHOD. (UN PROCEDIMENTO PER LA GENERAZION
AUTOMATICA DI RETICOLI TRIANGOLARI PIANI PER
APPLICAZIONE AL METODO DEGLI ELEMENTI FINITI)
ENERG. ELET. VOL. 48, NO. 7, 1971, PP. 449-454.

71-0367 MELOSH, R. J.
MANIPULATION ERRORS IN COMPUTER SOLUTION OF
CRITICAL SIZE STRUCTURAL EQUATIONS
NASA CONTRACT REP. CR-1784, JUNE 1971, 48 PP.

71-0368 NARAYANASWAMI, R.
HIGHER ORDER FINITE ELEMENT METHOD FOR STATIC AND
DYNAMIC ANALYSIS OF SHEAR WALLS
PH.D. DISSERTATION, TEXAS TECH. UNIVERSITY,
LUBBOCK, TEXAS, AUGUST 1971.

71-0369 NORRIE, D. H.
DE VRIES, G.
APPLICATION OF FINITE ELEMENT METHODS IN FLUID
DYNAMICS
AGARD LECTURES SERIES NO. 48, NUMERICAL METHODS IN
FLUID DYNAMICS, 1971.

71-0370 DE VRIES, G.
NORRIE, D. H.
THE APPLICATION OF THE FINITE ELEMENT TECHNIQUE TO
POTENTIAL FLOW PROBLEMS
TRANS. OF THE A.S.M.E., PAPER NO. 71-APM-22,
DECEMBER 1971, PP. 798-802.

71-0371 ATLURI, S.
A NEW ASSUMED STRESS HYBRID FINITE ELEMENT MODEL
FOR SOLID CONTINUA
J. AM. INST. AERON. ASTRON., VOL. 9, NO. 8, AUGUST
1971, PP. 1647-1649.

71-0372 KANG, C. M.
HANSEN, K. F.
FINITE ELEMENT METHODS FOR THE NEUTRON DIFFUSION
EQUATIONS
TRANS. AM. NUCL. SOC., VOL. 14, 1971, P. 199.

71-0373 SEMENZA, L. A.
LEWIS, E. E.
ROSSOW, E. C.
A FINITE ELEMENT TREATMENT OF NEUTRON DIFFUSION
TRANS. AM. NUCL. SOC., VOL. 14, 1971, P. 200.

71-0374 ZIENKIEWICZ, O. C.
TOO, J.
TAYLOR, R. L.
REDUCED INTEGRATION TECHNIQUE IN GENERAL ANALYSIS
OF PLATES AND SHELLS
INT. J. NUM. METH. ENG., VOL. 3, 1971.

71-0375 VERYARD, D. A.
PROBLEMS ASSOCIATED WITH THE CONVERGENCE OF
ISOPARAMETRIC AND MIXOPARAMETRIC FINITE ELEMENTS
M.SC. THESIS, UNIVERSITY OF WALES, 1971.

71-0376 GEORGE, J. A.
COMPUTER IMPLEMENTATION OF THE FINITE ELEMENT
METHOD
PH.D. THESIS, STANFORD UNIVERSITY, 1971.

71-0377 BIRKHOFF, G.
TRICUBIC POLYNOMIAL INTERPOLATION
PROC. NAT. ACAD. SCI., USA 68, 1971, PP.
1162-1164.

71-0378 DESAI, C. S.
SHERMAN, W. C.
UNCONFINED TRANSIENT SEEPAGE IN SLOPING BANKS
J. OF SOIL MECH. AND FOUND., PROC. ASCE, VOL. 99,
NO. SM2, FEBRUARY 1971, PP. 357-373.

71-0379 SCHOLES, A.
STROVER, E. M.
THE PIECEWISE LINEAR ANALYSIS OF TWO CONNECTED
STRUCTURES INCLUDING THE EFFECT OF CLEARANCE AT
THE CONNECTIONS
INT. J. NUM. METHODS IN ENG., VOL. 3, 1971, PP.
45-52.

71-0380 STRANG, G.
THE FINITE ELEMENT METHOD AND APPROXIMATION THEORY
IN NUMERICAL SOLUTION OF PARTIAL EQUATIONS - II B.
HUBBARD, EDITOR, ACADEMIC PRESS, NEW YORK, 1971,
PP. 547-584

71-0381 SHIBATA, H.
DYNAMIC ANALYSIS OF MECHANICAL STRUCTURES
PROC. U.S. - JAP. SEM. ON RECENT ADVANCES IN
MATRIX METHODS OF STRUCTURAL ANALYSIS AND DESIGN,
TOKYO, AUGUST 1969, PP. 515-544. (EDS) GALLAGHER,
R.H., YAMADA, Y. AND ODEN, J.T. (PUBL. BY UNIV. OF
ALABAMA PRESS, 1971)

71-0382 UEDA, Y.
YAMAKAWA, T.
ANALYSIS OF THERMAL ELASTIC PLASTIC STRESS AND
STRAIN DURING WELDING BY FINITE ELEMENT METHOD
TRANS. OF JAPAN WELDING SOCIETY, VOL. 2, NO. 2,
1971, PP. 90-100.

71-0383 ROBERTS, T. M.
ASHWELL, D. G.
THE USE OF FINITE ELEMENT MID-INCREMENT STIFFNESS
MATRICES IN THE POST BUCKLING ANALYSIS OF
IMPERFECT STRUCTURES
INT. J. SOLIDS STRUCT., VOL. 7, 1971, PP. 805-823.

71-0384 ANON
DEVELOPMENTS IN STRUCTURAL ANALYSIS AND SYNTHESIS
BY FUNCTION MINIMIZATION
AIR FORCE FLIGHT DYNAMICS, LAB., AIR FORCE SYSTEMS
COMMAND, WRIGHT-PATTERSON AIR FORCE BASE, OHIO,
TECH., REP. AFFDL-TR-70-175, APRIL 1971, 133 PP.

71-0385 ANON
STRUCTURAL VIBRATION - A MATRIX APPROACH
AUTOMOT. DES. ENG., VOL. 10, JULY-AUGUST 1971, PP.
14-20.

71-0386 ANON
STRESS ANALYSIS FOR CLEAN LEAN DESIGN
DES. ENG., VOL. 17, NO. 2, FEBRUARY 1971, PP.
32-33.

71-0387 ZIENKIEWICZ, O. C.
PAREKH, C. J.
THREE DIMENSIONAL ANALYSIS OF BUILDINGS COMPOSED
OF FLOOR AND WALL PANELS
PROC. INST. CIV. ENG., VOL. 49, JULY 1971, PP.
319-332.

71-0388 FUJINO, T.
SELF ADJOINT DIFFERENTIAL EQUATIONS AND
VARIATIONAL PRINCIPLE - 1.
MITSUBISHI HEAVY IND., MITSUBISHI TECH. BULL.,
NO. 74, AUGUST 1971, 12 PP.

71-0389 OHJI, K.
OGURA, K.
ANALYTICAL APPROACH TO THE NON-PROPAGATING CRACK
PROBLEM USING THE FINITE ELEMENT METHOD
PROC. OF 15TH JAP. CONGR. ON MATER. RES., TOKYO,
SEPTEMBER 1971. SOC. MATER., SCI., JAP., 1972, PP.
91-94.

71-0390 STUMPP, W.
BERECHNUNGSBEISPIELE UND ERFAHRUNGEN MIT DER
METHODE DER FINITEN ELEMENTE IN MASCHINENBAU.
(EXAMPLES OF CALCULATION AND EXPERIENCE WITH THE
METHOD OF FINITE ELEMENTS IN MACHINE CONSTRUCTION)
SCHWEZ. BAUZTG., VOL. 89, NO. 48, DECEMBER 2,
1971, PP. 1201-1205.

71-0391 FETTE, H.
MAY, B.
FINITE ELEMENTE FUR GEKRUEMMTE FLAECHENTRAGWERKE.
(FINITE ELEMENTS SOLUTIONS FOR CURVED SHELL
STRUCTURES)
KONSTR. INGENIEURBAU., BER. NO. 11, 1971, 80 PP.

71-0392 RYBICKI, E. E.
APPROXIMATE THREE-DIMENSIONAL SOLUTIONS FOR
SYMMETRIC LAMINATES UNDER INPLANE LOADING
J. COMPOS. MATER., VOL. 5, JULY 1971, PP. 354-360.

71-0393 DERNER, W. J.
HOLLOW-ENDED ROLLER. A SOLUTION FOR IMPROVING
FATIGUE LIFE IN ASYMMETRICALLY LOADED CYLINDRICAL
ROLLER BEARINGS
ASME, PAP. 71-LUB-14, FOR MEETING OCTOBER 5-7,
1971, 10 PP.

71-0394 HARKEGARD, G.
THEORETICAL STUDY OF THE INFLUENCE OF INCLUSIONS
UPON THE INITIATION AND GROWTH OF FATIGUE CRACKS
IN STEEL
JERNKONTERETS ANN., VOL. 155, NO. 6, 1971, PP.
289-294.

71-0395 DUMAS, G.
BARONET, C. N.
ELASTOPLASTIC INDENTATION OF HALF-SPACE BY AN
INFINITELY LONG RIGID CIRCULAR CYLINDER
INT. J. MECH. SCI., VOL. 13, NO. 6, JUNE 1971, PP.
519-530.

71-0396 PAWSEY, S. F.
CLOUGH, R. W.
IMPROVED NUMERICAL INTEGRATION OF THICK SHELL
FINITE ELEMENTS
INT. J. NUMER. METHODS ENG., VOL. 3, NO. 4,
OCTOBER-DECEMBER 1971, PP. 575-586.

71-0397 SWANSON, S. R.
FINITE-ELEMENT SOLUTION FOR A CRACKED TWO-LAYERED
ELASTIC CYLINDER
ENG. FRACT. MECH., VOL. 3, NO. 3, OCTOBER 1971,
PP. 283-289.

71-0398 SALONEN, E. M.
A RECTANGULAR PLATE BENDING ELEMENTS THE USE OF
WHICH IS EQUIVALENT TO THE USE OF THE FINITE
DIFFERENCE METHOD
INT. J. NUM. METHODS. ENGNG., VOL. 1, NO. 3, 1971,
PP. 261-274.

71-0399 BUSHNELL, D.
ALMROTH, B. O.
FINITE-DIFFERENCE ENERGY METHOD FOR NONLINEAR
SHELL ANALYSIS
COMPUT. STRUCT., VOL. 1, NO. 3, OCTOBER 1971, PP.
361-387.

71-0400 WHANG, B.
LAMINATED ORTHOTROPIC PLATES AND SHALLOW SHELLS
COMPUT. STRUCT., VOL. 1, NO. 3, OCTOBER 1971, PP.
465-493.

71-0401 TORIDIS, T. C.
KHOZEIMEH, K.
COMPUTER ANALYSIS OF RIGID FRAMES
COMPUT. STRUCT., VOL. 1, NO. 1-2, AUGUST 1971, PP.
193-221.

71-0402 MCCORMICK, C. W.
SHELL ANALYSIS WITH LARGE GENERAL PURPOSE PROGRAMS
COMPUT. STRUCT., VOL. 1, NO. 1-2, AUGUST 1971, PP.
323-332.

71-0403 KAMEL, H. A.
LIU, D.
APPLICATION OF THE FINITE ELEMENT METHOD TO SHIP
STRUCTURES
COMPUT. STRUCT., VOL. 1, NO. 1-2, AUGUST 1971, PP.
103-130.

71-0404 CHAN, S. K.
MANJOINE, M. J.
ANALYSIS OF STRESSES IN PRESSURIZED WELDED PIPE IN
THE CREEP RANGE
ASME PAP. 71-PVP-66 FOR MEETING MAY 10-12, 1971,
15 PP.

71-0405 POPPLEWELL, N.
VIBRATION OF A BOX-TYPE STRUCTURE - 1
J. SOUND VIBR., VOL. 14, NO. 3, FEBRUARY 8, 1971,
PP. 357-365.

71-0406 CICALA, P.
LINEAR SHELL THEORY FOR RAPID VARIATION SOLUTIONS
MECCANICA, VOL. 6, NO. 1, MARCH 1971, PP. 59-64.

71-0407 PISTER, K. S.
CONTINUUM MECHANICS - A BRIEF REVIEW
FINITE ELEMENT METHODS IN CONTINUUM MECHANICS,
(EDS.) ODEN, J.T., DE ARANTES E OLIVEIRA, E.R.,
THE ADVANCED STUDY INST., LABORATORIO NACIONAL DE
ENGENHARIA CIVIL, LISBON, PORTUGAL, SEPTEMBER
7-17, 1971, PP. 1-40.

71-0408 WOLF, J. P.
STRIP PROGRAM FOR DESIGNING LOAD BEARING SURFACE
STRUCTURES (PROGRAMME STRIP POUR LE CALCUL DES
STRUCTURES EN SURFACE PORTEUSE)
BULL. TECH. SUISSE ROMANDE, VOL. 97, NO. 17,
AUGUST 21, 1971, PP. 381-397.

71-0409 BUTTER, U.
FINITE ELEMENT METHOD FOR STRESS ANALYSIS OF
BODIES OF REVOLUTION SUBJECTED TO
AXISYMMETRIC LOADING
ING-ARCH., VOL. 40, NO. 5, 1971, PP. 281-303.

71-0410 GRIFFITHS, J. R.
OWEN, D. R. J.
ELASTIC-PLASTIC STRESS ANALYSIS FOR A NOTCHED BAR
IN PLANE STRAIN BENDING
J. MECH. PHYS. SOLIDS, VOL. 19, NO. 6, NOVEMBER
1971, PP. 419-431.

71-0411 COURBON, J.
CALCULATION OF THE BRIDGE DECKS BY USING NUMERICAL
METHODS. (LE CALCUL DES TABLIERS DE PONTS PAR DES
METHODS NUMERIQUES)
ANN. INST. TECH., BATIMENT, TRAV., PUB., NO. 288,
DECEMBER 1971, PP. 93-103.

71-0412 CHAN, A. S. L.
HENRYWOOD, R. K.
ANALYSIS OF RADIAL FLOW IMPELLERS BY THE MATRIX
FINITE ELEMENT METHOD
AERON. J., VOL. 75, NO. 732, DECEMBER 1971, PP.
850-860.

71-0413 THANGAM BABU, P. V.
REDDY, D. V.
FREQUENCY ANALYSIS OF SKEW ORTHOTROPIC PLATES BY
THE FINITE STRIP METHOD
J. SOUND VIBR., VOL. 18, NO. 4, OCTOBER 1971, PP.
465-474.

71-0414 OKUSHIMA, K.
KAKINO, Y.
RESIDUAL STRESS PRODUCED BY METAL CUTTING
ANN CIRP. VOL. 20, NO. 1, 1971, PP. 13-14.

71-0415 HRUDEY, T. M.
CREEP BENDING ANALYSIS OF PLATES BY THE FINITE
ELEMENT METHOD
NATL. RES. COUNC. CAN., AERONAUT REP. LR-552,
OCTOBER 1971, 43 PP. ALSO IN INT. J. OF
SOLIDS AND STRUCTURES, VOL. 9, 1973, PP. 291-305.

71-0416 SATA, T.
TAKASHIMA, N.
DYNAMIC ANALYSIS OF MACHINE TOOL STRUCTURES BY THE
FINITE ELEMENT METHOD
ANN CIRP. VOL. 20, NO. 1, 1971, PP. 75-76.

71-0417 CHEUNG, Y. K.
CHEUNG, M. S.
FLEXURAL VIBRATIONS OF RECTANGULAR AND OTHER
POLYGONAL PLATES
JOURNAL OF THE ENG. MECH. DIV., PROC. ASCE, APRIL
1971.

71-0418 MITCHELL, G. C.
ACCURACY AND OPTIMIZATION IN THE ANALYSIS OF SHIP
STRUCTURES. USING FINITE ELEMENT METHODS
EUR. SHIPBLDG., VOL. 20, NO. 4, 1971, PP. 16-28.

71-0419 RAO, A. K.
STRESS CONCENTRATIONS AND SINGULARITIES AT
INTERFACE CORNERS
Z. ANGEW. MATH. MECH., VOL. 51, NO. 5, AUGUST
1971, PP. 395-406.

71-0420 POPPLEWELL, N.
VIBRATION OF A BOX-TYPE STRUCTURE - 2.
J. SOUND VIBR., VOL. 18, NO. 4, OCROBER 22, 1971,
PP. 521-531.

71-0421 YANG, T. Y.
FINITE ELEMENT PROCEDURE FOR POSTBUCKLING ANALYSIS
OF INITIALLY CURVED PLATES
STRUCTURES, STRUCTURAL DYNAMICS, AND MATERIALS
CONF., AIAA/ASME 12TH, APRIL 19-21, 1971. AIAA,
1971, STRUCTURES AND MATERIALS, TECH. PAP. NO.
71-357, 10 PP.

71-0422 BROMBOLICH, L. J.
GOULD, P. L.
HIGH-PRECISION CURVED SHELL FINITE ELEMENT
PROC. 12TH CONF. ON STRUCTURES, STRUCTURAL
DYNAMICS AND MATERIALS, A.I.A.A./A.S.M.E., APRIL
19-21, 1971, PAPER NO. 71-360, 9 PP. ALSO IN
A.I.A.A. J., VOL. 10, NO. 6, JUNE 1972, PP.
727-728.

71-0423 SARGIOUS, M.
ROSS, G. A.
RIGID PAVEMENT STRESSES UNDER AIRCRAFT LOADING
ASCE TRANSP. ENG., J., VOL. 97, NO. TE 4, NOVEMBER
1971, PAPER 8492, PP. 579-590, UNIVERSITY OF
CALGARY, ALBERTA.

71-0424 WILSON, W. K.
THOMPSON, D. G.
FINITE ELEMENT METHOD FOR CALCULATING STRESS
INTENSITY FACTORS FOR CRACKED PLATES IN
BENDING
ENG. FRACT. MECH., VOL. 3, NO. 2, AUGUST 1971, PP.
97-102.

71-0425 NEILLE, D. S.
HELLINGS, J. E.
COMPARISON OF PRACTICAL METHODS OF ANALYSING
ELASTIC LATERALLY LOADED PLATES
CIV. ENG., S. AFR., VOL. 13, NO. 6, JUNE 1971, PP.
196-204.

71-0426 CORUM, J. M.
USE OF SMALL MODELS IN DESIGN AND ANALYSIS OF
PRESTRESSED-CONCRETE REACTOR VESSELS
NUCL. ENG., DES., VOL. 16, NO. 3, 1971, PP.
301-342.

71-0427 MARGOLIAS, D. S.
WEINGARTEN, V. I.
INEXTENSIONAL VIBRATIONS OF PARABOLOIDAL SHELLS OF
REVOLUTION, DOMES AND SHELLS
PROC. 12TH MIDWEST MECH. CONF. ON DEVELOPMENTS IN
MECHANICS, NOTRE DAME, ILL., AUGUST 16-18, 1971,
VOL. 6, PP. 995-1008. (PUBL. BY UNIV. NOTRE DAME
PRESS, 1971)

71-0428 CHEUNG, Y. K.
CHAKRABARTI, S.
ANALYSIS OF SIMPLY SUPPORTED THICK, LAYERED PLATES
JOURNAL OF THE ENG. MECH. DIV., PROC. ASCE, JUNE
1971.

71-0429 BIAREZ, J.
BOULIN, M.
APPLICATION DES CALCULUS DE FONDATIONS A
L'EVALUATION DU COMPORTEMENT DES CHAUSSEES EM DASH
1, 2. (APPLICATION OF FOUNDATION DESIGN METHOD TO
THE EVALUATION OF THE ROAD BEHAVIOR EM DASH 1, 2.)
GENIE CIV. VOL. 148, NO. 8-9, 10 AUGUST-SEPTEMBER,
1971, PP. 405-412, OCTOBER PP. 471-480.

71-0430 COWPER, G. R.
CURSHL: A HIGH-PRECISION FINITE ELEMENT FOR
SHELLS OF ARBITRARY SHAPE
NATL. RES. COUNCIL, CANADA, AERONAUT. REP. LR-560,
DECEMBER 1971, 52 PP.

71-0431 PLANEIX, J. M.
APPLICATION OF THE FINITE ELEMENT METHOD TO THE
CALCULATION OF SHIP HULLS
RFM, REV. FR. MECH., NO. 39, 1971, PP. 15-33.

71-0432 GUELLEC, P.
LES PLAQUES ET LES COQUES MINCES OU EPAISSES (THIN
AND THICK PLATES AND SHELLS)
RFM, REV. FR. MECH., NO. 39, 1971, PP. 35-42.

71-0433 BRADY, B. T.
EXACT SOLUTION TO THE RADIALLY END CONSTRAINED
CIRCULAR CYLINDER UNDER TRIAXIAL LOADING
INT. J. ROCK MECH. MINING SCI. VOL. 8, NO. 2,
MARCH 1971, PP. 165-178.

71-0434 DOKAINISH, M. A.
NEW APPROACH FOR PLATE VIBRATIONS. COMBINATION OF
TRANSFER MATRIX AND FINITE ELEMENT TECHNIQUE
A.S.M.E. MEETING, PAPER NO. 71-VIBR-85, SEPTEMBER
8-10, 1971, 5 PP. ALSO J. ENG. IND., TRANS.
A.S.M.E., VOL. 94, SER. B, NO. 2, MAY 1972, PP.
526-530.

71-0435 CHAN, S. K.
TUBA, I. S.
FINITE ELEMENT METHOD FOR CONTACT PROBLEMS OF
SOLID BODIES - 2.
INT. J. MECH. SCI., VOL. 13, NO. 7, JULY 1971, PP.
627-639.

71-0436 TRACEY, D. M.
FINITE ELEMENTS FOR DETERMINATION OF CRACK TIP
ELASTIC STRESS INTENSITY FACTORS
ENG. FRACT. MECH., VOL. 3, NO. 3, OCTOBER 1971,
PP. 255-265.

71-0437 GOULD, P. L.
SZABO, B. A.
HIGH-PRECISION PLATE AND SHELL FINITE ELEMENTS
DEVELOPMENTS IN MECHANICS, VOL. 6, PROC. 12TH
MIDWEST MECH. CONF., NOTRE DAME, ILL., AUGUST
16-18, 1971, UNIVERSITY OF NOTRE DAME PRESS, 1971,
PP. 801-814.

71-0438 DE ROUVRAY, A.
GOODMAN, R. E.
ANALYSIS AND MODEL STUDIES OF UNDERGROUND OPENINGS
IN JOINTED ROCK. RESEARCH ON ROCK BOLT
REINFORCEMENT
CORPS OF ENGINEERS, OMAHA DISTRICT NEB., TECH.
REP. 7, JUNE 1971, 210 PP.

71-0439 HEUZE, F. E.
GOODMAN, R. E.
NUMERICAL ANALYSES OF DEFORMABILITY TESTS IN
JOINTED ROCK. ' JOINT PERTURBATION ' AND ' NO
TENSION ' FINITE ELEMENT SOLUTIONS
ROCK MECH., FELSMECH, BEC. ROCHES, UNIV. OF
CALIF., BERKELEY, VOL. 3, NO. 1, MAY 1971, PP.
13-24.

71-0440 PAULSEN, W. C.
FINITE ELEMENT STRESS ANALYSIS PART 1
MACHINE DESIGN, VOL. 43, NO. 24, 1971, PP. 46-52.

71-0441 ODEN, J. T.
POE, J.
ON THE NUMERICAL SOLUTION OF A CLASS OF NONLINEAR
PROBLEMS IN DYNAMIC COUPLED THERMOELASTICITY
PROC. 5TH SOUTHEASTERN CONF. ON DEVELOPMENTS IN
THEORETICAL AND APPLIED MECHANICS, DUKE UNIV.,
NORTH CAROLINA, APRIL 16-17, 1970, VOL. 5. (PUBL.
BY UNIV. NORTH CAROLINA PRESS, CHAPEL HILL, 1971,
1046 PP.) ALSO AD-699603, OCTOBER 1969, 34 PP.

71-0442 VIGNON, B.
USE OF VARIATIONAL PRINCIPLE IN STUDY ON
RELATIVISTIC BOLTZMANN EQUATION
C. R. ACAD. SCI. SER A MATH., VOL. 273, NO. 25,
1971, PP. 1331-1335.

71-0443 VUJANOVIC, B.
STRAUSS, A. M.
HEAT TRANSFER WITH NONLINEAR BOUNDARY CONDITIONS
VIA A VARIATIONAL PRINCIPLE
AIAA, J. VOL. 9, NO. 2, 1971, PP. 327-330.

71-0444 YU, J. C. M.
APPLICATION OF CONFORMAL MAPPING AND VARIATIONAL
METHOD TO THE STUDY OF HEAT CONDUCTION IN
POLYGONAL PLATES WITH TEMPERATURE DEPENDENT
CONDUCTIVITY
INT. J. HEAT MASS TRANSFER, VOL. 14, NO. 1, 1971,
PP. 49-56.

71-0445 ZARGHAMER, M. S.
DIFFERENTIAL OF VARIATIONAL FORMS
SIAM J. APPL. MATH. VOL. 20, NO. 3, 1971, PP.
406-412.

71-0446 CIARLET, P. G.
WAGSCHAL, C.
MULTIPOINT TAYLOR FORMULAS AND APPLICATIONS TO THE
FINITE ELEMENT METHOD
NUMER. MATH. VOL. 17, NO. 1, 1971, PP. 84-100.

71-0447 CROLL, J. G. A.
WALKER, A. C.
THE FINITE DIFFERENCE AND LOCALIZED RITZ METHODS
INT. J. NUMER. METHODS ENG., VOL. 3, NO. 2, 1971,
PP. 155-160.

71-0448 FIX, G. J.
FINITE ELEMENT APPROXIMATIONS TO PARABOLIC
PROBLEMS
NOTICES AMER. MATH. SOC., VOL. 18, NO. 2, 1971,
PP. 380.

71-0449 FIX, G. J.
LARSEN, K.
ON THE CONVERGENCE OF SOR ITERATIONS FOR FINITE
ELEMENT APPROXIMATIONS TO ELLIPTIC BOUNDARY VALUE
PROBLEMS
SIAM J. NUMER. ANAL. VOL. 8, NO. 3, 1971, PP.
536-547.

71-0450 GONSHOR, H.
THE RING OF FINITE ELEMENTS IN A NON-STANDARD
MODEL OF THE REALS
J. LONDON MATH. SOC., VOL. 3, SER. 2, 1971, PP.
493-500.

71-0451 HUTTON, S. G.
ANDERSON, D. L.
FINITE ELEMENT METHOD: A GALERKIN APPROACH
ASCE J. ENG. MECH. DIV., VOL. 97, NO. EM5, PAPER
8448, 1971, PP. 1503-1520.

71-0452 MITCHELL, A. R.
VARIATIONAL PRINCIPLES AND FINITE ELEMENT METHOD
IN PARTIAL DIFFERENTIAL EQUATIONS
PROC. ROY. SOC. OF LONDON, SER. A 323, NO. 1553,
1971, PP. 211-217.

71-0453 NAKAMURA, S.
OHNISHI, T.
STUDY OF ITERATIVE SOLUTIONS FOR FINITE ELEMENT
METHOD
AMER. NUCL. SOC., WINTER MEETING, MIAMI BEACH,
FLORIDA, OCTOBER 17, 1971, ALSO PROC. OF SEM. ON
NUMER. REACT. CALCULATION, INST. OF ATOMIC ENERGY,
WIEN, AUSTRIA, JANUARY 1972.

71-0454 ODEN, J. T.
FINITE ELEMENT APPROXIMATION OF A CLASS OF
NONLINEAR OPERATORS
PAPER PRESENTED AT SIAM 1970 FALL MEETING, LENOX
HOTEL, BOSTON, MASS., OCTOBER 12-14, 1970.;
ABSTRACT IN SIAM REV. VOL. 13, NO. 2, 1971, P.
274.

71-0455 STRANG, G.
FINITE ELEMENT METHOD
NOTICES AMER. MATH. SOC. VOL. 18, NO. 2, 1971, 376
PP.

71-0456 WAIT, R.
MITCHELL, A. R.
CORNER SINGULARITIES IN ELLIPTIC PROBLEMS BY
FINITE ELEMENT METHODS
J. COMP. PHYS. VOL. 8, NO. 1, 1971, PP. 45-52.

71-0457 WALSH, J. E.
FINITE DIFFERENCE AND FINITE ELEMENT METHODS OF
APPROXIMATION
PROC. ROY. SOC. LONDON, SER. A., VOL. 323, NO.
1553, 1971, PP. 155-165.

71-0458 BRADLEY, R. G.
MILLER, R. D.
APPLICATION OF FINITE-ELEMENT THEORY TO AIRPLANE
CONFIGURATIONS
J. AIRCRAFT, VOL. 8, NO. 6, 1971, PP. 400-405.

71-0459 ODEN, J. T.
FINITE ELEMENT APPROXIMATIONS IN NONLINEAR
THERMOVISCOELASTICITY
FINITE ELEMENT METHODS IN CONTINUUM MECHANICS,
(EDS.) ODEN, J.T., DE ARANTES E OLIVEIRA, E.R.,
THE ADVANCED STUDY INST., LABORATORIO NACIONAL DE
ENGENHARIA CIVIL, LISBON, PORTUGAL, SEPTEMBER
7-17, 1971, PP. 77-119.

71-0460 CHARI, M. V.
SILVESTER, P.
ANALYSIS OF TURBOALTERNATOR MAGNETIC FIELDS BY
FINITE ELEMENTS
IEEE TRANS. ON POWER APPARATUS AND SYSTEMS, PA-90,
NO. 2, 1971, PP. 454-464.

71-0461 CHENG, R. T.
FINITE ELEMENT SOLUTION OF VISCOUS FLOW IN AN
ARBITRARY INTERNAL PASSAGE
AMER. PHYS. SOC. DIV., FLUID DYNAMICS MEETING, SAN
DIEGO, CALIF., NOVEMBER 22-24, 1971.

71-0462 COGGON, J. H.
ELECTROMAGNETIC AND ELECTRICAL MODELING BY THE
FINITE ELEMENT METHOD
GEOPHYSICS VOL. 36, NO. 1, 1971, PP. 132-155.

71-0463 CRAGGS, A.
TRANSIENT RESPONSE OF A COUPLED PLATE-ACOUSTIC
SYSTEM USING PLATE AND ACOUSTIC FINITE ELEMENTS
J. SOUND AND VIB. VOL. 15, NO. 4, 1971, PP.
509-528.

71-0464 CRASTAN, V.
APPLICATION OF AN ' ELEMENT-METHOD ' PROGRAM TO
SOLVING TEMPERATURE DISTRIBUTION PROBLEMS
(ANWENDUNG EINES ' ELEMENTENMETHODE ' - PROGRAMMES
ZUR LOSUNG VON TEMPERATURVERTEILUNSPROBLEMS)
NUCL. ENG. DES., VOL. 15, NO. 1, 1971, PP. 54-64.

71-0465 DALY, P.
HYBRID MODE ANALYSIS OF MICROSTRIP BY FINITE
ELEMENT METHODS
IEEE TRANS. MICROWAVE THEORY TECH., VOL. MTT-19,
NO. 1, 1971, PP. 19-25.

71-0466 DESAI, C. S.
NONLINEAR ANALYSES USING SPLINE FUNCTIONS
J. ASCE SOIL MECH. FOUND. DIV., VOL. 97, NO. SM10,
PAPER 8462, 1971, PP. 1461-1480.

71-0467 EMERY, A. F.
CARSON, W. W.
EVALUATION OF USE OF THE FINITE-ELEMENT METHOD IN
COMPUTATION OF TEMPERATURE
J. HEAT TRANSFER, TRANS. ASME VOL. 93, SER. C, NO.
2, 1971, PP. 136-145.

71-0468 EISENSTEIN, Z.
FINITE ELEMENTS METHOD IN SOIL MECHANICS (METODA
KONECNYCH PRVKU V MECHANICE SEMIN),
INZENYRSKE STAVBY, VOL. 19, NO. 2, 1971, PP.
64-67.

71-0469 FRANCE, P. W.
PAREKH, C. J.
PETERS, J. C.
TAYLOR, C.
NUMERICAL ANALYSIS OF FREE SURFACE SEEPAGE
PROBLEMS
J. ASCE IRRIG. DRAIN. DIV., VOL. 97, NO. IR1,
PAPER 7959, 1971, PP. 165-179.

71-0470 ODEN, J. T.
FINITE ELEMENT APPROXIMATIONS IN NONLINEAR
ELASTICITY
FINITE ELEMENT METHODS IN CONTINUUM MECHANICS,
(EDS.) ODEN, J.T., DE ARANTES E OLIVEIRA, E.R.,
THE ADVANCED STUDY INST., LABORATORIO NACIONAL DE
ENGENHARIA CIVIL, LISBON, PORTUGAL, SEPTEMBER
7-17, 1971, PP. 121-150.

71-0471 GEERTSMA, J.
FINITE-ELEMENT ANALYSIS OF SHALLOW TEMPERATURE
ANOMALIES
GEOPHYS. PROSPECT., VOL. 19, NO. 4, 1971, PP.
662-681.

71-0472 HARDER, R. L.
RODDEN, W. P.
KERNEL FUNCTION FOR NONPLANAR OSCILLATING SURFACES
IN SUPERSONIC FLOW
J. AIRCRAFT, VOL. 8, NO. 8, 1971, PP. 677-679.

71-0473 HOBBS, G. K.
METHODS OF MODELING AND ANALYZING VISCOELASTICALLY
DAMPED STRUCTURES
PAPER 71-VIBR-36 FOR ASME MEETING, SEPTEMBER 8-10,
1971, 8 PP.

71-0474 HSU, T. R.
TRANSIENT THERMAL STRESS TECHNIQUES IN NUCLEAR
REACTORS. APPLIED MECHANICS ASPECTS OF NUCLEAR
EFFECTS IN MATERIALS
ASME WINTER ANNUAL MEETING, WASHINGTON, D.C.,
DECEMBER 2, 1971, PP. 1-65.

71-0475 ISSACS, L. T.
FINITE ELEMENT METHODS: TWO-DIMENSIONAL SEEPAGE
WITH A FREE SURFACE
DEP. CIVIL ENG., BULL. 14, QUEENSLAND UNIV.,
JANUARY 1971, 55 PP.

71-0476 KAGAWA, Y.
NEW APPROACH TO ANALYSIS AND DESIGN OF
ELECTROMECHANICAL FILTERS BY FINITE-ELEMENT
TECHNIQUE
J. ACOUST. SOC. AMER, VOL. 49, NO. 5, 1971, PP.
1348-1356.

71-0477 KEALY, C. D.
BUSCH, R. A.
DETERMINING SEEPAGE CHARACTERISTICS OF
MILL-TAILING DAMS BY THE FINITE ELEMENT METHOD
U.S. BUR. MINES, REP. INVEST. 7477, JANUARY 1971,
113 PP.

71-0478 KONRAD, A.
SILVESTER, P.
SCALAR FINITE-ELEMENT PROGRAM PACKAGE FOR
TWO-DIMENSIONAL FIELD PROBLEMS
IEEE TRANS. MICROWAVE THEORY TECH. MTT-19, NO. 12,
1971, PP. 952-954.

71-0479 MALONE, D. W.
CONNOR, J. J.
FINITE ELEMENTS AND DYNAMIC VISCOELASTICITY
J. ASCE ENG. MECH. DIV., VOL. 97, NO. EM4, PAPER
8293, 1971, PP. 1145-1158.

71-0480 MILLER, D. W.
LEWIS, E. E.
ROSSOW, E. C.
TRANSPORT SOLUTIONS USING FINITE ELEMENTS IN
SPACE-ANGLE PHASE SPACE
WINTER MEETING OF AMER. NUCL. SOC., MIAMI BEACH,
FLORIDA, OCTOBER 17, 1971.

71-0481 MULLINS, W. M.
REYNOLDS, D. T.
STRESS ANALYSIS OF PARACHUTES USING FINITE
ELEMENTS
J. SPACECRAFT ROCKETS, VOL. 8, NO. 10, 1971, PP.
1068-1073.

71-0482 NEUMAN, S. P.
WITHERSPOON, P. A.
ANALYSIS OF NONSTEADY FLOW WITH A FREE SURFACE
USING FINITE ELEMENT METHOD
WATER RESOURCES RES., VOL. 7, NO. 3, 1971, PP.
611-623.

71-0483 OHNISHI, T.
FINITE ELEMENT METHOD APPLIED TO REACTOR PHYSICS
PROBLEMS
J. NUCL. SCI. TECHN. TOKYO, VOL. 8, NO. 12, 1971,
P. 717.

71-0484 PARISEAU, W. G.
GRAVITY FLOW OF POWDER IN A LUNAR ENVIRONMENT. PT.
2-ANALYSIS OF FLOW INITIATION
U.S. BUR. MINES, REP. INVEST. 7577, 1971, 20 PP.

71-0485 CHEUNG, M. S.
CHEUNG, Y. K.
ANALYSIS OF CURVED BOX GIRDER BRIDGES BY FINITE
STRIP METHOD
PUBLICATIONS, INTERNATIONAL ASSOCIATION OF BRIDGES
AND STRUCTURAL ENGINEERING, VOL. 31, NO. 1, 1971.

71-0486 RUHL, R. L.
PLECK, M. H.
GRAPHICAL DISPLAY OF COMPUTER SIMULATED UNBALANCED
ROTOR RESPONSE
PAPER 71-VIBR.-42, FOR ASME MEETING, SEPTEMBER
8-10, 1971, 8 PP.

71-0487 SCHACHT, C. A.
BOGDEN, F. J.
VERIFICATION AND QUALIFICATION OF A 3-D FINITE
ELEMENT COMPUTER ANALYSIS
ASME, SYMP. ON ENG. COMPUTER SOFTWARE, SAN
FRANCISCO, MAY 1971, PP. 88-95.

71-0488 SECREST, B. G.
BOYD, W. W.
SHAW, D. W.
APPLICATION OF FINITE ELEMENT METHOD TO MASS
TRANSPORT LIMITED EPITAXIAL GROWTH PROCESSES
J. CRYSTAL GROWTH, VOL. 10, NO. 3, 1971, PP.
251-259

71-0489 SEMENZA, L. A.
LEWIS, E. E.
ROSSOW, E. C.
DUAL FINITE ELEMENT METHODS FOR NEUTRON DIFFUSION
AMER. NUCL. SOC. MEETING, MIAMI BEACH, FLORIDA,
OCTOBER 17, 1971.

71-0490 SHUM, Y. M.
USE OF FINITE-ELEMENT METHOD IN THE SOLUTION OF
DIFFUSION-CONVECTION EQUATIONS
J. SOC. PETROL. ENG., VOL. 11, NO. 2, 1971, PP.
139-144.

71-0491 SILVESTER, P.
HSIEH, M. S.
FINITE ELEMENT SOLUTION OF TWO DIMENSIONAL
EXTERIOR FIELD PROBLEMS
PROC. INST. ELECT. ENG. LONDON, VOL. 118, NO. 12,
1971, PP. 1743-1747.

71-0492 TABARROK, B.
VARIATIONAL PRINCIPLE FOR THE DYNAMIC ANALYSIS OF
CONTINUA BY HYBRID FINITE ELEMENT METHOD
INT. J. SOLIDS STRUCTURE, VOL. 7, NO. 3, 1971, PP.
251-268.

71-0493 WANG, E. D.
SUN, M. C.
COMPUTER PROGRAM FOR GENERATING FINITE ELEMENT
MODELS OF MINE STRUCTURES
U.S. BUR. MINES, INFORM. CIRC. 8510, MAY 1971, 56
PP.

71-0494 WEXLER, A.
RICHARDS, D. J.
FINITE ELEMENT TECHNIQUES FOR SOLUTION OF
POISSON'S EQUATION
IEEE, G-MTT INT. MICROWAVE SYMP. WASHINGTON, D.C.,
DIGEST TECH. PAPER, MAY 16-19, 1971, PP. 132-133.

71-0495 JORDAN, S.
GALLO, A. M.
MAGIC II. AN AUTOMATED GENERAL PURPOSE SYSTEM FOR
STRUCTURAL ANALYSIS
TECH. REP. AFFDL-TR-71-1, (THREE VOLS.). AIRFORCE
FLIGHT DYNAMICS LAB., AIRFORCE SYSTEMS COMMAND,
WRIGHT-PATTERSON AIRFORCE BASE OHIO, MAY 1971.

71-0496 CHEUNG, Y. K.
CHEUNG, M. S.
HUIZER, A.
A THEORETICAL AND EXPERIMENTAL STUDY ON THE FREE
VIBRATION OF CURVED SLAB BRIDGES
PROCEEDINGS OF THE THIRD AUSTRALASIAN CONFERENCE
ON THE MECHANICS OF STRUCTURES AND MATERIALS,
UNIVERSITY OF AUCKLAND, AUGUST 1971.

71-0497 ODEN, J. T.
FINITE ELEMENT FORMULATIONS OF NONLINEAR BOUNDARY
VALUE PROBLEMS
LECTURES ON FINITE ELEMENT METHODS IN CONTINUUM
MECHANICS, EDITED BY J. T. ODEN AND E. P. DE
ARANTES E OLIVEIRA, LECTURES ON THE ADVANCED STUDY
INSTITUTE, LAB. NACIONAL DE ENGENHARIA CIVIL IN
LISBON, PORTUGAL, SEPTEMBER 7-17, 1971, PP.
187-208.

71-0498 DE ARANTES E OLIVEIRA, E. R.
MATHEMATICAL THEORY OF LINEAR STRUCTURES
LECTURES ON FINITE ELEMENT METHODS IN CONTINUUM
MECHANICS, EDITED BY J. T. ODEN AND E. R. DE
ARANTES E OLIVEIRA, LECTURES OF THE ADVANCED
STUDY INSTITUTE, LABORATORIO NACIONAL DE
ENGENHARIA CIVIL IN LISBON, PORTUGAL, SEPTEMBER
7-17, 1971, PP. 209-297.

71-0499 FRAEIJS DE VEUBEKE, B.
STATIC-GEOMETRIC ANALOGIES AND FINITE ELEMENT
MODELS
LECT. ON FINITE ELEMENT METH. IN CONTINUUM MECH.,
J.T. ODEN AND E. R. DE ARANTES E OLIVEIRA (EDS.),
LECTURES OF THE ADVANCED STUDY INSTITUTE,
LABORATORIO NACIONAL DE ENGENHARIA CIVIL IN
LISBON, PORTUGAL, SEPTEMBER 7-17, 1971, PP.
299-322.

71-0500 FRAEIJS DE VEUBEKE, B.
QUALITY IN STRUCTURAL ANALYSIS BY FINITE ELEMENTS
LECT. ON FINITE ELEMENT METH. IN CONTINUUM MECH.,
J. T. ODEN AND E. R. DE ARANTES E OLIVEIRA (EDS),
LECT OF THE ADVANCED STUDY INST., LABORATORIO
NACIONAL DE ENGENHARIA CIVIL, LISBON, PORTUGAL,
SEPTEMBER 7-17, 1971, PP. 323-355.

71-0501 FRAEIJS DE VEUBEKE, B.
THE DUAL PRINCIPLE OF ELASTODYNAMICS FINITE
ELEMENT APPLICATIONS
LECT. ON FINITE ELEMENT METH. IN CONTINUUM MECH.,
J.T. ODEN AND E.R. DE ARANTES E OLIVEIRA (EDS.),
LECT. OF THE ADVANCED STUDY INST., LABORATORIO
NACIONAL DE ENGENHARIA CIVIL, LISBON, PORTUGAL,
SEPTEMBER 7-17, 1971, PP. 357-377.

71-0502 ZIENKIEWICZ, O. C.
ISOPARAMETRIC ELEMENT FORMS IN FINITE ELEMENT
ANALYSIS
LECTURES ON FINITE ELEMENT METHODS IN CONTINUUM
MECHANICS, EDITED BY J. T. ODEN AND E. P. DE
ARANTES E OLIVEIRA, LECTURES OF THE ADVANCED
STUDY INSTITUTE, LABORATORIO NACIONAL DE
ENGENHARIA CIVIL IN LISBON, PORTUGAL, SEPTEMBER
7-17, 1971, PP. 379-414.

71-0503 ZIENKIEWICZ, O. C.
TAYLOR, C.
WEIGHTED RESIDUAL PROCESSES IN FINITE ELEMENT WITH
PARTICULAR REFERENCE TO SOME TRANSIENT AND COUPLED
PROBLEMS
LECTURES ON FINITE ELEMENT METHODS IN CONTINUUM
MECHANICS, EDITED BY J. T. ODEN, AND E. R. DE
ARANTES E OLIVEIRA, LECTURES OF THE ADVANCED
STUDY INSTITUTE, LABORATORIO NACIONAL DE
ENGENHARIA CIVIL IN LISBON, PORTUGAL, SEPTEMBER
7-17, 1971, PP. 415-458.

71-0504 ZIENKIEWICZ, O. C.
NAYLOR, D. J.
FINITE ELEMENT STUDIES OF SOILS AND POROUS MEDIA
LECTURES ON FINITE ELEMENT METHODS IN CONTINUUM
MECHANICS, EDITED BY J. T. ODEN AND E. R. DE
ARANTES E OLIVEIRA, LECTURES OF THE ADVANCED
STUDY INSTITUTE, LABORATORIO NACIONAL DE
ENGENHARIA CIVIL IN LISBON, PORTUGAL, SEPTEMBER
7-17, 1971, PP. 459-493.

71-0505 CLOUGH, R. W.
BASIC PRINCIPLES OF STRUCTURAL DYNAMICS
LECTURES ON FINITE ELEMENT METHODS IN CONTINUUM
MECHANICS, EDITED BY J. T. ODEN AND E. R. DE
ARANTES E OLIVEIRA, LECTURES OF THE ADVANCED
STUDY INSTITUTE, LABORATORIO NACIONAL DE
ENGENHARIA CIVIL IN LISBON, PORTUGAL, SEPTEMBER
7-17, 1971, PP. 495-511.

71-0506 CLOUGH, R. W.
VIBRATION ANALYSIS OF FINITE ELEMENT SYSTEMS
LECTURES ON FINITE ELEMENT METHODS IN CONTINUUM
MECHANICS, EDITED BY J. T. ODEN AND E. R. DE
ARANTES E OLIVEIRA, LECTURES OF THE ADVANCED
STUDY INSTITUTE, LABORATORIO NACIONAL DE
ENGENHARIA CIVIL IN LISBON, PORTUGAL, SEPTEMBER
7-17, 1971, PP. 513-523.

71-0507 CLOUGH, R. W.
NUMERICAL INTEGRATION OF THE EQUATION OF MOTION
LECTURES ON FINITE ELEMENT METHODS IN CONTINUUM
MECHANICS, EDITED BY J. T. ODEN AND E. R. DE
ARANTES E OLIVEIRA, LECTURES OF THE ADVANCED
STUDY INSTITUTE, LABORATORIO NACIONAL DE
ENGENHARIA CIVIL IN LISBON, PORTUGAL, SEPTEMBER
7-17, 1971, PP. 525-533.

71-0508 MARCAL, P. V.
LARGE STRAIN, LARGE DISPLACEMENT ANALYSIS
LECTURES ON FINITE ELEMENT METHODS IN CONTINUUM
MECHANICS, EDITED BY J. T. ODEN AND E. R. DE
ARANTES E OLIVEIRA, LECTURES OF THE ADVANCED
STUDY INSTITUTE, LABORATORIO NACIONAL DE
ENGENHARIA CIVIL IN LISBON, PORTUGAL, SEPTEMBER
7-17, 1971, PP. 535-543.

71-0509 CHEUNG, M. S.
CHEUNG, Y. K.
NATURAL VIBRATIONS OF THIN, FLAT WALLED STRUCTURES
WITH DIFFERENT BOUNDARY CONDITIONS
J. SOUND AND VIB., 1971, VOL. 18, PP. 325-337.

71-0510 AKIN, J. E.
FENTON, D. L.
STODDART, W. C. T.
THE FINITE ELEMENT METHOD - A BIBLIOGRAPHY OF ITS
THEORY AND APPLICATIONS
REPORT EM 72-1, DEPARTMENT OF ENGINEERING
MECHANICS, THE UNIVERSITY OF TENNESSEE,
KNOXVILLE, 1971.

71-0511 SHANNON, R. W. E.
AUSTIN, B. A.
APPLICATION OF FRACTURE MECHANICS TO FAILURE OF
THICK WALLED CYLINDERS
INST. MECH. ENG. PRACT. APPL. OF FRACT. MECH. TO
PRESSURE VESSEL TECHNOL., LONDON, 1971, CONF.
HELD IN LONDON, MAY 3-5, 1971, PP. 109-118.

71-0512 BABUSKA, I.
FINITE ELEMENT METHOD WITH PENALTY
TECHNICAL NOTE BN-710, UNIVERSITY OF MARYLAND,
INSTITUTE FORFLUID DYNAMICS AND APPLIED
MATHEMATICS, 1971.

71-0513 MIKKOLA, M.
ON THE CONVERGENCE OF THE FINITE ELEMENT METHOD
FOURTH SCANDINAVIAN MEETING ON STRENGTH OF
MATERIALS, HELSINKI, 1971.

71-0514 FRIED, I.
DISCRETIZATION AND ROUND-OFF ERROR IN THE FINITE
ELEMENT ANALYSIS OF ELLIPTIC BOUNDARY VALUE
PROBLEMS AND EIGENVALUE PROBLEMS
PH.D. DISSERTATION, MASSACHUSETTS INSTITUTE OF
TECHNOLOGY, CAMBRIDGE, 1971.

71-0515 MCLAY, R. W.
ON CERTAIN APPROXIMATIONS IN THE FINITE ELEMENT
METHOD
JOURNAL OF APPLIED MECHANICS, VOL. 38, NO. 4,
1971, PP 58-61

71-0516 KORNEEV, V. G.
CONCERNING THE FINITE ELEMENT METHOD FOR THE
SOLUTION OF PROBLEMS OF ELASTIC EQUILIBRIUM
STRUCTURAL MECHANICS OF BUILDING CONSTRUCTION (IN
RUSSIAN), LENINGRAD, U.S.S.R., 1971, PP. 28-46

71-0517 YAMAMOTO, Y.
TOKUDA, N.
THE ANALYSIS OF STRESS DISTRIBUTION AROUND THE TIP
OF A CRACK BY FINITE ELEMENT METHOD WITH THE
AID OF ANALYTIC SOLUTIONS (IN JAPANESE)
PROCEEDINGS OF THE FIFTH SYMPOSIUM ON MATRIX
METHODS OF STRUCTURAL ANALYSIS, SOCIETY OF
STEEL CONSTRUCTION OF JAPAN, JULY 1971, PP. 62-69.

71-0518 YOKOO, Y. ET AL
FINITE ELEMENT METHOD APPLIED TO BIOT'S
CONSOLIDATION THEORY
SOILS AND FOUNDATIONS, JAPANESE SOCIETY OF SOIL
MECHANICS AND FOUNDATION ENGINEERING, VOL. 11,
NO. 1, MARCH 1971. PP. 29-46.

71-0519
FRIED, I.
INFLUENCE OF POISSON'S RATIO ON THE CONDITION OF THE FINITE ELEMENT STIFFNESS MATRIX
BOSTON UNIVERSITY REPORT D71IF, 1971.

71-0520
FUJII, H.
FINITE ELEMENT GALERKIN METHOD FOR MIXED INITIAL-BOUNDARY VALUE PROBLEMS IN ELASTICITY THEORY
TECH. REP. CNA-34, CENTER FOR NUMERICAL ANALYSIS, THE UNIVERSITY OF TEXAS AT AUSTIN, 1971.

71-0521
FUJII, H.
HIBINO, S.
STABILITY OF FINITE ELEMENT SCHEMES FOR VIBRATION PROBLEMS OF THE EQUATION OF ELASTICITY THEORY
PROCEEDINGS OF THE 5TH JSSC CONFERENCE ON MATRIX STRUCTURAL ANALYSIS, 1971.

71-0522
YURKOVICH, R. N.
SCHMIDT, J. H.
DYNAMIC ANALYSIS OF STIFFENED PANEL STRUCTURES
J. AIRCRAFT, VOL. 8, NO. 4, MARCH 1971, PP. 149-155.

71-0523
YAMADA, Y.
NAGAI, Y.
ANALYSIS OF ONE-DIMENSIONAL STRESS WAVE BY THE FINITE ELEMENT METHOD
SEISAN KENKYU, J. OF INSTITUTE OF INDUST. SCI., UNIV. OF TOKYO, VOL. 23, NO. 5, 1971, PP. 186-189.

71-0524
BROMBOLICH, L. J.
GOULD, P. L.
HIGH PRECISION FINITE ELEMENT ANALYSIS OF SHELLS OF REVOLUTION
TRANSACTIONS OF JAPAN WELDING SOCIETY, VOL. 2, NO. 2, 1971, PP. 90-100.

71-0525
MCNAMARA, J. F.
MARCAL, P. V.
INCREMENTAL STIFFNESS METHOD FOR FINITE ELEMENT ANALYSIS OF THE NONLINEAR DYNAMIC PROBLEM
PROC. ONR SYMPOSIUM ON NUMERICAL AND COMPUTER METHODS IN STRUCTURAL MECHANICS, URBANA, ILLINOIS, SEPTEMBER 1971.

71-0526
YAMADA, Y.
IWATA, K.
FINITE ELEMENT ANALYSIS OF THERMO-VISCOELASTIC PROBLEMS
SEISAN KENKYU, MONTHLY J. OF INSTITUTE OF INDUST. SCI., UNIV. OF TOKYO, VOL. 24, NO. 4, 1972, PP. 165-170.

71-0527
ODEN, J. T.
ARMSTRONG, W. H.
ANALYSIS OF NONLINEAR DYNAMIC, COUPLED THERMOVISCOELASTICITY PROBLEMS BY THE FINITE ELEMENT METHOD
J. OF COMPUTERS AND STRUCTURES, VOL. 1, NO. 1, 1971.

71-0528
ARMSTRONG, W. H.
ANALYSIS OF NONLINEAR THERMOVISCOELASTICITY PROBLEMS BY THE FINITE ELEMENT METHOD
M.S. THESIS, UNIVERSITY OF ALABAMA IN HUNTSVILLE, HUNTSVILLE 1971.

71-0529
NAYAK, G. C.
PLASTICITY AND LARGE DEFORMATION PROBLEMS BY FINITE ELEMENT METHOD
PH.D. THESIS, UNIV. OF WALES, SWANSEA, 1971.

71-0530
BEST, B.
AN INVESTIGATION INTO THE USE OF FINITE ELEMENT METHODS FOR ANALYSING STRESS DISTRIBUTIONS IN BLOCK JOINTED MASSES
PH.D. THESIS, JAMES COOK UNIVERSITY, AUSTRALIA, 1971.

71-0531
IRONS, B. M.
QUADRATURE RULES FOR BRICK BASED FINITE ELEMENTS
INT. J. NUM. METH. ENG. VOL. 3, PP. 293-294, 1971.

71-0532
HUANG, C. T.
MORGENSTERN, N. R.
ON SOLUTION OF PLANE STRAIN CONSOLIDATION PROBLEMS BY FINITE ELEMENT METHOD
CANADIAN GEOTECHNICAL JOURNAL, VOL. 8, PP. 109-118, 1971.

71-0533
NAYLOR, D. J.
ZIENKIEWICZ, O. C.
SETTLEMENT ANALYSIS OF A STRIP FOOTING USING A CRITICAL STATE SOIL MODEL IN CONJUNCTION WITH FINITE ELEMENTS
SYMPOSIUM ON THE INTERACTION OF STRUCTURE AND FOUNDATION UNIVERSITY OF BIRMINGHAM, JULY 1971

71-0534
ZIENKIEWICZ, O. C.
NAYLOR, D. J.
ADAPTATION OF CRITICAL STATE SOIL MECHANICS FOR USE IN FINITE ELEMENTS
CONTRIBUTION TO ROSCOE MEMORIAL SYMPOSIUM, CAMBRIDGE, MARCH 1971.

71-0535
HARRISON, D. G.
A HIGH ORDER TRIANGULAR FIELD ELEMENTS
M.SC. THESIS, THE UNIVERSITY OF CALGARY, APRIL 1971.

71-0536
BOWLEY, W. W.
PRINCE, J. F.
FINITE ELEMENT ANALYSIS OF GENERAL FLUID FLOW PROBLEMS
AIAA 4TH FLUID AND PLASMA DYNAMICS CONFERENCE, JUNE 1971, PP. 1-11.

71-0537
WADA, S.
HAYASHI, H.
MIGITA, M.
APPLICATION OF FINITE ELEMENT METHOD TO HYDRODYNAMIC LUBRICATION PROBLEMS (PART 1, "INFINITE-WIDTH BEARINGS")
BULLETIN OF THE JAPAN SOCIETY OF MECHANICAL ENGINEERS, VOL. 14, NO. 77, NOV. 1971, P. 1222.

71-0538
WADA, S.
HAYASHI, H.
APPLICATION OF FINITE ELEMENT METHOD TO HYDRODYNAMIC LUBRICATION PROBLEMS (PART 2, " FINITE WIDTH BEARINGS ")
BULLETIN OF THE JAPAN SOCIETY OF MECHANICAL ENGINEERS, VOL. 14, NO. 77, NOVEMBER 1971, P. 1234.

71-0539
HUEBNER, K. H.
FINITE ELEMENT ANALYSIS OF CONTINUUM PROBLEMS AND ITS APPLICATION TO THE INCOMPRESSIBLE LUBRICATION PROBLEM
GENERAL MOTORS RESEARCH PUBLICATION GMR-1074, FEB. 1971.

71-0540
MCCORQUODALE, J. A.
LI, C. Y.
FINITE ELEMENT ANALYSIS OF SLUICE GATE FLOW
TRANSACTIONS OF THE ENGINEERING INSTITUTE OF CANADA, VOL. 14 NO. C-2, MARCH, 1971.

71-0541
EHLERS, K. D.
BERECHNUNG INSTATIONARER GRUND - UND SICKERWASSERSTROMUNGEN MIT FREIER OBERFLACHE NACH DER METHODE FINITER ELEMENTE
DISSERTATION, TECHNISCHE UNIVERSITAT HANNOVER, 1971.

71-0542
LUCKNER, L.
KADEN, ST.
GRUNDLAGEN DER DIGITALEN LOSUNG VON GRUNDWASSERSTROMUNGSPROBLEMEN NACH DER METHODE ENDLICHER ELEMENTE
FUNDAMENTALS OF THE DIGITAL SOLUTION OF GROUND-WATER FLOW PROBLEMS BY THE FINITE-ELEMENT-METHOD, NEUE BERGBAUTECHNIK, VOL. 1, NO. 9, 1971, PP. 665-676.

71-0543
LUCKNER, L.
KADEN, ST.
AUFBEREITUNG MATHEMATISCHER MODELLE FUR DIE DIGITALE BERECHNUNG VON GRUNDWASSERSTROMUNGSPROBLEMEN NACH DER METHODE ENDLICHER ELEMENTE
THE PREPARATION OF MATHEMATICAL MODELS FOR DIGITAL CALCULATION OF GROUND-WATER FLOW PROBLEMS BY THE FINITE- ELEMENT METHOD, NEUE BERGBAUTECHNIK, VOL. 1, NO. 12, 1971, PP. 897-904.

71-0544
UNNY, T. E.
WEAVER, D. S.
A FINITE ELEMENT SOLUTION FOR A CLASS OF TWO DIMENSIONAL VISCOUS FLUID FLOW DYNAMICS PROBLEMS
PROC. SYMPOSIUM HELD AT THE UNIVERSITY OF WATERLOO, 1971, PP. 493-508.

71-0545
KIKUCHI, F.
RECTANGULAR FINITE ELEMENT FOR PLATE BENDING ANALYSIS BASED ON HELLINGER-REISSNERS VARIATIONAL PRINCIPLE
JOURNAL OF NUCLEAR SCIENCE AND TECHNOLOGY, TOKYO, VOL. 9, NO 1, 1972, PP. 28-52.

71-0546
OLSON, M. D.
LINDBERG, G. M.
JET NOISE EXCITATION OF AN INTEGRALLY STIFFENED PANEL
JOUR. OF AIRCRAFT, VOL. 8, NO. 11, NOV. 1971.

71-0547
SAWKO, F.
MERRIMAN, P. A.
FINITE ELEMENT ANALYSIS OF BRIDGES CURVED IN PLAN
INT. SYMP. ON DEVELOPMENTS IN BRIDGE DESIGN AND CONSTR., CARDIFF, MARCH 1971.

71-0548
TREHARNE, G.
APPLICATIONS OF THE FINITE ELEMENT METHOD TO THE STRESS ANALYSIS OF MATERIALS SUBJECT TO CREEP
PH.D. THESIS, UNIVERSITY OF WALES, SWANSEA, 1971.

71-0549 GRIFFIN, D. S.
THE QUALIFICATION OF SOLUTION METHODS
SYMPOSIUM ON ENGINEERING COMPUTER SOFTWARE, NEW
YORK: AM.SOC. MECH. ENG., 1971, PP. 71-78.

71-0550 TOCHER, J. L.
A TECHNICAL EVALUATION OF THE NASTRAN COMPUTER
PROGRAM
TECH. NOTE #11, BOEING COMPUTER SERVICES, FEB.
1971, P. 43.

71-0551 MILES, G. A.
WRIGHT, G. C.
AN ECONOMICAL METHOD FOR DETERMINING THE SMALLEST
EIGENVALUES OF LARGE LINEAR SYSTEMS
INT. JOUR. FOR NUM. METH. IN ENG. VOL. 3, NO. 1,
1971, PP. 25-35.

71-0552 RADAJ, D.
COMPUTER-UNTERSTUTZTES KONSTRUIEREN, PRAKTISCHE
BEDEUTUNG DER FINIT-ELEMENT-VERFAHREN
KONSTRUKTION, VOL. 23, NO. 10, 1971, PP. 373-380.

71-0553 ROBINSON, J.
PETYT, M.
DYNAMIC ANALYSIS OF STRUCTURES USING THE RANK
FORCE METHOD
INT. JOUR. FOR NUM. METH. IN ENG. VOL. 3, NO. 1,
1971, PP. 103-119.

71-0554 AUTHORS
COMPUTATIONAL TECHNIQUES AS AN AID IN PHYSICAL
METALLURGY
PROC. CONF. COMPUT. TECH. AS AN AID IN PHYS.
METALL., UNIV. OF LEEDS, ENGLAND, JANUARY 4-5,
1971. (PUBL. BY IRON AND STEEL INST., LONDON,
ENGLAND, 1971, 73 PP.

71-0555 KIKUCHI, F.
TRIANGULAR FINITE ELEMENT FOR PLATE BENDING
ANALYSIS WITH TRANSVERSE SHEAR DEFORMATION
JOURNAL OF NUCLEAR SCIENCE AND TECHNOLOGY, TOKYO,
VOL. 8, NO. 12, 1971, P. 715.

71-0556 FELIPPA, C. A.
COMPUTATIONAL TECHNIQUES IN FINITE ELEMENT
ANALYSIS OF STRUCTURAL PROBLEMS
SIAM REVIEW VOL. 13, NO. 2, 1971, P. 266.

71-0557 GRIFFIS, C. A.
FINITE ELEMENT ANALYSIS OF NOTCHED TENSILE
SPECIMENS IN PLANE STRESS
REPORT OF NRL PROGRESS/U.S. NAVAL RESEARCH
LABORATORY VOL. 1971, MAY 1971, P. 30.

71-0558 KIKUCHI, F.
FINITE ELEMENT METHOD FOR PLATE BENDING ANALYSIS
BY DECOMPOSITION OF DIFFERENTIAL OPERATOR
JOURNAL OF NUCLEAR SCIENCE AND TECHNOLOGY, TOKYO,
VOL. 8, NO 10, 1971, P. 597.

71-0559 HARTUNG, R. F.
COMPUTER ORIENTED ANALYSIS OF SHELL STRUCTURES
US AIR FORCE SYST. COMMAND, AIR FORCE FLIGHT DYN.
LAB. TECH.REP., JUNE 1971, PP. 71-79.

71-0560 STRICKLIN, J. A.
HAISLER, W. E.
VON RIESEMANN, W. A.
GEOMETRICALLY NONLINEAR STRUCTURAL ANALYSIS BY THE
DIRECT STIFFNESS METHOD
J. OF STRUCT. DIV., ASCE, ST9, SEPT. 1971, PP.
2299-2314.

71-0561 SWANSON, J. A.
ANSYS USER'S MANUAL
SWANSON ANALYSIS SYSTEMS, INC., DEC. 1,1971.

71-0562 MARCAL, P. V.
MARC-CDC USER INFORMATION MANUALS
CONTROL DATA CORP., THREE VOLUMES, 1971.

71-0563 HUANG, T. C.
SACZALSKI, K. J.
ELASTODYNAMICS OF COMPLEX STRUCTURAL SYSTEMS
PROCEEDINGS OF THE 12TH MIDWESTERN MECHANICS
CONFERENCE, VOL. 6, AUG. 1971, PP. 675-688.

71-0564 RICE, J. R.
TRACEY, D. M.
COMPUTATIONAL FRACTURE MECHANICS
TECH. REPORT NASA NGL 40-002-08017, AUGUST 1971.

71-0565 ODEN, J. T.
BRAUCHLI, H. J.
A NOTE ON ACCURACY AND CONVERGENCE OF FINITE
ELEMENT APPROXIMATIONS
INTERNATIONAL JOUR. OF NUMERICAL METHODS IN ENG.,
VOL. 3, 1971, PP. 291-294.

71-0566 CHARI, M. V.
SILVESTER, P.
FINITE ELEMENT ANALYSIS OF MAGNETICALLY SATURATED
D-C MACHINES
IEE TRANS. ON POWER APPARATUS AND SYSTEMS, VOL.
PA90, NO. 5, 1971, P. 2362.

71-0567 HARRISON, W.
FANG, C. S.
WANG, S. N.
GROUNDWATER FLOWN IN A SANDY TIDAL BEACH .1.
ONE-DIMENSIONAL FINITE ELEMENT ANALYSIS
WATER RESOURCES RESEARCH, VOL. 7, NO. 5, 1971, P.
1313.

71-0568 KRAJCINOVIC, D.
COUPLED VIBRATIONS OF THIN-WALLED BEAMS OF OPEN
SECTION USING FINITE ELEMENT METHOD
INT. J. MECH. SCI., VOL. 13, NO. 8, 1971, P. 739.

71-0569 MAMET, J. C.
ANALYSIS OF SHEAR WALLS BY FINITE ELEMENTS
PROC. INST. OF CIVIL ENGINEERS, VOL. 50, NO. CT,
1971, P. 1774.

71-0570 GUILINGE, W. H.
FINITE ELEMENT METHOD
TRANS. AMERICAN NUCLEAR SOCIETY, VOL. 14, NO. 1,
1971, P. 199.

71-0571 BIRKHOFF, G.
FINITE ELEMENT METHODS IN REACTOR DESIGN
CALCULATIONS
TRANS. AMERICAN NUCLEAR SOCIETY, VOL. 14, NO. 1,
1971, P. 199.

71-0572 MOTE, C. D.
NONCONSERVATIVE STABILITY BY FINITE ELEMENT
ASCE J. ENG. MECH. DIV., VOL. 97, NO. EM3, JUNE
1971, PAPER 8204, PP. 645-656.

71-0573 HUTT, J. M.
SALAM, A. E.
DYNAMIC STABILITY OF PLATES BY FINITE ELEMENTS
ASCE J. ENG. MECH. DIV., VOL. 97, NO. EM3, JUNE
1971, PAPER 8211, PP. 879-899.

71-0574 BUELL, W. R.
SIMPSON, F. M.
APPLICATION OF THE FINITE ELEMENT METHOD TO
PREDICT STATIC AND DYNAMIC RESPONSE OF AN
UNSHROUDED CENTRIFUGAL COMPRESSOR BLADE
SAE PAP. 710554 FOR MEETING JUNE 7-11, 1971, 5 PP.

71-0575 BRAMLETTE, T. T.
PLANE POISEUILLE FLOW OF A RAREFIED GAS BASED ON
FINITE ELEMENT METHOD
PHYSICS OF FLUIDS, VOL. 14, NO. 2, 1971, P. 288.

71-0576 GANGAL, M. B.
PAUL, B.
FINITE ELEMENT METHOD IN STRESS ANALYSIS
COMPRESSED AIR, VOL. 76, NO. 2, 1971, P. 288.

71-0577 MCLAY, R. W.
CERTAIN APPROXIMATIONS IN FINITE ELEMENT METHOD
J. APPL. MECH., VOL. 38, NO. 1, 1971, P. 58.

71-0578 ZAPIR, Z.
STRUCTURAL ANALYSIS OF ARAVA FUSELAGE USING FINITE
ELEMENT METHOD
ISRAEL J. OF TECH., VOL. 9, NO. 1-2, 1971, P. 202.

71-0579 SMITH, I. M.
KAY, S.
STRESS ANALYSIS OF CONTRACTIVE OR DILATIVE SOIL
PROC. A.S.C.E., J. OF SOIL MECH. DIV., VOL. 97,
NO. SM7, 1971, PP. 981-997.

71-0580 KABAILA, A. P.
BECKERS, P.
COMPUTER INTEGRATION OF CONGRUENTLY TRANSFORMED
MATRICES
UNICIV REPORT R-75, UNIVERSITY OF NEW SOUTH WALES,
1971.

71-0581 BRAUN, K. A.
BRONLUND, O. E.
BUHLMEIER, J.
DIETRICH, G.
FRIK, G.
JOHNSEN, T. L.
KIESBAUER, H. T.
MALEJANNAKIS, G. A.
STRAUB, K.
VALLIANOS, G.
DYNAN LECTURE NOTES WITH COMPUTATIONAL EXAMPLES
ISD REPORT NO. 109, UNIVERSITY OF STUTTGART, 1971.

71-0582 BRAMBLE, J. H.
HILBERT, S. H.
BOUNDS FOR A CLASS OF LINEAR FUNCTIONALS WITH
APPLICATIONS TO HERMITE INTERPOLATION
NUME. MATH. VOL. 16, 1971, PP. 362-369.

71-0583 FRAEIJS DE VEUBEKE, B.
DUALITY IN STRUCTURAL ANALYSIS BY FINITE ELEMENTS,
STATIC GEOMETRIC ANALOGIES. THE DUAL PRINCIPLES OF
ELASTODYNAMICS
NATO ADVANCED STUDY INST., LECT. ON FINITE
ELEMENTS, UNIV. OF ALABAMA, HUNTSVILLE, 1971, PP.
299-377.

71-0584 CHENG, F. Y.
DYNAMIC RESPONSE OF NONLINEAR SPACE FRAMES BY
FINITE ELEMENT METHODS
PROC. OF SYMP. OF INT. ASSOC. FOR SHELL
STRUCTURES, TOKYO, KYOTO, JAPAN, 1971, PAPER 9-5.

71-0585 WEGMULLER, A. W.
FINITE ELEMENT ANALYSIS OF ELASTIC-PLASTIC PLATES
AND ECCENTRICALLY STIFFENED PLATES
PH.D. DISS., LEHIGH UNIVERSITY, BETHLEHEM, PENN.,
1971.

71-0586 COOLEY, R. L.
A FINITE DIFFERENCE METHOD FOR UNSTEADY FLOW IN
VARIABLY SATURATED POROUS MEDIA
WATER RES. RES., VOL. 7, NO. 6, 1971, PP.
1607-1625.

71-0587 AHMED, K. M.
ON THE ENERGY FUNCTIONS FOR CURVED SANDWICH PLATES
J. SOUND AND VIB., VOL. 1, NO. 16, 1971, PP.
461-463.

71-0588 AHMED, K. M.
THE EFFECT OF IN PLANE STATIC STRESSES ON THE
VIBRATION CHARACTERISTICS OF CURVED SANDWICH
PLATES
REPORT: ISVR-TR-50, JUNE 1971, UNIVERSITY OF
SOUTHAMPTON,

71-0589 BENNETT, J. G.
A STUDY OF A DYNAMICAL FINITE ELEMENT ANALYSIS FOR
APPLICATION TO AXIAL WAVE PROPAGATION PROBLEMS IN
SEMI-INFINITE AND FINITE MEMBRANES AND SHELLS OF
REVOLUTION
VA. POLYTECHNIC INST. AND STATE UNIV., 1971, 107
PP.

71-0590 BENZLEY, S. E.
TRANSIENT ANALYSIS OF THICK CYLINDRICAL SHELLS
UNIVERSITY OF CALIFORNIA, 1971, 138 PP.

71-0591 CHATANI, A.
SOME CONSIDERATIONS ON THE DYNAMIC STRESS ANALYSIS
OF A STRIP BY FINITE ELEMENT METHOD
MEM. FAC. TECHNOL. KANAZAWA UNIV., JAPAN, VOL. 6,
NO. 3, NOVEMBER 1971, PP. 253-262.

71-0592 DAMLE, S. K.
FEESER, L. J.
VIBRATION OF FOUR POINT-SUPPORTED PLATES BY A
FINITE ELEMENT METHOD
PROC. 23RD MEETING AERO. SOC. INDIA, INDIA INST.
OF TECH., KANPUR, INDIA, FEBRUARY 26-28, 1971.

71-0593 HARVEY, J.
KELSEY, S.
TRIANGULAR PLATE BENDING ELEMENT WITH ENFORCED
COMPATIBILITY
AIAA J., VOL. 9, 1971, PP. 1023-1026.

71-0594 GERADIN, M.
ERROR BOUNDS FOR EIGENVALUE ANALYSIS BY
ELIMINATION OF VARIABLES
ASPIRANT AU F.R.N.S., LAB. D'AERONAUTIQUE,
UNIVERSITE DE LIEGE, 4000 LIEGE, BELGIUM, J.
SOUND AND VIB., VOL. 19, NO. 2, NOVEMBER 22,
1971, PP. 111-132.

71-0595 GERADIN, M.
THE COMPUTATIONAL EFFICIENCY OF A NEW MINIMIZATION
ALGORITHM FOR EIGENVALUE ANALYSIS
J. SOUND AND VIB., VOL. 19, NO. 3, DECEMBER 8,
1971, PP. 319-331.

71-0596 HACKNEY, J. R.
PROBLEMS IN BUILDING AND MANIPULATING LARGE
STIFFNESS MATRICES
81ST MEETINGS OF THE ACOUSTICAL SOC. OF AMERICA,
APRIL 1971, PP. 20-23.

71-0597 HUNT, D. A.
DYNAMIC ANALYSIS OF A FLEXIBLE VEHICLE MOVING
ALONG A FLEXIBLE SUPPORT
J. SPACECRAFT ROCKETS, VOL. 8, NO. 6, JUNE 1971,
PP. 665-668.

71-0598 JAN, S. F.
IMPACT ON COMPLEX MECHANICAL STRUCTURES
AD-755 841, JULY 1971, 106 P.

71-0599 KANDIDOV, V. P.
CHESNOKOV, S. S.
APPLICATION OF FINITE ELEMENT METHODS FOR
CALUCLATION OF TRANSVERSE OSCILLATIONS IN ONE
DIMENSIONAL SYSTEMS
VESTN. MOSK. UNIV. FIZ. ASTRON. U.S.S.R., VOL. 1,
NO. 1, 1971, PP. 416-423.

71-0600 LYSMER, J.
DRAKE, L. A.
THE PROPAGATION OF LOVE WAVES ACROSS
NONHORIZONTALLY LAYERED STRUCTURES
BULL. SEISMOL. SOC. AM., U.S.A. VOL. 61, NO. 5,
OCTOBER 1971, PP. 1233-1251.

71-0601 MARGOLIAS, D. S.
WEINGARTEN, V. I.
FREE VIBRATIONS OF PRESSURE LOADED PARABOLOIDAL
SHELLS
UNIV. SOUTH CALIF., LOS ANGELES, AIAAA J. VOL. 9,
NO. 12, DECEMBER 1971, PP. 2339-2345.

71-0602 MIRZA, W. H.
PETYT, M.
ON THE VIBRATION OF POINT-SUPPORTED PLATES
J. SOUND AND VIB., VOL. 15, NO. 1, MARCH 1971, PP.
143-5.

71-0603 NEUBERT, V. H.
FINITE ELEMENT FOR SHOCK ANALYSIS
PENN. STATE UNIV., DEPT. ENGR. MECH., PROG. REP.
NO. 2, SEPTEMBER 1971, 100 PP.

71-0604 PAUL, B.
SHIFTING ROLES OF COMPUTERS, EXPERIMENTS AND
ANALYSIS IN APPLIED MECHANICS
EXP. MECH. VOL. 11, NO. 9, SEPTEMBER 1971, PP.
385-393.

71-0605 PERRELL, C. S.
DYNAMIC BEHAVIOUR OF ECCENTRICALLY STIFFENED
PLATES
UNIV. MISSOURI, ROLLA, MO., 1971, 129 PP.

71-0606 PETYT, M.
FLEISCHER, C. C.
FREE VIBRATION OF A CURVED BEAM
J. SOUND AND VIB., VOL. 18, SEPTEMBER 1971, PP.
17-30.

71-0607 RANGAIAH, V. P.
NEUBERT, V. H.
FINITE ELEMENT FOR TIMOSHENKO BEAM BASED ON
MECHANICAL IMPEDANCE
PA. STATE UNIV., UNIVERSITY PARK, PA., APRIL 1971,
PP. 133.

71-0608 SATO, H.
ON VIBRATION ANALYSIS OF RECTANGULAR CANTILEVER
PLATES WITH A STEPPED CROSS-SECTION BY THE
FINITE ELEMENT METHOD
MEM. FAC. TECHNOL. KANAZAWA UNIV., JAPAN, VOL. 6,
NO. 3, NOVEMBER 1971, PP. 227-239.

71-0609 THORNTON, E. A.
VIBRATION ANALYSIS OF A 1/15-SCALE DYNAMIC MODAL
OF A SPACE SHUTTLE CONFIGURATION
NASA-CR-111984, 1971, PP. 146.

71-0610 TONG, P.
PIAN, T. H. H.
BUCCIARELLI, L. L.
MODE SHAPES AND FREQUENCIES BY FINITE ELEMENT
METHOD USING CONSISTENT AND LUMPED MASSES
COMPUTERS AND STRUC. VOL. , NO. 4, DECEMBER 1971,
PP. 623- 638.

71-0611 TORIDIS, T. G.
KHOZEIMEH, K.
INELASTIC RESPONCE OF FRAMES TO DYNAMIC LOADS
ASCE J. ENG. MECH. DIV. VOL. 97, NO. 3, JUNE 1971,
PAPER 8205, PP. 847-863.

71-0612 ARALDSEN, P. O.
EGELAND, O.
GENERAL DESCRIPTION OF SESAM-69, SUPER ELEMENT
STRUCTURAL ANALYSIS (PROGRAM) MODULES
EUROPEAN SHIPBUILDING, NO. 2, 1971, PP. 21-35.

71-0613 ALDABBAGH, A.
GOODMAN, J. R.
BODIG, J.
FINITE ELEMENT METHOD FOR WOOD MECHANICS
J. STRUCT. DIV., A.S.C.E., VOL. 98, NO. ST3, 1971,
PP. 351-366.

71-0614 MEEK, J. L.
MATRIX STRUCTURAL ANALYSIS
NEW YORK: MCGRAW-HILL, 1971

71-0615 URAL, O.
MATRIX METHODS AND USE OF COMPUTERS IN STRUCTURAL
ENGINEERING
IN TEXT ED. NEW YORK AND LONDON, 1971.

71-0616 URAL, OKTAY
ANALYSIS OF TOTALLY SUBMERGED THIN SHEELS BY
FINITE ELEMENT TECHNIQUE
PROC. OF INTERNATIONAL ASSOCIATION OF SHELL
STRUCTURES, PACIFIC SYMPOSIUM, OCTOBER, 1971

71-0617 ARGYRIS, J. H.
SCHARPF, D. W.
METHODS OF ELASTO-PLASTIC ANALYSIS
RESEARCH REPORT NO. 105, INSTITUT FUR STATIK UND
DYNAMIK DER LUFTUND RAUMFAHRTKONSTRUKTIONEN, 1971.

71-0618 ARGYRIS, J. H.
MARECZEK, G.
THERMOMECHANICAL ANALYSIS OF STRUCTURES
RESEARCH REPORT NO. 113, INSTITUT FUR STATIK UND
DYNAMIK DER LUFTUND RAUMFAHRTKONSTRUKTIONEN,
OCTOBER 1971.

71-0619 TONG, P.
FUNG, Y. C.
SLOW VISCOUS FLOW AND ITS APPLICATION TO
BIOMECHANICS
J. APPLIED MECH., VOL. 38, SERIES E, NO. 4, 1971,
PP. 721-728.

71-0620 ALLWOOD, R. J.
MATRIX METHODS OF STRUCTURAL ANALYSIS
IN "LARGE SPARSE SETS OF LINEAR EQUATIONS", REID
(ED.), ACADEMIC PRESS, LONDON, 1971.

71-0621 ARGYRIS, J. H.
BUCK, K. E.
LOCHNER, N.
SCHARPF, D. W.
MATRIX DISPLACEMENT ANALYSIS OF PLATES AND SHELLS.
PART 11, GENERAL FORMULATION OF THE LINEAR THEORY
REPORT NO. 103, INSTITUT FUR STATIK UND DYNAMIK
DER LUFTUND RAUMFAHRTKONSTRUKTIONEN, 1971.

71-0622 MLEJNEK, H. P.
MAI, M.
TURBAN VERSION 1.0. EIN SPEZIELLES PROGRAMM ZUR
LINEAREN UND NICHTLINEAREN STATISCHEN UND
DYNAMISCHEN ANALYSE VON ROTIERENDEN
SCHALENTRAGWERKEN ANLEITUNG FUR BENUTZER
RESEARCH REPORT NO. 93, INSTITUT FUR STATIK UND
DYNAMIK DER LUFTUND RAUMFAHRTKONSTRUKTIONEN,
JANUARY 1971.

71-0623 ARGYRIS, J. H.
GRIEGER, I.
SCHREM, E.
STRUCTURAL ANALYSIS BY PROBLEM ORIENTED LANGUAGES
RESEARCH REPORT NO. 95, INSTITUT FUR STATIK UND
DYNAMIK DER LUFTUND RAUMFAHRTKONSTRUKTIONEN,
FEBRUARY 1971.

71-0624 ARGYRIS, J. H.
BRONLUND, O. E.
SCHREM, E.
ANWENDUNG DER METHODEN DER FINITEN ELEMENTE MIT
DEM PROGRAMMSYSTEM ASKA UNTER BESONDERER
BERUCKSICHTIGUNG DER DYNAMISCHEN PROBLEME
RESEARCH REPORT NO. 96, INSTITUT FUR STATIK UND
DYNAMIK DER LUFTUND RAUMFAHRTKONSTRUKTIONEN,
FEBRUARY 1971.

71-0625 JOHNSEN, TH.
STRAUB, K.
DYNAN USERS MANUEL
RESEARCH REPORT NO. 97, INSTITUT FUR STATIK UND
DYNAMIK DER LUFTUND RAUMFAHRTKONSTRUKTIONEN, MARCH
1971.

71-0626 SCHREM, E.
ROY, J.
AN AUTOMATIC SYSTEM FOR KINEMATIC ANALYSIS ASKA
PART 1
RESEARCH REPORT NO. 98, INSTITUT FUR STATIK UND
DYNAMIK DER LUFTUND RAUMFAHRTKONSTRUKTIONEN,
MARCH, 1971.

71-0627 MALEJANNAKIS, G. A.
ANWENDUNG DER MATRIZENVERSCHIEBUNGSMETHODE AUF
ERZWUNGENE SCHWINGUNGEN (SCHWINGUNGSPROBLEME)
(ARBEITSTITEL) PROPORTIONAL GEDAMPFTER
ELASTISCHER SYSTEME
RESEARCH REPORT NO. 99, INSTITUT FUR STATIK UND
DYNAMIK DER LUFTUND RAUMFAHRTKONSTRUKTIONEN,
SEPTEMBER 1971.

71-0628 ARGYRIS, J. H.
GRIEGER, I.
SORENSEN, M.
DIE METHODE DER ENDLICHEN ELEMENTE UND IHRE
ANWENDUNG IM MASCHINENBAU
RESEARCH REPORT NO. 101, INSTITUT FUR STATIK UND
DYNAMIK DERLUFTUND RAUMFAHRTKONSTRUKTIONEN, APRIL
1971.

71-0629 WILLAM, K.
WARNKE, P.
ROTATIONSSYMMETRISCHE BERECHNUNG DES 1:5 THTR
SPANNBETON-BEHAL-BEHAL TERMODELLS
RESEARCH REPORT NO. 102, INSTITUT FUR STATIK UND
DYNAMIK DER LUFTUND RAUMFAHRTKONSTRUKTIONEN, JULY
1971.

71-0630 WALLACE, D. B.
A COMPUTER BASED PROCEDURE FOR PREDICTING THE
TRANSIENT RESPONSE AND FAILURE OF A
TWO-DIMENSIONAL CONTINUUM WITH NONLINEAR MATERIAL
CHARACTERISTICS
UNIV. WISC., PH.D. THESIS, 1971, 424 PP.

71-0631 JENNEQUIN, G.
IS THE COMPUTATION OF NOISE LEVEL INSIDE A CAR
FEASIBLE
I. MECHE E/A.S.A.E. SYMP. ON VIB. AND NOISE IN
VEHICLES, PAPER C 108/71, 1971.

71-0632 ARGYRIS, J. H.
BRONLUND, O. E.
GRIEGER, I.
SORENSEN, M.
A SURVEY OF THE APPLICATION OF FINITE ELEMENT
METHODS TO STRESS ANALYSIS PROBLEMS WITH
PARTICULAR EMPHASIS ON THEIR APPLICATION TO
NUCLEAR ENGINEERING PROBLEMS
CONF. ON THE APPLICATION OF FINITE ELEMENT METHODS
IN NUCLEAR ENG. FIELD, OECD, ENEA, JUNE 29 - JULY
1, 1971. ISPRA. PUBLISHED IN ENEA NEWSLETTER NO.
12, OCT. 1971.

71-0633 SCHREM, E.
COMPUTER IMPLEMENTATION OF THE FINITE ELEMENT
PROCEDURE
RESEARCH REPORT NO. 111, INSTITUT FUR STATIK UND
DYNAMIK DER LUFTUND RAUMFAHRTKONSTRUKTIONEN,
SEPTEMBER 1971.

71-0634 DILWORTH, T. E. JR.
HINGE
DEPT. OF THE ARMY, WASHINGTON, D.C., REPORT NO:
PAT-APPL-173 181, PATENT-3 758 919, AUGUST 1971, 6
P. (N.T.I.S. - AD-164 805/4)

71-0635 ANDERHEGGEN, E.
STARR-PLASTISCHE TRAGLASTBERECHNUNGEN MITTELS DER
METHODE DER FINITEN ELEMENTE
INSTUT FUR BAUSTALIK, EIDGENOESSISCHE TECHNISCHE
HOCHSCHULE, ZURICH, REPORT NO. 0032, 1971.

71-0636 SNYDER, R. E.
A STABILITY ANALYSIS OF CYLINDRICAL PANELS USING A
FINITE ELEMENT FORMULATION
PH.D. THESIS, LANGLEY RESEARCH CENTER, NATIONAL
AERON. AND SPACE ADMIN., LANGLEY STATION, VA.,
REPORT NO: NASA-TM-X-69998, MAY 1971, 135 P.
(N.T.I.S. - N74-19548/8)

71-0637 ROSSI, F. A.
DELPUGLIA, A.
CALCULATION METHOD FOR THE STUDY OF CRACKED RIGID
STRUCTURES (METODO DI CALCOLO PER LO STUDIO DI
STRUTURE IRRIGIDITE FESSURATE)
PISA UNIV., ITALY, REPORT NO: REPT-1418, 1971, 96
P. (N.T.I.S. - N74-16590/3)

71-0638 GILLESPIE, D. C.
AN INVESTIGATION OF THE EFFECTS OF SQUARE CUTOUTS
ON THE NATURAL FREQUENCIES AND MODE SHAPES OF
RECTANGULAR PLATES
MASTER'S THESIS, SCHOOL OF ENGNG., AIR FORCE INST.
OF TECH., WRIGHT-PATTERSON AFB, OHIO, REPORT NO:
GAW/MC/71-8, JUNE 1971, 98 P. (N.T.I.S. - AD-769
915/0)

71-0639 ROYLANCE, D. K.
WAVE PROPAGATION IN A VISCOELASTIC FIBER SUBJECTED
TO TRANSVERSE IMPACT
PAPER-72-APM-27, APPLIED MECHANICS SUMMER CONF.,
CALIF. UNIV., LAJOLLA, JUNE 26-28, 1972, REPORT
NO: AMMRC-TR-73-33, NOVEMBER, 1971, 12 P. REVISION
OF REPORT AUGUST 5, 1971. (N.T.I.S - AD-766 380/0)
SEE ALSO J. APPLIED MECH., MARCH 1973, PP:
143-148.

71-0640 WU, C-H.
PERNG, D. Y. P.
ON THE ASYMPTOTICALLY SPHERICAL DEFORMATIONS OF
ARBITRARY MEMBRANES OF REVOLUTION FIXED ALONG AN
EDGE AND INFLATED BY LARGE PRESSURES - A NONLINEAR
BOUNDARY LAYER PHENOMENON
SIAM J. OF APPLIED MATH., VOL. 23, NO. 1, AUGUST
3, 1971, 23 P. (N.T.I.S. - AD-762 892)

71-0641 OSTERMANN, H.
FORM FACTORS OF FLAT BARS FOR FATIGUE STRENGTH
TESTS. (FORMZAHLEN VON FLACHSTAEBEN FUER
SCHWINGFESTIGKEITSVERSUCHE)
LABORATORIUM FUR BETRIEBSFESTIGKEIT, DARMSTADT,
WEST GERMANY. REPORT NO: LBF-TM-61/71, DECEMBER 8,
1971, 8 P. (N.T.I.S. -N73-15573)

71-0642 FITZGERALD, J. E.
HUFFERD, W. L.
HANDBOOK FOR THE ENGINEERING STRUCTURAL ANALYSIS
OF SOLID PROPELLANTS
JOHN HOPKINS UNIV., SILVER SPRING, MD., CHEMICAL
PROPULSION INFO. AGENCY, REPORT NO: CPIA-PUB-214,
REPORT NO: UTEC-CE-71-089, MAY 1971, 870 P.
(N.T.I.S. - AD-887 478)

71-0643 ARGYRIS, J. H.
WILLIAM, K.
CALCULATIONS FOR PRESTRESSED CONCRETE PRESSURE
VESSELS
INST. FUR STATIK UND DYNAMIK DER LUFT UND
RAUMFAHRTKONSTRUKTIONEN, UNIVERSITAT STUTTGART,
REPORT ISD-104, 1971.

71-0644 ASHWELL, D. G.
SABIR, A. B.
ROBERTS, T. M.
FURTHER STUDIES IN THE APPLICATION OF CURVED
FINITE ELEMENTS TO CIRCULAR ARCHES
INT. J. MECH. SCI., VOL. 13, 1971, PP. 507-517.

71-0645 BAZANT, E.
APPLICATION OF THE FINITE ELEMENT METHOD TO THE
CALCULATION OF REACTOR PRESSURE VESSELS
PROC. 1ST CONF. STRUCT. MECH. IN REACTOR TECH.,
VOL. 4, BERLIN, 1971, PP. 49-63.

71-0646 CRASTAN, V.
GENERALIZATION OF THE ELEMENT METHOD
NUCL. ENG. DES., VOL. 15, 1971, PP. 113-120.

71-0647 JENKINS, T. B.
EFFECT OF ELASTIC MODULAR RATIO AND ANISOTROPY ON
THE STRESS FIELD IN A TWO-DIMENSIONAL MATRIX INTO
WHICH A TWO-DIMENSIONAL RECTANGULAR INCLUSION HAS
BEEN EMBEDDED
MASTER'S THESIS, ORDNANCE RESEARCH LAB., PENN.
STATE UNIV., UNIV. PARK, REPORT NO. TM-71-73,
MARCH 23, 1971, 64 PP. (N.T.I.S. - AD-751 678)

71-0648 HARRISON, W.
BOON, J. D. 111
FANG, C. S.
FAUSAK, L. E.
WANG, S. N.
INVESTIGATION OF THE WATER TABLE IN A TIDAL BEACH
VIRGINIA INST. OF MARINE SCI., GLOUCESTER POINT,
REPORT NO: VIMS-SSR-60, OCTOBER 1971, 175 P.
(N.T.I.S.- AD-751 081)

71-0649 TADA, Y.
KUSAKA, K.
LARGE DEFLECTION OF CANTILEVER BEAMS
NATIONAL AEROSPACE LAB., TOKYO, JAPAN, REPORT NO:
NAL-TR-250, 1971, 28 P. (N.T.I.S. - N72-31914)

71-0650 SCHREM, E.
COMPUTER IMPLEMENTATION OF THE FINITE ELEMENT
PROCEDURE
O.N.R. SYMP. NUMER. AND COMPUT. METH. IN STRUCT.
MECH., UNIV. ILLINOIS AT URBANA-CHAMPAIGN, URBANA,
SEPTEMBER 1971.

71-0651 GHABOUSSI, J.
DYNAMIC STRESS ANALYSIS OF POROUS ELASTIC SOLIDS
SATURATED WITH COMPRESSIBLE FLUIDS
EARTHQUAKE ENGNG. RESEARCH CENTER, UNIV. OF
CALIF., BERKELEY, REPORT NO: EERC-71-6, AUGUST
1971, 108 P. (N.T.I.S. - PB-211 396)

71-0652 DUPONT, T.
ADI METHODS FOR FINITE ELEMENT SYSTEMS
NUMERICAL SOLUTION OF PARTIAL DIFFERENTIAL
EQUATIONS - II, SYNSPADE, 1970, B. E. HUBBARD
(ED.), ACADEMIC PRESS, NEW YORK, 1971.

71-0653 RIIKONEN, I.
ANALYSIS OF THICK-WALLED PRESSURE VESSELS USING
THE CONSTANT STRAIN AXISYMMETRIC FINITE ELEMENT
4TH NORDIC SYMPOSIUM ON THE STRENGTH OF MATERIALS,
OTANIEMI, FINLAND, AUGUST 5, 1971, 10 P. (N.T.I.S.
- CONF-710839) U.S. SALES ONLY

71-0654 WAIT, R.
A FINITE ELEMENT METHOD FOR THREE DIMENSIONAL
FUNCTION APPROXIMATION
DUNDEE CONF. ON APPLICS. OF NUMER. ANALYSIS,
SPRINGER-VERLAG, BERLIN, NO. 228, 1971.

71-0655 GLOSS, R. J.
TRANSPIRATION AND FILM COOLING BOUNDARY LAYER
COMPUTER PROGRAM. VOLUME 2, COMPUTER PROGRAM AND
USER'S MANUAL
DYNAMIC SCIENCE, IRVINE, CALIF., REPORT NO:
NASA-CR-126728, SN-230-VOL-2, JUNE 1971, 147 P.
(N.T.I.S. - N72-25301)

71-0656 HEUZE, F. E.
GOODMAN, R. E.
A DESIGN PROCEDURE FOR HIGH CUTS IN JOINTED HARD
ROCK. THREE DIMENSIONAL SOLUTIONS. VOLUME 1
UNIV. OF CALIF., BERKELEY, FINAL REPORT, MAY 1971,
143 P. (N.T.I.S. - PB-210 878) SEE ALSO VOL. 2,
PB-210 879.

71-0657 WADDOUPS, M. E.
MCCULLERS, L. A.
NABERHAUS, J. D.
COMPOSITE WING FOR TRANSONIC IMPROVEMENT. VOLUME
11. ADVANCED ANALYSIS EVALUATION
CONVAIR AEROSPACE DIV., GENERAL DYNAMICS, FORT
WORTH, TEX., TECHNICAL REPORT, NOVEMBER 1971, 173
P. (N.T.I.S. - AD-745 129)

71-0658 SOOSAAR, K.
DESIGN OF OPTICAL MIRROR STRUCTURES
CHARLES STARK DRAPER LAB., MASS. INST. OF TECH.,
CAMBRIDGE, REPORT NO: NASA-CR-122397, R-673,
JANUARY 1971, 148 P. (N.T.I.S. - N72-24495)

71-0659 SCORDELIS, A. C.
BOUWKAMP, J. G.
WASTI, S. T.
STRUCTURAL BEHAVIOR OF A TWO SPAN REINFORCED
CONCRETE BOX GIRDER BRIDGE MODEL. VOLUME 11.
REDUCTION, ANALYSIS AND INTERPRETATION OF RESULTS
STRUCT. ENGNG. LAB., UNIV. OF CALIFORNIA,
BERKELEY, REPORT NO: UCSESM-71-16, OCTOBER 1971,
198 P. (N.T.I.S. - PB-210 431) SEE ALSO VOL. 1,
PB-199 187.

71-0660 IZUMI, H.
ON ANALYSIS OF LARGE DEFORMATION PROBLEMS OF BEAM
NATIONAL AEROSPACE LAB., TOKYO, JAPAN, REPORT NO:
NAL-TR-246, NOVEMBER 1971, 10 P. (N.T.I.S. -
N72-22912)

71-0661 AUTHOR
THREE DIMENSIONAL SUPERTANKER COMPUTER PROGRAM
NAVAL CONST. RESEARCH ESTABLISHMENT, DUNFERMLINE,
SCOTLAND, REPORT NO: NCRE/R.564, APRIL 1971, 101
P. (N.T.I.S. - AD-743 743)

71-0662 BARTELDS, G.
OTTENS, H. H.
FINITE ELEMENT ANALYSIS OF SANDWICH PANELS
PROC. I.U.T.A.M. SYMP. ON HIGH SPEED COMPUTING OF
ELASTIC STRUCT., UNIV. OF LIEGE, VOL. 61, TOME 1,
SEPTEMBER 1971, PP. 357-382.

71-0663 LINK, M.
STABILITY INVESTIGATIONS OF PLATES USING FINITE
ELEMENT METHOD BASED ON AN EXPANDED VARIATION
PRINCIPLE (FINITE ELEMENTE ZUR
STABILITAETSUNTERSUCHUNG VON PLATTEN AUF DER
GRUNDLAGE EINES ERWEITERTEN VARIATIONSPRINZIPS)
FOURTH DGLR ANNUAL MEETING, BADEN-BADEN, WEST
GERMANY, OCTOBER 11-13, 1971, 22 P. (N.T.I.S. -
N72-21935)

71-0664 AUTHORS
JOURNAL OF RESEARCH OF THE NATIONAL BUREAU OF
STANDARDS: C. ENGINEERING AND INSTRUMENTATION
NATIONAL BUREAU OF STANDARDS, WASH. D.C., REPORT
NO: NBS-JRC-75-304, VOL. 75C, NO. 3, AND 4,
JULY-DECEMBER 1971, 71 PP. (N.T.I.S. -
COM-71-50314-0304)

71-0665 WEMPNER, G. A.
FINITE-DIFFERENCES VIA FINITE-ELEMENTS
RESEARCH INST., UNIV. OF ALABAMA, HUNTSVILLE,
REPORT NO: UARI-113, AUGUST 1971, 12 P. (N.T.I.S.
- AD-743 009)

71-0666 TILLERSON, J. R.
DYNASOR 11 - A FINITE ELEMENT PROGRAM FOR THE
DYNAMIC NONLINEAR ANALYSIS OF SHELLS OF REVOLUTION
(USER'S MANUAL)
DEPT. OF AEROSPACE ENGNG., TEXAS AGRI. AND MECH.
UNIV., COLLEGE STATION, OCTOBER 1971, 108 P.
(N.T.I.S. - SC-CR-70-6179) SEE ALSO NSA 2610, NO.
25112.

71-0667 GALLAGHER, R. H.
TRENDS AND DIRECTIONS IN THE APPLICATIONS OF
NUMERICAL ANALYSIS.
O.N.R. SYMP. ON NUMER. AND COMP. METH. IN STRUCT.
MECH., UNIV. OF ILLINOIS, 1971.

71-0668 CLOUGH, R. W.
RAPHAEL, J. M.
MOJTAHEDI, S.
FINITE ELEMENT ANALYSIS OF ARCH DAMS AND
FOUNDATIONS
UNIV. OF CALIF., BERKELEY, DECEMBER 1971, 113 P.
(N.T.I.S. - PB-208 698)

71-0669 GIENCKE, E.
SIMPLE FINITE ELEMENT METHOD FOR STRESS
CONCENTRATION PROBLEMS IN PRESSURE VESSELS
PROC. 1ST CONF. STRUCT. MECH. IN REACTOR TECH.,
BERLIN, VOL. 4, 1971, PP. 215-237.

71-0670 PIAN, T. H. H.
VARIATIONAL FORMULATIONS OF NUMERICAL METHODS IN
SOLID CONTINUA
PROC. OF SYMP. ON COMPUTER-AIDED ENGNG. SOLID
MECH. DIV., WATERLOO, ONT., MAY 11-13, 1971, PP.
421-448. (N.T.I.S. - AD-741 270)

71-0671 ATTAR-HASSAN, G. H.
A REFINED FINITE ELEMENT ANALYSIS OF FOLDED PLATE
STRUCTURES
THESIS, FOR MASTERS DEGREE OF APPL. SCI., DEPT. OF
CIVIL ENG., UNIV. OF OTTAWA, CANADA, AUGUST 1971.

71-0672 BUHLER, J.
LEVY, L. J.
EXPERIMENTAL AND THEORETICAL STUDIES OF
HULL-DECKHOUSE INTERACTION - 2-LEVEL DECKHOUSE
COLL. OF ENGNG., UNIV. OF CALIF., BERKELEY, REPORT
NO: NA-71-4, AUGUST 1971, 78 P. (N.T.I.S. - AD-740
818)

71-0673 KANG, C. M.
HANSEN, K. F.
FINITE ELEMENT METHODS FOR SPACE-TIME REACTOR
ANALYSIS
DEPT. OF NUCLEAR ENGNG., MASS. INST. OF TECH.,
CAMBRIDGE, REPORT NO: MITNE-135, NOVEMBER 1971,
192 P. (N.T.I.S. - MIT-3903-5)

71-0674 JEZERNIK, A.
ALUJEVIC, A.
HEAD, J. L.
FINITE ELEMENT MATRIX DISPLACEMENT METHOD FOR TIME
DEPENDENT STRESS ANALYSIS OF REACTOR MATERIALS IN
TWO AND THREE DIMENSIONS
PROC. 1ST CONF. STRUCT. MECH. IN REACTOR TECH.,
BERLIN, PAPER NO. M5/8, 1971.

71-0675 CHUNG, T. J.
BHUTWALA, J. N.
STEADY STATE AND TRANSIENT HEAT CONDUCTION AND
STRESS ANALYSES OF TWO AND THREE DIMENSIONAL
STRUCTURES BY ISOPARAMETRIC ELEMENTS
RESEARCH REPORT NO. 96, UNIV. OF ALABAMA,
HUNTSVILLE, 1971.

71-0676 KOLAR, V.
KRATOCHVIL, J.
ZLAMAL, M.
ZENISEK, A.
TECHNICAL, PHYSICAL AND MATHEMATICAL PRINCIPLES OF
THE FINITE ELEMENT METHOD
ACADEMIA, CZECHOSLOVAK ACADEMY OF SCI., PRAGUE,
1971.

71-0677 CANTIN, G.
AN EQUATION SOLVER OF VERY LARGE CAPACITY
INT. J. NUMER. METH. IN ENG., VOL. 3, NO. 3, 1971,
PP. 379-388.

71-0678 FRIED, I.
BASIC COMPUTATIONAL PROBLEMS IN THE FINITE ELEMENT
ANALYSIS OF SHELLS
INT. J. SOLIDS STRUCT., VOL. 7, 1971, PP.
1705-1715. REVISION OF REPORT, FEBRUARY 12, 1971.
(N.T.I.S. - AD-739 747)

71-0679 PIAN, T. H. H.
FORMULATIONS OF FINITE ELEMENT METHODS FOR SOLID
CONTINUA
AEROELASTIC AND STRUCT. RESEARCH LAB., MASS. INST.
OF TECH., CAMBRIDGE, DECEMBER 1971, 37 P.
(N.T.I.S. - AD-739 450)

71-0680 LINDLEY, P. B.
PLANE STRESS ANALYSIS OF RUBBER AT HIGH STRAINS
USING FINITE ELEMENTS
J. STRAIN ANALYSIS, VOL. 6, 1971, PP. 45-52.

71-0681 MACNEAL, R. H.
A HYBRID METHOD OF COMPONENT MODE SYNTHESIS
COMPUT. STRUCT., VOL. 1, 1971, PP. 581-602.

71-0682 MACNEAL, R. H.
MCCORMICK, C. W.
THE NASTRAN COMPUTER PROGRAM FOR STRUCTURAL
ANALYSIS
J. COMPUT. STRUCT. 1, 1971, PP. 389-412.

71-0683 REDDI, M. M.
FINITE ELEMENT SOLUTION OF LUBRICATION PROBLEMS
FRANKLIN INST., PHILADELPHIA, PA., REPORT NO:
NASA-CR-125069, F-C-2762-01, FEBRUARY 18, 1971, 65
P. (N.T.I.S. - N72-14487)

71-0684 FRANK, K. H.
THE FATIGUE STRENGTH OF FILLET WELDED CONNECTIONS
FRITZ ENGNG. LAB., LEHIGH UNIV., BETHLEHEM, PA.,
REPORT NO: FEL-358.37, OCTOBER 1971, 164 PP.
(N.T.I.S. - AD-739 341)

71-0685 SANTI, R. L.
A FINITE ELEMENT METHOD FOR OPTIMUM BEAM DESIGN
MASTER'S THESIS, NAVAL POSTGRADUATE SCHOOL,
MONTEREY, CALIF., DECEMBER 1971, 101 P. (N.T.I.S.
- AD-738 918)

71-0686 LEONIDAS, E.
A GENERAL PURPOSE THREE DIMENSIONAL STRESS
ANALYSIS PROGRAM
NAVAL POSTGRADUATE SCHOOL, MONTEREY, CALIF.,
DECEMBER 1971, 172 P. (N.T.I.S. - AD-738 906)

71-0687 MCCABE, J. T.
FULLER, D. D.
STABILITY OF TILTING-PAD JOURNAL BEARINGS
FRANKLIN INST. RESEARCH LABS., PHILADELPHIA, PA.,
REPORT NO: F-B-2131, NOVEMBER 1971, 17 P.
(N.T.I.S. - AD-738 608)

71-0688 SINGH, C.
TALL, L.
ON CRACKS IN WELDS AND WELDED STRUCTURES
FRITZ ENGNG. LAB., LEHIGH UNIV., BETHLEHEM, PA.,
REPORT NO: FEL-358.30, MAY 1971, 37 P. (N.T.I.S. -
AD-738 447)

71-0689 IYENGAR, N. G.
MURTHY, P. N.
FREE VIBRATION OF A SIMPLY SUPPORTED BEAM WITH
NONLINEAR MATERIAL PROPERTIES
DEPT. OF AERON. ENGNG., INDIAN INST. OF TECH.,
KANPUR, REPORT NO: AE-12/1971, 1971, 43 P.
(N.T.I.S. - N72-13879)

71-0690 MARGUERRE, K.
SCHALK, R.
WOELFEL, H.
CALCULATION OF VIBRATIONS OF STRUCTURES CAUSED BY
EARTHQUAKES USING THE FINITE ELEMENT METHOD
PROC. 1ST CONF. STRUCT. MECH. IN REACTOR TECH.,
BERLIN, VOL. 5, 1971, PP. 123-139.

71-0691 RUDDER, F. F. JR.
SKIN-STRINGER PANEL NORMAL MODE RESPONSE
EXPERIMENTAL DATA
LOCKHEED-GEORGIA CO., MARIETTA, REPORT NO:
NASA-CR-111988, NASA-CR-1959-SUPP, 1971, 194 PP.
(N.T.I.S. - 72-12905)

71-0692 WHETSONE, W. D.
JONES, C. E.
SUBSTRUCTURING TECHNIQUES
RESEARCH AND ENGNG. CENTER, LOCKHEED MISSILES AND
SPACE CO., HUNTSVILLE, ALA., REPORT NO:
NASA-CR-121051, LMSC/HREC-D162799-A, OCTOBER 1971,
25 P. (N.T.I.S. - N72-12123)

71-0693 AUTHOR
SINGLE-TRACK SUBWAY ENVIRONMENTAL SIMULATION
MODEL. PHASE 1
PARSONS, BRINCKERHOFF, QUADE AND DOUGLAS, N.Y.,
INTERIM TECHNICAL REPORT, AUGUST 1971, 175 P.
(N.T.I.S. - PB-206 895)

71-0694 TOKARZ, F. J.
INPUT SOIL PROPERTIES FOR THE FINITE ELEMENT CODE
NTEDYNS
LAWRENCE RADIATION LAB., UNIV. OF CALIF.,
LIVERMORE, MAY 3, 1971, 11 PP. (N.T.I.S. -
UCID-30028)

71-0695 TAKARZ, F. J.
USER'S MANUAL FOR NTEDYNS - A NONLINEAR FINITE
ELEMENT CODE FOR ANALYZING THE BLAST RESPONSE OF
UNDERGROUND STRUCTURES
LAWRENCE RADIATION LAB., UNIV. OF CALIF.,
LIVERMORE, JANUARY 7, 1971, 66 P. (N.T.I.S. -
UCID-30027)

71-0696 CROSE, J. G.
ASAAS: ASYMMETRIC STRESS ANALYSIS OF AXISYMMETRIC
SOLIDS WITH ORTHOTROPIC, TEMPERATURE-DEPENDENT
MATERIAL PROPERTIES THAT CAN VARY
CIRCUMFERENTIALLY
SAN BERNARDINO OPERATIONS, AEROSPACE CORP., SAN
BERNARDINO, CALIF., REPORT NO:
TR-0172(S2816-15)-1, DECEMBER 29, 1971, 275 P.
(N.T.I.S. - AD-737 357)

71-0697 SALONEN, E. M.
A FRAMEWORK METHOD FOR STRETCHED PLATES
FINNISH ACADEMY OF TECH. SCI., HELSINKI, REPORT
NO: APS-CI-6HNICA SCANDINAVICA PUB. OFFICE, BOX
5073, STOCKHOLM 5, SWEDEN

71-0698 TAKAHASHI, S. K.
MURTHA, R. N.
REDUCTION OF STRESSES IN BURIED STRUCTURES
NAVAL CIVIL ENGNG. LAB., PORT HUENEME, CALIF.,
REPORT NO: NCEL-TN-1199, DECEMBER 1971, 100 P.
(N.T.I.S. - AD-736 601)

71-0699 COVARRUBIAS, S. W.
CRACKING OF EARTH AND ROCKFILL DAMS: COMPARISON OF
OBSERVED AND THEORETICAL TENSILE STRAINS IN THE
CRESTS OF TWO EARTH AND ROCKFILL DAMS
HARVARD UNIV., CAMBRIDGE, MASS., FINAL REPORT,
APRIL 1971, 35 P. (N.T.I.S. - AD-736 372)

71-0700 CHEN, P. C. T.
CONSTITUTIVE MATRIX AND FINITE ELEMENT FORMULATION
FOR A DEFORMATION THEORY OF PLASTICITY
WATERVLIET ARSENAL, N.Y., REPORT NO: WVT-7110,
OCTOBER 1971, 20 P. (N.T.I.S. - AD-736 349)

71-0701 MASON, J. B.
FINITE ELEMENT ANALYSIS OF COUPLED IRREVERSIBLE
VECTOR PROCESSES
PH.D. THESIS, UNIV. OF MARYLAND, 1971.

71-0702 BERGAN, P. G.
NON-LINEAR ANALYSIS OF PLATES CONSIDERING
GEOMETRIC AND MATERIAL EFFECTS
STRUCT. ENGNG. LAB., UNIV. OF CALIF., BERKELEY,
REPORT NO: UCSESM-71-7, APRIL 1971, 184 P.
(N.T.I.S. - AD-735 937)

71-0703 KAVLIE, D.
CLOUGH, R. W.
A COMPUTER PROGRAM FOR ANALYSIS OF STIFFENED
PLATES UNDER COMBINED INPLANE AND LATERAL LOADS
DIV. OF STRUCT. ENGNG. AND STRUCT. MECH., UNIV. OF
CALIF., BERKELEY, REPORT NO: UCSESM-71-4, MARCH
1971, 71 P. (N.T.I.S. - AD-735 936)

71-0704 ANDERSEN, C. M.
NOOR, A. K.
USE OF GROUP-THEORETIC METHODS IN THE DEVELOPMENT
OF NONLINEAR SHELL FINITE ELEMENTS
PROC. INT. CONF. SYMMETRY, SIMILARITY AND
GROUP-THEORETIC METH. IN MECH., CALGARY, CANADA,
1974, PP. 533-558.

71-0705 MITCHELL, A. R.
PHILLIPS, G.
WACHSPRESS, E.
FORBIDDEN SHAPES IN THE FINITE ELEMENT METHOD
J. INST. MATH. APPL., VOL. 8, 1971, PP. 260-270.

71-0706 HARTUNG, R. F.
AN ASSESSMENT OF CURRENT CAPABILITY FOR COMPUTER
ANALYSIS OF SHELL STRUCTURES
PALO ALTO RESEARCH LAB., LOCKHEED MISSILES AND
SPACE CO., PALO ALTO, CALIF., FINAL REPORT, APRIL
1969 - APRIL 1971, 244 P. (N.T.I.S. - AD-735 726)

71-0707 AUTHORS
ARMY NUMERICAL ANALYSIS CONFERENCE
PROC. CONF. ARMY NUMER. ANALYSIS, REPORT NO:
AROD-71-4, DECEMBER 1971, 278 PP. (N.T.I.S. -
AD-735 690)

71-0708 HUNT, J. T.
BARACH, D.
JOHNSON, C.
VIBRATIONAL ANALYSIS OF THE FLEXING HEAD OF THE
BQS-6 TRANSDUCER
NAVAL UNDERSEA RES. AND DEV. CENTER, SAN DIEGO,
CALIF., REPORT NO. NUC-TP-239, NOVEMBER 1971, 33
PP. (N.T.I.S. -AD-735 459)

71-0709 SHINOZUKA, M.
PROBABILISTIC FORMULATION FOR ANALYTICAL MODELING
OC CONCRETE STRUCTURES
SHINOZUKA (MASANORU) RIDGEWOOD, N.J. REPORT NO:
TR-1, NOVEMBER 15, 1971, 50 P. (N.T.I.S. - AD-735
444)

71-0710 JUDD, W. R.
PERLOFF, W. H.
STRAIN DISTRIBUTION AROUND UNDERGROUND OPENINGS.
STATISTICAL RELATIONSHIPS FOR CERTAIN ROCK
PROPERTIES
SCHOOL OF CIVIL ENGNG., PURDUE UNIV., LAFAYETTE,
IND., REPORT NO: TR-6, OCTOBER 1971, 197 P.
(N.T.I.S. - AD-735 376) SEE ALSO TECH. REPORT NO.
5, AD-723 532.

71-0711 SHUKU, T.
ISHIHARA, K.
FINITE ELEMENT ANALYSIS OF IRREGULARLY SHAPED
ACOUSTIC FIELD
AUTUMN MEETING OF ACOUSTICAL SOC. OF JAPAN, 1971.

71-0712 BILLINGTON, A. E.
ON THE FICTION OF LEADING EDGE SUCTION
AERON. RESEARCH LABS., MELBOURNE, AUSTRALIA,
REPORT NO: ARL/SM-NOTE-360, JANUARY 1971, 37 PP.
(N.T.I.S. - N71-36399)

71-0713 AUTHORS
NASTRAN USERS EXPERIENCES
CONF. COLLOQ., HAMPTON, VA., SEPTEMBER 13-15,
1971, REPORT NO: NASA-TM-X-2378,L-7980, SEPTEMBER
1971, 785 P. (N.T.I.S. - N71-36253)

71-0714 ODEN, J. T.
LECTURES ON FINITE ELEMENT METHODS IN CONTINUUM
MECHANICS
NATO ADVANCED STUDY INST. ON FINITE ELEMENT METH.
IN CONTINUUM MECH., LISBON, 1971.

71-0715 COLLIER, W. D.
TRESS - A FINITE ELEMENT CODE FOR STRESS ANALYSIS
PROBLEMS
REACTOR GROUP, UNITED KINGDOM ATOMIC ENERGY
AUTHORITY, RISLEY, ENGLAND, 1971, 77 P. (N.T.I.S.
- TRG-REPORT-1813) SEE ALSO NSA 25 23, NO. 57285.

71-0716 LAKIS, A. A.
DYNAMICAL ANALYSIS OF AXIALLY NONUNIFORM, THIN
CYLINDRICAL SHELLS. PART 2 FREE VIBRATION
DEPT. OF MECH. ENGNG., MCGILL UNIV., MONTREAL,
QUE., REPORT NO: MERL-70-10, JANUARY 1971, 37 P.
(N.T.I.S. - N71-35138)

71-0717 OLSON, M.
ANALYSIS OF ARBITRARY SHELLS USING SHALLOW SHELL
FINITE ELEMENTS
DEPT. OF CIVIL ENG., UNIV. OF B.C., STRUCT.
RESEARCH SER. REPORT, 1971.

71-0718 SPREEUW, E.
FIESTA - FINITE ELEMENT STRESS AND TEMPERATURE
ANALYSIS. A DESCRIPTION OF THE THEORY USED FOR
THE CODE WITH SOME OTHER FINITE ELEMENT
APPLICATIONS
REACTOR CENTRUM, NEDERLAND, PETTEN, JUNE 1971, 65
P. (N.T.I.S. - RCN-149) SEE ALSO NSA 25 22, NO.
54377.

71-0719 SPANNAGEL, D. W.
TRANSVERSE REINFORCEMENT OF CONTINUOUS,
TROUGH-SHAPED CONCRETE RAILROAD STRUCTURES. AN
IMPLEMENTATION OF THE FOLDED PLATE ANALYSIS, MUPDI
AND THE FINITE ELEMENT ANALYSIS, FINPLA
BRIDGE DEPT., CALIF. STATE DIV. OF HIGHWAYS,
CALIF., REPORT NO: R/D-5-71, APRIL 1971, 252 P.
(N.T.I.S. - PB-204 542)

71-0720 POPPLEWELL, N.
MCDONALD, D.
CONFORMING RECTANGULAR AND TRIANGULAR PLATE
BENDING ELEMENTS
J. SOUND VIB., VOL. 19, 1971, PP. 333-347.

71-0721 YATES, D. N.
VISON, T. J.
SABLE, W. W.
THE DEVELOPMENT OF LARGE SCALE DIGITAL COMPUTER
CODES FOR PRODUCTION STRUCTURAL ANALYSIS
PROC. CONF. ON COMPUTER ORIENTED ANALYSIS OF SHELL
STRUCT., PALO ALTO, CALIF., REPORT AFFDL-TR-71-79,
AUGUST 1971, PP. 172-222.

71-0722 HORINO, F. G.
DUVALL, W. I.
BRADY, B. T.
THE USE OF ROCK BOLTS OR WIRE ROPE TO INCREASE THE
STRENGTH OF FRACTURED MODEL PILLARS
DENVER MINING RESEARCH CENTER, BUREAU OF MINES,
DENVER, COLO., REPORT NO: BM-R-7568, 1971, 30 P.
(N.T.I.S. - PB-204 153)

71-0723 ZLAMAL, M.
ZENISEK, A.
MATHEMATICAL ASPECTS OF THE FINITE ELEMENT METHOD
IN "TECHNICAL, PHYSICAL AND MATHEMATICAL
PRINCIPLES OF THE FINITE ELEMENT METHOD", (EDS)
KOLAR, V., KRATOCHVIL, J., ZLAMAL, M., AND
ZENISEK, A., ROSPRAVY CSAV, ACADEMIA, PRAGUE,
1971.

71-0724 SABIR, A. B.
ASHWELL, D. G.
FINITE ELEMENT ANALYSIS OF A BROAD ARCH BRIDGE
WITH A SLAB DECK
PROC. CONF. ON DEVELOPMENTS IN BRIDGE DESIGN,
UNIV. COLL., CARDIFF, 1971.

71-0725 KAHN, L. F.
INFLUENCE OF END-CLOSURE STIFFNESS ON BEHAVIOR OF
CONCRETE CYLINDRICAL HULLS SUBJECTED TO
HYDROSTATIC LOADING
NAVAL CIVIL ENGNG. LAB., PORT HUENEME, CALIF.,
REPORT NO: NCEL-TR-740, OCTOBER 1971, 57 P.
(N.T.I.S. - AD-732 363)

71-0726 TRATINA, G. G.
FRACTURE MECHANICS APPROACH TO ADHESIVE JOINTS
DEPT. OF THEORETICAL AND APPLIED MECH., UNIV. OF
ILLINOIS, URBANA, REPORT NO: T/AM-350,
UILU-ENG-71-60014, AUGUST 1971, 46 PP. (N.T.I.S. -
AD-731 992)

71-0727 BRANCA, T. R.
CREEP OF A UNIAXIAL METAL MATRIX COMPOSITE
SUBJECTED TO AXIAL AND NORMAL LATERAL LOADS
DEPT. OF THEORETICAL AND APPLIED MECH., UNIV. OF
ILLINOIS, URBANA, REPORT NO: T/AM-341,
UILU-ENG-71-6005, JUNE 1971, 64 P., (N.T.I.S. -
AD731 989)

71-0728 RINDAL, R. A.
WOOL, M. R.
POWARS, C. A.
BASELINE SOLUTIONS FOR THE SMOOTH WALL
THERMOCHEMICAL ABLATION RESPONSE OF GRAPHITE AND
CARBON PHENOLIC
AEROTHERM CORP., MOUNTAIN VIEW, CALIF., REPORT NO:
AEROTHERM-70-218 JULY 1971, 66 P. (N.T.I.S. -
AD-731 813)

71-0729 AHMED, K. M.
FREE VIBRATIONS OF CURVED SANDWICH BEAMS
INST. OF SOUND AND VIBRATION RESEARCH, SOUTHAMPTON
UNIV., ENGLAND, REPORT NO: ISVR-TR-45, MARCH 1971,
44 P. (N.T.I.S. - N71-33043)

71-0730 AHMED, K. M.
STATIC ANALYSIS OF SANDWICH STRUCTURES BY THE
FINITE ELEMENT METHOD
INST. OF SOUND AND VIBRATION RESEARCH, SOUTHAMPTON
UNIV., ENGLAND, REPORT NO: ISVR-TR-48, APRIL 1971,
52 P. (N.T.I.S. - N71-32860)

71-0731 LIN, F. T.
THE FINITE ELEMENT ANALYSIS OF LAMINATED
COMPOSITES
PH.D. THESIS, VIRGINIA POLYTECHNIC INST. AND STATE
UNIV., BLACKSBURG, VIRGINIA, 1971.

71-0732 COOK, W. A.
THREE-DIMENSIONAL GRAIN STRESS ANALYSIS USING THE
FINITE ELEMENT METHOD
ROCKET PROPULSION LAB., AIR FORCE SYSTEMS COMMAND,
EDWARDS AIR FORCE BASE, CALIF., AFRPL-TR-71-51,
1971.

71-0733 LEFEBVRE, G.
DUNCAN, J. M.
THREE-DIMENSIONAL FINITE ELEMENT ANALYSIS OF DAMS
OFFICE OF RESEARCH SERVICES, UNIV. OF CALIF.,
BERKELEY, REPORT NO: TE-71-5, MAY 1971, 68 P.
(N.T.I.S. - AD-731 231) SEE ALSO AD-713 900.

71-0734 ZIKE, J. H.
DESIGN OF PROPELLANT FINS SUBJECTED TO AXIAL
ACCELERATION
PROPULSION DIRECTORATE, ARMY MISSILE COMMAND,
REDSTONE ARSENAL, ALA., REPORT NO: RK-TR-71-15,
AUGUST 1971, 89 P. (N.T.I.S. - AD-731 208)

71-0735 GABRIELSEN, B.
MORRIS, P.
STRUCTURAL RESPONSE AND LOADING OF WALL PANELS
URS RESEARCH CO., SAN MATEO, CALIF., REPORT NO:
URS-709-11, JULY 1971, 196 P. (N.T.I.S. - AD-730
483)

71-0736 ELDER, A. S.
BURNS, B. P.
HURBAN, J. M.
STRESS ANALYSIS OF 175 MM PROJECTILE, HE, M437
BALLISTIC RESEARCH LABS., ABERDEEN PROVING GROUND,
MD., REPORT NO: BRL-MR-2113, JULY 1971, 41 P.
(N.T.I.S. AD-730 308)

71-0737 PEAKE, D. J.
TECHNICAL EVALUATION REPORT ON AGARD SPECIALISTS
MEETING ON AERODYNAMIC INTERFERENCE CONF. SILVER
SPRING, MD., NOVEMBER 28-30, 1970, MAY 1971, 18
PP., (N.T.I.S. - N71-31459).

71-0738 MASON, J. B.
THE NASTRAN HYDROELASTIC ANALYZER
GODDARD SPACE FLIGHT CENTER, NAT. AERON. AND SPACE
ADMIN., GREENBELT, MD., REPORT NO:
NASA-TM-X-656178X-321-71-174, MARCH 1971, 58 P.
(N.T.I.S. - N71-30609)

71-0739 JERGER, E. W.
HOGAN, J. C.
DEEP SEA ENGINEERING
COLL. OF ENGNG., UNIV. OF NOTRE DAME, IND.,
SEPTEMBER 1971, 48 P. (N.T.I.S. - AD-730 066)

71-0740 NELSON, I.
BARON, M. L.
SANDLER, I.
MATHEMATICAL MODELS FOR GEOLOGICAL MATERIALS FOR
WAVE PROPAGATION STUDIES
WEIDLINGER (PAUL) NEW YORK, TECHNICAL REPORT,
MARCH 1971, 191 P. (N.T.I.S. - AD-729 703)

71-0741 CROSE, J. G.
SAAS 111: FINITE ELEMENT STRESS ANALYSIS OF
AXISYMMETRIC AND PLANE SOLIDS WITH DIFFERENT
ORTHOTROPIC, TEMPERATURE-DEPENDENT MATERIAL
PROPERTIES IN TENSION AND COMPRESSION
SAN BERNARDINO OPERATIONS, AEROSPACE CORP., SAN
BERNARDINO, CALIF., REPORT NO:
TR-0059(S6816-53)-1, JUNE 22, 1971, 321 P.
(N.T.I.S. - AD-729 188) SEE ALSO AD-679 983,
SEPTEMBER, 1968.

71-0742 HEWITT, D. R.
PLANETARY EXPLORER STRUCTURAL CONCEPT EVALUATION
MODEL USING NASTRAN
GODDARD SPACE FLIGHT, NAT. AERON. AND SPACE
ADMIN., GREENBELT, MD., REPORT NO:
NASA-TM-X-65585, X-762-71-181, APRIL 1971, 42 P.
(N.T.I.S. - N71-28319)

71-0743 OPSTEEGH, J. D.
THE PREDICTION OF REGULAR WAVE FORCES ON FLOATING
BODIES WITH THE FINITE ELEMENT METHOD (IN DUTCH)
DEPT. OF APPL. MATH. TECH. UNIV., DELFT, 1971.

71-0744 HOWLETT, J. T.
APPLICATIONS OF NASTRAN TO COUPLED STRUCTURAL AND
HYDRODYNAMIC RESPONSES IN AIRCRAFT HYDRAULIC
SYSTEMS
NASTRAN COLLOQUIUM - USER'S EXPERIENCES, NASA,
LANGLEY, RES. CENTER, HAMPTON, VA., REPORT NO.
NASA TM-X-2378, SEPTEMBER 13-15, 1971, 19 PP.

71-0745 CHANG, C. H.
WEEKS, G. B.
PART 1 - TASK 3.2.9. BENDING AND BUCKLING OF A
ONE-DIMENSIONAL MODEL SIMULATING A CONICAL SHELL.
PART 11 - TASK 3.2.11. NONLINEAR ANALYSIS AND
COMPUTER PROGRAM FOR AXISYMMETRIC STATIC RESPONSE
OF SHELLS OF REVOLUTION
UNIV. BUREAU OF ENGNG. RESEARCH, UNIV. OF ALABAMA,
ALA., REPORT NO: BER-132-97, AUGUST 1971, 139 P.
(N.T.I.S. - AD-728 470)

71-0746 LANGLAND, R. T.
NONDETERMINISTIC PROBLEMS IN STRUCTURAL MECHANICS
NAVAL WEAPONS CENTER, CHINA LAKE, CALIF., REPORT
NO: NWC-TP-5207, JUNE 1971, 253 P. (N.T.I.S. -
AD-728 402)

71-0747 MCNEICE, G. M.
MARCAL, P. V.
OPTIMIZATION OF FINITE ELEMENT GRIDS BASED ON
MINIMUM POTENTIAL ENERGY
DIV. OF ENGNG., BROWN UNIV., PROVIDENCE, R.I.,
REPORT NO: TR-7, JUNE 1971, 25 P. (N.T.I.S. -
AD-727 993)

71-0748 CHEN, A. T. F.
KI, M. K.
MESH GENERATING COMPUTER PROGRAM FOR FINITE
ELEMENT ANALYSES OF SIMPLE SLOPES IN HORIZONTALLY
LAYERED GROUND
GEOLOGICAL SURVEY, MENLO PARK, CALIF., MARCH 1971,
43 P. (N.T.I.S. - PB-201 548)

71-0749 SAWYER, S. G.
BRLESC FINITE ELEMENT PROGRAM FOR AXISYMMETRIC,
PLANE STRAIN, AND PLANE STRESS, ORTHOTROPIC SOLIDS
WITH TEMPERATURE-DEPENDENT MATERIAL PROPERTIES
BALLISTIC RESEARCH LABS., ABERDEEN PROVING GROUND,
MD., REPORT NO: BRL-1539, MARCH 1971, 143 P.
(N.T.I.S. - AD-727 702)

71-0750 DE ROUVRAY, A.
GOODMAN, R. E.
DROZD, K.
HEUZE, F.
BREKKE, T. L.
ANALYSIS AND MODEL STUDIES OF UNDERGROUND OPENINGS
IN JOINTED ROCK. RESEARCH ON ROCK BOLT
REINFORCEMENT
UNIV. OF CALIF., BERKELEY, REPORT NO: TR-7, JUNE
1971, 225 P. (N.T.I.S. - AD-727 655) SEE ALSO
AD-723 397.

71-0751 HANSON, D. E.
THREE-DIMENSIONAL ELASTO-STATIC PROBLEMS USING
ISOPARAMETRIC FINITE ELEMENTS
MASTER'S THESIS, NAVAL POSTGRADUATE SCHOOL,
MONTEREY, CALIF., JUNE 1971, 94 P. (N.T.I.S. -
AD-727 117)

71-0752 FREMOND, M.
FORMULATIONS DUALES DES ENERGIES POTENTIELLES ET
COMPLEMENTAIRES. APPLICATION A LA METHODE DES
ELEMENTS FINIS
C.R. ACAD. SC., VOL. 273, SERIE A, 1971, PP.
775-777.

71-0753 DUPUIS, G. A.
INCREMENTAL FINITE ELEMENT ANALYSIS OF LARGE
ELASTIC DEFORMATION PROBLEMS
CONF. COMPUTER-AIDED ENGNG, WATERLOO UNIV., ONT.,
REPORT NO: TR-6, MAY 1971, 37 P. (N.T.I.S. -
AD-726 727)

71-0754 GEORGOPOULOS, G.
ROYSTER, L. H.
DEVELOPMENT OF A FINITE ELEMENT MODEL FOR THE
CLASS V FLEXTENSIONAL UNDERWATER TRANSDUCER SHELL
DEPT. OF MECH. AND AEROSPACE ENGNG., NORTH
CAROLINA STATE UNIV., RALEIGH, REPORT NO: TR-20,
JULY 1971, 23 P. (N.T.I.S. - AD-726 726)

71-0755 OWENS, R. H.
APPLICATIONS OF NUMERICAL ANALYSIS
CONF. ON APPLICATIONS OF NUM. ANALYSIS, DUNDEE
UNIV., REPORT NO: ONRL-C-12-71, JUNE 11, 1971, 20
P. (N.T.I.S. - AD-726 674)

71-0756 GALLO, A. M.
MAGIC 11: AN AUTOMATED GENERAL PURPOSE SYSTEM FOR
STRUCTURAL ANALYSIS. VOLUME 111. PROGRAMMER'S
MANUAL
BELL AEROSPACE CO., BUFFALO, N.Y. FINAL REPORT,
DECEMBER 2, 1968 - DECEMBER 2, 1970, MAY 1971, 568
P. (N.T.I.S. - AD-726 566) SEE ALSO VOL. 2, AD-726
565.

71-0757 JORDAN, S.
MAGIC 11: AN AUTOMATED GENERAL PURPOSE SYSTEM FOR
STRUCTURAL ANALYSIS. VOLUME 1. ENGINEER'S MANUAL
(ADDENDUM)
BELL AEROSPACE CO., BUFFALO, N.Y., FINAL REPORT,
DECEMBER 2, 1968 - DECEMBER 2, 1970, MAY 1971, 186
P. (N.T.I.S. - AD-726 564) SEE ALSO VOL. 2, AD-726
565

71-0758 PRAGER, W.
MARCAL, P. V.
OPTIMALITY CRITERIA IN STRUCTURAL DESIGN
DIV. OF ENGNG., BROWN UNIV., PROVIDENCE, R.I.,
TECHNICAL REPORT, JUNE 1969 - NOVEMBER 1970, MAY
1971, 123 P. (N.T.I.S. - AD-726 551)

71-0759 YALAMANCHILI, R. V. S.
CHU, S-C.
APPLICATION OF THE FINITE-ELEMENT METHOD TO
HEAT-TRANSFER PROBLEMS. PART 11. TRANSIENT
TWO-DIMENSIONAL HEAT TRANSFER WITH CONVECTION AND
RADIATION BOUNDARY CONDITIONS
RESEARCH DEVELOPMENT AND ENGNG. DIRECTORATE, ARMY
WEAPONS COMMAND, ROCK ISLAND, ILL., REPORT NO:
AMSWE-RE-71-41, JUNE 1971, 75 P. (N.T.I.S. -
AD-726 371) SEE ALSO PART 1, AD-726 371.

71-0760 CHU, S-C.
YALAMANCHILI, R. V. S.
APPLICATION OF THE FINITE-ELEMENT METHOD TO
HEAT-TRANSFER PROBLEMS. PART 1. FINITE-ELEMENT
METHOD APPLIED TO HEAT-CONDUCTION SOLIDS WITH
NONLINEAR BOUNDARY CONDITIONS
RESEARCH DEVELOPMENT AND ENGNG. DIRECTORATE, ARMY
WEAPONS COMMAND, ROCK ISLAND, ILL., REPORT NO:
ANSWE-RE-71-37, JUNE 1971, 51 P. (N.T.I.S. -
AD-726 370) SEE ALSO PART 2, AD-726 371.

71-0761 SCHIFFNER, K.
COMPARIOSN OF RESULTS OBTAINED FOR THE PROBLEM OF
SYMMETRIC FLUID OSCILLATIONS IN TANKS BY THE
FINITE DIFFERENCES TECHNIQUE AND THE FINITE
ELEMENT METHOD
ARBEITSGRUPPE THERMOFLASTIK, DEUTSCHE FORSCHUNGS
UND VERSUCHSANSTALT FUR LUFT UND RAUMFAHRT, PORZ,
WEST GERMANY, REPORT NO. DLR-FB-71-69, AUGUST 12,
1971, 61 PP. (N.T.I.S. - N72-21316).

71-0762 HARTUNG, R. F.
COMPUTER ORIENTED ANALYSIS OF SHELL STRUCTURES
PROC. CONF. ON COMPUTER ORIENTED ANALYSIS OF SHELL
STRUCTURES, PALO ALTO, CALIF., AUGUST 10-14, 1971,
JUNE 1971, 1301 P. (N.T.I.S. - AD-740 547)

71-0763 ZIENKIEWICZ, O. C.
ISOPARAMETRIC AND ALLIED NUMERICALLY INTEGRATED
ELEMENTS - A REVIEW
PROC. CONF. ON NUMERICAL AND COMPUTER METHODS IN
STRUCTURAL MECHANICS, UNIV., OF ILLINOIS, URBANA,
SEPTEMBER 1971.

71-0764 WILSON, W. K.
CRACK TIP FINITE ELEMENTS FOR PLANE ELASTICITY
WESTINGHOUSE RES. LAB., REPORT NO.
71-1E7-FMPWR-P2, JUNE 1971.

71-0765 CHAKRABARTI, S.
TRIGONOMETRIC FUNCTION REPRESENTATIONS FOR
RECTANGULAR PLATE BENDING ELEMENTS
INT. J. NUMER. METH. ENG., VOL. 3, 1971, PP.
261-273.

71-0766 DANHOF, R. H.
SECTOR FINITE ELEMENTS IN THE THEORY OF PLANE
ELASTICITY
DISSERTATION, UNIV. OF ILLINOIS, URBANA, 1971.

71-0767 FRANCE, P. W.
FINITE ELEMENT ANALYSIS OF UNCONFINED SEEPAGE
PROBLEMS
PH.D. THESIS, CIVIL ENG. DEPT., UNIV. COLLEGE OF
SWANSEA, WALES, 1971.

71-0768 HOFMANN, H. H.
LOAD LIMIT ANALYSIS, COMBINATION OF STATISTICAL
METHODS (MONTE CARLO METHOD) AND FINITE ELEMENT
METHOD FOR FRACTURE CALCULATIONS
PROC. 1ST CONF. STRUCT. MECH. IN REACTOR TECH.,
BERLIN, VOL. 4, 1971, PP. 253-272.

71-0769 KLEIN, S. A.
A STATIC AND DYNAMIC FINITE ELEMENT SHELL ANALYSIS
WITH EXPERIMENTAL VERIFICATION
INT. J. NUMER. METH. ENG., VOL. 3, 1971, PP.
299-316.

71-0770 THATCHER, P. W.
THE THEORY AND APPLICATION OF THE FINITE ELEMENT
METHOD
PH.D. THESIS, UNIV. LONDON, 1971.

71-0771 WOLF, J. P.
SYSTEMATIC ENFORCEMENT OF STRESS BOUNDARY
CONDITIONS IN THE ASSUMED STRESS HYBRID MODEL
BASED ON THE DEFORMATION METHOD
PROC. 1ST INTL. CONF. STRUCT. MECH. REACTOR TECH.,
BRUSSELS, PAPER M6/10, 1971, PP. 463-483.

71-0772 TOO, J. M.
TWO DIMENSION, PLATE, SHELL AND FINITE PRISM
ISOPARAMETRIC ELEMENTS AND THEIR APPLICATIONS
PH.D. THESIS, UNIV. OF WALES, 1971.

71-0773 ANON
CONFERENCE ON MATRIX METHODS IN STRUCTURAL
MECHANICS
PROC. 3RD CONF., WRIGHT-PATTERSON AFB, OHIO, 1971.

71-0774 ANON
CONFERENCE ON APPLICATION OF FINITE ELEMENT
METHODS TO STRESS ANALYSIS PROBLEMS IN NUCLEAR
ENGINEERING
PROC. CONF., ISPRA, ITALY, 1971.

71-0775 GRUBITZSCH, W.
APPROXIMATION METHOD FOR COMPUTATION OF NATURAL
FREQUENCIES OF SPIRAL SECTIONS
MASCHINENBAUTECHNIK, VOL. 20, NO. 11, NOVEMBER
1972, PP. 532-537.

71-0776 YAMADA, YOSHIAKI
YOKOUCHI, Y.
INCREMENTAL SOLUTION OF AXISYMMETRIC PLATE AND
SHELL FINITE DEFORMATION
PROC. I.U.T.A.M. HIGH SPEED COMPUT. OF ELASTIC
STRUCT., UNIV. OF LIEGE, 1971, PP. 663-682.

71-0777 YAMADA, YOSHIAKI
TAKATSUKA, K.
ELASTIC-PLASTIC ANALYSIS OF BEAMS WITH UNIFORM
CROSS-SECTION UNDER COMBINED LOADINGS
J. INST. OF IND. SCI., SEISAN-KENKYU, VOL. 23,
1971, PP. 531-535.

71-0778 YAMADA, YOSHIAKI
TAKATSUKA, K.
ELASTIC-PLASTIC ANALYSIS OF A BAR WITH UNIFORM
CROSS-SECTION UNDER COMBINED LOADINGS 1 -
COMBINATION OF AXIAL LOAD AND TORSION
PROC. 2ND NATL. SYMP. ON MATRIX METH. OF STRUCT.
ANALYSIS AND DESIGN, SOC. OF STEEL CONSTR. OF
JAPAN, 1971, PP. 252-259 (IN JAPANESE).

71-0779 YAMADA, YOSHIAKI
YOKOUCHI, Y.
ELASTIC-PLASTIC ANALYSIS OF AXISYMMETRIC PLATES
AND SHELLS
PROC. 2ND NATL. SYMP. ON MATRIX METH. OF STRUCT.
ANALYSIS AND DESIGN, SOC. OF STEEL CONSTR. OF
JAPAN, 1971, PP. 274-281 (IN JAPANESE).

71-0780 YAMADA, YOSHIAKI
NAGAI, Y.
HASHIMOTO, K.
ANALYSIS OF LONGITUDINAL WAVE IN A BAR EXHIBITING
VISCO-ELASTIC BEHAVIOR OR WITH MATERIAL DAMPING
PROC. 2ND NATL. SYMP. ON MATRIX METH. OF STRUCT.
ANALYSIS AND DESIGN, SOC. OF STEEL CONSTR. OF
JAPAN, 1971, PP. 338-345 (IN JAPANESE).

71-0781 YAMADA, YOSHIAKI
IWATA, K.
FINITE ELEMENT ANALYSIS OF THERMO-VISCOELASTIC
PROBLEMS
PROC. 2ND NATL. SYMP. ON MATRIX METH. OF STRUCT.
ANALYSIS AND DESIGN, SOC. OF STEEL CONSTR. OF
JAPAN, 1971, PP. 561-569 (IN JAPANESE).

71-0782 ZIENKIEWICZ, O. C.
STAGG, K. G.
PAREKH, C.
THERMAL STRESS AND TEMPERATURE FIELDS - A
FINITE-ELEMENT APPROACH
THERMAL STRESSES AND THERMAL FATIGUE, CENTRAL
ELECT. GEN. BRD. AND BERKELEY NUCL. LABS., 1971,
PP. 311-232.

71-0783 BACKLUND, J.
MIXED FINITE ELEMENT ANALYSIS OF PLATES IN
BENDING. SMALL DEFLECTION THEORY OF ELASTIC AND
ELASTO-PLASTIC PLATES
DEPT. OF STRUCT. MECH., CHALMERS UNIV. OF TECH.,
GOTEBORG, VOL. 71, NO. 4, 1971.

71-0784 HANAI, S.
AOKI, I.
HORITA, T.
KOMORIDA, H.
APPLICATION OF FINITE-ELEMENT METHOD AND MOIRE
TECHNIQUE TO SHEET-METAL FORMING PROBLEMS
PROC. INT. CONF. SCI. TECHNOL. IRON STEEL, TOKYO,
VOL. 11, 1971, PP. 894-896.

71-0785 ANON
FINITE ELEMENTS FOR DETERMINATION OF CRACK-TIP
ELASTIC STRESS INTENSITY FACTORS
ENG. FRACT. MECH., VOL. 3, NO. 3, OCTOBER 1971,
PP. 255-265.

71-0786 NAGAMATSU, A.
MUROTA, T.
NON-UNIFORM DEFORMATION OF A BLOCK IN PLANE-STRAIN
COMPRESSION CAUSED BY FRICTION. PART 4, ANALYSIS
BY FINITE ELEMENT METHOD IN GENERAL CONDITIONS
J.S.M.E. BULL., VOL. 14, NO. 70, 1971, PP.
322-330.

71-0787 BAKER, A. J.
A FINITE ELEMENT COMPUTATIONAL THEORY FOR THE
MECHANICS AND THERMODYNAMICS OF A VISCOUS,
COMPRESSIBLE MULTI-SPECIES FLUID
BELL AEROSPACE, RESEARCH REPORT, 9500-950200,
1971.

71-0788 YAMAMOTO, Y.
TOKUDA, N.
STRESS INTENSITY FACTORS IN PLATE STRUCTURES
CALCULATED BY THE FINITE ELEMENT METHOD (IN
JAPANESE)
J. SOC. NAVAL ARCH., JAPAN, VOL. 130, 1971, PP.
219-233.

71-0789 ABDELRAOUF, M. P. S.
FINITE ELEMENT ANALYSIS OF SHELL-TYPE STRUCTURES
PH.D., THE UNIV. OF TEXAS, AUSTIN, 1971.

71-0790 KAWAHARA, M.
HORII, K.
LARGE STRAIN, ELASTO-PLASTIC NUMERICAL ANALYSIS BY
MEANS OF FINITE ELEMENT METHOD
PROC. OF J. S. C. E., NO. 194, 1971, PP. 163-172.

71-0791 BAKER, A. J.
NUMERICAL SOLUTION TO THE DYNAMICS OF VISCOUS
FLUID FLOW BY A FINITE ELEMENT ALGORITHM: A FIRST
STEP TOWARDS COMPUTATIONAL CONTINUUM MECHANICS
PROC. INTL. ASSOC. FOR SHELL STRUCT., SYMP. ON
HYDROMECHANICALLY LOADED SHELLS, 1971.

71-0792 CASCIARO, R.
DI CARLO, A.
VALENTE, G.
DISCUSSION ON "PLANE STRESS LIMIT ANALYSIS BY
FINITE ELEMENTS"
PROC. A.S.C.E., J. ENG. MECH. DIV., VOL. 97, NO.
EM5, 1971, T. BELYTSCHKO, P. G. HODGE, JR.,
(EDS.), DECEMBER 1970.

71-0793 ANON
CALCULATION OF ELASTIC-PLASTIC DEFORMATION BY THE
FINITE ELEMENT METHOD OF NUMERICAL ANALYSIS
BERG HUTTENMANN MONATSH (GERMAN) VOL. 116, NO. 1,
JANUARY 1971, PP. 1-6.

71-0794 LEE, C. H.
ANALYSES OF AXYSYMMETRIC UPSETTING AND
PLANE-STRAIN SIDE-PRESSING OF SOLID CYLINDERS BY
THE FINITE ELEMENT METHOD
J. ENG. IND., TRANS. A.S.M.E., VOL. 93, NO. 2, MAY
1971, PP. 445-454.

71-0795 WILSON, I. H.
CRUCIFORM SPECIMENS FOR BIAXIAL FATIGUE TESTS - AN
INVESTIGATION USING FINITE-ELEMENT ANALYSIS AND
PHOTO-ELASTIC-COATING TECHNIQUES
J. STRAIN ANALYSIS, VOL. 6, NO. 1, JANUARY 1971,
PP. 27-37.

71-0796 FERRELL, C. S.
DYNAMIC BEHAVIOR OF ECCENTRICALLY STIFFENED PLATES
UNIV. MISSOURI, ROLLA, MO., 1971, 129 PP.

71-0797 ZANGHIRELLA, E.
PWR DEVELOPMENT FOR NUCLEAR PROPULSION: VIBRATION
ANALYSIS OF NUCLEAR POWER PLANT
CONF. NUCL. SHIPS, HAMBURG, GERMANY, CONF-710532,
VOL. 1, MAY 10, 1971, PP. 397-414.

71-0798 ANDERSON, P. V.
FUEL ROD FINITE ELEMENT STRESS ANALYSIS:
SEGMENTED FLIGHT REACTOR FUEL VERSUS S8DR FUEL
ATOMICS INT. DIV., CANOGA PARK, CALIF., MARCH 30,
1971, 42 PP. (N.T.I.S. - T1-759-240-035)

71-0799 DESALVO, G. J.
STRESS INTENSITY FACTORS FOR HEXAGONAL FRACTURE
TOUGHNESS SPECIMENS
ASTRONUCL. LAB., WESTINGHOUSE ELEC. CORP.,
PITTSBURGH, PA., APRIL 1971, 29 PP. (N.T.I.S. -
WANL-TME-2789)

71-0800 REMSON, I.
HORNBERGER, G. M.
MOLZ, F. J.
NUMERICAL METHODS IN SUBSURFACE HYDROLOGY: WITH AN
INTRODUCTION TO THE FINITE ELEMENT METHOD
WILEY-INTERSCI., NEW YORK, 1971, 389 PP.

71-0801 PERUMPRAL, J. V.
LILJEDAHL, J. B.
PERLOFF, W. H.
THE FINITE ELEMENT METHOD FOR PREDICTING STRESS
DISTRIBUTION AND SOIL DEFORMATION UNDER A TRACTIVE
DEVICE
TRANS. A.S.A.E., VOL. 14, NO. 6, NOVEMBER-DECEMBER
1971, PP. 1184-1188.

71-0802 AROCKIASAMY, M.
LIMIT ANALYSIS OF REINFORCED CONCRETE SLABS BY
FINITE ELEMENT METHOD
PH.D., THE UNIV. OF WISCONSIN, 1971.

71-0803 BASA, J. E.
STATIC ANALYSIS OF STIFFENED SHELLS BY THE FINITE
ELEMENT METHOD
PH.D., THE UNIV. OF WISCONSIN, 1971.

71-0804 CHOPRA, P. S.
FINITE ELEMENT INVESTIGATION OF SECONDARY CRACKING
DURING BRITTLE FRACTURE
PH.D., UNIV. OF WASHINGTON, 1971.

71-0805 CHUGH, Y. P.
FINITE ELEMENT STUDY OF PRESSURIZED CAVITIES IN
GEOLOGIC MEDIA
PH.D., THE PENNSYLVANIA STATE UNIV., 1971.

71-0806 CHAN, S. T. K.
FINITE ELEMENT ANALYSIS OF IRROTATIONAL FLOWS OF
AN IDEAL FLUID
PH.D., UNIV. OF CALIF., DAVIS, 1971.

71-0807 LUKAS, L.
STRESS SEPARATION IN PHOTOELASTICIMETRIC
MEASUREMENT BY THE FINITE ELEMENT METHOD
PROC. 4TH CONF. ON DIMENSION AND STRENGTH
CALCULATIONS, BUDAPEST, 1971.

71-0808 ECER, A.
FINITE ELEMENT ANALYSIS OF POST-BUCKLING BEHAVIOR
PH.D., UNIV. OF NOTRE DAME, 1971.

71-0809 EMERY, J. J.
FINITE ELEMENT ANALYSIS OF CREEP PROBLEMS IN SOIL
MECHANICS
PH.D., THE UNIV. OF BRITISH COLUMBIA, CANADA,
1971.

71-0810 FRANQUES, J. T. JR.
A FINITE ELEMENT MODEL FOR TWO-DIMENSIONAL STEADY
FLOW THROUGH CONTRACTIONS IN NATURAL CHANNELS
PH.D. THESIS, LOUISIANA STATE UNIV. AND AGRIC. AND
MECH. COLLEGE, 1971.

71-0811 GIBSON, W. C.
FESTRAN: A FINITE ELEMENT COMPUTER PROGRAM
PH.D., CASE WESTERN RESERVE UNIV., 1971.

71-0812 GORDON, J. L.
A FINITE ELEMENT ANALYSIS OF THE PLANE STRAIN
SHEET DRAWING PROBLEM
PH.D., CARNEGIE-MELLON UNIV., 1971.

71-0813 HAMMEL, D. J.
AN APPLICATION OF THE FINITE ELEMENT METHOD FOR
ROCK SLOPE STABILITY ANALYSIS
PH.D., THE UNIV. OF ARIZONA, 1971.

71-0814 HUANG, T. S.
CURVED TRIANGULAR FINITE ELEMENTS FOR THE ANALYSIS
OF THICK ELASTIC PLATES AND SHELLS
PH.D., UNIV. OF COLORADO, 1971.

71-0815 IGHARO, P.
ANALYSIS OF PLANE STRAIN CONSOLIDATION IN
SATURATED CLAYS BY FINITE ELEMENT METHOD
PH.D., UNIV. OF PITTSBURGH, 1971.

71-0816 JAIN, R. K.
NAVIGATION LOCK STRUCTURE ANALYSIS BY FINITE
ELEMENT METHOD
PH.D., TEXAS TECH. UNIV., 1971.

71-0817 KIRKLAND, J. L.
FINITE ELEMENT ANALYSIS OF BURIED CYLINDERS
PH.D., MISSISSIPPI STATE UNIV., 1971.

71-0818 KLEIN, S.
THE NONLINEAR DYNAMIC ANALYSIS OF SHELLS OF
REVOLUTION WITH ASYMMETRIC PROPERTIES BY THE
FINITE ELEMENT METHOD
PH.D., UNIV. OF SOUTHERN CALIF., 1971.

71-0819 KROUSKOP, T. A.
A FINITE ELEMENT MODEL FOR PREDICTING LONG BONE
FRACTURE FIXATION
PH.D., CARNEGIE-MELLON UNIV., 1971.

71-0820 LANGLAND, R. T.
PROBABILISTIC THEORY OF STRUCTURES USING FINITE
ELEMENTS
PH.D., UNIV. OF CALIF., DAVIS, 1971.

71-0821 LIEN, S-Y.
FINITE ELEMENT ELASTIC THIN SHELL PRE-BUCKLING AND
POST-BUCKLING ANALYSIS
PH.D., CORNELL UNIV., 1971.

71-0822 LI, C-T.
ANALYSIS OF INFLATABLE SHELLS BY THE FINITE
ELEMENT METHOD
PH.D., STATE UNIV. OF NEW YORK, BUFFALO, 1971.

71-0823 MASON, W. E. JR.
FINITE ELEMENT ANALYSIS OF CYLINDRICAL SHELLS
UTILIZING CURVED ELEMENTS CONTAINING RIGID BODY
DEFORMATIONS
PH.D., UNIV. OF CALIF., DAVIS, 1971.

71-0824 KING, G. J. W.
CHOWDHURY, R. N.
FINITE ELEMENT SOLUTION FOR QUANTITY OF STEADY
SEEPAGE
CIVIL ENG. AND PUBLIC WORKS REV., VOL. 66, 1971,
P. 785.

71-0825 NOORISHAD, J.
SEEPAGE IN FRACTURED ROCK: FINITE ELEMENT ANALYSIS
OF ROCK MASS BEHAVIOR UNDER COUPLED ACTION OF BODY
FORCES, FLOW FORCES, AND EXTERNAL LOADS
PH.D., UNIV. OF CALIF., BERKELEY, 1971.

71-0826 TAYLOR, R. L.
DISCUSSION OF "FINITE ELEMENT BENDING ANALYSIS OF
REISSNER PLATES"
A.S.C.E., J. ENG. MECH. DIV., VOL. 97, C. W. PRYOR
JR., R. M. BARKER AND D. FREDERICK (EDS.), 1971,
PP. 1571-1573.

534

71-0827 SACHS, K.
A FINITE ELEMENT ANALYSIS OF A DOUBLY CURVED
GOTHIC VAULT
PH.D., PRINCETON UNIV., 1971.

71-0828 STETSON, M.
FINITE ELEMENT STUDY OF THE ELASTIC BEHAVIOR OF
PLANE FRAMES WITH FILLER WALLS
PH.D., UNIV. OF ILLINOIS, URBANA, CHAMPAIGN, 1971.

71-0829 TU, R. K-C.
FREE BOUNDARY POTENTIAL FLOW USING FINITE ELEMENTS
PH.D., THE UNIV. OF ARIZONA, 1971.

71-0830 UCCIFERRO, J. J.
ELASTIC NONLINEAR FINITE ELEMENT ANALYSIS OF
PLANAR STRUCTURES
PH.D. THESIS, DREXEL UNIV., 1971.

71-0831 VERNER, E. A.
FINITE ELEMENT STRESS FORMULATION FOR WAVE
PROPAGATION
PH.D., THE UNIV. OF TEXAS, AUSTIN, 1971.

71-0832 WILSON, G. J.
KIRKHOPE, J.
DISCUSSION OF "FINITE ELEMENT BENDING ANALYSIS OF
REISSNER PLATES"
A.S.C.E., J. ENG. MECH. DIV., VOL. 97, C.W. PRYOR
JR., R. M. BARKER AND D. FREDERICK (EDS.), 1971,
PP. 1570-1571.

71-0833 KOVACS, W. D.
SEED, H. B.
IDRISS, I. H.
STUDIES OF SEISMIC RESPONSE OF CLAY BANKS
PROC. A.S.C.E., J. SOIL MECH. FOUNDATIONS DIV.,
VOL. 97, NO. SM2, FEBRUARY 1971, PP. 441-455.

71-0834 SCHOENSTER, J. A.
MEASURED AND CALCULATED VIBRATION PROPERTIES OF
RING-STIFFENED HONEY-COMB CYLINDERS
NASA, LANGLEY, JANUARY 1971, 53 PP.

71-0835 RAO, J. K. S.
PARAMASIVAN, P.
DISCRETE METHODS IN THE ANALYSIS OF PLATE SYSTEMS
J. AERON. SOC., INDIA, VOL. 23, NO. 1, FEBRUARY
1971, PP. 23-36.

71-0836 HOTOS, G. K.
METHODS OF MODELING AND ANALYZING VISCOELASTICALLY
DAMPED STRUCTURES
A.S.M.E., 1971.

71-0837 WILSON, R. R.
BREBBIA, C. A.
DYNAMIC BEHAVIOR OF STEEL FOUNDATIONS FOR
TURBO-ALTERNATORS
J. SOUND AND VIB., VOL. 18, NO. 3, OCTOBER 8,
1971, PP. 405-416.

71-0838 JONES, C. E.
ADVANCED SUBSTRUCTURING TECHNIQUES
NASA-CR-119927, JUNE 1971, 70 PP.

71-0839 LAKIS, A. A.
PAIDOUSSIS, M. P.
FREE VIBRATION OF CYLINDRICAL SHELLS PARTIALLY
FILLED WITH LIQUID
J. SOUND AND VIB., VOL. 19, NO. 1, NOVEMBER 8,
1971, PP. 1-15.

71-0840 GEORGOPOULOS, G.
ROYSTER, L. H.
DEVELOPMENT OF A FINITE ELEMENT MODEL FOR THE
CLASS 5 FLEXTENSIONAL UNDERWATER TRANSDUCER SHELL
MECH. AEROSPACE ENGR., N.C. STATE UNIV., RALEIGH,
N.C., JULY 1971, 23 PP.

71-0841 PRABHU, K. S.
AN IMPROVED MATRIX DISPLACEMENT METHOD FOR
VIBRATION OF NONUNIFORM BEAMS
PROC. 23RD MEET. AERONAUT. SOC., INDIA INST.
TECH., KANPUR, INDIA, FEBRUARY 26-28, 1971.

71-0842 WOOLEY, G. R.
CARVER, D. R.
STRESS CONCENTRATION FACTORS FOR BONDED LAP JOINTS
J. AIRCRAFT, VOL. 8, 1971, PP. 817-820.

71-0843 WOLDEMARIAM, Y.
DISTRIBUTION OF STRESS AROUND A TUNNEL CAVITY,
USING THE FINITE ELEMENT METHOD
PH.D., OKLAHOMA STATE UNIV., 1971.

71-0844 NEUMAN, S. P.
WITHERSPOON, P. A.
VARIATIONAL PRINCIPLES FOR FLUID FLOW IN POROUS
MEDIA
J. ENG. MECH. DIV., A.S.C.E., VOL. 97, 1971, PP.
359-374.

72-0001 LEONARD, J. W.
FINITE ELEMENT ANALYSIS OF PERTURBED COMPRESSIBLE
FLOW
INT. J. NUM. METHODS ENG. 4, NO. 1, 1972, PP.
123-132.

72-0002 SIMPSON, B.
THE USES OF FINITE ELEMENT TECHNIQUES IN SOIL
MECHANICS WITH PARTICULAR REFERENCE TO DEFORMATION
PLAIN STRAIN
PH.D. THESIS, ENG. DEPT., CAMBRIDGE UNIV., 1972.

72-0003 ALEWINE, R. W.
JUNGELS, P.
STUDY OF 1964 ALASKAN EARTHQUAKE USING FINITE
ELEMENT AND GENERALIZED INVERSION TECHNIQUES
TRANS. AMER. GEOPHYS. UNION 53, NO. 11, 1972, P.
1119.

72-0004 ARGYRIS, J. H.
MARECZEK, G.
POTENTIAL FLOW ANALYSIS BY FINITE ELEMENTS
INGENIEUR-ARCHIV 42, NO. 1, 1972, PP. 1-25.

72-0005 BEAUMONT, C.
LAMBERT, A.
CRUSTAL STRUCTURE FROM SURFACE LOAD TILTS USING A
FINITE ELEMENT MODEL
GEOPHYS. J. ROY. ASTRONOM. SOC. 29, NO. 2, 1972,
PP. 203-226

72-0006 BOOKER, J. F.
HUEBNER, K. H.
APPLICATION OF FINITE ELEMENT METHODS TO
LUBRICATION - ENGINEERING APPROACH
J. LUB. TECH. 94, NO. 4, 1972, PP. 313-323.

72-0007 BRISBANE, J. J.
ROHM AND HAAS THERMOSTRUCTURAL ANALYSIS SYSTEM.
THERMAL STRUCTURAL ANALYSIS PROGRAMS - A SURVEY
AND EVALUATION
ASME, NEW YORK, 1972, PP. 77-82.

72-0008 CHENG, R. T.
NUMERICAL SOLUTION OF NAVIER-STOKES EQUATIONS BY
FINITE ELEMENT METHOD
PHYS. FLUIDS 15, NO. 12, 1972, PP. 2098-2105.

72-0009 CHOWDHURY, P. C.
FLUID FINITE ELEMENTS FOR ADDED-MASS CALCULATIONS
INT. SHIPBUILD. PROG. 19, NO. 217, SEPTEMBER 1972,
PP. 303-309.

72-0010 CRAGGS, A.
USE OF SIMPLE THREE-DIMENSIONAL ACOUSTIC FINITE
ELEMENTS FOR DETERMINING NATURAL MODES AND
FREQUENCIES OF COMPLEX SHAPED ENCLOSURES
J. SOUND VIB. 23, NO. 3, 1972, PP. 331-339.

72-0011 CRASTAN, V.
DEVOS, J. E.
FINITE ELEMENT SOLUTION OF THE THREE DIMENSIONAL
FLOW PROBLEMS AND OF REYNOLD'S EQUATION FOR
INCOMPRESSIBLE FLUIDS
NUCL. ENG. DESIGN, VOL. 22, NO. 2, 1972, PP.
225-232

72-0012 CUNNINGHAM, A. M.
UNSTEADY SUBSONIC COLLOCATION METHOD FOR WINGS
WITH AND WITHOUT CONTROL SURFACES
J. AIRCR. 9, NO. 6, JUNE 1972, PP. 413-419.

72-0013 DAVIDS, N.
CHENG, R. C.
TRANSIENT LAMINAR FLOW IN DUCTS OF ARBITRARY
CROSS-SECTION BY FINITE ELEMENT METHODS
J. BIOMECH. 5, NO. 5, 1972, PP. 485-499.

72-0014 DESAI, C. S.
SEEPAGE ANALYSIS OF EARTH BANKS UNDER DRAWDOWN
J. SOIL MECH. FOUNDATIONS DIV., A.S.C.E. 98, NO.
SM-11, PAPER 9334, NOVEMBER 1972, PP. 1143-1162.

72-0015 DHIR, S. K.
SIKORA, J. P.
IMPROVED METHOD FOR OBTAINING THE
GENERAL-DISPLACEMENT FIELD FROM A HOLOGRAPHIC
INTERFEROGRAM
EXP. MECH. 12, NO. 7, JULY 1972, PP. 323-327.

72-0016 DRAKE, L. A.
RAYLEIGH WAVES AT A CONTINENTAL BOUNDARY BY FINITE
ELEMENT METHOD
BULL. SEISMOL. SOC. AMER. 62, NO. 5, 1972, PP.
1259-1268.

72-0017 GHABOUSSI, J.
WILSON, E. L.
VARIATIONAL FORMULATION OF DYNAMICS OF
FLUID-SATURATED POROUS ELASTIC SOLIDS
J. ENG. MECH. DIV., A.S.C.E. 98, NO. EM-4, PAPER
9152, AUGUST 1972, PP. 947-963.

72-0018 GUYMON, G. L.
FINITE ELEMENT SOLUTION OF DIFFUSION-CONVECTION
EQUATION
WATER RESOURCES RES. 8, NO. 5, 1972, PP.
1357-1360.

72-0019 HURST, C. J.
OLSON, D. R.
CONDUCTION THROUGH DROPLETS DURING DROPWISE
CONDENSATION
PAPER 72-HT-50, ASME MEETING, BLACKSBURG,
VIRGINIA, AUGUST 6-9, 1972.

72-0020 IWATA, K.
OSAKADA, K.
FUJINO, S.
ANALYSIS OF HYDROSTATIC EXTRUSION BY THE FINITE
ELEMENT METHOD
J. ENG. IND., TRANS. ASME 94, SER. B, NO. 2, PAPER
71-PROD-C, MAY 1972, PP. 697-703.

72-0021 SOLTIS, L. A.
CHRISTIANO, P.
FINITE DEFORMATION OF BIAXIALLY LOADED COLUMNS
J. STRUCTURAL DIV., ASCE., VOL. 98, NO. ST12,
PROC. PAPER 9407, 1972, PP. 2647-2661.

72-0022 LIKINS, P. W.
FINITE ELEMENT APPENDAGE EQUATIONS FOR HYBRID
COORDINATE DYNAMIC ANALYSIS
INT. J. SOLIDS STRUCT. 8, NO. 5, MAY 1972, PP.
709-731.

72-0023 LOZIUK, L. A.
ANDERSON, J. C.
HYDROTHERMAL ANALYSIS BY FINITE ELEMENT METHOD
J. HYD. DIV., A.S.C.E. 98, NO. HY-11, PAPER 9372,
NOVEMBER 1972, PP. 1983-1998.

72-0024 LUNG, M.
USE OF METHOD OF FINITE ELEMENTS FOR STUDY OF
STATIONARY METAL FORMING TECHNIQUES
Z. ANGEW. MATH. MECH., VOL. 52, NO. 4, 1972, PP.
63-65.

72-0025 MCDONALD, B. H.
WEXLER, A.
FINITE-ELEMENT SOLUTION OF UNBOUNDED FIELD
PROBLEMS
IEEE TRANS. MICROWAVE THEORY TECH. MT-20, NO. 12,
1972, PP. 841-847.

72-0026 MATTHEWS, F. L.
WEST, J. B.
FINITE ELEMENT DISPLACEMENT ANALYSIS OF A LUNG
J. BIOMECH. 15, NO. 6, 1972, PP. 591-600.

72-0027 MILLER, W. F.
LEWIS, E. E.
ROSSOW, E. C.
TWO-DIMENSIONAL TRANSPORT CALCULATIONS USING
PHASE-SPACE FINITE ELEMENTS
TRANS. AMER. NUCL. SOC. 15, NO. 2, 1972, P. 780.

72-0028 HOEG, K.
FINITE ELEMENT ANALYSIS OF STRAIN-SOFTENING CLAY
ASCE J. SOIL MECH. FOUND. DIV., VOL. 98, NO. SM1,
JANUARY 1972, PAPER 8650, PP. 43-58.

72-0029 MUKHERJI, B.
CRACK PROPAGATION IN SEA ICE - FINITE ELEMENT
APPROACH
TRANS. AMER. GEOPHYS. UNION 53, NO. 11, 1972, P.
1009.

72-0030 NALLUSWAMI, M.
LONGENBAUGH, R. A.
SUNADA, D. K.
FINITE ELEMENT METHOD FOR HYDRODYNAMIC DISPERISON
EQUATION WITH MIXED PARTIAL DERIVATIVES
WATER RESOURCES RES., VOL. 8, NO. 5, 1972, PP.
1247-1250.

72-0031 NELSON, J. M.
COOK, W. A.
STIBOR, G. S.
THREE-DIMENSIONAL PHOTOELASTIC AND FINITE-ELEMENT
ANALYSIS OF A PROPELLANT GRAIN
EXP. MECH. 12, NO. 9, 1972, PP. 436-440.

72-0032 ODEN, J. T.
WELLFORD, L. C.
ANALYSIS OF FLOW OF VISCOUS FLUIDS BY
FINITE-ELEMENT METHOD
AIAA J. 10, NO. 12, 1972, PP. 1590-1599.

72-0033 OH, K. P.
HUEBNER, K. H.
SOLUTION OF THE ELASTOHYDRODYNAMIC FINITE JOURNAL
BEARING PROBLEM
PAPER 72-LUB-26, ASME MEETING, WARREN, MICHIGAN,
OCTOBER 9-12, 1972.

72-0034 OWEN, D. R. J.
LYNESS, J. F.
INVESTIGATION OF BOND FAILURE IN FIBRE-REINFORCED
MATERIALS BY THE FINITE ELEMENT METHOD
FIBRE SCI. TECHNOL. 5, NO. 2, APRIL 1972, PP.
129-141.

72-0035 PALIT, K.
FENNER, R. T.
FINITE ELEMENT ANALYSIS OF TWO-DIMENSIONAL SLOW
NON-NEWTONIAN FLOWS
AICHE J. 18, NO. 6, 1972, PP. 1163-1170.

72-0036 PONTER, A. R. S.
APPLICATION OF DUAL MINIMUM THEOREMS TO THE FINITE
ELEMENT SOLUTION OF POTENTIAL PROBLEMS WITH
SPECIAL REFERENCE TO SEEPAGE
INT. J. NUM. METHODS ENG. 4, NO. 1,
JANUARY-FEBRUARY 1972, PP. 85-93.

72-0037 REINHOUDT, J. P.
ON THE STABILITY OF ROTOR-AND-BEARING SYSTEMS AND
ON THE CALCULATION OF SLIDING BEARINGS
PHILIPS RES. REP., SUPP. 1, 1972, 152 PP.

72-0038 RICHARDS, D. J.
WEXLER, A.
FINITE-ELEMENT SOLUTIONS WITHIN CURVED BOUNDARIES
IEEE TRANS. MICROWAVE THEORY TECH. MTT-20, NO. 10,
OCTOBER 1972, PP. 650-657.

72-0039 SEMENZA, L. A.
LEWIS, E. E.
ROSSOW, E. C.
FINITE ELEMENT NEUTRON DIFFUSION IN CURVILINEAR
COORDINATES
TRANS. AMER. NUCL. SOC. 15, NO. 2, 1972, P. 780.

72-0040 STONE, G. O.
COUPLING MATRICES FOR HIGH ORDER FINITE ELEMENT
ANALYSIS OF ACOUSTIC - WAVE PROPAGATION
ELECTRON LETT. VOL. 8, NO. 18, SEPTEMBER 7, 1972,
PP. 466-468.

72-0041 TAKEDA, T.
SHIMOMURA, Y.
OHTA, M.
YOSHIKAWA, M.
NUMERICAL ANALYSIS OF MAGNETOHYDRODYNAMIC
INSTABILITIES BY FINITE ELEMENT METHOD
PHYS. FLUIDS 15, NO. 12, 1972, PP. 2193-2201.

72-0042 TAYLOR, C.
O'CALLAGHAN, J. F.
NUMERICAL SOLUTION OF THE ELASTOHYDRODYNAMIC
LUBRICATION PROBLEM USING FINITE ELEMENTS
J. MECH. ENG. SCI. 14, NO. 4, AUGUST 1972, PP.
229-237.

72-0043 TEUSCHER, L. H.
ENGLAND, W. G.
HAUSER, L. E.
CALCULATION OF TEMPERATURE DISTRIBUTION WITHIN A
MELTING-FREEZING MATERIAL USING FINITE ELEMENT
TECHNIQUES
TRANS. AMER. GEOPHYS. UNION 53, NO. 11, 1972, P.
1020.

72-0044 TONG, P.
VAWTER, D.
ANALYSIS OF PERISTALTIC PUMPING
J. APPL. MECH., TRANS. A.S.M.E. 39, SER. E., NO.
4, PAPER 72-APM-19, DECEMBER 1972, PP. 857-862.

72-0045 VALATHUR, M.
DOVE, R. C.
ALBRECHT, B.
WAVE PROPAGATION IN A THIN HOLLOW CONE BY A FINITE
ELEMENT METHOD
J. SOUND VIB. 24, NO. 2, 1972, PP. 211-218.

72-0046 VERRUIJT, A.
SOLUTION OF TRANSIENT GROUNDWATER FLOW PROBLEMS BY
FINITE ELEMENT METHOD
WATER RESOURCES RES. 8, NO. 3, 1972, PP. 725-727.

72-0047 WADE, L. V.
KROKOSKY, E. M.
FINITE ELEMENT PREDICTION OF FIREPROOFING
EFFECTIVENESS
J. MATER. 7, NO. 4, 1972, PP. 496-500.

72-0048 ZIENKIEWICZ, O. C.
CORMEAU, I. C.
VISCO-PLASTICITY SOLUTION BY FINITE ELEMENT
PROCESS
ARCHIVES OF MECHANICS 24, NO. 5-6, 1972, PP.
873-889.

72-0049 ZULL, L.
CRAWFORD, P. B.
FINITE ELEMENT METHOD IN PETROLEUM RESERVOIR
ANALYSIS
TEXAS J. SCI. 24, NO. 3, 1972, P. 382.

72-0050 AKIN, J. E.
A FINITE ELEMENT COLLOCATION SOLUTION OF
DIFFERENTIAL EQUATIONS
INTERN. CONF. VARIATIONAL METHODS IN ENGINEERING,
SOUTHAMPTON UNIV., ENGLAND, SEPTEMBER 1972.

72-0051 SCORDELIS, A. C.
FINITE ELEMENT ANALYSIS OF REINFORCED CONCRETE
STRUCTURES
PROC. OF CONF. ON FINITE ELEMENT METH. IN CIVIL
ENG., MCGILL UNIV., MONTREAL, (EDS.) MCCUTCHEON,
J.O., MIRZA, M.S., MUFTI, A.A., JUNE 1-2, 1972,
PP. 71-114.

72-0052 AKIN, J. E.
SEN GUPTA, S. R.
FINITE ELEMENT APPLICATIONS OF THE LEAST SQUARES
METHOD
INTERN. CONF. VARIATIONAL METHODS IN ENGINEERING,
SOUTHAMPTON UNIV., ENGLAND, SEPTEMBER, 1972.

72-0053 ZIENKIEWICZ, O. C.
FINITE ELEMENTS - THE BACKGROUND STORY
THE MATHEMATICS OF FINITE ELEMENTS AND
APPLICATIONS, BRUNEL UNIVERSITY, ENGLAND, APRIL
18-20, 1972, PAPER NO. 1.

72-0054 MITCHELL, A. R.
INTRODUCTION TO THE MATHEMATICS OF THE FINITE
ELEMENT METHOD
THE MATHEMATICS OF FINITE ELEMENTS AND
APPLICATIONS, BRUNEL UNIVERSITY, ENGLAND, APRIL
18-20, 1972, PAPER NO.2.

72-0055 NEVILLE, A. M.
GHALI, A.
STRUCTURAL ANALYSIS: A UNIFIED CLASSICAL AND
MATRIX APPROACH
INTEXT PUBLS., SCRANTON, PENN., CHAPTERS 20 AND
24, 1972.

72-0056 BARNHILL, R. E.
WHITEMAN, J. R.
FUNCTIONAL ANALYSIS AND THE FINITE ELEMENT METHOD
APPLIED TO ELLIPTIC BOUNDARY VALUE PROBLEMS
THE MATHEMATICS OF FINITE ELEMENTS AND
APPLICATIONS, BRUNEL UNIVERSITY, ENGLAND, APRIL
18-20, 1972, PAPER NO. 4.

72-0057 CIARLET, P. G.
ORDERS OF CONVERGENCE IN FINITE ELEMENT METHODS
THE MATHEMATICS OF FINITE ELEMENTS AND
APPLICATIONS, BRUNEL UNIVERSITY, ENGLAND, APRIL
18-20, 1972, PAPER NO. 5.

72-0058 HOPPE, V.
FINITE ELEMENTS WITH HARMONIC INTERPOLATION
FUNCTIONS
THE MATHEMATICS OF FINITE ELEMENTS AND
APPLICATIONS, BRUNEL UNIVERSITY, ENGLAND, APRIL
18-20, 1972, PAPER NO. 6.

72-0059 NOBLE, B.
FINITE ELEMENT METHODS FOR INITIAL VALUE PROBLEMS
THE MATHEMATICS OF FINITE ELEMENTS AND
APPLICATIONS, BRUNEL UNIVERSITY, ENGLAND, APRIL
18-20, 1972, PAPER NO. 7.

72-0060 AKIN, J. E.
A LEAST SQUARES FINITE ELEMENT SOLUTION OF
NONLINEAR OPERATORS
THE MATHEMATICS OF FINITE ELEMENTS AND
APPLICATIONS, BRUNEL UNIVERSITY, ENGLAND, APRIL
18-20, 1972, PAPER NO. 8.

72-0061 FRIED, I.
THE L2 AND LOO CONDITION NUMBERS OF THE FINITE
ELEMENT MATRICES AND THE POINTWISE CONVERGENCE OF
THE METHOD
THE MATHEMATICS OF FINITE ELEMENTS AND
APPLICATIONS, BRUNEL UNIVERSITY, ENGLAND, APRIL
18-20, 1972, PAPER NO. 9.

72-0062 FREMOND, M.
DUAL FORMULATION FOR POTENTIAL ENERGY AND
COMPLEMENTARY ENERGY APPLICATION TO THE FINITE
ELEMENT METHOD
THE MATHEMATICS OF FINITE ELEMENTS AND
APPLICATIONS, BRUNEL UNIVERSITY, ENGLAND, APRIL
18-20, 1972, PAPER NO. 10.

72-0063 CZENDES, Z. J.
GOPINATH, A.
SILVER, R.
ON LOCAL APPROXIMATIONS OF OPERATOR EQUATIONS
THE MATHEMATICS OF FINITE ELEMENTS AND
APPLICATIONS, BRUNEL UNIVERSITY, ENGLAND, APRIL
18-20, 1972, PAPER NO. 11.

72-0064 BRUCH, J. C.
ZYVOLOSKI, G.
FINITE ELEMENT SOLUTION OF UNSTEADY AND
UNSATURATED FLOW IN POROUS MEDIA
THE MATHEMATICS OF FINITE ELEMENTS AND
APPLICATIONS, BRUNEL UNIVERSITY, ENGLAND, APRIL
18-20, 1972, PAPER NO. 12.

72-0065 PATTERSON, C.
SUFFICIENT CONDITIONS FOR CONVERGENCE IN THE
FINITE ELEMENT METHOD FOR ANY SOLUTION OF FINITE
ENERGY
THE MATHEMATICS OF FINITE ELEMENTS AND
APPLICATIONS, BRUNEL UNIVERSITY, ENGLAND, APRIL
18-20, 1972, PAPER NO. 13.

72-0066 HEISE, U.
COMBINED APPLICATION OF FINITE ELEMENT METHODS AND
RICHARDSON EXTRAPOLATION TO THE TORSION PROBLEM
THE MATHEMATICS OF FINITE ELEMENTS AND
APPLICATIONS, BRUNEL UNIVERSITY, ENGLAND, APRIL
18-20, 1972, PAPER NO. 14.

72-0067 ODEN, J. T.
FINITE ELEMENT APPLICATIONS IN MATHEMATICAL
PHYSICS
MATHEMATICS OF FINITE ELEMENTS WITH APPLICATIONS,
(ED. BY J.R. WHITEMAN), ACADEMIC PRESS, NEW
YORK-LONDON, 1973, PP. 239-282.

72-0068 CAMERON, I. G.
A DYNAMIC ELASTOPLASTIC ANALYSIS OF THIN SHELLS OF
REVOLUTION UNDER ASYMMETRIC MECHANICAL AND THERMAL
LOADINGS
THE MATHEMATICS OF FINITE ELEMENTS AND
APPLICATIONS, BRUNEL UNIVERSITY, ENGLAND, APRIL
18-20, 1972, PAPER NO. 16.

72-0069 HENSHELL, R. D.
ON HYBRID FINITE ELEMENTS
THE MATHEMATICS OF FINITE ELEMENTS AND
APPLICATIONS, BRUNEL UNIVERSITY, ENGLAND, APRIL
18-20, 1972, PAPER NO. 17.

72-0070 TAYLOR, C.
FRANCE, P. N.
ZIENKIEWICZ, O. C.
ON SOME FREE SURFACE ON TRANSIENT FLOW PROBLEMS OF
SEEPAGE AND IRROTATIONAL FLOW
THE MATHEMATICS OF FINITE ELEMENTS AND
APPLICATIONS, BRUNEL UNIVERSITY, ENGLAND, APRIL
18-20, 1972, PAPER NO. 18.

72-0071 BLACKBURN, W. S.
CALCULATION OF STRESS INTENSITY FACTORS AT CRACK
TIPS USING SPECIAL FINITE ELEMENTS
THE MATHEMATICS OF FINITE ELEMENTS AND
APPLICATIONS, BRUNEL UNIVERSITY, ENGLAND, APRIL
18-20, 1972, PAPER NO. 19.

72-0072 DALY, P.
SINGULARITIES IN ELECTROMAGNETIC TRANSMISSION
LINES
THE MATHEMATICS OF FINITE ELEMENTS AND
APPLICATIONS, BRUNEL UNIVERSITY, ENGLAND, APRIL
18-20, 1972, PAPER NO. 20.

72-0073 SCHMIDT, G. H.
APPLICATION OF THE FINITE ELEMENT METHOD TO THE
EXTENSIONAL VIBRATIONS OF PIEZOELECTRIC PLATES
THE MATHEMATICS OF FINITE ELEMENTS AND
APPLICATIONS, BRUNEL UNIVERSITY, ENGLAND, APRIL
18-20, 1972, PAPER NO. 21.

72-0074 PETYT, M.
FLEISCHER, C. C.
VIBRATION OF CURVED STRUCTURES USING QUADRILATERAL
FINITE ELEMENTS
THE MATHEMATICS OF FINITE ELEMENTS AND
APPLICATIONS, BRUNEL UNIVERSITY, ENGLAND, APRIL
18-20, 1972, PAPER NO. 22.

72-0075 KRISHNA MURTY, A. V.
VENKATESWARA RAO, G.
ASSESSMENT OF ACCURACIES OF FINITE ELEMENT
EIGENVALUES
THE MATHEMATICS OF FINITE ELEMENTS AND
APPLICATIONS, BRUNEL UNIVERSITY, ENGLAND, APRIL
18-20, 1972, PAPER NO. 23.

72-0076 BOLLAND, G. B.
FLEXIBILITY AND STIFFNESS MATRICES FOR AN
OPEN-TUBE WARPING CONSTRAINT FINITE ELEMENT
THE MATHEMATICS OF FINITE ELEMENTS AND
APPLICATIONS, BRUNEL UNIVERSITY, ENGLAND, APRIL
18-20, 1972, PAPER NO. 24.

72-0077 BARNARD, A. J.
A SANDWICH PLATE FINITE ELEMENT
THE MATHEMATICS OF FINITE ELEMENTS AND
APPLICATIONS, BRUNEL UNIVERSITY, ENGLAND, APRIL
18-20, 1972, PAPER NO. 25.

72-0078 ALUJEVIC, A.
HEAD, J. L.
APPLICATION OF THREE-DIMENSIONAL FINITE ELEMENTS
FOR ANALYSIS OF IRRADIATION INDUCED STRESSES IN
NUCLEAR REACTOR FUEL ELEMENTS
THE MATHEMATICS OF FINITE ELEMENTS AND
APPLICATIONS, BRUNEL UNIVERSITY, ENGLAND, APRIL
18-20, 1972, PAPER NO. 26.

72-0079 EVANS, D. J.
THE ANALYSIS AND APPLICATION OF SPARSE MATRIX
ALGORITHMS IN THE FINITE ELEMENT METHOD
THE MATHEMATICS OF FINITE ELEMENTS AND
APPLICATIONS, BRUNEL UNIVERSITY, ENGLAND, APRIL
18-20, 1972, PAPER NO. 27.

72-0080 JEZERNIK, A.
LEECH, A.
THE COMPARISON OF ITERATIVE AND DIRECT SOLUTION
TECHNIQUES IN THE ANALYSIS OF TIME-DEPENDENT
STRESS PROBLEMS, INCLUDING CREEP, BY THE FINITE
ELEMENT METHOD
THE MATHEMATICS OF FINITE ELEMENTS AND
APPLICATIONS, BRUNEL UNIVERSITY, ENGLAND, APRIL
18-20, 1972, PAPER NO. 28.

72-0081 TINGLEFF, O.
A METHOD FOR SIMULATING CONCENTRATED FORCES AND
LOCAL REINFORCEMENTS IN STRESS COMPUTATIONS
THE MATHEMATICS OF FINITE ELEMENTS AND
APPLICATIONS, BRUNEL UNIVERSITY, ENGLAND, APRIL
18-20, 1972, PAPER NO. 29.

72-0082 SEITELMAN, L. H.
SOME PRACTICAL SOLUTION TECHNIQUES FOR FINITE
ELEMENT ANALYSIS
THE MATHEMATICS OF FINITE ELEMENTS AND
APPLICATIONS, BRUNEL UNIVERSITY, ENGLAND, APRIL
18-20, 1972, PAPER NO. 30.

72-0083 LOCK, A. C.
SABIR, A. B.
ALGORITHM FOR THE LARGE DEFLECTION GEOMETRICALLY
NON-LINEAR BEHAVIOUR OF PLANE AND CURVED
STRUCTURES
THE MATHEMATICS OF FINITE ELEMENTS AND
APPLICATIONS, BRUNEL UNIVERSITY, ENGLAND, APRIL
18-20, 1972, PAPER NO. 31.

72-0084 CARPENTER, W. C.
GILL, P. A. T.
AUTOMATED SOLUTION OF TIME DEPENDENT PROBLEMS
THE MATHEMATICS OF FINITE ELEMENTS AND
APPLICATIONS, BRUNEL UNIVERSITY, ENGLAND, APRIL
18-20, 1972, PAPER NO. 32.

72-0085 ODEN, J. T.
REDDY, J. N.
MIXED CONJUGATE FINITE ELEMENT APPROXIMATIONS OF
LINEAR OPERATORS
INT. J. STRUCT. MECH., VOL. 1, NO. 1, 1972.

72-0086 KOBAYASHI, A. S.
WADE, B. G.
PHOTOELASTIC INVESTIGATION ON THE CRACK-ARREST
CAPABILITY OF A HOLE
EXP. MECH., VOL. 12, NO. 1, JANUARY 1972, PP.
32-37.

72-0087 MEHROTRA, B. L.
GOVIL, A. K.
SHEAR LAG ANALYSIS OF RECTANGULAR FULL-WIDTH TUBE
JUNCTIONS
A.S.C.E. J. STRUCT. DIV., VOL. 98, NO. ST1,
JANUARY 1972, PAAPER 8665, PP. 287-305.

72-0088 KAVANAGH, K. T.
EXTENSION OF CLASSICAL EXPERIMENTAL TECHNIQUES FOR
CHARACTERIZING COMPOSITE - MATERIAL BEHAVIOUR
EXP. MECH., VOL. 12, NO. 1, JANUARY 1972, PP.
50-56.

72-0089 HAY, J. K.
BLEW, J. M.
DYNAMIC TESTING AND COMPUTER ANALYSIS OF
AUTOMOTIVE FRAMES
SAE PAPER 720046 FOR MEETING JANUARY 10-14, 1972,
24 PP.

72-0090 HANDA, K. N.
ANALYSIS OF IN-PLANE VIBRATION OF SHEAR WALLS BY A
FINITE ELEMENT METHOD
J. SOUND AND VIBRATION, VOL. 21, NO. 2, 1972, PP.
169-180.

72-0091 HANDA, K. N.
ANALYSIS OF INPLANE VIBRATION OF BOX-TYPE
STRUCTURES BY A FINITE ELEMENT METHOD
J. SOUND AND VIBRATION, VOL. 21, NO. 1, 1972, PP.
107-114.

72-0092 TABARROK, B.
GASS, N.
NEW HYBRID CYLINDRICAL SHELL FINITE ELEMENT
J. SOUND AND VIBRATION, VOL. 21, NO. 3, 1972, PP.
369-378.

72-0093 ANDERSSON, H.
FINITE-ELEMENT ANALYSIS OF A FRACTURE TOUGHNESS
TEST SPECIMEN IN NON-LINEAR RANGE
J. MECH. AND PHYSICS OF SOLIDS, VOL. 20, NO. 1,
1972, PP. 33-51.

72-0094 ASHWELL, D. G.
SABIR, A. B.
NEW CYLINDRICAL SHELL FINITE ELEMENT BASED ON
SIMPLE INDEPENDENT STRAIN FUNCTIONS
INT. J. MECH. SC., VOL. 14, NO. 3, 1972, PP.
171-183.

72-0095 KALAJDZIC, M.
FINITE ELEMENT METHOD FOR DETERMINATION OF MACHINE
TOOL STRUCTURES
INST. ALATNE. MASINE. ALATE., MONOGR., NO. 4,
1972, P. 86.

72-0096 SABIR, A. B.
LOCK, A. C.
CURVED, CYLINDRICAL SHELL, FINITE ELEMENT
INT. J. MECH. SC., VOL. 14, NO. 2, 1972, PP.
125-135.

72-0097 HOMMEL, G.
COMMENT ON LIMITATIONS OF CERTAIN FINITE ELEMENTS
WHEN APPLIED TO ARCHES - FURTHER STUDIES IN
APPLICATION OF CURVED FINITE ELEMENTS TO CIRCULAR
ARCHES
INT. J. MECH. SC., VOL. 14, NO. 4, 1972, PP.
275-277.

72-0098 GUPTA, A. K.
MOHRAZ, B.
3-DIMENSIONAL ISOPARAMETRIC ELEMENT IN CURVILINEAR
COORDINATES
NUCL. ENG. AND DESIGN, VOL. 22, NO. 2, 1972, PP.
301-304.

72-0099 SCHIFFMAN, R. L.
THE EFFICIENT USE OF COMPUTER RESOURCES
PROC. SYMP. ON THE APPLICATIONS OF THE FINITE
ELEMENT METHOD IN GEOTECHNICAL ENG., VICKSBURG,
MISS., MAY 1-4, 1972, PP. 91-129. (PUBL. BY U.S.
ARMY ENG. WATERWAYS EXP. STATION, SEPPTEMBER 1972)

72-0100 BUELL, W. R.
BUSH, B. A.
MATHEMATICAL PROGRAMMING
PAPER NO. 72-WA/DE-2, A.S.M.E. MEETING, NOVEMBER
26-30, 1972, 7 PP.

72-0101 NICOLAIDES, R. A.
ON A CLASS OF FINITE ELEMENTS GENERATED BY
LAGRANGE INTERPOLATION.
SIAM J. NUMER. ANAL. VOL. 9, NO. 3, SEPTEMBER
1972, PP. 435-445.

72-0102 ARGYRIS, J. H.
GRIEGER, I.
DIE ANWENDUNG VON INTERAKTIVER COMPUTER-GRAPHIK
BEI DER BERECHNUNG VON TRAGWERKEN. (APPLICATION OF
INTERACTIVE COMPUTER GRAPHICS FOR THE ANALYSIS OF
STRUCTURES)
ANGEW INF., APPL. INF., VOL. 14, NO. 4, APRIL
1972, PP. 173-187.

72-0103 DESCLOUX, J.
ON FINITE ELEMENT MATRICES
SIAM J. NUMER. ANAL., VOL. 9, NO. 2, JUNE 1972,
PP. 260-265.

72-0104 FRIED, I.
CONDITION OF FINITE ELEMENT MATRICES GENERATED
FROM NONUNIFORM MESHES
AIAA J., VOL. 10, NO. 2, FEBRUARY 1972, PP.
219-221.

72-0105 TAYLOR, R. L.
ON COMPLETENESS OF SHAPE FUNCTIONS FOR FINITE
ELEMENT ANALYSIS
INT. J. NUMER. METHODS. ENG., VOL. 4, NO. 1,
JANUARY-FEBRUARY 1972, PP. 17-22.

72-0106 FUJINO, T.
SELF ADJOINT DIFFERENTIAL EQUATIONS AND
VARIATIONAL PRINCIPLE - 2.
MITSUBISHI HEAVY IND., MITSUBISHI, TECH. BULL.,
NO. 77, MARCH 1972, P. 26.

72-0107 FUJINO, T.
VARIATIONAL PRINCIPLE OF LINEAR DIFFERENTIAL
EQUATIONS
MITSUBISHI HEAVY IND., MITSUBISHI TECH. BULL., NO.
77, MARCH 1972, 26 PP.

72-0108 KIKUCHI, F.
ANDO, Y.
NEW VARIATIONAL FUNCTIONAL FOR THE FINITE-ELEMENT
METHOD AND ITS APPLICATION TO PLATE SHELL PROBLEMS
NUCL. ENG. DES., VOL. 21, NO. 1, 1972, PP. 95-113.
ALSO IN I.E.E.E. TRANS. MICROWAVE THEORY
TECH., VOL. MTT-19 NO. 12, 1971, PP. 952-954.

72-0109 MCNEICE, G. M.
MARCAL, P. V.
OPTIMIZATION OF FINITE ELEMENT GRIDS BASED ON
MINIMUM POTENTIAL ENERGY
PAPER NO. 72-PVP-3, A.S.M.E. MEETING, SEPTEMBER
17-21, 1972. ALSO IN J. ENG., IND., VOL. 95, NO.
1, 1973, PP. 186-190.

72-0110 CIARLET, P. G.
RAVIART, P. A.
GENERAL LAGRANGE AND HERMITE INTERPOLATION IN R**N
WITH APPLICATIONS TO FINITE ELEMENT METHODS
ARCH. RATION MACH. ANAL. VOL. 46, NO. 3, 1972, PP.
177-199.

72-0111 TRANTINA, G. G.
COMBINED MODE CRACK EXTENSION IN ADHESIVE JOINTS
J. COMPOS. MATER., VOL. 6, JULY 1972, PP. 371-385.

72-0112 DE, A. K.
INTRODUCTION TO FINITE ELEMENT METHOD AND ITS
APPLICATION TO MACHINE TOOL STRUCTURES
J. INST. ENG. (INDIA), MECH. ENG. DIV., VOL. 53,
PT ME 1, SEPTEMBER 1972, PP. 38-44.

72-0113 AYRES, D. J.
CULLEN, T. M.
CREEP DATA ACQUISITION AND APPLICATION TO REACTOR
COMPONENT DESIGN
NUCL. TECHNOL. VOL. 16, NO. 1, OCTOBER 1972, PP.
287-296.

72-0114 WALLACE, D. B.
SEIREG, A.
COMPUTER SIMULATION OF DYNAMIC STRESS,
DEFORMATION, AND FRACTURE OF GEAR TEETH
PAPER NO. 72-PTG-4, A.S.M.E. MEETING, OCTOBER
8-12, 1972, 7 PP.

72-0115 CHABERT, G.
EVALUATION OF STRESSES AND DEFLECTION OF SPUR GEAR
TEETH UNDER STRAIN
PAPER NO. 72-PTG-27, A.S.M.E. MEETING, OCTOBER
8-12, 1972, 16 PP.

72-0116 WILCOX, L.
COLEMAN, W.
APPLICATION OF FINITE ELEMENTS TO THE ANALYSIS OF
GEAR TOOTH STRESSES
PAPER NO. 72-PTG-30, A.S.M.E. MEETING, OCTOBER
8-12, 1972, 13 PP. ALSO IN J. OF ENG. FOR IND.,
VOL. 95, NO. 4, 1973, PP. 1139-1148.

72-0117 SHOTTER, B. A.
NEW APPROACH TO GEAR TOOTH ROOT STRESSES
PAPER NO. 72-PTG-42, A.S.M.E. MEETING, OCTOBER
8-12, 1972, 11 PP.

72-0118 WINTER, H.
HIRT, M.
MEASUREMENT OF ACTUAL DEDENDUM STRAINS, AND THE
INFLUENCE OF DEDENDUM RADIUS ON STRESSES AND
DEDENDUM STRENGTH
PAPER NO. 72-PTG-48, A.S.M.E. MEETING, OCTOBER
8-12, 1972, 23 PP.

72-0119 MEYERS, V. J.
RAMANI, D. T.
PLASTIC SANDWICH BUCKLE-SHELL STRUCTURES
ASCE J. STRUCT. DIV., VOL. 98, NO. ST11, NOVEMBER
1972, PAP. N. 9323, PP. 2381-2397.

72-0120 AALAMI, B.
LARGE DEFLECTION OF ELASTIC PLATES UNDER PATCH
LOADING
ASCE J. STRUCT. DIV., VOL. 98, NO. ST11, NOVEMBER
1972, PAP.N. 9359, PP. 2567-2586.

72-0121 BANAVALKAR, P. V.
GERGELY, P.
ANALYSIS OF THIN-SHELL HYPERBOLIC PARABOLOID
SHELLS
A.S.C.E. J. STRUCT. DIV., VOL. 98, NO. ST11, PAPER
NO. 9356, NOVEMBER 1972, PP. 2605-2621. (ALSO IN
ASCE, WATER RESOURCES ENG., MEETING AT PHOENIX,
ARIZONA, JANUARY 11-15, 1971, PAPER 1320, 30 PP.)

72-0122 SOKAL, Y. J.
NATURAL PERIODS OF INFILLED FRAMES
INST. ENG., AUST., CIV. ENG., TRANS., VOL. CE 14,
N. 1, APRIL 1972, PP. 13-18.

72-0123 WALLACE, D. B.
SEIREG, A.
FINITE ELEMENT BASED PROCEDURE FOR SIMULATING THE
TRANSIENT RESPONSE AND FAILURE OF A
TWO-DIMENSIONAL CONTINUUM WITH NONLINEAR MATERIAL
CHARACTERISTICS
PAPER NO. 72-WA/DE-4, A.S.M.E. MEETING, NOVEMBER
26-30, 1972, 8 PP. ALSO IN J. ENG. IND., VOL. 95,
NO. 1, 1973, PP. 345-352.

72-0124 GUPTA, K. K.
KNOELL, A. C.
MATHEMATICAL MODELLING AND STUCTURAL ANALYSIS OF
THE MANDIBLE
PAPER NO. 72-WA/BHF-4, A.S.M.E. MEETING, NOVEMBER
26-30, 1972, 5 PP.

72-0125 BELYTSCHKO, T.
KULAK, R. F.
NUMERICAL STRESS ANALYSIS OF INTERVERTEBRAL DISK
PAPER NO. 72-WA/BHF-12, A.S.M.E MEETING, NOVEMBER
26-30, 1972, 9 PP.

72-0126 BLAKE, W.
DESTRESSING TEST AT THE GALENA MINE, WALLACE,
IDAHO
TRANS. SOC. MIN. ENG. AIME. VOL. 252, NO. 3,
SEPTEMBER 1972, PP. 294-299.

72-0127 SABIR, A. B.
STABILITY INVESTIGATION OF A CYLINDRICAL SHELL
WITH FINITE ELEMENTS
DESTERR. ING-Z. VOL. 15, NO. 10, OCTOBER 1972, PP.
299-305.

72-0128 THOMPSON, E. G.
SAYLES, F. H.
IN SITU CREEP ANALYSIS OF ROOM IN FROZEN SOIL
ASCE J. SOIL MECH. FOUND. DIV., VOL. 9, NO. SM9,
SEPTEMBER 1972, PAP. N. 9202, PP. 899-915.

72-0129 WORKMAN, G. H.
DYNAMIC RESPONSE OF STRUCTURES SUBJECTED TO
TIME-DEPENDENT BOUNDARY CONDITIONS USING THE
FINITE ELEMENT METHOD
U. S. NAV. RES. LAB., SHOCK VIB. BULL. NO. 42, PT
2, JANUARY 1972, PP. 195-201.

72-0130 SISODIYA, R. G.
GHALI, A.
DIAPHRAGMS IN SINGLE- AND DOUBLE-CELL BOX GIRDER
BRIDGES WITH VARYING ANGLE OF SKEW
J. AM. CONCR. INST. VOL. 69, NO. 7, JULY 1972, PP.
415-419.

72-0131 WILLS, C. M. R.
NASTRAN - A FINITE ELEMENT PROGRAM FOR STRUCTURAL
ANALYSIS
COMPUT. AIDED DES. VOL. 4, NO. 4, JULY 1972, PP.
172-175.

72-0132 NOPPEN, R.
FINEL, UNIVERSELLES PROGRAMMSYSTEM ZUR BERECHNUNG
DER ELASTISCHEN EIGENSCHAFTEN VON
MASCHINENBAUTEILEN NACH DER METHODE FINITER
ELEMENTE. (FINEL. UNIVERSAL PROGRAMMING SYSTEM FOR
CALCULATING THE ELASTIC PROPERTIES OF STRUCTURAL
MACHINE PARTS ACCORDING TO THE METHOD OF FINITE
ELEMENTS)
IND-ANZ VOL. 94, N. 24, MARCH 21, 1972, PP.
483-486.

72-0133 RUDY DU PREEZ, M. S.
BERECHNUNG PRISMATISCHER FLAECHENTRAGWERKE MIT
HILFE VON FINITEN STREIFENELEMENTEN. (CALCULATION
OF PRISMATIC FOLDED PLATE STRUCTURES BY FINIE
ELEMENT METHOD)
BAUTECHNIK (AUSG A) VOL. 49, NO. 5, MAY 1972, PP.
151-162.

72-0134 KOBAYASHI, A. S.
NAKAGAWA, K.
THE EFFECT OF END RESTRAINT CONDITION ON THE
FRACTURE MECHANISM AND STRENGTH OF ROCK-LIKE
MATERIALS IN COMPRESSION
PROC. JAP. SOC. CIV. ENG. N. 197, JANUARY 1972,
PP. 125-134.

72-0135 BACKLUND, J.
MIXED FINITE ELEMENT ANALYSIS OF ELASTO-PLASTIC
PLATES IN BENDING
ARCH MECH. - ARCH MECH. STOSOW, VOL. 24, NO. 3,
1972, PP. 319-335.

72-0136 MIRSKY, I.
IN VIVO STRESSES IN THE HUMAN LEFT VENTRICULAR
WALL: ANALYSIS ACCOUNTING FOR THE IRREGULAR THREE
DIMENSIONAL GEOMETRY AND COMPARISON WITH IDEALIZED
GEOMETRY ANALYSIS
J. BIOMECH., VOL. 5, NO. 4, SEPTEMBER 1972, PP.
521-539.

72-0137 ANDO, Y.
IIDA, K.
FINITE ELEMENT LARGE DEFLECTION ANALYSIS OF
ELASTIC-PLASTIC SHELLS OF REVOLUTION SUBJECTED TO
AXISYMMETRIC LOADING
BULL. JSME, VOL. 15, NO. 85, JULY 1972, PP.
796-804.

72-0138 FRIED, I.
SCHMITT, K. H.
NUMERICAL RESULTS FROM THE APPLICATION OF GRADIENT
ITERATIVE TECHNIQUES TO THE FINITE ELEMENT
VIBRATION AND STABILITY ANALYSIS OF SKEW PLATES
AERONAUT. J. VOL. 76, NO. 735, MARCH 1972, PP.
166-169.

72-0139 CYR, N. A.
TETER, R. D.
FINITE ELEMENT THERMOELASTOPLASTIC ANALYSIS
ASCE J. STRUCT. DIV., VOL. 98, NO. ST7, JULY 1972,
PAPER 9068, PP. 1585-1603.

72-0140 CROSE, J. G.
STRESS ANALYSIS OF AXISYMMETRIC SOLIDS WITH
ASYMMETRIC PROPERTIES
AIAA J. VOL. 10, NO. 7, JULY 1972, PP. 866-871.

72-0141 SCHMIDT, G.
METHOD FOR FINITE ELEMENTS AS SPECIAL CASE OF
METHOD OF WEIGHTED RESIDUALS
ZEITSCHRIFT FUR ANGEWANDTE MATHEMATIK UND
MECHANIK, VOL. 52, NO. 9, 1972, PP. 461-469.

72-0142 JAENSSON, B. O.
SUNDSTROM, B. O.
DETERMINATION OF YOUNG'S MODULUS AND POISSON'S
RATIO FOR WC-CO ALLOYS BY THE FINITE ELEMENT
METHOD
MATER. SCI. ENG. VOL. 9, NO. 4, APRIL 1972, PP.
217-222

72-0143 MAU, S. T.
TONG, P.
FINITE ELEMENT SOLUTIONS FOR LAMINATED THICK PLATES
J. COMPOS. MATER. VOL. 6, APRIL 1972, PP. 304-311.

72-0144 WEBBER, J. P. H.
EXPERIMENTAL AND THEORETICAL BENDING DEFLECTIONS OF LAMINATED PLATES
J. STRAIN ANAL. VOL. 7, NO. 2, APRIL 1972, PP. 87-96.

72-0145 CUBAUD, J. C.
LEMAIRE, M.
APPLICATIONS DE LA METHODE DE GAUSS-SEIDEL A LA RESOLUTION DES PROBLEMS D'ELASTICITE. (APPLICATION OF THE GAUSS-SEIDEL METHOD FOR SOLVING ELASTICITY PROBLEMS)
CONSTR. MET. N. 1, MARCH 1972, PP. 15-20.

72-0146 HARRISON, N. L.
STRAIN ENERGY RELEASE RATES FOR TURNING CRACKS
FIBRE SCI. TECHNOL. VOL. 5, N. 3, JULY 1972, PP. 197-212.

72-0147 ARGYRIS, J. H.
CHAN, A. S. L.
APPLICATIONS OF FINITE ELEMENTS IN SPACE AND TIME
ING-ARCH. VOL. 41, NO. 4, 1972, PP. 235-257.
ALSO IN AERO. J., VOL. 71, DECEMBER 1969, PP. 1041-1044.

72-0148 MOHRAZ, B.
LUMPED PARAMETER ELEMENT FOR THE ANALYSIS OF HYPERBOLIC PARABOLOID SHELLS
INT. J. NUMER. METHODS ENG., VOL. 4, NO. 2, MARCH-APRIL 1972, PP. 235-249

72-0149 ELIAS, Z. M.
MIXED FINITE ELEMENT METHOD FOR AXISYMMETRIC SHELLS
INT. J. NUMER. METHODS ENG., VOL. 4, NO. 2, MARCH-APRIL 1972, PP. 261-277.

72-0150 WOLF, J. P.
DAS FLAECHENTRAGWERKSPROGRAM VON STRIP. (STRIP PROGRAM FOR PLATE STRUCTURES)
SCHWEIZ BAUZTG VOL. 90, NO. 3, JANUARY 20, 1972, PP. 41-52.

72-0151 HABIP, L. M.
STRUCTURAL EFFECTS OF CONFINED DYNAMIC LOADS
NUCL. SAFETY VOL. 13, NO. 2, MARCH-APRIL 1972, PP. 107-113.

72-0152 MASON, J. B.
NASTRAN STRUCTURAL/THERMAL ANALYZER
THERMAL STRUCTURAL ANALYSIS PROGRAMS--A SURVEY AND EVALUATION, ASME, NEW YORK, 1972, PP. 1-10.

72-0153 HURWITZ, M. M.
SCHROEDER, E. A.
THERMAL AND STRESS ANALYSIS BY NASTRAN
THERMAL STRUCTURAL ANALYSIS PROGRAMS--A SURVEY AND EVALUATION, ASME, NEW YORK, 1972, PP. 11-15.

72-0154 HIBBITT, H. D.
MARC SERIES OF PROGRAMS
THERMAL STRUCTURAL ANALYSIS PROGRAMS--A SURVEY AND EVALUATION, ASME, NEW YORK, 1972, PP. 17-28.

72-0155 SCHACHT, C. A.
THERMO-STRUCTURAL CAPABILITY OF ANSYS
THERMAL STRUCTURAL ANALYSIS PROGRAMS--A SURVEY AND EVALUATION, ASME, NEW YORK, 1972, PP. 29-36.

72-0156 SWANSON, J. A.
ANSYS (ENGINEERING ANALYSIS SYSTEM) PROGRAM
THERMAL STRUCTURAL ANALYSIS PROGRAMS--A SURVEY AND EVALUATION, ASME, NEW YORK, 1972, PP. 37-46.

72-0157 VON FUCHS, G.
ASKA SERIES OF PROGRAMS
THERMAL STRUCTURAL ANALYSIS PROGRAMS--A SURVEY AND EVALUATION, ASME, NEW YORK, 1972, PP. 47-57.

72-0158 OWEN, D. R. J.
FESS AND FINESSE PROGRAMS
THERMAL STRUCTURAL ANALYSIS PROGRAMS - A SURVEY AND EVALUATION, ASME, NEW YORK, 1972, PP. 59-69.

72-0159 WATWOOD, V. B.
THERMAL AND STRESS ANALYSIS PROGRAM - HYBOS
THERMAL STRUCTURAL ANALYSIS PROGRAMS - A SURVEY AND EVALUATION, ASME, NEW YORK, 1972, PP. 71-75.

72-0160 SUTPHIN, D. L.
VINSON, T. J.
MARK5C/DAISY SERIES OF PROGRAMS
THERMAL STRUCTURAL ANALYSIS PROGRAMS--A SURVEY AND EVALUATION, ASME, NEW YORK, 1972, PP. 83-90.

72-0161 HSU, M. B.
FETE PROGRAM PACKAGE
THERMAL STRUCTURAL ANALYSIS PROGRAMS--A SURVEY AND EVALUATION, ASME, NEW YORK, 1972, PP. 91-97.

72-0162 MARCAL, P. V.
THREE-DIMENSIONAL FINITE ELEMENT ANALYSIS FOR FRACTURE MECHANICS
SURFACE CRACK: PHYSICAL PROBLEMS AND COMPUTATIONAL SOLUTIONS, ASME, NEW YORK, 1972, PP. 187-202.

72-0163 CIARLET, P. G.
RAVIART, P. A.
LAGRANGE INTERPOLATION ON CURVED FINITE ELEMENTS
COMPTES RENDUS., ACADEMIE DES SCI. SER A MATH., VOL. 274, NO. 8, 1972, PP. 640-643.

72-0164 SMITH, C. S.
BUCKLING PROBLEMS IN THE DESIGN OF FIBREGLASS-REINFORCED PLASTIC SHIPS
J. SHIP RES. VOL. 16, NO. 3, SEPTEMBER 1972, PP. 174-190.

72-0165 LEVERENZ, R. K.
FINITE ELEMENT STRESS ANALYSIS OF A CRACK IN A BI-MATERIAL PLATE
INT. J. FRACT. MECH., VOL. 8, NO. 3, SEPTEMBER 1972, PP. 311-324.

72-0166 ARGYRIS, J. H.
ROY, J. R.
GENERAL TREATMENT OF STRUCTURAL MODIFICATIONS
ASCE J. STRUCT. DIV., VOL. 98, NO. ST2, FEBRUARY 1972, PAPER 8732, PP. 465-492.

72-0167 BELYTSCHKO, T.
VELEBIT, M.
FINITE ELEMENT METHOD FOR ELASTIC PLASTIC PLATES
ASCE J. ENG. MECH. DIV. VOL. 98, NO. EM1, FEBRUARY 1972, PAPER 8724, PP. 227-242.

72-0168 FRIED, I.
CONDENSATION OF FINITE ELEMENT EIGENPROBLEMS
A.I.A.A. J., VOL. 10, NO. 11, 1972, PP. 1529-1530.

72-0169 ARGYRIS, J. H.
SCHARPF, D. W.
LARGE DEFLECTION ANALYSIS OF PRESTRESSED NETWORKS
ASCE J. STRUCT. DIV. VOL. 98, NO. ST3, MARCH 1972, PAPER 8788, PP. 633-654.

72-0170 MAU, S. T.
GALLAGHER, R. H.
FINITE ELEMENT PROCEDURE FOR NONLINEAR PREBUCKLING AND INITIAL POSTBUCKLING ANALYSIS
NASA CONTRACT REP. CR-1936, JANUARY 1972, 56 PP.

72-0171 VOS, R. G.
GENERALIZATION OF PLATE FINITE ELEMENTS TO SHELLS
ASCE J. ENG. MECH. DIV., VOL. 98, NO. EM2, APRIL 1972, PAPER 8828, PP. 385-400.

72-0172 FU, C. C.
ON THE STABILITY OF EXPLICIT METHODS FOR THE NUMERICAL INTEGRATION OF THE EQUATIONS OF MOTION IN FINITE ELEMENT METHODS
INT. J. NUMER. METHODS ENG., VOL. 4, NO. 1, JANUARY - FEBRUARY 1972, PP. 95-107.

72-0173 CHERNUKA, M. W.
COWPER, G. R.
FINITE ELEMENT ANALYSIS OF PLATES WITH CURVED EDGES.
INT. J. NUMER. METHODS ENG., VOL. 4, NO. 1, JANUARY-FEBRUARY 1972, PP. 49-65.

72-0174 SINGH, S.
RAMASWAMY, G. S.
SECTOR ELEMENT FOR THIN PLATE FLEXURE
INT. J. NUMER. METHODS ENG., VOL. 4, NO. 1, JANUARY-FEBRUARY 1972, PP. 133-142.

72-0175 UEHARA, S.
COMPUTATIONAL ASPECTS OF FINITE ELEMENT METHOD FOR A FRAME STRUCTURE
PROC. JAP. SOC. CIV. ENG., NO. 201, MAY 1972, PP. 13-22.

72-0176 DEVERALL, L. I.
LINDSEY, G. H.
COMPARISON OF NUMERICAL METHODS FOR DETERMINING STRESS INTENSITY FACTORS
J. BASIC ENG., TRANS. ASME, VOL. 94, SER. D, NO. 2, JUNE 1972, PP. 508-509.

72-0177 HERAKOVICH, C. T.
PLASTIC STRAINS ASSOCIATED WITH TORSION OF NONCIRCULAR BARS
J. APPL. MECH., TRANS. ASME VOL. 38, SER. E, NO. 1, MARCH 1972, PP. 285-287.

72-0178 FOX, R. L.
MIURA, H.
AUTOMATED DESIGN OPTIMIZATION OF SUPERSONIC AIRPLANE WING STRUCTURES UNDER DYNAMIC CONSTRAINTS
AIAA/ASME/SAE, 13TH STRUCTURES, STRUCTURAL DYNAMICS, AND MATERIALS, CONF., SAN ANTONIO, TEX., APRIL 10-12, 1972, VOL. 3, STRUCTURES, 1972 PAPER NO. 71-333, 16 PP.

72-0179 PRZEMIENIECKI, J. S.
FINITE ELEMENT STRUCTURAL ANALYSIS OF LOCAL
INSTABILITY
AIAA/ASME/SAE, 13TH STRUCTURES, STRUCTURAL
DYNAMICS, AND MATERIALS, CONF., SAN ANTONIO, TEX.,
APRIL 10-12, 1972, VOL. 3, STRUCTURES, 1972 PAPER
NO. 72-354, 9 PP. ALSO IN AIAA
J., VOL. 11, NO. 1, 1973, PP. 33-39.

72-0180 SHARIFI, P.
POPOV, E. P.
NONLINEAR FINITE ELEMENT ANALYSIS OF SANDWICH
SHELLS OF REVOLUTION
AIAA/ASME/SAE, 13TH STRUCTURES, STRUCTURAL
DYNAMICS, AND MATERIALS, CONF., SAN ANTONIO, TEX.,
APRIL 10-12, 1972, VOL. 3, STRUCTURES, 1972 PAPER
NO. 72-356, 11 PP. ALSO IN
A.I.A.A. J., VOL. 11, NO. 5, 1973, PP. 715-722.

72-0181 TRANTINA, G. G.
FRACTURE MECHANICS APPROACH TO ADHESIVE JOINTS
J. COMPOS. MATER., VOL. 6, APRIL 1972, PP.
192-207.

72-0182 HASLUM, K.
PETTERSEN, E.
FINITE ELEMENTS IN TRANSVERSE FRAME ANALYSIS
EUR. SHIPBUILD. VOL. 21, NO. 1, 1972, PP. 2-8.

72-0183 DIXON, J. R.
STRANNIGAN, J. S.
DETERMINATION OF ENERGY RELEASE RATES AND STRESS -
INTENSITY FACTORS BY THE FINITE-ELEMENT METHOD
J. STRAIN ANAL. VOL. 7, NO. 2, APRIL 1972, PP.
125-131.

72-0184 JIROUSEK, J.
CALCUL DES PONTS BIAIS A POUTRES MULTIPLES SANS
ENTRETOISES PAR LA METHODE DES ELEMENTS
(CALCULATION OF SKEW MULTIBEAM BRIDGES WITHOUT
CROSS BRACINGS BY USING FINITE ELEMENT METHOD)
BULL. TECH. SUISSE ROMANDE, VOL. 9,, NO. 10, MAY
13, 1972, PP. 169-186.

72-0185 RADAJ, D.
ANWENDUNG DES FINITE ELEMENT VERFAHRENS SUR
BESTIMMUNG DER FESTIGKEIT VON
SCHWEISSKONSTRUKTIONEN. (FINITE ELEMENT METHOD OF
DETERMINING THE STRENGTH OF WELDED STRUCTURES)
CHEM-ING-TECH. VOL. 44, NO. 12, JUNE 1972, PP.
812-827.

72-0186 NEKI, I.
NAGAI, K.
GENERAL PURPOSE PROGRAM OF PLANE STRESS ANALYSIS
BY FINITE ELEMENT METHOD, AND ITS APPLICATIONS
IHI, ENG. REV., VOL. 5, NO. 1, MARCH 1972, PP.
10-24.

72-0187 GORHOLT, W.
NEYLAN, A. J.
DESIGN OF A COMPOSITE STEEL AND CONCRETE CLOSURE
FOR A PRESTRESSED CONCRETE REACTOR VESSEL.
ASME PAP. NO. 72-PVP-17 FOR MEET. SEPTEMBER 17-21,
1972, 8 PP.

72-0188 HENNING, G.
ZUR GENAUEN BERECHNUNG KONSTRUKTIV ORTHOTROPER
PLATTEN (EXACT CALCULATION OF STRUCTURALLY
ORTHOTROPIC PLATES)
STAHLBAU, VOL. 41, NO. 3, MARCH 1972, PP. 78-86.

72-0189 STACEY, T. R.
THREE DIMENSIONAL FINITE ELEMENT STRESS ANALYSIS
APPLIED TO TWO PROBLEMS IN ROCK MECHANICS
J. S. AFR. INST. MIN. METALL., VOL. 72, NO. 10,
MAY 1972, PP. 251-256.

72-0190 DEIST, F. H.
GEORGIADIS, E.
COMPUTER APPLICATIONS IN ROCK MECHANICS
J. S. AFR. INST. MIN. METALL. VOL. 72, NO. 10, MAY
1972, PP. 265-272.

72-0191 BELL, J. C.
ELMS, D. G.
NON-LINEAR ANALYSIS OF REINFORCED CONCRETE SLABS
MAG. CONCR. RES., VOL. 24, NO. 79, JUNE 1972, PP.
63-70.

72-0192 HARRYSSON, C.
EXPERIMENTELL UNDERSOKNING AV ELEMENTFOGARS
HALLFASTHETSEGENSKAPER (EXPERIMENTAL INVESTIGATION
OF STRENGTH PROPERTIES OF JOINTS BETWEEN PRECAST
CONCRETE ELEMENTS)
NORD. BETONG, VOL. 16, NO. 2, 1972, PP. 141-158.

72-0193 KUPTA, K. K.
DYNAMIC RESPONSE ANALYSIS OF GEOMETRICALLY
NON-LINEAR STRUCTURES SUBJECTED TO HIGH IMPACT
INT. J. NUMER. METHODS ENG., VOL. 4, NO. 2,
MARCH-APRIL 1972, PP. 163-174.

72-0194 KIRKHOPE, J.
WILSON, G. J.
VIBRATION OF CIRCULAR AND ANNULAR PLATES USING
FINITE ELEMENTS
INT. J. NUMER. METHODS ENG., VOL. 4, NO. 2,
MARCH-APRIL 1972, PP. 181-193.

72-0195 LYNN, P. P.
DHILLON, B. S.
CONVERGENCE OF EIGENVALUE SOLUTIONS IN CONFORMING
PLATE BENDING FINITE ELEMENTS
INT. J. NUMER. METHODS ENG., VOL. 4, NO. 2,
MARCH-APRIL 1972, PP. 217-234.

72-0196 BROWN, E. T.
HUDSON, J. A.
CONTROLLED FAILURE OF HOLLOW ROCK CYLINDERS IN
UNIAXIAL COMPRESSION
ROCK. MECH., FELSMECH., MEC. ROCHES, VOL. 4, NO. 1,
JUNE 1972, PP. 1-24.

72-0197 GHALI, A.
LE CALCUL DES TABLIERS DE PONTS PAR DES METHODES
NUMERIQUES
ANNALES ITBTP, DECEMBER 1971, PARIS, FRANCE, PP.
94-103 THIS PAPER IS TRANSLATED INTO JAPANESE,
DOBOKU-GIJUTSU (CIVIIL ENGINEERING TECHNICAL
JOURNAL), VOL. 27, NO. 8, 1972, PP.24-32

72-0198 BAZANT, Z. P.
MATRIX DIFFERENTIAL EQUATION AND HIGHER-ORDER
NUMERICAL METHODS FOR PROBLEMS OF NON-LINEAR
CREEP, VISCOELASTICITY AND ELASTO-PLASTICITY
INT. J. NUMER. METHODS ENG., VOL. 4, NO. 1,
JAN-FEB. 1972, PP. 11-15.

72-0199 BARTELDS, G.
APPLICATION OF THE FINITE ELEMENT DISPLACEMENT
METHOD TO PROBLEMS OF ELASTOPLASTIC DEFORMATION
CONTRIB. TO THE THEORY OF AIRCR. STRUCT., 1972,
PP. 99-115.

72-0200 KULHAWY, F. H.
DUNCAN, J. M.
STRESSES AND MOVEMENTS IN OROVILLE DAM
A.S.C.E. J., SOIL MECH. FOUND. DIV., VOL. 98, NO.
SM 7, PAPER NO. 9016, JULY 1972, PP. 653-665.

72-0201 WILLIAM, K. J.
SCORDELIS, A. C.
CELLULAR STRUCTURES OF ARBITRARY PLAN GEOMETRY
A.S.C.E. J., STRUCT. DIV., VOL. 98, NO. ST7, PAPER
9013, JULY 1972, PP. 1377-1394.

72-0202 KOENIG, H. A.
VOGEL, W.
ELASTO-PLASTIC CYCLIC ANALYSIS OF STRUCTURAL
MEMBERS
J. ENG. POWER, TRANS. ASME, VOL. 94, SER. A, NO.
3, JULY 1972, PP. 214-218.

72-0203 LIU, T. C. Y.
NILSON, A. H.
BIAXIAL STRESS-STRAIN RELATIONS FOR CONCRETE
A.S.C.E. J., STRUCT. DIV., VOL. 98, NO. ST5, PAPER
NO. 8905, MAY 1972, PP. 1025-1034.

72-0204 BURNETT, E. F. P.
RAJENDRA, R.
CLEMENT, S.
INFLUENCE OF JOINTS IN PANELIZED STRUCTURAL
SYSTEMS
A.S.C.E. J., STRUCT. DIV., VOL. 98, NO. ST9, PAPER
NO. 9207, SEPTEMBER 1972, PP. 1943-1955.

72-0205 BALUCH, M. H.
GOLDBERG, J. E.
FINITE ELEMENT APPROACH TO PLANE MICROELASTICITY
A.S.C.E. J., STRUCT. DIV., VOL. 98, NO. ST9, PAPER
NO. 9203, SEPTEMBER 1972, PP. 1957-1964.

72-0206 CANNON, M.
GREENFIELD, H.
ADAPTABILITY OF COMPUTER GRAPHICS TO STUDIES OF
ATHEROSCLEROSIS PATHOGENESIS
MED. INSTRUM., VOL. 6, NO. 3, MAY-JUNE 1972, PP.
250-255.

72-0207 RYBICKI, E. F.
SIMONEN, F. A.
ON THE MATHEMATICAL ANALYSIS OF STRESS IN THE
HUMAN FEMUR
J. BIOMECH VOL. 5, NO. 2, MARCH 1972, PP.
203-215.

72-0208 ARGYRIS, J. H.
RADAJ, D.
PARAMETRISCHE KERBSPANNUNGSUNTERSUCHUNG AM
ELASTISCHEN KERN.(PARAMETRICAL INVESTIGATION OF
NOTCH STRESSES AT AN ELASTIC CORE)
ACTA MECH. VOL. 13, NO. 3-4, 1972, PP. 303-314.

72-0209 SISODIYA, R. G.
CHEUNG, Y. K.
GHALI, A.
FINITE ELEMENT ANALYSIS OF SKEW BOX GIRDER BRIDGES
TRANSACTIONS EIC, 72-BR & STR 3, VOL. 15, NO. A-4,
MARCH 1972, THE ENGINEERING JOURNAL, PP. 1-6

72-0210 BYERS, N. R.
SCHULTZ, R. F.
ANALYSIS OF A STUB END BY FINITE ELEMENT METHOD
WELDING J., VOL. 51, NO. 1, 1972, PP. S 31 - S 35.

72-0211 RADAJ, D.
ZEHLEIN, H.
EINSATZ DES COMPUTERS BEI DER ENTWICKLUNG EINER
PANZERWANNE (THE USE OF A COMPUTER FOR DEVELOPING
A TANK HULL)
WEHRTECHNIK, NO. 4, APRIL 1972, PP. 172-177.

72-0212 SAUGY, B.
ZIMMERMANN, T. H.
APPLICATION OF NONLINEAR ANALYSIS TO SOLID
STRUCTURES
ANN. INST. TECH. BATIM TRAV., PUBLICS, NO. 292,
APRIL 1972, PP. 125-143.

72-0213 LEMAIRE, M. H.
DEPLACEMENTS DANS LES STRUCTURES MINCES A FAIBLE
RIGIDITE GEOMETRIQUE-DETERMINATION NUMERIQUE
(NUMERICAL DETERMINATION OF THE DISPLACEMENTS IN
THIN STRUCTURES WITH SMALL GEOMETRICAL STIFFNESS)
ANN. INST. TECH. BATIM TRAV., PUBL., NO. 293, MAY
1972, PP. 13-28.

72-0214 BOLLAND, G. B.
FLEXIBILITY MATRIX FOR AN EXACT TORSION-BENDING
FINITE ELEMENT
AERON. J., VOL. 76, NO. 742, 1972, PP. 613-615.

72-0215 BERG, C. A.
SALAMA, M.
FATIGUE OF PRENOTCHED GRAPHITE FIBER COMPOSITES IN
COMPRESSION
TEXT RES. J. VOL. 42, NO. 4, APRIL 1972, PP.
222-238.

72-0216 SCHNOBRICH, W. C.
ANALYSIS OF HIPPED ROOF HYPERBOLIC PARABOLOID
STRUCTURES
ASCE J. STRUCT. DIV. VOL. 98, NO. ST7, JULY 1972,
PAP. 9052, PP. 1575-1583.

72-0217 KOHNKE, P. C.
SCHNOBRICH, W. C.
ANALYSIS OF ECCENTRICALLY STIFFENED CYLINDRICAL
SHELLS
ASCE J. STRUCT. DIV., VOL. 98, NO. ST7, JULY 1972,
PAP. 9033, PP. 1493-1510.

72-0218 YANG, Y. C.
RECORD CONCRETE SHELL ROOF FEATURED IN COLLISEUM
CIV. ENG. VOL. 42, NO. 6, JUNE 1972, PP. 51-53.

72-0219 YEH, C.
NONLINEAR DYNAMIC ANALYSIS OF COOLING TOWER.
ASCE J. POWER DIV., VOL. 98, NO. POI, JUNE 1972,
PAP. 8983, PP. 49-63.

72-0220 MOTE, C. D.
MATSUMOTO, G. Y.
COUPLED, NONCONSERVATIVE STABILITY-FINITE ELEMENT
ASCE J. ENG. MECH. DIV., VOL. 98, NO. EM3, JUNE
1972, PAP. 8

72-0221 BUYUKOZTURK, O.
NILSON, A. H.
DEFORMATION AND FRACTURE OF PARTICULATE COMPOSITE
ASCE J. ENG. MECH. DIV., VOL. 98, NO. EM3, JUNE
1972, PAP. 8970, PP. 581-593.

72-0222 CHRISTIAN, J. T.
BOEHMER, J. W.
CONSOLIDATION OF A LAYER UNDER A STRIP LOAD
ASCE J. SOIL MECH. FOUND. DIV., VOL. 98, NO. SM7,
JULY 1972, PP. 693-707, PAPER 9030.

72-0223 HWANG, C. T.
PI, W. S.
NONLINEAR ACOUSTIC RESPONSE ANALYSIS OF PLATES
USING THE FINITE ELEMENT METHOD
AIAA J., VOL. 10, NO. 3, MARCH 1972, PP. 276-281.

72-0224 CHARNLEY, C. J.
SRINIVASAN, S.
INTERNATIONAL MACHINE TOOL DESIGN AND RESEARCH
CONFERENCE, 12TH PROCEEDINGS, 1971.
INT. MACH. TOOL DES. AND RES. CONF., 12TH PROC.,
MANCHESTER, ENGLAND, SEPTEMBER 15-17, 1971,
PUBLISHED BY MACMILLAN PRESS LTD., LONDON, ENGL.

72-0225 WANG, F. D.
SUN, M. C.
COMPUTER PROGRAM FOR PIT SLOPE STABILITY ANALYSIS
BY THE FINITE ELEMENT STRESS ANALYSIS AND LIMITING
EQUILIBRIUM METHOD
U. S. BUR. MINES, REP. INVEST., NO. 7685, 1972, 56
PP.

72-0226 BROWN, S. F.
BUSH, D. I.
DYNAMIC RESPONSE OF MODEL PAVEMENT STRUCTURE
ASCE TRANSP. ENG. J. VOL. 98, NO. TE4, NOVEMBER
1972   PAPER NO. 9863, PP. 1005-1022.

72-0227 KIM, Y. S.
DIMOCK, R. R.
CORRELATION BETWEEN MODEL STUDY AND FINITE ELEMENT
ANALYSIS IN AN INVESTIGATION OF SLOPE STABILITY
JOINT MEETING MMIJ-AIME, TOKYO, JAPAN, MAY 24-27,
1972, MIN AND METALL. INST. OF JAPAN, TOKYO, 1972,
5 VOLUMES, G. T-I, T-II, T-III, T-IV, VARIOUS
PAGINGS.

72-0228 GHALI, A.
FINITE ELEMENT ANALYSIS OF PERFORATED SHELLS
PROCEEDINGS, OF INTERNATIONAL ASSOCIATION FOR
SHELL STRUCTURES, JULY 3-6, 1972, PP. 405-413

72-0229 KOBAYASHI, A. S.
MEHTA, H. S.
FINITE ELEMENT ANALYSIS AND EXPERIMENTAL
INVESTIGATION OF SHEET METAL STRETCHING
ASME PAP. NO. 72-WA/PROD-11 FOR MEET. NOVEMBER
26-30, 1972, 7 PP.

72-0230 ROCKEY, K. C.
EL-GAALY, M. A.
FAILURE OF THIN-WALLED MEMBERS UNDER PATCH
LOADINGS
ASCE J. STRUCT. DIV., VOL. 98, NO. ST12. DECEMBER
1972, PAP. NO. 9409, PP. 2739-2752.

72-0231 NOVAK, M.
BEREDUGO, Y. O.
VERTICAL VIBRATION OF EMBEDDED FOOTINGS
ASCE J. SOUL MECH. FOUND. DIV., VOL. 98, NO.
SM12, DECEMBER 1972, PAPER NO. 9412, PP.
1291-1391.

72-0232 PICHARD, R. M.
MALVICK, A. J.
GALLAGHER, R. H.
MAU, S. T.
ELASTIC DEFORMATION OF LIGHTWEIGHT MIRRORS
METHOD OF LIMIT POINT CALCULATION IN
FINITE ELEMENT STRUCTURAL ANALYSIS
APPL. OPT. V. 12, N. 6, JUN. 1973, PP. 1220-1226.
NASA CONTRACT REP. CR-2115, SEPTEMBER
1972, 28 PP.

72-0233 CHENG, R. T.
NUMERICAL INVESTIGATIONS OF LAKE CIRCULATION
AROUND ISLANDS BY THE FINITE ELEMENT METHOD
INT. J. NUMER. METHODS IN ENGNG., VOL. 5, NO. 1,
SEPTEMBER-OCTOBER 1972, PP. 103-112.

72-0234 HOBBS, G. K.
TRANSIENT DYNAMIC RESPONSE OF VISCOELASTIC
STRUCTURES
SAE PAPER NO. 720812 FOR MEET., OCTOBER 2-5, 1972,
7 PP.

72-0235 CHELAPATI, C. V.
KENNEDY, R. P.
PROBABILISTIC ASSESSMENT OF AIRCRAFT HAZARD FOR
NUCLEAR POWER PLANTS
NUCL. ENG. DES. VOL. 19, NO. 2, 1972, PP. 333-364.

72-0236 SHINOZUKA, M.
PROBABILISTIC MODELING OF CONCRETE STRUCTURES
ASCE J. ENG. MECH. DIV., VOL. 98, NO. EM6,
DECEMBER 1972, PAPER NO. 9448, PP. 1433-1451.

72-0237 HEIFITZ, J. H.
CONSTANTINO, C. J.
DYNAMIC RESPONSE ON NONLINEAR MEDIA AT LARGE
STRAINS
ASCE J. ENG. MECH. DIV., VOL. 98, NO. EM 6,
DECEMBER 1972, PAPER NO. 9451, PP. 1511-1528.

72-0238 SAYEGH, A. F.
RUBINSTEIN, M. F.
ELASTIC-PLASTIC ANALYSIS BY QUADRATIC PROGRAMMING
ASCE J. ENG. MECH. DIV., VOL. 98, NO. EM6,
DECEMBER 1972, PAPER NO. 9435, PP. 1547-1572.

72-0239 DOKAINISH, M. A.
RAWTANI, S.
PSEUDO STATIC DEFORMATION AND FREQUENCIES OF
ROTATING TURBOMACHINERY BLADES
AIAA J., VOL. 10, NO. 11, NOVEMBER 1972, PP.
1397-1398.

72-0240 PARISEAU, W. G.
NICHOLSON, D. E.
GRAVITY FLOW BIN DESIGN FOR LUNAR SOIL
A.S.M.E. MEETING, PAPER NO. 72-MH-24, SEPTEMBER
17-20, 1972, 8 PP.

72-0241 LEE, C. H.
IWASAKI, H.
CALCULATION OF RESIDUAL STRESSES IN PLASTIC
DEFORMATION PROCESSES
ASME PAPER NO. 72-PROD-A, 1972, 9 PP.

72-0242 WILSON, E. A.
PARSONS, B.
RADIAL DISPLACEMENTS OF A HOLLOW SHAFT SUBJECTED
TO A UNIFORM BAND OF EXTERNAL PRESSURE
ASME PAPER NO. 72-DE-B, 1972, 6 PP.

72-0243 ODEN, J. T.
SOME ASPECTS OF THE MATHEMATICAL THEORY OF FINITE
ELEMENTS
IN ADVANCES IN COMPUTATIONAL METHODS IN STRUCTURAL
MECHANICS AND DESIGN, EDITED BY J. T. ODEN, R. W.
CLOUGH, AND Y. YAMAMOTO, PAPER PRESENTED AT THE
PROCEEDINGS OF THE 2ND U.S.- JAPAN SEMINAR ON
MATRIX METHODS OF STRUCTURAL ANALYSIS AND DESIGN,
AUGUST 1972, PP. 3-38.

72-0244 TURNER, C. E.
CHEUNG, J. S. T.
COMPUTATION OF POST-YIELD BEHAVIOUR IN NOTCH-BEND
AND TENSION TESTPIECES.
J. STRAIN ANAL. VOL. 7, NO. 4, OCTOBER 1972, PP.
303-312.

72-0245 MCNEICE, G. M.
HUNNISETT, S. F.
MIXED-DISPLACEMENT FINITE-ELEMENT ANALYSIS WITH
PARTICULAR APPLICATION USING PLANE-STRESS
TRIANGLES
J. STRAIN ANAL. VOL. 7, NO. 4, OCTOBER 1972, PP.
243-252.

72-0246 GAWRONSKI, W.
KRUSZWESKI, J.
ANALIZA ORGAN WYMUSZONYCH ZLOZONYCH UKLADOW
LINIOWYCH METODA SZTYWNYCH ELEMENTOW SKONCZONYCH
(ANALYSIS OF FORCED VIBRATIONS OF COMPOSITE LINEAR
SYSTEMS, BY THE STIFF FINITE ELEMENT METHOD)
ARCH. BUDOWY, MASZ., VOL. 19, NO. 4, 1972, PP.
623-641.

72-0247 TRIKHA, D. N.
EDWARDS, A. D.
ANALYSIS OF CONCRETE BOX GIRDERS BEFORE AND AFTER
CRACKING
PROC. INST. CIV. ENG. (LONDON) VOL. 53, PART 2,
DECEMBER 1972, PP. 515-528.

72-0248 GUPTA, A. K.
MOHRAZ, B.
ELASTO-PLASTIC ANALYSIS OF THREE-DIMENSIONAL
STRUCTURES USING THE ISOPARAMETRIC ELEMENT
NUCL. ENG. DES., VOL. 22, NO. 2, 1972, PP.
305-317. ALSO IN CIV. ENG. STUDIES,
STRUCT. RES. SER. NO. 381, 1971.

72-0249 OGLESBY, J. J.
LOMACKY, O.
EVALUATION OF FINITE ELEMENT METHODS FOR THE
COMPUTATION OF ELASTIC STRESS INTENSITY FACTORS
PAPER NO. 72-PVP-19, A.S.M.E. MEETING, SEPTEMBER
17-21, 1972, 9 PP., ALSO IN J. ENG. IND., VOL. 95,
NO. 1, 1973, PP. 177-185.

72-0250 RADAJ, D.
COMPARATIVE OBSERVATION ON PARAMETRIC STRESS
STUDIES WITH FINITE ELEMENTS WITH COMPLEX STRESS
FUNCTIONS
ZEITSCHRIFT FUR FLUGWISSENSCHAFTEN, VOL. 20, NO.
12, PP. 479-484, 1972.

72-0251 BELYTSCHKO, T.
FINITE ELEMENTS FOR AXISYMMETRIC SOLIDS UNDER
ARBITRARY LOADINGS WITH NODES ON ORIGIN,
AIAA J. VOL. 10, NO. 11, 1972, PP. 1532-1533.

72-0252 FRIED, I.
BEST FINITE ELEMENTS DISTRIBUTION AROUND A
SINGULARITY
AIAA J., VOL. 10, NO. 9, 1972, PP. 1244-1246.

72-0253 TABARROK, B.
GASS, N.
VIBRATION OF CYLINDRICAL SHELLS BY HYBRID FINITE
ELEMENT METHOD
AIAA J. VOL. 10, NO. 12, 1972, PP. 1553-1554.

72-0254 DAVIS, R.
HENSHELL, R. D.
WARBURTON, G. B.
CONSTANT CURVATURE BEAM FINITE ELEMENTS FOR
IN-PLANE VIBRATION
J. SOUND AND VIB., VOL. 25, NO. 4, 1972, PP.
561-576.

72-0255 GLADWELL, G. M.
TAHBILDAR, U. C.
FINITE ELEMENT ANALYSIS OF AXISYMMETRIC VIBRATIONS
OF CYLINDERS
J. SOUND AND VIB., VOL. 22, NO. 2, 1972, PP.
143-157.

72-0256 FRIED, I.
BOUNDS ON EXTERNAL EIGENVALUES OF FINITE ELEMENT
STIFFNESS AND MASS MATRICES AND THEIR SPECTRAL
CONDITION NUMBER
J. SOUND AND VIB., VOL. 22, NO. 4, 1972, PP.
407-418.

72-0257 WASHIZU, K.
SOME CONSIDERATIONS ON BASIC THEORY FOR THE FINITE
ELEMENT METHOD
IN ADVANCES IN COMPUTATIONAL METHODS IN STRUCTURAL
MECHANICS AND DESIGN, EDITED BY J. T. ODEN, R. W.
CLOUGH, AND Y. YAMAMOTO, PAPER PRESENTED AT THE
PROCEEDINGS OF THE 2ND U.S.- JAPAN SEMINAR ON
MATRIX METHODS OF STRUCTURAL ANALYSIS AND DESIGN,
AUGUST 1972, PP. 39-53.

72-0258 FIX, G. J.
ON THE EFFECTS OF QUADRATURE ERRORS IN THE FINITE
ELEMENT METHOD
IN ADVANCES IN COMPUTATIONAL METHODS IN STRUCTURAL
MECHANICS AND DESIGN, EDITED BY J. T. ODEN, R. W.
CLOUGH, AND Y. YAMAMOTO, PAPER PRESENTED AT THE
PROCEEDINGS OF THE 2ND U.S.- JAPAN SEMINAR ON
MATRIX METHODS OF STRUCTURAL ANALYSIS AND DESIGN,
AUGUST 1972, PP. 55-68.

72-0259 KANDA, T.
SAWADA, Y.
YASUNAMI, K.
STRESS ANALYSIS OF SELF-SEALING TYPE PISTON HEAD
BY FINITE-ELEMENT METHOD
HYDRAULICS AND PNEUMATICS, VOL. 25, NO. 9, 1972,
PP. 50.

72-0260 YAMAMOTO, Y.
SOME CONSIDERATIONS ON ROUND-OFF ERRORS OF THE
FINITE ELEMENT METHOD
IN ADVANCES IN COMPUTATIONAL METHODS IN STRUCTURAL
MECHANICS AND DESIGN, EDITED BY J. T. ODEN, R. W.
CLOUGH, AND Y. YAMAMOTO, PAPER PRESENTED AT THE
PROCEEDINGS OF THE 2ND U.S.- JAPAN SEMINAR ON
MATRIX METHODS OF STRUCTURAL ANALYSIS AND DESIGN,
AUGUST 1972, PP. 69-86.

72-0261 BABUSKA, I.
ROSENZWEIG, M. B.
FINITE ELEMENT SCHEME FOR DOMAINS WITH CORNERS
NUMERISCHE MATHEMATIK, VOL. 20, NO. 1, 1972, PP.
1-21. ALSO IN TECH. NOTE BN-720, U. OF
MARYLAND, INST. OF FLUID DYNAMICS AND APPLIED
MATHEMATICS, 1971.

72-0262 PIAN, T. H. H.
MAU, S. T.
SOME RECENT STUDIES IN ASSUMED STRESS HYBRID
MODELS
IN ADVANCES IN COMPUTATIONAL METHODS IN STRUCTURAL
MECHANICS AND DESIGN, EDITED BY J. T. ODEN, R. W.
CLOUGH, AND Y. YAMAMOTO, PAPER PRESENTED AT THE
PROCEEDINGS OF THE 2ND U.S.- JAPAN SEMINAR ON
MATRIX METHODS OF STRUCTURAL ANALYSIS AND DESIGN,
AUGUST 1972, PP. 87-106.

72-0263 KIKUCHI, F.
ANDO, Y.
LUMPED FINITE ELEMENT BASES FOR BEAMS AND PLATES
J. NUCL. SCI. AND TECHNOL., TOKYO, VOL. 9, NO. 12,
1972, PP. 749-751.

72-0264 NORVILLE, C. C.
MILLS, B.
APPLIED FINITE ELEMENT DISPLACEMENT ANALYSIS OF A
SIMPLE INTEGRAL CONSTRUCTION VEHICLE BODY SHELL
INT. J. MECH. SCI., VOL. 14, NO. 12, 1972, PP.
809-821.

72-0265 ANDERSON, D. T.
MILLS, B.
DYNAMIC ANALYSIS OF A CAR CHASSIS FRAME USING
FINITE ELEMENT METHOD
INT. J. MECH. SCI., VOL. 14, NO. 12, 1972, PP.
799-808.

72-0266 BELYTSCHKO, T.
PLANE STRESS SHAKEDOWN ANALYSIS BY FINITE ELEMENTS
INT. J. MECH. SCI., VOL. 14, NO. 9, 1972, PP.
619-625.

72-0267 DAWE, D. J.
RIGID-BODY MOTIONS AND STRAIN-DISPLACEMENT
EQUATIONS OF CURVED SHELL FINITE ELEMENTS
INT. J. MECH. SCI., VOL. 14, NO. 9, 1972, PP.
569-578.

72-0268 KIKUCHI, F.
ANDO, Y.
SOME FINITE ELEMENT SOLUTIONS FOR PLATE BENDING
PROBLEMS BY SIMPLIFIED HYBRID DISPLACEMENT METHOD
NUCL. ENG. AND DES., VOL. 23, NO. 2, 1972, PP.
155-178.

72-0269 FIX, G. F.
NASSIF, N.
FINITE ELEMENT APPROXIMATIONS TO TIME-DEPENDENT
PROBLEMS
NUMERISCHE MATHEMATIK, VOL. 19, NO. 2, 1972, PP.
127-135.

72-0270 YOKOUCHI, Y.
YAMADA, Y.
SANBONGI, S.
FINITE DIFFERENCE SOLUTIONS FOR LARGE DEFORMATIONS
OF CYLINDRICAL SHELLS - A COMPARISON WITH
FINITE ELEMENT SOLUTIONS
IN ADVANCES IN COMPUTATIONAL METHODS IN STRUCTURAL
MECHANICS AND DESIGN, EDITED BY J. T. ODEN, R. W.
CLOUGH, AND Y. YAMAMOTO, PAPER PRESENTED AT THE
PROCEEDINGS OF THE 2ND U.S.- JAPAN SEMINAR ON
MATRIX METHODS OF STRUCTURAL ANALYSIS AND DESIGN,
AUGUST 1972, PP. 107-125.

72-0271 ZIENKIEWICZ, O. C.
OWEN, D. R. J.
PHILLIPS, D. V.
NAYAK, G. C.
FINITE ELEMENT METHODS IN ANALYSIS OF REACTOR
VESSELS
NUCL. ENG. AND DES., VOL. 20, NO. 2, 1972, PP.
507-741.

72-0272 RADAJ, D.
FINITE-ELEMENT METHOD FOR DETERMINING STRENGTH OF
WELDED STRUCTURES
CHEMIE-INGENIEUR-TECHNIK, VOL. 44, NO. 12, 1972,
PP. 812-817

72-0273 FELIPPA, C. A.
BASIS FOR FORMULATION OF FINITE ELEMENT MODELS
IN ADVANCES IN COMPUTATIONAL METHODS IN STRUCTURAL
MECHANICS AND DESIGN, EDITED BY J. T. ODEN, R. W.
CLOUGH, AND Y. YAMAMOTO, PAPER PRESENTED AT THE
PROCEEDINGS OF THE 2ND U.S.- JAPAN SEMINAR ON
MATRIX METHODS OF STRUCTURAL ANALYSIS AND DESIGN,
AUGUST 1972, PP. 127-132.

72-0274 KITAGAWA, H.
SEGUCHI, Y.
TOMITA, Y.
INCREMENTAL THEORY OF LARGE STRAIN AND LARGE
DISPLACEMENT PROBLEMS AND ITS FINITE ELEMENT
FORMULATION
INGENIEUR-ARCHIV, VOL. 41, NO. 3, 1972, PP.
213-224.

72-0275 ZIENKIEWICZ, O. C.
TRANSIENT RESPONSE OF A COUPLED PLATE-ACOUSTIC
SYSTEM USING PLATE AND ACOUSTIC FINITE ELEMENTS
J. SOUND AND VIB., VOL. 20, NO. 2, 1972, PP. 253.

72-0276 MCLAY, R. W.
NATURAL BOUNDARY CONDITIONS IN FINITE-ELEMENT
METHOD
J. APPL. MECH., VOL. 39, NO. 4, 1972, PP.
1149-1150.

72-0277 FRIED, I.
PERTURBATION ERRORS IN FINITE ELEMENT METHOD
J. APPL. MECH. VOL. 39, NO. 2, 1972, PP. 629-631.

72-0278 HENSHELL, R. D.
WALTERS, D.
WARBURTON, G. B.
POSSIBLE LOSS OF ACCURACY IN CURVED FINITE
ELEMENTS
J. SOUND AND VIB., VOL. 23, NO. 4, 1972, PP.
510-513.

72-0279 FRIED, I.
COMMENT ON THE PAPER "POSSIBLE LOSS OF ACCURACY IN
CURVED (ISOPARAMETRIC) FINITE ELEMENTS" BY
HENSHELL, WALTERS, AND WARBURTON
J. SOUND AND VIB., VOL. 23, NO. 4, 1972, PP.
507-513.

72-0280 MEI, C.
FREE VIBRATIONS OF FINITE ELEMENT PLATES SUBJECTED
TO COMPLEX MIDDLE-PLANE FORCE SYSTEMS
J. SOUND AND VIB., VOL. 23, NO. 2, 1972, PP.
145-156.

72-0281 GOLDSBERRY, F.
CRAWFORD, P. B.
RELIABILITY STUDY OF LINEAR VARIATION TRIANGULAR
FINITE ELEMENT
TEXAS J. OF SCIENCE, VOL. 23, NO. 4, 1972, P. 561.

72-0282 GLOWINSKI, R.
METHOD FOR EXTERNAL APPROXIMATION USING FINITE
ELEMENTS FOR DELTA - 2 AND ITERATIVE METHOD FOR
SOLVING APPROXIMATION PROBLEM USING FINITE
ELEMENTS OF ORDER 2
COMPTES RENDUS... ACADEMIE DES SCIENCES SER A
MATH., VOL. 275, NO. 3, 1972, PP. 201-204.

72-0283 GLOWINSKI, R.
METHOD FOR EXTERNAL APPROXIMATION USING FINITE
ELEMENTS FOR DELTA -2 AND ITERATIVE METHOD FOR
SOLVING APPROXIMATION PROBLEM USING FINITE
ELEMENTS OF ORDER 2
COMPTES RENDUS... ACADEMIE DES SCIENCES SER A
MATH, VOL. 275, NO. 5, 1972, PP. 333-335.

72-0284 CROUZEIX, M.
NUMERICAL SOLUTION BY METHODS OF FINITE ELEMENTS
OF DEGENERATED ELLIPTICAL PROBLEMS
COMPTES RENDUS... ACADEMIE DES SCIENCES, SER. A
MATH., VOL. 275, NO. 21, 1972, PP. 1115-1118.

72-0285 WEINBERG, H. F.
VARIATIONAL PROPERTIES OF STEADY FALL IN STOKES
FLOW
J. FLUID MECH., VOL. 52, PT. 2, 1972, PP. 321-344.

72-0286 YOSHIKAWA, K. K.
VARIATIONAL SOLUTIONS BY USE OF STEPWISE CONSTANT
FUNCTIONS: ONE LINEAR CASE
AIAA J. VOL. 10, NO. 3, 1972, PP. 343-344.

72-0287 YOSHIDA, Y.
EQUIVALENT FINITE ELEMENTS ON DIFFERENT BASES
IN ADVANCES IN COMPUTATIONAL METHODS IN STRUCTURAL
MECHANICS AND DESIGN, EDITED BY J. T. ODEN, R. W.
CLOUGH, AND Y. YAMAMOTO, PAPER PRESENTED AT THE
PROCEEDINGS OF THE 2ND U.S.- JAPAN SEMINAR ON
MATRIX METHODS OF STRUCTURAL ANALYSIS AND
DESIGN, AUGUST 1972, PP. 133-149.

72-0288 BABUSKA, I.
FINITE ELEMENT METHOD FOR INFINITE DOMAINS
I. MATH. COMPUT., VOL. 26, NO. 117, 1972, PP.
1-11.

72-0289 DALY, P.
HELPS, J. D.
DIRECT METHOD OF OBTAINING CAPACITANCE FROM
FINITE-ELEMENT MATRICES
ELECTRON. LETT., VOL. 8, NO. 5, 1972, PP. 132-133.

72-0290 HOLEHOUSE, I.
SONIC FATIGUE OF DIFFUSION BOUNDED TITANIUM
SANDWICH STRUCTURES
AGARD CONF. PROC., NO. 113 FOR MEETING SEPTEMBER
26-27, 1972, 15 PP.

72-0291 FRIED, I.
ACCURACY OF COMPLEX FINITE ELEMENTS
AIAA J. VOL. 10, NO. 3, 1972, PP. 347-349.

72-0292 FRIED, I.
OPTIMAL GRADIENT MINIMIZATION SCHEME FOR FINITE
ELEMENT EIGENPROBLEMS
J. SOUND VIBRAT. VOL. 20, NO. 3, 1972, PP. 333.

72-0293 STRANG, G.
APPROXIMATION IN FINITE ELEMENT METHOD
NUMER. MATH. VOL. 19, NO. 1, 1972, 81 PP.

72-0294 WALLERSTEIN, D. V.
GENERAL LINEAR GEOMETRIC MATRIX FOR A FULLY
COMPATIBLE FINITE ELEMENT
AIAA J., VOL. 10, NO. 4, 1972, PP. 545-546.

72-0295 ALLAN, T.
APPLICATION OF FINITE ELEMENT ANALYSIS TO
HYDRODYNAMIC AND EXTERNALLY PRESSURIZED POCKET
BEARINGS
WEAR. USURE. VERSCHLEISS, VOL. 19, NO. 2, 1972, P.
169.

72-0296 CHACOUR, S.
DANUTA THREE DIMENSIONAL FINITE ELEMENT PROGRAM
USED IN ANALYSIS OF TURBOMACHINERY
J. BASIC ENG., VOL. 94, NO. 1, 1972, PP. 71-77.
(ALSO IN A.S.M.E., PAPER 71-WA/FE-29, MEETING,
NOVEMBER 28 - DECEMBER 2, 1971, 7 PP.

72-0297 CHENG, R. T.
TRANSIENT FREE-SURFACE FLOW IN POROUS MEDIA BY
FINITE ELEMENT METHOD
PAPER PRESENTED AT AMER. GEOPHYS. UN. 53RD ANN.
MEET., WASHINGTON, D.C., APRIL 17-21, 1972.

72-0298 YAMADA, YOSHIKAZU
DYNAMIC ANALYSIS OF STRUCTURES
IN "ADVANCES IN COMPUT. METH. IN STRUCT. MECH. AND
DESIGN", ODEN, J.T., CLOUGH, R.W., YAMAMOTO, Y.,
(EDS.), PAPER PRESENTED AT PROC. OF THE 2ND
U.S.-JAPAN SEM. ON MATRIX METH. OF STRUCT.
ANALYSIS AND DESIGN, AUGUST 1972, PP. 201-218.

72-0299 CZENDES, Z. J.
SILVESTER, P.
FINPLT; A FINITE-ELEMENT FIELD-PLOTTING PROGRAM
IEEE TRANS. MICROWAVE THEORY TECH. VOL. MTT-20,
NO. 4, 1972, PP. 294-295.

72-0300 FANG, C. S.
WANG, S. N.
GROUNDWATER FLOW IN A SANDY TIDAL BEACH: II.
2-DIMENSIONAL FINITE ELEMENT ANALYSIS
WATER RESOURCES RES., VOL. 8, NO. 1, 1972, PP.
121-128.

72-0301 HAYASHI, H.
WADA, S.
FINITE ELEMENT METHOD IN HYDRODYNAMIC LUBRICATION
J. JAP. SOC. LUBRICAT. ENG., VOL. 17, NO. 1, 1972,
3 PP.

72-0302 FUJII, H.
FINITE ELEMENT SCHEMES: STABILITY AND CONVERGENCE
IN ADVANCES IN COMPUTATIONAL METHODS IN STRUCTURAL
MECHANICS AND DESIGN, EDITED BY J. T. ODEN, R. W.
CLOUGH, AND Y. YAMAMOTO, PAPER PRESENTED AT THE
PROCEEDINGS OF THE 2ND U.S.- JAPAN SEMINAR ON
MATRIX METHODS OF STRUCTURAL ANALYSIS AND DESIGN,
AUGUST 1972, PP. 201-218.

72-0303 SHIMIZU, S.
FREE VIBRATION ANALYSIS OF STIFFENED PLATES
IN ADVANCES IN COMPUTATIONAL METHODS IN STRUCTURAL
MECHANICS AND DESIGN, EDITED BY J. T. ODEN, R. W.
CLOUGH, AND Y. YAMAMOTO, PAPER PRESENTED AT THE
PROCEEDINGS OF THE 2ND U.S.- JAPAN SEMINAR ON
MATRIX METHODS OF STRUCTURAL ANALYSIS AND DESIGN,
AUGUST 1972, PP. 219-236.

72-0304 JANZ, R. F.
GRIMM, A. F.
FINITE-ELEMENT MODEL FOR THE MECHANICAL BEHAVIOUR
OF THE LEFT VENTRICLE. PREDICTION OF DEFORMATION
IN THE POTASSIUM-ARRESTED RAT HEART
CIRCULAT. RES. VOL. 30, NO. 2, 1972, PP. 244-252.

72-0305 KREIG, R. D.
KEY, S. W.
TRANSIENT SHELL RESPONSE BY NUMERICAL TIME
INTEGRATION
IN ADVANCES IN COMPUTATIONAL METHODS IN STRUCTURAL
MECHANICS AND DESIGN, EDITED BY J. T. ODEN, R. W.
CLOUGH, AND Y. YAMAMOTO, PAPER PRESENTED AT THE
PROCEEDINGS OF THE 2ND U.S.- JAPAN SEMINAR ON
MATRIX METHODS OF STRUCTURAL ANALYSIS AND DESIGN,
AUGUST 1972, PP. 237-258.

72-0306 MORGAN, S.
DENNIS, R. W.
THEORETICAL PREDICTION OF DISC BRAKE TEMPERATURES
AND A COMPARISON WITH EXPERIMENTAL DATA
PAPER 720090 FOR SAE MEETING, JANUARY 10-14, 1972,
5 PP.

72-0307 SHIMOMURA, Y.
OHTA, M.
TAKEDA, T.
ANALYSIS OF HYDROMAGNETIC PLASMA STABILITY BY
FINITE-ELEMENT METHOD
NUCL. FUSION, VOL. 12, NO. 2, 1972, PP. 271-274.

72-0308 PALIT, K.
FENNER, R. T.
FINITE ELEMENT ANALYSIS OF SLOW NON-NEWTONIAN
CHANNEL FLOW
J. AICE, VOL. 18, NO. 3, 1972, PP. 628-633.

72-0309 RYU, J.
FINITE ELEMENT METHOD TO ELECTROMAGNETIC INDUCTION
PROBLEMS IN GEOPHYSICS
PAPER PRESENTED AT AMER. GEOPHYS. UN. 53RD ANN.
MEETING, WASHINGTON, D. C. APRIL 17-21, 1972.

72-0310 SEMENZA, L. A.
LEWIS, E. E.
ROSSOW, E. C.
APPLICATION OF FINITE ELEMENT METHOD TO MULTIGROUP
NEUTRON DIFFUSION EQUATION
NUCL. SCI. ENG. VOL. 47, NO. 3, 1972, PP. 302-310.

72-0311 SOLIMAN, J. I.
FAKHROO, E. A.
FINITE ELEMENT SOLUTION OF HEAT TRANSMISSION IN
STEEL INGOTS
J. MECH. SCI., VOL. 14, NO. 1, 1972, PP. 19-24.

72-0312 MATSUMOTO, K.
VIBRATION OF AN ELASTIC BODY IMMERSED IN A FLUID
IN ADVANCES IN COMPUTATIONAL METHODS IN STRUCTURAL
MECHANICS AND DESIGN, EDITED BY J. T. ODEN, R. W.
CLOUGH, AND Y. YAMAMOTO, PAPER PRESENTED AT THE
PROCEEDINGS OF THE 2ND U.S.- JAPAN SEMINAR ON
MATRIX METHODS OF STRUCTURAL ANALYSIS AND DESIGN,
AUGUST 1972, PP. 259-274.

72-0313 ELIAS, Z. M.
DYNAMIC ANALYSIS OF FRAME STRUCTURES BY THE FORCE
METHOD
IN ADVANCES IN COMPUTATIONAL METHODS IN STRUCTURAL
MECHANICS AND DESIGN, EDITED BY J. T. ODEN, R. W.
CLOUGH, AND Y. YAMAMOTO, PAPER PRESENTED AT THE
PROCEEDINGS OF THE 2ND U.S.- JAPAN SEMINAR ON
MATRIX METHODS OF STRUCTURAL ANALYSIS AND DESIGN,
AUGUST 1972, PP. 275-297.

72-0314 STRICKLIN, J. A.
VON RIESEMANN, W. A.
TILLERSON, J. R.
HAISLER, W. E.
STATIC GEOMETRIC AND MATERIAL NONLINEAR ANALYSIS
IN ADVANCES IN COMPUTATIONAL METHODS IN STRUCTURAL
MECHANICS AND DESIGN, EDITED BY J. T. ODEN, R. W.
CLOUGH, AND Y. YAMAMOTO, PAPER PRESENTED AT THE
PROCEEDINGS OF THE 2ND U.S.- JAPAN SEMINAR ON
MATRIX METHODS OF STRUCTURAL ANALYSIS AND DESIGN,
AUGUST 1972, PP. 301-324.

72-0315 YAMADA, Y.
INCREMENTAL FORMULATIONS FOR PROBLEMS WITH
GEOMETRIC AND MATERIAL NONLINEARITIES
IN ADVANCES IN COMPUTATIONAL METHODS IN STRUCTURAL
MECHANICS AND DESIGN, EDITED BY J. T. ODEN, R. W.
CLOUGH, AND Y. YAMAMOTO, PAPER PRESENTED AT THE
PROCEEDINGS OF THE 2ND U.S.- JAPAN SEMINAR ON
MATRIX METHODS OF STRUCTURAL ANALYSIS AND DESIGN,
AUGUST 1972, PP. 325-355.

72-0316 MALLETT, R. H.
HAFTKA, R. T.
PROGRESS IN NONLINEAR FINITE ELEMENT ANALYSIS
USING ASYMPTOTIC SOLUTION TECHNIQUES
IN "ADVANCES IN COMPUT. METH. IN STRUCT. MECH. AND
DESIGN",ODEN, J.T., CLOUGH, R.W., YAMAMOTO, Y.,
(EDS.), PAPER PRESENTED AT PROC. OF THE 2ND
U.S.-JAPAN SEM. ON MATRIX METH. OF STRUCT.
ANALYSIS AND DESIGN, AUGUST 1972, PP. 357-373.

72-0317 UEDA, Y.
YAMAKAWA, T.
THERMAL NONLINEAR BEHAVIOUR OF STRUCTURES
IN ADVANCES IN COMPUTATIONAL METHODS IN STRUCTURAL
MECHANICS AND DESIGN, EDITED BY J. T. ODEN, R. W.
CLOUGH, AND Y. YAMAMOTO, PAPER PRESENTED AT THE
PROCEEDINGS OF THE 2ND U.S.- JAPAN SEMINAR ON
MATRIX METHODS OF STRUCTURAL ANALYSIS AND DESIGN,
AUGUST 1972, PP. 375-392.

72-0318 ARMEN, H.
LEVINE, H. S.
PIFKO, A. B.
PLASTICITY THEORY AND FINITE ELEMENT APPLICATIONS
IN ADVANCES IN COMPUTATIONAL METHODS IN STRUCTURAL
MECHANICS AND DESIGN, EDITED BY J. T. ODEN, R. W.
CLOUGH, AND Y. YAMAMOTO, PAPER PRESENTED AT THE
PROCEEDINGS OF THE SECOND U.S.-JAPAN SEMINAR ON
MATRIX METHODS OF STRUCTURAL ANALYSIS AND DESIGN,
AUGUST 1972, PP. 393-437.

72-0319 OHTSUBO, H.
A METHOD OF ELASTIC-PLASTIC ANALYSIS OF LARGELY
DEFORMED PLATE PROBLEMS
IN ADVANCES IN COMPUTATIONAL METHODS IN STRUCTURAL
MECHANICS AND DESIGN, EDITED BY J. T. ODEN, R. W.
CLOUGH, AND Y. YAMAMOTO, PAPER PRESENTED AT THE
PROCEEDINGS OF THE 2ND U.S.- JAPAN SEMINAR ON
MATRIX METHODS OF STRUCTURAL ANALYSIS AND DESIGN,
AUGUST 1972, PP. 439-456.

72-0320 HARTZ, B. J.
CHOPRA, P. S.
A FINITE ELEMENT CONTRIBUTION TO FRACTURE
MECHANICS
IN ADVANCES IN COMPUTATIONAL METHODS IN STRUCTURAL
MECHANICS AND DESIGN, EDITED BY J. T. ODEN, R. W.
CLOUGH, AND Y. YAMAMOTO, PAPER PRESENTED AT THE
PROCEEDINGS OF THE SECOND U.S.-JAPAN SEMINAR ON
MATRIX METHODS OF STRUCTURAL ANALYSIS AND
DESIGN, AUGUST 1972, PP. 457-471.

72-0321 HANGAI, Y.
KAWAMATA, S.
PERTURBATION METHOD IN THE ANALYSIS OF
GEOMETRICALLY NON LINEAR AND STABILITY PROBLEMS
IN "ADVANCES IN COMPUT. METH. IN STRUCT. MECH. AND
DESIGN",ODEN, J.T., CLOUGH, R.W., YAMAMOTO, Y.,
(EDS.), PAPER PRESENTED AT PROC. OF THE 2ND
U.S.-JAPAN SEM. ON MATRIX METH. OF STRUCT.
ANALYSIS AND DESIGN, AUGUST 1972, PP. 473-489.

72-0322 KAWAI, T.
SOME PROGRAM DEVELOPMENTS OF MATRIX STRUCTURAL
ANALYSIS IN JAPAN
IN ADVANCES IN COMPUTATIONAL METHODS IN STRUCTURAL
MECHANICS AND DESIGN, EDITED BY J. T. ODEN, R. W.
CLOUGH, AND Y. YAMAMOTO, PAPER PRESENTED AT THE
PROCEEDINGS OF THE SECOND U.S.-JAPAN SEMINAR ON
MATRIX METHODS OF STRUCTURAL ANALYSIS AND
DESIGN, AUGUST 1972, PP. 493-516.

72-0323 MARCAL, P. V.
SURVEY OF GENERAL PURPOSE PROGRAMS FOR FINITE
ELEMENT ANALYSIS
IN ADVANCES IN COMPUTATIONAL METHODS IN STRUCTURAL
MECHANICS AND DESIGN, EDITED BY J. T. ODEN, R. W.
CLOUGH, AND Y. YAMAMOTO, PAPER PRESENTED AT THE
PROCEEDINGS OF THE SECOND U.S.-JAPAN SEMINAR ON
MATRIX METHODS OF STRUCTURAL ANALYSIS AND
DESIGN, AUGUST 1972, PP. 517-528.

72-0324 KIRIOKA, K.
HIRATA, T.
COMPUTING SYSTEM FOR STRUCTURAL ANALYSIS OF CAR
BODIES
IN ADVANCES IN COMPUTATIONAL METHODS IN STRUCTURAL
MECHANICS AND DESIGN, EDITED BY J. T. ODEN, R. W.
CLOUGH, AND Y. YAMAMOTO, PAPER PRESENTED AT THE
PROCEEDINGS OF THE 2ND U.S.- JAPAN SEMINAR ON
MATRIX METHODS OF STRUCTURAL ANALYSIS AND DESIGN,
AUGUST 1972, PP. 529-549.

72-0325 MCCORMICK, C. W.
THE NASTRAN PROGRAM FOR STRUCTURAL ANALYSIS
IN ADVANCES IN COMPUTATIONAL METHODS IN STRUCTURAL
MECHANICS AND DESIGN, EDITED BY J. T. ODEN, R. W.
CLOUGH, AND Y. YAMAMOTO, PAPER PRESENTED AT THE
PROCEEDINGS OF THE SECOND U.S.-JAPAN SEMINAR ON
MATRIX METHODS OF STRUCTURAL ANALYSIS AND
DESIGN, AUGUST 1972, PP. 551-571.

72-0326 NAKAO, Y.
KAWASHIMA, M.
PRACTICAL ANALYSIS OF STEEL STRUCTURES USING
HIGHER-ORDER ELEMENTS
IN ADVANCES IN COMPUTATIONAL METHODS IN STRUCTURAL
MECHANICS AND DESIGN, EDITED BY J. T. ODEN, R. W.
CLOUGH, AND Y. YAMAMOTO, PAPER PRESENTED AT THE
PROCEEDINGS OF THE 2ND U.S.- JAPAN SEMINAR ON
MATRIX METHODS OF STRUCTURAL ANALYSIS AND DESIGN,
AUGUST 1972, PP. 573-589.

72-0327 LOGCHER, R. D.
THE DEVELOPMENT OF ICES STRUDL
IN ADVANCES IN COMPUTATIONAL METHODS IN STRUCTURAL
MECHANICS AND DESIGN, EDITED BY J. T. ODEN, R. W.
CLOUGH, AND Y. YAMAMOTO, PAPER PRESENTED AT THE
PROCEEDINGS OF THE 2ND U.S.- JAPAN SEMINAR ON
MATRIX METHODS OF STRUCTURAL ANALYSIS AND DESIGN,
AUGUST 1972, PP. 591-606.

72-0328 SUHARA, J.
FUKUDA, J.
AUTOMATIC MESH GENERATION FOR FINITE ELEMENT
ANALYSIS
IN ADVANCES IN COMPUTATIONAL METHODS IN STRUCTURAL
MECHANICS AND DESIGN, EDITED BY J. T. ODEN, R. W.
CLOUGH, AND Y. YAMAMOTO, PAPER PRESENTED AT THE
PROCEEDINGS OF THE 2ND U.S.- JAPAN SEMINAR ON
MATRIX METHODS OF STRUCTURAL ANALYSIS AND DESIGN,
AUGUST 1972, PP. 607-624.

72-0329 WILSON, E. L.
SAP-A GENERAL STRUCTURAL ANALYSIS PROGRAM FOR
LINEAR SYSTEMS
IN ADVANCES IN COMPUTATIONAL METHODS IN STRUCTURAL
MECHANICS AND DESIGN, EDITED BY J. T. ODEN, R. W.
CLOUGH, AND Y. YAMAMOTO, PAPER PRESENTED AT THE
PROCEEDINGS OF THE 2ND U.S.- JAPAN SEMINAR ON
MATRIX METHODS OF STRUCTURAL ANALYSIS AND DESIGN,
AUGUST 1972, PP. 625-638. ALSO IN NUCL. ENG. DES.,
VOL. 25, NO. 2, JULY 1973, PP. 257-274.

72-0330 GALLAGHER, R. H.
APPLICATIONS OF FINITE ELEMENT ANALYSIS
IN ADVANCES IN COMPUTATIONAL METHODS IN STRUCTURAL
MECHANICS AND DESIGN, EDITED BY J. T. ODEN, R. W.
CLOUGH, AND Y. YAMAMOTO, PAPER PRESENTED AT THE
PROCEEDINGS OF THE 2ND U.S.- JAPAN SEMINAR ON
MATRIX METHODS OF STRUCTURAL ANALYSIS AND DESIGN,
AUGUST 1972, PP. 641-678.

72-0331 MIYAMOTO, H.
ISHIJIMA, Y.
SHIRATORI, M.
MIYOSHI, T.
THE APPLICATION OF THE FINITE ELEMENT METHOD TO
FRACTURE MECHANICS
IN ADVANCES IN COMPUTATIONAL METHODS IN STRUCTURAL
MECHANICS AND DESIGN, EDITED BY J. T. ODEN, R. W.
CLOUGH, AND Y. YAMAMOTO, PAPER PRESENTED AT THE
PROCEEDINGS OF THE 2ND U.S.- JAPAN SEMINAR ON
MATRIX METHODS OF STRUCTURAL ANALYSIS AND DESIGN,
AUGUST 1972, PP. 679-701.

72-0332 KAMEL, H. A.
LIU, D.
MCCABE, M. W.
PHILIPPOPOULOS, V.
SOME DEVELOPMENTS IN THE ANALYSIS OF COMPLEX SHIP
STRUCTURES
IN "ADVANCES IN COMPUT. METH. IN STRUCT. MECH. AND
DESIGN", ODEN, J.T., CLOUGH, R.W., YAMAMOTO, Y.,
(EDS.), PAPER PRESENTED AT PROC. OF THE 2ND
U.S.-JAPAN SEM. ON MATRIX METH. OF STRUCT.
ANALYSIS AND DESIGN, AUGUST 1972, PP. 703-726.

72-0333 KAWAI, T.
TADA, Y.
FINITE ELEMENT ANALYSIS OF A WING STRUCTURE
IN ADVANCES IN COMPUTATIONAL METHODS IN STRUCTURAL
MECHANICS AND DESIGN, EDITED BY J. T. ODEN, R. W.
CLOUGH, AND Y. YAMAMOTO, PAPER PRESENTED AT THE
PROCEEDINGS OF THE 2ND U.S.- JAPAN SEMINAR ON
MATRIX METHODS OF STRUCTURAL ANALYSIS AND DESIGN,
AUGUST 1972, PP. 727-744.

72-0334 MELOSH, R. J.
NUMERICAL ANALYSIS OF AUTOMOBILE STRUCTURES
IN ADVANCES IN COMPUTATIONAL METHODS IN STRUCTURAL
MECHANICS AND DESIGN, EDITED BY J. T. ODEN, R. W.
CLOUGH, AND Y. YAMAMOTO, PAPER PRESENTED AT THE
PROCEEDINGS OF THE 2ND U.S.- JAPAN SEMINAR ON
MATRIX METHODS OF STRUCTURAL ANALYSIS AND DESIGN,
AUGUST 1972, PP. 745-756.

72-0335 ENDA, Y.
ANALYSIS OF THIN-WALLED CURVED BEAMS BY THE
TRANSFER MATRIX METHOD
IN ADVANCES IN COMPUTATIONAL METHODS IN STRUCTURAL
MECHANICS AND DESIGN, EDITED BY J. T. ODEN, R. W.
CLOUGH, AND Y. YAMAMOTO, PAPER PRESENTED AT THE
PROCEEDINGS OF THE 2ND U.S.- JAPAN SEMINAR ON
MATRIX METHODS OF STRUCTURAL ANALYSIS AND DESIGN,
AUGUST 1972, PP. 757-774.

72-0336 KELSEY, S.
FINITE ELEMENT METHODS IN CIVIL ENGINEERING
IN ADVANCES IN COMPUTATIONAL METHODS IN STRUCTURAL
MECHANICS AND DESIGN, EDITED BY J. T. ODEN, R. W.
CLOUGH, AND Y. YAMAMOTO, PAPER PRESENTED AT THE
PROCEEDINGS OF THE SECOND U.S.-JAPAN SEMINAR ON
MATRIX METHODS OF STRUCTURAL ANALYSIS AND
DESIGN, AUGUST 1972, PP. 757-774.

72-0337 CHEUNG, R. T.
NUMERICAL INVESTIGATION OF LAKE CIRCULATION AROUND
ISLANDS BY THE FINITE ELEMENT METHOD
INT. J. NUM. METHODS IN ENG., VOL. 5, 1972, PP.
103-112.

72-0338 TOMITA, M.
KITAGAWA, H.
AN INCREMENTAL FINITE ELEMENT ANALYSIS OF TWO
DIMENSIONAL LARGE STRAIN AND LARGE DISPLACEMENT
PROBLEM FOR ELASTIC-PLASTIC MATERIAL
PROC. 21ST NATL. CONF. ON THEORETICAL AND APPLIED
MECH., UNIV. OF TOKYO PRESS, 1972.

72-0339 WHITEMAN, J. R.
A BIBLIOGRAPHY FOR FINITE ELEMENT METHODS
BRUNEL UNIVERSITY REPORT TR/9, DEPARTMENT OF
MATHEMATICS. BRUNEL UNIVERSITY, UXBRIDGE, 1972.

72-0340 REID, J. K.
ON THE CONSTRUCTION AND CONVERGENCE OF A FINITE
ELEMENT SOLUTION OF LAPLACE'S EQUATION
JOURNAL I.M.A., VOL. 9, 1972, PP. 1-13.

72-0341 BABUSKA, I.
ZLAMAL, M.
NONCONFORMING ELEMENTS IN THE FINITE ELEMENT
METHOD
TECHNICAL NOTE BN-729, INSTITUTE FOR FLUID
DYNAMICS AND APPLIED MATHEMATICS, MARCH, 1972,
COLLEGE PARK, MARYLAND.

72-0342 PIAN, T. H. H.
FINITE ELEMENT METHODS BY VARIATIONAL PRINCIPLES
WITH RELAXED CONTINUITY REQUIREMENTS
TO BE PRESENTED AT THE INTERNATIONAL CONFERENCE ON
VARIATIONAL METHODS IN ENGINEERING,
SOUTHAMPTON, ENGLAND, SEPTEMBER 25-29, 1972.

72-0343 FIX, G. J.
EFFECTS OF QUADRATURE ERRORS IN FINITE ELEMENT
APPROXIMATION OF STEADY STATE, EIGENVALUE, AND
PARABOLIC PROBLEMS
IN THE MATHEMATICAL FOUNDATIONS OF THE FINITE
ELEMENT METHOD WITH APPLICATIONS TO PARTIAL
DIFFERENTIAL EQUATIONS, EDITED BY A. K. AZIZ,
ACADEMIC PRESS, NEW YORK, 1972, PP. 525-556.

72-0344 HUTCHINSON, P.
WALLEY, W. J.
CALCULATION OF AREAL RAINFALL USING FINITE ELEMENT
TECHNIQUES WITH ALTITUDINAL CORRECTIONS
HYDROL. SCI. BULL. SCI. HYDROL., VOL. 17, NO. 3,
OCTOBER 1972, PP. 259-272.

72-0345 BARKER, R. M.
LIN, F. T.
DANA, J. R.
THREE-DIMENSIONAL FINITE ELEMENT ANALYSIS OF
LAMINATED COMPOSITES
NAT. SYMP. ON COMPUTERIZED STRUCT. ANALYSIS AND
DESIGN, GEORGE WASHINGTON UNIV., WASHINGTON, D.C.,
MARCH 27-29, 1972.

72-0346 STRICKLIN, J. A.
HAISLER, W. E.
VON RUSMAN, W. A.
FORMULATION, COMPUTATION, AND SOLUTION PROCEDURES
FOR MATERIAL AND/OR GEOMETRIC NONLINEAR STRUCTURAL
ANALYSIS BY THE FINITE ELEMENT METHOD
SANDIA CORP. CONTRACT REPORT SC-CR-72-3102,
JANUARY 1972.

72-0347 GALLAGHER, R. H.
GEOMETRICALLY NONLINEAR FINITE ELEMENT ANALYSIS
PROC. CONF. ON FINITE ELEMENT METH. IN CIVIL ENG.,
MCCUTCHEON, J.O., MIRZA, M.S., MUFTI, A.A.,
(EDS.), MCGILL UNIV., MONTREAL, JUNE 1-2, 1972,
PP. 3-34.

72-0348 CHU, T. C.
SCHNOBRICH, W.
FINITE ELEMENT ANALYSIS OF TRANSLATIONAL SHELLS
COMPUTERS AND STRUCTURES, 2, 1972, PP. 197-222.

72-0349 FORTIN
CALCUL NUMERIQUE DES ECOULEMENTS DES FLUIDES DE
BINGHAM ET DES FLUIDES NEWTONIENS
INCOMPRESSIBLES PAR LA METHODE DES ELEMENTS FINIS
THESIS, UNIVERSITE DE PARIS VI, 1972.

72-0350 MORANDI CECCHI, M.
CELLA, A.
AN EXTENDED THEORY FOR THE FINITE ELEMENT METHOD
PROC. INT. CONF. ON VARIATIONAL METHODS IN ENG.,
UNIV. OF SOUTHAMPTON, SEPTEMBER 1972, (EDS)
BREBBIA, C.A. AND TOTTENHAM, H., SOUTHAMPTON UNIV.
PRESS, 1973, VOL. 1, PP. 1/74-1/84.

72-0351 RAVIART, P. A.
THE USE OF NUMERICAL INTEGRATION IN FINITE ELEMENT
METHODS FOR SOLVING PARABOLIC EQUATIONS
CONFERENCE ON NUMERICAL ANALYSIS, ROYAL IRISH
ACADEMY, DUBLIN, AUGUST 1972, ALSO TOPICS ON
NUMERICAL ANALYSIS, MILLER, J. (ED.), ACADEMIC
PRESS, 1973.

72-0352 TONG, P.
FINITE ELEMENT METHODS FOR FLUID PROBLEMS
A SERIES OF TEN LECTURES GIVEN AT THE INTL. CENTER
FOR MECH.SCI., 1972. (PUBL. BY SPRINGER VERLAG).

72-0353 NEGRIN, M. P.
A FINITE ELEMENT APPROACH TO PHYSIOLOGICAL
TRANSPORT PHENOMENA BY CILIARY AND PERISTALTIC
PUMPING
MASTERS THESIS, MASSACHUSETTS INST. OF TECHNOLOGY,
CAMBRIDGE, JUNE 1972.

72-0354 OLSON, M. D.
FORMULATION OF A VARIATIONAL PRINCIPLE - FINITE
ELEMENT METHOD FOR VISCOUS FLOW
INT. CONF. ON VARIATIONAL METH. IN ENG.,
SOUTHAMPTON, ENGLAND, 1972.

72-0355 OLSON, M. D.
A VARIATIONAL FINITE ELEMENT METHOD FOR TWO
DIMENSIONAL STEADY VISCOUS FLOWS
PROC. MCGILL-EIC CONF. ON FINITE ELEMENT METH. IN
CIVIL ENG., MONTREAL, JUNE 1-2, 1972.

72-0356 CIARLET, P. G.
RAVIART, P. A.
INTERPOLATION THEORY OVER CURVED ELEMENTS WITH
APPLICATION TO THE FINITE ELEMENT METHOD
COMPUTER METHOD IN APPLIED MECHANICS AND
ENGINEERING, VOL. 1, 1972, PP. 217-249.

72-0357 CIARLET, P. G.
RAVIART, P. A.
THE COMBINED EFFECT TO CURVED BOUNDARIES AND
NUMERICAL INTEGRATION IN ISOPARAMETRIC FINITE
ELEMENT METHODS
IN THE MATHEMATICAL FOUNDATIONS OF THE FINITE
ELEMENT METHOD WITH APPLICATIONS TO PARTIAL
DIFFERENTIAL EQUATIONS, EDITED BY A. K. AZIZ,
ACADEMIC PRESS, NEW YORK, 1972, PP. 409-474.

72-0358 MARTIN, H. C.
CAREY, G. F.
INTRODUCTION TO FINITE ELEMENT ANALYSIS
MCGRAW-HILL, NEW YORK 1972.

72-0359 GROTKOP, G.
DIE BERECHNUNG VON FLACHWASSERWELLEN NACH DER
METHODE DER FINITEN ELEMENTE
PH. D. THESIS, TECHNISCHE UNIVERSITAT HANNOVER,
HANNOVER, 1972.

72-0360 WHITEMAN, J. R.
BARNHILL, R. E.
FINITE ELEMENT METHODS FOR ELLIPTIC BOUNDARY VALUE
PROBLEMS CONTAINING SINGULARITIES
PROC. EQUADIFF 3, CZECHOSLOVAK CONF. ON
DIFFERENTIAL EQUATIONS AND THEIR APPLICATIONS,
BRNO, CZECHOSLOVAKIA, 1972.

72-0361 BARNHILL, R. E.
GREGORY, J. A.
SARD KERNEL THEOREMS ON TRIANGULAR AND RECTANGULAR
DOMAINS WITH EXTENSIONS AND APPLICATIONS TO
FINITE ELEMENT ERROR BOUNDS
BRUNEL UNIVERSITY, DEPARTMENT OF MATHEMATICS,
REPORT TR/11, 1972.

72-0362 DESAI, C. S.
ABEL, J. F.
INTRODUCTION TO THE FINITE ELEMENT METHOD
VAN NOSTRAND REINHOLD CO., NEW YORK, 1972.

72-0363 LOZIUK, L. A.
ANDERSON, J. C.
BELYTSCHKO, T.
FINITE ELEMENT APPROACH TO HYDROTHERMAL ANALYSIS
OF SMALL LAKES
MTG. PREPRINT 1799. ASCE NAT. ENV. ENGRG. MTG.,
HOUSTON, OCT. 1972.

72-0364 NEUMAN, S. P.
FINITE ELEMENT COMPUTER PROGRAMS FOR FLOW IN
SATURATED UNSATURATED POROUS MEDIA
SECOND ANNUAL REPORT, PROJECT NO. A10-SWC-77,
HYDRAULIC ENGINEERING LABORATORY, TECHNION,
HAIFA, ISRAEL, 1972, P. 87

72-0365 IRONS, B. M.
RAZZAQUE, A.
EXPERIENCE WITH THE PATCH TEST FOR CONVERGENCE OF
FINITE ELEMENTS
IN THE MATHEMATICAL FOUNDATIONS OF THE FINITE
ELEMENT METHOD WITH APPLICATIONS TO PARTIAL
DIFFERENTIAL EQUATIONS, EDITED BY A. K. AZIZ,
ACADEMIC PRESS, NEW YORK, 1972, PP. 557-587.

72-0366 DESAI, C. S.
FINITE ELEMENT PROCEDURES FOR SEEPAGE ANALYSIS
USING AN ISOPARAMETRIC ELEMENT
PROC. SYMP. ON THE APPL. OF THE FINITE ELEMENT
METHOD IN GEOTECH. ENG., VICKSBURG, MISS., MAY
1-4, 1972, PP. 799-824. (PUBL. BY U.S. ARMY ENG.
WATERWAYS EXP. STATION, SEPTEMBER 1972)

72-0367 JOHNSON, C.
CONVERGENCE OF ANOTHER MIXED FINITE ELEMENT METHOD
FOR PLATE BENDING PROBLEMS
REPORT NO. 27, DEPARTMENT OF MATH., CHALMERS
INST., OF TECH. AND THE UNIVERSITY OF GOTEBORG,
1972.

72-0368 MEISSNER, U.
FEHLERBETRACHTUNGEN AUFGRUND DER METHODE DER
GEWICHTETEN RESIDUEN FUR DAS VERFAHREN DER
FINITEN ELEMENTE DEI GRUND UND
SICKERWASSERSTROMUNGEN
JAHRESBERICHT 1971 DES SONDERFORSCHUNGSBEREICHES
79 FUR WASSERFORSCHUNG IM KUSTENBEREICH DER
TECHNISCHEN UNIVERSITAT HANNOVER, NO. 2, 1972, PP.
16-38.

72-0369 BROWN, T. G.
GHALI, A.
FINITE STRIP ANALYSIS OF SKEW SLABS
PROCEEDINGS OF THE CONFERENCE ON FINITE ELEMENT
METHOD IN CIVIL ENGINEERING, MCGILL UNIVERSITY,
MONTREAL, JUNE 1972, PP.1141-1155

72-0370 CHEUNG, Y. K.
HARRISON, D. G.
A HIGHER ORDER TRIANGULAR FINITE ELEMENT FOR THE
SOLUTION OF FIELD PROBLEMS IN ORTHOTROPIC MEDIA
DEPARTMENT OF CIVIL ENGINEERING, UNIVERSITY OF
CALGARY, RESEARCH REPORT NO. CE 72-44, NOVEMBER
1972. ALSO IN INT. J. NUM. METHODS ENG., VOL. 7,
1972, PP. 287-295.

72-0371 WAGNER, R.
THERMAL STRESSES IN OPEN CYLINDRICAL SHELLS OF
ARBITRARY BOUNDARY CONDITIONS BY A FINITE STRIP
METHOD
PROC. INT. CONF. ON VARIATIONAL METH. IN ENG.,
UNIV. OF SOUTHAMPTON, SEPTEMBER 25, 1972, (EDS.)
BREBBIA, C.A. AND TOTTENHAM, H., SOUTHAMPTON UNIV.
PRESS, 1973, VOL. 1, PP. 6/87-6/99.

72-0372 STIANSEN, S. G.
ELBATOUTI, A.
FINITE ELEMENT ANALYSIS OF CONTAINER SHIPS
PRESENTED AT SYMPOSIUM ON THE COMPUTER IN FINITE
ELEMENT ANALYSIS OF SHIP STRUCTURES, UNIVERSITY
OF ARIZONA, TUCZON, ARIZONA, FEBRUARY 28-MARCH 6,
1972.

72-0373 DUNHAM, R. S.
WRIGHT, S. G.
BOTTOM STABILITY UNDER WATER INDUCED LOADING
PROC. 4TH ANN. OFFSHORE TECHNOL. CONF., HOUSTON,
TEXAS, MAY 1-3, 1972. PAPER 1603, VOL. 1, PP.
853-862.

72-0374 SCHIFF, M.
LAGNADO, I.
ANALYSIS AND REDESIGN OF SONAR PIEZOELECTRIC
RESONATORS USING THE FINITE ELEMENT METHOD.
IEEE REGION SIX (U S WESTERN REGION) CONFERENCE,
APRIL 19-21, 1972.

72-0375 YIH, C. S.
ADVANCES IN APPLIED MECHANICS, VOLUME 12.
ADV. APPL. MECH., V. 12, 250 P. PUBL BY ACADEMIC
PRESS, INC.NEW YORK, 1972.

72-0376 ANDERSEN, K. H.
FRIMANN CLAUSEN, C. J.
ANALYSIS OF GEOTECHNICAL PROBLEMS BY MEANS OF THE
FINITE ELEMENTS METHOD
NORG. GEOTEK. INST. PUBL., N. 94, 1972, 18 PP.

72-0377 PIAN, T. H. H.
TONG, P.
FINITE ELEMENT METHODS IN CONTINUUM MECHANICS.
ADV. APPL. MECH., V.12, 1972, PP. 1-58.

72-0378 MARTINEZ, J. E.
PETINER, A.
EFFICIENT COMPUTER ANALYSIS OF LARGE
THREE-DIMENSIONAL PIPING SYSTEMS BY THE FINITE
ELEMENT METHOD
PROC. 4TH ANN. OFFSHORE TECHNOL. CONF., HOUSTON,
TEXAS, MAY 1-38 1972, PAPER 1700, VOL. 2, PP.
753-758.

72-0379 LEONARD, J. W.
CURVED FINITE ELEMENT APPROXIMATION TO NONLINEAR
CABLES
PROC. 4TH ANN. OFFSHORE TECHNOL. CONF., HOUSTON,
TEXAS, MAY 1-3, 1972, PAPER 1533, VOL. 1, PP.
225-236.

72-0380 KWAN, C.
GRAFF, W. J.
ANALYSIS OF TUBULAR T-CONNECTIONS BY THE FINITE
ELEMENT METHOD: COMPARISON WITH EXPERIMENTS
PROC. 4TH ANN. OFFSHORE TECHNOL. CONF., HOUSTON,
TEXAS, MAY 1-3, 1972, PAPER 1669, VOL. 2, PP.
487-496.

72-0381 DAS, P. C.
CRISFIELD, M. A.
MULTI-CELL BRIDGE DECKS - MODEL TESTS AND
THEORETICAL ANALYSIS
TRANSP. RD. RES. LAB., (GB), TRRL. REP. LR. 493,
1972, 25 PP.

72-0382 BROWN, A. J.
STAUB, J. A.
FUNDAMENTAL STUDY OF UNDERWATER WELDING
PROC. 4TH ANN. OFFSHORE TECHNOL. CONF., HOUSTON,
TEXAS, MAY 1-3, 1972, PAPER 1621, VOL. 2, PP.
55-64.

72-0383 BERG, C. A.
SALAMA, M.
FATIGUE OF GRAPHITE FIBRE-REINFORCED EPOXY IN
COMPRESSION.
FIBRE SCI. TECHNOL., V. 6, N. 2, APR. 1973, PP.
79-188.

72-0384 YANG, T. Y.
FINITE DISPLACEMENT PLATE FLEXURE BY THE USE OF
MATRIX INCREMENTAL APPROACH
INTL. J. NUMER. METH. ENG., VOL. 4, NO. 3, MAY -
JUNE, 1972, PP. 415-432.

72-0385 YAMADA, Y.
NAKAGIRI, S.
TAKATSUKA, K.
ELASTIC PLASTIC ANALYSIS OF SAINT-VENANT TORSION
PROBLEM BY A HYBRID STRESS MODEL
INT. J. NUM. METH. ENG., VOL. 5, NO. 2, NOVEMBER -
DECEMBER, 1972, PP. 193-207.

72-0386 YAM, L. C. P.
CONFORMING FINITE ELEMENT WITH INTERNAL
EQUILIBRIUM FOR TWO DIMENSIONAL PROBLEMS
INTL. J. NUMER. METH. ENG., VOL. 4, NO. 3, MAY -
JUNE 1972, PP. 367-378.

72-0387 WITTKE, W.
APPLICATION OF THE FINITE ELEMENT METHOD TO THE
DESIGN OF UNDERGROUND BULKHEADS
ERZMETALL, V. 26, N. 2, FEB. 1973, PP.66-74.

72-0388 WEEKS, G. E.
FINITE ELEMENT MODEL FOR SHELLS BASED ON THE
DISCRETE   KIRCHHOFF HYPOTHESIS.
INT. J. NUMER. METHODS ENG., V. 5, N. 1, SEP-OCT.
1972, PP 3-16.

72-0389 LIN, C. W.
WANG, J. T. S.
SHELL STRUCTURES AND CLIMATIC INFLUENCES.
INT. ASSOC. FOR SHELL STRUCT., IASS CALGARY, SYMP.
PROC., UNIV OF CALGARY, ALTA., JUL 3-6 1972, 539
PP.

72-0390 VENKATESWARA RAO, G.
KRISHNA MURTY, A. V.
BOUNDS FOR EIGENVALUES IN SOME VIBRATION AND
STABILITY PROBLEMS.
INT. J.NUMER. METHODS ENG., V. 5, N. 2, NOV-DEC.
1972, PP. 237-242.

72-0391 TINAWI, R. A.
ANISOTROPIC TAPERED ELEMENTS USING DISPLACEMENT
MODELS
INT. J. NUMER. METH. ENG., VOL. 4, NO. 4, JULY -
AUGUST 1972, PP. 475-489.

72-0392 SILVESTER, P.
TETRAHEDRAL POLYNOMIAL FINITE ELEMENTS FOR THE
HELMHOLTZ   EQUATION.
INT. J.NUMER. METHODS ENG., V. 4, N. 3, MAY-JUN.
1972, PP. 405-413.

72-0393 SECREST, B. G.
BOYD, W. W.
FINITE ELEMENT METHOD APPLIED TO IDEAL GASEOUS
MIXTURES.
INT. J.NUMER. METHODS ENG., V. 4, N. 3, MAY-JUN.
1972, PP. 337-344.

72-0394 SCHNOBRICH, W. C.
STRESSES IN GABLE ROOF H. P. SHELLS.
BULL. INT. ASSOC. SHELL SPAT. STRUCT., N. 50, DEC.
1972, PP.21-25.

72-0395 RIGBY, G. L.
MCNEICE, G. M.
FURTHER LIMITATIONS ON GENERAL HEXAHEDRON FINITE
ELEMENTS   WITH DERIVATIVE DEGREES OF FREEDOM.
INT. J.NUMER. METHODS ENG., V. 5, N. 1, SEP-OCT.
1972, PP. 137-139.

72-0396 NICKEL, R. E.
SECOR, G. A.
CONVERGENCE OF CONSISTENTLY DERIVED TIMOSHENKO
BEAM FINITE ELEMENTS
INT. J. NUMER. METHODS ENGR., VOL. 5, NO. 2,
NOVEMBER-DECEMBER 1972, PP. 243-252.

72-0397 MAIER, G.
ZAVELANI-ROSSI, A.
FINITE ELEMENT APPROACH TO OPTIMAL DESIGN OF
PLASTIC STRUCTURES IN PLANE STRESS.
INT. J. NUMER. METHODS ENG., V. 4, N. 4, JUL-AUG.
1972, PP. 455-473.

72-0398 LEWIS, R. W.
GARNER, R. W.
FINITE ELEMENT SOLUTION OF COUPLED ELECTROKINETIC
AND HYDRODYNAMIC FLOW IN POROUS MEDIA
INT. J. NUMER. METHODS ENG., V. 5, N. 1, SEP-OCT.
1972, PP. 41-55.

72-0399 KORMAN, T.
SMITH, H. S.
SHELL STRUCTURES AND CLIMATIC INFLUENCES
PROC. OF INT. ASSOC. FOR SHELL STRUCT., CALGARY,
ALTA., JULY 3-6, 1972, 539 PP.

72-0400 GUPTA, K. K.
SOLUTION OF EIGENVALUE PROBLEMS BY STURM SEQUENCE
METHOD.
INT. J.NUMER. METHODS ENG., V. 4, N. 3, MAY-JUN.
1972, PP. 379-404.

72-0401 FRAEIJS DE VEUBEKE, B.
HOGGE, M. A.
DUAL ANALYSIS FOR HEAT CONDUCTION PROBLEMS BY
FINITE ELEMENTS
INT. J. NUMER. METH. ENG., VOL. 5, NO. 6,
SEPTEMBER-OCTOBER 1972, PP. 515-529.

72-0402 CHICUREL, R.
SPACE TIME FINITE ELEMENTS IN DYNAMIC PROBLEMS
PROC. 6TH SOUTHEAST CONF. ON THEORY AND APPL.
MECH., UNIV. OF SOUTH FLORIDA, TAMPA, MARCH 23-24,
1972, PP.515-529.

72-0403 COST, T. L.
HILL, J. L.
THERMOMECHANICAL   DISSIPATION   ANALYSIS   OF
THERMOVISCOELASTIC SOLIDS BY FINITE ELEMENTS
PROC. 6TH SOUTHEAST CONF. ON THEORY AND APPL.
MECH., UNIV. OF SOUTH FLORIDA, TAMPA, MARCH 23-24,
1972, PP. 727-759.

72-0404 DOBOVISEK, B.
DESIGN OF PRISMATIC SHELLS
BAUINGENIEUR, V. 47, N. 11, NOV. 1972. P.
393-400.   BANGAS

72-0405 CLEMENT, P.
DESCLOUX, J.
ON THE RIGID DISPLACEMENT CONDITION.
INT. J. NUMER. METHODS ENG., V. 4, N. 4, JUL-AUG.
1972, PP. 583-586.

72-0406 CHAN, H. S. Y.
COLLAPSE LOAD OF REINFORCED CONCRETE PLATES.
INT. J. NUMER. METHODS ENG., V. 5, N. 1, SEP-OCT.
1972, PP. 57-64.

72-0407 BUSBY, H. R. JR.
WEINGARTEN, V. I.
NON-LINEAR RESPONSE OF A BEAM TO PERIODIC LOADING.
INT. J. NON-LINEAR MECH., V. 7, N. 3, JUN. 1972,
PP. 289-303

72-0408 BRAMLETTE, T. TAZ
LEONARD, J. W.
FINITE ELEMENT ANALYSIS OF THE LINEARIZED BGK
BOLTZMANN   EQUATION.
INT. J. NUMER. METHODS ENG., V. 4, N. 4, JUL-AUG.
1972, PP. 501-507.

72-0409 RAZZAQUE, A.
FINITE ELEMENT ANALYSIS OF PLATES AND SHELLS
PH.D. THESIS, UNIVERSITY OF WALES, SWANSEA, 1972.

72-0410 DAVIDS, N.
MANI, M. K.
EFFECTS OF TURBULENCE ON FLOOD FLOW EXPLORED BY
FINITE ELEMENT ANALYSIS.
COMPUT. BIOL. MED., V. 2, N. 7, DEC. 1972, PP.
311-319.

72-0411 LANNOY, F. G.
DISPLACEMENT METHOD IN STRUCTURAL ANALYSIS AS
APPLIED TO RECTANGULAR SHELL BEAMS
RFM, REV. FR. MEC., N. 44, 1972, PP. 21-38.

72-0412 ANDO, Y.
KIKUCHI, F.
CONVERGENCE OF SIMPLIFIED HYBRID DISPLACEMENT
METHOD FOR PLATE BENDING
J. FAC. ENG. UNIV. TOKYO., SER. B., VOL. 31, NO.
4, DECEMBER 1972, PP. 692-713.

72-0413 MEEK, J. L.
DODDS, B. J.
COMPARISON OF SOME PLATE BENDING FINITE ELEMENTS
INST. ENG., AUST. CIV. ENG. TRANS., V. CE 14, N.
2, OCT. 1972, PP. 206-211.

72-0414 MALHOTRA, V. M.
HIBBERT, P. D.
IMPACT OF COMPUTERS ON THE PRACTICE OF STRUCTURAL
ENGINEERING IN CONCRETE
SYMP. ON IMPACT OF COMPUT. ON THE PRACTICE OF
STRUCT. ENG., AM. CONC. INST. CONVENTION, ST.
LOUIS, NOVEMBER 1-6, 1973. (SPEC. PUBL. 33, ACI,
1972)

72-0415 LOGCHER, R. D.
BIGGS, J. M.
IMPACT OF COMPUTERS ON THE PRACTICE OF STRUCTURAL
ENGINEERING IN CONCRETE
SYMP. ON IMPACT OF COMPUT. ON THE PRACTICE OF
STRUCT. ENG., AM. CONC. INST. CONVENTION, ST.
LOUIS, NOVEMBER 1 - 6, 1970. (SPEC. PUBL. 33, ACI,
1972)

72-0416 DWYER, W. J.
FINITE ELEMENT MODELING AND OPTIMIZATION OF
AEROSPACE STRUCTURES
US. AIR FORCE SYST. COMMAND, AIR FORCE FLIGHT DYN.
LAB., TECH. REP. NO. AFFDL-TR-72-59, AUGUST 1972,
50 PP.

72-0417 WANG, S. K.
CHEUNG, Y. K.
SARGIOUS, M.
AN ADVANCED ANALYSIS OF RIGID PAVEMENTS
TRANSPORTATION ENGINEERING JOURNAL, PROC. ASCE,
VOL. 98, TE1, FEBRUARY, 1972.

72-0418 UKHOV, S. B.
POSSIBILITY OF USING FINITE ELEMENT METHOD IN THE
ANALYSIS OF JOINT PERFORMANCE OF THE HYDRAULIC
STRUCTURES AND FOUNDATIONS
GIDROTEKH. STROIT.,N.11, NOV. 1972, PP. 29-35.

72-0419 TOLKACHNIK, S. V.
NATARIUS, Y. I.
INVESTIGATION OF THE EFFECT OF FOUNDATION YIELD ON
THE FREQUENCY AND SHAPE OF INHERENT VIBRATIONS OF
DAMS BY USING FINITE ELEMENT METHOD
GIDROTEKH. STROIT., NO. 11, NOVEMBER 1972, PP.
35-38.

72-0420 SANDHU, R. S.
FINITE ELEMENT ANALYSIS OF CONSOLIDATION IN CREEP
PROC. SYMP. ON THE APPLICATIONS OF THE FINITE
ELEMENT METH. IN GEOTECH. ENG., VICKSBURG, MISS.,
MAY 1-4, 1972, PP. 697-738. (PUBL. BY U.S. ARMY
ENG. WATERWAYS EXP. STATION, SEPTEMBER 1972)

72-0421 HWANG, C. T.
MORTENSTERN, N. R.
MURRAY, D. W.
APPLICATION OF THE FINITE ELEMENT METHOD TO
CONSOLIDATION PROBLEMS
PROC. SYMP. ON THE APPLICATIONS OF THE FINITE
ELEMENT METH. IN GEOTECH. ENG., VICKSBURG, MISS.,
MAY 1-4, 1972, PP. 739-765. (PUBL. BY U.S. ARMY
ENG. WATERWAYS EXP. STATION, SEPTEMBER 1972)

72-0422 REESE, L. C.
USE OF THE FINITE ELEMENT IN FOUNDATION
ENGINEERING
PROC. SYMP. ON THE APPLICATIONS OF THE FINITE
ELEMENT METH. IN GEOTECH. ENG., VICKSBURG, MISS.,
MAY 1-4, 1972, PP. 507-532. (PUBL. BY U.S. ARMY
ENG. WATERWAYS EXP. STATION, SEPTEMBER, 1972)

72-0423 RADAJ, D.
COMPARATIVE CONSIDERATION OF PARAMETRIC STRESS
CONCENTRATION INVESTIGATIONS WITH FINITE ELEMENTS
AND WITH COMPLEX STRESS FUNCTIONS
Z. FLUGWISS., VOL. 20, NO. 12, DECEMBER 1972, PP.
479-484.

72-0424 CHRISTIAN, J. T.
YATT, B. J.
UNDRAINED VISCO-ELASTIC ANALYSIS OF SOIL
DEFORMATION
PROC. SYMP. ON THE APPLICATIONS OF THE FINITE
ELEMENT METH. IN GEOTECH. ENG., VICKSBURG, MISS.,
MAY 1-4, 1972, PP. 533-579. (PUBL. BY U.S. ARMY
ENG. WATERWAYS EXP. STATION, SEPTEMBER, 1972)

72-0425 YANG, T. Y.
FINITE ELEMENT ANALYSIS OF PLATES ON A TWO
PARAMETER FOUNDATION MODEL.
COMPUT. STRUCT., V. 2, N. 4, SEP. 1972, PP.
593-614.

72-0426 VAN CAMPEN, D. H.
SPAAS, H.A.C.M.
ON THE STRESS DISTRIBUTION IN NOZZLE TO CYLINDER
CONNECTIONS FOR SMALL DIAMETER RATIOS
NUCL. ENG. DES., VOL. 21, NO. 3, 1972, PP.
368-395.

72-0427 THOMPSON, J. M. T.
LEWIS, G. M.
CONTINUUM AND FINITE ELEMENT BRANCHING STUDIES OF
THE CIRCULAR PLATE.
COMPUT. STRUCT., V. 2, N. 4, SEP. 1972, PP.
511-534.

72-0428 PARDOEN, G. C.
HAGEN, R. L.
SYMMETRICAL BENDING OF CIRCULAR PLATES USING
FINITE ELEMENTS
COMPUT. STRUCT., V. 2, N. 4, SEP. 1972, PP.
547-553.

72-0429 ODEN, J. T.
KEY, J. E.
ON THE EFFECT OF THE FORM OF THE STRAIN ENERGY
FUNCTION ON THE SOLUTION OF A BOUNDARY-VALUE
PROBLEM IN FINITE ELASTICITY.
COMPUT. STRUCT., V. 2, N. 4, SEP. 1972, PP.
585-592.

72-0430 SCHUMANN, W.
WUETHRICH, W.
SOME REMARKS CONCERNING HETEROGENEOUS ANISOTROPIC
PLATES.
J. COMPOS. MATER., V. , OCT 1972, PP. 536-546.

72-0431 SCHUMANN, W.
WUETHRICH, W.
ON SHELLS OF CONSTANT STRENGTH
ACTA. MECH., V. 15, N. 2-3, 1972, PP. 189-197.

72-0432 SATA, T.
SATO, N.
ANALYSIS OF THERMAL DEFORMATION OF MACHINE TOOL BY
THE FINITE ELEMENT METHOD.
ANN. CIRP., V. 21, N. 1, 1972, PP. 123-124.

72-0433 ROYLANCE, D.
WILDE, A.
BALLISTIC IMPACT OF TEXTILE STRUCTURES.
TEXT. RES. J., V. 43, N. 1, JAN. 1972, PP. 34-41.

72-0434 GHALI, A.
SISODIYA, R. G.
CHEUNG, Y. K.
FINITE ELEMENT OF SKEW VAULT BRIDGES
PUBLICATIONS, INTERNATIONAL ASSOCIATION OF BRIDGES
AND STRUCTURAL ENGINEERING, VOL. 32, NO. 2, 1972.

72-0435 PRETORIUS, P. C.
MONISMITH, C. L.
FATIGUE CRACK FORMATION AND PROPAGATION IN
PAVEMENTS CONTAINING SOIL-CEMENT BASES.
HIGHW. RES. REC., N. 407, 1972, PP. 102-115.

72-0436 NIELSON, R.
CHANG, P. Y.
STRUCTURAL ANALYSIS OF LONGITUDINALLY FRAMED SHIPS
U.S. NAV. SHIP ENG. CENT., SHIP STRUCT. COMM. REP.
N. SSC- 225, 1972, 65 PP.

72-0437 NICKELL, R. E.
USER EDUCATION: THE FINITE ELEMENT SHORT COURSE
A.S.M.E. MEETING ON SOFTWARE USER: EDUC. AND
QUALIFICATION, PRESSURE VESSEL AND PIPING DIV.,
NEW YORK, NOVEMBER 16, 1972, 38 PP.

72-0438 MUNROE, J. A.
CHEN, Y.
FINITE ELEMENT ANALYSIS OF AN AXISYMMETRICALLY
LOADED ORTHOTROPIC SHELL OF REVOLUTION.
COMPUT. STRUCT., V. 2, N. 5-6, DEC. 1972, PP.
723-732.

72-0439 MORLEY, L. S. D.
MERRIFIELD, B. C.
ON THE CONFORMING CUBIC TRIANGULAR ELEMENT FOR
PLATE BENDING
COMPUT. STRUCT., V. 2, N. 5-6, DEC. 1972, PP.
875-892.

72-0440 CHEUNG, Y. K.
CHAKRABARTI, S.
FREE VIBRATION OF THICK, LAYERED RECTANGULAR
PLATES BY FINITE LAYER METHOD
J. OF SOUND & VIBRATION, VOL. 21, NO. 3, 1972.

72-0441 KRAUS, H.
SOFTWARE USER: EDUCATION AND QUALIFICATION
A.S.M.E. MEETING ON SOFTWARE USER: EDUC. AND
QUALIFICATION, PRESSURE VESSEL AND PIPING DIV.,
NEW YORK, NOVEMBER 16, 1972, 38 PP.

72-0442 KRAFT, L. M.
KRISHNAMURTHY, N.
ANALYTICAL EVALUATION OF STRESS-STRAIN TEST DATA.
APPL. SCI. RES. (THE HAGUE), V. 27, N. 1, OCT.
1972, PP. 63-75.

72-0443 KEY, S. W.
ANALYSIS OF THIN SHELLS WITH A DOUBLY CURVED
ARBITRARY QUADRILATERAL FINITE ELEMENT.
COMPUT. STRUCT., V. 2, N. 4, SEP. 1972, PP.
637-673.

72-0444 KENNER, V. H.
DYNAMIC LOADING OF A FLUID-FILLED SPHERICAL SHELL.
INT. J. MECH. SCI., V. 14, N. 9, SEP. 1972, PP.
557-568.

72-0445 KAVANAGH, K. T.
APPROXIMATE ALGORITHM FOR THE REANALYSIS OF
STRUCTURES BY THE FINITE ELEMENT METHOD.
COMPUT. STRUCT., V. 2, N. 5-6, DEC. 1972, PP.
713-722.

72-0446 HORAK, M.
CALCULATION OF THE CRITICAL LOAD OF ELASTIC THIN
WALLED BARS BY THE METHOD OF FINITE ELEMENTS
STAVEBNICKY. CAS., VOL. 20, NO. 8, 1972, PP.
598-619.

72-0447 HODGE, P. G.
MCMAHON, A. A.
SIMPLE FINITE ELEMENT MODEL FOR ELASTIC-PLASTIC
PLATE BENDING.
COMPUT. STRUCT., V. 2, N. 5-6, DEC. 1972, PP.
841-854.

72-0448 HINDUJA, S.
COWLEY, A.
FINITE ELEMENT METHOD FOR THE ANALYSIS OF MACHINE
TOOL STRUCTURES.
ANN. CIRP., V. 21, N. 1, 1972, PP. 113-114.

72-0449 HAUPT, R. S.
OLSON, J. P.
PERFORMANCE OF EARTH AND EARTH-SUPPORTED
STRUCTURES
PROC. SPEC. CONF. ON PERFORMANCE OF EARTH AND
EARTH-SUPPORTED STRUCT., A.S.C.E., NEW YORK, JUNE
11-14, 1972, VOL. 1, 1555 PP. AND VOL. 2, 154 PP.

72-0450 GREEN, D. R.
INTERACTION OF SOLID SHEAR WALLS AND THEIR
SUPPORTING STRUCTURES.
BUILD. SCI., V. 7, N. 4, DEC. 1972, PP. 239-248.

72-0451 FRITZ, K.
JAHN, F.
NONSTEADY PYROMETRIC CALCULATIONS AS THE BASIS FOR
STRESS ANALYSIS OF STEAM GENERATORS AND HEAT
EXCHANGERS
NUCL. ENG. DES., V. 20, N. 1, 1972, PP. 5-29.

72-0452 EBNER, A. M.
UCCIFERRO, J. J.
THEORETICAL AND NUMERICAL COMPARISON OF ELASTIC
NONLINEAR FINITE ELEMENT METHODS.
COMPUT. STRUCT., V. 2, N. 5-6, DEC. 1972, PP.
1043-1061.

72-0453 DESAI, C. S.
JOHNSON, L. D.
EVALUATION OF TWO FINITE ELEMENT FORMULATIONS FOR
ONE-DIMENSIONAL CONSOLIDATION
COMPUT. STRUCT., VOL. 2, NO. 4, SEPTEMBER 1972,
PP. 469-486.

72-0454 FRAEIJS DE VEUBEKE, B.
NEW VARIATIONAL PRINCIPLE FOR FINITE ELASTIC
DISPLACEMENTS
INT. J. ENG. SCI., VOL. 10, NO. 9, SEPTEMBER 1972,
PP. 745-763.

72-0455 DE ROUVRAY, A. L.
GOODMAN, R. E.
FINITE ELEMENT ANALYSIS OF CRACK INITIATION IN A
BLOCK MODEL EXPERIMENT
ROCK. MECH., FELSMECH., MEC. ROCHES., VOL. 4, NO.
4, DECEMBER 1972, PP. 203-223.

72-0456 CROSE, J. G.
MCKINLEY, T. K.
THERMAL STRESS ANALYSIS OF REENTRY VEHICLE
NOSETIPS AT ANGLE OF ATTACK
COMPUT. STRUCT., V. 2, N. 5-6, DEC. 1972, PP.
1031-1042.

72-0457 BENTHIEN, G. W.
GURTIN, M. E.
ON THE SEMI-DISCRETE GALERKIN METHOD FOR
HYPERBOLIC PROBLEMS AND ITS APPLICATION TO
PROBLEMS IN ELASTODYNAMICS
ARCH. RATION. MECH. ANAL., VOL. 48, NO. 1, 1972,
PP. 51-63.

72-0458 CHEUNG, M. S.
CHEUNG, Y. K.
STATIC AND DYNAMIC BEHAVIOUR OF RECTANGULAR PLATES
USING HIGHER ORDER FINITE STRIPS
BUILDING SCIENCE, VOL. 7, NO. 3, SEPT. 1972.

72-0459 BALAKRISHNA RAO, H. A.
HARNAGE, D.
EVALUATION OF RIGID PAVEMENTS BY NONDESTRUCTIVE
TESTS.
HIGHW. RES. REC., N. 407, 1972, PP. 76-86.

72-0460 AYRES, D. J.
COMPANY TRAINING OF ENGINEERING IN STRUCTURAL
ANALYSIS COMPUTER PROGRAM USAGE
A.S.M.E. MEETING ON SOFTWARE USER:EDUC. AND
QUALIFICATIONS, NEW YORK, NOVEMBER 16, 1972, 38
PP.

72-0461 ANDERHEGGEN, E.
KNOEPFEL, H.
FINITE ELEMENT LIMIT ANALYSIS USING LINEAR
PROGRAMMING.
INT. J. SOLIDS STRUCT., V. 8, N. 12, DEC. 1972,
PP. 1413-1431.

72-0462 ALLGOOD, J. R.
TAKAHASHI, S. K.
BALANCED DESIGN AND FINITE-ELEMENT ANALYSIS OF
CULVERTS.
HIGHW. RES. REC., N. 413, 1972, PP. 45-56.

72-0463 ABE, T.
ELASTIC DEFORMATION OF POLYCRYSTALLINE METAL - 1.
INFLUENCE OF GRAIN SHAPE ON DISTRIBUTIONS OF
STRESS AND STRAIN IN GRAINS
BULL. JSME., V. 15, N. 86, AUG 1972, PP. 917-927.

72-0464 ZANGS, L.
CALCULATION OF THERMALLY INDUCED STRAINS IN
MACHINE TOOL ELEMENTS
IND-ANZ., V. 94, N. 105, DEC. 15, 1972, PP.
2514-2517.

72-0465 SISODIYA, R. G.
CHEUNG, Y. K.
NEW FINITE ELEMENTS WITH APPLICATION TO BOX GIRDER
BRIDGES.
PROC. INST. CIV. ENG., SUPPL. N. 10, 1972, PAP.
7479, PP. 207-225.

72-0466 OHTMER, O.
AUTOMATIC GENERATION AND CONDENSATION OF LARGE
MASS AND STIFFNESS MATRICES
ANGEW. INF., APPL. INFORMATICS., V. 14, N. 12, DEC.
1972, PP.569-574.

72-0467 GODFREY, D. A.
MOUSSA, M. M.
DYNAMIC ANALYSIS OF AXISYMMETRIC SHELLS UNDER
ARBITRARY TRANSIENT PRESSURES.
NUCL. ENG. DES., V. 23, N. 2, 1972, PP. 187-194.

72-0468 DUNN, J. C.
NICKELL, R. E.
RE-ENTRY THERMAL ANALYSIS OF VARIABLE THICKNESS
SPHERICAL VEHICLES.
PROG. ASTRONAUT. AERONAUT. FUNDAM OF SPACECR.
THERM. DES., V. 29, MIT. PRESS, CAMBRIDGE, MASS.,
1972, PP. 319-331.

72-0469 CAMBEFORT, H.
BRILLANT, J.
SPECIAL STUDY ON THE STABILITY OF A DAM SPILLWAY
TRAVAUX, N. 453, DEC 1972, PP. 3-8.

72-0470 ABSI, E.
THEORY OF EQUIVALENCE AND ITS APPLICATION TO
VARIOUS PROBLEMS IN ELASTICITY AND DESIGN OF SKEW
BRIDGES
ANN. TRAV. PUBLICS BELG. N. 3, 1971-1972, PP.
127-146.

72-0471 WOO, S. L. Y.
KOBAYASHI, A. S.
MATHEMATICAL MODEL OF THE CORNEO-SCLERAL SHELL AS
APPLIED TO INTRAOCULAR PRESSURE VOLUME RELATIONS
AND APPLANATION TONOMETRY
ANN. BIOMED. ENG., VOL. 1, NO. 1, 1972, PP. 87-98.

72-0472 IRONS, B. M.
RAZZAQUE, A.
SHAPE FUNCTION FORMULATIONS FOR ELEMENTS OTHER
THAN DISPLACEMENT MODELS
PROC. INT. CONF. ON VARIATIONAL METHODS IN ENG.,
UNIV. OF SOUTHAMPTON, SEPTEMBER 25, 1972, (EDS)
BREBBIA, C.A. AND TOTTENHAM, H. SOUTHAMPTON UNIV.
PRESS, 1973,VOL. 1, PP. 4/59-4/72.

72-0473 RAZZAQUE, A.
PROGRAM FOR TRIANGULAR BENDING ELEMENTS WITH
DERIVATIVES SMOOTHING
INT. J. NUM. METHODS IN ENG., VOL. 6, 1972, PP.
333-343.

72-0474 ANON
MACHINERY SPACE DEFLECTION AND VIBRATION ANALYSES
USING FINITE ELEMENT METHODS
MOT. SHIP, VOL. 53, NO. 621, APRIL 1972, PP.
24-27.

72-0475 ANON
REPORT ON POST CONF. MEETING ON COMPUTATIONAL
ASPECTS OF ASPECTS OF FINITE ELEMENT METHOD ISD
STUTTGART, SEPT. 27-28, 1971.
NUCL. ENG. AND DESIGN, VOL. 18, NO. 2, 1972, PP.
201-202.

2-0476 FRAEIJS DE VEUBEKE, B.
SANDER, G.
DUAL ANALYSIS BY FINITE ELEMENTS: LINEAR AND
NONLINEAR APPLICATIONS
U.S. AIR FORCE SYST. COMMAND, AIR FORCE FLIGHT
DYN. LAB., TECH. REP. NO. AFFDL-TR-72-93, DECEMBER
1972, 292 PP.

72-0477 VENKATESWARA, RAO, G.
FINITE ELEMENT ANALYSIS OF VIBRATION AND STABILITY
PROBLEMS
PH. D. THESIS, INDIAN INSTITUTE OF SCIENCE,
BANGALORE, 1972.

72-0478 PRASAD, K.S.R.K.
KRISHNA MURTY, A. V.
RAO, A. K.
A FINITE ELEMENT ANALOGUE OF THE MODIFIED
RAYLEIGH-RITZ METHOD FOR VIBRATION PROBLEMS
INT. J. NUM. METHODS IN ENG., VOL. 5, NO. 2, 1972,
PP. 163-169.

72-0479 MCCORMICK, C. W., (ED)
THE NASTRAN USER'S MANUAL
NASA SP-222, SEPTEMBER, 1970, REPRINTED JUNE,
1972.

72-0480 DOUGLAS, F.
THE NASTRAN PROGRAMMER'S MANUAL
NASA SP-223, SEPTEMBER, 1970, REPRINTED SEPTEMBER,
1972.

72-0481 EVERSTINE, G. C.
THE BANDIT COMPUTER PROGRAM FOR THE REDUCTION OF
MATRIX BANDWIDTH FOR NASTRAN
NSRDC REPORT 3827, MARCH, 1972.

72-0482 KABAILA, A. P.
PULMANO, V. A.
SOMERVAILLE, I. J.
A COMPATIBLE QUADRILATERAL PLATE BENDING ELEMENT
UNICIV REPORT R-86, JUNE, 1972.

72-0483 KABAILA, A. P.
PULMANO, V. A.
FRENCH, S. B.
A COMPATIBLE MEMBRANE QUADRILATERAL ELEMENT
UNICIV REPORT R-104, 1972.

72-0484 KABAILA, A. P.
SOMERVAILLE, I. J.
AN EQUILIBRIUM MEMBRANE QUADRILATERAL ELEMENT
UNICIV REPORT, R-95, 1972.

72-0485 CHEUNG, Y. K.
CHEUNG, M. S.
VIBRATION ANALYSIS OF CYLINDRICAL PANELS
J. OF SOUND & VIBRATION, VOL. 22, NO. 1, 1972.

72-0486 FUJINO, T.
APPLICATION OF FINITE ELEMENT METHOD TO THE
PROBLEM OF HEAT CONDUCTION AND FLUID MECHANICS
PREPRINT, 22ND NATL. CONGRESS FOR THEORETICAL AND
APPLIED MECHANICS, DECEMBER 1972, PP. 101-108. (IN
JAPANESE)

72-0487 CHEUNG, Y. K.
CHEUNG, M. S.
FREE VIBRATION OF CURVED AND STRAIGHT BEAM SLAB OR
BOX GIRDER BRIDGES
PUBLICATIONS, INTERNATIONAL ASSOCIATION FOR
BRIDGES AND STRUCTURAL ENGINEERING, VOL. 32, NO.
2, 1972.

72-0488 DOUGLAS, J.
DUPONT, T.
A FINITE ELEMENT COLLOCATION METHOD FOR THE HEAT
EQUATION
ISTITUTO NAZIONALE DI ALTA MATEMATICA, SYMPOSIA
MATHEMATICA, VOLUME 10, 1972.

72-0489 KHATUA, T. P.
CHEUNG, Y. K.
A TRIANGULAR ELEMENT FOR BENDING AND VIBRATION OF
MULTILAYER SANDWICH PLATES
J. ENG. MECH. DIV., PROC. ASCE., VOL. 98, NO. EM5,
OCTOBER 1972.

72-0490 CHEUNG, Y. K.
FINITE ELEMENT AND FINITE STRIP ANALYSES OF
BRIDGES
KEYNOTE ADDRESS, PROC. OF SPECIALTY CONFERENCE ON
THE FINITE ELEMENT METHOD IN CIVIL ENGINEERING,
MCGILL UNIVERSITY, MONTREAL, JUNE 1972.

72-0491 CHEUNG, Y. K.
WU, C. I.
FREE VIBRATION OF THICK, LAYERED CYLINDERS HAVING
FINITE LENGTH WITH VARIOUS BOUNDARY CONDITIONS
JOURNAL OF SOUND AND VIBRATION, VOL. 24, NO.
2, 1972.

72-0492 LICHARDUS, S.
CHEUNG, Y. K.
WATSON, M.
ORTHOTROPIC FLAT PLATES SUPPORTED ON COLUMNS
ELONGATED IN PLANE
ZAMM 52, T141-T142, 1972.

72-0493 CHAN, H. C.
CHEUNG, Y. K.
STATIC AND DYNAMIC ANALYSIS OF MULTILAYERED
SANDWICH PLATES
INT. J. MECH. SCI., VOL. 14, 1972.

72-0494 CHEUNG, Y. K.
KAJITA, T.
FINITE ELEMENT ANALYSIS OF CABLE STAYED BRIDGES
PUBLICATIONS, INTERNATIONAL ASSOCIATION FOR
BRIDGES AND STRUCTURAL ENGINEERING, VOL. 32, NO.
2, 1972.

72-0495 GLOWINSKI, R.
APPROXIMATIONS EXTERNES, PAR ELEMENTS FINIS DE
LAGRANGE D ORDRE, UN ET DEUX, DU PROBLEME DE
DIRICHLET POUR LA OPERATEUR BIHARMONIQUE, METHODE
ITERATIVE DE RESOLUTION DES PROBLEMES APPROCHES
CONF. NUMER. ANALYSIS, ROYAL IRISH ACADEMY, 1972.

72-0496 ATLURI, S.
PIAN, T. H. H.
THEORETICAL FORMULATION OF FINITE ELEMENT METHODS
IN LINEAR ELASTIC ANALYSIS OF GENERAL SHELLS
J. STRUCT. MECH., VOL. 1, NO. 1, PP. 1-41, 1972.

72-0497 BABUSKA, I.
AZIZ, A. K.
PRELIMINARY REMARKS
IN THE MATHEMATICAL FOUNDATIONS OF THE FINITE
ELEMENT METHOD WITH APPLICATIONS TO PARTIAL
DIFFERENTIAL EQUATIONS, EDITED BY A. K. AZIZ,
ACADEMIC PRESS, NEW YORK, 1972, PP. 5-14.

72-0498 AZIZ, A. K.
BABUSKA, I.
THE FUNDAMENTAL NOTIONS
IN THE MATHEMATICAL FOUNDATIONS OF THE FINITE
ELEMENT METHOD WITH APPLICATIONS TO PARTIAL
DIFFERENTIAL EQUATIONS, EDITED BY A. K. AZIZ,
ACADEMIC PRESS, NEW YORK, 1972, PP. 15-45.

72-0499 BABUSKA, I.
AZIZ, A. K.
PROPERTIES OF SOLUTIONS OF ELLIPTIC BOUNDARY VALUE
PROBLEMS
IN THE MATHEMATICAL FOUNDATIONS OF THE FINITE
ELEMENT METHOD WITH APPLICATIONS TO PARTIAL
DIFFERENTIAL EQUATIONS, EDITED BY A. K. AZIZ,
ACADEMIC PRESS, NEW YORK, 1972, PP. 47-81.

72-0500 BABUSKA, I.
AZIZ, A. K.
THEORY OF APPROXIMATION
IN THE MATHEMATICAL FOUNDATIONS OF THE FINITE
ELEMENT METHOD WITH APPLICATIONS TO PARTIAL
DIFFERENTIAL EQUATIONS, EDITED BY A. K. AZIZ,
ACADEMIC PRESS, NEW YORK, 1972, PP. 83-110.

72-0501 BABUSKA, I.
AZIZ, A. K.
VARIATIONAL PRINCIPLES
IN THE MATHEMATICAL FOUNDATIONS OF THE FINITE
ELEMENT METHOD WITH APPLICATIONS TO PARTIAL
DIFFERENTIAL EQUATIONS, EDITED BY A. K. AZIZ,
ACADEMIC PRESS, NEW YORK, 1972, PP. 111-184.

72-0502 BABUSKA, I.
AZIZ, A. K.
RATE OF CONVERGENCE OF THE FINITE ELEMENT METHOD
IN THE MATHEMATICAL FOUNDATIONS OF THE FINITE
ELEMENT METHOD WITH APPLICATIONS TO PARTIAL
DIFFERENTIAL EQUATIONS, EDITED BY A. K. AZIZ,
ACADEMIC PRESS, NEW YORK, 1972, PP. 185-241.

72-0503 BABUSKA, I.
AZIZ, A. K.
ONE PARAMETER FAMILIES OF VARIATIONAL PRINCIPLES
IN THE MATHEMATICAL FOUNDATIONS OF THE FINITE
ELEMENT METHOD WITH APPLICATIONS TO PARTIAL
DIFFERENTIAL EQUATIONS, EDITED BY A. K. AZIZ,
ACADEMIC PRESS, NEW YORK, 1972, PP. 243-263.

72-0504 BABUSKA, I.
AZIZ, A. K.
FINITE ELEMENT METHOD FOR NON SMOOTH DOMAINS AND
COEFFICIENTS
IN THE MATHEMATICAL FOUNDATIONS OF THE FINITE
ELEMENT METHOD WITH APPLICATIONS TO PARTIAL
DIFFERENTIAL EQUATIONS, EDITED BY A. K. AZIZ,
ACADEMIC PRESS, NEW YORK, 1972, PP. 265-283.

72-0505   BABUSKA, I.
AZIZ, A. K.
THE PROBLEMS OF PERTURBATIONS IN THE FINITE
ELEMENT METHOD
IN THE MATHEMATICAL FOUNDATIONS OF THE FINITE
ELEMENT METHOD WITH APPLICATIONS TO PARTIAL
DIFFERENTIAL EQUATIONS, EDITED BY A. K. AZIZ,
ACADEMIC PRESS, NEW YORK, 1972, PP. 285-301.

72-0506   BABUSKA, I.
AZIZ, A. K.
THE EIGENVALUE PROBLEM
IN THE MATHEMATICAL FOUNDATIONS OF THE FINITE
ELEMENT METHOD WITH APPLICATIONS TO PARTIAL
DIFFERENTIAL EQUATIONS, EDITED BY A. K. AZIZ,
ACADEMIC PRESS, NEW YORK, 1972, PP. 303-343.

72-0507   BABUSKA, I.
AZIZ, A. K.
THE FINITE ELEMENT METHOD FOR TIME DEPENDENT
PROBLEMS
IN THE MATHEMATICAL FOUNDATIONS OF THE FINITE
ELEMENT METHOD WITH APPLICATIONS TO PARTIAL
DIFFERENTIAL EQUATIONS, EDITED BY A. K. AZIZ,
ACADEMIC PRESS, NEW YORK, 1972, PP. 345-359.

72-0508   BIRKHOFF, G.
PIECEWISE ANALYTIC INTERPOLATION AND APPROXIMATION
IN TRIANGULATED POLYGONS
IN THE MATHEMATICAL FOUNDATIONS OF THE FINITE
ELEMENT METHOD WITH APPLICATIONS TO PARTIAL
DIFFERENTIAL EQUATIONS, EDITED BY A. K. AZIZ,
ACADEMIC PRESS, NEW YORK, 1972, PP. 363-385.

72-0509   BRAMBLE, J. H.
OSBORN, J. E.
APPROXIMATION OF STEKLOV EIGENVALUES OF
NON-SELFADJOINT SECOND ORDER ELLIPTIC OPERATORS
IN THE MATHEMATICAL FOUNDATIONS OF THE FINITE
ELEMENT METHOD WITH APPLICATIONS TO PARTIAL
DIFFERENTIAL EQUATIONS, EDITED BY A. K. AZIZ,
ACADEMIC PRESS, NEW YORK, 1972, PP. 387-408.

72-0510   DOUGLAS, J.
A SUPERCONVERGENCE RESULT FOR THE APPROXIMATE
SOLUTION OF THE HEAT EQUATION BY A COLLOCATION
METHOD
IN THE MATHEMATICAL FOUNDATIONS OF THE FINITE
ELEMENT METHOD WITH APPLICATIONS TO PARTIAL
DIFFERENTIAL EQUATIONS, EDITED BY A. K. AZIZ,
ACADEMIC PRESS, NEW YORK, 1972, PP. 475-490.

72-0511   DUPONT, T.
SOME L(2) ERROR ESTIMATES FOR PARABOLIC GALERKIN
METHODS
IN THE MATHEMATICAL FOUNDATIONS OF THE FINITE
ELEMENT METHOD WITH APPLICATIONS TO PARTIAL
DIFFERENTIAL EQUATIONS, EDITED BY A. K. AZIZ,
ACADEMIC PRESS, NEW YORK, 1972, PP. 491-504.

72-0512   EISENSTAT, S. C.
SCHULTZ, M. H.
COMPUTATIONAL ASPECTS OF THE FINITE ELEMENT METHOD
IN THE MATHEMATICAL FOUNDATIONS OF THE FINITE
ELEMENT METHOD WITH APPLICATIONS TO PARTIAL
DIFFERENTIAL EQUATIONS, EDITED BY A. K. AZIZ,
ACADEMIC PRESS, NEW YORK, 1972, PP. 505-522.

72-0513   KELLOGG, R. B.
HIGHER ORDER SINGULARITIES FOR INTERFACE PROBLEMS
IN THE MATHEMATICAL FOUNDATIONS OF THE FINITE
ELEMENT METHOD WITH APPLICATIONS TO PARTIAL
DIFFERENTIAL EQUATIONS, EDITED BY A. K. AZIZ,
ACADEMIC PRESS, NEW YORK, 1972, PP. 589-602.

72-0514   NITSCHE, J. A.
ON DIRICHLET PROBLEMS USING SUBSPACES WITH NEARLY
ZERO BOUNDARY CONDITIONS
IN THE MATHEMATICAL FOUNDATIONS OF THE FINITE
ELEMENT METHOD WITH APPLICATIONS TO PARTIAL
DIFFERENTIAL EQUATIONS, EDITED BY A. K. AZIZ,
ACADEMIC PRESS, NEW YORK, 1972, PP. 603-627.

72-0515   ODEN, J. T.
GENERALIZED CONJUGATE FUNCTIONS FOR MIXED FINITE
ELEMENT APPROXIMATIONS OF BOUNDARY VALUE PROBLEMS
IN THE MATHEMATICAL FOUNDATIONS OF THE FINITE
ELEMENT METHOD WITH APPLICATIONS TO PARTIAL
DIFFERENTIAL EQUATIONS, EDITED BY A. K. AZIZ,
ACADEMIC PRESS, NEW YORK, 1972, PP. 629-669.

72-0516   PIAN, T. H. H.
FINITE ELEMENT FORMULATION BY VARIATIONAL
PRINCIPLES WITH RELAXED CONTINUITY REQUIREMENTS
IN THE MATHEMATICAL FOUNDATIONS OF THE FINITE
ELEMENT METHOD WITH APPLICATIONS TO PARTIAL
DIFFERENTIAL EQUATIONS, EDITED BY A. K. AZIZ,
ACADEMIC PRESS, NEW YORK, 1972, PP. 671-687.

72-0517   STRANG, G.
VARIATIONAL CRIMES IN THE FINITE ELEMENT METHOD
IN THE MATHEMATICAL FOUNDATIONS OF THE FINITE
ELEMENT METHOD WITH APPLICATIONS TO PARTIAL
DIFFERENTIAL EQUATIONS, EDITED BY A. K. AZIZ,
ACADEMIC PRESS, NEW YORK, 1972, PP. 689-710. ALSO
IN ONR. SYMP. UNIVERSITY OF MARYLAND, BALTIMORE,
1972.

72-0518   THOMEE, V.
SPLINE APPROXIMATION AND DIFFERENCE SCHEMES FOR
THE HEAT EQUATION
IN THE MATHEMATICAL FOUNDATIONS OF THE FINITE
ELEMENT METHOD WITH APPLICATIONS TO PARTIAL
DIFFERENTIAL EQUATIONS, EDITED BY A. K. AZIZ,
ACADEMIC PRESS, NEW YORK, 1972, PP. 711-746.

72-0519   BARNHILL, R. E.
GREGORY, J. A.
WHITEMAN, J. R.
THE EXTENSION AND APPLICATION OF SARD KERNEL
THEOREMS TO COMPUTE FINITE ELEMENT ERROR BOUNDS
IN THE MATHEMATICAL FOUNDATIONS OF THE FINITE
ELEMENT METHOD WITH APPLICATIONS TO PARTIAL
DIFFERENTIAL EQUATIONS, EDITED BY A. K. AZIZ,
ACADEMIC PRESS, NEW YORK, 1972, PP. 749-755.

72-0520   BERGER, A. E..
TWO TYPES OF PIECEWISE QUADRATIC SPACES AND THEIR
ORDER OF ACCURACY FOR POISSON'S EQUATION
IN THE MATHEMATICAL FOUNDATIONS OF THE FINITE
ELEMENT METHOD WITH APPLICATIONS TO PARTIAL
DIFFERENTIAL EQUATIONS, EDITED BY A. K. AZIZ,
ACADEMIC PRESS, NEW YORK, 1972, PP. 757-761.

72-0521   DENDY, J. E. JR.
A METHOD TO GALERKIN TYPE ACHIEVING OPTIMUM L(2)
ACCURACY FOR FIRST ORDER HYPERBOLICS AND EQUATIONS
OF SCHRODINGER TYPE
IN THE MATHEMATICAL FOUNDATIONS OF THE FINITE
ELEMENT METHOD WITH APPLICATIONS TO PARTIAL
DIFFERENTIAL EQUATIONS, EDITED BY A. K. AZIZ,
ACADEMIC PRESS, NEW YORK, 1972, PP. 763-765.

72-0522   FAIRWEATHER, G.
JOHNSON, J. P.
RICHARDSON EXTRAPOLATION FOR PARABOLIC GALERKIN
METHODS
IN THE MATHEMATICAL FOUNDATIONS OF THE FINITE
ELEMENT METHOD WITH APPLICATIONS TO PARTIAL
DIFFERENTIAL EQUATIONS, EDITED BY A. K. AZIZ,
ACADEMIC PRESS, NEW YORK, 1972, PP. 767-768.

72-0523   GORDON, W. J.
HALL, C. A.
GEOMETRIC ASPECTS OF THE FINITE ELEMENT METHOD
IN THE MATHEMATICAL FOUNDATIONS OF THE FINITE
ELEMENT METHOD WITH APPLICATIONS TO PARTIAL
DIFFERENTIAL EQUATIONS, EDITED BY A. K. AZIZ,
ACADEMIC PRESS, NEW YORK, 1972, PP. 769-783.

72-0524   KAPER, H. G.
LEAF, G. K.
LINDEMAN, A. J.
THE USE OF INTERPOLATORY POLYNOMIALS FOR A FINITE
ELEMENT SOLUTION OF THE MULTIGROUP DIFFUSION
EQUATION
IN THE MATHEMATICAL FOUNDATIONS OF THE FINITE
ELEMENT METHOD WITH APPLICATIONS TO PARTIAL
DIFFERENTIAL EQUATIONS, EDITED BY A. K. AZIZ,
ACADEMIC PRESS, NEW YORK, 1972, PP. 785-789.

72-0525   MANSFIELD, L.
A " LOCAL " BASIS OF GENERALIZED SPLINES OVER
RIGHT TRIANGLES DETERMINED FROM A NONUNIFORM
PARTITIONING OF THE PLANE
IN THE MATHEMATICAL FOUNDATIONS OF THE FINITE
ELEMENT METHOD WITH APPLICATIONS TO PARTIAL
DIFFERENTIAL EQUATIONS, EDITED BY A. K. AZIZ,
ACADEMIC PRESS, NEW YORK, 1972, P. 791.

72-0526   PATENT, P. D.
LEAST SQUARE POLYNOMIAL SPLINE APPROXIMATION
IN THE MATHEMATICAL FOUNDATIONS OF THE FINITE
ELEMENT METHOD WITH APPLICATIONS TO PARTIAL
DIFFERENTIAL EQUATIONS, EDITED BY A.K. AZIZ,
ACADEMIC PRESS, NEW YORK, 1972, P. 791.

72-0527   SCOTT, R.
SUBSPACES WITH ACCURATELY INTERPOLATED BOUNDARY
CONDITIONS
IN THE MATHEMATICAL FOUNDATIONS OF THE FINITE
ELEMENT METHOD WITH APPLICATIONS TO PARTIAL
DIFFERENTIAL EQUATIONS, EDITED BY A.K. AZIZ,
ACADEMIC PRESS, NEW YORK, 1972, PP. 793-796.

72-0528   FIX, G. J.
NASSIF, N.
ON FINITE ELEMENT APPROXIMATIONS ON TIME DEPENDENT
PROBLEMS
NUMER. MATH., VOL. 19, 1972, PP. 127-135.

72-0529   TAYLOR, R. L.
IDING, R.
APPLICATION OF EXTENDED VARIATIONAL PRINCIPLES TO
FINITE ELEMENT ANALYSIS
PROC. INT. CONF. ON VARIATIONAL METH. IN ENG.,
UNIV. OF SOUTHAMPTON, SEPTEMBER 25, 1972, (EDS.)
BREBBIA, C.A. AND TOTTENHAM, H., SOUTHAMPTON UNIV.
PRESS, 1973, VOL. 1, PP. 2/54-2/67.

72-0530   GEORGE, J. A.
BLOCK ELIMINATION ON FINITE ELEMENT SYSTEMS OF
EQUATIONS
PROC. OF THE IBM SYMP. ON SPARSE MATRICES AND
THEIR APPLICATIONS, 1972.

72-0531 FAWKES, A. J.
SOME NONCONFORMING VARIANTS ON THE TRILINEAR
ISOPARAMETRIC BRICK FOR LINEAR ELASTIC PROBLEMS
M.SC. THESIS, UNIVERSITY OF WALES, 1972.

72-0532 IRONS, B. M.
AN ASSUMED STRESS VERSION OF THE WILSON 8-NODE
ISOPARAMETRIC BRICK
COMPUTER REPORT, CNME/CR/56, CIVIL ENGINEERING,
UNIVERSITY, SWANSEA, MAY 1972.

72-0533 GORDON, W. J.
HALL, C. A.
DISCRETIZATION ERROR BOUNDS FOR TRANSFINITE
ELEMENTS
GENERAL MOTORS RESEARCH REPORT GMR-1196, MAY 1972.

72-0534 GORDON, W. J.
HALL, C. A.
GEOMETRIC ASPECTS OF THE FINITE ELEMENT METHOD:
CONSTRUCTION OF CURVILINEAR COORDINATE SYSTEMS AND
THEIR APPLICATIONS TO MESH GENERATION
GENERAL MOTORS RESEARCH REPORT, OCTOBER 1972.

72-0535 KAPER, H. G.
LEAF, G. K.
LINDEMAN, A. J.
APPLICATIONS OF FINITE ELEMENT METHODS IN REACTOR
MATHEMATICS - NUMERICAL SOLUTION OF THE
NEUTRON DIFFUSION EQUATION
ARGONNE NATIONAL LAB., ANL-7925, 1972.

72-0536 DANIEL, I. M.
ROWLANDS, R. E.
DEFORMATION AND FAILURE OF BORON-EPOXY PLATE WITH
CIRCULAR HOLE
A.S.T.M. MEETING, SAN ANTONIO, TEXAS, APRIL 12-13,
1972, PAPER 521, PP. 143-164.

72-0537 KANDIDOV, V. P.
KHLYBOV, E. P.
ON THE CONVERGENCE OF THE METHOD OF FINITE
ELEMENTS IN THE ANALYSIS OF MEMBRANE DYNAMICS
APPL. MATH. MECH., VOL. 36, NO. 3, 1972, PP.
533-538.

72-0538 LINDBERG, G. M.
ACCURATE FINITE ELEMENT MODELLING OF FLAT AND
CURVED STIFFENED PANELS
PAPER #6, PROC. #113, AGARD STRUCTURES & MATERIALS
PANEL SYMPOSIUM ON ACOUSTIC FATIGUE, TOULOUSE,
FRANCE, SEPT. 25-28, 1 972.

72-0539 LINDLEY, P. B.
SOME APPLICATIONS OF FINITE ELEMENT ANALYSIS IN
RUBBER SPRING DESIGN
RUBBERCON 72, INT. RUBBER CONF., PROC., MAY 15-19,
1972. PUBLISHED BY INST. RUBBER IND., LONDON,
ENGLAND, 1972. 5 PP.

72-0540 MAJIDZADEH, K.
KAUFFMANN, E. M.
ANALYSIS OF FATIGUE OF PAVING MIXTURES FROM THE
FRACTURE MECHANICS VIEWPOINT
ASTM SPEC. TECH. PUBL. NO. 508, 1972 FOR MEETING
JUNE 27-JULY 2, 1971, PP. 67-84.

72-0541 ROSKAM, J.
STEADY STATE EQUATIONS OF MOTION, EQUILIBRIUM
SHAPE AND STABILITY DERIVATIVES OF ELASTIC
AIRPLANES EVALUATED WITH FINITE ELEMENT METHODS
CONTRIB. TO THE THEORY OF AIRCR. STRUCT. PP.
305-323, PUBL. FOR DELFT UNIV. PRESS, NETH., 1972.

72-0542 PRZEMIENIECKI, J. S.
MATRIX ANALYSIS OF LOCAL INSTABILITY IN PLATES,
STIFFENED PANELS AND COLUMNS
INT. J. NUM. METH. ENGNG., VOL. 5, NO. 2,
NOVEMBER-DECEMBER, 1972, PP. 209-216.

72-0543 FELIPPA, C. A.
AN ALPHANUMERIC FINITE ELEMENT MESH PLOTTER
INT. J. NUM. METH. ENGNG., VOL. 5, NO. 2,
NOVEMBER-DECEMBER, 1972, PP. 217-236.

72-0544 COOK, R. D.
TWO HYBRID ELEMENTS FOR ANALYSIS OF THICK, THIN
AND SANDWICH PLATES
INT. J. NUM. METH. ENGNG., VOL. 5, NO. 2,
NOVEMBER-DECEMBER, 1972, PP. 277-288.

72-0545 GEORGE, J. A.
ON THE DENSITY OF FINITE ELEMENT MATRICES
INT. J. NUM. METH. ENGNG., VOL. 5, NO. 2,
NOVEMBER-DECEMBER, 1972, PP. 297-300.

72-0546 GUPTA, A. K.
MOHRAZ, B.
A METHOD OF COMPUTING NUMERICALLY INTEGRATED
STIFFNESS MATRICES
INT. J. FOR NUMER. METHODS IN ENGNG., VOL. 5, NO.
1, SEPTEMBER-OCTOBER, 1972, PP. 83-89.

72-0547 NAYAK, G. C.
ZIENKIEWICZ, O. C.
ELASTO-PLASTIC STRESS ANALYSIS. A GENERALIZATION
FOR VARIOUS CONSTITUTIVE RELATIONS INCLUDING
STRAIN SOFTENING
INT. J. NUM. METH. ENGNG., VOL. 5, NO. 1,
SEPTEMBER-OCTOBER, 1972, PP. 113-135.

72-0548 COOK, R. D.
MORE ON REDUCED INTEGRATION AND ISOPARAMETRIC
ELEMENTS
INT. J. NUM. METH. ENGNG., VOL. 5, NO. 1,
SEPTEMBER-OCTOBER, 1972, P. 141-142.

72-0549 HOFMEISTER, L. D.
EVENSEN, D. A.
VIBRATION PROBLEMS USING ISOPARAMETRIC SHELL
ELEMENTS
INT. J. NUM. METH. ENGNG., VOL. 5, NO. 1,
SEPTEMBER-OCTOBER, 1972, 142-145.

72-0550 GANGAL, M. D.
FINITE ELEMENT ANALYSIS OF ELASTIC CONTACT PROBLEM
INT. J. NUM. METH. ENGNG., VOL. 5, NO. 1,
SEPTEMBER-OCTOBER, 1972, PP. 145-147.

72-0551 ZIENKIEWICZ, O. C.
CONSTRAINED VARIATIONAL PRINCIPLES AND PENALTY
FUNCTION METHODS IN FINITE ELEMENT ANALYSIS
CONF. ON NUM. SOLUTION OF DIFFERENTIAL EQUATIONS,
DUNDEE, 1972.

72-0552 ARALDSEN, P. O.
SESAM 69 A GENERAL PURPOSE FINITE ELEMENT METHOD
PROGRAM CDC/NASTRAN APPLICATIONS MANUAL
CYBERNET PUBL. DEPT., CONTROL DATA CORP.,
MINNEAPOLIS, 1972.

72-0553 PATTERSON, C.
A CLASS OF FUNCTIONALS GIVING IMPROVED CONVERGENCE
WITH FINITE ELEMENTS
INT. CONF. ON VARIATIONAL METHODS IN ENGINEERING,
UNIVERSITY OF SOUTHAMPTON, 1972.

72-0554 DIAB, B.
CONTRIBUTION A L'ETUDE DU COMPLEXE
BARRAGE-FONDATION ET EXPLICATION DES SEISMES DUS
AU REMPLISSAGE DE CERTAINS RESERVOIRS
(INVESTIGATION OF THE COMPLEX SYSTEM: GRAVITY DAM
AND FOUNDATION, AND EXPLICATION OF EARTHQUAKES
CAUSED BY FILLING OF SOME RESERVOIRS)
ANN. TRAV. PUBLICS BELG., NO. 6, 1972-1973, PP.
357-429.

72-0555 GUPTA, K. K.
DYNAMIC RESPONSE ANALYSIS OF GEOMETRICALLY
NON-LINEAR STRUCTURES SUBJECTED TO HIGH
IMPACT
INT. JOURNAL FOR NUM. METH. IN ENG. VOL. 4, NO. 2
MARCH- APR. 1972, PP. 163-174

72-0556 CHATTERJEE, A.
SETLUR, A. V.
A MIXED FINITE ELEMENT FORMULATION FOR PLATE
PROBLEMS
INT. JOUR. FOR NUM. METH. IN ENG. VOL. 4, NO. 1
PP. 67-84

72-0557 BRANDES, K.
ZUR SYSTEMATIK NUMERISCHER NAEHERUNGSVERFAHREN IN
DER STATIK UND DYNAMIK DER KONSTRUKTIONEN (METHODS
OF NUMERICAL APPROXIMATIONS IN THE STRUCTURAL
STATICS AND DYNAMICS)
NUCL. ENG. DES VOL. 18, NO. 3 1972, PP. 469-485.

72-0558 MCLEOD, R. Y. J.
MITCHELL, A. R.
THE CONSTRUCTION OF BASIC FUNCTIONS FOR CURVED
ELEMENTS IN THE FINITE ELEMENT METHOD
J. INST. MATHS. APPLICS, VOL. 10, 1972, P. 382.

72-0559 KARIAPPA, V.
SMITH, G. C. C.
FURTHER DEVELOPMENTS IN CONSISTENT UNSTEADY
SUPERSONIC AERODYNAMIC COEFFICIENTS
J. AIRCR VOL. 9, NO. 2 FEB 1972, PP. 157-161.

72-0560 KENNER, P. M.
STRUCTURAL ANALYSIS OF A PARAWING DURING
DEPLOYMENT
J. AIRCR VOL. 9, NO. 2 FEB 1972, PP. 168-177.

72-0561 CRISFIELD, M. A.
FINITE ELEMENT METHODS FOR ANALYSIS OF
MULTICELLULAR STRUCTURES
PROC. INSTIT. OF CIVIL ENGRS., VOL. 51, JANUARY
1972, P. 153.

72-0562 LAHELD, P.
DESIGN OF NEW SHIP TYPES BY MEANS OF FINITE
ELEMENT METHOD
NAVAL ENGINEERS JOURNAL, VOL. 84, NO. 5, 1972, P.
73.

72-0563 MITKEVICH, V.M.
CONSTRUCTION OF NORMAL DISPLACEMENT FUNCTIONS OF
TRIANGULAR FINITE ELEMENT OF PLATES AND SHELLS
DOPOVIDI AKADEMII NAUK UKRAINS'KOI R.S.R., SERIIA
A. VOL. 1972, NO. 10, 1972, P. 921.

72-0564 UKAI, S.
SOLUTION OF MULTI-DIMENSIONAL NEUTRON TRANSPORT
EQUATION BY FINITE ELEMENT METHOD
JOURNAL OF NUCLEAR SCIENCE AND TECHNOLOGY, TOKYO,
VOL. 9, NO. 6, 1972, P. 366.

72-0565 KIKUCHI, F.
FINITE ELEMENT ANALYSIS OF VIBRATION OF THIN
ELASTIC PLATES BY SIMPLIFIED HYBRID DISPLACEMENT
METHOD
JOURNAL OF NUCLEAR SCIENCE AND TECHNOLOGY, TOKYO,
VOL. 9, NO 3, 1972, P. 189.

72-0566 NAYAK, G. C.
ZIENKIEWICZ, O. C.
THE ALPHA CONSTANT-STIFFNESS METHOD FOR THE
ANALYSIS OF NONLINEAR PROBLEMS
INT. J. NUM. METHODS IN ENGNG., VOL. 4, 1972, PP.
579-582.

72-0567 JOSEPH, J. A.
MSC/NASTRAN APPLICATION MANUAL
MACNEAL-SCHWENDLER CORP., MSR-35, NOV., 1972.

72-0568 LINDLEY, P. B.
ENERGY FOR CRACK GROWTH IN MODEL RUBBER COMPONENTS
J. STRAIN ANALYSIS, 1972, P. 7, 132.

72-0569 YOUNG, J. W.
CRASH: A COMPUTER SIMULATION OF NONLINEAR
TRANSIENT RESPONSE OF STRUCTURES
PHILCO-FORD REPORT DOT-HS-09-1-125B, MARCH 1972.

72-0570 BRITTEN, S. S.
NELSON, M. F.
EIGENVALUE BOUNDS FOR FINITE ELEMENT MODELS
RESEARCH REPORT R72-29, DEPT. OF CIVIL ENG.,
M.I.T., CAMBRIDGE, MASSACHUSETTS, MAY 1972.

72-0571 STRICKLIN, J. A.
HAISLER, W. E.
VON RIESEMANN, W. A.
COMPUTATION AND SOLUTION PROCEDURE FOR NONLINEAR
ANALYSIS BY COMBINED FINITE ELEMENT-FINITE
DIFFERENCE METHODS
NATIONAL SYMPOSIUM ON COMPUTERIZED STRUCTURAL
ANALYSIS AND DESIGN, GEORGE WASHINGTON UNIV.,
1972.

72-0572 TURCKE, D. J.
MCNEICE, G. M.
A VARIATIONAL APPROACH TO GRID OPTIMIZATION IN THE
FINITE ELEMENT METHOD
CONF. ON VARIATIONAL METHODS IN ENGINEERING,
SOUTHHAMPTON UNIV. ENGLAND, SEPT. 1972.

72-0573 SKATTUM, K.S.
MODELING TECHNIQUES OF THIN-WALLED BEAMS WITH OPEN
CROSS SECTIONS
NASTRAN: USER'S EXPERIENCES, NASA TM X-2637, NASA
TECH. MEMORANDUM, WASHINGTON, D.C., 1972.

72-0574 LINDBERG, G. M.
COWPER, G. R.
HRUDEY, T. M.
REFINED FINITE ELEMENTS FOR FOLDED PLATE
STRUCTURES
PROC. OF THE MCGILL-E.I.C. FINITE ELEMENT CONF.,
JUNE 1-2, 1972.

72-0575 ODEN, J. T.
CLOUGH, R. W.
YAMAMOTO, Y.
RECENT ADVANCES IN COMPUTATIONAL METHODS IN
STRUCTURAL MECHANICS AND DESIGN.
PROC. OF THE 2ND JOINT U.S.-JAPAN SYMPOSIUM ON
MATRIX METHODS IN STRUCTURAL MECHANICS AND DESIGN,
BERKELEY, CALIFORNIA, AUGUST, 1972, ALSO IN THE
UAH PRESS, 1973.

72-0576 ODEN, J. T.
VARIATIONAL PRINCIPLES, NONLINEAR OPERATORS, AND
FINITE ELEMENT APPLICATIONS
ASCE NATIONAL STRUCTURAL ENGINEERING MEETING,
CLEVELAND, OHIO, APRIL 1972, PREPRINT 1704, 26 PP.

72-0577 ODEN, J. T.
AN IMBEDDING TECHNIQUE FOR THE GENERATION OF
WEAK-WEAK FINITE ELEMENT APPROXIMATIONS OF LINEAR
AND NONLINEAR OPERATORS
JOUR. OF THE ENGINEERING MECHANICS DIVISION, ASCE,
VOL. 98, NO. EM5, OCT., 1972, PP. 1315-1330.

72-0578 ODEN, J. T.
FINITE ELEMENT ANALYSIS OF FINITE DEFORMATIONS OF
PLATES, SHELLS, AND MEMBRANES
SHELL STRUCTURES AND CLIMATIC INFLUENCES, (ED. BY
P. G. GLOCKNER AND A. GHALI), UNIV. OF CALGARY,
ALBERTA, 1972, PP. 63-94.

72-0579 SMITH, E. T.
FAULKES, K. A.
KABAILA, A. P.
A FINITE ELEMENT PLATE BENDING PROGRAMME
UNICIV REPORT R-116, NEW SOUTH WALES UNIVERSITY,
1972.

72-0580 CUBITT, N. J.
ROY, S. K.
APPLICATION OF THE FINITE ELEMENT METHOD TO CASES
REQUIRING THE COMBINATION OF ELEMENTS POSSESSING
DIFFERENT NUMBERS OF DEGREES OF FREEDOM
VARIATIONAL METHODS IN ENGINEERING, CH. 4, C. A.
BREBBIA, H. TOTTENHAM, EDITORS, SOUTHAMPTON
UNIVERSITY PRESS, 1972.

72-0581 KHATUA, T. P.
CHEUNG, Y. K.
TRIANGULAR ELEMENT FOR MULTILAYER SANDWICH PLATES
J. ENGINEERING MECH. DIV., ASCE., 1972.

72-0582 LANSBERRY, C. R.
SHORE, S.
A FULLY COMPATIBLE ANNULAR SEGMENT FINITE ELEMENT
PROCEEDINGS OF THE SPECIALITY CONFERENCE ON FINITE
ELEMENT METHOD, MCGILL UNIVERSITY, CANADA, 1972.

72-0583 SPOKOWSKI, R. W.
FINITE ELEMENT ANALYSIS OF REINFORCED CONCRETE
MEMBERS
M. ENG. THESIS, MCGILL UNIVERSITY, MONTREAL 1972,
PP. 108.

72-0584 LUK, C. H.
ASSUMED STRESS HYBRID FINITE ELEMENT METHOD FOR
FRACTURE MECHANICS AND ELASTIC-PLASTIC ANALYSIS
TECH. REP. AFOSR-TR-73-0493, DEPT. OF AERONAUTICS
AND ASTRONAUTICS, MASS., INST. OF TECH.,
CAMBRIDGE, MASS., U.S.A., 1972.

72-0585 KONO, I.
ANALYSIS OF THE SEEPAGE THROUGH AN EARTH DAM BY
FINITE ELEMENT METHOD
J. JAP. SOC. OF SOIL MECHANICS AND FOUNDATION
ENGINEERING, VOL. 21, NO. 8, 1972, PP. 13-19.

72-0586 AHMED, K. M.
DYNAMIC ANALYSIS OF SANDWICH BEAMS
J. SOUND AND VIB., VOL. 21, NO. 3, APRIL 1972, PP.
263-276.

72-0587 ARCHAMBE, C. B.
JUNGELS, P.
DISPLACEMENT AND STRAIN FIELDS FROM UNDERGROUND
EXPLOSIONS USING ANALYTICAL AND FINITE ELEMENT
METHODS
TRANS. AMERICAN GEOPHYSICAL UNION, VOL. 53, NO.
11, 1972, P. 1118.

72-0588 BENZLEY, S. E.
TACOS: A FINITE ELEMENT COMPUTER PROGRAM FOR THE
TRANSIENT ANALYSIS OF CYLINDRICAL OBESE
SHELLS
N73-21852 SANDIA LABS., ALBUQUERQUE, NEW MEXICO,
SC-RR-72- 0454, OCTOBER 1972, 102 PP.

72-0589 CHOPRA, A. K.
CHAKRABARTI, P.
THE EARTHQUAKE EXPERIENCE AT KOYNA DAM AND
STRESSES IN CONCRETE GRAVITY DAMS
INTL. J. EARTHQUAKE ENGR., STRUC. DYN., VOL. 1,
NO. 2, OCTOBER/DECEMBER 1972, PP. 151-164.

72-0590 DAVIS, R. L.
HENSHELL, R. D.
WARBURTON, G. B.
A TIMOSHENKO BEAM ELEMENT
J. SOUND AND VIB., VOL. 1, NO. 22, JUNE 22, 1972,
PP. 475-487.

72-0591 EBERHARDT, A. C.
A FINITE ELEMENT APPROACH TO THE DYNAMIC ANALYSIS
OF CONTINUOUS HIGHWAY BRIDGES
UNIV. ILLINOIS AT URBANA-CHAMPAGN, PH.D. THESIS
1972, 232 PP.

72-0592 FERRITTO, J. M.
DYNAMIC RESPONSE OF A CYLINDER BURIED IN AN EARTH
BEAM - RESULTS OF A FINITE ELEMENT ANALYSIS
NAVAL CIVIL ENGR., LAB., PORT HUENEME, CALIF.,
NCEL-TR-764, APRIL 1, 1972, 77 PP.

72-0593 GEORGOPOULOS, G.
ROYSTER, L. H.
DEVELOPMENT OF A FINITE ELEMENT MODEL FOR THE
CLASS IV FLEXTENSIONAL UNDERWATER TRANSDUCER SHELL
GRE AD-748 187, SEPTEMBER 1972, 187 PP.

72-0594 GEORGOPOULOS, G.
ROYSTER, L. H.
A FORTRAN PROGRAM FOR CALCULATING THE FREQUENCIES
AND MODE SHAPES OF AN ARBITRARY SHAPE CYLINDRICAL
TYPE SHELL BY UTILIZING G.R.E.
AD-748 655, SEPTEMBER 1972, 36 PP.

72-0595 GEORGOPOULOS, G.
FREQUENCY COEFFICIENTS OF A SPHERICAL SHELL FOR
VARYING THICKNESS AND OPENING ANGLE WITH GUIDED
PINNED, CLAMPED AND PINNED BOUNDARY CONDITIONS
NORTH CAROLINA STATE UNIV., DEPT. OF MECH. AND
AEROSP. ENGR., RALEIGH, N.C., JANUARY 1972, 34 PP.

72-0596 NAVEH, A.
GELLERT, M.
GENERATION OF FINITE ELEMENTS BY EULER'S
VARIATIONAL METHOD
PROC. INT. CONF. ON VARIATIONAL METH. IN ENG.,
UNIV. OF SOUTHAMPTON, SEPTEMBER 25, 1972, (EDS.)
BREBBIA, C.A. AND TOTTENHAM, H., SOUTHAMPTON UNIV.
PRESS, 1973, VOL. 1, PP. 3/86-3/95.

72-0597 KANDIDOV, V. P.
CHESNOKOV, S. S.
COMPARISON OF CALCULATION BY THE FINITE ELEMENT
METHOD WITH EXPERIMENT AND AN EXAMPLE OF A DYNAMIC
STABILITY PROBLEM
VESTN. MOSK. UNIV., FIZ., ASTRON. U.S.S.R., VOL.
13, NO. 3, 1972, PP. 357-358.

72-0598 KANDIDOV, V. P.
CHESNOKOV, S. S.
A MODIFIED FINITE ELEMENT METHOD FOR CALCULATION
OF VIBRATION OF THIN PLATES
VESTN. MOSK. UNIV. FIZ. ASTRON. U.S.S.R., VOL. 13,
NO. 1, 192, PP. 44-51.

72-0599 KOBAYASHI, A. S.
WADE, B. G.
CRACK PROPAGATION AND ARREST IN IMPACTED PLATES
AD-745 792, JULY 1972, PP. 28.

72-0600 KUESSNER, H. G.
A COMPARISON OF METHODS USED IN FLUTTER RESEARCH
N72-33915 ADVISORY GROUP FOR AEROSP. RES. AND
DEV., PARIS, FRANCE, AGARD-R-592, AUGUST 1972,
141 PP.

72-0601 LEW, G. T.
A THREE DIMENSIONAL SOLUTION OF THE TRANSIENT
FIELD PROBLEM USING ISOPARAMETRIC FINITE ELEMENTS
AD-757 281, DECEMBER 1972, PP. 122.

72-0602 MAU, S. T.
WITMER, E. A.
STATIC, VIBRATION, AND THERMAL STRESS ANALYSES OF
LAMINATED PLATE AND SHELLS BY THE HYBRID-STRESS
FINITE-ELEMENT METHOD
AD-753 651, OCTOBER 1972, PP. 130.

72-0603 MCCOWAN, D. W.
SLAMCODE: FINITE ELEMENT STRESS WAVE ANALYSIS
AD-755 393, SEPTEMBER 14, 1972, 60 PP.

72-0604 PINSON, L. D.
VIBRATION CHARACTERISTICS OF THIN, PRESSURIZED
SHELLS OF REVOLUTION PARTIALLY FILLED WITH
LIQUID
VIRGINIA POLYTECH. INST. AND STATE UNIV., PH.D.
THESIS, 1972, PP. 128.

72-0605 RAO, S. S.
AUTOMATED OPTIMUM DESIGN OF AIRCRAFT WINGS TO
SATISFY STRENGTH, STABILITY, FREQUENCY AND
FLUTTER REQUIREMENTS
CASE WESTERN RESERVE UNIV., PHD THESIS, 1972, PP.
270

72-0606 SABIR, A. B.
LOCK, A. C.
LARGE AMPLITUDE, FREE VIBRATION GEOMETRICALLY
NONLINEAR CIRCULAR ARCHES
SYMP. ON NONLINEAR DYNAMICS, LOUGHBOROUGH UNIV.,
ENGLAND, MARCH 27-28, 1972, P. 21.

72-0607 SELNA, L.
SERPANOS, J. E.
FEATHER: FINITE ELEMENT ANALYSIS FOR
THREE-DIMENSIONAL ELASTIC RESPONSE
AD-753-211, AUGUST 1972, PP. 136.

72-0608 STRICKLIN, J. A.
HAISLER, W. E.
VON RIESEMANN, W. A.
LARGE DEFLECTION ELASTIC-PLASTIC DYNAMIC RESPONSE
OF STIFFENED SHELLS OF REVOLUTION
N73-24902, DECEMBER 1972, PP. 113.

72-0609 WU, R. W.
WITMER, E. A.
COMPUTER PROGRAM JET 3 TO CALCULATE THE LARGE
ELASTIC PLASTIC DYNAMICALLY INDUCED DEFORMATIONS
OF FREE AND RESTRAINED
N72-33917, AUGUST 1972, PP. 261.

72-0610 WU, J. J.
APPLICATION OF THE FINITE ELEMENT METHOD TO
NONCONSERVATIVE STABILITY PROBLEMS WITH DAMPING
AD-748086 WATERVLIET ARSENAL, BENET WEAPONS LAB.,
N.Y., AUGUST 1972, 30 PP.

72-0611 BLACKKETTER, D. O.
GARNER, F. R.
DYNAMIC RESPONSE OF THE HUMAN ARM AS A COMPOSITE
STRUCTURE
PROC. 6TH SYMP. COMPOS. MATER. ENG DES.,
WASHINGTON UNIV., ST. LOUIS, MAY 11-12, 1972, PP.
163-169. (AVAILABLE FROM ASM, METALS PART, OHIO,
1973.)

72-0612 PODATZ, W.
WITTKE, W.
WECHSELWIRKUNG ZWISCHEN DEFORMATION UND
DURCHSTROIMUNG IM KLUEFTIGEN, ANISOTROPEN GEBIRGE.
(INTERACTION BETWEEN DEFORMATION AND PERCOLATION
IN FISSURED ANISOTROPIC ROCK)
PROC. SYMP. ON PERCOLATION THROUGH FISSURED ROCK,
STUTTGART, GERMANY, PAPER T-2-1, SEPTEMBER 18-19,
1972, 8 PP. PUBL. BY DTSCH GES FUER ERD-UND
GRUNDBAU, ESSEN, GERMANY, 1972.

72-0613 KHOT, N. S.
VENKAYYA, V. B.
OPTIMUM DESIGN OF ADVANCED COMPOSITE STRUCTURES
FOR STATIC LOADS
PROC. 6TH SYMP. COMP. MATER. ENG. DES., WASHINGTON
UNIV., ST. LOUIS, MAY 11-12, 1972, PP. 417-428.
(AVAILABLE FROM ASM, METALS PARK, OHIO, 1973.)

72-0614 ROWLANDS, R. E.
DANIEL, I. M.
MECHANICAL BEHAVIOR OF A GRAPHITE-EPOXY LAMINATE
CONTAINING A HOLE
PROC. 6TH SYMP. COMPOS. MATER. ENG. DES.,
WASHINGTON UNIV., ST. LOUIS, MAY 11-12, 1972, PP.
365-376. (AVAILABLE FROM ASM, METALS PARK, OHIO,
1973.)

72-0615 COLE, R. T.
BIRCHFIELD, E. B.
EVALUATION OF COMPOSITE SHEAR PANELS WITH CUTOUTS
6TH PROC. SYMP., COMPOS. MATER. IN ENG. DES.,
WASHINGTON UNIV., ST. LOUIS, MAY 11-12, 1972, PP.
355-364. ALSO ASM, METALS PARK, OHIO, 1973.

72-0616 HYMAN, B. I.
NILFOROUSH, J. M.
EFFECT OF FIBRE DISTRIBUTION ON STRESS AND STRAIN
CONCENTRATION AT HOLES IN COMPOSITE PLATES
PROC. 6TH SYMP. COMPOS. MATER. ENG. DES.,
WASHINGTON UNIV., ST. LOUIS, MAY 11-12, 1972, PP.
621-635. (AVAILABLE FROM ASM, METALS PARK, OHIO,
1973.)

72-0617 KANDA, T.
SAWADA, Y.
YASUNAMI, K.
STRESS ANALYSIS OF THE "SELF-SEALING" TYPE PISTON
HEAD BY THE FINITE ELEMENT METHOD
PROC. 28TH ANNUAL NATL. CONF. ON FLUID POWER,
INST. OF TECH., CHICAGO, ILL., VOL. 26, SEPTEMBER
13-15, 1972. (AVAILABLE FROM FRAD. SCHOOL OF ILL.
INST. OF TECH., CHICAGO, 1972, P. 751.)

72-0618 THOMPSON, G. L.
BERT, C. W.
FINITE-ELEMENT ANALYSIS FOR FREE VIBRATION OF
GENERAL ANISOTROPIC LAMINATED THIN SHELLS
PROC. 6TH SYMP. COMPOS. MATER. ENG. DES.,
WASHINGTON UNIV., ST. LOUIS, MAY 11-12, 1972, PP.
72-80. (AVAILABLE FROM ASM, METALS PARK, OHIO,
1973.)

72-0619 MASER, K. R.
SOOSAAR, K.
ELASTIC ANALYSIS OF LARGE SPACEBOUND MIRRORS
PROC. 9TH INT. CONGR. COMM. FOR OPT., SANTA
MONICA, CALIF., OCTOBER 9-13, 1972, PP. 184-207.
(AVAILABLE FROM NAS, WASHINGTON, DC, 1974.)

72-0620 CHENG, W. K.
HOSAIN, M. U.
NEIS, V. V.
ANALYSIS OF CASTELLATED BEAMS BY THE FINITE
ELEMENT METHOD
PROC. CONF. ON FINITE ELEMENT METH. IN CIVIL ENG.,
MCGILL UNIV., MONTREAL, CANADA, 1972, PP. 152-163.

72-0621 RIGBY, G. L.
MCNEICE, G. M.
A STRAIN ENERGY BASIS FOR STUDIES OF ELEMENT
STIFFNESS MATRICES
A.I.A.A. J., VOL. 10, NO. 11, 1972, PP. 1490-1493.

72-0622 WALKER, R. E.
KIRKLAND, J. L.
DYNAMIC FINITE ELEMENT ANALYSIS OF AXI-SYMMETRIC
STRESS WAVE PROPAGATION IN SOIL-FILLED BIN
PROC. SYMP. ON THE APPLICATIONS OF THE FINITE
ELEMENT METHOD IN GEOTECH. ENG., VICKSBURG, MISS.,
MAY 1-4, 1972, PP. 1005-1026. (PUBL. BY U.S. ARMY
ENG. WATERWAYS EXP. STATION, SEPTEMBER 1972)

72-0623 BRON, J.
DHATT, G.
MIXED QUADRILATERAL ELEMENTS FOR BENDING
A.I.A.A. J., VOL. 10, NO. 10, OCT., 1972, PP:
1359-1361

72-0624 NOORISHAD, J.
WITHERSPOON, P.A.
INFLUENCE OF FLUID INJECTION ON THE STATE OF
STRESS IN THE EARTH'S CRUST
PROC. SYMP. ON PERCOLATION THROUGH FISSURED ROCK,
STUTTGART, GERMANY, PAPER T2-H, SEPT. 18-19, 1972,
11 PP. PUBL. BY DTSCH GES FUER ERD- UND GRUNDBAU,
ESSEN, GERMANY, 1972.

72-0625 ABEL, J.
DESAI, C.
COMPARISON OF FINITE ELEMENTS FOR PLATE BENDING
PROC. ASCE. J. STRUCT DIV., VOL. 98, NO. ST9,
SEPT. 1972. PP: 2143-2148

72-0626 COOK, R. D.
SOME ELEMENTS FOR ANALYSIS OF PLATE BENDING
PROC. ASCE. J. ENG. MECH. DIV., VOL. 98, NO. EM6,
1972. PP: 1452-1470

72-0627 URAL, O.
FINITE ELEMENT METHOD - A VERSATILE TOOL FOR CIVIL
ENGINEERING
PROC. CONF. OF MCGILL-EIC ON FINITE ELEMENT
METHOD, MONTREAL, JUNE 1972.

72-0628 LABRUJERE, TH. E.
MAARSINGH, R. A.
VOOREN, J. V. D.
PROPOSALS FOR THE APPLICATION OF THE FINITE
ELEMENT TECHNIQUE TO SOME AIRCRAFT AERODYNAMICAL
PROBLEMS (IN DUTCH)
NLR TR 72132 C, NOVEMBER, 1972

72-0629 MORGENSTERN, N. R.
GUTHER, H.
SEEPAGE INTO AN EXCAVATION IN A MEDIUM POSSESSING
STRESS-DEPENDENT PERMEABILITY
PROC. SYMP. ON PERCOLATION THROUGH FISSURED ROCK,
STUTTGART, GERMANY, PAPER T2-C, SEPTEMBER 18-19,
1972, 15 PP. PUBL. BY DTSCH GES FUER ERD-UND
GRUNDBAU, ESSEN, GERMANY, 1972.

72-0630 NAKAO, Y.
SASAKI, N.
KADO, F.
ON THE APPLICATION OF THE FINITE ELEMENT METHOD TO
THE VIBRATION ANALYSIS OF SPACE PLATE STRUCTURE
MITSUBISHIJUKO GIHO, VOL. 9, NO. 6, 1972.

72-0631 IRONS, B. M.
COMPUTER TECHNIQUE: THE FRONT SOLUTION
COURSE ON APPLICATIONS OF VARIATIONAL TECHNIQUES
IN ENG., UNIV. OF SOUTHAMPTON, ENGLAND, APRIL,
1972.

72-0632 BAKER, A. J.
FINITE ELEMENT COMPUTATIONAL THEORY FOR
THREE-DIMENSIONAL BOUNDARY LAYER FLOW
A.I.A.A. PAPER 72-108, 1972

72-0633 BAKER, A. J.
MANHARDT, P. D.
FINITE ELEMENT SOLUTION FOR ENERGY CONSERVATION
USING A HIGHLY STABLE EXPLICIT INTEGRATION
ALGORITHM
NASA CR-130149, 1972

72-0634 DIETERICH, J. H.
RALEIGH, C. B.
EARTHQUAKE TRIGGERING BY FLUID INJECTION AT
RANGELY, COLORADO
PROC. SYMP. ON PERCOLATION THROUGH FISSURED ROCK,
STUTTGART, GERMANY, PAPER T2-B, SEPT. 18-19, 1972,
12 PP. PUBL. BY DTSCH GES FUER ERD- UND GRUNDBAU,
ESSEN, GERMANY, 1972.

72-0635 DAILEY, J. E.
HARLEMAN, D. R. F.
NUMERICAL MODEL FOR THE PREDICTION OF TRANSIENT
WATER QUALITY IN ESTUARY NETWORKS
TECH. REP. NO. 158, RALPH M. PARSONS LAB., DEPT.
OF CIVIL ENGINEERING, MASSACHUSETTS INST. OF
TECH., OCTOBER, 1972

72-0636 PINDER, G. F.
FRIND, E. O.
APPLICATION OF GALERKIN'S PROCEDURE TO AQUIFER
ANALYSIS
WATER RESOURC. RES. VOL. 8, NO. 1, 1972, PP:
108-120

72-0637 HURR, R. T.
MODELLING GROUNDWATER FLOW BY THE FINITE ELEMENT
METHOD
INT. CONF. ON VARIATIONAL METHODS IN ENGRNG.,
SOUTHAMPTON UNIVERSITY, SEPTEMBER, 1972, PP:
5139-5149

72-0638 ARGYRIS, J. H.
GRIEGER, I.
PARAMETRISCHE UNTERSUCHUNGEN VON KERBEN MIT DER
MATRIZENVERSCHIEBUNGSMETHODE UND
BILDSCHIRMUNTERSTUTZUNG
TECHN. MITTEILUNGEN 65, NO. 7, 1972, PP: 301-309.
SEE ALSO INSTITUT FUR STATIK UND DYNAMIK DER
LUFTUND RAUMFAHRTKONSTRUTIONEN, REPORT NO: 122.

72-0639 ARGYRIS, J. H.
GRIEGER, I.
INTERACTIVE COMPUTER GRAPHICS IN STRUCTURAL
ANALYSIS: ONLINE 72
INTL. SYMP. OF ONLINE INTERACTIVE COMPUTING,
UXBRIDGE, ENGLAND, 1972.

72-0640 ARGYRIS, J. H.
ANGELOPOULOS, T.
THEORIE, PROGRAMMENTWICKLUNG UND ERFAHRUNG
ANVORGESPANNTEN NETZWERKKONSTRUKTIONEN
INTERNATION VEREINIGUNG FUR BRUCKENBAU UND
HOCHBAU, ZURICH SONDERDRUCK AUS DEM VORBERICHT D.
9 KONGRESSES, AMSTERDAM, MAY 8-13, 1972.

72-0641 ARGYRIS, J. H.
ANGELOPOULOS, T.
EIN VERFAHREN DUR DIE FORMFINDUNG VEON BELIEBIGEN
VORGESPANNTEN NETZWERKKONSTRUKTIONEN
INTERNATION. VEREINIGUNG FUR BRUCKENBAU UND
HOCHBAU, ZURICH SONDERDRUCK AUS DEM VORBERICHT D.
9 KONGRESSES AMSTERDAM MAY 8-13, 1972.

72-0642 ARGYRIS, J. H.
CHAN, A. S. L.
STATIC AND DYNAMIC ELASTO-PLASTIC ANALYSIS BY THE
METHOD OF FINITE ELEMENTS IN SPACE AND TIME
INGENIEUR ARCHIV. NO. 418 1972, PP: 235-257

72-0643 ARGYRIS, J. H.
SCHARPF, D. W.
MATRIX DISPLACEMENT ANALYSIS OF SHELLS AND PLATES
INCLUDING TRANSVERSE SHEAR STRAIN EFFECTS
COMPUTER METHODS IN APPLIED MECHANICS AND
ENGINEERING, NO. 1, 1972, PP: 81-139.

72-0644 MLEJNEK, H. P.
BERECHNUNG EINES EINACHSIGEN
STAHLDRAHT-KUNSTSTOFFVERBUNDES UNTER EINSATZ DER
FINIT-ELEMENT METHODE
RESEARCH REPORT NO. 131, INSTITUT FUR STATIK UND
DYNAMIK DER LUFT UND RAUMFAHTKONSTRUKTIONEN,
SEPTEMBER 1972.

72-0645 HENSHELL, R. D.
WALTERS, D.
WARBURTON, G. B.
A NEW FAMILY OF CURVILINEAR PLATE BENDING ELEMENTS
FOR VIBRATION AND STABILITY
J. SOUND AND VIB., VOL. 20, NO. 3, FEB. 8, 1972,
PP. 381-397.

72-0646 BRONLUND, O. E.
BUHLMEIER, J.
EINIGE VERFAHREN ZUR BERECHNUNG VON EIGENWERTEN
UND EIGENVEKTOREN VON NICHT HERMITESCHEN MATRIZEN
UNTER BESONDERER BERUCKSICHTIGUNG
STRUKTURDYNAMISCHER PROBLEME
INSTITUT FUR STATIK UND DYNAMIK DER LUFT UND
RAUMFAHRTKONSTRUKTIONEN, JULY 1972.

72-0647 BALMER, H.
DOLTSINIS, J.
MATERIAL NONLINEARITIES
INSTITUT FUR STATIK UND DYNAMIK DER LUFT UND
RAUMFAHRTKONSTRUKTIONEN, RESEARCH REPORT NO. 132,
NOVEMBER 1972. (SEE ALSO REPORT NO. 108)

72-0648 ARGYRIS, J. H.
BUCK, K. E.
GRIEGER, I.
WILLAM, K.
DISCUSSION ON FINITE ELEMENT ANALYSIS OF
PRESTRESSED CONCRETE REACTOR VESSELS
CONCRETE FOR NUCLEAR REACTORS, VOL. 1, 1972, P:
181.

72-0649 ARGYRIS, J. H.
DOLTSINIS, J. S.
GLOUDEMAN, J. F.
STRAUB, K.
WILLAM, K. J.
ASPECTS OF THE FINITE ELEMENT METHOD AS APPLIED TO
AERO-SPACE STRUCTURES
PROC. 23RD INTL. ASTRONAUTICAL CONGRESS, WIEN,
OCTOBER 1972, PP. 179-238, AND INSTITUT FUR STATIK
UND DYNAMIK DER LUFT UND RAUMFAHRTKONSTRUKTIONEN,
REPORT NO. 738.

72-0650 ARGYRIS, J. H.
ROY, J. R.
REANALYSIS FOR LIMITED STRUCTURAL DESIGN
MODIFICATIONS
AMER. SOC. CIVIL ENGRS. EM 6, 1972, PP: 1599-1601.

72-0651 ARGYRIS, J. H.
LOCHNER, N.
ON THE APPLICATION OF THE SHEBA SHELL ELEMENT
COMPUTER METHODS IN APPLIED MECH. AND ENG., VOL.
1, 1972, PP. 317-347. (SEE ALSO INSTITUT FUR
STATIK UND DYNAMIK DER LUFT UND
RAUMFAHRTKONSTRUKTIONEN, REPORT NO. 127)

72-0652 SINGLA, O. P.
ANALOG SUBROUTINES FOR THE FINITE ELEMENT METHOD
DOCTORAL THESIS, SCHOOL OF ENGRNG AND APPLIED
SCI., CALIFORNIA, LOS ANGELES, 1972, 201 P.
(N.T.I.S. / PB-237 527/7ST)

72-0653 DOMINQUEZ, R. F.
MUSKA, N. M.
MECHANICS OF CABLE MOORING SYSTEMS. VOLUME IV. A
COMPUTER PROGRAM FOR ANALYZING THE STEADY STATE
RESPONSE OF B1 AND QUAD CABLE ARRAYS
COASTAL AND OCEAN ENGNRG., TEXAS A AND M UNIV.,
COLLEGE STATION, REPORT NO: COE-160, DECEMBER
1972, 52 P. (N.T.I.S. - AD-786 184/2SL) SEE ALSO
VOL. 5, AD-786 185.

72-0654 DOMINQUEZ, R. F.
OWENS, G. E.
MECHANICS OF CABLE MOORING SYSTEMS. VOLUME 1.
THREE DIMENSIONAL RESPONSE OF DEEP WATER MOORING
LINES IN STEADY STATE FLOWS
COASTAL AND OCEAN ENGNRG., TEXAS A AND M UNIV.,
COLLEGE STATION, REPORT NO: COE-157, DECEMBER
1972, 243 P. (N.T.I.S. - AD-786 181/8SL) SEE ALSO
VOL. 2, AD-786 182.

72-0655 WEILER, F. C.
DERBIDGE, T. C.
POWERS, C. A.
ANALYSIS OF THERMAL STRESSES IN CORED CERAMIC
BRICKS USED IN HYPERSONIC WIND TUNNEL HEATERS
AEROTHERM DIV., ACUREX CORP., MOUNTAIN VIEW,
CALIF., REPORT NO: AEROTHERM-FR-72-61, OCTOBER
1972, 109 P. (N.T.I.S. - AD-783-359/3)

72-0656 DUNCAN, J. M.
WITHERSPOON, P. A.
MITCHELL, J. K.
WATKINS, D. J.
HARDCASTLE, J. H.
SEEPAGE AND GROUNDWATER EFFECTS ASSOCIATED WITH
EXPLOSIVE CRATERING
INST. OF TRANSP. AND TRAFFIC ENG., UNIV. OF
CALIF., BERKELEY, REPORT NO: TE-72-2, APRIL 1972,
207 PP. (N.T.I.S. - 782/356/0)

72-0657 ARALDSEN, P. O.
THE APPLICATION OF THE SUPER ELEMENT METHOD IN
ANALYSIS AND DESIGN OF SHIP STRUCTURES AND
MACHINERY COMPONENTS
NATL. SYMP. ON COMPUTERIZED STRUCT. ANALYSIS AND
DESIGN, WASH. D.C., 1972.

72-0658 EGGERS, H.
CONTINUUM CALCULATION FROM ELASTIC PLASTIC
REINFORCED MATERIAL USING A FINITE ELEMENT METHOD
(DIE BERECHNUNG VON KONTINUA AUS
ELASTISCH-PLASTISCHEM MATERIAL MIT VERFESTIGUNG
NACH EINER FINITEN ELEMENTMETHODE)
PH.D. THESIS, INSTITUT FUER STATIK, TECHNISCHE
UNIVERSITAET, BRUNSWICK, WEST GERMANY, REPORT NO:
REPT-72-3, 1972, 110 P. (N.T.I.S. - N74-13642/5)

72-0659 COMARTIN, C. D.
SCORDELIS, A. C.
ANALYSIS AND DESIGN OF SKEW BOX GIRDER BRIDGES
STRUCT. ENGNG. LAB., UNIV. OF CALIF., BERKELEY,
REPORT NO: UCSESM-72-14, DECEMBER 1972, 183 P.
(N.T.I.S. - PB-226 793/8)

72-0660 DE KONING, A. U.
A MATHEMATICAL THEORY OF PLASTICITY FOR STRAIN
HARDENING MATERIALS EXHIBITING INITIAL ANISOTROPY
DIV. STRUCT. AND MATER., NATL. AEROSPACE LAB.,
AMSTERDAM, REPORT NO: NLR-TR-72117-U, SEPTEMBER
20, 1972, 25 PP. (N.T.I.S.-N74-11719/3)

72-0661 WU, T. H.
INVESTIGATION OF SOIL-STRUCTURE INTERACTION:
FLEXIBLE CULVERT PIPES
ENGNG. EXPERIMENT STATION, OHIO STATE UNIV.,
COLUMBUS, REPORT NO: EES-279, MAY 1972, 82 P.
(N.T.I.S.-PB-224 974/6)

72-0662 RUNNER, T. A.
AN INVESTIGATION OF THE BENDING OF SYMMETRIC
PLATES UNDER SYMMETRIC TRANSVERSE LOADING
MASTER'S THESIS, SCHOOL OF ENGNG., AIR FORCE INST.
OF TECH., WRIGHT-PATTERSON AFB, OHIO, REPORT NO:
GAM/MC/73-4, DECEMBER 1972, 133 PP. (N.T.I.S. -
AD-769 928/3)

72-0663 SHORE, S.
LANSBERRY, C. R.
A FULLY COMPATIBLE ANNULAR SEGMENT FINITE ELEMENT
TOWNE SCHOOL OF CIVIL AND MECH. ENGNG.,
PHILADELPHIA, PA., INTERIM REPORT, FEBRUARY 1972,
200 PP. (N.T.I.S. - PB-223 620/6 SEE ALSO REPORT
NO. CURT-T0272 (CONSORTIUM OF UNIV. RESEARCH
TEAMS)

72-0664 BERGAN, P. G.
CLOUGH, R. W.
CONVERGENCE CRITERIA FOR ITERATIVE PROCESSES
A.I.A.A. J., VOL. 10, NO. 8, 1972, PP. 1107-1108.

72-0665 LOE, C. J.
FINITE ELEMENT CALCULATIONS OF THE COMPLIANCE OF A
TAPERED DOUBLE CANTILEVER BEAM SPECIMEN FOR
DIFFERENT CRACK CONFIGURATIONS
NATIONAL AEROSPACE LAB., AMSTERDAM, NETHERLANDS,
REPORT NO: NLR-TR-72083-1, JULY 1, 1972, 27 P.
(N.T.I.S. - N73-26941/7)

72-0666 RYAN, M. O.
SALEM, M. H.
GAMBLE, W. L.
MOHRAZ, B.
THICK WALLED MULTIPLE OPENING REINFORCED CONCRETE
CONDUITS
DEPT. OF CIVIL ENG., UNIV. OF ILLINOIS, URBANA,
REPORT NO: STRUCT. RESEARCH SER-390,
UILJ-ENG-72-2020, DECEMBER 1972, 199 P. (N.T.I.S.
- AD-766 813/0)

72-0667 KULICKI, J. M.
KOSTEM, C. N.
THE INELASTIC ANALYSIS OF REINFORCED AND
PRESTRESSED CONCRETE BEAMS
FRITZ ENGNG. LAB., LEHIGH UNIV., BETHLEHEM, PA.,
REPORT NO: FEL-378B.1, NOVEMBER 1972, 124 P.
(N.T.I.S. - PB-222 054/9)

72-0668 CHAN, S. S. M.
A CASE STUDY OF IN-SITU ROCK DEFORMATION BEHAVIOR
FOR THE DESIGN OF GROUND SUPPORT SYSTEM
BUREAU OF MINING RESEARCH, UNIV. OF IDAHO, MOSCOW,
RESEARCH REPORT, JULY 1973 - DECEMBER 1971,
JANUARY 31, 1972, 139 P. (N.T.I.S. - PB-221 880/8)

72-0669 ANON
NAVY-NASTRAN COLLOQUIM AT WASHINGTON, D. C.
PROC. 3RD CONF. NAVY-NASTRAN COLLOQ., NAVAL SHIP
RESEARCH AND DEVELOPMENT CENTER, BETHESDA, MD.,
MARCH 23, 1972, 228 PP. (N.T.I.S. -AD-764 299)

72-0670 BABUSKA, I.
AZIZ, A. K.
SURVEY LECTURES ON THE MATHEMATICAL FOUNDATIONS OF
THE FINITE ELEMENT METHOD
THE MATHEMATICAL FOUNDATIONS OF THE FINITE ELEMENT
METHOD WITH APPLICATIONS TO PARTIAL DIFFERENTIAL
EQUATIONS, A. K. AZIZ (ED.), ACADEMIC PRESS, NEW
YORK, 1972, PP. 1-345.

72-0671 SANDER, G.
BECKERS, P.
FRAEIJS DE VEUBEKE, B.
IMPROVEMENTS OF FINITE ELEMENT SOLUTIONS FOR
STRUCTURAL AND NONSTRUCTURAL APPLICATIONS
LABORATOIRE DE TECHNIQUES AERONAUTIQUES ET
SPATIALES, LIEGE UNIV., BELGIUM, FINAL REPORT,
JULY 1969 - MAY 1971, DECEMBER 1972, 48 PP.
(N.T.I.S. - AD-762 947)

72-0672 ABDELRAOUG, M. R.
MATLOCK, H.
FINITE-ELEMENT ANALYSIS OF BRIDGE DECKS
CENTER FOR HIGHWAY RESEARCH, TEXAS UNIV., AUSTIN,
REPORT NO: CFHR-3-5-63-56-28, AUGUST 1972, 136 PP.
(N.T.I.S. - PB-220 897/3)

72-0673 AUTHOR
M.I.T. TEST SECTION INSTRUMENTATION, MASSACHUSETTS
BAY TRANSPORTATION AUTHORITY, HAYMARKET-NORTH
EXTENSION. FINAL PROJECT REPORT
DEPT. OF CIVIL ENGNG., MASSACHUSETTS INST. OF
TECH., CAMBRIDGE, REPORT NO: R72-33, MARCH 1972,
359 P. (N.T.I.S. - PB-220 877/5) SEE ALSO NTIS,
PB-220 876-SET.

72-0674 HUDSON, W. R.
TREYBIG, H. J.
ABOU-AYYASH, A.
A SUMMARY OF DISCRETE-ELEMENT METHODS OF ANALYSIS
FOR PAVEMENT SLABS
CENTER FOR HIGHWAY RESEARCH, TEXAS UNIV., AUSTIN,
REPORT NO: CFHR-3-5-63-56-27, AUGUST 1972, 65 P.
(N.T.I.S. - PB-220 611/8)

72-0675 CHAKRABARTI, P.
CHOPRA, A. K.
EARTHQUAKE RESPONSE OF GRAVITY DAMS INCLUDING
RESERVOIR INTERACTION EFFECTS
EARTHQUAKE ENGNG. RESEARCH CENTER, UNIV. OF
CALIFORNIA, BERKELEY, REPORT NO: EERC-72-6,
DECEMBER 1972, 172 P. (N.T.I.S. - AD-762 330)

72-0676 BATOZ, J. L.
DHATT, G.
DEVELOPMENT OF TWO SIMPLE SHELL ELEMENTS
A.I.A.A. J., VOL. 10, 1972, PP. 237-238.

72-0677 BELL, K.
ON THE QUINTIC TRIANGULAR PLATE BENDING ELEMENT
DIV. OF STRUCT. MECH., UNIV. OF TRONDHEIM, REPORT
72-2, 1972.

72-0678 GIGNAC, D. A.
A COMPARATIVE STUDY OF SEVERAL CORE STORAGE
SCHEMES FOR LARGE SPARSE POSITIVE DEFINITE
MATRICES WITH REFERENCE TO THE CHOLESKY ALGORITHM
NAVAL SHIP RESEARCH AND DEVELOPMENT CENTER,
BETHESDA, MD., REPORT NO: NSRDC-4017, NOVEMBER
1972, 38 P. (N.T.I.S. - AD-760 669)

72-0679 HEUZE, F. E.
GOODMAN, R. E.
FINITE ELEMENT STUDIES OF PILE DRIVER TUNNELS
INCLUDING CONSIDERATIONS OF SUPPORT REQUIREMENTS
UNIV. OF CALIF., BERKELEY, INTERIM TECH. REPORT,
NOVEMBER 1972, 138 PP. (N.T.I.S. - AD-760 586)

72-0680 DI PASQUALE, S.
LEGGERI, B.
NENCIONI, S.
ENERGY FORMS IN THE FINITE ELEMENT TECHNIQUES
PROC. INT. CONF. ON VARIATIONAL METH. IN ENG.,
UNIV. OF SOUTHAMPTON, SEPTEMBER 25, 1972, (EDS.)
BREBBIA, C.A. AND TOTTENHAM, H., SOUTHAMPTON UNIV.
PRESS, 1973, VOL. 1, PP. 4/33-4/43.

72-0681 CLOUGH, R. W.
BATHE, K. J.
FINITE ELEMENT ANALYSIS OF DYNAMIC RESPONSE
ADVANCES IN COMPUTATIONAL METHODS IN STRUCT. MECH.
AND DESIGN. J. T. ODEN, R. W. CLOUGH, Y. YAMOMOTO
(EDS.), UNIV. OF ALABAMA PRESS, HUNTSVILLE, 1972,
PP. 153-180.

72-0682 OLINE, L.
MEDAGLIA, J.
ELASTIC PLATE SPALLATION
COLL. OF ENGNG., UNIV. OF SOUTH FLORIDA, TAMPA,
REPORT NO: NASA-CR-112294, MAY 31, 1972, 115 P.
(N.T.I.S. - N73-20911)

72-0683 WILSON, E. L.
BATHE, K. J.
DOVEY, H. H.
PETERSON, F. E.
COMPUTER PROGRAM FOR STATIC AND DYNAMIC ANALYSIS
OF LINEAR STRUCTURAL SYSTEMS
EARTHQUAKE ENGNG. RESEARCH CENTER, UNIV. OF
CALIF., BERKELEY, REPORT NO: EERC-72-10, NOVEMBER
1972, 62 PP. (N.T.I.S. - PB 220 437/8)

72-0684 NORRIS, D. M. JR.
TENSILE FIELD CALCULATION ABOUT A PRESSURIZED
CAVITY: MONERO
LAWRENCE LIVERMORE LAB., UNIV. OF CALIF.,
LIVERMORE, JULY 5, 1972, 9 PP. (N.T.I.S. -
UCID-16082)

72-0685 MIZUMACHI, W.
FINITE ELEMENT ANALYSIS IN NUCLEAR POWER PLANTS
(JAPANESE)
KARYOKU HATSUDEN, VOL. 23, 1972, PP. 505-514.

72-0686 BERGER, A. E.
ERROR ESTIMATES FOR THE FINITE ELEMENT METHOD
PH.D. THESIS, MASSACHUSETTS INST. OF TECH., 1972.

72-0687 AUTHOR
FINITE ELEMENT ANALYSIS OF MINE STRUCTURES
DEPT. OF CIVIL ENGNG. UNIV. OF CALIF., BERKELEY,
FINAL REPORT, SEPTEMBER 1972, 198 P. (N.T.I.S. -
PB-220 141/6)

72-0688 BOOKER, J. R.
A NUMERICAL METHOD FOR THE SOLUTION OF BIOT'S
CONSOLIDATION THEORY
CIVIL ENGNG. LABS., SYDNEY UNIV., AUSTRALIA,
REPORT NO: R-196, AUGUST 1972, 23 P. (N.T.I.S. -
PB-219 249/0)

72-0689 PETYT, M.
MIRZA, W. H.
DYNAMIC BEHAVIOR OF IN-LINE SHEAR WALLS CONNECTED
BY FLOOR SLABS
J. SOUND VIB., VOL. 25, NO. 3, DEC. 8, 1972, PP.
349-357.

72-0690 PANAK, J. J.
A DISCRETE-ELEMENT METHOD OF ANALYSIS FOR
ORTHOGONAL SLAB AND GRID BRIDGE FLOOR SYSTEMS
CENTER FOR HIGHWAY RESEARCH, UNIV. OF TEXAS,
AUSTIN, REPORT NO: 3-5063-56-25, MAY 1972, 129 P.
(N.T.I.S. - PB-219 231/8)

72-0691 DEVRIES, K. L.
WILLIAMS, M. L.
MECHANICS OF FRACTURE IN ADHESIVE JOINTS
COLL. OF ENGNG., UTAH UNIV., SALT LAKE CITY,
REPORT NO: UTEC-DO-72-163, OCTOBER 1972, 61 P.
(N.T.I.S. - PB-219 119/5)

72-0692 BURMAN, Z. I.
ON THE THEORY OF CALCULATION OF THE OVERALL
STRENGTH OF A FUSELAGE BY THE METHOD OF FINITE
ELEMENTS
IZV. VUZ. AVIATS. TEKH., VOL. 15, 1972, PP. 49-55.

72-0693 CAMBURN, G. L.
THE EFFECT OF GRAVITY UPON THE MELT-THROUGH TIME
OF A SOLID SUBJECTED TO A HIGH INTENSITY LASER
MASTER'S THESIS, SCHOOL OF ENGNG., AIR FORCE INST.
OF TECH., WRIGHT-PATTERSON AFB, OHIO, REPORT NO:
GAM/MC/73-6, MARCH 1972, 94 PP. (N.T.I.S. - AD-759
168)

72-0694 MERRIFIELD, B. C.
FORTRAN SUBROUTINES FOR FINITE ELEMENT ANALYSIS
STRUC. DEPT., ROYAL AIRCRAFT ESTABLISHMENT,
FARNBOROUGH, ENGLAND, REPORT NO: ARC-CP-1223,
RAE-TR-71156, 1972, 48 P. (N.T.I.S. - N73-18941)

72-0695 SIDNAY, S. E.
RODDIS, R.
GRAY, I.
THE APPLICATION OF FINITE ELEMENT ANALYSIS TO THE
STUDY OF TWO-DIMENSIONAL CRACKED BODIES
METALLURGY DEPT. BISRA-CORPORATE LABS. OF THE
BRITISH STEEL CORP., LONDON, ENGLAND, REPORT NO:
MG/21/72, NOVEMBER 1972, 15 P. (N.T.I.S. - PB-218
918/1)

72-0696 COOLEY, R. L.
HARSH, J. F.
LEWIS, D. C.
HYDROLOGIC ENGINEERING METHODS FOR WATER RESOURCES
DEVELOPMENT. VOLUME 10. PRINCIPLES OF GROUND-WATER
HYDROLOGY
HYDROLOGIC ENGNG. CENTER, DAVIS, CALIF., REPORT
NO: HEC-IHD-1000, APRIL 1972, 353 P. (N.T.I.S. -
AD-758 906) SEE ALSO VOL. 2, AD-758 905.

72-0697 BYKAT, A.
AUTOMATIC TRIANGULATION OF TWO DIMENSIONAL REGIONS
INST. OF COMPUTER SCI., UNIV. OF LONDON, REPORT
ICSI 420, 1972.

72-0698 CONLEY, W. H. JR.
CONSISTENT STRESSES FOR THE FINITE ELEMENT
STIFFNESS METHOD
MASTER'S THESIS, NAVAL POSTGRADUATE SCHOOL,
MONTEREY, CALIF., DECEMBER 1972, 102 PP. (N.T.I.S.
- AD-758 675)

72-0699 JONES, C. E.
FENG, G. C.
STUDY OF PROPELLANT DYNAMICS IN A SHUTTLE TYPE
LAUNCH VEHICLE
LOCKHEED MISSILES AND SPACE CO., HUNTSVILLE, ALA.,
REPORT NO: NASA-CR-124070, LMSC/HREC-D225823,
APRIL 1972, 29 P. (N.T.I.S. - N73-17801)

72-0700 CELLA, A.
APPROXIMATION TECHNIQUES IN THE FINITE ELEMENT
METHOD
CONSIGLIO NAZIONALE DELLE RICERCHE, INSTUIO DI
ELABORAZIONE DELLA INFORMAZIONE, PISA, REPORT NO.
I. 972-11, 1972.

72-0701 BAUER, A. M.
COMOC THERMAL ANALYSIS VARIANT, USER'S MANUAL
BELL AEROSPACE CO., BUFFALO, N.Y. REPORT NO:
NASA-CR-130148, REPT-9500-920256, OCTOBER 1972, 61
P. (N.T.I.S. - N73-16927)

72-0702 CROUZEIX, M.
THOMAS, J. M.
RESOLUTION NOMERIQUE PAR DES METHODES D ELEMENTS
FINIS DE PROBLEMES ELLIPTIQUES DEGENERES
COMPTES RENDUS DE LE ACAD. DES SCI., VOL. 275,
1972, PP. 1115-1118.

72-0703 KELLY, B. M.
THE NASTRAN CONTOUR PLOTTER
NAVAL SHIP RESEARCH AND DEVELOPMENT CENTER,
BETHESDA, MD., REPORT NO: NSRDC-3887, JUNE 1972,
24 P. (N.T.I.S. - AD-757 644)

72-0704 DESAI, C. S.
THEORY AND APPLICATION OF THE FINITE ELEMENT
METHOD IN GEOTECHNICAL ENGINEERING
PROC. SYMP. ON THE APPLICATIONS OF THE FINITE
ELEMENT METHOD IN GEOTECH. ENG., VICKSBURG, MISS.,
MAY 1-4, 1972, PP. 3-90. (PUBL. BY U.S. ARMY ENG.
WATERWAYS EXP. STATION, SEPTEMBER 1972)

72-0705 BOMBICH, A. A.
SULLIVAN, B. R.
ULTIMATE STRAIN CAPACITY AND TEMPERATURE RISE
STUDIES, TRUMBULL POND DAM
ARMY ENG. WATERWAYS EXPERIMENT STATION, VICKSBURG,
MISS., REPORT NO: AEWES-MISC-PAPER-C-72-20, AUGUST
1972, 85 P. (N.T.IS. - AD-757 379)

72-0706 DILLON, E. C.
O'BRIEN, M.
SOLUTION BOUNDS IN SOME STRESS PROBLEMS BY THE
HYPERCIRCLE AND FINITE ELEMENT METHODS
INT. CONF. VARIATIONAL METHODS IN ENG.,
SOUTHAMPTON UNIV., 1972.

72-0707 DYER, D. F.
FLUID MECHANICS AND FINITE ELEMENTS
OFFICE OF NAVAL RESEARCH, LONDON, ENGLAND, REPORT
NO: ONRL-R-21-72, DECEMBER 29, 1972, 16 P.
(N.T.I.S. - AD-757 266)

72-0708 TORVIK, P. J.
A NUMERICAL PROCEDURE FOR TWO DIMENSIONAL HEATING
AND MELTING CALCULATIONS, WITH APPLICATIONS TO
LASER EFFECTS
SCHOOL OF ENGNG., AIR FORCE INST. OF TECH.,
WRIGHT-PATTERSON AFB, OHIO, TECHNICAL REPORT,
MARCH 1972, 57 P. (N.T.I.S. -AD-757 189)

72-0709 EBNER, A. M.
UCCIFERRO, J. J.
A THEORETICAL AND NUMERICAL COMPARISON OF ELASTIC
NONLINEAR FINITE ELEMENT METHODS
NATL. SYMP. ON COMPUTERIZED STRUCT. ANAL. AND
DESIGN, GEORGE WASHINGTON UNIVERSITY, 1972.

72-0710 WEMPER, G.
DISCRETE APPROXIMATIONS OF ELASTIC-PLASTIC BODIES
BY VARIATIONAL METHODS
PROC. INT. CONF. ON VARIATIONAL METHODS IN ENGNG,
SOUTHAMPTON, ENGLAND, SEPTEMBER 1972, P: 1 - 35.
(N.T.I.S. - AD-756 979)

72-0711 DAME, R. E.
APPROXIMATION AND BOUNDS FOR EIGENVALUES OF
VARIOUS DOMAINS USING THE INTERMEDIATE PROBLEM
TECHNIQUE
INST. OF OCEAN SCI., AND ENGNG., CATHOLIC UNIV. OF
AMERICA, WASH., D.C., REPORT NO: 72-6, JUNE 1972,
130 P. (N.T.I.S. -AD-756 672)

72-0712 EDWARDS, B. A.
CHARACTERIZATION OF ELASTIC SOLIDS USING FINITE
ELEMENT METHODS
MASTER'S THESIS, NAVAL POSTGRADUATE SCHOOL,
MONTEREY, CALIF., DECEMBER 1972, 93 P. (N.T.I.S. -
AD-756 513)

72-0713 BARKER, R. M.
DANA, J. P.
PRYOR, C. W. JR.
THREE DIMENSIONAL ANALYSIS OF STRESS
CONCENTRATIONS NEAR HOLES IN LAMINATED COMPOSITES
CONF. ASCE SPECIALTY ON COMPOSITE MATERIALS,
PITTSBURGH, PA., NOVEMBER 13-14, 1972, REPORT NO:
VPI-E-72-27, DECEMBER 1972, 32 P. (N.T.I.S. -
AD-756 454)

72-0714 PALMERTON, J. B.
LEFEBVRE, G.
THREE-DIMENSIONAL BEHAVIOR OF A CENTRAL CORE DAM
ARMY ENG. WATERWAYS EXPERIMENT STATION, VICKSBURG,
MISS., REPORT NO: AEWES-RR-S-72-1, DECEMBER 1972,
74 P. (N.T.I.S. - AD-756 443)

72-0715 CARMIGNAMI, C.
CHARACTERISTICS OF A HYBRID CALCULATION SYSTEM
BASED ON THE FINITE ELEMENT METHOD.
(CARATTERISTICHE DI UNE SISTEMA DI CALCOLO IBRIDO
BASATO SUL METODO DEGLI ELEMENTI FINITI
PISA UNIV., ITALY, REPORT NO: RP-106(72), APRIL
1972, 29 P.)

72-0716 PAINE, A. A.
DEVELOPMENT AND APPLICATIONS OF SUPERSONIC
UNSTEADY CONSISTENT AERODYNAMICS FOR INTERFERING
PARALLEL WINGS USERS MANUAL
BELL AEROSPACE CO., BUFFALO, N.Y., REPORT NO:
NASA-CR-112184, REPT-2471-941001, AUGUST 1972, 54
PP. (N.T.I.S. N73-13290)

72-0717 MINAMI, T.
ELASTIC-PLASTIC EARTHQUAKE RESPONSE OF
SOIL-BUILDING SYSTEMS
DOCTORAL THESIS, EARTHQUAKE ENGNG. RESEARCH
CENTER, UNIV. OF CALIF., BERKELEY, REPORT NO:
EERC-72-3, AUGUST 1972, 164 P. (N.T.I.S. - PB-214
868/2)

72-0718 KEY, J. E.
USERS MANUAL FOR DIGITAL COMPUTER PROGRAM OK5
NUMERICAL ANALYSIS OF FINITE STRAINS AND
DISPLACEMENTS OF ELASTIC MEMBRANES
SCHOOL OF GRADUATE STUDIES AND RESEARCH, UNIV. OF
ALABAMA, HUNTSVILLE, REPORT NO: UAH-RR-134,
DECEMBER 1972, 78 P. (N.T.I.S. - AD-755 953)

72-0719 KEY, J. E.
COMPUTER PROGRAM FOR SOLUTION OF LARGE, SPARSE,
UNSYMMETRIC SYSTEMS OF LINEAR EQUATIONS
RESEARCH INST., UNIV. OF ALABAMA, HUNTSVILLE,
REPORT NO: UARI-RR-130, AUGUST 1972, 35 P.
(N.T.I.S. - AD-755 878)

72-0720 DAVIS, R.
HENSHELL, R. D.
WARBURTON, G. B.
CURVED BEAM FINITE ELEMENTS FOR COUPLED BENDING
AND TORSIONAL VIBRATION
INT. J. EARTHQUAKE ENG. STRUCT. DYN., VOL. 1, NO.
2, OCTOBER-DECEMBER, 1972, PP. 165-175.

72-0721 CANTIN, G.
THREE DIMENSIONAL FINITE ELEMENT STUDIES. PART
ONE: SERVICE ROUTINES
NAVAL POSTGRADUATE SCHOOL, MONTEREY, CALIF.,
REPORT NO: WPS-59C172121A, DECEMBER 1972, 69 P.
(N.T.I.S. - AD-755 539)

72-0722 HESS, J. L.
CALCULATION OF POTENTIAL FLOW ABOUT ARBITRARY
THREE-DIMENSIONAL LIFTING BODIES
DOUGLAS AIRCRAFT CO., LONG BEACH, CALIF., REPORT
NO: MDC-J5679-01, OCTOBER 1972, 165 P. (N.T.I.S. -
AD-755 480)

72-0723 BAI, K. J.
A VARIATIONAL METHOD IN POTENTIAL FLOWS WITH A
FREE SURFACE
COLL. OF ENGNG., UNIV. OF CALIFORNIA, BERKELEY,
REPORT NO: MA-72-2, SEPTEMBER 1972, 146 P.
(N.T.I.S. - AD-755 465)

72-0724 FREMOND, M.
LA METHODE DES ELEMENTS FINIS
ECOLE NATIONALE DES PONTS ET CHAUSSEES, REPORT,
PARIS, 1972.

72-0725 GALLAGHER, R. H.
THE FINITE ELEMENT METHOD OF THIN-SHELL STABILITY
ANALYSIS
NATL. SYMP. ON COMPUTERIZED STRUCT. ANALYSIS AND
DESIGN, GEORGE WASHINGTON UNIV., 1972

72-0726 GALLO, A. M.
MAGIC 111: AN AUTOMATED GENERAL PURPOSE SYSTEM FOR
STRUCTURAL ANALYSIS. VOLUME 111. PROGRAMMER'S
MANUAL
BELL AEROSPACE CO., BUFFALO, N.Y., FINAL REPORT,
MARCH 15, 1971 - MARCH 15, 1972, JULY 1972, 722 P.
(N.T.I.S. - AD-755 370) SEE ALSO VOL. 2, AD-755
369 AND NTIS REPORTS AD-755 368.

72-0727 BATT, J. R.
JORDAN, S.
MAGIC 111: AN AUTOMATED GENERAL PURPOSE SYSTEM FOR
STRUCTURAL ANALYSIS VOLUME 1. ENGINEER'S MANUAL
BELL AEROSPACE CO., BUFFALO, N.Y., FINAL REPORT,
MARCH 15, 1971 - MARCH 15, 1972, 182 PP. (N.T.I.S.
- AD-755 368) SEE ALSO VOL. 2, AD-755 369.

72-0728 WILLIAMS, R.
BRINSON, H. F.
APPLICATIONS OF TRILINEAR COORDINATES TO SOME
PROBLEMS IN PLANE ELASTICITY
COLL. OF ENGNG., VIRGINIA POLYTECHNIC INST. AND
STATE UNIV., BLACKSBURG, REPORT NO: VPI-E-72-22,
OCTOBER 1972, 107 PP. (N.T.I.S. - AD-755 181)

72-0729 HUNT, J. T.
SMITH, R. R.
BARACH, D.
MCCLEARY, L.
JOHNSON, C.
APPLICATIONS OF THE FINITE-ELEMENT METHOD AND
COMPUTER GRAPHICS. (A VIBRATIONAL ANALYSIS OF A
SONAR PROJECTOR TRANSDUCER ELEMENT)
NAVAL UNDERSEA CENTER, SAN DIEGO, CALIF., REPORT
NO: NUC-TP-321, DECEMBER 1972, 31 P. (N.T.I.S. -
AD-755 177)

72-0730 OHNISHI, T.
FINITE ELEMENT SOLUTION TECHNIQUE FOR NEUTRON
TRANSPORT EQUATION
NUMER. REACTOR CALCULATIONS, INT. ATOMIC ENERGY
AGENCY, VIENNA, 1972, PP. 629-638.

72-0731 FOST, R. B.
ODEN, J. T.
A FINITE ELEMENT ANALYSIS OF SHOCK AND
FINITE-AMPLITUDE WAVES IN ONE-DIMENSIONAL
HYPERELASTIC BODIES AT FINITE STRAIN
SCHOOL OF GRADUATE STUDIES AND RESEARCH, UNIV. OF
ALABAMA, HUNTSVILLE, REPORT NO: UAH-RR-135,
OCTOBER 1972, 44 P. (N.T.I.S. - AD-754 953)

72-0732 ODEN, J. T.
BHANDARI, D. R.
YAGAWA, G.
CHUNG, T. J.
A NEW APPROACH TO THE FINITE-ELEMENT FORMULATION
AND SOLUTION OF A CLASS OF PROBLEMS IN COUPLED
THERMOELASTOVISCOPLASTICITY OF CRYSTALLINE SOLIDS
RESEARCH REPORT: UARI-RR-131, SEPTEMBER 1972, 28
P. (N.T.I.S. - AD-754 952)

72-0733 GANGAL, M. D.
DIRECT FINITE ELEMENT ANALYSIS OF ELASTIC CONTACT
PROBLEMS
INT. J. NUMER. METH. ENG., VOL. 5, 1972, PP.
145-147.

72-0734 DWYER, W. J.
FINITE ELEMENT MODELING AND OPTIMIZATION OF
AEROSPACE STRUCTURES
GRUMMAN AEROSPACE CORP., BETHPAGE, N.Y., FINAL
REPORT, APRIL 15, 1971 - APRIL 15, 1972, AUGUST
1972, 58 P. (N.T.I.S. - AD-754 915)

72-0735 RICHARDSON, J. J.
A DERIVATION OF PERTURBED EQUATIONS OF MOTION AND
HEAT CONDUCTION OF A THERMOELASTIC SOLID
GROUND EQUIPMENT AND MATERIALS DIRECTORATE, ARMY
MISSILE COMMAND, REDSTONE ARSENAL, ALA., REPORT
NO: RL-TP-72-13, NOVEMBER 1, 1972, 34 P. (N.T.I.S.
- AD-754 542)

72-0736 CHENG, Y. F.
ON MAXIMUM FILLET STRESSES IN BREECH RING
WATERVLIET ARSENAL, N.Y., REPORT NO: WVT-7255,
OCTOBER 1972, 22 P. (N.T.I.S. - AD-754 531)

72-0737 JOHNSON, N. E.
GALLETLY, R. D.
THE COMPARISON OF THE RESPONSE OF A HIGHWAY BRIDGE
TO UNIFORM GROUND SHOCK AND MOVING GROUND
EXCITATION
U. S. NAVAL RES. LAB., SHOCK VIB. BULL. 42, NO. 2,
JAN. 1972, PP. 75-85.

72-0738 MCCOWAN, D. W.
NEWTON, C. A.
PAGE, E. A.
PETERSON, A. F.
FLINN, E. A.
STUDY OF NUMERICAL METHODS IN SEISMIC PROBLEMS
ALEXANDRIA LABS., TELEDYNE GEOTECH, ALEXANDRIA,
VA., REPORT NO: AL-72-2, SEPTEMBER 7, 1972, 61 P.
(N.T.I.S. - AD-754 212)

72-0739 FRYE, J. W.
TUCCHIO, M. A.
AN ANALYSIS OF SOME FACTORS THAT AFFECT HYDROPHONE
SENSITIVITY
NAVAL UNDERWATER SYSTEMS CENTER, NEWPORT, R.I.,
REPORT NO: NUSC-TR-4317, NOVEMBER 2, 1972, 77 P.
(N.T.I.S. - AD-754 099)

72-0740 LEONARD, J. W.
RECKER, W. W.
NONLINEAR DYNAMICS OF CABLES WITH LOW INITIAL
TENSION
A.S.C.E., J. ENG. MECH. DIV., VOL. 98, NO. EM2,
APR. 1972, PP. 293-309.

72-0741 GHISTA, D. N.
PAJU, I. S.
FINITE ELEMENT ANALYSES IN BIOMECHANICS
INT. CONF. VARIATIONAL METH. IN ENG., SOUTHAMPTON
UNIV., 1972.

72-0742 GLOWINSKI, R.
SURE UNE METHODE D'APPROXIMATION EXTERNE PAR
ELEMENTS FINIS D ORDRE DEUX
C.R. ACAD. SC., PARIS, A 275, 1972, PP. 201-204
AND PP. 333-335.

72-0743 PFEIFER, C. G.
EVALUATION OF A THREE-DIMENSIONAL STRESS ANALYSIS
PROGRAM
MASTER'S THESIS, NAVAL POSTGRADUATE SCHOOL,
MONTEREY, CALIF., SEPTEMBER 1972, 89 P. (N.T.I.S.
- AD-753 617)

72-0744 CHENG, Y. F.
FACTORS OF STRESS CONCENTRATION AT A HOLE IN
FINITE STRIPS UNDER BENDING
WATERVLIET ARSENAL, N.Y., REPORT NO: WVT-7240,
AUGUST 1972, 15 P. (N.T.I.S. - AD-753 574)

72-0745 WANG, M. C.
NACCI, V. A.
GROUNDWATER FLOW IN PARTIALLY SATURATED SOILS
DEPT. OF CIVIL AND ENVIRONMENTAL ENGNG., UNIV. OF
RHODE ISLAND, KINGSTON, COMPLETION REPORT, JULY 1,
1970 - JUNE 30 - 1972, AUGUST 10, 1972, 75 P.
(N.T.I.S. - PB-213 458/3)

72-0746 JERRAM, K.
HELLEN, T. K.
FINITE ELEMENT TECHNIQUES IN FRACTURE MECHANICS
INT. CONF. ON WELDING RESEARCH RELATED TO POWER
PLANTS, SOUTHAMPTON, U.K., 1972.

72-0747 VENKATESWARA RAO, G.
KRISHNA MURTY, A. V.
ASSESSMENT OF ACCURACIES OF FINITE ELEMENT
EIGENVALUES
IN "THE MATH. OF FINITE ELEMENTS AND APPL.", (ED)
WHITEMAN, J.R., ACADEMIC PRESS, LONDON, 1972, PP.
379-386.

72-0748 ORRINGER, O.
FRENCH, S. E.
FEABL (FINITE ELEMENT ANALYSIS BASIC LIBRARY)
USER'S GUIDE
AEROELASTIC AND STRUCT. RESEARCH LAB., MASS. INST.
OF TECH., CAMBRIDGE, REPORT NO: ASRL-TR-16203,
AUGUST 1972, 109 P. (N.T.I.S. - AD753 120)

72-0749 RUBINSTEIN, M. F.
ROBERTS, T. A.
DYNAMIC STABILITY ANALYSIS BY THE CONJUGATE
GRADIENT METHOD
J. FRANKLIN INST., VOL. 293, NO. 3, MAR. 1972, PP.
173-189.

72-0750 NIELSEN, R.
CHANG, P. Y.
DESCHAMPS, L. C.
TANKER TRANSVERSE STRENGTH ANALYSIS: USER'S MANUAL
COM/CODE CORP., ALEXANDRIA, VA., FINAL REPORT,
JULY 1972, 53 P. (N.T.I.S. - AD-752 771)

72-0751 CONNOR, J.
A SURVEY OF FINITE-ELEMENT METHODS IN CONTINUUM
MECHANICS
PROC. VAR. METH. IN ENG., SOUTHAMPTON UNIV.,
SEPTEMBER 1972, PP. 1 - 32.

72-0752 ARAI, K.
FINITE ELEMENT METHOD WITH CONSIDERATION TO THE
CONTINUITY OF NORMAL SLOPE
MITSUBISHI HEAVY-INDUSTRIES, LTD., TOKYO, JAPAN,
REPORT NO: TB-79, JULY 1972, 8 P. (N.T.I.S. -
N72-32908)

72-0753 AUTHOR
NASTRAN USERS EXPERIENCES
2ND NASTRAN USERS COLLOQ., HAMPTON, VA., SEPTEMBER
1972, 523 P. (N.T.I.S. - N72-32867)

72-0754 ROBERTS, E. JR.
RELABELING OF FINITE ELEMENT MESHES USING A RANDOM
PROCESS
LEWIS RESEARCH CENTER, NATIONAL AERON. AND SPACE
ADMIN., CLEVELAND, OHIO, REPORT NO. E-7001, 15,
1972, (N.T.I.S. -N72-32212)

72-0755 SANDHU, R. S.
STRIP FOOTING ANALYSIS
DEPT. OF CIVIL ENGNG., OHIO STATE UNIV., COLUMBUS,
FINAL REPORT NO: APRIL 1972, 62 P. (N.T.I.S. -
PB-213 149/8)

72-0756 THORNTON, E. A.
A NASTRAN CORRELATION STUDY FOR VIBRATIONS OF A
CROSS-STIFFENED SHIP'S DECK
NASTRAN: USER'S EXPERIENCES, SEPT. 1972, PP.
145-159.

72-0757 GOODMAN, R. E.
DUBOIS, J.
DUPLICATION OF DILATANT BEHAVIOR IN THE ANALYSIS
OF JOINTED ROCKS
UNIV. OF CALIF., BERKELEY, REPORT NO: TR-11,
OCTOBER 1972, 59 P. (N.T.I.S. - AD-752 195)

72-0758 RAMEY, M. R.
VIERENDEEL BENT CAPS
PREPARED IN COOPERATION WITH FEDERAL HWY. ADMIN.,
DEPT. OF CIVIL ENG., UNIV. OF CALIF., DAVIS, FINAL
REPORT, JULY 1972, 118 PP. (N.T.I.S. - PB-212
820/5)

72-0759 SCIARRA, J. J.
RICKS, R. G.
USE OF THE FINITE ELEMENT DAMPED FORCED RESPONSE
STRAIN ENERGY DISTRIBUTION FOR VIBRATION REDUCTION
PRESENTED AT ARO-D MILITARY THEME REVIEW, THE
HELICOPTER AND V/STOL AIRCRAFT RESEARCH CONF. U.S.
ARMY RESEARCH OFFICE, MOFFETT FIELD, CALIF.,
TECHNICAL REPORT, SEPTEMBER 26-27, 1972, 29 P.
(N.T.I.S. - AD-751 809)

72-0760 PETERS, D. H.
SHOCK LOADS ON PIPING SYSTEMS
MASTER'S THESIS, NAVAL POSTGRADUATE SCHOOL,
MONTEREY, CALIF., SEPTEMBER 1972, 87 P. (N.T.I.S.
- AD-751 649)

72-0761 BASAS, J.
HAYATDAVOUDI, A.
HEINS, R. W.
ASPECTS OF MECHANICAL BEHAVIOR OF ROCK UNDER
STATIC AND CYCLIC LOADING. PART A. MECHANICAL
BEHAVIOR OF ROCK UNDER STATIC LOADING
ENGNG. EXPERIMENT STATION, UNIV. OF WISCONSIN,
MADISON, REPORT NO: 144-C745-PT-A, SEPTEMBER 1972,
87 P. (N.T.I.S. - AD-751 566) SEE ALSO AD-751 572.

72-0762 NAIK, T.
LUNDQUIST, R. G.
BASA, J.
HEINS, R. W.
ASPECTS OF MECHANICAL BEHAVIOR OF ROCK UNDER
STATIC AND CYCLIC LOADING. PART A: MECHANICAL
BEHAVIOR OF ROCK UNDER STATIC LOADING
ENGNG. EXPERIMENT STATION, UNIV. OF WISCONSIN,
MADISON. REPORT NO: 144-B638, MARCH 1972, 103 P.
(N.T.I.S. - AD-751 051) SEE ALSO PART B, AD-751
050.

72-0763 NAIK, T. B.
ANALYSIS OF FOUNDATIONS ON A SEMI-INFINITE ELASTIC
MASS UNDER DYNAMIC LOADS
UNIV. WISCONSIN, 1972, 156 PP.

72-0764 HARTZMAN, M.
ELASTIC-PLASTIC STRESS-STRAIN RELATIONS FOR
MATERIALS WITH DIFFERENT YIELD STRENGTHS IN
TENSION AND COMPRESSION
LAWRENCE LIVERMORE LAB., UNIV. OF CALIF.,
LIVERMORE, FEBRUARY 10, 1972, 12 P.
(N.T.I.S.-UCID-16001) SEE ALSO NSA 3002, NO.
05807.

72-0765 RASHID, Y. R.
PART 1. THEORY REPORT FOR CREEP-PLAST COMPUTER
PROGRAM: ANALYSIS OF TWO-DIMENSIONAL PROBLEMS
UNDER SIMULTANEOUS CREEP AND PLASTICITY
NUCLEAR FUELS DEPT., GENERAL ELECTRIC CO., SAN
JOSE, CALIF., JANUARY 1972, 28 PP., (N.T.I.S. -
GEAP-10546) SEE ALSO NSA 2905, NO. 12417.

72-0766 THOMAS, J.
DOKUMACI, E.
CARNEGIE, W.
FINITE ELEMENT ANALYSIS OF COUPLED VIBRATION OF
TAPERED TWISTED BLADES
PROC. INT. CONF. ON VARIATIONAL METH. IN ENG.,
UNIV. OF SOUTHAMPTON, SEPTEMBER 25, 1972, (EDS.)
BREBBIA, C.A. AND TOTTENHAM, H., SOUTHAMPTON UNIV.
PRESS, 1973, VOL. 2, PP. 10/48-10/66.

72-0767 KAPER, H. G.
LEAF, G. K.
LINDEMAN, A. J.
TIMING COMPARISON STUDY FOR SOME HIGH ORDER FINITE
ELEMENT PROCEDURES AND A LOW ORDER FINITE
DIFFERENCE APPROXIMATION PROCEDURE FOR THE
NUMERICAL SOLUTION OF THE MULTIGROUP NEUTRON AND
DIFFUSION EQUATION
NUCL. SCI. ENG., VOL. 49, NO. 27, 1972.

72-0768 KAWAHARA, M.
LARGE STRAIN, VISCOELASTIC NUMERICAL ANALYSIS BY
MEANS OF FINITE ELEMENT METHOD
PROC. J.S.C.E., VOL. 204, 1972, PP. 141-149.

72-0769 LATHROP, K. D.
TRANSPORT THEORY NUMERICAL METHODS
LOS ALAMOS SCIENTIFIC LAB., N. MEX., REPORT NO:
CONF-730414-3, 1972, 10 P. (N.T.I.S. -
LA-UR-73-517) SEE ALSO NSA 2804, NO. 09509

72-0770 AGARWAL, B. D.
MICROMECHANICS ANALYSIS OF COMPOSITE MATERIALS
USING FINITE ELEMENT METHODS
ILLINOIS INST. OF TECH., CHICAGO, MAY 1972, 160 P.
(N.T.I.S.- COO-1794-16) SEE ALSO NSA 2804, NO.
10557.

72-0771 YAHR, G. T.
VALACHOVIC, R. S.
APPLICATION OF FRACTURE MECHANICS TO GRAPHITE
UNDER COMPLEX STRESS CONDITIONS
OAK RIDGE NATIONAL LAB., TENN., 1972, 11 P.
(N.T.I.S. -CONF-730707-2) SEE ALSO NSA 2806, NO.
13351.

72-0772 GANGADHARAN, A. C.
DESIGN AND ANALYSIS EXPERIENCE WITH A LIQUID METAL
HEAT EXCHANGER FOR PFTF SERVICE
FOSTER WHEELER CORP., NEW YORK, 1972, 26 P.
(N.T.I.S. - CONF-721112-11) SEE ALSO NSA 2806, NO.
14466.

72-0773 HAISLER, W. E.
STRICKLIN, J. A.
VON RIESEMANN, W. A.
DYNAPLAS: A FINITE ELEMENT PROGRAM FOR THE
DYNAMIC, LARGE DEFLECTION, ELASTIC-PLASTIC
ANALYSIS OF STIFFENED SHELLS OF REVOLUTION
DEPT. OF AEROSPACE ENGNG., TEXAS AGRICULTURAL AND
MECH. UNIV., COLLEGE STATION, REPORT NO:
TEES-72-27, DECEMBER 1, 1972, 160 P. (N.T.I.S. -
SLA-73-127) SEE ALSO NSA 2802, NO. 04988

72-0774 HAISLER, W. E.
STRICKLIN, J. A.
VON RIESEMANN, W. A.
SAMMSOR 111: A FINITE ELEMENT PROGRAM TO DETERMINE
STIFFNESS AND MASS MATRICES OF RING-STIFFENED
SHELLS OF REVOLUTION
TEXAS AGRI. AND MECH. UNIV., COLLEGE STATION,
REPORT NO: TEES-72-26, DECEMBER 1, 1972, 96 P.
(N.T.I.S. - SLA-73-126) SEE ALSO NSA 2803, NO.
07782.

72-0775 MASSINI, G.
PEPINO, A.
SERCHIA, L.
ZACCARELLI, P.
DESCRIPTION OF THE IBM-360 VERSION OF THE FINITE
ELEMENTS COMPUTER PROGRAM SAFE-3D
COMITATO NAZIONALE PER L'ENERGIA NUCLEARE, ROME,
ITALY, JULY 20, 1972, 84 P. (N.T.I.S. -
RT/ING-(72)16) SEE ALSO NSA 2801, NO. 00313.

72-0776 NEALE, B. K.
BERGEN, AN AUTOMATIC FINITE ELEMENT MESH
GENERATION PROGRAM FOR ARBITRARY STRUCTURES
BERKELEY NUCL. LABS., CENTRAL ELECT. GENERATING
BRD., BERKELEY, ENGLAND, OCTOBER 1972, 57 PP.
(N.T.I.S. - RD-B/N-2432) SEE ALSO NSA 2801, NO.
00312

72-0777 BABUSKA, I.
AZIZ, A.
LECTURES ON MATHEMATICAL FOUNDATIONS OF THE FINITE
ELEMENT METHOD
INST. FOR FLUID DYNAMICS AND APPLIED MATH., UNIV.
OF MARYLAND, COLLEGE PARK, REPORT NO: BM-748,
OCTOBER 1972, 347 P. (N.T.I.S. - ORO-3443-42) SEE
ALSO NSA 2708, NO. 19394.

72-0778 LEMMON, E. C.
COUPLE - A PSEUDO FINITE ELEMENT CONDUCTION CODE
FOR THE TRANSIENT THERMAL RESPONSE OF AXISYMMETRIC
OR PLANE ANISOTROPIC MATERIALS
SANDIA LABS., LIVERMORE, CALIF., DECEMBER 1972, 88
P. (N.T.I.S. - SCL-RR-720324) SEE ALSO NSA 2706,
NO. 14433.

72-0779 CALLABRESI, M. L.
HEIDELBERG, S. T. JR.
SASL - A FINITE ELEMENT CODE FOR THE STATIC
ANALYSIS OF AXISYMMETRIC AND PLANE SOLIDS
SUBJECTED TO AXISYMMETRIC AND PLANE LOADINGS
SANDIA LABS., LIVERMORE, CALIF., DECEMBER 1972,
116 P. (N.T.I.S. - SCL-DR-720061) SEE ALSO NSA
2704, NO. 09234.

72-0780 KIKUCHI, F.
ANDO, Y.
FINITE ELEMENT SOLUTIONS FOR PLATE BENDING
PROBLEMS BY SIMPLIFIED HYBRID DISPLACEMENT METHOD
J. NUCL. SCI. TECH., TOKYO, VOL. 15, 1972, PP.
155-178.

72-0781 KIKUCHI, F.
ANDO, Y.
SIMPLIFIED HYBRID DISPLACEMENT METHOD APPLIED TO
PLATE BUCKLING PROBLEMS
J. NUCL. SCI. TECH., VOL. 9, 1972, PP. 497-499.

72-0782 BLUM, R. E.
KIRKWOOD, W. F.
CALCULATING STRESSES IN A PRESSURE VESSEL WITH
FILL TUBE - COMPARISON OF THE FINITE ELEMENT AND
PHOTOELASTIC METHODS
LAWRENCE LIVERMORE LAB., UNIV. OF CALIF.,
LIVERMORE, JUNE 30, 1972, 22 P. (N.T.I.S. -
UCRL-51250) SEE ALSO NSA 2623 NO. 56034.

72-0783 STRICKLIN, J. A.
VON RIESEMANN, W. A.
TILLERSON, J. R.
HAISLER, W. E.
SURVEY OF STATIC GEOMETRIC AND MATERIAL NONLINEAR
ANALYSIS BY THE FINITE ELEMENT METHOD
DEPT. OF AEROSPACE ENGNG., TEXAS AGRI. AND MECH.
UNIV., COLLEGE STATION, REPORT NO: CONF-720812-2,
1972, 24 P. (N.T.I.S. - SC-DC-721391) SEE ALSO
MATRIX METHODS OF STRUCT. ANALYSIS, BERKELEY,
CALIF., AUGUST 14, 1972.

72-0784 AYALA-MILIAN, G.
BREBBIA, C. A.
SOLUTION OF WAVE PROPAGATION PROBLEMS IN SATURATED
MEDIUM
PROC. INT. CONF. ON VARIATIONAL METH. IN ENG.,
UNIV. OF SOUTHAMPTON, SEPTEMBER 25, 1972, (EDS.)
BREBBIA, C.A. AND TOTTENHAM, H., SOUTHAMPTON UNIV.
PRESS, 1973, VOL. 1, PP. 5/89-5/101.

72-0785 FRANCIS, P. H.
NAGY, A.
BEISSNER, R. E.
DEFORMATION MECHANICS OF DEEP SURFACE FLAW CRACKS
SOUTHWEST RESEARCH INST., SAN ANTONIO, TEX.,
REPORT NO: NASA-CR-128514, JULY 1972, 156 P.
(N.T.I.S. - N72-30921)

72-0786 GALLAGHER, R. H.
MAU, S. T.
A METHOD OF LIMIT POINT CALCULATION IN FINITE
ELEMENT STRUCTURAL ANALYSIS
CORNELL UNIV., ITHACA, N.Y., REPORT NO:
NASA-CR-2115, SEPTEMBER, 1972, 30 P. (N.T.I.S. -
N72-30918)

72-0787 PASSERELLO, C. E.
HUSTON, R. L.
AN ANALYSIS OF GENERAL CHAIN SYSTEMS
DEPT. OF ENGNG. ANALYSIS, UNIV. OF CINCINNATI,
OHIO, REPORT NO: NASA-CR-127924, 1972, 55 P.
(N.T.I.S. - N72-30532)

72-0788 MARR, W. A. JR.
FINITE ELEMENT ANALYSIS OF ELASTO-PLASTIC SOILS.
REPORT NO. 4, FINITE ELEMENT ANALYSIS OF
ELASTO-PLASTIC FRICTIONAL MATERIALS FOR
APPLICATION TO LUNAR EARTH SCIENCES
SCHOOL OF ENGNG., MASSACHUSETTS INST. OF TECH.,
CAMBRIDGE, REPORT NO: NASA-CR-128529,
SOILS-PUBL-301, JUNE 1972, 103 P. (N.T.I.S. -
N72-30353)

72-0789 GUPTA, K. K.
AKYUZ, F. A.
VISCEL A GENERAL PURPOSE COMPUTER PROGRAM FOR
ANALYSIS OF LINEAR VISCOELASTIC STRUCTURES, VOL. 2
INST. OF TECH., PASADENA, JET PROPULSION LAB.,
CALIF., REPORT NO: NASA-CR-127750,
JPL-TM-33-466-VOL-2, JULY 15, 1972, 50 P.
(N.T.I.S. - N72-29918)

72-0790 FERRISS, D. H.
SOLUTION OF MIXED BOUNDARY VALUE PROBLEMS IN
FINITE DOMAINS BY FOURIERS METHOD
DIV. OF NUMERICAL ANALYSIS AND COMPUTING, NATIONAL
PHYSICAL LAB., TEDDINGTON, ENGLAND, REPORT NO:
NPL-NAC-16, REPORT NO: NPL-NAC-16, APRIL 1972, 53
P. (N.T.I.S. - N72-29628)

72-0791 BARKER, R. M.
CARROLL, W. E.
A THEOREM FOR OPTIMUM IDEALIZATIONS IN
FINITE-ELEMENT ANALYSIS
FLORIDA TECHNOLOGICAL UNIV., ORLANDO, REPORT NO:
VPI-E-72-19, SEPTEMBER 7, 1972, 37 P. (N.T.I.S. -
AD-750 638)

72-0792 NEALE, B. K.
HENSHELL, R. D.
EDWARDS, G.
HYBRID PLATE BENDING ELEMENTS
J. SOUND VIB., VOL. 23, NO. 1, JULY 8, 1972, PP.
101-112.

72-0793 SAWYER, S. G.
HURBAN, J. M.
ANALYSIS OF BAND PRESSURE IN A GUN TUBE USING
FINITE ELEMENTS AND SHELL PUSHING DATA
BALLISTIC RESEARCH LABS., ABERDEEN PROVING GROUND,
MD., 1972, 14 P. (N.T.I.S. - AD-750 373)

72-0794 KOLAR, V.
KRATOCHVIL, J.
LEITNER, F.
ZENISEK, A.
STRESS ANALYSIS OF TWO AND THREE DIMENSIONAL
STRUCTURES BY THE FINITE ELEMENT METHOD
SPRINGER VERLAG, WIEN, (IN GERMAN) 1972.

72-0795 TABARROK, B.
VIBRATION OF CYLINDRICAL SHELLS BY A HYBRID FINITE
ELEMENT METHOD
DEPT. OF MECH. ENGNG., UNIV. OF TORONTO, ONTARIO,
JULY 5, 1972, 50 P. (N.T.I.S. - N72-27954) SEE
ALSO A.I.A.A. J., DECEMBER, 1971.

72-0796 OGAWA, A.
A STRUCTURAL ANALYSIS OF CYLINDER-CONE-CYLINDER
SHELLS BY THE FINITE ELEMENT METHOD
NATIONAL AEROSPACE LAB., TOKYO, JAPAN, REPORT NO:
NAL-TR-262, 1972, 19 P. (N.T.I.S. - N72-27952)

72-0797 MCLEAN, F. G.
WILLIAMS, R. D.
TURNBELL, R. C.
THE KANSAS TEST TRACK. NON-CONVENTIONAL TRACK
STRUCTURES. DESIGN REPORT
WESTENHOFF AND NOVICK, INC., CHICAGO, ILL.
SEPTEMBER 1972, 214 P. (N.T.I.S. - PB-212 358)

72-0798 DICKSON, J. N.
HSU, T-M.
MCKINNEY, J. M.
DEVELOPMENT OF AN UNDERSTANDING OF THE FATIQUE
PHENOMENA OF BONDED AND BOLTED JOINTS IN ADVANCED
FILAMENTARY COMPOSITE MATERIALS. VOLUME 1.
ANALYSIS METHODS
LOCKHEED-GEORGIA CO., MARIETTA, REPORT NO:
LGR-ER-11319-VOL-1, JUNE 1972, 134 P. (N.T.I.S. -
AD-750 132) SEE ALSO VOL. 2, AD-750 133.

72-0799 TAYLOR, M. A.
ROMSTAD, K. M.
HERRMANN, L. R.
RAMEY, M. R.
A FINITE ELEMENT COMPUTER PROGRAM FOR THE
PREDICTION OF THE BEHAVIOR OF REINFORCED AND
PRESTRESSED CONCRETE STRUCTURES SUBJECT TO
CRACKING
DEPT. OF CIVIL ENGNG., UNIV. OF CALIF., DAVIS,
FINAL REPORT, JUNE 30, 1970 - JUNE 1, 1972, 229 P.
(N.T.I.S. - AD-749 998)

72-0800 ZAR, J. L.
THE USE OF A LASER FOR ARPA MILITARY GEOPHYSICS
PROGRAM (ROCK MECHANICS AND RAPID EXCAVATION)
AVCO EVERETT RESEARCH LAB., EVERETT, MASS., FINAL
TECH. REPORT, APRIL 29, 1971 - SEPTEMBER 28, 1972,
90 P. (N.T.I.S. - AD -749 982)

72-0801 GOLDSMITH, W.
WAVE PROPAGATION IN ANISOTROPIC ROCKS
UNIV. OF CALIF., BERKELEY, SEMI-ANNUAL TECH.
REPORT NO. 1, FEBRUARY 23, - AUGUST 31, 1972,
SEPTEMBER 25, 1972, 96 P. (N.T.I.S. - AD-749 976)

72-0802 LOBITZ, D. W.
MESH GENERATION CODE AND A CONTOUR PLOTTING
ROUTINE
SANDIA LABORATORIES, ALBUQUERQUE, REPORT
SC-TM-710557, 1972.

72-0803 ZIMMERMAN, K. L.
PLOTS - THE PLOTTING SUBROUTINE INCORPORATED IN
THE BALLISTIC RESEARCH LABORATORIES ELECTRONIC
SCIENTIFIC COMPUTER (BRLESC) FINITE ELEMENT
COMPUTER PROGRAM
BALLISTIC RESEARCH LABS., ABERDEEN PROVING GROUND,
MD., REPORT NO: BRL-MR-2209, AUGUST 1972, 40 P.
(N.T.I.S. - AD-749 786)

72-0804 RIVELLO, R. M.
NASTRAN: USER EXPERIENCE WITH FOUR EXAMPLE
PROBLEMS
APPLIED PHYSICS LAB., JOHNS HOPKINS UNIV., SILVER
SPRING, MD., REPORT NO: APL-TG-1196, APRIL 1972,
132 P. (N.T.I.S. - AD-749 684)

72-0805 ODEN, J. T.
FINITE DEFORMATION OF ELASTIC PLATES, SHELLS, AND
MEMBRANES BY THE FINITE ELEMENT METHOD
PROC. SHELL STRUCT. AND CLIMATIC INFLUENCES,
CALGARY, JULY 3-6, 1972, PP. 63-89.(N.T.I.S. -
AD-749 565)

72-0806 WAAS, G.
EARTH VIBRATION EFFECTS AND ABATEMENT FOR MILITARY
FACILITIES. REPORT 3. ANALYSIS METHOD FOR
FOOTING VIBRATIONS THROUGH LAYERED MEDIA
COLL. OF ENGNG., UNIV. OF CALIF., BERKELEY,
RESEARCH REPORT, JUNE 1, 1970 - JUNE 30, 1972,
SEPTEMBER 1972, 195 P. (N.T.I.S. - AD-749 507) SEE
ALSO REPORT APRIL, 1972, AD-741 772.

72-0807 MCLEOD, R. J. Y.
BASIS FUNCTIONS FOR CURVED ELEMENTS IN THE FINITE
ELEMENT METHOD
PH.D. THESIS, UNIV. OF DUNDEE, 1972.

72-0808 ISENBERG, J.
ANALYTIC MODELING OF ROCK-STRUCTURE INTERACTION
ENGNG. AND APPLIED SCI. DIV., AGBABIAN ASSOC., LOS
ANGELES, CALIF., REPORT NO: AA-R-7215-2299, AUGUST
1972, 172 P. (N.T.I.S. - AD-749 373)

72-0809 KOBAYASHI, A. S.
WADE, B. G.
BRADLEY, W. B.
FRACTURE DYNAMICS OF HOMALITE-100 SHEETS
SHELL DEVELOPMENT CO., HOUSTON, TEXAS, REPORT NO:
TR-15, SEPTEMBER, 1972, 27 P. (N.T.I.S. - AD-749
098)

72-0810 MONFORTON, G. R.
MICHAIL, M. G.
FINITE ELEMENT ANALYSIS OF SKEW SANDWICH PLATES
A.S.C.E. J., VOL. 98, NO. EM 3, 1972, PP. 763-769.

72-0811 ODEN, J. T.
SOME ASPECTS OF RECENT CONTRIBUTIONS TO THE
MATHEMATICAL THEORY OF FINITE ELEMENTS
2ND. SEMINAR, U.S. - JAPAN, ON MATRIX METH. OF
STRUCT. ANALYSIS AND DESIGN, BERKELEY, CALIF.,
1972, PP. 1-38. (N.T.I.S. - AD-748 936)

72-0812 MIYOSHI, T.
CONVERGENCE OF FINITE ELEMENT SOLUTIONS
REPRESENTED BY A NON-CONFORMING BASIS
KUMAMOTO J. SCI. MATH., VOL. 9, 1972, PP. 11-20.

72-0813 KOICHI, A.
FINITE ELEMENT METHOD WITH CONSIDERATION TO THE
CONTINUITY OF NORMAL SLOPE
MITSUBISHI HEAVY IND. LTD., TOKYO, JAPAN, REPORT
NO: MTB-79, JULY 1972, 9 P. (N.T.I.S. - PB-211
759)

72-0814 MASAHIRO, M.
MASAAKI, M.
TETSURO, K.
MITSUAKI, N.
MAMORU, H.
APPLICATION OF PROGRAM FATIQUE TEST TO MEMBER
JOINTS OF HULLS
MITSUBISHI HEAVY IND. LTD., TOKYO, JAPAN, REPORT
NO: MTB-78, JULY 1972, 29 P. (N.T.I.S. - PB-211
758)

72-0815 SABIR, R. A.
LOCK, A. C.
THE APPLICATION OF FINITE ELEMENTS TO THE LARGE
DEFLECTION GEOMETRICALLY NONLINEAR BEHAVIOR OF
CYLINDRICAL SHELLS
PROC. INT. CONF. ON VARIATIONAL METH. IN ENG.,
UNIV. OF SOUTHAMPTON, SEPTEMBER 25, 1972, (EDS.)
BREBBIA, C.A. AND TOTTENHAM, H., SOUTHAMPTON UNIV.
PRESS, 1973, VOL. 2, PP. 7/66-7/75.

72-0816 MIZUMACHI, W.
APPLICATION OF FINITE ELEMENT METHOD TO BOILING
WATER REACTOR POWER PLANT (JAPANESE)
TOSHIBA REBYA, VOL. 27, 1972, PP. 1018-1022.

72-0817 ALLGOOD, J. R.
SUMMARY OF SOIL-STRUCTURE INTERACTION
NAVAL CIVIL ENGNG. LAB., PORT HUENEME, CALIF.,
REPORT NO: NCEL-TR-771, JULY 1972, 179 P.
(N.T.I.S. - AD-748 581)

72-0818 SPILKER, P. L.
A FINITE ELEMENT MODEL FOR LAMINATED PLATES
INCLUDING TRANSVERSE SHEAR DEFORMATION
AEROELASTIC AND STRUCT. RESEARCH LAB., MASS. INST.
OF TECH., CAMBRIDGE, REPORT NO: ASRL-TR-169-1,
JULY 1972, 58 P. (N.T.I.S. - AD-748 453)

72-0819 LIN, F-T.
THE ELASTIC STRESS ANALYSIS OF A BI-MATERIAL PLATE
WITH A CRACK NORMAL TO THE INTERFACES
DEPT. OF THEORETICAL AND APPLIED MECH., UNIV. OF
ILLINOIS, URBANA, REPORT NO: T/AM-356,
UILU-ENG-72-6004, MAY 1972, 134 P. (N.T.I.S. -
AD-748 288)

72-0820 NAKAMURA, S.
ITERATIVE SOLUTIONS FOR THE FINITE ELEMENT METHOD
NUMER. REACTOR CALCULATIONS, INTERNATIONAL ATOMIC
ENERGY AGENCY, VIENNA, 1972, PP. 639-656.

72-0821 NASSIF, N.
FINITE ELEMENT METHOD FOR TIME DEPENDENT PROBLEMS
PH.D. THESIS, HARVARD UNIV., 1972.

72-0822 KAHL, M. R.
REIFSNIDER, K. L.
EFFECTS OF LOCAL STRENGTH VARIATIONS ON PLASTIC
ENCLAVE DEVELOPMENT AT CRACK TIPS
DEPT. OF ENGNG. MECH., VIRGINIA POLYTECHNIC INST.
AND STATE UNIV., BLACKSBURG, REPORT NO:
VPI-E-72-15, JUNE 21, 1972, 76 P. (N.T.I.S. -
AD-747 783)

72-0823 BAKHSHANDEHPOUR, M.
FINITE ELEMENT SOLUTION FOR AXISYMMETRIC TRANSIENT
THERMAL STRESSES
MASTER'S THESIS, NAVAL POSTGRADUATE SCHOOL,
MONTEREY, CALIF., JUNE 1972, 17 P. (N.T.I.S. -
AD-747 523)

72-0824 BARNARD, A. J.
THE FINITE ELEMENT METHOD APPLIED TO THE ANALYSIS
OF SANDWICH PLATE AND SHELL STRUCTURES
M.SC. THESIS, LOUGHBOROUGH UNIV. OF TECH., 1972.

72-0825 HEAP, B. R.
ALGORITHMS FOR THE PRODUCTION OF CONTOUR MAPS OVER
AN IRREGULAR TRIANGULAR MESH
DIV. OF NUM. ANALYSIS AND COMPUTING, NAT. PHYSICAL
LAB., TEDDINGTON, ENGLAND, REPORT NO:
NPL-NAC-10, MARCH 1972, 31 P. (N.T.I.S. -
N72-25245)

72-0826 GUYMON, G. L.
APPLICATION OF THE FINITE-ELEMENT METHOD FOR
SIMULATION OF SURFACE WATER TRANSPORT PROBLEMS
COLLEGE INST. OF WATER RESOURCES, UNIV. OF ALASKA,
REPORT NO: IWR-21, JUNE 1972, 110 P. (N.T.I.S. -
PB-211 079)

72-0827 OUDENHOVEN, M. S.
BABCOCK, C. O.
BLAKE, W.
A METHOD FOR THE PREDICTION OF STRESSES IN AN
ISOTROPIC INCLUSION OR OREBODY OF IRREGULAR SHAPE
BUREAU OF MINES, WASH., D.C., REPORT NO:
BUMINES-RI-7645, JUNE 1972, 42 P. (N.T.I.S. -
PB-210 986)

72-0828 EHLE, P. E.
RAHE, A. E.
DEVELOPMENT OF A FINITE ELEMENT APPROACH FOR
APPROXIMATE ANALYSIS OF UNSTEADY COMPRESSIBLE
FLUID FLOW
WEAPONS LAB., ARMY WEAPONS COMMAND, ROCK ISLAND,
ILL., REPORT NO: SWERR-TR-72-36, JUNE 1972, 62 P.
(N.T.I.S. - AD-746 234)

72-0829 HODGE, P. G. JR.
COMPUTER SOLUTIONS OF PLASTICITY PROBLEMS
DEPT. OF AERO. ENGNG. AND MECH., UNIV. OF
MINNESOTA, MINN., REPORT NO: H1-3, 1972, 35 P.
(N.T.I.S. - AD-746 175)

72-0830 SVEC, O. J.
FINITE ELEMENT APPROACH TO THE CONTACT PROBLEM OF
A PLATE ON THE ELASTIC HALF SPACE
PH.D., UNIV. OF WATERLOO, CANADA, 1972.

72-0831 BHANDARI, D. R.
ODEN, J. T.
GENERAL MIXED FINITE ELEMENT METHODS OF NONLINEAR
CONTINUA
RESEARCH INST., UNIV. OF ALABAMA, HUNTSVILLE,
REPORT NO: UARI-RW-123, APRIL 1972, 32 P.
(N.T.I.S. - AD-745 994)

72-0832 SVALBONAS, V.
ARMENAKAS, A. E.
LINEAR AND NONLINEAR ANALYSIS OF SHELLS
DEPT. OF AERO. ENG. AND APPL. MECH., POLYTECHNIC
INST., BROOKLYN, N.Y., REPORT NO: PIBAL-72-20, MAY
1972, 213 PP. (N.T.I.S. - AD-745 967)

72-0833 NEY, R. A.
UTKU, S.
AN ALTERNATIVE FOR THE FINITE ELEMENT METHOD
PROC. SYMP. VARIATIONAL METHOD., UNIV. OF
SOUTHAMPTON, 1972.

72-0834 SHORE, S.
LANSBERRY, C. R.
HORIZONTALLY CURVED HIGHWAY BRIDGES. A FULLY
COMPATIBLE ANNULAR SEGMENT FINITE ELEMENT
TOWNE SCHOOL OF CIVIL AND MECH. ENGNG.,
PHILADELPHIA, PA., FEBRUARY 1972, 200 P. (N.T.I.S.
- PB-210 732)

72-0835 ROGERS, P. H.
SHIP (SIMPLIFIED-HELMHOLTZ-INTEGRAL PROGRAM): A
FAST COMPUTER PROGRAM FOR CALCULATING THE ACOUSTIC
RADIATION AND RADIATION IMPEDANCE FOR
FREE-FLOODED-RING AND FINITE-CIRCULAR-CYLINDER
SOURCES
NAVAL RESEARCH LAB., WASH., D.C., REPORT NO:
NRL-7240, JUNE 19, 1972, 64 P. (N.T.I.S. - AD-745
295)

72-0836 NOBARI, E. S.
DUNCAN, J. M.
EFFECT OF RESERVOIR FILLING ON STRESSES AND
MOVEMENT IN EARTH AND ROCKFILL DAMS. A REPORT OF
AN INVESTIGATION
COLL. OF ENG., UNIV. OF CALIF., BERKELEY, REPORT
NO: TE-72-1, JANUARY 1972, 184 PP. (N.T.I.S. -
AD-745 216) SEE ALSO REPORT MAY 1971, AD-731 231.

72-0837 ERICSON, W. A.
AN INVESTIGATION OF INITIAL YIELD SURFACES FOR
UNIDIRECTIONAL REINFORCED COMPOSITES
NAVAL POSTGRADUATE SCHOOL, MONTEREY, CALIF., MARCH
1972, 203 P. (N.T.I.S. - AD-745 176)

72-0838 PERLOFF, W. H.
STRAIN DISTRIBUTION AROUND UNDERGROUND OPENINGS.
FINITE ELEMENT ANALYSIS OF JOINTED SYSTEMS
SCHOOL OF CIVIL ENGNG., PURDUE UNIV., LAFAYETTE,
IND., REPORT NO: TR-7, MAY 1972, 152 P. (N.T.I.S.
- AD-745 139) SEE ALSO AD-735 376.

72-0839 PETYT, M.
MIRZA, W. H.
BREBBIA, C. A.
TOTTENHAM, H.
FINITE ELEMENT MODELLING TECHNIQUES FOR THE
ANALYSIS OF BUILDING VIBRATIONS
PROC. CONF. VARIATIONAL METH. ENG., UNIV. OF
SOUTHAMPTON, 1972.

72-0840 SALAMA, A. M.
ROWE, W. M.
YASUI, R. K.
STRESS ANALYSIS AND DESIGN OF SILICON SOLAR CELL
ARRAYS AND RELATED MATERIAL PROPERTIES
INST. OF TECH., JET PROPULSION LAB., PASADENA,
CALIF., REPORT NO: NASA-CR-126648 JPL-TR-32-1552,
MARCH 1, 1972, 35 P. (N.T.I.S. - N72-24041)

72-0841 RAJU, I. S.
FINITE ELEMENT ANALYSIS OF STRESS CONCENTRATIONS
AND SINGULARITIES
PH.D. THESIS, INDIAN INST. OF SCI., BANGALORE,
1972.

72-0842 ANDERSON, J. M.
CASE BOND STRESS CALCULATIONS FOR FLAPPED
CYLINDRICAL ANALOGS OF SOLID PROPELLANT ROCKET
MOTORS. BONDLINE PARAMETRIC STUDIES
BACCHUS WORKS, HERCULES INC., MAGNA, UTAH, INTERIM
REPORT ON TASK 1, MAY 1972, 92 P. (N.T.I.S. -
AD-744 901)

72-0843 KOBAYASHI, A. S.
WADE, B. G.
BRADLEY, W. B.
CHIU, S. T.
CRACK BRANCHING IN HOMALITE-100 SHEETS
DEPT. OF MECH. ENGNG., UNIV. OF WASH., SEATTLE,
REPORT NO: TR-13, JUNE 1972, 33 P. (N.T.I.S. -
AD-744 661)

72-0844 SMOLDEREN, J. J.
NUMERICAL METHODS IN FLUID DYNAMICS
ADVISORY GROUP FOR AEROSPACE RESEARCH AND
DEVELOPMENT, PARIS, FRANCE, REPORT NO:
AGARD-LS-48, MAY 1972, 333 P. (N.T.I.S. - AD-744
603)

72-0845 DOBYNS, A. L.
A TETRA-CORE STRESS ANALYSIS MODEL
BOEING CO., SEATTLE, WASH., REPORT NO:
D180-14257-1, APRIL 1972, 261 P. (N.T.I.S. -
AD-744 505)

72-0846 VAITYS, R. P.
HENZI, A. N.
LIS, S. J.
WEAPON-MOUNT INTERFACE STUDY - TRIPOD MOUNT M3
GENERAL AMERICAN RESEARCH DIV., GENERAL AMER.
TRANSP. CORP., NILES, ILL., REPORT NO: GARD-1504,
APRIL 1972, 193 P. (N.T.I.S. - AD-744 438)

72-0847 HIBBITT, H. D.
MARCAL, P. V.
A NUMERICAL THERMO-MECHANICAL MODEL FOR THE
WELDING AND SUBSEQUENT LOADING OF A FABRICATED
STRUCTURE
DIV. OF ENGNG., BROWN UNIV., PROVIDENCE, R.I.,
REPORT NO: TR-2, MARCH 1972, 62 P. (N.T.I.S. -
AD-744 212)

72-0848 PIAN, T. H. H.
TONG, P.
STATIC AND DYNAMIC NONLINEAR STRUCTURAL PROBLEMS
AEROELASTIC AND STRUCT. RESEARCH LAB., MASS. INST.
OF TECH., CAMBRIDGE. REPORT NO: ASRL-TR-144-6,
MARCH 1972, 12 P. (N.T.I.S. - AD-744 114)

72-0849 WU, R. W.
WITMER, E. A.
FINITE ELEMENT ANALYSIS OF LARGE TRANSIENT
ELASTIC-PLASTIC DEFORMATIONS OF SIMPLE STRUCTURES,
WITH APPLICATION TO THE ENGINE ROTOR FRAGMENT
CONTAINMENT/DEFLECTION PROBLEM
AEROELASTIC AND STRUCT. RESEARCH LAB., MASS. INST.
OF TECH., CAMBRIDGE. REPORT NO: NASA-CR-120886,
ASRL-TR-154-4, JANUARY, 1972, 326 P. (N.T.I.S. -
N72-22915)

72-0850 DENKMANN, W. J.
NICKELL, R. E.
STICKLER, D. C.
ANALYSIS OF STRUCTURAL-ACOUSTIC INTERACTIONS IN
METAL-CERAMIC TRANSDUCERS
DIV. OF ENGNG., BROWN UNIV., PROVIDENCE, R.I.,
REPORT NO: TR-10, APRIL 1972, 32 P. (N.T.I.S. -
AD-743 985)

72-0851 PECK, J. C.
PALACOL, E. L.
CHAO, H.
STRESS-WAVE PROPAGATION AND DYNAMIC EFFECTS IN
SHELLS (SPADES) MICROMECHANICAL STRESS-WAVE
RESPONSE OF 3-D QUARTZ PHENOLIC. BOOK 1:
DEVELOPMENT OF BLOCH-FOURIER ANALYSIS
MCDONNELL DOUGLAS ASTRONAUTICS, HUNTINGTON BEACH,
CALIF., FINAL TECH. RPT. JUNE 1970 - DECEMBER
1971, REPORT NO: MDC-G2709, JUNE 1972, 259 PP.
(N.T.I.S. - AD-743 898)

72-0852 CROSE, J. G.
MCKINLEY, T. K.
THERMAL STRESS ANALYSIS OF RE-ENTRY VEHICLE
NOSETIPS AT ANGLE OF ATTACK
SAN BERNARDINO OPERATIONS, AEROSPACE CORP., SAN
BERNARDINO, CALIF., REPORT NO: TR-072(S2816-15)-2,
APRIL 12, 1972, 27 PP. (N.T.I.S. - AD743 897)

72-0853 FOURNEY, W. L.
CRAWLEY, J. T.
PALMISANO, R. R.
NORMAL MODE ANALYSIS OF A FUZE SUPPORT STRUCTURE
USING NASTRAN. PART 1
HARRY DIAMOND LABS., WASH., D.C., REPORT NO:
HDL-TR-1581, APRIL 1972, 35 P. (N.T.I.S. - AD-743
879)

72-0854 KIRKLAND, J. L.
WALKER, R. E.
FUNDAMENTAL STUDIES OF MEDIUM-STRUCTURE
INTERACTION. REPORT 1. FINITE ELEMENT ANALYSIS OF
BURIED CYLINDERS
ARMY ENGINEER WATERWAYS EXPERIMENT STATION,
VICKSBURG, MISS., REPORT NO: AEWES-TR-N-72-7-1,
JUNE 1972, 63 P. (N.T.I.S. - AD-743 862)

72-0855 WILSON, E. L.
SOLID SAP. A STATIC ANALYSIS PROGRAM FOR THREE
DIMENSIONAL SOLID STRUCTURES
STRUCTURAL ENGNG. LAB., UNIV. OF CALIF., BERKELEY,
REPORT NO: UCSESM-71-19, MARCH 1972, 170 P.
(N.T.I.S. - PB-209 949)

72-0856 JENNINGS, P. C.
BIELAK, C.
DYNAMICS OF BUILDING - SOIL INTERACTION
EARTHQUAKE ENGNG. RESEARCH LAB., CALIF. INST. OF
TECH., PASADENA, REPORT NO: EERL-72-01, APRIL
1972, 81 P. (N.T.I.S. - PB-209 666)

72-0857 ODEN, J. T.
BASIC RESEARCH IN METHODS OF APPROXIMATION IN THE
NONLINEAR MECHANICS OF SOLIDS AND STRUCTURES
DIV. OF GRAD. PROGRAMS AND RESEARCH, UNIV. OF
ALABAMA, HUNTSVILLE, ANNUAL REPORT, JANUARY 1,
1971 - JANUARY 1, 1972, 88 P. (N.T.I.S. - AD-743
278)

72-0858 KOBAYASHI, A. S.
CHIU, S. T.
BEEUWKES, R.
A NUMERICAL AND EXPERIMENTAL INVESTIGATION OF THE
USE OF J-INTEGRALS
DEPT. OF MECH. ENGNG., UNIV. OF WASH., SEATTLE,
REPORT NO: TR-12, MAY 1972, 30 P. (N.T.I.S. -
AD-743 204)

72-0859 LUKAS, L.
CURVED BOUNDARY ELEMENTS FOR FINITE ELEMENT METHOD
PROC. 6TH I.K.M., WEIMAR, 1972.

72-0860 MEYER, P. M.
DATA GENERATOR FOR THE IDEALIZATION FOR FINITE
ELEMENT STRUCTURAL ANALYSIS OF NAVAL SHIP FLAT
PLATED GRILLAGES WITH MULTIPLE OPENINGS
NAVAL SHIP RESEARCH AND DEVELOPMENT CENTER,
BETHESDA, MD., REPORT NO: NSRDC-3807, MAY 1972, 49
P. (N.T.I.S. - AD-743 055)

72-0861 KEY, J. E.
USERS' MANUAL FOR DIGITAL COMPUTER PROGRAM OK1.
NUMERICAL ANALYSIS OF FINITE AXISYMMETRIC
DEFORMATIONS OF INCOMPRESSIBLE ELASTIC SOLIDS OF
REVOLUTION
RESEARCH INST., UNIV. OF ALABAMA, HUNTSVILLE,
REPORT NO: UARI-RR-121, APRIL 1972, 118 P.
(N.T.I.S. - AD-743 013)

72-0862 CHEN, K. C.
HIGH PRECISION FINITE ELEMENTS FOR PLANE ELASTIC
PROBLEMS
D.SC. DISSERTATION, DEPT. OF CIVIL ENG., WASH.
UNIV., ST. LOUIS, MO., MAY, 1972.

72-0863 WALKER, W. J.
VANDERLINDEN, J. W.
STRUCTURE- MEDIUM INTERACTION PHENOMENOLOGY
BOEING CO., SEATTLE, WASH., FINAL REPORT, APRIL 1,
1971 - JANUARY 31, 1972, 134 P. (N.T.I.S. - AD-742
839)

72-0864 RUDDER, F. F. JR.
STUDY OF EFFECTS OF DESIGN DETAILS ON STRUCTURAL
RESPONSE TO ACOUSTIC EXCITATION
LOCKHEED-GEORGIA CO., MARIETTA, REPORT NO:
NASA-CR-1959, MARCH 1972, 61 P. (N.T.I.S. -
N72-19921)

72-0865 KRAUSE, W.
THE MBB-UFE SUBSONIC PANEL METHOD. PART 2 - THE
LIFT-CONNECTED DISPLACEMENT PROBLEM IN
COMPRESSIBLE FLOW
NATL. AERON. SPACE ADMIN., WASH., D.C., REPORT NO:
NASA-TT-F-14117, UFE-633-70, MARCH 1972, 141 P.
(N.T.I.S. - N72-19325)

72-0866 SMITH, J. H.
NONLINEAR BEAM AND PLATE ELEMENTS
J. STRUCT. DIV., A.S.C.E., VOL. 98, NO. ST3, 1972,
PP. 553-568.

72-0867 FORREST, J. B.
KATONA, M. G.
GRIFFIN, D. F.
LAYERED PAVEMENT SYSTEMS. PART 1. LAYERED SYSTEM
DESIGN. PART 11. FATIGUE OF PLAIN CONCRETE
NAVAL CIVIL ENGNG. LAB., PORT HUENEME, CALIF.,
REPORT NO: NCEL-TR-763, APRIL 1972, 78 P.
(N.T.I.S. - AD-742 337)

72-0868 AUTHOR
ANALYSIS OF LAYERED SYSTEMS SUBJECTED TO WHEEL
LOADS
DEPT. OF CIVIL ENGNG., UNIV. OF CALIF., BERKELEY,
FINAL REPORT, AUGUST 1970 - JANUARY 1972, 18 P.
(N.T.I.S. - AD-742 147)

72-0869 WANG, C. Y-J.
FRACTURE MECHANICS FOR AN INTERFACIAL CRACK
BETWEEN ADHESIVELY BONDED DISSIMILAR MATERIALS
DEPT. OF THEORETICAL AND APPLIED MECH., UNIV. OF
ILL., URBANA, REPORT NO: T/AM-353,
UILU-ENG-72-6001, MARCH 1972, 101 P. (N.T.I.S. -
AD-742 124)

72-0870 MENDELSON, A.
GROSS, B.
SRAWLEY, J. E. 11
EVALUATION OF THE USE OF A SINGULARITY ELEMENT IN
FINITE ELEMENT ANALYSIS OF CENTER-CRACKED PLATES
LEWIS RESEARCH CENTER, NATL. AERON. SPACE ADMIN.,
CLEVELAND, OHIO. REPORT NO: E-6680, 1972.
(N.T.I.S. - N72-18910)

72-0871 FEIKEN, R. W.
FLEXURE OF A BEAM SUPPORTED ON AN ELASTIC
FOUNDATION
MASTER'S THESIS, SCHOOL OF ENGNG., AIR FORCE INST.
OF TECH., WRIGHT-PATTERSON AFB, OHIO, REPORT NO:
GSR/MC/72-4, MARCH 1972, 88 P. (N.T.I.S. - AD-741
453)

72-0872 MARTIN, M. S.
ROSE, D. J.
COMPLEXITY BOUNDS FOR REGULAR FINITE DIFFERENCE
AND FINITE ELEMENT GRIDS
DEPT. OF MATH., UNIV. OF DENVER, COLO., REPORT NO:
MS-R-7221, APRIL 1972, 18 P. (N.T.I.S. - AD-741
232)

72-0873 GOLDSMITH, W.
SACKMAN, J. L.
WAVE PROPAGATION IN ANISOTROPIC ROCKS
ANNUAL TECH. REPORT, UNIV. OF CALIF., BERKELEY,
NO. 1, FEBRUARY 5, 1971 - FEBRUARY 4, 1972, 91 PP.
(N.T.I.S. - AD-740 854)

72-0874 LASHKARI, M.
WEINGARTEN, V. I.
MARGOLIAS, D. S.
VIBRATIONS OF PRESSURE LOADED HYPERBOLOIDAL SHELLS
A.S.C.E. J. ENG. MECH. DIV., VOL. 98, NO. 5, OCT.
1972, PP. 1017-1030.

72-0875 SIMON, R. M.
LADD, C. C.
CHRISTIAN, J. T.
FINITE ELEMENT PROGRAM FEECON FOR UNDRAINED
DEFORMATION ANALYSES OF GRANULAR EMBANKMENTS ON
SOFT CLAY FOUNDATIONS
DEPT. OF CIVIL ENGNG., MASS. INST. OF TECH.,
CAMBRIDGE, REPORT NO: R72-9, SOILS PUB-294,
FEBRUARY 1972, 101 P. (N.T.I.S.- PB-208 407)

72-0876 NOSSEIR, S.
SHINOZUKA, M.
PROBABILISTIC MODEL FOR MATERIAL STRENGTH
VARIATION AND SIZE EFFECT
NAVAL CIVIL ENGNG. LAB., PORT HUENEME, CALIF.,
REPORT NO: NCEL-TR-756, FEBRUARY 1972, 34 P.
(N.T.I.S. - AD-740 752)

72-0877 CHANG, C-Y.
NAIR, K.
A THEORETICAL METHOD FOR EVALUATING STABILITY OF
OPENINGS IN ROCK
WOODWARD-LUNDGREN AND ASSOC., OAKLAND, CALIF.,
FINAL REPORT, MARCH 12, 1971 - MARCH 12, 1972, 148
P. (N.T.I.S. AD-740 341) SEE ALSO AD-733 500.

72-0878 MCKINSTRY, H. A.
SHULL, H. E.
BEUSSEM, W. R.
HIGH STRENGTH CERAMICS
UNIVERSITY PARK MATERIALS RESEARCH LAB., PENN.
STATE UNIV., FINAL REPORT, MARCH 30, 1972, 87 P.
(N.T.I.S. - AD-740 215)

72-0879 CHAMIS, C. C.
MECHANICS OF LOAD TRANSFER AT THE FIBER/MATRIX
INTERFACE
LEWIS RESEARCH CENTER, NAT. AERON. AND SPACE
ADMIN., CLEVELAND, OHIO., REPORT NO:
NASA-TN-D-6588, E-6447, FEBRUARY 1972, 64 P.
(N.T.I.S. - N72-16885)

72-0880 SVEC, O. J.
MCNEICE, G. M.
FINITE ELEMENT ANALYSIS OF FINITE SIZED PLATES
BOUNDED TO AN ELASTIC HALF SPACE
COMP. METH. APPL. MECH. ENG., VOL. 1, 1972, PP.
265-277.

72-0881 MASER, K. R.
AN ANALYSIS OF THE SMALL-SCALE STRENGTH TESTING OF
ICE
SEA GRANT PROJECT OFFICE, MASS. INST. OF TECH.,
CAMBRIDGE, REPORT NO: MITSG-72-6, JANUARY 25,
1972, 141 P. (N.T.I.S. - COM-72-10294)

72-0882 LEW, T. K.
EVALUATION OF A THREE-DIMENSIONAL STRESS CELL FOR
GRANULAR SOILS
NAVAL CIVIL ENGNG. LAB., PORT HUENEME, CALIF.,
REPORT NO: NCEL-TR-758, FEBRUARY 1972, 68 P.
(N.T.I.S. - AD-739 329)

72-0883 PERRONE, N.
KAO, R.
A GENERAL FINITE DIFFERENCE METHOD FOR ARBITRARY
MESHES
CATHOLIC UNIV. OF AMERICA, WASH. D.C., FEBRUARY
1972, 31 P. (N.T.I.S. - AD-738 982)

72-0884 KAO, R.
APPLICATIONS OF HILL FUNCTIONS (APPLIED
MATHEMATIC'S FINITE ELEMENT) TO APPLIED MECHANICS
PROBLEMS
DEPT. OF CIVIL AND MECH. ENGNG., CATHOLIC UNIV. OF
AMERICA, WASH. D.C., JANUARY 1972, 36 P. (N.T.I.S.
- AD-738 966)

72-0885 TAYLOR, R. L.
ON THE COMPLETENESS OF SHAPE FUNCTIONS FOR FINITE
ELEMENT ANALYSIS
INT. J. NUMER. METH. ENG., VOL. 4, 1972, PP.
17-22.

72-0886 GENALIS, P.
THE APPLICATION OF THE FINITE ELEMENT METHOD TO
MULTIHULLED SHIP STRUCTURAL ANALYSIS
NAVAL SHIP RESEARCH AND DEVELOPMENT CENTER,
BETHESDA, MD., REPORT NO: NSRDC-3780, JANUARY
1972, 28 P. (N.T.I.S. - AD-738 001)

72-0887 TEZCAN, S. S.
AGRAWAL, K. M.
KOSTRO, G.
FINITE ELEMENT ANALYSIS OF HYPERBOLIC PARABOLOID
SHELLS
J. STRUCT. DIV., A.S.C.E., VOL. 98, NO. ST3, 1972,
PP. 671-690.

72-0888 PALMERTON, J. B.
BANKS, D. C.
APPLICATION OF FINITE ELEMENT METHOD IN
DETERMINING STABILITY OF CRATER SLOPES
ARMY ENGR. WATERWAYS EXPERIMENT STATION,
VICKSBURG, MISS., REPORT NO:
AEWES-MISC-PAPER-S-72-2, JANUARY 1972, 131 P.
(N.T.I.S. - AD-737 178)

72-0889 MCMAHON, A.
HODGE, P. G.
A SIMPLE FINITE ELEMENT MODEL FOR ELASTIC-PLASTIC
PLATE BENDING
DEPT. OF MECH., ILL. INST. OF TECH., CHICAGO,
REPORT NO. DOMIIT-1-46, 1972, 66 PP. (N.T.I.S. -
AD-735 689)

72-0890 GEORGOPOULOS, G.
ROYSTER, L. H.
A FORTRAN PROGRAM FOR CALCULATING THE FREQUENCIES
AND MODE SHAPES OF A GUIDED CLAMPED, GUIDED
PINNED, CLAMPED, AND PINNED SPHERICAL SHELL
DEPT. OF MECH. AND AEROSPACE ENGNG., NORTH
CAROLINA STATE UNIV., RALEIGH, REPORT NO: TR-22,
JANUARY 1972, 45 P. (N.T.I.S. - AD-731 062)

72-0891 KANDIDOV, V. P.
CHESNOKOV, S. S.
MODIFICIROVANNYJ KONECHNYJ ELEMENT DIJA RASCHIOTA
KOLEBANIJ TONKIKH PLASTIN
VESTNIK MGU, FIZIKA ASTRONOMIJA, VOL. 13, 1972,
PP. 42-49.

72-0892 VAN CAMPEN, D. H.
SOME APPLICATIONS OF ENGINEERING MECHANICS IN
PRESSURE VESSEL ANALYSIS
LABORATORIUM VOOR ENERGIEVOORZIENING EN
KERNREACTOREN, TECHNISCHE HOGESCHOOL, DELFT,
NETHERLANDS, REPORT NO: WTHD-35, FEBRUARY 1972,
151 PP. (N.T.I.S. - N72-25901)

72-0893 TALBOT, R. J.
ANALYSIS OF THE DYNAMICS OF ELASTIC STRUCTURES BY
THE FINITE ELEMENT METHOD
DOCTORAL THESIS, SCHOOL OF ENGNG., AIR FORCE INST.
OF TECH., WRIGHT-PATTERSON AFB, OHIO, JUNE 1972,
182 P. (N.T.I.S. -AD-747 283)

72-0894 MAK, C. K.
KAO, D-W.
LEE, L. H. N.
INCREMENTAL VARIATIONAL METHOD FOR THE LARGE
DISPLACEMENT ANALYSIS OF SHELLS WITH GEOMETRIC
IMPERFECTIONS
COLL. OF ENGNG., UNIV. OF NOTRE DAME, IND., REPORT
NO: UND-M2-5, AUGUST 1972, 34 P. (N.T.I.S. -
AD-747 321)

72-0895 GROOMS, H. R.
ALGORITHM FOR MATRIX BANDWIDTH REDUCTION
PROC. A.S.C.E., STRUCT DIV., NO. ST1, 1972, PP.
203-214.

72-0896 OLSON, M. D.
A CONSISTENT FINITE ELEMENT METHOD FOR RANDOM
RESPONSE PROBLEMS
INT. J. COMPUT. AND STRUCT., VOL. 2, 1972, PP.
163-180.

72-0897 DE ARANTES E OLIVEIRA, E. R.
A GENERAL THEORY OF VARIATIONAL METHODS WITH
APPLICATION TO FINITE ELEMENTS
LABORATORIO NACIONAL DE ENGENHARIA CIVIL
(AVAILABLE FROM MINISTERIO DAS OBRAS PUBLICAS,
LISBON, PORTUGAL, 1972).

72-0898 WOOD, W. L.
SIMPLE TEACHING EXAMPLES FOR FINITE ELEMENT STRESS
ANALYSIS
INT. J. MATH. EDUC. SCI. TECHNOL., VOL. 3, 1972,
PP. 133-146.

72-0899 YANG, T. Y.
ELASTIC SNAP THROUGH ANALYSIS OF CURVED PLATES
USING DISCRETE ELEMENTS
A.I.A. A. J., VOL. 10, 1972, PP. 371-372.

72-0900 HILTON, P. D.
SIH, G. C.
APPLICATIONS OF THE FINITE ELEMENT METHOD TO THE
CALCULATIONS OF STRESS INTENSITY FACTORS
MECHANICS OF FRACTURE METHODS OF ANALYSIS AND
SOLUTION OF CRACK PROBLEMS, EDITED BY G.C. SIH,
NOORDHOFF INT. PUBL., LEYDEN, 1972, PP. 426-483.

72-0901 BABUSKA, I.
THE FINITE ELEMENT METHOD WITH LAGRANGIAN
MULTIPLIERS
INST. OF FLUID DYN. AND APPL. MATH., UNIV. OF
MARYLAND, TECH. NOTE BN-724, 1972.

72-0902 ANDERSON, J. C.
SEISMIC RESPONSE EFFECTS ON EMBEDDED STRUCTURES
BULL. SEISMOL. SOC. AM., VOL. 62, NO. 1, FEBRUARY
1972, PP. 177-194.

72-0903 KANDIDOV, V. P.
KIM, L. P.
RASCHIOT USTOICHIVOSTI IZGIBNO-KRUTILNYKH
KOLEBANIJ KRULA V DOZVUKOVOM POTOKE METODOM
KONECHNYKH ELEMENTOV
UCHENYE ZAPISKI CAGI, VOL. 3, 1972, PP. 111-117.

72-0904 NORRIS, V. A.
AN INTRODUCTION TO THE FINITE ELEMENT ANALYSIS OF
THE MOTION OF A GAS-LIQUID INTERFACE IN A POROUS
MEDIUM
M.SC. THESIS, CIVIL ENG. DEPT., UNIV. COLLEGE OF
SWANSEA, WALES, 1972.

72-0905 YAMADA, YOSHIAKI
PLASTICITY AND VISCOELASTICITY
BAIFUKAN PRESS, TOKYO, 1972. (IN JAPANESE)

72-0906 COALE, C. W.
KERTESZ, T. J.
INFLUENCE OF ASCENT HEATING ON THE SEPARATION
DYNAMICS OF A SPACECRAFT FAIRING
SHOCK AND VIB. BULL., VOL. 42, NO. 5, JANUARY
1972.

72-0907 GIORDANA, F.
CALCULATION OF THE NATURAL FREQUENCIES AND THE
PRINCIPAL MODES OF HELICOPTER BLADES
MECCANICA, VOL. 7, NO. 4, DECEMBER 1972, PP.
255-262.

72-0908 HAISLER, W. E.
STRICKLIN, J. A.
VON RIESEMANN, W. A.
A FINITE ELEMENT PROGRAM FOR THE DYNAMIC LARGE
DEFLECTION, ELASTIC-PLASTIC ANALYSIS OF STIFFENED
SHELLS OF REVOLUTION
AEROSP. ENG. DEPT., A AND M UNIV., COLLEGE
STATION, TEX., SLA-73-127, TEES-72-27, DECEMBER
1972, 160 PP.

72-0909 HAUG, E.
POWELL, G. H.
FINITE ELEMENT ANALYSIS OF NONLINEAR MEMBRANE
STRUCTURES
DEPT. OF CIVIL ENG., UNIV. OF CALIF., BERKELEY,
SESM REPORT NO. 72-7, 1972.

72-0910 HARRISON, N. L.
HARRISON, W. J.
THE STRESSES IN AN ADHESIVE LAYER
J. ADHESION, VOL. 3, 1972, PP. 195-212.

72-0911 ANON
SNAP DYNAMICS
LOCKHEED MISSILES AND SPACE CO., SUNNYVALE,
CALIF., MFS-21531, 1972.

72-0912 ANON
PROGRAM TO REDUCE THE SIZE OF STRUCTURAL MATRICES
NORTH AMER. ROCKWELL CORP., CANOGA PARK, SPACE
DIV., CALIF., MSC-17619, 1972.

72-0913 ROBINSON, J.
AUTOMATIC EXTRACTION OF RIGID BODY MODES FROM
STRESS AND STRAIN ELEMENTS
PROC. INT. CONF. ON VARIATIONAL METH. IN ENG.,
UNIV. OF SOUTHAMPTON, SEPTEMBER 25, 1972, (EDS.)
BREBBIA, C.A. AND TOTTENHAM, H., SOUTHAMPTON UNIV.
PRESS, 1973, VOL. 1, PP. 4/73-4/92.

72-0914 NEMAT-NASSER, S.
GENERAL VARIATIONAL PRINCIPLES IN NONLINEAR AND
LINEAR ELASTICITY WITH APPLICATIONS
NORTHWESTERN UNIV., EVANSTON, ILL., 1972, 51 PP.
(N.T.I.S. -AD-A007 895/6ST) PUBL. IN MECH. TODAY,
VOL. 1, 1972, PP. 21 4-261.

72-0915 KIMURA, L. A.
THOMAS, E. A.
YEE, F.
DIGITAL COMPUTER PROGRAMS FOR STEADY-STATE OR
TRANSIENT TEMPERATURE ANALYSIS OF PLANE AND
AXISYMMETRIC BODIES BY THE FINITE ELEMENT METHOD.
PROGRAMS E12202, E12205, E12206, E12207.
ENGINEERING OPERATIONS REPORT
AEROJET-GENERAL CORP., SACRAMENTO, CALIF., MAY 26,
1972, 253 PP. (N.T.I.S. - TID/SNA-2007)

72-0916 ROCK, A. O.
ENGINEERING OPERATIONS REPORT: NERVA THIN SHELL
FINITE ELEMENT STRESS ANALYSIS METHODOLOGY
COMPUTER PROGRAM E11101
AEROJET-GENERAL CORP., SACRAMENTO, CALIF., JUNE 2,
1972, 97 PP. (N.T.I.S. - TID-SNA-906)

72-0917 ROCK, A. O.
ENGINEERING OPERATIONS REPORT: NERVA AXISYMMETRIC
AND PLANE STRUCTURES FINITE ELEMENT STRESS
ANALYSIS METHODOLOGY. COMPUTER PROGRAMS E11401
AND E11405
AEROJET-GENERAL CORP., SACRAMENTO, CALIF., MAY 26,
1972, 190 PP. (N.T.I.S. - TID/SAN-903)

72-0918 REICH, M.
KOPLICK, B.
HENDRIE, J. M.
FINITE ELEMENT APPROACH TO POLYMER CONCRETE BRIDGE
DECK DESIGN ANALYSIS
REPORT ON POLYMER CONCRETE APPLICATIONS
DEVELOPMENT FOR THE BUREAU OF PUBLIC ROADS,
BROOKHAVEN NATL. LAB., UPTON, N.Y., MAY 1972, 104
PP. (N.T.I.S. - BNL-16890)

72-0919 TANAKA, K.
KINITAKE, M.
ANALYSIS OF SLOPE STABILITY BY FINITE ELEMENT
METHOD AND ITS APPLICATION PROGRAM. SOIL
MECHANICS
AGRIC. BULL., SAGA UNIV., VOL. 33, MAY 1972, PP.
41-58.

72-0920 SUGIYAMA, S.
NUMERICAL ANALYSIS ON STRESS DISTRIBUTION IN
PARTIAL COMPRESSED SPECIMEN (HARDBOARD) BY FINITE
ELEMENT METHOD
SCI. BULL., FAC. AGRIC., KYUSHU UNIV., VOL. 26,
NO. 1/4, MARCH 1972, PP. 489-504.

72-0921 ANON
THREE-DIMENSIONAL FINITE ELEMENT ANALYSIS FOR
FRACTURE MECHANICS
THE SURFACE CRACK, PHYS. PROBL. AND COMPUT.
SOLUTIONS, 1972, PP. 187-202.

72-0922 HENSHELL, R. D.
MAKOJU, J. O.
BOND, T. J.
RING FINITE ELEMENTS FOR AXISYMMETRIC AND
NON-AXISYMMETRIC THIN SHELL ANALYSIS
PROC. INT. CONF. ON VARIATIONAL METH. IN ENG.,
UNIV. OF SOUTHAMPTON, SEPTEMBER 25, 1972, (EDS.)
BREBBIA, C.A. AND TOTTENHAM, H., SOUTHAMPTON UNIV.
PRESS, 1973, VOL. 1, PP. 4/44-4/58.

72-0923 NAGAMATSU, A.
MUROTA, T.
JIMMA, T.
NON-UNIFORM DEFORMATION OF MATERIAL IN AXIALLY
SYMMETRIC COMPRESSION CAUSED BY FRICTION. PART 3,
THEORETICAL ANALYSIS OF COMPRESSION OF HOLLOW
CYLINDER BY FINITE-ELEMENT METHOD
BULL. JAPAN SOC. MECH. ENG., VOL. 15, NO. 89,
NOVEMBER 1972, PP. 1339-1347.

72-0924 GOULD, P. L.
SZABO, B. A.
SURYOUTOMO, H. B.
CURVED ROTATIONAL SHELL ELEMENTS BY THE CONSTRAINT
METHOD
PROC. INTL. CONF. VARIATIONAL METH. IN ENG., UNIV.
OF SOUTHAMPTON, 8/33-8/42, 1972.

72-0925 SMITH, F. W.
ELASTIC STRESSES IN LAYERED SNOW PACKS
GLACIOLOGY J., CAMBRIDGE, ENG., VOL. 11, NO. 63,
1972, PP. 407-414.

72-0926 JERRAM, K.
HELLEN, T. K.
THE USE OF FINITE ELEMENT TECHNIQUES IN FRACTURE
MECHANICS
C.E.G.B. RESEARCH DEPT., BERKELEY NUCL. LABS.,
KD/B/N2478, 1972.

72-0927 YANG, R. Y.
ELASTIC SNAP-TRHOUGH ANALYSIS OF CURVED PLATES
USING DISCRETE ELEMENTS
A.I.A.A. J., VOL. 10, 1972, PP. 371-373.

72-0928 DESAI, C. S.
JOHNSON, L. D.
SOME NUMERICAL PROCEDURES FOR ANALYSIS OF ONE
DIMENSIONAL CONSOLIDATION
PROC. SYMP. ON THE APPLICATIONS OF THE FINITE
ELEMENT METHOD IN GEOTECHNICAL ENG., VICKSBURG,
MISS., MAY 1-4, 1972, PP. 863-882. (PUBL. BY U.S.
ARMY ENG. WATERWAYS EXP. STATION, SEPTEMBER 1972)

72-0929 ABBASI, J. A.
REINFORCED CONCRETE - A FINITE-ELEMENT FORMULATION
PH.D., UNIV. OF MARYLAND, 1972.

72-0930 ARBIC, B. J.
NUMERICAL SOLUTION OF SOME NONSTATIONARY PARTIAL
DIFFERENTIAL EQUATIONS: THE FINITE ELEMENT METHOD
AND LINE METHOD
PH.D., UNIV. OF WYOMING, 1972.

72-0931 DESAI, C. S.
HOLLOWAY, M.
LOAD DEFORMATION ANALYSIS OF DEEP PILE FOUNDATION
PROC. SYMP. ON THE APPLICATIONS OF THE FINITE
ELEMENT METHOD IN GEOTECH. ENG., VICKSBURG, MISS.,
MAY 1-4, 1972, PP. 629-656. (PUBL. BY U.S. ARMY
ENG. WATERWAYS EXP. STATION, SEPTEMBER 1972)

72-0932 BRITTEN, S. S.
EIGENVALUE BOUNDS FOR FINITE ELEMENT MODELS
PH.D., MASSACHUSETTS INST. OF TECHNOL., 1972.

72-0933 CHI, H-M.
COMPUTATIONAL PROCEDURES FOR INELASTIC
FINITE-ELEMENT ANALYSIS
PH.D., UNIV. OF CALIF., BERKELEY, 1972.

72-0934 TOLMAN, F.
SPEELPENNING, B.
THE ELEMENT-CHAIN CONCEPT IN FINITE ELEMENT BASED
PROGRAMMING
INT. T.N.O. FOR BLDG. MATER. AND BLDG. STRUCT.,
DELFT, NETHERLANDS, REPORT B1-72-07.1.102, 1972.

72-0935 CHAR, C. V.
FINITE ELEMENT ANALYSIS OF CREEP IN ROCK-SALT
PILLAR MODELS
PH.D., THE LOUISIANA STATE UNIV. AND AGRICULTURAL
AND MECH. COL., 1972.

72-0936 CHENG, R. C-N.
FINITE-ELEMENT ANALYSIS OF FLOW IN BLOOD VESSEL
WITH ARBITRARY CROSS-SECTION
PH.D., THE PENNSYLVANIA STATE UNIV., 1972.

72-0937 CROSSMAN, F. W. JR.
COMPUTER SIMULATION OF THE DEFORMATION OF
COMPOSITE AND POLYCRYSTALLINE MATERIALS BY FINITE
ELEMENT ANALYSIS
PH.D., STANFORD UNIV., 1972.

72-0938 DESAI, S.
USE OF AUXILIARY FUNCTIONS IN THE FINITE ELEMENT
METHOD
PH.D., LEHIGH UNIV., 1972.

72-0939 FARAH, J. W.
STRESS ANALYSIS OF FIRST MOLARS WITH FULL CROWN
PREPARATIONS BY THREE-DIMENSIONAL PHOTOELASTICITY
AND THE FINITE-ELEMENT METHOD
PH.D., THE UNIV. OF MICHIGAN, 1972.

72-0940 FAHERTY, K. F.
AN ANALYSIS OF A REINFORCED AND A PRESTRESSED
CONCRETE BEAM BY FINITE-ELEMENT METHOD
PH.D., THE UNIV. OF IOWA, 1972.

72-0941 FORD, J. L.
A METHOD FOR MATERIAL CHARACTERIZATION EMPLOYING
THE FINITE-ELEMENT TECHNIQUE WITH APPLICATION TO
HOLOGRAPHY
PH.D., UNIV. OF ILLINOIS, URBANA, CHAMPAIGN, 1972.

72-0942 GANTAYAT, A. N.
STRESS ANALYSIS OF TWO JOINTS BY THE
FINITE-ELEMENT METHOD
PH.D., UNIV. OF CALIF., BERKELEY, 1972.

72-0943 GODBOLD, C. W.
CONVERGENCE CHARACTERISTICS OF CYLINDRICAL SHELL
FINITE ELEMENTS
SC.D., WASH. UNIV., 1972.

72-0944 HAUG, E.
FINITE ELEMENT ANALYSIS OF NONLINEAR MEMBRANE
STRUCTURES
PH.D., UNIV. OF CALIF., BERKELEY, 1972.

72-0945 HA, H. K.
ANALYSIS OF THREE-DIMENSIONAL ORTHOTROPIC SANDWICH
PLATE STRUCTURES BY FINITE-ELEMENT METHOD
PH.D., CONCORDIA UNIV., CANADA, 1972.

72-0946 HAND, F. R. JR.
NONLINEAR LAYERED FINITE-ELEMENT ANALYSIS OF
REINFORCED-CONCRETE PLATES AND SHELLS
PH.D., UNIV. OF ILLINOIS, URBANA, CHAMPAIGN, 1972.

72-0947 HEIFETZ, J. H.
FINITE DEFORMATIONS IN ELASTIC-PLASTIC MEDIA
SUBJECTED TO IMPULSIVE LOADINGS BY THE FINITE
ELEMENT METHOD
PH.D., THE CITY UNIVERSITY, NEW YORK, 1972.

72-0948 HUANG, S-W.
FINITE-ELEMENT ANALYSIS OF FRACTURE PROPAGATION IN
TWO-DIMENSIONAL ELASTIC BRITTLE SOLIDS
PH.D., THE OHIO STATE UNIV., 1972.

72-0949 IYER, M. S.
ANALYSIS OF A PRESSURE-VESSEL JUNCTION BY THE
FINITE-ELEMENT METHOD
PH.D., TEXAS TECH. UNIV., 1972.

72-0950 JOHNSON, J. A.
A TECHNIQUE FOR THE SETTLEMENT AND STRESS ANALYSIS
OF MAT FOUNDATIONS USING THE FINITE ELEMENT METHOD
PH.D., VANDERBILT UNIV., 1972.

72-0951 KAPPEL, V. V.
VIBRATION OF STIFFENED PLATES USING
FINITE-ELEMENTS
PH.D., UNIV. OF CALGARY, CANADA, 1972.

72-0952 KENNEDY, F. E. JR.
ANALYSIS OF NONLINEAR CONTACT PROBLEMS BY THE
FINITE-ELEMENT METHOD
PH.D., RENSSELAER POLYTECHNIC INST., 1972.

72-0953 KHATUA, T. P.
FINITE-ELEMENT ANALYSIS OF MULTILAYER SANDWICH
STRUCTURES
PH.D., UNIV. OF CALGARY, CANADA, 1972.

72-0954 LASSKER, A. J.
NONLINEAR BEHAVIOR OF REINFORCED CONCRETE BEAMS BY
THE FINITE-ELEMENT METHOD
PH.D., OKLAHOMA STATE UNIV., 1972.

72-0955 LIN, F. T.
THE FINITE ELEMENT ANALYSIS OF LAMINATED
COMPOSITES
PH.D., VIRGINIA POLYTECHNIC INST. AND STATE UNIV.,
1972.

72-0956 LORIOT DE ROUVRAY, A. D.
NONLINEAR FINITE ELEMENT ANALYSIS OF MULTI JOINTED
ROCK MASSES
PH.D., UNIV. OF CALIF., BERKELEY, 1972.

72-0957 MATSUMOTO, G. Y.
FINITE ELEMENT ANALYSIS OF PROBLEMS IN
NONCONSERVATIVE ELASTIC STABILITY
PH.D., UNIV. OF CALIF., BERKELEY, 1972.

72-0958 MASTROGEORGOPOULOS, S. C.
FINITE-ELEMENT ANALYSIS OF ORTHOTROPIC PLATES WITH
ECCENTRIC STIFFENERS
PH.D., VIRGINIA POLYTECHNIC INST. AND STATE UNIV.,
1972.

72-0959 MAMET, J-C.
FINITE-ELEMENT ANALYSIS OF TALL BUILDINGS
PH.D., MCGILL UNIV., CANADA, 1972.

72-0960 MCGEE, T. M.
AN INVESTIGATION OF MICRO-BUCKLING USING
PHOTOELASTIC AND FINITE-ELEMENT METHODS
PH.D., VANDERBILT UNIV., 1972.

72-0961 MELVILLE, J. G.
FINITE-ELEMENT ANALYSIS OF TAPERED VISCOUS FLOW
PH.D., THE PENNSYLVANIA STATE UNIV., 1972.

72-0962 MILLOY, J. A.
A VARIATIONAL FORMULATION FOR DISCONTINUOUS FIELDS
IN LINEARLY ELASTIC FINITE ELEMENT APPLICATIONS
PH.D., NORTHWESTERN UNIV., 1972.

72-0963 CAMPBELL, J. S.
FINESSE - A FINITE ELEMENT SYSTEM, PART V FINITE
ELEMENT DESIGN - DERIVATIVES FOR STRUCTURAL SHAPE
DESIGN
CIVIL ENG. DEPT., UNIV. COLLEGE OF SWANSEA, 1972.

72-0964 NEWSON, C. D.
FINITE-ELEMENT FORMULATION OF MIXED AND STRESS
VARIATIONAL PRINCIPLES FOR STRUCTURAL DYNAMIC
ANALYSIS
PH.D., THE UNIV. OF TEXAS, AUSTIN, 1972.

72-0965 OBAID, P.
THREE DIMENSIONAL FINITE ELEMENT ANALYSIS
PH.D., CASE WESTERN RESERVE UNIV., 1972.

72-0966 RAJASEKARAN, S.
FINITE ELEMENT ANALYSIS OF THIN-WALLED MEMBERS OF
OPEN SECTION
PH.D. THESIS, UNIV. OF CALGARY, CANADA, 1972.

72-0967 RAMEY, G. E.
A STUDY OF THE CONVERGENCE CHARACTERISTICS AND
ACCURACY OF CONFORMING FINITE ELEMENT SOLUTIONS
PH.D., UNIV. OF COLORADO, 1972.

72-0968 RAMANI, D. T.
THE FINITE ELEMENT ANALYSIS OF SANDWICH
BUCKLE-SHELL STRUCTURES
PH.D. THESIS, PURDUE UNIV., 1972.

72-0969 SADAT, M. M.
A QUADRILATERAL PLANE STRESS FINITE ELEMENT WITH
QUADRATIC PROPERTIES
PH.D., RUTGERS UNIV., THE STATE UNIV. OF NEW
JERSEY, 1972.

72-0970 SAMUELSON, A. C.
FINITE-ELEMENT ANALYSES OF INITIAL ELASTIC
DEFORMATION OF VERTICALLY DEFORMED CRYSTAL BLOCKS
PH.D., THE PENNSYLVANIA STATE UNIV., 1972.

72-0971 SEMENZA, L. A.
APPLICATION OF THE FINITE ELEMENT METHOD TO TWO
DIMENSIONAL MULTIGROUP NEUTRON DIFFUSION
PH.D., NORTHWESTERN UNIV., 1972.

72-0972 SIMON, R. M.
ANALYSIS OF EMBANKMENT CONSTRUCTION BY
FINITE-ELEMENTS
PH.D., MASSACHUSETTS INST. OF TECHNOLOGY, 1972.

72-0973 SISODIYA, R. G.
FINITE-ELEMENT ANALYSIS OF BRIDGES
PH.D., UNIV. OF CALGARY, CANADA, 1972.

72-0974 SINGH, R. D.
MECHANICAL CHARACTERIZATION AND FINITE-ELEMENT
ANALYSIS OF ELASTIC-PLASTIC, WORK-HARDENING SOILS
PH.D., THE OHIO STATE UNIV., 1972.

72-0975 SKJOLINGSTAD, L.
TRANSIENT AND NONLINEAR-ANALYSIS OF FIELD PROBLEMS
BY FINITE ELEMENTS
PH.D. THESIS, UNIV. OF CALGARY, CANADA, 1972.

72-0976 STRUIK, J. H. A.
APPLICATIONS OF FINITE-ELEMENT ANALYSIS TO
NONLINEAR PLANE STRESS PROBLEMS
PH.D., LEHIGH UNIV., 1972.

72-0977 STEPHENSON, J. W.
FINITE-ELEMENT PROCEDURES FOR PROBLEMS WITH CURVED
BOUNDARIES
PH.D., THE UNIV. OF SASKATCHEWAN, CANADA, 1972.

72-0978 TEBEDGE, N.
APPLICATIONS OF THE FINITE-ELEMENT METHOD TO
BEAM-COLUMN PROBLEMS
PH.D., LEHIGH UNIV., 1972.

72-0979 WANCHOO, M. K.
POST-ELASTIC FINITE-ELEMENT ANALYSIS OF REINFORCED
PLATES
PH.D., THE UNIV. OF NEW MEXICO, 1972.

72-0980 WALLACE, R. W.
A FINITE-ELEMENT, PLANAR-FLOW MODEL OF CAMAS
PRAIRIE, IDAHO
PH.D., UNIV. OF IDAHO, 1972.

72-0981 WOOLEY, G. R.
A FINITE-ELEMENT ANALYSIS OF STEADY,
TWO-DIMENSIONAL, INCOMPRESSIBLE, LAMINAR FLOW
PH.D., THE LOUISIANA STATE UNIV. AND AGRI. AND
MECH. COL., 1972.

72-0982 YOUNG, R. C.
LOCAL ERROR BOUNDS IN BOUNDARY VALUE PROBLEMS
USING FINITE ELEMENT DISCRETIZATION
PH.D., UNIV. OF CALIF., BERKELEY, 1972.

72-0983 YUZUGULLU, O.
FINITE-ELEMENT APPROACH FOR THE PREDICTION OF
INELASTIC BEHAVIOR OF SHEAR WALL-FRAME SYSTEMS
PH.D., UNIV. OF ILLINOIS, URBANA, CHAMPAIGN, 1972.

72-0984 HERRMANN, L. R.
INTERPRETATION OF FINITE ELEMENT PROCEDURE AS
STRESS ERROR MINIMIZATION PROCEDURE
A.S.C.E., J. ENG. MECH. DIV., VOL. 98, 1972, PP.
1330-1336.

72-0985 BACKLUND, J.
LIMIT ANALYSIS OF REINFORCED CONCRETE SLABS BY A
FINITE ELEMENT METHOD
PROC. CONF. FINITE ELEMENT METH., CIVIL ENG.,
MCGILL UNIV., MONTREAL, 1972, PP. 803-840.

72-0986 DAWE, D. J.
SHELL ANALYSIS USING A SIMPLE FACET ELEMENT
J. STRAIN ANALYSIS, VOL. 7, 1972, PP. 266-270.

72-0987 GAMBOLATI, G.
OPTIMIZZAZIONE DI RETICOLI TRIANGOLARI PER IL
METODO DEGLI ELEMENTI FINITI IN IDROLOGIA
PROC. A.I.C.A. CONGRESS, MILAN, NOVEMBER 1972, P.
87.

72-0988 DESAI, C. S.
OVERVIEW, TRENDS AND PROJECTIONS: THEORY AND
APPLICATIONS OF THE FINITE ELEMENT METHOD IN
GEOTECHNICAL ENGINEERING
PROC. SYMP. ON APPL. OF FINITE ELEMENT METHOD IN
GEOTECH. ENG., U.S.A. ENG. WATERWAYS EXP. STATION,
VICKSBURG, 1972.

72-0989 WILLOUGHBY, R.
A SURVEY OF SPARSE MATRIX TECHNOLOGY
IBM, YORKTOWN HEIGHTS, N.Y., REPT. RC3872,
(17585), 1972.

72-0990 ARANY, I.
SMYTH, W. F.
SZODA, L.
AN IMPROVED METHOD FOR REDUCING THE BANDWIDTH OF
SPARSE SYMMETRIC MATRICES
IN "INFORMATION PROCESSING 71", PROC. IFIP
CONGRESS, NORTH-HOLLAND, AMSTERDAM, 1972.

72-0991 SORENSEN, J. P.
SOLUTION OF TRANSPORT PROBLEMS BY USE OF TRIAL
FUNCTION TECHNIQUES
PH.D. THESIS, UNIV. OF WISCONSIN, 1972.

72-0992 ROBINSON, J.
THREE NODE ISOPARAMETRIC STRESS AXIAL ELEMENT
RA/ARD, REPORT NO. 23-3-72-6, MARCH 1972.

72-0993 DE KONING, A. U.
SANDERSE, A. A.
A COMPUTER PROGRAMME FOR ELASTIC-PLASTIC
CALCULATIONS WITH USE OF TRIM-6 FINITE ELEMENTS
NATL. AEROSPACE LAB., THE NETHERLANDS, REPORT
NLR-TR-72144 C., 1972.

72-0994 WHITEMAN, J. R.
AN INTRODUCTION TO VARIATIONAL AND FINITE ELEMENT
METHODS
DEPT. OF MATH., BRUNEL UNIV., FEBRUARY, 1972.

72-0995 BERGER, A.
RIDGWAY, S.
STRANG, G.
APPROXIMATE BOUNDARY CONDITIONS IN THE FINITE
ELEMENT METHOD
PROC. SYMP. ON NUMER. ANALYSIS, ISTITUTO NAZIONALE
DI ALTA MATICA, ROME, JANUARY 10-13, 1972, VOL.
10. (PUBL. BY ACADEMIC PRESS, NEW YORK, 1972)

72-0996 CUTHILL, E. H.
SEVERAL STRATEGIES FOR REDUCING THE BANDWIDTH OF
MATRICES
IN "SPARSE MATRICES AND THEIR APPLICATIONS", D. J.
ROSE, R. WILLOUGHBY, (EDS.), PLENUM PRESS, NEW
YORK, 1972, PP. 157-160.

72-0997 BOND, T. J.
SOME CONSIDERATIONS OF THE FINITE ELEMENT METHOD
IN STRESS ANALYSIS
PH.D. THESIS, NOTTINGHAM UNIV., 1972.

72-0998 GILL, J. A.
COMPUTER-AIDED DESIGN OF SHELL STRUCTURES USING
THE FINITE ELEMENT METHOD
PH.D. THESIS, UNIV. OF CAMBRIDGE, CAMBRIDGE,
ENGLAND, 1972.

72-0999 HENSHELL, R. D.
WALTERS, D.
WARBURTON, G. B.
A NEW FAMILY OF CURVILINEAR PLATE BENDING ELEMENTS
FOR VIBRATION AND STABILITY
J. SOUND VIB., VOL. 20, 1972, PP. 327-343.

72-1000 YAMOMOTO, Y.
YAGUCHI, M.
STRESS ANALYSIS BY SUPERPOSITION OF THE ANALYTICAL
SOLUTION AND FINITE ELEMENT SOLUTION
TRANS. JAPAN SOC. MECH. ENGRS., VOL. 306, 1972,
PP. 269-276.

72-1001 LUKAS, L.
CONTRIBUTION TO SOLUTION OF BODIES WITH CURVED
BOUNDARY BY FINITE ELEMENT METHOD
C.SC. THESIS, PLZEN, 1972.

72-1002 AL-HUSSAINI, M.M.
RADHAKRISHNAN, N.
ANALYSIS OF PLANE STRAIN TEST USING THE FINITE
ELEMENT METHOD
PROC. SYMP. ON THE APPLICATIONS OF THE FINITE
ELEMENT METHOD IN GEOTECH. ENG., VICKSBURG, MISS.,
MAY 1-4, 1972, PP. 215-257. (PUBL. BY U.S. ARMY
ENG. WATERWAYS EXP. STATION, SEPTEMBER, 1972)

72-1003 GROSS, H.
CLASSICAL AND NATURAL CONDUCTION AND DEFORMATION
MODES CONCEPTS FOR AXISYMMETRIC VARIATIONAL
METHODS
PROC. INT. CONF. ON VARIATIONAL METH. IN ENG.,
UNIV. OF SOUTHAMPTON, SEPTEMBER 25, 1972, (EDS.)
BREBBIA, C.A. AND TOTTENHAM, H., SOUTHAMPTON UNIV.
PRESS, 1973, VOL 2, PP. 10/114-10/129.

72-1004 WILSON, R. R.
BREBBIA, C. A.
RATE OF CHANGE OF DYNAMIC RESPONSE
PROC. INT. CONF. ON VARIATIONAL METH. IN ENG.,
UNIV. OF SOUTHAMPTON, SEPTEMBER 25, 1972, (EDS.)
BREBBIA, C.A. AND TOTTENHAM, H., SOUTHAMPTON UNIV.
PRESS, 1973, VOL. 2, PP. 10/67-10/76.

73-0001 NORRIE, D. H.
DE VRIES, G.
THE FINITE ELEMENT METHOD - FUNDAMENTALS AND
APPLICATIONS
ACADEMIC PRESS, NEW YORK, 1973.

73-0002 DE VRIES, G.
NORRIE, D. H.
APPLICATION OF THE PSEUDO-FUNCTIONAL FINITE
ELEMENT METHOD TO VISCO-PLASTIC TORSION
REPORT NO. 46, MECHANICAL ENGINEERING DEPARTMENT,
UNIVERSITY OF CALGARY, CANADA, OCTOBER 1973.

73-0003 NORRIE, D. H.
DE VRIES, G.
BERARD, G. P.
APPLICATION OF THE FINITE ELEMENT TECHNIQUE TO
COMPRESSIBLE FLOW PROBLEMS
MECH. ENG. REP. 18, DEPT. OF MECH. ENG., UNIV. OF
CALGARY.

73-0004 ANDERSEN, O. W.
TRANSFORMER LEAKAGE FLUX PROGRAM BASED ON FINITE
ELEMENT METHOD
IEEE TRANS. POWER APP. SYST. PA-92, NO. 2, 1973,
PP. 582-689.

73-0005 BECKETT, R. E.
CHU, S. C.
FINITE-ELEMENT METHOD APPLIED TO HEAT CONDUCTION
IN SOLIDS WITH NONLINEAR BOUNDARY CONDITIONS
J. HEAT TRANSFER 95, NO. 1, 1973, PP. 126-129.
ALSO IN ASME PAPER 69-WA/HT-39, NOVEMBER
1969.

73-0006 YOKOBORI, T.
KAMEI, A.
SIZE OF PLASTIC ZONE AT TIP OF A CRACK IN PLANE
STRAIN SLATE BY FINITE ELEMENT METHOD
INT. J. OF FRACTURES, VOL. 9, NO. 1, 1973, PP.
98-100.

73-0007 FLATABO, N.
TRANSIENT HEAT CONDUCTION PROBLEMS IN POWER CABLES
SOLVED BY FINITE ELEMENT METHOD
IEEE TRANS. POWER APP. SYST. PA-92, NO. 1, 1973,
PP. 56-63.

73-0008 KONRAD, A.
LINEAR ACCELERATOR CAVITY FIELD CALCULATION BY
FINITE ELEMENT
IEEE TRANS. NUCL. SCI., NS-20, NO. 1, 1973, PP.
802-808.

73-0009 LAGASSE, P. E.
FINITE ELEMENT ANALYSIS OF DISPERSION-FREE GUIDES
FOR ELASTIC SURFACE WAVES
IEEE TRANS. SONICS ULTRASONICS SU-20, NO. 1, 1973,
P. 56.

73-0010 MCKINSTRY, H. A.
SHULL, H. E.
BUESSEM, W. R.
TWO-DIMENSIONAL FINITE ELEMENT MODEL OF A CERAMIC
BODY
AMER. CERAM. SOC., BULL. 52, NO. 4, 1973, P. 349.

73-0011 MERCER, J. W.
PINDER, G. F.
FINITE ELEMENT APPROACH TO THE MODELING OF
HYDROTHERMAL SYSTEMS
TRANS. AMER. GEOPHYS. UNION, VOL. 54, NO. 4, 1973,
PP. 263. ALSO IN PH.D. THESIS, UNIVERSITY OF
ILLINOIS, URBANA, 1973.

73-0012 MINEAR, J. W.
FINITE ELEMENT MODELS OF FRACTURE
TRANS. AMER. GEOPHYS. UNION 54, NO. 4, 1973, P.
367.

73-0013 ODEN, J. T.
BHANDARI, D. R.
YAGAWA, G.
CHUNG, T. J.
NEW APPROACH TO FINITE ELEMENT FORMULATION AND
SOLUTION OF A CLASS OF PROBLEMS IN COUPLED
THERMOELASTOVISCOPLASTICITY OF CRYSTALLINE SOLIDS
NUCL. ENG. DESIGN, VOL. 24, NO. 3, 1973, PP.
420-430.

73-0014 PARRISH, D.
NONLINEAR FINITE ELEMENT FOLD MODEL
AMER. J. SCI. 273, NO. 4, 1973, PP. 318-334.

73-0015 PITKARANTA, J.
SILVENNOINEN, P.
COMPUTATIONAL EXPERIMENTATION ON FINITE ELEMENT
METHOD IN BARE SLAB CRITICALITY CALCULATIONS
NUCL. SCI. ENG. 50, NO. 3, 1973, PP. 297-300.

73-0016 SILVESTER, P.
FINITE ELEMENT ANALYSIS OF PLANAR MICROWAVE
NETWORKS
IEEE TRANS. MICROWAVE THEORY TECH. MT-21, NO. 2,
1973, PP. 104-108.

73-0017 SMITH, I. M.
FARRADAY, R. V.
O'CONNOR, B. A.
RAYLEIGH-RITZ AND GALERKIN FINITE ELEMENTS FOR
DIFFUSION-CONVECTION PROBLEMS
TRANS. AMER. GEOPHYS. UNION 54, NO. 1, 1973, P.
72. ALSO IN WATER RES. RES., VOL. 9, NO.
3, 1973, PP. 593-606.

73-0018 WAAS, G.
FINITE ELEMENT MODELING
GEOPHYSICS 38, NO. 1, 1973, P. 193.

73-0019 BARRACO, A.
APPLICATION DE LA METHODE DES ELEMENTS FINIS AU
CALCUL DES PLAQUES FLECHIES (APPLICATION OF THE
FINITE ELEMENTS TECHNIQUES IN CALCULATION OF PLATE
BENDING)
CONSTR. MET., NO. 3, SEPTEMBER 1973, PP. 27-33.

73-0020 WEXLER, A.
FINITE-ELEMENT FIELD ANALYSIS OF AN INHOMOGENEOUS
ANISOTROPIC RELUCTANCE MACHINE ROTOR
IEEE TRANS. POWER APP. SYST. PA-92, NO. 1, 1973,
PP. 145-149.

73-0021 SABIR, A. B.
LOCK, A. C.
LARGE DEFLECTION GEOMETRICALLY NON-LINEAR FINITE
ELEMENT ANALYSIS OF CIRCULAR ARCHES
INT. J. MECH. SCI., VOL. 15, NO. 1, 1973, PP.
37-47.

73-0022 MORIHARA, H.
CHENG, R. T.
NUMERICAL SOLUTION OF THE VISCOUS FLOW IN THE
ENTRANCE REGION OF PARALLEL PLATES
J. COMPUT. PHYSICS, VOL. 11, NO. 4, APRIL 1973.

73-0023 KIKUCHI, F.
ANDO, Y.
CONVERGENCE OF A MIXED FINITE ELEMENT SCHEME FOR
PLATE BENDING
NUCL. ENG. DESIGN, VOL. 24, NO. 3, 1973, PP.
357-373.

73-0024 ASHWELL, D. G.
SABIR, A. B.
CORRECTED ASSESSMENT OF CYLINDRICAL SHELL FINITE
ELEMENT OF BOGNER, FOX AND SCHMIT WHEN APPLIED TO
ARCHES
INT. J. MECH. SCIENCES, VOL. 15, NO. 4, 1973, PP.
325-327.

73-0025 HUNTER, A. R.
SKOGH, J.
PHOTOELASTIC AND FINITE ELEMENT INVESTIGATION OF A
NONSYMMETRICAL PLUG HATCH CONFIGURATION
EXP. MECH. VOL. 13, NO. 3, 1973, PP. 132-137.

73-0026 MURTY, A. V. K.
GALERKIN FINITE ELEMENT METHOD FOR VIBRATION
PROBLEMS
AIAA, J. VOL. 11, NO. 4, 1973, PP. 544-546.

73-0027 WOLF, J. A.
GENERALIZED HYBRID STRESS FINITE-ELEMENT MODELS
A.I.A.A. J., VOL. 11, NO. 3, 1973, PP. 385-388.

73-0028 MEI, C.
FINITE ELEMENT ANALYSIS OF NONLINEAR VIBRATION OF
BEAM COLUMNS
AIAA J. VOL. 11, NO. 1, 1973, PP. 115-117.

73-0029 CHANG, H. K.
CHENG, R. T.
FARHI, L. E.
A MODEL STUDY OF GAS DIFFUSION IN ALVEOLAR SACS
RESPIRATION PHYSIOLOGY 1973, 18, PP. 386-397.

73-0030 STRICKLIN, J. A.
TILLERSON, J. R.
CURVED FINITE ELEMENTS BY METHODS OF INITIAL
STRAINS
AIAA J. VOL. 11, NO. 2, 1973, PP. 249-250.

73-0031 HAFTKA, R. T.
ROBINSON, J.
EFFECT OF OUT-OF-PLANENESS OF MEMBRANE
QUADRILATERAL FINITE ELEMENTS
AIAA J., VOL. 11, NO. 5, 1973, PP. 742-744.

73-0032 THOMAS, D. L.
WILSON, J. M.
WILSON, R. R.
TIMOSHENKO BEAM FINITE ELEMENTS
J. OF SOUND AND VIBRATION 1973, VOL. 31, NO. 3,
PP. 315-330.

73-0033 THOMAS, J.
DOKUMACI, E.
IMPROVED FINITE ELEMENTS FOR VIBRATION ANALYSIS OF
TAPERED BEAMS
AERONAUTICAL QUARTERLY, VOL. 24, FEBRUARY 1973,
PP. 39-46.

73-0034 AKAGI, T.
OHNO, T.
VISCOELASTIC ANALYSIS OF P C CONTAINMENT USING
FINITE ELEMENT METHOD
PROC. JAP. SOC. CIV. ENG., JUNE 1973, PP. 1-8.

73-0035 ORRIS, R. M.
PETYT, M.
FINITE ELEMENT STUDY OF VIBRATION OF TRAPEZOIDAL
PLATES
J. SOUND AND VIBRATION, VOL. 27, NO. 3, 1973, PP.
325-344.

73-0036 YANG, T. Y.
SUN, C. T.
FINITE ELEMENTS FOR VIBRATION OF FRAMED SHEAR
WALLS
J. SOUND AND VIBRATION, VOL. 27, NO. 3, 1973, PP.
297-311.

73-0037 FONDER, G. A.
CLOUGH, R. W.
EXPLICIT ADDITION OF RIGID-BODY MOTIONS IN CURVED
FINITE ELEMENTS
AIAA J., VOL. 11, NO. 3, 1973, PP. 305-317.

73-0038 FRIED, I.
SHAPE FUNCTIONS AND ACCURACY OF ARCH FINITE
ELEMENTS
AIAA J., VOL. 11, NO. 3, 1973, PP. 287-281.

73-0039 FIX, G. J.
EIGENVALUE APPROXIMATION BY FINITE ELEMENT METHOD
ADVANCES IN MATH., VOL. 10, NO. 2, 1973, PP.
300-316.

73-0040 THOMAS, D. H.
WIXOM, J. A.
INTERACTIVE PROCEDURE FOR CURVILINEAR COORDINATION
OF A PLANAR REGION WITH APPLICATIONS TO
FINITE-ELEMENT METHODS
SIAM REVIEW, VOL. 15, NO. 1, 1973, 263 PP.

73-0041 STRANG, G.
PIECEWISE POLYNOMIALS AND FINITE ELEMENT METHOD
NOTICES OF THE AMER. MATH. SOC., VOL. 20, NO. 3,
1973, PP. A374-375. ALSO IN BULL. AM. MATH. SC.,
VOL. 79, NO. 6, 1973, PP. 1128-1137.

73-0042 GEORGE, A.
NESTED DISSECTION OF A REGULAR FINITE ELEMENT MESH
SIAM J. ON NUMERICAL ANALYSIS, VOL. 10, NO. 2,
1973, PP. 345-363.

73-0043 HOFFMAN, A. J.
MARTIN, M. S.
ROSE, D. J.
COMPLEXITY BOUNDS FOR REGULAR FINITE DIFFERENCE
AND FINITE ELEMENT GRIDS
SIAM J. ON NUMERICAL ANALYSIS, VOL. 10, NO. 2,
1973, PP. 364-369.

73-0044 NICOLAIDES, R. A.
CLASS OF FINITE ELEMENTS GENERATED BY LAGRANGE
INTERPOLATION - 2.
SIAM J. ON NUMERICAL ANALYSIS, VOL. 10, NO. 1,
1973, PP. 182-189.

73-0045 ZLAMAL, M.
CURVED ELEMENTS IN FINITE ELEMENT METHOD. 1.
SIAM J. ON NUMERICAL ANALYSIS, VOL. 10, NO. 1,
1973, PP. 229-240.

73-0046 BABUSKA, I.
FINITE ELEMENT METHOD WITH LAGRANGIAN MULTIPLIERS
NUMERISCHE MATHEMATIK, VOL. 20, NO. 3, 1973, PP.
179-192.

73-0047 DOUGLAS, J.
DUPONT, T.
FINITE ELEMENT COLLOCATION METHOD FOR QUASILINEAR
PARABOLIC EQUATIONS
MATHEMATICS OF COMPUTATION, VOL. 27, NO. 121,
1973, PP. 17-28.

73-0048 BUELL, W. R.
BUSH, B. A.
MESH GENERATION - SURVEY
J. OF ENGINEERING FOR INDUSTRY, VOL. 95, NO. 1,
1973, PP. 332-338.

73-0049 MCLAY, R. W.
BUTURLA, E. M.
ERROR BOUNDS IN AN OPTIMIZATION PROBLEM USING
FINITE-ELEMENT METHOD
J. APPL. MECH., VOL. 40, NO.1, 1973, PP. 204-208.
ALSO IN MECHANICAL ENGINEERING, VOL. 95,
NO. 6, 1973, P. 63.

73-0050 ODEN, J. T.
REDDY, J. N.
NOTE ON AN APPROXIMATE METHOD FOR COMPUTING
CONSISTENT CONJUGATE STRESSES IN ELASTIC
FINITE ELEMENTS
INTERNATIONAL JOURNAL FOR NUMERICAL METHODS IN
ENGINEERING (TO APPEAR) VOL. 6, NO. 1, 1973, PP.
55-61.

73-0051 YANG, T. Y.
SUN, C. T.
AXIAL-FLEXURAL VIBRATION OF FRAMEWORKS USING
FINITE-ELEMENT APPROACH
J. ACOUS. SOC. AMERICA., VOL. 53, NO. 1, 1973, PP.
137-146.

73-0052 MANSOUR, A.
FENTON, P. H.
STRUCTURAL ANALYSIS AND DESIGN OF A CATAMARAN
CROSS-STRUCTURE BY FINITE ELEMENT METHOD
NAV. ENG. J., VOL. 85, NO. 1, 1973, PP. 33-42.

73-0053 FARAH, J. W.
CRAIG, R. G.
FINITE ELEMENT STRESS ANALYSIS OF A RESTORED
AXISYMMETRIC FIRST MOLAR
J. DENTAL RES., VOL. 52, SER. 1, 1973, PP. 221.

73-0054 GERLACH, H. D.
FINITE ELEMENT METHOD IN ROLLING BEARING
ENGINEERING
BALL BEARING J., VOL. 1, NO. 175, 1973, PP. 1-10.

73-0055 BRUSA, L.
CIACCI, R.
TICOZZI, C.
TWO-DIMENSIONAL FINITE ELEMENT COMPUTER CODES FOR
STRESS ANALYSIS AND FIELD PROBLEMS
ENERGIA NUCLEARE, VOL. 20, NO. 3, 1973, PP.
154-170.

73-0056 BABUSKA, I.
ZLAMAL, M.
NONCONFORMING ELEMENTS IN FINITE ELEMENT METHOD
WITH PENALTY
SIAM JOURNAL ON NUMERICAL ANALYSIS 1973, VOL. 10,
NO. 5, PP. 863-875

73-0057 FRIED, I.
ACCURACY AND CONDITION OF CURVED (ISOPARAMETRIC)
FINITE ELEMENTS
JOURNAL OF SOUND AND VIBRATION 1973, VOL. 31, NO.
3, PP. 345-355

73-0058 PERRIAUX, J.
THREE DIMENSIONAL ANALYSIS OF COMPRESSIBLE
POTENTIAL FLOWS WITH THE FINITE ELEMENT METHOD
AVIONS MARCEL DASSAULT-BREGUET AVIATION, 1973, PP.
1-89.

73-0059 BIRD, T. S.
EVALUATION OF ATTENUATION FROM LOSSLESS
TRIANGULAR-FINITE-ELEMENT SOLUTIONS FOR
INHOMOGENEOUSLY FILLED GUIDING STRUCTURES
ELECTRONICS LETTERS 1973, VOL. 9, NO. 25, PP.
590-592.

73-0060 NAPOLITANO, L. G.
FINITE ELEMENT METHODS IN FLUID-DYNAMICS
AGARD-VKI LECTURE SERIES, ADVANCES IN NUMERICAL
FLUID DYNAMICS, VON KARMAN INSTITUTE FOR
FLUID DYNAMICS, BRUSSELS,MARCH 5-9, 1973.

73-0061 TAYLOR, C.
HOOD, P.
A NUMERICAL SOLUTION OF THE NAVIER-STOKES
EQUATIONS USING THE FINITE ELEMENT TECHNIQUE
INTER. J. OF COMPUTERS AND FLUIDS, VOL. 1, NO. 1,
1973, PP. 73-100.

73-0062 CHI CHEONG POON, D.
RESERVOIR SIMULATION BY FINITE ELEMENT METHOD
APRIL 1973, DEPARTMENT OF CHEMICAL ENGR.,
UNIVERSITY OF CALGARY, M.SC. THESIS.

73-0063 BAKER, A. J.
FINITE ELEMENT SOLUTION ALGORITHM FOR VISCOUS
INCOMPRESSIBLE FLUID DYNAMICS
INT. J. NUM. METH. ENG., VOL. 6, NO. 1, 1973, PP.
89-101.

73-0064 KING, I. P., ET AL
FINITE ELEMENT MODEL FOR TWO-DIMENSIONAL DENSITY
STRATIFIED FLOW
WATER RESOURCES ENGINEERS, INC., REPORT TO OWRR,
1973.

73-0065 NORTON, W. R.
FINITE ELEMENT MODEL FOR LOWER GRANITE RESERVOIR
U.S. ARMY CORPS. OF ENGINEERS, WATER RESOURCES
DIV., WALLA WALLA DISTRICT, 1973.

73-0066 ADAMEK, J. R.
AN AUTOMATIC MESH GENERATOR USING TWO AND THREE
DIMENSIONAL ISOPARAMETRIC ELEMENTS
M.S. THESIS, NAVAL POSTGRADUATE SCHOOL, MONTEREY,
CALIF., 1973.

73-0067 BRATANOW, T.
ECER, A.
KOBISKE, M.
FINITE ELEMENT ANALYSIS OF UNSTEADY INCOMPRESSIBLE
FLOW AROUND AN OSCILLATING OBSTACLE OF ARBITRARY
SHAPE
A.I.A.A. 11TH ANN. MEETING, WASHINGTON, D.C.,
JANUARY 1973, ALSO IN A.I.A.A. J., VOL. 11, NO.
11, NOVEMBER 1973, PP. 1471-1477.

73-0068 BRATANOW, T.
ECER, A.
COMPUTATIONAL CONSIDERATIONS IN APPLICATION OF THE
FINITE ELEMENT METHOD FOR ANALYSIS OF UNSTEADY
FLOW AROUND AIRFOILS
PRESENTED AT AIAA COMPUT. FLUID DYN. CONF., PALM
SPRINGS, CALIF., JULY 1973.

73-0069 BRATANOW, T.
ECER, A.
SUITABILITY OF THE FINITE ELEMENT METHOD FOR
ANALYSIS OF UNSTEADY FLOW AROUND OSCILLATING
AIRFOILS
INT. CONF. ON NUMER. METHODS IN FLUID DYN., UNIV.
OF SOUTHAMPTON, SEPTEMBER 26-28, 1973.

73-0070 TAYLOR, C.
DANTS, J.
FINITE ELEMENT NUMERICAL MODELLING OF FLOW AND
DISPERSION IN ESTUARIES
I.A.H.R. IN. SYMP. ON RIVER MECH., VOL. 3, 1973,
PP. 1023-1038.

73-0071 STRANG, G.
FIX, G. F.
AN ANALYSIS OF THE FINITE ELEMENT METHOD
PRENTICE HALL ENGLEWOOD CLIFFS, 1973.

73-0072 BARNHILL, R. E.
WHITEMAN, J. R.
ERROR ANALYSIS OF FINITE ELEMENT METHODS WITH
TRIANGLES FOR ELLIPTIC BOUNDARY VALUE PROBLEMS
THE MATHEMATICS OF FINITE ELEMENTS AND
APPLICATIONS, WHITEMAN, J. R. (EDITOR),
ACADEMIC PRESS, LONDON, 1973, PP. 83-112.

73-0073 GREGORY, J. A.
WHITEMAN, J. R.
LOCAL MESH REFINEMENT WITH FINITE ELEMENTS FOR
ELLIPTIC PROBLEMS
BRUNEL UNIV., DEPT. OF MATH., REPORT TR/24, 1973.

73-0074 HUEBNER, K. H.
APPLICATION OF FINITE ELEMENT METHODS TO
THERMOHYDRODYNAMIC LUBRICATION
GENERAL MOTORS RES. PUBL. GMR-1404, MAY 1973.

73-0075 SMITH, R. R.
HUNT, J. T.
BARACH, D. R.
FINITE ELEMENT ANALYSIS OF ACOUSTICALLY RADIATING
STRUCTURES WITH APPLICATIONS TO SONAR TRANSDUCERS
J. ACOUST. SOC. AM., VOL. 54, NO. 5, 1973, PP.
1277-1288.

73-0076 BELYTSCHKO, T.
KENNEDY, J. M.
RESPONSE OF REACTOR CORE SUBASSEMBLIES TO
IMPULSIVE LOADINGS
A.S.M.E. MEETING, PAPER NO. 73-DET-93, SEPTEMBER
9-12, 1973, 8 PP.

73-0077 PINDER, G. F.
FRIND, E. O.
PAPADOPULOUS, S. S.
FUNCTIONAL COEFFICIENTS IN THE ANALYSIS OF
GROUNDWATER FLOW USING FINITE ELEMENTS
WATER RESOURCES RESEARCH, VOL. 9, NO. 1, FEBRUARY
1973, PP. 222-226.

73-0078 MEISSNER, U.
A MIXED FINITE ELEMENT MODEL FOR USE IN POTENTIAL
FLOW PROBLEMS
INTERNATIONAL JOURNAL FOR NUMERICAL METHODS IN
ENGINEERING, VOL. 6, NO. 4, 1973, PP. 467-473.

73-0079 SCOTT, R.
FINITE ELEMENT CONVERGENCE FOR SINGULAR DATA
NUMERISCHE MATHEMATIK 1973, VOL. 21, NO. 4, PP.
317-327

73-0080 ZLAMAL, M.
THE FINITE ELEMENT METHOD IN DOMAINS WITH CURVED
BOUNDARIES
INT., J. NUM. METHODS IN ENGINEERING, VOL. 5.,
367-373 1973.

73-0081 KADEN, S. T.
LUCKNER, L.
QUAST, J.
PRAKTISCHE BEISPIELE DER DIGITALEN BERECHNUNG VON
GRUNDWASSERSTROMUNGS PROBLEMEN NACH DER METHODE
ENDLICHER ELEMENTE (DIGITAL COMPUTATION EXAMPLES
OF GROUND-WATER FLOW PROBLEMS BY THE FINITE
ELEMENT METHOD)
NEUE BERGBAUTECHNIK, VOL. 3, NO. 6, 1973, PP.
424-431.

73-0082 FENG, G. C.
JONES, C. E.
DYNAMICS OF A FLEXIBLE BULKHEAD AND CONTAINED
FLUID
FINAL REPORT, LOCKHEED MISSILES AND SPACE COMPANY,
MAY 1973,PP. 1-69.

73-0083 ALUJEVIC, A.
HEAD, J. L.
FUEL ELEMENT STRESSES AT DISCONTINUITIES AND
INTERACTIONS BY FINITE ELEMENTS IN TWO DIMENSIONS
ATOMKERNENERGIE, VOL. 21, NO. 2, 1973, PP. 75-80.

73-0084 ANDERSEN, D. W.
LAPLACIAN ELECTROSTATIC FIELD CALCULATIONS BY
FINITE ELEMENTS WITH AUTOMATIC GRID
GENERATION
IEEE TRANSACTIONS ON POWER APPARATUS AND SYSTEMS,
VOL. PA92,NO. 5, 1973, PP. 1485-1492.

73-0085 ANDERSON, G. P.
DE VRIES, K. L.
WILLIAMS, M. L.
FINITE ELEMENT IN NUMERICAL ANALYSIS OF ADHESIVE
FRACTURE
INTL. J. FRACT., VOL. 9, NO. 3, 1973, PP. 335-336.

73-0086 ANDERSSON, H.
FINITE ELEMENT REPRESENTATION OF STABLE
CRACK-GROWTH
J. MECHS. AND PHYSICS OF SOLIDS, VOL 21, NO. 5,
1973, PP. 337-356.

73-0087 ANDERSSON, H.
STEADILY GROWING ELASTIC-PLASTIC CRACK TIP IN A
FINITE ELEMENT TREATMENT
INT. J. OF FRACT., VOL. 9, NO. 2, 1973, PP.
231-233.

73-0088 ASHWELL, D. G.
BEHAVIOR WITH DIMINISHING CURVATURE OF
STRAIN-BASED ARCH FINITE ELEMENTS
J. SOUND AND VIBRATION, VOL. 28, NO. 1, 1973, PP.
133-137.

73-0089 ATLURI, S.
ASSUMED STRESS HYBRID FINITE ELEMENT MODEL FOR
LINEAR ELASTODYNAMIC ANALYSIS
A.I.A.A. J., VOL. 11, NO. 7, 1973, PP. 1028-1031.

73-0090 BELYTSCHKO, T.
KULAK, R. F.
FINITE ELEMENT METHOD FOR A SOLID ENCLOSING AN
INVISCID INCOMPRESSIBLE FLUID
J. APPL. MECH., VOL. 40, NO. 2, 1973, PP. 609-610.

73-0091 BHANDARI, D. R.
ODEN, J. T.
GENERAL MIXED FINITE ELEMENT METHODS OF NONLINEAR
CONTINUA
ZEITSCHRIFT FUR ANGEWANDTE MATHEMATIK UND MECHANIK
VOL. 53, ISS. N8, 1973, PP. 441-451.

73-0092 BOYD, T. J.
GARDNER, G. A.
GARDNER, L. R.
NUMERICAL STUDY OF HYDROGMAGNETIC STABILITY USING
FINITE ELEMENT METHOD
NUCLEAR FUSION, VOL. 13, NO. 5, 1973, PP. 764-766.

73-0093 PINDER, G. F.
GALERKIN-FINITE ELEMENT SIMULATION OF GROUNDWATER
CONTAMINATION ON LONG-ISLAND, NEW-YORK
WATER RESOURCES RESEARCH 1973, VOL. 9 NO. 6, PP.
1657-1669

73-0094 BUCK, K. E.
FINITE ELEMENTS FOR AXISYMMETRIC SOLIDS UNDER
ARBITRARY LOADINGS WITH NODES ON ORIGIN
A.I.A.A. J., VOL. 11, NO. 9, 1973, PP. 1357-1358.

73-0095 CARSLON, A. D.
KASPER, R. G.
TUCCHIO, M. A.
FINITE ELEMENT STRESS ANALYSIS OF A MULTICONDUCTOR
ARMOR CABLE
MARINE TECHNOL. SOC. J., VOL. 7, NO. 7, 1973, PP.
29-33.

73-0096 CRAGGS, A.
ACOUSTIC FINITE ELEMENT APPROACH FOR STUDYING
BOUNDARY FLEXIBILITY AND SOUND TRANSMISSION
BETWEEN IRREGULAR ENCLOSURES
JOURNAL OF SOUND AND VIBRATION, VOL. 30, ISS. N3,
1973, PP. 343-357.

73-0097 DAWE, D. J.
FREE VIBRATIONS OF FINITE ELEMENT PLATES SUBJECTED
TO COMPLEX MIDDLE-PLANE FORCE SYSTEMS
JOURNAL OF SOUND AND VIBRATION, VOL. 28, ISS. N4,
1973, PP. 759-762.

73-0098 DEPPE, L. O.
HANSEN, K. F.
APPLICATION OF FINITE ELEMENT METHOD TO
2-DIMENSIONAL DIFFUSION PROBLEMS
TRANSACTIONS OF THE AMERICAN NUCLEAR SOCIETY, VOL.
16, ISS. JUNE 1973, PP. 132-133.

73-0099 ECER, A.
FINITE ELEMENT ANALYSIS OF POSTBUCKLING BEHAVIOR
OF STRUCTURES
A.I.A.A. J., VOL. 11, NO. 11, 1973, PP.
1532-1538.

73-0100 EMERY, A. F.
USE OF SINGULARITY PROGRAMMING IN
FINITE-DIFFERENCE AND FINITE ELEMENT COMPUTATIONS
OF TEMPERATURE
J. OF HEAT TRANSFER, VOL. 95, NO. 3, 1973, PP.
344-351.

73-0101 ENGRAND, D.
BORDAS, J.
MATRIX ANALYSIS OF MULTILAYERED AND SANDWICH
SHELLS BY FINITE ELEMENT METHOD
RECHERCHE AEROSPATIALE, VOL. 1973, ISS. N2, 1973,
PP 109-118

73-0102 FIX, G. J.
GULATI, S.
WAKOFF, G. I.
USE OF SINGULAR FUNCTIONS WITH FINITE ELEMENT
APPROXIMATIONS
JOURNAL OF COMPUTATIONAL PHYSICS, VOL. 13, ISS.
N2, 1973, PP. 209-228.

73-0103 HARKEGARD, G.
APPLICATION OF FINITE ELEMENT METHOD ,TO CYCLIC
LOADING OF ELASTIC-PLASTIC STRUCTURES CONTAINING
EFFECTS
INTERNATIONAL JOURNAL OF FRACTURE, VOL. 9, ISS.
N3, 1973, PP. 322-323.

73-0104 HARTIG, D.
USE OF RECTANGULAR FINITE ELEMENTS
ZEITSCHRIFT FUR ANGEWANDTE MATHEMATIK UND
MECHANIK, VOL. 53, ISS. N4, 1973, PP. T135-136.

73-0105 HEISE, U.
FINITE-ELEMENT METHOD WITH ELEMENTS WITH CIRCULAR
RING   SECTOR FORM
ZEITSCHRIFT   FUR   ANGEWANDTE   MATHEMATIK   UND
MECHANIK, VOL. 53, ISS. N4, 1973, PP. T136-138.

73-0106 JOFRIET, J. C.
MCNEICE, G. M.
CSAGOLY, P.
FINITE ELEMENT ANALYSIS OF PRESTRESSED CONCRETE
VOIDED   BRIDGE DECKS
JOURNAL OF THE PRESTRESSED CONCRETE INSTITUTE,
VOL. 18,   ISS. N3, 1973, PP. 51-66.

73-0107 JOHNSON, C.
CONVERGENCE OF A MIXED FINITE-ELEMENT METHOD FOR
PLATE   BENDING PROBLEMS
NUMERISCHE MATHEMATIK, VOL. 21, ISS. N1, 1973, PP.
43-62.

73-0108 KANG, C. M.
HANSEN, K. F.
FINITE ELEMENT METHODS FOR REACTOR ANALYSIS
NUCLEAR SCIENCE AND ENGINEERING, VOL. 51, ISS. N4,
1973,   PP. 456-495.

73-0109 KARCHER, H. J.
TWO DUAL FINITE ELEMENT METHODS DERIVED FROM
EXTENDED VARIATIONAL PRINCIPLE
ZEITSCHRIFT   FUR   ANGEWANDTE   MATHEMATIK   UND
MECHANIK, VOL. 53, NO. 4, 1973, PP. T138-T139.

73-0110 KELLOGG, R. B.
SOME MATHEMATICAL ASPECTS OF FINITE ELEMENT
CALCULATIONS
TRANSACTIONS OF THE AMERICAN NUCLEAR SOCIETY, VOL.
16, ISS. JUN., 1973, P. 131.

73-0111 KIKUCHI, F.
OHYA, H.
ANDO, Y.
APPLICATION   OF   FINITE   ELEMENT   METHOD   TO
AXISYMMETRIC   BUCKLING   OF   SHALLOW   SPHERICAL
SHELLS UNDER EXTERNAL PRESSURE
JOURNAL OF NUCLEAR SCIENCE AND TECHNOLOGY, TOKYO,
VOL. 10,   ISS. N6, 1973, PP. 339-347.

73-0112 KITAGAWA, H.
ISIDA, M.
KURODA, M.
YUUKI, R.
SOME EXAMINATIONS ON AVAILABILITY OF FINITE
ELEMENT METHODS IN LINEAR FRACTURE MECHANICS
INTL. J. OF FRACT., VOL. 9, ISS. NO. 3, 1973,   PP.
337-338.

73-0113 KONRAD, A.
SILVESTER, P.
FINITE ELEMENT PROGRAM PACKAGE FOR AXISYMMETRIC
SCALAR   FIELD PROBLEMS
COMPUTER PHYSICS COMMUNICATIONS, VOL. 5, ISS. N6,
1973,   PP. 437-455.

73-0114 LAGASSE, P. E.
HIGHER-ORDER   FINITE-ELEMENT   ANALYSIS   OF
TOPOGRAPHIC GUIDES SUPPORTING ELASTIC SURFACE
WAVES
JOURNAL OF THE ACOUSTICAL SOCIETY OF AMERICA, V/I.
53, N4,   1973, PP. 1116-1122.

73-0115 LAGASSE, P. E.
FINITE ELEMENT ANALYSIS OF PIEZOELECTRIC ELASTIC
WAVEGUIDES
IEEE TRANSACTIONS ON SONICS AND ULTRASONICS, VOL.
SU20,   ISS. N4, 1973, PP. 354-359.

73-0116 LESAINT, P.
FINITE ELEMENT METHODS FOR SYMMETRIC HYPERBOLIC
EQUATIONS
NUMERISCHE MATHEMATIK, VOL. 21, ISS. N3, 1973, PP.
244-255.

73-0117 MAGHSOOD, J.
SCOTT, N. R.
FURRY, R. B.
SEXSMITH, R.
LINEAR AND NONLINEAR FINITE ELEMENT ANALYSIS OF
WOOD   STRUCTURAL MEMBERS
TRANS. OF AM. SOC. OF AGRICULTURAL ENGINEERS, VOL.
16,   ISS. N3, 1973, PP. 490-496.

73-0118 MALLETT, R. H.
DISNEY, R. K.
CURRENT FINITE ELEMENT ANALYSIS PRACTICES
TRANSACTIONS OF THE AMERICAN NUCLEAR SOCIETY, VOL.
16,   ISS. JUN, 1973, PP. 131-132.

73-0119 MARK, R.
ABEL, J. F.
ONEILL, K.
PHOTOELASTIC AND FINITE-ELEMENT ANALYSIS OF A
QUADRIPARTITE VAULT
EXPERIMENTAL MECHANICS, VOL. 13, ISS. N8, 1973,
PP. 322-329.

73-0120 PIAN, T. H. H.
MAU, S. T.
TONG, P.
VIBRATION ANALYSIS OF LAMINATED PLATES AND SHELLS
BY A   HYBRID STRESS ELEMENT
A.I.A.A., J., VOL. 11, NO. 10, 1973, PP.
1450-1452.

73-0121 DECRETON, M. C.
MCDONALD, B. H.
FRIEDMAN, M.
WEXLER, A.
INTEGRAL FINITE-ELEMENT APPROACH FOR SOLVING
LAPLACE   EQUATION
ELECTRONICS LETTERS, VOL. 9, ISS. N11, 1973, PP.
242-244.

73-0122 MEI, C.
YANG, T. Y.
FREE VIBRATIONS OF FINITE ELEMENT PLATES SUBJECTED
TO   COMPLEX   MIDDLE-PLANE   FORCE   SYSTEMS   -
REPLY
JOURNAL OF SOUND AND VIBRATION, VOL. 28, ISS. N4,
1973,   P. 763.

73-0123 MILLER, W. F.
LEWIS, E. E.
ROSSOW, E. C.
APPLICATION OF PHASE-SPACE FINITE ELEMENTS TO ONE
DIMENSIONAL NEUTRON TRANSPORT EQUATION
NUCLEAR SCIENCE AND ENGINEERING, VOL. 51, NO. 2,
1973, PP. 148-156.

73-0124 MORLEY, L. S. D.
FINITE ELEMENT SOLUTION OF BOUNDARY-VALUE PROBLEMS
WITH   NON-REMOVABLE SINGULARITIES
PHILOSOPHICAL TRANSACTIONS ROYAL SOCIETY LONDON,
SER A,   VOL. 275, ISS. N1252, 1973, PP.
463-488.

73-0125 ODEN, J. T.
AKAY, H. U.
JOHNSON, C. P.
EFFECT OF HIGHER ORDER TERMS IN CERTAIN NONLINEAR
FINITE   ELEMENT MODELS
A.I.A.A. J., VOL. 11, NO. 11, 1973, PP. 1589-1590.

73-0126 OHTA, M.
FUJISAWA, N.
ITOH, S.
ANALYSIS OF MAGNETIC FIELDS BY FINITE ELEMENT
METHOD
ELECTRICAL ENGINEERING IN JAPAN, VOL. 92, ISS. N3,
1973,   PP. 9-16.

73-0127 PATTERSON, C.
IMPROVED FINITE   ELEMENT   METHOD   FOR   STRESS
CONCENTRATIONS
INTERNATIONAL JOURNAL OF FRACTURE, VOL. 9, ISS.
N3, 1973, PP. 336-337.

73-0128 SILVENNOINEN, P.
PITKARANTA, J.
FINITE ELEMENT ANALYSIS OF SOME CRITICAL FAST
ASSEMBLIES
NUCLEAR SCIENCE AND ENGINEERING, VOL. 52, ISS. N4,
1973,   PP. 447-453.

73-0129 PRICE, H. L.
MURRAY, K. H.
FINITE ELEMENT ANALYSIS OF DIAMETRAL TEST OF
POLYMER   MOLDINGS
JOURNAL OF ENGINEERING MATERIALS AND TECHNOLOGY,
VOL. 95, ISS. NO. 3, 1973, PP. 186-191. ALSO
ASME PAPER NO. 73-MAT-H, 1973, 6 PP.

73-0130 REDDY, I. K.
RANKIN, D.
MAGNETOTELLURIC RESPONSE OF A 2-DIMENSIONAL
SLOPING CONTACT BY FINITE ELEMENT METHOD
PURE AND APPLIED GEOPHYSICS, VOL. 105, ISS. N4,
1973,   PP. 847-857.

73-0131 ROSSOW, M. P.
TAYLOR, J. E.
FINITE ELEMENT METHOD FOR OPTIMAL DESIGN OF
VARIABLE   THICKNESS SHEETS
A.I.A.A. J., VOL. 11, NO. 11, 1973, PP. 1566-1569.

73-0132 SEGAL, V. M.
SVIRID, G. P.
NUMERICAL   INVESTIGATION   OF   AXI-SYMMETRIC
VISCOPLASTIC FLOWS BY FINITE-ELEMENT PROCEDURE
DOKLADY AKADEMII ANUK BSSR, VOL. 17, ISS. N9,
1973,   PP. 819-821.

73-0133 SHUKU, T.
ISHIHARA, K.
ANALYSIS OF ACOUSTIC FIELD IN IRREGULARLY SHAPED
ROOMS BY FINITE ELEMENT METHOD
JOURNAL OF SOUND AND VIBRATION, VOL. 29, ISS. N1,
1973,   PP. 67-76.

73-0134 SILVESTER, P.
CABAYAN, H. S.
BROWNE, B. T.
EFFICIENT TECHNIQUES FOR FINITE ELEMENT ANALYSIS
OF ELECTRIC MACHINES
IEEE TRANSACTIONS ON POWER APPARATUS AND SYSTEMS,
VOL. PA92, NO. 4, 1973, PP. 1274-1281.

73-0135 SILVESTER, P.
KONRAD, A.
ANALYSIS OF TRANSFORMER LEAKAGE PHENOMENA BY
HIGH-ORDER FINITE ELEMENTS
IEEE TRANSACTIONS ON POWER APPARATUS AND SYSTEMS,
VOL. PA92,ISS. N6, 1973, PP. 1843-1855.

73-0136 SIMITSES, G. J.
KAMAT, M. P.
SMITH, C. V.
STRONGEST COLUMN BY FINITE ELEMENT DISPLACEMENT
METHOD
A.I.A.A. J., VOL. 11, NO. 9, 1973, PP. 1231-1232.

73-0137 STONE, G. O.
HIGH-ORDER FINITE ELEMENTS FOR INHOMOGENEOUS
ACOUSTIC GUIDING STRUCTURES
IEEE TRANSACTIONS ON MICROWAVE THEORY AND
TECHNIQUES, VOL. MT21, ISS. N8, 1973, PP.
538-542.

73-0138 WU, R. W.
WITMER, E. A.
NONLINEAR TRANSIENT RESPONSES OF STRUCTURES BY
SPATIAL FINITE-ELEMENT METHOD
A.I.A.A., J., VOL. 11, NO. 8, 1973, PP. 1110-1117.

73-0139 YALAMANCHILI, R.
CHU, S. C.
STABILITY AND OSCILLATION CHARACTERISTICS OF
FINITE ELEMENT, FINITE DIFFERENCE AND WEIGHTED
RESIDUAL METHODS
J. OF HEAT TRANSFER, VOL. 95, NO. 2, 1973, PP.
235-239.

73-0140 ZIEGLER, F.
PLANAR FINITE ELEMENT AT RANDOM TEMPERATURE
ZEITSCHRIFT FUR ANGEWANDTE MATHEMATIK UND
MECHANIK, VOL. 53,ISS. N4, 1973, PP. TU53-154.

73-0141 ZIENKIEWICZ, O. C.
PRACTICAL HINTS ON HOW TO UNDERSTAND, MODEL, USE
AND INTERPRET FINITE ELEMENTS
PROC INSTITUTION CIVIL ENGINEERS PT 2 RESEARCH &
THEORY, VOL. 55, ISS. JUN, 1973, PP. 549-551.

73-0142 FARAH, J. W.
CRAIG, R. G.
SIKARSKIE, D. L.
PHOTOELASTIC AND FINITE ELEMENT STRESS ANALYSIS OF
A RESTORED AXISYMMETRIC FIRST MOLAR
JOURNAL OF BIOMECHANICS, VOL. 6, ISS. N5, 1973,
PP. 511-520.

73-0143 WYNN, J. C.
STURGUL, J. R.
LAZEAR, G. D.
SOME APPLICATIONS OF FINITE ELEMENTS TO MODEL
GEOLOGICAL STRUCTURES
TRANSACTIONS, AMERICAN GEOPHYSICAL UNION, VOL. 54,
ISS. N11,1973, P. 1207.

73-0144 THOMPSON, E. G.
HAQUE, M. I.
HIGH ORDER FINITE ELEMENT FOR COMPLETELY
INCOMPRESSIBLE CREEPING FLOW
INT. J. NUMER. METHODS ENG., V. 6, N. 3, 1973, PP.
315-321.

73-0145 TIROSH, J.
EFFECT OF PLASTICITY AND CRACK BLUNTING ON THE
STRESS DISTRIBUTION IN ORTHOTROPIC COMPOSITE
MATERIALS
ASME PAPER N. 73-APMW-2 FOR MEET. SEP. 17-19 1973,
6 P.

73-0146 WOZNIAK, CZ.
BASIC CONCEPTS OF THE MECHANICS OF DISCRETIZED
BODIES WITH AN INTRODUCTION TO DISCRETE ELEMENT
CALCULUS.
ARCH MECH - ARCH. MECH. STOSOW., V. 25, N. 1,
1973, PP. 87-102.

73-0147 YOSHIDA, Y.
AMENIYA, E.
FLAT FINITE ELEMENT FOR THIN SHELL ANALYSIS
DERIVED BY ASSUMED STRESS APPROACH
PROC. JAP. SOC. CIV., ENG., NO. 211, MARCH 1973,
PP. 19-28.

73-0148 YAMAMOTO, Y.
TOKUDA, N.
DETERMINATION OF STRESS INTENSITY FACTORS IN
CRACKED PLATES BY THE FINITE ELEMENT METHOD
INT. J. NUMER. METHODS ENG., VOL. 6, NO. 3, 1973,
PP. 427-439.

73-0149 ATKINS, J. T.
PASHA, M. A.
CONTROLLING OPEN-PIT SLOPE FAILURES AT SHIRLEY
BASIN
MINING ENG. (N.Y.), V. 25, N. 6, JUN 1973, PP.
38-42.

73-0150 HOOPER, J. A.
FIELD INSTRUMENTATION FOR LONG TERM MEASUREMENT OF
PILE LOAD AND RAFT CONTACT PRESSURE
CIV. ENG. PUBLIC WORKS REV., VOL. 68, NO. 802, MAY
1973, PP. 438-446.

73-0151 KONDERLA, P.
SOLUTION OF A SINGLE HYPERBOLOID SHELL BY FINITE
ELEMENT METHOD
ARCH. INZ. LADOWEJ., V. 19, N. 1, 1973, PP.
181-191.

73-0152 SEN, S. K.
GOULD, P. L.
CRITERIA FOR FINITE ELEMENT DISCRETIZATION OF
SHELLS OF REVOLUTION
INT. J. NUMER. METHODS ENG., V. 6, N. 2, 1973, PP.
265-274.

73-0153 RYDZEWSKI, J. R.
HUMPHRIES, R. W.
APPROACH TO ANALYSIS OF ARCH DAMS IN WIDE VALLEYS
ASCE. J. POWER. DIV. V. 99. N. PO1, MAY 1973,
PAP. 9710, PP. 165-174.

73-0154 RAND, R. A.
SHEN, C. N.
OPTIMUM DESIGN OF COMPOSITE SHELLS SUBJECT TO
NATURAL FREQUENCY CONSTRAINTS
COMPUT. STRUCT., V. 3, N. 2, MAR. 1973, PP.
247-263.

73-0155 RADAJ, D.
ACCURACY OF THE FINITE ELEMENT ANALYSIS FOR THE
ELASTIC PLATE WITH A CIRCULAR HOLE
INT. J. NUMER. METHODS ENG., V. 6, N. 3, 1973, PP.
443-447.

73-0156 PAZ, M.
MATHEMATICAL OBSERVATIONS IN STRUCTURAL DYNAMICS
COMPUT. STRUCT., VOL. 3, NO. 2, MARCH 1973, PP.
385-396.

73-0157 PARDOEN, G. C.
STATIC, VIBRATION AND BUCKLING ANALYSIS OF
AXISYMMETRIC CIRCULAR PLATES USING FINITE
ELEMENTS
COMPUT. STRUCT., V. 3, N. 2, MAR. 1973, PP.
355-375.

73-0158 O'DONNELL, R.
COMPARISON OF STRUCTURAL TEST RESULTS WITH
PREDICTIONS OF FINITE ELEMENT ANALYSIS
PREPR. NO. 730340, S.A.E. MEETING, APRIL 3-6,
1973, 13 PP.

73-0159 ODEN, J. T.
FOST, R. B.
CONVERGENCE, ACCURACY AND STABILITY OF FINITE
ELEMENT APPROXIMATIONS OF A CLASS OF
NON-LINEAR HYPERBOLIC EQUATIONS
INT. J. NUMER. METHODS ENG., V. 6, N. 3, 1973, PP.
357-365.

73-0160 NEKI, I.
NAGAI, K.
3-DIMENSIONAL ANALYSIS OF HULL STRUCTURE BY FINITE
ELEMENT METHOD
IHI. ENG. REV., V. 6, N. 1, MAR. 1973, PP. 12-24.

73-0161 LASHKARI, M.
WEINGARTEN, V. I.
VIBRATIONS OF SEGMENTED SHELLS
EXP. MECH., V. 13, N. 3, MAR. 1973, PP. 120-125.

73-0162 KOENIG, H. A.
BERRY, G. F.
TRANSIENT RESPONSE OF NON-UNIFORM, NON-HOMOGENEOUS
BEAMS
INT. J. MECH. SCI., V. 15, N. 5, MAY 1973, PP.
339-413.

73-0163 JONES, R. E.
SHELL AND PLATE ANALYSIS BY FINITE ELEMENTS
ASCE. J. STRUCT DIV. V. 99, N. ST5, MAY 1973,
PAP. 9744, PP. 889-902.

73-0164 HELLAN, K.
ON THE UNITY OF THE CONSTANT STRAIN-CONSTANT
MOMENT FINITE ELEMENT METHODS
INT. J. NUMER. METHODS ENG., V. 6, N. 2, 1973, PP.
191-200.

73-0165 GRIEGER, I.
COMPUTER AIDED DESIGN AND CONSTRUCTION BY MEANS OF
CURVED SURFACE AND SOLID ELEMENTS
ANGEW. INF., APPL. INFORMATICS, VOL. 15, NO. 3,
MARCH 1973, PP. 101-113.

73-0166 GOULD, P. L.
CATALOGLU, A.
STRESS ANALYSIS OF THE HUMAN AORTIC VALVE
COMPUT. STRUCT., V. 3, N. 2, MAR. 1973, PP.
377-384.

73-0167 FUKII, T.
YUKI, T.
STUDY ON FINITE ELEMENT METHOD FOR STRUCTURAL
ANALYSIS - ISTRAN/PL (INH STRUCTURE ANALYSIS PLATE
STRUCTURE, LINEAR ANALYSIS) AND ITS APPLICATION
IHI. ENG. REV., VOL. 6, NO. 1, MARCH 1973, PP.
1-11.

73-0168 FOYE, R. L.
THEORETICAL POST YIELDING BEHAVIOUR OF COMPOSITE
LAMINATES - 1. INELASTIC MICROMECHANICS
J. COMPOS. MATER., VOL. 7, APRIL 1973, PP.
178-193.

73-0169 DVORAK, G. J.
RAO, M. S. M.
YIELDING IN UNIDIRECTIONAL COMPOSITES UNDER
EXTERNAL LOADS AND TEMPERATURE CHANGES.
J. COMPOS. MATER., V. 7, APR. 1973, PP. 194-216.

73-0170 DONEA, J.
GIULIANI, S.
CREEP ANALYSIS OF TRANSVERSELY ISOTROPIC BODIES
SUBJECTED TO TIME DEPENDENT LOADING
NUCL. ENG. DES., VOL. 24, NO. 3, MARCH 1973, PP.
410-419.

73-0171 DALY, P.
FINITE ELEMENTS FOR FIELD PROBLEMS IN CYLINDRICAL
CO-ORDINATES
INT. J. NUMER. METH. ENG., VOL. 6, NO. 2, 1973,
PP. 169-178.

73-0172 WETENKAMP, H. R.
THERMAL STRESS DEVELOPED IN S PLATE, STRAIGHT
PLATE, AND DEEP DISH WHEELS
ASME PAPER N. 73-RT-1 FOR MEET. APR. 11-12 1973, 8
P.

73-0173 TONG, P.
PIAN, T. H. H.
ON THE CONVERGENCE OF THE FINITE ELEMENT METHOD
FOR PROBLEMS WITH SINGULARITY
INT. J. SOLIDS STRUCT., VOL. 9, NO. 3, MARCH 1973,
PP. 313-321.

73-0174 SUN, C. T.
YANG, T. Y.
CONTINUUM APPROACH TOWARD DYNAMICS OF GRIDWORKS
J. APPL. MECH., TRANS. ASME., V. 40, SER. E, N. 1,
MAR. 1973, PP. 186-192.

73-0175 STOYKOVICH, M.
EVALUATION OF TORSIONAL EFFECTS IN THE SEISMIC
ANALYSIS OF SYMMETRIC AND UNSYMMETRIC NUCLEAR
POWER PLANT STRUCTURES
NUCL. ENG. DES., V. 24, N. 1, JAN. 1973, PP.
110-124.

73-0176 SKJOLINGSTAD, L.
CHEUNG, Y. K.
ANALYSIS OF ARCH DAMS USING A SPACE-FRAME MODEL
WATER POWER, V. 25, N. 2, FEB. 1973, PP. 51-55.

73-0177 SCAVUZZO, R. J.
ISENBERG, J.
EFFECTS OF FOUNDATION ROTATION ON SEISMIC INERTIA
FORCES OF NUCLEAR POWER PLANT STRUCTURES
NUCL. ENG. DES., V. 24, N. 2, FEB. 1973, PP.
195-202.

73-0178 SAW, C. B.
SMITH, R. G.
AUTOMATIC NODAL TRIANGULATION FOR FINITE ELEMENTS
COMPUT. AIDED DESIGN, VOL. 5, NO. 1, JANUARY 1973,
PP. 33-35.

73-0179 ROSEN, R.
INTRODUCTION TO FINITE ELEMENT ANALYSIS
ASME. PAPER N. 73-DE-17 FOR MEET. APR. 9-12 1973,
8 P.

73-0180 ROESSET, J. M.
HARMON, T. G.
SOME STRUCTURAL PROBLEMS - STANDARD OIL OF INDIANA
BUILDING
ASCE. J. STRUCT. DIV., VOL. 99, NO. ST4., APRIL
1973, PAPER NO. 964, PP. 637-654.

73-0181 PRASAD, K.S.R.K.
KRISHNA MURTY, A. V.
FINITE ELEMENT ANALOGUE OF THE MODIFIED RAYLEIGH
RITZ METHOD FOR VIBRATION PROBLEMS
INT. J. NUMER. METHODS ENG., VOL. 5, NO. 2,
NOVEMBER-DECEMBER 1972, PP. 163-169.

73-0182 PATEL, K. S.
LOGCHER, R. D.
COMPUTER - AIDED DESIGN - STANDARD OIL OF INDIANA
BUILDING
ASCE. J. STRUCT. DIV., V. 99, N. ST4, APR. 1973,
PAPER N. 9643, PP. 621-635.

73-0183 ODEN, J. T.
SOME RESULTS OF FINITE ELEMENT APPLICATIONS IN
FINITE ELASTICITY
COMPUT. STRUCT., V. 3, N. 1, JAN. 1973, PP.
175-194.

73-0184 MORRIS, A. J.
DEFICIENCY IN CURRENT FINITE ELEMENTS FOR THIN
SHELL APPLICATIONS
INT. J. SOLIDS STRUCT., V. 9, N. 3, MAR. 1973, PP.
331-346.

73-0185 MEI, C.
FINITE ELEMENT DISPLACEMENT METHOD FOR LARGE
AMPLITUDE FREE FLEXURAL VIBRATIONS OF BEAMS AND
PLATES
COMPUT. STRUCT., V. 3, N.1, JAN. 1973, PP. 163-174

73-0186 MAK, C. K.
KAO, D. W.
FINITE ELEMENT ANALYSIS OF BUCKLING AND
POST-BUCKLING BEHAVIORS OF ARCHES WITH
GEOMETRIC IMPERFECTIONS
COMPUT. STRUCT., V. 3, N. 1, JAN. 1973, PP.
149-161.

73-0187 KAMAT, M. P.
SIMITSES, G. J.
OPTIMAL BEAM FREQUENCIES BY THE FINITE ELEMENT
DISPLACEMENT METHOD
INT. J. SOLIDS STRUCT., V. 9, N. 3, MAR. 1973, PP.
415-429.

73-0188 JOHNS, J. T.
ECK, D. L.
ECONOMICAL DESIGN ANALYSIS OF PISTONS FOR INTERNAL
COMBUSTION ENGINES
ASME. PAPER NO. 73-DGP-16 FOR MEETING APRIL 1-4,
1973, 8 PP.

73-0189 HAMANN, W. C.
INTERFACING FINITE ELEMENT METHODS WITH PRODUCT
DESIGN ENGINEERING
ASME PAPER N. 73-DE-24 FOR MEET. APR. 9-12 1973,
11 P.

73-0190 AGARWAL, A. C.
HUGGINS, M. W.
DIFFERENTIAL EXPANSION IN ELASTIC LAMINATES
ASCE. J. STRUCT. DIV., V. 99, N. ST4, APR. 1973,
PAPER N. 9653, PP. 655-663.

73-0191 YAMAGAMI, T.
ANALYSIS OF QUICK SAND PHENOMENON BY FINITE
ELEMENT METHOD
PROC. JAP. SOC. CIV. ENG. N. 213, MAY 1973, PP.
29-40.

73-0192 WILSON, E. A.
PARSONS, B.
TRAPEZOIDAL FINITE ELEMENTS: THEIR DERIVATION AND
USE FOR AXISYMMETRIC ROTATING BODIES
ASME PAPER NO. 73 DET-47, FOR MEETING SEPTEMBER
9-12, 1973, 8 PP.

73-0193 WILSON, E. L.
FARHOOMAND, I.
NONLINEAR DYNAMIC ANALYSIS OF COMPLEX STRUCTURES
EARTHQUAKE ENG. STRUCT. DYN., V. 1, N. 3, JAN-MAR.
1973, PP. 241-252.

73-0194 BERGER, A. E.
L2 ERROR ESTIMATES FOR FINITE ELEMENTS WITH
INTERPOLATED BOUNDARY CONDITIONS
NUMERISCHE MATHEMATIK 1973, VOL. 21, NO. 4, PP.
345-348.

73-0195 TAYLOR, S.
COMPUTERS AND MECHANICAL DESIGN
COMPUT. AIDED. DES., V. 5, N. 3, JUL. 1973, PP.
174-176.

73-0196 SINGH, B.
CONTINUUM CHARACTERIZATION OF JOINTED ROCK MASSES
- 1, 2
INT. J. ROCK MECH. MIN. SCI., V. 10, N. 4, JULY
1973, PP. 311-349.

73-0197 SAINI, S. S.
KRISHNA, JAI
A SEISMIC STRENGTH OF KOLKEWADI DAM
EARTHQUAKE ENG. STRUCT. DYN., V. 1, N. 3, JAN-MAR.
1973, PP. 225-240.

73-0198 RAMEY, M. R.
TATTERSHALL, D. W.
REINFORCING REQUIREMENTS FOR CONCRETE BEAMS WITH
LARGE WEB OPENINGS
HIGHW. RES. REC. N. 728, 1973, PP. 5-21.

73-0199 PETERSON, W. S.
KOSTEM, C. N.
DYNAMIC ANALYSIS OF BEAM-SLAB HIGHWAY BRIDGES
HIGHW. RES. REC. N. 428, 1973, PP. 57-63.

73-0200 MERZ, K. L.
NONLINEAR SEISMIC ANALYSIS OF CURRENT DIVIDER
SUPPORT SYSTEM FOR PACIFIC DC INTERTIE
ASME. PAPER NO. 73-DET-63 FOR MEETING SEPTEMBER
9-12, 1973.

73-0201 KEMAIRE, M.
COURTADE, R. M.
AUTOMATIC CALCULATION OF THIN STRUCTURES
ANN. INST. TECH. BATIM TRAV. PUBLICS, N. 306, JUNE
1973,   PP. 71-81.

73-0202 LEFEBVRE, G.
DUNCAN, J. L.
THREE-DIMENSIONAL FINITE ELEMENT ANALYSES OF DAMS
ASCE. J. SOIL MECH. FOUND. DIV., V. 99, N. SM7,
JUL. 1973, PAP. N.9857, PP. 495-507.

73-0203 LARSSON, S. G.
CARLSSON, A. J.
INFLUENCE OF NON-SINGULAR STRESS TERMS AND
SPECIMEN GEOMETRY ON SMALL SCALE YIELDING AT CRACK
TIPS IN ELASTIC PLASTIC MATERIALS
J. MECH. PHYS. SOLIDS, V. 21, N. 4, JUL. 1973, PP.
263-277.

73-0204 KROENKE, W. C.
GUTZWILLER, M. J.
FINITE ELEMENT FOR REINFORCED CONCRETE FRAME STUDY
ASCE. J. STRUCT. DIV., V. 99, N. ST7, JUL. 1973,
PAP. N.   9835, PP. 1371-1390.

73-0205 HENRYWOOD, R. K.
DESIGN, DEVELOPMENT, DOCUMENTATION AND SUPPORT OF
A MAJOR FINITE ELEMENT SYSTEM
COMPUT. AIDED. DES., V. 5, N. 3, JUL. 1973, PP.
160-165.

73-0206 HASSLER, G. R.
RYBICKI, E. F.
FINITE ELEMENT MODEL SYSTEM OF A RABBIT CALVARIUM:
AN INVESTIGATION OF OPTIMAL COMPRESSIVE LOADING
BIOMECH. SYMP., A.S.M.E. JT. MEETING, GEORGIA
INST. OF TECHNOL., ATLANTA, JUNE 20-22, 1973, PP.
83-84.

73-0207 HANS, D.
ZUNDERDORP, H. J.
STRESS ANALYSIS OF TUBULAR JOINTS FOR OFFSHORE
STRUCTURES
SOC. PET. ENG. AIME. PREPR. N. SPE. 4342 FOR MEET.
APR 2-3 1973, 8 PP.

73-0208 HAND, F. R.
PECKNOLD, D. A.
NONLINEAR LAYERED ANALYSIS OF RC PLATES AND SHELLS
ASCE. J. STRUCT. DIV., V. 99, N. ST7, JUL. 1973,
PAP. N.   9860, PP. 1491-1505.

73-0209 FRIED, I.
BOUNDARY AND INTERIOR APPROXIMATION ERRORS IN THE
FINITE ELEMENT METHOD
ASME. PAPER NO.   73-APM-AAA, 1973, 5 PP.
ALSO IN J. OF APPL. MECH., VOL. 40, NO.
4, 1973, PP. 1113-1117.

73-0210 CROUCH, T. G.
FAIRBAIRN, I. S.
GRIP: AN INFORMATION SYSTEM FOR DESIGN AUTOMATION
ASME. PAPER. N. 73-DET-38 FOR MEET. SEP. 9-12
1973, 9 PP.

73-0211 BUTURLA, E. M.
MCLAY, R. W.
EVALUATION OF THE ERROR BOUNDS IN AN OPTIMIZATION
PROBLEM USING THE FINITE-ELEMENT METHOD
ASME. PAPER. N. 73-APMW-15 FOR MEET. SEP. 17-19
1973, 4 PP.

73-0212 BOECKER, H.
REICHERT, K.
DIGITAL CALCULATION OF ELECTRIC FIELDS IN METAL
CLAD   INSTALLATIONS
ELEKTROTECH. Z. AUSG. A., V. 94, N. 7, JUL. 1973,
PP 374-377

73-0213 BHATT, P.
EFFECT OF BEAM-SHEARWALL JUNCTION DEFORMATIONS ON
THE   FLEXIBILITY OF THE CONNECTING BEAMS
BUILD. SCI., V. 8, N. 2, JUN. 1973, PP. 149-151.

73-0214 ATLURI, S.
KOBAYASHI, A. S.
MECHANICS OF BRAIN TISSUE FRAGILITY
BIOMECH. SYMP., PRESENTED AT ASME JT. MEET.,
GEORGIA INST. OF TECHNOL., ATLANTA, JUNE 20-22,
1973, PP. 103-104.

73-0215 ABEL, J. F.
MARK, R.
STRESSES AROUND FLEXIBLE ELLIPTIC PIPES
ASCE. J. SOIL MECH. FOUND. DIV., V. 99, N. SM7,
JUL. 1973, PAP. N.9858, PP. 509-526.

73-0216 YUZUGULLU, O.
SCHNOBRICH, W. C.
NUMERICAL PROCEDURES FOR THE DETERMINATION OF THE
BEHAVIOUR OF A SHEAR WALL FRAME SYSTEM
J. AM. CONCRETE INST., VOL. 70, NO. 7, JULY 1973,
PP. 474-479.

73-0217 SWEDLOW, J. L.
PROCEDURE FOR SOLVING PROBLEMS OF ELASTO-PLASTIC
FLOW
COMPUT. STRUCT., V. 3, N. 4, JUL. 1973, PP.
879-998.

73-0218 SUTLIFF, D. R.
PIH, H.
THREE DIMENSIONAL SCATTERED-LIGHT STRESS ANALYSIS
OF DISCONTINUOUS FIBER-REINFORCED COMPOSITES
EXP. MECH., VOL. 13, NO. 7, 1973, PP. 294-298.

73-0219 OHTE, S.
FINITE ELEMENT ANALYSIS OF ELASTIC CONTACT
PROBLEMS
BULL. JSME., VOL. 16, NO. 95, MAY 1973, PP.
797-804.

73-0220 MARUYAMA, K.
STRESS ANALYSIS OF A BOLT-NUT JOINT BY THE FINITE
ELEMENT METHOD AND THE COPPER-ELECTROPLATING
METHOD - 1. STRESS AT THE ROOT OF A V-GROOVED ROD
UNDER A TENSILE LOAD
BULL JSME., VOL. 16, NO. 94, APRIL 1973, PP.
671-678.

73-0221 MCCULLERS, L. A.
NABERHAUS, J. D.
AUTOMATED STRUCTURAL DESIGN AND ANALYSIS OF
ADVANCED   COMPOSITE WING MODELS
COMPUT. STRUCT., V. 3, N. 4, JUL. 1973, PP.
925-935.

73-0222 KIRICHEVSKII, V. V.
SAKHAROV, A. S.
INVESTIGATION OF THE STRESS-STRAINED STATE OF
SPHERICAL SHELLS WITH AN ECCENTRIC CUTOUT ON
THE BASIS OF THE SPACE THEORY OF ELASTICITY BY
THE FINITE ELEMENT METHOD
PROBL. PROCHN., V. 5, N. 2, FEB. 1973, PP. 91-94.

73-0223 JACOBS, H. R.
TING, V. C.
INFLUENCE OF INTERLAMINATE FLAWS ON TRANSIENT
TEMPERATURE   DISTRIBUTIONS
ASME. PAPER N. 73-HT-48, FOR MEET AUG. 5-8 1973,
10 PP.

73-0224 GASCH, R.
UNBALANCE GENERATED VIBRATIONS AND STABILITY OF
TURBINE   ROTORS
KONSTR. MASCH. APP. GERAETEBAU., V. 25, N. 5, MAY.
1973,   PP. 161-168.

73-0225 FINN, W. D.
VAROGLU, E.
DYNAMICS OF GRAVITY DAM-RESERVOIR SYSTEMS
COMPUT. STRUCT., V. 3, N. 4, JUL. 1973, PP.
913-924.

73-0226 CYR, N. A.
TETER, R. D.
FINITE ELEMENT ELASTIC PLASTIC CREEP ANALYSIS OF
TWO DIMENSIONAL CONTINUUM WITH TEMPERATURE
DEPENDENT MATERIAL PROPERTIES
COMPUT. STRUCT., VOL. 3, NO. 4, JULY 1973, PP.
849-863.

73-0227 CLARKSON, B. L.
ASHIE, I.
COMPUTER BASED ANALYSES OF THE RESPONSE OF BOX
TYPE   STRUCTURES TO RANDOM PRESSURES
COMPUT. STRUCT., V. 3, N. 4, JUL. 1973, PP.
899-912.

73-0228 CARMICHAEL, G. D. T.
JERRAM, K.
APPLICATION OF FRACTURE MECHANICS TO PRESTRESSED
CONCRETE   PRESSURE VESSELS
CEM. CONCR. RES., V. 3, N. 4, JUL. 1973, PP.
459-467.

73-0229 WADE, L. V.
STAHL, P. A.
EFFECTS OF SPECIMEN RADIUS ON THE STRESS STATE
NEAR A ROOF BOLT ANCHOR: A FINITE ELEMENT
DETERMINATION
U.S. BUR. MINES, REP. INVEST. RI-7748, 1973, 28
PP.

73-0230 SIDDALL, J. N.
DOKAINISH, M. A.
MODEL TEST AND ANALYSIS OF OBLIQUE FOLDED PLATE
STRUCTURE
ASCE. J. STRUCT. DIV., V. 99, N. ST6, JUN. 1973,
PAP. N.   9772, PP. 1043-1049.

73-0231 SEGAL, V. M.
SVIRID, G. P.
INVESTIGATION OF THE STATIONARY FLOW OF A RIGID
PLASTIC MATERIAL BY THE NUMERICAL FINITE
ELEMENT METHOD
PRIKL. MEKH., V. 9, N. 4, APR. 1973, PP. 76-80.

73-0232 ADEY, R. A.
BREBBIA, C.
EFFICIENT METHOD FOR SOLUTION OF VISCOELASTIC
PROBLEMS
ASCE J. ENG. MECH. DIV., VOL. 99, NO. EM6,
DECEMBER 1973, PAPER NO. 10206, PP. 1119-1127.

73-0233 RAMESH, C. K.
DIXIT, V. D.
ANALYSIS OF A SKEW SLAB-BEAM SYSTEM BY FINITE
ELEMENT METHOD
J. INST. ENG. (INDIA), CIV. ENG. DIV., V. 53, PART
CI.5, MAY 1973, PP. 233-239.

73-0234 RAJASEKARAN, S.
MURRAY, D. W.
FINITE ELEMENT SOLUTION OF INELASTIC BEAM
EQUATIONS
ASCE. J. STRUCT. DIV., V. 99, N. ST6, JUN. 1973,
PAP. N.9773PP. 1025-1041.

73-0235 RAJASEKARAN, S.
MURRAY, D. W.
COUPLED LOCAL BUCKLING IN WIDE-FLANGE BEAM-COLUMNS
ASCE. J. STRUCT. DIV., V. 99, N. ST6, JUN. 1973,
PAP. N. 9774, PP. 1003-1023.

73-0236 RADAJ, D.
ZAFIRAKOPOULOS, N.
FORCES AND DEFORMATIONS AT THE LATERAL STIFFENER
OF THIN WALLED U-BARS
FORSCH. INGENIEURWES., VOL. 39, NO. 3, 1973, PP.
69-80

73-0237 OSTAF'EV, V. A.
DETERMINATION OF THE STRESS-STRAINED STATE WHEN
CUTTING METAL WITH A WEDGE
PROBL. PROCHN., V. 5, N. 1, JAN. 1973, PP. 66-69.

73-0238 NEEDLEMAN, A.
NUMERICAL STUDY OF UNIAXIAL COMPRESSION IN
CIRCULAR ELASTIC-PLASTIC COLUMNS
INT. J. SOLIDS STRUCT., V. 9, N. 8, AUG. 1973, PP.
981-998.

73-0239 MCCHESNEY, R. S.
IMPROVED FINITE-ELEMENT ANALYSIS OF
NUCLEAR-CONTAINMENT SHELL STRUCTURES
COMPUT. STRUCT., V. 3, N. 3, MAY 1973, PP.
619-637.

73-0240 MAMET, J. C.
MUFTI, A. A.
HYPAR DISPLACEMENT FIELDS FOR RECTANGULAR FINITE
ELEMENTS
ENG. J. (MONTREAL), V. 56, N. 4, APR. 1973, PP.
1-10.

73-0241 LI, C.
LEONARD, J. W.
FINITE ELEMENT ANALYSIS OF INFLATABLE SHELLS
ASCE. J. ENG. MECH. DIV., V. 99, N. EM3, JUN.
1973, PP. 495-514.

73-0242 LEVINE, H. W.
ARMEN, H.
NONLINEAR BEHAVIOR OF SHELLS OF REVOLUTION UNDER
CYCLIC LOADING
COMPUT. STRUCT., V. 3, N. 3, MAY 1973, PP.
589-617.

73-0243 LESTINGI, J.
PRACHUKTAM, S.
BLOCKING TECHNIQUE FOR LARGE SCALE STRUCTURAL
ANALYSIS
COMPUT. STRUCT., V. 3, N. 3, MAY 1973, PP.
669-714.

73-0244 LEONARD, J. W.
LI, C.
STRONGLY CURVED FINITE ELEMENT FOR SHELL ANALYSIS
ASCE. J. ENG. MECH. DIV., V. 99, N. EM3, JUN.
1973, PAPER 9805, PP. 515-535.

73-0245 KOBAYASHI, A. S.
CHIU, S. T.
NUMERICAL AND EXPERIMENTAL INVESTIGATION ON THE
USE OF J-INTEGRAL
ENG. FRACT. MECH., V. 5, N. 2, JUN. 1973, PP.
293-305.

73-0246 KIUSALAAS, J.
OPTIMAL DESIGN OF STRUCTURES WITH BUCKLING
CONSTRAINTS
INT. J. SOLIDS STRUCT., V. 9, N. 7, JUL. 1973, PP.
863-878.

73-0247 KIDYBINSKI, A.
BABCOCK, C. O.
STRESS DISTRIBUTION AND ROCK FRACTURE ZONES IN THE
ROOF OF LONGWALL FACE IN COAL MINE
ROCK MECH., FELSMECH., MEC. ROCHES, V. 5, N. 1, MAY
1973, PP. 1-19.

73-0248 KAWASHIMA, K.
NARUOKA, M.
ANALYSIS OF STRESSES IN SKEWED PLATES BY THE
METHOD OF FINITE ELEMENTS
BAUINGENIEUR, V. 48, N. 3, MAR. 1973, PP. 73-78.

73-0249 GUYMON, G. L.
FINITE ELEMENT SOLUTION FOR GENERAL FLUID MOTION
ASCE J. HYDRAUL DIV., VOL. 99, NO. HY6, JUNE 1973,
PAPER 9789, PP. 913-919.

73-0250 GALLAGHER, R. H.
FINITE ELEMENT METHOD IN SHELL STABILITY ANALYSIS
COMPUT. STRUCT., VOL. 3, NO. 3, MAY 1973, PP.
543-557.

73-0251 DORNQUAST, P.
PROBLEMS OF LONGITUDINAL STRENGTH IN THE
CONSTRUCTION OF VERY LARGE TANK TRUCKS
ALUMINIUM, V. 49, N. 3, MAR. 1973, PP. 239-243.

73-0252 CRUSE, T. A.
APPLICATION OF THE BOUNDARY-INTEGRAL EQUATION
METHOD TO THREE DIMENSIONAL STRESS ANALYSIS
COMPUT. STRUCT., V. 3, N. 3, MAY 1973, PP.
509-527.

73-0253 CHUNG, T. J.
EIDSON, R. L.
DYNAMIC ANALYSIS OF VISCOELASTOPLASTIC ANISOTROPIC
SHELLS
COMPUT. STRUCT., V. 3, N. 3, MAY 1973, PP.
483-496.

73-0254 CHAN, S. T. K.
LAROCK, B. E.
FREE-SURFACE IDEAL FLUID FLOWS BY FINITE ELEMENTS
ASCE. J. HYDRAUL. DIV., VOL. 99, NO. HY6, JUNE
1973, PAPER 9787, PP. 959-974.

73-0255 CARROLL, W. E.
BARKER, R. M.
THEOREM FOR OPTIMUM FINITE-ELEMENT IDEALIZATIONS
INT. J. SOLIDS STRUCT., V. 9, N. 7, JUL. 1973, PP.
883-895.

73-0256 BOROWSKI, V. J.
STEURY, R. L.
FINITE ELEMENT DYNAMIC ANALYSIS OF AN AUTOMOTIVE
FRAME
SAE. PREPR. N. 730506 FOR MEET., MAY 14-18 1973,
20 PP.

73-0257 BARTHOLOME, G.
KIMSCH, M.
FRACTURE AND SAFETY ANALYSIS OF NUCLEAR PRESSURE
VESSELS
ENG. FRACT. MECH., V. 5, N. 2, JUN. 1973, PP.
431-446.

73-0258 ANDERSON, D. M.
FRACTURE TOUGHNESS PARAMETERS AND ELASTIC-PLASTIC
ANALYSIS OF NON-MODERATE FRACTURE CONDITIONS
USING FINITE ELEMENT METHODS
ENG. FRACT. MECH., V. 5, N. 2, JUN. 1973, PP.
223-240.

73-0259 MAHOMED, M.
VENTER, R. D.
APPLICATION OF THE FINITE ELEMENT TECHNIQUE TO THE
STUDY OF AREAS OF HIGH STRESS GRADIENTS WITHIN A
GIVEN BODY
S. AFR. MECH. ENG., V. 23, N. 1, JAN. 1973, PP.
2-7.

73-0260 DAILEY, G.
CAYWOOD, W. C.
GENERAL PURPOSE COMPUTER PROGRAM FOR THE DYNAMIC
SIMULATION OF VEHICLE GUIDEWAY INTERACTIONS
A.I.A.A. J., VOL. 11, NO. 3, MARCH 1973, PP.
278-282. -

73-0261 CULVILLE, J.
BECKER, E. B.
LARGE DISPLACEMENT ANALYSIS OF THIN PLATES
A.S.C.E. J. STRUCT. DIV., PAPER NO. 9593, VOL. 99,
NO. ST3, MARCH 1973, PP. 349-364.

73-0262 BRUCH, J. C.
NONLINEAR EQUATION OF UNSTEADY GROUND-WATER FLOW
ASCE. J. HYDRAUL. DIV., V. 99, N. HY3, MAR. 1973,
PAP. N. 9589, PP. 395-403.

73-0263 WU, S. T.
UNSTEADY MHD DUCT FLOW BY THE FINITE ELEMENT
METHOD
INT. J. NUMER. METHODS ENG., V. 6, N. 1, 1973, PP.
3-10.

73-0264 UTKU, S.
FINITE ELEMENT METHOD FOR BLOCK ADJUSTMENT
PROBLEMS OF PHOTOGRAMMETRY
INT. J. NUMER. METH. ENG., VOL. 6, NO. 1, 1973,
PP. 33-38.

73-0265 UKHOV, S. B.
SEMENOV, V. V.
CALCULATION OF DISPLACEMENTS AND STRESSES IN
ANISOTROPIC ROCKS BY USING METHOD OF FINITE
ELEMENTS
GIDROTEKH. STROIT., N. 2, FEB. 1973, PP. 33-38.

73-0266 RADAJ, D.
KANDEL, W.
NOTCH STRESSES AT THE LOAD CARRYING ELASTIC CORE
FORSCH. INGENIEURWES, V. 39, N. 1, 1973, PP. 1-12.

73-0267 NAYAK, G. C.
OWEN, D. R. J.
KFOURI, A. P.
GRIFFITHS, J. R.
STRESSES IN A PARTLY YIELDED NOTCHED BAR - AN
ASSESSMENT OF THREE ALTERNATIVE PROGRAMS
INT. J. NUMER. METH. ENG., VOL. 6, NO. 1, 1973,
PP. 63-73.

73-0268 OWEN, D. R. J.
NAYAK, G. C.
STRESSES IN A PARTLY YIELDED NOTCHED BAR EM DASH
AN ASSESSMENT OF THREE ALTERNATIVE
PROGRAMS
INT. J. NUMER. METHODS ENG., V. 6, N. 1, 1973, PP.
63-73.

73-0269 MEISSNER, H.
WIBEL, A. P.
NUMERICAL COMPUTATION OF BOUNDARY VALUE PROBLEMS
IN SOIL MECHANICS
BAUTECHNIK. (AUSG A), V. 50, N. 2, FEB. 1973, PP.
65-69.

73-0270 LYNN, P. P.
ARYA, S. K.
USE OF THE LEAST SQUARES CRITERION IN THE FINITE
ELEMENT FORMULATION
INT. J. NUMER. METHOD ENGR., VOL. 6, NO. 1, 1973,
PP. 75-88.

73-0271 KIKUCHI, F.
FINITE ELEMENT METHOD FOR NON-SELF-ADJOINT
PROBLEMS
INT. J. NUMER. METHODS ENG., VOL. 6, NO. 1, 1973,
PP. 39-54.

73-0272 KHATUA, T. P.
CHEUNG, Y. K.
BENDING AND VIBRATION OF MULTILAYER SANDWICH BEAMS
AND PLATES
INT. J. NUMER. METHODS ENG., V. 6, N. 1, 1973, PP.
11-24.

73-0273 DUNCAN, J. M.
GATES, R. H.
APPLICATIONS OF THE FINITE ELEMENT METHOD IN
GEOTECHNICAL ENGINEERING, PROCEEDINGS OF THE
SYMPOSIUM
APPL. OF THE FINITE ELEM METHOD IN GEOTECH ENG.,
PROC. SYMP., VICKSBURG, MISS. MAY 1-4 1972, 1227P
PUBL BY U S ARMY ENG WATERW. EXP. STN. SOIL MECH
INF. ANAL. CENT, VICKSBURG, MISS., SEP. 1972.

73-0274 BABU, P. V.
REDDY, D. V.
FREQUENCY ANALYSIS OF ORTHOTROPIC CIRCULAR
CYLINDRICAL PANELS BY THE FINITE STRIP METHOD
BUILDING SCI. VOL. 8, PERMAGON PRESS 1973, PP.
229-241.

73-0275 ABE, T.
ELASTIC DEFORMATION OF POLYCRYSTALLINE METAL - 2
EFFECT OF SLENDERNESS OF GRAINS ON
DEFORMATION MODE
BULL. JSME, V. 16, N. 93, MAR. 1973, PP. 474-484.

73-0276 ROWLANDS, R. E.
DANIEL, I. M.
STRESS AND FAILURE ANALYSIS OF A GLASS-EPOXY
COMPOSITE PLATE WITH A CIRCULAR HOLE
EXP. MECH., V. 13, N. 1, JAN. 1973, PP. 31-37.

73-0277 JOFRIET, J. C.
KIKUCHI, F.
SHORT TERM DEFLECTIONS OF CONCRETE FLAT PLATES
ASCE. J. STRUCT. DIV., V. 99, N. ST1, JAN. 1973,
PAP. N. 9488, PP. 167-182.

73-0278 ALPAY, I. B.
UTKU, S.
ON RESPONSE OF INITIALLY STRESSED STRUCTURES TO
RANDOM EXCITATIONS
COMPUT. STRUCT., VOL. 3, NO. 5, SEPTEMBER 1973,
PP. 1079-1097.

73-0279 YANG, T. Y.
HIGH ORDER RECTANGULAR SHALLOW SHELL FINITE
ELEMENT
ASCE. J. ENG. MECH. DIV., V. 99, N. EM1, FEB.
1973, PAP. N. 9519, PP. 157-181.

73-0280 THOMPSON, J. E.
VEHICLE CRUSH PREDICTION USING FINITE-ELEMENT
TECHNIQUES
SAE. PREPR. N. 730157 FOR MEET., JAN 8-12, 1973,
15 PP.

73-0281 PENMAN, A.
CHARLES, A.
CONSTRUCTIONAL DEFORMATIONS IN ROCKFILL DAM
ASCE. J. SOIL. MECH. FOUND. DIV., V. 99, N. SM2,
FEB. 1973, PAP. N. 9560, PP. 139-163.

73-0282 JAEGER, L. G.
MUFTI, A. A.
STRUCTURAL ANALYSIS OF TALL BUILDINGS HAVING
IRREGULARLY POSITIONED SHEAR WALLS
BUILD. SCI., V. 8, N. 1, MAR. 1973, PP. 11-22.

73-0283 GODDEN, W. G.
STRUCTURAL MODEL STUDIES OF LARGE PIPE BIFURCATION
ASCE. TRANSP. ENG. J., V. 99, N. TE1, FEB. 1973,
PAP. N. 9549, PP. 93-109.

73-0284 BELL, C. G.
EARTHQUAKE-RESISTANT DESIGN OF ENGINEERING
STRUCTURES
NUCL. SAFETY, V. 14, N. 1, JAN-FEB 1973, PP. 1/5.

73-0285 BUSBY, H. R. JR.
WEINGARTEN, V. I.
RESPONSE OF NONLINEAR BEAM TO RANDOM EXCITATION
ASCE J. ENG. MECH. DIV., VOL. 99, NO. EM1,
FEBRUARY 1973, PAPER NO. 9530, PP. 55-66.

73-0286 BABUSKA, I.
AZIZ, A. K.
THE MATHEMATICAL FOUNDATIONS OF THE FINITE ELEMENT
METHOD - WITH APPLICATIONS TO PARTIAL DIFFERENTIAL
EQUATIONS
ACADEMIC PRESS, NEW YORK, 1973.

73-0287 ZLAMAL, M.
SOME RECENT ADVANCES IN THE MATHEMATICS OF FINITE
ELEMENTS
FROM "THE MATHEMATICS OF FINITE ELEMENTS AND
APPLICATIONS", J. WHITEMAN (ED.) ACADEMIC PRESS,
NEW YORK, 1973.

73-0288 WHITEMAN, J.
THE MATHEMATICS OF FINITE ELEMENTS AND
APPLICATIONS
ACADEMIC PRESS, NEW YORK, 1973.

73-0289 ODEN, J. T.
THE FINITE ELEMENT METHOD IN MATHEMATICAL PHYSICS
FROM "THE MATHEMATICS OF FINITE ELEMENTS AND
APPLICATIONS", J. WHITEMAN (ED.), ACADEMIC PRESS,
NEW YORK, 1973.

73-0290 BUTURLA, E. M.
MCLAY, R. W.
ON SPARSE MATRICES IN FINITE ELEMENT WAVE
PROPAGATION
ASME PAPER NO. 73-DET-142 FOR MEETING SEPTEMBER
9-12, 1973, 6 PP.

73-0291 PARK, K. C.
SACZALSKI, K. J.
TRANSIENT RESPONSE OF INELASTICALLY CONSTRAINED
RIGID BODY SYSTEMS
ASME J. ENGR. FOR INDUSTRY, 1973.

73-0292 MUTO, K.
ELASTIC-PLASTIC ANALYSIS OF REINFORCED CONCRETE
MEMBERS BY FEM
PROC. SYMP. OF THE JAPAN EARTHQUAKE ENG., TOKYO,
K. MUTO ET AL (EDS.), AUGUST 1973, PP. 349-356.

73-0293 NAKAO, Y.
SASAKI, N.
ABIRU, H.
DYNAMIC ANALYSIS OF THE GROUND USING THE THREE
DIMENSIONAL FINITE ELEMENT
MITSUBISHI JUKO GIHO, MITSUBISHI HEAVY IND. LTD.,
JAPAN, VOL. 10, NO. 4, 1973, PP. 43-51. (IN
JAPANESE)

73-0294 FAM, A. R. M.
TURKSTRA, C. J.
ANALYSIS OF CURVED BOX GIRDER BRIDGES
CAN. CONGR. OF APPL. MECH., 4TH PROC. PAPER EC
POLYTECH, MONTREAL, QUEBEC, MAY 28-JUNE 1, 1973,
PP. 297-298.

73-0295 KABAILA, A. P.
INTEGRATION FOR THE FINITE ELEMENTS - AN ALGORITHM
IN ORTHOGONAL AXES
UNICIV REPORT NO. R-99, THE UNIVERSITY OF NEW
SOUTH WALES, SYDNEY, AUSTRALIA, 1973.

73-0296 FUJITANI, Y.
KIHARA, T.
KAWAI, T.
A DIFFUSION ANALYSIS OF ION IN ELECTROLYSIS BY
FINITE ELEMENT METHOD
PROC. NAT. SYMP. ON MATRIX METHODS OF STRUCT.
ANALYSIS AND DES., SOC. OF STEEL CONST. OF JAPAN,
TOKYO, 1973, PP. 663-670. (IN JAPANESE)

73-0297 IWAKI, T.
ON THE PHYSICAL MEANINGS OF THE MATRICES DERIVED
FROM FINITE ELEMENT METHOD OF HEAT CONDUCTION
PROBLEMS
PROC. NATL. SYMP. ON MATRIX METH. OF STRUCT.
ANALYSIS AND DESIGN, SCC. OF STEEL CONSTRUCTION,
1973.

73-0298 USUKI, S.
KUDO, K.
LOCAL POTENTIAL APPROACH TO FINITE ELEMENT METHOD
IN UNSTEADY VISCOUS INCOMPRESSIBLE FLUID
FLOW
PROC. OF JSCE, NO. 216, AUGUST, 1973.

73-0299 YAMADA, Y.
ITO, K.
OHTSUBO, T.
FINITE ELEMENT FORMULATION FOR SOLUTION OF STEADY
FLOW OF INCOMPRESSIBLE VISCOUS FLUID
3RD NATL. SYMP. ON MATRIX METH. OF STRUCT.
ANALYSIS AND DESIGN, SOCIETY OF STEEL CONSTRUCTION
OF JAPAN, 1973, PP. 597-604. (IN JAPANESE)

73-0300 IKEGAWA, M.
WASHIZU, K.
FINITE ELEMENT METHOD APPLIED TO ANALYSIS OF FLOW
OVER A   SPILLWAY CREST
INT. J. NUM. METH. ENGNG., VOL. 6, PP. 179-189,
1973.

73-0301 TERAMAE, T.
APPLICATION OF FINITE ELEMENT METHOD TO
HYDRO-DYNAMICS
3RD CONF. ON MATRIX METH. AND STRUCT. ANALYSIS,
J.S.S.C., 1973. (IN JAPANESE)

73-0302 ODEN, J. T.
SOME CONTRIBUTIONS TO THE MATHEMATICAL THEORY OF
MIXED FINITE ELEMENT APPROXIMATIONS
IN "THEORY AND PRACTICE IN FINITE ELEMENT
STRUCTURAL ANALYSIS", EDITED BY Y. YAMADA AND R.
H. GALLAGHER, PROC. OF THE 1973 TOKYO SEMINAR ON
FINITE ELEMENT ANALYSIS, PP. 1-22. (THE UNIV. OF
TOKYO PRESS, NOVEMBER 1974)

73-0303 KIKUCHI, F.
SOME CONSIDERATIONS ON THE CONVERGENCE OF HYBRID
STRESS METHOD
IN "THEORY AND PRACTICE IN FINITE ELEMENT
STRUCTURAL ANALYSIS", EDITED BY Y. YAMADA AND R.H.
GALLAGHER, PROC. OF THE 1973 TOKYO SEMINAR ON
FINITE ELEMENT ANALYSIS, PP. 25-42.

73-0304 MOAN, T.
ON THE LOCAL DISTRIBUTION OF ERRORS BY FINITE
ELEMENT APPROXIMATIONS
IN "THEORY AND PRACTICE IN FINITE ELEMENT
STRUCTURAL ANALYSIS", EDITED BY Y. YAMADA AND R.H.
GALLAGHER, PROC. OF THE 1973 TOKYO SEMINAR ON
FINITE ELEMENT ANALYSIS, PP. 43-60.

73-0305 NIISEKI, S.
SATAKE, M.
SOME APPLICATIONS OF TOPOLOGICAL CONSIDERATION AND
WEIGHT MATRIX METHOD TO FINITE ELEMENT ANALYSIS
IN "THEORY AND PRACTICE IN FINITE ELEMENT
STRUCTURAL ANALYSIS", EDITED BY Y. YAMADA AND R.H.
GALLAGHER, PROC. OF THE 1973 TOKYO SEMINAR ON
FINITE ELEMENT ANALYSIS, PP. 61-73.

73-0306 YAMAMOTO, Y.
TOKUDA, N.
SUMI, Y.
FINITE ELEMENT TREATMENT OF SINGULARITIES OF
BOUNDARY VALUE PROBLEMS AND ITS APPLICATION TO
ANALYSIS OF STRESS INTENSITY FACTORS
IN "THEORY AND PRACTICE IN FINITE ELEMENT
STRUCTURAL ANALYSIS", EDITED BY Y. YAMADA AND R.H.
GALLAGHER, PROC. OF THE 1973 TOKYO SEMINAR ON
FINITE ELEMENT ANALYSIS, PP. 75-90.

73-0307 FUJII, H.
SOME REMARKS ON FINITE ELEMENT ANALYSIS OF TIME
DEPENDENT FIELD PROBLEMS
IN "THEORY AND PRACTICE IN FINITE ELEMENT
STRUCTURAL ANALYSIS", EDITED BY Y. YAMADA AND R.H.
GALLAGHER, PROC. OF THE 1973 TOKYO SEMINAR ON
FINITE ELEMENT ANALYSIS, PP. 91-106.

73-0308 GALLAGHER, R. H.
FINITE ELEMENT ANALYSIS OF GEOMETRICALLY NONLINEAR
PROBLEMS
IN "THEORY AND PRACTICE IN FINITE ELEMENT
STRUCTURAL ANALYSIS", EDITED BY Y. YAMADA AND R.H.
GALLAGHER, PROC. OF THE 1973 TOKYO SEMINAR ON
FINITE ELEMENT ANALYSIS, PP. 109-123.

73-0309 YAMADA, Y.
TAKATSUKA, K.
IWATA, K.
NONLINEAR ANALYSIS BY THE FINITE ELEMENT METHOD
AND SOME EXPOSITORY EXAMPLES
IN "THEORY AND PRACTICE IN FINITE ELEMENT
STRUCTURAL ANALYSIS", EDITED BY Y. YAMADA AND R.H.
GALLAGHER, PROC. OF THE 1973 TOKYO SEMINAR ON
FINITE ELEMENT ANALYSIS, PP. 125-138.

73-0310 SHIN, Y. K.
A COMPUTER METHOD FOR THE SECOND ORDER ELASTIC
ANALYSIS OF PLANE FRAMED STRUCTURES
IN "THEORY AND PRACTICE IN FINITE ELEMENT
STRUCTURAL ANALYSIS", EDITED BY Y. YAMADA AND R.
H. GALLAGHER, PROC. OF THE 1973 TOKYO SEMINAR ON
FINITE ELEMENT ANALYSIS, PP. 139-156.

73-0311 KAWAMATA, S.
MAGARA, K.
KUNITA, J.
ANALYSIS OF CABLE NETS IN MIXED FORMULATION
IN "THEORY AND PRACTICE IN FINITE ELEMENT
STRUCTURAL ANALYSIS", EDITED BY Y. YAMADA AND R.H.
GALLAGHER, PROC. OF THE 1973 TOKYO SEMINAR ON
FINITE ELEMENT ANALYSIS, PP. 157-175.

73-0312 TSUTA, T.
YAMAJI, S.
FINITE ELEMENT ANALYSIS OF CONTACT PROBLEMS
IN "THEORY AND PRACTICE IN FINITE ELEMENT
STRUCTURAL ANALYSIS", EDITED BY Y. YAMADA AND R.H.
GALLAGHER, PROC. OF THE 1973 TOKYO SEMINAR ON
FINITE ELEMENT ANALYSIS, PP. 177-194.

73-0313 MEEK, J. L.
EXCAVATION IN ROCK, AN APPRECIATION OF THE FINITE
ELEMENT METHOD OF ANALYSIS
IN "THEORY AND PRACTICE IN FINITE ELEMENT
STRUCTURAL ANALYSIS", EDITED BY Y. YAMADA AND R.H.
GALLAGHER, PROC. OF THE 1973 TOKYO SEMINAR ON
FINITE ELEMENT ANALYSIS, PP. 195-213.

73-0314 LEE, S. L.
ANAND, S. C.
ON TRESCA YIELD CRITERION IN FINITE ELEMENT
ANALYSIS OF PLANE STRESS PROBLEMS
IN "THEORY AND PRACTICE IN FINITE ELEMENT
STRUCTURAL ANALYSIS", EDITED BY Y. YAMADA AND R.H.
GALLAGHER, PROC. OF THE 1973 TOKYO SEMINAR ON
FINITE ELEMENT ANALYSIS, PP. 215-232.

73-0315 PURUSHOTHAMAN, P.
RAO, J. K. S.
MURTHY, P. N.
STUDIES ON TWO PHASE COMPOSITE MATERIALS THROUGH
COMPUTER SIMULATION
IN "THEORY AND PRACTICE IN FINITE ELEMENT
STRUCTURAL ANALYSIS", EDITED BY Y. YAMADA AND P.H.
GALLAGHER, PROC. OF THE 1973 TOKYO SEMINAR ON
FINITE ELEMENT ANALYSIS, PP. 233-245.

73-0316 GUPTA, K. K.
RECENT ADVANCES IN NUMERICAL ANALYSIS OF
STRUCTURAL EIGENVALUE PROBLEMS
IN "THEORY AND PRACTICE IN FINITE ELEMENT
STRUCTURAL ANALYSIS", EDITED BY Y. YAMADA AND R.H.
GALLAGHER, PROC. OF THE 1973 TOKYO SEMINAR ON
FINITE ELEMENT ANALYSIS, PP. 249-271.

73-0317 KAWAMO, K.
TAKETO, K.
A TEST ON CONVERGENCE OF EIGENSOLUTIONS BY
SIMULTANEOUS ITERATION METHOD
IN "THEORY AND PRACTICE IN FINITE ELEMENT
STRUCTURAL ANALYSIS", EDITED BY Y. YAMADA AND R.H.
GALLAGHER, PROC. OF THE 1973 TOKYO SEMINAR ON
FINITE ELEMENT ANALYSIS, PP. 273-288.

73-0318 SWANNELL, P.
THE AUTOMATIC COMPUTATION OF THE NATURAL
FREQUENCIES OF STRUCTURAL FRAMES USING AN EXACT
MATRIX TECHNIQUE
IN "THEORY AND PRACTICE IN FINITE ELEMENT
STRUCTURAL ANALYSIS", EDITED BY Y. YAMADA AND R.H.
GALLAGHER, PROC. OF THE 1973 TOKYO SEMINAR ON
FINITE ELEMENT ANALYSIS, PP. 289-304.

73-0319 RAO, A. K.
RAMAMURTHY, T. S.
RAO, G. V.
BOUNDS AND ERROR CONTROL FOR EIGENVALUES
IN "THEORY AND PRACTICE IN FINITE ELEMENT
STRUCTURAL ANALYSIS", EDITED BY Y. YAMADA AND R.H.
GALLAGHER, PROC. OF THE 1973 TOKYO SEMINAR ON
FINITE ELEMENT ANALYSIS, PP. 305-322.

73-0320 PRASAD, K.S.R.K.
MURTY, A.V.K.
RAO, A. K.
RAO, G. V.
FINITE ELEMENT MODELLING OF NATURAL VIBRATION
PROBLEMS
IN "THEORY AND PRACTICE IN FINITE ELEMENT
STRUCTURAL ANALYSIS", EDITED BY Y. YAMADA AND R.H.
GALLAGHER, PROC. OF THE 1973 TOKYO SEMINAR ON
FINITE ELEMENT ANALYSIS, PP. 323-338.

73-0321 KLOSTERMAN, A. L.
MCCLELLAND, W. A.
COMBINING EXPERIMENTAL AND ANALYTICAL TECHNIQUES
FOR DYNAMIC SYSTEM ANALYSIS
IN "THEORY AND PRACTICE IN FINITE ELEMENT
STRUCTURAL ANALYSIS", EDITED BY Y. YAMADA AND R.H.
GALLAGHER, PROC. OF THE 1973 TOKYO SEMINAR ON
FINITE ELEMENT ANALYSIS, PP. 339-356.

73-0322 RAMESH, C. K.
BELKUNE, R. M.
FREE VIBRATIONS OF PLATE-BEAM SYSTEMS
IN "THEORY AND PRACTICE IN FINITE ELEMENT
STRUCTURAL ANALYSIS", EDITED BY Y. YAMADA AND R.H.
GALLAGHER, PROC. OF THE 1973 TOKYO SEMINAR ON
FINITE ELEMENT ANALYSIS, PP. 357-370.

73-0323 MURTHY, P. N.
IYENGAR, N. G. R.
EFFECTS OF MATERIAL NONLINEARITIES ON VIBRATION
IN "THEORY AND PRACTICE IN FINITE ELEMENT
STRUCTURAL ANALYSIS", EDITED BY Y. YAMADA AND R.H.
GALLAGHER, PROC. OF THE 1973 TOKYO SEMINAR ON
FINITE ELEMENT ANALYSIS, PP. 371-398.

73-0324 MUTO, K.
OHMORI, N.
SUGANO, T.
MIYASHITA, T.
SHIMIZU, H.
NONLINEAR ANALYSIS OF REINFORCED CONCRETE
BUILDINGS
IN "THEORY AND PRACTICE IN FINITE ELEMENT
STRUCTURAL ANALYSIS", EDITED BY Y. YAMADA AND R.H.
GALLAGHER, PROC. OF THE 1973 TOKYO SEMINAR ON
FINITE ELEMENT ANALYSIS, PP. 371-398.

73-0325 NAKAO, Y.
TAKANO, S.
ELASTO-PLASTIC SEISMIC RESPONSE ANALYSIS OF FRAMED
STRUCTURES BY MODE SUPERPOSITION
IN "THEORY AND PRACTICE IN FINITE ELEMENT
STRUCTURAL ANALYSIS", EDITED BY Y. YAMADA AND R.H.
GALLAGHER, PROC. OF THE 1973 TOKYO SEMINAR ON
FINITE ELEMENT ANALYSIS, PP. 423-440.

73-0326 NAGY, L. I.
FINITE ELEMENT METHOD IN AUTOMOBILE DESIGN
IN "THEORY AND PRACTICE IN FINITE ELEMENT
STRUCTURAL ANALYSIS", EDITED BY Y. YAMADA AND R.H.
GALLAGHER, PROC. OF THE 1973 TOKYO SEMINAR ON
FINITE ELEMENT ANALYSIS, PP. 443-460.

73-0327 WOOD, R.
VALLIAPPAN, S.
SVENSSON, N. L.
STRESS ANALYSIS OF HUMAN FEMUR
IN "THEORY AND PRACTICE IN FINITE ELEMENT
STRUCTURAL ANALYSIS", EDITED BY Y. YAMADA AND R.H.
GALLAGHER, PROC. OF THE 1973 TOKYO SEMINAR ON
FINITE ELEMENT ANALYSIS, PP. 461-478.

73-0328 CHAN, Y. K.
KABAILA, A. P.
A COMPATIBLE ELEMENT FOR BUCKLING ANALYSIS OF SKEW
PLATES WITH PARALLEL EDGES
IN "THEORY AND PRACTICE IN FINITE ELEMENT
STRUCTURAL ANALYSIS", EDITED BY Y. YAMADA AND R.H.
GALLAGHER, PROC. OF THE 1973 TOKYO SEMINAR ON
FINITE ELEMENT ANALYSIS, PP. 479-497.

73-0329 KABAILA, A. P.
COMPUTER INTEGRATION FOR FINITE ELEMENTS
IN "THEORY AND PRACTICE IN FINITE ELEMENT
STRUCTURAL ANALYSIS", EDITED BY Y. YAMADA AND R.H.
GALLAGHER, PROC. OF THE 1973 TOKYO SEMINAR ON
FINITE ELEMENT ANALYSIS, PP. 499-509.

73-0330 NORRIE, D. H.
DE VRIES, G.
NON-STRUCTURAL APPLICATIONS OF THE FINITE ELEMENT
METHOD
IN "THEORY AND PRACTICE IN FINITE ELEMENT
STRUCTURAL ANALYSIS", EDITED BY Y. YAMADA AND R.H.
GALLAGHER, PROC. OF THE 1973 TOKYO SEMINAR ON
FINITE ELEMENT ANALYSIS, PP. 511-539.

73-0331 KAWAI, T.
ON THE FINITE ELEMENT ANALYSIS OF DIFFUSION
PROBLEMS
IN "THEORY AND PRACTICE IN FINITE ELEMENT
STRUCTURAL ANALYSIS", EDITED BY Y. YAMADA AND R.H.
GALLAGHER, PROC. OF THE 1973 TOKYO SEMINAR ON
FINITE ELEMENT ANALYSIS, PP. 541-555.

73-0332 KAWAHARA, M.
YOSHIMURA, N.
NAKAGAWA, K.
OHSAKA, H.
STEADY FLOW ANALYSIS OF INCOMPRESSIBLE VISCOUS
FLOW BY THE FINITE ELEMENT METHOD
IN "THEORY AND PRACTICE IN FINITE ELEMENT
STRUCTURAL ANALYSIS", EDITED BY Y. YAMADA AND R.H.
GALLAGHER, PROC. OF THE 1973 TOKYO SEMINAR ON
FINITE ELEMENT ANALYSIS, PP. 557-572.

73-0333 WASHIZU, K.
IKEGAWA, M.
LIFTING SURFACE PROBLEMS ANALYSIS
IN "THEORY AND PRACTICE IN FINITE ELEMENT
STRUCTURAL ANALYSIS", EDITED BY Y. YAMADA AND R.H.
GALLAGHER, PROC. OF THE 1973 TOKYO SEMINAR ON
FINITE ELEMENT ANALYSIS, PP. 573-582.

73-0334 MARCAL, P. V.
GENERAL PURPOSE PROGRAM FOR FINITE ELEMENT
ANALYSIS WITH REFERENCE TO INTERACTION BETWEEN
SOFTWARE AND HARDWARE
IN "THEORY AND PRACTICE IN FINITE ELEMENT
STRUCTURAL ANALYSIS", EDITED BY Y. YAMADA AND R.H.
GALLAGHER, PROC. OF THE 1973 TOKYO SEMINAR ON
FINITE ELEMENT ANALYSIS, PP. 585-596.

73-0335 MARTIN, C. W.
COMPUTER WORD LENGTH AND SOLUTION ACCURACY.
EXPERIENCE WITH ELAS AND THE IBM 360
IN "THEORY AND PRACTICE IN FINITE ELEMENT
STRUCTURAL ANALYSIS", EDITED BY Y. YAMADA AND R.H.
GALLAGHER, PROC. OF THE 1973 TOKYO SEMINAR ON
FINITE ELEMENT ANALYSIS, PP. 597-610.

73-0336 MCCORMICK, C. W.
SPARSE MATRIX OPERATIONS IN NASTRAN
IN "THEORY AND PRACTICE IN FINITE ELEMENT
STRUCTURAL ANALYSIS", EDITED BY Y. YAMADA AND R.H.
GALLAGHER, PROC. OF THE 1973 TOKYO SEMINAR ON
FINITE ELEMENT ANALYSIS, PP. 611-631.

73-0337 MCLAY, R. W.
KAWAHARA, M.
STEARNS, B. K.
BUTURLA, E. M.
SPARSE MATRICES IN FINITE ELEMENT PROGRAM
DEVELOPMENT
IN "THEORY AND PRACTICE IN FINITE ELEMENT
STRUCTURAL ANALYSIS", EDITED BY Y. YAMADA AND R.H.
GALLAGHER, PROC. OF THE 1973 TOKYO SEMINAR ON
FINITE ELEMENT ANALYSIS, PP. 633-650.

73-0338 CHULSOO, Y. J.
AUTOMATIC GENERATION OF FINITE ELEMENT MESH WITHIN
A BANDWIDTH AND AN EFFICIENT SOLUTION OF THE
BANDMATRIX
IN "THEORY AND PRACTICE IN FINITE ELEMENT
STRUCTURAL ANALYSIS", EDITED BY Y. YAMADA AND R.H.
GALLAGHER, PROC. OF THE 1973 TOKYO SEMINAR ON
FINITE ELEMENT ANALYSIS, PP. 651-670.

73-0339 YUKI, T.
FUJII, T.
AUTOMATED DATA GENERATION AND LARGE MATRIX
OPERATION FOR 3 DIMENSIONAL PLATE STRUCTURE
ANALYSIS
IN "THEORY AND PRACTICE IN FINITE ELEMENT
STRUCTURAL ANALYSIS", EDITED BY Y. YAMADA AND R.H.
GALLAGHER, PROC. OF THE 1973 TOKYO SEMINAR ON
FINITE ELEMENT ANALYSIS, PP. 671-686.

73-0340 SAGAWA, K.
AUTOMATIC MESH GENERATION FOR THREE DIMENSIONAL
STRUCTURES BASED ON THEIR THREE VIEWS
IN "THEORY AND PRACTICE IN FINITE ELEMENT
STRUCTURAL ANALYSIS", EDITED BY Y. YAMADA AND R.H.
GALLAGHER, PROC. OF THE 1973 TOKYO SEMINAR ON
FINITE ELEMENT ANALYSIS, PP. 687-703.

73-0341 KAMEL, H. A.
MCCABE, M. W.
AN INTERACTIVE GRAPHICS SHIP DESIGN ORIENTED
PROGRAM PACKAGE FOR TIME SHARING SYSTEMS
IN "THEORY AND PRACTICE IN FINITE ELEMENT
STRUCTURAL ANALYSIS", EDITED BY Y. YAMADA AND R.H.
GALLAGHER, PROC. OF THE 1973 TOKYO SEMINAR ON
FINITE ELEMENT ANALYSIS, PP. 705-719.

73-0342 YOSHIKI, M.
OKABE, T.
KAWAI, T.
HIDAKA, M.
ON THE DEVELOPMENT OF THE "PASSAGE" PROGRAM
IN "THEORY AND PRACTICE IN FINITE ELEMENT
STRUCTURAL ANALYSIS", EDITED BY Y. YAMADA AND R.H.
GALLAGHER, PROC. OF THE 1973 TOKYO SEMINAR ON
FINITE ELEMENT ANALYSIS, PP. 721-729.

73-0343 SCHIFFMAN, R. L.
PORTABILITY AND ADAPTABILITY OF CIVIL ENGINEERING
SOFTWARE
IN "THEORY AND PRACTICE IN FINITE ELEMENT
STRUCTURAL ANALYSIS", EDITED BY Y. YAMADA AND R.H.
GALLAGHER, PROC. OF THE 1973 TOKYO SEMINAR ON
FINITE ELEMENT ANALYSIS, PP. 731-733.

73-0344 WANG, S. K.
SARGIOUS, M. A.
CHEUNG, Y. K.
EFFECT OF OPENINGS ON STRESSES IN RIGID PAVEMENTS
TRANSPORTATION ENGINEERING, J., PROC. ASCE, VOL.
99, NO. TE2, MAY 1973.

73-0345 KHATUA, T. P.
CHEUNG, Y. K.
STABILITY ANALYSIS OF MULTILAYER SANDWICH
STRUCTRES
AIAA J., VOL. 2, NO. 9, SEPTEMBER 1973.

73-0346 SKJOLINGSTAD, L.
CHEUNG, Y. K.
CLARK, J. L.
GROUNDWATER PREDICTIONS FOR A HIGHWAY CUT USING
FINITE ELEMENTS
26TH CANADIAN GEOTECHNICAL CONF., TORONTO, OCTOBER
18, 1973.

73-0347 CHEUNG, Y. K.
FINITE ELEMENT METHOD AND FINITE STRIP METHOD
BRIDGES AND FOUNDATIONS. PART 1. SEPTEMBER . 1973,
PART 11. NOVEMBER 1973. (IN JAPANESE)

73-0348 DESAI, C. S.
APPROXIMATE SOLUTION FOR UNCONFINED SEEPAGE
J. IRRIG. DRAINAGE DIV., A.S.C.E., VOL. 99, NO.
11, PP. 71-8, PROC. PAPER 9607, MARCH 1973.

73-0349 DESAI, C. S.
JOHNSON, L. D.
EVALUATION OF SOME NUMERICAL SCHEMES FOR
CONSOLIDATION
INT. J. NUM. METHODS IN ENG., VOL. 7, 1973, PP.
243-254.

73-0350 CHAKRABARTI, P.
CHOPRA, A. K.
EARTHQUAKE ANALYSIS OF GRAVITY DAMS INCLUDING
HYDRODYNAMIC INTERACTION
EARTHQUAKE ENGR. STRUCT. DYN., VOL. 2, NO. 2,
OCTOBER-DECEMBER 1973, PP. 143-160.

73-0351 ROBINSON, J.
A GENERAL EQUILIBRIUM BEAM ELEMENT WITH OFFSETS
RA/ARD, REPORT NO. 8-1-74-21, JANUARY 1974.

73-0352 SERVICE, K.G.
DOUGLAS, A.
BOUNDARIES AND FRACTURES IN FINITE ELEMENT MODELS
OF GEOLOGICAL STRUCTURES
GEOPHYSICAL JOURNAL OF THE ROYAL ASTRONOMICAL
SOCIETY 1973, VOL. 32, NO. 1, PP. 1-14

73-0353 LAWRENCE, S. J.
SVED, G.
FINITE ELEMENT ANALYSIS OF CLAD STRUCTURES
ACTA TECHNICA ACADEMIAE SCIENTIARUM HUNGARICAE
1973, VOL. 75, NO. 1-4, PP. 261-275.

73-0354 TABARROK, B.
GASS, N.
NEW VARIATIONAL PRINCIPLE WITH APPLICATION TO
VIBRATIONS OF SHALLOW SHELLS USING FINITE ELEMENTS
METHOD
ZEITSCHRIFT FUR ANGEWANDTE MATHEMATIK UND MECHANIK
1973, VOL. 53, NO. 12, PP. 773-781

73-0355 ANDERSON, G. P.
DEVRIES, K. L.
WILLIAMS, M. L.
FINITE ELEMENT IN ADHESION ANALYSES
INT. J. OF FRACTURE, VOL. 9, NO. 4, 1973, PP.
421-436.

73-0356 NEWMAN, M.
PIPANO, A.
FAST MODAL EXTRACTION IN NASTRAN VIA THE FEER
COMPUTER PROGRAM
NASTRAN: USER'S EXPERIENCES, NASA TECH, MEMO.,
NASA TM X-2893, SEPTEMBER 1973, PP. 485-506.

73-0357 CROUZEIX, M.
THOMAS, J. M.
FINITE ELEMENTS AND DEGENERATED ELLIPTIC PROBLEMS
REVUE FRANC. D'AUTOMAT., INFORMAT. ET RECHERCHE
OPERAT., VOL. 7, DECEMBER 1973, PP. 77-104.

73-0358 HRUDEY, T. M.
APPLICATION OF A REFINED SHALLOW SHELL FINITE
ELEMENT TO GEOMETRICALLY NONLINEAR STRUCTURES
C.A.S.I. TRANSACTIONS, VOL. 6, NO. 2, 1973, PP.
92-96.

73-0359 NEDELEC, J. C.
PLANCHARD, J.
VARIATION METHOD ON FINITE ELEMENTS FOR NUMERICAL
RESOLUTION TO PROBLEMS WITH EXTERNAL R3
REVUE FRANC. D'AUTOMAT., INFORMAT. ET RECHERCHE
OPERAT., VOL. 7, DECEMBER 1973, PP. 105-129.

73-0360 AAMODT, B.
BERGAN, P. G.
CALCULATION OF STRESS INTENSITY FACTORS AND
FATIGUE CRACK PROPAGATION OF SEMI-ELLIPTICAL,
PART-THROUGH SURFACE CRACKS
PROC. 2ND INT. CONF. ON PRESSURE VESSEL TECHNOL.,
SAN ANTONIO, TEXAS, OCTOBER 1-4, 1973, PP.
911-921.

73-0361 ABE, T.
ELASTIC DEFORMATION OF POLYCRYSTALLINE METALS - 3
BULL. JSME., VOL. 16, NO. 100, OCTOBER 1973, PP.
1540-1549.

73-0362 STRICKLIN, J. A.
HAISLER, W. E.
EVALUATION OF SOLUTION PROCEDURES FOR MATERIAL
AND/OR GEOMETRICALLY NONLINEAR STRUCTURAL ANALYSIS
AIAA JOURNAL, VOL. 11, 1973, PP. 292-299.

73-0363 ALLEN, M.
ELLIS, J. A.
STRESS AND STRAIN DISTRIBUTION IN THE VICINITY OF
INTERFERENCE FIT FASTENERS
US AIR FORCE SYST. COMMAND, AIR FORCE FLIGHT DYN.
LAB., TECH. REP. NO. AFFDL-TR-72-153, JANUARY
1973, P. 573.

73-0364 ALMROTH, B. O.
BROGAN, F. A.
EXTENSIONS TO THE STAGS COMPUTER CODE
US AIR FORCE SYST. COMMAND, AIR FORCE FLIGHT DYN.
LAB., TECH. REP. NO. AFFDL-TR-73-14, MARCH 1973,
67 PP.

73-0365 ODEN, J. T.
WU, S. T.
CHUNG, T. J.
ANALYSIS OF RAREFIED GAS FLOW THROUGH AN ARBITRARY
CROSS SECTION BY THE FINITE ELEMENT METHOD
DEVELOPMENTS IN MECHANICS, VOL. 7, PROC. OF THE
13TH MIDWESTERN MECHANICS CONFERENCE, 1973, PP.
289-299.

73-0366 ANDO, Y.
YAGAWA, G.
LOW CYCLE FATIGUE STRENGTH OF PIPING COMPONENTS
PROC. 2ND INT. CONF. ON PRESSURE VESSEL TECHNOL.,
SAN ANTONIO, TEXAS, OCTOBER 1-4, 1973, PP.
723-742.

73-0367 ANON
RUBBER IN ENGINEERING
POLYM. AGE., VOL. 4, NO. 5, MAY 1973, PP. 171-173.

73-0368 ARGYRIS, J. H.
MARECZEK, G.
THERMOMECHANICAL ANALYSIS OF STRUCTURES
PROC. 4TH CONF. ON DIMENSIONING AND STRENGTH CALC.
HUNG. ACAD. OF SCI., BUDAPEST, OCTOBER 1971, PP.
267-285. (PUBL. BY AKAD KIADO, BUDAPEST, HUNGARY,
1973)

73-0369 ARMEN, H.
SALEME, E.
NONLINEAR CRACK ANALYSIS WITH FINITE ELEMENTS
ASME AMD., VOL. 6, 1973 FOR MEET., DETROIT, MICH.,
NOVEMBER 1973, PP. 171-209.

73-0370 ATKINSON, P.
PREDICTION OF MARINE PROPELLER DISTORTION AND
STRESSES USING A SUPERPARAMETRIC THICK SHELL
FINITE ELEMENT MODEL
R. INST. NAV. ARCHIT., SUPPL. PAPER, VOL. 115,
NOVEMBER 1973, PP. 359-375.

73-0371 ATLURI, S.
ON THE HYBRID STRESS FINITE ELEMENT MODEL FOR
INCREMENTAL ANALYSIS OF LARGE DEFLECTION PROBLEMS
INT. J. SOLIDS STRUCT., VOL. 9, NO. 10, OCTOBER
1973, PP. 1177-1191.

73-0372 BACK, N.
BURDEKIN, M.
PRESSURE DISTRIBUTION AND DEFORMATIONS OF MACHINED
COMPONENTS IN CONTACT.
INT. J. MECH. SCI., VOL. 15, NO. 12, DECEMBER
1973, PP. 993-1010.

73-0373 BALAS, J.
NASCH, L.
VELKE PRIEHYBY DOSIEK METODOU KONECNYCH PRVKOV
(LARGE DEFLECTIONS OF PLATES ANALYZED BY FINITE
ELEMENT METHOD)
STAVEBNICKY CAS., VOL. 21, NO. 3-5, 1973, PP.
156-168.

73-0374 BANDLER, J. W.
POPOVIC, J. R.
IEEE-G-MTT INTERNATIONAL MICROWAVE SYMPOSIUM,
1973: APPLICATIONS IN THE 70'S
IEEE-G-MTT INT. MICROWAVE SYMP., DIG. OF TECH.
PAP., UNIV. OF COLO., BOULDER, JUNE 4-6, 1973.

73-0375 BARKER, R. M.
HATT, F.
ANALYSIS OF BONDED JOINTS IN VEHICULAR STRUCTURES
AIAA J., VOL. 11, NO. 12, DECEMBER 1973, PP.
1650-1654.

73-0376 BARLA, G.
INNAURATO, N.
INDIRECT TENSILE TESTING OF ANISOTROPIC ROCKS
ROCK MECH., FELSMECH., MEC. ROCHES, VOL. 5, NO. 4,
DECEMBER 1973, PP. 215-230.

73-0377 BARRY, W. R.
STRUCTURAL ANALYSIS OF PETROCHEMICAL EQUIPMENT BY
THE FINITE ELEMENT METHOD
ASME PAPER NO. 73-PET-41 FOR MEET. SEPTEMBER
16-20, 1973, 15 PP.

73-0378 BELYTSCHKO, T.
HSIEH, B. J.
NON-LINEAR TRANSIENT FINITE ELEMENT ANALYSIS WITH
CONVECTED CO-ORDINATES
INT. J. NUMER. METHODS ENG., VOL. 7, NO. 3, 1973,
PP. 255-271.

73-0379 BENZLEY, S. E.
HUTCHINSON, J. R.
DYNAMIC SHELL THEORY COUPLING THICKNESS STRESS
WAVE EFFECTS WITH GROSS STRUCTURAL RESPONSE
J. APPL. MECH. TRANS. ASME, VOL. 40, SER. E NO. 3,
SEPTEMBER 1973, PP. 731-735.

580

73-0380 BERRY, G. F.
KOENIG, H. A.
TRANSIENT RESPONSE OF ELASTIC, VISCO-PLASTIC BEAMS
ARCH. MECH. - ARCH. MECH. STOSOW., VOL. 25, NO. 4,
1973, PP. 593-612.

73-0381 BLANC, G.
BARADAT, Y.
UTILISATION DE MODELES MATHEMATIQUES
MULTIDIMENSIONNELS POUR L'ETUDE DES ECOULEMENTS
DANS LES DIGUES ET BARRAGES EN TERRE (USE OF
MULTIDIMENSIONAL MATHEMATICAL MODELS FOR EARTH DAM
SEEPAGE INVESTIGATION)
HOUILLE BLANCHE, VOL. 28, NO. 5-6, 1973, PP.
435-441.

73-0382 BLOKHOV, V. G.
ISSLEDOVANIE PROCHNOSTI I ZHESTKOSTI PLASTIN,
NAGRUZHENNYKH V SVOEI PLOSKOSTI, METHODOM
KONECHNYKH ELEMENTOV (INVESTIGATION OF THE
STRENGTH AND RIGIDITY OF PLATES, LOADED IN THEIR
PLANE, BY THE FINITE ELEMENT METHOD)
IZV VYSSH UCHEBN ZAVED, MASHINOSTR, NO. 6, 1973,
PP. 11-15.

73-0383 BOND, T. J.
SWANNELL, J. H.
COMPARISON OF SOME CURVED TWO DIMENSIONAL FINITE
ELEMENTS
J. STRAIN ANAL., VOL. 8, NO. 3, JULY 1973, PP.
182-190.

73-0384 BOUSQUET, R. D.
APPLICATION OF INTERACTIVE GRAPHICS TO THE
NUMERICAL METHODS USED IN STRUCTURAL ANALYSIS
IEEE INTERCON TECH. PAP. 1973, FOR MEET., NEW
YORK, NEW YORK, MARCH 26-30, 1973, VOL. 4, PAPER
27/2, 6 PP.

73-0385 BRATANOW, T.
ECER, A.
FINITE ELEMENT ANALYSIS AND COMPUTER GRAPHICS
VISUALIZATION OF FLOW AROUND PITCHING AND PLUNGING
AIRFOILS
NASA CONTRACT REP. CR-2249, SEPTEMBER 1973, 71 PP.

73-0386 BROUTMAN, L. J.
AGARWAL, B. D.
THEORETICAL STUDY OF THE EFFECT OF THE INTERFACE
ON COMPOSITE TOUGHNESS
SPI. REINF. PLAST. DIV., PROC., ANNUAL TECH. CONF.
28TH, WASHINGTON, D. C., FEB. 6-9, 1973, PAPER
SECT. 5-B, 8 PP.

73-0387 BROZ, P.
PROCHAZKA, P.
STABILITA STEN V PLASTICKEM OBORU (STABILITY IN
WEBPLATES IN PLASTIC RANGE)
STAVEBNICKY CAS., VOL. 21, NO. 8, 1973, PP.
669-683.

73-0388 BRUCH, J. C.
ZYVOLOSKI, G.
FINITE ELEMENT WEIGHTED RESIDUAL SOLUTION TO ONE
DIMENSIONAL FIELD PROBLEMS
INT. J. NUMER. METHODS ENG., VOL. 6, NO. 4, 1973,
PP. 577-585.

73-0389 CALLADINE, C. R.
NEW FINITE ELEMENT METHOD FOR ANALYZING
SYMMETRICALLY LOADED THIN SHELLS OF REVOLUTION
INT. J. NUMER. METHODS ENG., VOL. 6, NO. 4, 1973,
PP. 475-487.

73-0390 CROUZIEX, M.
RAVIART, P. A.
CONFORMING AND NONCONFORMING FINITE ELEMENT
METHODS FOR SOLVING THE STATIONARY STOKES
EQUATIONS - 1
REV. FR. AUTOM. INF. RECH. OPER. VOL. 7, DEC 1973,
PP. 33-76.

73-0391 CHANG, J. C.
FORSYTH, R. A.
STRESSES AND DEFORMATIONS IN JAIL GULCH EMBANKMENT
HIGHW. RES. REC., NO. 457, 1973, PP. 51-59.

73-0392 CHAPLIN, E. C.
CARRETT, R. J.
DEVELOPMENT OF A DESIGN FOR A PRECAST CONCRETE
BRIDGE BEAM OF U-SECTION
STRUCT. ENG., VOL. 51, NO. 10, OCTOBER 1973, PP.
383-388.

73-0393 CORUM, J. R.
SARTORY, W. K.
ELASTIC-PLASTIC-CREEP ANALYSIS OF THERMAL
RATCHETTING IN STRAIGHT PIPE AND COMPARISONS WITH
TEST RESULTS
ASME PAP. NO. 73-WA/FE-15 FOR MEET., NOV. 11-15,
1973, 11 PP.

73-0394 CRISFIELD, M. A.
LARGE-DEFLECTION ELASTO-PLASTIC BUCKLING ANALYSIS
OF PLATES USING FINITE ELEMENTS
TRANSP. RD. RES. LAB., (GB), TRRL, REP. LR. 593,
1973, 47 PP.

73-0395 CROUCH, S. L.
FAIRHURST, C. F.
ANALYSIS OF ROCK MASS DEFORMATIONS DUE TO
EXCAVATIONS
ASME AMD VOL. 3, 1973 FOR MEET., DETROIT, MICH.,
NOV., 1973, PP. 25-40, UNIV. OF MINN.,
MINNEAPOLIS.

73-0396 CUNNINGHAM, F. M.
LEANHARDT, D. E.
VIBRATION CHARACTERISTICS OF FREE THIN CYLINDRICAL
SHELLS
ASME PAP. NO. 73-DET-141 FOR MEET. SEPT. 9-12,
1973, 5 PP.

73-0397 DASGUPTA, S.
KOENEMAN, J. B.
BENDING PROBLEM OF AXIALLY CONSTRAINED BEAMS
ASME PAP. NO. 73-DE-J 1973, 4 PP.

73-0398 DAVIS, M. E.
WILLIAMS, E. W.
FINITE BOUNDARY CORRECTIONS TO THE COPLANAR
WAVEGUIDE ANALYSIS
IEEE TRANS. MICROWAVE THEORY TECH., VOL. MTT-21,
NO. 9, SEPT. 1973, PP. 594-596.

73-0399 DECRETON, M. C.
CALCUL DES CHAMPS ELECTROMAGNETIQUES PAR LA
METHODE DES ELEMENTS FINIS (COMPUTATION OF
ELECTROMAGNETIC FIELDS BY THE METHOD OF FINITE
ELEMENTS)
BULL. ASSOC. SUISSE ELECTR., VOL. 64, NO. 19,
1973, PP. 1196-1203

73-0400 DE FRAINE, J.
PRAKTISCHE RESULTATEN, DOEL EN NUT VAN HET
ONTWERPEN VAN MECHANISCHE ELEMENTEN MET BEHULP VAN
COMPUTERS (PRACTICAL RESULTS, OBJECTIVES, AND
EFFICIENCY OF DESIGNING OF MECHANICAL ELEMENTS BY
MEANS OF COMPUTERS)
INGENIEURSBLAD, VOL. 42, NO. 23, DEC. 1. 1973, PP.
692-696.

73-0401 DENKMANN, W. J.
NICKELL, R. E.
ANALYSIS OF STRUCTURAL ACOUSTIC INTERACTIONS IN
METAL CERAMIC TRANSDUCERS
IEEE TRANS. AUDIO ELECTROACOUST., VOL. AU-21, NO.
4, AUGUST 1973, PP. 317-324.

73-0402 DONG, S. B.
WOLF, J. A.
EFFECT OF TRANSVERSE SHEAR DEFORMATION IN
VIBRATIONS OF PLANAR STRUCTURES COMPOSED OF BEAM
TYPE ELEMENTS
J. ACOUST. SOC. AM., VOL. 53, NO. 1, JANUARY 1973,
PP. 120-127.

73-0403 DUNCAN, J. M.
LEFEBVRE, G.
EARTH PRESSURES ON STRUCTURES DUE TO FAULT
MOVEMENT
ASCE J. SOIL MECH. FOUND. DIV., VOL. 99, NO. SM12,
DEC. 1973, PAPER 10237, PP. 1153-1156.

73-0404 DUNGLAS, J.
LOUDIERE, D.
NOUVELLE CONCEPTION DES DRAINS DANS LES BARRAGES
EN TERRE HOMOGENE DE PETITE ET MOYENNE DIMENSIONS
(NEW DRAIN DESIGN FOR MEDIUM SIZED AND FOR SMALL
HOMOGENEOUS EARTH DAMS)
HOUILLE BLANCH, VOL. 28, NO. 5-6, 1973, PP.
461-466.

73-0405 EBEL, H.
DIE BEULLASTEN DIAGONAL VERSTEIFTER
RECHTECKPLATTEN (BUCKLING LOADS IN DIAGONALLY
STIFFENED RECTANGULAR PANELS)
STAHLBAU, VOL. 42, NO. 8, AUGUST 1973, PP.
225-236.

73-0406 ELTER, C.
WALDNER, K.
STRESS ANALYSIS OF BWR PRESSURE VESSEL WITH
BERSAFE AND FLHE
PROC. 2ND INT. CONF. ON PRESSURE VESSEL TECHNOL.,
SAN ANTONIO, TEXAS, OCTOBER 1-4, 1973, PP.
221-238.

73-0407 FELIPPA, C. A.
SHARIFI, P.
COMPUTER IMPLEMENTATION OF NONLINEAR FINITE
ELEMENT ANALYSIS
A.S.M.E. A.M.D., VOL. 6, 1973, PP. 31-49.

73-0408 FOYE, R. L.
THEORETICAL POST YIELDING BEHAVIOUR OF COMPOSITE
LAMINATES -2.
J. COMPOS. MATER., VOL. 7, JULY 1973, PP. 310-319.

73-0409 FRIED, I.
BOUNDS ON THE SPECTRAL AND MAXIMUM NORMS OF THE
FINITE ELEMENT STIFFNESS, FLEXIBILITY AND MASS
MATRICES
INT. J. SOLIDS STRUCT., VOL. 9, NO. 9, SEPTEMBER
1973, PP. 1013-1034.

73-0410 GAMBOLATI, G.
DIAGONALLY DOMINANT MATRICES FOR THE FINITE
ELEMENT METHOD IN HYDROLOGY
INT. J. NUMER. METHODS ENG., VOL. 6, NO. 4, 1973,
PP. 587-594.

73-0411 GHABOUSSI, J.
WILSON, E. L.
FINITE ELEMENT FOR ROCK JOINTS AND INTERFACES
A.S.C.E. J., SOIL MECH. FOUND. DIV., VOL. 99, NO.
SM10, PAPER NO. 10095, OCTOBER 1973, PP. 833-884.

73-0412 GHABOUSSI, J.
WILSON, E. L.
SEISMIC ANALYSIS OF EARTH DAM RESERVOIR SYSTEMS
ASCE J. SOIL MECH. FOUND. DIV., VOL. 99, NO. SM10,
OCTOBER 1973, PAP. NO. 10053, PP. 849-886.

73-0413 GUPTA, K. K.
ON A COMBINED STURM SEQUENCE AND INVERSE ITERATION
TECHNIQUE FOR EIGENPROBLEM SOLUTION OF SPINNING
STRUCTURES
INT. J. NUMER. METHODS ENG., VOL. 7, NO. 4, 1973,
PP. 509-518.

73-0414 HAALAND, E.
TESTRAN PROGRAMS, A TOOL FOR TEMPERATURE AND
STRESS ANALYSIS
NOR. MARIT. RES. VOL. 1, NO. 4, 1973, PP. 12-20.

73-0415 HARDY, C. H.
MARCAL, P. V.
ELASTIC ANALYSIS OF A SKULL
ASME PAPER NO. 73-APMW-36 FOR MEET. SEPT. 17-19,
1973, 5 PP.

73-0416 HAWARD, R. N.
OWEN, D. R. J.
YIELDING OF A TWO DIMENSIONAL VOID ASSEMBLY IN AN
ORGANIC GLASS
J. MATER. SCI., VOL. 8, NO. 8, AUGUST 1973, PP.
1136-1144.

73-0417 HENRY, R.
LALANNE, M.
VIBRATION ANALYSIS OF ROTATING COMPRESSOR BLADES
ASME PAP. NO. 73-DET-140 FOR MEET. SEPT. 9-12,
1973, 8 PP.

73-0418 HIBBITT, H. D.
MARCAL, P. V.
NUMERICAL, THERMO-MECHANICAL MODEL FOR THE WELDING
AND SUBSEQUENT LOADING OF A FABRICATED STRUCTURE
COMPUT. STRUCT. VOL. 3, NO. 5, SEPT. 1973, PP.
1145-1174.

73-0419 HIBBITT, H. D.
SORENSEN, E. P.
ELASTIC PLASTIC AND CREEP ANALYSIS OF PIPELINES BY
FINITE ELEMENTS
INT. CONF. ON PRESSURE VESSEL TECHNOL. 2ND PROC.
PAP. SAN ANTONIO, TEXAS, OCTOBER 1-4, 1973, PP.
239-251.

73-0420 HINTON, E.
DAVIES, J. D.
LIMITATIONS OF THE FINITE ELEMENT METHOD IN
STRUCTURAL ANALYSIS
BUILD INT. (ENGL. ED), VOL. 6, NO. 3, MAY-JUNE
1973, PP. 299-319.

73-0421 HOOPER, J. A.
OBSERVATIONS ON THE BEHAVIOUR OF A PILED-RAFT
FOUNDATION ON LONDON CLAY.
PROC. INST. CIV. ENG (LONDON), VOL. 55, PART 2,
DECEMBER 1973, PP. 855-877.

73-0422 IMAMASA, J.
URAGAMI, K.
EXPERIMENTAL STUDY OF FLEXIBILITY FACTORS AND
STRESSES OF WELDING ELBOWS WITH END EFFECTS
PROC. 2ND. INT. CONF. ON PRESSURE VESSEL TECHNOL.,
SAN ANTONIO, TEXAS, OCTOBER 1-4, 1973, PP.
417-426.

73-0423 ISAACS, L. T.
CURVED CUBIC TRIANGULAR FINITE ELEMENT FOR
POTENTIAL FLOW PROBLEMS
INT. J. NUMER. METHODS ENG., VOL. 7, NO. 3, 1973,
PP. 337-344.

73-0424 ISENBERG, J.
LEE, L.
RESPONSE OF STRUCTURES TO COMBINED BLAST EFFECTS
ASCE TRANSP. ENG. J. VOL. 99, NO. TE4, NOVEMBER
1973, PAP. 10166, PP. 887-908.

73-0425 JONES, R. F.
COSTELLO, M. G.
SOLUTION PROCEDURE FOR NONLINEAR STRUCTURAL
PROBLEMS
ASME AMD, VOL. 6, 1973 FOR MEET., DETROIT, MICH.,
NOV. 1973, PP. 157-169.

73-0426 KALAJDZIC, M.
DOSTIGNUCA I TENDENCIJE U ISPITIVANJIMA ALATNIH
MASINA (ACHIEVEMENTS AND TENDENCIES IN MACHINE
TOOLS TESTING)
INST. ALATNE MASINE. ALATE. SAOPSTENJA. NO. 18,
1973, PP. 2393-2416.

73-0427 KARLSSON, B. I.
SOZEN, M. A.
PRESTRESSED CONCRETE DEEP SLABS WITH OPENINGS
NUCL. ENG. DES., VOL. 25, NO. 2, JULY 1973, PP.
290-300.

73-0428 SANDHU, R. S.
KARWOSKI, W.
NEW HORIZONS IN ROCK MECHANICS
PROC. 14TH SYMP. ON ROCK MECHANICS, PENN. STATE
UNIV., UNIV.PARK, JUNE 11-14, 1972, 785 PP. (PUBL.
BY A.S.C.E. NEW YORK, 1973)

73-0429 KAWAI, T.
APPLICATION OF FINITE ELEMENT METHODS TO SHIP
STRUCTURES
COMPUT. STRUCT., VOL. 3, NO. 5, SEPT. 1973, PP.
1175-1194.

73-0430 KENNEDY, F. E.
LING, F. S.
THERMAL, THERMOELASTIC, AND WEAR SIMULATION OF A
HIGH-ENERGY SLIDING CONTACT PROBLEM
PAPER NO. 73-LUB-6, A.S.M.E. MEETING, OCTOBER
14-19, 1973, 9 PP.

73-0431 KHOT, N. S.
VENKAYYA, V. B.
APPLICATION OF OPTIMALLY CRITERION TO
FIBER-REINFORCED COMPOSITES
US AIR FORCE SYST. COMMAND, AIR FORCE FLIGHT DYN.
LAB., TECH. REP. NO. AFFDL-TR-73-6, MAY 1973, 45
PP.

73-0432 KNUTELSKY, B. G.
DEMITRACK, G.
FINITE ELEMENT STRESS ANALYSIS OF ARTILLERY
PROJECTILES
PAPER NO. MS73-955, A.S.M.E. MEETING, OCTOBER 1-4,
1973, 14 PP.

73-0433 KOBAYASHI, A. S.
STABERG, L. G.
MECHANICS OF MACKAY-MARG TONOMETER
PAPER NO. 73-WA/BIO-21, A.S.M.E. MEETING, NOV.
11-15, 8 PP.

73-0434 KOBZA, W.
LIPINSKI, J.
METODA ELEMENTOW SKONSZONYCH ROZWIAZYWANIA
OSIOWO-SYMETRYCZNEGO PRZEWODNICTWA CIEPLA
(FINITE-ELEMENT METHOD FOR SOLVING THE PROBLEM OF
AXI-SYMMETRIC HEAT FLOW)
ARCH BUDOWY. MASZ., VOL. 20, NO. 3, 1973, PP.
491-501.

73-0435 KOBZA, W.
MULTIANGULAR ELEMENTS IN PLATE BENDING
INT. J. NUMER. METHODS ENG., VOL. 7, NO. 4, 1973,
PP. 545-551.

73-0436 KOLAR, V.
NEMEC, I.
EFFICIENT FINITE ELEMENT ANALYSIS OF RECTANGULAR
AND SKEW LAMINATED PLATES
INT. J. NUMER. METHODS ENG., VOL. 7, NO. 3, 1973,
PP. 309-323.

73-0437 KRASNIKOV, N. D.
TROITSKII, A. P.
RASCHET METODOM KONECHNYKH ELEMENTOV
DINAMICHESKIKH KHARAKTEERISTIK ZEMLYANOI PLOTINY
SOVMESTNO S OSNOVANIEM (FINITE ELEMENTS ANALYSIS
OF DYNAMIC CHARACTERISTICS OF AN EARTHFILL DAM
JOINTLY WITH ITS FOUNDATION)
GIDROTEKH STROIT, NO. 8, AUGUST 1973, PP. 19-23.

73-0438 KUESEL, T. R.
SCHMIDT, B.
SETTLEMENTS AND STRENGTHENING OF SOFT CLAY
ACCELERATED BY SAND DRAINS
HIGHW. RES. REC. NO. 457, 1973, PP. 18-26.

73-0439 LAM, P.
GREGORY, M. S.
MEASUREMENT OF SLOPE CHANGE ON CYLINDRICAL SHELL
SURFACES
INST. ENG. AUST. CIV. ENG. TRANS., VOL. CE-15, NO.
1-2, 1973, PP. 95-98.

73-0440 LAWTON, C. W.
STRAIN CONCENTRATION BEHAVIOUR IN A NOTCHED ROUND
BAR SUBJECTED TO CREEP
INT. CONF. ON PRESSURE VESSEL TECHNOL., 2ND PROC.
PAP., SAN ANTONIO, TEXAS, OCTOBER 1-4, 1973, PP.
253-262.

73-0441 LIKINS, P. W.
BARBERA, F. J.
MATHEMATICAL MODELING OF SPINNING ELASTIC BODIES
FOR MODAL ANALYSIS
AIAA J., VOL. 11, NO. 9, SEPT. 1973, PP.
1251-1258.

73-0442 LEMAIRE, M.
DUHAU, R.
MATRICE DE RIGIDITE D'UN ELEMENT DE POUTRE
SOLLICITE A LA TORSION-FLEXION (STIFFNESS MATRIX
OF A BEAM SECTION SUBJECTED TO TORSIONAL AND
BENDING STRESSES)
CONSTR. MET., NO. 3, SEPT., 1973, PP. 19-26.

73-0443 LEONE, S. G.
PERLMAN, A. B.
NUMERICAL STUDY OF DAMPING IN VISCOELASTIC
SANDWICH BEAMS
PAPER NO. 73-DET-73, A.S.M.E. MEETING, SEPTEMBER
9-12, 1973, 8 PP.

73-0444 LEWIS, R. W.
HUMPHESON, C.
NUMERICAL ANALYSIS OF ELECTRO-OSMOTIC FLOW IN
SOILS
ASCE J. SOIL. MECH. FOUND. DIV., VOL. 99, NO. SM8,
AUGUST 1973, PAPER 9912, PP. 603-616.

73-0445 LOZIUK, L. A.
ANDERSON, J. C.
TRANSIENT HYDROTHERMAL ANALYSIS OF SMALL LAKES
ASCE J. POWER DIV., VOL. 99, NO. PO2, NOVEMBER
1973, PAPER NO. 10150, PP. 349-364.

73-0446 LUNDQUIST, R. G.
HEINS, R. W.
NEW HORIZONS IN ROCK MECHANICS
PROC. 14TH SYMP. ON ROCK MECH., PENN. STATE UNIV.,
UNIV. PARK, JUNE 11-14, 1972, 785 PP. (PUBL. BY
A.S.C.E., NEW YORK, 1973)

73-0447 MAKSIMOVIC, M.
OPTIMUM POSITION OF THE CENTRAL CLAY CORE OF A
ROCKFILL DAM IN RESPECT TO ARCHING AND HYDRAULIC
FRACTURE
TRANS. 11TH INT. CONGR. ON LARGE DAMS, MADRID,
SPAIN, JUNE 11-15, 1973, VOL. 3, PP. 789-800.

73-0448 MAIER, G.
DRUCKER, D. C.
EFFECTS OF GEOMETRY CHANGE ON ESSENTIAL FEATURES
OF INELASTIC BEHAVIOUR
ASCE J. ENG. MECH. DIV., VOL. 99, NO. EM4, AUGUST
1973, PAP. 9914, PP. 819-834.

73-0449 MALNIG, H. W.
FINITE ELEMENTE ZUR BERECHNUNG DER
WAERMEUEBERTRAGUNG FUER EBENE UND AXISYMMETRISCHE
PROBLEME (FINITE ELEMENTS FOR CALCULATION OF THE
THERMAL CONDUCTIVITY FOR PLANE AND AXISYMMETRIC
PROBLEMS)
OESTERR. ING-Z., VOL. 16, NO. 9, SEPT. 1973, PP.
298-304.

73-0450 MARTIN, H.
ZUSAMMENHANG ZWISCHEN OBERFLAECHENBESCHAFFENHEIT,
VERBUND UND SPRENGWIRKUNG VON BEWEHRUNGSSTAEHLEN
UNTER KURZZEITBELASTUNG (RELATIONSHIP AMONG THE
SURFACE PROPERTIES, BOND AND TRANSVERSE CRACKING
OF REINFORCING STEELS UNDER SHORT-TERM LOAD)
DTSCH. AUSSCHUSS. STAHLBETON., NO. 228, 1973, 50
PP.

73-0451 MEEK, J. L.
MEMBRANE ANALYSIS OF UNFRAMED HIGH RISE BUILDINGS
INST. ENG. AUST. CIV. ENG. TRANS., VOL. CE-15, NO.
1-2, 1973, PP. 87-89.

73-0452 MELOSH, R. J.
INHERITED ERROR IN FINITE ELEMENT ANALYSES OF
STRUCTURES
COMPUT. STRUCT., VOL. 3, NO. 5, SEPT. 1973, PP.
1205-1217.

73-0453 MILLER, W. F.
LEWIS, E. E.
APPLICATION OF PHASE SPACE FINITE ELEMENTS TO THE
TWO DIMENSIONAL NEUTRON TRANSPORT EQUATION IN X-Y
GEOMETRY
NUCL. SCI. ENG., VOL. 52, NO. 1, SEPT. 1973, PP.
12-22.

73-0454 MIZERNYUK, G. N.
IVASHCHENKO, N. A.
OPREDELENIE STATSIONARNYKH TEMPERATURNYKH POLEI V
DETALYAKH DVIGATELEI VNUTRENNEGO SGORANIYA METODOM
ELEMENTA (DETERMINATION OF STATIONARY TEMPERATURE
FIELDS IN INTERNAL COMBUSTION ENGINE PARTS BY THE
ELEMENT METHOD)
IZV. VYSSH. UCHEBN. ZAVED., MASHINOSTR., NO. 6,
1973, PP. 112-115.

73-0455 MOAN, T.
ON SHELL EFFECTS ON FERRO-CEMENT VESSELS
NOR. MARIT. RES., VOL. 1, NO. 4, 1973, PP. 1-6.

73-0456 MUELLER, W.
DIE BERECHNUNG DER BEWEGUNGEN UND SPANNUNGEN DES
GEBIRGES VOM ABBAU BIS ZUR TAGESOBERFLAECHE NACH
DER METHODE DER ENGLISCHEN ELEMENTE (CALCULATION
OF DISPLACEMENTS AND STRESS IN ROCK STRATA BETWEEN
THE MINING EXPLOITATION AND THE SURFACE BY MEANS
OF FINITE ELEMENT METHOD)
GLUECKAUF-FORSCHUNGSH, VOL. 34, NO. 6, DEC. 1973,
PP. 228-236.

73-0457 BATHE, K. J.
RAMM, E.
WILSON, E. L.
FINITE ELEMENT FORMULATIONS FOR LARGE DISPLACEMENT
AND LARGE STRAIN ANALYSIS
SESM REPORT NO. 73-14, DEPT. OF CIVIL ENG. UNIV.
OF CALIF., BERKELEY, 1973.

73-0458 NAGAMATSU, A.
MUROTA, T.
ELASTIC, PLASTIC BEHAVIOUR OF ROTO - 1
BULL. JSME., VOL. 16, NO. 100, OCTOBER 1973, PP.
1532-1539.

73-0459 NEUMAN, S. P.
SATURATED-UNSATURATED SEEPAGE BY FINITE ELEMENTS
ASCE J. HYDRAUL. DIV., VOL. 99, NO. HY12, DECEMBER
1973, PAP. 10201, PP. 2233-2250.

73-0460 NICKELL, R. E.
MARCAL, P. V.
IN-VACUO MODAL DYNAMIC RESPONSE OF THE HUMAN SKULL
ASME PAPER NO. 73-DET-112 FOR MEET. SEPT. 6-12,
1973, 5 PP.

73-0461 NIKITENKO, V. I.
SOKOLOV, V. F.
CHISLENNYI METOD RASCHETA SOBSTVENNYKH I
VYNUZHDENNYKH KOLEBANII SOSTAVNYKH OBOLOCHECHNYKH
KONSTRUKTSII (NUMERICAL METHOD OF CALCULATION OF
NATURAL AND FORCED VIBRATIONS OF COMPOUND SHELL
STRUCTURES)
IZV. VYSSH. UCHEBN. ZAVED., MASHINOSTR., NO. 13,
1973, PP. 14-19.

73-0462 NIXON, C. D.
MANUEL, R. F.
FINITE ELEMENT ANALYSIS OF A DIAGONAL CRACKING
BEHAVIOUR IN REINFORCED CONCRETE
ENG. J. (MONTREAL), VOL. 56, NO. 7-8, JULY-AUGUST
1973, PP. I-V.

73-0463 OHJI, K.
OGURA, K.
FINITE ELEMENT APPROACH TO THE CRITICAL CYCLIC
STRESS REQUIRED TO PROPAGATE A CRACK
PROC. 16TH JAP. CONGR. ON MATER. RES., OSAKA,
JAPAN, AUGUST 1972, PP. 45-48. (PUBL. BY SOC. OF
MATER. SCI., KYOTO, JAPAN, 1973).

73-0464 OHNO, T.
AKAGI, T.
VISCOELASTIC ANALYSIS OF PRESTRESSED CONCRETE
SHELL OF REVOLUTION
PROC. 2ND INT. CONF. ON PRESSURE VESSEL TECHNOL.,
SAN ANTONIO, TEXAS, OCTOBER 1-4, 1973, PP.
469-483.

73-0465 ORLIN, A. S.
IVASHCHENKO, N. A.
RASCHET POLEI DEFORMATSII I NAPRYAZHENII V
DETALYAKH DVIGATELEI VNUTRENNEGO SGORANIYA
(CALCULATION OF THE DEFORMATION AND STRESS FIELDS
IN COMPONENTS OF INTERNAL COMBUSTION ENGINES)
IZV. VYSSH. UCHEBN. ZAVED, MASHINOSTR., NO. 12,
1973, PP. 87-91.

73-0466 OVEROYE, K. R.
APRAHAMIAN, R.
HOLOGRAPHIC INSTRUMENTATION OF TURBINE BLADES
HOLOG. AND OPT. FILTERING, CONF. PROC. HUNTSVILLE,
ALABAMA, MAY 24-25, 1971, PP. 15-26, NASA SP-299,
WASHINGTON, D.C., 1973.

73-0467 OWEN, D. R. J.
GRIFFITHS, J. R.
STRESS INTENSITY FACTORS FOR CRACKS IN A PLATE
CONTAINING A HOLE AND IN A SPINNING DISC
INT. J. FRACT., VOL. 9, NO. 4, DECEMBER 1973, PP.
471-476.

73-0468 PARDOEN, G. C.
DEFLECTION FUNCTION FOR THE ASYMMETRICAL BENDING
OF CIRCULAR PLATES
AIAA J., VOL. 11, NO. 9, SEPTEMBER 1973, PP.
1341-1342.

73-0469 PERSON, P.
CORRELATION BETWEEN PRESSURE AND THERMAL TRANSIENT
STRESS LEVELS DERIVED BY THEORY AND EXPERIMENT ON
AN ASYMMETRICAL PUMP CASING
PROC. 2ND INT. CONF. ON PRESSURE VESSEL TECHNOL.,
SAN ANTONIO, TEXAS, OCTOBER 1-4, 1973, PP.
611-622.

73-0470 GEORGESON, D. L.
PHILLIPS, R. V.
ENVIRONMENTAL CONSIDERATIONS OF DAM CONSTRUCTION
AND OPERATION IN SEISMICALLY ACTIVE URBAN AREAS
TRANS. 11TH INT. CONGR. ON LARGE DAMS, MADRID,
SPAIN, JUNE 11-15, 1973, VOL. 1, PP. 255-270.

73-0471 PIZIALI, R. L.
DYNAMIC TORSIONAL RESPONSE OF THE HUMAN LEG
RELATIVE TO SKIING INJURIES
A.S.M.E. MEETING, DETROIT, MICH., NOVEMBER 1973,
VOL. 4, PP. 305-315.

73-0472 POPE, G. G.
OPTIMUM DESIGN OF STRESSED SKIN STRUCTURES
AIAA J., VOL. 11, NO. 11, NOVEMBER 1973, PP.
1545-1552.

73-0473 PUGH, J. W.
ROSE, R. M.
STRUCTURAL MODEL FOR THE MECHANICAL BEHAVIOUR OF
TRABECULAR BONE
J. BIOMECH., VOL. 6, NO. 6, NOVEMBER 1973, PP.
657-670.

73-0474 RAESTAD, A. E.
SONTVEDT, T.
MARINE PROPELLER BLADES STRESSES AND DEFORMATIONS
WITH GIVEN, FROZEN, HYDRODYNAMIC LOADING
NORW. MARIT. RES., VOL. 1, NO. 3, 1973, PP. 1-12.

73-0475 RAJASEKARAN, S.
MURRAY, D. W.
INCREMENTAL FINITE ELEMENT MATRICES
ASCE J. STRUCT. DIV., VOL. 99, NO. ST12, DECEMBER
1973, PAPER NO. 10226, PP. 2423-2438.

73-0476 ROBINS, P. J.
KONG, F. K.
MODIFIED FINITE ELEMENT METHOD APPLIED TO R C DEEP
BEAMS
CIV. ENG. PUBLIC WORKS REV., VOL. 68, NO. 808,
NOVEMBER 1973, PP. 963-966.

73-0477 RUTLEDGE, P. C.
GOULD, J. P.
MOVEMENTS OF ARTICULATED CONDUITS UNDER EARTH DAMS
ON COMPRESSIBLE FOUNDATIONS
EMBANKMENT-DAM ENG., PP. 209-237, PUBL. BY
WILEY-INTERSCIENCE, DIVISION OF JOHN WILEY AND
SONS, INC., NEW YORK, 1973.

73-0478 SABIR, A. B.
APPLICATION OF THE FINITE ELEMENT METHOD TO THE
BUCKLING OF RECTANGULAR PLATES AND PLATES ON
ELASTIC FOUNDATION
STAVEBNICKY CAS., VOL. 21, NO. 10, 1973, PP.
689-711.

73-0479 SAITO, T.
DESIGN OF CONCRETE TRACK SLAB ON ELASTIC
FOUNDATION USING THE FINITE ELEMENT METHOD
PROC. JAP. SOC. CIV. ENG., NO. 219, NOVEMBER 1973,
PP. 83-93

73-0480 SAKAI, F.
EQUIVALENCY BETWEEN FINITE ELEMENT METHOD AND
FINITE DIFFERENCE METHOD AND A NEW DISCRETIZATION
APPROACH
PROC. JAP. SOC. CIV. ENG., NO. 220, DECEMBER 1973,
PP. 39-52

73-0481 SANDER, G.
BON, C.
FINITE ELEMENT ANALYSIS OF SUPERSONIC PANEL
FLUTTER
INT. J. NUMER METHODS ENG., VOL. 7, NO. 3, 1973,
PP. 379-394

73-0482 SATA, T.
OKUBO, N.
COMPUTERUNTERSTUETZTES SYSTEM FUER DIE
CONSTRUKTION DES WERKZEUGMASCHINENAUFBAUS
(COMPUTER-AIDED SYSTEM FOR MACHINE TOOL DESIGN)
WERKSTATT BETR., VOL. 106, NO. 9, SEPTEMBER 1973,
PP. 643-648.

73-0483 SATOH, K.
UEDA, Y.
RECENT TRENDS OF RESEARCH INTO RESTRAINT STRESSES
AND STRAINS IN RELATION TO WELD CRACKING
WELD WORLD, SOUDAGE MONDE., VOL. 11, NO. 5-6,
1973, PP. 133-156.

73-0484 SCHIFFNER, K.
KASTIL, H.
HYBRIDE FINITE VERFAHREN FUER LINEARE UND
NICHTLINEARE INSTATIONAERE WAERMELEITPROBLEME
(HYBRID FINITE-ELEMENT METHODS FOR LINEAR AND
NONLINEAR NON-STATIONARY HEAT CONDUCTION PROBLEMS)
Z FLUGWISS., VOL. 21, NO. 8, AUGUST 1973, PP.
283-293.

73-0485 SCHREYER, H. L.
SHIH, P.
LOWER BOUNDS TO COLUMN BUCKLING LOADS
ASCE J. ENG. MECH. DIV., VOL. 99, NO. EM5, OCTOBER
1973, PAPER 10064, PP. 1011-1022.

73-0486 SHANNON, R. W. E.
CRACK GROWTH MONITORING BY STRAIN SENSING
INT. J. PRESSURE VESSELS PIPING., VOL. 1, NO. 1,
JANUARY 1973, PP. 61-73.

73-0487 SHELLY, P. D.
ETTLES, C.
FINITE ELEMENT METHOD FOR THE CALCULATION OF LOCUS
PATHS IN DYNAMICALLY LOADED BEARINGS
INST. MECH. ENG. (LONDON), PROC. VOL. 187, NO. 5,
1973, PP. 79-86.

73-0488 SINGA-RAO, K.
AMBA-RAO, C. L.
VIBRATION OF BEAMS WITH OVERHANGS
AIAA J., VOL. 11, NO. 10, OCTOBER 1973, PP.
1455-1456

73-0489 SKOGH, J.
STERN, P.
POSTBUCKLING BEHAVIOUR OF A SECTION REPRESENTATIVE
OF THE B-1 AFT INTERMEDIATE FUSELAGE
US AIR FORCE SYSTEM COMMAND, AIR FORCE FLIGHT DYN.
LAB., TECH. REP. NO. AFFDL-TR-73-63, MAY 1973, 41
PP.

73-0490 SMITH, B. S.
RAHMAN, K. M. K.
THEORETICAL STUDY OF THE SEQUENCE OF FAILURE IN
PRECAST PANEL SHEAR WALLS
PROC. INST. CIV. ENG. (LONDON), VOL. 55, PART 2,
SEPTEMBER 1973, PP. 581-592.

73-0491 SPAAS, H. A. C. M.
DETERMINATION OF STRESS CONCENTRATIONS AT NOZZLE
TO CYLINDER INTERSECTIONS AND COMPARISON WITH
EXPERIMENTAL ANALYSIS
PROC. 2ND INT. CONF. ON PRESSURE VESSEL TECHNOL.,
SAN ANTONIO, TEXAS, OCTOBER 1-4, 1973, PP.
155-165.

73-0492 STACEY, T. R.
THREE-DIMENSIONAL CONSIDERATION OF THE STRESSES
SURROUNDING OPEN-PIT MINE SLOPES
INT. J. ROCK MECH. MIN. SCI., VOL. 10, NO. 6,
NOVEMBER 1973, PP. 523-533.

73-0493 STASENKO, I. V.
KULIKOV, Y. A.
TREUGOL'NY I ELEMENT DLYA KRUGOVOI
TSILINDRICHESKOI OBOLOCHKI (TRIANGULAR FINITE
ELEMENT FOR CIRCULAR CYLINDRICAL SHELLS)
IZV VYSSH UCHEBN ZAVED., MASHINOSTR., NO. 12,
1973, PP. 5-8.

73-0494 SUIDAN, M.
SCHNOBRICH, W. C.
FINITE ELEMENT ANALYSIS OF REINFORCED CONCRETE
ASCE J. STRUCT. DIV., VOL. 99, NO. ST10, OCTOBER
1973, PAPER NO. 10081, PP. 2109-2122.

73-0495 SUMPTER, J. D. G.
TURNER, C. E.
FRACTURE ANALYSIS IN AREAS OF HIGH NOMINAL STRAIN
PROC. 2ND INT. CONF. ON PRESSURE VESSEL TECHNOL.,
SAN ANTONIO, TEXAS, OCTOBER 1-4, 1973, PP.
1095-1103.

73-0496 SUNDARARAJAN, C.
RESPONSE OF RECTANGULAR ORTHOTROPIC PLATES TO
RANDOM EXCITATIONS
EARTHQUAKE ENG. STRUCT. DYN., VOL. 2, NO. 2,
OCTOBER-DECEMBEER 1973, PP. 161-170.

73-0497 TADROS, G. S.
GHALI, A.
CONVERGENCE OF SEMIANALYTICAL SOLUTION OF PLATES
ASCE J. ENG. MECH. DIV., VOL. 99, NO. EM5, OCTOBER
1973, PAPER 10054, PP. 1023-1035.

73-0498 TANNER, R. I.
DIE-SWELL RECONSIDERED: SOME NUMERICAL SOLUTIONS
USING A FINITE ELEMENT PROGRAM
APPL. POLY. SYMP., NO. 20, 1973, PP. 201-208.

73-0499 TEBEDGE, N.
TALL, L.
LINEAR STABILITY ANALYSIS OF BEAM-COLUMNS
ASCE J. STRUCT. DIV., VOL. 99, NO. ST12, DECEMBER
1973, PAPER NO. 10232, PP. 2439-2457.

73-0500 TELEGA, J. J.
METODA ELEMENTOW SKONSZONYCH W MECHANICE GRUNTOW I
MECHANICE GOROTWORU (FINITE ELEMENT METHOD IN SOIL
- AND ROCK MECHANICS)
MECH. TEOR. I. STOSOW., VOL. 11, NO. 3, 1973, PP.
195-210.

73-0501 TESK, J. A.
         WIDERA, O.
         STRESS DISTRIBUTION IN BONE ARISING FROM LOADING
         ON ENDOSTEAL DENTAL IMPLANTS
         J. BIOMED. MATER. RES., BIOMED. MATER. SYMP., SEP.
         4, VOL. 7, NO. 3, 1973, PP. 251-261.

73-0502 THAMM, B. R.
         ANWENDUNG   DER   FINITE-ELEMENT-METHODE   ZUR
         BERECHNUNG VON SPANNUNGEN IN WASSERGESAETTIGTEN
         BOEDEN (FINITE ELEMENT METHOD USED FOR
         DETERMINATION OF STRESSES IN SATURATED SOILS)
         BAUINGENIEUR., VOL. 48, NO. 10, OCTOBER 1973, PP.
         370-374.

73-0503 THELEN, J. F.
         THORNTON, C. H.
         COMPUTERIZED STRUCTURAL ANALYSIS OF THE WORLD'S
         LARGEST LIGHT-GAGE STEEL PRIMARY STRUCTURAL SYSTEM
         COMPUT. STRUCT. VOL. 3, NO. 5, SEPTEMBER 1973, PP.
         1251-1262

73-0504 THOMAS, M. D.
         KNIGHT, W. A.
         INTERNATIONAL MACHINE TOOL DESIGN AND RESEARCH
         CONFERENCE, 13TH PROCEEDINGS, 1972
         PROC. 13TH INT. MACH. TOOL DES. AND RES. CONF.,
         BIRMINGHAM, ENGLAND, SEPTEMBER 18-22, 1972. (PUBL.
         BY MACMILLAN PRESS LTD., LONDON, ENGLAND, 1973)

73-0505 THRESHER, R. W.
         SAITO, G. E.
         STRESS ANALYSIS OF HUMAN TEETH
         J. BIOMECH., VOL. 6, NO. 5, SEPTEMBER 1973, PP.
         443-449.

73-0506 TILLERSON, J. R.
         STRICKLIN, J. A.
         NUMERICAL METHODS FOR THE SOLUTION OF NONLINEAR
         PROBLEMS IN STRUCTURAL ANALYSIS
         A.S.M.E. A.M.D., VOL. 6, 1973, PP. 67-101.

73-0507 TOENSHOFF, H. K.
         SCHMOHL, H. P.
         MESSUERFAHREN RECHENVERFAHREN ZUR ERMITTLUNG
         DREIACHSIGER EIGENSPANNUNGSZUSTAENDE
         IND-ANZ., VOL. 95, NO. 76, SEPTEMBER 11, 1973, PP.
         1760-1764

73-0508 TONG, P.
         PIAN, T. H. H.
         HYBRID-ELEMENT APPROACH TO CRACK PROBLEMS IN PLANE
         ELASTICITY
         INT. J. NUMER. METHODS ENG., VOL. 7, NO. 3, 1973,
         PP. 297-308.

73-0509 UDOGUCHI, T.
         ASADA, Y.
         APPROACH TO INVESTIGATE NOTCH EFFECT ON LOW-CYCLE
         FATIGUE WITH FINITE ELEMENT
         PROC. 2ND INT. CONF. ON PRESSURE VESSEL TECHNOL.,
         SAN ANTONIO, TEXAS, OCTOBER 1-4, 1973, PP.
         785-800.

73-0510 UDOGUCHI, T.
         OKAMURA, H.
         ERROR ANALYSIS OF VARIOUS FINITE ELEMENT PATTERNS
         BULL. JSME., VOL. 16, NO. 102, DECEMBER 1973, PP.
         1803-1813.

73-0511 VINJE, T.
         CALCULATION OF THE CHARACTERISTIC FREQUENCIES OF
         FLUID-FILLED SPHERICAL SHELLS
         NORW. MARIT. RES., VOL. 1, NO. 1, 1973, PP. 15-26.

73-0512 WALSH, P. F.
         NUMERICAL ANALYSIS IN ORTHOTROPIC LINEAR FRACTURE
         MECHANICS
         INST. ENG., AUST., CIV. ENG. TRANS., VOL. CE-15,
         NO. 1-2, 1973, PP. 115-116.

73-0513 WALTON, D.
         WOODMAN, N. J.
         FINITE-ELEMENT METHOD APPLIED TO PREDICTING
         FATIGUE-CRACK GROWTH
         J. STRAIN ANAL., VOL. 8, NO. 4, OCTOBER 1973, PP.
         293-304.

73-0514 WANG, N.
         FINITE ELEMENT ANALYSIS OF CUT-GROWTH IN SHEETS OF
         HIGHLY ELASTIC MATERIALS
         INT. J. SOLIDS STRUCT., VOL. 9, NO. 10, OCTOBER
         1973, PP. 1211-1223.

73-0515 WEINGARTEN, V. I.
         VERONDA, D. R.
         BUCKLING OF SEGMENTS OF TOROIDAL SHELLS
         AIAA J., VOL. 11, NO. 10, OCTOBER 1973, PP.
         1422-1424.

73-0516 WELLS, A. A.
         FRACTURE CONTROL: PAST, PRESENT AND FUTURE
         EXP. MECH., VOL. 13, NO. 10, OCTOBER 1973, PP.
         401-410.

73-0517 WRIGHT, S. G.
         KULHAWY, F. H.
         ACCURACY OF EQUILIBRIUM SLOPE STABILITY ANALYSIS
         ASCE J. SOIL MECH. FOUND. DIV., VOL. 99, NO. SM10,
         OCTOBER 1973, PAPER NO. 10097, PP. 783-790.

73-0518 YAMAMOTO, Y.
         RATE OF CONVERGENCE FOR THE ITERATIVE APPROACH IN
         ELASTIC-PLASTIC ANALYSIS OF CONTINUA
         INT. J. NUMER. METHODS ENG., VOL. 7, NO. 4, 1973,
         PP. 497-508.

73-0519 HSIEH, B. J.
         BELYTSCHKO, T.
         NONLINEAR TRANSIENT ANALYSIS OF SHELLS AND SOLIDS
         OF REVOLUTION BY CONVECTED ELEMENTS
         AIAA/ASME/SAE STRUCTURES, STRUCTURAL DYNAMICS, AND
         MATERIALS CONFERENCE, WILLIAMSBURGH, VIRGINIA,
         MAR. 20-22, 1973.

73-0520 ZIENKIEWICZ, O. C.
         PAREKH, C. J.
         APPLICATION OF FINITE ELEMENTS TO HEAT CONDUCTION
         PROBLEMS INVOLVING LATENT HEAT
         ROCK MECH., FELSMECH, MEC. ROCHES., VOL. 5, NO. 2,
         AUGUST 1973, PP. 65-76.

73-0521 COLLATZ, L.
         METHODS FOR SOLUTION OF PARTIAL DIFFERENTIAL
         EQUATIONS
         IN NUMERICAL SOLUTION OF PARTIAL DIFFERENTIAL
         EQUATIONS, EDITED BY J. G. GRAM, PROC. OF NATO
         ADVANCED STUDY INSTITUTE, HELD AT KJELLER, NORWAY,
         AUGUST 20-24, 1973, PP. 1-16.

73-0522 MITCHELL, A. R.
         ELEMENT TYPES AND BASE FUNCTIONS
         IN NUMERICAL SOLUTION OF PARTIAL DIFFERENTIAL
         EQUATIONS, EDITED BY J. G. GRAM, PROC. OF NATO
         ADVANCED STUDY INSTITUTE, HELD AT KJELLER, NORWAY,
         AUGUST 20-24, 1973, PP 107-150.

73-0523 ZIENKIEWICZ, O. C.
         SOME LINEAR AND NON-LINEAR PROBLEMS IN FLUID
         MECHANICS. FINITE ELEMENT METHOD FORMULATION
         IN NUMERICAL SOLUTION OF PARTIAL DIFFERENTIAL
         EQUATIONS, EDITED BY J. G. GRAM, PROC. OF NATO
         ADVANCED STUDY INSTITUTE, HELD AT KJELLER, NORWAY,
         AUGUST 20-24, 1973, PP. 173-194.

73-0524 HOLAND, I.
         APPLICATION OF FINITE ELEMENT METHODS TO STRESS
         ANALYSIS
         IN NUMERICAL SOLUTION OF PARTIAL DIFFERENTIAL
         EQUATIONS, EDITED BY J. G. GRAM, PROC. OF NATO
         ADVANCED STUDY INSTITUTE, HELD AT KJELLER, NORWAY,
         AUGUST 20-24, 1973, PP. 195-221.

73-0525 SPREEUW, E.
         FINITE ELEMENT COMPUTER PROGRAMS FOR HEAT
         CONDUCTION PROBLEMS
         IN NUMERICAL SOLUTION OF PARTIAL DIFFERENTIAL
         EQUATIONS, EDITED BY J. G. GRAM, PROC. OF NATO
         ADVANCED STUDY INSTITUTE, HELD AT KJELLER, NORWAY,
         AUGUST 20-24, 1973, PP. 241-266.

73-0526 ONDRACEK, E.
         HADDOW, J. B.
         A NUMERICAL METHOD FOR THE ANALYSIS OF
         LONGITUDINAL ELASTIC PLASTIC STRESS WAVE
         PROPAGATION
         INT. J. NUMER. METH. IN ENG., VOL. 6, NO. 1, 1973,
         PP. 103-116.

73-0527 SOBIESZCZANSKI, J.
         FULTON, R. E.
         APPLICATION OF COMPUTER-AIDED AIRCRAFT DESIGN IN A
         MULTIDISCIPLINARY ENVIRONMENT
         14TH CONF. ON STRUCT. DYNAMICS AND MATER.,
         WILLIAMSBURG, VA., MARCH 20-22, 1973, AND SPEC.
         CONF. ON DYNAMICS, WILLIAMSBURG, VA., MARCH 19-20,
         1973. COLLECT. OF TECH. PAPERS, 2 VOLS. (PUBL. BY
         A.I.A.A., NEW YORK, 1973)

73-0528 BELL, K.
         HATLESTAD, B.
         HANSTEEN, O. E.
         ARALDSEN, P. O.
         ORSAM - A PROGRAMMING SYSTEM FOR THE FINITE
         ELEMENT METHOD
         USER'S MANUAL, FEBRUARY 1973.

73-0529 BERGAN, P. G.
         AAMODT, B.
         FINITE ELEMENT ANALYSIS OF CRACK PROPAGATION IN
         THREE-DIMENSIONAL SOLIDS UNDER CYCLING LOADING
         PROC. 2ND INT. CONF. ON STRUCT. MECH. IN REACTOR
         TECHNOL., BERLIN, SEPTEMBER 10-14, 1973.

73-0530 BERG, S.
         NONLINEAR FINITE ELEMENT ANALYSIS OF REINFORCED
         CONCRETE PLATES
         DIV. OF STUCT. MECH., NORWEGIAN INST. OF TECH.,
         REPORT NO. 73-1, FEBRUARY 1973.

73-0531 ROBINSON, J.
         INTEGRATED THEORY OF FINITE ELEMENT METHODS
         JOHN WILEY AND SONS, 1973.

73-0532 ROBINSON, J.
         BASIS FOR ISOPARAMETRIC STRESS ELEMENTS
         COMPUTER METHODS IN APPLIED MECHANICS AND
         ENGINEERING, VOL. 2, NO. 1, 1973.

585

73-0533 AHMED, S. R.
BERECHNUNG DES REIBUNGSLOSEN STROEMUNGSFELDES VON
DREIDIMENSIONALEN AUFTRIEBSBEHAFTETENTRAGFLUEGELN
RUEMPFEN UND FLUEGEL-RUMPF-KOMBINATIONEN NACH DEM
PANEL-VERFAHREN (CALCULATION OF THE INVISCID FLOW
FIELD AROUND THREE DIMENSIONAL LIFTING WINGS,
FUSELAGES AND WING-FUSELAGE COMBINATIONS BY THE
FINITE ELEMENT METHOD)
DTSCH LUFT RAUMFAHRT FUR SCHUNGSBER DLR-FB73-102,
1973, 67 PP.

73-0534 BIWAS, A. K.
REYNOLDS, P. J.
PROCEEDINGS OF INTERNATIONAL ASSOCIATION FOR
HYDRAULIC RESEARCH AND PROCEEDINGS OF
INTERNATIONAL SYMPOSIUM ON RIVER MECHANICS
PROC. INTL. ASSOC. FOR HYDRAUL. RES., INTL. SYMP.
ON RIVER MECH., BANGKOK, THAILAND, VOL. 1-3,
JANUARY 9-12, 1973, 1855 PP. (PUBL. BY ASIAN INST.
OF TECHNOL., BANGKOK, THAILAND, 1973)

73-0535 BUTMAN, Z. I.
LUKASHENKO, V. I.
SOME RESULTS OF FUSELAGE ANALYSIS BY THE FINITE
ELEMENT METHOD USING A COMPUTER
SOV. AERONAUT., VOL. 16, NO. 1, 1973, PP. 10-14.

73-0536 DLUGACH, M. I.
KOVAL'CHUK, N. V.
METOD KONECHNYKH ELEMENTOV V PRIMENENII K RASCHETU
TSILINDRICHESKIKH OBOLOCHEK S PRYAMOUGOL'NYMI
OTVERSTIYAMI (FINITE ELEMENT METHOD AS APPLIED TO
THE CALCULATION OF CYLINDRICAL SHELLS WITH
RECTANGULAR CUTOUTS)
PRIKL. MEKH., VOL. 9, NO. 11, NOVEMBER 1973, PP.
35-41.

73-0537 TURCKE, D. J.
MCNEICE, G. M.
PROCEDURES FOR SELECTING NEAR OPTIMUM FINITE
ELEMENT GRIDS
2ND INTERNATIONAL CONF. ON PRESSURE VESSEL AND
PIPING TECH.,SAN ANTONIO, TEXAS, OCT. 1973.

73-0538 HUANG, Y. H.
WANG, S. T.
FINITE-ELEMENT ANALYSIS OF CONCRETE SLABS AND ITS
IMPLICATIONS FOR RIGID PAVEMENT DESIGN
HIGHW. RES. REC., NO. 466, 1973, PP. 55-69.

73-0539 LUBKIN, J. L.
STRUCTURAL MODELING OF A WELDED JOINT TYPICAL FOR
VEHICLE FRAMES
DIV. OF ENGNG. RESEARCH, MICHIGAN STATE
UNIVERSITY, EAST LANSING, MICHIGAN, DECEMBER 1973.

73-0540 LAING, I. E.
TENSION ZONES IN DECKED ROCKFILL DAMS
CONF. ON STRESS AND STRAIN IN ENGLAND, BRISBANE,
AUST., AUGUST 23-24, 1973, PP. 151-155.

73-0541 ANON
MRI/STARDYNE USER INFORMATION MANUAL
CDC CYBERNET PUB., DEPT. PUB., NO. 86603500, 1973.

73-0542 MIURA, I.
AASOKA, K.
SIMULATION ON THE TENSILE BEHAVIOUR OF TWO-PHASE
ALLOY CONTAINING A FIBER UNDER TENSILE LOAD
J. JAP. INST. MET., VOL. 37, NO. 11, NOVEMBER
1973, PP. 1212-1216.

73-0543 ANON
USER INFORMATION MANUAL
CDC CYBERNET PUBLICATIONS, DEPT. PUBLICATION NO.
86616800, 1973.

73-0544 PARISEAU, W. G.
ROCK MECHANICS AND RISK IN OPEN PIT MINING
PROC. 11TH INT. SYMP. ON COMPUT. APPL. IN THE
MINER. IND., TUCSON, ARIZONA, VOL. 1, SECTION A,
APRIL 16-20, 1973, PP. 106-124.

73-0545 PRICE, J.
SPENCER, A. M.
ELASTIC DEFLECTION IN THIN WALLED CONES ROTATING
ABOUT THEIR AXES OF SYMMETRY
CONF. ON STRESS AND STRAIN IN ENG., BRISBANE,
AUST., AUGUST 23-24, 1973, PP. 146-150.

73-0546 RONEY, B. D.
STONE, B. J.
APPLICATION OF COMPUTER GRAPHICS TO FINITE ELEMENT
MESH GENERATION
CONF. ON STRESS AND STRAIN IN ENG., BRISBANE,
AUST., AUGUST 23-24, 1973, PP. 156-163.

73-0547 SABBAGH, H. A.
KRILE, T. F.
FINITE ELEMENT ANALYSIS OF ELASTIC WAVE SCATTERING
FROM DISCONTINUITIES
PROC. ULTRASONIC SYMP., I.E.E.E., NAVAL POSTGRAD.
SCHOOL, MONTEREY, CALIF., NOVEMBER 5-7, 1973, PP.
216-219.

73-0548 SALAM, Y. M.
MONISMITH, C. L.
FRACTURE (ULTIMATE STRENGTH) ANALYSES OF ASPHALT
PAVEMENT LAYERS RESULTING FROM TRAFFIC LOADING
HIGHW. RES. REC., VOL. 466, 1973, PP. 127-138.

73-0549 RAYMOND, G. P.
FOUNDATION FAILURE OF NEW LISKEARD EMBANKMENT
HIGHW. RES. REC., NO. 463, 1973, PP. 1-17.

73-0550 ARGYRIS, J. H.
FAUST, G.
FINITE ELEMENTS METHOD IN DESIGN OF PRESTRESSED
REACTOR CONCRETE CONSTRUCTION
DTSCH AUSSCHUSS. STAHLBETON, NO. 234, 1973, 69 PP.

73-0551 URAL, O.
FINITE ELEMENT METHOD AND ROCK MECHANICS
PROC. 8TH CAN. ROCK MECH. SYMP., UNIV. OF TORONTO,
ONT., NOVEMBER 30-DECEMBER 1, 1972, 299 PP. (PUBL.
BY MINES BRANCH, DEPT. OF ENERGY, MINES AND
RESOURCES, OTTAWA, 1973).

73-0552 BECK, H.
MEHLHORN, G.
ZUSAMMENWIRKEN VON EINZELNEN FERTIGTEILEN ALS
GROSSFLAECHIGE SCHEIBE (INTERACTION OF INDIVIDUAL
PREFABRICATED ELEMENTS AS A LARGE SHEAR PANEL)
DTSCH. AUSSCHUSS. STAHLBETON, NO. 224, 1973, 93
PP.

73-0553 BUSBY, H. R.
WEINGARTEN, V. I.
DYNAMIC STABILITY OF A NONLINEAR BEAM SUBJECTED TO
BOTH LONGITUDINAL AND TRANSVERSE EXCITATION
PROC. 13TH MIDWEST MECH. CONF., UNIV. OF
PITTSBURGH, PA., AUGUST 13-15, 1973, PAPER 27, PP.
367-382.

73-0554 CARMIGNANI, C.
CELLA, A.
COMPUTER SYSTEM FOR THE ANALYSIS OF LARGE SCALE
PROBLEMS IN SOLID, FLUID AND THERMODYNAMICS
SYST. APPROACHES TO DEV. CTRY., SYMP. PROC.,
ALGIERS, ALGERIA, MAY 28-31, 1973, PP. 81-87.

73-0555 CHUNG, T. J.
YAGAWA, G.
INCREMENTAL THEORY OF THREE DIMENSIONAL TRANSIENT
THERMOELASTOPLASTICITY - FORMULATION AND SOLUTION
PROC. 13TH MIDWEST MECH. CONF., UNIV. OF
PITTSBURGH, PA., AUGUST 13-15, 1973, PAPER 59, PP.
843-856.

73-0556 DOKAINISH, M. A.
GOSSAIN, D. M.
DEFLECTION AND VIBRATION ANALYSIS OF A FLEXIBLE
FAN BLADE
J. MECH. ENG. SCI., VOL. 15, NO. 6, DECEMBER 1973,
PP. 387-391.

73-0557 LAUDIERO, F.
TRALLI, A.
FINITE ELEMENT INCREMENTAL ANALYSIS OF
ELASTOPLASTIC PLATE BENDING
MECCANICA, VOL. 8, NO. 3, SEPTEMBER 1973, PP.
190-202.

73-0558 LINK, M.
ZUR BERECHNUNG VON PLATTEN NACH DER THEORIE
ZWEITER ORDNUNG MIT HILFE EINES HYBRIDEN
DEFORMATIONS MODELLS (CALCULATION OF PLATES BY
SECOND ORDER THEORY WITH THE AID OF A HYBRID
DISPLACEMENT MODEL - 2)
ING-ARCH., VOL. 42, NO. 6, 1973, PP. 381-394.

73-0559 NEMAT-NASSER, S.
LEE, K. N.
FINITE ELEMENT FORMULATIONS FOR ELASTIC PLATES BY
GENERAL VARIATIONAL STATEMENTS WITH DISCONTINUOUS
FIELDS
MIDWEST MECH. CONF., 13TH PROC. UNIV. OF
PITTSBURGH, PA., AUGUST 13-15, 1973, PAP. 68, PP.
979-995.

73-0560 IRONS, B. M.
THE CONJUGATE NEWTON METHOD
NATO ADVANCED STUDY INST., CALGARY, UNIVERSITY,
AUGUST 1973.

73-0561 ROGERS, C. R.
FINITE ELEMENT APPLICATION TO STRAIN GAGE DATA
REDUCTION
MIDWEST MECH. CONF., 13TH. PROC. UNIV. OF
PITTSBURGH, PA., AUGUST 13-15, 1973, PAPER 48, PP.
675-683.

73-0562 SANDIDGE, D. W.
ODEN, J. T.
STABILITY AND POSTBUCKLING BEHAVIOUR OF
HYPERELASTIC BODIES AT FINITE STRAIN BY THE FINITE
ELEMENT METHOD
MIDWEST MECH. CONF. 13TH PROC. UNIV. OF
PITTSBURGH, PA., AUGUST 13-15, 1973, PAP. 23, PP.
305-320.

73-0563 TABARROK, B.
GASS, N.
EIN NEUES VARIATIONSPRINZIP MIT ANWENDUNG AUF
SCHWINGUNGEN FLACHER SCHALEN NACH DER METHODE DER
FINITEN ELEMENTE (NEW VARIATIONAL PRINCIPLE AS
APPLIED TO VIBRATIONS OF PLANE SHELLS BY A FINITE
ELEMENT METHOD)
Z. ANGEW. MATH. MECH., VOL. 53, NO. 12, DEC. 1973,
PP. 773-781.

73-0564 TING, E. C.
YANG, H. T. Y.
NUMERICAL PROCEDURE FOR THE ANALYSIS OF STRUCTURES
SUPPORTED BY VISCOELASTIC FOUNDATIONS
PROC. 13TH MIDWEST MECH. CONF., UNIV. OF
PITTSBURGH, PA., AUGUST 13-15, 1973, PAPER 67, PP.
959-975.

73-0565 TOEWS, N. A.
YU, Y. S.
COMPUTER PROGRAM FOR AUTOMATIC MESH GENERATION FOR
THE 2-D FINITE ELEMENT PROGRAM WILAX
CAN. MINES BR. RESEARCH, REPORT R.272, NOVEMBER
1973, 109 PP.

73-0566 WU, J. J.
SOLUMN INSTABILITY UNDER NONCONSERVATIVE FORCES,
WITH INTERNAL AND EXTERNAL DAMPING FINITE ELEMENT
USING ADJOINT VARIATIONAL PRINCIPLES
PROC. 13TH MIDWEST MECH. CONF., UNIV. OF
PITTSBURGH, PA., AUGUST 13-15, 1973, PAPER 37, PP.
501-514.

73-0567 ARGYRIS, J. H.
DUNNE, P. C.
ANGELOPOULOS, T.
DYNAMIC RESPONSE BY LARGE STEP INTEGRATION
EARTHQUAKE ENGINEERING AND STRUCTURAL DYNAMICS,
VOL. 2, 1973, PP. 185-203.

73-0568 ARGYRIS, J. H.
DUNNE, P. C.
ANGELOPOULOS, T.
NONLINEAR OSCILLATIONS USING THE FINITE ELEMENT
TECHNIQUE
J. COMPUTER METHODS IN APPL. MECHS. AND ENGR.,
VOL. 2, 1973, PP. 203-250.

73-0569 TONG, P.
PIAN, T. H. H.
LASRY, S. J.
A HYBRID-ELEMENT APPROACH TO CRACK PROBLEMS IN
PLANE ELASTICITY
INT. J. NUMER. METH. IN ENGR., VOL. 7, 1973, PP.
297-308.

73-0570 FELIPPA, C. A.
FINITE ELEMENT AND FINITE DIFFERENCE ENERGY
TECHNIQUES FOR THE NUMERICAL SOLUTION OF PARTIAL
DIFFERENTIAL EQUATIONS
SUMMER COMPUT SIMULATION CONF. PROC. MONTREAL,
QUE. JUL 17-19 1973, VOL. 1, PP. 1-14.

73-0571 FULLARD, K.
CALCULATION OF THERMAL STRESSES IN A
CYLINDER-CYLINDER INTERSECTION BY MEANS OF FINITE
ELEMENTS
INT. J. PRESSURE VESSELS PIPING VOL. 1, NO. 3, JUL
1973, PP. 177-198.

73-0572 GEORGE, A.
SURVEY OF SPARSE MATRIX METHODS IN THE DIRECT
SOLUTION OF FINITE ELEMENT EQUATIONS
SUMMER COMPUT SIMULATION CONF. PROC. MONTREAL,
QUE. JUL 17-19 1973, VO. 1, PP. 15-20.

73-0573 GRAY, W. G.
PINDER, G. F.
GALERKIN APPROXIMATION OF THE TIME DERIVATIVE IN
THE FINITE ELEMENT ANALYSIS
SUMMER COMPUT SIMULATION CONF. PROC. MONTREAL,
QUE. JUL 17-19 1973, VOL. 1, PP. 21-24.

73-0574 KHANNA, S. K.
KACHROD, P. N.
MODIFICATION OF LAYERED THEORY FOR PAVEMENT DESIGN
ANALYSIS
AUST. ROAD RES., VOL. 5, NO. 4, DEC. 1973, PP.
67-85.

73-0575 KHOT, N. S.
VENKAYYA, V. B.
OPTIMIZATION OF STRUCTURES FOR STRENGTH AND
STABILITY REQUIREMENTS
US AIR FORCE SYST COMMAND AIR FORCE FLIGHT DYN LAB
TECH REP AFFDL-TR 73-98 DEC 1973, 48 PP.

73-0576 NEDELEC, J. C.
PLANCHARD, J.
UNE METHODE VARIATIONNELLE ELEMENTS FINIS POUR LA
RESULUTION NUMERIQUE UN PROBLEME EXTERIEUR DANS
R**3 (VARIATIONAL METHOD OF FINITE ELEMENTS FOR
THE NUMERICAL RESOLUTION OF AN EXTERIOR PROBLEM IN
3 DIMENSIONS)
REV. FR. AUTOM. INF. RECH. OPER., VOL.7, DEC.
1973, PP. 105-129.

73-0577 OPTIZ, H.
NOPPEN, R.
EVALUATION OF COMPUTER AIDED METHODS FOR THE
DESIGN OF MACHINE TOOL STRUCTURES
ANN CIRP VOL. 22, NO. 2 1973, PP. 227-231.

73-0578 SAHA, S. K.
DE, A. K.
SOME DESIGN ASPECTS OF IC ENGINE PISTON HEADS
J. INST. ENG (INDIA), MECH. ENG DIV. VOL. 54, PART
ME 1, SEP 1973, PP. 14-18.

73-0579 ZOROWSKI, C.F.
MATHEMATICAL PREDICTION OF DYNAMIC TIRE BEHAVIOR
TIRE SCI TECHNOL VOL. 1, NO. 1 FEB 1973, PP.
99-117.

73-0580 CHEN, P.M.M.
SOLVING A CLASS OF LARGE SPARSE LINEAR SYSTEMS OF
EQUATIONS BY PARTITIONING
IEEE INT. SYMP. ON CIRCUIT THEORY, PROC, DIG PAP,
TORONTO, ONT, APR 9-11 1973, PP. 223-226.

73-0581 BUCK, K. E.
SCHARPF, D. W.
SCHREM, E.
STEIN, E.
EINIGE ALLGEMEINE PROGRAMMSYSTEME FUR FINITE
ELEMENTE
IN: FINITE ELEMENTE IN DER STATIK, BERLIN,
MUNCHEN, DUSSELDORF: W. ERNST, 1973, PP. 389-454.

73-0582 CROUZEIX, M.
THOMAS, J. M.
ELEMENTS FINIS ET PROBLEMES ELLIPTIQUES DEGENERES
(FINITE ELEMENTS AND DEGENERATE ELLIPTIC PROBLEMS)
REV. FR. AUTOM. INF. RECH. OPER. VOL. 7, DEC 1973,
PP. 77-104.

73-0583 DE, A. K.
ANALYSIS OF CLAMPED CIRCULAR PLATES OF VARIABLE
THICKNESS SUBJECTED TO UNIFORMLY DISTRIBUTED LOAD
BY FINITE ELEMENT TECHNIQUE
J. INST. ENG. (INDIA), MECH. ENG. DIV. VOL. 54,
PART ME 1 SEPT 1973, PP. 19-21.

73-0584 PICKETT, R. M.
RUBINSTEIN, M. F.
AUTOMATED STRUCTURAL SYNTHESIS USING A PE NUMBER
OF DESIGN COORDINATES
AIAA/ASME/SAE, 14TH STRUCT. DYN. AND MATER. CONF.,
WILLIAMSBURG, VA., MARCH 20-22 1973, AND AIAA DYN.
SPEC. CONF., WILLIAMSBURG, VA., MARCH 19-20, 1973,
COLLECT. OF TECH. PAP., 2 VOLS., (PUBL. BY
A.I.A.A., NEW YORK, N.Y., 1973)

73-0585 GOULD, P. L.
STRESS RESULTANTS IN SHELLS BY AN AUGMENTED FINITE
ELEMENT METHOD
PROC. 4TH CAN. CONGR. OF APPL. MECH., EC.
POLYTECH., MONTREAL, QUE., MAY 28-JUNE 1, 1973,
PP. 93-94.

73-0586 HENNART, J. P.
COMPARISON OF EXTRAPOLATION TECHNIQUES WITH HIGH
ORDER FINITE ELEMENT METHODS FOR DIFFUSION
EQUATIONS WITH PIECEWISE CONTINUOUS MATERIAL
PROPERTIES
5TH ANS. TOP. MEET., UNIV. OF MICH., ANN ARBOR,
APRIL 9-11, 1973, PP. 95-119.

73-0587 HENSHAW, D. H.
TWO-DIMENSIONAL AIRFOIL ANALYSIS USING A REFINED
FINITE ELEMENT TECHNIQUE
PROC. 4TH CAN. CONGR. OF APPL. MECH., EC.
POLYTECH, MONTREAL, QUE., MAY 28-JUNE 1, 1973, PP.
721-722.

73-0588 HERSCOVITCH, P.
MUFTI, A. A.
FINITE ELEMENT CALCULATION OF LEFT VENTRICULAR
WALL STRESS IN NORMAL AND DISEASE STATES
PROC. 4TH CAN. CONGR. OF APPL. MECH., EC.
POLYTECH., MONTREAL, QUE., MAY 28-JUNE 1, 1973, PP.
843-844.

73-0589 HILL, J. L.
DAVIS, C. G.
EFFECT OF INITIAL STATE FORCES ON THE HYDROELASTIC
VIBRATION OF TUBES
PROC. 4TH CAN. CONGR. OF APPL. MECH., EC.
POLYTECH, MONTREAL, QUE., MAY 28-JUNE 1, 1973, PP.
563-564.

73-0590 HRUDEY, T. M.
LINDBERG, G. M.
FINITE ELEMENT VIBRATION ANALYSIS OF
MULTICONDUCTOR ELECTRICAL TRANSMISSION LINES
PROC. 4TH CAN. CONGR. OF APPL. MECH., EC.
POLYTECH, MONTREAL, QUE., MAY 28-JUNE 1, 1973, PP.
517-518.

73-0591 LEE, G. C.
STRAIN ENERGY FUNCTIONS AND MATHEMATICAL MODELS OF
LUNG ELASTICITY
PROC. 4TH CAN. CONGR. OF APPL. MECH., EC.
POLYTECH, MONTREAL, QUE., MAY 28-JUNE 1, 1973, PP.
847-848.

73-0592 LEWIS, R. W.
HUMPHESON, C.
FINITE ELEMENT ANALYSIS OF ELECTRO-OSMOTIC
DEWATERING
PROC. 4TH CAN. CONGR. OF APPL. MECH., EC.
POLYTECH, MONTREAL, QUE., MAY 28-JUNE 1, 1973, PP.
657-658.

73-0593   SODHI, D. S.
MIXED FINITE ELEMENT METHOD FOR VIBRATION ANALYSIS
OF SHALLOW SHELLS
PROC. 4TH CAN. CONGR. OF APPL. MECH., EC.
POLYTECH, MONTREAL, QUE., MAY 28-JUNE 1, 1973, PP.
527-528.

73-0594   SVEC, O. T.
REFINED HEAT CONDUCTION FINITE ELEMENT FOR
PERMAFROST ANALYSIS
PROC. 4TH CAN. CONGR. OF APPL. MECH., EC.
POLYTECH, MONTREAL, QUE., MAY 28-JUNE 1, 1973, PP.
775-776.

73-0595   HILDEBRAND, R. W.
KOTHAWALA, K. S.
FINAL REPORT ON NASTRAN/GRAPHICS, 1971 FORD DOOR
PROJECT
ENG. MECH. RESEARCH CORP., REPORT, FORD MOTOR CO.,
JUNE 1973.

73-0596   NAGY, L. I.
EVALUATION OF NASTRAN FROM THE THE USER'S POINT OF
VIEW
JSAE SYMP. ON ANALYSIS AND DESIGN OF AUTOMOBILE
BODY STRUCTURES, TOKYO, NOVEMBER 13, 1973.

73-0597   HENNRICH, C. H.
KONRATH, E. J.
SUBSTRUCTURE ANALYSIS TECHNIQUES AND AUTOMATION
NASTRAN USER'S EXPERIENCES COLLOQUIUM, NASA
TMX-2893, SEPT. 1973.

73-0598   ODEN, J. T.
SOME DISTRIBUTIONAL AND CONVERGENCE PROPERTIES OF
THE FINITE ELEMENT METHOD, WITH APPLICATIONS IN
NONLINEAR ELASTODYNAMICS
PROC., ACM NATIONAL CONFERENCE, (SIGNUM INVITED
LECTURE), ATLANTA, AUGUST, 1973.

73-0599   ODEN, J. T.
REDDY, J. N.
CONVERGENCE OF MIXED FINITE ELEMENT APPROXIMATIONS
OF A CLASS OF LINEAR-BOUNDARY VALUE PROBLEMS
INTERNATIONAL JOUR. OF STRUCT. MECH. VOL. 2, NO.
2, PP. 83-108.

73-0600   ODEN, J. T.
BHANDARI, D. R.
YAGAWA, G.
CHUNG, T. J.
A NEW APPROACH TO THE FINITE-ELEMENT FORMULATION
AND SOLUTION OF A CLASS OF PROBLEMS IN COUPLED
THERMOELASTOVISCOPLASTICITY OF CRYSTALLINE SOLIDS
JOUR. OF NUCLEAR ENG. AND DESIGN, VOL. 24, 1973,
PP. 420-430.

73-0601   ODEN, J. T.
CLOUGH, R. W.
YAMAMOTO, Y.
SOME ASPECTS OF RECENT CONTRIBUTIONS TO THE
MATHEMATICAL THEORY OF FINITE ELEMENTS
RECENT ADVANCES IN COMPUTATIONAL METHODS IN
STRUCTURAL MECHANICS AND DESIGN, THE UNIV. OF
ALABAMA IN HUNTSVILLE PRESS, HUNTSVILLE, 1973.

73-0602   ODEN, J. T.
BHANDARI, D. R.
KUBITZA, W. K.
A LARGE DEFORMATION ANALYSIS OF CRYSTALLINE
ELASTIC-VISCOPLASTIC MATERIALS
PROC. 2ND INT. CONF. ON STRUCT. MECH. IN REACTOR
TECH. (SMIRT), BERLIN, W. GERMANY, SEPT. 1973.

73-0603   ODEN, J. T.
AKAY, H. U.
JOHNSON, C. P.
ON THE EFFECT OF APPROXIMATING HIGHER ORDER TERMS
IN CERTAIN FINITE ELEMENT MODELS OF NONLINEAR
STRUCTURAL BEHAVIOR
AIAA JOUR. VOL. 11, NO. 11, 1973, PP. 1589-1590.

73-0604   ODEN, J. T.
CHUNG, T. J.
WU, S. T.
A FINITE ELEMENT ANALYSIS OF TRANSIENT RAREFIED
GAS FLOW
FINITE ELEMENT METHODS IN FLOW PROBLEMS (ED. BY
J.T. ODEN, O. C. ZIENKIEWICZ, R. H. GALLAGHER, AND
C. TAYLOR), UAH PRESS, HUNTSVILLE, 1973.

73-0605   ODEN, J. T.
SOME RESULTS OF FINITE ELEMENT APPLICATIONS ON
NONLINEAR ELASTICITY
INT. J. OF COMPUTERS AND STRUCTURES, VOL. 3, 1973,
PP. 175-194.

73-0606   ODEN, J. T.
KEY, J. E.
SOLUTION OF NONLINEAR EQUATIONS BY EXPLICIT TIME
INTEGRATION - A SHORT COMMUNICATION
INT. JOUR. OF NUMERICAL METHODS IN ENG., VOL. 7,
NO. 2, 1973, PP. 225-227.

73-0607   ZIENKIEWICZ, O. C.
LEWIS, R. W.
AN ANALYSIS OF VARIOUS TIME STEPPING SCHEMES FOR
INITIAL VALUE PROBLEMS
INT. J. FOR EARTHQUAKE ENGINEERING AND STRUCTURAL
DYNAMICS, VOL. 1, NO. 1, 1973.

73-0608   LEWIS, R. W.
NORRIS, V. A.
AN APPLICATION OF THE FINITE ELEMENT METHOD TO
TRANSIENT INTERFACE PROBLEMS IN POROUS MEDIA
FOURTH CANADIAN CONGRESS OF APPLIED MECHANICS,
MONTREAL, 1973.

73-0609   BRUCH, J. C.
LEWIS, R. W.
ELECTRO-OSMOSIS AND THE CONTAMINATION OF
UNDERGROUND FLUIDS
J. ENV. PLANNING AND POLLUTION CONTROL, VOL. 1,
NO. 4, 1973.

73-0610   ROUSSOPO, A. A.
SYRMAKEZ, C. A.
APPLICATION OF FINITE ELEMENT METHOD TO A
MICROZONATION STUDY OF SALONICA REGION
ANNALI DI GEOFISICA, VOL. 26, NO. 4, 1973, PP.
513-523.

73-0611   SMITH, I. M.
O'CONNOR, B. A.
FARRADAY, R. V.
NUMERICAL ANALYSIS OF PLASTICITY IN SOILS
PROC. SYMP. ON PLASTICITY AND SOIL MECHANICS,
CAMBRIDGE, ENGLAND, 1973, PP. 279-289.

73-0612   KABAILA, A. P.
HALL, A. S.
FINITE ELEMENT TECHNIQUES - A SHORT COURSE OF
FUNDAMENTALS AND APPLICATIONS
SYDNEY, UNISEARCH LTD., UNIVERSITY OF NEW SOUTH
WALES, FEBRUARY 1973.

73-0613   CHAN, Y. K.
KABAILA, A. P.
A CONFORMING QUADRILATERAL ELEMENT FOR BUCKLING
ANALYSIS OF STIFFENED PLATES
UNICIV REPORT R-121, UNIVERSITY OF NEW SOUTH
WALES, 1973, 42 PP.

73-0614   BASU, A. K.
NEW LIGHT ON THE NAYAK ALPHA TECHNIQUE
INT. J. NUMER. METHODS IN ENGNG., VOL. 6, 1973, P.
152.

73-0615   WOOD, W. L.
NOTE ON A MODIFIED CONJUGATE GRADIENT METHOD
INT. J. NUMER. METHOD. ENGNG., VOL. 7, 1973, PP.
228-232.

73-0616   THOMAS, D. L.
WILSON, R. R.
USE OF STRAIGHT BEAM FINITE ELEMENTS FOR ANALYSIS
OF VIBRATIONS OF CURVED BEAMS
J. OF SOUND AND VIBRATION, VOL. 26, NO. 1, 1973,
PP. 155-158

73-0617   IRONS, B. M.
RAZZAQUE, A.
INTRODUCTION TO SHEARS DEFORMATION INTO A THIN
PLATE DISPLACEMENTS FORMULATIONS
J. AIAA. 1973, P. 1438.

73-0618   BALDWIN, J.
RAZZAQUE, A.
IRONS, B. M.
SHAPE FUNCTION SUBROUTING FOR AN ISOPARAMETRIC
THIN PLATE ELEMENT
INT. J. NUMER. METH. ENG., VOL. 7, NO. 4, 1973,
PP. 431-440.

73-0619   IRONS, B. M.
RAZZAQUE, A.
A FURTHER MODIFICATION TO AHMAD'S SHELL ELEMENT
J. NUM. METHODS IN ENGNG., VOL. 5, 1973, PP.
588-589.

73-0620   ALBUQUERQUE, F.
A BEAM ELEMENT FOR USE WITH THE SEMILOOP SHELL
ELEMENT
M.SC. THESIS, UNIVERSITY OF WALES, SWANSEA, 1973.

73-0621   CIARLET, P. G.
CONFORMING AND NONCONFORMING FINITE ELEMENT
METHODS FOR SOLVING THE PLATE PROBLEM
CONF. ON NUMER. SOLUTION OF DIFFERENTIAL
EQUATIONS, UNIVER. OF DUNDEE, JULY 3-6, 1973.

73-0622   SCOTT, R.
FINITE ELEMENT TECHNIQUES FOR CURVED BOUNDARIES
PH. D. DISSERTATION, MASS INT. TECH., 1973.

73-0623 GERADIN, M.
ANALYSE DYNAMIQUE DUALE DES STRUCTURES PAR LA
METHODE DES ELEMENTS FINIS
COLLECTION DES PUBLICATIONS DE LA FACULTE DES
SCIENCES APPLIQUEES DE L'UNIVERSITE DE LIEGE, VOL.
36, 1973, PP. 1-173.

73-0624 IMBERT, J. F.
GIRARD, A.
GERADIN, M.
MODAL ANALYSIS OF A SATELLITE PRIMARY STRUCTURE
USING A FINITE ELEMENT PROCEDURE
SYMP. ON STRUCTURES OF SPACE VEHICLES AND
SPACECRAFT, UNIVERSITY COLLEGE, LONDON, 1973.

73-0625 GLOWINSKI, R.
MARROCCO, A.
ANALYSE NUMERIQUE DU CHAMP MAGNETIQUE D'UN
ALTERNATEUR PAR ELEMENTS FINIS ET SURRELAXATION
PONCTUELLE NON LINEAIRE
COMPUTING METHODS IN APPLIED SCIENCES AND
ENGINEERING, PART1, INT. SYMP. VERSAILLES,
DECEMBER 17-21, 1973, P. 409.

73-0626 TABARROK, B.
ON SOME ASPECTS OF CONTINUUM MECHANICS
INT. RES. SEMINAR ON THE THEORY AND APPLICATION OF
FINITE ELEMENT METHODS, THE UNIVERSITY OF
CALGARY, 1973.

73-0627 WATKINS, D.
ANALYSIS
INT. RES. SEMINAR ON THE THEORY AND APPLICATION OF
FINITE ELEMENT METHODS, THE UNIVERSITY OF
CALGARY, 1973.

73-0628 PHILLIPS, G.
NUMERICAL ANALYSIS
INT. RES. SEMINAR ON THE THEORY AND APPLICATION OF
FINITE ELEMENT METHODS, THE UNIVERSITY OF
CALGARY, 1973.

73-0629 SCHMIDT, E.
SPLINES AND FINITE ELEMENTS
INT. RES. SEMINAR ON THE THEORY AND APPLICATION OF
FINITE ELEMENT METHODS, THE UNIVERSITY OF
CALGARY, 1973.

73-0630 WESTBROOK, D. R.
VARIATIONAL AND WEIGHTED RESIDUAL METHODS
INT. RES. SEMINAR ON THE THEORY AND APPLICATION OF
FINITE ELEMENT METHODS, THE UNIVERSITY OF
CALGARY, 1973.

73-0631 FRAEIJS DE VEUBEKE, B.
ELASTIC EQUILIBRIUM MODELS, STRESS FUNCTIONS, AND
DUALITY NON LINEARITIES AND STABILITY, FIVE
LECTURES
INT. RES. SEMINAR ON THE THEORY AND APPLICATION OF
FINITE ELEMENT METH., THE UNIVERSITY OF CALGARY,
1973.

73-0632 NITSCHE, J. A.
NON PROJECTIONAL METHODS
INT. RES. SEMINAR ON THE THEORY AND APPLICATION OF
FINITE ELEMENT METHODS, THE UNIVERSITY OF
CALGARY, 1973.

73-0633 GALLAGHER, R. H.
PLATES AND SHELLS, INELASTIC ANALYSIS, TIME
DEPENDENT INELASTICITY, FIVE LECTURES
INT. RES. SEMINAR ON THE THEORY AND APPLICATION OF
FINITE ELEMENT METHODS, THE UNIVERSITY OF
CALGARY, 1973.

73-0634 IRONS, B. M.
COMPUTATIONAL TECHNIQUES, NONLINEAR EQUATIONS,
ISOPARAMETRIC ELEMENTS, FIVE LECTURES
INT. RES. SEMINAR ON THE THEORY AND APPLICATION OF
FINITE ELEMENT METHODS, THE UNIVERSITY OF
CALGARY, 1973.

73-0635 STRANG, W. G.
SURVEY OF UNDERLYING THEORY, APPROXIMATION BY
PIECEWISE POLYNOMIALS, ILLEGAL ELEMENTS AND
THE PATCH TEST, FIVE LECTURES
INT. RES. SEMINAR ON THE THEORY AND APPLICATION OF
FINITE ELEMENT METHODS, THE UNIVERSITY OF
CALGARY, 1973.

73-0636 ZLAMAL, M.
CURVED ELEMENTS, ERROR BOUNDS, TIME DEPENDENT
PROBLEMS   FIVE LECTURES
INT. RES. SEMINAR ON THE THEORY AND APPLICATION OF
FINITE ELEMENT METHODS, THE UNIVERSITY OF
CALGARY, 1973.

73-0637 ODEN, J. T.
SOME RECENT DEVELOPMENTS IN THE THEORY AND
APPLICATION OF FINITE ELEMENTS TO MIXED PROBLEMS
AND SHOCK WAVES IN NONLINEAR MATERIALS
INT. RES. SEMINAR ON THE THEORY AND APPLICATION OF
FINITE ELEMENT METHODS, UNIV. OF CALGARY, 1973.

73-0638 MITCHELL, A. R.
CURVED BOUNDARIES IN THE FINITE ELEMENT METHOD
INT. RES. SEMINAR ON THE THEORY AND APPLICATION OF
FINITE ELEMENT METHODS, THE UNIVERSITY OF
CALGARY, 1973.

73-0639 DE ARANTES E OLIVEIRA, E. R.
A THEORY OF VARIATIONAL METHODS AND ITS
APPLICATION TO FINITE ELEMENTS
INT. RES. SEMINAR ON THE THEORY AND APPLICATION OF
FINITE ELEMENT METHODS, THE UNIVERSITY OF
CALGARY, 1973.

73-0640 CIARLET, P. G.
NON-CONFORMING FINITE ELEMENT METHODS AND THE
PATCH TEST
INT. RES. SEMINAR ON THE THEORY AND APPLICATION OF
FINITE ELEMENT METHODS, THE UNIVERSITY OF
CALGARY, 1973.

73-0641 SCHATZ, A. H.
SOME INTERIOR SUPER CONVERGENCE ESTIMATES FOR
SPLINE   APPROXIMATION
INT. RES. SEMINAR ON THE THEORY AND APPLICATION OF
FINITE ELEMENT METHODS, THE UNIVERSITY OF
CALGARY, 1973.

73-0642 CHEUNG, Y. K.
THE FINITE STRIP METHOD AND ITS APPLICATIONS
INT. RES. SEMINAR ON THE THEORY AND APPLICATION OF
FINITE ELEMENT METHODS, THE UNIVERSITY OF
CALGARY, 1973.

73-0643 KOTERAS, J. R.
A HYBRID FINITE ELEMENT FOR DETERMINATION OF
STRESS INTENSITY FACTORS AT A CRACK TIP
TICOM REPORT 73-2, THE UNIVERSITY OF TEXAS AT
AUSTIN, 1973.

73-0644 GALLAGHER, R. H.
THE FINITE ELEMENT METHOD IN PLATE AND SHELL
STABILITY   ANALYSIS
PROC. 4TH AUSTRALASIAN CONFERENCE ON THE MECHANICS
OF STRUCTURES AND MATERIALS, UNIVERSITY OF
QUEENSLAND, 1973, PP. 77-86.

73-0645 HOUDE, J.
STUDY OF FORCE DISPLACEMENT RELATIONSHIPS FOR THE
FINITE ELEMENT ANALYSIS OF REINFORCED CONCRETE
PH.D. THESIS, UNIVERSITY OF MCGILL, MONTREAL,
1973, P. 326.

73-0646 URAL, O.
FINITE ELEMENT METHOD: BASIC CONCEPTS AND
APPLICATIONS
INTEXT EDUCATIONAL PUB., NEW YORK, 1973.

73-0647 VYSLOUKH, V. A.
KANDIDOV, V. P.
CHESNOKOV, S. S.
REDUCTION OF THE DEGREES OF FREEDOM IN SOLVING
DYNAMIC PROBLEMS BY THE FINITE ELEMENT METHOD
INT. J. FOR NUM. METHODS IN ENGNG., VOL. 7, 1973,
PP. 185-194.

73-0648 ARAL, M. M.
FINITE ELEMENT SOLUTION OF PARTIAL DIFFERENTIAL
EQUATIONS   FEMAC COMPUTER PROGRAM
MIDDLE EAST TECHNICAL UNIVERSITY PUBLICATION,
1973.

73-0649 TORIDIS, T. G.
DYNAMIC ANALYSIS OF FRAME AND PLATE STRUCTURES
WITH GEOMETRIC AND MATERIAL NONLINEARITIES
AD-762008, NAVAL SHIP RES. AND DEV. CTR.,
BETHESDA, MD., REPT. NO. NSRDC-3988, MAY 1973, PP.
136.

73-0650 BATHE, K. J.
WILSON, E. L.
PETERSON, F. E.
SAP IV, A STRUCTURAL ANALYSIS PROGRAM FOR STATIC
AND   DYNAMIC RESPONSE OF LINEAR SYSTEMS
PB-221 967/3, JUNE 1973, P. 182.

73-0651 CHUNG, T. J.
EIDSON, R. L.
STATIC AND DYNAMIC ANALYSIS OF VISCOELASTOPLASTIC
FIBER-REINFORCED COMPOSITE SHELLS IN MISSILE
STRUCTURES
ALABAMA UNIV., DEPT. OF MECH. ENGR., JANUARY 1973,
203 PP.

73-0652 DALY, P.
HELPS, J. D.
EXACT FINITE ELEMENT SOLUTIONS TO THE HELMHOLTZ
EQUATIONS
INT. J. NUM. METHODS ENGR., VOL. 6, NO. 4, 1973,
PP. 529- 542.

73-0653 WINFREY, R. C.
MULTIDEGREE-OF-FREEDOM ELASTIC SYSTEMS HAVING
MULTIPLE CLEARANCES
U.S. NAVAL RES. LAB., SHOCK VIB. BULL., NO. 43,
PART 3, JUNE 1973, PP. 23-30.

73-0654 HARTZMAN, M.
APPROXIMATE METHOD FOR CALCULATING THE RESPONSE OF
EMPLACEMENT STRUCTURES SUBJECTED TO GROUND SHOCK
FROM UNDERGROUND NUCLEAR DETONATIONS
U.S. NAVAL RES. LAB., SHOCK VIB. BULL., NO. 43,
PART 2, JUNE 1973, PP. 1-12.

73-0655 HUA, H. M.
A FINITE ELEMENT METHOD FOR CALCULATING
AERODYNAMIC COEFFICIENTS OF A SUBSONIC AIRPLANE
J. AIRCRAFT, VOL. 10, NO. 7, JULY 1973, PP.
422-426.

73-0656 KOOPMANN, G. H.
BUILDING VIBRATIONS AND ACOUSTICS
J. SOUND VIB., VOL. 28, NO. 3, JUNE 8, 1973, PP.
471-485.

73-0657 NAGARAJAN, S.
POPOV, E. P.
ELASTIC PLASTIC DYNAMIC ANALYSIS OF AXISYMMETRIC
SOLIDS
AD-764 244, JULY 1973, 86 PP.

73-0658 PICHUMANI, R.
FINITE ELEMENT ANALYSIS OF PAVEMENT STRUCTURES
USING AFPAV CODE LINEAR ELASTIC ANALYSIS
AD-761 524, MAY 1973, 90 PP.

73-0659 PRASAD, K. S. R. L.
KRISHNA MURTY, A. V.
GALERKIN FINITE ELEMENT METHOD FOR VIBRATION
PROBLEMS
AIAA J. VOL. 11, NO. 4, APRIL 1973, PP. 544-546.

73-0660 RAMAMURTI, V.
APPLICATION OF SIMULTANEOUS ITERATION METHOD TO
TORSIONAL VIBRATION PROBLEMS
J. SOUND VIB. VOL. 298 NO. 3, AUGUST 8, 1973, PP.
331-340

73-0661 TABARROK, B.
SODHI, D. S.
ON THE GENERALIZATION OF STRESS FUNCTION PROCEDURE
FOR DYNAMIC ANALYSIS OF PLATES
INTL. J. NUMER. METHODS VOL. 5, NO. 4, MARCH-APRIL
1973, PP. 523-542.

73-0662 MCMICHAEL, C. L.
THOMAS, G. W.
RESERVOIR SIMULATION BY GALERKIN'S METHOD
SOC. PET. ENG. J., VOL. 13, 1973, PP. 125-138.

73-0663 MATSUSHIMA, Y.
HYSTERETIC DAMPING SYSTEMS APPLIED TO SOIL
STRUCTURE INTERACTION BY FINITE ELEMENT METHOD
PROC. 5TH WORLD CONF. ON EARTHQUAKE ENG., JUNE
25-29 1973, ROME, ITALY, PAPER 3013, VOL. 2, 1974,
(AVAILABLE FROM EDIGRAF, ROME, ITALY)

73-0664 SCHUBERT, L.
ZUR BERECHNUNG, BEMESSUNG UND BEWEHRUNG
RECHTECKIGER PLATTEN KONSTNATER DICKE MIT
OEFFNUNGEN ODER AUSSPARUNGEN (DESIGN AND
REINFORCEMENT OF RECTANGULAR PLATES OF
CONSTANT THICKNESS WITH HOLES OR RECESSES)
WISS. Z. TECH. UNIV. DRESDEN, VOL. 22, NO. 6,
1973, PP: 1089-1097.

73-0665 ELLIOTT, G. L.
PICK, R. J.
CALCULATION OF NATURAL FREQUENCIES OF HEAT
EXCHANGER TUBES
PROC. INT. SYMP. ON VIB. PROBLEM IN IND., KESWICK,
ENGLAND, PAPER 422, APRIL 10-12, 1973, P. 16.
(AVAILABLE FROM U.K. ATOMIC ENERGY AUTH.,
WINDSCALE, ENGLAND)

73-0666 SEWELL, M. J.
GOVERNING EQUATIONS AND EXTREMUM PRINCIPLES OF
ELASTICITY AND PLASTICITY GENERATED FROM A
SINGLE FUNCTIONAL
J. STRUCT. MECH. VOL. 2, NO. 2, 1973, PP: 135-158.

73-0667 MASLENNIKOV, A. M.
VYUZITI METODY KONECNYCH PRVKU PRO PROJEKTOVANI
KOMBINOVANYCH KONSTRUCKCI (FINITE ELEMENT
METHOD IDEAS APPLIED TO THE COMPOSED
CONSTRUCTION DESIGN)
STAVEBNICKY CAS, VOL. 21, NO. 12, 1973, PP:
893-910.

73-0668 DEMS, K.
WIELOSTOPNIOWA SYNTEZA MACIERZY SZTYWNOSCI
(MULTI-STAGE SYNTHESIS OF THE STIFFNESS
MATRICES)
MECH. TEOR, 1, STOSOW, VOL. 11, NO. 4, 1973, PP:
407-415.

73-0669 DEMICHELE, D. W.
DARBY, J. B. JR.
THREE-DIMENSIONAL MECHANICAL STRESSES IN TOROIDAL
MAGNETS FOR CONTROLLED THERMONUCLEAR REACTORS
PROC. 5TH SYMP. ON ENG. PROBL. OF FUSION RES.,
PRINCETON UNIV., N. J., NOVEMBER 5-9, 1973, PP.
558-569. PUBL. BY IEEE NUCL. AND PLASMA SCI.
SOC., (AVAILABLE FROM IEEE (73CH0843-3-3-NPS), N.
Y. 1974).

73-0670 GEORGE, W.
VAUGHAN, G. N.
CALCULATION OF THERMAL STRESSES IN TUBULAR CERAMIC
COMPONENTS
PROC. 7TH INT. CONF. OF SCI. OF CERAM.,
JUAN-LES-PINS, FRANCE, SEPTEMBER 24-46 PAPER 6,
1973. (PUBL. BY SOC. FR. DE CERAAM, PARIS, 1973)

73-0671 MCCABE, M. W.
KAMEL, H. A.
GIFTS SYSTEM FOR THE PDP-15 AND OTHER
CONSIDERATIONS IN DEVELOPING GRAPHICS AND
CAPABILITIES FOR FINITE ELEMENT PROGRAMS
PROC. FALL SYMP. DIGITAL EQUIP. COMPUT. USERS
SOC., SAN FRANCISCO, CALIF., NOVEMBER 28-30, 1973,
PP. 69-76. (PUBL. BY DECUS, MAYNARD, MASS., 1974)

73-0672 COWDREY, D. R.
WILLIS, J. R.
FINITE ELEMENT CALCULATIONS RELEVANT TO AS-CUT
QUARTZ RESONATORS
PROC. 27TH SYMP. ON FREQ. CONTROL, CHERRY HILL,
N.J., JUNE 12-14 1973, PP. 7-10, (AVAILABLE FROM
EIA, WASHINGTON, D.C., 1973)

73-0673 REICH, M.
KOPLIK, B.
ANALYSIS OF BRIDGE DECKS USING POLYMER IMPREGNATED
CONCRETE
SYMP. POLYM. IN CONCR., AM. CONC. INST., FALL
CONV., HOLLYWOOD, FLA., 1972 AND 1973, SPRING
CONV., ATLANTIC CITY, N.J., PP. 247-281.
(AVAILABLE FROM ACI, PUBL. SP-40, DETROIT, MICH.,
1973)

73-0674 THURSTON, G. A.
ON THE INTERRELATION BETWEEN STABILITY AND
COMPUTATIONS
PROC. SYMP. NONLINEAR ELASTICITY, UNIV. OF
WISCONSIN, MADISON, APRIL 16-18, 1973, PP.
253-288. (PUBL. BY ACAD. PRESS, NEW YORK, 1973)

73-0675 BAGCI, C.
COMPUTER METHOD FOR COMPUTING CRITICAL SPEEDS OF
NON-UNIFORM SHAFTS GEARED SYSTEMS AND CURVED
ASSEMBLIES
PROC. 3RD APPL. MECH. CONF., OKLA. STATE UNIV.,
STILLWATER, PAPER 40, NOVEMBER 5-7, 1973, P. 15.
(AVAILABLE FROM OKLA. STATE UNIV., STILLWATER)

73-0676 BARLOW, J.
ON THE AUTOMATIC SELECTION OF INDEPENDENT
DISPLACEMENT VARIABLES IN GENERAL PURPOSE FINITE
ELEMENT CODES
INT. J. NUMER. METH. IN ENG., VOL. 6, NO. 1, 1973,
PP. 141-144.

73-0677 ALEXANDER, R. M.
LAWRENCE, K. L.
DYNAMIC STRAINS IN A FOUR BAR MECHANISM
PROC. 3RD CONF. APPL. MECH., OKLA. STATE UNIV.,
STILLWATER, PAPER NO. 26, NOVEMBER 5-7, 1973, P.
16. (AVAILABLE FROM OKLA. STATE UNIV., STILLWATER)

73-0678 NATH, B.
FINITE ELEMENT - ANALOGUE METHOD FOR DETERMINING
THE DYNAMIC CHARACTERISTICS OF AN ARCH
DAM-RESERVOIR SYSTEM
PROC. 5TH WORLD CONF. ON EARTHQUAKE ENG., ROME,
ITALY, VOL. 2, JUNE 25-29, 1973, P. 3013.
(AVAILABLE FROM EDIGRAF, ROME)

73-0679 EVERLING, G.
VORAUSSAGE DES GEBIRGSDRUCKES UND SEINER
AUSQIRKUNGEN AUF STREBEN UND STRECKEN IM
STEINKOHLENBERGBAU (FORE-CASTING THE ROCK
PRESSURE AND ITS EFFECTS ON LONGWALL FACES AND
ROADWAYS IN COAL MINES)
PROT. AGAINST ROCK FALL, SYMP., KATOWICE, POL.
OCTOBER 23-24, 1973, PAPER 1-4, P: 20.
(AVAILABLE FROM ASSOC. OF MIN. ENG. AND TECH.,
KATOWICE, POL. 1973)

73-0680 WATANABE, H.
ANALYSIS OF NONELASTIC AND NONLINEAR VIBRATION OF
ROCK DAMS USING FINITE ELEMENT METHOD
PROC. 5TH WORLD CONF. ON EARTHQUAKE ENG., ROME,
ITALY, VOL. 2, JUNE 25-29, 1973, P. 3013.
(AVAILABLE FROM EDIGRAF, ROME)

73-0681 REYES, F.
RUIZ, P.
FINITE ELEMENT FOR THE INTERACTION OF FRAMES AND
SHEAR WALLS
PROC. 5TH WORLD CONF. ON EARTHQUAKE ENG., ROME,
ITALY, JUNE 25-29, 1973, P. 3013, VOL. 2
(AVAILABLE FROM EDIGRAF, ROME, ITALY)

73-0682 PIAN, T. H.
HYBRID MODELS IN NUMERICAL AND COMPUTER METHODS
IN STRUCTURAL MECHANICS, S. J. FENVES, ET AL
(ED.), ACADEMIC PRESS, NEW YORK, 1973.

73-0683 EISENBERG, M. A.
MALVERN, L. E.
ON FINITE ELEMENT INTEGRATION IN NATURAL
COORDINATES
INT. J. NUM. METHODS ENG., VOL. 7, 1973, PP:
574-575.

73-0684 CARPENTER, W. C.
ANALYSIS OF PLATES ON ELASTIC FOUNDATIONS
INT. J. NUMER. METH. IN ENG., VOL. 7, NO. 3, 1973,
PP. 408-409.

73-0685 FRISCH-FAY, R.
LINEARIZED LARGE DEFLECTION PROBLEM
PROC. 4TH AUSTR. CONF. ON MECH. OF STRUCT. AND
MATER., UNIV. OF QUEENSLAND, BRISBANE, AUST.,
AUGUST 20-22, 1973.

73-0686 ARGYRIS, J. H.
GRIEGER, I.
WAGEMANN, G.
BERECHNUNG VON SPANNUNGSKONZENTRATIONEN AN
TRAGWERKEN UNTER THERMO-MECHANISCHER BELASTUNG
TECHNISCHE MITTEILUNGEN, REPORT NO. 66, INSTITUT
FUR STATIK UND DYNAMIK DER LUFT UND
RAUMFAHRTKONSTRUKTIONEN, AUGUST, 1973, PP.
401-013.

73-0687 DU PREEZ, R. J.
ANALYSIS OF SLAB BRIDGES BY FINITE STRIP METHOD
PROC. SYMP. ON BRIDGE DECK ANAL., JOHANNESBURG, S.
AFR., MAY 7-9, 1973.

73-0688 SILVESTER, P.
KONRAD, A.
AXISYMETRIC TRIANGULAR ELEMENT FOR THE SCALAR
HELMHOLTZ EQUATION
INT. J. NUM. METHODS ENG., VOL. 5, NO. 4, 1972,
PP: 481-498.

73-0689 BREBBIA, C. A.
CONNOR, J. J.
NUMERICAL METHODS IN FLUID DYNAMICS
PROCEEDINGS OF THE INTERNATIONAL CONF. HELD AT THE
U. OF SOUTHAMPTON, ENGLAND, SEPTEMBER 26 - 28,
1973

73-0690 GOPALACHARYULU, S.
A HIGHER ORDER CONFORMING RECTANGULAR ELEMENT
INT. J. NUM. METHODS ENG., VOL. 6, NO. 2, 1973,
PP: 305-308.

73-0691 WEGMULLER, A.
KOSTEM, C.
FINITE ELEMENT ANALYSIS OF PLATES AND
ECCENTRICALLY STIFFENED PLATES
REPORT NO. 378A, 3, FRITZ ENG. LAB., LEHIGH U.,
BETHLEHEM, PA., FEBRUARY 1973.

73-0692 FRIED, I.
SHEAR IN C AND C PLATE BENDING ELEMENTS
INT. J. SOLIDS STRUCT., VOL. 9, NO. 4, 1973, PP:
449-460.

73-0693 SVEC, O. J.
GLADWELL, G.
A TRIANGULAR PLATE BENDING ELEMENT FOR CONTACT
PROBLEMS
INT. J. SOLIDS STRUCT., VOL. 6, 1970, PP: 435-446.

73-0694 COOK, R. D.
COMMENT ON DISCRETE ELEMENT IDEALIZATION OF AN
INCOMPRESSIBLE LIQUID FOR VIBRATION ANALYSIS
A.I.A.A. JOL., VOL. 11, NO. 5, MAY 1973, PP:
766-767.

73-0695 MCMILLAN, C. M.
ANALYSIS OF SLAB BRIDGES BY FINITE ELEMENTS
PROC. SYMP. ON BRIDGE DECK ANAL., JOHANNESBURG, S.
AFR., MAY 7-9, 1973.

73-0696 CURTIN, J.
BILLERBECK, W. J.
ADVANCED INTERCONNECT SYSTEMS FOR LIGHTWEIGHT
SOLAR ARRAYS
PROC. INT. CONGRESS ON SUN IN THE SERVICE OF
MANKIND, PARIS, FRANCE, JULY 2-6, 1973. PP.
311-326. PUBL. BY CENT. NATL. D'ETUD. SPATIALES,
PARIS, FRANCE, 1973.

73-0697 DU PREEZ, R. J.
REVIEW OF BASIC THEORY AND METHODS OF ANALYSIS OF
CONTINUOUS STRUCTURES
PROC. SYMP. ON BRIDGE DECK ANAL., JOHANNESBURG, S.
AFR., MAY 7-9, 1973.

73-0698 BREBBIA, C. A.
SOME APPLICATIONS OF FINITE ELEMENTS FOR FLOW
PROBLEMS
IN 'VARIATIONAL METHODS IN ENGRNG', VOL. 1, C.
BREBBIA AND H. TOTTENHAM (EDS.) SOUTHAMPTON
UNIVERSITY PRESS, 1973.

73-0699 ROBINSON, J.
A GENERAL EQUILIBRIUM BENDING ELEMENT
RA/ARD, REPORT NO. 15-10-73-19, OCTOBER 1973.

73-0700 YAMADA, YOSHIAKI
GALLAGHER, R. H.
THEORY AND PRACTICE IN FINITE ELEMENT STRUCTURAL
ANALYSIS
PROC. TOKYO SEMIN. ON FINITE ELEM. ANALYSIS,
TOKYO, JAPAN, NOVEMBER 5-10, 1973, 733 PP. (PUBL.
BY UNIV. OF TOKYO PRESS, TOKYO, JAPAN, 1973).

73-0701 GRIEGER, I.
INGA - INTERAKTIVE GRAPHISCHE ANALYSE BENUTZERHAND
RESEARCH REPORT NO. 135, INSTITUT FUR STATIK UND
DYNAMIK DER LUFT UND RAUMFAHRTKONSTRUKTIONEN,
JANUARY 1973.

73-0702 ARGYRIS, J. H.
FAUST, G.
ROY, J. R.
SZIMMAT, J.
WARNKE, E. P.
WILLAM, K.
FINITE ELEMENTE ZUR BERECHNUNG VON
SPANNBETONREAKTORDRUCKBEHALTERN
RESEARCH REPORT NO. 137, INSTITUT FUR STATIK UND
DYNAMIK DER LUFT UND RAUMFAHRTKONSTRUKTIONEN, MAY
1973.

73-0703 ARGYRIS, J. H.
DOLTSINIS, J.
GLOUDEMAN, J. F.
STRAUB, K.
WILLAM, K.
ASPECTS OF FINITE ELEMENT METHOD APPLIED TO
AEROSPACE STRUCTURES
RESEARCH REPORT NO. 138, INSTITUT FUR STATIK UND
DYNAMIK DER LUFT UND RAUMFAHRTKONSTRUKTIONEN, JUNE
1973.

73-0704 ARGYRIS, J. H.
SCHREM, E.
BRONLUND, O. E.
ASKA, SOFTWARE ENGINEERING FOR THE SEVENTIES IN
STRUCTURAL ANALYSIS
PROC. INT. CONF. ON STRUCTURAL MECH. IN REACTOR
TECHNOLOGY, BERLIN, SEPTEMBER 1973, PP: 10-14.

73-0705 ARGYRIS, J. H.
DUNNE, P. C.
SOME CONTRIBUTIONS TO NONLINEAR SOLID MECHANICS
PROC. CONF. SYMP. ON COMPUTING METHODS IN APPLIED
SCIENCES AND ENG., VERSAILLES, DECEMBER 1973, PP:
17-21.

73-0706 ARGYRIS, J. H.
WILLAM, K.
SOME CONSIDERATIONS FOR THE EVALUATION OF FINITE
ELEMENT MODELS
RESEARCH REPORT NO. 147, INSTITUT FUR STATIK UND
DYNAMIK DER LUFT UND RAUMFAHRTKONSTRUKTIONEN,
DECEMBER 1973.

73-0707 ARGYRIS, J. H.
GREIGER, I.
KONSTRUKTION UND BERECHNUNG MIT DER METHODE FINITE
ELEMENTE
RESEARCH REPORT NO. 148, INSTITUT FUR STATIK UND
DYNAMIK DER LUFT UND RAUMFAHRTKONSTRUKTIONEN,
DECEMBER 1973.

73-0708 RASHID, Y. R.
WOOD, J. E.
ANALYSIS OF NUCLEAR FUEL ROD LOCALIZED STRESSES
PROC. INT. CONF. ON NUCL. FUEL PERFORMANCE,
LONDON, ENGL., SESS. 4, PAPER 69, OCTOBER 15-19, 5
PP. PUBL. BY BR. NUCL. ENERGY SOC., LONDON, ENGL.,
1973.

73-0709 ARGYRIS, J. H.
HAASE, M.
MALEJANNAKIS, G. A.
NATURAL GEOMETRY OF SURFACE WITH SPECIFIC
REFERENCE TO THE MATRIX DISPLACEMENT ANALYSIS
OF SHELLS. 1, 11, 111
NO. 5. KONINKLIJKE NEDERLANDSE AKADEMIE VAN
WETENSCHAPPEN PROCEEDINGS, SERIES B, VOL. 76,
1973.

73-0710 ARGYRIS, J. H.
WILLAM, K. J.
SOME CONSIDERATION FOR THE EVALUATION OF THE
FINITE ELEMENT MODELS
PROC. 2ND INT. CONF. ON STRUCT. MECH. IN REACTOR
TECH., BERLIN, SEPTEMBER 1973, PP: 10-14. (VOL. 6,
PART B SUPPLEMENT)

73-0711 CANTATORE, W. P.
THE NUMERICAL SOLUTION OF THE STEADY INTERFACE IN
A COASTAL CONFINED AQUIFER
THESIS, (M. ENG. SC.), JAMES COOK UNIV. OF NORTH
QUEENSLAND, 1973.

73-0712 ZYVOLOSKI, G.
FINITE ELEMENT WEIGHTED RESIDUAL SOLUTION OF
ONE-DIMENSIONAL, UNSTEADY AND UNSATURATED FLOWS IN
POROUS MEDIA
DEPT. OF MECH. ENG., UNIV., OF CALIFORNIA, SANTA
BARBARA. MASTER'S THESIS. DEPT. OF MECH. ENG.,
UNIV. OF CALIFORNIA, SANTA BARBARA, JUNE 1973, 74
P. (N.T.I.S.-PB-029/3ST)

73-0713 WU, T. H.
LONG-TERM DEFORMATION IN CLAY
ENGINEERING EXPERIMENT STATION, OHIO STATE UNIV.,
COLUMBUS, REPORT NO: EES-290, MAY 1973, 299P.
(N.T.I.S. - PB-239 078/ 9ST)

73-0714 KOSTEM, C. N.
USERS MANUAL FOR PROGRAM BEAM
FRITZ ENG. LAB., LEHIGH UNIV., BETHLEHEHEM, PA.,
REPORT NO: FEL-378B.2, FEBRUARY 1973, 125 P.
(N.T.I.S. - PB-237 959/2ST)

73-0715 LIPPMANN, M. J.
TWO DIMENSIONAL STOCHASTIC MODEL OF A
HETEROGENEOUS GEOLOGIC SYSTEM
DOCTORAL THESIS, WATER RESOURCES CENTER,
CALIFORNIA UNIV., BERKELEY, 1973, 150 P. (N.T.I.S.
- PB-237 682/0ST)

73-0716 MARLOWE, D. E.
HALSEY, N.
MITCHELL, R. A.
MORDFIN, L.
NON-METALLIC ANTENNA-SUPPORT MATERIALS
ENGINEERING MECH. SECTION, NATIONAL BUREAU OF
STANDARDS, WASHINGTON, D.C., REPORT NO:
NBSIR-73-233, APRIL 1973, 54 P. (N.T.I.S. -
COM-74-11770/6ST)

73-0717 GOLDHAMMER, M. I.
LOPEZ, M. L.
METHODS OF PREDICTING THE AERODYNAMIC AND
STABILITY AND CONTROL CHARACTERISTICS OF STOL
AIRCRAFT. VOLUME 111. ENGINEERING METHODS
DOUGLAS AIRCRAFT CO., LONG BEACH, CALIF., REPORT
NO: MDC-J5965-03, DECEMBER 1973, 258 P. (N.T.I.S.
- AD/A-001 582/6SL) SEE ALSO VOL. 2, AD/A-001 581
AND VOL. 1, AD/A-001 580.

73-0718 GOLDHAMMER, M. I.
WASSON, N. F.
METHODS FOR PREDICTING THE AERODYNAMIC AND
STABILITY AND CONTROL CHARACTERISTICS OF STOL
AIRCRAFT. VOLUME 11. STOL AERODYNAMIC METHODS
COMPUTER PROGRAM
DOUGLAS AIRCRAFT CO., LONG BEACH, CALIF., REPORT
NO: MDC-J5965-02, DECEMBER 1972, 231 P. (N.T.I.S.
- AD/A-001 581/8SL) SEE ALSO VOL. 1, AD/A-001 580
AND VOL. 3, AD/A-001 582.

73-0719 GOLDHAMMER, M. I.
LOPEZ, M. L.
SHEN, C. C.
METHODS FOR PREDICTING THE AERODYNAMIC AND
STABILITY AND CONTROL CHARACTERISTICS OF STOL
AIRCRAFT. VOLUME 1. BASIC THEORETICAL METHODS
DOUGLAS AIRCRAFT CO., LONG BEACH, CALIF., REPORT
NO: MDC-J5965-01, DECEMBER 1973, 301 P. (N.T.I.S.
- AD/A-001 580/0SL) SEE ALSO VOLUME 2, AD-A-001
581.

73-0720 DE KONING, A. U.
THEORETICAL AND EXPERIMENTAL ANALYSIS OF
ELASTOPLASTIC DEFORMATION IN A UNIAXIALLY LOADED
STRIP OF 2034 T3 ROLLED SHEET MATERIAL WEAKENED BY
A CENTRAL ROUND HOLE
NATL. AEROSPACE LAB., AMSTERDAM, NETHERLANDS,
REPORT NO: NLK-TR-73011-U, JANUARY 1973, 52 PP.
(N.T.I.S. - N74-33364/2SL)

73-C721 LACASSE, S. M.
LADD, C. C.
BEHAVIOR OF EMBANKMENTS ON NEW LISKHEARD VARVED
CLAY
DEPT. OF CIVIL ENG., INST. OF TECH., MASS., REPORT
NO: R73-44, SOILS-PUB-327, SEPTEMBER 1973, 271 P.
(PB-237 001/3SL)

73-0722 VACHARAJITTIPHAN, P.
TRAHAIR, N. S.
WARPING AND DISTORTION AT I-SECTION JOINTS
CIVIL ENGINEERING LABS., UNIV. OF SYDNEY,
AUSTRALIA, REPORT NO: R-211, MAY 1973, 38 P.
(N.T.I.S. - PB-236 097/2SL)

73-0723 BAKHREBAH, S. A.
PAUL, S. L.
SCHNOBRICH, W. C.
FINITE ELEMENT SOLUTION OF NORMALLY INTERSECTING
CYLINDERS
PROC. 2ND CONF. STRUCT. MECH. IN REACTOR TECH.,
PAPER G2/5A, BERLIN, 1973.

73-0724 DOMINQUEZ, R. F.
GREER, G. G.
LIAU, A. C.
MECHANICS OF CABLE MOORING SYSTEMS. VOLUME V11.
THE STEADY-STATE BEHAVIOR OF A PYRAMID ARRAY
SYSTEM
COASTAL AND OCEAN ENGRNG., TEXAS A AND M UNIV.,
COLLEGE STATION. REPORT NO: COE-163, FEBRUARY
1973, 60 P. (N.T.I.S. - AD-786 187/5SL) SEE ALSO
VOL. 8, AD-786 188.

73-0725 DOMINQUEZ, R. F.
GREER, G. G.
LIAU, A. C.
MECHANICS OF CABLE MOORING SYSTEMS. VOLUME V1. A
COMPUTER PROGRAM FOR ANALYZING THE STEADY STATE
CONFIGURATION OF A TRI-MOORED ARRAY WITH INCLUDED
RIGID AND DEFORMABLE MEMBERS
COASTAL AND OCEAN ENGRNG., TEXAS A AND M UNIV.,
COLLEGE STATION, REPORT NO: COE-162, JANUARY 1973,
51 P. (N.T.I. S. - AD-786 186/7SL) SEE ALSO VOL.
7, AD-786 187.

73-0726 BAKER, G. A.
FINITE ELEMENT - GALERKIN METHODS FOR HYPERBOLIC
EQUATIONS WITH DISCONTINUOUS COEFFICIENTS
DEPT. DE MATHEMATIQUES, ECOLE POLYTECHNIQUE
FEDERALE DE LAUSANNE, TECH. REPORT, 1973.

73-0727 GALLAGHER, R. H.
LIGGETT, J. A.
CHAN, S. T. K.
FINITE ELEMENT CIRCULATION ANALYSIS OF
VARIABLE-DEPTH SHALLOW LAKES
A.S.C.E. J., HYD. DIV., VOL. 99, NO. HY 7, 1973,
PP. 1083-1096.

73-0728 ANDO, Y.
YAGAWA, G.
OKABAYASHI, K.
THE APPLICATION OF THE FINITE ELEMENT METHOD TO
THE ANALYSIS OF FRACTURE OF CYLINDRICAL SHELLS
2ND INT. CONF. STRUCT. MECH. IN REACTOR TECH.,
BERLIN, VOL. 3, PAPER G5/4, SEPTEMBER 10-14, 1973.

73-0729 ROBINSON, J.
A GENERAL EQUILIBRIUM AXIAL ELEMENT
RA/ARD, REPORT NO. 5-10-73-18, OCTOBER 1973.

73-0730 FREDERCIK, D. L.
HIGGINS, J. F.
RAPID CASE DESIGN USING FINITE ELEMENT STRESS
ANALYSIS AND NASTRAN
FRANKFORD ARSENAL, PHILADELPHIA, PA., 1973, 16 P.
(N.T.I.S. - AD-785 622/2SL)

73-0731 BARNARD, AL. J.
SHARMAN, P. W.
THE ELASTO-PLASTIC AND LARGE-DISPLACEMENT RESPONSE
OF PLATES TO BLAST LOADING
DEPT. OF TRANSPORT TECHNOLOGY, UNIV. OF
TECHNOLOGY, LOUGHBOROUGH, ENGLAND, REPORT NO:
TT.7313, OCTOBER 1973, 210 P. (N.T.I.S. - AD-784
353/5)

73-0732 ALEXANDER, S. S.
MCCOWAN, D. W.
GLOVER, P.
EFFECTS OF SOURCE STRUCTURE AND SITE RESPONSE ON
GROUND MOTION IN THE NEAR FIELD OF EARTHQUAKES
COLL. OF EARTH AND MINERAL SCIENCES, PENNSYLVANIA
STATE UNIV., UNIVERSITY PARK, PENN., FINAL REPORT:
APRIL 1, 1972 - JUNE 30, 1973, 76 P. (N.T.I.S. -
COM-74-11416/6)

73-0733 NILSON, A. H.
ANALYSIS OF LIGHT GATE STEEL SHEAR DIAPHRAGMS
2ND SPEC. CONF. ON COLD-FORMED STEEL STRUCT., ST.
LOUIS, MO., OCTOBER 22-24, 1973, PP. 325-363,
PUBL. BY UNIV. OF MO., ROLLA AND CORNELL UNIV.,
ITHACA, N.Y., 1973.

73-0734 FREESE, C. E.
COLLOCATION AND FINITE ELEMENTS - A COMBINED
METHOD
ARMY MATERIALS AND MECHANICS RESEARCH CENTER,
WATERTOWN, MASS., REPORT NO: AMMRC-TR-73-28, JUNE
1973, 7 P. (N.T.I.S. - AD-782 945/0)

73-0735 WINK, J. H.
APPLICATION OF FINITE ELEMENT METHOD FOR THE
DETAILED ANALYSIS OF HATCH CORNER STRESSES
NEDERLANDS SCHEEPS-STUDIECENTRUM TNO, DELFT,
REPORT NO: NSR-MEDEDEL-37-S, TDCK-64236, DECEMBER
1973, 43 P. (N.T.I.S. - N74-26374/0)

73-0736 CARPENTER, L. R.
COWAN, W. C.
SPENCER, R. W.
POLYMER-IMPREGNATED CONCRETE TUNNEL SUPPORT AND
LINING
ENGINEERING RESEARCH CENTER, BUREAU OF
RECLAMATION, DENVER, COLO., REPORT NO:
REC-ERC-73-23, DECEMBER 1973, 165 P. (N.T.I. S. -
PB-233 963/8)

73-0737 FOTHERGILL, J. W.
LEE, H. Y.
FOTHERGILL, P. A.
PREDICTION OF LONG-TERM STRESS RANGES: USER'S
MANUAL - BRIDGE LOAD GENERATOR
INTEGRATED SYSTEMS, INC., CHEVY CHASE, MD., FINAL
REPORT, JUNE, 1973, 194 P. (N.T.I.S. - PB-233
490/2)

73-0738 GOLDEN, M. E.
HURWITZ, M. M.
DEFINIT - A NEW ELEMENT DEFINITION CAPABILITY FOR
NASTRAN: USER'S MANUAL
NAVAL SHIP RESEARCH AND DEVELOPMENT CENTER,
BETHESDA, MD., REPORT NO: NSRDC-4250, DECEMBER
1973, 200 P. (N.T.I.S. - AD- 782 513/6)

73-0739 SPANNAGEL, D. W.
DAVIS, R. E. SR.
BACHER, A. E. SR.
STRUCTURAL BEHAVIOR OF A FLEXIBLE METAL CULVERT
UNDER A DEEP EARTH EMBANKMENT USING METHOD A
BACKFILL
BRIDGE DEPT., CALIFORNIA STATE DIV. OF HIGHWAYS,
SACRAMENTO, REPORT NO: CA-HY-BD-624111-73-6, JUNE
1973, 220 P. (N.T.I.S. - PB-233 589/1)

73-0740 BATHE, K. J.
WILSON, E. L.
NONSAP - A GENERAL FINITE ELEMENT PROGRAM FOR
NONLINEAR DYNAMIC ANALYSIS OF COMPLEX STRUCTURES
PROC. 2ND CONF. STRUCT. MECH. IN REACTOR TECH.,
PAPER M3/1, BERLIN, 1973.

73-0741 KEAYS, R. H.
ANALYSIS OF AN ORTHOTROPIC PLATE WITH TWO OPPOSITE
EDGES SIMPLY SUPPORTED AND THE OTHER TWO FREE
AERONAUTICAL RESEARCH LABS., MELBOURNE, AUSTRALIA,
REPORT NO: ARL/SM-NOTE-398, SEPTEMBER 1973, 24 P.
(N.T.I.S. - N74-22536/8)

73-0742 MORINO, L.
A FINITE-ELEMENT FORMULATION FOR SUBSONIC FLOWS
AROUND COMPLEX CONFIGURATIONS
DEPT. OF AEROSPACE ENGINEERING, BOSTON UNIV.,
MASS., REPORT NO: NASA-CR-138142, TR-73-05,
DECEMBER 1973, 79 P. (N.T.I.S. - N74-21923/9)

73-0743 CONNELL, D. H.
GARLANGER, J. E.
LADD, C. C.
PERFORMANCE OF AN EMBANKMENT CONSTRUCTED ON VARVED
CLAY
DEPT. OF CIVIL ENG., MASSACHUSETTS INST. OF TECH.,
CAMBRIDGE, REPORT NO: R73-26, SOILS MECH. PUB-321,
MAY 1973, 141 P. (N.T.I.S.-PB-233 140/3) SEE ALSO
REPORT PB-191 402, OCT. 1969

73-0744 HARKRIDER, D. G.
SEISMOLOGY AND ACOUSTIC-GRAVITY WAVES
SEISMOLOGICAL LAB., CALIFORNIA INST. OF TECH.,
PASADENA, SEMI-ANNUAL TECH. REPORT, JULY 1, 1972 -
DECEMBER 31, 1973, 109P. (N.T.I.S. - AD-779 939/8)

73-0745 ARYGYRIS, J. H.
AICHER, W.
KIERNER, G.
OSCILLATION STABILITY OF INTERNALLY AND EXTERNALLY
NOTCHED PLANE BARS OF EQUAL FORM FACTOR FROM
39436497(7075-T6) (SCHWINGFESTIGKEIT INNENUND
AUSSENGEKERBTER FLACHSTAEBE GLEICHER FORMZAHL AUS)
INSTITUT FUR STATIK UND DYNAMIK DER LUFT UND
RAUMFAHRTKONSTRUKTIONEN, STUTTGART UNIV., WEST
GERMENY, REPORT NO. ISD-145, 1973, 30 PP.
(N.T.I.S. - N74-21554/2)

73-0746 APPERT, K.
BERGER, D.
GRUBER, R.
A USEFUL FORM OF THE VARIATIONAL PRINCIPLE OF
IDEAL MHD FOR ONE-DIMENSIONAL NUMERICAL STABILITY
CALCULATIONS
CENTRE DE RECHERCHES EN PHYSIQUE DES PLASMAS,
ECOLE POLYTECHNIQUE FEDERALE DE LAUSANNE,
SWITZERLAND, REPORT NO: LRP-76/73, DECEMBER 1973,
11 P. (N.T.I.S. - N74-21339/8)

73-0747 BOROWIEC, Z.
STATICAL CALCULATIONS OF PLANE AND SPACE BAR
STRUCTURES AS WELL AS STUMPY FRAMEWORKS USING A
MODIFIED BAR SCHEME
SYMP. ON IND. SPAT AND SHELL STRUCT., KIELCE,
POL., JUNE 18-23, 1973, PP. 137-152.

73-0748 VIDOUSE, F.
DETERMINATION OF THE TECHNICAL CONSTANTS OF
LAMINATES IN OBLIQUE DIRECTIONS. THEORY AND
EXPERIMENTAL RESULTS (DETERMINATION DES CONSTANTES
TECHNIQUES DES STRATIFIES DANS LES DIRECTIONS
OBLIQUES)
CENTRE DE RECHERCHES SCIENTIFIQUES ET TECHNIQUES
DE L'INDUSTRIE DES FABRICATIONS METALLIQUES,
BRUSSELS, BELGIUM, REPORT NO: CRIF-PL-4, NOVEMBER
1973, 42 PP. (N.T.I.S. - N74-21166/5)

73-0749 FAIZ, A.
PAVEMENT DESIGN AND CONTINUOUSLY REINFORCED
CONCRETE PAVEMENT PERFORMANCE
TRANSPORTATION RESEARCH BOARD, WASHINGTON, D.C.,
REPORT NO: TRN-485, ISBN-0-309-02271-1, 1973, 88
P. (N.T.I.S. - PB-232 201/4)

73-0750 RAO, M. S.
DVORAK, G. J.
PLASTICITY THEORY OF FIBROUS COMPOSITES - PART 11
DEPT. OF CIVIL ENGNG., DUKE UNIV., DURHAM, N.C.,
REPORT NO: SOLID MECHS. SER-9, DECEMBER 1973, 42
P. (N.T.I.S. - AD-779 481/1) SEE ALSO PART 1,
AD-779480.

73-0751 DVORAK, G. J.
RAO, M. S. M.
PLASTICITY THEORY OF FIBROUS COMPOSITES - PART 1
DEPT. OF CIVIL ENGNG., DUKE UNIV., DURHAM, N.C.,
REPORT NO: SOLID MECH., SER-8, DECEMBER 1973, 39
P. (N.T.I.S. - AD-779 480/3) SEE ALSO PART 2,
AD-779 481.

73-0752 ZAK, A. R.
DEVELOPMENT OF A GUN-TURRET MODEL AND OF THE
EQUATIONS FOR COUPLING IT WITH A HELICOPTER
FINITE-ELEMENT STRUCTURAL MODEL
ILLINOIS UNIV., URBANA, FINAL REPORT, JUNE -
OCTOBER, 1973, 60 P. (N.T.I.S. - AD-779 413/4)

73-0753 EIBL, J.
FINITE ELEMENTS IN CURVILINEAR COORDINATES -
SECOND ORDER SURFACES
SYMP. ON IND. SPAT AND SHELL STRUCT., KIELCE,
POL., JUNE 18-23, 1973, PP. 179-189. *

73-0754 TONG, P.
LASRY, S. J.
A SUPER-ELEMENT FOR CRACK ANALYSIS
INT. J. OF FRACTURE MECH., VOL. 9, 1973, PP:
234-236. (N.T.I.S-AD-779 187/4)

73-0755 PADUART, A.
CONTINUOUS CONCEPT OF A SPACE GRID
SYMP. ON IND. SPAT AND SHELL STRUCT., KIELCE,
POL., JUNE 18-23, 1973, PP. 351-362.

73-0756 MIYACHI, T.
OGAWA, A.
HOSHIYA, S.
SOFUE, Y.
FLEXUAL RIGIDITY OF A THIN WALLED BUILD-UP ROTOR
FOR JET ENGINE. MEASUREMENT BY STATIC LOAD TEST
AND VIBRATION TEST AND CALCULATION BY FINITE
ELEMENT METHOD
NATIONAL AEROSPACE LAB., TOKYO, JAPAN, REPORT NO:
NAL-TR-329, JULY 1973, 17 P. (N.T.I.S. -
N74-20572/5)

73-0757 SHAH, R. C.
EXPLORATORY INVESTIGATION OF RAPID CRACK
PROPAGATION AND CRACK ARREST
BOEING AEROSPACE CO., SEATTLE, WASH., REPORT NO:
D180-17529-1, AUGUST 1973, 104 P. (N.T.I.S. -
AD-778 822/7)

73-0758 SIMPSON, A.
KRON'S METHOD: AN ALGORITHM FOR THE EIGENVALUE
ANALYSIS OF LARGE-SCALE STRUCTURAL SYSTEMS
BRISTOL UNIV., ENGLAND, REPORT NO: ARC-R/M-3733,
ARC-34098, 1973, 16 P. (N.T.I.S. - N74-19564/5)

73-0759 BEECKMAN, R.
VAN LEEUWEN, D.
A FINITE ELEMENT ANALYSIS OF SHELLS OF REVOLUTION
CENTRE DE RECHERCHES SCIENTIFGUES DE L'INDUSTRIE
DES FABRICATIONS METALLIGUES, BRUSSELS, BELGIUM,
SECTION CONSTRUCTION METALLIQUE, REPORT NO.
CRIF-MT-84, OCTOBER 1973, 42 PP. (N.T.I.S. -
N74-19561/1)

73-0760 MILLER, K. J.
AN ELASTIC-PLASTIC FINITE ELEMENT ANALYSIS OF
CRACK TIP FIELDS UNDER BIAXIAL LOADING CONDITIONS
DEPT. OF ENGNG., CAMBRIDGE UNIV., ENGLAND, REPORT
NO: CUED/C-MAT-TR-7, 1973, 31 P. (N.T.I.S. -
N74-19547/0)

73-0761 THOMPSON, D. S.
FINITE ELEMENT ANALYSIS OF THE FLOW THROUGH A
CASCADE OF AEROFOILS
DEPT. OF ENGNG., CAMBRIDGE UNIV., ENGLAND, REPORT
NO: CUED/A-TURBO/TR-45, 1973, 48 P. (N.T.I.S. -
N74-18920/0)

73-0762 BIOT, M. A.
NONLINEAR THERMOELASTICITY, IRREVERSIBLE
THERMODYNAMICS AND ELASTIC INSTABILITY
MATHEMATICS J., INDIANA UNIV., INTERIM REPORT,
VOL. 23, NO. 4, 1973, PP: 309-335.
(N.T.I.S.-AD-778 211/3)

73-0763 JAMES, J. H.
ACOUSITC FINITE ELEMENT ANALYSIS OF AXISYMMETRIC
FLUID REGIONS
ADMIRALTY RESEARCH LAB., TEDDINGTON, ENGLAND,
REPORT NO: ARL/R/R/4, BR37560, AUGUST 1973, 21 P.
(N.T.I.S. - N74-18330/2)

73-0764 CLOUGH, G. W.
ANALYTICAL PROBLEMS IN MODELING SLURRY WALL
CONSTRUCTION
DEPT. OF CIVIL ENGNG., DUKE UNIV., DURHAM, N.C.,
FINAL REPORT, SEPTEMBER 1973, 63 P. (N.T.I.S. -
PB-230 940/9)

73-0765 DAVIS, R. E.
CASTLETON, G. A.
LOAD DISTRIBUTION IN A COMPOSITE STEEL BOX GIRDER
BRIDGE
BRIDGE DEPT., CALIF. STATE DIV. OF HIGHWAYS,
SACRAMENTO, REPORT NO: CA-HY-BD-4137-73-7, JUNE
1973, 195 P. (N.T.I.S. - PB-229 948/5)

73-0766 HAHN, G. T.
HOAGLAND, R. G.
KANNINEN, M. F.
ROSENFIELD, A. R.
SEJNOHA, R.
FAST FRACTURE RESISTANCE AND CRACK ARREST IN
STRUCTURAL STEELS
BATTELLE COLUMBUS LABS., OHIO, PROGRESS REPORT,
JUNE 1973, 129 P. (N.T.I.S. - AD-776 986/2)

73-0767 REPNAU, T.
ADAMS, D. F.
HIGH-PERFORMANCE COMPOSITE MATERIALS FOR VEHICLE
CONSTRUCTION: A FINITE ELEMENT COMPUTER PROGRAM
FOR THE ELASTOPLASTIC ANALYSIS OF CRACK
PROPAGATION IN A UNIDIRECTIONAL COMPOSITE
RAND CORP., SANTA MONICA, CALIF., REPORT NO:
R-1392-PR, OCTOBER 1973, 158 P. (N.T.I.S. - AD-776
822/9)

73-0768 KONIECZNY, S.
WOZNIAK, CZ.
MULTI-ELEMENT SPATIAL STRUCTURES
SYMP. ON IND. SPAT AND SHELL STRUCT., KIELCE,
POL., JUNE 18-23, 1973, PP. 411-421.

73-0769 HUNSAKER, B. JR.
VAUGHAN, D. K.
STRICKLIN, J. A.
HAISLER, W. E.
A COMPARISON OF CURRENT WORK-HARDENING MODELS USED
IN THE ANALYSIS OF PLASTIC DEFORMATIONS
TEXAS ENGNG. EXPERIMENT STATION, COLLEGE STATION,
REPORT NO: TEES/RPT-2926-73-0, OCTOBER 1973, 79 P.
(N.T.I.S. - AD-776 667/8)

73-0770 COLLINS, T. P.
WITMER, E. A.
APPLICATION OF THE COLLISION-IMPARTED VELOCITY
METHOD FOR ANALYZING THE RESPONSES OF CONTAINMENT
AND DEFLECTOR STRUCTURES TO ENGINE ROTOR FRAGMENT
IMPACT
AEROELASTIC AND STRUCT. RESEARCH LAB., MASS. INST.
OF TECH., CAMBRIDGE, REPORT NO:
NASA-CR-134494,ASRL-TR-154-8, AUGUST 1973, 250 P.
(N.T.I.S. - N74-16592/9)

73-0771 IDRISS, I. M.
LYSMER, J.
HWANG, R.
SEED, H. B.
QUAD-4: A COMPUTER PROGRAM FOR EVALUATING THE
SEISMIC RESPONSE OF SOIL STRUCTURES BY VARIABLE
DAMPING FINITE ELEMENT PROCEDURES
EARTHQUAKE ENGNG. RESEARCH CENTER, UNIV. OF
CALIF., BERKELEY, REPORT NO: EERC-73-16, JULY
1973, 80 P. (N.T.I.S. - PB-229 424/1)

73-0772 MAJIDZADEH, K.
KAUFFMANN, E. M.
CHANG, G. W.
VERIFICATION OF FRACTURE MECHANICS CONCEPTS TO
PREDICT CRACKING OF FLEXIBLE PAVEMENTS
DEPT. OF CIVIL ENGNG., OHIO STATE UNIV., COLUMBUS,
REPORT NO: RF-3200-FR, JUNE 1973, 359 P. (N.T.I.S.
- PB-228 159/0)

73-0773 DIXON, J. D.
STRUCTURAL DESIGN DATA FOR CONCRETE DRIFT LININGS
IN BLOCK CAVING
PREPARED IN CO-OPERATION WITH DENVER MINING
RESEARCH CENTER, COLO. REPORT NO: BUMINES-RI-7792,
1973, 84 P. (N.T.I.S. -PB-227 548/5) SEE ALSO
REPORT DATED OCTOBER 1969, PB-187 741.

73-0774 REIMER, R. B.
DECONVOLUTION OF SEISMIC RESPONSE FOR LINEAR
SYSTEMS
EARTHQUAKE ENGNG. RESEARCH CENTER, UNIV. OF
CALIF., BERKELEY, REPORT NO: EERC-73-10, OCTOBER
1973, 175 P. (N.T.I.S. - PB-227 179/9)

73-0775 SAXTON, H. J.
JONES, A. T.
WEST, A. J.
NAMAROS, T. C.
LOAD POINT COMPLIANCE OF THE CHARPY IMPACT
SPECIMEN
SANDIA LABS., LIVERMORE, CALIF., REPORT NO:
CONF-730690-1, RS-8232-2-39428, JULY 1973, 39 P.
(N.T.I.S. - SCL-DC-720372) SEE ALSO NSA 3004, NO.
12163.

73-0776 HUANG, P. C.
MATRA, J. P. JR.
PLANFORM INPUT GENERATOR (PING), A NASTRAN
PRE-PROCESSOR FOR LIFTING SURFACES-THEORETICAL
DEVELOPMENT
NAVAL ORDNANCE LAB., WHITE OAK, MD., REPORT NO:
NOLTR-73-228, DECEMBER 1973, 34 P. (N.T.I.S. -
AD-775 633/1)

73-0777 MURRAY, K. H.
FINITE ELEMENT ANALYSIS OF A COMPOSITE MATERIAL
INTERFACE
OLD DOMINION UNIV., NORFOLK, VA., REPORT NO:
NASA-CR-132341, AUGUST 1973, 44 P. (N.T.I.S. /
N74-15626/6)

73-0778 PASTERNACK, S.
ANALYTICAL FINITE ELEMENT SIMULATION MODEL FOR
STRUCTURAL CRASHWORTHINESS PREDICTION
TRANSPORTATION SYSTEMS CENTER, CAMBRIDGE, MASS.,
REPORT NO: DOT-TSC-NHTSA-73-12, OCTOBER 1973, 65
P. (N.T.I.S. - PB-228 136/8)

73-0779 SHEN, S. F.
CHAN, S. T. K.
VARIATIONAL APPROACH TO THE LIFTING SURFACE
PROBLEM
11TH A.I.A.A. AEROSPACE SCI. MEET., WASHINGTON,
D.C. PAPER NO: 73-87, JANUARY 10-12, 1973.

73-0780 HUANG, P. C.
MATRA, J. P. JR.
PLANFORM INPUT GENERATOR (PING), A NASTRAN
PROCESSOR FOR LIFTING SUPFACES-THEORETICAL
DEVELOPMENT, USER'S MANUAL, AND PROGRAM LISTING
NAVAL ORDNANCE LAB., WHITE OAK, MD., REPORT NO:
NOLTR-73-199, DECEMBER, 1973, 185 P. (N.T.I.S. -
AD-775 303/1)

73-0781 WOZNIAK, CZ.
EXACT THEORY OF SPATIAL STRUCTURES
SYMP. ON IND. SPAT AND SHELL STRUCT., KIELCE,
POL., JUNE 18-23, 1973, PP. 499-513.

73-0782 BOOKER, J. R.
EMBANKMENT DEFORMATIONS DUE TO WATER LOADS
CIVIL ENGNG. LABS., SYDNEY UNIV., AUSTRALIA,
REPORT NO: R-224, SEPTEMBER 1973, 40 P. (N.T.I.S.
- PB-228 454/5)

73-0783 ROBINSON, J.
A SIX NODE TRIANGULAR EQUILIBRIUM (CONSTANT
STRESS) MEMBRANE ELEMENT
RA/ARD, REPORT NO. 7-11-73-20, NOVEMBER 1973.

73-0784 HARTZMAN, M.
ELASTIC-PLASTIC STRESS-STRAIN RELATIONS USED IN
CODE EPSOLA
LAWRENCE LIVERMORE LAB., UNIV. OF CALIF.,
LIVERMORE, APRIL 20, 1973, 15 P.
(N.T.I.S.-UCID-16541) SEE ALSO NSA 3008, NO.
23488.

73-0785 WESTPHAL, J. L.
ADDITION OF AN ARBITRARY BODY ANALYSIS CAPABILITY
TO THE BOEING TEA 236 FINITE ELEMENT COMPUTER
PROGRAM
MASTER'S THESIS, SCHOOL OF ENGNG., AIR FORCE INST.
OF TECH., WRIGHT-PATTERSON AFB, OHIO. REPORT NO:
GAM/AE/73A-20, DECEMBER, 1973, 91 P. (N.T.I.S. -
AD-774 430/3)

73-0786 PIAN, T. H. H.
LINEAR DYNAMIC ANALYSES OF LAMINATED PLATES AND
SHELLS BY THE HYBRID-STRESS FINITE-ELEMENT METHOD
AEROELASTIC AND STRUCT. RESEARCH LAB., MASS. INST.
OF TECH., CAMBRIDGE, REPORT NO: ASRL-TR-172-2,
OCTOBER 1973, 232 P. (N.T.I.S. - AD-774-296/8)

73-0787 ABDEL-HAMIDABDELLAH, G.
A FINITE ELEMENT METHOD FOR RANDOM FOLDED PLATE
STRUCTURES (EINE FINITE ELEMENT METHODE ZUR
BERECHNUNG BELIEBIGER FALTWERK)
INSTITUT FUR STATIK, TECHNISCHE UNIV. BRUNSWICK,
WEST GERMANY, REPORT NO. REPT-73-10, 1973, 81 PP.
(N.T.I.S. - N74-13644/1)

73-0788 HARBORD, R.
COMPUTATION OF SHELLS WITH FINITE DISPLACEMENTS:
MIXED FINITE ELEMENTS (BERECHNUNG VON SCHALEN MIT
ENDLICHEN VERSCHIEBUNGEN: GEMISCHTE FINITE
ELEMENTE)
INSTITUT FUR STATIK, TECHNISCHE UNIV. BRUNSWICK,
WEST GERMANY, REPORT NO. REPT-72-7, 1973, 147 PP.
(N.T.I.S. - N74-13643/3)

73-0789 PUCCINELLI, E. F.
VIEW FACTOR COMPUTER PROGRAM (PROGRAM VIEW) USER'S
MANUAL
NATIONAL AERONAUTICS AND SPACE ADMIN., GODDARD
SPACE FLIGHT CENTER, GREENBELT, MD., REPORT NO:
NASA-TM-X-70538,X-324-73-272, JULY 1973, 127 P.
(N.T.I.S. - N74-13638/3)

73-0790 VANBUREN, W.
VISSER, C.
STRESS ANALYSIS OF A CERAMIC BLADE DESIGN
SUBJECTED TO CENTRIFUGAL AND THERMAL LOADS
PROC. 2ND. CONF., HYANNIS, MASS., NOVEMBER 13-16,
1973, PP. 157-178. PUBL. BY BROOK HILL PUBL. CO.,
CHESTNUT HILL, MASS., 1974.

73-0791 CHANG, C. Y.
NAIR, K.
SINGH, R. D.
DEVELOPMENT AND APPLICATIONS OF THEORETICAL
METHODS FOR EVALUATING STABILITY OF OPENINGS IN
ROCK
WOODWARD-LUNDGREN AND ASSOC., OAKLAND, CALIF.,
FINAL REPORT, MARCH 14, 1972 - DECEMBER 14, 1973,
208 PP. (N.T.I.S. -AD-773 858/6)

73-0792 LIU, Y.
KING, P. G.
A FINITE ELEMENT ANALYSIS OF WAVE PROPAGATION IN
HUMAN SPINE
BIOMECHANICS LAB., TULANE UNIV., NEW ORLEANS, LA.,
FINAL REPORT, DECEMBER 1, 1971 - APRIL 30, 1973,
89 P. (N.T.I.S. - AD-773 858/6)

73-0793 STOWE, R. L.
PROPAGATION OF FAILURE IN A CIRCULAR CYLINDER OF
ROCK SUBJECTED TO A COMPRESSIVE FORCE
ARMY ENGINEER WATERWAYS EXPERIMENT STATION,
VICKSBURG, MISS., REPORT NO:
AEWES-MISC-PAPER-C-73-12, DECEMBER 1973, 82 P.
(N.T.I.S. - AD-773 469/2)

73-0794 GEISS, R. H.
JOHNSON, P. W.
MECHANICS OF DEFORMABLE SOLIDS - EURASIAN
COMMUNIST COUNTRIES
ARMY FOREIGN SCI., AND TECH., CENTER,
CHARLOTTESVILLE, VA., REPORT NO:
FSTC-CS-01-100-74, SEPTEMBER 15, 1973, 75 P.
(N.T.I.S. - AD-773 334/8)

73-0795 YOUNG, R. C.
MOTE, C. D.
SOLUTION OF MIXED BOUNDARY VALUE PROBLEMS WITH
LOCAL ERROR BOUND BY THE FINITE ELEMENT METHOD
COMP. METH. APPL. MECH. ENG., VOL. 2, 1973, PP.
159-184.

73-0796 ARGYRIS, J. H.
GRIEGER, I.
WAGEMANNM G.
CALCULATION OF STRESS CONCENTRATIONS ON SUPPORT
STRUCTURES UNDER THERMOMECHANICAL LOADING
(BERECHNUNG VON SPANNUNGSKONZZENTRATIONEN UND
TRAGWERKEN UNTER THERMOMECHANISCHER BELASTUNG)
PROC. 7TH COLLOQ. ON SCHWEISSTECH., ESSEN, MARCH
23, 1973, (INSTITUT FUR STATIK UND DYNAMIK DER
LUFT UND RAUMFAHRTKONSTRUKTIONEN, STUTTGART UNIV.,
WEST GERMANY, REPORT NO. ISD-143, 1973, 29 PP.)
N.T.I.S.-N74-14567/5

73-0797 ARGYRIS, J. H.
FAUST, G.
ROY, J. R.
SZIMMAT, J.
WARNKE, E.
FINITE ELEMENTS FOR CALCULATION OF PRESTRESSED
CONCRETE REACTOR PRESSURE VESSELS (FINITE ELEMENTE
ZUR BERECHNUNG VON
SPANNBETON-REAKTORDRUCKBEHAELTERN)
INSTITUT FUER STATIK UND DYNAMIK DER LUFT- UND
RAUMFAHRTKONSTRUKTIONEN, STUTTGART UNIV., WEST
GERMANY, REPORT NO: ISD-137, APRIL 1973, 191 P.
(N.T.I.S. - N74-12566/1)

73-0798 ADELMAN, H. M.
LESTER, H. C.
ROGERS, J. L. JR.
A FINITE ELEMENT FOR THERMAL STRESS ANALYSIS OF
SHELLS OF REVOLUTION
LANGLEY RESEARCH CENTER, NATIONAL AERON. AND SPACE
ADMIN., LANGLEY STATION, VA., REPORT NO:
NASA-TN-D-7286,L-8670, DECEMBER 1973, 59 P.
(N.T.I.S. - N74-12551/9)

73-0799 PINKNEY, R. B.
CYCLIC PLASTIC ANALYSIS OF STRUCTURAL STEEL JOINTS
EARTHQUAKE ENG. RESEARCH CENTER, UNIV. OF CALIF.,
BERKELEY, REPORT NO. EERC-73-15, AUGUST 1973, 172
PP., (N.T.I.S. - PB-225 843/1).

73-0800 CONNOR, J. J.
WANG, J. D.
BRIGGES, D. A.
MADSEN, O. S.
MATHEMATICAL MODELS OF THE MASSACHUSETTS BAY. PART
1. FINITE ELEMENT MODELING OF TWO-DIMENSIONAL
HYDRODYNAMIC CIRCULATION IN SHALLOW WATER MASSES.
PART 11. ANALYTICAL MODELS FOR ONE-AND TWO LAYER
SYSTEMS IN RECTANGULAR BASINS
SEA GRANT PROJECT OFFICE, MASS. INST. OF TECH.,
CAMBRIDGE, REPORT NO: MITSG-74-4, OCTOBER 15,
1973, 160 P. (N.T.I.S. - COM-74-10190/8)

73-0801 SHIH, C. F.
SMALL SCALE YIELDING ANALYSIS OF MIXED MODE PLANE
STRAIN CRACK PROBLEMS
DIV. OF ENGINEERING AND APPLIED PHYSICS, HARVARD
UNIV., CAMBRIDGE, MASS., REPORT NO:
NASA-CR-138246, DEAP-S-1, 1973, 37 P. (N.T.I.S. -
N74-23451/9)

73-0802 VAN LAETHEM, M.
DE COEN, J.
INFLUENCE OF THE UNPRECISION OF EXECUTION ON THE
BEHAVIOUR OF A SPHERICAL SHELL WITH CONCENTRATED
LOAD P AT CROWN
PROC. INT. SYMP. ON PREFAB. SHELLS, HAIFA, ISR,
PAPER D-6, VOL. 2, SEPTEMBER 10-13, 1973, PP.
344-351. PUBL. BY INT. ASSOC. FOR SHELL AND SPAT
STRUCT., TEL AVIV, ISR, 1973.

73-0803 MERCIER, B.
NUMERICAL SOLUTION OF THE BIHARMONIC PROBLEM BY
MIXED FINITE OF CLASS C-0
REPORT, LABORATORIO DI ANALISI NUMERICA DEL
C.N.R., PAVIA, 1973.

73-0804 REDDY, J. N.
A MATHEMATICAL THEORY OF COMPLEMENTARY-DUAL
VARIATIONAL PRINCIPLES AND MIXED FINITE ELEMENT
APPROXIMATIONS OF LINEAR BOUNDARY VALUE PROBLEMS
IN CONTINUUM MECHANICS
PH.D. DISSERTATION, UNIV., OF ALABAMA, HUNTSVILLE,
1973.

73-0805 ARMAN, A.
MUNFAKH, G. A.
EFFECT OF DENSIFICATION ON THE ENGINEERING
CHARACTERISTICS OF ORGANIC SOILS - VOL. 1 AND 2.
BULL. ENG. RES., L. A. STATE UNIV., NO. 113,
DECEMBER 1973, 335 PP.

73-0806 BABUSKA, I.
KELLOGG, R. B.
MATHEMATICAL AND COMPUTATIONAL PROBLEMS IN REACTOR
CALCULATIONS
PROC. CONF. MATHEMATICAL MODELS AND COMPUTATIONAL
TECHNIQUES FOR ANALYSIS OF NUCLEAR SYSTEMS, USAEC
TECH. INF. CENTER, UNIV. OF ANN ARBOR, MICH.,
APRIL 9-11, 1973, PP. 67-94.

73-0807 TORBE, I.
CHURCH, K.
A GENERAL QUADRILATERAL PLATE ELEMENT
REPORT NO: AASU329, UNIV. OF SOUTHAMPTON, 1973.

73-0808 GLANVILLE, J. I.
SHAH, A. H.
PREFABRICATED PRETENSIONED LIGHTWEIGHT CONCRETE
FOLDED PLATE
PROC. INT. SYMP. ON PREFAB. SHELLS, HAIFA, ISR,
PAPER D-7, VOL. 2, SEPTEMBER 10-13, 1973, PP.
352-370. PUBL. BY INT. ASSOC. FOR SHELL AND SPAT
STRUCT., TEL AVIV, ISR, 1973.

73-0809 MARSHALL, G.
VAN SPIEGEL, E.
ON THE NUMERICAL TREATMENT OF THE NAVIER-STOKES
EQUATIONS FOR AN INCOMPRESSIBLE FLUID
J. ENGNG. MATHS., VOL. 7, 1973, 173 P.

73-0810 SONTVEDT, T.
JOHANSSON, P.
MARGASS, B.
VAAGE, B.
LOADS AND RESPONSE OF LARGE DUCTED PROPELLER
SYSTEMS
PROC. OF SYMP. ON DUCTED PROPELLERS, NATL.
PHYSICAL LAB., ENGLAND, MAY 3 - JUNE 1, 1973,
(PUBL. BY ROY. INST. OF NAVAL ARCHITECTS, 1973)

73-0811 HAYASHI, I.
IWAMOTO, J.
STRESS DISTRIBUTION NEAR A BROKEN POINT OF
FILAMENT IN UNIDIRECTIONALLY FIBER REINFORCED
COMPOSITES
PROC. 17TH JPN. CONGR. ON MATER. RES., TOKYO,
SEPTEMBER 1973, PP. 162-166. ALSO IN SOC. OF
MATER. SCI., JPN., 1974.

73-0812 SINGH, S.
SRINIVASAN, R.
STRUCTURAL DESIGN OF NUCLEAR PLANT FACILITIES
SPEC. CONF. STRUCT. DES. NUCL. PLANT FACIL.,
CHICAGO, ILL., VOL. 2, DECEMBER 17-18, 1973, PP.
1081.

73-0813 KREISS, H. O.
BURSTEIN, S. Z.
ADVANCES IN NUMERICAL FLUID DYNAMICS
AGARD LECT., BRUSSELS, BELG., NO. 64, MARCH 5-9,
1973.

73-0814 KEEGSTRA, P. N. R.
HEAD, J. L.
FINITE ELEMENT STRESS ANALYSIS OF INTERACTING FUEL
PELLET AND CANNING UNDER FAST REACTOR CONDITION
2ND INT. CONF. ON STRUCT. MECH. IN REACT.
TECHNOL., BERLIN, GERMANY, PAPER 3, VOL. 1,
SEPTEMBER 10-14, 1973, 10 PP. PUBL. BY COMM. OF
THE EUR. COMMUNITIES, CENT. FOR INF. AND DOC.,
LUXEMB., 1973.

73-0815 LEVY, S.
WILKINSON, J. P. D.
THREE-DIMENSIONAL STUDY OF NUCLEAR FUEL ROD
BEHAVIOUR DURING STARTUP
2ND INT. CONF. ON STRUCT. MECH. IN REACT.
TECHNOL., BERLIN, GERMANY, PAPER 5, VOL. 1,
SEPTEMBER 10-14, 1973, 12 PP. PUBL. BY COMM. OF
THE EUR. COMMUNITIES, CENT. FOR INF. AND DOC.,
LUXEMB., 1973.

73-0816 ALUJEVIC, A.
HEAD, J. L.
SOME MODELS FOR THE ANALYSIS OF STRESSES IN A
TUBULAR FUEL PIN FOR HIGH TEMPERATURE REACTORS BY
FINITE ELEMENTS
2ND INT. CONF. ON STRUCT. MECH. IN REACT.
TECHNOL., BERLIN, GERMANY, PAPER 5, VOL. 1,
SEPTEMBER 10-14, 1973, 10 PP. PUBL. BY COMM. OF
THE EUR. COMMUNITIES, CENT. FOR INF. AND DOC.,
LUXEMB., 1973.

73-0817 GROS, G.
GOLDSTEIN, S.
TENUE DES STRUCTURES DE GRAPHITE VIS-A-VIS DES
SOLLICITATIONS RENCONTREES DANS UN COEUR DE
REACTEUR A HAUTE TEMPERATURE. (GRAPHITE
STRUCTURES BEHAVIOR IN A HIGH TEMPERATURE REACTOR)
2ND INT. CONF. ON STRUCT. MECH. IN REACT.
TECHNOL., BERLIN, GERMANY, PAPER 5, VOL. 1,
SEPTEMBER 10-14, 1973, 12 PP. PUBL. BY COMM. OF
THE EUR. COMMUNITIES, CENT. FOR INF. AND DOC.,
LUXEMB., 1973.

73-0818 CARUSO, S.
GENCO, M.
INTERNAL SUPPORTING STRUCTURES OF THE (CIRENE)
REACTOR CORE - RESULTS OF STATIC STRESS ANALYSIS
FOR A NORMAL CONDITION
2ND INT. CONF. ON STRUCT. MECH. IN REACT.
TECHNOL., BERLIN, GERMANY, PAPER 8, VOL. 2,
SEPTEMBER 10-14, 1973, 10 PP. PUBL. BY COMM. OF
THE EUR. COMMUNITIES, CENT. FOR INF. AND DOC.,
LUXEMB., 1973.

73-0819 NICKELL, R. E.
HIBBITT, H. D.
THERMAL AND MECHANICAL ANALYSIS OF WELDED
STRUCTURES
2ND INT. CONF. ON STRUCT. MECH. IN REACT.
TECHNOL., BERLIN, GERMANY, PAPER 1, VOL. 2,
SEPTEMBER 10-14, 1973, 16 PP. PUBL. BY COMM. OF
THE EUR. COMMUNITIES, CENT. FOR INF. AND DOC.,
LUXEMB., 1973.

73-0820 WRIGHT, M. B.
THERMAL STRESSES IN FACT REACTOR BOILER TUBES AT
THE SODIUM/GAS INTERFACE
2ND INT. CONF. ON STRUCT. MECH. IN REACT.
TECHNOL., BERLIN, GERMANY, PAPER 3, VOL. 2,
SEPTEMBER 10-14, 1973, 10 PP. PUBL. BY COMM. OF
THE EUR. COMMUNITIES, CENT. FOR INF. AND DOC.,
LUXEMB., 1973.

73-0821 GOLDBERG, J. E.
YANG, H. T. Y.
ANALYSIS OF MODERATELY THICK FLANGED REACTOR
PRESSURE VESSELS
2ND INT. CONF. ON STRUCT. MECH. IN REACT.
TECHNOL., BERLIN, GERMANY, PAPER 3, VOL. 3,
SEPTEMBER 10-14, 1973, 12 PP. PUBL. BY COMM. OF
THE EUR. COMMUNITIES, CENT. FOR INF. AND DOC.,
LUXEMB., 1973.

73-0822 WILSON, J. D.
THERMAL STRESSES IN THIN CYLINDRICAL SHELLS
2ND INT. CONF. ON STRUCT. MECH. IN REACT.
TECHNOL., BERLIN, GERMANY, PAPER 10, VOL. 3,
SEPTEMBER 10-14, 1973, 12 P. PUBL. BY COMM. OF THE
EUR. COMMUNITIES, CENT. FOR INF. AND DOC.,
LUXEMB., 1973.

73-0823 PUTZEYS, J.
REYNEN, J.
STRESS ANALYSIS OF PRESSURE VESSEL NOZZLES
2ND INT. CONF. ON STRUCT. MECH. IN REACT.
TECHNOL., BERLIN, GERMANY, PAPER 11, VOL. 3,
SEPTEMBER 10-14, 1973, 8 PP. PUBL. BY COMM. OF THE
EUR. COMMUNITIES, CENT. FOR INF. AND DOC.,
LUXEMB., 1973.

73-0824 FULLARD, K.
THERMAL STRESSES IN A NOZZLE-DRUM INTERSECTION
2ND INT. CONF. ON STRUCT. MECH. IN REACT.
TECHNOL., BERLIN, GERMANY, PAPER 4, VOL. 3,
SEPTEMBER 10-14, 1973, 10 PP. PUBL. BY COMM. OF
THE EUR. COMMUNITIES, CENT. FOR INF. AND DOC.,
LUXEMB., 1973.

73-0825 HELLEN, T. K.
FINITE ELEMENT CALCULATION OF STRESS INTENSITY
FACTORS USING ENERGY TECHNIQUES
2ND INT. CONF. ON STRUCT. MECH. IN REACT.
TECHNOL., BERLIN, GERMANY, PAPER 3, VOL. 3,
SEPTEMBER 10-14, 1973, 13 PP. PUBL. BY COMM. OF
THE EUR. COMMUNITIES, CENT. FOR INF. AND DOC.,
LUXEMB., 1973.

73-0826 SAUGY, B.
ZIMMERMANN, T.
THREE DIMENSIONAL RUPTURE ANALYSIS OF A
PRESTRESSED CONCRETE PRESSURE VESSEL INCLUDING
CREEP EFFECTS
2ND INT. CONF. ON STRUCT. MECH. IN REACT.
TECHNOL., BERLIN, GERMANY, PAPER 5, VOL. 3,
SEPTEMBER 10-14, 1973, 11 PP. PUBL. BY COMM. OF
THE EUR. COMMUNITIES, CENT. FOR INF. AND DOC.,
LUXEM., 1973.

73-0827 STEFANOU, G. D.
STRESS DISTRIBUTION IN THE LIGAMENT ZONE OF
PERFORATED CIRCULAR DISCS WITH SPECIAL REFERENCE
TO THE STANDPIPE REGION OF THE PCPV FOR GAS-COOLED
REACTORS
2ND INT. CONF. ON STRUCT. MECH. IN REACT.
TECHNOL., BERLIN, GERMANY, PAPER 7, VOL. 3,
SEPTEMBER 10-14, 1973, 18 PP. PUBL. BY COMM. OF
THE EUR. COMMUNITIES, CENT. FOR INF. AND DOC.,
LUXEMB., 1973.

73-0828 ANDO, Y.
YAGAWA, G.
APPLICATION OF THE FINITE ELEMENT METHOD TO THE
ANALYSIS OF CYLINDRICAL SHELLS
2ND INT. CONF. ON STRUCT. MECH. IN REACT.
TECHNOL., BERLIN, GERMANY, PAPER 4, VOL. 3,
SEPTEMBER 10-14, 1973, 10 PP. PUBL. BY COMM. OF
THE EUR. COMMUNITIES, CENT. FOR INF. AND DOC.,
LUXEMB., 1973.

73-0829 KURODA, T.
OHNISHI, S.
EXPERIMENTAL AND THEORETICAL STUDY ON A 1/10TH
SCALE MODEL PCPV
2ND INT. CONF. ON STRUCT. MECH. IN REACT.
TECHNOL., BERLIN, GERMANY, PAPER 1, VOL. 3,
SEPTEMBER 10-14, 1973, 12 PP. PUBL. BY COMM. OF
THE EUR. COMMUNITIES, CENT. FOR INF. AND DOC.,
LUXEMB., 1973.

73-0830 BRUSA, L.
CIACCI, R.
COMPUTATIONAL METHOD FOR DYNAMIC ANALYSIS OF
STRUCTURES WITH LARGE SIZE STIFFNESS AND MASS
MATRICES. APPLICATIONS TO SEISMIC ANALYSIS
2ND INT. CONF. ON STRUCT. MECH. IN REACT.
TECHNOL., BERLIN, GERMANY, PAPER 3, VOL. 4,
SEPTEMBER 10-14, 1973, 16 PP. PUBL. BY COMM. OF
THE EUR. COMMUNITIES, CENT. FOR INF. AND DOC.,
LUXEMB., 1973.

73-0831 SZABO, B. A.
TSAI, C-T.
THE QUADRATIC PROGRAMMING APPROACH TO THE FINITE
ELEMENT METHOD
INT. J. NUMER. METHOD IN ENG., VOL. 5, 1973, PP.
375-381.

73-0832 KAWAI, T.
MURAKI, T.
ON THE FINITE ELEMENT ANALYSIS OF THIN-WALLED
STRUCTURES WITH APPLICATION TO STRUCTURAL PROBLEMS
IN REACTOR TECHNOLOGY
2ND INT. CONF. ON STRUCT. MECH. IN REACT.
TECHNOL., BERLIN, GERMANY, PAPER 3, VOL. 4,
SEPTEMBER 10-14, 1973, 15 PP. PUBL. BY COMM. OF
THE EUR. COMMUNITIES, CENT. FOR INF. AND DOC.,
LUXEMB., 1973.

73-0833 COSTANTINO, C. J.
MILLER, C. A.
INFLUENCE OF SOIL/STRUCTURE INTERACTION PARAMETERS
ON FLOOR RESPONSE SPECTRA
2ND INT. CONF. ON STRUCT. MECH. IN REACT.
TECHNOL., BERLIN, GERMANY, PAPER 3, VOL. 4,
SEPTEMBER 10-14, 1973, 9 PP. PUBL. BY COMM. OF THE
EUR. COMMUNITIES, CENT. FOR INF. AND DOC.,
LUXEMB., 1973.

73-0834 CHU, S. L.
AGRAWAL, P. K.
FINITE ELEMENT TREATMENT OF SOIL-STRUCTURE
INTERACTION PROBLEM FOR NUCLEAR POWER PLANT UNDER
SEISMIC EXCITATION
2ND INT. CONF. ON STRUCT. MECH. IN REACT.
TECHNOL., BERLIN, GERMANY, PAPER 4, VOL. 4,
SEPTEMBER 10-14, 1973, 14 PP. PUBL. BY COMM OF EUR
COMMUNITIES, CENT. FOR INF. AND DOC., LUXEMB.,
1973.

73-0835 ANDERSON, S. I.
OTTOSEN, N. S.
ULTIMATE LOAD BEHAVIOUR OF PCRV TOP CLOSURES.
THEORETICAL AND EXPERIMENTAL INVESTIGATIONS
2ND INT. CONF. ON STRUCT. MECH. IN REACT.
TECHNOL., BERLIN, GERMANY, PAPER 3, VOL. 3,
SEPTEMBER 10-14, 1973, 12 PP. PUBL. BY COMM. OF
THE EUR. COMMUNITIES, CENT. FOR INF. AND DOC.,
LUXEMB., 1973.

73-0836 CARMIGNANI, F.
CELLA, A.
STRUCTURAL DYNAMICS BY FINITE ELEMENTS: MODAL AND
FOURIER ANALYSIS
2ND INT. CONF. ON STRUCT. MECH. IN REACT.
TECHNOL., BERLIN, GERMANY, PAPER 2, VOL. 4,
SEPTEMBER 10-14, 1973, 14 PP. PUBL. BY COMM. OF
THE EUR. COMMUNITIES, CENT. FOR INF. AND DOC.,
LUXEMB., 1973.

73-0837 YAGAWA, G.
ANDO, Y.
THREE-DIMENSIONAL FINITE ELEMENT METHOD OF
THERMOELASTOPLASTICITY WITH CREEP EFFECT
2ND INT. CONF. ON STRUCT. MECH. IN REACT.
TECHNOL., BERLIN, GERMANY, PAPER 3, VOL. 5,
SEPTEMBER 10-14, 1973, 10 PP. PUBL. BY COMM. OF
THE EUR. COMMUNITIES, CENT. FOR INF. AND DOC.,
LUXEMB., 1973.

73-0838 IYENGAR, N. G. R.
MURTHY, P. N.
NON-LINEAR EFFECTS OF HIGH TEMPERATURE ON THE
VIBRATION OF A SIMPLY SUPPORTED BEAM WITH A
CENTRAL MASS BY MATHEMATICAL PROGRAMMING
TECHNIQUE.
2ND INT. CONF. ON STRUCT. MECH. IN REACT.
TECHNOL., BERLIN, GERMANY, PAPER 5, VOL. 5,
SEPTEMBER 10-14, 1973, 9 PP. PUBL. BY COMM. OF THE
EUR. COMMUNITIES, CENT. FOR INF. AND DOC.,
LUXEMB., 1973.

73-0839 REMEDIOS, E. E.
BELL, R.
ENGINEERING APPROACH TO FINITE ELEMENT ANALYSIS OF
NUCLEAR COMPONENTS
2ND INT. CONF. ON STRUCT. MECH. IN REACT.
TECHNOL., BERLIN, GERMANY, PAPER 2, VOL. 5,
SEPTEMBER 10-14, 1973, 12 PP. PUBL. BY COMM. OF
THE EUR. COMMUNITIES, CENT. FOR INF. AND DOC.,
LUXEMB., 1973.

73-0840 HASSELIN, G.
HELIOT, J.
USE OF THE ' TITUS ' FINITE ELEMENT CODE FOR THE
NUCLEAR STRUCTURES AND INTERNAL STRESS-ANALYSIS
2ND INT. CONF. ON STRUCT. MECH. IN REACT.
TECHNOL., BERLIN, GERMANY, PAPER 3, VOL. 5,
SEPTEMBER 10-14, 1973, 14 PP. PUBL. BY COMM. OF
THE EUR. COMMUNITIES, CENT FOR INF. AND DOC.,
LUXEMB., 1973.

73-0841 KIKUCHI, F.
ANDO, Y.
ON THE CONVERGENCE OF FINITE ELEMENT SCHEMES BASED
ON NON-CONFORMING AND SIMPLIFIED HYBRID
DISPLACEMENT METHODS
2ND INT. CONF. ON STRUCT. MECH. IN REACT.
TECHNOL., BERLIN, GERMANY, PAPER 6, VOL. 5,
SEPTEMBER 10-14, 1973, 11 PP. PUBL. BY COMM. OF
THE EUR. COMMUNITIES, CENT. FOR INF. AND DOC.,
LUXEMB., 1973.

73-0842 DEDONATO, O.
MAIER, G.
FINITE ELEMENT ELASTOPLASTIC ANALYSIS BY QUADRATIC
PROGRAMMING: THE MULTISTAGE METHOD
2ND INT. CONF. ON STRUCT. MECH. IN REACT.
TECHNOL., BERLIN, GERMANY, PAPER 8, VOL. 5,
SEPTEMBER 10-14, 1973, 12 PP. PUBL. BY COMM. OF
THE EUR. COMMUNITIES, CENT. FOR INF. AND DOC.,
LUXEMB., 1973.

73-0843 SHATOFF, H. D.
THREE-DIMENSIONAL ELASTIC FINITE ELEMENT ANALYSIS
USING GRADIENT DEGREES OF FREEDOM
2ND INT. CONF. ON STRUCT. MECH. IN REACT.
TECHNOL., BERLIN, GERMANY, PAPER 10, VOL. 5,
SEPTEMBER 10-14, 1973, 14 PP. PUBL. BY COMM. OF
THE EUR. COMMUNITIES, CENT. FOR INF. AND DOC.,
LUXEMB., 1973.

73-0844 NAGATO, K.
DYNAMIC ELASTIC-PLASTIC ANALYSIS OF AXISYMMETRIC
STRUCTURES BY FINITE ELEMENT METHOD COMPUTER CODE
2ND INT. CONF. ON STRUCT. MECH. IN REACT.
TECHNOL., BERLIN, GERMANY, PAPER 3, VOL. 5,
SEPTEMBER 10-14, 1973, 11 PP. PUBL. BY COMM. OF
THE EUR. COMMUNITIES, CENT. FOR INF. AND DOC.,
LUXEMB., 1973.

73-0845 BERG, S.
BERGAN, P. G.
NONLINEAR FINITE ELEMENT ANALYSIS OF REINFORCED
CONCRETE PLATES
2ND INT. CONF. ON STRUCT. MECH. IN REACT.
TECHNOL., BERLIN, GERMANY, PAPER 5, VOL. 5,
SEPTEMBER 10-14, 1973, 12 PP. PUBL. BY COMM. OF
THE EUR. COMMUNITIES, CENT. FOR INF. AND DOC.,
LUXEMB., 1973.

73-0846 PATIL, S. S.
RAO, A. K.
SPECIAL FINITE ELEMENTS FOR FLAW AND DISCONTINUITY
STRESSES IN COMPOSITES
2ND INT. CONF. ON STRUCT. MECH. IN REACT.
TECHNOL., BERLIN, GERMANY, PAPER 6, VOL. 5,
SEPTEMBER 10-14, 1973, 12 PP. PUBL. BY COMM. OF
THE EUR. COMMUNITIES, CENT. FOR INF. AND DOC.,
LUXEMB., 1973.

73-0847 BALUCH, M. H.
KORMAN, T.
FINITE ELEMENT ANALYSIS OF A THICK WALLED
MICROPOLAR CYLINDER LOADED AXISYMMETRICALLY
2ND INT. CONF. ON STRUCT. MECH. IN REACT.
TECHNOL., BERLIN, GERMANY, PAPER 8, VOL. 5,
SEPTEMBER 10-14, 1973, 10 PP. PUBL. BY COMM. OF
THE EUR. COMMUNITIES, CENT. FOR INF. AND DOC.,
LUXEMB., 1973.

73-0848 LESTINGI, J.
BROWN, S. JR.
COMPARISON OF THE NUMERICAL INTEGRATION TECHNIQUE
AND THE FINITE ELEMENT METHOD IN THE ANALYSIS OF
THIN-SHELL STRUCTURES
2ND INT. CONF. ON STRUCT. MECH. IN REACT.
TECHNOL., BERLIN, GERMANY, PAPER 6, VOL. 5,
SEPTEMBER 10-14, 1973, 11 P. PUBL. BY COMM. OF THE
EUR. COMMUNITIES, CENT. FOR INF. AND DOC.,
LUXEMB., 1973.

73-0849 BATOZ, J. L.
DHATT, G.
BUCKLING OF DEEP SHELLS
2ND INT. CONF. ON STRUCT. MECH. IN REACT. TECHNOL,
BERLIN, GERMANY, PAPER 7, VOL. 5, SEPTEMBER 10-14,
1973, 12 PP. PUBL. BY COMM. OF THE EUR.
COMMUNITIES, CENT. FOR INF. AND DOC., LUXEMB.,
1973.

73-0850 BELYTSCHKO, T.
HSIEH, B. J.
CONVECTED COORDINATES FOR TRANSIENT NONLINEAR
FINITE-ELEMENT ANALYSIS
2ND INT. CONF. ON STRUCT. MECH. IN REACT. TECHNOL,
BERLIN, GERMANY, PAPER 9, VOL. 5, SEPTEMBER 10-14,
1973, 13 PP. PUBL. BY COMM. OF THE EUR.
COMMUNITIES, CENT. FOR INF. AND DOC., LUXEMB.,
1973.

73-0851 MIYAMOTO, H.
FUKUDA, S.
COMPUTER PREDICTION OF FATIQUE CRACK PROPAGATION
UNDER RANDOM LOADING
2ND INT. CONF. ON STRUCT. MECH. IN REACT. TECHNOL,
BERLIN, GERMANY, PAPER 9, VOL. 5, SEPTEMBER 10-14,
1973, 10 P. PUBL. BY COMM. OF THE EUR.
COMMUNITIES, CENT. FOR INF. AND DOC., LUXEMB.,
1973.

73-0852 CIARLET, P. G.
RAVIART, P. A.
MAXIMUM PRINCIPLE AND UNIFORM CONVERGENCE FOR THE
FINITE ELEMENT METHOD
COMP. METH. IN APPL. MECH. ENG., VOL. 2, 1973, PP.
17-31.

73-0853 MIKSCH, M.
SCHMITT, W.
APPLICATION OF THE FINITE ELEMENT METHOD FOR THE
SAFETY EVALUATION OF REACTOR COMPONENTS
2ND INT. CONF. ON STRUCT. MECH. IN REACT.
TECHNOL., BERLIN, GERMANY, PAPER 8, VOL. 6A,
SEPTEMBER 10-14, 1973, 8 PP. PUBL. BY COMM. OF THE
EUR. COMMUNITIES, CENT. FOR INF. AND DOC.,
LUXEMB., 1973.

73-0854 HUBER, A. E.
HOFMANN, H. H.
EINFACHES NAEHERUNGSVERFAHREN FUR DIE
GRENZTRAGFAEHIGKEITSANLYSE VON
SPANNBETON-REAKTORDRUCKBEHAELTERN. (SIMPLE
APPROXIMATE METHOD FOR THE ANALYSIS OF THE
ULTIMATE CARRYING CAPACITY OF PRESTRESSED CONCRETE
REACTOR PRESSURE VESSELS)
2ND INT. CONF. ON STRUCT. MECH. IN REACT.
TECHNOL., BERLIN, GERMANY, PAPER 1, VOL. 6A,
SEPTEMBER 10-14, 1973, 10 PP. PUBL. BY COMM. OF
THE EUR. COMMUNITIES, CENT. FOR INF. AND DOC.,
LUXEMB., 1973.

73-0855 BINDSEIL, P.
UEBER DAS VERHALTEN VON
SPANNEBETONDRUCKBEHAELTRERN IM
GRENZTRAGFAEHIGKEITSZUSTAND UND DESSEN EINFULUSS
AUF BERECHUNG UND KONSTRUKTION (UNTERSPEZIELLER
BEZUGNAHME AUF DEN THTR BEHAELTER) (ON THE
BEHAVIOR OF PCPV'S UNDER ULTIMATE LOAD CONDITIONS
AND ITS INFLUENCE ON CALCULATION AND DESIGN
ESPECIALLY APPLIED TO THE THTR-VESSEL)
2ND INTL. CONF. ON STRUCT. MECH. IN REACT.
TECHNOL., BERLIN, GERMANY, PAPER 2, VOL. 6A,
SEPTEMBER 10-14, 1973, 20 PP. (PUBL. BY COMM. OF
THE EUR. COMM., CENT. FOR INF. AND DOC., LUXEMB.,
1973.

73-0856 FRUEHAUF, H.
MEERWALD, K.
ENTWURF UND SPANNUNGSBERECHNUNG EINES SPANNBETON
DRUCKBEHAELTERS AUS FERTIGTEILEN. (DESIGN AND
STRESS ANALYSIS OF A PREFABRICATED PRESTRESSED
CONCRETE PRESSURE VESSEL)
2ND. INT. CONF. ON STRUCT. MECH. IN REACT.
TECHNOL., GERMANY, PAPER 10, VOL. 6A, SEPTEMBER
10-14, 1973, 19 PP. (PUBL. BY COMM. OF THE EUR.
COMMUNITIES, CENT. FOR INF. AND DOC., LUXEMB.,
1973)

73-0857 SCOTTO, F. L.
THIN-WALLED PRESTRESSED CONCRETE PRESSURE VESSEL
FOR HIGH TEMPERATURE REACTORS: EXPERIMENTAL
INVESTIGATIONS ON THREE 1:20 SCALE PCPV MODELS AND
DESIGN PHILOSPHY PROPOSAL
2ND INT. CONF. ON STRUCT. MECH. IN REACT.
TECHNOL., BERLIN, GERMANY, PAPER 5, VOL. 6A,
SEPTEMBER 10-14, 1973, 12 PP. PUBL. BY COMM. OF
THE EUR. COMMUNITIES, CENT. FOR INF. AND DOC.,
LUXEMB., 1973.

73-0858 DIETRICH, R.
FUERSTE, W.
BERECHNUNGSMETHODEN ZUR ERMITTLUNG VON
BELASTUNGSBEDINGUNGEN UND
GRENZTRAGFAEHIGKEITSVERHALTEN VON KERNKRAFTWERKS
BAUWERKSSTRUKTUREN BEI AUSSERGEWEOHNLICHEN
AUSSEREN EINWIRKUNGEN WIE FLUGZENGASTURZ UND
GASDRUCKWELLE. (METHODS FOR ANALYZING LOADING
CONDITIONS AND LIMIT LOAD STRUCTURAL BEHAVIOUR OF
NUCLEAR POWER PLANT BUILDING STRUCTURES IN THE
EVENT OF EXCEPTIONAL EXTERNAL INFLUENCES AS
AIRCRAFT CRASHES AND GAS PRESSURE WAVES)
2ND INT. CONF. ON STRUCT. MECH. IN REACT.
TECHNOL., BERLIN, GERMANY, PAPER 1, VOL. 6B,
SEPTEMBER 10-14, 1973, 22 PP. PUBL. BY COMM. OF
THE EUR. COMMUNITIES, CENT. FOR INF. AND DOC.,
LUXEMB., 1973.

73-0859 POWELL, G. H.
CHI, H. M.
COMPUTATIONAL PROCEDURES FOR INELASTIC STRESS
ANALYSIS
2ND INT. CONF. ON STRUCT. MECH. IN REACT.
TECHNOL., BERLIN, GERMANY, PAPER 4, VOL. 6B,
SEPTEMBER 10-14, 1973, 12 PP. PUBL. BY COMM. OF
THE EUR. COMMUNITIES, CENT. FOR INF. AND DOC.,
LUXEMB., 1973.

73-0860 RAMANI, D. T.
TRIANGULAR THICK SHELL FINITE ELEMENT FOR THE
THREE-DIMENSIONAL STRESS ANALYSIS OF PRESSURE
VESSELS
2ND INT. CONF. ON STRUCT. MECH. IN REACT.
TECHNOL., BERLIN, GERMANY, PAPER 7, VOL. 6B,
SEPTEMBER 10-14, 1973, 16 PP. PUBL. BY COMM. OF
THE EUR. COMMNUNITIES, CENT. FOR INF. AND DOC.,
LUXEMB., 1973.

73-0861 CHUNG, T. J.
YAGAWA, G.
INCREMENTAL THERMOMECHANICAL THEORY OF
VISCOELASTOPLASTIC SOLIDS AND SOLUTION BY FINITE
ELEMENTS
2ND INT. CONF. ON STRUCT. MECH. IN REACT.
TECHNOL., BERLIN, GERMANY, PAPER L4/2, SEPTEMBER
10-14, 1973. PUBL. BY COMM. OF THE EUR.
COMMUNITIES , CENT. FOR INF. AND DOC., LUXEM.,
1973.

73-0862 MAK, C. K.
KELSEY, S.
PROBABILISTIC BUCKLING BEHAVIOR OF STRUCTURES WITH
RANDOM LACK-OF-FIT AND GEOMETRIC IMPERFECTIONS
2ND INT. CONF. ON STRUCT. MECH. IN REACT.
TECHNOL., BERLIN, GERMANY, PAPER 10, VOL. 5,
SEPTEMBER 10-14, 1973, 10 PP. PUBL. BY COMM. OF
THE EUR. COMMUNITIES, CENT. FOR INF. AND DOC.,
LUXEMB., 1973.

73-0863 CHOPRA, P. S.
KAEMPF, H.
INFLUENCE OF DEFECTS ON FUEL ELEMENT CLADDING
STRESSES
2ND INT. CONF. ON STRUCT. MECH. IN REACT.
TECHNOL., BERLIN, GERMANY, PAPER 3, VOL. 6A,
SEPTEMBER 10-14, 1973, 1 P. PUBL. BY COMM. OF THE
EUR. COMMUNITIES, CENT. FOR INF. AND DOC.,
LUXEMB., 1973.

73-0864 MACIOTTA, R.
SIGNIFICANT APPLICATIONS OF FINITE ELEMENTS AND
FINITE DIFFERENCES METHODS TO SOLVE PROBLEMS OF
THERMAL LOADS, STRESSES AND VIBRATIONS WITH
SPECIAL REGARD TO THE 550 BORE MEDIUM-SPEED ENGINE
PROC. INT. SYMP. ON MAR. ENG., SESS. 3-6, NOVEMBER
12-15, 1973, TOKYO, JAPAN, PP. 1-15 (PUBL. BY MAR.
ENG. SOC., JAPAN)

73-0865 CHOI, C-K.
SCHNOBRICH, W. C.
USE OF NONCONFORMING MODES IN FINITE ELEMENT
ANALYSIS OF PLATES AND SHELLS
DEPT. OF CIVIL ENGNG., UNIV. OF ILLINOIS, URBANA,
REPORT NO: STRUCT. RESEARCH SER-401,
UILU-ENG-73-2019, SEPTEMBER 1973. 124 P. (N.T.I.S.
- AD-767 169/6)

73-0866 BAKHREEAH, S. A.
SCHNOBRICH, W. C.
FINITE ELEMENT ANALYSIS OF INTERSECTING CYLINDERS
DEPT. OF CIVIL ENGNG., UNIV. OF ILLINOIS, URBANA,
REPORT NO: UILU-ENG-73-2018, STRUCT. RESEARCH
SER-400, SEPTEMBER 1973, 145 P. (N.T.I.S. - AD-767
168/8)

73-0867 HAFTKA, R. T.
AUTOMATED PROCEDURE FOR DESIGN OF WING STRUCTURES
TO SATISFY STRENGTH AND FLUTTER REQUIREMENTS
LANGLEY RESEARCH CENTER, NATIONAL AERON. AND SPACE
ADMIN., LANGLEY STATION, VA., REPORT NO:
NASA-TN-D-7264, L-8592, JULY 1973, 34 P. (N.T.I.S.
- N73-26927/6)

73-0868 FRAZIER, G. A.
ALEXANDER, J. H.
PETERSEN, C. M.
3-D SEISMIC CODE FOR ILLIAC-1V
SYSTEMS SCIENCE AND SOFTWARE, LA JOLLA, CALIF.,
REPORT NO: SSS-R-73-1506, FEBRUARY 9, 1973, 118
P. (N.T.I.S. / AD-767 010/2)

73-0869 EBERT, L. J.
GRIESBACH, T. J.
WRIGHT, P. K.
FINITE ELEMENT ANALYSIS SYSTEM FOR THE MECHANICAL
BEHAVIOR OF ORIENTED FIBER COMPOSITE MATERIALS
UNDER COMBINED STRESSES
DIV. OF METALLURGY AND MATERIALS SCI., CASE
WESTERN RESERVE UNIV., CLEVELAND, OHIO, INTERIM
REPORT, MAY 31, 1973, 120 P. (N.T.I.S. - AD-766
963/3)

73-0870 LASRY, S. J.
DERIVATION OF CRACK ELEMENT-STIFFNESS MATRIX BY
THE COMPLEX VARIABLE APPROACH
AEROELASTIC AND STRUCT. RESEARCH LAB., MASS. INST.
OF TECH., CAMBRIDGE, REPORT NO: ASRL-TR-170-2,
FEBRUARY 1973, 35 P. (N.T.I.S. - AD-766 922/9)

73-0871 SALAMON, P. F. JR.
LEAST SQUARE OPTIMAL APPROXIMATION SIMULATION
TECHNIQUES
SYSTEMS RESEARCH CENTER, CASE WESTERN RESERVE
UNIV., CLEVELAND, OHIO, REPORT NO: SRC-73-3,
FEBRUARY 16, 1973, 172 P. (N.T.I.S. - AD-766
854/4)

73-0872 HEUZE, F. E.
GOODMAN, R. E.
NUMERICAL AND PHYSICAL MODELLING OF REINFORCEMENT
SYSTEMS FOR TUNNELS IN JOINTED ROCK
ARMY ENGINEER DISTRICT, OMAHA, NEBRASKA, REPORT
NO: TR-16, AUGUST 1973, 116 P. (N.T.I.S. - AD-766
833/8)

73-0873 NOBARI, E. S.
LEE, K. K.
DUNCAN, J. M.
HYDRAULIC FRACTURING IN ZONED EARTH AND ROCKFILL
DAMS: A REPORT OF AN INVESTIGATION
COLL. OF ENGNG., UNIV. OF CALIFORNIA, BERKELEY,
REPORT NO: TE-73-1, JANUARY 1973, 78 P. (N.T.I.S.
- AD-766 728/0)

73-0874 WILSON, H. B.
ARMEN, H. JR.
HILL, J. H.
MATHEWS, D.
FEASIBILITY OF USING FINITE ELEMENTS IN ANALYSIS
OF SECOND BREAKDOWN IN SEMICONDUCTOR DEVICES
GUIDANCE AND CONTROL DIRECTORATE, ARMY MISSILE
COMMAND REDSTONE ARSENAL, ALA., REPORT NO:
RG-73-14, JUNE 8, 1973, 25 P. (N.T.I.S. - AD-766
688/6)

73-0875 SAKENEM E.
PIFKO, A.
LEVINE, H.
NONLINEAR CRACK ANALYSIS WITH FINITE ELEMENTS
RESEARCH DEPT., GRUMMAN AEROSPACE CORP., BETHPAGE,
N.Y., REPORT NO: RE-460J, JULY 1973, 49 P.
(N.T.I.S. - AD-766 558/1)

73-0876 CHOPRA, A. K.
VAISH, A. K.
EARTHQUAKE ANALYSIS OF STRUCTURE-FOUNDATION
SYSTEMS
EARTHQUAKE ENGNG. RESEARCH CENTER, UNIV. OF
CALIF., BERKELEY, REPORT NO: EERC-73-9, MAY 1973,
141 P. (N.T.I.S. - AD-766 272/9)

73-0877 CHEN, A. C. T.
CHEN, W. F.
CONSTITUTIVE RELATIONS OF CONCRETE AND
PUNCH-INDENTATION PROBLEMS
FRITZ ENGNG. LAB., LEHIGH UNIV., BETHLEHEM, PA.,
REPORT NO: FEL-370.11, MAY 1973, 160 P. (N.T.I.S.
- PB-222 615/7)

73-0878 MITCHELL, R. A.
WOOLEY, R. M.
CHWIRUT, D. J.
COMPOSITE-OVERLAY REINFORCEMENT OF CUTOUTS AND
CRACKS IN METAL SHEET
ENGNG. MECH. SECTION, NATIONAL BUREAU OF
STANDARDS, WASH., D.C., REPORT NO: NBSIR-73-201,
FEBRUARY 1973, 97 P. (N.T.I.S. - COM-73-11221/1)

73-0879 MA, J. H.
STRESS ANALYSIS OF COMPLEX SHIP COMPONENTS BY A
NUMERICAL PROCEDURE USING CURVED FINITE ELEMENTS
NAVAL SHIP RESEARCH AND DEVELOPMENT CENTER,
BETHESDA, MD., REPORT NO: NSPDC-4057, JULY 1973,
103 P. (N.T.I.S. - AD-765 712/5)

73-0880 FORREST, J. B.
LEW, T. K.
A COMPUTER MODEL FOR PREDICTING THE
LOAD-DEFLECTION RESPONSE OF EXPEDIENT SOIL
SURFACING
NAVAL CIVIL ENGNG. LAB., PORT HUENEME, CALIF.,
REPORT NO: NCEL-TN-1280, JULY 1973, 61 P.
(N.T.I.S. - AD-765 569/9)

73-0881 GARNET, H.
CROUZET-PASCAL, J.
DOUBLY CURVED TRIANGULAR FINITE ELEMENTS FOR
SHELLS OF ARBITRARY SHAPE
RESEARCH DEPT., GRUMMAN AEROSPACE CORP., BETHPAGE,
N. Y., REPORT NO. PE-453, APRIL 1973, 107 PP.,
(N.T.I.S. - AC-765 382/7).

73-0882 DANA, J. R.
THREE DIMENSIONAL FINITE ELEMENT ANALYSIS OF THICK
LAMINATED COMPOSITES - INCLUDING INTERLAMINAR AND
BOUNDARY EFFECTS NEAR CIRCULAR HOLES
PH.D. DISSERTATION, VIRGINIA POLYTECHNIC INST. AND
STATE UNIV., AUGUST 1973.

73-0883 VOS, R. G.
VANN, W. P.
A FINITE ELEMENT TENSOR APPROACH TO PLATE BUCKLING
AND POST BUCKLING
INT. J. NUMER. METH. ENG. VOL. 5, 1973, PP.
351-365.

73-0884 HASTING, R. F.
BALL, R. E.
A COMPARISON OF SEVERAL COMPUTER SOLUTIONS TO
THREE STRUCTURAL SHELL ANALYSIS PROBLEMS
AIR FORCE FLIGHT DYNAMICS LAB., WRIGHT-PATTERSON
AIR FORCE BASE. TECH. REPORT AFFDL-TR-73-15, 1973.

73-0885 BERGER, A. E.
(L SUP 2) ERROR ESTIMATES FOR FINITE ELEMENTS WITH
INTERPOLATED BOUNDARY CONDITIONS
NAVAL ORDNANCE LAB., WHITE OAK, MD., REPORT NO:
NOLTR-73-138, JULY 2, 1973, 13 P. (N.T.I.S. -
AD-764 883/5)

73-0886 GARG, V. K.
ANAND, S. C.
HODGE, P. G. JR.
STRUCTURAL INELASTICITY. V11. ELASTIC-PLASTIC
ANALYSIS OF A WHEEL ROLLING ON A RIGID TRACK
DEPT. OF AEROSPACE ENGNG. AND MECH. UNIV. OF
MINNESOTA, MINNEAPOLIS, REPORT NO: AEM-H1-7, MAY
1973, 33 P. (N.T.I.S. - AD-764 575/7) SEE ALSO
AD-764 574 AND AD-764 576.

73-0887 HODGE, P. G. JR.
GARG, V. K.
ANAND, S. C.
STRUCTURAL INELASTICITY. V1. A FINITE ELEMENT
METHOD FOR PLASTICITY PROBLEMS
DEPT. OF AEROSPACE ENGNG. AND MECH., MINNESOTA
UNIV., MINNEAPOLIS, REPORT NO: AEM-H1-6, MAY 1973,
22 P. (N.T.I.S. - AD-764 574/0) SEE ALSO AD-764
575.

73-0888 AUTHORS
PROCEEDINGS OF THE NAVY-NASTRAN COLLOQUIUM (4TH)
HELD AT NAVY SHIP RESEARCH AND DEVELOPMENT CENTER,
WASHINGTON, D.C. ON MARCH 27, 1973
NAVAL SHIP RESEARCH AND DEVELOPMENT CENTER,
BETHESDA, MD., MARCH 27, 1973, 181 P.
(N.T.I.S.-AD-764 508) SEE ALSO AD-764 299

73-0889 ARGYRIS, J. H.
DUNNE, P. C.
SHAPE DETERMINATION OF PREFABRICATED PRESTRESSED
NETWORKS
PROC. INT. SYMP. ON PREFAB. SHELLS, HAIFA, ISR.,
PAPER C-3, VOL. 1, SEPTEMBER 10-13, 1973, PP.
218-240. PUBL. BY INT. ASSOC. FOR SHELL AND SPAT
STRUCT., TEL AVIV, ISR, 1973.

73-0890 HALLWACHS, D. A.
AN INVESTIGATION INTO THE METHODS OF DETECTING
DEWETTED ZONES IN SOLID ROCKET PROPELLANTS USING
HOLOGRAPHIC INTERFEROMETRY
MASTER'S THESIS, NAVAL POSTGRADUATE SCHOOL,
MONTEREY, CALIF., JUNE 1973, 67 P. (N.T.I.S. -
AD-764 470)

73-0891 WHITESIDE, J. B. D.
THE BEHAVIOR OF ADVANCED FILAMENTARY COMPOSITE
PLATES WITH CUTOUTS
IIT RESEARCH INST. CHICAGO, ILL., FINAL TECH.
REPORT, MARCH 1970 - SEPTEMBER 1972, JUNE 1973,
170 P. (N.T.I.S. - AD-764 362)

73-0892 WILTON, C.
GABRIELSEN, B. L.
VERNER, E. A.
BECKER, E. B.
SHOCK TUNNEL TESTS OF PRELOADED AND ARCHED WALL
PANELS FINITE ELEMENT STRESS FORMULATION FOR WAVE
PROPAGATION
INT. J. NUMER. METH. ENG., VOL. 7, 1973, PP.
441-460.

73-0893 GOULD, PHILLIP L.
APPROXIMATE ANALYSIS OF PREFABRICATED TOWER
STRUCTURES USING THIN SHELL FINITE ELEMENTS
PROC. INT. SYMP. ON PREFAB. SHELLS, HAIFA, ISR.,
PAPER D-1, VOL. 2, SEPTEMBER 10-13, 1973, PP.
271-280. PUBL. BY INT. ASSOC. FOR SHELL AND SPAT
STRUCT., TEL AVIV, ISR. 1973.

73-0894 EBERHARDT, A. C.
AIRCRAFT-PAVEMENT INTERACTION STUDIES, PHASE 1: A
FINITE ELEMENT MODEL OF A JOINTED CONCRETE
PAVEMENT ON A NON-LINEAR VISCOUS SUBGRADE (DYNAMIC
INTERACTION OF AIRCRAFT-PAVEMENT SYSTEMS)
ARMY CONSTRUCTION ENGNG. RESEARCH LAB., CHAMPAIGN,
ILL., REPORT NO: CERL-PR-S-19, JUNE 1973, 30 P.
(N.T.I.S. - AD-764 243)

73-0895 KULHAWY, F. H.
ANALYSIS OF UNDERGROUND OPENINGS IN ROCK BY FINITE
ELEMENT METHODS
DEPT. OF CIVIL ENGNG., SYRACUSE UNIV., N. Y.,
FINAL TECH. REPORT. MARCH 11, 1971 - JULY 31,
1972, APRIL 1973, 310 P. (N.T.I.S. - AD - 764 123)

73-0896 BALIGH, M. M.
NUMERICAL STUDY OF UNIAXIAL AND TRIAXIAL ROCK
COMPRESSION TESTS
SYSTEMS SCI. AND SOFTWARE, LA JOLLA, CALIF.,
REPORT NO: SSS-R-73-1658, MAY 1973, 167 P.
(N.T.I.S. - AD-764 099)

73-0897 TAKAHASHI, S. K.
EFFECT OF BACKPACKING AND INTERNAL PRESSURIZATION
ON STRESSES TRANSMITTED TO BURIED CYLINDERS
NAVAL CIVIL ENGNG. LAB., PORT HUENEME, CALIF.,
REPORT NO: NCEL-TR-789, MAY 1973, 48 P. (N.T.I.S.
- AD-764 058)

73-0898 AUTHORS
SYMPOSIUM ON ACOUSTIC FATIGUE
PROC. 35TH MEET. OF STRUCT. AND MATER. PANEL,
TOULOUSE, FRANCE, SEPTEMBER 26-27, 1972, (REPORT
NO. AGARD-CP-113, MAY 1973, 271 PP.)
(N.T.I.S.-AD-763 918)

73-0899 AVENT, R. R.
DEAN, D. L.
FIELD ANALYSIS FOR PREFABRICATED RIBBED
CYLINDRICAL SHELLS
PROC. SYMP. ON PREFAB. SHELLS, HAIFA, ISR., PAPER
D-5, VOL. 2, SEPTEMBER 10-13, 1973, PP. 328-343,
PUBL. BY INT. ASSOC. FOR SHELL AND SPAT STRUCT.,
TEL AVIV, ISR., 1973.

73-0900 MERCER, J. W. JR.
PINDER, G. F.
GALERKIN FINITE-ELEMENT SIMULATION OF A GEOTHERMAL
RESERVOIR
GEOTHERMICS, VOL. 2, NO. 3-4, SEPTEMBER-DECEMBER,
1973, PP. 81-89.

73-0901 RICHARDS, B. G.
THEORETICAL TRANSIENT BEHAVIOUR OF SATURATED AND
UNSATURATED SOILS UNDER LOAD AND CHANGING MOISTURE
CONDITIONS
DIV. OF APPLIED GEOMECHANICS, COMMONWEALTH
SCIENTIFIC AND INDUSTRIAL RESEARCH ORGANIZATION,
MELBOURNE, AUSTRALIA, REPORT NO: TECH. PAPER -16,
1973, 26 P. (N.T.I.S. - PB-221 740/4)

73-0902 ABU-SHUMAYS, I. K.
BARFIS, E. H.
SINGULAR ELEMENTS IN VARIATIONAL AND FINITE
ELEMENT TRANSPORT COMPUTATIONS
TRANS. AMER. NUCL. SOC., VOL. 17, 1973, PP.
236-237.

73-0903 RADHAKRISHNAN, N.
ANALYSIS OF STRESS AND STRAIN DISTRIBUTIONS IN
TRIAXIAL TESTS USING THE METHOD OF FINITE ELEMENTS
ARMY ENG. WATERWAYS EXPERIMENT STATION, VICKSBURG,
MISS., REPORT NO: AEWES-TR-S-73-4, MAY 1973, 191
P. (N.T.I.S. - AD-763 181)

73-0904 MEIJERS, P.
OPTIMIZATION OF STRUCTURAL DESIGN 1
INSTITUTE FOR MECHANICAL CONSTRUCTIONS, DELFT,
NETHERLANDS, REPORT NO. REPT-10432, JANUARY 1973,
23 PP. (N.T.I.S. -N73-23911)

73-0905 LEIMBACH, K. R.
PROZAN, R. J.
STUDY OF HYPERVELOCITY METEOROID IMPACT ON ORBITAL
SPACE STATIONS
RESEARCH AND ENGNG. CENTER, LOCKHEED MISSILES AND
SPACE CO., HUNTSVILLE, ALA., REPORT NO:
NASA-CR-124241, HREC-8473-1, APRIL 1973, 39 P.
(N.T.I.S. - N73-23841)

73-0906 COWPER, G. R.
GAUSSIAN QUADRATURE FORMULAS FOR TRIANGLES
INT. J. NUMER. METH. ENG., VOL. 7, NO. 3, 1973,
PP. 405-407.

73-0907 HARTUNG, R. F.
BALL, R. E.
A COMPARISON OF SEVERAL COMPUTER SOLUTIONS TO
THREE STRUCTURAL SHELL ANALYSIS PROBLEMS
PALO ALTO RESEARCH LAB., LOCKHEED MISSILES AND
SPACE CO., INC., PALO ALTO, CALIF., FINAL REPORT,
OCTOBER 1970 - OCTOBER 1972, APRIL 1973, 110 P.
(N.T.I.S. - AD-762 946)

73-0908 DAOUD, N. A-H.
FORCE AND MOMENTS ON ASYMMETRIC AND YAWED BODIES
IN A FREE SURFACE
COLL. OF ENGNG., UNIV. OF CALIFORNIA, BERKELEY,
REPORT NO: NA-73-2, MAY 1973, 78 P. (N.T.I.S. -
AD-762 758)

73-0909 DESAI, C. S.
SEEPAGE IN MISSISSIPPI RIVER BANKS. REPORT 1.
ANALYSIS OF TRANSIENT SEEPAGE USING A VISCOUS-FLOW
MODEL AND THE FINITE DIFFERENCE AND FINITE ELEMENT
METHODS
ARMY ENG. WATERWAYS EXPERIMENT STATION, VICKSBURG,
MISS., REPORT NO. AEWES-TR-S-73-5-1, MAY 1973, 108
PP. (N.T.I.S. -AD-762 556)

73-0910 APPA, K.
SMITH, G. C.
FINITE ELEMENT APPROACH TO THE INTEGRATED
POTENTIAL FORMULATION OF GENERAL UNSTEADY
SUPERSONIC AERODYNAMICS
BELL AEROSPACE CO., BUFFALO, N.Y., REPORT NO.
NASA-CR-112296, 1973, 38 PP. (N.T.I.S. -
(73-21904)

73-0911 KING, I. P.
NORTON, W. R.
ORLOB, G. T.
A FINITE ELEMENT SOLUTION FOR TWO-DIMENSIONAL
DENSITY STRATIFIED FLOW
WATER RESOURCES ENG. INC., WALNUT CREEK, CALIF.,
REPORT NO. WRE-11360, MARCH 1973, 87 PP. (N.T.I.S.
PB-220 967/4)

73-0912 YEE, W. S.
LATERAL RESISTANCE AND DEFLECTION OF VERTICAL
PILES
BRIDGE DEPT., CALIF. STATE DIV. OF HIGHWAYS,
SACRAMENTO. REPORT NO: CA-HY-BD-4135-2-73-4,
JANUARY 1973, 111 P. (N.T.I.S. - PB-220 889/0)

73-0913 BIZRI, H.
STRUCTURAL CAPACITY OF REINFORCED CONCRETE COLUMNS
SUBJECTED TO FIRE INDUCED THERMAL GRADIENTS
DOCTORAL THESIS. STRUCT. ENGNG. LAB., UNIV. OF
CALIFORNIA, BERKELEY, REPORT NO: UCSESM-73-1,
JANUARY 1973, 206 P. (N.T.I.S. /PB-219 992/5)

73-0914 ALLEN, J. J.
THE EFFECT OF STRESS HISTORY ON THE RESILIENT
RESPONSE OF SOILS
ARMY CONSTRUCTION ENGNG. RESEARCH LAB., CHAMPAIGN,
ILL., REPORT NO: CERL-TR-M-49, JUNE 1973, 221 P.
(N.T.I.S. - AD-762 194)

73-0915 ACHUTARAMAYYA, G.
SCOTT, W. D.
FRACTURE SURFACE ENERGIES BY FINITE ELEMENT STRESS
ANALYSIS
AMER. CERAMIC SOC., BULL. 52, 1973, P. 709.

73-0916 MCKINNIS, J. A.
MEASUREMENT OF THERMAL EXPANSION IN SOLID
PROPELLANTS
AIR FORCE ROCKET PROPULSION LAB., EDWARDS AFB,
CALIF., REPORT NO: AFRPL-TR-72-109, MARCH 1973, 29
P. (N.T.I.S. - AD-761 815)

73-0917 ALMROTH, B. O.
BROGAN, F. A.
MELLER, E.
PETERSEN, H. T.
EXTENSIONS TO THE STAGS COMPUTER CODE
PALO ALTO RESEARCH LAB., LOCKHEED MISSILES AND
SPACE CO., INC., PALO ALTO, CALIF., FINAL REPORT,
OCTOBER 1970 - OCTOBER 1972, MARCH 1973, 75 P.
(N.T.I.S. - AD-761 806)

73-0918 ISENBERG, J.
ANALYTIC MODELING OF ROCK-STRUCTURE INTERACTION.
VOLUME 3. COMPUTER PROGRAM
ENGNG. AND APPLIED SCI., DIV., AGBABIAN
ASSOCIATES, EL SEGUNDO, CALIF., REPORT NO:
AA-R-7215-3-2701, APRIL 1973, 9 P. (N.T.I.S. -
AD-761 650) SEE ALSO VOL. 2, AD-761 649.

73-0919 ISENBERG, J.
ANALYTIC MODELING OF ROCK-STRUCTURE INTERACTION.
VOLUME 2. USERS GUIDE FOR A COMPUTER PROGRAM
ENGNG. AND APPLIED SCI., DIV., AGBABIAN
ASSOCIATES, EL SEGUNDO, CALIF., REPORT NO:
AA-R-7215-2-2701, APRIL 1, 1973, 105 P. (N.T.I.S.
- AD-761 649) SEE ALSO VOL. 1, AD-761 648 AND VOL.
3, AD-761 650.

73-0920 ISENBERG, J.
ANALYTIC MODELING OF ROCK-STRUCTURE INTERACTION.
VOLUME 1
ENGNG. AND APPLIED SCI., DIV., AGBABIAN
ASSOCIATES, EL SEGUNDO, CALIF., REPORT NO:
AA-R-7215-1-2701, APRIL 1973, 221 P. (N.T.I.S. -
AD-761 648) SEE ALSO VOL. 2, AD-761 649.

73-0921 AHN, C. S.
MATHEMATICAL MODEL OF HEATED EFFLUENTS IN COASTAL
WATER
TRANS. AMER. NUCL. SOC., VOL. 16, 1973, PP. 16-17.

73-0922 GOLDSMITH, W.
SACKMAN, J. L.
WAVE PROPAGATION IN ANISOTROPIC ROCKS
UNIV. OF CALIFORNIA, BERKELEY, TECH. REPORT NO.
4(FINAL), FEBRUARY 23, 1972 - JUNE 1, 1973, 260 P.
(N.T.I.S. - AD-761 203) SEE ALSO REPORT, SEPTEMBER
25, 1972, AD-749 976.

73-0923 ROCKWELL, R. D.
COMPUTER-AIDED INPUT/OUTPUT FOR THE USE WITH
FINITE ELEMENT STRUCTURAL ANALYSES
NAVAL SHIP RESEARCH AND DEVELOPMENT CENTER,
BETHESDA, MD., REPORT NO: NSRDC-3844, FEBRUARY
1973, 76 P. (N.T.I.S. - AD-761 154)

73-0924 CELESTINI, A. C.
APPLICATION OF FINITE ELEMENT ANALYSIS TO COPLANAR
WAVEGUIDE
MASTER'S THESIS. SCHOOL OF ENGNG., AIR FORCE INST.
OF TECH.,WRIGHT-PATTERSON AFB, OHIO, REPORT NO:
GE-EE-73-4, MARCH 1973, 80 P. (N.T.I.S.-AD-760
756)

73-0925 AUTHORS
TRANSACTIONS OF ARMY MATHEMATICIANS RESEARCH
CONFERENCE
PROC. 18TH CONF. ARMY MATHEMATICIANS, ARMY
RESEARCH OFFICE, DURHAM, N.C., REPORT NO.
AROD-73-1, JANUARY 1973, 809 PP. (N.T.I.S. -
AD-760 464) ALSO REPORT AD-481 004, MARCH 1966.

73-0926 CHOW, T. S.
KOWALIK, J. S.
COMPUTING WITH SPARSE MATRICES
INT. J. NUMER. METH. IN ENG., VOL. 7, NO. 2, 1973,
PP. 211-224.

73-0927 IDING, R. H.
IDENTIFICATION OF NONLINEAR MATERIALS BY FINITE
ELEMENT METHODS
STRUCT. ENGNG. LAB., UNIV. OF CALIF., BERKELEY,
REPORT NO: UCSESM-73-4, JANUARY 1973, 199 P.
(N.T.I.S. - PB-219 090/8)

73-0928 BATDORF, S. B.
CROSE, J. G.
A STATISTICAL THEORY FOR THE FRACTURE OF BRITTLE
STRUCTURES SUBJECTED TO NONUNIFORM POLYAXIAL
STRESSES
AEROSPACE CORP., EL SEGUNDO, CALIF., REPORT NO:
TR-0073(3450-76)-2, APRIL 26, 1973, 40 P.
(N.T.I.S. - AD-759 805)

73-0929 BIOT, M. A.
MECHANICS AND THERMOMECHANICS
BIOT (MAURICE - A), BRUSSELS, BELGIUM, INTERIM
SCIENTIFIC REPORT NO. 2, FEBRUARY 15 1972 -
FEBRUARY 14 1973, 17 PP. (N.T.I.S. - AD-758 884)

73-0930 HODGE, P. G. JR.
SOME APPLICATIONS OF THE PRINCIPLE OF VIRTUAL WORK
DEPT. OF AEROSPACE ENGNG. AND MECH., UNIV. OF
MINN. MINNEAPOLIS, REPORT NO: AEM-H1-5, MARCH
1973, 40 P. (N.T.I.S. - AD-758 727)

73-0931 LEHMAN, G. M.
PURDY, D. M.
ALLEN, F. C.
DIETZ, C. G.
DEVELOPMENT OF A GRAPHITE HORIZONTAL STABILIZER
DOUGLAS AIRCRAFT CO., LONG BEACH, CALIF., REPORT
NO: MDC-J5841, JANUARY 1973, 113 P. (N.T.I.S. -
AD-758 718)

73-0932 SEQUIT, W. T.
COMPUTER PROGRAMS FOR THE SOLUTION OF SYSTEMS OF
LINEAR ALGEBRAIC EQUATIONS
MEMPHIS STATE UNIV., TENN., REPORT NO.
NASA-CR-2173, JANUARY 1973, 251 PP., (N.T.I.S. -
N73-17181)

73-0933 POSTNOV, V. A.
KHARKHURIM, I. Y.
KHEGAZI, S. K.
APPLICATION OF THE METHOD OF FINITE ELEMENTS TO
THE CALCULATION OF SHIP STRUCTURES (ISPOLZOVANIE
METODA KONECHNYKH ELEMENTOV DLYA RASCHETA SUDOVYKH
PEREKRYTII)
TRANSLATION DIV., NAVAL INTELLIGENCE SUPPORT
CENTER, WASH. D.C., REPORT NO: NISC-TRANS-3420,
MARCH 19, 1973, 13 P. (N.T.I.S. - AD-758 036)

73-0934 MILEIKOVSKII, I. E.
RAIZER, V. D.
THE DEVELOPMENT OF APPLIED METHODS IN PROBLEMS OF
STATIC CALCULATION OF THIN-WALLED
THREE-DIMENSIONAL SYSTEMS (SHELLS AND FOLDS)
(RAZVITIE PRKLADNYKHH METODOV V ZADACHAKH
STATICHESK.GO RASCHETA TONKOSTENNYKH
PROSTRANSTVENNYKH SISTEM (OBLOCHKI I SKLADKI))
TRANSLATION DIV., NAVAL INTELLIGENCE SUPPORT
CENTER, WASH., D.C., REPORT NO: NISC-TRANS-3405,
FEBRUARY 21, 1973, 25 PP. (N.T.I.S. - AD-758 028)

73-0935 FAIRHURST, C.
ESTIMATION OF THE MECHANICAL PROPERTIES OF ROCK
MASSES
DEPT. OF CIVIL AND MINERAL ENGNG., MINN. UNIV.,
MINNEAPOLIS, FINAL REPORT, JANUARY 65. - DECEMBER
3, 1972, (N.T.I.S. - AD-757 768)

73-0936 CHOO, Y-I.
CHUNG, T. J.
CASARELLA, M. J.
A SURVEY OF ANALYTICAL METHODS FOR DYNAMIC
SIMULATION OF CAB
DEPT. OF CIVIL AND MECH. ENGNG., CATHOLIC UNIV. OF
AMERICA, WASH., D.C., REPORT NO: 73-1, MARCH 1973,
42 PP. (N.T.I.S. - AD-757 348)

73-0937 BATOZ, J. L.
DHATT, G. S.
LARGE DEFLECTION ANALYSIS OF ARBITRARY SHALLOW
SHELLS BY THE FINITE ELEMENT METHOD
DEPT. OF CIVIL ENG., LAVAL UNIV., QUEBEC, CANADA,
1973.

73-0938 MCCORMICK, C. W.
REVIEW OF NASTRAN DEVELOPMENT RELATIVE TO
EFFICIENCY OF EXECUTION
NASTRAN USER'S EXPERIENCES, NASA TM X-2893,
SEPTEMBER 1973, PP. 7-28.

73-0939 SLAGTER, W.
FAST REACTOR PROGRAMME. FINITE ELEMENT TEMPERATURE
ANALYSIS IN FUEL PINS
STICHTING REACTOR CENTRUM NEDERLAND, PETTEN,
REPORT NO: RCN-188, NOVEMBER 1973, 28 PP.
(N.T.I.S.-EURFNR-1158 AND RCN-188) SEE ALSO NSA
3003, NO. 08698.

73-0940 HOFFMANN, A.
LIVOLANT, M.
ROCHE, R.
ANALYSIS OF SHELLS OF ANY FORM IN THE PLASTIC
REGION BY THE FINITE ELEMENT METHOD
KRUPP (FRIEDR.) G.M.B.H., ESSEN (F.R. GERMANY),
FORSCHUNGSINSTITUT., REPORT NO: CONF-730942-5,
1973, 24 PP. (N.T.I.S. - CEA-CONF-2571) IN FRENCH.
SEE ALSO NSA 3003, NO. 06392.

73-0941 MELLO, R. M.
GRIFFIN, D. S.
PLASTIC COLLAPSE LOADS FOR PIPE ELBOWS USING
INELASTIC ANALYSIS
ADVANCED REACTORS DIV., WESTINGHOUSE ELECTRIC
CORP., PITTSBURGH, PA. AND MADISON, PA., REPORT
NO. CONF-740606-11, DECEMBER 1973, 32 PP.,
(N.T.I.S. - WARD-5224) SEE ALSO NSA 3002, NO.
05437.

73-0942 HARTZMAN, M.
NONLINEAR DYNAMIC ANALYSIS OF AXISYMMETRIC SOLIDS
BY THE FINITE ELEMENT METHOD
LAWRENCE LIVERMORE LAB., UNIV. OF CALIF.,
LIVERMORE. REPORT NO: CONF-740606-2, JULY 31,
1973, 27 P. (N.T.I.S. - UCRL-74978) SEE ALSO NSA
3002, NO. 03054.

73-0943 SCHAUER, D. A.
ELFED: A COMPUTER CODE TO GENERATE FINITE ELEMENT
MESH FOR PROBLEMS OF COMPLEX AXISYMMETRIC GEOMETRY
LAWRENCE LIVERMORE LAB., UNIV. OF CALIF.,
LIVERMORE, REPORT NO: CONF-740606-14, DECEMBER 3,
1973, 19 P. (N.T.I.S. - UCRL-74879) SEE ALSO NSA
3002, NO. 03053.

73-0944 BOLSTAD, J. H.
LEAF, G. K.
LINDEMAN, A. J.
KAPER, H. G.
EMPIRICAL INVESTIGATION OF REORDERING AND DATA
MANAGEMENT FOR FINITE ELEMENT SYSTEMS OF EQUATIONS
ARGONNE NATIONAL LAB., ILL. SEPTEMBER 1973, 51 P.
(N.T.I.S. -ANL-8056) SEE ALSO NSA 3001, NO. 02228.

73-0945 CESARI, F.
LABANTI, L.
PALMIERI, I.
DESCRIPTION AND INSTRUCTION FOR USE OF THE
SAFE-SHELL. B FINITE ELEMENT PROGRAM IN THE
IBM/360 VERSION
COMITATO NAZIONALE PER L'ENERGIA NUCLEARE, ROME,
DECEMBER 1973, 55 P. (N.T.I.S.-RT/ING-(73)19) SEE
ALSO NSA 2911 NO. 26789.

73-0946 APPERT, K.
BERGER, D.
GRUBER, R.
USEFUL FORM OF THE VARIATIONAL PRINCIPLE OF IDEAL
MHD FOR ONE-DIMENSIONAL NUMERICAL STABILITY
CALCULATIONS
ECOLE POLYTECHNIQUE FEDERALE, LAUSANNE,
SWITZERLAND, DECEMBER 1973, 11 P.
(N.T.I.S.-LRP-76-73) SEE ALSO NSA 2911, NO. 28432.

73-0947 ROBINSON, J.
SPOONER, J. B.
ASAS - A COMMERCIAL FINITE ELEMENT SYSTEM WITH
DISPLACEMENT AND FORCE CONTINUITY OPTIONS
PROC. OF THE CONGR. ON FINITE ELEMENT METH.,
BADEN-BADEN, WEST GERMANY, NOVEMBER 1973.

73-0948 WILSON, W.K.
BEGLEY, J.A.
HEAVY SECTION STEEL TECHNOLOGY PROGRAM TECHNICAL
OR PROGRAMMATIC MANUSCRIPT NO. 25. VARIABLE
THICKNESS STUDY OF THE EDGE CRACKED BEND SPECIMEN
WESTINGHOUSE ELECTRIC CORP., PITTSBURGH, PA.,
NOVEMBER 1973 46 PP. (N.T.I.S.-WCAP-8837) SEE ALSO
NSA 2910, NO. 24726.

73-0949 GABRIELSON, V. K.
FEMESH: A FINITE ELEMENT CODE PREPROCESSOR
SANDIA LABS., LIVERMORE, CALIF., REPORT NO:
CONF-730626-28 1973, (N.T.I.S.-SLL-73-5281) SEE
ALSO NSA 2910, NO. 24128.

73-0950 AXELSON, K.
ANISOTROPIC WORK HARDENING IN METALS. FINITE
ELEMENT APPLICATION TO PLANE STRESS AND PLANE
STRAIN
DEPT. OF STRUCT. MECH., CHALMERS UNIV. OF TECH.,
1973.

73-0951 RICCARDELLA, P. C.
SWEDLOW, J. L.
COMBINED ANALYTICAL – EXPERIMENTAL FRACTURE STUDY
OF THE TWO LEADING THEORIES OF ELASTIC-PLASTIC
FRACTURE (J-INTEGRAL AND EQUIVALENT ENERGY)
WESTINGHOUSE ELECTRIC CORP., PITTSBURGH, PA.,
REPORT NO: HSSTP-TR-33, OCTOBER 1973, 50 P.
(N.T.I.S. – WCAP-8284) SEE ALSO 2908, NO.
18259.

73-0952 MARCAL, P. V.
BETTES, R. S.
STUART, P. M.
ELASTIC-PLASTIC BEHAVIOR OF A LONGITUDINAL
SEMI-ELLIPTIC CRACK IN A THICK PRESSURE VESSEL
DIV. OF ENG., BROWN UNIV., PROVIDENCE, R.I., JUNE
1973, 38 PP. (N.T.I.S. –HSSTP-TR-28) ALSO NSA
2908, NO. 18252.

73-0953 TOKARZ, F. J.
MURRAY, R. C.
SEISMIC EFFECTS ON A PROPOSED UNDERGROUND REACTOR
FACILITY
LAWRENCE LIVERMORE LAB., UNIV. OF CALIF.,
LIVERMORE, FEBRUARY 26, 1973, 12 P.
(N.T.I.S.-UCID-16407) SEE ALSO NSA 2907, NO.
19684.

73-0954 ZEHLEIN, H.
ASSESSMENT OF COMMERCIALLY AVAILABLE FINITE
ELEMENT CODES FOR THE DYNAMIC ANALYSIS OF
STRUCTURES
INSTITUT FUR REAKTORENTWICKLUNG,
KERNFORSCHUNGSZENTRUM KARLSRUHE (F.R. GERMANY),
REPORT NO. KFK-1791, APRIL 1973, 88 PP. (N.T.I.S.
– EURFNR-1102) ALSO NSA 2906, NO. 14733.

73-0955 REED, W. H.
HILL, T. R.
BRINKLEY, F. W.
LATHROP, K. D.
TRIPLET: A TWO-DIMENSIONAL, MULTIGROUP, TRIANGULAR
MESH, PLANAR GEOMETRY, EXPLICIT TRANSPORT CODE
LOS ALAMOS SCIENTIFIC LAB., N. MEX., OCTOBER 1973,
69 P. (N.T.I.S. – LA-5428-MS) SEE ALSO NSA 2903,
NO. 06397.

73-0956 CALDER, C. A.
WILCOX, W. W.
APPLICATIONS OF PULSED LASER INDUCED STRESS WAVES
LAWRENCE LIVERMORE LAB., UNIV. OF CALIF.,
LIVERMORE, SEPTEMBER 27, 1973, 27 P.
(N.T.I.S.-UCID-16353) SEE ALSO NSA 2901, NO.
02406.

73-0957 SPEARS, R. K.
FINITE ELEMENT VISCOELASTIC STRESS ANALYSIS STUDY
USING THE GENERALIZED KELVIN SOLID MODEL
NEUTRON DEVICES DEPT., GENERAL ELECTRIC CO., ST.
PETERSBURG, SEE ALSO NSA 2902, NO. 03117.

73-0958 HARTZMAN, M.
STATIC STRESS ANALYSIS OF AXISYMMETRIC SOLIDS WITH
MATERIAL AND GEOMETRIC NONLINEARITIES BY THE
FINITE ELEMENT METHOD
LAWRENCE LIVERMORE LAB., UNIV. OF CALIF.,
LIVERMORE, JANUARY 19, 1973, 33 P.
(N.T.I.S.-UCRL-51390) SEE ALSO NSA 2911, NO.
29561.

73-0959 BENZLEY, S. E.
BEISINGER, Z. E.
CHILES: A FINITE ELEMENT COMPUTER PROGRAM THAT
CALCULATES THE INTENSITIES OF LINEAR ELASTIC
SINGULARITIES
SANDIA LABS., ALBUQUERQUE, N. MEX., SEPTEMBER
1973, 48 P. (N.T.I.S. – SLA-73-894) SEE ALSO NSA
2812, NO. 32132.

73-0960 FROEHLICH, R.
REVIEW OF CURRENT PROBLEMS FOR MULTIDIMENSIONAL
REACTOR STATICS CALCULATIONS
INSTITUT FUR NEUTRONENPHYSIK UND REAKTORTECHNIK,
KERNFORSCHUNGSZENTRUM, KARLSRUHE, WEST GERMANY,
JUNE 1973, 69 PP. (N.T.I.S. –KFK-1821) ALSO NSA
2811, NO. 29248.

73-0961 BAREISS, E. H.
ABU-SHUMAYS, I. K.
FINITE ELEMENTS IN NEUTRON TRANSPORT THEORY
NORTHWESTERN UNIV., EVANSTON, ILL. MAY 1973, 56 P.
(N.T.I.S. –COO-2280-1) SEE ALSO NSA 2811, NO.
28623.

73-0962 DE WINDT, P.
REYNEN, J.
EURCYL: A COMPUTER PROGRAM TO GENERATE FINITE
ELEMENT MESHES FOR CYLINDER-CYLINDER INTERSECTIONS
JOINT NUCLEAR RESEARCH CENTER, EUROPEAN ATOMIC
ENERGY COMMUNITY, ISPRA, ITALY, AUGUST 1973, 22
PP. (N.T.I.S. – EUR-5030) SEE ALSO NSA 2810, NO.
26829.

73-0963 LEMMON, E. C.
FINITE CONDUCTANCE ELEMENT METHOD OF CONDUCTION
HEAT TRANSFER
SANDIA LABS., LIVERMORE, CALIF., JUNE 1973, 103
PP. (N.T.I.S. – SLA-73-79) SEE ALSO NSA 2907, NO.
17702.

73-0964 KEY, S. W.
BEISINGER, Z. E.
SLADE D: A COMPUTER PROGRAM FOR THE DYNAMIC
ANALYSIS OF THIN SHELLS
SANDIA LABS., ALBUQUERQUE, N. MEX., JANUARY 1973,
155 P. (N.T.I.S. – SLA-73-79) SEE ALSO NSA 2807,
NO. 17702.

73-0965 SWARTZ, B.
WENDROFF, B.
COMPARATIVE EFFICIENCY OF CERTAIN FINITE ELEMENT
AND FINITE DIFFERENCE METHODS FOR A HYPERBOLIC
PROBLEM
LOS ALAMOS SCIENTIFIC LAB., N. MEX., REPORT NO:
CONF-730713-1, 1973, 13 P. (N.T.I.S.-LA-UR-73-918)
SEE ALSO NSA 2808, NO. 20577.

73-0966 NEALE, B. K.
FINITE ELEMENT MESH GENERATION PROGRAM FOR
ARBITRARY TWO-AND THREE-DIMENSIONAL STRUCTURES
PROC. 2N CONF. STRUCT. MECH. IN REACTOR TECH.,
BERLIN, PAPER M4/5. 1973.

73-0967 BABUSKA, I.
THE FINITE ELEMENT METHOD WITH PENALTY
MATH. COMP., VOL. 27, 1973, PP. 221-229.

73-0968 BABUSKA, I.
NUMERICAL SOLUTION OF PARTIAL DIFFERENTIAL
EQUATIONS
INST. FOR FLUID DYNAMICS AND APPLIED MATH., UNIV.
OF MARYLAND, COLLEGE PARK, REPORT NO:
CONF-730429-1, BN-761, MARCH 1973, 26 P. (N.T.I.S.
– ORO-3443-44) SEE ALSO NSA 2804, NO. 10505.

73-0969 HANSEN, K. F.
HENRY, A. F.
ANNUAL PROGRESS REPORT, FY1973
DEPT. OF NUCLEAR ENGNG., MASSACHUSETTS INST. OF
TECH., CAMBRIDGE, APRIL 1973, 37 P. (N.T.I.S. –
COO-2262-2) SEE ALSO NSA 2806, NO. 14666.

73-0970 HUTULA, D. N.
WIANCKO, B. E.
MATUS: A THREE-DIMENSIONAL FINITE ELEMENT PROGRAM
FOR SMALL-STRAIN ELASTIC ANALYSIS
BETTIS ATOMIC POWER LAB., PITTSBURG, P.A. MARCH
1973, 194 P. (N.T.I.S. – WAPD-TM-1081) SEE ALSO NSA
2802 NO. 04994.

73-0971 HUTULA, D. N.
WIANCKO, B. E.
ZEILER, S. M.
APACHE: A THREE-DIMENSIONAL FINITE ELEMENT PROGRAM
FOR STEADY-STATE OR TRANSIENT HEAT CONDUCTION
ANALYSIS
BETTIS ATOMIC POWER LAB., PITTSBURGH, PA., MARCH
1973, 111 P. (N.T.I.S.-WAPD-TM-1080) SEE ALSO NSA
2802, NO. 02510.

73-0972 HUTULA D. N.
ZEILER, S. M.
MESH 3D: A THREE-DIMENSIONAL FINITE ELEMENT MESH
GENERATOR PROGRAM FOR EIGHT-NODE ISOPARAMETRIC
ELEMENTS
BETTIS ATOMIC POWER LAB., PITTSBURGH, PA., MARCH
1973, 403 P. (N.T.I.S. – WAPD-TM-1079) SEE ALSO
NSA 2802, NO. 04993.

73-0973 BABUSKA, I.
KELLOGG, R. B.
MATHEMATICAL AND COMPUTATIONAL PROBLEMS IN REACTOR
CALCULATIONS
INST. FOR FLUID DYNAMICS AND APPLIED MATH., UNIV.
OF MARYLAND, COLLEGE PARK, REPORT NO: BN-760,
MARCH 1973, 29 P. (N.T.I.S. – ORO-3443-43) SEE
ALSO NSA 2803, NO. 07444.

73-0974 WACHSPRESS, E. L.
ALGEBRAIC-GEOMETRY FOUNDATIONS FOR FINITE-ELEMENT
COMPUTATION
KNOLLS ATOMIC POWER LAB., SCHENECTADY, N.Y., 1973,
11 P. (N.T.I.S. – KAPL-P-3962) SEE ALSO NSA 2803,
NO. 07635.

73-0975 BACKLUND, J.
FINITE ELEMENT ANALYSIS OF NONLINEAR STRUCTURES
PH.D. THESIS, DEPT. OF STRUCT. MECH., CHALMERS
INST. OF TECH., GOTEBORG, 1973.

73-0976 LASCAUX, P.
APPLICATION OF THE FINITE ELEMENTS METHOD IN
TWO-DIMENSIONAL HYDRODYNAMICS USING THE LAGRANGE
VARIABLES
CENTRE D'ETUDES, COMMISSARIAT A L'ENERGIE
ATOMIQUE, LIMEIL-BREVANNES, FRANCE, 1973, 23 P.
(N.T.I.S.-LA-TR-73-3) SEE ALSO NSA 2711, NO.
27178.

73-0977 DEPPE, L.
HANSEN, K.
FINITE ELEMENT METHOD APPLIED TO NEUTRON DIFFUSION
PROBLEMS
DEPT. OF NUCLEAR ENGNG., MASSACHUSETTS INST. OF
TECH., CAMBRIDGE. REPORT NO: MITNE-145, FEBRUARY
1973, 124 P. (N.T.I.S. - COO-2262-1) SEE ALSO NSA
2711, NO. 27134.

73-0978 GAGGERO, G.
CANDOLFO, G.
ON THE SOLUTION OF PLANE STRESS PROBLEMS BY FINITE
ELEMENTS COMPUTER PROGRAMS
JOINT NUCLEAR RESEARCH CENTER, EUROPEAN ATOMIC
ENERGY COMM., ISPRA, ITALY, FEBRUARY 1973, 30 P.
(N.T.I.S. - EUR-4928) SEE ALSO NSA 2708, NO.
19347.

73-0979 EMERY, A.
CALCULATION OF HARMONIC FIELDS IN WHICH
SINGULARITIES ARE PRESENT USING FINITE ELEMENT AND
FINITE DIFFERENCE TECHNIQUES BY INCORPORATING
SINGULARITY PROGRAMMING
SANDIA LABS., LIVERMORE, CALIF., JANUARY 1973, 36
P. (N.T.I.S. - SCL-RR-720093) SEE ALSO NSA 2707,
NO. 16924.

73-0980 VARGA, R. S.
EXTENSIONS OF THE SUCCESSIVE OVER-RELAXATION
THEORY WITH APPLICATIONS TO FINITE ELEMENT
APPROXIMATIONS
TOPICS IN NUMER. ANALYSIS, J. J. H. MILLER (ED.),
ACADEMIC PRESS, LONDON, 1973, PP. 329-343.

73-0981 RESTAD, K.
VOLCY, G. C.
MASSON, J. C.
INVESTIGATION ON FREE AND FORCED VIBRATIONS OF A
LONG TANKER WITH OVERLAPPING PROPELLER ARRANGEMENT
S.N.A.M.E., VOL. 81, 1973.

73-0982 DU PREEZ, R. J.
ANALYSIS OF BOX GIRDERS BY FOLDED PLATE AND STRIP
METHODS
PROC. SYMP. ON BRIDGE DECK ANAL., JOHANNESBURG, S.
AFR., MAY 7-9, 1973.

73-0983 GALLAGHER, R. H.
FINITE ELEMENT METHOD IN PLATE AND SHELL
INSTABILITY ANALYSIS
PROC. 4TH AUSTR. CONF. ON MECH. OF STRUCT. AND
MATER., UNIV. OF QUEENSLAND, BRISBANE, AUST.,
AUGUST 20-22, 1973.

73-0984 GODBOLE, P. N.
MEEK, J. L.
ZIENKIEWICZ, O. C.
ELASTOPLASTIC ANALYSIS OF A CANTILEVER BEAM
PROC. 4TH AUSTR. CONF. ON MECH. OF STRUCT. AND
MATER., UNIV. OF QUEENSLAND, BRISBANE, AUST.,
AUGUST 20-22, 1973.

73-0985 HOLMES, P.
ARNAOUTI, C.
YIELD CRITERION FOR REINFORCED CONCRETE SLABS
PROC. 4TH AUSTR. CONF. ON MECH. OF STRUCT. AND
MATER., UNIV. OF QUEENSLAND, BRISBANE, AUST.,
AUGUST 20-22, 1973.

73-0986 GRUNDY, P.
STATISTICAL ASPECTS OF THE STABILITY OF COMPOUND
SYSTEMS WITH BILINEAR ELEMENTS
PROC. 4TH AUSTR. CONF. ON MECH. OF STRUCT. AND
MATER., UNIV. OF QUEENSLAND, BRISBANE, AUST.,
AUGUST 20-22, 1973.

73-0987 WELLFORD, L. C.
ODEN, J. T.
ACCURACY AND CONVERGENCE OF FINITE ELEMENT
GALERKIN APPROXIMATIONS OF TIME DEPENDENT PROBLEMS
WITH EMPHASIS ON DIFFUSION AND CONVECTION
TICOM, REPORT 73-8, 1973.

73-0988 HERRMANN, L. R.
EFFICIENCY EVALUATION OF A TWO-DIMENSIONAL
INCOMPATIBLE FINITE ELEMENT
COMPUT. AND STRUCT., VOL. 3, 1973, PP. 1377-1395.

73-0989 BERGAN, P. G.
CLOUGH, R. W.
LARGE DEFLECTION ANALYSIS OF PLATES AND SHALLOW
SHELLS USING THE FINITE ELEMENT METHOD
INT. J. NUMER. METH. ENG., VOL. 5, 1973, PP.
543-556.

73-0990 BIFFLE, J. H.
FINITE ELEMENT ANALYSIS FOR WAVE PROPAGATION IN
ELASTIC-PLASTIC SOLIDS
TEXAS INST. FOR COMP. MECH., UNIV. OF TEXAS,
AUSTIN, REPORT 73-4, 1973.

73-0991 BLOKHOV, V. V.
INVESTIGATION OF FLEXURAL STIFFNESS OF RECTANGULAR
PLATES BY THE FINITE ELEMENT METHOD
IZV. VUZ. MASHINOSTR. (RUSSIAN), VOL. 1, 1973, PP.
5-9.

73-0992 GAMBOLATI, G.
MATHEMATICAL SIMULATION OF THE SUBSIDENCE OF
VENICE 1 THEORY
WATER RESOUR., VOL. 9, 1973, PP. 721-733.

73-0993 COLLINS, R. J.
BANDWIDTH REDUCTION BY AUTOMATIC RENUMBERING
INT. J. NUMER. METH. ENG., VOL. 6, 1973, PP.
345-356.

73-0994 COOK, R. D.
A NOTE ON CERTAIN INCOMPATIBLE ELEMENTS
INT. J. NUMER. METH. ENG., VOL. 6, 1973, PP.
146-147.

73-0995 WAIT, R.
FINITE-ELEMENT-TYPE SOLUTION OF INTEGRAL EQUATIONS
PROC. INT. COMPUT. SYMP., DAVOS, SWITZ., SEPTEMBER
4-7, 1973, PP. 344-347. (AVAILABLE FROM AMER.
ELSEVIER PUBL. CO., NEW YORK, N.Y.)

73-0996 HINTON, E.
RAZZAQUE, A.
FINITE ELEMENT ANALYSIS OF PLATES ALLOWING FOR
TRANSVERSE SHEAR DEFORMATION EFFECTS
CIVIL ENG. DEPT., UNIV. COLL. OF SWANSEA, INTERNAL
REPORT: C/R/192/73, 1973.

73-0997 DARIO, N. P.
BRADLEY, W. A.
A COMPARISON OF FIRST AND SECOND ORDER AXIALLY
SYMMETRIC FINITE ELEMENTS
INT. J. NUMER. METH. ENG., VOL. 5, 1973, PP.
573-583.

73-0998 DE DONATO, O.
FRANCHI, A.
A MODIFIED GRADIENT METHOD FOR FINITE ELEMENT
ELASTOPLASTIC ANALYSIS BY QUADRATIC PROGRAMMING
COMP. METH. IN APPL. MECH. ENG., VOL. 2, 1973, PP.
107-132.

73-0999 IRONS, B. M.
COMMENT ON A HIGHER ORDER CONFORMING RECTANGULAR
PLATE ELEMENT
INT. J. NUMER. METH. IN ENG., VOL. 6, NO. 2, 1973,
PP. 308-309.

73-1000 DOUGLAS, J.
DUPONT, T.
A FINITE ELEMENT COLLOCATION METHOD FOR
QUASILINEAR PARABOLIC EQUATIONS
MATH. COMP. VOL. 27, 1973, PP. 17-28.

73-1001 GALLAGHER, R. H.
COMPUTATIONAL METHODS IN NUCLEAR REACTOR
STRUCTURAL DESIGN FOR HIGH TEMPERATURE
APPLICATIONS: AN INTERPRETIVE REPORT
OAK RIDGE NATIONAL LAB., TENNESSEE, REPORT
ORNL-4756, 1973.

73-1002 GALLAGHER, R. H.
CHAN, S. T. K.
HIGHER-ORDER FINITE ELEMENT ANALYSIS OF LAKE
CIRCULATION
INT. J. COMPUT. AND FLUIDS. VOL. 1, 1973, PP.
119-132.

73-1003 GERIJ, J.
THERMO-ELASTIC-PLASTIC CYCLIC ANALYSIS BY FINITE
ELEMENT METHOD
PROC. 2ND CONF. STRUCT. MECH. IN REACTOR TECH.,
BERLIN, PAPER L7/7, 1973.

73-1004 GHABOUSSI, J.
WILSON, E. L.
FLOW OF COMPRESSIBLE FLUID IN POROUS ELASTIC MEDIA
INT. J. NUMER. METH. ENG., VOL. 5, 1973, PP.
419-442.

73-1005 GIENCKE, E.
A SIMPLE MIXED METHOD FOR PLATE AND SHELL PROBLEMS
PROC. 2ND CONF. STRUCT. MECH. IN REACTOR TECH.,
BERLIN, PAPER M2/5, 1973.

73-1006 GORDON, W. J.
HALL, C. A.
TRANSFINITE ELEMENT METHODS: BLENDING FUNCTION
INTERPOLATION OVER ARBITRARY CURVED ELEMENT
DOMAINS
NUMER. MATH., VOL. 21, 1973, PP. 109-129.

73-1007 GREENBAUM, G. A.
HOFMEISTER, L. D.
EVENSEN, D. A.
PURE MOMENT LOADING OF AXISYMMETRIC FINITE ELEMENT
MODELS
INT. J. NUMER. METH. ENG., VOL. 5, 1973, PP.
459-463.

73-1008 GROTKOP, G.
FINITE ELEMENT ANALYSIS OF LONG-PERIOD WATER WAVES
COMP. METH. IN APPL. MECH. ENG., VOL. 2, 1973, PP.
133-146.

73-1009 HARTIG, D.
HERMITEAN RECTANGLE FINITE ELEMENT FAMILY
PROC. 2ND CONF. STRUCT. MECH. IN REACTOR TECH.,
BERLIN, PAPER M2/4, 1973.

73-1010 COOK, R. D.
TAKEMOTO, H.
SOME MODIFICATIONS OF AN ISOPARAMETRIC SHELL
ELEMENT
INTL. J. NUMER. METH. ENG., VOL. 7, NO. 3, 1973,
PP. 401-405.

73-1011 ROBINSON, J.
A SIX NODE TRIANGULAR SHEAR PANEL FOR EQUILIBRIUM
MODELS
RA/ARD, REPORT NO. 25-4-73-14, APRIL, 1973.

73-1012 HEISE, U.
COMPILED APPLICATION OF FINITE ELEMENT METHODS AND
RICHARDSON EXTRAPOLATION TO THE TORSION PROBLEM
IN "THE MATH. OF FINITE ELEMENTS AND APPL.", J. P.
WHITEMAN (ED.), ACADEMIC PRESS, LONDON, 1973, PP.
225-237.

73-1013 ROBINSON, J.
A SYNTHESIZED WARPED QUADRILATERAL ELEMENT GIVING
SIMPLIFIED INPUT FOR EQUILIBRIUM MODELS
RA/ARD, REPORT NO. 2-2-73-11, FEBRUARY 1973.

73-1014 HENDRY, J. A.
DELVES, L. M.
VARIATIONAL SOLUTION OF TWO DIMENSIONAL FLUID FLOW
PROBLEMS
DEPT. OF COMPUT. AND STATIS. SCI., UNIV. OF
LIVERPOOL, INTERNAL REPORT CSS-73-4, 1973.

73-1015 HENNART, J. P.
SINGULARITIES IN THE FINITE ELEMENT APPROXIMATION
OF TWO DIMENSIONAL DIFFUSION PROBLEMS
TRANS. AMER. NUCL. SOC., VOL. 17, 1973, PP.
239-240.

73-1016 HENSHELL, R. D.
SHAW, K. G.
CRACK TIP FINITE ELEMENTS ARE UNNECESSARY
DEPT. OF MECH. ENG., UNIV. OF NOTTINGHAM, TECH.
REPORT, 1973.

73-1017 HOFFMANN, A.
LIVOLANT, M.
ROCHE, R.
PLASTIC ANALYSIS OF SHELLS BY FINITE ELEMENT
METHODS
PROC. 2ND CONF. STRUCT. MECH. IN REACTOR TECH.,
BERLIN, PAPER L6/2, 1973.

73-1018 BERKOVIC, M.
THE CALCULATION OF STATIC AND DYNAMIC
CHARACTERISTICS OF FLIGHT VEHICLE STRUCTURES BY
THE FINITE ELEMENT METHOD
1ST CONGR. YUGOSLAV AEROCOSMONAUTICS, BELGRADE,
1973. (IN SERBO-CROATIAN)

73-1019 CHENG, R. T.
LI, C. Y.
ON THE SOLUTION OF TRANSIENT FREE-SURFACE FLOW
PROBLEMS IN POROUS MEDIA BY THE FINITE ELEMENT
METHOD
J. OF HYDROLOGY, VOL. 20, 1973.

73-1020 TUFF, A. D.
JENNINGS, A.
AN ITERATIVE METHOD FOR LARGE SYSTEMS OF LINEAR
STRUCTURAL EQUATIONS
INTL. J. NUMER. METH. IN ENG., VOL. 7, NO. 2,
1973, PP. 175-184.

73-1021 JACQUIN, J. C.
HANSEN, K. F.
FINITE ELEMENT SOLUTIONS FOR MULTIREGION PROBLEMS
TRANS. AMER. NUCL. SOC., VOL. 17, 1973, PP.
238-239.

73-1022 JAMOVSKY, V.
ELLIPTIC BOUNDARY VALUE PROBLEMS WITH
NONVARIATIONAL PERTURBATION AND THE FINITE ELEMENT
METHOD
APLIKACE MATEMATIKY, VOL. 18, 1973, PP. 422-433.

73-1023 DOUGLAS, J.
DUPONT, T.
GALERKIN METHODS FOR PARABOLIC EQUATIONS WITH
NONLINEAR BOUNDARY CONDITIONS
NUMERISCHE MATHEMATIK, VOL. 20, 1973, PP. 213-237.

73-1024 ALLMAN, D. J.
CALCULATION OF THE ELASTIC BUCKLING LOADS OF THIN
FLAT REINFORCED PLATES USING TRIANGULAR FINITE
ELEMENTS
RAE, TECH. REPORT: 72215, 1973.

73-1025 SAUER, G.
LAMA, R. D.
APPLICATION OF NEW AUSTRIAN TUNNELLING METHOD IN
DIFFICULT BUILTOVER AREAS IN FRANKFURT/MAIN METRO
PROC. SYMP. ON ROCK MECH. AND TUNNELLING PROBLEMS,
REG. ENG. COLL., KURUKSHETRA, INDIA, VOL. 1,
DECEMBER 17-18, 1973, PP. 79-92. (PUBL. BY SARITA
PRAKASHAN, MEERUT, INDIA, 1973)

73-1026 KIKUCHI, F.
ANDO, Y.
CONVERGENCE OF LUMPED FINITE ELEMENT SCHEMES FOR
SELECTED INITIAL VALUE PROBLEMS
PROC. 2ND CONF. STRUCT. MECH. IN REACTOR TECH.,
BERLIN, PAPER M2/7, 1973.

73-1027 TSAI, C. T.
SZABO, B. A.
THE CONSTRAINT METHOD - A NEW FINITE ELEMENT
TECHNIQUE
NASA TECH. MEMORANDUM, NASA TM X-2893, 1973, PP.
551-558.

73-1028 KOUKAL, S.
PIECEWISE POLYNOMIAL INTERPOLATIONS IN THE FINITE
ELEMENT METHOD
APLIKACE MATEMATIKY, VOL. 18, 1973, PP. 146-160.

73-1029 WILSON, W. K.
FINITE ELEMENT METHODS FOR ELASTIC BODIES
CONTAINING CRACKS
METH. OF ANALYSIS AND SOLUTIONS OF CRACK PROBLEMS,
G. C. SIH (ED.), NOORDHOFF, NETHERLANDS, 1973, PP.
484-515.

73-1030 LESAINT, P.
RAYLEIGH RITZ GALERKIN METHODS FOR SYMMETRIC
POSITIVE DIFFERENTIAL EQUATIONS APPLICATION TO
FINITE ELEMENT METHODS
CENTRE D'ETUDES DE LIMEIL,
VILLENEUVE-SAINT-GEORGES, 1973.

73-1031 NEWTON, R. E.
DEGENERATION OF BRICK-TYPE ISOPARAMETRIC ELEMENTS
INTL. J. NUMER. METH. IN ENG., VOL. 7, NO. 4,
1973, PP. 579-584.

73-1032 LEWIS, E. E.
MILLER, W. E.
FINITE ELEMENT INTEGRAL NEUTRON TRANSPORT
TRANS. AMER. NUCL. SOC., VOL. 17, 1973, PP.
237-238.

73-1033 MACNEAL, R. H.
SOME ORGANIZATIONAL ASPECTS OF NASTRAN
PROC. 2ND CONF. STRUCT. MECH. IN REACTOR TECH.,
BERLIN, PAPER M1/3, 1973.

73-1034 MAREK, I.
NEDOMA, J.
FINITE ELEMENT IN THE THEORY OF SH-WAVE
PROPAGATION
IN "NUMERISCHE METHODEN IN DER GEOPHYSIK,",
K.A.P.G., VOL. 52, CZECHOSLOVAK ACADEMY, PRAGUE,
1973, PP. 31-37.

73-1035 MARSHALL, J. A.
MITCHELL, A. R.
AN EXACT BOUNDARY TECHNIQUE FOR IMPROVED ACCURACY
IN THE FINITE ELEMENT METHOD
J. INST. MATH. APPL., VOL. 12, 1973, PP. 355-362.

73-1036 MELOSH, R. J.
COMPUTATIONAL TECHNIQUES FOR FINITE ELEMENT
ANALYSIS
PROC. 2ND CONF. STRUCT. MECH. IN REACTOR TECH.,
BERLIN, PAPER M4/1, 1973.

73-1037 KIKUCHI, F.
ANDO, Y.
APPLICATION OF SIMPLIFIED HYBRID DISPLACEMENT
METHOD TO PLATE AND SHELL PROBLEMS
PROC. 2ND INT. CONF. STRUCT. MECH. IN REACTOR
TECH., VOL. 5, PART M, M5/5, 1973, PP. 1-13.

73-1038 VENKATESWARA RAO, G.
RAJU, I. S.
AN INHOMOGENEOUS, TAPERED FINITE ELEMENT
INTL. J. NUM. METH. IN ENG., VOL. 7, NO. 4, 1973, PP.
568-570.

73-1039 MILLER, W. F.
LEWIS, E. E.
QUADRATIC FINITE ELEMENT IN NEUTRON TRANSPORT
TRANS. AMER. NUCL. SOC., VOL. 17, 1973, P. 235.

73-1040 SEGUI, W. T.
COMPUTER PROGRAMS FOR THE SOLUTION OF SYSTEMS OF
LINEAR ALGEBRAIC EQUATIONS
INTL. J. NUMER. METH. ENG., VOL. 7, NO. 4, 1973,
PP. 479-490.

73-1041 MIYOSHI, T.
FINITE ELEMENT METHOD OF MIXED TYPE AND ITS
CONVERGENCE IN LINEAR SHELL PROBLEMS
KUMAMOTO J. SCI. (MATH) VOL. 10, 1973, PP. 35-38.

73-1042 MOAN, T.
FINITE ELEMENT STRESS FIELD SOLUTION OF THE
PROBLEM OF SAINT VENANT TORSION
INT. J. NUMER. METH. ENG., VOL. 5, 1973, PP.
455-458.

73-1043 VERNER, E. A.
BECKER, E. B.
FINITE ELEMENT STRESS FORMULATION FOR WAVE
PROPAGATION
INTL. J. NUMER. METH. IN ENG., VOL. 7, NO. 4,
1973, PP. 441-460.

73-1044 PEDERSEN, P.
SOME PROPERTIES OF LINEAR STRAIN TRIANGLES AND
OPTIMAL FINITE ELEMENT MODELS
INTL. J. NUMER. METH. IN ENG., VOL. 7, NO. 4,
1973, PP. 415-430.

73-1045 SINGH, B.
FAIRHURST, C.
CHRISTIANO, P. P.
COMPUTER SIMULATION OF LAMINATED ROOF REINFORCED
WITH GROUTED BOLTS
PROC. SYMP. ON ROCK MECH. AND TUNNELLING PROBL.,
ENG. COLL., KURUKSHETRA, INDIA, VOL. 1, DECEMBER
17-18, 1973, PP. 41-47. (PUBL. BY SARITA
PRAKASHAN, MEERUT, INDIA, 1973)

73-1046 NOPPEN, R.
BERECHNUNG DER ELASTIZITATSFIGENSCHAFTEN VON
MASCHINENBAUTEILEN NACH DER METHODE FINITER
ELEMENTE
PH.D. THESIS, TECHNISCHE HOCHSCHULE AACHEN, 1973.

73-1047 PALACOL, E. L.
STANTON, E. L.
ANISOTROPIC PARAMETRIC PLATE DISCRETE ELEMENTS
INT. J. NUMER. METH. ENG., VOL. 6, 1973, PP.
413-425.

73-1048 RAMANATHA, I.
ATHANASIU, C.
IN-SITU HORIZONTAL STRESSES IN LATERITE BY FINITE
ELEMENT ANALYSIS
PROC. SYMP. ON ROCK MECH. AND TUNNELLING PROBL.,
ENG. COLL., KURUKSHETRA, INDIA, VOL. 1, DECEMBER
17-18, 1973, PP. 248-253. (PUBL. BY SARITA
PRAKASHAN, MEERUT, INDIA, 1973)

73-1049 RAJU, I. S.
KRISHNA MURTY, A. V.
RAO, A. K.
SECTOR ELEMENTS FOR MATRIX DISPLACEMENT ANALYSIS
INT. J. NUMER. METH. ENG., VOL. 6, 1973, PP.
553-563.

73-1050 BREBBIA, C. A.
CONNOR, J. J.
FUNDAMENTALS OF FINITE ELEMENT TECHNIQUES FOR
STRUCTURAL ENGINEERS
PUBL. BY BUTTERWORTHS, LONDON, 1973.

73-1051 RAO, A. K.
REVIEW OF CONTINUUM, FINITE ELEMENT AND HYBRID
TECHNIQUES IN THE ANALYSIS OF STRESS
CONCENTRATIONS IN STRUCTURES
PROC. 2ND CONF. STRUCT. MECH. IN REACTOR TECH.,
BERLIN, PAPER M5/1, 1973.

73-1052 RAO, A. K.
DATTAGURU, B.
VENKATARAMAN, N. S.
RAJAIAH, K.
DETERMINATION OF STRESSES DUE TO DISCONTINUITIES
IN FINITE PLATES OF ISOTROPIC AND ORTHOTROPIC
MATERIALS
PROC. 2ND CONF. STRUCT. MECH. IN REACTOR TECH.,
PAPER M5/2, 1973.

73-1053 RAVIART, P. A.
METHODE DES ELEMENTS FINIS
LABORATOIRE ANALYSE NUMERIQUE, UNIVERSITE DE
PARIS, REPORT 73005, 1973.

73-1054 REMEDIOS, F. E.
BELL, R.
LOVATT, J. D.
AN ENGINEERING APPROACH TO FINITE ELEMENT ANALYSIS
OF NUCLEAR COMPONENTS
PROC. 2ND CONF. STRUCT. MECH. IN REACTOR TECH.,
BERLIN, PAPER M2/2, 1973.

73-1055 SCHMIDT, F. A. R.
FRANKE, H. P.
FINITE ELEMENTS VERSUS FINITE DIFFERENCES, A
COMPARISON OF THE TWO METHODS FOR THE SOLUTION OF
THE DIFFUSION EQUATION
VP. LEOPOLDSHAFEN GER., ZENTRALSTELLE FUF
ATONKERNENERGIE DUKUMENTATION, 1973.

73-1056 SCHOMBURG, U.
THE FINITE ELEMENT METHOD AND LOCAL BOUNDS FOR
BOUNDARY VALUE PROBLEMS OF ELASTIC STRUCTURES
PROC. 2ND CONF. STRUCT. MECH. IN REACTOR TECH.,
BERLIN, PAPER M2/9, 1973.

73-1057 BAUDENDISTEL, M.
INTERACTION OF ROCK AND TUNNEL LINING
PROC. SYMP. ON ROCK MECH. AND TUNNELLING PROBL.,
ENG. COLL., KURUKSHETRA, INDIA, VOL. 1, DECEMBER
17-18, 1973, PP. 1-11. (PUBL. BY SARITA PRAKASHAN,
MEERUT, INDIA, 1973)

73-1058 SOMERVAILLE, I. J.
A TECHNIQUE FOR MESH GRADING APPLIED TO CONFORMING
PLATE BENDING FINITE ELEMENTS
INT. J. NUMER. METH. ENG., VOL. 6, 1973, PP.
310-311.

73-1059 TAYLOR, C.
DAVIS, J. M.
TIDAL AND LONG WAVE PROPAGATION - A FINITE ELEMENT
APPROACH
DEPT. CIVIL ENG., UNIV. OF WALES, SWANSEA. REPORT
C/R/189, 1973.

73-1060 TEMAM, R.
NUMERICAL ANALYSIS
D. REIDEL PUBL. CO., DORDRECHT, 1973.

73-1061 THIERAUF, G.
ELASTIC-PLASTIC DEFORMATIONS OF FLEXURALLY STIFF
FRAMEWORKS FROM SECOND ORDER STRESS THEORY
ING. ARCH., VOL. 42, 1973, PP. 285-295.

73-1062 REED, W. H.
HILL, T. R.
TRIANGULAR MESH METHODS FOR THE NEUTRON TRANSPORT
EQUATION
LOS ALAMOS SCIENTIFIC LAB., REPORT LA-UR-73-479,
1973.

73-1063 TILLERSON, J. R.
A TREATISE OF NONLINEAR FINITE ELEMENT ANALYSIS
PH.D. THESIS, TEXAS A AND M UNIV., 1973.

73-1064 FACCIOLI, E.
VITIELLO, E.
A FINITE ELEMENT LINEAR PROGRAMMING METHOD FOR THE
LIMIT ANALYSIS OF THIN PLATES
INT. J. NUMER. METH. ENG., VOL. 5, 1973, PP.
311-325.

73-1065 HOPPE, V.
FINITE ELEMENTS WITH HARMONIC INTERPOLATION
FUNCTIONS
IN "THE MATH. OF FINITE ELEMENTS AND APPL.", J. R.
WHITEMAN (ED.), ACADEMIC PRESS, LONDON, 1973, PP.
131-142.

73-1066 FUHRING, H.
A DISCUSSION - ON THE DENSITY OF FINITE ELEMENT
MATRICES
INTL. J. NUMER. METH. IN ENG., VOL. 8, NO. 2,
1973, P. 432.

73-1067 TRACEY, D. M.
ON THE FRACTURE MECHANICS ANALYSIS OF
ELASTIC-PLASTIC MATERIALS USING THE FINITE ELEMENT
METHOD
PH.D. THESIS, BROWN UNIV., 1973.

73-1068 WILLIAMS, F. W.
COMPARISON BETWEEN SPARSE STIFFNESS MATRIX AND
SUB-STRUCTURE METHODS
INT. J. NUMER. METH. ENG., VOL. 5, 1973, PP.
383-394.

73-1069 YEO, M. F.
A MORE EFFICIENT FRONT SOLUTION: ALLOCATING
ASSEMBLY LOCATIONS BY LONGEVITY CONSIDERATIONS
INT. J. NUMER. METH. ENG., VOL. 7, 1973, PP.
570-573.

73-1070 YAMADA, YOSHIAKI
TAKATSUKA, K.
ELASTIC-PLASTIC ANALYSIS OF BEAMS WITH UNIFORM
CROSS-SECTION UNDER COMBINED LOADINGS (2)
J. INST. OF IND. SCI., SEISAN-KENKYU, VOL. 25,
1973, PP. 238-242.

73-1071 YAMADA, YOSHIAKI
NAGAI, Y.
FINITE ELEMENT ANALYSIS OF LOAD CELL RESPONSE IN
HIGH SPEED TENSILE TESTING
J. INST. OF IND. SCI., SEISAN-KENKYU, VOL. 25,
1973, PP. 259-262.

73-1072 YAMADA, YOSHIAKI
TAKABATAKE, H.
DYNAMIC RESPONSE OF VISCOELASTIC MATERIALS TO
SINUSOIDALLY VARYING LOADINGS
J. INST. OF IND. SCI., SEISAN-KENKYU, VOL. 25,
1973, PP. 293-296.

73-1073 YAMADA, YOSHIAKI
TAKATSUKA, K.
ANALYSIS OF ELASTIC-PLASTIC BEHAVIOR OF BEAMS AND
COLUMNS BY PLANE FINITE ELEMENT MODEL
PROC. 3RD NATL. SYMP. ON MATRIX METH. OF STRUCT.
ANALYSIS AND DESIGN, SOC. OF STEEL CONSTR. OF
JAPAN, 1973, PP. 307-316 (IN JAPANESE).

73-1074 YAMADA, YOSHIAKI
IWATA, K.
NONLINEAR ANALYSIS OF BEAMS AND SHELLS
PROC. 3RD NATL. SYMP. ON MATRIX METH. OF STRUCT.
ANALYSIS AND DESIGN, SOC. OF STEEL CONSTR. OF
JAPAN, 1973, PP. 331-340 (IN JAPANESE).

73-1075 YAMADA, YOSHIAKI
TAKABATAKE, H.
DYNAMIC RESPONSE OF VISCOELASTIC MATERIALS
PROC. 3RD NATL. SYMP. ON MATRIX METH. OF STRUCT.
ANALYSIS AND DESIGN, SOC. OF STEEL CONSTR. OF
JAPAN, 1973, PP. 497-504 (IN JAPANESE).

73-1076 WILSON, J. F.
THE FINITE ELEMENT ANALYSIS OF THIN ELASTIC SHELLS
PH.D. THESIS, ARIZONA STATE UNIV., 1973.

73-1077 GUPTA, K. K.
FREE VIBRATION ANALYSIS OF SPINNING STRUCTURAL
SYSTEMS
INT. J. NUMER. METHODS ENGR., VOL. 5, NO. 3, 1973,
PP. 395-418.

73-1078 BERNHARDT, K.
ROY, J. R.
DISCUSSION OF A PAPER BY F. W. WILLIAMS
INTL. J. NUMER. METH. IN ENG., VOL. 6, NO. 4,
1973, P. 601.

73-1079 CAPRILI, M.
CELLA, A.
GHERI, G.
SPLINE INTERPOLATION TECHNIQUES FOR VARIATIONAL
METHODS
INTL. J. NUMER. METH. IN ENG., VOL. 6, NO. 4,
1973, PP. 565-576.

73-1080 CAPURSO, M.
ON THE EXTREMAL PROPERTIES OF THE SOLUTION IN
DYNAMICS OF RIGID VISCOPLASTIC BODIES ALLOWING FOR
LARGE DISPLACEMENT EFFECTS
MECCANICA, VOL. 7, NO. 4, DECEMBER 1973, PP.
236-247.

73-1081 ZLAMAL, M.
A REMARK ON THE SERENDIPITY FAMILY
INTL. J. NUMER. METH. IN ENG., VOL. 7, NO. 1,
1973, PP. 98-100.

73-1082 HOPPE, V.
SPECIFICATION OF GEOMETRICAL PARAMETERS FOR
ELEMENTS WITH CUBIC MAPPING FUNCTIONS
INTL. J. NUMER. METH. IN ENG., VOL. 7, NO. 1,
1973, PP. 94-97.

73-1083 BATES, C. J.
A COMPUTATIONAL TECHNIQUE FOR THE EFFICIENT
HANDLING OF LARGE MATRICES
INTL. J. NUMER. METH. IN ENG., VOL. 7, NO. 1,
1973, PP. 85-93.

73-1084 OSTROFF, A. J.
MCCANN, M.
ANALYSIS OF THE DYNAMICS OF THIN PRIMARY MIRRORS
FOR LARGE ASTRONOMICAL TELESCOPES
LANGLEY STATION, VA., NASA-TM-X-2790, JULY 1973,
64 PP.

73-1085 BAZANT, Z. P.
EL NIMEIRI, M.
LARGE DEFLECTION SPATIAL BUCKLING OF THIN WALLED
BEAMS AND FRAMES
A.S.C.E. J. ENGR. MECH. DIV., VOL. 99, NO. EM6,
DECEMBER 1973, PP. 1259-1281.

73-1086 RAO, G. V.
RAJU, I. S.
MURTHY, T. V. G.
VIBRATION OF RECTANGULAR PLATES WITH MIXED
BOUNDARY CONDITIONS
J. SOUND AND VIB., VOL. 30, NO. 2, SEPTEMBER 1973,
PP. 257-260.

73-1087 GRAGGS, A.
AN ACOUSTIC FINITE ELEMENT APPROACH FOR STUDYING
BOUNDARY FLEXIBILITY AND SOUND TRANSMISSION
BETWEEN IRREGULAR ENCLOSURES
J. SOUND AND VIB., VOL. 30, NO. 3, NOVEMBER 1973,
PP. 343-357.

73-1088 RED, W. E.
MCMUNN, J. C.
FINITE ELEMENT ANALYSIS OF STATIC AND DYNAMIC
STRESSES AROUND SUDDENLY PUNCHED HOLES IN PLATES
AND SHELLS
COMPUTERS AND SCI., VOL. 3, NO. 6, NOVEMBER 1973,
PP. 1275-1292.

73-1089 VENANCIO-FILHO, F.
IGUTI, F.
VIBRATIONS OF GRIDS BY THE FINITE ELEMENT METHOD
COMPUT. AND STRUCT., VOL. 3, NO. 6, NOVEMBER 1973,
PP. 1331-1344.

73-1090 MURAKAMI, Y.
KUSUMOTO, S.
NOTCH EFFECT IN LOW CYCLE FATIGUE - REPORT 1:
LOW-CYCLE FATIGUE STRENGTH OF 0.48 PERCENT CARBON
STEEL UNDER CONSTANT LOAD
J.S.M.E. BULL., VOL. 16, NO. 101, NOVEMBER 1973,
PP. 1637-1647.

73-1091 HADEDANK, G.
FREQUENCY-DEPENDENT SHAPE FUNCTIONS FOR FREE AND
FORCED STRUCTURAL VIBRATION CALCULATION
44TH EUROMECH. COLLOQUIUM, DYNAMICS OF MACHINE
FOUNDATIONS, POLYTECH. INST., BUCHAREST, ROMANIA,
OCTOBER 29-31, 1973, PP. 35-45.

73-1092 PAWTANI, S.
THE EFFECT OF CAMBER ON THE NATURAL FREQUENCIES OF
LOW ASPECT RATIO TURBOMACHINERY BLADES
J. AERON. SOC., INDIA, VOL. 25, NO. 3, AUGUST
1973, PP. 119-126.

73-1093 KENINGSBERG, I. J.
CH-53A FLEXIBLE FRAME VIBRATION ANALYSIS/TEST
CORRELATION
UNITED AIRCRAFT CORP., SIKORSKY AIRCRAFT DIV.,
STRATFORD, CONN., REPT. NO. SER-651195, MARCH 28,
1973, 245 PP.

73-1094 KONRAD, A.
SILVESTER, P.
TRIANGULAR FINITE ELEMENTS FOR THE GENERALIZED
BESSEL EQUATION OF ORDER M
INTL. J. NUMER. METH. IN ENG., VOL. 7, NO. 1,
1973, PP. 43-56.

73-1095 BATHE, K. J.
WILSON, E. L.
SOLUTION METHODS FOR EIGENVALUE PROBLEMS IN
STRUCTURAL MECHANICS
INT. J. NUMER. METH. ENG., VOL. 6, 1973, PP.
213-226.

73-1096 CARPENTER, W. C.
FINITE ELEMENT ANALYSIS OF BONDED CONNECTIONS
INT. J. NUMER. METH. ENG., VOL. 6, 1973, PP.
450-451.

73-1097 FREMOND, M.
LA METHODE FRONTALE POUR LA RESOLUTION DES
SYSTEMES LINEARESS. (FRONTAL METHOD FOR SOLVING
LINEAR SYSTEMS)
PROC. INT. COMPUT. SYMP., DAVOS, SWITZ., SEPTEMBER
4-7, 1973, PP. 337-343. (AVAILABLE FROM AM.
ELSEVIER PUBL. CO., NEW YORK, N.Y.)

73-1098 CROLL, J. C. A.
THE TREATMENT OF NATURAL BOUNDARY CONDITIONS IN
THE FINITE ELEMENT AND FINITE DIFFERENCE METHODS
INT. J. NUMER. METH. ENG., VOL. 5, 1973, PP.
443-445.

73-1099 LIU, W-H.
SHERMAN, A. H.
COMPARATIVE ANALYSIS OF THE CUTHILL-MCKEE AND THE
REVERSE CUTHILL-MCKEE ORDERING ALGORITHMS FOR
SPARSE MATRICES
DEPT. OF COMPUT. SCI., YALE UNIV., NEW HAVEN,
CONN., REPORT NO: RR-28, 1973, 35 PP. (N.T.I.S. -
AD-A016 683/5ST)

73-1100 ANON
REPORT OF INTERAGENCY CONFERENCE ON COORDINATION
OF RESEARCH ACTIVITIES
ARMY ENG. WATERWAYS EXPERIMENT STATION, VICKSBURG,
MISS., NOVEMBER 13-15, 1973, 103 PP. (N.T.I.S. -
AD-A011 991/ST)

73-1101 PANAK, J. J.
DEVELOPMENT OF METHODS FOR COMPUTER SIMULATION OF
BEAM-COLUMNS AND GRID-BEAM AND SLAB SYSTEMS
CENTER FOR HIGHWAY RESEARCH, TEXAS UNIV., AUSTIN,
REPORT NO: CFHR-3-5-63-56-29F, AUGUST 1973, 86 PP.
(N.T.I.S. - PB-242 363/0ST)

73-1102 TONG, P.
ADACHI, J.
ON THE DYNAMIC BUCKLING OF SHELLS OF REVOLUTION
ARMY MATER. AND MECH. RESEARCH CENTER, WATERTOWN,
MASS., REPORT NO: AMMRC-TR-75-9, OCTOBER 5, 1973,
6 PP. (N.T.I.S. - AD-A011 300/1ST)

73-1103 BERNDT, P.
IMPORTANCE OF VARIATIONAL CALCULUS IN FLUID
MECHANICS AND ITS APPLICATION TO COMPRESSIBLE
POTENTIAL FLOWS (BEDEUTUNG DER VARIATIONSRECHNUNG
IN DER STROEMUNGSMECHANIK UND IHRE ANWENDUNG BEI
KOMPRESSIBLEM POTENTIALSTROEMUNGEN)
TECHNISCHE UNIV., MUNICH, WEST GERMANY, PH.D.
THESIS, JULY 26, 1973, 27 PP. (N.T.I.S. -
N75-15909/5ST)

73-1104 NAPOLITANO, L. G.
FUNCTIONAL ANALYSIS DERIVATION AND GENERALIZATION
OF HYBRID VARIATIONAL METHODS
ISTITUTO DI AERODINAMICA, NAPLES UNIV., ITALY,
REPORT NO: IA-218, DECEMBER 1973, 83 PP. (N.T.I.S.
- AD-A-007 068/0ST)

73-1105 KULICKI, J. M.
KOSTEM, C. N.
USER'S MANUAL FOR PROGRAM BEAM
FRITZ ENG. LAB., LEHIGH UNIV., BETHLEHEM, PA.,
REPORT NO: FEL-378B.2, FEBRUARY 1973, 125 PP.
(N.T.I.S. - PB-237 959/2ST)

73-1106 NEALE, B. K.
FINITE ELEMENT CRACK ANALYSIS USING THE J-INTEGRAL
METHOD
BERKELEY NUCL. LABS., CENTRAL ELECTRICITY
GENERATING BOARD, BERKELEY, (UK), SEPTEMBER 1973,
27 PP. (N.T.I.S. - RD/B/N-2785)

73-1107 CHENG, K. Y.
NOTE ON MINIMIZING THE BANDWIDTH OF SPARSE,
SYMMETRIC MATRICES
COMPUT. J., VOL. 11, 1973, PP. 27-30.

73-1108 SAMUELSON, A.
MIXED FINITE ELEMENT METHODS IN THEORY AND
APPLICATIONS
LECTURE NOTES ON THE FINITE ELEMENT COURSE IN
TIRRENIA, 1973.

73-1109 YOSHIDA, I.
EVALUATION OF STRESSES IN SOIL UNDER RIGID WHEELS
BY FINITE ELEMENT METHOD
J. SOC. AGRIC. MACHINERY, JAPAN, VOL. 34, NO. 4,
1973, PP. 312-317.

73-1110 SWADA, T.
YOSHITAKE, Y.
ANALYSIS OF SEEPAGE THROUGH DAMS AND IN FOUNDATION
BY FINITE ELEMENT METHOD
TRANS. JAP. SOC. IRRIG. DRAIN RECLAM. ENG., VOL.
47, OCTOBER 1973, PP. 16-21.

73-1111 GERIJ, J.
FINITE-ELEMENT METHOD IN CYCLIC
THERMO-ELASTOPLASTICITY
SCI. TECH. ARMEMENT, (FRENCH) VOL. 47, NO. 2,
1973, PP. 439-442.

73-1112 FRAEIJS DE VEUBEKE, B.
DIFFUSIVE EQUILIBRIUM MODELS
INTL. RESEARCH SEM. ON THE THEORY AND APPLICS. OF
THE FINITE ELEMENT METH., UNIV. OF CALGARY,
CANADA, 1973.

73-1113 BECKERS, P.
LES FONCTIONS DE TENSION DANS LA METHODE DES
ELEMENTS FINIS
PH.D., COLLECTION DES PUBLICATIONS DE LA FACULTE
DES SCIENCES APPLIQUEES DE L'UNIVERSITE DE LIEGE,
NO. 41, 1973.

73-1114 MOWBRAY, D. F.
MCCONNELEE, J. E.
APPLICATIONS OF FINITE ELEMENT STRESS ANALYSIS AND
STRESS-STRAIN PROPERTIES IN DETERMINING NOTCH
FATIGUE SPECIMEN DEFORMATION AND LIFE
CYCLIC STRESS-STRAIN BEHAVIOR ANALYSIS, EXP. AND
FAILURE PRED., 1973, PP. 151-169.

73-1115 HEISE, U.
EINE FINITE-ELEMENT-METHODE MIT
KREISRINGSEKTORFORMIGEN ELEMENTEN
ZAMM, VOL. 53, 1973, PP. 136-138.

73-1116 ROBINSON, J.
FLAT QUADRILATERAL EQUILIBRIUM MEMBRANE ELEMENTS
RA/ARD, REPORT NO. 28-9-73-17, SEPTEMBER 1973.

73-1117 BERGAN, P. G.
SOREIDE, T. H.
A COMPARATIVE STUDY OF DIFFERENT NUMERICAL
SOLUTION TECHNIQUES AS APPLIED TO A NONLINEAR
STRUCTURAL PROBLEM
COMPUT. METH. IN APPL. MECH. AND ENG., VOL. 2,
1973, PP. 185-201.

73-1118 SEWELL, E. G.
AUTOMATIC GENERATION OF TRIANGULATIONS FOR
PIECEWISE POLYNOMIAL APPROXIMATION
PH.D. THESIS, PURDUE UNIV., LAFAYETTE, INDIANA,
1973.

73-1119 O'BRIEN, M. A.
FITZGERALD, J. J.
DILLON, E. C.
APPLICATION OF THE FAST FOURIER TRANSFORM TO
LINEAR SYSTEMS IN CIVIL ENGINEERING
DEPT. OF CIVIL ENG., UNIV. COLLEGE, CORK, EIRE,
1973.

73-1120 REDDY, J. N.
ACCURACY AND CONVERGENCE OF MIXED FINITE-ELEMENT
APPROXIMATIONS OF THIN BARS, MEMBRANES, AND PLATES
ON ELASTIC FOUNDATIONS
PROC. CONF. GRAD. RESEARCH IN APPL. MECH., LAS
CRUCES, NEW MEXICO, PAPER 185, MARCH 1973.

73-1121 COLLEN, M. J. P.
A SIMPLE FINITE ELEMENT METHOD FOR METEOROLOGICAL
PROBLEMS
J. INST. OF MATH. AND APPL., ACADEMIC PRESS,
LONDON, VOL. 11, 1973, PP. 15-31.

73-1122 DE KONING, A. U.
RESULT OF CALCULATIONS WITH TRIM 6 AND TRIAX 6
ELASTIC-PLASTIC ELEMENTS
NATL. AEROSPACE LAB., THE NETHERLANDS, NLR
MP73010V, 1973.

73-1123 TONG, P.
PIAN, T. H. H.
APPLICATION OF FINITE ELEMENT METHOD TO MIXED-MODE
FRACTURE
10TH MEET. SOC. OF ENG. SCI., NORTH CAROLINA STATE
UNIV., RALEIGH, N.C., NOVEMBER 5-7, 1973.

73-1124 NEMAT-NASSER, S.
LEE, K. N.
FINITE ELEMENT FORMULATIONS FOR ELASTIC PLATES BY
GENERAL VARIATIONAL STATEMENTS WITH DISCONTINUOUS
FIELDS
COMP. METH. IN APPL. MECH. ENG., VOL. 2, NO. 1,
1973, PP. 33-41.

73-1125 WILSON, G. J.
APPLICATION OF THE FINITE ELEMENT METHOD TO THE
VIBRATION ANALYSIS OF AXIAL FLOW TURBINES
PH.D. THESIS, CARLETON UNIV., CANADA, 1973.

73-1126 ANON
FRACTURE TOUGHNESS PARAMETERS AND ELASTIC-PLASTIC
ANALYSIS OF NON-MODERATE FRACTURE CONDITIONS USING
FINITE-ELEMENT METHODS
ENG. FRACT. MECH., VOL. 5, NO. 2, JUNE 1973, PP.
223-240.

73-1127 FENTON, R. G.
FINITE ELEMENT SOLUTION OF EXTRUSION OF
ELASTO-PLASTIC WORK HARDENING MATERIALS
PROC. CONF. NORTH AMER. METALWORKING RESEARCH,
MCMASTER UNIV., HAMILTON, ONT., VOL. 1, MAY 1973,
PP. 49-61.

73-1128 TIEU, A. K.
OIL-FILM TEMPERATURE DISTRIBUTION IN AN INFINITELY
WIDE SLIDER BEARING: AN APPLICATION OF THE FINITE
ELEMENT METHOD
J. MECH. ENG. SCI., VOL. 15, NO. 4, 1973, P. 311.

73-1129 COOMER, J. T.
FINITE-ELEMENT ANALYSIS OF PERIODIC-STRUCTURES
PH.D., UNIV. OF CINCINNATI, 1973.

73-1130 DENKMANN, W. J.
FINITE-ELEMENT ANALYSIS OF SANDWICH SHELLS
PH.D. THESIS, STEVENS INST. OF TECHNOL., HOBOKEN,
N.J., 1973.

73-1131 DINYOVSZKY, P. P.
FINITE-ELEMENT TECHNIQUES FOR THE DYNAMIC ANALYSIS
OF UNSYMMETRICALLY LAMINATED PLATES
PH.D. THESIS, STEVENS INST. OF TECHNOL., HOBOKEN,
N.J., 1973.

73-1132 DIETRICH, D. E.
NONLINEAR ANALYSIS OF ARBITRARY, HYPERELASTIC
MEMBRANE SHELLS BY THE FINITE-ELEMENT METHOD
PH.D., STATE UNIV. OF NEW YORK, BUFFALO, 1973.

73-1133 GEORGOPOULOS, G.
VIBRATIONAL ANALYSIS OF GENERAL SHELLS UTILIZING
FINITE-ELEMENT TECHNIQUES
PH.D., NORTH CAROLINA STATE UNIV., RALEIGH, 1973.

73-1134 GLANCY, J. J.
DETERMINATION OF THE LOWEST CRITICAL BUCKLING
STRESS OF VARIOUS SHAPED PLATES BY FINITE-ELEMENTS
PH.D., ARIZONA STATE UNIV., 1973.

73-1135 HAGUE, M. I.
CONVERGENCE OF THE FINITE-ELEMENT METHOD UNDER THE
CONSTRAINT OF MATERIAL INCOMPRESSIBILITY
PH.D., COLORADO STATE UNIV., 1973.

73-1136 HAGEDOORN, A. H. J.
AN ELASTIC FINITE-ELEMENT ANALYSIS OF THE BUCKLING
OF ECCENTRICALLY STIFFENED PLATES AND SHELLS
PH.D., CORNELL UNIV., 1973.

73-1137 HSIUNG, H. C. H.
DYNAMIC ANALYSIS OF HYDROELASTIC SYSTEMS USING THE
FINITE-ELEMENT METHOD
PH.D., UNIV. OF SOUTHERN CALIF., 1973.

73-1138 JENSEN, F. R.
THREE-DIMENSIONAL FINITE-ELEMENT ELASTIC ANALYSIS
OF PRISMATIC SOLIDS
PH.D., BRIGHAM YOUNG UNIV., 1973.

73-1139 JONES, L. R.
UNIFICATION OF THE FINITE ELEMENT AND CLASSICAL
RITZ METHODS
PH.D., UNIV. OF CALIF., BERKELEY, 1973.

73-1140 JUDAH, O. M.
SIMULATION OF RUNOFF HYDROGRAPHS FROM NATURAL
WATERSHEDS BY FINITE-ELEMENT METHOD
PH.D., VIRGINIA POLYTECHNIC INST. AND STATE UNIV.,
1973.

73-1141 KETCHMAN, J.
APPLICATION OF THE FINITE-ELEMENT METHOD TO TOWED
CABLE DYNAMICS
ENG. SC.D., COLUMBIA UNIV., 1973.

73-1142 KHAN, A. Q.
GEOMETRICALLY NONLINEAR BEHAVIOUR OF THIN-WALLED
MEMBERS USING FINITE-ELEMENTS
PH.D., MCGILL UNIV., CANADA, 1973.

73-1143 KIM, H. W.
A STUDY OF FREE-VIBRATIONS OF THIN ELASTIC SHELLS
SUBJECTED TO INITIAL LOADS USING FINITE-ELEMENTS
PH.D., PURDUE UNIV., 1973.

73-1144 KISHORE, B. R.
CHARACTERISTICS OF FINITE-ELEMENTS FOR ANALYSIS OF
BEAMS SUBJECTED TO IMPACT LOAD
PH.D., OKLAHOMA STATE UNIV., 1973.

73-1145 KIM, Y. J.
AN EFFICIENT ITERATIVE PROCEDURE FOR USE WITH THE
FINITE-ELEMENT METHOD
PH.D., UNIV. OF ILLINOIS, URBANA-CHAMPAIGN, 1973.

73-1146 KLAMECKI, B. E.
INCIPIENT CHIP FORMATION IN METAL CUTTING — A
THREE- DIMENSION FINITE-ELEMENT ANALYSIS
PH.D., UNIV. OF ILLINOIS, URBANA-CHAMPAIGN, 1973.

73-1147 KNAPP, R. H.
FINITE-ELEMENT NONLINEAR BUCKLING ANALYSIS OF A
PSEUDO-CYLINDRICAL CONCAVE POLYHEDRAL SHELL UNDER
EXTERNAL PRESSURE
PH.D. THESIS, UNIV. OF HAWAII, HONOLULU, 1973.

73-1148 LAKSHMINARAYANAN, V.
ON THE DEVELOPMENT OF A REFINED FINITE-ELEMENT
FREE FROM DISCRETIZATION — DISPERSION FOR ELASTIC
WAVE PROPAGATION MODELLING
PH.D., MONTANA STATE UNIV., 1973.

73-1149 LIU, S-N.
AN ANALYSIS OF THE PHYSICAL BEHAVIOR OF NONWOVEN
FIBROUS MATERIAL BY THE FINITE-ELEMENT METHOD
PH.D., UNIV. OF WASHINGTON, 1973.

73-1150 MILLER, W. F. JR.
APPLICATION OF FINITE-ELEMENTS TO ONE-DIMENSIONAL
AND TWO-DIMENSIONAL NEUTRON-TRANSPORT
PH.D., NORTHWESTERN UNIV., 1973.

73-1151 MOHAMMED, M. A. A.
VIBRATION ANALYSIS OF PLATES, SHELLS AND BLADED
RIMMED DISCS USING THE
TRANSFER-MATRIX-FINITE-ELEMENT METHOD
PH.D. THESIS, MCMASTER UNIV., 1973.

73-1152 MOHR, C. L.
NONLINEAR STRESS ANALYSIS OF INDETERMINATE
STRUCTURAL ELEMENTS SUBJECTED TO BIAXIAL STRESS
STATES USING THE FINITE-ELEMENT TECHNIQUE
PH.D., WASHINGTON STATE UNIV., 1973.

73-1153 MOHAN, A.
FINITE-ELEMENT ANALYSIS OF HEAT FLOW AROUND BURIED
PIPES
PH.D., PURDUE UNIV., 1973.

73-1154 NAM, C. H.
A FINITE-ELEMENT STUDY ON THE NONLINEAR BEHAVIOR
OF REINFORCED CONCRETE FLEXURAL MEMBERS DUE TO
SHORT-TIME LOADING
PH.D., UNIV. OF WISCONSIN, 1973.

73-1155 NAY, R. A.
AN ALTERNATIVE FOR THE FINITE-ELEMENT METHOD
PH.D., DUKE UNIV., 1973.

73-1156 NATARAJA, M. S.
FINITE-ELEMENT SOLUTION OF STRESSES AND
DISPLACEMENTS IN A SOIL-CULVERT SYSTEM
PH.D., UNIV. OF PITTSBURGH, 1973.

73-1157 OZAWA, Y.
ELASTO-PLASTIC FINITE ELEMENT ANALYSIS OF SOIL
DEFORMATION
PH.D., UNIV. OF CALIFORNIA, BERKELEY, 1973.

73-1158 PANDARINATHAN, V. G.
COMPARISON OF SOME FINITE-ELEMENT MODELS FOR
CIRCULAR CYLINDRICAL SHELLS
PH.D., UNIV. OF NOTRE DAME, 1973.

73-1159 PATEL, M. R.
ROCK FRAGMENTATION BY SUBSURFACE THERMAL
INCLUSIONS — A FINITE ELEMENT STUDY
PH.D., UNIV. OF MISSOURI, ROLLA, 1973.

73-1160 PECQUET, R. A.
A FINITE-ELEMENT PROGRAM FOR THE ANALYSIS OF
EMBANKMENTS OVER SOFT, SATURATED SOILS INCLUDING
CONSOLIDATION AND CREEP EFFECTS
PH.D., THE LOUISIANA STATE UNIV. AND AGRIC. AND
MECH. COL., 1973.

73-1161 POLENSEK, A.
STATIC AND DYNAMIC ANALYSIS OF WOOD-JOIST FLOORS
BY THE FINITE-ELEMENT METHOD
PH.D., OREGON STATE UNIV., 1973.

73-1162 ROSSOW, M. P.
A FINITE-ELEMENT APPROACH TO OPTIMAL STRUCTURAL
DESIGN
PH.D., THE UNIV. OF MICHIGAN, 1973.

73-1163 SCHULTCHEN, E. G.
INTERACTIVE SOLUTION OF LINEAR EQUATIONS IN FINITE
ELEMENT ANALYSIS
PH.D. THESIS, LEHIGH UNIV., PENNSYLVANIA, 1973.

73-1164 SEN, S. K.
HIGH-PRECISION FINITE-ELEMENT ANALYSIS OF SHELLS
OF REVOLUTION
SC.D., WASHINGTON UNIV., 1973.

73-1165 SEMSARZADEH, G. A.
AN INTERACTIVE COMPUTER-AIDED MODELING SYSTEM FOR
THE `DISCRETIZATION OF BRIDGE STRUCTURES FOR
FINITE-ELEMENT ANALYSIS
PH.D., UNIV. OF PENNSYLVANIA, 1973.

73-1166 SHEFFLER, A. W.
A COMPACT FINITE-ELEMENT METHOD FOR THE ANALYSIS
OF ANISOTROPIC LAMINATED PLATES
PH.D., WEST VIRGINIA UNIV., 1973.

73-1167 CHEN, S. F.
THE AIRFOIL PROBLEM VIA THE FINITE ELEMENT METHOD
SYMP. APPLI. OF COMPUT. TO FLUID DYNA. ANALYSIS
AND DESIGN. POLY. INST., BROOKLYN, JANUARY 1973.

73-1168 STORM, J. H.
A FINITE-ELEMENT MODEL TO SIMULATE THE NONLINEAR
RESPONSE OF REINFORCED CONCRETE FRAMES WITH
MASONRY FILLER WALLS
PH.D., UNIV. OF ILLINOIS, URBANA, CHAMPAIGN, 1973.

73-1169 THOMAS, G. P.
NONLINEAR FINITE-ELEMENT ANALYSIS OF THIN SHELLS
PH.D., CORNELL UNIV., 1973.

74-0001 ZIENKIEWICZ, O. C.
FINITE ELEMENT METHODS IN FLOW PROBLEMS
IN FINITE ELEMENT METHOD IN FLOW PROBLEMS, EDITED
BY J. T. ODEN, O. C. ZIENKIEWICZ, R. H.
GALLAGHER, AND C. TAYLOR, PAPER PRESENTED AT
THE INTERNATIONAL SYMPOSIUM ON FINITE ELEMENT
METHODS IN FLOW PROBLEMS, HELD IN JANUARY 1974, IN
SWANSEA, UNITED KINGDOM, PP. 3-4.

74-0002 ODEN, J. T.
WELLFORD, L. C.
MATHEMATICAL FEATURES OF FINITE ELEMENT
APPROXIMATIONS OF CERTAIN FLOW PROBLEMS WITH
EMPHASIS ON CONVECTION AND DIFFUSION
IN FINITE ELEMENT METHOD IN FLOW PROBLEMS, EDITED
BY J. T. ODEN, O. C. ZIENKIEWICZ, R. H.
GALLAGHER, AND C. TAYLOR, PAPER PRESENTED AT
THE INTERNATIONAL SYMPOSIUM ON FINITE ELEMENT
METHODS IN FLOW PROBLEMS, HELD IN JANUARY 1974, IN
SWANSEA, UNITED KINGDOM, PP. 5-14.

74-0003 FINLAYSON, B. A.
WEIGHTED RESIDUAL METHODS AND THEIR RELATION TO
FINITE ELEMENT METHODS IN FLOW PROBLEMS
IN FINITE ELEMENT METHOD IN FLOW PROBLEMS, EDITED
BY J. T. ODEN, O. C. ZIENKIEWICZ, R. H.
GALLAGHER, AND C. TAYLOR, PAPER PRESENTED AT
THE INTERNATIONAL SYMPOSIUM ON FINITE ELEMENT
METHODS IN FLOW PROBLEMS, HELD IN JANUARY 1974, IN
SWANSEA, UNITED KINGDOM, PP. 13-19.

74-0004 NORRIE, D. H.
DE VRIES, G.
THE PSEUDO FUNCTIONAL FINITE ELEMENT METHOD
IN FINITE ELEMENT METHOD IN FLOW PROBLEMS, EDITED
BY J. T. ODEN, O. C. ZIENKIEWICZ, R. H.
GALLAGHER, AND C. TAYLOR, PAPER PRESENTED AT
THE INTERNATIONAL SYMPOSIUM ON FINITE ELEMENT
METHODS IN FLOW PROBLEMS, HELD IN JANUARY 1974, IN
SWANSEA, UNITED KINGDOM, PP. 21-28.

74-0005 CHENG, R. T.
ON THE STUDY OF CONVECTIVE DISPERSION EQUATION
IN FINITE ELEMENT METHOD IN FLOW PROBLEMS, EDITED
BY J. T. ODEN, O. C. ZIENKIEWICZ, R. H.
GALLAGHER, AND C. TAYLOR, PAPER PRESENTED AT
THE INTERNATIONAL SYMPOSIUM ON FINITE ELEMENT
METHODS IN FLOW PROBLEMS, HELD IN JANUARY 1974, IN
SWANSEA, UNITED KINGDOM, PP. 29-47.

74-0006 BAKER, A. J.
FINITE ELEMENT SOLUTION ALGORITHM FOR
INCOMPRESSIBLE FLUID DYNAMICS
IN FINITE ELEMENT METHOD IN FLOW PROBLEMS, EDITED
BY J. T. ODEN, O. C. ZIENKIEWICZ, R. H.
GALLAGHER, AND C. TAYLOR, PAPER PRESENTED AT
THE INTERNATIONAL SYMPOSIUM ON FINITE ELEMENT
METHODS IN FLOW PROBLEMS, HELD IN JANUARY 1974, IN
SWANSEA, UNITED KINGDOM, PP. 51-55.

74-C007 TONG, P.
ON THE SOLUTION OF THE NAVIER-STOKES EQUATION IN
TWO DIMENSIONAL AND AXIAL SYMMETRIC PROBLEMS
IN FINITE ELEMENT METHOD IN FLOW PROBLEMS, EDITED
BY J. T. ODEN, O. C. ZIENKIEWICZ, R. H.
GALLAGHER, AND C. TAYLOR, PAPER PRESENTED AT THE
INTERNATIONAL SYMPOSIUM ON FINITE ELEMENT METHODS
IN FLOW PROBLEMS, HELD IN JANUARY 1974, IN
SWANSEA, UNITED KINGDOM, PP. 57-66.

74-0008 HUTTON, A. G.
ON FLOW NEAR SINGULAR POINTS OF A WALL BOUNDARY
IN FINITE ELEMENT METHOD IN FLOW PROBLEMS, EDITED
BY J. T. ODEN, O. C. ZIENKIEWICZ, R. H.
GALLAGHER, AND C. TAYLOR, PAPER PRESENTED AT
THE INTERNATIONAL SYMPOSIUM ON FINITE ELEMENT
METHODS IN FLOW PROBLEMS, HELD IN JANUARY 1974, IN
SWANSEA, UNITED KINGDOM, PP. 67-83.

74-0009 LEIBER, P.
WEN, K. S.
ATTIA, A. V.
FINITE ELEMENT METHOD AS AN ASPECT OF THE
PRINCIPLE OF MAXIMUM UNIFORMITY: NEW
HYDRODYNAMICAL RAMIFACATIONS
IN FINITE ELEMENT METHOD IN FLOW PROBLEMS, EDITED
BY J. T. ODEN, O. C. ZIENKIEWICZ, R. H.
GALLAGHER, AND C. TAYLOR, PAPER PRESENTED AT
THE INTERNATIONAL SYMPOSIUM ON FINITE ELEMENT
METHODS IN FLOW PROBLEMS, HELD IN JANUARY 1974, IN
SWANSEA, UNITED KINGDOM, PP. 85-96.

74-C010 HUANG, A. B.
YOUNG, V.Y.C.
A NON VARIATIONAL FINITE ELEMENT ANALYSIS FOR THE
NAVIER STOKES EQUATIONS
IN FINITE ELEMENT METHOD IN FLOW PROBLEMS, EDITED
BY J. T. ODEN, O. C. ZIENKIEWICZ, R. H.
GALLAGHER, AND C. TAYLOR, PAPER PRESENTED AT
THE INTERNATIONAL SYMPOSIUM ON FINITE ELEMENT
METHODS IN FLOW PROBLEMS, HELD IN JANUARY 1974, IN
SWANSEA, UNITED KINGDOM, PP. 97-98.

74-0011 IKENOUCHI, M.
KIMURA, N.
AN APPROXIMATE NUMERICAL SOLUTION OF THE NAVIER
STOKES EQUATIONS BY GALERKIN METHOD
IN FINITE ELEMENT METHOD IN FLOW PROBLEMS, EDITED
BY J. T. ODEN, O. C. ZIENKIEWICZ, R. H.
GALLAGHER, AND C. TAYLOR, PAPER PRESENTED AT
THE INTERNATIONAL SYMPOSIUM ON FINITE ELEMENT
METHODS IN FLOW PROBLEMS, HELD IN JANUARY 1974, IN
SWANSEA, UNITED KINGDOM, PP. 99-100.

74-0012 OLSON, M. D.
VARIATIONAL FINITE ELEMENT METHODS FOR TWO
DIMENSIONAL AND AXISYMMETRIC NAVIER STOKES
EQUATIONS
IN FINITE ELEMENT METHOD IN FLOW PROBLEMS, EDITED
BY J. T. ODEN, O. C. ZIENKIEWICZ, R. H.
GALLAGHER, AND C. TAYLOR, PAPER PRESENTED AT
THE INTERNATIONAL SYMPOSIUM ON FINITE ELEMENT
METHODS IN FLOW PROBLEMS, HELD IN JANUARY 1974, IN
SWANSEA, UNITED KINGDOM, PP. 103-106.

74-0013 KAWAHARA, M.
YOSHIMURA, N.
NAKAGAWA, K.
ANALYSIS OF STEADY INCOMPRESSIBLE VISCOUS FLOW
IN FINITE ELEMENT METHOD IN FLOW PROBLEMS, EDITED
BY J. T. ODEN, O. C. ZIENKIEWICZ, R. H.
GALLAGHER, AND C. TAYLOR, PAPER PRESENTED AT
THE INTERNATIONAL SYMPOSIUM ON FINITE ELEMENT
METHODS IN FLOW PROBLEMS, HELD IN JANUARY 1974, IN
SWANSEA, UNITED KINGDOM, PP. 107-120.

74-0014 HOOD, P.
TAYLOR, C.
NAVIER STOKES EQUATIONS USING MIXED INTERPOLATION
IN FINITE ELEMENT METHOD IN FLOW PROBLEMS, EDITED
BY J. T. ODEN, O. C. ZIENKIEWICZ, R. H.
GALLAGHER, AND C. TAYLOR, PAPER PRESENTED AT
THE INTERNATIONAL SYMPOSIUM ON FINITE ELEMENT
METHODS IN FLOW PROBLEMS, HELD IN JANUARY 1974, IN
SWANSEA, UNITED KINGDOM, PP. 121-132.

74-0015 KING, I. P.
NORTON, W. R.
ICEMAN, K. R.
A FINITE ELEMENT MODEL FOR TWO DIMENSIONAL FLOW
IN FINITE ELEMENT METHOD IN FLOW PROBLEMS, EDITED
BY J. T. ODEN, O. C. ZIENKIEWICZ, R. H.
GALLAGHER, AND C. TAYLOR, PAPER PRESENTED AT
THE INTERNATIONAL SYMPOSIUM ON FINITE ELEMENT
METHODS IN FLOW PROBLEMS, HELD IN JANUARY 1974, IN
SWANSEA, UNITED KINGDOM, PP. 133-137.

74-0016 LASCAUX, P.
APPLICATION OF THE FINITE ELEMENT METHOD TO TWO
DIMENSIONAL LAGRANGIAN HYDRODYNAMICS
IN FINITE ELEMENT METHOD IN FLOW PROBLEMS, EDITED
BY J. T. ODEN, O. C. ZIENKIEWICZ, R. H.
GALLAGHER, AND C. TAYLOR PAPER PRESENTED AT
THE INTERNATIONAL SYMP. ON FINITE ELEMENT METHODS
IN FLOW PROBLEMS, HELD IN JANUARY 1974, IN
SWANSEA, UNITED KINGDOM, PP. 139-152.

74-0017 HIRIART, G.
SARPKAYA, T.
JET IMPINGEMENT ON AXISYMMETRIC CURVED DEFLECTORS
IN FINITE ELEMENT METHOD IN FLOW PROBLEMS, EDITED
BY J. T. ODEN, O. C. ZIENKIEWICZ, R. H.
GALLAGHER, AND C. TAYLOR, PAPER PRESENTED AT
THE INTERNATIONAL SYMPOSIUM ON FINITE ELEMENT
METHODS IN FLOW PROBLEMS, HELD IN JANUARY 1974, IN
SWANSEA, UNITED KINGDOM, PP. 153-157.

74-0018 NEWTON, R. E.
CHENAULT, D. W.
SMITH, D. A. JR.
FINITE ELEMENT SOLUTION FOR ADDED MASS AND DAMPING
IN FINITE ELEMENT METHOD IN FLOW PROBLEMS, EDITED
BY J. T. ODEN, O.C. ZIENKIEWICZ, R.H. GALLAGHER,
AND C. TAYLOR, PAPER PRESENTED AT THE INTL. SYMP.
ON FINITE ELEMENT METHODS IN FLOW PROBLEMS,
JANUARY 1974, SWANSEA, U.K., PP. 159-170.

74-0019 O'CARROLL, M. J.
MORGAN, L. A.
DIFFERENCE AND ELEMENT METHODS FOR POTENTIAL FLOW
IN FINITE ELEMENT METHOD IN FLOW PROBLEMS, EDITED
BY J. T. ODEN, O. C. ZIENKIEWICZ, R. H.
GALLAGHER, AND C. TAYLOR, PAPER PRESENTED AT
THE INTERNATIONAL SYMPOSIUM ON FINITE ELEMENT
METHODS IN FLOW PROBLEMS, HELD IN JANUARY 1974, IN
SWANSEA, UNITED KINGDOM, PP. 171-183.

74-0020 FISCHL, C. F.
FLOW BETWEEN LINES OF A GIVEN PRESSURE
IN FINITE ELEMENT METHOD IN FLOW PROBLEMS, EDITED
BY J. T. ODEN, O. C. ZIENKIEWICZ, R. H.
GALLAGHER, AND C. TAYLOR, PAPER PRESENTED AT
THE INTERNATIONAL SYMPOSIUM ON FINITE ELEMENT
METHODS IN FLOW PROBLEMS, HELD IN JANUARY 1974, IN
SWANSEA, UNITED KINGDOM, PP. 185-186.

74-0021 SHEN, S. F.
AN AERODYNAMICIST LOOKS AT THE FINITE ELEMENT
METHOD
IN FINITE ELEMENT METHOD IN FLOW PROBLEMS, EDITED
BY J. T. ODEN, O. C. ZIENKIEWICZ, R. H.
GALLAGHER, AND C. TAYLOR, PAPER PRESENTED AT
THE INTERNATIONAL SYMPOSIUM ON FINITE ELEMENT
METHODS IN FLOW PROBLEMS, HELD IN JANUARY 1974, IN
SWANSEA, UNITED KINGDOM, PP. 189-193.

74-0022 WASHIZU, K.
IKEGAWA, M.
FINITE ELEMENT TECHNIQUE IN LIFTING SURFACE
PROBLEMS
IN FINITE ELEMENT METHOD IN FLOW PROBLEMS, EDITED
BY J. T. ODEN, O. C. ZIENKIEWICZ, R. H.
GALLAGHER, AND C. TAYLOR, PAPER PRESENTED AT
THE INTERNATIONAL SYMPOSIUM ON FINITE ELEMENT
METHODS IN FLOW PROBLEMS, HELD IN JANUARY 1974, IN
SWNASEA, UNITED KINGDOM, PP. 195-207.

74-0023 AWBI, H. B.
SWANNELL, J. H.
A FINITE ELEMENT SOLUTION OF THE CONFINED FLOW
OVER A CIRCULAR CYLINDER
IN FINITE ELEMENT METHOD IN FLOW PROBLEMS, EDITED
BY J. T. ODEN, O. C. ZIENKIEWICZ, R. H.
GALLAGHER, AND C. TAYLOR, PAPER PRESENTED AT
THE INTERNATIONAL SYMPOSIUM ON FINITE ELEMENT
METHODS IN FLOW PROBLEMS, HELD IN JANUARY 1974, IN
SWANSEA, UNITED KINGDOM, PP. 209-224.

74-0024 BRATANOW, T.
EVER, A.
ANALYSIS OF MOVING BODY PROBLEMS IN AERODYNAMICS
IN FINITE ELEMENT METHOD IN FLOW PROBLEMS, EDITED
BY J. T. ODEN, O. C. ZIENKIEWICZ, R. H.
GALLAGHER, AND C. TAYLOR, PAPER PRESENTED AT
THE INTERNATIONAL SYMPOSIUM ON FINITE ELEMENT
METHODS IN FLOW PROBLEMS, HELD IN JANUARY 1974, IN
SWANSEA, UNITED KINGDOM, PP. 225-242.

74-0025 CARCY, G. F.
PERTURBATION VARIATIONAL SOLUTION OF COMPRESSIBLE
FLOWS
IN FINITE ELEMENT METHOD IN FLOW PROBLEMS, EDITED
BY J. T. ODEN, O. C. ZIENKIEWICZ, R. H.
GALLAGHER, AND C. TAYLOR, PAPER PRESENTED AT
THE INTERNATIONAL SYMPOSIUM ON FINITE ELEMENT
METHODS IN FLOW PROBLEMS, HELD IN JANUARY 1974, IN
SWANSEA, UNITED KINGDOM, PP. 243-248.

74-0026 PERRIAUX, J.
TWO AND THREE DIMENSIONAL ANALYSIS OF SUBSONIC
COMPRESSIBLE POTENTIAL FLOWS IN DUCTS AND AROUND
LIFTING BODIES WITH FINITE ELEMENT TECHNIQUES
IN FINITE ELEMENT METHOD IN FLOW PROBLEMS, EDITED
BY J. T. ODEN, O. C. ZIENKIEWICZ, R. H.
GALLAGHER, AND C. TAYLOR PAPER PRESENTED AT
THE INTERNATIONAL SYMPOSIUM ON FINITE ELEMENT
METHODS IN FLOW PROBLEMS, HELD IN JANUARY 1974 IN
SWANSEA, UNITED KINGDOM, PP. 249-250.

74-0027 CHUNG, T. J.
ODEN, J. T.
WU, S. T.
THE FINITE ELEMENT ANALYSIS OF TRANSIENT RAREFIED
GAS FLOW
IN FINITE ELEMENT METHOD IN FLOW PROBLEMS, EDITED
BY J. T. ODEN, O. C. ZIENKIEWICZ, R. H.
GALLAGHER, AND C. TAYLOR, PAPER PRESENTED AT
THE INTERNATIONAL SYMPOSIUM ON FINITE ELEMENT
METHODS IN FLOW PROBLEMS, HELD IN JANUARY 1974, IN
SWANSEA, UNITED KINGDOM, PP. 251-265.

74-0028 CALZOLARI, P. U.
GRAFFI, S.
PIERINI, G.
FINITE DIFFERENCE AND FINITE ELEMENT METHODS FOR
THE SEMICONDUCTOR FLOW EQUATIONS
IN FINITE ELEMENT METHOD IN FLOW PROBLEMS, EDITED
BY J. T. ODEN, O. C. ZIENKIEWICZ, R. H.
GALLAGHER, AND C. TAYLOR PAPER PRESENTED AT
THE INTERNATIONAL SYMPOSIUM ON FINITE ELEMENT
METHODS IN FLOW PROBLEMS, HELD IN JANUARY 1974 IN
SWANSEA, UNITED KINGDOM, PP. 267-268

74-0029 MILNE, R. D.
APPLICATION OF INTEGRAL EQUATIONS TO FLUID FLOWS
IN UNBOUNDED REGIONS
IN FINITE ELEMENT METHOD IN FLOW PROBLEMS, EDITED
BY J. T. ODEN, O. C. ZIENKIEWICZ, R. H.
GALLAGHER, AND C. TAYLOR PAPER PRESENTED AT
THE INTERNATIONAL SYMPOSIUM ON FINITE ELEMENT
METHODS IN FLOW PROBLEMS, HELD IN JANUARY 1974 IN
SWANSEA, UNITED KINGDOM, PP. 269-273

74-0030 VISSER, W.
VAN DER WILT, M.
A NUMERICAL APPROACH TO THE STUDY OF IRREGULAR
SHIP MOTIONS
IN FINITE ELEMENT METHOD IN FLOW PROBLEMS, EDITED
BY J. T. ODEN, O. C. ZIENKIEWICZ, R. H.
GALLAGHER, AND C. TAYLOR PAPER PRESENTED AT
THE INTERNATIONAL SYMPOSIUM ON FINITE ELEMENT
METHODS IN FLOW PROBLEMS, HELD IN JANUARY 1974 IN
SWANSEA, UNITED KINGDOM, PP. 277-281

74-0031 ALLOUARD, Y.
COUDERT, J. F.
NUMERICAL STUDY OF TRANSITORY OR NONLINEAR WAVES
IN FINITE ELEMENT METHOD IN FLOW PROBLEMS, EDITED
BY J. T. ODEN, O. C. ZIENKIEWICZ, R. H.
GALLAGHER, AND C. TAYLOR PAPER PRESENTED AT
THE INTERNATIONAL SYMPOSIUM ON FINITE ELEMENT
METHODS IN FLOW PROBLEMS, HELD IN JANUARY 1974 IN
SWANSEA, UNITED KINGDOM, PP. 283-288.

74-0032 BERKHOFF, J. C. W.
COMPUTATION OF WAVE PROPAGATION BY MEANS OF THE
FINITE ELEMENT METHOD
IN FINITE ELEMENT METHOD IN FLOW PROBLEMS, EDITED
BY J. T. ODEN, O. C. ZIENKIEWICZ, R. H.
GALLAGHER, AND C. TAYLOR PAPER PRESENTED AT
THE INTERNATIONAL SYMPOSIUM ON FINITE ELEMENT
METHODS IN FLOW PROBLEMS, HELD IN JANUARY 1974 AT
SWANSEA, UNITED KINGDOM, PP. 289-293.

74-0033 SCHULZE, K. W.
FINITE ELEMENT ANALYSIS OF LONG WAVES IN OPEN
CHANNEL SYSTEMS
IN FINITE ELEMENT METHOD IN FLOW PROBLEMS, EDITED
BY J. T. ODEN, O. C. ZIENKIEWICZ, R. H.
GALLAGHER, AND C. TAYLOR PAPER PRESENTED AT
THE INTERNATIONAL SYMPOSIUM ON FINITE ELEMENT
METHODS IN FLOW PROBLEMS, HELD IN JANUARY 1974 IN
SWANSEA, UNITED KINGDOM, PP. 295-303.

74-0034 TAYLOR, R. L.
DASGUPTA, G.
KHALIL, T. B.
WAVE PROPAGATION IN FLUIDS
IN "FINITE ELEMENT METHOD IN FLOW PROBLEMS", ODEN,
J.T., ZIENKIEWICZ, O.C., GALLAGHER, R.H., TAYLOR,
C., (EDS.), PAPER PRESENTED AT THE INTL. SYMP. ON
FINITE ELEMENT METH. IN FLOW PROBLEMS, SWANSEA,
U.K., JANUARY 1974, PP. 309-313.

74-0035 BOYD, T. J.
GARDNER, G. A.
GARDNER, L. R. T.
STUDIES OF HYDROMAGNETIC STABILITY
IN FINITE ELEMENT METHOD IN FLOW PROBLEMS, EDITED
BY J. T. ODEN, O. C. ZIENKIEWICZ, R. H.
GALLAGHER, AND C. TAYLOR, PAPER PRESENTED AT
THE INTERNATIONAL SYMPOSIUM ON FINITE ELEMENT
METHODS IN FLOW PROBLEMS, HELD IN JANUARY 1974, IN
SWANSEA, UNITED KINGDOM, PP. 309-313.

74-0036 WHITEMAN, J. R.
NUMERICAL SOLUTION OF STEADY STATE DIFFUSION
PROBLEMS CONTAINING SINGULARITIES
IN FINITE ELEMENT METHOD IN FLOW PROBLEMS, EDITED
BY J. T. ODEN, O. C. ZIENKIEWICZ, R. H.
GALLAGHER, AND C. TAYLOR PAPER PRESENTED AT
THE INTERNATIONAL SYMPOSIUM ON FINITE ELEMENT
METHODS IN FLOW PROBLEMS, HELD IN JANUARY 1974 IN
SWANSEA, UNITED KINGDOM, PP. 317-321.

74-0037 APOSTOL, R. T.
CHARLTON, J. A.
SOLUTION OF A TWO DIMENSIONAL DIFFUSION CONVECTION
EQUATION DESCRIBING THE MASS TRANSFER OF A
NEUTRALLY BUOYANT TRACER
IN FINITE ELEMENT METHOD IN FLOW PROBLEMS, EDITED
BY J. T. ODEN, O. C. ZIENKIEWICZ, R. H.
GALLAGHER, AND C. TAYLOR PAPER PRESENTED AT
THE INTERNATIONAL SYMPOSIUM ON FINITE ELEMENT
METHODS IN FLOW PROBLEMS, HELD IN JANUARY 1974 IN
SWANSEA, UNITED KINGDOM, PP. 323-332.

74-0038 USUKI, S.
FINITE ELEMENT METHOD VIA LOCAL POTENTIAL
IN FINITE ELEMENT METHOD IN FLOW PROBLEMS, EDITED
BY J. T. ODEN, O. C. ZIENKIEWICZ, R. H.
GALLAGHER, AND C. TAYLOR, PAPER PRESENTED AT
THE INTERNATIONAL SYMPOSIUM ON FINITE ELEMENT
METHODS IN FLOW PROBLEMS, HELD IN JANUARY 1974 IN
SWANSEA, UNITED KINGDOM, PP. 333-335.

74-0039 BARROW, H.
GILBERT, D. E.
MISTRY, J.
THE FINITE ELEMENT SOLUTION OF FLOW AND HEAT
TRANSFER IN ELLIPTICAL DUCTS
IN FINITE ELEMENT METHOD IN FLOW PROBLEMS, EDITED
BY J. T. ODEN, O. C. ZIENKIEWICZ, R. H.
GALLAGHER, AND C. TAYLOR, PAPER PRESENTED AT
THE INTERNATIONAL SYMPOSIUM ON FINITE ELEMENT
METHODS IN FLOW PROBLEMS, HELD IN JANUARY 1974 IN
SWANSEA, UNITED KINGDOM, PP. 337-338.

74-0040 MANGARELLA, P. A.
KNAPP, R.
NUMERICAL ANALYSIS OF COOLING TOWER PLUMES AND FOG
INCIDENCE
IN FINITE ELEMENT METHOD IN FLOW PROBLEMS, EDITED
BY J. T. ODEN, O. C. ZIENKIEWICZ, R. H.
GALLAGHER, AND C. TAYLOR, PAPER PRESENTED AT
THE INTERNATIONAL SYMPOSIUM ON FINITE ELEMENT
METHODS IN FLOW PROBLEMS, HELD IN JANUARY 1974 IN
SWANSEA, UNITED KINGDOM, PP. 339-340.

74-0041 HEUBNER, K. H.
FINITE ELEMENT ANALYSIS OF FLUID FILM LUBRICATION
- A SURVEY
IN FINITE ELEMENT METHOD IN FLOW PROBLEMS, EDITED
BY J.T. ODEN, O.C. ZIENKIEWICZ, R.H. GALLAGHER,
AND C. TAYLOR, PAPER PRESENTED AT THE
INTERNATIONAL SYMPOSIUM ON FINITE ELEMENT METHODS
IN FLOW PROBLEMS, HELD IN JANUARY 1974 IN SWANSEA,
UNITED KINGDOM, PP. 343-350.

74-0042 BARWELL, F. T.
THE ROLE OF NUMERICAL METHODS IN THE STUDY OF
LUBRICATION
IN FINITE ELEMENT METHOD IN FLOW PROBLEMS, EDITED
BY J. T. ODEN, O. C. ZIENKIEWICZ, R. H.
GALLAGHER, AND C. TAYLOR, PAPER PRESENTED AT
THE INTERNATIONAL SYMPOSIUM ON FINITE ELEMENT
METHODS IN FLOW PROBLEMS, HELD IN JANUARY 1974 IN
SWANSEA, UNITED KINGDOM, PP. 351-358.

74-0043 GALLAGHER, R. H.
FINITE ELEMENT LAKE CIRCULATION AND THERMAL
ANALYSIS
IN FINITE ELEMENT METHOD IN FLOW PROBLEMS, EDITED
BY J. T. ODEN, O. C. ZIENKIEWICZ, R. H.
GALLAGHER, AND C. TAYLOR, PAPER PRESENTED AT
THE INTERNATIONAL SYMPOSIUM ON FINITE ELEMENT
METHODS IN FLOW PROBLEMS, HELD IN JANUARY 1974 IN
SWANSEA, UNITED KINGDOM, PP. 361-369.

74-0044 TAYLOR, C.
DAVIS, J.
A FINITE ELEMENT MODEL OF TIDES IN ESTUARIES
IN FINITE ELEMENT METHOD IN FLOW PROBLEMS, EDITED
BY J. T. ODEN, O. C. ZIENKIEWICZ, R. H.
GALLAGHER, AND C. TAYLOR, PAPER PRESENTED AT
THE INTERNATIONAL SYMPOSIUM ON FINITE ELEMENT
METHODS IN FLOW PROBLEMS, HELD IN JANUARY 1974 IN
SWANSEA, UNITED KINGDOM, PP. 371-377.

74-0045 TAYLOR, C.
DAVIS, J.
A NUMERICAL MODEL OF DISPERSION IN ESTUARIES
IN FINITE ELEMENT METHOD IN FLOW PROBLEMS, EDITED
BY J. T. ODEN, O. C. ZIENKIEWICZ, R. H.
GALLAGHER, AND C. TAYLOR, PAPER PRESENTED AT
THE INTERNATIONAL SYMPOSIUM ON FINITE ELEMENT
METHODS IN FLOW PROBLEMS, HELD IN JANUARY 1974 IN
SWANSEA, UNITED KINGDOM, PP. 378-384.

74-0046 AL-MASHIDANI, G.
TAYLOR, C.
FINITE ELEMENT SOLUTIONS OF THE SHALLOW WATER
EQUATIONS SURFACE RUN OFF
IN FINITE ELEMENT METHOD IN FLOW PROBLEMS, EDITED
BY J. T. ODEN, O. C. ZIENKIEWICZ, R. G.
GALLAGHER, AND C. TAYLOR, PAPER PRESENTED AT THE
INTERNATIONAL SYMPOSIUM ON FINITE ELEMENT METHODS
IN FLOW PROBLEMS, HELD IN JANUARY 1974 IN SWANSEA,
UNITED KINGDOM, PP. 385-398.

74-0047 FARRADAY, R. V.
O'CONNOR, B. A.
SMITH, I. M.
GALERKIN FINITE ELEMENT SOLUTIONS FOR POLLUTION
PROBLEMS IN PARTIALLY MIXED ESTUARIES
IN FINITE ELEMENT METHOD IN FLOW PROBLEMS, EDITED
BY J. T. ODEN, O. C. ZIENKIEWICZ, R. H.
GALLAGHER, AND C. TAYLOR, PAPER PRESENTED AT
THE INTERNATIONAL SYMPOSIUM ON FINITE ELEMENT
METHODS IN FLOW PROBLEMS, HELD IN JANUARY 1974 IN
SWANSEA, UNITED KINGDOM, PP. 399-400.

74-0048 MERCER, J. W.
PINDER, G. F.
FINITE ELEMENT ANALYSIS OF HYDROTHERMAL SYSTEMS
IN FINITE ELEMENT METHOD IN FLOW PROBLEMS, EDITED
BY J. T. ODEN, O. C. ZIENKIEWICZ, R. H.
GALLAGHER, AND C. TAYLOR, PAPER PRESENTED AT
THE INTERNATIONAL SYMPOSIUM ON FINITE ELEMENT
METHODS IN FLOW PROBLEMS, HELD IN JANUARY 1974 IN
SWANSEA, UNITED KINGDOM, PP. 401-414.

74-0049 THIRRIOT, C.
GAUDU, R.
DARBOIS, F.
SALOMON, D.
USE OF FINITE ELEMENTS IN THE ANALYSIS OF A
HYDROTHERMAL DOUBLET
IN FINITE ELEMENT METHOD IN FLOW PROBLEMS, EDITED
BY J. T. ODEN, O. C. ZIENKIEWICZ, R. H.
GALLAGHER, AND C. TAYLOR, PAPER PRESENTED AT
THE INTERNATIONAL SYMPOSIUM ON FINITE ELEMENT
METHODS IN FLOW PROBLEMS, HELD IN JANUARY 1974 IN
SWANSEA, UNITED KINGDOM, PP. 415-426.

74-0050 HSU, M. B.
NICKELL, R. E.
COUPLED CONVECTIVE AND CONDUCTIVE HEAT TRANSFER BY
FINITE ELEMENT METHODS
IN FINITE ELEMENT METHOD IN FLOW PROBLEMS, EDITED
BY J. T. ODEN, O. C. ZIENKIEWICZ, R. H.
GALLAGHER, AND C. TAYLOR, PAPER PRESENTED AT
THE INTERNATIONAL SYMPOSIUM ON FINITE ELEMENT
METHODS IN FLOW PROBLEMS, HELD IN JANUARY 1974 IN
SWANSEA, UNITED KINGDOM, PP. 427-450.

74-0051 COLE, J. A.
DAVIS, J. M.
IDENTIFICATION OF AQUIFER CONSTANTS: TESTS OF TWO
INVERSE METHODS
IN FINITE ELEMENT METHOD IN FLOW PROBLEMS, EDITED
BY J. T. ODEN, O. C. ZIENKIEWICZ, R. H.
GALLAGHER, AND C. TAYLOR, PAPER PRESENTED AT
THE INTERNATIONAL SYMPOSIUM ON FINITE ELEMENT
METHODS IN FLOW PROBLEMS, HELD IN JANUARY 1974 IN
SWANSEA, UNITED KINGDOM, PP. 451-453.

74-0052 NARAYANAN, J.
SHANKAR, N. J.
A NUMERICAL MODEL FOR THE SIMULATION OF TWO
DIMENSIONAL CONVECTIVE DISPERSION IN SHALLOW
ESTUARIES
IN FINITE ELEMENT METHOD IN FLOW PROBLEMS, EDITED
BY J. T. ODEN, O. C. ZIENKIEWICZ, R. H.
GALLAGHER, AND C. TAYLOR, PAPER PRESENTED AT
THE INTERNATIONAL SYMPOSIUM ON FINITE ELEMENT
METHODS IN FLOW PROBLEMS, HELD IN JANUARY 1974 IN
SWANSEA, UNITED KINGDOM, PP. 455-458.

74-0053 ZIENKIEWICZ, O. C.
GODBOLE, A. N.
VISCOUS, INCOMPRESSIBLE FLOW WITH SPECIAL
REFERENCE TO NON-NEWTONIAN (PLASTIC) FLUIDS
IN FINITE ELEMENT METHOD IN FLOW PROBLEMS, EDITED
BY J. T. ODEN, O. C. ZIENKIEWICZ, R. H.
GALLAGHER, AND C. TAYLOR, PAPER PRESENTED AT
THE INTERNATIONAL SYMPOSIUM ON FINITE ELEMENT
METHODS IN FLOW PROBLEMS, HELD IN JANUARY 1974 IN
SWANSEA, UNITED KINGDOM, PP. 461-464.

74-0054 YAMADA, Y.
ITO, K.
YOKOUCHI, Y.
TAMANO, T.
OHISUBO, T.
FINITE ELEMENT ANALYSIS OF STEADY FLUID AND METAL
FLOW
IN FINITE ELEMENT METHOD IN FLOW PROBLEMS, EDITED
BY J. T. ODEN, O. C. ZIENKIEWICZ, R. H. GALLAGHER,
AND C. TAYLOR, PAPER PRESENTED AT THE
INTERNATIONAL SYMPOSIUM ON FINITE ELEMENT METHODS
IN FLOW PROBLEMS, JANUARY 1974, SWANSEA, U.K., PP.
465-469.

74-0055 BRISTEAU, M.
GLOWINSKI, R.
FINITE ELEMENT ANALYSIS OF THE UNSTEADY FLOW OF A
VISCOUS PLASTIC FLUID IN A CYLINDRICAL PIPE
IN FINITE ELEMENT METHOD IN FLOW PROBLEMS, EDITED
BY J. T. ODEN, O. C. ZIENKIEWICZ, R. H.
GALLAGHER, AND C. TAYLOR, PAPER PRESENTED AT THE
INTERNATIONAL SYMPOSIUM ON FINITE ELEMENT METHODS
IN FLOW PROBLEMS, HELD IN JANUARY 1974 IN SWANSEA,
UNITED KINGDOM, PP. 471-488.

74-0056 OWEN, D. R. J.
LYNESS, J. F.
ZIENKIEWICZ, O. C.
FINITE ELEMENT ANALYSIS OF THE STEADY FLOW OF NON
NEWTONIAN FLUIDS THROUGH PARALLEL SIDED CONDUITS
IN FINITE ELEMENT METHOD IN FLOW PROBLEMS, EDITED
BY J. T. ODEN, O. C. ZIENKIEWICZ, R. H.
GALLAGHER, AND C. TAYLOR, PAPER PRESENTED AT
THE INTERNATIONAL SYMPOSIUM ON FINITE ELEMENT
METHODS IN FLOW PROBLEMS, HELD IN JANUARY 1974 IN
SWANSEA, UNITED KINGDOM, PP. 489-503.

74-0057 DESAI, C. S.
FINITE ELEMENT METHODS FOR FLOW IN POROUS MEDIA
IN FINITE ELEMENT METHOD IN FLOW PROBLEMS, EDITED
BY J. T. ODEN, O. C. ZIENKIEWICZ, R. H.
GALLAGHER, AND C. TAYLOR, PAPER PRESENTED AT
THE INTERNATIONAL SYMPOSIUM ON FINITE ELEMENT
METHODS IN FLOW PROBLEMS, HELD IN JANUARY 1974 IN
SWANSEA, UNITED KINGDOM, PP. 511-515.

74-0058 NEUMAN, S. P.
GALERKIN APPROACH TO UNSATURATED FLOW IN SOILS
IN FINITE ELEMENT METHOD IN FLOW PROBLEMS, EDITED
BY J. T. ODEN, O. C. ZIENKIEWICZ, R. H.
GALLAGHER, AND C. TAYLOR, PAPER PRESENTED AT
THE INTERNATIONAL SYMPOSIUM ON FINITE ELEMENT
METHODS IN FLOW PROBLEMS, HELD IN JANUARY 1974 IN
SWANSEA, UNITED KINGDOM, PP. 517-522.

74-0059 CHOWDHURY, R. N.
GENERALIZED STEADY STATE FIELD PROBLEM AND ITS
SOLUTION
IN FINITE ELEMENT METHOD IN FLOW PROBLEMS, EDITED
BY J. T. ODEN, O. C. ZIENKIEWICZ, R. H.
GALLAGHER, AND C. TAYLOR, PAPER PRESENTED AT
THE INTERNATIONAL SYMPOSIUM ON FINITE ELEMENT
METHODS IN FLOW PROBLEMS, HELD IN JANUARY 1974 IN
SWANSEA, UNITED KINGDOM, PP. 523-534.

74-0060 VERNER, E. A.
LEWIS, R. W.
ZIENKIEWICZ, O. C.
FINITE ELEMENTS FOR TWO PHASE FLOW IN POROUS MEDIA
IN FINITE ELEMENT METHOD IN FLOW PROBLEMS, EDITED
BY J. T. ODEN, O. C. ZIENKIEWICZ, R. H.
GALLAGHER, AND C. TAYLOR, PAPER PRESENTED AT
THE INTERNATIONAL SYMPOSIUM ON FINITE ELEMENT
METHODS IN FLOW PROBLEMS, HELD IN JANUARY 1974 IN
SWANSEA, UNITED KINGDOM, PP. 535-540.

74-0061 MAYER, P. G.
SMITH, C. V.
THE FATE OF RADIONUCLIDES IN A GROUNDWATER
ENVIRONMENT
IN FINITE ELEMENT METHOD IN FLOW PROBLEMS, EDITED
BY J. T. ODEN, O. C. ZIENKIEWICZ, R. H.
GALLAGHER, AND C. TAYLOR, PAPER PRESENTED AT
THE INTERNATIONAL SYMPOSIUM ON FINITE ELEMENT
METHODS IN FLOW PROBLEMS, HELD IN JANUARY 1974 IN
SWANSEA, UNITED KINGDOM, PP. 541-544.

74-0062 MCCORQUODALE, J. A.
NASSER, M. S.
NUMERICAL METHODS FOR UNSTEADY NON-DARCY FLOW
IN FINITE ELEMENT METHOD IN FLOW PROBLEMS, EDITED
BY J. T. ODEN, O. C. ZIENKIEWICZ, R. H.
GALLAGHER, AND C. TAYLOR, PAPER PRESENTED AT
THE INTERNATIONAL SYMPOSIUM ON FINITE ELEMENT
METHODS IN FLOW PROBLEMS, HELD IN JANUARY 1974 IN
SWANSEA, UNITED KINGDOM, PP. 545-557.

74-0063 FRIND, E. O.
PINDER, G. F.
FINITE ELEMENTS IN THE SOLUTION OF THE INVERSE
PROBLEM IN GROUNDWATER FLOW
IN FINITE ELEMENT METHOD IN FLOW PROBLEMS, EDITED
BY J. T. ODEN, O. C. ZIENKIEWICZ, R. H.
GALLAGHER, AND C. TAYLOR, PAPER PRESENTED AT
THE INTERNATIONAL SYMPOSIUM ON FINITE ELEMENT
METHODS IN FLOW PROBLEMS, HELD IN JANUARY 1974 IN
SWANSEA, UNITED KINGDOM, PP. 559-572.

74-0064 SANDHU, R. S.
VARIABLE TIME STEP ANALYSIS OF UNCONFINED SEEPAGE
IN FINITE ELEMENT METHOD IN FLOW PROBLEMS. EDITED
BY J. T. ODEN, O. C. ZIENKIEWICZ, R. H.
GALLAGHER, AND C. TAYLOR, PAPER PRESENTED AT
THE INTERNATIONAL SYMPOSIUM ON FINITE ELEMENT
METHODS IN FLOW PROBLEMS, HELD IN JANUARY 1974 IN
SWANSEA, UNITED KINGDOM, PP. 573-579.

74-0065 GALE, J. E.
TAYLOR, R. L.
WITHERSPOON, P. A.
AYATOLLAHI, M. S.
FLOW IN ROCKS WITH DEFORMABLE FRACTURES
IN FINITE ELEMENT METHOD IN FLOW PROBLEMS. EDITED
BY J. T. ODEN, O. C. ZIENKIEWICZ, R. H.
GALLAGHER, AND C. TAYLOR, PAPER PRESENTED AT
THE INTERNATIONAL SYMPOSIUM ON FINITE ELEMENT
METHODS IN FLOW PROBLEMS, HELD IN JANUARY 1974 IN
SWANSEA, UNITED KINGDOM, PP. 583-598.

74-0066 DUGUID, J. O.
ABEL, J. F.
FINITE ELEMENT GALERKIN METHOD FOR ANALYSIS OF
FLOW IN FRACTURED POROUS MEDIA
IN FINITE ELEMENT METHOD IN FLOW PROBLEMS, J. T.
ODEN, O. C. ZIENKIEWICZ, R. H. GALLAGHER, AND C.
TAYLOR, (EDS.), PAPER PRESENTED AT THE INTL. SYMP.
ON FINITE ELEMENT METH. IN FLOW PROBLEMS, SWANSEA,
JANUARY 1974, PP. 599-615.

74-0067 MEISSNER, U.
GENERALIZED VARIATIONAL PRINCIPLES FOR USE IN FLOW
PROBLEMS THROUGH POROUS MEDIA
IN FINITE ELEMENT METHOD IN FLOW PROBLEMS. EDITED
BY J. T. ODEN, O. C. ZIENKIEWICZ, R. H.
GALLAGHER, AND C. TAYLOR, PAPER PRESENTED AT
THE INTERNATIONAL SYMPOSIUM ON FINITE ELEMENT
METHODS IN FLOW PROBLEMS, HELD IN JANUARY 1974 IN
SWANSEA, UNITED KINGDOM, PP. 617-626.

74-0068 THIRRIOT, C.
JEAN, C.
USE OF HYBRID COMPUTATION FOR SOLVING TRANSIENT
FLOW PROBLEM IN POROUS MEDIA BY FINITE ELEMENT
METHODS
IN FINITE ELEMENT METHOD IN FLOW PROBLEMS. EDITED
BY J. T. ODEN, O. C. ZIENKIEWICZ, R. H.
GALLAGHER, AND C. TAYLOR, PAPER PRESENTED AT THE
INTERNATIONAL SYMPOSIUM ON FINITE ELEMENT METHODS
IN FLOW PROBLEMS, HELD IN JANUARY 1974 IN SWANSEA,
UNITED KINGDOM, PP. 617-626.

74-0069 TROSCH, J.
TRANSIENT TWO DIMENSIONAL UNCONFINED AQUIFER
PROBLEMS
IN FINITE ELEMENT METHOD IN FLOW PROBLEMS. EDITED
BY J. T. ODEN, O. C. ZIENKIEWICZ, R. H.
GALLAGHER, AND C. TAYLOR, PAPER PRESENTED AT
THE INTERNATIONAL SYMPOSIUM ON FINITE ELEMENT
METHODS IN FLOW PROBLEMS, HELD IN JANUARY 1974 IN
SWANSEA, UNITED KINGDOM, PP. 633-640.

74-0070 CHAN, H. T.
KENNEY, T. C.
TEST OF VALIDITY OF DARCY'S LAW FOR A CLAY SOIL
IN FINITE ELEMENT METHOD IN FLOW PROBLEMS. EDITED
BY J. T. ODEN, O. C. ZIENKIEWICZ, R. H.
GALLAGHER, AND C. TAYLOR, PAPER PRESENTED AT
THE INTERNATIONAL SYMPOSIUM ON FINITE ELEMENT
METHODS IN FLOW PROBLEMS, HELD IN JANUARY 1974 IN
SWANSEA, UNITED KINGDOM, PP. 641-652.

74-0071 LEFEBVRE DU PREY, E. J.
WEILL, L. M.
FRONT DISPLACEMENT MODEL IN A FRACTURED RESERVOIR
IN FINITE ELEMENT METHOD IN FLOW PROBLEMS. EDITED
BY J. T. ODEN, O. C. ZIENKIEWICZ, R. H. GALLAGHER,
AND C. TAYLOR, PAPER PRESENTED AT THE
INTERNATIONAL SYMPOSIUM ON FINITE ELEMENT METHODS
IN FLOW PROBLEMS, JANUARY 1974, SWANSEA, U.K. PP.
653-671.

74-0072 BRUCH, J. C. JR.
AL-MASHIDANI, G.
TAYLOR, C.
DISPERSION IN STEADY UNSTEADY SEEPAGE FLOWFIELD
IN FINITE ELEMENT METHOD IN FLOW PROBLEMS. EDITED
BY J. T. ODEN, O. C. ZIENKIEWICZ, R. H. GALLAGHER,
AND C. TAYLOR, PAPER PRESENTED AT THE
INTERNATIONAL SYMPOSIUM ON FINITE ELEMENT METHODS
IN FLOW PROBLEMS, JANUARY 1974, SWANSEA, U.K., PP.
673-676.

74-0073 GRANEY, L.
SMITH, I. M.
WALSH, J. E.
RESPONSE OF UNCONFINED AQUIFERS TO PUMPING
IN FINITE ELEMENT METHOD IN FLOW PROBLEMS. EDITED
BY J. T. ODEN, O. C. ZIENKIEWICZ, R. H.
GALLAGHER, AND C. TAYLOR, PAPER PRESENTED AT
THE INTERNATIONAL SYMPOSIUM ON FINITE ELEMENT
METHODS IN FLOW PROBLEMS, HELD IN JANUARY 1974 IN
SWANSEA, UNITED KINGDOM, PP. 677-692.

74-0074 LUIR, H. P.
ORGANIZATION TECHNIQUES OF THE FINITE ELEMENT
METHOD FOR REDUCING CORE STORAGE REQUIREMENTS
IN FINITE ELEMENT METHOD IN FLOW PROBLEMS. EDITED
BY J. T. ODEN, O. C. ZIENKIEWICZ, R. H.
GALLAGHER, AND C. TAYLOR, PAPER PRESENTED AT
THE INTERNATIONAL SYMPOSIUM ON FINITE ELEMENT
METHODS IN FLOW PROBLEMS, HELD IN JANUARY 1974 IN
SWANSEA, UNITED KINGDOM, PP. 695-697.

74-0075 BEPLAMONT, J.
THE PROGRAM " ASSEN " FOR STEADY FREE SURFACE
FLOWS
IN FINITE ELEMENT METHOD IN FLOW PROBLEMS. EDITED
BY J. T. ODEN, O. C. ZIENKIEWICZ, R. H.
GALLAGHER, AND C. TAYLOR, PAPER PRESENTED AT
THE INTERNATIONAL SYMPOSIUM ON FINITE ELEMENT
METHODS IN FLOW PROBLEMS, HELD IN JANUARY 1974 IN
SWANSEA, UNITED KINGDOM, PP. 699-700.

74-0076 PERKINS, H. J.
HORLOCK, J. H.
COMPUTATION OF FLOWS IN TURBOMACHINES
IN FINITE ELEMENT METHOD IN FLOW PROBLEMS. EDITED
BY J. T. ODEN, O. C. ZIENKIEWICZ, R. H.
GALLAGHER, AND C. TAYLOR, PAPER PRESENTED AT
THE INTERNATIONAL SYMPOSIUM ON FINITE ELEMENT
METHODS IN FLOW PROBLEMS, HELD IN JANUARY 1974 IN
SWANSEA, UNITED KINGDOM, PP. 701-706.

74-0077 THOMPSON, D. S.
FLOW THROUGH A CASCADE OF AEROFOILS
IN FINITE ELEMENT METHOD IN FLOW PROBLEMS. EDITED
BY J. T. ODEN, O. C. ZIENKIEWICZ, R. H.
GALLAGHER, AND C. TAYLOR, PAPER PRESENTED AT
THE INTERNATIONAL SYMPOSIUM ON FINITE ELEMENT
METHODS IN FLOW PROBLEMS, HELD IN JANUARY 1974 IN
SWANSEA, UNITED KINGDOM, PP. 707-720.

74-0078 KADEN, ST.
LUCKNER, L.
QUAST, J.
APPLICATION OF FINITE ELEMENT METHODS IN
GEOHYDRAULICS
IN FINITE ELEMENT METHOD IN FLOW PROBLEMS. EDITED
BY J. T. ODEN, O. C. ZIENKIEWICZ, R. H. GALLAGHER,
AND C. TAYLOR, PAPER PRESENTED AT THE
INTERNATIONAL SYMPOSIUM ON FINITE ELEMENT METHODS
IN FLOW PROBLEMS, JANUARY 1974, SWANSEA, U.K., PP.
721-735.

74-0079 IKEGAWA, M.
FINITE ELEMENT ANALYSIS OF FLUID MOTION IN A
CONTAINER
IN FINITE ELEMENT METHOD IN FLOW PROBLEMS. EDITED
BY J. T. ODEN, O. C. ZIENKIEWICZ, R. H.
GALLAGHER, AND C. TAYLOR, PAPER PRESENTED AT
THE INTERNATIONAL SYMPOSIUM ON FINITE ELEMENT
METHODS IN FLOW PROBLEMS, HELD IN JANUARY 1974 IN
SWANSEA, UNITED KINGDOM, PP. 737-738.

74-0080 HAREN, R.
VALLIAPPAN, S.
FINITE ELEMENT ANALYSIS OF ELECTROMAGNETIC FIELD
PROBLEMS
IN FINITE ELEMENT METHOD IN FLOW PROBLEMS. EDITED
BY J. T. ODEN, O. C. ZIENKIEWICZ, R. H.
GALLAGHER, AND C. TAYLOR, PAPER PRESENTED AT
THE INTERNATIONAL SYMPOSIUM ON FINITE ELEMENT
METHODS IN FLOW PROBLEMS, HELD IN JANUARY 1974 IN
SWANSEA, UNITED KINGDOM, PP. 739-740.

74-0081 VALLIAPPAN, S.
LEE, I. K.
BOONLUALOHR, P.
FINITE ELEMENT ANALYSIS OF CONSOLIDATION PROBLEM
IN FINITE ELEMENT METHOD IN FLOW PROBLEMS. EDITED
BY J. T. ODEN, O. C. ZIENKIEWICZ, R. H.
GALLAGHER, AND C. TAYLOR, PAPER PRESENTED AT
THE INTERNATIONAL SYMPOSIUM ON FINITE ELEMENT
METHODS IN FLOW PROBLEMS, HELD IN JANUARY 1974 IN
SWANSEA, UNITED KINGDOM, PP. 741-755.

74-0082 IROBE, M.
AKAGI, T.
ITOH, Y.
FINITE ELEMENT ANALYSIS OF CONSOLIDATION OF
UNSATURATED SOIL
IN FINITE ELEMENT METHOD IN FLOW PROBLEMS. EDITED
BY J. T. ODEN, O. C. ZIENKIEWICZ, R. H.
GALLAGHER, AND C. TAYLOR, PAPER PRESENTED AT
THE INTERNATIONAL SYMPOSIUM ON FINITE ELEMENT
METHODS IN FLOW PROBLEMS, HELD IN JANUARY 1974 IN
SWANSEA, UNITED KINGDOM, PP. 757-767.

74-0083 WU, J. C.
DIVISION OF COMPUTATION FIELD FOR THE FINITE
ELEMENT METHOD
IN FINITE ELEMENT METHOD IN FLOW PROBLEMS. EDITED
BY J. T. ODEN, O. C. ZIENKIEWICZ, R. H.
GALLAGHER, AND C. TAYLOR, PAPER PRESENTED AT
THE INTERNATIONAL SYMPOSIUM ON FINITE ELEMENT
METHODS IN FLOW PROBLEMS, HELD IN JANUARY 1974 IN
SWANSEA, UNITED KINGDOM, PP. 769-770.

74-0084 KOSSOWSKI, R.
APPLICATION OF THE FINITE ELEMENT METHOD FOR
SOLVING ORDINARY LINEAR DIFFERENTIAL
EQUATION WITH BOUNDARY CONDITIONS
IN FINITE ELEMENT METHOD IN FLOW PROBLEMS, EDITED
BY J. T. ODEN, O. C. ZIENKIEWICZ, R. H.
GALLAGHER, AND C. TAYLOR. PAPER PRESENTED AT
THE INTERNATIONAL SYMPOSIUM ON FINITE ELEMENT
METHODS IN FLOW PROBLEMS, HELD IN JANUARY 1974 IN
SWANSEA, UNITED KINGDOM, PP. 771-772.

74-0085 YANG, T. Y.
KIM, H. W.
ASYMMETRICAL BENDING AND VIBRATION OF A CONICAL
SHELL FINITE ELEMENT
A.I.A.A. J., VOL. 12, NO. 3, 1974, PP. 257-258.

74-0086 EIDELBERG, B. E.
FINITE ELEMENT ANALYSIS OF LUBRICATION IN NATURAL
JOINTS
PH.D. THESIS, CORNELL UNIVERSITY, JAN. 1974.

74-0087 DAY, C. P.
TRANSIENT ELASTOHYDRODYNAMIC LUBRICATION BY FINITE
ELEMENT METHODS
M. S. THESIS, CORNELL UNIVERSITY, JAN. 1974.

74-0088 LUCKNER, L.
GRABER, P.-W.
KADEN, ST.
BEITRAG ZUR SIMULATION NICHTSTATIONARER
HORIZONTAL-EVENER GRUNDWASSERSTROMUNGSFELDER AUF
DER GRUNDLAGE UNREGELMAOIGER DREIECKSFORMIGER
FELDELEMENTE
A CONTRIBUTION TO THE SIMULATION OF NON-STATIONARY
GROUND- WATER FLOW FIELDS ON THE BASIS OF
IRREGULAR TRIANGLE-SHAPED FIELD ELEMENTS,
MITTEILUNGEN DES INSTITUTS FUR
WASSERWIRTSCHAFT, BERLIN, 1974.

74-0089 ANQUEZ, L.
BERGER, H.
OHAYON, R.
VALID, R.
VIBRATIONS OF TANKS PARTIALLY FILLED WITH LIQUIDS
PAPER PRESENTED AT INTERNATIONAL SYMPOSIUM ON THE
APPLICATION OF FINITE ELEMENT METHODS IN
FLOW PROBLEMS SWANSEA, WALES, U. K., JANUARY
7-11, 1974, PP 1-28.

74-0090 DE VRIES, G.
BERARD, G. P.
NORRIE, D. H.
APPLICATION OF THE FINITE ELEMENT TECHNIQUE TO
COMPRESSIBLE FLOW PROBLEMS
REPORT NO. 51, MECHANICAL ENGINEERING DEPARTMENT,
UNIVERSITY OF CALGARY, CANADA, JANUARY 1974.

74-0091 RAS, G. V.
RAJU, I. S.
RADHAMOH, S. K.
BUCKLING OF SHELLS BY FINITE ELEMENT METHOD
J. ENGR., MECH. DIV. ASCE., VOL. 100, NO. 5, 1974,
PP. 1092-1096.

74-0092 SHARIFI, P.
YATES, D. N.
NONLINEAR THERMO ELASTIC PLASTIC AND CREEP
ANALYSIS BY FINITE ELEMENT METHOD
A.I.A.A. J., VOL. 12, NO. 9, PP. 1210-1215.

74-0093 WEGMULLER, A. W.
ELASTIC PLASTIC FINITE ELEMENT ANALYSIS OF PLATES
PROC. INST. CIV. ENGRS., PART 2, RESEARCH AND
THEORY, VOL. 57, SEPTEMBER 1974, PP. 535-543.

74-0094 ASHWELL, D. G.
SABIR, A. B.
FINITE ELEMENT CALCULATION OF STRESS DISTRIBUTIONS
IN ARCHES
INT. J. OF MECH. SCI., VOL. 16, NO. 1, 1974, PP.
21-29.

74-0095 HUEBNER, K. H.
FINITE ELEMENT METHOD - STRESS ANALYSIS AND MUCH
MORE
MACHINE DESIGN 1974, VOL. 46, NO. 1, PP. 92-99

74-0096 THOMAS, J.
SOARES, C. A.
FINITE ELEMENT ANALYSIS OF ROTATING SHELLS
MECHANICAL ENGINEERING 1974, VOL. 96, NO. 1, PP.
60-61.

74-0097 RAO, G. V.
KADHAMOH, S. K.
RAJU, I. S.
REINVESTIGATION OF BUCKLING OF SHELLS OF
REVOLUTION BY A REFINED FINITE ELEMENT
AMERICAN INSTITUTE AERONAUTICS AND ASTRONAUTICS.
JOUR. 1974, VOL. 12, NO. 1, PP. 100-101

74-0098 WITTRICK, W.H.
FINITE ELEMENT STRUCTURAL ANALYSIS OF LOCAL
INSTABILITY
AIAA. JOUR. 1974, VOL. 12, NO. 1, PP. 123-124

74-0099 PRZEMIENIECKI, J.S.
FINITE ELEMENT STRUCTURAL ANALYSIS OF LOCAL
INSTABILITY - REPLY
AIAA. JOUR. 1974, VOL. 12, NO. 1, PP. 124

74-0100 ROSEN, R.
FINITE ELEMENT ANALYSIS - INTRODUCTION
MECHANICAL ENGINEERING 1974, VOL. 96, NO. 1, PP.
21-26

74-0101 KHATUA, T. P.
CHEUNG, Y. K.
FINITE ELEMENT ANALYSIS OF AXISYMMETRIC MULTILAYER
SANDWICH PLATES AND SHELLS
PAPER FOR CONFERENCE ON FINITE ELEMENT METHODS IN
ENGINEERING, SCHOOL OF CIVIL ENGINEERING,
UNIVERSITY OF NEW SOUTH WALES, SYDNEY, AUSTRALIA,
1974.

74-0102 CANTIN, G.
RIGID BODY MOTIONS AND EQUILIBRIUM IN FINITE
ELEMENTS
CONFERENCE ON FINITE ELEMENTS FOR THIN SHELLS AND
CURVED MEMBERS, CARDIFF, MAY 20-21, 1974.

74-0103 ANDERSON, C. M.
NOOR, A. K.
USE OF SYMBOLIC MANIPULATION IN DEVELOPMENT OF
TWO-DIMENSIONAL FINITE ELEMENTS
SIAM REVIEW, VOL. 16, NO. 1, 1974, PP. 115-116.

74-0104 ANON
USING FINITE ELEMENTS IN AUTOMOTIVE DESIGN
AUTOMOTIVE ENGINEERING, VOL. 82, NO. 4, 1974, PP.
25-31.

74-0105 BRATANOW, T.
ECER, A.
APPLICATIONS OF FINITE ELEMENT METHOD IN UNSTEADY
AERODYNAMICS
AIAA J., VOL. 12, NO. 4, 1974, PP. 503-510.

74-0106 CHARI, M. V.
FINITE-ELEMENT SOLUTION OF EDDY-CURRENT PROBLEM IN
MAGNETIC STRUCTURES
IEEE TRANS. ON POWER APPARATUS AND SYSTEMS, VOL.
PA93, NO. 1, 1974, PP. 62-72.

74-0107 CHOU, S. I.
MEAN-SQUARE ERRORS OF FINITE ELEMENT
APPROXIMATIONS ON LINEAR ELASTOSTATICS,
ELASTODYNAMICS, AND THERMOELASTO-DYNAMICS
SIAM REVIEW, VOL. 16, NO. 1, 1974, PP. 119.

74-0108 COLVILLE, J.
ABBASI, J.
PLANE STRESS REINFORCED CONCRETE FINITE ELEMENTS
J. STRUCT. DIV., ASCE., VOL. 100, NO. ST5, 1974,
PP. 1067-1083.

74-0109 DALY, P.
POLAR GEOMETRY WAVEGUIDES BY FINITE-ELEMENT METHOD
IEEE TRANS. ON MICROWAVE THEORY AND TECHNIQUES,
VOL. TT22, NO. 3, 1974, PP. 202-209.

74-0110 DAWKINS, W. P.
FINITE ELEMENT FOR REINFORCED CONCRETE FRAME STUDY
J. OF STRUCT. DIV., ASCE., VOL. 100, NO. ST4,
1974, PP. 822-824.

74-0111 DECRETON, M. C.
ANALYSIS OF OPEN STRUCTURES BY FINITE ELEMENT
RESOLUTION OF EQUIVALENT CLOSED-BOUNDARY PROBLEM
ELECTRONICS LETTERS, VOL. 10, NO. 4, 1974, PP.
43-44.

74-0112 DOUGLAS, J.
LOCATION OF POINTS AT WHICH FINITE ELEMENT
SOLUTIONS ARE VERY ACCURATE
SIAM REVIEW, VOL. 16, NO. 1, 1974, P. 121.

74-0113 FABIAN, R. J.
DESIGNING BY FINITE ELEMENTS CATCHING ON IN
AUTOMOTIVE ENGINEERING
AUTOMOTIVE ENGINEERING, VOL. 82, NO. 3, 1974, P.
5.

74-0114 FELIPPA, C. A.
DATA-STRUCTURES IN FINITE ELEMENT ANALYSIS
SIAM REVIEW, VOL. 16, NO. 1, 1974, PP. 122-123.

74-0115 GEORGE, A.
SPARSE MATRIX TECHNIQUES IN DIRECT SOLUTION OF
FINITE ELEMENT EQUATIONS
SIAM REVIEW, VOL. 16, NO. 1, 1974, PP. 124.

74-0116 GERARD, R.
FINITE ELEMENT SOLUTION FOR FLOW IN NONCIRCULAR
CONDUITS
J. HYDRAULICS DIV., ASCE., VOL. 100, NO. HY3,
1974, PP. 425-441.

74-0117 HAMANN, W. C.
FINITE ELEMENT METHODS IN PRODUCT DESIGN
MECHANICAL ENGINEERING, VOL. 96, NO. 2, 1974, PP.
30-36.

74-0118 HAYES, D. J.
TURNER, C. E.
APPLICATION OF FINITE ELEMENT TECHNIQUES TO
POST-YIELD ANALYSIS OF PROPOSED STANDARD 3 POINT
BEND FRACTURE TEST PIECES
INT. J. OF FRACT., VOL. 10, NO. 1, 1974, PP.
17-32.

74-0119 HINATA, M.
SHIMASAKI, M.
KIYONO, T.
NUMERICAL SOLUTION OF PLATEAUS PROBLEM BY A FINITE
ELEMENT METHOD
MATHEMATICS OF COMPUTATION, VOL. 28, NO. 125,
1974, PP. 45-60.

74-0120 HUANG, Y. H.
FINITE ELEMENT ANALYSIS OF SLABS ON ELASTIC SOLIDS
TRANS. ENG. J. OF ASC., VOL. 100, NO. TE2, 1974,
PP. 403-416

74-0121 BARACH, D. R.
HUNT, J. T.
KNITTEL, M. R.
FINITE ELEMENT APPROACH TO ACOUSTIC RADIATION FROM
ELASTIC STRUCTURES
J. ACOUSTICAL SOC. OF AMERICA, VOL. 55, NO. 2,
1974, PP. 269-280.

74-0122 JAMET, P.
RAVIART, P. A.
CONVERGENCE AND ERROR ESTIMATES FOR FINITE ELEMENT
APPROXIMATIONS OF STATIONARY NAVIER-STOKES
EQUATIONS
SIAM REVIEW, VOL. 16, NO. 1, 1974, PP. 128.

74-0123 JOHNSON, C. P.
WILL, K. M.
BEAM BUCKLING BY FINITE ELEMENT PROCEDURE
J. STRUCT. DIV., ASCE., VOL. 100, NO. ST3, 1974,
PP. 669-685

74-0124 KINSNER, W.
TORRE, E. D.
ITERATIVE APPROACH TO FINITE-ELEMENT METHOD IN
FIELD PROBLEMS
IEEE TRANS. ON MICROWAVE THEORY AND TECHNIQUES,
VOL TT22, NO. 3, 1974, PP. 221-228.

74-0125 KOTCHERGENKO, I. D.
DE AMORIM, A. M. A.
FREE-SURFACE IDEAL FLUID FLOWS BY FINITE ELEMENTS
J. HYDRAUL. DIV., ASCE., VOL. 100, NO. HY3, 1974,
PP. 497-498.

74-0126 MAU, S. T.
TONG, P.
CALCULATION OF MECHANICAL IMPEDANCE BY FINITE
ELEMENT HYBRID MODEL
AIAA J., VOL. 12, NO. 2, 1974, PP. 249-250.

74-0127 MILSTED, M. G.
HUTCHINSON, J. R.
USE OF TRIGONOMETRIC TERMS IN FINITE ELEMENT
METHOD WITH APPLICATION TO VIBRATING MEMBRANES
J. SOUND AND VIBRATION, VOL. 32, NO. 3, 1974, PP.
237-346.

74-0128 NILSON, A. H.
AMMAR, A. R.
FINITE ELEMENT ANALYSIS OF METAL DECK SHEAR
DIAPHRAGMS
J. STRUCT. DIV., ASCE., VOL. 100, NO. ST4, 1974,
PP. 711-726

74-0129 OH, K. P.
APPLICATION OF A GRADIENT METHOD IN FINITE ELEMENT
SOLUTION OF ELASTOHYDRODYNAMIC PROBLEMS
ASLE TRANSACTIONS, VOL. 17, NO. 2, 1974, PP.
111-116.

74-0130 ORRIS, R. M.
PETYT, M.
FINITE ELEMENT STUDY OF HARMONIC WAVE PROPAGATION
IN PERIODIC STRUCTURES
J. SOUND AND VIBRATION, VOL. 33, NO. 2, 1974, PP.
223-236.

74-0131 ROHDE, S. M.
FINITE ELEMENT OPTIMIZATION OF FINITE STEPPED
SLIDER BEARING PROFILES
ASLE TRANSACTIONS, VOL. 17, NO. 2, 1974, PP.
105-110.

74-0132 RYBICKI, E. F.
HOPPER, A. T.
TRANSFORMATIONS FOR ELLIPTICAL HOLES APPLIED TO
FINITE ELEMENT ANALYSIS
J. OF COMPOSITE MATERIALS, VOL. 8, JANUARY 1974,
PP. 93-96.

74-0133 FONDER, G.
STUDIES IN DOUBLY CURVED SHELL ELEMENTS
CONFERENCE ON FINITE ELEMENTS FOR THIN SHELLS AND
CURVED MEMBERS, CARDIFF, MAY 20-21, 1974.

74-0134 SCHULTZ, M. H.
COMPARISON OF FINITE ELEMENT METHODS FOR A MODEL
PROBLEM
SIAM REVIEW, VOL. 16, NO. 1, 1974, P. 138.

74-0135 SIMPSON, L. A.
HSU, T. R.
MERRETT, G.
ESTABLISHMENT OF A VALID FRACTURE TOUGHNESS
SPECIMEN GEOMETRY USING A FINITE ELEMENT ANALYSIS
AMERICAN CERAMIC SOCIETY BULLETIN, VOL. 53, NO. 4,
1974, PP. 318.

74-0136 TRACEY, D. M.
FINITE ELEMENTS FOR THREE-DIMENSIONAL ELASTIC
CRACK ANALYSIS
NUCLEAR ENGINEERING AND DESIGN, VOL. 26, NO. 2,
1974, PP. 282-290.

74-0137 AGARWAL, B. D.
BROUTMAN, L. J.
THREE-DIMENSIONAL FINITE ELEMENT ANALYSIS OF
SPHERICAL PARTICLE COMPOSITES
FIBRE SCI. TECHNOL., VOL. 7, NO. 1, JANUARY 1974,
PP. 63-77.

74-0138 AGARWAL, B. D.
LIFSHITZ, J. M.
ELASTIC-PLASTIC FINITE ELEMENT ANALYSIS OF SHORT
FIBRE COMPOSITES
FIBRE SCI. TECHNOL., VOL. 7, NO. 1, JANUARY 1974,
PP. 45-62.

74-0139 LEVERENZ, R. K.
INTERACTIVE GRAPHICS - AN AID FOR DEVELOPING
FINITE ELEMENT MODELS
INT. CONF. ON VEHICLE STRUCTURAL MECH., FINITE
ELEMENT APPLICATION TO VEHICLE DESIGN, DETROIT,
MICHIGAN, MARCH 26-28, 1974.

74-0140 CHANG, D. C.
EFFECTS OF FLEXIBLE CONNECTIONS ON BODY STRUCTURAL
RESPONSE
S.A.E. MEETING, PAPER 740041, FEBRUARY 25-MARCH 1,
1974, 12 PP.

74-0141 COLLINS, J. D.
HART, G. C.
STATISTICAL IDENTIFICATION OF STRUCTURES
AIAA J., VOL. 12, NO. 2, FEBRUARY 1974, PP.
185-190.

74-0142 ANAND, S. C.
STRESS DISTRIBUTIONS AROUND SHALLOW BURIED RIGID
PIPES
ASCE J. STRUCT. DIV., VOL. 100, NO. ST1, JANUARY
1974, PAPER 10258, PP. 161-174.

74-0143 IDRISS, I. M.
SEED, H. B.
SEISMIC RESPONSE BY VARIABLE DAMPING FINITE
ELEMENTS
ASCE J. GEOTECH. ENG. DIV., VOL. 100, NO. GT1,
JANUARY 1974, PAPER 10284, PP. 1-13.

74-0144 KATO, B.
MORITA, K.
STRENGTH OF TRANSVERSE FILLET WELDED JOINTS
WELD J., (FLA), VOL. 53, NO. 2, FEBRUARY 1974, PP.
59-64.

74-0145 RICE, J. R.
LIMITATIONS TO THE SMALL SCALE YIELDING
APPROXIMATION FOR CRACK TIP PLASTICITY
J. MECH. PHYS. SOLIDS., VOL. 22, NO. 1, JANUARY
1974, PP. 17-26

74-0146 SCHALLER, R. J.
VISSER, R. J.
CERAMIC ROTATING BLADES: SOME CRITICAL DESIGN
PARAMETERS FOR GAS TURBINE APPLICATIONS
PAPER NO. 74-GT-96, A.S.M.E. MEETING, MARCH
31-APRIL 4, 1974, 7 PP.

74-0147 STORAASLI, O. O.
SOBIESZCZANSKI, J.
ON THE ACCURACY OF THE TAYLOR APPROXIMATION FOR
STRUCTURE RESIZING
AIAA J., VOL. 12, NO. 2, FEBRUARY 1974, PP.
231-233.

74-0148 SWITZKY, H.
WANG, P.
MINIMUM WEIGHT DESIGN OF FINITE ELEMENT STRUCTURES
AIAA J., VOL. 12, NO. 2, FEBRUARY 1974, PP.
170-175.

74-0149 WEGMULLER, A. W.
FULL RANGE ANALYSIS OF ECCENTRICALLY STIFFENED
PLATES
ASCE J. STRUCT. DIV., VOL. 100, NO. ST1, JANUARY
1974, PAPER 10286, PP. 143-159.

74-0150 WU, R. W. H.
WITMER, E. A.
DYNAMIC RESPONSES OF CYLINDRICAL SHELLS INCLUDING
GEOMETRIC AND MATERIAL NONLINEARITIES
INT. J. SOLIDS STRUCT., VOL. 10, NO. 2, FEBRUARY
1974, PP. 243-260.

74-0151 MEI, C.
CHEN, H. S.
OSCILLATIONS AND WAVE FORCES IN A MAN-MADE HARBOUR
IN THE OPEN SEA
13TH NAVAL HYDRODYNAMICS SYMP., DEPT. OF CIVIL
ENG., M.I.T., CAMBRIDGE, MASS., JUNE 1974.

74-0152 MANSFIELD, L.
HIGHER ORDER COMPATIBLE TRIANGULAR FINITE ELEMENTS
NUMERISCHE MATHEMATIK, VOL. 22, NO. 2, 1974, PP.
89-97.

74-0153 MARUYAMA, K.
STRESS ANALYSIS OF A BOLT-NUT JOINT BY FINITE
ELEMENT METHOD AND COPPER-ELECTROPLATING METHOD -
2. STRESS AT ROOT OF A BOLT THREAD UNDER A TENSILE
LOAD
B. JSME, VOL. 17, NO. 106, 1974, PP. 442-450.

74-0154 ARGYRIS, J. H.
MARECZEK, G.
FINITE-ELEMENT ANALYSIS OF SLOW INCOMPRESSIBLE
VISCOUS FLUID MOTION
INGENIEUR-ARCHIV., VOL. 43, NO. 2-3, 1974, PP.
92-109.

74-0155 KAUFMAN, S.
LOVERHER, A.
FINITE ELEMENT METHOD FOR DETERMINATION OF THERMAL
STRESSES INTERACTIVE GRAPHICS FOR FINITE ELEMENT
MODEL DATA CHECKOUT IN ANISOTROPIC SOLIDS OF
REVOLUTION
PROC. S.A.E. INTL. CONF. ON VSM, DETROIT, MARCH
1974. (ALSO IN BELL SYSTEM TECH. J., VOL. 53, NO.
5, 1974)

74-0156 HARTLEY, G. A.
MCNEICE, G. M.
STENSCH, W.
VOGT BOUNDARY FOR FINITE ELEMENT ARCH DAM ANALYSIS
J. STRUCT. DIV., ASCE., VOL. 100, NO. ST1, 1974,
PP. 51-62.

74-0157 PICHUMANI, R.
CRAWFORD, J. E.
TRIANDAF, G. E.
FINITE ELEMENT ANALYSIS OF PILE-SUPPORTED PAVEMENT
STRUCTURES
TRANSPORTATION ENGINEERING J. VOL. 100, NO. TE2,
1974, PP. 293-303.

74-0158 DISTEFANO, N.
INVARIANT IMBEDDING AND SOLUTION OF FINITE ELEMENT
EQUATIONS
J. OF MATH. ANALYSIS AND APPLICATIONS, VOL. 46,
NO. 2, 1974, PP. 487-498.

74-0159 APPERT, K.
BERGER, D.
GRUBER, R.
RAPPAZ, J.
TROYON, F.
NUMERICAL STUDY OF EIGEN OSCILLATIONS OF A SCREW
PINCH WITH FINITE ELEMENTS
ZEITSCHRIFT FUR ANGEWANDTE MATHEMATIK UND PHYSIK,
VOL. 25, NO. 1, 1974, P. 116.

74-0160 APPERT, K.
BERGER, D.
GRUBER, R.
RAPPAZ, J.
TROYON, F.
NUMERICAL STUDY OF PROPER MODES OF A HOMOGENEOUS
CYLINDRICAL PLASMA BY METHOD OF FINITE ELEMENTS
ZEITSCHRIFT FUR ANGEWANDTE MATHEMATIK UND PHYSIK,
VOL. 25, NO. 1, 1974, PP. 229-240.

74-0161 ARGYRIS, J. H.
GRIEGER, I.
DESIGN AND CALCULATION WITH FINITE ELEMENTS
CHEMIE-INGENIEUR-TECHNIK, VOL. 46, NO. 10, 1974,
P. 451.

74-0162 BUCK, K. E.
PRACTICAL APPLICATION OF METHOD OF FINITE ELEMENTS
CHEMIE-INGENIEUR-TECHNIK, VOL. 46, NO. 10, 1974,
PP. 415-418

74-0163 FAZIO, P. P.
HA, K. H.
SANDWICH PLATE STRUCTURE ANALYSIS BY FINITE
ELEMENT
J. STRUCT. DIV., ASCE., VOL. 100, NO. ST6, 1974,
PP. 1243-1262.

74-0164 FOWLKES, C. W.
FINITE ELEMENT ANALYSIS OF CRACKS IN REGION OF A
BI-MATERIAL INTERFACE
J. OF ADHESION, VOL. 6, NO. 1-2, 1974, PP. 49-84.

74-0165 FRANECK, H.
RECKE, H. G.
PROBLEMS OF CROSS-LINKAGE IN USING METHOD OF
FINITE ELEMENTS
MASCHINENBAU TECHNIK, VOL. 23, NO. 6, 1974, PP.
279-282.

74-0166 GLOVER, P.
MCCOWAN, D. W.
ALEXANDER, S. S.
DYNAMIC FINITE ELEMENT METHOD CALCULATION OF
ELASTIC WAVE PROPAGATION IN TWO DIMENSIONAL
STRUCTURES
TRANSACTIONS, AMERICAN GEOPHYSICAL UNION, VOL. 55,
NO. 4, 1974, P. 351.

74-0167 GONG, C.
JACHENS, R. C.
KUO, J. T.
PSEUDO-THREE-DIMENSIONAL FINITE ELEMENT
FORMULATION FOR ELASTOSTATIC PROBLEMS
TRANSACTIONS, AMERICAN GEOPHYSICAL UNION, VOL. 55,
1974, NO. 4, P. 351.

74-0168 HABERL, G.
OCH, F.
FINITE ELEMENTS SOLUTION FOR TORSION RIGIDITY AND
SHEAR CENTER OF ARBITRARY CROSS-SECTIONS
ZEITSCHRIFT FUR FLUGWISSENSCHAFTEN, VOL. 22, NO.
4, 1974, PP. 115-119.

74-0169 KOLATA, G. B.
FINITE ELEMENT METHOD - MATHEMATICAL REVIVAL
SCIENCE, VOL. 184, NO. 4139, 1974, PP. 887-889.

74-0170 MCCOWAN, D. W.
GLOVER, P.
ALEXANDER, S. S.
INVERSION TECHNIQUE FOR STATIC FINITE ELEMENT
METHOD
TRANS., AMERICAN GEOPHYSICAL UNION, VOL. 55, NO.
4, 1974, PP. 353.

74-0171 WARZEE, G.
APPLICATION OF LAPLACE TRANSFORM AND FINITE
ELEMENT METHOD TO SOLUTION OF NONSTATIONARY
THERMAL CONDUCTION EQUATION
COMPTES RENDUS... ACADEMIE DES SCIENCES SER. A
MATH., VOL. 278, NO. 19, 1974, PP. 1265-1266.

74-0172 ZLAMAL, M.
CURVED ELEMENTS IN FINITE ELEMENT METHOD - 2
SIAM J. NUMERICAL ANALYSIS, VOL. 11, NO. 2, 1974,
PP. 347-362.

74-0173 ZLAMAL, M.
FINITE ELEMENT METHODS FOR PARABOLIC EQUATIONS
MATHEMATICS OF COMPUTATION, VOL. 28, NO. 126,
1974, PP. 393-404.

74-0174 APPERT, K.
BERGER, D.
GRUBER, R.
TROYON, F.
EIGENMODES OF CYLINDRICAL PLASMA USING FINITE
ELEMENT METHOD
ZEITSCHRIFT FUR ANGEWANDTE MATHEMATIK UND PHYSIK
1974, VOL. 25, NO. 2, PP. 229-240.

74-0175 ADAMS, D. G.
AUTOMOBILE PANEL SWEEP STIFFNESS ANALYSIS
PREPRINT NO. 740080, S.A.E. MEETING, FEBRUARY 25 -
MARCH 1, 1974, 12 PP.

74-0176 LEVERENZ, R. K.
NG, B. L.
BIRCHLER, W. D.
PERIARD, A. R.
ESSELINK, L.
USING INTERACTIVE GRAPHICS FOR THE PREPARATION AND
MANAGEMENT OF FINITE ELEMENT DATA
INT. CONF. ON VEHICLE STRUCTURAL MECHANICS: FINITE
ELEMENT APPLICATION TO VEHICLE DESIGN, DETROIT,
MICHIGAN, MAR. 26-28 1974, PP. 279-285.

74-0177 ARALDSEN, P. O.
EXAMPLE OF LARGE-SCALE STRUCTURAL ANALYSIS.
COMPARISON BETWEEN FINITE ELEMENT CALCULATION AND
FULL SCALE MEASUREMENTS ON THE OIL TANKER ESSO
NORWAY
COMPUT STRUCT., VOL. 4, NO. 1, JANUARY 1974, PP.
69-93.

74-0178 BAZANT, Z. P.
THREE-DIMENSIONAL HARMONIC FUNCTIONS NEAR
TERMINATION OR INTERSECTION OF GRADIENT
SINGULARITY LINES: A GENERAL NUMERICAL METHOD
INT. J. ENG. SCI., VOL. 12, NO. 3, MARCH 1974, PP.
221-243.

74-0179 CHAINOV, N. D.
IVASHCHENKO, N. A.
METODY RASCHETNOGO OPREDELENIYA TEMPERATURNYKH
NAPRYAZHENII V KRYSHKAKH TSILINDROV DVIGATELEI
VNUTRENNEGO SGORANIYA (METHODS OF CALCULATION OF
THERMAL STRESSES IN CLYINDER COVERS OF INTERNAL
COMBUSTION ENGINES)
IZV. VYSSH. UCHEBN. ZAVED., MASHINOSTR, NO. 1,
1974, PP. 81-84.

74-0180 CLOUGH, R. W.
AREAS OF APPLICATION OF THE FINITE ELEMENT METHOD
COMPUT. STRUCT. VOL. 4, NO. 1, JANUARY 1974, PP.
17-40.

74-0181 CORRADI, L.
ZAVELANI, A.
LINEAR PROGRAMMING APPROACH TO SHAKEDOWN ANALYSIS
OF STRUCTURES
COMPUT. METHODS APPL. MECH. ENG. VOL. 3, NO. 1,
JANUARY 1974, PP. 37-53.

74-0182 DONEA, J.
ON THE ACCURACY OF FINITE ELEMENT SOLUTIONS TO THE
TRANSIENT HEAT CONDUCTION EQUATION
INT. J. NUMER. METHODS ENG., VOL. 8, NO. 1, 1974,
PP. 103-110.

74-0183 EGELAND, O.
ARALDSEN, P. O.
SESAM-69 - A GENERAL PURPOSE FINITE ELEMENT METHOD
PROGRAM
COMPUT. STRUCT., VOL. 4, NO. 1, JANUARY 1974, PP.
41-68.

74-0184 FRANQUES, J. T.
YANNITELL, D. M.
TWO-DIMENSIONAL ANALYSIS OF BACKWATER AT BRIDGES
ASCE J. HYDRAUL. DIV., VOL. 100, NO. HY3, MARCH
1974, PAPER 10403, PP. 379-392.

74-0185 FRIED, I.
NOTES ON THE FINITE ELEMENT ANALYSIS OF THE
AXISYMMETRIC ELASTIC SOLID
INT. J. SOLIDS STRUCT., VOL. 10, NO. 3, MARCH
1974, PP. 383-386.

74-0186 SOBOLESKI, J. J.
NASTRAN PLOTTING AT A REMOTE TERMINAL
INT. CONF. ON VEHICLE STRUCTURAL MECHANICS: FINITE
ELEMENT APPLICATION TO VEHICLE DESIGN, DETROIT,
MICHIGAN, MAR. 26-28 1974, PP. 274-279.

74-0187 ROBINSON, J.
A GENERAL EQUILIBRIUM TORQUE ELEMENT
RA/ARD, REPORT NO. 16-4-74-23, APRIL 1974.

74-0188 GOULD, P. L.
SEN, S. K.
DYNAMIC ANALYSIS OF COLUMN-SUPPORTED HYPERBOLOIDAL
SHELLS
EARTHQUAKE ENG. STRUCT. DYN., VOL. 2, NO. 3,
JANUARY-MARCH 1974, PP. 269-280.

74-0189 HANSEN, H. R.
SOME EXAMPLES OF THE APPLICATION OF THE FINITE
ELEMENT METHOD SHIP STRUCTURES
COMPUT. STRUCT., VOL. 4, NO. 1, JANUARY 1974, PP.
205-212.

74-0190 HELLEN, T. K.
PROTHEROE, S. J.
BERSAFE FINITE ELEMENT SYSTEM
COMPUT. AIDED DES., VOL. 6, NO. 1, JANUARY 1974,
PP. 15-24.

74-0191 HOLAND, I.
FUNDAMENTALS OF THE FINITE ELEMENT METHOD
COMPUT. STRUCT., VOL. 4, NO. 1, JANUARY 1974, PP.
3-15.

74-0192 HOLZLOEHNER, U.
FINITE ELEMENT ANALYSIS FOR TIME-DEPENDENT
PROBLEMS
INT. J. NUMER. METHODS ENG., VOL. 8, NO. 1, 1974,
PP. 55-69

74-0193 HOOKER, R. J.
O'BRIEN, D. J.
NATURAL FREQUENCIES OF BOX-TYPE STRUCTURES BY A
FINITE-ELEMENT METHOD
ASME PAPER NO. 74-APM-F, 1974, 3 PP.

74-0194 FAWCETT, D. J.
GENERATION OF FINITE ELEMENT MODELS VIA COMPUTER
GRAPHICS
INT. CONF. ON VEHICLE STRUCTURAL MECHANICS: FINITE
ELEMENT APPLICATION TO VEHICLE DESIGN, DETROIT,
MICHIGAN, MAR. 26-28 1974, PP. 264-274.

74-0195 LOVERHER, A.
FINITE ELEMENT MODEL DATA CHECKOUT WITH
INTERACTIVE GRAPHICS
CONF. ON VEHICLE STRUCTURAL MECHANICS: FINITE
ELEMENT APPLICATION TO VEHICLE DESIGN, DETROIT,
MICHIGAN, MARCH 26-28, 1974, PP. 256-264.

74-0196 KIRIOKA, K.
HOTTA, Y.
ELASTO-PLASTIC ANALYSIS OF AUTOMOBILE BODY
STRUCTURE BY THE FINITE ELEMENT METHOD
S.A.E. MEETING, PAPER NO. 740039, FEBRUARY
25-MARCH 1, 1974, 9 PP.

74-0197 KORMAN, T.
GOLDBERG, J. E.
BOUSSINESQ PROBLEM OF PLANE MICROPOLAR ELASTICITY
INT. J. NUMER. METHODS ENG., VOL. 8, NO. 1, 1974,
PP. 45-54.

74-0198 LANGBALLE, M.
AASEN, E.
APPLICATION OF THE FINITE ELEMENT METHOD TO
MACHINERY
COMPUT. STRUCT., VOL. 4, NO. 1, JANUARY 1974, PP.
149-192.

74-0199 LATZKO, D. G. H.
DESIGN ASPECTS OF HIGH DUTY PRESSURE COMPONENTS
J. PRESSURE VESSEL TECHNOL., TRANS. ASME., VOL.
96, SER. J, NO. 1, FEBRUARY 1974, PP. 15-19.

74-0200 LYNN, P. P.
ARYA, S. K.
FINITE ELEMENTS FORMULATED BY THE WEIGHTED
DISCRETE LEAST SQUARES METHOD
INT. J. NUMER. METHODS ENG., VOL. 8, NO. 1, 1974,
PP. 71-90.

74-0201 NAIR, R. S.
BIRKEMOE, P. C.
HIGH STRENGTH BOLTS SUBJECT TO TENSION AND PRYING
ASCE J. STRUCT. DIV., VOL. 100, NO. ST2, FEBRUARY
1974, PAPER 10373, PP. 351-372.

74-0202 OSIAS, J. R.
SWEDLOW, J. L.
FINITE ELASTO-PLASTIC DEFORMATION - 1. THEORY AND
NUMERICAL EXAMPLES
INT. J. SOLIDS STRUCT., VOL. 10, NO. 3, MARCH
1974, PP. 321-339.

74-0203 OWEN, D. R. J.
ZIENKIEWICZ, O. C.
SHORT COMMUNICATIONS: TORSION OF AXI-SYMMETRIC
SOLIDS OF VARIABLE DIAMETER - INCLUDING
ACCELERATION EFFECTS
INT. J. NUMER. METHODS ENG., VOL. 8 NO. 1, 1974,
PP. 195-209

74-0204 ROSS, A. L.
DESIGNING WITH THREE DIRECTIONAL COMPOSITES
PAPER NO. 74-DE-25, A.S.M.E. MEETING, APRIL 1-4,
1974, 7 PP.

74-0205 SCHWARZ, H. R.
EIGENVALUE PROBLEMS (A - LAMBDA B) X EQUALS 0 FOR
SYMMETRIC MATRICES OF HIGH ORDER
COMPUT. METHODS APPL. MECH. ENG., VOL. 3, NO. 1,
JANUARY 1974, PP. 11-28.

74-0206 SLETTEN, R.
PEDERSEN, B.
APPLICATION OF THE FINITE ELEMENT METHOD TO
OFF-SHORE STRUCTURES
COMPUT. STRUCT., VOL. 4, NO. 1, JANUARY 1974, PP.
131-148.

74-0207 SONTVEDT, T.
PROPELLER BLADE STRESSES, APPLICATION OF FINITE
ELEMENT METHODS
COMPUT. STRUCT., VOL. 4, NO. 1, JANUARY 1974, PP.
193-204.

74-0208 SVEC, O. J.
UNBONDED CONTACT PROBLEM OF A PLATE ON THE ELASTIC
HALF SPACE
COMPUT. METHODS APPL. MECH. ENG., VOL. 3, NO. 1,
JANUARY 1974, PP. 105-113.

74-0209 LUBKIN, J. L.
THE FLEXIBILITY OF A TUBULAR WELDED JOINT IN A
VEHICLE FRAME
INT. CONF. ON VEHICLE STRUCTURAL MECHANICS:
FINITE ELEMENT APPLICATION TO VEHICLE DESIGN,
DETROIT, MICHIGAN, MARCH 26-28, 1974, PP. 250-256.

74-0210 ZIENKIEWICZ, O. C.
GODBOLE, P. N.
FLOW OF PLASTIC AND VISCO-PLASTIC SOLIDS WITH
SPECIAL REFERENCE TO EXTRUSION AND FORMING
PROCESSES
INT. J. NUMER. METHODS ENG., VOL. 8, NO. 1, 1974,
PP. 3-16.

74-0211 BRIDWELL, R. J.
NONLINEAR IN-PLANE BENDING OF A PLANE
STRESS-STRAIN FINITE ELEMENT MODEL
J. GEOPHYSICAL RESEARCH, VOL. 79, NO. 11, 1974,
PP. 1674-1678.

74-0212 BRIDWELL, R. J.
SWOLFS, H. S.
STABILITY ANALYSIS OF AN EXPERIMENTALLY DEFORMED
SINGLE LAYER OF INDIANA LIMESTONE USING FINITE
ELEMENTS
J. OF GEOPHYSICAL RESEARCH, VOL. 79, NO. 11, 1974,
PP. 1679-1686.

74-0213 BRILLA, J.
FINITE ELEMENT METHOD IN LINEAR VISCOELASTICITY
ZEITSCHRIFT FUR ANGEWANDTE MATHEMATIK UND
MECHANIK, VOL. 54, NO. 4, 1974, PP. T 47-48.

74-0214 KARCHER, H. J.
APPLICATION OF A CONJUGATED-GRADIENT-PROCESS IN
FINITE-ELEMENT-CALCULATION OF NON-LINEAR
STRUCTURES
ZEITSCHRIFT FUR ANGEWANDTE MATHEMATIK UND
MECHANIK, VOL. 54, NO. 4, 1974, PP. T 85-86.

74-0215 OHTMER, O.
CALCULATION OF TWO AND THREE DIMENSIONAL FLOW
PROBLEMS ACCORDING TO SUB-STRUCTURE THEORY WITH
ISOPARAMETRIC FINITE ELEMENTS
ZEITSCHRIFT FUR ANGEWANDTE MATHEMATIK UND
MECHANIK, VOL. 54, NO. 4, 1974, PP. T 139-141.

74-0216 MCRAE, G. J.
FINITE ELEMENT SOLUTIONS OF THE WAVE EQUATION
NOISE, SHOCK AND VIBRATION CONF., MONASH
UNIVERSITY, MELBOURNE, 1974, PP. 286-295.

74-0217 ADAMS, D. F.
ELASTOPLASTIC CRACK PROPAGATION IN A TRANSVERSELY
LOADED UNIDIRECTIONAL COMPOSITE
J. COMPOS. MATER., VOL. 8, JAN 1974, PP. 38-54.

74-0218 ALTMAN, W.
VENANCIO-FILHO, F.
STABILITY OF PLATES USING A MIXED FINITE ELEMENT
FORMULATION
COMPUT. STRUCT., VOL. 4, NO. 2, MAR 1974, PP.
437-443

74-0219 BABCOCK, C. O.
GEOMETRIC METHOD FOR THE PREDICTION OF STRESSES IN
INCLUSIONS, OREBODIES, AND MINING SYSTEMS
US. BUR. MINES. REP. INVEST., NO. 7838, 1974, 35
PP.

74-0220 BISWAS, J. K.
THREE-DIMENSIONAL ANALYSIS OF SHEAR WALL BUILDINGS
TO LATERAL LOAD
ASCE, J. STRUCT. DIV., VOL. 100, NO. ST5, MAY
1974, PAPER NO. 10537, PP. 1019-1036.

74-0221 FENNER, D.N.
DUGDALE MODEL SOLUTIONS FOR A SINGLE EDGE CRACKED
PLATE
INT. J. FRACT., VOL. 10, NO. 1, MAR 1974, PP.
71-76.

74-0222 LEE, G. C.
FRANKUS, A.
THEORY FOR DISTORTION STUDIES OF LUNG PARENCHYMA
BASED ON ALVEOLAR MEMBRANE PROPERTIES
J. BIOMECH., VOL. 7, NO. 1, JAN 1974, PP. 101-107

74-0223 GJELLEC, P.
RICARD, A.
CALCUL DE MASSIFS EN ELASTOPLASTICITE APPLICATIONS
A LA MECANIQUE DES ROCHES ET DES SOLS.
(APPLICATION OF THE RESULTS OF CALCULATION OF THE
ELASTOPLASTICITY TO SOIL AND ROCK MECHANICS)
LAB. CENT. PONTS. CHAUSSEES BULL. LIAISON LAB.
PONTS. CHAUSSEES, NO. 69, JAN-FEB 1974, PP.
115-123.

74-0224 CHENG, Y. K.
HOSAIN, M. U.
DEFLECTION ANALYSIS OF EXPANDED OPEN-WEB STEEL
BEAMS
COMPUT. STRUCT., VOL. 4, NO. 2, MAR 1974, PP.
327-336.

74-0225 KENNEDY, F. F.
LING, F. F.
ELASTO-PLASTIC INDENTATION OF A LAYERED MEDIUM
ASME. PAP. NO. 74-MAT-B, 1974, 7 PP.

74-0226 KOST, G.
WEAVER, W.
NONLINEAR DYNAMIC ANALYSIS OF FRAMES WITH FILLER
PANELS
ASCE, J. STRUCT. DIV., VOL. 100, NO. ST4, APR
1974, PAP. 10481, PP. 743-757.

74-0227 HARKEGARD, G.
LARSSON, S. G.
ON THE FINITE ELEMENT ANALYSIS OF CRACK AND
INCLUSION PROBLEMS IN ELASTIC-PLASTIC MATERIALS
COMPUT. STRUCT., VOL. 4, NO. 2, MAR 1974, PP.
293-305.

74-0228 LEVY, R.
MELOSH, R. J.
NONPROPORTIONAL LOADING LIMIT FOR STRUCTURES
COMPUT. STRUCT., VOL. 4, NO. 2, MAR 1974, PP.
453-465

74-0229 MILLER, C. J.
LIGHT GAGE STEEL INFILL PANELS IN MULTISTORY STEEL
FRAMES
ENG. J. AM. INST. STEEL CONSTR., VOL. 11, NO. 2,
SECOND Q., 1974, PP. 42-47.

74-0230 ODEN, J. T.
KEY, J. E.
NOTE ON THE ANALYSIS OF NONLINEAR DYNAMICS OF
ELASTIC MEMBRANES BY THE FINITE ELEMENT METHOD
COMPUT. STRUCT., VOL. 4, NO. 2, MAR 1974, PP.
445-452.

74-0231 RAWTANI, S.
NATURAL FREQUENCIES OF WIDE CHORD COMPRESSOR
BLADES
J. INST. ENG. (INDIA), MECH. ENG. DIV., VOL. 54,
PART ME. 4, MAR 1974, PP. 152-156

74-0232 SEN, S. K.
GOULD, P. L.
FREE VIBRATION OF SHELLS OF REVOLUTION USING
FINITE ELEMENT METHOD
ASCE. J. ENG. MECH. DIV., VOL. 100, NO. EM2, APR
1974, PAP. 10488, PP. 283-303.

74-0233 SINGH, B. P.
DHOOPAR, B. L.
MEMBRANE ANALOGY FOR ANISOTROPIC CABLE NETWORKS
ASCE J. STRUCT. DIV., VOL. 100, NO. ST5, MAY 1974,
PAPER NO. 10544, PP. 1053-1066.

74-0234 TAYLOR, C.
AL-MASHIDANI, G.
FINITE ELEMENT APPROACH TO WATERSHED RUNOFF
J. HYDROL, VOL. 21, NO. 3, MAR 1974, PP. 231-246.

74-0235 WANG, F. D.
ROPCHAN, D. M.
STRUCTURAL ANALYSIS OF A COAL MINE OPENING IN
ELASTIC, MULTILAYERED MATERIAL
US. BUR. MINES REP. INVEST., NO. 7845, 1974, 35
PP.

74-0236 YANG, H. T. Y.
BALDWIN, J. J.
ANALYSIS OF SPACE CLOSING SPRINGS IN ORTHODONTICS
J. BIOMECH., VOL. 7, NO. 1, JANUARY 1974, PP.
21-28.

74-0237 FLEMING, J. M.
PERCY, M. J.
DIESEL ENGINE COMPONENT DESIGN USING THE FINITE
ELEMENT METHOD AND INTERACTIVE GRAPHICS
INT. CONF. ON VEHICLE STRUCTURAL MECHANICS: FINITE
ELEMENT APPLICATION TO VEHICLE DESIGN, DETROIT,
MICHIGAN, MAR. 26-28 1974, PP. 217-228.

74-0238 BELYTSCHKO, T.
RADAJ, D.
ZIMMER, A.
GEISSLER, H.
FINITE ELEMENT ANALYSIS, AN AUTOMOBILE ENGINEER'S
TOOL
INT. CONF. ON VEHICLE STRUCTURAL MECHANICS:
FINITE ELEMENT APPLICATION TO VEHICLE DESIGN,
DETROIT, MICHIGAN, MARCH 26-28, 1974, PP. 228-244.

74-0239 THOMAS, J.
DOKUMACI, E.
SIMPLE FINITE ELEMENTS FOR PRE-TWISTED BLADING
VIBRATION
AERONAUTICAL QUARTERLY VOL. 25, 1974, MAY, PP.
109-118.

74-0240 FRIED, I.
FINITE ELEMENT METHOD - ACCURACY AT A POINT
QUARTERLY OF APPL. MATH., VOL. 32, NO. 2, 1974,
PP. 149-161.

74-0241 BOSSONEY, C.
FINITE ELEMENT METHOD APPLIED TO CALCULATIONS OF
NON-COPLANAR PLATE SYSTEMS
ZEITSCHRIFT FUR ANGEWANDTE MATHEMATIK UND PHYSIK
1974, VOL. 25, NO. 2, PP. 269-291.

74-0242 MELOSH, R. J.
STATUS REPORT ON COMPUTATIONAL TECHNIQUES FOR
FINITE ELEMENT ANALYSES
NUCL. ENG. AND DESIGN, VOL. 27, NO. 2, 1974, PP.
274-285.

74-0243 CHEUNG, Y. K.
WESTBROOK, D. R.
CHAKRABAS, S.
THREE DIMENSIONAL FINITE ELEMENT METHOD FOR PLATE
BENDING
INT. J. MECH. SCI., VOL. 16, NO. 7, 1974, PP.
479-487.

74-0244 BELYTSCHKO, T.
KULAK, R. F.
SCHULTZ, A. B.
GALANTE, J. O.
FINITE ELEMENT STRESS ANALYSIS OF AN
INTERVERTEBRAL DISC
J. BIOMECH., VOL. 7, NO. 3, 1974, PP. 277-285.

74-0245 GLOWINSKI, R.
MARROCCO, A.
APPROXIMATION BY FIRST ORDER FINITE ELEMENTS AND
SOLUTION BY PENALTY AND DUALITY OF NONLINEAR
DIRICHLET PROBLEMS
COMPTES RENDUS ACADEMIE DES SCIENCES SER A MATH.,
VOL. 278, NO. 26, 1974, PP. 1649-1652.

74-0246 PAGAY, S. N.
FINITE ELEMENT FOR REINFORCED CONCRETE FRAME STUDY
- SPACE DISCUSSION
J. STRUCT. DIV., A.S.C.E., VOL. 100, 1974, PP.
1714-1715.

74-0247 OJALVO, I. U.
IMPROVED THERMAL STRESS DETERMINATION BY FINITE
ELEMENT METHODS
AMERICAN INST. AERONAUTICS AND ASTRONAUTICS. JOUR.
1974, VOL. 12, NO. 8, PP. 1131-1132.

74-0248 ANDO, Y.
ANALYSIS FOR CRACKS IN SHIP STRUCTURE USING THE
FINITE ELEMENT METHOD WITH THE CONCEPT OF CRACK
SUB-STRUCTURE
MITSUBISHI HEAVY IND. TECH REV VOL. 11, NO. 1, FFB
1974, PP.18-27.

74-0249 BACKLUND, J.
WENNERSTROM, H.
FINITE ELEMENT ANALYSIS OF ELASTO-PLASTIC SHELLS
INT. J. NUMER METHODS ENG VOL. 8, NO. 2 1974, PP.
415-424.

74-0250 CULLIMORE, M. S. G.
ECKHART, J. B.
DISTRIBUTION OF THE CLAMPING PRESSURE IN
FRICTION-GRIP BOLTED JOINTS
STRUCT ENG VOL. 52, NO. 4, APP 1974, PP. 129-131.

74-0251 DEMIRCHYAN, K. S.
EFIMOV, Y. N.
REALIZATSIYA METODA KONECHNYKH ELEMENTOV NA EVM
DLYA RASCHETA DVUMERNYKH ELEKTRICHESKIKH I
MAGNITNYKH POLEI (COMPUTER AIDED REALIZATION OF
THE FINITE ELEMENT METHOD FOR THE CALCULATION OF
TWO-DIMENSIONAL ELECTRIC AND MAGNETIC FIELDS)
IZV AKAD NAUK (USSR) ENERG. TRANSP., NO. 1,
JANUARY - FEBRUARY, 1974, PP. 142-148.

74-0252 DONEA, J.
GIULIANI, S.
FINITE ELEMENTS IN THE SOLUTION OF ELECTROMAGNETIC
INDUCTION PROBLEMS
INT. J. NUMER. METHODS ENG. VOL. 8, NO. 2 1974,
PP. 359-367

74-0253 ELSTE, P.
SCHWINGUNGSMAESSIGE AUSLEGUNG DER BESCHAUFELUNG
AXIALER TURBOMASCHINEN (DESIGN OF
DISPLACEMENT-FREE BLADES IN AXIAL TURBOMACHINERY)
MASCHINENBAUTECHNIK, VOL. 23, NO. 4, APR 1974, PP
156-158.

74-0254 FRANCE, P. W.
FINITE ELEMENT ANALYSIS OF THREE-DIMENSIONAL
GROUNDWATER FLOW PROBLEMS
J. HYDROL VOL. 21, NO. 4, APR 1974, PP. 381-398.

74-0255 GOLDBERG, J. E.
BALUCH, M. H.
FINITE ELEMENT APPROACH TO BENDING OF MICROPOLAR
PLATES
INT. J. NUMER METHODS ENG VOL. 8, NO. 2 1974, PP.
311-321.

74-0256 HARBERL, G.
OCH, F.
EINE FINITE-ELEMENTE-LOESUNG FUR DIE
TORSIONSSTEIFIGKEIT UND DEN SCHUBMITTELPUNKT
BELIEBIGER QUERSCHNITTE (FINITE ELEMENTS SOLUTION
FOR TORSIONAL RIGIDITY AND SHEAR CENTER OF ANY
CROSS SECTION)
Z. FLUGWISS., VOL. 22, NO. 4, APRIL 1974, PP.
115-119.

74-0257 HILL, J. R. M.
MCDONALD, M. M.
SUPPORT PERFORMANCE OF HYDRAULIC BACKFILL
US BUR MINES REP INVEST NO. 7850 1974, 12 PP.

74-0258 RADAJ, D.
FESTIGKEITSNACHWEISE
DUSSELDORF DEUTSCHER VERLAG FUR SCHWEISSTECHNIK,
1974.

74-0259 HUNT, M. O.
SUDDARTH, S. K.
PREDICTION OF ELASTIC CONSTANTS OF PARTICLEBOARD
FOR PROD. J., VOL. 24, NO. 5, MAY 1974, PP. 52-57.

74-0260 LOMAX, R. J.
BARNES, J. J.
TWO-DIMENSIONAL FINITE ELEMENT SIMULATION OF
SEMICONDUCTOR DEVICES
ELECTRONICS LETTERS 1974, VOL. 10, NO. 16, PP.
341-343.

74-0261 JANSSEN, T. L.
SWANEY, T. G.
SHELL INSTABILITY ANALYSIS APPLIED TO A RADOME
A.I.A.A. J. VOL. 12, NO. 5, MAY 1974 PP. 714-716.

74-0262 KAUFMAN, S.
FINITE-ELEMENT METHOD FOR THE DETERMINATION OF
THERMAL STRESSES IN ANISOTROPIC SOLIDS OF
REVOLUTION
BELL SYST. TECH. J., VOL. 53, NO. 5, MAY-JUN.1974,
PP. 827-845.

74-0263 KORMAN, T.
MORGHEM, F.T.
BENDING OF MICROPOLAR PLATES
NUCL. ENG. DES. VOL. 26, NO. 3, FEB 1974, PP.
432-439.

74-0264 LO, K. Y.
EVALUATION OF THE STABILITY OF NATURAL SLOPES IN
PLASTIC CHAMPLAIN CLAYS
CAN GEOTECH J. VOL. 11, NO. 1 FEB 1974, PP.
165-181.

74-0265 MATSUOKA, T.
BABA, K.
STRESS ANALYSIS OF FRAME STRUCTURE CONNECTED WITH
PLATES
MITSUBISHI HEAVY IND. TECH REV VOL 11, NO, 1 FEB
1974, PP. 1 - 8.

74-0266 MAWENYA, A. S.
DAVIES, J. D.
FINITE ELEMENT BENDING ANALYSIS OF MULTILAYER
PLATES
INT. J. NUMER. METHODS ENG., VOL. 8, NO. 2, 1974,
PP. 215-225.

74-0267 MOAN, T.
NOTE ON THE CONVERGENCE OF FINITE ELEMENT
APPROXIMATIONS FOR PROBLEMS FORMULATED IN
CURVILINEAR COORDINATE SYSTEMS
COMPUT. METHODS APPL. MECH. ENG., VOL. 2, NO. 2,
MAY 1974. PP. 209-235.

74-0268 NAKAO, Y.
SASAKI, N.
THREE DIMENSIONAL SEISMIC ANALYSIS OF
STRUCTURE-GROUND SYSTEMS BY THE FINITE ELEMENT
METHOD
MITSUBISHI HEAVY IND. TECH REV VOL. 11, NO. 1 FEB
1974, PP. 9 - 17.

74-0269 OWEN, D. R. J.
PRAKASH, A.
FINITE ELEMENT ANALYSIS OF ELASTO-PLASTIC
MATERIALS BY USE OF DISLOCATION DIPOLE SYSTEMS
INT. J. NUMER METHODS ENG. VOL. 8, NO. 2 1974, PP.
277-288.

74-0270 PADOVAN, J.
SEMI-ANALYTICAL FINITE ELEMENT PROCEDURE FOR
CONDUCTION IN ANISOTROPIC AXISYMMETRIC SOLIDS
INT. J. NUMER METHODS ENG. VOL. 8, NO. 2 1974, PP.
295-310.

74-0271 SCHWERZLER, D. D.
CRAWFORD, J. E.
A TECHNIQUE FOR CONNECTING BEAM ELEMENTS TO A
PLATE MODEL OF A COMPLICATED BOX SECTION
INT. CONF. ON VEHICLE STRUCTURAL MECHANICS: FINITE
ELEMENT APPLICATION TO VEHICLE DESIGN, DETROIT,
MICHIGAN, MAR. 26-28 1974, PP. 244-250.

74-0272 RAO, S. S.
OPTIMIZATION OF COMPLEX STRUCTURES TO SATISFY
STATIC, DYNAMIC AND AEROELASTIC REQUIREMENTS
INT. J. NUMER METHODS ENG. VOL. 8, NO. 2 1974, PP.
249-269.

74-0273 SATOH, K.
UEDA, Y.
TENDANCES ACTUELLES DES RECHERCHES SUR LES
CONTRAINTES ET DEFORMATIONS DUES AU BRIDAGE EN
RELATION AVEC LA FISSURATION DES SOUDURES (RECENT
TRENDS OF RESEARCH INTO RESTRAINT STRESSES AND
STRAINS IN RELATION TO WELD CRACKING)
SOUDAGE TECH CONNEXES VOL. 28, NO. 1-2 JAN-FEB
1974, PP. 43-61.

74-0274 STOKER, J. R.
FINITE ELEMENTS
JOURNAL OF THE BRITISH NUCLEAR ENERGY SOCIETY
1974, VOL. 13, NO. 3, PP. 324-330.

74-0275 TAVENAS, F. A.
CHAPEAU, C.
IMMEDIATE SETTLEMENTS OF THREE TEST EMBANKMENTS ON
CHAMPLAIN CLAY
CAN GEOTECH J.VOL. 11, NO. 1 FEB 1974, PP.
109-141.

74-0276 VANDHAN, C. P.
KAPOOR, M. P.
INTEGRATED SEQUENTIAL SOLVER FOR LARGE MATRIX
EQUATIONS
INT. J. NUMER METHODS ENG. VOL. 8, NO. 2 1974, PP.
227-248.

74-0277 WARZEE, G.
FINITE ELEMENT ANALYSIS OF TRANSIENT HEAT
CONDUCTION. APPLICATION OF THE WEIGHTED RESIDUAL
PROCESS
COMPUT METHODS APPL MECH. ENG. VOL. 3, NO. 2 MAY
1974, PP. 255-268.

74-0278 WILSON, C. R.
WITHERSPOON, P. A.
STEADY STATE FLOW IN RIGID NETWORKS OF FRACTURES
WATER RESOURCES RESEARCH, VOL. 10, NO. 2, APRIL
1974, PP. 328-335.

74-0279 ZIENKIEWICZ, O. C.
OWEN, D. R. J.
LEAST SQUARE-FINITE ELEMENT FOR ELASTO-STATIC
PROBLEMS
INT. J. NUMER METHODS ENG. VOL. 8, NO. 2 1974, PP.
341-358.

74-0280 ANON
CATCHING WAVES WITH COMPUTERS
SURVEYOR, NEW YORK, VOL. 8, NO. 2, MAY 1974, PP.
3-13.

74-0281 ARAI, H.
APPLICATION OF FINITE ELEMENT METHOD TO THE STUDY
OF STRENGTH OF CAR BODY STRUCTURE AND WHEEL
Q. REP. RAILWAY TECH. RES. INST. TOKYO, VOL. 15,
NO. 1, MARCH 1974, PP. 25-43.

74-0282 NORRIE, D. H.
DE VRIES, G.
APPLICATION OF THE PSEUDO-FUNCTIONAL FINITE
ELEMENT METHOD TO NON-LINEAR PROBLEMS
IN FINITE ELEMENT METHODS IN FLOW PROBLEMS, ED.
GALLAGHER, ET. AL., JOHN WILEY, 1974.

74-0283 ROBINSON, J.
A MODE AMPLITUDE TECHNIQUE FOR GENERAL FINITE
ELEMENT EQUILIBRIUM
RA/ARD, REPORT NO. 23-1-74-22, JANUARY 1974.

74-0284 NORRIE, D. H.
DE VRIES, G.
A FINITE ELEMENT BIBLIOGRAPHY - PART I: AUTHOR
LISTING
REPORT NO. 57, DEPT. OF MECH. ENG., UNIV. OF
CALGARY, JUNE 1974

74-0285 NORRIE, D. H.
DE VRIES, G.
A FINITE ELEMENT BIBLIOGRAPHY - PART II: KEYWORD
LISTING
REPORT NO. 58, DEPT. OF MECH. ENG., UNIV. OF
CALGARY, JUNE 1974

74-0286 NORRIE, D. H.
DE VRIES, G.
A FINITE ELEMENT BIBLIOGRAPHY - PART III: CITATION
LISTING
REPORT NO. 59, DEPT. OF MECH. ENG., UNIV. OF
CALGARY, JUNE 1974

74-0287 NORRIE, D. H.
DE VRIES, G.
THE APPLICATION OF THE PSEUDO-FUNCTIONAL FINITE
ELEMENT METHOD TO VISCOUS FLOW PROBLEMS
REPORT NO. 63, DEPT. OF MECH. ENG. UNIV. OF
CALGARY, MAY 1974

74-0288 DAVIES, J. D.
DOYLE, W. S.
HARRISON, P. L.
MAWENYA, A. S.
EXPLICIT TRIANGULAR BENDING ELEMENT MATRIX
ASCE J. STRUCT. DIV. VOL. 100, NO. ST7, JULY 1974,
PP. 1459-1472.

74-0289 HAHN, O.
UNTERSUCHUNGEN ZUM FESTIGKEITSVERHALTEN VON
METALLKLEBUNGEN IM HINBLICK AUF DIE ERSTELLUNG
EINES PRAKTIKABLEN DIMENSIONIEUNGSVERFAHRENS
(INVESTIGATIONS OF THE STRENGTH BEHAVIOR OF BONDED
METAL JOINTS WITH RESPECT TO THE ESTABLISHMENT OF
A PRATICABLE DIMENSIONING METHOD)
IND-ANZ. VOL. 96, NO. 32, APR. 1974, PP. 709-712.

74-0290 KIRCHHOFF, L. W.
COMPUTER GRAPHICS FOR 3-D FINITE ELEMENT MODELS
PAP. NO. 74, A.S.M.E. MEETING, JUNE 24-28, 7 PP.

74-0291 KOBAYASHI, A. S.
WADE, B. G.
CRACK BRANCHING IN HOMALITE-100 SHEETS
ENG. FRACT. MECH. VOL. 6, NO. 1, MAR. 1974, PP.
81-92.

74-0292 LAURA, P. A.
REYES, J. A.
NUMERICAL EXPERIMENTS ON THE DETERMINATION OF
STRESS CONCENTRATION FACTORS
STRAIN. VOL. 10, NO. 2, APR. 1974, PP. 58-63.

74-0293 MANDELL, J. F.
MCGARRY, F.
STRESS INTENSITY FACTORS FOR ANISOTROPIC FRACTURE
TEST SPECIMENS OF SEVERAL GEOMETRIES
J. COMPOS. MATER. VOL. 8, APR. 1974, PP. 106-116.

74-0294 MARTZ, J. W.
MCCLELLAND, W. A.
IMPROVED TECHNIQUES FOR DYNAMIC ANALYSIS OF
EARTHMOVING EQUIPMENT
PREPR. NO. 740425, S.A.E. MEETING, APRIL 23-24,
1974, 13 PP.

74-0295 MEYER, T. O.
MCVEY, J. R.
NX BOREHOLE JACK MODULUS DETERMINATIONS IN
HOMOGENEOUS, ISOTROPIC, ELASTIC MATERIALS
US BUR. MINES REP. INVEST. NO. 7855, 1974, 50 P.

74-0296 MILLER, K. J.
KFOURI, A. P.
ELASTIC-PLASTIC FINITE ELEMENT ANALYSIS OF CRACK
TIP FIELDS UNDER BIAXIAL LOADING CONDITIONS
INT. J. OF FRACTURE. VOL. 10, NO. 3, 1974, PP.
393-404.

74-0297 PADOVAN, J.
QUASI-ANALYTICAL FINITE ELEMENT PROCEDURES FOR
AXISYMMETRIC ANISOTROPIC SHELLS AND SOLIDS
COMPUT. STRUCT. VOL. 4, NO. 3, MAY 1974, PP.
467-483.

74-0298 PLANEIX, J. M.
HUTHER. M.
DYNAMIC AND STOCHASTIC ASPECTS IN THE STRUCTURAL
ANALYSIS OF VLCCS AND ULCCS
TRANS. NORTH EAST COAST INST. ENG. SHIPBUILD VOL.
90, NO. 5,MAY 1974, PP. 133-150.

74-0299 SCHWING, H.
MEHLHORN, G.
ZUM TRAGVERHALTEN VON WAENDEN AUS FERTIGTEILTAFELN
(BEHAVIOR OF LARGE PANEL SHEAR WALLS)
BETONWERK FERTIGTEIL-TECH. VOL. 40, NO. 5, MAY
1974, PP. 313-324.

74-0300 SHANNON, R. W. E.
STRESS INTENSITY FACTORS FOR THICK-WALLED
CYLINDERS
INT. J. PRESSURE VESSELS PIPING VOL. 2, NO. 1, JAN
1974, PP.19-29.

74-0301 SMITH, H. W.
FINITE ELEMENT AIRPLANE COST ANALYSIS
PREPR. NO. 740390, S.A.E. MEETING, APRIL 2-5,
1974, 8 PP.

74-0302 SZABO, B. A.
CHEN, K. C.
CONFORMING FINITE ELEMENTS BASED ON COMPLETE
POLYNOMIALS
COMPUT. STRUCT., VOL. 4, NO. 3, MAY 1974, PP.
521-530.

74-0303 TURCKE, D. J.
MCNEICE, G. M.
GUIDELINES FOR SELECTING FINITE ELEMENT GRIDS
BASED ON AN OPTIMIZATION STUDY
COMPUT. STRUCT. VOL. 4, NO. 3, MAY 1974, PP.
499-519.

74-0304 WALLACE, J. A.
STRUCTURAL FINITE ELEMENT ANALYSIS AIDED BY
COMPUTER GRAPHICS
PREPR. NO. 740389, S.A.E. MEETING, APRIL 2-5,
1974, 11 PP.

74-0305 WINKEL, B. V.
STRESS AND DEFORMATION ANALYSIS OF IRRADIATION
INDUCED SWELLING
PAPER NO. 74, A.S.M.E. MEETING, JUNE 24-28, 1974,
8 PP.

74-0306 ZUDANS, Z.
REDDI M. M.
TSAI, H. C.
DYPLAS, A FINITE ELEMENT DYNAMIC ELASTIC-PLASTIC
LARGE DEFORMATION ANALYSIS PROGRAM
NUCL. ENG. AND DES., VOL. 27, NO. 3, 1974, PP.
398-412.

74-0307 STRICKLIN, J. A.
HAISLER, W. E.
SURVEY OF SOLUTION PROCEDURES FOR NONLINEAR STATIC
AND DYNAMIC ANALYSES
INT. CONF. ON VEHICLE STRUCTURAL MECHANICS: FINITE
ELEMENT APPLICATION TO VEHICLE DESIGN, DETROIT,
MICHIGAN, MAR. 26-28 1974, PP. 1-18.

74-0308 LARKIN, L. A.
ELASTIC-PLASTIC PLATE BENDING WITH CONSTANT
CURVATURE ELEMENTS
INT. CONF. ON VEHICLE STRUCTURAL MECHANICS:
FINITE ELEMENT APPLICATION TO VEHICLE DESIGN,
DETROIT, MICHIGAN, MARCH 26-28, 1974, PP. 18-26.

74-0309 MELOSH, R. J.
FINITE ELEMENT ANALYSIS OF AUTOMOBILE STRUCTURES
INT. CONF. ON VEHICLE STRUCTURAL MECHANICS: FINITE
ELEMENT APPLICATION TO VEHICLE DESIGN, DETROIT,
MICHIGAN, MAR. 26-28 1974, PP. 26-39.

74-0310 PARK, S. W.
DUVALL, F. W.
FINITE ELEMENT STRUCTURAL ANALYSIS AS APPLIED TO
AN AUTOMOTIVE DOOR STRUCTURE
INT. CONF. ON VEHICLE STRUCTURAL MECHANICS: FINITE
ELEMENT APPLICATION TO VEHICLE DESIGN, DETROIT,
MICHIGAN, MAR. 26-28 1974, PP. 39-58.

74-0311 FENSEL, P. A.
AN AXISYMMETRIC FINITE ELEMENT ANALYSIS OF THE
MECHANICAL AND THERMAL STRESSES IN BRAKE DRUMS
INT. CONF. ON VEHICLE STRUCTURAL MECHANICS: FINITE
ELEMENT APPLICATION TO VEHICLE DESIGN, DETROIT,
MICHIGAN, MAR. 26-28 1974, PP. 58-67.

74-0312 WADLEIGH, K. H.
APPLICATION OF FINITE ELEMENT METHODS TO COMPLETE
AUTOMOBILE STRUCTURAL DESIGN EVALUATION
INT. CONF. ON VEHICLE STRUCTURAL MECHANICS: FINITE
ELEMENT APPLICATION TO VEHICLE DESIGN, DETROIT,
MICHIGAN, MAR. 26-28 1974, PP. 67-73.

74-0313 NAGY, L. I.
STATIC ANALYSIS VIA SUBSTRUCTURING OF AN
EXPERIMENTAL VEHICLE FRONT-END BODY STRUCTURE
INT. CONF. ON VEHICLE STRUCTURAL MECHANICS: FINITE
ELEMENT APPLICATION TO VEHICLE DESIGN, DETROIT,
MICHIGAN, MAR. 26-28 1974, PP. 73-81.

74-0314 BARRON, G.E.
THE USE OF ELASTIC-PLASTIC FINITE ELEMENT ANALYSIS
IN THE CALCULATION OF CUMULATIVE FATIGUE DAMAGE
INT. CONF. ON VEHICLE STRUCTURAL MECHANICS: FINITE
ELEMENT APPLICATION TO VEHICLE DESIGN, DETROIT,
MICHIGAN, MAR. 26-28 1974, PP. 81-89.

74-0315 OH, H. L.
WANG, N.
THE COMPUTATION OF TEARING ENERGY OF NICKED RUBBER
STRIPS IN EXTENSION
INT. CONF. ON VEHICLE STRUCTURAL MECHANICS: FINITE
ELEMENT APPLICATION TO VEHICLE DESIGN, DETROIT,
MICHIGAN, MAR. 26-28 1974, PP. 89-95.

74-0316 MCCLELLAND, W. A.
KLOSTERMAN, A. L.
USING NASTRAN FOR DYNAMIC ANALYSIS OF VEHICLE
SYSTEMS
INT. CONF. ON VEHICLE STRUCTURAL MECHANICS: FINITE
ELEMENT APPLICATION TO VEHICLE DESIGN, DETROIT,
MICHIGAN, MAR. 26-28 1974, PP. 95-108.

74-0317 SACZALSKI, K. J.
PARK, K. C.
AN INTERACTIVE HYBRID TECHNIQUE FOR CRASHWORTHY
DESIGN OF COMPLEX VEHICULAR STRUCTURAL SYSTEMS
INT. CONF. ON VEHICLE STRUCTURAL MECHANICS: FINITE
ELEMENT APPLICATION TO VEHICLE DESIGN, DETROIT,
MICHIGAN, MAR. 26-28 1974, PP. 108-124.

74-0318 HOWELL, L. J.
POWER SPECTRAL DENSITY ANALYSIS OF VEHICLE
VIBRATION USING THE NASTRAN COMPUTER PROGRAM
INT. CONF. ON VEHICLE STRUCTURAL MECHANICS: FINITE
ELEMENT APPLICATION TO VEHICLE DESIGN, DETROIT,
MICHIGAN, MAR. 26-28 1974, PP. 124-134.

74-0319 VAIL, C.F.
A MODAL SYNTHESIS TECHNIQUE FOR DETERMINING
DYNAMIC PROPERTIES FOR A STRUCTURE FOR MASS AND
STIFFNESS CHANGES
INT. CONF. ON VEHICLE STRUCTURAL MECHANICS: FINITE
ELEMENT APPLICATION TO VEHICLE DESIGN, DETROIT,
MICHIGAN, MAR. 26-28 1974, PP. 134-145.

74-0320 NELSON, M. E.
THE USE OF CONDENSATION TECHNIQUES FOR SOLVING
DYNAMICS PROBLEMS
INT. CONF. ON VEHICLE STRUCTURAL MECHANICS: FINITE
ELEMENT APPLICATION TO VEHICLE DESIGN, DETROIT,
MICHIGAN, MAR. 26-28 1974, PP. 145-154.

74-0321 KARNES, R. N.
SEBASTIAN, J. D.
TOCHER, J. L.
TWIGG, D. W.
A USER-ORIENTED PROGRAM FOR CRASH DYNAMICS
INT. CONF. ON VEHICLE STRUCTURAL MECHANICS: FINITE
ELEMENT APPLICATION TO VEHICLE DESIGN, DETROIT,
MICHIGAN, MAR. 26-28 1974, PP. 154-164.

74-0322 JAEGER, L. G.
YOUSSEF, A.
THE ROLE OF FINITE DEFORMATION ANALYSIS IN PLANE
STRESS AND STRAIN FRACTURES
INT. CONF. ON VEHICLE STRUCTURAL MECHANICS: FINITE
ELEMENT APPLICATION TO VEHICLE DESIGN, DETROIT,
MICHIGAN, MAR. 26-28 1974, PP. 164-173.

74-0323 TANG, S. C.
PETROF, R. C.
ELASTO-PLASTIC ANALYSIS OF STRESS IN A GAS-TURBINE
WHEEL
INT. CONF. ON VEHICLE STRUCTURAL MECHANICS: FINITE
ELEMENT APPLICATION TO VEHICLE DESIGN, DETROIT,
MICHIGAN, MAR. 26-28 1974, PP. 173-188.

74-0324 BELYTSCHKO, T.
BRUCE, R. W.
WELCH, R. E.
LARGE DISPLACEMENT, NONLINEAR TRANSIENT ANALYSIS
BY FINITE ELEMENTS
INT. CONF. ON VEHICLE STRUCTURAL MECHANICS: FINITE
ELEMENT APPLICATION TO VEHICLE DESIGN, DETROIT,
MICHIGAN, MAR. 26-28 1974, PP. 188-198.

74-0325 CHEN, K.
FINE, D. S.
STIFFNESS ANALYSIS OF SHEET METAL SHELLS UNDER
CONCENTRATED LOADS
INT. CONF. ON VEHICLE STRUCTURAL MECHANICS: FINITE
ELEMENT APPLICATION TO VEHICLE DESIGN, DETROIT,
MICHIGAN, MAR. 26-28, 1974, PP. 198-205.

74-0326 TURCKE, D. J.
MCNEICE, G. M.
APPLICATION OF GRID SELECTION PROCEDURES FOR
IMPROVED FINITE ELEMENT STRESS ANALYSIS
INT. CONF. ON VEHICLE STRUCTURAL MECHANICS: FINITE
ELEMENT APPLICATION TO VEHICLE DESIGN, DETROIT,
MICHIGAN, MAR. 26-28 1974, PP. 205-217.

74-0327 ARAL, M. M.
FINITE ELEMENT SOLUTIONS OF SELECTED PARTIAL
DIFFERENTIAL EQUATIONS "FEMAC COMPUTER PROGRAM"
MIDDLE EAST TECH. UNIV. 1974, PP. 1-99.

74-0328 WILSON, E. L.
THE STATIC CONDENSATION ALGORITHM
INT. J. NUMER. METH. IN ENG., VOL. 8, NO. 1, 1974,
PP. 198-202.

74-0329 DANA, J. R.
STRESS CONCENTRATIONS NEAR HOLES IN LAMINATES
ASCE J. ENG. MECH. DIV. VOL. 100, NO. M-3, JUNE
1974, PAP 10590, PP. 477-488.

74-0330 BIRKHOFF, G.
COMPATIBLE TRIANGULAR FINITE ELEMENTS
JOURNAL OF MATHEMATICAL ANALYSIS AND APPLICATIONS
1974, VOL.47, NO. 3, PP. 531-553.

74-0331 BUSSE, L.
SCHWINGUNGEN ZYLINDRISCHER SCHRAUBENFEDERN
(VIBRATIONS OF CYLINDRICAL COIL SPRINGS)
KONSTR. MASCH. APP. GERAETEBAU VOL. 26, NO. 5, MAY
1974, PP.171-176.

74-0332 DAWE, D. J.
CURVED FINITE ELEMENTS FOR THE ANALYSIS OF SHALLOW
AND DEEP ARCHES
COMPUT. STRUCT. VOL. 4, NO. 3, MAY 1974, PP.
559-580.

74-0333 DESAI, C. S.
NUMERICAL DESIGN-ANALYSIS FOR PILES IN SANDS
ASCE J. GEOTECH. ENG. DIV. VOL. 100, NO. GT6, JUNE
1974, PP.613-635.

DJURIC, M. P.
NEW METHOD OF BULKHEAD ANALYSIS IN AIRCRAFT
STRUCTURES
PREPR. NO. 740388, S.A.E. MEETING, APRIL 2-5,
1974, 11 PP.

STRANG, G.
MOSCO, U.
ONE-SIDED   APPROXIMATION   AND   VARIATIONAL
INEQUALITIES
BULL. AMER. MATH. SOC., VOL. 80, 1974, PP.
308-312.

STRANG, G.
ONE-SIDED APPROXIMATION AND PLATE BENDING
COMPUTING   METHODS   IN   APPLIED   SCIENCE   AND
ENGINEERING, SPPINGER-VERLAG, BERLIN, 1974, PP.
140-155.

74-0337 STRANG, G.
THE FINITE ELEMENT METHOD-LINEAR AND NONLINEAR
APPLICATIONS
PROC. INTERNAT. CONG. OF MATHEMATICIANS 1974,
VANCOUVER, CANADA.

74-0338 COWPER, G. R.
LINDBERG, G. M.
STRESS ANALYSIS OF MULTI-CELLULAR CAISSONS
PROC. OF THE 2ND SYMPOSIUM ON APPLICATIONS OF
SOLID MECHANICS, MCMASTER UNIVERSITY, HAMILTON,
ONT., JUNE 17-18, 1974.

74-0339 IRONS, B. M.
THE PATCH TEST
FINITE ELEMENT SYMP., ATLAS COMPUT. LAB., CHILTON,
DIPCOT, BERKS., U.K., MARCH 1974.

74-0340 ODEN, J. T.
A THEORY OF MIXED FINITE ELEMENT APPROXIMATIONS OF
NON-SELF-ADJOINT BOUNDARY-VALUE PROBLEMS
PROC. OF THE 7TH U.S. NATIONAL CONGRESS OF APPLIED
MECHANICS, BOULDER, COLORADO, JUNE 1974.

74-0341 ODEN, J. T.
KEY, J. E.
FOST, R. B.
A NOTE ON THE ANALYSIS OF NONLINEAR DYNAMICS OF
ELASTIC MEMBRANES BY THE FINITE ELEMENT METHOD
COMPUTERS & STRUCTURES, VOL. 4, NO. 2, 1974, PP.
445-452.

74-0342 ODEN, J. T.
REDDY, J. N.
ON COMPLEMENTARY-DUAL VARIATIONAL PRINCIPLES IN
MATHEMATICAL PHYSICS
INT. JOUR. OF ENGINEERING SCIENCE, VOL. 12, NO. 1,
1974.

74-0343 ODEN, J. T.
APPROXIMATIONS AND NUMERICAL ANALYSIS OF FINITE
DEFORMATIONS OF ELASTIC SOLIDS
NONLINEAR ELASTICITY, ED. BY W. DICKEY, ACADEMIC
PRESS, NEW YORK, 1974, PP. 175-228.

74-0344 ODEN, J. T.
GODS, G.
HARMINAS, J.
FORMULATION AND APPLICATION OF CERTAIN PRIMAL AND
MIXED FINITE ELEMENT MODELS OF FINITE DEFORMATIONS
OF ELASTIC BODIES
LECTURE NOTES IN COMPUTER SCIENCE, VOL. 10,
COMPUTER   METHODS   IN   APPLIED   SCIENCES   AND
ENGINEERING, PART 1, SPRINGER-VERLAG,
BERLIN-HEIDELBERG, PP. 334-365.

74-0345 COWDREY, D. R.
WILLIS, J. R.
APPLICATION OF FINITE ELEMENT METHOD TO VIBRATIONS
OF QUARTZ PLATES
J. ACOUSTICAL SOC. OF AMERICA, VOL. 56, NO. 1,
1974, PP. 94-98.

74-0346 LEWIS, R. W.
NORRIS, V.
FRANCE, P. W.
THE DETERMINATION OF A WATER/GAS INTERFACE USING
FINITE ELEMENTS
INT. J. NUMER. METHODS IN ENGNG., VOL. 8, NO. 4,
1974.

74-0347 LEWIS, R. W.
BRUCH, J. C.
AN APPLICATION OF LEAST SQUARES TO ONE-DIMENSIONAL
TRANSIENT PROBLEMS
INT. J. FOR NUM. METH. IN ENG., VOL. 8, NO. 3,
1974.

74-0348 COMINI, C.
DEL GUIDICE, C.
LEWIS, R. W.
ZIENKIEWICZ, O. C.
FINITE ELEMENT SOLUTION OF NON-LINEAR HEAT
CONDUCTION PROBLEMS WITH SPECIAL REFERENCE TO
PHASE CHANGE
INT. J. FOR NUM. METH. IN ENG., VOL. 8, NO. 3,
1974.

74-0349 ZIENKIEWICZ, O. C.
LEWIS, R. W.
POROUS MEDIA - SOME PROBLEMS IN CIVIL AND
PETROLEUM ENGINEERING
SEMINAR ON NUMERICAL METHODS FOR PETROLEUM AND
CIVIL ENGINEERING, UNIVERSIDAD CENTRAL DE
VENEZUELA, CARACAS, FEBRUARY 1974.

74-0350 LEWIS, R. W.
A FINITE ELEMENT FORMULATION OF HEAT CONDUCTION
AND HEAT AND MASS TRANSFER PROBLEMS
FINITE ELEMENT SYMPOSIUM, ATLAS COMPUTER LAB.,
SCIENCE RESEARCH COUNCIL, MARCH 1974.

74-0351 HUANG, Y. H.
SONNENFE, S. L.
ANALYSIS OF UNSTEADY FLOW TOWARD AN ARTESIAN WELL
BY 3-DIMENSIONAL FINITE ELEMENTS
WATER RESOURCES RESFARCH, VOL. 10, NO. 3, 1974,
PP. 591-596.

74-0352 STEINCHE, W.
MENZEL, R.
PROCEDURE FOR SOLVING LINEAR EQUATION SYSTEMS WITH
SPARSELY OCCUPIED COEFFICIENT MATRICES ASSOCIATED
WITH METHOD OF FINITE ELEMENTS
ANGEWANDTE INFORMATIK, BRAUNSCHWEIG, NO. 7, 1974,
PP. 294-298.

74-0353 ZIENKIEWICZ, O. C.
ANDERSON, C. A.
SPONTANEOUS IGNITION - FINITE ELEMENT SOLUTIONS
FOR STEADY AND TRANSIENT CONDITIONS
J. OF HEAT TRANSFER, VOL. 96, NO. 3, 1974, PP.
398-404.

74-0354 BAZANT, Z. P.
INCREMENTAL FINITE ELEMENT MATRICES
J. STRUCT. DIV., ASCE., VOL. 100, NO. ST9, 1974,
PP. 1976-1977.

74-0355 COOK, R. D.
IMPROVED TWO-DIMENSIONAL FINITE ELEMENT
J. STRUCT. DIV., ASCE., VOL. 100, NO. ST9, 1974,
PP. 1851-1863.

74-0356 RAJASEKARAN, S.
FINITE ELEMENT ANALYSIS OF REINFORCED CONCRETE
J. STRUCT. DIV., ASCE., VOL. 100, NO. ST9, 1974,
PP. 1967-1968.

74-0357 NICKELL, R. E.
TANNER, R. I.
CASWELL, B.
SOLUTION OF VISCOUS INCOMPRESSIBLE JET AND
FREE-SURFACE FLOWS USING FINITE ELEMENT METHODS
J. FLUID MECHANICS, VOL. 65, NO. 12, 1974, PP.
189-206.

74-0358 HAREN, R. J.
THREE DIMENSIONAL FINITE ELEMENT APPROXIMATION IN
ELECTROMAGNETIC EXPLORATION
GEOEXPLORATION, VOL. 12, NO. 2-3, 1974, PP.
225-226.

74-0359 PADOVAN, J.
STEADY CONDUCTION OF HEAT IN LINEAR AND NONLINEAR
FULLY ANISOTROPIC MEDIA BY FINITE ELEMENTS
J. OF HEAT TRANSFER, VOL. 96, NO. 3, 1974, PP.
313-318.

74-0360 SCHAUER, D. A.
COMPUTER CODE TO GENERATE FINITE ELEMENT MESH FOR
PROBLEMS OF COMPLEX AXISYMMETRIC GEOMETRY
MECHANICAL ENGINEERING, VOL. 96, NO. 9, 1974, P.
64.

74-0361 ARAL, K.
DATA GENERATION FOR A 3-D FINITE ELEMENT SYSTEM
MECHANICAL ENGINEERING, VOL. 96, NO. 9, 1974, P.
60.

74-0362 DUNHAM, R. S.
BECKER, E. B.
GUERRA, F. M.
ORGANIZATION AND FUNCTIONAL PURPOSE OF A FINITE
ELEMENT   COMPUTER PROGRAM
MECHANICAL ENGINEERING, VOL. 96, NO. 9, 1974, P.
64.

74-0363 ARMEN, H.
LEVINE, H. S.
NONLINEAR ANALYSIS OF STRUCTURES
NASA CONTRACT REP. CR-2351, MARCH 1974, 128 PP.

74-0364 BRUCH, J. C.
ZYVOLOSKI, G.
TRANSIENT TWO DIMENSIONAL HEAT CONDUCTION PROBLEMS
SOLVED BY THE FINITE ELEMENT METHOD
INT. J. NUMER. METH. ENG., VOL. 8, NO. 3, 1974,
PP. 481-494.

621

74-0365 BURNHAM, M. W.
BENZLEY, S. E.
SHELL DEFORMATION UNDER TOOL LOADS REPRESENTATION
OF SINGULARITIES WITH ISOPARAMETRIC FINITE
ELEMENTS
TECH. PAPER NO. 1074-115, A.S.M.E. MEETING, APRIL
29-MAY 3, 1974, ALSO IN INT. J. NUMER. METHODS
ENG., VOL. 8, NO. 3, 1974, PP. 537-554.

74-0366 CARL, E. J.
HAMANN, W. C.
HOW FINITE ELEMENT METHODS ARE INTRODUCED IN LARGE
AND SMALL ORGANIZATIONS
SAE SPEC. PUBL. NO. 740006 1974, PP. 22-34.

74-0367 CHENG, R. T.
ON THE ACCURACY OF CERTAIN C DEGREE CONTINUOUS
FINITE ELEMENT REPRESENTATIONS
INT. J. NUMER. METH. ENG., VOL. 8, NO. 3, 1974,
PP. 649-657.

74-0368 CHEN, P. C.
FINITE ELEMENT ANALYSIS OF ELASTIC PLASTIC
THICK-WALLED TUBES
PROC. OF 3RD BIENN. ARMY SYMP. ON SOLID MECH.,
OCEAN CITY, MD., OCTOBER 3-5, 1972, PP. 243-253.
(PUBL. BY ARMY MATER. AND MECH. RES. CENTER,
WATERTOWN, MASS., 1974).

74-0369 SMITH, I. M.
BOORMAN, R.
THE ANALYSIS OF FLEXIBLE BULKHEADS IN SANDS
PROC. INSTITUTION OF CIVIL ENGINEERS, LONDON, VOL.
57, 1974, PP. 413-436.

74-0370 COOK, R. D.
LADKANY, S. G.
OBSERVATIONS REGARDING ASSUMED STRESS HYBRID PLATE
ELEMENTS
INT. J. NUMER. METH. ENG., VOL. 8, NO. 3, 1974,
PP. 513-519.

74-0371 CURTIN, D. J.
BILLERBECK, W. J.
DEVELOPMENT OF ADVANCED INTERCONNECTORS FOR SOLAR
CELLS
COMSAT. TECH. REV., VOL. 4, NO. 1, SPRING 1974,
PP. 53-68.

74-0372 DAVIS, P. L.
HOW FINITE ELEMENT METHODS IMPROVE THE DESIGN
CYCLE
SAE SPEC. PUBL. NO. 740003, 1974, PP. 8-10.

74-0373 DAVIS, C. S.
ILLUSTRATIONS OF AUTOMOTIVE FINITE ELEMENT MODELS
- STATICS
SAE SPEC. PUBL. NO. 740004, 1974, PP. 11-15.

74-0374 EVERLING, G.
FORECASTING THE ROCK PRESSURE AND ITS EFFECT ON
THE LONGWALL FACES AND ROADWAYS IN THE BITUMINOUS
COAL MINING
PRZEGL. CORN. VOL. 30, NO. 1, JANUARY 1974, PP.
20-26.

74-0375 GIRAULT, V.
THEORY OF A FINITE DIFFERENCE METHOD ON IRREGULAR
NETWORKS
SIAM J. NUMER. ANAL., VOL. 11, NO. 2, APRIL 1974,
PP. 260-282.

74-0376 GOLD, L. M.
FINITE ELEMENT DESIGN STUDY FOR 30-MM CARTRIDGE
CASE
PROC. OF 3RD BIENN. ARMY SYMP. ON SOLID MECH.,
OCEAN CITY, MD., OCTOBER 3-5, 1972, PP. 324-328.
(PUBL. BY ARMY MATER. AND MECH. RES. CENTER,
WATERTOWN, MASS., 1974).

74-0377 HECHT, M.
MARGERIE, J. C.
STUDY OF THE FLEXURAL BEHAVIOUR OF CASTINGS
FONDERIE, VOL. 29, NO. 334, MAY 1974, PP. 183-195.

74-0378 HINTON, E.
CAMPBELL, J. S.
LOCAL AND GLOBAL SMOOTHING OF DISCONTINUOUS FINITE
ELEMENT FUNCTIONS USING A LEAST SQUARES METHOD
INT. J. NUMER. METHODS ENG., VOL. 8, NO. 3, 1974,
PP. 461-480.

74-0379 KOEHLER, W.
PITTR, J.
CALCULATION OF TRANSIENT TEMPERATURE FIELDS WITH
FINITE ELEMENTS IN SPACE AND TIME DIMENSIONS
INT. J. NUMER. METHODS ENG., VOL. 8, NO. 3, 1974,
PP. 625-631.

74-0380 LUNG, M.
MAHRENHOLTZ, O.
FINITE ELEMENT PROCEDURE FOR ANALYSIS OF METAL
FORMING PROCESSES
TRANS. CAN. SOC. MECH. ENG., VOL. 2, NO. 1,
1973-1974, PP. 31-36.

74-0381 MAHATA, P. C.
NEW TRIANGULAR ELEMENT FOR FINITE DIFFERENCE
SOLUTION OF AXISYMMETRIC CONDUCTION PROBLEMS IN
CYLINDRICAL COORDINATES
INT. J. NUMER. METHODS ENGNG., VOL. 8, NO. 3,
1974, PP. 547-567.

74-0382 MEDAGLIA, J. M
OLINE, L. W.
INVESTIGATION OF ELASTIC PLATE SPALLATION BY
FINITE ELEMENTS
PROC. OF 3RD BIENN. ARMY SYMP. ON SOLID MECH.,
OCEAN CITY, MD., OCTOBER 3-5, 1972, PP. 350-356.
(PUBL. BY ARMY MATER. AND MECH. RES. CENTER,
WATERTOWN, MASS. 1974).

74-0383 NAYLOR, D. J.
STRESSES IN NEARLY INCOMPRESSIBLE MATERIALS BY
FINITE ELEMENTS WITH APPLICATION TO THE
CALCULATION OF EXCESS PORE PRESSURES
INT. J. NUMER. METHODS ENG., VOL. 8, NO. 3, 1974,
PP. 443-460.

74-0384 PISKUN, V. V.
SAVCHENKO, V. G.
SOLUTION OF A THREE DIMENSIONAL AXISYMMETRICAL
PROBLEM OF THERMOPLASTICITY AS APPLIED TO THICK
TURBINE DISKS
PROBL. PROCHN., VOL. 6, NO. 5, MAY 1974, PP. 8-13.

74-0385 ROUCH, K. E.
JOUNG, K. S.
STRUCTURAL ANALYSIS OF MAST BY FINITE ELEMENT
TECHNIQUE
ALLIS-CHALMERS ENG. REV., VOL. 39, NO. 1, 1974,
PP. 20-23.

74-0386 SCHACHT, C. A.
HRIBAR, J. A.
ELASTIC ANALYSIS OF A REFRACTORY SPRUNG ARCH
AM. CERAM SOC. BULL., VOL. 53, NO. 7, JULY 1974,
PP. 528-532

74-0387 SIEKMANN, J.
SCHILLING, U.
CALCULATION OF THE FREE OSCILLATIONS OF A LIQUID
IN AXISYMMETRIC MOTIONLESS CONTAINERS OF ARBITRARY
SHAPE
Z. FLUGWISS., VOL. 22, NO. 5, MAY 1974, PP.
168-173.

74-0388 SMITH, G. L.
AUTOMOTIVE USE OF FINITE ELEMENT METHODS -
INTRODUCTION AND OVERVIEW
SAE SPEC. PUBL. NO. SP-387, 1974, PP. 1-7.

74-0389 SVALBONAS, V.
KEY, J. E.
STATIC, STABILITY, AND DYNAMIC ANALYSIS OF SHELLS
OF REVOLUTION BY NUMERICAL INTEGRATION - A
COMPARISON
NUCL. ENG. DES., VOL. 27, NO. 1, MARCH 1974, PP.
30-45.

74-0390 THOMPSON, J. E.
FUTURE DEVELOPMENTS IN STRUCTURAL ANALYSIS
SAE SPEC. PUBL. 740008, 1974, PP. 35-40.

74-0391 VAIL, C. F.
ILLUSTRATIONS OF AUTOMOTIVE FINITE ELEMENT MODELS
- DYNAMICS
SAE SPEC. PUBL. NO. 740005, 1974, PP. 16-21.

74-0392 WU, R. W.
DYNAMIC PLASTIC LARGE DEFLECTIONS OF CLAMPED
RECTANGULAR PLATE
J. APPL. MECH. TRANS. ASME, VOL. 41, SER. E., NO.
2, JUNE 1974, PP. 531-533.

74-0393 SMITH, I. M.
HOBBS, R.
NUMERICAL AND PHYSICAL MODELLING
IN NUMERICAL METHODS IN GEOTECHNICAL ENGINEERING,
MCGRAW HILL, NEW YORK, 1974.

74-0394 PULMANO, V. A.
BLACK, D. C.
KABAILA, A. P.
SUBSTRUCTURE ANALYSIS OF MULTISTOREY FLAT SLAB
BUILDINGS
PROC. ASCE-IABSE CONF. ON TALL BUILDINGS, BANGKOK,
JANUARY 1974.

74-0395 BLACK, D. C.
PULMANO, V. A.
KABAILA, A. P.
CERTAIN FINITE ELEMENTS FOR PLATE BENDING PROBLEMS
UNICIV REPORT R-126, UNIVERSITY OF NEW SOUTH
WALES, 1974.

74-0396 FRENCH, S.
PULMANO, V. A.
KABAILA, A. P.
PLANE STRESS ELEMENT OF VARYING THICKNESS APPLIED
TO END BLOCK ANALYSIS
PROC. CONF. FINITE ELEMENT METHODS IN ENGINEERING,
COIMBATORE INST. OF TECHNOLOGY, INDIA,
1974.

74-0397 SOMERVAILLE, I. J.
KABAILA, A. P.
MESH GRADING TECHNIQUES FOR COMPATIBLE AND
EQUILIBRIUM ELEMENTS
FINITE ELEMENT METHODS IN ENGINEERING (EDS.
PULMANO, V. A. , KABAILA, A. P.), PROC. INT.
CONF., UNIV OF NEW SOUTH WALES, AUSTRALIA, 1974,
PP. 257-269.

74-0398 BLACK, D. C.
PULMANO, V. A.
KABAILA, A. P.
ON THE BENDING STIFFNESS OF SLABS IN CROSS-WALLED
STRUCTURES
UNICIV REPORT R-133, UNIVERSITY OF NEW SOUTH
WALES, 1974.

74-0399 BLACK, D. C.
PULMANO, V. A.
KABAILA, A. P.
DUAL ANALYSIS OF COLUMN SUPPORTED SLABS
PROC. CONF. FINITE ELEMENT METHODS IN ENGINEERING,
COIMBATORE INST. OF TECHNOLOGY, INDIA,
1974.

74-0400 PULMANO, V. A.
KABAILA, A. P.
FINITE ELEMENT METHODS IN ENGINEERING
PROC. OF THE INT. CONF. ON FINITE ELEMENT METHODS
IN ENGINEERING, EDS. PULMANO, V.A., KABAILA, A.P.,
UNIV. OF NEW SOUTH WALES, AUSTRALIA, AUGUST 1974,
840 PP.

74-0401 LOCKE, B.
AN ANALYSIS OF AN INTERSECTION PROBLEM USING THE
SEMILOOP SHELL ELEMENT
UNDERGRADUATE THESIS, CIVIL ENGINEERING,
UNIVERSITY OF WALES, SWANSEA, 1974.

74-0402 REID, J. K.
A DISCUSSION ON A MODIFIED CONJUGATE GRADIENT
METHOD BY W.L. WOOD
INT. J. NUMER. METH. IN ENG., VOL. 8, NO. 2, 1974,
PP. 431-432.

74-0403 IRONS, B. M.
A TECHNIQUE FOR DEGENERATING BRICK-TYPE
ISOPARAMETRIC ELEMENTS USING HIERARCHICAL MIDSIDE
NODES
INT. J. NUMER. METH. IN ENG., VOL. 8, NO. 1, 1974,
PP. 203-209.

74-0404 BOGNER, F. K.
DISCUSSION OF A SHORT COMMUNICATION BY S.
GOPALACHARYULU
INT. J. NUMER. METH. IN ENG., VOL. 8, NO. 1, 1974,
PP. 209-211.

74-0405 BOISSERIE, J. M.
APPLICATION DE LA METHODE DES ELEMENTS FINIS — UN
PROCEDE DE SOUS-ASSEMBLAGE
INT. SYMP. ON COMPUTING METHODS IN APPLIED
SCIENCES AND ENGINEERING, PART I, VERSAILLES,
DECEMBER 17-21, 1973, PP. 312-332.

74-0406 CIARLET, P. G.
QUELQUES METHODES D'ELEMENTS FINIS POUR DE
PROBLEME D'UNE PLAQUE ENCASTREE
INT. SYMP. ON COMPUTING METHODS IN APPLIED
SCIENCES AND ENGINEERING, PART I, VERSAILLES,
DECEMBER 17-21, 1973, PP. 156-176.

74-0407 IRONS, B. M.
UN NOVEL ELEMENT DE COQUES GENERALES
INT. SYMP. ON COMPUTING METHODS IN APPLIED
SCIENCES AND ENGINEERING, PART I, VERSAILLES,
DECEMBER 17-21, 1973, PP. 177-192.

74-0408 JAMET, P.
RAVIART, P. A.
NUMERICAL SOLUTION OF THE STATIONARY NAVIER-STOKES
EQUATIONS BY FINITE ELEMENT METHODS
INT. SYMP. ON COMPUTING METHODS IN APPLIED
SCIENCES AND ENGINEERING, PART I, VERSAILLES,
DECEMBER 17-21, 1973, PP. 193-223.

74-0409 FRAEIJS DE VEUBEKE, B.
FINITE ELEMENTS METHOD IN AEROSPACE ENGINEERING
PROBLEMS
INT. SYMP. ON COMPUTING METHODS IN APPLIED
SCIENCES AND ENGINEERING, PART I, VERSAILLES,
DECEMBER 17-21, 1973, PP. 224-258.

74-0410 ZIENKIEWICZ, O. C.
VISCO-PLASTICITY AND PLASTICITY-AN ALTERNATIVE FOR
FINITE ELEMENT SOLUTION OF MATERIAL NONLINEARITIES
INT. SYMP. ON COMPUTING METHODS IN APPLIED
SCIENCES AND ENGINEERING, PART I, VERSAILLES,
DECEMBER 17-21, 1973, PP. 259-287.

74-0411 DOUGLAS, J.
DUPONT, T.
WHEELER, M. F.
SOME SUPERCONVERGENCE RESULTS FOR AN H-GALERKIN
PROCEDURE FOR THE HEAT EQUATION
INT. SYMP. ON COMPUTING METHODS IN APPLIED
SCIENCES AND ENGINEERING, PART I, VERSAILLES,
DECEMBER 17-21, 1973, PP. 288-311.

74-0412 BEGIS, D.
GLOWINSKI, R.
APPLICATION DE LAT METHODE DES ELEMENTS FINIS A LA
RESOLUTION D'UN PROBLEME DE DOMAINE OPTIMAL
INT. SYMP. ON COMPUTING METHODS IN APPLIED
SCIENCES AND ENGINEERING, PART II, VERSAILLES,
DECEMBER 17-21, 1973, PP. 403-434.

74-0413 GLOWINSKI, P.
MARROCCO, A.
ETUDE NUMERIQUE DU CHAMP MAGNETIQUE DANS UN
ALTERNATEUR TETRAPOLAIRE PAR LA METHODE DES
ELEMENTS FINIS
INT. SYMP. ON COMPUTING METHODS IN APPLIED
SCIENCES AND ENGINEERING, VERSAILLES, DECEMBER
17-21, 1973.

74-0414 FENTON, D. L.
ECONOMICAL SOLUTION TECHNIQUES FOR LOAD DEFLECTION
EQUATIONS
FINITE ELEMENT METHOD IN STRUCTURAL ANALYSIS AND
USE OF COMPUTERS AS PROBLEM SOLVERS, OKTAY URAL,
UNIVERSITY OF MISSOURI-ROLLA, 1974, PP. 7.1-7.9.

74-0415 SMITH, I. M.
HOBBS, R.
FINITE ELEMENT ANALYSIS OF CENTRIFUGED AND
BUILT-UP SLOPES
GEOTECHNIQUE, VOL. 24, NO. 4, 1974.

74-0416 DE ARANTES E OLIVEIRA, E. R.
RESULTS ON THE CONVERGENCE OF THE FINITE ELEMENT
METHOD IN STRUCTURAL AND NONSTRUCTURAL CASES
PROC. INT. CONF. ON FINITE ELEMENT METHODS IN
ENG., (EDS.) PULMANO, V.A. AND KABAILA, A.P.,
UNIV. OF NEW SOUTH WALES, AUSTRALIA, AUGUST 28-30,
1974, PP. 3-14.

74-0417 ARGYRIS, J. H.
ROY, J. R.
ON THE REDUCTION OF NUMERICAL ERROR IN THE MATRIX
DISPLACEMENT METHOD
PROC. INT. CONF. ON FINITE ELEMENT METHODS IN
ENG., (EDS.) PULMANO, V.A. AND KABAILA, A.P.,
UNIV. OF NEW SOUTH WALES, AUSTRALIA, AUGUST 28-30,
1974, PP. 15-32.

74-0418 CAMPBELL, J. S.
A PENALTY FUNCTION APPROACH TO THE MINIMIZATION OF
QUADRATIC FUNCTIONALS IN FINITE ELEMENT ANALYSIS
PROC. INT. CONF. ON FINITE ELEMENT METHODS IN
ENG., (EDS.) PULMANO, V.A. AND KABAILA, A.P.,
UNIV. OF NEW SOUTH WALES, AUSTRALIA, AUGUST 28-30,
1974, PP. 33-54.

74-0419 CAREY, G. F.
BASIS FUNCTIONS IN FINITE ELEMENT THEORY AND
APPLICATION
PROC. INT. CONF. ON FINITE ELEMENT METHODS IN
ENG., (EDS.) PULMANO, V.A. AND KABAILA, A.P.,
UNIV. OF NEW SOUTH WALES, AUSTRALIA, AUGUST 28-30,
1974, PP. 55-74.

74-0420 O'CARROLL, M. J.
RASMUSSEN, H.
UNIFORM CONVERGENCE OF VARIATIONAL FINITE ELEMENT
AND RITZ METHODS
PROC. INT. CONF. ON FINITE ELEMENT METHODS IN
ENG., (EDS.) PULMANO, V.A. AND KABAILA, A.P.,
UNIV. OF NEW SOUTH WALES, AUSTRALIA, AUGUST 28-30,
1974, PP. 75-81.

74-0421 AHMAD, S.
IRONS, B. M.
AN ASSUMED STRESS APPROACH TO REFINED
ISOPARAMETRIC ELEMENTS IN THREE DIMENSIONS
PROC. INT. CONF. ON FINITE ELEMENT METHODS IN
ENG., (EDS.) PULMANO, V.A. AND KABAILA, A.P.,
UNIV. OF NEW SOUTH WALES, AUSTRALIA, AUGUST 28-30,
1974, PP. 85-100.

74-0422 BARRY, J. M.
HRDINA, J.
QUAASS, S. T.
THREE DIMENSIONAL STRESS ANALYSIS OF PENETRATIONS
IN THICK WALLED PRESSURE VESSELS
PROC. INT. CONF. ON FINITE ELEMENT METHODS IN
ENG., (EDS.) PULMANO, V.A. AND KABAILA, A.P.,
UNIV. OF NEW SOUTH WALES, AUSTRALIA, AUGUST 28-30,
1974, PP. 101-116.

74-0423 BECKER, E. B.
DUNHAM, R. S.
STERN, M.
SOME STRESS INTENSITY CALCULATIONS USING FINITE
ELEMENTS
PROC. INT. CONF. ON FINITE ELEMENT METHODS IN
ENG., (EDS.) PULMANO, V.A. AND KABAILA, A.P.,
UNIV. OF NEW SOUTH WALES, AUSTRALIA, AUGUST 28-30,
1974, PP. 117-138.

74-0424 HUGHES, O. F.
MISTREE, F.
SOME CONSIDERATIONS REGARDING STRUCTURAL
OPTIMIZATION AND FINITE ELEMENT ANALYSIS
PROC. INT. CONF. ON FINITE ELEMENT METHODS IN
ENG., (EDS.) PULMANO, V.A. AND KABAILA, A.P.,
UNIV. OF NEW SOUTH WALES, AUSTRALIA, AUGUST 28-30,
1974, PP. 139-159.

74-0425 KHATUA, T. P.
CHEUNG, Y. K.
FINITE ELEMENT ANALYSIS OF MULTILAYER SANDWICH
PLATES AND SHELLS
PROC. INT. CONF. ON FINITE ELEMENT METHODS IN
ENG., (EDS.) PULMANO, V.A. AND KABAILA, A.P.,
UNIV. OF NEW SOUTH WALES, AUSTRALIA, AUGUST 28-30,
1974, PP. 161-176.

74-0426 KWOK, W. L.
CHEUNG, Y. K.
ANALYSIS OF CIRCULAR AND ANGULAR LAMINATED THICK
PLATES
PROC. INT. CONF. ON FINITE ELEMENT METHODS IN
ENG., (EDS.) PULMANO, V.A. AND KABAILA, A.P.,
UNIV. OF NEW SOUTH WALES, AUSTRALIA, AUGUST 28-30,
1974, PP. 177-194.

74-0427 MARSHALL, J.
MISHU, F.
AN INVESTIGATION INTO THE ECONOMICS OF USING
COMPLEX ELEMENTS IN PLANE STRESS ANALYSIS
PROC. INT. CONF. ON FINITE ELEMENT METHODS IN
ENG., (EDS.) PULMANO, V.A. AND KABAILA, A.P.,
UNIV. OF NEW SOUTH WALES, AUSTRALIA, AUGUST 28-30,
1974, PP. 195-211.

74-0428 MARTIN, C. W.
AN ITERATIVE IMPROVEMENT FOR FINITE ELEMENT
ANALYSIS
PROC. INT. CONF. ON FINITE ELEMENT METHODS IN
ENG., (EDS.) PULMANO, V.A. AND KABAILA, A.P.,
UNIV. OF NEW SOUTH WALES, AUSTRALIA, AUGUST 28-30,
1974, PP. 213-217.

74-0429 PATTERSON, C.
AN EVALUATION OF FINITE ELEMENTS WITH IMPROVED
ASSURED CONVERGENCE FOR PLANE STRESS AND PLANE
STRAIN
PROC. INT. CONF. ON FINITE ELEMENT METHODS IN
ENG., (EDS.) PULMANO, V.A. AND KABAILA, A.P.,
UNIV. OF NEW SOUTH WALES, AUSTRALIA, AUGUST 28-30,
1974, PP. 219-230.

74-0430 PULMANO, V. A.
A COMPATIBLE PLATE BENDING ANNULAR SECTOR FINITE
ELEMENT OF NONUNIFORM THICKNESS
PROC. INT. CONF. ON FINITE ELEMENT METHODS IN
ENG., (EDS.) PULMANO, V.A. AND KABAILA, A.P.,
UNIV. OF NEW SOUTH WALES, AUSTRALIA, AUGUST 28-30,
1974, PP. 231-245.

74-0431 SOMERVAILLE, I. J.
A FAMILY OF EQUILIBRIUM PLATE BENDING ELEMENTS
PROC. INT. CONF. ON FINITE ELEMENT METHODS IN
ENG., (EDS.) PULMANO, V.A. AND KABAILA, A.P.,
UNIV. OF NEW SOUTH WALES, AUSTRALIA, AUGUST 28-30,
1974, PP. 247-256.

74-0432 VAN DER VOOREN, J.
LABRUJERE, TH.E.
FINITE ELEMENT SOLUTION OF THE INCOMPRESSIBLE FLOW
OVER AN AIRFOIL IN A NON UNIFORM STREAM
PROC. INT. CONF. ON NUMER. METH. IN FLUID
DYNAMICS, UNIV. OF SOUTHAMPTON, ENGLAND, SEPTEMBER
26-28, 1973, PP. 23-41.

74-0433 YOSHIDA, Y.
A HYBRID STRESS ELEMENT FOR THIN SHELL ANALYSIS
PROC. INT. CONF. ON FINITE ELEMENT METHODS IN
ENG., (EDS.) PULMANO, V.A. AND KABAILA, A.P.,
UNIV. OF NEW SOUTH WALES, AUSTRALIA, AUGUST 28-30,
1974, PP. 271-284.

74-0434 AL-HASHIMI, K.
FRACTURE IN SOIL CEMENT USING THE FINITE ELEMENT
METHOD
PROC. INT. CONF. ON FINITE ELEMENT METHODS IN
ENG., (EDS.) PULMANO, V.A. AND KABAILA, A.P.,
UNIV. OF NEW SOUTH WALES, AUSTRALIA, AUGUST 28-30,
1974, PP. 287-298.

74-0435 ARAI, H.
ANALYSIS ON THE LARGE DEFORMATION OF PLATE
STRUCTURES
PROC. INT. CONF. ON FINITE ELEMENT METHODS IN
ENG., (EDS.) PULMANO, V.A. AND KABAILA, A.P.,
UNIV. OF NEW SOUTH WALES, AUSTRALIA, AUGUST 28-30,
1974, PP. 299-313.

74-0436 BOOKER, J. R.
SMALL, J. C.
THE NUMERICAL SOLUTION OF VISCO-ELASTIC PROBLEMS
USING LAPLACE TRANSFORMS
PROC. INT. CONF. ON FINITE ELEMENT METHODS IN
ENG., (EDS.) PULMANO, V.A. AND KABAILA, A.P.,
UNIV. OF NEW SOUTH WALES, AUSTRALIA, AUGUST 28-30,
1974, PP. 315-326.

74-0437 DAVIS, E. H.
RING, G. J.
BOOKER, J. R.
THE SIGNIFICANCE OF THE RATE OF PLASTIC WORK IN
ELASTO-PLASTIC ANALYSIS
PROC. INT. CONF. ON FINITE ELEMENT METHODS IN
ENG., (EDS.) PULMANO, V.A. AND KABAILA, A.P.,
UNIV. OF NEW SOUTH WALES, AUSTRALIA, AUGUST 28-30,
1974, PP. 327-335.

74-0438 HAMID, M. S.
GHISTA, D. N.
FINITE ELEMENT ANALYSIS OF HUMAN CARDIAC
STRUCTURES
PROC. INT. CONF. ON FINITE ELEMENT METHODS IN
ENG., (EDS.) PULMANO, V.A. AND KABAILA, A.P.,
UNIV. OF NEW SOUTH WALES, AUSTRALIA, AUGUST 28-30,
1974, PP. 337-348.

74-0439 HAQUE, M. N.
VALLIAPPAN, S.
COOK D. J.
TENSILE CREEP ANALYSIS OF CONCRETE STRUCTURES
PROC. INT. CONF. ON FINITE ELEMENT METHODS IN
ENG., (EDS.) PULMANO, V.A. AND KABAILA, A.P.,
UNIV. OF NEW SOUTH WALES, AUSTRALIA, AUGUST 28-30,
1974, PP. 349-364.

74-0440 HENRICHSEN, L.
VISCOELASTIC FINITE ELEMENTS
PROC. INT. CONF. ON FINITE ELEMENT METHODS IN
ENG., (EDS.) PULMANO, V.A. AND KABAILA, A.P.,
UNIV. OF NEW SOUTH WALES, AUSTRALIA, AUGUST 28-30,
1974, PP. 365-377.

74-0441 MAJID, K. I.
THE EFFECT OF COMPOSITE ACTION ON THE
ELASTOPLASTIC ANALYSIS OF COMPLETE BUILDING
STRUCTURES
PROC. INT. CONF. ON FINITE ELEMENT METHODS IN
ENG., (EDS.) PULMANO, V.A. AND KABAILA, A.P.,
UNIV. OF NEW SOUTH WALES, AUSTRALIA, AUGUST 28-30,
1974, PP. 379-401.

74-0442 SAEED MIRZA, M.
NONLINEAR FINITE ELEMENT ANALYSIS OF REINFORCED
CONCRETE STRUCTURES
PROC. INT. CONF. ON FINITE ELEMENT METHODS IN
ENG., (EDS.) PULMANO, V.A. AND KABAILA, A.P.,
UNIV. OF NEW SOUTH WALES, AUSTRALIA, AUGUST 28-30,
1974, PP. 403-417.

74-0443 PIAN, T. H. H.
TONG, P.
LUK, C. H.
SPILKER, R. L.
ELASTIC PLASTIC ANALYSIS BY ASSUMED STRESS HYBRID
MODEL
PROC. INT. CONF. ON FINITE ELEMENT METHODS IN
ENG., (EDS.) PULMANO, V.A. AND KABAILA, A.P.,
UNIV. OF NEW SOUTH WALES, AUSTRALIA, AUGUST 28-30,
1974, PP. 419-434.

74-0444 RAJASEKARAN, S.
MURRAY, D. W.
FINITE ELEMENT LARGE DEFLECTION ANALYSIS OF
THIN-WALLED BEAMS
PROC. INT. CONF. ON FINITE ELEMENT METHODS IN
ENG., (EDS.) PULMANO, V.A. AND KABAILA, A.P.,
UNIV. OF NEW SOUTH WALES, AUSTRALIA, AUGUST 28-30,
1974, PP. 435-454.

74-0445 WORKMAN, G. H.
RODABAUGH, E. C.
SIMPLIFIED INELASTIC HIGH TEMPERATURE STRUCTURAL
ANALYSIS OF MODERATELY COMPLEX SPATIALLY THREE
DIMENSIONAL PIPING SYSTEM
PROC. INT. CONF. ON FINITE ELEMENT METHODS IN
ENG., (EDS.) PULMANO, V.A. AND KABAILA, A.P.,
UNIV. OF NEW SOUTH WALES, AUSTRALIA, AUGUST 28-30,
1974, PP. 455-475.

74-0446 BOONLUALOHR, P.
VALLIAPPAN, S.
LEE, I. K.
ELASTIC PLASTIC ANALYSIS OF SHALLOW FOUNDATIONS
PROC. INT. CONF. ON FINITE ELEMENT METHODS IN
ENG., (EDS.) PULMANO, V.A. AND KABAILA, A.P.,
UNIV. OF NEW SOUTH WALES, AUSTRALIA, AUGUST 28-30,
1974, PP. 479-492.

74-0447 KING, G. J. W.
CHANDRASEKARAN, V. S.
INTERACTIVE ANALYSIS OF A RAFTED MULTISTOREY SPACE
FRAME RESSTING ON AN INHOMOGENEOUS CLAY STRATUM
PROC. INT. CONF. ON FINITE ELEMENT METHODS IN
ENG., (EDS.) PULMANO, V.A. AND KABAILA, A.P.,
UNIV. OF NEW SOUTH WALES, AUSTRALIA, AUGUST 28-30,
1974, PP. 493-509.

74-0448 KAMESWARA RAO, N. S. V.
DASGUPTA, S. P.
FINITE ELEMENT SOLUTION TO SOME PLANE PROBLEMS IN
SOIL DYNAMICS
PROC. INT. CONF. ON FINITE ELEMENT METHODS IN
ENG., (EDS.) PULMANO, V.A. AND KABAILA, A.P.,
UNIV. OF NEW SOUTH WALES, AUSTRALIA, AUGUST 28-30,
1974, PP. 511-531.

74-0449 RICHARDS, B. G.
THE USE OF THE FINITE ELEMENT METHOD IN THE
SOLUTION OF THE FLOW EQUATIONS IN SOILS
PROC. INT. CONF. ON FINITE ELEMENT METHODS IN
ENG., (EDS.) PULMANO, V.A. AND KABAILA, A.P.,
UNIV. OF NEW SOUTH WALES, AUSTRALIA, AUGUST 28-30,
1974, PP. 533-549.

74-0450 URAL, O.
FINITE ELEMENT METHOD AND COMPUTER PROGRAMS
PROC. INT. CONF. ON FINITE ELEMENT METHODS IN
ENG., (EDS.) PULMANO, V.A. AND KABAILA, A.P.,
UNIV. OF NEW SOUTH WALES, AUSTRALIA, AUGUST 28-30,
1974, PP. 550-563.

74-0451 WARDLE, L. J.
FRASER, R. A.
FINITE ELEMENT ANALYSIS OF A PLATE ON A LAYERED
CROSS-ANISOTROPIC FOUNDATION
PROC. INT. CONF. ON FINITE ELEMENT METHODS IN
ENG., (EDS.) PULMANO, V.A. AND KABAILA, A.P.,
UNIV. OF NEW SOUTH WALES, AUSTRALIA, AUGUST 28-30,
1974, PP. 565-578.

74-0452 ATLURI, S. N.
FINITE ELEMENT PERTURBATION ANALYSIS OF NONLINEAR
DYNAMIC RESPONSE OF ELASTIC CONTINUUA
PROC. INT. CONF. ON FINITE ELEMENT METHODS IN
ENG., (EDS.) PULMANO, V.A. AND KABAILA, A.P.,
UNIV. OF NEW SOUTH WALES, AUSTRALIA, AUGUST 28-30,
1974, PP. 581-595.

74-0453 CHENG, F. Y.
FINITE ELEMENT ANALYSIS OF STRUCTURAL INSTABILITY
BY ASSOCIATION OF PULSATING EXCITATIONS
PROC. INT. CONF. ON FINITE ELEMENT METHODS IN
ENG., (EDS.) PULMANO, V.A. AND KABAILA, A.P.,
UNIV. OF NEW SOUTH WALES, AUSTRALIA, AUGUST 28-30,
1974, PP. 597-609.

74-0454 CHESNOKOV, S. S.
KANDIDOV, V. P.
VYSLOUKH, V. A.
STUDY ON THE VIBRATION OF THIN PLATES IN A GAS
STREAM BY THE FINITE ELEMENT METHOD
PROC. INT. CONF. ON FINITE ELEMENT METHODS IN
ENG., (EDS.) PULMANO, V.A. AND KABAILA, A.P.,
UNIV. OF NEW SOUTH WALES, AUSTRALIA, AUGUST 28-30,
1974, PP. 611-620.

74-0455 DAVIDSON, B. J.
MEDLAND, I. C.
A FINITE ELEMENT APPROACH TO STABILITY ANALYSIS IN
FRAMES
PROC. INT. CONF. ON FINITE ELEMENT METHODS IN
ENG., (EDS.) PULMANO, V.A. AND KABAILA, A.P.,
UNIV. OF NEW SOUTH WALES, AUSTRALIA, AUGUST 28-30,
1974, PP. 621-637.

74-0456 OVUNC, B. A.
ANALYSIS OF BUILDINGS ON ELASTIC MEDIUM UNDER
DYNAMIC OR SEISMIC LOADS
PROC. INT. CONF. ON FINITE ELEMENT METHODS IN
ENG., (EDS.) PULMANO, V.A. AND KABAILA, A.P.,
UNIV. OF NEW SOUTH WALES, AUSTRALIA, AUGUST 28-30,
1974, PP. 639-659.

74-0457 RUSSELL, J. J.
ANDERSON, W. J.
WHIRLING CABLE SUBJECTED TO VISCOUS DRAG
PROC. INT. CONF. ON FINITE ELEMENT METHODS IN
ENG., (EDS.) PULMANO, V.A. AND KABAILA, A.P.,
UNIV. OF NEW SOUTH WALES, AUSTRALIA, AUGUST 28-30,
1974, PP. 661-676.

74-0458 WEGMULLER, A. W.
INELASTIC RESPONSE OF PLATES AND ECCENTRICALLY
STIFFENED PLATES
PROC. INT. CONF. ON FINITE ELEMENT METHODS IN
ENG., (EDS.) PULMANO, V.A. AND KABAILA, A.P.,
UNIV. OF NEW SOUTH WALES, AUSTRALIA, AUGUST 28-30,
1974, PP. 677-701.

74-0459 AL-MASHIDANI, G.
TAYLOR, C.
FINITE ELEMENT SOLUTION OF BOUSSINESQ'S EQUATION
FOR UNSTEADY GROUNDWATER FLOW
PROC. INT. CONF. ON FINITE ELEMENT METHODS IN
ENG., (EDS.) PULMANO, V.A. AND KABAILA, A.P.,
UNIV. OF NEW SOUTH WALES, AUSTRALIA, AUGUST 28-30,
1974, PP. 705-717.

74-0460 ARAL, M. M.
YAZICI, A.
FINITE ELEMENT SOLUTION OF PROBLEMS WITH UNKNOWN
AND MOVING BOUNDARIES
PROC. INT. CONF. ON FINITE ELEMENT METHODS IN
ENG., (EDS.) PULMANO, V.A. AND KABAILA, A.P.,
UNIV. OF NEW SOUTH WALES, AUSTRALIA, AUGUST 28-30,
1974, PP. 719-727.

74-0461 BEER, G.
MEEK, J. L.
TRANSIENT HEAT FLOW IN SOLIDS
PROC. INT. CONF. ON FINITE ELEMENT METHODS IN
ENG., (EDS.) PULMANO, V.A. AND KABAILA, A.P.,
UNIV. OF NEW SOUTH WALES, AUSTRALIA, AUGUST 28-30,
1974, PP. 729-740.

74-0462 CHEUNG, Y. K.
SKJOLINGSTAD, L.
UNSTEADY RADIAL FLOW IN GAS RESERVOIRS BY FINITE
ELEMENTS
PROC. INT. CONF. ON FINITE ELEMENT METHODS IN
ENG., (EDS.) PULMANO, V.A. AND KABAILA, A.P.,
UNIV. OF NEW SOUTH WALES, AUSTRALIA, AUGUST 28-30,
1974, PP. 741-750.

74-0463 CHEUNG, Y. K.
SKJOLINGSTAD, L.
TWO AND THREE DIMENSIONAL GROUNDWATER SEEPAGE BY
FINITE ELEMENTS
PROC. INT. CONF. ON FINITE ELEMENT METHODS IN
ENG., (EDS.) PULMANO, V.A. AND KABAILA, A.P.,
UNIV. OF NEW SOUTH WALES, AUSTRALIA, AUGUST 28-30,
1974, PP. 751-766.

74-0464 FISHER, I.
MEDLAND, I. C.
THE MULTIDIMENSIONAL STEFAN PROBLEM: A FINITE
ELEMENT APPROACH
PROC. INT. CONF. ON FINITE ELEMENT METHODS IN
ENG., (EDS.) PULMANO, V.A. AND KABAILA, A.P.,
UNIV. OF NEW SOUTH WALES, AUSTRALIA, AUGUST 28-30,
1974, PP. 767-783.

74-0465 GODBOLE, P. N.
ZIENKIEWICZ, O. C.
FINITE ELEMENT ANALYSIS OF STEADY FLOW OF
NON-NEWTONIAN FLUIDS
PROC. INT. CONF. ON FINITE ELEMENT METHODS IN
ENG., (EDS.) PULMANO, V.A. AND KABAILA, A.P.,
UNIV. OF NEW SOUTH WALES, AUSTRALIA, AUGUST 28-30,
1974, PP. 785-798.

74-0466 HUYAKORN, P. S.
FINITE ELEMENT SOLUTIONS OF TRANSIENT, TWO-REGIME
FLOW TOWARDS WELLS
PROC. INT. CONF. ON FINITE ELEMENT METHODS IN
ENG., (EDS.) PULMANO, V.A. AND KABAILA, A.P.,
UNIV. OF NEW SOUTH WALES, AUSTRALIA, AUGUST 28-30,
1974, PP. 799-813.

74-0467 KONO, I.
ANALYSIS OF INTERFACE PROBLEM IN GROUNDWATER FLOW
BY FINITE ELEMENT METHOD
PROC. INT. CONF. ON FINITE ELEMENT METHODS IN
ENG., (EDS.) PULMANO, V.A. AND KABAILA, A.P.,
UNIV. OF NEW SOUTH WALES, AUSTRALIA, AUGUST 28-30,
1974, PP. 815-826.

74-0468 WU, J. C.
INTEGRAL REPRESENTATION OF FIELD VARIABLES FOR THE
FINITE ELEMENT SOLUTION OF VISCOUS FLOW PROBLEMS
PROC. INT. CONF. ON FINITE ELEMENT METHODS IN
ENG., (EDS.) PULMANO, V.A. AND KABAILA, A.P.,
UNIV. OF NEW SOUTH WALES, AUSTRALIA, AUGUST 28-30,
1974, PP. 827-840.

74-0469 ARGYRIS, J. H.
FAUST, G.
WILLAM, K.
RECENT DEVELOPMENTS IN FINITE ELEMENT ANALYSIS OF
PRESTRESSED CONCRETE REACTOR VESSELS
NUCLEAR ENGINEERING AND DESIGN, VOL. 28, NO. 1,
1974, PP. 42-75.

74-0470 ANDERSSON, H.
FINITE ELEMENT TREATMENT OF A UNIFORMLY MOVING
ELASTIC-PLASTIC CRACK TIP
J. MECHANICS AND PHYSICS OF SOLIDS, VOL. 22, NO.
4, 1974, PP. 285-308.

74-0471 ARGYRIS, J. H.
WILLAM, K.
SOME CONSIDERATIONS FOR EVALUATION OF FINITE
ELEMENT MODELS
NUCLEAR ENGINEERING AND DESIGN, VOL. 28, NO. 1,
1974, PP. 76-96.

74-0472 CHARI, M. V.
NONLINEAR FINITE ELEMENT SOLUTION OF ELECTRICAL
MACHINES UNDER NO-LOAD AND FULL-LOAD CONDITIONS
IEEE TRANS. ON MAGNETICS, VOL. MA10, NO. 3, 1974,
PP. 686-689.

74-0473 COOK, R. D.
EFFECT OF HIGHER ORDER TERMS IN CERTAIN NONLINEAR
FINITE ELEMENT MODELS - COMMENT
AIAA J., VOL. 12, NO. 8, 1974, P. 1290.

74-0474 LEE, C. H.
FINITE ELEMENT METHOD FOR TRANSIENT LINEAR VISCOUS
FLOW PROBLEMS
PROC. INT. CONF. ON NUMER. METH. IN FLUID
DYNAMICS, UNIV. OF SOUTHAMPTON, ENGLAND, SEPTEMBER
26-28, 1973, PP. 140-152.

74-0475 SALEM, M. H.
ARAL, K.
DATA GENERATION FOR A 3-D FINITE ELEMENT SYSTEM
PAPER NO. 74-PVP-23, A.S.M.E. MEETING, JUNE 24-28,
1974, P. 5.

74-0476 BELYTSCHKO, T.
HSIEH, B. J.
NONLINEAR TRANSIENT ANALYSIS OF SHELLS AND SOLIDS
OF REVOLUTION BY CONVECTED ELEMENTS
AIAA J. VOL. 12, NO. 8, AUGUST 1974, PP:
1031-1035.

74-0477 BRILLA, J.
NEMETHY, A.
RIESENIE VAZKOPRUZNYCH ANIZOTROPICKYCH DOASK
METODOU KONECNYCH PRVKOV (SOLUTION OF
VISCOELASTIC ANISOTROPIC PLATES BY FINITE
ELEMENT METHOD)
STAVEBNICKY CAS., VOL. 22, NO. 1, 1974, PP: 3-16.

74-0478 ▶ STOKER, J. R.
FINITE ELEMENTS
NUCLEAR ENG. AND DESIGN, VOL. 28, NO. 3, 1974, PP:
352-358

74-0479 BERGAN, P. G.
AAMODT, B.
FINITE ELEMENT ANALYSIS OF CRACK PROPAGATION IN
3-DIMENSIONAL SOLIDS UNDER CYCLIC LOADING
NUCLEAR ENG. AND DESIGN., VOL. 29, NO. 2, 1974,
PP: 180-188

74-0480 CAVENDISH, J. C.
AUTOMATIC TRIANGULATION OF ARBITRARY PLANAR
DOMAINS FOR THE FINITE ELEMENT METHOD
INT. J. NUMEP. METHODS ENG., VOL. 8, NO. 4, 1974,
PP: 679- 696.

74-0481 ADLER, D.
KRIMERMA, Y.
NUMERICAL CALCULATION OF MERIDIONAL FLOW FIELD IN
TURBOMACHINES USING FINITE ELEMENTS METHOD
ISRAEL J. OF TECH., VOL. 12, NO. 3-4, 1974, PP.
268-274.

74-0482 JEZERNIK, A.
MILLER, M. C.
LARGE USER-ORIENTATED SYSTEMS OF PROGRAMS FOR
STRUCTURAL ANALYSIS AND DESIGN
NUCL. ENG. DES., VOL. 27, NO. 2, MAY 1974, PP:
238-273.

74-0483 ADAMS, D. F.
PRACTICAL PROBLEMS ASSOCIATED WITH THE APPLICATION
OF THE FINITE ELEMENT METHOD TO COMPOSITE
MATERIAL MICROMECHANICAL ANALYSES
FIBRE SCI. TECHNOL. VOL. 7, NO. 2, APRIL 1974, PP:
111-122

74-0484 ROSSETTOS, J. N.
TONG, P.
FINITE ELEMENT ANALYSIS OF VIBRATION AND FLUTTER
OF CANTILEVER ANISOTROPIC PLATES
PAPER NO. 74-WA-APM-15, A.S.M.E. MEETING, NOVEMBER
17-21, 1974, P. 6.

74-0485 TABARROK, B.
HOA, V. S.
THERMAL STRESS ANALYSIS OF PLATES AND SHALLOW
SHELLS BY HYBRID FINITE-ELEMENT METHOD
J. STRAIN ANAL., VOL. 9, NO. 3, JULY 1974, PP:
152-158.

74-0486 KENINGSBERG, I. J.
DEAN, W.
CORRELATION OF FINITE ELEMENT STRUCTURAL DYNAMIC
ANALYSIS WITH MEASURED FREE VIBRATION
CHARACTERISTICS FOR A FULL-SCALE HELICOPTER
FUSELAGE
PAPER NO. 7 OF THE SPEC. MEET. ON ROTORCR. DYN.
PROC. SESS. NASA, AMES RES. CENT., MOFFETT FIELD,
CALIF., FEBRUARY 13-15, 1974, P. 14 (AVAILABLE
FROM JAMES C. BIGGERS, NASA AMES RES. CENT.
MOFFETT FIELD, CALIF., 1974)

74-0487 ELLIOTT, G. L.
PARAMETRIC ANALYSIS OF CONTACT CHATTER IN RELAYS
PROC. OF THE 3RD INT. AND 22ND NATL. RELAY CONF.,
OKLA STATE UNIV., STILLWATER, APRIL 30-MAY 1,
1974, PAPER NO. 21, P. 5. (PUBL. BY NATL. ASSOC.
OF RELAY MANUF., SCOTTSDALE ARIZONA, 1974)

74-0488 HOGGE, M.
FAMILLE D ELEMENTS FINIS DE COQUE PLAN
CINEMATIQUEMENT ADMISSIBLES POUR L'ANALYSE DES
STRUCTURES
AEROSPACE LAB., UNIV. OF LIEGE, REPORT SF-22,
1974.

74-0489 SUHARA, T.
TASAI, F.
STUDY OF MOTION AND STRENGTH OF FLOATING MARINE
STRUCTURES IN WAVES
PROC. OF THE 6TH ANNUAL OFFSHORE TECH. CONFERENCE,
HOUSTON, TEXAS, MAY 6-8, 1974, VOL. 2, PREPRINT OF
PAPER 2068, PP. 379-380. (PUBL. BY OFFSHORE
TECHNOL. CONF. C/O D.L. RILEY, DALLAS)

74-0490 PINCEMIN, M.
PLANFIX, J. M.
INTEGRATED PROGRAM FOR THE DYNAMIC STRUCTURAL
CALCULATION OF MOBIL OFFSHORE UNITS
PROC. OF THE 6TH ANNUAL OFFSHORE TECH. CONFERENCE,
HOUSTON, TEXAS, MAY 6-8, 1974, VOL. 2, PREPRINT OF
PAPER 2052, PP. 205-218. (PUBL. BY OFFSHORE
TECHNOL. CONF. C/O D.L. RILEY, DALLAS)

74-0491 BLACKSTONE, W. R.
DEHART, R. C.
DISCONTINUITY STRESS DECAY IN OFFSHORE PLATFORM
JOINTS
PROC. OF THE 6TH ANNUAL OFFSHORE TECH. CONFERENCE,
HOUSTON, TEXAS, MAY 6-8, 1974, VOL. 2, PREPRINT OF
PAPER 2102, PP. 691-700. (PUBL. BY OFFSHORE
TECHNOL. CONF. C/O D.L. RILEY, DALLAS)

74-0492 VISSER, W.
ON THE STRUCTURAL DESIGN OF TUBULAR JOINTS
PROC. OF THE 6TH ANNUAL OFFSHORE TECH. CONFERENCE,
HOUSTON, TEXAS, MAY 6-8, 1974, VOL. 2, PREPRINT OF
PAPER 2117, PP. 881-894. (PUBL. BY OFFSHORE
TECHNOL. CONF., C/O D.L. RILEY, DALLAS)

74-0493 KWAN, G.
LIMITATIONS ON THE APPLICATION OF EFFECTIVE
BREADTH CONCEPT IN BOX GIRDER DESIGN
PROC. OF THE 6TH ANNUAL OFFSHORE TECH. CONFERENCE,
HOUSTON, TEXAS, MAY 6-8, 1974, VOL. 2, PREPRINT OF
PAPER 2116, PP. 869-880. (PUBL. BY OFFSHORE
TECHNOL. CONF., C/O D.L. RILEY, DALLAS)

74-0494 SLOT, T.
BRANCA, T. R.
ON THE DETERMINATION OF EFFECTIVE ELASTIC-PLASTIC
PROPER TIES FOR THE EQUIVALENT SOLID PLATE
ANALYSIS OF TUBE SHEETS
PAPER NO. 74-PVP-18, A.S.M.E. MEETING, JUNE 24-28,
1974, P. 8.

74-0495 MATERNA, A.
SMIRAK, S.
RESENI MAZNIHO STAVU ZELEZOBETONOVYCH DESEK
(PLASTIC ANALYSIS OF REINFORCED CONCRETE PLATES)
STAVEBNICKY CAS, VOL. 22, NO. 12, DEC. 1974, PP.
848-860.

74-0496 STRICKLIN, J. A.
HAISLER, W. E.
LARGE DEFLECTION ELASTIC-PLASTIC DYNAMIC RESPONSE
OF STIFFENED SHELLS OF REVOLUTION
PAPER NO. 74-PVP-3, A.S.M.E. MEETING, JUNE 24-28,
1974, P. 9.

74-0497 KONO, I.
FINITE ELEMENT ANALYSIS OF INTERFACE PROBLEM IN
GROUND-WATER FLOW
PROC. JAP. SOC. CIV. ENG., NO. 228, AUGUST 1974,
PP: 109-115

74-0498 STALEY, J. A.
SCIARRA, J. J.
COUPLED ROTOR-AIRFRAME VIBRATION PREDICTION
METHODS
PAPER NO. 8 OF THE SPEC. MEET. CN ROTORCR. DYN.
PROC. SESS. NASA, AMES RES. CENT., MOFFETT FIELD,
CALIF., FEBRUARY 13-15, 1974, P. 10 (AVAILABLE
FROM JAMES C. BIGGERS, NASA AMES RES. CENT.
MOFFETT FIELD, CALIF., 1974)

74-0499 NAYFEH, A. H.
MOOK, D. T.
NUMERICAL-PERTURBATION METHOD FOR THE NONLINEAR
ANALYSIS OF STRUCTURAL VIBRATIONS
AIAA J., VOL. 12, NO. 9, SEPTEMBER 1974, PP:
1222-1228.

74-0500 HEISE, U.
FINITE ELEMENT ANALYSIS IN POLAR CO-ORDINATES OF
THE SAINT VENANT TORSICN PROBLEM
INT. J. NUMER METHODS ENG., VOL. 8, NO. 4, 1974,
PP: 713-729.

74-0501 GOTOH, M.
FINITE ELEMENT ANALYSIS OF GENERAL DEFORMATION OF
SHEET METALS
INT. J. NUMER. METHODS ENG., VOL. 8, NO. 4, 1974,
PP: 731-741.

74-0502 HIBBITT, H. D.
SPECIAL STRUCTURAL ELEMENTS FOR PIPING ANALYSIS
PROC. A.S.M.E. CONF., MIAMI BEACH, FLA., JUNE
24-28, 1974, PP. 1-110.

74-0503 DARWIN, D.
PECKNOLD, D. A. W.
INELASTIC MODEL FOR CYCLIC BIAXIAL LOADING OF
REINFORCED CONCRETE
ILL. UNIV. DEPT. CIV. ENG. STRUCT. RES., SERIES
NO. 409, JULY 1974, P: 169.

74-0504 DALEY, G. C.
OPTIMIZATION OF TENSION LEVEL AND STINGER LENGTH
FOR OFFSHORE PIPELINE INSTALLATION
ASME PAPER NO. 74-PET-A, 1974, P: 3.

74-0505 MCNAMARA, J. F.
SOLUTION SCHEMES FOR PROBLEMS OF NONLINEAR
STRUCTURAL DYNAMICS
PAPER NO. N74-PVP-30, A.S.M.E. MEETING, JUNE
24-28, 1974, P. 7.

74-0506 RICCARDELLA, P.C.
BAMFORD, W. H.
REACTOR COOLANT PUMP FLYWHEELS OVERSPEED
EVALUATION
PAPER NO. 74-PVP-25, A.S.M.E. MEETING, JUNE 24-28,
1974, P. 7.

74-0507 MELLO, R. M.
GRIFFIN, D. S.
PLASTIC COLLAPSE LOADS FOR PIPE ELBOWS USING
INELASTIC ANALYSIS
PAPER NO. 74-PVP-16, A.S.M.E. MEETING, JUNE 24-28,
1974, P. 7.

74-0508 FROST, G. R.
MODIFICATIONS TO ARL COMPUTER PROGRAMS USED FOR
DESIGN OF AXIAL COMPRESSOR AIRFOILS
U. S. AIR FORCE SYST. COMMAND, AEROSPACE RES. LAB.
TR., 74-0060, JUNE 1974, P: 479

74-0509 IDELSOHN, S.
ANALYSE STATIQUE ET DYNAMIQUE DES COQUES PAR LA
METHODE DES ELEMENTS FINIS
PH.D. THESIS, AEROSPACE LAB., UNIV. OF LIEGE,
REPORT SF-25, 1974.

74-0510 HARTZMAN, M.
NONLINEAR DYNAMIC ANALYSIS OF AXISYMMETRIC SOLIDS
BY THE FINITE ELEMENT METHOD
PAPER NO. 74-PVP-36, A.S.M.E. MEETING, JUNE 24-28,
1974, P. 10.

74-0511 FUNNELL, W. R.
LASZLO, C. A.
SIMULATING BEHAVIOR OF EARDRUM BY FINITE-ELEMENT
METHOD
J. OF THE ACOUSTICAL SOC. OF AMERICA, VOL. 56,
1974, P: 3.

74-0512 CLARK, J. N.
BARNHART, D. F.
STRESS ANALYSIS OF INDUSTRIAL COMPONENTS WITH
PLASTIC AND FINITE ELEMENT MODELS
PREPRINT NO. 740706, S.A.E. MEETING, SEPTEMBER
9-12, 1974, P. 7.

74-0513 KEALY, C.D.
BUSCH, R.A.
SEEPAGE-ENVIRONMENTAL ANALYSIS OF THE SLIME ZONE
OF A TAILINGS POND
U.S. BUR. MINES REP. INVEST. NO. 7439, 1974, 89
PP.

74-0514 MATERNA, A.
SMIRAK, S.
STATICKE RESENI FYZIKALNE NELINEARNIHO CHOVANI
ZELEZOBETO- NOVYCH DESEK (ANALYSIS OF PHYSICALLY
NONLINEAR BEHAVIOR OF REINFORCED CONCRETE
SLABS)
STAVEBNICKY, CAS., VOL. 22, NO. 3, 1974, PP:
616-175.

74-0515 POPOV, E. P.
SHARIFI, P.
INELASTIC BUCKLING ANALYSIS OF PIPES SUBJECTED TO
INTERNAL PRESSURE, FLEXURE AND AXIAL LOADING
PROC. A.S.M.E. CONF. ON PRESSURE VESSELS AND
PIPING, MIAMI BEACH, FLA., JUNE 24-28, 1974, PP.
11-23. (PUBL. BY A.S.M.E., NEW YORK, N.Y., AND
UNIV. OF CALIF., BERKELEY)

74-0516 JUNKINS, J. L.
FINITE ELEMENT MODEL OF EARTHS GRAVITY FIELD
TRANSACTIONS, AMERICAN GEOPHYSICAL UNION, VOL. 55,
NO. 12, 1974, P: 1104.

74-0517 ROCK, T.
HINTON, E.
FREE VIBRATION AND TRANSIENT RESPONSE OF THICK
AND THIN PLATES USING THE FINITE ELEMENT METHOD
EARTHQUAKE ENG. STRUCT. DYN., VOL. 3, NO. 1, JULY
- SEPTEMBER 1974, PP: 51-63.

74-0518 HEIDARI, M.
CARTWRIGHT, K.
ANALYSIS OF LIQUID-WASTE INJECTION WELLS IN
ILLINOIS BY MATHEMATICAL MODELS
ILL. UNIV. WATER RES. CENT., REPORT NO. 77,
FEBRUARY 1974, P. 114.

74-0519 BANGASH, Y.
RIDDLE OF BONDED AND UNBONDED TENDONS IN
PRESTRESSED CONCRETE REACTOR VESSELS
CONCRETE (LOND), VOL. 8, NO. 6, JUNE 1974, PP:
46-48.

74-0520 ANDERSEN, C. M.
USE OF COMPUTERIZED SYMBOLIC INTEGRATION IN FINITE
ELEMENT DEVELOPMENT
PROC. OF THE A.C.M. CONF., SAN DIEGO, CALIF.,
1974, PP. 554-562.

74-0521 BARSOUM, R. S.
APPLICATION OF QUADRATIC ISOPARAMETRIC FINITE
ELEMENTS IN LINEAR FRACTURE MECHANICS
INT. J. OF FRACTURE, VOL. 10, NO. 4, 1974, PP:
603-605.

74-0522 ALLIK, H.
WEBMAN, K. M.
HUNT, J. T.
VIBRATIONAL RESPONSE OF SONAR TRANSDUCERS USING
PIEZO-ELECTRIC FINITE ELEMENTS
J. OF THE ACOUSTICAL SOC. OF AMER., VOL. 56, NO.
6, 1974, PP: 1782-1791.

74-0523 AHMED, K. M.
APPLICATIONS OF CURVED FINITE ELEMENT TO SANDWICH
BEAMS USING SHEAR STRAIN FORMULATION
PROC. OF THE 8TH INT. CONGRESS ON ACOUSTICS, 1974.

74-0524 KNITTEL, M. R.
NICHOLS, C. S.
SMITH, R. R.
BARACH, D.
FINITE ELEMENT ANALYSIS OF ACOUSTICALLY RADIATING
STRUCTURES WITH APPLICATIONS TO SONAR TRANSDUCERS
- COMMENTS
J. OF THE ACOUSTICAL SOCIETY OF AMERICA, VOL. 56,
NO. 6, 1974, PP: 1905-1907.

74-0525 SEVAK, N. M.
MCLARNAN, C. W.
OPTIMAL SYNTHESIS OF FLEXIBLE LINK MECHANISMS WITH
LARGE STATIC DEFLECTIONS
ASME PAPER NO. 74-DET-83 FOR MEET OCTOBER 6-10,
1974, P: 7.

74-0526 ABE, T.
STRESS DISTRIBUTION UNDER UNIAXIAL ELASTIC PLANE
STRAIN COMPRESSION WITH FRICTION
BULL. JSME, VOL. 17, NO. 110, AUGUST 1974, PP:
1000-1008.

74-0527 KEY, S. W.
FINITE ELEMENT PROCEDURE FOR THE LARGE DEFORMATION
DYNAMIC RESPONSE OF AXISYMMETRIC SOLIDS
COMPUT. METHODS APPL. MECH. ENG., VOL. 4, NO. 2,
SEPTEMBER 1974, PP: 195-218.

74-0528 ARGYRIS, J. H.
DUNNE, P. C.
LARGE NATURAL STRAINS AND SOME SPECIAL
DIFFICULTIES DUE TO NONLINEARITY AND
INCOMPRESSIBILITY IN FINITE ELEMENTS
COMPUT. METHODS APPL. MECH. ENG., VOL. 4, NO. 2,
SEPTEMBER 1974, PP. 219-278.

74-0529 NAGTEGAAL, J. C.
PARKS, D. M.
ON NUMERICALLY ACCURATE FINITE ELEMENT SOLUTIONS
IN THE FULLY PLASTIC RANGE
COMPUT. METHODS APPL. MECH. ENG., VOL. 4, NO. 2,
SEPTEMBER 1974, PP: 153-177.

74-0530 IDING, R. H.
PISTER, K. S.
IDENTIFICATION OF NONLINEAR ELASTIC SOLIDS BY A
FINITE ELEMENT METHOD
COMPUT. METHODS APPL. MECH. ENG., VOL. 4, NO. 2,
SEPTEMBER 1974, PP: 121-142.

74-0531 BRASFIELD, R.L.
BAGCI, C.
MATRIX-DISPLACEMENT DIRECT-ELEMENT METHOD FOR
FORCE AND TORQUE ANALYSIS OF DETERMINATE AND
INDETERMINATE SPACE MECHANISMS WITH ARBITRARY SKEW
ANGLES
PAPER NO. 74-DET-66, A.S.M.E. MEETING, OCTOBER
6-10, 1974, P. 11.

74-0532 VOLCY, G.
GARNIER, H.
DEFORMABILITY OF THE HULL STEEL-WORK AND
DEFORMATIONS OF THE ENGINE-ROOM OF LARGE
TANKERS
INT. SHIPBUILD. PROG., VOL. 21, NO. 240, AUGUST
1974, PP: 229-251.

74-0533 SALEM, M. H.
MOHRAZ, B.
NONLINEAR ANALYSIS OF PLANAR REINFORCED CONCRETE
STRUCTURES
DEPT. CIV. ENG. STRUCT. RES., ILL. UNIV., SERIES
NO. 410, JULY 1974, P. 181.

74-0534 WRIGHT, W. B.
RODABAUGH, E. C.
INFLUENCE OF END-EFFECTS ON STRESSES AND
FLEXIBILITY OF A PIPING ELBOW WITH IN-PLANE MOMENT
PROC. OF A.S.M.E. CONF. ON PRESSURE VESSELS AND
PIPING, MIAMI BEACH, FLA., JUNE 24-28, 1974, PP.
95-106. (PUBL. BY A.S.M.E., NEW YORK, N.Y., 1974)

74-0535 GALLAGHER, J. P.
STALNAKER, H. D.
SPECTRUM TRUNCATION AND DAMAGE TOLERANCE STUDY
ASSOCIATED WITH THE C-5A OUTBOARD PYLON AFT
TRUSS LUGS
US AIR FORCE SYST. COMMAND AIR FORCE FLIGHT DYN.
LAB. TECH. REP. AFFDL-TR-74-5, MAY 1974, P:
52.

74-0536 FRANCIS, E. G.
DEVERALL, L. I.
FRACTURE OF ROCKET GRAIN MODELS SUBJECTED TO
COOLDOWN: A STUDY OF TEMPERATURE-INITIATED
FRACTURES OF TWO-DIMENSIONAL ROCKET GRAIN MODELS
J. TEST EVAL., VOL. 2, NO. 5, SEPTEMBER 1974, PP.
354-360.

74-0537 ORRIS, R. M.
A FINITE ELEMENT STUDY OF THE VIBRATION OF
SKIN-RIB STRUCTURES
PH.D. THESIS, UNIV. OF SOUTHAMPTON, 1974.

74-0538 SAUGY, B.
ZIMMERMANN, T.
THREE DIMENSIONAL RUPTURE ANALYSIS OF A
PRESTRESSED CONCRETE PRESSURE VESSEL INCLUDING
CREEP EFFECTS
NUCL. ENG. DES., VOL. 28, NO. 1, 1974, PP. 97-120.

74-0539 KANDAOUROFF, W.
GRAVAS, J.
TOUR MAINE-MONTPARNASSE QUELQUES PROBLEMES DE
STRUCTURE (SPECIFIC PROBLEMS ENCOUNTERED IN
THE CONSTRUCTION OF THE MAINE-MONTPARNASSE
HIGH-RISE)
ANN. INST. TECH. BATIM. TRAV. PUBLICS, VOL. 320,
SEPTEMBER 1974, PP: 1-14.

74-0540 BELYTSCHKO, T.
KENNEDY, J. M.
DYNAMIC RESPONSE OF FAST-REACTOR CORE
SUBASSEMBLIES
NUCL. ENG. DES. VOL. 28, NO. 1, 1974, PP: 31-41.

74-0541 SYKES, J. F.
LENNOX, W. C.
TWO-DIMENSIONAL HEATED PIPELINE IN PERMAFROST
PIPELINES
ASCE J. GEOTECH. ENG. DIV., VOL. 100, NO. 11,
NOVEMBER 1974, PP: 1203-1214.

74-0542 YUFIN, S. A.
RASCHET NAPRYAZHENIL L PEREMESHCHENII V SVODE
PODZEMNOGO MASHINNOGO ZALA GES I V
OKRUZHAYUSHCHEM SKALNOM MASSIVE SUCHETOM
POETAPNOSTI RAZRABOTKI (CALCULATION OF STRESSES
AND DISPLACEMENTS IN CONCRETE VAULT OF THE
UNDERGROUND POWERHOUSE AND IN SURROUNDING
ROCK MASSIF DURING STEP-BY STEP EXCAVATION BY
USING FINITE ELEMENT METHOD)
GIDROTEKH STROIT, NO. 9, SEPTEMBER 1974, PP:
16-21.

74-0543 SYKES, J. F.
LENNOX, W. C.
FINITE ELEMENT PERMAFROST THAW SETTLEMENT MODEL
ASCE J. GEOTECH. ENG. DIV., VOL. 100, NO. 11,
NOVEMBER 1974, PP: 1185-1201.

74-0544 LINZER, V.
LEMPP, W.
BERECHNUNG VON SPANNUNGEN UND TEMPERATURFELDERN IN
FLOSSENWAENDEN UND AUFGESCHWEISSIEN BAUTEILEN.
(CALCULATION OF STRESSES AND TEMPERATURE FIELDS IN
FINNED-TUBE WALLS AND WELDED-ON COMPONENTS)
VGB KRAFTSWERKSTECH, VOL. 54, NO. 9, SEPTEMBER
1974, PP: 595-601.

74-0545 OTTENS, H. H.
LOF, C. J.
COMPLIANCE OF A TAPERED OCB SPECIMEN -
CONFIGURATIONS BY A FINITE ELEMENT METHOD
ENG. FRACT. MECH. VOL. 6, NO. 3, OCTOBER 1974, PP:
573-585.

74-0546 SUNDSTROM, B.
ENERGY CONDITION FOR INITIATION OF INTERFACIAL
MICROCRACKS AT INCLUSIONS
ENG. FRACT. MECH., VOL. 6, NO. 3, OCTOBER 1974,
PP: 483-492.

74-0547 FRANECK, H.
RECKE, H. G.
BERECHNUNG VON LAUFRAEDERN MIT DER METHODE DER
FINITEN ELEMENTE (DESIGN OF ROTATING WHEELS
BY THE FINITE ELEMENT METHOD)
MASCHINENBAUTECHNIK, VOL. 23, NO. 9, SEPTEMBER
1974, PP: 423-426.

74-0548 ADAMS, R. D.
PEPPIATT, N. A.
STRESS ANALYSIS OF ADHESIVE-BONDED LAP JOINTS
J. STRAIN ANAL., VOL. 9, NO. 3, JULY 1974, PP:
185-196.

74-0549 BRINK, F. I. A.
VAN DER KROGT, A. H.
STRESS ANALYSIS OF A TUBULAR CROSS-JOINT WITHOUT
ANY INTERNAL STIFFENING FOR OFFSHORE STRUCTURES
MET. CONST. BR. WELD. J., VOL. 6, NO. 9, SEPTEMBER
1974, PP: 290-294.

74-0550 VEIDINGE, L.
ORDER OF CONVERGENCE OF A FINITE ELEMENT SCHEME
ACTA MATHEMATICA ACADEMIAE SCIENTIARUM HUNGARICAE,
VOL. 25, NO. 3-4, 1974, PP: 401-412.

74-0551 WASHIZU, K.
ON THE ROLE OF VARIATIONAL PRINCIPLES AND METHOD
OF WEIGHTED RESIDUALS IN THE FINITE ELEMENT
METHODS
PROC. OF THE J.S.S.C. SYMP. ON THE FINITE ELEMENT
METHOD, TOKYO, JAPAN, NOVEMBER 1974.

74-0552 VENKATES, G.
KANAKARA, K.
GALERKIN FINITE ELEMENT ANALYSIS OF A UNIFORM BEAM
CARRYING A CONCENTRATED MASS AND ROTARY
INERTIA WITH A SPRING HINGE
J. OF SOUND AND VIBRATION, VOL. 37, NO. 4, 1974,
PP: 567-569

74-0553 BERGER, D.
APPERT, K.
GRUBER, R.
DETERMINATION OF EIGENMODES OF HOMOGENEOUS PLASMA
WITH ELLIPTICAL SECTION USING BI-DIMENSIONAL
FINITE ELEMENT METHOD
ZEITSCHRIFT FUR ANGEWANDTE MATHEMATICK UND PHYSIK,
VOL. 25, NO. 5, 1974, P: 674.

74-0554 ALLEN, J. J.
THOMPSON, M. R.
SIGNIFICANCE OF VARIABLY CONFINED TRIAXIAL TESTING
ROADBUILDING MATERIALS
ASCE TRANSP. ENG. J., VOL. 100, NO. 4, NOVEMBER
1974, PP: 827-843.

74-0555 LAURA, P. A.
MAURIZI, M. J.
USE OF TRIGONOMETRIC TERMS IN FINITE ELEMENT
METHOD WITH APPLICATION TO VIBRATING MEMBRANES
J. OF SOUND AND VIBRATION, VOL. 37, NO. 1, 1974,
P: 137-139

74-0556 FELIPPA, C. A.
INCREMENTAL FINITE ELEMENT MATRICES
J. OF THE STRUCTURAL DIV. / ASCE, VOL. 100, NO.
12, 1974, PP: 2521-2523.

74-0557 BYSKOV, E.
NIELSEN, L. O.
FINITE ELEMENT METHOD AND DENDIXEN-OSTENFELD SLOPE
DEFLECTION METHOD
BYGNEINGSSTATISKE MEDDELELSER, VOL. 45, NO. 2,
1974, PP. 35-64.

74-0558 VAISH, A. K.
CHOPRA, A. K.
EARTHQUAKE FINITE ELEMENT ANALYSIS OF
STRUCTURE-FOUNDATION SYSTEMS
J. ENG. MECH. DIV., ASCE, VOL. 100, NO. 6, 1974,
PP: 1101-1116.

74-0559 HAUG, E.
FINITE ELEMENT ANALYSIS OF INFLATABLE SHELLS
J. ENG. MECH. DIV., ASCE, VOL. 100, NO. 6, 1974,
P: 1256.

74-0560 NAM, C. H.
SALMON, C. G.
FINITE ELEMENT ANALYSIS OF CONCRETE BEAMS
J. OF THE STRUCTURAL DIV. ASCE., VOL. 100, NO. 12,
1974, PP: 2419-2432.

74-0561 YAMAMOTO, Y.
SOME PROBLEMS ON THE BASIC THEORY OF THE FINITE
ELEMENT METHODS
PROC. OF THE J.S.S.C. SYMP. ON THE FINITE ELEMENT
METHOD, TOKYO, JAPAN, NOVEMBER 1974.

74-0562 SILVESTER, P.
RAFINEJA, P.
CURVILINEAR FINITE ELEMENTS FOR 2-DIMENSIONAL
SATURABLE MAGNETIC FIELDS
IEEE TRANSACTIONS ON POWER APPARATUS AND SYSTEMS,
VOL. 93, NO. 6, 1974, PP: 1861-1870.

74-0563 YAGAWA, G.
MIYAZAKI, N.
ANDO, Y.
ANALYSIS OF LARGE ELASTIC-PLASTIC CREEP
DEFORMATIONS OF AXISYMMETRIC SHELLS BY THE FINITE
ELEMENT METHOD
COLLOQUIUM ON THERMOPLASTICITY, WARSAW, 1974.

74-0564 SETO, H.
YAMAMOTO, Y.
FUNDAMENTAL STUDIES ON STEADY SHIP WAVE PROBLEMS
BY THE FINITE ELEMENT METHOD
J. SOC. OF NAVAL ARCH. OF JAPAN, VOL. 136,
DECEMBER 1974.

74-0565 ADAMS, D. F.
MICROMECHANICAL ANALYSIS OF CRACK PROPAGATION IN
AN ELASTOPLASTIC COMPOSITE MATERIAL
FIBRE SCI. TECHNOL. VOL. 7, NO. 4, OCTOBER 1974,
PP: 237-256

74-0566 HSU, T. R.
BERTELS, A. W. M.
IMPROVED APPROXIMATION OF CONSTITUTIVE
ELASTO-PLASTIC STRESS STRAIN RELATIONSHIP FOR
FINITE ELEMENT ANALYSIS
AIAA J. VOL. 12, NO. 10, OCTOBER 1974, PP:
1450-1452.

74-0567 NARAYANASWAMI, R.
DEPENDENCE OF PLATE-BENDING FINITE ELEMENT
DEFLECTIONS AND EIGENVALUES ON POISSONS RATIO
AIAA J. VOL. 12, NO. 10, OCTOBER 1974, PP:
1420-1421.

74-0568 MURTHY, K. N.
CHRISTIANO, P. P.
OPTIMAL DESIGN FOR PRESCRIBED BUCKLING LOADS
ASCE J. STRUCT. DIV., VOL. 100, NO. 11, NOVEMBER
1974, PP: 2175-2190

74-0569 MEHTA, Y. B.
BAGCI, C.
FORCE AND TORQUE ANALYSIS OF CONSTRAINED SPACE
MECHANISMS AND PLANE MECHANISMS WITH OFFSET LINKS
BY MATRIX DISPLACEMENT DIRECT ELEMENT METHOD
MECH. MACH. THEORY, VOL. 9, NO. 3-4, 1974, PP.
385-403.

74-0570 BARRY, J. M.
APPLICATION OF THE FINITE ELEMENT METHOD TO
ELLIPTIC PROBLEMS
AUST. COMPUT. J., VOL. 6, NO. 2, JULY 1974, PP:
51-60.

74-0571 GALLETLY, G. D.
MISTRY, J.
FREE VIBRATIONS OF CYLINDRICAL SHELLS WITH VARIOUS
END CLOSURES
NUCL. ENG. DES. VOL. 30, NO. 2, SEPTEMBER 1974,
PP: 249-268

74-0572 DONEA, J.
GIULIANI, S.
FINITE ELEMENT ANALYSIS OF STEADY-STATE NONLINEAR
HEAT TRANSFER PROBLEMS
NUCL. ENG. DES. VOL. 30, NO. 2, SEPTEMBER 1974,
PP: 205-213

74-0573 ZIENKIEWICZ, O. C.
OWEN, D. R. J.
ANALYSIS OF VISCOPLASTIC EFFECTS IN PRESSURE
VESSELS BY THE FINITE ELEMENT METHOD
NUC. ENG. DES., VOL. 28, NO. 2, SEPTEMBER 1974,
PP: 278-288.

74-0574 PARDOEN, G. C.
VIBRATION AND BUCKLING ANALYSIS OF AXISYMMETRIC
POLAR ORTHOTROPIC CIRCULAR PLATES
COMPUT. STRUCT., VOL. 4, NO. 5, OCTOBER 1974, PP:
951-960.

74-0575 WARZEE, G.
FINITE ELEMENT METHOD AND LAPLACE TRANSFORM
COMPARATIVE SOLUTIONS OF TRANSIENT HEAT CONDUCTION
PROBLEMS
COMPUT. STRUCT., VOL. 4, NO. 5, OCTOBER 1974, PP:
979-991.

74-0576 FRIED, I.
NUMERICAL INTEGRATION IN THE FINITE ELEMENT METHOD
COMPUT. STRUCT. VOL. 4, NO. 5, OCTOBER 1974, PP:
921-932.

74-0577 LEE, C. H.
CHENG, R. T.
ON SEAWATER ENCROACHMENT IN COASTAL AQUIFERS
WATER RESOURC. RES., VOL. 10, NO. 5, OCTOBER 1974,
PP: 1039-1043.

74-0578 GUYMON, G. L.
LUTHIN, J. N.
COUPLED HEAT AND MOISTURE TRANSPORT MODEL FOR
ARCTIC SOILS
WATER RESOUR. RES., VOL. 10, NO. 5, OCTOBER 1974,
PP: 995-1001.

74-0579 BAKER, A. J.
FINITE ELEMENT SOLUTION THEORY FOR
THREE-DIMENSIONAL BOUNDARY FLOWS
COMPUT. METHODS APPL. MECH. ENG., VOL. 4, NO. 3,
NOVEMBER 1974, PP: 367-386.

74-0580 BYSKOV, E.
NIELSEN, L. O.
ELEMENTMETODEN OG DEFORMATIONSMETODEN (FINITE
ELEMENT AND DEFORMATION METHODS IN STRESS
ANALYSIS)
DAN SELSK BYGNINGSSTAT MEDD. VOL. 45, NO. 2, JUNE
1974. PP: 35-64.

74-0581 WITTBRODT, E.
KRUSZEWSKI, J.
ALGORYTM OBLICZEN ORGAN WYMUSZONYCH METODA
SZTYWNYCH ELEMENTOW SKONCZONYCH (ALGORITHM
FOR CALCULATING FORCED VIBRATIONS BY MEANS OF
RIGID FINITE ELEMENT METHOD)
ARCH. INZ. LADOWE J., VOL. 20, NO. 3, 1974, PP:
477-489.

74-0582 ARGYRIS, J. H.
DOLTSINIS, J.
FAUST, G.
SZIMMAT, J.
WILLAM, K. J.
RECENT DEVELOPMENTS IN THE FINITE ELEMENT ANALYSIS
OF PRE-STRESSED CONCRETE REACTOR VESSELS
RESEARCH REPORT NO. 151, INSTITUT FUR STATIK UND
DYNAMIK DER LUFT UND RAUMFAHRTKONSTRUKTIONEN,
JANUARY 1974.

74-0583 NARAYANA, R.
DEPENDENCE OF PLATE-BENDING FINITE ELEMENT
DEFLECTIONS AND EIGENVALUES ON POISSONS RATIO
A.I.A.A. J. VOL. 12, NO. 10, 1974, PP: 1420-1421.

74-0584 ZINGONE, G.
MANCUSO, P.
LIMIT ANALYSIS OF FRAME SYSTEMS STIFFENED BY
PANELS
MECCANICA, VOL. 9, NO. 1, MARCH 1974, PP: 51-56.

74-0585 DESAI, C. S.
JOHNSON, L. D.
ANALYSIS OF PILE-SUPPORTED GRAVITY LOCK
ASCE J. GEOTECH. ENG. DIV., VOL. 100, NO. 9,
SEPTEMBER 1974, PP: 1009-1029.

74-0586 SCANLON, A.
MURRAY, D. W.
TIME DEPENDENT REINFORCED CONCRETE SLAB
DEFLECTIONS
ASCE J. STRUCT. DIV., VOL. 100, NO. 9, SEPTEMBER
1974, PP: 1911-1924.

74-0587 CORRADI, L.
DE DONATO, O.
INELASTIC ANALYSIS OF REINFORCED CONCRETE FRAMES
ASCE J. STRUCT. DIV., VOL. 100, NO. 9, SEPTEMBER
1974, PP: 1925-1942.

74-0588 ALSPAUGH, W.
KUNOO, K.
OPTIMUM CONFIGURATIONAL AND DIMENSIONAL DESIGN OF
TRUSS STRUCTURES
COMPUT. STRUCT., VOL. 4, NO. 4, AUGUST 1974, PP:
755-770.

74-0589 MATSUBARA, S.
KUDO, H.
STRESS ANALYSIS IN SOME COLD FORGING TOOL
COMPONENTS AND THEIR DESIGN GUIDES
ANN CORP., VOL. 23, NO. 1, 1974, PP. 53-51.

74-0590 SHELLY, P.
ETTLES, C.
APPLICATION OF A FINITE ELEMENT METHOD TO
EVALUATION OF OIL WHIRL CHARACTERISTICS
J. OF MECH. ENG. SCI., VOL. 16, NO. 2, 1974, PP:
101-108.

74-0591 SRIVASTAVA, N. K.
ANALYSIS OF PNEUMATIC SHELLS WITH OR WITHOUT CABLE
FORMULATION
COMPUT. STRUCT., VOL. 4, NO. 4, AUGUST 1974, PP.
813-828.

74-0592 GRINDLEY, R. E.
BESANT, C. B.
INEXPENSIVE, INTERACTIVE COMPUTER-AIDED DESIGN
APPLIED TO FINITE ELEMENT STRESS ANALYSIS
COMPUT. AIDED DES., VOL. 6, NO. 3, JULY 1974, PP:
125-131.

74-0593 VENKATESWARA, G.
CONICAL SHELL FINITE ELEMENT
COMPUT. STRUCT., VOL. 4, NO. 4, AUGUST 1974, PP:
901-915.

74-0594 FRIED, I.
RESIDUAL ENERGY BALANCING TECHNIQUE IN THE
GENERATION OF PLATE BENDING FINITE ELEMENTS
COMPUT. STRUCT. VOL. 4, NO. 4, AUGUST 1974, PP:
771-778.

74-0595 LINDEMAN, J.
LEAF, G. K.
COMPUTATIONAL ANALYSIS AND EVALUATION OF THE
FINITE ELEMENT METHOD FOR A CLASS OF NUCLEAR
REACTOR CONFIGURATIONS
COMPUT. METHODS APPL. MECH. ENG., VOL. 4, NO. 1,
JULY 1974, PP: 97-117.

74-0596 TOMLINSON, G. R.
LEONARD, R.
COMPUTER AIDED DESIGN OF THERMALLY LOADED
AXISYMMETRIC DIESEL ENGINE COMPONENTS
COMPUT. AIDED DES., VOL. 6, NO. 3, JULY 1974, PP:
132-596.

74-0597 GIMM, W.
HEINRICH, F.
BERECHNUNG UND BEURTEILUNG DER STANDSICHERHEIT VON
ABBAUHOHLRAEUMEN IN STEILEN GANGERZLAGERSTAETTEN
(CALCULATION AND EVALUATION OF THE STABILITY OF
MINING EXCAVATIONS IN THE STEEP-PITCH TYPE ORE
DEPOSITS)
NEUE BERGBAUTECH, VOL. 4, NO. 6, JUNE 1974, PP:
447-453.

74-0598 STEIN, E.
AHMAD, R.
ON THE STRESS COMPUTATION IN FINITE ELEMENT MODELS
BASED UPON DISPLACEMENT APPROXIMATIONS
COMPUT. METHODS APPL. MECH. ENG., VOL. 4, NO. 1,
JULY 1974, PP: 81-96

74-0599 CAREY, G. F.
UNIFIED APPROACH TO THREE FINITE ELEMENT THEORIES
FOR GEOMETRIC NONLINEARITY
COMPUT. METHODS APPL. MECH. ENG., VOL. 4, NO. 1,
JULY 1974, PP: 69-79.

74-0600 DAWE, D. J.
NUMERICAL STUDIES USING CIRCULAR ARCH FINITE
ELEMENTS
COMPUT STRUCT., VOL. 4, NO. 4, AUGUST 1974, PP:
729-740.

74-0601 WURMNEST, W.
ANTRAS COMPUTER PROGRAM FOR LINEAR STRESS ANALYSIS
BY FINITE ELEMENT METHOD
TECHNISCHE MITTEILUNGEN KRUPP. FORSCHUNGSBERICHTE,
VOL. 32, NO. 3, 1974, PP: 97-107.

74-0602 SMITH, R. A.
JERRAM, K.
EXPERIMENTAL AND THEORETICAL FATIGUE-CRACK
PROPAGATION LIVES OF VARIOUSLY NOTCHED PLATES
J. STRAIN ANAL., VOL. 9, NO. 2, APRIL 1974, PP:
61-66.

74-0603 WILSON, I. H.
TWO-DIMENSIONAL PHOTOELASTIC STUDY OF
STEAM-TURBINE CASING FLANGES WITH INTERFACE RELIEF
J. STRAIN ANAL., VOL. 9, NO. 2, APRIL 1974, PP:
109-117.

74-0604 CHANG, T. Y.
CHU, S. C.
ELASTIC PLASTIC DEFORMATION OF CYLINDRICAL
PRESSURE VESSELS UNDER CYCLIC LOADING
NUCL. ENG. DES., VOL. 27, NO. 2, MAY 1974, PP:
228-237.

74-0605 MITCHELL, R. A.
WOOLLEY, R. M.
HIGH STRENGTH END FITTINGS FOR FRP ROD AND ROPE
ASCE. J. ENG. MECH. DIV., VOL. 100, NO. EM4,
AUGUST 1974.   PAPER 10716, PP: 687-706.

74-0606 SKAAR, K. T.
ON THE FINITE ELEMENT STRESS ANALYSIS OF OIL
TANKER STRUCTURES
NOR. MARIT. RES., VOL. 2, NO. 2, 1974, PP: 2-11.

74-0607 ARGYRIS, J. H.
PISTER, K. S.
WILLIAM, K. J.
THERMOMECHANICAL CREEP OF AGEING CONCRETE - A
UNIFIED APPROACH
INSTITUT REPORT NO. 167, UNIVERSITAT STUTTGART,
1974.

74-0608 TIEU, A. K.
RESEARCH NOTE: A THREE-DIMENSIONAL OIL FILM
TEMPERATURE DISTRIBUTION IN TILTING THRUST
BEARINGS
J. MECH. ENG. SCI., VOL. 16, NO. 2, APRIL 1974,
PP: 121-124

74-0609 JINOCH, J.
BILEK, Z.
APLIKACE MKP NA RESENI PROBLEMU DIFUZE V KOVECH.
(APPLICATION OF THE FINITE ELEMENT METHOD
TO THE SOLUTION OF DIFFUSION PROBLEMS IN METALS)
KOVOVE MATER., VOL. 12, NO. 3, 1974, PP: 305-317

74-0610 KROENKE, W. C.
CLASSIFICATION OF FINITE ELEMENT STRESSES
ACCORDING TO ASME SECTION 111 STRESS CATEGORIES
PROC. OF A.S.M.E. CONF. ON PRESSURE VESSELS AND
PIPING. MIAMI BEACH, FLA., JUNE 24-28, 1974, PP.
107-140. (PUBL. BY A.S.M.E. NEW YORK, N.Y., 1974)

74-0611 BRATONOW, T.
ECER, A.
SENSITIVITY OF ROTOR BLADE VIBRATION
CHARACTERISTICS TO TORSIONAL OSCILLATIONS
J. AIRCR., VOL. 11, NO. 7, JULY 1974, PP: 375-381.

74-0612 SABOTA, Z.
VYPOCET KRUHOVYCH LOMENICOVYCH KONSTRUKCII METODOU
KONECNYCH PASOV. (CALCULATION OF CIRCULAR
FOLDED-PLATE STRUCTURES USING THE METHOD OF
FINITE STRIPS)
STAVEBNICKY CAS. VOL. 22, NO. 2, 1974, PP: 95-107.

74-0613 MCAULAY, A. D.
CHARAP, S. H.
NUMERICAL METHOD FOR INVESTIGATING THE PROPAGATION
OF SURFACE WAVES ON DISSIPATIVE GUIDES
PROC. IEEE., VOL. 62, NO. 3, MARCH 1974, PP:
402-403.

74-0614 KIKUCHI, F.
ANDO, R.
SIMPLIFIED HYBRID DISPLACEMENT METHOD FOR LINEAR
FINITE ELEMENT ANALYSIS OF GENERAL SHELLS
INT. J. PRESSURE VESSELS PIPING, VOL. 2, NO. 2,
APRIL 1974. PP: 155-164.

74-0615 BUTTSTAEDT, K. H.
THEIMERT, P. H.
DER EINSATZ DER FINIT-ELEMENT-METHODE IN
PRESSENBAU (APPLICATION OF THE FINITE
ELEMENT METHOD FOR THE CALCULATION IN
PRESS DESIGNING)
WERKSTATT BETR., VOL. 107, NO. 13, OCTOBER 1974,
PP: 589-594

74-0616 LYNN, P. P.
LEAST SQUARES FINITE ELEMENT ANALYSIS OF LAMINAR
BOUNDARY LAYER FLOWS
INT. J. NUM. METHODS ENG., VOL. 8, NO. 4, 1974,
PP: 865-876.

74-0617 SCHMIDT, W. F.
PROJECTIVE METHOD APPLIED TO THREE-DIMENSIONAL
ELASTICITY EQUATIONS
INT. J. NUM. METHODS ENG., VOL. 8, NO. 4, 1974,
PP: 697-711.

74-0618 ANON
TRANSVERSE STRENGTH AT FORE AND AFT SECTION OF
TANKER CARGO HOLD
IHI. ENG. REV., VOL. 7, NO. 3, SEPTEMBER 1974, PP:
30-38.

74-0619 SAKAI, K.
LINO, N.
STUDY ON FRACTURE TOUGHNESS EVALUATION BY
ELASTO-PLASTIC ANALYSIS AROUND NOTCHES
IHI. ENG. REV., VOL. 7, NO. 3, SEPTEMBER 1974, PP:
1-10.

74-0620 STEVEN, G. P.
STRESS DIFFUSION IN THICK STIFFENED PANELS
J. AIRCR., VOL. 11, NO. 12, DECEMBER 1974, PP:
779-781.

74-0621 KFOURI, A. P.
MILLER, K. J.
STRESS, DISPLACEMENT, LINE INTEGRAL AND CLOSURE
ENERGY DETERMINATIONS OF CRACK TIP STRESS
INTENSITY FACTORS
INT. J. PRESSURE VESSELS PIPING, VOL. 2, NO. 3,
JULY 1974. PP: 179-191.

74-0622 JOUNG, K. S.
HO, B. P. C.
SEISMIC ANALYSIS OF A THREE-PHASE BREAKER
INSTALLATION
ALLIS-CHALMERS ENG. REV., VOL. 39, NO. 2, 1974,
PP: 22-26.

74-0623 SIMPSON, L. A.
HSU, T. R.
APPLICATION OF THE SINGLE-EDGE NOTCHED BEAM TO
FRACTURE TOUGHNESS TESTING OF CERAMICS
J. TEST EVAL., VOL. 2, NO. 6, NOVEMBER 1974, PP:
503-509.

74-0624 SAKAI, F.
TSUKIOKA, Y.
FINITE ELEMENT ANALYSIS OF SURFACE WAVE PROBLEMS
USING INTEGRAL EQUATIONS
J. SOC. OF NAVAL ARCH. OF JAPAN, VOL. 136,
DECEMBER 1974.

74-0625 MORGAN, J. D.
KING, W. W.
ELASTODYNAMICS OF CRACKED STRUCTURES USING FINITE
ELEMENTS
AM. INST. AERON. AND ASTRON. J., VOL. 12, NO. 12,
1974,   PP: 1767-1769.

74-0626 WILSON, E. L.
BATHE, K. J.
PETERSON, F. E.
FINITE ELEMENT ANALYSIS OF LINEAR AND NONLINEAR
HEAT TRANSFER
NUCL. ENG. AND DESIGN, VOL. 29, NO. 1, 1974, PP:
110-124.

74-0627 NARAYANA, R.
NEW TRIANGULAR PLATE-BENDING FINITE ELEMENT WITH
TRANSVERSE SHEAR FLEXIBILITY
A.I.A.A. J., VOL. 12, NO. 12, 1974, PP: 1761-1763.

74-0628 DE ARANTES E OLIVEIRA, E. R.
TOVAR DE LEMOS, A.
CONVERGENCE OF FINITE ELEMENT SOLUTIONS IN
VISCOELASTICITY
IUTAM SYMP. ON THE MECH. OF VISCOELASTIC MEDIA AND
BODIES, GOTTENBURG, SEPTEMBER 1974.

74-0629 IRONS, B. M.
THE SEMI-LOOF SHELL ELEMENT
SYMP. ON FINITE ELEMENTS FOR THIN SHELLS AND
CURVED MEMBERS, UNIV. OF CARDIFF, MAY 1974.

74-0630 KULHAWY, F. H.
FINITE ELEMENT MODELING CRITERIA FOR UNDERGROUND
OPENINGS IN ROCK
INT. J. OF ROCK. MECH. AND MINING SCI., VOL. 2,
NO. 12, 1974 PP: 465-472.

74-0631 PIAN, T. H.
NONLINEAR CREEP ANALYSIS BY ASSUMED STRESS FINITE
ELEMENT METHODS
AM. INST. AERON. AND ASTRON. J. VOL. 12, NO. 12,
1974,   PP: 1756-1758.

74-0632 MORANDI CECCHI, M.
A NUMERICAL STUDY OF NONLINEAR INSTABILITY BY
MEANS OF AN EXTENDED FINITE ELEMENT METHOD
PROC. INT. CONF. IN NUMER. METH. IN FLUID
DYNAMICS, UNIV. OF SOUTHAMPTON, ENGLAND, SEPTEMBER
26-28, 1974, PP. 172-185.

74-0633 JESPERSE, D. C.
ARAKAWAS METHOD IS A FINITE-ELEMENT METHOD
J. OF COMPARATIVE PHYSIOLOGY, VOL. 16, NO. 4,
1974,   PP: 383-390

74-0634 RUMSEY, T. R.
ANALYSIS OF VISCOELASTIC CONTACT STRESSES IN
AGRICULTURAL PRODUCTS USING A FINITE ELEMENT
METHOD
PH.D., UNIV. OF CALIF., DAVIS, 1974.

74-0635 NICKELL, R. E.
FINITE-ELEMENT METHODS FOR COUPLED FIELD-STRESS
PROBLEMS
TRANSACTIONS OF THE AM. NUCLEAR SOC., VOL. 19,
OCTOBER 27, 1974, PP: 151-153.

74-0636 BECKER, E. B.
FINITE-ELEMENT ANALYSIS OF STRUCTURAL PROBLEMS
TRANSACTIONS OF THE AMER. NUCLEAR SOC., VOL. 19,
OCTOBER 27, 1974, PP: 151-152.

74-0637 JUIGNET, N.
CALCULATION BY FINITE ELEMENT TECHNIQUE OF
STEADY-STATE OR TRANSIENT HEAT-FLOW CONDUCTION
BULLETIN D'INFORMATIONS SCIENT. ET TECH. C.E.A.,
VOL. 1974, NO. 197, 1974, PP: 57-60.

74-0638 GOODELL, R. A.
ANALYTICAL AND EXPERIMENTAL INVESTIGATION OF A
NOZZLE TO CYLINDRICAL SHELL JUNCTION
PROC. OF A.S.M.E. CONF. ON PRESSURE VESSELS AND
PIPING, MIAMI BEACH, FLA., JUNE 24-28, 1974, PP.
141-157. (PUBL. BY A.S.M.E. NEW YORK, N.Y., 1974).

74-0639 WALKER, T. J.
QUANTITATIVE STRAIN-AND-STRESS STATE CRITERION FOR
FAILURE IN THE VICINITY OF SHARP CRACKS
NUCL. TECH., VOL. 23, NO. 2, AUGUST 1974, PP.
189-203.

74-0640 WOZNIAK, C. Z.
NONLINEAR MECHANICS OF CONSTRAINED MATERIAL
CONTINUA: FOUNDATIONS OF THE THEORY
ARCH MECH. ARCH MECH. STOSOW. VOL. 26, NO. 1,
1974,   PP: 105-118.

74-0641 ROMSTAD, K. M.
TAYLOR, M. A.
NUMERICAL BIAXIAL CHARACTERIZATION FOR CONCRETE
ASCE J. ENG. MECH. DIV., VOL. 100, NO. 5, OCTOBER
1974,   PP: 935-948.

74-0642 WANG, Y. S.
BILLINGTON, D. P.
BUCKLING OF CYLINDRICAL SHELLS BY WIND PRESSURE
A.S.C.E. J., ENG. MECH. DIV., VOL. 100, NO. 5,
OCTOBER 1974, PP. 1005-1024.

74-0643 HORLOCK, J. H.
FINITE ELEMENT METHODS IN FLOW PROBLEMS
J. OF FLUIDS ENG., VOL. 96, NO. 3, 1974, P. 197.

74-0644 YANG, T. Y.
LIANIS, G.
LARGE DISPLACEMENT ANALYSIS OF VISCOELASTIC BEAMS
AND FRAMES BY FINITE-ELEMENT METHOD
J. OF APPL. MECH., VOL. 41, NO. 3, 1974, PP.
635-640.

74-0645 MOCK, B.
LUMPING OF FINITE ELEMENT MATRICES
NOTICES OF THE AMER. MATH. SOC., VOL. 21, NO. 6,
1974, P: A550.

74-0646 FRANECK, H.
RECKE, H. G.
COMPUTATION OF IMPELLERS BY METHOD OF FINITE
ELEMENTS
MASCHINENBAU TECHNIK, VOL. 23, NO. 9, 1974, PP:
423-426.

74-0647 ARGYRIS, J. H.
BALMER, J. H.
BOLTSINIS, J. S.
MATERIAL NONLINEARITIES IN THE FINITE ELEMENT
ANALYSIS
PROC. OF INT. CONF. ON NUM. METHODS IN NONLINEAR
MECH., JABLONNA, POLEN, SEPTEMBER 1974.

74-0648 PITKARAN, J.
FINITE ELEMENT SOLUTION OF MULTIGROUP TRANSPORT
EQUATION IN 2 DIMENSIONAL GEOMETRIES
ACTA POLYTECHNICA SCANDINAVICA, PHYSICS INC.
NUCL., NO. 101, 1974, PP: 1-33.

74-0649 CULLEN, M. J.
INTEGRATIONS OF PRIMITIVE EQUATIONS ON A SPHERE
USING FINITE ELEMENT METHOD
QUARTERLY J. OF THE R. METEOROLOGICAL S., VOL.
100, NO. 426 1974, PP: 555-562.

74-0650 DANIEL, D. E.
OLSON, R. E.
STRESS-STRAIN PROPERTIES OF COMPACTED CLAYS
ASCE J. GEOTECH. ENG. DIV. VOL. 100, NO. 10,
OCTOBER 1974   PP: 1123-1136.

74-0651 AKTAN, A. E.
PECKNOLD, D. A.
RC COLUMN EARTHQUAKE RESPONSE IN TWO DIMENSIONS
ASCE J. STRUCT. DIV., VOL. 100, NO. 10, OCTOBER
1974,   PP: 1999-2015.

74-0652 BERKE, L.
VENKAYYA, V. B.
REVIEW OF OPTIMALITY CRITERIA APPROACHES TO
STRUCTURAL OPTIMIZATION
A.S.M.E. MEETING, NEW YORK, NOVEMBER 17-21, 1974.
VOL. 7, PP. 23-24.

74-0653 SETOGUCHI, K.
WADA, H.
STRUCTURAL ANALYSIS OF WELDED WALL FOR BOILER
FURNACE
MITSUBISHI HEAVY IND. TECH. REV., VOL. 11, NO. 2,
JUNE 1974, PP: 137-145.

74-0654 ANON
BOX CAR BODY BOLSTER PROBED IN STRESS STUDY
RAILW LOCOMOTIVE CARS, VOL. 148, NO. 6, AUGUST -
SEPTEMBER 1974, P: 24-25.

74-0655 LIU, S. C.
APPLICATION OF RESPONSE-BOUND METHOD IN SHOCK AND
VIBRATION ANALYSIS OF TELEPHONE STRUCTURES
BELL SYST. TECH. J. VOL. 53, NO. 7, SEPTEMBER
1974,   PP: 1403-1426.

74-0656 RYBICKI, E. F.
SIMONEN, F. A.
MATHEMATICAL AND EXPERIMENTAL STUDIES ON THE
MECHANICS OF PLATED TRANSVERSE FRACTURES
J. BIOMECH. VOL. 7, NO. 4, AUGUST 1974, PP:
377-384.

74-0657 MEISSNER, H.
WIBEL, A.
PARAMETER EINES ELASTO-PLASTISCHEN STOFFANSATZES
FUER KOERNIGE ERDSTOFFE (DETERMINATION OF
PARAMETER IN ELASTIC-PLASTIC PROBLEMS FOR GRANULAR
MATERIALS)
BAUTECHNIK (AUS G A) VOL. 51, NO. 8, AUGUST 1974,
PP: 263-269.

74-0658 GRAY, W. G.
PINDER, G. F.
GALERKIN APPROXIMATION OF THE TIME DERIVATIVE IN
THE FINITE ELEMENT ANALYSIS OF GROUNDWATER
FLOW
WATER RESOUR. RES., VOL. 10, NO. 4, AUGUST 1974,
PP: 821-828

74-0659 VOROSHKO, P. P.
KVITKA, A. L.
CHISLENNOE RESHENIE PLOSKIKH ZADACH
TEPLOPROVODNOSTI DLYA OBLASTEI SLOZHNOI
FORMY (NUMERICAL SOLUTION OF PLANE PROBLEMS OF
THERMAL CONDUCTIVITY FOR REGIONS OF COMPLEX
FORM)
PROBL. PROCHN. VOL. 6, NO. 6, JUNE 1974, PP: 3-7.

74-0660 MEHRAIN, M.
AALAMI, B.
ROTATIONAL STIFFNESS OF CONCRETE SLABS
J. AM. CONCR. INST. VOL. 71, NO. 9, SEPTEMBER
1974,   PP: 429-435.

74-0661 DITTMAR, S.
STRESS DETERMINATION IN 2 CONDENSER HEAT EXCHANGER
BAFFLE CAPS BY METHOD OF FINITE ELEMENTS AND A
STRENGTH ASSESSMENT OF THE SE COMPONENTS
CHEIMIE-INGENIEUR-TECHNIK, VOL. 46, NO. 21, 1974,
P: 915.

74-0662 NARAYANA, R.
ADELMAN, H. M.
INCLUSION OF TRANSVERSE SHEAR DEFORMATION IN
FINITE ELEMENT DISPLACEMENT FORMULATIONS
AMERICAN INST. AERONAUTICS AND ASTRONAUTICS, J.
VOL. 12, NO. 11, 1974, PP: 1613-1614.

74-0663 ANAND, S. C.
GARG, V. K.
SHAKEDOWN OF ROLLING WHEEL UNDER HUB LOADING
ASCE J. ENG. MECH. DIV., VOL. 100, NO. 6, DECEMBER
1974,   PP: 1237-1251.

74-0664 KRIZEK, R. J.
FARZIN, M. H.
EVALUATION OF STRESS CELL PERFORMANCE
ASCE J. GEOTECH. ENG. DIV., VOL. 100, NO. 12,
DECEMBER 1974, PP: 1275-1295.

74-0665 CLOUGH, G. W.
TSUI, Y.
PERFORMANCE OF TIED-BACK WALLS IN CLAY
ASCE J. GEOTECH. ENG. DIV., VOL. 100, NO. 12,
DECEMBER 1974, PP: 1259-1273.

74-0666 PRIETO, L.
AN INCORE FINITE ELEMENT PROGRAM FOR RESEARCH AND
TEACHING PURPOSE
RESEARCH REPORT NO. 162, INSTITUT FUR STATIK UND
DYNAMIK DER LUFTUND RAUMFAHRTKONSTRUKTIONEN,
OCTOBER 1974.

74-0667 LARRABEE, R. D.
BILLINGTON, D. P.
THERMAL LOADING OF THIN-SHELL CONCRETE COOLING
TOWERS
ASCE J. STRUCT. DIV., VOL. 100, NO. 12, DECEMBER
1974,   PP: 2367-2383.

74-0668 BAZANT, Z. P.
WU, S. T.
CREEP AND SHRINKAGE LAW FOR CONCRETE AT VARIABLE
HUMIDITY
ASCE J. ENG. MECH. DIV.,VOL. 100, NO. 6, DECEMBER
1974,   PP: 1183-1209

74-0669 PAO, Y. C.
WOODS, E. H.
RITMAN, E. L.
FINITE ELEMENT ANALYSIS OF LEFT VENTRICULAR
MYOCARDIAL STRESSES
J. OF BIOMECHANICS, VOL. 7, NO. 6, 1974, PP:
469-477.

74-0670 BELYTSCHKO, T.
MARCHERTAS, A. H.
NONLINEAR FINITE-ELEMENT METHOD FOR PLATES AND ITS
APPLICATION TO DYNAMIC RESPONSE OF REACTOR FUEL
SUBASSEMBLIES
J. PRESSURE VESSEL TECH. TRANS. ASME., VOL. 96,
NO. 4, NOVEMBER 1974, PP: 251-257.

74-0671 VARSHNEY, R. S.
DAMS ON ROCKS OF VARYING ELASTICITY
INT. J. ROCK. MECH. MIN. SCI., GEOMECH. ABSTR.,
VOL. 11, NO. 1, JANUARY 1974, PP: 1-12.

74-0672 ARGYRIS, J. H.
FAUST, G.
SZIMMAT, J.
WARNKE, E. P.
WILLAM, K. J.
FINITE ELEMENT ULTIMATE LOAD ANALYSIS OF THREE
DIMENSIONAL CONCRETE STRUCTURES
REPORT NO. 166, INSTITUT FUR STATIK UND DYNAMIK
DER LUFTUND RAUMFAHRTKONSTRUKTIONEN, 1974.

74-0673 ARGYRIS, J. H.
DUNNE, P. C.
ANGELOPOULDS, T.
DIE LOSUNG NICHT-LINEARER PROBLEME NACH DER
METHODE DER FINITEN ELEMENTE
PROC. DER CLAUSTHAL-CONF. (ERSCHEINT IM
BIRKHAUSER-VERLAG, BASEL), 1974.

74-0674 DUNNE, P. C.
SOME CONTRIBUTIONS TO NONLINEAR SOLID MECHANICS
IRIA LECTURE NOTES IN COMPUT. SCI., VOL. 10,
SPRINGER-VERLAG, BERLIN, PART 1, 1974, PP. 42-139.

74-0675 THOMPSON, G. V.
RAUSCHENPLAT, H. C.
CLOSURE ANALYSIS USING SELF ENERGIZING, DOUBLE
CONICAL GASKETS
PROC. OF CONF. ON PRESSURE VESSELS AND PIPING,
MIAMI BEACH, FLA., JUNE 24-28, 1974, PP. 159-170.
(PUBL. BY A.S.M.E., NEW YORK, N.Y., 1974)

74-0676 ARGYRIS, J. H.
DUNNE, P. C.
THE FINITE ELEMENT METHOD APPLIED TO FLUID
MECHANICS
PROC. OF CONF. ON COMPUTATIONAL METHODS AND
PROBLEMS IN AERONAUTICAL FLUID DYNAMICS, U. OF
MANCHESTER, ENGLAND, SEPTEMBER, 1974.

74-0677 ARGYRIS, J. H.
DUNNE, P. C.
ANGELOPOULOS, T.
BICHAT, B.
LINEAR AND NONLINEAR OSCILLATIONS IN MECHANICS
PROC. OF 9TH CONGR. OF THE INT. COUNCIL OF THE
AERON. SCI. HAIFA, ISRAEL, AUGUST 1974.

74-0678 ARGYRIS, J. H.
GRIEGER, I.
KONSTRUKTION UND BERECHNUNG MIT FINITEN ELEMENTEN
SONDERDRUCK CHEMIE-INGENIEUR-TECHNIK, ZEITSCHRIFT
FUR VERFAHRENSTECHNIK, VOL. 46, NO. 13, HEFT,
1974, P: 451.

74-0679 ARGYRIS, J. H.
KONIG, M.
NAGY, D. A.
HAASE, M.
MALEJANNAKIS, G.
GEOMETRIC NONLINEARITY AND THE FINITE ELEMENT
DISPLACEMENT METHOD
PROC. OF INT. CONF. ON NUM. METHODS IN NONLINEAR
MECHANICS, JABLONNA, POLEN, SEPTEMBER 1974.

74-0680 DIETRICH, G.
JOHNSEN, T. L.
SIMULTANEOUS EIGENVECTOR ITERATION APPLIED TO
LARGE SYSTEMS
RESEARCH REPORT NO. 149, INSTITUT FUR STATIK UND
DYNAMIK DER LUFT UND RAUMFAHRTKONSTRUKTIONEN,
JANUARY 1974.

74-0681 GREIGER, I.
INGA - INTERACTIVE GRAPHIC ANALYSIS
USERS REFERENCE MANUAL RESEARCH REPORT NO. 156,
INSTITUT FUR STATIK UND DYNAMIK DER LUFT UND
RAUMFAHRTKONSTRUKTIONEN, OCTOBER 1974.

74-0682 MCCARTHY, J. B.
CAMPBELL, R. D.
INELASTIC ANALYSIS OF A BRANCH SHELL JUNCTION
PROC. OF A.S.M.E. CONF. ON PRESSURE VESSELS AND
PIPING, MIAMI BEACH, FLA., JUNE 24-28, 1974, PP.
187-197. (PUBL. BY A.S.M.E., NEW YORK, N.Y., 1974)

74-0683 MLEJNEK, H. P.
BUHLMEIER, J.
MAI, M.
MALEN DIMENSIONIERUNG VON TRAGWERKEN
UNTERSUCHUNG UND WEITERENTWICKLUNG EINIGER
VERFAHREN ZUR OPTINSTITUT FUR STATIK UND DYNAMIK
DER LUFT UND RAUMFAHRTKONSTRUKTIONEN, RESEARCH
REPORT NO. 259, JULY 1974.

74-0684 HSU, M. B.
BERMAN, I.
EVALUATION OF THE COMPUTER PROGRAM CREEP-BLAST
(CDC VERSION)
PROC. OF A.S.M.E. CONF. ON PRESSURE VESSELS AND
PIPING, MIAMI BEACH, FLA., JUNE 24-28, 1974, PP.
199-219. (PUBL. BY A.S.M.E., NEW YORK, N.Y., 1974)

74-0685 DECRETON, M. C.
GARDIOL, F. E.
NUMERICAL ANALYSIS OF THE LINE CAPACITANCE AND
CROSSTALK FACTOR FOR INSULATED WIRE PAIRS
AEU. ARCH. ELEKTRON UEBERTRAG ELECTRON COMMUN.,
VOL. 28, NO. 10, OCTOBER 1974, PP: 415-420.

74-0686 SOVA, J. A.
CREWS, J. H.
A METHOD FOR DETERMINING LOCAL ELASTOPLASTIC
STRESS AND STRAIN IN METALLURGICALLY BONDED
NOTCHED LAMINATES SUBJECTED TO A LOADING CYCLE
NATIONAL AERON. AND SPACE ADMIN., LANGLEY RESEARCH
CENTER, LANGLEY STATION, VA., REPORT NO.
NASA-TN-D7766, L-9753, OCTOBER 1974, 23 PP.
(N.T.I.S. - N74-3530/9ST)

74-0687 CREWS, J. H. JR.
AN ELASTOPLASTIC ANALYSIS OF A UNIAXIALLY LOADED
SHEET WITH AN INTERFERENCE - FIT BOLT
NATIONAL AERON. AND SPACE ADMIN., LANGLEY,
RESEARCH CENTER, LANGLEY STATION, VA., REPORT NO:
NASA-TR-D-7748,L-9593, OCTOBER 1974, 25 P.
(N.T.I.S. - N74-35304/6ST)

74-0688 ZUK, J.
COMPRESSIBLE SEAL FLOW ANALYSIS USING THE FINITE
ELEMENT METHOD WITH GALERKIN SOLUTION TECHNIQUE
NATIONAL AERON. AND SPACE ADMIN., LEWIS RESEARCH
CENTER, CLEVELAND, OHIO. REPORT NO:
NASA-TM-X-71604, E-5116-2, 1974, 54 P. (N.T.I.S. -
N74-34698/2ST)

74-0689 REMINGTON, P. J.
O'CALLAGHAN, J. C.
MADDEN, R.
ANALYSIS OF STRESSES AND DEFLECTIONS IN FRAME
SUPPORTED TENTS
BOLT BERANEK AND NEWMAN INC., CAMBRIDGE, MASS.,
REPORT NO: BBN-2802, APRIL 1974, 131 PP. (N.T.I.S.
- AD/A-001 072/7ST)

74-0690 SHIH, C. F.
J INTEGRAL ESTIMATES FOR STRAIN HARDENING
MATERIALS USING FULLY PLASTIC SOLUTIONS
DIV. OF ENGRNG AND APPLIED PHYSICS, HARVARD UNIV.,
CAMBRIDGE, MASS., REPORT NO: DEAP-S-10, AUGUST
1974, 38 P. (N.T.I.S. - AD/A-001 939/8ST)

74-0691 DESILVA, B. M. E.
GRANT, G. N. C.
HYBRID FINITE ELEMENT OPTIMIZATION PROGRAMS FOR
DISCS USING A DYNAMICS DESIGN TECHNOLOGY
12TH A.I.A.A. AEROSPACE SCI. MEETING, WASHINGTON,
D.C., PAPER-74-103, JANUARY 30 - FEBRUARY 1, 1974,
REPORT NO: MATHS-RE-S-35, MARCH 1974, 31 P.
(N.T.I.S.- N74-34249/4SL)

74-0692 MILLER, R. W.
RETSCP: A COMPUTER PROGRAM FOR ANALYSIS OF ROCKET
ENGINE THERMAL STRAINS WITH CYCLIC PLASTICITY
ATKINS AND MERRILL, INC., ASHLAND, MASS., REPORT
NO: NASA-CR-134640, JUNE 1971, 168 P. (N.T.I.S. -
N74-34244/5SL)

74-0693 MILLER, R. W.
CYCLIC FATIGUE ANALYSIS OF ROCKET THRUST CHAMBERS.
VOLUME 2: ATTITUDE CONTROL THRUSTER HIGH CYCLE
FATIGUE
ATKINS AND MERRILL, INC., ASHLAND, MASS., REPORT
NO: NASA-CR-134641-VOL-2, JUNE 1974, 41 P.
(N.T.I.S. - N74-34243/7SL)

74-0694 MILLER, R. W.
CYCLIC FATIGUE ANALYSIS OF ROCKET THRUST CHAMBERS.
VOLUME 1: OFHC COPPER CHAMBER LOW CYCLE FATIGUE
ATKINS AND MERRILL, INC., ASHLAND, MASS., REPORT
NO: NASA-CR-134641-VOL-1, JUNE 1971 69 P.
(N.T.I.S. - N74-34242/9SL)

74-0695 MANDELL, J. F.
WANG, S. S.
MCGARRY, F. J.
FRACTURE OF GRAPHITE FIBER REINFORCED COMPOSITES
DEPT. OF CIVIL ENGNRG., INST. OF TECH., CAMBRIDGE,
MASS., FINAL TECH. REPORT, MAY 15, 1974 - NOVEMBER
15, 1974, 169 P. (N.T.I.S.- AD/A001 612/1SL) SEE
ALSO AD-235 927.

74-0696 LEFEBVRE, G. D.
DUNCAN, J. M.
FINITE ELEMENT ANALYSIS OF TRANSVERSE CRACKING IN
LOW-EMBANKMENT DAMS
OFFICE OF RESEARCH SERVICE, UNIV. OF CALIF.,
BERKELEY, FINAL REPORT NO. 5, OCTOBER 1974, 98 PP.
(N.T.I.S. - AD/A-001 523/0SL) SEE ALSO REPORT
AD-766 728, JANUARY 1973.

74-0697 WATKINSON, K. W.
THE MIDCOHV WEIGHT AND BALANCE COMPUTER PROGRAM
(WTBAL)
NAVAL COASTAL SYSTEMS LAB., PANAMA CITY, FLA.,
REPORT NO: NCSL-220-74, SEPTEMBER 1974, 30 P.
(N.T.I.S. - AD/A-001 279/9SL)

74-0698 COST, TH. L.
WEEKS, G. E.
AUTOMATED EVALUATION OF INTERCEPTOR ROCKET MOTOR
DESIGNS UNDER COMBINED OPERATIONAL AND NUCLEAR
EFFECTS LOADS
BUREAU OF ENG. RESEARCH, UNIV. OF ALABAMA, REPORT
NO: BER-176-97, JULY 1974, 65 P. (N.T.I.S. -
AD/A-000 915/9SL)

74-0699 DAVIDSON, H. L.
CHEN, W. F.
ELASTIC-PLASTIC LARGE DEFORMATION RESPONSE OF CLAY
TO FOOTING LOADS
FRITZ ENG. LAB., LEHIGH UNIV., BETHLEHEM, PA.,
REPORT NO: FEL-355.18, FEBRUARY 1974, 190 P.
(N.T.I.S. - PB-237 524/4ST)

74-0700 GARDNER, L. R. T.
GARDNER, G. A.
STUDY OF THE MHD INSTABILITIES OF CYLINDRICAL BELT
PINCH USING THE FINITE ELEMENT METHOD
INST. FOR THEORETICAL PHYSICS, INNSBRUCK UNIV.,
AUSTRIA, REPORT NO: UNICP-SR-103, JULY 25, 1974,
11 PP. (N.T.I.S. - AD/A-000 540/5ST)

74-0701 ODEN, J. T.
REDDY, J. N.
ON MIXED FINITE ELEMENT APPROXIMATIONS
INST. FOR COMPUTATIONAL MECH., AUSTIN, TEXAS,
REPORT NO: TICOM-74-3, MAY 1974, 26 P. (N.T.I.S. -
AD/A-000 386/3ST)

74-0702 NICHOLS, C. S.
KNITTEL, M. R.
BARACH, D.
CARLSON, M.
A FINITE ELEMENT APPROACH TO SCATTERING FROM
ELASTIC SPHERICAL SHELLS
NAVAL UNDERSEA CENTER, SAN DIEGO, CALIF., REPORT
NO: NUC-TP-425, SEPTEMBER 1974, 14 P. (N.T.I.S. -
AD/A-000 098/4ST)

74-0703 LEVY, A.
ZALESAK, J.
BERNSTEIN, M.
MASON, P. W.
DEVELOPMENT OF TECHNOLOGY FOR MODELING OF A
1/8-SCALE DYNAMIC MODEL OF THE SHUTTLE SOLID
ROCKET BOOSTER
GRUMMAN AEROSPACE CORP., BETHPAGE, N.Y., REPORT
NO: NASA-CR-132492, JULY 1974, 143 P. (N.T.I.S. -
N75-13926/1ST)

74-0704 MURAKI, T.
MASUBUCHI, K.
THERMAL ANALYSIS OF M552 EXPERIMENT FOR MATERIALS
PROCESSING IN SPACE
MASS. INST. OF TECH., CAMBRIDGE, REPORT NO:
NASA-CR-129043, OCTOBER 1974, 48 PP. (N.T.I.S. -
N75-13896/6ST)

74-0705 BATSON, G. B.
INFLATION FORMING OF STEEL FIBER-REINFORCED
CONCRETE DOMES
ARMY CONSTRUCTION ENG. RESEARCH LAB., CHAMPAIGN,
ILL. (405279), REPORT NO: CERL-IR-M/115, DECEMBER
1974, 23 PP. (N.T.I.S. D/A-005 046/8ST)

74-0706 BATHE, K. J.
PETERSON, F. E.
USERS MANUAL FOR CEL/NONSAP, A NONLINEAR
STRUCTURAL ANALYSIS PROGRAM
ENGINEERING/ANALYSIS CORP., BERKELEY, CALIF.,
REPORT NO: E/AC-74-2, OCTOBER 1974, 149 P.
(N.T.I.S. - AD/A-004 98802ST)

74-0707 BAYLOR, J. L.
BIENIEK, M. P.
WRIGHT, J. P.
TRANAL: A 3-D FINITE ELEMENT CODE FOR TRANSIENT
NONLINEAR ANALYSIS
WEIDLINGER ASSOCIATES, NEW YORK, REPORT: APRIL 1,
1973 - JUNE 30, 1974, 48 P. (N.T.I.S. - AD/A-004
979/1ST)

74-0708 OJALVO, I. U.
AUSTIN, F.
LEVY, A.
VIBRATION AND STRESS ANALYSIS OF SOFT-BONDED
SHUTTLE INSULATION TILES. MODAL ANALYSIS WITH
COMPACT WIDELY SPACED STRINGERS
GRUMMAN AEROSPACE CORP., BETHPAGE, N.Y., REPORT
NO: NASA-CR-132553, SEPTEMBER 1974, 136 P.
(N.T.I.S. - N75-13311/6ST)

74-0709 AUTHORS
PROCEEDINGS OF THE 5TH NAVY-NASTRAN COLLOQUIUM.
HELD AT NAVAL SHIP RESEARCH AND DEVELOPMENT
CENTER, BETHESDA, MD.
COMPUTATION AND MATHS. DEPT., NAVAL SHIP RESEARCH
AND DEVELOPMENT CENT., BETHESDA, MD., REPORT NO.
CMD-32-74, SEPTEMBER 1974, 140 PP. (N.T.I.S. -
AD/A-004 604/5ST).

74-0710 GEFFEN, N.
YANIV, S.
FINITE ELEMENTS FOR FLUID DYNAMICS. VARIATIONAL
FORMULATIONS FOR NONLINEAR, INITIAL-VALUE PROBLEMS
DEPT. OF APPL. MATH., TEL-AVIV UNIV., ISRAEL,
INTERIM REPORT, JUNE 1, 1973 - APRIL 30, 1974,
GRANT AF-AFOSR-2561-73. (N.T.I.S. AD/A-004
484/2ST)

74-0711 DOYLE, D. M.
STIFFNESS OF STEPPED BARS
MASTER'S THESIS, NAVAL POSTGRADUATE SCHOOL,
MONTEREY, CALIF., DECEMBER 1974, 81 P. (N.T.I.S. -
AD/A-004 251/5ST)

74-0712 RENICK, J. D.
ANALYSIS OF THE MUTUAL INDUCTANCE PARTICLE
VELOCIMETER
AFB N. MEX., AIR FORCE WEAPONS LAB., KIRTLAND,
REPORT NO: AFWL-TR-74-205, NOVEMBER 1974, 133 P.
(N.T.I.S. - AD/A-004 219/2ST)

74-0713 GROSSKURTH, J. F. JR.,
WHITE, R. N.
GALLAGHER, R. H.
THOMAS, G. R.
SHEAR BUCKLING OF SQUARE PERFORATED PLATES
CORNELL UNIV., ITHACA, N. Y. REPORT NO:
NASA-CR-132548, 1974, 47 P. (N.T.I.S. -
N75-12364/6ST)

74-0714 EVERSTINE, G. C.
MCKEE, J. M.
A SURVEY OF PRE AND POSTPROCESSORS FOR NASTRAN
NAVAL SHIP RESEARCH AND DEVELOPMENT CENTER,
BETHESDA, MD., REPORT NO: NSRDC-4391, JUNE 1974,
25 P. (N.T.I.S. - AD/A-004 177/2ST)

74-0715 ODEN, J. T.
A THEORY OF MIXED FINITE ELEMENT APPROXIMATIONS OF
NON-SELF-ADJOINT BOUNDARY VALUE PROBLEMS
UNIV. BUREAU OF ENG. RESEARCH, ALABAMA UNIV.,
REPORT NO: BER-184-07, OCTOBER 1974, 131 P.
(N.T.I.S. - AD/A-004 146/7ST)

74-0716 DIETRICH, D. E.
THREE DIMENSIONAL INELASTIC ANALYSIS OF A SHELL
APPURTENANCE PER SECTION 111 - APPENDIX F.
PROC. OF A.S.M.E. CONF. ON PRESSURE VESSELS AND
PIPING, MIAMI BEACH, FLA., JUNE 24-28, 1974, PP.
221-233. (PUBL. BY A.S.M.E., NEW YORK, N.Y.,
1974).

74-0717 LEE, S. W.
AN ASSUMED STRESS HYBRID FINITE ELEMENT FOR
THREE-DIMENSIONAL PLASTIC STRUCTURAL ANALYSIS
AEROELASTIC AND STRUCTURES RESEARCH LAB., INST. OF
CAMBRIDGE TECH., MASSACHUSETTS. REPORT NO:
ASRL-TR-170-3, MAY 1974, 111 PP. (N.T.I.S. -
AD/A-004 139/2ST)

74-0718 SPILKER, R. L.
A STUDY OF ELASTIC-PLASTIC ANALYSIS BY THE
ASSUMED-STRESS HYBRID FINITE-ELEMENT MODEL, WITH
APPLICATION TO THICK SHELLS OF REVOLUTION
AEROELASTIC AND STRUCTURES RESEARCH LAB., INST. OF
CAMBRIDGE TECH., MASSACHUSETTS, REPORT NO:
ASRL-TR-175-1, DECEMBER 1974, 385 P.
(N.T.I.S.-AD/A-004 027/9ST)

74-0719 NATH, B.
FUNDAMENTALS OF FINITE ELEMENT METHODS FOR
ENGINEERS
ATHLONE PRESS, LONDON, 1974.

74-0720 BILGEN, E.
TOO, J. J. M.
ON THE FINITE ELEMENT FORMULATION OF NAVIER-STOKES
EQUATIONS
TRANS. CAN. SOC. MECH. ENG., VOL. 2, NO. 4,
1973-1974, PP. 205-208.

74-0721 ODEN, J. T.
REDDY, J. N.
WELLFORD, L. C.
FOST, R. B.
KEY, J. E.
THEMIS FINAL REPORT, VOL. IV, RECENT DEVELOPMENTS
IN THE NUMERICAL SOLUTION OF CERTAIN PROBLEMS IN
NONLINEAR ELASTODYNAMICS AND IN MIXED FINITE
ELEMENT APPLICATIONS
SCHOOL OF GRADUATE STUDIES AND RESEARCH, UNIV. OF
ALABAMA, HUNTSVILLE, ALABAMA. SCIENTIFIC REPORT,
OCTOBER 1974, 343 PP. (N.T.I.S. - AD/A-003
728/3ST).

74-0722 CHUNG, T. J.
THEMIS FINAL REPORT, VOL. II, NONLINEAR MECHANICS
IN SOLIDS AND STRUCTURES
SCHOOL OF GRADUATE STUDIES AND RESEARCH, UNIV. OF
ALABAMA, HUNTSVILLE, ALABAMA. REPORT NO.
UAH-RR-162, OCTOBER 1974, 169 PP. (N.T.I.S. -
AD/A-003 714/3ST).

74-0723 BRAUCHLI, J. H.
CHUNG, T. J.
COST, T. L.
ODEN, J. T.
WEEKS, G. R.
THEMIS FINAL REPORT, VOL. 1, BASIC RESEARCH IN
METHODS OF APPROXIMATION IN THE NONLINEAR
MECHANICS OF SOLIDS AND STRUCTURES
SCHOOL OF GRADUATE STUDIES AND RESEARCH, UNIV. OF
ALABAMA, HUNTSVILLE, ALABAMA, REPORT NO.
UAH-RR-161, OCTOBER 1974, 112 PP. (N.T.I.S. -
AD/A-003 713/5ST).

74-0724 BALL, R. E.
AIRCRAFT FUEL TANK VULNERABILITY TO HYDRAULIC RAM:
MODIFICATTION OF THE NORTHROP FINITE ELEMENT
COMPUTER CODE BR-1 TO INCLUDE FLUID-STRUCTURE
INTERACTION-THEORY AND USERS MANUAL FOR BR-1HR
NAVAL POSTGRADUATE SCHOOL, MONTEREY, CALIF.,
REPORT NO: NPS-57BP74071, JULY 1974, 35 P.
(N.T.I.S. - AD/A-003 471/0ST)

74-0725 MURAKI, T.
MASUBUCHI, K.
THERMAL ANALYSIS OF M551 EXPERIMENT FOR MATERIALS
PROCESSING IN SPACE
MASSACHUSETTS INST. OF TECH., CAMBRIDGE, REPORT
NO: NASA-CR-120517, NOVEMBER 1974, 50 P. (N.T.I.S.
- N75-10966/0ST)

74-0726 HIRSCH, C.
WARZEE, G.
A FINITE ELEMENT METHOD FOR FLOW CALCULATIONS IN
TURBOMACHINES
DEPT. OF FLUID MECHANICS, BRUSSELS UNIV., BELGIUM,
REPORT NO: VUB-STR-5, JULY 1974, 76 P. (N.T.I.S. -
N75-10946/2ST)

74-0727 CHEN, Y-N.
PREDICTION OF LONG TERM DEFORMATION OF A COMPACTED
COHESIVE SOIL EMBANKMENT OVER A SOFT FOUNDATION
JOINT HIGHWAY RESEARCH PROJECT, PURDUE UNIV.,
LAFAYETTE, IND. REPORT NO: JHRP-74-4, MAY 1974,
202 P. (N.T.I.S. - PB-238 159/8ST)

74-0728 RAI, I.
SANDHU, R. S.
FINITE ELEMENT ANALYSIS OF ANISOTROPIC PLATES
USING Q-19 ELEMENT. PART 1. THEORETICAL
DEVELOPMENT
RESEARCH FOUNDATION, OHIO STATE UNIV., COLUMBUS,
REPORT NO. OSURF-3613-73-IF, OCTOBER 1974, 116 PP.
(N.T.I.S. AD/A-002 921/5ST)

74-0729 ATLURI, S. N.
KOBAYASHI, A. S.
NAKAGAKI, M.
APPLICATION OF AN ASSUMED DISPLACEMENT HYBRID
FINITE ELEMENT PROCEDURE TO TWO-DIMENSIONAL
PROBLEMS IN FRACTURE MECHANICS
PROC. OF 15TH A.I.A.A./A.S.M.E. CONF. ON STRUCT.
DYNAMICS AND MATER., LAS VEGAS, NEV., PAPER NO.
74-390, APRIL 17-19, 1974.

74-0730 VANDERBILT, M. D.
GOODMAN, J. R.
CRISWELL, M. E.
BODIG, J.
A RATIONAL ANALYSIS AND DESIGN PROCEDURE FOR WOOD
JOIST FLOOR SYSTEMS
COLORADO STATE UNIV., FORT COLLINS, REPORT NO:
CSU/CEWS/74-1, NOVEMBER 1974, 133 P. (N.T.I.S. -
PB-237 769/5ST)

74-0731 RAI, I.
IQBAL, S.
RANBIR, S.
FINITE ELEMENT ANALYSIS OF ANISOTROPIC PLATES
USING 0-19 ELEMENT: PART 11. INSTRUCTIONS FOR
USERS AND FORTRAN LISTING
RESEARCH FOUNDATION, OHIO STATE UNIV., COLUMBUS,
OHIO, REPORT NO: OSURF-3613-73-2F, OCTOBER 1974,
72 P. (N.T.I.S. - AD/A-002 689/8ST)

74-0732 DWYER, W. J.
AN IMPROVED AUTOMATED STRUCTURAL OPTIMIZATION
PROGRAM
GRUMMAN AEROSPACE CORP., BETHPAGE, N.Y., FINAL
REPORT: MARCH 27, 1973 - AUGUST 30, 1974, 158 P.
(N.T.I.S. - AD/A-002 688/0ST)

74-0733 WHITE, J. A.
OH - 58A PROPULSION SYSTEM VIBRATION INVESTIGATION
BELL HELICOPTER CO., FORT WORTH, TEX., AUGUST
1974, 193 P. (N.T.I.S. - AD/A-002 672/4ST)

74-0734 ELBATOUTI, A. M.
LIU, D.
JAN, H. Y.
STRUCTURAL ANALYSIS OF SL-7 CONTAINERSHIP UNDER
COMBINED LOADING OF VERTICAL, LATERAL AND
TORSIONAL MOMENTS USING FINITE ELEMENT TECHNIQUES
SHIP STRUCTURE COMMITTEE, WASHINGTON, DC., REPORT
NO: SSC-243, MAY 1974, 59 P. (N.T.I.S. - AD/A-002
620/3ST)

74-0735 ANDERSON, W. E.
FRACTURE MECHANICS OF TINY CRACKS NEAR FASTENERS
BATTELLE LABS., COLUMBUS, OHIO, JUNE 1974, 61 P.
(N.T.I.S - AD/A-002 554/4ST)

74-0736 KOBAYASHI, A. S.
CHAN, C. F.
A DYNAMIC PHOTOELASTIC ANALYSIS OF DYNAMIC TEAR
TEST SPECIMEN
DEPT. OF MECHANICAL ENG., UNIV. OF WASH., SEATTLE,
REPORT NO: TR-20, OCTOBER 1974, 29 P. (N.T.I.S. -
AD/A-002 431/5ST)

74-0737 VICK, J. L.
THE USE OF EIGEN-DERIVATIVES TO PREDICT THE EFFECT
OF DESIGN CHANGES UPON THE DYNAMIC CHARACTERISTICS
OF STRUCTURAL ELEMENTS
AIR FORCE INST. OF TECH., WRIGHT-PATTERSON AFB.,
OHIO SCHOOL OF ENG., REPORT NO: GA/MC/74-6,
DECEMBER 1974, 86 PP. (N.T.I.S. - AD-005 281/1ST)

74-0738 PEEL, M.
WOLBERG, J. R.
GLASER, E.
NICKERSON, E.
A FINITE ELEMENT PROGRAM FOR DESIGN AND ANALYSIS
OF THREE-DIMENSIONAL FLAT SPRINGS
PROC. INT. CONF. ON COMPUT. IN ENG. AND BUILD.
DES., IMP. COLL. OF SCI. AND TECH., LONDON, ENGL.,
SEPTEMBER 25-27, 1974. PUBL. BY IPC BUS. PRESS,
LTD., GUILDFORD, SURREY, ENGL. 1974.

74-0739 MEI, K. K.
KUO, W. C.
CHANG, S. K.
SCATTERING BY BURIED OBSTACLES
ELECTRONICS RESEARCH LAB., CALIFORNIA UNIV.,
BERKELEY. FINAL REPORT: APRIL 20, 1971 - JULY 31,
1974, 119P. (N.T.I.S. - AD/A-005 237/3ST)

74-0740 BATHE, K. J.
PETERSON, F. E.
THEORETICAL BASIS FOR CBL/NONSAP, A NONLINEAR
STRUCTURAL ANALYSIS PROGRAM
ENGINEERING ANALYSIS CORP., BERKELEY, CALIF.,
REPORT NO: E/AD/A-004 989/0ST

74-0741 AZIZ, A. K.
LEVENTHAL, S. H.
ON NUMERICAL SOLUTIONS OF EQUATIONS OF
HYPERBOLIC-ELLIPTIC TYPE
NAVAL ORDNANCE LAB., WHITE OAK, MD., REPORT NO:
NOLTR-74-144, AUGUST 1, 1974, 22 P.
(N.T.I.S.-AD/A-000 070/3ST)

74-0742 HUANG, Y. H.
WU, S. J.
ANALYSIS OF UNSTEADY FLOW TOWARD ARTESIAN WELLS BY
THREE-DIMENSIONAL FINITE ELEMENTS
KENTUCKY WATER RESOURCES RESEARCH INST.,
LEXINGTON, REPORT NO: RR-75, AUGUST 1974, 172 P.
(PB-236 925/4SL)

74-0743 WIGGERTS, D. C.
TWO-DIMENSIONAL FINITE ELEMENT MODELING OF
TRANSIENT FLOW IN REGIONAL AQUIFER SYSTEMS.
ANALYSIS OF GROUNDWATER FLOW IN RELATION TO WATER
QUALITY IMPROVEMENT
INST. OF WATER RESEARCH, MICHIGAN STATE UNIV.,
EAST LANSING. REPORT NO: TR-41, AUGUST 1974, 112
P. (N.T.I.S. - PB-236 855/3SL)

74-0744 LYSMER, J.
UDAKA, T.
SEED, H. B.
HWANG, R.
LUSH: A COMPUTER PROGRAM FOR COMPLEX RESPONSE
ANALYSIS OF SOIL-STRUCTURE SYSTEMS
EARTHQUAKE ENG. RESEARCH CENTER, UNIV. OF CALIF.,
BERKELEY, REPORT NO: EERC-74-4, APRIL 1974, 89 PP.
(N.T.I.S. - PB-236 796/9SL)

74-0745 SEED, H. B.
LYSMER, J.
EWANG, R.
SOIL-STRUCTURE INTERACTION ANALYSES FOR EVALUATING
SEISMIC RESPONSE
EARTHQUAKE ENGRNG. RESEARCH CENTER, UNIV., OF
CALIFORNIA, BERKELEY, REPORT NO: EERC-74-6, APRIL
1974, 51 P. (PB-236 519/5SL)

74-0746 BOOKER, J. R.
SMALL, J. C.
AN INVESTIGATION OF THE STABILITY OF NUMERICAL
SOLUTIONS OF BIOT'S EQUATIONS OF CONSOLIDATION
CIV. ENGINEERING LABS, UNIV. OF SYDNEY, AUSTRALIA,
REPORT NO: R-235, APRIL 1974, 23 P. (N.T.I.S. -
PB-236 064/2SL)

74-0747 BECKER, J. M.
BRESLER, B.
FIRES-RC, A COMPUTER PROGRAM FOR THE FIRE RESPONSE
OF STRUCTURES-REINFORCED CONCRETE FRAMES
FIRE RESEARCH GROUP, UNIV. OF CALIFORNIA,
BERKELEY. REPORT NO: UCB-FRG-74-3, JULY 1974, 171
P. (N.T.I.S. - PB-235 954/5SL)

74-0748 HABERCOM, G. E. JR.
TUNNEL CONSTRUCTION - A BIBLIOGRAPHY WITH
ABSTRACTS
NATIONAL TECH. INFORMATION SERVICE, SPRINGFIELD,
VA., REPORT FOR 1964 - SEPTEMBER, 1974, (N.T.I.S.
- /PS-74/096)

74-0749 DAVIES, L. G.
DURFEE, R. L.
FOURNEY, W. G.
LEAKAGE THROUGH CRACKS IN LNG TANKAGE
VERSAR, INC., SPRINGFIELD, VA., FINAL REPORT, MAY
31, 1974, 91 P. (N.T.I.S. - COM-74-11630/2SL)

74-0750 NAGARAJAN, S.
POPOV, E. P.
NONLINEAR ELASTIC-VISCOPLASTIC ANALYSIS USING THE
FINITE ELEMENT METHOD
PROC. CONF. COMP. METH. IN NONLINEAR MECH.,
AUSTIN, TEXAS, 1974.

74-0751 NOOR, A. K.
MATHERS, M. D.
NONLINEAR FINITE ELEMENT ANALYSIS OF LAMINATED
COMPOSITE SHELLS
PROC. CONF. COMP. METH. IN NONLINEAR MECH.,
AUSTIN, TEXAS, 1974.

74-0752 ODEN, J. T.
ZIENKIEWICZ, O. C.
GALLAGHER, R. H.
TAYLOR, C.
FINITE ELEMENT METHODS IN FLOW PROBLEMS
PROC. OF INT. SYMPOSIUM ON FINITE ELEMENT METHODS
IN FLOW PROBLEMS, SWANSEA, UNITED KINGDOM, PUBL.
BY UNIV. OF ALABAMA, HUNTSVILLE, ALABAMA, 1974,
743 P. (N.T.I.S. -AD-787 665/9SL)

74-0753 CHEN, C. J.
LI, P.
ERROR ANALYSIS OF A REMOTE TRANSIENT HEAT FLUX
SENSOR
INST. OF HYDRAULIC RESEARCH, IOWA CITY, IOWA,
REPORT NO: IIHHR-E-CJC-74-003, E-CJC-74-003,
AUGUST 1974, 67 P. (N.T.I.S. - AD-787 534/7SL)

74-0754 TS'AO, H. S.
REDDY, D. P.
ROSS, W.
MAGAZINE HEADWALL RESPONSE TO EXPLOSIVE BLAST
AGBABIAN ASSOCIATES, EL SEGUNDO, CALIF., REPORT
NO: AA-R-7336-3284, JANUARY 1974, 150 P. (N.T.I.S.
- AD-787 466/2SL)

74-0755 DANA, J. R.
BARKER, R. M.
THREE-DIMENSIONAL FINITE-ELEMENT COMPUTER PROGRAM
- USERS GUIDE
DEPT. OF ENG. SCI., AND MECH., POLYTECHNIC INST.
AND STATE UNIV., VIRGINIA, REPORT NO: VPT-E-74 19,
AUGUST 1974, 142 P. (N.T.I.S. - AD-787 422/5SL)

74-0756 GOLD, L. M.
ANALYTICAL PERFORMANCE OF STEEL, ALUMINUM AND
PLASTIC FOR A NEW DESIGN HIGH PRESSURE THIN-WALLED
CARTRIDGE CASE
FRANKFORD ARSENAL, PHILADELPHIA, PA., REPORT NO:
FA-M74-9-1, JUNE 1974, 14 P. (N.T.I.S. - AD-787
306/0SL)

74-0757 FESSLER, H.
SWANNELL, J. H.
PREDICTION OF THE CREEP BEHAVIOUR OF A FLANGED
JOINT
CONF. CREEP BEHAV. OF PIPING, LONDON, ENGLAND,
FEBRAURY 28, 1974, PAPER C51074, PP. 39-49. (PUBL.
BY INST. OF MECH. ENG., LONDON, ENG., 1974).

74-0758 KNOTHE, K.
VARIATIONAL PRINCIPLES OF ELASTOSTATICS WITH
RELAXED CONTINUITY REQUIREMENTS AND THEIR
APPLICABILITY TO THE FINITE ELEMENT METHOD
(VARIATIONSPRINZIPIEN DER ELASTOSTATIK MIT
GELOCKERTEN STETIGKEITSFORDERUNGEN UND IHRE
VERWENDBARKEIET IM RAHMEN DER METHODE DOER FINITEN
ELEMENTE)
INST. FUER LUFT UND RAUMFAHRT., TECH. UNIV.,
BERLIN. REPORT NO: ILR-13-1974, 75 PP. (N.T.I.S. -
N74-32360/1ST)

74-0759 NEWMAN, J. C. JR.
A FINITE ELEMENT ANALYSIS OF FATIGUE CRACK CLOSURE
8TH NATIONAL SYMP. ON FRACTURE MECH., PROVIDENCE,
AUGUST 26-28, 1974. NATIONAL AERON. AND SPACE
ADMIN., LANGLEY RESEARCH CENTER, LANGLEY STATION,
VA., REPORT NO: NASA-TM-X-72005, L-9715, SEPTEMBER
1974, 34 P. (N.T.I.S. - N74-32353/6ST)

74-0760 BUYUKOZTURK, O.
HIBBITT, H. D.
SORENSEN, E. P.
WATER IMPACT ANALYSIS OF SPACE SHUTTLE SOLID
ROCKET BY THE FINITE ELEMENT METHOD
MARC ANALYSIS RESEARCH CORP., PROVIDENCE, R.I.,
REPORT NO: NASA-CR-120319,TR73-7, MARCH 1974, 90
P. (N.T.I.S. - N74-32345/2ST)

74-0761 LAING, V. C.
DYNAMIC RESPONSE OF STRUCTURES IN LAYERED SOILS
DEPT. OF CIV. ENG., INST. OF TECH., CAMBRIDGE,
MASS. REPORT NO: R74-10, STRUCT. PUB-383, JANUARY
1974, 199 P. (N.T.I.S. -PB-236 449/5ST) ALSO PUB.
AS SOIL MECH. PUB-335, AND NONLINEAR AND COUPLED
SEISMIC EFFECTS-3. SEE ALSO PB-224 134.

74-0762 KAUSEL, E.
FORCED VIBRATIONS OF CIRCULAR FOUNDATIONS ON
LAYERED MEDIA
DEPT. OF CIVIL ENG., INST. OF TECH., CAMBRIDGE,
MASS., REPORT NO: NSF-GI-35139, JANUARY 1974, 242
P. (N.T.I.S. - PB-236 312/5ST) ALSO PUB. AS
STRUCT. PUB-384, AND NONLINEAR AND COUPLED SEISMIC
EFFECTS-4.

74-0763 BALAAM, N. P.
POULOS, H. G.
BOOKER, J. R.
FINITE ELEMENT ANALYSIS OF THE EFFECTS OF
INSTALLATION ON PILE LOAD-SETTLEMENT BEHAVIOUR
CIVIL ENG. LABS., SYDNEY UNIV., AUSTRALIA, REPORT
NO: R-246, JUNE, 1974, 23 P. (N.T.I.S. - PE-236
081/6ST)

74-0764 DAVIS, E. H.
BOOKER, J. R.
THE SIGNIFICANCE OF THE RATE OF PLASTIC WORK IN
ELASTO-PLASTIC ANALYSIS
CIVIL ENG. LABS., SYDNEY UNIV., AUSTRALIA, REPORT
NO: R-242, APRIL 1974, 19 P. (N.T.I.S. - PB-236
065/9ST)

74-0765 KITIPORNCHAI, S.
TRAHAIR, N. S.
BUCKLING OF INELASTIC I-BEAMS UNDER MOMENT
GRADIENT
CIVIL ENG. LABS., SYDNEY UNIV., AUSTRALIA, REPORT
NO: R-237, FEBRUARY 1974, 32 P. (N.T.I.S. - PB-235
834/9ST)

74-0766 KOBAYASHI, A. S.
AN INVESTIGATION ON THE PLASTICITY MAGNIFICATION
FACTOR IN DEEP SURFACE FLOWS
DEPT. OF MECH. ENG., UNIV. OF WASHINGTON, SEATTLE,
REPORT NO: 4, (FINAL), AUGUST 1974, 8 P. (N.T.I.S.
- AD-786 661/9ST)

74-0767 AUTHORS
THE ROLE OF MECHANICS IN DESIGN-STRUCTURAL JOINTS
PROC. OF THE ARMY SYMPOSIUM ON SOLID MECHANICS,
BASS RIVER, MASS. ARMY MATERIALS AND MECHANICS
RESEARCH CENTER, WATERTOWN, MASS., REPORT NO:
AMMRC-MS-74-8, SEPTEMBER 1974, 346 P. (N.I.T.S. -
AD-786 543/9ST)

74-0768 BERKE, L.
CHEN, P. C. T.
THE FINITE ELEMENT ANALYSIS OF ELASTIC-PLASTIC
THICK-WALLED TUBES
WATERVLIET ARSENAL, N.Y., REPORT NO: WVT-TR-74039,
SEPTEMBER 1974, 34 P. (N.T.I.S. - AD-786 533/0ST)

74-0769 CHEN, P. C. T.
ELASTIC-PLASTIC SOLUTION OF A TWO-DIMENSIONAL TUBE
PROBLEM BY THE FINITE ELEMENT METHOD
WATERVLIET ARSENAL, N.Y., REPORT NO: WVT-TR-74038,
SEPTEMBER 1974, 29 P. (N.T.I.S. - AD-786 532/2ST)

74-0770 AUTHOR
THE ROLE OF MECHANICS IN DESIGN - STRUCTURAL
JOINTS
IRIA LECTURE NOTES IN COMPUT. SCI.,
SPRINGER-VERLAG, BERLIN,ALSO SYMP. ON SOLID MECH.,
REPORT NO. AMMRC-MS-74-9, ARMY MATER. AND MECH.
RES. CENT., WATERTOWN MASS., VOL. 10, PART 1,
SEPTEMBER 1974, 26 PP. (N.T.I.S. - AD-786/9ST).

74-0771 LIKINS, P. W.
ANALYTICAL DYNAMICS AND NONRIGID SPACECRAFT
SIMULATION
JET PROPULSION LAB., CALIF. INST. OF TECH.,
PASADENA, REPORT NO: NASA/CR-139502, JPL/32-1593,
JULY 15, 1974, 147 P. (N.T. I.S. / N74-31333/9SL)

74-0772 ARMEN, H. JR.
APPLICATION OF A SUBSTRUCTURING TECHNIQUE TO THE
PROBLEM OF CRACK EXTENSION AND CLOSURE
RESEARCH DEPT. GRUMMAN AEROSPACE CORP., BETHPAGE,
N.Y., REPORT NO: NASA-CR-132458, RE-480, JULY
1974, 40 P. (N.T.I.S. - 30339/7SL)

74-0773 STEVEN, G. P.
FINITE ELEMENT ANALYSIS OF A HYDRODYNAMIC BEARING
DEPT. OF AERON. ENG., SYDNEY, AUSTRALIA, REPORT
NO. ATN-7402, MAY 1974, 39 PP. (N.T.I.S. -
N74-29639/3SL).

74-0774 DRNEVICH, V. P.
CONSTRAINED AND SHEAR MODULI FOR FINITE ELEMENTS
DEPT. CIV. ENG., KENTUCKY UNIV., LEXINGTON, REPORT
NO: SOIL MECH. - SER-18, UKY-TR85-74-CE22, JUNE
1971, 36 PP. (N.T.I.S. - PB-235 673/SL)

74-0775 NAGARAJAN, S.
POPOV, E. P.
NONLINEAR FINITE ELEMENT DYNAMIC ANALYSIS OF
AXISYMMETRIC SOLIDS
STRUCT. ENG., UNIV. OF CALIF., BERKELEY, REPORT
NO: UCSESM-74-9, JULY 1974, 54 PP. (N.T.I.S. -
AD-785 734/5SL)

74-0776 DESAI, C. S.
SANDHU, R. S.
FINITE ELEMENT ANALYSIS OF UNCONFINED FLOW WITH A
VARIABLE TIME-STEP PROCEDURE
ARMY ENGINEER WATERWAYS EXPERIMENT STATION,
VICKSBURG, MISS., 1974, 15 P. (N.T.I.S. - AD-785
616/4SL)

74-0777 PAUL, S. L.
KESLER, C. E.
GAYLORD, C. E.
MOHRA, B.
HENDRON, A. J.
RESEARCH TO IMPROVE TUNNEL SUPPORT SYSTEMS
DEPT. OF CIVIL ENGINEERING, UNIV., OF ILLINOIS,
URBANA, REPORT NO: UILU-ENG-74-2016, JUNE 1974,
316 P. (N.T.I.S. - PB-235 762/2)

74-0778 ARORA, J. S.
ON IMPROVING EFFICIENCY OF AN ALGORITHM FOR
STRUCTURAL OPTIMIZATION AND A USERS MANUAL FOR
PROGRAM TRUSSOPT2
DEPT. OF MECH. AND HYDRAULICS, UNIV., OF IOWA,
IOWA CITY, REPORT NO: TR-12, JUNE 1974, 96 P.
(N.T.I.S. - AD-785 564/6)

74-0779 LEVY, A.
A THREE DIMENSIONAL "VARIABLE NODE" ISOPARAMETRIC
SOLID ELEMENT
RESEARCH DEPT. GRUMMAN AEROSPACE CORP., BETHPAGE,
N.Y., REPORT NO: RE-587, JULY 1974, 24 P.
(N.T.I.S. - AD-785 353/4)

74-0780 FRAZIER, G. A.
PETERSEN, C. M.
3-D STRESS WAVE CODE FOR THE ILLTAC 1V
SYSTEMS SCIENCE AND SOFTWARE, LA JOLLA, CALIF.,
REPORT NO: SSS-R-2103, JULY 26, 1974, 99 P.
(N.T.I.S. - AD-785 260/1)

74-0781 MATTHEWS, A. T.
BLEICH, H. H.
COMPARISON OF THE RESULTS OF THE DYNAMIC RESPONSE
OF CYLINDRICAL SHELLS BY CHARACTERISTIC AND BY
FINITE ELEMENT METHODS: AXISYMMETRIC CASE
WEIDLINGER ASSOCIATES, NEW YORK, FINAL REPORT,
AUGUST 14, 1974, 44 P. (N.T.I.S. - AD-785 258/5)

74-0782 KARWOSKI, W. J.
THEORETICAL INVESTIGATION OF ROCK AND SUPPORT
INTERACTION - DEVELOP MORE RATIONAL DESIGN METHODS
AND DEVELOP NEW TYPES OF SUPPORT AND MINING
SYSTEMS. PART 1
SPOKANE MINING RESEARCH CENTER, BUREAU OF MINES,
SPOKANE, WASH., FINAL TECH. REPORT: FEBRUARY
1971-JUNE 1972, 92 PP. (N.T.I.S. - AD-784 987/0)

74-0783 ADLER, W. F.
DUPREE, D. M.
STRESS ANALYSIS OF COLDWORKED FASTENER HOLES
BELL AEROSPACE CO., BUFFALO, N.Y., TECHNICAL
SUMMARY REPORT: JUNE 1973 - MARCH 1974, 103 P.
(N.T.I.S. - AD-784 920/1)

74-0784 BAKER, A. J.
A FINITE ELEMENT SOLUTION ALGORITHM FOR THE
NAVIER-STOKES EQUATIONS
BELL AEROSPACE CO., BUFFALO, N.Y., REPORT NO:
NASA-CR-2391, D9198-950001, JUNE 1971, 77 P.
(N.T.I.S. - N74-29035/4)

74-0785 KLINEBERG, J. M.
STEGER, J. L.
THE NUMERICAL CALCULATION OF LAMINAR
BOUNDARY-LAYER SEPARATION
PROC. OF 12TH A.I.A.A. AEROSPACE SCI. MEETING,
WASHINGTON, D.C., JANUARY 30-FEBRUARY, 1974, AND
AMES RES. CENT., NAT. AERON. AND SPACE ADMIN.,
MOFFETT FIELD CALIF., REPORT NO.
NASA-TN-D-7732,A-5281, JULY 1974, 103 PP.
(N.T.I.S. - N74-28759/0).

74-0786 BOOKER, J. R.
SMALL, J. C.
THE ECONOMICAL SOLUTION OF ELASTIC PROBLEMS FOR A
RANGE OF POISSONS RATIO
CIVIL ENGRG. LABS., SYDNEY UNIV., AUSTRALIA,
REPORT NO: R-238, FEBRUARY 1974, 15 P. (N.T.I.S. -
PB-234 804/3)

74-0787 HALTINER, G. J.
WILLIAMS, R. T.
RECENT ADVANCES IN NUMERICAL WEATHER PREDICTION
PROC. OF 54TH ANN. MEET. OF THE AMER.
METEOROLOGICAL SOC., HONOLULU, HAWAII, JANUARY
8-11, 1974. NAVAL POSTGRADUATE SCHOOL, MONTEREY,
CALIF., REPORT NO. NPS-5UHAWU74081, AUGUST 1974,
65 PP. (N.T.I.S. AD-784/2).

74-0788 PLANK, R. J.
WITTRICK, W. H.
BUCKLING UNDER COMBINED LOADING OF THIN
FLAT-WALLED STRUCTURES BY A COMPLEX FINITE STRIP
METHOD
INT. J. NUMER. METH. ENG., VOL. 8, 1974, PP.
323-339.

74-0789 BRUMAN, Z. I.
METHODS OF FINITE ELEMENTS IN ANALYSES OF
CONTINUOUS AND COMBINED CONSTRUCTIONS
FOREIGN TECHNOLOGY DIV., WRIGHT-PATTERSON AFB,
OHIO. REPORT NO: FTD-MT-24-342-74, JULY 19 1971,
37 P. (N.T.I.S. - AD-783 969/9)

74-0790 ATCHISON, D. L.
FINITE ELEMENT SOLUTION OF THE INTERACTION OF A
PLANE ACOUSTIC BLAST WAVE AND A CYLINDRIC
STRUCTURE
MASTER'S THESIS, NAVAL POSTGRADUATE SCHOOL,
MONTEREY, CALIF., JUNE 1974, 80 P. (N.T.I.S. -
AD-783 861/8)

74-0791 SMITH, L. J.
STRESS CONCENTRATION FACTORS FOR CIRCULAR NOTCHES
MASTER'S THESIS, NAVAL POSTGRADUATE SCHOOL,
MONTEREY, CALIF., JUNE 1974, 165 P. (N.T.I.S. -
AD-783 784/2)

74-0792 GIBBS, N. E.
POOLE, W. G. JR.
STOCKMEYER, P. K.
AN ALGORITHM FOR REDUCING THE BANDWITH AND PROFILE
OF A SPARSE MATRIX
DEPT. OF MATHEMATICS, COLLEGE OF WILLIAM AND MARY,
WILLIAMSBURG, REPORT NO: TR-5, JULY 1974, 33 P.
(N.T.I.S. - AD-783 695/0)

74-0793 DANA, J. R.
BARKER, R. M.
THREE-DIMENSIONAL ANALYSIS FOR THE STRESS
DISTRIBUTION NEAR CIRCULAR HOLES IN LAMINATED
COMPOSITES
DEPT. OF ENGINEERING SCI., AND MECH., POLYTECHNIC
INST. AND STATE UNIV., BLACKSBURG, VIRGINIA,
REPORT NO: VPI-E-74-18, AUGUST 1974, 42 P.
(N.T.I.S. - AD-783 504/4)

74-0794 GOODHEIM, H.
O'HARA, G. P.
FINITE ELEMENT ANALYSIS OF SEVERAL SWAGE MANDREL
DESIGNS
WATERVLIET ARSENAL, N.Y., REPORT NO: WVT-TR-74011,
JUNE 1974, 50 P. (N.T.I.S. - AD-783 486/4)

74-0795 WU, J. J.
A UNIFIED FINITE ELEMENT APPROACH TO COLUMNS
SUBJECTED TO NONCONSERVATIVE LOADS, WITH DAMPING
EFFECTS AND VARIOUS END CONDITIONS
WATERVLIET ARSENAL, N.Y., REPORT NO: WVT-TR-74020,
JULY 1974, 20 P. (N.T.I.S. - AD-783-483/1)

74-0796 MCMEEKING, R. N.
RICE, J. R.
FINITE ELEMENT FORMULATIONS FOR PROBLEMS OF LARGE
ELASTIC-PLASTIC DEFORMATION
DIV. OF ENGRG., BROWN UNIV., PROVIDENCE, RHODE
ISLAND, REPORT NO: NASA-CR-138820, MAY 1974, 32
PP. (N.T.I.S.-74-28410/0)

74-0797 NAGTEGAAL, J. C.
PARKS, D. M.
RICE, J. R.
ON NUMERICALLY ACCURATE FINITE ELEMENTS
DIV. OF ENGINEERING, BROWN UNIV., PROVIDENCE,
R.I., REPORT NO: NASA-CR-138821, MARCH 1974, 49 P.
(N.T.I.S. - N74-28409/2)

74-0798 SPIER, E. E.
MILLER, M. F.
DESIGN, MANUFACTURE, DEVELOPMENT, TEST, AND
EVALUATION OF BORON/ALUMINUM STRUCTURAL COMPONENTS
FOR SPACE SHUTTLE. VOLUME 3: SHEAR BEAM COMPONENT
TEST AND ANALYSIS
GENERAL DYNAMICS/CONVAIR, SAN DIEGO, CALIF.,
REPORT NO: GDCA-DBG-73-006-VOL-3, FEBRUARY 1974,
56 P. (N.T.I.S. - N74-28323/5)

74-0799 OSIAS, J. R.
USER'S GUIDE FOR ANALYSIS OF FINITE ELASTOPLASTIC
DEFORMATION: THE FIPDEF AND FIPAX PROGRAMS FOR THE
CDC 6600
LEWIS RESEARCH CENTER, NATIONAL AERON. AND SPACE
ADMIN., CLEVELAND, OHIO., REPORT NO:
NASA-TM-X-2998,E-7610, JUNE 1974, 74 P. (N.T.I.S.
- N74-27411/9)

74-0800 ZAK, A. R.
FINITE ELEMENT COMPUTER PROGRAM FOR THE ANALYSIS
OF LAYERED ORTHOTROPIC STRUCTURES
DEPT. OF AERON. AND ASTRON. ENGINEERING, ILLINOIS
UNIV., URBBANA, CONTRACT REPORT: JANUARY 11, 1973
- JANUARY 31, 1974, 92 P. (N.T.I.S. - AD-783
412/0)

74-0801 HODGE, P. G.
FINITE ELEMENT METHODS IN PLASTICITY
DEPT. OF AERO. ENG. AND MECH., MINNESOTA UNIV.,
MINNEAPOLIS, REPORT NO. AEM-HI-9, JUNE 1974, 18
PP. (N.T.I.S. - AD-783 092/0).

74-0802 KNITTEL, M. R.
BARACH, D.
A FINITE-ELEMENT APPROACH TO THE ANALYSIS OF
TANGENTIALLY POLARIZED PIEZOELECTRIC-CERAMIC
FREE-FLOODED CYLINDER TRANSDUCERS
NAVAL UNDERSEA CENTER, SAN DIEGO, CALIF., REPORT
NO: NUC-TP-412, JULY 1974, 54 P. (N.T.I.S. -
AD-783 031/8)

74-0803 CHEBAN, W. G.
SABODASH, P. F.
EXCITATION OF PLANE THERMOELASTIC WAVES IN A LAYER
WITH A FINITE RATE OF HEAT PROPAGATION
FOREIGN TECHNOLOGY DIV., WRIGHT-PATTERSON AFB,
OHIO, REPORT NO: FTD-HT-23-1783-74, AUGUST 21,
1974, 12 P. (N.T.I.S. - AD-785 190/3)

74-0804 HOSKINS, E. R.
OSHIER, E. H.
A DEEP HOLE STRESS MEASUREMENT DEVICE
DEPT. OF MINING ENG., SOUTH DAKOTA SCHOOL OF MINES
AND TECHNOLOGY, RAPID CITY, REPORT NO: 102, JULY
10, 1974, 142 P. (N.T.I.S. - AD-783 304/9)

74-0805 HODGE, P. G.
A NEW MODEL FOR ELASTIC-PLASTIC TRUSSES
DEPT. OF AEROSPACE ENG. AMD MECH., MINNESOTA
UNIV., MINNEAPOLIS, REPORT NO. AEM-H1-10, JULY
1974, 48 PP. (N.T.I.S. - AD-783 093/8).

74-0806 HUGHES, T. J.
TAYLOR, R. L.
SACKMAN, J. L.
FINITE ELEMENT FORMULATION AND SOLUTION OF
CONTACT-IMPACT PROBLEMS IN CONTINUUM MECHANICS
STRUCT. ENG. LAB., CALIFORNIA UNIV., BERKELEY,
REPORT NO: UCSESM-74-8, MAY 1974, 102 P. (N.T.I.S.
- PB-233 888/1)

74-0807 FOTHERGILL, J. W.
LEE, H. Y.
FOTHERGILL, P. A.
PREDICTION OF LONG-TERM STRESS RANGES: USER'S
MANUAL - BRIDGE DYNAMIC STRESS ANALYSIS
INTEGRATED SYSTEMS, INC., CHEVY CHASE, MD., FINAL
REPORT, JUNE 1973, 214 P. (N.T.I.S. - PB-233
491/0)

74-0808 PILKEY, W. D.
HAVILAND, J. K.
A METHOD OF ANALYSIS OF LINE STRUCTURES BY
TRANSFER MATRICES DERIVED FROM FINITE ELEMENTS
APPLIED MECHANICS GROUP, UNIV. OF VIRGINIA,
CHARLOTTESVILLE. REPORT NO: TR-74-1, SEPTEMBER 1,
1974, 29 P. (N.T.I.S. - AD-785 001/9)

74-0809 DANIEU, D. J.
RUGGLES, V. L.
MINUTEMAN STV STRESS ANALYSIS AND TESTING
WASATCH DIV., THIOKOL CORP., BRIGHAM CITY, UTAH,
REPORT NO: TWR-7496, JUNE 1974, 37 P. (N.T.I.S. -
AD-781 381/9)

74-0810 COMSTOCK, C.
SOLVING THE LINEAR BALANCE EQUATION GLOBALLY
NAVAL POSTGRADUATE SCHOOL, MONTEREY, CALIF.,
REPORT NO: NPS-53ZK74061, JUNE 1974, 12 P.
(N.T.I.S. - AD-781 362/9)

74-0811 LASKER, G.
MALONEY, J. G.
SHELTON, M. T.
UNDERHILL, D. A.
STRUCTURAL DYNAMIC PROPERTIES OF TACTICAL MISSILE
JOINTS - PHASE 3
POMONA DIV., GENERAL DYNAMICS, POMONA, CALIF.,
REPORT NO. CR-6-348-945-003, MAY 1974, 186 PP.
(N.T.I.S. - AD-781 317/3).

74-0812 BAKER, A. J.
NONLINEAR INITIAL-BOUNDARY VALUE SOLUTIONS BY THE
FINITE ELEMENT METHOD
PROC. CONF. COMP. METH. IN NONLINEAR MECH.,
AUSTIN, TEXAS, 1974.

74-0813 MEACHAM, H. W.
PRAUSE, R. H.
WADDELL, J.
ASSESSMENT OF DESIGN TOOLS AND CRITERIA FOR URBAN
RAIL TRACK STRUCTURES. VOLUME 11. AT-GRADE SLAB
TRACK
BATTELLE COLUMBUS LABS., OHIO. FINAL REPORT, MARCH
1973 - MARCH 1974, 101 PP. (N.T.I.S. - PB-233
017/3).

74-0814 SMITH, D. A. JR.
FINITE ELEMENT ANALYSIS OF THE FORCED OSCILLATION
OF SHIP HULL FORMS
MASTER'S THESIS, NAVAL POSTGRADUATE SCHOOL,
MONTEREY, CALIF., JUNE 1974, 102 P. (N.T.I.S. -
AD-780 942/9)

74-0815 MARTIN, J. F.
CYCLIC MECHANICAL TESTS AND AN APPROPRIATE
ANALYTICAL STRESS-STRAIN MODEL FOR A36 STEEL
ARMY CONSTRUCTION ENGINEERING RESEARCH LAB.,
CHAMPAIGN, ILL., REPORT NO: CERL-TR-M-86, MAY
1974, 61 P. (N.T.I.S. - AD-780 802/5)

74-0816 PLUMMER, F. B. JR.
A NEW LOOK AT STRUCTURAL ENERGY DISSIPATION
ARMY CONSTRUCTION ENG. RESEARCH LAB., CHAMPAIGN,
ILL., REPORT NO: CERL-TM-M-82, MAY 1974, 63 P.
(N.T.I.S. - AD-780 801/1)

74-0817 FRANCIS, E. C.
DEVERALL, L. I.
CARLTON, C. H.
ZITZER, H. H.
KNAUSS, W. G.
CASE LINER BOND ANALYSIS
UNITED TECHNOLOGY CENTER, SUNNYVALE, CALIF.,
REPORT NO: UTC-2439-FR, JULY 1974, 470 P.
(N.T.I.S. - AD-780 642/5)

74-0818 AL-MAHAIDI, R. S.
NILSON, A. H.
COUPLED SHEAR WALLS. IMPROVED METHODS FOR ELASTIC
ANALYSIS
DEPT. OF STRUCT. ENG., CORNELL UNIV., ITHACA,
N.Y., REPORT NO: 355, FEBRUARY 1974, 147 P.
(N.T.I.S. - PB-232 519/9)

74-0819 TAYLOR, E. G.
ESTIMATIONS OF THE EFFECT OF DESIGN CHANGES ON THE
MODES AND FREQUENCIES OF VIBRATING STRUCTURAL
ELEMENTS
MASTER'S THESIS, SCHOOL OF ENG., AIR FORCE INST.
OF TECH., WRIGHT-PATTERSON AFB., OHIO, REPORT NO:
GAW/MC/74-17, MARCH 1974, 82 P. (N.T.I.S - AD-780
499/0)

74-0820 MARCUS, L. A.
STINCHCOMB, W. W.
TURGAY, H. M.
FATIGUE CRACK INITIATION IN A BORON EPOXY PLATE
WITH A CIRCULAR HOLE
DEPT. OF ENG. SCI. AND MECH., VIRGINIA POLYTECHNIC
INST. AND STATE UNIV., BLACKSBURG, REPORT NO.
VPT-E-74-6, FEBRUARY 1974, 68 PP. (N.T.I.S. -
AD-780 106/1)

74-0821 FORREST, J. B.
SHUGAR, T. A.
A STRUCTURAL EVALUATION OF RAPID METHODS OF
BACKFILLING FOR BOMB DAMAGE REPAIR
NAVAL CIVIL. ENG. LAB., PORT HUENEME, CALIF.,
FINAL REPORT, JULY 1971 - APRIL 1973, 94 P
(N.T.I.S. - AD-780 104/6)

74-0822 BABUSKA, I.
SOLUTION OF PROBLEMS WITH INTERFACES AND
SINGULARITIES
MATHEMATICAL ASPECTS OF FINITE ELEMENTS IN PARTIAL
DIFFERENTIAL EQUATIONS, ACADEMIC PRESS, 1974, PP.
213-277.

74-0823 LEE, E. H.
MALLETT, R. L.
TING, T. C. T.
YANG, W. H.
DYNAMIC ANALYSIS OF THE DEFORMATION OF STRUCTURES
AND OF METAL FORMING
DEPT. OF APPLIED MECH., STANFORD UNIV., CALIF.,
REPORT NO: SUDAM, 74-28 APRIL 1974, 30 P.
(N.T.I.S. - PB-232 552/0)

74-0824 MITCHELL, R. A.
WOOLLEY, R. M.
CHWIRUT, D. J.
ANALYSIS OF COMPOSITE REINFORCED CUTOUTS AND
CRACKS
PROC. 15TH CONF. OF STRUCTS, STRUCT. DYN. AND
MATER., LAS VEGAS, NEV., APRIL 17-19, 1974, PAPER
NO. 74-377, PP. 1-13. (N.T.I.S. COM-74-50478/8).

74-0825 WU, R. W. H.
WITMER, E. A.
FINITE-ELEMENT PREDICTIONS OF TRANSIENT
ELASTIC-PLASTIC LARGE DEFLECTIONS OF STIFFENED
AND/OR UNSTIFFENED RINGS AND CYLINDRICAL SHELLS
AEROELASTIC AND STRUCTURES RESEARCH LAB.,
MASSACHUSETTS INST. OF TECH., CAMBRIDGE, REPORT
NO: ASRL-TR-171-4, APRIL 1974, 191 P. (N.T.I.S. -
AD-779 453/0)

74-0826 BIRKHOFF, G.
FIX, G. J.
HIGHER-ORDER LINEAR FINITE ELEMENT METHODS
DEPT. OF MATHS., HARVARD UNIV., CAMBRIDGE, MASS.,
REPORT NO: TR-1, MARCH 1974, 37 P. (N.T.I.S. -
AD-779 341/7)

74-0827 USUKI, S.
APPLICATIONS OF FINITE ELEMENT METHODS TO UNSTEADY
VISCOUS FLOW AROUND A BOX-GIRDER BRIDGE
OSCILLATING IN UNIFORM FLOW
PROC. CONF. COMP. METH. IN NONLINEAR MECH.,
AUSTIN, TEXAS. 1974.

74-0828 FUNG, Y. C.
SECHLER, E. E.
THIN-SHELL STRUCTURES: THEORY, EXPERIMENT, AND
DESIGN
PROC. OF SYMP. ON THIN SHELL STRUCTURES, PASADENA,
CALIF., JUNE 29-30, 1972, 621 PP. (N.T.I.S. -
AD-779 088/4).

74-0829 MIYATA, H.
SHIDA, S.
KUSUMOTO, S.
THE SIMPLE METHOD OF EVALUATION OF STRESS
INTENSITY FACTOR USING THE FINITE ELEMENT METHOD
SYMP. ON MECH. BEHAVIOR OF MATER., KYOTO, AUGUST
21-24, 1974.

74-0830 LAKIS, A. A.
FREE VIBRATION AND RESPONSE TO RANDOM PRESSURE
FIELD OF ANISOTROPIC THIN CYLINDRICAL SHELLS
PROC. CONF. NOISE, SHOCK AND VIB., MONASH UNIV.,
MELBOURNE, VICTORIA, AUST., MAY 22-25, 1974, PP.
89-98, PUBL. BY MONASH UNIV., DEPT. MECH. ENG.,
CLAYTON, VICTORIA, AUST., 1974.

74-0831 FANELLI, M.
GIUSEPPETTI, G.
FINITE ELEMENT STUDY OF THE TRIAXIAL STRESS STATE
AROUND AN INSPECTION TUNNEL IN AN ARCH DAM
INT. ASSOC. FOR BRIDGE AND STRUCT. ENG., SEMIN. ON
CONCR. STRUCT. SUBJ. TO TRIAXIAL STRESSES,
BERGAMO, ITALY, MAY 17-19, 1974.

74-0832 SUNLEY, V. K.
TURNEY, M.
COMPUTER GRAPHICS FOR STRUCTURAL ANALYSIS
CONF. I.E.E., NO. 111, 1974, FOR MEET., UNIV. OF
SOUTHAMPTON, HAMPS., ENGL., APRIL 8-11, 1974, PP.
1-5. PUBL. (LONDON)

74-0833 NARAYANASWAMI, R.
NEW TRIANGULAR AND QUADRILATERAL PLATE-BENDING
FINITE ELEMENTS
NATIONAL AERON. AND SPACE ADMIN., LANGLEY RESEARCH
CENTRE, LANGLEY STATION, VA., REPORT NO:
NASA-TN-D-7407,L-9180, APRIL 1974, 74 P. (N.T.I.S.
- N74-19645/4)

74-0834 AUTHOR
CYCLIC-STRESS ANALYSIS OF NOTCHES FOR SUPERSONIC
TRANSPORT CONDITIONS
BOEING AEROSPACE CO., SEATTLE, WASH., REPORT NO:
NASA-CR-132387, FEBRUARY 1974, 51 P. (N.T.I.S. -
N74-19544/1)

74-0835 BERGER, A. E.
CIMENT, M.
ROGERS, J. C. W.
NUMERICAL SOLUTION OF A DIFFUSION CONSUMPTION
PROBLEM WITH A FREE BOUNDARY
NAVAL ORDNANCE LAB., WHITE OAK, MD., REPORT NO:
NOLTR-74-7, JANUARY 31, 1974, 61 P. (N.T.I.S. -
AD-778 309/5)

74-0836 HUCK, P. J.
LIBER, T.
CHIAPETTA, R. L.
THOMOPOULOS, N. T. JR.
SINGH, M. M.
DYNAMIC RESPONSE OF SOIL/CONCRETE INTERFACES AT
HIGH PRESSURE
IIT. RESEARCH INST., CHICAGO, ILL., FINAL REPORT,
NOVEMBER 17, 1972 - DECEMBER 17, 1973, APRIL 1974,
308 P. (N.T.I.S. - AD-778 101/6)

74-0837 ATCHLEY, C. E.
RAYMOND, L.
USE OF FINITE ELEMENT ANALYSIS IN FRACTURE
MECHANICS
LAB. OPERATIONS, AEROSPACE CORP., EL SEGUNDO,
CALIF., REPORT NO: TR-0074(4250-10)-6, APRIL 5,
1974, 68 P. (N.T.I.S. - AD-778 098/4)

74-0838 SANDIDGE, D. W.
STABILITY AND POSTBUCKLING BEHAVIOR OF
HYPERELASTIC BODIES AT FINITE STRAIN BY THE FINITE
ELEMENT METHOD
SYSTEMS ENGNG. DIRECTORATE, ARMY MISSILE COMMAND
REDSTONE ARSENAL, ALA., REPORT NO: RC-74-2,
FEBRUARY 25, 1974, 61 PP. (N.T.I.S. - AD-777
789/9)

74-0839 KENNEDY, F. E.
WU, J. J.
LING, F. F.
A THERMAL, THERMOELASTIC, AND WEAR ANALYSIS OF
HIGH-ENERGY DISK BRAKES
TRIBOLOGY LAB., RENSSELAER POLYTECHNIC INST.,
TROY, N.Y., REPORT NO: NASA-CR-134507, JANUARY
1974, 53 P. (N.T.I.S. - N74-18325/2)

74-0840 WANG, F. D.
ROPCHAN, D. M.
SUN, M. C.
STRUCTURAL ANALYSIS OF A COAL MINE OPENING IN
ELASTIC, MULTILAYERED MATERIAL
DENVER MINING RESEARCH CENTER, COLO., REPORT NO:
BUMINES-RI-7845, MARCH 1974, 42 P. (N.T.I.S. -
PB-231 298/1)

74-0841 LEITERER, R.
NUMERICAL INTEGRATION METHODS IN FINITE ELEMENTS
CEA CENTRE D'ETUDES DE LIMEIL, 94
VILLENEUVE-SAINT-GEORGES, FRANCE, NOVEMBER 1974,
36 PP. (N.T.I.S. -CEA-R-4611) IN FRENCH. SEE ALSO
NSA 3105, NO. 13458.

74-0842 GROOMS, D. W.
STRUCTURAL MECHANICS SOFTWARE. A BIBLIOGRAPHY WITH
ABSTRACTS
NATIONAL TECH. INF. SERVICE, SPRINGFIELD, VA.,
REPORT, MARCH 1971 - MARCH 1974, 130 P. (N.T.I.S.
- COM-74-10833/3)

74-0843 NEU, T. F.
FINITE-ELEMENT ANALYSIS OF EDGE EFFECTS IN
ANGLE-PLY COMPOSITE LAMINATES
INSTER, PA., REPORT NO: NADC-74051-30, MARCH 19,
1974, 15 P. (N.T.I.S. - AD-777 654/5)

74-0844 TONG, P.
MAU, S. T.
PIAN, T. H. H.
DERIVATION OF GEOMETRIC STIFFNESS AND MASS
MATRICES FOR FINITE ELEMENT HYBRID MODELS
INT. J. SOLIDS STRUCT. VOL. 10, 1974, PP. 919-932.

74-0845 BALSARA, J. P.
WALKER, R. E.
FOWLER, J.
VIBRATION CHARACTERISTICS OF THE NORTH FORK DAM
MODEL
ARMY ENG. WATERWAYS EXPERIMENT STATION, VICKSBURG,
MISS., REPORT NO: AWEES-TP-N-74-2, MARCH 1974, 79
P. (N.T.I.S. - AD-777 548/9)

74-0846 BRATANOW, T.
ECER, A.
SENSITIVITY ANALYSIS OF TORSIONAL VIBRATION
CHARACTERISTICS OF HELICOPTER ROTOR BLADES. PART
1: STRUCTURAL DYNAMICS ANALYSIS
WISCONSIN UNIV., MILWAUKEE, REPORT NO:
NASA-CR-2379, MARCH 1974, 48 P. (N.T.I.S. -
N74-17610/8)

74-0847 BRATANOW, T.
ECER, A.
KOBISKE, M.
NUMERICAL CALCULATIONS OF VELOCITY AND PRESSURE
DISTRIBUTION AROUND OSCILLATING AIRFOILS
WISCONSIN UNIV., MILWAUKEE, REPORT NO:
NASA-CR-2368, FEBRURAY 1974, 86 P. (N.T.I.S. -
N74-16704/0)

74-0848 BATHE, K. J.
WILSON, E. L.
IDING, R. H.
NONSAP: A STRUCTURAL ANALYSIS PROGRAM FOR STATIC
AND DYNAMIC RESPONSE OF NONLINEAR SYSTEMS
STRUCTURAL ENGNG. LAB., UNIV. OF CALIF., BERKELEY,
REPORT NO: UCSESM-74-3, FEBRUARY 1974, 172 P.
(N.T.I.S. - PB-231 112/4)

74-0849 DOVEY, H. H.
EXTENSION OF THREE DIMENSIONAL ANALYSIS TO SHELL
STRUCTURES USING THE FINITE ELEMENT IDEALIZATION
STRUCTURAL ENGNG. LAB., UNIV. OF CALIF., BERKELEY,
REPORT NO: UCSESM-74-2, JANUARY 1974, 186 P.
(N.T.I.S.-PB-230 353/5)

74-0850 BECKER, J. M.
BIZRI, H.
BRESLER, B.
FIRES-T. A COMPUTER PROGRAM FOR THE FIRE RESPONSE
OF STRUCTURES - THERMAL
FIRE RESEARCH GROUP, UNIV. OF CALIF., BERKELEY,
REPORT NO: UCB-FRG-74-1, JANUARY 1974, 76 P.
(N.T.I.S. - PB-230 004/4)

74-0851 BURMAN, Z. I.
PROBLEM OF USING THE METHOD OF FINITE ELEMENTS TO
CONSTRUCT AN ALGORITHM FOR CALCULATING A
THIN-WALLED REINFORCED SHELL OF AN EXTREMELY
IRREGULAR NATURE AND ADAPTING THIS PROBLEM TO THE
ELECTRONIC DIGITAL COMPUTER
FOREIGN TECHNOLOGY DIV., WRIGHT-PATTERSON AFB,
OHIO, REPORT NO: FTD-HT-23-359-74, MARCH 8, 1974,
25 P. (N.T.I.S. - AD-776 892/2)

74-0852 LIKINS, P. W.
DYNAMIC ANALYSIS OF A SYSTEM OF HINGE-CONNECTED
RIGID BODIES WITH NONRIGID APPENDAGES
INST. OF TECH., JET PROPULSION LAB., PASADENA,
CALIF., REPORT NO: NASA-CR-136627,JPL-TR-32-1576,
FEBRUARY 1, 1974, 29 P. (N.T.I.S. - N74-16588/1)

74-0853 KAPER, H. G.
LEAF, G. K.
LINDEMAN, A. J.
APPLICATIONS OF FINITE ELEMENT METHODS IN REACTOR
MATHEMATICS. NUMERICAL SOLUTION OF THE NEUTRON
TRANSPORT EQUATION
ARGONNE NATIONAL LAB., ILL., OCTOBER 1974, 94 P.
(N.T.I.S.-ANL-8126) SEE ALSO NSA 3105, NO. 12749

74-0854 PIAN, T. H. H.
INTERACTIVE PROGRAM IN DESIGN AND ANALYSIS OF
COMPOSITE MATERIALS
AEROELASTIC AND STRUCT. RESEARCH LAB., MASS. INST.
OF TECH., CAMBRIDGE, REPORT NO: ASRL-TR-163-1,
JANUARY 1974, 37 P. (N.T.I.S. - AD-775 730/5)

74-0855 WILSON, H. B.
RICHARDSON, J. J.
MODAL RESPONSE OF FREE ROCKETS UNDER THRUST
LOADING
GROUND EQUIPMENT AND MATERIALS DIRECTORATE, ARMY
MISSILE COMMAND REDSTONE ARSENAL, ALA., REPORT NO:
RL-TR-74-2, JANUARY 14, 1974, 64 P. (N.T.I.S. -
AD-775 377/5)

74-0856 BART, J.
KOMBINATION EINES INTEGRALGLEICHUNGS VERFAHRENS
MIT METHODE DER FINITEN ELEMENTEN ZUR BERECHNUNG
EBENER SPENNUNGS KONZENTRATIONS PROBLEME
PH.D. THESIS, UNIV. OF MUNCHEN, 1974.

74-0857 VENKATESWARA, R. G.
RAJU, I. S.
A COMPARATIVE STUDY OF VARIABLE AND CONSTANT
THICKNESS HIGH PRECISION TRIANGULAR PLATE BENDING
ELEMENTS IN THE ANALYSIS OF VARIABLE THICKNESS
PLATES
NUCL. ENG. DESIGN, VOL. 26, 1974, PP. 299-304.

74-0858 ARMEN, H. A.
GARNET, H.
EVALUATION OF NUMERICAL TIME INTEGRATION METHODS
AS APPLIED TO ELASTIC-PLASTIC DYNAMIC PROBLEMS
INVOLVING WAVE PROPAGATION
RESEARCH DEPT., GRUMMAN AEROSPACE CORP., BETHPAGE,
N.Y., REPORT NO: RE-475, MARCH 1974, 45 P.
(N.T.I.S. - AD-781 704/2)

74-0859 ARMEN, H. A. JR.
PLASTIC ANALYSIS
RESEARCH DEPT., GRUMMAN AEROSPACE CORP., BETHPAGE,
N.Y., REPORT NO: RE-476J, MAY 1974, 46 P.
(N.T.I.S. - AD-781 703/4)

74-0860 ROSE, D. J.
WHITTEN, G. F.
AUTOMATIC NESTED DISSECTION
CENTER FOR RESEARCH IN COMPUTING TECH., HARVARD
UNIV., CAMBRIDGE, MASS., REPORT NO: TR-15-74, MAY
1974, 13 P. (N.T.I.S. - AD-779 863/0)

74-0861 KATONA, M. G.
ICE ENGINEERING: VISCOELASTIC FINITE ELEMENT
FORMULATION
NAVAL CIVIL ENGNG. LAB., PORT HUENEME, CALIF.,
REPORT NO: NCEL-TR-803, JANUARY 1974, 60 P.
(N.T.I.S. - AD-774 482/4)

74-0862 JENSEN, P. S.
FELIPPA, C. A.
VARIABLE GRID FINITE DIFFERENCE-ELEMENT SOLUTION
OF ELLIPTIC PARTIAL DIFFERENTIAL EQUATIONS
LOCKHEED MISSILES AND SPACE CO. INC., SUNNYVALE,
CALIF., REPORT NO: LMSC-D401891, MAY 13, 1974, 56
P. (N.T.I.S. - AD-782 301/6)

74-0863 BALMER, H.
DOLTSINIS J.
KONIG. M.
ELASTOPLASTIC AND CREEP ANALYSIS WITH THE ASKA
PROGRAM SYSTEM
COMP. METH. APPL. MECH. ENG., VOL. 3, 1974, PP.
87-104.

74-0864 COOK, W. A.
THREE-DIMENSIONAL MESH GENERATOR FOR FINITE
ELEMENT COMPUTER CODES
LOS ALAMOS SCIENTIFIC LAB., N. MEX., REPORT NO:
CONF-741001-31, 1974, 8 P. (N.T.I.S. -
LA-UR-74-1693) SEE ALSO NSA 3205 NO. 13469

74-0865 SCHREM, E.
DEVELOPMENT AND MAINTENANCE OF LARGE FINITE
ELEMENT SOFTWARE SYSTEMS
IN "STRUCT. MECH. COMPUTER PROGRAMS", PILKEY, W.,
SACZALSKI, K., AND SCHAEFFER, H. (EDS), UNIV.
PRESS OF VIRGINIA, 1974, PP. 669-685.

74-0866 GRANTHAM, P.
CRONSHAGEN, E.
COMPUTER PROGRAMS FOR SIMULATING PHYSICAL
PHENOMENA IN THREE SPACE DIMENSIONS AND TIME
LAWRENCE LIVERMORE LAB., UNIV. OF CALIF.,
LIVERMORE, REPORT NO: CONF-741001-28, OCTOBER 8,
1974, 9 P. (N.T.I.S. - UCRL-76084) SEE ALSO NSA
3104, NO. 10453.

74-0867 BIFFLE, J. H.
INTERACTIVE GRAPHICS AIDED STRUCTURAL ANALYSIS
SANDIA LABS., ALBUQUERQUE, N. MEX., REPORT NO:
CONF-741001-26, 5 P. (N.T.I.S.-SAND-74-5570) SEE
ALSO NSA 3104, NO. 10448

74-0868 AUTHOR
USE OF SPARSITY IN THE SOLUTION OF FINITE ELEMENT
SYSTEMS OF EQUATIONS BY GAUSSIAN ELIMINATION-TYPE
METHODS
INST. OF MATH. SCI., NEW YORK UNIV., N.Y., REPORT
NO: CONF-740236-1, 1974, 35 P.
(N.T.I.S.-COO-3077-66) SEE ALSO 3104, NO. 10438

74-0869 CHOPRA, P. S.
SIGNIFICANCE OF SYMMETRY IN MULTIPLE FRACTURE ANALYSIS
ARGONNE NATIONAL LAB., ILL. 1974, 15 P. (N.T.I.S. - CONF-740846-1) SEE ALSO NSA 3104, NO. 10436.

74-0870 MALANG, S.
HETRAP: A HEAT TRANSFER ANALYSIS PROGRAM
OAK RIDGE NATIONAL LAB., TENN. SEPTEMBER 1974, 51 P. (N.T.I.S.-ORNL-TM-4555) SEE ALSO NSA 3103, NO. 07726.

74-0871 YOUNG, R. C.
CALLABRESI, M. L.
ORGANIZATION OF GNATS: A GENERAL NONLINEAR ANALYSIS COMPUTER PROGRAM FOR TWO-DIMENSIONAL STRUCTURES
SANDIA LABS., LIVERMORE, CALIF., OCTOBER 1974, 85 P. (N.T.I.S.-SLL-74-0022) SEE ALSO NSA 3102, NO. 05459.

74-0872 JONES, R. E.
QMESH: A SELF-ORGANIZING MESH GENERATION PROGRAM
SANDIA LABS., ALBUQUERQUE, N. MEX., REPORT NO: CONF-741001-19, 1974, 7 P. (N.T.I.S. - SAND-74-5486) SEE ALSO NSA 3102, NO. 05454.

74-0873 MASSINI, G.
ZACCARELLI, P.
VERIFICATION OF THE FINITE ELEMENT COMPUTER CODE SAFE-2D
COMITATO NAZIONALE PER L'ENERGIA NUCLEARE, ROME, ITALY, FEBRUARY 1974, 163 P. (N.T.I.S. - RT/ING-(74)10) IN ITALIAN. SEE ALSO NSA 3102, NO. 05452.

74-0874 MARCHERTAS, A. H.
BELYTSCHKO, T. B.
NONLINEAR FINITE-ELEMENT FORMULATION FOR TRANSIENT ANALYSIS OF THREE DIMENSIONAL THIN STRUCTURES
ARGONNE NATIONAL LAB., ILL. JUNE 1974, 49 P. (N.T.I.S. - ANL-8104) SEE ALSO NSA 3102, NO. 05334.

74-0875 YOUNG, R. C.
CALLABRESI, M. L.
GNATS: A FINITE ELEMENT COMPUTER PROGRAM FOR THE GENERAL NONLINEAR ANALYSIS OF TWO-DIMENSIONAL STRUCTURES
SANDIA LABS., LIVERMORE, CALIF., OCTOBER 1974, 102 PP. (N.T.I.S. - SLL-74-0023) SEE ALSO NSA 3101, NO. 02734.

74-0876 APPERT, K.
BERGER, D.
GRUBER, P.
NEW FINITE ELEMENT APPROACH TO THE NORMAL MODE ANALYSIS IN MAGNETOHYDRODYNAMICS
ECOLE POLYTECHNIQUE FEDERALE, LAUSANNE, SWITZERLAND. JUNE 1974, 27 PP. (N.T.I.S. - LRP-83/74) SEE ALSO NSA 3101, NO. 00295.

74-0877 BABUSKA, I.
CONNECTION BETWEEN THE FINITE DIFFERENCE LIKE METHODS AND THE METHODS BASED ON INITIAL VALUE PROBLEMS FOR ODE
COLL. PARK INST. FOR FLUID DYNAMICS AND APPLIED MATH., UNIV., OF MARYLAND, REPORT NO: CONF-740627-2, BN-796, JULY 1974, 30 P. (N.T.I.S. - ORO-3443-50) SEE ALSO NSA 3012, NO. 34502.

74-0878 LATHROP, K. D.
NEW TRANSPORT METHODS AND CODES
LOS ALAMOS SCIENTIFIC LAB., N. MEX., REPORT NO: CONF-740903-5, 1974, 9 P. (N.T.I.S. - LA-UR-74-1262) FOR ABSTRACT, SEE NSA 30 12, NO. 34032.

74-0879 REEVES, M.
DUGUID, J. O.
WATER FLOW THROUGH SATURATED-UNSATURATED POROUS MEDIA USING FINITE-ELEMENT-GALERKIN METHODS
OAK RIDGE NATL. LAB., TENN., 1974, 22 PP. (N.T.I.S. - CONF-740904-3) SEE ALSO NSA 3010, NO. 28896.

74-0880 JONES, R. E.
USERS MANUAL FOR QMESH. A SELF-ORGANIZING MESH GENERATION PROGRAM
SANDIA LABS., ALBUQUERQUE, N. MEX., JULY 1974, 66 P. (N.T.I.S. - SLA-74-0239) SEE ALSO NSA 3009, NO. 26154.

74-0881 ELLINGTON, J. P.
SIMPLE TRIANGULAR MEMBRANE FINITE ELEMENT
UKAEA REACTOR GROUP, RISLEY, APRIL 1974, 11 PP. (N.T.I.S. - TRG-REPORT-2504(R)), SEE ALSO NSA 3008, NO. 20931.

74-0882 WINKEL, B. V.
STRESS AND DEFORMATION ANALYSIS OF IRRADIATION INDUCED SWELLING
AEROJET NUCLEAR CO., IDAHO FALLS, IDAHO, 1974, 23 P. (N.T.I.S. - CONF-740606-19) SEE ALSO NSA 3009, NO. 24723.

74-0883 CESARI, E.
VERIFICATION OF THE PEC CONTAINMENT VESSEL FOR SOME OPERATING CONDITIONS BY USE OF THE FINITE ELEMENT CODE SAFE-SHELL. B
COMITATO NAZIONALE PER L'ENERGIA NUCLEARE, ROME, ITALY, JANUARY 1974, 160 PP. (N.T.I.S. - RT/ING-(74)3) IN ITALIAN. SEE ALSO NSA 3008, NO. 23104.

74-0884 LAWTON, R. G.
PLACID: A GENERAL FINITE ELEMENT COMPUTER PROGRAM FOR STRESS ANALYSIS OF PLANE AND AXISYMMETRIC SOLIDS
LOS ALAMOS SCI. LAB., NEW MEXICO, MAY 1974, 33 PP. (N.T.I.S.- LA-5621-MS) SEE ALSO NSA 3008, NO. 23318.

74-0885 LAWTON, R. G.
AYER HEAT CONDUCTION COMPUTER PROGRAM
LOS ALAMOS SCIENTIFIC LAB., N. MEX., MAY 1974, 27 P. (N.T.I.S. -LA-5613-MS) SEE ALSO NSA 3008, NO. 23317.

74-0886 BROWNING, R. V.
MILLER, D. G.
ANDERSON, C. A.
TSAAS: FINITE ELEMENT THERMAL AND STRESS ANALYSIS OF AXISYMMETRIC SOLIDS WITH ORTHOTROPIC TEMPERATURE-DEPENDENT MATERIALS PROPERTIES
LOS ALAMOS SCIENTIFIC LAB., N. MEX., MAY 1974, 52 P. (N.T.I.S. - LA-5599-MS) SEE ALSO NSA 3008, NO. 23418.

74-0887 STADTER, J. T.
WEISS, R. O.
CONPLOT: A GENERAL PURPOSE CONTOUR PLOTTING PROGRAM
APPLIED PHYSICS LAB., JOHNS HOPKINS UNIV., SILVER SPRING, MD., REPORT NO: ANSP-M-9, JUNE 1974, 78 P. (N.T.I.S. - AEC/SNS-3060-009) SEE ALSO NSA 3008, NO. 23291.

74-0888 SWARTZ, B K.
CREATION AND COMPARISON OF FINITE DIFFERENCE ANALOGS OF SOME FINITE ELEMENT SCHEMES
LOS ALAMOS SCIENTIFIC LAB., N. MEX., REPORT NO: CONF-740454-1, 1974, 40 P. (N.T.I.S.-LA-UR-74-771) SEE ALSO NSA 3006, NO. 18047.

74-0889 WATKINS, D. S.
BLENDING FUNCTIONS AND FINITE ELEMENTS
PH.D. THESIS, DEPT. OF MATH., STATISTICS AND COMP. SCI., UNIV. OF CALGARY, 1974.

74-0890 BABUSKA, I.
KELLOGG, R. B.
NONUNIFORM ERROR ESTIMATES FOR THE FINITE ELEMENT METHOD
COLL. PARK INST. FOR FLUID DYNAMICS AND APPLIED MATH., UNIV., OF MARYLAND, REPORT NO: BN-790, APRIL 1974, 16 P. (N.T.I.S. - ORO-3443-49) SEE ALSO NSA 3003, NO. 08963.

74-0891 BABUSKA, I.
SOLUTION OF PROBLEMS WITH INTERFACES AND SINGULARITIES
COL. PARK INST. FOR FLUID DYN. AND APPL. MATH., UNIV. OF MARYLAND, REPORT NO: BN-789, APRIL 1974, 75 PP. (N.T.I.S. - ORO-3443-48) SEE ALSO NSA 3003, NO. 08817.

74-0892 CAREY, J. J.
VALENTIN, R. A.
EVALUATION OF A CLASS OF METHODS FOR BOUNDING STEADY-STATE CREEP DEFORMATION
ARGONNE NATIONAL LAB., ILL., FEBRUARY 1974, 45 P. (N.T.I.S.-ANL-8016) SEE ALSO NSA 3002, NO. 03041.

74-0893 GABRIELSON, V. K.
GRAPHICS APPLICATIONS FOR FINITE ELEMENT CODE PROCESSING
SANDIA LABS., LIVERMORE, CALIF., REPORT NO: CONF-740512-4, APRIL 1974, 20 P. (N.T.I.S.-SLL-74-5218) SEE ALSO NSA 3001, NO. 02256.

74-0894 FRIEDMAN, E.
DIRECT ITERATION METHOD FOR THE INCORPORATION OF PHASE CHANGE IN FINITE ELEMENT HEAT CONDUCTION PROGRAMS
BETTIS ATOMIC POWER LAB., PITTSBURGH, PA., MARCH 1974, 27 P. (N.T.I.S. - WAPD-TM-1133) SEE ALSO NSA 2912, NO. 29247.

74-0895 UTKU, S.
SYSTEMATIC SUBSTRUCTURING
PROC. 6TH CONF. ELECTRONIC COMPUTATION, J. STRUCT. DIV., A.S.C.E., 1974, PP. 717-730.

74-0896 LEAF, G. K.
LINDEMAN, A. J.
KAPER, H. G.
CONSTRUCTION OF A FINITE ELEMENT APPROXIMATION WHICH CROSSES MATERIAL INTERFACES
ARGONNE NATIONAL LAB., ILL., JANUARY 1974, 24 P. (N.T.I.S. -ANL-8052) SEE ALSO NSA 2910, NO. 25469.

74-0897 GURLAND, J.
RICE, J. R.
COMBINED MACROSCOPIC AND MICROSCOPIC APPROACH TO
THE FRACTURE OF METALS
DIV. OF ENGNG., BROWN UNIV., PROVIDENCE, R.I.,
TECH. PROGRESS REPORT, JULY 1973 - JUNE 1974, 13
PP. (N.T.I.S. - COO-3084-29) SEE ALSO NSA 3009,
NO. 24585.

74-0898 MURFIN, W. B.
EFFECT OF GEOLOGIC IRREGULARITIES OF SEISMIC
RESPONSE
SANDIA LABS., ALBUQUERQUE, N. MEX., FEBRUARY 1974,
43 P. (N.T.I.S. - SLA-74-24) SEE ALSO NSA 2910,
NO. 24156.

74-0899 JONES, R. E.
QMESH: A SELF-ORGANIZING MESH GENERATION PROGRAM
SANDIA LABS., ALBUQUERQUE, N. MEX., JULY 1974, 59
P. (N.T.I.S. -SLA-73-1088) SEE ALSO NSA 3008 NO.
23328.

74-0900 YAMADA, YOSHIAKI
TAKABATAKE, H.
SATO, T.
EFFECT OF TIME-DEPENDENT MATERIAL PROPERTIES ON
DYNAMIC RESPONSE
INT. J. NUMER. METH. ENG., VOL. 8, 1974, PP.
403-424.

74-0901 DONEA, J.
GIULIANI, S.
CODE TAFEST: NUMERICAL SOLUTION TO TRANSIENT
HEAT-CONDUCTION PROBLEMS USING FINITE ELEMENTS IN
SPACE AND TIME
COMMISSION OF THE EUROPEAN COMM., LUXEM. AND
ISPRA, ITALY, JOINT NUCL. RES. CENTER, 1974, 41
PP. (N.T.I.S. - EUR-5049) SEE ALSO NSA 2909, NO.
20801.

74-0902 PILKEY, W.
SACZALSKI, K.
SCHAEFER, H.
STRUCTURAL MECHANICS COMPUTER PROGRAMS
UNIV. OF VIRGINIA, CHARLOTTESVILLE, 1974.

74-0903 AAMODT, B.
APPLICATION OF THE FINITE ELEMENT METHOD TO
PROBLEMS IN LINEAR FRACTURE MECHANICS
DISSERTATION, DIV. OF STRUCT. MECH., THE NORWEGIAN
INST. OF TECH., UNIV. OF TRONDHEIM, NORWAY, MAY
1974.

74-0904 SCHMID, G.
INCOMPRESSIBLE FLOW IN MULTIPLY CONNECTED REGIONS
PROC. INT. CONF. ON NUMER. METH. IN FLUID
DYNAMICS, UNIV. OF SOUTHAMPTON, ENGLAND, SEPTEMBER
26-28, 1973, PP. 153-171.

74-0905 TAYLOR, C.
DAVIS, J. M.
NUMERICAL MODEL OF DISPERSION IN ESTUARIES
PROC. INT. SYMP. ON FINITE ELEMENT METHODS IN FLOW
PROBLEMS, UNIV. OF SWANSEA, WALES, JANUARY 1974,
PP. 379-384, PUBL. BY UNIV. OF ALA., HUNTSVILLE,
1974.

74-0906 AAMODT, B.
BERGAN, P. G.
PROPAGATION OF ELLIPTICAL SURFACE CRACKS AND
NONLINEAR FRACTURE MECHANICS BY THE FINITE ELEMENT
METHOD
PROC. 5TH CONF. ON DIMENSIONING AND STRENGTH
CALCULATIONS, BUDAPEST, VOL. 1, OCTOBER 1974, PP.
31-42.

74-0907 BABUSKA, I.
AZIZ, A. K.
ON THE ANGLE CONDITION IN THE FINITE ELEMENT
METHOD
TECH. NOTE BN-808, INSTITUTE FOR FLUID DYNAMICS
AND APPLIED MATHEMATICS, UNIV. OF MARYLAND,
NOVEMBER 1974, PP. 1-23.

74-0908 COOK, W. A.
BODY ORIENTED (NATURAL) CO-ORDINATES FOR
GENERATING THREE-DIMENSIONAL MESHES
INT. J. NUMER. METH. IN ENG., VOL. 8, NO. 1, 1974,
PP. 27-44.

74-0909 NASRELDIN, H. A.
INTERACTIVE FINITE ELEMENT DATA GENERATION FOR
THREE-DIMENSIONAL PLATE STRUCTURES
INT. CONF. AND EXHIBITION ON COMPUTERS IN ENGNG.
AND BUILDING DESIGN, IMPERIAL COLLEGE, LONDON,
ENGLAND, SEPTEMBER 25-27, 1974.

74-0910 SWARTZ, B.
WENDROFF, B.
THE COMPARATIVE EFFICIENCY OF CERTAIN FINITE
ELEMENT AND FINITE DIFFERENCE SCHEMES FOR A
HYPERBOLIC PROBLEM
CONF. ON THE NUM. SOLUTION OF DIFFERENTIAL
EQUATIONS, (ED) WATSON, G.A., SPRINGER-VERLAG, NEW
YORK, PP. 153-163

74-0911 SWARTZ, B.
WENDROFF, B.
THE RELATIVE EFFICIENCY OF FINITE DIFFERENCE AND
FINITE ELEMENT METHODS, 1. HYPERBOLIC PROBLEMS
AND SPLINES
SIAM J. NUMER. ANAL. 11, 1974.

74-0912 SWARTZ, B.
WENDROFF, B.
THE RELATION BETWEEN THE GALERKIN AND COLLOCATION
METHODS USING SMOOTH SPLINES
SIAM J. NUMER. ANAL. 11, 1974.

74-0913 SAKAI, F.
KAWAI, S.
APPLICATION OF THE FINITE ELEMENT METHOD TO
SURFACE WAVE ANALYSIS
COASTAL ENG., JAPAN, VOL. 17, 1974, PP. 13-22.

74-0914 RAO, A. K.
REVIEW OF CONTINUUM, FINITE ELEMENT AND HYBRID
TECHNIQUES IN ANALYSIS OF STRESS CONCENTRATIONS IN
STRUCTURES
NUCL. ENG. DES., VOL. 31, NO. 3, 1974, PP.
427-433.

74-0915 GONCALVES MORGADO DE AZEVEDO, L.
SOBRE TORRES ESPIADAS - METODO DE CALCULO E UM
EXEMPLO DE APLICACAO. (METHOD AND AN EXAMPLE OF
CALCULATION OF GUYED TOWERS)
TECNICA, LISBON, VOL. 49, NO. 422, NOVEMBER 1974,
PP. 49-77.

74-0916 DIANA, G.
RUGGIERI, G.
SULLA DETERMINAZIONE DELLE FREQUENZE PROPRIE DI
ALBERI SU PIU SUPPORTI ELASTICI. (DETERMINATION
OF FREQUENCIES OCCURRING IN SHAFTS ON SEVERAL
ELASTIC BEARINGS)
ENERG. ELETTR., VOL. 51, NO. 2, FEBRUARY 1974, PP.
84-90.

74-0917 SIMS, F. A.
JONES, C. J. F.
COMPARISON BETWEEN THEORETICAL AND MEASURED EARTH
PRESSURES ACTING ON A LARGE MOTORWAY RETAINING
WALL
HIGHWAY ENG., VOL. 21, NO. 12, DECEMBER 1974, PP.
26-29.

74-0918 WURMNEST, W.
DAS PROGRAMMSYSTEM ANTRAS EDV-SYSTEM ZUR
BERECHNUNG BELIEBIGER STATISCHER SYSTEME NACH DER
METHODE DER FINITEN ELEMENTE (ANTRAS
COMPUTER-PROGRAM FOR LINEAR STRESS ANALYSIS BY THE
FINITE ELEMENT METHOD)
TECH. MITT. KRUPP. FORSCHUNGSBER, VOL. 32, NO. 3,
DECEMBER 1974, PP. 97-107.

74-0919 KULAK, R. F.
BELYTSCHKO, T. B.
NONLINEAR BEHAVIOR OF THE HUMAN INTERVERTEBRAL
DISC UNDER AXIAL LOADING
PROC. ASME ANNU. MEET., NEW YORK, N.Y., NOVEMBER
17-21, 1974, PP. 88-90.

74-0920 PLANT, R. E.
BARTEL, D. L.
FINITE ELEMENT ANALYSIS OF A BONE-PLATE-SCREW
SYSTEM
PROC. ASME ANNU. MEET., NEW YORK, N.Y., NOVEMBER
17-21, 1974, PP. 85-87.

74-0921 KUFEL, W.
PIETRAS, F.
OCENA DOKLADNOSCI ROZWIAZAN PROBLEMOW BRZEGOWYCH
SPREZYSTYCH CIAL DYSKERTYZOWANYCH. (ASSESSING THE
ACCURACY OF SOLVENCY BOUNDARY PROBLEMS OF
DISCRETE, ELASTIC BONDS)
ROZPR. INZ., VOL. 22, NO. 3, 1974, PP. 427-434.

74-0922 PIAN, T. H. H.
NONLINEAR CREEP ANALYSIS BY ASSUMED STRESS FINITE
ELEMENT METHODS
A.I.A.A. J., VOL. 12, NO. 12, DECEMBER 1974, PP.
1756-1758.

74-0923 KRUSZEWSKI, J.
GAWRONSKI, W.
METHOD SZTYWNYCH ELEMENTOW SKONCZONYCH W
OBLICZENIACH KONSTRUCJI OKRETOWYCH. (STIFF FINITE
ELEMENT METHOD IN COMPUTATION OF SHIP STRUCTURES)
ROZPR. INZ., VOL. 22, NO. 3, 1974, PP. 337-358.

74-0924 CIAVALDINI, J. F.
NEDELEC, J. C.
SUR L ELEMENT DE FRAEIJS DE VEUBEKE ET SANDER
(FRAEIJS DE VEUBEKE-SANDER ELEMENT)
REV. FR. AUTOM. INF. RECH. OPER., VOL. 8, AUGUST
1974, PP. 29-46.

74-0925 LESAINT, P.
FINITE ELEMENT METHODS FOR THE TRANSPORT EQUATION
REV. FR. AUTOM. INF. RECH. OPER., VOL. 8, AUGUST
1974, PP. 67-93.

74-0926 GILLHAM, R. W.
FARVOLDEN, R. N.
SENSITIVITY ANALYSIS OF INPUT PARAMETERS IN
NUMERICAL MODELING OF STEADY STATE REGIONAL
GROUNDWATER FLOW
WATER RESOUR. RES., VOL. 10, NO. 3, JUNE 1974, PP.
529-538.

74-0927 CLOUGH, G. W.
WILSON, J. E.
SUBTERRANEAN GUIDEWAY RESPONSES TO LEVITATION
VEHICLE LOADING
J. HIGH SPEED GROUND TRANSP., VOL. 8, NO. 1, 1974,
PP. 153-164.

74-0928 OWEN, D. R. J.
PRAKASH, A.
FINITE ELEMENT ANALYSIS OF NON-LINEAR COMPOSITE
MATERIALS BY USE OF OVERLAY SYSTEMS
COMPUT. STRUCT., VOL. 4, NO. 6, DECEMBER 1974, PP.
1251-1267.

74-0929 CHOPRA, P. S.
FINITE ELEMENT FRACTURE MECHANICS ANALYSIS OF
CREEP RUPTURE OF FUEL ELEMENT CLADDING
NUCL. ENG. DES., VOL. 29, NO. 1, 1974, PP. 7-21.

74-0930 RASHID, Y. R.
MATHEMATICAL MODELING AND ANALYSIS OF FUEL RODS
NUCL. ENG. DES., VOL. 29, NO. 1, 1974, PP. 22-32.

74-0931 COLEMAN, G. E. 111.
PERUMPRAL, J. V.
FINITE ELEMENT ANALYSIS OF SOIL COMPACTION
AMER. SOC. AGRIC. ENG. TRANS., VOL. 17, NO. 5,
SEPTEMBER-OCTOBER, 1974, PP. 856-960.

74-0932 HUDSPITH, I. B.
THE ASAS ANALYSIS OF THE LYNX
WESTLAND HELICOPTERS LTD., REPORT NO. SD968, JULY
1974.

74-0933 ROSSETTOS, J. N.
TONG, P.
FINITE-ELEMENT ANALYSIS OF VIBRATION AND FLUTTER
OF CANTILEVER ANISOTROPIC PLATES
ASME J. APPL. MECH., VOL. 41, NO. 4, DECEMBER
1974, PP. 1075-1080.

74-0934 CHAN, H. S.
MATHEMATICAL MODEL FOR CLOSED HEAD IMPACT
PROC. 15TH CONF. STAPP CAR CRASH, UNIV. OF MICH.,
ANN ARBOR, PAPER 741191, DECEMBER 4-5, 1974, PP.
557-578.

74-0935 PAYER, H. G.
BROELMANN, J.
TRAGLAST VON T-TRAEGERN. EXPERIMENT UND F.E.
BERECHNUNGEN (ULTIMATE STRENGTH OF T-BEAMS.
EXPERIMENTS AND COMPUTATIONS)
SCHIFFSTECHNIK, VOL. 21, NO. 106, NOVEMBER 1974,
PP. 117-128.

74-0936 KLOTZ, J. A.
KRUEGER, R. F.
EFFECT OF PERFORATION DAMAGE ON WELL PRODUCTIVITY
J. PET. TECHNOL., VOL. 26, NOVEMBER 1974 PP.
1303-1314.

74-0937 TABARROK, B.
FENTON, R. G.
APPLICATION OF A REFINED PLATE BENDING ELEMENT TO
BUCKLING PROBLEMS
COMPUT. STRUCT., VOL. 4, NO. 6, DECEMBER 1974, PP.
1313-1321.

74-0938 JAEGER, L. G.
NEAR-CRACK DEFORMATION FIELDS IN STRAIN-HARDENING
MATERIAL
TRANS. CAN. SOC., MECH. ENG., VOL. 2, NO. 2,
1973-1974, PP. 86-90.

74-0939 RAMEY, G. E.
SOME EFFECTS OF SYSTEM IDEALIZATIONS,
SINGULARITIES AND MESH PATTERNS ON FINITE ELEMENT
SOLUTIONS
COMPUT. STRUCT., VOL. 4, NO. 6, DECEMBER 1974, PP.
1173-1184.

74-0940 RAMEY, G. E.
KRISHNAMURTHY, N.
ERROR ESTIMATES FOR CONFORMING FINITE ELEMENT
SOLUTIONS
COMPUT. STRUCT., VOL. 4, NO. 6, DECEMBER 1974, PP.
1207-1222.

74-0941 RAMEY, G. E.
KRISHNAMURTHY, N.
THEORETICAL AND NUMERICAL CONVERGENCE RATES FOR
SOME CONFORMING FINITE ELEMENTS
COMPUT. STRUCT., VOL. 4, NO. 6, DECEMBER 1974, PP.
1185-1206.

74-0942 BON, C.
GERADIN, M.
ON THE NUMERICAL SOLUTION OF LARGE EIGENVALUE
PROBLEMS ARISING IN PANEL FLUTTER ANALYSIS BY THE
FINITE ELEMENT METHOD
COMPUT. STRUCT., VOL. 4, NO. 6, DECEMBER 1974, PP.
1223-1250.

74-0943 MCGEORGE, R.
SWEC, L. F. JR.
REFINED CRACKED CONCRETE ANALYSIS OF CONCRETE
CONTAINMENT STRUCTURES SUBJECT TO OPERATIONAL AND
ENVIRONMENTAL LOADINGS
NUCL. ENG. DES., VOL. 29, NO. 1, 1974, PP. 58-70.

74-0944 DLUGACH, M. I.
KOVAL'CHUK, N. V.
ISSLEDOVANIE NAPRYAZHENNOGO SOSTOYANIYA REBRISTYKH
ISILINDRICHESKIKH OBOLOCHEK S PRYAMOUGOL'NYMI
OTVERSTIYAMI METODOM KONECHNYKH ELEMENTOV.
(INVESTIGATION OF THE STRESSED STATE OF RIBBED
CYLINDRICAL SHELLS WITH RECTANGULAR CUTOUTS BY THE
FINITE ELEMENT METHOD)
PRIKL. MEKH., VOL. 10, NO. 10, OCTOBER 1974, PP.
22-30.

74-0945 YOON, K. E.
GOLEMBESKI, T. S.
STRUCTURAL ANALYSIS OF A THREE-DIMENSIONAL
PERFORATED PLATE BY THE FINITE ELEMENT METHOD WITH
TEST CORRELATION
NUCL. ENG. DES., VOL. 29, NO. 1, PP. 33-38.

74-0946 ABU-SHUMAYS, I. K.
BAREISS, E. H.
ADJOINING APPROPRIATE SINGULAR ELEMENTS TO
TRANSPORT THEORY COMPUTATIONS
J. MATH. ANAL. APPL., VOL. 48, NO. 10, OCTOBER
1974, PP. 200-222.

74-0947 MIYAMOTO, H.
ODA, J.
EFFECT OF GRAIN BOUNDARY FOR THE ELASTIC BEHAVIOR
OF AL AND CU CRYSTALS (ANALYSIS BY THE FINITE
ELEMENT METHOD)
PROC. 17TH CONGR. ON MATER. RES., JAPAN, SEPTEMBER
1973, PP. 53-55. ALSO IN SOC. OF MATER. SCI.,
TOKYO, JPN, 1974.

74-0948 LUTTRELL, A. L.
HENDERSON, H. L.
PRESSURE VESSEL CERTIFICATION BASED ON FRACTURE
MECHANICS TECHNOLOGY
J. SPACECR. ROCKETS, VOL. 11, NO. 12, DECEMBER
1974, PP. 838-844.

74-0949 RYNN, P. G.
PROGRESS IN COMPUTERIZED STRUCTURAL ANALYSIS OF
LARGE VESSELS
PROC. CONF. SUPER OCEAN CARRIER, NEW YORK, JANUARY
16-18, 1974, PP. 498-517.

74-0950 GOLDSMITH, W.
IMPACT ON A TRANSVERSELY ANISOTROPIC HALF-SPACE
INT. J. ROCK. MECH. MIN. SCI., GEOMECH. ABSTR.,
VOL. 11, NO. 11, NOVEMBER 1974, PP. 413-421.

74-0951 CRISFIELD, M. A.
COLLAPSE ANALYSIS OF BOX-GIRDER COMPONENTS USING
FINITE ELEMENTS
SYMP. ON NON-LINEAR TECHNIQUES AND BEHAVIOUR IN
STRUCT. ANALYSIS, TRANS. AND ROAD RESEARCH LAB.,
CROWTHORNE, BERKSHIRE, DECEMBER 1974.

74-0952 GIENCKE, E.
SIMPLE "MIXED" METHOD FOR PLATE AND SHELL PROBLEMS
NUCL. ENG. DES., VOL. 29, NO. 1, 1974, PP.
141-155.

74-0953 LUCO, J. E.
HADJIAN, A. H.
BOS, H. D.
DYNAMIC MODELING OF HALF-PLANE BY FINITE ELEMENTS
NUCL. ENG. DES., VOL. 31, NO. 2, 1974, PP.
184-194.

74-0954 HADJIAN, A. H.
LUCO, J. E.
TSAI, N. C.
SOIL-STRUCTURE INTERACTION — CONTINUUM OR FINITE
ELEMENT
NUCL. ENG. DES., VOL. 31, NO. 2, 1974, PP.
151-167.

74-0955 MIYAMOTO, H.
ODA, J.
SIMULATION OF ELASTIC MODULUS AND POISSON'S RATIO
OF SPHEROIDAL GRAPHITE CAST IRON
BULL. JSME., VOL. 17, NO. 112, OCTOBER 1974, PP.
1233-1239.

74-0956 PARKS, D. M.
STIFFNESS DERIVATIVE FINITE ELEMENT TECHNIQUE FOR
DETERMINATION OF CRACK TIP STRESS INTENSITY
FACTORS
INT. J. FRACT., VOL. 10, NO. 4, DECEMBER 1974, PP.
487-502.

74-0957 PAUL, S. L.
KESLER, C. E.
STUDIES TO IMPROVE TUNNEL SUPPORT SYSTEMS
FINAL REPORT FRA-ORDED 74-51, NTIS, SPRINGFIELD,
VA., JUNE 1974.

74-0958
ISEKI, H.
JIMMA, T.
FINITE ELEMENT METHOD OF ANALYSIS OF THE
HYDROSTATIC BULGING OF A SHEET METAL - 1.
BULL. JSME., VOL. 17, NO. 112, OCTOBER 1974, PP.
1240-1246.

74-0959
TROMPETTTE, P.
LALANNE, M.
VIBRATION ANALYSIS OF ROTATING TURBINE BLADES
PAPER NO 74-WA/DE-23, ASME MEETING, NOVEMBER 17-22
1974, 8 P.

74-0960
MATSUSHITA, T.
ANALYTICAL STUDY OF THE PROBLEM OF EVAPORATION AND
CONDENSATION USING THE FINITE ELEMENT METHOD
PROC. 9TH INT. SYMP. ON RAREFIED GAS DYN., DFVLR
PRESS, GOETTINGEN, GERMANY, VOL. 2, PAPER F. 3,
JULY, 1974, 12 P.

74-0961
BURMAN, B. C.
DEVELOPMENT OF A NUMERICAL MODEL FOR DISCONTINUA
AUST. GEOMECH. J., VOL. G4, NO. 1, 1974, PP.
13-22.

74-0962
TARNOWSKI, J.
ZASTOSOWANIE METODY SZTYWNYCH ELEMENTOW
SKONCZONYCH DO OBLICZEN ORGAN WALOW OKRETOYCH Z
UWZGLEDNIENIEM EFEKTOW GIROSKOPOWYCH.
(APPLICATION OF THE METHOD OF STIFF FINITE
ELEMENTS TO THE CALCULATION OF SHIP TRANSMISSION
SHAFTS, GYROSCOPIC EFFECTS BEING TAKEN INTO
CONSIDERATION)
ROZPR. INZ., VOL. 22, NO. 3, 1974, PP. 297-305.

74-0963
WITTBRODT, E.
HYBRYDOWA METODA ELEMENTOW SKONCZONYCH W
ZASTOSOWANIU DO OBLICZEN ORGAN URZADZEN
OKRETOWYCH. (HYBRID METHOD OF FINITE ELEMENTS
APPLIED TO THE CALCULATION OF VIBRATIONS OF SHIP
STRUCTURES)
ROZPR. INZ., VOL. 22, NO. 3, 1974, PP. 369-385.

74-0964
NARAYANASWAMI, R.
NEW TRIANGULAR PLATE-BENDING FINITE ELEMENT WITH
TRANSVERSE SHEAR FLEXIBILITY
A.I.A.A. J., VOL. 12, NO. 11, DECEMBER 1974, PP.
1761-1763.

74-0965
VOROSHKO, P. P.
KVITKA, A. L.
ISSLEDOVANIE NAPRYAZHENNO-DEFORMIROVANNOGO
SOSTOYANIYA SIMMETRICHNYKH DISKOV SO STUPITSEI.
(INVESTIGATION OF THE STRESS STRAINED STATE OF
SYMMETRICAL DISKS WITH A HUB)
PROBL. PROCHN., VOL. 6, NO. 11, NOVEMBER 1974, PP.
24-27.

74-0966
PIFKO, A.
LEVINE, H. S.
PLANS - A FINITE ELEMENT PROGRAM FOR NONLINEAR
ANALYSIS OF STRUCTURES
PAPER NO. 74-WA/PVP-6, ASME MEETING, NOVEMBER
17-22, 1974, 19 PP.

74-0967
KEY, S. W.
FINITE ELEMENT PROCEDURE FOR THE LARGE DEFORMATION
DYNAMIC RESPONSE OF AXISYMMETRIC SOLIDS
PAPER NO. 74-WA/PVP-7, ASME MEETING, NOVEMBER
17-22, 1974, 16 P.

74-0968
MINAMI, H. M.
VELJOVICH, W.
PRESSURE VESSEL WALL THICKNESS LIMITS BASED ON
THERMAL STRESS RACHETTING AND CREEP FATIGUE
INTERACTION
PAPER NO. 74-WA/PVP-11, ASME MEETING, NOVEMBER
17-22, 1974, 27 PP.

74-0969
OSTERBERG, J. O.
GILL, S. A.
LOAD TRANSFER MECHANISM FOR PIERS SOCKETTED IN
HARD SOILS OR ROCK
PROC. 9TH SYMP. CAN. ROCK MECH., MONTREAL, QUE.,
DECEMBER 13-15, 1973, PP. 235-262. PUBL. BY DEPT.
OF ENERGY, MINES AND RESOURCES, MINES BRANCH,
OTTAWA, ONT., 1974.

74-0970
BOEKENBRINK, D.
INFLUENCE OF COOLING, FREE-STREAM TURBULENCE AND
SURFACE ROUGHNESS ON THE AERODYNAMIC BEHAVIOR OF
CASCADES
PAPER NO. 74-WA/GT-9, ASME MEETING, NOVEMBER
17-22, 1974, 16 PP.

74-0971
FILSTRUP, A. W.
FINITE ELEMENT ANALYSIS OF A GAS TURBINE BLADE
PAPER NO. 74-WA/GT-11, ASME MEETING, NOVEMBER
17-22, 1974, 9 PP.

74-0972
TANG, S. C.
PETROF, R. C.
METHOD FOR THE COMPUTATION OF CYCLIC STRESSES IN A
GAS-TURBINE WHEEL
PAPER NO. 74-WA/GT-12, ASME MEETING, NOVEMBER
17-22, 1974, 16 PP.

74-0973
SHARIFI, P.
NONLINEAR BUCKLING ANALYSIS OF COMPOSITE SHELLS
PROC. 15TH STRUCT. CONF., A.I.A.A., LAS VEGAS,
NEV., PAPER NO. 74-411, APRIL 1974.

74-0974
JANZ, R. F.
KUBERT, B. R.
DEFORMATION OF THE DIASTOLIC LEFT VENTRICLE - 2.
NONLINEAR GEOMETRIC EFFECTS
J. BIOMECH., VOL. 7, NO. 6, 1974, PP. 509-516.

74-0975
HASSLER, C. R.
RYBICKI, E. F.
MEASUREMENTS OF HEALING AT AN OSTEOTOMY IN A
RABBIT CALVARIUM: THE INFLUENCE OF APPLIED
COMPRESSIVE STRESS ON COLLAGEN SYNTHESIS AND
CALCIFICATION
J. BIOMECH., VOL. 7, NO. 6, 1974, PP. 545-550.

74-0976
CHOWDHURY, P. C.
FREE VIBRATIONS OF FLUID-BORNE STRUCTURES:
INVESTIGATIONS ON A SIMPLE MODEL
INST. ENG. SHIPBUILD., TRANS NORTH EAST COAST,
VOL. 91, NO. 1, NOVEMBER 1974, PP. 15-28.

74-0977
MOAN, T.
EXPERIENCES WITH ORTHOGONAL POLYNOMIALS AND "BEST"
NUMERICAL INTEGRATION FORMULAS ON A TRIANGLE; WITH
PARTICULAR REFERENCE TO FINITE ELEMENT
APPROXIMATIONS
Z. ANGEW MATH. MECH., VOL. 54, NO. 8, AUGUST 1974,
PP. 501-508.

74-0978
MEHLHORN, G.
DOERR, K.
BERECHNUNG VON STAHLBETONSCHEIBEN IM ZUSTAND II
BEI ANNAHME EINES WIRKLICHKEITSNAHEN
WERKSTOFFVERHALTENS. (STRUCTURAL ANALYSIS OF
REINFORCED CONCRETE PANELS AT MORE REALISTIC
MATERIAL BEHAVIOR)
DTSCH. AUSSCHUSS STAHLBETON, NO. 238, 1974, PP.
29-102.

74-0979
BOND, D.
OPTIMUM DESIGN OF CONCRETE STRUCTURES
ENG. OPTIMIZATION, VOL. 1, NO. 1, 1974, PP. 17-28.

74-0980
POLIZZOTTO, C.
OPTIMUM PLASTIC DESIGN FOR MULTIPLE SETS OF LOADS
MECCANICA, VOL. 9, NO. 3, SEPTEMBER 1974, PP.
206-213.

74-0981
MEYERS, J. F.
BUILDING THREE NEW MILLION-HORSEPOWER HYDRAULIC
TURBINES FOR GRAND COULEE
ALLIS-CHALMERS ENG. REV., VOL. 39, NO. 3, 1974,
PP. 8-11.

74-0982
BRATANOW, T.
ECER, A.
ANALYSIS OF THREE-DIMENSIONAL UNSTEADY VISCOUS
FLOW AROUND OSCILLATING WINGS
A.I.A.A. J., VOL. 12, NO. 11, NOVEMBER 1974, PP.
1577-1584.

74-0983
MONDKAR, D. P.
POWELL, G. H.
LARGE CAPACITY EQUATION SOLVER FOR STRUCTURAL
ANALYSIS
J. COMPUT. AND STRUCT., VOL. 4, 1974, PP. 699-728.

74-0984
LARSEN, P. K.
POPOV, E. P.
NOTE ON INCREMENTAL EQUILIBRIUM EQUATIONS AND
APPROXIMATE CONSTITUTIVE RELATIONS IN LARGE
INELASTIC DEFORMATIONS
ACTA MECH., VOL. 19, NO. 1-2, 1974, PP. 1-14.

74-0985
RAI, I. S.
SANDHU, R. S.
FINITE ELEMENT ANALYSIS OF ANISOTROPIC PLATES
USING Q-19 ELEMENT - 1, 2
REPORT TR-74-120, US AIR FORCE SYST. COMMAND, AIR
FORCE FLIGHT DYN. LAB. TECH., PART 1, OCTOBER
1974, 108 P. PART 2, 63 P.

74-0986
SWARTZ, B.
WENDROFF, B.
RELATIVE EFFICIENCY OF FINITE DIFFERENCE AND
FINITE ELEMENT METHODS - 1. HYPERBOLIC PROBLEMS
AND SPLINES
SIAM J. NUMER. ANAL., VOL. 11, NO. 5, OCTOBER
1974, PP. 979-993.

74-0987
KOEHLER, W.
PITTR, J.
BERECHNUNG INSTATIONAERER TEMPERATUREFELDER MIT
FINITEN ELEMENTEN IN RAUM-UND ZEITDIMENSION.
(CALCULATION OF TRANSIENT TEMPERATURE FIELDS WITH
FINITE ELEMENTS IN SPACE AND TIME DIMENSIONS)
WAERME STOFFUEBERTRAG THERMO FLUID DYN., VOL. 7,
NO. 4, 1974, PP. 195-199.

74-0988 STRATONOVA, M. M.
TOLMACH, G. U.
RASCHET KILEBANII PLASTIN SLOZHNOI FORMY S
NEODNORODNYMI GRANICHNYMI USLOVIYAMI.
(CALCULATION OF VIBRATIONS OF PLATES OF COMPLEX
SHAPES WITH NONUNIFORM BOUNDARY CONDITIONS)
PROBL. PROCHN., VOL. 6, NO. 12, DECEMBER 1974, PP.
73-76.

74-0989 GOFFI, L.
BARLA, G.
DIRECT TENSILE TESTING OF ANISOTROPIC ROCKS
PROC. 3RD CONGR., INT. SOC. FOR ROCK MECH.,
DENVER, COLO., VOL. 2, PART A, SEPTEMBER 1-7,
1974, PP. 93-98.

74-0990 PELLEGRINO, A.
SURFACE FOOTINGS ON SOFT ROCKS
PROC. 3RD CONGR., INT. SOC. FOR ROCK MECH.,
DENVER, COLO., VOL. 2, PART B, SEPTEMBER 1-7,
1974, PP. 733-738.

74-0991 WALLNER, M.
WITTKE, W.
THEORIE UND EXPERIMENT ZU EINEM NEUEN
INJEKTIONSVERFAHREN. (THEORY AND EXPERIMENT OF A
NEW GROUTING PROCEDURE)
PROC. 3RD CONGR., INT. SOC. FOR ROCK MECH.,
DENVER, COLO., VOL. 2, PART B, SEPTEMBER 1-7,
1974, PP. 744-750.

74-0992 GOLDIN, A. L.
TROITSKY, A. P.
BEHAVIOUR OF THE ROCK MASS AT THE SITE OF A
ROCKFILL DAM
PROC. OF 3RD CONGR., INT. SOC. FOR ROCK MECH.,
DENVER, COLO., VOL. 2, PART B, SEPTEMBER 1-7,
1974, PP. 896-901.

74-0993 BARLA, G.
OTTOVIANI, M.
STRESSES AND DISPLACEMENTS AROUND TWO ADJACENT
CIRCULAR OPENINGS NEAR TO THE GROUND SURFACE
PROC. 3RD CONGR., INT. SOC. FOR ROCK MECH.,
DENVER, COLO., VOL. 2, PART B, SEPTEMBER 1-7,
1974, PP. 975-980.

74-0994 CHANG, C-Y.
NAIR, K.
DEVELOPMENT AND APPLICATIONS OF A GENERAL COMPUTER
PROGRAM FOR EVALUATING STABILITY OF OPENINGS IN
ROCK
PROC. 3RD CONGR., INT. SOC. FOR ROCK MECH.,
DENVER, COLO., VOL. 2, PART B, SEPTEMBER 1-7,
1974, PP. 981-989.

74-0995 DESCOEUDRES, F.
ANALYSE TRIDIMENSIONELLE DE LA STABILITE D'UN
TUNNEL AU VOISINAGE DU FRONT DE TAILLE DANS UNE
ROCHE ELASTO-PLASTIQUE. (THREE DIMENSIONAL
ANALYSIS OF TUNNEL STABILITY NEAR THE FACE IN AN
ELASTO-PLASTIC ROCK)
PROC. 3RD CONGR., INT. SOC. FOR ROCK MECH.,
DENVER, COLO., VOL. 2, PART B, SEPTEMBER 1-7,
1974, PP. 1130-1135.

74-0996 ORTLEPP, W. D.
FAILURE OF THE CONCRETE LINING IN AN INCLINED
SHAFT AT GREAT DEPTH.
PROC. 3RD CONGR., INT. SOC. FOR ROCK MECH.,
DENVER, COLO., VOL. 2, PART B, SEPTEMBER 1-7,
1974, PP. 1157-1162.

74-0997 SAGASETA, G.
ESCARIO, V.
INFLUENCE OF CONSTRUCTION PROCEDURES ON GROUND
PRESSURES AROUND A TUNNEL
PROC. OF 3RD CONGR., INT. SOC. FOR ROCK MECH.,
DENVER, COLO.VOL. 2, PART B, SEPTEMBER 1-7, 1974,
PP. 1318-1325.

74-0998 CLARK, G. B.
LEHNHOFF, T. F.
THERMAL-MECHANICAL FRAGMENTATION OF HARD ROCK -
FIELD TESTS
PROC. OF 3RD CONG., INT. SOC. FOR ROCK MECH.,
DENVER, COLO. VOL. 2, PART B, SEPTEMBER 1-7, 1974,
PP. 1428-1433.

74-0999 JONES, A. T.
FRACTURE TOUGHNESS TESTING WITH SECTIONS OF
CYLINDERS
ENG. FRACT. MECH., VOL. 6, NO. 4, DECEMBER 1974,
PP. 653-662.

74-1000 PAPAIOANNOU, S. G.
HILTON, P. D.
FINITE ELEMENT METHOD FOR CALCULATING STRESS
INTENSITY FACTORS AND ITS APPLICATION TO
COMPOSITES
ENG. FRACT. MECH., VOL. 6, NO. 4, DECEMBER 1974,
PP. 807-823.

74-1001 MERCKX, K. R.
CALCULATIONAL PROCEDURE FOR DETERMINING CREEP
COLLAPSE OF LWR FUEL RODS
NUCL. ENG. DES., VOL. 31, NO. 1, 1974, PP. 95-101.

74-1002 VENKATESWARA, R. G.
VENKATARAMANA, J.
VIBRATIONS OF THICK PLATES USING A HIGH PRECISION
TRIANGULAR ELEMENT
NUCL. ENG. DES., VOL. 31, NO. 1, 1974, PP.
102-105.

74-1003 LEVY, S.
WILKINSON, J. P. D.
THREE-DIMENSIONAL STUDY OF NUCLEAR FUEL ROD
BEHAVIOR DURING STARTUP
NUCL. ENG. DES., VOL. 29, NO. 2, 1974, PP.
157-166.

74-1004 PIAN, T. H. H.
NONLINEAR ANALYSIS BY ASSUMED STRESS HYBRID MODELS
24TH NATL. CONGRESS ON THEORETICAL AND APPL.
MECH., TOKYO, JAPAN, NOVEMBER 12, 1974.

74-1005 MACNEAL, R. H.
SOME ORGANIZATIONAL ASPECTS OF NASTRAN
NUCL. ENG. DES., VOL. 29, NO. 2, 1974, PP.
254-265.

74-1006 BATHE, K. J.
WILSON, E. L.
NONSAP - A NONLINEAR STRUCTURAL ANALYSIS PROGRAM
NUCL. ENG. DES., VOL. 29, NO. 2, 1974, PP.
266-293.

74-1007 CANTATORE, W. P.
VOLKER, R. E.
NUMERICAL SOLUTION OF THE STEADY INTERFACE IN A
CONFINED COASTAL AQUIFER
INSTITUTE ENG. AUST. CIV. ENG., VOL. 16, NO. 28,
1974, PP. 115-119.

74-1008 TEBEDGE, N.
ETUDES DES POUTRES ET POTEAUX PAR LA METHODE DES
ELEMENTS FINIS. (STABILITY ANALYSIS OF BEAMS AND
COLUMNS BY FINITE ELEMENT METHOD)
CONSTR. MET., VOL. 11, NO. 4, DECEMBER 1974, PP.
21-34.

74-1009 WURMNEST, W.
FESTIGKEITSUNTERSUCHUNG MIT DER METHODE DER
FINITEN ELEMENTE. (STRENGTH OF THE MATERIALS
INVESTIGATED BY MEANS OF THE FINITE ELEMENT
METHOD)
FOERDERN HEBEN, VOL. 24, NO. 17, DECEMBER 1974, P.
1632-1633.

74-1010 YOSHIDA, Y.
ON THE VARIATIONAL PRINCIPLES IN THE FINITE
ELEMENT METHOD
PROC. JAPAN SOC. CIV. ENG., NO. 232, DECEMBER
1974, PP. 25-36.

74-1011 LEE, H-P.
USING THE NASTRAN THERMAL ANALYZER TO SIMULATE A
FLIGHT SCIENTIFIC INSTRUMENT PACKAGE
PROC. 20TH ANNUAL MEETING, INST. OF ENVIRON. SCI.,
MT. PROSPECT, ILL, APRIL 28 - MAY 1, 1974, PP.
152-158.

74-1012 GRYCZMANSKI, M.
GENERATION OF SERENDIPEAN SHAPE FUNCTIONS IN
FINITE ELEMENT METHOD
BULLETIN DE L'ACADEMIE POL. DES SCI. SERIE
TECHNIQUES, VOL. 22, NO. 11, 1974, PP. 915-921.

74-1013 GRYCZMANSKI, M.
SCALAR ALGORITHM OF DETERMINATION OF STIFFNESS
MATRIX AND OF ANOTHER FINITE ELEMENT
CHARACTERISTIC
BULL. DE L'ACADEMIE POL. DES SCI. SERIE
TECHNIQUES, VOL. 22, NO. 11, PP. 909-914.

74-1014 KAGAWA, Y.
YAMABUCHI, T.
FINITE ELEMENT SIMULATION OF TWO-DIMENSIONAL
ELECTROMECHANICAL RESONATORS
IEEE TRANS. SONICS ULTRASON., VOL. SU-21, NO. 4,
OCTOBER 1974, PP. 275-283.

74-1015 SAW, C. B.
LINEAR ELASTIC FINITE ELEMENT ANALYSIS OF MASONRY
WALLS ON BUILDINGS
BUILD. SCI., VOL. 9, NO. 4, DECEMBER 1974, PP.
299-307.

74-1016 IMAMASA, J.
MIKI, T.
STRESS ANALYSIS PROGRAM BY FINITE ELEMENT METHOD
FOR RADIAL FLOW IMPELLERS
MITSUBISHI HEAVY IND. TECH. REV., VOL. 11, NO. 3,
1971, PP. 191-202.

74-1017 HUSTON, R. L.
PASSERELLO, C. E.
ON CONSTRAINT EQUATIONS - A NEW APPROACH
ASME J. APPLIED MECH., VOL. 41, NO. 4, DECEMBER
1974, PP. 1130-1131.

74-1018 BARNARD, A. J.
SHARMAN, P. W.
NONLINEAR RESPONSE OF PLATES TO IMPULSIVE LOADING
USING HYBRID FINITE ELEMENTS
PROC. CONF. DISCRETE METHODS IN ENG., MILAN,
SEPTEMBER 16-20, 1974.

74-1019 PAPANTONOPOULOS, C.
LADANYI, B.
ANALYSE DE LA STABILITY DES TALUS ROCHEUX PAR UNE
METHODE GENERALISEE DE L EQUILIBRE LIMITE (ROCK
SLOPE ANALYSIS BY A GENERALIZED LIMIT EQUILIBRIUM
METHOD)
PROC. 9TH SYMP. CAN. ROCK MECH., MONTREAL, QUE.,
DECEMBER 13-15, 1973, PP. 167-196. (PUBL. BY DEPT.
OF ENERGY MINES AND RESOURCES, MINES BRANCH,
OTTAWA, ONT., 1974)

74-1020 MEI, K. K.
UNIMOMENT METHOD OF SOLVING ANTENNA AND SCATTERING
PROBLEMS
IEEE TRANS. ANTENNAS PROPAG., NO. 6, NOVEMBER
1974, PP. 760-766.

74-1021 PARISEAU, W. G.
INFLUENCE OF ROCK PROPERTIES VARIABILITY ON MINE
OPENING STABILITY ANALYSIS
PROC. 9TH SYMP. CAN. ROCK MECH., MONTREAL, QUE.,
DEC. 13-15, 1973, PP. 141-165. PUBL. BY DEPT. OF
ENERGY, MINES AND RESOURCES, MINES BRANCH, OTTAWA,
ONT., 1974.

74-1022 CIARLET, P. G.
SUR L ELEMENT DE CLOUGH ET TOCHER (CLOUGH-TOCHER
ELEMENT)
REV. FR. AUTOM. INF. RECH. OPER., VOL. 8, AUGUST
1974, PP. 18-27.

74-1023 BRAMBLE, J. H.
THOMEE, V.
INTERIOR MAXIMUM NORM ESTIMATES FOR SOME SIMPLE
FINITE ELEMENT METHODS
REV. FR. AUTOM. INF. RECH. OPER., VOL. 8, AUGUST
1974, PP. 5-18.

74-1024 CHEN, H. S.
MEI, C. C.
OSCILLATIONS AND WAVE FORCES IN AN OFFSHORE HARBOR
- APPLICATIONS OF HYBRID FINITE ELEMENT METHOD TO
WATER-WAVE SCATTERING
MIT DEPT. CIV. ENG., RALPH M. PARSONS LAB. WATER
RESOURCE. HYDRODYN REPORT NO. 190, AUGUST 1974,
215 P.

74-1025 CHANG, S. K.
MEI, K. K.
SCATTERING BY DIELECTRIC CYLINDERS
IEE CONF., LONDON, ENGL., NO. 114, JULY 9-12,
1974, PP. 160-162.

74-1026 MURAKI, T.
BRYAN, J. J.
ANALYSIS OF THERMAL STRESSES AND METAL MOVEMENT
DURING WELDING 1. ANALYTICAL STUDY
PAPER NO. 74-WA/MAT-4, ASME MEETING, NOVEMBER
17-22, 1974, 4 PP.

74-1027 ARGYRIS, J. H.
DUNNE, P.C.
LARGE NATURAL STRAINS AND SOME SPECIAL
DIFFICULTIES DUE TO NONLINEARITY AND
INCOMPRESSIBILITY IN FINITE ELEMENTS
DTSCH LUFT RAUMFAHRT FORSCHUNGSBER, DLR-FB 74-62,
NO. 1, 1974, 79 PP.

74-1028 GERBERT, B. G.
PRESSURE DISTRIBUTION AND BELT DEFORMATION IN
V-BELT DRIVES
PAPER NO. 74-WA/DE-1, A.S.M.E. MEETING, NOVEMBER
17-22, 1974, 7 PP.

74-1029 CALZONA, R.
VESTRONI, F.
ANALISI NUMERICA DEL COMPORTAMENTO ELASTICO DI UNA
STRUTTURA BIDIMENSIONALE PIANA. (NUMERICAL
ANALYSIS OF THE ELASTIC BEHAVIOR OF A PLANE
TWO-DIMENSIONAL STRUCTURE)
G. GENIO CIV., VOL. 112, NO. 10-12, OCTOBER -
DECEMBER, 1974, PP. 387-403.

74-1030 KEVICZKY, L.
NONLINEAR STRUCTURES FOR SYSTEM IDENTIFICATION
PERIOD POLYTECH. ELECTR. ENG., VOL. 18, NO. 4,
1974, PP. 393-404.

74-1031 GOLAS, J.
KASPERSKI, Z.
ZASTOSOWANIE ITERACJI SEIDLA W METODZIE ELEMENTOW
SKONCZONYCCH NA PRZYKLADZIE OBLICZEN STATYCZNYCH
PLYT. (APPLICATION OF SEIDEL'S ITERATIONS FOR
FINITE ELEMENT METHOD IN EXAMPLE OF A STATICAL
CALCULATION OF PLATES)
MECH. TEOR I STODOW., VOL. 12, NO. 3, 1971, PP.
341-351.

74-1032 RAVINGER, J
DJUBEK, J.
GEOMETRICKY NELINEARNE ULOHY PLOCHYCH SKRUPIN
RIESENE PRIRASTKOVOU METODOU KONECNYCH PRVKOV.
(GEOMETRICALLY NONLINEAR PROBLEMS OF SHALLOW
SHELLS TREATED BY THE INCREMENTAL FINITE ELEMENT
METHOD)
STAVEBNICKY CAS., VOL. 22, NO. 8, 1974, PP.
488-498.

74-1033 NAPOLITANO, L. G.
MONTI, R.
PREPROCESSORS FOR GENERAL PURPOSE FINITE ELEMENT
PROGRAMS
PROC. SYMP. STRUCT. MECH. COMPUT. PROGRAMS, UNIV.
OF MD., COLLEGE PARK, JUNE 1974, PP. 807-823.
PUBL. BY UNIV. PRESS OF VA., CHARLOTTESVILLE,
1974.

74-1034 MIYOSHI, T.
SHIRATORI, M.
STUDY ON J//I//C FRACTURE CRITERION
PROC. SYMP. MECH. BEHAV. OF MATER., KYOTO, JAPAN,
VOL. 1, AUGUST 21-27, 1974, PP. 81-90.

74-1035 CHRISTIANSEN, H. N.
APPLICATIONS OF CONTINUOUS TONE COMPUTER-GENERATED
IMAGES IN STRUCTURAL MECHANICS
PROC. SYMP. STRUCT. MECH. COMPUT. PROGRAMS, UNIV.
OF MD., COLLEGE PARK, JUNE 1974, PP. 1003-1015.
PUBL. BY UNIV. PRESS OF VA., CHARLOTTESVILLE,
1974.

74-1036 SKALSKI, S. C.
OPTIM 11: (MAGIC) COMPATIBLE LARGE SCALE AUTOMATED
MINIMUM WEIGHT DESIGN PROGRAM - V 2. PROGRAMMER'S
MANUAL
REPORT AFFDL-TR-74-97, AIR FORCE FLIGHT DYN. LAB.
TECH., U.S. AIR FORCE SYST. COMMAND, JULY 1974, 84
PP.

74-1037 HWANG, J. H.
LORD, W.
FINITE ELEMENT ANALYSIS OF THE MAGNETIC FIELD
DISTRIBUTION INSIDE A ROTATING FERROMAGNETIC BAR
IEEE. TRANS. MAGN., VOL. MAG-10, NO. 4, DECEMBER
1974, PP. 1113-1118.

74-1038 SUGIYAMA, S.
NISHIMURA, M.
TRANSIENT TEMPERATURE RESPONSE OF COMPOSITE SLABS
INT. J. HEAT MASS. TRANSFER, VOL. 17, NO. 8,
AUGUST 1974, PP. 875-883.

74-1039 GURTIN, M. E.
RALSTON, T. D.
GALERKIN METHOD AS APPLIED TO PROBLEMS IN
VISCOELASTICITY
INT. J. SOLIDS STRUCT., VOL. 10, NO. 9, SEPTEMBER
1974, PP. 933-943.

74-1040 GARG, V. K.
ANAND, S. C.
ELASTIC-PLASTIC ANALYSIS OF A WHEEL ROLLING ON A
RIGID TRACK
INT. J. SOLIDS STRUCT., VOL. 10, NO. 9, SEPTEMBER
1974, PP. 945-956.

74-1041 FRIED, I.
FINITE ELEMENT ANALYSIS OF INCOMPRESSIBLE MATERIAL
BY RESIDUAL ENERGY BALANCING
INT. J. SOLIDS STRUCT., VOL. 10, NO. 9, SEPTEMBER
1974, PP. 993-1002.

74-1042 RISH, R. F.
ANALYSIS OF CYLINDRICAL SHELL ROOFS WITH POST
TENSIONED EDGEBEAMS
INT. J. SOLIDS STRUCT., VOL. 10, NO. 9, SEPTEMBER
1974, PP. 1035-1052.

74-1043 KRAJCINOVIC, D.
PRISMATIC SHELLS WITH HEXAGONAL CROSS-SECTION
INT. J. SOLIDS STRUCT., VOL. 10, NO. 10, OCTOBER
1974, PP. 1069-1089.

74-1044 BEVILACQUA, L.
FEIJOO, R.
VARIATIONAL PRINCIPLE FOR THE LAPLACE'S OPERATOR
WITH APPLICATION IN THE TORSION OF COMPOSITE RODS
INT. J. SOLIDS STRUCT., VOL. 10, NO. 10, OCTOBER
1974, PP. 1091-1102.

74-1045 WEGMULLER, A. W.
REFINED PLATE BENDING FINITE ELEMENT
INT. J. SOLIDS STRUCT., VOL. 10, NO. 11, NOVEMBER,
1974, PP. 1173-1178.

74-1046 COWDREY, D. R.
WILLIS, J. R.
VARIATION WITH TEMPERATURE OF RESONANT FREQUENCIES
OF ANISOTROPIC PLATES
J. ACOUST. SOC. AM., VOL. 56, NO. 4, OCTOBER 1974,
PP. 1153-1157.

74-1047 HITCHINGS, D.
DANCE, S. H.
RESPONSE OF NUCLEAR STRUCTURAL SYSTEMS TO
TRANSIENT AND RANDOM EXCITATIONS, USING BOTH
DETERMINISTIC AND PROBABILISTIC METHODS
NUCL. ENG. DES. VOL. 29, NO. 3, 1971, PP. 311-337.

74-1048 GELLATLY, R. A.
DUPREE, D. M.
OPTIM 11: A (MAGIC) COMPATIBLE LARGE SCALE
AUTOMATED MINIMUM WEIGHT DESIGN PROGRAM - 1.
ENGINEERS AND USERS MANUAL
REPORT NO: AFFDL-TR-74-97, AIR FORCE FLIGHT DYN.
LAB. TECH., U.S. AIR FORCE SYST. COMMAND, JULY
1974, 134 PP.

74-1049 SAITO, T.
STRESS ANALYSIS OF CONCRETE TRACK SLABS ON AN
ELASTIC FOUNDATION BY THE FINITE ELEMENT METHOD
RAILW. TECH. RES. INST., TOKYO, JPN., VOL. 15, NO.
4, DECEMBER 1974, PP. 186-190.

74-1050 GARTLING, D. K.
BECKER, E. B.
COMPUTATIONALLY EFFICIENT FINITE ELEMENT ANALYSIS
OF VISCOUS FLOW PROBLEMS
IN "COMPUTATIONAL METHODS IN NONLINEAR MECHANICS",
TICOM, J. T. ODEN ET AL (EDS.), 1974.

74-1051 JONES, R. E.
A SELF-ORGANIZING MESH GENERATION PROGRAM
TRANS. A.S.M.E., VOL. 96, SERIES J, AUGUST 1974,
PP. 193-199.

74-1052 HADJIAN, A. H.
NIEHOFF, D.
SIMPLIFIED SOIL-STRUCTURE INTERACTION ANALYSIS
WITH STRAIN DEPENDENT SOIL PROPERTIES
NUCL. ENG. DES., VOL. 31, NO. 2, 1974, PP.
218-233.

74-1053 MACCHI, A.
TEMPERATURES AND STRESSES IN A BOILER MEMBRANE
WALL TUBE
NUCL. ENG. DES., VOL. 31, NO. 2M, 1974, PP.
280-293.

74-1054 BETTESS, P.
A NOTE ON SOME VARIATIONAL STATEMENTS FOR THE SLOW
FLOW OF A NAVIER-POISSON FLUID
INT. J. NUMER. METH. IN ENG., VOL. 8, NO. 1, 1974,
PP. 17-26.

74-1055 DEB NATH, J. M.
USE OF HIGHER ORDER DISPLACEMENT FUNCTIONS IN THE
FREE VIBRATION ANALYSIS OF SHELLS OF REVOLUTION
HAVING MERIDIONAL SINGULARITIES
J. SOUND AND VIB., VOL. 36, NO. 2, SEPTEMBER 22,
1974, PP. 253-272.

74-1056 VENKATESWARA, R. G.
SUNDARARAMAIAH, V.
FINITE ELEMENT ANALYSIS OF VIBRATIONS OF INITIALLY
STRESSED THIN SHELLS OF REVOLUTION
J. SOUND AND VIB., VOL. 37, NO. 1, NOVEMBER 8,
1974, PP. 57-64.

74-1057 WU, J. J.
ON THE NUMERICAL CONVERGENCE OF MATRIX EIGENVALUE
PROBLEMS DUE TO CONSTRAINT CONDITIONS
J. SOUND AND VIB., VOL. 37, NO. 3, DECEMBER 8,
1974, PP. 349-358.

74-1058 KLEIN, L.
TRANSVERSE VIBRATIONS OF NON-UNIFORM BEAMS
J. SOUND AND VIB., VOL. 37, NO. 4, DECEMBER 22,
1974, PP. 491-505.

74-1059 BARTEL, L. C.
COMPUTER SIMULATION OF AN IN SITU SHALE-OIL RETORT
NTIS CONF. 741001-11, PAPER (SAND-74-5172), 1974,
8 P.

74-1060 KAMEL, H. A.
MCCABE, M. W.
GIFTS SYSTEM
PROC. SYMP. STRUCT. MECH. COMPUT. PROGRAMS, UNIV.
OF MD., COLLEGE PARK, JUNE 1974, PP. 969-990.
PUBL. BY UNIV. PRESS OF VA., CHARLOTTESVILLE,
1974.

74-1061 RATTI, U.
SU LA DETERMINAZIONE DEI CAMPI MAGNETICI LINEARI E
NON LINEARI MEDIANTE L APPLICAZIONE DEI METODI
DELLE DIFFERENZE FINITE (DETERMINATION OF LINEAR
AND NONLINEAR MAGNETIC FINITE DIFFERENCE METHODS)
ENERG. ELETTR., VOL. 51, NO. 8, AUGUST 1974, PP.
448-463.

74-1062 GHABOUSSI, J.
RANKEN, R. E.
TUNNEL DESIGN CONSIDERATIONS: ANALYSIS OF
MEDIUM-SUPPORT INTERACTION
U.S. FED. RAILROAD ADMIN., REPORT 75-24, NOVEMBER
1974, 84 PP.

74-1063 BABCOCK, C. D.
A NEW METHOD OF ANALYSIS TO OBTAIN EXACT SOLUTIONS
FOR STRESSES AND STRAINS IN CIRCULAR INCLUSIONS
DENVER MINING RESEARCH CENTER, BUREAU OF MINES,
DENVER, COLO., FINAL REPT. OF INVESTIGATIONS,
REPORT NO: BUMINES RI 7967, DECEMBER 1974, 45 PP.
(N.T.I.S. - PB-240 142/0ST)

74-1064 LABRUJERE, TH. E.
VAN DER VOOREN, J.
FINITE ELEMENT CALCULATION OF AXISYMMETRIC
SUBCRITICAL COMPRESSIBLE FLOW
NATL. AEROSPACE LAB., THE NETHERLANDS, NLR TR
74162 U, 1974, 21 PP.

74-1065 SALAAM, U.
FINITE ELEMENTS WITH RELAXED CONTINUITY - AN
INVESTIGATION OF THE VARIATIONAL BASIS AND
NUMERICAL PERFORMANCE
PH.D. THESIS, OHIO STATE UNIV., 1974.

74-1066 ROZENDAL, D. B.
FINITE ELEMENT ANALYSIS OF PRESTRESSED FOLDED
PLATE STRUCTURES
PH.D. THESIS, PURDUE UNIV., 1974.

74-1067 WOLF, J. P.
GENERALIZED STRESS MODELS FOR FINITE ELEMENT
ANALYSIS
PH.D. THESIS, SWISS FEDERAL INST. OF TECH.,
ZURICH, SWITZERLAND, 1974. ALSO INST. OF STRUCT.
ENG., BIRKHAUSER, BASEL, SWITZERLAND, REPORT NO.
52, 1974.

74-1068 SOGGE, R. L.
FINITE ELEMENT ANALYSIS OF ANCHORED BULKHEAD
BEHAVIOR
PH.D. THESIS, UNIV. OF ARIZONA, 1974.

74-1069 TANIGUCHI, S.
A THERMO-ELASTIC ANALYSIS OF THE PARALLEL SURFACE
THRUST BEARING
MITSUBISHI HEAVY-INDUSTRIES LTD., NAGASAKI, JAPAN,
REPORT NO: MTB-95, DECEMBER 1974, 16 PP. (N.T.I.S.
N75-16856/7ST)

74-1070 TSOU, H. M.
STRAIN DISTRIBUTION IN AN ALUMINA BICRYSTAL
STUDIES BY THE FINITE ELEMENT METHOD AND X-RAY
DIFFRACTION
PH.D. THESIS, PENNYSYLAVANIA STATE UNIV., 1974.

74-1071 BERKOVIC, M.
HYBRID FINITE ELEMENTS IN THE PLANE STRESS PROBLEM
12TH YUGOSLAV CONGR. OF RATIONAL AND APPL. MECH.,
OHRID, 1974. (IN SERBO-CROATION)

74-1072 VALLIAPPAN, P.
NONLINEAR STRESS DEFORMATION ANALYSES OF LAKE
AGASSIZ CLAYS USING FINITE ELEMENT METHOD
PH.D. THESIS, UNIV. OF MANITOBA, CANADA, 1974.

74-1073 VERMA, V. K.
FINITE ELEMENT ANALYSIS OF NONLINEAR CABLE
REINFORCED MAMBRANES
PH.D. THESIS, ILLINOIS INST. OF TECHNOLOGY, 1974.

74-1074 VIGNESWARAN, K.
THE FINITE-ELEMENT METHOD AS APPLIED TO SHALLOW
FOUNDATIONS
PH.D., UNIV. OF WATERLOO, CANADA, 1974.

74-1075 BENZLEY, S. E.
REPRESENTATION OF SINGULARITIES WITH ISOPARAMETRIC
FINITE ELEMENTS
INTL. J. NUMER. METH. IN ENG., VOL. 8, NO. 3,
1974, PP. 537-546.

74-1076 WARD, C. C.
A DYNAMIC FINITE ELEMENT MODEL OF THE HUMAN BRAIN
PH.D. THESIS, UNIV. OF CALIFORNIA, LOS ANGELES,
1974.

74-1077 WAGNER, P. J.
A FINITE ELEMENT DISPLACEMENT APPROACH FOR THE
ELASTO-PLASTIC ANALYSIS OF THIN CYLINDRICAL SHELLS
PH.D. THESIS, PURDUE UNIV., 1974.

74-1078 YANG, S. K.
CONDITION OF FINITE ELEMENT MATRICES
PH.D. THESIS, BOSTON UNIV. GRAD. SCHOOL, 1974.

74-1079 LARDER, R. A.
APPLICATION OF THREE-DIMENSIONAL FINITE ELEMENT
ANALYSES TO THE MICROMECHANICS OF FIBROUS
COMPOSITE MATERIALS
LAWRENCE LIVERMORE LAB., UNIV. OF CALIF.,
LIVERMORE, REPORT NO. CONF-7410001-43, OCTOBER 23,
1974, 6 PP. (N.T.I.S. - UCRL-76463) SEE ALSO NSA
3109, NO. 25178.

74-1080 KURIBAYA, E.
IIDA, Y.
APPLICATION OF FINITE ELEMENT METHOD TO
SOIL-FOUNDATION INTERACTION ANALYSES
5TH SYMP. ON EARTHQUAKE ENG., UNIV. OF ROORKEE,
INDIA, VOL. 1, NOVEMBER 9-11, 1974, PP. 151-159.
(PUBL. BY SARITA PRAKASAN, NAUCHANDI, INDIA, 1974)

74-1081 HORIKAMI, K.
NAKAHARA, Y.
FUJIMURA, T.
OHNISHI, T.
FINITE ELEMENT METHOD FOR SOLVING NEUTRON
TRANSPORT PROBLEMS IN TWO-DIMENSIONAL CYLINDRICAL
GEOMETRY
JAPAN ATOMIC ENERGY RESEARCH INST., TOKAI,
IBARAKI, JULY 1974, 48 PP. (N.T.I.S. -
JAERI-M-5793) SEE ALSO NSA 3110, NO. 27511.

74-1082 ANDERSON, C. A.
CAVITY STABILITY: FINITE ELEMENT ANALYSIS FOR
STEADY AND TRANSIENT CREEP
LOS ALAMOS SCIENTIFIC LAB., N. MEX., OCTOBER 1974,
20 PP. (N.T.I.S. - LA-5769) SEE ALSO NSA 3110, NO.
25497.

74-1083 BABUSKA, I.
AZIZ, A. K.
ANGLE CONDITION IN THE FINITE ELEMENT METHOD
TECHNICAL NOTE BN-808, COLLEGE PARK INST. FOR
FLUID DYNAMICS AND APPLIED MATH., MARYLAND UNIV.,
NOVEMBER 1974, 25 PP. (N.T.I.S. - ORO-3443-51)
ALSO NSA 3110, NO. 28894.

74-1084 BABUSKA, I.
METHOD OF WEAK ELEMENTS
TECHNICAL NOTE BN-809, COLLEGE PARK INST. FOR
FLUID DYNAMICS AND APPLIED MATH., MARYLAND UNIV.,
DECEMBER 1974, 68 PP. (N.T.I.S. - ORO-3443-52)
ALSO NSA 3110, NO. 28895.

74-1085 APPERT, K.
BERGER, D.
GRUBER, R.
TROYON, F.
ROBERTS, K. V.
THALIA: A ONE-DIMENSIONAL MAGNETOHYDRODYNAMIC
STABILITY PROGRAM USING THE METHOD OF FINITE
ELEMENTS
ECOLE POLYTECHNIQUE FEDERALE, LAUSANNE,
SWITZERLAND, OCTOBER 1974, 35 PP. (N.T.I.S. -
LRP-87-74) SEE ALSO NSA 3201, NO. 00302.

74-1086 DE WINDT, P.
REYNEN, J.
EURCYL: A PROGRAM TO GENERATE FINITE ELEMENT
MESHES FOR PRESSURE VESSEL NOZZLES
JOINT RESEARCH CENTRE, COMMISSION OF THE EUROPEAN
COMMUNITIES, ISPRA, ITALY, 1974, 40 PP. (N.T.I.S.
- EUR-5257E) SEE ALSO NSA 3202, NO. 05319.

74-1087 CLARKE, P. W.
CONIDA: A FINITE ELEMENT PROGRAM FOR THE STRESS
ANALYSIS OF AXISYMMETRIC THIN SHELLS
UKAEA REACTOR GROUP, RISLEY, DECEMBER 1974, 31 PP.
(N.T.I.S. -TRG-2382(R)) SEE ALSO NSA 3203, NO.
06239.

74-1088 CAREY, G. F.
PERTURBATION-VARIATIONAL SOLUTION OF COMPRESSIBLE
FLOWS
PROC. INT. SYMP. FINITE ELEMENT METHODS IN FLOW
PROBL., UNIV. OF SWANSEA, WALES, JANUARY 1974, PP.
243-248. (PUBL. BY UNIV. OF ALA., HUNTSVILLE,
1974)

74-1089 FRANKE, H. P.
SCHMIDT, F. A. R.
FEM2D: PROGRAM FOR SOLVING THE TWO-DIMENSIONAL
DIFFUSION EQUATION BY THE METHOD OF FINITE
ELEMENTS
JULY 1, 1974, 79 PP. (N.T.I.S. - ORNL-TR-2971) SEE
ALSO NSA 3207, NO. 19157.

74-1090 DONG, S. B.
GLOBAL-LOCAL FINITE ELEMENT STRUCTURAL ANALYSIS,
ANALYTICAL CONSIDERATIONS/ELEMENT FORMULATION
SCHOOL OF ENG. AND APPLIED SCI., CALIF. UNIV., LOS
ANGELES, REPORT NO: UCLA-ENG-7466, OCTOBER 1974,
53 PP. (N.T.I.S. - AD/A-003 936/2ST)

74-1091 COST, T. L.
WEEKS, G. E.
BASIC RESEARCH IN METHODS OF APPROXIMATION IN THE
NONLINEAR MECHANICS OF SOLIDS AND STRUCTURES.
VOLUME 111. TRANSIENT ANALYSES
UNIV. BUREAU OF ENG. RESEARCH, UNIV. OF ALABAMA,
REPORT NO: BER-184-07, OCTOBER 1974, 131 PP.
(N.T.I.S. - AD/A-004 146/7ST) SEE ALSO VOL 1,
AD/A-003 713.

74-1092 HILTON, P. D.
GIFFORD, L. N. JR.
LOMACKY, O.
FINITE ELEMENT FRACTURE MECHANICS ANALYSIS OF
TWO-DIMENSIONAL AND AXISYMMETRIC ELASTIC AND
ELASTIC-PLASTIC CRACKED STRUCTURES
NAVAL SHIP RESEARCH AND DEVELOPMENT CENTER,
BETHESDA, MD., REPORT NO: NSRDC-4493, NOVEMBER
1974, 119 PP. (N.T.I.S. - AD/A-005 880/0ST)

74-1093 SKALSKI, S. C.
OPTIM 11: MAGIC COMPATIBLE LARGE SCALE AUTOMATED
MINIMUM WEIGHT DESIGN PROGRAM. VOLUME 11.
PROGRAMMER'S MANUAL
BELL AEROSPACE CO., BUFFALO, N.Y., FINAL REPORT,
JUNE 1972 - APRIL 1974, REPORT NO: D2459-9500001,
JULY 1974, 93 PP. (N.T.I.S. - AD-A006 412/1ST)

74-1094 LINDEMAN, R. A.
FINITE ELEMENT COMPUTER PROGRAM FOR THE SOLUTION
OF NONLINEAR AXISYMMETRIC CONTACT PROBLEMS WITH
INTERFERENCE FITS
NAVAL WEAPONS LAB., DAHLGREN, VA., TECH. REPT. NO:
NWL-TR-3148, JUNE 1974, 59 PP. (N.T.I.S. - AD-920
158/3ST)

74-1095 ANON
DESIGN OPTIMIZATION IN UNDERGROUND COAL SYSTEMS
DIV. OF MINERALS ENG., VIRGINIA POLYTECHNIC INST.
AND STATE UNIV., BLACKSBURG, INTERIM REPT. NO. 1,
DECEMBER 1973 - JULY 1974, 89 PP. (N.T.I.S. -
PB-239 075/5ST) SEE ALSO GPO 163.10:91/INT.1.

74-1096 GELLATLY, R. A.
DUPREE, D. M.
BERKE, L.
OPTIM 11: A MAGIC COMPATIBLE LARGE-SCALE AUTOMATED
MINIMUM WEIGHT DESIGN PROGRAM. VOLUME 1. ENGINEERS
AND USERS MANUAL
BELL AEROSPACE CO., BUFFALO, N.Y., FINAL REPORT,
JUNE 1972 - APRIL 1974, REPORT NO: D2459-950001,
JULY 1974, 146 PP. (N.T.I.S. - AD/A-005 747/0ST)
SEE ALSO VOL. 2, AD/A-006 412.

74-1097 AKAY, H. U.
LINEAR BUCKLING AND GEOMETRICALLY NONLINEAR
ANALYSIS OF PLANAR PLATE-STIFFENER TYPE STRUCTURES
BY THE FINITE ELEMENT METHOD
PH.D. THESIS, UNIV. OF TEXAS, AUSTIN, 1974.

74-1098 ALLEN, V. D.
DIRECT AND INVERSE HEAT CONDUCTION IN ROCK
MATERIALS USING THE FINITE ELEMENT METHOD
PH.D. THESIS, UNIV. OF MISSOURI, ROLLA, 1974.

74-1099 APOSTAL, M. C.
DEVELOPMENT OF AN ANISOTROPIC SINGULARITY FINITE
ELEMENT UTILIZING THE HYBRID DISPLACEMENT METHOD
PH.D. THESIS, STATE UNIV. OF NEW YORK, BUFFALO,
1974.

74-1100 HUEBNER, K. H.
FINITE ELEMENT ANALYSIS OF FLUID FILM LUBRICATION
- A SURVEY
PROC. INT. SYMP. FINITE ELEMENT METHODS IN FLOW
PROBL., UNIV. OF SWANSEA, WALES, JANUARY 1974, PP.
343-350. (PUBL. BY UNIV. OF ALA., HUNTSVILLE,
1974)

74-1101 CHANDRASEKARAN, A. R.
BICKOVSKI, V.
PETROVSKI, J.
PASKALOV, T.
STATIC AND DYNAMIC ANALYSIS OF AN ARCH DAM
PROC. 5TH SYMP. ON EARTHQUAKE ENG., UNIV. OF
ROORKEE, INDIA, NOVEMBER 9-11, 1974, VOL. 1, PP.
47-54. (PUBL. BY SARITA PRAKASHAN, NAUCHANDI,
INDIA, 1974).

74-1102 ARIATHURAI, C. R.
A FINITE ELEMENT MODEL FOR SEDIMENT TRANSPORT IN
ESTUARIES
PH.D., UNIV. OF CALIF., DAVIS, 1974.

74-1103 AUGUSTUS, K. J.
A FINITE ELEMENT SOLUTION OF THE TRANSIENT
MAGNETOHYDRODYNAMIC COUETTE FLOW
D.SC. THESIS, NEW MEXICO STATE UNIV., 1974.

74-1104 ARRIYAVAT, P.
ANALYSIS OF NONLINEAR REINFORCED CONCRETE FRAME BY
FINITE ELEMENT METHOD
PH.D. THESIS, PURDUE UNIV., 1974.

74-1105 REIGSTAD, G. H.
FINITE ELEMENT ANALYSIS OF DRIVEN PILES
PH.D. THESIS, UNIV. OF MINNESOTA, 1974.

74-1106 AGRAWAL, P. K.
GUPTAK, D. C.
KUMAR, V.
FINITE ELEMENT ANALYSIS FOR THE SEISMIC STABILITY
OF EARTH STRUCTURES
5TH SYMP. ON EARTHQUAKE ENG., UNIV. OF ROORKEE,
INDIA, VOL. 1, NOVEMBER 9-11, 1974, PP. 189-198.
(PUBL. BY SARITA PRAKASHAN, NAUCHANDI, INDIA,
1974)

74-1107 BAHGAT, B. M.
GENERAL FINITE ELEMENT VIBRATIONAL ANALYSIS OF
PLANAR MECHANISMS
PH.D. THESIS, CLARKSON COLLEGE OF TECH., 1974.

74-1108 ANON
BULLETIN DE L ASSOCIATION TECHNIQUE MARITIME ET
AERONAUTIQUE (BULLETIN OF THE TECHNICAL MARITIME
AND AERONAUTICS)
BULL. ASSOC. TECH. MARIT. AERONAUT., PARIS, NO.
74, MAY 6-10, 1974, 755 PP.

74-1109 PHU, N. D.
A FINITE ELEMENT METHOD FOR THREE DIMENSIONAL
ANALYSIS OF METAL FORMING
PH.D. THESIS, UNIV. OF MISSOURI, ROLLA, 1974.

74-1110 PALKA, L. C.
INVESTIGATION OF VENEER CUTTING BY ELASTIC FINITE
ELEMENT MODELS
PH.D. THESIS, UNIV. OF CALIF., BERKELEY, 1974.

74-1111 OLIVER, W. A.
FINITE ELEMENT ANALYSIS OF THE PIEZOELECTRIC
TOPOGRAPHIC WAVEGUIDE
PH.D. THESIS, UNIV. OF CALIF., LOS ANGELES, 1974.

74-1112 NEWMAN, J. C. R.
FINITE ELEMENT ANALYSIS OF FATIGUE CRACK
PROPAGATION - INCLUDING THE EFFECTS OF CRACK
CLOSURE
PH.D. THESIS, VIRGINIA POLYTECHNIC INST. AND STATE
UNIV., 1974.

74-1113 MURNEN, G. J.
ANALYSIS AND DESIGN OF AXISYMMETRIC REINFORCEMENT
AROUND A CIRCULAR HOLE IN A THIN FLAT PLATE BY THE
FINITE ELEMENT METHOD
PH.D. THESIS, UNIV. OF NOTRE DAME, 1974.

74-1114 MORRIS, W. F.
A FINITE ELEMENT APPROACH TO THE DETERMINATION OF
THE DYNAMIC RESPONSE OF BEAMS SUPPORTING MOVING
MASSES
PH.D. THESIS, VIRGINIA POLYTECHNIC INST. AND STATE
UNIV., 1974.

74-1115 MORGAN, J. D.
DYNAMICS OF CRACKED STRUCTURES USING FINITE
ELEMENTS
PH.D. THESIS, GEORGIA INST. OF TECH., 1974.

74-1116 MANG, H. A.
ANALYSIS OF DOUBLE CORRUGATED SHELL STRUCTURES BY
THE FINITE ELEMENT METHOD
PH.D. THESIS, TEXAS TECH. UNIV., 1974.

74-1117 MA, P. S.
APPLICATION OF THE FINITE ELEMENT METHOD TO
PROBLEMS OF TEMPERATURE DISTRIBUTION ANALYSIS FOR
BODIES OF WATER
PH.D. THESIS, CORNELL UNIV., 1974.

74-1118 LEE, K. N.
FINITE ELEMENT METHODS IN ELASTICITY
PH.D. THESIS, NORTHWESTERN UNIV., 1974.

74-1119 LASKARIS, T. E.
FINITE ELEMENT ANALYSIS OF SEVERAL COMPRESSIBLE
AND INCOMPRESSIBLE VISCOUS FLOW PROBLEMS
PH.D. THESIS, RENSSELAER POLYTECHNIC INST., 1974.

74-1120 LUEHR, H-P.
ORGANIZATION TECHNIQUES OF THE FINITE ELEMENT
METHOD FOR REDUCING CORE STORAGE REQUIREMENTS
PROC. INT. SYMP. FINITE ELEMENT METHODS IN FLOW
PROBL., UNIV. OF SWANSEA, WALES, JANUARY 1974, PP.
695-697. (PUBL. BY UNIV. OF ALA., HUNTSVILLE,
1974)

74-1121 KULAK, R. F.
A STUDY OF INTERVERTEBRAL DISC MECHANICS BY THE
FINITE ELEMENT METHOD
PH.D. THESIS, UNIV. OF ILLINOIS AT CHICAGO CIRCLE,
1974.

74-1122 KAMAL, S. A.
ANALYSIS OF MULTI-AXIAL ANISOTROPIC CREEP, ROOF
RESIN BOLTS, AND BED SEPARATION IN ROCK STRUCTURES
USING THE FINITE ELEMENT METHOD
PH.D. THESIS, MICHIGAN TECH. UNIV., 1974.

74-1123 KANG, S. I.
ON THE DETERMINATION OF EFFECTIVE MODULI OF
COMPOSITE MATERIALS BY A THREE-DIMENSIONAL FINITE
ELEMENT METHOD
PH.D. THESIS, GEORGIA INST. OF TECH., 1974.

74-1124 HOSSAIN, Q. A.
NONLINEAR FINITE ELEMENT ANALYSIS OF REINFORCED
CONCRETE FRAMES
PH.D. THESIS, UNIV. OF CALIF., DAVIS, 1974.

74-1125 HOUSTIS, E. N.
FINITE ELEMENT METHODS FOR SOLVING INITIAL
BOUNDARY VALUE PROBLEMS
PH.D. THESIS, PURDUE UNIV., 1974.

74-1126 HATT, F.
THREE-DIMENSIONAL FINITE ELEMENT ANALYSIS
INCLUDING ELASTIC-PLASTIC MATERIAL RESPONSE
PH.D. THESIS, VIRGINIA POLYTECHNIC INST. AND STATE
UNIV., 1974.

74-1127 HARTMAN, J. P.
FINITE ELEMENT PARAMETRIC STUDY OF VERTICAL STRAIN
INFLUENCE FACTORS AND THE PRESSUREMETER TEST TO
ESTIMATE THE SETTLEMENT OF FOOTINGS IN SAND
PH.D THESIS, UNIV. OF FLORIDA, 1974.

74-1128 HARDY, R. H.
A HIGH-ORDER FINITE ELEMENT FOR TWO-DIMENSIONAL
CRACK PROBLEMS
PH.D. THESIS, GEORGIA INST. OF TECH., 1974.

74-1129 VENKATASUBBU, S.
DESROCHERS, P.
THERMAL SLEEVE DESIGN FOR PRESSURE VESSELS
TRANS. CAN. SOC. MECH. ENG., VOL. 2, NO. 4, 1974,
PP. 229-234.

74-1130 BEAVERS, J. E.
BEAUFAIT, F. W.
HIGH-ORDER FINITE ELEMENT TECHNIQUE FOR PRISMATIC
AND NONPRISMATIC FOLDED PLATE STRUCTURES
PROC. SYMP. ON FOLDED PLATES AND SPAT PANEL
STRUCT., UDINE, ITALY, SEPTEMBER 23-27, 1974, 18
PP. (PUBL. BY INT. CENT. FOR MECH. SCI., UDINE,
ITALY, 1974)

74-1131 KOSTEM, C. N.
ANALYTICAL MODELING OF BEAM-SLAB BRIDGES
PROC. SYMP. ON FOLDED PLATES AND SPAT PANEL
STRUCT., UDINE, ITALY, SEPTEMBER 23-27, 1974, 11
PP. (PUBL. BY INT. CENT. FOR MECH. SCI., UDINE,
ITALY, 1974)

74-1132 HA, K. H.
FINITE ELEMENT ANALYSIS OF SPATIAL SANDWICH PLATE
STRUCTURES
PROC. SYMP. ON FOLDED PLATES AND SPAT PANEL
STRUCT., UDINE, ITALY, SEPTEMBER 23-27, 1974, 18
PP. (PUBL. BY INT. CENT. FOR MECH. SCI., UDINE,
ITALY, 1974)

74-1133 KNAPP, R. H.
NUMERICAL AND EXPERIMENTAL ANALYSIS OF A
PSEUDOCYLINDRICAL SHELL
PROC. SYMP. ON FOLDED PLATES AND SPAT PANEL
STRUCT., UDINE, ITALY, SEPTEMBER 23-27, 1974, 15
P. (PUBL. BY INT. CENT. FOR MECH. SCI., UDINE,
ITALY, 1974)

74-1134 SCORDELIS, A. C.
ANALYTICAL AND EXPERIMENTAL STUDIES OF MULTI-CELL
CONCRETE BOX GIRDER BRIDGES
PROC. SYMP. ON FOLDED PLATES AND SPAT PANEL
STRUCT., UDINE, ITALY, SEPTEMBER 23-27, 1974, 19
PP. (PUBL. BY INT. CENT. FOR MECH. SCI., UDINE,
ITALY, 1974)

74-1135 FISCHER, F. D.
ZUR LOESUNG DES KONTAKTPROBLEMS ELASTISCHER
KOERPER MIT AUSGEDEHNTER KONTAKTFLAECHE DURCH
QUADRATISCHE PROGRAMMIERUNG (SOLUTION OF THE
CONTACT PROBLEM OF ELASTIC BODIES WITH EXTENDED
CONTACT AREAS BY QUADRATIC PROGRAMMING)
COMPUT. (VIENNA/N.Y.), VOL. 13, NO. 3-4, 1974, PP.
353-384.

74-1136 BUGROV, A. K.
SOLUTION OF THE MIXED PROBLEM OF THE THEORY OF
ELASTICITY AND THE THEORY OF PLASTICITY OF SOILS
SOIL MECH. FOUND. ENG., VOL. 11, NO. 6,
NOVEMBER-DECEMBER 1974, PP. 392-398.

74-1137 SUBE, R. M.
CHATTOT, J. J.
COUPLAGE ENTRE L ECOULEMENT AUTOUR D UN
ARRIERE-CORPS ET LE JET PROPULSIF EN THEORIE DE
FLUIDE PARFAIT (COUPLING BETWEEN THE FLOW AROUND
AFTERBODIES AND PROPULSIVE JETS IN THE INVISCID
FLOW THEORY)
PROC. CONF. FOR AGARD MEET., ROME, ITALY, PAPER 8,
SEPTEMBER 3-6, 1974, 12 PP.

74-1138 BJERTNAES, L. T.
TVEIT, A.
MEASUREMENT OF THE WATER DIFFUSIVITY OF POROUS
MATERIALS
2ND SYMP. ON MOIST PROBL. IN BUILD., ROTTERDAM,
NETH., PAPER 5. 2. 3, SEPTEMBER 10-12, 1974, 9 PP.
(SPONSORED BY ORGAN. FOR APPL. SCI. RES. (TNO),
THE HAGUE, NETH., 1974)

74-1139 FOSTER, E. P.
EXPERIMENTAL AND FINITE ELEMENT ANALYSIS OF CABLE
ROOF STRUCTURES INCLUDING PRECAST PANELS
PH.D. THESIS, VANDERBILT UNIV., 1974.

74-1140 FOSTER, W. A.
A METHOD FOR THE TRANSIENT NONLINEAR COUPLED
THERMOELASTIC ANALYSIS OF AXISYMMETRIC BODIES
USING FINITE ELEMENTS
PH.D. THESIS, AUBURN UNIV., 1974.

74-1141 PALEKAR, M. G.
NUMERICAL SOLUTION OF 2-POINT BOUNDARY VALUE
PROBLEMS
INT. J. COMPUT. MATH., VOL. 4, NO. 2, 1974, PP.
191-195.

74-1142 MCEVILY, A. J. JR.
ON THE ROLE OF DEFECTS IN CRACK INITIATION IN
WELDED STRUCTURES
PROC. JPN-US SEMINAR ON SIGNIFICANCE OF DEFECTS IN
WELDED STRUCT., TOKYO, OCTOBER 15-19, 1973, PP.
96-104. (PUBL. BY UNIV. OF TOKYO PRESS, 1974)

74-1143 ELTAWILA, F. M.
AN OPTIMUM IDEALIZATION IN FINITE ELEMENT METHOD
FOR A TWO-GROUP NEUTRON DIFFUSION PROBLEM
PH.D. THESIS, VIRGINIA POLYTECHNIC INST. AND STATE
UNIV., 1974.

74-1144 MIYAMOTO, H.
MIYOSHI, T.
ANALYSIS OF CRACK PROPAGATION IN WELDED STRUCTURES
PROC. JPN-US SEMINAR ON SIGNIFICANCE OF DEFECTS IN
WELDED STRUCT., TOKYO, OCTOBER 15-19, 1973, PP.
189-202. (PUBL. BY UNIV. OF TOKYO PRESS, 1974.)

74-1145 CHANG, C. Y.
NAIR, K.
ANALYTICAL METHODS FOR PREDICTING SUBSIDENCE ABOVE
SOLUTION-MINED CAVITIES
4TH PROC. SYMP. ON SALT, HOUSTON, TEX., VOL. 2,
APRIL 8-12, 1973, PP. 101-117. (PUBL. BY NORTH
OHIO GEOL. SOC., CLEVELAND, 1974)

74-1146 PRIVITZER, E.
WIDERA, O.
SOME FACTORS AFFECTING DENTAL IMPLANT DESIGN
6TH ANNUAL SYMP. FOR INT. BIOMATER, CLEMSON UNIV.,
APRIL 20-24, 1974, PP. 251-255.

74-1147 HAWS, E. T.
LIPPARD, D. C.
FOUNDATION INSTRUMENTATION FOR THE NATIONAL
WESTMINISTER BANK TOWER
PROC. SYMP. ON FIELD INSTRUM. IN GEOTECH. ENG.,
LONDON, ENGLAND, MAY 30 - JUNE 1, 1973, PP.
180-193. (PUBL. BY HALSTED PRESS, DIV. OF JOHN
WILEY AND SONS, N.Y.)

74-1148 DWYER, M. G.
THOMS, R. L.
FINITE ELEMENT ANALYSES OF SALT DOMES WITH STORED
HOT WASTES
PROC. 4TH SYMP. ON SALT, HOUSTON, TEX., VOL. 2,
APRIL 8-12, 1973, PP. 343-348. (PUBL. BY NORTH
OHIO GEOL. SOC., CLEVELAND, 1974)

74-1149 NAIR, K.
CHANG, C. Y.
TIME-DEPENDENT ANALYSIS TO PREDICT CLOSURE IN SALT
CAVITIES
PROC. 4TH SYMP. ON SALT, HOUSTON, TEX., VOL. 2,
APRIL 8-12, 1973, PP. 129-139. (PUBL. BY NORTH
OHIO GEOL. SOC., CLEVELAND, 1974)

74-1150 CHAO, R.
LONG TERM CREEP CLOSURE OF SOLUTION CAVITY SYSTEM
PROC. 4TH SYMP. ON SALT, HOUSTON, TEX., VOL. 2,
APRIL 8-12, 1973, PP 119-127. (PUBL. BY NORTH
OHIO GEOL. SOC., CLEVELAND, 1974)

74-1151 DANA, J. R.
THREE-DIMENSIONAL FINITE ELEMENT ANALYSIS OF THICK
LAMINATED COMPOSITES - INCLUDING INTERLAMINAR AND
BOUNDARY EFFECTS NEAR CIRCULAR HOLES
PH.D. THESIS, VIRGINIA POLYTECHNIC INST. AND STATE
UNIV., 1974.

74-1152 HOOPER, J. A.
ANALYSIS OF A CIRCULAR RAFT IN ADHESIVE CONTACT
WITH A THICK ELASTIC LAYER
GEOTECHNIQUE, VOL. 24, NO. 4, DECEMBER 1974, PP.
561-580.

74-1153 MOHITPOUR, M.
LENGYEL, B.
TEMPERATURE RISE IN THE HIGH SPEED COMPRESSION OF
RIGHT CYLINDRICAL BILLETS
PROC. 2ND CONF. NORTH AM. METALWORK RES., MADISON,
WIS., MAY 20-22, 1974, PP. 48-59. (PUBL. BY SOC.
OF MANUF. ENG., DEARBORN, MICH. 1974)

74-1154 GOPDON, J. L.
WEINSTEIN, A. S.
FINITE ELEMENT ANALYSIS OF THE PLANE STRAIN
DRAWING PROBLEM
PROC. 2ND CONF. NORTH AM. METALWORK RES., MADISON,
WIS., MAY 20-22, 1974, PP. 194-208. (PUBL. BY SOC.
OF MANUF. ENG., DEARBORN, MICH. 1974)

74-1155 BERKOWITZ, L.
KUHN, H. A.
SIMPLIFIED MINIMUM INTERNAL ENERGY OF DEFORMATION
POSTULATE AND ITS APPLICATION TO METALFORMING
PROCESSES
PROC. 2ND CONF. ON NORTH AM. METALWORK RES.,
MADISON, WIS. MAY 20-22, 1974, PP. 336-349.
(PUBL. BY SOC. OF MANUF. ENG., DEARBORN, MICH.)

74-1156 RAMSDEN, V. S.
APPLICATION OF THE FINITE ELEMENT METHOD TO EDDY
CURRENT PROBLEMS
PROC. CONF. ON COMPUT. IN ENG., SYDNEY, AUST., MAY
16-17, 1974, PP. 102-106. (PUBL. BY INST. OF ENG.,
SYDNEY, AUST., 1974)

74-1157 HARRISON, H. B.
NON-LINEAR ELASTIC ANALYSIS OF UNGUYED TOWERS AND
STACKS
PROC. CONF. ON COMPUT. IN ENG., SYDNEY, AUST., MAY
16-17, 1974, PP. 178-182. (PUBL. BY INST. OF
ENG., SYDNEY, AUST.)

74-1158 LYALL, J. S.
MODELLING HEAT FLOW FROM UNDERGROUND POWER CABLES
PROC. CONF. ON COMPUT. IN ENG., SYDNEY, AUST., MAY
16-17, 1974, PP. 107-110. (PUBL. BY INST. OF ENG.,
SYDNEY)

74-1159 GILL, J. I.
SHELLS, PLATES AND MEMBRANES - A COMPUTER-AIDED
DESIGN SYSTEM
PROC. CONF. ON COMPUT. IN ENG., SYDNEY, AUST., MAY
16-17, 1974, PP. 183-187. (PUBL. BY INST. OF
ENG., SYDNEY)

74-1160 VALLIAPPAN, S.
WILSON, M.
USE OF COMPUTERS IN WELDING TECHNOLOGY
PROC. CONF. ON COMPUT. IN ENG., SYDNEY, AUST., MAY
16-17, 1974, PP. 188-192. (PUBL. BY INST. OF
ENG., SYDNEY)

74-1161 YUNG, C.
FRIDLEY, R. B.
COMPUTER SIMULATION OF VIBRATION OF WHOLE TREE
SYSTEMS USING FINITE ELEMENT METHODS
67TH ASAE ANNU. MEET., OKLA STATE UNIV.,
STILLWATER, JUNE 23-26, 1974, AND WINTER MEET.,
CHICAGO, ILL., DECEMBER 10-13, 1974. PAPER
74-1034, 30 PP. (PUBL. BY ASAE, ST. JOSEPH,
MICH.,)

74-1162 YUNG, C.
FRIDLEY, R. B.
COMPUTER ANALYSIS OF FRUIT DETACHMENT DURING TREE
SHAKING
67TH A.S.A.E. ANN. MEETING, OKL. STATE UNIV.,
STILLWATER, JUNE 23-26, 1974, AND PAPER NO.
74-3009, WINTER MEETING, CHICAGO, ILL., DECEMBER
10-13, 1974, 30 PP. (PUBL. BY A.S.A.E., ST.
JOSEPH, MICH.).

74-1163 CHU, S-C.
LEECH, W. J.
DISCRETE VARIABLE METHOD APPLIED TO TRANSIENT HEAT
CONDUCTION PROBLEMS
PROC. 5TH. CONF., INT. HEAT TRANSFER, TOKYO, JPN.,
PAPER CU2-4, VOL. 1, SEPTEMBER 3-7, 1974, PP.
184-187. (PUBL. BY JPN. SOC. OF MECH. ENG., TOKYO)

74-1164 HALTEMAN, E. K.
GERRISH, R. W.
PERIODIC HEAT CONDUCTION IN A TWO PHASE, TWO
DIMENSIONAL SOLID DOMAIN
PROC. 5TH CONF. INT. HEAT TRANSFER, TOKYO, JAPAN,
SEPTEMBER 3-7, 1974, PAPER CU3-2, VOL. 1, PP.
216-219. (PUBL. BY JPN. SOC. OF MECH. ENG.,
TOKYO).

74-1165 BAINBRIDGE, C. A.
SMEDLEY, G. P.
DESIGN APPRAISAL OF OFFSHORE PLATFORMS OF WELDED
TUBULAR CONSTRUCTION
MET. CONSTR. BR. WELD., VOL. 6, NO. 11, NOVEMBER
1974, PP. 337-341.

74-1166 CORP, E. L.
A FINITE ELEMENT MODEL FOR STABILITY ANALYSIS OF
MINE WASTE EMBANKMENTS SUBJECTED TO SEEPAGE
PH.D. THESIS, UNIV. OF IDAHO, 1974.

74-1167 CHAKRABARTI, S.
APPROXIMATIONS IN FINITE-ELEMENT HEAT CONDUCTION
ANALYSIS
PH.D., UNIV. OF PITTSBURGH, 1974.

74-1168 PARMELEE, R. A.
COROTIS, R. B.
ANALYTICAL AND EXPERIMENTAL EVALUATION OF MODULUS
OF SOIL REACTION
TRANSP. RES. BRD., NO. 518, 1974, PP. 29-38.

74-1169 FERREIRA, S.
TRANSIENT NATURAL CONVECTION COOLING OF A VERTICAL
CIRCULAR CYLINDER
NUCL. ENG. DES., VOL. 31, NO. 3, 1974, PP.
346-350.

74-1170　LAURA, P. A. A.
REYES, J. A.
COMPARISON OF ANALYTICAL AND NUMERICAL SOLUTIONS
IN HEAT CONDUCTION PROBLEMS
NUCL. ENG. DES., VOL. 31, NO. 3, 1974, PP.
379-382.

74-1171　BERGER, H.
OHAYON, R.
UNE METHODE DE CALCUL PAR ELEMENTS FINIS DES
MOUVEMENTS DE LIQUIDES DANS DES RESERVOIRS RIGIDES
OU DEFORMABLES
BULL. ASSOC. TECH. MARIT. AERONAUT., PARIS,
FRANCE, MAY 6-10, 1974, PP. 241-262.

74-1172　TANIGUCHI, S.
THERMO-ELASTIC ANALYSIS OF THE PARALLEL SURFACE
THRUST BEARING
MITSUBISHI HEAVY IND., TECH. BULL. NO. 95,
DECEMBER 1974, 15 PP.

74-1173　DALY, P.
PROPAGATION IN ELLIPTICAL AND PARABOLIC WAVEGUIDES
5TH PROC. COLLOQ. ON MICROWAVE COMMUN., BUDAPEST,
HUNGARY, VOL. 3, JUNE 24-30, 1974, PP. 65-74.
(PUBL. BY AKAD KIADO, BUDAPEST, HUNGARY, 1974)

74-1174　CHOU, Y. T.
HUTCHINSON, R. L.
DESIGN METHOD FOR FLEXIBLE AIRFIELD PAVEMENTS
TRANSP. RES. BOARD, NO. 521, 1974, PP. 1-13.

74-1175　OWEN, M. J.
JOHNSON, K.
STRESS ANALYSIS OF LAMINATED CFRP PLATES
CONTAINING A HOLE
PROC. 2ND INT. CONF. ON CARBON FIBRES, LONDON,
ENGLAND, FEBRUARY 18-20, 1974, PAPER NO. 30, 11
PP. (PUBL. BY PLAST. INST., LONDON).

74-1176　SHARPE, W. N. JR.
MUHA, T. J. JR.
COMPARISON OF THEORETICAL AND EXPERIMENTAL SHEAR
STRESS IN THE ADHESIVE LAYER OF A LAP JOINT MODEL
PROC. ARMY SYMP. ON SOLID MECH., BASS RIVER,
MASS., SEPTEMBER 10-12, 1974, PP. 23-44. (PUBL.
BY ARMY MATER. AND MECH., WATERTOWN, MASS.)

74-1177　GHADIALI, N D.
HOPPER, A. T.
ELASTIC-PLASTIC ANALYSIS OF INTERFERENCE FIT
FASTENERS
PROC. ARMY SYMP. ON SOLID MECH., BASS RIVER,
MASS., SEPTEMBER 10-12, 1974, PP. 61-74. (PUBL. BY
ARMY MATER AND MECH., WATERTOWN, MASS.)

74-1178　ADAMS, F. D.
GRIFFITH, W. I.
INTERFERENCE-FIT FASTENER DISPLACEMENT MEASUREMENT
BY SPECKLE PHOTOGRAPHY
PROC. ARMY SYMP. ON SOLID MECH., BASS RIVER,
MASS., SEPTEMBER 10-12, 1974, PP. 75-98. (PUBL.
BY ARMY MATER. AND MECH., WATERTOWN, MASS.)

74-1179　O'HARA, P.
FINITE ELEMENT ANALYSIS OF THREADED CONNECTIONS
PROC. ARMY SYMP. ON SOLID MECH., BASS RIVER,
MASS., SEPTEMBER 10-12, 1974, PP. 99-119. (PUBL.
BY ARMY MATER. AND MECH. WATERTOWN, MASS.)

74-1180　GOLD, L. M.
STOWELL, N. A.
FINITE ELEMENT ANALYSIS OF A MULTI-COMPONENT
KINETIC ENERGY PROJECTILE
PROC. ARMY SYMP. ON SOLID MECH., BASS RIVER,
MASS., SEPTEMBER 10-12, 1974, PP. 329-342. (PUBL.
BY ARMY MATER. AND MECH., WATERTOWN, MASS.)

74-1181　VOLKER, R. E.
NUMERICAL DETERMINATION OF SEA WATER INTRUSION IN
AQUIFERS
PROC. 5TH AUSTRALAS CONF. ON HYDRAUL. AND FLUID
MECH., UNIV., OF CANTERBURY, CHRISTCHURCH, N.Z.,
VOL. 2, DECEMBER 9-13, 1974, PP. 171-178.

74-1182　DUDGEON, C. R.
KESHAVARZ, M. H.
OPTIMAL DESIGN OF WATER WELLS IN UNCONSOLIDATED
SEDIMENTS
PROC. 5TH AUSTRALAS CONF. ON HYDRAUL. AND FLUID
MECH., UNIV. OF CANTERBURY, CHRISTCHURCH, N.Z.,
VOL. 2, DECEMBER 9-13, 1974, PP. 195-201.

74-1183　DE VRIES, G.
BALASUBRAMANIAN, R.
NORRIE, D. H.
APPLICATION OF THE PSEUDO-FUNCTIONAL FINITE
ELEMENT METHOD TO VISCOUS FLOW PROBLEMS
PROC. 5TH AUSTR. CONF. ON HYDRAUL. AND FLUID
MECH., UNIV. OF CANTERBURY, CHRISTCHURCH, N.Z.,
DECEMBER 9-13, 1974, VOL. 2, PP. 561-571.

74-1184　STEVEN, G. P.
SOLUTION OF FLUID DYNAMIC PROBLEMS BY FINTE
ELEMENTS
PROC. 5TH AUSTR. CONF. ON HYDRAUL. AND FLUID
MECH., UNIV. OF CANTERBURY, CHRISTCHURCH, N.Z.,
DECEMBER 9-13, 1974, VOL. 2, PP. 572-279.

74-1185　MIHALCEA, A.
MOROIANU, A.
COMPORTAREA ELASTICA A DIGURILOR DE PAMINT
SOLICIATE SEISMIC (ELASTIC BEHAVIOR OF THE EARTH
DAMS SEISMICALLY STRESSED)
STUD. GEOTEH. FUND CONSTRU. HIDROTEH., VOL. 18,
1974, PP. 99-146.

74-1186　MASON, P. W.
HARRIS, H. G.
ZALESAK, J.
BERNSTEIN, M.
ANALYTICAL AND EXPERIMENTAL INVESTIGATION OF A 1/8
SCALE DYNAMIC MODEL OF THE SHUTTLE ORBITER. VOLUME
1: SUMMARY REPORT
GRUMMAN AEROSPACE CORP., BETHPAGE, N.Y., REPORT
NO.: NASA-CR-132488, MAY 1974, 47 PP. (N.T.I.S. -
N74-15680/2ST)

74-1187　TRAUBENIK, M.
SILVER, M. L.
BELYTSCHKO, T. B.
STATIC THREE DIMENSIONAL ANALYSIS OF ELEVATED
STEEL TRANSPORTATION STRUCTURES
DEPT. OF MATERIALS ENG., UNIV. OF ILLINOIS,
CHICAGO. INTERIM REPORT, JUNE 1973 - MAY 1974, 64
PP. (N.T.I.S. - PB-239 870/9ST)

74-1188　MCLAY, R. W.
ABSHER, R. G.
ANALYSIS OF FEEDBACK CONTROL STABILIZED VIBRATION
BY THE FINITE ELEMENT METHOD
PROC. INTL. CONF. ON PROD. ENG., TOKYO, JPN., PART
1, AUGUST 26-29, 1974, PP. 169-171. (PUBL. BY JPN.
SOC. OF PRECIS. ENG., TOKYO, 1974)

74-1189　OKUBO, N.
TAKEUCHI, Y.
COMPUTER AIDED DESIGN OF MACHINE TOOL STRUCTURE
PROC. INTL. CONF. ON PROD. ENG., TOKYO, JAPAN,
PART 1, AUGUST 26-29, 1974, PP. 220-225. (PUBL. BY
JPN. SOC. OF PRECIS. ENG., TOKYO, 1974)

74-1190　SETHURATHNAM, A.
DAYARATNAM, P.
INTERACTION OF FRAME AND SHEAR WALL WITH OPENINGS
PROC. REG. CONF. ON TALL BUILDINGS, BANGKOK,
THAILAND, JANUARY 23-25, 1974, PP. 301-313.

74-1191　CHAKRABARTI, G. S.
A DISCRETE LEAST-SQUARES FINITE ELEMENT METHOD FOR
TRANSIENT HEAT CONDUCTION PROBLEMS
PH.D. THESIS, UNIV. OF COLORADO, 1974.

74-1192　WANG, J. D.
CONNOR, J. J.
FINITE ELEMENT MODEL OF TWO LAYER COASTAL
CIRCULATION
PROC. 14TH COASTAL ENG. CONF., COPENHAGEN, DEN.,
VOL. 3, JUNE 24-28, 1974, PP. 2401-2420. (PUBL. BY
ASCE, N.Y., 1975)

74-1193　CAREY, G. F.
STUDIES IN FINITE-ELEMENT ANALYSIS
PH.D., UNIV. OF WASHINGTON, 1974.

74-1194　PUTTER, S.
MANOR, H.
USE OF INTEGRAL PARAMETERS IN ROD PROBLEMS
ISR. J. TECHNOL., VOL. 12, NO. 5-6, 1974, PP.
352-355.

74-1195　CHOUDHURY, J. R.
SMITH, B. S.
ELASTIC ANALYSIS OF SPATIAL SYSTEMS OF
INTERCONNECTED SHEAR WALLS AND FRAMES
PROC. REG. CONF. ON TALL BUILD., BANGKOK,
THAILAND, JANUARY 23-25, 1974, PP. 461-476.

74-1196　PEYROT, A. H.
SAUL, W. E.
MULTIDEGREE DYNAMIC ANALYSIS OF TALL BUILDINGS
SUBJECTED TO WIND AS A STOCHASTIC PROCESS
PROC. REG. CONF. ON TALL BUILD., BANGKOK,
THAILAND, JANUARY 23-25, 1974, PP. 555-569.

74-1197　MAK, C. K.
CHENG, W. C.
CREEP DEFORMATION OF A MULTI-STOREY STRUCTURE IN A
FIRE
PROC. REG. CONF. ON TALL BUILD., BANGKOK,
THAILAND, JANUARY 23-25, 1974, PP. 815-828.

74-1198　ZIENKIEWICZ, O. C.
WHY FINITE ELEMENTS
PROC. INT. CONF. ON FINITE ELEMENT METH. FLOW
PROBLEMS, UNIV. OF WALES, SWANSEA, WALES, 1974.

74-1199　RUMSEY, T. R.
FRIDLEY, R. B.
ANALYSIS OF VISCOELASTIC CONTACT STRESSES IN
AGRICULTURAL PRODUCTS USING A FINITE ELEMENT
METHOD
PROC. 67TH A.S.A.E. ANN. MEETING, OKLA. STATE
UNIV., STILLWATER, JUNE 23-26, 1974, AND WINTER
MEETING, CHICAGO, ILL., DECEMBER 10-13, 1974,
PAPER 74-3513, 31 PP. (PUBL. BY A.S.A.E. ST.
JOSEPH, MICH., 1974).

74-1200 SINGH, R. P.
SEGERLIND, L. J.
FINITE ELEMENT METHOD IN FOOD ENGINEERING
PROC. 67TH A.S.A.E. ANN. MEETING, OKLA. STATE
UNIV., STILLWATER, JUNE 23-26, 1974, AND WINTER
MEETING, CHICAGO, ILL., DECEMBER 10-13, 1974,
PAPER 74-6015, 32 PP. (PUBL. BY A.S.A.E. ST.
JOSEPH, MICH., 1974).

74-1201 CATALOGLU, A.
THE FINITE ELEMENT STRESS ANALYSIS OF THE HUMAN
AORTIC VALVE TOWARD THE DEVELOPMENT OF A
PROSTHETIC TRILEAFLET AORTIC VALVE
D.SC. THESIS, WASHINGTON UNIV., 1974.

74-1202 BANGASH, Y.
CIRCUMFERENTIAL PRESTRESSING LOAD ANALYSIS DUE TO
WIRE/STRAND WINDING SYSTEMS ON PRESTRESSED
CONCRETE REACTOR VESSELS
PROC. INST. CIV. ENG., LONDON, VOL. 57, PART 2,
SEPTEMBER 1974, PP. 437-450.

74-1203 LAU, P. C. M.
PRESCOTT, P.
INTERPRETATION OF COMPLEX STRESS ANALYSIS BY
MULTIPLE REGRESSION
PROC. INST. CIV. ENG., LONDON, VOL. 57, PART 2,
SEPTEMBER 1974, PP. 513-522.

74-1204 CHRISTIANO, P. P.
RIZZO, P. C.
COMPLIANCES OF LAYERED ELASTIC SYSTEMS
PROC. INST. CIV. ENG., LONDON, VOL. 57, PART 2,
DECEMBER 1974, PP. 673-683.

74-1205 ZIENKIEWICZ, O. C.
CORMEAN, I. C.
VISCOPLASTICITY, PLASTICITY AND CREEP. A UNIFIED
NUMERICAL SOLUTION APPROACH
IN "ADVANCED TOPICS IN FINITE ELEMENT ANALYSIS",
I.C.C.A.D. LECTURE SERIES N. 1/74, GENOVA, 1974.

74-1206 ZAVELANI-ROSSI, A.
FINITE ELEMENT TECHNIQUES IN PLANE LIMIT PROBLEMS
MECCANICA, VOL. 9, NO. 4, DECEMBER 1974, PP.
312-324.

74-1207 POBEDRIA, B. E.
SOME METHODS OF SOLVING PROBLEMS OF NON-LINEAR
THERMOVISCOELASTICITY
ROZPR. INZ., VOL. 22, NO. 4, 1974, PP. 545-563.

74-1208 CROSSMAN, F. W.
KARLAK, R. F.
CREEP OF B/AL COMPOSITES AS INFLUENCED BY RESIDUAL
STRESSES, BOND STRENGTH, AND FIBER PACKING
GEOMETRY
PROC. 6TH AIME ANNUAL SPRING MEET., 2ND SYMP. ON
FAILURE MODES IN COMPOS., UNIV. OF PITTSBURGH,
PA., MAY 19-23, 1974, PP. 8-31. (PUBL. BY METALL.
SOC. OF AIME, N.Y.)

74-1209 KARLAK, R. F.
CROSSMAN, F. W.
INTERFACE FAILURES IN COMPOSITES
PROC. 6TH AIME ANNUAL SPRING MEET., 2ND SYMP. ON
FAILURE MODES IN COMPOS., UNIV. OF PITTSBURGH,
PA., MAY 19-23, 1974, PP. 119-130. (PUBL. BY
METALL. SOC. OF AIME, N.Y., 1974)

74-1210 MELOSH, R. J.
UNBOUNDED ERRORS IN NUMERICAL ANALYSIS OF
STRUCTURES
IND. MATH., VOL. 24, PART 1, 1974, PP. 1-8.

74-1211 DEHART, R. C.
USE OF FRACTURE MECHANICS AND FINITE ELEMENT
TECHNIQUES FOR DETERMINING THE SAFETY OF OFFSHORE
STRUCTURES
PROC. 2ND COLLOQ. INT. SUR L'EXPLOIT DES OCEANS,
PARIS, FRANCE, OCTOBER 1-4, 1974, PAPER ? 303,
VOL. 3, 22 PP. (PUBL. BY ASSOC. POUR L'ORGAN DE
COLLOQ. OCEANOL., BORDEAUX, PARIS).

74-1212 TAY, A. O.
STEVENSON, M. G.
USING THE FINITE ELEMENT METHOD TO DETERMINE
TEMPERATURE DISTRIBUTIONS IN ORTHOGONAL MACHINING
INST. MECH. ENG., LONDON, VOL. 188, NO. 55, 1974,
PP. 627-638.

74-1213 HIRAI, T.
TANAKA, T.
HYBRID EFFECT ON BENDING BEHAVIORS OF CFRP-AL
LAMINATE
PROC. 18TH JPN. CONGR. ON MATER. RES., KYOTO,
JPN., SEPTEMBER 1974, PP. 15-18. (PUBL. BY SOC. OF
MATER. SCI., JAPAN)

74-1214 CANDOGAN, A.
CYCLIC LOADING AND CRACK PROPAGATION — AN
ELASTOPLASTIC FINITE ELEMENT STUDY
PH.D. THESIS, VIRGINIA POLYTECH. INST. AND STATE
UNIV., 1974.

74-1215 VAN DER REE, H.
BASTING, W. J.
PREDICTION OF TEMPERATURE DISTRIBUTIONS IN CARGOES
WITH THE AID OF A COMPUTER PROGRAM USING THE
METHOD OF FINITE ELEMENT
MEETING, INST. INT. FROID ANNEXE 1974-2,
WAGENINGEN, NETH., APRIL 22-26, 1974, PP. 195-220.

74-1216 ANAND, S. C.
SHAW, H.
NONLINEAR ANALYSIS OF A ROLLING CONTACT PROBLEM BY
FINITE ELEMENTS
PROC. CONF. COMP. METH. IN NONLINEAR MECH.,
AUSTIN, TEXAS, 1974.

74-1217 ARGYRIS, J. H.
ANGELOPOULOS, T.
BICHAT, B.
A GENERAL METHOD FOR THE SHAPE FINDING OF
LIGHTWEIGHT TENSION STRUCTURES
COMP. METH. APPL. MECH. ENG., VOL. 3, 1974, PP.
135-149.

74-1218 STECCO, S.
GUERRINI, M.
BOVE, B.
A FINITE ELEMENT PROCEDURE FOR NUMERICAL
OPTIMIZATION IN MECHANICAL DESIGN
INT. CONF. OF C.S.E.I., NAPLES, ITALY, DECEMBER
1974.

74-1219 BARNHILL, R. E.
WHITEMAN, J. R.
COMPUTABLE ERROR BOUNDS FOR THE FINITE ELEMENT
METHOD FOR ELLIPTIC BOUNDARY VALUE PROBLEMS
IN "NUMERISCHE METHODEN BEI
DIFFERENTIALGLEICHUNGEN UND MIT
FUNKTIONALANALYTISCHEN HILFSMITTEIN", J. ALBRECHT
AND L. COLLATZ (EDS.), I.S.N.M., VOL. 19,
BIRKHAUSER VERLAG, BASEL, 1974, PP. 9-28.

74-1220 HOOD, P.
THE SOLUTION OF THE NAVIER STOKES EQUATIONS
PH.D. THESIS, UNIV. OF WALES, SWANSEA, 1974.

74-1221 ALAYLIOGLU, H.
DEVELOPMENT AND APPLICATION OF HYBRID STRESS
FINITE ELEMENTS TO AN AUTOMOTIVE STRUCTURE
PH.D. THESIS, LOUGHBOROUGH UNIV. OF TECH., 1974.

74-1222 BONNEROT, R.
JAMET, P.
A SECOND ORDER FINITE ELEMENT METHOD FOR THE
ONE-DIMENSIONAL STEFAN PROBLEM
INT. J. NUMER. METH. ENG., VOL. 8, 1974, PP.
811-820.

74-1223 BRASHERS, M. R.
CHAN, S. T. K.
YOUNG, V. Y. C.
FINITE ELEMENT ANALYSIS OF TRANSONIC FLOW
PROC. CONF. COMP. METH. IN NONLINEAR MECH.,
AUSTIN, TEXAS, 1974.

74-1224 BUTURLA, E. M.
COTTRELL, P. E.
TWO-DIMENSIONAL FINITE ELEMENT ANALYSIS OF
SEMI-CONDUCTOR STEADY STATE TRANSPORT EQUATIONS
PROC. CONF. COMP. METH. IN NONLINEAR MECH.,
AUSTIN, TEXAS, 1974.

74-1225 HODGE, P. G. JR.
GARG, V. K.
ANAND, S. C.
A FINITE ELEMENT MODEL FOR PLASTICITY PROBLEMS
PROC. 7TH SOUTHEASTERN CONF. ON THEORETICAL AND
APPL. MECH., M. HELLER AND M. CHI (EDS.), THE
CATHOLIC UNIV. OF AMER., WASH., VOL. 7, 1974, PP.
369-383.

74-1226 CECCI, M. M.
CELLA, A.
A FINITE ELEMENT SOLUTION OF THE STABILITY OF
SUPERPOSED FLUIDS
PROC. SYMP. FINITE ELEMENT METH. IN FLOW PROBLEMS,
SWANSEA, 1974.

74-1227 CIARLET, P. G.
RAVIART, P. A.
A MIXED FINITE ELEMENT METHOD FOR THE BIHARMONIC
EQUATION
SYMP. MATH. ASPECTS OF FINTIE ELEMENT IN PARTIAL
DIFFERENTIAL EQUATIONS, MATH RESEARCH CENTER,
UNIV. OF WISCONSIN, MADISON, 1974.

74-1228 CIARLET, P. G.
RAVIART, P. A.
LA METHODE DES ELEMENTS FINIS POUR LES PROBLEMES
AUX LIMITES ELLIPTIQUES
LABORATOIRE ANALYSE NUMERIQUE, UNIVERSITE PARIS,
REPORT 740006, 1974.

74-1229 CLEMENT, PH.
PINI, F.
APPROXIMATION BY FINITE ELEMENT FUNCTIONS USING
REGULARIZATION
DEPARTEMENT DE MATHEMATIQUES, ECOLE POLYTECHNIQUE
FEDERALE DE LAUSANNE, TECHNICAL REPORT, 1974.

74-1230 CRISFIELD, M. A.
A NOTE ON THE ELASTIC ANALYSIS OF BOX-GIRDER
DIAPHRAGMS USING FINITE ELEMENTS
DEPT. ENVIRONMENT, TRANSPORT AND RD. RESEARCH
LAB., CROWTHORNE, REPORT: SR118 UC, 1974.

74-1231 CRISFIELD, M. A.
SOME APPROXIMATIONS IN THE NONLINEAR ANALYSIS OF
RECTANGULAR PLATES USING FINITE ELEMENTS
DEPT. OF THE ENVIRONMENT, TRANSPORT AND ROAD
RESEARCH LAB., CROWTHORNE, TRRL REPORT: SR51 UC,
1974.

74-1232 ODEN, J. T.
COMPUTATIONAL METHODS IN NONLINEAR MECHANICS
PROC. INTL. CONF. ON COMPUT. METH. IN NONLINEAR
MECH., UNIV. OF TEXAS, AUSTIN, TEXAS, SEPTEMBER
23-25, 1974.

74-1233 DESAI, C. S.
A CONSISTENT FINITE ELEMENT TECHNIQUE FOR
WORKSOFTENING BEHAVIOUR
PROC. CONF. COMP. METH. IN NONLINEAR MECH.,
AUSTIN, TEXAS, 1974.

74-1234 DESAI, C. S.
ON APPLICATION OF THE FINITE ELEMENT METHOD TO
LIMIT ANALYSIS CONSISTENT FINITE ELEMENT TECHNIQUE
FOR WORK-SOFTENING BEHAVIOR
PROC. INT. CONF. ON COMPUT. METH. IN NONLINEAR
MECH., UNIV. OF TEXAS, AUSTIN, SEPTEMBER 23-25,
1974.

74-1235 DI CARLO, A.
PIVA, R.
FINITE ELEMENT SIMULATION OF THERMALLY INDUCED
FLOW FIELDS
PROC. CONF. COMP. METH. IN NONLINEAR MECH.,
AUSTIN, TEXAS, 1974.

74-1236 DLUGACH, M. I.
KOVALCHUK, N. V.
AN APPLICATION OF THE FINITE ELEMENT METHOD TO THE
CALCULATION OF CYLINDRICAL SHELLS WITH RECTANGULAR
OPENINGS
PRIKLADY MEKH. (RUSSIAN), VOL. 9, 1974, PP. 35-41.

74-1237 WICKS, T. M.
BECKER, E. B.
YEW, C. H.
DUNHAM, R. S.
ON APPLICATION OF THE FINITE ELEMENT METHOD TO
LIMIT ANALYSIS
PROC. INTL. CONF. ON COMPUT. METH. IN NONLINEAR
MECH., UNIV. OF TEXAS, AUSTIN, SEPTEMBER 23-25,
1974.

74-1238 FIX, G. J.
FINITE ELEMENT METHOD FOR NONLINEAR HYPERBOLIC
EQUATIONS
PROC. INTL. CONF. ON COMPUT. METH. IN NONLINEAR
MECH., UNIV. OF TEXAS, AUSTIN, SEPTEMBER 23-25,
1974.

74-1239 CAREY, G. F.
VARANASI, S. R.
EXPERIENCE WITH FINITE ELEMENT COMPUTATIONS FOR
NONLINEAR STRUCTURAL PROBLEMS
PROC. INTL. CONF. ON COMPUT. METH. IN NONLINEAR
MECH., UNIV. OF TEXAS, AUSTIN, SEPTEMBER 23-25,
1974.

74-1240 GALLAGHER, R. H.
GENERAL POTENTIAL ENERGY AND COMPLEMENTARY ENERGY
MODELS BASED ON STRESS PARAMETERS
LECTURES AT ICCAD COURSE ON ADVANCED TOPICS IN
FINITE ELEMENT ANALYSIS, SANTA MARGHERITA, ITALY,
1974.

74-1241 GALLAGHER, R. H.
PERTURBATION PROCEDURES IN NONLINEAR FINITE
ELEMENT ANALYSIS
PROC. INTL. CONF. ON COMPUT. METH. IN NONLINEAR
MECH., UNIV. OF TEXAS, AUSTIN, SEPTEMBER 23-25,
1974.

74-1242 ARGYRIS, J. H.
DUNNE, P. C.
ANGELOPOULOS, T.
BICHAT, B.
FINITE ELEMENT DIFFICULTIES IN NONLINEAR AND
INCOMPRESSIBLE MATERIAL
PROC. INTL. CONF. ON COMPUT. METH. IN NONLINEAR
MECH., UNIV. OF TEXAS, AUSTIN, SEPTEMBER 23-25,
1974.

74-1243 GODBOLE, P. N.
MEEK, J. L.
APPLICATION OF FINITE ELEMENT METHOD TO PLANE
STRAIN EXTRUSION PROCESSES
PROC. INTL. CONF. ON COMPUT. METH. IN NONLINEAR
MECH., UNIV. OF TEXAS, AUSTIN, SEPTEMBER 23-25,
1974.

74-1244 HERRERA, I.
BIELAK, J.
A SIMPLIFIED VERSION OF GURTIN'S VARIATIONAL
PRINCIPLE
ARCH. RAT. MECH. ANAL., VOL. 53, 1974, PP.
131-149.

74-1245 HOOD, P.
FINITE ELEMENT FORMULATION WITH REFERENCE TO FLUID
DYNAMICS
PROC. INTL. CONF. ON COMPUT. METH. IN NONLINEAR
MECH., UNIV. OF TEXAS, AUSTIN, SEPTEMBER 23-25,
1974.

74-1246 HSU, T. R.
BERTELS, A. W. M.
ARYA, B.
BANERJEE, S.
APPLICATION OF THE FINITE ELEMENT METHOD TO THE
NONLINEAR ANALYSIS OF NUCLEAR REACTOR FUEL
BEHAVIOR
PROC. INTL. CONF. ON COMPUT. METH. IN NONLINEAR
MECH., UNIV. OF TEXAS, AUSTIN, SEPTEMBER 23-25,
1974.

74-1247 JAMET, P.
ESTIMATIONS D'ERREUR POUR DES ELEMENTS FINIS
DROITS PRESQUE DEGENERES
CENTRE DE ETUDES DE LIMEIL, VILLENEUVE,
SAINT-GEORGES, REPORT CRM-447, 1974.

74-1248 REDDY, J. N.
SOME MATHEMATICAL PROPERTIES OF CERTAIN MIXED
GALERKIN APPROXIMATIONS IN NONLINEAR ELASTICITY
PROC. INTL. CONF. ON COMPUT. METH. IN NONLINEAR
MECH., UNIV. OF TEXAS, AUSTIN, SEPTEMBER 23-25,
1974.

74-1249 JONES, R. E.
A SELF-ORGANIZING MESH GENERATION PROGRAM
CONF. PRESSURE VESSELS AND PIPING, NUCL. AND
MATER. DIV., A.S.M.E., MIAMI, 1974.

74-1250 KALKANI, E. C.
MESH GENERATION PROGRAM FOR HIGHWAY EXCAVATION
CUTS
INT. J. NUMER. METH. ENG., VOL. 8, 1974, PP.
369-394.

74-1251 CARMIGNANI, C.
CELLA, A.
DEPAULIS, A.
FUNCTIONAL MINIMIZATION IN NONLINEAR SOLID
MECHANICS
PROC. INTL. CONF. ON COMPUT. METH. IN NONLINEAR
MECH., UNIV. OF TEXAS, AUSTIN, SEPTEMBER 23-25,
1974.

74-1252 LAKSHMIKANTHAM, C.
TONG, P.
STRESSES AROUND HOLES IN STIFFENED COMPOSITE
PANELS USING LAURENT-SERIES AND FINITE ELEMENT
METHODS
5TH INT. CONF. EXP. STRESS ANALYSIS, UDINE, 1974.

74-1253 LAKSHMINARAYANAN, V.
LANG, T. E.
DISCRETIZATION-DISPERSION IN FINITE ELEMENT
MODELLING OF WAVE PROPAGATION IN SOLIDS
PROC. CONF. COMP. METH. IN NONLINEAR MECH.,
AUSTIN, TEXAS, 1974.

74-1254 LEE, K. N.
NEMAT-NASSER, S.
MIXED VARIATIONAL PRINCIPLES, FINITE ELEMENTS AND
FINITE ELASTICITY
PROC. CONF. COMP. METH. IN NONLINEAR MECH.,
AUSTIN, TEXAS, 1974.

74-1255 LESAINT, P.
RAVIART, P. A.
ON A FINITE ELEMENT METHOD FOR SOLVING THE NEUTRON
TRANSPORT EQUATION
LABORATOIRE ANALYSE NUMERIQUE, UNIV. OF PARIS,
REPORT 74008, 1974.

74-1256 LUKAS, I. L.
CURVED BOUNDARY ELEMENTS. GENERAL FORMS OF
POLYNOMIAL MAPPINGS
PROC. CONF. COMP. METH. IN NONLINEAR MECH.,
AUSTIN, TEXAS, 1974.

74-1257 LYNN, P. P.
A LEAST SQUARES FINITE ELEMENT ANALYSIS OF
NONLINEAR BIOPHYSICAL DIFFUSION-KINETICS EQUATIONS
PROC. CONF. COMP. METH. IN NONLINEAR MECH.,
AUSTIN, TEXAS, 1974.

74-1258 GARTLING, D. K.
BECKER, E. B.
COMPUTATIONALLY EFFICIENT FINITE ELEMENT ANALYSIS
OF VISCOUS FLOW PROBLEMS
PROC. INTL. CONF. ON COMPUT. METH. IN NONLINEAR
MECH., UNIV. OF TEXAS, AUSTIN, SEPTEMBER 23-25,
1974.

74-1259 MITCHELL, A. R.
THE FINITE ELEMENT METHOD
BULL. INST. MATH. APPL., VOL. 10, 1974, PP. 76-79.

74-1260 MITCHELL, A. R.
MARSHALL, J. A.
MATCHING OF ESSENTIAL BOUNDARY CONDITIONS IN THE
FINITE ELEMENT METHOD
PROC. CONF. ON NUMER. ANALYSIS, J. MILLER (ED.),
DUBLIN, 1974.

74-1261 MITCHELL, A. R.
MCLEOD, R.
CURVED ELEMENTS IN THE FINITE ELEMENT METHOD
PROC. CONF. ON NUMER. SOLUTION OF DIFFERENTIAL
EQUATION, LECTURE NOTES IN MATH., G. A. WATSON
(ED.), SPRINGER-VERLAG, BERLIN, NO. 363, 1974, PP.
89-104.

74-1262 MITTELMANN, H. D.
FINITE ELEMENT VERFAHREN BEI QUASILINEAREN
ELLIPTISCHEN RANDWERTPROBLEM
FACHBEREICH MATHEMATICK, TECHNISCHE HOCHSCHULE
DARMSTADT, PREPRINT 133, 1974.

74-1263 MITTELMANN, H. D.
NICHTLINEARE DIRICHTLET PROBLEME UND EINFACHE
FINITE ELEMENT VERFAHREN
FACHBEREICH MATHEMATIK, TECHNISCHE HOCHSCHULE
DARMSTADT, REPORT 157, 1974.

74-1264 NORRIE, D. H.
DE VRIES, G.
A LAGRANGIAN FINITE ELEMENT SOLUTION TO UNSTEADY
FLOW
DEPT. MECH. ENG., UNIV. OF CALGARY, REPORT 17,
1974.

74-1265 FROIER, M.
NILSSON, L.
SAMUELSSON, A.
THE RECTANGULAR PLANE STRESS ELEMENT BY TURNER,
PIAN AND WILSON
INT. J. NUMER. METH. ENG., VOL. 8, NO. 2, 1974,
PP. 433-437.

74-1266 PRAKASO, B.
VENKATESWARA, G.
AXISYMMETRIC VIBRATIONS OF LINEARLY TAPERED
ANNULAR PLATES
J. SOUND AND VIB., VOL. 32, 1974, PP. 507-512.

74-1267 RAVIART, P. A.
HYBRID FINITE ELEMENT METHODS FOR SOLVING 2ND
ORDER ELLIPTIC EQUATIONS
LABORATOIRE ANALYSE NUMERIQUE, UNIVERSITY OF
PARIS, REPORT 74015, 1974.

74-1268 SALINAS, E.
NGUYEN, D. H.
FINITE ELEMENT SOLUTIONS OF A NONLINEAR REACTOR
DYNAMICS PROBLEM
PROC. CONF. COMP. METH. IN NONLINEAR MECH.,
AUSTIN, TEXAS, 1974.

74-1269 SCAIFFE, B. K. P.
STUDIES IN NUMERICAL ANALYSIS
ACADEMIC PRESS, NEW YORK, 1974.

74-1270 SCHAUER, D. A.
ELFED A COMPUTER CODE TO GENERATE A FINITE ELEMENT
MESH FOR PROBLEMS OF COMPLEX AXISYMMETRIC GEOMETRY
LAWRENCE LIVERMORE LAB., UNIV. OF CALIF., REPORT
UCRL-74879, 1974.

74-1271 KAWAHARA, M.
KAMEMURA, K.
ELASTO-VISCOPLASTIC FINITE ELEMENT ANALYSIS BY
PERTURBATION METHOD
PROC. INTL. CONF. ON COMPUT. METH. IN NONLINEAR
MECH., UNIV. OF TEXAS, AUSTIN, SEPTEMBER 23-25,
1974.

74-1272 MCLAY, R. W.
BULLIS, S.
ON THE CONVERGENCE OF THE FINITE ELEMENT SOLUTION
IN A NONLINEAR HEAT CONDUCTION PROBLEM
PROC. INTL. CONF. ON COMPUT. METH. IN NONLINEAR
MECH., UNIV. OF TEXAS, AUSTIN, SEPTEMBER 23-25,
1974.

74-1273 VOS, R. G.
FINITE ELEMENT SOLUTION OF NONLINEAR STRUCTURES BY
PERTURBATION TECHNIQUE
PROC. INTL. CONF. ON COMPUT. METH. IN NONLINEAR
MECH., UNIV. OF TEXAS, AUSTIN, SEPTEMBER 23-25,
1974.

74-1274 WASHIZU, K.
IKEGAWA, M.
SOME APPLICATIONS OF THE FINITE ELEMENT METHOD TO
FLUID MECHANICS
THEORETICAL AND APPLIED MECH., UNIV. OF TOKYO,
VOL. 22, 1974, PP. 143-154.

74-1275 FELIPPA, C. A.
FINITE ELEMENT ANALYSIS OF THREE-DIMENSIONAL CABLE
STRUCTURES
PROC. INTL. CONF. ON COMPUT. METH. IN NONLINEAR
MECH., UNIV. OF TEXAS, AUSTIN, SEPTEMBER 23-25,
1974.

74-1276 WHITEMAN, J. R.
LAGRANGIAN FINITE ELEMENT AND FINITE DIFFERENCE
METHODS FOR POISSON PROBLEMS
IN "NUMERISCHE BEHANDLUNG VON
DIFFERENTIALGLEICHYNGEN", I.S.M. M., L. COLLATZ
(ED.), BIRKHAUSER VERLAG, BASEL, 1974.

74-1277 BAKER, A. J.
A HIGHLY STABLE EXPLICIT INTEGRATION TECHNIQUE FOR
COMPUTATIONAL CONTINUUM MECHANICS
PROC. INT. CONF. ON NUMER. METH. IN FLUID
DYNAMICS, UNIV. OF SOUTHAMPTON, ENGLAND, SEPTEMBER
26-28, 1973, PP. 99-120.

74-1278 ZIENKIEWICZ, O. C.
VISCOPLASTICITY, PLASTICITY AND PLASTIC FLOW
PROC. CONF. COMP. METH. IN NONLINEAR MECH.,
AUSTIN, TEXAS, 1974.

74-1279 ZIENKIEWICZ, O. C.
CORMEAU, I. C.
VISCO-PLASTICITY, PLASTICITY AND CREEP IN ELASTIC
SOLIDS. A UNIFIED NUMERICAL SOLUTION APPROACH
INT. J. NUMER. METH. ENG., VOL. 8, 1974, PP.
821-845.

74-1280 ZLAMAL, M.
FINITE ELEMENT METHODS FOR PARABOLIC EQUATIONS
PROC. CONF. NUMER. SOLUTION OF DIFFERENTIAL
EQUATIONS, LECTURE NOTES IN MATH., G. A. WATSON
(ED.), SPRINGER-VERLAG, BERLIN, NO. 363, 1974, PP.
215-221.

74-1281 DALY, P.
FINITE ELEMENT SOLUTIONS FOR AN EQUILATERAL
TRIANGLE
INT. J. NUMER. METH. ENG., VOL. 8, 1974, PP.
495-531.

74-1282 FRAEIJS DE VEUBEKE, B.
VARIATIONAL PRINCIPLES AND THE PATCH TEST
INT. J. NUMER. METH. ENG., VOL. 8, 1974, PP.
783-801.

74-1283 MITTELMANN, H. D.
STABILITAT BEI DER METHODE DER FINITEN ELEMENTE
FUR QUASILINEARE ELLIPTISCHE RANDWERTPROBLEME
FACHBEREICH MATHEMATIK, TECHNISCHE HOCHSCHULE
DARMSTADT, REPORT 154, 1974.

74-1284 SYNGE, J. L.
THE HYPERCIRCLE METHOD
IN "STUDIES IN NUMERICAL ANALYSIS", B.K.P. SCAIFF
(ED.), ACADEMIC PRESS, NEW YORK, 1974.

74-1285 WESTBROOK, D. R.
A VARIATIONAL PRINCIPLE WITH APPLICATIONS IN
FINITE ELEMENTS
J. INST. MATH. APPL., VOL. 14, 1974, PP. 79-82.

74-1286 ZLAMAL, M.
UNCONDITIONALLY STABLE FINITE ELEMENT SCHEMES FOR
PARABOLIC EQUATIONS
PROC. 2ND INT. CONF. NUMER. ANALYSIS, DUBLIN,
1974.

74-1287 ZIENKIEWICZ, O. C.
PHILLIPS, D. V.
OWENS, D. R. J.
FINITE ELEMENT ANALYSIS OF SOME CONCRETE
NONLINEARITIES. THEORY AND EXAMPLES
INT. ASSOC. FOR BRIDGE AND STRUCT. ENG., SEMIN. ON
CONCR. STRUCT. SUBJ. TO TRIAXIAL STRESSES,
BERGAMO, ITALY, MAY 17-19, 1974.

74-1288 DE DONATO, O.
FRANCHI, A.
GIDDA, G.
FINITE ELEMENT ELASTOPLASTIC ANALYSIS OF
UNDERGROUND OPENINGS BY QUADRATIC PROGRAMMING
INT. ASSOC. FOR BRIDGE AND STRUCT. ENG., SEMIN. ON
CONCR. STRUCT. SUBJ. TO TRIAXIAL STRESSES,
BERGAMO, ITALY, MAY 17-19, 1974.

74-1289 FANELLI, M.
RICCIONI, R.
ROBUTTI, G.
FINITE ELEMENT ANALYSIS OF PRESTRESSED CONCRETE
PRESSURE VESSELS
INT. ASSOC. FOR BRIDGE AND STRUCT. ENG., SEMIN. ON
CONCR. STRUCT. SUBJ. TO TRIAXIAL STRESSES,
BERGAMO, ITALY, MAY 17-19, 1974.

74-1290 WILLIAMS, G. M. J.
SURVEY OF STRUCTURAL ANALYSIS SYSTEMS
PROC. INT. CONF. ON COMPUT. IN ENG. AND BUILD.
DES., IMP. COLL. OF SCI. AND TECH., LONDON, ENGL.,
SEPTEMBER 25-27, 1974. PUBL. BY IPC BUS. PRESS,
LTD., GUILDFORD, SURREY, ENGL., 1974.

74-1291 NASRELDIN, H. A.
INTERACTIVE FINITE ELEMENT DATA GENERATION USING
THE SUB-STRUCTURING TECHNIQUE
PROC. INT. CONF. ON COMPUT. IN ENG. AND BUILD.
DES., IMP. COLL. OF SCI. AND TECH., LONDON, ENGL.,
SEPTEMBER 25-27, 1974. PUBL. BY IPC BUS. PRESS,
LTD., GUILDFORD, SURREY, ENGL., 1974.

74-1292 SPOONER, J. B.
LAWRENCE, D. J.
FINITE ELEMENT ANALYSIS, MODELLING AND COMPUTER
GRAPHICS
PROC. INT. CONF. ON COMPUT. IN ENG. AND BUILD.
DES., IMP. COLL. OF SCI. AND TECH., LONDON, ENGL.,
SEPTEMBER 25-27, 1974. PUBL. BY IPC BUS. PRESS,
LTD., GUILDFORD, SURREY, ENGL., 1974.

74-1293 YAMADA, YOSHIAKI
IWATA, K.
KAKIMI, T.
HOSOMURA, T.
LARGE DEFORMATION AND CRITICAL LOADS ANALYSIS OF
FRAMED STRUCTURES
PROC. INT. CONF. ON COMPUT. METH. IN NONLINEAR
MECH., UNIV. OF TEXAS, AUSTIN, 1974, PP. 819-828.

74-1294 YAMADA, YOSHIAKI
SAKURAI, T.
TAKE, H.
A FINITE ELEMENT SIMULATION OF MECHANICAL
PROPERTIES OF COMPOSITE MATERIALS
COMPOSITE MATER. AND STRUCT., FUKUGO ZAIRYO
KENKYU, VOL. 3, NO. 4, 1974, PP. 7-11.

74-1295 GORDON, W. J.
HALL, C. A.
CONSTRUCTION OF CURVILINEAR COORDINATE SYSTEMS AND
APPLICATIONS TO MESH GENERATION
INTL. J. NUMER. METHODS ENGR., VOL. 7, NO. 4,
JANUARY 1974, PP. 461-477.

74-1296 LIKINS, P. W.
GEOMETRIC STIFFNESS CHARACTERISTICS OF A ROTATING
ELASTIC APPENDAGE
INTL. J. SOLIDS STRUCT., VOL. 10, NO. 2, FEBRUARY
1974, PP. 161-167.

74-1297 PETYT, M.
FLEISCHER, C. C.
VIBRATION OF MULTISUPPORTED CURVED BEAMS
J. SOUND AND VIB., VOL. 32, NO. 3, FEBRUARY 1974,
PP. 359-365.

74-1298 BELYTSCHKO, T.
FINITE ELEMENT APPROACH TO HYDRODYNAMICS AND MESH
STABILIZATION
PROC. INTL. CONF. ON COMPUT. METH. IN NONLINEAR
MECH., UNIV. OF TEXAS, AUSTIN, SEPTEMBER 23-25,
1974.

74-1299 SMITH, G. L.
USING FINITE ELEMENTS IN AUTOMOTIVE DESIGN
AUTOMOT. ENGR., VOL. 82, NO. 4, APRIL 1974, PP.
25-31.

74-1300 VERMA, M. K.
MURTHY, A. V. K.
NONLINEAR VIBRATIONS OF NONUNIFORM BEAMS WITH
CONCENTRATED MASSES
J. SOUND AND VIB., VOL. 33, NO. 1, MARCH 1974, PP.
1-12.

74-1301 STROEBEL, G. J.
DYNAMIC COUPLING IN AN UNSYMMETRICALLY TAPERED
BEAM
J. SOUND AND VIB., VOL. 34, NO. 2, MAY 22, 1974,
PP. 275-283.

74-1302 BATHE, K. J.
OZDEMIR, H.
WILSON, E. L.
STATIC AND DYNAMIC GEOMETRIC AND MATERIAL
NONLINEAR ANALYSIS
STRUCT. ENG. LAB., UNIV. OF CALIF., BERKELEY,
CALIF., REPORT NO. UCSESM 74-4, FEBRUARY 1974, 178
PP.

74-1303 SCIARRA, J. J.
VIBRATION REDUCTION BY USING BOTH THE FINITE
ELEMENT STRAIN ENERGY DISTRIBUTION AND MOBILITY
TECHNIQUES
SHOCK AND VIB. BULL., U.S. NAVAL RES. LAB., VOL.
44, NO. 2, AUGUST 1974, PP. 193-199.

74-1304 SNYDER, R. E.
VIKING DYNAMICS - AN OVERVIEW
SHOCK AND VIB. BULL., U.S. NAVAL RES. LAB., VOL.
44, NO. 2, AUGUST 1974, PP. 19-23.

74-1305 CHRISTIANSEN, H. N.
COMPUTER GENERATED DISPLAYS OF STRUCTURES IN
VIBRATION
SHOCK AND VIB. BULL., U.S. NAVAL RES. LAB., VOL.
44, NO. 2, AUGUST 1974, PP. 185-192.

74-1306 HOUDE, J.
MIRZA, M. S.
FINITE ELEMENT ANALYSIS OF SHEAR STRENGTH OF
REINFORCED CONCRETE BEAMS
SHEAR IN REINF. CONCR., VOL. 2, 1974. (PUBL. BY
AMER. CONCRETE. INST., DETROIT, MICH.).

74-1307 MASTERSON, D. M.
LONG, A. E.
PUNCHING STRENGTH OF SLABS, A FLEXURAL APPROACH
USING FINITE ELEMENTS
SHEAR IN REINF. CONCR., VOL. 2, 1974. (PUBL. BY
AMER. CONCR. INST., DETROIT, MICH.).

74-1308 FRANKLIN, H. A.
FINITE ELEMENT STUDY OF REINFORCED CONCRETE BEAMS
WITH DIAGONAL TENSION CRACKS
SHEAR IN REINF. CONCR., VOL. 2, 1974. (PUBL. BY
AMER. CONCR. INST., DETROIT, MICH.).

74-1309 LEIBOWITZ, R. C.
VIBROACOUSTIC RESPONSE OF TURBULENCE EXCITED THIN
RECTANGULAR FINITE PLATES IN HEAVY AND FLUID MEDIA
DAVID W. TAYLOR NAVAL SHIP RESEARCH AND DEVELOP.
CENTER, BETHESDA, MD., REPORT NO: DTNSRDC-2976G,
JUNE 3, 1974, 67 PP. (N.T.I.S. - AD-A016 540/7ST)
SEE ALSO J. SOUND AND VIB., VOL. 40, NO. 4, 1975,
PP. 441-495.

74-1310 MASSINI, G.
ZOLTI, E.
REBO: COMPUTER CODE FOR THE GENERATION OF A
THREE-DIMENSIONAL FINITE-ELEMENT GRID OF THE
NOZZLE-CYLINDRICAL VESSEL JUNCTION ZONE
COMITATO NAZIONALE PER L'ENERGIA NUCLEARE, ROME,
ITALY, DECEMBER 1974, 154 PP. (N.T.I.S. -
RT/ING-(74)35)

74-1311 STUEMKE, H.
CONTRIBUTIONS TO FLUID MECHANICS AND FLIGHT
MECHANICS
DEUTSCHE GESELLSCHAFT FUER LUFT- UND RAUMFAHRT,
COLOGNE, WEST GERMANY, REPORT NO: DLR-FB-74-62,
JULY 25, 1974, 234 PP. (N.T.I.S. - N75-30082/2ST)

74-1312 KASHIWAGI, T.
A RADIATIVE IGNITION MODEL OF A SOLID FUEL
NATL. BUREAU OF STANDARDS, WASH., D.C., FINAL
REPORT, 1974, 12 PP. (N.T.I.S. - COM-75-50574/3ST)
PUBL. IN COMB. SCI. AND TECH., VOL. 8, 1974, PP.
225-236.

74-1313 POWELL, G. H.
FINITE ELEMENT ANALYSIS OF ELASTO-PLASTIC TEE
JOINTS
COLL. OF ENG., UNIV. OF CALIF., BERKELEY,
SEPTEMBER 1974, 107 PP. (N.T.I.S.
-ORNL-SUB-3193-2)

74-1314 CASCIARO, R.
DI CARLO, A.
MIXED FINITE ELEMENT MODELS IN LIMIT ANALYSIS
PROC. INTL. CONF. ON COMPUT. METH. IN NONLINEAR
MECH., UNIV. OF TEXAS, AUSTIN, SEPTEMBER 23-25,
1974.

74-1315 KUNOO, K.
OPTIMUM CONFIGURATION OF TRUSS STRUCTURES
NATL. AEROSPACE LAB., TOKYO, JAPAN, REPORT NO:
NAL-TR-388T, SEPTEMBER 1974, 12 PP. (N.T.I.S. -
N75-29471/0ST)

74-1316 MORINO, L.
CHEN, L. T.
A FINITE-ELEMENT ANALYSIS FOR STEADY-AND
OSCILLATORY SUPERSONIC FLOWS AROUND COMPLEX
CONFIGURATIONS
DEPT. OF AEROSPACE ENG., BOSTON UNIV., MASS.,
REPORT NO: NASA-CR-143099, TR-74-03, DECEMBER
1974, 68 PP. (N.T.I.S. - N75-27306/0ST)

74-1317 CHEN, L. T.
SUCIU, E. O.
MORINO, L.
A FINITE ELEMENT ANALYSIS FOR STEADY AND
OSCILLATORY SUBSONIC FLOW AROUND COMPLEX
CONFIGURATIONS
DEPT. OF AEROSPACE ENG., BOSTON UNIV., MASS.,
REPORT NO: NASA-CR-143103, TR-74-02, DECEMBER
1974, 55 PP. (N.T.I.S. - N75-27297/1ST)

74-1318 WRIGHT, F. D.
DESIGN OF ROOF BOLT PATTERNS FOR JOINTED ROCK
ROCK MECH. LAB., KENTUCKY UNIV., LEXINGTON,
OCTOBER 1974, 132 PP. (N.T.I.S. - PB-243 899/2ST)

74-1319 BEECKMAN, R.
VANLEEUWEN, D.
A FINITE ELEMENT ANALYSIS OF BODIES OF REVOLUTION.
A COMPARISON BETWEEN CURRENTLY USED TYPES OF
ELEMENTS
CENTRE DE RECH. SCI. ET TECH. DE L'IND. DES FAB.
METAL., BRUSSELS, BELGIUM, REPORT NO: CRIF-MT-102,
DECEMBER 1974, 56 PP. (N.T.I.S. - N75-24422/8ST)

74-1320 MORINO, L.
SUBSONIC AND SUPERSONIC INDICIAL AERODYNAMICS AND
AERODYNAMIC TRANSFER FUNCTION FOR COMPLEX
CONFIGURATIONS
DEPT. OF AEROSPACE ENG., BOSTON UNIV., MASS.,
REPORT NO. NASA-CR-142818, TN-74-01, SEPTEMBER
1974, 33 PP. (N.T.I.S. N75-23478/1ST)

74-1321 MORINO, L.
SUCIU, E.
A FINITE-ELEMENT METHOD FOR LIFTING SURFACES IN
STEADY INCOMPRESSIBLE SUBSONIC FLOW
DEPT. AEROSPACE ENG., BOSTON UNIV., MASS., REPORT
NO: NASA -CR-142811, TR-74-05, DECEMBER 1974, 38
PP. (N.T.I.S. - N75-23477/3ST)

74-1322 SILVENNOINEN, P.
TUOMINEN, J.
CURVED LOW ORDER ANGULAR ELEMENTS IN TWO
DIMENSIONAL NEUTRON TRANSPORT PROBLEMS
NUCL. ENG. LAB., TECH. RESEARCH CENTRE, FINLAND,
HELSINKI, REPORT NO: PUBL-8, 1974, 14 PP.
(N.T.I.S. - N75-23262/9ST)

74-1323 HYLARIDES, S.
TRANSVERSE VIBRATIONS OF SHIP'S PROPULSION
SYSTEMS. PART 1: THEORETICAL ANALYSIS
SCHEEPSBOUW AFDELING TECHNIEK, NEDERLANDS
SCHEEPS-STUDIECENTRUM TNO, DELFT, REPORT NO.
REPT-197-M, TDCK-65871, OCTOBER 1974, 20 PP.
(N.T.I.S. - N75-22823/9ST).

74-1324 WEVERS, L. J.
TRANSVERSE VIBRATIONS OF SHIP'S PROPULSION
SYSTEMS. PART 2: EXPERIMENTAL ANALYSIS
SCHEEPSBOUW AFDELING TECHNIEK, NEDERLANDS
SCHEEPS-STUDIECENTRUM TNO, DELFT, REPORT NO.
REPT-203-M, TDCK-65875, DECEMBER 1974, 20 PP.
(N.T.I.S. - N75-22824/7ST).

74-1325 MEIJERS, P.
NUMERICAL HULL VIBRATION ANALYSIS OF A FAR EAST
CONTAINER SHIP
SCHEEPSBOUW AFDELING, NEDERLANDS
SCHEEPS-STUDIECENTRUM TNO, DELFT, REPORT NO:
REPT-195-S, TDCK-65495, JULY 1974, 28 PP.
(N.T.I.S. - N75-22822/1ST)

74-1326 CARMIGNANI, C.
CELLA, A.
DEPAULIS, A.
ON THE RESOLUTION OF NONLINEAR ELASTIC AND
ELASTOPLASTIC PROBLEMS BY THE METHOD OF FINITE
ELEMENTS. (SULLA RISOLUZIONE DEI PROBLEMI DI
ELASTICITA NON LINEARE E DI ELASTO-PLASTICITA
ATTRAVERSO IL METODO DEGLI ELEMENTI FINITI)
PISA UNIV., ITALY, REPORT NO: REPT-39, 1974, 14
PP. (N.T.I.S. - N75-22816/3ST)

74-1327 ANON
LARGE STRUCTURES FOR MANNED SPACECRAFT:
MATHEMATICAL ANALYSIS, DESIGN, CONSTRUCTION AND
TESTS
EUROPEAN SPACE RESEARCH ORGANIZATION, PARIS,
FRANCE, REPORT NO: ESRO-SP-99, MARCH 1974, 618 PP.
(N.T.I.S. - N75-22504/5ST)

74-1328 NERLI, G.
CALCULATION OF AXISYMMETRICAL OSCILLATION OF
FLEXIBLE TURBINE DISCS. (CALCOLO DELLE
OSCILLAZIONI ASSIALSIMMETRICHE FLESSIONALI DEI
DISCHI DELLE TURBINE)
IST. DI AERONAUTICA, PISA UNIV., ITALY, REPORT NO:
REPT-41, 1974, 25 PP. (N.T.I.S. N75-22327/1ST)

74-1329 CARTER, J. P.
POULOS, H. G.
BOOKER, J. R.
EFFECT OF SEEPAGE ON EMBANKMENT DEFORMATIONS DUE
TO WATER LOADING
CIVIL ENG. LABS., SYDNEY UNIV., AUSTRALIA, REPORT
NO: R-256, DECEMBER 1974, 19 PP. (N.T.I.S. -
PB-242 450/5ST)

74-1330 RAMEY, G. E.
THE BEHAVIOR OF STATISTICALLY HETEROGENEOUS
EXCAVATED EARTH SLOPES
DEPT. OF CIVIL ENG., AUBURN UNIV., ALA., REPORT
NO: HPR-67-B, JUNE 1974, 181 PP. (N.T.I.S. -
PB-241 917/4ST)

74-1331 CARMIGNANI, C.
FOURIER ANALYSIS OF TRANSIENT PROBLEMS BY THE
FINITE ELEMENT METHOD
IST. DI IMPIANTI NUCLEARI, PISA UNIV., ITALY,
REPORT NO: RP-170(74), 1974, 30 PP. (N.T.I.S. -
N75-20763/9ST)

74-1332 POCHA, J. J.
STUDY OF THE ACOUSTIC EXCITATION OF STRUCTURES
SPACE DIV., HAWKER SIDDELEY DYNAMICS LTD.,
HATFIELD, ENGLAND, REPORT NO: HSD-TP-7508,
ESRO-CR-(P)-596, DECEMBER 1974, 176 PP. (N.T.I.S.
- N75-20464/4ST)

74-1333 PETERSON, W. S.
KOSTEM, C. N.
KULICKI, J. M.
THE INELASTIC ANALYSIS OF REINFORCED CONCRETE
SLABS
FRITZ ENG. LAB., LEHIGH UNIV., BETHLEHEM, PA.,
REPORT NO: FEL-378B.3, MAY 1974, 171 PP.
(N.T.I.S. - PB-241 634/5ST)

74-1334 LAI, N. W.
DOMINGUEZ, R. F.
DUNLAP, W. A.
NUMERICAL SOLUTIONS FOR DETERMINING WAVE-INDUCED
PRESSURE DISTRIBUTIONS AROUND BURIED PIPELINES
COASTAL AND OCEAN ENG. DIV., A AND M UNIV.,
COLLEGE STATION, REPORT NO: TAMU-SG-75-205,
DECEMBER 1974, 105 PP. (N.T.I.S.-COM-75-10503/1ST)

74-1335 STEVEN, G. P.
AUTOMATIC GENERATION OF FINITE ELEMENTS
DEPT. AERON. ENG., SYDNEY UNIV., REPORT NO:
ATN-7405, SEPTEMBER 1974, 28 PP. (N.T.I.S. -
N75-20012/1ST)

74-1336 TAYLOR, C.
AL-MASHIDANI, G.
AN ANALYSIS OF TWO-DIMENSIONAL SURFACE RUN-OFF BY
FINITE ELEMENTS
PROC. INTL. CONF. ON COMPUT. METH. IN NONLINEAR
MECH., UNIV. OF AUSTIN, SEPTEMBER 23-25, 1974.

74-1337 MATSUSUE, K.
NAGAMATSU, A.
HASHIMOTO, R.
YONAIYAMA, M.
STRESSES IN A ROTATING DISK WITH CENTRAL AND
NONCENTRAL HOLES (STRESS ANALYSIS BY THE FINITE
ELEMENT METHOD)
NAT. AEROSPACE LAB., TOKYO, JAPAN, REPORT NO:
NAL-TR-390, OCTOBER 1974, 15 PP. (N.T.I.S. -
N75-19735/0ST)

74-1338 CRONKHITE, J. D.
BERRY, V. L.
BRUNKEN, J. E.
A NASTRAN VIBRATION MODEL OF THE AH-IG HELICOPTER
AIRFRAME. VOLUME 1
BELL HELICOPTER CO., FORT WORTH, TEXAS, REPORT NO.
BHC-209-099-432 VOL. 1, JUNE 1974, 425 PP.
(N.T.I.S. - AD-A009 482/1ST).

74-1339 CRONKHITE, J. D.
BERRY, V. L.
BRUNKEN, J. E.
A NASTRAN VIBRATION MODEL OF THE AH-IG HELICOPTER
AIRFRAME. VOLUME 11
BELL HELICOPTER CO., FORTH WORTH, TEXAS, REPORT
NO. BHC-209-099-432, VOL 2, JUNE 1974, 639 PP.
(N.T.I.S. - AD-A009 483/9ST).

74-1340 BERKHOFF, J. C. W.
LINEAR WAVE PROPAGATION PROBLEMS AND THE FINITE
ELEMENT METHOD
WATERLOOPKUNDIG LABORATORIUM, DELFT, NETHERLANDS,
REPORT NO. PUBL-124, MARCH 1974, 31 PP. (N.T.I.S.
N75-17929/1ST). ALSO IN PROC. INT. SYMP. ON FINITE
ELEMENT METHODS IN FLOW PROBLEMS, SWANSEA, WALES,
JANUARY 7-11, 1974.

74-1341 COOLEY, R. L.
FINITE ELEMENT SOLUTIONS FOR THE EQUATIONS OF
GROUND-WATER FLOW
DESERT RESEARCH INST., NEVADA UNIV., RENO, REPORT
NO: H-W-PUB-18, JANUARY 1974, 141 PP. (N.T.I.S. -
PB-240 813/6ST)

74-1342 MULINAZZI, T. E.
SATTERLY, G. T. JR.
GEOMETRIC HIGHWAY AND CULVERT DESIGN
TRANSPORTATION RESEARCH BOARD, WASH., D.C., REPORT
NO: TRB-TRR-518, ISBM-/-309-02363-7, 1974, 72 PP.
(N.T.I.S. - PB-240 192/5ST)

74-1343 OATES, G. C.
CAREY, G. F.
A VARIATIONAL FORMULATION OF THE COMPRESSIBLE
THROUGHFLOW PROBLEM
AEROSPACE RESEARCH LAB., UNIV. OF WASHINGTON,
SEATTLE, FINAL REPORT, MARCH 1973-APRIL 1974, 93
PP. (N.T.I.S. - AD-A007 979/8ST)

74-1344 DEWAAL, J. F.
STRESS CONCENTRATION FACTORS FOR A FATIGUE TEST
SPECIMEN WITH A U-TYPE NOTCH IN ONE EDGE BY FINITE
ELEMENT ANALYSIS
NATL. AERON. ESTABL., OTTAWA, ONTARIO, REPORT NO:
LTR-ST-721, AUGUST 2, 1974, 27 PP. (N.T.I.S. -
N75-16876/5ST)

74-1345 CULLEN, M. J. P.
SOLUTION OF WATER PREDICTION PROBLEMS BY FINITE
ELEMENT METHODS
PROC. INTL. CONF. ON COMPUT. METH. IN NONLINEAR
MECH., UNIV. OF TEXAS, AUSTIN, SEPTEMBER 23-25,
1974.

74-1346 TURCKE, D. J.
OPTIMUM MESH CONFIGURATIONS IN THE FINITE ELEMENT
METHOD
PH.D. THESIS, UNIV. OF WATERLOO, WATERLOO, ONT.,
AUGUST 1974.

74-1347 CULLEN, M. J. P.
INTEGRATIONS OF THE PRIMITIVE EQUATIONS ON A
SPHERE USING THE FINITE ELEMENT METHOD
QUART. J. ROYAL METEOR. SOC., BRACKNELL, ENGLAND,
VOL. 100(426), OCTOBER 1974, PP. 555-562.

74-1348 NISHIYAMA, S.
SASAKI, T.
KUROKAWA, Y.
STUDIES ON THE DETERMINATION OF SEEPAGE-LINE OF
FILL DAM BY THE METHOD OF FINITE ELEMENT. WATER
FLOW
TECH. BULL., FAC. AGRIC., KAGAWA UNIV., VOL. 26,
NO. 1, OCTOBER 1974, PP. 47-51.

74-1349 OHYA, H. G.
YAGAWA, G
ANDO, Y.
SNAP-THROUGH BUCKLING ANALYSIS OF SHELLS OF
REVOLUTION USING INCREMENTAL STIFFNESS MATRICES
PROC. INTL. CONF. ON COMPUT. METH. IN NONLINEAR
MECH., UNIV. OF TEXAS, AUSTIN, SEPTEMBER 23-25,
1974.

74-1350 DEAK, A. L.
ATLURI, S.
NONLINEAR HYBRID STRESS FINITE ELEMENT ANALYSIS OF
LAMINATED SHELLS
PROC. INTL. CONF. ON COMPUT. METH. IN NONLINEAR
MECH., UNIV. OF TEXAS, AUSTIN, SEPTEMBER 23-25,
1974.

74-1351 HUANG, Y. H.
SONNENFELD, S. L.
ANALYSIS OF UNSTEADY FLOW TOWARD AN ARTESIAN WELL
BY THREE-DIMENSIONAL FINITE ELEMENTS
WATER RESOURCES RESEARCH, VOL. 10, NO. 3, JUNE
1974, PP. 591-596.

74-1352 HUYAKORN, P.
FINITE ELEMENT SOLUTION OF TWO-REGIME FLOW TOWARDS
WELLS
WATER RES. LAB., NEW SOUTH WALES, REPORT NO. 137,
1974, PP. 167-175.

74-1353 BALASUBRAMANIAN, R.
AN ITERATIVE FINITE ELEMENT METHOD OF VISCOUS FLOW
M.SC. THESIS, UNIV. OF CALGARY, JANUARY 1974.

74-1354 ANON
SOLVING THE PROBLEM OF PROCESSING METALS BY
PRESSURE THROUGH THE METHOD OF FINITE ELEMENTS
IZVEST. VUZ CHERNAYA MET., VOL. 5, 1974, PP.
74-78.

74-1355 LESAINT, P.
RAVIART, P. A.
ON A FINITE ELEMENT METHOD FOR SOLVING THE NEUTRON
TRANSPORT EQUATION
IN "MATHEMATICAL ASPECTS OF FINITE ELEMENTS IN
PARTIAL DIFFERENTIAL EQUATIONS", (ED) CARL DE
BOOR, ACADEMIC PRESS, N.Y., PP. 89-123.

74-1356 NAIR, P.
UNIMOD - APPLICATIONS ORIENTED FINITE ELEMENT
SCHEME FOR THE ANALYSIS OF FRACTURE MECHANICS
PROBLEMS
"FRACTURE ANALYSIS", A.S.T.M., PHILADELPHIA, PA.,
1974, PP. 211-225.

74-1357 OHNAKA, I.
CALCULATION OF SOLIDIFICATION OF MOLTEN METAL BY
FINITE-ELEMENT METHOD
TECHNOL. REP., OSAKA UNIV., VOL. 24, OCTOBER 1974,
PP. 461-475.

74-1358 ANON
STIFFNESS DERIVATIVE FINITE ELEMENT TECHNIQUE FOR
DETERMINATION OF CRACK TIP STRESS INTENSITY
FACTORS
INT. J. FRACT., VOL. 10, NO. 4, DECEMBER 1974, PP.
487-502.

74-1359 ANON
FINITE ELEMENT METHOD FOR LINEAR FRACTURE
MECHANICS
CONF. AUSTRALIAN FRACT. GROUP., AUSTRALIAN INST.
OF METALS, PARKVILLE, VICTORIA, 1974, PP. 78-86.

74-1360 DE VRIES, G.
NORRIE, D. H.
APPLICATION OF THE PSEUDO-FUNCTIONAL FINITE
ELEMENT METHOD TO COMPRESSIBLE FLOW PROBLEMS
MECH. ENG. DEPT., UNIV. OF CALGARY, REPORT NO: 51,
JANUARY 1974.

74-1361 WHITE, J. K.
A FINITE ELEMENT DETERMINISTIC CATCHMENT MODEL
PROC. INTL. CONF. IN NUMER. METH. IN FLUID
DYNAMICS, UNIV. OF SOUTHAMPTON, ENGLAND, SEPTEMBER
26-28, 1974, PP. 533-543.

74-1362 PINDER, G. F.
SIMULATION OF GROUND WATER CONTAMINATION USING A
GALERKIN FINITE ELEMENT TECHNIQUE
PROC. INTL. CONF. IN NUMER. METH. IN FLUID
DYNAMICS, UNIV. OF SOUTHAMPTON, ENGLAND, SEPTEMBER
26-28, 1974, PP. 512-532.

74-1363 NYASHIN, YU. I.
ANANEV, I. N.
SOLUTION OF THE PROBLEM OF PROCESSING METALS BY
PRESSURE USING THE METHOD OF FINITE ELEMENTS
IZVEST. VUZ CHERNAYA MET., VOL. 5, 1974, PP.
74-78.

74-1364 KOLAR, V.
KRATOCHVIL, J.
LEITNER, F.
ZENISEK, A.
DESIGN OF TWO AND THREE DIMENSIONAL STRUCTURES BY
THE FINITE ELEMENT METHOD (IN GERMAN)
PUBL. BY SPRINGER VERLAG, WIEN AND NEW YORK, 1974.

74-1365 MUYAMOTO, H.
EFFECTS OF A PEAK OVERLOAD ON FATIGUE CRACK
PROPAGATION RATE-PLANE STRESS ANALYSIS OF CRACKED
PLATE BY FINITE ELEMENT METHOD
MECH. BEHAV. MATER., SOC. MATER. SCI., KYOTO,
JAPAN, 1974, PP. 149-158.

74-1366 JINOCH, J.
BILEK, Z.
APPLICATION OF THE FINITE ELEMENT METHOD IN METAL
DIFFUSION
KOVOVE MAT., VOL. 12, NO. 3, 1974, PP. 305-317.

74-1367 ANON
ANALYSIS FOR CRACKS IN SHIP STRUCTURES USING THE
FINITE-ELEMENT METHOD WITH THE CONCEPT OF CRACK
SUB-STRUCTURE
MITSUBISHI TECH. REV., VOL. 11, NO. 1, FEBRUARY
1974, PP. 18-27.

74-1368 TONG, P.
FINITE ELEMENT SOLUTION OF THE WIND DRIVEN
CURRENTS AND ITS MASS TRANSPORT IN LAKES
PROC. INTL. CONF. IN NUMER. METH. IN FLUID
DYNAMICS, UNIV. OF SOUTHAMPTON, ENGLAND, SEPTEMBER
26-28, 1974, PP. 440-453.

74-1369 DAILEY, J. E.
HARLEMAN, D. R. F.
A NUMERICAL MODEL OF TRANSIENT WATER QUALITY IN A
ONE-DIMENSIONAL ESTUARY BASED ON THE FINITE
ELEMENT METHOD
PROC. INTL. CONF. IN NUMER. METH. IN FLUID
DYNAMICS, UNIV. OF SOUTHAMPTON, ENGLAND, SEPTEMBER
26-28, 1974, PP. 412-439.

74-1370 CONNOR, J. J.
WANG, J.
FINITE ELEMENT MODELLING OF HYDRODYNAMIC
CIRCULATION
PROC. INTL. CONF. IN NUMER. METH. IN FLUID
DYNAMICS, UNIV. OF SOUTHAMPTON, ENGLAND, SEPTEMBER
26-28, 1974, PP. 355-387.

74-1371 ADEY, R. A.
BREBBIA, C. A.
FINITE ELEMENT SOLUTION FOR EFFLUENT DISPERSION
PROC. INTL. CONF. IN NUMER. METH. IN FLUID
DYNAMICS, UNIV. OF SOUTHAMPTON, ENGLAND,
SEPTEMBER 26-28, 1974, PP. 325-354.

74-1372 BEDFORD, K.
A NUMERICAL INVESTIGATION OF STABLY STRATIFIED,
WIND DRIVEN CAVITY FLOW BY THE FINITE ELEMENT
METHOD
PH.D. DISSERTATION, SCHOOL OF CIVIL AND
ENVIRONMENTAL ENG., CORNELL UNIV., 1974.

74-1373 COOK, R. D.
CONCEPTS AND APPLICATIONS OF FINITE ELEMENT
ANALYSIS
JOHN WILEY, NEW YORK, 1974.

74-1374 OLSON, M. D.
ANALYSIS OF ARBITRARY SHELLS USING SHALLOW SHELL
FINITE ELEMENTS
IN "THIN SHELL STRUCTURES", (EDS.) FUNG, Y.C., AND
SECHLER, E.E., PP. 403-434, (PUBL. BY PRENTICE
HALL, N.J., 1974).

74-1375 KONO, I.
FINITE ELEMENT ANALYSIS OF A NONSTEADY GROUNDWATER
SEEPAGE PROBLEM DAM, DRAINAGE
SOILS FOUND., VOL. 14, NO. 4, DECEMBER 1974, PP.
75-85.

74-1376 SMITH, I. M.
HOBBS, R.
FINITE ELEMENT ANALYSIS OF CENTRIFUGED AND BUILT
UP SLOPES
J. BIOMECH., VOL. 7, NO. 1, JANUARY 1974, PP.
21-28.

74-1377 STEINMUELLER, G.
RESTRICTIONS IN THE APPLICATION OF AUTOMATIC
GENERATION SCHEMES BY ISOPARAMETRIC CO-ORDINATES
INTL. J. NUMER. METH. IN ENG., VOL. 8, NO. 2,
1974, PP. 289-294.

75-0001 ROSSOW, M. P.
LEE, J. C.
CHEN, K. C.
COMPUTER IMPLEMENTATION OF THE CONSTRAINT METHOD
WASH. UNIV., ST. LOUIS, MO. AND AMCAR DIV. ACF
IND., ST. CHARLES, MO., REPORT DOT-OS-30108-3,
1975.

75-0002 KATZ, I. M.
PEANO, A. G.
SZABO, B. A.
NODAL VARIABLES FOR ARBITRARY ORDER FINITE
ELEMENTS
WASH. UNIV., ST. LOUIS, MO., REPORT
DOT-OS-30108-5, 1975.

75-0003 BAKER, A. J.
PREDICTIONS IN ENVIRONMENTAL HYDRODYNAMICS USING
FINITE ELEMENT METHOD 1, THEORETICAL DEVELOPMENT
A.I.A.A. J., VOL. 13, NO. 1, 1975, PP: 36-42.

75-0004 RAO, S. S.
FINITE ELEMENT FLUTTER ANALYSIS OF MULTI-WEB WING
STRUCTURES
J. OF SOUND AND VIB., VOL. 38, NO. 2, 1975, PP:
233-244.

75-0005 ASKAR, A.
FINITE ELEMENT METHOD FOR BOUND STATE CALCULATIONS
IN QUANTUM MECHANICS
J. OF CHEMICAL PHYSICS, VOL. 62, NO. 2, 1975, PP:
732-734.

75-0006 WELLFORD, L. C.
ODEN, J. T.
ACCURACY AND CONVERGENCE OF FINITE
ELEMENT/GALERKIN APPROXIMATIONS OF TIME DEPENDENT
PROBLEMS WITH EMPHASIS ON DIFFUSION
IN "FINITE ELEMENTS IN FLUIDS", GALLAGHER, R.H.,
ET AL (EDS), VOL. 2, 1975, PP. 31-54. (PUBL. BY
JOHN WILEY)

75-0007 TAYLOR, R. L.
IMPACT OF FINITE ELEMENT METHOD ON STRUCTURAL
ENGINEERING
NOTICES OF THE AMER. MATH. SOC., VOL. 22, NO. 1,
1975, PP:A264-265.

75-0008 ROBINSON, J.
EQUILIBRIUM MODELS
WORLD CONGRESS ON FINITE ELEMENT METH. IN STRUCT.
MECH., BOURNEMOUTH, DORSET, ENGLAND, OCTOBER
12-17, 1975, PP. I.1-I.41.

75-0009 HENNART, J. P.
FINITE-ELEMENT METHODS FOR REACTOR ANALYSIS -
COMMENT
NUCLEAR SCIENCE AND ENG., VOL. 56, NO. 2, 1975, P:
225-226.

75-0010 SCHMITT, K.
APPLICATIONS OF VARIATIONAL EQUATIONS TO ORDINARY
AND PARTIAL DIFFERENTIAL EQUATIONS - MULTIPLE
SOLUTIONS OF BOUNDARY VALUE PROBLEMS
J. OF DIFFERENTIAL EQUATIONS, VOL. 17, NO. 1,
1975, PP: 154-186.

75-0011 PIFKO, A.
LEVINE, H. S.
ARMEN, H.
LEVY, A.
PLANS - A FINITE-ELEMENT PROGRAM FOR NONLINEAR
ANALYSIS OF STRUCTURES
MECHANICAL ENG. VOL. 97, NO. 2, 1975, P: 50.

75-0012 AAMODT, B.
MO, O.
ELASTO-PLASTIC ANALYSIS USING AN EFFICIENT
FORMULATION OF THE FINITE ELEMENT METHOD
PROC. 3RD INT. CONF. ON STRUCT. MECH. IN REACTOR
TECH., LONDON, VOL. 7, SEPTEMBER 1975.

75-0013 KEY, S. W.
FINITE ELEMENT PROCEDURE FOR LARGE DEFORMATION
DYNAMIC RESPONSE OF AXISYMMETRIC SOLIDS
MECHANICAL ENG. VOL. 97, NO. 2, 1975, P: 51.

75-0014 NICKELL, R. E.
COMPUTER PROGRAM CONSTRUCTION AND MAINTENANCE -
FUTURE OF CENTRALIZED FINITE ELEMENT ACTIVITY
NOTICES OF THE AMERICAN MATH. SOCI., VOL. 22, NO.
1, 1975, P: A265-266.

75-0015 DUPONT, T.
MODELING WAVE PROPAGATION WITH FINITE ELEMENT
METHODS
NOTICES OF THE AMER. MATH. SOC., VOL. 22, NO. 1,
1975, P. A265.

75-0016 FRAEIJS DE VEUBEKE, B.
STRESS FUNCTION APPROACH
WORLD CONGRESS ON FINITE ELEMENT METH. IN STRUCT.
MECH., BOURNEMOUTH, DORSET, ENGLAND, OCTOBER
12-17, 1975, PP. J.1-J.51.

75-0017 CRAWFORD, J. E.
JOHNSON, F. R.
FINITE ELEMENT ANALYSES OF QUICK CAMP MODULE
CIV. ENG. LAB., PORT HUENEME, CALIF., REPORT NO:
CEL-TN-1369 JANUARY 1975, 72 P. (N.T.I.S. -
AD/A/004 932/OST)

75-0018 WARD, G.
WILLSHARE, G. T.
PROPELLER-EXCITED VIBRATION WITH PARTICULAR
REFERENCE TO FULL-SCALE MEASUREMENTS
PAPER NO. 4, MEETING OF THE ROYAL INSTITUTION OF
NAVAL ARCHITECTS, LONDON, APRIL 23, 1975, PP.
1-16.

75-0019 ROCKEY, K. C. ET AL
THE FINITE ELEMENT METHOD: A BASIC INTRODUCTION
CROSBY LOCKWOOD STAPLES, LONDON, 1975, 239 PP.

75-0020 HANSEN, K. F.
KANG, C. M.
FINITE ELEMENT METHODS IN REACTOR PHYSICS ANALYSIS
ADVANCES IN NUCLEAR SCIENCE AND TECHNOLOGY, EDITED
BY E. J. HENLEY, J. LEWINS, ACADEMIC PRESS, VOL.
8, FEBRUARY 1975, PP. 362.

75-0021 OATES, G. C.
KNIGHT, C. J.
CAREY, G. F.
A VARIATIONAL FORMULATION OF THE COMPRESSIBLE
THROUGHFLOW PROBLEM
PREPR. PAPER NO. 75-GT-32, A.S.M.E. GAS TURBINE
CONF., HOUSTON, TEXAS, MARCH 2-6, 1975.

75-0022 SALAAM, U.
SANDHU, R. S.
NUMERICAL PERFORMANCE OF A FINITE ELEMENT GALERKIN
PROCEDURE
PROC. 5TH CAN. CONGRESS OF APPLIED MECH., UNIV. OF
NEW BRUNSWICK, FREDERICTON, N.B., MAY 26-30, 1975,
PP. 657-658.

75-0023 SCHMIDT, W. F.
A COMPARISON AND EVALUATION OF INCREMENTAL FINITE
ELEMENT FORMULATIONS
FIFTH CAN. CONGRESS OF APPLIED MECH., UNIV. OF NEW
BRUNSWICK, FREDERICTON, N.B., MAY 26-30, 1975.

75-0024 ELSAIE, A. M.
FENTON, R. G.
TABARROK, B.
VIBRATION OF SQUARE PLATES HAVING VARIOUS EDGE
CONDITIONS AND LOADINGS USING A REFINED FINITE
ELEMENT
FIFTH CAN. CONGRESS OF APPLIED MECH., UNIV. OF NEW
BRUNSWICK, FREDERICTON, N.B., MAY 26-30, 1975.

75-0025 LEWIS, R. W.
A TWO-DIMENSIONAL FINITE ELEMENT - LEAST SQUARES
SOLUTION
FIFTH CAN. CONGRESS OF APPLIED MECH., UNIV. OF NEW
BRUNSWICK, FREDERICTON, N.B., MAY 26-30, 1975.

75-0026 IRONS, B. M.
THE SUPERPATCH THEOREM AND OTHER PROPOSITIONS
RELATING TO THE PATCH TEST
FIFTH CAN. CONGRESS OF APPLIED MECH., UNIV. OF NEW
BRUNSWICK, FREDERICTON, N.B., MAY 26-30, 1975

75-0027 MIRZA, F. A.
OLSON, M. D.
A HYBRID FINITE ELEMENT METHOD FOR TWO-DIMENSIONAL
NAVIER STOKES EQUATIONS
FIFTH CAN. CONGRESS OF APPLIED MECH., UNIV. OF NEW
BRUNSWICK, FREDERICTON, N.B., MAY 26-30, 1975.

75-0028 TURCKE, D. J.
FURTHER DEVELOPMENTS IN GRID SELECTION PROCEDURES
IN THE FINITE ELEMENT METHOD
FIFTH CAN. CONGRESS OF APPLIED MECH., UNIV. OF NEW
BRUNSWICK, FREDERICTON, N.B., MAY 26-30, 1975.

75-0029 DEAK, A. L.
ATLURI, S.
NONLINEAR HYBRID STRESS FINITE ELEMENT ANALYSIS OF
PNEUMATIC TIRES
FIFTH CAN. CONGRESS OF APPLIED MECH., UNIV. OF NEW
BRUNSWICK, FREDERICTON, N.B., MAY 26-30, 1975.

75-0030 HRUDEY, T. M.
A FINITE ELEMENT PROCEDURE FOR PLATES WITH CURVED
BOUNDARIES
FIFTH CAN. CONGRESS OF APPLIED MECH., UNIV. OF NEW
BRUNSWICK, FREDERICTON, N.B., MAY 26-30, 1975.

75-0031 THOMAS, G. R.
COMPUTATIONAL EFFICIENCY ASPECTS OF THE FINITE
ELEMENT NONLINEAR ANALYSIS OF SHELLS
FIFTH CAN. CONGRESS OF APPLIED MECH., UNIV. OF NEW
BRUNSWICK, FREDERICTON, N.B., MAY 26-30, 1975.

75-0032 YALAMANCHILI, R.
AN EXTENSION OF FINITE-ELEMENT METHOD TO INCLUDE
NONLINEAR BOUNDARY CONDITIONS
FIFTH CAN. CONGRESS OF APPLIED MECH., UNIV. OF NEW
BRUNSWICK, FREDERICTON, N.B., MAY 26-30, 1975.

75-0033 TINAWI, R.
NESSIM, A.
A COMPARISON BETWEEN TWO-HIGH-PRECISION FINITE
ELEMENTS
FIFTH CAN. CONGRESS OF APPLIED MECH., UNIV. OF NEW
BRUNSWICK, FREDERICTON, N.B., MAY 26-30, 1975.

75-0034 FRIED, I.
FINITE ELEMENT ANALYSIS OF THIN ELASTIC SHELLS
WITH RESIDUAL ENERGY BALANCING AND THE ROLE OF THE
RIGID BODY MODES
PAPER NO. 75-APM-6, A.S.M.E, MEETING, JUNE 23-25,
1975, 6 PP.

75-0035 CLOUGH, R. W.
APPLICATION OF THE FINITE ELEMENT METHOD
PROC. CONF. ON THE MATHS. OF FINITE ELEMENTS AND
APPL., BRUNEL UNIV., APRIL 7-10, 1975.

75-0036 TORBE, I.
CHURCH, K.
IMPROVED GENERAL QUADRILATERAL PLATE ELEMENTS
J. ROY. INST. NAVAL ARCHITECTS, JANUARY 1975, P.
14.

75-0037 STURGUL, J. R.
GRINSHPAN, Z.
FINITE-ELEMENT MODEL FOR POSSIBLE ISOSTATIC
REBOUND IN GRAND CANYON
GEOLOGY, VOL. 3, NO. 4, 1975, PP. 169-171.

75-0038 VYSLOUKH, V. A.
KANDIDOV, V. P.
CHESNOKOV, S. S.
MODEL OF ELASTIC PLATE FROM FINITE ELEMENTS IN A
SUPERSONIC FLOW
PRIKLADNAYA MATEMATIKA I MEKHANIKA, VOL. 39, NO.
1, 1975, PP. 86-94.

75-0039 NOOR, A. K.
FULTON, R. E.
IMPACT OF CDC STAR-100 COMPUTER ON FINITE ELEMENT
SYSTEMS
ASCE J. STRUCT. DIV., VOL. 101, NO. 4, 1975, PP.
731-750.

75-0040 VISSER, C.
LIEN, S.
SCHALLER, R. J.
APPLICATION OF FINITE ELEMENT ANALYSIS TO CERAMIC
COMPONENTS
J. AMER. CERAMIC SOC., VOL. 58, NO. 3-4, 1975, PP.
131-135.

75-0041 MERCIER, B.
FINITE ELEMENT APPROXIMATION, AND SOLUTION BY A
PENALTY-DUALITY ALGORITHM OF AN ELASTO-PLASTIC
PROBLEM
COMPTES RENDUS, ACADEMIE DES SCIENCES SER A MATH.,
VOL. 280, NO. 5, 1975, PP. 287-290.

75-0042 FIX, G. J.
FINITE ELEMENTS AND SCIENTIFIC COMPUTING
SIAM REVIEW, VOL. 17, NO. 2, 1975, P. 378.

75-0043 VANDERVORST, M. J.
BERGER, A. E.
USE OF FINITE ELEMENT METHOD IN A MARKER AND CELL
PROGRAM FOR SOLVING TIME-DEPENDENT FLUID FLOW
PROBLEMS WITH FREE SURFACES
SIAM REVIEW, VOL. 17, NO. 2, 1975, P. 390.

75-0044 HENNART, J. P.
EXTRAPOLATED FINITE ELEMENT PROCEDURES OVER
CARTESIAN PRODUCT DOMAINS
SIAM REVIEW, VOL. 17, NO. 2, 1975, PP. 380-381.

75-0045 MARGOLIS, S. G.
FINITE ELEMENT METHODS FOR UNSTEADY COMPRESSIBLE
GAS DYNAMICS
SIAM REVIEW, VOL. 17, NO. 2, 1975, P. 385.

75-0046 LORD, W.
HWANG, J. H.
PREDICTING MAGNETIC LEAKAGE FIELDS BY FINITE
ELEMENT TECHNIQUES
SIAM REVIEW, VOL. 17, NO. 2, 1975, P. 384.

75-0047 FUNNELL, W. R. J.
LASZLO, C. A.
MODELING EARDRUM AS A DOUBLY CURVED SHELL USING
FINITE-ELEMENT METHOD
J. ACOUST. SOC. AM., VOL. 57, NO. 1, 1975, P. 72.

75-0048 FRIED, I.
FINITE-ELEMENT ANALYSIS OF THIN ELASTIC SHELLS
WITH RESIDUAL ENERGY BALANCING AND ROLE OF RIGID
BODY MODES
J. APPLIED MECH., VOL. 42, NO. 1, 1975, PP.
99-104.

75-0049 MURPHY, D. J.
CLOUGH, G. W.
TEMPORARY EXCAVATION IN VARVED CLAY
ASCE J. GEOTECH. ENG. DIV., VOL. 101, NO. 3, MARCH
1975, PP. 279-295.

75-0050 YUDHBIR,
VARADARAJAN, A.
STRESS-PATH DEPENDENT DEFORMATION MODULI OF CLAY
ASCE J. GEOTECH. ENG. DIV., VOL. 101, NO. 3, MARCH
1975, PP. 315-327.

75-0051 MOWATT, G. A.
STRENGTH OF SHIP STRUCTURAL ELEMENTS
TRANS. NORTH EAST COAST INST. ENG. SHIPBUILD.,
VOL. 91, NO. 3, FEBRUARY 1975, PP. 85-104.

75-0052 MOHRAZ, B.
HENDRON, A. J. JR.
LINER-MEDIUM INTERACTION IN TUNNELS
ASCE J. CONSTR. DIV., VOL. 101, NO. 1, MARCH 1975,
PP. 127-141.

75-0053 SCORDELIS, A. C.
LIN, C-S.
NONLINEAR ANALYSIS OF RC SHELLS OF GENERAL FORM
ASCE J. STRUCT. DIV., VOL. 101, NO. 3, MARCH 1975,
PP. 523-538.

75-0054 SKERMER, N. A.
MICA DAM EMBANKMENT STRESS ANALYSIS
ASCE J. GEOTECH. ENG. DIV., VOL. 101, NO. 3, MARCH
1975, PP. 229-242.

75-0055 KULHAWY, F. H.
FLANAGAN, R. F.
ANALYSIS OF BEHAVIOR OF EDWARD HYATT POWER PLANT
ASCE J. GEOTECH. ENG. DIV., VOL. 101, NO. 3, MARCH
1975, PP. 243-257.

75-0056 PREVOST, J. H.
HOEG, K.
EFFECTIVE STRESS-STRAIN-STRENGTH MODEL FOR SOILS
ASCE J. GEOTECH. ENG. DIV., VOL. 101, NO. 3, MARCH
1975, PP. 259-278.

75-0057 ARGYRIS, J. H.
BRONLUND, O. E.
NATURAL FACTOR FORMULATION OF THE STIFFNESS FOR
THE MATRIX DISPLACEMENT METHOD
COMPUTER METHODS APPL. MECH. ENG., VOL. 5, NO. 1,
JANUARY 1975, PP. 97-119.

75-0058 LEE, E. H.
MALLETT, R. L.
DYNAMIC ANALYSIS OF STRUCTURAL DEFORMATION AND
METAL FORMING
COMPUTER METHODS APPL. MECH. ENG., VOL. 5, NO. 1,
JANUARY 1975, PP. 69-82.

75-0059 DISTEFANO, N.
SAMARTIN, A.
DYNAMIC PROGRAMMING APPROACH TO THE FORMULATION
AND SOLUTION OF FINITE ELEMENT EQUATIONS
COMPUTER METH. APPL. MECH. ENG., VOL. 5, NO. 1,
JANUARY 1975, PP. 37-52.

75-0060 KNUDSON, R. M.
SCHNIEWIND, A. P.
PERFORMANCE OF STRUCTURAL WOOD MEMBERS EXPOSED TO
FIRE
PROD. J., VOL. 25, NO. 2, FEBRUARY 1975, PP.
23-32.

75-0061 HAY, J. K.
MARTZ, J. W.
AVOIDING DANGEROUS AND COSTLY FAN FAILURES
MACH. DES., VOL. 47, NO. 4, FEBRUARY 20, 1975, PP.
113-119.

75-0062 SATAKE, M.
NIISEKI, S.
STUDY ON WEIGHT MATRIX IN STRUCTURAL ANALYSIS
PROC. JAP. SOC. CIV. ENG., NO. 233, JANUARY 1975,
PP. 25-34.

75-0063 JONES, R. M.
NELSON, D. A. R. JR.
NEW MATERIAL MODEL FOR THE NONLINEAR BIAXIAL
BEHAVIOR OF ATJ-S GRAPHITE
J. COMPOS. MATER., VOL. 9, NO. 1, JANUARY 1975,
PP. 10-27.

75-0064 ZELAZNY, S. W.
BAKER, A. J.
PREDICTIONS IN ENVIRONMENTAL HYDRODYNAMICS USING
THE FINITE ELEMENT METHOD 2. APPLICATIONS
A.I.A.A. J., VOL. 13, NO. 1, JANUARY 1975, PP.
43-46.

75-0065 MOKRY, M.
INTEGRAL EQUATION METHOD FOR SUBSONIC FLOW PAST
AIRFOILS IN VENTILATED WIND TUNNELS
A.I.A.A. J., VOL. 13, NO. 1, JANUARY 1975, PP.
47-53.

75-0066 BRANDES, H.
MARTIN, H.
SPANNUNGEN IN MEMBRANWAENDEN DURCH
WAERMEBEANSPRUCHUNG UND MECHANISCHE BELASTUNG.
(STRESSES IN MEMBRANE WALLS DUE TO THERMAL STRESS
AND MECHANICAL LOADING)
VGB KRAFTSWERKSTECH, VOL. 55, NO. 1, JANUARY 1975,
PP. 12-20.

75-0067 RAJU, K. K.
RAO, G. V.
CALCULATION OF NON-LINEAR AXISYMMETRIC VIBRATIONS
OF THIN SHELLS OF REVOLUTION BY A FINITE ELEMENT
METHOD
J. SOUND AND VIB., VOL. 38, NO. 4, 1975, PP.
505-509.

75-0068 HWANG, J. H.
LORD, W.
FINITE ELEMENT MODELING OF MAGNETIC FIELD DEFECT
INTERACTIONS
J. TEST EVAL., VOL. 3, NO. 1, JANUARY 1975, PP.
21-25.

75-0069 HUNT, J. T.
KNITTEL, M. R.
NICHOLS, C. S.
BARACH, D.
FINITE-ELEMENT APPROACH TO ACOUSTIC SCATTERING
FROM ELASTIC STRUCTURES
J. ACOUST. SOC. AM., VOL. 57, NO. 2, 1975, PP.
287-299.

75-0070 BRUSA, L.
CIACCI, R.
GRECO, A.
TICOZZI, C.
USE OF PAS-1 FINITE ELEMENT COMPUTER SYSTEM FOR
STATIC AND DYNAMIC ANALYSIS OF 2-DIMENSIONAL AND
3-DIMENSIONAL STRUCTURES
ENERGIA NUCLEARE, VOL. 22, NO. 1, 1975, PP. 41-50.

75-0071 DOUGLAS, J.
DUPONT, T.
WAHLBIN, L.
STABILITY IN LQ OF L2-PROJECTION INTO FINITE
ELEMENT FUNCTION SPACES
NUMER. MATH., VOL. 23, NO. 3, 1975, PP. 193-197.

75-0072 PITKARANTA, J.
NON-SELF-ADJOINT VARIATIONAL PROCEDURE FOR
FINITE-ELEMENT APPROXIMATION OF TRANSPORT EQUATION
TRANSPORT THEORY AND STATISTICAL PHYSICS. VOL. 4,
NO. 1, 1975, PP. 1-24.

75-0073 WEINSTEIN, A. M.
KLAWITTER, J. J.
ANAND, S. C.
ADAMS, R. R.
FINITE ELEMENT STRESS ANALYSIS OF PROSTHETIC TOOTH
IMPLANTS
J. DENTAL RESEARCH, VOL. 54, 1975, P. 115.

75-0074 WANG, M-H.
CHENG, R. T-S.
STUDY OF CONVECTIVE-DISPERSION EQUATION BY
ISOPARAMETRIC FINITE ELEMENTS
J. HYDROL., VOL. 24, NO. 1-2, 1975, PP. 45-56.

75-0075 MATHES, H. B.
PRICE, E. W.
METHODS FOR DETERMINING CHARACTERISTICS OF
ACOUSTIC WAVES IN ROCKET MOTORS
J. SPACECRAFT ROCKETS, VOL. 12, NO. 1, JANUARY
1975, PP. 39-43.

75-0076 LIAW, C. Y.
CHOPRA, A. K.
EARTHQUAKE ANALYSIS OF AXISYMMETRIC TOWERS
PARTIALLY SUBMERGED IN WATER
EARTHQUAKE ENG. STRUCT. DYN., VOL. 3, NO. 3,
JANUARY - MARCH 1975, PP. 233-248.

75-0077 RODRIGUES, J. S. N.
NODE NUMBERING OPTIMIZATION IN STRUCTURAL ANALYSIS
ASCE J. STRUCT. DIV., VOL. 101, NO. 2, FEBRUARY
1975, PP. 361-376.

75-0078 AZAR, J. J.
BOMBA, J. G.
PIPELINE STRESS ANALYSIS
ASCE TRANSP. ENG. J., VOL. 101, NO. 1, FEBRUARY
1975, PP. 163-170.

75-0079 GANGARAO, H. V. S.
ELMEGED, A. B.
MACRO-APPROACH FOR RIBBED AND GRID PLATE SYSTEMS
ASCE J. ENG. MECH. DIV., VOL. 101, NO. 1, FEBRUARY
1975, PP. 25-43.

75-0080 MUKHERJI, B.
NOTE ON RECTANGULAR FINITE ELEMENTS WITH IN-PLANE
FORCES
A.I.A.A. J., VOL. 13, NO. 2, 1975, PP. 240-241.

75-0081 LENCOVA, B.
FINITE ELEMENT METHOD IN ELECTRON OPTICS
CESKOSLOVENSKY CASOPIS PRO FYSIKU, VOL. 25, NO. 1,
1975, PP. 57-61.

75-0082 SCHMIT, L. C. JR.
FRAME OPTIMIZATION INCLUDING FREQUENCY CONSTRAINTS
ASCE J. STRUCT. DIV., VOL. 101, NO. 1, JANUARY
1975, PP. 283-293.

75-0083 CHENG, R. T-S.
HU, M-H.
STUDY OF FLUID MOVEMENTS THROUGH CAUSEWAY
ASCE J. HYDRAUL. DIV., VOL. 101, NO. 1, JANUARY
1975, PP. 155-165.

75-0084 GREEN, R.
POST-TENSIONED SLAB BRIDGES ON ISOLATED SUPPORTS
ASCE J. STRUCT. DIV., VOL. 101, NO. 1, JANUARY
1975, PP. 233-247.

75-0085 WANCHOO, M. K.
MAY, G. W.
CRACKING ANALYSIS OF REINFORCED CONCRETE PLATES
ASCE J. STRUCT. DIV., VOL. 101, NO. 1, JANUARY
1975, PP. 201-215.

75-0086 RICHARDS, J. A.
KERFOOT, R. P.
WOOD SHEAR PANELS BONDED WITH FLEXIBLE ADHESIVES
ASCE J. STRUCT. DIV., VOL. 101, NO. 1, JANUARY
1975, PP. 131-149.

75-0087 KROENKE, W. C.
GUTZWILLER, M. J.
LEE, R. H.
FINITE ELEMENT FOR REINFORCED CONCRETE FRAME STUDY
- CLOSURE
ASCE J. STRUCT. DIV., VOL. 101, NO. 3, 1975, P.
605.

75-0088 WADE, B. G.
KOBAYASHI, A. S.
PHOTOELASTIC INVESTIGATION ON THE CRACK-ARREST
CAPABILITY OF A PRETENSIONED STIFFENED PLATE
EXP. MECH., VOL. 15, NO. 1, JANUARY 1975, PP. 1-9.

75-0089 KIKUCHI, F.
THE HYBRID DISPLACEMENT METHOD APPLIED TO PLATE
AND SHELL PROBLEMS
PROC. NATL. SYMP. OF THE SOC. OF STEEL CONSTR. OF
JAPAN ON THE MATRIX METHODS OF STRUCT. ANALYSIS,
JUNE 1975.

75-0090 WHITEMAN, J. R.
SOME ASPECTS OF THE MATHEMATICS OF FINITE ELEMENTS
PROC. CONF. ON THE MATHS. OF FINITE ELEMENTS AND
APPL., BRUNEL UNIV., APRIL 7-10, 1975.

75-0091 MITCHELL, A. R.
BASIC FUNCTIONS FOR CURVED ELEMENTS IN THE
MATHEMATICAL THEORY OF FINITE ELEMENTS
PROC. CONF. ON THE MATHS. OF FINITE ELEMENTS AND
APPL., BRUNEL UNIV., APRIL 7-10, 1975.

75-0092 MCLEOD, R.
OVERCOMING LOSS OF ACCURACY WHEN USING CURVED
FINITE ELEMENTS
PROC. CONF. ON THE MATHS. OF FINITE ELEMENTS AND
APPL., BRUNEL UNIV., APRIL 7-10, 1975.

75-0093 BARNHILL, R. E.
BLENDING FUNCTION FINITE ELEMENTS FOR CURVED
BOUNDARIES
PROC. CONF. ON THE MATHS. OF FINITE ELEMENTS AND
APPL., BRUNEL UNIV., APRIL 7-10, 1975.

75-0094 WATKINS, D. S.
SOME CONFORMING, RECTANGULAR, PLATE ELEMENTS
PROC. CONF. ON THE MATHS. OF FINITE ELEMENTS AND
APPL., BRUNEL UNIV., APRIL 7-10, 1975.

75-0095 ZLAMAL, M.
FINITE ELEMENT METHODS IN HEAT CONDUCTION PROBLEMS
PROC. CONF. ON THE MATHS. OF FINITE ELEMENTS AND
APPL., BRUNEL UNIV., APRIL 7-10, 1975.

75-0096   CIARLET, P. G.
CONFORMING FINITE ELEMENT METHOD FOR THE SHELL
PROBLEM
PROC. CONF. ON THE MATHS. OF FINITE ELEMENTS AND
APPL., BRUNEL UNIV., APRIL 7-10, 1975.

75-0097   BABUSKA, I.
PROBLEMS OF SELF ADAPTIVENESS IN THE FINITE
ELEMENT METHOD
PROC. CONF. ON THE MATHS. OF FINITE ELEMENTS AND
APPL., BRUNEL UNIV., APRIL 7-10, 1975.

75-0098   SCOTT, R.
ACCURATE FINITE ELEMENT APPROXIMATION FOR PROBLEMS
WITH SINGULAR DATA
PROC. CONF. ON THE MATHS. OF FINITE ELEMENTS AND
APPL., BRUNEL UNIV., APRIL 7-10, 1975.

75-0099   LESAINT, P.
CONTINUOUS AND DISCONTINUOUS FINITE ELEMENT
METHODS FOR SOLVING THE TRANSPORT EQUATION
PROC. CONF. ON THE MATHS. OF FINITE ELEMENTS AND
APPL., BRUNEL UNIV., APRIL 7-10, 1975.

75-0100   GREGORY, J. A.
FINITE ELEMENT ERROR BOUNDS
PROC. CONF. ON THE MATHS. OF FINITE ELEMENTS AND
APPL., BRUNEL UNIV., APRIL 7-10, 1975.

75-0101   MEYER, G. H.
SEWELL, E. G.
FINITE ELEMENT SOLUTION OF DEGENERATE INTERFACE
PROBLEMS
PROC. CONF. ON THE MATHS. OF FINITE ELEMENTS AND
APPL., BRUNEL UNIV., APRIL 7-10, 1975.

75-0102   CELLA, A.
ON THE ACCURACY AND NUMERICAL STABILITY OF THE
FINITE ELEMENT APPROXIMATION FOR PARABOLIC AND
HYPERBOLIC OPERATORS
PROC. CONF. ON THE MATHS. OF FINITE ELEMENTS AND
APPL., BRUNEL UNIV., APRIL 7-10, 1975.

75-0103   NASSIF, N. R.
ON THE DISCRETIZATION OF THE TIME VARIABLE FOR
PARABOLIC PARTIAL DIFFERENTIAL EQUATIONS
PROC. CONF. ON THE MATHS. OF FINITE ELEMENTS AND
APPL., BRUNEL UNIV., APRIL 7-10, 1975.

75-0104   CECCHI, M. M.
ERROR ESTIMATES FOR FINITE ELEMENT METHOD FOR HEAT
TRANSFER PROBLEMS IN THE SPACE-TIME DOMAIN
PROC. CONF. ON THE MATHS. OF FINITE ELEMENTS AND
APPL., BRUNEL UNIV., APRIL 7-10, 1975.

75-0105   O'CARROLL, M. J.
HARRISON, H. T.
A VARIATIONAL METHOD FOR FREE BOUNDARY PROBLEMS
PROC. CONF. ON THE MATHS. OF FINITE ELEMENTS AND
APPL., BRUNEL UNIV., APRIL 7-10, 1975.

75-0106   BRILLA, J.
A GENERALIZATION OF FINITE ELEMENT METHOD TO
DYNAMIC VISCOELASTIC ANALYSIS
PROC. CONF. ON THE MATHS. OF FINITE ELEMENTS AND
APPL., BRUNEL UNIV., APRIL 7-10, 1975.

75-0107   FROIDEVAUX, H.
A FINITE ELEMENT METHOD FOR THE RESOLUTION OF
NON-LINEAR BOUNDARY-VALUE PROBLEMS
PROC. CONF. ON THE MATHS. OF FINITE ELEMENTS AND
APPL., BRUNEL UNIV., APRIL 7-10, 1975.

75-0108   ARAL, M. M.
ADALI, S.
FINITE ELEMENT SOLUTION OF A NON-LINEAR
DIFFERENTIAL EQUATION USING KACHANOV'S AND
SUCCESSIVE APPROXIMATION METHODS
PROC. CONF. ON THE MATHS. OF FINITE ELEMENTS AND
APPL., BRUNEL UNIV., APRIL 7-10, 1975.

75-0109   ZIENKIEWICZ, O. C.
NEWTONIAN AND NON-NEWTONIAN VISCOUS FLOW
PROC. CONF. ON THE MATHS. OF FINITE ELEMENTS AND
APPL., BRUNEL UNIV., APRIL 7-10, 1975.

75-0110   ODEN, J. T.
FINITE ELEMENT ANALYSIS OF WAVES IN NONLINEAR
MATERIALS
PROC. CONF. ON THE MATHS. OF FINITE ELEMENTS AND
APPL., BRUNEL UNIV., APRIL 7-10, 1975.

75-0111   ARGYRIS, J. H.
BRONLUND, O.
THE NATURAL FACTOR APPROACH IN THE DISPLACEMENT
METHOD - FOUNDATION AND FURTHER DEVELOPMENTS
PROC. CONF. ON THE MATHS. OF FINITE ELEMENTS AND
APPL., BRUNEL UNIV., APRIL 7-10, 1975.

75-0112   GUREGHIAN, A. B.
YOUNGS, E.
THE USE OF THE FINITE ELEMENT METHOD IN OBTAINING
SOIL-WATER PRESSURE AND SOIL-WATER CONTENT
DISTRIBUTIONS IN DRAINED LAND WITH A UNIFORM
RAINFALL INCIDENT ON THE SURFACE
PROC. CONF. ON THE MATHS. OF FINITE ELEMENTS AND
APPL., BRUNEL UNIV., APRIL 7-10, 1975.

75-0113   BRUCH, J. C.
ZYVOLOSKI, G.
FINITE ELEMENT SOLUTION OF TWO DIMENSIONAL
UNSTEADY AND UNSATURATED FLOW IN POROUS MEDIA
PROC. CONF. ON THE MATHS. OF FINITE ELEMENTS AND
APPL., BRUNEL UNIV., APRIL 7-10, 1975.

75-0114   BLACKBURN, W. S.
AN APPLICATION OF LEAST SQUARES FINITE ELEMENT
METHODS TO TWO DIMENSIONAL STEADY INVISCID
COMPRESSIBLE FLOW WITHOUT CONSERVATION OF MASS OR
VORTICITY
PROC. CONF. ON THE MATHS. OF FINITE ELEMENTS AND
APPL., BRUNEL UNIV., APRIL 7-10, 1975.

75-0115   HIRSCH, CH.
WARZEE, G.
THE MERIDIONAL THROUGH-FLOW CALCULATION IN AN
AXIAL FLOW MACHINE BY THE FINITE ELEMENT METHOD
PROC. CONF. ON THE MATHS. OF FINITE ELEMENTS AND
APPL., BRUNEL UNIV., APRIL 7-10, 1975.

75-0116   TAYLOR, C.
IJAM, A.
COUPLED CONVECTIVE/CONDUCTIVE HEAT TRANSFER
INCLUDING VELOCITY FIELD EVALUATION
PROC. CONF. ON THE MATHS. OF FINITE ELEMENTS AND
APPL., BRUNEL UNIV., APRIL 7-10, 1975.

75-0117   BARROW, H.
MISTRY, J.
THE PREDICTION OF FULLY DEVELOPED TURBULENT FLOW
IN DUCTS BY THE FINITE ELEMENT METHOD
PROC. CONF. ON THE MATHS. OF FINITE ELEMENTS AND
APPL., BRUNEL UNIV., APRIL 7-10, 1975.

75-0118   TORBE, I.
A CRUCIFORM ELEMENT FOR SURFACES OF INFLATABLE
STRUCTURES
PROC. CONF. ON THE MATHS. OF FINITE ELEMENTS AND
APPL., BRUNEL UNIV., APRIL 7-10, 1975.

75-0119   CHEN, C. H. S.
APPLICATION OF FINITE ELEMENT METHOD TO PNEUMATIC
TYRE
PROC. CONF. ON THE MATHS. OF FINITE ELEMENTS AND
APPL., BRUNEL UNIV., APRIL 7-10, 1975.

75-0120   CORDS, H.
DIEMONT, W.
THERMAL AND STRESS ANALYSIS ON PRISMATIC NUCLEAR
FUEL ELEMENTS
PROC. CONF. ON THE MATHS. OF FINITE ELEMENTS AND
APPL., BRUNEL UNIV., APRIL 7-10, 1975.

75-0121   ENGLAND, R.
HENNART, J. P.
MARTIN, J. G.
MELENDEZ, L.
SEMIDISCRETE GALERKIN TECHNIQUES WITH TIME
INTERPOLATION AND SPLITTING UP FOR PLASMA
SIMULATION
PROC. CONF. ON THE MATHS. OF FINITE ELEMENTS AND
APPL., BRUNEL UNIV., APRIL 7-10, 1975.

75-0122   SCHMIDT, F. A. R.
FRANKE, H. P.
SAPPER, E.
ON THE APPLICATION OF THE FINITE ELEMENT METHOD IN
REACTOR PHYSICS
PROC. CONF. ON THE MATHS. OF FINITE ELEMENTS AND
APPL., BRUNEL UNIV., APRIL 7-10, 1975.

75-0123   PIVA, R.
DI CARLO, A.
NUMERICAL TECHNIQUES FOR CONVECTION/DIFFUSION
PROBLEMS
PROC. CONF. ON THE MATHS. OF FINITE ELEMENTS AND
APPL., BRUNEL UNIV., APRIL 7-10, 1975.

75-0124   SABIR, A. B.
STIFFNESS MATRICES FOR THE GENERAL DEFORMATION
(OUT-OF-PLANE AND INPLANE) OF CURVED BEAM MEMBERS
BASED ON INDEPENDENT STRAIN FUNCTIONS
PROC. CONF. ON THE MATHS. OF FINITE ELEMENTS AND
APPL., BRUNEL UNIV., APRIL 7-10, 1975.

75-0125   HINTON, E.
OWEN, D. R. J.
SHANTARAM, D.
DYNAMIC TRANSIENT NON-LINEAR BEHAVIOUR OF THICK
AND THIN PLATES
PROC. CONF. ON THE MATHS. OF FINITE ELEMENTS AND
APPL., BRUNEL UNIV., APRIL 7-10, 1975.

75-0126   YOUNG, D. M.
ITERATIVE SOLUTION OF LINEAR SYSTEMS ARISING FROM
FINITE ELEMENT TECHNIQUES
PROC. CONF. ON THE MATHS. OF FINITE ELEMENTS AND
APPL., BRUNEL UNIV., APRIL 7-10, 1975.

75-0127   RHEINBOLDT, W. C.
ON THE SOLUTION OF SETS OF NONLINEAR EQUATIONS
ARISING IN THE APPLICATION OF FINITE ELEMENT
METHODS
PROC. CONF. ON THE MATHS. OF FINITE ELEMENTS AND
APPL., BRUNEL UNIV., APRIL 7-10, 1975.

75-0128 BERGAN, P. G.
HANSSEN, L.
A NEW APPROACH FOR DERIVING GOOD ELEMENT STIFFNESS
MATRICES
PROC. CONF. ON THE MATHS. OF FINITE ELEMENTS AND
APPL., BRUNEL UNIV., APRIL 7-10, 1975.

75-0129 TAGNFORS, H.
WIBERG, N.
GENERAL SOLUTION ROUTINES FOR SYMMETRIC EQUATION
SYSTEMS
PROC. CONF. ON THE MATHS. OF FINITE ELEMENTS AND
APPL., BRUNEL UNIV., APRIL 7-10, 1975.

75-0130 HELLEN, T. K.
NUMERICAL INTEGRATION CONSIDERATIONS IN TWO AND
THREE DIMENSIONAL ISOPARAMETRIC FINITE ELEMENTS
PROC. CONF. ON THE MATHS. OF FINITE ELEMENTS AND
APPL., BRUNEL UNIV., APRIL 7-10, 1975.

75-0131 MARTIN, W. C.
HARROLD, A. J.
REMOVAL OF TRUNCATION ERROR IN FINITE ELEMENT
ANALYSIS
PROC. CONF. ON THE MATHS. OF FINITE ELEMENTS AND
APPL., BRUNEL UNIV., APRIL 7-10, U975.

75-0132 AKIN, J. E.
PARDUE, R. M.
ELEMENT RESEQUENCING FOR FRONTAL SOLUTIONS
PROC. CONF. ON THE MATHS. OF FINITE ELEMENTS AND
APPL., BRUNEL UNIV., APRIL 7-10, 1975.

75-0133 SEWELL, G.
AN ADAPTIVE COMPUTER PROGRAM FOR DIV. (P(X,Y) GRAD
U) EQUALS F(X,Y,U) IN A POLYGONAL REGION
PROC. CONF. ON THE MATHS. OF FINITE ELEMENTS AND
APPL., BRUNEL UNIV., APRIL 7-10, 1975.

75-0134 SMITH, S. L.
BREBBIA, C. A.
FINITE-ELEMENT SOLUTION OF NAVIER-STOKES EQUATIONS
FOR TRANSIENT 2-DIMENSIONAL INCOMPRESSIBLE FLOW
J. COM. PHYSICS., VOL. 17, NO. 3, 1975, PP.
235-245.

75-0135 SEGOL, G.
PINDER, G. F.
GRAY, W. G.
GALERKIN-FINITE ELEMENT TECHNIQUE FOR CALCULATING
TRANSIENT POSITION OF SALTWATER FRONT
WATER RESOURCES RESEARCH, VOL. 11, NO. 2, 1975,
PP. 343-347.

75-0136 BARTELDS, G.
SANDWICH AND COMPOSITE ELEMENTS
WORLD CONGRESS ON FINITE ELEMENT METH. IN STRUCT.
MECH., BOURNEMOUTH, DORSET, ENGLAND, OCTOBER
12-17, 1975, 27 PP.

75-0137 NEUMAN, S. P.
FEDDES, R. A.
BRESLER, E.
FINITE ELEMENT ANALYSIS OF 2-DIMENSIONAL FLOW IN
SOILS CONSIDERING WATER UPTAKE BY ROOTS .1. THEORY
PROC. SOIL SCI. SOC. OF AMER., VOL. 39, NO. 2,
1975, PP. 224-230.

75-0138 KONRAD, A.
SILVESTER, P.
FINITE ELEMENT PROGRAM PACKAGE FOR AXISYMMETRIC
VECTOR FIELD PROBLEMS
COMP. PHYSICS COMM., VOL. 9, NO. 3, 1975, PP.
193-204.

75-0139 WEGMULLER, A. W.
CRISFIELD, M. A.
ELASTIC-PLASTIC FINITE ELEMENT ANALYSIS OF PLATES
- DISCUSSION
PROC. INST. CIV. ENGRS., PART 2, VOL. 59, MARCH
1975, PP. 219-221.

75-0140 HINTON, E.
RAZZAQUE, A.
ZIENKIEWICZ, O. C.
DAVIES, J. D.
SIMPLE FINITE ELEMENT SOLUTION FOR PLATES OF
HOMOGENEOUS, SANDWICH AND CELLULAR CONSTRUCTION
PROC. INST. CIV. ENGRS., PART 2, VOL. 59, MARCH
1975, PP. 43-65.

75-0141 ROBINSON, J.
A WARPED QUADRILATERAL STRAIN MEMBRANE ELEMENTS
COMPUT. METH. IN APPL. MECH. AND ENG.,
NORTH-HOLLAND PUBL. CO. AMSTERDAM, NETHERLANDS,
1975.

75-0142 KAGAWA, Y.
ARAI, H.
OKUDA, S.
SHIRAI, K.
FINITE ELEMENT SIMULATION OF ENERGY-TRAPPED
ELECTROMECHANICAL RESONATORS
J. SOUND AND VIB., VOL. 39, NO. 3, 1975, PP.
317-335.

75-0143 CARPENTER, C. J.
FINITE-ELEMENT NETWORK MODELS AND THEIR
APPLICATION TO EDDY-CURRENT PROBLEMS
PROC. INST. ELECTRICAL ENG., LONDON, VOL. 122, NO.
4, 1975, PP. 455-462.

75-0144 GONTAROVSKII, V. P.
KOZLOV, I. A.
METHOD OF FINITE ELEMENTS FOR CALCULATION OF
HETEROGENEOUS BODIES OF REVOLUTION IN
ELASTO-PLASTIC MEDIUM
DOPOVIDI AKADEMII NAUK UKRAINS'KOI R.S.R., NO. 3,
1975, PP. 212-215.

75-0145 POULOS, H. G.
BROWN, P. T.
WIESNER, T. J.
FINITE ELEMENT ANALYSIS OF PILE-SUPPORTED PAVEMENT
STRUCTURES
ASC J. TRANSP. ENGNG., VOL. 101, NO. 2, 1975, PP.
399-401.

75-0146 ATLURI, S. N.
KOBAYASHI, A. S.
NAKAGAKI, N.
ASSUMED DISPLACEMENT HYBRID FINITE ELEMENT MODEL
FOR LINEAR FRACTURE MECHANICS
INT. J. FRACT., VOL. 11, NO. 2, 1975, PP. 257-272.

75-0147 FEDDES, R. A.
NEUMAN, S. P.
BRESLER, E.
FINITE ELEMENT ANALYSIS OF 2-DIMENSIONAL FLOW IN
SOILS CONSIDERING WATER UPTAKE BY ROOTS .2. FIELD
APPLICATIONS
PROC. SOIL SCI. SOC. OF AMER., VOL. 39, NO. 2, PP.
231-237.

75-0148 PARIS, P. C.
MCMEEKING, R. M.
EFFICIENT FINITE ELEMENT METHODS FOR STRESS
INTENSITY FACTOR
INT. J. FRACT., VOL. 11, NO. 2, 1975, PP. 354-358.

75-0149 NETHERCOT, D. A.
INELASTIC BUCKLING OF STEEL BEAMS UNDER NON
UNIFORM MOMENT
STRUCT. ENG., VOL. 53, NO. 2, FEBRUARY 1975, PP.
73-78.

75-0150 SANDHU, R. S.
HUANG, S. W.
APPLICATION OF GRIFFITH'S THEORY TO ANALYSIS OF
PROGRESSIVE FRACTURE
INT. J. FRACT., VOL. 11, NO. 1, FEBRUARY 1975, PP.
107-121.

75-0151 DENKE, P. H.
EIDE, G. R.
MATRIX DIFFERENCE EQUATION ANALYSIS OF VIBRATING
PERIODIC STRUCTURES
A.I.A.A. J., VOL. 13, NO. 2, FEBRUARY 1975, PP.
160-166.

75-0152 HIRIART, G.
SARPKAYA, T.
ANALYSIS OF CURVED TARGET-TYPE THRUST REVERSERS
A.I.A.A. J., VOL. 13, NO. 2, FEBRUARY 1975, PP.
185-192.

75-0153 BROERE, R.
USE OF THE FINITE ELEMENT METHOD IN EXAMINING
ELASTIC BUCKLING OF THIN PLATES OF ANY SHAPE
ANN. INST. TECH. BATIM. TRAV. PUBLICS., NO. 325,
FEBRUARY 1975, PP. 73-78.

75-0154 BERWANGER, C.
SYMKO, Y.
THERMAL STRESSES IN STEEL-CONCRETE COMPOSITE
BRIDGES
CAN. J. CIV. ENG., VOL. 2, NO. 1, MARCH 1975, PP.
66-84.

75-0155 SAKAI, T.
A UNIFIED THEORY OF VARIATIONAL PRINCIPLES IN
MECHANICS
PROC. NATL. SYMP. OF THE SOC. OF STEEL CONSTR. OF
JAPAN ON THE MATRIX METHODS OF STRUCT. ANALYSIS,
JUNE 1975.

75-0156 BRAUER, J. R.
SIMPLE EQUATIONS FOR THE MAGNETIZATION AND
RELUCTIVITY CURVES OF STEEL
IEEE TRANS. MAGN., VOL. MAG-11, NO. 1, JANUARY
1975, P. 81.

75-0157 DICELLO, J. A.
ADAMS, D. G.
HIGH STRENGTH MATERIALS AND VEHICLE WEIGHT
REDUCTION ANALYSIS
PREPR. NO. 750221, SAE MEET., FEBRUARY 24-28,
1975, 10 PP.

75-0158 YAMAGAMI, T.
ODA, E.
METHOD OF ANALYZING NON-DARCY FLOW PROBLEMS BY
FINITE ELEMENTS
PROC. JAP. SOC. CIV. ENG., NO. 234, FEBRUARY 1975,
PP. 111-120.

75-0159 YAGAWA, G.
MIYAZAKI, N.
ANDO, Y.
FINITE ELEMENT CREEP, BUCKLING ANALYSIS OF
VISCO-ELASTIC SHALLOW SPHERICAL SHELL
PROC. NATL. SYMP. OF THE SOC. OF STEEL CONSTR. OF
JAPAN ON THE MATRIX METHODS OF STRUCT. ANALYSIS,
JUNE 1975.

75-0160 YOUNG, C-I. J.
CROCKER, M. J.
PREDICTION OF TRANSMISSION LOSS IN MUFFLERS BY THE
FINITE ELEMENT METHOD
J. ACOUST. SOC. AM., VOL. 57, NO. 1, JANUARY 1975,
PP. 144-148.

75-0161 YAMAMOTO, Y.
URA, T.
SIMULATION OF FINITE ELEMENT METHOD FOR LOADING
PROCESS OF GRAIN CARGO AND INTERACTION WITH SHIP'S
STRUCTURE
PROC. NATL. SYMP. OF THE SOC. OF STEEL CONSTR. OF
JAPAN ON THE MATRIX METHODS OF STRUCT. ANALYSIS,
JUNE 1975.

75-0162 ATLURI, S. N.
KOBAYASHI, A. S.
BRAIN TISSUE FRAGILITY - A FINITE STRAIN ANALYSIS
BY A HYBRID FINITE-ELEMENT METHOD
PAPER NO. 75-APMW-46, ASME MEETING, MARCH 25-27,
1975, 5 PP.

75-0163 BARRON, G. E.
FINITE ELEMENT AND CUMULATIVE DAMAGE ANALYSIS OF A
KEYHOLE TEST SPECIMEN
PREPR. NO. 750041, SAE MEETING, FEBRUARY 24-28
1975, 8 PP.

75-0164 SOEDEL, W.
CONTACT OF AN INFLATED TOROIDAL MEMBRANE WITH A
FLAT SURFACE AS AN APPROACH TO THE TIRE DEFLECTION
PROBLEM
TIRE SCI. TECHNOL., VOL. 3, NO. 1, FEBRUARY 1975,
PP. 43-61.

75-0165 HELLEN, T. K.
DOWLING, A. R.
THREE-DIMENSIONAL CRACK ANALYSIS APPLIED TO AN LWR
NOZZLE-CYLINDER INTERSECTION
INT. J. PRESSURE VESSELS PIPING, VOL. 3, NO. 1,
JANUARY 1975, PP. 57-74.

75-0166 MIKSCH, M.
MERA, A.
STATIC AND HEAT TRANSFER ANALYSIS OF A SPHERE-CONE
INTERSECTION IN A NUCLEAR CONTAINMENT VESSEL
INT. J. PRESSURE VESSELS PIPING, VOL. 3, NO. 1,
JANUARY 1975, PP. 27-42.

75-0167 GARGESA, G.
NEW DESIGN FACTOR FOR A SHRINK FITTED ASSEMBLY
PAPER NO. 75-DE-50, A.S.M.E. MEETING, APRIL 21-24,
1975, 8 PP.

75-0168 KULHAWY, F. H.
STRESSES AND DISPLACEMENTS AROUND OPENINGS IN
HOMOGENEOUS ROCK
INT. J. ROCK MECH. MIN. SCI., GEOMECH. ABSTR.,
VOL. 12, NO. 3, MARCH 1975, PP. 43-57.

75-0169 KULHAWY, F. H.
STRESSES AND DISPLACEMENTS AROUND OPENINGS IN ROCK
CONTAINING AN ELASTIC DISCONTINUITY
INT. J. ROCK. MECH. MIN. SCI., GEOMECH. ABSTR.,
VOL. 12, NO. 3, MARCH 1975, PP. 59-72.

75-0170 KULHAWY, F. H.
STRESS AND DISPLACEMENTS AROUND OPENINGS IN ROCK
CONTAINING AN INELASTIC DISCONTINUITY
INT. J. ROCK MECH., MIN. SCI., GEOMECH. ABSTR.,
VOL. 12, NO. 3, MARCH 1975, PP. 73-78.

75-0171 ARCHER, J. S.
DYNAMIC ANALYSIS
WORLD CONGRESS ON FINITE ELEMENT METH. IN STRUCT.
MECH., BOURNEMOUTH, DORSET, ENGLAND, OCTOBER
12-17, 1975, PP. K.1-K.34.

75-0172 BROWN, R. C. JR.
BURNS, N. H.
COMPUTER ANALYSIS OF SEGMENTALLY ERECTED BRIDGES
ASCE J. STRUCT. DIV., VOL. 101, NO. 4, APRIL 1975,
PP. 761-778.

75-0173 CHENG, W-C.
COMPUTER ANALYSIS OF STEEL FRAME IN FIRE
ASCE J. STRUCT. DIV., VOL. 101, NO. 4, APRIL 1975,
PP. 855-867.

75-0174 WHITEMAN, J. R.
CONFORMING FINITE ELEMENT METHODS FOR THE CLAMPED
PLATE PROBLEM
IN "FINITE ELEMENTE UND DIFFERENZENVERFAHREN",
COLLATZ, L. (ED) I.S.N.M., BIRKHAUSER VERLAG,
BASEL, 1975.

75-0175 PARDOEN, G. C.
ASYMMETRIC BENDING OF CIRCULAR PLATES USING THE
FINITE ELEMENT METHOD
COMPUT. STRUCT., VOL. 5, NO. 2-3, JUNE 1975, PP.
197-202.

75-0176 FARNUM, P.
CAREY, G. F.
COMPUTER MODELLING OF PLUG SEEDLING SURVIVAL
12TH MEETING, SOC. ENG. SCI., UNIV. TEXAS, AUSTIN,
OCTOBER 20-22, 1975.

75-0177 ALLMAN, D. J.
CALCULATION OF THE ELASTIC BUCKLING LOADS OF THIN
FLAT REINFORCED PLATES USING TRIANGULAR FINITE
ELEMENTS
INT. J. NUMER. METH. ENG., VOL. 9, NO. 2, 1975,
PP. 415-432.

75-0178 YAMADA, YOSHIAKI
ITO, K.
YOKOUCHI, Y.
TAMANO, T.
OHTSUBO, T.
FINITE ELEMENT ANALYSIS OF STEADY FLUID AND METAL
FLOW
CHAPTER 4, FINITE ELEMENTS IN FLUIDS, VOL. 1:
VISCOUS FLOW AND HYDRODYNAMICS, JOHN WILEY, 1975,
PP. 73-94.

75-0179 FERRING, M.
HALL, U.
PERSSON, B.
THE FINITE ELEMENT METHOD APPLIED TO COMPRESSIBLE
CASCADE FLOW
VOLVO FLYGMOTOR AB, TROLLHATTAN, SWEDEN, REPORT
NO. 340 331 331B 331M 331U FMV-F:MO(3) FMV-BNLD
F:TSAAB/L (2) 333L 392, JUNE 17 1975, 21 PP.

75-0180 YAMADA, YOSHIAKI
IWATA, K.
INSTABILITY ANALYSIS BY THE FINITE ELEMENT METHOD
PROC. 4TH NATL. SYMP. ON MATRIX METHOH. OF STRUCT.
ANALYSIS AND DESIGN, SOC. OF STEEL CONSTR. OF
JAPAN, 1975, PP. 57-64. (IN JAPANESE)

75-0181 YAMADA, YOSHIAKI
YOKOUCHI, Y.
NISHIMURA, Y.
FINITE ELEMENT ANALYSIS OF COMPOSITE MATERIAL
BEHAVIOURS FROM MECHANICAL PROPERTIES OF
CONSTITUENT MATERIALS
PROC. 4TH NATL. SYMP. ON MATRIX METH. OF STRUCT.
ANALYSIS AND DESIGN, SOC. OF STEEL CONSTR. OF
JAPAN, 1975, PP. 471-477. (IN JAPANESE)

75-0182 CARROLL, W. E.
RAMIFICATIONS OF OPTIMUM IDEALIZATION GEOMETRY IN
DISCRETE ELEMENT ANALYSIS
IN "WORLD CONGRESS ON FINITE ELEMENT METH. IN
STRUCT. MECH.", BOURNEMOUTH, DORSET, ENGLAND,
OCTOBER 12-17, 1975, 15 PP.

75-0183 MORTON, K. W.
STABILITY AND ACCURACY OF NUMERICAL APPROXIMATIONS
TO TIME DEPENDENT FLOWS
AGARD MEETING, BRUSSELS, BELG., NO. 73, PAPER 5,
FEBRUARY 17-22, 1975, 12 PP.

75-0184 BELLEVAUX, C.
MAILLE, M.
APPLICATIONS OF FINITE ELEMENT METHODS IN FLUID
DYNAMICS
AGARD MEETING, BRUSSELS, BELG., NO. 73, PAPER 7,
FEBRUARY 17-22, 1975, 28 PP.

75-0185 NG, S. S. F.
KWOK, W. L.
FINITE-ELEMENT ANALYSIS OF SKEW SANDWICH PLATES
A.S.M.E., PAPER NO. 75-APM-Q, 1975, 8 PP.

75-0186 RAQUET, E.
EINSATZ MODERNER RECHENVERFAHREN ZUR ERMITTLUNG
DER SPANNUNGEN UND VERFORMUNGEN BEI RADSAETZEN.
(USE OF MODERN COMPUTER CALCULATION METHODS TO
DETERMINE STRESSES AND DEFORMATIONS OF WHEEL SETS)
Z EISENBAHNWES VERKEHRSTECH GLASERS ANN, VOL. 99,
NO. 9, SEPTEMBER 1975, PP. 249-255.

75-0187 ANDRESEN, K.
BERECHNUNG DES KEGELDRUCKVERSUCHES MIT EINER
FINITE-ELEMENT-METHODE. (CALCULATION OF THE CONE
INDENTATION TEST USING A FINITE-ELEMENT METHOD)
ARCH EISENHUETTENWES, VOL. 46, NO. 9, SEPTEMBER
1975, PP. 571-574.

75-0188 OHJI, K.
OGURA, K.
CYCLIC ANALYSIS OF A PROPAGATING CRACK AND ITS
CORRELATION WITH FATIGUE CRACK GROWTH
ENG. FRACT. MECH., VOL. 7, NO. 3, SEPTEMBER 1975,
PP. 457-464.

75-0189 AOYAMA, T.
INASAKI, I.
DYNAMIC BEHAVIOR OF HYDROSTATIC THRUST BEARING
PAPER NO. 75-DET-2, A.S.M.E. MEETING, SEPTEMBER
17-19, 1975, 8 PP.

75-0190 BOTKIN, M. E.
LUBKIN, J. L.
WELDED JOINT STIFFNESS OBTAINED USING SOLID FINITE
ELEMENTS
PAPER NO. 75-DET-3, A.S.M.E. MEETING, SEPTEMBER
17-19, 1975, 8 PP.

75-0191 DIETERICH, D. A.
FINITE ELEMENT COMPUTER PROGRAM FOR PREDICTING THE
NONLINEAR STATIC AND DYNAMIC BEHAVIOR OF
VISCOELASTIC COMPONENTS
PAPER NO. 75-DET-7, A.S.M.E. MEETING, SEPTEMBER
17-19, 1975, 7 PP.

75-0192 CURTIS, A. J.
ON THE APPLICATION OF RESPONSE LIMITING TO FINITE
ELEMENT STRUCTURAL ANALYSIS PROGRAMS
PAPER NO. 75-DET-12, A.S.M.E. MEETING, SEPTEMBER
17-19, 1975, 5 PP.

75-0193 VON RIESEMANN, W. A.
GUBBELS, M. H.
ANALYSIS OF AN IN-PILE REACTOR TUBE
PAPER NO. 75-DET-43, A.S.M.E. MEETING, SEPTEMBER
17-19, 1975, 8 PP.

75-0194 THOMAS, J.
ABBAS, B. A. H.
DYNAMIC STABILITY OF TIMOSHENKO BEAMS BY FINITE
ELEMENT METHOD
PAPER NO. 75-DET-78, A.S.M.E. MEETING, SEPTEMBER
17-19, 1975, 5 PP.

75-0195 WILSON, G. J.
KIRKHOPE, J.
VIBRATION ANALYSIS OF AXIAL FLOW TURBINE DISKS
USING FINITE ELEMENTS
PAPER NO. 75-DET-80, A.S.M.E. MEETING, SEPTEMBER
17-19, 1975, 6 PP.

75-0196 CONWAY, J. C.
FINITE ELEMENT TECHNIQUES APPLIED TO CRACKS
INTERACTING WITH SELECTED SINGULARITIES
J. AMER. CERAM. SOC., VOL. 58, NO. 9-10,
SEPTEMBER-OCTOBER 1975, PP. 402-405.

75-0197 POPPLEWELL, N.
RESPONSE OF BOX-LIKE STRUCTURES TO WEAK EXPLOSIONS
J. SOUND AND VIB., VOL. 42, NO. 1, SEPTEMBER 8,
1975, PP. 65-84.

75-0198 CHEUNG, Y. K.
KWOK, W. L.
DYNAMIC ANALYSIS OF CIRCULAR AND SECTOR THICK,
LAYERED PLATES
J. SOUND AND VIB., VOL. 42, NO. 2, SEPTEMBER 22,
1975, PP. 147-158.

75-0199 GLADWELL, G. M. L.
VIJAY, D. K.
VIBRATION ANALYSIS OF AXISYMMETRIC RESONATORS
J. SOUND AND VIB., VOL. 42, NO. 2, SEPTEMBER 22,
1975, PP. 137-145.

75-0200 FANELLI, M.
IL METODO DEGLI ELEMENTI FINITI: POSSIBILITA DI
APPLICAZIONE A PROBLEMI DI INTERESSE DEGLI
ELETTROTECNICI. (FINITE ELEMENT METHOD:
POSSIBILITIES OF ITS USE FOR SOLVING
ELECTROTECHNICAL PROBLEMS)
ELETTROTECNICA, VOL. 62, NO. 6, JUNE 1975, PP.
513-520.

75-0201 PURDY, D. M.
DISCRETE ELEMENT ANALYSIS OF COMPOSITE STRUCTURES
COMPOS. MATER., VOL. 8, 1975, PP. 1-31.

75-0202 AL-MAHAIDI, R. S.
NILSON, A. H.
COUPLED SHEAR WALL ANALYSIS BY LAGRANGE
MULTIPLIERS
A.S.C.E. J. STRUCT. DIV., VOL. 101, NO. 11,
NOVEMBER, 1975, PP. 2359-2366.

75-0203 GLADWELL, G. M. L.
VIJAY, D. K.
ERRORS IN SHELL FINITE-ELEMENT MODELS FOR
VIBRATION OF CIRCULAR-CYLINDERS
J. SOUND AND VIB., VOL. 43, NO. 3, 1975, PP.
511-528.

75-0204 EHRLICH, L. W.
GUPTA, M. M.
SOME DIFFERENCE SCHEMES FOR THE BIHARMONIC
EQUATION
SIAM J. NUMER. ANAL., VOL. 12, NO. 5, OCTOBER
1975, PP. 773-790.

75-0205 HELLEN, T. K.
BLACKBURN, W. S.
CALCULATION OF STRESS INTENSITY FACTORS FOR
COMBINED TENSILE AND SHEAR LOADING
INT. J. FRACT., VOL. 11, NO. 4, AUGUST 1975, PP.
605-617.

75-0206 FIX, G. J.
FINITE ELEMENT MODELS FOR OCEAN CIRCULATION
PROBLEMS
SIAM J. APPL. MATH., VOL. 29, NO. 3, NOVEMBER
1975, PP. 371-387.

75-0207 DI MONACO, A.
GIUSEPPETTI, G.
STUDIO DI CAMPI ELETTRICI E MAGNETICI STANZIONARI
CON IL METODO DEGLI ELEMENTI FINITI. APPLICAZIONE
AI TRASFORMATORI. (ANALYSIS OF STEADY ELECTRICAL
AND MAGNETIC FIELDS BY THE FINITE ELEMENT METHOD.
APPLICATION TO TRANSFORMERS)
ELETTROTECNICA, VOL. 62, NO. 7, JULY 1975, PP.
585-598.

75-0208 NIELSEN, L. O.
SPAENDINGSHYBRIDE FINITE ELEMENTER TIL
SVINGNINGSPROBLEMER. (STRESS HYBRID FINITE ELEMENT
METHOD FOR THE SOLUTION OF VIBRATION PROBLEMS)
STRUCT. RES. LAB., TECH. UNIV. OF DEN., LYNGBY,
REPORT NO. 61, 1975, 19 PP.

75-0209 KLEIBER, M.
LAGRANGIAN AND EULERIAN FINITE ELEMENT FORMULATION
FOR LARGE STRAIN ELASTO-PLASTICITY
BULL. POL. ACAD. OF SCI., WARSAW, VOL. 23, NO. 3,
1975, PP. 209-218.

75-0210 MCNAMARA, J. F.
SHARMA, S. K.
ISOTROPIC-KINEMATIC HARDENING MODEL FOR
ELASTIC-PLASTIC CYCLIC STRUCTURAL ANALYSIS
CONSTR. ENG. RES. LAB. TECH., REPORT NO. M-148,
AUGUST 1975, 16 PP.

75-0211 DACKO, M.
GAJOWNICZEK, S.
OBLICZANIE KONSTRUKCJI TOROIDALNEGO ZBIORNIKA NA
WODE. (STATICAL ANALYSIS OF A TOROIDAL WATER
RESERVOIR)
ARCH INZ LADOWEJ, VOL. 21, NO. 3, 1975, PP.
441-456.

75-0212 ATLURI, S. N.
KOBAYASHI, A. S.
FRACTURE MECHANICS APPLICATION OF AN ASSUMED
DISPLACEMENT HYBRID FINITE ELEMENT PROCEDURE
A.I.A.A. J., VOL. 13, NO. 6, JUNE 1975, PP.
734-739.

75-0213 FRIEDMAN, E.
THERMOMECHANICAL ANALYSIS OF THE WELDING PROCESS
USING THE FINITE ELEMENT METHOD
A.S.M.E., J. PRESSURE VESSEL TECH., VOL. 97, NO.
3, AUGUST 1975, PP. 206-213.

75-0214 SHARIFI, P.
NONLINEAR BUCKLING ANALYSIS OF COMPOSITE SHELLS
A.I.A.A. J., VOL. 13, NO. 6, JUNE 1975, PP.
729-734.

75-0215 BELYTSCHKO, T.
KENNEDY, J. M.
FINITE ELEMENT STUDY OF PRESSURE WAVE ATTENUATION
BY REACTOR FUEL SUBASSEMBLIES
A.S.M.E., J. PRESSURE VESSEL TECH., VOL. 97, NO.
3, AUGUST 1975, PP. 173-177.

75-0216 JONES, D. P.
FINITE ELEMENT ANALYSIS OF PERFORATED PLATES
CONTAINING TRIANGULAR PENETRATION PATTERNS OF 5
AND 10 PERCENT LIGAMENT EFFICIENCY
A.S.M.E., J. PRESSURE VESSEL TECH., VOL. 97, NO.
3, AUGUST 1975, PP. 199-205.

75-0217 KLEIN, S.
NONLINEAR DYNAMIC ANALYSIS OF SHELLS OF REVOLUTION
WITH ASYMMETRIC PROPERTIES BY THE FINITE ELEMENT
METHOD
A.S.M.E., J. PRESSURE VESSEL TECH., VOL. 97, NO.
3, AUGUST 1975, PP. 163-171.

75-0218 CHERN, J. M.
PAI, D. H.
INELASTIC ANALYSIS OF A STRAIGHT TUBE UNDER
COMBINED BENDING, PRESSURE AND THERMAL LOADS
A.S.M.E., J. PRESSURE VESSEL TECH., VOL. 97, NO.
3, AUGUST 1975, PP. 155-162.

75-0219 BRUCH, J. C. JR.
         LEWIS, R. W.
         TRANSIENT TWO-DIMENSIONAL PROBLEMS UTILIZING THE
         LEAST SQUARES ALGORITHM
         A.S.M.E., J. HEAT TRANSFER, VOL. 97, NO. 3, AUGUST
         1975, PP. 467-469.

75-0220 MOAN, T.
         HAVER, S.
         STOCHASTIC DYNAMIC RESPONSE ANALYSIS OF OFFSHORE
         PLATFORMS, WITH PARTICULAR REFERENCE TO
         GRAVITY-TYPE PLATFORMS
         PROC. 7TH ANNUAL CONF., OFFSHORE TECHNOL.,
         HOUSTON, TEXAS, MAY 5-8, 1975, PAPER OTC 2407,
         VOL. 3, PP. 707-720.

75-0221 BROUGHTON, P.
         OFFSHORE GRAVITY BASED OIL PRODUCTION PLATFORM
         INTERACTION WITH THE SEA BED
         PROC. 7TH ANN. CONF., OFFSHORE TECHNOL., HOUSTON,
         TEXAS, MAY 5-8, 1975, PAPER OTC 2372, VOL. 3, PP.
         387-398.

75-0222 LEHNHOFF, T. F.
         SCHELLER, J. D.
         INFLUENCE OF TEMPERATURE DEPENDENT PROPERTIES ON
         THERMAL ROCK FRAGMENTATION
         INT. J. ROCK MECH. MIN. SCI. GEOMECH., VOL. 12,
         NO. 8, AUGUST 1975, PP. 255-260.

75-0223 GRAY, W. H.
         SCHNURR, N. M.
         COMPARISON OF THE FINITE ELEMENT AND FINITE
         DIFFERENCE METHODS FOR THE ANALYSIS OF STEADY TWO
         DIMENSIONAL HEAT CONDUCTION PROBLEMS
         COMPUT. METH. APPL. MECH. ENG., VOL. 6, NO. 2,
         SEPTEMBER 1975, PP. 243-245.

75-0224 TANNER, R. I.
         NICKELL, R. E.
         FINITE ELEMENT METHODS FOR THE SOLUTION OF SOME
         INCOMPRESSIBLE NON-NEWTONIAN FLUID MECHANICS
         PROBLEMS WITH FREE SURFACES
         COMPUT. METH. APPL. MECH. ENG., VOL. 6, NO. 2,
         SEPTEMBER 1975, PP. 155-174.

75-0225 MCKEOWN, J. J.
         QUASI-LINEAR PROGRAMMING ALGORITHM FOR OPTIMIZING
         FIBRE-REINFORCED STRUCTURES OF FIXED STIFFNESS
         COMPUT. METH. APPL. MECH. ENG., VOL. 6, NO. 2,
         SEPTEMBER 1975, PP. 123-154.

75-0226 KAWAI, T.
         RECENT DEVELOPMENTS OF FINITE ELEMENT METHODS IN
         JAPAN
         IN "WORLD CONGRESS ON FINITE ELEMENT METH. IN
         STRUCT. MECH.", BOURNEMOUTH, DORSET, ENGLAND,
         OCTOBER 12-17, 1975, 20 PP.

75-0227 BLAKER, B.
         VIBRATION OF SUBMERGED STRUCTURES AS COMPUTED BY
         THE FINITE ELEMENT METHOD
         IN "WORLD CONGRESS ON FINITE ELEMENT METH. IN
         STRUCT. MECH.", BOURNEMOUTH, DORSET, ENGLAND,
         OCTOBER 12-17, 1975, 23 PP.

75-0228 DRNEVICH, V. P.
         CONSTRAINED AND SHEAR MODULI FOR FINITE ELEMENTS
         A.S.C.E. J. GEOTECH. ENG. DIV., VOL. 101, NO. 5,
         MAY 1975, PP. 459-473.

75-0229 POCESKI, A.
         MIXED FINITE ELEMENT METHOD FOR BENDING OF PLATES
         INT. J. NUMER. METH. ENG., VOL. 9, NO. 1, 1975,
         PP. 3-15.

75-0230 PETYT, M.
         FINITE ELEMENT METHODS FOR THE RESPONSE OF
         STRUCTURES TO RANDOM EXCITATIONS
         IN "WORLD CONGRESS ON FINITE ELEMENT METH. IN
         STRUCT. MECH.", BOURNEMOUTH, DORSET, ENGLAND,
         OCTOBER 12-17, 1975, 20 PP.

75-0231 KAMAT, M. P.
         EFFECT OF SHEAR DEFORMATIONS AND ROTARY INERTIA ON
         OPTIMUM BEAM FREQUENCIES
         INT. J. NUMER. METHODS ENG., VOL. 9, NO. 1, 1975,
         PP. 51-62.

75-0232 MURPHY, W. D.
         CUBIC SPLINE GALERKIN APPROXIMATIONS TO PARABOLIC
         SYSTEMS WITH COUPLED NON-LINEAR BOUNDARY
         CONDITIONS
         INT. J. NUMER. METHODS ENG., VOL. 9, NO. 1, 1975,
         PP. 63-71.

75-0233 HORNBUCKLE, J. C.
         NEVILL, G. E.
         STRUCTURAL OPTIMIZATION USING THE FINITE ELEMENT
         METHOD APPLIED TO A BEAM
         INT. J. NUMER. METHODS ENG., VOL. 9, NO. 1, 1975,
         PP. 101-107.

75-0234 KAERCHER, H. J.
         FINITE ELEMENTS ON THE BASIS OF CONTINUUM
         MECHANICS
         INT. J. NUMER. METHODS ENG., VOL. 9, NO. 1, 1975,
         PP. 129-147.

75-0235 CHUNG, T. J.
         THERMOMECHANICAL RESPONSE OF INELASTIC FIBRE
         COMPOSITES
         INT. J. NUMER. METHODS ENG., VOL. 9, NO. 1, 1975,
         PP. 169-185.

75-0236 OWEN, D. R. J.
         SALONEN, E. M.
         THREE-DIMENSIONAL ELASTO-PLASTIC FINITE ELEMENT
         ANALYSIS
         INT. J. NUMER. METHODS ENG., VOL. 9, NO. 1, 1975,
         PP. 209-218.

75-0237 HELLEN, T. K.
         ON THE METHOD OF VIRTUAL CRACK EXTENSIONS
         INT. J. NUMER. METHODS ENG., VOL. 9, NO. 1, 1975,
         PP. 187-207.

75-0238 JARAUSCH, R.
         ALBRECHT, H.
         FESTIGKEITSUNTERSUCHUNGEN UND-BERECHNUNGEN AN
         PUMPENSPIRALGEHAEUSEN. (INVESTIGATIONS AND
         CALCULATIONS OF THE STRENGTH BEHAVIOR OF PUMP
         SPIRAL CASINGS)
         IND-ANZ, VOL. 97, NO. 47, JUNE 11, 1975, PP.
         977-981.

75-0239 OTTAVIANI, M.
         THREE-DIMENSIONAL FINITE ELEMENT ANALYSIS OF
         VERTICALLY LOADED PILE GROUPS
         GEOTECHNIQUE, VOL. 25, NO. 2, JUNE 1975, PP.
         159-174.

75-0240 RODRIGUES, J. S. N.
         SIMONS, N. E.
         FINITE ELEMENT ANALYSIS OF THE SURFACE DEFORMATION
         DUE TO A UNIFORM LOADING ON A LAYER OF GIBSON SOIL
         RESTING ON A SMOOTH RIGID BASE
         GEOTECHNIQUE, VOL. 25, NO. 2, JUNE 1975, PP.
         375-379.

75-0241 MORAND, H. J. P.
         OHAYON, R.
         INTERNAL PRESSURE EFFECTS ON THE VIBRATION OF
         PARTIALLY FILLED ELASTIC TANKS
         IN "WORLD CONGRESS ON FINITE ELEMENT METH. IN
         STRUCT. MECH.", BOURNEMOUTH, DORSET, ENGLAND,
         OCTOBER 12-17, 1975, 18 PP.

75-0242 WITTRICK, W. H.
         STRESS SINGULARITIES AT DIAPHRAGM-WEB-FLANGE
         JUNCTIONS IN BOX GIRDERS
         PROC. INST. CIV. ENG., LONDON, VOL. 59, PART 2,
         MARCH 1975, PP. 79-89.

75-0243 ATLURI, S. N.
         KOBAYASHI, A. S.
         BRAIN TISSUE FRAGILITY - A FINITE STRAIN ANALYSIS
         BY A HYBRID FINITE-ELEMENT METHOD
         A.S.M.E., J. APPL. MECH. TRANS., VOL. 42, NO. 2,
         JUNE 1975, PP. 269-273.

75-0244 MOLLENDORF, J. C.
         APPLICABILITY OF APPROXIMATE AND EXACT TRANSIENT
         HEAT TRANSFER ANALYSES TO HEATING PROCESSES USED
         TO SOLDER MULTILAYER CIRCUIT BOARDS
         I.E.E.E. TRANS., VOL. PHP-11, NO. 2, JUNE 1975,
         PP. 96-104.

75-0245 SHEN, C. C.
         LOPEZ, M. L.
         JET-WING LIFTING - SURFACE THEORY USING ELEMENTARY
         VORTEX DISTRIBUTIONS
         J. AIRCR., VOL. 12, NO. 5, MAY 1975, PP. 448-456.

75-0246 NAGARAJAN, S.
         POPOV, E. P.
         PLASTIC AND VISCOPLASTIC ANALYSIS OF AXISYMMETRIC
         SHELLS
         INT. J SOLIDS STRUCT., VOL. 11, NO. 1, JANUARY
         1975, PP. 1-19.

75-0247 TALBOT, R. J.
         PRZEMIENIECKI, J. S.
         FINITE ELEMENT ANALYSIS OF FREQUENCY SPECTRA FOR
         ELASTIC WAVEGUIDES
         INT. J. SOLIDS STRUCT., VOL. 11, NO. 1, JANUARY
         1975, PP. 115-138.

75-0248 FRISCH-FAY, R.
         STABILITY OF MASONRY PIERS
         INT. J. SOLIDS STRUCT., VOL. 11, NO. 2, FEBRUARY
         1975, PP. 187-198.

75-0249 BRADFORD, L. G.
         DONG, S. B.
         ELASTODYNAMIC BEHAVIOR OF LAMINATED ORTHOTROPIC
         PLATES UNDER INITIAL STRESS
         INT. J. SOLIDS STRUCT., VOL. 11, NO. 2, FEBRUARY
         1975, PP. 213-230.

75-0250 SABIR, A. B.
APPLICATION OF THE FINITE ELEMENT METHOD OF
ANALYSIS TO HYPERBOLIC PARABOLOIDS WITH EDGE BEAMS
STAVEBNICKY CAS., VOL. 23, NO. 5, MAY 1975, PP.
268-281.

75-0251 YAGAWA, G.
NISHIOKA, T.
FINITE ELEMENT CALCULATION OF STRESS INTENSITY
FACTORS USING SUPERPOSITION
2ND. NATL. CONGR. ON PRESSURE VESSELS AND PIPING,
COMPUT. FRACT. MECH., SAN FRANCISCO, CALIF., JUNE
23-27, 1975, PP. 21-34. PUBL. BY ASME, NEW YORK,
N.Y., 1975.

75-0252 EMERY, A. F.
CUPPS, F. J.
USE OF SINGULARITY PROGRAMMING IN FINITE ELEMENT
CALCULATIONS OF ELASTIC STRESS INTENSITY FACTORS.
PLANE AND AXISYMMETRIC - APPLIED TO THERMAL STRESS
FRACTURE
COMPUT. FRACT. MECH., 2ND NATL. CONGR. ON PRESSURE
VESSELS AND PIPING, SAN FRANCISCO, CALIF., JUNE
23-27, 1975, PP. 35-48. (PUBL. BY ASME, N.Y.,
1975)

75-0253 WALSH, P. F.
CRACKS, NOTCHES AND FINITE ELEMENTS
COMPUT. FRACT. MECH., 2ND NATL. CONGR. ON PRESSURE
VESSELS AND PIPING, SAN FRANCISCO, CALIF., JUNE
23-27, 1975, PP. 49-61. (PUBL. BY ASME, N.Y.,
1975)

75-0254 HELLEN, T. K.
BLACKBURN, W. S.
CALCULATION OF STRESS INTENSITY FACTORS IN TWO AND
THREE DIMENSIONS USING FINITE ELEMENTS
COMPUT. FRACT. MECH., 2ND NATL. CONGR. ON PRESSURE
VESSELS AND PIPING, SAN FRANCISCO, CALIF., JUNE
23-27, 1975, PP. 103-120. (PUBL. BY ASME, N.Y.,
1975)

75-0255 KOBAYASHI, A. S.
POLVANICH, N.
STRESS INTENSITY FACTOR OF A SURFACE CRACK IN A
PRESSURIZED CYLINDER
COMPUT. FRACT. MECH., 2ND NATL. CONGR. ON PRESSURE
VESSELS AND PIPING, SAN FRANCISCO, CALIF., JUNE
23-27, 1975, PP. 121-132. (PUBL. BY ASME, N.Y.,
1975)

75-0256 AYRES, D. J.
THREE-DIMENSIONAL ELASTIC ANALYSIS OF
SEMI-ELLIPTICAL SURFACE CRACKS SUBJECT TO THERMAL
SHOCK
COMPUT. FRACT. MECH., 2ND NATL. CONGR. ON PRESSURE
VESSELS AND PIPING, SAN FRANCISCO, CALIF., JUNE
23-27, 1975, PP. 133-143. (PUBL. BY ASME, N.Y.,
1975)

75-0257 STERN, M.
CALCULATION OF STRESS INTENSITY FACTORS IN
ANISOTROPIC MATERIALS BY A CONTOUR INTEGRAL METHOD
COMPUT. FRACT. MECH., 2ND NATL. CONGR. ON PRESSURE
VESSELS AND PIPING, SAN FRANCISCO, CALIF., JUNE
23-27, 1975, PP. 161-171. (PUBL. BY ASME, N.Y.,
1975)

75-0258 ANDERSON, J. M.
FINITE ELEMENT ANALYSIS OF CRACKED STRUCTURES
SUBJECTED TO SHOCK LOADS
COMPUT. FRACT. MECH., 2ND NATL. CONGR. ON PRESSURE
VESSELS AND PIPING, SAN FRANCISCO, CALIF., JUNE
23-27, 1975, PP. 173-184. (PUBL. BY ASME, N.Y.,
1975)

75-0259 SCHREM, E.
FINITE ELEMENT SOFTWARE IN THE NEXT DECADE
WORLD CONGRESS ON FINITE ELEMENT METH. IN STRUCT.
MECH., BOURNEMOUTH, DORSET, ENGLAND, OCTOBER
12-17, 1975, PP. R.1-R.18.

75-0260 AAMODT, B.
BERGAN, P. G.
NUMERICAL TECHNIQUES IN LINEAR AND NONLINEAR
FRACTURE MECHANICS
COMPUT. FRACT. MECH., 2ND NATL. CONGR. ON PRESSURE
VESSELS AND PIPING, SAN FRANCISCO, CALIF., JUNE
23-27, 1975, PP. 199-216. (PUBL. BY ASME, N.Y.,
1975)

75-0261 PAYER, H. G.
LONGREE, W. D.
DESIGN, ANALYSIS AND CONSTRUCTION OF THE RESEARCH
PLATFORM (NORDSEE)
PROC. 7TH ANNU. CONF., OFFSHORE TECHNOL., HOUSTON,
TEX., PAPER OTC 2168, VOL. 1, MAY 5-8, 1975, PP.
223-233.

75-0262 ZIENKIEWICZ, O. C.
THE FINITE ELEMENT METHOD AND BOUNDARY SOLUTION
PROCEDURES AS GENERAL APPROXIMATION METHODS FOR
FIELD PROBLEMS
WORLD CONGRESS ON FINITE ELEMENT METH. IN STRUCT.
MECH., BOURNEMOUTH, DORSET, ENGLAND, OCTOBER
12-17, 1975, PP. S.1-S.31.

75-0263 DOROSZ, S.
KLEIBER, M.
FUNKCJA KSZTALTU NIESPELNIAPACA WARUNKOW CIAGLOSCI
W ANALIZIE PLYT METODA ELEMENTOW SKONCZONYCH.
(SHAPE FUNCTION VIOLATING CONTINUITY REQUIREMENTS
AS APPLIED TO THE FINITE ELEMENT ANALYSIS OF
PLATES)
ARCH. INZ. LADOWEJ., VOL. 21, NO. 2, 1975, PP.
275-286.

75-0264 FEDOROVSKUU, V. G.
KAGANOVSKAYA, S. E.
RIGID PLATE ON NONLINEARLY DEFORMABLE COHESIVE
BASE (PLANE PROBLEM)
SOIL MECH. FOUND. ENG., VOL. 12, NO. 1, JANUARY -
FEBRUARY 1975, PP. 67-72.

75-0265 DHALLA, A. K.
ROCHE, R. V.
INELASTIC ANALYSIS AND SATISFACTION OF DESIGN
CRITERIA OF A HIGH TEMPERATURE COMPONENT
ADV. IN DES. FOR ELEVATED TEMP. ENVIRON., 2ND
NATL. CONGR. ON PRESSURE VESSELS AND PIPING, SAN
FRANCISCO, CALIF., JUNE 23-27, 1975, PP. 83-92.
(PUBL. BY ASME, N.Y., 1975)

75-0266 BUCHTA, K.
DEICKE, K.
KOLBENKUEHLUNG (PISTON COOLING)
MTZ. MOTORTECH. Z., VOL. 36, NO. 7, JULY - AUGUST,
1975, PP. 200-205.

75-0267 BERGER, H.
BOUJOT, J.
ON A SPECTRAL PROBLEM IN VIBRATION MECHANICS:
COMPUTATION OF ELASTIC TANKS PARTIALLY FILLED WITH
LIQUIDS
J. MATH. ANAL. APPL., VOL. 51, NO. 2, AUGUST 1975,
PP. 272-298.

75-0268 GELLERT, M.
LAURSEN, M. E.
FINITE ELEMENT EQUILIBRIUM MODEL FOR GENERAL BODY
FORCE DISTRIBUTION
DAN. SELSK. BYGNINGSSTAT. MEDD., VOL. 46, NO. 1,
MARCH 1975, 24 PP.

75-0269 JOHNSON, R. L.
COLLINS, A. G.
FINITE-ELEMENT METHOD FOR WATER-DISTRIBUTION
NETWORKS
J. AMER. WATER WORKS ASSOC., VOL. 67, NO. 7, JULY
1975, PP. 385-389.

75-0270 EIGENHEER, LUIZ PAULO, Q. T.
SOUTO SILVEIRA, E. B.
EMBANKMENT DAM FOR THE MARIMBONDO HYDRO PROJECT
INT. WATER POWER DAM CONSTR., VOL. 6-7, JUNE -
JULY, 1975, PP. 217-2219

75-0271 RADAJ, D.
BERECHNUNG DER SCHWEISSEIGENSPANNUNGEN UND
SCHWEISSFORMAENDERUNGEN MIT ELASTISCH-PLASTISCHEN
FINITEN ELEMENTEN. (CALCULATION OF INTERNAL
STRESSES AND DISTORTIONS CAUSED BY WELDING, USING
THE ELASTO-PLASTIC FINITE ELEMENT METHOD)
SCHWEISSEN SCHNEIDEN, VOL. 27, NO. 7, JULY 1975,
PP. 245-250.

75-0272 COURTADE, R. M.
HAMELIN, P.
CHARACTERISATION ELASTIQUE D'UN STRATIFIE
RESINE-VERRE PAR LA METHODE DES DEPLACEMENTS
(DETERMINATION OF ELASTIC PROPERTIES OF LAMINATED
RESIN-GLASS MATERIALS BY DISPLACEMENT METHOD)
FIBRE SCI. TECHNOL., VOL. 8, NO. 3, JULY 1875, PP.
207-219.

75-0273 PEYRONNE, C.
ETUDE THEORIQUE DU COMPORTEMENT D'UNE DALLE DE
BETON. (THEORETICAL STUDY OF THE BEHAVIOR OF A
CONCRETE SLAB)
PONTS ET CHAUSSEES, LAB CENT., PARIS, SUPPL. TO
NO. 77, MAY - JUNE, 1975, PP. 7-29.

75-0274 ARMIT, A. P.
LEMKE, H. U.
ICON: THE INTERACTIVE CREATION OF NASTRAN DATA. A
SYSTEM DESCRIPTION
COMPUT. AIDED DES., VOL. 7, NO. 3, JULY 1975, PP.
145-150.

75-0275 ABBAS, B. A. H.
THOMAS, J.
FINITE ELEMENT MODEL FOR DYNAMIC ANALYSIS OF
TIMOSHENKO BEAM
J. SOUND AND VIB., VOL. 41, NO. 3, AUGUST 8, 1975,
PP. 291-299.

75-0276 CARPENTER, S. H.
LYTTON, R. L.
PAVEMENT CRACKING IN WEST TEXAS DUE TO FREEZE-THAW
CYCLING
TRANSP. RES. BOARD. NO. 532, 1975, PP. 1-13.

75-0277 SEED, H. B.
IDRISS, I. M.
DYNAMIC ANALYSIS OF THE SLIDE IN THE LOWER SAN
FERNANDO DAM DURING THE EARTHQUAKE OF FEBRUARY 9,
1971
ASCE J. GEOTECH. ENG. DIV., VOL. 101, NO. 9,
SEPTEMBER 1975, PP. 889-911.

75-0278 LEE, K. L.
IDRISS, I. M.
STATIC STRESSES BY LINEAR AND NONLINEAR METHODS
ASCE J. GEOTECH. ENG. DIV., VOL. 101, NO. 9,
SEPTEMBER 1975, PP. 871-887.

75-0279 FRENCH, S.
KABAILA, A. P.
SINGLE ELEMENT PANEL FOR FLAT PLATE STRUCTURES
ASCE J. STRUCT. DIV., VOL. 101, NO. 9, SEPTEMBER
1975, PP. 1801-1812.

75-0280 RABIZADEH, R. O.
SHORE, S.
DYNAMIC ANALYSIS OF CURVED BOX-GIRDER BRIDGES
ASCE. J. STRUCT. DIV., VOL. 101, NO. 9, SEPTEMBER
1975, PP. 1899-1912.

75-0281 GINESU, F.
PICASSO, B.
ELASTO-PLASTIC ANALYSIS OF A ROTATING MODEL
CONICAL TURBINE DISC
J. STRAIN ANAL., VOL. 10, NO. 3, JULY 1975, PP.
167-171.

75-0282 ZIENKIEWICZ, O. C.
GODBOLE, P. N.
PENALTY FUNCTION APPROACH TO PROBLEMS OF PLASTIC
FLOW OF METALS WITH LARGE SURFACE DEFORMATIONS
J. STRAIN ANAL., VOL. 10, NO. 3, JULY 1975, PP.
180-183.

75-0283 MAK, S. W.
BOTMAN, M.
SUBSTRUCTURES ANALYSIS OF IMPELLER VIBRATION MODES
PAPER NO. 75-DET-112, A.S.M.E. MEETING, SEPTEMBER
17-19, 1975, 8 PP.

75-0284 FOYE, R. L.
INELASTIC MICROMECHANICS OF CURING STRESSES IN
COMPOSITES
A.S.M.E. MEETING, HOUSTON, TEXAS, NOVEMBER 30 -
DECEMBER 5, 1975, AMD VOL. 13, PP. 177-211.

75-0285 ANON
SHIP STRUCTURAL ANALYSIS IN LLOYD'S REGISTER OF
SHIPPING
NAV. ARCHIT., NO. 3, JULY 1975, PP. 89-90.

75-0286 CHEN, A. C. T.
CHEN, W. F.
CONSTITUTIVE EQUATIONS AND PUNCH-INDENTATION OF
CONCRETE
A.S.C.E. J. ENG. MECH. DIV., VOL. 101, NO. 6,
DECEMBER 1975, PP. 889-906.

75-0287 JAGANNATHAN, D. S.
EPSTEIN, H. I.
NONLINEAR ANALYSIS OF RETICULATED SPACE TRUSSES
A.S.C.E. J. STRUCT. DIV., VOL. 101, NO. 12,
DECEMBER 1975, PP. 2641-2658.

75-0288 THOMPSON, E. G.
GOODMAN, J. R.
FINITE ELEMENT ANALYSIS OF LAYERED WOOD SYSTEMS
A.S.C.E. J. STRUCT. DIV., VOL. 101, NO. 12,
DECEMBER 1975, PP. 2659-2672.

75-0289 BEDFORD, K. W.
LIGGETT, J. A.
CONVECTIVE TRANSPORT FINITE ELEMENT ANALOG
A.S.C.E. J. ENG. MECH. DIV., VOL. 101, NO. 6,
DECEMBER 1975, PP. 803-818.

75-0290 JUNKINS, J. L.
INVESTIGATION OF FINITE ELEMENT REPRESENTATIONS OF
THE GEOPOTENTIAL
CONF. AM. ASTRON. SOC., A.I.A.A., PAPER AAS,
NASSAU, BAHAMAS, JULY 28-30, 1975, 30 PP.

75-0291 HYLARIDES, S.
TRANSVERSE VIBRATIONS OF SHIP'S PROPULSION SYSTEMS
- 1 THEORETICAL ANALYSIS
INT. SHIPBUILD. PROG., VOL. 22, NO. 252, AUGUST
1975, PP. 275-288.

75-0292 ANTES, H.
UEBER PAUSCHALE FEHLERSCHRANKEN FUER
APPROXIMATIONEN VON SCHALENVERFORMUNGEN. (GLOBAL
BOUNDARY CONDITIONS FOR APPROXIMATION OF SHELL
DEFORMATIONS)
Z. ANGEW MATH. MECH., VOL. 55, NO. 9, SEPTEMBER
1975, PP. 491-501.

75-0293 BUYUKOZTUPK, O.
MARCAL, P. V.
STRENGTH OF REINFORCED CONCRETE CHAMBERS UNDER
EXTERNAL PRESSURE
A.S.M.E., J. PRESSURE VESSEL TECH. TRANS., VOL.
97, NO. 4, NOVEMBER 1975, PP. 309-314.

75-0294 GASS, N.
SOME TWO-FIELD VARIATIONAL PRINCIPLES FOR
NONLINEAR DEFORMATION ANALYSIS OF SHELLS
Z. ANGEW MATH. MECH., VOL. 55, NO. 9, SEPTEMBER
1975, PP. 515-521.

75-0295 JABLON, C.
TOURNAIRE, M.
COMPARAISON DES DIVERSES MODELISATIONS NUMERIQUES
D'UN SYSTEME MAGNETIQUE SATURABLE. (COMPARISON OF
DIFFERENT METHODS OF MODELING A SATURABLE MAGNETIC
SYSTEM)
REV. GEN. ELEC., VOL. 84, NO. 9, SEPTEMBER 1975,
PP. 615-618.

75-0296 GOLAS, J.
KASPERSKI, Z.
NUMERYCZNE ROZWIAZANIE PROBLEMU ORGAN WLASNYCH NA
PRZYKLADZIE PLYT. (NUMERICAL SOLUTION OF THE
PROBLEM OF FREE VIBRATIONS OF PLATES)
MECH. TEOR. I STOSOW, VOL. 13, NO. 1, 1975, PP.
85-94.

75-0297 BELYTSCHKO, T.
OSIAS, J. R.
FINITE ELEMENT ANALYSIS OF TRANSIENT NONLINEAR
STRUCTURAL BEHAVIOR
A.S.M.E. MEETING, HOUSTON, TEXAS, NOVEMBER 30 -
DECEMBER 5, 1975, AMD VOL. 14, 191 PP.

75-0298 BELYTSCHKO, T.
HOLMES, N.
EXPLICIT INTEGRATION - STABILITY, SOLUTION
PROPERTIES, COST
A.S.M.E. MEETING, HOUSTON, TEXAS, NOVEMBER 30 -
DECEMBER 5, 1975, AMD VOL. 14, PP. 1-21.

75-0299 NEMAT-NASSER, S.
CONTINUUM BASES FOR CONSISTENT NUMERICAL
FORMULATIONS OF FINITE STRAINS IN ELASTIC AND
INELASTIC STRUCTURES
A.S.M.E. MEETING, HOUSTON, TEXAS, NOVEMBER 30 -
DECEMBER 5, 1975, AMD VOL. 14, PP. 85-89.

75-0300 SACZALSKI, K. J.
MODELING AND ANALYSIS TECHNIQUES FOR PREDICTION OF
STRUCTURAL AND BIODYNAMIC CRASH-IMPACT RESPONSE
A.S.M.E. MEETING, HOUSTON, TEXAS, NOVEMBER 30 -
DECEMBER 5, 1975, AMD VOL. 14, PP. 99-117.

75-0301 WELCH, R. E.
BRUCE, R. W.
NONLINEAR TRANSIENT ANALYSIS OF AUTOMOTIVE
STRUCTURES BY FINITE ELEMENTS
A.S.M.E. MEETING, HOUSTON, TEXAS, NOVEMBER 30 -
DECEMBER 5, 1975, AMD VOL. 14, PP. 119-132.

75-0302 MESSIER, R. H.
MARCAL, P. V.
FINITE ELEMENT ALGORITHM FOR THE DETERMINATION OF
DYNAMIC BUCKLING
A.S.M.E. MEETING, HOUSTON, TEXAS, NOVEMBER 30 -
DECEMBER 5, 1975, AMD VOL. 14, PP. 133-156.

75-0303 REEFMAN, R. J. B.
SPREEUW, E.
PRINCIPLES OF THE FINITE ELEMENT METHOD
HOLECTECHNIEK, VOL. 5, NO. 2, SEPTEMBER 1975, PP.
56-60.

75-0304 ROHWER, K.
PIENING, M.
DESH, EIN EINFACHES FLACHES SCHALENELEMENT UEBER
DREIECKIGEM GRUNDRISS. (DESH, A SIMPLE TRIANGULAR
SHALLOW SHELL ELEMENT)
DTSCH LUFT RAUMFAHRT MITT., NO. 75-13, 1975, 36 PP.

75-0305 LARDER, R. A.
BEADLE, C. W.
STRENGTH DISTRIBUTIONS OF SINGLE FILAMENTS
J. COMPOS. MATER., VOL. 9, JULY 1975, PP. 241-243.

75-0306 ANDERSON, J. C.
GURFINKEL, G.
SEISMIC BEHAVIOUR OF FRAMED TUBES
EARTHQUAKE ENG. STRUCT. DYN., VOL. 4, NO. 2,
OCTOBER-DECEMBER, 1975, PP. 145-162.

75-0307 VOLCY, G. C.
BUDD, W. I. H.
REDUCTION GEAR DAMAGES RELATED TO EXTERNAL
INFLUENCES
MAR. TECH., VOL. 12, NO. 4, OCTOBER 1975, PP.
335-366.

75-0308 RUTLEDGE, J. C.
TIED-BACK RETAINING WALLS
NZ ENG., VOL. 30, NO. 8, AUGUST 15, 1975, PP.
242-247.

75-0309 FRANK, R.
ETUDE THEORIQUE DU COMPORTEMENT DES PIEUX SOUS
CHARGE VERTICALE. INTRODUCTION DE LA DILATANCE.
(THEORETICAL RESEARCH ON THE BEHAVIOR OF PILES
UNDER VERTICAL LOADING. INTRODUCING DILATANCY)
LAB. CENT., PONTS ET CHAUSSEEES, PARIS, FRANCE,
RAPP. RECH., NO. 46, JULY 1975, 238 PP.

75-0310 MOFFATT, K. R.
DOWLING, P. J.
SHEAR LAG IN STEEL BOX GIRDER BRIDGES
STRUCT. ENG., VOL. 53, NO. 10, OCTOBER 1975, PP.
439-448.

75-0311 HORI, Y.
HASUIKE, A.
STUDY ON FOIL BEARINGS - AN APPLICATION TO TAPE
MEMORY DEVICES
J.S.M.E./A.S.M.E. CONF. ON APPL. MECH., HONOLULU,
HAWAII, MARCH 24-27, 1975, PP. 121-128. (PUBL. BY
J.S.M.E., TOKYO, JAPAN, 1975)

75-0312 GOTOH, M.
PLASTICITY OF POLYCRYSTALS AND GENERAL DISCUSSIONS
ON PLASTICITY THEORY WITH IT
J.S.M.E./A.S.M.E. CONF. ON APPL. MECH., HONOLULU,
HAWAII, MARCH 24-27, 1975, PP. 239-246. (PUBL. BY
J.S.M.E., TOKYO, JAPAN, 1975)

75-0313 SHINDO, A.
SEGUCHI, Y.
ELASTIC-PLASTIC PROBLEMS OF COMPRESSIBLE STRAIGHT
BAR UNDER COMBINED LOADING CONDITION
J.S.M.E./A.S.M.E. CONF. ON APPL. MECH., HONOLULU,
HAWAII, MARCH 24-27, 1975, PP. 267-274. (PUBL. BY
J.S.M.E., TOKYO, JAPAN, 1975)

75-0314 YAGAWA, G.
NISHIOKA, T.
ELASTIC-PLASTIC FINITE ELEMENT ANALYSIS USING
SUPERPOSITION
J.S.M.E./A.S.M.E. CONF. ON APPL. MECH., HONOLULU,
HAWAII, MARCH 24-27, 1975, PP. 367-374. (PUBL. BY
J.S.M.E., TOKYO, JAPAN, 1975)

75-0315 SHIRATORI, M.
MIYOSHI, T.
NONLINEAR ANALYSIS OF CRACKED BODY BY FINITE
ELEMENT METHOD
J.S.M.E./A.S.M.E. CONF. ON APPL. MECH., HONOLULU,
HAWAII, MARCH 24-27, 1975, PP. 375-382. (PUBL. BY
J.S.M.E., TOKYO, JAPAN, 1975)

75-0316 MIYATA, H.
KUSUMOTO, S.
METHOD OF EVALUATION OF THREE-DIMENSIONAL STRESS
INTENSITY FACTOR USING THE FINITE ELEMENT METHOD
J.S.M.E./A.S.M.E. CONF. ON APPL. MECH., HONOLULU,
HAWAII, MARCH 24-27, 1975, PP. 383-390. (PUBL. BY
J.S.M.E., TOKYO, JAPAN, 1975)

75-0317 LIN, C-S.
SCORDELIS, A. C.
FINITE ELEMENT STUDY OF A REINFORCED CONCRETE
CYLINDRICAL SHELL THROUGH ELASTIC, CRACKING, AND
ULTIMATE RANGES
J. AM. CONCR. INST., VOL. 72, NO. 11, NOVEMBER
1975, PP. 628-633.

75-0318 ISIDA, M.
ARBITRARY SYMMETRIC LOADING PROBLEMS OF CENTRALLY
CRACKED RECTANGULAR PLATES
J.S.M.E./A.S.M.E. CONF. ON APPL. MECH., HONOLULU,
HAWAII, MARCH 24-27, 1975, PP. 399-406. (PUBL. BY
J.S.M.E., TOKYO, JAPAN, 1975)

75-0319 PETRASOVITS, G.
SETTLEMENT ANALYSIS OF DRIVEN PILES
A.C.T.A. TECH. (BUDAP), VOL. 80, NO. 3-4, 1975,
PP. 343-351.

75-0320 GOLDBERG, J. E.
PATHAK, D. V.
ANALYSIS OF MODERATELY THICK FLANGED REACTOR
VESSELS
NUCL. ENG. DES., VOL. 32, NO. 3, JULY 1975, PP.
363-381.

75-0321 SCHAEFER, H.
CONTRIBUTION TO THE SOLUTION OF CONTACT PROBLEMS
WITH THE AID OF BOND ELEMENTS
COMPUT. METHODS APPL. MECH. ENG., VOL. 6, NO. 3,
NOVEMBER 1975, PP. 335-353.

75-0322 TANIGUCHI, S.
ETTLES, C. M. M.
MECHANISM OF PARALLEL SURFACE THRUST BEARING
J. JAPAN SOC. LUBR. ENG., VOL. 20, NO. 1, 1975,
PP. 45-53.

75-0323 YACIUK, G.
MUIR, W. E.
SELECTION OF LOCATIONS FOR LONG TERM STORAGE OF
WHEAT
A.S.A.E. 68TH ANN. MEETING, UNIV. OF CALIF.,
DAVIS, JUNE 22-25, 1975, PAPER NO. 75-4057, 21 PP.

75-0324 LASCAUX, P.
LESAINT, P.
SOME NONCONFORMING FINITE ELEMENTS FOR THE PLATE
BENDING PROBLEM
REV. FR. AUTOM. INF. RECH. OPER., VOL. 9, R-1,
APRIL 1975, PP. 9-53.

75-0325 MASSMANN, J.
STRUCTURAL RESPONSE TO IMPACT DAMAGE
AGARD, REPORT NO. 633, SEPTEMBER 1975, 22 PP.

75-0326 OCHIAI, S-I.
YOSHINAGA, S.
FORMULATION OF STRESS STRAIN INDUCED DIFFUSION OF
HYDROGEN AND ITS SOLUTION BY COMPUTER-AIDED
FINITE ELEMENT METHOD
TRANS. IRON STEEL INST., JAPAN, VOL. 15, NO. 10,
1975, PP. 503-507.

75-0327 MOAN, T.
SOREIDE, T.
ANALYSIS OF STIFFENED PLATES CONSIDERING NONLINEAR
MATERIAL AND GEOMETRIC BEHAVIOUR
IN "WORLD CONGRESS ON FINITE ELEMENT METH. IN
STRUCT. MECH.", BOURNEMOUTH, DORSET, ENGLAND,
OCTOBER 12-17, 1975, 28 PP.

75-0328 MCNAMARA, J. F.
SHARMA, S. K.
ANALYTICAL MODEL FOR DETERMINING ENERGY
DISSIPATION IN DYNAMICALLY LOADED STRUCTURES
CONSTR. ENG. RES. LAB., TECH. REPORT M-165,
OCTOBER 1975, 35 PP.

75-0329 FOWLER, J. H.
FINITE-ELEMENT ANALYSIS OF HIGH-PRESSURE, LARGE
VALVE GATES
PAPER NO. 75-PET-6, A.S.M.E. MEETING, SEPTEMBER
21-25, 1975, 7 PP.

75-0330 BARRY, W. R.
FINITE ELEMENT ANALYSIS OF CASING THREADS
PAPER NO. 75-PET-9, A.S.M.E. MEETING, SEPTEMBER
21-25, 1975, 9 PP.

75-0331 KRAUS, H.
NELSON, N. W.
ANALYSIS OF BOTTOM ENTRY NOZZLES FOR ATMOSPHERIC
STORAGE TANKS
PAPER NO. 75-PET-23, A.S.M.E. MEETING, SEPTEMBER
21-25, 1975, 7 PP.

75-0332 NERLI, G.
STUDY OF UMBRELLA OSCILLATIONS OF TURBINE DISKS OF
ARBITRARY PROFILE BY THE FINITE ELEMENT METHOD
COMPUT. STRUCT., VOL. 5, NO. 4, NOVEMBER 1975, PP.
233-239.

75-0333 PEDERSEN, P.
MEGAHED, M. M.
AXISYMMETRIC ELEMENT ANALYSIS USING ANALYTICAL
COMPUTING
COMPUT. STRUCT., VOL. 5, NO. 4, NOVEMBER 1975, PP.
241-247.

75-0334 VENKATESWARA RAO, G.
BENKATARAMANA, J.
STABILITY OF MODERATELY THICK RECTANGULAR PLATES
USING A HIGH PRECISION TRIANGULAR FINITE ELEMENT
COMPUT. STRUCT., VOL. 5, NO. 4, NOVEMBER 1975, PP.
257-259.

75-0335 GOPALACHARYULU, S.
ALI, R.
SHARMAN, P. W.
THE DEVELOPMENT OF A SERIES OF HYBRID-STRESS
FINITE ELEMENTS
IN "WORLD CONGRESS ON FINITE ELEMENT METH. IN
STRUCT. MECH.", BOURNEMOUTH, DORSET, ENGLAND,
OCTOBER 12-17, 1975, 27 PP.

75-0336 HUMPHESON, C.
LEWIS, R. W.
APPLICATIONS OF ELECTRO-OSMOSIS TO GROUND-WATER
FLOW PROBLEMS
GROUND WATER, VOL. 13, NO. 6, NOVEMBER-DECEMBER
1975, PP. 484-491.

75-0337 DE ARANTES E OLIVEIRA, E. R.
CONVERGENCE OF FINITE ELEMENT SOLUTIONS IN VISCOUS
FLOW PROBLEMS
INT. J. NUMER. METHODS ENG., VOL. 9, NO. 4, 1975,
PP. 739-763.

75-0338 MEEK, J. L.
REDUNDANT STRUCTURE CONCEPTS IN FINITE ELEMENT
ANALYSES
INT. J. NUMER. METHODS ENG., VOL. 9, NO. 4, 1975,
PP. 765-773.

75-0339 TORBE, I.
CHURCH, K.
GENERAL QUADRILATERAL PLATE ELEMENT
INT. J. NUMER. METH. ENG., VOL. 9, NO. 4, 1975,
PP. 855-868.

75-0340 ROSS, C. T. F.
FINITE ELEMENTS FOR THE VIBRATION OF CONES AND
CYLINDERS
INT. J. NUMER. METH. ENG., VOL. 9, NO. 4, 1975,
PP. 833-845.

75-0341 ROHDE, S. M.
HIGHER ORDER FINITE ELEMENT METHODS FOR THE
SOLUTION OF COMPRESSIBLE POROUS BEARING PROBLEMS
INT. J. NUMER. METH. ENG., VOL. 9, NO. 4, 1975,
PP. 903-911.

75-0342 THOMPSON, E. G.
AVERAGE AND COMPLETE INCOMPRESSIBILITY IN THE
FINITE ELEMENT METHOD
INT. J. NUMER. METH. ENG., VOL. 9, NO. 4, 1975,
PP. 925-932.

75-0343 UDOGUCHI, T.
NOZUE, Y.
NOTCH EFFECT IN LOW-CYCLE FATIGUE. CONSIDERATIONS
ON NOTCH EFFECT IN LOW-CYCLE FATIGUE WITH
FINITE-ELEMENT METHOD
BULL. J.S.M.E., VOL. 18, NO. 126, 1975, PP.
1355-1364.

75-0344 NAHAVANDI, A. N.
BOHM, G. J.
PEDRIDO, R. R.
STRUCTURALLY COMPATIBLE FLUID FINITE-ELEMENT FOR
SOLID FLUID INTERACTION STUDIES
NUCL. ENG. DES., VOL. 35, NO. 3, 1975, PP.
335-347.

75-0345 MARCHUK, G. I.
ZALESNYI, V. B.
KUZIN, V. I.
FINITE-DIFFERENCE AND FINITE-ELEMENT METHODS IN
GLOBAL WIND-DRIVEN OCEAN CIRCULATION
IZVESTIYA AKADEMII NAUK SSSR., FIZIKA ATMOS., VOL.
11, NO. 12, 1975, PP. 1294-1300.

75-0346 SRINATHA, H. R.
RAO, G. V.
RAJU, I. S.
DISCONTINUITY STRESS ANALYSIS OF PRESSURE-VESSELS
USING FINITE-ELEMENT METHOD
NUCL. ENG. DES., VOL. 35, NO. 2, 1975, PP.
309-314.

75-0347 LASKARIS, T. E.
FINITE-ELEMENT ANALYSIS OF COMPRESSIBLE AND
INCOMPRESSIBLE VISCOUS-FLOW AND HEAT-TRANSFER
PROBLEMS
PHYSICS OF FLUIDS, VOL. 18, NO. 12, 1975, PP.
1639-1648.

75-0348 SCHRODER, J. J.
COMPARISON OF FINITE-DIFFERENCE AND FINITE-ELEMENT
SOLUTION TECHNIQUE IN TRANSIENT HEAT-CONDUCTION
FORSCHUNG IM INGENIEURWESEN, VOL. 41, NO. 6, 1975,
PP. 169-173.

75-0349 BEGIS, D.
GLOWINSKI, R.
APPLICATION OF FINITE-ELEMENT METHOD TO
APPROXIMATION OF OPTIMAL DOMAIN PROBLEM - SOLUTION
METHODS FOR APPROXIMATED PROBLEMS
APPL. MATH. AND OPTIMIZATION, VOL. 2, NO. 2, 1975,
PP. 130-169.

75-0350 HENRY, M. F.
BUCHAKJIAN, L.
GIGLIOTTI, M. F. X.
FINITE-ELEMENT MODELING OF DIRECTIONAL
SOLIDIFICATION OF SUPER ALLOY EUTECTICS
J. METALS, VOL. 27, NO. 128 1975, P. 43.

75-0351 BARLOW, J.
INERTIA LOADING IN FINITE-ELEMENT ANALYSIS OF
STRUCTURES SUBJECT TO COMPOUND MOTION
INT. J. NUMER. METH. ENG., VOL. 10, NO. 1, 1975,
PP. 197-209.

75-0352 BARSOUM, R. S.
USE OF ISOPARAMETRIC FINITE-ELEMENTS IN LINEAR
FRACTURE MECHANICS
INT. J. NUMER. METH. ENG., VOL. 10, NO. 1, 1975,
PP. 25-37.

75-0353 KIKUCHI, F.
THEORY AND EXAMPLES OF PARTIAL APPROXIMATION IN
FINITE-ELEMENT METHOD
INT. J. NUMER. METH. ENG., VOL. 10, NO. 1, 1975,
PP. 115-122.

75-0354 GERINROZE, J.
LESAINT, P.
ISOPARAMETRIC FINITE-ELEMENT METHODS FOR
2-DIMENSIONAL TRANSPORT CALCULATIONS
INT. J. NUMER. METH. ENG., VOL. 10, NO. 1, 1975,
PP. 171-183.

75-0355 YAGAWA, G.
MIYAZAKI, N.
ANDO, Y.
ANALYSIS OF ELASTIC-PLASTIC CREEP BUCKLING OF
AXISYMMETRIC SHELLS BY FINITE-ELEMENT METHOD
ARCHIVES OF MECH., VOL. 27, NO. 5-6, 1975, PP.
869-882.

75-0356 BATOZ, J. L.
CHATTOPADHYAY, A.
FINITE ELEMENT LARGE DEFLECTION ANALYSIS OF
SHALLOW SHELLS
INT. J. NUMER. METH. ENG., VOL. 10, NO. 1, 1975,
PP. 39-58.

75-0357 MATSUI, T.
MATSUOKA, O.
NEW FINITE-ELEMENT SCHEME FOR INSTABILITY ANALYSIS
OF THIN SHELLS
INT. J. NUMER. METH. ENG., VOL. 10, NO. 1, 1975,
PP. 145-170.

75-0358 WEISSHAAR, T. A.
PANEL FLUTTER OPTIMIZATION - REFINED
FINITE-ELEMENT APPROACH
INT. J. NUMER. METH. ENG., VOL. 10, NO. 1, 1975,
PP. 77-91.

75-0359 HIRSCH, C.
WARZEE, G.
FINITE-ELEMENT METHOD FOR AXISYMMETRIC FLOW
COMPUTATION IN A TURBOMACHINE
INT. J. NUMER. METH. ENG., VOL. 10, NO. 1, 1975,
PP. 93-113.

75-0360 HARTMANN, B.
SCHRADER, K. H.
MESY - EIN PROGRAMMSYSTEM ZUR UNTERSUCHUNG VON
TRAGWERKEN. (STRUCTURAL ANALYSIS BY MESY
PROGRAMMING SYSTEM)
KONSTR. INGENIEURBAU BER., NO. 22, 1975, 106 PP.

75-0361 FREESE, C. E.
BOWIE, O. L.
STRESS ANALYSIS OF CONFIGURATIONS INVOLVING SMALL
FILLETS
J. STRAIN ANALY., VOL. 10, NO. 1, JANUARY 1975,
PP. 53-58.

75-0362 LINDLEY, P. B.
FINITE-ELEMENT PROGRAMME FOR THE PLANE-STRAIN
ANALYSIS OF RUBBER
J. STRAIN ANAL., VOL. 10, NO. 1, JANUARY 1975, PP.
25-31.

75-0363 WARREN, G. E.
PHOTOELASTIC TESTS ON COMPLEX INTERSECTIONS OF
REINFORCED THIN-SHELL CYLINDERS
EXP. MECH., VOL. 15, NO. 5, MAY 1975, PP. 201-208.

75-0364 CAREY, G. F.
FINLAYSON, B. A.
ORTHOGONAL COLLOCATION ON FINITE ELEMENTS
CHEM. ENG. SCI., VOL. 30, NO. 5-6, MAY - JUNE,
1975, PP. 587-596.

75-0365 GUY, A. W.
LIN, J. C.
EFFECT OF 2450-MHZ RADIATION ON THE RABBIT EYE
I.E.E.E. TRANS. MICROWAVE THEORY TECH., VOL. 23,
NO. 6, JUNE 1975, PP. 492-498.

75-0366 SANDAY, S. C.
LAM, T. L.
DESIGN AND ANALYSIS OF A CERAMIC STATOR VANE
PAPER NO. 75-GY-100, A.S.M.E. MEETING, MARCH 2-6,
1975, 11 PP.

75-0367 ALLEN, J. M.
ERICKSON, L. B.
NASTRAN ANALYSIS OF A TURBINE BLADE AND COMPARISON
WITH TEST AND FIELD DATA
PAPER NO. 75-GI-77, A.S.M.E. MEETING, MARCH 2-6,
1975, 8 PP.

75-0368 KUO, P. S.
FINITE ELEMENT APPROACH TO THE VIBRATION ANALYSIS
OF ELASTIC DISKS ON A FLEXIBLE SHAFT
PAPER NO. 75-GT-57, A.S.M.E. MEETING, MARCH 2-6,
1975, 7 PP.

75-0369 BOLLETER, U.
BEGLINGER, V.
EFFECTS OF SHEAR DEFORMATION, ROTARY INERTIA, AND
ELASTICITY OF THE SUPPORT ON THE RESONANCE
FREQUENCIES OF SHORT CANTILEVER BEAMS
PAPER NO. 75-GT-37, A.S.M.E. MEETING, MARCH 2-6,
1975, 9 PP.

75-0370 LACHAT, J. C.
WATSON, J. O.
ANALYSIS OF CONTINUA BY THE FINITE ELEMENT AND
INTEGRAL EQUATION METHOD
IN "WORLD CONGRESS ON FINITE ELEMENT METH. IN
STRUCT. MECH.", BOURNEMOUTH, DORSET, ENGLAND,
OCTOBER 12-17, 1975, 25 PP.

75-0371 FILCEK, H.
HALAT, W.
DER SPANNUNGS UND BEANSPRUCHUNGSZUSTAND DES
BODENMATERIALS IN DER NACHBARSCHAFT EINER
BODENBOESCHUNG. (STATE OF STRESS AND STRAIN OF
SOIL MATERIAL IN THE VICINITY OF A SLOPE)
ARCH. GORN., VOL. 20, NO. 1, 1975, PP. 3-15.

75-0372 YEH, C.
DONG, S. B.
ARBITRARILY SHAPED INHOMOGENEOUS OPTICAL FIBER OR
INTEGRATED OPTICAL WAVEGUIDES
J. APPL. PHYS., VOL. 46, NO. 5, MAY 1975, PP.
2125-2129.

75-0373 CHEW, B.
KYTE, W. S.
DIFFUSION OF HYDROGEN IN FILLET WELDS
MET. TECHNOL., VOL. 2, PART 2, FEBRUARY 1975, PP.
66-72.

75-0374 HERTLING, J.
NUMERICAL TREATMENT OF ALGEBRAIC INTEGRAL
EQUATIONS BY VARIATIONAL METHODS
S.I.A.M. J. NUMER. ANAL., VOL. 12, NO. 2, APRIL
1975, PP. 203-212.

75-0375 STOVALL, R. E.
MEI, K. K.
APPLICATION OF A UNIMOMENT TECHNIQUE TO A
BICONICAL ANTENNA WITH INHOMOGENEOUS DIELECTRIC
LOADING
I.E.E.E. TRANS. ANTENNAS PROPAG., VOL. AP-23, NO.
3, MAY 1975, PP. 335-342.

75-0376 LACHAT, J. E.
WATSON, J. O.
SECOND GENERATION BOUNDARY INTEGRAL EQUATION
PROGRAM FOR THREE-DIMENSIONAL ELASTIC ANALYSIS
A.S.M.E./AMD MEETING, TROY, N.Y., JUNE 23-25,
1975, VOL. 11, PP. 85-100.

75-0377 BESUNER, P. M.
SNOW, D. W.
APPLICATION OF THE TWO-DIMENSIONAL INTEGRAL
EQUATION METHOD TO ENGINEERING PROBLEMS
A.S.M.E./AMD MEETING, TROY, N.Y., JUNE 23-25,
1975, VOL. 11, PP. 101-117.

75-0378 KHALIL, T. B.
HUBBARD, R. P.
DYNAMIC RESPONSE OF ELASTIC HEAD MODELS
A.S.M.E./AMD MEETING, TROY, N.Y., JUNE 23-25,
1975, VOL. 10, PP. 145-148.

75-0379 HIGHT, T. K.
NAGEL, D. A.
DETAILED STRUCTURAL ANALYSIS OF THE HUMAN TIBIA
A.S.M.E./AMD MEETING, TROY, N.Y., JUNE 23-25,
1975, VOL. 10, PP. 97-99.

75-0380 SIMON, B. R.
WOO, S. L. Y.
FINITE ELEMENT AND EXPERIMENTAL ANALYSIS OF AN
INTERNAL FIXATION PLATE MODEL
A.S.M.E./AMD MEETING, TROY, N.Y., JUNE 23-25,
1975, VOL. 10, PP. 93-95.

75-0381 MCNEICE, G. M.
FINITE ELEMENT STUDIES OF FEMORAL ENDOPROSTHESES
FOR HIP RECONSTRUCTION
A.S.M.E./AMD MEETING, TROY, N.Y., JUNE 23-25,
1975, VOL. 10, PP. 89-92.

75-0382 VOROSHKO, P. P.
KVITKA, A. L.
K VOPROSU OB AVTOMATIZATSII ZADANIYA INFORMATSII V
METODE KONECHNYKH ELEMENTOV. (PROBLEM OF
AUTOMATION OF INFORMATION SETTING IN THE FINITE
ELEMENT METHOD)
PROBL. PROCHN., VOL. 7, NO. 3, MARCH 1975, PP.
42-46.

75-0383 COOK, R. D.
AVOIDANCE OF PARASITIC SHEAR IN PLANE ELEMENT
A.S.C.E. J. STRUCT. DIV., VOL. 101, NO. 6, JUNE
1975, PP. 1239-1253.

75-0384 COLE, P. P.
ABEL, J. F.
BUCKLING OF COOLING-TOWER SHELLS: BIFURCATION
RESULTS
A.S.C.E. J. STRUCT. DIV., VOL. 101, NO. 6, 1975,
PP. 1205-1222.

75-0385 COLE, P. P.
ABEL, J. F.
BUCKLING OF COOLING-TOWER SHELLS: STATE-OF-THE-ART
A.S.C.E. J. STRUCT. DIV., VOL. 101, NO. 6, JUNE
1975, PP. 1185-1203.

75-0386 SHARIFI, P.
YATES, D. N.
NONLINEAR FINITE ELEMENT AND ITS APPLICATION TO
LARGE SCALE STRUCTURAL PROBLEMS
IN "WORLD CONGRESS ON FINITE ELEMENT METH. IN
STRUCT. MECH.", BOURNEMOUTH, DORSET, ENGLAND,
OCTOBER 12-17, 1975, 26 PP.

75-0387 CIAVALDINI, J. F.
ANALYSE NUMERIQUE D'UN PROBLEME DE STEFAN A DEUX
PHASES PAR UNE METHODE D'ELEMENTS FINIS.
(NUMERICAL ANALYSIS OF A TWO-PHASE STEFAN PROBLEM
BY THE FINITE ELEMENT METHOD)
S.I.A.M. J. NUMER. ANAL., VOL. 12, NO. 3, JUNE
1975, PP. 464-487.

75-0388 SCOTT, R.
INTERPOLATED BOUNDARY CONDITIONS IN THE FINITE
ELEMENT METHOD
S.I.A.M. J. NUMER. ANAL., VOL. 12, NO. 3, JUNE
1975, PP. 404-427.

75-0389 NAGARAJ, V. T.
SHANTHAKUMAR, P.
ROTOR BLADE VIBRATIONS BY GALERKIN FINITE-ELEMENT
METHOD
J. SOUND AND VIB., VOL. 43, NO. 3, 1975, PP.
575-577.

75-0390 TAYLOR, C.
DAVIS, J. M.
TIDAL PROPAGATION AND DISPERSION IN ESTUARIES
IN "FINITE ELEMENTS IN FLUIDS", R. GALLAGHER, ET
AL. (EDS.), VOL. 1, 1975, PP. 95-118. (PUBL. BY
WILEY, 1975.)

75-0391 KING, I. P.
NORTON, W. R.
ICEMAN, K. R.
A FINITE ELEMENT SOLUTION FOR TWO-DIMENSIONAL
STRATIFIED FLOW PROBLEMS
IN "FINITE ELEMENTS IN FLUIDS", R. GALLAGHER, ET
AL. (EDS.), VOL. 1, 1975, PP. 133-156. (PUBL. BY
WILEY, 1975.)

75-0392 LEWIS, R. W.
VERNER, E.
ZIENKIEWICZ, O. C.
A FINITE ELEMENT APPROACH TO TWO-PHASE FLOW IN
POROUS MEDIA
IN "FINITE ELEMENTS IN FLUIDS", R. GALLAGHER, ET
AL. (EDS.), VOL. 1, 1975, PP. 183-200. (PUBL. BY
WILEY, 1975.)

75-0393 NEUMAN, S. P.
GALERKIN APPROACH TO SATURATED-UNSATURATED FLOW IN
POROUS MEDIA
IN "FINITE ELEMENTS IN FLUIDS", R. GALLAGHER, ET
AL. (EDS.), VOL. 1, 1975, PP. 201-218. (PUBL. BY
WILEY, 1975.)

75-0394 NEWTON, R. E.
FINITE ELEMENT ANALYSIS OF TWO-DIMENSIONAL ADDED
MASS AND DAMPING
IN "FINITE ELEMENTS IN FLUIDS", R. GALLAGHER, ET
AL. (EDS.), VOL. 1, 1975, PP. 219-232. (PUBL. BY
WILEY, 1975.)

75-0395 BERKHOFF, J. C. M.
LINEAR WAVE PROPAGATION PROBLEMS AND THE FINITE
ELEMENT METHOD
IN "FINITE ELEMENTS IN FLUIDS", R. GALLAGHER, ET
AL. (EDS.), VOL. 1, 1975, PP. 251-264. (PUBL. BY
WILEY, 1975.)

75-0396 SARPKAYA, T.
HIRIART, G.
FINITE ELEMENT ANALYSIS OF JET IMPINGEMENT ON
AXISYMMETRIC CURVED DEFLECTORS
IN "FINITE ELEMENTS IN FLUIDS", R. GALLAGHER, ET
AL. (EDS.), VOL. 1, 1975, PP. 265-280. (PUBL. BY
WILEY, 1975.)

75-0397 CRISFIELD, M. A.
COMBINED MATERIAL AND GEOMETRIC NONLINEARITY FOR
THIN STEEL PLATES
IN "WORLD CONGRESS ON FINITE ELEMENT METH. IN
STRUCT. MECH.", BOURNEMOUTH, DORSET, ENGLAND,
OCTOBER 12-17, 1975, 25 PP.

75-0398 SPOONER, J. B.
A HISTORY OF THE FINITE ELEMENT METHOD
WORLD CONGRESS ON FINITE ELEMENT METH. IN STRUCT.
MECH., BOURNEMOUTH, DORSET, ENGLAND, OCTOBER
12-17, 1975, PP. A.1-A.22.

75-0399 TAIG, I. C.
MODELLING AND INTERPRETATION OF RESULTS
WORLD CONGRESS ON FINITE ELEMENT METH. IN STRUCT.
MECH., BOURNEMOUTH, DORSET, ENGLAND, OCTOBER
12-17, 1975, P. B1.

75-0400 JANSEN, R.
FINITE-ELEMENT CURRENT-DENSITY REPRESENTATION IN
NUMERICAL-SOLUTION OF MICROSTRIP PROBLEMS
AEU-ARCHIV FUR ELEKTRONIK VBERTRAGUNSTECH, ELECTRO
AND COM., VOL. 29, NO. 11, 1975, PP. 477-480.

75-0401 JAMET, P.
ERROR ESTIMATE FOR QUADRILATERAL ISOPARAMETRIC
FINITE-ELEMENTS OF TYPE Q1 WHICH CAN DEGENERATE
INTO TRIANGLES
COMPTES RENDUS, ACADEMIE DES SCI. SER A MATH.,
VOL. 281, NO. 22, 1975, PP. 983-984.

75-0402 WHITE, J. K.
JAYAWARDENA, A. W.
FINITE-ELEMENT APPROACH TO WATERSHED RUNOFF-
DISCUSSION
J. OF HYDROLOGY, VOL. 27, NO. 3-4, 1975, PP.
357-358.

75-0403 AAMODT, B.
EFFICIENT FORMULATIONS OF THE FINITE ELEMENT
METHOD IN LINEAR AND NONLINEAR FRACTURE MECHANICS
IN "WORLD CONGRESS ON FINITE ELEMENT METH. IN
STRUCT. MECH.", BOURNEMOUTH, DORSET, ENGLAND,
OCTOBER 12-17, 1975, 24 PP.

75-0404 TOMLINSON, E. T.
ROBINSON, J. C.
VONDY, D. R.
SOLUTION OF DIFFUSION AND P1 FINITE-ELEMENT
EQUATIONS BY ITERATION
AMER. NUCL. SOC. TRANS., VOL. 22, NO. 16, 1975, P.
248.

75-0405 HOLAND, I.
MEMBRANE AND PLATE BENDING ELEMENTS
WORLD CONGRESS ON FINITE ELEMENT METH. IN STRUCT.
MECH., BOURNEMOUTH, DORSET, ENGLAND, OCTOBER
12-17, 1975, PP. C.1-C.42.

75-0406 BARUCH, M.
INCLUSION OF TRANSVERSE-SHEAR DEFORMATION IN
FINITE-ELEMENT DISPLACEMENT FORMULATIONS - COMMENT
A.I.A.A. J., VOL. 13, NO. 9, 1975, PP. 1253-1254.

75-0407 KENNEDY, J. M.
BELYTSCHKO, T.
FINITE-ELEMENT ANALYSIS OF STRUCTURAL RESPONSE IN
HCDA
AMER. NUCL. SOC. TRANS., VOL. 22, NO. 16, 1975,
PP. 439-440.

75-0408 MARTIN, W. R.
CONVERGENCE OF FINITE-ELEMENT METHOD IN
NEUTRON-TRANSPORT
AMER. NUCL. SOC. TRANS., VOL. 22, NO. 16, 1975,
PP. 251-252.

75-0409 ENGLAND, R.
HENNART, J. P.
MARTIN, J. G.
MELENDEZ, L.
FINITE-ELEMENT CODE TO SIMULATE ANISOTROPIC-PLASMA
DIFFUSION IN LINEAR MULTIPOLES
AMER. NUCL. SOC. TRANS., VOL. 22, NO. 16, 1975,
PP. 271-272.

75-0410 KAUSEL, E.
CHANGLIANG, V.
EARTHQUAKE FINITE-ELEMENT ANALYSIS OF
STRUCTURE-FOUNDATION SYSTEMS
A.S.C.E., J. ENG. MECH. DIV., VOL. 101, NO. 6,
1975, PP. 909-911.

75-0411 LABBENS, R. C.
EFFICIENT FINITE-ELEMENT METHODS FOR STRESS
INTENSITY FACTORS USING WEIGHT FUNCTIONS
INT. J. OF FRACT., VOL. 11, NO. 6, 1975, PP.
1057-1058.

75-0412 ROSSOW, M. P.
LEAST-SQUARES VARIATIONAL PRINCIPLE FOR
FINITE-ELEMENT APPLICATIONS
J. APPLIED MECH., VOL. 42, NO. 4, 1975, PP.
900-901.

75-0413 FOWLER, J. H.
FINITE-ELEMENT ANALYSIS OF HIGH-PRESSURE LARGE
VALVE GATES
MECH. ENG., VOL. 97, NO. 12, 1975, P. 89.

75-0414 CRAGGS, A.
SOME PRACTICAL APPLICATIONS OF ACOUSTIC FINITE
ELEMENTS
IN "WORLD CONGRESS ON FINITE ELEMENT METH. IN
STRUCT. MECH.", BOURNEMOUTH, DORSET, ENGLAND,
OCTOBER 12-17, 1975, 20 PP.

75-0415 WILSON, G. J.
KIRKHOPE, J.
VIBRATION ANALYSIS OF AXIAL-FLOW TURBINE DISKS
USING FINITE-ELEMENTS
MECH. ENG., VOL. 97, NO. 12, 1975, PP. 95-96.

75-0416 LIGHT, M. F.
LUXMOORE, A.
EVANS, W. T.
PREDICTION OF SLOW CRACK GROWTH BY A
FINITE-ELEMENT METHOD
INT. J. FRACTURE, VOL. 11, NO. 6, 1975, PP.
1045-1046.

75-0417 BARRY, W. R.
FINITE-ELEMENT ANALYSIS OF CASING THREADS
MECH. ENG., VOL. 97, NO. 12, 1975, P. 89.

75-0418 ANSOURIAN, P.
APPERT, K.
APPLICATION OF METHOD OF FINITE ELEMENTS TO
ANALYSIS OF COMPOSITE FLOOR SYSTEMS
PROC. INST. CIVIL ENG., PART 2, RESEARCH AND
THEORY, DECEMBER 1975, VOL. 59, PP. 699-726.

75-0419 MACNEAL, R. H.
NCCORMICK, C. W.
COMPUTERIZED SUBSTRUCTURE ANALYSIS
IN "WORLD CONGRESS ON FINITE ELEMENT METH. IN
STRUCT. MECH.", BOURNEMOUTH, DORSET, ENGLAND,
OCTOBER 12-17, 1975, PP. Q.1-Q.29.

75-0420 GRUBER, R.
BERGER, D.
BIDIMENSIONAL COMPUTER-PROGRAM FOR CALCULATION OF
GROWTH-RATES OF INCLUDED AXIALLY SYMMETRICAL
PLASMA USING FINITE-ELEMENTS
ZEITSCHRIFT FUR ANGEWANDTE MATHEMATIK UND PHYSIK,
VOL. 26, NO. 5, 1975, P. 660.

75-0421 MCMEEKING, R. M.
PARIS, P. C.
EFFICIENT FINITE-ELEMENT METHODS FOR STRESS
INTENSITY FACTORS USING WEIGHT FUNCTIONS - REPLY
INT. J. FRACTURE, VOL. 11, NO. 6, 1975, P. 1059.

75-0422 THOMAS, J.
ABBAS, B. A. H.
DYNAMIC STABILITY OF TIMOSHENKO BEAMS BY FINITE
ELEMENT METHOD
MECH. ENG., VOL. 97, NO. 12, 1975, P. 95.

75-0423 SCHMIDT, F. A. R.
FRANKE, H. P.
SAPPER, E.
2-DIMENSIONAL AND 3-DIMENSIONAL REACTOR-PHYSICS
CALCULATIONS WITH FINITE-ELEMENT METHOD
ATOMKERNENERGIE, VOL. 26, NO. 3, 1975, PP.
158-162.

75-0424 CURTIS, A. J.
APPLICATION OF RESPONSE LIMITING TO FINITE-ELEMENT
STRUCTURAL-ANALYSIS PROGRAMS
J. ENG. FOR INDUSTRY, VOL. 97, NO. 4, 1975, PP.
1194-1198.

75-0425 IRONS, B.
RAZZAQUE, A.
THE EVOLUTION OF THE ISOPARAMETRIC ELEMENTS
WORLD CONGRESS ON FINITE ELEMENT METH. IN STRUCT.
MECH., BOURNEMOUTH, DORSET, ENGLAND, OCTOBER
12-17, 1975, PP. D.1-D.27.

75-0426 ORRIS, R. M.
PETYT, M.
RANDOM RESPONSE OF PERIODIC STRUCTURES BY A
FINITE-ELEMENT TECHNIQUE
J. SOUND AND VIB., VOL. 43, NO. 1, 1975, PP. 1-8.

75-0427 KING, J. T.
QUASIOPTIMAL FINITE-ELEMENT METHOD FOR ELLIPTIC
INTERFACE PROBLEMS
COMPUTING, VOL. 15, NO. 2, 1975, PP. 127-135.

75-0428 MARIASUBE, R.
FINITE ELEMENT METHOD FOR AXISYMMETRIC JETS OF
FLUID WITH GRAVITY FORCE
COMPTES RENDUS - ACADEMIE DES SCIENCES SER B.
PHYSIQUE, VOL. 281, NO. 16, 1975, PP. 333-336.

75-0429 YAGAWA, G.
NISHIOKA, T.
ELASTIC-PLASTIC FINITE-ELEMENT ANALYSIS USING
SUPERPOSITION
NUCL. ENG. DES., VOL. 34, NO. 2, 1975, PP.
247-254.

75-0430 ATTEIA, M.
SPLINE FUNCTIONS AND FINITE-ELEMENT METHOD
REVUE FRANC. D'AUTOMAT., INFORMAT ET RECHERCHE
OPERAT., VOL. 9, NO. 2, 1975, PP. 13-40.

75-0431 BOLT, B. A.
STIFLER, J. F.
LOVE WAVE DISPERSION THROUGH IRREGULAR STRUCTURES
USING FINITE-ELEMENT ALGORITHMS
TRANS. AM. GEOPHYSICAL UNION, VOL. 56, NO. 12,
1975, P. 1026.

75-0432 KATZ, I. N.
PEANO, A. G.
ROSSOW, M. P.
SZABO, B. A.
THE CONSTRAINT METHOD FOR FINITE ELEMENT STRESS
ANALYSIS
IN "WORLD CONGRESS ON FINITE ELEMENT METH. IN
STRUCT. MECH.", BOURNEMOUTH, DORSET, ENGLAND,
OCTOBER 12-17, 1975, 26 PP.

75-0433 AHMED, K. M.
APPLICATIONS OF CURVED FINITE ELEMENTS TO SANDWICH
SHELLS USING SHEAR STRAIN FORMULATION
IN "WORLD CONGRESS ON FINITE ELEMENT METH. IN
STRUCT. MECH.", BOURNEMOUTH, DORSET, ENGLAND,
OCTOBER 12-17, 1975, 21 PP.

75-0434 FRAZIER, G.
DAY, S. M.
FINITE-ELEMENT TREATMENT OF QUADRUPOLE GROUND
MOTIONS IN AXISYMMETRIC EARTH STRUCTURES
TRANS. AMER. GEOPHYSICAL UNION, VOL. 56, NO. 12,
1975, P. 1026.

75-0435 KAWAHARA, M.
LARGE STRAIN, VISCOELASTIC AND ELASTO-VISCOPLASTIC
NUMERICAL-ANALYSIS BY MEANS OF FINITE-ELEMENT
METHOD
NUCL. ENG. DES., VOL. 34, NO, 2, 1975, PP.
233-246.

75-0436 CLEMENT, P.
APPROXIMATION BY FINITE-ELEMENT FUNCTIONS USING
LOCAL REGULARIZATION
REVUE FRANC. D'AUTOMAT., INFORMAT ET RECHERCHE
OPERAT., VOL. 9, NO. 2, 1975, PP. 77-84.

75-0437 HSU, C. C.
GALERKIN METHOD FOR A CLASS OF STEADY,
2-DIMENSIONAL, INCOMPRESSIBLE, LAMINAR
BOUNDARY-LAYER FLOWS
J. FLUID MECH., VOL. 69, JUNE 24, 1975, PP.
783-802.

75-0438 ANDRESEN, K.
CALCULATION OF CONE INDENTATION TEST USING A
FINITE-ELEMENT METHOD
ARCHIV FUR DAS EISENHUTTENWESEN, VOL. 46, NO. 9,
1975, PP. 571-574.

75-0439 CIAVALDINI, J. F.
TOURNEMINE, G.
FINITE-ELEMENT METHOD TO COMPUTE STATIONARY
SUBCRITICAL FLOWS
COMPTES RENDUS - ACADEMIE DES SCIENCES SER. A
MATH., VOL. 281, NO. 12, 1975, PP. 487-490.

75-0440 THIERAUF, G.
TOPCU, A.
STRUCTURAL OPTIMIZATION USING THE FORCE METHOD
IN "WORLD CONGRESS ON FINITE ELEMENT METH. IN
STRUCT. MECH.", BOURNEMOUTH, DORSET, ENGLAND,
OCTOBER 12-17, 1975, 20 PP.

75-0441 CHUNG, T. J.
CONVERGENCE AND STABILITY OF NONLINEAR
FINITE-ELEMENT EQUATIONS
A.I.A.A. J., VOL. 13, NO. 7, 1975, PP. 963-966.

75-0442 BERKOVIC, M.
GENERAL MEMBRANE ISOPARAMETRIC ELEMENTS
IN "WORLD CONGRESS ON FINITE ELEMENT METH. IN
STRUCT. MECH.", BOURNEMOUTH, DORSET, ENGLAND,
OCTOBER 12-17, 1975, 22 PP.

75-0443 LEWIS, E. E.
MILLER, W. F.
HENRY, T. P.
2-DIMENSIONAL FINITE-ELEMENT METHOD FOR INTEGRAL
NEUTRON-TRANSPORT CALCULATIONS
NUCL. SCI. AND ENG., VOL. 58, NO. 2, 1975, PP.
203-212.

75-0444 WYATT, M. J.
DAVIES, G.
SNELL, C.
NEW DIFFERENCE BASED FINITE-ELEMENT METHOD
PROC. INST. CIVIL ENG., PT. 2 RESEARCH AND THEORY,
VOL. 59, SEPTEMBER 1975, PP. 395-409.

75-0445 FIX, G.
HYBRID FINITE-ELEMENT METHODS
AMER. MATH. SOC., VOL. 22, NO. 6, 1975, P. A659.

75-0446 SMITH, W. D.
APPLICATION OF FINITE-ELEMENT ANALYSIS TO BODY
WAVE-PROPAGATION PROBLEMS
GEOPHYSICAL J. ROY. ASTRO. SOC., VOL. 42, NO. 2,
1975, PP. 747-768.

75-0447 HAGER, W. W.
STRANG, G.
FREE BOUNDARIES AND FINITE-ELEMENTS IN ONE
DIMENSION
AMER. MATH. SOC., VOL. 22, NO. 6, 1975, P. A641.

75-0448 MARTIN, B. W.
CONTRIBUTION TO NUMERICAL-SOLUTION OF DEVELOPING
LAMINAR-FLOW IN ENTRANCE REGION OF CONCENTRIC
ANNULI WITH ROTATING INNER WALLS - COMMENT
J. FLUIDS ENG., VOL. 97, NO. 3, 1975, P. 394.

75-0449 ORR, F. M.
SCRIVEN, L. E.
RIVAS, A. P.
MENISCI IN ARRAYS OF CYLINDERS -
NUMERICAL-SIMULATION BY FINITE-ELEMENTS
J. COLLOID AND INTERFACE SCI., VOL. 52, NO. 3,
1975, PP. 602-610.

75-0450 NOOR, A. K.
ANDERSON, C. M.
MIXED ISOPARAMETRIC LAMINATED COMPOSITE SHELL
ELEMENTS
IN "WORLD CONGRESS ON FINITE ELEMENT METH. IN
STRUCT. MECH.", BOURNEMOUTH, DORSET, ENGLAND,
OCTOBER 12-17, 1975, 33 PP.

75-0451 WELLFORD, L. C.
ODEN, J. T.
DISCONTINUOUS FINITE-ELEMENT APPROXIMATIONS FOR
ANALYSIS OF SHOCK-WAVES IN NONLINEARLY
ELASTIC-MATERIALS
J. COMPUT. PHYSICS, VOL. 19, NO. 2, 1975, PP.
179-210.

75-0452 CAVENDISH, J. C.
LOCAL MESH REFINEMENT USING RECTANGULAR BLENDED
FINITE-ELEMENTS
J. COMPUT. PHYSICS, VOL. 19, NO. 2, 1975, PP.
211-228.

75-0453 REDDY, J. N.
ODEN, J. T.
MIXED FINITE-ELEMENT APPROXIMATIONS OF LINEAR
BOUNDARY-VALUE PROBLEMS
QUARTERLY APPL. MATH., VOL. 33, NO. 3, 1975, PP.
255-280.

75-0454 DEMEY, G.
FINITE-ELEMENT METHOD FOR POTENTIAL CALCULATIONS
IN A HALL PLATE
RADIO AND ELECTRONIC ENG., VOL. 45, NO. 9, 1975,
PP. 472-474.

75-0455 RIVIER, M. L.
FINITE-ELEMENT SOLUTIONS OF PLASMA EQUILIBRIUM
BULL. AMER. PHYSICAL SOC., VOL. 20, NO. 10, 1975,
P. 1278.

75-0456 SANKAR, A.
TONG, T. C.
CURRENT COMPUTATION ON COMPLEX STRUCTURES BY
FINITE-ELEMENT METHOD
ELECTRONICS LETTER, VOL. 11, NO. 20, 1975, PP.
481-482.

75-0457 GUREGHIAN, A. B.
YOUNGS, E. G.
CALCULATION OF STEADY-STATE WATER-TABLE HEIGHTS IN
DRAINED SOILS BY MEANS OF FINITE-ELEMENT METHOD
J. HYDROLOGY, VOL. 27, NO. 1-2, 1975, PP. 15-32.

75-0458 APPERT, K.
BERGER, D.
GRUBER, R.
TROYON, F.
ROBERTS, K. V.
THALIA - ONE-DIMENSIONAL MAGNETOHYDRODYNAMIC
STABILITY PROGRAM USING METHOD OF FINITE-ELEMENTS
COMPUT. PHYSICS COMM., VOL. 10, NO. 1, 1975, PP.
11-29.

75-0459 CARNAHAN, N. F.
ZIMMER, R. A.
ANCHOR-PILE DESIGN FOR OCEAN-FLOOR ENVIRONMENTS
USING FINITE-ELEMENT ANALYSIS
PROC. 7TH ANNUAL CONF. OFFSHORE TECH., HOUSTON,
TEXAS, MAY 5-8, 1975, PAPER OTC 2308, VOL. 2, PP.
625-632.

75-0460 WEBSTER, R. L.
NONLINEAR STATIC AND DYNAMIC RESPONSE OF
UNDERWATER CABLE STRUCTURES USING THE FINITE
ELEMENT METHOD
PROC. 7TH ANNUAL CONF. OFFSHORE TECH., HOUSTON,
TEXAS, MAY 5-8, 1975, PAPER OTC 2322, VOL. 2, PP.
753-764.

75-0461 PEARCE, B. R.
PAGENKOPF, J. R.
NUMERICAL CALCULATION OF STORM SURGES: AN
EVALUATION OF TECHNIQUES
PROC. 7TH ANNUAL CONF. OFFSHORE TECH., HOUSTON,
TEXAS, MAY 5-8, 1975, PAPER OTC 2333, VOL. 2, PP.
887-901.

75-0462 PUTTER, S.
MANOR, H.
FLEXURE ANALYSIS OF AXIALLY-LOADED BEAM BY
FINITE-ELEMENT METHOD WITH INTEGRAL PARAMETERS
ISR J. TECH., TEL AVIV, ISR. AND HAIFA, ISR, VOL.
13, NO. 1-2, MAY 21-22, 1975, PP. 134-142.

75-0463 BAGUELIN, F.
BUSTAMANTE, M.
LA CAPACITE PORTANTE DES PIEUX. (BEARING CAPACITY
OF PILES)
ANN. INST. TECH. BATIM TRAV. PUBLICS, NO. 330,
JULY-AUGUST 1975, PP. 1-22.

75-0464 KAMMINGA, W.
FINITE-ELEMENT SOLUTIONS FOR DEVICES WITH
PERMANENT MAGNETS
J. APPL. PHYS., VOL. 8, NO. 7, MAY 11, 1975, PP.
841-855.

75-0465 WATANABE, M.
HARLEMAN, D. R. F.
FINITE ELEMENT MODEL FOR TRANSIENT TWO-LAYER
COOLING POND BEHAVIOR
YDRODYN., NO. 202, JULY 1975, 284 PP.

75-0466 SHAUG, J. C.
BRUCH, J. C. JR.
SOLUTION OF A FREE-SURFACE BOUNDARY VALUE PROBLEM
USING AN INVERSE FORMULATION AND THE FINITE
ELEMENT METHOD
J. HYDROL., VOL. 26, NO. 1-2, JULY 1975, PP.
141-152.

75-0467 GELLATLY, R. A.
BERKE, L.
A REVIEW OF SOME ASPECTS OF STRUCTURAL
OPTIMIZATION
IN "WORLD CONGRESS ON FINITE ELEMENT METHOD IN
STRUCT. MECH.", BOURNEMOUTH, DORSET, ENGLAND,
OCTOBER 12-17, 1975, PP. N.6-N.52.

75-0468 KASHEF, A-A. I.
SAFAR, M. M.
COMPARATIVE STUDY OF FRESH-SALT WATER INTERFACES
USING FINITE ELEMENT AND SIMPLE APPROACHES
WATER RESOUR. BULL., VOL. 11, NO. 4, AUGUST 1975,
PP. 651-665.

75-0469 BUGROV, A. K.
METOD KONECHNYKH ELEMENTOV V RASCHETAKH
KONSOLIDATSII VODONASYSHCHENNYKH GRUNTOV. (FINITE
ELEMENT METHOD USED IN CALCULATION OF WATER
SATURATED SOILS CONSOLIDATION)
GIDROTEKH STROIT. NO. 7, JULY 1975, PP. 35-38.

75-0470 NAKAD, Y.
SASAKI, N.
CONSIDERATION ON THE SEISMIC RESPONSE ANALYSIS OF
UNDERGROUND STRUCTURES
MITSUBISHI HEAVY IND. TECH. REV., VOL. 12, NO. 2,
JUNE 1975, PP. 86-94.

75-0471 GOLIA, C.
LA TECNICA DEGLI ELEMENTI FINITI PER PROBLEMI DI
LUBRIFICAZIONE INCOMPRESSIBILE. (FINITE ELEMENT
METHOD FOR HYDRODYNAMIC LUBRICATION PROBLEMS)
TERMOTECNICA, MILAN, VOL. 29, NO. 6, JUNE 1975,
PP. 317-324.

75-0472 HENSHELL, R. D.
SHAW, K. G.
CRACK TIP FINITE ELEMENTS ARE UNNECESSARY
INT. J. NUMER. METH. ENG., VOL. 9, NO. 3, 1975,
PP. 495-507.

75-0473 CSENDES, Z. J.
FINITE ELEMENT METHOD FOR THE GENERAL SOLUTION OF
ORDINARY DIFFERENTIAL EQUATIONS
INT. J. NUMER. METH. ENG., VOL. 9, NO. 3, 1975,
PP. 551-561.

75-0474 NAGARAJAN, S.
POPOV, E. P.
NON-LINEAR DYNAMIC ANALYSIS OF AXISYMMETRIC SHELLS
INT. J. NUMER. METH. ENG., VOL. 9, NO. 3, 1975,
PP. 535-550.

75-0475 SZABO, B. A.
KASSOS, T.
LINEAR EQUALITY CONSTRAINTS IN FINITE ELEMENT
APPROXIMATION
INT. J. NUMER. METH. ENG., VOL. 9, NO. 3, 1975,
PP. 563-580.

75-0476 WOLF, J. P.
ALTERNATE HYBRID STRESS FINITE ELEMENT MODELS
INT. J. NUMER. METH. ENG., VOL. 9, NO. 3, 1975,
PP. 601-615.

75-0477 WOOD, W. L.
LEWIS, R. W.
COMPARISON OF TIME MARCHING SCHEMES FOR THE
TRANSIENT HEAT CONDUCTION EQUATION
INT. J. NUMER. METH. ENG., VOL. 9, NO. 3, 1975,
PP. 679-689.

75-0478 SANDER, G.
BECKERS, P.
DELINQUENT FINITE ELEMENTS FOR SHELL IDEALIZATION
IN "WORLD CONGRESS ON FINITE ELEMENT METH. IN
STRUCT. MECH.", BOURNEMOUTH, DORSET, ENGLAND,
OCTOBER 12-17, 1975, 31 PP.

75-0479 KIKUCHI, F.
ON THE VALIDITY OF THE FINITE ELEMENT ANALYSIS OF
CIRCULAR ARCHES REPRESENTED BY AN ASSEMBLAGE OF
BEAM ELEMENTS
COMPUT. METH. APPL. MECH. ENG., VOL. 5, NO. 3, MAY
1975, PP. 253-276.

75-0480 CIARLET, P. G.
GLOWINSKI, R.
DUAL ITERATIVE TECHNIQUES FOR SOLVING A FINITE
ELEMENT APPROXIMATION OF THE BIHARMONIC EQUATION
COMPUT. METH. APPL. MECH. ENG., VOL. 5, NO. 3, MAY
1975, PP.277-295.

75-0481 BUERHOP, H.
ZUR BERECHNUNG DER BIEGESTEIFIGKEIT ABGESETZTER
STAEBE UND WELLEN UNTER ANWENDUNG VON FINITEN
ELEMENTEN. (CALCULATION OF BENDING STRENGTH OF
FIXED RODS AND SHAFTS USING FINITE ELEMENTS)
FORTSCHR BER VDI Z., PART 1, NO. 36, MAY 1975, PP.
1-161.

75-0482 DEMS, K.
LIPINSKI, J.
APPLICATION OF FINITE DIFFERENCES FOR SOLVING THE
TWO-DIMENSIONAL ELASTICITY PROBLEM BY MEANS OF THE
FINITE ELEMENT METHOD
COMPUT. METH. APPL. MECH. ENG., VOL. 6, NO. 1,
JULY 1975, PP. 49-58.

75-0483 PRAGER, W.
ABSI, E.
COMPARISON OF EQUIVALENCE AND FINITE ELEMENT
METHODS
COMPUT. METH. APPL. MECH. ENG., VOL. 6, NO. 1,
JULY 1975, PP. 59-64.

75-0484 BIFFLE, J. H.
BECKER, E. B.
FINITE ELEMENT STRESS FORMULATION FOR DYNAMIC
ELASTIC-PLASTIC ANALYSIS
COMPUT. METH. APPL. MECH. ENG., VOL. 6, NO. 1,
JULY 1975, PP. 101-119.

75-0485 SALAH EL-DIN, A. S.
EL-ADAWY NASSEF, M. M.
MODIFIED APPROACH FOR ESTIMATING THE CRACKING
MOMENT OF REINFORCED CONCRETE BEAMS
J. AMER. CONCR. INST., VOL. 72, NO. 7, JULY 1975,
PP. 356-360.

75-0486 DOMASCHUK, L.
VALLIAPPAN, P.
NONLINEAR SETTLEMENT ANALYSIS BY FINITE ELEMENT
J. GEOTECH. ENG. DIV., A.S.C.E., VOL. 101, NO. 7,
JULY 1975, PP. 601-614.

75-0487 MARSHALL, J.
JATEGAONKAR, R. P.
SHEAR STRESSES IN BEAMS WITH DISCONTINUITY IN
DEPTH
J. STRUCT. DIV., A.S.C.E., VOL. 101, NO. 7, JULY
1975, PP. 1393-1402.

75-0488 VERONDA, D. R.
WEINGARTEN, V. I.
STABILITY OF HYPERBOLOIDAL SHELLS
J. STRUCT. DIV., A.S.C.E., VOL. 101, NO. 7, JULY
1975, PP. 1585-1602.

75-0489 HENSHELL, R. D.
THEORETICAL BASIS FOR HYBRID FINITE ELEMENTS IN
DYNAMIC PROBLEMS
J. SOUND AND VIB., VOL. 39, NO. 4, 1975, PP.
520-522.

75-0490 WELLFORD, L. C. JR.
ODEN, J. T.
NOTE ON THE ACCURACY AND CONVERGENCE OF FINITE
ELEMENT APPROXIMATIONS OF THE CONVECTION EQUATION
COMPUT. METH. APPL. MECH. ENG., VOL. 5, NO. 1,
JANUARY 1975, PP. 83-96.

75-0491 JOHNSON, C.
THOMEE, V.
ERROR ESTIMATES FOR A FINITE-ELEMENT APPROXIMATION
OF A MINIMAL SURFACE
MATH. OF COMPUT., VOL. 29, NO. 130, 1975, PP.
343-349.

75-0492 KAGAWA, Y.
YAMABUCHI, T.
FINITE ELEMENT SIMULATION OF ELECTROMECHANICAL
RESONATORS WITH ARBITRARY ELECTRODES
PROC. NATL. SYMP. ON THE MATRIX METH. OF STRUCT.
ANALYSIS, SOCIETY OF STEEL CONSTRUCTION OF JAPAN,
TOKYO, JUNE 1975.

75-0493 TANAKA, T.
KATAYAMA, D.
FINITE ELEMENT APPLICATION IN LAKE CIRCULATION
PROC. NATL. SYMP. ON THE MATRIX METH. OF STRUCT.
ANALYSIS, SOCIETY OF STEEL CONSTRUCTION OF JAPAN,
TOKYO, JUNE 1975.

75-0494 CIARLET, P. G.
CONVERGENCE OF CONFORMING FINITE-ELEMENT METHODS
FOR SHELL PROBLEMS
COMPTES RENDUS - ACADEMIE DES SCIENCES SER. A
MATH., VOL. 280, NO. 19, 1975, PP. 1299-1301.

75-0495 ORRINGER, O.
LIN, Y. K.
STALK, G.
TONG, P.
MAR, G. W.
APPLICATIONS OF THE ASSUMED-STRESS HYBRID METHOD
TO FINITE ELEMENT ANALYSIS IN FRACTURE MECHANICS
PROC. NAT. CONF. ON STRUCT. ENG., NEW ORLEANS,
LA., APRIL 1975.

75-0496 BRIGGS, L. L.
MILLER, W. F.
LEWIS, E. E.
RAY-EFFECT MITIGATION IN DISCRETE ORDINATE-LIKE
ANGULAR FINITE-ELEMENT APPROXIMATIONS IN
NEUTRON-TRANSPORT
NUCL. SCI. AND ENG., VOL. 57, NO. 3, 1975, PP.
205-217.

75-0497 HUANYEN, L.
ZUNGUO, H.
FINITE ELEMENT ANALYSIS OF TIME-DEPENDENT PUMPING
TEST DATA IN AN UNCONFINED AQUIFER
SCIENTIA GEOLOGICA SINICA, NO. 1, 1975, PP. 82-90.

75-0498 HILL, T. R.
LINEAR DISCONTINUOUS FINITE-ELEMENT SCHEME IN
TIME-DEPENDENT TRANSPORT CALCULATIONS
TRANS. AMER. NUCL. SOC., VOL. 21, JUNE 1975, PP.
228-230.

75-0499 WALTERS, W. F.
KOMORIYA, H.
ENERGY-DEPENDENT FINITE-ELEMENT METHOD FOR
FEW-GROUP DIFFUSION EQUATIONS
TRANS. AMER. NUCL. SOC., VOL. 21, JUNE 1975, PP.
224-225.

75-0500 HENNART, J. P.
MARTIN, J. G.
SANCHEZGUTIERREZ, J.
FINITE-ELEMENT PLASMA-DIFFUSION CODE FOR ARBITRARY
TOROIDAL GEOMETRY
TRANS. AMER. NUCL. SOC., VOL. 21, JUNE 1975, PP.
225-226.

75-0501 ZLAMAL, M.
FINITE-ELEMENT MULTISTEP DISCRETIZATIONS OF
PARABOLIC BOUNDARY-VALUE PROBLEMS
MATH. OF COMPUT., VOL. 29, NO. 130, 1975, PP.
350-359.

75-0502 HANSEN, K. F.
KANG, C. M.
FINITE-ELEMENT METHODS IN REACTOR-PHYSICS ANALYSIS
ADV. NUCL. SCI., VOL. 8, 1975, PP. 173-253.

75-0503 NGUYEN, D. H.
SPACE-TIME SOLUTIONS OF NONLINEAR REACTOR DYNAMICS
BY FINITE-ELEMENT METHOD
TRANS. AMER. NUCL. SOC., VOL. 21, JUNE 1975, PP.
475-477.

75-0504 YUAN, Y. C.
LEWIS, E. F.
SPACE FINITE-ELEMENTS 2-DIMENSIONAL MULTIGROUP
TRANSPORT CALCULATIONS USING PHASE-SPACE
FINITE-ELEMENTS
TRANS. AMER. NUCL. SOC., VOL. 21, JUNE 1975, P.
227.

75-0505 BESSELING, J. F.
POST-BUCKLING AND NON-LINEAR ANALYSIS BY FINITE
ELEMENT METHOD AS A SUPPLEMENT TO A LINEAR
ANALYSIS
ZEITSCHRIFT FUR ANGEWANDTE MATHEMATIK UND
MECHANIK, VOL. 55, NO. 4, 1975, PP. 3-16.

75-0506 SHUGAR, T. A.
KATONA, M. G.
DEVELOPMENT OF FINITE ELEMENT HEAD INJURY MODEL
J. ENG. MECH DIV., A.S.C.E., VOL. 101, NO. 3,
1975, PP. 223-239.

75-0507 MACBAIN, J. C.
VIBRATORY BEHAVIOR OF TWISTED CANTILEVERED PLATES
J. AIRCR., VOL. 12, NO. 4, APRIL 1975, PP.
343-349.

75-0508 ANDERSON, D. A.
MODELING OF GAS TURBINE ENGINE COMPRESSOR BLADES
FOR VIBRATION ANALYSIS
J. AIRCR., VOL. 12, NO. 4, APRIL 1975, PP.
357-359.

75-0509 SATTAR, S. A.
SUNDT, C. V.
GAS TURBINE ENGINE DISK CYCLIC LIFE PREDICTION
J. AIRCR., VOL. 12, NO. 4, APRIL 1975, PP.
360-365.

75-0510 LINK, M.
ZUR BERECHNUNG EINFACH-SYMMETRISCHER I-TRAEGER
NACH DER THEORIE - 2. ORDNUNG UNTER
BERUECKSICHTIGUNG DER QUERSCHNITTSVERFORMIUNG MIT
HILFE HYBRIDER FINITER ELEMENTE. (SECOND-ORDER
THEORY CALCULATION OF SINGLE-SYMMETRIC I-SHAPED
PROFILE BEAM WITH HYBRID FINITE ELEMENTS AND
CONSIDERING CROSS-SECTION DEFORMATION - 2)
STAHLBAU, VOL. 44, NO. 5, MAY 1975, PP. 140-146.

75-0511 ANON
GENERAL PURPOSE PROGRAM FOR SHELL STRUCTURES:
ISTRAN-S
IHI. ENG. REV., VOL. 8, NO. 2, MAY 1975, PP. 1-16.

75-0512 BUYUKOZTURK, O.
MARCAL, P. V.
STRENGTH OF REINFORCED CONCRETE CHAMBERS UNDER
EXTERNAL PRESSURE
PAPER NO. 75-PVP-7, A.S.M.E. MEETING, JUNE 23-27,
1975, 6 PP.

75-0513 CHERN, J. M.
PAI, D. H.
INELASTIC ANALYSIS OF A STRAIGHT TUBE UNDER
COMBINED BENDING, PRESSURE AND THERMAL LOADS
PAPER NO. 75-PVP-19, A.S.M.E. MEETING, JUNE 23-27,
1975, 8 PP.

75-0514 REYNEN, J.
ON THE USE OF FINITE ELEMENTS IN FRACTURE ANALYSIS
OF PRESSURE VESSEL COMPONENTS
PAPER NO. 75-PVP-20, A.S.M.E. MEETING, JUNE 23-27,
1975, 9 PP.

75-0515 ANDERSON, J. C.
SINGH, A. K.
INELASTIC RESPONSE OF NUCLEAR PIPING SUBJECTED TO
RUPTURE FORCES
PAPER NO. 75-PVP-21, A.S.M.E. MEETING, JUNE 23-27,
1975, 7 PP.

75-0516 HSU, T. R.
BERTELS, A. W. M.
PROPAGATION AND OPENING OF A THROUGH CRACK IN PIPE
SUBJECT TO COMBINED CYCLIC THERMOMECHANICAL
LOADING
PAPER NO. 75-PVP-23, A.S.M.E. MEETING, JUNE 23-27,
1975, 9 PP.

75-0517 SIMONEN, F. A.
HENDERSON, N. C.
ANALYSIS OF STRENGTH AND RESIDUAL STRESSES IN
FILAMENT REINFORCED ALUMINUM CYLINDERS
PAPER NO. 75-PVP-24, A.S.M.E. MEETING, JUNE 23-27,
1975, 7 PP.

75-0518 FRIEDMAN, E.
THERMOMECHANICAL ANALYSIS OF THE WELDING PROCESS
USING THE FINITE ELEMENT METHOD
PAPER NO. 75-PVP-27, A.S.M.E. MEETING, JUNE 23-27,
1975, 8 PP.

75-0519 HILTON, P. D.
ELASTIC-PLASTIC ANALYSIS FOR CRACKED MEMBERS
PAPER NO. 75-PVP-34, A.S.M.E. MEETING, JUNE 23-27,
1975, 9 PP.

75-0520 JONES, D. P.
FINITE ELEMENT ANALYSIS OF PERFORATED PLATES
CONTAINING TRIANGULAR PENETRATION PATTERNS OF 5
AND 10 PERCENT LIGAMENT EFFICIENCY
PAPER NO. 75-PVP-35, A.S.M.E. MEETING, JUNE 23-27,
1975, 7 PP.

75-0521 BERMAN, I.
HENSCHEL, R.
CREEP BENDING TESTS OF LARGE WELD OVERLAY AND BASE
METAL SPECIMENS AT 1000 DEGREE F. (537.7 DEG. C)
PAPER NO. 75-PVP-37, A.S.M.E. MEETING, JUNE 23-27,
1975, 6 PP.

75-0522 BELYTSCHKO, T.
KENNEDY, J. M.
FINITE ELEMENT APPROACH TO PRESSURE WAVE
ATTENUATION BY REACTOR FUEL SUBASSEMBLIES
PAPER NO. 75-PVP-43, A.S.M.E. MEETING, JUNE 23-27,
1975, 9 PP.

75-0523 HUNSAKER, B.
VAUGHAN, D. K.
COMPARISON OF THE CAPABILITY OF OUR HARDENING
RULES TO PREDICT A MATERIAL'S PLASTIC BEHAVIOR
PAPER NO. 75-PVP-43, A.S.M.E. MEETING, JUNE 23-27,
1975, 6 PP.

75-0524 CARLSEN, C. A.
KAVLIE, D.
INDETS - INTEGRATED DESIGN OF TANKER STRUCTURES
NOR. MARIT. RES., VOL. 3, NO. 2, 1975, PP. 2-18.

75-0525 SELNA, L. G.
SHILLINGBURG, H. T. JR.
FINITE ELEMENT ANALYSIS OF DENTAL STRUCTURES -
AXISYMMETRIC AND PLANE STRESS IDEALIZATIONS
J. BIOMED. MATER. RES., VOL. 9, NO. 2, MARCH 1975,
PP. 237-252.

75-0526 KAWAHARA, M.
OKAMOTO, T.
FINITE ELEMENT ANALYSIS OF VISCOUS FLOW BY STREAM
FUNCTION
PROC. NATL. SYMP. S.S.C.J. ON THE MATRIX METH. OF
STRUCT. ANALYSIS, JUNE 1975.

75-0527 VALLABHAN, C. V. G.
DUCKER, W. L.
ANALYSIS FOR THE DESIGN OF BEAM TESTING APPARATUS
J. TEST EVAL., VOL. 3, NO. 3, MAY 1975, PP.
163-166.

75-0528 DIMAROGONAS, A. D.
GENERAL METHOD FOR STABILITY ANALYSIS OF ROTATING
SHAFTS
ING. ARCH., VOL. 44, NO. 1, 1975, PP. 9-20.

75-0529 NISHINO, F.
KURAKATA, Y.
FINITE DISPLACEMENT BEAM THEORY
PROC. JAP. SOC. CIV. ENG., NO. 237, MAY 1975, PP.
11-26.

75-0530 GIMM, W.
HEINRICH, F.
NEUARTIGE ANWENDUNG GEOMECHANISCHER
UNTERSUCHUNGSMETHODEN ZUR BEURTEILUNG DER
ABBAUWIRKUNGEN IM GANGERZBERGBAU. (NEW
APPLICATIONS OF GEOMECHANICAL INVESTIGATION
METHODS TO EVALUATING EFFECTS OF VEIN ORE MINING)
NEUE BERGBAUTECH., VOL. 5, NO. 5, MAY 1975, PP.
386-397.

75-0531 ROHDE, S. M.
MCALLISTER, G. T.
VARIATIONAL FORMULATION FOR A CLASS OF FREE
BOUNDARY PROBLEMS ARISING IN HYDRODYNAMIC
LUBRICATION
INT. J. ENG. SCI., VOL. 13, NO. 9-10, SEPTEMBER -
OCTOBER, 1975, PP. 841-850.

75-0532 PALMBERG, J. O.
ON THERMO-ELASTO-HYDRODYNAMIC FLUID FILM BEARINGS
CHALMERS TEK., HOGSK. DOKTORSAVH, NO. 149, 1975,
170 PP.

75-0533 MELOSH, R. J.
LOBITZ, D. W.
ON A NUMERICAL SUFFICIENCY TEST FOR MONOTONIC
CONVERGENCE OF FINITE ELEMENT MODELS
A.I.A.A. J., VOL. 13, NO. 5, MAY 1975, PP.
675-678.

75-0534 VENKATESWARA, R. G.
NARASIMHA, R. R. V.
GALERKIN FINITE ELEMENT SOLUTION FOR THE STABILITY
OF CANTILEVER COLUMNS SUBJECTED TO TANGENTIAL
LOADS
A.I.A.A. J., VOL. 13, NO. 5, MAY 1975, PP.
690-691.

75-0535 JAMET, P.
BONNEROT, R.
NUMERICAL-SOLUTION OF EULERIAN EQUATIONS OF
COMPRESSIBLE FLOW BY A FINITE-ELEMENT METHOD WHICH
FOLLOWS FREE BOUNDARY AND INTERFACES
J. COMPUT. PHYSICS, VOL. 18, NO. 1, 1975, PP.
21-45.

75-0536 IRONS, B. M.
THE SEMILOOF SHELL ELEMENT
REPORT NO. CE 75-5, DEPT. CIVIL ENG., UNIV. OF
CALGARY, CANADA, JULY 1975, 45 PP.

75-0537 MURAKAMI, Y.
NISITANI, H.
NOTCH EFFECT IN LOW-CYCLE FATIGUE - 2. EFFECT OF
STRESS CHANGE AND MEAN STRESS
BULL. J.S.M.E., VOL. 18, NO. 119, MAY 1975, PP.
465-472.

75-0538 PULMANO, V. A.
TAPERED PLANE STRESS ANNULAR SECTOR FINITE ELEMENT
J. ENG. MECH. DIV., A.S.C.E., VOL. 101, NO. 4,
AUGUST 1975, PP. 349-360.

75-0539 CHOI, C-K.
SCHNOBRICH, W. C.
NONCONFORMING FINITE ELEMENT ANALYSIS OF SHELLS
J. ENG. MECH. DIV., A.S.C.E., VOL. 101, NO. 4,
AUGUST 1975, PP. 447-464.

75-0540 JOHNSON, G. R.
CHRISTIANO, P.
STIFFNESS COEFFICIENTS FOR EMBEDDED FOOTINGS
J. GEOTECH. ENG. DIV., A.S.C.E., VOL. 101, NO. 8,
AUGUST 1975, PP. 789-800.

75-0541 CHEN, A. C. T.
CHEN, W-F.
CONSTITUTIVE RELATIONS FOR CONCRETE
J. ENG. MECH. DIV., A.S.C.E., VOL. 101, NO. 4,
AUGUST 1975, PP. 465-481.

75-0542 KEY, J. E.
ON THE NUMERICAL SOLUTION OF CERTAIN PROBLEMS IN
FINITE ELASTICITY BY THE FINITE ELEMENT METHOD
INTL. J. NUM. METH. ENG., VOL. 9, NO. 2, 1975, PP.
483-487.

75-0543 PAZ, M.
DUNG, L.
POWER SERIES EXPANSION OF THE GENERAL STIFFNESS
MATRIX FOR BEAM ELEMENTS
INTL. J. NUM. METH. ENG., VOL. 9, NO. 2, 1975, PP.
449-460.

75-0544 KIKUCHI, F.
FINITE-ELEMENT SCHEME BASED ON DISCRETE KIRCHHOFF
ASSUMPTION
NUMERISCHE MATHEMATIK, VOL. 24, NO. 3, 1975, PP.
211-231.

75-0545 GALLAGHER, R. H.
SHELL ELEMENTS
WORLD CONGRESS ON FINITE ELEMENT METH. IN STRUCT.
MECH., BOURNEMOUTH, DORSET, ENGLAND, OCTOBER
12-17, 1975, PP. E.1-E.35.

75-0546 MCVERRY, G. H.
BRADFORD, E.
WOODING, R. A.
FINITE-ELEMENT CALCULATION OF PRESSURE HISTORY IN
A GAS FIELD
NEW ZEALAND, J. OF SCI., VOL. 18, NO. 3, 1975, PP.
345-360.

75-0547 STELZER, F.
PRACTICAL HANDLING OF FINITE-ELEMENT METHOD 2
KERNTECHNIK, VOL. 17, NO. 8, 1975, PP. 364-370.

75-0548 BARNES, J.
LOMAX, R.
2-DIMENSIONAL SIMULATION OF SEMICONDUCTOR-DEVICES
USING FINITE ELEMENT METHODS
AMER. MATH. SOC., VOL. 22, NO. 5, 1975, PP.
A595-596.

75-0549 MARCAL, P. V.
SOME CURRENT FINITE-ELEMENT MODELS FOR MATERIAL
BEHAVIOR
INT. J. OF FRACTURE, VOL. 11, NO. 4, 1975, PP.
675-676.

75-0550 DIMONACO, A.
GIUSEPPETTI, G.
TONTINI, G.
ANALYSIS OF STEADY ELECTRICAL AND MAGNETIC-FIELDS
BY FINITE-ELEMENT METHOD - APPLICATION TO
TRANSFORMERS
ELETTROTECNICA, VOL. 62, NO. 7, 1975, PP. 585-598.

75-0551 STELZER, F.
PRACTICAL HANDLING OF FINITE-ELEMENT METHOD 1
KERNTECHNIK, VOL. 17, NO. 7, 1975, PP. 324-328.

75-0552 DI NAPOLI, A.
RATTI, U.
EDDY CURRENTS SCREENING PROBLEMS SOLVED BY MEANS
OF FINITE ELEMENT DIFFERENCE AND FINITE ELEMENT
METHODS
ACTA TECH. CSAV., VOL. 20, NO. 2, 1975, PP.
147-177.

75-0553 ROEHRLE, M. D.
DETERMINATION OF STRESSES AND DEFORMATIONS ON
PISTONS BY MEANS OF COMPUTER PROGRAMS AND
PHOTOELASTICITY
PAPER NO. 75-DGP-18, A.S.M.E. MEETING, APRIL 6-10,
1975, 8 PP.

75-0554 PAGENKOPF, J. R.
PEARCE, B. R.
EVALUATION OF TECHNIQUES FOR NUMERICAL CALCULATION
OF STORM SURGES
RALPH M. PARSONS LAB. WATER RESOUR. HYDRODYN., MIT
DEPT. CIV. ENG., NO. 199, FEBRUARY 1975, 120 PP.

75-0555 NAYAK, A. L.
CHENG, P.
FINITE ELEMENT ANALYSIS OF LAMINAR CONVECTIVE HEAT
TRANSFER IN VERTICAL DUCTS WITH ARBITRARY
CROSS-SECTIONS
INTL. J. HEAT MASS TRANSFER, VOL. 18, NO. 2,
FEBRUARY 1975, PP. 227-236.

75-0556 HANSEN, H. R.
SKAAR, K. T.
HULL AND SUPERSTRUCTURE VIBRATIONS DESIGN
CALCULATIONS BY FINITE ELEMENTS
NOR VERITAS PUBL., NO. 86, JANUARY 1975, PP. 5-15.

75-0557 HENSHELL, R. D.
ONG, J. H.
AUTOMATIC MASTERS FOR EIGENVALUE ECONOMIZATION
EARTHQUAKE ENG. STRUCT. DYN., VOL. 3, NO. 4, APRIL
- JUNE, 1975, PP. 375-383.

75-0558 PIAN, T. H. H.
CRACK ELEMENTS
WORLD CONGRESS ON FINITE ELEMENT METH. IN STRUCT.
MECH., BOURNEMOUTH, DORSET, ENGLAND, OCTOBER
12-17, 1975, PP. F.1-F.39.

75-0559 MCHENRY, H. I.
HENSLEY, E. K.
EVALUATION OF DAMAGE TOLERANCE IN AIRCRAFT
STRUCTURES
J. AIRCR., VOL. 12, NO. 2, FEBRUARY 1975, PP.
93-99.

75-0560 MODENA, C.
DISTRIBUZIONI OTTIMALI DI ARMATURE DI
PRECOMPRESSIONE IN TRAVI-PARETE DI CALCESTRUZZO.
(OPTIMAL DISTRIBUTIONS OF PRESTRESSING FORCES IN
CONCRETE WALL)
G. GENIO CIV., VOL. 113, NO. 1-2-3,
JANUARY-FEBRUARY-MARCH, 1975, PP. 59-66.

75-0561 ODLAND, J. R.
SIMPLIFIED CALCULATION OF THE INTERACTION FORCES
BETWEEN HULL AND TANK SYSTEM FOR A MOSS ROSENBERG
TYPE LNG CARRIER
NOR. MARIT. RES., VOL. 3, NO. 1, 1975, PP. 2-10.

75-0562 CROSSMAN, F. W.
ASHBY, M. F.
NON-UNIFORM FLOW OF POLYCRYSTALS BY GRAIN-BOUNDARY
SLIDING ACCOMMODATED BY POWER-LAW CREEP
ACTA. METALL., VOL. 23, NO. 4, APRIL 1975, PP.
425-440.

75-0563 KAWAHARA, M.
LARGE STRAIN, VISCOELASTIC AND ELASTO VISCOPLASTIC
NUMERICAL ANALYSIS BY MEANS OF THE FINITE ELEMENT
METHOD
ARCH. MECH., STOSOW, VOL. 27, NO. 3, 1975, PP.
417-443.

75-0564 BATHE, K. M.
RAMM, E.
WILSON, E. L.
FINITE ELEMENT FORMULATIONS FOR LARGE DEFORMATION
DYNAMIC ANALYSIS
INTL. J. NUM. METH. ENG., VOL. 9, NO. 2, 1975, PP.
353-386.

75-0565 HAYASHI, N.
ON THE EXPLICIT REPRESENTATION OF
TRANSFER-MATRICES OF CURVED BEAMS AND THEIR
APPLICATIONS
BULL. J.S.M.E., VOL. 18, NO. 122, AUGUST 1975, PP.
797-806.

75-0566 PARMERTER, R. R.
MUKHERJI, B.
STRESS INTENSITY FACTORS FOR A NOTCHED BEAM ON AN
ELASTIC FOUNDATION
ENG. FRACT. MECH., VOL. 7, NO. 2, JUNE 1975, PP.
291-298.

75-0567 ITO, H.
MORIGUCHI, K.
ACTUAL STRESS OF HIGH HEAD PUMP-TURBINE RUNNER
HITACHI REV., VOL. 24, NO. 7, JULY 1975, PP.
293-300.

75-0568 DESAI, C. S.
LYTTON, R. L.
STABILITY CRITERIA FOR TWO FINITE ELEMENT SCHEMES
FOR PARABOLIC EQUATION
INT. J. NUMER. METH. ENG., VOL. 9, NO. 3, 1975,
PP. 721-726.

75-0569 GODBOLE, P. N.
CREEPING FLOW IN RECTANGULAR DUCTS BY FINITE
ELEMENT METHOD
INT. J. NUMER. METH. ENG., VOL. 9, NO. 3, 1975,
PP. 727-731.

75-0570 SAKAI, F.
KAWAI, S.
FINITE ELEMENT ANALYSIS OF WAVE PROPAGATION IN
OPEN CHANNELS
PROC. J.S.C.E., NO. 239, JULY 1975, PP. 25-36.

75-0571 LAEVASTU, T.
MATTHEWS, J. B.
HYDRODYNAMICAL-NUMERICAL MODELS FOR COASTAL WATERS
AND OPEN OCEAN AREAS
HN MODEL WORKSHOP AND AGU TOPICAL CONF. MONTEREY,
CALIF., DECEMBER 2-11, 1974, ALSO IN TRANS. AMER.
GEOPHYS. UNION, VOL. 56, NO. 9, SEPTEMBER 1975,
PP. 580-583.

75-0572 KAO, R.
APPLICATION OF HILL FUNCTIONS TO CIRCULAR PLATE
PROBLEMS
Q. APPL. MATH., VOL. 33, NO. 1, APRIL 1975, PP.
63-72.

75-0573 TANG, H. T.
RASHID, Y. R.
SAFE-COLAPS: A FINITE ELEMENT PROGRAM FOR THE
CREEP COLLAPSE ANALYSIS OF NUCLEAR FUEL RODS
2ND NATL. CONGR. PRESSURE VESSELS AND PIPING, FUEL
ELEM. ANAL., SAN FRANCISCO, CALIF., JUNE 23-27,
1975, PP. 1-11. (PUBL. BY A.S.M.E., NEW YORK,
N.Y., 1975.)

75-0574 TOO, J. J. M.
HSU, T. R.
FULMOD - AN INELASTIC ANALYSIS PROGRAM TO PREDICT
THE OPERATING BEHAVIOUR OF CANDU FUEL ELEMENTS
2ND NATL. CONGR. PRESSURE VESSELS AND PIPING, FUEL
ELEM. ANAL., SAN FRANCISCO, CALIF., JUNE 23-27,
1975, PP. 23-35. (PUBL. BY A.S.M.E., NEW YORK,
N.Y., 1975.)

75-0575 KENDALL, J. M.
HSU, M. B.
BLOWDOWN EFFECTS ON A FUEL ROD
2ND NATL. CONGR. PRESSURE VESSELS AND PIPING, FUEL
ELEM. ANAL., SAN FRANCISCO, CALIF., JUNE 23-27,
1975, PP. 37-51. (PUBL. BY A.S.M.E., NEW YORK,
N.Y., 1975.)

75-0576 PAO, Y. C.
ON COMPUTATIONS INVOLVING STIFFNESS MATRICES
STORED IN RECTANGULAR FORM
INTL. J. NUM. METH. ENG., VOL. 9, NO. 1, 1975, PP.
250-251.

75-0577 GEORGE, J.
DAMLE, P. S.
ON THE NUMERICAL SOLUTION OF FREE BOUNDARY
PROBLEMS
INTL. J. NUM. METH. ENG., VOL. 9, NO. 1, 1975, PP.
245-249.

75-0578 BROWN, T. G.
GHALI, A.
SEMI-ANALYTIC SOLUTION OF SKEW BOX GIRDER BRIDGES
PROC. INST. CIVIL ENG. (LONDON), VOL. 59, PT. 2,
SEPTEMBER 1975, PP. 487-500.

75-0579 VENTER, R.
MULLINS, M.
PROBABILISTIC EVALUATION OF THE RELATIVE BEHAVIOUR
OF TUNGSTEN CARBIDE COMPONENTS IN STRUCTURAL
APPLICATIONS
ANN. CIRP., VOL. 24, NO. 1, 1975, PP. 147-151.

75-0580 WECK, M.
MIESSEN, W.
ANWENDUNG DER METHODE FINITER ELEMENTE BEI DER
ANALYSE DES DYNAMISCHEN VERHALTENS GEDAEMPFTER
WERKZEUGMASCHINENSTRUKTUREN. (APPLICATION OF THE
FINITE ELEMENT METHOD FOR THE ANALYSIS OF THE
DYNAMIC BEHAVIOR OF DAMPED MACHINE TOOLS.)
ANN. CIRP., VOL. 24, NO. 1, 1975, PP. 303-307.

75-0581 GIESELBERG, K.
LANGE, K.
ELASTIC DEFORMATION OF SQUARE EXTRUSION DIES DUE
TO LOAD AND TEMPERATURE
ANN. CIRP., VOL. 24, NO. 1, 1975, PP. 167-171.

75-0582 HIJINK, J. A. W.
VAN DER WOLF, A. C. H.
ON THE DESIGN OF DIE-SETS
ANN. CIRP., VOL. 24, NO. 1, 1975, PP. 357-360.

75-0583 SEIDE, P.
ACCURACY OF SOME NUMERICAL METHODS FOR COLUMN
BUCKLING
J. ENG. MECH. DIV., A.S.C.E., VOL. 101, NO. 5,
OCTOBER 1975, PP. 549-560.

75-0584 MURRAY, D. W.
RAJASEKARAN, S.
TECHNIQUE FOR FORMULATING BEAM EQUATIONS
J. ENG. MECH. DIV., A.S.C.E., VOL. 101, NO. 5,
OCTOBER 1975, PP. 561-573.

75-0585 VERONDA, D. R.
WEINGARTEN, V. I.
STABILITY OF PRESSURIZED HYPERBOLOIDAL SHELLS
J. ENG. MECH. DIV., A.S.C.E., VOL. 101, NO. 5,
OCTOBER 1975, PP. 663-678.

75-0586 KAUSEL, E.
ROESSET, J. M.
DYNAMIC ANALYSIS OF FOOTINGS ON LAYERED MEDIA
J. ENG. MECH. DIV., A.S.C.E., VOL. 101, NO. 5,
OCTOBER 1975, PP. 679-693.

75-0587 HAGER, W. W.
STRANG, G.
FREE BOUNDARIES AND FINITE-ELEMENTS IN ONE
DIMENSION
MATH. COMPUT., VOL. 29, NO. 132, 1975, PP.
1020-1031.

75-0588 VENKATARAMANA, J.
RAO, G. V.
FINITE-ELEMENT ANALYSIS OF MODERATELY THICK SHELLS
NUCL. ENG. DES., VOL. 33, NO. 3, 1975, PP.
398-402.

75-0589 MCLEOD, R. J. Y.
MITCHELL, A. R.
USE OF PARABOLIC ARCS IN MATCHING CURVED
BOUNDARIES IN FINITE-ELEMENT METHOD
J. INST. MATH. APPL., VOL. 16, NO. 2, 1975, PP.
239-246.

75-0590 BAKER, G. A.
FINITE-ELEMENT METHOD FOR 1ST ORDER HYPERBOLIC
EQUATIONS
MATHS. COMPUT., VOL. 29, NO. 132, 1975, PP.
995-1006.

75-0591 SABBAGH, H. A.
KRILE, T. F.
3-DIMENSIONAL FINITE-ELEMENT ANALYSIS OF
ELECTROELASTIC NORMAL MODES OF A TANGENTIALLY
POLARIZED, SEGMENTED FERROELECTRIC SHELL
J. ACOUSTICAL SOC. AMER., VOL. 58, SER. 1, 1975.
P. 80.

75-0592 WARD, C.
MODES OF MOTION OF A FINITE-ELEMENT BRAIN MODEL
J. ACOUSTICAL SOC. AMER., VOL. 58, SER. 1, 1975,
P. 14.

75-0593 HINTON, E.
SCOTT, F. C.
RICKETTS, R. E.
LOCAL LEAST SQUARES STRESS SMOOTHING FOR PARABOLIC
ISOPARAMETRIC ELEMENTS
INTL. J. NUM. METH. ENG., VOL. 9, NO. 1, 1975, PP.
235-238.

75-0594 KRILE, T. F.
SABBAGH, H. A.
2-DIMENSIONAL FINITE-ELEMENT ANALYSIS OF
ELECTROELASTIC NORMAL MODES OF FERROELECTRIC DISKS
AND SHELLS
J. ACOUSTICAL SOC. AMER., VOL. 58, SER. 1, 1975,
P. 80.

75-0595 WINNICKI, R. T.
AUYER, S. E.
FINITE-ELEMENT MODELING OF PIEZOELECTRIC CERAMIC
HYDROPHONES
J. ACOUSTICAL SOC. AMER., VOL. 58, SER. 1, 1975,
P. 79.

75-0596 DIETERICH, J. H.
DECKER, R. W.
FINITE-ELEMENT MODELING OF SURFACE DEFORMATION
ASSOCIATED WITH VOLCANISM
J. GEOPHYS. RESEARCH, VOL. 80, NO. 29, 1975, PP.
4094-4102.

75-0597 GONG, C.
JACHENS, R. C.
KUO, J. T.
PSEUDO 3-DIMENSIONAL FINITE-ELEMENT FORMULATION
FOR ELASTOSTATIC PROBLEMS AND ITS GEOPHYSICAL
APPLICATIONS
J. GEOPHYS. RESEARCH, VOL. 80, NO. 29, 1975, PP.
4103-4110.

75-0598 CIARLET, P. G.
USE OF FINITE ELEMENT METHOD FOR SHELL PROBLEMS
COMPTES RENDUS - ACADEMIE DES SCIENCES SER. A
MATH., VOL. 280, NO. 18, 1975, PP. 1229-1232.

75-0599 WEGMULLER, A. W.
ELASTIC-PLASTIC FINITE ELEMENT ANALYSIS OF PLATES
INGENIEUR-ARCHIV, VOL. 44, NO. 2, 1975, PP. 63-77.

75-0600 FOGGIA, A.
SABONNADIERE, J. C.
SILVESTER, P.
FINITE ELEMENT SOLUTION OF SATURATED TRAVELLING
MAGNETIC FIELD PROBLEMS
I.E.E.E. TRANS. POWER APPAR. AND SYSTEMS, VOL.
PA94, NO. 3, 1975, PP. 866-871.

75-0601 MERCER, J. W.
PINDER, G. F.
DONALDSON, I. G.
GALERKIN-FINITE ELEMENT ANALYSIS OF HYDROTHERMAL
SYSTEM AT WAIRAKEI, NEW-ZEALAND
J. GEOPHYS. RESEARCH, VOL. 80, NO. 17, 1975, PP.
2608-2621.

75-0602 THOMAS, G. R.
GALLAGHER, R. H.
TRIANGULAR THIN SHELL FINITE ELEMENT: LINEAR
ANALYSIS
NASA, CONTRACT REPORT CR-2482, JULY 1975, 45 PP.

75-0603 FUEHRING, H.
PARAMETRISCHE KERBSPANNUNGSUNTERSUCHUNGEN AN DER
LOCHSCHEIBE MIT DER METHODE DER FINITEN ELEMENTE.
(FINITE ELEMENT METHOD PARAMETRIC STRESS
CONCENTRATION TESTS IN CIRCULAR HOLE STRIPS)
STAHLBAU, VOL. 44, NO. 9, SEPTEMBER 1975, PP.
272-279.

75-0604 GUREGHIAN, A. B.
STUDY BY THE FINITE-ELEMENT METHOD OF THE
INFLUENCE OF FRACTURES IN CONFINED AQUIFERS.
SOC. PET. ENG., A.I.M.E. J., VOL. 15, NO. 2, APRIL
1975, PP. 181-191.

75-0605 BAI, K. J.
DIFFRACTION OF OBLIQUE WAVES BY AN INFINITE
CYLINDER
J. FLUID MECH., VOL. 68, PART 3, APRIL 15, 1975,
PP. 513-535.

75-0606 GAMBOLATI, G.
USE OF THE OVER-RELAXATION TECHNIQUE IN THE
SIMULATION OF LARGE GROUNDWATER BASINS BY THE
FINITE ELEMENT METHOD
INTL. J. NUM. METH. ENG., VOL. 9, NO. 1, 1975, PP.
219-234.

75-0607 BAZANT, Z. P.
ELNIMEIRI, M.
FINITE-ELEMENT FOR BUCKLING OF CURVED BEAMS AND
SHELLS WITH SHEAR
J. STRUCT. DIV., A.S.C.E., VOL. 101, NO. 9, 1975,
PP. 1997-2004.

75-0608 UTKU, S.
SOLUTION OF LARGE SYSTEMS OF EQUATIONS
IN "WORLD CONGRESS ON FINITE ELEMENT METH. IN
STRUCT. MECH.", BOURNEMOUTH, DORSET, ENGLAND,
OCTOBER 12-17, 1975, PP. 1-44.

75-0609 WEISSGERBER, V.
HYBRIDE SCHALENELEMENTE MIT EINER ANWENDUNG AUF
WENDELSCHALEN. (HYBRID SHELL ELEMENTS IN
APPLICATION TO HELICAL SHELLS)
ING. ARCH., VOL. 44, NO. 2, 1975, PP. 103-112.

75-0610 HAJIDJAFARI, S.
WIGGERT, D. C.
NUMERICAL PREDICTION OF TRACER CONCENTRATION IN
AQUIFERS WITH TRANSIENT GROUNDWATER FLOW BY
FINITE-ELEMENT METHOD
TRANS. AMER. GEOPHYSICAL UNION, VOL. 56, NO. 9,
1975, P. 598.

75-0611 KEUNING, D. H.
APPLICATION OF FINITE ELEMENT METHOD WITH
SECTIONAL LINEARIZATION TO FLOW PROBLEMS
J. ENG. MATH., VOL. 9, NO. 3, 1975, PP. 251-260.

75-0612 BRAMBLE, J. H.
SURVEY OF SOME FINITE ELEMENT METHOD PROPOSED FOR
TREATING DIRICHLET PROBLEM
ADVANCES IN MATH., VOL. 16, NO. 2, 1975, PP.
187-196.

75-0613 COOK, R. D.
FINITE ELEMENT BUCKLING ANALYSIS OF HOMOGENEOUS
AND SANDWICH PLATES
INTL. J. NUM. METH. ENG., VOL. 9, NO. 1, 1975, PP.
39-50.

75-0614 FUHRING, H.
THE APPLICATION OF NODE-ELEMENT RULES FOR
FORECASTING PROBLEMS IN THE GENERATION OF FINITE
ELEMENT MESHES
INTL. J. NUM. METH. ENG., VOL. 9, NO. 3, 1975, PP.
617-631.

75-0615 MYERS, G. E.
STABILITY AND OSCILLATION CHARACTERISTICS OF
FINITE ELEMENT, FINITE-DIFFERENCE, AND
WEIGHTED-RESIDUALS METHODS FOR TRANSIENT
TWO-DIMENSIONAL HEAT CONDUCTION IN SOLIDS -
COMMENT
J. HEAT TRANSFER, VOL. 97, NO. 2, 1975, P. 320.

75-0616 CIAVALDINI, J. F.
NUMERICAL ANALYSIS OF 2-PHASE STEFAN PROBLEM BY
METHOD OF FINITE ELEMENTS
J. NUMER. ANALYSIS, S.I.A.M., VOL. 12, NO. 3,
1975, PP. 464-487.

75-0617 ODEN, J. T.
REDDY, J. N.
SOME OBSERVATIONS ON PROPERTIES OF CERTAIN MIXED
FINITE ELEMENT APPROXIMATIONS
INT. J. NUMER. METH. ENG., VOL. 9, NO. 4, 1975,
PP. 933-938.

75-0618 RAO, G. V.
RAO, R. V. N.
GALERKIN FINITE-ELEMENT SOLUTION FOR STABILITY OF
CANTILIVER COLUMNS SUBJECTED TO TANGENTIAL LOADS
A.I.A.A. J., VOL. 13, NO. 5, 1975, P. 690-691.

75-0619 BUCHANAN, G. R.
STRESS-HYBRID FINITE-ELEMENT
J. STRUCT. DIV., A.S.C.E., VOL. 101, NO. 7, 1975,
PP. 1629-1631.

75-0620 MELOSH, R. J.
LOBITZ, D. W.
NUMERICAL SUFFICIENCY TEST FOR MONOTONIC
CONVERGENCE OF FINITE ELEMENT MODELS
A.I.A.A. J., VOL. 13, NO. 5, 1975, PP. 675-678.

75-0621 FRANCAVILLA, A.
ZIENKIEWICZ, O. C.
A NOTE ON NUMERICAL COMPUTATION OF ELASTIC CONTACT
PROBLEMS
INT. J. NUMER. METH. ENG., VOL. 9, NO. 4, 1975,
PP. 913-925.

75-0622 SHORE, S.
WILSON, J. L.
INTERACTIVE TECHNIQUES WITH GRAPHICAL OUTPUT FOR
BRIDGE ANALYSES
COMPUT. METH. APPL. MECH. ENG., VOL. 5, NO. 2,
MARCH 1975, PP. 197-209.

75-0623 TIEU, A. K.
NUMERICAL SIMULATION OF FINITE-WIDTH THRUST
BEARINGS, TAKING INTO ACCOUNT VISCOSITY VARIATION
WITH TEMPERATURE AND PRESSURE
J. MECH. ENG. SCI., VOL. 17, NO. 1, FEBRUARY 1975,
PP. 1-10.

75-0624　BAGUELIN, F.
FRANK, R.
ETUDE DE LA CAPACITE PORTANTE DES PIEUX PAR LA
METHODE DES ELEMENTS FINIS. INFLUENCE DE LA
DILATANCE DU MILIEU. (STUDY OF THE BEARING
CAPACITY OF PILES BY THE FINITE ELEMENT METHOD.
INFLUENCE OF THE MEDIUM DILATANCY)
BULL. LIAISON LAB. CENT. PONT CHAUSSEES, NO. 75,
JANUARY - FEBRUARY, 1975, PP. 157-164.

75-0625　PERIAUX, J.
THREE DIMENSIONAL ANALYSIS OF COMPRESSIBLE
POTENTIAL FLOWS
NUMER. METH. ENG., VOL. 9, NO. 4, 1975, PP.
775-833.

75-0626　MEAR, J. L.
REDUNDANT STRUCTURE CONCEPTS IN FINITE ELEMENT
ANALYSES
INTL. J. NUM. METH. ENG., VOL. 9, NO. 4, 1975 PP.
765-775.

75-0627　BESSELING, J. F.
NONLINEAR ANALYSIS
WORLD CONGRESS ON FINITE ELEMENT METH. IN STRUCT.
MECH., BOURNEMOUTH, DORSET, ENGLAND, OCTOBER
12-17, 1975, PP. L.1-L.47.

75-0628　HANEKE, R.
TEMPERATURVERLAUF IM QUERSCHNITT GESCHWEISSTER
ROHRWAENDE BEI UEBERWIEGEND KONVEKTIVEM
WAERMEUEBERGANG AUF DER RAUCHGASSEITE.
(TEMPERATURE VARIATION IN THE CROSS-SECTION OF
WELDED TUBE WALLS WITH PREDOMINANTLY CONVECTIVE
HEAT TRANSFER ON THE FLUE GAS SIDE)
VGB KRAFTWERKSTECH, VOL. 55, NO. 11, NOVEMBER
1975, PP. 758-764.

75-0629　JONES, C. J. F. P.
SIMS, F. A.
EARTH PRESSURES AGAINST THE ABUTMENTS AND WING
WALLS OF STANDARD MOTORWAY BRIDGES
GEOTECHNIQUE, VOL. 25, NO. 4, DECEMBER 1975, PP.
731-742.

75-0630　DE ARANTES E OLIVEIRA, E. R.
CONVERGENCE AND ACCURACY IN THE FINITE ELEMENT
METHOD
IN "WORLD CONGRESS ON FINITE ELEMENT METH. IN
STRUCT. MECH.", BOURNEMOUTH, DORSET, ENGLAND,
OCTOBER 12-17, 1975, PP. 0.1-0.24.

75-0631　ATLURI, S. N.
KOBAYASHI, A. S.
CHENG, J. S.
BRAIN-TISSUE FRAGILITY - FINITE STRAIN ANALYSIS BY
A HYBRID FINITE-ELEMENT METHOD
MECH. ENG., VOL. 97, NO. 7, 1975, P. 102.

75-0632　JUDAH, O. M.
SHANHOLTZ, V. O.
CONTRACTOR, D. N.
FINITE-ELEMENT SIMULATION OF FLOOD HYDROGRAPHS
TRANS. AMER. SOC. OF AGRI. ENG., VOL. 18, NO. 3,
1975, PP. 518-522.

75-0633　DAVIS, R. E.
CASTLETON, G. A.
FIELD TESTS OF A STEEL-COMPOSITE BOX-GIRDER BRIDGE
TRANS. RES. BOARD, NO. 547, 1975, PP. 47-54.

75-0634　SPIER, E. E.
WANG, G.
ON BUCKLING OF UNIDIRECTIONAL BORON-ALUMINUM
STIFFENERS - A CAUTION TO DESIGNERS
J. COMPOS. MATER., VOL. 9, NO. 4, OCTOBER 1975,
PP. 347-360.

75-0635　RAJASEKARAN, S.
MURRAY, D. W.
INCREMENTAL FINITE-ELEMENT MATRICES
J. STRUCT. DIV., A.S.C.E., VOL. 101, NO. 8, 1975,
PP. 1701-1702.

75-0636　BRAUER, J. R.
SATURATED MAGNETIC ENERGY FUNCTIONAL FOR
FINITE-ELEMENT ANALYSIS OF ELECTRIC MACHINES
TRANS. POWER APPAR. AND SYSTEMS, I.E.E.E., VOL.
94, NO. 4, 1975, P. 1099.

75-0637　NEALE, B. K.
FINITE ELEMENT CRACK TIP MODELING WHEN EVALUATING
J-INTEGRAL
INT. J. OF FRACTURE, VOL. 11, NO. 1, 1975, PP.
177-178.

75-0638　PENMAN, A. D. M.
CHARLES, J. A.
NASH, J. K. T. L.
HAMPHREYS, J. D.
PERFORMANCE OF CULVERT UNDER WINSCAR DAM
GEOTECHNIQUE, VOL. 25, NO. 4, DECEMBER 1975, PP.
713-730.

75-0639　BARSOUM, R. S.
FURTHER APPLICATION OF QUADRATIC ISOPARAMETRIC
FINITE-ELEMENTS TO LINEAR FRACTURE MECHANICS OF
PLATE BENDING AND GENERAL SHELLS
INT. J. OF FRACTURE, VOL. 11, NO. 1, 1975, PP.
167-169.

75-0640　JAGANNATHAN, D. S.
EPSTEIN, H. I.
CHRISTIANO, P. P.
SNAP-THROUGH BUCKLING OF RETICULATED SHELLS
PROC. INST. CIVIL ENG., LONDON, VOL. 59, PART 2,
DECEMBER 1975, PP. 727-742.

75-0641　GARNET, H.
ARMEN, H.
VARIABLE TIME STEP METHOD FOR DETERMINING PLASTIC
STRESS REFLECTIONS FROM BOUNDARIES
A.I.A.A. J., VOL. 13, NO. 4, APRIL 1975, PP.
532-534.

75-0642　REEFMAN, R. J. B.
TEMPERATURE CALCULATIONS USING THE NASTRAN FINITE
ELEMENT PROGRAMME
HOLECTECHNIEK, VOL. 5, NO. 3, DECEMBER 1975, PP.
82-91.

75-0643　OWEN, M. J.
BISHIO, P. T.
PREDICTION OF STATIC AND FATIGUE DAMAGE AND CRACK
PROPAGATION IN COMPOSITE MATERIALS
PROC. AGARD CONF., MUNICH, GERMANY, PAPER NO. 1,
OCTOBER 13-19, 1974, 12 PP.

75-0644　HENZE, E.
ROTH, S.
PRACTICAL FINITE ELEMENT METHOD OF FAILURE
PREDICTION FOR COMPOSITE MATERIAL STRUCTURES
PROC. AGARD CONF., MUNICH, GERMANY, PAPER 5, NO.
163, OCTOBER 13-19, 1974, 11 PP.

75-0645　ANSOURIAN, P.
ROTH, W.
LEE, K. L.
FACTOR OF SAFETY APPROACH FOR EVALUATING SEISMIC
STABILITY OF SLOPES
PROC. U.S. NATL. CONF. ON EARTHQUAKE ENG., ANN
ARBOR, MICH., JUNE 18-20, 1975, PP. 156-165.
(PUBL. BY EARTHQUAKE ENG. RES. INST., OAKLAND,
CALIF., 1975)

75-0646　MARINSHAW, S. A.
LINDSEY, G. H.
PLANE STRESS CRACK TRAJECTORIES IN LOW MODULUS
MATERIALS
INT. J. FRACT., VOL. 11, NO. 2, APRIL 1975, PP.
273-282.

75-0647　SEGERLIND, L. J.
FINITE ELEMENTS AND THE AG ENGINEER
AGRIC. ENG., VOL. 56, NO. 6, JUNE 1975, PP. 14-16.

75-0648　NANDAKUMAR, K.
MASLIYAH, J. H.
FULLY DEVELOPED VISCOUS FLOW IN INTERNALLY FINNED
TUBES
CHEM. ENG. J., VOL. 10, NO. 2, OCTOBER 1975, PP.
113-120.

75-0649　KRUSZEWSKI, J.
WITTBRODT, E.
OBLICZANIE ORGAN METODA SZTYWNYCH ELEMENTOW
SKONCZONYCH. (CALCULATING VIBRATIONS BY THE
RIGID-FINITE-ELEMENT METHOD)
PRZEGL. MECH., VOL. 34, NO. 4, FEBRUARY 28, 1975,
PP. 113-117.

75-0650　ANDERSON, T. L.
PAN, Y. S.
THREE-DIMENSIONAL STRESS ANALYSES OF MITERED
ELBOWS
A.S.M.E. MEETING, PAPER NO. 75-PVP-54, JUNE 23-27,
1975, 12 PP.

75-0651　WU, J. N. C.
HECHMER, J. L.
THERMAL MECHANICAL ANALYSIS PROCEDURE FOR A
SEMICIRCULAR PLATE WITH CLAMPED STRAIGHT EDGE
PAPER NO. 75-PVP-55, A.S.M.E. MEETING, JUNE 23-27,
1975, 13 PP.

75-0652　PAI, D. H.
HSU, M. B.
INELASTIC ANALYSIS OF TUBESHEETS BY THE FINITE
ELEMENT METHOD
PAPER NO. 75-PVP-57, A.S.M.E. MEETING, JUNE 23-27,
1975, 8 PP.

75-0653　BERTSCH, O. L.
SUN, C. L.
THREE-DIMENSIONAL FINITE ELEMENT ANALYSIS OF A
STEAM GENERATOR CHANNEL-HEAD-COMPLEX - COMPARISON
OF THEORY AND EXPERIMENT
PAPER NO. 75-PVP-59, A.S.M.E. MEETING, JUNE 23-27,
1975, 8 PP.

75-0654 MELLO, R. M.
SCHELLER, J. D.
SIMPLIFIED INELASTIC (PLASTIC AND CREEP) ANALYSIS
OF PIPE ELBOWS SUBJECTED TO INPLANE AND
OUT-OF-PLANE BENDING
PAPER NO. 75-PVP-62, A.S.M.E. MEETING, JUNE 23-27,
1975, 12 PP.

75-0655 KROENKE, W. C.
ADDICOTT, G. W.
INTERPRETATION OF FINITE ELEMENT STRESSES
ACCORDING TO ASME SECTION 111
PAPER NO. 75-PVP-63, A.S.M.E. MEETING, JUNE 23-27,
1975, 12 PP.

75-0656 JONES, R. F.
RODERICK, J. E.
INCREMENTAL SOLUTION PROCEDURE FOR NONLINEAR
STRUCTURAL DYNAMICS PROBLEMS
PAPER NO. 75-PVP-4, A.S.M.E. MEETING, JUNE 23-27,
1975, 6 P.

75-0657 KLEIN, S.
NONLINEAR DYNAMIC ANALYSIS OF SHELLS OF REVOLUTION
WITH ASYMMETRIC PROPERTIES BY THE FINITE ELEMENT
METHOD
PAPER NO. 75-PVP-8, A.S.M.E. MEETING, JUNE 23-27,
1975, 9 PP.

75-0658 HARTLEY, G. A.
MCNEICE, G. M.
ARCH DAM CONSTRUCTION STRESSES DUE TO HYDROSTATIC
AND GRAVITY LOADS
PROC. INST. CIV. ENG., LONDON, VOL. 59, PART 2,
JUNE 1975, PP. 223-235.

75-0659 EVANS, H. R.
AL-RIFAIE, W. N.
EXPERIMENTAL AND THEORETICAL INVESTIGATION OF THE
BEHAVIOR OF BOX GIRDERS CURVED IN PLAN
PROC. INST. CIV. ENG., LONDON, VOL. 59, PART 2,
JUNE 1975, PP. 323-352.

75-0660 FRANKUS, A.
LEE, G. C.
LOCAL COMPLIANCE PROBLEM OF LUNG ELASTICITY
PAPER NO. 75-APMB-6, A.S.M.E. MEETING, JUNE 23-27,
1975, 12 PP.

75-0661 GAWRONSKI, W.
EFFECT OF SYSTEM PARAMETER VARIATIONS ON NATURAL
FREQUENCIES
COMPUT. STRUCT., VOL. 5, NO. 1, APRIL 1975, PP.
31-43.

75-0662 FELIPPA, C. A.
SOLUTION OF LINEAR EQUATIONS WITH SKYLINE-STORED
SYMMETRIC MATRIX
COMPUT. STRUCT., VOL. 5, NO. 1, APRIL 1975, PP.
13-29.

75-0663 KIKUCHI, F.
ON THE VALIDITY OF AN APPROXIMATION AVAILABLE IN
THE FINITE ELEMENT SHELL ANALYSIS
COMPUT. STRUCT., VOL. 5, NO. 1, APRIL 1975, PP.
1-8.

75-0664 LYNESS, J. F.
OWEN, D. R. J.
FINITE ELEMENT ANALYSIS OF ENGINEERING SYSTEMS
GOVERNED BY A NON-LINEAR QUASI-HARMONIC EQUATION
COMPUT. STRUCT., VOL. 5, NO. 1, APRIL 1975, PP.
65-79.

75-0665 TAHIANI, C.
LACHANCE, L.
LINEAR AND NON-LINEAR ANALYSIS OF THIN SHALLOW
SHELLS BY MIXED FINITE ELEMENTS
COMPUT. STRUCT., VOL. 5, NO. 2-3, JUNE 1975, PP.
167-177.

75-0666 TURKSTRA, C.
FAM, A.
FINITE ELEMENT SCHEME FOR BOX BRIDGE ANALYSIS
COMPUT. STRUCT., VOL. 5, NO. 2-3, JUNE 1975, PP.
179-186.

75-0667 NATARAJAN, R.
BLOMFIELD, J. A.
STRESS ANALYSIS OF CURVED PIPES WITH END
RESTRAINTS
COMPUT. STRUCT., VOL. 5, NO. 2-3, JUNE 1975, PP.
187-196.

75-0668 THOMAS, G. R.
GALLAGHER, R. H.
TRIANGULAR THIN SHELL FINITE ELEMENT: NONLINEAR
ANALYSIS
NASA, REPORT: CR-2483, JULY 1975, 69 PP.

75-0669 HORWATH, J. K.
STRUCTURAL AND SYSTEM MODELS
SAE SPEC. PUBL. SP-392, PAPER 750135, FEBRUARY
1975, PP. 21-30.

75-0670 WALLERSTEIN, D. V.
THERMAL DEFORMATION VECTOR FOR A BILINEAR
TEMPERATURE DISTRIBUTION IN AN ANISOTROPIC
QUADRILATERAL MEMBRANE ELEMENT
INT. J. NUMER. METH. ENG., VOL. 9, NO. 2, 1975,
PP. 325-336.

75-0671 EZZUDDIN, Z. Y.
NUMERICAL SOLUTIONS OF NONLINEAR PLASMA EQUATIONS
BY THE FINITE-ELEMENT METHOD
PH.D., UNIV. OF CALIF., LOS ANGELES, 1975.

75-0672 LEWIS, R. W.
NORRIS, V. A.
FINITE ELEMENT ANALYSIS OF THE MOTION OF A
GAS-LIQUID INTERFACE IN A POROUS MEDIUM
INT. J. NUMER. METH. ENG., VOL. 9, NO. 2, 1975,
PP. 433-448.

75-0673 RADAJ, D.
KERBSPANNUNGSANALYSE NACH DER
FINITE-ELEMENT-METHODE AN KEHLNAEHTEN. (NOTCH
EFFECT ANALYSIS USING THE FINITE ELEMENT METHOD ON
FILLET WELDS)
SCHWEISSEN SCHNEIDEN, VOL. 27, NO. 3, MARCH 1975,
PP. 86-89.

75-0674 GARTLING, D. K.
FINITE ELEMENT ANALYSIS OF VISCOUS, INCOMPRESSIBLE
FLUID FLOW
PH.D., THE UNIV. OF TEXAS, AUSTIN, 1975.

75-0675 ANON
NASTRAN: USER'S EXPERIENCES
NATL. AERON. SPACE ADMIN., LANGLEY RESEARCH
CENTER, LANGLEY STATION, REPORT NO:
NASA-TM-X-3278, L-10436, SEPTEMBER 1975, 621 PP.
(N.T.I.S. - N75-31485/6ST)

75-0676 MARION, C.
ISOFINEL: ISOPARAMETRIC FINITE ELEMENT CODE FOR
ELASTIC ANALYSIS OF TWO-DIMENSIONAL BODIES
DEPT. OF MECH. ENG., CARNEGIE-MELLON UNIV.,
PITTSBURGH, PA., REPORT NO: NASA-CR-143498,
SM-75-4, SEPTEMBER 1975, 67 PP. (N.T.I.S. -
N75-31484/9ST)

75-0677 ZALESAK, J.
MODAL COUPLING PROCEDURES ADAPTED TO NASTRAN
ANALYSIS OF THE 1/8 - SCALE SHUTTLE STRUCTURAL
DYNAMICS MODEL. VOLUME 1: TECHNICAL REPORT
GRUMMAN AEROSPACE CORP., BETHPAGE, N.Y., REPORT
NO: NASA-CR-132666, JULY 1975, 175 PP. (N.T.I.S. -
N75-31480/7ST)

75-0678 ZALESAK, J.
MODAL COUPLING PROCEDURES ADAPTED TO NASTRAN
ANALYSIS OF THE 1/8-SCALE SHUTTLE STRUCTURAL
DYNAMICS MODEL. VOLUME 2: SUPPORTING DATA
GRUMMAN AEROSPACE CORP., BETHPAGE, N.Y., REPORT
NO: NASA-CR-132667, JULY 1975, 392 PP. (N.T.I.S. -
N75-31481/5ST)

75-0679 MCNAMARA, J. F.
NUMERICAL SOLUTION SCHEMES FOR HIGHLY NONLINEAR
STATIC STRUCTURAL BEHAVIOR
ARMY CONST. ENG. RES. LAB., CHAMPAIGN, ILL.,
REPORT NO: CERL-TM-M-102, OCTOBER 1975, 28 PP.
(N.T.I.S. - AD-A016 985/4ST).

75-0680 HUANG, J-C.
ELDRIDGE, C. M.
ANALYSIS OF SHELLS OF REVOLUTION BY A CURVED
FINITE ELEMENT
ARMY MISSILE RES. DEVEL. ENG. LAB., REDSTONE
ARSENAL, ALA., REPORT NO.: RL-75-10, MAY 30, 1975,
46 PP. (N.T.I.S. - AD-A016 877/3ST)

75-0681 BUBECK, R. B.
TRANSIENT THERMAL STRESS ANALYSIS OF COMPOSITE
STRUCTURES INCLUDING CONTINUOUSLY VARYING
PROPERTIES
MASTER'S THESIS, NAVAL POSTGRADUATE SCHOOL,
MONTEREY, CALIF., SEPTEMBER 1975, 344 PP.
(N.T.I.S. - AD-A016 572/0ST)

75-0682 CRAWFORD, J.
FORREST, J. B.
A STRUCTURAL EVALUATION OF RAPID METHODS OF
BACKFILLING FOR BOMB DAMAGE REPAIR - PHASE 11
NAVAL CIV. ENG. LAB., PORT HUENEME, CALIF., FINAL
REPT., APRIL 1973 - AUGUST 1974, AUGUST 1975, 88
PP. (N.T.I.S. - AD-A016 498/8ST)

75-0683 OH, S. I.
KOBAYASHI, S.
DEFORMATION MODE OF VOID-GROWTH AND COALESCENCE IN
THE PROCESS OF DUCTILE FRACTURE
DEPT. OF MECH. ENG., UNIV. OF CALIF., BERKELEY,
FINAL REPORT, MAY 1974 - MARCH 1975, 82 PP.
(N.T.I.S. - AD-A016 473/1ST).

75-0684 NIELSEN, J. P.
AFPAV COMPUTER CODE FOR STRUCTURAL ANALYSIS OF
AIRFIELD PAVEMENTS
ERIC H. WANG CIVIL ENG. RESEARCH FACILITY, UNIV.
OF NEW MEXICO, ALBUQUERQUE, FINAL REPORT, OCTOBER
1975, 75 PP. (N.T.I.S. - AD-A016 428/5ST)

75-0685 IVANOV, Y. I.
CALCULATION OF THE REINFORCED THIN-WALLED
STRUCTURES BY THE METHOD OF FINITE ELEMENTS
(RASCHET PODKREPLENNYKH TONKOSTENNYKH KONSTRUKTSII
METODOM KONECHNOGO ELEMENTA)
TRANSLATION DIV., NAVAL INTELLIGENCE SUPPORT
CENTER, WASHINGTON, D.C., REPORT NO:
NISC-TRANS-3692, SEPTEMBER 30, 1975, 15 PP.
(N.T.I.S. - AD-A016 387/3ST)

75-0686 WANG, H-P
MODELING AN OCEAN POND. A TWO-DIMENSIONAL, FINITE
ELEMENT HYDRODYNAMIC MODEL OF NINIGRET POND,
CHARLESTOWN, RHODE ISLAND
DEPT. OF MECH. ENG. AND APPL. MECH., RHODE ISLAND
UNIV, KINGSTON, REPORT NO: MARINE TR-40, 1975, 66
PP. (N.T.I.S. - COM -75-11393/6ST)

75-0687 DAEMEN, J. J. K.
RATIONAL DESIGN OF TUNNEL SUPPORTS: TUNNEL
SUPPORT LOADING CAUSED BY ROCK FAILURE
DEPT. OF CIVIL AND MINING ENG., MINNESOTA UNIV.,
MINNEAPOLIS, FINAL REPORT, MAY 1975, 440 PP.
(N.T.I.S. - AD-A016 155/4ST)

75-0688 KARLAK, R. F.
CROSSMAN, F. W.
FAILURE MECHANISMS IN COMPOSITE SYSTEMS
PALO ALTO RESEARCH LAB., LOCKHEED MISSILES AND
SPACE CO. INC., PALO ALTO, CALIF., REPORT NO:
LMSC-D457462, AUGUST 1975, 133 PP. (N.T.I.S. -
AD-A015 885/7ST)

75-0689 EBERT, L. J.
FLYNN, P. L.
AN ANALYTIC METHOD FOR PREDICTING THE FLOW AND
FRACTURE IN NOTCHED FIBER COMPOSITE MATERIALS
DEPT. OF METALLURGY AND MATER. SCI., CASE WESTERN
RESERVE UNIV., CLEVELAND, OHIO, FINAL REPORT, JUNE
1, 1974 - MAY 31, 1975, 129 PP. (N.T.I.S. -
AD-A015 182/9ST)

75-0690 GWALTNEY, R. C.
BOLT, S. E.
CORUM, J. M.
BRYSON, J. W.
THEORETICAL AND EXPERIMENTAL STRESS ANALYSES OF
ORNL THIN-SHELL CYLINDER-TO-CYLINDER MODEL 3
OAK RIDGE NATL. LAB., TENN., JUNE 1975, 179 PP.
(N.T.I.S. - ORNL-5020)

75-0691 YU, Y. Y.
DYNAMIC INSTABILITY OF DUCTS CONVEYING FLUID
DEPT. OF MECH. ENG., POLYTECHNIC INST. OF NEW
YORK, BROOKLYN, REPORT NO: NASA-CR-143205, AUGUST
1975, 13 PP. (N.T.I.S. - N75-28372/1ST)

75-0692 KASHEF, A. I.
MANAGEMENT OF RETARDATION OF SALT WATER INTRUSION
IN COASTAL AQUIFERS
DEPT. OF CIVIL ENG., NORTH CAROLINA STATE UNIV.,
RALEIGH, COMPLETION REPORT, JULY 1975, 295 PP.
(N.T.I.S. - PB-244 721/7ST)

75-0693 SIMKINS, T. E.
STRUCTURAL RESPONSE TO MOVING PROJECTILE MASS BY
THE FINITE ELEMENT METHOD
BENET WEAPONS LAB., WATERVLIET ARSENAL, N.Y.,
REPORT NO: WVTT-TR-75044, JULY 1975, 49 PP.
(N.T.I.S. - AD-A/14 988/0ST)

75-0694 SUITS, K. W.
NONLINEAR ANALYSES OF THE STRESS FIELD NEAR A
CRACK OR HOLE IN A PLATE
ARMY MISSILE RESEARCH DEV. AND ENG. LAB., REDSTONE
ARSENAL, ALA., REPORT NO: RL-76-3, AUGUST 18,
1975, 125 PP. (N.T.I.S. - AD-A014 982/3ST)

75-0695 GWALTNEY, R. C.
BOLT, S. E.
BRYSON, J. W.
THEORETICAL AND EXPERIMENTAL STRESS ANALYSES OF
ORNL THIN-SHELL CYLINDER-TO-CYLINDER MODEL 4
OAK RIDGE NATL. LAB., TENN., JUNE 1975, 181 PP.
(N.T.I.S. - ORNL-5019)

75-0696 CLINARD, J. A.
CORUM, J. M.
SARTORY, W. K.
COMPARISON OF TYPICAL INELASTIC ANALYSIS
PREDICTIONS WITH BENCHMARK PROBLEM EXPERIMENTAL
RESULTS
OAK RIDGE NATL. LAB., TENN., 1975, 39 PP.
(N.T.I.S. - CONF-750617-2)

75-0697 THOMAS, G. R.
GALLAGHER, R. H.
A TRIANGULAR THIN SHELL FINITE ELEMENT: NONLINEAR
ANALYSIS
CORNELL UNIV., ITHACA, N.Y., REPORT NO:
NASA-CR-2483, JULY 1975, 69 PP. (N.T.I.S. -
N75-27430/8ST)

75-0698 THOMAS, G. R.
GALLAGHER, R. H.
A TRIANGULAR THIN SHELL FINITE ELEMENT: LINEAR
ANALYSIS
CORNELL UNIV., ITHACA, N. Y., REPORT NO:
NASA-CR-2482, JULY 1975, 45 PP. (N.T.I.S. -
N75-07429/0ST)

75-0699 VOS, R. G.
STRAAYER, J. W.
BOPACE 3-D (THE BOEING PLASTIC ANALYSIS CAPABILITY
FOR 3-DIMENSIONAL SOLIDS USING ISOPARAMETRIC
FINITE ELEMENTS)
RESEARCH AND ENG. DIV., BOEING AEROSPACE CO.,
SEATTLE, WASH., REPORT NO: NASA-CR-143891,
D180-18677-1, APRIL 15, 1975, 360 PP. (N.T.I.S. -
N75-27426/6ST)

75-0700 KAUFMAN, A.
ANALYSIS OF THE EFFECTS ON LIFE OF LEADING-EDGE
HOLES IN AN AIRFOIL SUBJECTED TO ARBITRARY
SPANWISE AND CHORDWISE TEMPERATURE DISTRIBUTIONS
NATL. AERO. AND SPACE ADMIN., LEWIS RESEARCH
CENTER, CLEVELAND, OHIO, REPORT NO:
NASA-TM-X-3257, E-8277, JULY 1975, 17 PP.
(N.T.I.S. - N75-27421/7ST)

75-0701 CREWS, J. H. JR.
ANALYTICAL AND EXPERIMENTAL INVESTIGATION OF
FATIGUE IN A SHEET SPECIMEN WITH AN
INTERFERENCE-FIT BOLT
NATL. AERO. AND SPACE ADMIN., LANGLEY RESEARCH
CENTER, LANGLEY STATION, VA., REPORT NO:
NASA-TN-D-7926, L-10067, JULY 1975, 29 PP.
(N.T.I.S. - N75-27418/3ST)

75-0702 BAKER, A. J.
MANHARDT, P. D.
THE FINITE ELEMENT METHOD IN LOW SPEED
AERODYNAMICS
SCHOOL OF ENG., OLD DOMINION UNIV., NORFOLK, VA.,
REPORT NO: NASA-CR-143190, TR-75-T5, MAY 1975, 73
PP. (N.T.I.S. - N75-26972/0ST)

75-0703 WIGHT, R. B.
SELBY, R. A.
THREE-DIMENSIONAL THERMAL AND STRESS ANALYSIS OF A
PIPING TEE
STRUCT. ENG. SERVICES, CONTROL DATA CORP., PALO
ALTO. CALIF., FINAL REPORT, MAY 1975, 128 PP.
(N.T.I.S. - PB-244 432/1ST)

75-0704 MISRA, N.
RAMAKRISHNAN, V.
BEHAVIOR OF ISOLATED SHEAR WALLS SUBJECTED TO
LATERAL LOADS
DEPT. OF CIVIL ENG., SOUTH DAKOTA SCHOOL OF MINES
AND TECH.,RAPID CITY, REPORT NO: SDSMT-CNSF-7502,
MAY 1975, 36 PP. (N.T.I.S. - PB-243 945/3ST)

75-0705 RAMAKRISHNAN, V.
COMPARISON OF THE THREE-DIMENSIONAL ANALYSIS OF A
SHEAR WALL BUILDING BY TWO DIFFERENT METHODS
DEPT. OF CIVIL ENG., SOUTH DAKOTA SCHOOL OF MINES
AND TECH., RAPID CITY, REPORT NO: SDSMT-CNSF-7501,
MAY 1975, 53 PP. (N.T.I.S. - PB-243 944/6ST)

75-0706 CHENG, R. T-S.
FINITE ELEMENT MODELING OF FLOW THROUGH POROUS
MEDIA
DEPT. OF CIVIL ENG., STATE UNIV. OF NEW YORK,
BUFFALO, REPORT NO: RR-75-2, MARCH 1975, 93 PP.
(N.T.I.S. - PB-243 750/7ST)

75-0707 MEYER, P. M.
DATA GENERATION FOR FINITE-ELEMENT STRUCTURAL
ANALYSIS OF THREE-DIMENSIONAL NAVAL SHIP
STRUCTURES
DAVID W. TAYLOR NAVAL SHIP RESEARCH AND DEV.
CENTER, BETHESDA, MD., REPORT NO: NSRDC-4637, JUNE
1975, 84 PP. (N.T.I.S. - AD-A014 125/9ST)

75-0708 SOOSAAR, K.
GRIN, R.
FUREY, M.
HAMILTON, J.
STRUCTURAL EVALUATION OF CANDIDATE DESIGNS FOR THE
LARGE SPACE TELESCOPE PRIMARY MIRROR
DRAPER (CHARLES STARK) LAB., INC., CAMBRIDGE,
MASS., REPORT NO: NASA-CR-143880, R-874, APRIL
1975, 173 PP. (N.T.I.S. - N75-26416/8ST)

75-0709 MERCHANT, D. H.
GATES, R. M.
STRAAYER, J. W.
EFFECT OF DAMPING ON EXCITABILITY OF HIGH-ORDER
NORMAL MODES
RESEARCH AND ENG. DIV., BOEING AEROSPACE CO.,
SEATTLE, WASH., REPORT NO: NASA-CR-143884,
D180-18835-1, MAY 30, 1975, 231 PP. (N.T.I.S. -
N75-26000/0ST)

75-0710 CREGO, H. L.
FUNDAMENTAL VIBRATIONAL FREQUENCY BY A 3D — FEM
PROCEDURE
MASTER'S THESIS, NAVAL POSTGRAD. SCHOOL, MONTEREY,
CALIF., JUNE 1975, 68 PP. (N.T.I.S. — AD-A013
622/6ST).

75-0711 JOHNSON, E. H.
OPTIMIZATION OF STRUCTURES UNDERGOING HARMONIC OR
STOCHASTIC EXCITATION
DEPT. OF AERON. AND ASTRON., STANFORD UNIV.,
CALIF., REPORT NO: NASA-CR-142936, JUNE 1975, 190
PP. (N.T.I.S. — N75-2522/05ST)

75-0712 BABUSKA, I.
RHEINBOLDT, W.
MESZTENYI, C.
SELF-ADAPTIVE REFINEMENTS IN THE FINITE ELEMENT
METHOD
COMPUTER SCI. CENTER, MARYLAND UNIV., COLLEGE
PARK, REPORT NO: TR-375, MAY 1975, 48 PP.
(N.T.I.S. — PB-243 251/6ST)

75-0713 HUS, M.
LEE, W.
THERMAL AND STRESS ANALYSIS OF A FUEL ROD
STRUCTURAL ENG. SERVICES, CONTROL DATA CORP., PALO
ALTO, CALIF., REPORT NO: SES/PSD-74/017, APRIL
1975, 185 PP. (N.T.I.S. — PB-243 219/3ST)

75-0714 GROOMS, D. W.
NUMERICAL METHODS IN FLUID FLOW PROBLEMS. VOL. 2.
NATL. TECH. INFORMATION SERVICE, SPRINGFIELD, VA.,
JANAURY —AUGUST 1975, 101 PP. (N.T.I.S. —
NTIS/PS-75/643/7ST) SEE ALSO NTIS/PS-75/642.

75-0715 GROOMS, D. W.
NUMERICAL METHODS IN FLUID FLOW PROBLEMS. VOL. 1
NATL. TECH. INFOR. SERVICE, SPRINGFIELD, VA.,
REPORT FOR 1970-1974, (N.T.I.S. — PS-75/642/9ST)
SEE ALSO 75/643

75-0716 ANON
COMPUTATIONAL METHODS FOR INVISCID AND VISCOUS
TWO-AND-THREE DIMENSIONAL FLOW FIELDS
ADVISORY GROUP FOR AEROSPACE RESEARCH AND
DEVELOPMENT, PARIS, LECTURE SERIES, REPORT NO:
AGARD-LS-73, 1975, 200 PP. (N.T.I.S. — AD-A013
269/6ST)

75-0717 HUANG, P. C.
PAO, C.
MATRA, J. P. JR.
MISSILE BODY INPUT GENERATOR (BING), A NASTRAN
PRE-PROCESSOR; THEORETICAL DEVELOPMENT, USER'S
MANUAL, AND PROGRAM LISTING
NAVAL SURFACE WEAPONS CENTER, WHITE OAK LAB.,
SILVER SPRING, MD., REPORT NO: NSWC/WOL/TR-75-9,
MARCH 1975, 333 PP. (N.T.I.S. — AD-A012 472/7ST)

75-0718 MONDKAR, D. P.
POWELL, G. H.
STATIC AND DYNAMIC ANALYSIS OF NONLINEAR
STRUCTURES
EARTHQUAKE ENG. RESEARCH CENTER., UNIV. OF CALIF.,
BERKELEY, REPORT NO: EERC-75-10, JUNE 11, 1975,
168 PP. (N.T.I.S. — PB-242 434/9ST)

75-0719 THOMS, R. L.
PECQUET, R. A.
ARMAN, A.
NUMERICAL ANALYSIS OF EMBANKMENTS OVER SOFT SOILS
DIV. ENG. RESEARCH, LOUISIANA STATE UNIV., BATON
ROUGE, REPORT NO: BULL-112, JUNE 1975, 212 PP.
(N.T.I.S. — PB-242 420/8ST)

75-0720 WANG, J. D.
MATHEMATICAL MODELING OF NEAR COASTAL CIRCULATION
SEA GRANT PROJECT OFFICE, MASS. INST. OF TECH.,
CAMBRIDGE, REPORT NO: MITSG-75-13, APRIL 20, 1975,
274 PP. (N.T.I.S. — COM-75-10889/4ST)

75-0721 HU, W-L.
LIU, H. W.
CRACK TIP STRAIN — A COMPARISON OF FEM
CALCULATIONS AND MEASUREMENTS
GEORGE SACHS FRACT. AND FATIGUE RESEARCH LAB.,
SYRACUSE UNIV., N.Y., TECH. REPORT, APRIL 1975, 24
PP. (N.T.I.S. — AD-A012 217/6ST)

75-0722 FARRIS, R. J.
HERRMANN, L. R.
HUTCHINSON, J. R.
SCHAPERY, R. A.
DEVELOPMENT OF A SOLID ROCKET PROPELLANT NONLINEAR
CONSTITUTTIVE THEORY
AEROJET SOLID PROPULSION CO., SACRAMENTO, CALIF.,
REPORT NO:ASPC-1074-26F, MAY 1975, 404 PP.
(N.T.I.S. — AD-A012 213/5ST) SEE ALSO VOL. 1,
JUNE, 1973, AD-769 264. VOL. 2, JUNE 1973, AD-769
263

75-0723 BARKER, R. M.
MELOSH, R. J.
HOLZER, S. M.
SINGER: A COMPUTER CODE FOR GENERAL ANALYSIS OF
TWO-DIMENSIONAL CONCRETE STRUCTURES. VOL. 11.
PROGRAM DOCUMENT
DEPT. OF CIV. ENG., VIRGINIA POLYTECHNIC INST. AND
STATE UNIV., BLACKSBURG, REPORT NO:
AFWL-TR-74-228-V2, MAY 1975, 430 PP. (N.T.I.S. —
AD-A012 211/9ST) SEE ALSO VOL. 1, AD-A011 942 AND
VOL. 3, AD-A011 585

75-0724 HOLZER, S. M.
MELOSH, R. J.
BARKER, R. M.
SOMERS, A. E.
SINGER: A COMPUTER CODE FOR GENERAL ANALYSIS OF
TWO-DIMENSIONAL REINFORCED CONCRETE STRUCTURES.
VOLUME 1. SOLUTION PROCESS
DEPT. OF CIVIL ENG., VIRGINIA POLYTECHNIC INST.
AND STATE UNIV., BLACKSBURG, FINAL REPORT AUGUST
1972 — AUGUST 1974, MAY 1975, 210 PP. (N.T.I.S. —
AD-A011 942/0ST) SEE ALSO VOL. 3, AD-A011 585.

75-0725 JENSEN, P. S.
BROGAN, F. A.
FELIPPA, C. A.
PROBLEM CONTROLLED GRID GENERATION FOR THE
NUMERICAL SOLUTION OF PARTIAL DIFFERENTIAL
EQUATIONS
LOCKHEED MISSILES AND SPACE CO. INC., SUNNYVALE,
CALIF., REPORT NO: LMSC-D469784, MAY 27, 1975, 81
PP. (N.T.I.S. — AD-A011 769/7ST)

75-0726 RAMSEY, J. W. JR.
STRESS CONCENTRATION FACTORS FOR CIRCULAR,
REINFORCED PENETRATIONS IN PRESSURIZED CYLINDRICAL
SHELLS
NATL. AERON. AND SPACE ADMIN., LANGLEY RESEARCH
CENTER, LANGLEY STATION, VA., REPORT NO:
NASA-TM-X-68733, MAY 1975, 205 PP. (N.T.I.S. —
N75-22813/0ST)

75-0727 HOPKINS, A. S.
THE MOTION OF INTERCONNECTED FLEXIBLE BODIES
SCHOOL OF ENG. AND APPL. SCI., CALIF. UNIV., LOS
ANGELES, REPORT NO: NASA-CR-120740,
UCLA-ENG-7513-VOL-3, FEBRUARY 1975, 233 PP.
(N.T.I.S. — N75-22811/4ST)

75-0728 WARNER, D. M.
WOHLEN, R. L.
PARK, A. C.
FINITE ELEMENT SOLUTION OF LOW BOND NUMBER
SLOSHING
MARTIN MARIETTA CORP., DENVER, COLO., REPORT NO:
NASA-CR-120299, MCR-75-139, APRIL 1975, 123 PP.
(N.T.I.S. —N75-22624/74ST)

75-0729 TEMAM, R.
FINITE ELEMENT METHODS IN FLUID FLOW
DEPT. OF MECH. ENG., UNIV. OF CALIF., BERKELEY,
REPORT NO: FM-75-2, FEBRUARY 1975, 32 PP.
(N.T.I.S. — AD-A011 671/5ST)

75-0730 BARKER, R. M.
MELOSH, R. J.
HOLZER, S. M.
BRADSHAW, J. C.
SINGER: A COMPUTER CODE FOR GENERAL ANALYSIS TO
TWO-DIMENSIONAL CONCRETE STRUCTURES. VOLUME 4.
DEMONSTRATION PROBLEMS
DEPT. OF CIVIL ENG., VIRGINIA POLYTECHNIC INST.
AND STATE UNIV., BLACKSBURG, FINAL REPORT AUGUST
1972 - AUGUST 1974, MAY 1975, 49 PP. (N.T.I.S. —
AD-A011 586/5ST) SEE ALSO VOL. 3, AD-A011 585.

75-0731 MELOSH, R. J.
BARKER, R. M.
HOLZER, S. M.
SINGER: A COMPUTER CODE FOR GENERAL ANALYSIS OF
TWO-DIMENSIONAL REINFORCED CONCRETE STRUCTURES.
VOLUME 3. USER'S GUIDE
DEPT. OF CIVIL ENG., VIRGINIA POLYTECHNIC INST.
AND STATE UNIV., BLACKSBURG, FINAL REPORT AUGUST
1972 - AUGUST 1974, MAY 1975, 89 PP. (N.T.I.S. —
AD-A011 585/7ST) SEE ALSO VOL. 4, AD-A011 586.

75-0732 SIH, G. C.
HILTON, P. D.
HARTRANFT, R. J.
KIEFER, B. V.
THREE-DIMENSIONAL STRESS ANALYSIS OF A FINITE SLAB
CONTAINING A TRANSVERSE CENTRAL CRACK
INST. OF FRACTURE AND SOLID MECH., LEHIGH UNIV.,
BETHLEHEM, PA., REPORT NO: IFSM-75-69, JUNE 1975,
62 PP. (N.T.I.S. — AD-A011 357/1ST)

75-0733 LAKSHMIKANTHAM, C.
TONG, P.
STRESSES AROUND HOLES IN STIFFENED COMPOSITE
PANELS USING LAURENT-SERIES AND FINITE ELEMENT
METHODS
ARMY MATER. AND MECH. RESEARCH CENTER, WATERTOWN,
MASS., REPORT NO: AMMRC-TR-75-7, APRIL 1975, 12
PP. (N.T.I.S. - AD-A011 256/5ST)

75-0734 HURWITZ, M. M.
PROGRAMMER'S MANUAL ADDITIONS AND DEMONSTRATION
PROBLEMS FOR A THERMOSTRUCTURAL CAPABILITY FOR
NASTRAN USING ISOPARAMETRIC FINITE ELEMENTS
NAVAL SHIP RESEARCH AND DEVELOPMENT CENTER,
BETHESDA, MD., REPORT NO: NSRDC-4656, MARCH 1975,
155 PP. (N.T.I.S. - AD-A011 126/0ST)

75-0735 JONES, D. P.
SWEDLOW, J. L.
INFLUENCE OF CRACK CLOSURE AND ELASTO-PLASTIC FLOW
ON THE BENDING OF A CRACKED PLATE
INTL. J. FRACT., VOL. 11, NO. 6, DECEMBER 1975,
PP. 897-914.

75-0736 MINICH, M. D.
CHAMIS, C. C.
ANALYTICAL DISPLACEMENTS AND VIBRATIONS OF
CANTILEVERED UNSYMMETRIC FIBER COMPOSITE LAMINATES
NATL. AERON. AND SPACE ADMINISTRATION, LEWIS
RESEARCH CENTER, CLEVELAND, OHIO, REPORT NO:
NASA-TM-X-71699, E-8306, 1975, 13 PP. (N.T.I.S. -
N75-21373/6ST)

75-0737 TONG, P.
ROSSETTOS, J. N.
MODULAR APPROACH TO STRUCTURAL SIMULATION FOR
VEHICLE CRASHWORTHINESS PREDICTION
TRANSPORTATION SYSTEMS CENTER, CAMBRIDGE, MASS.,
REPORT NO: DOT-TSC-NHTSA-74-7, MARCH 1975, 37 PP.
(N.T.I.S. - PB-241 784/8ST)

75-0738 SWIFT, G. W.
PLETTA, D. H.
HURST, H. T.
RELATIONSHIP BETWEEN FINITE ELEMENT AND MEASURED
JOINT-FLOOR CONNECTION RIGIDITIES
COLL. OF ENG., VIRGINIA POLYTECHNIC INST. AND
STATE UNIV., BLACKSBURG, REPORT NO: VPI-E-75-10,
MAY 1975, 22 PP. (N.T.I.S. - PB-241 741/8ST)

75-0739 GROOMS, D. W.
FINITE ELEMENTS IN STRUCTURAL ANALYSIS
NATL. TECH. INFO. SERVICE, SPRINGFIELD, VA., JUNE
1975, 75 PP. (N.T.I.S. - NTIS/PS-75/496/0ST)

75-0740 HUGHES, T. J.
TAYLOR, R. L.
SACKMAN, J. L.
FINITE ELEMENT FORMULATION AND SOLUTION OF
CONTACT-IMPACT PROBLEMS IN CONTINUUM MECHANICS-II
STRUCT. ENG. LAB., UNIV. OF CALIF., BERKELEY,
REPORT NO: UCSESM-75-3, JANUARY 1975, 54 PP.
(N.T.I.S. - AD-A011 103/9ST)

75-0741 HAZZENZAHL, W. W.
GRAY, W. H.
THIN WINDOWS FOR GASEOUS AND LIQUID TARGETS: AN
OPTIMIZATION PROCEDURE
CRYOGENICS, VOL. 15, NO. 11, NOVEMBER 1975, PP.
627-638.

75-0742 HABERCOM, G. E. JR.
MATHEMATICAL ANALYSIS OF STRESS CRACKS. VOLUME 2.
NATL. TECH. INFO. SERVICE, SPRINGFIELD, VA., JUNE
1975, 114 PP. (N.T.I.S. - NTIS/PS-75/467/1ST)
SEE ALSO VOL. 1, NTIS/PS-75/466.

75-0743 HABERCOM, G. E. JR.
MATHEMATICAL ANALYSIS OF STRESS CRACKS. VOLUME 1
NATL. TECH. INFO. SERVICE, SPRINGFIELD, VA., JUNE
1975, 212 PP. (N.T.I.S. - NTIS/PS-75/466/3ST)
SEE ALSO NTIS/PS-75/465 AND VOL. 2.
NTIS/PS-75/467.

75-0744 HABERCOM, G. E. JR.
STRESS CRACK PHENOMENA
NAT. TECH. INFO. SERVICE, SPRINGFIELD, VA., REPORT
FOR 1964 TO APRIL, 1975, 191 PP.
(NTIS/PS-75/465/5ST) SEE ALSO 75/466

75-0745 DESAI, C. S.
ODEN, J. T.
EVALUATION AND ANALYSES OF SOME FINITE ELEMENT AND
FINITE DIFFERENCE PROCEDURES FOR TIME-DEPENDENT
PROBLEMS
ARMY ENG. WATERWAYS EXP. STATION, VICKSBURG,
MISS., REPORT NO: AEWES-MISC-PAPER-S-75-7, APRIL
1975, 110 PP. (N.T.I.S. -AD-A009 739/4ST)

75-0746 FOOS, R. C.
MODELING AN INPUT-OUTPUT GEOKINETIC SYSTEM
UTILIZING A FINITE ELEMENT APPROACH
MASTER'S THESIS, NAVAL POSTGRADUATE SCHOOL,
MONTEREY, CALIF., MARCH 1975, 79 PP. (N.T.I.S. -
AD-A009 567/9ST)

75-0747 PICHUMANI, R.
DUNPHEY, E. P.
IUZZOLINI, H. J.
PARAMETRIC STUDIES OF LAYERED PAVEMENT SYSTEMS
USING AFPAV CODE WITH FAST EQUATION SOLVER
ERIC H. WANG CIV. ENG. RESEARCH FACIL., NEW MEXICO
UNIV., ALBUQUERQUE, FINAL REPORT, NOVEMBER 15,
1973 - JUNE 30, 1974, 54 PP. (N.T.I.S. - AD-A009
479/7ST)

75-0748 CRAWFORD, J.
PICHUMANI, R.
FINITE-ELEMENT ANALYSIS OF PAVEMENT STRUCTURES
USING AFPAV CODE (NONLINEAR ELASTIC ANALYSIS)
ERIC H. WANG CIV. ENG. RESEARCH FACIL., NEW MEXICO
UNIV., ALBUQUERQUE, FINAL REPORT, AUGUST 1972 -
NOVEMBER 1973, 87 PP. (N.T.I.S. - AD-A009 478/9ST)

75-0749 NEWMAN, M.
ZAPHIR, Z.
BODNER, S. R.
FINITE ELEMENT ANALYSIS FOR THE TIME-DEPENDENT
INELASTIC MATERIAL BEHAVIOR
MATER. MECH., TECHNION - ISRAEL INST. OF TECH.,
HAIFA, REPORT NO: SCIENTIFIC-5, FEBRUARY 1975, 28
PP. (N.T.I.S. - AD-A009 164/5ST)

75-0750 HRUDEY, T. M.
A FINITE ELEMENT PROCEDURE FOR PLATES WITH CURVED
BOUNDARIES
NAT. AERON. ESTABL., OTTAWA, REPORT NO:
NRC-LR-584, JANUARY 1975, 62 PP. (N.T.I.S. -
AD-A009 085/2ST).

75-0751 EVANS, J. D.
INVESTIGATION OF THE FINITE ELEMENT METHOD FOR
DETERMINING GREEN'S FUNCTIONS
MASTER'S THESIS, SCHOOL OF ENG., AIR FORCE INST.
OF TECH., WRIGHT-PATTERSON AFB, OHIO, REPORT NO.
GEP/PH/75-4, MARCH 1975, 65 PP. (N.T.I.S. -
AD-A008 670/2ST)

75-0752 CARSON, J.
DRYSDALE, W. H.
AN ANALYSIS OF FINITE ELEMENT CODE TORT 11
BALLISTIC RESEARCH LABS., ABERDEEN PROVING GROUND,
MD., REPORT NO: BRL-MR-2463, MARCH 1975, 18 PP.
(N.T.I.S. - AD-A008 319/6ST)

75-0753 ADHAM, S.
BHAUMIK, A.
TSENBERG, J.
REINFORCED CONCRETE ON CONSTITUTIVE RELATIONS
AGBABIAN ASSOC., EL SEGUNDO, CALIF., FINAL REPORT
MARCH 6, 1973 - MARCH 8, 1974, 358 PP. (N.T.I.S. -
AD-007 886/5ST)

75-0754 GOLD, L. M.
STOWELL, N. A.
FINITE ELEMENT ANALYSIS OF A MULTI-COMPONENT
KINETIC ENERGY PROJECTILE
FRANKFORD ARSENAL, PHILADELPHIA, PA., REPORT NO:
FA-TM-75010, FEBRUARY 1975, 21 PP. (N.T.I.S. -
AD-A007 805/5ST)

75-0755 KANDIDOV, V. P.
KIM, L. P.
STABILITY ANALYSIS OF THE TORSIONAL BENDING
VIBRATIONS OF WING IN A SUBSONIC FLOW BY METHOD OF
THE FINITE ELEMENTS
FOREIGN TECH. DIV., WRIGHT-PATTERSON AFB, OHIO,
REPORT NO: FFTD-MT-24-0435-75, FEBRUARY 14, 1975,
33 PP. (N.T.I.S. - AD-A007 450/0ST)

75-0756 IVANOV, Y. I.
CALCULATION OF THE REINFORCED THIN-WALLED
CONSTRUCTION BY THE FINITE ELEMENT
FOREIGN TECH. DIV., WRIGHT-PATTERSON AFB., OHIO,
REPORT NO: FTD-MT-24-0427-75, FEBRUARY 14, 1975,
36 PP. (N.T.I.S. - AD-A007 307/2ST)

75-0757 SINGH, J. N.
FINITE ELEMENT ANALYSIS OF MERCURY SLOSH IN THE
SOLAR ELECTRIC PROPULSION STAGE
MARTIN MARIETTA CORP., DENVER, COLO., REPORT NO:
NASA-CR-120597, MCR-74-464, JANUARY 1975, 136 PP.
(N.T.I.S. N75-15737/0ST)

75-0758 JOBSON, D. A.
PIECE-WISE CONTINUOUS FINITE ELEMENT FOR THE
STRESS ANALYSIS OF PLATES AND SHELLS
UKAEA REACTOR GROUP, RISLEY, MARCH 1975, 22 PP.
(N.T.I.S. - TRG-REPORT-2414(R))

75-0759 FRIEDRICH, C. M.
BESTRAN: A TECHNIQUE FOR PERFORMING STRUCTURAL
ANALYSES
BETTIS ATOMIC POWER DIV., WESTINGHOUSE ELEC.
CORP., PITTSBURGH, PA., FEBRUARY 1975, 79 PP.
(N.T.I.S. - WAPD-TM-1140)

75-0760   GARTLING, D. K.
BECKER, E. B.
FINITE ELEMENT ANALYSIS OF VISCOUS, INCOMPRESSIBLE
FLUID FLOW
SANDIA LABS., ALBUQUERQUE, N. MEX., FEBRUARY,
1975, 16 PP. (N.T.I.S. - SAND-75-5171)

75-0761   REEVES, M.
DUGUID, J. O.
WATER   MOVEMENT   THROUGH   SATURATED-UNSATURATED
POROUS MEDIA: A FINITE-ELEMENT GALERKIN MODEL
OAK RIDGE NATL. LAB., TENN., FEBRUARY 1975, 236
PP. (N.T.I.S. - ORNL-4927)

75-0762   LARDER, R. A.
STOCHASTIC   FINITE   ELEMENT   SIMULATION   OF   THE
NONLINEAR STRUCTURAL RESPONSE OF FIBROUS COMPOSITE
MATERIALS
LAWRENCE LIVERMORE LAB., CALIF. UNIV., LIVERMORE,
JANUARY 15, 1975, 98 PP. (N.T.I.S. - UCRL-51717)

75-0763   AKHRAS, G.
DHATT, G.
AN AUTOMATIC RELABELING ALGORITHM FOR BANDWIDTH
MINIMIZATION
PROC.   5TH CANADIAN CONGRESS APPL. MECH.,
FREDERICTON, MAY 26-30, 1975.

75-0764   EIDELBERG, B. E.
BOOKER, J. F.
APPLICATION   OF   FINITE   ELEMENT   METHODS   TO
LUBRICATION: SQUEEZE FILMS BETWEEN POROUS SURFACES
PAPER NO. 75-LUB-35, A.S.M.E. MEETING, OCTOBER
21-23, 1975, 5 PP.

75-0765   YUNG, C.
FRIDLEY, R. B.
SIMULATION OF VIBRATION OF WHOLE TREE SYSTEMS
USING FINITE ELEMENTS
TRANS. A.S.A.E., GEN. ED., VOL. 18, NO. 3,
MAY-JUNE 1975, PP. 475-481.

75-0766   DEMIRCHYAN, K. S.
SOLNYSHKIN, N. I.
RASHET TREKHMERNYKH MAGNITNYKE POLEI METODOM
KONECHNYKH   ELEMENTOV.   (CALCULATION   OF
THREE-DIMENSIONAL MAGNETIC FIELDS BY THE FINITE
ELEMENT METHOD)
IZV AKAD NAUK (SSSR), ENERGY TRANS., NO. 5,
SEPTEMBER-OCTOBER, 1975, PP. 39-49.

75-0767   AUCONI, F.
MODAL   RESPONSE   ANALYSIS   OF   CONSTRAINED
FINITE-ELEMENTS ELASTIC CONTINUUM
AEROTECNICA MISSILI E SPAZIO, VOL. 54, NO. 5-6,
1975, PP. 265-272.

75-0768   LAPSHIN, IU. G.
PINTUS, L. V.
APPLICATION OF THE FINITE ELEMENT METHOD TO
CALCULATE WOOD TENSION
LESN ZH IZV VYSSH UCHEBN ZAVED, VOL. 1, 1975, PP.
88-92.

75-0769   DUKES, W. H.
BRITTLE MATERIALS: A DESIGN CHALLENGE
MECH. ENG., VOL. 97, NO. 11, NOVEMBER 1975, PP.
42-47.

75-0770   ALEXANDER, J. M.
TURNER, T. W.
PRELIMINARY INVESTIGATION OF THE DIE-LESS DRAWING
OF TITANIUM AND SOME STEELS
PROC. 15TH INT. CONF. ON MACHINE TOOL DES. AND
RES., BIRMINGHAM, ENGLAND, SEPTEMBER 18-20, 1974,
PP. 525-537. (PUBL. BY HALSTED PRESS, DIV. OF JOHN
WILEY, NEW YORK, N.Y., 1975).

75-0771   SCHMUGAR, K. L.
VIBRATORY WEAR OF FUEL RODS
PAPER NO. 75-WA/HT-79, A.S.M.E. MEETING, NOVEMBER
30 - DECEMBER 4, 1975, 8 PP.

75-0772   GALLAGHER, R. H.
FINITE ELEMENT ANALYSIS: FUNDAMENTALS
PUBL. BY PRENTICE-HALL, ENGLEWOOD CLIFFS, N.J.,
1975, 420 PP.

75-0773   WANG, M. S.
CHENG, R. T. S.
A STUDY OF CONVECTIVE-DISPERSION EQUATION BY
ISOPARAMETRIC FINITE ELEMENTS. TRANSPORT OF
POLLUTANTS IN GROUND-WATER
J. HYDROL., VOL. 24, NO. 1/2, JANUARY 1975, PP.
45-56.

75-0774   ANON
FINITE ELEMENT AND CUMULATIVE DAMAGE ANALYSIS OF A
KEYHOLE TEST SPECIMEN
SOC. AUTO. ENG. INC., TWO PENN. PLAZA, N.Y.,
FEBRUARY 1975, 8 PP.

75-0775   GONTAROVSKII, V. P.
KOZLOV, I. A.
APPLICATION OF THE METHOD OF FINITE ELEMENTS TO
CALCULATIONS OF THE STRESSED AND STRAINED STATE OF
INHOMOGENEOUS SOLIDS OF REVOLUTION
PROB. PROCHN., VOL. 8, 1975, PP. 72-76.

75-0776   ANON.
THERMOMECHANICAL ANALYSIS OF THE WELDING PROCESS
USING THE FINITE ELEMENT METHOD
TRANS. A.S.M.E., J. PRESS. VESSEL TECH., VOL. 97,
NO. 3, AUGUST 1975, PP. 206-213.

75-0777   UEDA, Y.
ANALYSIS OF WELDING STRESS RELIEVING BY ANNEALING
BASED ON FINITE ELEMENT METHOD
TRANS. J.W.R.I., VOL. 4, NO. 1, 1975, PP. 39-45.

75-0778   ANON
SOME ASPECTS OF NON-LINEAR FINITE ELEMENT COMPUTER
PROGRAMS
PROC. 3RD CONF. NORTH AMER. METALWORKING RESEARCH,
PITTSBURGH, PA., 1975, PP. 115-126.

75-0779   BUCHALET, C.
FINITE-ELEMENT   ELASTIC-PLASTIC   ANALYSIS   OF
RESIDUAL STRESSES DUE TO CLAD WELDING IN REACTOR
VESSELS
CENTRAL   ELECTRICITY   GENERATING   BOARD,   ENGLAND,
1975, PP. 137-142.

75-0780   JERRAM, K.
FINITE-ELEMENT TECHNIQUES IN FRACTURE MECHANICS
CENTRAL   ELECTRICITY   GENERATING   BRD.,   ENGLAND,
1975, PP. 165-178.

75-0781   OLSON, M. D.
COMPATIBILITY
WORLD CONGRESS ON FINITE ELEMENT METH. IN STRUCT.
MECH.,   BOURNEMOUTH,   DORSET,   ENGLAND,   OCTOBER
12-17, 1975, PP. H.1-H.33.

75-0782   GUPTA, S. K. S.
A FINITE SOLUTION OF PUCHER'S EQUATION
PH.D., THE UNIVERSITY OF TENNESSEE, 1975, 141 PP.

75-0783   MURTY, V. V. N.
A FINITE ELEMENT MODEL FOR MISCIBLE DISPLACEMENT
IN GROUNDWATER AQUIFERS
PH.D., UNIV. OF CALIFORNIA, DAVIS, 1975, 212 PP.

75-0784   WANG, R. C-J.
REFINED FINITE ELEMENT STABILITY ANALYSIS OF
THIN-WALLED MEMBERS
PH.D., STATE UNIV. OF NEW YORK, BUFFALO, 1975, 130
PP.

75-0785   MCCOWAN, D. W.
DYNAMIC FINITE ELEMENT ANALYSIS WITH APPLICATIONS
TO SEISMOLOGICAL PROBLEMS
PH.D., THE PENNSYLVANIA STATE UNIV., 1975, 197 PP.

75-0786   TAKACS, R. R.
A FINITE ELEMENT PROGRAM FOR FREE FORM SHELLS
PH.D., WEST VIRGINIA UNIV., 1975, 331 PP.

75-0787   MCKNIGHT, R. L.
FINITE ELEMENT CYCLIC THERMOPLASTICITY ANALYSIS BY
THE METHOD OF SUBVOLUMES
PH.D., UNIV. OF CINCINNATI, 1975, 143 PP.

75-0788   FRILEY, J. R.
FINITE   ELEMENT   APPLICATIONS   OF   A   PRECISION
TRIANGULAR   PLATE   BENDING   ELEMENT   TO   LINEAR
BOUNDARY VALUE PROBLEMS GOVERNED BY EXTREMUM
PRINCIPLES
PH.D., PURDUE UNIV., 1975, 312 PP.

75-0789   GHAZVINIAN, B.
NONCONSERVATIVE STABILITY OF PLATES BY FINITE
ELEMENT METHOD
PH.D., ILLINOIS INST. OF TECHNOLOGY, 1975, 71 PP.

75-0790   YUNG, CH.
VIBRATION STUDY OF COMPLETE TREE SYSTEMS USING
FINITE ELEMENT METHOD
PH.D., UNIV. OF CALIFORNIA, DAVIS, 1975, 169 PP.

75-0791   CHOU, C-K.
IMPROVEMENT IN PLATE MODELING AND PLATE DESIGN BY
THE FINITE ELEMENT METHOD
PH.D., MICHIGAN STATE UNIV., 1975, 259 PP.

75-0792   MAHMOUD, M. H.
SILAGE - SILO INTERACTION USING MATERIAL
CHARACTERIZATION AND FINITE ELEMENT ANALYSIS
PH.D., THE OHIO STATE UNIV., 1975, 173 PP.

75-0793   HABASHI, W. G.
A STUDY OF THE FINITE ELEMENT METHOD FOR
AERODYNAMIC APPLICATIONS
PH.D., CORNELL UNIV., 1975, 180 PP.

75-0794 YILMAZ, C.
ULTIMATE STRENGTH OF BOX GIRDERS BY FINITE ELEMENT
METHOD
PH.D., LEHIGH UNIV., 1975, 134 PP.

75-0795 AMEND, J. H. III
A FINITE ELEMENT ANALYSIS OF DISSOLVED OXYGEN
DRAWDOWN AND SULFATE PRODUCTION IN STRIP MINE
SPOIL DAMS DUE TO PYRITIC CHEMICAL REACTION
PH.D., VIRGINIA POLYTECHNIC INST. AND STATE UNIV.,
1975, 122 PP.

75-0796 ALANI, K. G.
LEAST SQUARES FINITE ELEMENT ANALYSIS OF
TWO-DIMENSIONAL LAMINAR BOUNDARY-LAYER EQUATION
WITH SEPARATION
PH.D., UNIV. OF COLORADO, 1975, 195 PP.

75-0797 EVERETT, J.
MCMILLAN, C. M.
BENTONITE PILES IN DURBAN
PROC. 6TH REG. CONF. FOR AFR. CN SOIL MECH. AND
FOUND. ENG., DURBAN, SOUTH AFRICA, VOL. 1,
SEPTEMBER 1975, PP. 155-159. (PUBL. BY BALKEMA,
A.A., CAPE TOWN, SOUTH AFRICA, 1975).

75-0798 LARDER, R. A.
THE STOCHASTIC FINITE ELEMENT SIMULATION OF THE
NONLINEAR STRUCTURAL RESPONSE OF FIBROUS COMPOSITE
MATERIALS
D. ENG., UNIV. OF CALIFORNIA, DAVIS, 1975, 97 PP.

75-0799 AKBARZADEH, A.
STRESS ANALYSIS OF COMPOSITE MATERIALS BY THE
HYBRID FINITE ELEMENT METHOD
PH.D., UNIV. OF WYOMING, 1975, 203 PP.

75-0800 WANG, J-K.
BIT PENETRATION INTO ROCK - A FINITE ELEMENT STUDY
PH.D., UNIV. OF MISSOURI, ROLLA, 1975, 111 PP.

75-0801 DUNN, P. J. JR.
AN ERROR EXPANSION FOR A FINITE-ELEMENT SOLUTION
OF A TWO-POINT BOUNDARY-VALUE PROBLEM
PH.D., RICE UNIV., 1975, 103 PP.

75-0802 TSIPOURAS, P.
RECTANGULAR AND TRIANGULAR FINITE ELEMENTS FOR
POISSON'S EQUATION
PH.D., BOSTON UNIV. GRADUATE SCHOOL, 1975, 159 PP.

75-0803 CRISFIELD, M. A.
FULL-RANGE ANALYSIS OF STEEL PLATES AND STIFFENED
PLATING UNDER UNIAXIAL COMPRESSION
PROC. INST. CIVIL ENG., LONDON, VOL. 59, NO. 2,
DECEMBER 1975, PP. 595-624.

75-0804 MIRZAD, S. S.
CONTRIBUTION A L'ANALYSE DES VIBRATIONS LIBRES
D'UNE STRUCTURE. APPLICATION AUX AUBAGES DE
TURBOMACHINES. (CONTRIBUTION TO THE ANALYSIS OF
FREE VIBRATIONS OF A STRUCTURE AND ITS APPLICATION
TO TURBOMACHINE BLADINGS)
ELEC. FR. BULL. DIR ETUD. RECH. SER., A NUCL.
HYDRAUL. THERM., 1974, 139 PP.

75-0805 REYER, E.
UEBER DIE PRAKTISCHE ANWENDUNG VON FINITEN
ELEMENTEN IN DER PLATTENSTATIK (APPLICATION OF THE
FINITE ELEMENT METHOD IN STRUCTURAL ANALYSIS OF
PLATES)
BAUTECHNIK (AUSG. A), VOL. 52, NO. 1, JANUARY
1975, PP. 15-2

75-0806 HUYAKORN, P. S.
DUDGEON, C. R.
TECHNIQUES FOR HANDLING INPUT DATA FOR FINITE
ELEMENT ANALYSIS OF REGIONAL GROUNDWATER FLOW
HYDROL. SYMP., ARMIDALE, N.S.W., AUST., MAY 18-21,
1975, PP. 21-25. (PUBL. BY INST. OF ENG., AUST.,
(NATL. CONF. PUBL. N 75/3), SYDNEY, 1975).

75-0807 RADAJ, D.
VOLLSTAENDIGE
SPANNUNGS-DEHNUNGS-TEMPERATURAENDERUNGSBEZIEHUNG
FUER DIE SCHWEISSEIGENSPANNUNGSBERECHNUNG MIT
FINITEN ELEMENTEN. (COMPLETE
STRESS-STRAIN-TEMPERATURE CHANGE RELATION FOR THE
CALCULATION BY FINITE ELEMENTS OF THE RESIDUAL
STRESSES CAUSED BY WELDING)
SCHWEISSEN SCHNEIDEN, VOL. 27, NO. 10, OCTOBER
1975, PP. 394-396.

75-0808 MEI, C. C.
CHEN, H. S.
HYBRID-ELEMENT METHOD FOR WATER WAVES
PROC. 2ND ANN. SYMP. MODEL TECH., SAN FRANCISCO,
CALIF., VOL. 1, SEPTEMBER 3-5, 1975, PP. 63-81.
(PUBL. BY A.S.C.E., NEW YORK, N.Y., 1975).

75-0809 KETCHMAN, J.
LOU, Y. K.
APPLICATION OF THE FINITE ELEMENT METHOD TO TOWED
CABLE DYNAMICS
I.E.E.E. CONF. ON ENG. IN THE OCEAN ENVIRON. AND
11TH ANNUAL MEET. MAR. TECH. SOC., SAN DIEGO,
CALIF., SEPTEMBER 22-25, 1975, PP. 98-107. (PUBL.
BY I.E.E.E. (75 CHO 995-1 OEC), NEW YORK, N. Y.,
1975)

75-0810 RAQUET, E.
TACKE, G.
UNTERSUCHUNGEN ZUR SPURMASSTABILITAET BEI
RADSAETZEN MIT KLOTZGEBREMSTEN VOLLRAEDERN
(INVESTIGATIONS OF THE TRACK GAGE STABILITY OF
WHEEL SETS WITH BLOCK-BRAKED SOLID WHEELS)
Z EISENBAHNWES VERKEHRSTECH GLASERS ANN, VOL. 99,
NO. 11, NOVEMBER 1975, PP. 311-316.

75-0811 UEDA, Y.
APPLICATION OF FINITE ELEMENT METHOD TO PROBLEMS
OF NON-LINEAR BEHAVIOR WITH CREEP
PROC. NATL. SYMP. OF THE SOC. OF STEEL CONSTR. OF
JAPAN ON THE MATRIX METHODS OF STRUCT. ANALYSIS,
TOKYO, JAPAN, JUNE 1975.

75-0812 GOWAN, M. J.
LLOYD, T.
EFFECT OF THE VARIATION OF CONSISTENCY IN RESIDUAL
SOILS ON LOAD TRANSFER BY BORED PILES
PROC. 6TH REG. CONF. FOR AFR. ON SOIL MECH. AND
FOUND. ENG., DURBAN, SOUTH AFRICA, VOL. 1,
SEPTEMBER 1975, PP. 177-182. (PUBL. BY BALKEMA,
A.A., CAPE TOWN, SOUTH AFRICA, 1975).

75-0813 ANON
FINITE-ELEMENT METHOD FOR CALCULATING THE
CONE-INDENTATION TEST
ARCH. EISENHUTTENWES., VOL. 46, NO. 9, SEPTEMBER
1975, PP. 571-574.

75-0814 YANG, T. Y.
FLUTTER OF FLAT FINITE-ELEMENT PANELS IN A
SUPERSONIC POTENTIAL FLOW
A.I.A.A. J., VOL. 13, NO. 11, 1975, PP. 1502-1507.

75-0815 ROWAN, J. C.
BURNS, T. A.
AEROELASTIC LOADS PREDICTIONS USING FINITE-ELEMENT
AERODYNAMICS
J. AIRCRAFT, VOL. 12, NO. 11, 1975, PP. 890-898.

75-0816 STERN, M.
STRONGLY PLANAR AC-LATTICES IN WHICH IDEAL OF
FINITE-ELEMENTS IS STANDARD
ACTA MATHEMATICA ACADEMIAE SCIENTIARUM HUNGARICAE,
VOL. 26, NO. 3-4, 1975, PP. 229-232.

75-0817 VEIDINGER, L.
ORDER OF CONVERGENCE OF A FINITE-ELEMENT METHOD IN
REGIONS WITH CURVED BOUNDARIES
ACTA MATHEMATICA ACADEMIAE SCIENTIARUM HUNGARICAE,
VOL. 26, NO. 3-4, 1975, PP. 419-431.

75-0818 NOOR, A. K.
VOIGT, S. J.
HYPERMATRIX SCHEME FOR FINITE-ELEMENT SYSTEMS ON
CDC STAR-100 COMPUTER
COMPUT. AND STRUCT., VOL. 5, NO. 5-6, 1975, PP.
287-296.

75-0819 SUN, C. T.
HUANG, S. N.
TRANSVERSE IMPACT PROBLEMS BY HIGHER-ORDER BEAM
FINITE-ELEMENTS
COMPUT. AND STRUCT., VOL. 5, NO. 5-6, 1975, PP.
297-303.

75-0820 FRIED, I.
MONOTON FINITE-ELEMENT MATRICES AND THEIR COMPUTED
CONDITION NUMBERS
COMPUT. AND STRUCT., VOL. 5, NO. 5-6, 1975, PP.
317-319.

75-0821 NATTERER, F.
POINTWISE CONVERGENCE OF FINITE-ELEMENTS
NUMERISCHE MATHEMATIK, VOL. 25, NO. 1, 1975, PP.
67-77.

75-0822 CURTIS, A. J.
ON THE APPLICATION OF RESPONSE LIMITING TO FINITE
ELEMENT STRUCTURAL ANALYSIS PROGRAMS
TRANS. A.S.M.E., J. ENG. IND., VOL. 97, SER. 8,
NO. 4, NOVEMBER 1975, PP. 1194-1198.

75-0823 KACZKOWSKI, Z.
METHOD OF FINITE SPACE-TIME ELEMENTS IN DYNAMICS
OF STRUCTURES
J. TECH. PHYS., VOL. 16, NO. 1, 1975, PP. 69-84.

75-0824 HINTZ, R. M.
ANALYTICAL METHODS IN COMPONENT MODAL SYNTHESIS
A.I.A.A. J., VOL. 13, NO. 8, AUGUST 1975, PP.
1007-1016.

75-0825 BELYTSCHKO, T.
SCHOEBERLE, D. F.
ON THE UNCONDITIONAL STABILITY OF AN IMPLICIT
ALGORITHM FOR NONLINEAR STRUCTURAL DYNAMICS
PAPER NO. 75-WA/APM-14, A.S.M.E. MEETING, NOVEMBER
30 - DECEMBER 4, 1975, 5 PP.

75-0826 LAGUE, G.
BALDUR, R.
AXISYMMETRIC FINITE ELEMENT IN NATURAL COORDINATES
PAPER NO. 75-WA/PVP-24, A.S.M.E. MEETING, NOVEMBER
30 - DECEMBER 4, 1975, 9 PP.

75-0827 KUMBLE, R. G.
BERRY, J. T.
APPLICATION OF THE DOUBLE LIGAMENT TENSILE TEST IN
DEFORMATION PROCESSING
PAPER NO. 75-WA/PROD-17, A.S.M.E. MEETING,
NOVEMBER 30 - DECEMBER 4, 1975, 6 PP.

75-0828 HOOPER, J. A.
ELASTIC SETTLEMENT OF A CIRCULAR RAFT IN ADHESIVE
CONTACT WITH A TRANSVERSELY ISOTROPIC MEDIUM
GEOTECHNIQUE, VOL. 25, NO. 4, DECEMBER 1975, PP.
691-711.

75-0829 HUNT, C.
NASSIF, N. R.
ON A VARIATIONAL INEQUALITY AND ITS APPROXIMATION,
IN THE THEORY OF SEMICONDUCTORS
S.I.A.M., J. NUMER. ANAL., VOL. 12, NO. 6,
DECEMBER 1975, PP. 938-950.

75-0830 WU, R. W. H.
WITMER, E. A.
THEORETICAL AND EXPERIMENTAL STUDIES OF TRANSIENT
ELASTIC-PLASTIC LARGE DEFLECTIONS OF GEOMETRICALLY
STIFFENED RINGS
PAPER NO. 75-APM-Z, A.S.M.E. MEETING, NOVEMBER 30
- DECEMBER 4, 1975, 7 PP.

75-0831 WU, R. W. H.
WITMER, E. A.
THEORETICAL AND EXPERIMENTAL STUDIES OF TRANSIENT
ELASTIC-PLASTIC LARGE DEFLECTIONS OF GEOMETRICALLY
STIFFENED RINGS
TRANS. A.S.M.E., J. APPL. MECH., VOL. 42, SEP. E,
NO. 4, DECEMBER 1975, PP. 793-799.

75-0832 SCHAEFER, H.
LINK, J.
MEHLORN, G.
ZUR WIRKLICHKEITSNAHEN BERECHNUNG VON
STAHLBETONPLATTEN MIT DER FINITE-ELEMENT-METHODE.
(APPROXIMATE CALCULATION OF REINFORCED CONCRETE
PLATES BY USING FINITE-ELEMENT-METHOD)
BETON STAHLBETONBAU, VOL. 70, NO. 11, NOVEMBER
1975, PP. 265-273.

75-0833 HAYHURST, D. R.
DIMMER, P. R.
CHERNUKA, M. W.
ESTIMATES OF THE CREEP RUPTURE LIFETIME OF
STRUCTURES USING THE FINITE ELEMENT METHOD
J. MECH. PHYS. SOLIDS, VOL. 23, NO. 4-5,
AUGUST-OCTOBER 1975, PP. 335-355.

75-0834 JULLIEN, J. F.
CUBAUD, J. C.
MOREL, A.
PHOTOELASTIC DETERMINATION OF THE STRESSES IN THE
FOUNDATION OF THE NUCLEAR-POWER STATION BUGEY 2
EXP. MECH., VOL. 15, NO. 12, DECEMBER 1975, PP.
471-475.

75-0835 CORP, E. L.
SCHUSTER, R. L.
MCDONALD, M. M.
ELASTIC-PLASTIC STABILITY ANALYSIS OF MINE-WASTE
EMBANKMENTS
U.S. BUR. MINES, REPORT NO: 8069, 1975, 106 PP.

75-0836 MIURA, I.
ASAOKA, K.
SIMULATION ON THE INITIATION AND GROWTH OF DUCTILE
FRACTURE VOIDS
J. JAPAN INST. METALS, VOL. 39, NO. 10, OCTOBER
1975, PP. 1025-1032.

75-0837 EGGERS, H.
VARIATIONAL PRINCIPLES FOR ELASTOPLASTIC CONTINUA
J. STRUCT. MECH., VOL. 3, NO. 4, 1974-1975, PP.
345-358.

75-0838 OLIVER, L. R.
JOHNSON, C. O.
BREIG, W. F.
V-BELT LIFE PREDICTION AND POWER RATING
PAPER NO. 75-WA/DE-26, A.S.M.E. MEETING, NOVEMBER
30 - DECEMBER 4, 1975, 8 PP.

75-0839 TENERELLI, D. J.
THERMOSTRUCTURAL DESIGN CONSIDERATIONS TO ACHIEVE
THE LARGE SPACE TELESCOPE LINE-OF-SIGHT
REQUIREMENTS
PROC. 21ST ANN. MEETING AMER. ASTRONAUT. SOC.,
DENVER, COLO., PAPER NO. AAS 75-190, AUGUST 26-28,
1975, 17 PP.

75-0840 LAGER, J. R.
DESIGN OF LOW-THERMAL-DISTORTION LST METERING
STRUCTURE
PROC. 21ST ANN. MEETING AMER. ASTRONAUT. SOC.,
DENVER, COLO., PAPER NO. AAS 75-191, AUGUST 26-28,
1975, 29 PP.

75-0841 WAGNER, R. J.
YANG, T. Y.
ELASTIC-PLASTIC FORMULATION FOR A
CYLINDRICAL-SHELL FINITE-ELEMENT
A.I.A.A. J., VOL. 13, NO. 12, 1975, PP. 1545-1546.

75-0842 GALLAGHER, R. H.
ENHANCEMENTS OF THE FINITE ELEMENT METHOD THROUGH
MULTIDISCIPLINARY APPLICATIONS
PROC. 5TH CAN. CONGRESS OF APPL. MECH., UNIV. OF
NEW BRUNSWICK, FREDERICTON, MAY 26-30, 1975, PP.
G35-G47.

75-0843 FAIRCLOTH, J. M.
GAZDA, I. W.
SWITZER, C. Z.
SIGVALDASON, O. T.
FINITE-ELEMENT ANALYSIS OF AN ELECTRODE-NIPPLE
JOINT
CARBON, VOL. 13, NO. 6, 1975, P. 549.

75-0844 CRISFIELD, M. A.
AN ITERATIVE IMPROVEMENT FOR NON-CONFORMING
BENDING ELEMENTS
INTL. J. NUMER. METH. IN ENG., VOL. 9, NO. 3, JULY
1975.

75-0845 GLOWINSKI, R.
MARROCO, A.
SUR L APPROXIMATION, PAR ELEMENTS FINIS D ORDRE
UN, ET LA RESOLUTION, PAR PENALISATION-DUALITE, D
UNE CLASSE DE PROBLEMES DE DIRICHLET NON LINEAIRES
(APPROXIMATION BY FINITE ELEMENTS OF ORDER ONE AND
SOLUTION BY PENALIZATION-DUALITY OF A CLASS OF
NONLINEAR DIRICHLET PROBLEMS)
REV. FR. AUTOM. INF. RECH. OPER., VOL. 9, NO. R-2,
AUGUST 1975, PP. 41-76.

75-0846 ATTEIA, M.
FONCTIONS SPLINES ET METHODE D ELEMENTS FINIS
(SPLINE FUNCTIONS AND FINITE ELEMENT METHOD)
REV. FR. AUTOM. INF. RECH. OPER., VOL. 9, NO. R-2,
AUGUST 1975, PP. 13-40.

75-0847 KAHLERT, F.
BASSENGE, CH.
NUMERISCHE BERECHNUNG INSTATIONAERER
TEMPERATURFELDER UND THERMISCH BEDINGTER
VERFORMUNGEN VON WERKZEUGMASCHINENBAUTEILEN.
(NUMERICAL ANALYSIS OF UNSTEADY TEMPERATURE FIELDS
AND DEFORMATIONS ON MACHINE TOOL PARTS)
MASCHINENBAUTECHNIK, VOL. 24, NO. 11, NOVEMBER
1975, PP. 491-497.

75-0848 SCHNEIDER, G. E.
FINITE ELEMENT FORMULATION OF THE HEAT CONDUCTION
EQUATION IN GENERAL ORTHOGONAL CURVILINEAR
COORDINATES
PAPER NO. 75-WA-HT-95, A.S.M.E. MEETING, NOVEMBER
30 - DECEMBER 4, 1975, 9 PP.

75-0849 YALAMANCHILI, R.
ACCURACY, STABILITY, AND OSCILLATION
CHARACTERISTICS OF TRANSIENT TWO-DIMENSIONAL HEAT
CONDUCTION
PAPER NO. 75-WA/HT-85, A.S.M.E. MEETING, NOVEMBER
30 - DECEMBER 4, 1975, 9 PP.

75-0850 REDDY, I. K.
RANKIN, D.
MAGNETOTELLURIC RESPONSE OF LATERALLY
INHOMOGENEOUS AND ANISOTROPIC MEDIA
GEOPHYS., VOL. 40, NO. 6, DECEMBER 1975, PP.
1035-1045.

75-0851 TSENG, M. T.
FINITE ELEMENT MODEL FOR BACKWATER COMPUTATION
PROC. 2ND ANN. SYMP. ON MODEL TECH., SAN
FRANCISCO, CALIF., VOL. 2, SEPTEMBER 3-5, 1975,
PP. 1448-1466. (PUBL. BY A.S.C.E., NEW YORK, N.Y.,
1975)

75-0852 ODEN, J. T.
MATHEMATICAL ASPECTS OF FINITE ELEMENT
APPROXIMATIONS IN THEORETICAL MECHANICS
MECH. TODAY - 1973. S. NEMMAT-NASSER, (ED.),
PERGAMON PRESS, OXFORD, 1975.

75-0853 HUTTON, M.
RABINS, M. J.
SIMPLIFICATION OF HIGH ORDER MECHANICAL SYSTEMS
USING THE ROUTH APPROXIMATION
PAPER NO. 75-WA/AUT-10, A.S.M.E. MEETING, NOVEMBER
30 - DECEMBER 4, 1975, 10 PP.

75-0854 VOIGT, S.
PROGRAM DESIGN BY A MULTIDISCIPLINARY TEAM
TRANS. I.E.E.E. SOFTWARE ENG., VOL. SE-1, NO. 4,
DECEMBER 1975, PP. 370-376.

75-0855 KOBAYASHI, A. S.
BROWN, C. A.
EMERY, A. F.
VISCOELASTIC PRESPONSE OF THE CORNEO-SCLERAL SHELL
UNDER TONOMETER LOADING
PAPER NO. 75-WA/BIO-2, A.S.M.E. MEETING, NOVEMBER
30 - DECEMBER 4, 1975, 8 PP.

75-0856 AYORINDE, O. A.
KOBAYASHI, A. S.
CHEN, J. C. K
MERATI, J. K.
FINITE ELEMENT LARGE DEFORMATION ANALYSIS OF A
TAPERED AORTA
PAPER NO. 75-WA/BIO-5, A.S.M.E. MEETING, NOVEMBER
30 - DECEMBER 4, 1975, 4 PP.

75-0857 NEWMAN, J. C. JR.
ARMEN, H. JR.
ELASTIC-PLASTIC ANALYSIS OF A PROPAGATING CRACK
UNDER CYCLIC LOADING
A.I.A.A. J., VOL. 13, NO. 8, AUGUST 1975, PP.
1017-1023.

75-0858 WEGNER, R.
DUDDECK, H.
DER GERISSENE ZUSTAND ZWEISEITIG GELAGERTER
PLATTEN UNTER EINZELLASTEN - NICHTLINEARE
BERECHNUNG MIT FINITEN ELEMENTEN. (NONLINEAR
CALCULATION OF THE FLAWED CONDITION OF BILATERALLY
SUPPORTED PLATES UNDER CONCENTRATED LOAD BY USING
FINITE ELEMENT METHOD)
BETON STAHLBETONBAU, VOL. 70, NO. 11, NOVEMBER
1975, PP. 257-262.

75-0859 KATO, S.
NONLINEAR DYNAMIC ANALYSIS OF TORSIONAL SHELLS BY
COMBINED USE OF FINITE ELEMENT AND MODE
SUPERPOSITION METHOD
PROC. NATL. SYMP. OF THE SOC. OF STEEL CONSTR. OF
JAPAN, ON THE MATRIX METH. OF STRUCT. ANALYSIS,
JUNE 1975.

75-0860 KOBORI, K.
DYNAMIC RESPONSE ANALYSIS OF THE IRREGULAR GROUND
BY MEANS OF THE FINITE ELEMENT METHOD
PROC. 24TH JAPAN NATL. CONGRESS FOR THEORETICAL
AND APPL. MECH., 1975.

75-0861 YAGAWA, G.
APPLICATION OF FINITE ELEMENT DYNAMIC ANALYSIS TO
FRACTURE MECHANICS
3RD INT. CONF. ON STRUCT. MECH. IN REACTOR TECH.,
IMPERIAL COLLEGE, LONDON, SEPTEMBER 1-5, 1975.

75-0862 SHIRATORI, M.
KINOSHITA, H.
ANALYSIS OF DISLOCATION STRESS FIELD BY FINITE
ELEMENT METHOD
PROC. NAT. SYMP. OF THE SOC. OF STEEL CONSTR. OF
JAPAN, ON THE MATRIX METH. OF STRUCT. ANALYSIS,
TOKYO, JAPAN, JUNE 1975.

75-0863 KOUNO, T.
SHIMIZU, S.
A CONSIDERATION ON THE FINITE ELEMENT METHOD FOR
CALCULATING STRESS INTENSITY FACTOR KI USING THE J
INTEGRAL METHOD
PROC. NATL. SYMP. OF THE SOC. OF STEEL CONSTR. OF
JAPAN, ON THE MATRIX METH. OF STRUCT. ANALYSIS,
TOKYO, JAPAN, JUNE 1975.

75-0864 KLOSTERMAN, A. L.
MCCLELLAND, W. A.
SHERLOCK, J. E.
DYNAMIC SIMULATION OF COMPLEX SYSTEMS UTILIZING
EXPERIMENTAL AND ANALYTICAL TECHNIQUES
PAPER NO. 75-WA/AERO-9, A.S.M.E. MEETING, NOVEMBER
30-DECEMBER 4, 1975, 12 PP.

75-0865 KAWAHARA, M.
TAKEUCHI, N.
ANALYSIS OF VISCOELASTIC FLUID BY FINITE ELEMENT
METHOD
PROC. NATL. SYMP. OF THE SOC. OF STEEL CONSTR. OF
JAPAN, ON THE MATRIX METH. OF STRUCT. ANALYSIS,
TOKYO, JAPAN, JUNE 1975.

75-0866 KAWAHARA, M.
FINITE ELEMENT ANALYSIS OF INCOMPRESSIBLE
MICROPOLAR VISCOUS FLUID FLOW
PROC. NATL. SYMP. OF THE SOC. OF STEEL CONSTR. OF
JAPAN, ON THE MATRIX METH. OF STRUCT. ANALYSIS,
TOKYO, JAPAN, JUNE 1975.

75-0867 YOSHIMURA, N.
YAGISHITA, S.
TIDAL SIMULATION GY TIME SPACE ELEMENT
PROC. NATL. SYMP. OF THE SOC. OF STEEL CONSTR. OF
JAPAN, ON THE MATRIX METH. OF STRUCT. ANALYSIS,
TOKYO, JAPAN, JUNE 1975.

75-0868 HASHIMOTO, M.
THE APPLICATION OF FINITE ELEMENT METHOD TO THE
LIFTING SURFACE PROBLEM
PROC. NATL. SYMP. OF THE SOC. OF STEEL CONSTR. OF
JAPAN, ON THE MATRIX METH. OF STRUCT. ANALYSIS,
TOKYO, JAPAN, JUNE 1975.

75-0869 IKEGAWA, M.
FINITE ELEMENT ANALYSIS OF POTENTIAL FLOW PROBLEMS
J.S.S.C. SYMP. ON THE APPL. OF FINITE ELEMENT
METHOD, TOKYO, MARCH 1975.

75-0870 YAMAMOTO, Y.
SETO, H.
ANALYSIS OF WATER WAVE PROBLEMS BY THE METHOD OF
SUPERPOSITION OF ANALYTICAL AND FINITE ELEMENT
SOLUTIONS
PROC. NATL. SYMP. OF THE SOC. OF STEEL CONSTR. OF
JAPAN, ON THE MATRIX METH. OF STRUCT. ANALYSIS,
TOKYO, JAPAN, JUNE 1975.

75-0871 KAGAWA, Y.
OMOTE, T.
FINITE ELEMENT SIMULATION OF ACOUSTIC FILTERS WITH
VARIABLE CIRCULAR CROSS SECTION
PROC. NATL. SYMP. OF THE SOC. OF STEEL CONSTR. OF
JAPAN, ON THE MATRIX METH. OF STRUCT. ANALYSIS,
TOKYO, JAPAN, JUNE 1975.

75-0872 SAKAI, F.
TUKIOKA, Y.
FINITE ELEMENT METHOD FOR WAVE GUIDE PROBLEMS
PROC. NATL. SYMP. OF THE SOC. OF STEEL CONSTR. OF
JAPAN, ON THE MATRIX METH. OF STRUCT. ANALYSIS,
TOKYO, JAPAN, JUNE 1975.

75-0873 KAWAI, T.
FUJITANI, Y.
SHEAR DEFORMATION ANALYSIS OF BEAMS AND PLATES BY
FINITE ELEMENT METHOD
PROC. NATL. SYMP. OF THE SOC. OF STEEL CONSTR. OF
JAPAN, ON THE MATRIX METH. OF STRUCT. ANALYSIS,
TOKYO, JAPAN, JUNE 1975.

75-0874 YAMADA, Y.
FINITE ELEMENT ANALYSIS OF COMPOSITE MATERIAL
BEHAVIORS FROM MECHANICAL PROPERTIES OF
CONSTITUTENT MATERIALS
PROC. NATL. SYMP. OF THE SOC. OF STEEL CONSTR. OF
JAPAN, ON THE MATRIX METH. OF STRUCT. ANALYSIS,
TOKYO, JAPAN, JUNE 1975.

75-0875 UEDA, Y.
A NEW MEASURING METHOD OF RESIDUAL STRESSES WITH
THE AID OF FINITE ELEMENT METHOD AND RELIABILITY
OF ESTIMATED VALUES
PROC. NATL. SYMP. OF THE SOC. OF STEEL CONSTR. OF
JAPAN, ON THE MATRIX METH. OF STRUCT. ANALYSIS,
TOKYO, JAPAN, JUNE 1975.

75-0876 ANON
FINITE ELEMENT METHODS
J. MATH. SCI., THE SCI. CO., TOKYO, NO. 144, JUNE
1975.

75-0877 CAREY, G. F.
AUTOMATED MESH REFINEMENT STRATEGIES FOR FINITE
ELEMENT COMPUTATIONS
12TH MEET., SOC. OF ENG. SCI., UNIV. OF TEXAS,
AUSTIN, OCTOBER 20-22, 1975.

75-0878 BOYD, T. J. M.
GARDNER, L. R. T.
HYDROMAGNETIC STABILITY STUDIES USING THE FINITE
ELEMENT METHOD
IN "FINITE ELEMENTS IN FLUIDS", R. H. GALLAGHER ET
AL. (EDS.), VOL. 2, 1975, PP. 255-274. (PUBL. BY
JOHN WILEY)

75-0879 PENZIEN, J.
CHEN, M-C.
TSENG, W-S.
SEISMIC RESPONSE OF HIGHWAY BRIDGES
PROC. U.S. NATL. CONF. ON EARTHQUAKE ENG., ANN
ARBOR, MICH., JUNE 18-20, 1975, PP. 176-185.
(PUBL. BY EARTHQUAKE ENG. RES. INST., OAKLAND,
CALIF., 1975.)

75-0880 HUEBNER, K. H.
FINITE ELEMENT ANALYSIS OF FLUID FILM LUBRICATION
SURVEY
IN "FINITE ELEMENTS IN FLUIDS", R. H. GALLAGHER ET
AL. (EDS.), VOL. 2, 1975, PP. 225-254. (PUBL. BY
JOHN WILEY)

75-0881   BARWELL, F. T.
LUBRICATION PROBLEMS - THE SELECTION OF
MATHEMATICAL MODELS
IN "FINITE ELEMENTS IN FLUIDS", R. H. GALLAGHER ET
AL. (EDS.), VOL. 2, 1975, PP. 205-224. (PUBL. BY
JOHN WILEY)

75-0882   CAREY, G. F.
A COMPUTER PROGRAMME FOR ELASTIC PLASTIC
CALCULATIONS WITH DUAL PERTURBATION EXPANSION AND
VARIATIONAL SOLUTION FOR COMPRESSIBLE FLOWS USING
FINITE ELEMENTS
IN "FINITE ELEMENTS IN FLUIDS", R. H. GALLAGHER ET
AL. (EDS.), VOL. 2, 1975, PP. 159-178. (PUBL. BY
JOHN WILEY)

75-0883   ROBINSON, J.
A TRIANGULAR EQUILIBRIUM (LINEAR STRESS) MEMBRANE
ELEMENT
RA/ARD, REPORT NO: 8-4-75-28, APRIL 1975.

75-0884   ATLURI, S. N.
KATHIRESAN, K.
KOBAYASHI, A. S.
THREE-DIMENSIONAL LINEAR FRACTURE MECHANICS
ANALYSIS BY A DISPLACEMENT-HYBRID FINITE-ELEMENT
MODEL
3RD INTL. CONF. ON STRUCT. MECH. IN REACTOR TECH.,
LONDON, PAPER NO. L/3, SEPTEMBER 1-5, 1975.

75-0885   ROBINSON, J.
A WARPED QUADRILATERAL EQUILIBRIUM MEMBRANE
ELEMENT
RA/ARD, REPORT NO: 27-2-75-27, FEBRUARY 1975.

75-0886   YAGAWA, G.
NISHIOKA, T.
ANDO, Y.
OGURA, N.
THE FINITE ELEMENT CALCULATION OF STRESS INTENSITY
FACTORS USING SUPERPOSITION
2ND CONF. ON PRESSURE VESSELS AND PIPING,
A.S.M.E., SAN FRANCISCO, CALIF., JUNE 23-27, 1975.

75-0887   HUEBNER, K. H.
THE FINITE ELEMENT METHOD FOR ENGINEERS
JOHN WILEY, NEW YORK, 1975.

75-0888   HOGGE, M. A.
COMPARISON OF SOME FORMULATIONS IN THERMAL
ANALYSIS OF STRUCTURES
IN "WORLD CONGRESS ON FINITE ELEMENT METH. IN
STRUCT. MECH.", BOURNEMOUTH, DORSET, ENGLAND,
OCTOBER 12-17, 1975, 30 PP.

75-0889   BEAVERS, J. E.
A HIGH-ORDER FINITE ELEMENT FOR PRISMATIC AND
NONPRISMATIC FOLDED PLATE STRUCTURES
IN "WORLD CONGRESS ON FINITE ELEMENT METH. IN
STRUCT. MECH.", BOURNEMOUTH, DORSET, ENGLAND,
OCTOBER 12-17, 25 PP.

75-0890   ROHDE, S. M.
OH, K. P.
A UNIFIED TREATMENT OF THICK AND THIN FILM
ELASTOHYDRODYNAMIC PROBLEMS USING HIGHER ORDER
ELEMENT PROBLEMS
PROC. ROY. SOC. LONDON, VOL. 343, SER. A, 1975,
PP. 315-331.